MW00838009

The Electrical Engineering Handbook Series

Series Editor
Richard C. Dorf
University of California, Davis

Titles Included in the Series

The Avionics Handbook, Cary R. Spitzer
The Biomedical Engineering Handbook, 2nd Edition, Joseph D. Bronzino
The Circuits and Filters Handbook, Wai-Kai Chen
The Communications Handbook, Jerry D. Gibson
The Control Handbook, William S. Levine
The Digital Signal Processing Handbook, Vijay K. Madisetti & Douglas Williams
The Electrical Engineering Handbook, 2nd Edition, Richard C. Dorf
The Electric Power Engineering Handbook, Leo L. Grigsby
The Electronics Handbook, Jerry C. Whitaker
The Engineering Handbook, Richard C. Dorf
The Handbook of Formulas and Tables for Signal Processing, Alexander D. Poularikas
The Industrial Electronics Handbook, J. David Irwin
The Measurement, Instrumentation, and Sensors Handbook, John G. Webster
The Mechanical Systems Design Handbook, Osita D.I. Nwokah and Yidirim Hurmuzlu
The RF and Microwave Handbook, Mike Golio
The Mobile Communications Handbook, 2nd Edition, Jerry D. Gibson
The Ocean Engineering Handbook, Ferial El-Hawary
The Technology Management Handbook, Richard C. Dorf
The Transforms and Applications Handbook, 2nd Edition, Alexander D. Poularikas
The VLSI Handbook, Wai-Kai Chen
The Mechatronics Handbook, Robert H. Bishop
The Computer Engineering Handbook, Vojin G. Oklobdzija

Forthcoming Titles

The Circuits and Filters Handbook, 2nd Edition, Wai-Kai Chen
The Handbook of Ad hoc Wireless Networks, Mohammad Ilyas
The Handbook of Optical Communication Networks, Mohammad Ilyas
The Handbook of Nanoscience, Engineering, and Technology, William A. Goddard, Donald W. Brenner, Sergey E. Lyshevski, and Gerald J. Iafrate
The Communications Handbook, 2nd Edition, Jerry Gibson

THE CONTROL HANDBOOK

EDITOR
WILLIAM S. LEVINE

A CRC Handbook Published in Cooperation with IEEE Press

Library of Congress Cataloging-in-Publication Data

The control handbook / edited by William S. Levine.
p. cm.—(The electrical engineering handbook series)
Includes bibliographical references and index.
ISBN 0-8493-8570-9 (alk. paper)
1. Automatic control—Handbooks, manuals, etc. 2. Control theory—Handbooks, manuals, etc. I. Levine, William S. II. Series.
TJ213.C6593 1995
629.8—dc20 95-25551
CIP

Direct all inquiries to CRC Press LLC, 2000 N.W. Corporate Blvd., Boca Raton, Florida 33431.

No claim to original U.S. Government works
International Standard Book Number 0-8493-8570-9
Library of Congress Card Number 95-25551
Printed in the United States of America 8 9 0
Printed on acid-free paper

Preface

The purpose of *The Control Handbook* is to put the tools of control theory and practice into the hands of the reader. This means that the tools are not just described. Their use is explained and illustrated. Of course, one cannot expect to become an expert on a subject as vast and complicated as control from just one book, no matter how large. References are given to more detailed and specialized works on each of the tools.

One of the major challenges in compiling this book is the breadth and diversity of the subject. Control technology is remarkably varied. Control system implementations range from float valves to microprocessors. Control system applications include regulating the amount of water in a toilet tank, controlling the flow and generation of electrical power over huge geographic regions, regulating the behavior of gasoline engines, controlling the thickness of rolled products as varied as paper and sheet steel, and hundreds of controllers hidden in consumer products of all kinds. The different applications often require unique sensors and actuators. It quickly became obvious that it would be impossible to include a thorough and useful description of actuation and sensing in this handbook. Sensors and actuators are covered in another handbook in this series, the *Measurement and Instrumentation Handbook. The Control Handbook* thoroughly covers control theory and implementations from the output of the sensor to the input to the actuator—those aspects of control that are universal.

The book is organized in three major sections, Fundamentals, Advanced Methods, and Applications. The Fundamentals are just what the name implies, the basics of control engineering. Note that this section includes major subsections on digital control and modeling of dynamical systems. There are also chapters on specification of control systems, techniques for dealing with the most common and important control system nonlinearities, and digital implementation of control systems.

The section on Advanced Methods consists of chapters dealing with more difficult and more specialized control problems. Thus, this section contains subsections devoted to the analysis and design of multiple-input multiple-output systems, adaptive control, nonlinear control, stochastic control, and the control of distributed parameter systems.

The Applications section is included for several reasons. First, these chapters illustrate the diversity of control systems. Second, they provide examples of how the theory can be applied to specific practical problems. Third, they contain important information about aspects of control that are not fully captured by the theory, such as techniques for protecting against controller failure and the role of cost and complexity in specifying controller designs.

The Control Handbook is designed to be used as a traditional handbook. That is, if you have a question about some topic in control you should be able to find an article dealing with that topic in the book. However, I believe the handbook can also be used in several other ways. It is a picture of the present state-of-the-art. Browsing through it is a way to discover a great deal about control. Reading it carefully is a way to learn the subject of control.

Acknowledgments

I want to thank, first of all, Professor Richard C. Dorf, Editor-in Chief of the Electrical Engineering Handbook Series, for inviting me to edit *The Control Handbook.*

Several people helped make the job of editor much easier and more pleasant than I expected it to be. I cannot imagine how the book could have been completed without Joel Claypool, Engineering Publisher for CRC Press. He had a good solution to every problem and a calm confidence in the ultimate completion of the book that was very comforting and ultimately justified. His assistants, Michelle Veno and Marlaine Beratta could not have been more efficient or more helpful. Susan Fox did an excellent job as production editor. My editorial assistant, and daughter, Eleanor J. Levine proved to be both gifted at her job and fun to work with. Mrs. Patricia Keehn did the typing quickly, accurately and elegantly — as she always does.

Control is an extremely broad and diverse subject. No one person, and certainly not this one, could possibly have the breadth and depth of knowledge necessary to organize this handbook. The Advisory Board provided sound advice on every aspect of the book. Professor Mark Spong volunteered to organize the section on robotics and did a sterling job.

My wife, Shirley Johannesen Levine, deserves a substantial share of the credit for everything I have done.

Last, but most important, I would like to thank the authors of the chapters in this book. Only well respected experts were asked to write articles. Such people are always overworked. I am very grateful to all of them for finding the time and energy to contribute to the handbook.

Advisory Board

Contributors

Eyad H. Abed
Department of Electrical Engineering and the Institute for Systems Research, University of Maryland, College Park, MD

Anders Ahlén
Systems and Control Group, Department of Technology, Uppsala University, Uppsala, Sweden

Albert N. Andry, Jr.
Teledyne Electronic Devices, Marina del Rey, CA

Panos J. Antsaklis
Department of Electrical Engineering, University of Notre Dame, Notre Dame, IN

Brian Armstrong
Department of Electrical Engineering and Computer Science, University of Wisconsin—Milwaukee, Milwaukee, WI

Karl J. Åström
Department of Automatic Control, Lund Institute of Technology, Lund, Sweden

Michael Athans
Massachusetts Institute of Technology, Cambridge, MA

Derek P. Atherton
School of Engineering, The University of Sussex

David M. Auslander
Mechanical Engineering Department, University of California at Berkeley

J. Baillieul
Boston University

V. Balakrishnan
Purdue University

Gary J. Balas
Aerospace Engineering and Mechanics, University of Minnesota, Minnesota, MN

Maria Domenica Di Benedetto
Dipartimento di Ingegneria Elettrica, Università de L'Aquila, Monteluco di Roio (L'Aquila)

W.L. Bialkowski
EnTech Control Engineering Inc.

Robert H. Bishop
The University of Texas at Austin

F. Blanchini
Dipartimento di Matematica e Informatica, Università di Udine, Udine, Italy

Okko H. Bosgra
Mechanical Engineering Systems and Control Group, Delft University of Technology, Delft, The Netherlands

S. Boyd
Department of Electrical Engineering, Stanford University, Stanford, CA

Richard D. Braatz
University of Illinois, Department of Chemical Engineering, Urbana, IL

Herman Bruyninckx
Katholieke Universiteit Leuven, Department of Mechanical Engineering, Leuven, Belgium

Christopher I. Byrnes
Department of Systems Sciences and Mathematics, Washington University, St.Louis, MO

François E. Cellier
Department of Electrical and Computer Engineering, The University of Arizona, Tucson, AZ

Alan Chao
Laboratory for Information and Decision Systems, Massachusetts Institute of Technology, Cambridge, MA

Y. Cho
Department of Electrical Engineering, Stanford University, Stanford, CA

David W. Clarke
Department of Engineering Science, Parks Road, Oxford, UK

Charles M. Close
Electrical, Computer, and Systems Engineering Department, Rensselaer Polytechnic Institute, Troy, NY

J. A. Cook
Ford Motor Company, Scientific Research Laboratory, Control Systems Department, Dearborn, MI

Vincent T. Coppola
Department of Aerospace Engineering, The University of Michigan, Ann Arbor, MI

Bruce G. Coury
The Johns Hopkins University, Applied Physics Laboratory, Laurel, MD

John J. D'Azzo
Air Force Institute of Technology

Munther A. Dahleh
Lab. for Information and Decision Systems, M.I.T., Cambridge, MA

C. Davis
Semiconductor Process and Design Center, Texas Instruments, Dallas, TX

Edward J. Davison
Department of Electrical & Computer Engineering, University of Toronto, Toronto, Ontario, Canada

R. A. DeCarlo
School of Electrical and Computer Engineering, Purdue University, West Lafayette, IN

David F. Delchamps
Cornell University, Ithaca, NY

Bradley W. Dickinson
Princeton University

Rik W. De Doncker
Silicon Power Corporation, Malvern, PA

Richard C. Dorf
University of California, Davis

Joel Douglas
Department of Electrical Engineering and Computer Science, Massachusetts Institute of Technology, Cambridge, MA

S. V. Drakunov
Department of Electrical Engineering, Tulane University, New Orleans, LA

T.E. Duncan
Department of Mathematics, University of Kansas, Lawrence, KS

John M. Edmunds
UMIST, Manchester, England

Hilding Elmqvist
Dynasim AB, Research Park Ideon, Lund, Sweden

Jay A. Farrell
College of Engineering, University of California, Riverside

Clifford C. Federspiel
Johnson Controls, Inc., Milwaukee, WI

Xiangbo Feng
Department of Systems Engineering, Case Western Reserve University, Cleveland, OH

A. Feuer
Electrical Engineering Department, Technion-Israel Institute of Technology, Haifa, Israel

Bruce A. Francis
Department of Electrical and Computer Engineering, University of Toronto, Toronto, Ontario, Canada

G. Franklin
Department of Electrical Engineering, Stanford University, Stanford, CA

Dean K. Frederick
Electrical, Computer, and Systems Engineering Department, Rensselaer Polytechnic Institute, Troy, NY

Randy A. Freeman
University of California, Santa Barbara

James S. Freudenberg
Dept. Electrical Engineering & Computer Science, University of Michigan, Ann Arbor, MI

Bernard Friedland
Department of Electrical and Computer Engineering, New Jersey Institute of Technology, Newark, NJ

T.T. Georgiou
Department of Electrical Engineering, University of Minnesota

James T. Gillis
The Aerospace Corp., Los Angeles, CA

G.C. Goodwin
Department of Electrical and Computer Engineering, University of Newcastle, Newcastle, Australia

Stefan F. Graebe
PROFACTOR GmbH, Steyr, Austria

C. W. Gray
The Aerospace Corporation, El Segundo, CA

M.J. Grimble
Industrial Control Centre, University of Strathclyde, Glasgow, Scotland, U.K.

J. W. Grizzle
Department of EECS, Control Systems Laboratory, University of Michigan, Ann Arbor, MI

Simon Grocott
Space Engineering Research Center, Massachusetts Institute of Technology, Cambridge, MA

John A. Gubner
University of Wisconsin–Madison

P. Gyugyi
Department of Electrical Engineering, Stanford University, Stanford, CA

David Haessig
GEC-Marconi Systems Corporation, Wayne, NJ

Tore Hägglund
Department of Automatic Control, Lund Institute of Technology, Lund, Sweden

Fumio Hamano
California State University, Long Beach

R. A. Hess
University of California, Davis

Gene H. Hostetter

Constantine H. Houpis
Air Force Institute of Technology, Wright-Patterson AFB, OH

Petros Ioannou
University of Southern California, EE-Systems, MC-2562, Los Angeles, CA

Alberto Isidori
Dipartimento di Informatica e Sistemistica, Università di Roma "La Sapienza", Rome, and Department of Systems Sciences and Mathematics, Washington University, St.Louis, MO

Thomas M. Jahns
GE Corporate R&D, Schenectady, NY

Hodge Jenkins
The George W. Woodruff School of Mechanical Engineering, The Georgia Institute of Technology, Atlanta, GA

Christopher P. Jobling
Department of Electrical and Electronic Engineering, University of Wales, Swansea, Singleton Park, Wales, UK

M.A. Johnson
Industrial Control Centre, University of Strathclyde, Glasgow, Scotland, U.K.

Jason C. Jones
Mechanical Engineering Department, University of California at Berkeley

S. M. Joshi
NASA Langley Research Center

V. Jurdjevic
Department of Mathematics, University of Toronto, Ontario, Canada

T. Kailath
Department of Electrical Engineering, Stanford University, Stanford, CA

Edward W. Kamen
School of Electrical and Computer Engineering, Georgia Institute of Technology, Atlanta, GA

M. R. Katebi
Industrial Control Centre, Strathclyde University, Glasgow, Scotland

A. G. Kelkar
NASA Langley Research Center

Hassan K. Khalil
Michigan State University

Petar V. Kokotović
University of California, Santa Barbara

Karlene A. Kosanovich
Department of Chemical Engineering, University of South Carolina, Columbia, SC

Miroslav Krstić
Department of Mechanical Engineering, University of Maryland, College Park, MD

Vladimír Kučera
Institute of Information Theory and Automation, Prague, Academy of Sciences of the Czech Republic

P. R. Kumar
Department of Electrical and Computer Engineering and Coordinated Science Laboratory, University of Illinois, Urbana, IL

Thomas R. Kurfess
The George W. Woodruff School of Mechanical Engineering, The Georgia Institute of Technology, Atlanta, GA

Harry G. Kwatny
Drexel University

J. E. Lagnese
Department of Mathematics, Georgetown University, Washington, DC

Françoise Lamnabhi–Lagarrigue
Laboratoire des Signaux et Systèmes CNRS, Supélec, Gif–sur–Yvette, France

Einar V. Larsen
GE Power Systems Engineering, Schenectady, NY

B.P. Lathi
California State University, Sacramento

A.J. Laub
Department of Electrical and Computer Engineering, University of California, Santa Barbara, CA

B. Lehman
Northeastern University

G. Leugering
Fakultät für Mathematik und Physik, University of Bayreuth, Postfach Bayreuth, Germany

William S. Levine
Department of Electrical Engineering, University of Maryland, College Park, MD

F. L. Lewis
Automation and Robotics Research Institute, The University of Texas at Arlington, Ft. Worth, TX

M. K. Liubakka
Advanced Vehicle Technology, Ford Motor Company, Dearborn, MI

Lennart Ljung
Department of Electrical Engineering, Linköping University, Sweden

Douglas P. Looze
Dept. Electrical and Computer Engineering, University of Massachusetts, Amherst, MA

Kenneth A. Loparo
Department of Systems Engineering, Case Western Reserve University, Cleveland, OH

Leonard Lublin
Space Engineering Research Center, Massachusetts Institute of Technology, Cambridge, MA

Claudio Maffezzoni
Politecnico Di Milano

Mohamed Mansour
Swiss Federal Institute of Technology (ETH)

N. Harris McClamroch
Department of Aerospace Engineering, The University of Michigan, Ann Arbor, MI

R. H. Middleton
Department of Electrical and Computer Engineering, University of Newcastle, NSW, Australia

M. Moslehi
Semiconductor Process and Design Center, Texas Instruments, Dallas, TX

Neil Munro
UMIST, Manchester, England

Norman S. Nise
California State Polytechnic University, Pomona

S. Norman
Department of Electrical Engineering, Stanford University, Stanford, CA

Katsuhiko Ogata
University of Minnesota

Gustaf Olsson
Dept. of Industrial Electrical Engineering and Automation, Lund Institute of Technology, Lund, Sweden

A.W. Ordys
Industrial Control Centre, University of Strathclyde, Glasgow, Scotland, U.K.

Martin Otter
Institute for Robotics and System Dynamics, German Aerospace Research Establishment Oberpfaffenhofen (DLR), Wessling, Germany

M. Pachter
Department of Electrical and Computer Engineering, Air Force Institute of Technology, Wright-Patterson AFB, OH

Andy Packard
Mechanical Engineering, University of California, Berkeley, CA

Z.J. Palmor
Faculty of Mechanical Engineering, Technion - Israel Institute of Technology, Haifa, Israel

P. Park
Department of Electrical Engineering, Stanford University, Stanford, CA

John J. Paserba
GE Power Systems Engineering, Schenectady, NY

B. Pasik-Duncan
Department of Mathematics, University of Kansas, Lawrence, KS

Kevin M. Passino
Department of Electrical Engineering, Ohio State University, Columbus, OH

Stephen D. Patek
Laboratory for Information and Decision Systems, Massachusetts Institute of Technology, Cambridge, MA

R.V. Patel
Department of Electrical and Computer Engineering, Concordia University, Montreal, Quebec, Canada

A.W. Pike
Industrial Control Centre, University of Strathclyde, Glasgow, Scotland, U.K.

Michael J. Piovoso
DuPont Central Science & Engineering, Wilmington, DE

L. Praly
Centre Automatique et Systèmes, École Des Mines de Paris

Jörg Raisch
Institut für Systemdynamik und Regelungstechnik, Universität Stuttgart, Stuttgart, FR Germany

D.S. Rhode
Advanced Vehicle Technology, Ford Motor Company, Dearborn, MI

John R. Ridgely
Mechanical Engineering Department, University of California at Berkeley

C. Magnus Rimvall
ABB Corporate Research and Development, Heidelberg, Germany

Charles E. Rohrs
Tellabs, Mishawaka, IN

David L. Russell
Department of Mathematics, Virginia Tech, Blacksburg, VA

Juan J. Sanchez-Gasca
GE Power Systems Engineering, Schenectady, New York

Mohammed S. Santina
The Aerospace Corporation, Los Angeles, CA

K. Saraswat
Department of Electrical Engineering, Stanford University, Stanford, CA

C. Schaper
Department of Electrical Engineering, Stanford University, Stanford, CA

Gerrit Schootstra
Philips Research Laboratories, Eindhoven, The Netherlands

Joris De Schutter
Katholieke Universiteit Leuven, Department of Mechanical Engineering, Leuven, Belgium

John E. Seem
Johnson Controls, Inc., Milwaukee, WI

Thomas I. Seidman
Department of Mathematics and Statistics, University of Maryland Baltimore County, Baltimore, MD

M. E. Sezer
Bilkent University, Ankara, Turkey

S. Shakoor
Industrial Control Centre, University of Strathclyde, Glasgow, Scotland, U.K.

Jeff S. Shamma
Center for Control and Systems Research, Department of Aerospace Engineering and Engineering Mechanics, The University of Texas at Austin, Austin, TX

Eliezer Y. Shapiro
HR Textron, Valencia, CA

F. Greg Shinskey
Process Control Consultant, North Sandwich, NH

Adam Shwartz
Electrical Engineering, Technion—Israel Institute of Technology, Haifa, Israel

D. D. Šiljak
Santa Clara University, Santa Clara, CA

Kenneth M. Sobel
Department of Electrical Engineering, The City College of New York, New York, NY

Torsten Söderström
Systems and Control Group, Uppsala University, Uppsala, Sweden

E. Sontag
Department of Mathematics, Rutgers University

Mark W. Spong
The Coordinated Science Laboratory, University of Illinois at Urbana-Champaign

Raymond T. Stefani
Electrical Engineering Department, California State University, Long Beach

Maarten Steinbuch
Philips Research Laboratories, Eindhoven, The Netherlands

Allen R. Stubberud
University of California, Irvine, Irvine, CA

J. Sun
Ford Motor Company, Scientific Research Laboratory, Control Systems Department, Dearborn, MI

Jacob Tal
Galil Motion Control, Inc.

David G. Taylor
Georgia Institute of Technology, School of Electrical and Computer Engineering, Atlanta, GA

A.R. Teel
Department of Electrical Engineering, University of Minnesota

R. Tempo
CENS-CNR, Politecnico di Torino, Torino, Italy

Alberto Tesi
Dipartimento di Sistemi e Informatica, Università di Firenze, Firenze, Italy

A. L. Tits
University of Maryland

P.M. Van Dooren
Department of Mathematical Engineering, Université Catholique de Louvain, Belgium

George C. Verghese
Massachusetts Institute of Technology

Hua O. Wang
United Technologies Research Center, East Hartford, CT

John Ting-Yung Wen
Department of Electrical, Computer, and Systems Engineering, Rensselaer Polytechnic Institute

Trevor Williams
Department of Aerospace Engineering and Engineering Mechanics, University of Cincinnati, Cincinnati, OH

J. R. Winkelman
Advanced Vehicle Technology, Ford Motor Company, Dearborn, MI

Carlos Canudas de Wit
Laboratoire d'Automatique de Grenoble, ENSIEG, Grenoble, France

William A. Wolovich
Brown University

Jiann-Shiou Yang
Department of Electrical and Computer Engineering, University of Minnesota, Duluth, MN

Stephen Yurkovich
Department of Electrical Engineering, The Ohio State University, Columbus, OH

S. H. Żak
School of Electrical and Computer Engineering, Purdue University, West Lafayette, IN

Contents

PART A FUNDAMENTALS OF CONTROL

SECTION I Mathematical Foundations

SECTION II Models for Dynamical Systems

SECTION III Analysis and Design Methods for Continous-Time Systems

SECTION IV Digital Control

SECTION V Analysis and Design Methods for Nonlinear Systems

SECTION VI Software for Control System Analysis and Design

PART B ADVANCED METHODS OF CONTROL

SECTION VII Analysis Methods for MIMO Linear Systems

SECTION VIII Kalman Filter and Observers

SECTION IX Design Methods for MIMO LTI Systems

SECTION X Adaptive Control

SECTION XI Analysis and Design of Nonlinear Systems

SECTION XII System Identification

SECTION XIII Stochastic Control

SECTION XIV Control of Distributed Parameter Systems

PART C APPLICATIONS OF CONTROL

SECTION XV Process Control

SECTION XVI Mechanical Control Systems

SECTION XVII Electrical and Electronic Control Systems

SECTION XVIII Control Systems Including Humans

PART A

FUNDAMENTALS OF CONTROL

SECTION I

Mathematical Foundations

1

Ordinary Linear Differential and Difference Equations

B.P. Lathi
California State University, Sacramento

1.1 Differential Equations

A function containing variables and their derivatives is called a *differential expression*, and an equation involving differential expressions is called a *differential equation*. A differential equation is an *ordinary* differential equation if it contains only one independent variable; it is a *partial* differential equation if it contains more than one independent variable. We shall deal here only with ordinary differential equations.

In the mathematical texts, the independent variable is generally x, which can be anything such as time, distance, velocity, pressure, and so on. In most of the applications in control systems, the independent variable is time. For this reason we shall use here independent variable t for time, although it can stand for any other variable as well.

The following equation

$$\left(\frac{d^2y}{dt^2}\right)^4 + 3\frac{dy}{dt} + 5y^2(t) = \sin t$$

is an ordinary differential equation of second *order* because the highest derivative is of the second order. An nth-order differential equation is *linear* if it is of the form

$$a_n(t)\tfrac{d^ny}{dt^n} + a_{n-1}(t)\tfrac{d^{n-1}y}{dt^{n-1}} + \cdots + a_1(t)\tfrac{dy}{dt} + a_0(t)y(t) = r(t) \tag{1.1}$$

where the coefficients $a_i(t)$ are not functions of $y(t)$. If these coefficients (a_i) are constants, the equation is linear with *constant coefficients*. Many engineering (as well as nonengineering) systems can be modeled by these equations. Systems modeled by these equations are known as *linear time-invariant* (LTI) systems. In this chapter we shall deal exclusively with linear differential equations with constant coefficients. Certain other forms of differential equations are dealt with elsewhere in this volume.

Role of Auxiliary Conditions in Solution of Differential Equations

We now show that a differential equation does not, in general, have a unique solution unless some additional constraints (or conditions) on the solution are known. This fact should not come as a surprise. A function $y(t)$ has a unique derivative dy/dt, but for a given derivative dy/dt there are infinite possible functions $y(t)$. If we are given dy/dt, it is impossible to determine $y(t)$ uniquely unless an additional piece of information about $y(t)$ is given. For example, the solution of a differential equation

$$\tfrac{dy}{dt} = 2 \tag{1.2}$$

obtained by integrating both sides of the equation is

$$y(t) = 2t + c \tag{1.3}$$

for any value of c. Equation 1.2 specifies a function whose slope is 2 for all t. Any straight line with a slope of 2 satisfies this equation. Clearly the solution is not unique, but if we place an additional constraint on the solution $y(t)$, then we specify a unique solution. For example, suppose we require that $y(0) = 5$; then out of all the possible solutions available, only one function has a slope of 2 and an intercept with the vertical axis at 5. By setting $t = 0$ in Equation 1.3 and substituting $y(0) = 5$ in the same equation, we obtain $y(0) = 5 = c$ and

$$y(t) = 2t + 5$$

which is the unique solution satisfying both Equation 1.2 and the constraint $y(0) = 5$.

In conclusion, differentiation is an irreversible operation during which certain information is lost. To reverse this operation, one piece of information about $y(t)$ must be provided to restore the original $y(t)$. Using a similar argument, we can show that,

0-8493-8570-9/96/$0.00+$.50

given d^2y/dt^2, we can determine $y(t)$ uniquely only if two additional pieces of information (constraints) about $y(t)$ are given. In general, to determine $y(t)$ uniquely from its nth derivative, we need n additional pieces of information (constraints) about $y(t)$. These constraints are also called *auxiliary conditions*. When these conditions are given at $t = 0$, they are called *initial conditions*.

We discuss here two systematic procedures for solving linear differential equations of the form in Equation 1.1. The first method is the *classical method*, which is relatively simple, but restricted to a certain class of inputs. The second method (the convolution method) is general and is applicable to all types of inputs. A third method (Laplace transform) is discussed elsewhere in this volume. Both the methods discussed here are classified as *time-domain* methods because with these methods we are able to solve the above equation directly, using t as the independent variable. The method of Laplace transform (also known as the *frequency-domain* method), on the other hand, requires transformation of variable t into a frequency variable s.

In engineering applications, the form of linear differential equation that occurs most commonly is given by

$$\begin{aligned}&\tfrac{d^ny}{dt^n} + a_{n-1}\tfrac{d^{n-1}y}{dt^{n-1}} + \cdots + a_1\tfrac{dy}{dt} + a_0y(t)\\ &= b_m\tfrac{d^mf}{dt^m} + b_{m-1}\tfrac{d^{m-1}f}{dt^{m-1}} + \cdots + b_1\tfrac{df}{dt} + b_0f(t) \quad (1.4a)\end{aligned}$$

where all the coefficients a_i and b_i are constants. Using operational notation D to represent d/dt, this equation can be expressed as

$$\begin{aligned}&(D^n + a_{n-1}D^{n-1} + \cdots + a_1D + a_0)y(t)\\ &= (b_mD^m + b_{m-1}D^{m-1} + \cdots + b_1D + b_0)f(t) \quad (1.4b)\end{aligned}$$

or

$$Q(D)y(t) = P(D)f(t) \quad (1.4c)$$

where the polynomials $Q(D)$ and $P(D)$, respectively, are

$$\begin{aligned}Q(D) &= D^n + a_{n-1}D^{n-1} + \cdots + a_1D + a_0\\ P(D) &= b_mD^m + b_{m-1}D^{m-1} + \cdots + b_1D + b_0\end{aligned}$$

Observe that this equation is of the form of Equation 1.1, where $r(t)$ is in the form of a linear combination of $f(t)$ and its derivatives. In this equation, $y(t)$ represents an output variable, and $f(t)$ represents an input variable of an LTI system. Theoretically, the powers m and n in the above equations can take on any value. Practical noise considerations, however, require [1] $m \leq n$.

1.1.1 Classical Solution

When $f(t) \equiv 0$, Equation 1.4 is known as the *homogeneous* (or complementary) equation. We shall first solve the homogeneous equation. Let the solution of the homogeneous equation be $y_c(t)$, that is,

$$Q(D)y_c(t) = 0$$

or

$$(D^n + a_{n-1}D^{n-1} + \cdots + a_1D + a_0)y_c(t) = 0$$

We first show that if $y_p(t)$ is the solution of Equation 1.4, then $y_c(t) + y_p(t)$ is also its solution. This follows from the fact that

$$Q(D)y_c(t) = 0$$

If $y_p(t)$ is the solution of Equation 1.4, then

$$Q(D)y_p(t) = P(D)f(t)$$

Addition of these two equations yields

$$Q(D)\left[y_c(t) + y_p(t)\right] = P(D)f(t)$$

Thus, $y_c(t) + y_p(t)$ satisfies Equation 1.4 and therefore is the general solution of Equation 1.4. We call $y_c(t)$ the *complementary* solution and $y_p(t)$ the *particular* solution. In system analysis parlance, these components are called the *natural* response and *the forced* response, respectively.

Complementary Solution (The Natural Response)

The complementary solution $y_c(t)$ is the solution of

$$Q(D)y_c(t) = 0 \quad (1.5a)$$

or

$$\left(D^n + a_{n-1}D^{n-1} + \cdots + a_1D + a_0\right)y_c(t) = 0 \quad (1.5b)$$

A solution to this equation can be found in a systematic and formal way. However, we will take a short cut by using heuristic reasoning. Equation 1.5b shows that a linear combination of $y_c(t)$ and its n successive derivatives is zero, not at *some* values of t, but for all t. This is possible *if and only if* $y_c(t)$ and all its n successive derivatives are of the same form. Otherwise their sum can never add to zero for all values of t. We know that only an exponential function $e^{\lambda t}$ has this property. So let us assume that

$$y_c(t) = ce^{\lambda t}$$

is a solution to Equation 1.5b. Now

$$\begin{aligned}Dy_c(t) &= \frac{dy_c}{dt} = c\lambda e^{\lambda t}\\ D^2y_c(t) &= \frac{d^2y_c}{dt^2} = c\lambda^2e^{\lambda t}\\ \cdots\cdots &\ \cdots \ \cdots\cdots\\ D^ny_c(t) &= \frac{d^ny_c}{dt^n} = c\lambda^ne^{\lambda t}\end{aligned}$$

Substituting these results in Equation 1.5b, we obtain

$$c\left(\lambda^n + a_{n-1}\lambda^{n-1} + \cdots + a_1\lambda + a_0\right)e^{\lambda t} = 0$$

For a nontrivial solution of this equation,

$$\lambda^n + a_{n-1}\lambda^{n-1} + \cdots + a_1\lambda + a_0 = 0 \quad (1.6a)$$

This result means that $ce^{\lambda t}$ is indeed a solution of Equation 1.5 provided that λ satisfies Equation 1.6a. Note that the polynomial in Equation 1.6a is identical to the polynomial $Q(D)$ in Equation 1.5b, with λ replacing D. Therefore, Equation 1.6a can be expressed as

$$Q(\lambda) = 0 \tag{1.6b}$$

When $Q(\lambda)$ is expressed in factorized form, Equation 1.6b can be represented as

$$Q(\lambda) = (\lambda - \lambda_1)(\lambda - \lambda_2)\cdots(\lambda - \lambda_n) = 0 \tag{1.6c}$$

Clearly λ has n solutions: $\lambda_1, \lambda_2, \ldots, \lambda_n$. Consequently, Equation 1.5 has n possible solutions: $c_1e^{\lambda_1 t}, c_2e^{\lambda_2 t}, \ldots, c_ne^{\lambda_n t}$, with $c_1, c_2, \ldots, c_n$ as arbitrary constants. We can readily show that a general solution is given by the sum of these n solutions[1], so that

$$y_c(t) = c_1e^{\lambda_1 t} + c_2e^{\lambda_2 t} + \cdots + c_ne^{\lambda_n t} \tag{1.7}$$

where $c_1, c_2, \ldots, c_n$ are arbitrary constants determined by n constraints (the auxiliary conditions) on the solution.

The polynomial $Q(\lambda)$ is known as the *characteristic polynomial.* The equation

$$Q(\lambda) = 0 \tag{1.8}$$

is called the *characteristic* or auxiliary equation. From Equation 1.6c, it is clear that $\lambda_1, \lambda_2, \ldots, \lambda_n$ are the roots of the characteristic equation; consequently, they are called the *characteristic roots.* The terms *characteristic values, eigenvalues,* and *natural frequencies* are also used for characteristic roots [2]. The exponentials $e^{\lambda_i t}$ $(i = 1, 2, \ldots, n)$ in the complementary solution are the *characteristic modes* (also known as *modes* or *natural modes*). There is a characteristic mode for each characteristic root, and the *complementary solution is a linear combination of the characteristic modes.*

Repeated Roots

The solution of Equation 1.5 as given in Equation 1.7 assumes that the n characteristic roots $\lambda_1, \lambda_2, \ldots, \lambda_n$ are distinct. If there are repeated roots (same root occurring more than once), the form of the solution is modified slightly. By direct substitution we can show that the solution of the equation

$$(D - \lambda)^2 y_c(t) = 0$$

is given by

$$y_c(t) = (c_1 + c_2t)e^{\lambda t}$$

In this case the root λ repeats twice. Observe that the characteristic modes in this case are $e^{\lambda t}$ and $te^{\lambda t}$. Continuing this pattern, we can show that for the differential equation

$$(D - \lambda)^r y_c(t) = 0 \tag{1.9}$$

the characteristic modes are $e^{\lambda t}, te^{\lambda t}, t^2e^{\lambda t}, \ldots, t^{r-1}e^{\lambda t}$, and the solution is

$$y_c(t) = \left(c_1 + c_2t + \cdots + c_rt^{r-1}\right)e^{\lambda t} \tag{1.10}$$

Consequently, for a characteristic polynomial

$$Q(\lambda) = (\lambda - \lambda_1)^r(\lambda - \lambda_{r+1})\cdots(\lambda - \lambda_n)$$

the characteristic modes are $e^{\lambda_1 t}, te^{\lambda_1 t}, \ldots, t^{r-1}e^{\lambda t}, e^{\lambda_{r+1}t}, \ldots, e^{\lambda_n t}$. and the complementary solution is

$$y_c(t) = (c_1 + c_2t + \cdots + c_rt^{r-1})e^{\lambda_1 t} + c_{r+1}e^{\lambda_{r+1}t} + \cdots + c_ne^{\lambda_n t}$$

Particular Solution (The Forced Response): Method of Undetermined Coefficients

The particular solution $y_p(t)$ is the solution of

$$Q(D)y_p(t) = P(D)f(t) \tag{1.11}$$

It is a relatively simple task to determine $y_p(t)$ when the input $f(t)$ is such that it yields only a finite number of independent derivatives. Inputs having the form $e^{\zeta t}$ or t^r fall into this category. For example, $e^{\zeta t}$ has only one independent derivative; the repeated differentiation of $e^{\zeta t}$ yields the same form, that is, $e^{\zeta t}$. Similarly, the repeated differentiation of t^r yields only r independent derivatives. The particular solution to such an input can be expressed as a linear combination of the input and its independent derivatives. Consider, for example, the input $f(t) = at^2 + bt + c$. The successive derivatives of this input are $2at + b$ and $2a$. In this case, the input has only two independent derivatives. Therefore the particular solution can be assumed to be a linear combination of $f(t)$ and its two derivatives. The suitable form for $y_p(t)$ in this case is therefore

$$y_p(t) = \beta_2t^2 + \beta_1t + \beta_0$$

The undetermined coefficients β_0, β_1, and β_2 are determined by substituting this expression for $y_p(t)$ in Equation 1.11 and then equating coefficients of similar terms on both sides of the resulting expression.

Although this method can be used only for inputs with a finite number of derivatives, this class of inputs includes a wide variety of the most commonly encountered signals in practice. Table 1.1 shows a variety of such inputs and the form of the particular solution corresponding to each input. We shall demonstrate this procedure with an example.

Note: By definition, $y_p(t)$ cannot have any characteristic mode terms. If any term $p(t)$ shown in the right-hand column for the

---

[1]To prove this fact, assume that $y_1(t), y_2(t), \ldots, y_n(t)$ are all solutions of Equation 1.5. Then

$$\begin{aligned} Q(D)y_1(t) &= 0 \\ Q(D)y_2(t) &= 0 \\ \cdots\cdots \quad \cdots & \quad \cdots\cdots \\ Q(D)y_n(t) &= 0 \end{aligned}$$

Multiplying these equations by $c_1, c_2, \ldots, c_n$, respectively, and adding them together yields

$$Q(D)\left[c_1y_1(t) + c_2y_2(t) + \cdots + c_ny_n(t)\right] = 0$$

This result shows that $c_1y_1(t) + c_2y_2(t) + \cdots + c_ny_n(t)$ is also a solution of the homogeneous Equation 1.5

[2]The term *eigenvalue* is German for characteristic value.

TABLE 1.1

Input $f(t)$	Forced Response
1. $e^{\zeta t}$ $\zeta \neq \lambda_i$ $(i = 1, 2, \cdots, n)$	$\beta e^{\zeta t}$
2. $e^{\zeta t}$ $\zeta = \lambda_i$	$\beta t e^{\zeta t}$
3. k (a constant)	β (a constant)
4. $\cos(\omega t + \theta)$	$\beta \cos(\omega t + \phi)$
5. $(t^r + \alpha_{r-1}t^{r-1} + \cdots + \alpha_1 t + \alpha_0)\, e^{\zeta t}$	$(\beta_r t^r + \beta_{r-1}t^{r-1} + \cdots + \beta_1 t + \beta_0) e^{\zeta t}$

particular solution is also a characteristic mode, the correct form of the forced response must be modified to $t^i p(t)$, where i is the smallest possible integer that can be used and still can prevent $t^i p(t)$ from having a characteristic mode term. For example, when the input is $e^{\zeta t}$, the forced response (right-hand column) has the form $\beta e^{\zeta t}$. But if $e^{\zeta t}$ happens to be a characteristic mode, the correct form of the particular solution is $\beta t e^{\zeta t}$ (see Pair 2). If $t e^{\zeta t}$ also happens to be a characteristic mode, the correct form of the particular solution is $\beta t^2 e^{\zeta t}$, and so on.

EXAMPLE 1.1:

Solve the differential equation

$$\left(D^2 + 3D + 2\right) y(t) = Df(t) \tag{1.12}$$

if the input

$$f(t) = t^2 + 5t + 3$$

and the initial conditions are $y(0^+) = 2$ and $\dot{y}(0^+) = 3$.

The characteristic polynomial is

$$\lambda^2 + 3\lambda + 2 = (\lambda + 1)(\lambda + 2)$$

Therefore the characteristic modes are e^{-t} and e^{-2t}. The complementary solution is a linear combination of these modes, so that

$$y_c(t) = c_1 e^{-t} + c_2 e^{-2t} \qquad t \geq 0$$

Here the arbitrary constants c_1 and c_2 must be determined from the given initial conditions.

The particular solution to the input $t^2 + 5t + 3$ is found from Table 1.1 (Pair 5 with $\zeta = 0$) to be

$$y_p(t) = \beta_2 t^2 + \beta_1 t + \beta_0$$

Moreover, $y_p(t)$ satisfies Equation 1.11, that is,

$$\left(D^2 + 3D + 2\right) y_p(t) = Df(t) \tag{1.13}$$

Now

$$\begin{aligned} Dy_p(t) &= \frac{d}{dt}\left(\beta_2 t^2 + \beta_1 t + \beta_0\right) = 2\beta_2 t + \beta_1 \\ D^2 y_p(t) &= \frac{d^2}{dt^2}\left(\beta_2 t^2 + \beta_1 t + \beta_0\right) = 2\beta_2 \end{aligned}$$

and

$$Df(t) = \frac{d}{dt}\left[t^2 + 5t + 3\right] = 2t + 5$$

Substituting these results in Equation 1.13 yields

$$2\beta_2 + 3(2\beta_2 t + \beta_1) + 2(\beta_2 t^2 + \beta_1 t + \beta_0) = 2t + 5$$

or

$$2\beta_2 t^2 + (2\beta_1 + 6\beta_2)t + (2\beta_0 + 3\beta_1 + 2\beta_2) = 2t + 5$$

Equating coefficients of similar powers on both sides of this expression yields

$$\begin{aligned} 2\beta_2 &= 0 \\ 2\beta_1 + 6\beta_2 &= 2 \\ 2\beta_0 + 3\beta_1 + 2\beta_2 &= 5 \end{aligned}$$

Solving these three equations for their unknowns, we obtain $\beta_0 = 1$, $\beta_1 = 1$, and $\beta_2 = 0$. Therefore,

$$y_p(t) = t + 1 \qquad t > 0$$

The total solution $y(t)$ is the sum of the complementary and particular solutions. Therefore,

$$\begin{aligned} y(t) &= y_c(t) + y_p(t) \\ &= c_1 e^{-t} + c_2 e^{-2t} + t + 1 \qquad t > 0 \end{aligned}$$

so that

$$\dot{y}(t) = -c_1 e^{-t} - 2c_2 e^{-2t} + 1$$

Setting $t = 0$ and substituting the given initial conditions $y(0) = 2$ and $\dot{y}(0) = 3$ in these equations, we have

$$\begin{aligned} 2 &= c_1 + c_2 + 1 \\ 3 &= -c_1 - 2c_2 + 1 \end{aligned}$$

The solution to these two simultaneous equations is $c_1 = 4$ and $c_2 = -3$. Therefore,

$$y(t) = 4e^{-t} - 3e^{-2t} + t + 1 \qquad t \geq 0$$

The Exponential Input $e^{\zeta t}$

The exponential signal is the most important signal in the study of LTI systems. Interestingly, the particular solution for an exponential input signal turns out to be very simple. From Table 1.1 we see that the particular solution for the input $e^{\zeta t}$ has the form $\beta e^{\zeta t}$. We now show that $\beta = Q(\zeta)/P(\zeta)$ [3]. To determine

[3] This is true only if ζ is not a characteristic root.

the constant β, we substitute $y_p(t) = \beta e^{\zeta t}$ in Equation 1.11, which gives us

$$Q(D)\left[\beta e^{\zeta t}\right] = P(D)e^{\zeta t} \tag{1.14a}$$

Now observe that

$$\begin{aligned} De^{\zeta t} &= \frac{d}{dt}\left(e^{\zeta t}\right) = \zeta e^{\zeta t} \\ D^2 e^{\zeta t} &= \frac{d^2}{dt^2}\left(e^{\zeta t}\right) = \zeta^2 e^{\zeta t} \\ \cdots\cdots &\;\cdots\;\cdots\cdots \\ D^r e^{\zeta t} &= \zeta^r e^{\zeta t} \end{aligned}$$

Consequently,

$$Q(D)e^{\zeta t} = Q(\zeta)e^{\zeta t} \quad \text{and} \quad P(D)e^{\zeta t} = P(\zeta)e^{\zeta t}$$

Therefore, Equation 1.14a becomes

$$\beta Q(\zeta)e^{\zeta t} = P(\zeta)e^{\zeta t} \tag{1.14b}$$

and

$$\beta = \frac{P(\zeta)}{Q(\zeta)}$$

Thus, for the input $f(t) = e^{\zeta t}$, the particular solution is given by

$$y_p(t) = H(\zeta)e^{\zeta t} \qquad t > 0 \tag{1.15a}$$

where

$$H(\zeta) = \frac{P(\zeta)}{Q(\zeta)} \tag{1.15b}$$

This is an interesting and significant result. It states that for an exponential input $e^{\zeta t}$ the particular solution $y_p(t)$ is the same exponential multiplied by $H(\zeta) = P(\zeta)/Q(\zeta)$. The total solution $y(t)$ to an exponential input $e^{\zeta t}$ is then given by

$$y(t) = \sum_{j=1}^{n} c_j e^{\lambda_j t} + H(\zeta)e^{\zeta t}$$

where the arbitrary constants $c_1, c_2, \ldots, c_n$ are determined from auxiliary conditions.

Recall that the exponential signal includes a large variety of signals, such as a constant ($\zeta = 0$), a sinusoid ($\zeta = \pm j\omega$), and an exponentially growing or decaying sinusoid ($\zeta = \sigma \pm j\omega$). Let us consider the forced response for some of these cases.

The Constant Input f(t) = C

Because $C = Ce^{0t}$, the constant input is a special case of the exponential input $Ce^{\zeta t}$ with $\zeta = 0$. The particular solution to this input is then given by

$$\begin{aligned} y_p(t) &= CH(\zeta)e^{\zeta t} \qquad \text{with} \quad \zeta = 0 \\ &= CH(0) \end{aligned} \tag{1.16}$$

The Complex Exponential Input $e^{j\omega t}$

Here $\zeta = j\omega$, and

$$y_p(t) = H(j\omega)e^{j\omega t} \tag{1.17}$$

The Sinusoidal Input f(t) = cos ω_0t

We know that the particular solution for the input $e^{\pm j\omega t}$ is $H(\pm j\omega)e^{\pm j\omega t}$. Since $\cos \omega t = (e^{j\omega t} + e^{-j\omega t})/2$, the particular solution to $\cos \omega t$ is

$$y_p(t) = \frac{1}{2}\left[H(j\omega)e^{j\omega t} + H(-j\omega)e^{-j\omega t}\right]$$

Because the two terms on the right-hand side are conjugates,

$$y_p(t) = \text{Re}\left[H(j\omega)e^{j\omega t}\right]$$

But

$$H(j\omega) = |H(j\omega)|e^{j\angle H(j\omega)}$$

so that

$$\begin{aligned} y_p(t) &= \text{Re}\left\{|H(j\omega)|e^{j[\omega t + \angle H(j\omega)]}\right\} \\ &= |H(j\omega)| \cos\left[\omega t + \angle H(j\omega)\right] \end{aligned} \tag{1.18}$$

This result can be generalized for the input $f(t) = \cos(\omega t + \theta)$. The particular solution in this case is

$$y_p(t) = |H(j\omega)| \cos\left[\omega t + \theta + \angle H(j\omega)\right] \tag{1.19}$$

EXAMPLE 1.2:

Solve Equation 1.12 for the following inputs:
(a) $10e^{-3t}$ **(b)** 5 **(c)** e^{-2t} **(d)** $10\cos(3t + 30°)$.
The initial conditions are $y(0^+) = 2,\ \dot{y}(0^+) = 3$.

The complementary solution for this case is already found in Example 1.1 as

$$y_c(t) = c_1 e^{-t} + c_2 e^{-2t} \qquad t \geq 0$$

For the exponential input $f(t) = e^{\zeta t}$, the particular solution, as found in Equation 1.15 is $H(\zeta)e^{\zeta t}$, where

$$H(\zeta) = \frac{P(\zeta)}{Q(\zeta)} = \frac{\zeta}{\zeta^2 + 3\zeta + 2}$$

(a) For input $f(t) = 10e^{-3t}$, $\zeta = -3$, and

$$\begin{aligned} y_p(t) &= 10H(-3)e^{-3t} \\ &= 10\left[\frac{-3}{(-3)^2 + 3(-3) + 2}\right]e^{-3t} \\ &= -15e^{-3t} \qquad t > 0 \end{aligned}$$

The total solution (the sum of the complementary and particular solutions) is

$$y(t) = c_1 e^{-t} + c_2 e^{-2t} - 15e^{-3t} \qquad t \geq 0$$

and

$$\dot{y}(t) = -c_1e^{-t} - 2c_2e^{-2t} + 45e^{-3t} \qquad t \geq 0$$

The initial conditions are $y(0^+) = 2$ and $\dot{y}(0^+) = 3$. Setting $t = 0$ in the above equations and substituting the initial conditions yields

$$c_1 + c_2 - 15 = 2 \qquad \text{and} \qquad -c_1 - 2c_2 + 45 = 3$$

Solution of these equations yields $c_1 = -8$ and $c_2 = 25$. Therefore,

$$y(t) = -8e^{-t} + 25e^{-2t} - 15e^{-3t} \qquad t \geq 0$$

(b) For input $f(t) = 5 = 5e^{0t}$, $\zeta = 0$, and

$$y_p(t) = 5H(0) = 0 \qquad t > 0$$

The complete solution is $y(t) = y_c(t) + y_p(t) = c_1e^{-t} + c_2e^{-2t}$. We then substitute the initial conditions to determine c_1 and c_2 as explained in Part a.

(c) Here $\zeta = -2$, which is also a characteristic root. Hence (see Pair 2, Table 1.1, or the comment at the bottom of the table),

$$y_p(t) = \beta te^{-2t}$$

To find β, we substitute $y_p(t)$ in Equation 1.11, giving us

$$\left(D^2 + 3D + 2\right) y_p(t) = Df(t)$$

or

$$\left(D^2 + 3D + 2\right)\left[\beta te^{-2t}\right] = De^{-2t}$$

But

$$\begin{aligned} D\left[\beta te^{-2t}\right] &= \beta(1 - 2t)e^{-2t} \\ D^2\left[\beta te^{-2t}\right] &= 4\beta(t - 1)e^{-2t} \\ De^{-2t} &= -2e^{-2t} \end{aligned}$$

Consequently,

$$\beta(4t - 4 + 3 - 6t + 2t)e^{-2t} = -2e^{-2t}$$

or

$$-\beta e^{-2t} = -2e^{-2t}$$

This means that $\beta = 2$, so that

$$y_p(t) = 2te^{-2t}$$

The complete solution is $y(t) = y_c(t) + y_p(t) = c_1e^{-t} + c_2e^{-2t} + 2te^{-2t}$. We then substitute the initial conditions to determine c_1 and c_2 as explained in Part a.

(d) For the input $f(t) = 10\cos(3t + 30°)$, the particular solution [see Equation 1.19] is

$$y_p(t) = 10|H(j3)|\cos\left[3t + 30° + \angle H(j3)\right]$$

where

$$\begin{aligned} H(j3) &= \frac{P(j3)}{Q(j3)} = \frac{j3}{(j3)^2 + 3(j3) + 2} \\ &= \frac{j3}{-7 + j9} = \frac{27 - j21}{130} = 0.263e^{-j37.9°} \end{aligned}$$

Therefore,

$$|H(j3)| = 0.263, \qquad \angle H(j3) = -37.9°$$

and

$$\begin{aligned} y_p(t) &= 10(0.263)\cos(3t + 30° - 37.9°) \\ &= 2.63\cos(3t - 7.9°) \end{aligned}$$

The complete solution is $y(t) = y_c(t) + y_p(t) = c_1e^{-t} + c_2e^{-2t} + 2.63\cos(3t - 7.9°)$. We then substitute the initial conditions to determine c_1 and c_2 as explained in Part a.

1.1.2 Method of Convolution

In this method, the input $f(t)$ is expressed as a sum of impulses. The solution is then obtained as a sum of the solutions to all the impulse components. The method exploits the superposition property of the linear differential equations. From the sampling (or sifting) property of the impulse function, we have

$$f(t) = \int_0^t f(x)\delta(t - x)\,dx \qquad t \geq 0 \tag{1.20}$$

The right-hand side expresses $f(t)$ as a sum (integral) of impulse components. Let the solution of Equation 1.4 be $y(t) = h(t)$ when $f(t) = \delta(t)$ and all the initial conditions are zero. Then use of the linearity property yields the solution of Equation 1.4 to input $f(t)$ as

$$y(t) = \int_0^t f(x)h(t - x)\,dx \tag{1.21}$$

For this solution to be general, we must add a complementary solution. Thus, the general solution is given by

$$y(t) = \sum_{j=1}^{n} c_je^{\lambda_jt} + \int_0^t f(x)h(t - x)\,dx \tag{1.22}$$

where the lower limit 0 is understood to be 0^- in order to ensure that impulses, if any, in the input $f(t)$ at the origin are accounted for. side of (1.22) is well known in the literature as the *convolution integral.* The function $h(t)$ appearing in the integral is the solution of Equation 1.4 for the impulsive input $[f(t) = \delta(t)]$. It can be shown that [1]

$$h(t) = P(D)[y_o(t)u(t)] \tag{1.23}$$

where $y_o(t)$ is a linear combination of the characteristic modes subject to initial conditions

$$\begin{aligned} &y_o^{(n-1)}(0) = 1 \\ &y_o(0) = y_o^{(1)}(0) = \cdots = y_o^{(n-2)}(0) = 0 \end{aligned} \tag{1.24}$$

The function $u(t)$ appearing on the right-hand side of Equation 1.23 represents the unit step function, which is unity for $t \geq 0$ and is 0 for $t < 0$.

The right-hand side of Equation 1.23 is a linear combination of the derivatives of $y_o(t)u(t)$. Evaluating these derivatives is clumsy and inconvenient because of the presence of $u(t)$. The derivatives will generate an impulse and its derivatives at the origin [recall that $\frac{d}{dt}u(t) = \delta(t)$]. Fortunately when $m \leq n$ in Equation 1.4, the solution simplifies to

$$h(t) = b_n\delta(t) + [P(D)y_o(t)]u(t) \qquad (1.25)$$

EXAMPLE 1.3:

Solve Example 1.2, Part a using method of convolution.

We first determine $h(t)$. The characteristic modes for this case, as found in Example 1.1, are e^{-t} and e^{-2t}. Since $y_o(t)$ is a linear combination of the characteristic modes

$$y_o(t) = K_1e^{-t} + K_2e^{-2t} \qquad t \geq 0$$

Therefore,

$$\dot{y}_o(t) = -K_1e^{-t} - 2K_2e^{-2t} \qquad t \geq 0$$

The initial conditions according to Equation 1.24 are $\dot{y}_o(0) = 1$ and $y_o(0) = 0$. Setting $t = 0$ in the above equations and using the initial conditions, we obtain

$$K_1 + K_2 = 0 \qquad \text{and} \qquad -K_1 - 2K_2 = 1$$

Solution of these equations yields $K_1 = 1$ and $K_2 = -1$. Therefore,

$$y_o(t) = e^{-t} - e^{-2t}$$

Also in this case the polynomial $P(D) = D$ is of the first-order, and $b_2 = 0$. Therefore, from Equation 1.25

$$\begin{aligned} h(t) &= [P(D)y_o(t)]u(t) = [Dy_o(t)]u(t) \\ &= \left[\frac{d}{dt}(e^{-t} - e^{-2t})\right]u(t) \\ &= (-e^{-t} + 2e^{-2t})u(t) \end{aligned}$$

and

$$\begin{aligned} \int_0^t f(x)h(t-x)\,dx &= \int_0^t 10e^{-3x}[-e^{-(t-x)} \\ &\quad + 2e^{-2(t-x)}]\,dx \\ &= -5e^{-t} + 20e^{-2t} - 15e^{-3t} \end{aligned}$$

The total solution is obtained by adding the complementary solution $y_c(t) = c_1e^{-t} + c_2e^{-2t}$ to this component. Therefore,

$$y(t) = c_1e^{-t} + c_2e^{-2t} - 5e^{-t} + 20e^{-2t} - 15e^{-3t}$$

Setting the conditions $y(0^+) = 2$ and $y(0^+) = 3$ in this equation (and its derivative), we obtain $c_1 = -3$, $c_2 = 5$ so that

$$y(t) = -8e^{-t} + 25e^{-2t} - 15e^{-3t} \qquad t \geq 0$$

which is identical to the solution found by the classical method.

Assessment of the Convolution Method

The convolution method is more laborious compared to the classical method. However, in system analysis, its advantages outweigh the extra work. The classical method has a serious drawback because it yields the total response, which cannot be separated into components arising from the internal conditions and the external input. In the study of systems it is important to be able to express the system response to an input $f(t)$ as an explicit function of $f(t)$. This is not possible in the classical method. Moreover, the classical method is restricted to a certain class of inputs; it cannot be applied to any input [4].

If we must solve a particular linear differential equation or find a response of a particular LTI system, the classical method may be the best. In the theoretical study of linear systems, however, it is practically useless. General discussion of differential equations can be found in numerous texts on the subject [3].

1.2 Difference Equations

The development of difference equations is parallel to that of differential equations. We consider here only linear difference equations with constant coefficients. An nth-order difference equation can be expressed in two different forms; the first form uses delay terms such as $y[k-1]$, $y[k-2]$, $f[k-1]$, $f[k-2]$, ..., etc., and the alternative form uses advance terms such as $y[k+1], y[k+2]$, ..., etc. Both forms are useful. We start here with a general nth-order difference equation, using advance operator form

$$\begin{aligned} y[k+n] + a_{n-1}y[k+n-1] + \cdots + a_1y[k+1] + a_0y[k] \\ = b_mf[k+m] + b_{m-1}f[k+m-1] + \cdots \\ + b_1f[k+1] + b_0f[k] \end{aligned} \qquad (1.26)$$

Causality Condition

The left-hand side of Equation 1.26 consists of values of $y[k]$ at instants $k+n$, $k+n-1$, $k+n-2$, and so on. The right-hand side of Equation 1.26 consists of the input at instants $k+m$, $k+m-1$, $k+m-2$, and so on. For a causal equation, the solution cannot depend on future input values. This shows that when the equation is in the advance operator form of the Equation 1.26, causality requires $m \leq n$. For a general causal case, $m = n$, and Equation 1.26 becomes

$$y[k+n] + a_{n-1}y[k+n-1] + \cdots + a_1y[k+1] + a_0y[k]$$

[4]Another minor problem is that because the classical method yields total response, the auxiliary conditions must be on the total response, which exists only for $t \geq 0^+$. In practice we are most likely to know the conditions at $t = 0^-$ (before the input is applied). Therefore, we need to derive a new set of auxiliary conditions at $t = 0^+$ from the known conditions at $t = 0^-$. The convolution method can handle both kinds of initial conditions. If the conditions are given at $t = 0^-$, we apply these conditions only to $y_c(t)$ because by its definition the convolution integral is 0 at $t = 0^-$.

$$= b_n f[k+n] + b_{n-1} f[k+n-1] + \cdots + b_1 f[k+1] + b_0 f[k] \qquad (1.27a)$$

where some of the coefficients on both sides can be zero. However, the coefficient of $y[k+n]$ is normalized to unity. Equation 1.27a is valid for all values of k. Therefore, the equation is still valid if we replace k by $k-n$ throughout the equation. This yields the alternative form (the delay operator form) of Equation 1.27a

$$\begin{aligned} y[k] + a_{n-1} y[k-1] + \cdots + a_1 y[k-n+1] + a_0 y[k-n] \\ = b_n f[k] + b_{n-1} f[k-1] + \cdots \\ + b_1 f[k-n+1] + b_0 f[k-n] \qquad (1.27b) \end{aligned}$$

We designate the form of Equation 1.27a the *advance operator form*, and the form of Equation 1.27b the *delay operator form*.

1.2.1 Initial Conditions and Iterative Solution

Equation 1.27b can be expressed as

$$\begin{aligned} y[k] = -a_{n-1} y[k-1] - a_{n-2} y[k-2] - \cdots \\ - a_0 y[k-n] + b_n f[k] + b_{n-1} f[k-1] + \cdots \\ + b_0 f[k-n] \qquad (1.27c) \end{aligned}$$

This equation shows that $y[k]$, the solution at the kth instant, is computed from $2n+1$ pieces of information. These are the past n values of $y[k]$: $y[k-1]$, $y[k-2]$, ..., $y[k-n]$ and the present and past n values of the input: $f[k]$, $f[k-1]$, $f[k-2]$, ..., $f[k-n]$. If the input $f[k]$ is known for $k = 0, 1, 2, \ldots$, then the values of $y[k]$ for $k = 0, 1, 2, \ldots$ can be computed from the $2n$ initial conditions $y[-1]$, $y[-2]$, ..., $y[-n]$ and $f[-1]$, $f[-2]$, ..., $f[-n]$. If the input is causal, that is, if $f[k] = 0$ for $k < 0$, then $f[-1] = f[-2] = \ldots = f[-n] = 0$, and we need only n initial conditions $y[-1]$, $y[-2]$, ..., $y[-n]$. This allows us to compute iteratively or recursively the values $y[0]$, $y[1]$, $y[2]$, $y[3]$, ..., and so on.[5] For instance, to find $y[0]$ we set $k = 0$ in Equation 1.27c. The left-hand side is $y[0]$, and the right-hand side contains terms $y[-1]$, $y[-2]$, ..., $y[-n]$, and the inputs $f[0]$, $f[-1]$, $f[-2]$, ..., $f[-n]$. Therefore, to begin with, we must know the n initial conditions $y[-1]$, $y[-2]$, ..., $y[-n]$. Knowing these conditions and the input $f[k]$, we can iteratively find the response $y[0]$, $y[1]$, $y[2]$, ..., and so on. The following example demonstrates this procedure. This method basically reflects the manner in which a computer would solve a difference equation, given the input and initial conditions.

[5]For this reason Equation 1.27 is called a *recursive difference equation*. However, in Equation 1.27 if $a_0 = a_1 = a_2 = \cdots = a_{n-1} = 0$, then it follows from Equation 1.27c that determination of the present value of $y[k]$ does not require the past values $y[k-1]$, $y[k-2]$, ..., etc. For this reason when $a_i = 0$, $(i = 0, 1, \ldots, n-1)$, the difference Equation 1.27 is *nonrecursive*. This classification is important in designing and realizing digital filters. In this discussion, however, this classification is not important. The analysis techniques developed here apply to general recursive and nonrecursive equations. Observe that a nonrecursive equation is a special case of recursive equation with $a_0 = a_1 = \ldots = a_{n-1} = 0$.

EXAMPLE 1.4:

Solve iteratively

$$y[k] - 0.5y[k-1] = f[k] \qquad (1.28a)$$

with initial condition $y[-1] = 16$ and the input $f[k] = k^2$ (starting at $k = 0$). This equation can be expressed as

$$y[k] = 0.5y[k-1] + f[k] \qquad (1.28b)$$

If we set $k = 0$ in this equation, we obtain

$$\begin{aligned} y[0] &= 0.5y[-1] + f[0] \\ &= 0.5(16) + 0 = 8 \end{aligned}$$

Now, setting $k = 1$ in Equation 1.28b and using the value $y[0] = 8$ (computed in the first step) and $f[1] = (1)^2 = 1$, we obtain

$$y[1] = 0.5(8) + (1)^2 = 5$$

Next, setting $k = 2$ in Equation 1.28b and using the value $y[1] = 5$ (computed in the previous step) and $f[2] = (2)^2$, we obtain

$$y[2] = 0.5(5) + (2)^2 = 6.5$$

Continuing in this way iteratively, we obtain

$$\begin{aligned} y[3] &= 0.5(6.5) + (3)^2 = 12.25 \\ y[4] &= 0.5(12.25) + (4)^2 = 22.125 \\ &\cdots\cdots\cdots\cdots \end{aligned}$$

This iterative solution procedure is available only for difference equations; it cannot be applied to differential equations. Despite the many uses of this method, a closed-form solution of a difference equation is far more useful in the study of system behavior and its dependence on the input and the various system parameters. For this reason we shall develop a systematic procedure to obtain a closed-form solution of Equation 1.27.

Operational Notation

In difference equations it is convenient to use operational notation similar to that used in differential equations for the sake of compactness and convenience. For differential equations, we use the operator D to denote the operation of differentiation. For difference equations, we use the operator E to denote the operation for advancing the sequence by one time interval. Thus,

$$\begin{aligned} Ef[k] &\equiv f[k+1] \\ E^2 f[k] &\equiv f[k+2] \\ \cdots\cdots & \cdots \cdots\cdots \\ E^n f[k] &\equiv f[k+n] \qquad (1.29) \end{aligned}$$

A general nth-order difference Equation 1.27a can be expressed as

$$\begin{aligned} (E^n + a_{n-1} E^{n-1} + \cdots + a_1 E + a_0) y[k] \\ = (b_n E^n + b_{n-1} E^{n-1} + \cdots + b_1 E + b_0) f[k] \qquad (1.30a) \end{aligned}$$

or

$$Q[E]y[k] = P[E]f[k] \quad (1.30b)$$

where $Q[E]$ and $P[E]$ are nth-order polynomial operators, respectively,

$$Q[E] = E^n + a_{n-1}E^{n-1} + \cdots + a_1 E + a_0 \quad (1.31a)$$

$$P[E] = b_n E^n + b_{n-1}E^{n-1} + \cdots + b_1 E + b_0 \quad (1.31b)$$

1.2.2 Classical Solution

Following the discussion of differential equations, we can show that if $y_p[k]$ is a solution of Equation 1.27 or Equation 1.30, that is,

$$Q[E]y_p[k] = P[E]f[k] \quad (1.32)$$

then $y_p[k] + y_c[k]$ is also a solution of Equation 1.30, where $y_c[k]$ is a solution of the homogeneous equation

$$Q[E]y_c[k] = 0 \quad (1.33)$$

As before, we call $y_p[k]$ the particular solution and $y_c[k]$ the complementary solution.

Complementary Solution (The Natural Response)

By definition

$$Q[E]y_c[k] = 0 \quad (1.33a)$$

or

$$(E^n + a_{n-1}E^{n-1} + \cdots + a_1 E + a_0)y_c[k] = 0 \quad (1.33b)$$

or

$$y_c[k+n] + a_{n-1}y_c[k+n-1] + \cdots + a_1 y_c[k+1] + a_0 y_c[k] = 0 \quad (1.33c)$$

We can solve this equation systematically, but even a cursory examination of this equation points to its solution. This equation states that a linear combination of $y_c[k]$ and delayed $y_c[k]$ is zero *not for some values of k, but for all k.* This is possible *if and only if* $y_c[k]$ and delayed $y_c[k]$ have the same form. Only an exponential function γ^k has this property as seen from the equation

$$\gamma^{k-m} = \gamma^{-m}\gamma^k$$

This shows that the delayed γ^k is a constant times γ^k. Therefore, the solution of Equation 1.33 must be of the form

$$y_c[k] = c\gamma^k \quad (1.34)$$

To determine c and γ, we substitute this solution in Equation 1.33. From Equation 1.34, we have

$$\begin{aligned} Ey_c[k] &= y_c[k+1] = c\gamma^{k+1} = (c\gamma)\gamma^k \\ E^2 y_c[k] &= y_c[k+2] = c\gamma^{k+2} = (c\gamma^2)\gamma^k \\ \cdots & \quad \cdots \quad \cdots\cdots\cdots \\ E^n y_c[k] &= y_c[k+n] = c\gamma^{k+n} = (c\gamma^n)\gamma^k \quad (1.35) \end{aligned}$$

Substitution of this in Equation 1.33 yields

$$c(\gamma^n + a_{n-1}\gamma^{n-1} + \cdots + a_1\gamma + a_0)\gamma^k = 0 \quad (1.36)$$

For a nontrivial solution of this equation

$$(\gamma^n + a_{n-1}\gamma^{n-1} + \cdots + a_1\gamma + a_0) = 0 \quad (1.37a)$$

or

$$Q[\gamma] = 0 \quad (1.37b)$$

Our solution $c\gamma^k$ [Equation 1.34] is correct, provided that γ satisfies Equation 1.37. Now, $Q[\gamma]$ is an nth-order polynomial and can be expressed in the factorized form (assuming all distinct roots):

$$(\gamma - \gamma_1)(\gamma - \gamma_2)\cdots(\gamma - \gamma_n) = 0 \quad (1.37c)$$

Clearly γ has n solutions $\gamma_1, \gamma_2, \cdots, \gamma_n$ and, therefore, Equation 1.33 also has n solutions $c_1\gamma_1^k, c_2\gamma_2^k, \cdots, c_n\gamma_n^k$. In such a case we have shown that the general solution is a linear combination of the n solutions. Thus,

$$y_c[k] = c_1\gamma_1^k + c_2\gamma_2^k + \cdots + c_n\gamma_n^k \quad (1.38)$$

where $\gamma_1, \gamma_2, \cdots, \gamma_n$ are the roots of Equation 1.37 and c_1, c_2, ..., c_n are arbitrary constants determined from n auxiliary conditions. The polynomial $Q[\gamma]$ is called the *characteristic polynomial*, and

$$Q[\gamma] = 0 \quad (1.39)$$

is the *characteristic equation*. Moreover, $\gamma_1, \gamma_2, \cdots, \gamma_n$, the roots of the characteristic equation, are called *characteristic roots* or *characteristic values* (also *eigenvalues*). The exponentials $\gamma_i^k (i = 1, 2, \ldots, n)$ are the *characteristic* modes or *natural* modes. A characteristic mode corresponds to each characteristic root, and the complementary solution is a linear combination of the characteristic modes of the system.

Repeated Roots

For repeated roots, the form of characteristic modes is modified. It can be shown by direct substitution that if a root γ repeats r times (root of multiplicity r), the characteristic modes corresponding to this root are $\gamma^k, k\gamma^k, k^2\gamma^k, \ldots, k^{r-1}\gamma^k$. Thus, if the characteristic equation is

$$Q[\gamma] = (\gamma - \gamma_1)^r(\gamma - \gamma_{r+1})(\gamma - \gamma_{r+2})\cdots(\gamma - \gamma_n) \quad (1.40)$$

the complementary solution is

$$y_c[k] = (c_1 + c_2k + c_3k^2 + \cdots + c_rk^{r-1})\gamma_1^k + c_{r+1}\gamma_{r+1}^k + c_{r+2}\gamma_{r+2}^k + \cdots + c_n\gamma_n^k \tag{1.41}$$

Particular Solution

The particular solution $y_p[k]$ is the solution of

$$Q[E]y_p[k] = P[E]f[k] \tag{1.42}$$

We shall find the particular solution using the method of undetermined coefficients, the same method used for differential equations. Table 1.2 lists the inputs and the corresponding forms of solution with undetermined coefficients. These coefficients can be determined by substituting $y_p[k]$ in Equation 1.42 and equating the coefficients of similar terms.

TABLE 1.2

	Input $f[k]$	Forced Response $y_p[k]$
1.	$r^k \quad r \neq \gamma_i \; (i = 1, 2, \cdots, n)$	βr^k
2.	$r^k \quad r = \gamma_i$	$\beta k r^k$
3.	$\cos(\Omega k + \theta)$	$\beta \cos(\Omega k + \phi)$
4.	$\left(\sum_{i=0}^{m} \alpha_i k^i\right) r^k$	$\left(\sum_{i=0}^{m} \beta_i k^i\right) r^k$

Note: By definition, $y_p[k]$ cannot have any characteristic mode terms. If any term $p[k]$ shown in the right-hand column for the particular solution should also be a characteristic mode, the correct form of the particular solution must be modified to $k^i p[k]$, where i is the smallest integer that will prevent $k^i p[k]$ from having a characteristic mode term. For example, when the input is r^k, the particular solution in the right-hand column is of the form cr^k. But if r^k happens to be a natural mode, the correct form of the particular solution is $\beta k r^k$ (see Pair 2).

EXAMPLE 1.5:

Solve

$$(E^2 - 5E + 6)y[k] = (E - 5)f[k] \tag{1.43}$$

if the input $f[k] = (3k + 5)u[k]$ and the auxiliary conditions are $y[0] = 4$, $y[1] = 13$.

The characteristic equation is

$$\gamma^2 - 5\gamma + 6 = (\gamma - 2)(\gamma - 3) = 0$$

Therefore, the complementary solution is

$$y_c[k] = c_1(2)^k + c_2(3)^k$$

To find the form of $y_p[k]$ we use Table 1.2, Pair 4 with $r = 1$, $m = 1$. This yields

$$y_p[k] = \beta_1 k + \beta_0$$

Therefore,

$$y_p[k+1] = \beta_1(k+1) + \beta_0 = \beta_1 k + \beta_1 + \beta_0$$

$$y_p[k+2] = \beta_1(k+2) + \beta_0 = \beta_1 k + 2\beta_1 + \beta_0$$

Also,

$$f[k] = 3k + 5$$

and

$$f[k+1] = 3(k+1) + 5 = 3k + 8$$

Substitution of the above results in Equation 1.43 yields

$$\beta_1 k + 2\beta_1 + \beta_0 - 5(\beta_1 k + \beta_1 + \beta_0) + 6(\beta_1 k + \beta_0) = 3k + 8 - 5(3k + 5)$$

or

$$2\beta_1 k - 3\beta_1 + 2\beta_0 = -12k - 17$$

Comparison of similar terms on two sides yields

$$\left.\begin{aligned} 2\beta_1 &= -12 \\ -3\beta_1 + 2\beta_0 &= -17 \end{aligned}\right\} \Longrightarrow \begin{aligned} \beta_1 &= -6 \\ \beta_2 &= -\tfrac{35}{2} \end{aligned}$$

This means

$$y_p[k] = -6k - \frac{35}{2}$$

The total response is

$$\begin{aligned} y[k] &= y_c[k] + y_p[k] \\ &= c_1(2)^k + c_2(3)^k - 6k - \frac{35}{2} \qquad k \geq 0 \end{aligned} \tag{1.44}$$

To determine arbitrary constants c_1 and c_2 we set $k = 0$ and 1 and substitute the auxiliary conditions $y[0] = 4$, $y[1] = 13$ to obtain

$$\left.\begin{aligned} 4 &= c_1 + c_2 - \tfrac{35}{2} \\ 13 &= 2c_1 + 3c_2 - \tfrac{47}{2} \end{aligned}\right\} \Longrightarrow \begin{aligned} c_1 &= 28 \\ c_2 &= \tfrac{-13}{2} \end{aligned}$$

Therefore,

$$y_c[k] = 28(2)^k - \tfrac{13}{2}(3)^k \tag{1.45}$$

and

$$y[k] = \underbrace{28(2)^k - \frac{13}{2}(3)^k}_{y_c[k]} \underbrace{- 6k - \frac{35}{2}}_{y_p[k]} \tag{1.46}$$

A Comment on Auxiliary Conditions

This method requires auxiliary conditions $y[0], y[1], \ldots, y[n-1]$ because the total solution is valid only for $k \geq 0$. But if we are given the initial conditions $y[-1], y[-2], \ldots, y[-n]$, we can derive the conditions $y[0], y[1], \ldots, y[n-1]$ using the iterative procedure discussed earlier.

Exponential Input

As in the case of differential equations, we can show that for the equation

$$Q[E]y[k] = P[E]f[k] \tag{1.47}$$

the particular solution for the exponential input $f[k] = r^k$ is given by

$$y_p[k] = H[r]r^k \qquad r \neq \gamma_i \tag{1.48}$$

where

$$H[r] = \frac{P[r]}{Q[r]} \tag{1.49}$$

The proof follows from the fact that if the input $f[k] = r^k$, then from Table 1.2 (Pair 4), $y_p[k] = \beta r^k$. Therefore,

$$E^i f[k] = f[k+i] = r^{k+i} = r^i r^k \text{ and } P[E]f[k] = P[r]r^k$$

$$E^j y_p[k] = \beta r^{k+j} = \beta r^j r^k \text{ and } Q[E]y[k] = \beta Q[r]r^k$$

so that Equation 1.47 reduces to

$$\beta Q[r]r^k = P[r]r^k$$

which yields $\beta = P[r]/Q[r] = H[r]$.

This result is valid only if r is not a characteristic root. If r is a characteristic root, the particular solution is $\beta k r^k$ where β is determined by substituting $y_p[k]$ in Equation 1.47 and equating coefficients of similar terms on the two sides. Observe that the exponential r^k includes a wide variety of signals such as a constant C, a sinusoid $\cos(\Omega k + \theta)$, and an exponentially growing or decaying sinusoid $|\gamma|^k \cos(\Omega k + \theta)$.

A Constant Input $f(k) = C$

This is a special case of exponential Cr^k with $r = 1$. Therefore, from Equation 1.48 we have

$$y_p[k] = C\frac{P[1]}{Q[1]}(1)^k = CH[1] \tag{1.50}$$

A Sinusoidal Input

The input $e^{j\Omega k}$ is an exponential r^k with $r = e^{j\Omega}$. Hence,

$$y_p[k] = H[e^{j\Omega}]e^{j\Omega k} = \frac{P[e^{j\Omega}]}{Q[e^{j\Omega}]}e^{j\Omega k}$$

Similarly for the input $e^{-j\Omega k}$

$$y_p[k] = H[e^{-j\Omega}]e^{-j\Omega k}$$

Consequently, if the input

$$f[k] = \cos \Omega k = \frac{1}{2}(e^{j\Omega k} + e^{-j\Omega k})$$

$$y_p[k] = \frac{1}{2}\left\{H[e^{j\Omega}]e^{j\Omega k} + H[e^{-j\Omega}]e^{-j\Omega k}\right\}$$

Since the two terms on the right-hand side are conjugates

$$y_p[k] = \text{Re}\left\{H[e^{j\Omega}]e^{j\Omega k}\right\}$$

If

$$H[e^{j\Omega}] = |H[e^{j\Omega}]|e^{j\angle H[e^{j\Omega}]}$$

then

$$\begin{aligned} y_p[k] &= \text{Re}\left\{\left|H[e^{j\Omega}]\right| e^{j(\Omega k + \angle H[e^{j\Omega}])}\right\} \\ &= |H[e^{j\Omega}]| \cos\left(\Omega k + \angle H[e^{j\Omega}]\right) \end{aligned} \tag{1.51}$$

Using a similar argument, we can show that for the input

$$f[k] = \cos(\Omega k + \theta)$$

$$y_p[k] = |H[e^{j\Omega}]| \cos\left(\Omega k + \theta + \angle H[e^{j\Omega}]\right) \tag{1.52}$$

EXAMPLE 1.6:

Solve

$$(E^2 - 3E + 2)y[k] = (E + 2)f[k]$$

for $f[k] = (3)^k u[k]$ and the auxiliary conditions $y[0] = 2$, $y[1] = 1$.

In this case

$$H[r] = \frac{P[r]}{Q[r]} = \frac{r+2}{r^2 - 3r + 2}$$

and the particular solution to input $(3)^k u[k]$ is $H[3](3)^k$; that is,

$$y_p[k] = \frac{3+2}{(3)^2 - 3(3) + 2}(3)^k = \frac{5}{2}(3)^k$$

The characteristic polynomial is $(\gamma^2 - 3\gamma + 2) = (\gamma - 1)(\gamma - 2)$. The characteristic roots are 1 and 2. Hence, the complementary solution is $y_c[k] = c_1 + c_2(2)^k$ and the total solution is

$$y[k] = c_1(1)^k + c_2(2)^k + \frac{5}{2}(3)^k$$

Setting $k = 0$ and 1 in this equation and substituting auxiliary conditions yields

$$2 = c_1 + c_2 + \frac{5}{2} \qquad \text{and} \qquad 1 = c_1 + 2c_2 + \frac{15}{2}$$

Solution of these two simultaneous equations yields $c_1 = 5.5$, $c_2 = -5$. Therefore,

$$y[k] = 5.5 - 6(2)^k + \frac{5}{2}(3)^k \qquad k \geq 0$$

1.2.3 Method of Convolution

In this method, the input $f[k]$ is expressed as a sum of impulses. The solution is then obtained as a sum of the solutions to all the impulse components. The method exploits the superposition property of the linear difference equations. A discrete-time unit impulse function $\delta[k]$ is defined as

$$\delta[k] = \begin{cases} 1 & k = 0 \\ 0 & k \neq 0 \end{cases} \tag{1.53}$$

Hence, an arbitrary signal $f[k]$ can be expressed in terms of impulse and delayed impulse functions as

$$\begin{aligned} f[k] = f[0]\delta[k] + f[1]\delta[k-1] + f[2]\delta[k-2] + \cdots \\ + f[k]\delta[0] + \cdots \qquad k \geq 0 \end{aligned} \tag{1.54}$$

The right-hand side expresses $f[k]$ as a sum of impulse components. If $h[k]$ is the solution of Equation 1.30 to the impulse input $f[k] = \delta[k]$, then the solution to input $\delta[k-m]$ is $h[k-m]$. This follows from the fact that because of constant coefficients, Equation 1.30 has time invariance property. Also, because Equation 1.30 is linear, its solution is the sum of the solutions to each of the impulse components of $f[k]$ on the right-hand side of Equation 1.54. Therefore,

$$\begin{aligned} y[k] = f[0]h[k] + f[1]h[k-1] + f[2]h[k-2] + \cdots \\ + f[k]h[0] + f[k+1]h[-1] + \cdots \end{aligned}$$

All practical systems with time as the independent variable are causal, that is $h[k] = 0$ for $k < 0$. Hence, all the terms on the right-hand side beyond $f[k]h[0]$ are zero. Thus,

$$\begin{aligned} y[k] &= f[0]h[k] + f[1]h[k-1] + f[2]h[k-2] + \cdots \\ &\quad + f[k]h[0] \\ &= \sum_{m=0}^{k} f[m]h[k-m] \end{aligned} \tag{1.55}$$

The general solution is obtained by adding a complementary solution to the above solution. Therefore, the general solution is given by

$$y[k] = \sum_{j=1}^{n} c_j \gamma_j^k + \sum_{m=0}^{k} f[m]h[k-m] \tag{1.56}$$

The last sum on the right-hand side is known as the *convolution sum* of $f[k]$ and $h[k]$.

The function $h[k]$ appearing in Equation 1.56 is the solution of Equation 1.30 for the impulsive input ($f[k] = \delta[k]$) when all initial conditions are zero, that is, $h[-1] = h[-2] = \cdots = h[-n] = 0$. It can be shown that [3] $h[k]$ contains an impulse and a linear combination of characteristic modes as

$$h[k] = \tfrac{b_0}{a_0}\delta[k] + A_1\gamma_1^k + A_2\gamma_2^k + \cdots + A_n\gamma_n^k \tag{1.57}$$

where the unknown constants A_i are determined from n values of $h[k]$ obtained by solving the equation $Q[E]h[k] = P[E]\delta[k]$ iteratively.

EXAMPLE 1.7:

Solve Example 1.5 using convolution method. In other words solve

$$(E^2 - 3E + 2)y[k] = (E + 2)f[k]$$

for $f[k] = (3)^k u[k]$ and the auxiliary conditions $y[0] = 2,\ y[1] = 1$.

The unit impulse solution $h[k]$ is given by Equation 1.57. In this case $a_0 = 2$ and $b_0 = 2$. Therefore,

$$h[k] = \delta[k] + A_1(1)^k + A_2(2)^k \tag{1.58}$$

To determine the two unknown constants A_1 and A_2 in Equation 1.58, we need two values of $h[k]$, for instance $h[0]$ and $h[1]$. These can be determined iteratively by observing that $h[k]$ is the solution of $(E^2 - 3E + 2)h[k] = (E + 2)\delta[k]$, that is,

$$h[k+2] - 3h[k+1] + 2h[k] = \delta[k+1] + 2\delta[k] \tag{1.59}$$

subject to initial conditions $h[-1] = h[-2] = 0$. We now determine $h[0]$ and $h[1]$ iteratively from Equation 1.59. Setting $k = -2$ in this equation yields

$$h[0] - 3(0) + 2(0) = 0 + 0 \Longrightarrow h[0] = 0$$

Next, setting $k = -1$ in Equation 1.59 and using $h[0] = 0$, we obtain

$$h[1] - 3(0) + 2(0) = 1 + 0 \Longrightarrow h[1] = 1$$

Setting $k = 0$ and 1 in Equation 1.58 and substituting $h[0] = 0,\ h[1] = 1$ yields

$$0 = 1 + A_1 + A_2 \qquad \text{and} \qquad 1 = A_1 + 2A_2$$

Solution of these two equations yields $A_1 = -3$ and $A_2 = 2$. Therefore,

$$h[k] = \delta[k] - 3 + 2(2)^k$$

and from Equation 1.56

$$\begin{aligned} y[k] &= c_1 + c_2(2)^k + \sum_{m=0}^{k}(3)^m[\delta[k-m] - 3 + 2(2)^{k-m}] \\ &= c_1 + c_2(2)^k + 1.5 - 4(2)^k + 2.5(3)^k \end{aligned}$$

The sums in the above expression are found by using the geometric progression sum formula

$$\sum_{m=0}^{k} r^m = \frac{r^{k+1} - 1}{r - 1} \qquad r \neq 1$$

Setting $k = 0$ and 1 and substituting the given auxiliary conditions $y[0] = 2,\ y[1] = 1$, we obtain

$$2 = c_1 + c_2 + 1.5 - 4 + 2.5 \quad \text{and} \quad 1 = c_1 + 2c_2 + 1.5 - 8 + 7.5$$

Solution of these equations yields $c_1 = 4$ and $c_2 = -2$. Therefore,

$$y[k] = 5.5 - 6(2)^k + 2.5(3)^k$$

which confirms the result obtained by the classical method.

Assessment of the Classical Method

The earlier remarks concerning the classical method for solving differential equations also apply to difference equations. General discussion of difference equations can be found in texts on the subject [2].

References

[1] Birkhoff, G. and Rota, G.C., *Ordinary Differential Equations*, 3rd ed., John Wiley & Sons, New York, 1978.

[2] Goldberg, S., *Introduction to Difference Equations*, John Wiley & Sons, New York, 1958.

[3] Lathi, B.P., *Linear Systems and Signals*, Berkeley-Cambridge Press, Carmichael, CA, 1992.

2

The Fourier, Laplace, and z-Transforms

Edward W. Kamen

School of Electrical and Computer Engineering, Georgia Institute of Technology, Atlanta, GA

2.1 Introduction

The study of signals and systems can be carried out in terms of either a time domain or a transform domain formulation. Both approaches are often used together in order to maximize our ability to deal with a particular problem arising in applications. This is very much the case in controls engineering where both time domain and transform domain techniques are extensively used in analysis and design. The transform domain approach to signals and systems is based on the transformation of functions using the Fourier, Laplace, and z-transforms. The fundamental aspects of these transforms are presented in this section along with some discussion on the application of these constructs.

The development in this chapter begins with the Fourier transform (FT), which can be viewed as a generalization of the Fourier series representation of a periodic function. The Fourier transform and Fourier series are named after Jean Baptiste Joseph Fourier (1768-1830), who first proposed in a 1807 paper that a series of sinusoidal harmonics could be used to represent the temperature distribution in a body. In 1822 Fourier wrote a book on his work, which was translated into English many years later (see [2]). It was also during the first part of the 1800s that Fourier was successful in constructing a frequency domain representation for aperiodic (nonperiodic) functions. This resulted in the FT, which provides a representation of a function $f(t)$ of a real variable t in terms of the frequency components comprising the function. Much later (in the 1900s), a FT theory was developed for functions $f(k)$ of an integer variable k. This resulted in the discrete-time Fourier transform (DTFT) and the N-point discrete Fourier transform (N-point DFT), both of which are briefly considered in this section.

Also during the early part of the 1800s, Pierre Simon Laplace (1749-1827) carried out his work on the generalization of the FT, which resulted in the transform that now bears his name. The Laplace transform can be viewed as the FT with the addition of a real exponential factor to the integrand of the integral operation. This results in a transform that is a function of a complex variable $s = \sigma + j\omega$. Although the modification to the FT may not seem to be very major, in fact the Laplace transform is an extremely powerful tool in many application areas (such as controls) where the utility of the FT is somewhat limited. In this section, a brief presentation is given on the one-sided Laplace transform with much of the focus on rational transforms.

The discrete-time counterpart to the Laplace transform is the z-transform which was developed primarily during the 1950s (e.g., see [1], [3], and [5]). The one-sided z-transform is considered, along with the connection to the DTFT.

Applications and examples involving the Fourier, Laplace, and z-transforms are given in the second part of this section. There the presentation centers on the relationship between the pole locations of a rational transform and the frequency spectrum of the transformed function; the numerical computation of the FT; and the application of the Laplace and z-transforms to solving differential and difference equations. The application of the transforms to systems and controls is pursued in other chapters in this handbook.

2.2 Fundamentals of the Fourier, Laplace, and z-Transforms

Let $f(t)$ be a real-valued function of the real-valued variable t; that is, for any real number t, $f(t)$ is a real number. The function $f(t)$ can be viewed as a signal that is a function of the continuous-time variable t (in units of seconds) and where t takes values from

0-8493-8570-9/96/$0.00+$.50

$-\infty$ to ∞. The FT $F(\omega)$ of $f(t)$ is defined by

$$F(\omega) = \int_{-\infty}^{\infty} f(t)e^{-j\omega t}dt, \; -\infty < \omega < \infty \tag{2.1}$$

where ω is the frequency variable in radians per second (rad/s), $j = \sqrt{-1}$ and $e^{-j\omega t}$ is the complex exponential given by Euler's formula

$$e^{-j\omega t} = \cos(\omega t) - j\sin(\omega t) \tag{2.2}$$

Inserting Equation 2.2 into Equation 2.1 results in the following expression for the FT:

$$F(\omega) = R(\omega) + jI(\omega) \tag{2.3}$$

where $R(\omega)$ and $I(\omega)$ are the real and imaginary parts, respectively, of $F(\omega)$ given by

$$\begin{aligned} R(\omega) &= \int_{-\infty}^{\infty} f(t)\cos(\omega t)dt \\ I(\omega) &= -\int_{-\infty}^{\infty} f(t)\sin(\omega t)dt \end{aligned} \tag{2.4}$$

From Equation 2.3, it is seen that in general the FT $F(\omega)$ is a complex-valued function of the frequency variable ω. For any value of ω, $F(\omega)$ has a magnitude $|F(\omega)|$ and an angle $\angle F(\omega)$ given by

$$\begin{aligned} |F(\omega)| &= \sqrt{R^2(\omega) + I^2(\omega)} \\ \angle F(\omega) &= \tan^{-1}\left(\frac{I(\omega)}{R(\omega)}\right) \end{aligned} \tag{2.5}$$

where again $R(\omega)$ and $I(\omega)$ are the real and imaginary parts defined by Equation 2.4. The function $|F(\omega)|$ represents the magnitude of the frequency components comprising $f(t)$, and thus the plot of $|F(\omega)|$ vs. ω is called the **magnitude spectrum** of $f(t)$. The function $\angle F(\omega)$ represents the phase of the frequency components comprising $f(t)$, and thus the plot of $\angle F(\omega)$ vs. ω is called the **phase spectrum** of $f(t)$. Note that $F(\omega)$ can be expressed in the polar form

$$F(\omega) = |F(\omega)| \exp[j\angle F(\omega)] \tag{2.6}$$

whereas the rectangular form of $F(\omega)$ is given by Equation 2.3.

The function (or signal) $f(t)$ is said to have a FT in the ordinary sense if the integral in Equation 2.1 exists for all real values of ω. Sufficient conditions that ensure the existence of the integral are that $f(t)$ have only a finite number of discontinuities, maxima, and minima over any finite interval of time and that $f(t)$ be absolutely integrable. The latter condition means that

$$\int_{-\infty}^{\infty} |f(t)|dt < \infty \tag{2.7}$$

There are a number of functions $f(t)$ of interest for which the integral in Equation 2.1 does not exist; for example, this is the case for the constant function $f(t) = c$ for $-\infty < t < \infty$, where c is a nonzero real number. Since the integral in Equation 2.1 obviously does not exist in this case, the constant function does not have a FT in the ordinary sense, but it does have a FT in the generalized sense, given by

$$F(\omega) = 2\pi c\delta(\omega) \tag{2.8}$$

where $\delta(\omega)$ is the impulse function. If Equation 2.8 is inserted into the inverse FT given by Equation 2.11, the result is the constant function $f(t) = c$ for all t. This observation justifies taking Equation 2.8 as the definition of the (generalized) FT of the constant function.

The FT defined by Equation 2.1 can be viewed as an operator that maps a time function $f(t)$ into a frequency function $F(\omega)$. This operation is often given by

$$F(\omega) = \Im[f(t)] \tag{2.9}$$

where $\Im$ denotes the FT operator. From Equation 2.9, it is clear that $f(t)$ can be recomputed from $F(\omega)$ by applying the inverse FT operator denoted by $\Im^{-1}$; that is,

$$f(t) = \Im^{-1}[F(\omega)] \tag{2.10}$$

The inverse operation is given by

$$f(t) = \frac{1}{2\pi}\int_{-\infty}^{\infty} F(\omega)e^{j\omega t}d\omega \tag{2.11}$$

The FT satisfies a number of properties that are very useful in applications. These properties are listed in Table 2.1 in terms of functions $f(t)$ and $g(t)$ whose transforms are $F(\omega)$ and $G(\omega)$, respectively. Appearing in this table is the convolution $f(t)*g(t)$ of $f(t)$ and $g(t)$, defined by

$$f(t) * g(t) = \int_{-\infty}^{\infty} f(\tau)g(t-\tau)d\tau \tag{2.12}$$

Also in Table 2.1 is the convolution $F(\omega) * G(\omega)$ given by

$$F(\omega) * G(\omega) = \int_{-\infty}^{\infty} F(\lambda)G(\omega-\lambda)d\lambda \tag{2.13}$$

From the properties in Table 2.1 and the generalized transform given by Equation 2.8, it is possible to determine the FT of many common functions. A list of FT of some common functions is given in Table 2.2.

2.2.1 Laplace Transform

Given the real-valued function $f(t)$, the **two-sided** (or **bilateral**) Laplace transform $F(s)$ of $f(t)$ is defined by

$$F(s) = \int_{-\infty}^{\infty} f(t)e^{-st}dt \tag{2.14}$$

where s is a complex variable. The **one-sided** (or **unilateral**) Laplace transform of $f(t)$ is defined by

$$F(s) = \int_{0}^{\infty} f(t)e^{-st}dt \tag{2.15}$$

Note that if $f(t) = 0$ for all $t < 0$, the one-sided and two-sided Laplace transforms of $f(t)$ are identical. In controls engineering, the one-sided Laplace transform is primarily used, and thus our presentation focuses on only the one-sided Laplace transform, which is referred to as the Laplace transform.

TABLE 2.1 Properties of the Fourier Transform.

Property	Transform/Property
Linearity	$\Im[af(t)+bg(t)] = aF(\omega)+bG(\omega)$ for any scalars a, b
Right or left shift in t	$\Im[f(t-t_o)] = F(\omega)\exp(-j\omega t_o)$ for any t_o
Time scaling	$\Im[f(at)] = (1/a)F(\omega/a)$ for any real number $a > 0$
Time reversal	$\Im[f(-t)] = F(-\omega) = \overline{F(\omega)}$ = complex conjugate of $F(\omega)$
Multiplication by a power of t	$\Im[t^n f(t)] = j^n \frac{d^n}{d\omega^n} F(\omega), n = 1, 2, \ldots$
Multiplication by $\exp(j\omega_0 t)$	$\Im[f(t)\exp(j\omega_0 t)] = F(\omega-\omega_0)$ for any real number ω_0
Multiplication by $\sin(\omega_0 t)$	$\Im[f(t)\sin(\omega_0 t)] = (j/2)[F(\omega+\omega_0) - F(\omega-\omega_0)]$
Multiplication by $\cos(\omega_0 t)$	$\Im[f(t)\cos(\omega_0 t)] = (1/2)[F(\omega+\omega_0) + F(\omega-\omega_0)]$
Differentiation in the time domain	$\Im\left[\frac{d^n}{dt^n} f(t)\right] = (j\omega)^n F(\omega), n = 1, 2, \ldots$
Multiplication in the time domain	$\Im[f(t)g(t)] = \frac{1}{2\pi}[F(\omega) * G(\omega)]$
Convolution in the time domain	$\Im[f(t) * g(t)] = F(\omega)G(\omega)$
Duality	$\Im[F(t)] = 2\pi f(-\omega)$
Parseval's theorem	$\int_{-\infty}^{\infty} f(t)g(t)dt = \frac{1}{2\pi}\int_{-\infty}^{\infty} F(-\omega)G(\omega)d\omega$
Special case of Parseval's theorem	$\int_{-\infty}^{\infty} f^2(t)dt = \frac{1}{2\pi}\int_{-\infty}^{\infty} \lvert F(\omega)\rvert^2 d\omega$

TABLE 2.2 Common Fourier Transforms.

$f(t)$	$F(\omega)$
$\delta(t)$ = unit impulse	1
$c, -\infty < t < \infty$	$2\pi c\delta(\omega)$
$f(t) = \begin{cases} 1, & -T/2 \le t \le T/2 \\ 0, & \text{all other } t \end{cases}$	$\frac{2}{\omega}\sin\left(\frac{T\omega}{2}\right)$
$\frac{\sin(at)}{t}$	$\begin{cases} \pi, & -a < \omega < a \\ 0, & \text{all other } \omega \end{cases}$
$e^{-b\lvert t\rvert}$, any $b > 0$	$\frac{2b}{\omega^2+b^2}$
e^{-bt^2} any $b > 0$	$\sqrt{\frac{\pi}{b}}e^{-\omega^2/4b}$
$e^{j\omega_0 t}$	$2\pi\delta(\omega-\omega_0)$
$\cos(\omega_0 t+\theta)$	$\pi\left[e^{-j\theta}\delta(\omega+\omega_0) + e^{j\theta}\delta(\omega-\omega_0)\right]$
$\sin(\omega_0 t+\theta)$	$j\pi\left[e^{-j\theta}\delta(\omega+\omega_0) - e^{j\theta}\delta(\omega-\omega_0)\right]$

Given a function $f(t)$, the set of all complex numbers s such that the integral in Equation 2.15 exists is called the **region of convergence** of the Laplace transform of $f(t)$. For example, if $f(t)$ is the unit-step function $u(t)$ given by $u(t) = 1$ for $t \ge 0$ and $u(t) = 0$ for $t < 0$, the integral in Equation 2.15 exists for any $s = \sigma + j\omega$ with real part $\sigma > 0$. Hence, the region of convergence is the set of all complex numbers s with positive real part, and, for any such s, the transform of the unit-step function $u(t)$ is equal to $1/s$.

Given a function $f(t)$, if the region of convergence of the Laplace transform $F(s)$ includes all complex numbers $s = j\omega$ for ω ranging from $-\infty$ to ∞, then $F(j\omega) = F(s)|_{s=j\omega}$ is well defined (i.e., exists) and is given by

$$F(j\omega) = \int_0^{\infty} f(t)e^{-j\omega t}dt \tag{2.16}$$

Then if $f(t) = 0$ for $t < 0$, the right-hand side of Equation 2.16 is equal to the FT $F(\omega)$ of $f(t)$ (see Equation 2.1). Hence, the FT of $f(t)$ is given by

$$F(\omega) = F(s)|_{s=j\omega} \tag{2.17}$$

(Note that we are denoting $F(j\omega)$ by $F(\omega)$.) This fundamental result shows that the FT of $f(t)$ can be computed directly from the Laplace transform $F(s)$ if $f(t) = 0$ for $t < 0$ and the region of convergence includes the imaginary axis of the complex plane (all complex numbers equal to $j\omega$).

The Laplace transform defined by Equation 2.15 can be viewed as an operator, denoted by $F(s) = L[f(t)]$ that maps a time function $f(t)$ into the function $F(s)$ of the complex variable s. The inverse Laplace transform operator is often denoted by L^{-1}, and is given by

$$f(t) = L^{-1}[F(s)] = \frac{1}{2\pi j}\int_{c-j\infty}^{c+j\infty} X(s)e^{st}ds \tag{2.18}$$

The integral in Equation 2.18 is evaluated along the path $s = c + j\omega$ in the complex plane from $c - j\infty$ to $c + j\infty$, where c is any real number for which the path $c + j\omega$ lies in the region of convergence of the transform $F(s)$. It is often possible to determine $f(t)$ without having to use Equation 2.18; for example, this is the case when $F(s)$ is a rational function of s. The computation of the Laplace transform or the inverse transform is often facilitated by using the properties of the Laplace transform, which are listed in Table 2.3. In this table, $f(t)$ and $g(t)$ are two functions with Laplace transforms $F(s)$ and $G(s)$, respectively, and $u(t)$ is the unit-step function defined by $u(t) = 1$ for $t \ge 0$ and $u(t) = 0$ for $t < 0$. Using the properties in Table 2.3, it is possible to determine the Laplace transform of many common functions without having to use Equation 2.15. A list of common Laplace transforms is given in Table 2.4.

TABLE 2.3 Properties of the (One-Sided) Laplace Transform.

Property	Transform/Property
Linearity	$L[af(t)+bG(t)] = aF(s)+bG(s)$ for any scalars a, b
Right shift in t	$L[f(t-t_0)u(t-t_0)] = F(s)\exp(-st_0)$ for any $t_0 > 0$
Time scaling	$L[f(at)] = (1/a)F(s/a)$ for any real number $a > 0$
Multiplication by a power of t	$L[t^n f(t)] = (-1)^n \frac{d^n}{ds^n} F(s), n = 1, 2, \ldots$
Multiplication by $e^{\alpha t}$	$L[f(t)e^{\alpha t}] = F(s-\alpha)$ for any real or complex number α
Multiplication by $\sin(\omega_0 t)$	$L[f(t)\sin(\omega_0 t)] = (j/2)[F(s+j\omega_0) - F(s-j\omega_0)]$
Multiplication by $\cos(\omega_0 t)$	$L[f(t)cos(\omega_0 t)] = (1/2)[F(s+j\omega_0) + F(s-j\omega_0)]$
Differentiation in the time domain	$L\left[\frac{d}{dt}f(t)\right] = sF(s) - f(0)$
Second derivative	$L\left[\frac{d^2}{dt^2}f(t)\right] = s^2F(s) - sf(0) - \frac{d}{dt}f(0)$
nth derivative	$L\left[\frac{d^n}{dt^n}f(t)\right] = s^nF(s) - s^{n-1}f(0) - s^{n-2}\frac{d}{dt}f(0) - \cdots - \frac{d^{n-1}}{dt^{n-1}}f(0)$
Integration	$L\left[\int_0^t f(\tau)d\tau\right] = \frac{1}{s}F(s)$
Convolution in the time domain	$L[f(t) * g(t)] = F(s)G(s)$
Initial-value theorem	$f(0) = \lim_{s\to\infty} sF(s)$
Final-value theorem	If $f(t)$ has a finite limit $f(\infty)$ as $t \to \infty$, then $f(\infty) = \lim_{s\to 0} sF(s)$

TABLE 2.4 Common Laplace Transforms.

$f(t)$	Laplace Transform $F(s)$
$u(t)$ = unit-step function	$\frac{1}{s}$
$u(t) - u(t-T)$ for any $T > 0$	$\frac{1-e^{-Ts}}{s}$
$\delta(t)$ = unit impulse	1
$\delta(t-t_0)$ for any $t_0 > 0$	$e^{-t_0 s}$
$t^n, t \geq 0$	$\frac{n!}{s^{n+1}}, n = 1, 2, \ldots$
e^{-at}	$\frac{1}{s+a}$
$t^n e^{-at}$	$\frac{n!}{(s+a)^{n+1}}, n = 1, 2, \ldots$
$\cos(\omega t)$	$\frac{s}{s^2+\omega^2}$
$\sin(\omega t)$	$\frac{\omega}{s^2+\omega^2}$
$\cos^2 \omega t$	$\frac{s^2+2\omega^2}{s(s^2+4\omega^2)}$
$\sin^2 \omega t$	$\frac{2\omega^2}{s(s^2+4\omega^2)}$
$\sinh(at)$	$\frac{a}{s^2-a^2}$
$\cosh(at)$	$\frac{s}{s^2-a^2}$
$e^{-at}\cos(\omega t)$	$\frac{s+a}{(s+a)^2+\omega^2}$
$e^{-at}\sin(\omega t)$	$\frac{\omega}{(s+a)^2+\omega^2}$
$t\cos(\omega t)$	$\frac{s^2-\omega^2}{(s^2+\omega^2)^2}$
$t\sin(\omega t)$	$\frac{2\omega s}{(s^2+\omega^2)^2}$
$te^{-at}\cos(\omega t)$	$\frac{(s+a)^2-\omega^2}{\left[(s+a)^2+\omega^2\right]^2}$
$te^{-at}\sin(\omega t)$	$\frac{2\omega(s+a)}{\left[(s+a)^2+\omega^2\right]^2}$

2.2.2 Rational Laplace Transforms

The Laplace transform $F(s)$ of a function $f(t)$ is said to be a **rational function** of s if it can be written as a ratio of polynomials in s; that is,

$$F(s) = \frac{N(s)}{D(s)} \tag{2.19}$$

where $N(s)$ and $D(s)$ are polynomials in the complex variable s given by

$$N(s) = b_m s^m + b_{m-1}s^{m-1} + \cdots + b_1 s + b_0 \tag{2.20}$$

$$D(s) = s^n + a_{n-1}s^{n-1} + \cdots + a_1 s + a_0 \tag{2.21}$$

In Equations 2.20 and 2.21, m and n are positive integers and the coefficients $b_m, b_{m-1}, \ldots, b_1, b_0$ and $a_{n-1}, \ldots, a_1, a_0$ are real numbers. In Equation 2.19, it is assumed that $N(s)$ and $D(s)$ do not have any common factors. If there are common factors, they should always be cancelled. Also note that the polynomial $D(s)$ is monic; that is, the coefficient of s^n is equal to 1. A rational function $F(s)$ can always be written with a monic denominator polynomial $D(s)$. The integer n, which is the degree of $D(s)$, is called the order of the rational function $F(s)$. It is assumed that $n \geq m$, in which case $F(s)$ is said to be a **proper rational function.** If $n > m$, $F(s)$ is said to be **strictly proper.**

Given a rational transform $F(s) = N(s)/D(s)$ with $N(s)$ and $D(s)$ defined by Equations 2.20 and 2.21, let $z_1, z_2, \ldots, z_m$ denote the roots of the polynomial $N(s)$, and let $p_1, p_2, \ldots, p_n$ denote the roots of $D(s)$; that is, $N(z_i) = 0$ for $i = 1, 2, \ldots, m$ and $D(p_i) = 0$ for $i = 1, 2, \ldots, n$. In general, the z_i and the p_i may be real or complex numbers, but if any are complex, they must appear in complex conjugate pairs. The numbers $z_1, z_2, \ldots, z_m$ are called the **zeros** of the rational function $F(s)$ since $F(s) = 0$ when $s = z_i$ for $i = 1, 2, \ldots, m$; and the numbers $p_1, p_2, \ldots, p_n$ are called the **poles** of $F(s)$ since the magnitude $|F(s)|$ becomes infinite as s approaches p_i for $i = 1, 2, \ldots, n$.

If $F(s)$ is strictly proper ($n > m$) and the poles $p_1, p_2, \ldots, p_n$ of $F(s)$ are distinct (nonrepeated), then $F(s)$ has the partial fraction expansion

$$F(s) = \frac{c_1}{s-p_1} + \frac{c_2}{s-p_2} + \cdots + \frac{c_n}{s-p_n} \tag{2.22}$$

where the c_i are the **residues** given by

$$c_i = [(s-p_i)F(s)]_{s=p_i}, \quad i = 1, 2, \ldots, n \tag{2.23}$$

For a given value of i, the residue c_i is real if and only if the corresponding pole p_i is real, and c_i is complex if and only if p_i is complex.

From Equation 2.22 we see that the inverse Laplace transform $f(t)$ is given by the following sum of exponential functions:

$$f(t) = c_1 e^{p_1 t} + c_2 e^{p_2 t} + \cdots + c_n e^{p_n t} \quad (2.24)$$

If all the poles $p_1, p_2, \ldots, p_n$ of $F(s)$ are real numbers, then $f(t)$ is a sum of real exponentials given by Equation 2.24. If $F(s)$ has a pair of complex poles $p = \sigma \pm j\omega$, then $f(t)$ contains the term

$$c e^{(\sigma + j\omega)t} + \bar{c} e^{(\sigma - j\omega)t} \quad (2.25)$$

where $\bar{c}$ is the complex conjugate of c. Then writing c in the polar form $c = |c|e^{j\theta}$, we have

$$\begin{aligned} c e^{(\sigma + j\omega)t} + \bar{c} e^{(\sigma - j\omega)t} &= |c|e^{j\theta} e^{(\sigma + j\omega)t} + \\ &\quad |\bar{c}|e^{-j\theta} e^{(\sigma - j\omega)t} \\ &= |c|e^{\sigma t}\left[e^{j(\omega t + \theta)} + \right. \\ &\quad \left. e^{-j(\omega t + \theta)}\right] \end{aligned} \quad (2.26)$$

Finally, using Euler's formula, Equation 2.26 can be written in the form

$$c e^{(\sigma + j\omega)t} + \bar{c} e^{(\sigma - j\omega)t} = 2|c|e^{\sigma t}\cos(\omega t + \theta) \quad (2.27)$$

From Equation 2.27 it is seen that if $F(s)$ has a pair of complex poles, then $f(t)$ contains a sinusoidal term with an exponential amplitude factor $e^{\sigma t}$. Note that if $\sigma = 0$ (so that the poles are purely imaginary), Equation 2.27 is purely sinusoidal.

If one of the poles (say p_1) is repeated r times and the other $n - r$ poles are distinct, $F(s)$ has the partial fraction expansion

$$\begin{aligned} F(s) &= \frac{c_1}{s - p_1} + \frac{c_2}{(s - p_1)^2} + \cdots + \frac{c_r}{(s - p_1)^r} \\ &\quad + \frac{c_{r+1}}{s - p_{r+1}} + \cdots + \frac{c_n}{s - p_n} \end{aligned} \quad (2.28)$$

In Equation 2.28, the residues $c_{r+1}, c_{r+2}, \ldots, c_n$ are calculated as in the distinct-pole case; that is,

$$c_i = [(s - p_i)F(s)]_{s = p_i}, \quad i = r+1, r+2, \ldots, n \quad (2.29)$$

and the residues $c_1, c_2, \ldots, c_r$ are given by

$$c_i = \frac{1}{(r-i)!}\left\{\frac{d^{r-i}}{ds^{r-i}}\left[(s - p_1)^r F(s)\right]\right\}_{s = p_1} \quad (2.30)$$

Then, taking the inverse transform of Equation 2.28 yields

$$\begin{aligned} f(t) &= c_1 e^{p_1 t} + c_2 t e^{p_1 t} + \cdots + \frac{c_r}{(r-1)!} t^{r-1} e^{p_1 t} \\ &\quad + c_{r+1} e^{p_{r+1} t} + \cdots + c_n e^{p_n t} \end{aligned} \quad (2.31)$$

The above results reveal that the analytical form of the function $f(t)$ depends directly on the poles of $F(s)$. In particular, if $F(s)$ has a nonrepeated real pole p, then $f(t)$ contains a real exponential term of the form ce^{pt} for some real constant c. If a real pole p is repeated r times, then $f(t)$ contains terms of the form $c_1 e^{pt}, c_2 t e^{pt}, \ldots, c_r t^{r-1} e^{pt}$ for some real constants $c_1, c_2, \ldots, c_r$. If $F(s)$ has a nonrepeated complex pair $\sigma \pm j\omega$ of poles, then $f(t)$ contains a term of the form $ce^{\sigma t}\cos(\omega t + \theta)$ for some real constants c and θ. If the complex pair $\sigma \pm j\omega$ is repeated r times, $f(t)$ contains terms of the form $c_1 e^{\sigma t}\cos(\omega t + \theta_1), c_2 t e^{\sigma t}\cos(\omega t + \theta_2), \ldots, c_r t^{r-1} e^{\sigma t}\cos(\omega t + \theta_r)$ for some real constants $c_1, c_2, \ldots, c_r$ and $\theta_1, \theta_2, \ldots, \theta_r$. These results are summarized in Table 2.5.

If $F(s)$ is proper, but not strictly proper (so that $n = m$ in Equations 2.20 and 2.21), then using long division $F(s)$ can be written in the form

$$F(s) = b_n + \frac{R(s)}{D(s)} \quad (2.32)$$

where the degree of $R(s)$ is strictly less than n. Then $R(s)/D(s)$ can be expanded via partial fractions as was done in the case when $F(s)$ is strictly proper. Note that for $F(s)$ given by Equation 2.32, the inverse Laplace transform $f(t)$ contains the impulse $b_n\delta(t)$. Hence, having $n = m$ in $F(s)$ results in an impulsive term in the inverse transform.

From the relationship between the poles of $F(s)$ and the analytical form of $f(t)$, it follows that $f(t)$ converges to zero as $t \to \infty$ if and only if all the poles $p_1, p_2, \ldots, p_n$ of $F(s)$ have real parts that are strictly less than zero; that is, $Re(p_i) < 0$ for $i = 1, 2, \ldots, n$. This condition is equivalent to requiring that all the poles be located in the **open left half-plane (OLHP)**, which is the region in the complex plane to the left of the imaginary axis.

It also follows from the relationship between the poles of $F(s)$ and the form of $f(t)$ that $f(t)$ has a finite limit $f(\infty)$ as $t \to \infty$ if and only if all the poles of $F(s)$ have real parts that are less than zero, except that $F(s)$ may have a nonrepeated pole at $s = 0$. In mathematical terms, the conditions for the existence of a finite limit $f(\infty)$ are

$$Re(p_i) < 0 \quad \text{for all poles } p_i \neq 0 \quad (2.33)$$

$$\text{If} \quad p_i = 0 \quad \text{is a pole of } F(s), \text{ then } p_i \text{ is nonrepeated} \quad (2.34)$$

If the conditions in Equations 2.33 and 2.34 are satisfied, the limiting value $f(\infty)$ is given by

$$f(\infty) = [sF(s)]_{s=0} \quad (2.35)$$

The relationship in Equation 2.35 is a restatement of the final-value theorem (given in Table 2.3) in the case when $F(s)$ is rational and the poles of $F(s)$ satisfy the conditions in Equations 2.33 and 2.34.

2.2.3 Irrational Transforms

The Laplace transform $F(s)$ of a function $f(t)$ is said to be an **irrational function** of s if it is not rational; that is, $F(s)$ cannot be expressed as a ratio of polynomials in s. For example, $F(s) = e^{-t_0 s}/s$ is irrational since the exponential function $e^{-t_0 s}$ cannot be expressed as a ratio of polynomials in s. In this case, the inverse transform $f(t)$ is equal to $u(t - t_0)$ where $u(t)$ is the unit-step function.

Given any function $f(t)$ with transform $F(s)$ and given any real number $t_0 > 0$, the transform of the time-shifted (or time-

TABLE 2.5 Relationship Between the Poles of $F(s)$ and the Form of $f(t)$.

Pole Locations of $F(s)$	Corresponding Terms in $f(t)$
Nonrepeated real pole at $s = p$	ce^{pt}
Real pole at $s = p$ repeated r times	$\sum_{i=1}^{r} c_i t^{i-1} e^{pt}$
Nonrepeated complex pair at $s = \sigma \pm j\omega$	$ce^{\sigma t}\cos(\omega t + \theta)$
Complex pair at $s = \sigma \pm j\omega$ repeated r times	$\sum_{i=1}^{r} c_i t^{i-1} e^{\sigma t}\cos(\omega t + \theta_i)$

delayed) function $f(t-t_0)u(t-t_0)$ is equal to $F(s)e^{-t_0 s}$. Time-delayed signals arise in systems with time delays, and thus irrational transforms appear in the study of systems with time delays. Also, any function $f(t)$ that is of finite duration in time has a transform $F(s)$ that is irrational. For instance, suppose that

$$f(t) = \gamma(t)\,[u(t-t_0) - u(t-t_1)]\,,\quad 0 \le t_0 < t_1 \tag{2.36}$$

so that $f(t) = \gamma(t)$ for $t_0 \le t < t_1$, and $f(t) = 0$ for all other t. Then $f(t)$ can be written in the form

$$f(t) = \gamma_0(t-t_0)u(t-t_0) - \gamma_1(t-t_1)u(t-t_1) \tag{2.37}$$

where $\gamma_0(t) = \gamma(t+t_0)$ and $\gamma_1(t) = \gamma(t+t_1)$. Taking the Laplace transform of Equation 2.37 yields

$$F(s) = \Gamma_0(s)e^{-t_0 s} - \Gamma_1(s)e^{-t_1 s} \tag{2.38}$$

where $\Gamma_0(s)$ and $\Gamma_1(s)$ are the transforms of $\gamma_0(t)$ and $\gamma_1(t)$, respectively. Note that by Equation 2.38, the transform $F(s)$ is an irrational function of s.

To illustrate the above constructions, suppose that

$$f(t) = e^{-at}\,[u(t-1) - u(t-2)] \tag{2.39}$$

Writing $f(t)$ in the form of Equation 2.37 gives

$$f(t) = e^{-a}e^{-a(t-1)}u(t-1) - e^{-2a}e^{-a(t-2)}u(t-2) \tag{2.40}$$

Then, transforming Equation 2.40 yields

$$F(s) = \left[e^{-(s+a)} - e^{-2(s+a)}\right]\frac{1}{s+a} \tag{2.41}$$

Clearly, $F(s)$ is an irrational function of s.

2.2.4 Discrete-Time Fourier Transform

Let $f(k)$ be a real-valued function of the integer-valued variable k. The function $f(k)$ can be viewed as a discrete-time signal; in particular, $f(k)$ may be a sampled version of a continuous-time signal $f(t)$. More precisely, $f(k)$ may be equal to the sample values $f(kT)$ of a signal $f(t)$ with t evaluated at the sample times $t = kT$, where T is the sampling interval. In mathematical terms, the sampled signal is given by

$$f(k) = f(t)|_{t=kT} = f(kT),\quad k = 0, \pm 1, \pm 2, \ldots \tag{2.42}$$

Note that we are denoting $f(kT)$ by $f(k)$. The FT of a function $f(k)$ of an integer variable k is defined by

$$F(\Omega) = \sum_{k=-\infty}^{\infty} f(k)e^{-j\Omega k},\quad -\infty < \Omega < \infty \tag{2.43}$$

where Ω is interpreted as the real frequency variable. The transform $F(\Omega)$ is called the discrete-time Fourier transform (DTFT) since it can be viewed as the discrete-time counterpart of the FT defined above. The DTFT is directly analogous to the FT, so that all the properties of the FT discussed above carry over to the DTFT. In particular, as is the case for the FT, the DTFT $F(\Omega)$ is in general a complex-valued function of the frequency variable Ω, and thus $F(\Omega)$ must be specified in terms of a magnitude function $|F(\Omega)|$ and an angle function $\angle F(\Omega)$. The magnitude function $|F(\Omega)|$ (respectively, the angle function $\angle F(\Omega)$) displays the magnitude (respectively, the phase) of the frequency components comprising $f(k)$. All of the properties of the FT listed in Table 2.1 have a counterpart for the DTFT, but this will not be pursued here.

In contrast to the FT, the DTFT $F(\Omega)$ is always a periodic function of the frequency variable Ω with period 2π; that is,

$$F(\Omega + 2\pi) = F(\Omega)\ \text{for} - \infty < \Omega < \infty \tag{2.44}$$

As a result of the periodicity property in Equation 2.44, it is necessary to specify $F(\Omega)$ over a 2π interval only, such as 0 to 2π or $-\pi$ to π. Given $F(\Omega)$ over any 2π interval, $f(k)$ can be recomputed using the inverse DTFT. In particular, if $F(\Omega)$ is specified over the interval $-\pi < \Omega < \pi$, $f(k)$ can be computed from the relationship

$$f(k) = \frac{1}{2\pi}\int_{-\pi}^{\pi} F(\Omega)e^{jk\Omega}d\Omega \tag{2.45}$$

In practice, the DTFT $F(\Omega)$ is usually computed only for a discrete set of values of the frequency variable Ω. This is accomplished by using the N-point discrete Fourier transform (N-point DFT) of $f(k)$ given by

$$F_n = \sum_{k=0}^{N-1} f(k)e^{-j2\pi kn/N},\ n = 0, 1, \ldots, N-1 \tag{2.46}$$

where N is a positive integer. If $f(k) = 0$ for $k < 0$ and $k \ge N$, comparing Equations 2.46 and 2.43 reveals that

$$F_n = F\left(\tfrac{2\pi n}{N}\right),\qquad n = 0, 1, \ldots, N-1 \tag{2.47}$$

Hence, the DFT F_n is equal to the values of the DTFT $F(\Omega)$ with Ω evaluated at the discrete points $\Omega = 2\pi n/N$ for $n = 0, 1, 2, \ldots, N-1$.

The computation of the DFT F_n given by Equation 2.46 can be carried out using a fast algorithm called the Fast Fourier transform (FFT). The inverse FFT can be used to compute $f(k)$ from F_n. A development of the FFT is beyond the scope of this section (see "Further Reading").

2.2.5 z-Transform

Given the function $f(k)$, the **two-sided** (or **bilateral**) z-transform $F(z)$ of $f(k)$ is defined by

$$F(z) = \sum_{k=-\infty}^{\infty} f(k)z^{-k} \tag{2.48}$$

where z is a complex variable. The **one-sided** (or **unilateral**) z-transform of $f(k)$ is defined by

$$F(z) = \sum_{k=0}^{\infty} f(k)z^{-k} \tag{2.49}$$

Note that if $f(k) = 0$ for $k = -1, -2, \ldots$, the one-sided and two-sided z-transforms of $f(k)$ are the same. As is the case with the Laplace transform, in controls engineering the one-sided version is the most useful, and thus the development given below is restricted to the one-sided z-transform, which is referred to as the z-transform.

Given $f(k)$, the set of all complex numbers z such that the summation in Equation 2.49 exists is called the region of convergence of the z-transform of $f(k)$. If the region of convergence of the z-transform includes all complex numbers $z = e^{j\Omega}$ for Ω ranging from $-\infty$ to ∞, then $F(e^{j\Omega}) = F(z)|_{z=e^{j\Omega}}$ is well defined (i.e., exists) and is given by

$$F(e^{j\Omega}) = \sum_{k=0}^{\infty} f(k)\left(e^{j\Omega}\right)^{-k} \tag{2.50}$$

But $(e^{j\Omega})^{-k} = e^{-j\Omega k}$, and thus Equation 2.50 can be rewritten as

$$F(e^{j\Omega}) = \sum_{k=0}^{\infty} f(k)e^{-j\Omega k} \tag{2.51}$$

Then if $f(k) = 0$ for all $k < 0$, the right-hand side of Equation 2.51 is equal to the DTFT $F(\Omega)$ of $f(k)$ (see Equation 2.43). Therefore, the DTFT of $f(k)$ is given by

$$F(\Omega) = F(z)|_{z=e^{j\Omega}} \tag{2.52}$$

This result shows that the DTFT of $f(k)$ can be computed directly from the z-transform $F(z)$ if $f(k) = 0$ for all $k < 0$ and the region of convergence includes all complex numbers $z = e^{j\Omega}$ with $-\infty < \Omega < \infty$. Note that since $|e^{j\Omega}| = 1$ for any value of Ω and $\angle e^{j\Omega} = \Omega$, the set of complex numbers $z = e^{j\Omega}$ comprises the unit circle of the complex plane. Hence, the DTFT of $f(k)$ is equal to the values of the z-transform on the unit circle of the complex plane, assuming that the region of convergence of $F(z)$ includes the unit circle.

The z-transform defined by Equation 2.49 can be viewed as an operator, denoted by $F(z) = Z[f(k)]$, that maps a discrete-time function $f(k)$ into the function $F(z)$ of the complex variable z. The inverse z-transform operation is denoted by $f(k) = Z^{-1}[F(z)]$. As discussed below, when $F(z)$ is a rational function of z, the inverse transform can be computed using long division or by carrying out a partial fraction expansion of $F(z)$. The computation of the z-transform or the inverse z-transform is often facilitated by using the properties of the z-transform given in Table 2.6. In this table, $f(k)$ and $g(k)$ are two functions with z-transforms $F(z)$ and $G(z)$, respectively, and $u(k)$ is the unit-step function defined by $u(k) = 1$ for $k \geq 0$ and $u(k) = 0$ for $k < 0$. A list of common z-transforms is given in Table 2.7. In Table 2.7, the function $\delta(k)$ is the unit pulse defined by $\delta(0) = 1$, $\delta(k) = 0$ for $k \neq 0$.

2.2.6 Rational z-Transforms

As is the case for the Laplace transform, the z-transform $F(z)$ is often a rational function of z; that is, $F(z)$ is given by

$$F(z) = \frac{N(z)}{D(z)} \tag{2.53}$$

where $N(z)$ and $D(z)$ are polynomials in the complex variable z given by

$$N(z) = b_m z^m + b_{m-1}z^{m-1} + \cdots + b_1 z + b_0 \tag{2.54}$$

$$D(z) = z^n + a_{n-1}z^{n-1} + \cdots + a_1 z + a_0 \tag{2.55}$$

It is assumed that the order n of $F(z)$ is greater than or equal to m, and thus $F(z)$ is proper. The poles and zeros of $F(z)$ are defined in the same way as given above for rational Laplace transforms.

When the transform $F(z)$ is in the rational form in Equation 2.53, the inverse z-transform $f(k)$ can be computed by expanding $F(z)$ into a power series in z^{-1} by dividing $D(z)$ into $N(z)$ using long division. The values of the function $f(k)$ are then read off from the coefficients of the power series expansion. The first few steps of the process are carried out below:

$$\begin{array}{r|ll}
 & b_m z^{m-n} + (b_{m-1} - a_{n-1}b_m)z^{m-n-1} & +\cdots \\ \hline
z^n + a_{n-1}z^{n-1} + \cdots + a_1 z + a_0 & b_m z^m + b_{m-1}z^{m-1} & +\cdots \\
 & \underline{b_m z^m + a_{n-1}b_m z^{m-1}} & \underline{+\cdots} \\
 & (b_{m-1} - a_{n-1}b_m)z^{m-1} & +\cdots \\
 & \underline{(b_{m-1} - a_{n-1}b_m)z^{m-1}} & \underline{+\cdots} \\
 & & \vdots
\end{array}$$

Since the value of $f(k)$ is equal to the coefficient of z^{-k} in the power series expansion of $F(z)$, it follows from the above division process that $f(n-m) = b_m$, $f(n-m+1) = b_{m-1} - a_{n-1}b_m$, and so on.

To express the inverse z-transform $f(k)$ in closed form, it is necessary to expand $F(z)$ via partial fractions. It turns out that the form of the inverse z-transform $f(k)$ is simplified if $F(z)/z = N(z)/D(z)z$ is expanded by partial fractions. Note that $F(z)/z$ is strictly proper since $F(z)$ is assumed to be proper.

Letting $p_1, p_2, \ldots, p_n$ denote the poles of $F(z)$, if the p_i are distinct and are nonzero, then $F(z)/z$ has the partial fraction expansion

$$\frac{F(z)}{z} = \frac{c_0}{z} + \frac{c_1}{z-p_1} + \frac{c_2}{z-p_2} + \cdots + \frac{c_n}{z-p_n} \tag{2.56}$$

TABLE 2.6 Properties of the (One-Sided) z-Transform.

Property	Transform /Property
Linearity	$Z[af(k)+bg(k)] = aF(z)+bG(z)$ for any scalars a, b
Right shift of $f(k)u(k)$	$Z[f(k-q)u(k-q)] = z^{-q}F(z)$ for any integer $q \geq 1$
Right shift of $f(k)$	$Z[f(k-1)] = z^{-1}F(z)+f(-1)$
	$Z[f(k-2)] = z^{-2}F(z)+f(-2)+z^{-1}f(-1)$
	$\vdots$
	$Z[f(k-q)] = z^{-q}F(z)+f(-q)+z^{-1}f(-q+1)+\cdots+z^{-q+1}f(-1)$
Left shift in time	$Z[f(k+1)] = zF(z)-f(0)z$
	$Z[f(k+2)] = z^2F(z)-f(0)z^2-f(1)z$
	$\vdots$
	$Z[f(k+q)] = z^qF(z)-f(0)z^q-f(1)z^{q-1}-\cdots-f(q-1)z$
Multiplication by k	$Z[kf(k)] = -z\frac{d}{dz}F(z)$
Multiplication by k^2	$Z[k^2f(k)] = z\frac{d}{dz}F(z)+z^2\frac{d^2}{dz^2}F(z)$
Multiplication by a^k	$Z[a^kf(k)] = F(\frac{z}{a})$
Multiplication by $\cos(\Omega k)$	$Z[\cos(\Omega k)f(k)] = \frac{1}{2}\left[F(e^{j\Omega}z)+F(e^{-j\Omega}z)\right]$
Multiplication by $\sin(\Omega k)$	$Z[\sin(\Omega k)f(k)] = \frac{j}{2}\left[F(e^{j\Omega}z)-F(e^{-j\Omega}z)\right]$
Summation	$\left[\sum_{i=0}^{k} f(i)\right] = \frac{z}{z-1}F(z)$
Convolution	$Z[f(k)*g(k)] = F(z)G(z)$
Initial-value theorem	$f(0) = \lim_{z\to\infty} F(z)$
Final-value theorem	If $f(k)$ has a finite limit $f(\infty)$ as $k\to\infty$, then $f(\infty) = \lim_{z\to 1}(z-1)F(z)$

where the residues are given by

$$c_0 = F(0) \tag{2.57}$$

$$c_i = \left[(z-p_i)\frac{F(z)}{z}\right]_{z=p_i}, i = 1,2,\ldots,n \tag{2.58}$$

Then multiplying both sides of Equation 2.56 by z gives

$$F(z) = c_0 + \frac{c_1 z}{z-p_1} + \frac{c_2 z}{z-p_2} + \cdots + \frac{c_n z}{z-p_n} \tag{2.59}$$

and taking the inverse z-transform gives

$$f(k) = c_0\delta(k) + c_1 p_1^k + c_2 p_2^k + \cdots + c_n p_n^k \tag{2.60}$$

If the poles $p_1, p_2, \ldots, p_n$ of $F(z)$ are real numbers, then from Equation 2.60 it is seen that $f(k)$ is the sum of geometric functions of the form cp^k, plus a pulse function $c_0\delta(k)$ if $c_0 \neq 0$. If $F(z)$ has a pair of complex poles given in polar form by $\sigma e^{\pm j\Omega}$, then it can be shown that $f(k)$ contains a sinusoidal term of the form

$$c\sigma^k \cos(\Omega k + \theta) \tag{2.61}$$

for some constants c and θ.

If $F(z)$ has a real pole p that is repeated r times, then $f(k)$ contains the terms $c_1 p^k, c_2 kp^k, \ldots, c_r k^{r-1}p^k$; and if $F(z)$ has a pair of complex poles given by $\sigma e^{\pm j\Omega}$ that is repeated r times, then $f(k)$ contains terms of the form $c_1\sigma^k\cos(\Omega k+\theta_1)$, $c_2 k\sigma^k\cos(\Omega k+\theta_2), \ldots, c_r k^{r-1}\sigma^k\cos(\Omega k+\theta_r)$ for some constants $c_1, c_2, \ldots, c_r$ and $\theta_1, \theta_2, \ldots, \theta_r$. These results are summarized in Table 2.8.

From the relationship between the poles of $F(z)$ and the analytical form of $f(k)$, it follows that $f(k)$ converges to zero as $k \to \infty$ if and only if all the poles $p_1, p_2, \ldots, p_n$ of $F(z)$ have magnitudes that are strictly less than one; that is, $|p_i| < 1$ for $i = 1, 2, \ldots, n$. This is equivalent to requiring that all the poles be located on the **open unit disk** of the complex plane, which is the region of the complex plane consisting of all complex numbers whose magnitude is strictly less than one.

TABLE 2.7 Common z-Transform Pairs.

$f(k)$	z-Transform $F(z)$
$u(k)$ = unit-step function	$\frac{z}{z-1}$
$u(k)-u(k-N)$, $N = 1,2,\ldots$	$\frac{z^N-1}{z^{N-1}(z-1)}$, $N = 1,2,\ldots$
$\delta(k)$ = unit pulse	1
$\delta(k-q)$, $q = 1,2,\ldots$	$\frac{1}{z^q}$, $q = 1,2,\ldots$
a^k, a real or complex	$\frac{z}{z-a}$
k	$\frac{z}{(z-1)^2}$
$k+1$	$\frac{z^2}{(z-1)^2}$
k^2	$\frac{z(z+1)}{(z-1)^3}$
ka^k	$\frac{az}{(z-a)^2}$
k^2a^k	$\frac{az(z+a)}{(z-a)^3}$
$k(k+1)a^k$	$\frac{2az^2}{(z-a)^3}$
$\cos(\Omega k)$	$\frac{z^2-(\cos\Omega)z}{z^2-(2\cos\Omega)z+1}$
$\sin(\Omega k)$	$\frac{(\sin\Omega)z}{z^2-(2\cos\Omega)z+1}$
$a^k\cos(\Omega k)$	$\frac{z^2-(a\cos\Omega)z}{z^2-(2a\cos\Omega)z+a^2}$
$a^k\sin(\Omega k)$	$\frac{(a\sin\Omega)z}{z^2-(2a\cos\Omega)z+a^2}$

It also follows from the relationship between pole locations

TABLE 2.8 Relationship Between the Poles of $F(z)$ and the Form of $f(k)$.

Pole Locations of $F(z)$	Corresponding Terms in $f(k)$
Nonrepeated real pole at $z = p$	cp^k
Real pole at $z = p$ repeated r times	$\sum_{i=1}^{r} c_i k^{i-1} p^k$
Nonrepeated complex pair at $z = \sigma e^{\pm j\Omega}$	$c\sigma^k \cos(\Omega k + \theta)$
Complex pair at $z = \sigma e^{\pm j\Omega}$ repeated r times	$\sum_{i=1}^{r} c_i k^{i-1} \sigma^k \cos(\Omega k + \theta_i)$

and the form of the function that $f(k)$ has a finite limit $f(\infty)$ as $k \to \infty$ if and only if all the poles of $F(z)$ have magnitudes that are less than one, except that $F(z)$ may have a nonrepeated pole at $z = 1$. In mathematical terms, the conditions for the existence of a finite limit $f(\infty)$ are:

$$|p_i| < 1 \quad \text{for all poles } p_i \neq 0 \tag{2.62}$$

$$\text{If} \quad p_i = 1 \quad \text{is a pole of } F(z), \text{ then } p_i \text{ is nonrepeated} \tag{2.63}$$

If the conditions in Equations 2.62 and 2.63 are satisfied, the limiting value $f(\infty)$ is given by

$$f(\infty) = [(z-1)F(z)]_{z=1} \tag{2.64}$$

The relationship in Equation 2.64 is a restatement of the final-value theorem (given in Table 2.6) in the case when $F(z)$ is rational and the poles of $F(z)$ satisfy the conditions in Equations 2.62 and 2.63.

2.3 Applications and Examples

Given a real-valued signal $f(t)$ of the continuous-time variable t, the FT $F(\omega)$ reveals the frequency spectrum of $f(t)$; in particular, the plot of $|F(\omega)|$ vs. ω is the magnitude spectrum of $f(t)$, and the plot of $\angle F(\omega)$ versus ω is the phase spectrum of $f(t)$. The magnitude function $|F(\omega)|$ is sometimes given in decibels (dB) defined by

$$|F(\omega)|_{\text{dB}} = 20\log_{10}|F(\omega)| \tag{2.65}$$

Given a signal $f(t)$ with FT $F(\omega)$, if there exists a positive number B such that $|F(\omega)|$ is zero (or approximately zero) for all $\omega > B$, the signal $f(t)$ is said to be **band limited** or to have a finite bandwidth; that is, the frequencies comprising $f(t)$ are limited (for the most part) to a finite range from 0 to B rad/s. The **3-dB bandwidth** of such a signal is the smallest positive value $B_{3\text{dB}}$ such that

$$|F(\omega)| \leq .707 F_{\max} \quad \text{for all} \quad \omega > B_{3\text{dB}} \tag{2.66}$$

where $F_{\max}$ is the maximum value of $|F(\omega)|$. The inequality in Equation 2.66 is equivalent to requiring that the magnitude $|F(\omega)|_{\text{dB}}$ in decibels be down from its peak value by 3dB or more. For example, suppose that $f(t)$ is the T-second rectangular pulse defined by

$$f(t) = \begin{cases} 1, & -T/2 \leq t \leq T/2 \\ 0, & \text{all other } t \end{cases}$$

From Table 2.2, the FT is

$$F(\omega) = \frac{2}{\omega}\sin\left(\frac{T\omega}{2}\right) \tag{2.67}$$

Note that by l'Hôpital's rule, $F(0) = T$. A plot of $|F(\omega)|$ vs. ω is given in Figure 2.1.

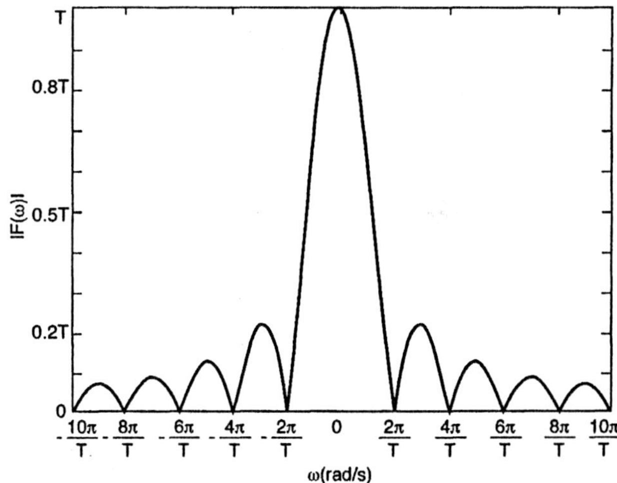

Figure 2.1 Magnitude spectrum of the T-second rectangular pulse.

From Figure 2.1, it is seen that most of the frequency content of the rectangular pulse is contained in the main lobe, which runs from $-2\pi/T$ to $2\pi/T$ rad/s. Also the plot shows that there is no finite positive number B such that $|F(\omega)|$ is zero for all $\omega > B$. However, $|F(\omega)|$ is converging to zero as $\omega \to \infty$, and thus this signal can still be viewed as being bandlimited. Since the maximum value of $|F(\omega)|$ is $F_{\max} = T$, the 3-dB bandwidth of the T-second rectangular pulse is the smallest positive number $B_{3\text{dB}}$ for which

$$\left|\frac{2}{\omega}\sin\left(\frac{T\omega}{2}\right)\right| \leq .707T \quad \text{for all} \quad \omega > B_{3\text{dB}} \tag{2.68}$$

From Figure 2.1 it is clear that if the duration T of the rectangular pulse is decreased, the magnitude spectrum spreads out, and thus the 3-dB bandwidth increases. Hence, a shorter duration pulse has a wider 3-dB bandwidth. This result is true in general; that is, signals with shorter time durations have wider bandwidths than signals with longer time durations.

2.3.1 Spectrum of a Signal Having a Rational Laplace Transform

Now suppose that the signal $f(t)$ is zero for all $t < 0$, and that the Laplace transform $F(s)$ of $f(t)$ is rational in s; that is $F(s) =$

$N(s)/D(s)$ where $N(s)$ and $D(s)$ are polynomials in s given by Equations 2.20 and 2.21. It was noted in the previous section that if the region of convergence of $F(s)$ includes the imaginary axis ($j\omega$-axis) of the complex plane, then the FT $F(\omega)$ is equal to the Laplace transform $F(s)$ with $s = j\omega$. When $F(s)$ is rational, it turns out that the region of convergence includes the $j\omega$-axis if and only if all the poles of $F(s)$ lie in the OLHP; thus, in this case, the FT is given by

$$F(\omega) = F(s)|_{s=j\omega} \tag{2.69}$$

For example, if $f(t) = ce^{-at}$ for $t \geq 0$ with $a > 0$, then $F(s) = c/(s+a)$ which has a single pole at $s = -a$. Since the point $-a$ lies in the OLHP, the FT of the exponential function ce^{-at} is

$$F(\omega) = \frac{c}{j\omega + a} \tag{2.70}$$

It follows from Equation 2.70 that the 3-dB bandwidth is equal to a. Hence, the farther over in the OLHP the pole $-a$ is (i.e., the larger a is), the larger the bandwidth of the signal. Since the rate of decay to zero of ce^{-at} increases as a is increased, this result again confirms the property that shorter duration time signals have wider bandwidths. In the case when $c = a = 1$, a plot of the magnitude spectrum $|F(\omega)|$ is shown in Figure 2.2. For any real values of a and c, the magnitude spectrum rolls off to zero at the rate of 20 dB/decade where a decade is a factor of ten in frequency.

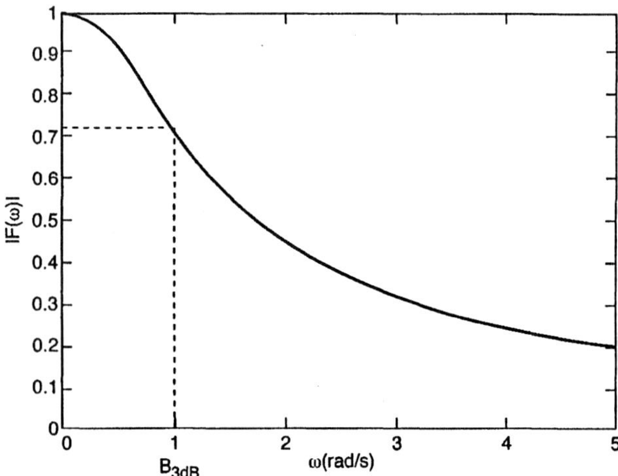

Figure 2.2 Magnitude spectrum of the exponential function e^{-t}.

As another example, consider the signal $f(t)$ whose Laplace transform is

$$F(s) = \frac{c}{s^2 + 2\zeta\omega_n s + \omega_n^2} \tag{2.71}$$

where ω_n is assumed to be strictly positive ($\omega_n > 0$). In this case, $F(s)$ has two poles p_1 and p_2 given by

$$\begin{aligned} p_1 &= -\zeta\omega_n + \omega_n\sqrt{\zeta^2 - 1} \\ p_2 &= -\zeta\omega_n - \omega_n\sqrt{\zeta^2 - 1} \end{aligned} \tag{2.72}$$

When $\zeta > 1$, both poles are real, nonrepeated, and lie in the OLHP (assuming that $\omega_n > 0$). As $\zeta \to \infty$, the pole p_1 moves along the negative real axis to the origin of the complex plane and the pole p_2 goes to $-\infty$ along the negative axis of the complex plane. For $\zeta > 1$, $F(s)$ can be expanded by partial fractions as follows:

$$F(s) = \frac{c}{(s-p_1)(s-p_2)} = \frac{c}{p_1 - p_2}\left[\frac{1}{s-p_1} - \frac{1}{s-p_2}\right] \tag{2.73}$$

Taking the inverse Laplace transform gives

$$f(t) = \frac{c}{p_1 - p_2}\left[e^{p_1 t} - e^{p_2 t}\right] \tag{2.74}$$

and thus $f(t)$ is a sum of two decaying real exponentials. Since both poles lie in the OLHP, the FT $F(\omega)$ is given by

$$F(\omega) = \frac{c}{\omega_n^2 - \omega^2 + j(2\zeta\omega_n\omega)} \tag{2.75}$$

For the case when $c = \omega_n^2 = 100$ and $\zeta = 2$, the plot of the magnitude spectrum $|F(\omega)|$ is given in Figure 2.3. In this case, the spectral content of the signal $f(t)$ rolls off to zero at the rate of 40dB/decade, starting with the peak magnitude of 1 at $\omega = 0$.

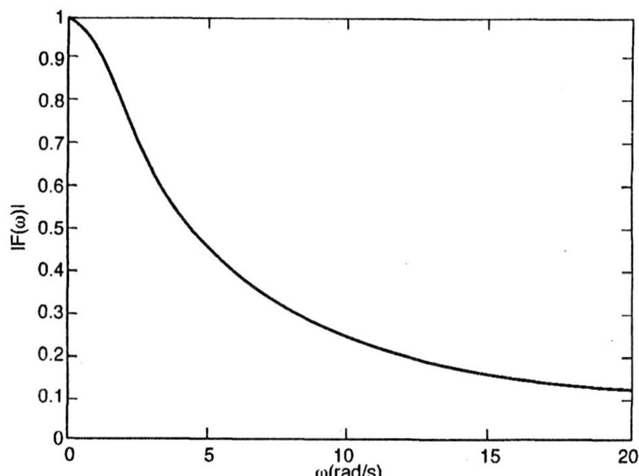

Figure 2.3 Magnitude spectrum of the signal with transform $F(s) = 100/(s^2 + 40s + 100)$.

When $\zeta = 1$, the poles p_1 and p_2 of $F(s)$ are both equal to $-\omega_n$, and $F(s)$ becomes

$$F(s) = \frac{c}{(s + \omega_n)^2} \tag{2.76}$$

Taking the inverse transform gives

$$f(t) = cte^{-\omega_n t} \tag{2.77}$$

Since ω_n is assumed to be strictly positive, when $\zeta = 1$ both the poles are in the OLHP; in this case, the FT is

$$F(\omega) = \frac{c}{(j\omega + \omega_n)^2} \tag{2.78}$$

As ζ varies from 1 to -1, the poles of $F(s)$ trace out a circle in the complex plane with radius ω_n. The loci of pole locations is shown in Figure 2.4. Note that the poles begin at $-\omega_n$ when $\zeta = 1$, then split apart and approach the $j\omega$-axis at $\pm j\omega_n$ as $\zeta \to 0$ and then move to ω_n as $\zeta \to -1$. For $-1 < \zeta < 1$, the inverse transform of $F(s)$ can be determined by first completing the square in the denominator of $F(s)$:

$$F(s) = \frac{c}{(s+\zeta\omega_n)^2+\omega_d^2} \tag{2.79}$$

where

$$\omega_d = \omega_n\sqrt{1-\zeta^2} > 0 \tag{2.80}$$

Note that ω_d is equal to the imaginary part of the pole p_1 given by Equation 2.72. Using Table 2.4, we have that the inverse transform of $F(s)$ is

$$f(t) = \frac{c}{\omega_d}e^{-\zeta\omega_n t}\sin\omega_d t \tag{2.81}$$

From Equation 2.81, it is seen that $f(t)$ now contains a sinusoidal factor. When $0 < \zeta < 1$, the poles lie in the OLHP, and the signal is a decaying sinusoid. In this case, the FT is

$$F(\omega) = \frac{c}{(j\omega+\zeta\omega_n)^2+\omega_d^2} \tag{2.82}$$

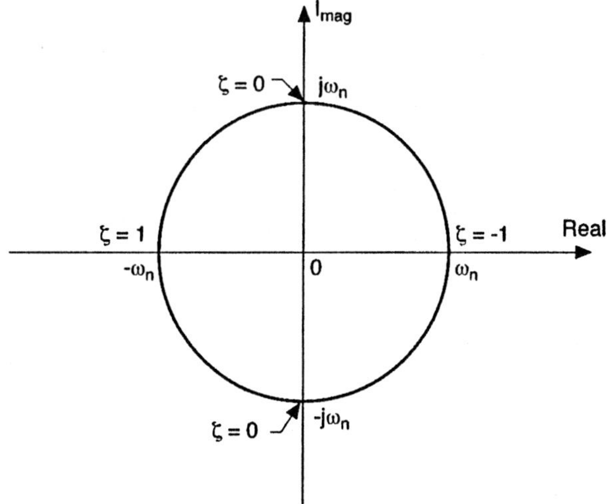

Figure 2.4 Loci of poles of $F(s)$ as ζ varies from 1 to -1.

The magnitude spectrum $|F(\omega)|$ is plotted in Figure 2.5 for the values $c = \omega_n^2 = 100$ and $\zeta = 0.3, 0.5, 0.7$. Note that for $\zeta = 0.5$ and 0.3, a peak appears in $|F(\omega)|$. This corresponds to the sinusoidal oscillation resulting from the $\sin(\omega_d t)$ factor in $f(t)$. Also note that as ζ is decreased from 0.5 to 0.3, the peak increases in magnitude, which signifies a longer duration oscillation in the signal $f(t)$. This result is expected since the poles are approaching the $j\omega$-axis of the complex plane as $\zeta \to 0$. As $\zeta \to 0$, the peak in $|F(\omega)|$ approaches ∞, so that $|F(\omega)|$ does not exist (in the ordinary sense) in the limit as $\zeta \to 0$. When $\zeta \to 0$, the signal $f(t)$ is purely oscillatory and does not have a FT in the ordinary sense. In addition, when $\zeta < 0$ there is no FT (in any sense) since there is a pole of $F(s)$ in the open right half-plane (ORHP).

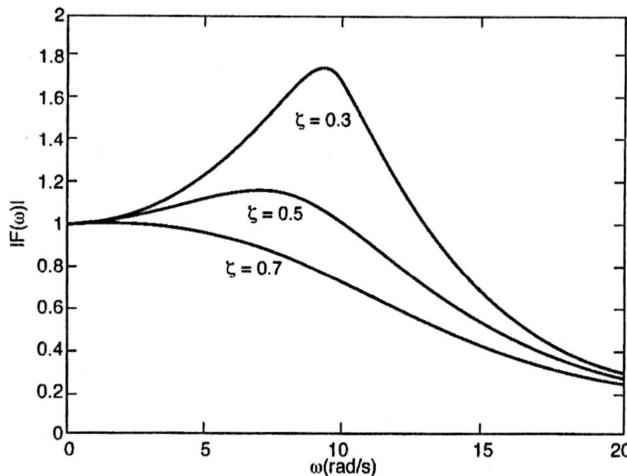

Figure 2.5 Magnitude spectrum of the signal with transform $F(s) = 100/(s^2 + 20\zeta s + 100)$ and with $\zeta = 0.7, 0.5, 0.3$.

The above results lead to the following generalized properties of the magnitude spectrum $|F(\omega)|$ of a signal $f(t)$ whose Laplace transform $F(s)$ is rational with all poles in the OLHP:

- If the poles of $F(s)$ are real, the magnitude spectrum $|F(\omega)|$ simply rolls off to zero as $\omega \to \infty$, starting with a peak value at $\omega = 0$ of $F(0)$.
- If $F(s)$ has a complex conjugate pair of poles at $s = \sigma \pm j\omega_d$ and if the ratio σ/ω_d is sufficiently small, then $|F(\omega)|$ will have a peak located approximately at $\omega = \omega_d$.

2.3.2 Numerical Computation of the Fourier Transform

In many applications, the signal $f(t)$ cannot be given in function form; rather, all one has are a set of sample values $f(k) = f(kT)$, where k ranges over a subset of integers and T is the sampling interval. Without loss of generality, we can assume that k starts with $k = 0$. Also, since all signals arising in practice are of finite duration in time, we can assume that $f(k)$ is zero for all $k \geq N$ for some positive integer N. The problem is then to determine the FT of $f(t)$ using the sample values $f(k) = f(kT)$ for $k = 0, 1, 2, \ldots, N-1$.

One could also carry out a discrete-time analysis by taking the N-point DFT F_n of the sampled signal $f(k)$. In the previous section, it was shown that F_n is equal to $F(\frac{2\pi n}{N})$ for $n = 0, 1, 2, \ldots, N-1$, where $F(\frac{2\pi n}{N})$ is the DTFT $F(\Omega)$ of $f(k)$ with the frequency variable Ω evaluated at $2\pi n/N$. Hence the discrete-time counterpart of the frequency spectrum can be determined from F_n. For details on this, see "Further Reading."

Again letting $F(\omega)$ denote the FT of $f(t)$, we can carry out a numerical computation of the FT as follows. First, since $f(t)$ is zero for $t < 0$ and $t \geq NT$, from the definition of the FT in Equation 2.1 we have

$$F(\omega) = \int_0^{NT} f(t)e^{-j\omega t}dt \tag{2.83}$$

Assuming that $f(t)$ is approximately constant over each T-second interval $[(k-1)T, kT]$, we obtain the following approximation to Equation 2.83:

$$F(\omega) = \sum_{k=0}^{N-1} \left[\int_{kT}^{kT+T} e^{-j\omega t}\right] f(k) \tag{2.84}$$

Then carrying out the integration in the right-hand side of Equation 2.84 gives

$$F(\omega) = \frac{1-e^{-j\omega T}}{j\omega} \sum_{k=0}^{N-1} e^{-j\omega kT} f(k) \tag{2.85}$$

Finally, setting $\omega = 2\pi n/NT$ in Equation 2.85 yields

$$F\left(\frac{2\pi n}{NT}\right) = \frac{1-e^{-j2\pi n/N}}{j2\pi n/NT} F_n \tag{2.86}$$

where F_n is the DFT of $f(k)$ given by Equation 2.46.

It should be stressed that the relationship in Equation 2.86 is only an approximation; that is, the right-hand side of Equation 2.86 is an approximation to $F(2\pi n/NT)$. In general, the approximation is more accurate the larger N is and/or the smaller T is. For a good result, it is also necessary that $f(t)$ be suitably small for $t < 0$ and $t \geq NT$.

As an example, let $f(t)$ be the 2-second pulse given by $f(t) = 1$ for $0 \leq t \leq 2$ and $f(t) = 0$ for all other t. Using the time shift property in Table 2.1 and the FT of a rectangular pulse given in Table 2.2, we have that the FT of $f(t)$ is

$$F(\omega) = \frac{2}{\omega}\sin(\omega)e^{-j\omega} \tag{2.87}$$

A MATLAB program (adapted from [4]) for computing the exact magnitude spectrum $|F(\omega)|$ and the approximation based on Equation 2.86 is given in Figure 2.6. The program was run for the case when $N = 128$ and $T = 0.1$, with the plots shown in Figure 2.7. Note that the approximate values are fairly close, at least for frequencies over the span of the main lobe. A better approximation can be achieved by increasing N and/or decreasing T. The reader is invited to try this using the program in Figure 2.6.

```
N = input('Input N: ');
T = input('Input T: ');
t = 0:T:2;
f = [ones(1,length(t))zeros(1,N-length(t))];
Fn = fft(f);
gam = 2*pi/N/T;
n = 0:10/gam;
Fapp = (1-exp(-j*n*gam*T))/j/n/gam*Fn;
w = 0:.05:10;
Fexact = 2*sin(w)./w;
plot (n*gam,abs(Fapp(1:length(n))),'og',w,abs(Fexact),'b')
```

Figure 2.6 MATLAB program for computing the Fourier transform of the 2-second pulse.

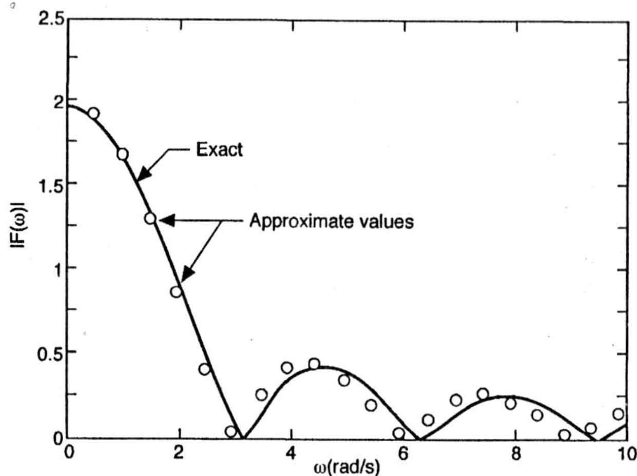

Figure 2.7 Exact and approximate magnitude spectra of the 2-second pulse.

2.3.3 Solution of Differential Equations

One of the major applications of the Laplace transform is in solving linear differential equations. To pursue this, we begin by considering the first-order linear constant-coefficient differential equation given by

$$\dot{f}(t) + af(t) = w(t), \quad t \geq 0 \tag{2.88}$$

where $\dot{f}(t)$ is the derivative of $f(t)$ and $w(t)$ is an arbitrary real-valued function of t. To solve Equation 2.88, we apply the Laplace transform to both sides of the equation. Using linearity and the derivative properties of the Laplace transform given in Table 2.3, we have

$$sF(s) - f(0) + aF(s) = W(s) \tag{2.89}$$

where $F(s)$ is the transform of $f(t)$, $W(s)$ is the transform of $w(t)$, and $f(0)$ is the initial condition. Then solving Equation 2.89 for $F(s)$ gives

$$F(s) = \frac{1}{s+a}\left[f(0) + W(s)\right] \tag{2.90}$$

Taking the inverse Laplace transform of $F(s)$ then yields the solution $f(t)$. For example, if $w(t)$ is the unit-step function $u(t)$ and $a \neq 0$, then $W(s) = 1/s$ and $F(s)$ becomes

$$F(s) = \frac{1}{s+a}\left[f(0) + \frac{1}{s}\right] = \frac{f(0)s+1}{(s+a)s} = \frac{f(0)-\frac{1}{a}}{s+a} + \frac{\frac{1}{a}}{s} \tag{2.91}$$

Taking the inverse transform gives

$$f(t) = \left[f(0) - \frac{1}{a}\right]e^{-at} + \frac{1}{a}, \quad t \geq 0 \tag{2.92}$$

Now consider the second-order differential equation

$$\ddot{f}(t) + a_1\dot{f}(t) + a_0 f(t) = w(t) \tag{2.93}$$

Again using the derivative property of the Laplace transform, taking the transform of both sides of Equation 2.93 we obtain

$$s^2F(s) - sf(0) - \dot{f}(0) + a_1\left[sF(s) - f(0)\right] + a_0F(s) = W(s) \tag{2.94}$$

where $f(0)$ and $\dot{f}(0)$ are the initial conditions. Solving Equation 2.94 for $F(s)$ yields

$$F(s) = \frac{1}{s^2+a_1s+a_0}\left[f(0)s + \dot{f}(0) + a_1f(0) + W(s)\right] \tag{2.95}$$

For example, if $a_0 = 2$, $a_1 = 3$, and $w(t) = u(t)$, then

$$F(s) = \frac{1}{(s+1)(s+2)}\left[f(0)s + \dot{f}(0) + 3f(0) + \frac{1}{s}\right]$$

$$F(s) = \frac{f(0)s^2 + \left[\dot{f}(0) + 3f(0)\right]s + 1}{(s+1)(s+2)s}$$

$$F(s) = \frac{2f(0) + \dot{f}(0) - 1}{s+1} + \frac{\frac{-5}{2}f(0) - \dot{f}(0) + \frac{1}{2}}{s+2} + \frac{\frac{1}{2}}{s} \quad (2.96)$$

Inverse transforming Equation 2.96 gives

$$f(t) = \left[2f(0) + \dot{f}(0) - 1\right]e^{-t} + \left[-2.5f(0) - \dot{f}(0) + 0.5\right]e^{-2t} + 0.5, \quad t \geq 0 \quad (2.97)$$

For the general case, consider the nth-order linear constant-coefficient differential equation

$$f^{(n)}(t) + \sum_{i=0}^{n-1} a_i f^{(i)}(t) = w(t) \quad (2.98)$$

where $f^{(i)}(t)$ is the ith derivative of $f(t)$. Given $w(t)$ and the initial conditions $f(0), \dot{f}(0), \ldots, f^{(n-1)}(0)$, the solution $f(t)$ to Equation 2.98 is unique. The solution can be determined by taking the transform of both sides of Equation 2.98 and solving for $F(s)$. This yields

$$F(s) = \frac{1}{D(s)}\left[N(s) + W(s)\right] \quad (2.99)$$

where $D(s)$ is the polynomial

$$D(s) = s^n + a_{n-1}s^{n-1} + \cdots + a_1 s + a_0 \quad (2.100)$$

and $N(s)$ is a polynomial in s of the form

$$N(s) = b_{n-1}s^{n-1} + b_{n-2}s^{n-2} + \cdots + b_1 s + b_0 \quad (2.101)$$

The coefficients $b_0, b_1, \ldots, b_{n-1}$ of $N(s)$ depend on the values of the n initial conditions $f(0), \dot{f}(0), \ldots, f^{(n-1)}(0)$. The relationship between the b_i and the initial conditions is given by the matrix equation

$$b = Px \quad (2.102)$$

where b and x are the column vectors

$$b = \begin{bmatrix} b_0 \\ b_1 \\ \vdots \\ b_{n-2} \\ b_{n-1} \end{bmatrix} \qquad x = \begin{bmatrix} f(0) \\ \dot{f}(0) \\ \vdots \\ f^{(n-2)}(0) \\ f^{(n-1)}(0) \end{bmatrix} \quad (2.103)$$

and P is the n-by-n matrix given by

$$P = \begin{bmatrix} a_1 & a_2 & \cdots & a_{n-2} & a_{n-1} & 1 \\ a_2 & a_3 & \cdots & a_{n-1} & 1 & 0 \\ \vdots & \vdots & & \vdots & \vdots & \vdots \\ a_{n-1} & 1 & \cdots & 0 & 0 & 0 \\ 1 & 0 & \cdots & 0 & 0 & 0 \end{bmatrix} \quad (2.104)$$

The matrix P given by Equation 2.104 is invertible for any values of the a_i, and thus there is a one-to-one and onto correspondence between the set of initial conditions and the coefficients of the polynomial $N(s)$ in Equation 2.101. In particular, this implies that for any given vector b of coefficients of $N(s)$, there is a vector x of initial conditions that results in the polynomial $N(s)$ with the given coefficients. From Equation 2.102, it is seen that $x = P^{-1}b$ where P^{-1} is the inverse of P.

Once $N(s)$ is computed using Equation 2.102, the solution $f(t)$ to Equation 2.98 can then be determined by inverse transforming Equation 2.99. If $W(s)$ is a rational function of s, then the right-hand side of Equation 2.99 is rational in s and thus, in this case, $f(t)$ can be computed via a partial fraction expansion.

An interesting consequence of the above constructions is the following characterization of a real-valued function $f(t)$ whose Laplace transform $F(s)$ is rational. Suppose that

$$F(s) = \frac{N(s)}{D(s)} \quad (2.105)$$

where $D(s)$ and $N(s)$ are given by Equations 2.100 and 2.101, respectively. Then comparing Equations 2.105 and 2.99 shows that $f(t)$ is the solution to the nth-order homogeneous equation

$$f^{(n)}(t) + \sum_{i=0}^{n-1} a_i f^{(i)}(t) = 0 \quad (2.106)$$

with the initial conditions given by $x = P^{-1}b$, where x and b are defined by Equation 2.103. Hence, any function $f(t)$ having a rational Laplace transform is the solution to a homogeneous differential equation. This result is of fundamental importance in the theory of systems and controls.

2.3.4 Solution of Difference Equations

The discrete-time counterpart to the solution of differential equations using the Laplace transform is the solution of difference equations using the z-transform. We begin by considering the first-order linear constant-coefficient difference equation

$$f(k+1) + af(k) = w(k), \quad k \geq 0 \quad (2.107)$$

where $w(k)$ is an arbitrary real-valued function of the integer variable k. Taking the z-transform of Equation 2.107 using the linearity and left shift properties given in Table 2.6 yields

$$zF(z) - f(0)z + aF(z) = W(z) \quad (2.108)$$

where $F(z)$ is the z-transform of $f(k)$ and $f(0)$ is the initial condition. Then solving Equation 2.108 for $F(z)$ gives

$$F(z) = \frac{1}{z+a}\left[f(0)z + W(z)\right] \quad (2.109)$$

For example, if $w(k)$ is the unit-step function $u(k)$ and $a \neq 1$, then $W(z) = z/(z-1)$ and $F(z)$ becomes

$$F(z) = \frac{1}{z+a}\left[f(0)z + \frac{z}{z-1}\right]$$

$$F(z) = \frac{f(0)z(z-1) + z}{(z+a)(z-1)} \quad (2.110)$$

Then

$$\frac{F(z)}{z} = \frac{f(0)(z-1)+1}{(z+a)(z-1)} \tag{2.111}$$

and expanding by partial fractions gives

$$\frac{F(z)}{z} = \frac{f(0)-\frac{1}{1+a}}{z+a} + \frac{\frac{1}{1+a}}{z-1} \tag{2.112}$$

Thus

$$F(z) = \frac{\left[f(0)-\frac{1}{1+a}\right]z}{z+a} + \frac{\frac{1}{1+a}z}{z-1} \tag{2.113}$$

and taking the inverse z-transform gives

$$f(k) = \left[f(0) - \tfrac{1}{1+a}\right](-a)^k + \tfrac{1}{1+a}, \quad k \geq 0 \tag{2.114}$$

For the general case, consider the nth-order linear constant-coefficient difference equation

$$f(k+n) + \sum_{i=0}^{n-1} a_i f(k+i) = w(k) \tag{2.115}$$

The initial conditions for Equation 2.115 may be taken to be the n values $f(0), f(1), \ldots, f(n-1)$. Another choice is to take the initial values to be $f(-1), f(-2), \ldots, f(-n)$. We prefer the latter choice since the initial values are given for negative values of the time index k. In this case, the use of the z-transform to solve Equation 2.115 requires that the equation be time shifted. This is accomplished by replacing k by $k-n$ in Equation 2.115, which yields

$$f(k) + \sum_{i=1}^{n} a_{n-i} f(k-i) = w(k-n) \tag{2.116}$$

Then using the right-shift property of the z-transform and transforming Equation 2.116 yields

$$\begin{aligned} F(z) &= \frac{1}{D(z^{-1})}\Bigg[N(z^{-1}) + z^{-n}W(z) \\ &\quad + \sum_{i=1}^{n} w(-i)z^{-n+i}\Bigg] \end{aligned} \tag{2.117}$$

where $D(z^{-1})$ and $N(z^{-1})$ are polynomials in z^{-1} given by

$$\begin{aligned} D(z^{-1}) &= 1 + a_{n-1}z^{-1} + \cdots + a_1 z^{-n+1} \\ &\quad + a_0 z^{-n} \end{aligned} \tag{2.118}$$

$$\begin{aligned} N(z^{-1}) &= b_{n-1} + b_{n-2}z^{-1} + \cdots + b_1 z^{-n+2} \\ &\quad + b_0 z^{-n+1} \end{aligned} \tag{2.119}$$

The coefficients b_i of $N(z^{-1})$ are related to the initial values by the matrix equation

$$b = Q\phi \tag{2.120}$$

where b and ϕ are the column vectors

$$b = \begin{bmatrix} b_0 \\ b_1 \\ \vdots \\ b_{n-2} \\ b_{n-1} \end{bmatrix} \qquad \phi = \begin{bmatrix} f(-1) \\ f(-2) \\ \vdots \\ f(-n+1) \\ f(-n) \end{bmatrix} \tag{2.121}$$

and Q is the n-by-n matrix given by

$$Q = \begin{bmatrix} a_0 & 0 & \cdots & 0 & 0 & 0 \\ a_1 & a_0 & \cdots & 0 & 0 & 0 \\ \vdots & \vdots & & \vdots & \vdots & \vdots \\ a_{n-2} & a_{n-3} & \cdots & a_1 & a_0 & 0 \\ a_{n-1} & a_{n-2} & \cdots & a_2 & a_1 & a_0 \end{bmatrix} \tag{2.122}$$

The matrix Q given by Equation 2.122 is invertible for any values of the a_i as long as $a_0 \neq 0$, and thus for any given vector b of coefficients of the polynomial $N(z^{-1})$, there is a vector ϕ of initial conditions that results in the polynomial $N(z^{-1})$ with the given coefficients. Clearly, if $a_0 \neq 0$, then $\phi = Q^{-1}b$ where Q^{-1} is the inverse of Q.

Once $N(z^{-1})$ is computed using Equation 2.120, the solution $f(k)$ to Equation 2.115 or Equation 2.116 can then be determined by inverse transforming Equation 2.117. If $W(z)$ is a rational function of z, then the right-hand side of Equation 2.117 is rational in z, and in this case, $f(k)$ can be computed via a partial fraction expansion.

Finally, it is worth noting (in analogy with the Laplace transform) that any function $f(k)$ having a rational z-transform $F(z)$ is the solution to a homogeneous difference equation of the form

$$f(k+n) + \sum_{i=0}^{n-1} a_i f(k+i) = 0 \tag{2.123}$$

where the initial conditions are determined using Equations 2.120 to 2.122.

2.3.5 Defining Terms

3-dB bandwidth: For a bandlimited signal, this is the smallest value $B_{3\text{dB}}$ for which the magnitude spectrum $|F(\omega)|$ is down by 3 dB or more from the peak magnitude for all $\omega > B_{3\text{dB}}$.

Bandlimited signal: A signal $f(t)$ whose Fourier Transform $F(\omega)$ is zero (or approximately zero) for all $\omega > B$, where B is a finite positive number.

Irrational function: A function $F(s)$ of a complex variable s that cannot be expressed as a ratio of polynomials in s.

Magnitude spectrum: The magnitude $|F(\omega)|$ of the Fourier Transform of a function $f(t)$.

One-sided (or unilateral) transform: A transform that operates on a function $f(t)$ defined for $t \geq 0$.

Open left half-plane (OLHP): The set of all complex numbers having negative real part.

Open unit disk: The set of all complex numbers whose magnitude is less than 1.

Phase spectrum: The angle $\angle F(\omega)$ of the FT of a function $f(t)$.

Poles of a rational function N(s)/D(s): The values of s for which $D(s) = 0$, assuming that $N(s)$ and $D(s)$ have no common factors.

Proper rational function: A rational function $N(s)/D(s)$ where the degree of $N(s)$ is less than or equal to the degree of $D(s)$.

Rational function: A ratio of two polynomials $N(s)/D(s)$ where s is a complex variable.

Region of convergence: The set of all complex numbers for which a transform exists (i.e., is well defined) in the ordinary sense.

Residues: The values of the numerator constants in a partial fraction expansion of a rational function.

Strictly proper rational function: A rational function $N(s)/D(s)$ where the degree of $N(s)$ is strictly less than the degree of $D(s)$.

Two-sided (or bilateral) transform: A transform that operates on a function $f(t)$ define for $-\infty < t < \infty$.

Zeros of a rational function $N(s)/D(s)$: The values of s for which $N(s) = 0$, assuming that $N(s)$ and $D(s)$ have no common factors.

References

[1] Barker, R.H., The pulse transfer function and its applications to sampling servo systems, *Proc. IEEE,* 99, Part IV, 302–317, 1952.

[2] Fourier, J.B.J., *The Analytical Theory of Heat,* Cambridge, (transl. A. Freeman) 1878.

[3] Jury, E.I., Analysis and synthesis of sampled-data control systems, *Communications and Electronics,* 1954, 1–15.

[4] Kamen, E.W. and Heck, B.S., *Fundamentals of Signals and Systems with MATLAB,* Prentice Hall, Englewood Cliffs, NJ, 1996.

[5] Ragazzini, J.R. and Zadeh, L. A., The analysis of sampled-data systems, *Trans. AIEE,* 71, Part II:225–232, 1952.

Further Reading

A mathematically rigorous development of the FT can be found in Papoulis, A. 1962. *The Fourier Integral and Its Applications,* McGraw-Hill, New York.
Bracewell, R.M. 1965. *The Fourier Transform and Its Applications,* McGraw-Hill, New York.

An in-depth treatment on the use of the Fast Fourier transform in Fourier analysis can be found in
Brigham, E.O. 1988. *The Fast Fourier Transform and Its Applications,* Prentice Hall, Englewood Cliffs, NJ.

For a detailed development of the Laplace transform, see Rainville, E.D. 1963. *The Laplace Transform: An Introduction,* Macmillan, New York.

For a thorough development of the z-transform, see Jury, E.I. 1964. *Theory and Application of the z-Transform Method,* Wiley, New York.

Treatments of the Fourier, Laplace, and z-transforms can be found in textbooks on signals and systems, such as

Oppenhein, A.V. and Willsky, A.S., 1983. *Signals and Systems,* Prentice Hall, Englewood Cliffs, NJ.

Ziemer, R.E., Tranter, W.H. and Fannin, D.R., 1993. 3rd ed., Macmillan, New York.

For a development that is integrated together with the MATLAB software package, see [4].

3

Matrices and Linear Algebra

Bradley W. Dickinson
Princeton University

3.1 Introduction

Matrices and linear algebra are indispensable tools for analysis and computation in problems involving systems and control. This chapter presents an overview of these subjects that highlights the main concepts and results.

3.2 Matrices

To introduce the notion of a *matrix*, we start with some notation that serves as the framework for describing matrices and the rules for manipulating them in algebraic expressions.

Let $\mathcal{R}$ be a *ring*, a set of quantities together with definitions for addition and multiplication operations. Standard notations for some examples of rings arising frequently in control systems applications include $\mathbb{R}$ (the real numbers), $\mathbb{C}$ (the complex numbers), $\mathbb{R}[s]$ (the set of polynomials in the variable s having real coefficients), and $\mathbb{R}(s)$ (the set of rational functions, i.e., ratios of polynomials). Each of these rings has distinguished elements 0 and 1, the identity elements for addition and multiplication, respectively. Rings for which addition and multiplication are commutative operations and for which multiplicative inverses of all nonzero quantities exist are known as *fields*; in the examples given, the real numbers, the complex numbers, and the rational functions are fields.

A *matrix* (more descriptively a $\mathcal{R}$-matrix, e.g., a complex matrix or a real matrix) is a rectangular array of *matrix elements* that belong to the ring $\mathcal{R}$. When $\mathbf{A}$ is a matrix with m rows and n columns, denoted $\mathbf{A} \in \mathcal{R}^{m\times n}$, $\mathbf{A}$ is said to be "an m by n (written $m \times n$) matrix," and its matrix elements are indexed with a double subscript, the first indicating the row position and the second indicating the column position. The notation used is

$$\mathbf{A} = \begin{bmatrix} a_{11} & a_{12} & \cdots & a_{1n} \\ a_{21} & a_{22} & \cdots & a_{2n} \\ \vdots & \vdots & & \vdots \\ a_{m1} & a_{m2} & \cdots & a_{mn} \end{bmatrix} \quad \text{and} \quad (\mathbf{A})_{ij} = a_{ij} \qquad (3.1)$$

Three special "shapes" of matrices commonly arise and are given descriptive names. $\mathcal{R}^{m\times 1}$ is the set of column matrices, also known as *column m-vectors*, *m-vectors*, or when no ambiguity results simply as *vectors*. Similarly, $\mathcal{R}^{1\times n}$ is the set of *row vectors*. Finally, $\mathcal{R}^{n\times n}$ is the set of *square matrices* of size n.

0-8493-8570-9/96/$0.00+$.50

3.2.1 Matrix Algebra

Since matrix elements belong to a ring $\mathcal{R}$, they may be combined in algebraic expressions involving addition and multiplication operations. This provides the means for defining algebraic operations for matrices. The usual notion of equality is adopted: two $m \times n$ matrices are equal if and only if they have the same elements.

Scalar Multiplication

The product of a matrix $\mathbf{A} \in \mathcal{R}^{m\times n}$ and a *scalar* $z \in \mathcal{R}$ may always be formed. The resulting matrix, also in $\mathcal{R}^{m\times n}$, is obtained by element-wise multiplication:

$$(z\mathbf{A})_{ij} = za_{ij} = (\mathbf{A}\,z)_{ij} \tag{3.2}$$

Matrix Addition

Two matrices, both in $\mathcal{R}^{m\times n}$, say $\mathbf{A}$ and $\mathbf{B}$, may be added to produce a third matrix $\mathbf{C} \in \mathcal{R}^{m\times n}$, $\mathbf{A} + \mathbf{B} = \mathbf{C}$, where $\mathbf{C}$ is the matrix of elementwise sums

$$(\mathbf{C})_{ij} = c_{ij} = a_{ij} + b_{ij} \tag{3.3}$$

Matrix Multiplication

Two matrices, say $\mathbf{A}$ and $\mathbf{B}$, may be multiplied with $\mathbf{A}$ as the left factor and $\mathbf{B}$ as the right factor if and only if their sizes are compatible: if $\mathbf{A} \in \mathcal{R}^{m_\mathrm{A}\times n_\mathrm{A}}$ and $\mathbf{B} \in \mathcal{R}^{m_\mathrm{B}\times n_\mathrm{B}}$, then it is required that $n_\mathrm{A} = m_\mathrm{B}$. That is, the number of columns of the left factor must equal the number of rows of the right factor. When this is the case, the product matrix $\mathbf{C} = \mathbf{A}\,\mathbf{B}$ is $m_\mathrm{A} \times n_\mathrm{B}$, that is, the product has the same number of rows as the left factor and the same number of columns as the right factor. With simpler notation, if $\mathbf{A} \in \mathcal{R}^{m\times n}$ and $\mathbf{B} \in \mathcal{R}^{n\times p}$, then the product matrix $\mathbf{C} \in \mathcal{R}^{m\times p}$ is given by

$$(\mathbf{C})_{ij} = c_{ij} = \sum_{k=1}^{n} a_{ik} b_{kj} \tag{3.4}$$

Using the interpretation of the rows and the columns of a matrix as matrices themselves, several important observations follow from the defining equation for the elements of a matrix product.

1. The columns of the product matrix $\mathbf{C}$ are obtained by multiplying the matrix $\mathbf{A}$ times the corresponding columns of $\mathbf{B}$.
2. The rows of the product matrix $\mathbf{C}$ are obtained by multiplying the corresponding rows of $\mathbf{A}$ times the matrix $\mathbf{B}$.
3. The (i, j)th element of the product matrix, $(\mathbf{C})_{ij}$, is the product of the ith row of $\mathbf{A}$ times the jth column of $\mathbf{B}$.
4. The product matrix $\mathbf{C}$ may be expressed as the sum of products of the columns of $\mathbf{A}$ times the rows of $\mathbf{B}$.

Unlike matrix addition, matrix multiplication is generally not commutative. If the definition of matrix multiplication allows for both of the products $\mathbf{A}\,\mathbf{B}$ and $\mathbf{BA}$ to be formed, the two products are square matrices but they are not necessarily equal nor even the same size.

The addition and multiplication operations for matrices obey familiar rules of associativity and distributivity: (a) $(\mathbf{A} + \mathbf{B}) + \mathbf{C} = \mathbf{A} + (\mathbf{B} + \mathbf{C})$; (b) $(\mathbf{A}\,\mathbf{B})\mathbf{C} = \mathbf{A}\,(\mathbf{BC})$; (c) $(\mathbf{A} + \mathbf{B})\mathbf{C} = \mathbf{A}\,\mathbf{C} + \mathbf{BC}$; and (d) $\mathbf{A}\,(\mathbf{B} + \mathbf{C}) = \mathbf{A}\,\mathbf{B} + \mathbf{A}\,\mathbf{C}$.

The Zero Matrix and the Identity Matrix

The *zero matrix*, denoted $\mathbf{0}$, is any matrix whose elements are all zero:

$$(\mathbf{0})_{ij} = 0 \tag{3.5}$$

Usually the numbers of rows and columns of $\mathbf{0}$ will be understood from context; $\mathbf{0}_{m\times n}$ will specifically denote the $m \times n$ zero matrix. Clearly $\mathbf{0}$ is the additive identity element for matrix addition: $\mathbf{0} + \mathbf{A} = \mathbf{A} = \mathbf{A} + \mathbf{0}$; indeed, $\mathcal{R}^{m\times n}$ with the operation of matrix addition is a group. For matrix multiplication, if $\mathbf{A}$ is $m \times n$, then $\mathbf{0}_{m\times m}\mathbf{A} = \mathbf{0}_{m\times n} = \mathbf{A}\,\mathbf{0}_{n\times n}$.

The *identity matrix*, denoted $\mathbf{I}$, is a square matrix whose only nonzero elements are the ones along its main diagonal:

$$(\mathbf{I})_{ij} = \begin{cases} 1, & \text{for } i = j \\ 0, & \text{for } i \neq j \end{cases} \tag{3.6}$$

Again, the dimensions of $\mathbf{I}$ are usually obtained from context; the $n \times n$ identity matrix is specifically denoted $\mathbf{I}_n$. The identity matrix serves as an identity element for matrix multiplication: $\mathbf{I}_m\mathbf{A} = \mathbf{A} = \mathbf{A}\,\mathbf{I}_n$ for any $\mathbf{A} \in \mathcal{R}^{m\times n}$. This has an important implication for square matrices: $\mathcal{R}^{n\times n}$, with the operations of matrix addition and matrix multiplication, is a (generally noncommutative) ring.

3.2.2 Matrix Inverse

Closely related to matrix multiplication is the notion of *matrix inverse.* If $\mathbf{A}$ and $\mathbf{X}$ are square matrices of the same size and they satisfy $\mathbf{A}\,\mathbf{X} = \mathbf{I} = \mathbf{X}\,\mathbf{A}$, then $\mathbf{X}$ is called the matrix inverse of $\mathbf{A}$, and is denoted by $\mathbf{A}^{-1}$. The inverse matrix satisfies

$$\mathbf{A}\,\mathbf{A}^{-1} = \mathbf{I} = \mathbf{A}^{-1}\mathbf{A} \tag{3.7}$$

For a square matrix $\mathbf{A}$, if $\mathbf{A}^{-1}$ exists it is unique and $(\mathbf{A}^{-1})^{-1} = \mathbf{A}$. If $\mathbf{A}$ has a matrix inverse, $\mathbf{A}$ is said to be *invertible*; the terms *nonsingular* and *regular* are also used as synonyms for invertible. If $\mathbf{A}$ has no matrix inverse, it is said to be *noninvertible* or *singular*.

The invertible matrices in $\mathcal{R}^{n\times n}$, along with the operation of matrix multiplication, form a group, the *general linear group*, denoted by $GL(\mathcal{R}, n)$; $\mathbf{I}_n$ is the identity element of the group.

If $\mathbf{A}$ and $\mathbf{B}$ are square, invertible matrices of the same size, then their products are invertible also and

$$(\mathbf{A}\,\mathbf{B})^{-1} = \mathbf{B}^{-1}\mathbf{A}^{-1}\,;\ (\mathbf{BA})^{-1} = \mathbf{A}^{-1}\mathbf{B}^{-1} \tag{3.8}$$

This extends to products of more than two factors, giving the *product rule for matrix inverses*: The inverse of a product of square matrices is the product of their inverses taken in reverse order, provided the inverses of all of the factors exist.

Some Useful Matrix Inversion Identities

1. If the $n \times n$ matrices $\mathbf{A}$, $\mathbf{B}$, and $\mathbf{A}+\mathbf{B}$ are all invertible, then

$$(\mathbf{A}^{-1}+\mathbf{B}^{-1})^{-1} = \mathbf{A}(\mathbf{A}+\mathbf{B})^{-1}\mathbf{B} \qquad (3.9)$$

2. Assuming that the matrices have suitable dimensions and that the indicated inverses all exist, then

$$(\mathbf{A}+\mathbf{BCD})^{-1} = \mathbf{A}^{-1} - \mathbf{A}^{-1}\mathbf{B}(\mathbf{C}^{-1} + \mathbf{D}\mathbf{A}^{-1}\mathbf{B})^{-1}\mathbf{D}\mathbf{A}^{-1} \qquad (3.10)$$

This simplifies in the important special case when $\mathbf{C} = 1$, $\mathbf{B}$ is a column vector, and $\mathbf{D}$ is a row vector.

Determining whether a square matrix is invertible and, if so, finding its matrix inverse, is important in a variety of applications. The determinant is introduced as a means of characterizing invertibility.

3.2.3 The Determinant

The *determinant* of a square matrix $\mathbf{A} \in \mathcal{R}^{n\times n}$, denoted $\det \mathbf{A}$, is a scalar function taking the form of a sum of signed products of n matrix elements. While an explicit formula for $\det \mathbf{A}$ can be given [5], it is common to define the determinant inductively as follows. For $\mathbf{A} = [a_{11}] \in \mathcal{R}^{1\times 1}$, $\det \mathbf{A} = a_{11}$. For $\mathbf{A} \in \mathcal{R}^{n\times n}$, with $n > 1$,

$$\det \mathbf{A} = \sum_{k=1}^{n} (-1)^{i+k} a_{ik}\Delta_{ik} \quad \text{or} \quad \det \mathbf{A} = \sum_{k=1}^{n} (-1)^{i+k} a_{ki}\Delta_{ki} \qquad (3.11)$$

These are the *Laplace expansions* for the determinant corresponding to the ith row and ith column of $\mathbf{A}$, respectively. In these formulas, the quantity Δ_{ik} is the determinant of the $(n-1)\times(n-1)$ square matrix obtained by deleting the ith row and kth column of $\mathbf{A}$, and similarly for Δ_{ki}.

The quantities Δ_{ik} and Δ_{ki} are examples of $(n-1)\times(n-1)$ *minors* of $\mathbf{A}$; for any k, $1 \le k \le n-1$, an $(n-k)\times(n-k)$ minor of $\mathbf{A}$ is the determinant of an $(n-k)\times(n-k)$ square matrix obtained by deleting some set of k rows and k columns of $\mathbf{A}$.

For $\mathbf{A} \in \mathcal{R}^{2\times 2}$, the Laplace expansion corresponding to any row or column leads to the well-known formula: $\det \mathbf{A} = a_{11}a_{22} - a_{12}a_{21}$. For any n, $\det \mathbf{I}_n = 1$.

Properties of the Determinant

Many properties of determinants can be verified directly from the Laplace expansion formulas. For example: replacing any row of a matrix by its sum with another row does not change the value of the determinant; replacing a row (or a column) of a matrix with a multiple of itself changes the determinant by the same factor; interchanging two rows (or columns) of a matrix changes only the sign of the determinant. (These operations are known as the *elementary row and column operations.*)

If $\mathbf{A} \in \mathcal{R}^{n\times n}$ and $z \in \mathcal{R}$, then $\det(z\mathbf{A}) = z^n \det \mathbf{A}$. If $\mathbf{A}$ and $\mathbf{B}$ are matrices for which both products $\mathbf{AB}$ and $\mathbf{BA}$ are defined, then $\det(\mathbf{AB}) = \det(\mathbf{BA})$. If, in addition, both matrices are square, then

$$\det(\mathbf{AB}) = \det(\mathbf{BA}) = \det \mathbf{A}\ \det \mathbf{B} = \det \mathbf{B}\ \det \mathbf{A} \qquad (3.12)$$

This is the *product rule for determinants.*

3.2.4 Determinants and Matrix Inverses

Characterization of Invertibility

The determinant of an invertible matrix and the determinant of its inverse are reciprocals. If $\mathbf{A}$ is invertible, then

$$\det(\mathbf{A}^{-1}) = 1/\det \mathbf{A} \qquad (3.13)$$

This result indicates that invertibility of matrices is related to existence of multiplicative inverses in the underlying ring $\mathcal{R}$. In ring-theoretic terminology, the *units* of $\mathcal{R}$ are those ring elements having multiplicative inverses. When $\mathcal{R}$ is a field, all nonzero elements are units, but for $\mathcal{R} = \mathbb{R}[s]$ (or $\mathbb{C}[s]$), the ring of polynomials with real (or complex) coefficients, only the nonzero constants (i.e., the nonzero polynomials of degree 0) are units.

Determinants provide a characterization of invertibility as follows:

> The matrix $\mathbf{A} \in \mathcal{R}^{n\times n}$ is invertible if and only if $\det \mathbf{A}$ is a unit in $\mathcal{R}$.

When $\mathcal{R}$ is a field, all nonzero ring elements are units and the criterion for invertibility takes a simpler form:

> When $\mathcal{R}$ is a field, the matrix $\mathbf{A} \in \mathcal{R}^{n\times n}$ is invertible if and only if $\det \mathbf{A} \neq 0$.

Cramer's Rule and PLU Factorization

Cramer's rule provides a general formula for the elements of $\mathbf{A}^{-1}$ in terms of a ratio of determinants:

$$(\mathbf{A}^{-1})_{ij} = (-1)^{i+j}\Delta_{ji}/\det \mathbf{A} \qquad (3.14)$$

where Δ_{ji} is the $(n-1)\times(n-1)$ minor of $\mathbf{A}$ in which the jth row and ith column of $\mathbf{A}$ are deleted.

If $\mathbf{A}$ is a 1×1 matrix over R, then it is invertible if and only if it is a unit; when $\mathbf{A}$ is invertible, $\mathbf{A}^{-1} = 1/\mathbf{A}$. (The 1×1 matrix s over the ring of polynomials, $\mathbb{R}[s]$, is not invertible; however as a matrix over $\mathbb{R}(s)$, the field of rational functions, it is invertible with inverse $1/s$.)

If $\mathbf{A} \in \mathcal{R}^{2\times 2}$, then $\mathbf{A}$ is invertible if and only if $\det \mathbf{A} = \Delta = a_{11}a_{22} - a_{21}a_{12}$ is a unit. When $\mathbf{A}$ is invertible,

$$\mathbf{A} = \begin{bmatrix} a_{11} & a_{12} \\ a_{21} & a_{22} \end{bmatrix} \quad \text{and} \quad \mathbf{A}^{-1} = \begin{bmatrix} a_{22}/\Delta & -a_{12}/\Delta \\ -a_{21}/\Delta & a_{11}/\Delta \end{bmatrix} \qquad (3.15)$$

A 2×2 polynomial matrix has a polynomial matrix inverse just in case Δ equals a nonzero constant.

Cramer's rule is almost never used for computations because of its computational complexity and numerical sensitivity. When a matrix of real or complex numbers needs to be inverted, certain matrix factorization methods are employed; such factorizations also provide the best methods for numerical computation of determinants.

Inversion of upper and lower triangular matrices is done by a simple process of back-substitution; the inverses have the same triangular form. This may be exploited in combination with the product rule for inverses (and for determinants) since any invertible matrix $\mathbf{A} \in \mathbb{R}^{n \times n}$ ($\mathbb{R}$ can be replaced by another field $\mathcal{F}$) can be factored into the form

$$\mathbf{A} = \mathbf{P}\,\mathbf{L}\,\mathbf{U} \tag{3.16}$$

where the factors on the right side are, respectively, a permutation matrix, a lower triangular matrix, and an upper triangular matrix. The computation of this *PLU factorization* is equivalent to the process of Gaussian elimination with pivoting [4]. The resulting expression for the matrix inverse (usually kept in its factored form) is

$$\mathbf{A}^{-1} = \mathbf{U}^{-1}\mathbf{L}^{-1}\mathbf{P}^{-1} \tag{3.17}$$

whereas $\det \mathbf{A} = \det \mathbf{P}\ \det \mathbf{L}\ \det \mathbf{U}$. ($\det \mathbf{P} = \pm 1$, since $\mathbf{P}$ is a permutation matrix.)

3.2.5 Equivalence for Polynomial Matrices

Multiplication by $\mathbf{A}^{-1}$ transforms an invertible matrix $\mathbf{A}$ to a simple form: $\mathbf{A}\mathbf{A}^{-1} = \mathbf{I}\mathbf{A}\mathbf{A}^{-1} = \mathbf{I}$. For $\mathbf{A} \in \mathcal{R}^{n \times n}$ with $\det \mathbf{A} \neq 0$ but $\det \mathbf{A}$ not equal to a unit in $\mathcal{R}$, transformations of the form $\mathbf{A} \mapsto \mathcal{P}\mathbf{A}\,\mathcal{Q}$, where $\mathcal{P}, \mathcal{Q} \in \mathcal{R}^{n \times n}$ are invertible matrices, produce $\det \mathbf{A} \mapsto \det \mathcal{P}\ \det \mathbf{A}\ \det \mathcal{Q}$, i.e., the determinant is multiplied by the invertible element $\det \mathcal{P}\ \det \mathcal{Q} \in \mathcal{R}$. Thus, invertible matrices $\mathcal{P}$ and $\mathcal{Q}$ can be sought to bring the product $\mathcal{P}\mathbf{A}\,\mathcal{Q}$ to some simplified form even when $\mathbf{A}$ is not invertible; $\mathcal{P}\mathbf{A}\,\mathcal{Q}$ and $\mathbf{A}$ are said to be related by *$\mathcal{R}$-equivalence*.

For equivalence of polynomial matrices (see [3] for details), where $\mathcal{R} = \mathbb{R}[s]$ (or $\mathbb{C}[s]$), let $\mathcal{P}(s)$ and $\mathcal{Q}(s)$ be invertible $n \times n$ polynomial matrices. Such matrices are called *unimodular*; they have constant, nonzero determinants. Let $\mathcal{A}(s)$ be an $n \times n$ polynomial matrix with nonzero determinant. Then, for the equivalent matrix $\bar{\mathcal{A}}(s) = \mathcal{P}(s)\mathcal{A}(s)\mathcal{Q}(s)$, $\det \bar{\mathcal{A}}(s)$ differs from $\det \mathcal{A}(s)$ only by a constant factor; with no loss of generality, $\bar{\mathcal{A}}(s)$ may be assumed to be scaled so that $\det \bar{\mathcal{A}}(s)$ is a *monic* polynomial, i.e., so that the coefficient of the highest power of s in $\det \bar{\mathcal{A}}(s)$ is 1.

In forming $\bar{\mathcal{A}}(s)$, the multiplication of $\mathcal{A}(s)$ on the left by unimodular $\mathcal{P}(s)$ corresponds to performing a sequence of elementary row operations on $\mathcal{A}(s)$, and the multiplication of $\mathcal{A}(s)$ on the right by unimodular $\mathcal{Q}(s)$ corresponds to performing a sequence of elementary column operations on $\mathcal{A}(s)$. By suitable choice of $\mathcal{P}(s)$ and $\mathcal{Q}(s)$, $\mathcal{A}(s)$ may be brought to the *Smith canonical form*, $\mathcal{A}_S(s)$, a diagonal polynomial matrix whose diagonal elements are monic polynomials $\{\phi_i(s) : 1 \leq i \leq n\}$ satisfying the following divisibility conditions: $\phi_k(s)$ is a factor of $\phi_{k-1}(s)$, for $1 < k \leq n$.

The polynomials in the Smith canonical form $\mathcal{A}_S(s)$ are the *invariant polynomials* of $\mathcal{A}(s)$, and they may be obtained from $\mathcal{A}(s)$ as follows. Let $\epsilon_0(s) = 1$, and for $1 \leq i \leq n$, let $\epsilon_i(s)$ be the monic greatest common divisor of all nonzero $i \times i$ minors of $\mathcal{A}(s)$. Then, the invariant polynomials are given by $\phi_i(s) = \epsilon_i(s)/\epsilon_{i-1}(s)$. It follows that $\det \mathcal{A}(s)$ is a constant multiple of the polynomial $\epsilon_n(s) = \phi_1(s)\phi_2(s)\cdots\phi_n(s)$.

As an example of the Smith canonical form, consider

$$\mathcal{A}(s) = \begin{bmatrix} s(s+1)(s^2+s+1) & s^2(s+1)^2 \\ s(s+1)^2/3 & s(s+1)^2/3 \end{bmatrix} \tag{3.18}$$

The invariant polynomials are found from $\epsilon_1(s) = s(s+1)$ and $\epsilon_2(s) = s^2(s+1)^3$, giving $\phi_1(s) = s(s+1)$ and $\phi_2(s) = s(s+1)^2$. The corresponding Smith canonical form is indeed equivalent to $\mathcal{A}(s)$:

$$\begin{aligned} \mathcal{A}_S(s) &= \begin{bmatrix} s(s+1) & 0 \\ 0 & s(s+1)^2 \end{bmatrix} = \\ & \begin{bmatrix} 1 & -s \\ 0 & 3 \end{bmatrix} \mathcal{A}(s) \begin{bmatrix} 1 & 0 \\ -1 & 1 \end{bmatrix} \end{aligned} \tag{3.19}$$

3.2.6 Matrix Transposition

Another operation on matrices that is useful in a number of applications is *matrix transposition*. If $\mathbf{A}$ is an $m \times n$ matrix with $(\mathbf{A})_{ij} = a_{ij}$, the *transpose* of $\mathbf{A}$, denoted $\mathbf{A}^{\mathrm{T}}$, is the $n \times m$ matrix given by

$$(\mathbf{A}^{\mathrm{T}})_{ij} = a_{ji} \tag{3.20}$$

Thus, the transpose of a matrix is formed by interchanging its rows and columns.

If a square matrix $\mathbf{A}$ satisfies $\mathbf{A}^{\mathrm{T}} = \mathbf{A}$, it is called a *symmetric matrix*. If a square matrix $\mathbf{A}$ satisfies $\mathbf{A}^{\mathrm{T}} = -\mathbf{A}$, it is called a *skew-symmetric matrix*.

For matrices whose elements may possibly be complex numbers, a generalization of transposition is often more appropriate. The *Hermitian transpose* of a matrix $\mathbf{A}$, denoted $\mathbf{A}^{\mathrm{H}}$, is formed by interchanging rows and columns and replacing each element by its complex conjugate:

$$(\mathbf{A}^{\mathrm{H}})_{ij} = a_{ji}^{*} \tag{3.21}$$

The matrix $\mathbf{A}$ is *Hermitian symmetric* if $\mathbf{A}^{\mathrm{H}} = \mathbf{A}$.

Properties of Transposition

Several relationships between transposition and other matrix operations are noteworthy. For any matrix, $(\mathbf{A}^{\mathrm{T}})^{\mathrm{T}} = \mathbf{A}$; for $\mathbf{A} \in \mathcal{R}^{m \times n}$ and $z \in \mathcal{R}$, $(z\mathbf{A})^{\mathrm{T}} = z\mathbf{A}^{\mathrm{T}}$. With respect to algebraic operations, $(\mathbf{A} + \mathbf{B})^{\mathrm{T}} = \mathbf{A}^{\mathrm{T}} + \mathbf{B}^{\mathrm{T}}$ and $(\mathbf{A}\,\mathbf{B})^{\mathrm{T}} = \mathbf{B}^{\mathrm{T}}\mathbf{A}^{\mathrm{T}}$. (The products $\mathbf{A}\,\mathbf{A}^{\mathrm{T}}$ and $\mathbf{A}^{\mathrm{T}}\mathbf{A}$ are always defined.) With respect to determinants and matrix inversion, if $\mathbf{A}$ is a square matrix, $\det(\mathbf{A}^{\mathrm{T}}) = \det \mathbf{A}$, and if $\mathbf{A}$ is an invertible matrix, $\mathbf{A}^{\mathrm{T}}$ is also invertible, with $(\mathbf{A}^{\mathrm{T}})^{-1} = (\mathbf{A}^{-1})^{\mathrm{T}}$. A similar list of properties holds for Hermitian transposition.

Even for 2×2 matrices, transposition appears to be a much simpler operation than inversion. Indeed, the class of matrices for which $\mathbf{A}^T = \mathbf{A}^{-1}$ is quite remarkable. A real matrix whose transpose is also its inverse is known as an *orthogonal matrix.* The set of $n \times n$ orthogonal matrices, along with the operation of matrix multiplication, is a group; it is a subgroup of the group of invertible matrices, $GL(\mathbb{R}, n)$. For complex matrices, when $\mathbf{A}$ satisfies $\mathbf{A}^H = \mathbf{A}^{-1}$, it is called a *unitary matrix*; the unitary matrices form a subgroup of $GL(\mathbb{C}, n)$.

3.2.7 Block Matrices

It is sometimes convenient to partition the rows and columns of a matrix so that the matrix elements are grouped into submatrices. For example, a matrix $\mathbf{A} \in \mathcal{R}^{m\times n}$ may be partitioned into n columns (submatrices in $\mathcal{R}^{m\times 1}$) or into m rows (submatrices in $\mathcal{R}^{1\times n}$). More generally

$$\mathbf{A} = \begin{bmatrix} \mathbf{A}_{11} & \cdots & \mathbf{A}_{1q} \\ \vdots & & \vdots \\ \mathbf{A}_{p1} & \cdots & \mathbf{A}_{pq} \end{bmatrix} \tag{3.22}$$

where all submatrices in each block row have the same number of rows and all submatrices in each block column have the same number of columns; i.e., submatrix $\mathbf{A}_{ij}$ is $m_i \times n_j$, with $m_1 + \cdots + m_p = m$ and $n_1 + \cdots + n_q = n$. Such a matrix $\mathbf{A}$ is said to be a $p \times q$ *block matrix*, and it is denoted by $\mathbf{A} = (\mathbf{A}_{ij})$ for simplicity.

Matrix addition can be carried out block-wise for $p \times q$ block matrices with *conformable* partitions, where the corresponding submatrices have the same numbers of rows and columns. Matrix multiplication can also be carried out block-wise provided the column partition of the left factor is *compatible* with the row partition of the right factor: it is required that if $\mathbf{A} = (\mathbf{A}_{ij})$ is a $p_A \times q_A$ block matrix with block column i having n_i columns and $\mathbf{B} = (\mathbf{B}_{ij})$ is a $p_B \times q_B$ block matrix with block row j having m_j rows, then when $q_A = p_B$ and, in addition, $n_i = m_i$ for each i, the product matrix $\mathbf{C} = \mathbf{A}\,\mathbf{B}$ is a $p_A \times q_B$ block matrix $\mathbf{C} = (\mathbf{C}_{ij})$, where block $\mathbf{C}_{ij}$ is given by

$$\mathbf{C}_{ij} = \sum_{k=1}^{r} \mathbf{A}_{ik}\mathbf{B}_{kj} \tag{3.23}$$

where $r = q_A = p_B$.

For square matrices written as $p \times p$ block matrices having square "diagonal blocks" $\mathbf{A}_{ii}$, the determinant has a block-wise representation. For a square 2×2 block matrix,

$$\det \mathbf{A} = \det \begin{bmatrix} \mathbf{A}_{11} & \mathbf{A}_{12} \\ \mathbf{A}_{21} & \mathbf{A}_{22} \end{bmatrix} = \det \mathbf{A}_{11} \det(\mathbf{A}_{22} - \mathbf{A}_{21}\mathbf{A}_{11}^{-1}\mathbf{A}_{12}) \tag{3.24}$$

provided $\det \mathbf{A}_{11} \neq 0$. If this block matrix is invertible, its inverse may be expressed as a conformable block matrix:

$$\mathbf{A}^{-1} = \begin{bmatrix} \mathbf{A}_{11} & \mathbf{A}_{12} \\ \mathbf{A}_{21} & \mathbf{A}_{22} \end{bmatrix}^{-1} = \begin{bmatrix} \mathbf{S}_{11} & \mathbf{S}_{12} \\ \mathbf{S}_{21} & \mathbf{S}_{22} \end{bmatrix} \tag{3.25}$$

and assuming $\mathbf{A}_{11}$ is invertible, the blocks of the inverse matrix are: $\mathbf{S}_{11} = \mathbf{A}_{11}^{-1} + \mathbf{A}_{11}^{-1}\mathbf{A}_{12}\Phi^{-1}\mathbf{A}_{21}\mathbf{A}_{11}^{-1}$; $\mathbf{S}_{21} = -\Phi^{-1}\mathbf{A}_{21}\mathbf{A}_{11}^{-1}$; $\mathbf{S}_{12} = -\mathbf{A}_{11}^{-1}\mathbf{A}_{12}\Phi^{-1}$; $\mathbf{S}_{22} = \Phi^{-1} = (\mathbf{A}_{22} - \mathbf{A}_{21}\mathbf{A}_{11}^{-1}\mathbf{A}_{12})^{-1}$.

3.2.8 Matrix Powers and Polynomials

If $\mathbf{A} \in \mathcal{R}^{n\times n}$, define $\mathbf{A}^0 = \mathbf{I}_n$, and $\mathbf{A}^r$ equal to the product of r factors of $\mathbf{A}$, for integer $r \geq 1$. When $\mathbf{A}$ is invertible, $\mathbf{A}^{-1}$ has already been introduced as the notation for the inverse matrix. Nonnegative powers of $\mathbf{A}^{-1}$ provide the means for defining $\mathbf{A}^{-r} = (\mathbf{A}^{-1})^r$.

For any polynomial, $p(s) = p_0 s^k + p_1 s^{k-1} + \cdots + p_{k-1}s + p_k$, with coefficients $p_i \in \mathcal{R}$, the *matrix polynomial* $p(\mathbf{A})$ is defined as $p(\mathbf{A}) = p_0\mathbf{A}^k + p_1\mathbf{A}^{k-1} + \cdots + p_{k-1}\mathbf{A} + p_k\mathbf{I}$. When the ring of scalars, $\mathcal{R}$, is a field (and in some more general cases), $n \times n$ matrices obey certain polynomial equations of the form $p(\mathbf{A}) = \mathbf{0}$; such a polynomial $p(s)$ is an *annihilating polynomial* of $\mathbf{A}$. The monic annihilating polynomial of least degree is called the *minimal polynomial* of $\mathbf{A}$; the minimal polynomial is the (monic) greatest common divisor of all annihilating polynomials. The degree of the minimal polynomial of an $n \times n$ matrix is never larger than n because of the remarkable *Cayley-Hamilton Theorem.*

> Let $\mathbf{A} \in \mathcal{R}^{n\times n}$, where $\mathcal{R}$ is a field. Let $\chi(s)$ be the nth degree monic polynomial defined by
>
> $$\chi(s) = \det(s\mathbf{I} - \mathbf{A}) \tag{3.26}$$
>
> Then, $\chi(\mathbf{A}) = \mathbf{0}$.

The polynomial $\chi(s) = \det(s\mathbf{I} - \mathbf{A})$ is called the *characteristic polynomial* of $\mathbf{A}$.

3.3 Vector Spaces

3.3.1 Definitions

A *vector space* consists of an ordered tuple $(\mathcal{V}, \mathcal{F}, +, \cdot)$ having the following list of attributes:

1. $\mathcal{V}$ is a set of elements called vectors, containing a distinguished vector $\mathbf{0}$, the zero vector.
2. $\mathcal{F}$ is a field of scalars; most commonly $\mathcal{F} = \mathbb{R}$ or $\mathbb{C}$, the real or complex numbers.
3. The $+$ operation is a vector addition operation defined on $\mathcal{V}$. For all $\mathbf{v}_1, \mathbf{v}_2, \mathbf{v}_3 \in \mathcal{V}$, the following properties must hold: (a) $\mathbf{v}_1 + \mathbf{v}_2 = \mathbf{v}_2 + \mathbf{v}_1$, (b) $\mathbf{v}_1 + \mathbf{0} = \mathbf{v}_1$, and (c) $(\mathbf{v}_1 + \mathbf{v}_2) + \mathbf{v}_3 = \mathbf{v}_1 + (\mathbf{v}_2 + \mathbf{v}_3)$.
4. The $\cdot$ operation is a scalar multiplication of vectors (and usually the $\cdot$ is not written explicitly). For all $\mathbf{v}_1, \mathbf{v}_2 \in \mathcal{V}$, and $\alpha_1, \alpha_2 \in \mathcal{F}$, the following properties must hold: (a) $0\mathbf{v}_1 = \mathbf{0}$, (b) $1\mathbf{v}_1 = \mathbf{v}_1$, (c) $\alpha_1(\mathbf{v}_1 + \mathbf{v}_2) = \alpha_1\mathbf{v}_1 + \alpha_1\mathbf{v}_2$, (d) $(\alpha_1 + \alpha_2)\mathbf{v}_1 = \alpha_1\mathbf{v}_1 + \alpha_2\mathbf{v}_1$, and (e) $\alpha_1(\alpha_2\mathbf{v}_1) = (\alpha_1\alpha_2)\mathbf{v}_1$.

These conditions formalize the idea that a vector space is a set of elements closed under the operation of taking linear combinations.

3.3.2 Examples and Fundamental Properties

The conventional notation for the vector space $\mathcal{V}$ consisting of (column) n-vectors of elements of $\mathcal{F}$ is $\mathcal{F}^n$; thus $\mathbb{C}^n$ and $\mathbb{R}^n$ denote the spaces of complex and real n-vectors, respectively. To show that the theory is widely applicable, some other examples of vector spaces will be mentioned. Still others will arise later on.

1. The set of $m \times n$ matrices over a field $\mathcal{F}$, with the usual rules for scalar multiplication and matrix addition, forms a vector space, denoted $\mathcal{F}^{m \times n}$.
2. The set of polynomial functions of a complex variable $\mathcal{P} = \{p(s) : p(s) = p_0 s^k + p_1 s^{k-1} + \cdots + p_{k-1}s + p_k\}$ is a vector space over $\mathbb{C}$ because addition of two polynomials produces another polynomial, as does multiplication of a polynomial by a complex number.
3. The set $C[0, T]$ of real-valued continuous functions defined on the closed interval $0 \le t \le T$ is a real vector space because the sum of two continuous functions is another continuous function and scalar multiplication also preserves continuity.

A number of important concepts from the theory of vector spaces will now be introduced.

Subspaces

If $\mathcal{V}$ is a vector space and $\mathcal{W}$ is a subset of vectors from $\mathcal{V}$, then $\mathcal{W}$ is called a *subspace* of $\mathcal{V}$ if $\mathbf{0} \in \mathcal{W}$ and if $\alpha_1\mathbf{v}_1 + \alpha_2\mathbf{v}_2 \in \mathcal{W}$ for all $\mathbf{v}_1$ and $\mathbf{v}_2 \in \mathcal{W}$ and all α_1 and $\alpha_2 \in \mathcal{F}$. Notice that this means that $\mathcal{W}$ is a vector space itself. The set $\mathcal{W} = \{\mathbf{0}\}$ is always a subspace, and $\mathcal{V}$ is a subspace of itself. If $\mathbf{v}$ is a nonzero vector in a vector space $\mathcal{V}$, then the set $\{\alpha\mathbf{v} : \alpha \in \mathcal{F}\}$ is a subspace of $\mathcal{V}$. For two subspaces $\mathcal{W}_1$ and $\mathcal{W}_2$, $\mathcal{W}_1 \cap \mathcal{W}_2$ is a subspace, and $\mathcal{W}_1 + \mathcal{W}_2 = \{\mathbf{w}_1 + \mathbf{w}_2 : \mathbf{w}_1 \in \mathcal{W}_1 \text{ and } \mathbf{w}_2 \in \mathcal{W}_2\}$ is a subspace. The geometric intuition of subspaces is that they consist of "lines" or "planes" (often called "hyperplanes" in spaces of high dimension) passing through the origin $\mathbf{0}$.

Linear Independence

A set of vectors $\{\mathbf{v}_1, \mathbf{v}_2, \ldots, \mathbf{v}_k\}$ is called *linearly independent* when the equation

$$\textstyle\sum_{i=1}^{k} \alpha_i \mathbf{v}_i = \mathbf{0} \tag{3.27}$$

is satisfied only by the trivial choice of the scalars: $\alpha_1 = \alpha_2 = \cdots = \alpha_k = 0$. No nontrivial linear combination of linearly independent vectors equals the zero vector. A set of vectors that is not linearly independent is called *linearly dependent.* Any set containing $\mathbf{0}$ is linearly dependent.

Spanning Set

If every vector $\mathbf{v} \in \mathcal{V}$ can be written as a linear combination of the vectors from some set, then that set is called a *spanning set* for $\mathcal{V}$; and the vectors of the set *span* $\mathcal{V}$. For any set of vectors, $\{\mathbf{v}_1, \mathbf{v}_2, \ldots, \mathbf{v}_k\}$, the *span* of the set is the subspace of elements $\{\mathbf{v} = \sum_{i=1}^{k} \alpha_i \mathbf{v}_i : \alpha_i \in \mathcal{F}, 1 \le i \le k\}$, denoted $\mathrm{sp}\{\mathbf{v}_1, \mathbf{v}_2, \ldots, \mathbf{v}_k\}$.

Basis

A *basis* for a vector space $\mathcal{V}$ is any spanning set for $\mathcal{V}$ consisting of linearly independent vectors.

Dimension

If a vector space has a basis with finitely many vectors, then the number of vectors in every basis is the same and this number is the *dimension* of the vector space. A vector space having no basis with finitely many vectors is called infinite-dimensional.

Coordinates

If $\mathcal{S}$ is a linearly independent set of vectors, and $\mathbf{v} \in \mathrm{sp}\,\mathcal{S}$, then there is a *unique* way of expressing $\mathbf{v}$ as a linear combination of the vectors in $\mathcal{S}$. Thus, given a basis, every vector in a vector space has a unique representation as a linear combination of the vectors in the basis. If $\mathcal{V}$ is a vector space of dimension n with basis $\mathcal{B} = \{\mathbf{b}_1, \mathbf{b}_2, \ldots, \mathbf{b}_n\}$, there is a natural correspondence between $\mathcal{V}$ and the vector space $\mathcal{F}^n$ defined by

$$\mathbf{v} \mapsto \mathbf{v}_{\mathcal{B}} = \begin{bmatrix} \alpha_1 \\ \alpha_2 \\ \vdots \\ \alpha_n \end{bmatrix} \tag{3.28}$$

where the elements of the *coordinate vector* $\mathbf{v}_{\mathcal{B}}$ give the representation of $\mathbf{v} \in \mathcal{V}$ with respect to the basis $\mathcal{B}$:

$$\mathbf{v} = \sum_{i=1}^{n} \alpha_i \mathbf{b}_i \tag{3.29}$$

In particular, the basis vectors of $\mathcal{V}$ correspond to the *standard basis vectors* of $\mathcal{F}^n$: $\mathbf{b}_i \mapsto \mathbf{e}_i$, where $\mathbf{e}_i \in \mathcal{F}^n$ is the vector whose ith element is 1 and all of its other elements are 0; the ith element of the standard basis of $\mathcal{F}^n$, $\mathbf{e}_i$, is called the ith *unit vector* or ith *principal axis vector.*

3.3.3 Linear Functions

Let $\mathcal{X}$ and $\mathcal{Y}$ be two vector spaces over a common field, $\mathcal{F}$. A *linear function* (sometimes called a linear transformation, linear operator, or linear mapping), denoted $f : \mathcal{X} \to \mathcal{Y}$, assigns to each $\mathbf{x} \in \mathcal{X}$ an element $\mathbf{y} \in \mathcal{Y}$ so as to make

$$f(\alpha_1\mathbf{x}_1 + \alpha_2\mathbf{x}_2) = \alpha_1 f(\mathbf{x}_1) + \alpha_2 f(\mathbf{x}_2) \tag{3.30}$$

for every $\mathbf{x}_1$ and $\mathbf{x}_2 \in \mathcal{X}$ (the domain of the function) and for every choice of scalars α_1 and $\alpha_2 \in \mathcal{F}$.

Let $\mathcal{L}(\mathcal{X}, \mathcal{Y})$ denote the set of all linear functions from $\mathcal{X}$ to $\mathcal{Y}$; $\mathcal{L}(\mathcal{X}, \mathcal{Y})$ is a vector space over $\mathcal{F}$. If $\mathcal{X}$ has dimension n and $\mathcal{Y}$ has dimension m, then $\mathcal{L}(\mathcal{X}, \mathcal{Y})$ has dimension mn. If $\{\mathbf{x}_1, \ldots, \mathbf{x}_n\}$ is a basis for $\mathcal{X}$ and $\{\mathbf{y}_1, \ldots, \mathbf{y}_m\}$ is a basis for $\mathcal{Y}$, then a basis for $\mathcal{L}(\mathcal{X}, \mathcal{Y})$ is the set of functions $\{f_{ij}(\mathbf{x}) : 1 \le i \le n,\ 1 \le j \le m\}$, where basis function $f_{ij}(\mathbf{x})$ takes the value $\mathbf{0}$ for all $\mathbf{x}$ except when $\mathbf{x} = \alpha\mathbf{x}_i$, $\alpha \in \mathcal{F}$, in which case $f_{ij}(\alpha\mathbf{x}_i) = \alpha\mathbf{y}_j$. $\mathcal{L}(\mathcal{X}, \mathcal{F})$ is known as the space of *linear functionals* on $\mathcal{X}$ or the *dual space* of $\mathcal{X}$.

If f is a linear function, then it necessarily maps subspaces of $\mathcal{X}$ to subspaces of $\mathcal{Y}$; if $\mathcal{W}$ is a subspace of $\mathcal{X}$, then the function $f|_{\mathcal{W}} : \mathcal{W} \to \mathcal{Y}$, called the *restriction of f to $\mathcal{W}$* and defined by $f|_{\mathcal{W}}(\mathbf{w}) = f(\mathbf{w})$, is a linear function from $\mathcal{W}$ to $\mathcal{Y}$. There are two subspaces of particular importance associated with a linear function $f : \mathcal{X} \to \mathcal{Y}$. The *nullspace* or *kernel* of f is the subspace

$$\ker f = \{\mathbf{x} \in \mathcal{X} | f(\mathbf{x}) = \mathbf{0} \in \mathcal{Y}\} \tag{3.31}$$

The *range* or *image* of f is the subspace

$$\operatorname{im} f = \{\mathbf{y} \in \mathcal{Y} | \mathbf{y} = f(\mathbf{x}) \text{ for some } \mathbf{x} \in \mathcal{X}\} \tag{3.32}$$

When f is a linear function from a vector space $\mathcal{X}$ to itself, $f : \mathcal{X} \to \mathcal{X}$, any subspace mapped into itself is said to be *f-invariant.* For a subspace $\mathcal{W} \subseteq \mathcal{X}$, $\mathcal{W}$ is f-invariant if $f(\mathbf{w}) \in \mathcal{W}$ for all $\mathbf{w} \in \mathcal{W}$. $\ker f$ is f-invariant, and for any $\mathbf{x} \in \mathcal{X}$, the subspace spanned by the vectors $\{\mathbf{x}, f(\mathbf{x}), f(f(\mathbf{x})), \ldots\}$ is f-invariant. When $\mathcal{W}$ is f-invariant, its restriction, $f|_{\mathcal{W}}$ is a linear function from $\mathcal{W}$ to $\mathcal{W}$.

Matrix Representations

Every linear function is uniquely determined by its values on any basis for its domain. This fact leads to the use of matrices for representing linear functions. Let $\mathbf{A}$ be an $m \times n$ matrix over $\mathcal{F}$. Then the function $f : \mathcal{F}^n \to \mathcal{F}^m$, defined in terms of matrix multiplication $\mathbf{x} \mapsto \mathbf{A}\mathbf{x}$, is a linear function. Indeed, every linear function $f : \mathcal{F}^n \to \mathcal{F}^m$ takes this form for a unique matrix $\mathbf{A}_f \in \mathcal{F}^{m\times n}$. Specifically, for $1 \le i \le n$ the ith column of $\mathbf{A}_f$ is defined to be the vector $f(\mathbf{e}_i) \in \mathcal{F}^m$, where $\mathbf{e}_i$ is the ith unit vector in $\mathcal{F}^n$. With this definition, for any $\mathbf{x} \in \mathcal{F}^n$, $f(\mathbf{x}) = \mathbf{A}_f\mathbf{x}$.

This same idea can be extended using bases and coordinate vectors to provide a matrix representation for any linear function. Let f be a linear function mapping $\mathcal{X}$ to $\mathcal{Y}$, let $\mathcal{B}_X$ be a basis for $\mathcal{X}$, and let $\mathcal{B}_Y$ be a basis for $\mathcal{Y}$. Suppose $\mathcal{X}$ has dimension n and $\mathcal{Y}$ has dimension m. Then there is a unique matrix $\mathbf{A}_f \in \mathcal{F}^{m\times n}$ giving $\mathbf{A}_f\mathbf{x}_{\mathcal{B}_X} = \mathbf{y}_{\mathcal{B}_Y}$ if and only if $f(\mathbf{x}) = \mathbf{y}$, where $\mathbf{x}_{\mathcal{B}_X} \in \mathcal{F}^n$ is the coordinate vector of $\mathbf{x}$ with respect to the basis $\mathcal{B}_X$ and $\mathbf{y}_{\mathcal{B}_Y} \in \mathcal{F}^m$ is the coordinate vector of $\mathbf{y}$ with respect to the basis $\mathcal{B}_Y$. Thus, the ith column of $\mathbf{A}_f$ is the coordinate vector (with respect to the basis $\mathcal{B}_Y$) of the vector $f(\mathbf{b}_i) \in \mathcal{Y}$, where $\mathbf{b}_i \in \mathcal{X}$ is the ith vector of the basis $\mathcal{B}_X$ for $\mathcal{X}$. $\mathbf{A}_f$ is called the *matrix representation of f with respect to the bases $\mathcal{B}_X$ and $\mathcal{B}_Y$.*

Composition of linear functions preserves linearity. If $h : \mathcal{X} \to \mathcal{Z}$ is defined as $h(\mathbf{x}) = g(f(\mathbf{x}))$, the composition of two other linear functions, $f : \mathcal{X} \to \mathcal{Y}$ and $g : \mathcal{Y} \to \mathcal{Z}$, then h is a linear function. When the three vector spaces are finite dimensional and bases are chosen, the corresponding relationship between the matrix representations of the linear functions is given by $\mathbf{A}_h = \mathbf{A}_g\mathbf{A}_f$, and so composition of linear functions corresponds to matrix multiplication.

On the one hand, bases provide a means for representing linear functions with matrices. On the other hand, linear functions provide a means of relating different bases. If $\mathcal{B} = \{\mathbf{b}_1, \mathbf{b}_2, \ldots, \mathbf{b}_n\}$ and $\widehat{\mathcal{B}} = \{\widehat{\mathbf{b}}_1, \widehat{\mathbf{b}}_2, \ldots, \widehat{\mathbf{b}}_n\}$ are two bases for a vector space $\mathcal{V}$, a linear function $t : \mathcal{X} \to \mathcal{X}$ is defined by its values on the basis $\mathcal{B}$: $\mathbf{b}_i \mapsto \widehat{\mathbf{b}}_i$ for $1 \le i \le n$; t is clearly invertible and its inverse, t^{-1}, is the linear function defined by its values on the basis $\widehat{\mathcal{B}}$: $\widehat{\mathbf{b}}_i \mapsto \mathbf{b}_i$ for $1 \le i \le n$. The matrix representation of t with respect to basis $\mathcal{B}$ for $\mathcal{X}$ (with basis $\mathcal{B}$ used to define coordinates of both $\mathbf{x}$ and $t(\mathbf{x})$) is the $n \times n$ matrix $\mathbf{T}$ whose ith column is the coordinate vector of $\widehat{\mathbf{b}}_i$ and, when $t(\mathbf{x}) = \bar{\mathbf{x}}$, the corresponding equation for coordinate vectors is

$$\mathbf{T}\mathbf{x}_{\mathcal{B}} = \bar{\mathbf{x}}_{\mathcal{B}} \tag{3.33}$$

The matrix $\mathbf{T}$ is invertible because the function t is invertible, and $\mathbf{T}^{-1}$ is the matrix representation of t^{-1} with respect to the basis $\widehat{\mathcal{B}}$. Specifically, the ith column of $\mathbf{T}^{-1}$ is the coordinate vector of $\mathbf{b}_i$ with respect to the $\widehat{\mathcal{B}}$ basis.

Now consider, along with bases $\mathcal{B}$ and $\widehat{\mathcal{B}}$ for $\mathcal{X}$ and the associated linear functions t and t^{-1} defined by the mappings between the basis vectors, a linear function $f : \mathcal{X} \to \mathcal{X}$. Let $\mathbf{A}_f$ denote the matrix representation of f with respect to basis $\mathcal{B}$, and let $\bar{\mathbf{A}}_f$ denote the matrix representation of f with respect to basis $\widehat{\mathcal{B}}$. The two matrix representations of f are related by the equation

$$\mathbf{T}^{-1}\mathbf{A}_f\mathbf{T} = \bar{\mathbf{A}}_f \tag{3.34}$$

or equivalently

$$\mathbf{A}_f = \mathbf{T}\bar{\mathbf{A}}_f\mathbf{T}^{-1} \tag{3.35}$$

Two $n \times n$ matrices $\mathbf{A}$ and $\hat{\mathbf{A}}$ related by the equation $\mathbf{T}^{-1}\mathbf{A}\,\mathbf{T} = \hat{\mathbf{A}}$, for some invertible matrix $\mathbf{T}$, are said to be *similar matrices,* with $\mathbf{T}$ being called a *similarity transformation.* Similarity is an equivalence relation on $\mathcal{F}^{n\times n}$, and a complete characterization of the similarity equivalence classes is of major importance.

3.3.4 Norms and Inner Products

Vectors in $\mathbb{R}^2$ and $\mathbb{R}^3$ are often viewed as points in two- and three-dimensional Euclidean space, respectively. The resulting geometric intuition may be extended to other vector spaces by developing more general notions of length and angle.

Vector Norms

The notion of vector norm is introduced to play the role of length. For a vector space $\mathcal{V}$ over the real or complex numbers, the notation $\|\mathbf{v}\|$ is used to denote the *norm* of vector $\mathbf{v}$; $\|\mathbf{v}\| \in \mathbb{R}$. To qualify as a norm, three properties must hold:

N1. For all $\mathbf{v} \in \mathcal{V}$, $\|\mathbf{v}\| \ge 0$ with equality holding only for $\mathbf{v} = \mathbf{0}$.

N2. For all $\mathbf{v} \in \mathcal{V}$ and all $\alpha \in \mathcal{F}$, $\|\alpha\mathbf{x}\| = |\alpha|\|\mathbf{x}\|$.

N3. (*Triangle inequality*) For all $\mathbf{v}_1$ and $\mathbf{v}_2 \in \mathcal{V}$, $\|\mathbf{v}_1 + \mathbf{v}_2\| \le \|\mathbf{v}_1\| + \|\mathbf{v}_2\|$

In N2, $|\alpha|$ denotes the absolute value when the field of scalars $\mathcal{F} = \mathbb{R}$, and it denotes the modulus (or magnitude) when $\mathcal{F} = \mathbb{C}$.

The *Euclidean norm* on $\mathbb{R}^n$ is given by

$$\|\mathbf{v}\| = (\mathbf{v}^T\mathbf{v})^{1/2} = \left(\sum_{i=1}^{n} v_i^2\right)^{1/2} \tag{3.36}$$

It corresponds to Euclidean length for vectors in $\mathbb{R}^2$ and $\mathbb{R}^3$. Other norms for $\mathbb{R}^n$ are the *uniform norm*, which will be denoted by $\|\mathbf{v}\|_\infty$, with

$$\|\mathbf{v}\|_\infty = \max\{|v_i| : 1 \le i \le n\} \tag{3.37}$$

and the family of p-norms, defined for real numbers $1 \le p < \infty$ by

$$\|\mathbf{v}\|_p = \left(\sum_{i=1}^{n} |v_i|^p\right)^{1/p} \tag{3.38}$$

The Euclidean norm is the p-norm for $p = 2$.

Various norms turn out to be appropriate for applications involving vectors in other vector spaces; as an example, a suitable norm for $C[0, T]$, the space of real-valued continuous functions on the interval $0 \le t \le T$ is the uniform norm:

$$\|c(t)\|_\infty = \max\{|c(t)| : 0 \le t \le T\} \tag{3.39}$$

The name and notation are the same as used for the uniform norm on $\mathbb{R}^n$ since the analogy is apparent. A notion of p-norm for vector spaces of functions can also be established. The 2-norm is the natural generalization of Euclidean norm:

$$\|c(t)\|_2 = \left(\int_0^T |c(t)|^2 dt\right)^{1/2} \tag{3.40}$$

As a final example, the *Frobenius norm* of a matrix $\mathbf{A} \in \mathbb{R}^{m\times n}$, denoted $\|\mathbf{A}\|_F$, is the Euclidean norm of the nm-vector consisting of all of the elements of $\mathbf{A}$:

$$\|\mathbf{A}\|_F = \left(\sum_{i=1}^{m}\sum_{j=1}^{n} a_{ij}^2\right)^{1/2} \tag{3.41}$$

Norms of Linear Functions

When matrices are viewed as representations of linear functions, it is more appropriate to employ a different kind of norm, one that arises from the role of a linear function as a mapping between vector spaces. For example, consider the linear function $f : \mathbb{R}^n \to \mathbb{R}^n$ given by $f(\mathbf{v}) = \mathbf{A}\mathbf{v}$. When $\mathbb{R}^n$ is equipped with the Euclidean norm, the *induced Euclidean norm* of $\mathbf{A}$, is defined by

$$\|\mathbf{A}\| = \max\{\|\mathbf{A}\mathbf{x}\| : \|\mathbf{x}\| = 1\} \tag{3.42}$$

This is easily generalized. For any linear function $f : \mathcal{X} \to \mathcal{Y}$ and norms $\|\cdot\|_\mathcal{X}$ and $\|\cdot\|_\mathcal{Y}$, the induced norm of f takes the form:

$$\|f\| = \max\{\|f(\mathbf{x})\|_\mathcal{Y} : \|\mathbf{x}\|_\mathcal{X} = 1\} \tag{3.43}$$

The subscripts are commonly suppressed when the choice of norms is readily determined from context. The induced matrix norm $\|\mathbf{A}\|$ for $\mathbf{A} \in \mathcal{F}^{n\times m}$ is simply the induced norm of the linear function $\mathbf{A}\mathbf{x}$.

A consequence of the definition of induced norm is the inequality

$$\|\mathbf{A}\mathbf{x}\| \le \|\mathbf{A}\|\,\|\mathbf{x}\| \tag{3.44}$$

which holds for all vectors $\mathbf{x}$. This inequality also implies the following inequality for the induced norm of a matrix product:

$$\|\mathbf{A}\mathbf{B}\| \le \|\mathbf{A}\|\,\|\mathbf{B}\| \tag{3.45}$$

Explicit expressions for three of the most important induced matrix norms can be determined. Suppose $\mathbf{A} \in \mathbb{R}^{n\times n}$. For the induced Euclidean norm,

$$\|\mathbf{A}\| = \sigma_1(\mathbf{A}) \tag{3.46}$$

which is the largest *singular value* of the matrix. (Singular values are discussed later.) For the induced uniform norm,

$$\|\mathbf{A}\|_\infty = \max_{1\le i\le n} \sum_{j=1}^{n} |a_{ij}| \tag{3.47}$$

which is the largest of the absolute row-sums of $\mathbf{A}$. For the induced 1-norm,

$$\|\mathbf{A}\|_1 = \max_{1\le j\le n} \sum_{i=1}^{n} |a_{ij}| \tag{3.48}$$

which is the largest of the absolute column-sums of $\mathbf{A}$.

Inner Products and Orthogonality

For nonzero vectors $\mathbf{v}_1$ and $\mathbf{v}_2$ in $\mathbb{R}^2$ or $\mathbb{R}^3$, the Euclidean geometric notion of angle is easily expressed in terms of the dot product. With $\mathbf{v}_1 \cdot \mathbf{v}_2 = \mathbf{v}_1^T\mathbf{v}_2$, the angle between $\mathbf{v}_1$ and $\mathbf{v}_2$, $\theta_{x,y}$, satisfies

$$\cos(\theta_{x,y}) = \frac{\mathbf{v}_1^T\mathbf{v}_2}{((\mathbf{v}_1^T\mathbf{v}_1)(\mathbf{v}_2^T\mathbf{v}_2))^{1/2}} = \frac{\mathbf{v}_1^T\mathbf{v}_2}{\|\mathbf{v}_1\|\,\|\mathbf{v}_2\|} \tag{3.49}$$

where the Euclidean norm is used in the second expression.

The notion of inner product of vectors is used to obtain a generalization of the dot product in order to provide a geometric interpretation for angles between vectors in other vector spaces. If $\mathcal{V}$ is a vector space over $\mathbb{R}$, the mapping from the Cartesian product (ordered pairs of vectors) $\mathcal{V} \times \mathcal{V}$ to $\mathbb{R}$ defined by $(\mathbf{v}_1, \mathbf{v}_2) \mapsto \langle\mathbf{v}_1, \mathbf{v}_2\rangle$ is an inner product if the following properties are satisfied:

I1. For all $\mathbf{v} \in \mathcal{V}$, $\langle\mathbf{v}, \mathbf{v}\rangle \ge 0$ with equality holding only for $\mathbf{v} = \mathbf{0}$.

I2. For all $\mathbf{v}_1$ and $\mathbf{v}_2 \in \mathcal{V}$, $\langle\mathbf{v}_1, \mathbf{v}_2\rangle = \langle\mathbf{v}_2, \mathbf{v}_1\rangle$

I3. For all $\mathbf{v}_1$ and $\mathbf{v}_2 \in \mathcal{V}$ and $\alpha \in \mathbb{R}$, $\langle\alpha\mathbf{v}_1, \mathbf{v}_2\rangle = \alpha\langle\mathbf{v}_1, \mathbf{v}_2\rangle$.

I4. For all $\mathbf{v}_1, \mathbf{v}_2$, and $\mathbf{v}_3 \in \mathcal{V}$, $\langle\mathbf{v}_1 + \mathbf{v}_2, \mathbf{v}_3\rangle = \langle\mathbf{v}_1, \mathbf{v}_3\rangle + \langle\mathbf{v}_2, \mathbf{v}_3\rangle$

Inner products for complex vector spaces are complex-valued and satisfy similar properties.

For the vector space $\mathbb{R}^n$, the definition $\langle \mathbf{v}_1, \mathbf{v}_2 \rangle = \mathbf{v}_1^T \mathbf{v}_2$, provides a generalization of the dot product. Furthermore, the Euclidean norm on $\mathbb{R}^n$ is *compatible* with this inner product, meaning that $\|\mathbf{v}\| = (\langle \mathbf{v}, \mathbf{v} \rangle)^{1/2}$. Finally, the interpretation

$$\cos(\theta_{x,y}) = \frac{\langle \mathbf{v}_1, \mathbf{v}_2 \rangle}{\|\mathbf{v}_1\| \, \|\mathbf{v}_2\|} \tag{3.50}$$

is appropriate because of the *Schwarz inequality*

$$|\langle \mathbf{v}_1, \mathbf{v}_2 \rangle| \leq \|\mathbf{v}_1\| \, \|\mathbf{v}_2\| \tag{3.51}$$

With the notion of angle now defined in terms of inner product, two vectors are said to be *orthogonal* if their inner product is zero.

For the vector space $\mathcal{C}[0, T]$ of real-valued continuous functions on the interval $0 \leq t \leq T$, the inner product of two functions $c_1(t)$ and $c_2(t)$ is defined by

$$\langle c_1(t), c_2(t) \rangle = \int_0^T c_1(t) c_2(t) dt \tag{3.52}$$

and the Euclidean norm on $\mathcal{C}[0, T]$ defined earlier is compatible with this inner product.

There are many cases of vector spaces having norms that are not compatible with any definition of an inner product. For example, on $\mathbb{R}^n$ the uniform norm does not correspond to any inner product.

Inner Product Spaces

Let $\mathcal{V}$ be a vector space with an inner product and compatible norm. A set of mutually orthogonal vectors is known as an orthogonal set, and a basis consisting of mutually orthogonal vectors is known as an *orthogonal basis*. An orthogonal basis consisting of vectors whose norms are all one (i.e., consisting of vectors having unit length) is called an *orthonormal basis*.

Given an orthonormal basis, any vector is easily expressed as a linear combination of the orthonormal basis vectors. If $\{\mathbf{w}_1, \ldots, \mathbf{w}_k\}$ is an orthonormal basis, the vector $\mathbf{v}$ is given by

$$\mathbf{v} = \sum_{i=1}^{k} \alpha_i \mathbf{w}_i \tag{3.53}$$

where $\alpha_i = \langle \mathbf{v}, \mathbf{w}_i \rangle$. Also, as a generalization of the Pythagorean theorem,

$$\|\mathbf{v}\|^2 = \sum_{i=1}^{k} \alpha_i^2 \tag{3.54}$$

Gram-Schmidt Orthogonalization and QR Factorization

There is a constructive procedure for obtaining an orthonormal basis starting from an arbitrary basis, the *Gram-Schmidt procedure*. Starting with a basis of k vectors $\{\mathbf{v}_1, \mathbf{v}_2, \ldots, \mathbf{v}_k\}$, the orthonormal basis $\{\mathbf{w}_1, \mathbf{w}_2, \ldots, \mathbf{w}_k\}$ is constructed sequentially according to the following steps:

1. $\mathbf{w}_1 = \mathbf{v}_1 / \|\mathbf{v}_1\|$
2. For $2 \leq i \leq k$, $\mathbf{w}_i = \mathbf{z}_i / \|\mathbf{z}_i\|$, where $\mathbf{z}_i = \mathbf{v}_i - \sum_{j=1}^{i-1} \langle \mathbf{v}_i, \mathbf{w}_j \rangle \mathbf{w}_j$

For k vectors in $\mathbb{R}^m$, $k \leq m$, take the vectors as columns of $\mathbf{V} \in \mathbb{R}^{m \times k}$. Then, the Gram-Schmidt procedure produces the matrix factorization $\mathbf{V} = \mathbf{W}\mathbf{U}$, where $\mathbf{W} \in \mathbb{R}^{m \times k}$ matrix, and $\mathbf{U} \in \mathbb{R}^{k \times k}$ is an upper triangular matrix. The columns of the matrix $\mathbf{W}$ are orthogonal so that $\mathbf{W}^T\mathbf{W} = \mathbf{I}_k$.

The factorization of $\mathbf{V}$ into a product of a matrix with orthogonal columns times an upper triangular matrix, $\mathbf{WU}$, is traditionally known as the QR factorization [4]. It is rarely computed column-by-column because better numerical accuracy can be achieved by taking a different approach. For simplicity, assume that $\mathbf{V}$ is $m \times m$. If any sequence of orthogonal matrices $\mathbf{W}_1, \mathbf{W}_2, \ldots, \mathbf{W}_j$ can be found so that $\mathbf{V}$ is transformed to an upper triangular matrix,

$$\mathbf{W}_j \cdots \mathbf{W}_2 \mathbf{W}_1 \mathbf{V} = \mathbf{U} \tag{3.55}$$

then multiplying both sides of this equation by $\mathbf{W} = \mathbf{W}_1^T \mathbf{W}_2^T \cdots$, $\mathbf{W}_j^T$ produces the QR factorization.

A commonly applied computational procedure for QR factorization involves a certain sequence of $j = m$ symmetric orthogonal matrices known as Householder transformations, matrices of the form $\mathbf{W}(\mathbf{y}) = \mathbf{I} - 2\mathbf{y}\mathbf{y}^T / \|\mathbf{y}\|^2$. The matrix $\mathbf{W}_i$ is chosen to be the Householder transformation that produces all subdiagonal elements of the ith column of $\mathbf{W}_i \cdots \mathbf{W}_1\mathbf{V}$ equal to zero without changing any of the zeros that are subdiagonal elements of the first $i - 1$ columns of $\mathbf{W}_{i-1} \cdots \mathbf{W}_1 \mathbf{V}$.

Orthogonal Projections and Orthogonal Functions

Suppose $\mathcal{V}$ is a vector space with an inner product, and let $\mathcal{W}$ be a subspace of $\mathcal{V}$. The subspace $\mathcal{W}^\perp = \{\mathbf{v} : \langle \mathbf{v}, \mathbf{w} \rangle = 0 \text{ for all } \mathbf{w} \in \mathcal{W}\}$ is called the *orthogonal complement of* $\mathcal{W}$ *in* $\mathcal{V}$. $\mathcal{W} \cap \mathcal{W}^\perp = \{0\}$, and $\mathcal{V} = \mathcal{W} \oplus \mathcal{W}^\perp$ is an orthogonal direct sum decomposition of $\mathcal{V}$: every $\mathbf{v} \in \mathcal{V}$ can be written uniquely in the form $\mathbf{v} = \mathbf{w} + \mathbf{w}^\perp$, where $\mathbf{w} \in \mathcal{W}$ and $\mathbf{w}^\perp \in \mathcal{W}^\perp$. The linear function $p_\mathcal{W} : \mathcal{V} \to \mathcal{V}$ defined in terms of the unique decomposition, $\mathbf{v} \mapsto \mathbf{w}$, is called the *orthogonal projection of* $\mathcal{V}$ *onto* $\mathcal{W}$. The orthogonal direct sum decomposition of $\mathcal{V}$ may be rewritten as $\mathcal{V} = \operatorname{im} p_\mathcal{W} \oplus \ker p_\mathcal{W}$. The *complementary orthogonal projection* of $p_\mathcal{W}$ is the orthogonal projection of $\mathcal{V}$ onto $\mathcal{W}^\perp$, or $p_{\mathcal{W}^\perp}$; its kernel and image provide another orthogonal direct sum representation of $\mathcal{V}$, whose terms correspond to the image and kernel of $p_\mathcal{W}$, respectively.

The orthogonal projection of a vector $\mathbf{v}$ onto a subspace $\mathcal{W}$, $p_\mathcal{W}(\mathbf{v})$, provides the solution to a *linear least squares approximation problem* since $\|p_{\mathcal{W}^\perp}(\mathbf{v})\| = \|\mathbf{v} - p_\mathcal{W}(\mathbf{v})\| \geq \|\mathbf{v} - \widehat{\mathbf{w}}\|$ for any choice of $\widehat{\mathbf{w}} \in \mathcal{W}$.

A linear function $f : \mathcal{V} \to \mathcal{V}$, where $\mathcal{V}$ is a real vector space with an inner product and compatible norm, is an *orthogonal function* if it maps an orthonormal basis to an orthonormal basis. If $\mathcal{V}$ is n-dimensional, the matrix representation of f with respect to an orthonormal basis is an *orthogonal matrix*, an $n \times n$ matrix

$\mathbf{O}$ satisfying $\mathbf{O}^{-1} = \mathbf{O}^{\mathrm{T}}$. The columns of an orthogonal matrix form an orthonormal basis for $\mathbb{R}^n$ (with the usual inner product, $\langle \mathbf{x}_1, \mathbf{x}_2 \rangle = \mathbf{x}_1^{\mathrm{T}} \mathbf{x}_2$). For an orthogonal function, $\|f(\mathbf{v})\| = \|\mathbf{v}\|$, and for an orthogonal matrix $\|\mathbf{O}\,\mathbf{x}\| = \|\mathbf{x}\|$, for all $\mathbf{x}$. Any orthogonal matrix has induced Euclidean norm equal to 1.

3.4 Linear Equations

For a linear function $f : \mathcal{X} \to \mathcal{Y}$, it is frequently of interest to find a vector $\mathbf{x} \in \mathcal{X}$ whose image under f is some given vector $\mathbf{y} \in \mathcal{Y}$, i.e., to *solve* the equation $f(\mathbf{x}) = \mathbf{y}$ for $\mathbf{x}$. By resorting to the matrix representation of f if necessary, there is no loss of generality in assuming that the problem is posed in the framework of matrices and vectors, a framework that is suited to numerical computation as well as to theoretical analysis using matrix algebra.

3.4.1 Existence and Uniqueness of Solutions

Let $\mathbf{A} \in \mathcal{F}^{m \times n}$, $\mathbf{x} \in \mathcal{F}^n$, and $\mathbf{y} \in \mathcal{F}^m$. Then the equation

$$\mathbf{A}\,\mathbf{x} = \mathbf{y} \tag{3.56}$$

specifies m *linear equations* in n unknowns (the elements of $\mathbf{x}$). From the definition of matrix multiplication, the left side of the equation is some linear combination of the n columns of $\mathbf{A}$, the unknown coefficients of the linear combination being the elements of the vector $\mathbf{x}$. Thus, for a given $\mathbf{y} \in \mathcal{F}^m$, there will be a solution if and only if $\mathbf{y} \in \operatorname{im} \mathbf{A}\,\mathbf{x}$; since this subspace, the image or range of the linear function $\mathbf{A}\,\mathbf{x}$, is spanned by the columns of $\mathbf{A}$, it is conventionally called the *range* of $\mathbf{A}$, denoted by $\mathsf{R}(\mathbf{A})$. In order that a solution $\mathbf{x}$ can be found for every possible choice of $\mathbf{y}$, it is necessary and sufficient that $\mathsf{R}(\mathbf{A})$ has dimension m, or equivalently that there are m linearly independent columns among the n columns of $\mathbf{A}$.

Solutions to linear equations are not necessarily unique; uniqueness holds if and only if the columns of $\mathbf{A}$ are linearly independent vectors. In the case $n > m$, uniqueness never holds. In the case $n = m$, the uniqueness condition coincides with the existence condition: the matrix $\mathbf{A}$ must be invertible, and this is equivalent to the condition that the determinant of $\mathbf{A}$ be nonzero.

If a linear equation has two distinct solutions, $\mathbf{x}_1$ and $\mathbf{x}_2$, then the difference vector $\mathbf{x} = \mathbf{x}_1 - \mathbf{x}_2$ is a *nontrivial* (nonzero) solution to the related homogeneous equation

$$\mathbf{A}\,\mathbf{x} = \mathbf{0} \tag{3.57}$$

This equation shows that a nontrivial solution to the homogeneous equation may be found if and only if the columns of $\mathbf{A}$ are not linearly independent. For $\mathbf{A} \in \mathcal{F}^{n \times n}$ this is equivalent to the condition $\det \mathbf{A} = 0$.

The set of all solutions to the homogeneous equation forms the subspace $\ker \mathbf{A}\,\mathbf{x}$, the kernel or nullspace of the linear function $\mathbf{A}\,\mathbf{x}$. It is conventionally called the *nullspace* of $\mathbf{A}$, denoted $\mathsf{N}(\mathbf{A})$. It plays an important role in specifying the entire set of solutions to underdetermined sets of equations. If $\mathbf{A}\,\mathbf{x} = \mathbf{y}$ has a solution, say $\mathbf{x}_0$, the set of all solutions can be expressed as

$$S_{\mathbf{x}} = \{\mathbf{x} : \mathbf{x} = \mathbf{x}_0 + \sum_{i=1}^{k} \alpha_i \mathbf{x}_i\} \tag{3.58}$$

where the α_i are arbitrary scalars and $\{\mathbf{x}_1, \ldots, \mathbf{x}_k\}$ is a basis for $\mathsf{N}(\mathbf{A})$. Thus, the condition for uniqueness of solutions is $\mathsf{N}(\mathbf{A}) = \{\mathbf{0}\}$.

The dimension of $\mathsf{R}(\mathbf{A})$ (a subspace of $\mathcal{F}^m$) is called the *rank* of $\mathbf{A}$. Thus the rank of $\mathbf{A}$ is the number of its linearly independent columns, and it is equal to the number of its linearly independent rows. The dimension of $\mathsf{N}(\mathbf{A})$ (a subspace of $\mathcal{F}^n$) is called the *nullity* of $\mathbf{A}$. These quantities are related by the equation $\operatorname{rank}(\mathbf{A}) + \operatorname{nullity}(\mathbf{A}) = n$.

3.4.2 Solution of Linear Equations

For $\mathbf{A} \in \mathcal{F}^{n \times n}$, when $\mathbf{A}$ is invertible the solution of $\mathbf{A}\,\mathbf{x} = \mathbf{y}$ can be written explicitly as a linear function of $\mathbf{y}$, $\mathbf{y} = \mathbf{A}^{-1}\mathbf{x}$; this relation also shows that the ith column of $\mathbf{A}^{-1}$ is the solution to the linear equation $\mathbf{A}\,\mathbf{x} = \mathbf{e}_i$, where $\mathbf{e}_i$ is the ith unit vector.

For numerical computation of the solution of $\mathbf{A}\,\mathbf{x} = \mathbf{y}$, the PLU factorization of $\mathbf{A}$ (i.e., Gaussian elimination) may be used. Given the factorization $\mathbf{A} = \mathbf{P}\,\mathbf{L}\,\mathbf{U}$, the solution $\mathbf{x}$ may be obtained from solving two triangular sets of linear equations: solve $\mathbf{P}\,\mathbf{L}\,\mathbf{z} = \mathbf{y}$ (which is triangular after a reordering of the equations), and then solve $\mathbf{U}\,\mathbf{x} = \mathbf{z}$.

For the case of linear equations over the field of real numbers, additional results may be developed by employing geometric concepts. Let $\mathbf{A} \in \mathbb{R}^{m \times n}$, $\mathbf{x} \in \mathbb{R}^n$, and $\mathbf{y} \in \mathbb{R}^m$. The usual inner products will be used for $\mathbb{R}^n$ and $\mathbb{R}^m$: $\langle \mathbf{x}_1, \mathbf{x}_2 \rangle = \mathbf{x}_1^{\mathrm{T}} \mathbf{x}_2$ and $\langle \mathbf{y}_1, \mathbf{y}_2 \rangle = \mathbf{y}_1^{\mathrm{T}} \mathbf{y}_2$. In this framework, the matrix $\mathbf{A}^{\mathrm{T}}\mathbf{A} \in \mathbb{R}^{n \times n}$ is called the *Gram matrix* associated with $\mathbf{A}$; the Gram matrix is symmetric and its (i, j)th element is the inner product of the ith and jth columns of $\mathbf{A}$. The Gram matrix is invertible if and only if $\mathbf{A}$ has linearly independent columns. So, to test for uniqueness of solutions to consistent linear equations, it suffices to verify that $\det \mathbf{A}^{\mathrm{T}}\mathbf{A}$ is nonzero. In this case, premultiplying both sides of the linear equation $\mathbf{A}\,\mathbf{x} = \mathbf{y}$ by $\mathbf{A}^{\mathrm{T}}$ produces the equation (the *normal equations* for the components of $\mathbf{x}$), $\mathbf{A}^{\mathrm{T}}\mathbf{A}\,\mathbf{x} = \mathbf{A}^{\mathrm{T}}\mathbf{y}$, which has the solution $\mathbf{x} = (\mathbf{A}^{\mathrm{T}}\mathbf{A})^{-1}\mathbf{A}^{\mathrm{T}}\mathbf{y}$. An alternative approach with better inherent numerical accuracy is to use the QR factorization $\mathbf{A} = \mathbf{W}\mathbf{U}$, premultiplying both sides of the linear equation by $\mathbf{W}^{\mathrm{T}}$ to give an easily solved triangular system of linear equations, $\mathbf{U}\mathbf{x} = \mathbf{W}^{\mathrm{T}}\mathbf{y}$.

Numerical Conditioning of Linear Equations

Geometric methods are also useful in sensitivity analysis for linear equations. For $\mathbf{A} \in \mathbb{R}^{n \times n}$, consider the linear equations $\mathbf{A}\,\mathbf{x} = \mathbf{y}$, and suppose that the vector $\mathbf{y}$ is perturbed to become $\mathbf{y} + \Delta\mathbf{y}$. Then $\mathbf{A}(\mathbf{x} + \Delta\mathbf{x}) = \mathbf{y} + \Delta\mathbf{y}$, where $\Delta\mathbf{x} = \mathbf{A}^{-1}\Delta\mathbf{y}$. Using norms to quantify the relative change in $\mathbf{x}$ arising from the relative change in $\mathbf{y}$, leads to the inequality

$$\frac{\|\Delta\mathbf{x}\|}{\|\mathbf{x}\|} \le \kappa(\mathbf{A}) \frac{\|\Delta\mathbf{y}\|}{\|\mathbf{y}\|} \tag{3.59}$$

where $\kappa(\mathbf{A})$ denotes the *condition number* of $\mathbf{A}$, defined as

$$\kappa(\mathbf{A}) = \|\mathbf{A}\|\|\mathbf{A}^{-1}\| \tag{3.60}$$

Since $\kappa(\mathbf{A}) = \|\mathbf{A}\|\|\mathbf{A}^{-1}\| \geq \|\mathbf{A}\mathbf{A}^{-1}\| = 1$, when $\kappa(\mathbf{A}) \approx 1$, the matrix $\mathbf{A}$ is well-conditioned, but when $\kappa(\mathbf{A}) \gg 1$, the matrix $\mathbf{A}$ is ill-conditioned. The condition number of $\mathbf{A}$ also serves as the multiplier scaling relative errors in $\mathbf{A}$, measured by the induced norm, to relative errors in $\mathbf{x}$ [4].

3.4.3 Approximate Solutions and the Pseudo-Inverse

A geometric approach also provides a means of circumventing complicated issues of existence and uniqueness of solutions by replacing the linear equation $\mathbf{A}\mathbf{x} = \mathbf{y}$, with the following more general problem formulation: among the vectors $\widehat{\mathbf{x}}$ that minimize the Euclidean norm of the error vector $\mathbf{A}\widehat{\mathbf{x}} - \mathbf{y}$, find that vector $\mathbf{x}$ of smallest Euclidean norm. A unique solution to this problem always exists and takes the form

$$\mathbf{x} = \mathbf{A}^{\dagger}\mathbf{y} \tag{3.61}$$

where the matrix $\mathbf{A}^{\dagger}$ is called the *pseudo-inverse* of $\mathbf{A}$ [4] because it coincides with $\mathbf{A}^{-1}$ when $\mathbf{A}$ is square and nonsingular. $\mathbf{A}^{\dagger} \in \mathbb{R}^{n\times m}$, and it is the unique solution to the following set of matrix equations:

$$\mathbf{A}\mathbf{A}^{\dagger}\mathbf{A} = \mathbf{A} \tag{3.62a}$$

$$\mathbf{A}^{\dagger}\mathbf{A}\mathbf{A}^{\dagger} = \mathbf{A}^{\dagger} \tag{3.62b}$$

$$(\mathbf{A}\mathbf{A}^{\dagger}) \text{ is symmetric} \tag{3.62c}$$

$$(\mathbf{A}^{\dagger}\mathbf{A}) \text{ is symmetric} \tag{3.62d}$$

These equations define the matrix $\mathbf{A}\mathbf{A}^{\dagger}$ as the orthogonal projection onto $\mathsf{R}(\mathbf{A})$, and they define the matrix $(\mathbf{I} - \mathbf{A}\mathbf{A}^{\dagger})$ as the orthogonal projection onto $\mathsf{N}(\mathbf{A}^{\dagger})$. Furthermore, the matrix $(\mathbf{I} - \mathbf{A}\mathbf{A}^{\dagger})$ is the complementary orthogonal projection; i.e., $\mathsf{N}(\mathbf{A}^{\dagger})$ is the orthogonal complement of $\mathsf{R}(\mathbf{A})$ in $\mathbb{R}^m$.

If $\mathbf{A}$ has linearly independent columns, then the Gram matrix $\mathbf{A}^T\mathbf{A}$ is invertible and $\mathbf{A}^{\dagger} = (\mathbf{A}^T\mathbf{A})^{-1}\mathbf{A}^T$; for this case, $\mathbf{A}^{\dagger}$ may also be expressed in terms of the QR factorization of $\mathbf{A} = \mathbf{W}\mathbf{U}$: $\mathbf{A}^{\dagger} = \mathbf{U}^{-1}\mathbf{W}^T$. If $\mathbf{A}^T$ has linearly independent columns, then the Gram matrix associated with $\mathbf{A}^T$, $\mathbf{A}\mathbf{A}^T$, is invertible and $\mathbf{A}^{\dagger} = \mathbf{A}^T(\mathbf{A}\mathbf{A}^T)^{-1}$. When neither $\mathbf{A}$ nor $\mathbf{A}^T$ has linearly independent columns, no simple expression for $\mathbf{A}^{\dagger}$ is available. However, it will be given in terms of a matrix factorization of $\mathbf{A}$ known as the Singular Value Decomposition that is introduced later.

3.5 Eigenvalues and Eigenvectors

Scalar multiplication, $\mathbf{v} \mapsto \alpha\mathbf{v}$, is the simplest kind of linear function that maps a vector space into itself. The zero function, $\mathbf{v} \mapsto \mathbf{0}$, and the identity function, $\mathbf{v} \mapsto \mathbf{v}$, are two special cases. For a general linear function, $f : \mathcal{V} \to \mathcal{V}$, it is natural to investigate whether or not there are vectors, and hence subspaces of vectors, on which f is equivalent to scalar multiplication.

3.5.1 Definitions and Fundamental Properties

If $\mathcal{W}_\lambda$ is a nonzero subspace such that $f(\mathbf{w}) = \lambda\mathbf{w}$ for all $\mathbf{w} \in \mathcal{W}_\lambda$, then it is called an *eigenspace* of f corresponding to *eigenvalue* λ. The nonzero vectors in $\mathcal{W}_\lambda$ are called *eigenvectors* of f corresponding to eigenvalue λ.

To study eigenvalues, eigenvectors, and eigenspaces it is customary to use matrix representations and coordinate spaces. In this framework, the equation determining an eigenvector and its corresponding eigenvalue takes the form

$$\mathbf{A}\mathbf{u} = \lambda\mathbf{u} \quad \text{for } \mathbf{u} \neq \mathbf{0} \tag{3.63}$$

where $\mathbf{A} \in \mathcal{F}^{n\times n}$, $\mathbf{u} \in \mathcal{F}^n$, and $\lambda \in \mathcal{F}$. Equivalently

$$(\lambda\mathbf{I} - \mathbf{A})\mathbf{u} = \mathbf{0} \tag{3.64}$$

A nontrivial solution of this homogeneous linear equation will exist if and only if $\det(\lambda\mathbf{I} - \mathbf{A}) = 0$. This equation is called the *characteristic equation* of the matrix $\mathbf{A}$, since it involves the monic nth degree characteristic polynomial of $\mathbf{A}$,

$$\det(\lambda\mathbf{I} - \mathbf{A}) = \chi(\lambda) = \lambda^n + \chi_1\lambda^{n-1} + \ldots + \chi_{n-1}\lambda + \chi_n \tag{3.65}$$

Eigenvalues are zeros of the characteristic polynomial, i.e., roots of the characteristic equation.

Depending on the field $\mathcal{F}$, roots of the characteristic equation may or may not exist; i.e., $(\lambda\mathbf{I} - \mathbf{A})$ may be invertible for all $\lambda \in \mathcal{F}$. For a characteristic polynomial such as $\lambda^2 + 1$, there are no real zeros even though the polynomial has real coefficients; on the other hand, this polynomial has two complex zeros. Indeed, by the Fundamental Theorem of Algebra, every nth degree polynomial with complex coefficients has n complex zeros, implying that

$$\det(\lambda\mathbf{I} - \mathbf{A}) = (\lambda - \lambda_1)(\lambda - \lambda_2)\cdots(\lambda - \lambda_n) \tag{3.66}$$

for some set of complex numbers $\lambda_1, \ldots, \lambda_n$, not necessarily distinct. Thus, for finding eigenvalues and eigenvectors of $\mathbf{A} \in \mathbb{R}^{n\times n}$ it is sometimes convenient to regard $\mathbf{A}$ as an element of $\mathbb{C}^{n\times n}$.

The eigenvalues and eigenvectors of real matrices are constrained by conjugacy conditions. If $\mathbf{A}$ is real and λ is an eigenvalue with nonzero imaginary part, then λ^*, the complex conjugate of λ, is also an eigenvalue of $\mathbf{A}$. (The characteristic polynomial of a real matrix has real coefficients and its complex zeros occur in conjugate pairs.) If $\mathbf{u}$ is an eigenvector of the real matrix $\mathbf{A}$ corresponding to eigenvalue λ having nonzero imaginary part, then $\mathbf{u}^*$ (component-wise conjugation) is an eigenvector of $\mathbf{A}$ corresponding to eigenvalue λ^*.

Some classes of real matrices have real eigenvalues. Since the diagonal elements of any upper triangular matrix are its eigenvalues, every real upper triangular matrix has real eigenvalues. The same is true of lower triangular matrices and diagonal matrices. More surprisingly, any *normal matrix*, a matrix $\mathbf{A} \in \mathbb{C}^{n\times n}$ with $\mathbf{A}^H\mathbf{A} = \mathbf{A}\mathbf{A}^H$, has real eigenvalues. A matrix $\mathbf{A} \in \mathbb{R}^{n\times n}$ is normal when $\mathbf{A}^T\mathbf{A} = \mathbf{A}\mathbf{A}^T$, and thus any real symmetric matrix, $\mathbf{Q} \in \mathbb{R}^{n\times n}$ with $\mathbf{Q}^T = \mathbf{Q}$, has real eigenvalues.

3.5.2 Eigenvector Bases and Diagonalization

When λ is an eigenvalue of $\mathbf{A} \in \mathbb{C}^{n\times n}$, the subspace of $\mathbb{C}^n$ given by $\mathcal{W}_\lambda = \mathsf{N}(\lambda\mathbf{I} - \mathbf{A})$ is the associated *maximal eigenspace*; it has dimension greater than zero. If λ_1 and λ_2 are two eigenvalues of $\mathbf{A}$ with $\lambda_1 \neq \lambda_2$, corresponding eigenvectors $\mathbf{u}(\lambda_1) \in \mathcal{W}_{\lambda_1}$ and $\mathbf{u}(\lambda_2) \in \mathcal{W}_{\lambda_2}$ are linearly independent. This leads to a sufficient condition for existence of a basis of $\mathbb{C}^n$ consisting of eigenvectors of $\mathbf{A}$. If $\mathbf{A}$ has n distinct eigenvalues, the set of n corresponding eigenvectors forms a basis for $\mathbb{C}^n$.

More generally, if $\mathbf{A}$ has $r \leq n$ distinct eigenvalues, $\{\lambda_1, \ldots, \lambda_r\}$, with associated maximal eigenspaces $\mathcal{W}_1, \mathcal{W}_2, \ldots, \mathcal{W}_r$ having dimensions $d_1, d_2, \ldots, d_r$ equal to the algebraic multiplicities of the eigenvalues (as zeros of the characteristic polynomial), respectively, then $d_1 + \cdots + d_r = n$ and $\mathbb{C}^n$ has a basis consisting of eigenvectors of $\mathbf{A}$. This case always holds for real symmetric matrices.

Let $\mathbf{A} \in \mathbb{C}^{n\times n}$ be a matrix whose eigenvectors $\{\mathbf{u}_1, \mathbf{u}_2, \ldots, \mathbf{u}_n\}$ form a basis $\mathcal{B}$ for $\mathbb{C}^n$; let $\{\lambda_1, \lambda_2, \ldots, \lambda_n\}$ be the corresponding eigenvalues. Let $\mathbf{T}$ be the invertible $n \times n$ matrix whose ith column is $\mathbf{u}_i$. Then $\mathbf{A}\,\mathbf{T} = \mathbf{T}\Lambda$, where Λ is a diagonal matrix formed from the eigenvalues:

$$\Lambda = \begin{bmatrix} \lambda_1 & 0 & \cdot & \cdot & \cdot & 0 \\ 0 & \lambda_2 & & & & \cdot \\ \cdot & & \cdot & & & \cdot \\ \cdot & & & \cdot & & \cdot \\ \cdot & & & & \cdot & 0 \\ 0 & \cdot & \cdot & \cdot & 0 & \lambda_n \end{bmatrix} \tag{3.67}$$

Solving for Λ gives

$$\mathbf{T}^{-1}\mathbf{A}\,\mathbf{T} = \Lambda \tag{3.68}$$

Thus, $\mathbf{A}$ is similar to the diagonal matrix of its eigenvalues, Λ, and $\mathbf{T}\Lambda\mathbf{T}^{-1} = \mathbf{A}$. Also, Λ is the matrix representation of the linear function $f(\mathbf{x}) = \mathbf{A}\,\mathbf{x}$ with respect to the eigenvector basis $\mathcal{B}$ of $\mathbb{C}^n$.

For any matrix whose eigenvectors form a basis of $\mathbb{C}^n$, the similarity equation $\mathbf{A} = \mathbf{T}\Lambda\mathbf{T}^{-1}$ may be rewritten using the definition of matrix multiplication, giving

$$\mathbf{A} = \mathbf{T}\Lambda\mathbf{T}^{-1} = \sum_{i=1}^{n} \lambda_i \mathbf{u}_i \mathbf{v}_i \tag{3.69}$$

where $\mathbf{v}_i$ is the ith row of $\mathbf{T}^{-1}$. This is the called the *spectral representation* of $\mathbf{A}$. The row vector $\mathbf{v}_i$ is called a *left eigenvector* of $\mathbf{A}$ since it satisfies $\mathbf{v}_i\mathbf{A} = \lambda_i\mathbf{v}_i$.

Symmetric Matrices

For a real symmetric matrix $\mathbf{Q}$, all eigenvalues are real and the corresponding eigenvectors may be chosen with real components. For this case, if λ_1 and λ_2 are two eigenvalues with $\lambda_1 \neq \lambda_2$, the corresponding real eigenvectors $\mathbf{u}(\lambda_1) \in \mathcal{W}_{\lambda_1}$ and $\mathbf{u}(\lambda_2) \in \mathcal{W}_{\lambda_2}$ are not only linearly independent, they are also orthogonal, $\langle \mathbf{u}(\lambda_1), \mathbf{u}(\lambda_2)\rangle = 0$. Further, each maximal eigenspace has dimension equal to the algebraic multiplicity of the associated eigenvalue as a zero of the characteristic polynomial, and each maximal eigenspace has an orthogonal basis of eigenvectors. Thus, for any real symmetric matrix $\mathbf{Q}$, there is an orthogonal basis for $\mathbb{R}^n$ consisting of eigenvectors; by scaling the lengths of the basis vectors to one, an orthonormal basis of eigenvectors is obtained. Thus, $\Lambda = \mathbf{O}^{\mathrm{T}}\mathbf{Q}\mathbf{O}$, where $\mathbf{O}$ is an orthogonal matrix. (This may be generalized. If $\mathbf{A}$ is a complex Hermitian matrix, i.e., $\mathbf{A}^{\mathrm{H}} = \mathbf{A}$ where $\mathbf{A}^{\mathrm{H}}$ denotes the combination of conjugation and transposition: $\mathbf{A}^{\mathrm{H}} = (\mathbf{A}^*)^{\mathrm{T}}$. Then $\mathbf{A}$ has real eigenvalues and there is a basis of $\mathbb{C}^n$ comprised of normalized eigenvectors so that $\Lambda = \mathbf{U}^{\mathrm{H}}\mathbf{A}\,\mathbf{U}$, where $\mathbf{U}$ is a unitary matrix.)

For symmetric $\mathbf{Q}$, when the eigenvectors are chosen to be orthonormal, the spectral representation simplifies to

$$\mathbf{Q} = \mathbf{O}\,\Lambda\mathbf{O}^{\mathrm{T}} = \sum_{i=1}^{n} \lambda_i \mathbf{u}_i \mathbf{u}_i^{\mathrm{T}} \tag{3.70}$$

A real symmetric matrix $\mathbf{Q}$ is said to be *positive definite* if the inequality $\mathbf{x}^{\mathrm{T}}\mathbf{Q}\mathbf{x} > 0$ holds for all $\mathbf{x} \in \mathbb{R}^n$ with $\mathbf{x} \neq \mathbf{0}$; $\mathbf{Q}$ is said to be *nonnegative definite* if $\mathbf{x}^{\mathrm{T}}\mathbf{Q}\mathbf{x} \geq 0$ for all $\mathbf{x} \in \mathbb{R}^n$. When all eigenvalues of $\mathbf{Q}$ are positive (nonnegative), $\mathbf{Q}$ is positive definite (nonnegative definite). When $\mathbf{Q}$ is nonnegative definite, let $\mathbf{V} = \lambda^{1/2}\mathbf{O}$, where $\lambda^{1/2}$ denotes the diagonal matrix of nonnegative square roots of the eigenvalues; then $\mathbf{Q}$ may be written in factored form as $\mathbf{Q} = \mathbf{V}^{\mathrm{T}}\mathbf{V}$. This shows that if $\mathbf{Q}$ is positive definite, it is the Gram matrix of a set of linearly independent vectors, the columns of $\mathbf{V}$. If the QR factorization of $\mathbf{V}$ is $\mathbf{V} = \mathbf{W}\mathbf{U}$, then the *Cholesky factorization* of $\mathbf{Q}$ is $\mathbf{Q} = \mathbf{U}^{\mathrm{T}}\mathbf{U}$ [4], which is also seen to be the (symmetric) PLU factorization of $\mathbf{Q}$ (i.e., $\mathbf{P} = \mathbf{I}$ and $\mathbf{L} = \mathbf{U}^{\mathrm{T}}$).

Conversely, for any matrix $\mathbf{H} \in \mathbb{R}^{m\times n}$, the symmetric matrices $\mathbf{Q}_{n\times n} = \mathbf{H}^{\mathrm{T}}\mathbf{H}$ (the Gram matrix of the columns of $\mathbf{H}$) and $\mathbf{Q}_{m\times m} = \mathbf{H}\mathbf{H}^{\mathrm{T}}$ (the Gram matrix of the columns of $\mathbf{H}^{\mathrm{T}}$) are both nonnegative definite since $\mathbf{x}^{\mathrm{T}}\mathbf{Q}_{m\times m}\mathbf{x} = \langle \mathbf{H}^{\mathrm{T}}\mathbf{x}, \mathbf{H}^{\mathrm{T}}\mathbf{x}\rangle \geq 0$ for all $\mathbf{x} \in \mathbb{R}^m$ and $\mathbf{x}^{\mathrm{T}}\mathbf{Q}_{n\times n}\mathbf{x} = \langle \mathbf{H}\mathbf{x}, \mathbf{H}\mathbf{x}\rangle \geq 0$ for all $\mathbf{x} \in \mathbb{R}^n$.

3.5.3 More Properties of Eigenvalues

From the factored form of the characteristic polynomial it follows that $\det\mathbf{A} = \lambda_1\lambda_2\cdots\lambda_n$, the product of the eigenvalues. Thus, $\mathbf{A}$ will be invertible if and only if it has no zero eigenvalue. If $\lambda = 0$ is an eigenvalue of $\mathbf{A}$, then $\mathsf{N}(\mathbf{A})$ is the associated maximal eigenspace.

Let λ be an eigenvalue of $\mathbf{A}$ with corresponding eigenvector $\mathbf{u}$. For integer $k \geq 0$, λ^k is an eigenvalue of $\mathbf{A}^k$ with corresponding eigenvector $\mathbf{u}$; more generally, for any polynomial $\alpha(s) = \alpha_0 s^d + \cdots + \alpha_{d-1}s + \alpha_d$, $\alpha(\lambda)$ is an eigenvalue of $\alpha(\mathbf{A})$ with corresponding eigenvector $\mathbf{u}$. If $\mathbf{A}$ is invertible, $1/\lambda$ is an eigenvalue of $\mathbf{A}^{-1}$ with corresponding eigenvector $\mathbf{u}$.

The *spectral radius* of a square matrix (over $\mathbb{C}$ or $\mathbb{R}$) is the magnitude of its largest eigenvalue: $\rho(\mathbf{A}) = \max\{|\lambda| : \det(\lambda\mathbf{I} - \mathbf{A}) = 0\}$. All eigenvalues of $\mathbf{A}$ lie within the disk $\{s \in \mathbb{C} : |s| \leq \rho(\mathbf{A})\}$ in the complex plane. If $\mathbf{A}$ is invertible, all eigenvalues of $\mathbf{A}$ lie within the annulus $\{s \in \mathbb{C} : \rho(\mathbf{A}^{-1}) \leq |s| \leq \rho(\mathbf{A})\}$ in the complex plane.

3.5.4 Eigenvectors, Commuting Matrices, and Lie Brackets

For two square matrices of the same size, $\mathbf{A}_1$ and $\mathbf{A}_2$, the eigenvalues of $\mathbf{A}_1\mathbf{A}_2$ are the same as the eigenvalues of $\mathbf{A}_2\mathbf{A}_1$. However, the difference matrix, $\mathbf{A}_1\mathbf{A}_2-\mathbf{A}_2\mathbf{A}_1$ is zero only when the matrices *commute*, i.e., when $\mathbf{A}_1\mathbf{A}_2 = \mathbf{A}_2\mathbf{A}_1$. If $\mathbf{A}_1$ and $\mathbf{A}_2$ are $n \times n$ and have a common set of n linearly independent eigenvectors, then they commute. Conversely, if $\mathbf{A}_1$ and $\mathbf{A}_2$ commute and if $\mathbf{A}_1$ has distinct eigenvalues, then $\mathbf{A}_1$ has n linearly independent eigenvectors that are also eigenvectors of $\mathbf{A}_2$.

Restated in terms of linear functions, the matrices $\mathbf{A}_1$ and $\mathbf{A}_2$ commute if and only if the linear function $f_{[\mathbf{A}_1,\mathbf{A}_2]}(\mathbf{x}) = (\mathbf{A}_1\mathbf{A}_2-\mathbf{A}_2\mathbf{A}_1)\mathbf{x} = \mathbf{0}$. For ease of notation, denote $[\mathbf{A}_1, \mathbf{A}_2] = \mathbf{A}_1\mathbf{A}_2 - \mathbf{A}_2\mathbf{A}_1$; $[\mathbf{A}_1, \mathbf{A}_2]$ is the *Lie product* of the matrices $\mathbf{A}_1$ and $\mathbf{A}_2$. $[\mathbf{A}_1, \mathbf{A}_2]\mathbf{x}$ is the *Lie bracket* of the linear functions $\mathbf{A}_1\mathbf{x}$ and $\mathbf{A}_2\mathbf{x}$.

Just as an invariant subspace of a linear function $f(\mathbf{x}) = \mathbf{A}\mathbf{x}$ provides a generalization of ker f useful for studying single linear functions, a Lie algebra of linear functions $f_i(\mathbf{x}) = \mathbf{A}_i\mathbf{x}$ provides a framework for the study of classes of matrices that do not necessarily commute. Let $\mathcal{L}(n)$ denote $\mathcal{L}(\mathbb{R}^n, \mathbb{R}^n)$, the real vector space of all linear functions mapping $\mathbb{R}^n$ to itself. The Lie bracket defines a mapping from $\mathcal{L}(n)\times\mathcal{L}(n)$ to $\mathcal{L}(n)$. A subspace $\mathcal{A}$ of $\mathcal{L}(n)$ is a *Lie algebra* of linear functions if the Lie bracket maps $\mathcal{A} \times \mathcal{A}$ into $\mathcal{A}$. In this sense, $\mathcal{A}$ is invariant under the Lie bracket. The dimension of $\mathcal{A}$ is its dimension as a subspace of the n^2-dimensional vector space $\mathcal{L}(n)$.

A Lie algebra of functions on $\mathbb{R}^n$ also associates with every point of $\mathbb{R}^n$ a subspace of $\mathbb{R}^n$ given by $\widehat{\mathbf{x}} \mapsto \{\mathbf{A}\widehat{\mathbf{x}} \in \mathbb{R}^n : \mathbf{A}\mathbf{x} \in \mathcal{A}\}$. Under certain conditions, this subspace is the space of tangent vectors at the point $\widehat{\mathbf{x}}$ to a manifold ("surface") in $\mathbb{R}^n$ described implicitly as the solution space of an associated nonlinear equation $\mathbf{F}_{\mathcal{A}}(\mathbf{x}) = \mathbf{0}$, with the normalization condition $\mathbf{F}_{\mathcal{A}}(\widehat{\mathbf{x}}) = \mathbf{0}$.

If $\mathbf{A} \neq \alpha\mathbf{I}$ for $\alpha \in \mathbb{R}$, then $\mathcal{A} = \{(\alpha_1\mathbf{A} + \alpha_2\mathbf{I})\mathbf{x} : \alpha_i \in \mathbb{R}\}$ is a 2-dimensional Lie algebra; every Lie bracket of functions in $\mathcal{A}$ is the zero function. When $\widehat{\mathbf{x}}$ is an eigenvector of $\mathbf{A}$, the tangent space at $\widehat{\mathbf{x}}$ is the one-dimensional space spanned by $\widehat{\mathbf{x}}$; at all other points $\widehat{\mathbf{x}}$, the tangent vectors form a two-dimensional subspace of $\mathbb{R}^n$.

$\mathcal{A}_{\text{skew}} = \{\mathbf{S}\mathbf{x} : \mathbf{S}^{\mathrm{T}} = -\mathbf{S}\}$ is the $n(n-1)/2$-dimensional Lie algebra of skew symmetric linear functions. Any two linearly independent functions in $\mathcal{A}_{\text{skew}}$ have a nonzero Lie bracket. For $n = 3$, the tangent vectors at each point $\widehat{\mathbf{x}}$ form a subspace of $\mathbb{R}^3$ having dimension 2, orthogonal to the vector $[\hat{x}_1\ \hat{x}_2\ \hat{x}_3]^{\mathrm{T}}$, the normal vector to the manifold given as the solution set of $\mathbf{F}(\mathbf{x}) = x_1^2 + x_2^2 + x_3^2 - (\hat{x}_1^2 + \hat{x}_2^2 + \hat{x}_3^2) = 0$. This manifold is recognized as the surface of the sphere whose squared radius is $\hat{x}_1^2 + \hat{x}_2^2 + \hat{x}_3^2$.

3.6 The Jordan Form and Similarity of Matrices

If the matrix $\mathbf{A}$ does not have a set of n linearly independent eigenvectors, then it is not similar to a diagonal matrix, but eigenvalues and eigenvectors still play a role in providing various useful representations; for at least one of its eigenvalues λ, the dimension of $\mathcal{W}_\lambda$ is smaller than the algebraic multiplicity of λ as a zero of the characteristic polynomial. For ease of notation let $\mathbf{A}_\lambda = (\lambda\mathbf{I}-\mathbf{A})$ so that $\mathcal{W}_\lambda = \mathsf{N}(\mathbf{A}_\lambda)$. Then, for $k \geq 1$, $\mathsf{N}(\mathbf{A}_\lambda^k) \subseteq \mathsf{N}(\mathbf{A}_\lambda^{k+1})$. Let $\mathrm{I}(\mathbf{A}_\lambda)$ be the *index* of $\mathbf{A}_\lambda$, the smallest positive integer k such that $\mathsf{N}(\mathbf{A}_\lambda^k) = \mathsf{N}(\mathbf{A}_\lambda^{k+1})$. Then the subspace $\mathcal{W}_{\lambda,\mathrm{I}} = \mathsf{N}(\mathbf{A}_\lambda^{\mathrm{I}})$ has dimension equal to the algebraic multiplicity of λ as a zero of the characteristic polynomial; when $\mathrm{I}(\mathbf{A}_\lambda) = 1$, $\mathcal{W}_{\lambda,\mathrm{I}} = \mathcal{W}_\lambda$, the associated maximal eigenspace.

For each eigenvalue λ, $\mathcal{W}_{\lambda,\mathrm{I}}$ is an $\mathbf{A}$-invariant subspace (i.e., f-invariant for $f(\mathbf{x}) = \mathbf{A}\mathbf{x}$). For eigenvalues $\lambda_1 \neq \lambda_2$, the corresponding "generalized eigenspaces" are independent, i.e., for nonzero vectors $\mathbf{v}_1 \in \mathcal{W}_{\lambda_1,\mathrm{I}_1}$ and $\mathbf{v}_2 \in \mathcal{W}_{\lambda_2,\mathrm{I}_2}$, $\mathbf{v}_1$ and $\mathbf{v}_2$ are linearly independent. The vectors obtained by choosing bases of all of the generalized eigenspaces, $\{\mathcal{W}_{\lambda_i,\mathrm{I}_i}\}$, may be collected together to form a basis of $\mathbb{C}^n$ consisting of eigenvectors and "generalized eigenvectors", and the general form of the matrix representation of the linear function $\mathbf{A}\mathbf{x}$ with respect to such a basis is a block diagonal matrix called the *Jordan form* [3]. Using the basis vectors as the columns of $\mathbf{T}$, the Jordan form of $\mathbf{A}$ is obtained by a similarity transformation,

$$\mathbf{T}^{-1}\mathbf{A}\,\mathbf{T} = \begin{bmatrix} \mathbf{M}(\lambda_1, d_1) & 0 & \cdot\ \cdot\ \cdot & 0 \\ 0 & \mathbf{M}(\lambda_2, d_2) & & \cdot \\ \cdot & & \cdot & \cdot \\ \cdot & & & \cdot \\ \cdot & & \cdot & 0 \\ 0 & \cdot & \cdot\ \cdot\ 0 & \mathbf{M}(\lambda_r, d_r) \end{bmatrix} \tag{3.71}$$

where block $\mathbf{M}(\lambda_i, d_i)$ has dimension $d_i \times d_i$ and d_i is the algebraic multiplicity of eigenvalue λ_i, assuming there are r distinct eigenvalues.

The matrix $\mathbf{M}(\lambda_i, d_i)$ is the matrix representation of $(\mathbf{A}\mathbf{x})|_{\mathcal{W}_{\lambda_i,\mathrm{I}_i}}$ with respect to the basis chosen for $\mathcal{W}_{\lambda_i,\mathrm{I}_i}$. If there are e_i linearly independent eigenvalues corresponding to eigenvalue λ_i, then the basis vectors can be chosen so that $\mathbf{M}(\lambda_i, d_i)$ takes the block diagonal form

$$\mathbf{M}(\lambda_i, d_i) = \begin{bmatrix} \mathbf{J}_1(\lambda_i) & 0 & \cdot\ \cdot\ \cdot & 0 \\ 0 & \mathbf{J}_2(\lambda_i) & & \cdot \\ \cdot & & \cdot & \cdot \\ \cdot & & & \cdot \\ \cdot & & \cdot & 0 \\ 0 & \cdot & \cdot\ \cdot\ 0 & \mathbf{J}_{e_i}(\lambda_i) \end{bmatrix} \tag{3.72}$$

where the *Jordan blocks* $\mathbf{J}_k(\lambda_i)$ take one of two forms: if the kth block has size $\delta_{i,k} = 1$, then $\mathbf{J}_k(\lambda_i) = [\lambda_i]$; if the kth block has size $\delta_{i,k} > 1$, $\mathbf{J}_k(\lambda_i)$ is given elementwise as

$$(\mathbf{J}_k(\lambda_i))_{p,q} = \begin{cases} \lambda_i, & \text{for } p = q \\ 1, & \text{for } p = q + 1 \\ 0, & \text{otherwise} \end{cases} \tag{3.73}$$

The sizes of the Jordan blocks can be expressed in terms of the dimensions of the subspaces $\mathcal{W}_{\lambda_i,j}$ for $1 \leq j \leq I_i$. The largest block size is equal to $I_i = I(\mathbf{A}_{\lambda_i})$. Of course $\sum_{k=1}^{e_i} \delta_{i,k} = d_i$. If every Jordan block has size 1, $e_i = d_i$ and the Jordan form is a diagonal matrix.

3.6.1 Invariant Factors and the Rational Canonical Form

Eigenvalues are invariant under similarity transformation. For any invertible matrix $\mathbf{T}$, when $\mathbf{T}^{-1}\mathbf{A}_1\mathbf{T} = \mathbf{A}_2$, $\mathbf{A}_1$ and $\mathbf{A}_2$ have the same eigenvalues. Let λ be an eigenvalue of $\mathbf{A}_1$ and of $\mathbf{A}_2$. If $\mathbf{u}$ is a corresponding eigenvector of $\mathbf{A}_1$, then $\mathbf{T}^{-1}\mathbf{u}$ is a corresponding eigenvector of $\mathbf{A}_2$. Likewise, $\mathbf{T}^{-1}$ transforms generalized eigenvectors of $\mathbf{A}_1$ to generalized eigenvectors of $\mathbf{A}_2$. Given a convention for ordering the Jordan blocks, the Jordan form is a canonical form for matrices under the equivalence relation of similarity, and two matrices are similar if and only if they have the same Jordan form.

The use of eigenvectors and generalized eigenvectors, and related methods involving invariant subspaces, is one approach to studying similarity of matrices. Polynomial matrix methods offer a second approach, based on the transformation of the polynomial matrix $(s\mathbf{I} - \mathbf{A})$ corresponding to similarity: $(s\mathbf{I} - \mathbf{A}) \mapsto \mathbf{T}^{-1}(s\mathbf{I} - \mathbf{A})\mathbf{T}$. This is a special case of the general equivalence transformation for polynomial matrices $\mathcal{A}(s) \mapsto \mathcal{P}(s)\mathcal{A}(s)\mathcal{Q}(s)$, whereby unimodular matrices $\mathcal{P}(s)$ and $\mathcal{Q}(s)$ transform $\mathcal{A}(s)$; two polynomial matrices are equivalent if and only if they have the same invariant polynomials [9].

Two real or complex matrices $\mathbf{A}_1$ and $\mathbf{A}_2$ are similar if and only if the polynomial matrices $(s\mathbf{I} - \mathbf{A}_1)$ and $(s\mathbf{I} - \mathbf{A}_2)$ have the same invariant polynomials [9]. The invariant polynomials of $(s\mathbf{I} - \mathbf{A})$ are also known as the *invariant factors* of $\mathbf{A}$ since they are factors of the characteristic polynomial $\det(s\mathbf{I} - \mathbf{A})$: $\det(s\mathbf{I} - \mathbf{A}) = \phi_1(s)\phi_2(s)\cdots\phi_n(s)$. Also, $\phi_{i+1}(s)$ is a factor of $\phi_i(s)$, and $\phi_1(s)$ is the *minimal polynomial* of $\mathbf{A}$, the monic polynomial $p(s)$ of least degree such that $p(\mathbf{A}) = 0$.

If $\mathbf{A}$ has q nontrivial invariant factors, then by similarity transformation it can be brought to the *rational canonical form* by a suitable choice of $\mathbf{T}$ [3]:

$$\mathbf{T}^{-1}\mathbf{A}\mathbf{T} = \begin{bmatrix} \mathbf{C}(\phi_1(s)) & 0 & \cdot & \cdot & \cdot & 0 \\ 0 & \mathbf{C}(\phi_2(s)) & & & & \cdot \\ \cdot & & \cdot & & & \cdot \\ \cdot & & & \cdot & & \cdot \\ \cdot & & & & \cdot & 0 \\ 0 & & \cdot & \cdot & 0 & \mathbf{C}(\phi_q(s)) \end{bmatrix} \tag{3.74}$$

where the ith diagonal block of this matrix, $\mathbf{C}(\phi_i(s))$, is a *companion matrix* associated with ith invariant factor: for a polynomial

$$\pi(s) = s^m + \pi_1 s^{m-1} + \cdots + \pi_{m-1}s + \pi_m,$$

$$\mathbf{C}(\pi(s)) = \begin{bmatrix} 0 & 0 & \cdot & \cdot & \cdot & 0 & -\pi_m \\ 1 & 0 & \cdot & \cdot & \cdot & 0 & -\pi_{m-1} \\ 0 & 1 & & & & \cdot & \cdot \\ \cdot & & \cdot & & & \cdot & \cdot \\ \cdot & & & \cdot & & \cdot & \cdot \\ \cdot & & & & 1 & 0 & -\pi_2 \\ 0 & \cdot & \cdot & \cdot & 0 & 1 & -\pi_1 \end{bmatrix} \tag{3.75}$$

The characteristic polynomial of $\mathbf{C}(\pi(s))$ is $\pi(s)$.

3.7 Singular Value Decomposition

Another approach may be taken to the transformation of a matrix to a diagonal form. Using two orthonormal bases obtained from eigenvectors of the symmetric nonnegative definite matrices $\mathbf{A}^T\mathbf{A}$ and $\mathbf{A}\mathbf{A}^T$ provides a representation known as the *singular value decomposition (SVD)* of $\mathbf{A}$ [4]:

$$\mathbf{A} = \mathbf{V}\Sigma\mathbf{U}^T \tag{3.76}$$

In this equation, $\mathbf{V}$ is an orthogonal matrix of eigenvectors of the product $\mathbf{A}\mathbf{A}^T$ and $\mathbf{U}$ is an orthogonal matrix of eigenvectors of the product $\mathbf{A}^T\mathbf{A}$. Σ is a diagonal matrix of nonnegative quantities $\sigma_1 \geq \sigma_2 \geq \cdots \geq \sigma_n \geq 0$ known as the *singular values* of $\mathbf{A}$. The singular values are obtained from the eigenvalues of the nonnegative definite matrix $\mathbf{A}\mathbf{A}^T$ (or those of $\mathbf{A}^T\mathbf{A}$) by taking positive square roots and reordering if necessary.

While not a similarity invariant, singular values are invariant under left and right orthogonal transformations. For any orthogonal matrices $\mathbf{O}_1$ and $\mathbf{O}_2$, when $\mathbf{O}_1\mathbf{A}_1\mathbf{O}_2 = \mathbf{A}_2$, $\mathbf{A}_1$ and $\mathbf{A}_2$ have the same singular values. The singular value decomposition holds for rectangular matrices in exactly the same form as for square matrices: $\mathbf{A} = \mathbf{V}\Sigma\mathbf{U}^T$, where $\mathbf{V}$ is an orthogonal matrix of eigenvectors of the product $\mathbf{A}\mathbf{A}^T$ and $\mathbf{U}$ is an orthogonal matrix of eigenvectors of the product $\mathbf{A}^T\mathbf{A}$. For $\mathbf{A} \in \mathbb{R}^{m\times n}$, the matrix $\Sigma \in \mathbb{R}^{m\times n}$ contains the singular values (the positive square roots of the common eigenvalues of $\mathbf{A}\mathbf{A}^T$ and $\mathbf{A}^T\mathbf{A}$) as diagonal elements of a square submatrix of size $\min(m, n)$ located in its upper left corner, with all other elements of Σ being zero.

The singular value decomposition is useful in a number of applications. If $\text{rank}(\mathbf{A}) = k$ (i.e., $\mathbf{A}$ has k linearly independent columns or rows), then k is the number of its nonzero singular values. When Σ must be determined by numerical techniques and is therefore subject to computational inaccuracies, judging the size of elements of Σ in comparison to the "machine accuracy" of the computer being used provides a sound basis for computing rank.

The SVD can be used to generate "low rank" approximations to the matrix $\mathbf{A}$. If $\text{rank}(\mathbf{A}) = k$, then for $\kappa < k$ the best rank κ approximation, i.e., the one minimizing the induced Euclidean norm $\|\mathbf{A} - \widehat{\mathbf{A}}\|$ over rank κ matrices $\widehat{\mathbf{A}}$, is obtained by setting the smallest $k - \kappa$ nonzero elements of Σ to zero and multiplying by $\mathbf{V}$ and $\mathbf{U}^T$ as in the defining equation for the SVD.

Finally, the SVD also provides a way of computing the pseudo-inverse:

$$\mathbf{A}^{\dagger} = \mathbf{U}\mathbf{D}^{\dagger}\mathbf{V}^{T} \tag{3.77}$$

where $\mathbf{D}^{\dagger}$, the pseudo-inverse of Σ, is obtained from Σ^T by inverting its nonzero elements.

The largest singular value of $\mathbf{A} \in \mathbb{R}^{n\times n}$ equals $\|\mathbf{A}\|$, its induced Euclidean norm. When $\mathbf{A}$ is invertible, its smallest singular value gives the distance of $\mathbf{A}$ from the set of singular matrices, and the ratio of its largest and smallest singular values is equal to its condition number, $\kappa(\mathbf{A})$.

The singular values of a real symmetric matrix $\mathbf{Q}$ are equal to the absolute values of its eigenvalues and, in particular, the spectral radius $\rho(\mathbf{Q}) = \|\mathbf{Q}\|$, its induced Euclidean norm.

3.8 Matrices and Multivariable Functions

It has already been noted that the most general linear function $f : \mathbb{R}^n \to \mathbb{R}^m$ can be expressed in terms of matrix multiplication as $\mathbf{y} = \mathbf{A}\mathbf{x}$, for $\mathbf{y} \in \mathbb{R}^m$, $\mathbf{x} \in \mathbb{R}^n$, and where $\mathbf{A} \in \mathbb{R}^{m\times n}$ is the matrix representation of $f(\mathbf{x})$ with respect to the standard bases of $\mathbb{R}^n$ and $\mathbb{R}^m$. Matrices are also useful in expressing much more general *nonlinear* functions.

3.8.1 Polynomial Functions

A function $h : \mathbb{R}^n \to \mathbb{R}^m$ satisfying $h(\alpha\mathbf{x}) = \alpha^k h(\mathbf{x})$ for all $\alpha \in \mathbb{R}$, and for all $\mathbf{x} \in \mathbb{R}^n$ is *homogeneous of order k*. Constant functions are homogeneous of order 0, and linear functions are homogeneous of order 1. For $\mathbf{x} = [x_1\ x_2\ \cdots\ x_n]^T$, a product of the form $x_1^{i_1} x_2^{i_2} \cdots x_n^{i_n}$, is homogeneous of order k, where k is the sum of the (nonnegative) powers: $i_1 + i_2 + \cdots + i_n = k$. Such *monomials* are the building blocks of (multivariable) polynomial functions.

Kronecker Products

The *Kronecker product*, or *tensor product*, for vectors and matrices provides a useful, systematic means of expressing general products of elements [5]. For any two matrices $\mathbf{A} \in \mathcal{R}^{m_A\times n_A}$ and $\mathbf{B} \in \mathcal{R}^{m_B\times n_B}$, the Kronecker product, denoted $\mathbf{A} \otimes \mathbf{B}$, is the $m_A m_B \times n_A n_B$ matrix written in block-matrix form as

$$\mathbf{A} \otimes \mathbf{B} = \begin{bmatrix} a_{11}\mathbf{B} & \cdots & a_{1n_A}\mathbf{B} \\ \vdots & & \vdots \\ a_{m_A 1}\mathbf{B} & \cdots & a_{m_A n_A}\mathbf{B} \end{bmatrix} \tag{3.78}$$

and with this definition, it satisfies (1) associativity: $(\mathbf{A} \otimes \mathbf{B}) \otimes \mathbf{C} = \mathbf{A} \otimes (\mathbf{B} \otimes \mathbf{C})$, so that $\mathbf{A} \otimes \mathbf{B} \otimes \mathbf{C}$ is unambiguous; (2) $(\mathbf{A}+\mathbf{B})\otimes(\mathbf{C}+\mathbf{D}) = (\mathbf{A}\otimes\mathbf{C})+(\mathbf{A}\otimes\mathbf{D})+(\mathbf{B}\otimes\mathbf{C})+(\mathbf{B}\otimes\mathbf{D})$; and (3) $(\mathbf{A}\mathbf{B})\otimes(\mathbf{C}\mathbf{D}) = (\mathbf{A}\otimes\mathbf{C})(\mathbf{B}\otimes\mathbf{D})$. (It is assumed that the numbers of rows and columns of the various matrices allow the matrix additions and matrix multiplications to be carried out.)

Some further properties of the Kronecker product include: (1) if $\mathbf{A}$ and $\mathbf{B}$ are invertible, then $(\mathbf{A} \otimes \mathbf{B})^{-1} = \mathbf{A}^{-1} \otimes \mathbf{B}^{-1}$; (2) $\mathbf{A} \otimes \mathbf{B} = \mathbf{0}$ if and only if $\mathbf{A} = \mathbf{0}$ or $\mathbf{B} = \mathbf{0}$; and (3) $\text{rank}(\mathbf{A} \otimes \mathbf{B}) = \text{rank}(\mathbf{A})\text{rank}(\mathbf{B})$.

Representation of Polynomial Functions

To represent a general homogeneous polynomial function of order k, $h : \mathbb{R}^n \to \mathbb{R}^m$, define $\mathbf{x}^{(k)} = \mathbf{x} \otimes \mathbf{x} \otimes \cdots \otimes \mathbf{x}$, where there are k factors; let $\mathbf{x}^{(1)} = \mathbf{x}$. Then $h(\mathbf{x}) = \mathbf{A}_k \mathbf{x}^{(k)}$ for some matrix $\mathbf{A}_k \in \mathbb{R}^{m\times n^k}$. (The matrix is not uniquely determined by h because for $\mathbf{x} \in \mathbb{R}^n$, a mixed product of elements such as $x_1 x_2^{k-1}$ is of course equal to $x_2^{k-1} x_1$.)

The composition of homogeneous polynomial functions with linear functions preserves the order of homogeneity. If $\mathbf{x} = \mathbf{T}\mathbf{z}$ for some matrix $\mathbf{T}$, then $\mathbf{x}^{(2)} = (\mathbf{T}\mathbf{z}) \otimes (\mathbf{T}\mathbf{z}) = \mathbf{T}^{(2)}\mathbf{z}^{(2)}$, and similarly $\mathbf{x}^{(k)} = \mathbf{T}^{(k)}\mathbf{z}^{(k)}$ for any k. If $\mathbf{T}$ is invertible and $h : \mathbb{R}^n \to \mathbb{R}^n$ is homogeneous of order k with $\mathbf{x} \mapsto \widehat{\mathbf{x}} = h(\mathbf{x}) = \mathbf{A}_k \mathbf{x}^{(k)}$, then with $\widehat{\mathbf{x}} = \mathbf{T}\widehat{\mathbf{z}}$, $\widehat{\mathbf{z}} = \mathbf{T}^{-1}\mathbf{A}_k\mathbf{T}^{(k)}\mathbf{z}^{(k)}$.

A general polynomial function $p : \mathbb{R}^n \to \mathbb{R}^m$ is a linear combination of homogeneous polynomials and can be represented using Kronecker products as

$$p(\mathbf{x}) = \mathbf{P}_0 + \mathbf{P}_1\mathbf{x} + \mathbf{P}_2\mathbf{x}^{(2)} + \cdots + \mathbf{P}_M\mathbf{x}^{(M)} \tag{3.79}$$

where $\mathbf{P}_k \in \mathbb{R}^{m\times n^k}$ is the kth coefficient matrix. $\mathbf{P}_0 \in \mathbb{R}^m$ is the constant term of the polynomial; $p(\mathbf{0}) = \mathbf{P}_0$.

3.8.2 Quadratic Forms

For scalar homogeneous polynomials of order 2, also known as *quadratic forms* or homogeneous scalar quadratic functions, an alternate matrix representation is commonly used. $Q : \mathbb{R}^n \to \mathbb{R}$ given by

$$Q(\mathbf{x}) = \mathbf{x}^T\mathbf{Q}\mathbf{x} \tag{3.80}$$

is a homogeneous polynomial of order 2, and any such function can be put in this form by suitable choice of the matrix $\mathbf{Q} \in \mathbb{R}^{n\times n}$. Without loss of generality, it may be assumed that $\mathbf{Q}$ is symmetric, since the symmetric part of $\mathbf{Q}$, $(\mathbf{Q}+\mathbf{Q}^T)/2$ yields the same quadratic function.

Since every symmetric matrix has real eigenvalues and can be diagonalized by an orthogonal similarity transformation, let $\mathbf{O}^T\mathbf{Q}\mathbf{O} = \Lambda$. Taking $\mathbf{y} = \mathbf{O}^T\mathbf{x}$ gives $Q(\mathbf{x}) = \mathbf{x}^T\mathbf{Q}\mathbf{x} = \mathbf{y}^T\Lambda\mathbf{y} = \sum_{i=1}^{n} \lambda_i y_i^2$. Thus, the quadratic function may be expressed as a weighted sum of squares of certain linear functions of $\mathbf{x}$. The quadratic function $Q(\mathbf{x})$ is positive (for all nonzero $\mathbf{x}$), or equivalently the matrix $\mathbf{Q}$ is positive definite, when all of the eigenvalues of $\mathbf{Q}$ are positive.

Another characterization of positive definiteness is given in terms of determinants. A set of principal minors of $\mathbf{Q}$ consists of the determinants of a nested set of n submatrices of $\mathbf{Q}$ formed as follows. Let $(\pi_1, \pi_2, \ldots, \pi_n)$ be a permutation of $(1, 2, \ldots, n)$. Let $\Delta_0 = \det \mathbf{Q}$, and for $1 \le i < n$ let Δ_i be the $(n-i)\times(n-i)$ minor given by the determinant of the submatrix of $\mathbf{Q}$ obtained by deleting rows and columns $\pi_1, \ldots, \pi_i$. $\mathbf{Q}$ is positive definite if and only if any set of principal minors has all positive elements.

A final characterization of positive definiteness is expressed in terms of Gram matrices. $\mathbf{Q}$ is positive definite if and only if it can be written as the Gram matrix of a set of n linearly independent vectors; taking such a set of vectors to be columns of a matrix $\mathbf{H}$, $\mathbf{Q} = \mathbf{H}^T\mathbf{H}$ and thus $Q(\mathbf{x}) = (\mathbf{Hx})^T(\mathbf{Hx})$, which expresses $Q(\mathbf{x})$ as $\|\mathbf{Hx}\|^2$, the squared Euclidean norm of the vector $\mathbf{Hx}$.

Starting with the quadratic form $\mathbf{x}^T\mathbf{Qx}$, an invertible linear change of variables $\mathbf{x} = \mathbf{T}^T\mathbf{y}$ produces another quadratic form, $\mathbf{y}^T\mathbf{TQT}^T\mathbf{y}$. The corresponding *congruence* transformation of symmetric matrices, whereby $\mathbf{Q} \mapsto \mathbf{TQT}^T$, does not necessarily preserve eigenvalues; however, the signs of eigenvalues are preserved. The number of positive, negative, and zero eigenvalues of $\mathbf{Q}$ characterizes its equivalence class under congruence transformations.

3.8.3 Matrix-Valued Functions

The algebra of matrices provides a direct means for defining a polynomial function of a square matrix. The Cayley-Hamilton Theorem is evidence that this is a fruitful concept and is one motivation for consideration of matrix-valued functions in a general setting. Such functions can be defined explicitly, as in the case of polynomial functions, or implicitly as the solution to some matrix equation(s).

Explicit Matrix Functions

Let $p : \mathbb{C} \to \mathbb{C}$, be a polynomial function, with $p(s) = p_0s^m + p_1s^{m-1} + \cdots + p_{m-1}s + p_m$, $p_i \in \mathbb{C}$, $0 \le i \le m$. When $\mathbf{A} \in \mathbb{C}^{n\times n}$, $p(\mathbf{A}) = p_0\mathbf{A}^m + p_1\mathbf{A}^{m-1} + \cdots + p_{m-1}\mathbf{A} + p_m\mathbf{I}$. For a function $f : \mathbb{C} \to \mathbb{C}$ given by a power series $f(s) = \sum_0^\infty f_i s^i$, convergent in some region $\{s \in \mathbb{C} : |s| < R\}$, the corresponding matrix function $f(\mathbf{A}) = \sum_0^\infty f_i\mathbf{A}^i$ is defined for matrices $\mathbf{A}$ with $\rho(\mathbf{A}) < R$, where $\rho(\mathbf{A})$ denotes the spectral radius of $\mathbf{A}$. Under this condition, the sequence of partial sums, $\mathbf{S}_n = \sum_0^n f_i\mathbf{A}^i$, converges to a limiting matrix $\mathbf{S}_\infty$, meaning that $\lim_{n\to\infty} \|\mathbf{S}_\infty - \mathbf{S}_n\| = 0$.

As a simple example, for $s \in \mathbb{C}$ with $|s| < 1$, the familiar formula for the geometric series defines the function

$$f(s) = (1-s)^{-1} = \sum_{i=0}^{\infty} s^i \tag{3.81}$$

and leads to the matrix function

$$f(\mathbf{A}) = (\mathbf{I}-\mathbf{A})^{-1} = \sum_{i=0}^{\infty} \mathbf{A}^i \tag{3.82}$$

which is defined for all $\mathbf{A} \in \mathbb{C}^{n\times n}$ satisfying $\rho(\mathbf{A}) < 1$.

Many interesting examples are obtained by considering functions $f(s)$ that involve a real or complex parameter and thereby produce parametric matrix functions. For example, consider the exponential function $f(s) = \exp(st)$, with parameter $t \in \mathbb{R}$, defined by the power series

$$\exp(st) = \sum_{i=0}^{\infty} \frac{t^i}{i!} s^i \tag{3.83}$$

For all $t \in \mathbb{R}$, this series converges for all $s \in \mathbb{C}$ so that the matrix exponential function $\exp(\mathbf{A}t)$ is given by

$$\exp(\mathbf{A}t) = \sum_{i=0}^{\infty} \frac{t^i}{i!} \mathbf{A}^i \tag{3.84}$$

for all $\mathbf{A} \in \mathbb{C}^{n\times n}$.

For any $\mathbf{A} \in \mathbb{C}^{n\times n}$, the power series also defines $\exp(\mathbf{A}t)$ as a matrix-valued function of t, $\exp(\mathbf{A}t) : \mathbb{R} \to \mathbb{C}^{n\times n}$. Analogs of the usual properties of the exponential function include $\exp(\mathbf{0}) = \mathbf{I}$; $\exp(\mathbf{A}(t+\tau)) = \exp(\mathbf{A}t)\exp(\mathbf{A}\tau)$; $(\exp(\mathbf{A}t))^{-1} = \exp(-\mathbf{A}t)$. However, $\exp((\mathbf{A}_1 + \mathbf{A}_2)t) \neq \exp(\mathbf{A}_1t)\exp(\mathbf{A}_2t)$, unless $\mathbf{A}_1$ and $\mathbf{A}_2$ commute; indeed, $\exp(-\mathbf{A}_2t)\exp((\mathbf{A}_1 + \mathbf{A}_2)t)\exp(-\mathbf{A}_1t) = \mathbf{I} + [\mathbf{A}_1, \mathbf{A}_2]t^2/2 + \cdots$, where $[\mathbf{A}_1, \mathbf{A}_2] = \mathbf{A}_1\mathbf{A}_2 - \mathbf{A}_2\mathbf{A}_1$, the Lie product of $\mathbf{A}_1$ and $\mathbf{A}_2$.

Similarity transformations can simplify the evaluation of matrix functions defined by power series. Suppose $\mathbf{A}$ is similar to a diagonal matrix of its eigenvalues, $\Lambda = \mathbf{T}^{-1}\mathbf{A}\mathbf{T}$; then, $\mathbf{A}^i = \mathbf{T}\Lambda^i\mathbf{T}^{-1}$ and for a function $f(\mathbf{A})$ defined as a power series, $f(\mathbf{A}) = \mathbf{T}f(\Lambda)\mathbf{T}^{-1}$. Since $f(\Lambda)$ is the diagonal matrix of values $f(\lambda_i)$, $f(\mathbf{A})$ is determined by: (a) the values of $f(s)$ on the eigenvalues of $\mathbf{A}$; and (b) the similarity transformation $\mathbf{T}$ whose columns are linearly independent eigenvectors of $\mathbf{A}$. When $\mathbf{A}$ is not similar to a diagonal matrix, $f(\mathbf{A})$ may still be evaluated using a similarity transformation to Jordan form.

When the eigenvalues of $\mathbf{A}$ are distinct, $f(\mathbf{A})$ can be obtained by finding an interpolating polynomial. Denote the characteristic polynomial by $\chi(s) = \det(s\mathbf{I} - \mathbf{A})$; its zeros are the eigenvalues, $\{\lambda_i, 1 \le i \le n\}$. Define polynomials $\chi_k(s) = \chi(s)/(s - \lambda_k)$, $1 \le k \le n$. Then, each $\chi_k(s)$ has degree $n-1$ and zeros $\{\lambda_j : j \neq k\}$; let $\chi_k = \chi_k(\lambda_k) \neq 0$. Then, the polynomial $L(s) = \sum_{k=1}^n (f(\lambda_k)/\chi_k)\chi_k(s)$ is the unique polynomial of degree $< n$ interpolating the function values; $L(\lambda_i) = f(\lambda_i)$, $1 \le i \le n$. Thus, $f(\mathbf{A}) = L(\mathbf{A})$, since $\mathbf{T}^{-1}f(\mathbf{A})\mathbf{T} = f(\Lambda) = L(\Lambda) = \mathbf{T}^{-1}L(\mathbf{A})\mathbf{T}$, where $\Lambda = \mathbf{T}^{-1}\mathbf{A}\mathbf{T}$.

Another important parametric matrix function is obtained from the power series for $f(s) = (\lambda - s)^{-1}$,

$$(\lambda - s)^{-1} = \sum_{i=0}^{\infty} \lambda^{-(i+1)} s^i \tag{3.85}$$

which, given $\lambda \in \mathbb{C}$, converges for those $s \in \mathbb{C}$ satisfying $|s| < |\lambda|$. The resulting matrix function is

$$(\lambda\mathbf{I} - \mathbf{A})^{-1} = \sum_{i=0}^{\infty} \lambda^{-(i+1)} \mathbf{A}^i \tag{3.86}$$

which is defined for $\mathbf{A} \in \mathbb{C}^{n\times n}$ satisfying $\rho(\mathbf{A}) < |\lambda|$.

For any given $\mathbf{A} \in \mathbb{C}^{n\times n}$, the power series also defines $(\lambda\mathbf{I} - \mathbf{A})^{-1}$ as a matrix-valued function of λ known as the *resolvent matrix* of $\mathbf{A}$, $(\lambda\mathbf{I} - \mathbf{A})^{-1} : \mathcal{D} \to \mathbb{C}^{n\times n}$ with the domain $\mathcal{D} = \{\lambda \in \mathbb{C} : |\lambda| > \rho(\mathbf{A})\}$. Additional properties of the resolvent matrix arise because $(\lambda\mathbf{I} - \mathbf{A})$ is a matrix of rational functions, i.e., $(\lambda\mathbf{I} - \mathbf{A}) \in \mathcal{F}^{n\times n}$, where $\mathcal{F}$ is the field $\mathbb{C}(\lambda)$ of rational functions of the complex variable λ with complex coefficients. Over $\mathbb{C}(\lambda)$, $(\lambda\mathbf{I} - \mathbf{A})$ is invertible because its determinant is the characteristic polynomial of $\mathbf{A}$, $\chi(\lambda)$, and is therefore a nonzero rational function. By Cramer's rule, the form of $(\lambda\mathbf{I} - \mathbf{A})^{-1}$ is

$$(\lambda\mathbf{I} - \mathbf{A})^{-1} = \frac{\Psi(\lambda)}{\det(\lambda\mathbf{I} - \mathbf{A})} \tag{3.87}$$

where $\Psi(\lambda)$ is a matrix whose elements are polynomials having degree $< n$. Multiplying both sides of this equation by $(\det(\lambda\mathbf{I} - \mathbf{A}))(\lambda\mathbf{I} - \mathbf{A})$ and equating coefficients of powers of λ leads to an explicit form for $\Psi(\lambda)$:

$$\begin{aligned}\Psi(\lambda) &= \mathbf{I}\lambda^{n-1} + (\mathbf{A} + \chi_1\mathbf{I})\lambda^{n-2} \\ &\quad + (\mathbf{A}^2 + \chi_1\mathbf{A} + \chi_2\mathbf{I})\lambda^{n-3} + \cdots \\ &\quad + (\mathbf{A}^{n-1} + \chi_1\mathbf{A}^{n-2} + \chi_2\mathbf{A}^{n-3} + \cdots \\ &\quad + \chi_{n-1}\mathbf{I})\end{aligned} \tag{3.88}$$

Expressing $(\lambda\mathbf{I} - \mathbf{A})^{-1}$ as a matrix of rational functions thus provides a means of defining it as a matrix-valued function for all $\lambda \in \mathbb{C}$ except for the zeros of $\det(\lambda\mathbf{I} - \mathbf{A})$, i.e., except for the eigenvalues of $\mathbf{A}$.

Solution of Matrix Equations

Matrix functions are not always given by explicit formulae; they can be defined implicitly as solutions to algebraic equations. For example, if $\mathbf{A} \in \mathcal{F}^{n\times n}$, the equation $\mathbf{A}\mathbf{X} = \mathbf{I}_n$ has a solution if and only if $\det\mathbf{A} \neq 0$; in this case the unique solution is $\mathbf{X} = \mathbf{A}^{-1}$. In fact, the Cayley-Hamilton Theorem may be used to express the solution as a polynomial matrix function of $\mathbf{A}$. If the characteristic polynomial of $\mathbf{A}$ is $s^n + \chi_1 s^{n-1} + \cdots + \chi_{n-1}\mathbf{A} + \chi_n\mathbf{I}$, then

$$\mathbf{A}^{-1} = \chi_n^{-1}(-\mathbf{A}^{n-1} - \chi_1\mathbf{A}^{n-2} - \cdots - \chi_{n-1}\mathbf{I}) \tag{3.89}$$

An important result for the study of matrix equations is the *Contraction Mapping Theorem* [8]. With $\|\cdot\|$ a norm on $\mathbb{R}^{n\times n}$, suppose $g : \mathbb{R}^{n\times n} \to \mathbb{R}^{n\times n}$, and suppose that $\mathcal{W}$ is a closed g-invariant subset (not necessarily a subspace) so that when $\mathbf{X} \in \mathcal{W}$ then $g(\mathbf{X}) \in \mathcal{W}$. If $\|g(\mathbf{X})\| < \gamma\|\mathbf{X}\|$ for some γ with $0 < \gamma < 1$, then g is called a contraction mapping on $\mathcal{W}$ with contraction constant γ. If g is a contraction mapping on $\mathcal{W}$, then a solution to the *fixed-point equation* $\mathbf{X} = g(\mathbf{X})$ exists and is unique in $\mathcal{W}$. The solution may be found by the method of successive approximation: for an arbitrary $\mathbf{X}_0 \in \mathcal{W}$ let

$$\mathbf{X}_i = g(\mathbf{X}_{i-1}) \qquad \text{for} \quad i > 0 \tag{3.90}$$

Then $\mathbf{X}_\infty = \lim_{i\to\infty}\mathbf{X}_i$ exists and satisfies $\mathbf{X}_\infty = g(\mathbf{X}_\infty)$.

Linear Matrix Equations

For solving a linear equation of the form $f(\mathbf{X}) = \mathbf{Y}$, where f is a linear function and $\mathbf{X}$ and $\mathbf{Y} \in \mathcal{F}^{n\times n}$, the selection of bases leads to a matrix representation $\mathbf{A}_f \in \mathcal{F}^{n^2\times n^2}$, and hence to a corresponding linear equation involving coordinate vectors: $\mathbf{A}_f\mathbf{x} = \mathbf{y}$, with $\mathbf{x}$ and $\mathbf{y} \in \mathcal{F}^{n^2}$.

When the linear function takes the form $f(\mathbf{X}) = \mathbf{A}_1\mathbf{X}\mathbf{A}_2$, for $\mathbf{A}_1$ and $\mathbf{A}_2 \in \mathcal{F}^{n\times n}$, an equivalent linear equation for coordinate vectors can be expressed concisely using Kronecker products. First form $\mathbf{x}^T$ by concatenating the rows of $\mathbf{X}$; similarly for $\mathbf{y}^T$. Then, the matrix equation $\mathbf{A}_1\mathbf{X}\mathbf{A}_2 = \mathbf{Y}$ is transformed to the form $(\mathbf{A}_1 \otimes \mathbf{A}_2^T)\mathbf{x} = \mathbf{y}$.

When the linear function $f(\mathbf{X})$ takes a more complicated form, which may always be expressed as a sum of such terms,

$$f(\mathbf{X}) = \sum_i \mathbf{A}_{1,i}\mathbf{X}\mathbf{A}_{2,i} \tag{3.91}$$

the Kronecker product approach may provide additional insight. For example, the linear matrix equation $\mathbf{A}_1\mathbf{X} - \mathbf{X}\mathbf{A}_2 = \mathbf{Y}$ becomes $(\mathbf{A}_1 \otimes \mathbf{I}_n - \mathbf{I}_n \otimes \mathbf{A}_2^T)\mathbf{x} = \mathbf{y}$. To characterize invertibility of the resulting $n^2 \times n^2$ matrix, it is most convenient to use the condition that it has no zero eigenvalue. Its eigenvalues are expressed in terms of the eigenvalues of $\mathbf{A}_1$ and those of $\mathbf{A}_2$ by the sums $\lambda_i(\mathbf{A}_1) - \lambda_j(\mathbf{A}_2)$; thus there will be no zero eigenvalue unless some eigenvalue of $\mathbf{A}_1$ is also an eigenvalue of $\mathbf{A}_2$. As a second example, the linear matrix equation $\mathbf{X} - \mathbf{A}_1\mathbf{X}\mathbf{A}_2 = \mathbf{Y}$ becomes $(\mathbf{I} - \mathbf{A}_1 \otimes \mathbf{A}_2^T)\mathbf{x} = \mathbf{y}$, and the resulting $n^2 \times n^2$ matrix is invertible unless some eigenvalue of $\mathbf{A}_1$ is the multiplicative inverse of some eigenvalue of $\mathbf{A}_2$.

References

[1] Barnett, S., *Matrices: Methods and Applications*, Oxford University Press, New York, NY, 1990.

[2] Bellman, R., *Introduction to Matrix Analysis*, Society for Industrial and Applied Mathematics, Philadelphia, PA, 1995.

[3] Gantmacher, F.R., *The Theory of Matrices*, 2 vols., Chelsea Publishing Co., New York, NY, 1959.

[4] Golub, G.H. and Van Loan, C.F., *Matrix Computations*, 2nd ed., The Johns Hopkins University Press, Baltimore, MD, 1989.

[5] Halmos, P.R., *Finite-Dimensional Vector Spaces*, Springer-Verlag, New York, NY, 1974.

[6] Householder, A.S., *The Theory of Matrices in Numerical Analysis*, Dover Publications, New York, NY, 1975.

[7] Lancaster, P. and Tismenetsky, M., *The Theory of Matrices*, 2nd ed., Academic Press, Orlando, FL, 1985.

[8] Luenberger, D.G., *Optimization by Vector Space Methods*, John Wiley & Sons, New York, NY, 1969.

[9] MacDuffee, C.C., *The Theory of Matrices*, Chelsea Publishing Co., New York, NY, 1946.

[10] Strang, G., *Linear Algebra and Its Applications*, 3rd ed., Harcourt, Brace, Jovanovich, San Diego, CA, 1988.

Further Reading

My undergraduate textbook (B. W. Dickinson, *Systems: Analysis, Design, and Computation*, Prentice Hall, 1991) covers material on linear algebra and matrices, thereby drawing on the applications to linear systems, nonlinear systems, and optimization for context and motivation. Most books on linear systems at the introductory graduate level cover basic material on linear algebra and matrices.

Thomas Kailath's *Linear Systems* (Prentice Hall, 1980) is a personal favorite that includes detailed coverage of polynomial matrices (and, for me, evokes nostalgia about my Stanford "roots" in the 1970s).

Other recommended books are Roger Brockett's *Finite Dimensional Linear Systems* (John Wiley, 1970), David Delchamps' *State Space and Input-Output Linear Systems* (Springer-Verlag, 1988), W.J. (Jack) Rugh's *Linear Systems* (Prentice Hall, 1993), and Eduardo Sontag's *Mathematical Control Theory: Deterministic Finite Dimensional Systems* (Springer-Verlag, 1990).

For coverage of current research on matrices and linear algebra, the following journals are recommended: *Linear Algebra and Its Applications* (Elsevier Science, Inc., New York, NY) and *SIAM Journal on Matrix Analysis and Applications* (Society for Industrial and Applied Mathematics, Philadelphia, PA).

A wealth of information is accessible in electronic form via the World Wide Web. Some `http` resources of particular interest are the following:

`math.technion.ac.il/iic`

Home page of the ILAS Information Center, primary contact point for the International Linear Algebra Society. This site has archives of tables of contents of recent issues of journals.

`www.mathworks.com`

Home page of The MathWorks, producers of the MATLAB software package.

`www-control.eng.cam.ac.uk/extras/Virtual_Library/Control_VL.html`

Home page of the WWW Virtual Library on Systems and Control, maintained at Cambridge University, England.

4

Complex Variables

C. W. Gray
The Aerospace Corporation, El Segundo, CA

4.1 Complex Numbers

From elementary algebra, the reader should be familiar with the *imaginary* number i where

$$i^2 = -1 \tag{4.1}$$

Historically in engineering mathematics, the square root of -1 is often denoted by j to avoid notational confusion with current i.

Every new number system in the history of mathematics created cognitive problems which were often not resolved for centuries. Even the terms for the *irrational number* $\sqrt{2}$, *transcendental number* π, and the *imaginary number* $i = \sqrt{-1}$ bear witness to the conceptual difficulties. Each system was encountered in the solution or completeness of a classical problem. Solutions to the quadratic and cubic polynomial equations were presented by Cardan in 1545, who apparently regarded the complex numbers as fictitious but used them formally. Remarkably, A. Girald (1590-1633) conjectured that any polynomial of the nth degree would have n roots in the complex numbers. This conjecture which is known as the *fundamental theorem of algebra* become famous and withstood false proofs from d'Alembert (1746) and Euler (1749). In fact, the dissertation of Gauss (1799) contains five different proofs of the conjecture, two of which are flawed [2].

4.1.1 The Algebra of Complex Numbers

A complex number is formed from a pair of real numbers (x, y) where the complex number is

$$z = x + iy \tag{4.2}$$

and where

$$x = \Re(z)\,,\, y = \Im(z) \tag{4.3}$$

are called the *real* and *imaginary* parts of z. A complex number z which has no real part $\Re(z) = 0$ is called *purely imaginary.* Two complex numbers are said to be *equal* if both their real and imaginary parts are equal. Assuming the ordinary rules of arithmetic, one derives the rules for addition and multiplication of complex numbers as follows:

$$(x + iy) + (u + iv) = (x + u) + i(y + v) \tag{4.4}$$

$$(x + iy) \cdot (u + iv) = (xu - yv) + i(xv + yu) \tag{4.5}$$

Complex numbers form a *field* satisfying the *commutative, associative,* and *distributive laws.* The real numbers 0 and 1 are the additive and multiplicative *identities.* Assuming these laws, the rule for division is easily derived:

$$\frac{x + iy}{u + iv} = \frac{x + iy}{u + iv} \cdot \frac{u - iv}{u - iv} = \frac{(xu + yv) + i(yu - xv)}{u^2 + v^2} \tag{4.6}$$

4.1.2 Conjugation and Modulus

The formula for complex multiplication employs the fact that $i^2 = -1$. The transformation,

$$z = x + iy \to \bar{z} = x - iy \tag{4.7}$$

is called *complex conjugation* and has the fundamental properties associated with an isomorphism

$$\overline{a + b} = \bar{a} + \bar{b} \tag{4.8}$$

$$\overline{ab} = \bar{a} \cdot \bar{b} \tag{4.9}$$

0-8493-8570-9/96/$0.00+$.50

The formulas

$$\Re(z) = \frac{z+\bar{z}}{2} \text{ and } \Im(z) = \frac{z-\bar{z}}{2i} \quad (4.10)$$

express the real and imaginary parts of z terms of conjugation. Consider the polynomial equation

$$a_n z^n + a_{n-1} z^{n-1} + \cdots + a_1 z + a_0 = 0.$$

Taking the complex conjugate of both sides,

$$\bar{a}_n \bar{z}^n + \bar{a}_{n-1} \bar{z}^{n-1} + \cdots + \bar{a}_1 \bar{z} + \bar{a}_0 = 0.$$

If the coefficients $a_i = \bar{a}_i$ are real, ξ and $\bar{\xi}$ are roots of the same equation, and hence, the nonreal roots of a polynomial with real coefficients occur in conjugate pairs. The product $z\bar{z} = x^2+y^2$ is always positive if $z \neq 0$. The *modulus* or *absolute value* is defined as

$$|z| = \sqrt{z\bar{z}} = \sqrt{x^2+y^2}. \quad (4.11)$$

Properties of conjugation can be employed to obtain

$$|ab| = |a| \cdot |b| \text{ and } |\frac{a}{b}| = \frac{|a|}{|b|}. \quad (4.12)$$

Formulas for the sum and difference follow from expansion:

$$|a+b|^2 = (a+b)\cdot(\bar{a}+\bar{b}) = a\bar{a} + (a\bar{b}+b\bar{a}) + b\bar{b},$$
$$|a+b|^2 = |a|^2 + |b|^2 + 2\Re(a\bar{b}). \quad (4.13)$$

The fact

$$\Re(a\bar{b}) \leq |ab|$$

can be combined with Equation 4.13

$$|a+b|^2 \leq (|a|+|b|)^2,$$

to yield the *triangle inequality*

$$|a+b| \leq |a| + |b|. \quad (4.14)$$

Cauchy's inequality is true for complex numbers:

$$|\sum_{i=1}^{n} a_i b_i|^2 \leq \left(\sum_{i=1}^{n} |a_i|^2\right)\cdot\left(\sum_{i=1}^{n} |b_i|^2\right). \quad (4.15)$$

4.1.3 Geometric Representation

A complex number $z = x + iy$ can be represented as a pair (x, y) on the *complex plane*. The x-axis is called the *real axis* and the y-axis the *imaginary axis*. The addition of complex numbers can be viewed as vector addition in the plane. The *modulus* or absolute value $|z|$ is interpreted as the *length* of the vector. The product of two complex numbers can be evaluated geometrically if we introduce polar coordinates:

$$x = r\cos\theta \quad (4.16)$$

and

$$y = r\sin\theta. \quad (4.17)$$

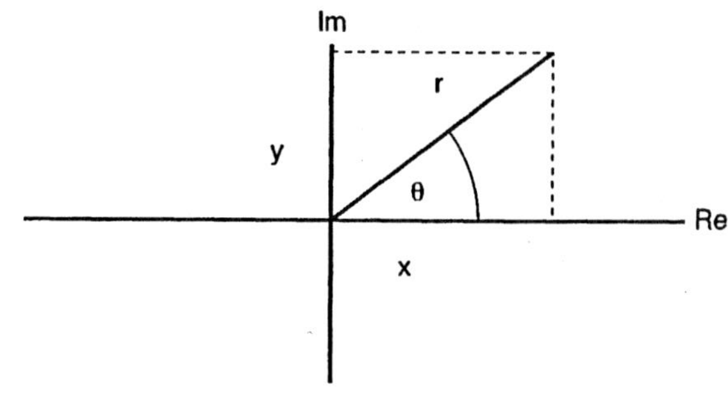

Figure 4.1 Polar representation.

Hence, $z = r(\cos\theta + i\sin\theta)$. This trigonometric form has the property that $r = |z|$ is the modulus and θ is called the *argument*,

$$\theta = \arg z. \quad (4.18)$$

Consider two complex numbers z_1, z_2 where, for $k = 1, 2$,

$$z_k = r_k(\cos\theta_k + i\sin\theta_k).$$

The product is easily computed as

$$\begin{aligned} z_1 z_2 &= r_1 r_2[(\cos\theta_1\cos\theta_2 - \sin\theta_1\sin\theta_2) \\ &\quad + i(\sin\theta_1\cos\theta_2 + \cos\theta_1\sin\theta_2)]. \end{aligned}$$

The standard addition formulas yield

$$\begin{aligned} z &= z_1 z_2 = r_1 r_2[\cos(\theta_1+\theta_2) \\ &\quad + i\sin(\theta_1+\theta_2)] = r(\cos\theta + i\sin\theta). \quad (4.19) \end{aligned}$$

The geometric interpretation of the product $z = z_1 z_2$ of complex numbers can be reduced to the dilation or stretch/contraction given by the product $r = r_1 r_2$ and the sum of the rotations

$$\arg(z_1 z_2) = \arg z_1 + \arg z_2. \quad (4.20)$$

The argument of the product is equal to the sum of the arguments. The argument of 0 is not defined and the polar angle θ in Equations 4.16 and 4.17 is only defined to a multiple of 2π.

The trigonometric form for the division $z = z_1/z_2$ can be derived by noting that the modulus is $r = r_1/r_2$ and

$$\arg\frac{z_1}{z_2} = \arg z_1 - \arg z_2. \quad (4.21)$$

From the preceding discussion, we can derive the powers of $z = r(\cos\theta + i\sin\theta)$ given by

$$z^n = r^n(\cos n\theta + i\sin n\theta). \quad (4.22)$$

For a complex number on the unit circle $r = 1$, we obtain *de Moivre's formula* (1730):

$$(\cos\theta + i\sin\theta)^n = \cos n\theta + i\sin n\theta. \quad (4.23)$$

The above formulas can be applied to find the roots of the equation $z^n = a$ where

$$a = r(\cos\theta + i\sin\theta)$$

and

$$z = \rho(\cos\phi + i\sin\phi).$$

Then Equation 4.22 yields

$$\rho^n(\cos n\phi + i\sin n\phi) = r(\cos\theta + i\sin\theta)$$

or

$$\rho = \sqrt[n]{r} \tag{4.24}$$

and

$$\phi = \frac{\theta}{n} + k \cdot \frac{2\pi}{n}, \tag{4.25}$$

for $k = 0, 1, \ldots, n-1$. We have found n roots to the equation $z^n = a$. If $a = 1$, then all of the roots lie on the unit circle and we can define the *primitive n^{th} root of unity* ξ:

$$\xi = \cos\frac{2\pi}{n} + i\sin\frac{2\pi}{n} \tag{4.26}$$

The roots of the equation $z^n = 1$ are easily expressed as $1, \xi, \xi^2, \ldots, \xi^{n-1}$.

4.2 Complex Functions

Let $\Omega \subseteq C$ be a subset of the complex plane. A rule of correspondence which associates each element $z = x + iy \in \Omega$ with a unique $w = f(z) = u(x, y) + iv(x, y)$ is called a *single-valued* complex function. Functions like $f(z) = \sqrt{z}$ are called *multiple-valued* and can be considered as a collection of single-valued functions. Definitions of the concepts of limit and continuity are analogous to those encountered in the functions of a real variable. The modulus function is employed as the *metric*.

DEFINITION 4.1 Open Region A subset $\Omega \subseteq C$ of the complex plane is called an open region or domain if, for every $z_0 \in \Omega$, there exists a $\delta > 0$ exists so that the circular disk $| z - z_0 | < \delta$, centered at z_0, is contained in Ω.

DEFINITION 4.2 Limit

$$\lim_{z \to z_0} f(z) = w_0 \tag{4.27}$$

if, for every $\epsilon > 0$, a $\delta > 0$ exists so that $| f(z) - w_0 | < \epsilon$ for all z satisfying $0 < | z - z_0 | < \delta$

DEFINITION 4.3 Continuity The function $f(z)$ is continuous at the point z_0 if

$$\lim_{z \to z_0} f(z) = f(z_0). \tag{4.28}$$

The function is said to be continuous in a region Ω if it is continuous at each point $z_0 \in \Omega$.

DEFINITION 4.4 Derivative If $f(z)$ is a single-valued complex function in some region Ω of the complex plane, the derivative of $f(z)$ is

$$f'(z) = \lim_{\Delta z \to 0} \frac{f(z + \Delta z) - f(z)}{\Delta z}. \tag{4.29}$$

The function is called differentiable provided that the limit exists and is the same regardless of the manner in which the complex number $\Delta z \to 0$.

A point where $f(z)$ is not differentiable is called a *singularity*. As in the theory of real valued functions, the sums, differences, products, and quotients (provided the divisor is not equal to zero) of continuous or differentiable complex functions are continuous or differentiable. It is important to note that the function $f(z) = \bar{z}$ is an example of a function which is continuous but nowhere differentiable.

$$\lim_{\Delta z \to 0} \frac{f(z + \Delta z) - f(z)}{\Delta z} = \pm 1 \tag{4.30}$$

depending upon whether the limit is approached through purely real or imaginary sequences Δz.

4.2.1 Cauchy-Riemann Equations

A function $f(z)$ is said to be *analytic* in a region Ω if it is differentiable and $f'(z)$ is continuous at every point $z \in \Omega$. Analytic functions are also called *regular* or *holomorphic*. A region Ω, for which a complex valued function is analytic, is called *a region of analyticity*. The previous example showed that the function $\bar{z}$ is not an analytic function. The requirement that a function be analytic is extremely strong. Consider an analytic function $w = f(z) = u(x, y) + iv(x, y)$. The derivative $f'(z)$ can be found by computing the limit through real variations $\Delta z = \Delta x \to 0$:

$$f'(z) = \frac{\partial f}{\partial x} = \frac{\partial u}{\partial x} + i\frac{\partial v}{\partial x} \tag{4.31}$$

or through purely imaginary variations $\Delta z = i\Delta y$:

$$f'(z) = \frac{1}{i}\frac{\partial f}{\partial y} = -i\frac{\partial f}{\partial y} = \frac{\partial v}{\partial y} - i\frac{\partial u}{\partial y} \tag{4.32}$$

Since the function is differentiable,

$$\frac{\partial f}{\partial x} = -i\frac{\partial f}{\partial y}. \tag{4.33}$$

Equating the expressions (4.31) and (4.32), one obtains the *Cauchy-Riemann* differential equations

$$\frac{\partial u}{\partial x} = \frac{\partial v}{\partial y}, \frac{\partial u}{\partial y} = -\frac{\partial v}{\partial x}. \tag{4.34}$$

Conversely, if the partial derivatives in Equation 4.34 are continuous and u, v satisfy the Cauchy-Riemann equations, then the function $f(z) = u(x, y) + iv(x, y)$ is analytic. If the second derivatives of u and v relative to x, y exist and are continuous, then, by differentiation and use of Equation 4.34,

$$\frac{\partial^2 u}{\partial x^2} + \frac{\partial^2 u}{\partial y^2} = 0, \frac{\partial^2 v}{\partial x^2} + \frac{\partial^2 v}{\partial y^2} = 0. \tag{4.35}$$

The real part u and imaginary part v satisfy *Laplace's equation* in two dimensions. Functions satisfying Laplace's equation are called *harmonic functions.*

4.2.2 Polynomials

The constant functions and the function $f(z) = z$ are analytic functions. Since the product and sum of analytic functions are analytic, it follows that any polynomial,

$$p(z) = a_n z^n + a_{n-1} z^{n-1} + \cdots + a_1 z + a_0, \tag{4.36}$$

with complex coefficients a_i is also an analytic function on the entire complex plane. If $a_n \neq 0$, the polynomial $p(z)$ is said to be of degree n. If $a_n = 1$ then $p(z)$ is called a *monic* polynomial.

THEOREM 4.1 **Fundamental Theorem of Algebra** *Every polynomial equation $p(z) = 0$ of degree n has exactly n complex roots $\xi_i, i = 1, \ldots, n$. The polynomial $p(z)$ can be uniquely factored as*

$$p(z) = \prod_{i=1}^{n} (z - \xi_i). \tag{4.37}$$

The roots ξ are not necessarily distinct. Roots of $p(z)$ are commonly called *zeros*. If the root ξ_i appears k times in the factorization, it is called a *zero of order k*.

Bernoulli's Method

The following numerical method, attributed to Bernoulii, can be employed to find the dominant (largest in modulus) root of a polynomial. The method can be employed as a quick numerical method to check if a discrete time system is stable (all roots of the characteristic polynomial lie in the unit circle). If there are several roots of the same modulus, then the method is modified and shifts are employed.

Given $a \in C^n$, define the monic polynomial:

$$p_a(z) = z^n + a_{n-1} z^{n-1} + \cdots + a_0.$$

Let $\{x_k\}$ be a nonzero solution to the difference equation

$$x_k = -a_{n-1} x_{k-1} - \cdots - a_0 x_{k-n}.$$

If $p_a(z)$ has a single largest dominant root r, then in general

$$r = \lim_{k \to \infty} \frac{x_{k+1}}{x_k}.$$

If a complex conjugate pair of roots r_1, r_2 is dominant and the coefficients are real $a \in R^{n+1}$, then $r_1, r_2 = r(\cos\theta \pm \sin\theta)$ where

$$r^2 = \lim_{k\to\infty} \frac{x_k^2 - x_{k+1}x_{k-1}}{x_{k-1}^2 - x_k x_{k-2}}$$

and

$$2r\cos\theta = \lim_{k\to\infty} \frac{x_{k+1}x_{k-1} - x_{k-1}x_k}{x_k x_{k-2} - x_{k-1}^2}.$$

We sketch the proof of Bernoulli's method for a single real dominant root. The typical response of the difference equation to a set of initial conditions can be written as

$$x_k = c_1 r_1^k + c_2 r_2^k + \cdots + c_n r_n^k$$

where $r_1, \ldots r_n$ are roots of the characteristic equation $p_a(x)$ with $| r_1 |>| r_2 |\geq \cdots \geq| r_n |$. If the initial conditions are selected properly, $c_1 \neq 0$ and

$$\frac{x_{k+1}}{x_k} = r_1 \frac{1 + (c_2/c_1)(r_2/r_1)^{k+1} + \cdots (c_n/c_1)(r_n/r_1)^{k+1}}{1 + (c_2/c_1)(r_2/r_1)^k + \cdots (c_n/c_1)(r_n/r_1)^k}.$$

If r_1 is dominant, then $| r_j/r_1 |< 1$ and the fractional expression tends toward 1. Hence x_{k+1}/x_k tends toward r_1. The proof of the complex dominant root formula is a slight generalization.

Genji's Formula

The following polynomial root perturbation formula can be employed with the root locus method to adjust or *tweak* the gains of a closed loop system.

Let $a \in C^{n+1}$ be a vector. Define the polynomial $p_a(z) = a_n z^n + a_{n-1} z^{n-1} + \cdots + a_0$. If $r \in C$ is a root $p_a(r) = 0$, then the following formula relates a perturbation of the root dr to a perturbation of the coefficients $da \in C^{n+1}$:

$$dr = -\frac{p_{da}(r)}{p_a'(r)}. \tag{4.38}$$

The formula follows from taking the total differential of the expression $p_a(r) = 0$,

$$[da_n r^n + da_{n-1} r^{n-1} + \cdots + da_0] + [n a_n r^{n-1} dr + (n-1) a_{n-1} r^{n-2} dr + \cdots + a_1 dr]$$

Hence,

$$p_{da}(r) + p_a'(r) dr = 0.$$

Lagrange's Interpolation Formula

Suppose that $z_0, z_1, \ldots z_n$ are $n+1$ distinct complex numbers. Given w_i where $0 \leq i \leq n$, we wish to find the polynomial $p(z)$ of degree n so that $p(z_i) = w_i$. The polynomial $p(z)$ can be employed as a method of interpolation. For $0 \leq i \leq n$, define

$$p_i(z) = \prod_{j \neq i} \left(\frac{z - z_j}{z_i - z_j} \right). \tag{4.39}$$

Clearly, $p_i(z_i) = 1$, and $p_i(z_j) = 0$, for $i \neq j$. Hence the interpolating polynomial can be found by

$$p(z) = \sum_{i=0}^{n} w_i p_i(z). \tag{4.40}$$

4.2.3 Zeros and Poles

The notion of repeated root can be generalized:

DEFINITION 4.5 **Zeros** An analytic function $f(z)$ has a zero at $z = a$ of order $k > 0$ if the following limit exists and is nonzero:

$$\lim_{z \to a} \frac{f(z)}{(z-a)^k} \neq 0. \tag{4.41}$$

A *singular point* of a function $f(z)$ is a value of z at which $f(z)$ fails to be analytic. If $f(z)$ is analytic in a region Ω, except at an *interior point* $z = a$, the point $z = a$ is called an *isolated singularity*. For example,

$$f(z) = \frac{1}{z-a}.$$

The concept of a *pole* is analogous to that of a zero.

DEFINITION 4.6 Poles A function $f(z)$ with an isolated singularity at $z = a$ has a pole of order $k > 0$ if the following limit exists and is nonzero:

$$\lim_{z\to a} f(z)(z-a)^k \neq 0. \tag{4.42}$$

A pole of order 1 is called a *simple pole*.

Clearly, if $f(z)$ has a pole of order k at $z = a$, then

$$f(z) = \frac{g(z)}{(z-a)^k}. \tag{4.43}$$

and,if $f(z)$ is analytic and has a zero of order k, then

$$f(z) = (z-a)^k g(z), \tag{4.44}$$

where $g(z)$ is analytic in a region including $z = a$ and $g(a) \neq 0$.

The function

$$f(z) = \frac{\sin z}{z}$$

is not defined at $z = 0$, but could be extended to an analytic function which takes the value 1 at $z = 0$. If a function can be extended to be analytic at a point $z = a$, then $f(z)$ is said to have a *removable singularity*. Hence, if a function $f(z)$ has a pole of order k at $z = a$, then the function $f(z)(z-a)^k$ has a removable singularity at $z = a$. A singularity at a point $z = a$, which is neither removable nor a pole of finite order k, is called *an essential singularity*.

4.2.4 Rational Functions

A *rational* function $H(z)$ is a quotient of two polynomials $N(z)$ and $D(z)$.

$$H(z) = \frac{N(z)}{D(z)} = \frac{b_m z^m + b_{m-1}z^{m-1} + \ldots b_1 z + b_0}{z^n + a_{n-1}z^{n-1} + \ldots + a_1 z + a_0}. \tag{4.45}$$

We shall assume in the discussion that the quotient is in *reduced form* and there are no common factors and hence no common zeros. A rational function is called *proper* if $m \leq n$. If the degrees satisfy $m < n$, then $H(z)$ is called *strictly proper*. In control engineering, rational functions most commonly occur as the transfer functions of linear systems. Rational functions $H(z)$ of a variable z denote the transfer functions of *discrete* systems and transfer functions $H(s)$ of the variable s are employed for *continuous* systems. Strictly proper functions have the property that

$$\lim_{z\to\infty} H(z) = 0$$

and hence *roll off* the power at high frequencies. Roots of the numerator and denominator are the zeros and poles of the corresponding rational function, respectively.

Partial Fraction Expansion

Consider a rational function $H(s) = N(s)/D(s)$ where the denominator $D(s)$ is a polynomial with distinct zeros $\xi_1, \xi_2, \ldots, \xi_n$. $H(s)$ can be expressed in a partial fraction expansion as

$$\frac{N(s)}{D(s)} = \frac{A_1}{s-\xi_1} + \frac{A_2}{s-\xi_2} + \cdots + \frac{A_n}{s-\xi_n}. \tag{4.46}$$

Multiplying both sides of the equation by $s - \xi_i$ and letting $s \to \xi_i$,

$$A_i = \lim_{s\to\xi_i} (s-\xi_i)H(s). \tag{4.47}$$

Applying L'Hospital's rule,

$$A_i = \lim_{s\to\xi_i} N(s)\frac{(s-\xi_i)}{D(s)} = N(\xi_i) \lim_{s\to\xi_i} \frac{1}{D'(s)} = \frac{N(\xi_i)}{D'(\xi_i)}.$$

Thus

$$\frac{N(s)}{D(s)} = \frac{N(\xi_1)}{D'(\xi_1)}\cdot\frac{1}{s-\xi_1} + \frac{N(\xi_2)}{D'(\xi_2)}\cdot\frac{1}{s-\xi_2} + \cdots + \frac{N(\xi_n)}{D'(\xi_n)}\cdot\frac{1}{s-\xi_n} \tag{4.48}$$

This formula is commonly called *Heaviside's expansion formula*, and it can be employed for computing the inverse Laplace transform of rational functions when the roots of $D(s)$ are distinct.

In general, any strictly proper rational function $H(s)$ can be written as a sum of the strictly proper rational functions

$$\frac{A_{\xi,r}}{(s-\xi)^r} \tag{4.49}$$

where ξ is a zero of $D(s)$ of order k where $r \leq k$. If ξ is a repeated zero of $D(s)$ of order k, the coefficient $A_{\xi,r}$ corresponding to the power $r \leq k$ can be found,

$$A_{\xi,r} = \lim_{s\to\xi_i} \frac{1}{(k-r)!}\frac{d^{k-r}}{ds^{k-r}}[(s-\xi_i)^k H(s)]. \tag{4.50}$$

Lucas' Formula

The *Nyquist stability criterion* or the *principle of the argument* relies upon a generalization of Lucas's formula. The derivative of a factored polynomial of the form,

$$P(s) = a_n(s-\xi_1)(s-\xi_2)\cdots(s-\xi_n),$$

yields Lucas' formula,

$$\frac{P'(s)}{P(s)} = \frac{1}{s-\xi_1} + \frac{1}{s-\xi_2} + \cdots + \frac{1}{s-\xi_n}. \tag{4.51}$$

Let $z = a$ be a zero of order k of the function $f(z)$. Application of Equation 4.44

$$\begin{aligned} f(z) &= (z-a)^k g(z), \\ f'(z) &= k(z-a)^{k-1}g(z) + (z-a)^k g'(z), \end{aligned}$$

gives

$$\frac{f'(z)}{f(z)} = \frac{k}{z-a} + \frac{g'(z)}{g(z)} \tag{4.52}$$

where $g(a) \neq 0$ and $g(z)$ is analytic at $z = a$. For a pole at $z = a$ of order k of a function $f(z)$, Equation 4.43 yields

$$\begin{aligned} f(z) &= (z-a)^{-k} g(z), \\ f'(z) &= -k(z-a)^{-k-1} g(z) + (z-a)^{-k} g'(z). \end{aligned}$$

This gives

$$\frac{f'(z)}{f(z)} = -\frac{k}{z-a} + \frac{g'(z)}{g(z)} \tag{4.53}$$

where $g(a) \neq 0$ and $g(z)$ is analytic around $z = a$. Inductive use of the above expressions results in a generalization of Lucas' formula (4.51). For a rational function with zeros α_j and poles ξ_i,

$$\begin{aligned} H(s) &= \frac{N(s)}{D(s)} = \frac{\prod_{j=1}^{m}(s-\alpha_j)}{\prod_{i=1}^{n}(s-\xi_i)} \\ &= \frac{(s-\alpha_1)(s-\alpha_2)\cdots(s-\alpha_m)}{(s-\xi_1)(s-\xi_2)\cdots(s-\xi_n)}, \end{aligned} \tag{4.54}$$

and

$$\frac{H'(s)}{H(s)} = \sum_{j=1}^{m} \frac{1}{s-\alpha_j} - \sum_{i=1}^{n} \frac{1}{s-\xi_i} \tag{4.55}$$

Rational functions can be generalized:

DEFINITION 4.7 Meromorphic function A function $f(z)$, which is analytic in an open region Ω and whose every singularity is an isolated pole, is said to be meromorphic.

The transfer function of every continuous time-invariant linear system is meromorphic in the complex plane. Systems or block diagrams which employ the Laplace transform for a delay e^{-sT} result in meromorphic transfer functions. If the meromorphic function $f(z)$ has a finite number of zeros α_j and poles ξ_i in a region Ω, then Equations 4.55, 4.52, and 4.53 yield a generalized Lucas formula,

$$\frac{f'(z)}{f(z)} = \sum_{j=1}^{m} \frac{1}{z-\alpha_j} - \sum_{i=1}^{n} \frac{1}{z-\xi_i} + \frac{g'(z)}{g(z)} \tag{4.56}$$

where $g(z) \neq 0$ is analytic in Ω.

4.2.5 Power Series Expansions

A *power series* is of the form

$$\begin{aligned} f(z) &= a_0 + a_1 z + a_2 z^2 + \ldots \\ &\quad + a_n z^n + \ldots = \sum_{n=0}^{\infty} a_n z^n. \end{aligned} \tag{4.57}$$

In general, a power series can be expanded around a point $z = z_0$,

$$f(z) = \sum_{n=0}^{\infty} a_n (z-z_0)^n \tag{4.58}$$

Series expansions do not always converge for all values of z. For example, the *geometric series*

$$\frac{1}{1-z} = 1 + z + z^2 + \ldots + z^n + \ldots \tag{4.59}$$

converges when $|z| < 1$. Every power series has a *radius of convergence* ρ. In particular, Equation 4.58 converges for all $|z - z_0| < \rho$ where, by *Hadamard's formula*,

$$\frac{1}{\rho} = \lim_{n\to\infty} \sup \sqrt[n]{|a_n|}. \tag{4.60}$$

Historically, two different approaches have been taken to set forth the fundamental theorems in the theory of analytic functions of a single complex variable. Cauchy's approach (1825) defines an analytic function as in Subsection 4.2.1 employing the Cauchy-Riemann equations (4.34) and Green's theorem in the plane to derive the famous integral formulas (Section 4.3.1). The existence of a power series expansion follows directly from the integral formulas.

***THEOREM 4.2* Taylor's Series** *Let $f(z)$ be an analytic function on a circular region Ω centered at $z = z_0$. For all points in the circle,*

$$f(z) = \sum_{n=0}^{\infty} a_n (z-z_0)^n \text{ where } a_n = \frac{f^{(n)}(z_0)}{n!}. \tag{4.61}$$

This expansion agrees with the form for the Taylor series expansion of a function of a real variable. Most texts base their exposition of the theory of a complex variable on Cauchy's approach incorporating a slight weakening of the definition due to Goursat. Weierstrauss' approach defines a function as analytic at a point z_0, if there is a convergent power series expansion (4.58). If one accepts the relevant theorems concerning the ability to move integrals and derivatives through power series expansions, the Cauchy integral formulas are easily demonstrated.

The Exponential Function

The exponential function is defined by the power series

$$e^z = 1 + \frac{z}{1!} + \frac{z^2}{2!} + \ldots + \frac{z^n}{n!} + \ldots \tag{4.62}$$

which converges for all complex values of z. Familiar identities of the form $e^{a+b} = e^a \cdot e^b$ are true by virtue of the formal power series expansion. Euler's formula (1749)

$$e^{i\theta} = \cos\theta + i\sin\theta \tag{4.63}$$

is easily derived from substitution in (4.62),

$$e^{i\theta} = (1 - \frac{\theta^2}{2!} + \cdots) + i(\frac{\theta}{1!} - \frac{\theta^3}{3!} + \ldots).$$

Thus, the polar form $z = r(\cos\theta + i\sin\theta)$ can be compactly expressed as $z = re^{i\theta}$. De Moivre's formula (4.23) states the obvious relationship

$$(e^{i\theta})^n = e^{in\theta}.$$

The unit circle $|z| = 1$ can be parameterized as $z = e^{i\theta}$ where $0 \leq \theta < 2\pi$. Substituting in a power series expansion of the form (4.57) yields a Fourier series expansion,

$$f(z) = \sum_{n=0}^{\infty} a_n z^n = \sum_{n=0}^{\infty} a_n e^{in\theta}.$$

Curves of the form,

$$\gamma(t) = r_1 e^{i\omega_1 t} + r_2 e^{i\omega_2 t} + \ldots + r_m e^{i\omega_m t},$$

are epicycles and examples of almost periodic functions. The ancient approach of employing epicycles to describe the motions of the planets can be viewed as an exercise in Fourier approximation. If $z = x + iy$, then $e^z = e^x(\cos y + i \sin y)$.

The multiple-valued *logarithm* function $\ln z$ is defined as the inverse of the exponential e^z. Hence, if $z = re^{i\theta}$ is in polar form and n is an integer,

$$\ln z = \ln r + i(\theta + 2\pi n) \tag{4.64}$$

The imaginary part of the logarithm is the same as the argument function. The addition theorem of the exponential implies

$$\ln(z_1 z_2) = \ln z_1 + \ln z_2 \tag{4.65}$$

which makes sense only if both sides of the equation represent the same infinite set of complex numbers.

Trigonometric Functions

The trigonometric functions are defined as

$$\cos z = \frac{e^{iz} + e^{-iz}}{2} \text{ and } \sin z = \frac{e^{iz} - e^{-iz}}{2i}. \tag{4.66}$$

Note that the sine and cosine functions are periodic,

$$\sin(z + 2\pi n) = \sin z \text{ and } \cos(z + 2\pi n) = \cos z.$$

The expressions for the trigonometric functions (4.66) can be employed to deduce the addition formulas

$$\cos(z_1 + z_2) = \cos z_1 \cos z_2 - \sin z_1 \sin z_2, \tag{4.67}$$

$$\sin(z_1 + z_2) = \cos z_1 \sin z_2 + \sin z_1 \cos z_2, \tag{4.68}$$

and the modulation formula

$$\cos \omega_1 t \cos \omega_2 t = \frac{1}{2}[\cos(\omega_1 + \omega_2)t + \cos(\omega_1 - \omega_2)t]. \tag{4.69}$$

If signals of frequencies ω_1, ω_2 are modulated with each other, they produce energy at the sum $\omega_1 + \omega_2$ and difference frequencies $\omega_1 - \omega_2$.

4.3 Complex Integrals

If $f(z) = u(x, y) + iv(x, y)$ is defined and continuous in a region Ω, we define the *integral of* $f(z)$ along some curve $\gamma \subseteq \Omega$ by

$$\begin{aligned} \int_\gamma f(z)dz &= \int_\gamma (u + iv)(dx + idy) \\ &= \int_\gamma udx - vdy + i\int_\gamma vdx + udy. \end{aligned} \tag{4.70}$$

These expressions depend upon line integrals for real valued functions. If the curve $\gamma(t)$ is a piecewise differentiable arc $\gamma(t)$ for $a \leq t \leq b$, then Equation 4.70 is equivalent to

$$\int_\gamma f(z)dz = \int_a^b f(\gamma(t))\gamma'(t)dt. \tag{4.71}$$

The most important property of the line integral (4.71) is its invariance with a change of parameter. Hence, if two curves start and end at the same points and trace out the same curve γ, the value of the integrals (4.71) will be the same. Distinctions are made in terms of the direction of travel,

$$\int_{-\gamma} f(z)dz = -\int_\gamma f(z)dz.$$

A curve or arc $\gamma(t)$ is said to be *closed* if the endpoints coincide $\gamma(a) = \gamma(b)$. A closed curve is called *simple* if it does not intersect itself.

All points to the left of a curve as it is traversed are said to be *enclosed* by it. A counterclockwise (CCW) traverse around a contour is said to be *positive*. A closed curve $\gamma(t)$ is said to make n *positive encirclements* of the origin $z = 0$ if vector $\gamma(t)$ rotates in a counterclockwise (CCW) direction and completes n rotations. A *negative encirclement* is obtained if the path is traversed in a clockwise (CW) directions.

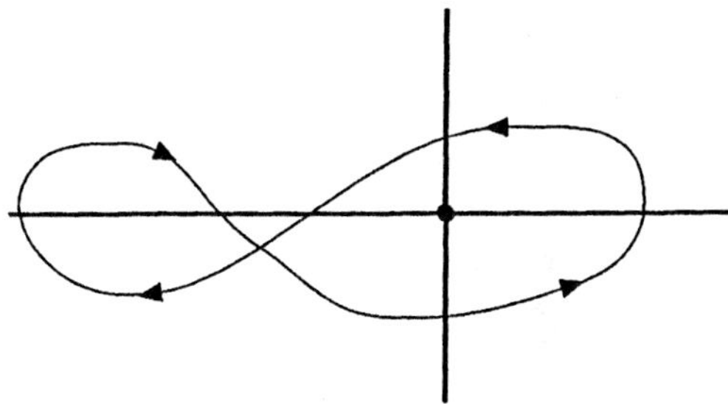

Figure 4.2 A single positive encirclement of the origin.

The notions of enclosement or encirclement have different conventions in the mathematical literature and in engineering expositions of classical control theory. Most texts in classical control state that a point is enclosed or encircled by a contour if it lies to the right of the curve as it is traversed, and clockwise (CW) contours and rotations are called positive.

4.3.1 Integral Theorems

Cauchy's Theorem

Suppose $f(z)$ is analytic in a region Ω bounded by a simple closed curve γ. *Cauchy's theorem* states that

$$\int_\gamma f(z)dz = \oint_\gamma f(z)dz = 0. \tag{4.72}$$

This equation is equivalent to saying that $\int_{z_1}^{z_2} f(z)dz$ is unique and is *independent of the path joining* z_1 *and* z_2.

Let $\gamma(t)$ for $0 \leq t \leq 1$ be a closed curve which does not pass through the origin $z = 0$. Consider the line integral for an integer n,

$$\int_\gamma z^n dz \text{ for } n \neq -1.$$

By Cauchy's theorem, this integral is zero if $n \geq 0$. By computation,

$$\int_\gamma z^n dz = \left.\frac{z^{n+1}}{n+1}\right|_\gamma = \frac{\gamma(1)^{n+1}}{n+1} - \frac{\gamma(0)^{n+1}}{n+1}.$$

Because the curve is closed, $\gamma(0) = \gamma(1)$, and

$$\int_\gamma z^n dz = 0 \text{ for } n \neq -1. \tag{4.73}$$

This argument can be generalized: for any closed curve $\gamma(t)$ not passing through the point $z = a$,

$$\int_\gamma (z-a)^n dz = 0 \text{ for } n \neq -1. \tag{4.74}$$

Let $f(z)$ be a power series expansion of the form

$$f(z) = \sum_{n=0}^{\infty} a_n (z-a)^n$$

and let $\gamma(t)$ lie within the radius of convergence. Applying (4.74) and moving the integration through the expansion

$$\int_\gamma f(z)dz = \sum_{n=0}^{\infty} \int_\gamma a_n (z-a)^n dz = 0 \tag{4.75}$$

gives a version of Cauchy's theorem.

Cauchy's Integral Formulas

Consider the closed curve $\gamma(t) = a + e^{it}$ for $0 \leq t \leq 2\pi k$. The curve lies on a unit circle centered at $z = a$ and completes k counterclockwise positive encirclements of $z = a$. Consider the line integral

$$\int_\gamma \frac{1}{z-a} dz. \tag{4.76}$$

By computation,

$$\int_\gamma \frac{1}{z-a} dz = \ln(z-a)|_\gamma$$

is a multivalued function. To obtain the integral, one must consider the expression

$$\ln e^{it} - \ln e^0 = it$$

for $0 \leq t \leq 2\pi k$. Thus

$$\int_\gamma \frac{1}{z-a} dz = 2\pi k i \tag{4.77}$$

The equation can be generalized as

$$n(\gamma, a) = \frac{1}{2\pi i} \int_\gamma \frac{1}{z-a} dz \tag{4.78}$$

where $n(\gamma, a)$ is called *the winding number of* γ *around* $z = a$. The integral counts the number of counterclockwise (CCW) encirclements of the point $z = a$.

If $f(z)$ is analytic within and on a region Ω bounded by a simple closed curve γ and $a \in \Omega$ is a point interior to γ, then

$$f(a) = \frac{1}{2\pi i} \oint_\gamma \frac{f(z)}{z-a} dz \tag{4.79}$$

where γ is traversed in the counterclockwise (CCW) direction. Higher order derivatives $f^{(r)}(a)$ can be expressed as

$$f^{(r)}(a) = \frac{r!}{2\pi i} \oint_\gamma \frac{f(z)}{(z-a)^{r+1}} dz. \tag{4.80}$$

Equations 4.79 and 4.80 are known as the *Cauchy integral formulas*. The formulas imply that, if the analytic function $f(z)$ is known on a simple closed curve γ, then its value (and, its higher derivatives) in the interior of γ are preordained by the behavior of the function along γ. This quite remarkable fact is contrary to any intuition that one might infer from real-valued functions.

If $f(z)$ has a Taylor power series expansion around the point $z = a$,

$$f(z) = \sum_{n=0}^{\infty} a_n (z-a)^n \text{ where } a_n = \frac{f^{(n)}(a)}{n!}, \tag{4.81}$$

and the closed curve γ is contained in the radius of convergence, then

$$\frac{1}{2\pi i} \int_\gamma \frac{f(z)}{(z-a)} dz = \frac{1}{2\pi i} \sum_{n=0}^{\infty} \int_\gamma a_n (z-a)^{n-1} dz.$$

The terms corresponding to $n > 0$ are zero by Cauchy's theorem (4.74), and the use of Equation 4.78, yields a version of the Cauchy integral formula (4.79):

$$n(\gamma, a) f(a) = n(\gamma, a) a_0 = \frac{1}{2\pi i} \int_\gamma \frac{f(z)}{z-a} dz. \tag{4.82}$$

Formal division of the power series expansion (4.81) by the term $(z-a)^{r+1}$ yields the higher derivative formulas

$$n(\gamma, a) f^{(r)}(a) = \frac{1}{2\pi i} \int_\gamma \frac{f(z)}{(z-a)^{r+1}} dz. \tag{4.83}$$

4.3.2 The Argument Principle

Every rational transfer function $H(s)$ is meromorphic. The transfer function of a time-invariant linear system is meromorphic, even when it employs delays of the form e^{-sT}. Let $f(z)$ be a meromorphic function in a region Ω which contains a finite number of zeros α_j and poles ξ_k. The generalized Lucas formula (4.56) gives

$$\frac{f'(z)}{f(z)} = \sum_{j=1}^{m} \frac{1}{z-\alpha_j} - \sum_{k=1}^{n} \frac{1}{z-\xi_k} + \frac{g'(z)}{g(z)} \tag{4.84}$$

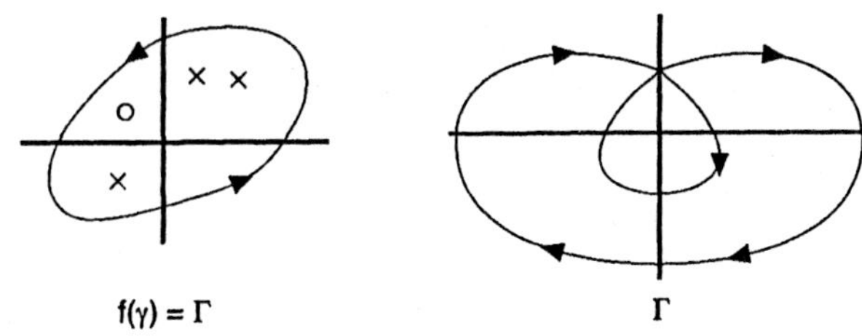

Figure 4.3 The number of encirclements of the origin by Γ is equal to the difference between the number encirclements of the zeros and poles.

where $g(z) \neq 0$ is analytic in Ω. Since $g'(z)/g(z)$ is analytic in Ω, one can apply Cauchy's theorem (4.72) to deduce

$$\frac{1}{2\pi i}\int_\gamma \frac{g'(z)}{g(z)}dz = 0$$

By Equations 4.78 and 4.84

$$\frac{1}{2\pi i}\int_\gamma \frac{f'(z)}{f(z)}dz = \sum_j^m n(\gamma,\alpha_j) - \sum_{k=1}^n n(\gamma,\xi_k). \quad (4.85)$$

The function $w = f(z)$ maps γ onto a closed curve $\Gamma(t) = f(\gamma(t))$ and, by a change in variables,

$$\tfrac{1}{2\pi i}\int_\Gamma \tfrac{1}{w}dw = \tfrac{1}{2\pi i}\int_\gamma \tfrac{f'(z)}{f(z)}dz \quad (4.86)$$

hence

$$n(\Gamma,0) = \textstyle\sum_j n(\gamma,\alpha_j) - \sum_k n(\gamma,\xi_k). \quad (4.87)$$

The left hand side of (4.85) can be viewed as the number of counterclockwise (CCW) encirclements of the origin $n(\Gamma, 0)$. If γ is a simple closed curve then Equation 4.85 computes the difference $m - n$ between the number of zeros and number of poles. Equation 4.87 is known as the *principle of the argument.*

For example, consider the simple closed curve γ of Figure 4.3. If a function $f(z)$ has three poles and a single zero enclosed by the curve γ, then the argument principle states that the curve $\Gamma = f(\gamma)$ must make two negative encirclements of the origin.

The argument principle (4.85) can be generalized. If $g(z)$ is analytic in a region Ω and $f(z)$ is meromorphic in Ω with a finite number of zeros and poles, then, for any closed curve γ,

$$\begin{aligned}\frac{1}{2\pi i}\int_\gamma g(z)\frac{f'(z)}{f(z)}dz &= \sum_{j=1}^m n(\gamma,\alpha_j)g(\alpha_j)\\ &\quad -\sum_{k=1}^n n(\gamma,\xi_k)g(\xi_k). \quad (4.88)\end{aligned}$$

The case $g(z) = z$ is of interest. Suppose that $f(z)$ is analytic in a circular region Ω of radius $r > 0$ around a. Ω is bounded by the simple closed curve $\gamma(t) = a + re^{it}$ for $0 \leq t \leq 2\pi$. Suppose the function $f(z)$ has an inverse in Ω, then $f(z) - w$ has only a single zero in Ω, and Equation 4.88 yields the *inversion formula*

$$f^{-1}(w) = \frac{1}{2\pi i}\oint_\gamma \frac{zf'(z)}{f(z)-w}dz. \quad (4.89)$$

Other Important Theorems

The following theorems are employed in H^∞ control theory.

1. **Liouville's Theorem:** If $f(z)$ is analytic and $| f(z) | < M$ is bounded in the entire complex plane, then $f(z)$ must be a constant.
2. **Cauchy's Estimate:** Suppose the analytic function $f(z)$ is bounded, $| f(z) |< M$, on and inside a circular region of radius r centered at $z = a$, then the kth derivative satisfies
$$| f^{(k)}(a) |\leq \frac{Mk!}{r^k}. \quad (4.90)$$
3. **Maximum Modulus Theorem:** If $f(z)$ is a nonconstant analytic function inside and on a simple closed curve γ, then the maximum value of $| f(z) |$ occurs on γ and is not achieved on the interior.
4. **Minimum Modulus Theorem:** If $f(z)$ is a nonzero analytic function inside and on a simple closed curve γ, then the minimum value of $| f(z) |$ occurs on γ.
5. **Rouche's Theorem:** If $f(z)$, $g(z)$ are analytic on a simple closed curve γ, then $f(z)$ and the sum $f(z)+g(z)$ have the same number of zeros inside γ.
6. **Gauss' Mean Value Theorem:** If $f(z)$ is analytic inside and on the circle of radius r centered at $z = a$, then $f(a)$ is the average value of $f(z)$ along the circle,
$$f(a) = \frac{1}{2\pi}\int_0^{2\pi} f(a + re^{i\theta})d\theta. \quad (4.91)$$

4.3.3 The Residue Theorem

Let $f(z)$ be analytic in a region Ω except at a pole at $z = a \in \Omega$ of order k. By Equation 4.43,

$$g(z) = (z-a)^k f(z)$$

has a removable singularity at $z = a$ and can be viewed as analytic over Ω. Thus, $g(z)$ may be expanded in a Taylor series about $z = a$. Dividing by $(z - a)^k$ yields the *Laurent expansion*

$$\begin{aligned}f(z) &= \frac{a_{-k}}{(z-a)^k} + \ldots + \frac{a_{-1}}{z-a} + a_0\\ &\quad + a_1(z-a) + a_2(z-a)^2 + \ldots. \quad (4.92)\end{aligned}$$

In general, a series of the form

$$\sum_{r=-\infty}^{\infty} a_r(z-a)^r$$

is called a *Laurent series.* Because power series expansions have a radius of convergence, a Laurent series can be viewed as the expansion of two analytic functions $h_1(z)$, $h_2(z)$ where

$$H(z) = h_1(\frac{1}{z-a}) + h_2(z-a) = \sum_{r=-\infty}^{\infty} a_r(z-a)^r.$$

The series converges for values of z which lie in an annular region $\rho_1 < \mid z-a \mid < \rho_2$ where ρ_1 can be zero and ρ_2 could be infinite. The *principal part* $h_1(1/(z-a))$ corresponds to the coefficients a_r, where $r < 0$ and the *analytic part* $h_2(z)$ corresponds to the coefficients a_r where $r \geq 0$. If the principal part has infinitely many nonzero terms $a_r \neq 0$, then $z = a$ is said to be an *essential singularity* of the function $H(z)$. The coefficient a_{-1} is called the *residue of $H(z)$ at the point $z = a$*.

If $f(z)$ is analytic within and on a simple closed curve γ except for an isolated singularity at $z = a$, then it has a Laurent series expansion around $z = a$ where, by Equation 4.74 and the Cauchy integral formula (4.79),

$$f(z) = \sum_{r=-\infty}^{\infty} a_r(z-a)^r, \text{ where } a_{r-1} = \frac{1}{2\pi i}\oint_\gamma \frac{f(z)}{(z-a)^r}dz. \tag{4.93}$$

The *residue* is defined as

$$\text{Res}(f, a) = a_{-1} = \frac{1}{2\pi i}\oint_\gamma f(z)dz, \tag{4.94}$$

and, for an arbitrary curve γ where $z = a$ is the only singularity enclosed by γ,

$$n(\gamma, a)\text{Res}(f, a) = \frac{1}{2\pi i}\int_\gamma f(z)dz.$$

At a simple pole of $f(z)$ at $z = a$,

$$\text{Res}(f, a) = \lim_{z\to a}(z-a)f(z), \tag{4.95}$$

and, at a pole of order k,

$$\text{Res}(f, a) = \lim_{z\to a}\frac{1}{(k-1)!}\frac{d^{k-1}}{dz^{k-1}}[(z-a)^k f(z)]. \tag{4.96}$$

For a simple pole Equation 4.95 is identical to Equation 4.47 and, for a pole of order k Equation 4.96 is identical to Equation 4.50 with $r = 1$.

The *residue theorem* states that, if $f(z)$ is analytic within and on a region Ω defined by a simple closed curve γ except at a finite number of isolated singularities $\xi_1, \ldots, \xi_k$, then,

$$\oint_\gamma f(z)dz = 2\pi i[\text{Res}(f, \xi_1) + \ldots + \text{Res}(f, \xi_k)]. \tag{4.97}$$

Cauchy's theorem (4.72) and the integral theorems can be viewed as special cases of the residue theorem.

The residue theorem can be employed to find the values of various integrals. For example, consider the integral

$$\int_{-\infty}^{\infty}\frac{1}{1+z^4}dz.$$

The poles of the function $f(z) = 1/(1+z^4)$ occur at the points

$$e^{i\pi/4}, \quad e^{i3\pi/4}, \quad e^{i5\pi/4}, \quad e^{i7\pi/4}.$$

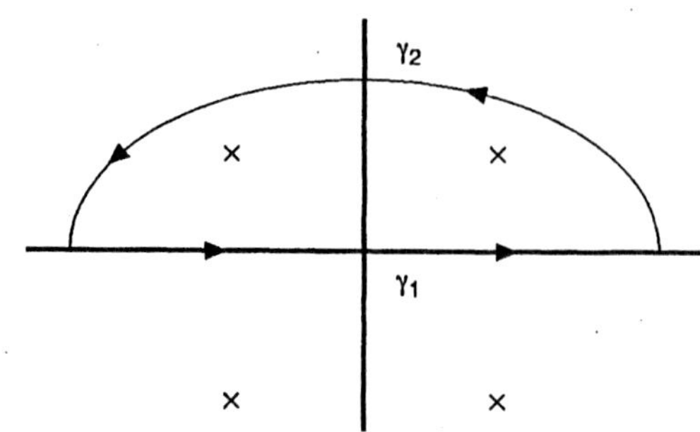

Figure 4.4 Residue theorem example.

Two of the poles lie in the upper half plane and two lie in the lower half plane. Employing Equation 4.95, one can compute the residues of the poles in the upper half plane

$$\text{Res}(f, e^{i\pi/4}) = \frac{e^{-i3\pi/4}}{4}$$

and

$$\text{Res}(f, e^{i3\pi/4}) = \frac{e^{-i\pi/4}}{4}$$

and the sum of the residues in the upper half plane is $-i\sqrt{2}/4$. Consider the contour integral of Figure 4.4. The curve γ consists of two curves γ_1 and γ_2, and hence,

$$\oint_\gamma f(z)dz = \int_{\gamma_1} f(z)dz + \int_{\gamma_2} f(z)dz.$$

By the residue theorem (4.97),

$$\oint_\gamma \frac{1}{1+z^4}dz = 2\pi i(-i\sqrt{2}/4) = \frac{\pi\sqrt{2}}{2}.$$

One can show that the limit of the line integral

$$\int_{\gamma_2} f(z)dz \to 0$$

as the radius of the semicircle approaches infinity and the curve γ_1 approaches the interval $(-\infty, \infty)$. Thus

$$\int_{-\infty}^{\infty}\frac{1}{1+z^4}dz = \frac{\pi\sqrt{2}}{2}.$$

4.4 Conformal Mappings

Every analytic function $w = f(z)$ can be viewed as a mapping from the z plane to the w plane. Suppose $\gamma(t)$ is a differentiable curve passing through a point z_0 at time $t = 0$. The curve $\Gamma = f(\gamma)$ is a curve passing through the point $w_0 = f(z_0)$. An application of the chain rule gives

$$\Gamma'(0) = f'(z_0)\gamma'(0). \tag{4.98}$$

Taking the argument and assuming $f'(z_0) \neq 0$ and $\gamma'(0) \neq 0$, then

$$\arg\Gamma'(0) = \arg f'(z_0) + \arg\gamma'(0)$$

Hence the angle between the directed tangents of γ and Γ at the point z_0 is the angle $\arg f'(z_0)$. Thus, if $f(z_0) \neq 0$, two curves γ_1, γ_2 which intersect at angle are mapped by $f(z)$ to two curves which intersect at the same angle. A mapping with this property is called *conformal*, and hence, if $f(z)$ is analytic and $f'(z) \neq 0$ in a region Ω, then $f(z)$ is conformal on Ω.

Equation 4.98 has an additional geometric interpretation. The quantity $| f'(z_0) |^2$ can be viewed as the *area dilation factor* at z_0. Infinitesimal area elements $dxdy$ around z_0 are expanded (or contracted) by a factor of $| f'(z_0) |^2$.

4.4.1 Bilinear or Linear Fractional Transformations

A transformation of the form,

$$w = T(z) = \frac{az + b}{cz + d} \tag{4.99}$$

where $ad - bc \neq 0$ is called a *linear fractional or bilinear transformation*. This important class of transformations occurs in control theory in developing Padé delay approximations, transformations between continuous s domain and discrete z domain realizations, and lead-lag compensators. There are four *fundamental transformations:*

1. Translation. $w = z + b$.
2. Rotation. $w = az$ where $a = e^{i\theta}$.
3. Dilation: $w = az$ where $a = r$ is real. If $a < 1$, the mapping contracts; if $a > 1$, its expands.
4. Inversion: $w = 1/z$.

Every fractional transformation can be decomposed into a combination of translations, rotations, dilations and inversions. In fact, every linear fractional transformation of the form (4.99) can be associated with a 2×2 complex matrix A_T,

$$A_T = \begin{bmatrix} a & b \\ c & d \end{bmatrix} \text{ where } \det A_T = ad - bc \neq 0. \tag{4.100}$$

By direct substitution, one can show that, if T, S are two bilinear transformations, then the composition $T \cdot S = T(S(z))$ is bilinear and

$$A_{T \cdot S} = A_T A_S \tag{4.101}$$

holds for the corresponding 2×2 matrix multiplication. The fundamental transformations correspond to the matrices

$$\begin{bmatrix} 1 & b \\ 0 & 1 \end{bmatrix}, \begin{bmatrix} e^{i\theta} & 0 \\ 0 & 1 \end{bmatrix}, \begin{bmatrix} r & 0 \\ 0 & 1 \end{bmatrix}, \text{ and } \begin{bmatrix} 0 & 1 \\ 1 & 0 \end{bmatrix}.$$

For a fractional transformation $w = T(z)$, if γ is a curve which describes a circle or a line in the z plane, then $\Gamma = T(\gamma)$ is a circle or a line in the w plane. This follows from the fact that it is valid for the fundamental transformations and, hence, for any composition.

Any scalar multiple αA, where $\alpha \neq 0$, corresponds to the same linear fractional transformation as A. Hence, one could assume that the matrix A_T is *unimodular* or $\det A_T = ad - bc = 1$; if $\alpha = \sqrt{\det A_T}$, then the linear transformation given by (a, b, c, d) is identical to one given by $(a, b, c, d)/\alpha$.

Every linear fractional transformation T (4.99) has an inverse which is a linear fractional transformation

$$z = T^{-1}(w) = \frac{dw - b}{-cw + a}. \tag{4.102}$$

If A is unimodular, then A^{-1} is unimodular and

$$A^{-1} = \begin{bmatrix} d & -b \\ -c & a \end{bmatrix}.$$

For example, for a sample time Δt, the *Tustin or bilinear transformation* is the same as the Padé approximation,

$$z^{-1} = e^{-s\Delta t} \approx \frac{1 - s\Delta t/2}{1 + s\Delta t/2},$$

or

$$z = T(s) = \frac{1 + s\Delta t/2}{1 - s\Delta t/2}, \tag{4.103}$$

and Equation 4.102 yields

$$s = T^{-1}(z) = \frac{z - 1}{z\Delta t/2 + \Delta t/2} = \frac{2}{\Delta t} \cdot \frac{z - 1}{z + 1}. \tag{4.104}$$

The Tustin transformation conformally maps the left half plane of the s-domain onto the unit disk in the z-domain. Thus if one designs a stable system with a transfer function $H(s)$ and discretizes the system by the Tustin transformation, one obtains a stable z-domain system with transfer function $G(z) = H[T^{-1}(z)]$.

4.4.2 Applications to Potential Theory

The real and the imaginary parts of an analytic function $f(z)$ satisfy *Laplace's equation,*

$$\nabla^2 \Phi = \frac{\partial^2 \Phi}{\partial x^2} + \frac{\partial^2 \Phi}{\partial y^2} = 0. \tag{4.105}$$

Solutions to Laplace's equation are called *harmonic.* Laplace's equation occurs in electromagnetics and the velocity potential of stationary fluid flow. An equation of the form

$$\nabla^2 \Phi = f(x, y) \tag{4.106}$$

is called *Poisson's equation* commonly occurring in problems solving for the potential derived from Gauss' law of electrostatics. Let Ω be a region bounded by a simple closed curve γ. Two types of *boundary-value problem* are commonly associated with Laplace's equation:

1. **Dirichlet's Problem**: Determine a solution to Laplace's equation subject to a set of prescribed values along the boundary γ.
2. **Neumann's Problem**: Determine a solution to Laplace's equation so that the derivative normal to the curve $\partial\Phi/\partial n$ takes prescribed values along γ.

Conformal mapping can be employed to find a solution of Poisson's or Laplace's equation. In general, one attempts to find an analytic or meromorphic function $w = f(z)$ which maps the region Ω to the interior of the unit circle or the upper half plane. The mapped boundary-valued problem is then solved on the w plane for the unit circle or upper half plane and is then transformed via $f^{-1}(w)$ to solve the problem on Ω.

Let $f(z) = u(x, y) + iv(x, y)$ be analytic on a region Ω. Both $u(x, y)$ and $u(x, y)$ and $v(x, y)$ satisfy (4.105). The function $v(x, y)$ is called a conjugate harmonic function to $u(x, y)$. Since the mapping $f(z)$ is conformal, the curves

$$u(x, y) = a\ ,\ v(x, y) = b$$

for a fixed a, b are orthogonal. The first curve $u(x, y) = a$ is often called the *equipotential line* and the curve $v(x, y) = b$ is called the *streamline* of the flow.

References

[1] Ahlfors, *Complex Analysis*, McGraw-Hill.

[2] Bell, E. T., *The Development of Mathematics*, Dover, New York, 1940.

[3] Cartan, H., *Elementary Theory of Analytic Functions of One or Several Complex Variables*, Hermann, Addison-Wesley, 1963.

[4] Churchill, R. V., *Introduction to Complex Variables and Applications*, McGraw-Hill, 2nd ed., 1962.

[5] Knopp, K. *Theory of Functions*, Dover, New York, 1945.

[6] Marsden, J. and Hoffman, M. *Complex Analysis*, W. H. Freeman, 2nd ed., 1987.

SECTION II

Models for Dynamical Systems

5

Standard Mathematical Models

William S. Levine
Department of Electrical Engineering, University of Maryland, College Park, MD

James T. Gillis
The Aerospace Corp., Los Angeles, CA

5.1 Input-Output Models

William S. Levine, *Department of Electrical Engineering, University of Maryland, College Park, MD*

5.1.1 Introduction

A fundamental problem in science and engineering is to predict the effect a particular action will have on a physical system. This problem can be posed more precisely as follows. What will be the response, $y(t)$, for all times t in the interval $t_0 \leq t < t_f$, of a specified system to an arbitrary input $u(t)$ over the same time interval ($t_0 \leq t < t_f$)? Because the question involves the future behavior of the system, its answer requires some sort of model of the system.

Engineers use many different kinds of models to predict the results of applying inputs to physical systems. One extreme example is the 15-acre scale model, scaled at one human stride to the mile, of the drainage area of the Mississippi river that the U.S. Corps of Engineers uses to predict the effect of flood control actions [1]. Such models, although interesting, are not very general. It is more useful, in a chapter such as this one, to concentrate on classes of models that can be used for a wide variety of problems.

The input-output models form just such a class. The fundamental idea behind input-output models is to try to model only the relation between inputs and outputs. No attempt is made to describe the "internal" behavior of the system. For example, electronic amplifiers are often described only by input-output models. The many internal voltages and currents are ignored by the model. Because of this concentration on the external behavior, many different physical systems can, and do, have the same input-output models. This is a particular advantage in design. Given that a particular input-output behavior is required and specified, the designer can choose the most advantageous physical implementation.

This chapter is restricted to input-output models. Chapter 5.2, deals with state-space models. A complete discussion of all types of input-output models would be virtually impossible. Instead, this chapter concentrates on an exemplary subclass of input-output models, those that are linear and time-invariant (LTI). Although no real system is either linear or time-invariant, many real systems are well approximated by LTI models within the time duration and range of inputs over which they are used. Even when LTI models are somewhat inaccurate, they have so many advantages over more accurate models that they are often still used, albeit cautiously. These advantages will be apparent from the ensuing discussion.

LTI ordinary differential and difference equation (ODE) models will be introduced in Section 5.1.2. The same acronym is used for both differential and difference equations because they are very similar, have analogous properties, and it will be clear from the context which is meant. ODE LTI models are very often obtained from the physics of a given system. For example, ODE models for electrical circuits and many mechanical systems can be deduced directly from the physics. The section concludes with a brief introduction to nonlinear and time-varying ODE input-output models.

Section 5.1.3 deals with continuous-time and discrete-time impulse response models. These are slightly more general than the ODE models. Such models are primarily used for LTI systems. An introduction to impulse response models for time-varying linear systems concludes the section.

Section 5.1.4 describes transfer function models of LTI systems. Transfer functions are very important in classical control theory and practice. They have the advantage of being directly measurable. That is, given a physical system that is approximately LTI, its transfer function can be determined experimentally.

The chapter concludes with Section 5.1.5, in which the equivalence among the different descriptions of LTI models is discussed.

0-8493-8570-9/96/$0.00+$.50

5.1.2 Ordinary Differential and Difference Equation Models

Consider a simple example of a system, such as an electronic amplifier. Such a system normally has an input terminal pair and an output terminal pair, as illustrated in Figure 5.1. Also, there is often a line cord that must be connected to a wall outlet to provide power for the amplifier. Typically, the amplifier is designed to have very high impedance at the input and very low impedance at the output. Because of this, the input voltage is not affected by the amplifier and the output voltage is determined only by the amplifier. Thus, in normal operation, the amplifier can be regarded as a system with input $u(t) = v_{in}(t)$ and output $y(t) = v_0(t)$. The relationship between $y(t)$ and $u(t)$ is designed to be approximately $y(t) = au(t)$, where a is some real constant. Notice that the power supply is ignored, along with the currents, in this simplified model of the amplifier. Furthermore, the facts that the amplifier saturates and that the gain, a, generally depends on the input frequency have also been ignored. This illustrates a fundamental aspect of modeling. Those features of the real system that are deemed unimportant should be left out of the model. This requires considerable judgement. The best models include exactly those details that are essential and no others. This is context dependent. Much of the modeling art is in deciding which details to include in the model.

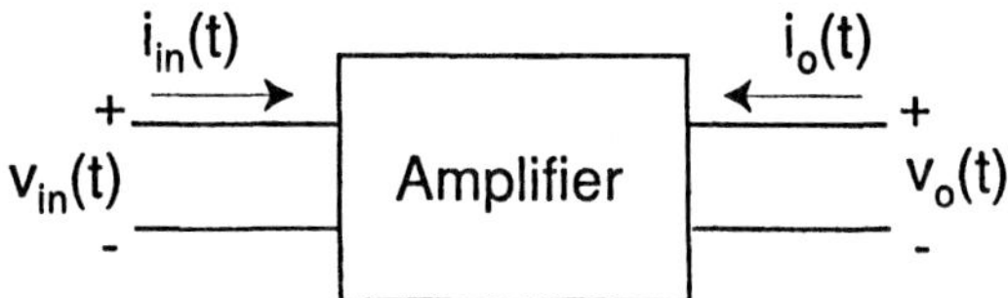

Figure 5.1 A representation of an electronic amplifier.

It is useful to generalize this simple electronic example to a large class of single-input single-output (SISO) models. Consider the structure depicted in Figure 5.2, where $u(t)$ and $y(t)$ are both scalars. In many physical situations, the relation between $u(t)$ and $y(t)$ is a function of the derivatives of both functions. For example, consider the RC circuit of Figure 5.3, where $u(t) = v(t)$ and $y(t) = i(t)$. It is well known that a mathematical model for this circuit is [2]

$$v(t) = Ri(t) + \frac{1}{C}\int_{-\infty}^{t} i(\tau)d\tau \tag{5.1}$$

Differentiating both sides once with respect to t, replacing $v(t)$ by $u(t)$, $i(t)$ by $y(t)$ and dividing by R gives

$$\frac{1}{R}\dot{u}(t) = \dot{y}(t) + \frac{1}{RC}y(t). \tag{5.2}$$

This example illustrates two important points. First, Equations 5.1 and 5.2 are approximations to reality. Real RC circuits behave linearly only if the input voltage is not too large. Real capacitors include some leakage (large resistor in parallel with the capacitance) and real resistors include some small inductance. The conventional model is a good model in the context of inputs,

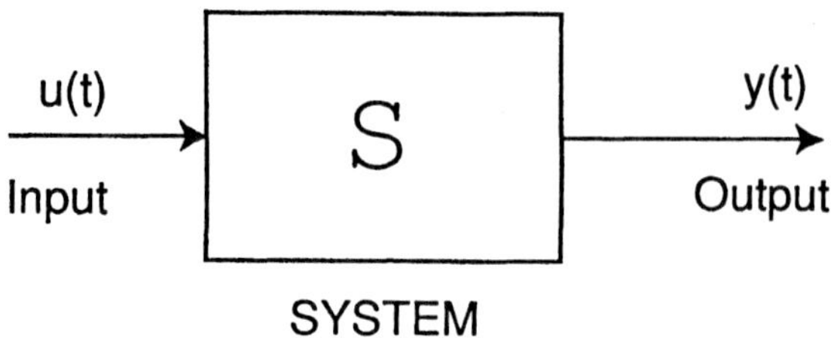

Figure 5.2 A standard input-out representation of a continuous-time system.

$v(t)$, that are not too large (so the nonlinearity can be ignored) and, not too high frequency (so the inductance can be ignored). The leakage current can be ignored whenever the capacitor is not expected to hold its charge for a long time.

Second, Equation 5.1 implicitly contains an assumption that the input and output are defined for past times, τ, $-\infty < \tau < t$. Otherwise, the integral in Equation 5.1 would be meaningless. This has apparently disappeared in Equation 5.2. However, in order to use Equation 5.2 to predict the response of the system to a given input, one would also need to know an "initial condition", such as $y(t_0)$ for some specific t_0. In the context of input-output models it is preferable not to have to specify separately the initial conditions and the input. The separate specification of initial conditions can be avoided, for systems that are known to be stable, by assuming the system is "initially at rest"— that is, by assuming the input is zero prior to some time t_0 (which may be $-\infty$) and that the initial conditions are all zero up to t_0 ($y(t)$ and all its derivatives are zero prior to t_0). If the response to nonzero initial conditions is important, as it is in many control systems, nonzero initial conditions can be specified. Given a complete set of initial conditions, the input prior to t_0 is irrelevant.

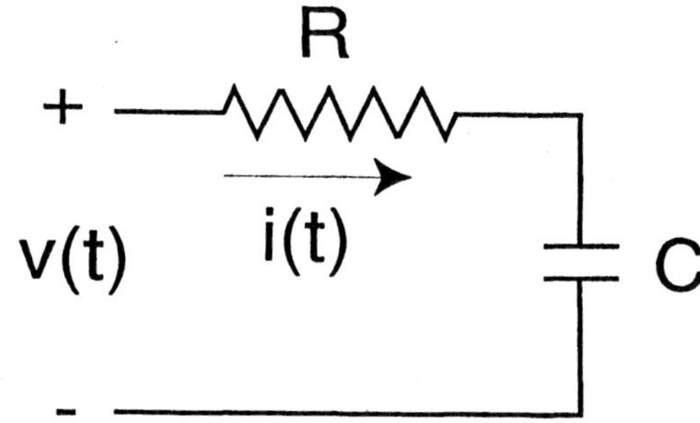

Figure 5.3 A series RC circuit.

The RC circuit is an example of a stable system. If the input is zero for a time duration longer than 5RC the charge on the capacitor and the current, $y(t)$, will decay virtually to zero. The choice of 5RC is somewhat arbitrary. The time constant of the transient response of the RC circuit is RC and 5 time constants is commonly used as the time at which the response is approximately zero. If the input subsequently changes from zero, say, at time t_0, the RC circuit can be modeled by a system that is "initially at rest" even though it may have had a nonzero input at some earlier time.

A simple generalization of the RC circuit example provides a large, and very useful class of input-output models for systems. This is the class of models of the form

$$\frac{d^n y(t)}{dt^n} + a_{n-1}\frac{d^{n-1}y(t)}{dt^{n-1}} + \ldots + a_0 y(t) = b_m \frac{d^m u(t)}{dt^m} + \ldots + b_0 u(t) \qquad (5.3)$$

where the a_i, $i = 0, 1, 2, \ldots, n-1$, and the b_j, $j = 0, 1, 2, \ldots, m$ are real numbers.

The reader is very likely to have seen such models before because they are common in many branches of engineering and physics. Both of the previous examples are special cases of Equation 5.3. Equations of this form are also studied extensively in mathematics.

Several features of Equation 5.3 are important. Both sides of Equation 5.3 could be multiplied by any nonzero real number without changing the relation between $y(t)$ and $u(t)$. In order to eliminate this ambiguity, the coefficient of $dy^n(t)/dt^n$ is always made to be one by convention.

Models of the form of Equation 5.3 are known as linear systems for the following reason. Assume, in addition to Equation 5.3, that the system is at rest prior to some time, t_0. Assume also that input $u_i(t)$, $t_0 \le t < t_f$, produces the response $y_i(t)$, $t_0 \le t < t_f$, for $i = 1, 2$. That is,

$$\frac{d^n y_i(t)}{dt^n} + a_{n-1}\frac{d^{n-1}y_i(t)}{dt^{n-1}} t \ldots + a_0 y_i(t) = b_m \frac{d^m u_i(t)}{dt^m} + \ldots + b_0 u_i(t) \qquad (5.4)$$

for $i = 1, 2$.

Then, if α and β are arbitrary constants (physically, α and β must be real, but mathematically they can be complex), then the input

$$u_s(t) = \alpha u_1(t) + \beta u_2(t) \qquad (5.5)$$

to the system (Equation 5.3) produces the response

$$y_s(t) = \alpha y_1(t) + \beta y_2(t). \qquad (5.6)$$

A proof is elementary. Substitute Equations 5.5 and 5.6 into 5.3 and rearrange the terms [3].

The mathematical definition of linearity requires that the superposition property described by Equations 5.5 and 5.6 be valid. More precisely, a system is linear if and only if the system's response, $y_s(t)$, to any linear combination of inputs ($u_s(t) = \alpha u_1(t) + \beta u_2(t)$) is the same linear combination of the responses to the inputs taken one at a time ($y_s(t) = \alpha y_1(t) + \beta y_2(t)$).

If nonzero initial conditions are included with Equation 5.3, as is often the case in control, the input $u_k(t)$, $k = 1, 2$, will produce output

$$y_k(t) = y_{ic}(t) + y_{u_k}(t) \qquad k = 1, 2 \qquad (5.7)$$

where $y_{ic}(t)$ denotes the response to initial conditions with $u(t) = 0$, and $y_{u_k}(t)$ denotes the response to $u_k(t)$ with zero initial conditions. When the input $u_s(t)$ in Equation 5.5 is applied to Equation 5.3 and the initial conditions are not zero, the resulting output is

$$y_s(t) = y_{ic}(t) + \alpha y_{u_1}(t) + \beta y_{u_2}(t) \qquad (5.8)$$

When $y_s(t)$ is computed by means of Equations 5.6 and 5.7, the result is

$$y_s(t) = (\alpha + \beta) y_{ic}(t) + \alpha y_{u_1}(t) + \beta y_{u_2}(t) \qquad (5.9)$$

The fact that Equation 5.8, the correct result, and Equation 5.9 are different proves that nonzero initial conditions invalidate the strict mathematical linearity of Equation 5.3. However, systems having the form of Equation 5.3 are generally known as linear systems, even when they have nonzero initial conditions.

Models of the form of Equation 5.3 are known as time-invariant systems for the following reason. Assume, in addition to Equation 5.3, the system is at rest prior to the time, t_0, at which the input is applied. Assume also that the input $u(t)$, $t_0 \le t < \infty$, produces the output $y(t)$, $t_0 \le t < \infty$. Then, applying the same input shifted by any amount T produces the same output shifted by an amount T. More precisely, letting (remember that $u(t) = 0$ for $t < t_0$ as part of the "at rest" assumption)

$$u_d(t) = u(t - T), \; t_0 + T < t < \infty \qquad (5.10)$$

Then, the response, $y_d(t)$, to the input $u_d(t)$ is

$$y_d(t) = y(t - T), \qquad t_0 + T < t < \infty. \qquad (5.11)$$

A proof that systems described by Equation 5.3 are time-invariant is simple; substitute $u(t - T)$ into Equation 5.3 and use the uniqueness of the solution to linear ODEs to show that the resulting response must be $y(t - T)$.

Of course, many physical systems are neither linear nor time-invariant. A simple example of a nonlinear system can be obtained by replacing the resistor in the RC circuit example by a nonlinear resistance, a diode for instance. Denoting the resistor current-voltage relationship by $v(t) = f(i(t))$, where $f(\cdot)$ is some differentiable function, Equation 5.1 becomes

$$v(t) = f(i(t)) + \frac{1}{C}\int_{-\infty}^{t} i(\tau)d\tau \qquad (5.12)$$

Differentiating both sides with respect to t, replacing $v(t)$ by $u(t)$, and $i(t)$ by $y(t)$ gives

$$\dot{u}(t) = \frac{df}{dy}(y(t))\dot{y}(t) + \frac{1}{C}y(t) \qquad (5.13)$$

The system in Equation 5.12 is not linear because an input of the form (Equation 5.5) would not produce an output of the form (Equation 5.6).

One could also allow the coefficients in Equation 5.3 (the a_is and b_js) to depend on time. The result would still be linear but would no longer be time-invariant. There will be a brief discussion of such systems in the following section.

Before introducing the impulse response, the class of discrete-time models will be introduced. Consider the ordinary difference equation

$$\begin{aligned} y(k+n) &+ a_{n-1}y(k+n-1)+\ldots+a_0y(k) \\ &= b_mu(k+m)+\ldots+b_0u(k) \qquad (5.14)\end{aligned}$$

where the $a_i, i = 0, 1, 2, \ldots, n-1$ and $b_j, j = 0, 1, 2, \ldots, m$ are real numbers and $k = k_0, k_0+1, k_0+2, \ldots, k_f$ are integers. Such models commonly arise from either sampling a continuous-time physical system or as digital simulations of physical systems. The properties of Equation 5.14 are similar to those of Equation 5.3. The leading coefficient (the coefficient of $y(k+n)$) is conventionally taken to be one. The computation of $y(k)$ given $u(k)$, $k = k_0, k_0+1, \ldots, k_f$ also requires n initial conditions. The need for initial conditions can be addressed, for stable systems, by assuming the system is "initially at rest" prior to the instant, k_0, at which an input is first applied. This initially "at rest" assumption means that (1) $u(k) = 0$ for $k < k_0$, and (2) that the initial conditions ($y(0), y(-1), \ldots, y(-n+1)$ for example) are all zero. When the system is initially at rest, analogous arguments to those given for Equation 5.3 show that Equation 5.14 is linear and time-invariant. Generally, even when the system is not initially at rest, systems of the form of Equation 5.14 are known as linear time-invariant (LTI) discrete-time systems.

5.1.3 Impulse Response

The starting point for discussion of the impulse response is not the system but the signal, specifically the input $u(t)$ or $u(k)$. Generally, inputs and outputs, as functions of time, are called signals. The discrete-time case, $u(k)$, will be described first because it is mathematically much simpler. The first step is to raise an important question that has been heretofore ignored. What is the collection of possible inputs to a system? This collection will be a set of signals. In Section 5.1.2, it was assumed that any $u(k)$, such that $u(k) = 0$ for $k < k_0$, could be an input. In reality this is not so. For example, it would be physically impossible to create the following signal

$$u(k) = \begin{cases} 0 & k \le 0 \\ k^2 & k \ge 0. \end{cases} \qquad (5.15)$$

Some energy is required to produce any physical signal; the energy needed for the signal in Equation 5.14 would be infinite.

There was a second assumption about the collection of signals buried in Section 5.1.2. Equation 5.5 and the definition of linearity assume that $u_s(t)$ ($u_s(k)$ gives the discrete-time version), defined only as a linear combination of possible inputs, is also a possible input. Mathematically, this amounts to assuming that the collection of possible input signals forms a vector space. For engineering purposes the requirement is the following.

If $u_i(k), i = 1, 2, 3, \ldots$ belong to a collection of possible input signals and $\alpha_i, i = 1, 2, 3, \ldots$ are real numbers, then

$$u_t(k) = \sum_{i=1}^{\infty} \alpha_i u_i(k) \qquad (5.16)$$

also belongs to the collection of possible input signals.

Equation 5.16 provides the first key to an economical description of LTI discrete-time systems. The second key is the signal known as the unit impulse or the unit pulse,

$$\delta(k) = \begin{cases} 1 & k = 0 \\ 0 & \text{all other integers.} \end{cases} \qquad (5.17)$$

Using $\delta(k)$ and Equation 5.16, any signal

$$u(k) = u_k \ (u_k \text{ a real number}) \quad -\infty < k < \infty \qquad (5.18)$$

can be rewritten as a sum of unit pulses

$$u(k) = \sum_{i=-\infty}^{\infty} u_i \delta(k-i) \qquad (5.19)$$

This initially seems to be a ridiculous thing to do. Equation 5.19 is just a complicated way to write Equation 5.18. However, suppose you are given a LTI discrete-time system, S, and that the response of this system to an input $\delta(k)$ is $h(k)$, as illustrated in Figure 5.4. Because the system is time-invariant, its response to an input $\delta(k-i)$ is just $h(k-i)$. Because the u_i in Equation 5.19 are constants, like α and β in Equations 5.5 and 5.6, because the system is linear, and because Equations 5.5 and 5.6 can be extended to infinite sums by induction, the following argument is valid. Denote the action of S on $u(k)$, $-\infty < k < \infty$, by

$$y(k) = \mathcal{S}(u(k))$$

Then,

$$\begin{aligned} y(k) &= \mathcal{S}\Big(\sum_{i=-\infty}^{\infty} u_i\delta(k-i)\Big) \\ &= \sum_{i=-\infty}^{\infty} u_i\mathcal{S}(\delta(k-i)) \\ y(k) &= \sum_{i=-\infty}^{\infty} u_i h(k-i) \qquad -\infty < k < \infty \qquad (5.20)\end{aligned}$$

Equation 5.20 demonstrates that the response of the LTI discrete-time system, $\mathcal{S}$, to any possible input, $u(k)$, $-\infty < k < \infty$, can be computed from *one* output signal, $h(k)$, $-\infty < k < \infty$, known as the impulse (or unit pulse) response of the system. Thus, the impulse response, $h(k)$, $-\infty < k < \infty$, is an input-output model of $\mathcal{S}$.

The main uses of impulse response models are theoretical. This is because using Equation 5.20 involves a very large amount of computation and there are several better ways to compute $y(k)$, $-\infty < k < \infty$, when the impulse response and the input are given. One example of the use of the impulse response is in the determination of causality.

A system is said to be causal if and only if the output at any time k, $y(k)$, depends only on the input at times up to and including time k, that is, on the set of $u(\ell)$ for $-\infty < \ell \le k$. Real systems must be causal. However, it is easy to construct mathematical models that are not causal.

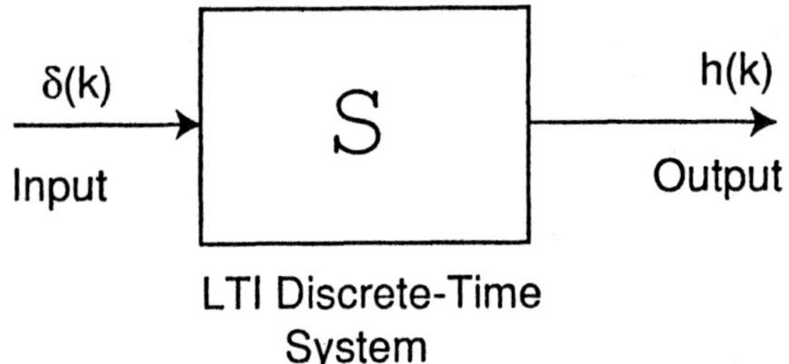

Figure 5.4 An input-output representation of a discrete-time LTI system showing a unit pulse as input and the discrete-time impulse response as output.

It is easy to see from Equation 5.20 that an impulse response, $h(k)$, is causal if and only if

$$h(k) = 0 \text{ for all } k < 0 \qquad (5.21)$$

For causal impulse responses, Equation 5.20 becomes

$$y(k) = \sum_{i=-\infty}^{k} u_i h(k-i) \qquad (5.22)$$

because $h(k-i) = 0$ by Equation 5.21 for $i > k$.

The development of the continuous-time impulse response as an input-output model for LTI continuous-time systems is analogous to that for discrete-time. The ideas are actually easy to understand, especially after seeing the discrete-time version. However, the underlying mathematics are very technical. Thus, proofs will be omitted.

A natural beginning is with the second key to an economical description of LTI systems, the definition of the unit impulse. For reasons that will be explained below, the continuous-time unit impulse is usually defined by the way it operates on smooth functions, rather than as an explicit function such as Equation 5.17.

DEFINITION 5.1 Let $f(t)$ be any function that is continuous on the interval $-\epsilon < t < \epsilon$ for every $0 < \epsilon < \epsilon_m$ and some ϵ_m. Then the unit impulse (also known as the Dirac delta function) $\delta(t)$ satisfies

$$f(0) = \int_{-\infty}^{\infty} f(\tau)\delta(\tau)d\tau \qquad (5.23)$$

To see why Equation 5.23 is used and not something like Equation 5.17, try to construct a $\delta(t)$ that would satisfy Equation 5.23. The required function must be zero everywhere but at $t = 0$ and its integral over any interval that includes $t = 0$ must be one. Good luck.

Given a signal, $u(t)\; -\infty < t < \infty$, the analog to Equation 5.19 then becomes, using Equation 5.23,

$$u(t) = \int_{-\infty}^{\infty} u(\tau)\delta(t-\tau)d\tau. \qquad (5.24)$$

Equation 5.24 can then be used analogously to Equation 5.19 to derive the continuous-time equivalent of Equation 5.20. That is, suppose you are given an LTI continuous-time system, $\mathcal{S}$, and that the response of this system to a unit impulse applied at $t = 0$, $\delta(t)$, is $h(t)$ for all t, $-\infty < t < \infty$. In other words,

$$h(t) = \mathcal{S}(\delta(t)) \qquad -\infty < t < \infty.$$

To compute the response of $\mathcal{S}$, $y(t)(-\infty < t < \infty)$, to an input $u(t)(-\infty < t < \infty)$, proceed as follows.

$$\begin{aligned} y(t) &= \mathcal{S}(u(t)) \\ &= \mathcal{S}\left(\int_{-\infty}^{\infty} u(\tau)\delta(t-\tau)d\tau\right) \end{aligned}$$

Because $\mathcal{S}(\cdot)$ acts on signals (functions of time, t), because $u(\tau)$ acts as a constant (not a function of t), and because integration commutes with the action of linear systems (think of integration as the limit of a sequence of sums)

$$\begin{aligned} y(t) &= \int_{-\infty}^{\infty} u(\tau)\mathcal{S}(\delta(t-\tau))d\tau \\ y(t) &= \int_{-\infty}^{\infty} u(\tau)h(t-\tau)d\tau \qquad -\infty < t < \infty \qquad (5.25) \end{aligned}$$

As in the discrete-time case, the primary use of Equation 5.25 is theoretical. There are better ways to compute $y(t)$ than by direct computation of the integral in Equation 5.25. Specifically, the Laplace or Fourier transform provides an efficient means to compute $y(t)$ from knowledge of $h(t)$ and $u(t)$, $-\infty < t < \infty$. The transforms also provide a good vehicle for the discussion of physical signals that can be used as approximations to the unit impulse. These transforms will be discussed in the next section.

Two apparently different classes of input-output models have been introduced for both continuous-time and discrete-time LTI systems. A natural question is whether the ODE models are equivalent to the impulse response models. It is easy to see that, in continuous-time, there are impulse responses for which there are not equivalent ordinary differential equations. The simplest such example is a pure delay. That is

$$y(t) = u(t - t_d) \qquad -\infty < t < \infty \qquad (5.26)$$

where $t_d > 0$ is a fixed time delay.

The impulse response for a pure delay is

$$h(t) = \delta(t - t_d) \qquad -\infty < t < \infty \qquad (5.27)$$

but there is no ODE that exactly matches Equation 5.26 or 5.27. Note that there are real systems for which a pure delay is a good model. Electronic signals travel at a finite velocity. Thus, long transmission paths correspond to pure delays.

The converse is different. Every ODE has a corresponding impulse response. It is easy to demonstrate this in discrete time. Simply let $\delta(k)$ be the input in Equation 5.14 with $n = 1$ and $m = 0$. Assuming Equation 5.14 is initially at rest results in a recursive calculation for $h(k)$. For example, let

$$y(k+1) + ay(k) = bu(k) \qquad -\infty < k < \infty$$

Replacing $u(k)$ by $\delta(k)$ gives an input that is zero prior to $k = 0$. Assuming the system is initially at rest makes $y(k) = 0$ for $k < 0$.

Then

$$y(0) + ay(-1) = bu(-1)$$

gives

$$\begin{aligned} y(0) &= h(0) = 0, \\ y(1) + ay(0) &= bu(0) = b \end{aligned}$$

gives

$$\begin{aligned} y(1) &= h(i) = b, \\ y(2) + ay(1) &= bu(1) = 0 \end{aligned}$$

gives

$$y(2) = h(2) = -ab,$$

etc.

The result in the continuous-time case is the same—every ODE has a corresponding impulse response—but the mathematics is more complicated unless one uses transforms.

The response of a system that is linear but time-varying to a unit impulse depends on the time at which the impulse is applied. Thus, the impulse response of a linear time-varying system must be denoted, in the continuous-time case, by $h(t, \tau)$ where τ is the time at which the impulse is applied and t is the time at which the impulse response is recorded. Because the system is linear, the argument that produced Equation 5.25 also applies in the time-varying case. The result is

$$y(t) = \int_{-\infty}^{\infty} u(\tau)h(t,\tau)d\tau \qquad -\infty < t < \infty. \tag{5.28}$$

The analogous result holds in discrete time.

$$y(k) = \sum_{i=-\infty}^{\infty} u(i)h(k,i) \qquad -\infty < k < \infty. \tag{5.29}$$

Please see Chapter 25 for more information about linear time-varying systems.

There are forms of impulse response models that are useful in the study of nonlinear systems. See Chapter 55.3.

5.1.4 Transfer Functions

The other common form for LTI input-output models is the transfer function. The transfer function, as an input-output model, played a very important role in communications and in the development of feedback control theory in the 1930s. Transfer functions are still very important and useful. One reason is that for asymptotically stable, continuous-time LTI systems the transfer function can be measured easily. To see this, suppose that such a system is given. Assume this system has an impulse response, $h(t)$, and that it is causal ($h(t) = 0$ for $t < 0$). Suppose this system is excited with the input

$$u(t) = \cos wt \qquad -\infty < t < \infty \tag{5.30}$$

Equation 5.30 is a mathematical idealization of a situation where the input cosinusoid started long enough in the past that all the transients have decayed to zero. The corresponding output is,

$$\begin{aligned} y(t) &= \int_{-\infty}^{\infty} h(t-\tau)\cos w\tau d\tau \qquad (5.31) \\ &= \int_{-\infty}^{\infty} h(t-\tau)\left(\frac{e^{jwt}+e^{-jw\tau}}{2}\right)d\tau \\ &= \frac{1}{2}\int_{-\infty}^{\infty} h(t-\tau)e^{jw\tau}d\tau \\ &\quad + \frac{1}{2}\int_{-\infty}^{\infty} h(t-\tau)e^{-jw\tau}d\tau \\ &= \frac{1}{2}\int_{-\infty}^{\infty} h(\sigma)e^{jw(t-\sigma)}d\sigma \\ &\quad + \frac{1}{2}\int_{-\infty}^{\infty} h(\sigma)e^{-jw(t-\sigma)}d\sigma \\ y(t) &= \left(\int_{-\infty}^{\infty} h(\sigma)e^{-jw\sigma}d\sigma\right)\frac{e^{jwt}}{2} \\ &\quad + \left(\frac{1}{2}\int_{-\infty}^{\infty} h(\sigma)e^{jw\sigma}d\sigma\right)\frac{e^{-jwt}}{2} \qquad (5.32) \end{aligned}$$

Define for all real w, $-\infty < w < \infty$,

$$H(jw) = \int_{-\infty}^{\infty} h(\sigma)e^{-jw\sigma}d\sigma \tag{5.33}$$

Notice that $H(jw)$ is a complex number for every real w and that the complex conjugate of $H(jw)$, denoted $H^*(jw)$, is $H(-jw)$. Then, Equation 5.32 becomes

$$\begin{aligned} y(t) &= \frac{H(jw)e^{jwt} + H^*(jw)e^{-jwt}}{2} \\ \text{or, } y(t) &= |H(jw)|\cos(wt + \measuredangle H(jw)) \qquad (5.34) \end{aligned}$$

where

$$\begin{aligned} |H(jw)| &= \text{magnitude of } H(jw) \\ \measuredangle H(jw) &= \text{angle of } H(jw). \end{aligned}$$

Of course, $H(jw)$, for all real w, $-\infty < w < \infty$, is the transfer function of the given LTI system. It should be noted that some authors call $H(jw)$ the frequency response of the system and reserve the term transfer function for $H(s)$ (to be defined shortly). Both $H(jw)$ and $H(s)$ will be called transfer functions in this article. Equation 5.34 shows how to measure the transfer function; evaluate Equation 5.34 experimentally for every value of w. Because it is impossible to measure $H(jw)$ for every w, $-\infty < w < \infty$, what is actually done is to measure $H(jw)$ for a finite collection of w's and interpolate.

Suppose the object of the experiment was to measure the impulse response. Recognize that Equation 5.33 defines $H(jw)$ to be the Fourier transform of $h(t)$. The inverse of the Fourier transform is given by

$$x(t) \triangleq \int_{-\infty}^{\infty} X(jw)e^{jwt}\frac{dw}{2\pi} \tag{5.35}$$

where $X(jw)$ is a function of w, $-\infty < w < \infty$.

Applying Equation 5.35 to Equation 5.33 gives

$$h(t) = \int_{-\infty}^{\infty} H(jw)e^{jwt}\frac{dw}{2\pi} \tag{5.36}$$

Equation 5.36 provides a good way to determine $h(t)$, for asymptotically stable continuous-time LTI systems. Measure $H(jw)$ and then compute $h(t)$ from Equation 5.36. Of course, it is not possible to measure $H(jw)$ for all w, $-\infty < w < \infty$. It is possible to measure $H(jw)$ for enough values of w to compute a good approximation to $h(t)$. In many applications, control design using Bode, Nichols, or Nyquist plots for example, knowing $H(jw)$ is sufficient.

Having just seen that the transfer function can be measured when the system is asymptotically stable, it is natural to ask what can be done when the system is unstable. The integral in Equation 5.33 blows up; because of this the Fourier transform of $h(t)$ does not exist. However, the Laplace transform of $h(t)$ is defined for unstable as well as stable systems and is given by

$$H(s) = \int_{-\infty}^{\infty} h(t)e^{-st}dt \tag{5.37}$$

for all complex s such that the integral in Equation 5.37 is finite.

Transfer functions have several important and useful properties. For example, it is easy to prove that the transfer function for a continuous-time LTI system satisfies

$$Y(s) = H(s)U(s) \tag{5.38}$$

where $Y(s)$ is the Laplace transform of the output $y(t)$, $-\infty < t < \infty$, and $U(s)$ is the Laplace transform of the input $u(t)$, $-\infty < t < \infty$.

To prove Equation 5.38 take the Laplace transform of both sides of Equation 5.25 to obtain

$$\begin{aligned} Y(s) &= \int_{-\infty}^{\infty}\left(\int_{-\infty}^{\infty} u(\tau)h(t-\tau)d\tau\right)e^{-st}dt \\ &= \int_{-\infty}^{\infty}\int_{-\infty}^{\infty} u(\tau)h(t-\tau)e^{st}dtd\tau \end{aligned}$$

Make the change of variables $\sigma = t - \tau$

$$\begin{aligned} &= \int_{-\infty}^{\infty}\int_{-\infty}^{\infty} u(\tau)h(\sigma)e^{-s(\sigma+\tau)}d\sigma d\tau \\ Y(s) &= \int_{-\infty}^{\infty} h(\sigma)e^{-s\sigma}d\sigma \int_{-\infty}^{\infty} u(\tau)e^{-st}d\tau \\ &= H(s)U(s) \end{aligned} \tag{5.39}$$

The Laplace transform provides an easy means to demonstrate the relationship between transfer functions, $H(s)$ or $H(jw)$, and ordinary differential equation models of LTI continuous-time systems. Take the Laplace transform of both sides of Equation 5.3, assuming that the system is initially at rest. The result is

$$\begin{aligned} &(s^n + a_{n-1}s^{n-1} + \ldots + a_0)Y(s) \\ &= (b_m s^m + \ldots + b_0)U(s) \end{aligned} \tag{5.40}$$

where the fact that the Laplace transform of $\dot{y}(t)$ is $sY(s)$ has been used repeatedly. Dividing through gives

$$Y(s) = \frac{b_m s^m + b_{m-1}s^{m-1} + \ldots + b_0}{s^n + a_{n-1}s^{n-1} + \ldots + a_0}U(s) \tag{5.41}$$

Equation 5.41 shows that, for a continuous-time ODE of the form of Equation 5.3,

$$H(s) = \frac{b_m s^m + b_{m-1}s^{m-1} + \ldots + b_0}{s^n + a_{n-1}s^{n-1} + \ldots + a_0} \tag{5.42}$$

The discrete-time case is very similar. The discrete-time analog of the Fourier transform is the discrete Fourier transform. The discrete-time analog of the Laplace transform is the Z-transform. The results and their derivations parallel those for continuous-time systems. See almost any textbook on signals and systems such as [3] or [4] for details.

5.1.5 Conclusions

Although this chapter has treated the ODE, impulse response, and transfer function descriptions of LTI systems separately, it should be apparent that they are equivalent descriptions for a large class of systems. A demonstration that the impulse response and transfer function descriptions are more general than the ODE descriptions has already been given; there is no continuous-time ODE corresponding to $H(s) = e^{-sT}$. However, all three descriptions are equivalent whenever $H(s)$ can be written as a rational function, that is, as a ratio of polynomials in s.

There is a result, known as Runge's theorem [5, p. 258], that proves that any $H(s)$ that is analytic in a region of the s-plane can be approximated to uniform accuracy, in that region, by a rational function. A family of such approximations is known as the Padé approximants [6]. The basic idea is to expand the given analytic function in a Taylor series (this is always possible) and then choose the coefficients of the rational function so as to match as many terms of the series as possible. For example,

$$\begin{aligned} e^{-sT} &= 1 - Ts + \frac{T^2}{2!}s^2 + \ldots + \frac{T^2}{n!}s^n \\ &+ \ldots \approx \frac{b_m s^m + b_{m-1}s^{m-1} + \ldots + b_0}{s^m + a_{m-1}s^{m-1} + \ldots + a_0} \end{aligned}$$

The $2m - 1$ coefficients $(b_0, b_1, \ldots, b_m, a_0, \ldots, a_{m-1})$ can then be selected to match the first $2m - 1$ coefficients of the Taylor series. The result is known as the Padé approximation to a pure delay of duration T in the control literature [7, p. 301]. The approximation improves with increasing m.

There are still examples for which approximation by ODEs is problematic. An example is the flow of heat in a long solid rod. Letting x denote the displacement along the rod and assuming that the input is applied at $x = 0$, that the initial temperature of the rod is zero, and that the output is the temperature of the rod at point x, then the transfer function observed at x is [8, pp. 182–184] and [9, pp. 145–150]

$$H(s,x) = e^{-x\sqrt{s/a}} \tag{5.43}$$

where a is the thermal diffusivity.

Even for a fixed x, this is an example of a transfer function to which Runge's theorem does not apply in any region of the complex plane that includes the origin. The reason is that $H(s,x)$

is not differentiable at $s = 0$ for any $x > 0$ and is therefore not analytic in any region containing the origin. In many applications it is nonetheless adequate to approximate this transfer function by a simple low pass filter,

$$H(s) = \frac{b_0}{s + a_0} \tag{5.44}$$

This example emphasizes the difficulty of making general statements about modeling accuracy. Deciding whether a given model is adequate for some purpose requires a great deal of expertise about the physical system and the intended use of the model. The decision whether to use an input-output model is somewhat easier. Input-output models are appropriate whenever the internal operation of the physical system is irrelevant to the problem of interest. This is true in many systems problems, including many control problems.

References

[1] McPhee, J., *The Control of Nature*, The Noonday Press, Farrar Strauss, and Giroux, 1989, 50.

[2] Bose, A.G. and Stevens, K.N., *Introductory Network Theory*, Harper and Row, 1965.

[3] Oppenheim, A.V. and Willsky, A.S., with Young, I.T., *Signals and Systems*, Prentice Hall, 1983.

[4] Lathi, B.P., *Linear Systems and Signals*, Berkeley-Cambridge Press, 1992.

[5] Rudin, W., *Real and Complex Analysis*, McGraw-Hill, 1966.

[6] Baker, Jr., G.A., *Essentials of Padé Approximants*, Academic Press, 1975.

[7] Franklin, G.F., Powell, J.D., and Emami-Naeni, A., *Feedback Control of Dynamic Systems*, 3rd ed., Addison-Wesley, 1994.

[8] Aseltine, J.A., *Transform Method in Linear Systems Analysis*, McGraw-Hill, 1958.

[9] Yosida, K., *Operational Calculus: A Theory of Hyperfunctions*, Springer-Verlag, 1984.

Further Reading

There are literally hundreds of textbooks on the general topic of signals and systems. Most have useful sections on input-output descriptions of LTI systems. References [3] and [4] are good examples. Reference [3] has a particularly good and well-organized bibliography.

A particularly good book for those interested in the mathematical technicalities of Fourier transforms and the impulse response is

[1] Lighthill, M.J., *Introduction to Fourier Analysis and Generalized Functions*, Cambridge Monographs on Mechanics and Applied Mathematics, 1958.

Those interested in nonlinear input-output models of systems should read Chapter 55.3. Those interested in linear time-varying systems should see Chapter 25. Chapters 7.2 and 58 are good starting points for those interested in the experimental determination of LTI input-output models.

5.2 State Space

James T. Gillis, The Aerospace Corp., Los Angeles, CA

5.2.1 Introduction

This chapter introduces the state space methods used in control systems; it is an approach deeply rooted in the techniques of differential equations, linear algebra, and physics.

The Webster's Ninth College Dictionary [12] defines a "state" as a "mode or condition of being" (1a), and "a condition or state of physical being of something" (2a). By a *state space approach* one means a description of a system in which the "state" gives a complete description of the system at a given time; it implies that there are orderly rules for the transition from one state to another. For example, if the system is a particle governed by Newton's Law: $F = ma$, then the state could be the position and velocity of the particle or the position and momentum of the particle. These are both state descriptions of such a system. Thus, state space descriptions of a system are not unique.

5.2.2 States

Basic Explanation

In this section the concepts of state and state space are introduced in an intuitive manner, then formally. Several examples of increasing complexity are discussed. A method for conversion of an ordinary differential equation (ODE) state space model to a transfer function model is discussed along with several conversions of a rational transfer function to a state space model.

The key concept is the *state* of a system, which is a set of variables which, along with the current time, summarizes the current configuration of a system. While some texts require it, there are good reasons for not requiring the variables to be a minimal set.[1]

---

[1] A typical example is the evolution of a direction cosine matrix. This is a three by three matrix that gives the orientation of one coordinate system with respect to another. Such matrices have two restrictions on them; they have determinant 1, and their transpose is their inverse. This is also called SO(3), the special orthogonal group of order 3. A smaller set of variables is pitch, roll, yaw ($\mathbb{R}^3$); however, this description is only good for small angles as $\mathbb{R}^3$ is commutative and rotations are not. When the relationship is a simple rotation with angular momentum $\omega = [\omega_x, \omega_y, \omega_z]$, the dynamics can be described with a state space in $\mathbb{R}^9 \sim \mathbb{R}^{3\times 3}$, as:

$$\frac{d}{dt}A = -\begin{bmatrix} 0 & -\omega_z & \omega_y \\ \omega_z & 0 & -\omega_x \\ -\omega_y & \omega_x & 0 \end{bmatrix} A$$

It is often desirable to work with a minimal set of variables, e.g., to improve numerical properties or to minimize the number of components used to build a system.

Given the state of a system at a given time, the prior history is of no additional help in determining the future behavior of the system. The state summarizes all the past behavior for the purposes of determining future behavior. The *state space* is the set of allowable values. The state space defines the topological, algebraic, and geometric properties associated with the evolution of the system over time. The state space description carries an internal model of the system dynamics. For example, one familiar equation is Newton's equation: $F(t) = ma(t) = m\ddot{x}(t)$. The Hamiltonian formulation for this problem, yields a set of coupled set of first order equations for position ($q = x$) and momentum ($p = m\dot{x}$):

$$\frac{d}{dt}\begin{bmatrix} q \\ p \end{bmatrix} = \begin{bmatrix} \frac{p}{m} \\ F(t) \end{bmatrix} \tag{5.45}$$

$$\begin{bmatrix} q(0) & p(0) \end{bmatrix}^T = \begin{bmatrix} x(0) & m\dot{x}(0) \end{bmatrix}^T. \tag{5.46}$$

This is indeed a state space description of Newton's equations. The state at a given time is the vector $\begin{bmatrix} q(t) & p(t) \end{bmatrix}^T$. The state space is $\mathbb{R}^2$. Examples of topological properties of the state space are: it is continuous, it has dimension two, etc. Of course, $\mathbb{R}^2$ enjoys many algebraic and geometric properties also. One could also integrate Newton's equations twice and get:

$$x(t) = x(0) + t\dot{x}(0) + \frac{1}{m}\int_0^t \int_0^s F(\tau)d\tau ds. \tag{5.47}$$

While very useful, this is an example of an "input/output" model of the system. The single variable "$x(t)$" is not sufficient to characterize the future behavior; clearly one needs $\dot{x}(t)$. Many facts about Equation 5.47 can be deduced by examining the solution; however, methods such as phase portraits (plot of q -vs- p) are frequently helpful in elucidating information about the behavior of the system without solving it. See [1, Chapter 2, section 1].

Often, the structure of the state space can be guessed by the structure of the initial conditions for the problem. This is because the initial conditions summarize the behavior of the system up to the time that they are given. In most mechanical systems, the state space is twice the number of the *degrees of freedom*; this assumes that the dynamics are second order. The degrees of freedom are positions,"x_i," and the additional variables needed to make up a state description are the velocities, "$\dot{x}_i$."

State space descriptions can be quite complicated, as is the case when two connected bodies separate (or collide), in which case the initial dimension of the state space would double (or half). Such problems occur in the analysis of the motions of launch vehicles, such as the space shuttle which uses, and ejects, solid rocket motors as well as a large liquid tank on its ascent into orbit. In such a case two different models are often created, and an attempt is made to capture and reconcile the forces of separation. This approach is not feasible for some systems, such as gimballed arms in robotics, which experience gimbal lock or the dropping or catching of gripped objects. Such problems are, usually, intrinsically difficult.

Reduction to First Order

In the case of Newton's laws, the state space description arose out of a reduction to a system of first order differential equations. This technique is quite general. Given a higher order differential equation:[2]

$$y^{(n)} = f(t, u(t), y, \dot{y}, \ddot{y}, ..., y^{(n-1)}) \tag{5.48}$$

with initial conditions:

$$y(t_0) = y_0,\ \dot{y}(t_0) = y_1,\ \cdots,\ y^{(n-1)}(t_0) = y_{n-1}. \tag{5.49}$$

Consider the vector $x \in \mathbb{R}^n$ with $x_1 = y(t)$, $x_2 = \dot{y}(t)$, ... , $x_n = y^{(n-1)}(t)$. Then Equation 5.48 can be written as:

$$\frac{d}{dt}x = \begin{bmatrix} x_2 \\ x_3 \\ \vdots \\ x_n \\ f(t, u(t), x_1, x_2, x_3, ..., x_n) \end{bmatrix} \tag{5.50}$$

with initial conditions:

$$x(t_0) = [y_0,\ y_1,\ \cdots,\ y_{n-1}]^T. \tag{5.51}$$

Here x is the state and the state space is $\mathbb{R}^n$. This procedure is known as *reduction to first order*. It is the center of discussion for most of what follows.

Another case that arises in electric circuits and elsewhere is the integro-differential equation, such as that associated with Figure 5.6:

$$e(t) = RI(t) + L\frac{d}{dt}I(t) + \frac{1}{C}\int_0^t I(\tau)d\tau. \tag{5.52}$$

Letting $x_1(t) = \int_0^t I(\tau)d\tau$, $x_2(t) = I(t)$, so that $\dot{x}_1 = x_2$. Then the state space formulation is:

$$\frac{d}{dt}x = \begin{bmatrix} 0 & 1 \\ -\frac{1}{LC} & -\frac{R}{L} \end{bmatrix}x + \begin{bmatrix} 0 \\ \frac{1}{L} \end{bmatrix}e(t) \tag{5.53}$$

$$I(t) = \begin{bmatrix} 0 & 1 \end{bmatrix}x. \tag{5.54}$$

This is a typical linear system with an output map.

Reduction to first order can also handle f (Equation 5.48) which depends on integrals of the inputs, $\int_0^t u(\tau)\,d\tau$, by setting $x_{n+1} = \int_0^t u(\tau)\,d\tau$ so that $\dot{x}_{n+1}(t) = u(t)$, etc. However,

---

which is not a minimal representation, but is simple to deal with. There is no worry about large angles in this representation.

[2] In this section superscripts in parenthesis represent derivatives, as do over–dots, e.g., $\ddot{y} = y^{(2)} = \frac{d^2}{dt^2}y$.

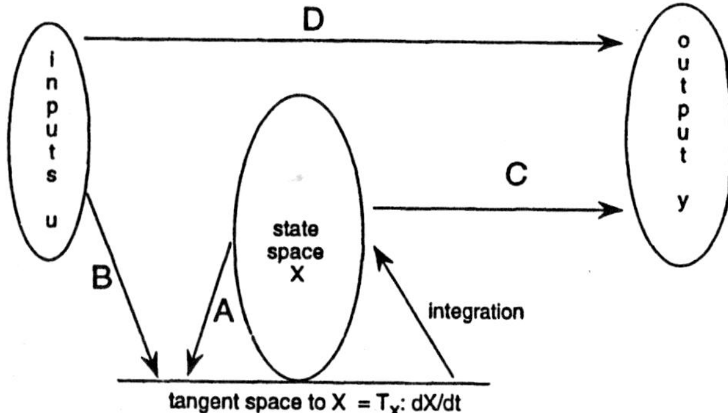

Figure 5.5 A pictorial view of the state space model for a system with dynamics: $\dot{x} = A(x) + B(u)$ and measurement equation $y = C(x) + D(u)$. Note that the tangent space for $\mathbb{R}^n$ is identified with $\mathbb{R}^n$ itself, but this is not true for other objects. For example, SO(3) is a differentiable manifold, and its tangent space is composed of skew symmetric 3×3 matrices noted earlier.

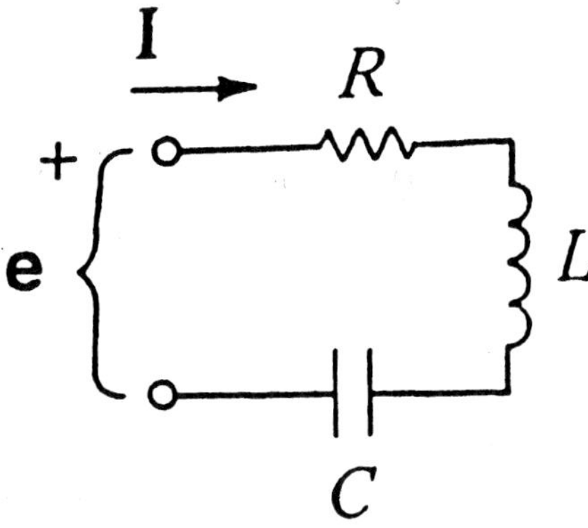

Figure 5.6 An RLC circuit, with impressed voltage $e(t)$ (input function), yields an equation for the current, $I(t)$.

unless $\dot{u}$ can be considered as the input, it cannot be handled in the reduction to first order. The sole exception is if f is linear, in which case the equation can be integrated until no derivatives of the input appear in it. This procedure will fail if there are more derivatives of the input than the original variable since there will not be enough initial conditions to be compatible with the integrations! That is differential equations of the form $\sum_0^N \frac{d^i}{dt^i} y_i(t) = \sum_0^M \frac{d^i}{dt^i} u_i(t)$, discussed in Section 5.2.2, must have $N \geq M$ in order to develop a state space model for the system.

Equivalent reduction to first order works for higher order difference equations (see Section 5.2.2).

Several other forms are commonly encountered. The first is the delay, e.g., $\dot{x}(t) = ax(t) + bx(t-T)$ for some fixed T. This problem is discussed in Section 5.5. Another form, the transfer function, is discussed in Section 5.2.2.

ARMA

The standard ARMA (auto regressive moving average) model is given by:

$$y_t = \sum_{l=1}^{n} a_l y_{t-l} + \sum_{l=0}^{n} b_l u_{t-l}. \tag{5.55}$$

The "$a_j y_{t-j}$" portion of this is the auto regressive part and the "$b_k u_{t-k}$" portion is the moving average part. This can always be written as a state–space model,

$$\begin{aligned} x_{t+1} &= Ax_t + Bu_t && (5.56)\\ y_t &= Cx_t + b_0 u_t, && (5.57) \end{aligned}$$

with $x \in \mathbb{R}^n$, as follows.

Using the "Z" transform, where $y_{t-1} \rightarrow z^{-1}Y(z)$, Equation 5.55 transforms to

$$\begin{aligned} [1 - a_1 z^{-1} \cdots &- a_n z^{-n}]Y(z) \\ &= [b_0 + b_1 z^{-1} \cdots + b_n z^{-n}]U(z). \\ \frac{Y(z)}{U(z)} &= \frac{b_0 + b_1 z^{-1} \cdots + b_n z^{-n}}{1 - a_1 z^{-1} \cdots - a_n z^{-n}} \\ &= b_0 + \frac{c_1 z^{-1} \cdots + c_{n-1} z^{-(n-1)}}{1 - a_1 z^{-1} \cdots - a_n z^{-n}} \end{aligned}$$

The last step is by division (multiply top and bottom by z^n, do the division, and multiply top and bottom by z^{-n}). This is the transfer function representation; then, apply the process directly analogous to the one explained in Section 5.2.2 for continuous systems to get the state space model. The resulting state space model is:

$$A = \begin{bmatrix} 0 & 1 & 0 & 0 & \cdots & 0 & 0 \\ 0 & 0 & 1 & 0 & \cdots & 0 & 0 \\ 0 & 0 & 0 & 1 & \cdots & 0 & 0 \\ \vdots & & & \vdots & & & \vdots \\ 0 & 0 & 0 & 0 & \cdots & 0 & 1 \\ a_n & a_{n-1} & & \cdots & & a_2 & a_1 \end{bmatrix}$$

$$B = [0\ 0 \cdots 0\ 1]^T$$

$$C = [c_1\ c_2 \cdots c_{n-1}]$$

The solution operator for this equation is given by the analog of the variation of constants formula:

$$x_t = A^t x_0 + \sum_{l=0}^{t-1} A^{t-1-l} B u_l \tag{5.58}$$

Ordinary Differential Equation

An ordinary differential equation, such as: $\ddot{y} + 2\omega\xi\dot{y} + \omega^2 y = u(t)$, with $y(0) = y_0$; $\dot{y}(0) = \dot{y}_0$ is actually the prototype for this model. Here $0 < \xi < 1$ and $\omega > 0$. Consider $x = \begin{bmatrix} y & \dot{y} \end{bmatrix}^T$, then the equation could be written:

$$\begin{aligned} \dot{x} &= \begin{bmatrix} 0 & 1 \\ -\omega^2 & -2\omega\xi \end{bmatrix} x + \begin{bmatrix} 0 \\ 1 \end{bmatrix} u(t) \\ y(t) &= \begin{bmatrix} 1 & 0 \end{bmatrix} x \\ x|_{t=0} &= x_0 = \begin{bmatrix} y_0 & \dot{y}_0 \end{bmatrix}^T. \end{aligned}$$

The state transition function is given by the variation of constants formula and the matrix exponential:[3]

$$\begin{aligned} x(t) &= \exp\left\{\begin{bmatrix} 0 & 1 \\ -\omega^2 & -2\omega\xi \end{bmatrix} t\right\} x_0 \\ &+ \int_0^t \exp\left\{\begin{bmatrix} 0 & 1 \\ -\omega^2 & -2\omega\xi \end{bmatrix}(t-\tau)\right\} \\ &\begin{bmatrix} 0 \\ 1 \end{bmatrix} u(\tau)\,d\tau \qquad (5.59) \\ &= \Gamma(0, t, x_0, u) \qquad (5.60) \\ w(t) = y(t) &= \begin{bmatrix} 1 & 0 \end{bmatrix} x(t). \end{aligned}$$

It is easy to confirm that the variation of constants formula meets the criterion to be a state transition function. The system is linear and time invariant.

This is a specific case of the general first order, linear time invariant vector system which is written:

$$\dot{x}(t) = Ax(t) + Bu(t) \qquad (5.61)$$

$$y(t) = Cx(t) + Du(t) \qquad (5.62)$$

where $x(t) \in \mathbb{R}^k$, $u(t) \in \mathbb{R}^p$, $y(t) \in \mathbb{R}^m$, and the quadruple $[A, B, C, D]$ are compatible constant matrices.

Transfer Functions and State Space Models

This section is an examination of linear time invariant differential equations (of finite order). All such systems can be reduced to first order, as already explained. Given a state space description with $x(t) \in \mathbb{R}^k$, $u(t) \in \mathbb{R}^p$, $y(t) \in \mathbb{R}^m$ and:

$$\dot{x} = Ax + Bu \quad \text{With initial conditions } x(t_0) = x_0 \qquad (5.63)$$

$$y = Cx + Du, \qquad (5.64)$$

where $[A, B, C, D]$ are constant matrices of appropriate dimensions, the transfer function is given by: $G(s) = C(sI - A)^{-1}B + D$. This is arrived at by taking the Laplace transform of the equation and substituting—using initial conditions of zero. While this expression is simple, it can be deceiving as numerically stable computation of $(sI - A)^{-1}$ may be difficult. For small systems, one often computes $(sI - A)^{-1}$ by the usual methods (cofactor expansion, etc.). Other methods are given in [4]. One consequence of this expression is that state space models are proper ($\lim_{s\to\infty} G(s) = D$).

Given a transfer function $G(s)$, which is a rational function, how does one develop a state space realization of the system? First the single–input–single–output (SISO) case is treated, then extend it to the multi–input–multi–output (MIMO) case (where $G(s)$ is a matrix of rational functions dimension m (outputs) by p (inputs)). The idea is to seek a quadruple $[A, B, C, D]$, representing the state equations as in Equations 5.63 and 5.64, so that $G(s) = C(sI - A)^{-1}B + D$, where A maps $\mathbb{R}^k \to \mathbb{R}^k$; C maps $\mathbb{R}^k \to \mathbb{R}^m$; B maps $\mathbb{R}^p \to \mathbb{R}^k$; D maps $\mathbb{R}^p \to \mathbb{R}^m$. Here the state space is $\mathbb{R}^k$.

Let

$$G(s) = n(s)/d(s) + e(s) \qquad (5.65)$$

where n, d, and e are polynomials with $deg(n) < deg(d)$, and this can be constructed by the Euclidean algorithm [13]. The coefficient convention is $d(s) = d_0 + d_1 s + \cdots + s^k$ for $deg(d) = k$. Notice that the leading order term of the denominator is normalized to one, and n is a polynomial of degree $k-1$. The transfer function is said to be *proper* if $e(s)$ is constant and *strictly proper* if $e(s)$ is zero. If a transfer function is strictly proper, then the input does not directly appear in the output $D \equiv 0$; if it is proper then D is a constant. If the transfer function is not proper then derivatives of the input appear in the output.

Given the transfer function 5.65, by taking the inverse Laplace transform of $\hat{y}(s) = G(s)\hat{u}(s)$, and substituting, one has the dynamics: [4]

$$d\left(\frac{d}{dt}\right)y(t) = \left(n\left(\frac{d}{dt}\right) + e_0 d\left(\frac{d}{dt}\right)\right)u(t) \qquad (5.66)$$

It is possible to work with this expression directly; however, the introduction of an auxiliary variable leads to two standard state space representations for transfer functions. Introduce the variable $z(t)$ as:

$$d\left(\frac{d}{dt}\right)z(t) = u(t) \qquad (5.67)$$

$$y(t) = n\left(\frac{d}{dt}\right)z + e_0 u \qquad (5.68)$$

or:

$$d\left(\frac{d}{dt}\right)z(t) = n\left(\frac{d}{dt}\right)u \qquad (5.69)$$

$$y(t) = z(t) + e_0 u. \qquad (5.70)$$

Both of these expressions can be seen to be equivalent to 5.66, by substitution.

Equations 5.67 and 5.68 can be reduced to first order by writing:

$$\frac{d^k}{dt^k}(z) + \frac{d^{k-1}}{dt^{k-1}}(d_{k-1}z) + \cdots + d_0 z = u.$$

---

[3]Since it is useful, let $C(t) = \cos\omega\sqrt{1-\xi^2}t$ and $S(t) = \sin\omega\sqrt{1-\xi^2}t$ then:

$$\exp\left\{\begin{bmatrix} 0 & 1 \\ -\omega^2 & -2\omega\xi \end{bmatrix} t\right\} = \exp\{-\xi\omega t\}\begin{bmatrix} C(t) + \frac{\xi}{\sqrt{1-\xi^2}}S(t) & \frac{1}{\omega\sqrt{1-\xi^2}}S(t) \\ -\frac{\omega}{\sqrt{1-\xi^2}}S(t) & C(t) - \frac{\xi}{\sqrt{1-\xi^2}}S(t) \end{bmatrix}$$

[4]Here the initial conditions are taken as zero when taking the inverse Laplace transform to get the dynamics; they must be incorporated as the initial conditions of the state space model. As a notational matter, by $d(\frac{d}{dt})y(t)$ one means $d_0 y(t) + d_1\dot{y}(t) + \cdots + d_{k-1}y^{(k-1)} + y^{(k)}$. Thus, $d(\frac{d}{dt})$ is a polynomial in the operator $\frac{d}{dt}$ and the same is true for $n(\frac{d}{dt})$.

Then let $x_1 = z, x_2 = \dot{z}$, so that $\frac{d}{dt}x_1 = x_2, \frac{d}{dt}x_2 = x_3$, etc. This can be written in first order matrix form as:

$$\dot{x} = \begin{bmatrix} 0 & 1 & 0 & \cdots & 0 & 0 \\ 0 & 0 & 1 & \cdots & 0 & 0 \\ 0 & 0 & 0 & \cdots & 0 & 0 \\ \vdots & \vdots & \vdots & \vdots & \vdots & \vdots \\ 0 & 0 & 0 & \cdots & 1 & 0 \\ 0 & 0 & 0 & \cdots & 0 & 1 \\ -d_0 & -d_1 & -d_2 & \cdots & -d_{k-2} & -d_{k-1} \end{bmatrix} x + \begin{bmatrix} 0 \\ 0 \\ 0 \\ \vdots \\ 0 \\ 0 \\ 1 \end{bmatrix} u(t)$$

$$y(t) = [n_0\, n_1 \cdots n_{k-2}\, n_{k-1}]\, x + e_0\, u(t).$$

This is known as the *controllable canonical form.*

The second pair of equations can be reduced to first order by writing Equation 5.69 as:

$$0 = d_0 z - n_0 u + \frac{d}{dt}(d_1 z - n_1 u + \frac{d}{dt}(d_2 z - n_2 u \frac{d}{dt} (\cdots d_{k-2} z - n_{k-2} u + \frac{d}{dt}(d_{k-1} z - n_{k-1} u + \frac{d}{dt}(z))\cdots))). \quad (5.71)$$

Let x_1 = everything past the first $\frac{d}{dt}$, then the complete equation is $\dot{x}_1 = -d_0 z + n_0 u$. Let x_2 = everything past the second $\frac{d}{dt}$, then $x_1 = d_1 z - n_1 u + \dot{x}_2$, which is the same as: $\dot{x}_2 = x_1 - d_1 z + n_1 u$. Proceeding in this fashion, $\dot{x}_l = x_{l-1} - d_{l-1} z + n_{l-1} u$; however, the interior of the last $\frac{d}{dt}$ is the variable z so that $x_k = z$! Substituting and writing in matrix form:

$$\dot{x} = \begin{bmatrix} 0 & 0 & 0 & \cdots & 0 & 0 & -d_0 \\ 1 & 0 & 0 & \cdots & 0 & 0 & -d_1 \\ 0 & 1 & 0 & \cdots & 0 & 0 & -d_2 \\ \vdots & & \vdots & \vdots & \vdots & \vdots & \\ 0 & 0 & 0 & \cdots & 1 & 0 & -d_{k-2} \\ 0 & 0 & 0 & \cdots & 0 & 1 & -d_{k-1} \end{bmatrix} x + \begin{bmatrix} n_0 \\ n_1 \\ n_2 \\ \vdots \\ n_{k-1} \end{bmatrix} u(t)$$

$$y(t) = [0\,0 \cdots 0\,1]\, x + e_0\, u(t). \quad (5.72)$$

This is known as *observable canonical form.* This form is usually reduced from a block diagram making the choice of variables more obvious.

Notice that Controllable Form is equal to Observable Form transposed,

$$[A, B, C, D] = [A^T, C^T, B^T, D^T],$$

which is the beginning of duality for controllability and observability. The A matrix in these two examples is in companion form, the negative of the coefficients of the characteristic equation on the outside edge, with ones off the diagonal. There are two other companion forms, and these are also related to controllability and observability.

If $G(s)$ is strictly improper, that is $deg(e) \geq 1$, there is no state space model; however, there is at least one trick used to circumvent this. As an example, one might deal with this by modeling $e(s) = s^2$ with $\frac{s^2}{s^2+as+b}$, moving the poles outside the range of the system. Since $\frac{s^2}{s^2+as+b}$ is proper, it has a state space model; then adjoin this model to the original, feeding its output into the output original system. One effect is the bandwidth (dynamic range) of the system is greatly increased. This can result in "stiff" ODEs which have numerical problems associated with them. Obviously, this technique could be applied to more complex transfer functions $e(s)$. However, improper systems should be closely examined as such a system has an increasingly larger response to higher frequency inputs!

Miscellanea

Given a state space representation of the form $\dot{x} = Ax + Bu$; $y = Cx + Du$, one could introduce a change of variables $z = Px$, with P^{-1} existing. Then the system can be written, by substitution: $\dot{z} = PAP^{-1}x + PBu$; $y = CP^{-1}z + Du$. Hence $Px \to z$ induces $[A, B, C, D] \to [PAP^{-1}, PB, CP^{-1}, D]$. Note that such a transformation does not change the characteristic equation; and hence the eigenvalues associated with A, since:

$$\begin{aligned} det(sI - A) &= det(P)det(P^{-1})det(sI - A) \\ &= det(P)det(sI - A)det(P^{-1}) \\ &= det(P(sI - A)P^{-1}) \\ &= det(sPP^{-1} - PAP^{-1}) \\ &= det(sI - PAP^{-1}). \end{aligned}$$

A little more algebra shows that the transfer function associated with the two systems is the same. The transformation $A \to P^{-1}AP$ is known as a *similarity transformation.*

Finding a state space representation of smaller dimension has two guises: the first is the search for a minimal realization, touched on in Example 5.4. The second is the problem of model reduction.

Usual operations on transfer functions involve combining them, usually adding or cascading them. In both of these cases the state space model can be determined from the state space models of the component transfer functions. The total state space size is usually the sum of the dimensions of the component state space models. Here x_i is taken as the state (subvector) corresponding to transfer function G_i, rather than a single component of the state vector. The exception to this rule about component size has to do with repeated roots and will be illustrated below. The size of the state space is the degree of the denominator polynomial of the combined system—note that cancellation can occur between the numerator and the denominator of the combined system.

When two transfer functions are cascaded (the output of the first is taken as input to the second), the resulting transfer function is the product of the two transfer functions: $H(s) = G_2(s)G_1(s)$. To build the corresponding state space model, one could simply generate a state space model for $H(s)$; however, if state space models exist for G_1 and G_2, there are sound reasons for proceeding to work in the state space domain. This is because higher order transfer functions are more likely to lead to poor numerical conditioning (see the discussion at the end of Section 5.5). Given $G_1 \sim [A_1, B_1, C_1, D_1]$ and $G_2 \sim [A_2, B_2, C_2, D_2]$, then simple algebra shows that:

$$H = G_2G_1 \sim \tag{5.73}$$

$$\dot{x} = \begin{bmatrix} A_1 & 0 \\ B_2C_1 & A_2 \end{bmatrix} x(t) + \begin{bmatrix} B_1 \\ B_2D_1 \end{bmatrix} u(t) \tag{5.74}$$

$$y(t) = [D_2C_1 \ C_2]\, x(t) + D_2D_1u(t). \tag{5.75}$$

The state space used in this description is $\begin{bmatrix} x_1 & x_2 \end{bmatrix}^T$, where $x_1 \sim G_1$ and $x_2 \sim G_2$.

When two transfer functions are added, $H(s) = G_1(s) + G_2(s)$, and state space models are available the equivalent construction yields:

$$H = G_1 + G_2 \sim \tag{5.76}$$

$$\dot{x} = \begin{bmatrix} A_1 & 0 \\ 0 & A_2 \end{bmatrix} x(t) + \begin{bmatrix} B_1 \\ B_2 \end{bmatrix} u(t) \tag{5.77}$$

$$y(t) = \begin{bmatrix} C_1 & C_2 \end{bmatrix} x(t) + [D_1 + D_2]\, u(t). \tag{5.78}$$

If state feedback is used, that is $u = Fx(t) + v(t)$, then,

$$[A, B, C, D] \rightarrow [A + BF, B, C + DF, D]$$

is the state space system acting on $v(t)$. If output feedback is used, that is $u = Fy(t) + v(t)$ then,

$$[A, B, C, D] \rightarrow [A + BFC, B, C + DFC, D]$$

is the state space system acting on $v(t)$. Both of these are accomplished by simple substitution. Clearly, there is a long list of such expressions, each one equivalent to a transfer function manipulation.

To illustrate these ideas, consider the system:

$$G(s) = \frac{2(s^2+4)(s-2)}{(s+3)(s+2)^2}$$

$$= \frac{2s^3 - 4s^2 + 8s - 16}{s^3 + 7s^2 + 16s + 12} \tag{5.79}$$

$$= 2 + \frac{-18s^2 - 24s - 40}{s^3 + 7s^2 + 16s + 12} \tag{5.80}$$

$$= 2 - \frac{130}{s+3} + \frac{112}{s+2} - \frac{64}{(s+2)^2}. \tag{5.81}$$

This can be realized as a state space model using four techniques. The last expression is the partial fractions expansion of $G(s)$ (see Chapter 4).

EXAMPLE 5.1:

Controllable canonical form; $G(s) \sim [A_c, B_c, C_c, D_c]$ where:

$$A_c = \begin{bmatrix} 0 & 1 & 0 \\ 0 & 0 & 1 \\ -12 & -16 & -7 \end{bmatrix},$$

$$B_c = \begin{bmatrix} 0 \\ 0 \\ 1 \end{bmatrix}, C_c^T = \begin{bmatrix} -40 \\ -24 \\ -18 \end{bmatrix}, \ D_c = [2].$$

EXAMPLE 5.2:

Observable canonical form; $G(s) \sim [A_o, B_o, C_o, D_o]$ where:

$$A_o = \begin{bmatrix} 0 & 0 & -12 \\ 1 & 0 & -16 \\ 0 & 1 & -7 \end{bmatrix},$$

$$B_o = \begin{bmatrix} -40 \\ -24 \\ -18 \end{bmatrix}, \ C_o^T = \begin{bmatrix} 0 \\ 0 \\ 1 \end{bmatrix}, \ D_o = [2].$$

Note that $[A_c, B_c, C_c, D_c] = [A_o^T, C_o^T, B_o^T, D^T]$, verifying the Controllable-Observable duality in this case.

EXAMPLE 5.3:

As a product, $G = G_2G_1$, $G_1(s) = \frac{s-2}{s+3} = 1 + \frac{-5}{s+3}$, $G_2(s) = 2\frac{s^2+4}{(s+2)^2} = 2 + \frac{-8s}{(s+2)^2}$. $G_1(s) \sim [A_1, B_1, C_1, D_1] = [-3, 1, -5, 1]$; $G_2(s) \sim [A_2, B_2, C_2, D_2]$ where:

$$A_2 = \begin{bmatrix} 0 & 1 \\ -4 & -4 \end{bmatrix},$$

$$B_2 = \begin{bmatrix} 0 \\ 1 \end{bmatrix}, \ C_2^T = \begin{bmatrix} 0 \\ 8 \end{bmatrix}, \ D_2 = [2].$$

The combined system is:

$$A_{21} = \begin{bmatrix} -3 & 0 & 0 \\ 0 & 0 & 1 \\ -5 & -4 & -4 \end{bmatrix},$$

$$B_{21} = \begin{bmatrix} 1 \\ 0 \\ 1 \end{bmatrix}, \ C_{21}^T = \begin{bmatrix} -10 \\ 0 \\ 8 \end{bmatrix}, \ D_{21} = [2].$$

EXAMPLE 5.4:

As a sum, $G = G_1 + G_2 + G_3$, where the summands are derived by the partial fractions method (see Chapter 8). This case is used to illustrate the *Jordan* representation of the state space. If one were to proceed hastily, using the addition method, then the resulting systems would have $1+1+2=4$ states. The other realizations have three states. Why? Since the term $s+2$ appears in the last two transfer functions, it is possible to "reuse" it. The straightforward combination is not wrong, it simply fails to be minimal. The transfer function for $\frac{1}{(s+a)^2}$ is given by the product rule, Equations 5.74 and 5.75: Let $G^2(s) = G(s)G(s)$, where $G(s) = \frac{1}{s+a} \sim [-a, 1, 1, 0]$, so that $G^2 \sim [A, B, C, D]$ with:

$$A = \begin{bmatrix} -a & 0 \\ 1 & -a \end{bmatrix} \quad B = \begin{bmatrix} 1 \\ 0 \end{bmatrix}$$
$$C^T = \begin{bmatrix} 0 \\ 1 \end{bmatrix} \quad D = [0].$$

The third power, $G^3 = GG^2\ [A, B, C, D]$, with:

$$A = \begin{bmatrix} -a & 0 & 0 \\ 1 & -a & 0 \\ 0 & 1 & -a \end{bmatrix}, \quad B = \begin{bmatrix} 1 \\ 0 \\ 0 \end{bmatrix},$$
$$C^T = \begin{bmatrix} 0 \\ 0 \\ 1 \end{bmatrix}, \quad D = [0].$$

This can be continued, and the A matrix will have $-a$ on the diagonal and 1 on the subdiagonal; $B = [1\ 0\ \cdots\ 0]^T$, $C = [0\ 0\ \cdots\ 0\ 1]$, $D = 0$. This is very close to the standard matrix form known as Jordan form. For G^2 a change of variables $x' = \begin{bmatrix} 0 & 1 \\ 1 & 0 \end{bmatrix} x$ takes the system to:

$$A = \begin{bmatrix} -a & 1 \\ 0 & -a \end{bmatrix} \quad B = \begin{bmatrix} 0 \\ 1 \end{bmatrix}$$
$$C^T = \begin{bmatrix} 1 \\ 0 \end{bmatrix} \quad D = [0]. \tag{5.82}$$

So that this is the system in Jordan form (this is just the exchange of variables $x' = [x_2\ x_1]^T$). Notice that if $C = \begin{bmatrix} 0 & 1 \end{bmatrix}$ in system 5.82, the transfer function is $\frac{1}{s+a}$! Thus if $G(s) = \frac{\alpha_1}{s+a} + \frac{\alpha_2}{(s+a)^2}$, use system 5.82 with $C = \begin{bmatrix} \alpha_2 & \alpha_1 \end{bmatrix}$.
In general, consider $G^k \sim [A, B, C, D]$, obtained by the product rules Equations 5.74 and 5.75, apply the conversion is $x' = Px$, with $P = antidiagonal(1)$ (note that $P = P^{-1}$). The resulting system is:

$$\dot{x} = \begin{bmatrix} -a & 1 & 0 & \cdots & 0 & 0 & 0 \\ 0 & -a & 1 & \cdots & 0 & 0 & 0 \\ 0 & 0 & -a & \cdots & 0 & 0 & 0 \\ \vdots & & \vdots & \vdots & \vdots & \vdots & \\ 0 & 0 & 0 & \cdots & 0 & -a & 1 \\ 0 & 0 & 0 & \cdots & 0 & 0 & -a \end{bmatrix} x + \begin{bmatrix} 0 \\ 0 \\ 0 \\ \vdots \\ 0 \\ 1 \end{bmatrix} u(t)$$

$$y(t) = [1\,0\,\cdots\,0\,0]\,x.$$

And if the desired transfer function is $G(s) = \sum_{l=1}^{k} \frac{\alpha_l}{(s+a)^l}$, then $C = [\alpha_k\ \alpha_{k-2}\ \cdots\ \alpha_1]$.
Returning to the example, Equation 5.81, $G_1(s) = 2 - \frac{130}{s+3} \sim [-3, 1, -130, 2]$ where:
The last two transfer functions are in Jordan form: $G_2(s)+G_3(s) = \frac{112}{s+2} - \frac{64}{(s+2)^2} \sim [A_2, B_2, C_2, D_2]$ where:
$G(s) \sim [A, B, C, D]$ where:

$$A = \begin{bmatrix} -2 & 1 \\ 0 & -2 \end{bmatrix} \quad B = \begin{bmatrix} 0 \\ 1 \end{bmatrix}$$
$$C^T = \begin{bmatrix} -64 \\ 112 \end{bmatrix} \quad D = [0].$$

Since G_1 is already in Jordan form (trivially), the combined system is $G_{321}(s) \sim [A_{321}, B_{321}, C_{321}, D_{321}]$ where:

$$A_{321} = \begin{bmatrix} -2 & 1 & 0 \\ 0 & -2 & 1 \\ 0 & 0 & -2 \end{bmatrix},$$
$$B_{321} = \begin{bmatrix} 1 \\ 0 \\ 1 \end{bmatrix}, \ C_{321}^T = \begin{bmatrix} -64 \\ 112 \\ -130 \end{bmatrix}, \ D_{321} = [2].$$

Here A is also in Jordan form; this representation is very nice as it displays the eigenvalues and their multiplicities.

MIMO Transfer Functions to State Space

One of the historical advantages of the state space methods was the ability to deal with multi-input-multi-output systems. The frequency domain methods have matured to deal with this case also, and the following section deals with how to realize a state space model from a transfer function $G(s)$, which is a matrix with entries that are rational functions.

The methods discussed here are straightforward generalizations of Equations 5.69 and 5.70 and Equations 5.67 and 5.68.

A variety of other methods are discussed in [4] and [7], where numerical considerations are considered. The reader is referred to the references for more detailed information, including the formal proofs of the methods being presented.

A useful approach is to reduce the multi-input-multi-output system to a series of single-input-multi-output systems and then combine the results. This means treating the columns of $G(s)$ one at a time. If $[A_i, B_i, C_i, D_i]$ are the state space descriptions associated with the $i-th$ column of $G(s)$, denoted $G_i(s)$, then the state space description of $G(s)$ is

$$[diag(A_1, A_2, \cdots, A_n), [B_1, B_2, \cdots, B_n],$$
$$[C_1 C_2 \cdots C_n], [D_1 D_2 \cdots D_n]] .$$

Where:

$$diag(A_1, A_2) = \begin{bmatrix} A_1 & 0 \\ 0 & A_2 \end{bmatrix},$$

etc. The input to this system is the vector $[u_1\ u_2\ \cdots\ u_n]^T$. With this in mind, consider the development of a single-input-multi-output transfer function to a state space model.

Given $G(s)$, a column vector, first subtract off the vector $E = G(s)|_{s=\infty}$ from G leaving strictly proper rational functions as entries, then find the least common denominator of all of the entries, $d(s) = s^k + \sum_0^{k-1} d_l s^l$, and factor it out. This leaves a vector of polynomials $n_i(s)$ as entries of the vector. Thus $G(s)$ has been decomposed as:

$$G(s) = \begin{bmatrix} g_1(s) \\ g_2(s) \\ \cdots \\ g_q(s) \end{bmatrix} = \begin{bmatrix} e_1 \\ e_2 \\ \cdots \\ e_q \end{bmatrix} + \frac{1}{d(s)} \begin{bmatrix} n_1(s) \\ n_2(s) \\ \cdots \\ n_q(s) \end{bmatrix}.$$

Writing $n_j(s) = \sum_{l=0}^{k-1} \nu_{j,l} s^l$, then the state space realization is:

$$\frac{d}{dt}x = \begin{bmatrix} 0 & 1 & 0 & \cdots & 0 & 0 \\ 0 & 0 & 1 & \cdots & 0 & 0 \\ 0 & 0 & 0 & \cdots & 0 & 0 \\ \vdots & \vdots & \vdots & \vdots & \vdots & \vdots \\ 0 & 0 & 0 & \cdots & 1 & 0 \\ 0 & 0 & 0 & \cdots & 0 & 1 \\ -d_0 & -d_1 & -d_2 & \cdots & -d_{k-2} & -d_{k-1} \end{bmatrix} x + \begin{bmatrix} 0 \\ 0 \\ 0 \\ \vdots \\ 0 \\ 0 \\ 1 \end{bmatrix} u(t)$$

$$\begin{bmatrix} y_1 \\ y_2 \\ \cdots \\ y_q \end{bmatrix} = \begin{bmatrix} \nu_{1,k-1} & \nu_{1,k-2} & \cdots & \nu_{1,0} \\ \nu_{2,k-1} & \nu_{2,k-2} & \cdots & \nu_{2,0} \\ \cdots & \cdots & \cdots & \cdots \\ \nu_{q,k-1} & \nu_{q,k-2} & \cdots & \nu_{q,0} \end{bmatrix} x + \begin{bmatrix} e_1 \\ e_2 \\ \cdots \\ e_q \end{bmatrix} u(t).$$

This a controllable realization of the transfer function. In order to get an observable realization, treat the transfer function by rows and proceed in a similar manner (this is multi-input-single output approach, see [4]). The duality of the two realizations still holds.

Additional realizations are possible, perhaps the most important being Jordan form. This is handled like the single-input-single-output case, using partial fractions with column coefficients rather than scalars. That is, expanding the entries of $G(s)$, a single-input-multi-output (column) transfer function, as partial fractions with (constant) vector coefficients and then using Equations 5.77 and 5.78 for dealing with the addition of transfer functions. Multiple inputs are handled one at a time and stacked appropriately.

EXAMPLE 5.5:

Let

$$\begin{aligned} G(s) &= \begin{bmatrix} \frac{s+3}{s^2+2s+2} \\ \frac{s^2+4}{(s+1)^2} \end{bmatrix} \\ &= \begin{bmatrix} \frac{2}{s+1} - \frac{1}{s+2} \\ 1 - \frac{2}{s+1} + \frac{5}{(s+1)^2} \end{bmatrix} \\ &= \begin{bmatrix} 0 \\ 1 \end{bmatrix} + \frac{1}{s+1}\begin{bmatrix} 2 \\ -2 \end{bmatrix} \\ &\quad - \frac{1}{s+2}\begin{bmatrix} 1 \\ 0 \end{bmatrix} + \frac{1}{(s+1)^2}\begin{bmatrix} 0 \\ 5 \end{bmatrix}. \end{aligned}$$

This results in the two systems:

$$\begin{aligned} \dot{x_1} &= \begin{bmatrix} -1 & 1 \\ 0 & -1 \end{bmatrix} x_1 + \begin{bmatrix} 0 \\ 1 \end{bmatrix} u(t) \\ y_1(t) &= \begin{bmatrix} -1 & 0 \\ 0 & 5 \end{bmatrix} x_1 + \begin{bmatrix} 0 \\ 1 \end{bmatrix} u \\ \dot{x_2} &= -2x_2 + u \\ y_2(t) &= \begin{bmatrix} 2 \\ -2 \end{bmatrix} x_2. \end{aligned}$$

Which combine to form:

$$\begin{aligned} \frac{d}{dt}\begin{bmatrix} x_1 \\ x_2 \end{bmatrix} &= \begin{bmatrix} -1 & 1 & 0 \\ 0 & -1 & 0 \\ 0 & 0 & -2 \end{bmatrix} \begin{bmatrix} x_1 \\ x_2 \end{bmatrix} + \begin{bmatrix} 0 \\ 1 \\ 1 \end{bmatrix} u(t) \\ y(t) &= \begin{bmatrix} -1 & 0 & 2 \\ 0 & 5 & -2 \end{bmatrix} \begin{bmatrix} x_1 \\ x_2 \end{bmatrix} + \begin{bmatrix} 0 \\ 1 \end{bmatrix} u(t). \end{aligned}$$

Padè Approximation of Delay

The most common nonrational transfer function model encountered is that resulting from a delay in the system:[5] $e^{-sT}\hat{f}(s) = L\{f(t-T)\}$. The function e^{-sT} is clearly transcendental, and this changes the fundamental character of the system if treated rigorously. However, it is fortunate that there is an often useful approximation of the transfer function which is rational and hence results in the addition of finite states to the differential equation.

The general idea is to find a rational function approximation to a given function, in this case e^{-sT}, and replace the function with the rational approximation. Then use the methods of Section 5.2.2 to construct a state space description. Specifically, given $f(s) = \sum_0^\infty a_n s^n$ find $P_M^N(s)$:

$$P_M^N(s) = \frac{\sum_0^N A_n s^n}{1+\sum_1^M B_n s^n} \tag{5.83}$$

so that the coefficients of the first N+M+1 terms of the two Taylor series match.[6] It is usual to look at the diagonal approximate $P_N^N(s)$. As will be shown, the diagonal Padé approximation of e^{-sT} shares with e^{-sT} the fact that it has unit magnitude along the imaginary axis. Thus, the Bode magnitude plot (log magnitude of the transfer function vs. log frequency) is the constant zero. However, the phase approximation deviates dramatically, asymptotically. Put another way: the approximation is perfect in magnitude; all of the error is in phase. The difference in the phase (or angle) part of the Bode plot (which is phase vs. log frequency) is displayed in Figure 5.7, which compares the pure delay, e^{-sT}, with several Padé approximations.

The coefficients of Padé approximation to the delay can be in calculated in the following way which takes advantage of the properties of the exponential function. Here it is easier to allow the B_0 term to be something other than unity; it will work out to be one:

$$e^{-sT} = \frac{e^{-s\frac{T}{2}}}{e^{s\frac{T}{2}}} \tag{5.84}$$

Therefore:

$$\frac{e^{-s\frac{T}{2}}}{e^{s\frac{T}{2}}} = \frac{\sum_0^N A_n s^n}{\sum_0^M B_n s^n} + O(M+N+1) \tag{5.85}$$

Hence:

$$e^{-s\frac{T}{2}} \times \sum_0^M B_n s^n = e^{s\frac{T}{2}} \times \left(\sum_0^N A_n s^n + O(M+N+1)\right). \tag{5.86}$$

Equality will hold if $\sum_0^M B_n s^n$ is the first M+1 terms in the Taylor expansion of $e^{s\frac{T}{2}}$ and $\sum_0^N A_n s^n$ is the first N+1 terms in the Taylor expansion of $e^{-s\frac{T}{2}}$. This yields the expression usually found in texts:

$$P_M^N(s) = \frac{\sum_0^N (-1)^n \frac{\left(\frac{sT}{2}\right)^n}{n!}}{\sum_0^M \frac{\left(\frac{sT}{2}\right)^n}{n!}}. \tag{5.87}$$

Notice that $P_N^N(s)$ is of the form $p(-s)/p(s)$ so that it is an *all pass filter*; that is, it has magnitude one for all $j\omega$ on the imaginary axis. To see this, let $s = j\omega$ and multiply the numerator and denominator by their conjugates. Then note that $(-j\omega)^n = (-1)^n(j\omega)^n$ and that $(-1)^{2n} = 1$. Hence the diagonal approximation has magnitude one for all $s = j\omega$ on the imaginary axis, just as e^{-sT} does. This is one of the reasons that this approximation is so useful for delays.

The approximation in Equation 5.87, has a zero-th order term of 1 and the highest order term contains $\frac{1}{n!}$; thus, numerical stability of the approximation is a concern, especially for high order (meaning over several decades of frequency). It has already been noted that $e^{-sT} = e^{-s\frac{T}{2}}e^{-s\frac{T}{2}}$, so that an approximation is possible as cascaded lower order approximations. In practice it is best not to symmetrically divide the delay, thus avoiding repeated roots in the system. For example:

$$e^{-sT} = e^{-s\frac{1.1T}{3}}e^{-s\frac{T}{3}}e^{-s\frac{.99T}{3}} \tag{5.88}$$

and approximate each of the delays using a fourth order approximation. This gives a twelve state model, using cascaded state space systems (see Equations 5.74 and 5.75). This model is compared in Figure 5.7, where it is seen to be not as good as an order twelve Padé approximation; however, the condition number[7] of the standard twelfth order Padé approximation is on the order of 10^{14}. Breaking the delay into three unequal parts resulting in no repeated roots and a condition number for the A matrix of about 10^4, quite a bit more manageable numerically. The resulting twelve state model is about as accurate as the tenth order approximation, which has a condition of about 10^{10}.

A nice property of the Padé approximation of a delay is that the poles are all in the left half-plane; however, the zeros are all in the right half-plane which is difficult (Problem 8.61 [3]). This does reflect the fact that delay is often destabilizing in feedback loops.

[5] $L\{\cdot\}$ is used to denote the Laplace transform of the argument with respect to the variable t.

[6] There is an extensive discussion of Padé approximations in [3]. This includes the use of Taylor expansions at two points to generate Padé approximates. While the discussion is not aimed at control systems, it is wide ranging and otherwise complete. Let: $f(s) \sim \sum_0^\infty a_n(s-s_0)^n$ as $s \to s_0$ and $f(s) \sim \sum_0^\infty b_n(s-s_1)^n$ as $s \to s_1$ then choose $P_M^N(s)$, as before a rational function, so that the first J terms of the Taylor series match the s_0 representation and so the first K terms of the s_1 representation likewise agree, where $J + K = N + M + 1$. The systems of equations used to solve for A_n and B_n are given in [3] in both the simple and more complex circumstance. In the case of matching a single Taylor series, the relationship between Padé approximations and continued fractions is exploited for a recursive method of computing the Padé coefficients. A simpler trick will suffice for the case at hand.

[7] The condition number is the ratio of the largest to the smallest magnitude of the eigenvalue, see [9] for more on condition numbers.

There are other common non–rational transfer functions, such as those that arise from partial differential equations (PDEs) and from the study of turbulence (e.g., the von Karman model for turbulence, which involves fractional powers of s). In such cases, use the off diagonal approximates to effect an appropriate roll-off at high frequency (for the exponential function this forces the choice of the diagonal approximate).

Remarks on Padé Approximations The Padé approximation is one of the most common approximation methods; however, in control systems, it has several drawbacks that should be mentioned. It results in a finite dimensional model, which may hide fundamental behavior of a PDE. While accurately curve fitting the transfer function in the s domain, it is not an approximation that has the same pole-zero structure as the original transfer function. Therefore, techniques that rely on pole–zero cancellation will not result in effective control. In fact, the Padé approximation of a stable transfer function is not always stable!

A simple way to try to detect the situation where pole–zero cancellation is being relied upon is to use different models for design and for validation of the control system. The use of several models provides a simple, often effective method for checking the robustness of a control system design.

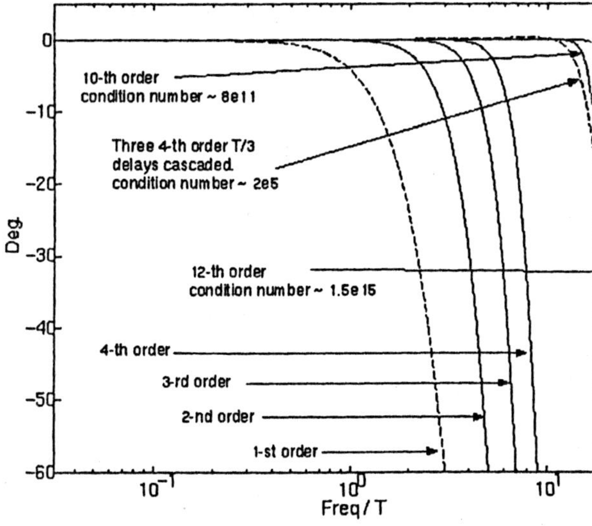

Figure 5.7 Comparison of e^{-sT} and various order Padé approximations.

5.2.3 Linearization

Preliminaries

This section is a brief introduction to linearization of nonlinear systems. The point of the section is to introduce enough information to allow linearization, when possible, in a way that permits for stabilization of a system by feedback. The proofs of the theorems are beyond the scope of this section, since they involve finding generalized energy functions, known as Lyapunov[8] functions. A good reference for this information is [11], Chapter 5.

Let $x = [x_1 x_2 \cdots x_n]^T \in \mathbb{R}^n$ then the Jacobian of a function $\mathbf{f}(\mathbf{x}, t) = [f_1 f_2 \cdots f_m]^T$ is the $m \times n$ matrix of partials:

$$\left[\frac{\partial \mathbf{f}}{\partial \mathbf{x}}\right] = \begin{bmatrix} \frac{\partial f_1}{\partial x_1} & \frac{\partial f_1}{\partial x_2} & \cdots & \frac{\partial f_1}{\partial x_n} \\ \frac{\partial f_2}{\partial x_1} & \frac{\partial f_2}{\partial x_2} & \cdots & \frac{\partial f_2}{\partial x_n} \\ \vdots & \vdots & \vdots & \vdots \\ \frac{\partial f_m}{\partial x_1} & \frac{\partial f_m}{\partial x_2} & \cdots & \frac{\partial f_m}{\partial x_n} \end{bmatrix} \tag{5.89}$$

$$= \left[\frac{\partial \mathbf{f}}{\partial \mathbf{x}}\right]. \tag{5.90}$$

So that the Taylor expansion of $\mathbf{f}(\mathbf{x}, t)$ about $\mathbf{x}_0$ is

$$\mathbf{f}(\mathbf{x}, t) = \mathbf{f}(\mathbf{x}_0, t) + \left[\frac{\partial f}{\partial x}\right]\Bigg|_{\mathbf{x}=\mathbf{x}_0} \mathbf{x} + O(|\mathbf{x} - \mathbf{x}_0|^2).$$

The stability of a system of differential equations is discussed at length in Chapter 1. A brief review is provided here for ease of reference. A differential equation, $\frac{d}{dt}\mathbf{x} = \mathbf{f}(\mathbf{x}, t)$, is said to have an *equilibrium point* at $\mathbf{x}_0$ if $\mathbf{f}(\mathbf{x}_0, t) = \mathbf{0}$ for all t, thus $\frac{d}{dt}\mathbf{x}\Big|_{\mathbf{x}_0} = \mathbf{0}$ and, the differential equation would have solution $\mathbf{x}(t) \equiv \mathbf{x}_0$.

A differential equation, $\frac{d}{dt}\mathbf{x} = \mathbf{f}(\mathbf{x}, t)$, with an equilibrium point at the origin is said to be *stable* if for every $\epsilon > 0$ and $t > 0$ there is a $\delta(\epsilon, t_0)$ so that $|\mathbf{x}_0| < \delta(\epsilon, t_0)$ implies that $|\mathbf{x}(t)| < \epsilon$ for all t. If δ depends only on ϵ, then the system is said to be *uniformly stable.* The equation is said to be *exponentially stable* if there is a ball $B_r = |\mathbf{x}| \leq r$ so that for all $\mathbf{x}_0 \in B_r$ the solution obeys the following bound for some $a, b > 0$: $|\mathbf{x}_0| < a|\mathbf{x}_0| \exp\{-bt\}$ for all t. This is a strong condition. For autonomous linear systems, all of these forms collapse to exponential stability, for which all of the eigenvalues having negative real parts is necessary and sufficient. However, as Examples 5.1 and 5.2 show, this simple condition is not universal. For time varying systems, checking exponential stability can be quite difficult.

It is true that if the system is "slowly varying", things do work out. That is, if $\mathbf{x}_0$ is an equilibrium point and $\left[\frac{\partial \mathbf{f}}{\partial \mathbf{x}}\right]$ has eigenvalues with negative real parts bounded away from the imaginary axis and additional technical conditions hold, then the differential equation is exponentially stable. See Theorem 15, Chapter 5, Section 8 in [11]. Note that Example 5.2 is *not* slowly varying. The physical example of a pendulum in which the length is varied (even slightly) close to twice the period (which is a model of a child on a swing), shows that rapidly varying dynamics can take an exponentially stable system (a pendulum

[8]The methods used here were developed in Lyapunov's monograph of 1892. There are several spellings of his name, benign variations on the transliteration of a Russian name. The other common spelling is Liapunov, and this is used in the translations of V. I. Arnold's text [1] (Arnold's name is also variously transliterated!).

with damping) and change it into an unstable system. See Section 25 in [1], where there is an analysis of Mathieu's equation: $\ddot{q} = -\omega_0(1 + \epsilon\cos(t))q$, $\epsilon \ll 1$, which is unstable for $\omega_0 = \frac{k}{2}$, $k = 1, 2, \cdots$. This instability continues in the presence of a little damping. This phenomenon is known as *parametric resonance.*

Lyapunov's Linearization Method

Consider $\frac{d}{dt}\mathbf{x} = \mathbf{f}(\mathbf{x}, t)$, with an equilibrium point at $\mathbf{x}_0$ and $\mathbf{f}$ continuously differentiable in both arguments. Then:

$$\frac{d}{dt}\mathbf{x} = \mathbf{f}(\mathbf{x}, t) \tag{5.91}$$

$$= \left\{\left[\frac{\partial \mathbf{f}}{\partial \mathbf{x}}\right]\bigg|_{\mathbf{x}=\mathbf{x}_0}\right\}\mathbf{x} + \mathbf{f}_1(\mathbf{x}, t) \tag{5.92}$$

$$= \mathbf{A}(t)\mathbf{x} + \mathbf{f}_1(\mathbf{x}, t). \tag{5.93}$$

Here $\frac{d}{dt}\mathbf{x} = \mathbf{A}(t)\mathbf{x}$ is the candidate for the linearization of the system. The additional condition which must hold is that the approximation be uniform in t, that is:

$$\lim_{|\mathbf{x}|\to\mathbf{x}_0} \sup_{t\geq 0} \frac{|\mathbf{f}_1(\mathbf{x}, t)|}{|\mathbf{x}|} = 0. \tag{5.94}$$

Note that if $\mathbf{f}_1(\mathbf{x}, t) = \mathbf{f}_1(\mathbf{x})$ then the definition of the Jacobian guarantees that condition 5.94 holds. (see Example 5.1 for an example where uniform convergence fails.)

Under these conditions, the system $\frac{d}{dt}\mathbf{x} = \mathbf{A}(t)\mathbf{x}$ represents a linearization of system about $\mathbf{x}_0$.

If it is desired to hold the system at $\mathbf{x}_0$ which is not an equilibrium point, then the system must be modified to subtract off $\mathbf{f}(\mathbf{x}_0, t)$. That is, $\tilde{\mathbf{f}}(\mathbf{x}, t) = \mathbf{f}(\mathbf{x}, t) - \mathbf{f}(\mathbf{x}_0, t)$. Thus, $\tilde{\mathbf{f}}$ has an equilibrium point at $\mathbf{x}_0$, so the theorems apply.

THEOREM 5.1 *Consider $\frac{d}{dt}\mathbf{x} = \mathbf{f}(\mathbf{x}, t)$, with an equilibrium point at the origin, and*

1. *$\mathbf{f}(\mathbf{x}, t) - \mathbf{A}(t)\mathbf{x}$ converges uniformly to zero (i.e., that 5.94 holds).*
2. *$\mathbf{A}(t)$ is bounded.*
3. *The system $\frac{d}{dt}\mathbf{x} = \mathbf{A}(t)\mathbf{x}$ is exponentially stable.*

Then $\mathbf{x}_0$ is an exponentially stable equilibrium of $\frac{d}{dt}\mathbf{x} = \mathbf{f}(\mathbf{x}, t)$.

If the system does not depend explicitly on time (i.e., $\mathbf{f}(\mathbf{x}, t) = \mathbf{f}(\mathbf{x})$) it is said to be *autonomous.* For autonomous systems, conditions 1 and 2 are automatically true, and exponential stability of 3 is simply the condition that all of the eigenvalues of $\mathbf{A}$ have negative real parts.

THEOREM 5.2 *Suppose $\frac{d}{dt}\mathbf{x} = \mathbf{f}(\mathbf{x}, t)$ has an equilibrium point at $\mathbf{x}_0$ and $\mathbf{f}(\mathbf{x}, t)$ is continuously differentiable, further that the Jacobian $\mathbf{A}(t) = \mathbf{A}$ is constant, and that the uniform approximation condition Equation 5.94 holds. Then, the equilibrium is unstable if $\mathbf{A}$ has at least one eigenvalue with positive real part.*

Failures of Linearization

EXAMPLE 5.6:

This example shows what happens in the case that there is not uniform approximation (i.e., that condition 5.94 fails).

$$\frac{dx}{dt} = -x + tx^2 \tag{5.95}$$

The right hand side of this equation is clearly differentiable. The candidate for linearization is the first term of Equation 5.95, *it is not a linearization*—this system has no linearization. It is clear as t grows large the quadratic term will dominate the linear part. Thus, the system will grow rather than having zero as a stable equilibrium, as would be predicted by Theorem 5.1, if the uniform approximation condition were not necessary.

EXAMPLE 5.7:

This example has fixed eigenvalues with negative real parts; nonetheless it is unstable.

$$\frac{dx}{dt}\begin{bmatrix} x_1(t) \\ x_2(t) \end{bmatrix} = \begin{bmatrix} -1 & 0 \\ e^{at} & -2 \end{bmatrix}\begin{bmatrix} x_1(t) \\ x_2(t) \end{bmatrix},$$

$$\mathbf{x} = \begin{bmatrix} 1 \\ 1 \end{bmatrix} \tag{5.96}$$

The solution to this equation is: $x_1(t) = exp\{-t\}$ and $\dot{x}_2(t) = -2x_2(t) + \exp\{(a-1)t\}$. So that $x_2(t) = \exp\{-2t\} + \int_0^t \exp\{-2(t-s)\}\exp\{(a-1)s\}ds$, or $x_2(t) = \exp\{-2t\} + \exp\{-2t\}\int_0^t \exp\{(a+1)s\}ds$. Clearly for $a > 1$ the system is unstable. This example is simple. Other slightly more complex examples show that systems with bounded $A(t)$, can have fixed eigenvalues and still result in an unstable system (see Example 90, Chapter 5, Section 4 in [11]).

Example of Linearization

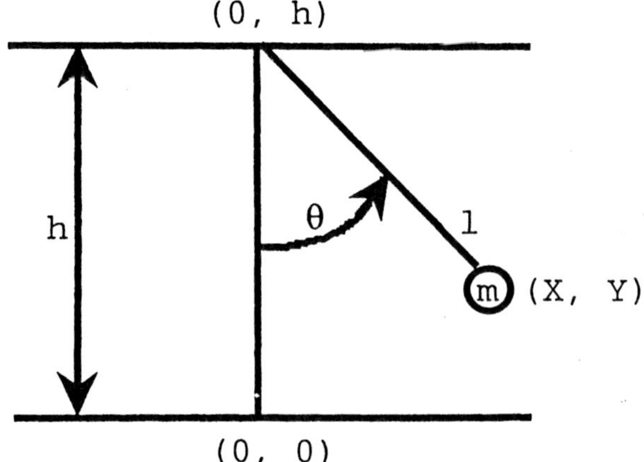

Figure 5.8 Simple pendulum.

Consider a simple pendulum and allow the length of the pendulum to vary (see Figure 5.8). Let (X, Y) be the position of the pendulum at time t. Then, $X = l\sin\theta$, $Y = h - l\cos\theta$. Further $\dot{X} = \dot{l}\sin\theta + l\dot{\theta}\cos\theta$, $\dot{Y} = -\dot{l}\cos\theta + l\dot{\theta}\sin\theta$. The kinetic energy (K.E.) is given by $(1/2)m(\dot{X}^2 + \dot{Y}^2)$, or where m is the mass at the end of the pendulum (and this is the only mass). The potential energy (P.E.) is: $mgY = mg(h - l\cos\theta)$, g being the gravitational constant (about 9.8 m/sec^2 at sea level). So, both g and l are nonnegative. The Lagrangian is given by $L(\theta, l, \dot{\theta}, \dot{l}) = L(\mathbf{x}, \dot{\mathbf{x}}) = K.E. - P.E.$; and the Euler–Lagrange equations:

$$0 = \frac{d}{dt}\left[\frac{\partial L}{\partial \dot{\mathbf{x}}}\right] - \left[\frac{\partial L}{\partial \mathbf{x}}\right], \tag{5.97}$$

give the motions of the system (See [1]).

Here

$$\left[\frac{\partial L}{\partial \dot{\mathbf{x}}}\right] = \begin{bmatrix} ml^2\dot{\theta} \\ m\dot{l} \end{bmatrix}, \tag{5.98}$$

so

$$\frac{d}{dt}\left[\frac{\partial L}{\partial \dot{\mathbf{x}}}\right] = \begin{bmatrix} 2ml\dot{l}\dot{\theta} + ml^2\ddot{\theta} \\ m\ddot{l} \end{bmatrix}, \tag{5.99}$$

and

$$\left[\frac{\partial L}{\partial \mathbf{x}}\right] = \begin{bmatrix} -mgl\sin\theta \\ m\dot{\theta}^2 l + mg\cos\theta \end{bmatrix}. \tag{5.100}$$

Combined, these yield:

$$\mathbf{0} = \begin{bmatrix} 2ml\dot{l}\dot{\theta} + ml^2\ddot{\theta} \\ m\ddot{l} \end{bmatrix} - \begin{bmatrix} -mgl\sin\theta \\ m\dot{\theta}^2 l + mg\cos\theta \end{bmatrix}. \tag{5.101}$$

After a bit of algebraic simplification:

$$0 = \ddot{\theta} + 2\frac{\dot{l}}{l}\dot{\theta} + \frac{g}{l}\sin\theta \tag{5.102}$$

$$0 = \ddot{l} - \dot{\theta}^2 l - g\cos\theta. \tag{5.103}$$

Remarks If the motion for l is prescribed, then the first equation of the Lagrangian is correct, but l and $\dot{l}$ are not degrees of freedom in the Lagrangian (i.e., $L = L(\theta, \dot{\theta})$). This happens because the re-derivation of the system yields the first equation of our system. Using the above derivation and setting l, $\dot{l}$ to their prescribed values *will not* result in the correct equations. Consider Case I below, if $l = l_0$, $\dot{l} = \ddot{l} = 0$ is substituted, then the second equation becomes $\dot{\theta}^2 l_0 + g\cos\theta = 0$. Taking $\frac{d}{dt}$ of this results in: $\dot{\theta}(\ddot{\theta} + \frac{g}{2l_0}\sin\theta) = 0$—which is inconsistent with the first equation: $\ddot{\theta} + \frac{g}{l_0}\sin\theta = 0$.

Case I Let $\dot{l} \equiv 0$, then $\ddot{l} \equiv 0$, and let l be a positive constant, equations of motion are (see the above remark):

$$0 = \ddot{\theta} + \frac{g}{l}\sin\theta \tag{5.104}$$

Writing Equation 5.104 in first order form, by letting $\mathbf{x} = \begin{bmatrix} \theta & \dot{\theta} \end{bmatrix}^T$, then:

$$\dot{\mathbf{x}} = \mathbf{f}(\mathbf{x}) = \begin{bmatrix} x_2 \\ \frac{g}{l}\sin x_1 \end{bmatrix}. \tag{5.105}$$

It is easy to check that $\mathbf{f}(\mathbf{0}) = \mathbf{0}$, and that $\mathbf{f}$, otherwise meets the linearization criterion, so that a linearization exists; it is given by:

$$\dot{\mathbf{x}} = \begin{bmatrix} 0 & 1 \\ -\frac{g}{l} & 0 \end{bmatrix}\mathbf{x}. \tag{5.106}$$

Case II Let $\mathbf{x} = [\theta, l, \dot{\theta}, \dot{l}]^T$, then Equations 5.102 and 5.103 are given by:

$$\dot{\mathbf{x}} = \begin{bmatrix} x_3 \\ x_4 \\ -2\frac{x_4}{x_2}x_3 - \frac{g}{x_2}\sin x_1 \\ x_3^2 x_2 + g\cos x_1 \end{bmatrix}. \tag{5.107}$$

Clearly, $\mathbf{0}$ is not an equilibrium point of this system. Physically, $l = 0$ doesn't make much sense anyway. Let us seek an equilibrium at $\mathbf{x}_0 = [0, l_0, 0, 0]$, which is a normal pendulum of length l_0, at rest. Note $\mathbf{f}(\mathbf{x}_0) = [0, 0, 0, g]^T$; hence open loop feedback of the form $[0, 0, 0, -g]^T$ will give the equilibrium at the desired point. Let $\mathbf{F}(\mathbf{x}) = \mathbf{f}(\mathbf{x}) - [0, 0, 0, g]^T$, then $\mathbf{F}(\mathbf{x}_0) = \mathbf{0}$, and it can be verified that $\dot{\mathbf{x}} = \mathbf{F}(\mathbf{x})$ meets all the linearization criteria at $\mathbf{x}_0$. Thus, $-\mathbf{f}(\mathbf{x}_0)$ is the open loop control that must be applied to the system. The linearized system is given by $\mathbf{z} = \mathbf{x} - \mathbf{x}_0$, $\dot{\mathbf{z}} = \left[\frac{\partial \mathbf{F}}{\partial \mathbf{x}}\right]|_{\mathbf{x}_0}\mathbf{z}$.

$$\left[\frac{\partial \mathbf{F}}{\partial \mathbf{x}}\right] = \begin{bmatrix} 0 & 0 & 1 & 0 \\ 0 & 0 & 0 & 1 \\ -\frac{g}{x_2}\cos x_1 & \frac{2x_3x_4 + g\sin x_1}{x_2^2} & -2\frac{x_4}{x_2} & -2\frac{x_3}{x_2} \\ -g\sin x_1 & x_3^2 & 2x_3x_2 & 0 \end{bmatrix} \tag{5.108}$$

$$\left[\frac{\partial \mathbf{F}}{\partial \mathbf{x}}\right]|_{\mathbf{x}_0} = \begin{bmatrix} 0 & 0 & 1 & 0 \\ 0 & 0 & 0 & 1 \\ -\frac{g}{l_0} & 0 & 0 & 0 \\ 0 & 0 & 0 & 0 \end{bmatrix} \tag{5.109}$$

To determine the eigenvalues, examine $det(sI - A)$; in this case expansion is easy around the last row and $det(sI - A) = s^2(s^2 + \frac{g}{l_0})$. Thus, the system is not exponentially stable.

References

[1] Arnold, V.I., *Mathematical Methods of Classical Mechanics,* Second Edition, Springer-Verlag, 1989.

[2] Balakrishnan, A.V., *Applied Functional Analysis,* Second Edition, Springer-Verlag, 1981.

[3] Bender C. and Orzag, S., *Nonlinear Systems Analysis,* Second Edition, McGraw-Hill, 1978.

[4] Chen, C.T., *Linar Systems Theory and Design,* Holt, Rinehart and Wiston, 1984.

[5] Dorf, R.C., *Modern Control Systems*, 3rd ed., Addison-Wesley, 1983.

[6] Fattorini, H., *The Cauchy Problem*, Addison-Wesley, 1983.

[7] Kailith, T., *Linear Systems Theory*, Prentice Hall, 1980.

[8] Kalman, R.E., Falb, P.E., and Arbib, M.A., *Topics in Mathematical Systems Theory*, McGraw-Hill, 1969.

[9] Press, W., et. al., *Numerical Recipes in C*, Cambridge, 1988.

[10] Reed, M. and Simon, B., *Functional Analysis*, Academic Press, 1980.

[11] Vidyasagar, M., *Nonlinear Systems Analysis*, Second ed., Prentice Hall, 1993.

[12] *Webster's Ninth New Collegiate Dictionary, First Digital Edition*, Merriam-Webster and NeXT Comuter, Inc., 1988.

[13] Wolovich, W., *Automatic Control Systems*, Harcourt Brace, 1994.

6

Graphical Models

Dean K. Frederick
Electrical, Computer, and Systems Engineering Department, Rensselaer Polytechnic Institute, Troy, NY

Charles M. Close
Electrical, Computer, and Systems Engineering Department, Rensselaer Polytechnic Institute, Troy, NY

Norman S. Nise
California State Polytechnic University, Pomona

6.1 Block Diagrams

Dean K. Frederick, Electrical, Computer, and Systems Engineering Department, Rensselaer Polytechnic Institute, Troy, NY
Charles M. Close, Electrical, Computer, and Systems Engineering Department, Rensselaer Polytechnic Institute, Troy, NY

6.1.1 Introduction

A block diagram is an interconnection of symbols representing certain basic mathematical operations in such a way that the overall diagram obeys the system's mathematical model. In the diagram, the lines interconnecting the blocks represent the variables describing the system behavior, such as the input and state variables. Inspecting a block diagram of a system may provide new insight into the system's structure and behavior beyond that available from the differential equations themselves.

Throughout most of this chapter we restrict the discussion to fixed linear systems that contain no initial stored energy. After we transform the equations describing such a system, the variables that we use are the Laplace transforms of the corresponding functions of time. The parts of the system can then be described by their transfer functions. Recall that transfer functions give only the zero-state response. However, the steady-state response of a stable system does not depend on the initial conditions, so in that case there is no loss of generality in using only the zero-state response.

After defining the components to be used in our diagrams, we develop rules for simplifying block diagrams, emphasizing those diagrams that represent feedback systems. The chapter concludes by pointing out that graphical models can be used for more general systems than those considered here, including nonlinear blocks, multi-input/multi-output (MIMO) systems, and discrete-time systems.

Computer programs for the analysis and design of control systems exist that allow the entry of block diagram models in graphical form. These programs are described in another section of this book.

6.1.2 Diagram Blocks

The operations that we generally use in block diagrams are summation, gain, and multiplication by a transfer function. Unless otherwise stated, all variables are Laplace-transformed quantities.

Summer

The addition and subtraction of variables is represented by a *summer*, or *summing junction*. A summer is represented by a circle that has any number of arrows directed toward it (denoting inputs) and a single arrow directed away from it (denoting the output). Next to each entering arrowhead is a plus or minus symbol indicating the sign associated with the variable that the particular arrow represents. The output variable, appearing as the one arrow leaving the circle, is defined to be the sum of all the incoming variables, with the associated signs taken into account. A summer having three inputs $X_1(s)$, $X_2(s)$, and $X_3(s)$ appears in Figure 6.1.

0-8493-8570-9/96/$0.00+$.50

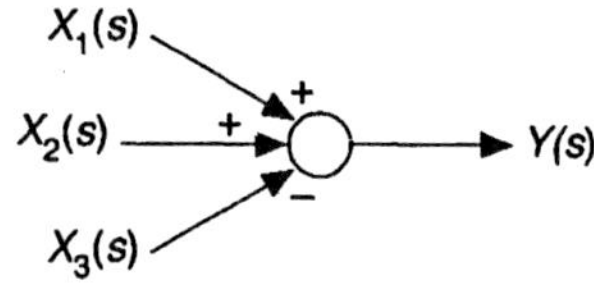

Figure 6.1 Summer representing $Y(s) = X_1(s) + X_2(s) - X_3(s)$.

Gain

The multiplication of a single variable by a constant is represented by a *gain* block. We place no restriction on the value of the gain, which may be positive or negative. It may be an algebraic function of other constants and/or system parameters. Several self-explanatory examples are shown in Figure 6.2.

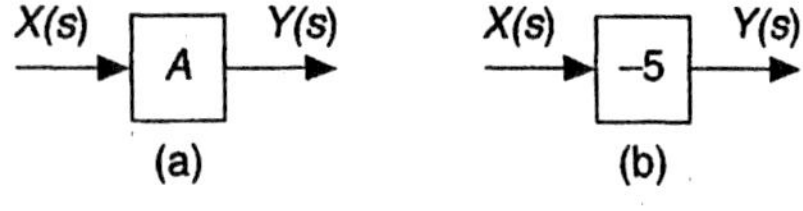

X(s) $\frac{K}{M}$ Y(s)

(c)

Figure 6.2 Gains. (a) $Y(s) = AX(s)$; (b) $Y(s) = -5X(s)$; (c) $Y(s) = (K/M)X(s)$.

Transfer Function

For a fixed linear system with no initial stored energy, the transformed output $Y(s)$ is given by

$$Y(s) = H(s)U(s)$$

where $H(s)$ is the transfer function and $U(s)$ is the transformed input. When dealing with parts of a larger system, we often use $F(s)$ and $X(s)$ for the transfer function and transformed input, respectively, of an individual part. Then

$$Y(s) = F(s)X(s) \tag{6.1}$$

Any system or combination of elements can be represented by a block containing its transfer function $F(s)$, as indicated in Figure 6.3(a). For example, the first-order system that obeys the input-output equation

$$\dot{y} + \frac{1}{\tau}y = Ax(t)$$

has as its transfer function

$$F(s) = \frac{A}{s + \frac{1}{\tau}}$$

Thus, it could be represented by the block diagram shown in Figure 6.3(b). Note that the gain block in Figure 6.2(a) can be considered as a special case of a transfer function block, with $F(s) = A$.

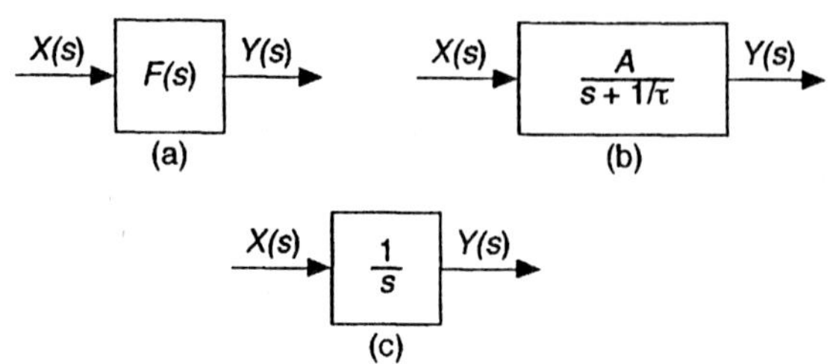

Figure 6.3 Basic block diagrams. (a) Arbitrary transfer function; (b) first-order system; (c) integrator.

Integrator

Another important special case of a general transfer function block, one that appears frequently in our diagrams, is the *integrator* block. An integrator that has an input $x(t)$ and an output $y(t)$ obeys the relationship

$$y(t) = y(0) + \int_0^t x(\lambda)d\lambda$$

where λ is the dummy variable of integration. Setting $y(0)$ equal to 0 and transforming the equation give

$$Y(s) = \frac{1}{s}X(s)$$

Hence, the transfer function of the integrator is $Y(s)/X(s) = 1/s$, as shown in Figure 6.3(c).

Because a block diagram is merely a pictorial representation of a set of algebraic Laplace-transformed equations, it is possible to combine blocks by calculating equivalent transfer functions and thereby to simplify the diagram. We now present procedures for handling series and parallel combinations of blocks. Methods for simplifying diagrams containing feedback paths are discussed in the next section.

Series Combination

Two blocks are said to be in *series* when the output of one goes only to the input of the other, as shown in Figure 6.4(a). The transfer functions of the individual blocks in the figure are $F_1(s) = V(s)/X(s)$ and $F_2(s) = Y(s)/V(s)$.

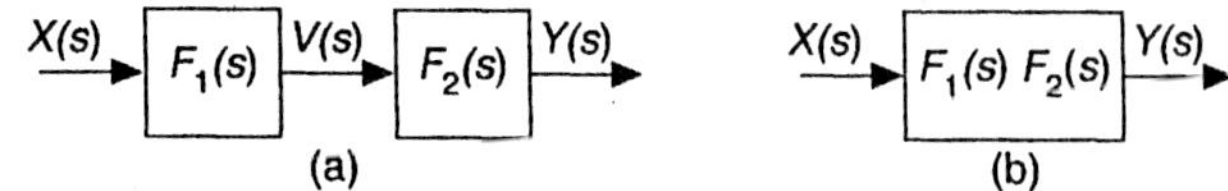

Figure 6.4 (a) Two blocks in series; (b) equivalent diagram.

When we evaluate the individual transfer functions, it is essential that we take any *loading effects* into account. This means that $F_1(s)$ is the ratio $V(s)/X(s)$ when the two subsystems are connected, so any effect the second subsystem has on the first is accounted for in the mathematical model. The same statement holds for calculating $F_2(s)$. For example, the input-output relationship for a linear potentiometer loaded by a resistor connected from its wiper to the ground node differs from that of the unloaded potentiometer.

In Figure 6.4(a), $Y(s) = F_2(s)V(s)$ and $V(s) = F_1(s)X(s)$. It follows that

$$\begin{aligned} Y(s) &= F_2(s)[F_1(s)X(s)] \\ &= [F_1(s)F_2(s)]X(s) \end{aligned}$$

Thus, the transfer function relating the input transform $X(s)$ to the output transform $Y(s)$ is $F_1(s)F_2(s)$, the product of the individual transfer functions. The equivalent block diagram is shown in Figure 6.4(b).

Parallel Combination

Two systems are said to be in *parallel* when they have a common input and their outputs are combined by a summing junction. If, as indicated in Figure 6.5(a), the individual blocks have the transfer functions $F_1(s)$ and $F_2(s)$ and the signs at the summing junction are both positive, the overall transfer function $Y(s)/X(s)$ is the sum $F_1(s) + F_2(s)$, as shown in Figure 6.5(b). To prove this statement, we note that

$$Y(s) = V_1(s) + V_2(s)$$

where $V_1(s) = F_1(s)X(s)$ and $V_2(s) = F_2(s)X(s)$. Substituting for $V_1(s)$ and $V_2(s)$, we have

$$Y(s) = [F_1(s) + F_2(s)]X(s)$$

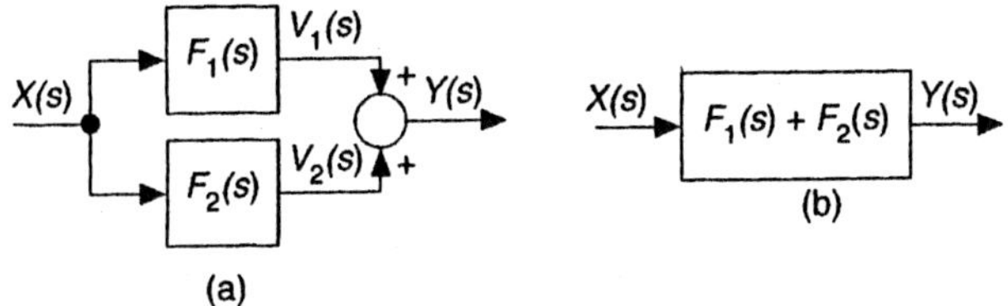

Figure 6.5 (a) Two blocks in parallel; (b) equivalent diagram.

If either of the summing-junction signs associated with $V_1(s)$ or $V_2(s)$ is negative, we must change the sign of the corresponding transfer function in forming the overall transfer function. The following example illustrates the rules for combining blocks that are in parallel or in series.

EXAMPLE 6.1:

Evaluate the transfer functions $Y(s)/U(s)$ and $Z(s)/U(s)$ for the block diagram shown in Figure 6.6, giving the results as rational functions of s.

Solution Because $Z(s)$ can be viewed as the sum of the outputs of two parallel blocks, one of which has $Y(s)$ as its output, we first evaluate the transfer function $Y(s)/U(s)$. To do this, we observe that $Y(s)$ can be considered the output of a series combination of two parts, one of which is a parallel combination of two blocks. Starting with this parallel combination, we write

$$\frac{2s+1}{s+4} + \frac{s-2}{s+3} = \frac{3s^2+9s-5}{s^2+7s+12}$$

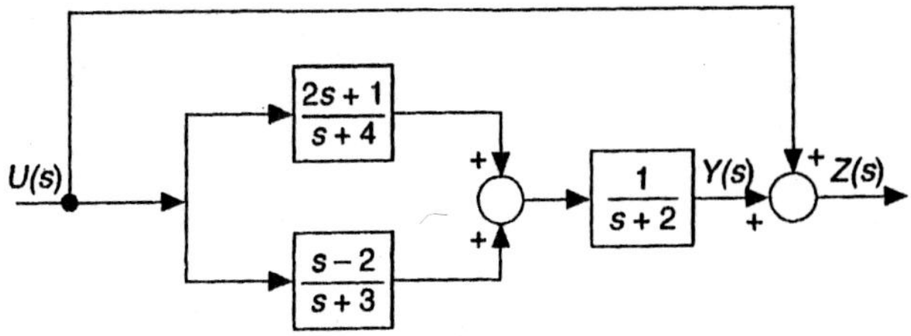

Figure 6.6 Block diagram for Example 6.1.

and redraw the block diagram as shown in Figure 6.7(a). The series combination in this version has the transfer function

$$\begin{aligned} \frac{Y(s)}{U(s)} &= \frac{3s^2+9s-5}{s^2+7s+12} \cdot \frac{1}{s+2} \\ &= \frac{3s^2+9s-5}{s^3+9s^2+26s+24} \end{aligned}$$

which leads to the diagram shown in Figure 6.7(b). We can reduce the final parallel combination to the single block shown in Figure 6.7(c) by writing

$$\begin{aligned} \frac{Z(s)}{U(s)} &= 1 + \frac{Y(s)}{U(s)} \\ &= 1 + \frac{3s^2+9s-5}{s^3+9s^2+26s+24} \\ &= \frac{s^3+12s^2+35s+19}{s^3+9s^2+26s+24} \end{aligned}$$

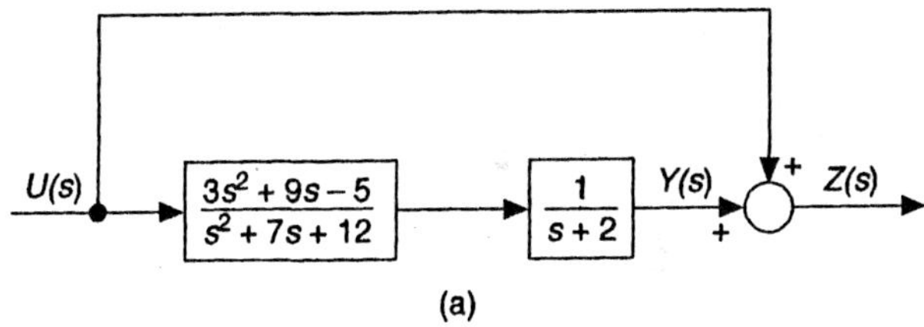

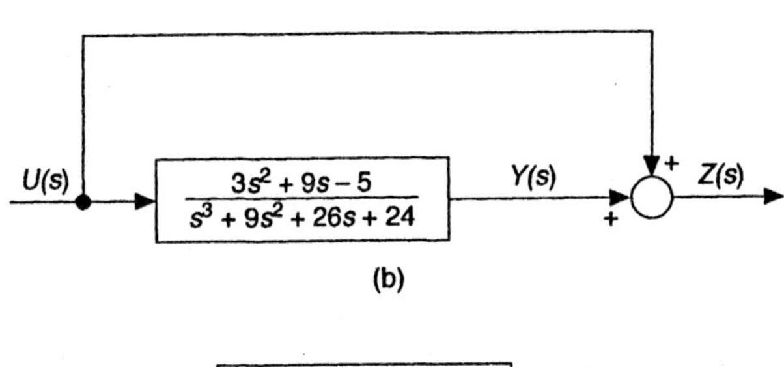

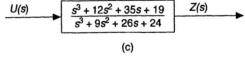

Figure 6.7 Equivalent block diagrams for the diagram shown in Figure 6.6.

In general, it is desirable to reduce the transfer functions of combinations of blocks to rational functions of s in order to simplify the subsequent analysis. This will be particularly important in the following section when we are reducing feedback loops to obtain an overall transfer function.

6.1.3 Block Diagrams of Feedback Systems

Figure 6.8(a) shows the block diagram of a general feedback system that has a forward path from the summing junction to the output and a feedback path from the output back to the summing junction. The transforms of the system's input and output are $U(s)$ and $Y(s)$, respectively. The transfer function $G(s) = Y(s)/V(s)$ is known as the *forward transfer function*, and $H(s) = Z(s)/Y(s)$ is called the *feedback transfer function*. We must evaluate both of these transfer functions with the system elements connected in order properly to account for the loading effects of the interconnections. The product $G(s)H(s)$ is referred to as the *open-loop transfer function*. The sign associated with the feedback signal from the block $H(s)$ at the summing junction is shown as minus because a minus sign naturally occurs in the majority of feedback systems, particularly in control systems.

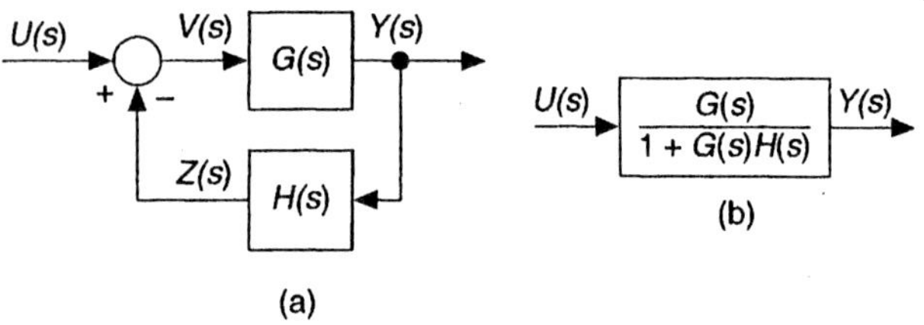

Figure 6.8 (a) Block diagram of a feedback system; (b) equivalent diagram.

Given the model of a feedback system in terms of its forward and feedback transfer functions $G(s)$ and $H(s)$, it is often necessary to determine the *closed-loop transfer function* $T(s) = Y(s)/U(s)$. We do this by writing the algebraic transform equations corresponding to the block diagram shown in Figure 6.8(a) and solving them for the ratio $Y(s)/U(s)$. We can write the following transform equations directly from the block diagram.

$$\begin{aligned} V(s) &= U(s) - Z(s) \\ Y(s) &= G(s)V(s) \\ Z(s) &= H(s)Y(s) \end{aligned}$$

If we combine these equations in such a way as to eliminate $V(s)$ and $Z(s)$, we find that

$$Y(s) = G(s)[U(s) - H(s)Y(s)]$$

which can be rearranged to give

$$[1 + G(s)H(s)]Y(s) = G(s)U(s)$$

Hence, the closed-loop transfer function $T(s) = Y(s)/U(s)$ is

$$T(s) = \frac{G(s)}{1 + G(s)H(s)} \tag{6.2}$$

where it is implicit that the sign of the feedback signal at the summing junction is negative. It is readily shown that when a plus sign is used at the summing junction for the feedback signal, the closed-loop transfer function becomes

$$T(s) = \frac{G(s)}{1 - G(s)H(s)} \tag{6.3}$$

A commonly used simplification occurs when the feedback transfer function is unity, that is, when $H(s) = 1$. Such a system is referred to as a *unity-feedback system*, and Equation 6.2 reduces to

$$T(s) = \frac{G(s)}{1 + G(s)} \tag{6.4}$$

We now consider three examples that use Equations 6.2 and 6.3. The first two illustrate determining the closed-loop transfer function by reducing the block diagram. They also show the effects of feedback gains on the closed-loop poles, time constant, damping ratio, and undamped natural frequency. In Example 6.4, a block diagram is drawn directly from the system's state-variable equations and then reduced to give the system's transfer functions.

EXAMPLE 6.2:

Find the closed-loop transfer function for the feedback system shown in Figure 6.9(a), and compare the locations of the poles of the open-loop and closed-loop transfer functions in the s-plane.

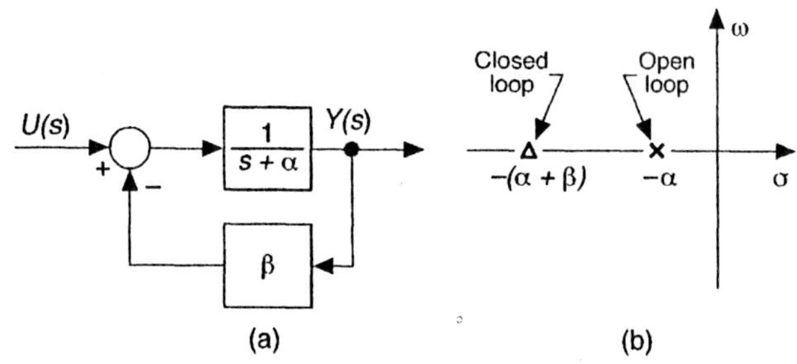

Figure 6.9 Single-loop feedback system for Example 6.2.

Solution By comparing the block diagram shown in Figure 6.9(a) with that shown in Figure 6.8(a), we see that $G(s) = 1/(s + \alpha)$ and $H(s) = \beta$. Substituting these expressions into Equation 6.2 gives

$$T(s) = \frac{\dfrac{1}{s+\alpha}}{1+\left(\dfrac{1}{s+\alpha}\right)\beta}$$

which we can write as a rational function of s by multiplying the numerator and denominator by $s + \alpha$. Doing this, we obtain the closed-loop transfer function

$$T(s) = \frac{1}{s+\alpha+\beta}$$

This result illustrates an interesting and useful property of feedback systems: the fact that the poles of the closed-loop transfer function differ from the poles of the open-loop transfer function $G(s)H(s)$. In this case, the single open-loop pole is at $s = -\alpha$, whereas the single closed-loop pole is at $s = -(\alpha + \beta)$. These pole locations are indicated in Figure 6.9(b) for positive α and β. Hence, in the absence of feedback, the pole of the transfer function $Y(s)/U(s)$ is at $s = -\alpha$, and the free response is of the form $\epsilon^{-\alpha t}$. With feedback, however, the free response is $\epsilon^{-(\alpha+\beta)t}$. Thus, the time constant of the open-loop system is $1/\alpha$, whereas that of the closed-loop system is $1/(\alpha + \beta)$.

EXAMPLE 6.3:

Find the closed-loop transfer function of the two-loop feedback system shown in Figure 6.10. Also express the damping ratio ζ and the undamped natural frequency ω_n of the closed-loop system in terms of the gains a_0 and a_1.

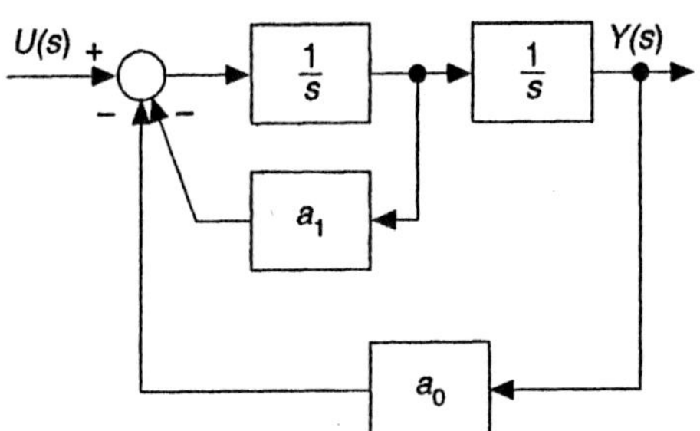

Figure 6.10 System with two feedback loops for Example 6.3.

Solution Because the system's block diagram contains one feedback path inside another, we cannot use Equation 6.2 directly to evaluate $Y(s)/U(s)$. However, we can redraw the block diagram such that the summing junction is split into two summing junctions, as shown in Figure 6.11(a). Then it is possible to use Equation 6.2 to eliminate the inner loop by calculating the transfer function $W(s)/V(s)$. Taking $G(s) = 1/s$ and $H(s) = a_1$ in Equation 6.2, we obtain

$$\frac{W(s)}{V(s)} = \frac{\frac{1}{s}}{1+\frac{a_1}{s}} = \frac{1}{s+a_1}$$

Redrawing Figure 6.11(a) with the inner loop replaced by a block having $1/(s + a_1)$ as its transfer function gives Figure 6.11(b). The two blocks in the forward path of this version are in series and can be combined by multiplying their transfer functions, which gives the block diagram shown in Figure 6.11(c). Then we can apply Equation 6.2 again to find the overall closed-loop transfer function $T(s) = Y(s)/U(s)$ as

$$T(s) = \frac{\frac{1}{s(s+a_1)}}{1 + \frac{1}{s(s+a_1)} \cdot a_0} = \frac{1}{s^2 + a_1 s + a_0} \tag{6.5}$$

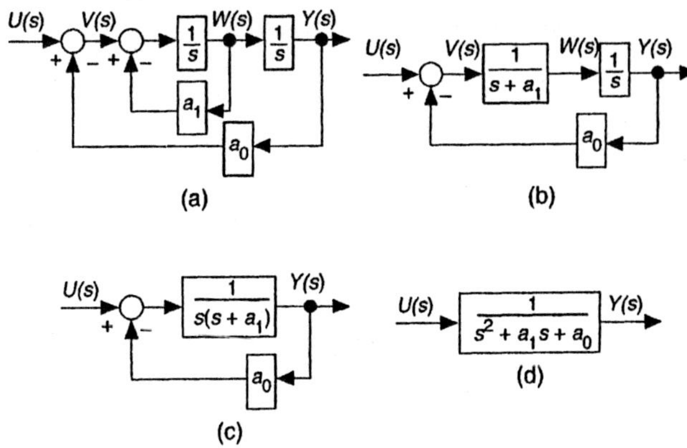

Figure 6.11 Equivalent block diagrams for the system shown in Figure 6.10.

The block diagram representation of the feedback system corresponding to Equation 6.5 is shown in Figure 6.11(d).

The poles of the closed-loop transfer function are the roots of the equation

$$s^2 + a_1 s + a_0 = 0 \tag{6.6}$$

which we obtain by setting the denominator of $T(s)$ equal to zero and which is the characteristic equation of the closed-loop system. Equation 6.6 has two roots, which may be real or complex, depending on the sign of the quantity $a_1^2 - 4a_0$. However, the roots of Equation 6.6 will have negative real parts and the closed-loop system will be stable provided that a_0 and a_1 are both positive.

If the poles are complex, it is convenient to rewrite the denominator of $T(s)$ in terms of the damping ratio ζ and the undamped natural frequency ω_n. By comparing the left side of Equation 6.6 with the characteristic polynomial of a second-order system written as $s^2 + 2\zeta\omega_n s + \omega_n^2$, we see that

$$a_0 = \omega_n^2 \text{ and } a_1 = 2\zeta\omega_n$$

Solving the first of these equations for ω_n and substituting it into the second gives the damping ratio and the undamped natural frequency of the closed-loop system as

$$\zeta = \frac{a_1}{2\sqrt{a_0}} \text{ and } \omega_n = \sqrt{a_0}$$

We see from these expressions that a_0, the gain of the outer feedback path in Figure 6.10, determines the undamped natural frequency ω_n and that a_1, the gain of the inner feedback path, affects only the damping ratio. If we can specify both a_0 and a_1 at will, then we can attain any desired values of ζ and ω_n for the closed-loop transfer function.

EXAMPLE 6.4:

Draw a block diagram for the translational mechanical system shown in Figure 6.12, whose state-variable equations can be written as

$$\dot{x}_1 = v_1$$

$$\dot{v}_1 = \frac{1}{M}[-K_1x_1 - B_1v_1 + K_1x_2 + f_a(t)]$$
$$\dot{x}_2 = \frac{1}{B_2}[K_1x_1 - (K_1 + K_2)x_2]$$

Reduce the block diagram to determine the transfer functions $X_1(s)/F_a(s)$ and $X_2(s)/F_a(s)$ as rational functions of s.

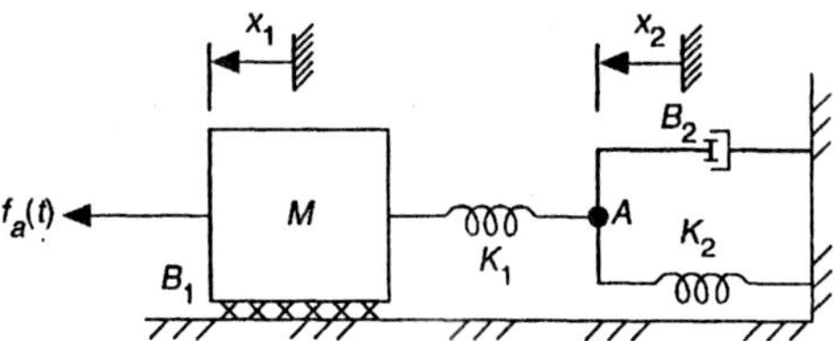

Figure 6.12 Translational system for Example 6.4.

Solution Transforming the three differential equations with zero initial conditions, we have

$$sX_1(s) = V_1(s)$$
$$MsV_1(s) = -K_1X_1(s) - B_1V_1(s) + K_1X_2(s) + F_a(s)$$
$$B_2sX_2(s) = K_1X_1(s) - (K_1 + K_2)X_2(s)$$

We use the second of these equations to draw a summing junction that has $MsV_1(s)$ as its output. After the summing junction, we insert the transfer function $1/Ms$ to get $V_1(s)$, which, from the first equation, equals $sX_1(s)$. Thus, an integrator whose input is $V_1(s)$ has $X_1(s)$ as its output. Using the third equation, we form a second summing junction that has $B_2sX_2(s)$ as its output. Following this summing junction by the transfer function $1/B_2s$, we get $X_2(s)$ and can complete the four feedback paths required by the summing junctions. The result of these steps is the block diagram shown in Figure 6.13(a).

To simplify the block diagram, we use Equation 6.2 to reduce each of the three inner feedback loops, obtaining the version shown in Figure 6.13(b). To evaluate the transfer function $X_1(s)/F_a(s)$, we can apply Equation 6.3 to this single-loop diagram because the sign associated with the feedback signal at the summing junction is positive rather than negative. Doing this with

$$G(s) = \frac{1}{Ms^2 + B_1s + K_1}$$

and

$$H(s) = \frac{K_1^2}{B_2s + K_1 + K_2}$$

we find

$$\frac{X_1(s)}{F_a(s)} = \frac{\dfrac{1}{Ms^2 + B_1s + K_1}}{1 - \dfrac{1}{Ms^2 + B_1s + K_1} \cdot \dfrac{K_1^2}{B_2s + K_1 + K_2}}$$
$$= \frac{B_2s + K_1 + K_2}{(Ms^2 + B_1s + K_1)(B_2s + K_1 + K_2) - K_1^2}$$
$$= \frac{B_2s + K_1 + K_2}{P(s)} \tag{6.7}$$

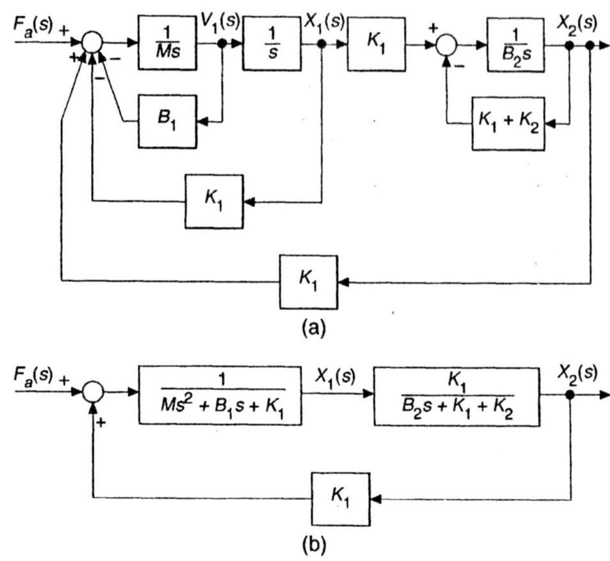

Figure 6.13 Block diagrams for the system in Example 6.4. (a) As drawn from the differential equations; (b) with the three inner feedback loops eliminated.

where

$$P(s) = MB_2s^3 + [(K_1 + K_2)M + B_1B_2]s^2 + [B_1(K_1 + K_2) + B_2K_1]s + K_1K_2$$

To obtain $X_2(s)/F_a(s)$, we can write

$$\frac{X_2(s)}{F_a(s)} = \frac{X_1(s)}{F_a(s)} \cdot \frac{X_2(s)}{X_1(s)}$$

where $X_1(s)/F_a(s)$ is given by Equation 6.7 and, from Figure 6.13(b),

$$\frac{X_2(s)}{X_1(s)} = \frac{K_1}{B_2s + K_1 + K_2} \tag{6.8}$$

The result of multiplying Equations 6.7 and 6.8 is a transfer function with the same denominator as Equation 6.7 but with a numerator of K_1.

In the previous examples, we used the rules for combining blocks that are in series or in parallel, as shown in Figures 6.4 and 6.5. We also repeatedly used the rule for simplifying the basic feedback configuration given in Figure 6.8(a). A number of other operations can be derived to help simplify block diagrams. To conclude this section, we present and illustrate two of these additional operations.

Keep in mind that a block diagram is just a means of representing the algebraic Laplace-transformed equations that describe a system. Simplifying or reducing the diagram is equivalent to manipulating the equations. In order to prove that a particular operation on the block diagram is valid, we need only show that the relationships among the transformed variables of interest are left unchanged.

Moving a Pick-Off Point

A *pick-off point* is a point where an incoming variable in the diagram is directed into more than one block. In the partial

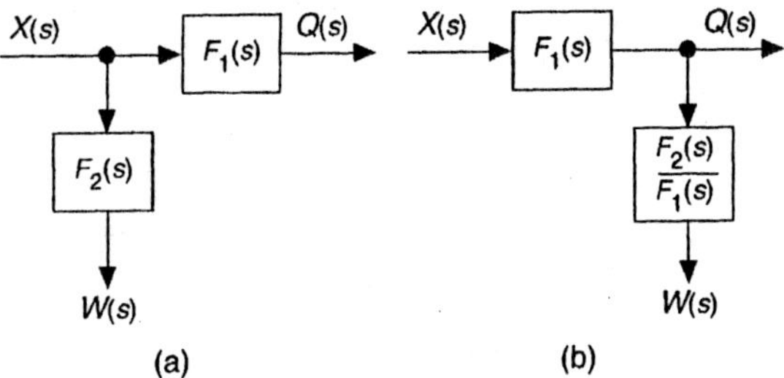

Figure 6.14 Moving a pick-off point.

diagram of Figure 6.14(a), the incoming signal $X(s)$ is used not only to provide the output $Q(s)$ but also to form the signal $W(s)$, which in practice might be fed back to a summer that appears earlier in the complete diagram. The pick-off point can be moved to the right of $F_1(s)$ if the transfer function of the block leading to $W(s)$ is modified as shown in Figure 6.14(b). Both parts of the figure give the same equations:

$$Q(s) = F_1(s)X(s)$$
$$W(s) = F_2(s)X(s)$$

EXAMPLE 6.5:

Use Figure 6.14 to find the closed-loop transfer function for the system shown in Figure 6.10.

Solution The pick-off point leading to the gain block a_1 can be moved to the output $Y(s)$ by replacing a_1 by $a_1 s$, as shown in Figure 6.15(a). Then the two integrator blocks, which are now in series, can be combined to give the transfer function $G(s) = 1/s^2$. The two feedback blocks are now in parallel and can be combined into the single transfer function $a_1 s + a_0$, as shown in Figure 6.15(b). Finally, by Equation 6.2,

$$T(s) = \frac{Y(s)}{U(s)} = \frac{1/s^2}{1+(a_1 s+a_0)/s^2} = \frac{1}{s^2+a_1 s+a_0}$$

which agrees with Equation 6.5, as found in Example 6.3.

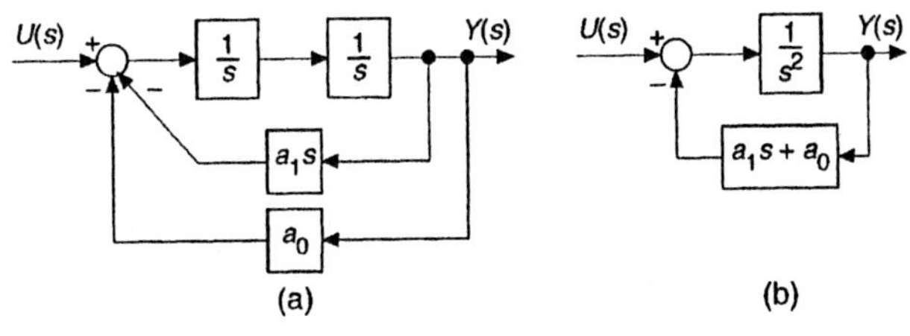

Figure 6.15 Equivalent block diagrams for the system shown in Figure 6.10.

Moving a Summing Junction

Suppose that, in the partial diagram of Figure 6.16(a), we wish to move the summing junction to the left of the block that has the transfer function $F_2(s)$. We can do this by modifying the transfer function of the block whose input is $X_1(s)$, as shown in part (b) of the figure. For each part of the figure,

$$Q(s) = F_1(s)X_1(s) + F_2(s)X_2(s)$$

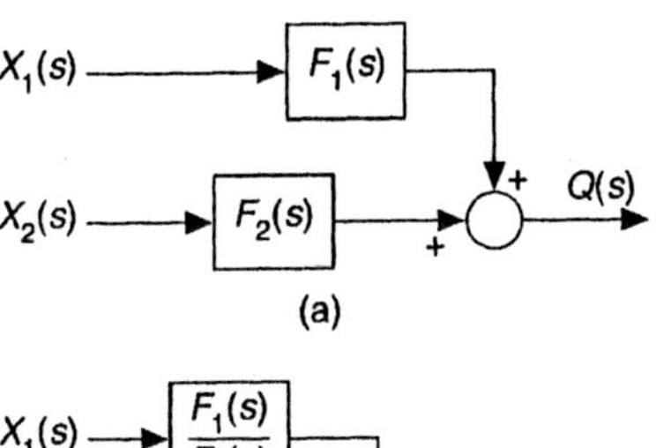

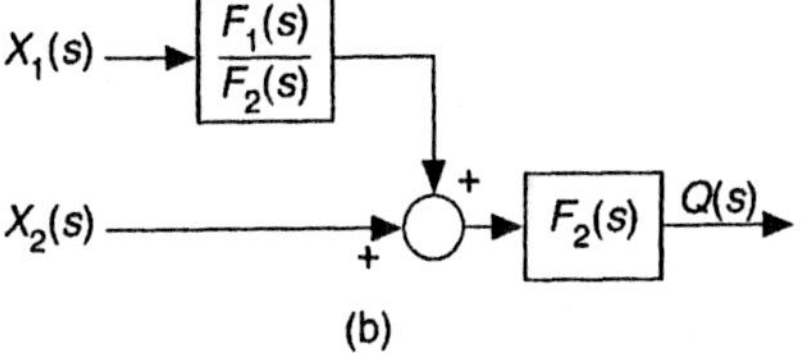

Figure 6.16 Moving a summing junction.

EXAMPLE 6.6:

Find the closed-loop transfer function $T(s) = Y(s)/U(s)$ for the feedback system shown in Figure 6.17(a).

Solution We cannot immediately apply Equation 6.2 to the inner feedback loop consisting of the first integrator and the gain block a_1 because the output of block b_1 enters a summer within that loop. We therefore use Figure 6.16 to move this summer to the left of the first integrator block, where it can be combined with the first summer. The resulting diagram is given in Figure 6.17(b).

Now Equation 6.2 can be applied to the inner feedback loop to give the transfer function

$$G_1(s) = \frac{1/s}{1+a_1/s} = \frac{1}{s+a_1}$$

The equivalent block with the transfer function $G_1(s)$ is then in series with the remaining integrator, which results in a combined transfer function of $1/[s(s+a_1)]$. Also, the two blocks with gains of sb_1 and 1 are in parallel and can be combined into a single block. These simplifications are shown in Figure 6.17(c).

We can now repeat the procedure and move the right summer to the left of the block labeled $1/[s(s+a_1)]$, where it can again be combined with the first summer. This is done in part (d) of the figure. The two blocks in parallel at the left can now be combined by adding their transfer functions, and Equation 6.2 can be applied to the right part of the diagram to give

$$\frac{\dfrac{1}{s(s+a_1)}}{1+\dfrac{a_0}{s(s+a_1)}} = \frac{1}{s^2+a_1 s+a_0}$$

These steps yield Figure 6.17(e), from which we see that

$$T(s) = \frac{b_2 s^2 + (a_1 b_2 + b_1)s + 1}{s^2 + a_1 s + a_0} \tag{6.9}$$

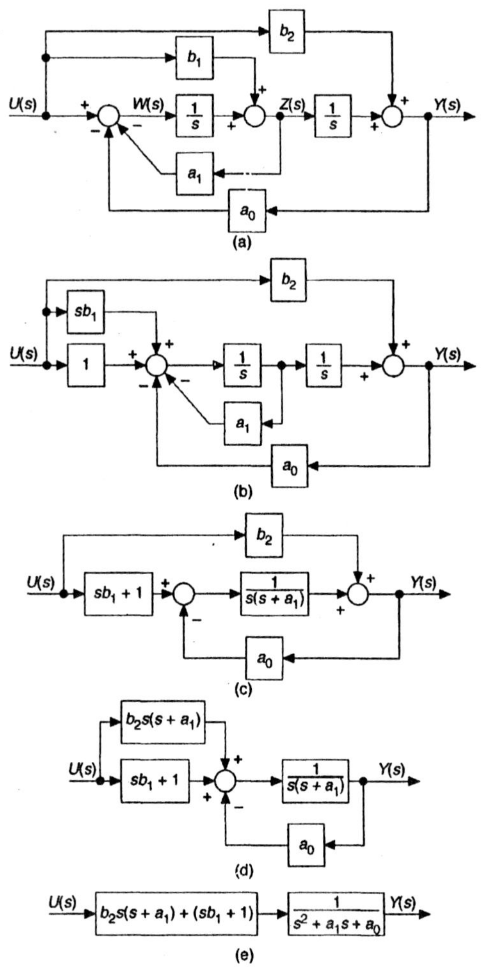

Figure 6.17 (a) Block diagram for Example 6.6; (b)–(e) equivalent block diagrams.

Because performing operations on a given block diagram is equivalent to manipulating the algebraic equations that describe the system, it may sometimes be easier to work with the equations themselves. As an alternative solution to the last example, suppose that we start by writing the equations for each of the three summers in Figure 6.17(a):

$$\begin{aligned} W(s) &= U(s) - a_0 Y(s) - a_1 Z(s) \\ Z(s) &= \frac{1}{s} W(s) + b_1 U(s) \\ Y(s) &= \frac{1}{s} Z(s) + b_2 U(s) \end{aligned}$$

Substituting the expression for $W(s)$ into the second equation, we see that

$$Z(s) = \tfrac{1}{s}[U(s) - a_0 Y(s) - a_1 Z(s)] + b_1 U(s)$$

from which

$$Z(s) = \frac{1}{s + a_1}[-a_0 Y(s) + (b_1 s + 1) U(s)] \qquad (6.10)$$

Substituting Equation 6.10 into the expression for $Y(s)$ gives

$$Y(s) = \tfrac{1}{s(s+a_1)}[-a_0 Y(s) + (b_1 s + 1) U(s)] + b_2 U(s)$$

Rearranging this equation, we find that

$$\frac{Y(s)}{U(s)} = \frac{b_2 s^2 + (a_1 b_2 + b_1) s + 1}{s^2 + a_1 s + a_0}$$

which agrees with Equation 6.9, as found in Example 6.6.

6.1.4 Summary

Block diagrams are an important way of representing the structure and properties of fixed linear systems. We start the construction of a block diagram by transforming the system equations, assuming zero initial conditions.

The blocks used in diagrams for feedback systems may contain transfer functions of any degree of complexity. We developed a number of rules, including those for series and parallel combinations and for the basic feedback configuration, for simplifying block diagrams.

6.1.5 Block Diagrams for Other Types of Systems

So far, we have considered block diagrams of linear, continuous-time models having one input and one output. Because of this restriction, we can use the system's block diagram to develop an overall transfer function in terms of the complex Laplace variable s. For linear discrete-time models we can draw block diagrams and do corresponding manipulations to obtain the closed-loop transfer function in terms of the complex variable z, as used in the z–transform. We can also construct block diagrams for MIMO systems, but we must be careful to obey the rules of matrix algebra when manipulating the diagrams to obtain equivalent transfer functions.

System models that contain nonlinearities, such as backlash, saturation, or dead band, can also be represented in graphical form. Typically, one uses transfer-function blocks for the linear portion of the system and includes one or more special blocks to represent the specific nonlinear operations. While such diagrams are useful for representing the system and for preparing computer simulations of the nonlinear model, they cannot be analyzed in the ways that we have done in the previous two sections. Similar comments can be made for time-varying elements.

Acknowledgments

This chapter is closely based on Sections 13.1 and 13.5 of *Modeling and Analysis of Dynamic Systems, Second Edition*, 1993, by Charles M. Close and Dean K. Frederick. It is used with permission of the publisher, Houghton Mifflin Company.

6.2 Signal-Flow Graphs[1]

Norman S. Nise, California State Polytechnic University, Pomona

6.2.1 Introduction

Signal-flow graphs are an alternate system representation. Unlike **block diagrams** of **linear systems,** which consist of blocks, signals, summing junctions, and pickoff points, signal-flow graphs of linear systems consist of only **branches,** which represent systems, and **nodes,** which represent signals. These elements are shown in Figure 6.18a and Figure 6.18b respectively. A system (Figure 6.18a) is represented by a line with an arrow showing the

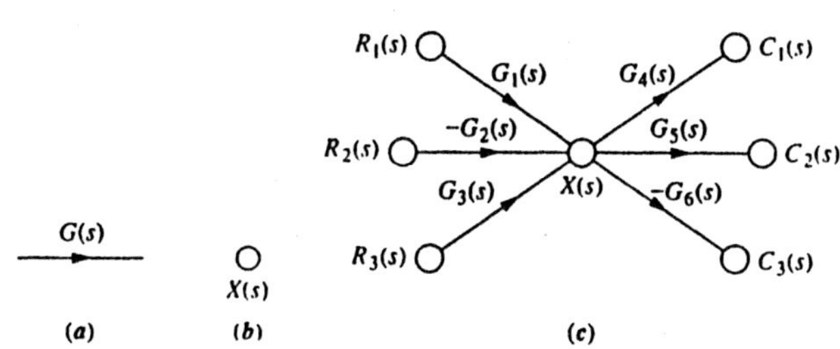

Figure 6.18 Signal-flow graph component parts: (a) system; (b) signal; (c) interconnection of systems and signals.

direction of signal flow through the system. Adjacent to the line we write the **transfer function.** A signal (Figure 6.18b) is a node with the signal name written adjacent to the node.

Figure 6.18c shows the interconnection of the systems and the signals. Each signal is the sum of signals flowing into it. For example, in Figure 6.18c the signal $X(s) = R_1(s)G_1(s) - R_2(s)G_2(s)+R_3(s)G_3(s)$. The signal, $C_3(s) = -X(s)G_6(s) = -R_1(s)G_1(s)G_6(s) + R_2(s)G_2(s)G_6(s) - R_3(s)G_3(s)G_6(s)$.

Notice that the summing of negative signals is handled by associating the negative sign with the system and not with a summing junction as in the case of block diagrams.

6.2.2 Relationship Between Block Diagrams and Signal-Flow Graphs

To show the parallel between block diagrams and signal-flow graphs, we convert some block diagram forms to signal-flow graphs. In each case, we first convert the signals to nodes and then interconnect the nodes with systems.

Let us convert the **cascaded, parallel,** and **feedback** forms of the block diagrams shown in Figures 6.19 through 6.21, respectively, into signal-flow graphs.

In each case, we start by drawing the signal nodes for that

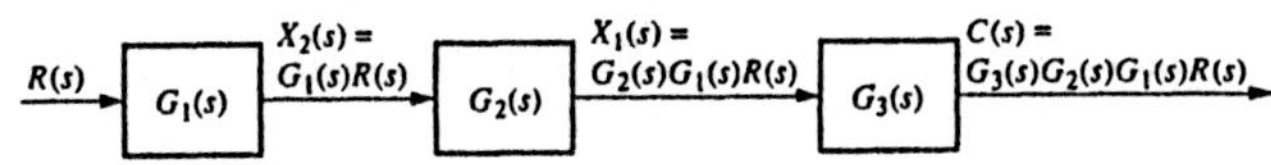

Figure 6.19 Cascaded subsystems.

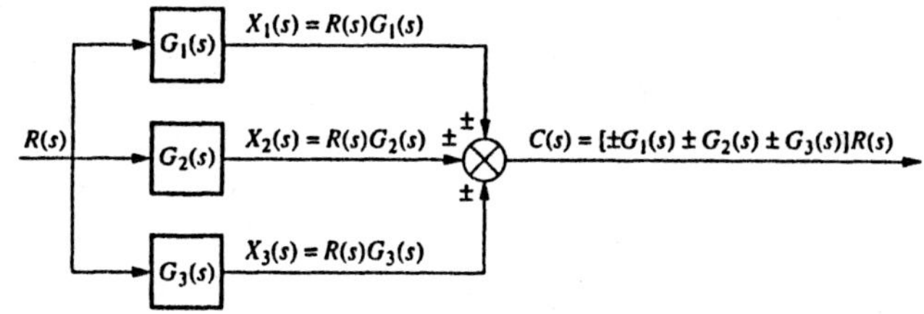

Figure 6.20 Parallel subsystems.

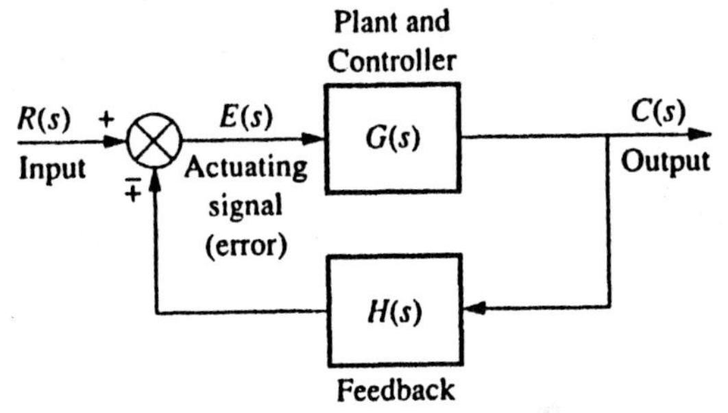

Figure 6.21 Feedback control system.

system. Next, we interconnect the signal nodes with system branches. Figures 6.22a, 6.22c, and 6.22e show the signal nodes for the cascaded, parallel, and feedback forms, respectively. Next, interconnect the nodes with branches that represent the subsystems. This is done in Figures 6.22b, 6.22d, and 6.22f for the cascaded, parallel, and feedback forms, respectively. For the parallel form, positive signs are assumed at all inputs to the summing junction; for the feedback form, a negative sign is assumed for the feedback.

In the next example we start with a more complicated block diagram and end with the equivalent signal-flow graph. Convert the block diagram of Figure 6.23 to a signal-flow graph. Begin by drawing the signal nodes as shown in Figure 6.24a.

Next, interconnect the nodes showing the direction of signal flow and identifying each transfer function. The result is shown in Figure 6.24b. Notice that the negative signs at the summing junctions of the block diagram are represented by the negative transfer functions of the signal-flow graph.

Finally, if desired, simplify the signal-flow graph to the one shown in Figure 6.24c by eliminating signals that have a single flow in and a single flow out such as $V_2(s)$, $V_6(s)$, $V_7(s)$, and $V_8(s)$.

6.2.3 Mason's Rule

Block diagram reduction requires successive application of fundamental relationships in order to arrive at the system transfer function. On the other hand, Mason's Rule for the reduction of signal-flow graphs to a transfer function relating the output to the input requires the application of a single formula. The formula was derived by S. J. Mason when he related the signal-flow

[1]Adapted from *Control Systems Engineering*, second edition, by Norman S. Nise.

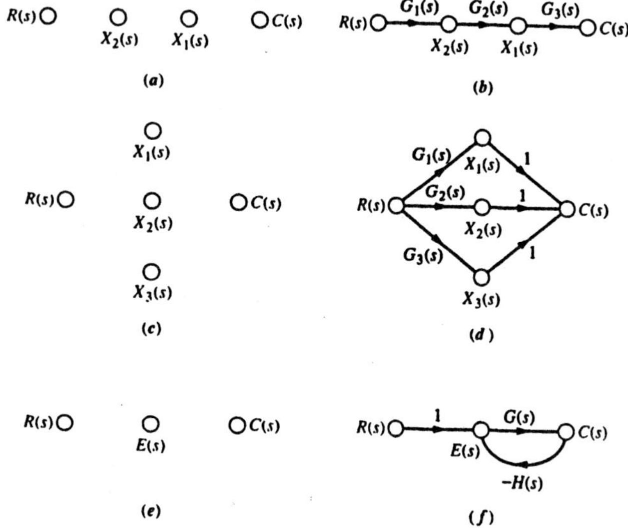

Figure 6.22 Building signal-flow graphs: (a) cascaded system: nodes (from Figure 6.19); (b) cascaded system: signal-flow graph; (c) parallel system: nodes (from Figure 6.20); (d) parallel system: signal-flow graph; (e) feedback system: nodes (from Figure 6.21); (f) feedback system: signal-flow graph.

graphs to the simultaneous equations that can be written from the graph ([1], [2]).

In general, it can be complicated to implement the formula without making mistakes. Specifically, the existence of what we later call nontouching loops increases the complexity of the formula. However, many systems do not have nontouching loops and thus lend themselves to the easy application of Mason's gain formula. For these systems, you may find Mason's Rule easier to use than block diagram reduction.

The formula has several component parts that must be evaluated. We must first be sure that the definitions of the component parts are well understood. Then, we must exert care in evaluating the component parts of Mason's formula. To that end, we now discuss some basic definitions applicable to signal-flow graphs. Later we will state Mason's Rule and show an example.

Definitions

Loop Gain: Loop gain is the product of branch gains found by traversing a path that starts at a node and ends at the same node without passing through any other node more than once and following the direction of the signal flow. For examples of loop gains, look at Figure 6.25. There are four loop gains as follows:

1. $G_2(s)H_1(s)$ (6.11a)
2. $G_4(s)H_2(s)$ (6.11b)
3. $G_4(s)G_5(s)H_3(s)$ (6.11c)
4. $G_4(s)G_6(s)H_3(s)$ (6.11d)

Forward-Path Gain: Forward-path gain is the product of gains found by traversing a path from the input node to the output node of the signal-flow graph in the direction of signal flow. Examples of forward-path gains are also shown in Figure 6.25. There are two forward-path gains as follows:

$$G_1(s)G_2(s)G_3(s)G_4(s)G_5(s)G_7(s) \quad (6.12a)$$

$$G_1(s)G_2(s)G_3(s)G_4(s)G_6(s)G_7(s) \quad (6.12b)$$

Nontouching Loops: Nontouching loops are loops that do not have any nodes in common. In Figure 6.25, loop $G_2(s)H_1(s)$ does not touch loop $G_4(s)H_2(s)$, loop $G_4(s)G_5(s)H_3(s)$, or loop $G_4(s)G_6(s)H_3(s)$.

Nontouching-Loop Gain: Nontouching-loop gain is the product of loop gains from nontouching loops taken two, three, four, etc. at a time. In Figure 6.25, the product of loop gain $G_2(s)H_1(s)$ and loop gain $G_4(s)H_2(s)$ is a nontouching-loop gain taken two at a time. In summary, all three of the nontouching-loop gains taken two at a time are

1. $[G_2(s)H_1(s)][G_4(s)H_2(s)]$ (6.13a)
2. $[G_2(s)H_1(s)][G_4(s)G_5(s)H_3(s)]$ (6.13b)
3. $[G_2(s)H_1(s)][G_4(s)G_6(s)H_3(s)]$ (6.13c)

The product of loop gains $[G_4(s)G_5(s)H_3(s)][G_4(s)G_6(s)H_3(s)]$ is not a nontouching-loop gain since these two loops have nodes in common. In our example there are no nontouching-loop gains taken three at a time since three nontouching loops do not exist in the example.

We are now ready to state Mason's Rule.

Mason's Rule: The transfer function, $C(s)/R(s)$, of a system represented by a signal-flow graph is

$$G(s) = \frac{C(s)}{R(s)} = \frac{\sum_k T_k \Delta_k}{\Delta} \quad (6.14)$$

where		
$\sum$		denotes summation
k	=	number of forward paths
T_k	=	the kth forward-path gain
Δ	=	$1 - \sum$ loop gains $+ \sum$ nontouching-loop gains taken two at a time $- \sum$ nontouching-loop gains taken three at a time $+ \sum$ nontouching-loop gains taken four at a time $- \ldots$
Δ_k	=	$\Delta - \sum$ loop gain terms in Δ touching the kth forward path. In other words, Δ_k is formed by eliminating from Δ those loop gains that touch the kth forward path.

Notice the alternating signs for each component part of Δ. The following example will help clarify Mason's Rule.

Find the transfer function, $C(s)/R(s)$, for the signal-flow graph in Figure 6.26. First, identify the **forward-path gains.**

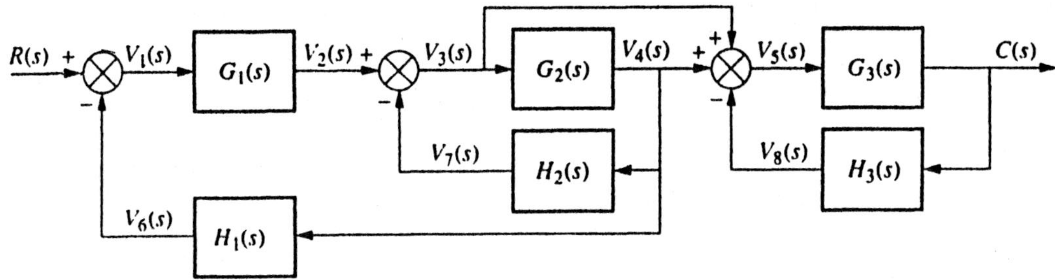

Figure 6.23 Block diagram.

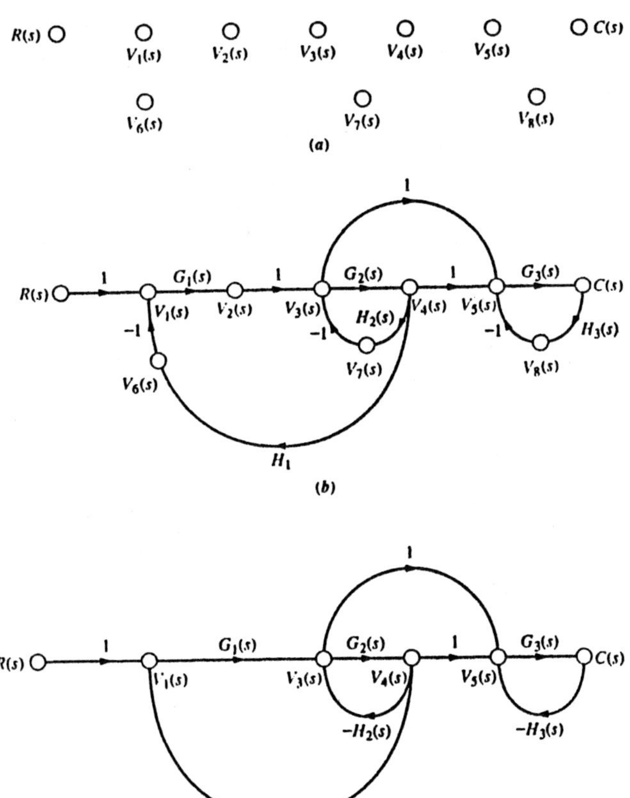

Figure 6.24 Signal-flow graph development: (a) signal nodes; (b) signal-flow graph; (c) simplified signal-flow graph.

In this example, there is only one:

$$G_1(s)G_2(s)G_3(s)G_4(s)G_5(s) \tag{6.15}$$

Second, identify the **loop gains**. There are four loops as follows:

1. $G_2(s)H_1(s)$ (6.16a)
2. $G_4(s)H_2(s)$ (6.16b)
3. $G_7(s)H_4(s)$ (6.16c)
4. $G_2(s)G_3(s)G_4(s)G_5(s)G_6(s)G_7(s)G_8(s)$ (6.16d)

Third, identify the **nontouching-loop gains taken two at a time.** From Equations 6.16a to 6.16d and Figure 6.26, we see that loop 1 does not touch loop 2, loop 1 does not touch loop 3, and loop 2

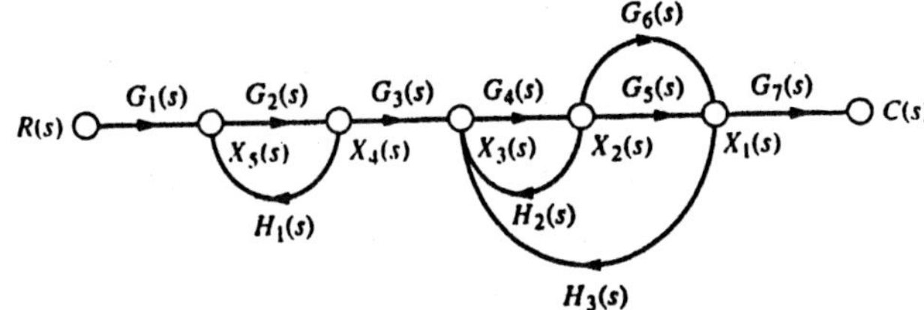

Figure 6.25 Sample signal-flow graph for demonstrating Mason's Rule.

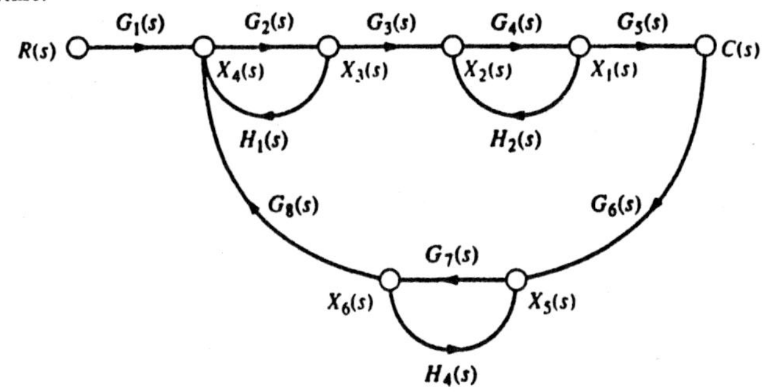

Figure 6.26 Signal-flow graph.

does not touch loop 3. Notice that loops 1, 2, and 3 all touch loop 4. Thus, the combinations of nontouching-loop gains taken two at a time are as follows:

Loop 1 and loop 2:	$G_2(s)H_1(s)G_4(s)H_2(s)$	(6.17a)
Loop 1 and loop 3:	$G_2(s)H_1(s)G_7(s)H_4(s)$	(6.17b)
Loop 2 and loop 3:	$G_4(s)H_2(s)G_7(s)H_4(s)$	(6.17c)

Finally, the **nontouching-loop gains taken three at a time** are as follows:

$$\text{Loops 1, 2, and 3:} \quad G_2(s)H_1(s)G_4(s)H_2(s)G_7(s)H_4(s) \tag{6.18}$$

Now, from Equation 6.14 and its definitions, we form Δ and Δ_k:

$$\begin{aligned}\Delta = {} & 1 - [G_2(s)H_1(s) + G_4(s)H_2(s) + G_7(s)H_4(s) \\ & + G_2(s)G_3(s)G_4(s)G_5(s)G_6(s)G_7(s)G_8(s)] \\ & + [G_2(s)H_1(s)G_4(s)H_2(s) \\ & + G_2(s)H_1(s)G_7(s)H_4(s) \\ & + G_4(s)H_2(s)G_7(s)H_4(s)] \\ & - [G_2(s)H_1(s)G_4(s)H_2(s)G_7(s)H_4(s)]\end{aligned} \tag{6.19}$$

We form Δ_k by eliminating from Δ those loop gains that touch the kth forward path:

$$\Delta_1 = 1 - G_7(s)H_4(s) \tag{6.20}$$

Equations 6.15, 6.19, and 6.20 are substituted into Equation 6.14 yielding the transfer function

$$\begin{aligned} G(s) &= \frac{T_1\Delta_1}{\Delta} \\ &= \frac{[G_1(s)G_2(s)G_3(s)G_4(s)G_5(s)][1 - G_7(s)H_4(s)]}{\Delta} \end{aligned} \tag{6.21}$$

Since there is only one forward path, $G(s)$ consists only of one term rather than the sum of terms each coming from a forward path.

6.2.4 Signal-Flow Graphs of Differential Equations and State Equations

In this section we show how to convert a differential equation or state-space representation to a signal-flow graph. Consider the differential equation

$$\frac{d^3c}{dt^3} + 9\frac{d^2c}{dt^2} + 26\frac{dc}{dt} + 24c = 24r \tag{6.22}$$

Converting to the **phase-variable** representation in state-space [2]

$$\dot{x}_1 = x_2 \tag{6.23a}$$
$$\dot{x}_2 = x_3 \tag{6.23b}$$
$$\dot{x}_3 = -24x_1 - 26x_2 - 9x_3 + 24r \tag{6.23c}$$
$$y = x_1 \tag{6.23d}$$

To draw the associated signal-flow graph, first identify three nodes, as in Figure 6.27a, to be the three state variables, x_1, x_2, and x_3. Also identify a node as the input, r, and another node as the output, y. The first of the three state equations, $\dot{x}_1 = x_2$, is modeled in Figure 6.27b by realizing that the derivative of state variable x_1, which is x_2, would appear to the left at the input to an integrator. Remember, division by s in the frequency domain is equivalent to integration in the time domain. Similarly, the second equation, $\dot{x}_2 = x_3$, is added in Figure 6.27c. The last of the state equations, $\dot{x}_3 = -24x_1 - 26x_2 - 9x_3 + 24r$, is added in Figure 6.27d by forming $\dot{x}_3$ at the input of an integrator whose output is x_3. Finally, the output, $y = x_1$, is also added in Figure 6.27d completing the signal-flow graph. Notice that the state variables are outputs of the integrators.

6.2.5 Signal-Flow Graphs of Transfer Functions

To convert transfer functions to signal-flow graphs we first convert the transfer function to a state-space representation. The signal-flow graph then follows from the state equations as in the preceding section. Consider the transfer function

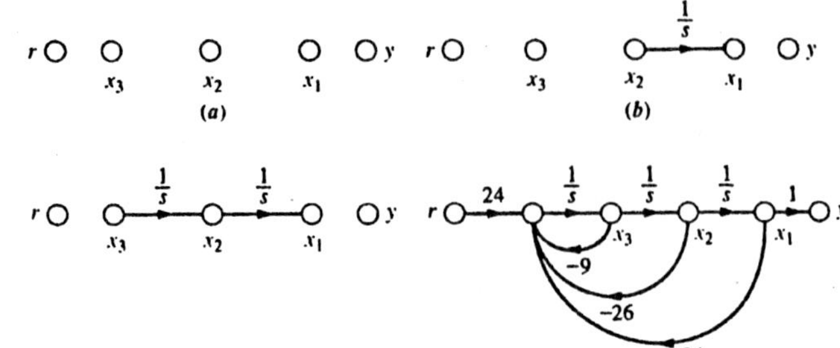

Figure 6.27 Stages in the development of a signal-flow graph in phase-variable form for the system of Equation 6.22 or Equations 6.23.

$$G(s) = \frac{s^2+7s+2}{s^3+9s^2+26s+24} \tag{6.24}$$

Converting to the phase-variable representation in state-space [3]

$$\dot{x}_1 = x_2 \tag{6.25a}$$
$$\dot{x}_2 = x_3 \tag{6.25b}$$
$$\dot{x}_3 = -24x_1 - 26x_2 - 9x_3 + 24r \tag{6.25c}$$
$$y = 2x_1 + 7x_2 + x_3 \tag{6.25d}$$

Following the same procedure used to obtain Figure 6.27d, we arrive at Figure 6.28. Notice that the denominator of the transfer function is represented by the feedback paths, while the numerator of the transfer function is represented by the **linear combination** of state variables forming the output.

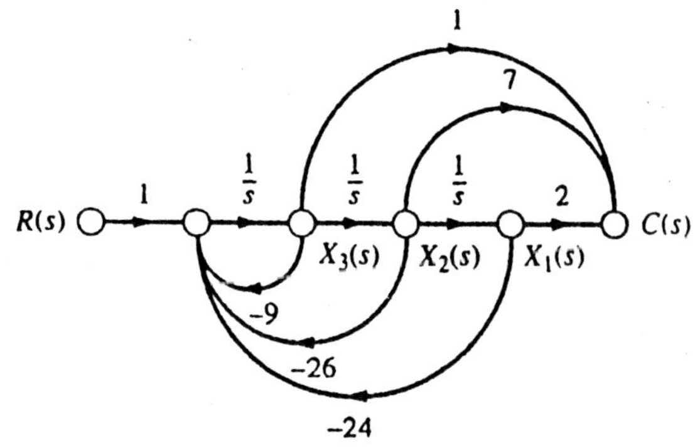

Figure 6.28 Signal-flow graph in phase-variable form for Equation 6.24.

[2] See [3] for description of how to convert differential equations into the phase-variable representation in state space.

[3] See [3] for a description of how to convert transfer functions into the phase-variable representation in state space.

6.2.6 A Final Example

We conclude this chapter with an example that demonstrates the application of signal-flow graphs and the previously discussed forms to represent in state space the feedback control system shown in Figure 6.29. We first draw the signal-flow diagram for the forward transfer function, $G(s) = 100(s+5)/[(s+2)(s+3)]$, and then add the feedback path. In many physical systems, the forward transfer function consists of several systems in cascade. Thus, for this example, instead of representing $G(s)$ in phase-variable form using the methods previously described, we arrive at the signal-flow graph by considering $G(s)$ to be the following terms in cascade:

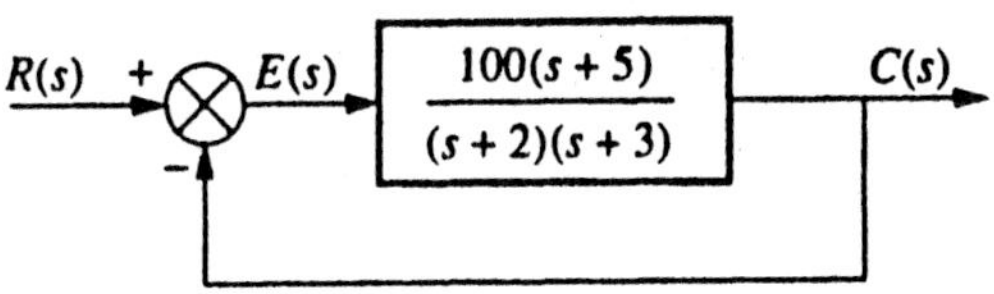

Figure 6.29 Feedback control system.

$$G(s) = 100 * \frac{1}{(s+2)} * \frac{1}{(s+3)} * (s+5) \tag{6.26}$$

Each first-order term is of the form

$$\frac{C_i(s)}{R_i(s)} = \frac{1}{(s+a_i)} \tag{6.27}$$

Cross multiplying,

$$(s+a_i)C_i(s) = R_i(s) \tag{6.28}$$

Taking the inverse Laplace transform with zero initial conditions,

$$\frac{dc_i(t)}{dt} + a_i c_i(t) = r_i(t) \tag{6.29}$$

Solving for $\frac{dc_i(t)}{dt}$,

$$\frac{dc_i(t)}{dt} = -a_i c_i(t) + r_i(t) \tag{6.30}$$

Figure 6.30 shows Equation 6.30 as a signal-flow graph. Here again, a node is assumed for $C_i(s)$ at the output of an integrator, and its derivative formed at the input. Using Figure 6.30 as a model, we represent the first three terms on the right of Equation 6.26 as shown in Figure 6.31a. To cascade the zero, $(s+5)$, we identify the output of $100/[(s+2)(s+3)]$ as $X_1(s)$. Cascading $(s+5)$ yields

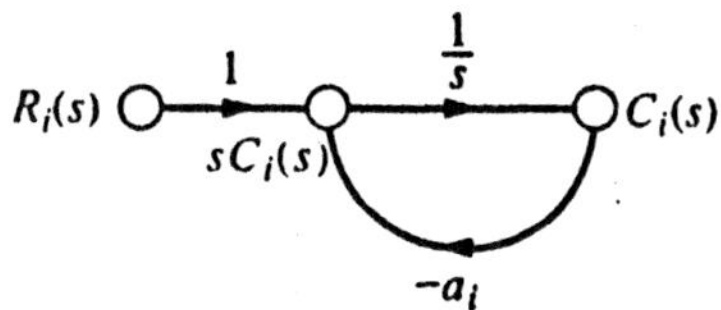

Figure 6.30 First-order subsystem.

$$C(s) = (s+5)X_1(s) = sX_1(s) + 5X_1(s) \tag{6.31}$$

Thus, $C(s)$ can be formed from a linear combination of previously derived signals as shown in Figure 6.31a. Finally, add the feedback and input paths as shown in Figure 6.31b.

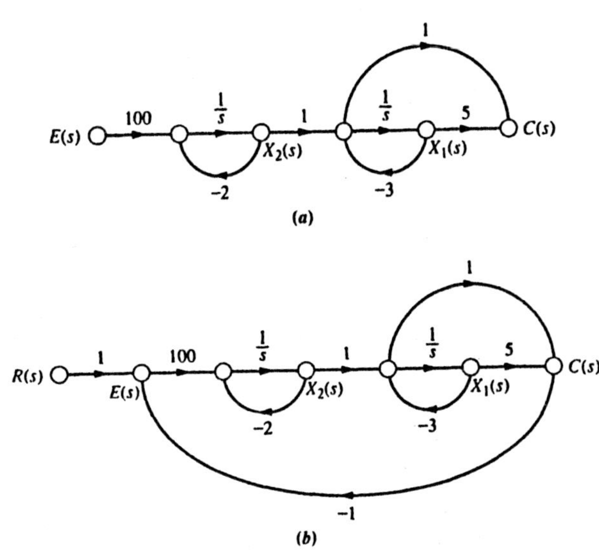

Figure 6.31 Steps in drawing the signal-flow graph for the feedback system of Figure 6.29: (a) forward transfer function; (b) closed-loop system.

Now, by inspection, write the state equations.

$$\dot{x}_1 = -3x_1 + x_2 \tag{6.32a}$$

$$\dot{x}_2 = -2x_2 + 100(r - c) \tag{6.32b}$$

But from Figure 6.31b,

$$c = 5x_1 + (x_2 - 3x_1) = 2x_1 + x_2 \tag{6.32c}$$

Substituting Equation 6.32c into Equation 6.32b, the state equations for the system are

$$\dot{x}_1 = -3x_1 + x_2 \tag{6.33a}$$

$$\dot{x}_2 = -200x_1 - 102x_2 + 100r \tag{6.33b}$$

The output equation is the same as Equation 6.32c, or

$$y = c = 2x_1 + x_2 \tag{6.33c}$$

In vector-matrix form,

$$\dot{\mathbf{x}} = \begin{bmatrix} -3 & 1 \\ -200 & -102 \end{bmatrix}\mathbf{x} + \begin{bmatrix} 0 \\ 100 \end{bmatrix} r \tag{6.34a}$$

$$y = \begin{bmatrix} 2 & 1 \end{bmatrix}\mathbf{x} \tag{6.34b}$$

In this chapter we discussed signal-flow graphs. We defined them and related them to block diagrams. We showed that signals are represented by nodes, and systems by branches. We showed how to draw signal-flow graphs from state equations and how to use signal-flow graphs as an aid to obtaining state equations.

6.2.7 Defining Terms

Block diagram: A representation of the interconnection of subsystems. In a linear system, the block diagram consists of blocks representing subsystems; arrows representing signals; summing junctions, which show the algebraic summation of two or more signals; and pickoff points, which show the distribution of one signal to multiple subsystems.

Branches: Lines that represent subsystems in a signal-flow diagram.

Cascaded form: An interconnection of subsystems, where the output of one subsystem is the input of the next. For linear systems with real and distinct poles (eigenvalues), this model leads to a triangular system matrix in state space with the poles along the diagonal.

Companion matrix: A system matrix that contains the coefficients of the characteristic equation along a row or column. If the first row contains the coefficients, the matrix is an upper companion matrix; if the last row contains the coefficients, the matrix is a lower companion matrix; if the first column contains the coefficients, the matrix is a left companion matrix; and if the last column contains the coefficients, the matrix is a right companion matrix.

Feedback form: An interconnection of two subsystems: forward-path and feedback. The input to the forward-path subsystem is the algebraic sum of two signals: (1) the system input and (2) the system output operated on by the feedback subsystem.

Forward-path gain: The product of gains found by traversing a path from the input node to the output node of a signal-flow graph in the direction of signal flow.

Linear combination: A linear combination of n variables, x_i, for $i = 1$ to n, is given by the sum, $S = K_n X_n + K_{n-1} X_{n-1} + \cdots K_1 X_1$, where each K_i is a constant.

Linear system: A system possessing the properties of superposition and homogeneity.

Loop gain: The product of branch gains found by traversing a path that starts at a node and ends at the same node without passing through any other node more than once and following the direction of the signal flow.

Mason's Rule: The transfer function, $C(s)/R(s)$, of a system represented by a signal-flow graph is

$$G(s) = \frac{C(s)}{R(s)} = \frac{\sum_k T_k \Delta_k}{\Delta}$$

where $\sum$ denotes summation

k = number of forward paths

T_k = the kth forward-path gain

Δ = $1 - \sum$ loop gains $+ \sum$ nontouching-loop gains taken two at a time $- \sum$ nontouching-loop gains taken three at a time $+ \sum$ nontouching-loop gains taken four at a time $- \ldots$

Δ_k = $\Delta - \sum$ loop gain terms in Δ touching the kth forward path. In other words, Δ_k is formed by eliminating from Δ those loop gains that touch the kth forward path.

Nodes: Points in a signal-flow diagram that represent signals.

Nontouching-loop gain: The product of loop gains from nontouching loops taken two, three, four, etc. at a time.

Nontouching loops: Loops that do not have any nodes in common.

Parallel form: An interconnection of subsystems, where the input is common to all subsystems and the output is an algebraic sum of the outputs of the subsystems.

Phase-variable form: A system representation where the state variables are successive derivatives and the system matrix is a lower **companion matrix**.

Signal-flow graphs: A representation of the interconnection of subsystems that form a system. The representation consists of nodes representing signals, and lines with arrows representing subsystems.

Transfer function: The ratio of the Laplace transform of the output of a system to the Laplace transform of the input.

References

[1] Mason, S., Feedback theory—some properties of signal-flow graphs, *Proc. IRE*, September 1953, 1144–1156.

[2] Mason, S., Feedback theory—further properties of signal-flow graphs, *Proc. IRE*, July 1956, 920–926.

[3] Nise, N.S., *Control Systems Engineering*, 2nd ed., Benjamin/Cummings, Redwood City, CA, 1995, 134–139.

Further Reading

[1] Dorf, R. and Bishop, R. 1995. *Modern Control Systems*, 7th ed., Addison-Wesley, Reading, MA.

[2] Hostetter, G., Savant, C., Jr., and Stefani, R. 1994. *Design of Feedback Control Systems*, 3rd ed., Holt Rinehart and Winston, New York.

[3] Kuo, B. 1995. *Automatic Control Systems*, 7th ed., Prentice Hall, Englewood Cliffs, NJ.

[4] Timothy, L. and Bona, B. 1968. *State Space Analysis: An Introduction*, McGraw-Hill, New York.

7

Determining Models

François E. Cellier
Department of Electrical and Computer Engineering, The University of Arizona, Tucson, AZ

Hilding Elmqvist
Dynasim AB, Research Park Ideon, Lund, Sweden

Martin Otter
Institute for Robotics and System Dynamics, German Aerospace Research Establishment Oberpfaffenhofen (DLR), Wessling, Germany

William S. Levine
Department of Electrical Engineering, University of Maryland, College Park, MD

7.1 Modeling from Physical Principles

François E. Cellier, Department of Electrical and Computer Engineering, The University of Arizona, Tucson, AZ
Hilding Elmqvist, Dynasim AB, Research Park Ideon, Lund, Sweden
Martin Otter, Institute for Robotics and System Dynamics, German Aerospace Research Establishment Oberpfaffenhofen (DLR), Wessling, Germany

7.1.1 Introduction

Modeling and simulation are central to the design of control systems, since, for complex, large-scale, nonlinear industrial plants, there usually don't exist, or at least are not known, any suitable analytical design approaches. Simulation may often be the only resort short of building the physical plant itself and experimenting with the real system.

Chapter 22 deals with a wide palette of issues relating to the simulation of control systems. This chapter, on the other hand, shall concentrate on issues relating to modeling the physical plant to be controlled.

Modeling physical systems seems to be a straightforward task. Since physical systems and experiments are often reproducible in a reliable fashion, since measurements from physical systems are frequently available in abundance and of high quality, since the meta-laws of physics are mostly well understood, it seems to be a particularly easy task to come up with accurate mathematical descriptions of most physical plants.

Yet, there are some typical pitfalls and frequent misconceptions about the modeling of physical systems, especially among control engineers. These are illustrated, and a sound methodological basis for modeling from physical principles is then created.

7.1.2 Common Misconceptions

1st Misconception: State-Space Models Form the Basis of Physics

Control engineers are used to dealing with state-space models of the form:

$$\begin{aligned} \dot{\mathbf{x}} &= \mathbf{f}(\mathbf{x}, \mathbf{u}, t) \\ \mathbf{y} &= \mathbf{g}(\mathbf{x}, \mathbf{u}, t) \end{aligned} \tag{7.1}$$

Consequently, they look at the physical world as if it consisted of these types of equations, and, when modeling, try to start out with system descriptions that come as close as possible to the familiar state-space form.

This concept is explained by means of a simple lunar lander module, as shown in Figure 7.1. Only the vertical movement of the space craft is considered. The modeler starts out with Newton's laws (translational motion) or the d'Alembert principle, since Newton's laws come very close, in description, to the desired state-space formulation. Since the mass of the space craft is time-varying (due to fuel consumption), the resulting equation takes the following form:

$$\frac{d(m \cdot v)}{dt} = \frac{dm}{dt} \cdot v + m \cdot \frac{dv}{dt} = thrust - m \cdot g \tag{7.2}$$

0-8493-8570-9/96/$0.00+$.50

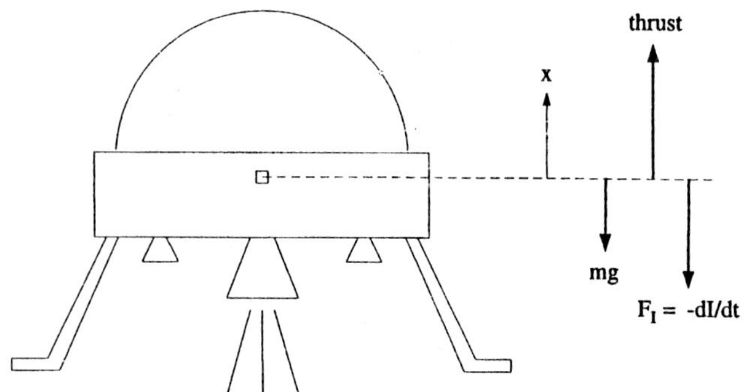

Figure 7.1 Lunar lander module.

Making the reasonable assumption that the change in mass is proportional to the magnitude of the thrust:

$$\frac{dm}{dt} = -k \cdot |thrust| \tag{7.3}$$

and assuming that the thrust is always positive, Equation 7.2 can be rewritten as:

$$\frac{dv}{dt} = \frac{1}{m} \cdot ((1 + k \cdot v) \cdot thrust - m \cdot g) \tag{7.4}$$

which is already very close to the desired state-space model.

There is only one major drawback of this model — it is, unfortunately, incorrect. This can be seen easily. For simplicity, it will be assumed that the space craft is far away from any source of gravity. Thus, $g \approx 0$, and Equation 7.4 can be simplified to:

$$\frac{dv}{dt} = \frac{1}{m} \cdot (1 + k \cdot v) \cdot thrust \tag{7.5}$$

Making the additional assumption that the initial velocity of the spacecraft is

$$v_0 = -\frac{1}{k} \tag{7.6}$$

it follows that the acceleration is exactly equal to zero, irrespective of the thrust, and the doomed crew will not be able to either accelerate or decelerate their space craft ever again. Luckily for them, physics doesn't work that way.

Where is the bug in the model? Had the modeler, instead of blindly and unthinkingly applying Newton's laws to the space craft, formulated either the principle of energy conservation or, alternatively, that of momentum conservation (both concepts work equally well in this simple example) to a piece of space surrounding the rocket, he or she would have noticed that the plume of exhaust of the space craft also carries mass and momentum and energy, and the disaster could have been prevented. More precisely:

$$\mathcal{I}(t + \Delta t) = \mathcal{I}(t) + \Delta\mathcal{I}(t \rightarrow t + \Delta t) \tag{7.7}$$

i.e., the total momentum $\mathcal{I}$ of a system at time $t + \Delta t$ equals the total momentum at time t plus the (positive or negative) momentum added to (subtracted from) the system between time t and time $t + \Delta t$. Applied to the space craft:

$$(m - \Delta m) \cdot (v + \Delta v) + \Delta m \cdot v = m \cdot v + thrust \cdot \Delta t \tag{7.8}$$

The first term on the left-hand side of Equation 7.8 denotes the momentum of the space craft at time $t + \Delta t$. The second term denotes the momentum of the cloud of exhaust at the same time. The first term on the right-hand side denotes the momentum of the space craft at time t, and the second term denotes the added momentum due to the drive of the space craft. Notice that the exhaust must be included. Either the cloud of exhaust is considered a part of the system by adding it to the left-hand side of Equation 7.8, as done above, or the cloud must be considered as leaving the system between time t and time $t + \Delta t$, and then the same term must be subtracted from the right-hand side of Equation 7.8.

Neglecting terms in Equation 7.8 that are of second-order small leads to:

$$m \cdot \Delta v = thrust \cdot \Delta t \tag{7.9}$$

or by dividing through Δt and letting Δt go to zero:

$$m \cdot \frac{dv}{dt} = thrust \tag{7.10}$$

Thus, although the mass of the space craft is undeniably changing with time, the more familiar form of Newton's laws must be used in this case.

This is of course a very simple example, and few control engineers would fall into this trap. However, such errors indeed do happen, and often, it is because of the control engineers' infatuation with state-space models.

2nd Misconception: Signals Capture Physics

The second misconception has to do with treating individual signals as portraying physical principles. This idea is illustrated by means of a simple thermal model describing, as a function of time, the temperature distribution along a perfectly insulated rod.

Most physics textbooks (e.g., [10]) can be consulted to provide the following model:

$$\frac{\partial T}{\partial t} = \frac{\lambda}{c \cdot \rho} \cdot \frac{\partial^2 T}{\partial x^2} \tag{7.11}$$

where t denotes time, x denotes the location along the rod, T is the temperature measured in Kelvin, λ is the specific thermal conductance of the material, c represents the specific thermal capacitance of the material, and ρ stands for its density.

Control engineers don't usually like partial differential equations, so they conceptually cut the rod into slices of length Δx, approximate the spacial derivative by third-order accurate finite differences, and, introducing the abbreviation:

$$\sigma = \frac{\lambda}{c \cdot \rho} \tag{7.12}$$

end up with the state-space model:

$$\frac{dT_i}{dt} = \sigma \cdot \frac{T_{i+1} - 2T_i + T_{i-1}}{\Delta x^2} \tag{7.13}$$

Is there anything wrong with this model? Indeed, this model describes rather accurately what happens to the temperature at each point along the rod and at each moment in time, if, starting out from an initial spatial temperature distribution, the system is allowed to settle into steady state without external influences, which is a fine physical experiment, but doesn't represent what most engineers might like to do with most rods most of the time.

The problem is that the state-space model of Equation 7.13 is autonomous. There is no input at all to this model. If an engineer, for example, decides to let a current flow through the rod and observe how the rod heats up as a consequence, there is no way that the above model will be able to help him or her describe the desired effect.

The problem with the above model is that it is expressed in terms of *temperature*. Temperature is not what drives the physical phenomenon. It is only an observational quantity. What really goes on in the rod is a phenomenon of *heat flow*. Thus, it would have been much more sensible to describe the heat flow through the rod, rather than focusing on the temperature distribution.

In essence, the two misconceptions mentioned so far boil down to the same thing. By focusing on heat flow, the modeler is thinking in terms of energy. By focusing on temperature, he or she is modeling in terms of a phenomenological signal.

Every first-year undergraduate student of electrical engineering knows that the relationship between voltage and current in an electrical capacitor is described by the equation:

$$i_C = C \cdot \frac{du_C}{dt} \tag{7.14}$$

and the students are told early on in their careers that the behavior of electrical circuits is governed by two fundamental variables: *voltage* and *current*. Equation 7.14 doesn't tell why current is flowing through the capacitor. Yet students will rarely ask that question. To them, the correct circuit behavior follows directly from a systematic application of Kirchhoff's laws. But what precisely do these laws express in terms of physical, rather than mathematical, knowledge?

The product of voltage and current comes into the picture only as an afterthought, if at all. It rarely sinks in with students that what the circuit is really doing is, starting out from an initial state, trying to balance energy at all times.

Most state-space models are expressed in terms of phenomenological variables that are related by equations representing remote consequences of the basic physical laws, not the laws themselves. It is then the modeler's obligation to check whether all the conditions, all the silent assumptions, that went into these equations are indeed satisfied in the situation at hand, and, in order to make this assertion, the modeler often needs to revert to the physical laws that form the basis of these equations. Yet, engineers often don't do this. In fact, they may have forgotten the basic laws altogether and only remember the derived laws that they use frequently in their everyday work. And this is where the problems start.

Might it not make more sense to provide a modeling tool that can start out from the basic physical principles directly and derive the final state-space model on its own?

3rd Misconception: Causality Forms the Basis of Physics

The third misconception has to do with the notion of causality in physical systems. Many engineers believe that physics is essentially causal in nature. Someone takes a conscious decision to affect the world in a particular way, thereby *causing* the world to react to his or her actions.

Sir Isaac Newton followed the same line of reasoning when he formulated his famous law about *action* being equal to *reaction*. If someone or something exerts a force (the action) on a body that cannot move, then the system has to react by creating a counterforce (the reaction) of equal magnitude and opposite direction, such that the two forces annihilate each other.

Yet, the distinction between action and reaction is a deeply human and moral concept, not a physical one. There is no physical experiment in the world that can distinguish between action and reaction. It is clear that, if a drunk driver smashes his or her car into a tree, this is the fault of the driver and not of the tree. Yet, this is a purely human concept. Physics can't tell the difference.

The relationship between voltage across and current through an electrical resistor can be described by Ohm's law:

$$u_R = R \cdot i_R \tag{7.15}$$

yet, whether it is the current flowing through the resistor that causes a voltage drop, or whether it is the difference between the electrical potentials on the two wires that causes current to flow is, from a physical perspective, a meaningless question.

Physics is essentially acausal. Even where time is involved, physics locally only obeys conservation laws that, by nature, again are acausal. In spite of the recent efforts of science philosophers such as Roger Penrose, physics has still not discovered the true nature of the conscious mind. Maybe, consciousness is in fact a mere illusion, the true nature of which, however, is impossible to determine, due to Gödel's theorem?

Luckily, being engineers, we don't have to concern ourselves with these philosophical questions. Yet, the acausal nature of physics is indeed of much concern. State-space models are written in assignment statement form. There is always exactly one variable to the left of the equal sign, and the model implies that the expression to the right of the equal sign is evaluated, and the result of this evaluation is assigned as a new value to the variable to the left.

Consequently, the modeler needs two different models to describe an electrical resistor: a *voltage-drop-causer* model and a *current-flow-causer* model. As mentioned earlier, from a physical perspective, this makes no sense whatsoever.

Would it not be more meaningful to have a modeling tool available that allows the formulation of models in a declarative (acausal) form, and let the software worry about establishing the appropriate *computational causality* of each equation before generating simulation run-time code?

7.1.3 Energy Modeling and Bond Graphs

For a long time, mechanical engineers have used the Lagrangian or Hamiltonian equations to model mechanical systems, i.e., the modeling is based on the total *energy* contained in a mechanical system. However, the approach has two major drawbacks: (1) If non-conservative forces are present, the Lagrangian and Hamiltonian equations are no longer based solely on the energy of the system, but additionally on the virtual work of the non-conservative forces. (2) The equations are based on the total energy of the complete system, i.e., it is not possible to use the energy of subsystems, and then connect these subsystems together in a *modular* fashion. Similar problems occur in other engineering fields if modeling is based on energy.

To overcome these two deficiencies, Henry Paynter invented, in the early 1960s, the *bond graph* [18]. By computing the time derivative of the energy:

$$P(t) = \frac{dE(t)}{dt} \tag{7.16}$$

and using the power P instead of the energy E for modeling, both of the aforementioned difficulties vanish. In this new approach, *power continuity* equations are formulated instead of *energy conservation* laws. It turns out that, in any physical system, the power balance is a local property, i.e., the modeler can express power balance equations for each subsystem separately, and then connect all the subsystems, as long as he or she makes sure that the power is also balanced at all the interfaces between submodels [12].

This very same property also makes it possible to use power balance equations to describe dissipative systems. Resistors simply become two-port elements where free energy "opts out" and decides to henceforth be called heat.

As a bonus, power in any physical system can be written as the product of two adjugate variables, called the *effort* and the *flow* in bond graph terminology. In a bond graph, energy flow from one point of a system to another is denoted by a harpoon (a semi-arrow), as shown in Figure 7.2.

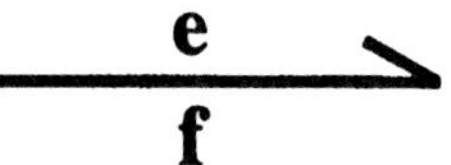

Figure 7.2 The bond.

Power is the product of effort and flow:

$$P = e \cdot f \tag{7.17}$$

In an electrical system, it is customary to select the voltage (or electrical potential), u, as the effort variable, and the current, i, as the flow variable. In a translational mechanical system, the force, f, will be treated as effort, and the velocity, v, as flow. In a rotational system, the torque, τ, assumes the role of the effort, and the angular velocity, ω, that of the flow. However, the assignment is arbitrary. Effort and flow are dual variables, and the assignment could just as well be done the other way around.

When power splits, or is combined, in a so-called *junction*, the power continuity equation dictates that the sum of incoming power streams must equal the sum of outgoing power streams. This requirement can be satisfied in many different ways, but the two easiest ones are certainly to keep one of the two adjugate variables constant around the junction, and formulate the balance equation for the other. In bond graph terminology, these two junction types are called the *0-junction* and the *1-junction*. Their constitutive equations are shown in Figure 7.3.

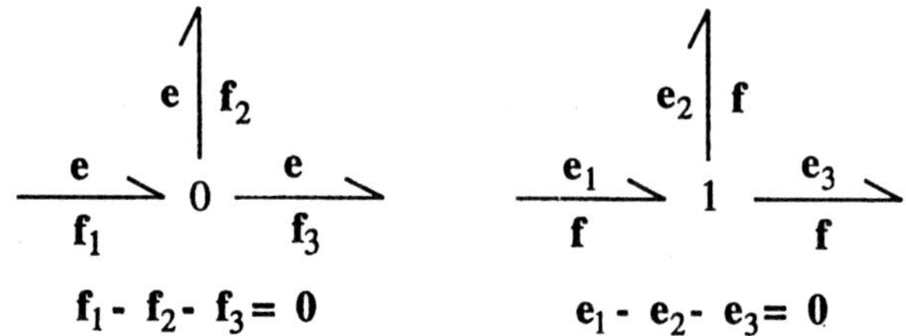

Figure 7.3 The two basic junction types.

Luckily for us engineers, physics seems to have a preference for simple solutions, and therefore, these two basic junction types are very commonly found in all kinds of physical systems. For example, one recognizes at once Kirchhoff's current law formulated for any node or cutset (the 0-junction), and Kirchhoff's voltage law formulated for any mesh or loop (the 1-junction).

The concept and benefits of bond graph modeling is demonstrated by means of a model describing an armature-controlled DC motor. A schematic diagram of the motor is shown in Figure 7.4 and the corresponding bond graph is shown in Figure 7.5.

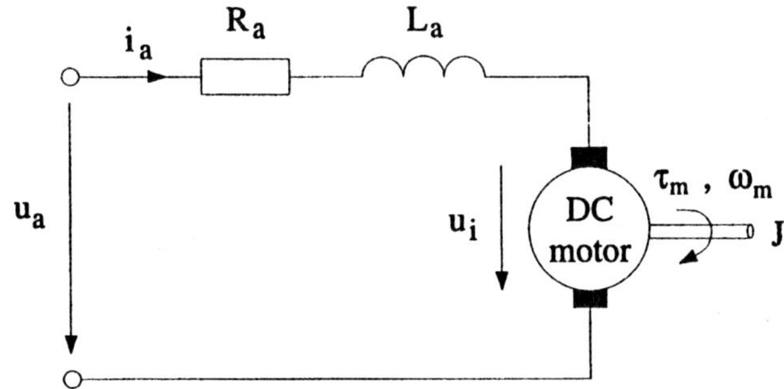

Figure 7.4 Model of a DC motor.

The 0-junction to the left of the bond graph depicts the electrical power port. Through it, the power:

$$P_{\text{elect}} = u_a \cdot i_a \tag{7.18}$$

enters the subsystem containing the DC motor. The 0-junction is the interface to the next subsystem to the left. Only a portion of the incoming power is available for conversion to mechanical energy. Some of the power gets stored in the armature inductance L_a, and some gets irreversibly converted to heat in the armature resistance R_a. Since the armature circuit is a mesh, the distribution of energy is described by a 1-junction. The remaining power

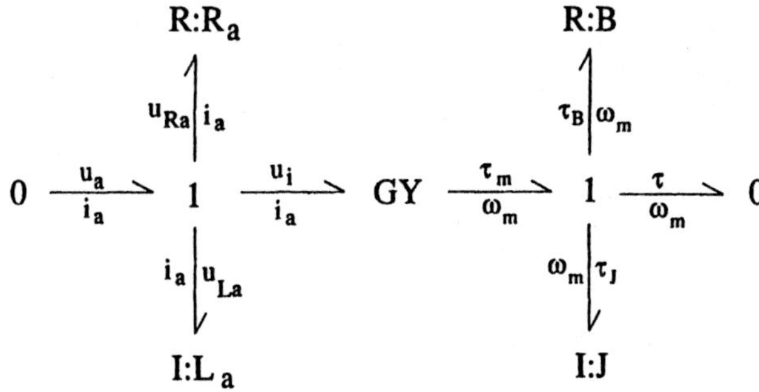

Figure 7.5 Bond graph of a DC-motor.

is available for conversion to mechanical energy, thus:

$$P_{\text{mot}} = u_i \cdot i_a = \tau_m \cdot \omega_m \tag{7.19}$$

Again, power continuity across the converter can be satisfied in many different ways, yet, there are only two simple solutions, as shown in Figure 7.6, namely, the *transformer (TF)* and the *gyrator*

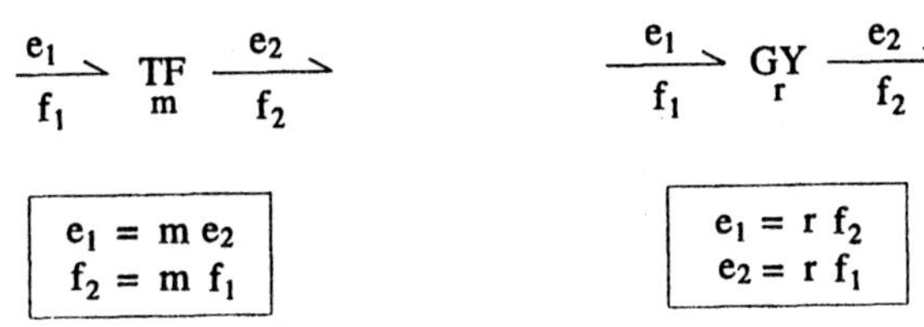

$e_1 \xrightarrow[f_1]{} \text{TF}_m \xrightarrow[f_2]{e_2}$ $\qquad$ $e_1 \xrightarrow[f_1]{} \text{GY}_r \xrightarrow[f_2]{e_2}$

$$e_1 = m\, e_2 \qquad f_2 = m\, f_1$$

$$e_1 = r\, f_2 \qquad e_2 = r\, f_1$$

Figure 7.6 The two basic converter types.

(GY). In fact, these two converters are dual to each other. Had the convention for efforts and flows been chosen dually on either the electrical or the mechanical side, the gyrator of Figure 7.5 would have become a transformer.

The right-hand part of Figure 7.5 shows the mechanical side of the DC motor. Some of the converted power gets stored in the rotor J (another inductance), and some of it gets dissipated in the bearings B (another resistor). Since the angular velocity is the same for all these elements, the power distribution is described by another 1-junction. The remaining power:

$$P_{\text{mech}} = \tau \cdot \omega_m \tag{7.20}$$

is available to drive a rotational load. It leaves the subsystem through the second power port, which is represented in the bond graph as a second 0-junction.

The reader may notice that the flux constant, ψ_{Flux}:

$$\tau_m = \psi_{\text{Flux}} \cdot i_a \tag{7.21}$$

expressed in Nm/A, and the electromotive force constant, ψ_{EMF}:

$$u_i = \psi_{\text{EMF}} \cdot \omega_m \tag{7.22}$$

expressed in Vs/rad, are physically two different aspects of one and the same phenomenon, and the numerical values of these two constants must, hence, be the same:

$$\psi_{\text{EMF}} = \psi_{\text{Flux}} \tag{7.23}$$

otherwise, the gyrator will either lose or miraculously generate energy. The consequent application of energy-flow modeling (here expressed in terms of a bond graph) makes it possible to discover such types of modeling constraints. Signal-flow modeling (e.g., expressed in terms of a block diagram) will not reveal this relationship.

Does the bond graph approach to power modeling help resolve some of the previously mentioned riddles? In fact, it does. To clarify this, the problem of temperature distribution along an ideally insulated rod is revisited.

It has been known since the time of Gabriel Kron [13] that the finite difference approximation of Equation 7.13 can be interpreted as the electrical circuit shown in Figure 7.7. A bond graph

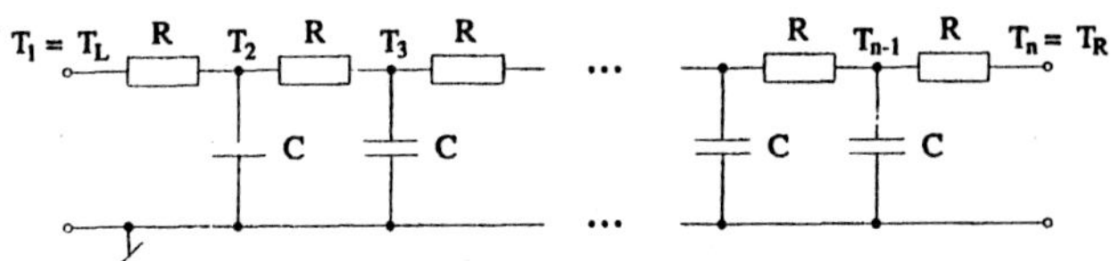

Figure 7.7 Electrical circuit representation of a diffusion chain.

representation of this circuit is given in Figure 7.8. In bond graph

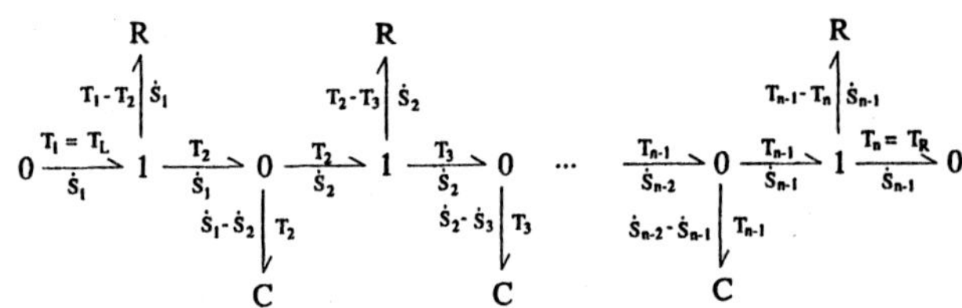

Figure 7.8 Bond graph representation of a diffusion chain.

notation, it is common to write heat flow as the product of temperature and entropy flow:

$$\frac{dQ}{dt} = T \cdot \frac{dS}{dt} \tag{7.24}$$

where the temperature, T, is used as the effort variable, and the entropy flow, dS/dt, is interpreted as the flow variable. Thus, it seems that the bond graph model is providing more than just an implementation of the circuit diagram, since it not only computes temperature values, but also provides a heat flow interpretation.

Unfortunately, the bond graph of Figure 7.8 is certainly incorrect. The problem with this bond graph has to do with the dissipated heat in the resistive elements. It was okay to ignore the generated heat in the armature resistance and bearings of the DC motor model. It simply means that the modeler chose not to include the thermal variables and equations in his or her model. However, the temperature dissipation model is already in the thermal domain. There is no way that the modeler can ignore what happens to the generated entropy and get away with it. Since the rod is thermally insulated, the generated entropy cannot escape and has to stay in the rod.

Jean Thoma proposed to represent resistors with entropy generation as *resistive source (RS)* elements [19]. The modified bond graph takes the form of Figure 7.9. The bond graph of Figure 7.9

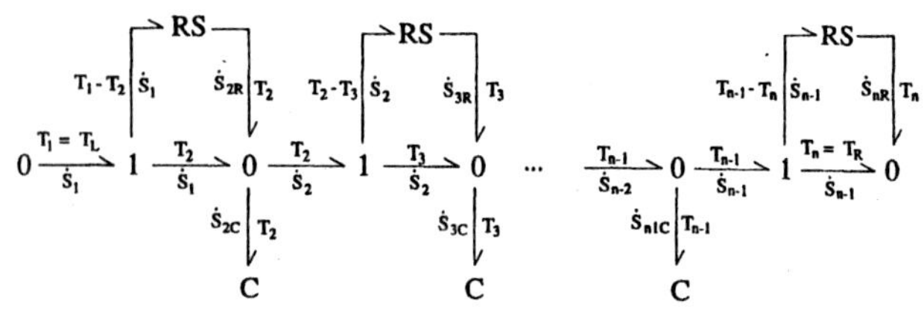

Figure 7.9 Corrected bond graph representation of a diffusion chain.

correctly represents both the entropy generation and flow in the rod, as well as the temperature distribution. Now, it has become easy to model also the electrically heated rod. Each resistor element is represented in the bond graph through an additional small source of entropy, as shown in Figure 7.10.

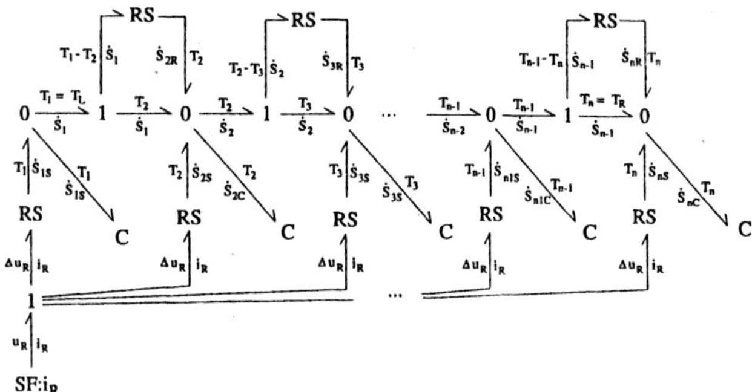

Figure 7.10 Bond graph representation of an electrically heated rod.

7.1.4 Object-Oriented Modeling

It was shown that using bond graphs for representing physical systems constitutes a safe way of modeling. However, in some cases, bond graphs may be unnecessarily constraining. What is important is not the bond graph syntax per se, but some of the properties inherent in bond graphs.

The most important feature of a bond graph is that it operates on energy flows rather than on individual signals. Thereby, it is guaranteed that no component and no interface will ever generate or lose energy in an uncontrolled fashion. Yet, other modeling formalisms may offer this same feature.

Another important facet of bond graphs is the fact that they are inherently acausal. Looking at the bond graphs presented in the previous section of this article, nothing indicates whether the resistive elements are of the voltage-drop-causer or of the current-flow-causer variety. The equations that are derived from a bond graph are declarative in nature. It is correct that some researchers have introduced so-called "causality strokes" into the bond graph methodology, turning the formerly acausal bond graphs into a causal variety. However, as shown in [6], causality strokes are not necessarily helpful, and, in the case of switching circuits, they are even harmful. Hence, it was decided not to augment the bond graphs shown in this article by causality strokes.

A third, very important, feature of bond graphs is that they are modular and hierarchical [4]. This makes it possible to construct bond graphs of submodels and connect them topologically to more complex bond graph elements that can then be connected further in a hierarchical manner. Since the interface points between bond graph submodels can be restricted to be always 0-junctions, power continuity at the interface between submodels is automatically guaranteed.

However, a bond graph is a fairly low-level modeling interface. It may not always be convenient or efficient to model down to that interface. Other modeling methodologies share some of the benefits of bond graphs but offer either a more convenient or higher-level interface. For example, if a modeler wants to describe an electrical circuit, there is nothing wrong with using the circuit diagram as a modeling tool. Circuit diagrams are modular and hierarchical, and power continuity at the interface between submodels is guaranteed by systematically applying Kirchhoff's laws to nodes and/or meshes [11]. The bond graph has the advantage of being domain independent, but a circuit diagram may be a more natural and equally powerful tool as far as electrical circuits are concerned.

The case of three-dimensional mechanical devices, so-called multibody systems (MBS) [15], is considerably worse. An important subclass of MBS devices are tree-structured robots. The use of bond graphs to describe the motion of such robots will, within the constraints of today's software technology, almost invariably lead to very inefficient simulation code. The problem is that bond graphers always express the motion of bodies in terms of absolute velocities [3]. This leads, in the generated equations, necessarily to large algebraically coupled higher index equation systems that are very hard to break. The trick in generating efficient simulation code is always to express the motion of each joint relative to the motion of the previous one [16]. It is evidently true that the bond graph contains complete information about the system, and a smart preprocessor could convert the model to a form that might be used to generate efficient run-time code, but this would necessarily call for specialized translation algorithms that only work in the case of MBS problems. It is not useful to develop such a preprocessor, because it is perfectly feasible to come up with a modular hierarchical description of MBS components that can be connected in a safe fashion and which generates efficient simulation code directly.

The object-oriented modeling paradigm, introduced by Hilding Elmqvist in the late 1970s [7], provides for a platform that allows the modeler to implement all of the above modeling formalisms including electrical circuit diagrams [5], bond graphs [4] and MBS [16]. Use of this methodology is a little less safe than using a more restricted modeling tool, such as a circuit diagram editor or a bond graph editor, because it provides more flexibility. It is the user's and/or model library developer's responsibility to ensure that the power continuity equation is satisfied at the interfaces between models. Modelers are thus advised not to define the *cuts* (the interface points) of their models arbitrarily, but to restrain themselves, and only use proven connection mechanisms. The question that the modeler should *always* ask him- or herself when designing these model interfaces is whether connecting such models in an arbitrary fashion will always ensure that power, momentum, and mass are balanced at the interfaces.

It is the authors' experience that many modeling errors are introduced while interconnecting models that violate balance equations at their interfaces. Hence, it is important to systematize the approach to interfacing models, and the large-scale example shown later in this chapter explains one way of accomplishing this.

An object-oriented modeling environment for physical system modeling should offer at least the following features:

- *Encapsulation of knowledge*: The modeler must be able to encode all knowledge related to a particular object in a compact fashion in one place with well-defined interface points to the outside.
- *Topological interconnection capability*: The modeler should be able to interconnect objects in a topological fashion, plugging together component models in the same way as an experimenter would plug together real equipment in a laboratory. This requirement entails that the equations describing the models must be declarative in nature, i.e., they must be acausal.
- *Hierarchical modeling*: The modeler should be able to declare interconnected models as new objects, making them indistinguishable from the outside from the basic equation models. Models can then be built up in a hierarchical fashion.
- *Object instantiation*: The modeler should have the possibility to describe generic object classes, and instantiate actual objects from these class definitions by a mechanism of model invocation.
- *Class inheritance*: A useful feature is class inheritance, since it allows the encapsulation of knowledge even below the level of a physical object. The so-encapsulated knowledge can then be distributed through the model by an inheritance mechanism, which ensures that the same knowledge will not have to be encoded several times in different places of the model separately.
- *Generalized Networking Capability*: A useful feature of a modeling environment is the capability to interconnect models through *nodes*. Nodes are different from regular models (objects) in that they offer a variable number of connections to them. This feature mandates the availability of *across* and *through* variables, so that power continuity across the nodes can be guaranteed.

In the following section, one such object-oriented modeling environment, Dymola [9], is described. The language Omola with its environment Omsim [1] represents another research effort with similar aims and properties.

7.1.5 Dymola: An Object-Oriented Modeling Tool

Dymola offers all of the above features plus a few more. In Dymola, an electrical resistor (a simple object) can be described as:

```
model class Resistor
  cut WireA(vA/iA), WireB(vB/iB)
  main cut TwoWires [ WireA , WireB ]
  main path Orientation < WireA , WireB >
  local u, i
  parameter R
    u = vA − vB
    i = iA
    iA + iB = 0
    u = R * i
end
```

A *cut* is a mechanism used for interconnection of models. It describes an interface point of the model. The resistor has two such interface points, namely its two pins, called *WireA* and *WireB*, respectively. Each pin carries two variables, an electrical potential (an across variable), and the current *into* the device (a through variable).

In Dymola, both the across variables and the through variables are lists of variables separated by a comma; the two lists themselves are separated by a slash. The difference between across and through variables becomes apparent in a connection only: *across variables* are set equal at a connection point, whereas *through variables* are summed up to zero. Note that, in the above example, the product of the two *cut* variables (e.g., $v_A \cdot i_A$) is the power flowing into the element. In this way, it is guaranteed that the energy is balanced at the interfaces of the resistor. In other words, across and through variables can be interpreted as the effort and flow variables of the bond graphs, whereby connection points are always 0-junctions.

The third and *main cut* is a hierarchical cut. It consists of the two previously defined basic cuts. This declaration enables the modeler to bend the legs of the resistor and plug the resistor as a whole with a single motion into a socket attached to a circuit board. The *path* declaration provides a logical orientation of the device. If resistors are connected in parallel or in series, Dymola uses the *path* declaration to determine which way they are connected into the circuit. In the case of a resistor, this may not really matter, but in other cases, it might. The four equations formulating the relationships between the terminal (interconnect) variables, the local variables, the parameters, and the constants are declarative in nature. Dymola will solve each of them for the appropriate variable, once it has access to all the equations.

The reader may notice that many of the equations described in the above model are shared by all one-port devices. Thus, it makes sense to encapsulate this knowledge in a separate model class, and then migrate the knowledge to the resistor class by means of inheritance. In Dymola, this is done in the following way:

```
model class OnePort
  cut WireA(vA/i), WireB(vB/ − i)
  main cut TwoWires [ WireA , WireB ]
```

```
    main path Orientation < WireA , WireB >
    local u
        u = vA − vB
end

model class Resistor
    inherit OnePort
    parameter R
        u = R * i
end
```

The *inherit* statement tells Dymola to copy all the declarations and equations from the parent class to the inheriting class. The current equations were simplified slightly to avoid having to carry around unnecessary aliases.

An interesting one-port is the electrical switch:

```
model class Switch
    inherit OnePort
    terminal open
        0 = if open then i else u
end
```

If the switch is open, then $i = 0$, else $u = 0$. The reader may remember that equations are declarative in nature, and that the computational causality will only be determined later. It turns out that Dymola can turn around *if* expressions as easily as algebraic expressions. A *terminal* is another interface point. It is a simplified *cut* that contains only a single across variable. Whereas *WireA* and *WireB* are power interfaces, *open* is a signal interface. The variable *open* is a control input to the model. It could have been declared as *input*, but there were good reasons for not doing so and using the direction-neutral *terminal* declaration instead, as will soon become clear. By using a signal interface, the modeler acknowledges that he or she considers the energy flow needed to turn the switch negligible in comparison with other power paths in the system. It is also acknowledged that both the time needed to deliver the energy to the switch as well as the time needed for the switch to react can be ignored. Models never reflect all facets of reality, and simplifications are okay, as long as the modeler makes them consciously, and knows what he or she is doing.

An ideal diode can be described in the following way:

```
model class Diode
    inherit Switch
        open = u <= 0 and not i > 0
end
```

An ideal diode is a switch. A logical equation is added that determines the value of the Boolean variable *open*. The diode goes into its *on* state (switch closed), when the voltage becomes positive, and it goes into its *off* state when the current becomes negative.

Evidently, Dymola's inheritance option is hierarchical, since diodes are switches which in turn are one-ports.

It becomes evident now why *open* had not been explicitly declared as an *input*. In the diode model, the computational causality of the signal port is turned around. Now, *open* is no longer a control input, but merely a measurement variable that is reported to the outside world through the interface. It is hardly ever justified to declare a variable explicitly as *input* or *output*, except at the top level, i.e., in the main model.

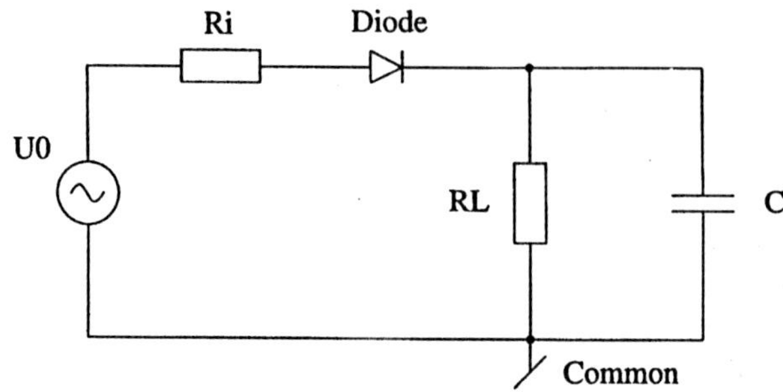

Figure 7.11 Rectifier with load.

Examine the simple electrical circuit shown in Figure 7.11: This circuit can be modeled in Dymola as follows:

```
model Rectifier
    submodel (VSource)   U0
    submodel (Resistor)  Ri(R = 10), RL(R = 50)
    submodel (Capacitor) C(C = 0.001)
    submodel (Diode)     D
    submodel Common

    parameter f = 50

    connect Common − ((U0 − Ri − D)//C//RL)

    U0.u0 = sin(2 * 3.14159 * f * Time)
end
```

Instantiations of models from model classes are obtained using the *submodel* statement. The *connect* statement plugs the circuit together. Since this circuit was sufficiently simple, no *nodes* were explicitly introduced, and the interconnections between the models were simply accomplished through series (−) and parallel (//) connections[1]. The dot-notation is used to access individual variables of submodels directly rather than through port connections. $U0.u0$ is the variable $u0$ of the model $U0$, which in turn is of model class *VSource*. This is a case of back-door programming, and the feature should be used sparingly.

Object-oriented modeling can be used as an alternative to bond graph modeling, provided the model classes are designed in such a way that the power balance is satisfied at the model interfaces. Object-oriented modeling has the additional benefit that it more closely resembles the physical reality (e.g., electrical components are represented in their familiar form rather than through more abstract domain-independent bond graph elements). Object-oriented models are more general than bond graphs, since they can just as easily be used to represent block diagrams or signal flow graphs when the need arises.

7.1.6 Object-Oriented Model of a Steam Power Plant

Figure 7.12 shows a top-level object-diagram of a steam power plant originally modeled by Lindahl [14], consisting of a drum

[1] For larger models it may be more convenient to make the connections using Dymodraw, a graphical object-diagram editor that generates Dymola code.

system, superheaters, preheaters, several turbines operating at different pressure levels, attemperators, deaerators, condensors, a feedwater pump, a combustion chamber, and a few other components.

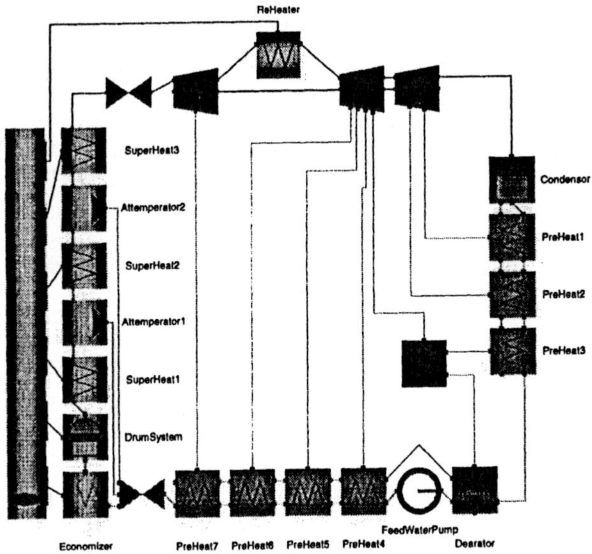

Figure 7.12 Steam power plant modeled in Dymodraw.

The drum can, for example, be described in the following manner. The model given here has been derived from first principles by Åström and Bell [2]. Total energy and mass balances are stated, as well as energy and mass balances for the risers. The variation of the steam-water mass ratio along the risers is complex. A linear variation is assumed, which, after integration along the risers, gives the average steam-water volume ratio, a_m, as shown below.

```
model class Drum
  { Including risers and downcomers }

  parameter Adrum = 20 { Drum wet area [m^2] },
            Vdrum = 40 { Drum volume [m^3] },
            Vr = 37    { Riser volume [m^3] },
            Vdc = 19   { Downcomer volume [m^3] },
            k = 0.01   { Friction coefficient },
            cp = 360   { Specific heat of metal [J kg^-1 K^-1] },
            m = 0      { Mass of metal [kg] },

  cut InWater  (p, Hfw/qfw)
  { Pressure, enthalpy and flow rate of feedwater. }
  cut InPower  ( /Pow) { Power from fuel. }
  cut OutSteam (p, Hs/ - qs)
  { Pressure, enthalpy and flow rate of steam. }

  local dl    { Drum level },
        am    { Steam quality volume ratio },
        Vw    { Drum water volume },
        Vwt   { Total water volume },
        xr    { Steam quality at riser outlet },
        Vst   { Total steam volume },
        Tm    { Metal temperature },
        qdc   { Downcomer flow },
        qr    { Riser flow },
        rs    { Steam density },
        Hw    { Water enthalpy },
        rw    { Water density },
        Ts    { Steam temperature }
  local Ed, Md, Er, Mr { Energies and masses }

    { Total energy balance. }
    Ed = rs * Vst * Hs + rw * Vwt * Hw + m * cp * Tm
    der(Ed) = Pow + qfw * Hfw - qs * Hs

    { Total mass balance. }
    Md = rs * Vst + rw * Vwt
    der(Md) = qfw - qs

    { Energy balance in risers. }
    Er = rs * am * Vr * Hs + rw * (1 - am) * Vr * Hw
    der(Er) = Pow + qdc * Hw - xr * qr * Hs
      - (1 - xr) * qr * Hw

    { Mass balance in risers. }
    Mr = am * rs * Vr + (1 - am) * rw * Vr
    der(Mr) = qdc - qr

    { Momentum balance for downcomers. }
    am * Vr * (rw - rs) = k * qdc * *2/2

    { Average steam-water volume ratio in the risers. }
    am = rw/(rw - rs) * (1 - rs/(rw - rs)/xr
      *ln(1 + (rw/rs - 1) * xr))

    { Total steam volume. }
    Vst = Vdrum - Vw + am * Vr

    { Total water volume. }
    Vwt = Vw + Vdc + (1 - am) * Vr

    { Drum water level. }
    dl = (Vw + am * Vr)/Adrum

    { Steam and water properties. }
    Hs = H2P(p)
    rs = 1/V2P(p)
    Hw = H1P(p)
    rw = 1/V1P(p)
    Ts = TP(p)
    Tm = Ts
end
```

The model of the drum is in itself interesting. It is of index two because of the constraint that the sum of the water volume and steam volume must be equal to the drum volume including risers and downcomers. This means that water and steam volumes cannot both be chosen as state variables. the index can be reduced by differentiating certain equations [17]. The index reduction technique also makes it possible to choose more appropriate state variables p, V_w, and x_r than the variables appearing differentiated E_d, M_d, E_r, and M_r. When solving for the corresponding derivatives, a system of 13 linear simultaneous equations is detected. It can be reduced to a 3×3 system by use of tearing, which can then be solved either symbolically or numerically [8]. In [2], the determination of which equations to differentiate, the differentiation, elimination of variables, and the solution of the linear system of equations were done manually. Dymola is capable of performing all these steps automatically.

7.1.7 Conclusions

This article has shown energy flow modeling to be a corner stone in safe descriptions of physical systems. Many potential problems can be avoided by systematically applying this approach to modeling. However, energy flow modeling puts heavy demands on the modeling software. Only a full-fledged object-oriented modeling tool, such as Dymola, can satisfy these demands in all circumstances.

A very important problem has not been dwelled upon in this article. Evidently, it is never possible to consider all aspects of a system in a model. Any model must always represent an idealization, a simplification of reality. How this is done, which facets of reality are left out from the model, is a question that needs to be addressed by the modeler. Finding the answer to this question is not something that can be automated in general. It requires intuition into how the system works, and a lot of experience on the side of the *human* modeler.

Yet, energy balances can assist the user also in this respect. It is a sound proposition to request that even after the model simplifications have become effective, all energy flows internal to the model are still balanced, and also that energy can enter or leave the model only through a limited number of well-defined and carefully monitored ports. Although it may not always be possible to decide on the basis of energy considerations alone, which facets of reality to keep in the model and which other facets to throw out, energy modeling *will* support the user in ensuring that simplifications, once decided upon, are implemented in a consistent fashion.

References

[1] Andersson, M. 1994. *Object-Oriented Modeling and Simulation of Hybrid Systems*, Ph.D. thesis TFRT-1043, Dept. of Automatic Control, Lund Inst. of Technology, Lund, Sweden.

[2] Åström, K.J. and Bell, R.D. 1988. Simple drum-boiler models. *Proc. IFAC Symp. Power Systems Modelling and Control Applications*, Brussels, Belgium.

[3] Bos, A.M. 1986. *Modelling Multibody Systems in Terms of Multibond Graphs with Application to a Motorcycle*, Ph.D. dissertation, Technical University Twente, Enschede, The Netherlands.

[4] Cellier, F.E. 1992. Hierarchical non-linear bond graphs: a unified methodology for modeling complex physical systems. *Simulation*, 58(4):230–248.

[5] Cellier, F.E. and Elmqvist, H. 1993. Automated formula manipulation supports object-oriented continuous-system modeling. *IEEE Control Systems*, 13(2):28–38.

[6] Cellier, F.E., Otter, M., and Elmqvist, H. 1995. Bond graph modeling of variable structure systems. *Proc. ICBGM '95, 2nd Int. Conf. on Bond Graph Modeling and Simulation*, Las Vegas, NV, pp. 49–55.

[7] Elmqvist, H. 1978. *A Structured Model Language for Large Continuous Systems*, Ph.D. dissertation. Report CODEN:LUTFD2/(TFRT-1015), Dept. of Automatic Control, Lund Inst. of Technology, Lund, Sweden.

[8] Elmqvist, H., and Otter, M. 1994. Methods for Tearing Systems of Equations in Object-Oriented Modeling. *Proc. ESM '94, European Simulation Multiconference*, Barcelona, Spain, pp.326–332.

[9] Elmqvist, H. 1995. *Dymola — User's Manual*, Dynasim AB, Research Park Ideon, Lund, Sweden.

[10] Holman, J.P. 1992. *Heat Transfer*, 7th ed., McGraw-Hill, New York.

[11] Huelsman, L.P. 1984. *Basic Circuit Theory* 2nd ed., Prentice Hall, Englewood Cliffs, N.J.

[12] Karnopp, D.C., Margolis, D.L., and Rosenberg, R.C. 1990. *System Dynamics: A Unified Approach*, 2nd ed., John Wiley & Sons, New York.

[13] Kron, G. 1962. *Diakoptics — The Piecewise Solution to Large-Scale Systems*, MacDonald & Co., London.

[14] Lindahl, S. 1976. *A Non-linear Drum Boiler — Turbine Model*, Report TFRT-3132, Dept. of Automatic Control, Lund Inst. of Technology, Lund, Sweden.

[15] Nikravesh, P.E. 1988. *Computer-Aided Analysis of Mechanical Systems*, Prentice Hall, Englewood Cliffs, N.J.

[16] Otter, M., Elmqvist, H., and Cellier, F.E. 1993. Modeling of multibody systems with the object-oriented modeling language Dymola. *Proc. NATO/ASI, Computer-Aided Analysis of Rigid and Flexible Mechanical Systems*, Troia, Portugal, Vol. 2, pp. 91–110.

[17] Pantelides, C.C. 1988. The consistent initialization of differential-algebraic systems. *SIAM J. Sci. Stat. Comput.*, 9:213–231.

[18] Paynter, H.M. 1961. *Analysis and Design of Engineering Systems*, MIT Press, Cambridge, MA.

[19] Thoma, J.U. 1990. *Simulation by Bondgraphs*, Springer-Verlag, Berlin.

7.2 System Identification When Noise Is Negligible

William S. Levine, Department of Electrical Engineering, University of Maryland, College Park, MD

7.2.1 Introduction

Suppose a system is handed to you and you are asked to "identify" it. To make the problem vivid and precise, suppose the system is encased in a black box that allows access only to the system's input and output terminals. It is then impossible to discover what system is actually inside the box because many physical systems have identical input-output behavior. What is possible, in some cases, is to determine a model of the input-output behavior that is adequate to predict the response of the system to any possible input. Once this input-output model is known, the system is "identified".

Such black box identification problems fall into one of two categories, those for which the observations are corrupted by noise (unpredictable disturbances) and those for which the observations are perfect (correspond exactly to the true input and output signals). Of course, real observations always include some noise but the perfect observation case models the situation where the noise is negligible. It is this case that is treated here for continuous-time single-input single-output systems. The results can be applied to multi-input multi-output systems by identifying one input-output pair at a time. The identification of discrete-time systems in the presence of noise is discussed in Chapter 58.

If one really knows nothing about the system, the identification problem is still impossible. For example, if the system is rapidly and unpredictably time-varying, then identification experiments performed at one time interval provide no information about the system's behavior at other times. Thus, in order for the identification problem to be meaningful it is necessary to assume that the unknown system is time-invariant. As usual, this is an idealization of the common reality that the time variation of the system is negligible.

The identification of arbitrary nonlinear time-invariant noiseless systems is, at least, difficult. Adding the assumption that the system in the black box is linear makes the identification problem tractable. Thus, this chapter deals with the identification (in the sense described in the previous paragraphs) of linear time-invariant (LTI) systems.

There are basically two methods. Identification by means of a set of sinusoidal signals will be described first, in Section 7.2.2. Section 7.2.3 describes the second method, identification by means of signals that approximate steps, impulses, or white noise. The chapter concludes with a brief admonition.

7.2.2 Identification by Sinusoidal Inputs

As demonstrated in Chapter 5.1 the input-output behavior of any stable continuous-time LTI system can be completely described by its transfer function, $H(jw)$ for all real w ($-\infty < w < \infty$). The transfer function can be measured for a given fixed input frequency, w_0, as follows. Apply the input

$$u(t) = a\cos w_0 t \qquad -\infty < t < \infty, \quad \text{where } a \text{ is real and } > 0, \quad \text{and } w_0 \text{ is real and } \geq 0. \tag{7.25}$$

It is, of course, not possible to apply an input starting at $-\infty$ and extending to $+\infty$ to a real system. Equation 7.25 is an idealization of an input $a\cos w_o t$ that is applied for a long enough time for the system to reach steady state. Because the system is stable, all the effects of initial conditions become negligible in finite time. Once all these transients have decayed to zero, the system is in steady state and its output, $y(t)$, is given by

$$y(t) = a|H(jw_0)|\cos(w_0 t + \measuredangle H(jw_0)) \qquad -\infty < t < \infty \tag{7.26}$$

The transfer function at w_0,

$$H(jw_0) = |H(jw_0)|e^{j\measuredangle H(jw_0)} \tag{7.27}$$

can be read from an oscilloscope display of $y(t)$. For real systems, knowing $H(jw_0)$ for any positive w_0 implies knowledge of $H(-jw_o)$ because

$$H(-jw_0) = |H(jw_0)|e^{-j\measuredangle H(jw_0)} \tag{7.28}$$

In practice, this measurement can be made in a reasonable length of time. It is thus possible to repeat the measurement for other values of w until $H(jw)$ is known for a collection of n values of w, as illustrated in Figure 7.13. Is it possible to reconstruct $H(jw)$, $-\infty < w < \infty$, from a collection of $H(jw_k)$, $k = 0, 1, 2, \ldots, n-1$?

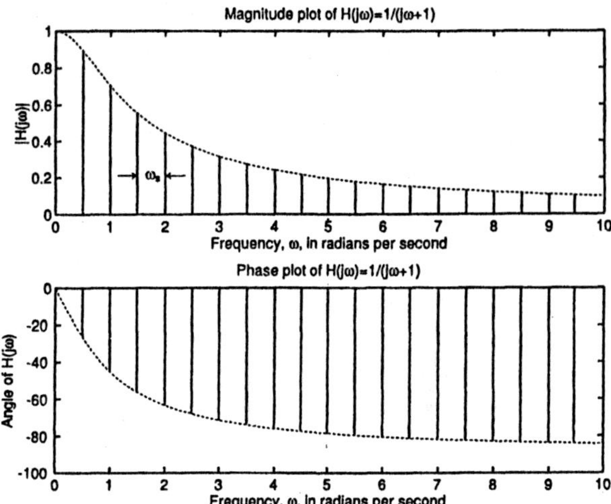

Figure 7.13 $|H(jw)|$ and $\measuredangle\ H(jw)$ vs. w for $H(jw) = 1/(jw+1)$. The dotted lines are the values at every w, $0 \leq w \leq 10$. The vertical lines are the frequency-sampled values, at $w = 0, w_s, 2w_s, \ldots, 20w_s$.

A precise answer can be obtained under the following assumptions.

A1. The impulse response, $h(t)$, satisfies

$$h(t) = 0 \qquad \text{for all } |t| \geq T \tag{7.29}$$

A2. $H(jw)$ is measured at $w = w_k$, $k = 0, 1, 2, \ldots$ and

$$w_k = kw_s \tag{7.30}$$

(In other words, $H(jw)$ is sampled at every frequency that is an integer multiple of w_s).

A3.

$$\frac{2\pi}{w_s} > 2T \tag{7.31}$$

If these assumptions are satisfied, $h(t)$ (and hence $H(jw)$) can be reconstructed perfectly from the infinite set of measurements

$$H(jw_k), \quad w_k = kw_s, \quad k = 0, 1, 2, \ldots \tag{7.32}$$

A proof follows immediately from the dual of the sampling theorem [1, pp. 540–543]. As a practical matter, $H(jw)$ goes to zero as w increases toward infinity so one can set $H(jw_k) =$

0 for $k > N$ for some finite N. Theoretically, there is also a problem with T. The impulse response of a stable LTI system generally approaches zero exponentially as $t \to \infty$. Thus, A1 is not precisely true. As a practical matter, one can always choose a reasonable finite value for T and use a smaller value of w_s than the maximum given by Equation 7.31.

The reconstruction procedure based on the dual of the sampling theorem is, in most cases, not the method used to interpolate between the data points in applications. The main reason for this is that the resulting transfer function, $H(jw)$, is not a rational function. That is, $H(jw)$ cannot be written as a ratio of polynomials in jw. Many techniques for analyzing and designing control systems, such as root locus plots, require that $H(jw)$ be a rational function. However, it is important to be aware that control system design by means of Bode, Nichols, or Nyquist plots can be accomplished directly from measured values of $H(jw)$, $-w_m < w < w_m$. There is no need to write an explicit formula for $H(jw)$ or to find the poles and zeros of $H(jw)$.

Why not just connect the data points by straight line segments? First, the result is not a rational function. More importantly, straight line interpolation can underestimate the maximum of $|H(jw)|$ and this can be dangerous. An example is given shortly. Thus, the best approach to interpolation of the measurements $H(jkw_s)$, $k = 0, 1, 2, \ldots, w_N$, is to fit a rational function to the data. Chapter 58 contains a nice introduction to curve-fitting by rational functions. Ljung's discussion is for discrete-time systems in the time domain. But, with minor adjustments the ideas he presents apply equally well in the frequency domain.

What happens to this identification procedure if the system is not really linear? One of the strengths of identification by means of pure cosinusoidal inputs is that nonlinearity is easily detected. Nonlinearity will distort the output signal so that $y(t)$ no longer has exactly the form of Equation 7.26. For large amounts of nonlinearity, the distortion of $y(t)$ would be visible to the naked eye from an oscilloscope display. For example, an input that caused the system to saturate would produce a $y(t)$ that was flattened at its peaks. More generally, a spectral analysis of $y(t)$ [2] could be used to determine whether frequencies other than w_0 were present. The presence in $y(t)$ of other frequencies than w_0 is proof that the system is not linear.

To summarize, direct measurement of the transfer function by means of pure cosinusoidal inputs is conceptually easy to do and has a convenient built-in way to check that the system is approximately linear. The only real objection to the use of this method is that it can be time-consuming. One has to repeat the measurements at a large number of frequencies and each measurement requires enough time for the system to reach steady state. There are several ways to speed up the measurement process.

Several companies sell spectrum and network analyzers that can be used to measure $H(jw)$ and to determine the frequency content of $y(t)$ when $u(t) = a\cos w_o t$. To speed up the identification process, many network analyzers can input a swept sinusoidal input, that is, an input that is approximately cosinusoidal with slowly increasing frequency. Such an input signal can determine $H(jw)$, $-w_m < w < w_m$ (w_m is some value of w beyond which $|H(jw)|$ is very small ($\ll 1$)) from one sweep. See [2], [3], or [4] for more details. Note also that modern spectrum and network analyzers often contain software for fitting a rational function to the data they measure. See [3] or [4] for an example.

As with all measurements on real systems it is important to be observant and careful in measuring $H(jw)$. It is easy to construct examples of systems for which the measurement of $H(jw)$ is tricky. One such is the following. Suppose you are given a system

$$H(jw) = \frac{w_n^2}{(jw)^2 + 2\zeta w_n(jw) + w_n^2} \tag{7.33}$$

Assume also that ζ is very small so the system is very lightly damped. The corresponding impulse response is [1]

$$h(t) = \begin{cases} \frac{w_n e^{-\zeta w_n t}}{\sqrt{1-\zeta^2}} \sin(w_n\sqrt{1-\zeta^2}t) & t > 0 \\ 0 & t \le 0. \end{cases} \tag{7.34}$$

The time constant is $1/\zeta w_n$ so the duration of $h(t)$, T in (7.29), can be taken to be $5/\zeta w_n$. The choice of 5 time constants is common, but arbitrary. Both larger and smaller values are plausible. This forces $w_s < \pi\zeta w_n/5$, as required by (7.31). The exact maximum of $|H(jw)|$ occurs at $w_{\max} = \sqrt{1-2\zeta^2}w_n$ and has magnitude $|H(jw_{\max})| = \frac{1}{2\zeta\sqrt{1-\zeta^2}}$. If $\zeta = .01$, this maximum of $|H(jw)|$ is approximately 50. This much amplification may drive the system into saturation even though the input amplitude is small enough to avoid saturation at most other frequencies. In addition, the peak in $|H(jw)|$ is very narrow. Even with a small value of w_s it is unlikely that the exact maximum of $H(jw)$ will be found at kw_s for any integer k. Because exact reconstruction of $H(jw)$ for samples w_s apart is complicated most people simply interpolate between the measured values by straight lines. In other words, they connect the data points by straight lines. In this example straight line interpolation or unnoticed saturation would result in seriously over estimating the damping coefficient, ξ, unless you were very lucky.

It is dangerous to overestimate the damping coefficient of a lightly damped pair of poles. Thus, it is important to check for saturation whenever peaks occur in $|H(jw)|$. In the vicinity of such peaks, one should fit a second-order transfer function to the data [3, 4].

7.2.3 Identification in the Time Domain

Because continuous-time LTI systems are completely characterized by their impulse response, $h(t)$, one naturally thinks of trying to measure $h(t)$ directly. The obvious thing to do is to input an impulse, $\delta(t)$, and measure the response. It is equally obvious that this is not possible. You cannot create an input $\delta(t)$. The best you could do would be some sort of approximation. For example, people frequently suggest and occasionally use the signal ($\epsilon > 0$ is a small number)

$$\hat{\delta}_\epsilon(t) = \begin{cases} 1/\epsilon & 0 \le t < \epsilon \\ 0 & \text{otherwise} \end{cases} \tag{7.35}$$

as an approximation to $\delta(t)$. There are two practical problems with Equation 7.35. First, if ϵ is small, the magnitude of $\hat{\delta}_\epsilon(0)$ is large—probably large enough to drive the system into saturation. Second, if one constrains the maximum amplitude of $\hat{\delta}_\epsilon(t)$, the result is a signal containing relatively little energy. As a result, the response of the system to $\hat{\delta}_\epsilon(t)$ is usually small and difficult to measure.

Because of these two problems, the impulse response is rarely measured by means of an approximation to $\delta(t)$. There are two alternatives that are used much more commonly. The integral from $-\infty$ to t of the unit impulse is known as the unit step. It will be denoted by

$$u_{-1}(t) = \begin{cases} 1 & t > 0 \\ 0 & t \leq 0. \end{cases} \tag{7.36}$$

The input $au_{-1}(t)$, where a is a real number > 0, is often used for identifying and characterizing systems, even nonlinear ones. There are several chapters in this handbook that illustrate the use of step inputs to identify and characterize systems. See particularly Chapters 52 and 10.1. A step input has the advantage that it supplies infinite energy (arbitrarily large energy in practice) for any value of a. Another advantage of the step input is that a step is a cosinusoid for which $w = 0$. Thus, the step gives an accurate measurement of $H(j0)$ provided you wait long enough for the transients to die out.

The major disadvantage of using a step input to identify a continuous-time LTI system is that the fast transients are not easily seen. It is easiest to demonstrate this by means of the transfer function, $H(jw)$. Denote the response of an LTI system having impulse response $h(t)$ to a unit step by $y_s(t)$.

$$y_s(t) = \int_{-\infty}^{\infty} h(\tau)u_{-1}(t-\tau)d\tau = \int_{-\infty}^{t} h(\tau)d\tau, \; t > 0 \tag{7.37}$$

Taking Fourier transforms of both sides gives

$$Y_s(jw) = \frac{H(jw)}{jw} \quad \text{for all } w \neq 0 \tag{7.38}$$

For large w, $|Y_s(jw)|$ will be very small. Any measurement error, or noise, will effectively hide whatever information that is present in $Y_s(jw)$ about $H(jw)$. Thus, although the measured step response is a good descriptor of the low frequency (slow) components of the response of an LTI system, the fast, high frequency, aspects of the system's behavior are not easily visible in the step response.

A very useful alternative to a step input for identifying LTI systems is white noise. See Chapter 34 for the basics of white noise. It is again easiest to explain the use of white noise in the identification of continuous-time LTI systems by means of a frequency domain (Fourier transform) analysis. It is shown, (in Chapter 34) that white noise has the spectral density function

$$s(w) = a \quad \text{for } a > 0 \quad \text{for all } -\infty < w < \infty. \tag{7.39}$$

It is also shown there that the spectral density of the output, $s_y(\omega)$, of a LTI system driven by a white noise input is

$$s_y(w) = a|H(jw)|^2 \qquad -\infty < w < \infty \tag{7.40}$$

where $H(jw)$ is the transfer function of the LTI system.

Thus, $|H(jw)|$ can be obtained from $s_y(w)$ by dividing by a and taking the positive square root of the result. The $\measuredangle H(jw)$ can be obtained from $|H(jw)|$, provided $H(jw)$ has no poles or zeros in the closed right half-plane (that is, no poles or zeros with real part greater than or equal to zero), by means of Bode's gain-phase relation (see Chapter 31).

Modern spectrum and network analyzers are generally capable of producing signals that are good approximations to white noise for frequencies below some value, say w_m. This signal can be used as a test input to an LTI system and the corresponding response, $y(t)$, can be recorded. If the duration of the white noise test input is given by $2T$, then w_s in Equation 7.31 is the resolution of the measured frequency response. Thus, Equation 7.31 and the desired frequency resolution can be used to determine the required duration of the test input. These same spectrum and network analyzers usually contain hardware and software that compute good estimates of $s_y(w)$ and $|H(jw)|$ for $|w| < w_m$ from $y(t), 0 \leq 2T$. See [2]–[4] for information about the details of the calculations. In fact, today these computations are done digitally on sampled time signals. Chapter 58 describes the discrete-time calculations that need to be performed to compute $s_y(w)$, $s_{yu}(w)$, and $s_u(w)$ in the continuous-time case. Note that the author uses Φ instead of s to denote spectral density.

White noise is a good input signal for the accurate determination of the resonant peaks of $|H(jw)|$ although one must be careful to keep the system in its linear range. Thus, the use of white noise for system identification complements the use of cosinusoidal signals. However, there is a serious potential difficulty with identification by means of white noise. It is impossible to detect the presence of nonlinearity, or measurement noise (as opposed to the white *input* noise) from $s_y(w)$, $|w| < w_m$, alone. Thus, it is extremely important to have a means of detecting nonlinearities and measurement noise.

The cross spectral density, $s_{yu}(w)$, provides a test for nonlinearity of the system and for the presence of measurement noise. A precise definition of $s_{yu}(w)$ can be found in [5, pp. 346–347]. There are three essential points.

1. $s_{yu}(w)$ can be computed from $y(t)$ and $u(t)$, $0 \leq t < T_s$, by virtually the same methods as are used to compute $s_y(w)$.
2. For an LTI system with transfer function $H(jw)$,

$$s_y(w) = s_{yu}(w)H(jw) \tag{7.41}$$
$$s_{yu}(w) = s_u(w)H^*(jw) \tag{7.42}$$

where $*$ denotes complex conjugate and $s_u(w)$ is the spectral density of the input, $u(t)$. Equations 7.41 and 7.42 can be combined to give

$$\frac{|s_{yu}(w)|^2}{|s_y(w)| \cdot |s_u(w)|} = 1 \tag{7.43}$$

Note that the magnitude signs in the denominator are unnecessary; $s_y(w)$ and $s_u(w)$ are real numbers and greater than zero.

3. When $y(t)$ and $u(t)$ are completely unrelated, as would be the case if $y(t)$ were entirely measurement noise, then

$$|s_{yu}(w)| = 0 \quad (7.44)$$

The ratio

$$\frac{|s_{yu}(w)|^2}{|s_y(w)| \cdot |s_u(w)|} \quad |w| < w_m \quad (7.45)$$

is sometimes called the coherence [3], [4] in the spectrum and network analyzer literature. The coherence is bounded, for every w, between 0 and 1. A coherence value of 1 for all $w, 0 \leq |w| < w_u$, means that the system being tested is LTI and there is no measurement noise. Values of the coherence that are much less than one for any $w, 0 \leq |w| < w_m$, are warnings of nonlinearity and/or measurement noise. Certainly, one needs to be very cautious when the coherence is not close to one.

As was noted earlier, white noise and cosinusoidal input signals are complimentary for purposes of identification. It is a good idea to use both whenever possible.

7.2.4 Conclusions

This chapter has addressed the identification of continuous-time LTI systems when the identification can be performed separately from control. In adaptive control systems the system identification has to occur in real time. Real time, recursive, identification is a different problem from the one discussed here. Chapter 58 is a good introduction to recursive identification.

Off-line, nonrecursive, identification has the advantage that the measurements can continue until one is satisfied with the accuracy of system identification. However, as was mentioned earlier, the user needs to be aware of the identification errors that can result from nonlinearities and measurement noise. Even when the system is truly LTI, the accuracy of the identification is dependent on many measurement details including the extent to which the spectrum of the input signal matches the assumptions implicit in this chapter (perfect cosinusoid, step, or white noise), and the precision with which the spectral densities are computed. See [2]–[4] for information on the accuracy with which $H(jw)$ can be measured. However, with care and modern equipment, the methods described in this chapter can be very effective.

References

[1] Oppenheim, A.V. and Willsky. A.S., with Yeung, I.T., *Signals and Systems*, Prentice Hall, 1983.

[2] Witte, R.A., *Spectrum and Network Measurements*, Prentice Hall, 1991.

[3] HP 35665A Dynamic Signal Analyzer Concepts Guide, Hewlett Packard Co., 1991.

[4] HP 35665A Operator's Reference, Hewlett Packard Co., 1991.

[5] Papoulis, A., *Probability, Random Variables, and Stochastic Processes*, McGraw-Hill, 1965.

SECTION III

Analysis and Design Methods for Continuous-Time Systems

8

Analysis Methods

Raymond T. Stefani
Electrical Engineering Department, California State University, Long Beach

William A. Wolovich
Brown University

8.1 Time Response of Linear Time-Invariant Systems

Raymond T. Stefani, *Electrical Engineering Department, California State University, Long Beach*

8.1.1 Introduction

Linear time-invariant systems[1] may be described by either a scalar nth order linear differential equation with constant coefficients or a coupled set of n first-order linear differential equations with constant coefficients using state variables. The solution in either case may be separated into two components: the **zero-state response**, found by setting the initial conditions to zero; and the **zero-input response**, found by setting the input to zero. Another division is into the **forced response** (having the form of the input) and the **natural response** due to the characteristic polynomial.

8.1.2 Scalar Differential Equation

Suppose the system has an input (forcing function) $r(t)$ with the resulting output being $y(t)$. As an example of a second-order linear differential equation with constant coefficients

$$\frac{d^2y}{dt} + 6\frac{dy}{dt} + 8y = \frac{dr}{dt} + 8r \tag{8.1}$$

[1]This section includes excerpts from *Design of Feedback Control Systems*, Third Edition by Raymond T. Stefani, Clement J. Savant, Barry Shahian, and Gene H. Hostetter, copyright ©1994 by Saunders College Publishing, reprinted by permission of the publisher.

The Laplace transform may be evaluated where the initial conditions are taken at time 0^- and $y'(0^-)$ means the value of dy/dt at time 0^-.

$$\begin{aligned} s^2Y(s) - sy(0^-) &- y'(0^-) + 6[sY(s) - y(0^-)] + 8Y(s) \\ &= sR(s) - r(0^-) + 8R(s) \\ Y(s)[s^2 + 6s + 8] &= (s+8)R(s) + sy(0^-) \\ &\quad + y'(0^-) + 6y(0^-) - r(0^-) \end{aligned} \tag{8.2}$$

In Equation 8.2 the quadratic $s^2 + 6s + 8$ is the **characteristic polynomial** while the first term on the right hand side is due to the input $R(s)$ and the remaining terms are due to initial conditions (the initial state). Solving for $Y(s)$

$$Y(s) = \underbrace{\left[\frac{s+8}{s^2+6s+8}\right]R(s)}_{\text{zero-state response}} + \underbrace{\frac{sy(0^-) + y'(0^-) + 6y(0^-) - r(0^-)}{s^2+6s+8}}_{\text{zero-input response}} \tag{8.3}$$

In Equation 8.3 the zero-state response results by setting the initial conditions to zero while the zero-input response results from setting the input $R(s)$ to zero. The system transfer function results from

$$T(s) = \left.\frac{Y(s)}{R(s)}\right|_{\text{initial conditions}=0} = \frac{s+8}{s^2+6s+8} \tag{8.4}$$

Thus, the zero-state response is simply $T(s)R(s)$. The denominator of $T(s)$ is the characteristic polynomial which, in this case, has roots at -2 and -4.

0-8493-8570-9/96/$0.00+$.50

To solve Equation 8.4, values must be established for the input and the initial conditions. With a unit-step input and choices for $y(0^-)$ and $y'(0^-)$

$$\begin{aligned} r(0^-) &= 0 \quad R(s) = 1/s \quad y(0^-) = 10 \quad y'(0^-) = -4 \\ Y(s) &= \underbrace{\left[\frac{s+8}{s^2+6s+8}\right]\frac{1}{s}}_{\text{zero-state response}} + \underbrace{\frac{10s+56}{s^2+6s+8}}_{\text{zero-input response}} \end{aligned} \quad (8.5)$$

Next, the zero-state and zero-input responses can be expanded into partial fractions for the poles of $T(s)$ at -2 and -4 and for the pole of $R(s)$ at $s = 0$.

$$Y(s) = \underbrace{\overbrace{\frac{1}{s}}^{\text{forced response}} \overbrace{- \frac{1.5}{s+2} + \frac{0.5}{s+4}}_{\text{zero-state response}} + \underbrace{\frac{18}{s+2} - \frac{8}{s+4}}^{\text{natural response}}}_{\text{zero-input response}} \quad (8.6)$$

In this case, the forced response is the term with the pole of $R(s)$, and the natural response contains the terms with the poles of $T(s)$, since there are no common poles between $R(s)$ and $T(s)$. When there are common poles, those multiple poles are usually assigned to the forced response.

The division of the total response into zero-state and zero-input components is a rather natural and logical division because these responses can easily be obtained empirically by setting either the initial conditions or the input to zero and then obtaining each response. The forced response cannot be obtained separately in most cases, so the division into forced and natural components is more mathematical than practical.

The inverse transform of Equation 8.6 is

$$y(t) = \underbrace{\overbrace{1}^{\text{forced response}} - 1.5e^{-2t} + 0.5e^{-4t}}_{\text{zero-state response}} + \underbrace{18e^{-2t} - 8e^{-4t}}_{\text{zero-input response}} \quad (8.7)$$

(natural response: $-1.5e^{-2t} + 0.5e^{-4t} + 18e^{-2t} - 8e^{-4t}$)

The total response is therefore

$$y(t) = 1 + 16.5e^{-2t} - 7.5e^{-4t} \quad (8.8)$$

Figure 8.1 contains a plot of $y(t)$ and its components.

8.1.3 State Variables

A linear system may also be described by a coupled set of n first-order linear differential equations, in this case having constant coefficients.

$$\begin{aligned} \frac{dx}{dt} &= Ax + Br \\ y &= Cx + Dr \end{aligned} \quad (8.9)$$

where x is an nx1 column vector. If r is a scalar and there are m outputs, then A is nxn, B is $nx1$, C is mxn and D is $mx1$.

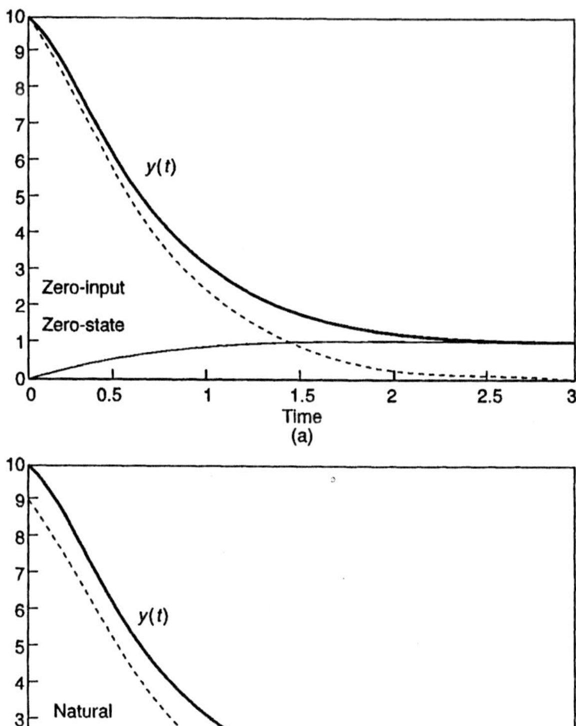

Figure 8.1 $y(t)$ and components. (a) Zero-state and zero-input components. (b) Forced and natural components.

In most practical systems, D is zero because there is not usually an instantaneous output response due to an applied input. The Laplace transform can be obtained in vector form, where I is the identity matrix

$$\begin{aligned} sIX(s) - x(0^-) &= AX(s) + BR(s) \\ Y(s) &= CX(s) + DR(s) \end{aligned} \quad (8.10)$$

Solving

$$\begin{aligned} X(s) &= \underbrace{(sI-A)^{-1}BR(s)}_{\text{zero-state response}} + \underbrace{(sI-A)^{-1}x(0^-)}_{\text{zero-input response}} \\ Y(s) &= \underbrace{[C(sI-A)^{-1}B + D]R(s)}_{\text{zero-state response}} \\ &\quad + \underbrace{C(sI-A)^{-1}x(0^-)}_{\text{zero-input response}} \end{aligned} \quad (8.11)$$

Thus, the transfer function $T(s)$ becomes $[C(sI-A)^{-1}B+D]$.

The time response can be found in two ways. The inverse Laplace transform of Equation 8.11 can be taken, or the response can be calculated by using the **state transition matrix** $\Phi(t)$, which is the inverse Laplace transform of the **resolvant matrix** $\Phi(s)$.

$$\Phi(s) = (sI-A)^{-1} \quad \Phi(t) = L^{-1}\{\Phi(s)\} \quad (8.12)$$

Figure 8.2 contains a second-order system in state-variable form. The system is chosen to have the dynamics of Equation 8.1. From Figure 8.2

$$\begin{bmatrix} dx_1/dt \\ dx_2/dt \end{bmatrix} = \begin{bmatrix} 0 & 1 \\ -8 & -6 \end{bmatrix}\begin{bmatrix} x_1 \\ x_2 \end{bmatrix} + \begin{bmatrix} 0 \\ 1 \end{bmatrix} r$$
$$y = \begin{bmatrix} 8 & 1 \end{bmatrix}\begin{bmatrix} x_1 \\ x_2 \end{bmatrix} \tag{8.13}$$

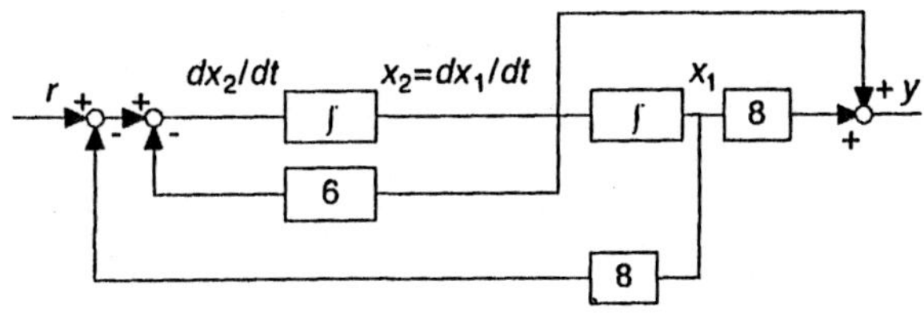

Figure 8.2 Second order system in state variable form.

Thus, D is zero. The resolvant matrix is

$$\begin{aligned} \Phi(s) &= (sI - A)^{-1} = \begin{bmatrix} s & -1 \\ 8 & s+6 \end{bmatrix}^{-1} \\ &= \frac{1}{s^2+6s+8}\begin{bmatrix} s+6 & 1 \\ -8 & s \end{bmatrix} \end{aligned} \tag{8.14}$$

8.1.4 Inverse Laplace Transform Approach

First, the time response for $y(t)$ is calculated using the inverse Laplace transform approach. The transfer function is as before since D is zero and

$$\begin{aligned} C(sI - A)^{-1}B &= \frac{\begin{bmatrix} 8 & 1 \end{bmatrix}}{s^2+6s+8}\begin{bmatrix} s+6 & 1 \\ -8 & s \end{bmatrix}\begin{bmatrix} 0 \\ 1 \end{bmatrix} \\ &= \frac{s+8}{s^2+6s+8} \end{aligned} \tag{8.15}$$

Suppose a unit-step input is chosen so that $R(s)$ is $1/s$. It follows that the zero-state response of Equation 8.11 is the same as in Equations 8.5 to 8.7 since both $T(s)$ and $R(s)$ are the same. Suppose $x_1(0^-)$ is 1 and $x_2(0^-)$ is 2. The zero-input response becomes

$$\begin{aligned} C(sI - A)^{-1}x(0^-) &= \frac{\begin{bmatrix} 8 & 1 \end{bmatrix}}{s^2+6s+8}\begin{bmatrix} s+6 & 1 \\ -8 & s \end{bmatrix}\begin{bmatrix} 1 \\ 2 \end{bmatrix} \\ &= \frac{10s+56}{s^2+6s+8} \end{aligned} \tag{8.16}$$

The zero-input response is also the same as in Equations 8.5 to 8.7 because the initial conditions on the state variables cause the same initial conditions as were used for y; that is,

$$\begin{aligned} y(0^-) &= 8x_1(0^-) + x_2(0^-) = 8 + 2 = 10 \\ y'(0^-) &= 8dx_1/dt(0^-) + dx_2/dt(0^-) \\ &= 8x_2(0^-) + [-8x_1(0^-) - 6x_2(0^-) + r(0^-)] \\ &= 16 + [-8 - 12 + 0] = -4 \end{aligned}$$

8.1.5 State Transition Matrix Approach

The second procedure for calculating the time response is to use the state transition matrix. $\Phi(s)$ in Equation 8.14 may be expanded in partial fractions to obtain $\Phi(t)$

$$\begin{aligned} \Phi(s) &= \begin{bmatrix} (2/s+2 - 1/s+4) & (.5/s+2 - .5/s+4) \\ (-4/s+2 + 4/s+4) & (-1/s+2 + 2/s+4) \end{bmatrix} \\ \Phi(t) &= \begin{bmatrix} (2e^{-2t} - e^{-4t}) & (.5e^{-2t} - .5e^{-4t}) \\ (-4e^{-2t} + 4e^{-4t}) & (-e^{-2t} + 2e^{-4t}) \end{bmatrix} \end{aligned} \tag{8.17}$$

Using $\Phi(t)$, the solution to Equation 8.11 is

$$\begin{aligned} x(t) &= \underbrace{\int_0^t \Phi(t-\tau)Br(\tau)d\tau}_{\text{zero-state response}} + \underbrace{\Phi(t)x(0^-)}_{\text{zero-input response}} \\ y(t) &= \underbrace{\int_0^t C\Phi(t-\tau)Br(\tau)dt + Dr(t)}_{\text{zero-state response}} \\ &+ \underbrace{C\Phi(t)x(0^-)}_{\text{zero-input response}} \end{aligned} \tag{8.18}$$

For this example, the zero-input response for $y(t)$ is

$$\begin{aligned} y_{zi}(t) &= \begin{bmatrix} 8 & 1 \end{bmatrix} \\ &\quad \begin{bmatrix} (2e^{-2t} - e^{-4t}) & (0.5e^{-2t} - 0.5e^{-4t}) \\ (-4e^{-2t} + 4e^{-4t}) & (-e^{-2t} + 2e^{-4t}) \end{bmatrix} \\ &\quad \begin{bmatrix} 1 \\ 2 \end{bmatrix} \\ &= 18e^{-2t} - 8e^{-4t} \end{aligned} \tag{8.19}$$

which agrees with Equation 8.7. The form for the zero-state response in Equation 8.18 is called a convolution integral. It is therefore necessary to evaluate the integral of

$$\begin{aligned} C\Phi(t-\tau)Bu(\tau) &= \begin{bmatrix} 8 & 1 \end{bmatrix} \\ &\quad \begin{bmatrix} \Phi_{11}(t-\tau) & \Phi_{12}(t-\tau) \\ \Phi_{21}(t-\tau) & \Phi_{22}(t-\tau) \end{bmatrix} \\ &\quad \begin{bmatrix} 0 \\ 1 \end{bmatrix} (1) \\ &= 8\Phi_{12}(t-\tau) + \Phi_{22}(t-\tau) \\ &= 3e^{-2(t-\tau)} - 2e^{-4(t-\tau)} \\ &= 3e^{-2t}e^{2\tau} - 2e^{-4t}e^{4\tau} \end{aligned} \tag{8.20}$$

After integrating with respect to τ

$$\begin{aligned} y_{zs}(t) &= (3/2)e^{-2t}[e^{2\tau}]\Big|_0^t - (2/4)e^{-4t}[e^{4\tau}]\Big|_0^t \\ &= 1.5e^{-2t}[e^{2t} - 1] - 0.5e^{-4t}[e^{4t} - 1] \quad (8.21) \\ &= 1.5[1 - e^{-2t}] - 0.5[1 - e^{-4t}] \\ &= 1 - 1.5e^{-2t} + 0.5e^{-4t} \end{aligned}$$

which agrees with Equation 8.7.

8.1.6 Computer Algorithms

Both the resolvant matrix and the state transition matrix can be calculated from A by a computer algorithm. The resolvant matrix is computable by an algorithm due to Fadeeva. The resolvant is written in the form

$$\begin{aligned} \Phi(s) &= N(s)/d(s) \\ N(s) &= s^{n-1}N_0 + s^{n-2}N_1 + \cdots + sN_{n-2} + N_{n-1} \\ d(s) &= s^n + d_1 s^{n-1} + \cdots + d_{n-1}s + d_n \end{aligned} \tag{8.22}$$

where $d(s)$ is the characteristic polynomial, $N(s)$ is a matrix, and the system is of order n. The matrices and coefficients can be found by a series of steps

$$\begin{array}{ll} N_0 = I & d_1 = -\text{trace}(A) \\ N_1 = N_0 A + d_1 I & d_2 = -(1/2)\text{trace}(N_1 A) \\ \cdot & \cdot \\ N_{n-1} = N_{n-2}A + d_{n-1}I & d_n = -(1/n)\text{trace}(N_{n-1}A) \end{array}$$

For the system of Equation 8.13, the calculations are

$$\begin{aligned} N_0 &= \begin{bmatrix} 1 & 0 \\ 0 & 1 \end{bmatrix}, \; d_1 = 6, \; N_1 = \begin{bmatrix} 6 & 1 \\ -8 & 0 \end{bmatrix}, \; d_2 = 8 \\ \Phi(s) &= \left\{ s\begin{bmatrix} 1 & 0 \\ 0 & 1 \end{bmatrix} + \begin{bmatrix} 6 & 1 \\ -8 & 0 \end{bmatrix} \right\} / (s^2 + 6s + 8) \end{aligned}$$

which agrees with Equation 8.14.

The state transition matrix can be calculated using a Taylor series for the matrix exponential

$$\begin{aligned} \Phi(t) = e^{At} \approx P(t) &= I + At + (1/2)A^2t^2 + \cdots \\ &+ (1/m!)A^m t^m \end{aligned} \tag{8.23}$$

See Chapter 21 in this handbook and [2] for a detailed discussion of algorithms for computing $\Phi(t) = e^{At}$ and related computational difficulties that may occur. One procedure is to choose a small value of t so the series of Equation 8.23 converges, then the time response is found by iterating in steps of t seconds. A good selection of t is

$$t \le 0.1/\textit{largest}\,|\textit{real part (eigenvalue of } A)| \tag{8.24}$$

In the example, the eigenvalues are -3 and -4 so t would be no more than $0.1/4$ or 0.025. If 0.02 is selected, the true value of $\Phi(t)$ and the value of the series for an increasing number of terms is as follows for four decimal digits:

$$\begin{bmatrix} 0.9985 & 0.0188 \\ -0.1507 & 0.8854 \end{bmatrix}$$

$$\begin{bmatrix} 1 & 0 \\ 0 & 1 \end{bmatrix}, \begin{bmatrix} 1 & 0.02 \\ -0.16 & 0.88 \end{bmatrix},$$

$$\begin{bmatrix} 0.9984 & 0.0188 \\ -0.1504 & 0.8856 \end{bmatrix}, \begin{bmatrix} 0.9985 & 0.0188 \\ -0.1507 & 0.8854 \end{bmatrix}$$

Thus, the series converges to four-decimal-digit accuracy in four terms.

For a step input, the convolution integral can be found by replacing $\Phi(t)$ in Equation 8.18 with $P(t)$ and removing the constant $Br(t)$ from the integral. By changing variables it can be shown that

$$\begin{aligned} \int_0^t \Phi(t-\tau)d\tau &= \int_0^t \Phi(t)dt = E(t) \\ &\approx It + (1/2)At^2 + (1/6)A^2t^3 + \cdots \\ &\quad + [1/(m+1)!]A^m t^{m+1} \end{aligned} \tag{8.25}$$

The small value of t from Equation 8.24 can be used to ensure convergence of Equation 8.25. The time response can then be found by iterating in steps of t seconds. Thus, Equation 8.18 becomes (where k is an integer 1, 2, 3, and so on)

$$\begin{aligned} x(kt) &= \underbrace{E(t)Br([k-1]t)}_{\text{zero-state response}} + \underbrace{P(t)x([k-1]t)}_{\text{zero-input response}} \\ y(kt) &= \underbrace{[CE(t)B + D]r([k-1]t)}_{\text{zero-state response}} + \underbrace{CP(t)x([k-1]t)}_{\text{zero-input response}} \end{aligned} \tag{8.26}$$

If t is 0.02, the series for $E(t)$ is as follows for an increasing number of terms until the first four decimal digits converge:

$$\begin{bmatrix} 0.0200 & 0 \\ 0 & 0.0200 \end{bmatrix}, \begin{bmatrix} 0.0200 & 0.0002 \\ -0.0016 & 0.0188 \end{bmatrix},$$

$$\begin{bmatrix} 0.0200 & 0.0002 \\ -0.0015 & 0.0188 \end{bmatrix}$$

Since D is 0 here, if k is 1 and t is 0.02, then from Equation 8.26

$$\begin{aligned} y(0.02) &= CE(0.02)Br(0) + CP(0.02)x(0) \\ &= \begin{bmatrix} 8 & 1 \end{bmatrix}\begin{bmatrix} 0.0200 & 0.0002 \\ -0.0015 & 0.0188 \end{bmatrix}\begin{bmatrix} 0 \\ 1 \end{bmatrix}(1) \\ &\quad + \begin{bmatrix} 8 & 1 \end{bmatrix}\begin{bmatrix} 0.9985 & 0.0188 \\ -0.1507 & 0.8854 \end{bmatrix}\begin{bmatrix} 1 \\ 2 \end{bmatrix} \\ &= 9.93 \end{aligned}$$

This result agrees with Equation 8.8, using $t = 0.02$.

8.1.7 Eigenvalues, Poles, and Zeros

It has been noted that the denominator of a transfer function $T(s)$ is the characteristic polynomial. In the previous examples, that denominator was $s^2 + 6s + 8$ with roots at -2 and -4. The characteristic polynomial may also be found from the system matrix since the characteristic polynomial is $|sI - A|$, where $|.|$ means the determinant. The eigenvalues of a matrix are those s values satisfying $|sI - A| = 0$; hence, the eigenvalues of A are the same as the poles of the transfer function. As in Equations 8.6 and 8.7, the poles of $T(s)$ establish terms present in the natural response. The coefficients of the partial fraction expansion (and, thus, the shape of the response) depend on the numerator of $T(s)$ which, in turn, depends on the zeros of $T(s)$. In fact, some zeros of $T(s)$ can cancel poles of $T(s)$, eliminating terms from the natural response.

To simplify this discussion, suppose the initial conditions are set to zero, and interest is focused on the zero-state response for a unit-step input. As one example, consider a system with a closed-loop transfer function $T_1(s)$

$$T_1(s) = \frac{6s^2+10s+2}{(s+1)(s+2)} = \frac{6(s+0.232)(s+1.434)}{(s+1)(s+2)}$$

The zero-state response for a unit-step input is

$$Y_1(s) = \frac{6s^2+10s+2}{s(s+1)(s+2)} = \frac{1}{s} + \frac{2}{s+1} + \frac{3}{s+2}$$

If the denominator of the transfer function (the characteristic polynomial) remains the same but the numerator changes to $3s^2 + 7s + 2$, the zero-state response changes considerably due to a cancellation of the pole at -2 by a zero at the same location:

$$\begin{aligned} T_2(s) &= \frac{3s^2+7s+2}{(s+1)(s+2)} = \frac{3(s+1/3)(s+2)}{(s+1)(s+2)} \\ Y_2(s) &= \frac{3s^2+7s+2}{s(s+1)(s+2)} = \frac{1}{s} + \frac{2}{s+1} \end{aligned}$$

Similarly, if the numerator changes to $4s^2 + 6s + 2$ there is a cancellation of the pole at -1 by a zero at -1.

$$\begin{aligned} T_3(s) &= \frac{4s^2+6s+2}{(s+1)(s+2)} = \frac{4(s+0.5)(s+1)}{(s+1)(s+2)} \\ Y_3(s) &= \frac{4s^2+6s+2}{s(s+1)(s+2)} = \frac{1}{s} + \frac{3}{s+2} \end{aligned}$$

The time responses $y_1(t)$, $y_2(t)$, and $y_3(t)$ are shown in Figure 8.3.

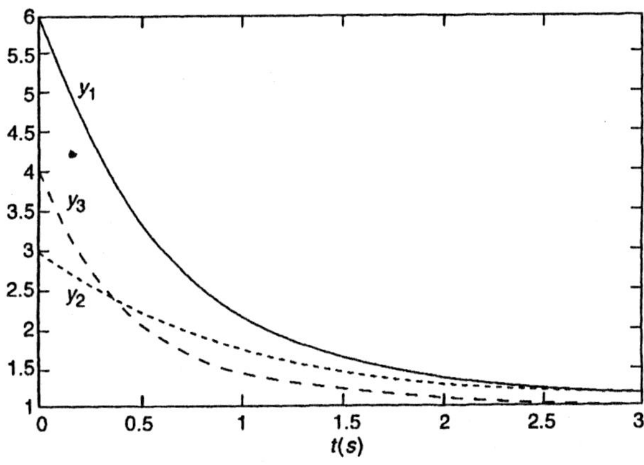

Figure 8.3 Time response example.

In summary, the terms in the time response are determined by the poles of the transfer function (eigenvalues of the system matrix) while the relative excitation of each term is dictated by the zeros of the transfer function.

8.1.8 Second-Order Response

Many systems are describable by a second-order linear differential equation with constant coefficients while many other higher-order systems have complex conjugate dominant roots that cause the response to be nearly second order. Thus, the study of second-order response characteristics is important in understanding system behavior. The standard form for a second-order transfer function is

$$T(s) = \frac{\omega_n^2}{s^2+2\zeta\omega_n+\omega_n^2} \tag{8.27}$$

where ζ is called the damping ratio and ω_n is called the undamped natural frequency. The roots of the characteristic polynomial depend on ζ as shown in Table 8.1. When the damping ratio exceeds one (overdamped response), there are two real roots, which are distinct; hence, the natural response contains two exponentials with differing time constants. When the damping ratio equals one (critically damped response), there are two equal real roots and the natural response contains one term $K_1 exp(-\omega_n t)$ and a second term $K_2 t exp(-\omega_n t)$. When $0 \leq \zeta < 1$, the resulting oscillatory response is called underdamped. The zero-state response for a unit-step input is

TABLE 8.1 Roots of a Second-Order Characteristic Polynomial.

Range for ζ	Type of Response	Root Locations
$\zeta > 1$	Overdamped	$-\zeta\omega_n \pm \omega_n(\zeta^2 - 1)^{1/2}$
$\zeta = 1$	Critically damped	$-\omega_n, \quad -\omega_n$
$0 \leq \zeta < 1$	Underdamped	$-\zeta\omega_n \pm j\omega_n(1-\zeta^2)^{1/2}$

$$\begin{aligned} y(t) &= 1 - (1/k)e^{-at}\cos(\omega t - \Theta) \\ k &= (1-\zeta^2)^{1/2} \quad \Theta = \tan^{-1}(\zeta/k) \\ a &= \zeta\omega_n \quad \omega = \omega_n(1-\zeta^2)^{1/2} \end{aligned} \tag{8.28}$$

When the damping ratio ζ is zero (undamped behavior), the system is marginally stable and there are complex roots $\pm j\omega_n$; hence, the radian frequency of the sinusoid becomes ω_n, explaining the term undamped natural frequency. When $0 < \zeta < 1$, the system is underdamped-stable and the radian frequency becomes $\omega = \omega_n(1-\zeta^2)^{1/2}$, called the damped natural frequency.

Figure 8.4 shows the unit-step response for various values of ζ from 0 to 1. To normalize (generalize) these plots, the horizontal axis is $\omega_n t$. Associated with this type of response are three figures of merit: percent overshoot, rise time, and settling time.

Percent overshoot is defined by

$$\%\ \text{overshoot} = 100\frac{\text{max value} - \text{steady-state value}}{\text{steady-state value}}$$

From Figure 8.4, notice that percent overshoot varies from 0 to 100%.

Rise time, Tr, is defined as the time required for the unit-step response to rise from 10% of the steady-state value to 90% of the steady-state value. Alternatively, rise time may be defined from 5% of the steady-state value to 95% of the steady-state value, but the 10 to 90% range is used here.

Settling time, Ts, is defined as the minimum time after which the response remains within ±5% of the steady-state value.

Figure 8.5a shows the product $\omega_n T_r$ vs. damping ratio ζ. Figure 8.5b shows percent overshoot vs. ζ. Figure 8.5c shows the

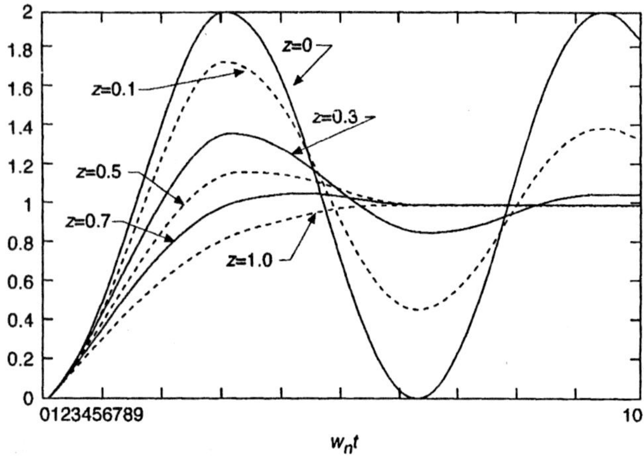

Figure 8.4 Second order zero-state unit step responses.

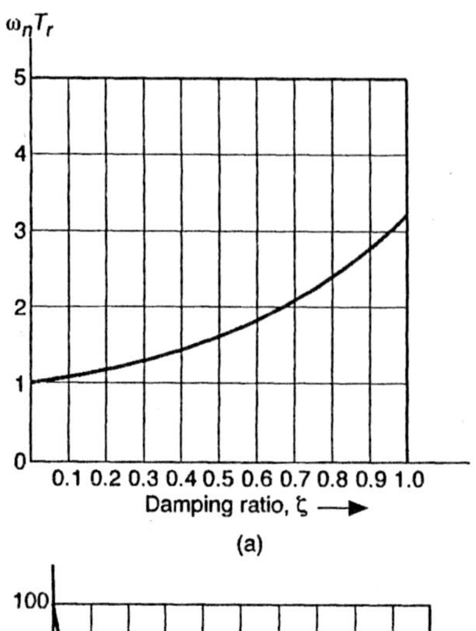

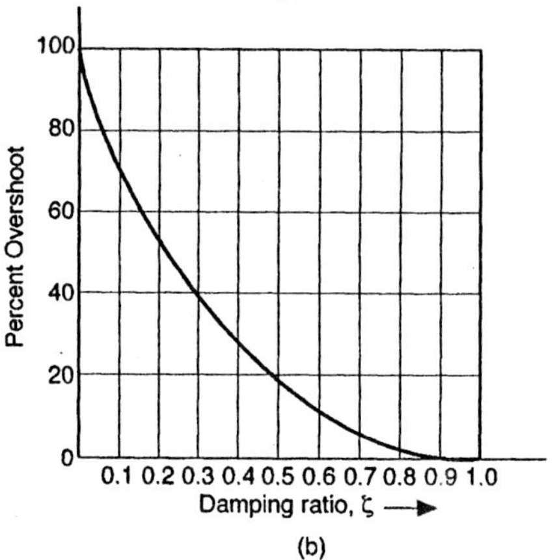

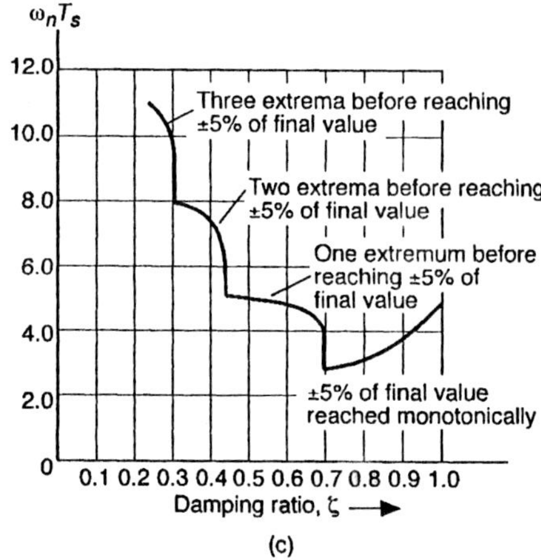

Figure 8.5 Figures of merit for second order zero-state unit step responses. (a) Rise time. (b) Percent overshoot. (c) Settling time.

product $\omega_n T_s$ vs. ζ. Notice that Figures 8.5a and 8.5b have opposite slopes. As ζ diminishes from 1 to 0, the rise time drops while the percent overshoot increases. The settling time curve of Figure 8.5c provides a trade-off between percent overshoot and rise time. As ζ drops from 1 to 0.7, the product $\omega_n T_s$ drops monotonically toward a minimum value. For that range of ζ values, the time response enters a value 5% below the steady state-value at $\omega_n T_s$ and does not exceed the upper limit 5% above the steady-state value after $\omega_n T_s$. Near $\zeta = 0.7$, the percent overshoot is near 5%, so a small reduction in ζ below 0.7 causes $\omega_n T_s$ to jump upward to a value at which the response peaks and then enters the ±5% boundary. The segment of Figure 8.5c, as ζ drops from about 0.7 to about 0.43, is a plot of increasing values of $\omega_n T_s$, corresponding to response curves in Figure 8.4, which go through a peak value where the derivative is zero (extremum point) prior to entering and staying within the ±5% boundary. Additional $\omega_n T_s$ curve segments correspond to regions where the unit-step response curve goes through an integer number of extrema prior to entering the ±5% boundary, which is entered alternatively from above and below. For a damping ratio of zero, the value of $\omega_n T_s$ is infinite since the peak values are undamped.

8.1.9 Defining Terms

Characteristic polynomial: Denominator of the transfer function. The roots determine terms in the **natural response.**

Forced response: The part of the response of the form of the forcing function.

Natural response: The part of the response whose terms follow from the roots of the **characteristic polynomial.**

Percent overshoot: 100 (max value−steady-state value)/ steady-state value.

Resolvant matrix: $\Phi(s) = [sI - A]^{-1}$.

Rise time: The time required for the unit-step response to rise from 10% of the steady-state value to 90% of the steady-state value.

Settling time: The minimum time after which the response remains within ±5% of the steady-state value.

State transition matrix: $\Phi(t)$ = The inverse Laplace transform of the **resolvant matrix.**

Zero-input response: The part of the response found by setting the input to zero.

Zero-state response: The part of the response found by setting the initial conditions to zero.

References

[1] Stefani, R.T., Savant, C.J., Shahian, B., and Hostetter, G.H., *Design of Feedback Control Systems,* 3rd ed., Saunders College Publishing, Boston, MA, 1994.

[2] Moler, C. and Van Loan, C., Nineteen dubious ways to compute the exponential of a matrix, *SIAM Review,* 20 (4), 801–836, 1978.

Further Reading

Additional information may be found in *IEEE Control Syst. Mag.*; *IEEE Trans. Autom. Control;* and *IEEE Trans. Syst., Man, and Cybern.*

8.2 Controllability and Observability

William A. Wolovich, Brown University

8.2.1 Introduction

The ultimate objective of any control system[2] is to improve and often to optimize the performance of a given dynamical system. Therefore, an obvious question that should be addressed is *how* do we design an appropriate controller? Before we can resolve this question, however, it is usually necessary to determine whether or not an appropriate controller exists; i.e., *can* we design a satisfactory controller?

Most physical systems are designed so that the control input does affect the complete system and, as a consequence, an appropriate controller does exist. However, this is not always the case. Moreover, in the *multi-input/multi-output* (MIMO) cases, certain control inputs may affect only part of the dynamical behavior. For example, the steering wheel of an automobile does not affect its speed, nor does the accelerator affect its heading; i.e., the speed of an automobile is uncontrollable via the steering wheel, and the heading is uncontrollable via the accelerator. In certain cases it is important to determine whether or not complete system control is possible if one or more of the inputs (actuators) or outputs (sensors) fails to perform as expected.

The primary purpose of this chapter is to introduce two fundamental concepts associated with dynamical systems, namely controllability and observability, which enable us to resolve the "can" question for a large class of dynamical systems. These dual concepts, which were first defined by R. E. Kalman [4] using state-space representations, are by no means restricted to systems described in state-space form. Indeed, problems associated with analyzing and/or controlling systems that were either uncontrollable or unobservable were encountered long before the state-space approach to control system analysis and design was popularized in the early 1960s.

In those (many) cases where state-space equations can be employed to define the behavior of a dynamical system, controllability implies an ability to transfer the entire state of the system from any initial state $x(t_0)$ to any final state $x(t_f)$ over any arbitrary time interval $t_f - t_0 > 0$ through the employment of an appropriate control input $u(t)$ defined over the time interval. The concept of observability implies an ability to determine the entire initial state of the system from knowledge of the input and the output $y(t)$ over any arbitrary time interval $t_f - t_0 > 0$. These dual concepts play a crucial role in many of the control system design methodologies that have evolved since the early 1960s, such as pole placement, LQG, H_∞ and minimum time optimization, realization theory, adaptive control and system identification.

These dual concepts are not restricted to linear and/or time-invariant systems, and numerous technical papers have been directed at extending controllability and observability to other classes of systems, such as nonlinear, distributed-parameter, discrete-event and behavioral systems. This chapter, however, focuses on the class of linear time-invariant systems whose dynamical behavior can be described by finite dimensional state-space equations or (equivalently) by one or more ordinary linear differential equations, since a fundamental understanding of these two concepts should be obtained before any extensions can be undertaken.

8.2.2 State-Space Controllability and Observability

This section deals with the controllability and observability properties of systems described by linear, time-invariant state-space representations. In particular, consider a *single-input/single-output* (SISO) linear, time-invariant system defined by the *state-space representation*:

$$\dot{\mathbf{x}}(t) = A\mathbf{x}(t) + Bu(t); \quad y(t) = C\mathbf{x}(t) + Eu(t), \tag{8.29}$$

whose state matrix A has (n) distinct eigenvalues, $\lambda_1, \lambda_2, \ldots, \lambda_n$, which define the *poles* of the system and the corresponding *modes* $e^{\lambda_i t}$. Such an A can be diagonalized by any one of its eigenvector matrices V. More specifically, there exists a state transformation matrix $Q = V^{-1}$ that diagonalizes A, so that if

$$\hat{\mathbf{x}}(t) = Q\mathbf{x}(t) = V^{-1}\mathbf{x}(t), \tag{8.30}$$

the dynamical behavior of the equivalent system in *modal canonical form* then is defined by the state-space representation:

$$\dot{\hat{\mathbf{x}}}(t) = \begin{bmatrix} \dot{\hat{x}}_1(t) \\ \dot{\hat{x}}_2(t) \\ \vdots \\ \dot{\hat{x}}_n(t) \end{bmatrix} = \underbrace{\begin{bmatrix} \lambda_1 & 0 & 0 & \ldots \\ 0 & \lambda_2 & 0 & \ldots \\ \vdots & & \ddots & \\ 0 & 0 & \ldots & \lambda_n \end{bmatrix}}_{QAQ^{-1} = \hat{A}} \begin{bmatrix} \hat{x}_1(t) \\ \hat{x}_2(t) \\ \vdots \\ \hat{x}_n(t) \end{bmatrix} +$$

[2]Excerpts and figures from *Automatic Control Systems, Basic Analysis and Design,* by William A. Wolovich, copyright ©1994 by Saunders College Publishing, reproduced by permission of the publisher.

$$\underbrace{\begin{bmatrix} \hat{B}_{11} \\ \hat{B}_{21} \\ \vdots \\ \hat{B}_{n1} \end{bmatrix}}_{QB = \hat{B}} u(t);$$

$$y(t) = \underbrace{[\hat{C}_{11}, \hat{C}_{12}, \ldots \hat{C}_{1n}]}_{CQ^{-1} = \hat{C}} \begin{bmatrix} \hat{x}_1(t) \\ \hat{x}_2(t) \\ \vdots \\ \hat{x}_n(t) \end{bmatrix} + Eu(t), \quad (8.31)$$

as depicted in Figure 8.6.

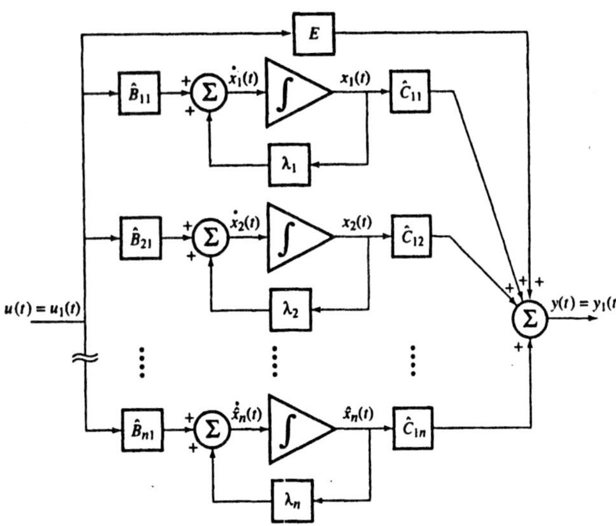

Figure 8.6 A state-space system in modal canonical form. (From Wolovich, William A., *Automatic Control Systems, Basic Analysis and Design*, copyright 1994 by Saunders College Publishing. Reproduced by permission of the publisher.)

Controllability

If $\hat{B}_{k1} = 0$ for any $k = 1, 2, \ldots, n$, then the state $\hat{x}_k(t)$ is *uncontrollable* by the input $u(t) = u_1(t)$, since its time behavior is characterized by the mode $e^{\lambda_k t}$, independent of $u(t)$; i.e.,

$$\hat{x}_k(t) = e^{\lambda_k(t-t_0)}\hat{x}_k(t_0). \quad (8.32)$$

The lack of controllability of the state $\hat{x}_k(t)$ (or the mode $e^{\lambda_k t}$) by $u(t)$ is reflected by a completely zero kth row of the so-called *controllability matrix* of the system, namely the $(n \times n)$ matrix

$$\hat{\mathcal{C}} \stackrel{\text{def}}{=} [\hat{B}, \hat{A}\hat{B}, \ldots, \hat{A}^{n-1}\hat{B}] = \begin{bmatrix} \hat{B}_{11} & \lambda_1\hat{B}_{11} & \ldots & \lambda_1^{n-1}\hat{B}_{11} \\ \hat{B}_{21} & \lambda_2\hat{B}_{21} & \ldots & \lambda_2^{n-1}\hat{B}_{21} \\ \vdots & \vdots & \ddots & \vdots \\ \hat{B}_{n1} & \lambda_n\hat{B}_{n1} & \ldots & \lambda_n^{n-1}\hat{B}_{n1} \end{bmatrix}, \quad (8.33)$$

because $\hat{A}^m = \Lambda^m = diag[\lambda_i{}^m]$, a diagonal matrix for all integers $m \geq 0$. Therefore, each zero kth row element $\hat{B}_{k1}$ of $\hat{B}$ implies an *uncontrollable state* $\hat{x}_k(t)$, whose time behavior is characterized by the *uncontrollable mode* $e^{\lambda_k t}$, as well as a completely zero kth row of the controllability matrix $\hat{\mathcal{C}}$.[3]

On the other hand, each nonzero kth row element of $\hat{B}$ implies a direct influence of $u(t)$ on $\hat{x}_k(t)$, hence a *controllable* state $\hat{x}_k(t)$ (or mode $e^{\lambda_k t}$) and a corresponding nonzero kth row of $\hat{\mathcal{C}}$ defined by $\hat{B}_{k1}[1, \lambda_k, \lambda_k^2, \ldots, \lambda_k^{n-1}]$. In the case (assumed here) of distinct eigenvalues, each such nonzero row of $\hat{B}$ increases the rank of $\hat{\mathcal{C}}$ by one. Therefore, the rank of $\hat{\mathcal{C}}$ corresponds to the total number of states or modes that are controllable by the input $u(t)$, which is termed the *controllability rank* of the system.

Fortunately, it is not necessary to transform a given state-space system to modal canonical form in order to determine its controllability rank. In particular, Equation 8.31 implies that $B = Q^{-1}\hat{B}$, $AB = Q^{-1}\hat{A}QQ^{-1}\hat{B} = Q^{-1}\hat{A}\hat{B}$, or that $A^mB = Q^{-1}\hat{A}^m\hat{B}$ in general, which defines the *controllability matrix* of the system defined by Equation 8.29, namely,

$$\mathcal{C} \stackrel{\text{def}}{=} [B, AB, \ldots, A^{n-1}B] = Q^{-1}\hat{\mathcal{C}}, \quad (8.34)$$

with $Q^{-1} = V$ nonsingular. Therefore, *the rank of* $\mathcal{C}$ (which is equal to the rank of $\hat{\mathcal{C}}$) *equals the controllability rank of the system.* It is important to note that this result holds in the case of nondistinct eigenvalues, as well the multi-input case where B has m columns, so that

$$B = [B_1, B_2, \ldots, B_m], \quad (8.35)$$

and the controllability matrix $\mathcal{C}$, as defined by Equation 8.34, is an $(n \times nm)$ matrix.

In light of the preceding, any state-space system defined by Equation 8.29 is said to be *completely (state or modal) controllable* if its $(n \times nm)$ controllability matrix $\mathcal{C}$ has full rank n. Otherwise, the system is said to be *uncontrollable*, although some ($< n$) of its states generally will be controllable. Note that for a general state-space system, the rank of $\mathcal{C}$ tells us only the *number* of controllable (and uncontrollable) modes, and not their identity, an observation that holds relative to the observability properties of a system as well.

We finally observe that there are several alternative ways of establishing state-space controllability. In particular, it is well known [6], [3] that *a state-space system defined by* Equation 8.29 *is controllable if and only if any one of the following (equivalent) conditions is satisfied:*

- The (n) rows of $e^{At}B$, where e^{At} represents the (unique) *state transition matrix of the system*, are linearly independent over the real field $\mathcal{R}$ for all t.
- The *controllability grammian*

$$G_c(t_0, t_f) \stackrel{\text{def}}{=} \int_{t_0}^{t_f} e^{-A\tau}BB^Te^{-A^T\tau}d\tau$$

[3]The reader should be careful not to confuse the controllability matrix $\hat{\mathcal{C}}$ of Equation 8.33 with the output matrix $\hat{C}$ of Equation 8.31.

is nonsingular for all $t_f > t_0$.

- The controllability matrix $\mathcal{C}$ defined by Equation 8.34 has full rank n.
- The $n \times (n+m)$ matrix $[\lambda I - A, \quad B]$ has rank n at all eigenvalues λ_i of A or, equivalently, $\lambda I - A$ and B are *left coprime*[4] polynomial matrices.

Since the solution to Equation 8.29 is given by

$$\mathbf{x}(t) = e^{A(t-t_0)}\mathbf{x}(t_0) + \int_{t_0}^{t} e^{A(t-\tau)} B\mathbf{u}(\tau)d\tau, \qquad (8.36)$$

it follows that the controllability grammian-based control input

$$\mathbf{u}(t) = B^T e^{-A^T t} G_c^{-1}(t_0, t_f)\left[e^{-At_f}\mathbf{x}(t_f) - e^{-At_0}\mathbf{x}(t_0)\right] \qquad (8.37)$$

transfers any initial state $\mathbf{x}(t_0)$ to any arbitrary final state $\mathbf{x}(t_f)$ at any arbitrary $t_f > t_0$, an observation that is consistent with the more traditional definition of controllability.

Observability

We next note, in light of Figure 8.6, that if $\hat{C}_{1i} = 0$ for any $i = 1, 2, \ldots n$, then the state $\hat{x}_i(t)$ is *unobservable* at the output $y(t) = y_1(t)$, in the sense that the mode $e^{\lambda_i t}$, which defines the time behavior of

$$\hat{x}_i(t) = e^{\lambda_i(t-t_0)}\hat{x}_i(t_0), \qquad (8.38)$$

does not appear at the output $y(t)$. This lack of observability of the state $\hat{x}_i(t)$ (or the mode $e^{\lambda_i t}$) at $y(t)$ is reflected by a completely zero (ith) column of the so-called *observability matrix* of the system, namely, the $(n \times n)$ matrix

$$\hat{\mathcal{O}} \overset{\text{def}}{=} \begin{bmatrix} \hat{C} \\ \hat{C}\hat{A} \\ \vdots \\ \hat{C}\hat{A}^{n-1} \end{bmatrix} = \begin{bmatrix} \hat{C}_{11} & \hat{C}_{12} & \ldots & \hat{C}_{1n} \\ \lambda_1\hat{C}_{11} & \lambda_2\hat{C}_{12} & \ldots & \lambda_n\hat{C}_{1n} \\ \vdots & \vdots & \ddots & \vdots \\ \lambda_1^{n-1}\hat{C}_{11} & \lambda_2^{n-1}\hat{C}_{12} & \ldots & \lambda_n^{n-1}\hat{C}_{1n} \end{bmatrix}, \qquad (8.39)$$

analogous to a completely zero (kth) row of $\hat{\mathcal{C}}$ in Equation 8.33.

On the other hand, each nonzero ith column element $\hat{C}_{1i}$ of $\hat{C}$ implies a direct influence of $\hat{x}_i(t)$ on $y(t)$, hence an *observable* state $\hat{x}_i(t)$ or mode $e^{\lambda_i t}$, and a corresponding nonzero ith column of $\hat{\mathcal{O}}$ defined by $[1, \lambda_i, \lambda_i^2, \ldots, \lambda_i^{n-1}]^T \hat{C}_{1i}$. In the case (assumed here) of distinct eigenvalues, each such nonzero element of $\hat{C}$ increases the rank of $\hat{\mathcal{O}}$ by one. Therefore, the rank of $\hat{\mathcal{O}}$ corresponds to the total number of states or modes that are observable at the output $y(t)$, which is termed the *observability rank* of the system.

As in the case of controllability, it is not necessary to transform a given state-space system to modal canonical form in order to determine its observability rank. In particular, Equation 8.31 implies that $C = \hat{C}Q, CA = \hat{C}QQ^{-1}\hat{A}Q = \hat{C}\hat{A}Q$, or that $CA^m = \hat{C}\hat{A}^m Q$ in general, which defines the *observability matrix* of the system defined by Equation 8.29, namely,

$$\mathcal{O} \overset{\text{def}}{=} \begin{bmatrix} C \\ CA \\ \vdots \\ CA^{n-1} \end{bmatrix} = \hat{\mathcal{O}}Q, \qquad (8.40)$$

with $Q = V^{-1}$ nonsingular. Therefore, *the rank of* $\mathcal{O}$ (which is equal to the rank of $\hat{\mathcal{O}}$) *equals the observability rank of the system.* It is important to note that this result holds in the case of nondistinct eigenvalues, as well as the multi-output case where C has p rows, so that

$$C = \begin{bmatrix} C_1 \\ C_2 \\ \vdots \\ C_p \end{bmatrix}, \qquad (8.41)$$

and the observability matrix $\mathcal{O}$, as defined by Equation 8.40, is a $(pn \times n)$ matrix. In view of the preceding, a state-space system defined by Equation 8.29 is said to be *completely (state or modal) observable* if its $(pn \times n)$ observability matrix $\mathcal{O}$ has full rank n. Otherwise, the system is said to be *unobservable*, although some $(< n)$ of its states generally are observable.

As in the case of controllability, there are several alternative ways of establishing state-space observability. In particular, it is well known [6] [3] that *a state-space system defined by* Equation 8.29 *is observable if and only if any one of the following (equivalent) conditions is satisfied:*

- The (n) columns of Ce^{At} are linearly independent over $\mathcal{R}$ for all t.
- The *observability grammian*

$$G_o(t_0, t_f) \overset{\text{def}}{=} \int_{t_0}^{t_f} e^{A^T\tau} C^T C e^{A\tau} d\tau$$

  is nonsingular for all $t_f > t_0$.
- The observability matrix $\mathcal{O}$ defined by Equation 8.40 has full rank n.
- The $(n+p) \times n$ matrix $\begin{bmatrix} \lambda I - A \\ C \end{bmatrix}$ has rank n at all eigenvalues λ_i of A or, equivalently, $\lambda I - A$ and C are *right coprime* polynomial matrices.

If a state-space system is observable, and if

$$f(t) \overset{\text{def}}{=} y(t) - C\int_{t_0}^{t} e^{A(t-\tau)} Bu(\tau)d\tau - Eu(t), \qquad (8.42)$$

it then follows that its initial state can be determined via the relation

$$\mathbf{x}(t_0) = e^{At_0} G_o^{-1}(t_0, t_f) e^{A^T t_0} \int_{t_0}^{t_f} e^{A^T(t-t_0)} C^T f(t)dt, \qquad (8.43)$$

[4]Two polynomials are called *coprime* if they have no common roots. Two polynomial matrices $P(\lambda)$ and $R(\lambda)$, which have the same number of rows, are left coprime if the rank of the composite matrix $[P(\lambda), \quad R(\lambda)]$ remains the same for all (complex) values of λ. Right coprime polynomial matrices, which have the same number of columns, are defined in an analogous manner.

which is consistent with the more traditional definition of observability.

Component Controllability and Observability

In the multi-input and/or multi-output cases, it often is useful to determine the controllability and observability rank of a system relative to the individual components of its input and output. Such a determination would be important, for example, if one or more of the actuators or sensors were to fail.

In particular, suppose the system defined by Equation 8.29 has $m > 1$ inputs, $u_1(t), u_2(t), \ldots, u_m(t)$, so that the input matrix B has m columns, as in Equation 8.35. If we disregard all inputs except $u_j(t)$, the resulting *controllability matrix associated with input* $u_j(t)$ is defined as the $(n \times n)$ matrix

$$\mathcal{C}_j \stackrel{\text{def}}{=} [B_j, AB_j, \ldots, A^{n-1}B_j]. \tag{8.44}$$

The rank of each such $\mathcal{C}_j$ would determine the number of states or modes that are controllable by input component $u_j(t)$.

In a dual manner, suppose the given state-space system has $p > 1$ outputs, $y_1(t), y_2(t), \ldots y_p(t)$, so that the output matrix C has p rows, as in Equation 8.41. If we disregard all outputs except $y_q(t)$, the resulting *observability matrix associated with output* $y_q(t)$ is defined as the $(n \times n)$ matrix

$$\mathcal{O}_q \stackrel{\text{def}}{=} \begin{bmatrix} C_q \\ C_qA \\ \vdots \\ C_qA^{n-1} \end{bmatrix}. \tag{8.45}$$

As in the case of controllability, the rank of each such $\mathcal{O}_q$ determines the number of states or modes that are observable by output component $y_q(t)$.

EXAMPLE 8.1:

To illustrate the preceding, we next note that the linearized equations of motion of an orbiting satellite can be defined by the state-space representation [2]

$$\underbrace{\begin{bmatrix} \dot{x}_1(t) \\ \dot{x}_2(t) \\ \dot{x}_3(t) \\ \dot{x}_4(t) \end{bmatrix}}_{\dot{\mathbf{x}}(t)} = \underbrace{\begin{bmatrix} 0 & 1 & 0 & 0 \\ 3\omega^2 & 0 & 0 & 2\omega \\ 0 & 0 & 0 & 1 \\ 0 & -2\omega & 0 & 0 \end{bmatrix}}_{A} \underbrace{\begin{bmatrix} x_1(t) \\ x_2(t) \\ x_3(t) \\ x_4(t) \end{bmatrix}}_{\mathbf{x}(t)} + \underbrace{\begin{bmatrix} 0 & 0 \\ 1 & 0 \\ 0 & 0 \\ 0 & 1 \end{bmatrix}}_{B} \underbrace{\begin{bmatrix} u_1(t) \\ u_2(t) \end{bmatrix}}_{\mathbf{u}(t)},$$

with a defined output

$$\underbrace{\begin{bmatrix} y_1(t) \\ y_2(t) \end{bmatrix}}_{\mathbf{y}(t)} = \underbrace{\begin{bmatrix} 1 & 0 & 0 & 0 \\ 0 & 0 & 1 & 0 \end{bmatrix}}_{C} \mathbf{x}(t).$$

The reader can verify that the $(n \times nm = 4 \times 8)$ controllability matrix $\mathcal{C} = [B, AB, A^2B, A^3B]$ has full rank $4 = n$ in this case, so that the entire state is controllable using both inputs. However, since

$$\mathcal{C}_1 = [B_1, AB_1, A^2B_1, A^3B_1] = \begin{bmatrix} 0 & 1 & 0 & -\omega^2 \\ 1 & 0 & -\omega^2 & 0 \\ 0 & 0 & -2\omega & 0 \\ 0 & -2\omega & 0 & 2\omega^3 \end{bmatrix}$$

is singular (i.e., the determinant of $\mathcal{C}_1$, namely $| \mathcal{C}_1 |= 4\omega^4 - 4\omega^4 = 0$, with rank $\mathcal{C}_1 = 3 < 4 = n$,) it follows that one of the "states" cannot be controlled by the radial thruster $u_1(t)$ alone, which would be unfortunate if the tangential thruster $u_2(t)$ were to fail.

We next note that

$$\mathcal{C}_2 = [B_2, AB_2, A^2B_2, A^3B_2] = \begin{bmatrix} 0 & 0 & 2\omega & 0 \\ 0 & 2\omega & 0 & -2\omega^3 \\ 0 & 1 & 0 & -4\omega^2 \\ 1 & 0 & -4\omega^2 & 0 \end{bmatrix}$$

is nonsingular, since $| \mathcal{C}_2 |= 4\omega^4 - 16\omega^4 = -12\omega^4 \neq 0$, so that complete state control is possible by the tangential thruster $u_2(t)$ alone if the radial thruster $u_1(t)$ were to fail.

Insofar as observability is concerned, $y_1(t) = C_1\mathbf{x}(t) = x_1(t) = r(t) - d$ represents the radial deviation of $r(t)$ from a nominal radius $d = 1$, while output $y_2(t) = C_2\mathbf{x}(t) = x_3(t) = \alpha(t) - \omega t$ represents the tangential deviation of $\alpha(t)$ from a nominal angular position defined by ωt. The reader can verify that the $(pn \times n = 8 \times 4)$ observability matrix $\mathcal{O}$ has full rank $n = 4$ in this case, so that the entire state is observable using both outputs. However, since

$$\mathcal{O}_1 = \begin{bmatrix} C_1 \\ C_1A \\ C_1A^2 \\ C_1A^3 \end{bmatrix} = \begin{bmatrix} 1 & 0 & 0 & 0 \\ 0 & 1 & 0 & 0 \\ 3\omega^2 & 0 & 0 & 2\omega \\ 0 & -\omega^2 & 0 & 0 \end{bmatrix}$$

is clearly singular (because its third column is zero), with rank $\mathcal{O}_1 = 3 < 4 = n$, it follows that one of the "states" cannot be observed by $y_1(t)$ alone.

We finally note that

$$\mathcal{O}_2 = \begin{bmatrix} C_2 \\ C_2A \\ C_2A^2 \\ C_2A^3 \end{bmatrix} = \begin{bmatrix} 0 & 0 & 1 & 0 \\ 0 & 0 & 0 & 1 \\ 0 & -2\omega & 0 & 0 \\ -6\omega^3 & 0 & 0 & -4\omega^2 \end{bmatrix}$$

is nonsingular, since $|\mathcal{O}_2| = -12\omega^4 \neq 0$, so that the entire state can be observed by $y_2(t)$ alone.

MIMO Case

In the general MIMO case, the explicit modal controllability and observability properties of a system with distinct eigenvalues can be determined by transforming the system to modal canonical form. In particular, a zero in any kth row of column $\hat{B}_j$ of the input matrix $\hat{B}$ implies the uncontrollability of state $\hat{x}_k(t)$ (or the mode $e^{\lambda_k t}$) by $u_j(t)$. Furthermore, a completely zero kth row of $\hat{B}$ implies the complete uncontrollability of state $\hat{x}_k(t)$ (or the mode $e^{\lambda_k t}$) with respect to the entire vector input $\mathbf{u}(t)$. Each such zero row of $\hat{B}$ implies a corresponding zero row of $\hat{C}$, thereby reducing the rank of the $(n \times nm)$ controllability matrices $\hat{C}$ and C by one. The number of controllable modes therefore is given by the rank of $\hat{C}$ or C, the controllability rank of the system.

Dual results hold with respect to the observability properties of a system. In particular, a zero in any ith column of row $\hat{C}_q$ of the output matrix $\hat{C}$ implies the unobservability of state $\hat{x}_i(t)$ (or the mode $e^{\lambda_i t}$) by $y_q(t)$. Furthermore, a completely zero ith column of $\hat{C}$ implies the complete unobservability of state $\hat{x}_i(t)$ (or the mode $e^{\lambda_i t}$) with respect to the entire vector output $\mathbf{y}(t)$. Each such zero column of $\hat{C}$ implies a corresponding zero column of $\hat{O}$, thereby reducing the rank of the $(pn \times n)$ observability matrices $\hat{O}$ and O by one. The number of observable modes therefore is given by the rank of $\hat{O}$ or O, the observability rank of the system. Section 2.6 of [7] contains a MIMO example that illustrates the preceding.

8.2.3 Differential Operator Controllability and Observability

Suppose the defining differential equations of a dynamical system are in the *differential operator form*

$$a(D)z(t) = b(D)u(t);$$
$$y(t) = c(D)z(t) + e(D)u(t), \tag{8.46}$$

where $a(D)$, $b(D)$, $c(D)$ and $e(D)$ are polynomials[5] in the *differential operator* $D = \frac{d}{dt}$, with $a(D)$ a monic polynomial of degree n, which defines the *order* of this representation, and $z(t)$ is a single-valued function of time called the *partial state.* We often find it convenient to "abbreviate" Equation 8.46 by the polynomial quadruple $\{a(D), b(D), c(D), e(D)\}$; i.e., $\{a(D), b(D), c(D), e(D)\} \Longleftrightarrow$ Equation 8.46.

An Equivalent State-Space Representation

We first show that Equation 8.46 has an *equivalent state-space representation* that can be determined directly by inspection of $a(D)$ and $b(D)$ when $deg[b(D)] < n = deg[a(D)]$. In particular, suppose we employ the coefficients of

$$a(D) = D^n + a_{n-1}D^{n-1} + \ldots + a_1 D + a_0$$

and

$$b(D) = b_{n-1}D^{n-1} + \ldots + b_1 D + b_0,$$

in order to define the following state-space system:

$$\underbrace{\begin{bmatrix} \dot{x}_1(t) \\ \dot{x}_2(t) \\ \vdots \\ \dot{x}_n(t) \end{bmatrix}}_{\dot{\mathbf{x}}(t)} = \underbrace{\begin{bmatrix} 0 & 0 & 0 & \ldots & -a_0 \\ 1 & 0 & 0 & \ldots & -a_1 \\ 0 & 1 & 0 & \ldots & \\ \vdots & \vdots & \ddots & & \vdots \\ 0 & 0 & \ldots & 1 & -a_{n-1} \end{bmatrix}}_{A} \underbrace{\begin{bmatrix} x_1(t) \\ x_2(t) \\ \vdots \\ x_n(t) \end{bmatrix}}_{\mathbf{x}(t)} + \underbrace{\begin{bmatrix} b_0 \\ b_1 \\ \vdots \\ b_{n-1} \end{bmatrix}}_{B} u(t), \tag{8.47}$$

with

$$z(t) \stackrel{\text{def}}{=} x_n(t) = [0\ 0\ \ldots 0\ 1]\mathbf{x}(t) \stackrel{\text{def}}{=} C_z\mathbf{x}(t). \tag{8.48}$$

Since A is a (right column) companion matrix, the *characteristic polynomial of A* is given by

$$|\lambda I - A| = \lambda^n + a_{n-1}\lambda^{n-1} + \ldots + a_1\lambda + a_0 = a(\lambda). \tag{8.49}$$

Therefore, *the n zeros of $a(\lambda)$ correspond to the n eigenvalues λ_i of A, which define the system modes $e^{\lambda_i t}$*. As in the previous section, we assume that these n eigenvalues of A are distinct.

In terms of the differential operator D, Equation 8.47 can be written as

$$\underbrace{\begin{bmatrix} D & 0 & 0 & \ldots & a_0 \\ -1 & D & 0 & \ldots & a_1 \\ 0 & -1 & D & \ldots & a_2 \\ \vdots & & \ddots & & \vdots \\ 0 & 0 & \ldots & -1 & D + a_{n-1} \end{bmatrix}}_{(DI - A)} \underbrace{\begin{bmatrix} x_1(t) \\ x_2(t) \\ \vdots \\ x_n(t) \end{bmatrix}}_{\mathbf{x}(t)} = \underbrace{\begin{bmatrix} b_0 \\ b_1 \\ \vdots \\ b_{n-1} \end{bmatrix}}_{B} u(t). \tag{8.50}$$

If we now premultiply Equation 8.50 by the row vector $[1\ D\ D^2 \ldots D^{n-1}]$, noting that $x_n(t) = z(t)$, we obtain the relation

$$[0\ 0\ \ldots\ 0\ a(D)] \begin{bmatrix} x_1(t) \\ x_2(t) \\ \vdots \\ x_n(t) \end{bmatrix} = a(D)z(t) = b(D)u(t), \tag{8.51}$$

thereby establishing the equivalence of the state-space system defined by Equation 8.47 and the partial state/input relation $a(D)z(t) = b(D)u(t)$ of Equation 8.46.

Since $x_n(t) = z(t)$, in light of Equation 8.48, Equation 8.47 implies that

$$Dz(t) = \dot{x}_n(t) = x_{n-1}(t) - a_{n-1}x_n(t) + b_{n-1}u(t),$$

[5]We will later allow both $b(D)$ and $c(D)$ to be polynomial vectors, thereby enlarging the class of systems considered beyond the SISO case defined by Equation 8.46.

$$D^2z(t) = \dot{x}_{n-1}(t) - a_{n-1}\dot{x}_n(t) + b_{n-1}\dot{u}(t)$$

$$= x_{n-2}(t) - a_{n-2}x_n(t) + b_{n-2}u(t)$$

$$-a_{n-1}[x_{n-1}(t) - a_{n-1}x_n(t) + b_{n-1}u(t)] + b_{n-1}\dot{u}(t),$$

etc., which enables us to express the output relation of Equation 8.46, namely,

$$y(t) = c(D)z(t) + e(D)u(t) = c(D)x_n(t) + e(D)u(t)$$

as a function of $\mathbf{x}(t)$ and $u(t)$ and its derivatives. As a consequence,

$$y(t) = C\mathbf{x}(t) + E(D)u(t) \tag{8.52}$$

for some constant $(1 \times n)$ vector C and a corresponding polynomial $E(D)$.

We have therefore established a *complete equivalence relationship* between the differential operator representation of Equation 8.46 and the state-space representation defined by Equations 8.47 and 8.52, with E expanded to $E(D)$ (if necessary) to include derivatives of the input. We denote this equivalence relationship as

$$\underbrace{\{A, B, C, E(D)\}}_{\text{of Equations 8.47 and 8.52}} \overset{\text{equiv}}{\Longleftrightarrow} \underbrace{\{a(D), b(D), c(D), e(D)\}}_{\text{of Equation 8.46}} \tag{8.53}$$

Observable Canonical Forms

If $c(D) = 1$ in Equation 8.46, so that

$$a(D)z(t) = b(D)u(t); \quad y(t) = z(t) + e(D)u(t), \tag{8.54}$$

the equivalent state-space system defined by Equations 8.47 and 8.52 is characterized by an output matrix $C = C_z = [0\ 0\ \ldots 0\ 1]$ and an $E(D) = e(D)$ in Equation 8.52; i.e.,

$$y(t) = \underbrace{[0\ 0\ \ldots 0\ 1]}_{C = C_z}\mathbf{x}(t) + \underbrace{E(D)}_{e(D)}u(t). \tag{8.55}$$

Therefore, Equations 8.47 and 8.55 represent a state-space system equivalent to the differential operator system defined by Equation 8.54. We denote this equivalence relationship as

$$\underbrace{\{A, B, C, E(D)\}}_{\text{of Equations 8.47 and 8.55}} \overset{\text{equiv}}{\Longleftrightarrow} \underbrace{\{a(D), b(D), c(D) = 1, e(D)\}}_{\text{of Equation 8.54}} \tag{8.56}$$

Moreover, *both of these representations are completely observable.* In particular, the differential operator representation is observable because $a(D)$ and $c(D) = 1$ are coprime,[6] and the state-space representation is observable because its observability matrix

$$\mathcal{O} = \begin{bmatrix} C \\ CA \\ CA^2 \\ \vdots \\ CA^{n-1} \end{bmatrix} = \begin{bmatrix} 0 & \ldots & 0 & 1 \\ 0 & \ldots & 1 & * \\ \vdots & & * & \vdots \\ 0 & 1 & \ldots & * \\ 1 & * & \ldots & * \end{bmatrix} \tag{8.57}$$

(where $*$ denotes an irrelevant scalar) is nonsingular.

Note further that the $\{A, C\}$ pair of Equations 8.47 and 8.55 is in a special canonical form. In particular, A is a right column companion matrix and C is identically zero except for a 1 in its right-most column. In light of these observations, we say that both of the representations defined by Equation 8.56 are in *observable canonical form.* Figure 8.7 depicts a block diagram of a state-space system in observable canonical form, as defined by Equations 8.47 and 8.55.

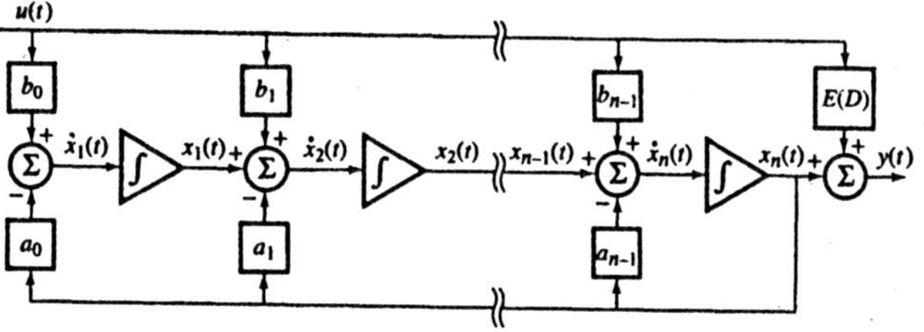

Figure 8.7 A State-Space System in Observable Canonical Form. (From Wolovich, William A., *Automatic Control Systems, Basic Analysis and Design,* copyright 1994 by Saunders College Publishing. Reproduced by permission of the publisher.)

Differential Operator Controllability

Because of the right column companion form structure of A in Equation 8.47, it follows that in the case (assumed here) of distinct eigenvalues,[7] the vector $[1\ \lambda_i\ \lambda_i^2 \ldots \lambda_i^{n-1}]$ is a row eigenvector of A in the sense that

$$[1\ \lambda_i\ \lambda_i^2 \ldots \lambda_i^{n-1}]A = \lambda_i[1\ \lambda_i\ \lambda_i^2 \ldots \lambda_i^{n-1}], \tag{8.58}$$

for each $i = 1, 2, \ldots n$. Therefore, the transpose of a Vandermonde matrix V of n column eigenvectors of A, namely,

$$V^T = \begin{bmatrix} 1 & 1 & \ldots & 1 \\ \lambda_1 & \lambda_2 & \ldots & \lambda_n \\ \lambda_1^2 & \lambda_2^2 & \ldots & \lambda_n^2 \\ \vdots & \vdots & & \vdots \\ \lambda_1^{n-1} & \lambda_2^{n-1} & \ldots & \lambda_n^{n-1} \end{bmatrix}^T$$

[6] We formally establish this condition for differential operator observability later in this section.

[7] Although the results presented hold in the case of nondistinct eigenvalues as well.

$$= \begin{bmatrix} 1 & \lambda_1 & \lambda_1^2 & \dots & \lambda_1^{n-1} \\ 1 & \lambda_2 & \lambda_2^2 & \dots & \lambda_2^{n-1} \\ \vdots & \vdots & \vdots & & \vdots \\ 1 & \lambda_n & \lambda_n^2 & \dots & \lambda_n^{n-1} \end{bmatrix} \quad (8.59)$$

diagonalizes A. Otherwise stated, a transformation of state defined by $\hat{\mathbf{x}}(t) = V^T\mathbf{x}(t)$ reduces the state-space system defined by Equation 8.47 to the modal canonical form

$$\begin{bmatrix} \dot{\hat{x}}_1(t) \\ \dot{\hat{x}}_2(t) \\ \vdots \\ \dot{\hat{x}}_n(t) \end{bmatrix} = \underbrace{\begin{bmatrix} \lambda_1 & 0 & 0 & \dots \\ 0 & \lambda_2 & 0 & \dots \\ \vdots & & \ddots & \\ 0 & 0 & \dots & \lambda_n \end{bmatrix}}_{V^TAV^{-T} = \hat{A} = diag[\lambda_i]} \begin{bmatrix} \hat{x}_1(t) \\ \hat{x}_2(t) \\ \vdots \\ \hat{x}_n(t) \end{bmatrix} + \underbrace{\begin{bmatrix} b(\lambda_1) \\ b(\lambda_2) \\ \vdots \\ b(\lambda_n) \end{bmatrix}}_{V^TB} u(t), \quad (8.60)$$

with the elements of V^TB given by $b(\lambda_i)$ because

$$[1\ \lambda_i\ \lambda_i^2 \dots \lambda_i^{n-1}]B = b(\lambda_i), \quad \text{for } i = 1, 2, \dots, n. \quad (8.61)$$

In light of the results presented in the previous section, and Figure 8.6 in particular, each $\hat{x}_i(t)$ is controllable if and only if $b(\lambda_i) \neq 0$ when $a(\lambda_i) = 0$. Therefore, *the state-space system defined by* Equation 8.47 *is completely (state or modal) controllable if and only if the polynomials $a(\lambda)$ and $b(\lambda)$, or the differential operator pair $a(D)$ and $b(D)$, are coprime.*

When this is not the case, every zero λ_k of $a(\lambda)$, which also is a zero of $b(\lambda)$, implies an uncontrollable state $\hat{x}_k(t) = [1\ \lambda_k\ \lambda_k^2 \dots \lambda_k^{n-1}]\mathbf{x}(t)$, characterized by an uncontrollable mode $e^{\lambda_k t}$. Moreover, each such λ_k reduces the controllability rank of the system by one. The controllability properties of a dynamical system in differential operator form therefore can be completely specified by the (zeros of the) polynomials $a(D)$ and $b(D)$ of Equation 8.46, independent of any state-space representation.

Controllable Canonical Forms

When $b(D) = 1$ and $deg[c(D)] < n = deg[a(D)]$, the differential operator system defined by Equation 8.46, namely,

$$\underbrace{(D^n + a_{n-1}D^{n-1} + \dots + a_1D + a_0)}_{a(D)} z(t) = u(t);$$
$$y(t) = \underbrace{(c_{n-1}D^{n-1} + \dots + c_1D + c_0)}_{c(D)} z(t) + e(D)u(t), \quad (8.62)$$

has an alternative, *equivalent state-space representation*, which can be determined directly by inspection of $a(D)$ and $c(D)$.

In particular, suppose we employ the coefficients of $a(D)$ and $c(D)$ to define the following state-space system:

$$\underbrace{\begin{bmatrix} \dot{x}_1(t) \\ \dot{x}_2(t) \\ \vdots \\ \dot{x}_n(t) \end{bmatrix}}_{\dot{\mathbf{x}}(t)} = \underbrace{\begin{bmatrix} 0 & 1 & 0 & \dots & 0 \\ 0 & 0 & 1 & \dots & 0 \\ \vdots & \vdots & & \ddots & \vdots \\ -a_0 & -a_1 & & \dots & -a_{n-1} \end{bmatrix}}_{A} \underbrace{\begin{bmatrix} x_1(t) \\ x_2(t) \\ \vdots \\ x_n(t) \end{bmatrix}}_{\mathbf{x}(t)} + \underbrace{\begin{bmatrix} 0 \\ \vdots \\ 0 \\ 1 \end{bmatrix}}_{B} u(t);$$

$$y(t) = \underbrace{[c_0\ c_1\ \dots\ c_{n-1}]}_{C} \begin{bmatrix} x_1(t) \\ x_2(t) \\ \vdots \\ x_n(t) \end{bmatrix} + \underbrace{E(D)}_{e(D)} u(t). \quad (8.63)$$

Since A is a (bottom row) companion matrix, the characteristic polynomial of A is given by

$$|\lambda I - A| = \lambda^n + a_{n-1}\lambda^{n-1} + \dots + a_1\lambda + a_0 = a(\lambda), \quad (8.64)$$

as in Equation 8.49. Therefore, the n zeros of $a(\lambda)$ correspond to the n eigenvalues λ_i of A, which define the system modes $e^{\lambda_i t}$.

If $z(t) \stackrel{\text{def}}{=} x_1(t)$ in Equation 8.63, it follows that $Dz(t) = \dot{x}_1(t) = x_2(t)$, $D^2z(t) = \dot{x}_2(t) = x_3(t)$, ... $D^{n-1}z(t) = \dot{x}_{n-1}(t) = x_n(t)$, or that

$$\begin{bmatrix} 1 \\ D \\ \vdots \\ D^{n-1} \end{bmatrix} z(t) = \begin{bmatrix} x_1(t) \\ x_2(t) \\ \vdots \\ x_n(t) \end{bmatrix} = \mathbf{x}(t). \quad (8.65)$$

The substitution of Equation 8.65 for $\mathbf{x}(t)$ in Equation 8.63 therefore implies that

$$\begin{bmatrix} D & -1 & 0 & \dots & 0 \\ 0 & D & -1 & \dots & 0 \\ \vdots & \vdots & & \ddots & \vdots \\ a_0 & a_1 & \dots & & D + a_{n-1} \end{bmatrix} \begin{bmatrix} 1 \\ D \\ \vdots \\ D^{n-1} \end{bmatrix} z(t) = \begin{bmatrix} 0 \\ \vdots \\ 0 \\ 1 \end{bmatrix} u(t),$$

or that

$$a(D)z(t) = u(t);$$
$$y(t) = C\mathbf{x}(t) + E(D)u(t) = c(D)z(t) + e(D)u(t), \quad (8.66)$$

thus establishing the equivalence of the two representations. We denote this *equivalence relationship* as

$$\underbrace{\{A, B, C, E(D)\}}_{\text{of Equation 8.63}} \stackrel{\text{equiv}}{\Longleftrightarrow} \underbrace{\{a(D), b(D) = 1, c(D), e(D)\}}_{\text{of Equation 8.62}} \quad (8.67)$$

Note that *both of the representations defined by* Equation 8.67 *are completely controllable.* In particular, the differential operator representation is controllable because $a(D)$ and $b(D) = 1$ clearly are coprime, and the state-space representation is controllable because its controllability matrix, namely,

$$\mathcal{C} = [B, AB, \ldots, A^{n-1}B] = \begin{bmatrix} 0 & \ldots & 0 & 1 \\ 0 & \ldots & 1 & * \\ \vdots & & * & \vdots \\ 0 & 1 & \ldots & * \\ 1 & * & \ldots & * \end{bmatrix} \quad (8.68)$$

(where $*$ denotes an irrelevant scalar) is nonsingular.

Furthermore, the $\{A, B\}$ pair of Equation 8.63 is in a special canonical form. In particular, A is a bottom row companion matrix and B is identically zero except for the 1 in its bottom row. In light of these observations, we say that both of the representations defined by Equation 8.67 are in *controllable canonical form.* Figure 8.8 depicts a block diagram of a state-space system in controllable canonical form, as defined by Equation 8.63.

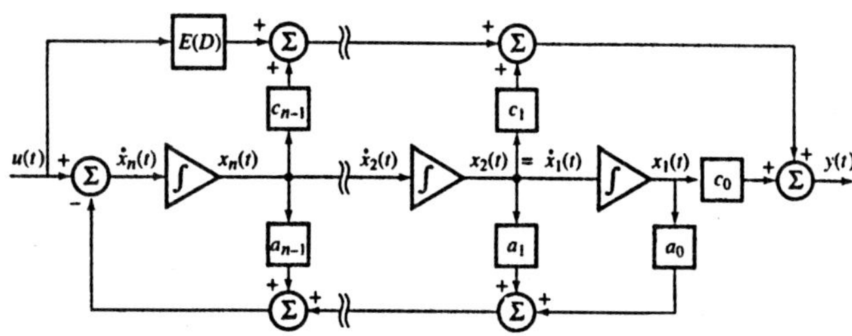

Figure 8.8 A State-Space System in Controllable Canonical Form. (From Wolovich, William A., *Automatic Control Systems, Basic Analysis and Design,* copyright 1994 by Saunders College Publishing. Reproduced by permission of the publisher.)

Differential Operator Observability

Because of the bottom row companion form structure of A in Equation 8.63, it follows that for each $i = 1, 2, \ldots n$, $\begin{bmatrix} 1 \\ \lambda_i \\ \vdots \\ \lambda_i^{n-1} \end{bmatrix}$ is a column eigenvector of A in the sense that

$$A \begin{bmatrix} 1 \\ \lambda_i \\ \vdots \\ \lambda_i^{n-1} \end{bmatrix} = \begin{bmatrix} 1 \\ \lambda_i \\ \vdots \\ \lambda_i^{n-1} \end{bmatrix} \lambda_i. \quad (8.69)$$

Therefore, if V is a Vandermonde matrix of n column eigenvectors of A, as in Equation 8.59, it follows that its inverse V^{-1} diagonalizes A. Otherwise stated, a transformation of state defined by $\hat{\mathbf{x}}(t) = V^{-1}\mathbf{x}(t)$ reduces the state-space system defined by Equation 8.63 to the following modal canonical form:

$$\begin{bmatrix} \dot{\hat{x}}_1(t) \\ \dot{\hat{x}}_2(t) \\ \vdots \\ \dot{\hat{x}}_n(t) \end{bmatrix} = \underbrace{\begin{bmatrix} \lambda_1 & 0 & 0 & \ldots \\ 0 & \lambda_2 & 0 & \ldots \\ \vdots & & \ddots & \\ 0 & 0 & \ldots & \lambda_n \end{bmatrix}}_{V^{-1}AV = \hat{A} = diag[\lambda_i]} \begin{bmatrix} \hat{x}_1(t) \\ \hat{x}_2(t) \\ \vdots \\ \hat{x}_n(t) \end{bmatrix} + \underbrace{\begin{bmatrix} \hat{b}_0 \\ \hat{b}_1 \\ \vdots \\ \hat{b}_{n-1} \end{bmatrix}}_{V^{-1}B} u(t);$$

$$y(t) = \underbrace{[c(\lambda_1),\ c(\lambda_2),\ \ldots c(\lambda_n)]}_{CV} \begin{bmatrix} \hat{x}_1(t) \\ \hat{x}_2(t) \\ \vdots \\ \hat{x}_n(t) \end{bmatrix} + E(D)u(t), \quad (8.70)$$

with the elements of CV given by $c(\lambda_i)$ because

$$C \begin{bmatrix} 1 \\ \lambda_i \\ \lambda_i^2 \\ \vdots \\ \lambda_i^{n-1} \end{bmatrix} = c(\lambda_i), \quad \text{for } i = 1, 2, \ldots n. \quad (8.71)$$

In light of the results presented in the previous section, and Figure 8.6 in particular, each $\hat{x}_i(t)$ is observable if and only if $c(\lambda_i) \neq 0$. Therefore, *the state-space system defined by* Equations 8.47 and 8.52 *is completely (state or modal) observable if and only if the polynomials $a(\lambda)$ and $c(\lambda)$, or the differential operator pair $a(D)$ and $c(D)$, are coprime.*

When this is not the case, every zero λ_k of $a(\lambda)$, which is also a zero of $c(\lambda)$, implies an unobservable state $\hat{x}_k(t)$, characterized by an uncontrollable mode $e^{\lambda_k t}$. Moreover, each such λ_k reduces the observability rank of the system by one. The observability properties of a dynamical system in differential operator form therefore can be completely specified by the (zeros of the) polynomials $a(D)$ and $c(D)$ of Equation 8.46, independent of any state-space representation.

The MIMO Case

Although we initially assumed that Equation 8.46 defines a SISO system, it can be modified to include certain MIMO systems as well. In particular, a vector input

$$\mathbf{u}(t) = \begin{bmatrix} u_1(t) \\ u_2(t) \\ \vdots \\ u_m(t) \end{bmatrix} \quad (8.72)$$

can be accommodated by allowing the polynomial $b(D)$ in Equation 8.46 to be a row vector of polynomials, namely,

$$b(D) = [b_1(D), b_2(D), \ldots, b_m(D)]. \quad (8.73)$$

Each polynomial element of $b(D)$ then defines a corresponding real $(n \times 1)$ column of the input matrix B of an equivalent state-space system, analogous to that defined by Equation 8.47.

In a dual manner, a vector output

$$\mathbf{y}(t) = \begin{bmatrix} y_1(t) \\ y_2(t) \\ \vdots \\ y_p(t) \end{bmatrix} \quad (8.74)$$

can be accommodated by allowing the polynomial $c(D)$ in Equation 8.46 to be a column vector of polynomials, namely,

$$c(D) = \begin{bmatrix} c_1(D) \\ c_2(D) \\ \vdots \\ c_p(D) \end{bmatrix}. \quad (8.75)$$

Of course, $e(D)$ also is a vector or matrix of polynomials in these cases. Each polynomial element of $c(D)$ then defines a corresponding real $(1 \times n)$ row of the output matrix C of an equivalent state-space system, analogous to that defined by Equation 8.52. A block diagram of such a MIMO system is depicted in Figure 8.9.

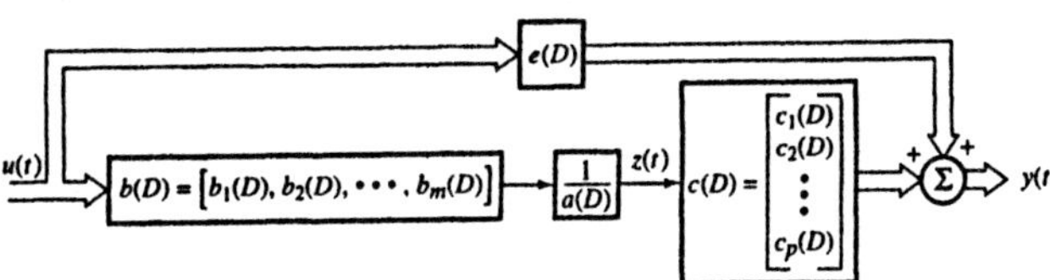

Figure 8.9 A MIMO Differential Operator System. (From Wolovich, William A., *Automatic Control Systems, Basic Analysis and Design*, copyright 1994 by Saunders College Publishing. Reproduced by permission of the publisher.)

EXAMPLE 8.2:

To illustrate the preceding, consider a dynamical system defined by the (two-input/two-output) differential equation

$$\frac{d^4z(t)}{dt^4} + 2\frac{d^3z(t)}{dt^3} - 6\frac{d^2z(t)}{dt^2} - 22\frac{dz(t)}{dt} - 15z(t) =$$

$$\frac{d^2u_1(t)}{dt^2} + 4\frac{du_1(t)}{dt} + 5u_1(t) + \frac{d^2u_2(t)}{dt^2} - u_2(t),$$

with

$$y_1(t) = -2\frac{dz(t)}{dt} + 6z(t)$$

and

$$y_2(t) = -\frac{dz(t)}{dt} - z(t).$$

This system can readily be placed in a MIMO differential operator form analogous to that defined by Equation 8.46, namely,

$$\underbrace{(D^4 + 2D^3 - 6D^2 - 22D - 15)}_{a(D)} z(t) = \underbrace{[D^2 + 4D + 5,\ D^2 - 1]}_{b(D) = [b_1(D),\ b_2(D)]} \underbrace{\begin{bmatrix} u_1(t) \\ u_2(t) \end{bmatrix}}_{\mathbf{u}(t)};$$

$$\mathbf{y}(t) = \begin{bmatrix} y_1(t) \\ y_2(t) \end{bmatrix} = \begin{bmatrix} c_1(D) \\ c_2(D) \end{bmatrix} z(t) = \underbrace{\begin{bmatrix} -2D + 6 \\ -D - 1 \end{bmatrix}}_{c(D)} z(t).$$

Since $a(D)$ can be factored as

$$\begin{aligned} a(D) &= (D+1)(D-3)(D^2+4D+5) \\ &= (D+1)(D-3)(D+2-j)(D+2+j), \end{aligned}$$

the system modes are defined by the $(n = 4)$ zeros of $a(D)$, namely, -1, $+3$, and $-2 \pm j$.

We next note that $b_1(D) = D^2 + 4D + 5$, which is a factor of $a(D)$ as well. Therefore, the modes $e^{(-2+j)t}$ and $e^{(-2-j)t}$, which imply the real-valued modes $e^{-2t} \sin t$ and $e^{-2t} \cos t$, are uncontrollable by $u_1(t)$. Moreover, since $b_2(D) = (D+1)(D-1)$, the mode e^{-t} is uncontrollable by $u_2(t)$. Therefore, the remaining mode e^{3t} is the only one that is controllable by both inputs. Since all of the modes are controllable by at least one of the inputs, the system is completely (state or modal) controllable by the vector input $\mathbf{u}(t)$. This latter observation also holds because $a(D)$ and the polynomial vector $b(D) = [b_1(D),\ b_2(D)]$ are coprime; i.e., none of the zeros of $a(D)$ are also zeros of *both* $b_1(D)$ and $b_2(D)$.

We further note that $c_1(D) = -2(D-3)$ while $c_2(D) = -(D+1)$. Therefore, the mode e^{3t} is unobservable by $y_1(t)$, while e^{-t} is unobservable by $y_2(t)$. Since all of the modes are observable by at least one of the outputs, the system is completely (state or modal) observable by the vector output $\mathbf{y(t)}$. This latter observation also holds because $a(D)$ and the polynomial vector $c(D)$ are coprime; i.e., none of the zeros of $a(D)$ are also zeros of *both* $c_1(D)$ and $c_2(D)$.

In the general p-output, m-input differential operator case, $a(D)$ in Equation 8.46 could be a $(q \times q)$ polynomial matrix, with $\mathbf{z}(t)$ a q-dimensional partial state vector [6] [3]. The zeros of the determinant of $a(D)$ would then define the (n) poles of the MIMO system and its corresponding modes. Moreover, $b(D)$, $c(D)$ and $e(D)$ would be polynomial matrices in D of dimensions $(q \times m)$, $(p \times q)$ and $(p \times m)$, respectively. In such cases, the controllability and observability properties of the system can be determined directly in terms of the defining polynomial matrices.

In particular, as shown in [6], $a(D)$ and $b(D)$ always have a *greatest common left divisor*, namely a nonsingular polynomial matrix $g_l(D)$ that is a left divisor of both $a(D)$ and $b(D)$ in the sense that $a(D) = g_l(D)\hat{a}(D)$ and $b(D) = g_l(D)\hat{b}(D)$, for some appropriate pair of polynomial matrices, $\hat{a}(D)$ and $\hat{b}(D)$. Furthermore, *the determinant of $g_l(D)$ is a polynomial of maximum degree whose zeros define all of the uncontrollable modes of the system.*

If the degree of $|g_l(D)|$ is zero, then the following (equivalent) conditions hold:

- $g_l(D)$ is a *unimodular matrix*[8].
- $a(D)$ and $b(D)$ are *left coprime.*
- The differential operator system is controllable.

The astute reader will note that a non-unimodular $g_l(D)$ implies a lower order differential operator representation between $\mathbf{z}(t)$ and $\mathbf{u}(t)$ than that defined by Equation 8.46, namely, $\hat{a}(D)\mathbf{z}(t) = \hat{b}(D)\mathbf{u}(t)$, which implies a corresponding pole-zero "cancellation" relative to the transfer function matrix relationship between the partial state and the input.

By duality, $a(D)$ and $c(D)$ always have a *greatest common right divisor* $g_r(D)$, *whose determinant defines all of the unobservable modes of the system.* If the degree of $|g_r(D)|$ is zero, then the following (equivalent) conditions hold:

- $g_r(D)$ is a *unimodular matrix.*
- $a(D)$ and $c(D)$ are *right coprime.*
- The differential operator system is observable.

The astute reader will note that a non-unimodular $g_l(D)$ or $g_r(D)$ implies a lower order differential operator representation between $\mathbf{z}(t)$ and $\mathbf{u}(t)$ or $\mathbf{y}(t)$ than that defined by Equation 8.46. For example, if $g_l(D)$ is non-unimodular, then $\hat{a}(D)\mathbf{z}(t) = \hat{b}(D)\mathbf{u}(t)$, which implies a corresponding pole-zero "cancellation" relative to the transfer function matrix relationship between the partial state and the input. A dual observation holds when $g_r(D)$ is non-unimodular.

The preceding observations, which extend the notions of controllability and observability to a more general class of differential operator systems, are fully developed and illustrated in a number of references, such as [6] and [3].

References

[1] Athans, Michael and Falb, Peter L., *Optimal Control: An Introduction to the Theory and Its Applications,* McGraw-Hill Book Company, New York,1966.

[2] Brockett, Roger W., *Finite Dimensional Linear Systems,* John Wiley & Sons, New York, 1970.

[3] Chen, Chi-Tsong, *Linear System Theory and Design,* Holt, Rinehart & Winston, New York, 1984.

[4] Kalman, R. E., Contributions to the theory of optimal control, *Bol. Sociedad Mat. Mex.,* 1960.

[5] *The MATLAB User's Guide,* The Math Works, Inc., South Natick, MA

[6] Wolovich, W. A., *Linear Multivariable Systems,* Springer-Verlag, 1974.

[7] Wolovich, W. A., *Automatic Control Systems, Basic Analysis and Design,* Saunders College Publishing, Boston, MA, 1994.

[8] A polynomial matrix whose determinant is a real scalar, independent of D, so that its inverse is also a unimodular matrix.

9

Stability Tests

Robert H. Bishop
The University of Texas at Austin

Richard C. Dorf
University of California, Davis

Charles E. Rohrs
Tellabs, Mishawaka, IN

Mohamed Mansour
Swiss Federal Institute of Technology (ETH)

Raymond T. Stefani
Electrical Engineering Department, California State University, Long Beach

9.1 The Routh-Hurwitz Stability Criterion

Robert H. Bishop, The University of Texas at Austin
Richard C. Dorf, University of California, Davis

9.1.1 Introduction

In terms of linear systems, we recognize that the stability requirement may be defined in terms of the location of the poles of the closed-loop transfer function. Consider a single-input, single-output closed-loop system transfer function given by

$$\begin{aligned} T(s) &= \frac{p(s)}{q(s)} \\ &= \frac{K\prod_{i=1}^{M}(s+z_i)}{\prod_{k=1}^{Q}(s+\sigma_k)\prod_{m=1}^{R}\left(s^2+2\alpha_m s+\alpha_m^2+\omega_m^2\right)}, \end{aligned} \tag{9.1}$$

where $q(s)$ is the characteristic equation whose roots are the poles of the closed-loop system. The output response for an impulse function input is then

$$c(t) = \sum_{k=1}^{Q} A_k e^{-\sigma_k t} + \sum_{m=1}^{R} B_m \left(\frac{1}{\omega_m}\right) e^{-\alpha_m t} \sin \omega_m t\ . \tag{9.2}$$

To obtain a bounded response to a bounded input, the poles of the closed-loop system must be in the left-hand portion of the s-plane (i.e., $\sigma_k > 0$ and $\alpha_m > 0$). *A necessary and sufficient condition that a feedback system be stable is that all the poles of the system transfer function have negative real parts.* We will call a system not stable if not all the poles are in the left half-plane. If the characteristic equation has simple roots on the imaginary axis ($j\omega$-axis) with all other roots in the left half-plane, the steady-state output is sustained oscillations for a bounded input, unless the input is a sinusoid (which is bounded) whose frequency is equal to the magnitude of the $j\omega$-axis roots. For this case, the output becomes unbounded. Such a system is called marginally stable, since only certain bounded inputs (sinusoids of the frequency of the poles) cause the output to become unbounded. For an unstable system, the characteristic equation has at least one root in the right half of the s-plane or repeated $j\omega$-axis roots; for this case, the output becomes unbounded for any input.

9.1.2 The Routh-Hurwitz Stability Criterion

The discussion and determination of **stability** has occupied the interest of many engineers. Maxwell and Vishnegradsky first considered the question of stability of dynamic systems. In the late 1800s, A. Hurwitz and E. J. Routh published independently a method of investigating the stability of a linear system [1] and [2]. The Routh-Hurwitz stability method provides an answer to the question of stability by considering the characteristic equation of

0-8493-8570-9/96/$0.00+$.50

the system. The characteristic equation in Equation 9.1 can be written as

$$q(s) = a_n s^n + a_{n-1} s^{n-1} + \cdots + a_1 s + a_o = 0 . \tag{9.3}$$

We require that all the coefficients of the polynomial must have the same sign if all the roots are in the left half-plane. Also, it is necessary that all the coefficients for a stable system be nonzero. However, although necessary, these requirements are not sufficient. That is, we immediately know the system is unstable if they are not satisfied; yet if they are satisfied, we must proceed to ascertain the stability of the system. The **Routh-Hurwitz criterion** is a necessary and sufficient criterion for the stability of linear systems. The method was originally developed in terms of determinants, but here we utilize the more convenient array formulation. The Routh-Hurwitz criterion is based on ordering the coefficients of the characteristic equation in Equation 9.3 into an array or schedule as follows [3]:

$$\begin{array}{c|cccc} s^n & a_n & a_{n-2} & a_{n-4} & \cdots \\ s^{n-1} & a_{n-1} & a_{n-3} & a_{n-5} & \cdots \end{array} .$$

Further rows of the schedule are then completed as follows:

$$\begin{array}{c|ccc} s^n & a_n & a_{n-2} & a_{n-4} \\ s^{n-1} & a_{n-1} & a_{n-3} & a_{n-5} \\ s^{n-2} & b_{n-1} & b_{n-3} & b_{n-5} \\ s^{n-3} & c_{n-1} & c_{n-3} & c_{n-5} \\ \cdot & \cdot & \cdot & \cdot \\ \cdot & \cdot & \cdot & \cdot \\ \cdot & \cdot & \cdot & \cdot \\ s^o & h_{n-1} & h_{n-3} & \end{array}$$

where

$$\begin{aligned} b_{n-1} &= \frac{(a_{n-1})(a_{n-2}) - a_n(a_{n-3})}{a_{n-1}} \\ &= -\frac{1}{a_{n-1}} \begin{vmatrix} a_n & a_{n-2} \\ a_{n-1} & a_{n-3} \end{vmatrix} , \\ b_{n-3} &= -\frac{1}{a_{n-1}} \begin{vmatrix} a_n & a_{n-4} \\ a_{n-1} & a_{n-5} \end{vmatrix} , \end{aligned}$$

and

$$c_{n-1} = -\frac{1}{b_{n-1}} \begin{vmatrix} a_{n-1} & a_{n-3} \\ b_{n-1} & b_{n-3} \end{vmatrix} ,$$

and so on. The algorithm for calculating the entries in the array can be followed on a determinant basis or by using the form of the equation for b_{n-1}.

The Routh-Hurwitz criterion states that the number of roots of $q(s)$ with positive real parts is equal to the number of changes of sign in the first column of the array. This criterion requires that there be no changes in sign in the first column for a stable system. This requirement is both necessary and sufficient.

Four distinct cases must be considered and each must be treated separately:

1. No element in the first column is zero.
2. There is a zero in the first column, but some other elements of the row containing the zero in the first column are nonzero.
3. There is a zero in the first column, and the other elements of the row containing the zero are also zero.
4. As in case 3 with *repeated* roots on the $j\omega$-axis.

Case 1. *No element in the first column is zero.*
EXAMPLE: The characteristic equation of a third-order system is

$$q(s) = a_3 s^3 + a_2 s^2 + a_1 s + a_o . \tag{9.4}$$

The array is written as

$$\begin{array}{c|cc} s^3 & a_3 & a_1 \\ s^2 & a_2 & a_o \\ s^1 & b_1 & 0 \\ s^o & c_1 & 0 \end{array}$$

where

$$b_1 = \frac{a_2 a_1 - a_o a_3}{a_2} \quad \text{and} \quad c_1 = \frac{b_1 a_o}{b_1} = a_0 .$$

For the third-order system to be stable, it is necessary and sufficient that the coefficients be positive and $a_2 a_1 > a_o a_3$. The condition $a_2 a_1 = a_o a_3$ results in a marginal stability case, and one pair of roots lies on the imaginary axis in the s-plane. This marginal stability case is recognized as Case 3 because there is a zero in the first column when $a_2 a_1 = a_o a_3$. It is discussed under Case 3.

Case 2. *Zeros in the first column while some other elements of the row containing a zero in the first column are nonzero.* If only one element in the array is zero, it may be replaced with a small positive number ϵ that is allowed to approach zero after completing the array.

EXAMPLE: Consider the characteristic equation

$$q(s) = s^4 + s^3 + s^2 + s + K , \tag{9.5}$$

where it is desired to determine the gain K that results in marginal stability. The Routh-Hurwitz array is then

$$\begin{array}{c|ccc} s^4 & 1 & 1 & K \\ s^3 & 1 & 1 & 0 \\ s^2 & \epsilon & K & 0 \\ s^1 & c_1 & 0 & 0 \\ s^o & K & 0 & 0 \end{array}$$

where

$$c_1 = \frac{\epsilon - K}{\epsilon} \to \frac{-K}{\epsilon} \quad \text{as } \epsilon \to 0.$$

Therefore, for any value of K greater than zero, the system is unstable (with $\epsilon > 0$). Also, because the last term in the first column is equal to K, a negative value of K results in an unstable system. Therefore, the system is unstable for all values of gain K.

Case 3. *Zeros in the first column, and the other elements of the row containing the zero are also zero.* Case 3 occurs when all the elements in one row are zero or when the row consists of a single element that is zero. This condition occurs when the characteristic polynomial contains roots that are symmetrically located about the origin of the s-plane. Therefore, Case 3 occurs when factors such as $(s+\sigma)(s-\sigma)$ or $(s+j\omega)(s-j\omega)$ occur. This problem is circumvented by utilizing the *auxiliary* polynomial, $U(s)$, which is formed from the row that immediately precedes the zero row in the Routh array. The order of the auxiliary polynomial is always even and indicates the number of symmetrical root pairs.

To illustrate this approach, let us consider a third-order system with a characteristic equation

$$q(s) = s^3 + 2s^2 + 4s + K\,, \tag{9.6}$$

where K is an adjustable loop gain. The Routh array is then

s^3	1	4
s^2	2	K
s^1	$\frac{8-K}{2}$	0
s^o	K	0

Therefore, for a stable system, we require that

$$0 < K < 8\,.$$

When $K = 8$, we have two roots on the $j\omega$-axis and a marginal stability case. Note that we obtain a row of zeros (Case 3) when $K = 8$. The auxiliary polynomial, $U(s)$, is formed from the row preceding the row of zeros which, in this case, is the s^2 row. We recall that this row contains the coefficients of the even powers of s and therefore, in this case, we have

$$U(s) = 2s^2 + Ks^0 = 2s^2 + 8 = 2(s^2+4) = 2(s+j2)(s-j2)\,. \tag{9.7}$$

Case 4. *Repeated roots of the characteristic equation on the $j\omega$-axis.* If the roots of the characteristic equation on the $j\omega$-axis are simple, the system is neither stable nor unstable; it is instead called marginally stable, since it has an undamped sinusoidal mode. If the $j\omega$-axis roots are repeated, the system response will be unstable, with a form $t\,(\sin(\omega t + f))$. The Routh-Hurwitz criterion does not reveal this form of instability [4]. Consider the system with a characteristic equation

$$\begin{aligned} q(s) &= (s+1)(s+j)(s-j)(s+j)(s-j) \\ &= s^5 + s^4 + 2s^3 + 2s^2 + s + 1\,. \end{aligned} \tag{9.8}$$

The Routh array is

s^5	1	2	1
s^4	1	2	1
s^3	ϵ	ϵ	0
s^2	1	1	
s^1	ϵ	0	
s^o	1		

where $\epsilon \to 0$. Note the absence of sign changes, a condition that falsely indicates that the system is marginally stable. The impulse response of the system increases with time as $t\sin(t+f)$. The auxiliary equation at the s^2 line is (s^2+1) and the auxiliary equation at the s^4 line is $(s^4+2s^2+1) = (s^2+1)^2$, indicating the repeated roots on the $j\omega$-axis.

9.1.3 Design Example: Tracked Vehicle Turning Control

Using Routh-Hurwitz methods, the design of a turning control system for a tracked vehicle (which can be modeled as a two-input, two-output system [5]) is considered. As shown in Figure 9.1a, the system has throttle and steering inputs and vehicle heading and track speed differences as outputs. The two vehicle tracks are operated at different speeds in order to turn the vehicle. The two-input, two-output system model can be simplified to two independent single-input, single-output systems for use in the control design phase. The single-input, single-output vehicle heading feedback control system is shown in Figure 9.1b. For purposes of discussion, the control problem is further simplified to the selection of two parameters. Our objective is to select the parameters K and a so that the system is stable and the steady-state error for a ramp command is less than or equal to 24% of the magnitude of the command. The characteristic equation of the feedback system is

$$1 + G_cG(s) = 0 \tag{9.9}$$

or

$$1 + \frac{K(s+a)}{s(s+1)(s+2)(s+5)} = 0\,. \tag{9.10}$$

Therefore, we have

$$q(s) = s(s+1)(s+2)(s+5) + K(s+a) = 0 \tag{9.11}$$

or

$$s^4 + 8s^3 + 17s^2 + (K+10)s + Ka = 0\,. \tag{9.12}$$

To determine the stable region for K and a, we establish the Routh array as

s^4	1	17	Ka
s^3	8	$K+10$	0
s^2	b_3	Ka	
s^1	c_3		
s^o	Ka		

where

$$b_3 = \frac{126-K}{8} \quad \text{and} \quad c_3 = \frac{b_3(K+10) - 8Ka}{b_3}\,.$$

For the elements of the first column to be positive, we require that Ka, b_3, and c_3 be positive. We therefore require

$$\begin{aligned} &K < 126 \\ &Ka > 0 \\ &(K+10)(126-K) - 64Ka > 0\,. \end{aligned} \tag{9.13}$$

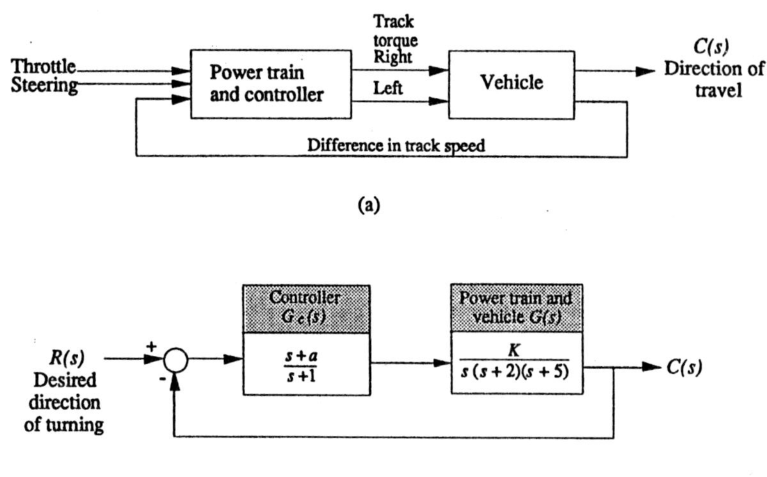

Figure 9.1 (a) Turning control for a two-track vehicle; (b) block diagram. (From Dorf, R. C. and Bishop, R. H., *Modern Control Systems*, 7th ed., Addison-Wesley, Reading, MA, 293, 1995. With permission.)

The region of stability for $K > 0$ is shown in Figure 9.2. The steady-state error to a ramp input $r(t) = At, t > 0$ is

$$e_{ss} = \frac{A}{K_v}, \tag{9.14}$$

where K_v is the *velocity error constant*, and in this case $K_v = Ka/10$. Therefore, we have

$$e_{ss} = \frac{10A}{Ka}. \tag{9.15}$$

When e_{ss} is equal to 23.8% of A, we require that $Ka = 42$. This can be satisfied by the selected point in the stable region when $K = 70$ and $a = 0.6$, as shown in Figure 9.2. Of course, another acceptable design is attained when $K = 50$ and $a = 0.84$. We can calculate a series of possible combinations of K and a that can satisfy $Ka = 42$ and that lie within the stable region, and all will be acceptable design solutions. However, not all selected values of K and a will lie within the stable region. Note that K cannot exceed 126.

The corresponding unit ramp input response is shown in Figure 9.3. The steady-state error is less than 0.24, as desired.

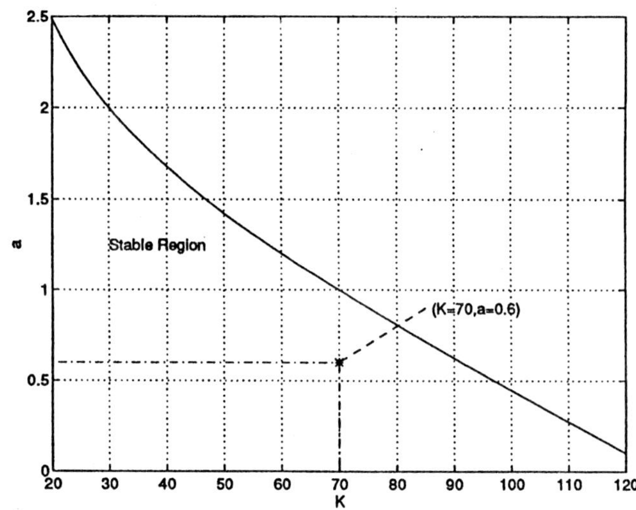

Figure 9.2 The stability region. (From Dorf, R. C. and Bishop, R. H., *Modern Control Systems*, 7th ed., Addison-Wesley, Reading, MA, 299, 1995. With permission.)

9.1.4 Conclusions

In this chapter, we have considered the concept of the stability of a feedback control system. A definition of a stable system in terms of a bounded system response to a bounded input was outlined and related to the location of the poles of the system transfer function in the s-plane.

The Routh-Hurwitz stability criterion was introduced, and several examples were considered. The relative stability of a feedback control system was also considered in terms of the location of the poles and zeros of the system transfer function in the s-plane.

9.1.5 Defining Terms

Stability: A performance measure of a system. A system is stable if all the poles of the transfer function have negative real parts.

Routh-Hurwitz criterion: A criterion for determining the stability of a system by examining the characteristic equation of the transfer function.

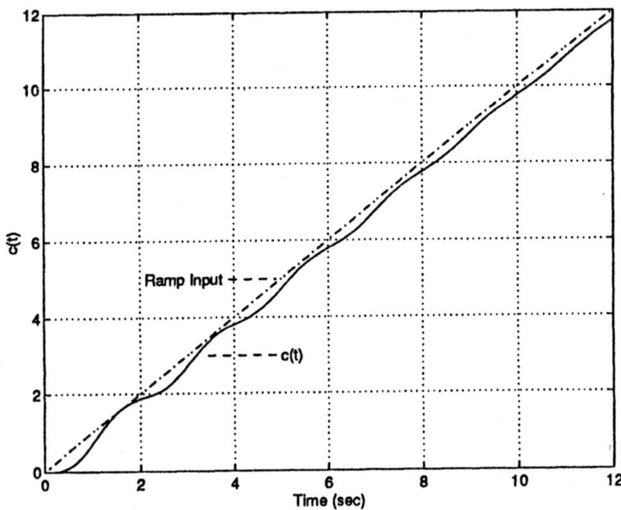

Figure 9.3 Ramp response for $a = 0.6$ and $K = 70$ for two-track vehicle turning control. (From Dorf, R. C. and Bishop, R. H., *Modern Control Systems,* 7th ed., Addison-Wesley, Reading, MA, 300, 1995. With permission.)

References

[1] Hurwitz, A., On the conditions under which an equation has only roots with negative real parts, *Mathematische Annalen,* 46, 273–284, 1895. Also in *Selected Papers on Mathematical Trends in Control Theory,* Dover, New York, 70–82, 1964.

[2] Routh, E. J., *Dynamics of a System of Rigid Bodies,* Macmillan, New York, 1892.

[3] Dorf, R. C. and Bishop, R. H., *Modern Control Systems,* 7th ed., Addison-Wesley, Reading, MA, 1995.

[4] Clark, R. N., The Routh-Hurwitz stability criterion, revisited, *IEEE Control Syst. Mag.,* 12 (3), 119–120, 1992.

[5] Wang, G. G., Design of turning control for a tracked vehicle, *IEEE Control Syst. Mag.,* 10 (3), 122–125, 1990.

9.2 The Nyquist Stability Test[1]

Charles E. Rohrs, *Tellabs, Mishawaka, IN*

9.2.1 The Nyquist Criterion

Development of the Nyquist Theorem

The Nyquist criterion is a graphical method and deals with the loop gain transfer function, i.e., the open-loop transfer function. The graphical character of the Nyquist criterion is one of its most appealing features.

Consider the controller configuration shown in Figure 9.4. The loop gain transfer function is given simply by $G(s)$. The closed-loop transfer function is given by

$$\begin{aligned}\frac{Y(s)}{R(s)} &= \frac{G(s)}{1+G(s)} = \frac{K_G N_G(s)/D_G(s)}{1+K_G N_G(s)/D_G(s)} \\ &= \frac{K_G N_G(s)}{D_G(s)+K_G N_G(s)} = \frac{K_G N_G(s)}{D_k(s)}\end{aligned}$$

where $N_G(s)$ and $D_G(s)$ are the numerator and denominator, respectively, of $G(s)$, K_G is a constant gain, and $D_k(s)$ is the denominator of the closed-loop transfer function. The closed-loop poles are equal to the zeros of the function

$$1+G(s) = 1+\frac{K_G N_G(s)}{D_G(s)} = \frac{D_G(s)+K_G N_G(s)}{D_G(s)} \tag{9.16}$$

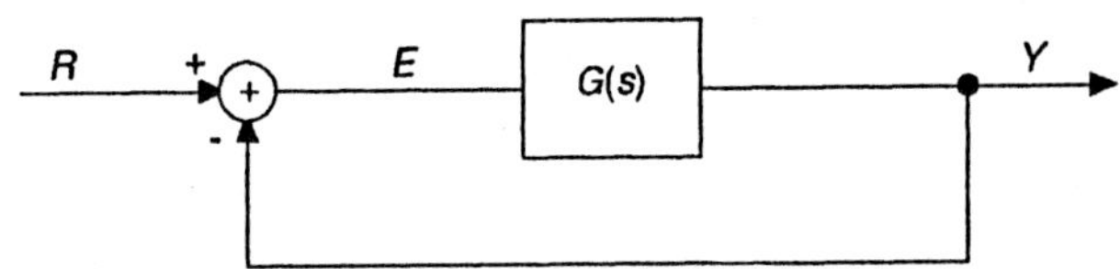

Figure 9.4 A control loop showing the loop gain $G(s)$.

Of course, the numerator of Equation 9.16 is just the closed-loop denominator polynomial, $D_k(s)$, so that

$$1+G(s) = \frac{D_k(s)}{D_G(s)} \tag{9.17}$$

In other words, we can determine the stability of the closed-loop system by locating the zeros of $1 + G(s)$. This result is of prime importance in the following development.

For the moment, let us assume that $1 + G(s)$ is known in factored form so that we have

$$1+G(s) = \frac{(s+\lambda_{k1})(s+\lambda_{k2})\cdots(s+\lambda_{kn})}{(s+\lambda_1)(s+\lambda_2)\cdots(s+\lambda_n)} \tag{9.18}$$

Obviously, if $1+G(s)$ were known in factored form, there would be no need for the use of the Nyquist criterion, since we could simply observe whether any of the zeros of $1 + G(s)$ [which are the poles of $Y(s)/R(s)$], lie in the right half of the s plane. In fact, the primary reason for using the Nyquist criterion is to avoid this factoring. Although it is convenient to think of $1 + G(s)$ in factored form at this time, no actual use is made of that form.

Let us suppose that the pole-zero plot of $1+G(s)$ takes the form shown in Figure 9.5a. Consider next an arbitrary closed contour, such as that labeled Γ in Figure 9.5a, which encloses one and only one zero of $1+G(s)$ and none of the poles. Associated with each point on this contour is a value of the complex function $1+G(s)$. The value of $1+G(s)$ for any value of s on Γ may be found analytically by substituting the appropriate complex value of s into the function. Alternatively, the value may be found graphically by considering the distances and angles from s on Γ to the zeros and poles.

If the complex value of $1 + G(s)$ associated with every point on the contour Γ is plotted, another closed contour Γ' is created

[1]Much of the material of this section is taken from Rohrs, Charles E., Melsa, James L., and Schultz, Donald G., *Linear Control Systems,* McGraw-Hill, New York, 1993. It is used with permission.

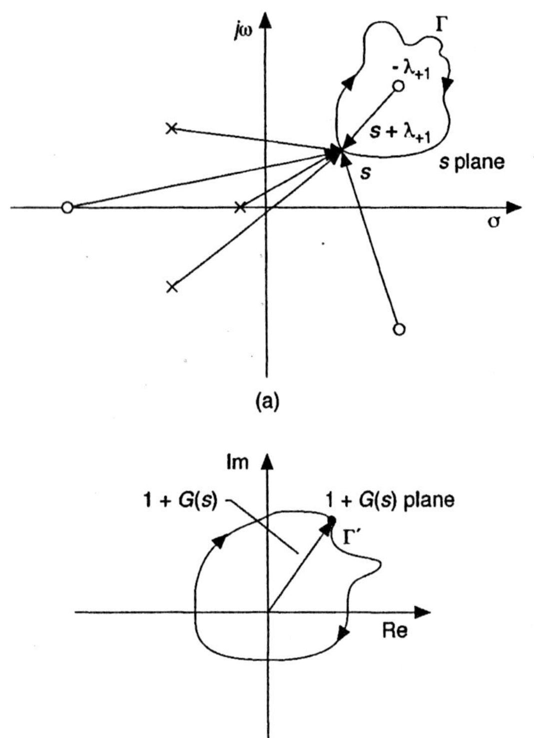

Figure 9.5 (a) Pole-zero plot of $1 + G(s)$ in the s plane; (b) plot of the Γ contour in the $1 + G(s)$ plane.

in the complex $1 + G(s)$ plane, as shown in Figure 9.5b. The function $1 + G(s)$ is said to map the contour Γ in the s plane into the contour Γ' in the $1 + G(s)$ plane. What we wish to demonstrate is that, if a zero is enclosed by the contour Γ, as in Figure 9.5a, the contour Γ' encircles the origin of the $1 + G(s)$ plane in the same sense that the contour Γ encircles the zero in the s plane. In the s plane, the zero is encircled in the clockwise direction; hence we must show that the origin of the $1 + G(s)$ plane is also encircled in the clockwise direction. This result is known as the *Principle of the Argument.*

The key to the Principle of the Argument rests in considering the value of the function $1 + G(s)$ at any point s as simply a complex number. This complex number has a magnitude and a phase angle. Since the contour Γ in the s plane does not pass through a zero, the magnitude is never zero. Now we consider the phase angle by rewriting Equation 9.18 in polar form:

$$\begin{aligned}
& 1 + G(s) \\
&= \frac{|s+\lambda_{k1}| \underline{/\arg(s+\lambda_{k1})} \cdots |s+\lambda_{kn}| \underline{/\arg(s+\lambda_{kn})}}{|s+\lambda_1| \underline{/\arg(s+\lambda_1)} \cdots |s+\lambda_n| \underline{/\arg(s+\lambda_n)}} \\
&= \frac{|s+\lambda_{k1}| \cdots |s+\lambda_{kn}|}{|s+\lambda_1| \cdots |s+\lambda_n|} \underline{/\arg(s+\lambda_{k1}) + \cdots} \\
&\quad \underline{+ \arg(s+\lambda_{kn}) - \arg(s+\lambda_1) - \cdots - \arg(s+\lambda_n)}
\end{aligned} \tag{9.19}$$

We assume that the zero encircled by Γ is at $s = -\lambda_{k1}$. Then the phase angle associated with this zero changes by a full $-360°$ as the contour Γ is traversed clockwise in the s plane. Since the argument or angle of $1 + G(s)$ includes the angle of this zero, the argument of $1 + G(s)$ also changes by $-360°$. As seen from Figure 9.5a, the angles associated with the remaining poles and zeros make no net change as the contour Γ is traversed. For any fixed value of s, the vector associated with each of these other poles and zeros has a particular angle associated with it. Once the contour has been traversed back to the starting point, these angles return to their original value; they have not been altered by plus or minus 360° simply because these poles and zeros are not enclosed by Γ.

In a similar fashion, we could show that, if the Γ contour were to encircle two zeros of $1+G(s)$ in the clockwise direction on the s plane, the Γ' contour would encircle the origin of the $1 + G(s)$ plane twice in the clockwise direction. On the other hand, if the Γ contour were to encircle only one pole and no zero of $1+G(s)$ in the *clockwise* direction, then the contour Γ' would encircle the origin of the $1 + G(s)$ plane once in the *counterclockwise* direction. This change in direction comes about because angles associated with poles are accompanied by negative signs in the evaluation of $1 + G(s)$, as indicated by Equation 9.19. In general, the following conclusion can be drawn: *The net number of clockwise encirclements by Γ' of the origin in the $1 + G(s)$ plane is equal to the difference between the number of zeros n_z and the number of poles n_p of $1 + G(s)$ encircled in the clockwise direction by Γ.*

This result means that the difference between the number of zeros and the number of poles enclosed by *any* closed contour Γ may be determined simply by counting the net number of clockwise encirclements of the origin of the $1 + G(s)$ plane by Γ'. For example, if we find that Γ' encircles the origin three times in the clockwise direction and once in the counterclockwise direction, then $n_z - n_p$ must be equal to $3 - 1 = 2$. Therefore, in the s plane, Γ must encircle two zeros and no poles, three zeros and one pole, or any other combination such that $n_z - n_p$ is equal to 2.

In terms of stability analysis, the problem is to determine the number of zeros of $1 + G(s)$, i.e., the number of poles of $Y(s)/R(s)$, that lie in the right half of the s-plane. Accordingly, the contour Γ is chosen as the entire $j\omega$ axis and an infinite semicircle enclosing the right half-plane as shown in Figure 9.6a. This contour is known as the *Nyquist D contour* as it resembles the capital letter D.

In order to avoid any problems in plotting the values of $1+G(s)$ along the infinite semicircle, let us assume that

$$\lim_{|s|\to\infty} G(s) = 0$$

This assumption is justified since, in general, the loop gain transfer function $G(s)$ is strictly proper. With this assumption, the entire infinite semicircle portion of Γ maps into the single point $s = +1 + j0$ on the $1 + G(s)$ plane.

The mapping of Γ therefore involves simply plotting the complex values of $1 + G(s)$ for $s = j\omega$ as ω varies from $-\infty$ to $+\infty$. For $\omega \geq 0$, Γ' is nothing more than the polar plot of the frequency response of the function $1 + G(s)$. The values of $1 + G(j\omega)$ for negative values of ω are the mirror image of the values of $1 + G(j\omega)$ for positive values of ω reflected about the real axis.

The Γ' contour may therefore be found by plotting the frequency response $1+G(s)$ for positive ω and then reflecting this plot about the real axis to find the plot for negative ω. The Γ' plot is always symmetrical about the real axis of the $1+G$ plane. Care must be taken to establish the direction that the Γ' plot is traced as the D-contour moves up the $j\omega$ axis, around the infinite semicircle and back up the $j\omega$ axis from $-\infty$ towards 0.

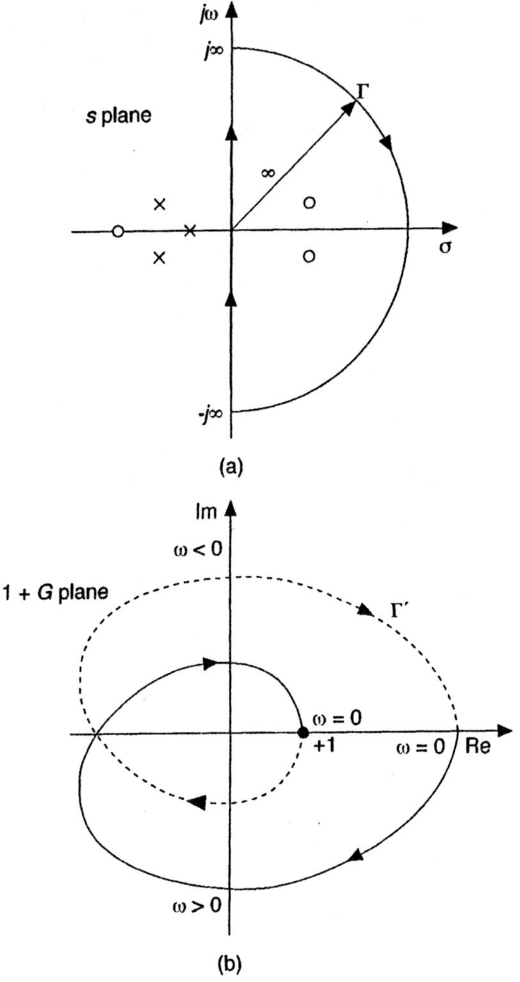

Figure 9.6 (a) Γ contour in the s plane—the Nyquist contour; (b) Γ' contour in the $1 + G$ plane.

From the Γ' contour in the $1 + G(s)$ plane, as shown in Figure 9.6b, the number of zeros of $1 + G(s)$ in the right half of the $s-$plane may be determined by the following procedure. The net number of clockwise encirclements of the origin by Γ' is equal to the number of zeros minus the number of poles of $1+G(s)$ in the right half of the $s-$plane. Note that we must know the number of poles of $1 + G(s)$ in the right half-plane if we are to be able to ascertain the exact number of zeros in the right half-plane and therefore determine stability. This requirement usually poses no problem since the poles of $1+G(s)$ correspond to the poles of the loop gain transfer function. In Equation 9.17 the denominator of $1 + G(s)$ is just $D_G(s)$, which is usually described in factored form. Hence, the number of zeros of $1 + G(s)$ or, equivalently, the number of poles of $Y(s)/R(s)$ in the right half-plane may be found by determining the net number of clockwise encirclements of the origin by Γ' and then adding the number of poles of the loop gain located in the right-half $s-$plane.

At this point the reader may revolt. Our plan for finding the number of poles of $Y(s)/R(s)$ in the right-half $s-$plane involves counting encirclements in the $1 + G(s)$ plane and observing the number of loop gain poles in the right-half $s-$plane. Yet we were forced to start with the assumption that all the poles and zeros of $1+G(s)$ are known, so that the Nyquist contour can be mapped by the function of $1 + G(s)$. Admittedly, we know the poles of this function because they are the poles of the loop gain, but we do not know the zeros; in fact, we are simply trying to find how many of these zeros lie in the right-half $s-$plane.

What we do know are the poles and zeros of the loop gain transfer function $G(s)$. Of course, this function differs from $1+G(s)$ only by unity. Any contour that is chosen in the $s-$plane and mapped through the function $G(s)$ has exactly the same shape as if the contour were mapped through the function $1 + G(s)$ except that it is displaced by one unit. Figure 9.7 is typical of such a situation. In this diagram the -1 point of the $G(s)$ plane is the origin of the $1+G(s)$ plane. If we now map the boundary of the right-half s-plane through the mapping function $G(s)$, which we often know in pole-zero form, information concerning the zeros of $1+G(s)$ may be obtained by counting the encirclements of the -1 point. The important point is that, by plotting the open-loop frequency-response information, we may reach stability conclusions regarding the closed-loop system.

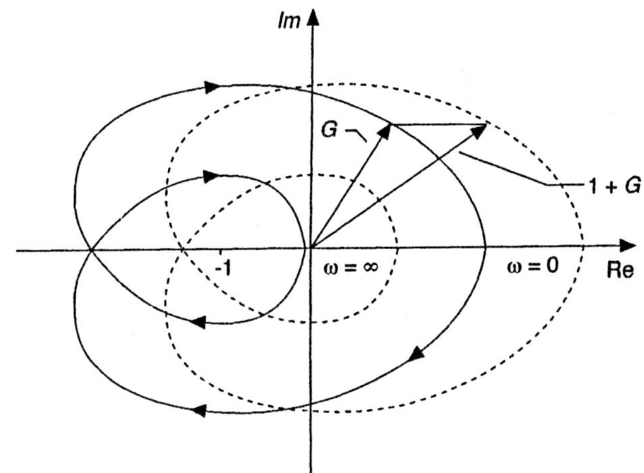

Figure 9.7 Comparison of the $G(s)$ and $1 + G(s)$ plots.

As mentioned previously contour Γ of Figure 9.6a is referred to as the Nyquist D-contour. The map of the Nyquist D-contour through $G(s)$ is called the *Nyquist diagram of* $G(s)$. There are three parts to the Nyquist diagram. The first part is the polar plot of the frequency response of $G(s)$ from $\omega = 0$ to $\omega = \infty$. The second part is the mapping of the infinite semicircle around the right half-plane. If $G(s)$ is strictly proper, this part maps entirely into the origin of the $G(s)$ plane. The third part is the polar plot of the negative frequencies, $\omega = -\infty$ to $\omega = 0$. The map of these frequencies forms a mirror image in the $G(s)$ plane about the real axis of the first part.

In terms of the Nyquist diagram of $G(s)$, the Nyquist stability criterion may be stated as follows:

THEOREM 9.1 **The Nyquist Theorem.** *The closed-loop system is stable if and only if the net number of clockwise encirclements of the points $s = -1 + j0$ by the Nyquist diagram of $G(s)$ plus the number of poles of $G(s)$ in the right half-plane is zero.*

Notice that while the *net* number of clockwise encirclements are counted in the first part of the Nyquist criterion, only the number of right half-plane *poles* of $G(s)$ are counted in the second part. Right half-plane zeros of $G(s)$ are not part of the formula in determining stability using the Nyquist criterion.

Because the Nyquist diagram involves the loop gain transfer function $G(s)$, a good approximation of the magnitude and phase of the frequency response plot can be obtained by using the Bode diagram straight-line approximations for the magnitude and for the phase. The Nyquist plot can be obtained by transferring the magnitude and phase information to a polar plot. If a more accurate plot is needed, the exact magnitude and phase may be determined for a few values of ω in the range of interest. However, in most cases, the approximate plot is accurate enough for practical problems.

An alternative procedure for obtaining the Nyquist diagram is to plot accurately the poles and zeros of $G(s)$ and obtain the magnitude and phase by graphical means. In either of these methods, the fact that $G(s)$ is known in factored form is important. Even if $G(s)$ is not known in factored form, the frequency-response plot can still be obtained by simply substituting the values $s = j\omega$ into $G(s)$ or by frequency-response measurements on the actual system.

Of course, computer programs that produce Nyquist plots are generally available. However, the ability to plot Nyquist plots by hand helps designers know how they can affect such plots by adjusting compensators.

It is also important to note that the information required for a Nyquist plot may be obtainable by measuring the frequency response of a stable plant directly and plotting this information. Thus, Nyquist ideas can be applied even if the system is a "black box" as long as it is stable.

Examples of the Nyquist Theorem

EXAMPLE 9.1:

To illustrate the use of the Nyquist[2] criterion, let us consider the simple first-order system shown in Figure 9.8a. For this system the loop gain transfer function takes the form

$$G(s) = KG_P(s) = \tfrac{K}{s+10} = \tfrac{50}{s+10}$$

The magnitude and phase plots of the frequency response of $KG_P(s)$ are shown. From these plots the Nyquist diagram $KG_P(s)$ may be easily plotted, as shown in Figure 9.9. For example, the point associated with $\omega = 10$ rad/s is found to have a magnitude of $K/(10\sqrt{2})$ and a phase angle of $-45°$. The point at $\omega = -10$ rad/s is just the mirror image of the value at $\omega = 10$ rad/s.

From Figure 9.9 we see that the Nyquist diagram can never encircle the $s = -1 + j0$ point for any positive value of K, and therefore the closed-loop system is stable for all positive values of K. In this simple example, it is easy to see that this result is correct since the closed-loop transfer function is given by

$$\frac{Y(s)}{R(s)} = \frac{K}{s + 10 + K}$$

For all positive values of K, the pole of $Y(s)/R(s)$ is in the left half-plane.

In this example, $G(s)$ remains finite along the entire Nyquist contour. This is not always the case even though we have assumed that $G(s)$ approaches zero as $|s|$ approaches infinity. If a pole of $G(s)$ occurs on the $j\omega$ axis, as often happens at the origin because of an integrator in the plant, a slight modification of the Nyquist contour is necessary. The method of handling the modification is illustrated in the Example 9.2.

EXAMPLE 9.2:

Consider a system whose loop transfer function is given by

$$G(s) = \frac{(2K/7)\left[(s + 3/2)^2 + \left(\sqrt{5/2}\right)^2\right]}{s(s+2)(s+3)}$$

The pole-zero plot of $G(s)$ is shown in Figure 9.10a. Since a pole occurs on the standard Nyquist contour at the origin, it is not clear how this problem should be handled. As a beginning, let us plot the Nyquist diagram for $\omega = +\varepsilon$ to $\omega = -\varepsilon$, including the infinite semicircle; when this is done, the small area around the origin is avoided. The resulting plot is shown as the solid line in Figure 9.10b with corresponding points labeled.

From Figure 9.10b we cannot determine whether the system is stable until the Nyquist diagram is completed by joining the points at $\omega = -\varepsilon$ and $\omega = +\varepsilon$. In order to join these points, let us use a semicircle of radius ε to the right of the origin, as shown in Figure 9.10a. Now $G(s)$ is finite at all points on the contour in the s plane, and the mapping to the G plane can be completed as shown by the dashed line in Figure 9.10b. The small semicircle

[2]Throughout the examples of this section, we assume that the gain $K > 0$. If $K < 0$, all the theory holds with the critical point shifted to $+1 + j0$.

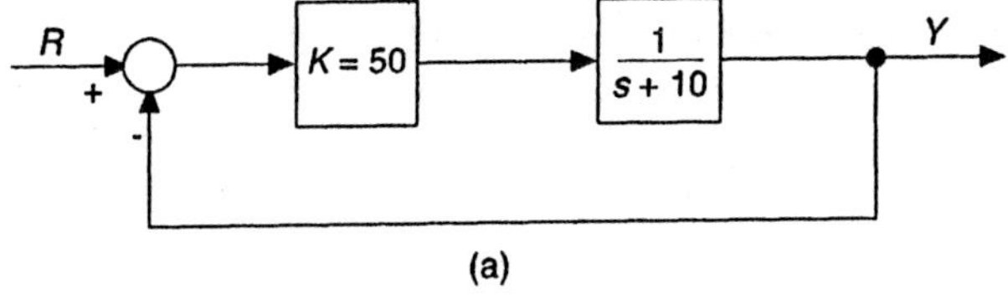

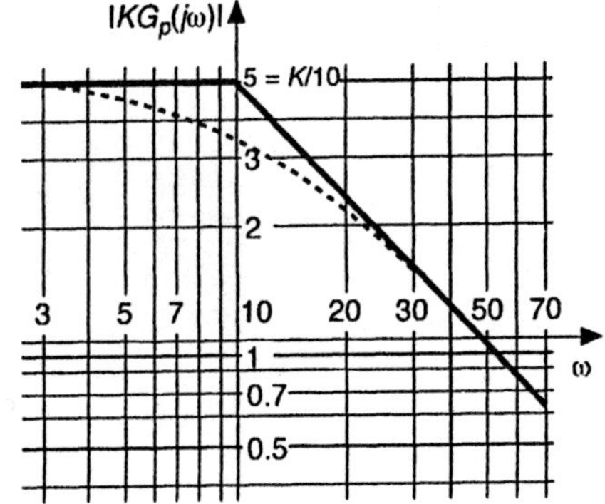

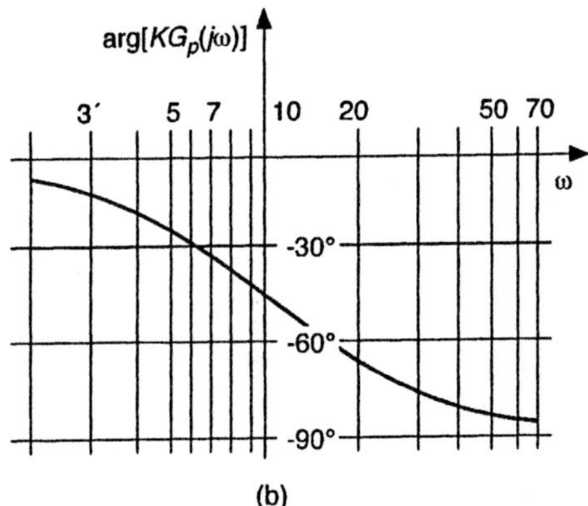

Figure 9.8 Simple first-order example. (a) Block diagram; (b) magnitude and phase plots.

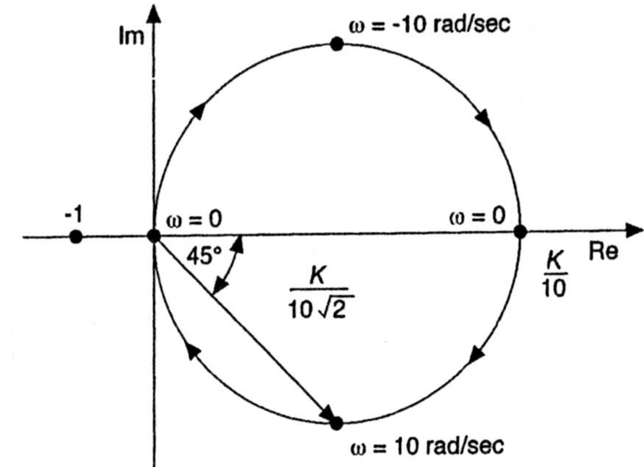

Figure 9.9 Nyquist diagram for Example 9.3.

used to avoid the origin in the s plane maps into a large semicircle in the G plane.

It is important to know whether the large semicircle in the G plane swings to the right around positive real values of s or to the left around negative real values of s. There are two ways to determine this. The first way borrows a result from complex variable theory, which says that the Nyquist diagram is a *conformal map* and for a conformal map right turns in the s plane correspond to right turns in the $G(s)$ plane. Likewise, left turns in the s plane correspond to left turns in the $G(s)$ plane.

The second method of determining the direction of the large enclosing circle on the $G(s)$ plane comes from a graphical evaluation of $G(s)$ on the circle of radius ε in the s plane. The magnitude is very large here due to the proximity of the pole. The phase at $s = -\varepsilon$ is slightly larger than $+90°$ as seen from the solid line of the Nyquist plot. The phase contribution from all poles and zeros except the pole at the origin does not change appreciably as the circle of the radius ε is traversed. The angle from the pole at the origin changes from $-90°$ through $0°$ to $+90°$. Since angles from poles contribute in a negative manner, the contribution from the pole goes from $+90°$ through $0°$ to $-90°$. Thus, as the semicircle of radius ε is traversed in the s-plane, a semicircle moving in a clockwise direction through about $180°$ is traversed in the $G(s)$ plane. The semicircle is traced in the clockwise direction as the angle associated with $G(s)$ becomes more negative. Notice that this is consistent with the conformal mapping rule, which matches right turns of $90°$ at the top and bottom of both circles.

In order to ensure that no right half-plane zeros of $1+G(s)$ can escape discovery by lying in the ε-radius semicircular indentation in the s plane, ε is made arbitrarily small, with the result that the radius of the large semicircle in the G plane approaches infinity. As $\varepsilon \to 0$, the shape of the Nyquist diagram remains unchanged, and we see that there are no encirclements of the $s = -1 + j0$ point. Since there are no poles of $G(s)$ in the right half-plane, the system is stable. In addition, since changing the magnitude of K can never cause the Nyquist diagram to encircle the -1 point, the closed-loop system must be stable for all values of positive K.

We could just as well close the contour with a semicircle of radius ε into the left half-plane. Note that if we do this, the contour encircles the pole at the origin and this pole is counted as a right half-plane pole of $G(s)$. In addition, by applying either the conformal mapping with left turns or the graphical evaluation, we close the contour in the $G(s)$ plane by encircling the negative real axis. There is 1 counterclockwise encirclement (-1 clockwise encirclement) of the -1 point. The Nyquist criterion says that -1 clockwise encirclement plus 1 right half-plane pole of $G(s)$ yield zero closed-loop right half-plane poles. The result that the closed-loop system is stable for all positive values of K remains unchanged, as it must. The two approaches are equally good although philosophically the left turn contour, which places the pole on the $j\omega$ axis in the right half-plane, is more in keeping with the convention of poles on the $j\omega$ axis being classified as unstable.

In each of the two preceding examples, the system was open-loop stable; that is, all the poles of $G(s)$ were in the left half-plane. The next example illustrates the use of the Nyquist criterion when the system is open-loop unstable.

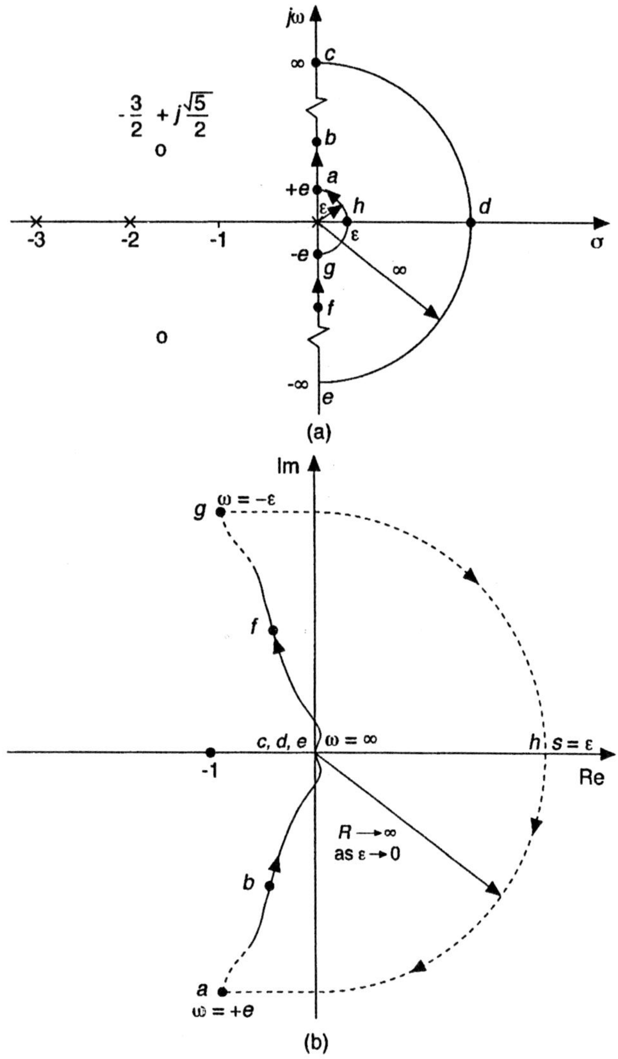

Figure 9.10 Example 9.2. (a) Pole-zero plot; (b) Nyquist diagram.

EXAMPLE 9.3:

This example is based on the system shown in Figure 9.11. The loop gain transfer function for this system is

$$G(s) = \frac{K(s+1)}{(s-1)(s+2)}$$

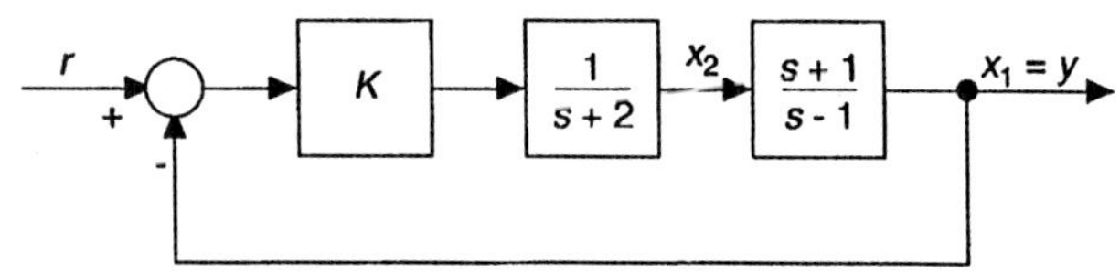

Figure 9.11 Example 9.3.

We use the Bode diagrams of magnitude and phase as an assistance in plotting the Nyquist diagram. The magnitude and phase plots are shown in Figure 9.12a.

The Nyquist diagram for this system is shown in Figure 9.12b. Note that the exact shape of the plot is not very important since the only information we wish to obtain at this time is the number of encirclements of the $s = -1 + j0$ point. It is easy to see that the Nyquist diagram encircles the -1 point once in the *counterclockwise* direction if $K > 2$ and has no encirclements if $K < 2$. Since this system has one right half-plane pole in $G(s)$, it is necessary that there be one counterclockwise encirclement if the system is to be stable. Therefore, this system is stable if and only if $K > 2$.

Besides providing simple yes/no information about whether a closed-loop system is stable, the Nyquist diagram also provides a clear graphical image indicating how close to instability a system may be. If the Nyquist diagram passes close to the -1 point and there is some mismodeling of the plant so that the characteristics of the plant are slightly different from those plotted in the Nyquist plot, then the true Nyquist characteristic may encircle the -1 point more or fewer times than the nominal Nyquist plot. The actual closed-loop system may be unstable. The Nyquist plot gives direct visual evidence of the frequencies where the plant's nominal Nyquist plot passes near the -1 point. At these frequencies great care must be taken to be sure that the nominal Nyquist plot accurately represents the plant transfer characteristic, or undiagnosed instability may result. These ideas are formalized by a theory that goes under the name of stability robustness theory.

9.2.2 Closed-Loop Response and Nyquist Diagrams

The Nyquist diagram has another important use. There are many possible designs that result in closed-loop systems that are stable but have highly oscillatory and thus unsatisfactory responses to inputs and disturbances. Systems that are oscillatory are often said to be relatively less stable than systems that are more highly damped. The Nyquist diagram is very useful in determining the relative stability of a closed-loop system.

For this development we must start with a system in the G configuration (Figure 9.4). The key for extracting information about the closed-loop system is to determine the frequency-response function of the closed-loop system, often referred to as the M−curve. The M−curve is, of course, a function of frequency and may be determined analytically as

$$M(j\omega) = \frac{Y(j\omega)}{R(j\omega)} = \frac{G(j\omega)}{1+G(j\omega)} \tag{9.20}$$

Figure 9.13 illustrates how the value of $M(j\omega_1)$ may be determined directly from the Nyquist diagram of $G(j\omega)$ at one particular frequency, ω_1. In this figure the vectors -1 and $G(j\omega_1)$ are indicated, as is the vector $(G(j\omega_1) - (-1)) = 1 + G(j\omega_1)$. The length of the vector $G(j\omega_1)$ divided by the length of $1+G(j\omega_1)$ is thus the value of the magnitude $M(j\omega_1)$. The arg $M(j\omega_1)$ is determined by subtracting the angle associated with the $1+G(j\omega_1)$ vector from that of $G(j\omega_1)$. The complete M−curve may be found by repeating this procedure over the range of frequencies of interest.

In terms of the magnitude portion of the $M(j\omega)$ plot, the point-by-point procedure illustrated above may be considerably simplified by plotting contours of constant $|M(j\omega)|$ on the

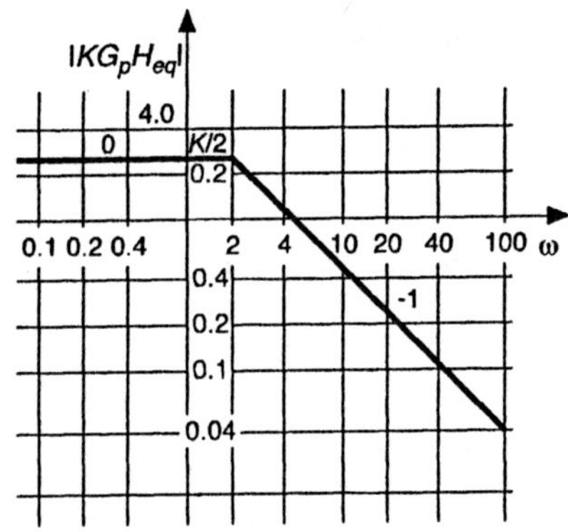
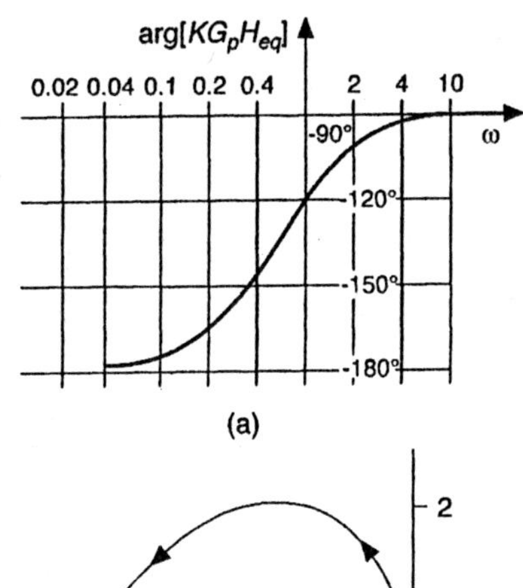

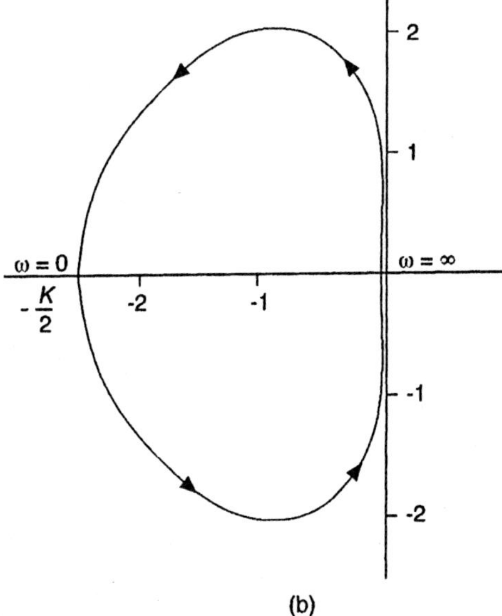

Figure 9.12 Example 9.3. (a) Magnitude and phase plots; (b) Nyquist diagram.

Nyquist plot of $G(s)$. The magnitude plot of $M(j\omega)$ can then be read directly from the Nyquist diagram of $G(s)$. Fortunately, these contours of constant $|M(j\omega)|$ have a particularly simple form. For $|M(j\omega)| = M$, the contour is simply a circle. These circles are referred to as constant M-circles or simply M-circles.

If these constant M-circles are plotted together with the Nyquist diagram of $G(s)$, as shown in Figure 9.14 for the system $G(s) = \frac{42}{s(s+2)(s+15)}$, the values of $|M(j\omega)|$ may be read directly from the plot. Note that the $M = 1$ circle degenerates to the straight line $X = -0.5$. For $M < 1$, the constant M-circles lie to the right of this line, whereas, for $M > 1$, they lie to the left.

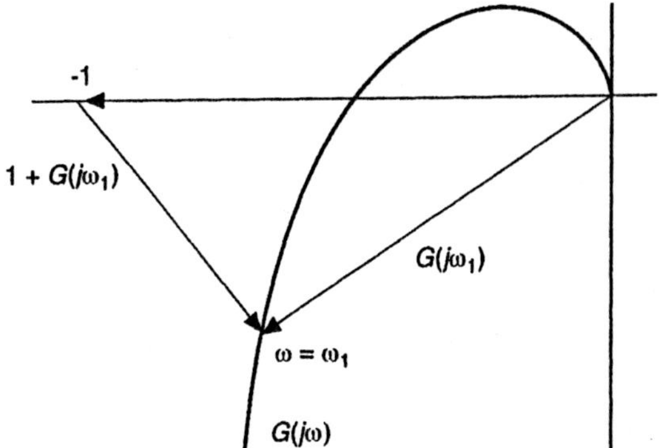

Figure 9.13 Graphical determination of $M(j\omega)$.

In addition, the $M = 0$ circle is the point $0 + j0$, and $M = \infty$ corresponds to the point $-1.0 + j0$.

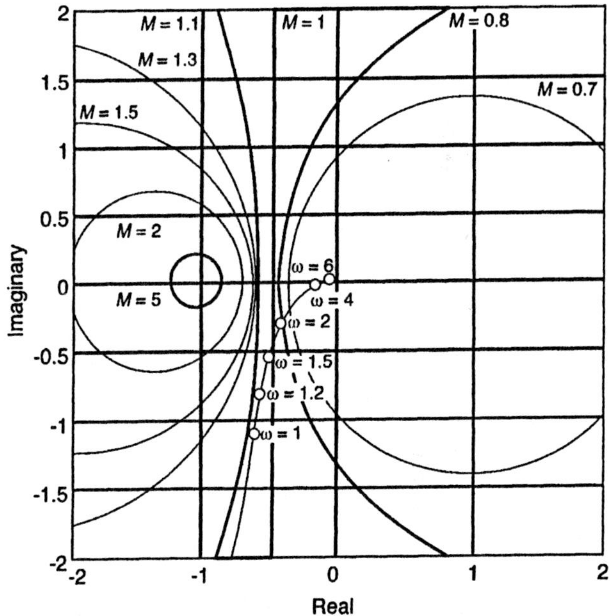

Figure 9.14 Constant M-contours.

In an entirely similar fashion, the contours of constant $\arg(M(j\omega))$ can be found. These contours turn out to be segments of circles. The circles are centered on the line $X = -\frac{1}{2}$. The contour of the $arg(M(j\omega)) = \beta$ for $0 < \beta < 180°$ is the upper half-plane portion of the circle centered at $-\frac{1}{2} + j1/(2\tan\beta)$ with a radius $|1/(2\sin\beta)|$. For β in the range $-180° < \beta < 0°$, the portions of the same circles in the lower half-plane are used. Figure 9.15 shows the plot of the constant-phase contours for some values of β. Notice that one circle represents $\beta = 45°$ above the real axis while the same circle represents $\beta = -135°$ below the real axis.

By using these constant-magnitude and constant-phase contours, it is possible to read directly the complete closed-loop frequency response from the Nyquist diagram of $G(s)$. In practice it

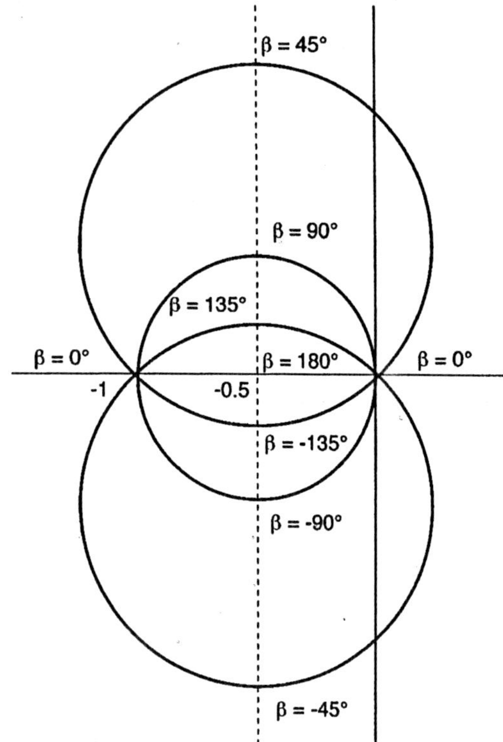

Figure 9.15 Constant-phase contours.

is common to dispense with the constant-phase contours, since it is the magnitude of the closed-loop frequency response that provides the most information about the transient response of the closed-loop system. In fact, it is common to simplify the labor further by considering only one point on the magnitude plot, namely, the point at which M is maximum. This point of peak magnitude is referred to as M_p, and the frequency at which the peak occurs is ω_p. The point M_p may be easily found by considering the contours of larger and larger values of M until the contour is found that is just tangent to the plot of $G(s)$. The value associated with this contour is then M_p, and the frequency at which the M_p contour and $G(s)$ touch is ω_p. In the plot of $G(s)$ shown in Figure 9.14, for example, the value of M_p is 1.1 at the frequency $\omega_p \approx 1.1$ rad/s.

One of the primary reasons for determining M_p and ω_p, in addition to the obvious saving of labor as compared with the determination of the complete frequency response, is the close correlation of these quantities with the behavior of the closed-loop system. In particular, for the simple second-order closed-loop system,

$$\frac{Y(s)}{R(s)} = \frac{\omega_n^2}{s^2+2s\zeta\omega_n+\omega_n^2} \tag{9.21}$$

the values of M_p and ω_p completely characterize the system. In other words, for this second-order system, M_p and ω_p specify the damping, ζ , and the natural frequency, ω_n, the only parameters of the system. The following equations relate the maximum point of the frequency response of Equation 9.21 to the values of ζ and ω_n;

$$\omega_p = \omega_n\sqrt{1-2\zeta^2} \tag{9.22}$$

$$M_p = \frac{1}{2\zeta\sqrt{1-\zeta^2}} \text{ for } \zeta \leq 0.707 \tag{9.23}$$

From these equations one may determine ζ and ω_n if M_p and ω_p are known, and vice versa. Figure 9.16 graphically displays the relations between M_p and ω_p and ζ and ω_n for a second-order system. Once ζ and ω_n are known, we may determine the time behavior of this second-order system.

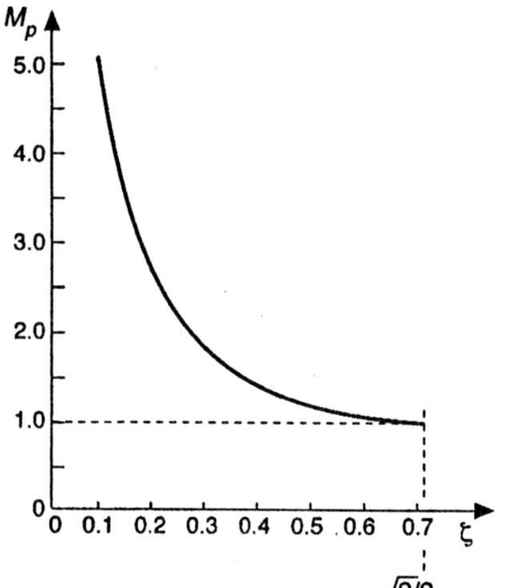

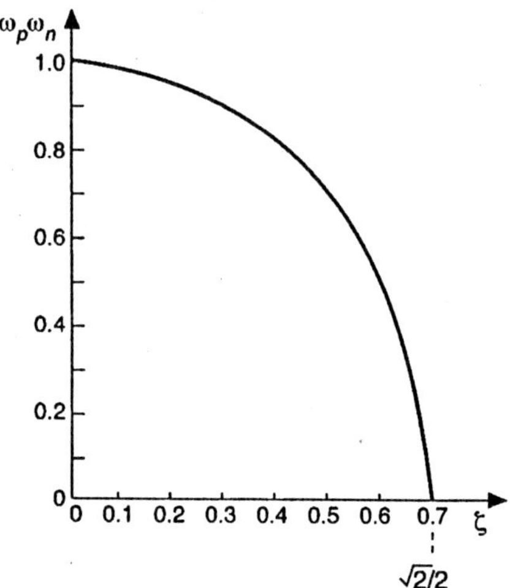

Figure 9.16 Plots of M_p and ω_p/ω_n vs. ζ for a simple second-order system.

Not all systems are of a simple second-order form. However, it is common practice to assume that the behavior of many high-order systems is closely related to that of a second-order system with the same M_p and ω_p.

Two other measures of the qualitative nature of the closed-loop response that may be determined from the Nyquist diagram of $G(s)$ are the phase margin and crossover frequency. The crossover frequency ω_c is the positive value of ω for which the magnitude of $G(j\omega)$ is equal to unity, that is,

$$|G(j\omega_c)| = 1 \tag{9.24}$$

The phase margin ϕ_m is defined as the difference between the argument of $G(j\omega_c)$ (evaluated at the crossover frequency) and $-180°$. In other words, if we define β_c as

$$\beta_c = \arg(G(j\omega_c)) \tag{9.25}$$

the phase margin is given by

$$\phi_m = \beta_c - (-180°) = 180° + \beta_c \tag{9.26}$$

While it is possible for a complicated system to possess more than one crossover frequency, most systems are designed to possess just one. The phase margin takes on a particularly simple and graphic meaning in the Nyquist diagram of $G(s)$. Consider, for example, the Nyquist diagram shown in Figure 9.17. In that diagram, we see that the phase margin is simply the angle between the negative real axis and the vector $G(j\omega_c)$. The vector $G(j\omega_c)$ may be found by intersecting the $G(s)$ locus with the unit circle. The frequency associated with the point of intersection is ω_c.

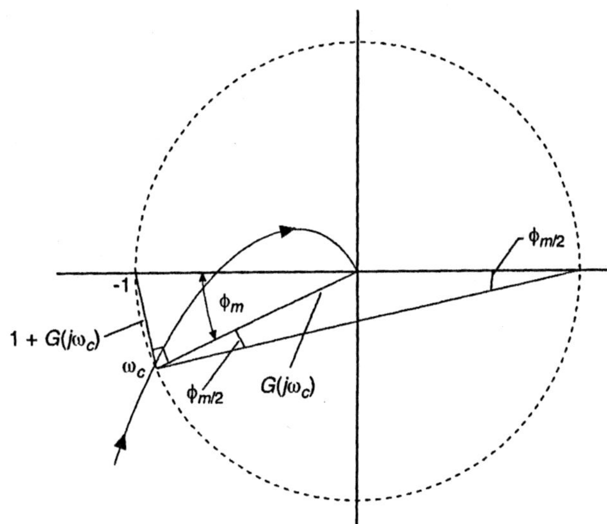

Figure 9.17 Definition of phase margin.

It is possible to determine ϕ_m and ω_c more accurately directly from the Bode plots of the magnitude and phase of $G(s)$. The value of ω for which the magnitude crosses unity is ω_c. The phase margin is then determined by inspection from the phase plot by noting the difference between the phase shift at ω_c and $-180°$. Consider, for example, the Bode magnitude and phase plots shown in Figure 9.18 for the $G(s)$ function of Figure 9.14. In time-constant form this transfer function is

$$G(s) = \frac{1.4}{s(1+s/2)(1+s/15)}$$

From this figure we see that $\omega_c = 1.4$ and $\phi_m = 60°$.

The value of the magnitude of the closed-loop frequency response at ω_c can be derived from ϕ_m. We shall call this value M_c. Often the closest point to the -1 point on a Nyquist plot occurs at a frequency that is close to ω_c. This means that M_c is often a good approximation to M_p. A geometric construction shown in Figure 9.17 shows that a right triangle exists with a hypotenuse

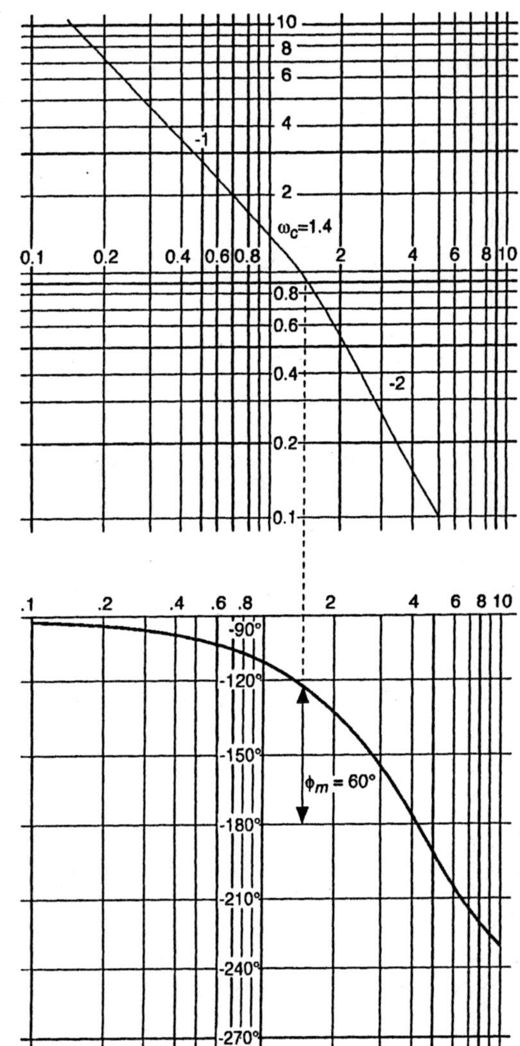

Figure 9.18 Magnitude and phase plots of $G(s)$.

of 2, one side of length $|1 + G(j\omega_c)|$, and the opposite angle of $\phi_m/2$ where ϕ_m is the phase margin. From this construction, we see

$$\sin \phi_m/2 = \frac{|1+G(j\omega_c)|}{2} \tag{9.27}$$

Since at $\omega = \omega_c$,

$$|G(j\omega_c)| = 1 \tag{9.28}$$

$$M_c = \frac{|G(j\omega_c)|}{|1+G(j\omega_c)|} = \frac{1}{2\sin \phi_m/2} \tag{9.29}$$

An oscillatory characteristic in the closed-loop time response can be identified by a large peak in the closed-loop frequency response which, in turn, can be identified by a small phase margin and the corresponding large value of M_c. Unfortunately, the correlation between response and phase margin is somewhat poorer than the correlation between the closed-loop time response and the peak M. This lower reliability of the phase margin measure is a direct consequence of the fact that ϕ_m is determined by considering only one point, ω_c on the G plot, whereas M_p is found by examining the entire plot, to find the maximum M. Consider, for example, the two Nyquist diagrams shown in Figure 9.19. The phase margin for these two diagrams is identical; however, it can be seen from the above discussion that the closed-loop step response resulting from closing the loop gain of Figure 9.19b is far more oscillatory and underdamped then the closed-loop step response resulting from closing the loop gain of Figure 9.19a.

In other words, the relative ease of determining ϕ_m as compared with M_p has been obtained only by sacrificing some of the reliability of M_p. Fortunately, for many systems the phase margin provides a simple and effective means of estimating the closed-loop response from the $G(j\omega)$ plot.

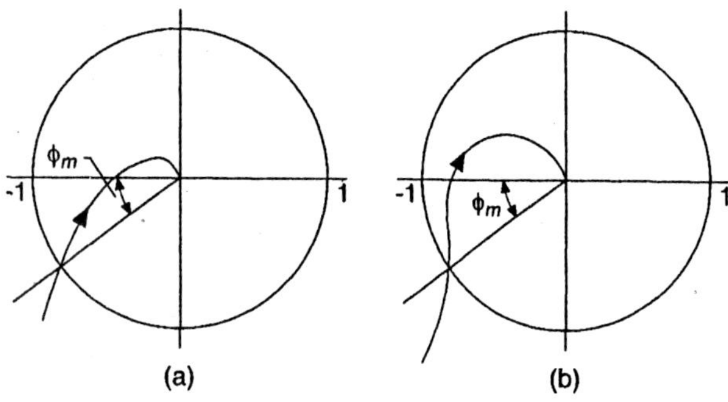

Figure 9.19 Two systems with the same phase margin but different M_p.

A system such as that shown in Figure 9.19b can be identified as a system having a fairly large M_p by checking another parameter, the gain margin. The gain margin is easily determined from the Nyquist plot of the system. The gain margin is defined as the ratio of the maximum possible gain for stability to the actual system gain. If a plot of $G(s)$ for $s = j\omega$ intercepts the negative real axis at a point $-a$ between the origin and the critical -1 point, then the gain margin is simply

$$\text{Gain margin} = GM = \frac{1}{a}$$

If a gain greater than or equal to $1/a$ were placed in series with $G(s)$, the closed-loop system would be unstable.

While the gain margin does not provide very complete information about the response of the closed-loop system, a small gain margin indicates a Nyquist plot that approaches the -1 point closely at the frequency where the phase shift is 180°. Such a system has a large M_p and an oscillatory closed-loop time response independent of the phase margin of the system.

While a system may have a large phase margin and a large gain margin and still get close enough to the critical point to create a large M_p, such phenomena can occur only in high-order loop gains. However, one should never forget to check any results obtained by using phase margin and gain margin as indicators of the closed-loop step response, lest an atypical system slip by. A visual check to see if the Nyquist plot approaches the critical -1 point too closely should be sufficient to determine if the resulting closed-loop system may be too oscillatory.

Using the concepts that give rise to the M-circles, a designer can arrive at a pretty good feel for the nature of the closed-loop transient response by examining the loop gain Bode plots. The chain of reasoning is as follows: From the loop gain Bode plots, the shape of the Nyquist plot of the loop gain can be envisioned. From the shape of the loop gain Nyquist plot, the shape of the Bode magnitude plot of the closed-loop system can be envisioned using the concepts of this section. Certain important points are evaluated by returning to the loop gain Bode plots. From the shape of the Bode magnitude plot of the closed-loop system, the dominant poles of the closed-loop transfer function are identified. From the knowledge of the dominant poles, the shape of the step response of the closed-loop system is determined. Example 9.4 illustrates this chain of thought.

EXAMPLE 9.4:

Consider the loop gain transfer function

$$G(s) = \frac{80}{s(s+1)(s+10)}$$

The Bode plots for this loop gain are given in Figure 9.20.

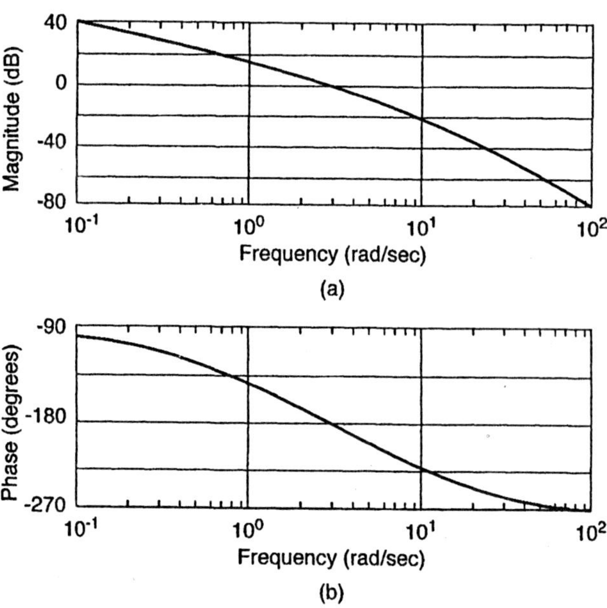

Figure 9.20 Bode plots of loop gain. (a) Magnitude plot; (b) Phase plot.

From the Bode plots the Nyquist plot can be envisioned. The Nyquist plot begins far down the negative imaginary axis since the Bode plot has large magnitude and $-90°$ phase at low frequency. It swings to the left as the phase lag increases and then spirals clockwise towards the origin, cutting the negative real axis and approaching the origin from the direction of the positive imaginary axis, i.e., from the direction associated with $-270°$ phase. From the Bode plot it is determined that the Nyquist plot does not encircle the -1 point since the Bode plot shows that the magnitude crosses unity ($0dB$) before the phase crosses $-180°$.

From the Bode plot it can be seen that the Nyquist plot passes very close to the -1 point near the crossover frequency. In this case $\omega_p \approx \omega_c$ and the phase margin is a key parameter to establish how large M_p, the peak in the closed-loop frequency magnitude plot, is. The crossover frequency is read from the Bode magnitude plot as $\omega_c = 2.5$ rad/s and the phase margin is read from the Bode phase plot as $\phi_m = 6°$.

Our visualization of the Nyquist plot is confirmed by the diagram of the actual Nyquist plot shown in Figure 9.21.

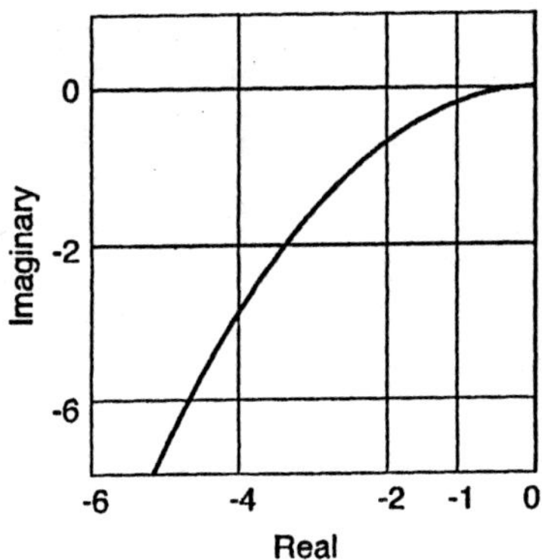

Figure 9.21 Nyquist plot of loop gain.

The magnitude of the closed-loop frequency response for this system can be envisioned using the techniques learned in this section. At low frequencies $G(s)$ is very large; the distance from the origin to the Nyquist plot is very nearly the same as the distance from the -1 point to the Nyquist plot and the closed-loop frequency response has magnitude near one. As the Nyquist plot of the loop gain approaches -1, the magnitude of the closed-loop frequency-response function increases to a peak. At higher frequencies the loop gain becomes small and the closed-loop frequency response decreases with the loop gain since the distance from -1 point to the loop gain Nyquist plot approaches unity. Thus, the closed-loop frequency response starts near $0dB$, peaks as the loop gain approaches -1 and then falls off.

The key point occurs when the loop gain approaches the -1 point and the closed-loop frequency response peaks. The closest approach to the -1 point occurs at a frequency very close to the crossover frequency, which has been established as $\omega_c = 2.5$ rad/s. The height of the peak can be established using the phase margin which has been established as $\phi_m = 6°$, and Equation 9.29. The height of the peak should be very close to $(2\sin(\phi_m/2))^{-1} = 9.5 = 19.6$ dB.

Our visualization of the magnitude of the closed-loop frequency response is confirmed by the actual plot shown in Figure 9.22.

From the visualization of the closed-loop frequency response function and the information about the peak of the frequency response, it is possible to identify the dominant closed-loop poles. The frequency of the peak identifies the natural frequency of a pair of complex poles, and the height of the peak identifies the damping ratio. More precisely,

$$M_p = \frac{1}{2\zeta\sqrt{1-\zeta^2}} \approx \frac{1}{2\zeta} \text{ for } \zeta \text{ small}$$

and

$$\omega_p = \omega_n\sqrt{1-2\zeta^2} \approx \omega_n \text{ for } \zeta \text{ small}$$

Using the approximations for ω_p and M_p that are obtained from the loop gain crossover frequency and phase margin, respectively, the following values are obtained: $\zeta \approx 1/(2M_p) \approx 0.05$ and $\omega_n \approx \omega_p \approx \omega_c \approx 2.5$ rad/s.

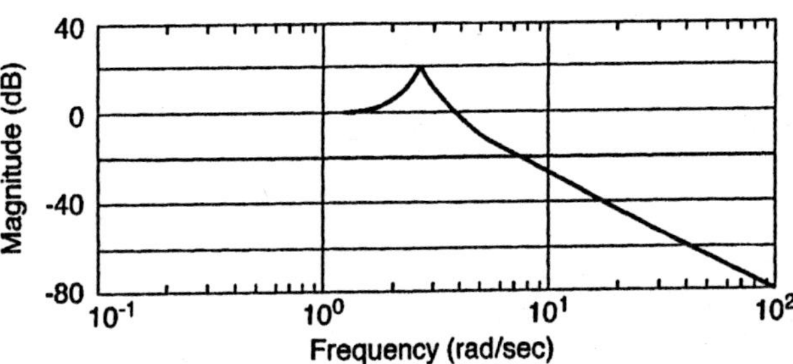

Figure 9.22 Magnitude plot of closed-loop frequency response.

If the Nyquist plot of a loop gain does not pass too closely to the -1 point, the closed-loop frequency response does not exhibit a sharp peak. In this case, the dominant poles are well damped or real. The distance of these dominant poles from the origin can be identified by the system's bandwidth, which is given by the frequency at which the closed-loop frequency response begins to decrease. From the $M-$circle concept it can be seen that the frequency at which the closed-loop frequency response starts to decrease is well approximated by the crossover frequency of the loop gain.

Having established the position of the dominant closed-loop poles, it is easy to describe the closed-loop step response. The step response has a percent overshoot given by

$$PO = 100e^{-\left(\frac{\zeta\pi}{\sqrt{1-\zeta^2}}\right)} \approx 85\%$$

The period of the oscillation is given

$$T_d = \frac{2\pi}{\omega_d} = \frac{2\pi}{\omega_n\sqrt{1-\zeta^2}} \approx 2.5\text{s}$$

The first peak in the step response occurs at a time equal to half of the period of oscillations, or about 1.25 s. The envisioned step response is confirmed in the plot of the actual closed-loop step response shown in Figure 9.23.

The method of the previous example may seem a long way to go in order to get an approximation to the closed-loop step response. Indeed, it is much simpler to calculate the closed-loop transfer function directly from the loop gain transfer function. The importance of the logic in the example is not to create a computational method; the importance lies in the insight that is achieved in predicting problems with the closed-loop transient response by examining the Bode plot of the loop gain. The essence of the insight can be summarized in a few sentences: Assume that the Nyquist plot of the loop gain indicates a stable closed-loop system. *If the Nyquist plot of the loop gain approaches the -1 point*

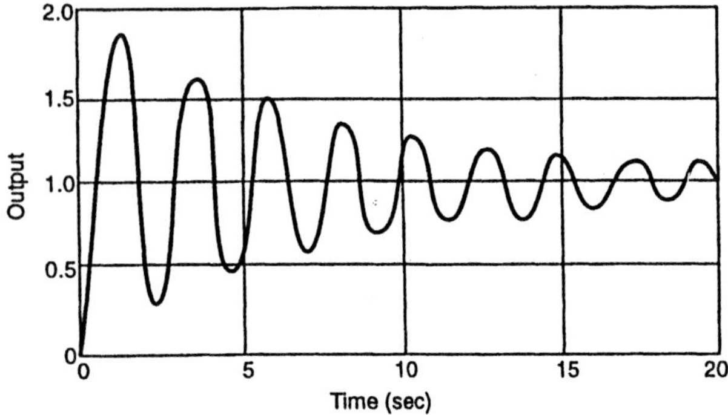

Figure 9.23 Closed-loop step response.

too closely, the transient response characteristics of the closed-loop system are oscillatory. The speed of the transient response of the closed-loop system is usually indicated by the loop gain crossover frequency. Detailed information about the loop gain Nyquist plot is available in the loop gain Bode plots. In particular, the crossover frequency and the phase margin can be read from the Bode plots.

Any information that can be wrenched out of the Bode plots of the loop gain is critically important for two reasons. First, the Bode plots are a natural place to judge the properties of the feedback loop. When the magnitude of the loop gain is large, positive feedback properties such as good disturbance rejection and good sensitivity reduction are obtained. When the magnitude of the loop gain is small, these properties are not enhanced. The work of this chapter completes the missing information about transient response that can be read from the loop gain Bode plots. Second, it is the Bode plot that we are able to manipulate directly using series compensation techniques. It is important to be able to establish the qualities of the Bode plots that produce positive qualities in a control system because only then can the Bode plots be manipulated to attain the desired qualities.

References

[1] Bode, H.W., *Network Analysis and Feedback Amplifier Design,* Van Nostrand, New York, 1945.

[2] Churchill, R.V., Brown, J.W., and Verhey, R.F., *Complex Variables and Applications,* McGraw-Hill, New York, 1976.

[3] Horowitz, I.M., *Synthesis of Feedback Systems,* Academic, New York, 1963.

[4] Nyquist, H., Regeneration Theory, *Bell System Tech. J.,* 11, 126–147, 1932.

[5] Rohrs, C.E., Melsa, J.L., and Schultz, D.G., *Linear Control Systems,* McGraw-Hill, 1993.

9.3 Discrete-Time and Sampled-Data Stability Tests

Mohamed Mansour, *Swiss Federal Institute of Technology (ETH)*

9.3.1 Introduction

Discrete-time dynamic systems are described by difference equations. Economic systems are examples of these systems where the information about the system behavior is known only at discrete points of time.

On the other hand, in sampled-data systems some signals are continuous and others are discrete in time. Some of the discrete-time signals come from continuous signals through sampling. An example of a sampled-data system is the control of a continuous process by a digital computer. The digital computer only accepts signals at discrete points of time so that a sampler must transform the continuous time signal to a discrete time signal.

Stability is the major requirement of a control system. For a linear discrete-time system, a necessary and sufficient condition for stability is that all roots of the characteristic polynomial using the z-transform lie inside the unit circle in the complex plane. A solution to this problem was first obtained by Schur [1].

The stability criterion in table and determinant forms was published by Cohn [2]. A symmetrix matrix form was obtained by Fujiwara [3]. Simplifications of the table and the determinant forms were obtained by Jury [4] and Bistritz [5]. A Markov stability test was introduced by Nour Eldin [6]. It is always possible to solve the stability problem of a discrete-time system by reducing it to the stability problem of a continuous system with a bilinear transformation of the unit circle to the left half-plane. For sampled-data systems, if the z-transform is used, then the same criteria apply. If the δ-transform is used, a direct solution of the stability problem, without transformation to the z- or s-plane, is given by Mansour [7] and Premaratne and Jury [8].

9.3.2 Fundamentals

Representation of a Discrete-Time System

A linear discrete-time system can be represented by a difference equation, a system of difference equations of first order, or a transfer function in the z-domain.

Difference equation:

$$y(k+n) + a_1 y(k+n-1) + \ldots + a_n y(k) = b_1 u(k+n-1) + \ldots + b_n u(k) \qquad (9.30)$$

If z is the shift operator, i.e., $zy(k) = y(k+1)$ then the difference equation can be written as

$$z^n y(k) + a_1 z^{n-1} y(k) + \ldots + a_n y(k) = b_1 z^{n-1} u(k) + \ldots + b_n u(k) \qquad (9.31)$$

System of difference equations of first order: Equation 9.30 can be decomposed in the following n difference equations using the state variables:

$$
\begin{aligned}
x_1(k) &\quad \dots \quad x_n(k) \\
x_1(k+1) &= x_2(k) \\
x_2(k+1) &= x_3(k) \\
&\vdots \\
x_{n-1}(k+1) &= x_n(k) \\
x_n(k+1) &= -a_n x_1(k) - \dots - a_1 x_n(k) + u(k) \\
y(k) &= b_n x_1(k) + b_{n-1} x_2(k) + \dots + b_1 x_n(k)
\end{aligned}
$$

This can be written as

$$x(k+1) = Ax(k) + bu(k), \quad y(k) = c^T x(k) \tag{9.32}$$

where A is in the companion form.

Transfer function in the z-domain: The z transform of Equation 9.30 gives the transfer function

$$G(z) = \frac{Y(z)}{U(z)} = \frac{b_1 z^{n-1} + \dots + b_n}{z^n + a_1 z^{n-1} + \dots + a_n} \tag{9.33}$$

Representation of a Sampled-Data System

The digital controller is a discrete-time system represented by a difference equation, a system of difference equations or a transfer function in the z-domain. The continuous process is originally represented by a differential equation, a system of differential equations or a transfer function in the s-domain. However, because the input to the continuous process is normally piecewise constant (constant during a sampling period) then the continuous process can be represented in this special case by difference equations or a transfer function in the z-domain. Figure 9.24 shows a sampled-data control system represented as a discrete time system.

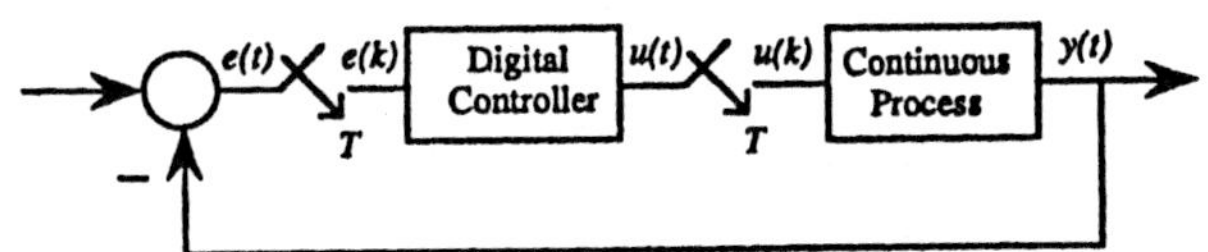

Figure 9.24 Sampled-Data control system as a discrete time system.

Representation in the δ-domain: Use the δ-operator which is related to the z-operator by

$$\delta = \frac{z-1}{\Delta}, \tag{9.34}$$

where Δ is the sampling period.

In this case, δ corresponds to the differentiation operator in continuous systems and tends to it if Δ goes to zero. The continuous system is the limiting case of the sampled-data system when the sampling period becomes very small.

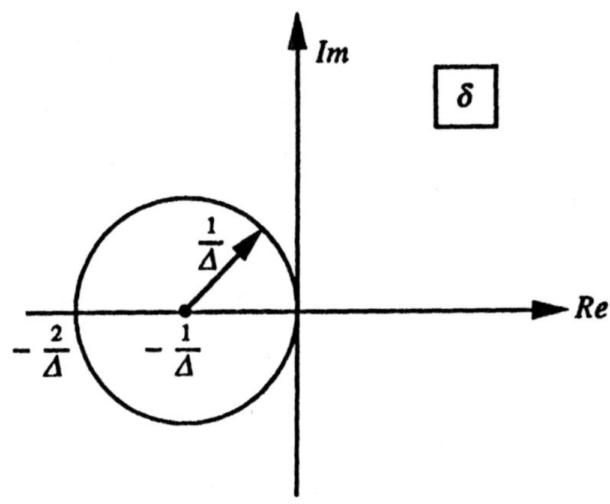

Figure 9.25 The stability region in the δ-domain.

For stability, the characteristic equation with the δ-operator should have all its roots inside the circle of radius $1/\Delta$ in the left half-plane of Figure 9.25.

Representing sampled-data systems with the δ-operator has numerical advantages [9].

Definition of Stability

The output of a SISO discrete system is given by

$$y(k) = \sum_{i=0}^{\infty} g(i)u(k-i), \tag{9.35}$$

where $g(i)$ is the impulse response sequence. This system is BIBO-stable if, and only if, a real number $P > 0$ exists so that $\sum_{i=0}^{\infty} |g(i)| \leq P < \infty$.

Basics of Stability Criteria for Linear Systems

Stability criteria for linear systems are obtained by a simple idea: an n-degree polynomial is reduced to an (n-1)-degree polynomial insuring that no root crosses the stability boundary. This can be achieved for discrete-time and sampled-data systems using the Rouché [1] or Hermite-Bieler theorem [10]. This reduction operation can be continued, thus obtaining a table form for checking stability, i.e., the impulse response is absolutely summable. This is achieved if all roots of the characteristic equation (or the eigenvalues of the system matrix) lie inside the unit circle. For sampled-data systems represented by the δ-operator, all roots of the characteristic equation must lie inside the circle in Figure 9.25.

9.3.3 Stability Criteria

Necessary Conditions for Stability of Discrete-Time Systems

[11]

Consider the characteristic polynomial,

$$f(z) = a_0 z^n + a_1 z^{n-1} + \dots + a_n. \tag{9.36}$$

The following conditions are necessary for the roots of Equation 9.36 to lie inside the unit circle (with $a_0 = 1$):

$$0 < f(1) < 2^n,$$
$$0 < (-1)^n f(-1) < 2^n. \quad (9.37)$$
$$|a_n| < 1. \quad (9.38)$$

The ranges of the coefficients $a_1, a_2, \ldots a_n$ are given by the following table.

	a_1	a_2	a_3	a_4	a_5	a_6	a_7	...
$n=2$	2	1						
	−2	−1						
$n=3$	3	3	−1					
	−3	−1	1					
$n=4$	4	6	4	1				
	−4	−2	−4	−1				
$n=5$	5	10	10	5	1			
	−5	−2	−10	−3	−1			
$n=6$	6	15	20	15	6	1		
	−6	−3	−20	−5	−6	−1		
$n=7$	7	21	35	35	21	7	1	
⋮	−7	−3	−35	−5	−21	−5	−1	

Thus for example, a necessary condition for the stability of the characteristic polynomial, $f(z) = z^2 + a_1 z + a_2$, is that $-2 < a_1 < 2$ and $-1 < a_2 < 1$. This table can detect instability without calculations. It is analogous to the positivity of the characteristic polynomial coefficients for a continuous system, but not equivalent to it.

Sufficient Conditions (with $a_0 > 0$)

6.1

$$a_0 > \sum |ak| \quad (9.39)$$

[2].

6.2

$$0 < a_n < a_{n-1} < \ldots < a_1 < a_0 \quad (9.40)$$

[12].

Necessary and Sufficient Conditions ($a_0 > 0$)

Frequency domain criterion 1: "$f(z)$ has all roots inside the unit circle of the z-plane if, and only if, $f(e^{j\theta})$ has a change of argument of $n\pi$ when θ changes from 0 to π." The proof of this criterion is based on the principle of the argument.

Frequency domain criterion 2: "$f(z)$ has all its roots inside the unit circle of the z-plane if, and only if, $f^* = h^* + jg^*$ has a change of argument of $n\pi/2$ when θ changes from 0 to π."

$$f(z) = h(z) + g(z), \quad (9.41)$$

where

$$h(z) = 1/2[f(z) + z^n f(1/z)] \quad (9.42)$$

and

$$g(z) = 1/2[f(z) - z^n f(1/z)] \quad (9.43)$$

are the symmetric and antisymmetric parts of $f(z)$ respectively. Also

$$f(e^{j\theta}) = 2e^{jn\theta/2}[h^* + jg^*]. \quad (9.44)$$

For n even, $n = 2m$,

$$h^*(\theta) = \alpha_0 \cos m\theta + \alpha_1 \cos(m-1)\theta + \ldots + \alpha_{m-1}\cos\theta + \alpha_m/2 \quad (9.45)$$
$$g*(\theta) = \beta_0 \sin m\theta + \beta_1 \sin(m-1)\theta + \ldots + \beta_{m-1}\sin\theta \quad (9.46)$$

and for n odd, $n = 2m-1$

$$h^*(\theta) = a_0 \cos(m-1/2)\theta + a_1\cos(m-3/2)\theta + \ldots + \alpha_{m-1}\cos\theta/2 \quad (9.47)$$
$$g^*(\theta) = \beta_0 \sin(m-1/2)\theta + \beta_1 \sin(m-3/2)\theta + \ldots + \beta_{m-1}\sin\theta/2 \quad (9.48)$$

$h^*(x)$ and $g^*(x)$ are the projections of $h(z)$ and $g(z)$ on the real axis with $x = \cos\theta$.

$$\text{For } n = 2m, \quad h^*(x) = \sum_0^{m-1} \alpha_i T_{m-1} + \frac{\alpha_m}{2} \quad (9.49)$$
$$g^*(x) = \sum_0^{m-1} \beta_i U_{m-i} \quad (9.50)$$

where T_k, U_k are Tshebyshef polynomials of the first and second kind, respectively. T_k and U_k can be obtained by the recursions

$$T_{k+1}(x) = 2xT_k(x) - T_{k-1}(x) \quad (9.51)$$
$$U_{k+1}(x) = 2xU_k(x) - U_{k-1}(x) \quad (9.52)$$

where

$$T_0(x) = 1, \quad T_1(x) = x, \quad U_0(x) = 0, U_1(x) = 1$$

Discrete Hermite-Bieler theorem: "$f(z)$ has all its roots inside the unit circle of the z-plane if, and only if, $h(z)$ and $g(z)$ have simple alternating roots on the unit circle and $\left|\frac{a_n}{a_0}\right| < 1$ [13]. The necessary condition $\left|\frac{a_n}{a_0}\right| < 1$ distinguishes between $f(z)$ and its inverse which has all roots outside the unit circle.

Schur-Cohn criterion: "$f(z)$ with $|a_n/a_0| < 1$ has all roots inside the unit circle of the z-plane if, and only if, the polynomial $\frac{1}{z}\left[f(z) - \frac{a_n}{a_0}z^n f(1/z)\right]$ has all roots inside the unit circle."

The proof of this theorem is based on the Rouché theorem [1] or on the discrete Hermite-Bieler theorem [10]. This criterion can be translated in table form as follows:

a_0	$a_1 \quad a_2 \ldots a_n$
a_n	$a_{n-1} \quad a_{n-2} \ldots a_0$
b_0	$b_1 \ldots b_{n-1}$
b_{n-1}	$b_{n-2} \ldots b_0$
c_0	$c_1 \ldots c_{n-2}$
c_{n-2}	$c_{n-3} \ldots c_0$
	⋱
g_0	g_1
g_1	g_0
h_0	

where

$$b_0 = \begin{vmatrix} a_0 & a_n \\ a_n & a_0 \end{vmatrix}, \quad b_1 = \begin{vmatrix} a_0 & a_{n-1} \\ a_n & a_1 \end{vmatrix}, \quad \ldots,$$

$$b_{n-1} = \begin{vmatrix} a_0 & a_1 \\ a_n & a_{n-1} \end{vmatrix},$$

$$c_0 = \begin{vmatrix} b_0 & b_{n-1} \\ b_{n-1} & b_0 \end{vmatrix}, \quad \ldots$$

The necessary and sufficient condition for stability is that

$$b_0, c_0, \ldots, g_0, h_0 > 0. \tag{9.53}$$

Jury [4] has replaced the last condition by the necessary conditions

$$f(1) > 0, \text{ and } (-1)^n f(-1) > 0. \tag{9.54}$$

Example: $f(z) = 6z^2 + z - 1$.

6	1	-1	
-1	1	6	
35	7		35 > 0
7	35		
1176			1176 > 0

Hence there is stability.

Bistritz table [5] Bistritz used a sequence of symmetric polynomials $T_i(z)$ $i = 0, 1, \ldots, n$,

$$T_n(z) = 2h(z) \tag{9.55}$$

$$T_{n-1}(z) = 2g(z)/(z-1) \tag{9.56}$$

$$T_i(z) = [\delta_{i+2}(z+1)T_{i+1}(z) - T_{i+2}(z)]/z \qquad i = n-2, n-3, \ldots, 0 \tag{9.57}$$

where

$$\delta_{i+2} = [T_{i+2}(0)]/[T_{i+1}(0)].$$

"The polynomial $f(z)$ is stable if and only if

i) $T_i(0) \neq 0$, $i = n-1, n-2, \ldots, 0$, and
ii) $T_n(1), T_{n-1}(1), \ldots, T_0(1)$

have the same sign."

Example: $n = 2$ $\quad a_0 > 0$.

$$\begin{aligned} T_2(z): &\quad (a_0 + a_2)z^2 + 2a_1 z + a_0 + a_2 \\ T_1(z): &\quad (a_0 - a_2)z + (a_0 - a_2) \\ T_0(z): &\quad 2(a_0 + a_2 - a_1) \end{aligned}$$

$$\begin{aligned} T_2(1) &= 2(a_0 + a_1 + a_2) = 2f(1) \\ T_1(1) &= 2(a_0 - a_2) \\ T_0(1) &= 2(a_0 + a_2 - a_1) = 2f(-1) \end{aligned}$$

A necessary and sufficient condition for stability of a second-order system of characteristic equation $f(z)$ is

$$\begin{aligned} f(1) &> 0 \\ f(-1) &> 0 \\ a_0 - a_2 &> 0 \end{aligned}$$

Determinant criterion [2]

$$f(z) = a_0 z^n + a_1 z^{n-1} + \ldots + a_n$$

has all roots inside the unit circle if, and only if,

$$\begin{aligned} \Delta_k &< 0 \qquad k \text{ odd} \\ \Delta_k &> 0 \qquad k \text{ even} \qquad k = 1, 2, \ldots, n \end{aligned} \tag{9.58}$$

where

$$\Delta_k = \begin{bmatrix} A_k & B_k^T \\ B_k & A_k^T \end{bmatrix} \tag{9.59}$$

$$A_k = \begin{bmatrix} a_n & & & 0 \\ a_{n-1} & a_n & & \\ \vdots & \ddots & \ddots & \\ a_{n+k-1} & \cdots & a_{n-1} & a_n \end{bmatrix}$$

$$B_k = \begin{bmatrix} a_0 & & & 0 \\ a_1 & a_0 & & \\ \vdots & \ddots & \ddots & \\ a_{k-1} & \cdots & a_1 & a_0 \end{bmatrix}$$

Jury simplified this criterion so that only determinants of dimension $n-1$ are computed [4]. The necessary conditions $f(1) > 0$ and $(-1)f(-1) > 0$ replace the determinants of dimension n. Example: $f(z) = 6z^2 + z - 1$.

$$\Delta_1 = \begin{vmatrix} -1 & 6 \\ 6 & -1 \end{vmatrix} = -35 < 0$$

$$\Delta_2 = \begin{vmatrix} -1 & 0 & 6 & 1 \\ 1 & -1 & 0 & 6 \\ 6 & 0 & -1 & 1 \\ 1 & 6 & 0 & -1 \end{vmatrix} = 161 > 0.$$

Hence there is stability.

Positive definite symmetric matrix [3] "$f(z)$ has all its roots inside the unit circle, if and only if, the symmetric matrix $C = [c_{ij}]$ is positive definite."

C is given by

$$c_{ij} = \sum_{p=1}^{\min(i,j)} (a_{i-p} - a_{j-p} a_{n-i+p} a_{n-j+p}). \tag{9.60}$$

For $n = 2$,

$$C = \begin{bmatrix} a_0^2 - a_2^2 & a_0 a_1 - a_1 a_2 \\ a_0 a_1 - a_1 a_2 & a_0^2 - a_2^2 \end{bmatrix}. \tag{9.61}$$

For $n = 3$,

$$C = \begin{bmatrix} a_0^2 - a_3^2 & a_0 a_1 - a_2 a_3 & a_0 a_2 - a_1 a_3 \\ a_0 a_1 - a_2 a_3 & a_0^2 + a_1^2 - a_2^2 - a_3^2 & a_0 a_1 - a_2 a_3 \\ a_0 a_2 - a_1 a_3 & a_0 a_1 - a_2 a_3 & a_0^2 - a_3^2 \end{bmatrix}. \tag{9.62}$$

For $n = 4$,

$$C = \begin{bmatrix} a_0^2 - a_4^2 & a_0 a_1 - a_3 a_4 \\ a_0 a_1 - a_3 a_4 & a_0^2 + a_1^2 - a_3^2 - a_4^2 \\ a_0 a_2 - a_2 a_4 & a_0 a_1 + a_1 a_2 - a_2 a_3 - a_3 a_4 \\ a_0 a_3 - a_1 a_4 & a_0 a_2 - a_2 a_4 \end{bmatrix}$$

$$\begin{matrix} a_0a_2 - a_2a_4 & a_0a_3 - a_1a_4 \\ a_0a_1 + a_1a_2 - a_2a_3 - a_3a_4 & a_0a_2 - a_2a_4 \\ a_0^2 + a_1^2 - a_3^2 - a_4^2 & a_0a_1 - a_3a_4 \\ a_0a_1 - a_3a_4 & a_0^2 - a_4^2 \end{matrix}\Bigg] \quad (9.63)$$

Jury simplified this criterion. See, for example, [14] and [15].
Example: $f(z) = 6z^2 + z - 1$.

$$C = \begin{bmatrix} 35 & 7 \\ 7 & 35 \end{bmatrix} > 0.$$

Markov stability criterion [6] If

$$g^*(x)/h^*(x) = s_0/x + s_1/x^2 + s_2/x^3 + \ldots, \quad (9.64)$$

then, for $n = 2m$, the polynomial $f(z)$ has all roots inside the unit circle if, and only if, the Hankel matrices

$$S_a = \begin{bmatrix} s_0 + s_1 & s_1 + s_2 & \cdots & s_{m-1} + s_m \\ s_1 + s_2 & s_2 + s_3 & \cdots & s_m + s_{m+1} \\ \vdots & \vdots & & \vdots \\ s_{m-1} + s_m & s_m + s_{m+1} & \cdots & s_{2m-2} + s_{2m-1} \end{bmatrix} \quad (9.65)$$

and

$$S_b = \begin{bmatrix} s_0 - s_1 & s_1 - s_2 & \cdots & s_{m-1} - s_m \\ s_1 - s_2 & s_2 - s_3 & \cdots & s_m - s_{m+1} \\ \vdots & \vdots & & \vdots \\ s_{m-1} - s_m & s_m - s_{m+1} & \cdots & s_{2m-2} - s_{2m-1} \end{bmatrix} \quad (9.66)$$

are positive definite.

For n odd, $zf(z)$ is considered instead of $f(z)$. Simplifications of the above critrion can be found in [16] and [17].

Stability of discrete-time systems using Lyapunov theory: Given the discrete system,

$$x(k+1) = Ax(k) \quad (9.67)$$

and using a quadratic form as a Lyapunov function

$$V(k) = x(k)^T Px(k), \quad (9.68)$$

the change in $V(K)$

$$\Delta V(k) = x(k)^T [A^T PA - P]x(k). \quad (9.69)$$

For stability, $V(k) > 0$, and $\Delta V(k) < 0$.

This is achieved by solving the matrix equation

$$A^T PA - P = -Q \quad (9.70)$$

P and Q are symmetric matrices.

Q is chosen positive definite, e.g., the unity matrix, and P is determined by solving a set of algebraic equations.

Necessary and sufficient conditions for the asymptotic stability of Equation 9.67 are that P is positive definite.

Stability conditions of low order polynomials [14]

$n = 2:$

$$f(z) = a_0z^2 + a_1z + a_2, \quad a_0 > 0$$

a) $f(1) > 0, \quad f(-1) > 0$
b) $a_0 - a_2 > 0$

$n = 3:$

$$f(z) = a_0z^3 + a_1z^2 + a_2z + a_3, \quad a_0 > 0$$

a) $f(1) > 0, \quad f(-1) < 0$
b) $|a_3| < a_0$
c) $a_3^2 - a_0^2 < a_3a_1 - a_0a_2$

$n = 4:$

$$f(z) = a_0z^4 + a_1z^3 + a_2z^2 + a_3z + a_4, \quad a_0 > 0$$

a) $f(1) > 0, \quad f(-1) > 0$
b) $a_4^2 - a_0^2 - a_4a_1 + a_3a_0 < 0$
c) $a_4^2 - a_0^2 + a_4a_1 - a_3a_0 < 0$
d) $a_4^3 + 2a_4a_2a_0 + a_3a_1a_0 - a_4a_0^2 - a_2a_0^2 - a_4a_1^2 - a_4^2a_0 - a_4^2a_2 - a_3^2a_0 + a_0^3 + a_4a_3a_1 > 0$

$n = 5:$

$$f(z) = a_0z^2 + a_1z^4 + a_2z^3 + a_3z^2 + a_4z + a_5, \quad a_0 > 0$$

a) $f(1) > 0, \quad f(-1) < 0$
b) $|a_5| < a_0$
c) $a_4a_1a_0 - a_5a_0^2 - a_3a_0^2 + a_5^3 + a_5a_2a_0 - a_5a_1^2 - (a_4^2a_0 - a_5a_3a_0 - a_0^3 + a_5^2a_2 + a_5^2a_0 - a_5a_4a_1) > 0$
d) $a_4a_1a_0 - a_5a_0^2 - a_3a_0^2 + a_5^3 + a_5a_2a_0 - a_5a_1^2 + (a_4^2a_0 - a_5a_3a_0 - a_0^3 + a_5^2a_2 + a_5^2a_0 - a_5a_4a_1) < 0$
e) $(a_5^2 - a_0^2)^2 - (a_5a_1 - a_4a_0)^2 + (a_5a_1 - a_4a_0)(a_1^2 + a_4a_2 - a_4^2 - a_3a_1 - a_5^2 + a_0^2) + (a_5a_2 - a_3a_0)(a_5a_4 - a_1a_0 - a_5a_2 + a_3a_0) - (a_5a_3 - a_2a_0)[(a_5^2 - a_0^2) - 2(a_5a_1 - a_4a_0)] > 0$

The critical stability constraints that determine the boundary of the stability region in the coefficient space are given by the first condition

a) $f(1) > 0$ and $(-1)^n f(-1) > 0$

and the last condition of the above conditions, i.e., condition b) for $n = 2$, condition c) for $n = 3$ and so on.

Stability criteria for delta-operator polynomials: The characteristic equation of a sampled-data system, whose characteristic equation is in the δ-domain, is given by

$$f(\delta) = a_0\delta^n + a_1\delta^{n-1} + \ldots + a_n, \quad (9.71)$$

where Δ is the sampling period. For stability, the roots of $f(\delta)$ must lie inside the circle in Figure 9.25.

Necessary conditions for stability: [10]

i) $a_1, a_2, \ldots a_n > 0,$ (9.72)

ii) $(-1)^n f(-2/\Delta) > 0,$ and (9.73)

iii) $a_i < \binom{n}{i} (2/\Delta)^i \quad i = 1, 2, \ldots, n.$ (9.74)

Necessary and sufficient conditions: The stability of Equation 9.71 can be checked by one of the following methods:

1. transforming Equation 9.71 to the s-domain by the transformation

$$\delta = \frac{2s}{2 - \Delta s} \tag{9.75}$$

   and applying the Routh-Hurwitz criterion
2. transforming Equation 9.71 to the z-domain by the transformation

$$\delta = \frac{z-1}{\Delta} \tag{9.76}$$

   and applying Schur-Cohn criterion
3. using a direct approach such as the one given in [8]

This direct approach is as follows: Let

$$f^*(\delta) = (1 + \Delta\delta)^n f\left(\frac{\delta}{1 + \Delta\delta}\right). \tag{9.77}$$

Consider the sequence of polynomials,

$$T_n(\delta) = f(\delta) + f^*(\delta) \tag{9.78}$$

$$T_{n-1}(\delta) = 1/\delta[f(\delta) - f^*(\delta)] \tag{9.79}$$

$$T_j(\delta) = \frac{1}{1 + \Delta\delta}\left[\delta_{j+2}(2 + \Delta\delta)T_{j+1}(\delta) - T_{j+2}(\delta)\right] \tag{9.80}$$

with

$$\delta_{j+2} = \frac{T_{j+2}\left(-\frac{1}{\Delta}\right)}{T_{j+1}\left(-\frac{1}{\Delta}\right)}, \quad j = n-2, n-3, \ldots, 0, \tag{9.81}$$

where

$$T_{j-1}(-1/\Delta) \neq 0, \quad j = 1, 2, \ldots, n,$$

Stability is concluded if

$$\text{var}\left\{T_j(0)\right\}_{j=0}^{n} = 0 \tag{9.82}$$

(no change of sign).

References

[1] Schur, I., Ueber Potenzreihen die in Innern des Einheitskreises beschränkt sind. *S. Fuer Math., 147*, 205–232, 1917.

[2] Cohn, A., Ueber die Anzahl der Wurzeln einer algebraischen Gleichung in einem Kreise. *Math. Z., 14*, 110–148, 1922.

[3] Fujiwara, M., Ueber die algebraischen Gleichung deren Wurzeln in einem Kreise oder in einer Halbebene liegen. *Math. Z. 24*, 160–169, 1962.

[4] Jury, E.I., A simplified Stability Criterion for Linear Discrete Systems. *IRE Proc. 50(6)*, 1493–1500, 1962.

[5] Bistritz, Y., Zero location with respect to the unit circle of discrete-time linear system polynomials. *Proc. IEEE, 72*, 1131–1142, 1984.

[6] Nour Eldin, H.A., Ein neues Stabilitaets kriterium fuer abgetastete Regelsysteme. *Regelungstechnik u. Prezess-Daten verabeitung, 7*, 301–307, 1971.

[7] Mansour, M., Stability and Robust Stability of Discrete-time Systems in the δ-Transform. In *Fundamentals of Discrete Time Systems*, M. Jamshidi et al., Eds., TSI Press, Albuquerque, NM, 1993, 133–140.

[8] Premaratne, K. and Jury, E.I., Tabular Method for Determining Root Distribution of Delta-Operator Formulated Real Polynomials, *IEEE Trans. AC., 1994, 39(2)*, 352–355, 1994.

[9] Middleton, R.H. and Goodwin, G.C., *Digital Control and Estimation. A Unified Approach.* Prentice Hall, Englewood Cliffs, NJ, 1990.

[10] Mansour, M., Robust Stability in Systems Described by Rational Functions. In *Control and Dynamic Systems*, Leondes, Ed., 79–128, Academic Press, 1992, Vol. 51.

[11] Mansour, M., Instability Criteria of Linear Discrete Systems. *Automatica, 2*, 1985, 167–178, 1965.

[12] Ackerman, J., *Sampled-Data Control Systems.* Springer, Berlin, 1985.

[13] Scuessler, H.W., A stability theorem for discrete systems. *IEEE Trans, ASSP. 24*, 87–89, 1976.

[14] Jury, E.I., *Theory and Applications of the Z-Transform Method.* Robert E. Krieger, Huntingdon, NY, 1964.

[15] Jury, E.I., *Inners and Stability of Dynamic Systems.* John Wiley & Sons, New York, 1974.

[16] Anderson, B.D.O., Jury, E.I. and Chaparro, L.F., On the root distribution of a real polynomial with respect to the unit circle. *Regelungstechnik, 1976, 3*, 101–102, 1976.

[17] Mansour, M. and Anderson, B.D.O, On the Markov Stability Criterion for Discrete Systems. *IEEE Trans. CAS, 37(12)*, 1576–1578, 1990.

[18] Astroem, K.J. and Wittenmark, B., *Computer Controlled Systems*, Prentice Hall, Englewood Cliffs, NJ, 1984.

Further Reading

For comprehensive discussions of the stability of linear discrete-time and sampled-data systems, see [14] and [15]. The Delta-operator approach and its advantages is dealt with in [9]. The application of the Nyquist stability criterion, using the discrete frequency response, is given in [12] and [18].

9.4 Gain Margin and Phase Margin

Raymond T. Stefani, Electrical Engineering Department, California State University, Long Beach

9.4.1 Introduction

According[3] to Nyquist plot stability evaluation methods, a system with no open-loop right half-plane (RHP) poles should have no clockwise (CW) encirclements of the -1 point for stability, and a system with open-loop RHP poles should have as many counter-clockwise (CCW) encirclements as there are open-loop RHP poles. It is often possible to determine whether the -1 point is encircled by looking at only that part of the Nyquist plot (or Bode plot) that identifies the presence of an encirclement, that is, the part of the Nyquist plot near the -1 point (the part of the Bode plot near 0 dB for magnitude and $-180°$ for phase).

Similarly, it is possible to examine part of the Nyquist plot (or Bode plot) to determine the factor by which the system magnitude can be changed to make the system marginally stable. That factor is called the **gain margin.** It is also possible to examine part of the Nyquist plot (or Bode plot) to determine the amount of phase shift required to make the system marginally stable. The negative of that phase shift is called the **phase margin.** Both margins are discussed in this chapter.

A polynomial is called **minimum phase** when all the roots are in the left half-plane (LHP) and **non-minimum** phase when there are RHP roots. This means that stability is relatively easy to determine when $G(s)H(s)$ has minimum-phase poles, since there should be no encirclements of -1 for stability (a requirement that is easy to verify), but special care must be taken for the nonminimum-phase RHP pole case where stability demands CCW encirclements.

9.4.2 Gain Margin

In general, a Nyquist plot establishes the stability of a system of the form of Figure 9.26 with $K = 1$ by mapping $G(s)H(s)$ for s along the RHP boundary. One measure of stability arises from use of the **phase crossover frequency,** denoted ω_{PC} and defined as the frequency at which the phase of $G(s)H(s)$ is $-180°$, that is,

$$\text{phase } G(j\omega_{PC})H(j\omega_{PC}) = -180° = \Phi(\omega)$$

The magnitude of $G(s)H(s)$ at the phase crossover frequency is denoted $A(\omega_{PC})$. The gain margin, GM, is defined to be $1/A(\omega_{PC})$. Suppose that the gain K in Figure 9.26 is not selected to be one; rather K is selected to be $K = GM$. Then

$$KG(j\omega_{PC})H(j\omega_{PC}) = KA(\omega_{PC})\angle - 180°$$

$$= 1\angle - 180°$$

and the system becomes marginally stable.

For example, a system with $G(s)H(s) = 4/s(s+1)(s+2)$ has the Nyquist plot of Figure 9.27. Table 9.1 contains magnitude and phase data spanning the frequency range from zero to infinity.

[3]This section includes excerpts from *Design Feedback Control Systems,* Third Edition by Raymond T. Stefani, Clement J. Savant, Barry Shahian, and Gene H. Hostetter, copyright ©1994 by Saunders College Publishing, reprinted by permission of the publisher.

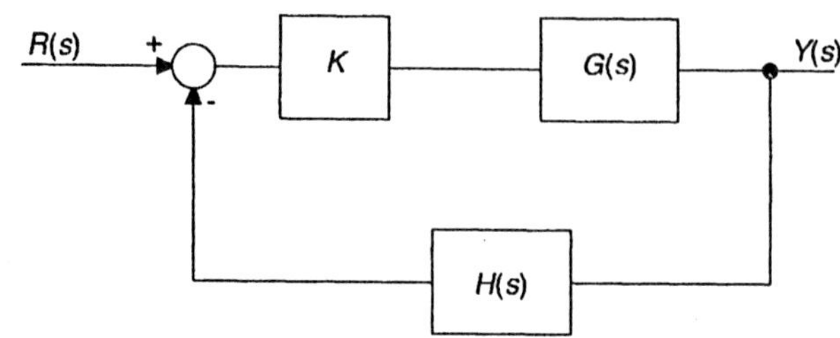

Figure 9.26 Closed-loop system. (From Stefani, Raymond T., Shahian, Barry, Savant, Clement J., and Hostetter, Gene H., *Design of Feedback Control Systems,* 3rd ed., Saunders College Publishing, Boston, MA, 109, 1994. With permission.)

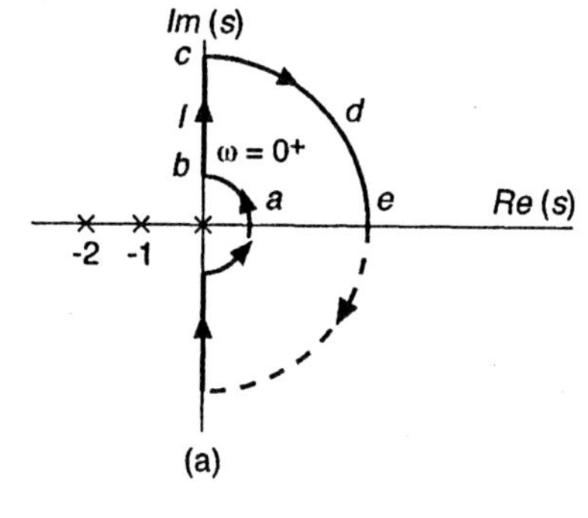

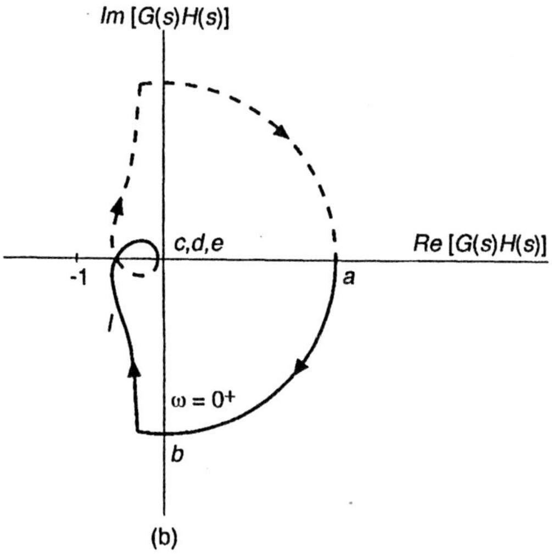

Figure 9.27 Nyquist plot for $G(s)H(s) = 4/s(s+1)(s+2)$. (a) RHP boundary; (b) Nyquist plot. (Not to scale.) (From Stefani, Raymond T., Shahian, Barry, Savant, Clement J., and Hostetter, Gene H., *Design of Feedback Control Systems,* 3rd ed., Saunders College Publishing, Boston, MA, 109, 1994. With permission.)

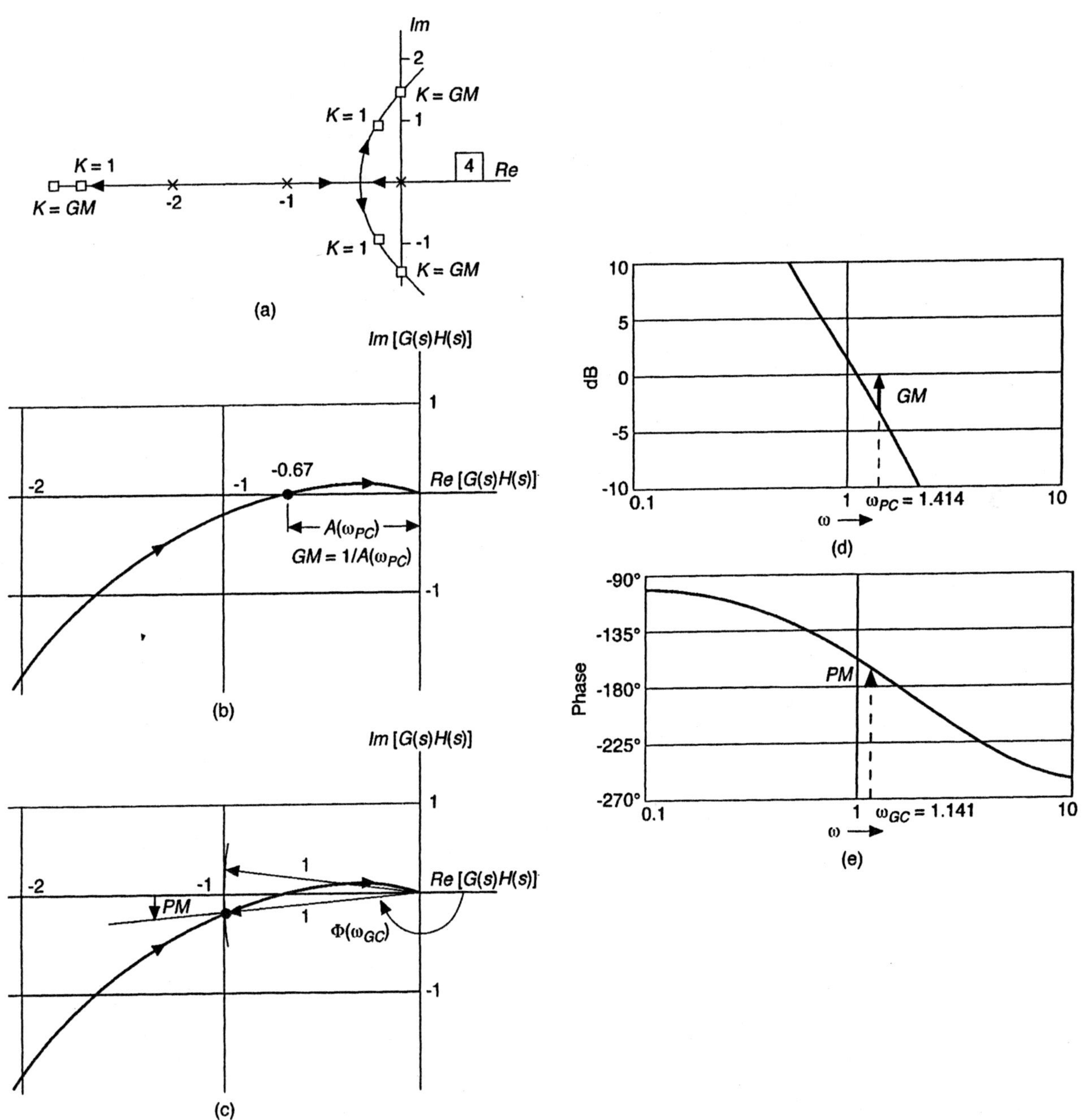

Figure 9.28 Stability of the system $G(s)H(s) = 4/s(s+1)(s+2)$. (a) Root locus for variable K; (b) Nyquist plot of $G(s)H(s)$ showing gain margin; (c) Nyquist plot of $G(s)H(s)$ showing phase margin. (d) Bode magnitude plot of $G(s)H(s)$; (e) Bode phase plot of $G(s)H(s)$. (From Stefani, Raymond T., Shahian, Barry, Savant, Clement J., and Hostetter, Gene H., *Design of Feedback Control Systems*, 3rd ed., Saunders College Publishing, Boston, MA, 109, 1994. With permission.)

TABLE 9.1 Evaluation for $G(s)H(s)$ for $s = j\omega$.

ω	0	.5	1	1.141	1.414	2	10	∞
$A(\omega)$	∞	3.47	1.26	1.00	0.67	0.31	0.004	0
$\Phi(\omega)$	$-90°$	$-131°$	$-162°$	$-169°$	$-180°$	$-198°$	$-253°$	$-270°$

Figure 9.28a shows the root locus with variable K for the same system as Figure 9.27. Figures 9.28b and 9.28c contain part of the Nyquist plot of $G(s)H(s)$ and Figures 9.28d and 9.28e show the Bode plot of $G(s)H(s)$. The gain margin GM for the system of Figure 9.26 with K nominally equal to 1 is the K for marginal stability on the root locus of Figure 9.28a. More generally, GM is the gain at which the system becomes marginally stable divided by the nominal gain. That ratio can be expressed as a base 10 number or in dB.

Viewed from the perspective of a Nyquist plot, $A(\omega_{PC})$ is the distance from the origin to where the Nyquist plot crosses the negative real axis in Figure 9.28b, which occurs for $A(\omega_{PC}) = 0.67$. The gain margin is thus measured at $\omega_{PC} = 1.414$ rad/sec. The gain margin is $1/0.67 = 1.5$ as a base 10 number while

$$dB(GM) = -dB[A(\omega_{PC})] = 20\log_{10}(1.5) = 3.5\,dB$$

The GM in dB is the distance on the Bode magnitude plot from the amplitude at the phase crossover frequency up to the 0-dB point (see Figure 9.28d).

When a system is stable for all positive K, the phase crossover frequency is generally infinite: $A(\omega_{PC})$ is zero; and GM is infinite. Conversely, when a system is unstable for all positive K, the phase crossover frequency is generally at 0 rad/s; $A(\omega_{PC})$ is infinite; and the GM is zero.

Gain margin can be interpreted in two ways. First, the designer can purposely vary K to some value other than one and $K = GM$ represents an upper bound on the value of K for which the closed-loop system remains stable. Second, the actual system open-loop transmittance may not actually be $G(s)H(s)$. When the uncertainty in $G(s)H(s)$ is only in the magnitude, the gain margin is a measure of the allowable margin of error in knowing $|G(s)H(s)|$ before the system moves to marginal stability.

As an open-loop RHP pole example, consider the Nyquist and Bode plots of Figure 9.29. The Nyquist plot of Figure 9.29a has one CCW encirclement of -1 so the number of closed-loop RHP poles is -1 (due to the CCW encirclement) $+1$ (due to the open-loop RHP pole), which equals zero indicating stability. Here the phase is $-180°$ at $\omega_{PC} = 0$ rad/s and $G(0)H(0)$ is -1.5 so that $A(\omega_{PC})$ is 1.5 and GM is 0.67 or -3.5 dB. In this case, the system is stable with a $dB(GM)$ that is negative.

Figure 9.30a shows part of the Nyquist plot for $G(s)H(s) = 0.75(s+2)^2/s^2(s+0.5)$. From the complete Nyquist plot it is easy to show that there are no CW encirclements of the -1 point so the system is stable with $K = 1$. The phase crossover frequency is 1.41 rad/s with a gain margin of $1/1.5 = 2/3 = 0.67(-3.53dB)$. The system is stable for a negative value of $dB(GM)$. If the $G(s)H(s)$ of Figure 9.30a is divided by three, as in Figure 9.30b, the complete Nyquist plot indicates that the system is unstable. Predictably, the GM of Figure 9.30b is three times that of Figure 9.30a; thus, the GM of Figure 9.30b is 2.0

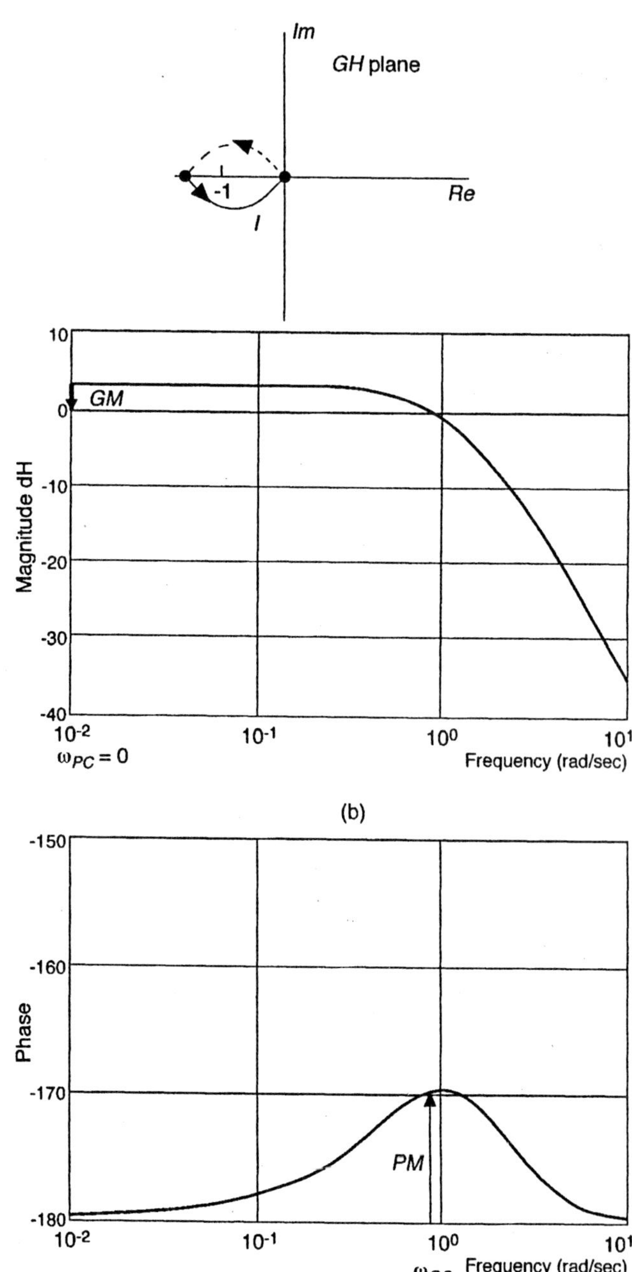

Figure 9.29 Stability of the system $G(s)H(s) = 2(s+3)/(s+2)^2(s-1)$. (a) Nyquist plot; (b) Bode magnitude plot of $G(s)H(s)$ showing gain margin; (c) Bode phase plot of $G(s)H(s)$ showing phase margin. (Figure 9.29a from Stefani, Raymond T., Shahian, Barry, Savant, Clement J., and Hostetter, Gene H., *Design of Feedback Control Systems*, 3rd ed., Saunders College Publishing, Boston, MA, 109, 1994. With permission.)

(6 dB) with a phase crossover frequency of 1.41 rad/s. Here the system is unstable with a positive $dB(GM)$ value.

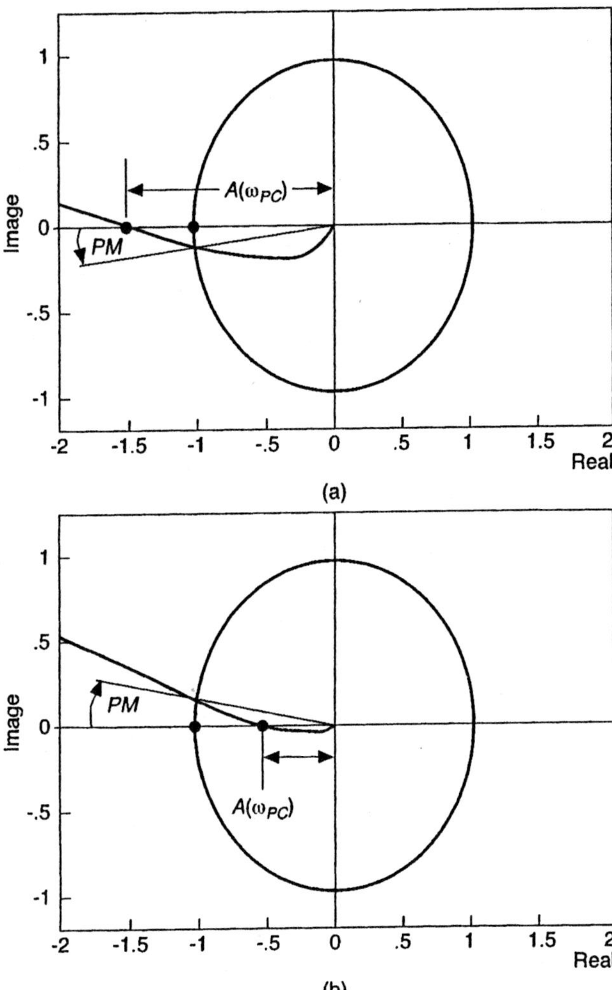

Figure 9.30 Partial Nyquist plots.
(a) $G(s)H(s) = 0.75(s+2)^2/s^2(s+0.5)$, $GM = 2/3(-3.53dB)$, $PM = 7.3°$, stable system;
(b) $G(s)H(s) = 0.25(s+2)^2/s^2(s+0.5)$, $GM = 2(6dB)$, $PM = -9.2°$, unstable system.

To be sure of stability, it is good practice to examine the Nyquist plot and the root locus plot. When there is more than one GM value (due to more than one phase crossover frequency) for a stable system, it is good practice to select the smallest GM value to ensure stability.

9.4.3 Phase Margin

In contrast to gain margin, when only the magnitude of $KG(s)H(s)$ is changed compared to that of $G(s)H(s)$, suppose instead that the gain K has unit magnitude and only the phase of $KG(s)H(s)$ is changed compared to that of $G(s)H(s)$.

It is useful to define the **gain crossover frequency** ω_{GC} as the frequency at which the magnitude of $G(s)H(s)$ is one (0 dB). Thus $A(\omega_{GC}) = 1$. The phase of $G(s)H(s)$ at the gain crossover frequency is denoted by $\Phi(\omega_{GC})$. The phase margin, PM, is defined by

$$PM = 180° + \Phi(\omega_{GC})$$

Suppose the gain K in Figure 9.26 is selected to be $K = 1\angle - PM$. Then at the gain crossover frequency

$$\begin{aligned} KG(j\omega_{GC})H(j\omega_{GC}) &= [1\angle - PM][1\angle\Phi(\omega_{GC})] \\ |KG(j\omega_{GC})H(j\omega_{GC})| &= 1 \\ \text{phase } KG(j\omega_{GC})H(j\omega_{GC}) &= -PM + \Phi(\omega_{GC}) \\ &= -180° - \Phi(\omega_{GC}) \\ &\quad + \Phi(\omega_{GC}) \\ &= -180° \end{aligned}$$

and the system is marginally stable.

For example, consider again the system with $G(s)H(s) = 4/s(s+1)(s+2)$. From Table 9.1 the gain crossover frequency is at 1.141 rad/s so that $A(\omega_{GC}) = 1$ while $\Phi(\omega_{GC}) = \Phi(1.141) = -169°$. Therefore, PM is $180° - 169° = 11°$. The phase margin PM is the angle in the Nyquist plot of Figure 9.28c drawn from the negative real axis to the point at which the Nyquist plot penetrates a circle of unit radius (called the unit circle). On the Bode phase plot of Figure 9.28e, the PM is the distance from $-180°$ to the phase at the gain crossover frequency. *The phase margin is therefore the negative of the phase through which the Nyquist plot can be rotated, and similarly the Bode plot can be shifted, so that the closed-loop system becomes marginally stable.*

In order to properly calculate PM, it is generally best to define $\Phi(\omega_{GC})$ as $-270° \leq \Phi(\omega_{GC}) \leq 90°$. For example, a third quadrant $\Phi(\omega_{GC})$ would be written as $-160°$ so PM would be $+20°$.

For a nonminimum-phase example, consider again the system of Figure 9.29. The Nyquist plot indicates stability. Here the gain crossover frequency is at 0.86 rad/s, $\Phi(\omega_{GC})$ is $-170°$ and the PM is $180° - 170°$ or $10°$; hence, the upward-directed arrow on the phase plot of Figure 9.29c.

For the example of Figure 9.30a, which is a stable system, ω_{GC} is 1.71 rad/s and the PM is 7.3°. For the unstable system of Figure 9.30b, ω_{GC} is 1.71 rad/s and the PM is $-9.2°$. It should be noted that there is a positive phase margin for all the stable systems just examined and a negative phase margin for all the unstable systems. This sign-stability relationship holds for the phase margin of most systems, while no such relationship holds for the sign of $dB(GM)$ in the examples just examined.

If there is more than one gain crossover frequency, there is more than one phase margin. For a stable system, the smallest candidate phase margin should be chosen.

As noted earlier, when $K = 1\angle - PM$, the system becomes marginally stable. That fact can be interpreted in two ways. First, the designer can purposely vary K away from one and then $1\angle - PM$ represents one extreme of selection of K. Second, the actual system open-loop transmittance may not actually be $G(s)H(s)$. When the uncertainty in $G(s)H(s)$ affects only the phase, the phase margin is the allowable margin of error in knowing phase $G(s)H(s)$ before the system moves to marginal stability. In most systems, there is uncertainty in both the magnitude and phase of $G(s)H(s)$ so that substantial gain and phase margins are required to assure the designer that imprecise knowledge of $G(s)H(s)$ does not necessarily cause instability. In fact, there are examples of systems that have large gain and phase margins, but small variations in gain *and* phase cause instability. It is of course important to check the complete Nyquist plot when there is any question about stability.

Suppose in Figure 9.26 that $K = 1$, $H(s) = 1$ and $G(s) = \omega_n^2/s(s + 2\zeta\omega_n)$. The closed-loop transfer function is $T(s) = \omega_n^2/(s^2 + 2\zeta\omega_n s + \omega_n^2)$, the standard form for a second-order system with damping ratio ζ and undamped natural frequency ω_n. For this system, the gain crossover frequency can be found in closed form and the phase margin follows from a trigonometric identity.

$$\begin{aligned} \omega_{GC} &= k\omega_n \\ k &= ((4\zeta^4 + 1)^{0.5} - 2\zeta^2)^{0.5} \\ PM &= \tan^{-1}(2\zeta/k) \end{aligned}$$

Table 9.2 shows values of gain crossover frequency and phase margin for this standard-form second-order system.

TABLE 9.2 Phase Margin for a Standard-Form Second-Order System.

Damping Ratio ζ	k ω_{GC}/ω_n	Phase Margin Degrees
0.0	1	0
0.1	0.99	11
0.2	0.96	23
0.3	0.91	33
0.4	0.85	43
0.5	0.79	52
0.6	0.72	59
0.7	0.65	65
0.8	0.59	70
0.9	0.53	74
1.0	0.49	76

For other systems with two dominant closed-loop underdamped poles, Table 9.2 is often a good approximation. Thus, the phase margin is approximately 100ζ degrees for damping ratios from zero to about 0.7.

Defining Terms

Gain crossover frequency (rad/s): Frequency at which the magnitude of GH is one (zero dB).

Gain margin: Negative of the dB of GH measured at the phase crossover frequency (inverse of the base 10 magnitude). When $K = GM$, the system becomes marginally stable.

Phase crossover frequency (rad/s): Frequency at which the phase of GH is $-180°$.

Phase margin: $180° +$ phase of GH measured at the gain cross-over frequency. When $K = 1\angle - PM$ the system becomes marginally stable.

Reference

[1] Stefani, R.T., Savant, C.J., Shahian, B., and Hostetter, G.H., *Design of Feedback Control Systems,* 3rd ed., Saunders College Publishing, Boston, MA, 1994.

Further Reading

Additional information may be found in *IEEE Control Syst. Mag.; IEEE Trans. Autom. Control;* and *IEEE Trans. Syst., Man, and Cybern.*

10

Design Methods

Jiann-Shiou Yang
Department of Electrical and Computer Engineering, University of Minnesota, Duluth, MN

William S. Levine
Department of Electrical Engineering, University of Maryland, College Park, MD

Richard C. Dorf
University of California, Davis

Robert H. Bishop
The University of Texas at Austin

John J. D'Azzo
Air Force Institute of Technology

Constantine H. Houpis
Air Force Institute of Technology

Karl J. Åström
Department of Automatic Control, Lund Institute of Technology, Lund, Sweden

Tore Hägglund
Department of Automatic Control, Lund Institute of Technology, Lund, Sweden

Katsuhiko Ogata
University of Minnesota

Richard D. Braatz
University of Illinois, Department of Chemical Engineering, Urbana, IL

Z.J. Palmor
Faculty of Mechanical Engineering, Technion - Israel Institute of Technology, Haifa, Israel

0-8493-8570-9/96/$0.00+$.50

10.1 Specification of Control Systems

Jiann-Shiou Yang, Department of Electrical and Computer Engineering, University of Minnesota, Duluth, MN

William S. Levine, Department of Electrical Engineering, University of Maryland, College Park, MD

10.1.1 Introduction

Generally, control system specifications can be divided into two categories, performance specifications and robustness specifications. Although the boundaries between the two can be fuzzy, the performance specifications describe the desired response of the nominal system to command inputs. The robustness specifications limit the degradation in performance due to variations in the system and disturbances. Section 10.1.2 of this chapter describes the classical performance specifications for single-input single-output (SISO) linear time-invariant (LTI) systems. This is followed by a discussion of the classical robustness specifications for SISO LTI systems. The fourth section gives some miscellaneous classical specifications. The fifth section describes performance specifications that are unique to multi-input multi-output (MIMO) systems. This is followed by a section on robustness specifications for MIMO systems. The final section contains conclusions.

10.1.2 Performance Specifications for SISO LTI Systems

Transient Response Specifications

In many practical cases, the desired performance characteristics of control systems are specified in terms of time-domain quantities, and frequently, in terms of the transient and steady-state response to a unit step input. The unit step signal, one of the three most commonly used test signals (the other two are ramp and parabolic signals), is often used because there is a close correlation between a system response to a unit step input and the system's ability to perform under normal operating conditions. And many control systems experience input signals very similar to the standard test signals. Note that if the response to a unit step input is known, then it is mathematically possible to compute the response to any input. We discuss the transient and steady-state response specifications separately in this section and Section 10.1.2. We emphasize that both the transient and steady-state specifications require that the closed-loop system is stable.

The transient response of a controlled system often exhibits damped oscillations before reaching steady state. In specifying the transient response characteristics, it is common to specify the following quantities:

1. Rise time (t_r)
2. Percent overshoot (PO)
3. Peak time (t_p)
4. Settling time (t_s)
5. Delay time (t_d)

The rise time is the time required for the response to rise from x% to y% of its final value. For overdamped second-order systems, the 0% to 100% rise time is normally used, and for underdamped systems (see Figure 10.1), the 10% to 90% rise time is commonly used.

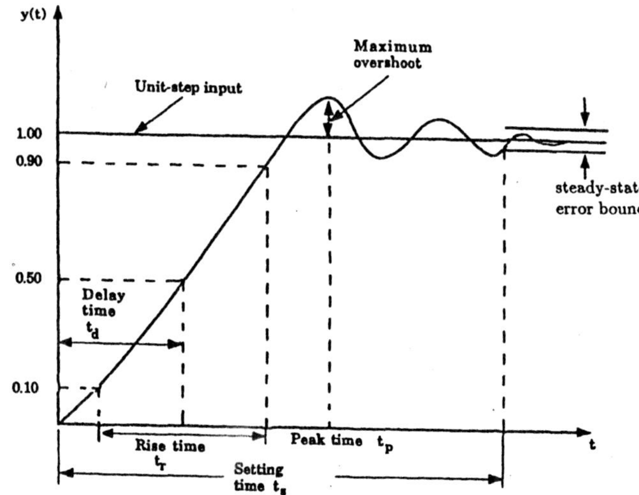

Figure 10.1 Typical underdamped unit-step response of a control system. An overdamped unit-step response would not have a peak.

The peak time is the time required for the response to reach the first (or maximum) peak. The settling time is defined as the time required for the response to settle to within a certain percent of its final value. Typical percentage values used are 2% and 5%. The settling time is related to the largest time constant of the controlled system. The delay time is the time required for the response to reach half of its final value for the very first time. The percent overshoot represents the amount that the response overshoots its steady-state (or final) value at the peak time, expressed as a percentage of the steady-state value. Figure 10.1 shows a typical unit step response of a second-order system

$$G(s) = \frac{\omega_n^2}{s^2 + 2\zeta\omega_n s + \omega_n^2}$$

where ζ is the damping ratio and ω_n is the undamped natural frequency. For this second-order system with $0 \le \zeta < 1$ (an underdamped system), we have the following properties:

$$\begin{aligned} PO &= e^{-\zeta\pi/\sqrt{1-\zeta^2}} \\ t_p &= \frac{\pi}{\omega_n\sqrt{1-\zeta^2}} \\ t_s &= \frac{4}{\zeta\omega_n} \end{aligned}$$

where the 2% criterion is used for the settling time t_s. If 5% is used, then t_s can often be approximated by $t_s = 3/\zeta\omega_n$. A precise formula for rise time t_r and delay time t_d in terms of damping ratio ζ and undamped natural frequency ω_n cannot be found. But useful approximations are

$$t_d \cong \frac{1.1+0.125\zeta+0.469\zeta^2}{\omega_n}$$
$$t_r \cong \frac{1-0.4167\zeta+2.917\zeta^2}{\omega_n}$$

Note that the above expressions are only accurate for a second-order system. Many systems are more complicated than the pure second-order system. Thus, when using these expressions, the designer should be aware that they are only rough approximations. The time-domain specifications are quite important because most control systems must exhibit acceptable time responses. If the values of t_s, t_d, t_r, t_p, and PO are specified, then the shape of the response curve is virtually determined. However, not all these specifications necessarily apply to any given case. For example, for an overdamped system ($\zeta > 1$) or a critically damped system ($\zeta = 1$), t_p and PO are not useful specifications. Note that the time-domain specifications, such as PO, t_r, ζ, etc., can be applied to discrete-time systems with minor modifications.

Quite often the transient response requirements are described in terms of pole-zero specifications instead of step response specifications. For example, a system may be required to have its poles lying to the left of some constraint boundary in the s-plane, as shown in Figure 10.2.

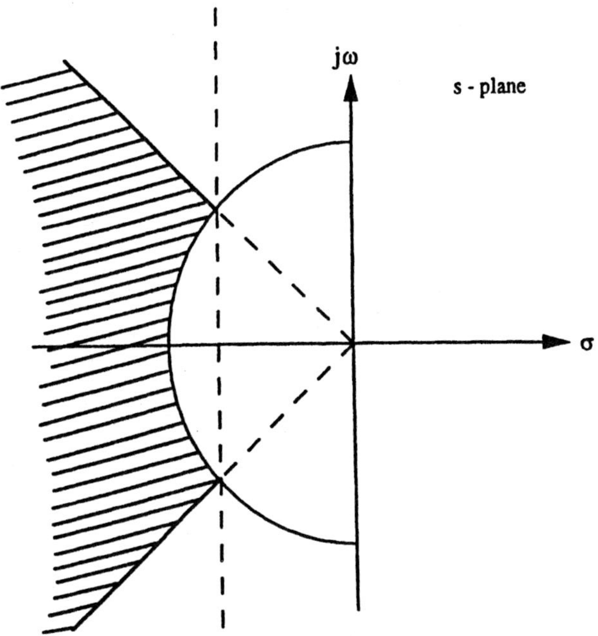

Figure 10.2 The shaded region is the allowable region in the s-plane.

The parameters given above can be related to the pole locations of the transfer function. For example, certain transient requirements on the rise time, settling time, and percent overshoot (e.g., t_r, t_s, and PO less than some particular values) may restrict poles to a region of the complex plane. The poles of a second-order system with $\zeta < 1$ are given in terms of ζ and ω_n by

$$p_1 = -\zeta\omega_n + j\omega_n\sqrt{1-\zeta^2}$$
$$p_2 = -\zeta\omega_n - j\omega_n\sqrt{1-\zeta^2}$$

Notice that $|p_1| = |p_2| = \omega_n$ and $\angle p_1 = -\angle p_2 = 180^o - tan^{-1}\sqrt{1-\zeta^2}/\zeta$. Thus, contours of constant ω_n are circles in the complex plane, while contours of constant ζ are straight lines ($0 \leq \zeta \leq 1$), as shown in Figure 10.3.

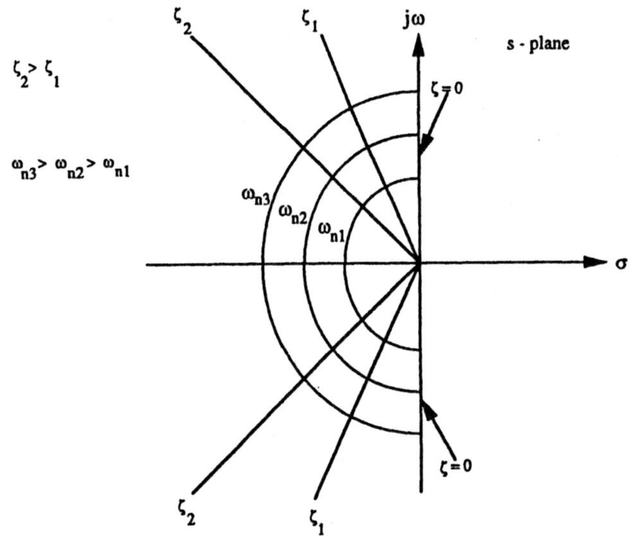

Figure 10.3 Contours of constant ζ and ω_n for a second-order system.

Because high-order systems can always be decomposed into a parallel combination of first- and second-order subsystems, the parameters related to the time response of these high-order systems can be estimated using the expressions given above. For instance, t_s can be approximated by four (or three) times the slowest time constant (i.e., the slowest subsystem normally determines the system settling time).

It is known that the location of the poles of a transfer function in the s-plane has great effects on the transient response of the system. The poles that are close to the $j\omega$-axis in the left half s-plane give transient responses that decay relatively slowly, whereas those poles far away from the $j\omega$-axis (relative to the dominant poles) correspond to more rapidly decaying time responses. The relative dominance of poles is determined by the ratio of the real parts of the poles, and also by the relative magnitudes of the residues evaluated at these poles (the magnitudes of the residues depend on both the poles and zeros). It has been recognized in practice that if the ratios of the real parts exceed five, then the poles nearest the $j\omega$-axis will dominate in the transient response behavior. The poles that have dominant effects on the transient response behavior are called dominant poles. And those poles with the magnitudes of their real parts at least five times greater than the dominant poles may be regarded as insignificant (as far as the transient response is concerned). Quite often the dominant closed-loop poles occur in the form of a complex-conjugate pair. It is not uncommon that some high-order systems can be approximated by low-order systems. In other words, they contain insignificant poles that have little effect on the transient response, and may be approximated by dominant poles only. If this is the case, then the parameters described above can still be used in specifying the system dynamic behavior, and we can use dominant poles to control the dynamic performance, whereas the insignificant poles are used for ensuring the controller designed can be physically realized.

Although, in general, high-order systems may not have dominant poles (and thus, we can no longer use ζ, ω_n, etc., to specify the design requirements), the time domain requirements on transient and steady-state performance may be specified as bounds on the command step response, such as that shown in Figure 10.4 (i.e., the system has a step response inside some constraint boundaries).

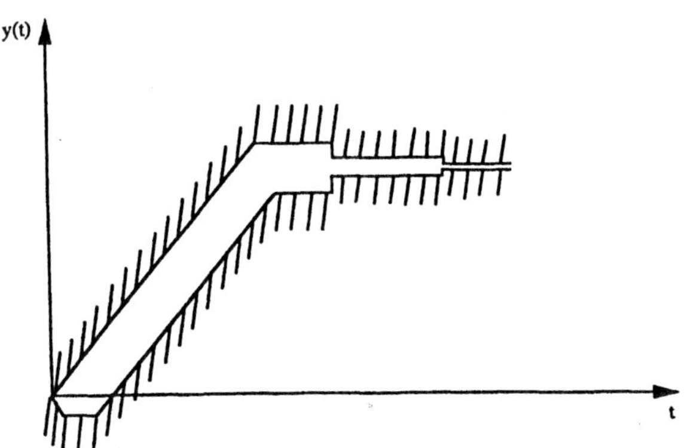

Figure 10.4 General envelope specification on a step response. The step response is required to be inside the region indicated.

Steady-State Accuracy

If the output of a system at steady state does not exactly agree with the input, the system is said to have steady-state error. This error is one measure of the accuracy of the system. Since actual system inputs can frequently be considered combinations of step, ramp, and parabolic types of signals, control systems may be classified according to their ability to follow step, ramp, parabolic inputs, etc. In general, the steady-state error depends not only on the inputs but also on the "type" of control system. Let $G_o(s)$ represent the open-loop transfer function of a stable unity feedback control system (see Figure 10.5), then $G_o(s)$ can generally be expressed as

$$G_o(s) = \frac{k(s - z_1)(s - z_2)\cdots(s - z_m)}{s^N(s - p_1)(s - p_2)\cdots(s - p_n)}$$

where z_i ($\neq 0$, $i = 1, 2, \cdots, m$) are zeros, p_j ($\neq 0$, $j = 1, 2, \cdots, n$) and 0 (a pole at the origin with multiplicity N) are poles, and $m < n + N$.

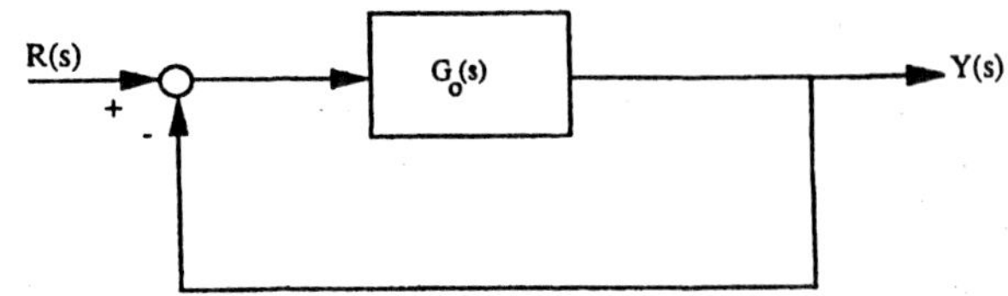

Figure 10.5 A simple unity feedback system with open-loop transfer function $G_o(s)$ and closed-loop transfer function $Y(s)/R(s) = G_o(s)/[1 + G_o(s)]$.

The type of feedback system refers to the order of the pole of the open-loop transfer function $G_o(s)$ at $s = 0$ (i.e., the value of the exponent N of s in $G_o(s)$). In other words, the classification is based on the number of pure integrators in $G_o(s)$. A system is called type 0, type 1, type 2, $\cdots$ if $N = 0, 1, 2, \cdots$, respectively. Note that a nonunity feedback system can be mathematically converted to an equivalent unity feedback system from which its "effective" system type and static error constants (to be defined later) can be determined. For instance, for a control system with forward path transfer function $G(s)$ and feedback transfer function $H(s)$, the equivalent unity feedback system has the forward path transfer function $G_o(s) = G(s)/(1 + G(s)[H(s) - 1])$.

Static error constants describe the ability of a system to reduce or eliminate steady-state errors. Therefore, they can be used to specify the steady-state performance of control systems. For a stable unity feedback system with open-loop transfer function $G_o(s)$, the position error constant K_p, velocity error constant K_v, and acceleration error constant K_a are defined, respectively, as

$$\begin{aligned} K_p &= lim_{s \to 0}\, G_o(s) \\ K_v &= lim_{s \to 0}\, sG_o(s) \\ K_a &= lim_{s \to 0}\, s^2 G_o(s) \end{aligned}$$

In terms of K_p, K_v, and K_a, the system's steady-state error for the three commonly used test signals, i.e., a unit step input ($u(t)$), a ramp input ($tu(t)$), and a parabolic input ($\frac{1}{2}t^2u(t)$), can be expressed, respectively, as

$$\begin{aligned} e(\infty) &= \frac{1}{1 + K_p} \\ e(\infty) &= \frac{1}{K_v} \\ e(\infty) &= \frac{1}{K_a} \end{aligned}$$

where the error $e(t)$ is the difference between the input and output, and $e(\infty) = lim_{t \to \infty}\, e(t)$. Therefore, the value of the steady-state error decreases as the error constants increase. Just as damping ratio (ζ), settling time (t_s), rise time (t_r), delay time

(t_d), peak time (t_p), and percent overshoot (PO) are used as specifications for a control system's transient response, so K_p, K_v, and K_a can be used as specifications for a control system's steady-state errors.

To increase the static error constants (and hence, improve the steady-state performance), we can increase the type of the system by adding integrator(s) to the forward path. For example, provided the system is stable, a type 1 system has no steady-state error for a constant input, a type 2 system has no steady-state error for a constant or ramp input, a type 3 system has no steady-state error for a constant, a ramp, or a parabolic input, and so forth. Clearly, as the type number is increased, accuracy is improved. However, increasing the type number aggravates the stability problem. It is desirable to increase the error constants, while maintaining the transient response within an acceptable range. A compromise between steady-state accuracy and relative stability is always necessary.

Step response envelope specifications, similar to that of Figure 10.4, are often used as the time-domain specifications for control system design. These specifications cover both the transient and steady-state performance requirements. For MIMO systems, the diagonal entries of the transfer function matrix represent the functions from the commanded inputs to their associated commanded variables, and the off-diagonal entries are the transfer functions from the commands to other commanded variables. It is generally desirable to require that the diagonal elements lie within bounds such as those shown in Figure 10.4 , while the off-diagonal elements are reasonably small.

Frequency-Domain Performance Specifications

A stable LTI system $G(s)$ subjected to a sinusoidal input will, at steady state, have a sinusoidal output of the same frequency as the input. The amplitude and phase of the system output will, in general, be different from those of the system input. In fact, the amplitude of the output is given by the product of the amplitude of the input and $|G(j\omega)|$, while the phase angle differs from the phase angle of the input by $\angle G(j\omega)$. In other words, if the input is $r(t) = A\ sin\omega t$, the output steady-state response $y(t)$ will be of the form $AR(\omega)\ sin[\omega t + \phi(\omega)]$, where $R(\omega) = |G(s)|_{s=j\omega} = |G(j\omega)|$ and $\phi(\omega) = \angle G(s)|_{s=j\omega} = \angle G(j\omega)$ vary as the input frequency ω is varied. Therefore, the magnitude and phase of the output for a sinusoidal input may be found by simply determining the magnitude and phase of $G(j\omega)$. The complex function $G(j\omega) = G(s)|_{s\to j\omega}$ is referred to as the frequency response of the system $G(s)$. It is understood that the system is required to be stable. In control system design by means of frequency-domain methods, the following specifications are often used in practice.

1. Resonant peak (M_p)
2. Bandwidth (ω_b)
3. Cutoff rate

The resonant peak M_p is defined as the maximum magnitude of the closed-loop frequency response, and the frequency at which M_p occurs is called the resonant frequency (ω_p). More precisely, if we consider a LTI open-loop system described by its transfer function $G_o(s)$ and the unity feedback closed-loop system pictured in Figure 10.5, then the closed-loop transfer function $G_{cl}(s)$ will have transfer function

$$G_{cl}(s) = \frac{G_o(s)}{1 + G_o(s)}$$

and

$$\begin{aligned} M_p &= max_{\omega\geq 0}\,|G_{cl}(j\omega)| \\ \omega_p &= arg\,\{max_{\omega\geq 0}\,|G_{cl}(j\omega)|\} \end{aligned}$$

In general, the magnitude of M_p gives an indication of the relative stability of a stable system. Normally, a large M_p corresponds to a large maximum overshoot of the step response in the time domain. For most control systems, it is generally accepted in practice that the desirable M_p should lie between 1.1 and 1.5. The bandwidth, ω_b, is defined as the frequency at which the magnitude of the closed-loop frequency response drops to 0.707 of its zero-frequency value. In general, the bandwidth of a controlled system gives a measure of the transient response properties, in that a large bandwidth corresponds to a faster response. Conversely, if the bandwidth is small, only signals of relatively low frequencies are passed, and the time response will generally be slow and sluggish. Bandwidth also indicates the noise-filtering characteristics and the robustness of the system. Often, bandwidth alone is not adequate as an indication of the characteristics of the system in distinguishing signals from noise. Sometimes it may be necessary to specify the cutoff rate of the frequency response, which is the slope of the closed-loop frequency response at high frequencies. The performance criteria defined above are illustrated in Figure 10.6.

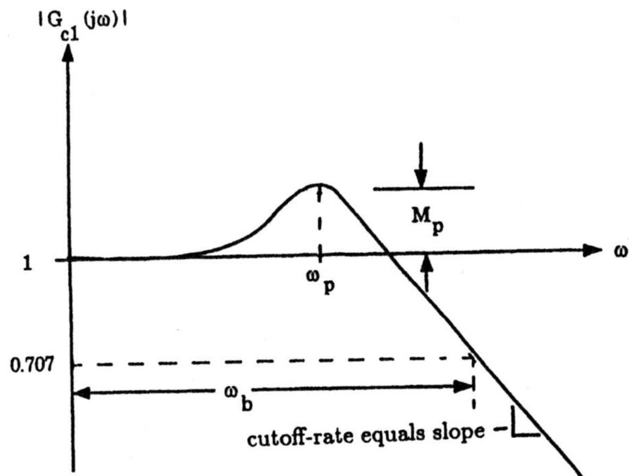

Figure 10.6 Frequency response specification.

The closed-loop time response is related to the closed-loop frequency response. For example, overshoot in the transient response is related to resonance in the closed-loop frequency response. However, except for first- and second-order systems, the exact relationship is complex and is generally not used. For the standard second-order system, the resonant peak M_p, the resonant frequency ω_p, and the bandwidth ω_b are uniquely related to the damping ratio ζ and undamped natural frequency ω_n. The relations are given by the following equations:

$$\begin{aligned} \omega_p &= \omega_n\sqrt{1-2\zeta^2} \quad \text{for } \zeta \leq 0.707 \\ M_p &= \frac{1}{2\zeta\sqrt{1-\zeta^2}} \quad \text{for } \zeta \leq 0.707 \\ \omega_b &= \omega_n[\,(1-2\zeta^2)+\sqrt{4\zeta^4-4\zeta^2+2}\,]^{1/2} \end{aligned}$$

Like the general envelope specifications on a step response (e.g., Figure 10.4), the frequency-domain requirements may also be given as constraint boundaries similar to those shown in Figure 10.7. That is, the closed-loop frequency response of the designed system should lie within the specified bounds. This kind of specification can be applied to many different situations.

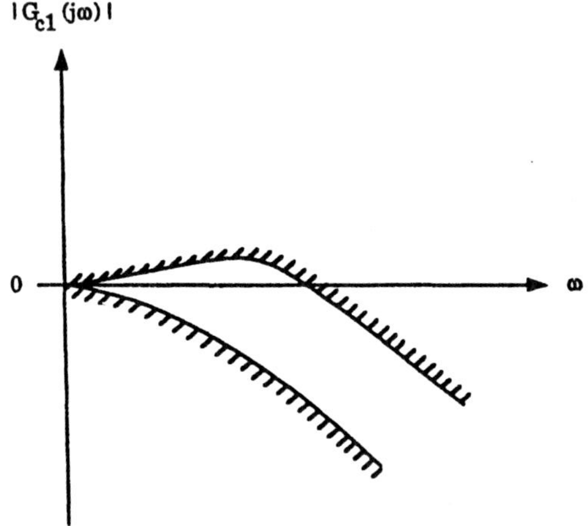

Figure 10.7 General envelope specification on the closed-loop frequency response.

10.1.3 Robustness Specifications for SISO LTI Systems

Relative Stability — Gain and Phase Margins

In control system design, in general, we require the designed system to be not only stable, but to have a certain guarantee of stability. In the time domain, relative stability is measured by parameters such as the maximum overshoot and the damping ratio. In the frequency domain, the resonant peak M_p can be used to indicate relative stability. Gain margin (G.M.) and phase margin (P.M.) are two design criteria commonly used to measure the system's relative stability. They provide an approximate indication of the closeness of the Nyquist plot of the system's open-loop frequency response $L(j\omega)$. ($L(j\omega)$ is also often called the loop transfer function) to the critical point, -1, in the complex plane. The open-loop frequency response, $L(j\omega)$, is obtained by connecting all of the elements in the loop in series and not closing the loop. For example, $L(j\omega) = G_o(j\omega)$ for the unity feedback system of Figure 10.5, but $L(j\omega) = H(j\omega)G_p(j\omega)G_c(j\omega)$ for the more complex closed-loop system of Figure 10.8.

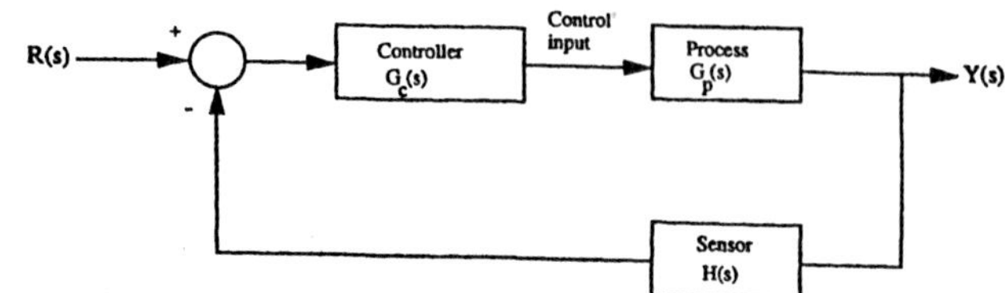

Figure 10.8 A typical closed-loop control system.

The decibel (abbreviated dB) is a commonly used unit for the frequency response magnitude. The magnitude of $L(j\omega)$ is then $20\,log_{10}\,|L(j\omega)|\,dB$. For example, if $|L(j\omega_o)| = 5$, $|L(j\omega_o)| = 20\,log_{10}\,5\,dB$. The gain margin is the amount of gain in dB that can be inserted in the loop before the closed-loop system reaches instability. The phase margin is the change in open-loop phase shift required at unity gain to make the closed-loop system unstable, or the magnitude of the minimum angle by which the Nyquist plot of the open-loop transfer function must be rotated in order to intersect the -1 point (for a stable closed-loop system). The Nyquist plot showing definitions of the gain and phase margins is given in Figure 10.9.

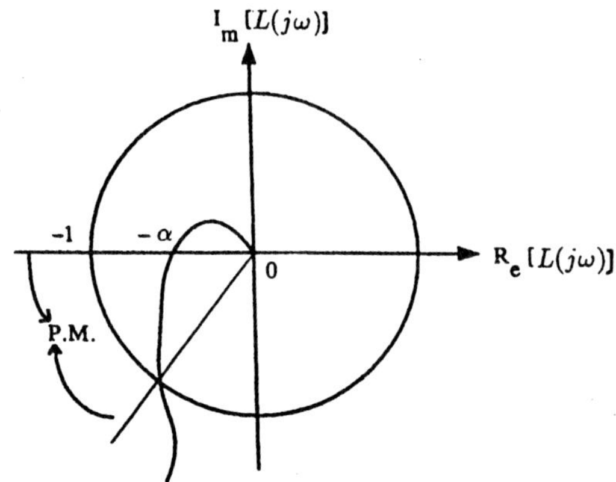

Figure 10.9 Definitions of the gain and phase margins. The gain margin is 20 log $\frac{1}{\alpha}$.

Although G.M. and P.M. can be obtained directly from a Nyquist plot, they are more often determined from a Bode plot of the open-loop transfer function. The Bode plot (which includes Bode magnitude plot and phase plot) and Nyquist plot (i.e., $L(s)|_{s=j\omega}$ drawn in polar coordinates) differ only in the coordinates and either one can be obtained from the other. The gain crossover frequency ω_c is the frequency at which $|L(j\omega_c)| = 1$ (i.e., the frequency at which the loop gain crosses the 0 dB line). Therefore, comparing both Bode magnitude and phase plots, the distance in degrees between the -180° line and $\angle\, L(j\omega_c)$ is the P.M. If $\angle\, L(j\omega_c)$ lies above the -180° line, P.M. is positive, and if it lies below, P.M. is negative. The phase crossover frequency ω_p is the frequency at which $\angle\, L(j\omega_p) = -180^o$ (i.e., the frequency at which $\angle L(j\omega_p)$ crosses the -180° line). Thus, the distance in decibels between the 0 dB line and $|L(j\omega_p)|$ is the G.M. (i.e., $-20\ log_{10}\ |L(j\omega_p)|\ dB$). If $|L(j\omega_p)|$ lies below the 0 dB line, G.M. is positive. Otherwise, it is negative. A proper transfer function is called minimum-phase if all its poles and zeros lie in the open left half s-plane. In general, for a minimum-phase $L(s)$, the system is stable if G.M. (in dB) > 0 and unstable if G.M. (in dB) < 0. And generally, a minimum-phase system has a positive P.M., and it becomes unstable if P.M. < 0. For non-minimum phase systems, care must be taken in interpreting stability based on the sign of G.M. (dB) and P.M. In this case, the complete Nyquist plot or root locus must be examined for relative stability. For a system in which multiple -180° crossovers occur, the simplest approach is to convert the Bode plot to the corresponding Nyquist plot to determine stability and the frequencies at which the stability margins occur. Obviously, systems with greater gain and phase margins can withstand greater changes in system parameter variations before becoming unstable.

It should be noted that neither the gain margin alone or the phase margin alone gives a sufficient indication of the relative stability. For instance, the G.M. does not provide complete information about the system response; a small G.M. indicates the Nyquist plot $|L(j\omega_p)|$ is very close to the -1 point. Such a system will have a large M_p (which means an oscillatory closed-loop time response) independent of the P.M. A Nyquist plot approaching the -1 point too closely also indicates possible instability in the presence of modeling error and uncertainty. Therefore, both G.M. and P.M. should be given in the determination of relative stability. These two values bound the behavior of the closed-loop system near the resonant frequency. Since for most systems there is uncertainty in both the magnitude and phase of the open-loop transfer function $L(s)$, a substantial amount of G.M. and P.M. is required in the control design to assure that the possible variations of $L(s)$ will not cause instability of the closed-loop system. For satisfactory performance, the phase margin should lie between 30° and 60°, and the gain margin should be greater than 6 dB. For an underdamped second-order system, P.M. and ζ are related by

$$P.M. = tan^{-1}\ \frac{2\zeta}{\sqrt{\sqrt{4\zeta^4+1}-2\zeta^2}}$$

This gives a close connection between performance and robustness, and allows one to fully specify such a control system by means of phase margin and bandwidth alone. Note that for first- and second-order systems, the Bode phase plot never crosses the -180° line. Thus, G.M. $= \infty$. In some cases the gain and phase margins are not helpful indicators of stability. For example, a high-order system may have large G.M. and P.M. (both positive); however, its Nyquist plot may still get close enough to the -1 point to incur a large M_p. Such a phenomenon can only occur in high-order systems. Therefore, the designer should check any results obtained by using G.M. and P.M. as indicators of relative stability.

Sensitivity to Parameters

During the design process, the engineer may want to consider the extent to which changes in system parameters affect the behavior of a system. One of the main advantages of feedback is that it can be used to make the response of a system relatively independent of certain types of changes or inaccuracies in the plant model. Ideally, parameter changes due to heat, humidity, age, or other causes should not appreciably affect a system's performance. The degree to which changes in system parameters affect system transfer functions, and hence performance, is called sensitivity. The greater the sensitivity, the worse is the effect of a parameter change.

A typical closed-loop control system may be modeled as shown in Figure 10.8, where $G_p(s)$ represents the plant or process to be controlled, $G_c(s)$ is the controller, and $H(s)$ may represent the feedback sensor dynamics. The model $G_p(s)$ is usually an approximation to the actual plant dynamic behavior, with parameters at nominal values and high-frequency dynamics neglected. The parameter values in the model are often not precisely known and may also vary widely with operating conditions. For the system given in Figure 10.8, the closed-loop transfer function $G_{cl}(s)$ is

$$G_{cl}(s) = \frac{C(s)}{R(s)} = \frac{G_c(s)G_p(s)}{1+G_c(s)G_p(s)H(s)}$$

If the loop gain $|L| = |G_cG_pH| >> 1$, C/R depends almost entirely on the feedback H alone and is virtually independent of the plant and other elements in the forward path and their parameter variations. This is because $|1+G_cG_pH| \approx |G_cG_pH|$ and $|G_cG_p/G_cG_pH| \approx 1/|H|$. Therefore, the sensitivity of the closed-loop performance to the elements in the forward path reduces as the loop gain increases. This is a major reason for using feedback. With open-loop control (i.e., $H = 0$), $C/R = G_pG_c$. Choice of G_c on the basis of an approximate plant model or a model for which the parameters are subjected to variations will cause errors in C proportional to those in G_p. With feedback, the effects due to approximations and parameter variations in G_p can be greatly reduced. Note that, unlike G_p, the feedback element H is usually under the control of the designer.

The sensitivity of the closed-loop transfer function G_{cl} to changes in the forward path transfer function, especially the plant transfer function G_p, can be defined by

$$S = \frac{\partial G_{cl}}{\partial G_p}\frac{G_p}{G_{cl}} = \frac{1}{1+L}$$

We can plot $S(j\omega)$ as a function of ω, and such a plot shows how sensitivity changes with the frequency of a sinusoidal input R. Obviously, for the sensitivity to be small over a given frequency band, the loop gain L over that band should be large. Generally, for good sensitivity in the forward path of the control system, the loop gain (by definition, the loop gain is $|L(j\omega)|$) is made large over as wide a band of frequencies as possible.

The extent to which the plant model is unknown will be called uncertainty. Uncertainty is another important issue which designers might need to face. It is known that some uncertainty is always present, both in the environment of the system and in the system itself. We do not know in advance exactly what disturbance and noise signals the system will be subjected to. In the system itself, we know that no mathematical expressions can exactly model a physical system. The uncertainty may be caused by parameter changes, neglected dynamics (especially, high-frequency unmodeled dynamics), or other unspecified effects, which might adversely affect the performance of a control system. Figure 10.10 shows a possible effect of high-frequency plant uncertainty due to dynamics in the high-frequency range being neglected in the nominal plant.

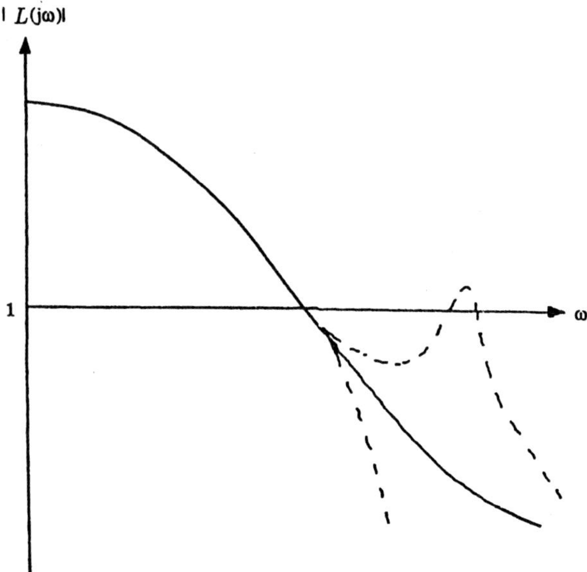

Figure 10.10 Effect of high-frequency unmodeled plant uncertainty.

In Figure 10.10, instability can result if an unknown high-frequency resonance causes the magnitude to rise above 1. The likelihood of an unknown resonance in the plant G_p rising above 1 can be reduced if we can keep the loop gain small in the high-frequency range.

In summary, to reduce the sensitivity we need to increase the loop gain. But, in general, increasing the loop gain degrades the stability margins. Hence, we usually have a trade-off between low sensitivity and adequate stability margins.

Disturbance Rejection and Noise Suppression

All physical systems are subjected to some types of extraneous signals or noise during operation. External disturbances, such as a wind gust acting on an aircraft, are quite common in controlled systems. Therefore, in the design of a control system, consideration should be given so that the system is insensitive to noise and disturbance. The effect of feedback on noise and disturbance depends greatly on where these extraneous signals occur in the system. But in many situations, feedback can reduce the effect of noise and disturbance on system performance. To explain these effects, let's consider a closed-loop unity feedback system as shown in Figure 10.11, where a disturbance $d(t)$ and a sensor noise $n(t)$ have been added to the system.

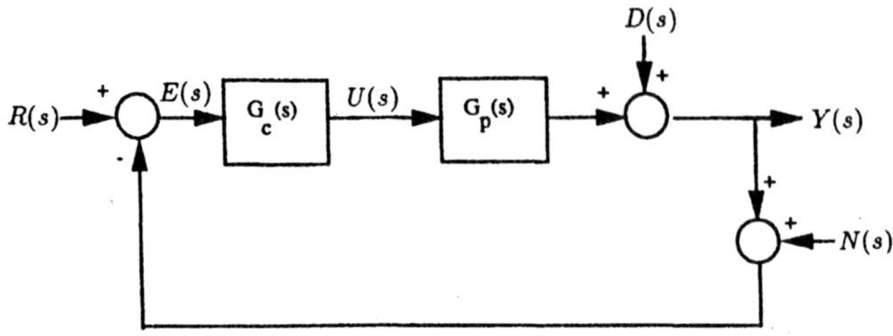

Figure 10.11 A unity feedback control system showing sources of noise, $N(s)$, and disturbance, $D(s)$.

For simplicity, we assume that the effect of external disturbances is collected and presented at the plant output. The sensor noise, $n(t)$, is introduced into the system via sensors. Both disturbances and sensor noise usually include random high-frequency signals. Let $D(s)$, $N(s)$, $R(s)$, and $Y(s)$ be, respectively, the Laplace transform of the disturbance $d(t)$, sensor noise $n(t)$, system input $r(t)$, and system output $y(t)$. It is easy to find that, by superposition, the total output $Y(s)$ is

$$\begin{aligned} Y(s) &= \frac{G_c(s)G_p(s)}{1+G_c(s)G_p(s)}R(s) + \frac{1}{1+G_c(s)G_p(s)}D(s) \\ &\quad - \frac{G_c(s)G_p(s)}{1+G_c(s)G_p(s)}N(s) \end{aligned}$$

and the tracking error $e(t)$, defined as $e(t) = r(t) - c(t)$ with its corresponding Laplace transform $E(s)$, becomes

$$\begin{aligned} E(s) &= \frac{1}{1+G_c(s)G_p(s)}R(s) - \frac{1}{1+G_c(s)G_p(s)}D(s) \\ &\quad + \frac{G_c(s)G_p(s)}{1+G_c(s)G_p(s)}N(s) \end{aligned}$$

In terms of the sensitivity function S and closed-loop transfer function G_{cl} defined in Section 10.1.3, the output $Y(s)$ and tracking error $E(s)$ become

$$\begin{aligned} Y(s) &= G_{cl}(s)R(s) + S(s)D(s) - G_{cl}(s)N(s) \\ E(s) &= S(s)R(s) - S(s)D(s) + G_{cl}(s)N(s) \end{aligned}$$

Note that the transfer function G_{cl} is also called the complementary sensitivity function because S and G_{cl} are related by $S(s) + G_{cl}(s) = 1$ for all frequencies. It is clear that $S(s)$ must be kept small to minimize the effects of disturbances. From the definition of S, this can be achieved if the loop gain (i.e., $|L(j\omega)| = |G_c(j\omega)G_p(j\omega)|$) is large. $|G_{cl}(j\omega)|$ must be kept small to reduce the effects of sensor noise on the system's output, and this can be achieved if the loop gain is small. For good tracking, $S(s)$ must be small, which implies that the loop gain should be large over the frequency band of the input signal $r(t)$.

Tracking and disturbance rejection require small S, while noise suppression requires small G_{cl}. From the relation between S and G_{cl}, clearly, we cannot reduce both functions simultaneously. However, in practice, disturbances and commands are often low-frequency signals, whereas sensor noises are often high-frequency signals. Therefore, we can still meet both objectives by keeping S small in the low-frequency range and G_{cl} small in the high-frequency range.

Putting together the requirements of reducing the sensitivity to parameters, disturbance rejection, and noise suppression we arrive at a general desired shape for the open-loop transfer function, which is as shown in Figure 10.12.

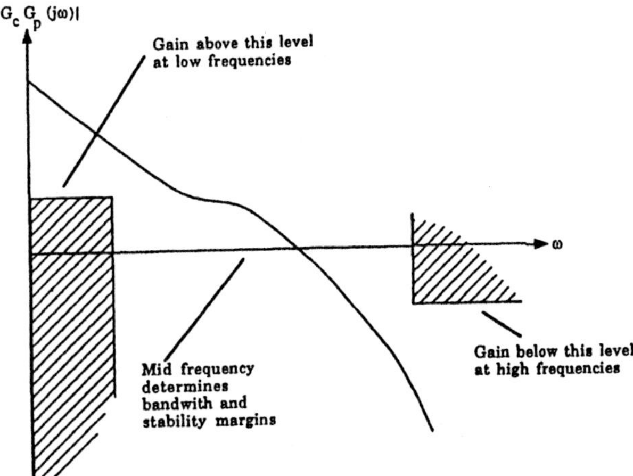

Figure 10.12 Desirable shape for the open-loop frequency response.

The general features of the open-loop transfer function are that the gain in the low-frequency region should be large enough, and in the high-frequency region, the gain should be attenuated as much as possible. The gain at intermediate frequencies typically controls the gain and phase margins. Near the gain crossover frequency ω_c, the slope of the log-magnitude curve in the Bode plot should be close to $-20\,dB/decade$ (i.e., the transition from the low- to high-frequency range must be smooth). Note that the phase margin (P.M.) of the feedback system is $180^o + \phi_c$ with $\phi_c = \angle\, G_c(j\omega_c)G_p(j\omega_c)$. If the loop transfer function $L(j\omega) = G_c(j\omega)G_p(j\omega)$ is stable, proper, and minimum phase, then ϕ_c is uniquely determined from the gain plot of G_cG_p (i.e., $|G_c(j\omega)G_p(j\omega)|$). Bode actually showed that ϕ_c is given, in terms of the weighted average attenuation rate of $|G_cG_p|$, by

$$\phi_c = \frac{1}{\pi}\int_{-\infty}^{\infty} \frac{d\,ln\,|G_c(j\omega(\mu))G_p(j\omega(\mu))|}{d\mu} (ln\,coth\,\frac{|\mu|}{2})\,d\mu$$

where $\mu = ln\,(\omega/\omega_c)$. And ϕ_c is large if $|G_cG_p|$ attenuates slowly and small if it attenuates rapidly. Therefore, a rapid attenuation of $|G_cG_p|$ at or near crossover frequency will decrease the P.M. and a more than $-20\ dB/decade$ slope near ω_c indicates that P.M. is inadequate. Controlling ϕ_c is important because it is related to the system stability and performance measure.

Based on the loop gain shown in Figure 10.12, therefore, the desirable shapes for the sensitivity and complementary sensitivity functions of a closed-loop system should be similar to those shown in Figure 10.13. That is, the sensitivity function S must

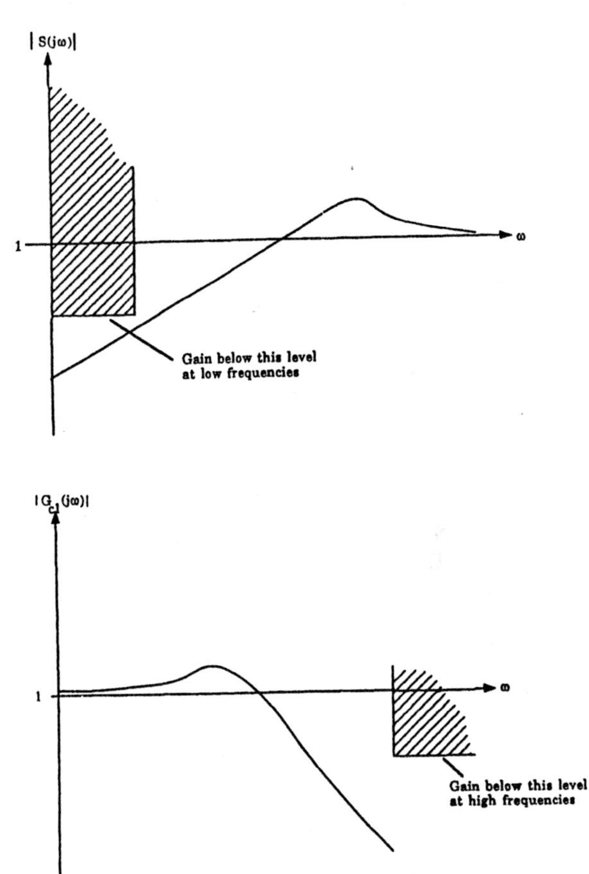

Figure 10.13 Desirable shape for the sensitivity and complementary sensitivity functions.

be small at low frequencies and roll off to 1 (i.e., $0\ dB$) at high frequencies, whereas G_{cl} must be 1 ($0\ dB$) at low frequencies and get small at high frequencies.

Notice that Figures 10.12 and 10.13 can be viewed as specifications for a control system. Such graphical specifications of the "loop shape" are suitable for today's computer-aided design packages, which have extensive graphical capabilities. As will be seen shortly, these specifications on the loop shape are particularly easy to adapt to MIMO systems.

In order to achieve the desired performance shown in Figures 10.12 and 10.13, a "loop shaping" method, which presents a graphical technique for designing a controller to achieve robust performance, may be considered. The idea of loop shaping is to design the Bode magnitude plot of the loop transfer function $L(j\omega) = G_c(j\omega)G_p(j\omega)$ to achieve (or at least approximate) the requirements shown in Figure 10.12, and then to back-solve for the controller from the loop transfer function. In other words, we first convert performance specifications on $|S(j\omega)|$ and $|G_{cl}(j\omega)|$ (as given in Figure 10.13) into specifications on $|L(j\omega)|$ (as shown in Figure 10.12). We then shape $|L(j\omega)|$ to make it lie above the first constraint curve ($|L(j\omega)| >> 1$) at low frequencies, lie below the second constraint curve ($|L(j\omega)| << 1$), and roll off as fast as possible in the high-frequency range, and make a smooth transition from low to high frequency, i.e., keep the slope as gentle as possible (about $-20\,dB/decade$) near the crossover frequency ω_c. In loop shaping, the resulting controller has to be checked in the closed-loop system to see whether a satisfactory trade-off between $|S(j\omega)|$ and $|G_{cl}(j\omega)|$ has been reached. Note that it is also possible to directly shape $|S(j\omega)|$ and/or $|G_{cl}(j\omega)|$.

Control Effort

It is important to be aware of the limits on actuator signals in specifying controllers. Most actuators have upper limits on their magnitudes and on their rates of change. These limits can severely constrain the performance and robustness of the closed-loop system. For example, actuator saturation can cause instability in conditionally stable systems and very poor transient response because of integral windup (see Chapter 20.1).

The problem should be addressed in two places. The actuator specifications should state the allowable limits on the magnitudes and rates of actuator signals. Proper specification of the actuators greatly facilitates the rest of the design. Secondly, the response of the closed-loop system to "large" inputs should be specified. Such specifications ensure that, if it is necessary to saturate the actuators, the performance and robustness of the closed-loop system remain satisfactory.

10.1.4 Miscellaneous Specifications

There are many other aspects of a control system that are often specified. There are usually constraints on the allowable cost of the controller. In some applications, especially spacecraft, the size, weight, and power required for the controller's operation are restricted.

Control system reliability is also often specified. The simplest such specification is the life expectancy of the controller. This is usually given as the mean time before failure (MTBF). The allowable ways in which a controller may fail are also often specified, especially in applications involving humans. For example, a control system may be required to "fail safe". A stronger requirement is "fail soft".

A good example of a control system that is designed to "fail soft" is the controller that regulates the height of the water in a toilet tank. Sooner or later the valve that stops the flow of water when the tank is full gets stuck open. This would cause the tank to overflow, creating a mess. The MTBF is a few years, not very long in comparison to the life of the whole system. This is acceptable because the control system includes an overflow tube that causes the overflow to go into the toilet and, from there, into the sewer. This is a soft failure. The controller is not working properly. Water is being wasted. But, the failure is not creating a serious problem.

10.1.5 Performance Specifications for MIMO LTI Systems

From the discussion given for SISO LTI systems, it is clear that the open-loop transfer function $L(j\omega)$ plays an essential role in determining various performance and robustness properties of the closed-loop system. For MIMO systems, the inputs and outputs are generally interacting. Due to such interactions, it can be difficult to control a MIMO system. However, the classical Bode gain/phase plots can be generalized for MIMO systems. In the following, we describe performance specifications for MIMO systems.

The responses of a MIMO system are generally coupled. That is, every input affects more than one output, and every output is influenced by more than one input. If a controller can be found such that every input affects one and only one output, then we say the MIMO system is decoupled. Exact decoupling can be difficult, if not impossible, to achieve in practice. There are various ways to specify approximate decoupling. Because decoupling is most obvious for square transfer functions, let $G(j\omega)$ be a strictly proper $m \times m$ transfer function and let $g_{ij}(j\omega)$ denote the ijth element of the matrix $G(j\omega)$. Then the requirement that

$$\begin{aligned} |g_{ij}(j\omega)| &< \delta \quad \text{for all } 0 \le \omega < \infty; \\ i, j &= 1, 2, \cdots, m; \;\; i \neq j \end{aligned}$$

would force $G(j\omega)$ to be approximately decoupled provided δ were small enough. Such a specification might be defective because it allows the diagonal transfer functions, the $g_{ii}(j\omega)$, $i = 1, 2, \cdots, m$, to be arbitrary. Typically, one wants $|g_{ii}(j\omega)|$, $i = 1, 2, \cdots, m$ close to one, at least in some range $\omega_1 \le \omega \le \omega_2$. The requirement that

$$\begin{aligned} \frac{|g_{ij}(j\omega)|}{|g_{ii}(j\omega)|} &< \delta \quad \text{for all } 0 \le \omega_1 \le \omega \le \omega_2; \\ i, j &= 1, 2, \cdots, m; \;\; i \neq j \end{aligned}$$

forces the $g_{ij}(j\omega)$ to be small relative to the $g_{ii}(j\omega)$. Of course, nothing prevents adding SISO specifications on the $g_{ii}(j\omega)$ to the decoupling specifications.

Another useful decoupling specification is diagonal dominance. A strictly proper square $m \times m$ transfer function $G(j\omega)$ is said to be row diagonal dominant if

$$\sum_{j=1, j\neq i}^{m} |g_{ij}(j\omega)| < |g_{ii}(j\omega)| \quad \text{for all}$$
$$i = 1, 2, \cdots, m \text{ and all } 0 \le \omega < \infty$$

Column diagonal dominance is defined in the obvious way. Decoupling specifications can also be written in the time domain as limits on the impulse or step responses of the closed-loop system.

Decoupling is not always necessary or desirable. Thus, it is necessary to have other ways to specify the performance of MIMO controlled systems. One effective way to do this is by means of the singular value decomposition (SVD). It is known that the SVD is a useful tool in linear algebra, and it has found many applications in control during the past decade. Let G be an $m \times n$ complex (constant) matrix. Then the positive square roots of the eigenvalues of G^*G (where G^* means the complex conjugate transpose of G) are called the singular values of G. These square roots are always real numbers because the eigenvalues of G^*G are always real and ≥ 0. The maximum and minimum singular values of G are denoted by $\bar{\sigma}(G)$ and $\underline{\sigma}(G)$, respectively; and they can also be expressed by

$$\begin{aligned} \bar{\sigma}(G) &= max_{||u||=1} \, ||Gu|| \\ \underline{\sigma}(G) &= min_{||u||=1} \, ||Gu|| \end{aligned}$$

where $u \in C^n$ and the vector norm $|| \cdot ||$ is the Euclidean norm. That is, $\bar{\sigma}(G)$ and $\underline{\sigma}(G)$ are the maximum and minimum gains of the matrix G. For a square G, $\underline{\sigma}(G)$ is a measure of how far G is from singularity, and $\bar{\sigma}(G)/\underline{\sigma}(G)$ is the condition number which is a measure of the difficulty of inverting G. The best way to compute the SVD is by means of an algorithm also known as the singular value decomposition (SVD) (see Chapter 21).

For MIMO systems, the transfer function matrices evaluated at $s = j\omega$ have proven useful in resolving the complexities of MIMO design. The idea is to reduce the transfer function matrices to two critical gains versus frequency, that is, the maximum and minimum singular values of the transfer function matrix. Consider a MIMO system represented by a transfer matrix $G(s)$. Similar to the constant matrix case discussed, if we let $s = j\omega$ $(0 \leq \omega < \infty)$, then the singular values $\sigma_i(G(j\omega))$ will be functions of ω, and a plot of $\sigma_i(G(j\omega))$ is called a singular value plot (or σ-plot) which is analogous to a Bode magnitude plot of a SISO transfer function. The maximum and minimum singular values of $G(j\omega)$ are defined, respectively, as

$$\begin{aligned} \bar{\sigma}(G(j\omega)) &= \sqrt{\lambda_{max}[G(j\omega)^*G(j\omega)]} \\ \underline{\sigma}(G(j\omega)) &= \sqrt{\lambda_{min}[G(j\omega)^*G(j\omega)]} \end{aligned}$$

where $\lambda[\cdot]$ denotes eigenvalues. Note that the H_∞ norm of a stable transfer matrix G is the maximum of $\bar{\sigma}(G)$ over all frequencies, i.e., $||G||_\infty = sup_{\omega \geq 0}\, \bar{\sigma}(G(j\omega))$. For the performance and robustness measures of MIMO systems, we can examine $\bar{\sigma}$ and $\underline{\sigma}$ in a manner similar to that used to examine the frequency-response magnitude of a SISO transfer function. If $\bar{\sigma}(G)$ and $\underline{\sigma}(G)$ are very close to each other, then we can simply treat the system like a SISO system. In general, they are not close and $\bar{\sigma}$ is important to bound the performance requirements.

Without loss of generality, consider the MIMO unity feedback system shown in Figure 10.11. Like the SISO case, define the sensitivity and complementary sensitivity matrices as $S(s) = [I + G_p(s)G_c(s)]^{-1}$ (where $(\cdot)^{-1}$ means the matrix inverse) and $G_{cl}(s) = I - S(s)$, respectively. Note that $G_{cl}(s) = [I + G_p(s)G_c(s)]^{-1}G_p(s)G_c(s)$ is the closed-loop transfer function matrix which describes the system's input-output relationship. From Figure 10.11, we have

$$\begin{aligned} Y(s) &= G_{cl}(s)R(s) + S(s)D(s) - G_{cl}(s)N(s) \\ E(s) &= S(s)R(s) - S(s)D(s) + G_{cl}(s)N(s) \end{aligned}$$

Similar to the arguments stated for SISO systems, for disturbance rejection, tracking error reduction, and insensitivity to plant parameter variations, we need to make $S(j\omega)$ "small" over the frequency range (say, $0 \leq \omega \leq \omega_0$) where the commands, disturbances, and parameter changes of G_p are significant. That is, to keep $\bar{\sigma}[(I + G_pG_c)^{-1}]$ as small as possible, or $\underline{\sigma}[I + G_pG_c]$ as large as possible $\forall\ \omega \leq \omega_0$ since $\bar{\sigma}[(I + G_pG_c)^{-1}] = 1/\underline{\sigma}[I + G_pG_c]$. The inequalities $max\ (0, \underline{\sigma}[G_pG_c] - 1) \leq \underline{\sigma}[I + G_pG_c] \leq \underline{\sigma}[G_pG_c] + 1$ further imply that the loop gain $\underline{\sigma}[G_pG_c]$ should be made as large as possible $\forall\ \omega \leq \omega_0$.

From the equations shown, it is also clear that for sensor noise reduction, the transfer matrix $G_{cl}(j\omega)$ should be "small" over the frequency range (say, $\omega \geq \omega_1$) where the noise is significant. Using the equalities $(I + X)^{-1}X = (I + X^{-1})^{-1}$ and $\bar{\sigma}(X)\underline{\sigma}(X^{-1}) = 1$ (assume X^{-1} exists), this implies that $\bar{\sigma}[(I + G_pG_c)^{-1}G_pG_c] = \bar{\sigma}[(I + (G_pG_c)^{-1})^{-1}] = 1/\underline{\sigma}[I + (G_pG_c)^{-1}]$ should be small, or $\underline{\sigma}[I + (G_pG_c)^{-1}]$ should be large $\forall\ \omega \geq \omega_1$. Since $\underline{\sigma}[I + (G_pG_c)^{-1}] \geq \underline{\sigma}[(G_pG_c)^{-1}]$, we should make $\underline{\sigma}[(G_pG_c)^{-1}]$ large, which further means that $\bar{\sigma}[G_pG_c]$ should be kept as small as possible $\forall\ \omega \geq \omega_1$.

In summary, disturbance rejection, tracking error, and sensitivity (to plant parameter variations) reduction require large $\underline{\sigma}[G_pG_c]$, while sensor noise reduction requires small $\bar{\sigma}[G_pG_c]$. Since $\bar{\sigma}[G_pG_c] \geq \underline{\sigma}[G_pG_c]$, a conflict between the requirements of $S(j\omega)$ and $G_{cl}(j\omega)$ exists, which is exactly the same as that discussed for SISO systems. Therefore, a performance tradeoff for command tracking and disturbance reduction versus sensor noise reduction is unavoidable. Typical specifications for MIMO system loop gain requirements will be similar to those of Figure 10.12, with the "low-frequency boundary (or constraint)" being the lower boundary for $\underline{\sigma}[G_pG_c]$ when ω is small, and the "high-frequency boundary" representing the upper boundary for $\bar{\sigma}[G_pG_c]$ when ω becomes large. In other words, we need "loop shaping" of the singular values of the plant transfer function matrix G_p by using G_c (i.e., the design of controller G_c) so that the nominal closed-loop system is stable, $\underline{\sigma}[G_pG_c]$ (thus, $\bar{\sigma}[G_pG_c]$) at low frequencies lies above the low-frequency boundary, and $\bar{\sigma}[G_pG_c]$ (thus, $\underline{\sigma}[G_pG_c]$) at high frequencies lies below the high-frequency boundary. That is, we want to increase the low-frequency value of $\underline{\sigma}$ (thus, $\bar{\sigma}$) of G_pG_c to ensure adequate attenuation of (low-frequency) disturbances and better command tracking, and roll-off $\bar{\sigma}$ and $\underline{\sigma}$ in the high-frequency range to ensure robust stability. Note that the "forbidden" areas shown in Figure 10.12 are problem dependent, and they may be constructed from the design specifications.

It is generally desirable to require that the gap between $\bar{\sigma}[G_pG_c]$ and $\underline{\sigma}[G_pG_c]$ be fairly small and that their slope be close to $-20\ dB/decade$ near the gain crossover frequencies. Here, there are a range of gain crossover frequencies from the frequency $\underline{\omega}_c$ at which $\underline{\sigma}[G_p(j\omega)G_c(j\omega)] = 1$ to the frequency $\bar{\omega}_c$ at which $\bar{\sigma}[G_p(j\omega)G_c(j\omega)] = 1$. As for SISO systems, the requirements near the crossover frequencies primarily address robustness.

10.1.6 Robustness Specifications for MIMO LTI Systems

It is known that, in control system design, the plant model used is only an approximate representation of the physical system. The discrepancies between a system and its mathematical representation (model) may lead to a violation of some performance specification, or even to closed-loop instability. We say the system is robust if the design performs satisfactorily under variations in the dynamics of the plant (including parameter variations and various possible uncertainties). Stability and performance robustness are two important issues that should be considered in control design. Generally, the form of the plant uncertainty can be parametric, nonparametric, or both. Typical sources of uncertainty include unmodeled high-frequency dynamics, neglected nonlinearities, plant parameter variations (due to changes of environmental factors), etc. The parametric uncertainty in the plant model can be expressed as $G_p(s,\gamma)$, a parameterized model of the nominal plant $G_p(s)$ with the uncertain parameter γ. The three most commonly used models to represent unstructured uncertainty are additive, input multiplicative, and output multiplicative types; and can be represented, respectively, as

$$\begin{aligned}\tilde{G}_p(s) &= G_p(s) + \Delta_a(s)\\ \tilde{G}_p(s) &= G_p(s)[I + \Delta_i(s)]\\ \tilde{G}_p(s) &= [I + \Delta_o(s)]G_p(s)\end{aligned}$$

where $\tilde{G}_p(s)$ is the plant transfer matrix as perturbed from its nominal model $G_p(s)$ due to the uncertainty Δ (Δ_a or Δ_i or Δ_o) with $\bar{\sigma}[\Delta(j\omega)]$ bounded above. That is, we have $\bar{\sigma}[\Delta(j\omega)] \le l(\omega)$, where $l(\omega)$ is a known positive real scalar function. In general, $l(\omega)$ is small at low frequencies because we can model the plant more accurately in the low-frequency range. The plant high-frequency dynamics are less known, resulting in large $l(\omega)$ at high frequencies. Note that Δ_i (Δ_o) assumes all the uncertainty occurred at the plant input (output). Both Δ_i and Δ_o represent relative deviation while Δ_a represents absolute deviation from $G_p(s)$. Of course, there are many possible ways to represent uncertainty, for example, $\tilde{G}_p(s) = [I + \Delta_o(s)]G_p(s)[I + \Delta_i(s)]$, $\tilde{G}_p(s) = [N(s) + \Delta_N(s)][D(s) + \Delta_D(s)]^{-1}$ (a matrix fractional representation with $G_p(s) = N(s)D^{-1}(s)$), etc. We may even model the plant by combining both parametric and unstructured uncertainties in the form of $\tilde{G}_p(s) = [I + \Delta_o(s)]G_p(s,\gamma)[I + \Delta_i(s)]$.

In the following, we examine the robustness of performance together with robust stability under output multiplicative uncertainty Δ_o. Similar results can be derived if other types of unstructured uncertainty are used. Via a multivariable version of the standard Nyquist criterion, the closed-loop system will remain stable under the uncertainty Δ_o if

$$\bar{\sigma}[G_pG_c(I + G_pG_c)^{-1}(j\omega)] < \frac{1}{\bar{\sigma}(\Delta_o)} \quad \forall\,\omega \ge 0$$

where we assume that the nominal closed-loop system is stable, and $G_p(s)$, $\tilde{G}_p(s)$ have the same number of unstable poles. The above condition is actually necessary and sufficient for robust stability. Note that the expression inside $\bar{\sigma}[\cdot]$ in the left-hand side of the inequality is the complementary sensitivity matrix $G_{cl}(j\omega)$. Thus, the model uncertainty imposes an upper bound on the singular values of $G_{cl}(s)$ for robust stability. Rewriting the inequality, we get

$$\begin{aligned}\bar{\sigma}(\Delta_o) &< \frac{1}{\bar{\sigma}[G_pG_c(I + G_pG_c)^{-1}(j\omega)]}\\ &= \underline{\sigma}[I + (G_pG_c)^{-1}(j\omega)] \quad \forall\,\omega \ge 0\end{aligned}$$

where we assume that G_pG_c is invertible. Obviously, $\underline{\sigma}[I + (G_pG_c)^{-1}]$ can be used as a measure of the degree of stability of the feedback system. This is a multivariable version of SISO stability margins (i.e., gain and phase margins) because it allows gain and phase changes in each individual output channel and/or simultaneous gain and phase changes in several channels. The extent to which these changes are allowed is determined by the inequality shown above. Therefore, we can use the singular values to define G.M. for MIMO systems.

In the high-frequency range (where the loop gain is small), from the inequality given above, we have

$$\bar{\sigma}[G_p(j\omega)G_c(j\omega)] < \frac{1}{\bar{\sigma}(\Delta_o)}$$

This is a constraint on the loop gain G_pG_c for robust stability. This constraint implies that at high frequencies the loop gain on G_pG_c (i.e., $\bar{\sigma}[G_pG_c]$) should lie below a certain limit for robust stability (the same argument described in the previous section). Obviously, satisfaction of the high-frequency boundary shown in Figure 10.12 is mandatory for robust stability (not just desired for sensor noise reduction !).

For robust performance (i.e., good command tracking, disturbance reduction, and insensitivity to plant parameter variations (i.e., structured uncertainty) under all possible $\tilde{G}_p(s)$), we should keep $\bar{\sigma}[(I + \tilde{G}_pG_c)^{-1}]$ of the "perturbed" sensitivity matrix $\tilde{S}(s)$ as small as possible in the low-frequency range. Following the same argument given in the previous section, this further implies that $\underline{\sigma}[\tilde{G}_pG_c]$ should be large at low frequencies. Making $\underline{\sigma}[G_pG_c]$ large enough ensures that $\underline{\sigma}[\tilde{G}_pG_c]$ is large. Clearly, high loop gain can compensate for model uncertainty. Therefore, for robust performance the low-frequency boundary given in Figure 10.12 should be satisfied, although this is not mandatory.

For MIMO systems, as for the requirements on the loop gain in SISO control design, any violation of the low-frequency boundary in Figure 10.12 constitutes a violation of the robust performance specifications, while a violation of the upper boundary (in the high-frequency range) leads to a violation of the robust stability specifications. The main distinction between MIMO and SISO design is the use of singular values of the transfer function matrix to express the "size" of functions.

10.1.7 Conclusions

Writing specifications for control systems is not easy. The different aspects of controller performance and robustness are interrelated and, in many cases, competing. While a good controller can compensate for some deficiencies in the plant (system to be controlled), the plant implies significant limits on controller performance. It is all too easy to impose unachievable specifications on a controller.

For these reasons, it is important to have the plant designer, the person responsible for the controller specifications, and the control designer work together from the beginning to the end of any project with demanding control requirements.

Further Reading

Most undergraduate texts on control systems contain useful descriptions of performance and robustness specifications for SISO LTI systems. Three examples are

[1] Franklin, G.F, Powell, J.D., and Emami-Naeini, A., *Feedback Control of Dynamic Systems*, 3rd Edition, Addison-Wesley, Reading, MA, 1994.
[2] Nise, N.S., *Control Systems Engineering*, 2nd Edition, Benjamin/Cummings, Redwood City, CA, 1995.
[3] Dorf, R.C. and Bishop, R.H., *Modern Control Systems*, 7th Edition, Addison-Wesley, Reading, MA, 1995.

Several textbooks have been published recently that include a more loop shaping-oriented discussion of control specifications. These include

[4] Boyd, S.P. and Barratt, C.H., *Linear Controller Design - Limits of Performance*, Prentice Hall, Englewood Cliffs, NJ, 1991.
[5] Doyle, J.C., Francis, B.A., and Tannenbaum, A.R., *Feedback Control Theory*, Macmillan, New York, NY, 1992.
[6] Belanger, P.R., *Control Engineering - A Modern Approach*, Saunders College Publishing, New York, 1995.
[7] Wolovich, W.A., *Automatic Control Systems - Basic Analysis and Design*, Saunders College Publishing, New York, 1994.

Specifications for MIMO LTI systems are covered in

[8] Maciejowski, J.M., *Multivariable Feedback Design*, Addison-Wesley, Reading, MA, 1989.
[9] Green, M. and Limebeer, D.J.N., *Linear Robust Control*, Prentice Hall, Englewood Cliffs, NJ, 1995.

The limitations on control systems, including Bode's results on the relation between gain and phase, are described in

[10] Freudenberg, J.S. and Looze, D.P., *Frequency Domain Properties of Scalar and Multivariable Feedback Systems*, Springer-Verlag, New York, NY, 1988.

A good, detailed example of the specifications for a complex MIMO control system is

[11] Hoh, R.H., Mitchell, D.G., and Aponso, B.L., Handling Qualities Requirements for Military Rotorcraft, Aeronautical Design Standard, ADS-33C, US Army ASC, Aug. 1989.
[12] Hoh, R.H., Mitchell, D.G., Aponso, B.L., Key, D.L., and Blanken, C.L., Background Information and User's Guide for Handling Quantities Requirements for Military Rotorcraft, USAAVSCOM TR89-A-8, US Army ASC, Dec. 1989.

Finally, all three publications of the IEEE Control Systems Society, the Transactions on Automatic Control, the Transactions on Control Systems Technology, and the Control Systems Magazine regularly contain articles on the design of control systems.

10.2 Design Using Performance Indices

Richard C. Dorf, University of California, Davis
Robert H. Bishop, The University of Texas at Austin

10.2.1 Introduction

Modern control theory assumes that the systems engineer can specify quantitatively the required system performance. Then a **performance index** can be calculated or measured and used to evaluate the system's performance. A quantitative measure of the performance of a system is necessary for automatic parameter optimization of a control system and for the design of optimum systems.

We consider a feedback system as shown in Figure 10.14 where the closed-loop transfer function is

$$\frac{C(s)}{R(s)} = T(s) = \frac{G_c(s)G(s)}{1 + G_c(s)G(s)} . \qquad (10.1)$$

Whether the aim is to improve the design of a system or to design a control system, a performance index may be chosen [1]:

A performance index is a quantitative measure of the performance of a system and is chosen so that emphasis is given to the important system specifications.

The system is considered an *optimum control system* when the system parameters are adjusted so that the index reaches an extremum value, commonly a minimum value. A performance index, to be useful, must be a number that is always positive or zero. Then the best system is defined as the system that minimizes this index.

It is also possible to design an optimum system to achieve a *deadbeat response*, which is characterized by a fast response with minimal overshoot. The desired closed-loop system characteristic equation coefficients are selected to minimize settling time and rise time.

10.2.2 The ISE Index

One performance index is the integral of the square of the error (ISE), which is defined as

$$ISE = \int_0^T e^2(t)dt\,. \qquad (10.2)$$

The upper limit T is a finite time chosen somewhat arbitrarily so that the integral approaches a steady-state value. It is usually convenient to choose T as the settling time, T_s. This criterion will discriminate between excessively overdamped and excessively underdamped systems. The minimum value of the integral occurs for a compromise value of the damping. The performance index of Equation 10.2 is easily adapted for practical measurements because a squaring circuit is readily obtained. Furthermore, the squared error is mathematically convenient for analytical and computational purposes.

10.2.3 The ITAE Index

To reduce the contribution of the relatively large initial error to the value of the performance integral, as well as to emphasize errors occurring later in the response, the following index has been proposed [2]:

$$ITAE = \int_0^T t|e(t)|dt\,, \qquad (10.3)$$

where ITAE is the integral of time multiplied by the absolute magnitude of the error.

The coefficients that will minimize the ITAE performance criterion for a step input have been determined for the general closed-loop transfer function [2]

$$T(s) = \frac{C(s)}{R(s)} = \frac{b_0}{s^n + b_{n-1}s^{n-1} + \cdots + b_1 s + b_0}\,. \qquad (10.4)$$

This transfer function has a steady-state error equal to zero for a step input. Note that the transfer function has n poles and no zeros. The optimum coefficients for the ITAE criterion are given in Table 10.1 for a step input. The transfer function, Equation 10.4, implies the plant and controller $G_c(s)G(s)$ have one or more pure integrations to provide zero steady-state error. The responses using optimum coefficients for a step input are given in Figure 10.15 for ITAE. The responses are provided for normalized time, $\omega_n t$. Other standard forms based on different performance indices are available and can be useful in aiding the designer to determine the range of coefficients for a specific problem.

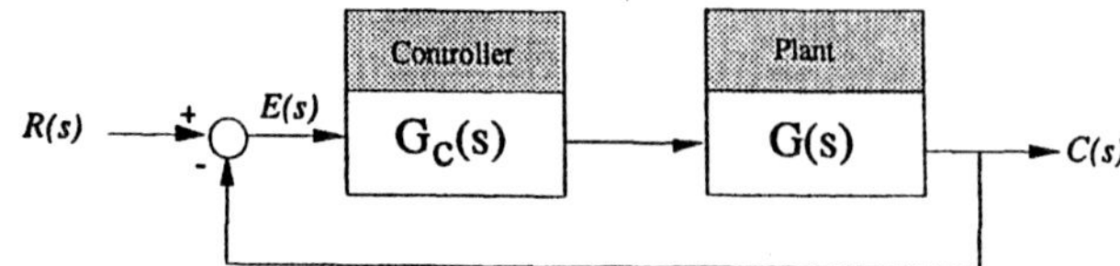

Figure 10.14 Feedback control system.

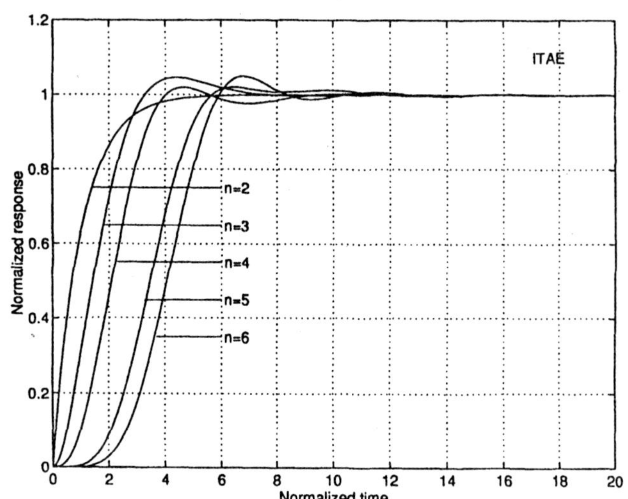

Figure 10.15 The step response of a system with a transfer function satisfying the ITAE criterion.

For a ramp input, the coefficients have been determined that minimize the ITAE criterion for the general closed-loop transfer function [2]:

$$T(s) = \frac{b_1 s + b_0}{s^n + b_{n-1}s^{n-1} + \cdots + b_1 s + b_0}\,. \qquad (10.5)$$

This transfer function has a steady-state error equal to zero for a ramp input. The optimum coefficients for this transfer function are given in Table 10.2. The transfer function, Equation 10.5, implies that the plant and controller $G_c(s)G(s)$ have two or more pure integrations, as required to provide zero steady-state error.

10.2.4 Normalized Time

We consider the transfer function of a closed-loop system, $T(s)$. To determine the coefficients that yield the optimal **deadbeat response**, the transfer function is first normalized. An example of this for a third-order system is

$$T(s) = \frac{\omega_n^3}{s^3 + \alpha\omega_n s^2 + \beta\omega_n^2 s + \omega_n^3}\,. \qquad (10.6)$$

TABLE 10.1 The Optimum Coefficients of $T(s)$ Based on the ITAE Criterion for a Step Input.

$s + \omega_n$
$s^2 + 1.4\omega_n s + \omega_n^2$
$s^3 + 1.75\omega_n s^2 + 2.15\omega_n^2 s + \omega_n^3$
$s^4 + 2.1\omega_n s^3 + 3.4\omega_n^2 s^2 + 2.7\omega_n^3 s + \omega_n^4$
$s^5 + 2.8\omega_n s^4 + 5.0\omega_n^2 s^3 + 5.5\omega_n^3 s^2 + 3.4\omega_n^4 s + \omega_n^5$
$s^6 + 3.25\omega_n s^5 + 6.6\omega_n^2 s^4 + 8.6\omega_n^3 s^3 + 7.45\omega_n^4 s^2 + 3.95\omega_n^5 s + \omega_n^6$

TABLE 10.2 The Optimum Coefficients of $T(s)$ Based on the ITAE Criterion for a Ramp Input.

$s^2 + 3.2\omega_n s + \omega_n^2$
$s^3 + 1.75\omega_n s^2 + 3.25\omega_n^2 s + \omega_n^3$
$s^4 + 2.41\omega_n s^3 + 4.39\omega_n^2 s^2 + 5.14\omega_n^3 s + \omega_n^4$
$s^5 + 2.19\omega_n s^4 + 6.5\omega_n^2 s^3 + 6.3\omega_n^3 s^2 + 5.24\omega_n^4 s + \omega_n^5$

Dividing the numerator and denominator by ω_n^3 yields

$$T(s) = \frac{1}{\frac{s^3}{\omega_n^3} + \alpha\frac{s^2}{\omega_n^2} + \beta\frac{s}{\omega_n} + 1} . \tag{10.7}$$

Let $\bar{s} = s/\omega_n$ to obtain

$$T(s) = \frac{1}{\bar{s}^3 + \alpha\bar{s}^2 + \beta\bar{s} + 1} . \tag{10.8}$$

Equation 10.8 is the normalized third-order closed-loop transfer function. For a higher-order system, the same method is used to derive the normalized equation. When we let $\bar{s} = s/\omega_n$, this has the effect in the time-domain of normalizing time, $\omega_n t$. The step response for a normalized system is as shown in Figure 10.15.

10.2.5 Deadbeat Response

Often the goal for a control system is to achieve a fast response to a step command with minimal overshoot. We define a **deadbeat response** as a response that proceeds rapidly to the desired level and holds at that level with minimal overshoot. We use the $\pm 2\%$ band at the desired level as the acceptable range of variation from the desired response. Then, if the response enters the band at time T_s, it has satisfied the settling time T_s upon entry to the band, as illustrated in Figure 10.16. A deadbeat response has the following characteristics:

1. Zero steady-state error
2. Fast response $\rightarrow$ minimum rise time and settling time
3. $0.1\% \leq$ percent overshoot $< 2\%$
4. Percent undershoot $< 2\%$

Characteristics 3 and 4 require that the response remain within the $\pm 2\%$ band so that the entry to the band occurs at the settling time.

A more general normalized transfer function of a closed-loop system may be written as

$$T(s) = \frac{1}{\bar{s}^n + \alpha\bar{s}^{n-1} + \beta\bar{s}^{n-2} + \gamma\bar{s}^{n-3} + \cdots + 1} . \tag{10.9}$$

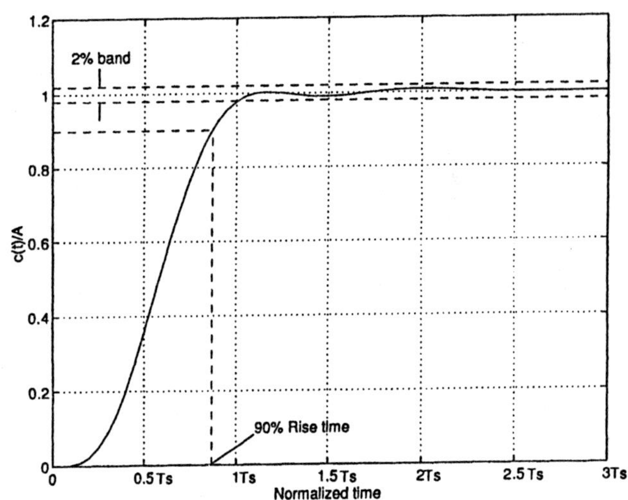

Figure 10.16 The deadbeat response, where A is the magnitude of the step input.

The coefficients of the denominator equation (α, β, γ, and so on) are then assigned the values necessary to meet the requirement of deadbeat response. The coefficients recorded in Table 10.3 were selected to achieve deadbeat response and to minimize settling time and rise time to 100% of the desired command. The form of Equation 10.9 is normalized since $\bar{s} = s/\omega_n$. Thus, we choose ω_n based on the desired settling time or rise time. For example, if we have a third-order system with a required settling time of 1.2 seconds, we note from Table 10.3 that the normalized settling time is

$$\omega_n T_s = 4.04 .$$

Therefore, we require

$$\omega_n = \frac{4.04}{T_s} = \frac{4.04}{1.2} = 3.37 .$$

Once ω_n is chosen, the complete third-order closed-loop transfer function is known, having the form of Equation 10.6, where $\alpha = 1.9$ and $\beta = 2.2$. When designing a system to obtain a deadbeat

TABLE 10.3 Coefficients and Response Measures of a Deadbeat System.

System Order	Coefficients α	β	γ	δ	ϵ	Percent Overshoot P.O.	Percent Undershoot P.U.	90% Rise Time $T_{r_{90}}$	100% Rise Time T_r	Settling Time T_s
2nd	1.82					0.10%	0.00%	3.47	6.58	4.82
3rd	1.90	2.20				1.65%	1.36%	3.48	4.32	4.04
4th	2.20	3.50	2.80			0.89%	0.95%	4.16	5.29	4.81
5th	2.70	4.90	5.40	3.40		1.29%	0.37%	4.84	5.73	5.43
6th	3.15	6.50	8.70	7.55	4.05	1.63%	0.94%	5.49	6.31	6.04

Note: All time is normalized.

response, the compensator is chosen and the closed-loop transfer function is found. This compensated transfer function is then set equal to Equation 10.9 and the required compensator parameters can be determined.

EXAMPLE 10.1:

Consider a sample plant

$$G(s) = \frac{1}{s(s+p)} \tag{10.10}$$

and a controller $G_c(s) = K$. The goal is to select the parameters p and K to yield (1) an ITAE response, and alternatively (2) a deadbeat response. In addition, the settling time for a step response is specified as less than 1 second. The closed-loop transfer function is

$$T(s) = \frac{K}{s^2 + ps + K} . \tag{10.11}$$

For the ITAE system we examine Figure 10.15 and note that for $n = 2$ we have $\omega_n T_s = 8$. Then, for $T_s = 0.8$, we use $\omega_n = 10$. From Table 10.1, we require

$$s^2 + 1.4\omega_n s + \omega_n^2 = s^2 + ps + K .$$

Therefore, we have $K = \omega_n^2 = 100$ and $p = 1.4\omega_n = 14$.

If we seek a deadbeat response, we use Table 10.3 and note that the normalized settling time is

$$\omega_n T_s = 4.82 .$$

In order to obtain T_s less than 1 second, we use $\omega_n = 6$ and $T_s = 0.8$. Then we use Table 10.3 to obtain $\alpha = 1.82$. The closed-loop transfer function is

$$T(s) = \frac{\omega_n^2}{s^2 + \alpha\omega_n s + \omega_n^2} = \frac{36}{s^2 + 10.9s + 36} . \tag{10.12}$$

Then we require $K = 36$ and $p = 10.9$. The deadbeat and the ITAE step response are shown in Figure 10.17.

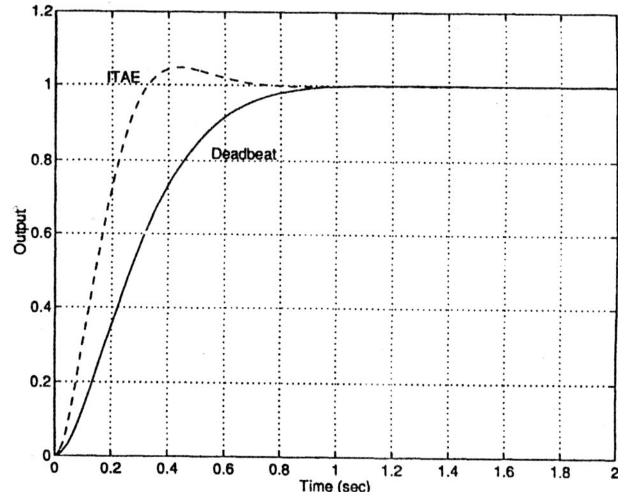

Figure 10.17 ITAE ($p = 14$ and $K = 100$) and deadbeat ($p = 10.9$ and $K = 36$) system response.

10.2.6 Conclusions

The design of a feedback system using performance indices or deadbeat control leads to predictable system responses. This permits the designer to select system parameters to achieve desired performance.

10.2.7 Defining Terms

Performance index: A quantitative measure of the performance of a system.

ISE: Integral of the square of the error.

ITAE: Integral of time multiplied by the absolute error.

Deadbeat response: A system with a rapid response, minimal overshoot, and zero steady-state error for a step input.

References

[1] Dorf, R.C. and Bishop, R.H., *Modern Control Systems*, 7th ed., Addison-Wesley, Reading, MA, 1995.

[2] Graham, D. and Lathrop, R.C., The synthesis of optimum response: criteria and standard forms. II, *Trans. AIEE 72*, 273–288, November 1953.

10.3 Nyquist, Bode, and Nichols Plots

John J. D'Azzo, Air Force Institute of Technology

Constantine H. Houpis, Air Force Institute of Technology

10.3.1 Introduction

The frequency-response[1] method of analysis and design is a powerful technique for the comprehensive study of a system by conventional methods. Performance requirements can be readily expressed in terms of the frequency response. Since noise, which is always present in any system, can result in poor overall performance, the frequency-response characteristics of a system permit evaluation of the effect of noise. The design of a passband for the system response may result in excluding the noise and therefore improving the system performance as long as the dynamic tracking performance specifications are met. The frequency response is also useful in situations for which the transfer functions of some or all of the components in a system are unknown. The frequency response can be determined experimentally for these situations, and an approximate expression for the transfer function can be obtained from the graphical plot of the experimental data. The frequency-response method is also a very powerful method for analyzing and designing a robust multi-input/multi-output (MIMO) system with structured uncertain plant parameters. In this chapter two graphical representations of transfer functions are presented: the logarithmic plot and the polar plot. These plots are used to develop Nyquist's stability criterion [1], [2], [6], [8] and closed-loop design procedures. The plots are also readily obtained by use of computer-aided design (CAD) packages like MATLAB^R or TOTAL-PC (see [11]). The closed-loop feedback response $M(j\omega)$ is obtained as a function of the open- loop transfer function $G(j\omega)$. Design methods for adjusting the open-loop gain are developed and demonstrated. They are based on the polar plot of $G(j\omega)$ and the Nichols plot. Both methods achieve a peak value M_m and a resonant frequency ω_m of the closed-loop frequency response. A correlation between these frequency-response characteristics and the time response is developed.

[1]The material contained in this chapter is based on Chapters 8 and 9 from the text *Linear Control System Analysis & Design — Conventional and Modern*, 4th ed., McGraw-Hill, New York, 1995.

10.3.2 Correlation of the Sinusoidal and Time Responses

Once the frequency response [2] of a system has been determined, the time response can be determined by inverting the corresponding Fourier transform. The behavior in the frequency domain for a given driving function $r(t)$ can be determined by the Fourier transform as

$$R(j\omega) = \int_{-\infty}^{\infty} r(t)e^{-j\omega t}dt \tag{10.13}$$

For a given control system, the frequency response of the controlled variable is

$$C(j\omega) = \frac{G(j\omega)}{1+G(j\omega)H(j\omega)}R(j\omega) \tag{10.14}$$

By use of the inverse Fourier transform, the controlled variable as a function of time is

$$c(t) = \frac{1}{2\pi}\int_{-\infty}^{\infty} C(j\omega)e^{j\omega t}d\omega \tag{10.15}$$

Equation 10.15 can be evaluated by numerical or graphical integration or by reference to a table of definite integrals. This is necessary if $C(j\omega)$ is available only as a curve and cannot be simply expressed in analytical form, as is often the case. The procedure is described in several books [4]. In addition, methods have been developed based on the Fourier transform and a step input signal, relating $C(j\omega)$ qualitatively to the time solution without actually taking the inverse Fourier transform. These methods permit the engineer to make an approximate determination of the system response through the interpretation of graphical plots in the frequency domain.

Elsewhere [11, sect. 4.12] it is shown that the frequency response is a function of the pole-zero pattern in the s−plane. It is therefore related to the time response of the system. Two features of the frequency response are the maximum value M_m and the resonant frequency ω_m. Section 10.3.8 describes the qualitative relationship between the time response and the values M_m and ω_m. Since the location of the poles can be determined from the root locus, there is a direct relationship between the root-locus and frequency-response methods.

Frequency-Response Curves

The frequency domain plots belong to two categories. The first category is the plot of the magnitude of the output-input ratio vs. frequency in rectangular coordinates, as illustrated in [11, sect. 4.12]. In logarithmic coordinates these are known as *Bode plots.* Associated with this plot is a second plot of the corresponding phase angle vs. frequency. In the second category the output-input ratio may be plotted in polar coordinates with frequency as a parameter. Direct polar plots are generally used only for the open-loop response and are commonly referred to as *Nyquist plots.* [7] The plots can be obtained experimentally or by CAD packages. [3], [4] When a CAD program is not available, the Bode plots are easily obtained by a graphical procedure. The other plots can then be obtained from the Bode plots.

For a given sinusoidal input signal, the input and steady-state output are of the following forms:

$$r(t) = R \sin \omega t \quad (10.16)$$

$$c(t) = C \sin(\omega t + \alpha) \quad (10.17)$$

The closed-loop frequency response is given by

$$\frac{C(j\omega)}{R(j\omega)} = \frac{G(j\omega)}{1+G(j\omega)H(j\omega)} = M(\omega)\angle[\alpha(\omega)] \quad (10.18)$$

For each value of frequency, Equation 10.18 yields a phasor quantity whose magnitude is M and whose phase angle α is the angle between $C(j\omega)$ and $R(j\omega)$. An ideal system may be defined as one where $\alpha = 0°$ and $R(j\omega) = C(j\omega)$ for $0 < \omega < \infty$ (see curves 1 in Figures 10.18a and 10.18b). However, this definition implies an instantaneous transfer of energy from the input to the output. Such a transfer cannot be achieved in practice since any physical system has some energy dissipation and some energy-storage elements. Curves 2 and 3 in Figures 10.18a and 10.18b represent the frequency responses of practical control systems. The passband, or bandwidth, of the frequency response is defined as the range of frequencies from 0 to the frequency ω_b, where $M = 0.707$ of the value at $\omega = 0$. However, the frequency ω_m is more easily obtained than ω_b. The values M_m and ω_m are often used as figures of merit (F.O.M.).

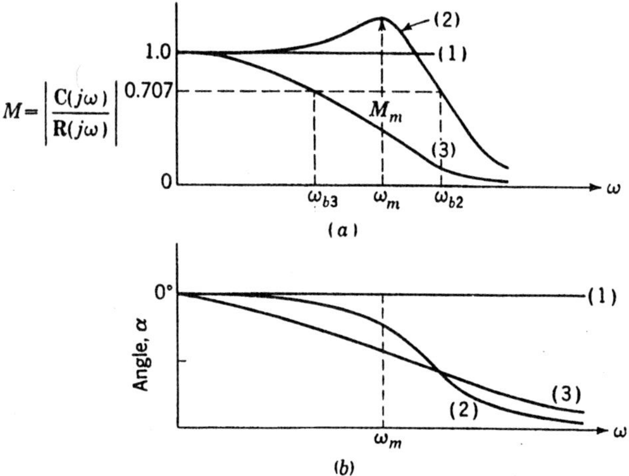

Figure 10.18 Frequency-response characteristics of $C(j\omega)/R(j\omega)$ in rectangular coordinates.

In any system the input signal may contain spurious noise signals in addition to the true signal input, or there may be sources of noise within the closed-loop system. This noise may be in a band of frequencies above the dominant frequency band of the true signal. In that case, in order to reproduce the true signal and attenuate the noise, feedback control systems are designed to have a definite passband. In certain cases the noise frequency may exist in the same frequency band as the true signal. However, when this occurs, the problem of estimating the desired signal is more complicated. Therefore, even if the ideal system were possible, it would not be desirable.

Bode Plots (Logarithmic Plots)

The use of semilog paper eliminates the need to take logarithms of very many numbers and expands the low-frequency range, which is of primary importance. The basic factors of the transfer function fall into three categories, and these can easily be plotted by means of straight-line asymptotic approximations. The straight-line approximations are used to obtain approximate performance characteristics very quickly or to check values obtained from the computer. As the design becomes more firmly specified, the straight-line curves can be corrected for greater accuracy.

Some basic definitions of logarithmic terms follow.

Logarithm. The logarithm of a complex number is itself a complex number. The abbreviation log is used to indicate the logarithm to the base 10:

$$\log|G(j\omega)|\,e^{j\phi(\omega)} := \log|G(j\omega)| + \log e^{j\phi(\omega)} = \log|G(j\omega)| + j0.434\phi(\omega) \quad (10.19)$$

The real part is equal to the logarithm of the magnitude, $\log|G(j\omega)|$, and the imaginary part is proportional to the angle, $0.434\phi(\omega)$. In the rest of this chapter, the factor 0.434 is omitted and only the angle $\phi(\omega)$ is used.

Decibels. The unit commonly used for the logarithm of the magnitude is the *decibel* (dB). When logarithms of transfer functions of physical systems are used, the input and output variables are not necessarily in the same units; e.g., the output may be speed in radians per second (rad/s), and the input may be voltage in volts (V).

Log magnitude. The logarithm of the magnitude of a transfer function $G(j\omega)$ expressed in decibels is

$$20\log|G(j\omega)| \text{ dB} \quad (10.20)$$

This quantity is called the *log magnitude*, abbreviated Lm. Thus,

$$\text{Lm}G(j\omega) = 20\log|G(j\omega)| \text{ dB} \quad (10.21)$$

Since the transfer function is a function of frequency, the Lm is also a function of frequency.

Octave and decade. Two units used to express frequency bands or frequency ratios are the octave and the decade. An octave is a frequency band from f_1 to f_2, where $f_2/f_1 = 2$. Thus, the frequency band from 1 to $2Hz$ is 1 octave in width, and the frequency band from 17.4 to $34.8Hz$ is also 1 octave in width. Note that 1 octave is not a fixed frequency bandwidth but depends on the frequency range being considered. The number of octaves in the frequency range from f_1 to f_2 is

$$\frac{\log(f_2/f_1)}{\log 2} = 3.32\log\frac{f_2}{f_1} \text{ octaves} \quad (10.22)$$

There is an increase of 1 decade from f_1 to f_2 when $f_2/f_1 = 10$. The frequency band from 1 to $10Hz$ or from 2.5 to $25Hz$ is 1 decade in width. The number of decades from f_1 to f_2 is given by

$$\log \frac{f_2}{f_1} \text{ decades} \qquad (10.23)$$

The dB values of some common numbers are given in Table 10.4. Note that the reciprocals of numbers differ only in sign. Thus, the dB value of 2 is +6 dB and the dB value of 1/2 is −6 dB. The following two properties are illustrated in Table 10.4:

TABLE 10.4 Decibel Values of Some Common Numbers.

Number	Decibels
0.01	−40
0.1	−20
0.5	−6
1.0	0
2.0	6
10.0	20
100.0	40
200.0	46

Property 1. As a number doubles, the decibel value increases by 6 dB. The number 2.70 is twice as large as 1.35, and its decibel value is 6 dB greater. The number 200 is twice as large as 100, and its decibel value is 6 dB greater.

Property 2. As a number increases by a factor of 10, the decibel value increases by 20 dB. The number 100 is 10 times as large as the number 10, and its decibel value is 20 dB greater. The number 200 is 100 times as large as the number 2, and its decibel value is 40 dB greater.

General Frequency Transfer Function Relationships

The frequency transfer function can be written in generalized form as the ratio of polynomials

$$G(j\omega) = \frac{K_m(1+j\omega T_1)(1+j\omega T_2)^r\cdots}{(j\omega)^m(1+j\omega T_a)\left[1+(2\zeta/\omega_n)j\omega+(1/\omega^2)(j\omega)^2\right]\cdots} \qquad (10.24)$$

where K_m is the gain constant. The logarithm of the transfer function is a complex quantity; the real portion is proportional to the Lm, and the complex portion is proportional to the angle. Two separate equations are written, one for the Lm and one for the angle, respectively:

$$\begin{aligned} \mathrm{Lm}G(j\omega) &= \mathrm{Lm}K_m + \mathrm{Lm}(1+j\omega T_1) + r\mathrm{Lm}(1+j\omega T_2) \\ &\quad + \cdots - m\mathrm{Lm}j\omega - \mathrm{Lm}(1+j\omega T_a) \\ &\quad - \mathrm{Lm}\left[1+\frac{2\zeta}{\omega_n}j\omega + \frac{1}{\omega^2}(j\omega)^2\right] - \cdots \end{aligned} \qquad (10.25)$$

$$\begin{aligned} \angle[G(j\omega)] &= \angle[K_m] + \angle[1+j\omega T_1] + r\angle[1+j\omega T_2] \\ &\quad + \cdots - m\angle[j\omega] - \angle[1+j\omega T_a] \\ &\quad - \angle\left[1+\frac{2\zeta}{\omega_n}j\omega + \frac{1}{\omega^2}(j\omega)^2\right] - \cdots \end{aligned} \qquad (10.26)$$

The angle equation may be rewritten as

$$\begin{aligned} \angle[G(j\omega)] &= \angle[K_m] + \tan^{-1}\omega T_1 + r\tan^{-1}\omega T_2 \\ &\quad + \cdots - m90^\circ - \tan^{-1}\omega T_a \\ &\quad - \tan^{-1}\frac{2\zeta\omega/\omega_n}{1-\omega^2/\omega_n^2} - \cdots \end{aligned} \qquad (10.27)$$

The gain K_m is a real number but may be positive or negative; therefore, its angle is correspondingly 0° or 180°. Unless otherwise indicated, a positive value of gain is assumed in this chapter. Both the Lm and the angle given by these equations are functions of frequency. When the Lm and the angle are plotted as functions of the log of frequency, the resulting curves are referred to as the Bode plots or the *Lm diagram and the phase diagram.* Equations 10.25 and 10.26 show that the resultant curves are obtained by the addition and subtraction of the corresponding individual terms in the transfer function equation. The two curves can be combined into a single curve of Lm vs. angle, with frequency ω as a parameter. This curve is called the Nichols or the *log magnitude-angle diagram.*

Drawing the Bode Plots

The properties of frequency-response plots are presented in this section, but the data for these plots usually are obtained from a CAD program. The generalized form of the transfer function as given by Equation 10.24 shows that the numerator and denominator have four basic types of factors:

$$K_m \qquad (10.28)$$

$$(j\omega)^{\pm m} \qquad (10.29)$$

$$(1+j\omega T)^{\pm r} \qquad (10.30)$$

$$\left[1+\frac{2\zeta}{\omega_n}j\omega + \frac{1}{\omega_n^2}(j\omega)^2\right]^{\pm p} \qquad (10.31)$$

Each of these terms except K_m may appear raised to an integral power other than 1. The curves of Lm and angle vs. the log of frequency can easily be drawn for each factor. Then these curves for each factor can be added together graphically to get the curves for the complete transfer function. The procedure can be further simplified by using asymptotic approximations to these curves, as shown in the following pages.

Constants Since the constant K_m is frequency invariant, the plot of

$$\mathrm{Lm}K_m = 20\log K_m \text{ dB}$$

is a horizontal straight line. The constant raises or lowers the Lm curve of the complete transfer function by a fixed amount. The angle, of course, is zero as long as K_m is positive.

$j\omega$ Factors The factor $j\omega$ appearing in the denominator has an Lm

$$\text{Lm}(j\omega)^{-1} = 20\log\left|(j\omega)^{-1}\right| = -20\log\omega \quad (10.32)$$

When plotted against $log\ \omega$, this curve is a straight line with a negative slope of 6 dB/octave or 20 dB/decade. Values of this function can be obtained from Table 1 for several values of ω. The angle is constant and equal to $-90°$. When the factor $j\omega$ appears in the numerator, the Lm is

$$\text{Lm}(j\omega) = 20\log|\omega| = 20\log\omega \quad (10.33)$$

This curve is a straight line with a positive slope of 6 dB/octave or 20 dB/decade. The angle is constant and equal to $+90°$. Notice that the only difference between the curves for $j\omega$ and for $1/j\omega$ is a change in the sign of the slope of the Lm and a change in the sign of the angle. Both Lm curves go through the point 0 dB at $\omega = 1$. For the factor $(j\omega)^{\pm m}$ the Lm curve has a slope of $\pm 6m$ dB/octave or $\pm 20m$ dB/decade, and the angle is constant and equal to $\pm m90°$.

$1+j\omega T$ Factors The factor $1 + j\omega T$ appearing in the denominator has an Lm

$$\begin{aligned}\text{Lm}(1+j\omega T)^{-1} &= 20\log|1+j\omega T|^{-1} \\ &= -20\log\sqrt{\left[1+\omega^2T^2\right]} \quad (10.34)\end{aligned}$$

For very small values of ω, that is, $\omega T \ll 1$,

$$\text{Lm}(1+j\omega T)^{-1} \approx \log 1 = 0\text{dB} \quad (10.35)$$

Thus, the plot of the Lm at small frequencies is the 0 – dB line. For very large values of ω, that is, $\omega T \gg 1$,

$$\text{Lm}(1+j\omega T)^{-1} \approx 20\log|j\omega T|^{-1} = -20\log\omega T \quad (10.36)$$

The value of Equation 10.36 at $\omega = 1/T$ is 0. For values of $\omega > 1/T$ this function is a straight line with a negative slope of $6dB/octave$. Therefore, the asymptotes of the plot of $\text{Lm}(1 + j\omega T)^{-1}$ are two straight lines, one of zero slope below $\omega = 1/T$ and one of $-6dB/octave$ slope above $\omega = 1/T$. These asymptotes are drawn in Figure 10.19.

The frequency at which the asymptotes to the Lm curve intersect is defined as the corner frequency ω_{cf}. The value $\omega_{cf} = 1/T$ is the corner frequency for the function

$$(1+j\omega T)^{\pm r} = (1+j\omega/\omega_{cf})^{\pm r}$$

The exact values of Lm $(1 + j\omega T)^{-1}$ are given in Table 10.5 for several frequencies in the range a decade above and below the corner frequency. The exact curve is also drawn in Figure 10.19 The error, in dB, between the exact curve and the asymptotes is approximately as follows:

- At the corner frequency: 3 dB
- One octave above and below the corner frequency: 1 dB
- Two octaves from the corner frequency: 0.26 dB

TABLE 10.5 Values of $\text{Lm}(1 + j\omega T)^{-1}$ for Several Frequencies.

$\frac{\omega}{\omega_{cf}}$	Exact value, dB	Value of the asymptote, dB	Error, dB
0.1	−0.04	0.	−0.04
0.25	−0.26	0.	−0.26
0.5	−0.97	0.	−0.97
0.76	−2.00	0.	−2.00
1	−3.01	0.	−3.01
1.31	−4.35	−2.35	−2.00
2	−6.99	−6.02	−0.97
4	−12.30	−12.04	−0.26
10	−20.04	−20.0	−0.04

Preliminary design studies are often made by using the asymptotes only. The correction to the straight-line approximation to yield the true Lm curve is shown in Figure 10.20. The phase curve for this function is also plotted in Figure 10.19. At zero frequency the angle is 0°; at the corner frequency $\omega = \omega_{cf}$ the angle is $-45°$; and at infinite frequency the angle is $-90°$. The angle curve is symmetrical about the corner frequency value when plotted against log (ω/ω_{cf}) or $log\ \omega$. Since the abscissa of the curves in Figure 10.19 is ω/ω_{cf}, the shapes of the angle and Lm curves are independent of the time constant T. Thus, when the curves are plotted with the abscissa in terms of ω, changing T just "slides" the Lm and the angle curves left or right so that the -3 dB and the $-45°$ points occur at the frequency $\omega = \omega_{cf}$. The approximation of the phase curve is a straight line drawn through the following three points:

ω/ω_{cf}	0.1	1.0	10
Angle	0°	−45°	−90°

The maximum error resulting from this approximation is about $\pm 6°$. For greater accuracy, a smooth curve is drawn through the points given in Table 10.6.

TABLE 10.6 Angles of $(1 + j\omega/\omega_{cf})^{-1}$ for Key Frequency Points

$\frac{\omega}{\omega_{cf}}$	Angle, deg
0.1	−5.7
0.5	−26.6
1.0	−45.0
2.0	−63.4
10.0	−84.3

The factor $1 + j\omega T$ appearing in the numerator has the Lm

$$\text{Lm}(1+j\omega T) = 20\log\sqrt{(1+\omega^2T^2)}$$

This is the same function as its inverse Lm $(1 + j\omega T)^{-1}$ except that it is positive. The corner frequency is the same, and the angle varies from 0 to 90° as the frequency increases from zero to infinity. The Lm and angle curves for the function $(1 + j\omega T)$ are symmetrical about the abscissa to the curves for $(1 + j\omega T)^{-1}$.

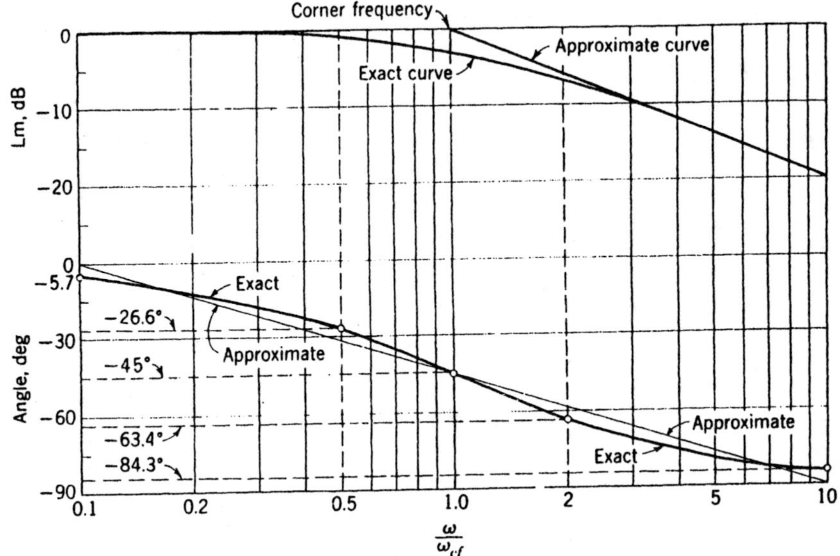

Figure 10.19 Log magnitude and phase diagram for $(1 + j\omega T)^{-1} = [1 + j(\omega/\omega_{cf})]^{-1}$.

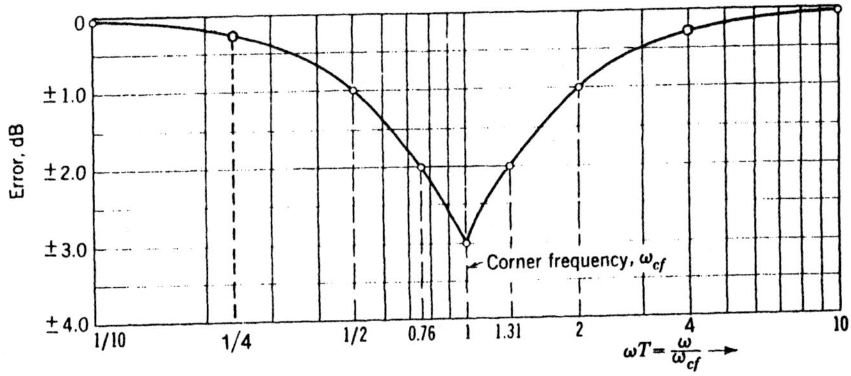

Figure 10.20 Log magnitude correction for $(1 + j\omega T)^{\pm 1}$.

Quadratic Factors Quadratic factors in the denominator of the transfer function have the form

$$\left[1 + \frac{2\zeta}{\omega_n} j\omega + \frac{1}{\omega_n^2}(j\omega)^2\right]^{-1} \tag{10.37}$$

For $\zeta > 1$, the quadratic can be factored into two first-order factors with real zeros which can be plotted in the manner shown previously. But for $\zeta < 1$ Equation 10.37 contains conjugate-complex factors, and the entire quadratic is plotted without factoring:

$$\text{Lm}\left[1 + \frac{2\zeta}{\omega_n} j\omega + \frac{1}{\omega_n^2}(j\omega)^2\right]^{-1} =$$

$$-20 \log\left[\left(1 - \frac{\omega^2}{\omega_n^2}\right)^2 + \left(\frac{2\zeta\omega}{\omega_n}\right)^2\right]^{1/2} \tag{10.38}$$

$$\angle\left[1 + \frac{2\zeta}{\omega_n} j\omega + \frac{1}{\omega_n^2}(j\omega)^2\right]^{-1} = -\tan^{-1}\frac{2\zeta\omega/\omega_n}{1-\omega^2/\omega_n^2} \tag{10.39}$$

From Equation 10.38 it is seen that for very small values of ω, the low-frequency asymptote is represented by Lm= 0 dB. For very high values of frequency, the high-frequency asymptote has a slope of $-40dB/decade$. The asymptotes cross at the corner frequency $\omega_{cf} = \omega_n$.

From Equation 10.38 it is seen that a resonant condition exists in the vicinity of $\omega = \omega_n$, where the peak value of the Lm > 0 dB. Therefore there may be a substantial deviation of the Lm curve from the straight-line asymptotes, depending on the value of ζ. A family of Lm curves of several values of $\zeta < 1$ is plotted in Figure 10.21. For the appropriate ζ, the Lm curve can be drawn by selecting sufficient points from Figure 10.21 or computed from Equation 10.38.

The phase-angle curve for this function also varies with ζ. At zero frequency the angle is 0°; at the corner frequency the angle is −90°; and at infinite frequency the angle is −180°. A family of phase-angle curves for various values of $\zeta < 1$ is plotted in Figure 10.21. Enough values to draw the appropriate phase-angle curve can be taken from Figure 10.21 or computed from Equation 10.39. When the quadratic factor appears in the numerator, the magnitudes of the Lm and phase angle are the same as those in Figure 10.21, except that they are changed in sign.

The $\text{Lm}[1 + j2\zeta/\omega_n + (j\omega/\omega_n)^2]^{-1}$ with $\zeta < 0.707$ has a peak value. The magnitude of this peak value and the frequency at which it occurs are important terms. These values, derived in

Section 10.3.8 [see 11, Sect. 9.3], are repeated here:

$$M_m = \frac{1}{2\zeta\sqrt{(1-\zeta^2)}} \tag{10.40}$$

$$\omega_m = \omega_n\sqrt{(1-2\zeta^2)} \tag{10.41}$$

Note that the peak value M_m depends only on the damping ratio ζ. Since Equation 10.41 is meaningful only for real values of ω_m, the curve of M vs. ω has a peak value greater than unity only for $\zeta < 0.707$. The frequency at which the peak value occurs depends on both the damping ratio ζ and the undamped natural frequency ω_n. This information is used when adjusting a control system for good response characteristics. These characteristics are discussed in Sections 10.3.8 and 10.3.9.

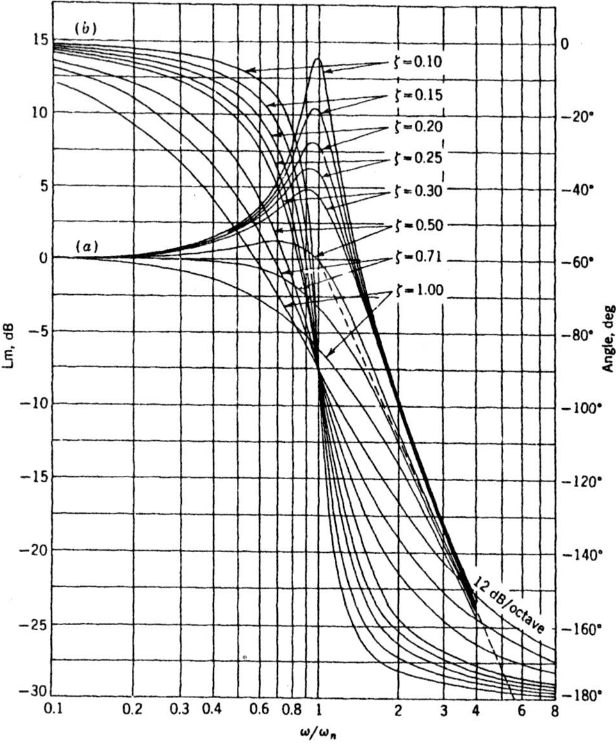

Figure 10.21 Log magnitude and phase diagram for $[1 + j2\zeta\omega/\omega_n + (j\omega/\omega_n)^2]^{-1}$.

The Lm curves for poles and zeros lying in the right-half (RH) s−plane are the same as those for poles and zeros located in the left-half (LH) s−plane. However, the angle curves are different. For example, the angle for the factor $(1 - j\omega T)$ varies from 0 to $-90°$ as ω varies from zero to infinity. Also, if ζ is negative, the quadratic factor of Equation 10.37 contains RH s−plane poles or zeros. Its angle varies from $-360°$ at $\omega = 0$ to $-180°$ at $\omega = \infty$. This information can be obtained from the pole-zero diagram [11, Sect. 4.12] with all angles measured in a *counter-clockwise* (CCW) direction. Some CAD packages do not consistently use a CCW measurement direction, thus resulting in inaccurate angle values.

10.3.3 System Type and Gain as Related to Lm Curves

The steady-state error of a closed-loop system depends on the system type and the gain. The system error coefficients are determined by these two characteristics [11, Chap. 6]. For any given Lm curve, the system type and gain can be determined. Also, with the transfer function given so that the system type and gain are known, they can expedite drawing the Lm curve. This is described for Type 0, 1, and 2 systems.

Type 0 System

A first-order Type 0 system has a transfer function of the form

$$G(j\omega) = \frac{K_0}{1 + j\omega T_a}$$

At low frequencies, $\omega < 1/Ta$, $\text{Lm}G(j\omega) \approx 20 \log K_0$, which is a constant. The slope of the Lm curve is zero below the corner frequency $\omega_1 = 1/T_a$ and −20 dB/decade above the corner frequency. The Lm curve is shown in Figure 10.22.

For a Type 0 system the characteristics are as follows:

1. The slope at low frequencies is zero.
2. The magnitude at low frequencies is $20 \log K_0$.
3. The gain K_0 is the steady-state step error coefficient.

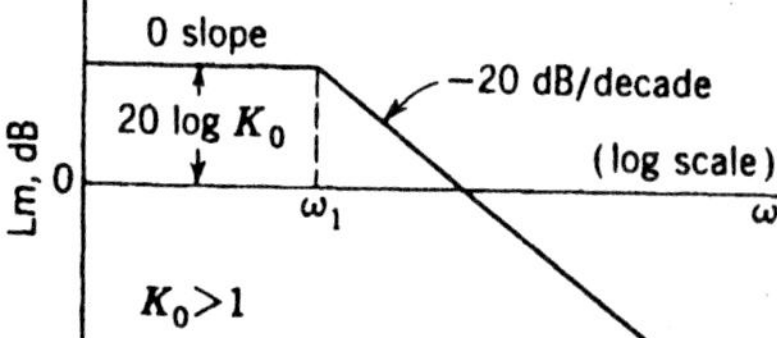

Figure 10.22 Log magnitude plot for $G(j\omega) = K_0/(1 + j\omega T_a)$.

Type 1 System

A second-order Type 1 system has a transfer function of the form

$$G(j\omega) = \frac{K_1}{j\omega(1 + j\omega T_a)}$$

At low frequencies, $\omega < 1/T_a$, $\text{Lm}[G(j\omega)] \approx \text{Lm}(K_1/j\omega) = \text{Lm } K_1 - \text{Lm } j\omega$, which has a slope of −20 dB/decade. At $\omega = K_1$, $\text{Lm}(K1/j\omega) = 0$. If the corner frequency $\omega_1 = 1/T_a$ is greater than K_1, the low-frequency portion of the curve of slope −20 dB/decade crosses the 0−dB axis at a value of $\omega_x = K_1$, as shown in Figure 10.23a. If the corner frequency is less than K_1, the low-frequency portion of the curve of slope −20 dB/decade may be extended until it does cross the 0−dB axis. The value of the frequency at which the extension crosses the 0−dB axis is $\omega_x = K_1$. In other words, the plot Lm $(K_1/j\omega)$ crosses the 0−dB value at $\omega_x = K_1$, as illustrated in Figure 10.23b.

At $\omega = 1$, Lm $j\omega = 0$; therefore, Lm $(K_1/j\omega)_{\omega=1} = 20 \log(K_1)$. For $T_a < 1$ this value is a point on the slope of −20 dB/decade. For $T_a > 1$ this value is a point on the extension of

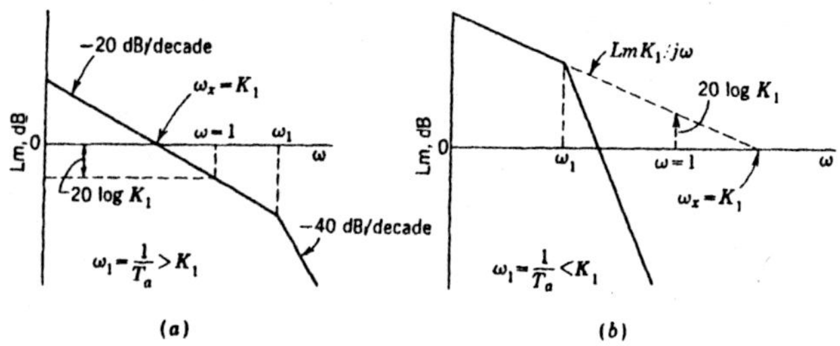

Figure 10.23 Log magnitude plot for $G(j\omega) = K_1/j\omega(1 + j\omega T_a)$.

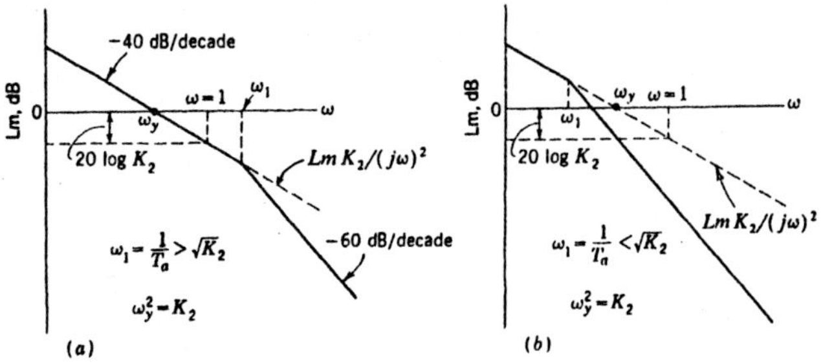

Figure 10.24 Log magnitude plot for $G(j\omega) = K_2/(j\omega)^2(1 + j\omega T_a)$.

the initial slope, as shown in Figure 10.23b. The frequency ω_x is smaller or larger than unity according as K_1 is smaller or larger than unity. For a Type 1 system the characteristics are as follows:

1. The slope at low frequencies is -20 dB/decade.
2. The intercept of the low-frequency slope of -20 dB/decade (or its extension) with the 0–dB axis occurs at the frequency ω_x, where $\omega_x = K_1$.
3. The value of the low-frequency slope of -20 dB/decade (or its extension) at the frequency $\omega = 1$ is equal to $20\log(K_1)$.
4. The gain K_1 is the steady-state ramp error coefficient.

Type 2 System

A Type 2 system has a transfer function of the form

$$G(j\omega) = \frac{K_2}{(j\omega)^2(1 + j\omega T_a)}$$

At low frequencies, $\omega < 1/T_a$, $\text{Lm}\,[G(j\omega)] = \text{Lm}\,[K_2/(j\omega)^2] = \text{Lm}[K_2] - \text{Lm}[j\omega]^2$, for which the slope is $-40dB/decade$. At $\omega^2 = K_2$, Lm $[K_2/(j\omega)^2] = 0$; therefore the intercept of the initial slope of $-40dB/decade$ (or its extension, if necessary) with the 0–dB axis occurs at the frequency ω_y, where $\omega_y^2 = K_2$.

At $\omega = 1$, Lm $[j\omega]^2 = 0$; therefore, Lm $[K_2/(\omega)^2]_{\omega=1} = 20\log[K_2]$. This point occurs on the initial slope or on its extension, according as $\omega_1 = 1/T_a$ is larger or smaller than $\sqrt{K_2}$. If $K_2 > 1$, the quantity $20\log[K_2]$ is positive, and if $K_2 < 1$, the quantity $20\log[K_2]$ is negative.

The Lm curve for a Type 2 transfer function is shown in Figure 10.24. The determination of gain K_2 from the graph is shown. For a Type 2 system the characteristics are as follows:

1. The slope at low frequencies is -40 dB/decade.
2. The intercept of the low-frequency slope of -40 dB/decade (or its extension, if necessary) with the 0–dB axis occurs at a frequency ω_y, where $\omega_y^2 = K_2$.
3. The value on the low-frequency slope of -40 dB/decade (or its extension) at the frequency $\omega = 1$ is equal to $20\log[K_2]$.
4. The gain K_2 is the steady-state parabolic error coefficient.

10.3.4 Experimental Determination of Transfer Functions

The magnitude and angle of the ratio of the output to the input can [4, 8] be obtained experimentally for a steady-state sinusoidal input signal at a number of frequencies. For stable plants, the Bode data for the plant is used to obtain the exact Lm and angle diagram. Asymptotes are drawn on the exact Lm curve, using the fact that their slopes must be multiples of ± 20 dB/decade. From these asymptotes and their intersections, the system type and the approximate time constants are determined. In this manner, the transfer function of the system can be synthesized. Care must be exercised in determining whether any zeros of the transfer function are in the RH s–plane. A system that has no open-loop zeros in the RH s–plane is defined as a *minimum-phase* system, [5], [11], [12] and all factors have the form $(1 + Ts)$ and/or $(1 + As + Bs^2)$. A system that has open-loop zeros in the RH s–plane is a *nonminimum-phase* system. The stability is determined by the location of the poles and does not affect the designation of minimum or nonminimum phase. The angular variation for poles or zeros in the RH s–plane is different from those in the LH plane [5]. For this situation, one or more terms in the transfer function have the form $(1 - Ts)$ and/or $(1 \pm As \pm Bs^2)$. Care must be exercised in interpreting the angle plot to determine whether any factors of the transfer function lie in the RH s–plane. Many practical systems are minimum phase. Unstable plants must be handled with care. That is, first a stabilizing compensator must be added to form a stable closed-loop system. From the experimental data for the stable closed-loop system, the plant transfer function is determined by using the known compensator transfer function.

10.3.5 Direct Polar Plots

The magnitude and angle of $G(j\omega)$, for sufficient frequency points, are readily obtainable from the $\text{Lm}[G(j\omega)]$ and $\angle[G(j\omega)]$ vs. $\log[\omega]$ curves or by the use of a CAD program. It is also possible to visualize the complete shape of the frequency-response curve from the pole-zero diagram because the angular contribution of each pole and zero is readily apparent. The polar plot of $G(j\omega)$ is called the *direct polar plot*. The polar plot of $[G(j\omega)]^{-1}$ is called the *inverse polar plot* [10].

Lag-Lead Compensator [11]

The compensator transfer function is

$$G_c(s) = \frac{1+(T_1+T_2)s+T_1T_2s^2}{1+(T_1+T_2+T_{12})s+T_1T_2s^2} \quad (10.42)$$

As a function of frequency, the transfer function is

$$G_c(j\omega) = \frac{E_0(j\omega)}{E_i(j\omega)} = \frac{(1-\omega^2T_1T_2)+j\omega(T_1+T_2)}{(1-\omega^2T_1T_2)+j\omega(T_1+T_2+T_{12})} \quad (10.43)$$

By the proper choice of the time constants, the compensator acts as a lag network [i.e., the output signal $E_0(j\omega)$ lags the input signal $E_i(j\omega)$] in the lower-frequency range of 0 to ω_x and as a lead network [i.e., the output signal leads the input signal] in the higher-frequency range of ω_x to ∞. The polar plot of this transfer function is a circle with its center on the real axis and lying in the first and fourth quadrants. Its properties are

1. $\lim_{\omega\to 0}[G(j\omega T_1)] \to 1\angle 0^\circ$
2. $\lim_{\omega\to \infty}[G(j\omega T_1)] \to 1\angle 0^\circ$
3. At $\omega = \omega_x$, for which $\omega_x^2 T_1T_2 = 1$, Equation 10.43 yields the value

$$\begin{aligned} G(j\omega_xT_1) &= \frac{T_1+T_2}{T_1+T_2+T_{12}} \\ &= |G(j\omega_xT_1)|\angle 0^\circ \end{aligned} \quad (10.44)$$

Note that Equation 10.44 represents the minimum value of the transfer function in the whole frequency spectrum. For frequencies below ω_x the transfer function has a negative or lag angle. For frequencies above ω_x it has a positive or lead angle.

Type 0 Feedback Control System

The field-controlled servomotor [10] illustrates a typical Type 0 device. It has the transfer function

$$G(j\omega) = \frac{C(j\omega)}{E(j\omega)} = \frac{K_0}{(1+j\omega T_f)(1+j\omega T_m)} \quad (10.45)$$

$$\text{Note: } G(j\omega) \to \begin{pmatrix} K_0\angle 0^\circ & \text{as} & \omega\to 0^+ \\ 0\angle -180^\circ & \text{as} & \omega\to\infty \end{pmatrix} \quad (10.46)$$

Also, for each term in the denominator the angular contribution to $G(j\omega)$, as ω goes from 0 to ∞, goes from 0 to -90°. Thus, the polar plot of this transfer function must start at $G(j\omega) = K_0\angle 0^\circ$ for $\omega = 0$ and proceed first through the fourth and then through the third quadrants to $\lim_{\omega\to\infty} G(j\omega) = 0\angle -180^\circ$ as the frequency approaches infinity. In other words, the angular variation of $G(j\omega)$ is continuously decreasing, going in a clockwise (CW) direction from 0° to -180°. The exact shape of this plot is determined by the particular values of the time constants T_f and T_m.

Consider the transfer function

$$G(j\omega) = \frac{K_0}{(1+j\omega T_f)(1+j\omega T_m)(1+j\omega T)} \quad (10.47)$$

In this case, when $\omega\to\infty$, $G(j\omega)\to 0\angle -270^\circ$. Thus, the curve crosses the negative real axis at a frequency ω_x for which the imaginary part of the transfer function is zero. When a term of the form $(1+j\omega T)$ appears in the numerator, the transfer function experiences an angular variation of 0 to 90° (a CCW rotation) as the frequency is varied from 0 to ∞. Thus, the angle of $G(j\omega)$ may not change continuously in one direction. Also, the resultant polar plot may not be as smooth as the one for Equations 10.45 and 10.47.

In the same manner, a quadratic in either the numerator or the denominator of a transfer function results in an angular contribution of 0 to $\pm 180^\circ$, respectively, and the polar plot of $G(j\omega)$ is affected accordingly. It can be seen from the examples that the polar plot of a Type 0 system always starts at a value K_0 (step error coefficient) on the positive real axis for $\omega = 0$ and ends at zero magnitude (for $n > \omega$) and tangent to one of the major axes at $\omega = \infty$. The final angle is -90° times the order n of the denominator minus the order w of the numerator of $G(j\omega)$.

Type 1 Feedback Control System

A typical Type 1 system containing only poles is

$$G(j\omega) = \frac{C(j\omega)}{E(j\omega)} = \frac{K_1}{j\omega(1+j\omega T_m)(1+j\omega T_c)(1+j\omega T_q)} \quad (10.48)$$

$$\text{Note: } G(j\omega) \to \begin{pmatrix} \infty\angle -90^\circ & \text{as} & \omega\to 0^+ \\ 0\angle -360^\circ & \text{as} & \omega\to\infty \end{pmatrix} \quad (10.49)$$

Note that the $j\omega$ term in the denominator contributes the angle -90° to the total angle of $G(j\omega)$ for all frequencies. Thus, the basic difference between Equations 10.47 and 10.48 is the presence of the term $j\omega$ in the denominator of the latter equation. Since all the $(1+j\omega T)$ terms of Equation 10.48 appear in the denominator, the angle of the polar plot of $G(j\omega)$ decreases continuously (CW) in the same direction from -90 to -360° as ω increases from 0 to ∞. The frequency of the crossing point on the negative real axis of the $G(j\omega)$ function is that value of frequency ω_x for which the imaginary part of $G(j\omega)$ is equal to zero. The real-axis crossing point is very important because it determines closed-loop stability, as described in later sections dealing with system stability.

Type 2 Feedback Control System

The transfer function of a Type 2 system is

$$G(j\omega) = \frac{C(j\omega)}{E(j\omega)} = \frac{K_2}{(j\omega)^2(1+j\omega T_f)(1+j\omega T_m)} \quad (10.50)$$

Its properties are

$$G(j\omega) \to \begin{pmatrix} \infty\angle -180^\circ & \text{as} & \omega\to 0^+ \\ 0\angle -360^\circ & \text{as} & \omega\to +\infty \end{pmatrix} \quad (10.51)$$

The presence of the $(j\omega)^2$ term in the denominator contributes -180° to the total angle of $G(j\omega)$ for all frequencies. The polar plot for the transfer function of Equation 10.50 is a smooth curve whose angle $\phi(\omega)$ decreases continuously from -180 to -360°. The introduction of an additional pole and a zero can alter the shape of the polar plot. It can be shown that as $\omega \rightarrow 0^+$, the polar plot of a Type 2 system is below the real axis if

$$\sum T_{numerator} - \sum T_{denominator} \quad (10.52)$$

is a positive value, and above the real axis if it is a negative value.

Summary: Direct Polar Plots

To obtain the direct polar plot of a system's forward transfer function, the following steps are used to determine the key parts of the curve. Figure 10.25 shows the typical polar plot shapes for different system types.

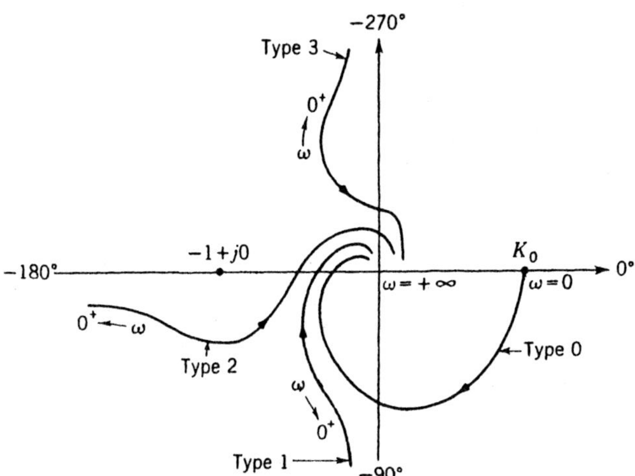

Figure 10.25 A summary of direct polar plots of different types of systems.

Step 1. The forward transfer function has the general form

$$G(j\omega) = \frac{K_m(1+j\omega T_a)(1+j\omega T_b)\cdots(1+j\omega T_w)}{(j\omega)^m(1+j\omega T_1)(1+j\omega T_2)\cdots(1+j\omega T_u)} \quad (10.53)$$

For this transfer function, the system type is equal to the value of m and determines the portion of the polar plot representing the $\lim_{\omega\rightarrow 0}[G(j\omega)]$. The low-frequency polar-plot characteristics (as $\omega \rightarrow 0$) of the different system types are determined by the angle at $\omega = 0$, i.e., $\angle G(j0) = m(-90°)$.
Step 2. The high-frequency end of the polar plot can be determined as follows:

$$\lim_{\omega\rightarrow +\infty}[G(j\omega)] = 0\angle[(w - m - u)90°] \quad (10.54)$$

Note that since the degree of the denominator of Equation 10.53 is always greater than the degree of the numerator, the angular condition of the high-frequency point ($\omega = \infty$) is approached in the CW sense. The plot ends at the origin and is tangent to the axis determined by Equation 10.54. Tangency may occur on either side of the axis.
Step 3. The asymptote that the low-frequency end approaches, for a Type 1 system, is determined by taking the limit as $\omega \rightarrow 0$ of the real part of the transfer function.
Step 4. The frequencies at the points of intersection of the polar plot with the negative real axis and the imaginary axis are determined, respectively, by

$$Im[G(j\omega)] = 0 \quad (10.55)$$

$$Re[G(j\omega)] = 0 \quad (10.56)$$

Step 5. If there are no frequency-dependent terms in the numerator of the transfer function, the curve is a smooth one in which the angle of $G(j\omega)$ continuously decreases as ω goes from 0 to ∞. With time constants in the numerator, and depending upon their values, the angle may not change continuously in the same direction, thus creating "dents" in the polar plot.
Step 6. As is seen later, it is important to know the exact shape of the polar plot of $G(j\omega)$ in the vicinity of the $-1 \pm j0$ point and the crossing point on the negative real axis.

10.3.6 Nyquist's Stability Criterion

The Nyquist stability criterion [2], [6], [8] provides a simple graphical procedure for determining closed-loop stability from the frequency-response curves of the open-loop transfer function $G(j\omega)H(j\omega)$ (for the case of no poles or zeros on the imaginary axis, etc.). The application of this method in terms of the polar plot is covered in this section; application in terms of the log magnitude-angle (Nichols) diagram is covered in a later section.

For a stable closed-loop system, the roots of the characteristic equation

$$B(s) = 1 + G(s)H(s) = 0 \quad (10.57)$$

cannot be permitted to lie in the RH s−plane or on the $j\omega$ axis. In terms of $G = N_1/D_1$ and $H = N_2/D_2$, Equation 10.57 becomes

$$\begin{aligned} B(s) &= 1 + \frac{N_1N_2}{D_1D_2} = \frac{D_1D_2 + N_1N_2}{D_1D_2} \\ &= \frac{(s-Z_1)(s-Z_2)\cdots(s-Z_n)}{(s-p_1)(s-p_2)\cdots(s-p_n)} \end{aligned} \quad (10.58)$$

Note that the numerator and denominator of $B(s)$ have the same degree and the *poles of the open-loop transfer function* $G(s)H(s)$ *are the poles of* $B(s)$. The closed-loop transfer function of the system is

$$\frac{C(s)}{R(s)} = \frac{G(s)}{1+G(s)H(s)} = \frac{N_1D_2}{D_1D_2+N_1N_2} \quad (10.59)$$

The denominator of $C(s)/R(s)$ is the same as the numerator of $B(s)$. The condition for stability may therefore be restated as: For a stable system none of the zeros of $B(s)$ can lie in the RH s−plane or on the imaginary axis. Nyquist's stability criterion relates the number of zeros and poles of $B(s)$ that lie in the RH s−plane to the polar plot of $G(s)H(s)$.

Limitations

In this analysis it is assumed that all the control systems are inherently linear or that their limits of operation are confined to give a linear operation. This yields a set of linear differential equations that describe the dynamic performance of the systems. Because of the physical nature of feedback control systems, the degree of the denominator $D_1 D_2$ is equal to or greater than the degree of the numerator $N_1 N_2$ of the open-loop transfer function $G(s)H(s)$. Mathematically, this means that $\lim_{s\to\infty}[G(s)H(s)] \to 0$ or a constant. These two factors satisfy the necessary limitations to the generalized Nyquist stability criterion.

Generalized Nyquist's Stability Criterion

Consider a closed contour Q such that the whole RH s-plane is encircled (see Figure 10.26a with $\epsilon \to 0$), thus enclosing all zeros and poles of $B(s)$ that have positive real parts. The theory of complex variables used in the rigorous derivation requires that the contour Q must not pass *through any poles or zeros of* $B(s)$. When these results are applied to the contour Q, the following properties are noted:

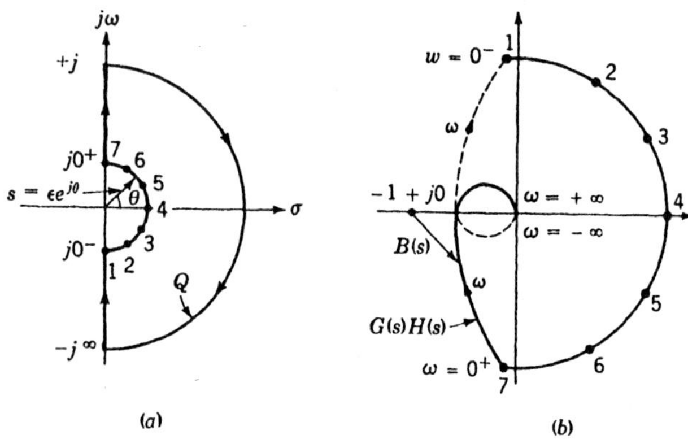

Figure 10.26 (a) The contour Q, which encircles the right-half s-plane; (b) complete plot for $G(s)H(s) = \frac{K_1}{s(1+T_1 s)(1+T_2 s)}$.

1. The total number of CW rotations of $B(s)$ due to its zeros is equal to the total number of zeros Z_R in the RH s-plane.
2. The total number of CCW rotations of $B(s)$ due to its poles is equal to the total number of poles P_R in the RH s-plane.
3. The *net* number of rotations N of $B(s) = 1 + G(s)H(s)$ about the origin is equal to the total number of poles P_R minus the total number of zeros Z_R in the RH s-plane. N may be positive (CCW), negative (CW), or zero.

The essence of these three conclusions can be represented by

$$N = \frac{\text{phase change of } [1+G(s)H(s)]}{2\pi} = P_R - Z_R \quad (10.60)$$

where CCW rotation is defined as being positive and CW rotation is negative. In order for $B(s)$ to realize a net rotation N, the directed line segment representing $B(s)$ (see Figure 10.27a) must rotate about the origin $360N°$, or N complete revolutions. Solving for Z_R in Equation 10.60 yields $Z_R = P_R - N$. Since $B(s)$ can have no zeros Z_R in the RH s-plane for a stable system, it is therefore concluded that, *for a stable system, the net number of rotations of* $B(s)$ *about the origin must be CCW and equal to the number of poles* P_R *that lie in the RH plane.* In other words, if $B(s)$ experiences a net CW rotation, this indicates that $Z_R > P_R$, where $P_R \geq 0$, and thus the closed-loop system is unstable. If there are zero net rotations, then $Z_R = P_R$ and the system may or may not be stable, according as $P_R = 0$ or $P_R > 0$.

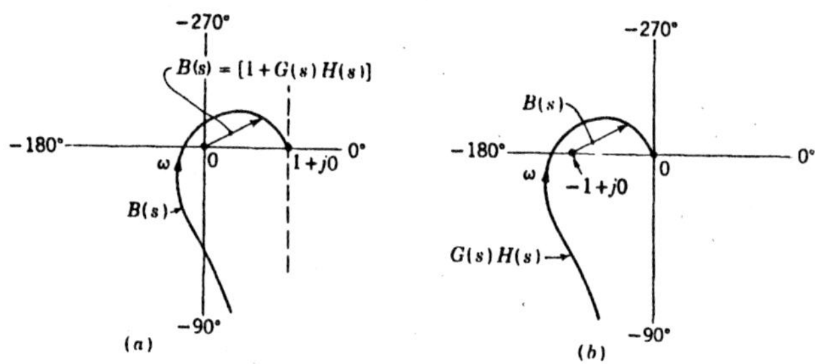

Figure 10.27 A change of reference for $B(s)$.

Obtaining a Plot of $B(s)$

Figures 10.27a and 10.27b show a plot of $B(s)$ and a plot of $G(s)H(s)$, respectively. By moving the origin of Figure 10.27b to the $-1 + j0$ point, the curve is now equal to $1 + G(s)H(s)$, which is $B(s)$. Since $G(s)H(s)$ is known, this function is plotted and then the origin is moved to the -1 point to obtain $B(s)$. In general, the open-loop transfer functions of many physical systems do not have any poles P_R in the RH s-plane. In this case, $Z_R = N$. *Thus, for a stable system the net number of rotations about the* $-1 + j0$ *point must be zero when there are no poles of* $G(s)H(s)$ *in the RH* s*-plane.*

Analysis of Path Q

In applying Nyquist's criterion, the whole RH s-plane must be encircled to ensure the inclusion of all poles or zeros in this portion of the plane. In Figure 10.26 the entire RH s-plane is enclosed by the closed path Q which is composed of the following four segments:

1. One segment is the imaginary axis from $-j\infty$ to $j0^-$.
2. The second segment is the semicircle of radius $\epsilon \to 0$.
3. The third segment is the imaginary axis from $j0^+$ to $+j\infty$.
4. The fourth segment is a semicircle of infinite radius that encircles the entire RH s-plane.

The portion of the path along the imaginary axis is represented mathematically by $s = j\omega$. Thus, replacing s by $j\omega$ in Equation 10.58 and letting ω take on all values from $-\infty$ to $+\infty$ gives the portion of the $B(s)$ plot corresponding to that portion of the closed contour Q that lies on the imaginary axis.

One of the requirements of the Nyquist criterion is that $\lim_{s\to\infty}[G(s)H(s)] \to 0$ or a constant. Thus, $\lim_{s\to\infty}[B(s)] = \lim_{s\to\infty}[1 + G(s)H(s)] \to 1$ or 1 plus the constant. As a consequence, the segment of the closed contour represented by the semicircle of infinite radius, the corresponding portion of the $B(s)$ plot is a fixed point. As a result, the movement along only the imaginary axis from $-j\infty$ to $+j\infty$ results in the same net rotation of $B(s)$ as if the whole contour Q were considered. *In other words, all the rotation of $B(s)$ occurs while the point O, in Figure 10.26a, goes from $-j\infty$ to $+j\infty$ along the imaginary axis.* More generally, this statement applies only to those transfer functions $G(s)H(s)$ that conform to the limitations stated earlier in this section [6].

Effect of Poles at the Origin on the Rotation of B(s)

The manner in which the $\omega = 0^-$ and $\omega = 0^+$ portions of the plot in Figure 10.26a are joined is now investigated for those transfer functions $G(s)H(s)$ that have s^m in the denominator. Consider the transfer function with positive values of T_1 and T_2 :

$$G(s)H(s) = \frac{K_1}{s(1+T_1s)(1+T_2s)} \tag{10.61}$$

The direct polar plot of $G(j\omega)H(j\omega)$ of this function is obtained by substituting $s = j\omega$ into Equation 10.61, as shown in Figure 10.26b. The plot is drawn for both positive and negative frequency values. *The polar plot drawn for negative frequencies* $(0^- > \omega > -\infty)$ is the conjugate of the plot drawn for positive frequencies. This means that the curve for negative frequencies is symmetrical to the curve for positive frequencies, with the real axis as the axis of symmetry.

The closed contour Q of Figure 10.26a, in the vicinity of $s = 0$, has been modified as shown. In other words, the point O is moved along the negative imaginary axis from $s = -j\infty$ to a point where $s = -j\epsilon = 0^-\angle - \pi/2$ becomes very small. Then the point O moves along a semicircular path of radius $s = \epsilon e^{j\theta}$ in the RH $s-$plane with a very small radius ϵ until it reaches the positive imaginary axis at $s = +j\epsilon = j0^+ = 0^+\angle\pi/2$. From here the point O proceeds along the positive imaginary axis to $s = +j\infty$. Then, letting the radius approach zero, $\epsilon \to 0$, for the semicircle around the origin ensures the inclusion of all poles and zeros in the RH $s-$plane. To complete the plot of $B(s)$ in Figure 10.27, the effect of moving point O on this semicircle around the origin must be investigated. For the semicircular portion of the path Q represented by $s = \epsilon e^{j\theta}$, where $\epsilon \to 0$ and $-\pi/2 \le \theta \le \pi/2$, Equation 10.61 becomes

$$G(s)H(s) = \frac{K_1}{s} = \frac{K_1}{\epsilon e^{j\theta}} = \frac{K_1}{\epsilon}e^{-j\theta} = \frac{K_1}{\epsilon}e^{j\psi} \tag{10.62}$$

where $K_1/\epsilon \to \infty$ as $\epsilon \to 0$, and $\psi = -\theta$ goes from $\pi/2$ to $-\pi/2$ as the directed segment s goes CCW from $\epsilon\angle - \pi/2$ to $\epsilon\angle + \pi/2$. Thus, in Figure 10.26b, the end points from $\omega \to 0^-$ and $\omega \to 0^+$ are joined by a semicircle of infinite radius in the first and fourth quadrants. Figure 10.26b shows the completed contour of $G(s)H(s)$ as the point O moves along the contour Q in the $s-$plane in the CW direction. When the origin is moved to the $-1 + j0$ point, the curve becomes $B(s)$. The plot of $B(s)$ in Figure 10.26b does not encircle the $-1 + j0$ point; therefore, the encirclement N is zero. From Equation 10.61 there are no poles within Q; that is, $P_R = 0$. Thus, when Equation 10.60 is applied, $Z_R = 0$ and the closed-loop system is stable.

Transfer functions that have the term s^m in the denominator have the general form, with $s = \epsilon e^{j\theta}$ as $\epsilon \to 0$,

$$\begin{aligned} G(s)H(s) &= \frac{K_m}{s^m} = \frac{K_m}{(\epsilon^m)e^{jm\theta}} \\ &= \frac{K_m}{\epsilon^m}e^{-jm\theta} = \frac{K_m}{\epsilon^m}e^{jm\psi} \end{aligned} \tag{10.63}$$

where $m = 1, 2, 3, 4, \ldots$. It is seen from Equation 10.63 that, as s moves from 0^- to 0^+, the plot of $G(s)H(s)$ traces m CW semicircles of infinite radius about the origin. Since the polar plots are symmetrical about the real axis, all that is needed is to determine the shape of the plot of $G(s)H(s)$ for a range of values of $0 < \omega < +\infty$. The net rotation of the plot for the range of $-\infty < \omega < +\infty$ is twice that of the plot for the range of $0 < \omega < +\infty$.

When G(jω)H(jω) Passes through the Point -1 + j0

When the curve of $G(j\omega)H(j\omega)$ passes through the $-1 + j0$ point, the number of encirclements N is indeterminate. This corresponds to the condition where $B(s)$ has zeros on the imaginary axis. A necessary condition for applying the Nyquist criterion is that the path encircling the specified area must not pass through any poles or zeros of $B(s)$. When this condition is violated, the value for N becomes indeterminate and the Nyquist stability criterion cannot be applied. Simple imaginary zeros of $B(s)$ mean that the closed-loop system will have a continuous steady-state sinusoidal component in its output that is independent of the form of the input. Unless otherwise stated, this condition is considered unstable.

Nyquist's Stability Criterion Applied to Systems Having Dead Time

The transfer function representing transport lag (dead time) is

$$G_\tau(s) = e^{-\tau s} \to G_\tau(j\omega) = e^{-j\omega\tau} = 1\angle - \omega\tau \tag{10.64}$$

It has a magnitude of unity and a negative angle whose magnitude increases directly in proportion to frequency. The polar plot of Equation 10.64 is a unit circle that is traced indefinitely, as shown in Figure 10.28a. The corresponding Lm and phase-angle diagram shows a constant value of 0dB and a phase angle that decreases with frequency. When the contour Q is traversed and the polar-plot characteristic of dead time, shown in Figure 10.28, is included, the effects on the complete polar plot are as follows:

1. In traversing the imaginary axis of the contour Q between $0^+ < \omega < +\infty$, the portion of the polar plot of $G(j\omega)H(j\omega)$ in the third quadrant is shifted CW, closer to the $-1 + j0$ point (see Figure 10.28c). Thus, if the dead time is increased sufficiently, the $-1 + j0$ point is enclosed by the polar plot and the system becomes unstable.

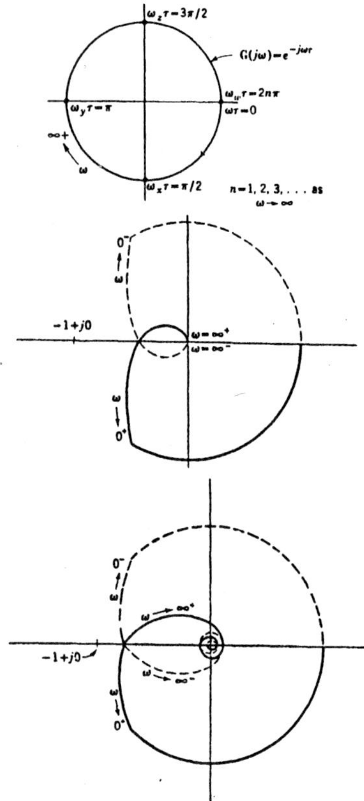

Figure 10.28 (a) Polar plot characteristic for transport lag, Equation 10.64; (b) polar plot for Equation 10.64a without transport lag ($\tau = 0$); (c) destabilizing effect of transport lag in Equation 10.64a;

2. As $\omega \to +\infty$, the magnitude of the angle contributed by the transport lag increases indefinitely. This yields a spiraling curve as $|G(j\omega)H(j\omega)| \to 0$.

A transport lag therefore tends to make a system less stable. This is illustrated for the transfer function

$$G(s)H(s) = \frac{K_1 e^{-\tau s}}{s(1+T_1 s)(1+T_2 s)} \tag{10.64a}$$

Figure 10.28b shows the polar plot without transport lag; Figure 10.28c shows the destabilizing effect of transport lag.

10.3.7 Definitions of Phase Margin and Gain Margin and Their Relation to Stability

The stability and approximate degree of stability [11] can be determined from the Lm and phase diagram. The stability characteristic is specified in terms of the following quantities:

- **Gain crossover.** This is the point on the plot of the transfer function at which the magnitude of $G(j\omega)$ is unity [$\text{Lm}G(j\omega) = 0$ dB]. The frequency at gain crossover is called the *phase-margin frequency* ω_ϕ.
- **Phase margin angle** γ. This angle is 180° plus the negative trigonometrically considered angle of the transfer function at the gain crossover point. It is designated as the angle γ, which can be expressed as $\gamma = 180° + \phi$, where $\angle[G(j\omega_\phi)] = \phi$ is negative.
- **Phase crossover.** This is the point on the plot of the transfer function at which the phase angle is $-180°$. The frequency at which phase crossover occurs is called the *gain-margin frequency* ω_c.
- **Gain margin.** The gain margin is the factor a by which the gain must be changed in order to produce instability., i.e.,

$$\begin{aligned} |G(j\omega_c)|a &= 1 \to |G(j\omega)| = \frac{1}{a} \\ \to \text{Lm}\, a &= -\text{Lm}G(j\omega_c) \end{aligned} \tag{10.65}$$

These quantities are illustrated in Figure 10.29 on both the Lm and the polar curves. Note the algebraic sign associated with these two quantities as marked on the curves. Figures 10.29a and 10.29b represent a stable system, and Figures 10.29c and 10.29d represent an unstable system. The phase margin angle is the amount of phase shift at the frequency ω_ϕ that would just produce instability. The γ for minimum-phase systems must be positive for a stable system, whereas a negative γ means that the system is unstable.

It is shown later that γ is related to the effective damping ratio ζ of the system. Satisfactory response is usually obtained in the range of $40° \le \gamma \le 60°$. As an individual acquires experience, the value of γ to be used for a particular system becomes more evident. This guideline for system performance applies only to those systems where behavior is that of an equivalent second-order system. The gain margin must be positive when expressed in decibels (greater than unity as a numeric) for a stable system. A negative gain margin means that the system is unstable. The damping ratio ζ of the system is also related to the gain margin. However, γ gives a better estimate of damping ratio, and therefore of the transient overshoot of the system, than does the gain margin.

The values of ω_ϕ, γ, ω_c, and Lm a are also readily identified on the Nichols plot as shown in Figure 10.30 and described in the next section. Further information about the speed of response of the system can be obtained from the maximum value of the control ratio and the frequency at which this maximum occurs. The relationship of stability and gain margin is modified for a conditionally stable system. [6] Instability can occur with both an increase or a decrease in gain. Therefore, both "upper" and "lower" gain margins must be identified, corresponding to the upper crossover frequency ω_{cu} and the lower crossover frequency ω_{cl}.

Stability Characteristics of the Lm and Phase Diagram

The total phase angle of the transfer function at any frequency is closely related to the slope of the Lm curve at that frequency. A slope of −20 dB/decade is related to an angle of −90°; a slope of −40 dB/decade is related to an angle of −180°; a slope of −60 dB/decade is related to an angle of −270°; etc. Changes of slope at higher and lower corner frequencies, around

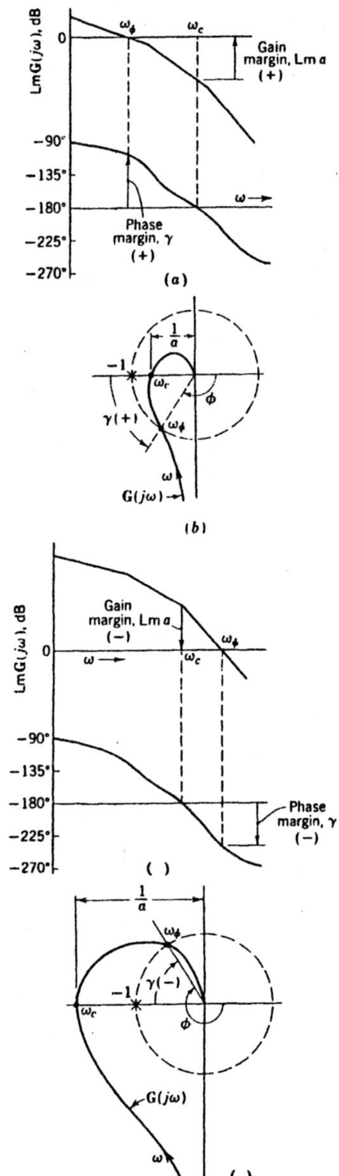

Figure 10.29 Log magnitude and phase diagram and polar plots of $G(j\omega)$, showing gain margin and phase margin: (a) and (b) stable; (c) and (d) unstable.

the particular frequency being considered, contribute to the total angle at that frequency. The farther away the changes of slope are from the particular frequency, the less they contribute to the total angle at that frequency. The stability of a minimum-phase system requires that $\gamma > 0$. For this to be true, the angle at the gain crossover [Lm $G(j\omega) = 0$ dB] must be greater than $-180°$. This places a limit on the slope of the Lm curve at the gain crossover. *The slope at the gain crossover should be more positive than −40 dB/decade if the adjacent corner frequencies are not close.* A slope of −20 dB/decade is preferable. This is derived from the consideration of a theorem by Bode. Thus, the Lm and phase diagram reveals some pertinent information. For example, the gain can be adjusted (this raises or lowers the Lm curve) to achieve the desirable range of $45° \leq \gamma \leq 60°$. The shape of the

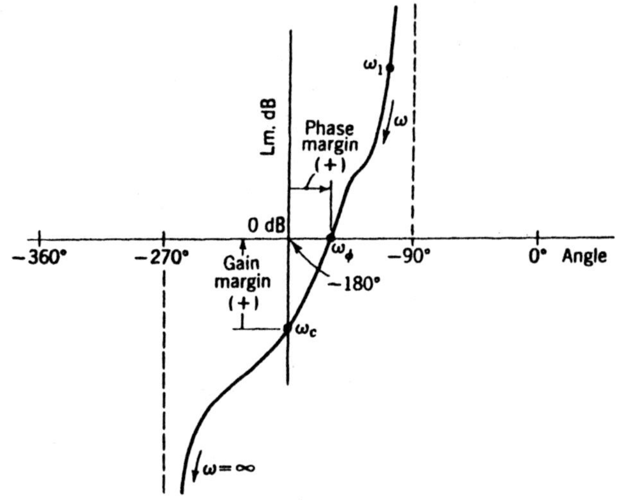

Figure 10.30 Typical Log magnitude-angle diagram for $G(j\omega) = \frac{4(1+j0.5\omega)}{j\omega(1+j2\omega)[1+j0.5\omega+(j0.125\omega)^2]}$.

low-frequency portion of the curve determines system type and therefore the degree of steady-state accuracy. The system type and the gain determine the error coefficients and therefore the steady-state error. The value of ω_ϕ gives a qualitative indication of the speed of response of a system.

Stability from the Nichols Plot (Lm-Angle Diagram)

The Lm-angle diagram is drawn by picking for each frequency the values of Lm and angle from the Lm and phase diagrams vs. ω (Bode plot). The resultant curve has frequency as a parameter. The curve for the example shown in Figure 10.30, shows a positive gain margin and phase margin angle; therefore, this represents a stable system. Changing the gain raises or lowers the curve without changing the angle characteristics. Increasing the gain raises the curve, thereby decreasing the gain margin and phase margin angle, with the result that the stability is decreased. Increasing the gain so that the curve has a positive Lm at $-180°$ results in negative gain margin and phase margin angle; therefore, an unstable system results. Decreasing the gain lowers the curve and increases stability. However, a large gain is desired to reduce steady-state errors [11].

The Lm-angle diagram for $G(s)H(s)$ can be drawn for all values of s on the contour Q of Figure 10.26a. The resultant curve for minimum-phase systems is a closed contour. Nyquist's criterion can be applied to this contour by determining the number of points (having the values 0 dB and odd multiples of 180°) enclosed by the curve of $G(s)H(s)$. This number is the value of N that is used in the equation $Z_R = N - P_R$ to determine the value of Z_R. An example for $G(s) = K_1/[s(1+Ts)]$ is shown in Figure 10.31. The Lm-angle contour for a nonminimum-phase system does not close [11]; thus, it is more difficult to determine the value of N. For these cases the polar plot is easier to use to determine stability.

It is not necessary to obtain the complete Lm-angle contour to determine stability for minimum-phase systems. Only that portion of the contour is drawn representing $G(j\omega)$ for the range of values $0^+ < \omega < \infty$. The stability is then determined from the position of the curve of $G(j\omega)$ relative to the (0 dB, $-180°$) point. In other words, the curve is traced in the direction of increasing frequency, i.e., walking along the curve in the direction of increasing frequency. The system is stable if the (0 dB, $-180°$) point is to the right of the curve. This is a simplified rule of thumb, which is based on Nyquist's stability criterion for a minimum-phase system.

A conditionally stable system is one in which the curve crosses the $-180°$ axis at more than one point. The gain determines whether the system is stable or unstable. That is, instability (or stability) can occur with both an increase (or a decrease) in gain.

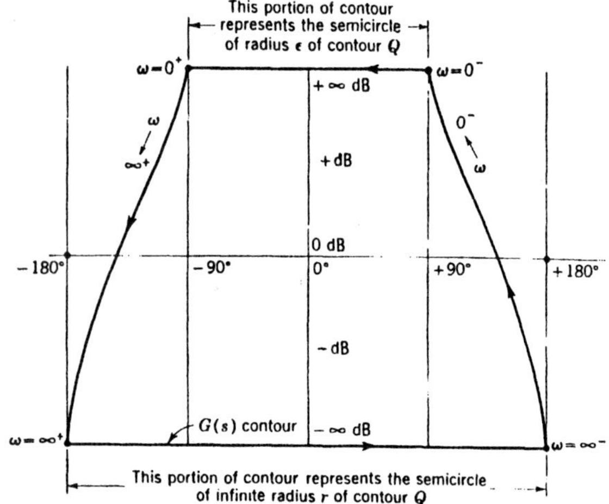

Figure 10.31 The log magnitude-angle diagram for $G(s) = \frac{K_1}{s(1+Ts)}$.

10.3.8 Closed-Loop Tracking Performance Based on the Frequency Response

A correlation between the frequency and time responses of a system, leading to a method of gain setting in order to achieve a specified closed-loop frequency response, is now developed [9]. The closed-loop frequency response is obtained as a function of the open-loop frequency response. Although the design is performed in the frequency domain, the closed-loop responses in the time domain are also obtained. Then a "best" design is selected by considering both the frequency and the time responses. Both the polar plot and the Lm-angle diagram (Nichols plot) are used.

Direct Polar Plot

The frequency control ratio $C(j\omega)/R(j\omega)$ for a unity feedback system is given by

$$\frac{C(j\omega)}{R(j\omega)} = \frac{A(j\omega)}{B(j\omega)} = \frac{|A(j\omega)|e^{j\phi(\omega)}}{|B(j\omega)|e^{j\lambda(\omega)}} = \frac{G(j\omega)}{1+G(j\omega)}$$

$$\frac{C(j\omega)}{R(j\omega)} = \left|\frac{A(j\omega)}{B(j\omega)}\right| e^{j(\phi-\lambda)} = M(\omega)e^{j\alpha(\omega)} \quad (10.66)$$

where $A(j\omega) = G(j\omega)$ and $B(j\omega) = 1 + G(j\omega)$. Since the magnitude of the angle $\phi(\omega)$, as shown in Figure 10.32, is greater than the magnitude of the angle $\lambda(\omega)$, the value of the angle $\alpha(\omega)$ is negative. Remember that CCW rotation is taken as positive. The error control ratio $E(j\omega)/R(j\omega)$ is given by

$$\frac{E(j\omega)}{R(j\omega)} = \frac{1}{1+G(j\omega)} = \frac{1}{|B(j\omega)|e^{j\lambda}} \quad (10.67)$$

From Equation 10.67 and Figure 10.32 it is seen that the greater the distance from the $-1 + j0$ point to a point on the $G(j\omega)$ locus, for a given frequency, the smaller the steady-state sinusoidal error for a stated sinusoidal input. Thus, the usefulness and importance of the polar plot of $G(j\omega)$ have been enhanced.

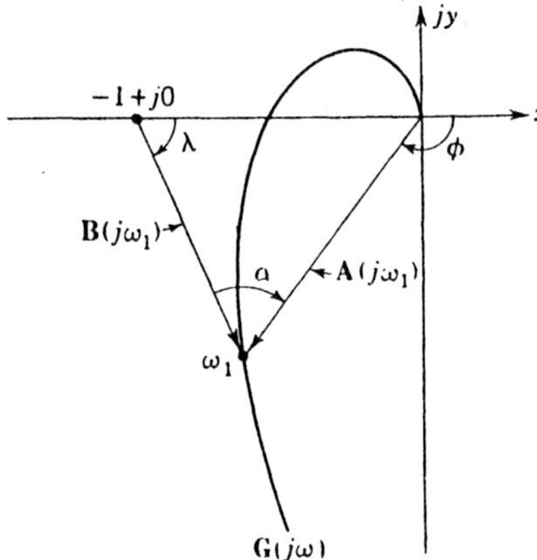

Figure 10.32 Polar plot of $G(j\omega)$ for a unity-feedback system.

Determination of M_m and ω_m for a Simple Second-Order System

The frequency at which the maximum value of $|C(j\omega)/R(j\omega)|$ occurs (see Figure 10.33) is referred to as the *resonant frequency* ω_m. The maximum value is labeled M_m. These two quantities are figures of merit (F.O.M.) of a system. Compensation to improve system performance is based upon a knowledge of ω_m and M_m. For a *simple second-order system* a direct and simple relationship can be obtained for M_m and ω_m in terms of the system parameters [11, Sect. 9.3]. These relationships are

$$\omega_m = \omega_n\sqrt{1-2\zeta^2} \quad M_m = \frac{1}{2\zeta\sqrt{1-\zeta^2}} \quad (10.68)$$

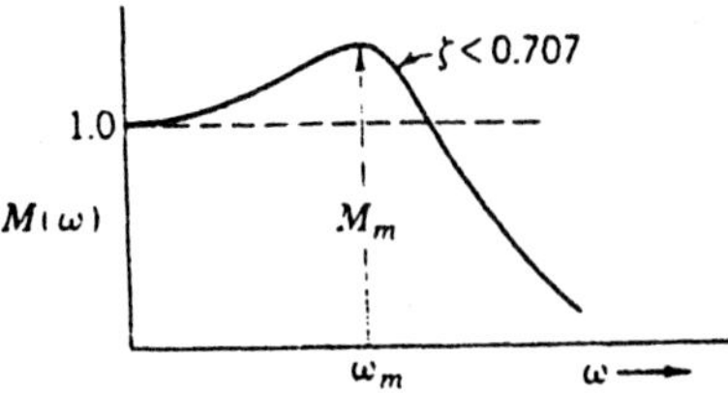

Figure 10.33 A closed-loop frequency-response curve indicating M_m and ω_m.

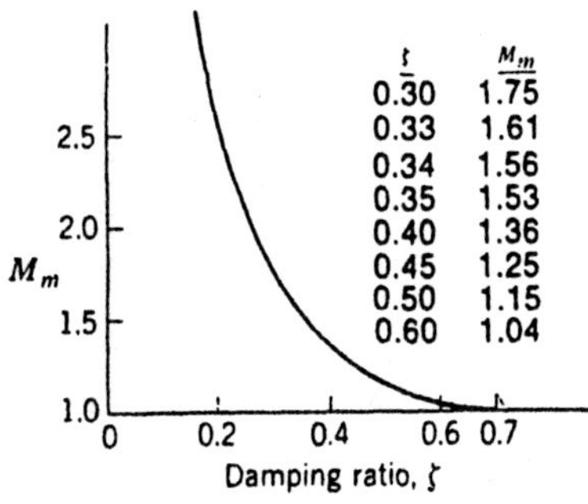

Figure 10.34 A plot of M_m vs. ζ for a simple second-order system.

From these equations it is seen that the curve of M vs. ω has a peak value, other than at $\omega = 0$, for only $\zeta < 0.707$. Figure 10.34 shows a plot of M_m *vs*. ζ for a simple second-order system. It is seen for values of $\zeta < 0.4$ that M_m increases very rapidly in magnitude; the transient oscillatory response is therefore excessively large and might damage the physical system. The correlation between the frequency and time responses is shown qualitatively in Figure 10.35.

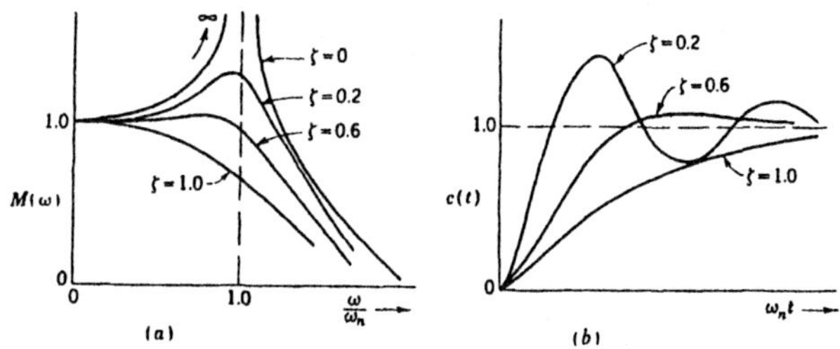

Figure 10.35 (a) Plots of M vs. ω/ω_n for a simple second-order system; (b) corresponding time plots for a step input.

The corresponding time domain F.O.M [11] are

- The damped natural frequency ω_d
- The peak value M_p
- The peak time T_p,
- The settling time $t_s (\pm 2\%)$.

For a unit-step forcing function, these F.O.M. for the transient of a simple second-order system are

$$M_p = 1 + e^{-\zeta\pi\sqrt{1-\zeta^2}} \quad (10.69)$$

$$T_p = \pi/\omega_d \quad (10.70)$$

$$\omega_d = \omega_n\sqrt{1-\zeta^2} \quad (10.71)$$

$$t_s = 4/\zeta\omega_n \quad (10.72)$$

Therefore, for a simple second-order system the following conclusions are obtained in correlating the frequency and time responses:

1. Inspection of Equation 10.68 reveals that ω_m is a function of both ω_n and ζ. Thus, for a given ζ, the larger the value of ω_m, the larger ω_n, and the faster the transient time of response for this system given by Equation 10.72.
2. Inspection of Equations 10.68 and 10.69 shows that both M_m and M_p are functions of ζ. The smaller ζ becomes, the larger in value M_m and M_p become. Thus, it is concluded that the larger the value of M_m, the larger the value of M_p. For values of $\zeta < 0.4$, the correspondence between M_m and M_p is only qualitative for this simple case. In other words, for $\zeta = 0$ the time domain yields $M_p = 2$, whereas the frequency domain yields $M_m = \infty$. When $\zeta > 0.4$, there is a close correspondence between M_m and M_p.
3. Note that the shorter the distance between the $-1 + j0$ point and a particular $G(j\omega)$ plot (see Figure 10.36), the smaller the damping ratio. Thus, M_m is larger and consequently M_p is also larger.

From these characteristics, a designer can obtain a good approximation of the time response of a simple second-order system by knowing only the M_m and ω_m of its frequency response. A corresponding correlation for ω_m and M_m becomes tedious for more complex systems. Therefore, a graphic procedure is generally used, as shown in the following sections [13], [14].

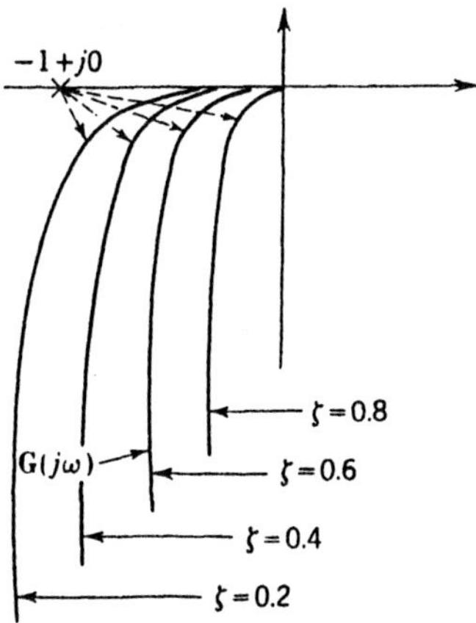

Figure 10.36 Polar plots of $G(j\omega) = K_1/[j\omega(1 + j\omega T)]$ for different values of K_1 and the resulting closed-loop damping ratios.

Correlation of Sinusoidal and Time Response

It has been found by experience [11] that M_m is also a function of the *effective* ζ and ω_n for higher-order systems. The effective ζ and ω_n of a higher-order system is dependent upon the ζ and ω_n of each second-order term, the zeros of $C(s)/R(s)$, and the values of the real roots in the characteristic equation of $C(s)/R(s)$. Thus, in order to alter the M_m, the location of some of the roots must be changed. Which ones should be altered depends on which are dominant in the time domain. From the analysis for a simple second-order system, whenever the frequency response has the shape shown in Figure 10.33, the following correlation exists between the frequency and time responses for systems of any order:

1. The larger ω_m is made, the faster the time of response for the system.
2. The value of M_m gives a qualitative measure of M_p within the acceptable range of the *effective* damping ratio $0.4 < \zeta < 0.707$. In terms of M_m, the acceptable range is $1 < M_m < 1.4$.
3. The closer the $G(j\omega)$ curve comes to the $-1 + j0$ point, the larger the value of M_m.

The larger K_p, K_v, or K_a is made, the greater the steady-state accuracy for a step, a ramp, and a parabolic input, respectively. In terms of the polar plot, the farther the point $G(j\omega)|_{\omega=0} = K_0$ is from the origin, the more accurate is the steady-state time response for a step input. For a Type 1 system, the farther the low-frequency asymptote (as $\omega \to 0$) is from the imaginary axis, the more accurate is the steady-state time response for a ramp input. All the factors mentioned above are merely *guideposts* in the *frequency domain* to assist the designer in obtaining an *approximate* idea of the time response of a system. They serve as "stop-and-go signals" to indicate if one is headed in the right direction in achieving the desired time response. If the desired performance specifications are not satisfactorily met, compensation techniques must be used.

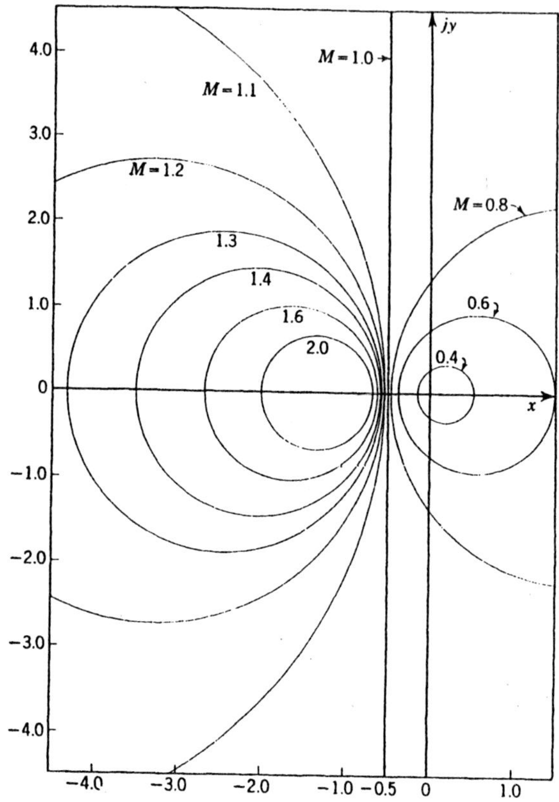

Figure 10.37 Constant M contours.

Constant $M(\omega)$ and $\alpha(\omega)$ Contours of $C(j\omega)/R(j\omega)$ on the Complex Plane (Direct Plot)

The contours of constant values of M drawn in the complex plane yield a rapid means of determining the values of M_m and ω_m and the value of gain required to achieve a desired value of M_m. In conjunction with the contours of constant values of $\alpha(\omega)$, also drawn in the complex plane, the plot of $C(j\omega)/R(j\omega)$ can be obtained rapidly. The M and α contours are developed only for unity-feedback systems by inserting $G(j\omega) = x + jy$ into $M(j\omega)$ [11]. The derivation of the M and α contours yields the equation of a circle with its center at the point (a, b) and having radius r. The location of the center and the radius for a specified value of M are given by

$$x_0 = -\frac{M^2}{M^2 - 1} \tag{10.73}$$

$$y_0 = 0 \tag{10.74}$$

$$r_0 = \left|\frac{M}{M^2 - 1}\right| \tag{10.75}$$

This circle is called a *constant M contour* for $M = M_a$. Figure 10.37 shows a family of circles in the complex plane for different values of M. Note that the larger the value M, the smaller its corresponding M circle. A further inspection of Figure 10.37 and Equation 10.73 reveals the following:

1. For $M \to \infty$, which represents a condition of oscillation ($\zeta \to 0$), the center of the M circle $x_0 \to -1 + j0$ and the radius $r_0 \to 0$. Thus, as the $G(j\omega)$ plot comes closer to the $-1 + j0$ point, the effective ζ becomes smaller and the degree of stability decreases.
2. For $M(\omega) = 1$, which represents the condition where $C(j\omega) = R(j\omega)$, $r_0 \to \infty$ and the M contour becomes a straight line perpendicular to the real axis at $x = -1/2$.
3. For $M \to 0$, the center of the M circle $x_0 \to 0$ and the radius $r_0 \to 0$.
4. For $M > 1$, the centers x_0 of the circles lie to the left of $x = -1 + j0$; and for $M < 1$, x_0 of the circles lie to the right of $x = 0$. All centers are on the real axis.

$\alpha(\omega)$ Contours The $\alpha(\omega)$ contours, representing constant values of phase angle $\alpha(\omega)$ for $C(j\omega)/R(j\omega)$, can also be determined in the same manner as for the M contours [11]. The derivation results in the equation of a circle, with $N = \tan\alpha$ as a parameter, given by

$$\left(x + \tfrac{1}{2}\right)^2 + \left(y - \tfrac{1}{2N}\right)^2 = \tfrac{1}{4}\tfrac{N^2+1}{N^2} \tag{10.76a}$$

whose center is located at $x_q = -\frac{1}{2}$, $y_q = \frac{1}{2N}$ with a radius

$$r_q = \tfrac{1}{2}\left(\tfrac{N^2+1}{N^2}\right)^{1/2} \tag{10.76b}$$

Different values of α result in a family of circles in the complex plane with centers on the line represented by $(-1/2, y)$, as illustrated in Figure 10.38.

Tangents to the M Circles [11] The line drawn through the origin of the complex plane and tangent to a given M circle plays

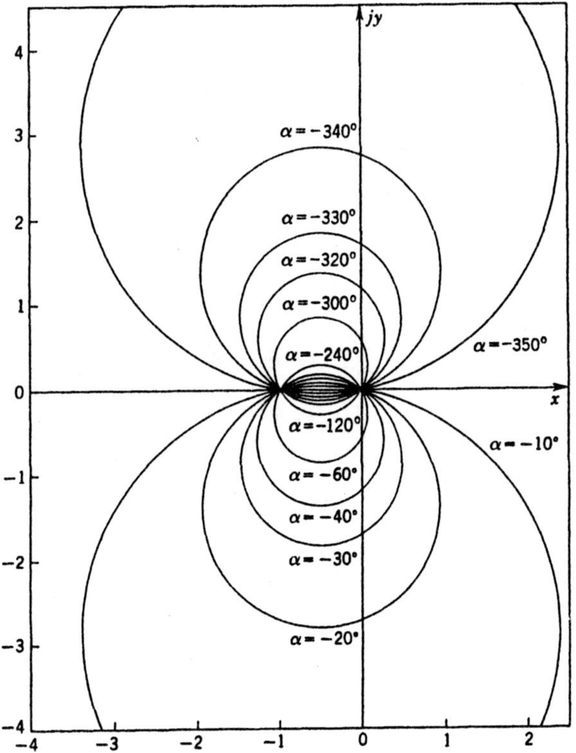

Figure 10.38 Constant α contours.

an important part in setting the gain of $G(j\omega)$. Referring to Figure 10.39 and recognizing that $bc = r_0$ is the radius and $ob = x_0$ is the distance to the center of the particular M circle yields $\sin\psi = 1/M$ and $oa = 1$.

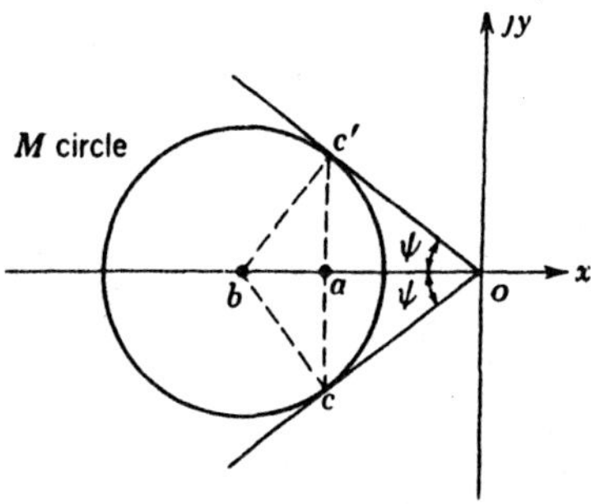

Figure 10.39 Determination of $\sin\psi$.

10.3.9 Gain Adjustment for a Desired M_m of a Unity-Feedback System (Direct Polar Plot)

Gain adjustment is the first step in adjusting the system for the desired performance. The procedure for adjusting the gain is outlined in this section. Figure 10.40a shows $G_x(j\omega)$ with its respective M_m circle in the complex plane. Since

$$G_x(j\omega) = x + jy = K_x G'_x(j\omega) = K_x(x' + jy') \quad (10.77)$$

then

$$x' + jy' = \frac{x}{K_x} + j\frac{y}{K_x}$$

where $G'_x(j\omega) = G_x(j\omega)/K_x$ is defined as the frequency-

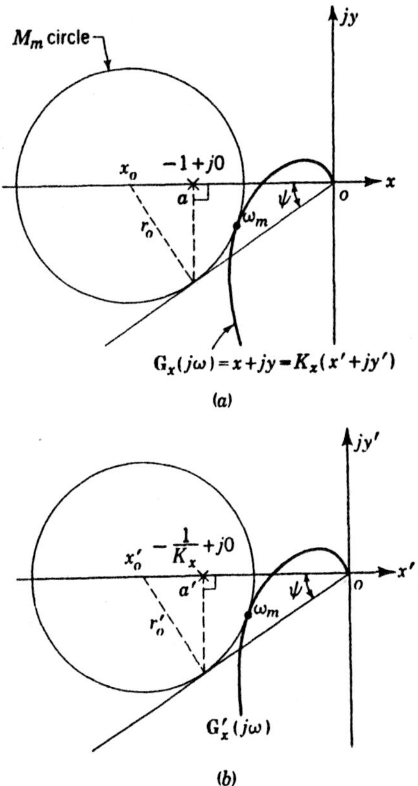

Figure 10.40 (a) Plot of $G_x(j\omega)$ with respective M_m circle; (b) circle drawn tangent to both the plot of $G'_x(j\omega)$ and the line representing the angle $\psi = \sin^{-1}(1/M_m)$.

sensitive portion of $G_x(j\omega)$ with unity gain. Note that changing the gain merely changes the amplitude and not the angle of the locus of points of $G_x(j\omega)$. Thus, if in Figure 10.40a a change of scale is made by dividing the x, y coordinates by K_x so that the new coordinates are x', y', the following are true:

1. The $G_x(j\omega)$ plot becomes the $G'_x(j\omega)$ plot.
2. The M_m circle becomes a circle that is simultaneously tangent to $G'_x(j\omega)$ and the line representing $\sin\psi = 1/M_m$.
3. The $-1 + j0$ point becomes the $-1/K_x + j0$ point.
4. The radius r_0 becomes $r'_0 = r_0/K_x$.

It is possible to determine the required gain to achieve a desired M_m for a given system by using the following graphical procedure:

Step 1. If the original system has a transfer function

$$\begin{aligned} G_x(j\omega) &= K_x G'_x(j\omega) \\ &= \frac{K_x(1 + j\omega T_1)(1 + j\omega T_2)\cdots}{(j\omega)^m(1 + j\omega T_a)(1 + j\omega T_b)(1 + j\omega T_c)\cdots} \end{aligned} \quad (10.78)$$

with an original gain K_x, only the frequency-sensitive portion $G'_x(j\omega)$ is plotted.

Step 2. Draw a straight line through the origin at the angle $\psi = \sin^{-1}(1/M_m)$, measured from the negative real axis.

Step 3. By trial and error, find a circle whose center lies on the negative real axis and is simultaneously tangent to both the $G'_x(j\omega)$ plot and the line drawn at the angle ψ.

Step 4. Having found this circle, locate the point of tangency on the ψ−angle line. Draw a vertical line from this point of tangency perpendicular to the real axis. Label the point where this line intersects the real axis as a'.

Step 5. For this circle to be an M circle representing M_m, the point a' must be the $-1 + j0$ point. Thus, the x', y' coordinates must be multiplied by a gain factor K_m in order to convert this plot into a plot of $G(j\omega)$. From the graphical construction the gain value is $K_m = 1/oa'$.

Step 6. The original gain must be changed by a factor $A = K_m/K_x$.

Note that if $G_x(j\omega)$, which includes a gain K_x, is already plotted, it is possible to work directly with the plot of the function $G_x(j\omega)$. Following the procedure just outlined results in the determination of the *additional* gain required to produce the specified M_m; that is, the additional gain is

$$A = \frac{K_m}{K_x} = \frac{1}{oa'} \tag{10.79}$$

10.3.10 Constant M and α Curves on the Lm-Angle Diagram (Nichols Chart)

The transformation of the constant M curves (circles) [11], [14] on the polar plot to the Lm-angle diagram is done more easily by starting from the inverse polar plot since all the M^{-1} circles have the same center at $-1 + j0$. Also, the constant α contours are radial lines drawn through this point [11]. There is a change of sign of the Lm and angle obtained, since the transformation is from the inverse transfer function on the inverse polar plot to the direct transfer function on the Lm vs. ϕ plot. This transformation results in constant M and α curves that have symmetry at every 180° interval. An expanded 300° section of the constant M and α graph is shown in Figure 10.41. This graph is commonly referred to as the *Nichols chart.* Note that the $M = 1$ (0 dB) curve is asymptotic to $\phi = -90°$ and $\phi = -270°$ and the curves for $M < 1/2(-6dB)$ are always negative. The curve for $M = \infty$ is the point at 0 dB, −180°, and the curves for $M > 1$ are closed curves inside the limits $\phi = -90°$ and $\phi = -270°$. These loci for constant M and α on the Nichols Chart apply only for stable unity-feedback systems.

The Nichols Chart has the Cartesian coordinates of dB vs. phase angle ϕ. Standard graph paper with loci of constant M and α for the closed-loop transfer function is available. The open-loop frequency response $G(j\omega)$, with the frequencies noted along the plot, is superimposed on the Nichols Chart as shown in Figure 10.41. The intersections of the $G(j\omega)$ plot with the M and α contours yield the closed-loop frequency response $M\angle\alpha$.

By plotting Lm $G'_x(j\omega)$ vs. $\angle G'_x(j\omega)$ on the Nichols Chart, the value of K_x required to achieve a desired value of M_m can be determined. The amount Δ dB required to raise or lower this plot of $G_x(j\omega)$ vs. ϕ in order to make it just tangent to the desired $M = M_m$ contour yields $\text{Lm}K_x = \Delta$. The frequency value at the point of tangency, i.e., $\text{Lm}G_x(j\omega_m)$, yields the value of the resonant frequency $\omega = \omega_m$.

10.3.11 Correlation of Pole-Zero Diagram with Frequency and Time Responses

Whenever the closed-loop control ratio $M(j\omega)$ has the characteristic [11] form shown in Figure 10.33, the system may be approximated as a simple second-order system. This usually implies that the poles, other than the dominant complex pair, are either far to the left of the dominant complex poles or are close to zeros. When these conditions are not satisfied, the frequency response may have other shapes. This can be illustrated by considering the following three control ratios:

$$\frac{C(s)}{R(s)} = \frac{1}{s^2 + s + 1} \tag{10.80}$$

$$\frac{C(s)}{R(s)} = \frac{0.313(s + 0.8)}{(s + 0.25)(s^2 + 0.3s + 1)} \tag{10.81}$$

$$\frac{C(s)}{R(s)} = \frac{4}{(s^2 + s + 1)(s^2 + 0.4s + 4)} \tag{10.82}$$

The pole-zero diagram, the frequency response, and the time response to a step input for each of these equations are shown in Figure 10.42. For Equation 10.80 the following characteristics are noted from Figure 10.42a:

1. The control ratio has only two complex dominant poles and no zeros.
2. The frequency-response curve has the following characteristics:
   a. A single peak $M_m = 1.157$ at $\omega_m = 0.7$.
   b. $1 < M < M_m$ in the frequency range $0 < \omega < 1$.
3. The time response has the typical waveform for a simple underdamped second-order system. That is, the first maximum of $c(t)$ due to the oscillatory term is greater than $c(t)_{ss}$, and the $c(t)$ response after this maximum oscillates around the value of $c(t)_{ss}$.

For Equation 10.81 the following characteristics are noted from Figure 10.42b:

1. The control ratio has two complex poles and one real pole, all dominant, and one real zero.
2. The frequency-response curve has the following characteristics:

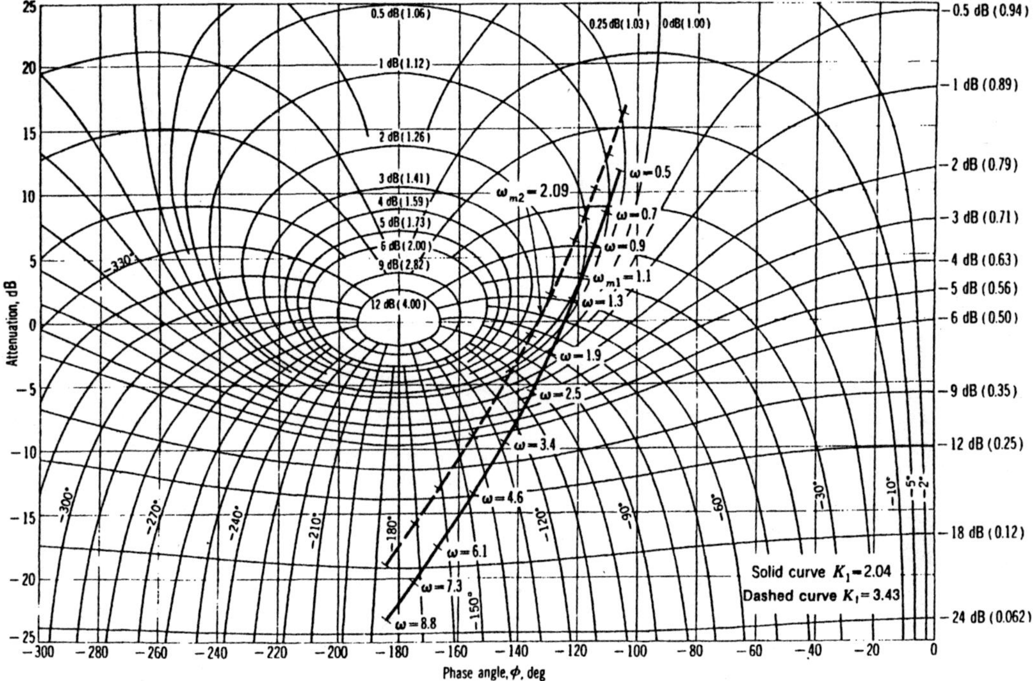

Figure 10.41 Use of the Log magnitude-angle diagram (Nichols Chart) for $G_x(j\omega) = K_x G'_x(j\omega)$.

a. A single peak, $M_m = 1.27$ at $\omega_m = 0.95$.

b. $M < 1$ in the frequency range $0 < \omega < \omega_x$.

c. The peak M_m occurs at $\omega_m = 0.95 > \omega_x$.

3. The time response does not have the conventional waveform. That is, the first maximum of $c(t)$ due to the oscillatory term is less than $c(t)_{ss}$ because of the transient term $A_3 e^{-0.25t}$.

For Equation 10.82 the following characteristics are noted from Figure 10.42c:

1. The control ratio has four complex poles, all dominant, and no zeros.
2. The frequency-response curve has the following characteristics:

a. There are two peaks, $M_{m1} = 1.36$ at $\omega_{m1} = 0.81$ and $M_{m2} = 1.45$ at $\omega_{m2} = 1.9$.

b. $1 < M < 1.45$ in the frequency range $0 < \omega < 2.1$.

c. The time response does not have the simple second-order waveform. That is, the first maximum of $c(t)$ in the oscillation is greater than $c(t)_{ss}$, and the oscillatory portion of $c(t)$ does not oscillate about a value of $c(t)_{ss}$. This time response can be predicted from the pole locations in the s-plane and from the two peaks in the plot of M vs. ω.

10.3.12 Summary

The different types of frequency-response plots are presented in this chapter. All of these plots indicate the type of system under consideration. Both the polar plot and the Nichols plot can be used to determine the necessary gain adjustment that must be made to improve its response. The methods presented for obtaining the Lm frequency-response plots stress graphical techniques. For greater accuracy a CAD program should be used to calculate this data. This chapter shows that the polar plot of the transfer function $G(s)$, in conjunction with Nyquist's stability criterion, gives a rapid means of determining whether a system is stable or unstable. The phase margin angle and gain margin are also used as a means of measuring stability. This is followed by the correlation between the frequency and time responses. The F.O.M. M_m and ω_m are established as guideposts for evaluating the tracking performance of a system. The addition of a pole to an open-loop transfer function produces a CW shift of the direct polar plot, which results in a larger value of M_m. The time response also suffers because ω_m becomes smaller. The reverse is true if a zero is added to the open-loop transfer function. This agrees with the analysis using the root locus, which shows that the addition of a pole or zero results in a less stable or more stable system, respectively. Thus, the qualitative correlation between the root locus and the frequency response is enhanced. The M and α contours are an aid in adjusting the gain to obtain a desired M_m. The methods described for adjusting the gain for a desired M_m are based on the fact that generally the desired values of M_m are slightly greater than 1. This yields a time response having an underdamped response, with a small amplitude of oscillation that reaches steady state rapidly. When the gain adjustment does not

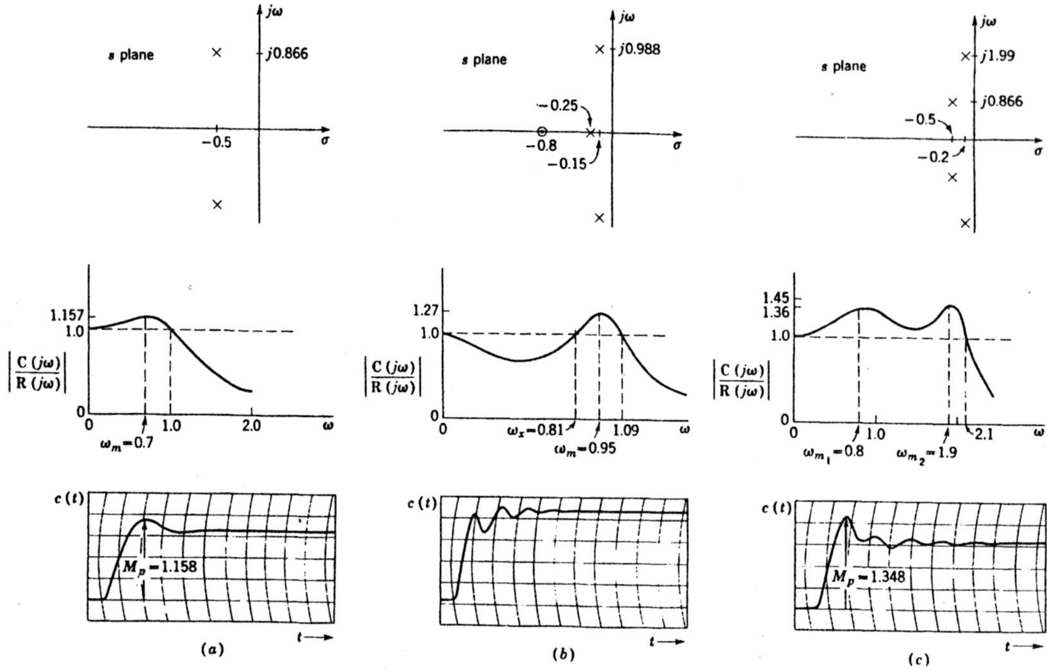

Figure 10.42 Comparison of frequency and time responses for three pole-zero patterns.

yield a satisfactory value of ω_m, the system must be compensated in order to increase ω_m without changing the value of M_m.

References

[1] Maccoll, L. A., *Fundamental Theory of Servomechanisms*, Van Nostrand, Princeton, NJ, 1945.

[2] James, H. M., Nichols, N.B., and Phillips, R.S., *Theory of Servomechanisms*, McGraw-Hill, New York, 1947.

[3] Nyquist, H., Regeneration theory, *Bell Syst. Tech. J.*, 11, 126–147, 1932.

[4] Bruns, R. A. and Saunders, R.M., *Analysis of Feedback Control Systems*, McGraw-Hill, New York, 1955, chap. 14.

[5] Balabanian, N. and LePage, W.R., What is a minimum-phase network?, *Trans. AIEE*, 74, pt. II, 785–788, 1956.

[6] Bode, H. W., *Network Analysis and Feedback Amplifier Design*, Van Nostrand, Princeton, NJ, 1945, chap. 8.

[7] Chestnut, H. and Mayer, R.W., *Servomechanisms and Regulating System Design*, Vol. 1, 2nd ed., Wiley, New York, 1959.

[8] Sanathanan, C. K. and Tsukui, H., Synthesis of transfer function from frequency response data, *Int. J. Syst. Sci.*, 5(1), 41–54, 1974.

[9] Brown, G. S. and Campbell, D.P., *Principles of Servomechanisms*, Wiley, New York, 1948.

[10] D'Azzo, J. J. and Houpis, C.H., *Feedback Control System Analysis and Synthesis*, 2nd ed., McGraw-Hill, New York, 1966.

[11] D'Azzo, J. J. and Houpis, C.H., *Linear Control System Analysis and Design: Conventional and Modern*, 4th ed., McGraw-Hill, New York, 1995.

[12] Freudenberg, J. S. and Looze, D.P., Right half-plane poles and zeros and design trade-offs in feedback systems, *IEEE Trans. Autom. Control*, AC-30, 555–565, 1985.

[13] Chu, Y., Correlation between frequency and transient responses of feedback control systems, *Trans. AIEE*, 72, pt. II, 81–92, 1953.

[14] James, H. M., Nichols, N.B., and Phillips, R.S., *Theory of Servomechanisms*, McGraw-Hill, New York, 1947, chap. 4.

[15] Higgins, T. J., and Siegel, C.M., Determination of the maximum modulus, or the specified gain of a servomechanism by complex variable differentiation, *Trans. AIEE*, 72, pt. II, 467, 1954.

10.4 The Root Locus Plot

William S. Levine, *Department of Electrical Engineering, University of Maryland, College Park, MD*

10.4.1 Introduction

The root locus plot was invented by W. R. Evans around 1948 [1, 2]. This is somewhat surprising because the essential ideas behind the root locus were available many years earlier. All that is really needed is the Laplace transform, the idea that the poles of a linear time-invariant system are important in control design, and the geometry of the complex plane. One could argue that the essentials were known by 1868 when Maxwell published his paper "On Governors" [3]. It is interesting to speculate on why it took so long to discover such a natural and useful tool.

It has become much easier to produce root locus plots in the last few years. Evans's graphical construction has been superseded by computer software. Today it takes just a few minutes to input the necessary data to the computer. An accurate root locus plot is available seconds later [4]. In fact, the computer makes it possible to extend the basic idea of the root locus to study graphically almost any property of a system that can be parameterized by a real number.

The detailed discussion of root locus plots and their uses begins with an example and a definition. This is followed by a description of the original rules and procedures for constructing root locus plots. Using the computer introduces different questions. These are addressed in Section 4. The use of root locus plots in the design of control systems is described in Section 5. In particular, the design of lead, lag, and lead/lag compensators, as well as the design of notch filters, is described. This is followed by a brief introduction to other uses of the basic idea of the root locus. The final section summarizes and mentions some limitations.

10.4.2 Definition

The situation of interest is illustrated in Figure 10.43, where $G(s)$ is the transfer function of a single-input single-output linear time-invariant system and k is a real number. The closed-loop system has the transfer function

$$G_{cl}(s) = \frac{y(s)}{r(s)} = \frac{kG(s)}{1+kG(s)} \tag{10.83}$$

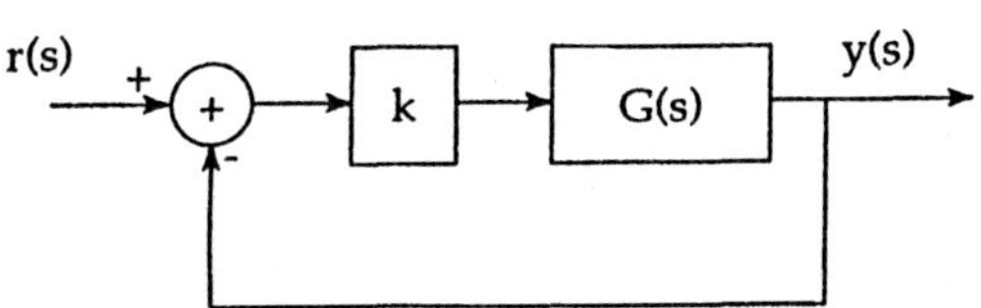

Figure 10.43 Block diagram for a simple unity feedback control system.

The standard root locus only applies to the case where $G(s)$ is a rational function of s. That is,

$$G(s) = n(s)/d(s) \tag{10.84}$$

and $n(s)$ and $d(s)$ are polynomials in s with real coefficients. If this is true then it is easy to show that

$$G_{cl}(s) = \frac{kn(s)}{d(s)+kn(s)} \tag{10.85}$$

Notice that the numerators of $G(s)$ and of $G_{cl}(s)$ are identical. We have just proven that, except possibly for pole-zero cancellations, the open-loop system, $G(s)$, and the closed-loop system, $G_{cl}(s)$, have exactly the same zeros regardless of the value of k.

What happens to the poles of $G_{cl}(s)$ as k varies? This is precisely the question that is answered by the root locus plot. By definition, the root locus plot is a plot of the poles of $G_{cl}(s)$ in the complex plane as the parameter by k varies. It is very easy to generate such plots for simple systems. For example, if

$$G(s) = \frac{3}{s^2+4s+3}$$

then

$$G_{cl}(s) = \frac{3k}{s^2+4s+(3+3k)}$$

The poles of $G_{cl}(s)$, denoted by p_1 and p_2, are given by

$$\begin{aligned} p_1 &= -2+\sqrt{1-3k} \\ p_2 &= -2-\sqrt{1-3k} \end{aligned} \tag{10.86}$$

It is straightforward to plot p_1 and p_2 in the complex plane as k varies. This is done for $k \geq 0$ in Figure 10.44. Note that, strictly speaking, $G_{cl}(s) = 0$ when $k = 0$. However, the denominators of Equations 10.84 and 10.85 both give the same values for the closed-loop poles when $k = 0$, namely -1 and -3. Those values are the same as the open-loop poles. By convention, all root locus plots use the open-loop poles as the closed-loop poles when $k = 0$.

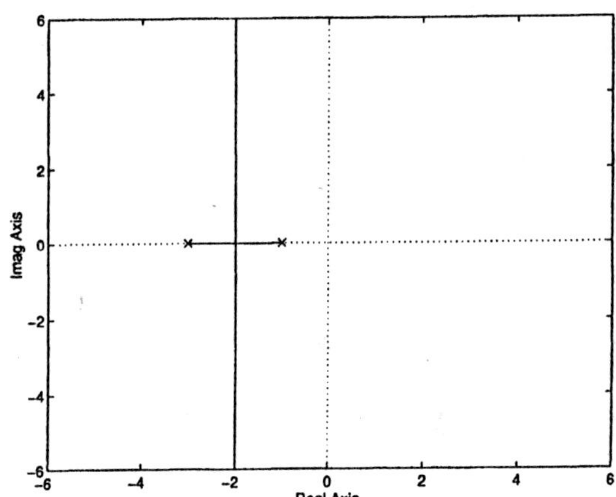

Figure 10.44 Root locus plot for $G(s) = \frac{3}{s^2+4s+3}$ and $k \geq 0$.

The plot provides a great deal of useful information. First, it gives the pole locations for every possible closed-loop system that can be created from the open-loop plant and any positive gain k. Second, if there are points on the root locus for which the closed-loop system would meet the design specifications, then simply applying the corresponding value of k completes the design. For example, if a closed-loop system with damping ratio $\zeta = .707$ is desired for the system whose root locus is plotted in Figure 10.44, then simply choose the value of k that puts the poles of $G_{cl}(s)$ at $-2 \pm j2$. That is, from Equation 10.86, choose $k = 5/3$.

The standard root locus can be easily applied to non-unity feedback control systems by using block diagram manipulations to put the system in an equivalent unity feedback form (see Chapter 6). Because the standard root locus depends only on properties of polynomials it applies equally well to discrete-time systems. The only change is that $G(s)$ is replaced by $G(z)$, the z-transform transfer function.

10.4.3 Some Construction Rules

Evans's procedure for plotting the Root Locus consists of a collection of rules for determining if a test point, s_t, in the complex plane is a pole of $G_{cl}(s)$ for some value of k. The first such rule has already been explained.

Rule 1: The open-loop poles, i.e., the roots of $d(s) = 0$, are all points in the root locus plot corresponding to $k = 0$.

The second rule is also elementary. Suppose that

$$\begin{aligned} d(s) &= s^n + a_{n-1}s^{n-1} + a_{n-2}s^{n-2} + \ldots + a_0 \\ n(s) &= b_m s^m + b_{m-1}s^{m-1} + \ldots + b_0 \end{aligned} \tag{10.87}$$

For physical systems it is always true that $n > m$. Although it is possible to have reasonable mathematical models that violate this condition, it will be assumed that $n > m$. The denominator of Equation 10.85 is then

$$d_{cl}(s) = s^n + a_{n-1}s^{n-1} + \ldots + (a_{n-m} + kb_m)s^m + \ldots + (a_0 + kb_0) \tag{10.88}$$

Rule 2 is an obvious consequence of Equation 10.88, the assumption that $n > m$, and the fact that a polynomial of degree n has exactly n roots.

Rule 2: The root locus consists of exactly n branches.

The remaining rules are derived from a different form of the denominator of $G_{cl}(s)$. Equation 10.83 shows that the denominator of $G_{cl}(s)$ can be written as $1 + kG(s)$. Even though $1 + kG(s)$ is not a polynomial it is still true that the poles of $G_{cl}(s)$ must satisfy the equation

$$1 + kG(s) = 0 \tag{10.89}$$

Because s is a complex number, $G(s)$ is generally complex and Equation 10.89 is equivalent to two independent equations. These could be, for instance, that the real and imaginary parts of Equation 10.89 must separately and independently equal zero. It is more convenient, and equivalent, to use the magnitude and angle of Equation 10.89. That is, Equation 10.89 is equivalent to the two equations

$$\begin{aligned} |kG(s)| &= 1 \\ \angle(kG(s)) &= \pm(2h+1)180^\circ, \text{ where } h = 0, 1, 2, \ldots \end{aligned} \tag{10.90}$$

The first equation explicitly states that, for Equation 10.89 to hold, the magnitude of $kG(s)$ must be one. The second equation shows that the phase angle of $kG(s)$ must be $\pm 180^\circ$, or $\pm 540^\circ$, etc. It is possible to simplify Equations 10.90 somewhat because k is a real number. Thus, for $k \geq 0$ Equations 10.90 become

$$\begin{aligned} |G(s)| &= 1/k \\ \angle(G(s)) &= \pm(2h+1)180^\circ, \text{where } h = 0, 1, 2, \ldots \end{aligned} \tag{10.91}$$

The form for $k \leq 0$ is the same for the magnitude of $G(s)$, except for a minus sign ($|G(s)| = -\frac{1}{k}$), but the angle condition becomes integer multiples of 360°.

Equations 10.91 are the basis for plotting the root locus. The first step in producing the plot is to mark the locations of the open-loop poles and zeros on a graph of the complex plane. The poles are denoted by $\times$s, as in Figure 10.44, while the zeros are denoted by os. If the poles and zeros are accurately plotted it is then possible to measure $|G(s_t)|$ and $\angle(G(s_t))$ for any given test point s_t.

For example, suppose $G(s) = 10\frac{(s+4)}{(s+3+j4)(s+3-j4)}$. The poles and zeros of this transfer function are plotted in Figure 10.45. Notice that the plot does not depend, in any way, on the gain 10. It is generally true that pole-zero plots are ambiguous with respect to pure gain. Figure 10.45 contains a plot of the complex number $(s+4)$ for the specific value $s_t = -1 + j3$. It is exactly the same length as, and parallel to, the vector drawn from the zero to the point $s = -1 + j3$, also shown. The same is true of the vectors corresponding to the two poles. To save effort, only the vectors from the poles to the test point $s_t = -1 + j3$ are drawn. Once the figure is drawn, simple measurements with a ruler and a protractor provide

$$\begin{aligned} |G(-1+j3)| &= l_1/l_2 l_3 \\ \angle(G(-1+j3)) &= \phi_1 - \phi_2 - \phi_3 \end{aligned}$$

One can then check the angle condition in Equations 10.91 to see if $s_t = -1 + j3$ is a point on the root locus for this $G(s)$.

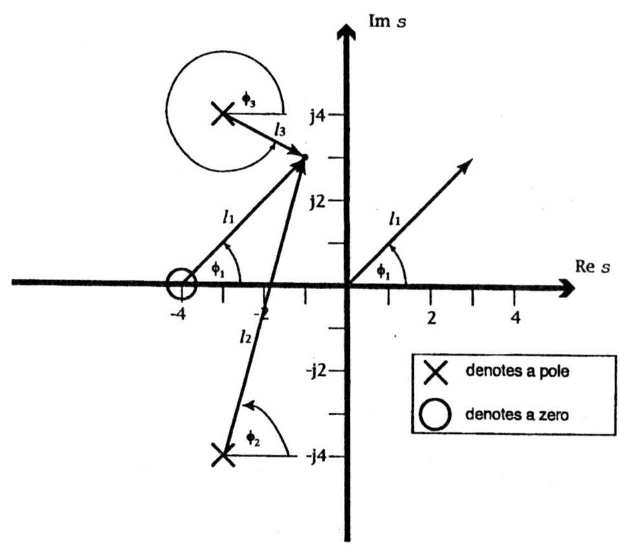

Figure 10.45 Poles and zeros of $G(s) = \frac{10(s+4)}{(s+3+j4)(s+3-j4)}$. The vectors from each of the singularities to the test point $s_t = -1 + j3$ are also shown, as is the vector $s + 4|_{s_t=-1+j3}$.

Of course, it would be tedious to check every point in the complex plane. This is not necessary. There is a collection of rules for finding points on the root locus plot. A few of these are developed below. The others can be found in most undergraduate textbooks on control, such as [5], [6].

Rule 3: For $k \geq 0$, any point on the real axis that lies to the left of an odd number of singularities (poles plus zeros) on the real axis is a point on the root locus. Any other point on the real axis is not. (Change "odd" to "even" for negative k).

A proof follows immediately from applying the angle condition in Equations 10.91 to test points on the real axis. The angular contributions due to poles and zeros that are not on the real axis cancel as a result of symmetry. Poles and zeros to the left of the test point have angles equal to zero. Poles and zeros to the right of the test point contribute angles of $-180°$ and $+180°$, respectively. In fact, a fourth rule follows easily from the symmetry.

Rule 4: The root locus is symmetric about the real axis.

We already know that all branches start at the open-loop poles. Where do they end?

Rule 5: If $G(s)$ has n poles and m finite zeros ($m \leq n$) then exactly m branches terminate, as $k \to \infty$, on the finite zeros. The remaining $n-m$ branches go to infinity as $k \to \infty$.

The validity of the rule can be proved by taking limits as $k \to \infty$ in the magnitude part of Equations 10.91. Doing so gives

$$\lim_{k\to\infty} |G(s)| = \lim_{k\to\infty} \frac{1}{k} = 0$$

Thus, as $k \to \infty$, it must be true that $|G(s)| \to 0$. This is true when s coincides with any finite zero of $G(s)$. From Equations 10.84 and 10.87,

$$\begin{aligned}\lim_{s\to\infty} G(s) &= \lim_{s\to\infty} \frac{b_m s^m + b_{m-1}s^{m-1} + \ldots + b_0}{s^n + a_{n-1}s^{n-1} + \ldots + a_0} \\ &= \lim_{s\to\infty} \frac{s^{m-n}(b_m + b_{m-1}s^{-1} + \ldots + b_0 s^{-m})}{(1 + a_{n-1}s^{-1} + \ldots + a_0 s^{-n})}\end{aligned}$$

Finally,

$$\lim_{s\to\infty} G(s) = \lim_{s\to\infty} b_m s^{m-n} = 0 \qquad (10.92)$$

The b_m factors out and the fact that $|G(s)| \to 0$ as $s \to \infty$ with multiplicity $n-m$ is apparent. One can think of this as a demonstration that $G(s)$ has $n-m$ zeros at infinity. Equation 10.92 plays an important role in the proof of the next rule as well.

Rule 6: If $G(s)$ has n poles and m finite zeros ($n \geq m$) and $k \geq 0$ then the $n-m$ branches that end at infinity asymptotically approach lines that intersect the real axis at a point σ_0 and that make an angle γ with the real axis, where

$$\gamma = \pm\frac{(1+2h)180°}{n-m} \quad \text{where } h = 0, 1, 2, \ldots$$

and

$$\sigma_0 = \frac{\sum_{i=1}^{n} Re(p_i) - \sum_{l=1}^{m} Re(z_l)}{n-m}$$

A proof of the formula for γ follows from applying the angle condition of Equations 10.91 to Equation 10.92. That is,

$$\angle(s^{m-n}) = \pm(1+2h)180°$$

so

$$\gamma = \angle(s) = \frac{\mp(1+2h)180°}{n-m}$$

A proof of the equation for σ_0 can be found in [5], pp. 252–254. Most textbooks include around a dozen rules for plotting the root locus; see [5]–[7] for example. These are much less important today than they were just a few years ago because good, inexpensive software for plotting root loci is now widely available.

10.4.4 Use of the Computer to Plot Root Loci

There are many different software packages that can be used to plot the root locus. Particularly well-known examples are MATLAB, Matrix-X, and Control-C. To some extent the software is foolproof. If the data are input correctly, the resulting root locus is calculated correctly. However, there are several possible pitfalls. For example, the software automatically scales the plot. The scaling can obscure important aspects of the root locus, as is described in the next section. This, and other possible problems associated with computer-generated root locus plots, is discussed in detail in [4].

10.4.5 Uses

The root locus plot can be an excellent tool for designing single-input single-output control systems. It is particularly effective when the open-loop transfer function is accurately known and is, at least approximately, reasonably low order. This is often the case in the design of servomechanisms. The root locus is also very useful as an aid to understanding the effect of feedback and compensation on the closed-loop system poles. Some ways to use the root locus are illustrated below.

Design of a Proportional Feedback Gain

Consider the open-loop plant $G(s) = 1/s(s+1)(0.1s+1)$. This is a typical textbook example. The plant is third order and given in factored form. The plant has been normalized so that $\lim_{s\to 0} sG(s) = 1$. This is particularly helpful in comparing different candidate designs. A servo motor driving an inertial load would typically have such a description. The motor plus load would correspond to the poles at 0 and -1. The electrical characteristics of the motor add a pole that is normally fairly far to the left, such as the pole at -10 in this example.

The simplest controller is a pure gain, as in Figure 10.43. Suppose, for illustration, that the specifications on the closed-loop system are that ζ, the damping ratio, must be exactly 0.707 and the natural frequency, ω_n, must be as large as possible. Because the closed-loop system is actually third order, both damping ratio and natural frequency are not well defined. However, the open-loop system has a pair of dominant poles, those at 0 and -1. The closed-loop system will also have a dominant pair of poles because the third pole will be more than a factor of ten to the left of the complex conjugate pair of poles. The damping ratio and natural frequency will then be defined by the dominant poles.

A root locus plot for this system, generated by MATLAB, is shown in Figure 10.46. (Using the default scaling it was difficult to see the exact intersection of the locus with the $\zeta = 0.707$ line because the automatic scaling includes all the poles in the visible plot. A version of the same plot with better scaling was created and is the one shown.) The diagonal lines on the plot correspond to $\zeta = 0.707$. The four curved lines are lines of constant ω_n, with values 1, 2, 3, 4. The value of gain k corresponding to $\zeta = 0.707$ is approximately 5, and the closed-loop poles resulting from this gain are roughly -10, $-1/2 \pm j/2$. These values are easily obtained from MATLAB [4] or any of the other software packages mentioned earlier. The bandwidth of the closed-loop system is fairly small because, as you can see from Figure 10.46, the dominant poles have $\omega_n < 1$. This indicates that the transient response will be slow. If a closed-loop system that responds faster is needed, two approaches could be used. The plant could be changed. For instance, if the plant is a motor then a more powerful motor could be obtained. This would move the pole at -1 to the left, say to -3. A faster closed-loop system could then be designed using only a feedback gain. The other option is to

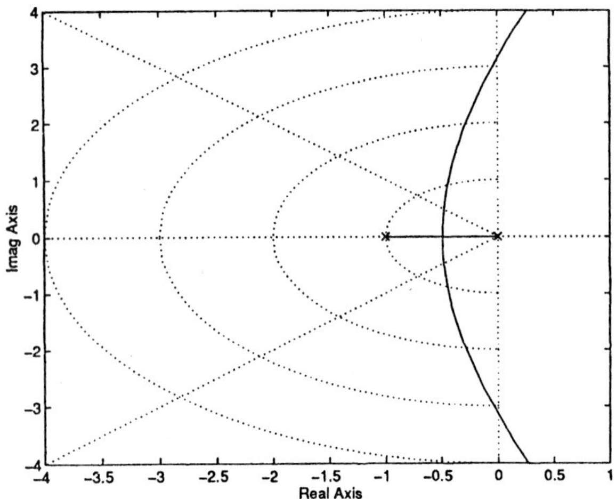

Figure 10.46 Root locus plot of $\frac{1}{s(s+1)(s+10)}$. The dotted diagonal lines correspond to $\zeta = .707$. The dotted elliptical lines correspond to $\omega_n = 1, 2, 3, 4$, respectively.

introduce a lead compensator.

Design of Lead Compensation

A lead compensator is a device that can be added in series with the plant and has a transfer function $G_c(s) = (s - z)/(s - p)$. Both z and p are real and negative. The zero, z, lies to the right of the pole, p ($z > p$). Because the magnitude of a lead compensator will increase with frequency between the zero and the pole, some combination of actuator limits and noise usually forces $p/z < 10$. Lead compensators are always used in conjunction with a gain, k.

The purpose of a lead compensator is to speed up the transient response. The example we have been working on is one for which lead compensation is easy and effective. For a system with three real poles and no zeros one normally puts the zero of the lead compensator close to the middle pole of the plant. The compensator pole is placed as far to the left as possible. The result of doing this for our example is shown in Figure 10.47. Comparison of the root loci in Figures 10.46 and 10.47 shows that the lead compensator has made it possible to find a gain, k, for which the closed-loop system has $\zeta = .707$ and $w_n > 4$. This approach to designing lead compensators basically maximizes the bandwidth of the closed-loop system for a given damping ratio. If the desired transient response of the closed-loop system is specified in more detail, there are other procedures for placing the compensator poles and zeros [7], pp. 439–445.

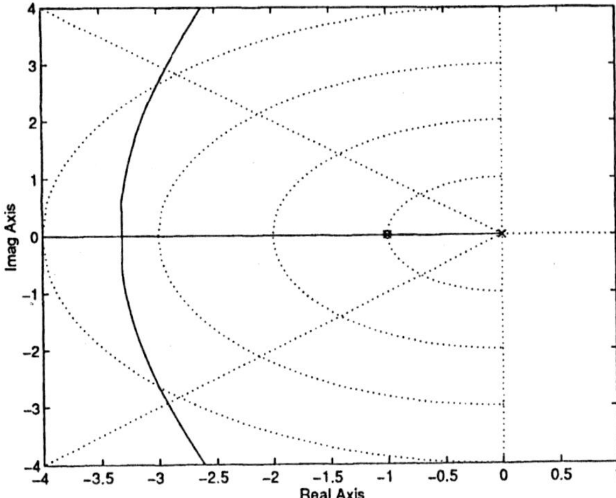

Figure 10.47 Root locus plot of $G_c(s)G(s) = \frac{(s+1)}{(s+10)}\frac{1}{s(s+1)(s+10)}$. The dotted diagonal lines correspond to $\zeta = .707$. The dotted elliptical curves correspond to $\omega_n = 1, 2, 3, 4$.

Design of Lag Compensation

A lag compensator is a device that can be added in series with the plant and has a transfer function $G_c(s) = (s - z)/(s - p)$. Both z and p are real and negative. The zero, z, lies to the left of the pole, p ($z < p$). Lag compensators are always used in series with a gain, k. Again, it is usually not feasible to have the pole and the zero of the compensator be different by more than a factor of ten. One important reason for this is explained below.

The purpose of a lag compensator is to improve the steady-state response of the closed-loop system. Again, the example we have been working on illustrates the issues very well. Because our example already has an open-loop pole at the origin it is a type 1 system. The closed-loop system will have zero steady-state error in response to a unit step input. Adding another open-loop pole close to the origin would make the steady-state error of the closed-loop system in response to a unit ramp smaller. In fact, if we put the extra pole at the origin we would reduce this error to zero. Unfortunately, addition of only an open-loop pole close to the origin will severely damage the transient response. No choice of gain will produce a closed-loop system with a fast transient response.

The solution is to add a zero to the left of the lag compensator's pole. Then, if the gain, k, is chosen large enough, the compensator's pole will be close to the compensator's zero which will approximately cancel the pole in the closed-loop system. The result is that the closed-loop transient response will be nearly unaffected by the lag compensator while the steady-state error of the closed-loop system is reduced. Note that this will not be true if the gain is too low. In that case the closed-loop transient response will be slowed by the compensator pole.

Design of Lead/Lag Compensation

Conceptually, the lead and lag compensators are independent. One can design a lead compensator so as to produce a closed-loop system that satisfies the specifications of the transient response while ignoring the steady-state specifications. One can then design a lag compensator to meet the steady-state requirements knowing that the effect of this compensator on the transient response will be negligible. As mentioned above, this does require that the gain is large enough to move the pole of the lag compensator close to its zero. Otherwise, the lag compensator will slow down the transient response, perhaps greatly.

There are a number of different ways to implement lead/lag compensators. One relatively inexpensive implementation is shown in [7], p. 473. It has the disadvantage that the ratio of the lead compensator zero to the lead compensator pole must be identical to the ratio of the lag compensator pole to the lag compensator zero. This introduces some coupling between the two compensators which complicates the design process. See [7], pp. 590–595 for a discussion.

Design of a Notch Filter

The notch filter gets its name from the appearance of a notch in the plot of the magnitude of its transfer function versus frequency. Nonetheless, there are aspects of the design of notch filters that are best understood by means of the root locus plot. It is easiest to begin with an example where a notch filter would be appropriate. Such an example would have open-loop transfer function $G(s) = 1/s(s+1)(s+.1+j5)(s+.1-j5)$. This transfer function might correspond to a motor driving an inertial load at the end of a long and fairly flexible shaft. The flexure of the shaft introduces the pair of lightly damped poles and greatly complicates the design of a good feedback controller for this plant. While it is fairly rare that a motor can only be connected to its load by a flexible shaft, problems where the open-loop system includes a pair of lightly damped poles are reasonably common.

The obvious thing to do is to add a pair of zeros to cancel the offending poles. Because the poles are stable, although lightly damped, it is feasible to do this. The only important complication is that you cannot implement a compensator that has two zeros and no poles. In practice one adds a compensator consisting of the desired pair of zeros and a pair of poles. The poles are usually placed as far to the left as feasible and close to the real axis. Such a compensator is called a notch filter.

One rarely knows the exact location of the lightly damped poles. Thus, the notch filter has to work well, even when the poles to be cancelled are not exactly where they were expected to be. Simply plotting the root locus corresponding to a plant plus notch filter for which the zeros are above the poles, as is done in Figure 10.48, shows that such a design is relatively safe. The root locus lies to the left of the lightly damped poles and zeros. The root locus plot corresponding to the situation where the compensator zeros are below the poles curves the opposite way, showing that this is a dangerous situation, in the sense that such a system can easily become unstable as a result of small variations in the plant gain.

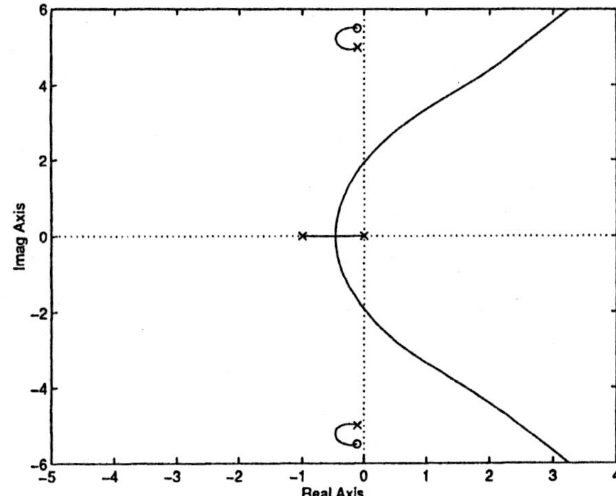

Figure 10.48 Root locus plot for the system $G_c(s)G(s) = \frac{(s+.11+j5.5)(s+.11-j5.5)}{(s+8)^2}\frac{1}{s(s+1)(s+.1+j5)(s+.1-j5)}$.

Other Uses of the Root Locus Plot

Any time the controller can be characterized by a single parameter it is possible to plot the locus of the closed-loop poles as a function of that parameter. This creates a kind of root locus that, although often very useful and easy enough to compute, does not necessarily satisfy the plotting rules given previously. An excellent example is provided by a special case of the optimal linear quadratic regulator. Given a single-input single-output (SISO) linear time-invariant system described in state space form by

$$\begin{aligned}\dot{x}(t) &= Ax(t) + bu(t)\\ y(t) &= cx(t)\end{aligned}$$

find the control that minimizes

$$J = \int_0^\infty (y^2(t) + ru^2(t))dt$$

The solution, assuming that the state vector $x(t)$ is available for feedback, is $u(t) = kx(t)$, where k is a row vector containing n elements. The vector k is a function of the real number r so it is possible to plot the locus of the closed-loop poles as a function of r. Under some mild additional assumptions this locus demonstrates that these poles approach a Butterworth pattern as r goes to zero. The details can be found in [9], pp. 218–233.

10.4.6 Conclusions

The root locus has been one of the most useful items in the control engineer's toolbox since its invention. Modern computer software for plotting the root locus has only increased its utility. Of course, there are situations where it is difficult or impossible to use it. Specifically, when the system to be controlled is not accurately known or cannot be well approximated by a rational transfer function, then it is better to use other tools.

References

[1] Evans, W.R., Control System Synthesis by Root Locus Method, *AIEE Trans.*, 69, 66–69, 1950.

[2] Evans, W.R., Graphical Analysis of Control Systems, *AIEE Trans.*, 67, 547–551, 1948.

[3] Maxwell, J.C., On Governors, Proc. R. Soc. London, 16, 270–283, 1868. (Reprinted in Selected Papers on Mathematical Trends in Control Theory, R. Bellman and R. Kalaba, Eds., Dover Publishing, 1964).

[4] Leonard, N.E. and Levine, W.S., Using MATLAB to Analyze and Design Control Systems, 2nd ed., Benjamin/Cummings, Menlo Park, CA, 1995.

[5] Franklin, G.F., Powell, J.D., and Emami-Naeni, A., Feedback Control of Dynamic Systems, 3rd ed., Addison-Wesley, Reading, MA, 1994.

[6] D'Azzo, J.J. and Houpis, C.H., Linear Control System Analysis and Design, 2nd ed., McGraw-Hill, New York, NY, 1981.

[7] Nise, N.S., Control Systems Engineering, 2nd ed., Benjamin/Cummings, Menlo Park, CA, 1995.

[8] Kuo, B.C., Automatic Control Systems, 7th ed., Prentice Hall, Englewood Cliffs, NJ, 1995.

[9] Kailath, T., Linear Systems, Prentice Hall, Englewood Cliffs, NJ, 1980.

10.5 PID Control

Karl J. Åström, Department of Automatic Control, Lund Institute of Technology, Lund, Sweden
Tore Hägglund, Department of Automatic Control, Lund Institute of Technology, Lund, Sweden

10.5.1 Introduction

The proportional-integral-derivative (PID) controller is by far the most commonly used controller. About 90 to 95% of all control problems are solved by this controller, which comes in many forms. It is packaged in standard boxes for process control and in simpler versions for temperature control. It is a key component of all distributed systems for process control. Specialized controllers for many different applications are also based on PID control. The PID controller can thus be regarded as the "bread and butter" of control engineering. The PID controller has gone through many changes in technology. The early controllers were based on relays and synchronous electric motors or pneumatic or hydraulic systems. These systems were then replaced by electronics and, lately, microprocessors.

Much interest was devoted to PID control in the early development of automatic control. For a long time researchers paid very little attention to the PID controller. Lately there has been a resurgence of interest in PID control because of the possibility of making PID controllers with automatic tuning, automatic generation of gain schedules and continuous adaptation. See the chapter "Automatic Tuning of PID Controllers" in this handbook.

Even if PID controllers are very common, they are not always used in the best way. The controllers are often poorly tuned. It is quite common that derivative action is not used. The reason is that it is difficult to tune three parameters by trial and error.

In this chapter we will first present the basic PID controller in Section 10.5.2. When using PID control it is important to be aware of the fact that PID controllers are parameterized in several different ways. This means for example that "integral time" does not mean the same thing for different controllers. PID controllers cannot be understood from linear theory. Amplitude and rate limitations in the actuators are key elements that lead to the windup phenomena. This is discussed in Section 10.5.4 where different ways to avoid windup are also discussed. Mode switches also are discussed in the same section.

Most PID controllers are implemented as digital controllers. In Section 10.5.5 we discuss digital implementation. In Section 10.5.6 we discuss uses of PID control, and in Section 10.5.7 we describe how complex control systems are obtained in a "bottom up" fashion by combining PID controllers with other simple systems.

We also refer to the companion chapter "Automatic Tuning of PID Controllers" in this handbook, which treats design and tuning of PID controllers. Examples of industrial products are also given in that chapter.

10.5.2 The Control Law

In a PID controller the control action is generated as a sum of three terms. The control law is thus described as

$$u(t) = u_P(t) + u_I(t) + u_D(t) \tag{10.93}$$

where u_P is the proportional part, u_I the integral part and u_D the derivative part.

Proportional Control

The proportional part is a simple feedback

$$u_P(t) = Ke(t) \tag{10.94}$$

where e is the control error, and K is the controller gain. The error is defined as the difference between the set point y_{sp} and the process output y, i.e.,

$$e(t) = y_{sp}(t) - y(t) \tag{10.95}$$

The modified form,

$$u_P(t) = K(by_{sp}(t) - y(t)) \tag{10.96}$$

where b is called *set point weighting*, admits independent adjustment of set point response and load disturbance response.

Integral Control

Proportional control normally gives a system that has a steady-state error. Integral action is introduced to remove this. Integral action has the form

$$u_I(t) = k_i \int^t e(s)ds = \frac{K}{T_i} \int^t e(s)ds \qquad (10.97)$$

The idea is simply that control action is taken even if the error is very small provided that the average of the error has the same sign over a long period.

Automatic Reset

A proportional controller often gives a steady-state error. A manually adjustable reset term may be added to the control signal to eliminate the steady-state error. The proportional controller given by Equation 10.94 then becomes

$$u(t) = Ke(t) + u_b(t) \qquad (10.98)$$

where u_b is the reset term. Historically, integral action was the result of an attempt to obtain automatic adjustment of the reset term. One way to do this is shown in Figure 10.49.

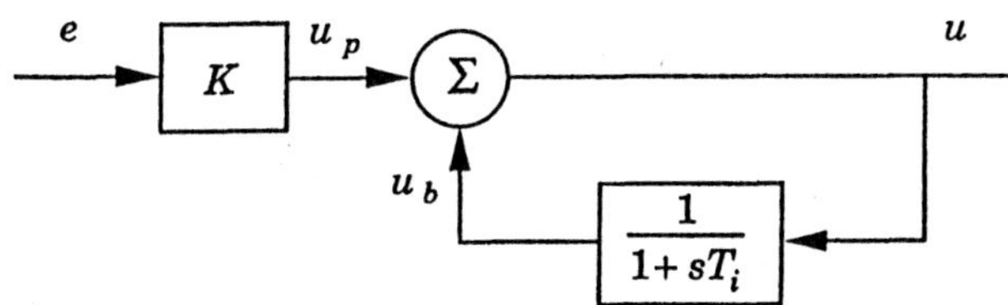

Figure 10.49 Controller with integral action implemented as automatic reset.

The idea is simply to filter out the low frequency part of the error signal and add it to the proportional part. Notice that the closed loop has positive feedback. Analyzing the system in the figure we find that

$$U(s) = K(1 + \frac{1}{sT_i})E(s)$$

which is the input-output relation of a proportional-integral (PI) controller. Furthermore, we have

$$u_b(t) = \frac{K}{T_i} \int^t e(s)ds = u_I(t)$$

The automatic reset is thus the same as integral action.

Notice, however, that set point weighting is not obtained when integral action is obtained as automatic reset.

Derivative Control

Derivative control is used to provide anticipative action. A simple form is

$$u_D(t) = k_d \frac{de(t)}{dt} = KT_d \frac{de(t)}{dt} \qquad (10.99)$$

The combination of proportional and derivative action is then

$$u_P(t) + u_D(t) = K[e(t) + T_d \frac{de(t)}{dt}]$$

This means that control action is based on linear extrapolation of the error T_d time units ahead. See Figure 10.50. Parameter T_d, which is called derivative time, thus has a good intuitive interpretation.

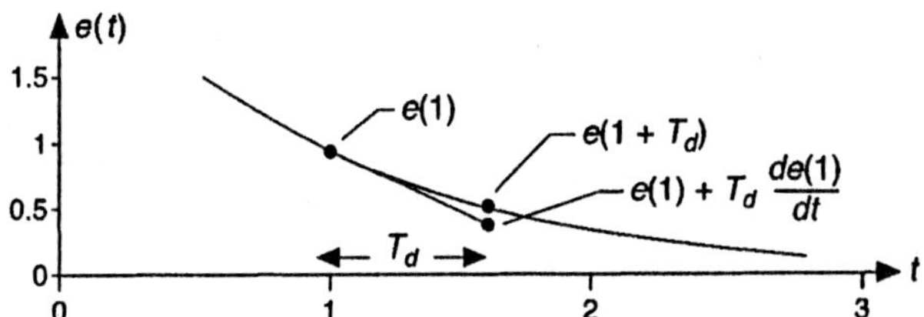

Figure 10.50 Interpretation of derivative action as prediction.

The main difference between a PID controller and a more complex controller is that a dynamic model admits better prediction than straight-line extrapolation.

In many practical applications the set point is piecewise constant. This means that the derivative of the set point is zero except for those time instances when the set point is changed. At these time instances the derivative becomes infinitely large. Linear extrapolation is not useful for predicting such signals. Also, linear extrapolation is inaccurate when the measurement signal changes rapidly compared to the prediction horizon T_d.

A better realization of derivative action is, therefore,

$$U_D(s) = \frac{KT_d s}{1+sT_d/N}(cY_{sp}(s) - Y(s)) \qquad (10.100)$$

The signals pass through a low-pass filter with time constant T_d/N. Parameter c is a set point weighting, which is often set to zero.

Filtering of Process Variable

The process output can sometimes be quite noisy. A first-order filter with the transfer function

$$G_f(s) = \frac{1}{1+sT_f} \qquad (10.101)$$

is often used to filter the signal. For PID controllers that are implemented digitally, the filter can be combined with the antialiasing filter as discussed in Section 10.5.5.

Set Point Weighting

The PID controller introduces extra zeros in the transmission from set point to output. From Equations 10.96, 10.97, and 10.99, the zeros of the PID controller can be determined as the roots of the equation

$$cT_iT_ds^2 + bT_is + 1 = 0 \qquad (10.102)$$

There are no extra zeros if $b = 0$ and $c = 0$. If only $c = 0$, then there is one extra zero at

$$s = -\frac{1}{bT_i} \qquad (10.103)$$

This zero can have a significant influence on the set point response. The overshoot is often too large with $b = 1$. It can be reduced substantially by using a smaller value of b. This is a much better solution than the traditional way of detuning the controller.

This is illustrated in Figure 10.51, which shows PI control of a system with the transfer function

$$G_p(s) = \frac{1}{s+a} \quad (10.104)$$

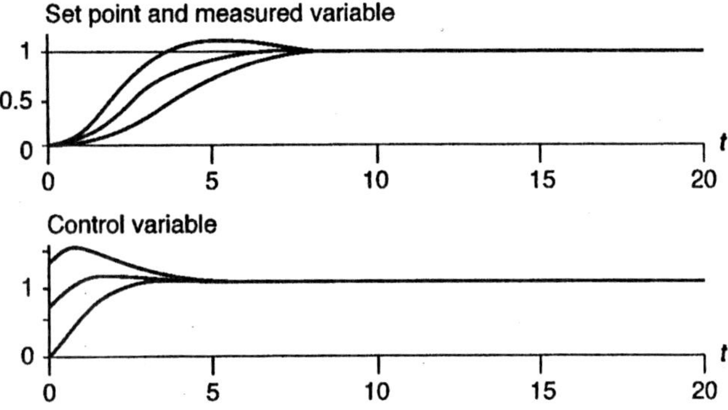

Figure 10.51 The usefulness of set point weighting. The values of the set point weighting parameter are 0, 0.5 and 1.

10.5.3 Different Representations

The PID controller discussed in the previous section can be described by

$$U(s) = G_{sp}(s)Y_{sp}(s) - G_c(s)Y(s) \quad (10.105)$$

where

$$\begin{aligned} G_{sp}(s) &= K(b + \frac{1}{sT_i} + c\frac{sT_d}{1 + sT_d/N}) \\ G_c(s) &= K(1 + \frac{1}{sT_i} + \frac{sT_d}{1 + sT_d/N}) \end{aligned} \quad (10.106)$$

The linear behavior of the controller is thus characterized by two transfer functions: $G_{sp}(s)$, which gives the signal transmission from the set point to the control variable, and $G_c(s)$, which describes the signal transmission from the process output to the control variable.

Notice that the signal transmission from the process output to the control signal is different from the signal transmission from the set point to the control signal if either set point weighting parameter $b \neq 1$ or $c \neq 1$. The PID controller then has two degrees of freedom.

Another way to express this is that the set point parameters make it possible to modify the zeros in the signal transmission from set point to control signal.

The PID controller is thus a simple control algorithm that has seven parameters: controller gain K, integral time T_i, derivative time T_d, maximum derivative gain N, set point weightings b and c, and filter time constant T_f. Parameters K, T_i and T_d are the primary parameters that are normally discussed. Parameter N is a constant, whose value typically is between 5 and 20. The set point weighting parameter b is often 0 or 1, although it is quite useful to use different values. Parameter c is mostly zero in commercial controllers.

The Standard Form

The controller given by Equations 10.105 and 10.106 is called the *standard form*, or the ISA (Instrument Society of America) form. The standard form admits complex zeros, which is useful when controlling systems with oscillatory poles. The parameterization given in Equation 10.106 is the normal one. There are, however, also other parameterizations.

The Parallel Form

A slight variation of the standard form is the *parallel form,* which is described by

$$\begin{aligned} U(s) &= k[bY_{sp}(s) - Y(s)] + \frac{k_i}{s}[Y_{sp}(s) - Y(s)] \\ &+ \frac{k_d s}{1 + sT_{df}}[cY_{sp}(s) - Y(s)] \end{aligned} \quad (10.107)$$

This form has the advantage that it is easy to obtain pure proportional, integral or derivative control simply by setting appropriate parameters to zero. The interpretation of T_i and T_d as integration time and prediction horizon is, however, lost in this representation. The parameters of the controllers given by Equations 10.105 and 10.107 are related by

$$\begin{aligned} k &= K \\ k_i &= \frac{K}{T_i} \\ k_d &= KT_d \end{aligned} \quad (10.108)$$

Use of the different forms causes considerable confusion, particularly when parameter $1/k_i$ is called integral time and k_d derivative time.

The form given by Equation 10.107 is often useful in analytical calculations because the parameters appear linearly. However, the parameters do not have nice physical interpretations.

Series Forms

If $T_i > 4T_d$ the transfer function $G_c(s)$ can be written as

$$G_c'(s) = K'\left(1 + \frac{1}{sT_i'}\right)(1 + sT_d') \quad (10.109)$$

This form is called the series form. If $N = 0$ the parameters are related to the parameters of the parallel form in the following way:

$$\begin{aligned} K &= K'\frac{T_i' + T_d'}{T_i'} \\ T_i &= T_i' + T_d' \end{aligned}$$

$$T_d = \frac{T_i' T_d'}{T_i' + T_d'} \tag{10.110}$$

The inverse relation is

$$\begin{aligned} K' &= \frac{K}{2}\left(1+\sqrt{1-4T_d/T_i}\right) \\ T_i' &= \frac{T_i}{2}\left(1+\sqrt{1-4T_d/T_i}\right) \\ T_d' &= \frac{T_i}{2}\left(1-\sqrt{1-4T_d/T_i}\right) \end{aligned} \tag{10.111}$$

Similar, but more complicated, formulas are obtained for $N \neq 0$. Notice that the parallel form admits complex zeros while the series form has real zeros.

The parallel form given by Equations 10.105 and 10.106 is more general. The series form is also called the classical form because it is obtained naturally when a controller is implemented as automatic reset. The series form has an attractive interpretation in the frequency domain because the zeros of the feedback transfer function are the inverse values of T_i' and T_d'. Because of tradition, the form of the controller remained unchanged when technology changed from pneumatic via electric to digital.

It is important to keep in mind that different controllers may have different structures. This means that if a controller in a certain control loop is replaced by another type of controller, the controller parameters may have to be changed. Note, however, that the series and parallel forms differ only when both the integral and the derivative parts of the controller are used.

The parallel form is the most general form because pure proportional or integral action can be obtained with finite parameters. The controller can also have complex zeros. In this way it is the most flexible form. However, it is also the form where the parameters have little physical interpretation. The series form is the least general because it does not allow complex zeros in the feedback path.

Velocity Algorithms

The PID controllers given by Equations 10.105, 10.107 and 10.109 are called positional algorithms because the output of the algorithms is the control variable. In some cases it is more natural to let the control algorithm generate the rate of change of the control signal. Such a control law is called a velocity algorithm. In digital implementations, velocity algorithms are also called incremental algorithms.

Many early controllers that were built around motors used velocity algorithms. Algorithms and structure were often retained by the manufacturers when technology was changed in order to have products that were compatible with older equipment. Another reason is that many practical issues, like windup protection and bumpless parameter changes, are easy to implement using the velocity algorithm.

A velocity algorithm cannot be used directly for a controller without integral action because such a controller cannot keep the stationary value. The system will have an unstable mode, an integrator, that is canceled. Special care must therefore be exercised for velocity algorithms that allow the integral action to be switched off.

10.5.4 Nonlinear Issues

So far we have discussed only the linear behavior of the PID controller. There are several nonlinear issues that also must be considered. These include effects of actuator saturation, mode switches, and parameter changes.

Actuator Saturation and Windup

All actuators have physical limitations, a control valve cannot be more than fully open or fully closed, a motor has limited velocity, etc. This has severe consequences for control. Integral action in a PID controller is an unstable mode. This does not cause any difficulties when the loop is closed. The feedback loop will, however, be broken when the actuator saturates because the output of the saturating element is then not influenced by its input. The unstable mode in the controller may then drift to very large values. When the actuator desaturates it may then take a long time for the system to recover. It may also happen that the actuator bounces several times between high and low values before the system recovers.

Integrator windup is illustrated in Figure 10.52, which shows simulation of a system where the process dynamics is a saturation at a level of ± 0.1 followed by a linear system with the transfer function

$$G(s) = \frac{1}{s(s+1)}$$

The controller is a PI controller with gain $K = 0.27$ and $T_i = 7.5$. The set point is a unit step. Because of the saturation in the

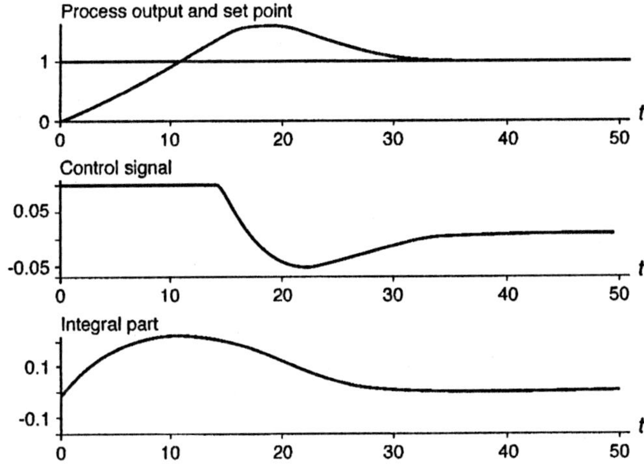

Figure 10.52 Simulation that illustrates integrator windup.

actuator, the control signal saturates immediately when the step is applied. The control signal then remains at the saturation level and the feedback is broken. The integral part continues to increase because the error is positive. The integral part starts to decrease when the output equals the set point, but the output remains saturated because of the large integral part. The output finally decreases around time $t = 14$ when the integral part has decreased sufficiently. The system then settles. The net effect is that there is a large overshoot. This phenomenon, which was observed experimentally very early, is called "integrator windup." Many so-called anti-windup schemes for avoiding windup have been developed; conditional integration and tracking are two common methods.

Conditional Integration

Integrator windup can be avoided by using integral action only when certain conditions are fulfilled. Integral action is thus switched off when the actuator saturates, and it is switched on again when it desaturates. This scheme is easy to implement, but it leads to controllers with discontinuities. Care must also be exercised when formulating the switching logic so that the system does not come to a state where integral action is never used.

Tracking

Tracking or back calculation is another way to avoid windup. The idea is to make sure that the integral is kept at a proper value when the actuator saturates so that the controller is ready to resume action as soon as the control error changes. This can be done as shown in Figure 10.53. The actuator output is measured

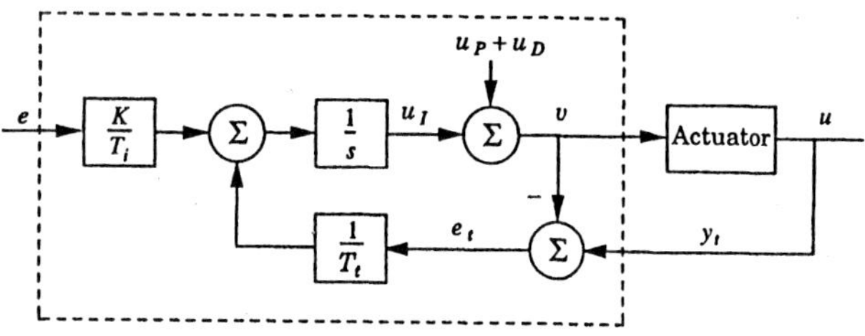

Figure 10.53 PID controller that avoids windup by tracking.

and the signal e_t, which is the difference between the input v and the output u of the actuator, is formed. The signal e_t is different from zero when the actuator saturates. The signal e_t is then fed back to the integrator. The feedback does not have any effect when the actuator does not saturate because the signal e_t is then zero. When the actuator saturates, the feedback drives the integrator output to a value such that the error e_t is zero.

Figure 10.54 illustrates the effect of using the anti-windup scheme. The simulation is identical to the one in Figure 10.52, and the curves from that figure are copied to illustrate the properties of the system. Notice the drastic difference in the behavior of the system. The control signal starts to decrease before the output reaches the set point. The integral part of the controller is also initially driven towards negative values.

The signal y_t may be regarded as an external signal to the controller. The PID controller can then be represented as a block with three inputs, y_{sp}, y and y_t, and one output v, and the anti-windup scheme can then be shown as in Figure 10.55. Notice that tracking is disabled when the signals y_t and v are the same.

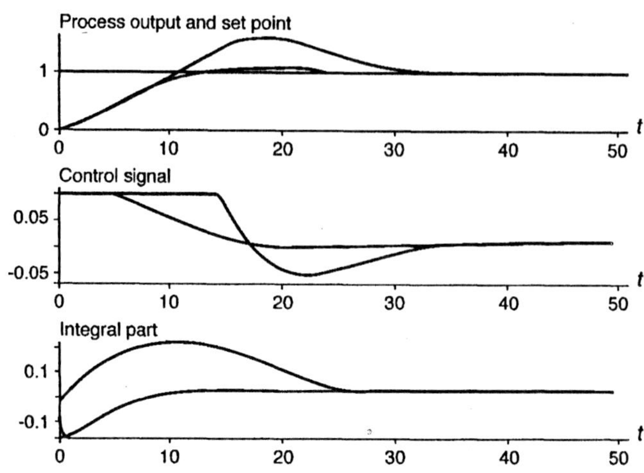

Figure 10.54 Simulation of PID controller with tracking. For comparison, the response for a system without windup protection is also shown. Compare with Figure 10.52.

The signal y_t is called the tracking signal because the output of

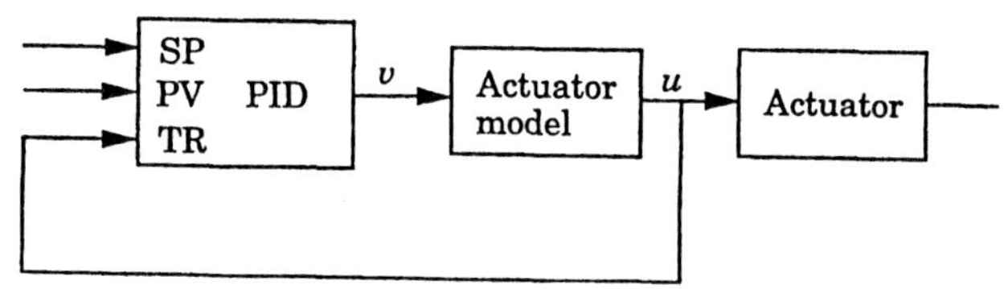

Figure 10.55 Anti-windup in PID controller with tracking input.

the controller tracks this signal. The time constant T_t is called the tracking time constant.

The configuration with a tracking input is very useful when several different controllers are combined to build complex systems. One example is when controllers are coupled in parallel or when selectors are used.

The tracking time constant influences the behavior of the system as shown in Figure 10.56. The values of the tracking constant are 1, 5, 20, and 100. The system recovers faster with smaller

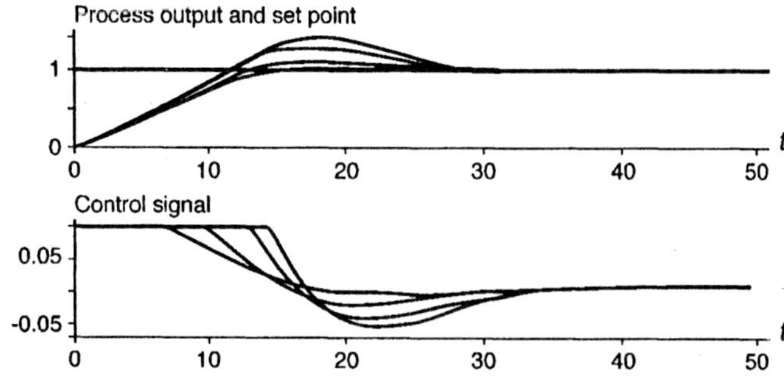

Figure 10.56 Effect of the tracking time constant on the anti-windup. The values of the tracking time constant are 1, 5, 20 and 100.

tracking constants. It is, however, not useful to make the time constant too small because tracking may then be introduced accidentally by noise. It is reasonable to choose $T_t < T_i$ for a PI controller and $T_d < T_t < T_i$ for a PID controller.

The Proportional Band

Let u_{max} and u_{min} denote the limits of the control variable. The proportional band K_p of the controller is then

$$K_p = \frac{u_{max} - u_{min}}{K}$$

This is sometimes used instead of the gain of the controller; the value is often expressed in percent (%).

For a PI controller, the values of the process output that correspond to the limits of the control signal are given by

$$\begin{aligned} y_{max} &= by_{sp} + \frac{u_I - u_{max}}{K} \\ y_{min} &= by_{sp} + \frac{u_I - u_{min}}{K} \end{aligned}$$

The controller operates linearly only if the process output is in the range (y_{min}, y_{max}). The controller output saturates when the predicted output is outside this band. Notice that the proportional band is strongly influenced by the integral term. A good insight into the windup problem and anti-windup schemes is obtained by investigating the proportional band. To illustrate this, Figure 10.57 shows the same simulation as Figure 10.52, but the proportional band is now also shown. The figure shows that

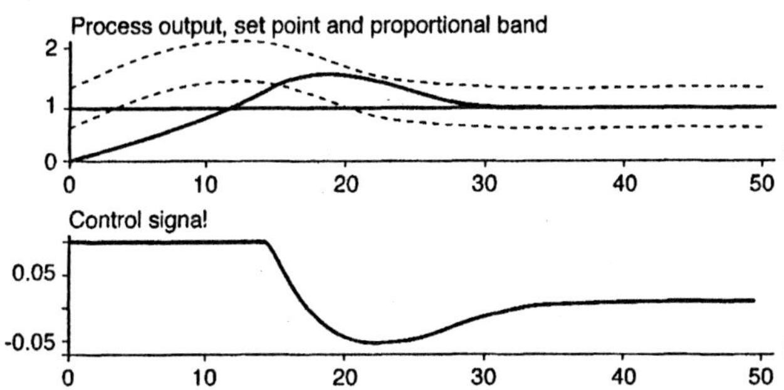

Figure 10.57 Proportional band for simulation in Figure 10.52.

the output is outside the proportional band initially. The control signal is thus saturated immediately. The signal desaturates as soon as the output leaves the proportional band. The large overshoot is caused by windup, which increases the integral when the output saturates.

Anti-Windup in Controller on Series Form

A special method is used to avoid windup in controllers with a series implementation. Figure 10.58 shows a block diagram of the system. The idea is to make sure that the integral term that represents the automatic reset is always inside the saturation limits. The proportional and derivative parts do, however, change the output directly. It is also possible to treat the input to the saturation as an external tracking signal.

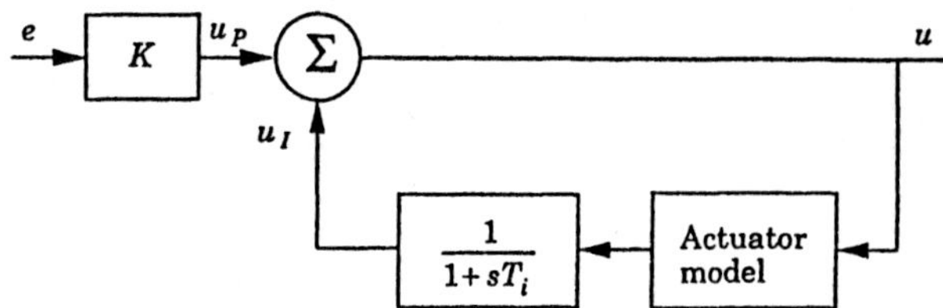

Figure 10.58 A scheme for avoiding windup in a controller with a series implementation.

Notice that the tracking time constant in the controller in Figure 10.58 is equal to the integration time. Better performance can often be obtained with smaller values. This is a limitation of the scheme in Figure 10.58.

Anti-Windup in Velocity Algorithms

In a controller that uses a velocity algorithm we can avoid windup simply by limiting the input to the integrator. The behavior of the system is then similar to a controller with conditional integration.

Mode Switches

Most PID controllers can be operated in one of two modes, manual or automatic. So far we have discussed the automatic mode. In the manual mode the controller output is manipulated directly. This is often done by two buttons labeled "increase" and "decrease". The output is changed with a given rate when a button is pushed. To obtain this function the buttons are connected to the output via an integrator.

It is important that the system be implemented in such a way that there are no transients when the modes are switched. This is very easy to arrange in a controller based on a velocity algorithm, where the same integrator is used in both modes.

It is more complicated to obtain bumpless parameter changes in the other implementations. It is often handled via the tracking mode.

Parameter Changes

Switching transients may also occur when parameters are changed. Some transients cannot be avoided, but others are implementation dependent. In a proportional controller it is unavoidable to have transients if the gain is changed when the control error is different from zero.

For controllers with integral action, it is possible to avoid switching transients even if the parameters are changed when the error is not zero, provided that the controller is implemented properly.

If integral action is implemented as

$$\begin{aligned} \frac{dx}{dt} &= e \\ I &= \frac{K}{T_i}x \end{aligned}$$

there will be a transient whenever K or T_i is changed when $x \neq 0$.

If the integral part is realized as

$$\begin{aligned}\frac{dx}{dt} &= \frac{K}{T_i}e \\ I &= x\end{aligned}$$

we find that the transient is avoided. This is a manifestation that linear time-varying systems do not commute.

10.5.5 Digital Implementation

Most controllers are implemented using digital controllers. In this handbook several chapters deal with these issues. Here we will summarize some issues of particular relevance to PID control. The following operations are performed when a controller is implemented digitally:

- Step 1. Wait for clock interrupt.
- Step 2. Read analog input.
- Step 3. Compute control signal.
- Step 4. Set analog output.
- Step 5. Update controller variables.
- Step 6. Go to 1.

To avoid unnecessary delays, it is useful to arrange the computations so that as many as possible of the calculations are performed in Step 5. In Step 3 it is then sufficient to do two multiplications and one addition.

When computations are based on sampled data, it is good practice to introduce a prefilter that effectively eliminates all frequencies above the Nyquist frequency, $f_N = \pi/h$, where h is the sampling period. If this is not done, high-frequency disturbances may be aliased so that they appear as low-frequency disturbances. In commercial PID controllers this is often done by a first-order system.

Discretization

So far we have characterized the PID controller as a continuous time system. To obtain a computer implementation we have to find a discrete time approximation. There are many ways to do this. Refer to the section on digital control for a general discussion; here we do approximations specifically for the PID controller. We will discuss discretization of the different terms separately. The sampling instants are denoted as t_k where $k = 0, 1, 2, \ldots$. It is assumed that the sampling instants are equally spaced. The sampling period is denoted by h. The proportional action, which is described by

$$u_p = K(by_{sp} - y)$$

is easily discretized by replacing the continuous variables with their sampled versions. This gives

$$u_p(t_k) = K\left(by_{sp}(t_k) - y(t_k)\right) \tag{10.112}$$

The integral term is given by

$$u_I(t) = \frac{K}{T_i}\int_0^t e(\tau)d\tau$$

Differentiation with respect to time gives

$$\frac{du_I}{dt} = \frac{K}{T_i}e$$

There are several ways to discretize this equation. Approximating the derivative by a forward difference gives

$$u_I(t_{k+1}) = u_I(t_k) + \tfrac{Kh}{T_i}\,e(t_k) \tag{10.113}$$

If the derivative is approximated by a backward difference we get instead

$$u_I(t_k) = u_I(t_{k-1}) + \tfrac{Kh}{T_i}\,e(t_k) \tag{10.114}$$

Another possibility is to approximate the integral by the trapezoidal rule, which gives

$$u_I(t_{k+1}) = u_I(t_k) + \tfrac{Kh}{T_i}\,\tfrac{e(t_{k+1})+e(t_k)}{2} \tag{10.115}$$

Yet another method is called ramp equivalence. This method gives exact outputs at the sampling instants if the input signal is continuous and piecewise linear between the sampling instants. In this particular case, the ramp equivalence method gives the same approximation of the integral term as the Tustin approximation. The derivative term is given by

$$\frac{T_d}{N}\frac{du_D}{dt} + u_D = -KT_d\frac{dy}{dt}$$

This equation can be approximated in the same way as the integral term.

The forward difference approximation is

$$\begin{aligned}u_D(t_{k+1}) &= \left(1 - \frac{Nh}{T_d}\right)u_D(t_k) - KN\left(y(t_{k+1})\right. \\ &\quad \left. -y(t_k)\right)\end{aligned} \tag{10.116}$$

The backward difference approximation is

$$\begin{aligned}u_D(t_k) &= \frac{T_d}{T_d + Nh}u_D(t_{k-1}) \\ &\quad - \frac{KT_dN}{T_d + Nh}\left(y(t_k) - y(t_{k-1})\right)\end{aligned} \tag{10.117}$$

Tustin's approximation gives

$$\begin{aligned}u_D(t_k) &= \frac{2T_d - Nh}{2T_d + Nh}u_D(t_{k-1}) \\ &\quad - \frac{2KT_dN}{2T_d + Nh}\left(y(t_k) - y(t_{k-1})\right)\end{aligned} \tag{10.118}$$

The ramp equivalence approximation gives

$$\begin{aligned}u_D(t_k) &= e^{-Nh/T_d}u_D(t_{k-1}) \\ &\quad - \frac{KT_d(1 - e^{-Nh/T_d})}{h} \\ &\quad \left(y(t_k) - y(t_{k-1})\right)\end{aligned} \tag{10.119}$$

Unification

The approximations of the integral and derivative terms have the same form, namely

$$\begin{aligned} u_I(t_k) &= u_I(t_{k-1}) + b_{i1}e(t_k) + b_{i2}e(t_{k-1}) \\ u_D(t_k) &= a_d u_D(t_{k-1}) - b_d(y(t_k) - y(t_{k-1})) \end{aligned} \tag{10.120}$$

The parameters for the different approximations are given in Table 10.7.

TABLE 10.7 Parameters for the Different Approximations.

	Forward	Backward	Tustin	Ramp Equivalence
b_{i1}	0	$\frac{Kh}{T_i}$	$\frac{Kh}{2T_i}$	$\frac{Kh}{2T_i}$
b_{i2}	$\frac{Kh}{T_i}$	0	$\frac{Kh}{2T_i}$	$\frac{Kh}{2T_i}$
a_d	$1-\frac{Nh}{T_d}$	$\frac{T_d}{T_d+Nh}$	$\frac{2T_d-Nh}{2T_d+Nh}$	e^{-Nh/T_d}
b_d	KN	$\frac{KT_dN}{T_d+Nh}$	$\frac{2KT_dN}{2T_d+Nh}$	$\frac{KT_d(1-e^{-Nh/T_d})}{h}$

The controllers obtained can be written as

$$\begin{aligned} u(t_k) &= t_0 y_{sp}(t_k) + t_1 y_{sp}(t_{k-1}) + t_2 y_{sp}(t_{k-2}) \\ &\quad - s_0 y(t_k) - s_1 y(t_{k-1}) - s_2 y(t_{k-2}) \\ &\quad + (1+a_d)u(t_{k-1}) - a_d u(t_{k-2}) \end{aligned} \tag{10.121}$$

where
$$\begin{aligned} s_0 &= K + b_{i1} + b_d \\ s_1 &= -K(1+a_d) - b_{i1}a_d + b_{i2} - 2b_d \\ s_2 &= Ka_d - b_{i2}a_d + b_d \\ t_0 &= Kb + b_{i1} \\ t_1 &= -Kb(1+a_d) - b_{i1}a_d + b_{i2} \\ t_2 &= Kba_d - b_{i2}a_d \end{aligned}$$

Equation 10.121 gives the linear behavior of the controller. To obtain the complete controller we have to add the anti-windup feature and facilities for changing modes and parameters.

Discussion

There is no significant difference between the different approximations of the integral term. The approximations of the derivative term have, however, quite different properties.

The approximations are stable when $|a_d| < 1$. For the forward difference approximation, this implies that $T_d > Nh/2$. The approximation is thus unstable for small values of T_d. The other approximations are stable for all values of T_d. Tustin's approximation and the forward difference method give negative values of a_d if T_d is small. This is undesirable because the approximation then exhibits ringing. The backward difference approximation gives good results for all values of T_d.

Tustin's approximation and the ramp equivalence approximation give the best agreement with the continuous time case; the backward approximation gives less phase advance; and the forward approximation gives more phase advance. The forward approximation is seldom used because of the problems with instability for small values of derivative time T_d. Tustin's algorithm has the ringing problem for small T_d. Ramp equivalence requires evaluation of an exponential function. The backward difference approximation is used most commonly. The backward difference is well behaved.

Computer Code

As an illustration we give the computer code for a reasonably complete PID controller that has set point weighting, limitation of derivative gain, bumpless parameter changes and anti-windup protection by tracking.

Code

```
Compute controller coefficients
bi=K*h/Ti
ad=(2*Td-N*h)/(2*Td+N*h)
bd=2*K*N*Td/(2*Td+N*h)
a0=h/Tt
Bumpless parameter changes
uI=uI+Kold*(bold*ysp-y)-Knew*(bnew*ysp-y)
Read set point and process output from AD converter
ysp=adin(ch1)
y=adin(ch2)
Compute proportional part
uP=K*(b*ysp-y)
Update derivative part
uD=ad*uD-bd*(y-yold)
Compute control variable
v=uP+uI+uD
u=sat(v,ulow,uhigh)
Command analog output
daout(ch1)
Update integral part with windup protection
uI=uI+bi*(ysp-y)+ao*(u-v)
yold=y
```

Precomputation of the controller coefficients *ad, ao, bd* and *bi* in Equation 10.121 saves computer time in the main loop. These computations are made only when the controller parameters are changed. The main program is called once every sampling period. The program has three states: *yold, uI,* and *uD.* One state variable can be eliminated at the cost of a less readable code.

PID controllers are implemented in many different computers, standard processors as well as dedicated machines. Word length is usually not a problem if general-purpose machines are used. For special-purpose systems, it may be possible to choose word length. It is necessary to have sufficiently long word length to properly represent the integral part.

Velocity Algorithms

The velocity algorithm is obtained simply by taking the difference of the position algorithm

$$\Delta u(t_k) = u(t_k) - u(t_{k-1}) = \Delta u_P(t_k) + \Delta I(t_k) + \Delta D(t_k)$$

The differences are then added to obtain the actual value of the control signal. Sometimes the integration is done externally. The differences of the proportional, derivative and integral terms are obtained from Equations 10.112 and 10.120.

$$\begin{aligned}
\Delta u_P(t_k) &= u_P(t_k) - u_P(t_{k-1}) \\
&= K\left(by_{sp}(t_k) - y(t_k) - by_{sp}(t_{k-1}) + y(t_{k-1})\right) \\
\Delta u_I(t_k) &= u_I(t_k) - u_I(t_{k-1}) \\
&= b_{i1}\, e(t_k) + b_{i2}\, e(t_{k-1}) \\
\Delta u_D(t_k) &= u_D(t_k) - u_D(t_{k-1}) \\
&= a_d \Delta u_D(t_{k-1}) \\
&\quad - b_d\left(y(t_k) - 2y(t_{k-1}) + y(t_{k-2})\right)
\end{aligned}$$

One advantage with the incremental algorithm is that most of the computations are done using increments only. Short word-length calculations can often be used. It is only in the final stage where the increments are added that precision is needed. Another advantage is that the controller output is driven directly from an integrator. This makes it very easy to deal with windup and mode switches. A problem with the incremental algorithm is that it cannot be used for controllers with P or proportional-derivative (PD) action only. Therefore, Δu_P has to be calculated in the following way when integral action is not used:

$$\Delta u_P(t_k) = K\left(by_{sp}(t_k) - y(t_k)\right) + u_b - u(t_{k-1})$$

where u_b is the bias term. When there is no integral action, it is necessary to adjust this term to obtain zero steady-state error.

10.5.6 Uses of PID Control

The PID controller is by far the control algorithm that is most commonly used. It is interesting to observe that in order to obtain a functional controller it is necessary to consider linear and nonlinear behavior of the controller as well as operational issues such as mode switches and tuning. For a discussion of tuning we refer to the chapter "Automatic Tuning of PID Controllers" in this handbook. These questions have been worked out quite well for PID controllers, and the issues involved are quite well understood.

The PID controller in many cases gives satisfactory performance. It can often be used on processes that are difficult to control provided that extreme performance is not required. There are, however, situations when it is possible to obtain better performance by other types of controllers. Typical examples are processes with long relative dead times and oscillatory systems.

There are also cases where PID controllers are clearly inadequate. If we consider the fact that a PI controller always has phase lag and that a PID controller can provide a phase lead of at most 90°, it is clear that neither will work for systems that require more phase advance. A typical example is stabilization of unstable systems with time delays.

A few examples are given as illustrations.

Systems with Long Time Delays

Processes with long time delays are difficult to control. The loop gain with proportional control is very small so integral action is necessary to get good control. Such processes can be controlled by PI controllers, but the performance can be increased by more sophisticated controllers. The reason derivative action is not so useful for processes of this type is that prediction by linear extrapolation of the output is not very effective. To make a proper prediction, it is necessary to take account of the past control signals that have not yet shown up in the output. To illustrate this we consider a process with the transfer function

$$G(s) = \frac{e^{-10s}}{(s+1)^3}$$

The dynamics of this process is dominated by the time delay. A good PI controller that gives a step response without overshoot has a gain $K = 0.27$ and $T_i = 4.8$. The response to set point changes and load disturbances of the system is shown in Figure 10.59. This figure shows the response to a step in the set point at time $t = 0$ and a step at the process input at time $t = 50$.

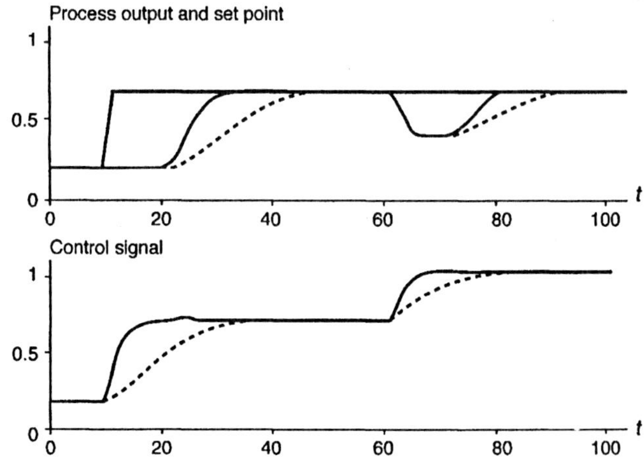

Figure 10.59 Control of a process with long time delays with a PI controller (dashed lines) and a Smith predictor (solid lines).

One way to obtain improved control is to use a controller with a Smith predictor. This controller requires a model of the process. If a model in the form of a first-order system with gain K_p, time constant T, and a time delay L is used, the controller becomes

$$\begin{aligned}
U(s) &= K\left(1 + \frac{1}{sT_i}\right)\Big(E(s) \\
&\quad - \frac{K_p}{1+sT}\left(1 - e^{-sL}\right)U(s)\Big) \qquad (10.122)
\end{aligned}$$

The controller can predict the output better than a PID controller because of the internal process model. The last term in the right-hand side of Equation 10.122 can be interpreted as the effect on the output of control signals that have been applied in the time interval $(t - T, t)$. Because of the time delay the effect of these signals has not appeared in the output at time t. The improved performance is seen in the simulation in Figure 10.59.

If load disturbance response is evaluated with the integrated absolute error (IAE), we find that the Smith predictor is about 30% better than the PI controller. There are situations when the increased complexity is worth while.

Systems with Oscillatory Modes

Systems with poorly damped oscillatory modes are another case where more complex controllers can outperform PID control. The reason for this is that it pays to have a more complex model in the controller. To illustrate this we consider a system with the transfer function

$$G(s) = \frac{25}{(s+1)(s^2+25)}$$

This system has two complex undamped poles.

The system cannot be stabilized with a PI controller with positive coefficients. To stabilize the undamped poles with a PI controller, it is necessary to have controllers with a zero in the right half-plane. Some damping of the unstable poles can be provided in this way. It is advisable to choose set point weighting $b = 0$ in order to avoid unnecessary excitation of the modes. The response obtained with such a PID controller is shown in Figure 10.60. In this figure a step change in the set point has been introduced at time $t = 0$, and a step change in the load disturbance has been applied at time $t = 20$. The set point weighting b is zero. Because of this we avoid a right half-plane zero in the transfer function from set point to output, and the oscillatory modes are not excited much by changes in the set point. The oscillatory modes are, however, excited by the load disturbance.

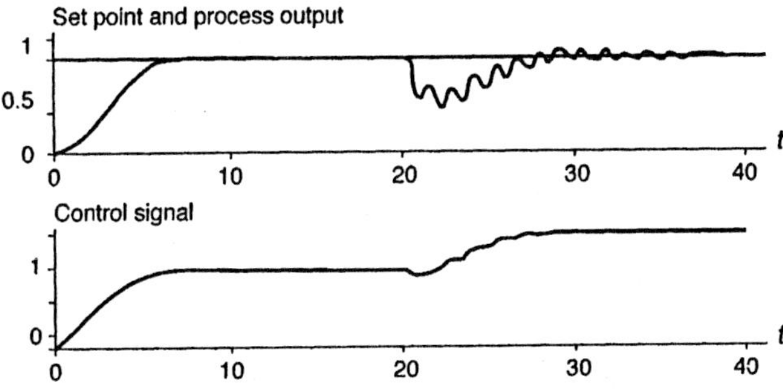

Figure 10.60 Control of an oscillatory system with PI control. The controller parameters are $K = -0.25$, $T_i = -1$ and $b = 0$.

By using a controller that is more complex than a PID controller it is possible to introduce damping in the system. This is illustrated by the simulation in Figure 10.61. The controller has the transfer functions

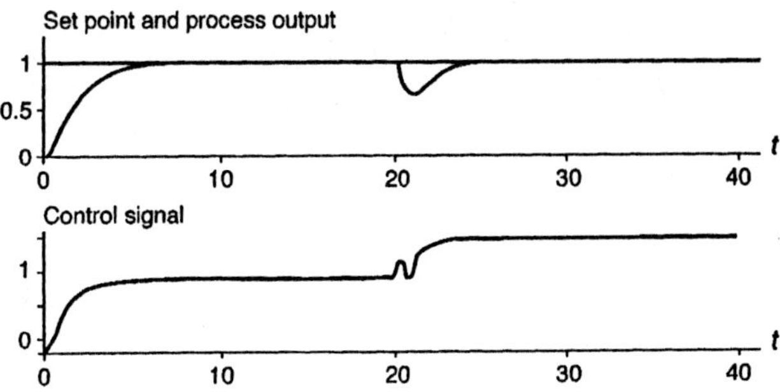

Figure 10.61 Control of the system in Figure 10.60 with a third-order controller.

$$\begin{aligned} G_c(s) &= \frac{21s^3 - 14s^2 + 65s + 100}{s(s^2 + 16s + 165)} \\ G_{sp}(s) &= \frac{100}{s(s^2 + 16s + 165)} \end{aligned}$$

The transfer function $G_c(s)$ has poles at 0 and $-8 \pm 10.05i$ and zeros at -1 and $0.833 \pm 2.02i$. Notice that the controller has two complex zeros in the right half-plane. This is typical for controllers of oscillatory systems. The controller transfer function can be written as

$$\begin{aligned} G_c(s) &= 0.6061(1 + \frac{1}{s}) \frac{1 - 0.35s + 0.21s^2}{1 + 0.0970s + 0.00606s^2} \\ G_{sp} &= \frac{0.6061}{s} \cdot \frac{1}{1 + 0.0970s + 0.00606s^2} \end{aligned}$$

The controller can thus be interpreted as a PI controller with an additional compensation. Notice that the gain of the controller is 2.4 times larger than the gain of the PI controller used in the simulation in Figure 10.60. This gives faster set point response and a better rejection of load disturbances.

10.5.7 Bottom-Up Design of Complex Systems

Control problems are seldom solved by a single controller. Many control systems are designed using a "bottom up" approach where PID controllers are combined with other components, such as filters, selectors and others.

Cascade Control

Cascade control is used when there are several measured signals and one control variable. It is particularly useful when there are significant dynamics (e.g., long dead times or long time constants) between the control variable and the process variable. Tighter control can then be achieved by using an intermediate measured signal that responds faster to the control signal. Cascade control is built up by nesting the control loops, as shown in Figure 10.62. The system in this figure has two loops. The inner loop is called *the secondary loop*; the outer loop is called *the primary loop*. The reason for this terminology is that the outer loop controls the signal we are primarily interested in. It is also possible to have a cascade control with more nested loops.

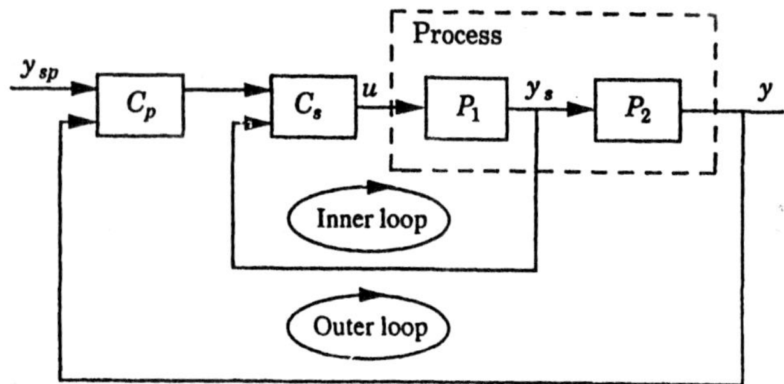

Figure 10.62 Block diagram of a system with cascade control.

The performance of a system can be improved with a number of measured signals, up to a certain limit. If all state variables are measured, it is often not worthwhile to introduce other measured variables. In such a case the cascade control is the same as state feedback.

Feedforward Control

Disturbances can be eliminated by feedback. With a feedback system it is, however, necessary that there be an error before the controller can take actions to eliminate disturbances. In some situations it is possible to measure disturbances before they have influenced the processes. It is then natural to try to eliminate the effects of the disturbances before they have created control errors. This control paradigm is called *feedforward*. The principle is illustrated simply in Figure 10.63. Feedforward can be used for

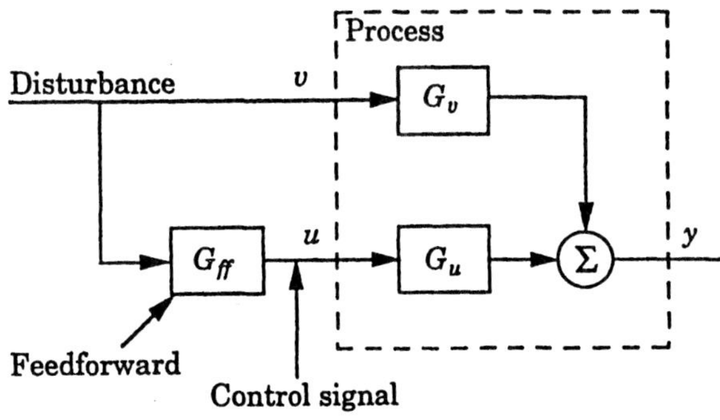

Figure 10.63 Block diagram of a system with feedforward control from a measurable disturbance.

both linear and nonlinear systems. It requires a mathematical model of the process.

As an illustration we consider a linear system that has two inputs, the control variable u and the disturbance v, and one output y. The transfer function from disturbance to output is G_v, and the transfer function from the control variable to the output is G_u. The process can be described by

$$Y(s) = G_u(s)U(s) + G_v(s)V(s)$$

where the Laplace transformed variables are denoted by capital letters. The feedforward control law

$$U(s) = -\frac{G_v(s)}{G_u(s)}V(s)$$

makes the output zero for all disturbances v. The feedforward transfer function thus should be chosen as

$$G_{ff}(s) = -\frac{G_v(s)}{G_u(s)}$$

10.5.8 Selector Control

Selector control can be viewed as the inverse of split range control. In split range, there is one measured signal and several actuators. In selector control, there are many measured signals and only one actuator. A selector is a static device with many inputs and one output. There are two types of selectors: *maximum* and *minimum*. For a maximum selector, the output is the largest of the input signals.

There are situations where several controlled process variables must be taken into account. One variable is the primary controlled variable, but it is also required that other process variables remain within given ranges. Selector control can be used to achieve this. The idea is to use several controllers and to have a selector that chooses the controller that is most appropriate. For example, selector control is used when the primary controlled variable is temperature and we must ensure that pressure does not exceed a certain range for safety reasons.

The principle of selector control is illustrated in Figure 10.64. The primary controlled variable is the process output y. There

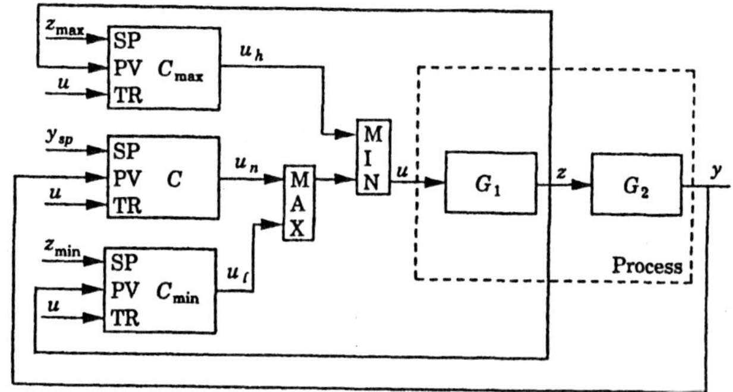

Figure 10.64 Control system with selector control.

is an auxiliary measured variable z that should be kept within the limits z_{min} and z_{max}. The primary controller C has process variable y, setpoint y_{sp} and output u_n. There are also secondary controllers with measured process variables that are the auxiliary variable z and with set points that are bounds of the variable z. The outputs of these controllers are u_h and u_l. The controller C is an ordinary PI or PID controller that gives good control under normal circumstances. The output of the minimum selector is the smallest of the input signals; the output of the maximum selector is the largest of the inputs.

Under normal circumstances the auxiliary variable is larger than the minimum value z_{min} and smaller than the maximum value z_{max}. This means that the output u_h is large and the output u_l is small. The maximum selector, therefore, selects u_n and the minimum selector also selects u_n. The system acts as if the maximum and minimum controller were not present. If the variable z reaches its upper limit, the variable u_h becomes small and is selected by the minimum selector. This means that the control system now attempts to control the variable z and drive it towards its limit. A similar situation occurs if the variable z becomes smaller than z_{min}. To avoid windup, the finally selected control u is used as a tracking signal for all controllers.

References

[1] Åström, K. J. and Hägglund, T., *PID Control—Theory, Design and Tuning*, 2nd ed., Instrument Society of America, Research Triangle Park, NC, 1995.

[2] Åström, K. J., Hägglund, T., Hang, C.C., and Ho, W. K., Automatic tuning and adaptation for PID controllers—a survey, *Control Eng. Pract.*, 1(4), 699–714, 1993.

[3] Åström, K. J., Hang, C.C., Persson, P., and Ho, W. K., Towards intelligent PID control, *Automatica*, 28(1), 1–9, 1992.

[4] Fertik, H. A. Tuning controllers for noisy processes, *ISA Trans.*, 14, 292–304, 1975.

[5] Fertik, H. A. and Ross, C.W., Direct digital control algorithms with anti-windup feature, *ISA Trans.*, 6(4), 317–328, 1967.

[6] Ross, C. W., Evaluation of controllers for deadtime processes, *ISA Trans.*, 16(3), 25–34, 1977.

[7] Seborg, D. E., Edgar, T.F., and Mellichamp, D.A., *Process Dynamics and Control*, Wiley, New York, 1989.

[8] Shinskey, F. G. *Process-Control Systems. Application, Design, and Tuning*, 3rd ed., McGraw-Hill, New York, 1988.

[9] Smith, C. L. and Murrill, P.W., A more precise method for tuning controllers, *ISA Journal*, May, 50–58, 1966.

10.6 State Space – Pole Placement

Katsuhiko Ogata, University of Minnesota

10.6.1 Introduction

In this chapter[2] we present a design method commonly called the pole placement or pole assignment technique. We assume that all state variables are measurable and are available for feedback. It will be shown that if the system considered is completely state controllable, then poles of the closed-loop system may be placed at any desired locations by means of state feedback through an appropriate state feedback gain matrix.

The present design technique begins with a determination of the desired closed-loop poles based on the transient-response and/or frequency-response requirements, such as speed, damping ratio, or bandwidth, as well as steady-state requirements.

Let us assume that we decide that the desired closed-loop poles are to be at $s = \mu_1, s = \mu_2, \ldots, s = \mu_n$. By choosing an appropriate gain matrix for state feedback, it is possible to force the system to have closed-loop poles at the desired locations, provided that the original system is completely state controllable.

In what follows, we treat the case where the control signal is a scalar and prove that a necessary and sufficient condition that the closed-loop poles can be placed at any arbitrary locations in the s plane is that the system be completely state controllable. Then we discuss three methods for determining the required state feedback gain matrix.

It is noted that when the control signal is a vector quantity, the state feedback gain matrix is not unique. It is possible to choose freely more than n parameters; that is, in addition to being able to place n closed-loop poles properly, we have the freedom to satisfy some of the other requirements, if any, of the closed-loop system. This chapter, however, discusses only the case where the control signal is a scalar quantity. (For the case where the control signal is a vector quantity, refer to MIMO LTI systems in this handbook.)

10.6.2 Design via Pole Placement

In the conventional approach to the design of a single-input, single-output control system, we design a controller (compensator) such that the dominant closed-loop poles have a desired damping ratio ζ and undamped natural frequency ω_n. In this approach, the order of the system may be raised by 1 or 2 unless pole-zero cancellation takes place. Note that in this approach we assume the effects on the responses of nondominant closed-loop poles to be negligible.

Different from specifying only dominant closed-loop poles (conventional design approach), the present pole placement approach specifies all closed-loop poles. (There is a cost associated with placing all closed-loop poles, however, because placing all

[2] Most of the material presented here is from [1].

closed-loop poles requires successful measurements of all state variables or else requires the inclusion of a state observer in the system.) There is also a requirement on the part of the system for the closed-loop poles to be placed at arbitrarily chosen locations. The requirement is that the system be completely state controllable.

Consider a control system

$$\dot{x} = Ax + Bu \tag{10.123}$$

where x	=	state vector (n-vector)
u	=	control signal (scalar)
A	=	$n \times n$ constant matrix
B	=	$n \times 1$ constant matrix

We shall choose the control signal to be

$$u = -Kx \tag{10.124}$$

This means that the control signal is determined by instantaneous state. Such a scheme is called *state feedback*. The $1 \times n$ matrix K is called the state feedback gain matrix. In the following analysis we assume that u is unconstrained.

Substituting Equation 10.124 into Equation 10.123 gives

$$\dot{x}(t) = (A - BK)x(t)$$

The solution of this equation is given by

$$x(t) = e^{(A-BK)t}x(0) \tag{10.125}$$

where $x(0)$ is the initial state caused by external disturbances. The stability and transient response characteristics are determined by the eigenvalues of matrix $A - BK$. If matrix K is chosen properly, then matrix $A - BK$ can be made an asymptotically stable matrix, and for all $x(0) \neq 0$ it is possible to make $x(t)$ approach 0 as t approaches infinity. The eigenvalues of matrix $A - BK$ are called the regulator poles. If these regulator poles are located in the left half of the s plane, then $x(t)$ approaches 0 as t approaches infinity. The problem of placing the closed-loop poles at the desired location is called a pole placement problem.

Figure 10.65(a) shows the system defined by Equation 10.123. It is an open-loop control system because the state x is not fed back to the control signal u. Figure 10.65(b) shows the system with state feedback. This is a closed-loop control system because the state x is fed back to the control signal u.

In what follows, we prove that arbitrary pole placement for a given system is possible if and only if the system is completely state controllable.

10.6.3 Necessary and Sufficient Condition for Arbitrary Pole Placement

Consider the control system defined by Equation 10.123. We assume that the magnitude of the control signal u is unbounded. If the control signal u is chosen as

$$u = -Kx$$

where K is the state feedback gain matrix ($1 \times n$ matrix), then the system becomes a closed-loop control system as shown in Figure 10.65(b) and the solution to Equation 10.123 becomes as given by Equation 10.125, or

$$x(t) = e^{(A-BK)t}x(0)$$

Note that the eigenvalues of matrix $A - BK$ (which we denote $\mu_1, \mu_2, \ldots, \mu_n$) are the desired closed-loop poles.

We now prove that a necessary and sufficient condition for arbitrary pole placement is that the system be completely state controllable. We first derive the necessary condition. We begin by proving that if the system is not completely state controllable, then there are eigenvalues of matrix $A - BK$ that cannot be controlled by state feedback.

Suppose the system of Equation 10.123 is not completely state controllable. Then the rank of the controllability matrix is less than n, or

$$\text{rank}[B|AB|\cdots|A^{n-1}B] = q < n$$

This means that there are q linearly independent column vectors in the controllability matrix. Let us define such q linearly independent column vectors as $f_1, f_2, \ldots, f_q$. Also, let us choose $n - q$ additional n-vectors $v_{q+1}, v_{q+2}, \ldots, v_n$ such that

$$P = [f_1|f_2|\cdots|f_q|v_{q+1}|v_{q+2}|\cdots|v_n]$$

is of rank n. Then it can be shown that

$$\hat{A} = P^{-1}AP = \left[\begin{array}{c|c} A_{11} & A_{12} \\ \hline 0 & A_{22} \end{array}\right], \quad \hat{B} = P^{-1}B = \left[\begin{array}{c} B_{11} \\ \hline 0 \end{array}\right]$$

Define

$$\hat{K} = KP = [k_1 \mid k_2]$$

Then we have

$$\begin{aligned}
|sI - A + BK| &= |P^{-1}(sI - A + BK)P| \\
&= |sI - P^{-1}AP + P^{-1}BKP| \\
&= |sI - \hat{A} + \hat{B}\hat{K}| \\
&= \left| sI - \left[\begin{array}{c|c} A_{11} & A_{12} \\ \hline 0 & A_{22} \end{array}\right] \right. \\
&\quad \left. + \left[\begin{array}{c} B_{11} \\ \hline 0 \end{array}\right][k_1 \mid k_2] \right| \\
&= \begin{vmatrix} sI_q - A_{11} + B_{11}k_1 & -A_{12} + B_{11}k_2 \\ 0 & sI_{n-q} - A_{22} \end{vmatrix} \\
&= |sI_q - A_{11} + B_{11}k_1| \cdot |sI_{n-q} - A_{22}| \\
&= 0
\end{aligned}$$

where I_q is a q-dimensional identity matrix and I_{n-q} is an $(n - q)$-dimensional identity matrix.

Notice that the eigenvalues of A_{22} do not depend on K. Thus, if the system is not completely state controllable, then there are eigenvalues of matrix A that cannot be arbitrarily placed. Therefore, to place the eigenvalues of matrix $A - BK$ arbitrarily, the

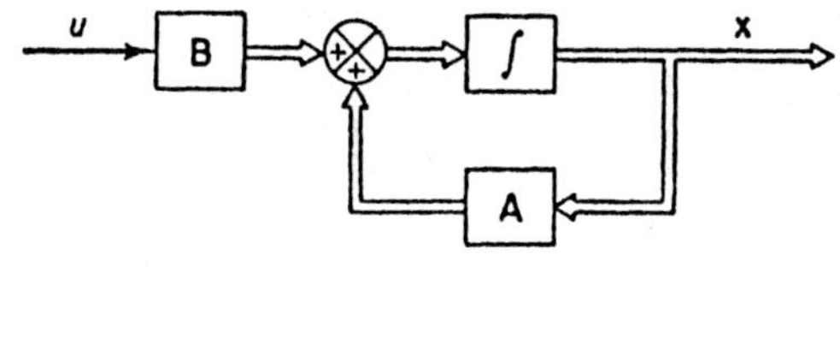

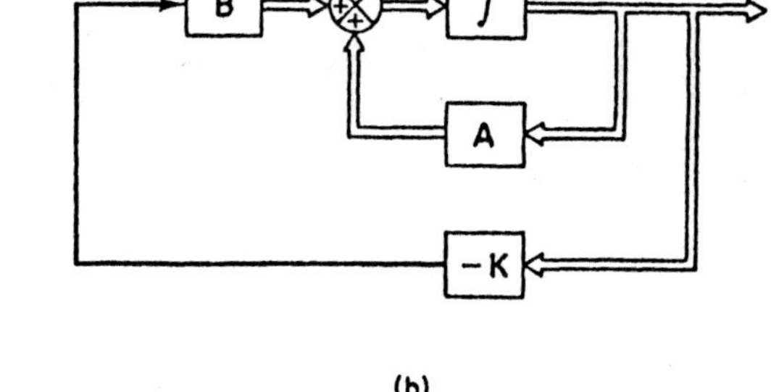

Figure 10.65 (a) Open-loop control system; (b) closed-loop control system with $u = -Kx$. (From Ogata, Katsuhiko, *Modern Control Engineering*, 2nd ed., Prentice Hall, Inc., Englewood Cliffs, NJ, 1990, 777. With permission.)

system must be completely state controllable (necessary condition).

Next we prove a sufficient condition: that is, if the system is completely state controllable (meaning that matrix M given by Equation 10.127 has an inverse), then all eigenvalues of matrix A can be arbitrarily placed.

In proving a sufficient condition, it is convenient to transform the state equation given by Equation 10.123 into the controllable canonical form.

Define a transformation matrix T by

$$T = MW \tag{10.126}$$

where M is the controllability matrix

$$M = [B|AB|\cdots|A^{n-1}B] \tag{10.127}$$

and

$$W = \begin{bmatrix} a_{n-1} & a_{n-2} & \cdots & a_1 & 1 \\ a_{n-2} & a_{n-3} & \cdots & 1 & 0 \\ \vdots & \vdots & & \vdots & \vdots \\ a_1 & 1 & \cdots & 0 & 0 \\ 1 & 0 & \cdots & 0 & 0 \end{bmatrix} \tag{10.128}$$

where each a_i is a coefficient of the characteristic polynomial

$$|sI - A| = s^n + a_1 s^{n-1} + \cdots + a_{n-1}s + a_n$$

Define a new state vector $\hat{x}$ by

$$x = T\hat{x}$$

If the rank of the controllability matrix M is n (meaning that the system is completely state controllable), then the inverse of matrix T exists and Equation 10.123 can be modified to

$$\dot{\hat{x}} = T^{-1}AT\hat{x} + T^{-1}Bu \tag{10.129}$$

where

$$T^{-1}AT = \begin{bmatrix} 0 & 1 & 0 & \cdots & 0 \\ 0 & 0 & 1 & \cdots & 0 \\ \vdots & \vdots & \vdots & & \vdots \\ 0 & 0 & 0 & \cdots & 1 \\ -a_n & -a_{n-1} & -a_{n-2} & \cdots & -a_1 \end{bmatrix} \tag{10.130}$$

$$T^{-1}B = \begin{bmatrix} 0 \\ 0 \\ \vdots \\ 0 \\ 1 \end{bmatrix} \tag{10.131}$$

Equation 10.129 is in the controllable canonical form. Thus, given a state equation, Equation 10.123, it can be transformed into the controllable canonical form if the system is completely state controllable and if we transform the state vector x into state vector $\hat{x}$ by use of the transformation matrix T given by Equation 10.126.

Let us choose a set of the desired eigenvalues as $\mu_1, \mu_2, \ldots, \mu_n$. Then the desired characteristic equation becomes

$$(s - \mu_1)(s - \mu_2)\cdots(s - \mu_n) = s^n + \alpha_1 s^{n-1} + \cdots + \alpha_{n-1}s + \alpha_n = 0 \tag{10.132}$$

Let us write

$$\hat{K} = KT = [\delta_n\ \delta_{n-1}\ \cdots\ \delta_1] \tag{10.133}$$

When $u = -\hat{K}\hat{x} = -KT\hat{x}$ is used to control the system given by Equation 10.129, the system equation becomes

$$\dot{\hat{x}} = T^{-1}AT\hat{x} - T^{-1}BKT\hat{x}$$

The characteristic equation is

$$|sI - T^{-1}AT + T^{-1}BKT| = 0$$

This characteristic equation is the same as the characteristic equation for the system, defined by Equation 10.123, when $u = -Kx$ is used as the control signal. This can be seen as follows: Since

$$\dot{x} = Ax + Bu = (A - BK)x$$

the characteristic equation for this system is

$$\begin{aligned} |sI - A + BK| &= |T^{-1}(sI - A + BK)T| \\ &= |sI - T^{-1}AT + T^{-1}BKT| = 0 \end{aligned}$$

Now let us simplify the characteristic equation of the system in the controllable canonical form. Referring to Equations 10.130,

10.131 and 10.133, we have

$$
\begin{aligned}
&|sI - T^{-1}AT + T^{-1}BKT| \\
&= \left| sI - \begin{bmatrix} 0 & 1 & \cdots & 0 \\ \vdots & \vdots & & \vdots \\ 0 & 0 & \cdots & 1 \\ -a_n & -a_{n-1} & \cdots & -a_1 \end{bmatrix} \right. \\
&\quad \left. + \begin{bmatrix} 0 \\ \vdots \\ 0 \\ 1 \end{bmatrix} [\delta_n \; \delta_{n-1} \cdots \delta_1] \right| \\
&= \begin{vmatrix} s & -1 & \cdots & 0 \\ 0 & s & \cdots & 0 \\ \vdots & \vdots & & \vdots \\ a_n + \delta_n & a_{n-1} + \delta_{n-1} & \cdots & s + a_1 + \delta_1 \end{vmatrix} \\
&= s^n + (a_1 + \delta_1)s^{n-1} + \cdots + (a_{n-1} + \delta_{n-1})s \\
&\quad + (a_n + \delta_n) = 0 \qquad (10.134)
\end{aligned}
$$

This is the characteristic equation for the system with state feedback. Therefore, it must be equal to Equation 10.132, the desired characteristic equation. By equating the coefficients of like powers of s, we get

$$
\begin{aligned}
a_1 + \delta_1 &= \alpha_1 \\
a_2 + \delta_2 &= \alpha_2 \\
&\vdots \\
a_n + \delta_n &= \alpha_n
\end{aligned}
$$

Solving the preceding equations for each δ_i and substituting them into Equation 10.133, we obtain

$$
\begin{aligned}
K = \hat{K}T^{-1} &= [\delta_n \; \delta_{n-1} \cdots \delta_1]\, T^{-1} \\
&= [\alpha_n - a_n | \alpha_{n-1} - a_{n-1} | \\
&\quad \cdots | \alpha_2 - a_2 | \alpha_1 - a_1]\, T^{-1} \qquad (10.135)
\end{aligned}
$$

Thus, if the system is completely state controllable, all eigenvalues can be arbitrarily placed by choosing matrix K according to Equation 10.135 (sufficient condition).

We have thus proved that the necessary and sufficient condition for arbitrary pole placement is that the system be completely state controllable.

10.6.4 Design Steps for Pole Placement

Suppose that the system is defined by

$$\dot{x} = Ax + Bu$$

and the control signal is given by

$$u = -Kx$$

The feedback gain matrix K that forces the eigenvalues of $A - BK$ to be $\mu_1, \mu_2, \ldots, \mu_n$ (desired values) can be determined by the following steps. (If μ_i is a complex eigenvalue, then its conjugate must also be an eigenvalue of $A - BK$.)

Step 1. Check the controllability condition for the system. If the system is completely state controllable, then use the following steps.

Step 2. From the characteristic polynomial for matrix A :

$$|sI - A| = s^n + a_1 s^{n-1} + \cdots + a_{n-1}s + a_n$$

determine the values of $a_1, a_2, \ldots, a_n$.

Step 3. Determine the transformation matrix T that transforms the system state equation into the controllable canonical form. (If the given system equation is already in the controllable canonical form, then $T = I$.) It is not necessary to write the state equation in the controllable canonical form. All we need here is to find the matrix T. The transformation matrix T is given by Equation 10.126, or

$$T = MW$$

where M is given by Equation 10.127 and W is given by Equation 10.128.

Step 4. Using the desired eigenvalues (desired closed-loop poles), write the desired characteristic polynomial

$$
\begin{aligned}
&(s - \mu_1)(s - \mu_2) \cdots (s - \mu_n) = \\
&\quad s^n + \alpha_1 s^{n-1} + \cdots + \alpha_{n-1}s + \alpha_n
\end{aligned}
$$

and determine the values of $\alpha_1, \alpha_2, \ldots, \alpha_n$.

Step 5. The required state feedback gain matrix K can be determined from Equation 10.135, rewritten thus:

$$K = [\alpha_n - a_n | \alpha_{n-1} - a_{n-1} | \cdots | \alpha_2 - a_2 | \alpha_1 - a_1]\, T^{-1}$$

10.6.5 Comments

Note that if the system is of lower order ($n \leq 3$), then direct substitution of matrix K into the desired characteristic polynomial may be simpler. For example, if $n = 3$, then write the state feedback gain matrix K as

$$K = [k_1 \; k_2 \; k_3]$$

Substitute this K matrix into the desired characteristic polynomial $|sI - A + BK|$ and equate it to $(s - \mu_1)(s - \mu_2)(s - \mu_3)$, or

$$|sI - A + BK| = (s - \mu_1)(s - \mu_2)(s - \mu_3)$$

Since both sides of this characteristic equation are polynomials in s, by equating the coefficients of the like powers of s on both sides it is possible to determine the values of k_1, k_2, and k_3. This approach is convenient if $n = 2$ or 3. (For $n = 4, 5, 6, \ldots$, this approach may become very tedious.)

There are other approaches for the determination of the state feedback gain matrix K. In what follows, we present a well-known formula, known as Ackermann's formula, for the determination of the state feedback gain matrix K.

10.6.6 Ackermann's Formula

Consider the system given by Equation 10.123, rewritten thus:

$$\dot{x} = Ax + Bu$$

We assume that the system is completely state controllable. We also assume that the desired closed-loop poles are at $s = \mu_1, s = \mu_2, \ldots, s = \mu_n$.

Use of the state feedback control

$$u = -Kx$$

modifies the system equation to

$$\dot{x} = (A - BK)x \tag{10.136}$$

Let us define

$$\tilde{A} = A - BK$$

The desired characteristic equation is

$$\begin{aligned} |sI - A + BK| &= |sI - \tilde{A}| \\ &= (s-\mu_1)(s-\mu_2)\cdots(s-\mu_n) \\ &= s^n + \alpha_1 s^{n-1} + \cdots + \alpha_{n-1}s + \alpha_n = 0 \end{aligned}$$

Since the Cayley-Hamilton theorem states that $\tilde{A}$ satisfies its own characteristic equation, we have

$$\phi(\tilde{A}) = \tilde{A}^n + \alpha_1 \tilde{A}^{n-1} + \cdots + \alpha_{n-1}\tilde{A} + \alpha_n I = 0 \tag{10.137}$$

We utilize Equation 10.137 to derive Ackermann's formula. To simplify the derivation, we consider the case where $n = 3$. (For any other positive integer n, the following derivation can be easily extended.)

Consider the following identities:

$$\begin{aligned} I &= I \\ \tilde{A} &= A - BK \\ \tilde{A}^2 &= (A-BK)^2 = A^2 - ABK - BK\tilde{A} \\ \tilde{A}^3 &= (A-BK)^3 = A^3 - A^2BK - ABK\tilde{A} - BK\tilde{A}^2 \end{aligned}$$

Multiplying the preceding equations in order by $\alpha_3, \alpha_2, \alpha_1, \alpha_0$ (where $\alpha_0 = 1$), respectively, and adding the results, we obtain

$$\begin{aligned} \alpha_3 I &+ \alpha_2\tilde{A} + \alpha_1\tilde{A}^2 + \tilde{A}^3 \\ &= \alpha_3 I + \alpha_2(A - BK) + \alpha_1(A^2 - ABK - BK\tilde{A}) \\ &\quad + A^3 - A^2BK - ABK\tilde{A} - BK\tilde{A}^2 \\ &= \alpha_3 I + \alpha_2 A + \alpha_1 A^2 + A^3 - \alpha_2 BK \\ &\quad - \alpha_1 ABK - \alpha_1 BK\tilde{A} - A^2BK - ABK\tilde{A} - BK\tilde{A}^2 \end{aligned} \tag{10.138}$$

Referring to Equation 10.137, we have

$$\alpha_3 I + \alpha_2\tilde{A} + \alpha_1\tilde{A}^2 + \tilde{A}^3 = \phi(\tilde{A}) = 0$$

Also, we have

$$\alpha_3 I + \alpha_2 A + \alpha_1 A^2 + A^3 = \phi(A) \neq 0$$

Substituting the last two equations into Equation 10.138, we have

$$\begin{aligned} \phi(\tilde{A}) &= \phi(A) - \alpha_2 BK - \alpha_1 BK\tilde{A} - BK\tilde{A}^2 \\ &\quad - \alpha_1 ABK - ABK\tilde{A} - A^2BK \end{aligned}$$

Since $\phi(\tilde{A}) = 0$, we obtain

$$\begin{aligned} \phi(A) &= B(\alpha_2 K + \alpha_1 K\tilde{A} + K\tilde{A}^2) \\ &\quad + AB(\alpha_1 K + K\tilde{A}) + A^2BK \qquad (10.139) \\ &= [B \mid AB \mid A^2B] \begin{bmatrix} \alpha_2 K + \alpha_1 K\tilde{A} + K\tilde{A}^2 \\ \alpha_1 K + K\tilde{A} \\ K \end{bmatrix} \end{aligned}$$

Since the system is completely state controllable, the inverse of the controllability matrix

$$[B \mid AB \mid A^2B]$$

exists. Premultiplying the inverse of the controllability matrix to both sides of Equation 10.139, we obtain

$$[B \mid AB \mid A^2B]^{-1}\phi(A) = \begin{bmatrix} \alpha_2 K + \alpha_1 K\tilde{A} + K\tilde{A}^2 \\ \alpha_1 K + K\tilde{A} \\ K \end{bmatrix}$$

Premultiplying both sides of this last equation by $[0\,0\,1]$, we obtain

$$[0\,0\,1][B \mid AB \mid A^2B]^{-1}\phi(A) = [0\,0\,1] \begin{bmatrix} \alpha_2 K + \alpha_1 K\tilde{A} + K\tilde{A}^2 \\ \alpha_1 K + K\tilde{A} \\ K \end{bmatrix} = K$$

which can be rewritten as

$$K = [0\,0\,1][B \mid AB \mid A^2B]^{-1}\phi(A)$$

This last equation gives the required state feedback gain matrix K.

For an arbitrary positive integer n, we have

$$\begin{aligned} K &= [0\,0\cdots0\,1] \\ &\quad [B \mid AB \mid \cdots \mid A^{n-1}B]^{-1}\phi(A) \end{aligned} \tag{10.140}$$

Equations 10.136 to 10.140 collectively are known as Ackermann's formula for the determination of the state feedback gain matrix K.

EXAMPLE 10.2:

Consider the system defined by

$$\dot{x} = Ax + Bu$$

where

$$A = \begin{bmatrix} 0 & 1 & 0 \\ 0 & 0 & 1 \\ -1 & -5 & -6 \end{bmatrix}, \quad B = \begin{bmatrix} 0 \\ 0 \\ 1 \end{bmatrix}$$

By using the state feedback control $u = -Kx$, it is desired to have the closed-loop poles at $s = -2 \pm j4$, and $s = -10$. Determine the state feedback gain matrix K.

First, we need to check the controllability of the system. Since the controllability matrix M is given by

$$M = [B \mid AB \mid A^2B] = \begin{bmatrix} 0 & 0 & 1 \\ 0 & 1 & -6 \\ 1 & -6 & 31 \end{bmatrix}$$

we find that det $M = -1$ and therefore rank $M = 3$. Thus, the system is completely state controllable and arbitrary pole placement is possible.

Next, we solve this problem. We demonstrate each of the three methods presented in this chapter.

Method 1. The first method is to use Equation 10.125. The characteristic equation for the system is

$$\begin{aligned} |sI - A| &= \begin{vmatrix} s & -1 & 0 \\ 0 & s & -1 \\ 1 & 5 & s+6 \end{vmatrix} \\ &= s^3 + 6s^2 + 5s + 1 \\ &= s^3 + a_1 s^2 + a_2 s + a_3 = 0 \end{aligned}$$

Hence,

$$a_1 = 6, \quad a_2 = 5, \quad a_3 = 1$$

The desired characteristic equation is

$$\begin{aligned} &(s + 2 - j4)(s + 2 + j4)(s + 10) \\ &= s^3 + 14s^2 + 60s + 200 \\ &= s^3 + \alpha_1 s^2 + \alpha_2 s + \alpha_3 = 0 \end{aligned}$$

Hence,

$$\alpha_1 = 14, \quad \alpha_2 = 60, \quad \alpha_3 = 200$$

Referring to Equation 10.135 we have

$$K = [\alpha_3 - a_3 \mid \alpha_2 - a_2 \mid \alpha_1 - a_1]T^{-1}$$

where $T = I$ for this problem because the given state equation is in the controllable canonical form. Then we have

$$\begin{aligned} K &= [200 - 1 \mid 60 - 5 \mid 14 - 6] \\ &= [199 \;\; 55 \;\; 8] \end{aligned}$$

Method 2. By defining the desired state feedback gain matrix K as

$$K = [k_1 \, k_2 \, k_3]$$

and equating $|sI - A + BK|$ with the desired characteristic equation, we obtain

$$\begin{aligned} |sI - A + BK| &= \left| \begin{bmatrix} s & 0 & 0 \\ 0 & s & 0 \\ 0 & 0 & s \end{bmatrix} - \begin{bmatrix} 0 & 1 & 0 \\ 0 & 0 & 1 \\ -1 & -5 & -6 \end{bmatrix} + \begin{bmatrix} 0 \\ 0 \\ 1 \end{bmatrix} [k_1 \, k_2 \, k_3] \right| \\ &= \begin{vmatrix} s & -1 & 0 \\ 0 & s & -1 \\ 1 + k_1 & 5 + k_2 & s + 6 + k_3 \end{vmatrix} \\ &= s^3 + (6 + k_3)s^2 + (5 + k_2)s + 1 + k_1 \\ &= s^3 + 14s^2 + 60s + 200 \end{aligned}$$

Thus,

$$6 + k_3 = 14, \quad 5 + k_2 = 60, \quad 1 + k_1 = 200$$

from which we obtain

$$k_1 = 199, \quad k_2 = 55, \quad k_3 = 8$$

or

$$K = [199 \quad 55 \quad 8]$$

Method 3. The third method is to use Ackermann's formula. Referring to Equation 10.140 we have

$$K = [0\,0\,1][B \mid AB \mid A^2B]^{-1}\phi(A)$$

Since

$$\begin{aligned} \phi(A) &= A^3 + 14A^2 + 60A + 200I \\ &= \begin{bmatrix} 0 & 1 & 0 \\ 0 & 0 & 1 \\ -1 & -5 & -6 \end{bmatrix}^3 + 14 \begin{bmatrix} 0 & 1 & 0 \\ 0 & 0 & 1 \\ -1 & -5 & -6 \end{bmatrix}^2 \\ &\quad + 60 \begin{bmatrix} 0 & 1 & 0 \\ 0 & 0 & 1 \\ -1 & -5 & -6 \end{bmatrix} \\ &\quad + \begin{bmatrix} 200 & 0 & 0 \\ 0 & 200 & 0 \\ 0 & 0 & 200 \end{bmatrix} \\ &= \begin{bmatrix} 199 & 55 & 8 \\ -8 & 159 & 7 \\ -7 & -43 & 117 \end{bmatrix} \end{aligned}$$

and

$$[B \mid AB \mid A^2B] = \begin{bmatrix} 0 & 0 & 1 \\ 0 & 1 & -6 \\ 1 & -6 & 31 \end{bmatrix}$$

we obtain

$$\begin{aligned} K &= [0\,0\,1] \begin{bmatrix} 0 & 0 & 1 \\ 0 & 1 & -6 \\ 1 & -6 & 31 \end{bmatrix}^{-1} \begin{bmatrix} 199 & 55 & 8 \\ -8 & 159 & 7 \\ -7 & -43 & 117 \end{bmatrix} \\ &= [0\,0\,1] \begin{bmatrix} 5 & 6 & 1 \\ 6 & 1 & 0 \\ 1 & 0 & 0 \end{bmatrix} \begin{bmatrix} 199 & 55 & 8 \\ -8 & 159 & 7 \\ -7 & -43 & 117 \end{bmatrix} \\ &= [199 \;\; 55 \;\; 8] \end{aligned}$$

As a matter of course, the feedback gain matrix K obtained by the three methods are the same. With this state feedback, the closed-loop poles are located at $s = -2 \pm j4$ and $s = -10$, as desired.

It is noted that if the order n of the system is 4 or higher, methods 1 and 3 are recommended, since all matrix computations can be carried by a computer. If method 2 is used, hand computations become necessary because a computer may not handle the characteristic equation with unknown parameters $k_1, k_2, \ldots, k_n$.

10.6.7 Comments

It is important to note that matrix K is not unique for a given system, but depends on the desired closed-loop pole locations (which determine the speed and damping of the response) selected. Note that the selection of the desired closed-loop poles or the desired characteristic equation is a compromise between the rapidity of the response of the error vector and the sensitivity to disturbances and measurement noises. That is, if we increase the speed of error response, then the adverse effects of disturbances and measurement noises generally increase. If the system is of second order, then the system dynamics (response characteristics) can be precisely correlated to the location of the desired closed-loop poles and the zero(s) of the plant. For higher-order systems, the location of the closed-loop poles and the system dynamics (response characteristics) are not easily correlated. Hence, in determining the state feedback gain matrix K for a given system, it is desirable to examine by computer simulations the response characteristics of the system for several different matrices K (based on several different desired characteristic equations) and to choose the one that gives the best overall system performance.

References

[1] Ogata, Katsuhiko, *Modern Control Engineering,* 2nd ed., Prentice Hall, Inc., Englewood Cliffs, NJ, 1990.

[2] Ogata, Katsuhiko, *Designing Linear Control Systems with MATLAB,* Prentice-Hall, Inc., Englewood Cliffs, NJ, 1994.

[3] Ogata, Katsuhiko, *Discrete-Time Control Systems,* 2nd ed., Prentice-Hall, Inc., Englewood Cliffs, NJ, 1995.

[4] Willems, J. C. and Mitter, S. K., Controllability, observability, pole allocation, and state reconstruction, *IEEE Trans. Autom. Control,* 16, 582–595, 1971.

[5] Wonham, W. M., On pole assignment in multi-input controllable linear systems, *IEEE Trans. Autom. Control,* 12, 660–665, 1967.

10.7 Internal Model Control

Richard D. Braatz, University of Illinois, Department of Chemical Engineering, Urbana, IL

10.7.1 Introduction

The field of process control experienced a surge of interest during the 1960s as engineers worked to apply the newly developed state-space optimal control theory to chemical processes. Though these methods had been applied successfully to the control of many mechanical and electrical systems, applications to chemical processes were not so forthcoming. By the 1970s, both industrialists and academicians began to realize that the theory would not be applied to chemical processes to any significant extent. It also began to be understood that certain characteristics of chemical processes make it very difficult (perhaps impossible) to directly apply this theory in a consistent and reproducible manner.

One characteristic of chemical processes is that unknown disturbances, inaccurate values for the physical parameters of the process, and lack of complete understanding of the underlying physical phenomena (for example, of the kinetics in an industrial polymerization reactor) make it impossible to generate a highly accurate model for most chemical processes, either phenomenologically or via input-output identification. Another characteristic that makes chemical processes especially difficult and interesting to control is the overwhelming importance of constraints on the manipulated variables (e.g., valve positions, pump and compressor throughput) and on the controlled variables (e.g., pressure, temperature, or capacity limits). Both model uncertainty and process constraints were not explicitly addressed by the state-space optimal control theory of the 1960s, and this to a large part explains the difficulties in applying this theory to the control of chemical processes.

Methods for explicitly addressing model uncertainty and process constraints began to coalesce in the late 1970s. Collectively, these methods came to be referred to as **internal model control** (IMC). After more than a decade in development, IMC is now widely used in the chemical industries, mostly in the form of proportional integral derivative (PID) tuning rules, in which a single parameter provides a clear tradeoff between closed-loop performance and robustness to model uncertainty. Besides its industrial importance, IMC also provides a convenient theoretical framework for understanding Smith predictors, multiple-degree-of-freedom problems, and the performance limitations due to nonminimum-phase behavior and model uncertainty. Here we describe the framework of IMC for the case where the process is stable. Since most chemical processes are stable, the greatest strengths of the IMC framework occur in this case and this will simplify the presentation. The results will be developed in continuous time, with comments on the discrete time case given at the end of the chapter.

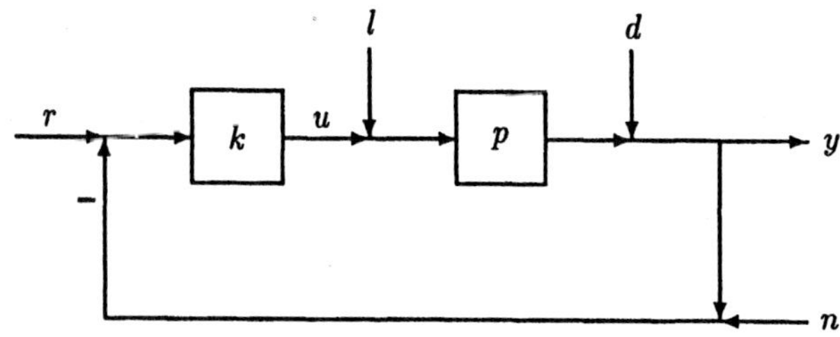

Figure 10.66 Classical control structure.

10.7.2 Fundamentals

Here the IMC and classical **control structures** are compared, which will illustrate the advantages of IMC in terms of addressing actuator constraints and model uncertainty in the control design. Then the IMC design procedure is presented.

Classical Control Structure

Before describing the IMC structure, let us consider the classical control structure used for the feedback control of single-input, single-output (SISO) processes (shown in Figure 10.66). Here p refers to the transfer function of the process; d and l refer to the output and load disturbances, respectively; y refers to the controlled variable; n refers to measurement noise; r refers to the setpoint; and u refers to the manipulated variable specified by the controller k. The controlled variable is related to the setpoint, measurement noise, and unmeasured disturbances by

$$\begin{aligned} y &= \frac{pk}{1+pk}(r-n) + \frac{1}{1+pk}(d+pl) \\ &= T(r-n) + S(d+pl), \end{aligned} \tag{10.141}$$

where

$$S = \frac{1}{1+pk}; \qquad T = 1 - S = \frac{pk}{1+pk} \tag{10.142}$$

are the **sensitivity** and **complementary sensitivity** functions, respectively.

A well-known requirement of any closed-loop system is that it be internally stable; that is, that bounded signals injected at any point in the control system generate bounded signals at any other point. From the viewpoint of **internal stability**, only the boundedness of outputs y and u need to be considered, since all other signals in the system are bounded provided u, y, and all inputs are bounded. Similarly, in terms of internal stability only the inputs r and l need to be considered. Thus, the classic control structure is internally stable if and only if all elements in the following 2×2 transfer matrix are stable (that is, have all their poles in the open left half-plane)

$$\begin{pmatrix} y \\ u \end{pmatrix} = \begin{pmatrix} pS & T \\ -T & kS \end{pmatrix} \begin{pmatrix} l \\ r \end{pmatrix}. \tag{10.143}$$

The closed-loop system is internally stable if and only if the transfer functions pS, T, and kS are stable. The stability of these three transfer functions is implied by the stability of only one transfer function, kS (the stability of p and kS implies the stability of $T = pkS$; the stability of 1 and T implies the stability of $S = 1 - T$; and the stability of p and S implies the stability of pS). Thus, for a stable process p, the closed-loop system is internally stable if and only if

$$kS = \frac{k}{1+pk} \tag{10.144}$$

is stable.

For good setpoint tracking, it would be desirable in Equation 10.141 to have $T(j\omega) \approx 1$; and for good disturbance rejection, it would be desirable to have $S(j\omega) \approx 0$ for all frequencies. These performance requirements are commensurate, since $S + T = 1$. On the other hand, to avoid magnifying measurement noise at high frequencies, it is desirable to have $T(j\omega)$ roll off here. Thus, there is a fundamental tradeoff between system performance (which corresponds to $S \approx 0$) and insensitivity of the closed-loop system to measurement noise (which corresponds to $T \approx 0$).

To explicitly account for model uncertainty, it is necessary to quantify the accuracy of the process model $\tilde{p}$ used in the control design procedure. A natural and convenient method for quantifying model uncertainty is as a frequency-dependent bound on the difference between the process model $\tilde{p}$ and the true plant p

$$\left| \frac{p(j\omega) - \tilde{p}(j\omega)}{\tilde{p}(j\omega)} \right| \le l_m(\omega); \qquad \forall\omega. \tag{10.145}$$

It should be expected that the inaccuracy of the model described by $l_m(\omega)$ would increase with frequency and eventually exceed 1, as it would be difficult to ascertain whether the true process has unmodeled zeros on the imaginary axis at sufficiently high frequencies (this would happen, for example, if there were any uncertainty in a process time delay), and these zeros would give $|(p(j\omega) - \tilde{p}(j\omega))/\tilde{p}(j\omega)| = |0 - \tilde{p}(j\omega)/\tilde{p}(j\omega)| = 1$.

The Nyquist Theorem can be used to show that the closed-loop system is internally stable for all plants defined by Equation 10.145 if and only if the nominal closed-loop system is internally stable and

$$|\tilde{T}(j\omega)| = \left| \frac{\tilde{p}(j\omega)k(j\omega)}{1 + \tilde{p}(j\omega)k(j\omega)} \right| < \frac{1}{l_m(\omega)}; \qquad \forall\omega. \tag{10.146}$$

As $l_m(\omega)$ is expected to be greater than one at high frequencies, it is necessary for $\tilde{T}(j\omega)$ to be detuned at high frequencies to prevent the control system from being sensitive to model uncertainty. Thus, there is a fundamental tradeoff between nominal system performance (which corresponds to $\tilde{S} \approx 0$) and insensitivity of the closed-loop system to model uncertainty (which corresponds to $\tilde{T} \approx 0$). For chemical processes, designing the controller to be insensitive to model uncertainty usually provides a greater limitation on closed-loop performance than measurement noise.

One disadvantage of the classical control structure is that the controller k enters the stability and performance specifications of Equations 10.144 and 10.146 in an inconvenient manner. It is also not clear how to address process constraints in a manner that ensures internal stability of the closed-loop system. It is well known, for example, that a controller implemented using the classical control structure can give arbitrarily poor performance or even instability when the control action becomes limited.

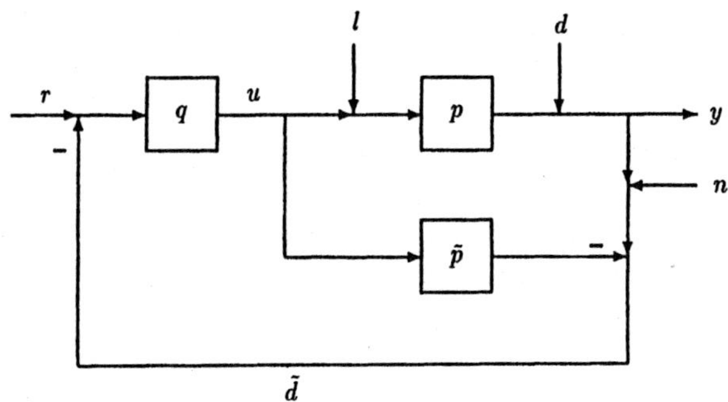

Figure 10.67 IMC structure.

IMC Structure

The IMC structure is shown in Figure 10.67, where $\tilde{p}$ refers to a model of the true process p, and q refers to the IMC controller. Simple block diagram manipulations show that the IMC structure is equivalent to the classical control structure provided

$$k = \frac{q}{1 - \tilde{p}q}, \quad \text{or, equivalently,} \quad q = \frac{k}{1 + \tilde{p}k}. \tag{10.147}$$

This control structure is referred to as *internal model control*, because the process *model* $\tilde{p}$ is explicitly an *internal* part of the controller k.

In terms of the IMC controller q, the transfer functions between the controlled variable and the setpoint, measurement noise, and unmeasured disturbances are given by

$$\begin{aligned} y &= T(r-n) + S(d+pl) \\ &= \frac{pq}{1+q(p-\tilde{p})}(r-n) \\ &\quad + \frac{1-\tilde{p}q}{1+q(p-\tilde{p})}(d+pl). \end{aligned} \tag{10.148}$$

When the process model is not equal to the true plant, then the closed-loop transfer functions S and T in Equation 10.148 do not appear to be any simpler for the IMC structure than for the classical control structure in Equation 10.141. However, when the process model $\tilde{p}$ is equal to the true process p, then Equation 10.148 simplifies to

$$\begin{aligned} y &= T(r-n) + S(d+pl) \\ &= \tilde{p}q(r-n) + (1-\tilde{p}q)(d+\tilde{p}l), \end{aligned} \tag{10.149}$$

and the IMC controller is related to the classical controller by Equation 10.147

$$q = \frac{k}{1+\tilde{p}k} = \frac{k}{1+pk} = kS. \tag{10.150}$$

For stable plants, Equation 10.150 is exactly the condition for internal stability derived for the classical control structure. This replaces the somewhat inconvenient task of selecting a controller k to stabilize $k/(1+pk)$ with the simpler task of selecting any stable transfer function q. Also, the IMC controller q enters the closed-loop transfer functions T and S in Equation 10.149 in an affine manner, that is

$$T = \tilde{p}q; \qquad S = 1 - \tilde{p}q. \tag{10.151}$$

This makes the tradeoff between nominal performance and model uncertainty very simple and is exploited in the IMC design procedure described later.

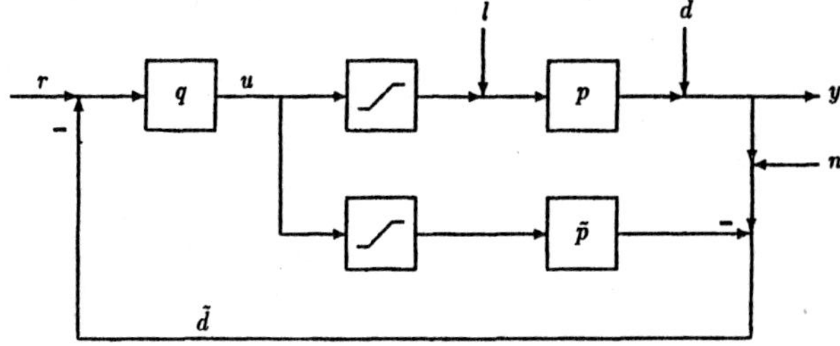

Figure 10.68 IMC implementation with actuator constraints.

Another advantage of the IMC structure over the classical control structure is that the explicit consideration of the process model provides a convenient means for understanding the role of model uncertainty in the control system design. To see this, let us interpret the feedback signal $\tilde{d}$ in Figure 10.67 for the case where the process model is not an exact representation of the true plant ($\tilde{p} \neq p$):

$$\tilde{d} = (p - \tilde{p})u + n + d + pl. \tag{10.152}$$

When there are no unknown disturbances or measurement noise ($n = d = l = 0$) and no model uncertainty ($\tilde{p} = p$), then the feedback signal $\tilde{d}$ is zero and the control system is open loop, that is, no feedback is necessary. If there are disturbances, measurement noise, or model uncertainty, then the feedback signal $\tilde{d}$ is not equal to zero. This illustrates clearly that $\tilde{d}$ expresses all that is unknown about the process, with the magnitude of $\tilde{d}$ directly related to the magnitude of the unknown process characteristics. This motivates the idea of placing a filter on $\tilde{d}$ to reduce the ability of the deleterious signals to destabilize the system; this is an important step in the IMC design procedure discussed in the next section.

In addition, if model uncertainty is ignored ($p = \tilde{p}$), then actuator constraints do not destabilize the closed-loop system provided that the constrained plant input is sent to the model $\tilde{p}$ rather than the output of the IMC controller q (see Figure 10.68). To see this, calculate $\tilde{d}$ from Figure 10.68 to be

$$\tilde{d} = n + d + pl. \tag{10.153}$$

When the process model is equal to the true plant, then the control system is open loop, and the system is internally stable if and only if all the blocks in series are stable. In this case, internal stability is implied by the stability of the true plant p, the IMC controller q, and the actuator nonlinearity. When model uncertainty is taken into account, then

$$\tilde{d} = (p\alpha - \tilde{p}\alpha)u + n + d + pl = (p - \tilde{p})\alpha u + n + d + pl, \tag{10.154}$$

where α represents a stable nonlinear operator, in this case, the static actuator limitation nonlinearity. We again see that $\tilde{d}$ represents all that is unknown about the process.

IMC Design Procedure

The objectives of setpoint tracking and disturbance rejection are to minimize the error $e = y - r$. When the process model $\tilde{p}$ is equal to the true process p, then the error is given by Equation 10.149

$$\begin{aligned} e &= y - r = -pqn + (1 - pq)(d + pl - r) \\ &= -Tn + S(d + pl - r). \end{aligned} \tag{10.155}$$

The error e is an affine function of the IMC controller q. The preceding section discussed how introducing a filter in the IMC feedback path (that is, in q) would reduce the ability of model uncertainty and measurement noise to destabilize the system. This motivates the IMC design procedure, which consists of designing the IMC controller q in two steps:

Step 1. Nominal Performance. A nominal IMC controller $\tilde{q}$ is designed to yield optimal tracking and disturbance rejection, ignoring measurement noise, model uncertainty, and constraints on the manipulated variable.

Step 2. Robust Stability and Performance. An IMC filter f is used to detune the controller $\tilde{q}$ (that is, $q = \tilde{q} f$), to trade off performance with smoothness of the control action and robustness to measurement noise and model uncertainty.

These steps are detailed below.

Performance Measures Almost any reasonable performance measure can be used in the design of the nominal IMC controller. For fixed inputs (that is, disturbances and/or setpoint), two of the most popular performance measures are the **integral absolute error** (IAE) and the **integral square error** (ISE):

$$\mathrm{IAE}\{e\} \equiv \int_0^\infty |e(t)|\,\mathrm{d}t; \tag{10.156}$$

$$\mathrm{ISE}\{e\} \equiv \int_0^\infty e^2(t)\mathrm{d}t. \tag{10.157}$$

When the inputs are best described by a set representation, two of the most popular performance measures are the **error variance** (EV) and the **worst case error** (WCE). The EV is appropriate for stochastic inputs (for example, filtered white noise with zero mean and specified variance) and given by

$$\mathrm{EV}\{e\} \equiv \text{Expected Value}\left\{\int_0^\infty e^2(t)\mathrm{d}t\right\}. \tag{10.158}$$

The WCE is the worst-case ISE for sets of inputs whose ISE is bounded

$$\mathrm{WCE}\{e\} \equiv \sup_{\int_0^\infty v^2(t)\mathrm{d}t \le 1} \int_0^\infty e^2(t)\mathrm{d}t, \tag{10.159}$$

and is commonly used for loopshaping closed-loop transfer functions to have desired frequency domain properties.

It is straightforward (using the Parseval's Lemma) to show that minimizing the EV for a given set of stochastic inputs is mathematically equivalent to minimizing the ISE for a corresponding fixed input signal. Since the ISE performance measure for fixed inputs is the most popular in IMC, it is used in what follows.

Irrespective of the closed-loop performance measure that a control engineer may prefer, it is usually important that the closed-loop system satisfy certain steady-state properties. For example, a common control system requirement is that the error signal resulting from step inputs approaches zero at steady-state. It can be shown from the final value theorem applied to Equation 10.148 that this is equivalent to

$$q(0) = \tilde{p}^{-1}(0). \tag{10.160}$$

Another typical requirement is that the error signal resulting from ramp inputs approaches zero at steady-state. The final value theorem can be used to show that this requirement is equivalent to having both of the following conditions satisfied:

$$q(0) = \tilde{p}^{-1}(0); \qquad \frac{\mathrm{d}(\tilde{p}q)}{\mathrm{d}s}(0) = 0. \tag{10.161}$$

These conditions are used when selecting the IMC filter.

ISE Optimal Performance The ISE optimal controller can be solved via a simple analytical procedure when the process is stable and has no zeros on the imaginary axis. The first step in this procedure is a factorization of the process model $\tilde{p}$ into an allpass portion $\tilde{p}_A$ and a minimum-phase portion $\tilde{p}_M$

$$\tilde{p} = \tilde{p}_A \tilde{p}_M, \tag{10.162}$$

where $\tilde{p}_A$ includes all the open right half-plane zeros and delays of $\tilde{p}$ and has the form

$$\tilde{p}_A = e^{-s\theta} \prod_i \frac{-s + z_i}{s + \bar{z}_i}; \qquad \mathrm{Re}\{z_i\}, \theta > 0, \tag{10.163}$$

θ is the time delay, z_i is a right half-plane zero in the process model, and $\bar{z}_i$ is the complex conjugate of z_i. We use v to refer to the fixed unmeasured input in Equation 10.155, namely, $v = d + pl - r$. The controller that minimizes the ISE for these inputs is given in the following theorem.

THEOREM 10.1 *Assume that the process model $\tilde{p}$ is stable. Factor $\tilde{p}$ and the input v into allpass and minimum-phase portions*

$$\tilde{p} = \tilde{p}_A \tilde{p}_M; \qquad v = v_A v_M. \tag{10.164}$$

The controller that minimizes the ISE is given by

$$\tilde{q} = (\tilde{p}_M v_M)^{-1} \left\{\tilde{p}_A^{-1} v_M\right\}_* \tag{10.165}$$

where the operator $\{\cdot\}_*$ denotes that after a partial faction expansion of the operand all terms involving the poles of $\tilde{p}_A^{-1}$ are omitted.

Provided that the input v have been chosen to be of the appropriate type (for example, step or ramp), the ISE optimal controller $\tilde{q}$ satisfies the appropriate asymptotic steady-state performance requirements of Equations 10.160 and 10.161. In general, the $\tilde{q}$ given by Theorem 10.1 will not be proper, and the complementary sensitivity function $\tilde{T} = \tilde{p}\tilde{q}$ will have undesirable high-frequency behavior. The nominal IMC controller $\tilde{q}$ is augmented by a low-pass filter f (that is, $q = \tilde{q}f$) to provide desirable high-frequency behavior, to prevent sensitivity to model uncertainty and measurement noise, and to avoid overly rapid or large control actions. This filter f provides the compromise between performance and robustness, and its selection is described next.

IMC Filter Forms The IMC filter f should be chosen so that the closed-loop system retains its asymptotic properties as $\tilde{q}$ is detuned for robustness. In particular, for the error signal resulting from step inputs to approach zero at steady-state, the filter f should satisfy

$$f(0) = 1. \tag{10.166}$$

Filters that satisfy this form include

$$f(s) = \frac{1}{(\lambda s + 1)^n}, \tag{10.167}$$

and

$$f(s) = \frac{\beta s + 1}{(\lambda s + 1)^n}, \tag{10.168}$$

where λ is an adjustable filter parameter that provides the tradeoff between performance and robustness, n is selected large enough to make q proper, and β is another free parameter that can be useful for some applications (its use is described in Example 10.8). For the error signal resulting from ramp inputs to approach zero at steady-state, the filter f must satisfy

$$f(0) = 1, \quad \text{and} \quad \frac{df}{ds}(0) = 0. \tag{10.169}$$

A filter form that satisfies these conditions is

$$f(s) = \frac{n\lambda s + 1}{(\lambda s + 1)^n}. \tag{10.170}$$

Notice that the parameter β, which is free in Equation 10.168, becomes fixed in Equation 10.170 to satisfy the additional condition in Equation 10.169.

The IMC controller q is calculated from $q = \tilde{q}f$, and the adjustable parameters are tuned to arrive at the appropriate tradeoff between performance and robustness. The corresponding classical controller, if desired, can be calculated by substituting q into Equation 10.147.

Rapid changes in the control action are generally undesirable, as they waste energy and may cause the control actuators to wear out prematurely. The IMC filter allows the control engineer to directly detune the control action, as can be seen from (ignoring model uncertainty in Figure 10.67)

$$u = \tilde{q}f(r - n - d - pl). \tag{10.171}$$

10.7.3 Applications

The IMC design procedure is applied to examples of processes that are common in the chemical industries.

EXAMPLE 10.3: Distillation Column Reboiler

Processes with inverse response are common in the chemical industries. In the single-loop case, these correspond to plants with nonminimum-phase (right half-plane) zeros. For example, a model of the level in a reboiler (located at the base of a distillation column) to a change in steam duty could be given by

$$\tilde{p} = \frac{-3s + 1}{s(s + 1)}. \tag{10.172}$$

Developing an accurate model for the level in a reboiler is difficult because the level depends on frothing, which does not respond in completely reproducible manner. The uncertainty in this model of the process can be described by a frequency-dependent bound on the difference between the process model $\tilde{p}$ and the true plant p

$$\left|\frac{p(j\omega) - \tilde{p}(j\omega)}{\tilde{p}(j\omega)}\right| \le l_m(\omega); \qquad \forall\omega, \tag{10.173}$$

where $l_m(\omega) = |(2j\omega + 0.2)/(j\omega + 1)|$ is shown in Figure 10.69. This uncertainty covers up to 20% error in the steady-state gain and up to 200% error at high frequencies.

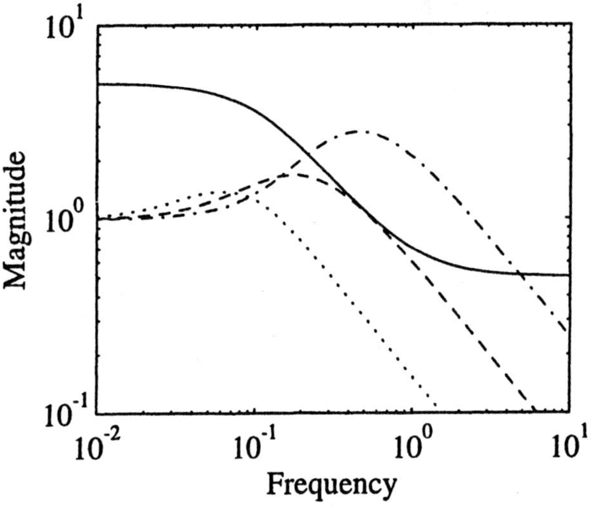

Figure 10.69 IMC controller design for robustness: $1/|l_m|$ (—); and $\tilde{T}$ for $\lambda = 2.7$ (- · -), $\lambda = 5.4$ (---), and $\lambda = 10.8$ (· · ·).

The performance specification is to minimize the LSE in rejecting ramp output disturbances, $d = 1/s^2$. Because the plant is not stable, a controller implemented using the IMC structure (in Figure 10.67) would not be internally stable, as bounded **load disturbances** would lead to unbounded plant outputs. On the other hand, it turns out that Theorem 10.1 can still be used to design the ISE optimal controller, as long as the controller is implemented using the classical control structure (in Figure 10.66) and the integrators in the plant also appear in the input v [6].

The first step in calculating the nominal IMC controller $\tilde{q}$ is to factor the process model $\tilde{p}$ and the input $v = d$ into allpass and minimum-phase portions as in Equation 10.164

$$\tilde{p}_A = \tfrac{-3s+1}{3s+1}; \qquad \tilde{p}_M = \tfrac{3s+1}{s(s+1)};$$

$$v_A = 1; \qquad v_M = \frac{1}{s^2}. \tag{10.174}$$

The ISE optimal controller is given by Equation 10.165

$$\begin{aligned}\tilde{q} &= (\tilde{p}_M v_M)^{-1}\left\{\tilde{p}_A^{-1} v_M\right\}_* &(10.175)\\ &= \left(\frac{1}{s}\cdot\frac{3s+1}{s+1}\cdot\frac{1}{s^2}\right)^{-1}\\ &\quad\cdot\left\{\left(\frac{-3s+1}{3s+1}\right)^{-1}\cdot\frac{1}{s^2}\right\}_* &(10.176)\\ &= \frac{s^3(s+1)}{3s+1}\cdot\left\{\frac{6}{s}+\frac{1}{s^2}+\frac{18}{-3s+1}\right\}_* &(10.177)\\ &= \frac{s^3(s+1)}{3s+1}\cdot\left(\frac{6}{s}+\frac{1}{s^2}\right) &(10.178)\\ &= \frac{s(s+1)(6s+1)}{3s+1}. &(10.179)\end{aligned}$$

This is augmented with a filter form appropriate for ramp inputs (Equation 10.170), where the order of the denominator is chosen so that the IMC controller

$$\begin{aligned}q &= \tilde{q}f = \frac{s(s+1)(6s+1)}{3s+1}\cdot\frac{3\lambda s+1}{(\lambda s+1)^3}\\ &= \frac{s(s+1)(6s+1)(3\lambda s+1)}{(3s+1)(\lambda s+1)^3} &(10.180)\end{aligned}$$

is proper. The value of λ is selected just large enough that the inequality described by Equation 10.146

$$\begin{aligned}|\tilde{T}(j\omega)| &= |\tilde{p}(j\omega)q(j\omega)|\\ &= \left|\frac{(6j\omega+1)(-3j\omega+1)(3\lambda j\omega+1)}{(3j\omega+1)(\lambda j\omega+1)^3}\right|\\ &< \left|\frac{j\omega+1}{2j\omega+0.2}\right| &(10.181)\end{aligned}$$

is satisfied for all frequencies. A value of $\lambda = 5.4$ is adequate (from Figure 10.69) for the closed-loop system to be robust to model uncertainty (any larger would result in overly sluggish performance). As stated earlier, this controller cannot be implemented using the IMC structure, but must be implemented using the classical control structure with Equation 10.147

$$\begin{aligned}k &= \frac{q}{1-\tilde{p}q}\\ &= \frac{(s+1)(6s+1)(3\lambda s+1)}{s(3\lambda^3 s^2+(\lambda^2+9\lambda+54)\lambda s+3\lambda^2+18)}. &(10.182)\end{aligned}$$

Closed-loop responses to ramp output disturbances for several plants included in the uncertainty description are shown in Figure 10.70 (two have the same dynamics but different steady-state error; the other two were chosen so that the inequality in Equation 10.173 is satisfied as an equality). The closed-loop systems are stable as desired.

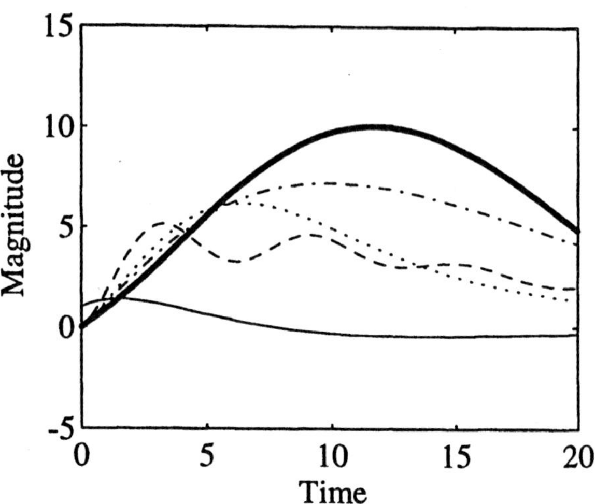

Figure 10.70 Closed-loop responses y for a ramp output disturbance: $\tilde{p}$ (—), $1.2\tilde{p}$ (···), $0.8\tilde{p}$ (-·-), $(-3s+1)(3s+1.2)/s(s+1)^2$ (---), $(-s+.8)(-3s+1)/s(s+1)^2$ (**—**).

EXAMPLE 10.4: Minimum-Phase Process Models

Although most chemical processes have some nonminimum-phase character, a large number of processes can be approximated by a minimum-phase model. In this case, $\tilde{p}_A = 1$ and the ISE optimal controller is given by Equation 10.165

$$\begin{aligned}\tilde{q} &= (\tilde{p}_M v_M)^{-1}\left\{\tilde{p}_A^{-1} v_M\right\}_* &(10.183)\\ &= (\tilde{p}_M v_M)^{-1}\{v_M\}_* &(10.184)\\ &= \tilde{p}_M^{-1}. &(10.185)\end{aligned}$$

The value for $\tilde{q}$ is the inverse of the process model (for this reason, IMC is often referred to as being **model inverse based**). In this case, the ISE for the unfiltered nominal system is zero, irrespective of the characteristics of the inputs because the nominal error from (Equation 10.155)

$$\begin{aligned}e &= y - r = (1-\tilde{p}\tilde{q})(d+pl-r)\\ &= (1-\tilde{p}_M\tilde{p}_M^{-1})(d+pl-r) = 0. &(10.186)\end{aligned}$$

is zero. Augmenting the ISE optimal controller with an IMC filter gives the controlled output of Equation 10.149

$$\begin{aligned}y &= T(r-n)+S(d+pl)\\ &= \tilde{p}q(r-n)+(1-\tilde{p}q)(d+pl)\\ &= f(r-n)+(1-f)(d+pl). &(10.187)\end{aligned}$$

If the process has relative degree one, then the IMC filter can be chosen to be relative degree one to give

$$\begin{aligned}y &= T(r-n)+S(d+pl)\\ &= \frac{1}{\lambda s+1}(r-n)+\frac{\lambda s}{\lambda s+1}(d+pl). &(10.188)\end{aligned}$$

We see that the bandwidth of T is very nearly equal to the bandwidth of S, and λ specifies the exact location of these bandwidths. When the process model is not minimum phase, then the bandwidths of T and S can be far apart, although the IMC filter parameter will still provide the tradeoff between nominal performance ($S \approx 0$) and robustness ($T \approx 0$).

EXAMPLE 10.5: Processes With Common Stochastic Disturbances

Although the IMC design procedure was presented in terms of minimizing the ISE for a fixed input, disturbances in the chemical industries are often more naturally modeled as stochastic inputs. Fortunately, Parseval's Lemma informs us that Theorem 10.1 provides the minimum variance controller for stochastic inputs, if v is chosen correctly. For example, the minimum variance controller for integrated white noise inputs is equal to the ISE optimal controller designed for a step input v. This and other equivalences are provided in Table 10.8.

Since most chemical process disturbances are represented well by one of these stochastic descriptions, for convenience the simplified expressions for the minimum variance (or ISE optimal) controller are given in the third column of Table 10.8. These expressions follow by analytically performing the partial fraction expansion and applying the $\{\cdot\}_*$ operator in Theorem 10.1. For example, for $v = 1/s$, the ISE optimal controller is from Equation 10.165

$$\tilde{q} = (\tilde{p}_M v_M)^{-1} \left\{\tilde{p}_A^{-1} v_M\right\}_* \tag{10.189}$$

$$= \left(\tilde{p}_M \cdot \frac{1}{s}\right)^{-1} \cdot \left\{\tilde{p}_A^{-1} \cdot \frac{1}{s}\right\}_* \tag{10.190}$$

$$= s \cdot \tilde{p}_M^{-1} \cdot \left\{\frac{\tilde{p}_A^{-1}(0)}{s} + \cdots\right\}_* \tag{10.191}$$

$$= \tilde{p}_M^{-1}. \tag{10.192}$$

EXAMPLE 10.6: Processes With Time Delay

It is common for chemical process models to include time delays. This may be due to transport delays in reactors or process piping, or to approximating high-order dynamics, such as is commonly done when modeling distillation column dynamics. Let us design an IMC controller for the following process

$$\tilde{p} = \overline{p} e^{-s\theta}, \tag{10.193}$$

where θ is the time delay and $\overline{p}$ is the delay-free part of the plant and can include nonminimum-phase zeros. Irrespective of the assumptions on the nature of the inputs, the IMC controller will have the form

$$q = \tilde{q} f. \tag{10.194}$$

From Equation 10.147, the corresponding classical controller has the form

$$k = \frac{\tilde{q} f}{1 - \tilde{q} f \ \overline{p} e^{-\theta s}}. \tag{10.195}$$

Some readers may notice a strong familiarity between the form of this controller and the structure of the well-known **Smith predictor** controller (shown in Figure 10.71) given by Smith [8]

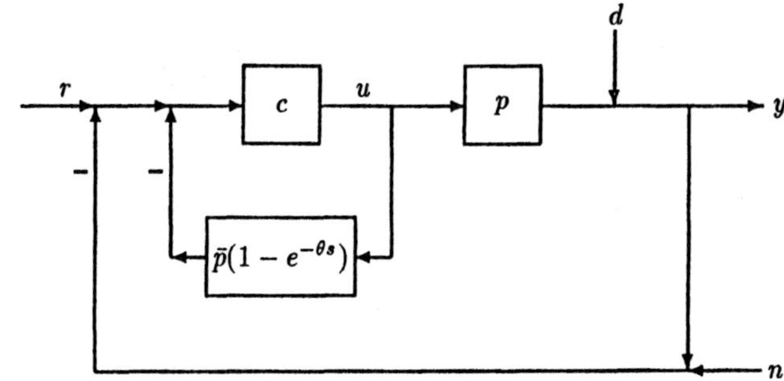

Figure 10.71 Smith predictor control structure.

$$k = \frac{c}{1 + c\overline{p}(1 - e^{-\theta s})}. \tag{10.196}$$

Actually, setting Equation 10.195 equal to Equation 10.196 and rearranging gives the Smith predictor controller in terms of the IMC controller and vice versa

$$c = \frac{\tilde{q} f}{1 - \overline{p} \tilde{q} f}; \qquad \tilde{q} f = \frac{c}{1 + \overline{p} c}. \tag{10.197}$$

This implies the Smith predictor control structure and the IMC structure are completely equivalent (this was first noticed by Brosilow [2]).

Smith states in his original manuscript that the Smith predictor control structure in Figure 10.71 allows the controller c to be designed via any optimal controller design method applied to the delay-free plant. This seems to be confirmed by Equation 10.197, where c would be the form of the classical controller designed via IMC applied to the delay-free plant. Although it is true that c could be designed by ignoring the delay in the plant, the nominal closed-loop performance depends on the sensitivity

$$\tilde{S} = 1 - \tilde{p}\tilde{q} = 1 - \overline{p} e^{-\theta s} \tilde{q} f, \tag{10.198}$$

which is a function of the time delay, and thus its effect should be considered in the controller design. An appropriate method of designing the controller c would be to design $\tilde{q}$ based on the process model with delay (as in Equation 10.194) and to tune the IMC filter based on $\tilde{S}$ and $\tilde{T}$ taking performance and robustness into account. Thus, IMC provides a transparent method for designing Smith predictor controllers.

EXAMPLE 10.7: PID Tuning Rules for Low-Order Processes

In IMC, the resulting controller is of an order roughly equivalent to that of the process model. Because most models for SISO chemical processes are low order, IMC controllers based on these models will be of low order and can be exactly or approximately described as **proportional-integral-derivative** (PID) controllers

$$k = k_c \left(1 + \tau_D s + \frac{1}{\tau_I s}\right), \tag{10.199}$$

where k_c is the gain, τ_I is the integral time constant, and τ_D is the derivative time constant. PID controllers are the most popular

TABLE 10.8 Minimum Variance Controllers for Common Stochastic Inputs

v	Stochastic Inputs	Minimum Variance $\tilde{q}$
$\frac{1}{s}$	Integrated white noise	$\tilde{p}_M^{-1}$
$\frac{1}{\tau s+1}$	Filtered white noise	$p_M^{-1} p_A^{-1}(-1/\tau)$
$\frac{1}{s(\tau s+1)}$	Filtered integrated white noise	$p_M^{-1}\left(1+\left(1-p_A^{-1}(-1/\tau)\right)\tau s\right)$
$\frac{1}{s^2}$	Double integrated white noise	$p_M^{-1}\left(1-s\cdot\frac{dp_A}{ds}(0)\right)$

and reliable controllers in the process industries. To a large part, this explains why the largest number of industrial applications of IMC to SISO processes is for the tuning of PID controllers.

To provide an example of the derivation of IMC PID tuning rules, consider a first-order process model

$$\tilde{p} = \tilde{p}_M = \frac{k_p}{\tau s + 1}, \tag{10.200}$$

where k_p is the steady-state gain and τ is the time constant. Table 1 gives the IMC controller (with a first-order filter) for step inputs as

$$q = \tilde{p}_M^{-1} = \frac{\tau s + 1}{k_p(\lambda s + 1)}. \tag{10.201}$$

The corresponding classical controller is given by Equation 10.147

$$k = \frac{\tau s + 1}{k_p \lambda s}. \tag{10.202}$$

This can be rearranged to be in the form of an ideal proportional-integral (PI) controller

$$k = k_c\left(1 + \frac{1}{\tau_I s}\right) \tag{10.203}$$

with

$$k_c = \frac{\tau}{k_p \lambda}; \qquad \tau_I = \tau. \tag{10.204}$$

An advantage of designing the PID controllers via IMC is that only one parameter is required to provide a clear tradeoff between robustness and performance whereas PID has three parameters that do not provide this clear tradeoff. IMC PID tuning rules for low-order process models and the most common disturbance and setpoint model (step) in the chemical industries are provided by Rivera, Skogestad, and Morari [7] and listed in Table 10.9.

TABLE 10.9 IMC PID Tuning Rules

Process Model $\tilde{p}$	k_c	τ_I	τ_D
$\frac{k_p}{\tau s+1}$	$\frac{\tau}{\lambda k_p}$	τ	-
$\frac{k_p}{\tau^2 s^2+2\zeta\tau s+1}$	$\frac{2\zeta\tau}{\lambda k_p}$	$2\zeta\tau$	$\frac{\tau}{2\zeta}$
$\frac{k_p}{s}$	$\frac{1}{\lambda k_p}$	-	-
$\frac{k_p}{s(\tau s+1)}$	$\frac{1}{\lambda k_p}$	-	τ

EXAMPLE 10.8: Processes With a Single Dominant Lag

Many chemical processes have a time lag that is substantially slower than the other time lags and the time delay. Several researchers over the last 10 years have claimed that IMC gives poor rejection of load disturbances for these processes. To aid in the understanding of this claim, consider a process that is modeled by a dominant lag

$$\tilde{p} = \frac{1}{100s + 1}. \tag{10.205}$$

An IMC controller designed for this process model is (see Example 10.4)

$$q = \frac{100s + 1}{\lambda s + 1}. \tag{10.206}$$

For simplicity of presentation only, let $r = n = 0$ and ignore model uncertainty in what follows. The controlled output is related to the output disturbance d and load disturbance l by (Equation 10.188)

$$\begin{aligned} y &= S(d + pl) = (1 - \tilde{p}q)d + p(1 - \tilde{p}q)l && (10.207)\\ &= \frac{\lambda s}{\lambda s + 1}d + \frac{\lambda s}{(\lambda s + 1)(100s + 1)}l. && (10.208)\end{aligned}$$

The value for λ is chosen to be 20 to provide nominal performance approximately five times faster than open loop. The closed-loop responses to unit-step load and output disturbances are shown in Figure 10.72. As expected, the control system rejects the unit-step output disturbance d with a time constant of approximately 20 time units. On the other hand, the control system rejects the load disturbance l very slowly. This difference in behavior is easily understood from Equation 10.208, since the slow process time lag appears in the transfer function between the load disturbance and the controlled output, irrespective of the magnitude of the filter parameter λ (as long as $\lambda \neq 0$). The open-loop dynamics appear in the closed-loop dynamics, resulting in the long tail in Figure 10.72.

Several researchers have proposed ad hoc fixes for this problem. Perhaps the simplest solution is presented here. The closed-loop dynamics are poor because the IMC filter forms in common use are designed for output disturbances, not load disturbances. Thus, a simple "fix" is to design the correct filter for the load disturbance. Consider the IMC filter given in Equation 10.168 which provides an extra degree of freedom (β) over the other filter forms of Equations 10.167 and 10.170. The order n of the filter is chosen equal to 2 so that the IMC controller will be proper. Then the controlled output is related to the output disturbance

d and load disturbance l by Equation 10.207

$$\begin{aligned} y &= Sd + pSl = (1 - \tilde{p}q)d + p(1 - \tilde{p}q)l \quad (10.209) \\ &= \frac{\lambda^2 s^2 + (2\lambda - \beta)s}{(\lambda s + 1)^2} d \\ &+ \frac{\lambda^2 s^2 + (2\lambda - \beta)s}{(\lambda s + 1)^2 (100s + 1)} l. \quad (10.210) \end{aligned}$$

Since the sluggish response to load disturbances is due to the open-loop pole at $s = -1/100$, select the extra degree of freedom β to cancel this pole

$$\beta = 2\lambda - \frac{\lambda^2}{100}. \quad (10.211)$$

Then the controlled output in Equation 10.210 is given by

$$\begin{aligned} y &= \frac{\lambda^2 s^2 + \lambda^2 s/100}{(\lambda s + 1)^2} d + \frac{\lambda^2 s^2 + \lambda^2 s/100}{(\lambda s + 1)^2 (100s + 1)} l \quad (10.212) \\ &= \frac{s\lambda^2 (s + 1/100)}{(\lambda s + 1)^2} d + \frac{s\lambda^2/100}{(\lambda s + 1)^2} l. \quad (10.213) \end{aligned}$$

The closed-loop responses to unit-step load and output disturbances are shown in Figure 10.72 (with $\lambda = 20$). This time the undesirable open-loop time constant does not appear in the controlled variable.

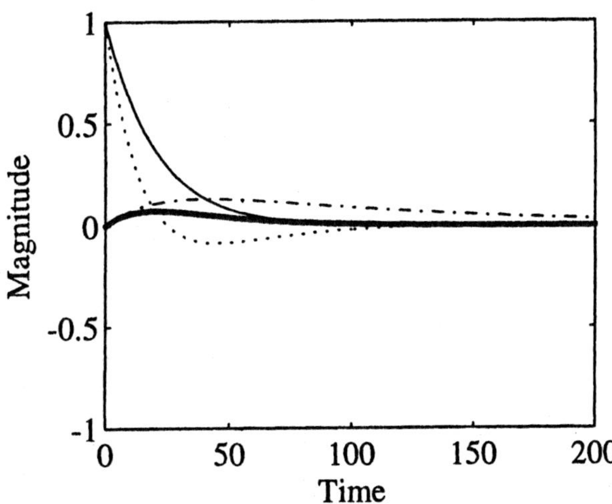

Figure 10.72 Closed-loop response y with common filter design (Equation 10.208) for unit step output (—) and load (- · -) disturbances; and closed-loop response y with correct filter design (Equation 10.213) for unit step output (· · ·) and load (**—**) disturbances.

10.7.4 Defining Terms

Complementary sensitivity: The transfer function T between the setpoint r and the controlled variable y. For the classical control structure, this transfer function is given by $T = pk/(1 + pk)$, where p is the plant and k is the controller.

Control structure: A method of placing control blocks and connecting these blocks to the process.

Error variance: A closed-loop performance measure that is appropriate for stochastic inputs, defined by

$$\text{EV}\{e\} \equiv \text{Expected Value}\left\{\int_0^{\infty} e^2(t)\text{d}t\right\}, \quad (10.214)$$

where e is the closed-loop error.

Internal model control (IMC): A method of implementing and designing controllers in which the process model is explicitly an internal part of the controller.

Integral square error (ISE): A closed-loop performance measure that is appropriate for fixed inputs, defined by

$$\text{ISE}\{e\} = \int_0^{\infty} e^2(t)\text{d}t, \quad (10.215)$$

where e is the closed-loop error.

Internal stability: The condition where bounded signals injected at any point in the control system generate bounded signals at any other point.

Inverse-based control: Any control design method in which an explicit inverse of the model is used in the design procedure.

Load disturbance: A disturbance that enters the input of the process.

Proportional-integral-derivative (PID) controller: The most common controller in the chemical process industries. The ideal form of this controller is given by

$$k = k_c\left(1 + \tau_D s + \frac{1}{\tau_I s}\right). \quad (10.216)$$

Sensitivity: The transfer function S between disturbances d at the output of the plant and the controlled variable y. For the classical control structure, this transfer function is given by $S = 1/(1 + pk)$, where p is the plant and k is the controller.

Smith predictor: A strategy for designing controllers for processes with significant time delay in which a predictor in the control structure allows the controller to be designed ignoring the time delay.

References

[1] Braatz, R.D., Ogunnaike, B.A., and Schwaber, J.S., Failure tolerant globally optimal linear control via parallel design, Paper 232b, in AIChE Annu. Meet., San Francisco, 1994.

[2] Brosilow, C.B., The structure and design of Smith predictors from the viewpoint of inferential control, *Proc. Jt. Autom. Control Conf.*, 288–288, 1979.

[3] Frank, P.M., Entwurf von Regelkreisen mit vorgeschriebenem Verhalten, Ph.D. thesis, G. Braun, Karlsruhe, 1974.

[4] Garcia, C.E., Prett, D.M., and Morari, M., Model predictive control: theory and practice—a survey, *Automatica*, 25, 335–348, 1989.

[5] Henson, M.A. and Seborg, D.E., An internal model control strategy for nonlinear systems, *AIChE J.*, 37, 1065–1081, 1991.

[6] Morari, M. and Zafiriou, E., *Robust Process Control*, Prentice Hall, Englewood Cliffs, NJ, 1989.

[7] Rivera, D.E., Skogestad, S., and Morari, M., Internal model control 4: PID controller design. *Ind. Eng. Chemical Process Design Dev.*, 25, 252–265, 1986.

[8] Smith, O.J.M., Closer control of loops with dead time, *Chem. Eng. Prog.*, 53, 217–219, 1957.

Further Reading

A review of the origins of the IMC structure is provided in the Ph.D. thesis of P.M. Frank [3].

Coleman B. Brosilow and Manfred Morari popularized the structure in a series of conference presentations and journal articles published in the chemical engineering literature in the late 1970s and the 1980s (AIChE Annu. Meet., *AIChE J.*, *Chemical Eng. Sci.*, and *Ind. Eng. Chemical Process Design Dev.*). A more thorough description of IMC, covering many of the topics in this chapter, is provided in the research monograph by Manfred Morari and Evanghelos Zafiriou [6].

A control system is said to have *multiple degrees of freedom* if it has more than one controller block, with each block having different input signals. Examples common in the process industries include reference prefiltering, cascade control, feedforward-feedback control, and inferential control. Strategies for placing and designing the control blocks in an IMC setting for the simple cases above are provided in P.M. Frank's thesis [3] and summarized in Morari and Zafiriou's research monograph [6]. A general method for constructing the optimal and most general multiple-degree-of-freedom control strategies, in which each control block is designed independently, was presented at the 1994 AIChE Annu. Meet. and is described in a technical report by Richard D. Braatz, Babatunde A. Ogunnaike, and James S. Schwaber [1]. For a copy of this report, contact: Large-Scale Systems Research Laboratory, University of Illinois, 600 South Mathews Avenue, Box C-3, Urbana, IL, 61801-3792. Phone (217) 333-5073.

A survey of efforts to generalize IMC to nonlinear processes is provided by Michael A. Henson and Dale E. Seborg [5].

When there are multiple process inputs and/or outputs, IMC is usually treated in discrete time, and the performance objective is optimized on-line subject to the constraints. This method of control is referred to by many names, including *model predictive control, model predictive heuristic control, generalized predictive control, dynamic matrix control,* and *IDCOM,* and is the most popular multivariable control method used by chemical process control engineers. A survey of these methods is provided by Carlos E. Garcia, David M. Prett, and Manfred Morari [4].

10.8 Time-Delay Compensation — Smith Predictor and its Modifications

Z.J. Palmor, *Faculty of Mechanical Engineering, Technion - Israel Institute of Technology, Haifa, Israel*

10.8.1 Introduction

Time delays or dead times (DTs) between inputs and outputs are common phenomena in industrial processes, engineering systems, economical and biological systems. Transportation and measurement lags, analysis times, computation and communication lags all introduce DTs into control loops. DTs are also inherent in distributed parameter systems and frequently are used to compensate for model reduction where high-order systems are represented by low-order models with delays. The presence of DTs in the control loops has two major consequences: 1. It greatly complicates the analysis and the design of feedback controllers for such systems. 2. It makes satisfactory control more difficult to achieve.

In 1957, O.J.M. Smith [1] presented a control scheme for single-input single-output (SISO) systems, which has the potential of improving the control of loops with DTs. This scheme became known as the Smith predictor (SP) or Smith dead-time compensator (DTC). It can be traced back to optimal control [2]. Early attempts to apply the SP demonstrated that classical design methods were not suitable for the SP or similar schemes. Theoretical investigations performed in the late 1970s and early 1980s clarified the special properties of the SP and provided tools for understanding and designing such algorithms. Over the years, numerous studies on the properties of the SP have been performed, both in academia and in industry. Many modifications have been suggested, and the SP was extended to multi-input and multi-output (MIMO) cases with multiple DTs.

The SP contains a model of the process with a DT. Its implementation on analog equipment was therefore difficult and inconvenient. When digital process controllers began to appear in the marketplace at the beginning of the 1980s, it became relatively easy to implement the DTC algorithms. Indeed, in the early 1980s some microprocessor-based industrial controllers offered the DTC as a standard algorithm like the PID.

It is impossible to include all the available results on the topic and the many modifications and extensions in a single chapter. Hence, in this chapter the SISO continuous case is treated. This case is a key to understanding the sampled-data and the multivariable cases. Attention is paid to both theoretical and practical aspects. To make the reading more transparent, proofs are omitted but are referenced.

10.8.2 Control Difficulties Due to Time Delays

A linear time-invariant (LTI) SISO plant with an input delay is represented in the state space as follows:

$$\dot{x}(t) = Ax(t) + Bu(t-\theta)$$
$$y(t) = Cx(t) \qquad (10.217)$$

where $x \in R^n$ is the state vector, $u \in R$ is the input and $y \in R$ is the output. A, B, C are matrices of appropriate dimensions and θ is the time delay (or DT). Similarly, an LTI SISO plant with an output delay is given by:

$$\dot{x}(t) = Ax(t) + Bu(t)$$
$$y(t) = Cx(t-\theta) \qquad (10.218)$$

The transfer function of both Equations 10.217 and 10.218 is

$$y(s)/u(s) = P(s) = P_o(s)e^{-\theta s} \qquad (10.219)$$

where

$$P_o(s) = C(sI - A)^{-1}B \qquad (10.220)$$

$P_o(s)$ is seen to be a rational transfer function of order n. The presence of a DT in the control loop complicates the stability analysis and the control design of such systems. Furthermore, it degrades the quality of control due to unavoidable reduction in control gains as is demonstrated by the following simple example.

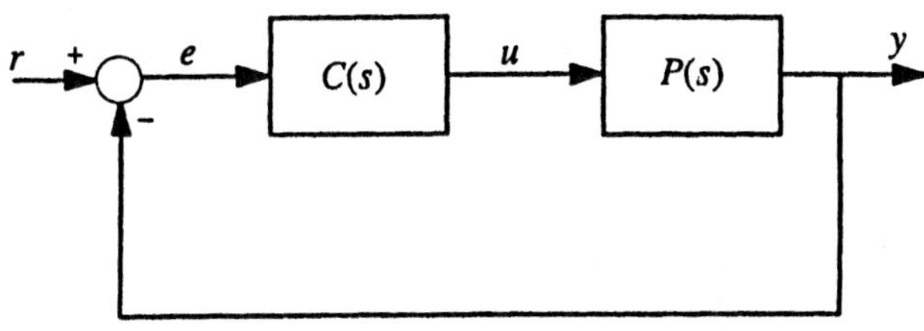

Figure 10.73 A feedback control system with plant P and controller C.

Assume that in the feedback control loop shown in Figure 10.73, the controller, $C(s)$, is a proportional (P) controller [i.e., $C(s) = K$,] and that $P_o(s)$ is a first-order filter [i.e., $P_o(s) = 1/(\tau s + 1)$]. $P(s)$ is thus given by:

$$P(s) = e^{-\theta s}/(\tau s + 1) \qquad (10.221)$$

The transfer function of the closed-loop system relating the output, $y(s)$, to the setpoint (or reference), $r(s)$, is

$$\frac{y(s)}{r(s)} = \frac{Ke^{-\theta s}}{\tau s+1+Ke^{-\theta s}} \qquad (10.222)$$

First, note that the characteristic equation contains $e^{-\theta s}$. Hence, it is a transcendental equation in s, which is more difficult to analyze than a polynomial equation. Second, the larger the ratio between the DT, θ, and the time constant, τ, the smaller becomes the maximum gain, K_{max}, for which stability of the closed loop holds. When $\theta/\tau = 0$ (that is, the process is DT free), then $K_{max} \to \infty$, at least theoretically. When $\theta/\tau = 1$ (that is, the DT equals the time constant), the maximum gain reduces drastically, from ∞ to about 2.26, and when $\theta/\tau \to \infty$, $K_{max} \to 1$.

The preceding example demonstrates clearly that when DTs are present in the control loop, controller gains have to be reduced to maintain stability. The larger the DT is relative to the time scale of the dynamics of the process, the larger the reduction required. Under most circumstances this results in poor performance and sluggish responses. One of the first control schemes aimed at improving the closed-loop performance for systems with DTs was that proposed by Smith [1]. This scheme is discussed next.

10.8.3 Smith Predictor (DTC)

a) Structure and Basic Properties

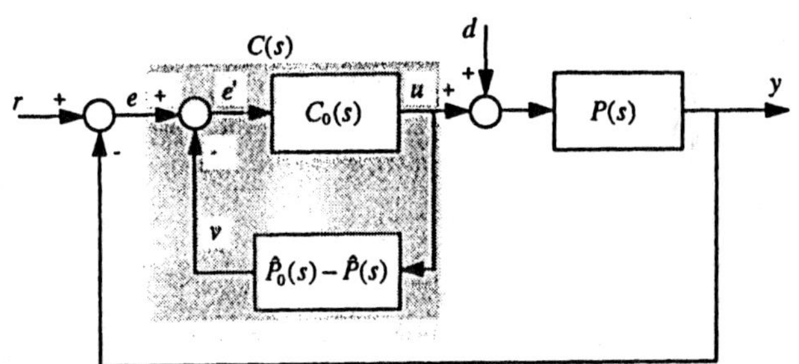

Figure 10.74 Classical configuration of a system incorporating SP.

The classical configuration of a system containing an SP is depicted in Figure 10.74. $P(s)$ is the transfer function of the process, which consists of a stable rational transfer function $P_o(s)$ and a DT as in Equation 10.219. $\hat{P}_o(s)$ and $\hat{P}(s)$ are models, or nominal values, of $P_o(s)$ and $P(s)$, respectively. The shaded area in Figure 10.74 is the SP, or the DTC. It consists of a *primary controller*, $C_o(s)$, which in industrial controllers is usually the conventional proportional-integral (PI) controller or proportional-integral-derivative (PID) controller, and a minor feedback loop, which contains the model of the process with and without the DT. The overall transfer function of the DTC is given by:

$$C(s) = C_o(s)/[1 + C_o(s)(\hat{P}_o(s) - \hat{P}(s))] \quad (10.223)$$

The underlying idea of the SP is clear, if one notices that the signal $v(t)$ (see Figure 10.74), contains a prediction of $y(t)$ DT units of time into the future. For that reason the minor feedback around the primary controller is called a "predictor." It is noted that $e' = r - P_o u$, whereas $e = r - Pu$. Therefore the "adjusted" error, $e'(t)$, which is fed into the primary controller, carries that part of the error that is "directly" caused by the primary controller. This eliminates the overcorrections associated with conventional controllers that require significant reductions in gains as was discussed earlier. Thus, it is seen that the SP should permit higher gains to be used.

The above qualitative arguments can be supported analytically. Assuming perfect model matching (which is called in the sequel the *ideal case*), i.e., $\hat{P}(s) = P(s)$, the transfer function of the closed loop in Figure 10.74 from the setpoint to the output is

$$G_{r(s)} = \frac{y(s)}{r(s)} = \frac{C_o P}{1+C_o P_o} \tag{10.224}$$

where the arguments have been dropped for convenience. It is observed that the DT has been removed from the denominator of Equation 10.224. This is a direct consequence of using the predictor. In fact, the denominator of Equation 10.224 is the same as the one of a feedback system with the DT-free process, P_o, and the controller C_o, without a predictor. Furthermore, Equation 10.224 is also the transfer function of the system shown in Figure 10.75, which contains neither DT nor DTC inside the closed loop.

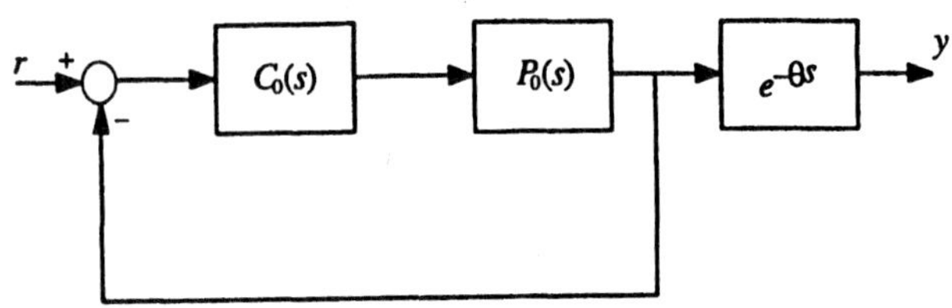

Figure 10.75 An input-output equivalent system.

The input-output equivalence of the two systems in Figures 10.74 and 10.75 may lead to the conclusion that one can design the primary controller in the SP by considering the system in Figure 10.75 as if DT did not exist. While the elimination of the DT from the characteristic equation is the main source for the potential improvement of the SP, its design cannot be based on Equation 10.224 or on the system in Figure 10.75. The reason is that the two input-output equivalent systems possess completely different sensitivity and robustness properties. It turns out that, under certain circumstances (to be discussed in the next section), an asymptotically stable design, which is based upon Equation 10.224, with seemingly large stability margins may in fact be *practically unstable*. That is, the overall system with an SP may lose stability under infinitesimal model mismatchings.

An alternative way from which it can be concluded that the design and tuning of the SP should not rely on the equivalent system in Figure 10.75 is to write down G_r, the closed-loop transfer function from r to y, for the more practical situation, in which mismatchings, or uncertainties, are taken into account. This transfer function is denoted G'_r. Thus, when $\hat{P} \neq P$, G'_r takes the following form:

$$G'_r(s) = \frac{y(s)}{r(s)} = \frac{C_o P}{1+C_o \hat{P}_o - C_o \hat{P} + C_o P} \tag{10.225}$$

It is evident from Equation 10.225 that when mismatching exists, the DT is not removed in its totality from the denominator and therefore affects the stability. It is therefore more appropriate to state that the SP minimizes the effect of the DT on stability, thereby allowing tighter control to be used. Also, note that the DT has not been removed from the numerator of G_r (and G'_r). Consequently, the SP tracks reference variations with a time delay.

The transfer function from the input disturbance d (see Figure 10.74) to the output y, in the ideal case, is denoted G_d and is given by:

$$G_d(s) = \frac{y(s)}{d(s)} = P\left[1 - \frac{C_o P}{1+C_o P_o}\right] \tag{10.226}$$

It is seen that the closed-loop poles consist of the zeros of $1+C_o P_o$ and the poles of P, the open-loop plant. Consequently, the "classical" SP, shown in Figure 10.74, can be used for stable plants only. In section 10.10, modified SP schemes for unstable plants are presented. In addition, the presence of the open-loop poles in G_d strongly influences the regulatory capabilities of the SP. This is discussed further in Section 10.8.3c.

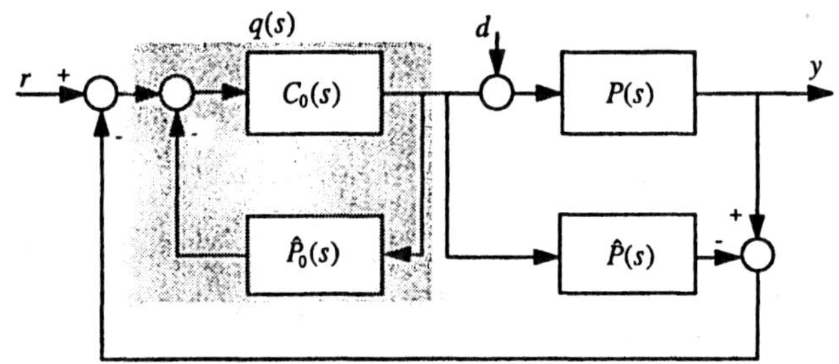

Figure 10.76 SP in IMC form.

An equivalent configuration of the SP is shown in Figure 10.76. Since the scheme in Figure 10.76 results from a simple rearrangement of the block diagram in Figure 10.74, it is apparent that it leaves all input-output relationships unaffected. Although known long ago, this scheme has become to be known as the IMC (internal model control) form of the SP [4, 5]. The dashed area in Figure 10.76 is the IMC controller, $q(s)$, which is related to the primary controller $C_o(s)$ via the following relationship:

$$q(s) = C_o(s)/[1 + C_o(s)\hat{P}_o(s)] \tag{10.227}$$

The controller $q(s)$ is usually cascaded with a filter $f(s)$. The filter parameters are adjusted to comply with robustness requirements. Thus, the overall IMC controller, $\overline{q}(s)$, is

$$\overline{q}(s) = f(s)q(s) \tag{10.228}$$

The IMC parameterization is referred to in Section 10.8.3d.

b) Practical, Robust and Relative Stability

Several stability results that are fundamental to understanding the special stability properties of the SP are presented. Among other things, they clarify why the design of the SP cannot be based on the ideal case. To motivate the development to follow let us examine the following simple example.

EXAMPLE 10.9:

Let the process in Figure 10.74 be given by $P(s) = e^{-s}/(s+1)$ and the primary controller by $C_o(s) = 4(0.5s + 1)$, an ideal proportional-derivative (PD) controller. In the ideal case (i.e., perfect matching), the overall closed loop including the SP has, according to Equation 10.224, a single pole at $s = -5/3$. The system not only is asymptotically stable, but possesses a gain margin and a phase margin of approximately 2 and 80°, respectively, as indicated by the Nyquist plot (the solid line) in Figure 10.77.

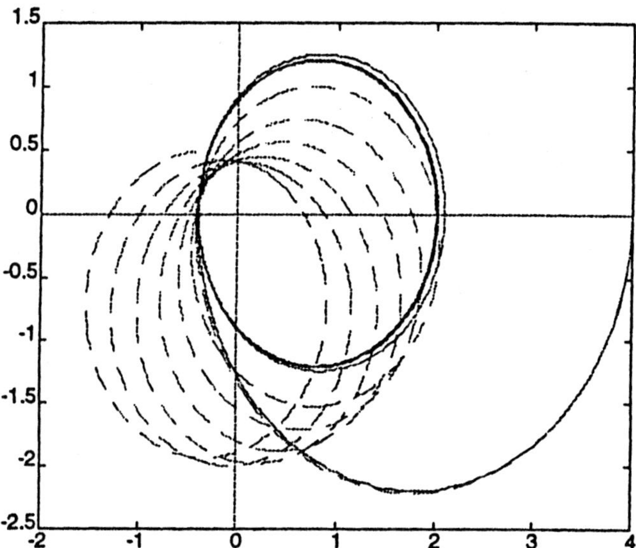

Figure 10.77 Nyquist plots of the system in Example 10.9. Solid line-ideal case; dashed line, nonideal case (5% mismatch in DT).

However, for a slight mismatch in the DT the overall system goes unstable, as is clearly observed from the dashed line in Figure 10.77, which shows the Nyquist plot for the nonideal case with 5% mismatch in the DT. In other words, the system is practically unstable.

For methodological reasons, the next definition of *practical instability* is presented in rather nonrigorous fashion.

DEFINITION. A system that is asymptotically stable in the ideal case but becomes unstable for infinitesimal modeling mismatches is called a practically unstable system.

A necessary condition for practical stability of systems with SP is developed next. To this end, the following quantity is defined:

$$Q(s) = C_o\hat{P}_o/(1 + C_o\hat{P}_o) \qquad (10.229)$$

It is noticed that $Q(s)$ is $G_r(s)/e^{-\theta s}$ where G_r, defined in Equation 10.224, is the transfer function of the closed loop in the ideal case. Hence, it is assumed in the sequel that $Q(s)$ is *stable*. Denoting $Im(s)$ by ω we have the following theorem:

THEOREM 10.2 *[3] For the system with an SP to be closed-loop practically stable, it is necessary that*

$$\lim_{\omega\to\infty} |Q(j\omega)| < 1/2 \qquad (10.230)$$

REMARK 10.1 If only mismatches in DT are considered, then it can be shown that the condition in Equation 10.230 is sufficient as well.

REMARK 10.2 For $Q(s)$ to satisfy Equation 10.230, it must be at least proper. If $Q(s)$ is strictly proper, then the system is practically stable.

EXAMPLE 10.10:

Equation 10.230 is applied to Example 10.9. It is easily verified that $Q(s)$ in that case is $Q(s) = (2s + 4)/(3s + 5)$. Hence, $\lim_{s\to\infty} |Q(s)| = 2/3 > 1/2$ and the system is practically unstable as was confirmed in Example 10.9.

Unless stated otherwise, it is assumed in all subsequent results that $Q(s)$ satisfies Equation 10.230. When the design complies with the condition in Theorem 10.2, one still may distinguish between two possible cases: one in which the design is, stability-wise, completely insensitive to mismatches in the DT; and the second, where there is a finite maximum mismatch in the DT below which the system remains stable. $\Delta\theta$ denotes the mismatch in DT and is given by:

$$\Delta\theta = \theta - \hat{\theta} \qquad (10.231)$$

where $\hat{\theta}$ is the estimated DT used in the SP. In the following theorem it is assumed that $P_o = \hat{P}_o$, i.e., mismatches may exist only in the DTs.

THEOREM 10.3 *[3] (a) The closed-loop system is asymptotically stable for any $\Delta\theta$ if*

$$|Q(j\omega)| < 1/2 \;\; \forall\omega \geq 0 \qquad (10.232)$$

(b) If

$$|Q(j\omega)| \leq 1 \;\; \forall\omega \geq 0 \text{ and } \lim_{\omega\to\infty} |Q(j\omega)| < 1/2 \qquad (10.233)$$

then there exists a finite positive $(\Delta\theta)_m$ such that the closed loop is asymptotically stable for all $|\Delta\theta| < (\Delta\theta)_m$.

REMARK 10.3 In [3] it is shown that a rough (and frequently conservative) estimate of $(\Delta\theta)_m$ is given by:

$$(\Delta\theta a)_m = \pi/(3\omega_o) \qquad (10.234)$$

where ω_o is the frequency above which $|Q(j\omega)| < 1/2$.

The next example demonstrates the application of Theorem 10.3a.

EXAMPLE 10.11:

If in Example 10.9 the gain of the primary controller is reduced such that $C_o = 0.9(0.5s + 1)$, then the corresponding $Q(s)$ satisfies the condition in Equation 10.232. Consequently, the system with the above primary controller not only is practically stable but also maintains stability for any mismatch in DT.

The conditions for robust stability presented so far were associated with uncertainties just in the DT. While the SP is largely sensitive to mismatches in DTs (particularly when the DTs are large as compared to the time constants of the process), conditions for robust stability of the SP for simultaneous uncertainties in all parameters, or even for structural differences between the model used in SP and the plant, may be derived. When modeling error is represented by uncertainty in several parameters, it is often mathematically convenient to approximate the uncertainty with a single multiplicative perturbation. Multiplicative perturbations on a nominal plant are commonly represented by:

$$P(s) = \hat{P}(s)\,[1 + \ell_m(s)] \qquad (10.235)$$

where, as before, $\hat{P}(s)$, is the model used in the SP. Hence, $\ell_m(s)$, the multiplicative perturbation, is given by:

$$\ell_m(s) = (P(s) - \hat{P}(s))/\hat{P}(s) = \frac{P_o(s)}{\hat{P}_o(s)} e^{-\Delta\theta s} - 1 \qquad (10.236)$$

A straightforward application of the well-known robust stability theorem (see [4], for example) leads to the following result:

THEOREM 10.4 *[3] Assume that $Q(s)$ is stable. Then the closed-loop system will be asymptotically stable for any multiplicative perturbation satisfying the following condition:*

$$|Q(j\omega)\ell_m(j\omega)| < 1 \qquad \forall\omega \geq 0 \qquad (10.237)$$

REMARK 10.4 It is common, where possible, to norm bound the multiplicative error $\ell_m(j\omega)$. If that bound is denoted by $\ell(\omega)$, then, the condition in Equation 10.237 can be restated as:

$$|Q(j\omega)|\,\ell(\omega) < 1 \qquad \forall\omega \geq 0 \qquad (10.238)$$

In [5], for example, the smallest possible $\ell(\omega)$ for the popular first-order with DT model:

$$P(s) = k_p e^{-\theta s}/(\tau s + 1) \qquad (10.239)$$

was found for simultaneous uncertainties in gain, k_p, time-constant, τ, and DT, θ. When $\ell(\omega)$ is available or can be determined, it is quite easy to check whether the SP design complies with Equation 10.238. This can be done by plotting the amplitude Bode diagram of $|Q(j\omega)|$ and verifying that it stays below $1/\ell(\omega)$ for all frequencies.

Considering further the properties of the rational function $Q(s)$, conditions under which the closed-loop system containing the SP possesses some attractive relative stability properties may be derived. The following result is due to [3]:

THEOREM 10.5 *[3] Let $Q(s)$ be stable. If*

$$|Q(j\omega)| \leq 1 \qquad \forall\omega \geq 0 \qquad (10.240)$$

then the closed-loop system has:

a. A minimum gain margin of 2

b. A minimum phase margin of 60°.

REMARK 10.5 It should be emphasized that unless the design is practically stable, the phase margin property may be misleading. That is to say that a design may satisfy Equation 10.240 but not Equation 10.230. Under such circumstances the system will go unstable for an infinitesimal mismatch in DT despite the 60° phase margin.

REMARK 10.6 Note that the gain margin property relates to the overall gain of the loop and not to the gain of the primary controller.

c) Performance

Several aspects related to the performance of the SP are briefly discussed. The typical improvements in performance, due to the SP, that can be expected are demonstrated. Both reference tracking and disturbance attenuation are considered. It is shown that while the potential improvement in reference tracking is significant, the SP is less effective in attenuating disturbances. The reasons for that are clarified and the feedforward SP is presented.

First, the steady-state errors to step changes in the reference (r) and in the input disturbance (d) (see Figure 10.74) are examined. The following assumptions are made:

1. $P(s)$, the plant, is asymptotically stable.
2. $P(s)$ does not contain a pure differentiator.
3. $C_o(s)$, the primary controller, contains an integrator.
4. $Q(s)$ is asymptotically stable and practically stable.

With the above assumptions it is quite straightforward to prove the following theorem by applying the final-value theorem to G_r in Equation 10.224 and to G_d in Equation 10.226.

THEOREM 10.6 *Under assumptions 1 to 4, the SP system will display zero steady-state errors to step reference inputs and to step disturbances.*

REMARK 10.7 Theorem 10.6 remains valid for all uncertainties (mismatchings) for which the closed-loop stability is maintained.

The next example is intended to demonstrate the typical improved performance to be expected by employing the SP and to motivate further discussion on one of the structural properties of the SP that directly affects its regulation capabilities.

EXAMPLE 10.12:

The SP is applied to the following second-order with DT process:

$$P(s) = e^{-.5s}/(s+1)^2 \qquad (10.241)$$

The primary controller, $C_o(s)$, is chosen to be the following ideal PID controller:

$$C_o(s) = K(s+1)^2/s \quad ; \quad K = 6 \qquad (10.242)$$

This choice of $C_o(s)$ is discussed in more detailed form in Section 10.8.3d. It should be emphasized that the value of the gain $K = 6$ is quite conservative. This can be concluded from inspection of $Q(s)$ in Equation 10.229 (or equivalently, G_r in Equation 10.224), which in this case is

$$Q(s) = K/(s+K) \qquad (10.243)$$

It is seen that one may employ K as large as desired in the ideal case, without impairing stability. K will have, however, to be reduced to accommodate mismatching. For reasons to be elaborated in Section 10.8.3d the value of 6 was selected for the gain. Note that without the SP the maximum gain allowed would be 4.7 approximately.

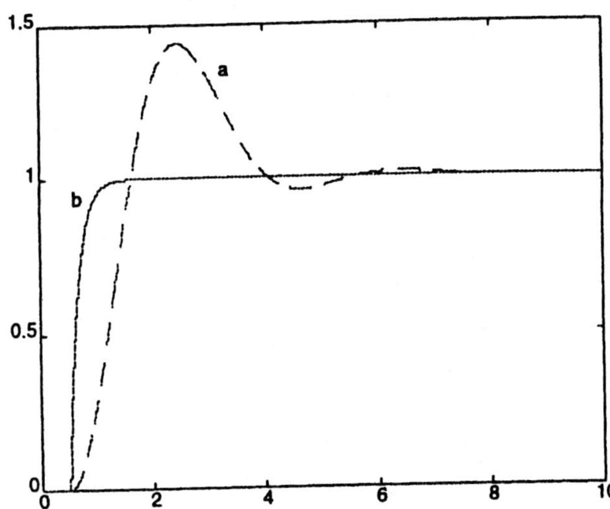

Figure 10.78 Responses to step change in setpoint, the ideal case: (a) PID; (b) SP.

In Figure 10.78 the time responses, in the ideal case, of the SP and a conventional PID controller to a step change in reference are compared. The PID controller settings were determined via the well known Ziegler-Nichols rules, which can be found in another chapter in this handbook. It is evident that despite the relatively conservative tuning of the SP it outperformed the PID in all respects: better rise-time, better settling time, no overshoot, etc. One may try to optimize the PID settings, but the response of the SP is hard to beat.

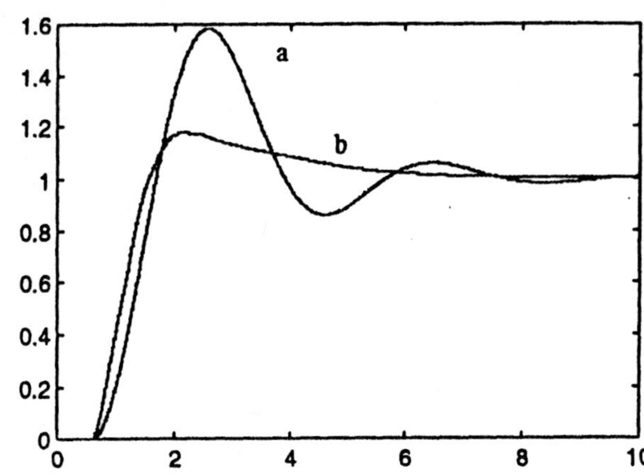

Figure 10.79 Setpoint step responses, nonideal case: (a) PID; (b) SP.

In Figure 10.79 the same comparison of responses is made, but with a mismatch of 20% in the DT, namely, $\hat{\theta} = 0.5$, but the DT has been changed to 0.6. The effect of the mismatch on the responses of both the SP and the PID is noticeable. However, the response of the SP is still considerably better.

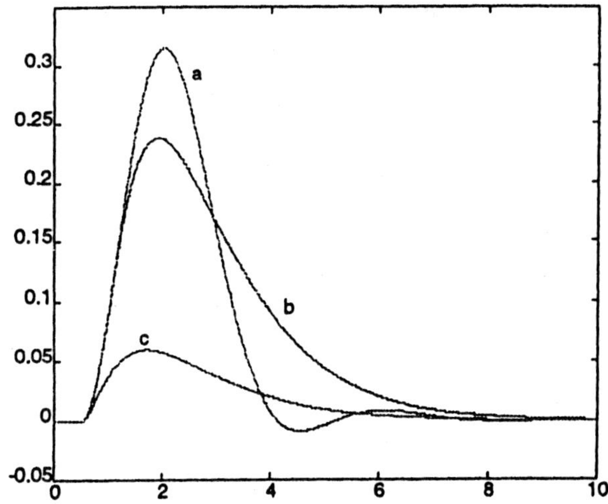

Figure 10.80 Responses to a unit-step input disturbance: (a) PID; (b) SP; (c) feedforward SP.

In Figure 10.80 the corresponding responses to a unit input disturbance are depicted. While the response of the SP has a smaller overshoot, its settling time is inferior to that of the PID. This point is elaborated upon next.

Example 10.12 demonstrates that, on one hand, the SP provides significant improvements in tracking properties over conventional controllers, but on the other hand its potential enhancements in regulatory capabilities are not as apparent. The reason for this has been pointed out in Section 10.8.3a, where it was shown that the open-loop poles are present in the transfer function G_d. These poles are excited by input disturbances but not by the reference. Depending on their locations relative to the closed-loop poles, these poles may dominate the response. The slower the open-loop poles are, the more sluggish the response to input disturbances will be. This is exactly the situation in Example 10.12: the closed-loop pole (the zero of $1 + C_o P_o$) is $s = -6$, while the two open-loop poles are located at $s = -1$. The presence of the open-loop poles in G_d is a direct consequence of the structure of the SP, and many modifications aimed at improving that shortcoming of the SP were proposed. Several modifications are presented in Section 10.9. It is worthwhile noting, however, that the influence of the open-loop poles on the response to disturbances is less pronounced in cases with large DTs. In such a circumstance, the closed-loop poles cannot usually be shifted much to the left, mainly due to the effect of model uncertainties. Hence, in such situations the closed-loop poles do not differ significantly from the open-loop ones, and their influence on the response to disturbances is less prominent.

For the reasons mentioned above, the SP is usually designed for tracking, and if necessary, a modification aimed at improving the disturbance rejection properties is added. When other than step inputs are considered, C_o should and can be designed to accommodate such inputs. For a given plant and inputs, an H_2−optimal design in the framework of IMC (see Section 10.8.3a) was proposed. The interested reader is referred to [4] and [5] for details. A DTC of a special structure that can accommodate various disturbances is presented in Section 10.9.3.

Another way to improve on the regulation capabilities of the SP is to add a standard feedforward controller, which requires online measurements of the disturbances. However, if disturbances are measurable, the SP can provide significant improvements in disturbance attenuation in a direct fashion. This may be achieved by transmitting the measured disturbance into the predictor. The general idea, in the most simple form, is shown in Figure 10.81.

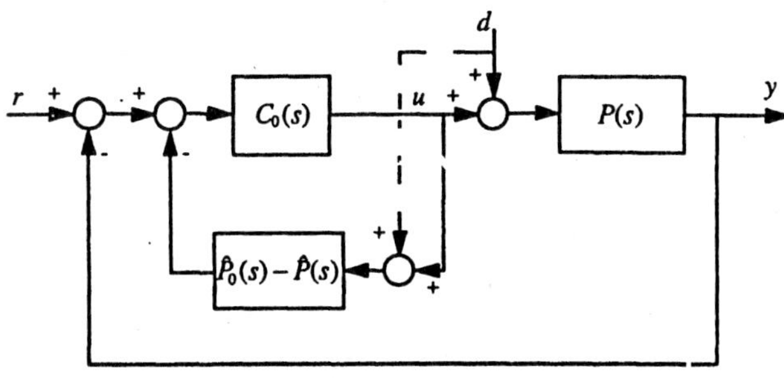

Figure 10.81 A simple form of the feedforward SP for measurable disturbances.

In this fashion the plant model in the predictor is further exploited to predict the effect of the disturbance on the output. The advantage of this scheme may be better appreciated by comparing the closed-loop relations between the control signal, u, and the disturbance, d, in both the "conventional" SP, Figure 10.74, and the one in Figure 10.81. In the conventional SP scheme (Fig.10.74) that relation is given by:

$$u(s)/d(s) = -C_o P_c e^{-\theta s}/(1 + C_o P_o) \quad (10.244)$$

The corresponding relation in the scheme in Figure 10.81 is:

$$u(s)/d(s) = -C_o P_o/(1 + C_o P_o) \quad (10.245)$$

and it is evident that the DT by which the control action is delayed in the conventional SP is effectively canceled in the scheme in Figure 10.81. By counteracting the effect of the disturbance before it can appreciably change the output, the scheme in Figure 10.81 behaves in a manner similar to that of a conventional feedforward controller. For this reason, the scheme in Figure 10.81 is called *feedforward SP*. In fact, the feedforward SP eliminates the need for a separate feedforward controller in many circumstances under which it would be employed. The advantage of the scheme is demonstrated in Figure 10.80. It should be noted that the feedforward SP in Figure 10.81 is presented in its most simplistic form. More realistic forms and other related issues can be found in [7].

Finally, it is worth noting that it has been found from practical experience that a properly tuned SP performs better than a PID in many loops typical to the process industries, even when the model used in the SP is of lower order than the true behavior of the loop. This applies in many circumstances even to DT-free loops. In those cases, the DT in the model is used to compensate for the order reduction.

d) Tuning Considerations

Since the SP is composed of a primary controller and a model of the plant, its tuning, in practice, involves the determination of the parameters of the model and the settings of the primary controller. In this chapter, however, it is assumed that the model is available, and the problem with which we are concerned is the design and setting of the primary controller. From the preceding sections it is clear that the tuning should be related to the stability and robustness properties of the SP.

A simple tuning rule for simple primary controllers, $C_o(s)$, is presented. For low-order plant models with poles and zeros in the left half-plane (LHP), a simple structure for $C_o(s)$, which can be traced back to optimal control [2], is given by

$$C_o(s) = \tfrac{K}{s} \hat{P}_o(s)^{-1} \quad (10.246)$$

When $\hat{P}_o(s)$ is a first-order filter or a second-order one, then the resulting $C_o(s)$ is the classical PI or PID controller, respectively. More specifically, in the first-order and the second-order cases, the corresponding $\hat{P}_o$ is given as in Equations 10.247a and 10.247b, respectively:

$$\hat{P}_o(s) = k_p/(\tau s + 1) \quad (10.247a)$$

$$\hat{P}_o(s) = k_p/(\tau^2 s^2 + 2\tau\xi s + 1) \quad (10.247b)$$

The "textbook" transfer functions of the PI and PID controllers are as follows:

$$K_c\left(1 + \tfrac{1}{\tau_i s}\right) \quad (10.248a)$$

$$K_c\left(1 + \tfrac{1}{\tau_i s} + \tau_D s\right) \quad (10.248b)$$

If Equation 10.247a is substituted into Equation 10.246, the resulting primary controller, C_o, is equivalent to the PI controller in Equation 10.248a if

$$K_c = K\tau/k_p \quad ; \quad \tau_i = \tau \tag{10.249}$$

Similarly, for the second-order case (Equation 10.247b), C_o will be equivalent to Equation 10.248b if:

$$K_c = K(2\tau\xi)/k_p; \quad \tau_i = 2\tau\xi; \quad \tau_D = \tau/2\xi \tag{10.250}$$

Commercially available SPs offer PI and PID controllers for C_o. Thus, the structure in Equation 10.246 is well suited to SP with simple plant models. It should be noted that if the pole excess in $\hat{P}_o$ is larger than 2, then C_o in Equation 10.246 must be supplemented by an appropriate filter to make it proper.

The particular choice in Equation 10.246 leads to a $Q(s)$ with the following simple form:

$$Q(s) = K/(s + K) \tag{10.251}$$

and it is evident that stability of the closed loop, in the ideal case, is maintained for any positive K. However, model uncertainty imposes an upper limit on K. Since the SP is mostly sensitive to mismatches in the DT, a simple rule for the setting of K, the single tuning parameter of C_o in Equation 10.246, can be obtained by calculating K_{max} [6] for stability as a function of δ_θ, the relative error in DT.

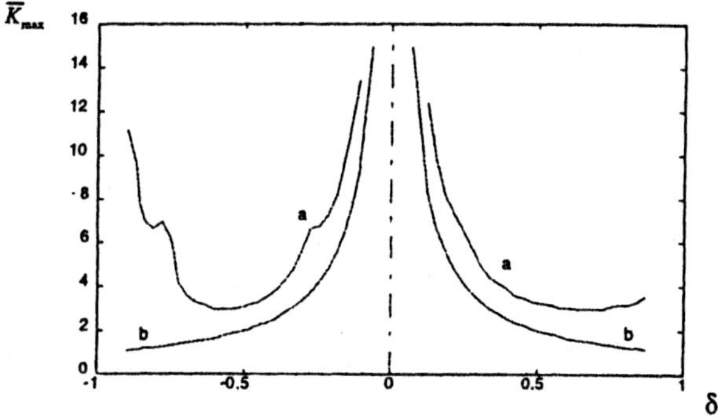

Figure 10.82 (a) Maximum $\overline{K}$ for stability vs. δ_θ; (b) the rule of thumb suggested in [4].

Curve (a) in Figure 10.82 depicts $\overline{K}_{max}$ as a function of δ_θ. $\overline{K}$ and δ_θ are defined as follows:

$$\overline{K} \triangleq K\hat{\theta} \quad ; \quad \delta_\theta \triangleq \Delta\theta/\hat{\theta} \tag{10.252}$$

Figure 10.82 reveals an interesting property, namely, that the choice of $\overline{K} < 3$ assures stability for $\pm100\%$ mismatches in DT. Thus, a simple tuning rule is to set K as follows:

$$K = 3/\hat{\theta} \tag{10.253}$$

This rule was applied in Example 10.12 in section 10.8.3c. While curve (a) in Figure 10.82 displays the exact $\overline{K}_{max}$ for all cases for which C_o is given by Equation 10.246, some caution must be exercised in using it, or Equation 10.253, since it takes into account mismatches in DT only. Curve (b) in Figure 10.82 displays a rule of thumb suggested in [4] where $\overline{K}$ is chosen as $1/\delta_\theta$. This latter rule was developed for first-order systems (see Equation 10.239) and mismatches in the DT only.

A method for tuning the primary controller of the SP for robust performance in the presence of simultaneous uncertainties in parameters was developed in [5] in the framework of IMC. For the three parameters first-order with DT model given in Equation 10.239, the overall H_2−optimal IMC controller (see Equation 10.228) for step references and equipped with a filter, is given by:

$$\bar{q}(s) = (\tau s + 1)/\left[k_p(\lambda s + 1)\right] \tag{10.254}$$

where λ is the time constant of the filter $f(s)$. In Equation 10.254 $\bar{q}(s)$ is equivalent to C_o in Equation 10.246, for this case, with $\lambda = 1/K$. Using various robust performance criteria, tuning tables for λ for various simultaneous uncertainties in the three parameters have been developed and can be found in [5].

10.9 Modified Predictors

In the previous section, it was seen that the improvements in disturbance attenuation offered by the SP are not as good as for tracking. The reasons for that were discussed and a "remedy" in the form of the feedforward SP was presented. The feedforward SP is applicable, however, only if disturbances can be measured on-line. This may not be possible in many cases. Quite a number of researchers have recognized this fact and proposed modifications aimed at improving on the regulatory capabilities of the SP. Three such designs are briefly described in this section. In the first two, the structure of the SP is kept, but a new component is added or an existing one is modified. In the third one, a new structure is proposed.

10.9.1 Internal Cancellation of Model Poles

The scheme suggested in [8] has the same structure as the SP in Figure 10.74, but $\hat{P}_1$ replaces $\hat{P}_o$ in the minor feedback around the primary controller. Hence, the minor feedback consists of $(\hat{P}_1 - \hat{P})$ instead of $(\hat{P}_o - \hat{P})$. $\hat{P}_1$ may by considered to be a modified plant model without a DT. $\hat{P}_1$ is the nominal model of P_1, which is given by:

$$P_1(s) = Ce^{-A\theta}(sI - A)^{-1}B - \int_0^\theta Ce^{-A\tau}Bd\tau \tag{10.255}$$

where A, B, and C are the "true" matrices in the state-space representation of the plant given in Equation 10.217. The role of P_1 will be clarified later. Note that $Q(s)$ may be defined in a similar fashion to the one defined in Equation 10.229 for the conventional SP. For the scheme under consideration, it is given by:

$$Q(s) = C_o\hat{P}_o/(1 + C_o\hat{P}_1) \tag{10.256}$$

and all the previous results on practical and robust stability of Section 10.8.3b apply to this case as well.

In [8] some general results, applicable to the scheme considered here with any stable $\hat{P}_l$ were stated and proven. Under assumptions 1 to 4 of Section10.8.3d, and for the particular $\hat{P}_l$ in Equation 10.255, the application of the general results yields the following theorem:

THEOREM 10.7 *Under assumptions 1 to 4, the modified SP, with $\hat{P}_o$ replaced by $\hat{P}_l$ in the minor feedback, has the following properties:*

a. *A zero steady-state error to step reference.*
b. *A zero steady-state error to step disturbance.*
c. *The poles of $(\hat{P}_l - \hat{P})$, the minor feedback, are canceled with its zeros.*

The following remarks explain and provide some insight into the properties stated in Theorem 10.7.

REMARK 10.8 Property (a) holds for any P_l satisfying $\lim_{s\to 0} P_o(s)/P_l(s) = 1$. It can be verified that the P_l in Equation 10.255 satisfies the latter condition.

REMARK 10.9 Property (b) holds for any stable P_l satisfying $\lim_{s\to 0}(P_l(s) - P(s)) = 0$. It is easy to show that the P_l in Equation 10.255 satisfies that condition.

REMARK 10.10 Property (c) represents the major advantage of the scheme discussed here and is the source for its potential improvement in regulatory capabilities. Due to the pole-zero cancellation in the minor feedback, G_d (see Equation 10.226), the transfer function relating y to d, no longer contains the open-loop poles. The response to disturbances, under these circumstances, is governed by the zeros of $(1 + C_o P_l)$, after θ units of time from the application time of the disturbances. That is, the error decay can be made as fast as desired (in the ideal case) after the DT has elapsed. See the discussion following Example 10.13.

REMARK 10.11 An equivalent way to express properties (b) and (c) is to say that the states of the minor feedback, as well as the state of the integrator of the primary controller, are unobservable in v (see Figure 10.74).

REMARK 10.12 Note that for given A, B, C and θ, the integral in P_l (Equation 10.255) is a constant scalar. Its sole purpose is to make the state of the integrator unobservable in v.

REMARK 10.13 If the minor feedback, $\hat{P}_l - \hat{P}$, is realized as a dynamical system (i.e., by Equations 10.255 and 10.220), then this scheme is not applicable to unstable plants. However, with a different realization, to be discussed in Section 10.10, it can be applied to unstable plants.

EXAMPLE 10.13:

Two cases are considered. In both, the plant is a first-order with DT (see Equation 10.239). In the first case the plant parameters are $k_p = 1$, $\tau = 1$, $\theta = 0.2$. In the second case θ, the DT, is increased to $\theta = 1$. Primary controllers, C_o, are designed for the SP and for the modified predictor, and the responses to input disturbances are compared. For a fair comparison both designs are required to guarantee stability for $\pm 60\%$ mismatching in the DT. C_o for the SP is taken to be as in Equation 10.246 with K according to Equation 10.253. Thus, the C_o for the SP is a PI controller, which clearly satisfies the above stability requirement. A PI controller is also selected for the modified predictor. K_c and τ_i, the parameters of the PI controller (Equation 10.248a), for the modified predictor are determined such that the stability requirement is satisfied and the response optimized. The resulting parameters of the PI controllers for both schemes and for the two cases are as follows: 1.SP a.Case 1 $-K_c = 15$, $\tau_i = 1/15$ b. Case 2 $-K_c = 3$, $\tau_i = 0.33$.
2. Modified predictor a. Case 1 $-K_c = 3$, $\tau_i = 0.14$ b. Case 2 $-K_c = 0.46$, $\tau_i = 1.92$.
Notice the substantial reduction in the gains of the modified predictor, relative to the SP, required to guarantee stability for the same range of mismatches in the DT. The responses to a unit-step input disturbance of the SP [curve (1)] and the modified predictor [curve (2)] are compared in Figure 10.83a for Case 1 and in Figure 10.83b for Case 2.

While the improvement achieved by the modified predictor is evident in Case 1, it is insignificant in Case 2. This point is elaborated on next.

Example 10.13 demonstrates the potential improvement in the regulatory performance of the modified scheme. However, practical considerations reduce the effectiveness of this scheme in certain cases. First, note that P_l is a proper transfer function and frequently is non-minimum phase. Hence, the design of C_o is usually more involved. Simple primary controllers, like those in Equation 10.246, are not applicable, and in many cases the conventional PI and PID controllers may not stabilize the system even in the ideal case. Second, when it is designed for robust stability or robust performance, the resulting gains of the primary controllers are considerably lower than those allowed in the conventional SP for the same range of uncertainties. Therefore, the improvements in disturbance attenuation are usually less prominent. It turns out, as is also evident in Example 10.13, that the modified scheme is advantageous in cases where DTs are small relative to the time constants of the plant, or when uncertainties are small. The chief reason for this is that it is possible to improve on the conventional SP only if the closed-loop poles (i.e., zeros of $1 + C_o P_l$) can be made considerably faster than the open-loop ones. This can be achieved if relatively high gains are allowed.

It was pointed out in the previous paragraph that the design of C_o in the modified scheme is considerably more involved than in the conventional SP, more so when plant models are of order

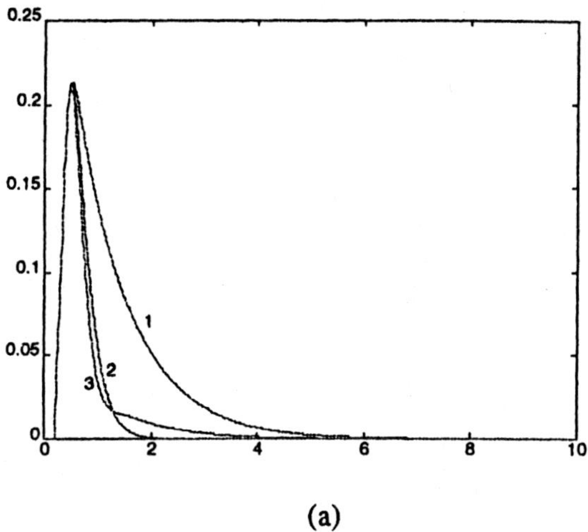

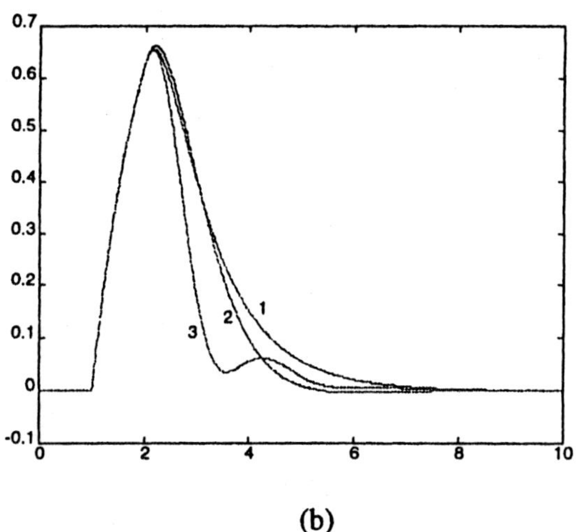

Figure 10.83 Responses to a unit-step input disturbance: (a) Case 1 ($\theta/\tau = 0.2$); (b) Case 2 ($\theta/\tau = 1$). (1) SP. (2) Modified predictor with internal cancellations. (3) SP with approximate inverse of DT.

higher than two. In those cases, C_o may be determined by input-output pole placement. A realizable C_o and of low order, if possible, is sought such that the closed-loop poles in the ideal case (i.e., the zeros of $1 + C_o \hat{P}_1$) are placed in predetermined locations. Then the design is checked for robust stability and robust performance. If the design fails the robustness tests, then C_o is modified in a trial-and-error fashion until it complies with the robustness requirements.

10.9.2 An Approximate Inverse of DT

A simple modification was suggested in [9]. It consists of a simple predictor, $M(s)$, which is placed in the major feedback of the SP as shown in Figure 10.84.

It is desired to have $M(s)$ equal to the inverse of the DT, i.e., $M(s) = e^{\hat{\theta}s}$. In this fashion the output, y, is forecast one DT into the future. This in turn eliminates the DT between the dis-

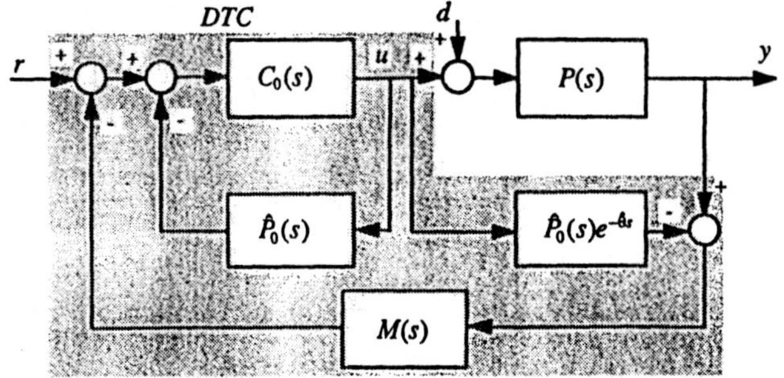

Figure 10.84 Modified SP with an approximate inverse of DT.

turbance, d, and the control, u, in a similar fashion to the feed-forward SP (see Figure 10.81). Indeed, with $M(s)$ as above, the transfer function $u(s)/d(s)$ is exactly the one in Equation 10.245. It is clear, however, that it is impossible to realize an inverse of the DT. Hence, a realizable approximation of the inverse of the DT is employed. In [9], the following $M(s)$ is suggested:

$$M(s) = (1 + B(s))/[1 + B(s)e^{-Ls}] \qquad (10.257)$$

If $B(s)$ is a high-gain low-pass filter given by:

$$B(s) = K_m/(\tau_m s + 1) \qquad (10.258)$$

then $M(s)$ in Equation 10.257 approximates e^{Ls} at low frequencies.

A method for the design of $M(s)$ is suggested in [9]. It consists of two steps. First, an SP is designed based on methods like the one in [5], or the one described in Section 10.8.3d. In the second step, the $M(s)$ in Equations 10.257 and 10.258 is designed to cope with uncertainties. With $M(s)$ in the major feedback, the condition for stability under multiplicative error, which corresponds to the one in Equation 10.238, is easily shown to be

$$|M(j\omega)Q(j\omega)|\ell(\omega) < 1 \qquad \forall\omega \geq 0$$

or equivalently,

$$|M(j\omega)| < [|Q(j\omega)|\ell(\omega)]^{-1} \qquad \forall\omega \geq 0 \quad (10.259)$$

For good performance it is desired to have $|M(j\omega)Q(j\omega)|$ close to one at frequencies below the bandwidth of the closed loop. Thus, the design of $M(s)$ is to choose the three parameters, k_m, τ_m and L, such that the magnitude curve of $|M(j\omega)|$ lies as close as possible to $|Q(j\omega)|^{-1}$ and beneath the curve $[|Q(j\omega)|\ell(\omega)]^{-1}$.

The three parameters of $M(s)$ optimizing the regulatory properties of the SP with the primary controller in Equation 10.246 tuned according to Equation 10.253 were determined experimentally in [6] for the first-order case (Equation 10.239) and are summarized in Table 10.10.

It was found that for $\hat{\theta}/\hat{\tau}$ up to 2, the inclusion of the simple filter $M(s)$ enhances the disturbance attenuation properties of the SP. However, for $\hat{\theta}/\hat{\tau}$ above 2, the improvement via $M(s)$ is minor and the use of $M(s)$ is not justified in those cases. In Figure 10.83, the responses to a step input disturbance for the two cases considered in Example 10.13, with $M(s)$ in place, are

TABLE 10.10 Parameters of $M(s)$ for the First-Order with DT Case with the C_o in Equation 10.246 and tuning rule in Equation 10.253.

$\hat{\theta}/\hat{\tau}$	k_m	$L/\hat{\theta}$	L/τ_m
0.3	10	0.75	0.05
0.6	8	0.6	0.1
1.0	4	0.45	0.2
2.0	2	0.27	0.3

shown [curve (3)], and compared to those of the two predictors discussed in Example 10.13. The primary controllers are the same as those used in the conventional SP, and the parameters of $M(s)$ were determined from Table 10.10. The improvement achieved by $M(s)$ in both cases shown in Figure 10.83 is evident.

10.9.3 Observer-Predictor

The structure of the observer-predictor (OP) is depicted in Figure 10.85. It consists of

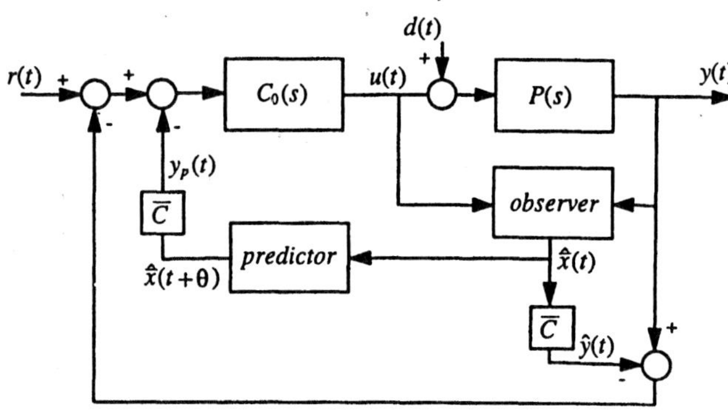

Figure 10.85 The structure of the observer-predictor (OP).

- An asymptotic observer that estimates the states of both the plant and the disturbance
- A predictor that uses the estimated state to forecast the output one DT into the future
- A primary controller, C_o

The basic structure of the OP is not a new one. It was suggested in [10], where a static-state feedback was employed and where the main concern was the stability properties of the scheme. No attention was paid to tracking and regulation capabilities. Indeed, the performance of the OP in [10] is poor. The OP outlined in this section was developed in [11] and contains several modifications aimed at improving its regulatory properties. First, it contains a dynamical primary controller that operates on the forecast output. Second, a model of the dynamics of the disturbance is incorporated in the observer [12]. It enables the on-line estimation of the disturbance. Third, an additional feedback, similar to the one used in the SP, which carries the difference between the measured and the estimated outputs, is introduced. The main objective of that feedback is to compensate for uncertainties and disturbances. The equations of the OP in Figure 10.85, are given next, followed by brief comments on several properties of the OP. An example demonstrating the effectiveness of the OP in attenuating disturbances concludes this section.

As in Equation 10.217, the plant is given by:

$$\begin{aligned} \dot{x}(t) &= Ax(t) + B_1u(t-\theta) + B_2d(t-\theta) \\ y(t) &= Cx(t) \end{aligned} \quad (10.260)$$

The plant model in Equation 10.260 is slightly more general than before, as the control, u, and the disturbance, d, go through different input matrices. The disturbance is assumed to obey the following model:

$$\begin{aligned} \dot{z}(t) &= Dz(t) \\ d(t) &= Hz(t) \end{aligned} \quad (10.261)$$

where $z \epsilon R^m$ is the state vector of the disturbance model. It is further assumed that the pairs (A, C) and (D, H) are observable. Substitution of d from Equation 10.261 into Equation 10.260 yields:

$$\dot{x}(t) = Ax(t) + B_1u(t-\theta) + B_2He^{-D\theta}z(t) \quad (10.262)$$

Next, an augmented state vector is defined:

$$\overline{x}(t)^T = (x(t)^T, z(t)^T) \quad (10.263)$$

By means of Equation 10.263, the plant and the disturbance models are combined:

$$\begin{aligned} \dot{\overline{x}}(t) &= \overline{A}\overline{x}(t) + \overline{B}u(t-\theta) \\ y(t) &= \overline{C}\overline{x}(t) \end{aligned} \quad (10.264)$$

where

$$\overline{A} = \begin{pmatrix} A & B_2He^{-D\theta} \\ 0 & D \end{pmatrix}; \quad \overline{B} = \begin{pmatrix} B_1 \\ 0 \end{pmatrix}; \quad \overline{C} = (C \quad 0) \quad (10.265)$$

The observer is given by:

$$\dot{\hat{\overline{x}}}(t) = \overline{A}\hat{\overline{x}}(t) + \overline{B}u(t-\theta) - L(y(t) - \overline{C}\hat{\overline{x}}(t)) \quad (10.266)$$

where $\hat{\overline{x}}$ is the estimate of $\overline{x}$, and L is the vector of gains of the observer. The predictor is given by:

$$\begin{aligned} y_p(t) &\triangleq \hat{y}(t+\theta) \\ &= Ce^{A\theta}\hat{x}(t) + C\int_{-\theta}^{0} e^{-Ah} \\ &\quad \left[B_1u(t+h) + B_2\hat{d}(t+h)\right]dh \end{aligned} \quad (10.267)$$

where $y_p(t)$ is the forecast of the estimated output θ units of time ahead, and $\hat{x}(t)$ and $\hat{d}(t)$ are the estimates of the state and of the disturbance, respectively, both generated by the observer.

Equation 10.267 presents the *integral form* of the predictor. If Equation 10.267 is Laplace transformed, then the *dynamical form* of the predictor is obtained:

$$y_p(s) = Ce^{A\theta}\hat{x}(s) + C(I - e^{-(sI-A)\theta})(sI - A)^{-1} \left[B_1 u(s) + B_2 \hat{d}(s)\right] \quad (10.268)$$

Finally, the control signal $u(s)$ is

$$u(s) = C_o(s)\left[r(s) - y_p(s) - (y(s) - \hat{y}(s))\right] \quad (10.269)$$

The design of the OP consists of the selection of L, the observer's gain vector, and of $C_o(s)$. L may be determined by the well-known pole-placement techniques. If, in addition to the observability assumptions stated above, it is further assumed that no pole-zero cancellations between the models of the plant and the disturbances occur, then the pair $(\overline{A}, \overline{C})$ is observable and the observer poles can be placed at will. The design of C_o is referred to in the subsequent remarks.

The main properties of the OP are summarized in the following remarks:

REMARK 10.14 In the ideal case the DT is eliminated from the characteristic equation of the closed loop.

REMARK 10.15 The overall closed-loop transfer function relating r to y is identical, in the ideal case, to that of the SP given in Equation 10.224.

REMARK 10.16 If the predictor is realized in the integral form of Equation 10.267, then the closed-loop poles consist of the observer poles (i.e., the zeros of $det[sI - \overline{A} - L\overline{C}]$) and of the zeros of $1 + C_o P_o$.

REMARK 10.17 According to the *internal model principle* (see [12], for example), C_o should contain the disturbance poles in order to reject disturbances. Note that the poles of the disturbance in Equation 10.261 are the zeros of $det(sI - D)$. With the latter requirement the design of C_o may be carried out by appropriate placement of the zeros of $1 + C_o P_o$.

REMARK 10.18 While the OP has far better disturbance rejection properties than the SP, as demonstrated in Example 10.14, it is, in general, considerably more sensitive to uncertainties.

Example 10.14 demonstrates the improved capability of the OP to reject disturbances.

EXAMPLE 10.14:

The plant is the one used in Example 10.12, Equation 10.241. The disturbance, however, in this case is given by:

$$d(t) = \sin 2t \quad (10.270)$$

The D in Equation 10.261 is therefore:

$$D = \begin{pmatrix} 0 & 1 \\ -4 & 0 \end{pmatrix}$$

and $det(sI - D) = s^2+4$. The latter factor is included (according to Remark 10.17) in the denominator of C_o.

The observer gains were selected such that all the observer poles are placed at $s = -2$. In addition to containing the disturbance poles, C_o was required to have an integrator. The rest of the design of C_o was based on placing the zeros of $1 + C_o P_o$ at $s = -2$. No claim is made that this is an optimal design. For comparison purposes the same C_o was used in the SP. The responses of the OP and the SP to the sinusoidal disturbance in Equation 10.270 are shown in Figure 10.86.

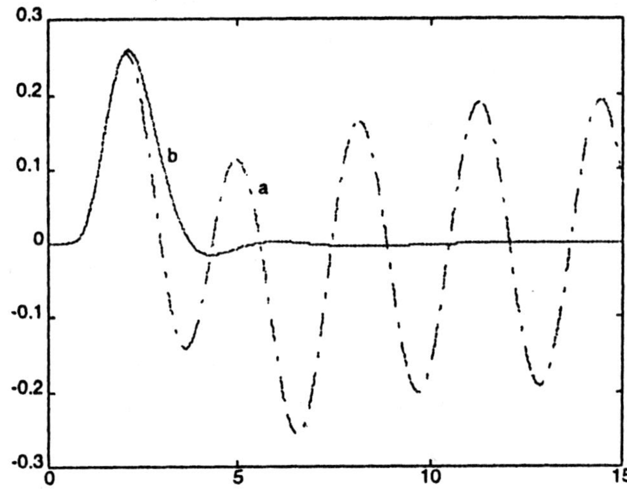

Figure 10.86 Response of sinusoidal input disturbance: (a) SP; (b) OP.

It is apparent that while the SP cannot handle such a disturbance, the OP does a remarkable job.

10.10 Time-Delay Compensation for Unstable Plants

In this section we briefly discuss which of the schemes presented in the preceding sections is applicable to unstable plants with DTs and under what conditions.

In section 10.8.3a it was pointed out that the SP cannot be applied to unstable plants. It has been shown that the plant models in the minor feedback of the SP are the cause for the appearance of the poles of the plant as poles of the closed loop. This fact was evident in the transfer function G_d in Equation 10.226. A straightforward calculation of the closed-loop poles of a system with an SP shows that they consist of the zeros of $1+C_o P_o$ and the poles of the open-loop plant. Hence, it is concluded that the SP is internally unstable if the plant is unstable [13]. An alternative way to arrive at the same conclusion is to look at the IMC form of the SP in Figure 10.76. It is seen that the control signal, u, is fed in parallel to the plant and to the model. Such a structure is clearly uncontrollable. The above conclusion applies to the modified SP with the approximate inverse of the DT as well.

The modified predictor with internal pole-zero cancellations in section 10.9.1 is in the same situation. Although the poles of the minor feedback are canceled with its zeros and therefore do not show up in G_d, for example, these poles are clearly internal modes of the closed loop. Upon noting that the poles of the minor feedback are those of the plant, we may conclude that the modified predictor with the internal cancellation will be internally unstable for unstable plants. It is possible, however, to make the modified predictor cope with unstable plants. Fortunately enough, the minor feedback in the modified predictor can be realized in an equivalent, but nondynamical form, which does not possess poles. The derivation of the *nondynamical form* of the minor feedback is outlined next. Recall that the minor feedback is

$$P_1 - P \tag{10.271}$$

Upon substitution of the P_1 in Equation 10.255 and P in Equations 10.219 and 10.220, Equation 10.271 can be written explicitly as follows:

$$\begin{aligned} P_1 - P &= Ce^{-A\theta}(sI - A)^{-1}B - \int_0^\theta Ce^{-A\tau}Bd\tau \\ &\quad - C(sI - A)^{-1}Be^{-\theta s} \end{aligned} \tag{10.272}$$

With the aid of the following, easily verified identity:

$$\begin{aligned} &Ce^{-A\theta}(sI - A)^{-1}B - C(sI - A)^{-1}Be^{-\theta s} = \\ &\quad Ce^{-A\theta}\textstyle\int_{-\theta}^{0} e^{\tau(sI-A)}Bd\tau \end{aligned} \tag{10.273}$$

Equation 10.272 becomes:

$$\begin{aligned} P_1 - P &= Ce^{-A\theta}\int_{-\theta}^{0} e^{\tau(sI-A)}Bd\tau \\ &\quad - \int_0^\theta Ce^{-A\tau}Bd\tau \end{aligned} \tag{10.274}$$

Finally, inverse Laplace transformation of Equation 10.272 yields:

$$\begin{aligned} v(t) &= Ce^{-A\theta}\int_{-\theta}^{0} e^{-A\tau}Bu(t+\tau)d\tau \\ &\quad - \left[\int_0^\theta Ce^{-A\tau}Bd\tau\right]u(t) \end{aligned} \tag{10.275}$$

where $v(t)$ is the output variable of the minor feedback (see Figure 10.74). Due to the finite limits of the integral, the right-hand side of Equation 10.275 is an entire function that does not possess singularities. Consequently, if the minor feedback is realized via Equation 10.275, the modified predictor with the internal cancellations is applicable to unstable plants. By applying exactly the same arguments to the case of the OP of section 10.9.3 it may be concluded, at once, that the OP can be applied to unstable plants if the predictor is realized in the integral form given in Equation 10.267.

10.10.1 Concluding Remarks

In this chapter we have presented the basic features and the special properties of the SP. The advantages and drawbacks of the SP were discussed. For the benefit of potential users, attention was paid to both theoretical and practical aspects. Several modifications and alternative schemes that were developed over the years to improve, in certain cases, on some of the properties of the SP were presented also. Due to lack of space, however, we confined our attention to the continuous SISO case. As mentioned in the introduction, a vast material exists on this topic and it was impossible to cover many additional contributions and extensions in a single chapter.

References

[1] Smith, O.J.M., *Chem. Eng. Prog.*, 53, 217, 1959.

[2] Palmor, Z.J., *Automatica*, 18(1), 107–116, 1982.

[3] Palmor, Z.J., *Int. J. Control*, 32(6), 937–949, 1980.

[4] Morari, M. and Zafiriou, E., *Robust Process Control*, Prentice Hall, Englewood Cliffs, NJ, 1989.

[5] Laughlin, D.L., Rivera, D.E., and Morari, M., *Int. J. Control*, 46(2), 477–504, 1987.

[6] Palmor, Z.J. and Blau, M., *Int. J. Control*, 60(1), 117–135, 1994.

[7] Palmor, Z.J. and Powers, D.V., *AIChE J.*, 31(2), 215–221, 1985.

[8] Watanabe, K. and Ito, M., *IEEE Trans. Autom. Control*, 26(6), 1261–1269, 1981.

[9] Huang, H.P., Chen, C.L., Chao, Y.C., and Chen, P.L., *AIChE J.*, 36(7), 1025–1031, 1990.

[10] Furakawa, T. and Shimemura, E., *Int. J. Control*, 37(2), 399–412, 1983.

[11] Stein, A., M.Sc. thesis, Faculty of Mech. Eng., Technion, Haifa, 1994.

[12] Johnson, C.D., *Control and Dynamic Systems*, Vol. 12, Leondes, C.T., Eds. Academic Press, New York, 1976, 389–489.

[13] Palmor, Z.J. and Halevi, Y., *Automatica*, 26(3), 637–640, 1990.

[14] Ogunnaike, B.A. and Ray, W.H., *AIChE J.*, 25(6), 1043, 1979.

[15] Jerome, N.F. and Ray, W.H., *AIChE J.*, 32(6), 914–931, 1986.

[16] Palmor, Z.J. and Halevi, Y., *Automatica*, 19(3), 255–264, 1983.

[17] Astrom, K.J., Hang, C.C., and Lim, B.C., *IEEE Trans. Autom. Control*, 39(2), 343–345, 1994.

Further Reading

For the interested reader we mention just a few references that contain additional results and extensions.

Results on the sampled-data version of SP can be found in [13], where it was shown that while some of the properties of the continuous SP carry over to its discrete counterpart, there are properties unique to the sampled-data case.

The extension of the SP to MIMO plants with multiple delays was given in [14] and in more general form in [15].

The stability properties of the multivariable SP were analyzed in [16].

An SP-like scheme, specific for plants with an integral mode, which decouples tracking from regulation, has recently been presented in [17].

Finally, a simple automatic tuner for SP with simple models and that simultaneously identifies the model and tunes the primary controller can be found in [6].

SECTION IV

Digital Control

11

Discrete-Time Systems

Mohammed S. Santina
The Aerospace Corporation, Los Angeles, CA

Allen R. Stubberud
University of California, Irvine, Irvine, CA

Gene H. Hostetter

11.1 Discrete-Time Systems

11.1.1 Introduction to Digital Control

Rapid advances in digital system technology have radically altered the control system design options. It has become routinely practicable to design very complicated digital controllers and to carry out the extensive calculations required for their design. These advances in implementation and design capability can be obtained at low cost because of the widespread availability of inexpensive and powerful digital computers and their related devices.

A *digital control system* uses digital hardware, usually in the form of a programmed digital computer, as the heart of the controller. A typical digital controller has analog components at its periphery to interface with the plant. It is the processing of the controller equations that distinguishes analog from digital control.

In general, digital control systems have many advantages over analog control systems. Some of the advantages are

1. Low cost, low weight, and low power consumption
2. Zero drift of system parameters despite wide variations in temperature, humidity, and component aging
3. High accuracy
4. High reliability and ease of making software and design changes

The signals used in the description of digital control systems are termed *discrete-time signals.* Discrete-time signals are defined only for discrete instants of time, usually at evenly spaced time steps. Discrete-time computer-generated signals have discrete (or *quantized*) amplitudes and thus attain only discrete values. Figure 11.1 shows a continuous amplitude signal that is represented by a 3-bit binary code at evenly spaced time instants. In general, an n-bit binary code can represent only 2^n different values. Because of the complexity of dealing with quantized signals, digital control system design proceeds as if the signals involved are not of discrete amplitude. Further analysis usually must be performed to determine whether the proposed level of quantization is acceptable.

A discrete-time system is said to be *linear* if it satisfies the principle of *superposition.* Any linear combination of inputs produces the same linear combination of corresponding output components. If a system is not linear, then it is termed *nonlinear.* A discrete-time system is *step invariant* if its properties do not change with time step. Any time shift of the inputs produces an equal time shift of every corresponding output signal.

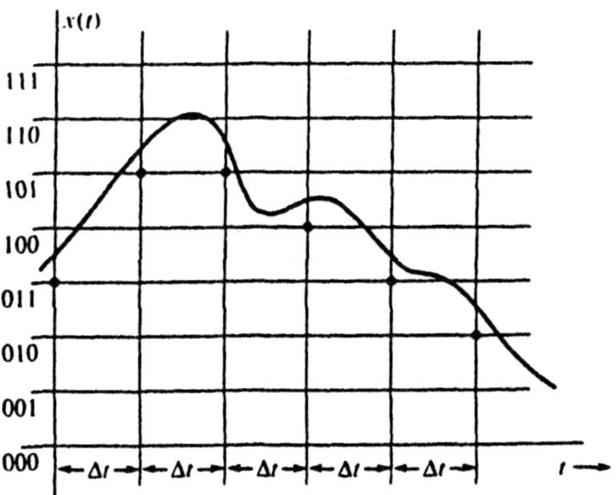

Figure 11.1 An example of a 3-bit quantized signal.

Figure 11.2 shows a block diagram of a typical digital control system for a continuous-time plant. The system has two reference inputs and five outputs, two of which are measured directly by analog sensors. The *analog-to-digital converters* (A/D) *sample* the analog sensor signals and produce equivalent binary representations of these signals. The sampled sensor signals are then modified by the digital controller algorithms, which are designed to produce the necessary digital control inputs $u_1(k)$ and $u_2(k)$.

0-8493-8570-9/96/$0.00+$.50

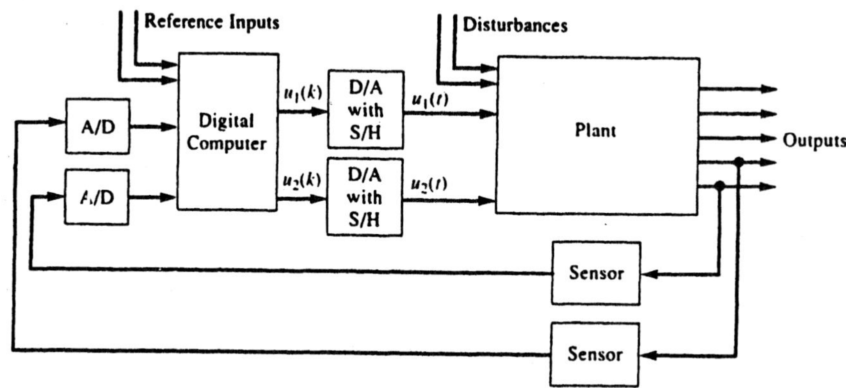

Figure 11.2 A digital control system controlling a continuous-time plant.

Consequently, the control inputs $u_1(k)$ and $u_2(k)$ are converted to analog signals $u_1(t)$ and $u_2(t)$ using *digital-to-analog converters* (D/A). The D/A transforms the digital codes to signal *samples* and then produces *step reconstruction* from the signal samples by transforming the binary-coded digital input to voltages. These voltages are held constant during the *sampling period* T until the next sample arrives. This process of holding each of the samples is termed *sample and hold.* Then the analog signals $u_1(t)$ and $u_2(t)$ are applied to control the behavior of the plant. Not shown in Figure 11.2 is a real-time clock that synchronizes the actions of the A/D, D/A, and shift registers.

Of course, there are many variations on this basic theme, including situations where the signals of the analog sensors are sampled at different sampling periods and where the system has many controllers with different sampling periods. Other examples include circumstances where (1) the A/D and D/A are not synchronized; (2) the sampling rate is not fixed; (3) the sensors produce digital signals directly; (4) the A/D conversion is different from sample and hold; and (5) the actuators accept digital commands.

11.1.2 Discrete-Time Signals and Systems

A *discrete-time signal* $f(k)$ is a sequence of numbers called samples. It is a function of the discrete variable k, termed the *step index.* Figure 11.3 shows some fundamental sequences, all having samples that are zero prior to $k = 0$. In the figure, the step and the ramp sequences consist of samples that are values of the corresponding continuous-time functions at evenly spaced points in time. But the unit pulse sequence and the unit impulse function are not related this way because the pulse has a unit sample at $k = 0$ while the impulse is infinite at $t = 0$.

Z-Transformation

The one-sided z-transform of a sequence $f(k)$ is defined by the equation

$$\mathcal{Z}[f(k)] = F(z) = \sum_{k=0}^{\infty} f(k)z^{-k}$$

It is termed one-sided because samples before step zero are not included in the transform. The z-transform plays much the same role in the analysis of discrete-time systems as the Laplace transform does with continuous-time systems. Important sequences and their z-transforms are listed in Table 11.1, and properties of the z-transform are summarized in Table 11.2.

TABLE 11.1 Z-Transform Pairs

$f(k)$	$F(z)$
$\delta(k)$, unit pulse	1
$u(k)$, unit step	$\frac{z}{z-1}$
$ku(k)$	$\frac{z}{(z-1)^2}$
$c^k u(k)$	$\frac{z}{z-c}$
$kc^k u(k)$	$\frac{cz}{(z-c)^2}$
$u(k) \sin \Omega k$	$\frac{z \sin \Omega}{z^2-2z\cos\Omega+1}$
$u(k) \cos \Omega k$	$\frac{z(z-\cos\Omega)}{z^2-2z\cos\Omega+1}$
$u(k)c^k \sin \Omega k$	$\frac{z(c\sin\Omega)}{z^2-(2c\cos\Omega)z+c^2}$
$u(k)c^k \cos \Omega k$	$\frac{z(z-c\cos\Omega)}{z^2-(2c\cos\Omega)z+c^2}$

A sequence that is zero prior to $k = 0$ is recovered from its z-transform via the inverse z-transform

$$f(k) = \frac{1}{2\pi j} \oint F(z)z^{k-1}dz$$

in which the integration is performed in a counterclockwise direction along a closed contour on the complex plane. In practice, the integrals involved are often difficult, so other methods of inversion have been developed to replace the inverse transform calculations. For rational $F(z)$, sequence samples can be obtained from $F(z)$ by long division. Another method of recovering the sequence from its z-transform is to expand $F(z)/z$ in a partial fraction expansion and use appropriate transform pairs from Table 11.1. Rather than expanding a z-transform $F(z)$ directly in a partial fraction, the function $F(z)/z$ is expanded so that terms with a z in the numerator result. Yet another method of determining the inverse z-transform that is well suited to dig-

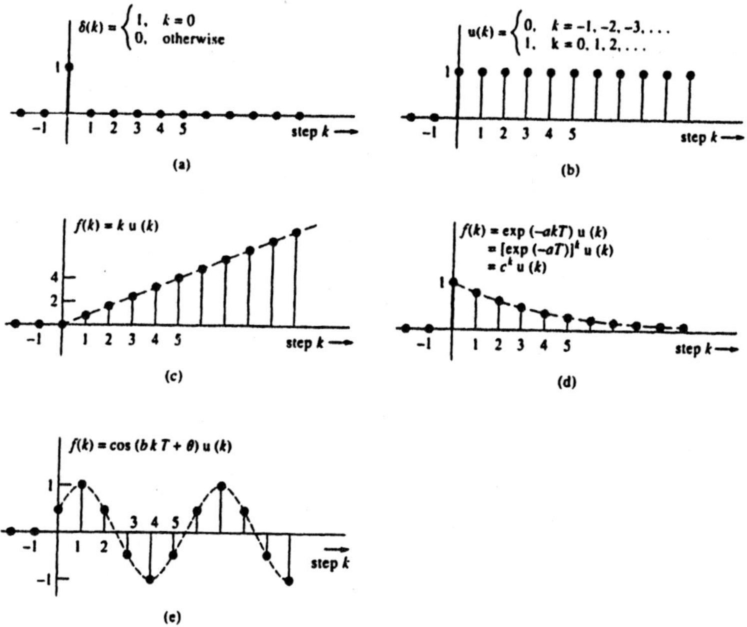

Figure 11.3 Some fundamental sequences. (a) unit pulse; (b) unit step; (c) unit ramp; (d) geometric (or exponential); (e) sinusoidal.

TABLE 11.2 Z-Transform Properties

$\mathcal{Z}[f(k)] = \sum_{k=0}^{\infty} f(k)z^{-k} = F(z)$
$\mathcal{Z}[cf(k)] = cF(z)$ c a constant
$\mathcal{Z}[f(k) + g(k)] = F(z) + G(z)$
$\mathcal{Z}[kf(k)] = -z\frac{dF(z)}{dz}$
$\mathcal{Z}[c^k f(k)] = F(z/c)$ c a constant
$\mathcal{Z}[f(k-1)] = f(-1) + z^{-1}F(z)$
$\mathcal{Z}[f(k-2)] = f(-2) + z^{-1}f(-1) + z^{-2}F(z)$
$\mathcal{Z}[f(k-n)] = f(-n) + z^{-1}f(1-n) + z^{-2}f(2-n)$
$\quad + \cdots + z^{1-n}f(-1) + z^{-n}F(z)$
$\mathcal{Z}[f(k+1)] = zF(z) - zf(0)$
$\mathcal{Z}[f(k+2)] = z^2F(z) - z^2f(0) - zf(1)$
$\mathcal{Z}[f(k+n)] = z^nF(z) - z^nf(0) - z^{n-1}f(1)$
$\quad - \cdots - z^2f(n-2) - zf(n-1)$
$f(0) = \lim_{z\to\infty} F(z)$
If $\lim_{k\to\infty} f(k)$ exists and is finite,
$\quad \lim_{k\to\infty} f(k) = \lim_{z\to 1}\left[\frac{z-1}{z}F(z)\right]$
$\mathcal{Z}\left[\sum_{i=0}^{k} f_1(k-i)f_2(i)\right] = F_1(z)F_2(z)$

ital computation is to construct a difference equation from the rational function and then solve the difference equation recursively.

Difference Equations

Analogous to the differential equations that describe continuous-time systems, the input-output behavior of a discrete-time system can be described by difference equations. Linear discrete-time systems are described by linear difference equations. If the coefficients of a difference equation are constant, the system is step invariant and the difference equation has the form

$$\begin{aligned} &y(k+n) + a_{n-1}y(k+n-1) + a_{n-2}y(k+n-2) \\ &\quad + \cdots + a_1y(k+1) + a_0y(k) \\ &= b_mr(k+m) + b_{m-1}r(k+m-1) \\ &\quad + \cdots + b_1r(k+1) + b_0r(k) \end{aligned} \tag{11.1}$$

where r is the input and y is the output. The order of the difference equation is n, which is the number of past output steps that are involved in calculating the present output:

$$\begin{aligned} &y(k+n) = \\ &\underbrace{-a_{n-1}y(k+n-1) - a_{n-2}y(k+n-2) - \cdots}_{n \text{ terms}} \\ &\underbrace{\cdots - a_1y(k+1) - a_0y(k)} \\ &\underbrace{+b_mr(k+m) + b_{m-1}r(k+m-1) + \cdots}_{m+1 \text{ terms}} \\ &\underbrace{\cdots + b_1r(k+1) + b_0r(k)} \end{aligned}$$

Returning to Equation 11.1, a discrete-time system is said to be *causal* if $m \leq n$ so that only past and present inputs, not future ones, are involved in the calculation of the present output. An alternative equivalent form of the difference equation is obtained by replacing k by $k-n$ in Equation 11.1.

Difference equations can be solved recursively using the equation and solutions at past steps to calculate the solution at the next step. For example, consider the difference equation

$$y(k) - y(k-1) = 2u(k) \tag{11.2}$$

with the initial condition $y(-1) = 0$ and $u(k) = 1$ for all k. Letting $k = 0$ and substituting into the difference equation gives

$$\begin{aligned} y(0) - y(-1) &= 2u(0) \\ y(0) &= 2 \end{aligned}$$

Letting $k = 1$ and substituting

$$\begin{aligned} y(1) - y(0) &= 2u(1) \\ y(1) &= 2 + 2 = 4 \end{aligned}$$

At step 2,

$$\begin{aligned} y(2) - y(1) &= 2 \\ y(2) &= 6 \end{aligned}$$

and so on.

A difference equation can be constructed using a computer by programming its recursive solution. A digital hardware realization of the difference equation can also be constructed by coding the signals as binary words, storing present and past values of the input and output in registers, and using binary arithmetic devices to multiply the signals by the equation coefficients and add them to form the output.

11.1.3 Z-Transfer Function Methods

Solutions of linear step-invariant difference equations can be found using z-transformation. For example, consider the single-input, single-output system described by Equation 11.2. Using z-transformation,

$$\begin{aligned} Y(z) - z^{-1}Y(z) &= \frac{2z}{z-1} \\ \frac{Y(z)}{z} &= \frac{2z}{(z-1)^2} = \frac{k_1}{(z-1)} + \frac{k_2}{(z-1)^2} \\ Y(z) &= \frac{2z}{(z-1)} + \frac{2z}{(z-1)^2} \end{aligned}$$

and

$$y(k) = 2u(k) + 2ku(k)$$

Checking,

$$\begin{aligned} y(0) &= 2 \\ y(1) &= 2 + 2 = 4 \\ y(2) &= 2 + 4 = 6 \end{aligned}$$

which agrees with the recursive solution in the previous example.

In general, an nth-order linear discrete-time system is modeled by a difference equation of the form

$$\begin{aligned} &y(k+n) + a_{n-1}y(k+n-1) + \cdots + a_1 y(k+1) + a_0 y(k) \\ &\quad = b_m r(k+m) + \cdots + b_1 r(k+1) + b_0 r(k) \end{aligned}$$

or

$$\begin{aligned} &y(k) + a_{n-1}y(k-1) + \cdots + a_1 y(k-n+1) + a_0 y(k-n) \\ &\quad = b_m r(k+m-n) + \cdots + b_1 r(k-n+1) + b_0 r(k-n) \end{aligned}$$

which has a z-transform given by

$$\begin{aligned} &Y(z) + a_{n-1}[z^{-1}Y(z) + y(-1)] + \cdots \\ &\quad + a_1[z^{-n+1}Y(z) + z^{-n+2}y(-1) + \cdots + y(-n+1)] \\ &\quad + a_0[z^{-n}Y(z) + z^{-n+1}y(-1) + \cdots + y(-n)] \\ &= b_m[z^{-n+m}R(z) + z^{-n+m-1}r(-1) + \cdots \\ &\quad + r(-n+m-1)] + \cdots \\ &\quad + b_0[z^{-n}R(z) + z^{-n+1}r(-1) + \cdots + r(-n)] \end{aligned}$$

$$Y(z) = \underbrace{\frac{b_m z^m + b_{m-1}z^{m-1} + \cdots + b_1 z + b_0}{z^n + a_{n-1}z^{n-1} + \cdots + a_1 z + a_0} R(z)}_{\text{Zero-state component}}$$

$$+ \underbrace{\frac{\text{(Polynomial in } z \text{ of degree } n \text{ or less with coefficients dependent on initial conditions)}}{z^n + a_{n-1}z^{n-1} + \cdots + a_1 z + a_0}}_{\text{Zero-input component}}$$

If all the initial conditions are zero, the zero-input component of the response is zero. The zero-state component of the response is the product of the system z-transfer function

$$T(z) = \frac{b_m z^m + b_{m-1}z^{m-1} + \cdots + b_1 z + b_0}{z^n + a_{n-1}z^{n-1} + \cdots + a_1 z + a_0}$$

and the z-transform of the system input:

$$Y_{\text{zero state}}(z) = T(z)R(z)$$

Analogous to continuous-time systems, the *transfer function* of a linear step-invariant discrete-time system is the ratio of the z-transform of the output to the z-transform of the input when all initial conditions are zero.

It is also common practice to separate the system response into *natural* (or *transient*) and *forced* (or *steady-state*) components. The natural response is a solution of the *homogeneous* difference equation. This is the solution of the difference equation due to initial conditions only with all independent inputs set to zero. The remainder of the response, which includes a term in a form dependent on the specific input, is the forced response component.

Stability and Response Terms

As shown in Figure 11.4, when the input to a linear step-invariant discrete-time system is the unit pulse $\delta(k)$ and all initial conditions are zero, the response of the system is given by

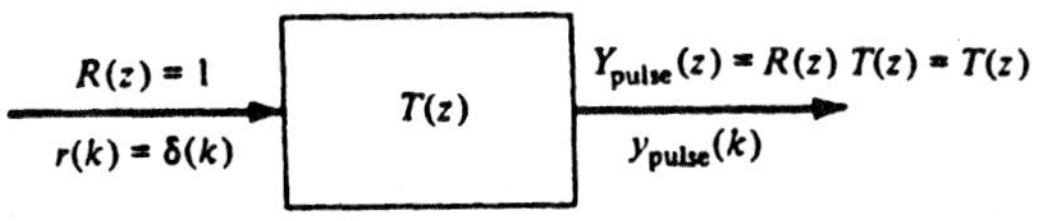

$R(z) = 1$, $r(k) = \delta(k)$ → $T(z)$ → $Y_{\text{pulse}}(z) = R(z)\,T(z) = T(z)$, $y_{\text{pulse}}(k)$

Figure 11.4 Unit pulse response of a discrete-time system.

$$Y_{\text{pulse}}(z) = R(z)T(z) = T(z)$$

A linear step-invariant discrete-time system is said to be *input-output stable* if its pulse response decays to zero asymptotically.

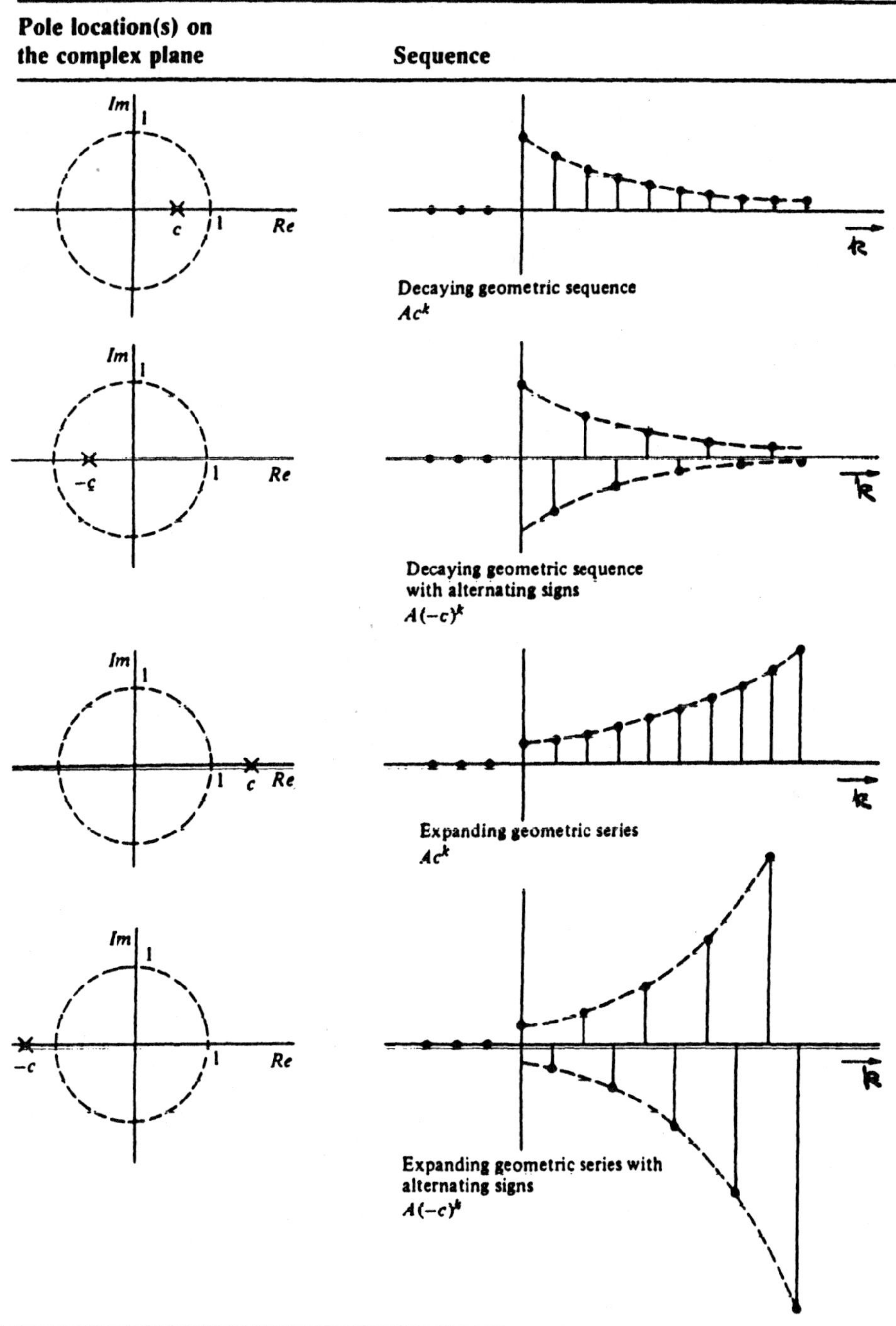

Figure 11.5 Sequences corresponding to various z-transform pole locations.

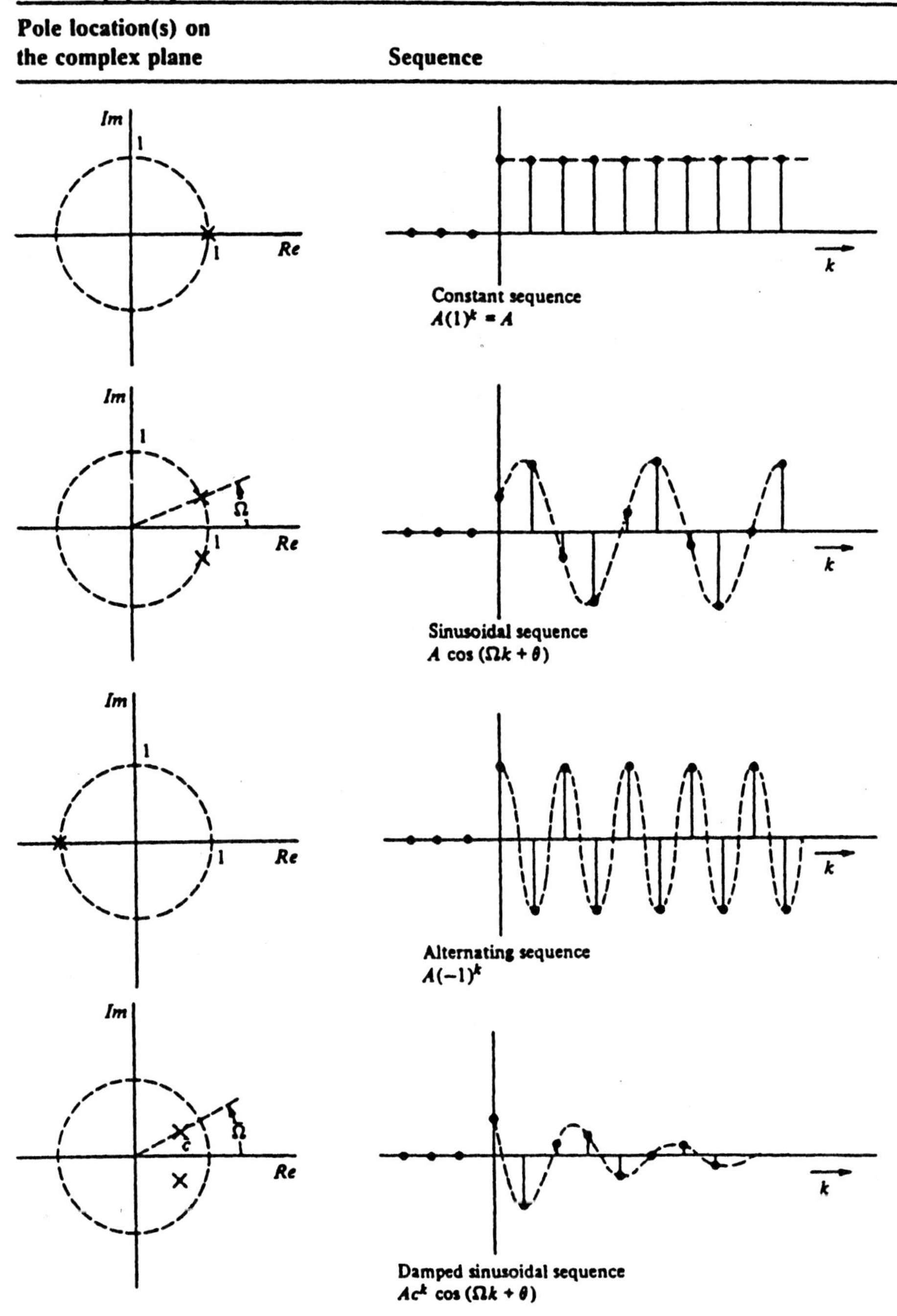

Figure 11.5 *(Continued)* Sequences corresponding to various z-transform pole locations.

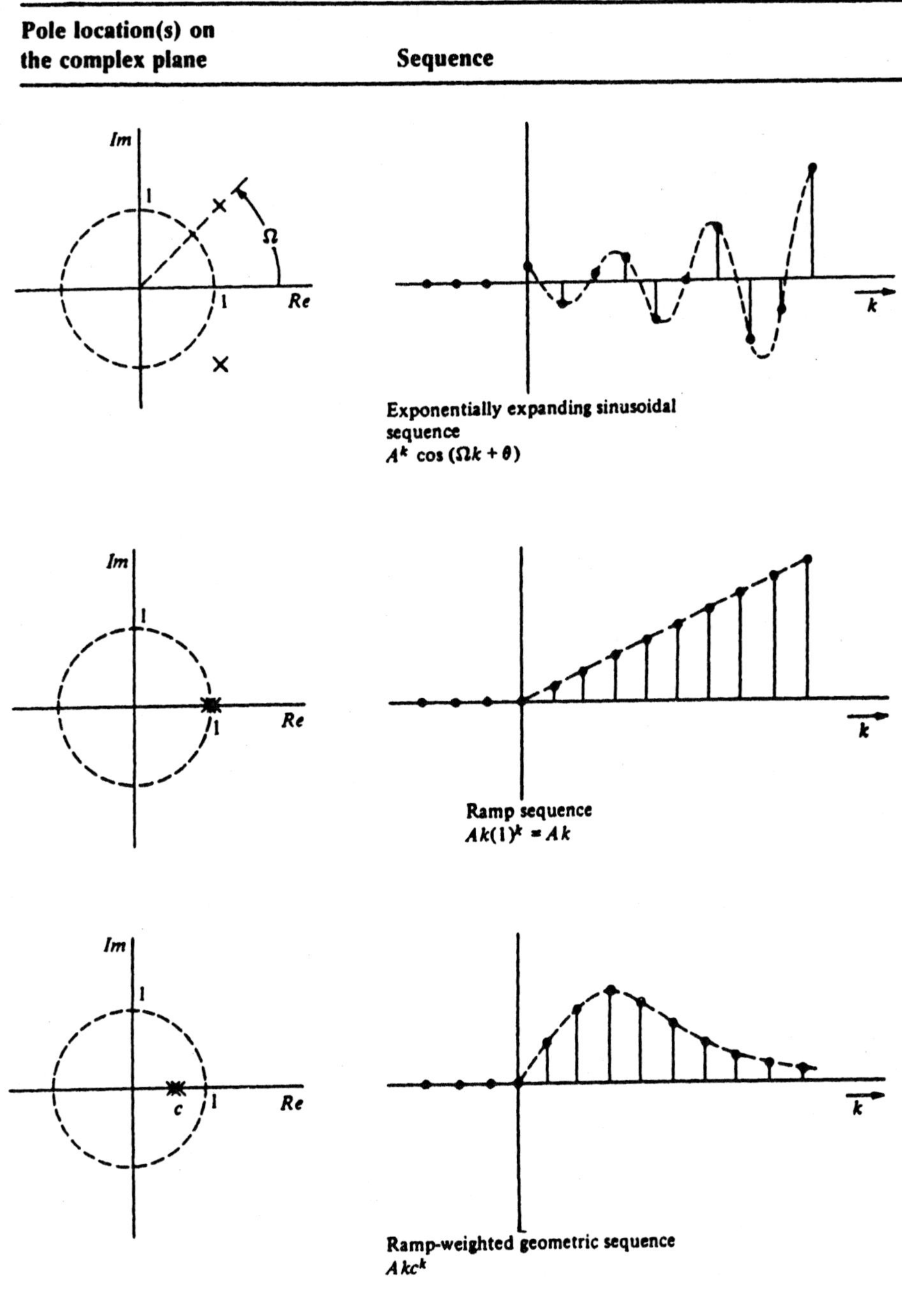

Figure 11.5 *(Continued)* Sequences corresponding to various z-transform pole locations.

This occurs if and only if *all* the roots of the denominator polynomial of the transfer function are inside the unit circle in the complex plane.

Figure 11.5 shows pulse responses corresponding to various *pole* (denominator root) locations. Sequences corresponding to pole locations inside the unit circle decay to zero asymptotically so they are stable. Systems with poles that are outside the unit circle or repeated on the unit circle have outputs that expand with step and are thus *unstable*. Systems with nonrepeated poles on the unit circle have responses that neither decay nor expand with step and are termed *marginally stable*. Methods for testing the location of the roots of the denominator polynomial of a transfer function are presented in Chapter 7 of this handbook.

A pole-zero plot of a z-transfer function consists of Xs denoting poles and Os denoting zeros in the complex plane. The z-transfer function

$$
\begin{aligned}
T(z) &= \frac{3z + 3z^3}{4z^4 + 6z^3 - 4z^2 + z + 2} \\
&= \left(\frac{3}{4}\right) \frac{z(z+j)(z-j)}{(z+2)(z+\frac{1}{2})(z-\frac{1}{2}+j\frac{1}{2})(z-\frac{1}{2}-j\frac{1}{2})}
\end{aligned}
$$

has the pole-zero plot shown in Figure 11.6. It represents an unstable discrete-time system because it has a pole at $z = -2$, which is outside the unit circle.

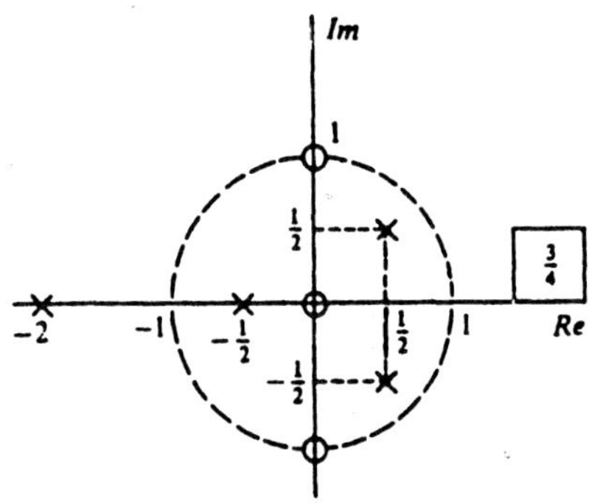

Figure 11.6 An example.

Block Diagram Algebra

The rules of block diagram algebra for linear time-invariant continuous-time systems apply to linear step-invariant discrete-time systems as well. Combining blocks in cascade or in tandem or moving a pick-off point in front of or behind a block, etc. with discrete-time systems is done the same way as with continuous-time systems. However, as we see in Chapter 12, these rules do not necessarily apply for sampled data systems containing discrete-time as well as continuous-time components.

Similar to a continuous-time system, when a discrete-time system has several inputs and/or outputs, there is a z-transfer function relating each one of the inputs to each one of the outputs,

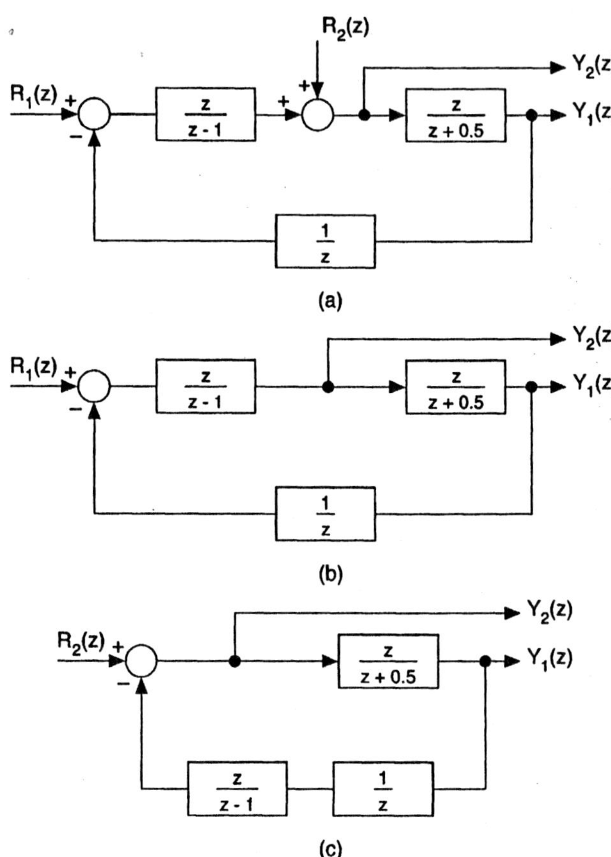

Figure 11.7 Multiple-input, multiple-output block diagram reduction. (a) two-input, two-output system; (b) block diagram reduction to determine $T_{11}(z)$ and $T_{12}(z)$; (c) block diagram reduction to determine $T_{21}(z)$ and $T_{22}(z)$.

with all other inputs set to zero:

$$
T_{ij}(z) = \left. \frac{Y_i(z)}{R_j(z)} \right| \begin{array}{l} \text{When all initial conditions} \\ \text{are zero and when all inputs} \\ \text{except } R_j \text{ are zero} \end{array}
$$

In general, when all the initial conditions of a system are zero, the outputs of the system are given by

$$
\begin{aligned}
Y_1(z) &= T_{11}(z)R_1(z) + T_{12}(z)R_2(z) + T_{13}(z)R_3(z) + \cdots \\
Y_2(z) &= T_{21}(z)R_1(z) + T_{22}(z)R_2(z) + T_{23}(z)R_3(z) + \cdots \\
Y_3(z) &= T_{31}(z)R_1(z) + T_{32}(z)R_2(z) + T_{33}(z)R_3(z) + \cdots \\
&\vdots
\end{aligned}
$$

For example, the four transfer functions of the two-input, two-output system shown in Figure 11.7 are as follows:

$$
\begin{aligned}
T_{11}(z) &= \frac{Y_1(z)}{R_1(z)} = \frac{z^2}{z^2 + 0.5z - 0.5} \\
T_{21}(z) &= \frac{Y_2(z)}{R_1(z)} = \frac{z^2(z+0.5)}{z^2 + 0.5z - 0.5}
\end{aligned}
$$

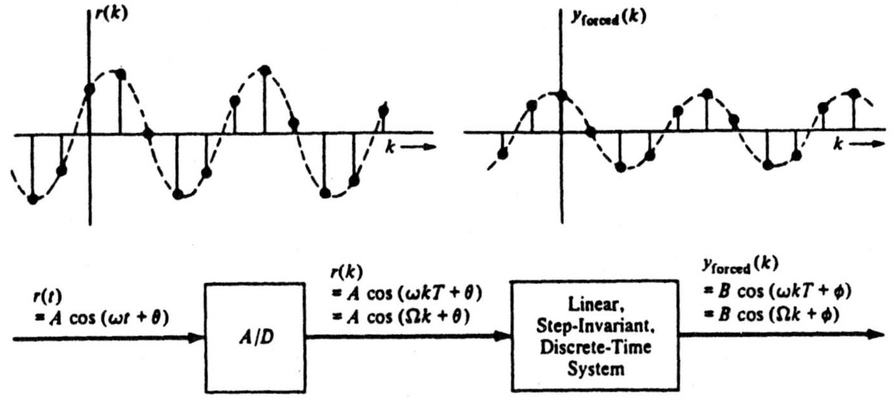

Figure 11.8 Discrete-time system with a sinusoidal input sequence and the sinusoidal forced output sequence.

$$T_{12}(z) = \frac{Y_1(z)}{R_2(z)} = \frac{z(z-1)}{z^2+0.5z-0.5}$$
$$T_{22}(z) = \frac{Y_2(z)}{R_2(z)} = \frac{(z-1)(z+0.5)}{z^2+0.5z-0.5}$$

A linear step-invariant discrete-time multiple-input, multiple-output system is input-output stable if, and only if, *all* the poles of *all* its z-transfer functions are inside the unit circle on the complex plane.

Discrete-Frequency Response

As shown in Figure 11.8, when the input to a linear step-invariant discrete-time system is a sinusoidal sequence of the form

$$r(k) = A\cos(\Omega k + \theta)$$

the forced output $y(k)$ of the system includes another sinusoidal sequence with the same frequency, but generally with different amplitude B and different phase ϕ.

If a discrete-time system is described by the difference equation

$$y(k+n) + a_{n-1}y(k+n-1) + \cdots + a_1 y(k+1) + a_0 y(k)$$
$$= b_m r(k+m) + b_{m-1} r(k+m-1) + \cdots + b_0 r(k)$$

its transfer function is given by

$$T(z) = \frac{b_m z^m + b_{m-1}z^{m-1} + \cdots + b_1 z + b_0}{z^n + a_{n-1}z^{n-1} + \cdots + a_1 z + a_0}$$

The *magnitude* of the z-transfer function, evaluated at $z = \exp(j\Omega)$, is the ratio of the amplitude of the forced output to the amplitude of the input:

$$\frac{B}{A} = |T(z = e^{j\Omega})|$$

The angle of the z-transfer function, evaluated at $z = \exp(j\Omega)$, is the phase difference between the input and output:

$$\phi - \theta = \underline{/T(z = e^{j\Omega})}$$

These results are similar to the counterpart for continuous-time systems, in which the transfer function is evaluated at $s = j\omega$.

Frequency response plots for a linear step-invariant discrete-time system are plots of the magnitude and angle of the z-transfer function, evaluated at $z = \exp(j\Omega)$, vs. Ω. The plots are periodic because $\exp(j\Omega)$ is periodic in Ω with period 2π. This is illustrated in Figure 11.9, which gives the frequency response for the z-transfer function in the accompanying pole-zero plot. In general, the frequency response plots for discrete-time systems are symmetric about $\Omega = \pi$, as shown. The amplitude ratio is even symmetric while the phase shift is odd symmetric about $\Omega = \pi$. Therefore, the frequency range of Ω from 0 to π is adequate to completely specify the frequency response of a discrete-time system. Logarithmic frequency response plots for the system given in Figure 11.9 are shown in Figure 11.10.

11.1.4 Discrete-Time State Equations and System Response

We now make the transition from classical system description and analysis methods to state variable methods. System response is expressed in terms of discrete convolution and in terms of z-transforms. z-Transfer function matrices of multiple-input, multiple-output systems are found in terms of the state equations. The state equations and response of step-varying systems are also discussed.

State Variable Models of Linear Step-Invariant Discrete-Time Systems

An nth-order linear discrete-time system can be modeled by a state equation of the form

$$x(k+1) = Ax(k) + Bu(k) \tag{11.3a}$$

where x is the n-vector state of the system, u is an r-vector of input signals, the state coupling matrix A is $n \times n$, and the input coupling matrix B is $n \times r$. The m-vector of system measurement outputs y is related to the state and inputs by a measurement equation of the form

$$y(k) = Cx(k) + Du(k) \tag{11.3b}$$

where the output coupling matrix C is $m \times n$, and the input-to-output coupling matrix D is $m \times r$. A block diagram showing how the various quantities of the state and output equations are related is shown in Figure 11.11. In the diagram, wide arrows

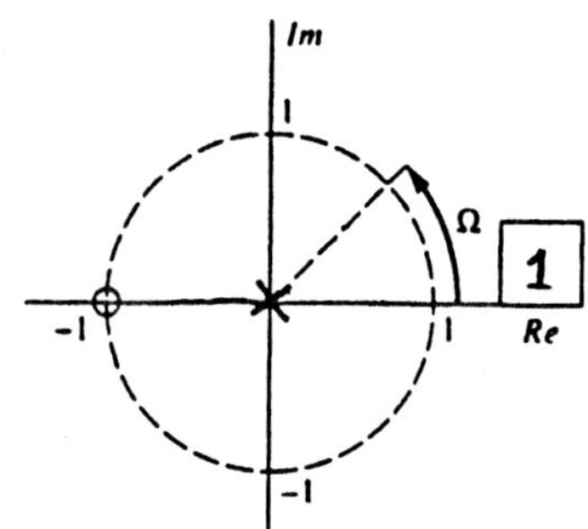

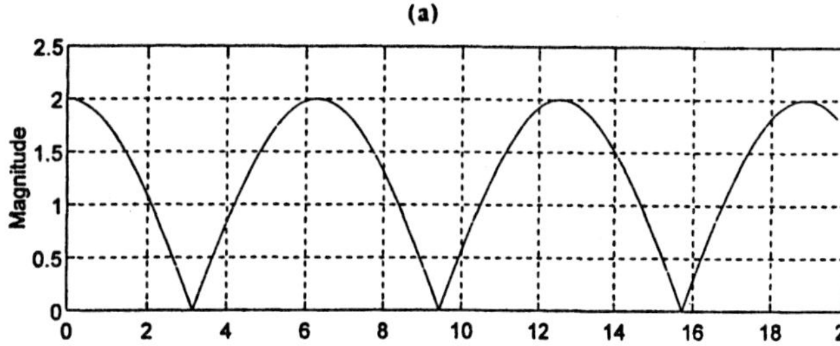

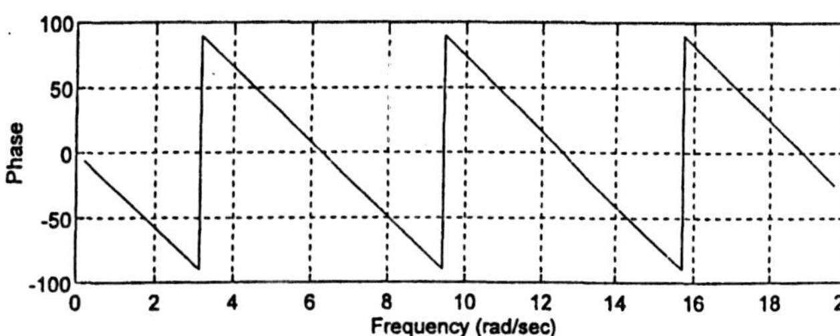

Figure 11.9 Periodicity of the frequency response of a discrete-time system. (a) Pole-zero plot of a transfer function $T(z)$; (b) frequency response plots for $T(z)$.

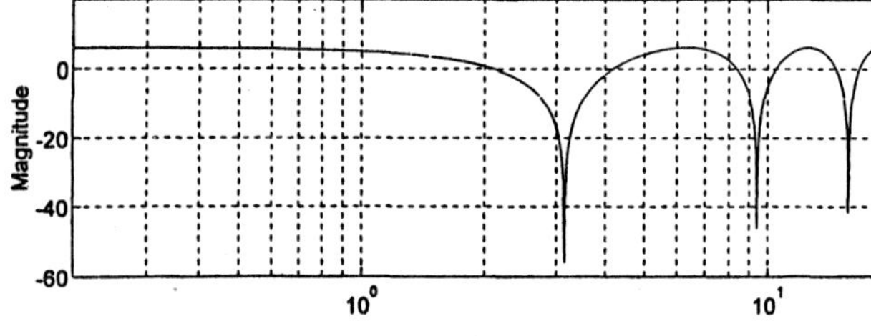

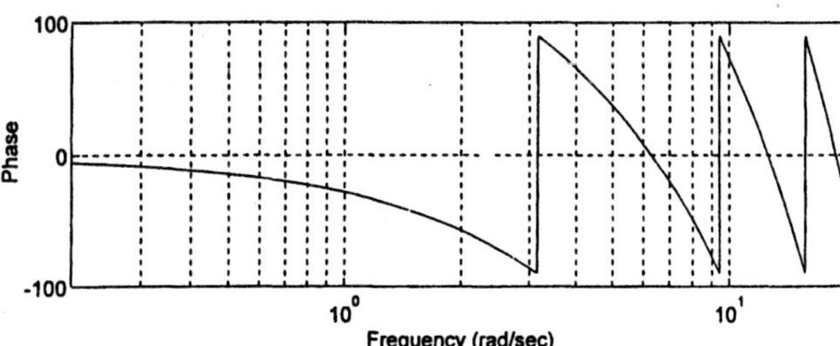

Figure 11.10 Logarithmic frequency response plots for the system shown in Figure 11.9.

represent vectors of signals. A system given by Equations 11.3a, b is termed *step invariant* if the matrices A, B, C, D do not change with step.

Response in Terms of Discrete Convolution

In terms of the initial state $x(0)$ and the inputs $u(k)$ at step zero and beyond, the solution for the state after step zero can be calculated recursively. From $x(0)$ and $u(0)$, $x(1)$ can be calculated

$$x(1) = Ax(0) + Bu(0)$$

Then, using $x(1)$ and $u(1)$

$$x(2) = Ax(1) + Bu(1) = A^2x(0) + ABu(0) + Bu(1)$$

From $x(2)$ and $u(2)$

$$x(3) = Ax(2)+Bu(2) = A^3x(0)+A^2Bu(0)+ABu(1)+Bu(2)$$

and in general

$$x(k) = \underbrace{A^k x(0)}_{\text{Zero-input component}} + \underbrace{\textstyle\sum_{i=0}^{k-1} A^{k-1-i} Bu(i)}_{\text{Zero-state component}} \tag{11.4}$$

and the solution for the output $y(k)$ in Equation 11.3 is

$$y(k) = CA^k x(0) + \left\{ \sum_{i=0}^{k-1} CA^{k-i-1} Bu(i) \right\} + Du(k)$$

The system output when all initial conditions are zero is termed the *zero-state* response of the system. When the system initial conditions are not all zero, the additional components in the outputs are termed the *zero-input* response components.

Response in Terms of Z-Transform

The response of a discrete-time system described by the state Equations 11.3a, b can be found by z-transforming the state equations

$$x(k+1) = Ax(k) + Bu(k)$$

That is,

$$zX(z) - zx(0) = AX(z) + BU(z)$$

or

$$(zI - A)X(z) = zx(0) + BU(z)$$

Hence,

$$X(z) = z(zI - A)^{-1}x(0) + (zI - A)^{-1}BU(z)$$

and

$$Y(z) = Cz(zI - A)^{-1}x(0) + \{C(zI - A)^{-1}B + D\}U(z) \tag{11.5}$$

The solution for the state is then

$$x(k) = \mathcal{Z}^{-1}[z(zI - A)^{-1}]x(0) + \mathcal{Z}^{-1}[(zI - A)^{-1}BU(z)]$$

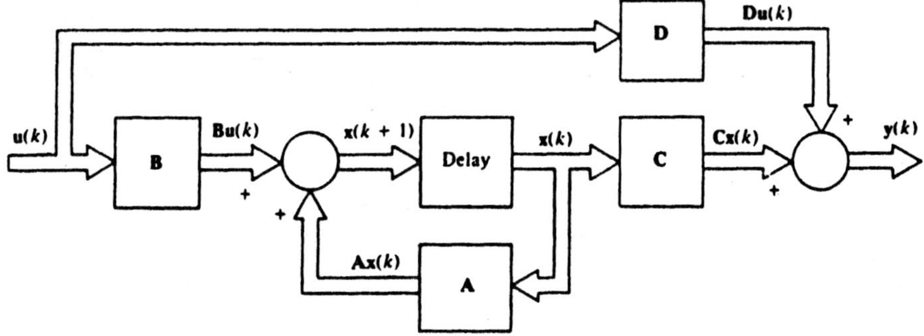

Figure 11.11 Block diagram showing the relations between signal vectors in a discrete-time state-variable model.

Comparing this result with Equation 11.4 shows that

$$A^k = \mathcal{Z}^{-1}[z(zI - A)^{-1}]$$

which is analogous to the continuous-state transition matrix. Setting the initial conditions in Equation 11.5 to zero gives the $m \times r$ transfer function matrix

$$T(z) = C(zI - A)^{-1}B + D$$

where m is the number of outputs in $y(k)$ and r is the number of inputs in $u(k)$. The elements of $T(z)$ are functions of the variable z, and the element in the ith row and jth column of $T(z)$ is the transfer function relating the ith output to the jth input:

$$T_{ij}(z) = \left.\frac{Y_i(z)}{U_j(z)}\right| \begin{array}{l}\text{Zero initial conditions} \\ \text{and all other inputs zero}\end{array}$$

For an $n \times n$ matrix A,

$$(zI - A)^{-1} = \frac{\text{adj}(zI - A)}{|zI - A|},$$

where $|zI - A|$ is the determinant of $zI - A$, and hence each element of $T(z)$ is a ratio of polynomials in z that shares the denominator polynomial

$$q(z) = |zI - A|$$

which is the characteristic polynomial of the matrix A. Each transfer function of $T(z)$ then has the same poles, although there may be pole-zero cancellation. Stability requires that all the system poles (or *eigenvalues*) be within the unit circle on the complex plane.

As an example, consider the second-order three-input, two-output system described by

$$\begin{aligned}
\begin{bmatrix} x_1(k+1) \\ x_2(k+1) \end{bmatrix} &= \begin{bmatrix} 2 & -5 \\ \frac{1}{2} & -1 \end{bmatrix}\begin{bmatrix} x_1(k) \\ x_2(k) \end{bmatrix} \\
&\quad + \begin{bmatrix} 1 & -2 & 0 \\ 0 & 1 & 3 \end{bmatrix}\begin{bmatrix} u_1(k) \\ u_2(k) \\ u_3(k) \end{bmatrix} \\
&= Ax(k) + Bu(k) \\
\begin{bmatrix} y_1(k) \\ y_2(k) \end{bmatrix} &= \begin{bmatrix} 2 & 0 \\ 1 & -1 \end{bmatrix}\begin{bmatrix} x_1(k) \\ x_2(k) \end{bmatrix} \\
&\quad + \begin{bmatrix} 0 & 4 & 0 \\ 0 & 0 & -2 \end{bmatrix}\begin{bmatrix} u_1(k) \\ u_2(k) \\ u_3(k) \end{bmatrix} \\
&= Cx(k) + Du(k)
\end{aligned}$$

This system has the characteristic equation

$$\begin{aligned}
|zI - A| &= \begin{vmatrix} (z-2) & 5 \\ -\frac{1}{2} & (z+1) \end{vmatrix} = z^2 - z + \frac{1}{2} \\
&= (z - \frac{1}{2} - j\frac{1}{2})(z - \frac{1}{2} + j\frac{1}{2}) = 0
\end{aligned}$$

Its six transfer functions all share the poles

$$z_1 = \frac{1}{2} + j\frac{1}{2} \qquad z_2 = \frac{1}{2} - j\frac{1}{2}$$

The transfer function matrix for this system, which is stable, is given by

$$\begin{aligned}
T(z) &= C(zI - A)^{-1}B + D \\
&= \begin{bmatrix} 2 & 0 \\ 1 & -1 \end{bmatrix}\begin{bmatrix} (z-2) & 5 \\ -\frac{1}{2} & (z+1) \end{bmatrix}^{-1} \\
&\quad \begin{bmatrix} 1 & -2 & 0 \\ 0 & 1 & 3 \end{bmatrix} + \begin{bmatrix} 0 & 4 & 0 \\ 0 & 0 & -2 \end{bmatrix} \\
&= \begin{bmatrix} \frac{2z+2}{z^2-z+\frac{1}{2}} & 4 + \frac{-4z-14}{z^2-z+\frac{1}{2}} & \frac{-30}{z^2-z+\frac{1}{2}} \\ \frac{z+\frac{1}{2}}{z^2-z+\frac{1}{2}} & \frac{-3z-4}{z^2-z+\frac{1}{2}} & -2 + \frac{-3z-9}{z^2-z+\frac{1}{2}} \end{bmatrix}
\end{aligned}$$

Linear step-invariant discrete-time causal systems must have transfer functions with numerator polynomials of an order less than or equal to that of the denominator polynomials. Only causal systems can be represented by the standard state-variable models.

State Equations and Response of Step-Varying Systems

A linear step-varying discrete-time system has a state equation of the form

$$\begin{aligned}
x(k+1) &= A(k)x(k) + B(k)u(k) \\
y(k) &= C(k)x(k) + D(k)u(k)
\end{aligned}$$

In terms of the initial state $x(0)$ and the inputs $u(k)$ at step zero and beyond, the solution for the state after step zero is

$$x(k) = \Phi(k, 0)x(0) + \sum_{i=0}^{k-1} \Phi(k, i+1)B(i)u(i)$$

where the state transition matrices, $\Phi(.,.)$, are the $n \times n$ products of state coupling matrices

$$\begin{aligned} \Phi(k, j) &= A(k-1)A(k-2)\ldots A(j-1)A(j) \quad k > j \\ \Phi(i, i) &= I \end{aligned}$$

where I is the $n \times n$ identity matrix.

A linear step-varying discrete-time system of the form

$$x(k+1) = A(k)x(k) + B(k)u(k)$$

is said to be *zero-input stable* if, and only if, for every set of initial conditions $x_{\text{zero-input}}(0)$, the zero-input component of the state, governed by

$$x_{\text{zero-input}}(k+1) = A(k)x_{\text{zero-input}}(k)$$

approaches zero with step. That is,

$$\lim_{k \to \infty} \|x_{\text{zero-input}}(k)\| = 0$$

where the symbol $\|.\|$ denotes the Euclidean norm of the quantity.

The system is *zero-state stable* if, and only if, for zero initial conditions and every bounded input

$$\|u(k)\| < \delta \qquad k = 0, 1, 2, \ldots$$

the zero-state component of the state, governed by

$$\begin{cases} x_{\text{zero-state}}(k+1) = A(k)x_{\text{zero-state}}(k) + B(k)u(k) \\ x_{\text{zero-state}}(0) = 0 \end{cases}$$

is bounded:

$$\|x_{\text{zero-state}}(k)\| < \sigma \qquad k = 0, 1, 2, \ldots$$

Change of Variables

A nonsingular change of state variables

$$x(k) = P\bar{x}(k) \quad \bar{x}(k) = P^{-1}x(k)$$

in discrete-time state-variable equations

$$\begin{aligned} x(k+1) &= Ax(k) + Bu(k) \\ y(k) &= Cx(k) + Du(k) \end{aligned}$$

gives new equations of the same form

$$\begin{aligned} \bar{x}(k+1) &= (P^{-1}AP)\bar{x}(k) + (P^{-1}B)u(k) = \bar{A}\bar{x}(k) + \bar{B}u(k) \\ y(k) &= (CP)\bar{x}(k) + Du(k) = \bar{C}\bar{x}(k) + Du(k) \end{aligned}$$

The system transfer function matrix is unchanged by a nonsingular change of state variables

$$\begin{aligned} \bar{T}(z) &= \bar{C}(zI - \bar{A})^{-1}\bar{B} + D \\ &= CP(zP^{-1}P - P^{-1}AP)^{-1}P^{-1}B + D \\ &= CP[P^{-1}(zI - A)P]^{-1}P^{-1}B + D \\ &= CPP^{-1}(zI - A)^{-1}PP^{-1}B + D \\ &= C(zI - A)^{-1}B + D = T(z) \end{aligned}$$

Each different set of state-variable equations having the same z-transfer function matrix is termed a *realization* of the z-transfer functions. The transformation

$$\bar{A} = P^{-1}AP$$

is called a *similarity transformation*. The transformation matrix P can be selected to take the system to a convenient realization such as the *controllable form, observable form, diagonal form, block Jordan form*, etc. Using these forms, it is especially easy to synthesize systems having desired transfer functions. For example, if the eigenvalues of the A matrix are distinct, there exists a nonsingular matrix P such that the state coupling matrix A of the new system

$$\begin{aligned} \bar{x}(k+1) &= \bar{A}\bar{x}(k) + \bar{B}u(k) \\ y(k) &= \bar{C}\bar{x}(k) + Du(k) \end{aligned}$$

is diagonal with the eigenvalues of A as the diagonal elements. The new state equations are decoupled from one another, and each equation involves only one state variable. In this example, the columns of the P matrix are the eigenvectors of the A matrix. It should be noted that taking a system from one realization to another may not always be possible. This depends on the characteristics of the system.

Controllability and Observability

A discrete-time system is said to be *completely controllable* if, by knowing the system model and its state $x(k)$ at any specific step k, a control input sequence $u(k), u(k+1), \ldots, u(k+i-1)$ can be determined that it will take the system to any desired later state x in a finite number of steps. For a step-invariant system, if it is possible to move the state at any step, say $x(0)$, to an arbitrary state at a later step, then it is possible to move it to an arbitrary desired state starting with any beginning step.

For an nth-order step-invariant system with r inputs

$$\begin{aligned} x(k+1) &= Ax(k) + Bu(k) \\ y(k) &= Cx(k) + Du(k) \end{aligned}$$

and a desired state δ, the system state at step n, in terms of the initial state $x(0)$ and the inputs, is

$$\delta = x(n) = A^n x(0) + \sum_{i=0}^{n-1} A^{n-1-i}Bu(i)$$

or

$$\begin{aligned} &Bu(n-1) + ABu(n-2) + \cdots + A^{n-2}Bu(1) \\ &+ A^{n-1}Bu(0) = \delta - A^n x(0) \end{aligned}$$

where the terms on the right-hand side are known. These equations have a solution for the inputs $u(0), u(1), \ldots, u(n-1)$ if, and only if the $n \times (rn)$ array of coefficients

$$M_c = [B \,|\, AB \,|\, \cdots \,|\, A^{n-2}B \,|\, A^{n-1}B]$$

called the *controllability matrix* of the system, is of full rank. Additional steps, giving additional equations with coefficients A^nB, $A^{n+1}B$, and so on, do not affect this result because, by the Cayley-Hamilton theorem, these equations are linearly dependent on the others.

For a multiple-input system, the smallest possible integer η for which the matrix

$$M_c(\eta) = [B|AB|A^2B|\cdots|A^{\eta-1}B]$$

has full rank is called the *controllability index* of the system. It is the minimum number of steps needed to control the system state.

A discrete-time system is said to be *completely observable* if its state $x(k)$ at any specific step k can be determined from the system model and its inputs and measurement outputs for a finite number of steps. For a step-invariant system, if it is possible to determine the state at any step, $x(0)$, then with a shift of step, the state at any other step can be determined in the same way.

For an nth-order step-invariant system with m outputs,

$$\begin{aligned} x(k+1) &= Ax(k) + Bu(k) \\ y(k) &= Cx(k) + Du(k) \end{aligned}$$

the initial state $x(0)$, in terms of the outputs and inputs, is given by

$$\begin{cases} y(0) = Cx(0) + Du(0) \\ y(1) = Cx(1) + Du(1) = CAx(0) \\ \quad +CBu(0) + Du(1) \\ y(2) = Cx(2) + Du(2) \\ \quad = CA^2x(0) + CABu(0) + CBu(1) + Du(2) \\ \vdots \\ y(n-1) = Cx(n-1) + Du(n-1) \\ \quad = CA^{n-1}x(0) + CA^{n-2}Bu(0) + CA^{n-3}Bu(1) + \cdots \\ \qquad + CBu(n-2) + Du(n-1) \end{cases}$$

Collecting the $x(0)$ terms on the left

$$\begin{cases} Cx(0) = y(0) - Du(0) \\ CAx(0) = y(1) - CBu(0) - Du(1) \\ CA^2x(0) = y(2) - CABu(0) - CBu(1) - Du(2) \\ \vdots \\ CA^{n-1}x(0) = y(n-1) - CA^{n-2}Bu(0) - \cdots \\ \quad -CBu(n-2) - Du(n-1) \end{cases}$$

where the terms on the right are known and $x(0)$ is unknown. This set of linear algebraic equations can be solved for $x(0)$ only if the array of coefficients

$$M_0 = \begin{bmatrix} C \\ \hline CA \\ \hline CA^2 \\ \hline \vdots \\ \hline CA^{n-1} \end{bmatrix}$$

is of full rank. Additional outputs are of no help because they give additional equations with coefficients CA^n, CA^{n+1}, ..., which are linearly dependent on the others.

For a multiple-output system, the smallest possible integer ν for which

$$M_0 = \begin{bmatrix} C \\ \hline CA \\ \hline CA^2 \\ \hline \vdots \\ \hline CA^{\nu-1} \end{bmatrix}$$

has full rank is called the *observability index* of the system. It is the minimum number of steps needed to determine the system state.

The replacements

$$\begin{cases} A \to A^\dagger \\ B \to C^\dagger \\ C \to B^\dagger \end{cases}$$

where $\dagger$ denotes matrix transposition, creates a system with a controllability matrix that is the observability matrix of the original system and an observability matrix that is the controllability matrix of the original system. Every controllability result has a corresponding observability result and vice versa, a concept termed *duality*.

References

[1] Santina, M.S., Stubberud, A.R., and Hostetter, G.H., *Digital Control System Design*, 2nd ed., Saunders College Publishing, Philadelphia, PA, 1994.

[2] DiStefano, J.J.,III, Stubberud, A.R., and Williams, I.J., *Feedback and Control Systems (Schaum's Outline)*, 2nd ed., McGraw-Hill, New York, NY, 1990.

[3] Jury, E.I., *Theory and Application of the z-Transform Method*, John Wiley & Sons, New York, NY, 1964.

[4] Oppenheim, A.V. and Willsky, A.W., *Signals and Systems*, Prentice Hall, Englewood Cliffs, NJ, 1983.

[5] Papoulis, A., *Circuits and Systems: A Modern Approach*, Saunders College Publishing, Philadelphia, PA, 1980.

[6] Cadzow, J.A. and Martens, H.R., *Discrete-Time and Computer Control Systems*, Prentice Hall, Englewood Cliffs, NJ, 1970.

[7] McGillem, C.D. and Cooper, G.R., *Continuous and Discrete Signal and System Analysis*, 3rd ed., Saunders Publishing, Philadelphia, PA, 1991.

[8] Chen, C.T., *System and Signal Analysis*, 2nd ed., Saunders College Publishing, Philadelphia, PA, 1980.

[9] Kailath, T., *Linear Systems*, Prentice Hall, Englewood Cliffs, NJ, 1980.

[10] DeRusso, P.M., Roy, R.J., and Close, C.M., *State Variables for Engineers*, Wiley, New York, NY, 1965.

12

Sampled Data Systems

A. Feuer
Electrical Engineering Department, Technion-Israel Institute of Technology, Haifa, Israel

G.C. Goodwin
Department of Electrical and Computer Engineering, University of Newcastle, Newcastle, Australia

12.1 Introduction and Mathematical Preliminaries

The advances in digital computer technology have led to its application in a very wide variety of areas. In particular, it has been used to replace the analog controller in many control systems. However, to use the digital computer as a controller one has to overcome the following problem: The input and output signals of the physical plant are analog, namely, continuous-time signals, and the digital computer can only accept and generate sequences of numbers, namely, discrete-time signals (we do not discuss here the quantization problem).

This problem was solved by developing two types of interfacing units: a sampler, which transforms the analog signal to a discrete one, and a hold unit, which transforms the discrete signal to an analog one. A typical configuration of the resulting system, Figure 12.1, shows that the control system contains continuous-time signals and discrete-time signals (drawn in full and broken lines, respectively).

Because mathematical tools were available for analyzing systems with either continuous-time signals or discrete-time signals, the approach to controller design evolved accordingly. One approach is to design a continuous-time controller and then approximate it by a discrete-time system. Another approach is to look at the system, from the input to the hold unit to the output of the sampler, as a discrete-time system. Then, use discrete-time control design methods (the development of which was prompted by that approach) to design the controller.

Both approaches are acceptable when the sampling is fast enough. However, in many applications the sampling rate may be constrained and as a result the approaches above may prove inappropriate. This realization prompted many researchers to develop tools to analyze systems containing both continuous- and discrete-time signals, referred to as sampled data systems.

The purpose of this chapter is to introduce one such tool which is based on frequency domain considerations and seems to be a very natural approach. Because we will heavily rely on Fourier transforms, it will be helpful to review some of the definitions and properties:

Fourier Transform (FT):

$$X(\omega) = \mathcal{F}\{x(t)\} = \int_{-\infty}^{\infty} x(t)e^{-j\omega t}\,dt, \qquad (12.1)$$

$x(t)$, a continuous-time signal,
ω, angular frequency (in rad/sec).

Inverse Fourier transform (IFT):

$$x(t) = \frac{1}{2\pi}\int_{-\infty}^{\infty} X(\omega)e^{-j\omega t}\,d\omega. \qquad (12.2)$$

Discrete-Time Fourier transform (DTFT):

$$X^s(\omega) = \Delta \sum_{k=-\infty}^{\infty} x[k]e^{-j\omega\Delta k}, \qquad (12.3)$$

$x[k]$, a discrete-time signal obtained by sampling $x(t)$ at the instants $t = k\Delta \quad k = 0, 1, 2, \ldots,$
Δ, the associated sampling time interval.

Inverse Discrete Time Fourier transform (IDTFT):

$$x[k] = \frac{1}{2\pi}\int_{-\pi/\Delta}^{\pi/\Delta} X^s(\omega)e^{j-\omega\Delta k}\,d\omega. \qquad (12.4)$$

0-8493-8570-9/96/$0.00+$.50

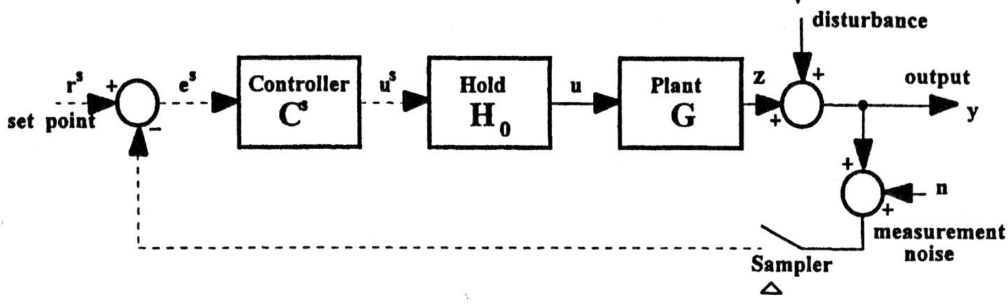

Figure 12.1 Configuration of a typical sampled data control system.

Given that $x[k]$ are samples of $x(t)$, namely,

$$x[k] = x(k\Delta), \tag{12.5}$$

$$X^s(\omega) = \sum_{k=-\infty}^{\infty} X\left(\omega - k\frac{2\pi}{\Delta}\right) \tag{12.6}$$

(see the chapter "Sampled-Rate Selection" in this Handbook). $X^d(\omega)$ results from 'folding' $X(\omega)$ every $\frac{2\pi}{\Delta}$ and repeating it periodically (this is sometimes referred to as 'aliasing').

Following we adopt the notation

$$[X]^s \triangleq \sum_{k=-\infty}^{\infty} X\left(\omega - k\frac{2\pi}{\Delta}\right) \tag{12.7}$$

and readily observe the following properties:

$$[X_1 + X_2]^s = [X_1]^s + [X_2]^s \tag{12.8}$$

$$[C^s X]^s = C^s [X]^s \tag{12.9}$$

where C^s is the frequency response of a discrete-time system.

We should also point out that the most commonly used hold unit is the Zero Order Hold (ZOH) given by

$$H_o(\omega) = \frac{1 - e^{-j\omega\Delta}}{j\omega\Delta} \quad . \tag{12.10}$$

12.2 'Sensitivity Functions' in Sampled Data Control Systems

It is well-known that sensitivity functions play a key role in control design, be it a continuous-time controller for a continuous-time system or a discrete-time controller for a discrete-time system. Let us start our discussion with a brief review of commonly known facts about sensitivity functions. Consider the system in Figure 12.2.

Denoting by capital letters the Fourier transforms of their lower case counterparts in the time domain, from Figure 12.2:

$$Y = TR + SV - TN, \tag{12.11}$$

where

$$S = \frac{1}{1+GC} \tag{12.12}$$

is the sensitivity function, and

$$T = \frac{GC}{1+GC} \tag{12.13}$$

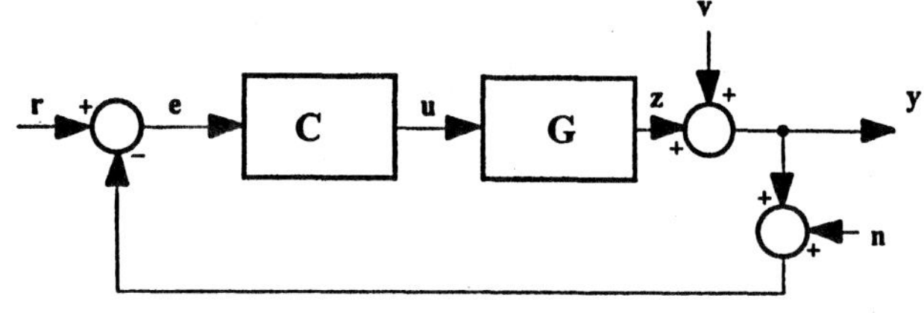

Figure 12.2 Control system with uniform type of signals (either all continuous-time or all discrete-time).

is the complementary sensitivity function as well as the closed-loop transfer function. Here are some facts regarding these functions:

1. Let ΔG denote the change in the open-loop transfer function G, and let ΔT denote the corresponding change in the closed-loop transfer function. Then
$$\frac{\Delta T}{T} \simeq S\frac{\Delta G}{G}. \tag{12.14}$$
2. Clearly
$$S + T = 1. \tag{12.15}$$
3. The zeros of T are the open-loop zeros. Hence, if z_o is an open-loop zero, $S(z_o) = 1$.
4. The zeros of S are the open-loop poles. Hence if p_o is an open-loop pole, $T(p_o) = 1$.
5. $|T|$ is usually made to approach 1 at low frequencies to give zero steady-state errors at d.c.
6. $|T|$ is usually made small at high frequencies to give insensitivity to high frequency noise n (this implies $|S| \approx 1$ at high frequencies).
7. Because of (3) and (5), to avoid peaks in $|T|$ it is desirable that, moving from low to high frequencies, we meet a closed-loop pole before we meet each open-loop zero.
8. Because of (4) and (6), to avoid large peaks in $|S|$ it is desirable that, moving from high to low frequencies, we meet a closed-loop pole before we meet each open-loop pole.
9. For stable, well-damped open-loop poles and zeros, (7) and (8) can easily be achieved by cancellation. However, open-loop unstable poles and zeros place fundamental limitation on the desirable closed-loop

bandwidth. In particular, the following bandwidth limitations are necessary to avoid peaks in either S or T:

$$\left.\begin{array}{l}\text{bandwidth} < \text{open-loop unstable zeros}\\ \text{bandwidth} > \text{open-loop unstable poles}\end{array}\right\} \quad (12.16)$$

With the above in mind, one may adopt the approach mentioned in Section 1. View the system of Figure 12.1 as a discrete-time system by looking at the discrete-time equivalent of the ZOH, the plant, and the sampler. Then, using the above, one gets a discrete-time system for which a discrete-time controller can be designed. The problem is that this approach guarantees desired behavior only *at the sampled outputs.* However, there is no a priori reason to presume that the response between samples would not deviate significantly from what is observed at the sample points. Indeed we shall see later that it is quite possible for the intersample response to be markedly different from the sampled response. It is then clear that the sensitivity functions calculated for the discrete equivalents are unsatisfactory tools for the sampled data system. Following we will develop equivalent functions for the sampled data system.

Let us again consider the system in Figure 12.1. We have the following key result describing the *continuous output* response under the digital control law:

THEOREM 12.1 *Subject to closed-loop stability, the Fourier transform of the continuous-time output of the single-input, single-output system in Figure 12.1 is given by:*

$$\begin{aligned} Y(\omega) &= P(\omega)(R^s(\omega) - N^s(\omega)) + D(\omega)V(\omega) \\ &\quad - P(\omega) \sum_{\substack{k=-\infty \\ k\neq 0}}^{\infty} V\left(\omega - k\frac{2\pi}{\Delta}\right), \\ &= P(\omega)(R^s(\omega) - N^s(\omega)) \\ &\quad + V(\omega) - P(\omega)[V]^s. \end{aligned} \qquad (12.17)$$

$P(\omega)$ and $D(\omega)$ are frequency response functions given, respectively, by

$$P(\omega) \triangleq C^s(\omega)G(\omega)H_o(\omega)S^s(\omega) \qquad (12.18)$$

and

$$D(\omega) = 1 - P(\omega) \qquad (12.19)$$

where $S^s(\omega)$ is the usual discrete sensitivity calculated for the discrete equivalent system given by

$$S^s(\omega) \triangleq \frac{1}{1+C^s(\omega)[GH_o]^s}. \qquad (12.20)$$

Note that $[GH_o]^s$ is the frequency response of the discrete equivalent of GH_o.

PROOF 12.1 Observing Figure 12.1 we have (using Equation 12.6)

$$\begin{aligned} Y^s(\omega) &= \sum_{k=-\infty} Y\left(\omega - k\frac{2\pi}{\Delta}\right), \\ &= \sum_{k=-\infty} \left(Z\left(\omega - k\frac{2\pi}{\Delta}\right) + V\left(\omega - k\frac{2\pi}{\Delta}\right)\right), \\ &= \sum_{k=-\infty} G\left(\omega - k\frac{2\pi}{\Delta}\right) H_o\left(\omega - k\frac{2\pi}{\Delta}\right) C^s \\ &\quad \left(\omega - k\frac{2\pi}{\Delta}\right)\left(R^s\left(\omega - k\frac{2\pi}{\Delta}\right)\right. \\ &\quad \left. - Y^s\left(\omega - k\frac{2\pi}{\Delta}\right) - N^s\left(\omega - k\frac{2\pi}{\Delta}\right)\right) \\ &\quad + V\left(\omega - k\frac{2\pi}{\Delta}\right). \end{aligned}$$

Because, C^s, R^s, Y^s and N^s are periodic functions of ω,

$$\begin{aligned} Y^s(\omega) &= C^s(\omega) \sum_{k=-\infty} G\left(\omega - k\frac{2\pi}{\Delta}\right) H_o\left(\omega - k\frac{2\pi}{\Delta}\right) \\ &\quad \left[R^s(\omega) - Y^s(\omega) - N^s(\omega)\right] + V^s(\omega) \\ &= C^s(\omega)[GH_o]^s\left[R^s(\omega) - Y^s(\omega) - N^s(\omega)\right] \\ &\quad + V^s(\omega). \end{aligned}$$

Hence

$$\begin{aligned} Y^s(\omega) &= [1 - S^s(\omega)][R^s(\omega) - N^s(\omega)] \\ &\quad + S^s(\omega)V^s(\omega). \end{aligned} \qquad (12.21)$$

Now, from Figure 12.1,

$$Y(\omega) = G(\omega)H_o(\omega)C^s(\omega)(R^s(\omega) - Y^s(\omega) - N^s(\omega)) + V(\omega)$$

and substituting Equation 12.21 results in

$$\begin{aligned} Y(\omega) &= G(\omega)H_o(\omega)C^s(\omega)S^s(\omega)[R^s(\omega) - N^s(\omega)] \\ &\quad - G(\omega)H_o(\omega)C^s(\omega)S^s(\omega)V^s(\omega) + V(\omega) \end{aligned}$$

which, by substituting Equation 12.18 and 12.19 leads to Equation 12.17.

Comparing Equation 12.17 and 12.11, the roles that D and P play in a sampled data system are very similar to the roles that sensitivity and complementary sensitivity play in Figure 12.2.

We also note that the functions $P(\omega)$ and $D(\omega)$ allow computing the continuous output in the frequency domain using the input $R^s(\omega)$, the disturbance $V(\omega)$, and the noise $N^s(\omega)$. We will thus refer to $P(\omega)$ and $D(\omega)$ as the *reference* and *disturbance gain functions*, respectively. Observe that the infinite sum defining $[GH_o]^s$ is convergent provided the transfer function GH_o is strictly proper.

The result in Theorem 12.1 holds for general reference, noise, and disturbance inputs. However, it is insightful to consider the special case of sinusoidal signals. In this case, the result simplifies as follows. Let $r[k]$ be a sampled sinewave

$$r[k] = \cos(\omega_o k\Delta) \quad .$$

Then

$$R^s(\omega) = \pi \sum_{k=-\infty}^{\infty} \left[\delta\left(\omega - \omega_o - k\frac{2\pi}{\Delta}\right) + \delta\left(\omega + \omega_o - k\frac{2\pi}{\Delta}\right)\right]. \quad (12.22)$$

Hence, using Equation 12.17 (assuming $V(\omega) = N^s(\omega) = 0$),

$$Y(\omega) = \pi \sum_{k=-\infty}^{\infty} P(\omega)\left[\delta\left(\omega - \omega_o - k\frac{2\pi}{\Delta}\right) + \delta\left(\omega + \omega_o - k\frac{2\pi}{\Delta}\right)\right]. \quad (12.23)$$

Thus, the continuous-time output in this case, is *multifrequency* with corresponding magnitudes and phases determined by the *reference gain function* $P(\omega)$. In particular, for a sinusoidal reference signal as above, where $0 < \omega_o < \frac{\pi}{\Delta}$, the first two components in the output are at frequencies ω_o and $\frac{2\pi}{\Delta} - \omega_o$ and have amplitudes $|P(\omega)|$ and $|P(\frac{2\pi}{\Delta} - \omega_o)|$, respectively. Similar observations can be made for $N^s(\omega)$ and $V(\omega)$.

In the next section we will show that the connections of P and D with T^s and S^s go beyond the apparent similarity in roles.

12.3 Sensitivity Consideration

In the previous section we found that the reference gain function $P(\omega)$ and the disturbance gain function $D(\omega)$ allow computing the *continuous-time* output response in a sampled data system, namely, a *digital* controller in a closed-loop, with a *continuous-time* plant. We recall the definitions for convenience

$$P(\omega) = C^s(\omega)G(\omega)H_o(\omega)S^s(\omega) \quad (12.24)$$

$$D(\omega) = 1 - P(\omega) \quad (12.25)$$

$$S^s(\omega) = \frac{1}{1 + C^s(\omega)[GH_o]^s} \quad (12.26)$$

$$T^s(\omega) = \frac{C^s(\omega)[GH_o]^s}{1 + C^s(\omega)[GH_o]^s} \quad (12.27)$$

where C^s, G and H_o are the frequency responses of the controller, plant, and ZOH, respectively.

First we note that, as in 12.15 for S^s and T^s,

$$P + D = 1. \quad (12.28)$$

Next it is interesting to note that the *open-loop continuous-time* zeros of the plant appear as zeros of $P(\omega)$. Thus, *irrespective* of any *discrete-time consideration*, the locations of the continuous-time plant zeros are of concern because they affect the continuous-time output responses. Specifically, the existence of a nonminimum phase zero in the plant results in performance constraints for any type of controller, continuous time or discrete time.

The magnitude of the ratio between the P and T^s at frequency ω_o is

$$\left|\frac{P(\omega_o)}{T^s(\omega_o)}\right| = \left|\frac{G(\omega_o)H_o(\omega_o)}{[GH_o]^s}\right|. \quad (12.29)$$

Hence, to avoid large peaks in $|P(\omega)|$, $|T^s(\omega)|$ must not be near 1 at any frequency where the gain of the composite continuous-time transfer function GH_o is significantly greater than the gain of its discrete-time equivalent $[GH_o]^s$. Otherwise, as Equation 12.28 indicates, a large $|P(\omega)|$ and a large $|D(\omega)|$ result, showing large sensitivity to disturbances. There are several reasons why the gain of GH_o might be significantly greater than that of $[GH_o]^s$. Two common reasons are

1. For continuous plants having a relative degree exceeding one, there is usually a discrete zero near the point $z = -1$. Thus, the gain of $[GH_o]^s$ typically falls near $\omega = \pi/\Delta$ (i.e., the folding frequency). Hence, it is rarely desirable to have a discrete closed-loop bandwidth approaching the folding frequency as will be demonstrated in later examples.
2. Sometimes high frequency resonances can perturb the discrete transfer function away from the continuous-time transfer function by folding effects leading to differences between GH_o and $[GH_o]^s$.

One needs to be careful about the effect these factors have on the differences between P and T^s. In particular, the bandwidth must be kept well below any frequency where folding effects reduce $[GH_o]^s$ relative to the continuous plant transfer function. This will also be illustrated in the examples presented later.

Finally, we look at the sensitivity of the closed-loop system to changes in the open-loop plant transfer function.

Recall, using Equation 12.15 and Equation 12.26 that

$$T^s(\omega) = \frac{C^s(\omega)[GH_o]^s}{1+C^s(\omega)[GH_o]^s} \quad (12.30)$$

and

$$P(\omega) = C^s(\omega)G(\omega)H_o(\omega)S^s(\omega). \quad (12.31)$$

Clearly,

$$T^s(\omega) = [P]^s. \quad (12.32)$$

Furthermore, we have the following result which extends Equation 12.14 to the case of mixed continuous and discrete signals:

LEMMA 12.1 The relative changes in the *reference gain function* and the closed-loop *discrete-time* transfer function are

$$\frac{\Delta P}{P} = D\frac{\Delta G}{G} - \sum_{\substack{k=-\infty \\ k\neq 0}}^{\infty} P\left(\omega - k\frac{2\pi}{\Delta}\right)\frac{\Delta G(\omega - k\frac{2\pi}{\Delta})}{G(\omega - k\frac{2\pi}{\Delta})} \quad (12.33)$$

and

$$\frac{\Delta T^s}{T^s} = \frac{S^s(\omega)}{[GH_o]^s}\sum_{k=-\infty}^{\infty} G\left(\omega - k\frac{2\pi}{\Delta}\right) \cdot H\left(\omega - k\frac{2\pi}{\Delta}\right)\frac{\Delta G(\omega - k\frac{2\pi}{\Delta})}{G(\omega - k\frac{2\pi}{\Delta})}. \quad (12.34)$$

PROOF 12.2 By differentiating Equation 12.30 and 12.31 with respect to G, we see that, up to first order,

$$\Delta P \simeq \frac{C^s(\omega)H_o(\omega)}{(1+C^s(\omega)[GH_o]^s)^2}\left[(1+C^s(\omega)[H_oG]^s)\Delta G(\omega) - G(\omega)C^s(\omega)[H_o\Delta G]^s\right]$$

$$\Delta T^s \simeq \frac{C^s(\omega)}{(1+C^s(\omega)[GH_o]^s)^2}[H_o\Delta G]^s.$$

Dividing by P and T^s, respectively, and recalling Equation 12.7 leads to Equation 12.33 and 12.34.

Again note the similarity of the roles of P and D in Equation 12.33 to those of T and S in Equation 12.14. Typically the magnitude of $(\frac{\Delta G}{G})$ approaches unity at high frequencies. Hence, Equation 12.33 and 12.34 show that a sensitivity problem will exist even for the discrete-time transfer function, if $[GH_o]^s$ is small at a frequency where GH_o is large, unless, of course, S^s is small at the same frequency. This further reinforces the claim that the bandwidth must be kept well below any frequencies where folding effects reduce $[GH_o]^s$ relative to the continuous plant transfer function GH_o.

12.4 Examples

In this section, we present some simple examples which illustrate the application of Theorem 12.1 in computing the intersample behavior of continuous-time systems under the action of sampled data control. The examples are not intended to illustrate good sampled data control design but have been chosen to show the utility of the functions $P(\omega)$ and $D(\omega)$ in giving correct qualitative and quantitative understanding of the continuous time response in a difficult situation (when it is significantly different from that indicated by the sampled response).

For each example we give G, Δ and the desired T^s. Then we show the resulting functions, P, D and S^s and make comparisons enabling us to predict the effects of various test signals, the results of which will also be shown.

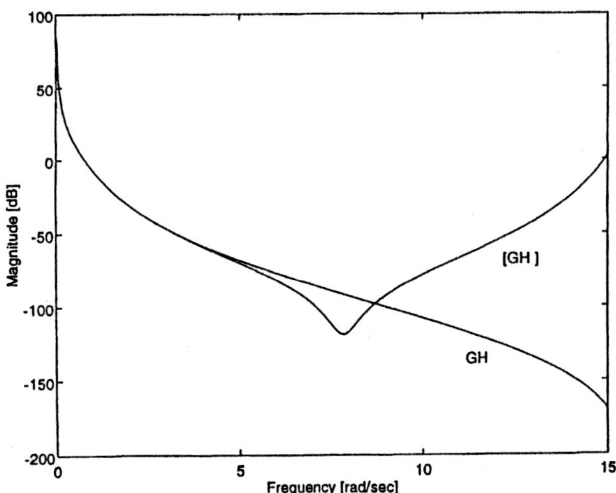

Figure 12.3 Continuous and discrete frequency response.

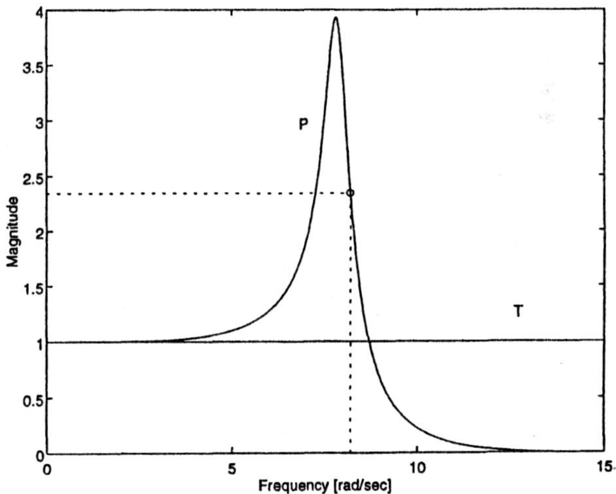

Figure 12.4 Complementary sensitivity and reference gain functions.

12.4.1 Example 1

$$G(\omega) = \frac{1}{j\omega(j\omega+1)} \qquad \Delta = 0.4 \text{ sec.}$$

C^s is chosen so that

$$T^s(\omega) = e^{-0.4j\omega}.$$

Figure 12.3 shows $|G(\omega)H_o(\omega)|$ and $|[GH_o]^s|$. Over the range $(0, \pi/\Delta)$ the two are very nearly equal. Only near $\omega = \pi/\Delta = 7.85$ is there 'some' discrepancy due to the sampling zero of $[GH_o]^s$. Figure 12.4 shows $|T^s|$ and $|P|$. Although T^s seems ideal, the graph for P indicates 'trouble' around $\omega = \pi/\Delta$ rad/sec. The peak in P results from the discrepancy in Figure 12.3 around the same frequency. Similar 'trouble' is, naturally, observed in Figure 12.5 from the graph of D. These peaks indicate that a large continuous-time response is to be expected whenever reference input, noise, or disturbance have frequency content around that frequency. This is demonstrated in Figure 12.6 for a step reference input and in Figure 12.7 for a sinusoidal disturbance of frequency $3/\Delta$ rad/sec and unit amplitude. The exact expressions for both responses can be derived from Equation 12.17. In particular, as marked in Figures 12.4 and 12.5 the disturbance response of the two dominant frequencies will be $3/\Delta$ with amplitude $|D(3/\Delta)| \approx 4$ and $(2\pi-3)/\Delta$ with amplitude $\left|P\left(\frac{2\pi-3}{\Delta}\right)\right| \approx 2.4$. Adding the two, with the appropriate phase shift, will give the signal in Figure 12.7. Note that the sampled responses for both cases (marked in both figures) are very misleading.

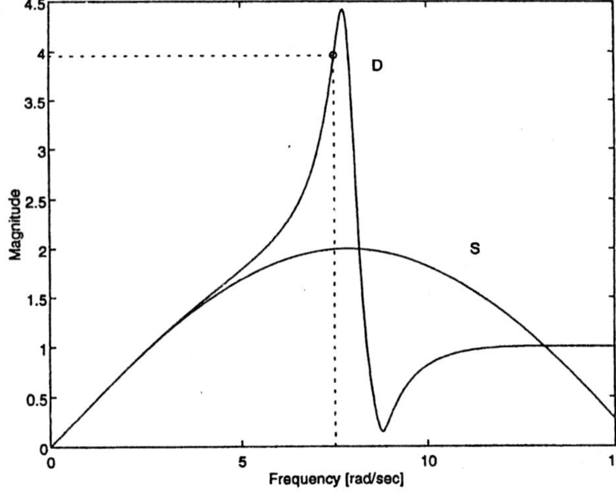

Figure 12.5 Sensitivity and disturbance gain functions.

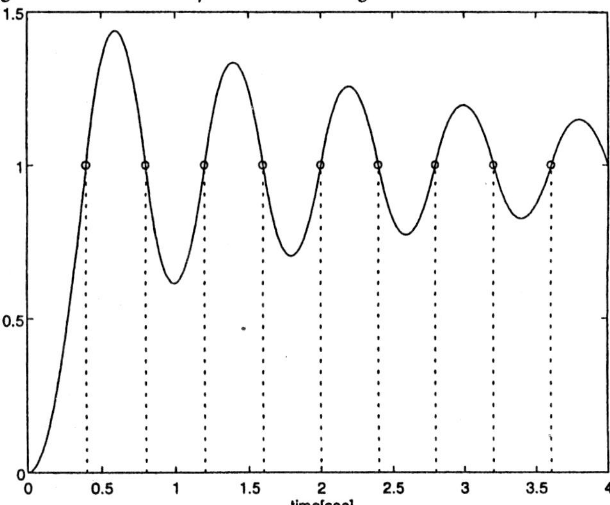

Figure 12.6 Step response.

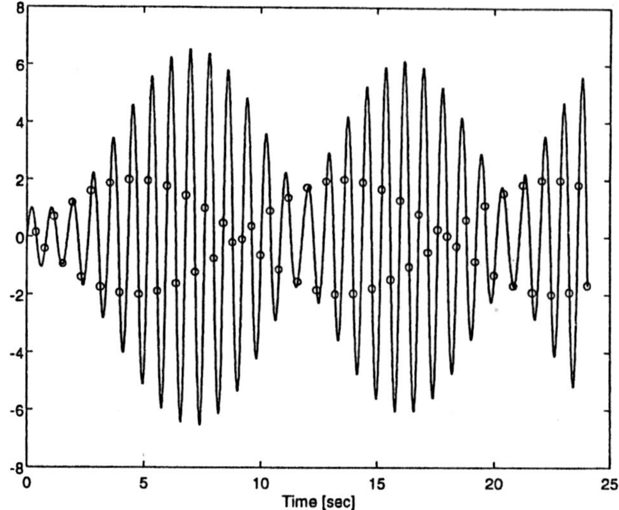

Figure 12.7 Disturbance response.

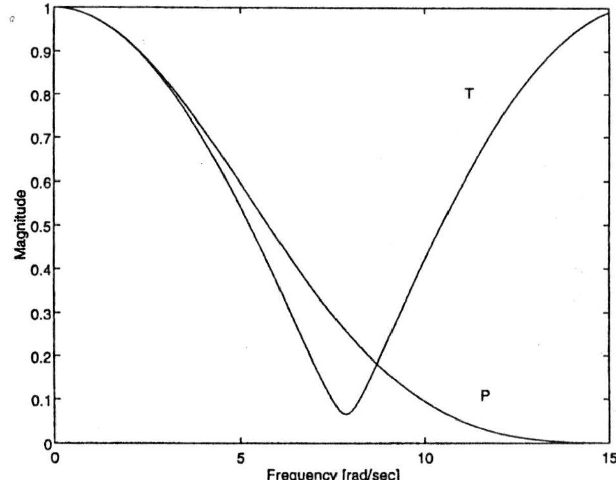

Figure 12.8 Complementary sensitivity and reference gain functions.

12.4.2 Example 2

The plant and Δ are the same as in Example 12.4.1. However, C^s is chosen so that

$$T^s(\omega) = \frac{B(\omega)}{B(0)}\, e^{-0.8j\omega}$$

where $B(\omega)$ is the numerator of $[GH_o]^s$ (this is a deadbeat control). In this case both P and T^s and D and S^s are very close in the range $(0, \pi/\Delta)$ as in Figures 12.8 and 12.9. Predictably, the same test signal as in Example 12.4.1 produces a sampled response more indicative of the continuous-time response. This is clearly noted in Figures 12.10 and 12.11.

Examples 12.4.1 and 12.4.1 represent two controller designs, the design in 12.4.2 is more along the recommendations in the previous section with clearly superior *continuous-time* performance.

12.4.3 Example 3

$$G(\omega) = \frac{100}{(j\omega)^2 + 2j\omega + 100} \qquad \Delta = 0.5 \text{ sec.}$$

C^s is chosen so that

$$T^s(\omega) = \frac{0.5}{e^{0.5j\omega} - 0.5}.$$

Figure 12.12 compares the $|[GH_o]^s|$ and $|GH_o|$. The resonant peak has been folded into the low frequency range in the discrete frequency response. Figure 12.13 shows $|P|$ and $|T^s|$. $P(\omega)$ has a significant peak at $\omega = \pi/\Delta$, reflected in the intersample behavior of the step response in Figure 12.15. However, the sampled response in Figure 12.15 is a simple exponential as could be predicted from T^s in Figure 12.13.

Significant differences in S^s and D can also be observed in Figure 12.14. When a sinusoidal disturbance of frequency $1/\Delta$ rad/sec was applied, we observe a multifrequency continuous time response, but the sampled response is a single sinusoid of frequency $1/\Delta$ shown in Figure 12.16. This result can again be predicted from Equation 12.17 and Figures 12.13 and 12.14.

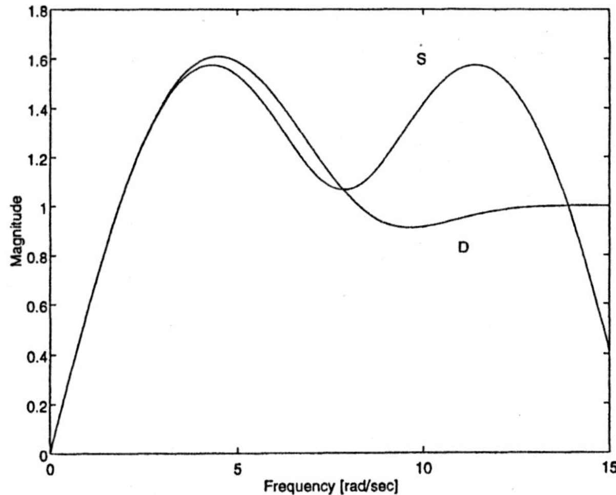

Figure 12.9 Sensitivity and disturbance gain functions.

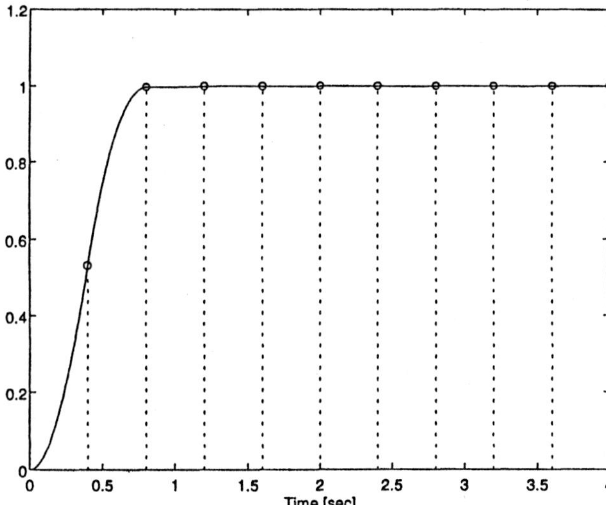

Figure 12.10 Step response.

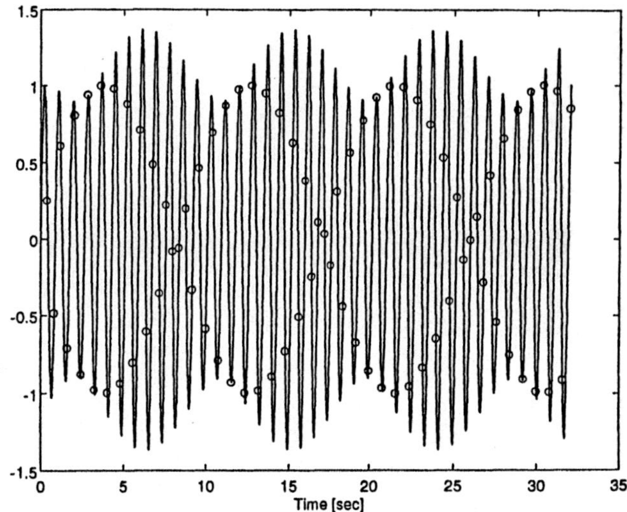

Figure 12.11 Disturbance response.

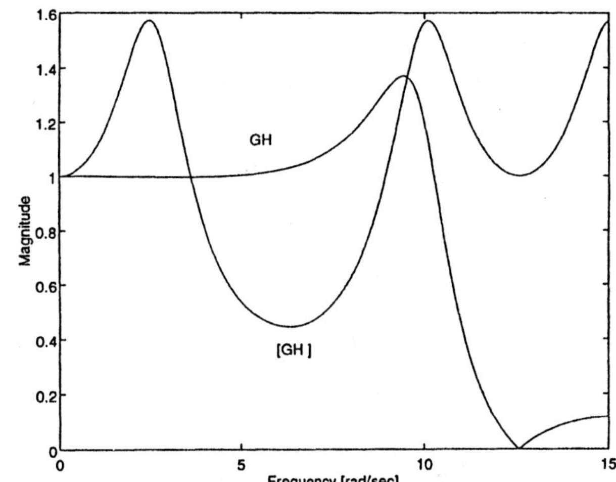

Figure 12.12 Continuous and discrete frequency responses.

12.4.4 Observations and Comments from Examples

1. The reference and disturbance gain functions, $P(\omega)$ and $D(\omega)$, give qualitative and quantitative information about the true continuous-time response resulting from reference, noise, or disturbance input.
2. In many cases, the first two components in the multifrequency output response suffice for an accurate qualitative description of the response to a sinewave disturbance input.
3. For a sinewave disturbance of frequency $\omega_o \in [0, \pi/\Delta)$, the continuous-time response will (if we consider only the first two frequency components) be an amplitude modulated waveform of carrier frequency π/Δ and modulating frequency of $(\pi/\Delta - \omega_o)$.
4. Significant resonant peaks in the system frequency response outside the range $[0, \pi/\Delta)$ can be folded back into the frequency range of interest even with antialiasing filters. These are a potential source of difficulty in sampled data design.
5. All continuous-time systems of relative degree greater than one, lead to discrete-time models with extra zeros (the "sampling zeros") in the vicinity of -1. Sampled data control systems which attempt to cancel these zeros will necessarily have high gain at frequencies near π/Δ (see Example 1). This is a potential source of poor intersample behavior. Control laws with this property include those based on discrete Loop Transfer Recovery (LTR) designs. These are frequently advocated for sampled data systems because of the recovery of the discrete-time sensitivity performance. However, such designs are not recommended due to the inevitable poor intersample behavior when applied to continuous-time plants.

As might be expected, the differences between P, D and T^s, S^s can be made small if one abides by the usual rules of thumb for discrete control, i.e., sample 10 times faster than the open- or closed-loop rise time, use antialiasing filters and keep the closed-loop response time well above the sample period. However, there

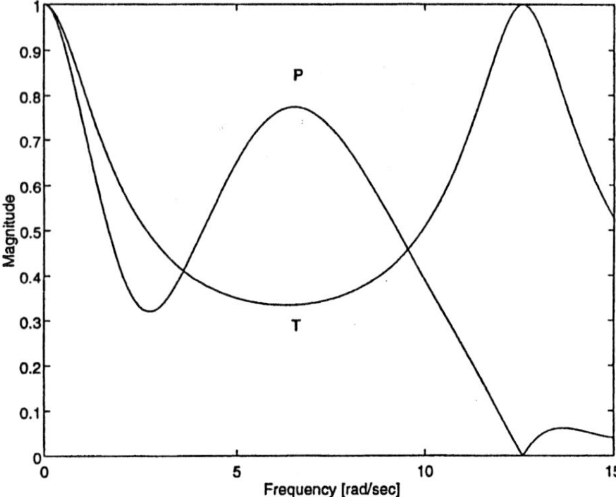

Figure 12.13 Complementary sensitivity and reference gain functions.

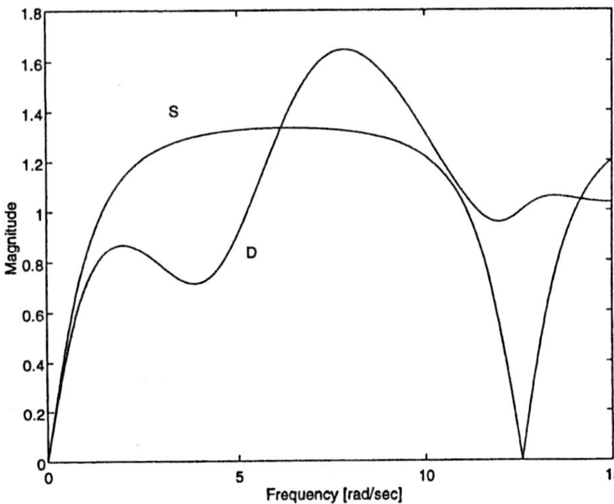

Figure 12.14 Sensitivity and disturbance gain functions.

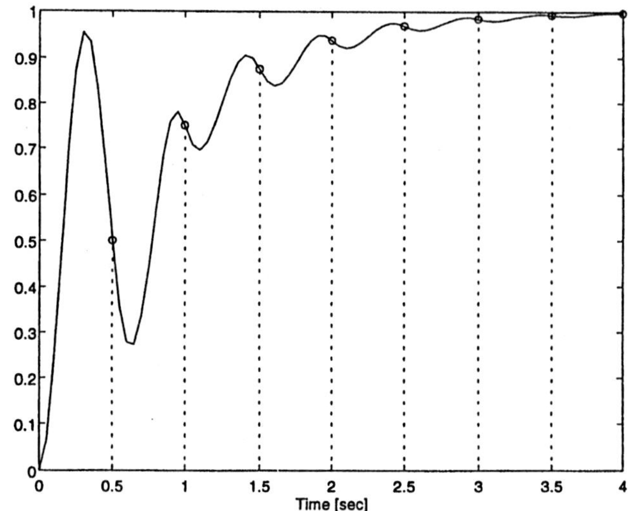

Figure 12.15 Step response.

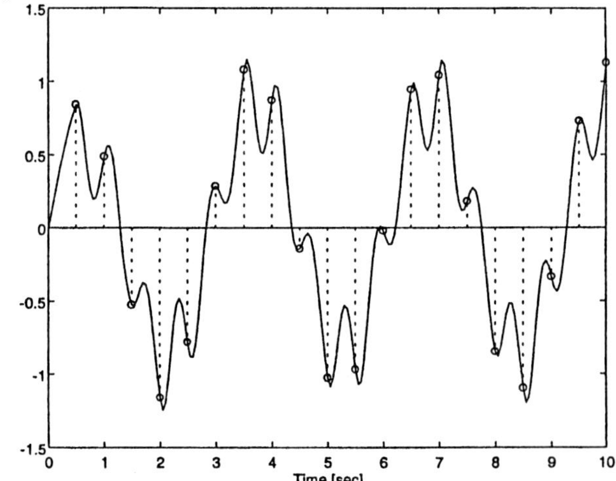

Figure 12.16 Disturbance response.

are situations where these rules cannot be obeyed due to hardware limitations and, in these cases, the gain functions described here can be used to predict the correct continuous-time output behavior. Hence, they are a useful tool for designing sampled data control systems.

12.5 Linear Quadratic Design of Sampled-Data Controllers

The linear quadratic optimal regulator for discrete systems, if applied in the sampled data configuration, will *optimize* the *at sample response*. However, in view of our discussion in earlier sections, we may somtimes get rather poor intersample performance. We therefore proceed in this section to investigate how one might *optimize* the *continuous response* using a discrete-time control law.

Let us suppose that the plant (including anti-aliasing filter if needed) has a minimal *continuous-time* state-space representation of the form

$$\frac{d}{dt}x(t) = Ax(t) + Bu(t); \qquad x(0) = x_o. \tag{12.35}$$

We assume that (A, B) is stabilizable and that the sampling rate avoids pathological loss of stabilizability in the discrete-time system.

Now we assume that we want to choose a discrete-time control law (using samples of x to generate u via a ZOH) so as to minimize a continuous-time quadratic cost function of the form:

$$\begin{aligned} J &= x(t_1)^T P_o x(t_1) \\ &\quad + \int_{t_o}^{t_1} \left(x(t)^T Q_c x(t) + u(t)^T R_c u(t) \right) dt \end{aligned} \tag{12.36}$$

The control must be

$$u(t) = u[k] \quad \text{for} \quad k\Delta \leq t < (k+1)\Delta \tag{12.37}$$

where

$$u[k] = -K[k]x[k] \quad . \tag{12.38}$$

We assume that the reader is familiar with finding $K[k]$ when the system and criterion are both in discrete form (see e.g., [14]). The discrete equivalent of Equation 12.35 is well-known. Hence, rewriting Equation 12.36 in a discrete form, we have a solution for the sampled data control problem.

Note first that Equation 12.35 and 12.37 can be used to write

$$\frac{d}{dt}\begin{bmatrix} x(t) \\ u(t) \end{bmatrix} = \begin{bmatrix} A & B \\ 0 & 0 \end{bmatrix}\begin{bmatrix} x(t) \\ u(t) \end{bmatrix} \quad \text{for} \quad k\Delta \le t < (k+1)\Delta \tag{12.39}$$

So, in this time interval,

$$\begin{bmatrix} x(t) \\ u(t) \end{bmatrix} = e^{\bar{A}(t-k\Delta)}\begin{bmatrix} x[k] \\ u[k] \end{bmatrix} \tag{12.40}$$

where

$$\bar{A} = \begin{bmatrix} A & B \\ 0 & 0 \end{bmatrix} \tag{12.41}$$

and

$$x[k] = x(k\Delta)$$

Rewriting Equation 12.36 and substituting Equation 12.40 (assuming w.l.o.g that $t_o = k_o\Delta$ and $t_1 = k_1\Delta$)

$$\begin{aligned} J &= x[k_1]^T P_o x[k_1] + \sum_{k=k_o}^{k_1-1}\int_{k\Delta}^{(k+1)\Delta}\left[x(t)^T u(t)^T\right] \\ &\quad \begin{bmatrix} Q_c & 0 \\ 0 & R_c \end{bmatrix}\begin{bmatrix} x(t) \\ u(t) \end{bmatrix} dt \\ &= x[k_1]^T P_o x[k_1] + \sum_{k=k_o}^{k_1-1}\int_{k\Delta}^{(k+1)\Delta}\left[x[k]^T u[k]^T\right] \\ &\quad e^{\bar{A}^T(t-k\Delta)}\begin{bmatrix} Q_c & 0 \\ 0 & R_c \end{bmatrix} \\ &\quad e^{\bar{A}(t-k\Delta)}\begin{bmatrix} x[k] \\ u[k] \end{bmatrix} dt \end{aligned}$$

or

$$\begin{aligned} J &= x[k_1]^T P_o x[k_1] + \sum_{k=k_o}^{k_1-1}\left[x[k]^T u[k]^T\right] \\ &\quad \begin{bmatrix} Q_d & S_d^T \\ S_d & R_d \end{bmatrix}\begin{bmatrix} x[k] \\ u[k] \end{bmatrix} \end{aligned} \tag{12.42}$$

where

$$\begin{bmatrix} Q_d & S_d^T \\ S_d & R_d \end{bmatrix} = \int_0^{\Delta} e^{\bar{A}^T\tau} \begin{bmatrix} Q_C & 0 \\ 0 & R_C \end{bmatrix} e^{\bar{A}^T} d\tau. \tag{12.43}$$

Recall that the discrete equivalent of Equation 12.35 is

$$x[k+1] = A_s x[k] + B_s u[k], \qquad x[k_o] = x_o \tag{12.44}$$

where

$$\begin{aligned} A_s &= e^{A\Delta} \\ B_s &= \int_0^{\Delta} e^{A\sigma} B d\sigma, \end{aligned}$$

and we recognize that Equation 12.44 with Equation 12.43 is a standard discrete-time linear quadratic optimal control problem. Thus we can conclude that a continuous-time linear quadratic problem for sampled data systems can be solved via discrete-time linear quadratic design tools with a properly modified cost function. This remarkable property will be illustrated by a simple example.

12.5.1 Example 1

Consider the continuous-time system in state-space form,

$$\begin{aligned} \dot{x}(t) &= Ax(t) + Bu(t) \\ y(t) &= Cx(t) \end{aligned}$$

where

$$\begin{aligned} A &= \begin{bmatrix} 0 & 0 \\ 0 & -1 \end{bmatrix},\ B = \begin{bmatrix} 1 \\ 1 \end{bmatrix}, \\ C &= [1, -1],\ x(0) = \begin{bmatrix} 1 \\ -1 \end{bmatrix}. \end{aligned}$$

Design an optimal time-invariant discrete-time regulator for this system of the form

$$u[k] = -Kx[k]$$

where $x[k] = x(k\Delta)$, $u(t) = u[k]$ for $k\Delta \le t < (k+1)\Delta$ and $\Delta = 0.1$ sec, with the cost function

$$J = \int_0^{\infty} y^2(t)\,dt = \int_0^{\infty} x(t)^T \begin{bmatrix} 1 & -1 \\ -1 & 1 \end{bmatrix} x(t)\,dt \quad .$$

Solution:

Initially suppose we take the discrete equivalent of the system and approximate J by

$$J_1 = \Delta \sum_{k=0}^{\infty} y[k]^2$$

where $y[k] = y(k\Delta)$.

This leads to the feedback gain matrix

$$K_1 = [206.6263,\ -186.9588] \quad .$$

This feedback will move one closed-loop discrete pole to the origin and cancel the sampling zero with the other. This guarantees that the discrete output goes to zero in one step as in Figure 12.17. However, we warned against the sampling zero cancellation in previous sections. When we look at the *continuous-time* response in Figure 12.17, we see a completely

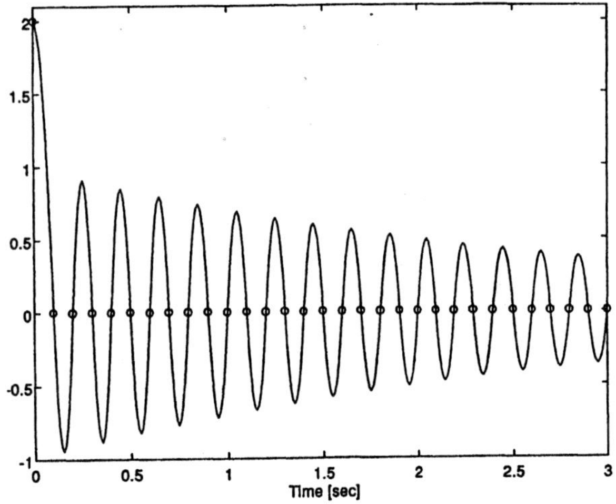

Figure 12.17 Sampled and continuous response for first design (Example 12.5.1).

different story. Furthermore, calculating the corresponding J and J_1,

$$J_1 = 0.4, \text{ and } J = 0.9227.$$

Using the methodology described here for properly designing sampled data optimal regulator,

$$\bar{A} = \begin{bmatrix} 0 & 0 & 1 \\ 0 & -1 & 1 \\ 0 & 0 & 0 \end{bmatrix}$$

$$\longrightarrow e^{\bar{A}t} = \begin{bmatrix} 1 & 0 & t \\ 0 & e^{-t} & 1-e^{-t} \\ 0 & 0 & 1 \end{bmatrix},$$

$$\longrightarrow \begin{bmatrix} Q_d & S_d^T \\ S_d & R_d \end{bmatrix} = \int_0^{\Delta} e^{\bar{A}^T\tau} \begin{bmatrix} 1 & -1 & 0 \\ -1 & 1 & 0 \\ 0 & 0 & 0 \end{bmatrix} e^{\bar{A}\tau}\, d\tau,$$

$$= \begin{bmatrix} 0.1 & -0.0952 & 0.0002 \\ -0.0952 & 0.0906 & -0.0002 \\ 0.0002 & -0.0002 & 0 \end{bmatrix}.$$

Then, for

$$J = \sum_{k=o}^{\infty} \left[x[k]^T u[k] \right] \begin{bmatrix} Q_d & S_d^T \\ S_d & R_d \end{bmatrix} \begin{bmatrix} x[k] \\ u[k] \end{bmatrix},$$

the feedback gain matrix,

$$K_2 = [156.7922 \quad -139.7698].$$

The resulting trajectories are shown in Figure 12.18. Clearly, the sampled response is not as good as in Figure 12.17 but, what really counts, the continuous time response is significantly improved.

This is also reflected in the corresponding cost functions,

$$J_1 = 0.4305, \qquad J = 0.22531.$$

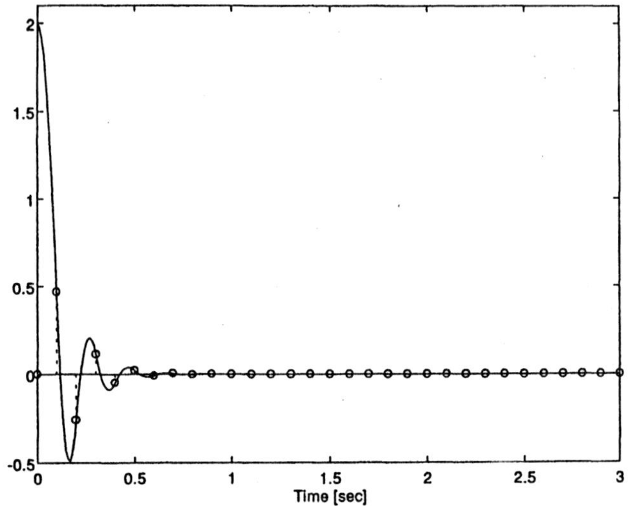

Figure 12.18 Sampled and continuous response for second design (Example 12.5.1).

Further Reading

The analysis of sampled data control systems is still a continuing research topic. In this article we have pointed out some potential problems if the sampled data system is treated in the discrete context only. We have presented some tools (the reference and disturbance gain functions) useful in analyzing sampled data control systems. We have also presented the LQ optimal control modified for the sampled data control system.

The material presented here is based on a recent book [10] and relies on a paper [11]. Closely related work appears in [1], [2], [3]. A comprehensive survey of recent development in sampled data control with many references is contained in [4].

The linear quadratic optimization results are based on [8]. Related work also appears in [7], [12].

Treatments of other optimization criteria in sampled data control systems appear in [6], [12], [18], [9], [17], [13], [5], [19].

References

[1] Araki, M. and Y. Ito. Frequency Response of Sampled Data Systems I: Open Loop Considerations, Technical Report 92-04, Department of Electrical Engineering II, Kyoto University, 1992.

[2] Araki, M. and Y. Ito. Frequency Response of Sampled Data Systems II: Closed Loop Considerations, Technical Report 92-05, Department of Electrical Engineering II, Kyoto University, 1992.

[3] Araki, M. and Y. Ito. On Frequency Response of Sampled-Data Control Systems, SICE Symp. Control Theory, Kariya, 1992.

[4] Araki, M. Recent Developments in Digital Control Theory, IFAC World Congress, Sydney, 1993.

[5] Bamieh, B., M.A. Dahleh and J.B. Pearson. Minimization of the L^{∞}-Induced Norm for Sampled-data Sys-

tems, *IEEE Trans. Automat. Contr. 38,* (5), 717–732, 1993.

[6] Bamieh, B.A and J.B. Pearson, Jr. A General Framework for Linear Periodic Systems with Application to H^∞ Sampled-Data Control, *IEEE Trans. Automat. Contr. 37,* (4), 1418–1435, 1992.

[7] Bamieh, B.A and J.B. Pearson. The H^2 Problem for Sampled-Data Systems, *Sys. Contr. Lett., 19,* (1), 1–12, 1992.

[8] Chen, T. and B.A. Francis. H_2-Optimal Sampled-Data Control, *IEEE Trans. Automat. Contr., 36,* (4), 387–397, 1991.

[9] Dullerud, G.E. and B.A. Francis. L_1 Analysis and Design of Sampled-Data Systems, *IEEE Trans. Automat. Contr., 37,* (4), 436–446, 1992.

[10] Feuer, A. and G.C. Goodwin. *Sampled Data Systems,* CRC, Boca Raton, FL, (to appear).

[11] Goodwin, G.C. and M.E. Salgado. Frequency Domain Sensitivity Functions for Continuous Time Systems Under Sampled Data Control, 1993, to appear Automatica.

[12] Kabamba, P. and Hara, S. On Computing the Induced Norm of Sampled Data Feedback Systems, Proc. Am. Contr. Conf., San Diego, 1990, 319–320.

[13] Linnemann, A. L_∞-Induced Optimal Performance in Sampled-Data Systems, *Syst. Contr. Lett. 18,* 265–275, 1992.

[14] Middleton, R.H. and G.C. Goodwin. *Digital Control and Estimation: A Unified Approach,* Prentice Hall, Englewood Cliffs, NJ, 1990.

[15] Osborn, S.L. and D.S. Bernstein. An Exact Treatment of the Achievable Closed Loop H_2 Performance of Sampled Data Controllers: From Continous Time to Open Loop. Proc. IEEE CDC Conference San Antonio, Texas, 1993.

[16] Sivashankar, N. and P.P. Khargonekar. L^∞-Induced Norm of Sampled-Data Systems, Proc. Am. Contr. Conf., Boston, MA., 1991, 167–172.

[17] Sivashankar, N. and P.P. Khargonekar. Robust Stability and Performance of Sampled-Data Systems, Proc. 30th CDC Brighton, U.K., 1992, 881–886.

[18] Toivonen, H.T. Sampled Data Control of Continuous Time System with an H^∞ Optimality Criterion, Rep. 90-1 Department of Chemical Engineering, Abo Akademic, Finland, 1990.

[19] Yamamoto, Y. New Approach to Sampled-Data Control Systems – A Function Space Method, Proc., 29th CDC, Honolulu, 1990, 1882–1887.

13

Discrete-Time Equivalents to Continuous-Time Systems

Mohammed S. Santina
The Aerospace Corporation, Los Angeles, CA

Allen R. Stubberud
University of California, Irvine, Irvine, CA

Gene H. Hostetter

13.1 Introduction

The traditional approach to designing digital control systems for continuous-time plants is first to design an analog controller for the plant, then to derive a digital counterpart that closely approximates the behavior of the original analog controller. This approach is particularly useful when the designer is replacing all or part of an existing analog controller with a digital controller. However, even for small sampling periods, the digital approximation usually performs less well than the analog controller from which it was derived.

The other approach to designing digital controllers for continuous-time plants is first to derive a discrete-time equivalent of the plant and then to design a digital controller directly to control the discretized plant. Classical and modern state-space design methods for discrete-time systems parallel the classical and modern control methods for continuous-time systems. These methods will be discussed in the following article.

In this section, we first present several methods for discretizing analog controllers. Next, the relationship between continuous-time state variable plant models and their discrete counterparts is derived to use these results for designing digital controllers for discrete-time systems.

13.2 Digitizing Analog Controllers

Consider the situation in Figure 13.1 where a digital compensator is wanted which approximates the behavior of the analog controller described by $G_c(s)$. As shown in the figure, the digital controller consists of an A/D converter driving the discrete-time system described by the z-transfer function $H(z)$ followed by a D/A converter with sample and hold. This configuration is called a *digital filter.* Specifically, the problem can be stated as follows: Given $G_c(s)$ of the analog controller, what is the corresponding $H(z)$ so that the digital controller approximates the behavior of the analog one? To answer this question, several methods for discretizing analog controllers are now presented. These methods are

1. Numerical approximation of differential equations.
2. Matching step and other responses.
3. Pole-zero matching.

In the material to follow, the ^ over a symbol denotes an approximation of the quantity.

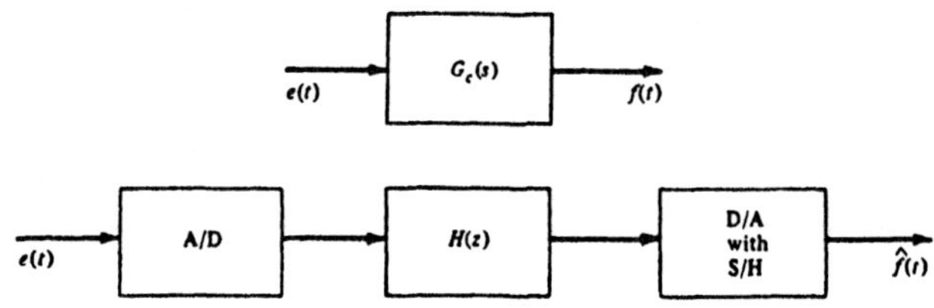

Figure 13.1 Digitizing an analog controller.

13.2.1 Numerical Approximation of Differential Equations

The problem of approximating an analog controller with a digital one can be viewed as converting the analog controller transfer function $G_c(s)$ to a differential equation and then obtaining a numerical approximation to the solution of the differential equation. There are basically two methods for numerically approximating the solution of differential equations, numerical integra-

0-8493-8570-9/96/$0.00+$.50

tion and numerical differentiation. We first discuss numerical integration and then summarize numerical differentiation.

Approximate numerical calculation of integrals is an important computational problem fundamental to numerical differential equation solution. The most common approach to numerical integration is to divide the interval of integration into many T subintervals and to approximate the contribution to the integral in each T strip by the integral of a *polynomial approximation* to the integrand in that strip.

Consider the transfer function

$$G_c(s) = \frac{F(s)}{E(s)} = \frac{1}{s} \tag{13.1}$$

which has the corresponding differential equation

$$\frac{df}{dt} = e(t). \tag{13.2}$$

Integrating both sides of Equation 13.2 from t_0 to t,

$$f(t) = f(t_0) + \int_{t_0}^{t} e(t)dt, \quad t \geq t_0.$$

For evenly spaced sample times $t = kT, k = 0, 1, 2, \ldots$ and during one sampling interval $t_0 = kT$ to $t = kT + T$, the solution is

$$f(kT + T) = f(kT) + \int_{kT}^{kT+T} e(t)dt. \tag{13.3}$$

Euler's Forward Method (One Sample)

The simplest approximation of the integral in Equation 13.3 is simply to approximate the integrand by a constant equal to the value of the integrand at the *left* endpoint of each T subinterval and multiply by the sampling interval T as in Figure 13.2(a). Thus,

$$\hat{f}(kT + T) = \hat{f}(kT) + Te(kT). \tag{13.4}$$

Z-transforming both sides of Equation 13.4,

$$z\hat{F}(z) - \hat{F}(z) = TE(z),$$

and, therefore,

$$H(z) = \frac{\hat{F}(z)}{E(z)} = \frac{T}{z-1}. \tag{13.5}$$

Comparing Equation 13.5 with the analog controller transfer function Equation 13.1 implies that a discrete-time equivalence of an analog controller can be determined with Euler's forward method by simply replacing each s in the analog controller transfer function with $(z - 1)/T$, that is,

$$H(z) = G_c(s)|_{s=\frac{z-1}{T}}.$$

Euler's Backward Method (One Sample)

Instead of approximating the integrand in Equation 13.3 during one sampling interval by its value at the left endpoint, *Euler's backward* method approximates the integrand by its value at the *right* endpoint of each T subinterval and multiplies by the sampling interval, as in Figure 13.2(b). Then, Equation 13.3 becomes

$$\hat{f}(kT + T) = \hat{f}(kT) + Te(kT + T) \tag{13.6}$$

or, using the z-transformation,

$$H(z) = \frac{\hat{F}(z)}{E(z)} = \frac{Tz}{z-1} \tag{13.7}$$

Comparing Equation 13.6 with Equation 13.1 shows that the equivalent discrete-time transfer function of the analog controller can be obtained by replacing each s in $G_c(s)$ with $(z - 1)/Tz$, that is,

$$H(z) = G_c(s)|_{s=\frac{z-1}{Tz}}.$$

For the analog controller,

$$G_c(s) = \frac{a}{s+a}$$

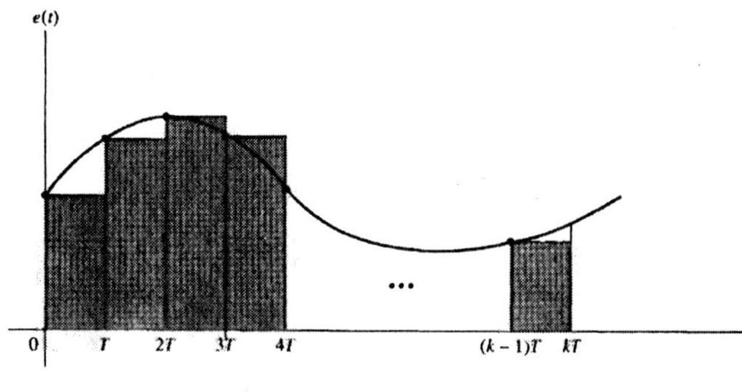

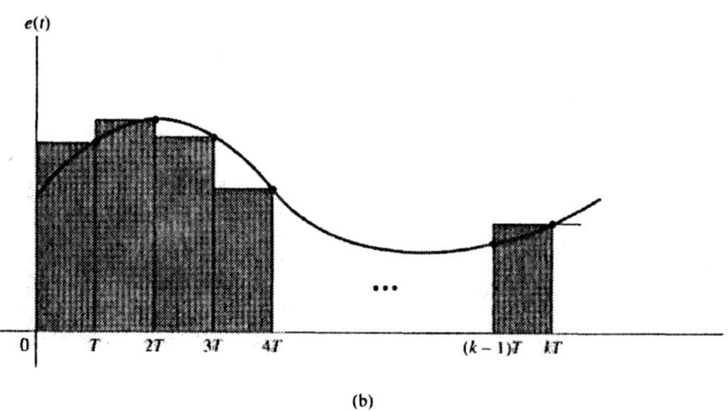

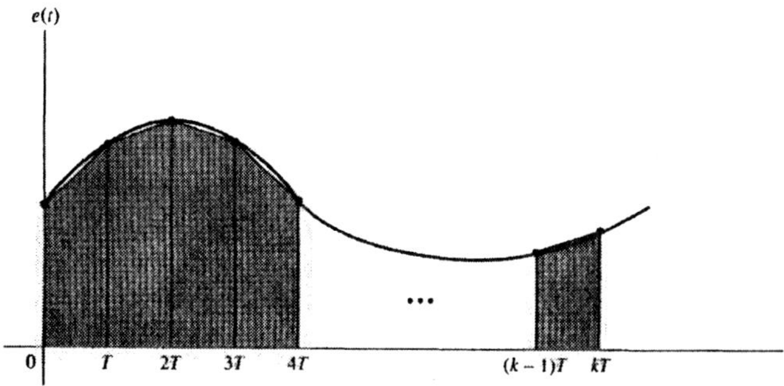

Figure 13.2 Comparing Euler's and trapezoidal integration approximations. (a) approximation using Euler's forward method, (b) approximation using Euler's backward method, (c) approximation using the trapezoidal method.

for example, the discrete-time equivalent controller using Euler's backward method is

$$H(z) = \frac{a}{(z-1)/Tz + a} = \frac{aTz}{(1+aT)z - 1}$$

and the discrete-time equivalent controller using Euler's forward method is

$$H(z) = \frac{aT}{z - 1 + aT}.$$

Trapezoidal Method (Two Samples)

Euler's forward and backward methods are sometimes called *rectangular* methods because, during the sampling interval, the area under the curve is approximated with a rectangle. Additionally, Euler's methods are also called *first order* because they use one sample during each sampling interval.

The performance of the digital controller can be improved over the simpler approximation by either Euler's forward or backward methods if more than one sample is used to update the approximation of the analog controller transfer function during a sampling interval. As in Figure 13.2(c), the trapezoidal approximation approximates the integrand with a straight line. Applying the trapezoidal rule to the integral in Equation 13.3 gives:

$$\hat{f}(kT+T) = \hat{f}(kT) + \frac{T}{2}\{e(kT) + e(kT+T)\}$$

which has a corresponding z-transfer function,

$$(z-1)\hat{F}(z) = \frac{T}{2}(z+1)E(z),$$

or

$$H(z) = \frac{\hat{F}(z)}{E(z)} = \frac{T}{2}\left(\frac{z+1}{z-1}\right). \qquad (13.8)$$

Comparing Equation 13.8 with Equation 13.1, the digital controller transfer function can be obtained by simply replacing each s in the analog controller transfer function with $\frac{2}{T}\frac{z-1}{z+1}$.

That is,

$$H(z) = G_c(s)|_{s=\frac{2}{T}\frac{z-1}{z+1}}$$

The trapezoidal method is also called *Tustin's* method, or the *bilinear* transformation. Higher order polynomial integrals can be approximated in the same way, but for a recursive numerical solution, an integral approximation should involve only present and past values of the integrand, not future ones.

A summary of some common approximations for integrals, along with the corresponding z-transfer function of each integral approximation, is shown in Table 13.1. Higher order approximations result in digital controllers of progressively higher order. The higher the order of the approximation, the better the approximation to the analog integrations and the more accurately the digital controller output tends to track samples of the analog controller output for any input. The digital controller, however, probably has a sample-and-held output between samples, so that accurate tracking of samples is of less concern to the designer than the sample rate.

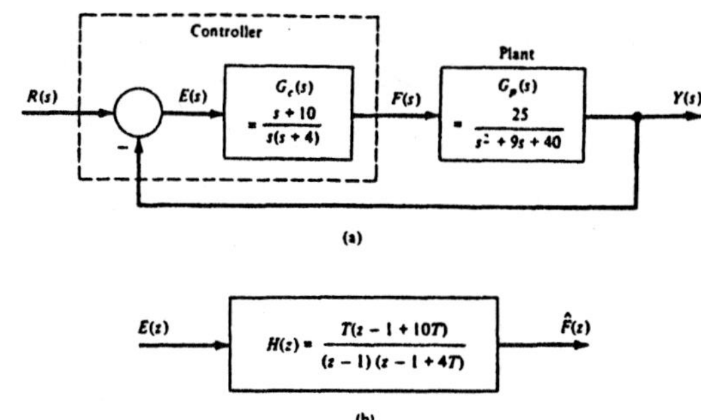

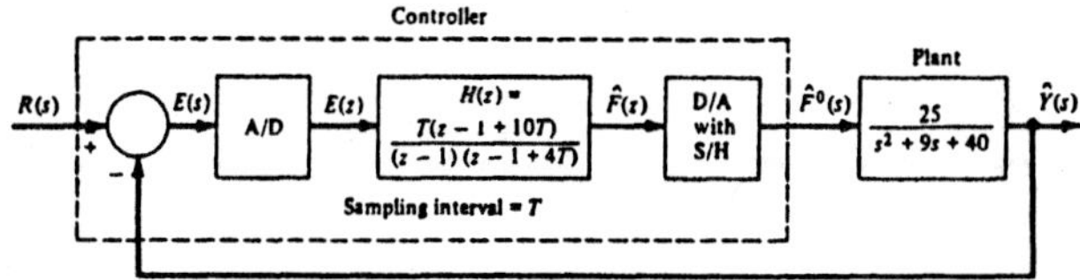

Figure 13.3 Obtaining the z-transfer function associated with Euler's forward approximation of an analog controller. (a) analog control system, (b) digital controller transfer function using Euler's forward method, (c) plant with digital controller.

An Example

As an example of designing a digital controller for a continuous-time plant, consider the system shown in Figure 13.3(a), where it is assumed that the analog controller

$$G_c(s) = \frac{s+10}{s(s+4)}$$

has been designed to control the continuous-time plant with the transfer function,

$$G_p(s) = \frac{25}{s^2 + 9s + 40}.$$

The feedback system has the overall transfer function,

$$T(s) = \frac{G_c(s)G_p(s)}{1 + G_c(s)G_p(s)} = \frac{25s + 250}{s^4 + 13s^3 + 76s^2 + 185s + 250},$$

with all of its poles to the left of $s = -1.5$ and zero steady-state error to a step input.

Using Euler's forward method, the digital controller transfer function,

$$H(z) = G_c(s)|_{s=\frac{z-1}{T}} = \frac{(z-1)/T + 10}{[(z-1)/T][(z-1)/T + 4]} = \frac{T(z - 1 + 10T)}{(z-1)(z-1+4T)},$$

as shown in Figure 13.3(b). The plant and the digital controller are shown in Figure 13.3(c). The step response of this digital controller for various sampling intervals T is shown in Figure 13.4. When $T = 0.2$ sec., which is relatively large compared to the

TABLE 13.1 Some Integral Approximations Using Present and Past Integrand Samples

Approximation to the integral over one step	Difference equation for the approximate integral	Z-Transmittance of the approximate integrator
One-Sample		
$\int_{kT}^{kT+T} e(t)dt \cong Te(kT)$	$\hat{f}[(k+1)T] = \hat{f}(kT) + Te(kT)$	$\frac{T}{z-1}$
$\int_{kT}^{kT+T} e(t)dt \cong Te(kT+T)$	$\hat{f}[(k+1)T] = \hat{f}(kT) + Te(kT+T)$	$\frac{Tz}{z-1}$
Two-Sample (Tustin approximation)		
$\int_{kT}^{kT+T} e(t)dt \cong T\left\{\frac{1}{2}e[(k+1)T] + \frac{1}{2}e(kT)\right\}$	$\hat{f}[(k+1)T] = \hat{f}(kT) + \frac{T}{2}e[(k+1)T] + \frac{T}{2}e(kT)$	$\frac{T(z+1)}{2(z-1)}$
Three-Sample		
$\int_{kT}^{kT+T} e(t)dt \cong T\left\{\frac{5}{12}e[(k+1)T] + \frac{8}{12}e(kT) - \frac{1}{12}e[(k-1)T]\right\}$	$\hat{f}[(k+1)T] = \hat{f}(kT) + \frac{5T}{12}e[(k+1)T] + \frac{8T}{12}e(kT) - \frac{T}{12}e[(k-1)T]$	$\frac{T[(5/12)z^2+(8/12)z-(1/12)]}{z(z-1)}$

analog controller's fastest mode e^{-4t}, the digital controller approximation deviates significantly from the analog controller. As shown in Figures 13.4(b) and 13.4(c), as the sampling interval is decreased so that there are several steps during each time constant of the analog controller's mode e^{-4t}, the step responses of the analog and digital controllers are nearly the same. The same can be said for the response of the overall feedback system. Step responses of the feedback system for various sampling intervals T are shown in Figure 13.5.

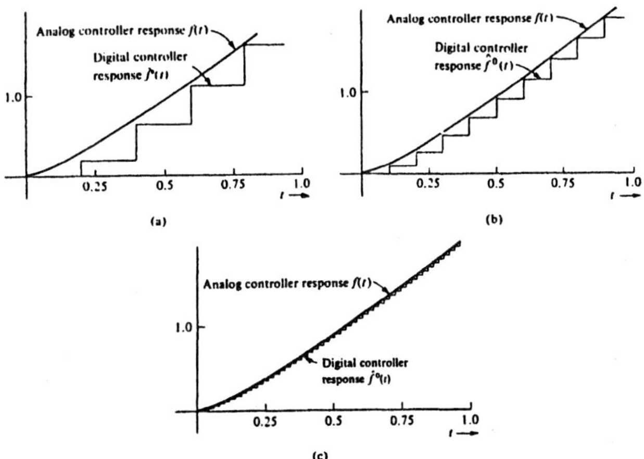

Figure 13.4 Step response of the Euler's forward approximation digital controller. (a) sampling interval $T = 0.2$, (b) $T = 0.1$, (c) $T = 0.02$.

If, on the other hand, we use the trapezoidal method, the transfer function of the digital controller becomes

$$H(z) = \frac{T(z+1)[(5T+1)z + (5T-1)]}{(z-1)[(4T+2)z + (4T-2)]}.$$

The character of the response of this controller is similar to that of the controller response shown in Figure 13.4. For the same sampling interval, the controller using Tustin's method tends to track the analog controller output more accurately at the sample times because the approximations to the analog integrations are better. The step response of the overall feedback system using Tustin's and Euler's forward methods is shown in Figure 13.6 for various sampling intervals. Obviously, Tustin's method gives results better than Euler's for the same sampling intervals.

Because Euler's and Tustin's methods result in controllers of the same order, the designer usually opts for Tustin's approximation. In general, if the sampling interval is sufficiently small and the approximation is sufficiently good, the behavior of the digital controller, designed by any one of the approximation methods, will be nearly indistinguishable from that of the analog controller. One should not be too hasty in abandoning the simple Euler approximation for higher order approximation. In modeling physical systems, the poor accuracy of the Euler approximation with a very small sampling interval indicates an underlying lack of physical robustness that ought to be carefully examined.

Warning:

The approximation methods summarized in Table 13.1 apply by replacing each s in the analog controller transfer function with the corresponding z-transmittance. Every z-transmittance is a mapping from the s-plane to the z-plane. As shown in Figure 13.7(b), Euler's forward method has the potential of mapping poles in the left half of the s-plane to poles outside the unit circle on the complex plane. Then, some stable analog controllers may produce unstable digital controllers. Euler's backward rule maps the left hand of the s-plane to a region inside the unit circle as shown in Figure 13.7(c). The trapezoidal rule, however, maps the left half of the s-plane to the interior of the unit circle on the z-plane, the right half of the s-plane to the exterior of the unit circle, and the imaginary axis of the s-plane to the boundaries of the unit circle as shown in Figure 13.7(d).

Bilinear Transformation with Frequency Prewarping

In many digital control and digital signal processing applications, it is desired to design a digital filter $G(z)$ that closely approximates the frequency response of a continuous-time filter $G(s)$ within the bandlimited range

$$G(z = e^{j\omega T}) \cong G(s = j\omega) \qquad 0 \le \omega < \omega_0 = \frac{\pi}{T}$$

The bilinear(trapezoidal) method applies but with minor modifications. If the frequency response of the digital filter $G(z)$ is to approximate the frequency response of the analog controller

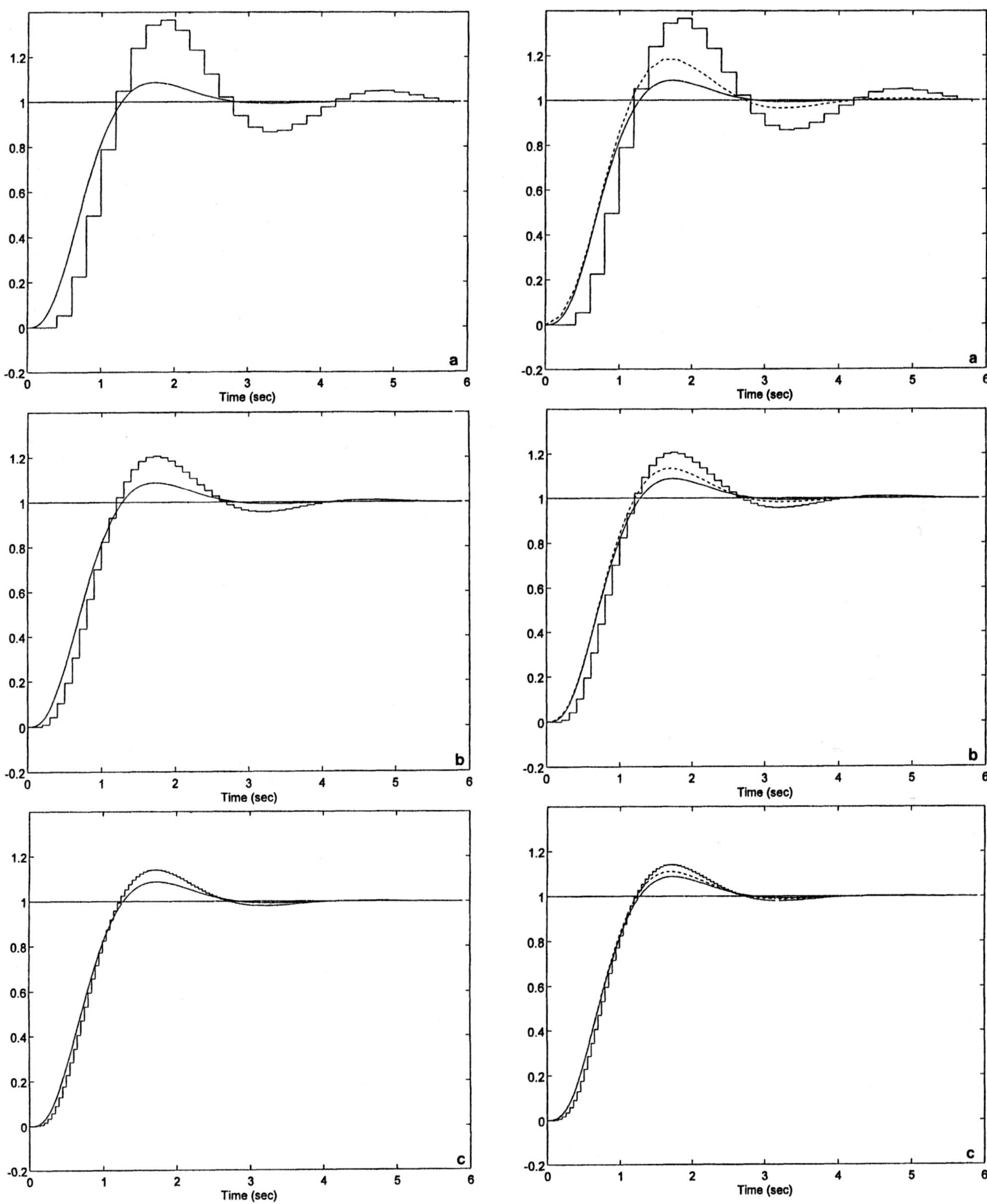

Figure 13.5 Step response of the feedback system with an Euler's forward approximation digital controller (staircase: Euler, solid: continuous). (a) sampling interval $T = 0.4$, (b) $T = 0.2$, (c) $T = 0.1$.

Figure 13.6 Step response of the feedback system using Euler's and Tustin's approximation digital controller (staircase: Euler, dotted: Tustin, solid: continuous). (a) sampling interval $T = 0.4$, (b) $T = 0.2$, (c) $T = 0.1$.

$G(s)$, then

$$\begin{aligned} G(z) &= G\left(s = \frac{2}{T}\frac{z-1}{z+1}\right) \\ G(z = e^{j\omega_d T}) &= G\left(j\omega_c = \frac{2}{T}\frac{e^{j\omega_d T}-1}{e^{j\omega_d T}+1}\right) \\ &= G\left[j\omega_c = \frac{2}{T}\left(\frac{e^{(j\omega_d T/2)} - e^{(-j\omega_d T/2)}}{e^{(j\omega_d T/2)} + e^{(-j\omega_d T/2)}}\right)\right] \\ &= G\left[j\omega_c = j\frac{2}{T}\frac{\sin\omega_d T/2}{\cos\omega_d T/2}\right] \\ &= G\left[j\omega_c = j\frac{2}{T}\tan\omega_d T\right] \end{aligned}$$

and, therefore,

$$\omega_c = \frac{2}{T}\tan\frac{\omega_d T}{2} \tag{13.9}$$

where ω_c is the continuous frequency and ω_d is the discrete frequency. This nonlinear relationship arises because the *entire* $j\omega$-axis of the s-plane is mapped into one complete revolution of the unit circle in the z-plane.

For relatively small values of ω_d as compared to the *folding frequency* π/T, then

$$j\omega_c \cong j\frac{2}{T}\frac{\omega_d T}{2} = j\omega_d$$

and the behavior of the discrete-time filter closely approximates the frequency response of the corresponding continuous-time filter. When ω_d approaches the folding frequency π/T,

$$\begin{aligned} j\omega_c &= j\frac{2}{T}\tan\frac{\omega_d T}{2} \\ &\to j\infty \end{aligned}$$

the continuous frequency approaches infinity, and distortion becomes evident. However, if the bilinear transformation is applied together with Equation 13.9 near the frequencies of interest, the frequency distortion can be reduced considerably.

The general design procedure for discretizing a continuous-time filter using the bilinear transformation with frequency *prewarping* is as follows:

1. Beginning with the continuous-time filter $G(s)$, obtain a new continuous-time filter with transfer function $G'(s)$ whose poles and zeros with critical frequencies $(s+\alpha')$ are related to those of the original $G(s)$ by

$$(s+\alpha) \to (s+\alpha')\big|_{\alpha'=2/T\tan\alpha T/2}$$

in the case of real roots, and by

$$s^2 + 2\zeta\omega_n s + \omega_n^2 \to s^2 + 2\zeta\omega'_n s + \omega'^2_n\Big|_{\omega'_n=2/T\tan\omega_n T/2}$$

in the case of complex roots.

2. Apply the bilinear transformation to $G'(s)$ by replacing each s in $G'(s)$ with

$$s = \frac{2}{T}\frac{z-1}{z+1}$$

3. Scale the multiplying constant of $G(z)$ to match the multiplying constant of the continuous-time filter $G(s)$ at a specific frequency.

To illustrate the steps above, consider the second-order lowpass filter described by the transfer function,

$$G(s) = \frac{(200\pi)^2}{s^2 + 200\pi s + (200\pi)^2}.$$

This filter has a unity D.C. gain, undamped natural frequency $f = 100$ Hz, and a damping ratio $\zeta = 0.5$. For a sampling interval $T = 0.002$ sec., the folding frequency is $f_0 = 250$ Hz, which is above the 100 Hz cutoff of the filter. At $\omega_n = 200\pi$,

$$\omega'_n = \tfrac{2}{T}\tan\tfrac{\omega_n T}{2} = 1000\tan\tfrac{200\pi}{1000} = 726.54 \quad \text{rad/sec.}$$

and hence the warped transfer function is

$$G'(s) = \frac{K}{s^2 + 726.54s + (726.54)^2}.$$

Therefore,

$$\begin{aligned} G(z) &= \frac{K}{[1000(z-1)/(z+1)]^2 + 726542(z-1)/(z+1) + (726.54)^2} \\ &= \frac{K(z+1)^2}{2254405z^2 - 944273z + 801321}. \end{aligned}$$

For a unity D.C. gain of the continuous-filter,

$$G(1) = \frac{4K}{2111453.1} = 1.$$

Then

$$K = 527863.28,$$

and hence

$$G(z) = \frac{0.2341(z+1)^2}{z^2 - 0.4189z + 0.3554}. \tag{13.10}$$

which is the required filter.

If the example above is repeated for a sampling interval $T = 0.001$ sec., the resulting digital filter transfer function becomes

$$G'(z) = \frac{0.0738(z+1)^2}{z^2 - 1.2505z + 0.5457}. \tag{13.11}$$

Figure 13.8 shows the frequency responses of the continuous-time filter and the digital filters given by Equations 13.10 and 13.11. As expected, the filter with the sampling interval $T = 0.001$ sec. tends to approximate the frequency response of the continuous-time filter more accurately than the filter with $T = 0.002$ sec.

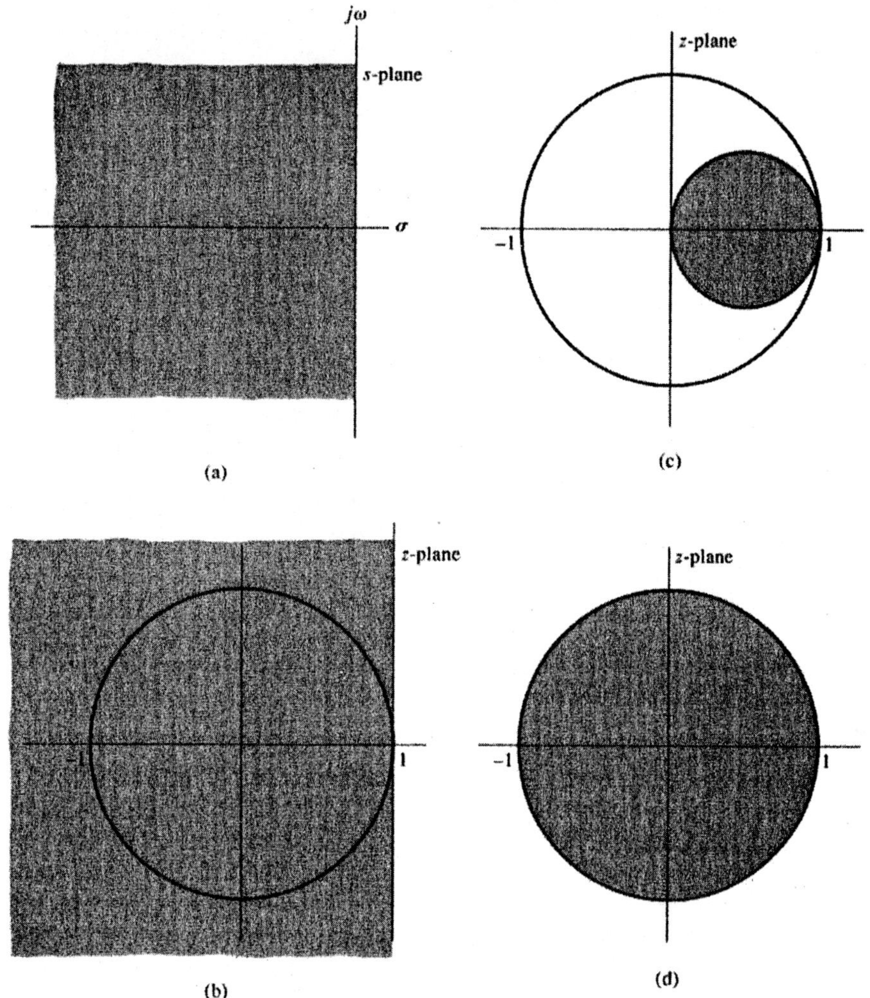

Figure 13.7 Mappings between the s-plane and the z-plane using Euler's and Tustin's methods. (a) stability region in the s-plane, (b) corresponding stability region in the z-plane using Euler's forward method, (c) corresponding region in the z-plane using Euler's backward method, (d) corresponding region in the z-plane using Tustin's method.

Numerical Differentiation

The other main approach to approximating the solution of a differential equation is *numerical differentiation.* The approximate solution of the differential equation is obtained by replacing the derivative terms in the differential equation with finite difference approximations. The resulting difference equation can then be solved numerically. Three methods of first- and second-order derivative approximations listed in Table 13.2 are called finite-difference approximations of derivatives. The corresponding z-transmittance of each of these finite difference approximations is also listed in Table 13.2.

As an example, for the analog controller

$$G_c(s) = \frac{K(s+a)}{(s+b)},$$

the forward-difference approximation gives the digital controller,

$$H(z) = \frac{K[(z-1)/T+a]}{(z-1)/T+b} = \frac{K(z-1+aT)}{z-1+bT},$$

and the backward-difference approximation yields the digital controller

$$\begin{aligned} H(z) &= \frac{K[(z-1)/Tz+a]}{(z-1)/Tz+b} \\ &= \frac{K(1+aT)}{1+bT}\left[\frac{z-\frac{1}{1+aT}}{z-\frac{1}{1+bT}}\right]. \end{aligned}$$

Similar to integral approximations, higher order derivative approximations can be generated in the same way, but for a recursive numerical solution, a derivative approximation should involve only present and past values of the input, not future ones.

13.2.2 Matching Step and Other Responses

Another way of approximating the behavior of the analog controller with a digital controller is to require that, at the sampling times, the step response of the digital controller *matches* the analog controller step response. Consider the unit step response $f_{\text{step}}(t)$ of the analog controller with transfer function

TABLE 13.2 Finite Difference Approximations of Derivatives

	Derivative approximation	Z-Transmittance of the approximate differentiator
First-Order derivative		
Forward difference	$\frac{f[(k+1)T]-f(kT)}{T}$	$\frac{z-1}{T}$
Backward difference	$\frac{f(kT)-f[(k-1)T]}{T}$	$\frac{z-1}{Tz}$
Central difference	$\frac{f[(k+1)T]-f[(k-1)T]}{2T}$	$\frac{z^2-1}{2Tz}$
Second-Order derivative		
Forward difference	$\frac{f[(k+2)T]-2f[(k+1)T]+f(kT)}{T^2}$	$\frac{z^2-2z+1}{T^2}$
Backward difference	$\frac{f(kT)-2f[(k-1)T]+f[(k-2)T]}{T^2}$	$\frac{z^2-2z+1}{T^2z^2}$
Central difference	$\frac{f[(k+1)T]-2f(kT)+f[(k-1)T]}{T^2}$	$\frac{z^2-2z+1}{T^2z}$

$G_c(s)$ shown Figure 13.9(a). Our objective is to design a discrete-portion $H(z)$ of the digital controller, as in Figure 13.9(b), such that its step response $f_{\text{step}}(k)$ to a unit step input consists of samples of $f_{\text{step}}(t)$. Then, as in Figure 13.9(c), the digital controller has a step response that equals the step response of the analog controller at the sample times. This method is termed a *step-invariant approximation* of the analog system by a digital system.

As an example, consider the analog controller with the transfer function,

$$G_c(s) = \frac{4s^2+17s+12}{s^2+5s+6}.$$

This controller has the unit step response,

$$\begin{aligned} F_{\text{step}}(s) &= \frac{1}{s}G_c(s) = \frac{4s^2+17s+12}{s(s+2)(s+3)} \\ &= \frac{2}{s}+\frac{3}{s+2}+\frac{-1}{s+3}. \end{aligned}$$

For a sampling interval $T = 0.2$ sec., the samples of $f_{\text{step}}(t)$ have the z-transform,

$$F_{\text{step}}(z) = \frac{2z}{z-1}+\frac{3z}{z-e^{-0.4}}-\frac{z}{z-e^{-0.6}}.$$

Taking these samples as the output of a discrete-time system $H(z)$ driven by a unit step sequence,

$$\begin{aligned} F_{\text{step}}(z) &= \left(\frac{z}{z-1}\right)H(z) \\ &= \frac{2z}{z-1}+\frac{3z}{z-0.67}-\frac{z}{z-0.54}, \end{aligned}$$

the step-invariant approximation is

$$\begin{aligned} H(z) &= 2+\frac{3(z-1)}{z-0.6703}-\frac{(z-1)}{z-0.5488} \\ &= \frac{4z^2-5.4143z+1.7118}{(z-0.6703)(z-0.5488)}. \end{aligned}$$

Figure 13.10(a) shows the step response of the analog controller and the step response of the step-invariant approximation. Reducing the sampling interval to $T = 0.1$ sec., the step-invariant approximation becomes

$$H(z) = \frac{4z^2-6.5227z+2.6167}{z^2-1.5595z+0.6065}.$$

The step response of this controller and the analog one are shown in Figure 13.10(b).

It is occasionally desirable to design digital controllers so that their response to some input other than a step consists of a sampled-and-held version of an analog controller's response to that input. A *ramp-invariant approximation*, for example, has a discrete-time response to a unit ramp sequence that consists of samples of the unit ramp response of the continuous-time system. The *impulse-invariant approximation* is another possibility.

13.2.3 Pole-Zero Matching

Yet another method of approximating an analog controller by a digital one is to map the poles and zeros of the analog controller transfer function $G_c(s)$ to those of the corresponding digital controller $H(z)$ as follows:

$$(s+a) \to z-e^{-aT},$$

for real roots, and

$$(s+a)^2+b^2 \to z^2-2(e^{-aT}\cos bT)z+e^{-2aT},$$

for complex conjugate pairs.

Usually, an analog controller has more finite poles than zeros. In this case its high frequency response tends to zero as ω_c approaches infinity. Because the entire $j\omega$ axis of the s-plane is mapped into one complete revolution of the unit circle in the z-plane, the highest possible frequency on the $j\omega$-axis is at $\omega_c = \pi/T$. Hence,

$$z = e^{sT} = e^{j(\pi/T)T} = -1,$$

and, therefore, infinite zeros of the analog controller map into finite zeros located at $z = -1$ in the corresponding digital equivalence. The resulting transfer function of the digital controller will always have the number of poles equal to the number of zeros.

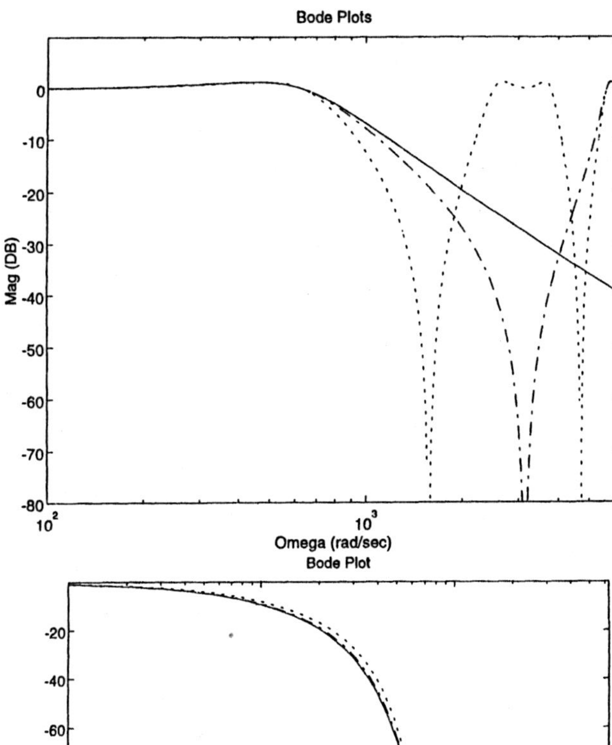

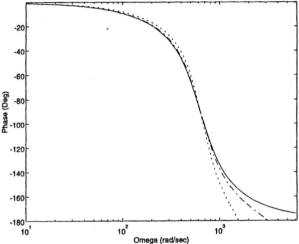

Figure 13.8 Frequency response plots of the second-order low-pass filter using bilinear transformation with frequency prewarping (solid: continuous, dash: $T = 0.001$, dotted: $T = 0.002$). (a) magnitude, (b) phase.

For example, the analog controller,

$$G_c(s) = \frac{6s+10}{s^2+2s+5} = \frac{6(s+\frac{5}{3})}{(s+1)^2+2^2},$$

has two finite poles and one finite zero. For a sampling interval $T = 0.1$ sec.,

$$\begin{aligned} H(z) &= \frac{K(z+1)(z-e^{-(5/30)})}{z^2-2(e^{-0.1}\cos 2T)z+e^{-0.2}} \\ &= \frac{K(z+1)(z-0.85)}{z^2-1.773z+0.818}. \end{aligned}$$

The D.C. gain of the analog controller is

$$G_c(s=j0) = 2.$$

For the identical D.C. gain of the digital controller

$$H(z=1) = \frac{K(2)(0.15)}{1-1.773+0.818} = 2$$

then $K = 0.3$. Hence,

$$H(z) = \frac{0.3(z+1)(z-0.85)}{z^2-1.773z+0.818}.$$

If the analog controller has poles or zeros at the origin of the s-plane, the multiplying constant of the digital controller is selected to match the gain of the analog controller at a specified frequency. At low frequency,

$$H(z)|_{z=1} = G_c(s)|_{s=0},$$

and at high frequency,

$$H(z)|_{z=-1} = G_c(s)|_{s=\infty}.$$

All of the approximation methods presented thus far in this section for digitizing analog controllers tend to perform well for sufficiently short sampling intervals. For longer sampling intervals, dictated by design and cost constraints, one approximation method or another might perform the best in a given situation.

Digitizing an analog controller is not a good general design technique, although it is very useful when replacing all or part of an existing analog controller with a digital one. The technique requires beginning with a good analog design, which is probably as difficult as creating a good digital design. The digital design usually performs less well than the analog counterpart from which it was originally derived. Furthermore, the step-invariant and other approximations are not easily extended to systems with multiple inputs and outputs. When the resulting feedback system performance is inadequate, the designer may have few options besides raising the sampling rate.

13.3 Discretization of Continuous-Time State Variable Models

We now discuss the relationship between continuous-time state variable plant models and discrete-time models of plant signal samples.

13.3.1 Discrete-Time Models of Continuous-Time Systems

Consider an nth order continuous-time system described by the state variable equation

$$\begin{aligned} \dot{x}(t) &= \mathcal{A}x(t) + \mathcal{B}u(t) \\ y(t) &= Cx(t) + Du(t) \end{aligned} \tag{13.12}$$

Script symbols are now used for state and input coupling matrices to distinguish between them and the corresponding discrete-time models. The time t is a continuous variable, x is the n-vector state of the system, u is an r-vector of system inputs, and y is an m-vector of system outputs. The remaining matrices in Equation 13.12 are of appropriate dimensions.

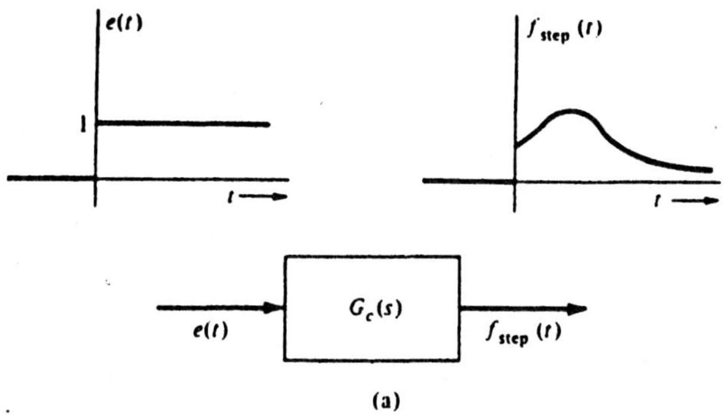

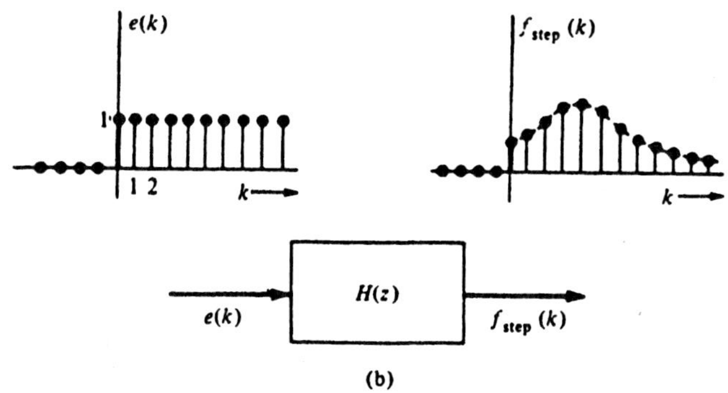

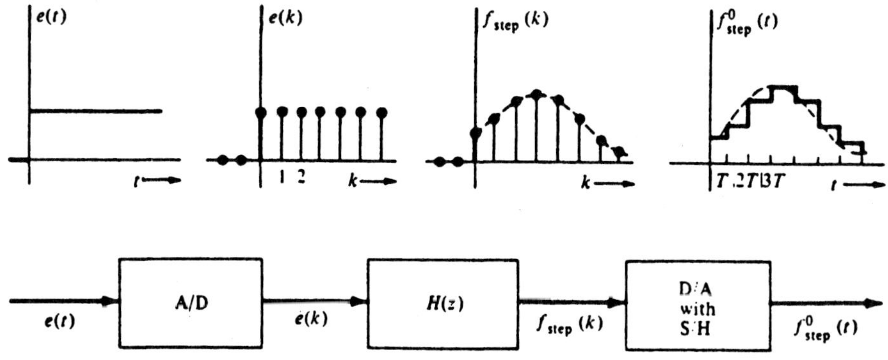

Figure 13.9 Finding a step-invariant approximation of a continuous-time controller. (a) analog controller step response, (b) discrete step response consisting of samples of the analog controller step response, (c) step-invariant digital controller.

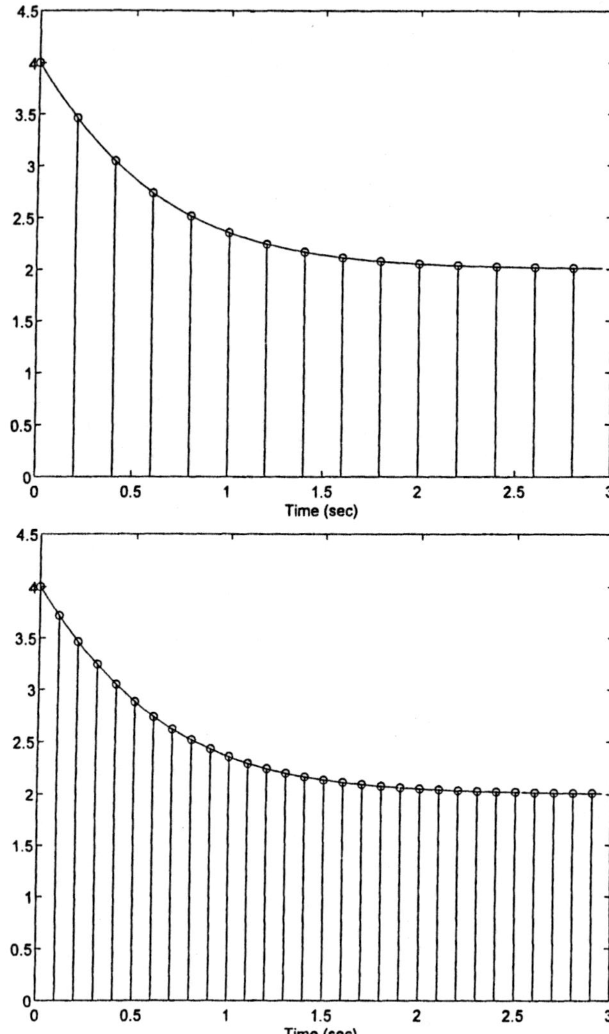

Figure 13.10 Step responses of a continuous-time controller and step-invariant approximation. (a) $T = 0.2$, (b) $T = 0.1$.

The solution for the state for $t \geq 0$ is given by the convolution,

$$x(t) = e^{\mathcal{A}t}x(0) + \int_0^t e^{\mathcal{A}(t-\tau)}\mathcal{B}u(\tau)d\tau.$$

At the sample times kT, $k = 0, 1, 2, \ldots$, the state is

$$x(kT) = e^{\mathcal{A}kT}x(0) + \int_0^{kT} e^{\mathcal{A}(kT-\tau)}\mathcal{B}u(\tau)d\tau.$$

The state at the $(k+1)$th step can be expressed in terms of the state at the kth step as follows:

$$\begin{aligned}
x(kT+T) &= e^{\mathcal{A}(kT+T)}x(0) \\
&\quad + \int_0^{kT+T} e^{\mathcal{A}(kT+T-\tau)}\mathcal{B}u(\tau)d\tau \\
&= e^{\mathcal{A}T}e^{\mathcal{A}kT}x(0) + \int_0^{kT} e^{\mathcal{A}(kT+T-\tau)}\mathcal{B}u(\tau)d\tau \\
&\quad + \int_{kT}^{kT+T} e^{\mathcal{A}(kT+T-\tau)}\mathcal{B}u(\tau)d\tau \\
&= e^{\mathcal{A}T}\left[e^{\mathcal{A}kT}x(0) + \int_0^{kT} e^{\mathcal{A}(kT-\tau)}\mathcal{B}u(\tau)d\tau\right] \\
&\quad + \int_{kT}^{kT+T} e^{\mathcal{A}(kT+T-\tau)}\mathcal{B}u(\tau)d\tau \\
&= e^{\mathcal{A}T}x(kT) + \quad \text{(input term)}
\end{aligned}$$

When the input $u(t)$ is constant during each sampling interval, as it is when driven by sample-and-hold, then the input term becomes

$$\int_{kT}^{kT+T} e^{\mathcal{A}(kT+T-\tau)}\mathcal{B}u(\tau)d\tau = \left[\int_{kT}^{kT+T} e^{\mathcal{A}(kT+T-\tau)}d\tau\right]\mathcal{B}u(kT)$$

The discrete-time input coupling matrix is

$$B = \left[\int_{kT}^{kT+T} e^{\mathcal{A}(kT+T-\tau)}d\tau\right]\mathcal{B}$$

Letting

$$\gamma = kT + T - \tau, \quad d\gamma = -d\tau,$$

then

$$B = \left[\int_0^T e^{\mathcal{A}\gamma}d\gamma\right]\mathcal{B}.$$

Expanding the integrand into a power series,

$$e^{\mathcal{A}\gamma} = I + \frac{\mathcal{A}\gamma}{1!} + \frac{\mathcal{A}^2\gamma^2}{2!} + \cdots + \frac{\mathcal{A}^i\gamma^i}{i!} + \cdots,$$

and integrating term by term results in

$$\begin{aligned}
B &= \left\{\int_0^T \left[I + \frac{\mathcal{A}\gamma}{1!} + \frac{\mathcal{A}^2\gamma^2}{2!} + \cdots \right.\right. \\
&\quad \left.\left. + \frac{\mathcal{A}^i\gamma^i}{i!} + \cdots\right]d\gamma\right\}\mathcal{B} \\
&= \left[IT + \frac{\mathcal{A}T^2}{2!} + \frac{\mathcal{A}^2T^3}{3!} + \cdots + \frac{\mathcal{A}^iT^{i+1}}{(i+1)!} + \cdots\right]\mathcal{B}.
\end{aligned}$$

Because

$$\begin{aligned}
\mathcal{A}B &= \left[\frac{\mathcal{A}T}{1!} + \frac{\mathcal{A}^2T^2}{2!} + \cdots + \frac{\mathcal{A}^{i+1}T^{i+1}}{(i+1)!} + \cdots\right]\mathcal{B} \\
&= (e^{\mathcal{A}T} - I)\mathcal{B}
\end{aligned}$$

and, if $\mathcal{A}$ is nonsingular, then

$$B = \mathcal{A}^{-1}(e^{\mathcal{A}T} - I)\mathcal{B} = (e^{\mathcal{A}T} - I)\mathcal{A}^{-1}\mathcal{B}.$$

The discrete-time model of Equation 13.12 is then

$$\begin{aligned}
x[(k+1)T] &= Ax(kT) + Bu(kT) \\
y(kT) &= Cx(kT) + Du(kT)
\end{aligned}$$

or

$$\begin{aligned}
x(k+1) &= Ax(k) + Bu(k) \\
y(k) &= Cx(k) + Du(k)
\end{aligned}$$

where

$$A = e^{\mathcal{A}T} = I + \frac{\mathcal{A}T}{1!} + \frac{\mathcal{A}^2T^2}{2!} + \cdots + \frac{\mathcal{A}^iT^i}{i!} + \cdots \quad (13.13)$$

$$B = \left[IT + \frac{\mathcal{A}T^2}{2!} + \frac{\mathcal{A}^2T^3}{3!} + \cdots + \frac{\mathcal{A}^iT^{i+1}}{(i+1)!} + \cdots\right]\mathcal{B} \quad (13.14)$$

and where

$$B = \mathcal{A}^{-1}[e^{\mathcal{A}T} - I]\mathcal{B} = [e^{\mathcal{A}T} - I]\mathcal{A}^{-1}\mathcal{B} \quad (13.15)$$

when $\mathcal{A}$ is nonsingular.

As a numerical example, consider the continuous-time system,

$$\begin{bmatrix} \dot{x}_1(t) \\ \dot{x}_2(t) \end{bmatrix} = \begin{bmatrix} -2 & 2 \\ 1 & -3 \end{bmatrix}\begin{bmatrix} x_1(t) \\ x_2(t) \end{bmatrix} + \begin{bmatrix} -1 \\ 5 \end{bmatrix} u(t)$$

$$y(t) = [2 \quad -4]\begin{bmatrix} x_1(t) \\ x_2(t) \end{bmatrix} + 6u(t)$$

with a sampling interval $T = 0.2$. The matrix exponential is

$$e^{\mathcal{A}T} = e^{0.2\mathcal{A}} = \begin{bmatrix} 0.696 & 0.246 \\ 0.123 & 0.572 \end{bmatrix}$$

which can be calculated by truncating the power series

$$e^{\mathcal{A}T} \cong I + \mathcal{A}T + \frac{(\mathcal{A}T)^2}{2!} + \frac{(\mathcal{A}T)^3}{3!} + \cdots + \frac{(\mathcal{A}T)^i}{i!}.$$

By examining the finite series as more and more terms are added, it can be decided when to truncate the series. However, there are pathologic matrices for which the series converges slowly, for which the series seems to converge first to one matrix then to another, and for which numerical rounding can give misleading results. See "Numerical and Computational Issues in Linear Control" in this Handbook.

Continuing with the example, if the input $u(t)$ is constant in each interval from kT to $kT + T$, then

$$\begin{aligned} b &= [e^{\mathcal{A}T} - I]\mathcal{A}^{-1}\mathcal{B} \\ &= \begin{bmatrix} -0.304 & 0.246 \\ 0.123 & -0.428 \end{bmatrix}\begin{bmatrix} -\frac{3}{4} & -\frac{1}{2} \\ -\frac{1}{4} & -\frac{1}{2} \end{bmatrix}\begin{bmatrix} -1 \\ 5 \end{bmatrix} \\ &= \begin{bmatrix} -0.021 \\ 0.747 \end{bmatrix} \end{aligned}$$

which could also have been found using a truncated series in Equation 13.13. The discrete-time model of the continuous-time system is then

$$\begin{bmatrix} x_1(kT+T) \\ x_2(kT+T) \end{bmatrix} = \begin{bmatrix} 0.696 & 0.246 \\ 0.123 & 0.572 \end{bmatrix}\begin{bmatrix} x_1(kT) \\ x_2(kT) \end{bmatrix} + \begin{bmatrix} -0.021 \\ 0.747 \end{bmatrix} u(kT)$$

$$y(kT) = [2 \quad -4]\begin{bmatrix} x_1(kT) \\ x_2(kT) \end{bmatrix} + 6u(kT) \quad (13.16)$$

13.3.2 Approximation Methods

Another method for finding discrete-time equivalents of continuous-time systems described by state variable equations is to integrate Equation 13.12 as follows:

$$x(t) = x(t_0) + \int_{t_0}^{t} [\mathcal{A}x(t) + \mathcal{B}u(t)]dt \quad (13.17)$$

For evenly spaced samples, at $t = kT$, $k = 0, 1, 2, \ldots$

$$x(kT+T) = x(kT) + \int_{kT}^{kT+T} [\mathcal{A}x(t) + \mathcal{B}u(t)]dt$$

Applying Euler's forward rectangular approximation of the integral,

$$x(kT+T) \cong x(kT) + [\mathcal{A}x(kT) + \mathcal{B}u(kT)]T$$

or

$$x(kT+T) \cong [I + \mathcal{A}T]x(kT) + \mathcal{B}Tu(kT)$$

For the previous example, Euler's forward rectangular rule gives

$$\begin{bmatrix} x_1(kT+T) \\ x_2(kT+T) \end{bmatrix} \cong \left\{\begin{bmatrix} 1 & 0 \\ 0 & 1 \end{bmatrix} + \begin{bmatrix} -2 & 2 \\ 1 & -3 \end{bmatrix}(0.2)\right\} \begin{bmatrix} x_1(kT) \\ x_2(kT) \end{bmatrix} + \begin{bmatrix} -1 \\ 5 \end{bmatrix}(0.2)u(kT)$$

or

$$\begin{bmatrix} x_1(kT+T) \\ x_2(kT+T) \end{bmatrix} \cong \begin{bmatrix} 0.6 & 0.4 \\ 0.2 & 0.4 \end{bmatrix}\begin{bmatrix} x_1(kT) \\ x_2(kT) \end{bmatrix} + \begin{bmatrix} -0.2 \\ 1 \end{bmatrix} u(kT)$$

and

$$y(kT) = [2 \quad -4]\begin{bmatrix} x_1(kT) \\ x_2(kT) \end{bmatrix} + 6u(kT)$$

which does not match well with the result given by Equation 13.16.

Reducing the sampling interval to $T = 0.01$ sec., Euler's approximation gives

$$\begin{bmatrix} x_1(kT+T) \\ x_2(kT+T) \end{bmatrix} \cong \begin{bmatrix} 0.98 & 0.02 \\ 0.01 & 0.97 \end{bmatrix}\begin{bmatrix} x_1(kT) \\ x_2(kT) \end{bmatrix} + \begin{bmatrix} -0.01 \\ 0.05 \end{bmatrix} u(kT)$$

$$y(kT) = [2 \quad -4]\begin{bmatrix} x_1(kT) \\ x_2(kT) \end{bmatrix} + 6u(kT)$$

Taking the first two terms in the series in Equations 13.13 and 13.14 results in

$$\begin{bmatrix} x_1(kT+T) \\ x_2(kT+T) \end{bmatrix} \cong \begin{bmatrix} 0.9803 & 0.0195 \\ 0.0098 & 0.9706 \end{bmatrix}\begin{bmatrix} x_1(kT) \\ x_2(kT) \end{bmatrix} + \begin{bmatrix} -0.0094 \\ 0.0492 \end{bmatrix} + u(kT)$$

which is in close agreement with Euler's result.

Euler's backward approximation of the integral in Equation 13.17 gives

$$x(kT+T) \cong x(kT) + [\mathcal{A}x(kT+T) + \mathcal{B}u(kT+T)]T$$

or

$$[I-\mathcal{A}T]x(kT+T) \cong x(kT) + \mathcal{B}Tu(kT+T)$$

Hence

$$x(kT+T) \cong [I-\mathcal{A}T]^{-1}x(kT) + [I-\mathcal{A}T]^{-1}\mathcal{B}Tu(kT+T)$$

Letting

$$\bar{x}(kT+T) = x(kT)$$

then

$$[I-\mathcal{A}T]x(kT+T) \cong \bar{x}(kT+T) + \mathcal{B}Tu(kT+T)$$

or

$$[I-\mathcal{A}T]x(kT) \cong \bar{x}(kT) + \mathcal{B}Tu(kT)$$

Hence

$$x(kT) \cong [I-\mathcal{A}T]^{-1}\bar{x}(kT) + [I-\mathcal{A}T]^{-1}\mathcal{B}Tu(kT)$$

and, therefore,

$$\bar{x}(kT+T) \cong [I-\mathcal{A}T]^{-1}\bar{x}(kT) + [I-\mathcal{A}T]^{-1}\mathcal{B}Tu(kT)$$

The output equation,

$$y(kT) = Cx(kT) + Du(kT)$$

in terms of the new variable $\bar{x}$, becomes

$$y(kT) \cong C[I-\mathcal{A}T]^{-1}\bar{x}(kT) + C[I-\mathcal{A}T]^{-1}\mathcal{B}Tu(kT) + Du(kT)$$

or

$$y(kT) \cong C[I-\mathcal{A}T]^{-1}\bar{x}(kT) + \left\{C[I-\mathcal{A}T]^{-1}\mathcal{B}T + D\right\}u(kT)$$

Some formulas for discretizing continuous-time state variable equations using numerical integration are listed in Table 13.3. Derivative approximations, such as those listed in Table 13.2 are also possibilities for discretizing continuous-time state equations. Improved accuracy and a reduced sampling interval may also result from using more involved approximations, such as predictor-corrector or Runge-Kutta methods.

One of the most important control system design tools is simulation, computer modeling of the plant and controller to verify the properties of a preliminary design and to test its performance under conditions (e.g., noise, disturbances, parameter variations, and nonlinearities) that might be difficult or cumbersome to study analytically. Through simulation, difficulties with between-sample plant response are discovered and solved.

When a continuous-time plant is simulated on a digital computer, its response is computed at closely spaced discrete times. It is plotted by joining the closely spaced calculated response values with straight line segments approximating a continuous curve. Digital computer simulation of discrete-time control of a continuous-time system involves at least two sets of discrete-time calculations. One runs at high rate to simulate the continuous-time plant. The other runs at a lower rate (say once every 10 to 50 of the former calculations) to generate new control signals at each discrete control step.

TABLE 13.3 Some Formulas for Discretizing Continuous-Time State Variable Models

For the nth order continuous-time plant described by

$$\begin{aligned} \dot{x}(t) &= \mathcal{A}x(t) + \mathcal{B}u(t) \\ y(t) &= Cx(t) + Du(t) \end{aligned}$$

its discrete-time equivalence is given by

1. Zero-Order Hold Method

$$\begin{aligned} x[(k+1)T] &= Ax(kT) + Bu(kT) \\ y(kT) &= Cx(kT) + Du(kT) \end{aligned}$$

where

$$\begin{aligned} A &= e^{\mathcal{A}T} = I + \frac{\mathcal{A}T}{1!} + \frac{\mathcal{A}^2T^2}{2!} + \cdots + \frac{\mathcal{A}^iT^i}{i!} + \cdots \\ B &= \left[IT + \frac{\mathcal{A}T^2}{2!} + \frac{\mathcal{A}^2T^3}{3!} + \cdots + \frac{\mathcal{A}^iT^{i+1}}{(i+1)!} + \cdots\right]\mathcal{B} \end{aligned}$$

or

$$B = \mathcal{A}^{-1}[\exp(\mathcal{A}T) - I]\mathcal{B} = [\exp(\mathcal{A}T) - I]\mathcal{A}^{-1}\mathcal{B}$$

when $\mathcal{A}$ is nonsingular.

2. Euler's Forward Rectangular Method

$$\begin{aligned} x(kT+T) &\cong x(kT) + [\mathcal{A}x(kT) + \mathcal{B}u(kT)]T \\ y(kT) &= Cx(kT) + Du(kT) \end{aligned}$$

3. Euler's Backward Rectangular Method

$$\begin{aligned} \bar{x}(kT+T) &\cong [I - \mathcal{A}T]^{-1}\bar{x}(kT) + [I - \mathcal{A}T]^{-1}\mathcal{B}Tu(kT) \\ y(kT) &\cong C[I - \mathcal{A}T]^{-1}\bar{x}(kT) + \left\{C[I - \mathcal{A}T]^{-1}\mathcal{B}T + D\right\}u(kT) \end{aligned}$$

where

$$x(kT) = \bar{x}(kT+T)$$

4. Trapezoidal Method

$$\begin{aligned} \bar{x}(kT+T) &= (I - \mathcal{A}T/2)^{-1}(I + \mathcal{A}T/2)\bar{x}(kT) \\ &\quad + \frac{T}{2}(I - \mathcal{A}T/2)^{-1}\mathcal{B}u(kT-T) + \frac{T}{2}(I - \mathcal{A}T/2)^{-1}\mathcal{B}u(kT) \\ y(kT) &= C(I - \mathcal{A}T/2)^{-1}(I + \mathcal{A}T/2)\bar{x}(kT) \\ &\quad + \frac{T}{2}C(I - \mathcal{A}T/2)^{-1}\mathcal{B}u(kT-T) + \left[\frac{T}{2}C(I - \mathcal{A}T/2)^{-1}\mathcal{B} + D\right]u(kT) \end{aligned}$$

where

$$x(kT) = \bar{x}(kT+T)$$

References

[1] Santina, M.S., Stubberud, A.R., and Hostetter, G.H., *Digital Control System Design,* 2nd ed., Saunders College Publishing, Philadelphia, 1994.

[2] DiStefano III, J.J., Stubberud, A.R., and Williams, I.J., *Feedback and Control Systems (Schaum's Outline),* 2nd ed., McGraw-Hill, New York, 1990.

[3] Kuo, B.C., *Digital Control Systems,* 2nd ed., Saunders Publishing, Philadelphia, 1992.

[4] Franklin, G.F., Powell, J.D., and Workman, M.L., *Digital Control of Dynamic Systems,* 2nd ed., Addison-Wesley, Reading, MA, 1990.

[5] Phillips, C.L. and Nagle, Jr., H.T., *Digital Control System Analysis and Design,* Prentice Hall, Englewood Cliffs, NJ, 1990.

[6] Ogata, K., *Discrete-Time Control Systems,* Prentice Hall, Englewood Cliffs, NJ, 1987.

[7] Åström, K.J. and Wittenmark, B., *Computer Controlled Systems,* Prentice Hall, Englewood Cliffs, NJ, 1987.

[8] VanLandingham, H.F., *Introduction to Digital Control Systems,* Macmillan, New York, 1985.

[9] Houpis, C.H. and Lamont, G.B., *Digital Control Systems: Theory, Hardware, Software,* 2nd ed., McGraw-Hill, New York, 1985.

[10] Katz, P., *Digital Control Using Microprocessors,* Prentice Hall, Englewood Cliffs, NJ, 1987.

14

Design Methods for Discrete-Time, Linear Time-Invariant Systems

Mohammed S. Santina
The Aerospace Corporation, Los Angeles, CA

Allen R. Stubberud
University of California, Irvine, Irvine, CA

Gene H. Hostetter

14.1 An Overview

The starting point of most beginning studies of classical and state space control is with control of a linear, single-input/single-output, time-invariant plant. The tools of classical discrete-time linear control system design, which parallel the tools for continuous-time systems, are the z-transform, stability testing, root locus and frequency response methods.

As in the classical approach to designing analog controllers, one begins with simple digital controllers, increasing their complexity until the performance requirements are met. The digital controller parameters are chosen to give feedback system pole locations that result in acceptable zero-input (transient) response. At the same time, requirements are placed on the overall system's zero-state response components for representative discrete-time reference inputs, such as steps or ramps.

Extending classical single-input/single-output control system design methods to the design of complicated feedback structures involving many loops, each including a compensator, is not easy. Put another way, modern control systems require the design of compensators having multiple inputs and multiple outputs. Design *is* iterative, involving considerable trial and error. Therefore with many design variables, it is important to deal efficiently with those design decisions that need not be iterative. The powerful methods of state space offer insights about what is possible and what is not. They also provide an excellent framework for general methods of approaching and accomplishing design objectives.

A *tracking system* is one in which the plant outputs are controlled so that they become and remain nearly equal to externally applied *reference inputs.* A special case of a tracking system in which the desired tracking system output is zero is termed a *regulator.* In general, tracking control system design has two basic concerns:

1. obtaining acceptable zero-input response
2. obtaining acceptable zero-state system response to reference inputs

In addition, if the plant is continuous-time, and the controller is discrete-time, it is necessary to

3. obtain acceptable between-sample response of the continuous-time plant

Through superposition, the zero-input response due to initial conditions and the individual zero-state response contributions of each input can be dealt with separately. The character of a system's zero-input response is determined by its pole locations, so that the first concern of tracking system design is met by selecting a controller that places all of the overall system poles in

0-8493-8570-9/96/$0.00+$.50

acceptable locations. Having designed a feedback structure placing all of the overall system poles to achieve the desired character of zero-input response, additional design freedom can then be used to obtain good tracking of reference inputs.

The above three concerns of digital tracking system design are the subject of this section.

14.2 Classical Control System Design Methods

A typical classical control system design problem is to determine and specify the transfer function $G_c(z)$ of a cascade compensator that results in a feedback tracking system with prescribed performance requirements. This is only a part of complete control system design, of course. It applies after a suitable model has been found and the performance requirements are quantified. In describing solution methods for idealized problems such as these, we separate general design principles from the highly specialized details of a particular application.

The basic system configuration for this problem is shown in Figure 14.1. There are, of course, many variations on this basic theme, including situations where the system structure is more involved, where there is a feedback transmittance $H(z)$, and where there are disturbance inputs to be considered. Usually, these disturbances are undesirable inputs that the plant *should not* track.

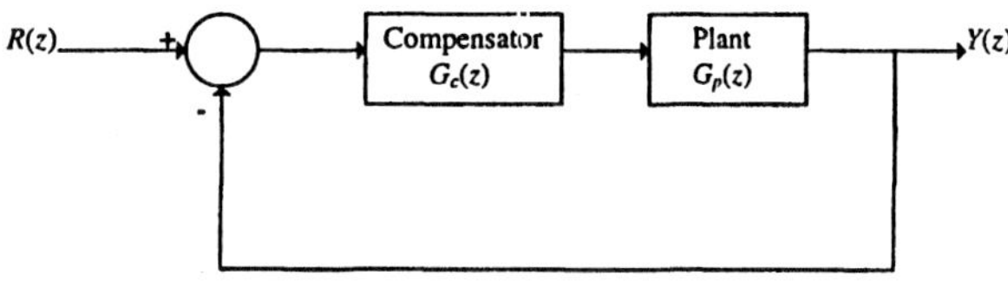

Figure 14.1 Cascade compensation of a unity feedback system.

The first concern of tracking system design is met by choosing the compensator $G_c(z)$ that results in acceptable pole locations for the overall transfer function

$$T(z) = \frac{Y(z)}{R(z)} = \frac{G_c(z)G_p(z)}{1 + G_c(z)G_p(z)}$$

Root locus is an important design tool because, with it, the effects on closed-loop system pole location of varying design parameters are quickly and easily visualized.

The second concern of tracking system design is obtaining acceptable closed-loop zero-state response to reference inputs. Zero-state performance is simple to deal with if it can be expressed as a maximum steady-state error to a power-of-time input. For the discrete-time system shown in Figure 14.2, the open-loop transfer function may have the form

$$\begin{aligned} KG(z)H(z) &= \frac{K(z+\alpha_1)(z+\alpha_2)\cdots(z+\alpha_l)}{(z-1)^n(z+\beta_1)(z+\beta_2)\cdots(z+\beta_m)} \\ &= \frac{KN(z)}{(z-1)^n D(z)} \end{aligned} \tag{14.1}$$

If n is nonnegative the system is said to be *type n*.

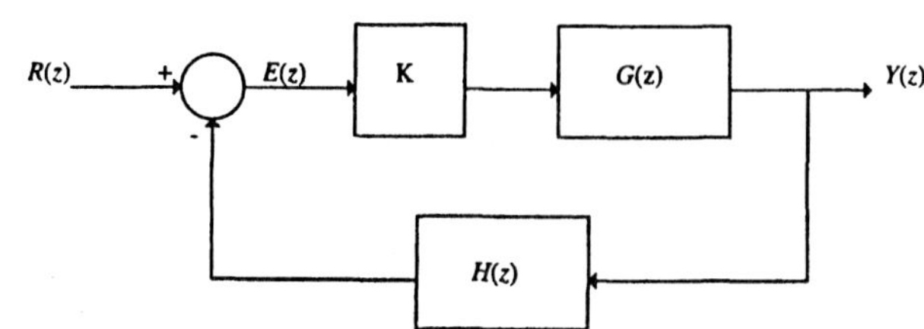

Figure 14.2 A discrete-time control system.

The error between the input and the output of the system is

$$E(z) = R(z) - Y(z)H(z)$$

but

$$Y(z) = KE(z)G(z)$$

Then

$$T_E(z) = \frac{E(z)}{R(z)} = \frac{1}{1 + KG(z)H(z)}.$$

The steady-state error to a power-of-time input is given by the final value theorem

$$\begin{aligned} \lim_{k\to\infty} e(k) &= \lim_{z\to 1}(1 - z^{-1})E(z) \\ &= \lim_{z\to 1}\frac{(1 - z^{-1})R(z)}{1 + KG(z)H(z)} \end{aligned} \tag{14.2}$$

provided that the limit exists. A necessary condition for the limit to exist and be finite is that all the closed-loop poles of the system be inside the unit circle on the z-plane.

Similar to continuous-time systems, there are three reference inputs for which steady-state errors are commonly defined. They are the step (position), ramp (velocity), and parabolic (acceleration) inputs. The step input has the form

$$r(kT) = Au(kT)$$

or, in the z-domain,

$$R(z) = \frac{Az}{z-1}$$

and the ramp input is

$$r(kT) = AkTu(kT)$$

or

$$R(z) = \frac{ATz}{(z-1)^2}.$$

The parabolic input is

$$r(kT) = \frac{1}{2}A(kT)^2u(kT)$$

or

$$R(z) = \frac{T^2}{2}\frac{Az(z+1)}{(z-1)^3}.$$

Table 14.1 summarizes steady-state errors using Equations 14.1 and 14.2 for various system types for power-of-time inputs.

We now present an overview of classical discrete-time control system design using an example.

TABLE 14.1 Steady State Errors to Power-of-Time Inputs

System type	Steady-state error to step input $R(z)=\frac{Az}{z-1}$	Steady-state error to ramp input $R(z)=\frac{ATz}{(z-1)^2}$	Steady-state error to parabolic input $R(z)=\frac{T^2}{2}\frac{Az(z+1)}{(z-1)^3}$
0	$\frac{A}{1+K\frac{N(1)}{D(1)}}$	∞	∞
1	0	$\frac{AT}{K\frac{N(1)}{D(1)}}$	∞
2	0	0	$\frac{AT^2}{K\frac{N(1)}{D(1)}}$
$\vdots$	$\vdots$	$\vdots$	$\vdots$
n	0	0	0

14.2.1 Root Locus Design Methods

Similar to continuous-time systems, the root locus plot of a discrete-time system consists of the loci of the poles of a transfer function as some parameter is varied. For the configuration shown in Figure 14.2 where the constant gain K is the parameter of interest, the overall transfer function of this system is

$$T(z)=\frac{KG(z)}{1+KG(z)H(z)}$$

and the poles of the overall system are the roots of

$$1+KG(z)H(z)=0$$

which depend on the parameter K. The rules for constructing the root locus of discrete-time systems are identical to the rules for plotting the root locus of continuous-time systems (see Chapter 10.4). The root locus plot, however, must be interpreted relative to the z-plane.

Consider the block diagram of the commercial broadcast videotape-positioning system shown in Figure 14.3(a). The transfer function $G(s)$ relates the applied voltage to the drive motor armature and the tape speed at the recording and playback heads. The delay term accounts for the propagation of speed changes along the tape over the distance physically separating the tape drive mechanism and the recording and playback heads. The pole term in $G(s)$ represents the dynamics of the motor and tape drive capstan. Tape position is sensed by a recorded signal on the tape itself.

It is desired to design a digital controller that results in zero steady-state error to any step change in desired tape position. Also, the system should have a zero-input (or transient) response that decays to no more than 10% of any initial value within a 1/30 sec. interval, the video frame rate. The sampling period of the controller is chosen as $T=1/120$ sec. to synchronize the tape motion control with the 1/60 sec. field rate (each frame consists of two fields of the recorded video). In Figure 14.3(b) the diagram of Figure 14.3(a) has been rearranged to emphasize the discrete-time input $R(z)$ and the discrete-time samples $P(z)$ of the tape position.

The open-loop transfer function of the system is

$$\begin{aligned}
G_p(z) &= \underset{T=1/120}{\text{sample at}}\left[\left(\frac{1-e^{-(1/120)s}}{s}\right)\left(\frac{40e^{-(1/120)s}}{s+40}\right)\left(\frac{1}{s}\right)\right]\\
&= \underset{T=1/120}{\text{sample at}}\left\{[1-e^{-(1/120)s}]e^{-(1/120)s}\left[\frac{-(1/140)}{s}+\frac{1}{s^2}+\frac{1/40}{s+40}\right]\right\}\\
&= (1-z^{-1})z^{-1}\left[\frac{-z/40}{z-1}+\frac{z/120}{(z-1)^2}+\frac{z/40}{z-0.72}\right]\\
&= \frac{0.00133(z+0.75)}{z(z-1)(z-0.72)}
\end{aligned}$$

The position error signal, in terms of the compensator's z-transfer function $G_c(z)$, is

$$\begin{aligned}
E(z) &= R(z)-Y(z)\\
&= \left[1-\frac{G_c(z)G_p(z)}{1+G_c(z)G_p(z)}\right]R(z)\\
&= \frac{1}{1+G_c(z)G_p(z)}R(z)
\end{aligned}$$

For a unit step input sequence

$$E(z)=\frac{1}{1+G_c(z)G_p(z)}\left(\frac{z}{z-1}\right)$$

Assuming that the feedback system is stable

$$\lim_{k\to\infty}e(k)=\lim_{z\to 1}\left[\frac{z-1}{z}E(z)\right]=\lim_{z\to 1}\left[\frac{1}{1+G_c(z)G_p(z)}\right]$$

Provided that the compensator does not have a zero at $z=1$, the system type is 1 and, therefore according to Table 14.1, the steady-state error to a step input is zero. For the feedback system transient response to decay by at least a factor 1/10 within 1/30 sec., the desired closed-loop poles must be located so that a decay of at least this amount occurs every 1/120 sec. step. This implies that the closed-loop poles must lie within a radius c of the origin on the z-plane, where

$$c^4=0.1, \quad c=0.56$$

Similar to continuous-time systems, one usually begins with the simplest compensator consisting of only a gain K. The feedback system is stable for

$$0<K<95$$

but, as shown in Figure 14.4, this compensator is inadequate because there are always poles at distances from the origin greater than the required $c=0.56$ regardless of the value of K. As shown in Figure 14.5(a), another compensator with z-transfer function,

$$G_c(z)=\frac{K(z-0.72)}{z}$$

which cancels the plant pole at $z=0.72$, is tried. The root locus plot for this system is shown in Figure 14.5(b). For $K=90$, the design is close to meeting the requirements, but it is not quite

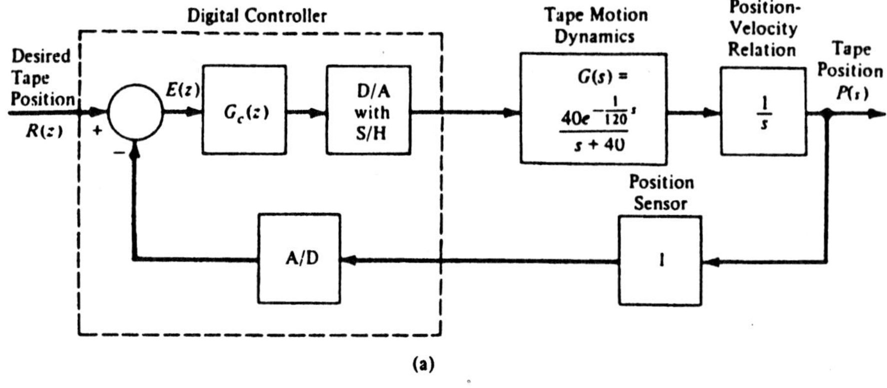

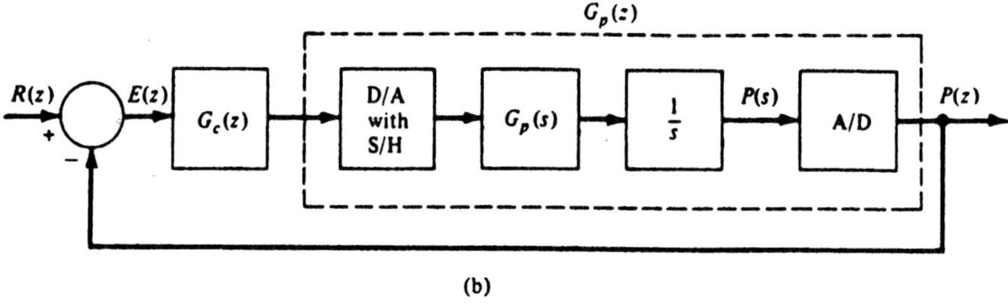

Figure 14.3 Videotape-positioning system. (a) block diagram. (b) relation between discrete-time signals.

good enough. However, if the compensator pole is moved from the origin to the left as shown in Figure 14.6, the root locus is pulled to the left and the performance requirements are met.

For the compensator with z-transfer function,

$$G_c(z) = \frac{150(z-0.72)}{z+0.4} \tag{14.3}$$

the feedback system z-transfer function is

$$\begin{aligned} T(z) &= \frac{G_c(z)G_p(z)}{1+G_c(z)G_p(z)} = \frac{0.2(z+0.75)}{z^3-0.6z^2-0.2z+0.15} \\ &= \frac{0.2(z+0.75)}{(z-0.539-j0.155)(z-0.539+j0.155)} \\ &\times \frac{1}{(z+0.477)} \end{aligned}$$

As expected, the steady-state error to a step input is zero,

$$\begin{aligned} &\lim_{z\to 1}\left\{\left(\frac{z-1}{z}\right)[1-T(z)]\left(\frac{z}{z-1}\right)\right\} \\ &= \lim_{z\to 1}\frac{z^3-0.6z^2-0.4z}{z^3-0.6z^2-0.2z+0.15} \\ &= 0 \end{aligned}$$

The steady state error to a unit ramp input is

$$\begin{aligned} &\lim_{z\to 1}\left\{\left(\frac{z-1}{z}\right)[1-T(z)]\left[\frac{Tz}{(z-1)^2}\right]\right\} \\ &= \lim_{z\to 1}\frac{\frac{1}{120}(z^2+0.4z)}{z^3-0.6z^2-0.2z+0.15} \\ &= \frac{1}{30} \end{aligned}$$

For a compensator with a z-transfer function of the form,

$$G_c(z) = \frac{150(z-0.72)}{z+a}$$

the feedback system has the z-transfer function,

$$\begin{aligned} T(z) &= \frac{G_c(z)G_p(z)}{1+G_c(z)G_p(z)} \\ &= \frac{0.2(z+0.75)}{(z+a)(z^2-z)+0.2(z+0.75)} \\ &= \frac{0.2(z+0.75)}{z^3-z^2+0.2z+0.15+a(z^2-z)} \\ &= \frac{0.2(z+0.75)/[z^3-z^2+0.2z+0.15]}{1+az(z-1)/[z^3-z^2+0.2z+0.15]} \\ &= \frac{\text{numerator}}{1+az(z-1)/[(z-0.637-j0.378)\cdot(z-0.637+j0.378)(z+0.274)]} \end{aligned}$$

A root locus plot in terms of positive a in Figure 14.7, shows that choices of a between 0.4 and 0.5 give a controller that meets the performance requirements.

Classical discrete-time control system design is an iterative process just like its continuous-time counterpart. Increasingly complicated controllers are tried until both the steady-state error and transient performance requirements are met. Root locus is an important tool because it easily indicates qualitative closed-loop system pole locations as a function of a parameter. Once feasible controllers are selected, root locus plots are refined to show quantitative results.

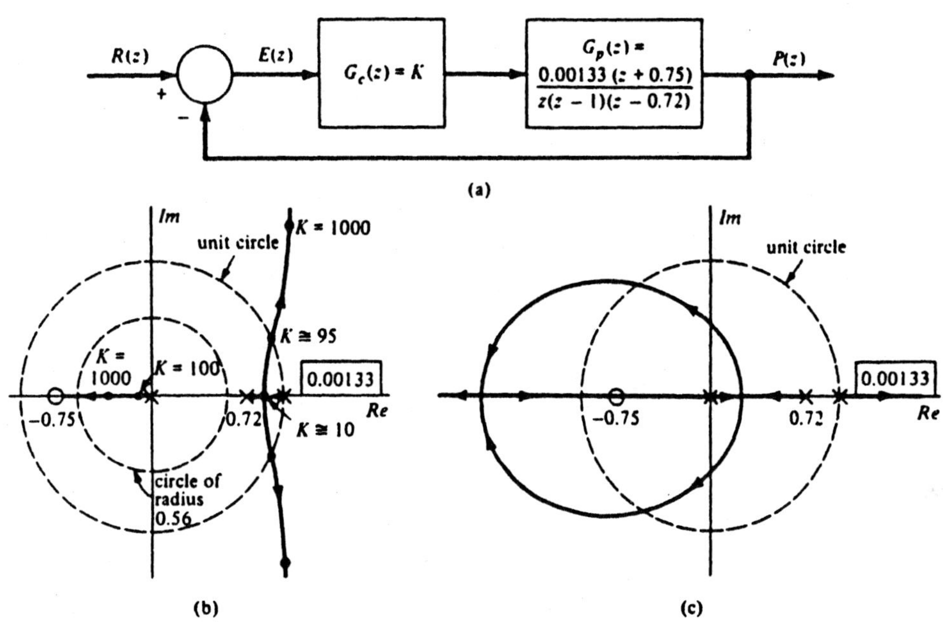

Figure 14.4 Constant-gain compensator. (a) block diagram. (b) root locus for positive K. (c) root locus for negative K.

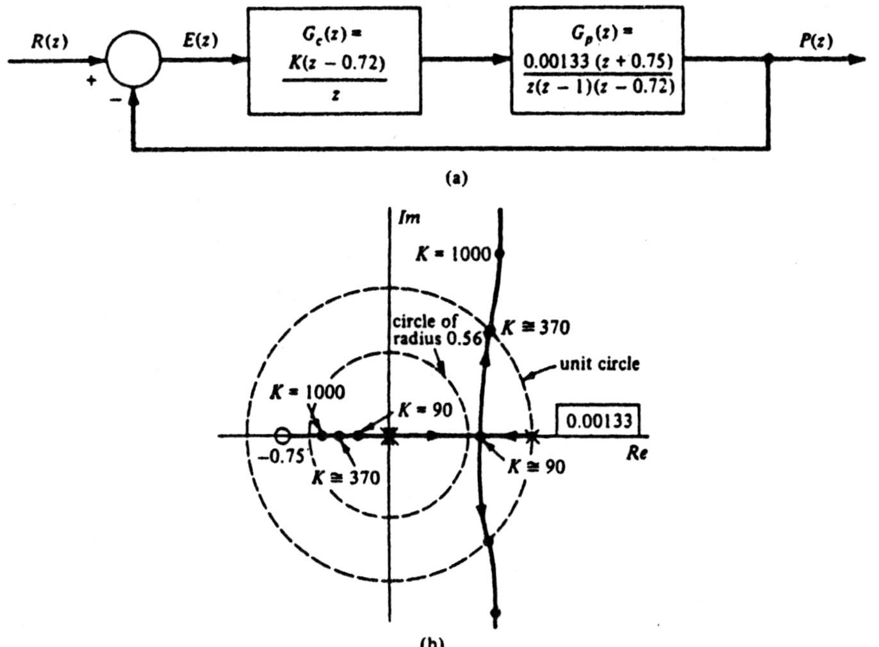

Figure 14.5 Compensator with zero at $z = 0.72$ and pole at $z = 0$. (a) block diagram. (b) root locus for positive K.

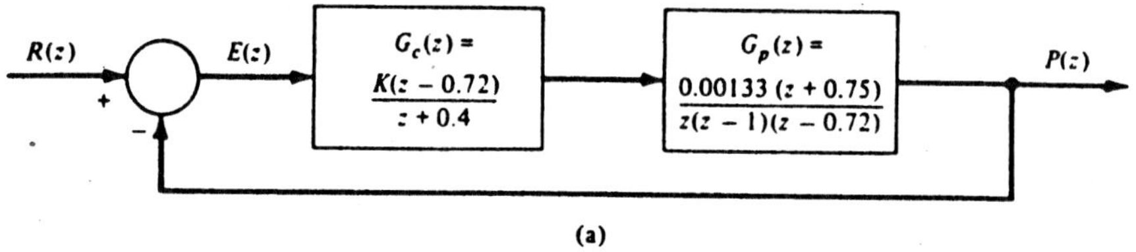

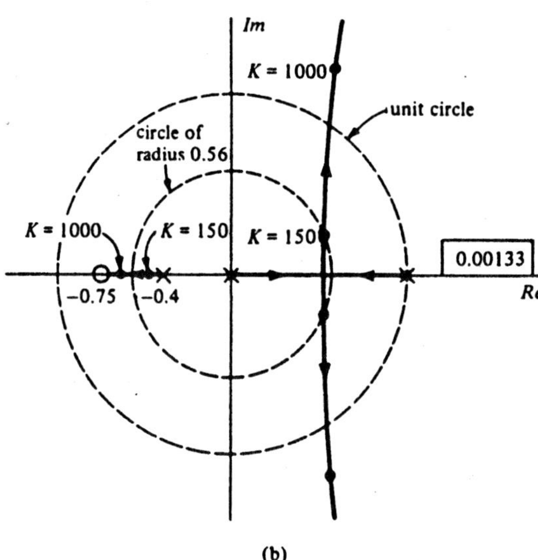

Figure 14.6 Compensator with zero at $z = 0.72$ and pole at $z = -0.4$. (a) block diagram. (b) root locus for positive K.

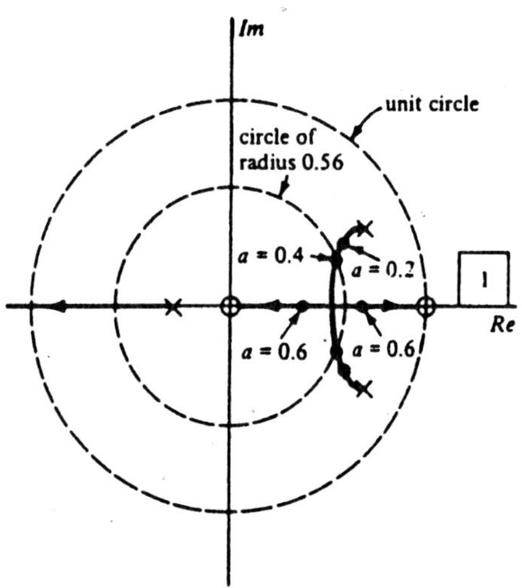

Figure 14.7 Root locus plot as a function of the compensator pole location.

14.2.2 Frequency Domain Methods

Frequency response characterizations of systems have long been popular because steady-state sinusoidal response methods are easy and practical. Furthermore, frequency response methods do not require explicit knowledge of system transfer function models.

For the videotape positioning system, the open loop z-transfer function which includes the compensator given by Equation 14.3 is

$$G_c(z)G_p(z) = \frac{(150)(0.00133)(z+0.75)}{z(z+0.4)(z-1)}$$

Substituting $z = e^{j\omega T}$, then

$$G_c(e^{j\omega T})G_p(e^{j\omega T}) = \frac{0.1995(e^{j\omega T}+0.75)}{e^{j\omega T}(e^{j\omega T}+0.4)(e^{j\omega T}-1)} \qquad (14.4)$$

which has frequency response plots shown in Figure 14.8. At the *phase crossover frequency* (114.2 rad/sec.), the *gain margin* is 11.48 dB, and, at the *gain crossover frequency* (30 rad/sec.), the *phase margin* is about 66.5 degrees.

For ease in generating frequency response plots and gaining greater insight into the design process, frequency domain methods such as Nyquist, Bode, Nichols, etc. for discrete-time systems are best developed with the *w-transform*. In the w-plane, the wealth of tools and techniques developed for continuous-time systems are directly applicable to discrete-time systems as well.

The w-transform is

$$w = \frac{z-1}{z+1}, \qquad z = \frac{w+1}{1-w}$$

which is a bilinear transformation between the w-plane and the z-plane.

The general procedure for analyzing and designing discrete-time systems with the w-transform is summarized as follows:

1. Apply the w-transform to the open-loop transfer function $G(z)H(z)$ by replacing each z in $G(z)H(z)$

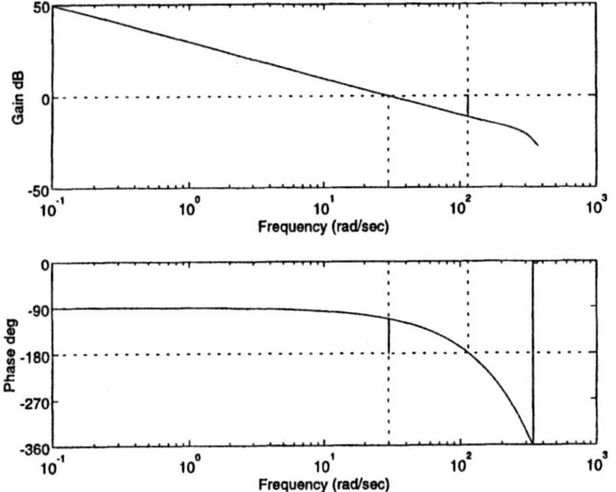

Figure 14.8 Frequency response plots of the videotape positioning system.

with

$$z = \frac{w+1}{1-w}$$

to obtain $G(w)H(w)$. Note that the functions G and H actually are different after the substitution.

2. Visualizing the w-plane as if it were the s-plane, substitute $w = j\nu$ into $G(w)H(w)$ and generate frequency response plots in terms of the real frequency ν, such as Nyquist, Bode, Nichols, etc.
3. Determine the gain margin, phase margin, crossover frequencies, bandwidth, closed-loop frequency response or any other desired frequency response characteristics.
4. If it is necessary, design an appropriate compensator $G_c(w)$ to satisfy the frequency domain performance requirements.
5. Convert critical frequencies ν in the w-plane to frequencies ω in the z-domain according to

$$\nu = \tan\frac{\omega T}{2}$$

or

$$\omega = \frac{2}{T}\tan^{-1}\nu.$$

6. Transform the controller $G_c(w)$ to $G_c(z)$ according to

$$w = \frac{z-1}{z+1}.$$

Control system design for discrete-time systems using Bode, Nyquist or Nichols methods can be found in [2] and [3]. Frequency response methods are most useful in developing models from experimental data, in verifying the performance of a system designed by other methods, and in dealing with those systems and situations in which rational transfer function models are not adequate.

14.3 Eigenvalue Placement with State Feedback

All of the results for eigenvalue placement with state feedback for continuous-time systems carry over to discrete-time systems. For a linear, step-invariant nth order system, described by the state equations

$$x(k+1) = Ax(k) + Bu(k),$$

consider the state feedback arrangement,

$$u(k) = Ex(k) + \rho(k),$$

where $\rho(k)$ is a vector of external inputs, as shown in Figure 14.9. Provided that the plant is completely controllable, and that the state is accessible for feedback,[1] the feedback gain matrix E can always be chosen so that each of the eigenvalues of the feedback system,

$$x(k+1) = (A + BE)x(k) + B\rho(k)$$

is at an arbitrary desired location selected by the designer. This is to say that the designer can freely choose the character of the overall system's transient performance.

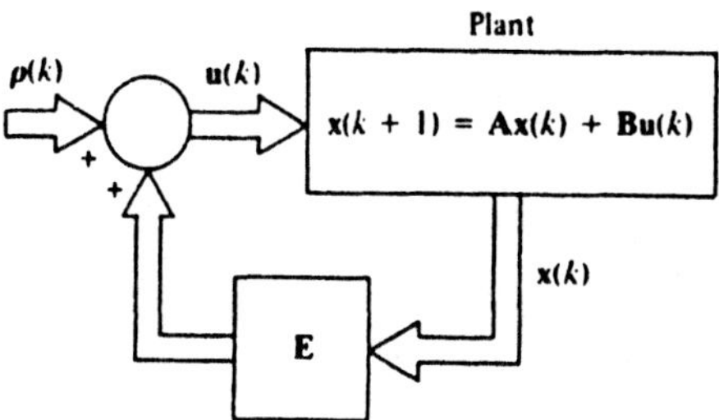

Figure 14.9 State feedback.

14.3.1 Eigenvalue Placement for Single-Input Systems

There are a number of methods for finding the state feedback gain vector of single-input plants, one summarized below, and additional ones can be found in [1] and [7] and in Chapters 5.2 and 38.

One method for calculating the state feedback gain vector is given by *Ackermann's formula:*

$$e^{\dagger} = -j_n^{\dagger} M_c^{-1} \Delta_c(A)$$

where $j_n^{\dagger}$ is the transpose of the nth-unit coordinate vector

$$j_n^{\dagger} = [0 \quad 0 \quad \cdots \quad 0 \quad 1],$$

[1]When the plant state vector is not available for feedback, as is usually the case, an *observer* is designed to estimate the state vector. The observer state estimate is used for feedback in place of the state itself.

M_c is the controllability matrix of the system, and $\Delta_c(A)$ is the desired characteristic equation with the matrix A substituted for the variable z.

For example, for the completely controllable system,

$$\begin{bmatrix} x_1(k+1) \\ x_2(k+1) \end{bmatrix} = \begin{bmatrix} 1 & -1 \\ 3 & 0 \end{bmatrix} \begin{bmatrix} x_1(k) \\ x_2(k) \end{bmatrix} + \begin{bmatrix} 1 \\ 2 \end{bmatrix} u(k),$$

it is desired to place the feedback system eigenvalues at $z = 0, -0.5$. Then,

$$\Delta_c(z) = z(z+0.5) = z^2 + 0.5z$$

and

$$\Delta_c(A) = A^2 + 0.5A.$$

Using Ackermann's formula, the state feedback gain vector is

$$\begin{aligned} e^\dagger &= -[0 \quad 1] \begin{bmatrix} 1 & -1 \\ 2 & 3 \end{bmatrix}^{-1} \left\{ \begin{bmatrix} -2 & -1 \\ 3 & -3 \end{bmatrix} \right. \\ &\quad \left. + 0.5 \begin{bmatrix} 1 & -1 \\ 3 & 0 \end{bmatrix} \right\} \\ &= [-1.5 \quad 0]. \end{aligned}$$

14.3.2 Eigenvalue Placement with Multiple-Inputs

If the plant has multiple inputs and if it is completely controllable from one of the inputs, then that one input alone can be used for feedback. If the plant is not completely controllable from a single input, a single input can usually be distributed to the multiple ones so that the plant is completely controllable from the single input.

For example, for the system

$$\begin{bmatrix} x_1(k+1) \\ x_2(k+1) \\ x_3(k+1) \end{bmatrix} = \begin{bmatrix} -0.5 & 0 & 1 \\ 0 & 0.5 & 2 \\ 1 & -1 & 0 \end{bmatrix} \begin{bmatrix} x_1(k) \\ x_2(k) \\ x_3(k) \end{bmatrix} + \begin{bmatrix} 1 & 0 \\ 0 & -2 \\ -1 & 1 \end{bmatrix} \begin{bmatrix} u_1(k) \\ u_2(k) \end{bmatrix}.$$

Letting

$$u_1(k) = 3\mu(k)$$

and

$$u_2(k) = \mu(k)$$

then

$$\begin{bmatrix} x_1(k+1) \\ x_2(k+1) \\ x_3(k+1) \end{bmatrix} = \begin{bmatrix} -0.5 & 0 & 1 \\ 0 & 0.5 & 2 \\ 1 & -1 & 0 \end{bmatrix} \begin{bmatrix} x_1(k) \\ x_2(k) \\ x_3(k) \end{bmatrix} + \begin{bmatrix} 3 \\ -2 \\ -2 \end{bmatrix} \mu(k)$$

which is a controllable single input system. If the desired eigenvalues are located at $z_1 = -0.1$, $z_2 = -0.15$, and $z_3 = 0.1$, Ackermann's formula gives

$$e^\dagger = [0.152 \quad 0.0223 \quad 0.2807]$$

and hence, the feedback gain matrix for the multiple input system is

$$E = \begin{bmatrix} 0.4559 & 0.0669 & 0.8420 \\ 0.1520 & 0.0223 & 0.2807 \end{bmatrix}$$

The structure of this system is shown in Figure 14.10.

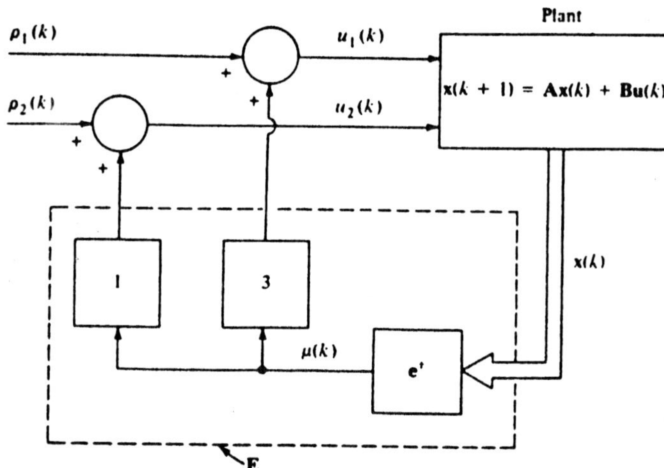

Figure 14.10 State feedback to a plant with multiple inputs.

14.3.3 Eigenvalue Placement with Output Feedback

It is the measurement vector of a plant, not the state vector, that is available for feedback. For the nth order plant with state and output equations

$$\begin{aligned} x(k+1) &= Ax(k) + Bu(k) \\ y(k) &= Cx(k) + Du(k) \end{aligned}$$

if the output coupling matrix C has n linearly independent rows, then the plant state can be recovered from the plant inputs and the measurement outputs and the method of the previous section applied:

$$x(k) = C^{-1}\{y(k) - Du(k)\}$$

When the nth order plant does not have n linearly independent measurement outputs, it still might be possible to select a feedback matrix E in

$$\begin{aligned} u(k) &= E\{y(k) - Du(k)\} + \rho(k) \\ &= ECx(k) + \rho(k) \end{aligned}$$

to place all of the feedback system eigenvalues, those of $(A + BEC)$, acceptably. Generally, however, measurement feedback alone is insufficient for arbitrary eigenvalue placement.

14.3.4 Pole Placement with Feedback Compensation

Similar to output feedback, pole placement with feedback compensation assumes that the measurement outputs of a plant, not the state vector, are available for feedback.

Consider the single-input/single-output, nth order, linear, step-invariant, discrete-time system described by the transfer function $G(z)$. Arbitrary pole placement of the feedback system can be accomplished with an mth order feedback compensator as shown in Figure 14.11.

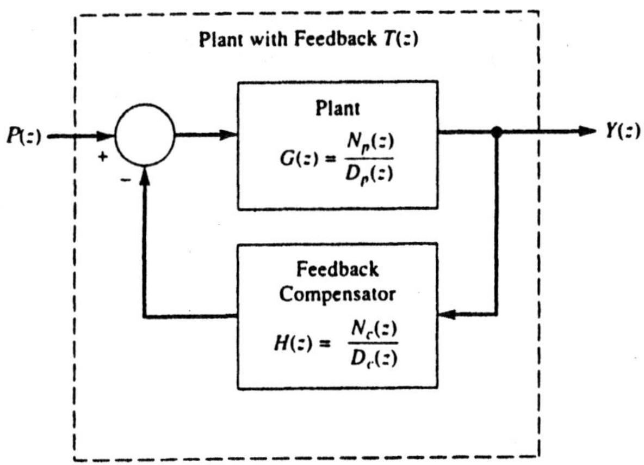

Figure 14.11 Pole placement with feedback compensation.

Let the numerator and denominator polynomials of $G(z)$ be $N_p(z)$ and $D_p(z)$, respectively. Also, let the numerator and denominator of the compensator transfer function $H(z)$ be $N_c(z)$, and $D_c(z)$, respectively. Then, the overall transfer function of the system is

$$\begin{aligned} T(z) &= \frac{G(z)}{1 + G(z)H(z)} \\ &= \frac{N_p(z)/D_p(z)}{1 + [N_p(z)/D_p(z)][N_c(z)/D_c(z)]} \\ &= \frac{N_p(z)D_c(z)}{D_p(z)D_c(z) + N_p(z)N_c(z)} = \frac{P(z)}{Q(z)} \end{aligned}$$

which has closed-loop zeros in $P(z)$ that are those of the plant, in $N_p(z)$, together with zeros that are the poles of the feedback compensator, in $D_c(z)$.

For a desired set of poles of $T(z)$, given with an unknown multiplicative constant by the polynomial $Q(z)$,

$$D_p(z)D_c(z) + N_p(z)N_c(z) = Q(z)$$

The desired polynomial $Q(z)$ has the form

$$Q(z) = \alpha_0(z^{n+m} + b_{n+m-1}z^{n+m-1} + \cdots + \beta_1 z + \beta_0)$$

where the βs are known coefficients, but the α_0 is unknown. In general, for a solution to exist, there must be at least as many unknowns as equations:

$$n + m + 1 \leq 2m + 2$$

or

$$m \geq n - 1 \qquad (14.5)$$

where n is the order of the plant and m is the order of the compensator. Equation 14.5 states that the order of the feedback controller is at least one less than the plant order. If the plant transfer function has *coprime* numerator and denominator polynomials (that is, plant pole-zero cancellations have been made), then a solution is guaranteed to exist.

For example, consider the second-order plant

$$G(z) = \frac{(z+1)(z+0.5)}{z(z-1)} = \frac{N_p(z)}{D_p(z)} \qquad (14.6)$$

According to Equation 14.5, a first-order feedback compensator of the form,

$$H(z) = \frac{\alpha_1 z + \alpha_2}{z + \alpha_3} = \frac{N_c(z)}{D_c(z)},$$

places the three closed-loop poles of the feedback system at any desired location in the z-plane by appropriate choice of α_1, α_2, and α_3. Let the desired poles of the plant with feedback be at $z = 0.1$. Then,

$$Q(z) = \alpha_0(z-0.1)^3 = \alpha_0(z^3 - 0.3z^2 + 0.03z - 0.001). \quad (14.7)$$

In terms of the compensator coefficients, the characteristic equation of the feedback system is

$$\begin{aligned} & D_p(z)D_c(z) + N_p(z)N_c(z) \\ = \; & z(z-1)(z+\alpha_3) + (z+1)(z+0.5)(\alpha_1 z + \alpha_2), \\ = \; & (\alpha_1 + 1)z^3 + (1.5\alpha_1 + \alpha_2 + \alpha_3 - 1)z^2 \\ & + (0.5\alpha_1 + 1.5\alpha_2 - \alpha_3)z + 0.5\alpha_2. \end{aligned} \qquad (14.8)$$

Equating coefficients in Equations 14.7 and 14.8 and solving for the unknowns,

$$\alpha_0 = 1.325 \quad \alpha_1 = 0.325 \quad \alpha_2 = -0.00265 \quad \alpha_3 = 0.1185$$

Therefore, the compensator

$$H(z) = \frac{0.325z - 0.00265}{z + 0.1185}$$

will place the closed-loop poles where desired.

As far as feedback system pole placement is concerned, a feedback compensator of order $n - 1$, where n is the order of the plant, can always be designed. It is possible, however, that a lower order feedback controller may give acceptable feedback pole locations even though those locations are constrained and not completely arbitrary. This is the thrust of classical control system design, in which increasingly higher order controllers are tested until satisfactory results are obtained.

For the plant given by Equation 14.6, for example, a *zero*th order feedback controller of the form

$$H(z) = K$$

gives overall closed-loop poles at $z = 0.1428$ and $z = 0.5$ for $K = 1/6$ which might be an adequate pole placement design.

14.4 Step-Invariant Discrete-Time Observer Design

When the plant state vector is not entirely accessible, as is usually the case, the state is *estimated* with an observer, and the estimated state is used in place of the actual state for feedback. See the article entitled "Observers".

14.4.1 Full-Order Observers

A full-order state observer of an nth order step-invariant discrete-time plant,

$$\begin{aligned} x(k+1) &= Ax(k) + Bu(k) \\ y(k) &= Cx(k) + Du(k) \end{aligned} \tag{14.9}$$

is another nth order system of the form,

$$\xi(k+1) = F\xi(k) + Gy(k) + Hu(k) \tag{14.10}$$

driven by the inputs and outputs of the plant so that the error between the plant state and the observer state,

$$\begin{aligned} & x(k+1) - \xi(k+1) \\ &= Ax(k) + Bu(k) - F\xi(k) - Gy(k) - Hu(k) \\ &= Ax(k) + Bu(k) - F\xi(k) - GCx(k) \\ &\quad - GDu(k) - Hu(k) \\ &= (A - GC)x(k) - F\xi(k) + (B - GD - H)u(k) \end{aligned}$$

is governed by an autonomous equation. This requires that

$$F = A - GC \tag{14.11}$$

$$H = B - GD \tag{14.12}$$

so that the error satisfies

$$x(k+1) - \xi(k+1) = (A - GC)[x(k) - \xi(k)]$$

or

$$\begin{aligned} x(k) - \xi(k) &= (A - GC)^k [x(0) - \xi(0)] \\ &= F^k [x(0) - \xi(0)] \end{aligned}$$

The eigenvalues of $F = A - GC$ can be placed arbitrarily by the choice of G, provided that the system is completely observable. The observer error, then, approaches zero with step regardless of the initial values of $x(0)$ and $\xi(0)$, that is, the observer state $\xi(k)$ will approach the plant state $x(k)$. The full-order observer relations are summarized in Table 14.2. If all n of the observer eigenvalues (eigenvalues of F) are selected to be zero, then the characteristic equation of F is

$$\lambda^n = 0$$

and, because every matrix satisfies its own characteristic equation, then

$$F^n = 0.$$

At the nth step, the error between the plant state and the observer state is

$$x(n) - \xi(n) = F^n[x(0) - \xi(0)]$$

so that

$$x(n) = \xi(n)$$

and the observer state equals the plant state. Such an observer is termed *deadbeat*. In subsequent steps, the observer state continues to equal the plant state.

TABLE 14.2 Full-Order State Observer Relations

Plant model
$x(k+1) = Ax(k) + Bu(k)$
$y(k) = Cx(k) + Du(k)$
Observer
$\xi(k+1) = F\xi(k) + Gy(k) + Hu(k)$
where
$F = A - GC$
$H = B - GD$
Observer error
$x(k+1) - \xi(k+1) = F[x(k) - \xi(k)]$
$x(k) - \xi(k) = F^k[x(0) - \xi(0)]$

There are a number of ways for calculating the observer gain matrix g for single-output plants. Similar to the situation with state feedback, the eigenvalues of $F = A - gc^\dagger$ can be placed arbitrarily by choice of g as given by Ackermann's formula:

$$g = \Delta_0(A) M_0^{-1} j_n \tag{14.13}$$

provided that $(A, c^\dagger)$ is completely observable. In Equation 14.13, $\Delta_0(A)$ is the desired characteristic equation of the observer eigenvalues with the matrix A substituted for the variable z, M_0 is the observability matrix

$$M_0 = \begin{bmatrix} c^\dagger \\ \hline c^\dagger A \\ \hline c^\dagger A^2 \\ \hline \vdots \\ c^\dagger A^{n-1} \end{bmatrix}$$

and j_n is the nth-unit coordinate vector

$$j_n = \begin{bmatrix} 0 \\ 0 \\ 0 \\ \vdots \\ 1 \end{bmatrix}$$

It is enlightening to express the full-order observer equations given by Equations 14.10, 14.11, and 14.12 in the form

$$\begin{aligned} \xi(k+1) &= (A - GC)\xi(k) + Gy(k) + (B - GD)u(k) \\ &= A\xi(k) + Bu(k) + G[y(k) - w(k)] \end{aligned}$$

where

$$w(k) = C\xi(k) + Du(k)$$

The observer consists of a model of the plant driven by the input $u(k)$ and the error between the plant output $y(k)$ and the plant output that is estimated by the model $w(k)$.

14.4.2 Reduced-Order State Observers

Rather than estimating the entire state vector of a plant, if a completely observable nth order plant has m linearly independent outputs, a *reduced-order* state observer, of order $n - m$, having an output that observes the plant state can be constructed. See the article entitled "Observers".

When an observer's state,

$$\xi(k+1) = F\xi(k) + Gy(k) + Hu(k)$$

estimates a linear combination $Mx(k)$ of the plant state rather than the state itself, the error between the observer state and the plant state transformation is given by

$$\begin{aligned} &Mx(k+1) - \xi(k+1) \\ &= MAx(k) + MBu(k) - F\xi(k) - Gy(k) \\ &\quad - Hu(k) \\ &= (MA - GC)x(k) - F\xi(k) + (MB - GD - H)u(k) \end{aligned}$$

where M is $(n - m)xn$. For the observer error system to be autonomous,

$$\begin{aligned} FM &= MA - GC \\ H &= MB - GD \end{aligned} \tag{14.14}$$

so that the error is governed by

$$Mx(k+1) - \xi(k+1) = F[Mx(k) - \xi(k)]$$

For a completely observable plant, the observer gain matrix g can always be chosen so that all of the eigenvalues of F are inside the unit circle on the complex plane. Then the observer error

$$Mx(k) - \xi(k) = F^k[Mx(0) - \xi(0)]$$

will approach zero asymptotically with step, and then

$$\xi(k) \to Mx(k)$$

If the plant outputs, which also involve linear transformation of the plant state, are used in the formulation of a state observer, the dynamic order of the observer can be reduced. For the nth order plant given by Equation 14.9 with the m rows of C linearly independent, an observer of order $n - m$ with n outputs,

$$w'(k) = \begin{bmatrix} 0 \\ I \end{bmatrix} \xi(k) + \begin{bmatrix} I \\ 0 \end{bmatrix} y(k) - \begin{bmatrix} D \\ 0 \end{bmatrix} u(k)$$

observes

$$w'(k) \to \begin{bmatrix} C \\ M \end{bmatrix} x(k) = Nx(k)$$

Except in special cases, the rows of M and the rows of C are linearly independent. If they are not so, slightly different observer eigenvalues can be chosen to give linear independence. Therefore,

$$w(k) = N^{-1}w'(k)$$

observes $x(k)$.

14.4.3 Eigenvalue Placement with Observer Feedback

When observer feedback is used in place of plant state feedback, the eigenvalues of the feedback system are those the plant would have, if the state feedback were used, and those of the observer. This result is known as the *separation theorem* for observer feedback. For a completely controllable and completely observable plant, an observer of the form

$$\xi(k+1) = F\xi(k) + Gy(k) + Hu(k) \tag{14.15}$$

$$w(k) = L\xi(k) + N[y(k) - Du(k)] \tag{14.16}$$

with feedback to the plant given by

$$u(k) = Ew(k) \tag{14.17}$$

can thus be designed so that the overall feedback system eigenvalues are specified by the designer. The design procedure can proceed in two steps. First, the state feedback is designed to place the n state feedback system eigenvalues at desired locations as if the state vector is accessible. Second, the state feedback is replaced by feedback of an observer estimate of the same linear transformations of the state. As an example of eigenvalue placement with observer feedback, Figure 14.12 shows eigenvalue placement with full order state observer. The eigenvalues of the overall system are those of the state feedback and those of the full-order observer.

14.5 Tracking System Design

The second concern of tracking system design, obtaining acceptable zero-state system response to reference inputs, is now discussed.

A tracking system is one in which the plant's outputs are controlled so that they become and remain nearly equal to externally applied reference signals $r(k)$ as shown in Figure 14.13(a). The outputs $\bar{y}(k)$ are said to *track* or *follow* the reference inputs.

As shown in Figure 14.13(b), a linear, step-invariant controller of a multiple-input/multiple-output plant is described by two transfer function matrices: one relating the reference inputs to the plant inputs and the other relating the output feedback vector to the plant inputs. The feedback compensator is used for shaping the plant's zero-input response by placing the feedback

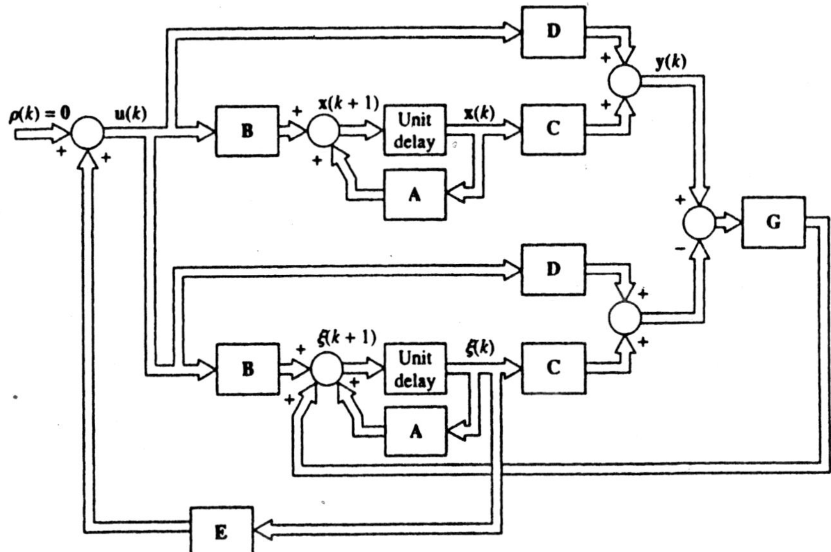

Figure 14.12 Eigenvalue placement with full-order state observer feedback.

system eigenvalues at desired locations as discussed in the previous subsections. The input compensator, on the other hand, is designed to achieve good tracking of the reference inputs by the system outputs.

The output of any linear system can always be decomposed into two parts: the zero-input component due to the initial conditions alone and the zero-state component due to the input alone, that is

$$\bar{y}(k) = \bar{y}_{\text{zero-input}}(k) + \bar{y}_{\text{zero-state}}(k)$$

Basically, there are three methods for tracking system design:

1. ideal tracking system design
2. response model design
3. reference model design

14.5.1 Ideal Tracking System Design

In this first method, *ideal tracking* is obtained if the measurement output equals the tracking input,

$$\bar{y}_{\text{zero-state}}(k) = r(k)$$

The tracking outputs $\bar{y}(k)$ have initial transient error due to any nonzero plant initial conditions, after which they are equal to the reference inputs $r(k)$, no matter what these inputs are.

As shown in Figure 14.13(c), if the plant with feedback has the z-transfer function matrix $T(z)$ relating the tracking output to the plant inputs, then

$$\bar{Y}(z) = T(z)\rho(z)$$

An input compensator or a *reference input filter*, as shown in Figure 14.13(d), with transfer function matrix $G_r(z)$, for which

$$\rho(z) = G_r(z)R(z)$$

gives

$$\bar{Y}(z) = T(z)G_r(z)R(z)$$

Ideal tracking is achieved if

$$T(z)G_r(z) = I$$

where I is the identity matrix with dimensions equal to the number of reference inputs and tracking outputs. This is to say that ideal tracking is obtained if the reference input filter is an *inverse filter* of the plant with feedback. Reference input filters do not change the eigenvalues of the plant with feedback which are assumed to have been previously placed with output or observer feedback.

When a solution exists, ideal tracking system design achieves *exact* zero-state tracking of any reference input. Because it involves constructing inverse filters, ideal tracking system design may require unstable or noncausal filters. An ideal tracking solution can also have other undesirable properties, such as unreasonably large gains, highly oscillatory plant control inputs, and the necessity of canceling plant poles and zeros when the plant model is not known accurately.

14.5.2 Response Model Design

When ideal tracking is not possible or desirable, the designer can elect to design *response model* tracking, for which

$$T(z)G_r(z) = \Omega(z)$$

where the response model z-transfer function matrix $\Omega(z)$ characterizes an acceptable relation between the tracking outputs and the reference inputs. Clearly, the price one pays for the added freedom designing a reference model can be degraded tracking performance. However, performance can be improved by increasing the order of the reference input filter. Response model design is a generalization of the classical design technique of imposing requirements for a controller's steady state response to power-of-time inputs.

The difficulty with the response model design method is in selecting suitable model systems. For example, when two or

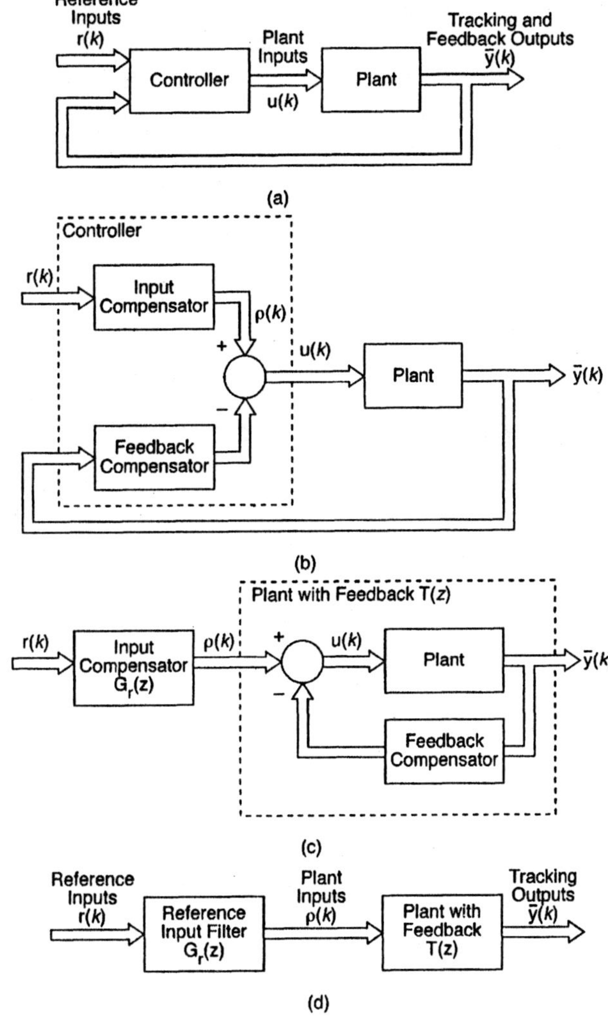

Figure 14.13 Controlling a multiple-input/multiple-output plant. The output $\bar{y}(k)$ is to track the reference input $r(k)$. (a) A tracking system using the reference inputs and plant outputs. (b) Representing a controller with a feedback compensator and an input compensator. (c) Feedback compensator combined with plant to produce a plant-with-feedback transfer function matrix $T(z)$. (d) Using a reference input filter for tracking.

more reference input signals are to be tracked simultaneously, the response model z-transfer functions selected include those relating plant tracking outputs and the reference inputs they are to track, and those relating unwanted coupling between each tracking output and the other reference inputs.

14.5.3 Reference Model Tracking System Design

The practical response model performance is awkward to design because it is difficult to relate performance criteria to the z-transfer functions of response models. An alternative design method models the reference input signals $r(k)$ instead of the system response. This method, termed reference model tracking system design, allows the designer to specify a class of representative reference inputs that are to be tracked exactly, rather than having to specify acceptable response models for all the possible inputs.

In reference model tracking system design, additional external input signals $r(k)$ to the composite system are applied to the original plant inputs and to the observer state equations so that the feedback system, instead of Equations 14.15, 14.16, and 14.17, is described by Equation 14.9 and

$$\begin{aligned} \xi(k+1) &= F\xi(k) + Gy(k) + Hu(k) + Jr(k) \quad (14.18) \\ w(k) &= L\xi(k) + N[y(k) - Du(k)] \quad (14.19) \end{aligned}$$

with

$$u(k) = Ew(k) + Pr(k) \quad (14.20)$$

Then, the overall composite system has the state equations

$$\begin{aligned} \begin{bmatrix} x(k+1) \\ \xi(k+1) \end{bmatrix} &= \begin{bmatrix} A + BENC & BEL \\ GC + \bar{H}ENC & F + \bar{H}EL \end{bmatrix} \begin{bmatrix} x(k) \\ \xi(k) \end{bmatrix} + \begin{bmatrix} BP \\ \bar{H}P + J \end{bmatrix} r(k) \\ &= \bar{A}\bar{x}(k) + \bar{B}r(k) \end{aligned}$$

and the output equation

$$\begin{aligned} \bar{y}(k) &= [C + DENC \quad DEL]\bar{x}(k) + DPr(k) \\ &= \bar{C}\bar{x}(k) + \bar{D}r(k) \end{aligned}$$

where

$$\bar{H} = H + GD$$

The composite state coupling matrix $\bar{A}$ above shows that the coupling of external inputs $r(k)$ to the feedback system does not affect its eigenvalues. The input coupling matrix $\bar{B}$ has matrices P and J which are entirely arbitrary and thus can be selected by the designer. Our objective is to select P and J so that the system output $\bar{y}(k)$ tracks the reference input $r(k)$.

Consider the class of reference signals, generated by the autonomous state variable model of the form,

$$\begin{aligned} \sigma(k+1) &= \Psi\sigma(k) \\ r(k) &= \Theta\sigma(k) \quad (14.21) \end{aligned}$$

The output of this *reference input model system* may consist of step, ramp, parabolic, exponential, sinusoidal, and other common sequences. For example, the model,

$$\begin{aligned} \begin{bmatrix} \sigma_1(k+1) \\ \sigma_2(k+1) \end{bmatrix} &= \begin{bmatrix} 2 & 1 \\ -1 & 0 \end{bmatrix} \begin{bmatrix} \sigma_1(k) \\ \sigma_2(k) \end{bmatrix} \\ r(k) &= [1 \quad 0] \begin{bmatrix} \sigma_1(k) \\ \sigma_2(k) \end{bmatrix} \end{aligned}$$

has an arbitrary constant plus an arbitrary ramp,

$$r(k) = \beta_1 + \beta_2 k$$

In reference model tracking system design, the concept of an observer is used in a new way; it is the plant with feedback that is an observer of the *fictitious* reference input model system in Figure 14.14. When driven by $r(k)$, the state of the composite system observes

$$\bar{x}(k) \to M\sigma(k)$$

where M, according to Equation 14.14, satisfies

$$M\Psi - \bar{A}M = \bar{b}\Theta \tag{14.22}$$

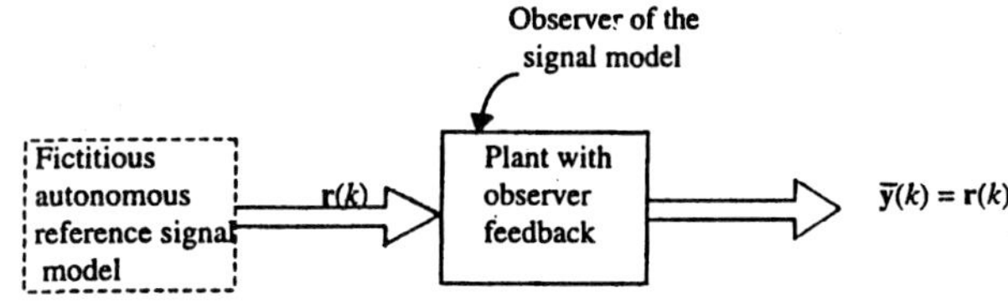

Figure 14.14 Observing a reference signal model.

The plant tracking output $\bar{y}(k)$ observes

$$\bar{y}(k) = \bar{C}\bar{x}(k) + \bar{D}r(k) \quad \to \quad \bar{C}M\sigma(k) + \bar{D}r(k)$$

and, for

$$\bar{y}(k) \quad \to \quad r(k)$$

it is necessary that

$$\bar{C}M\sigma(k) + \bar{D}r(k) = r(k) \tag{14.23}$$

Equations 14.22 and 14.23 constitute a set of linear algebraic equations where the elements of M, P, and J are unknowns. If, for an initial problem formulation, there is no solution to the equations, one can reduce the order of the reference signal model and/or raise the order of the observer used for plant feedback until an acceptable solution is obtained.

The autonomous reference input model has no physical existence; the actual reference input $r(k)$ likely deviates somewhat from the prediction of the model. The designer deals with representative reference inputs, such as constants and ramps, and, by designing for exact tracking of these, obtains acceptable tracking performance for other reference inputs.

14.6 Designing Between-Sample Response

The first two design problems for a tracking system, obtaining acceptable zero-input response and zero-state response, were discussed and solved in the previous subsections.

When a digital controller is to control a continuous-time plant, a third design problem is achieving good between-sample response of the continuous-time plant. A good discrete-time design will insure that samples of the plant response are well-behaved, but satisfactory response between the discrete-time steps is also necessary. Signals in a continuous-time plant can fluctuate wildly, even though discrete-time samples of those signals are well-behaved. The basic problem is illustrated in Figure 14.15 with the zero-input continuous-time system

$$\begin{aligned}
\begin{bmatrix} \dot{x}_1(t) \\ \dot{x}_2(t) \end{bmatrix} &= \begin{bmatrix} -0.2 & 1 \\ -1.01 & 0 \end{bmatrix} \begin{bmatrix} x_1(t) \\ x_2(t) \end{bmatrix} = Ax(t) \\
y(t) &= [1 \quad 0] \begin{bmatrix} x_1(t) \\ x_2(t) \end{bmatrix} = c^{\dagger}x(t)
\end{aligned}$$

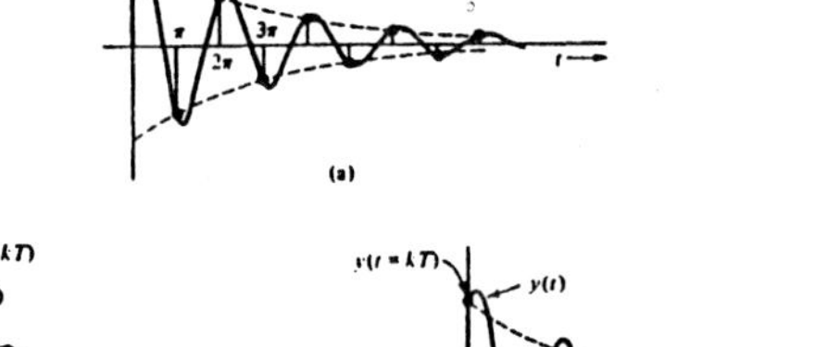

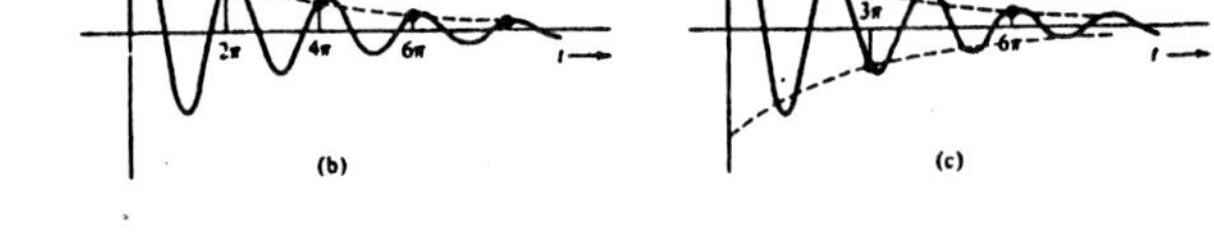

Figure 14.15 Hidden oscillations in a sampled continuous-time signal. (a) $T = \pi$, (b) $T = 2\pi$, (c) $T = 3\pi$.

This system has a response of the form

$$y(t) = Me^{-0.1t}\cos(t + \theta)$$

where M and θ depend on the initial conditions $x(0)$. When the output of this system is sampled with $T = \pi$, the output samples are

$$\begin{aligned}
y(k) &= y(t = k\pi) = Me^{-0.1k\pi}\cos(k\pi + \theta) \\
&= M(e^{-0.1\pi})^k(-1)^k\cos\theta = M\cos\theta(-0.73)^k
\end{aligned}$$

as shown in Figure 14.15(a). But these samples are the response of a first-order discrete-time system with a single geometric series model. The wild fluctuations of $y(t)$ between sampling times, termed *hidden oscillations*, cannot be determined from the samples $y(k)$.

As one might expect, the discrete-time model of this continuous-time system

$$\begin{aligned}
x(k+1) &= [\exp(\mathcal{A}T)]x(k) \\
y(k) &= c^{\dagger}x(k)
\end{aligned}$$

or

$$\begin{bmatrix} x_1(k+1) \\ x_2(k+1) \end{bmatrix} = \begin{bmatrix} -0.73 & 0 \\ 0 & -0.73 \end{bmatrix} \begin{bmatrix} x_1(k) \\ x_2(k) \end{bmatrix} = Ax(k)$$

$$y(t) = [1 \quad 0] \begin{bmatrix} x_1(k) \\ x_2(k) \end{bmatrix} = c^{\dagger}x(k)$$

is not completely observable in this circumstance because

$$M_0 = \begin{bmatrix} c^{\dagger} \\ \hline c^{\dagger}A \end{bmatrix} = \begin{bmatrix} 1 & 0 \\ -0.73 & 0 \end{bmatrix}$$

This phenomenon is called *loss of observability due to sampling.* The discrete-time system would normally have two modes, those given by its characteristic equation

$$\begin{bmatrix} (z+0.73) & 0 \\ 0 & (z+0.73) \end{bmatrix} = (z+0.73)^2 = 0$$

which are $(-0.73)^k$ and $k(-0.73)^k$. Only the $(-0.73)^k$ mode appears in the output, however.

Hidden oscillations occur at any other integer multiple of the same sampling period $T = \pi$. Figure 14.15(b) shows sampling with $T = 2\pi$, for which only a $[(-0.73)^2]^k = (0.533)^k$ mode is observable from $y(k)$. In Figure 14.15(c), with $T = 3\pi$, only a $[(-0.73)^3]^k = (-0.39)^k$ is observable from $y(k)$. For a slightly different sampling interval, for example $T = 3$ sec., there are no hidden oscillations, and the discrete-time model is completely observable.

Hidden oscillations in a continuous-time system and the accompanying loss of observability of the discrete-time model occur when the sampling interval is half the period of oscillation of an oscillatory mode or an integer multiple of that period. Although it is very unlikely that the sampling interval chosen for a plant control system would be precisely one resulting in hidden oscillations, intervals close to these result in modes in discrete-time models that are "almost unobservable". With limited numerical precision in measurements and in controller computations, these modes can be difficult to detect and control.

Between input changes, the continuous-time plant behaves as if it has a constant input *without* feedback. Often, the abrupt changes in the plant inputs at the sample times are the major cause of poor between-sample plant response. If the between-sample response of a continuous-time plant is not acceptable, there are a number of ways of improving the between-sample response.

The first approach is to *increase the controller sampling rate* so that each step change in the plant input is of smaller amplitude. Higher sample rates are often expensive, however, because the amount of computation that must be performed by the controller is proportional to the sample rate.

The second approach is to *change the plant*, perhaps by adding continuous-time feedback. The continuous-time plant with feedback, rather than the original plant, then becomes the plant to be controlled digitally. This too is often undesirable because of its susceptibility to noise and drift at the expense of routing analog signals.

The third approach is to *change the shape of the plant input signals* from having step changes at the controller sample rate to a shape that gives improved plant response. This third approach is now examined. First, input signal shaping with analog filters is considered, then input signal shaping with high-speed dedicated digital filters is discussed.

14.6.1 Analog Plant Input Filtering

Analog filters between the D/A converters and the plant inputs are usually acceptable in a controller design. As indicated in Figure 14.16, the idea is to use a filter or filters to smooth or shape the plant inputs so that the undesirable modes of the plant's open- loop response are not excited as much as they would be with abrupt changes in the plant inputs at each step. Figure 14.17 shows simulation results for the continuous-time plant

$$\begin{aligned} \begin{bmatrix} \dot{x}_1(t) \\ \dot{x}_2(t) \end{bmatrix} &= \begin{bmatrix} -0.6 & 1 \\ -9 & 0 \end{bmatrix} \begin{bmatrix} x_1(t) \\ x_2(t) \end{bmatrix} + \begin{bmatrix} -1 \\ 1 \end{bmatrix} u(t) \\ &= \mathcal{A}x(t) + \mathcal{B}u(t) \\ \bar{y}(t) &= [1 \quad 0] \begin{bmatrix} x_1(t) \\ x_2(t) \end{bmatrix} = \bar{c}^{\dagger}x(t) \end{aligned} \tag{14.24}$$

driven by a discrete-time control system

$$u(k) = -9r(k)$$

where $r(k)$ is the reference input and $T = 5$. The highly undamped zero-input plant response results in large fluctuations ("ringing") of the tracking output each time there is a step change in the plant input by the controller. The plant response is improved considerably by the insertion of a plant input analog filter with the transfer function

$$G(s) = \frac{1/3}{s + 1/3}$$

as shown in Figure 14.17(b). The filter was designed with a 3-sec. time constant to smooth the plant input waveform during each 5-sec. sampling interval, resulting in much less ringing of the tracking output.

In general, insertion of a plant input filter with the transfer function matrix $G(s)$ before a plant with transfer function matrix $T(s)$ results in a composite continuous-time *model plant* with the transfer function matrix

$$M(s) = G(s)T(s)$$

Designing analog plant input filters is quite similar to designing discrete-time reference input filters. The objective of the design is to improve the between-sample plant response by obtaining a model plant with acceptable step response. As an example, consider again the continuous-time plant Equation 14.24 with the transfer function

$$T(s) = \bar{c}^{\dagger}(sI - \mathcal{A})^{-1}\mathcal{B} = \frac{-s+1}{s^2 + 0.6s + 9}$$

A plant input filter with the transfer function,

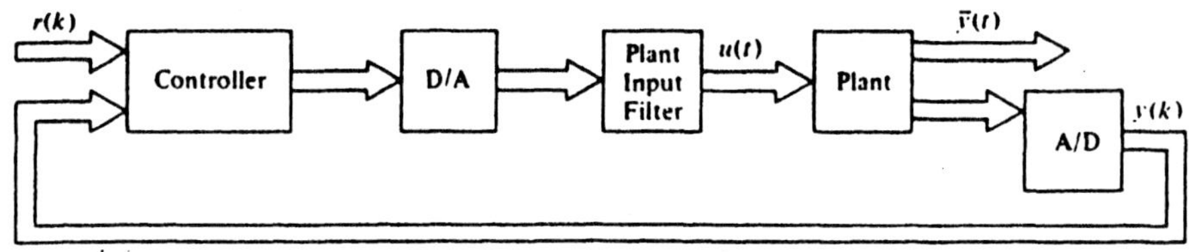

Figure 14.16 Use of an analog plant input filter to improve between-sample response.

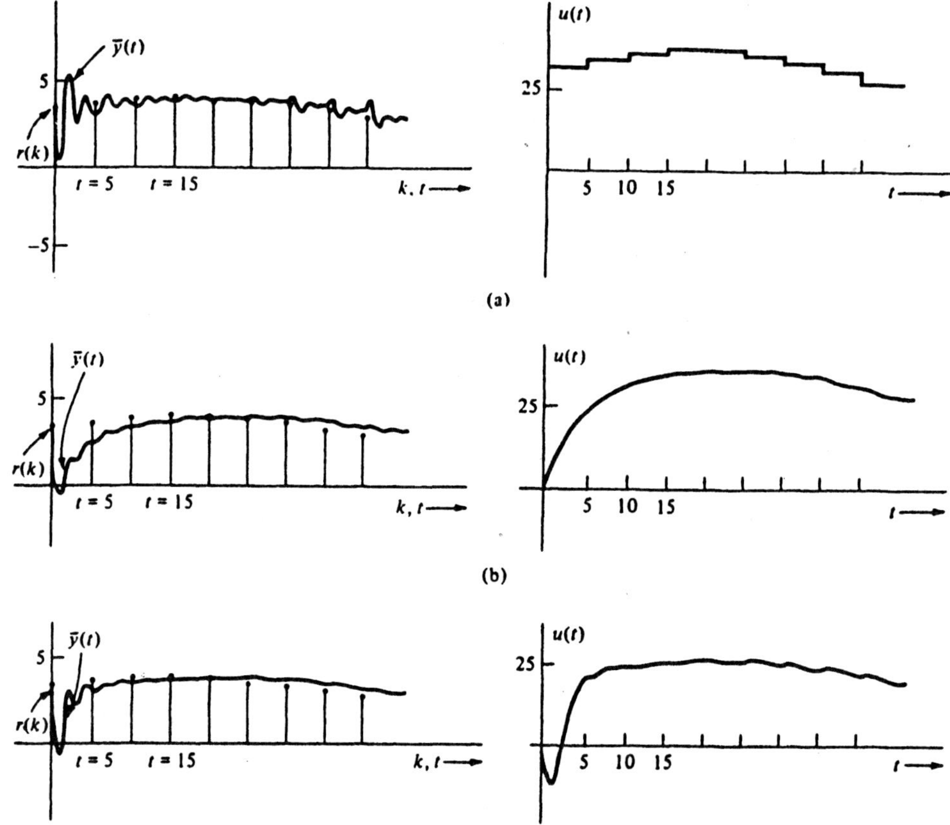

Figure 14.17 Improvement of between-sample response with analog plant input filtering. (a) response of a system without a plant input filter. (b) response with an added first-order analog plant input filter. (c) response with a better analog plant input filter.

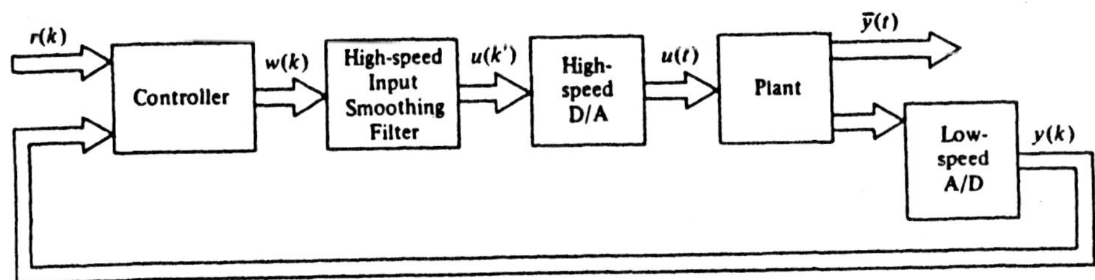

Figure 14.18 Use of high-speed digital plant input filtering to improve between-sample response.

$$G(s) = \frac{s^2 + 0.6s + 9}{(s+1)^2} = 1 + \frac{-1.4s + 8}{s^2 + 2s + 1},$$

cancels the plant poles and results in a model plant transfer function,

$$M(s) = G(s)T(s) = \frac{-s+1}{(s+1)^2}.$$

The resulting response is shown in Figure 14.17(c). The original plant's zero-input response, excited by nonzero plant initial conditions, is apparent at first but eventually decays to zero.

14.6.2 Higher Rate Digital Filtering and Feedback

Another method for performing plant input filtering is to use a digital filter that operates at many times the rate of the controller, as in Figure 14.18. Analog-to-digital conversion is usually accomplished with repeated approximations, one for each bit. With a given technology, one can perform several D/A conversions as fast as one A/D conversion. The cost of a digital plant input filter is thus relatively low because it requires higher D/A, not A/D, speed.

As an example, consider again the continuous-time plant given by Equation 14.24. The discrete-time controller for this plant operates with a sampling interval $T = 5$ sec.

It was found earlier that an analog filter with transfer function,

$$G(s) = \frac{1/3}{s + 1/3},$$

improves the plant's between-sample response. The step response of the plant alone and the response of the plant with this input filter are shown in Figure 14.19(a).

A state variable realization of the analog filter is

$$\begin{aligned} \dot{\varepsilon}(t) &= -\frac{1}{3}\varepsilon(t) + \frac{1}{3}w(t), \\ u(t) &= \varepsilon(t). \end{aligned}$$

A discrete-time model of this filter with sampling interval $\Delta t = 1$, one-fifth the controller interval, is

$$\begin{aligned} \varepsilon(k'+1) &= 0.717\varepsilon(k') + 0.283w(k') \\ u(k') &= \varepsilon(k') \end{aligned}$$

where k' is the index for steps of size Δt. The step response of the plant with a digital filter that approximates the continuous-time filter is shown in Figure 14.19(b). Figure 14.19(c) is a tracking response plot for the combination of plant, high-speed digital plant input filter, and lower speed digital controller.

Higher rate digital plant feedback can improve a plant's between-sample response, but this requires high-rate A/D as well as high-rate D/A conversion. Another possibility is to sample the plant measurement outputs at the lower rate, but estimate the plant state at a high rate, feeding the state estimate back at high rate.

14.6.3 Higher Order Holds

A traditional approach to improving reconstruction is to employ holds of higher order than the zero-order one. An nth-order hold produces a piecewise nth-degree polynomial output that passes through the most recent $n + 1$ input samples. As the order of the hold is increased, a well-behaved signal is reconstructed with increasing accuracy. Several holds and their typical responses are shown in Figure 14.20.

Although higher-order holds do have a smoothing effect on the plant inputs, the resulting improvement of plant between-sample response is generally poor compared with that possible with a conventional filter of comparable complexity. Hardware for holds of higher than zero order (which is the sample-and-hold operation) is not routinely available. One approach is to employ high speed digital devices and D/A conversion, as in the technique of Figure 14.18, but where the high speed input-smoothing filter performs the interpolation calculations for a hold.

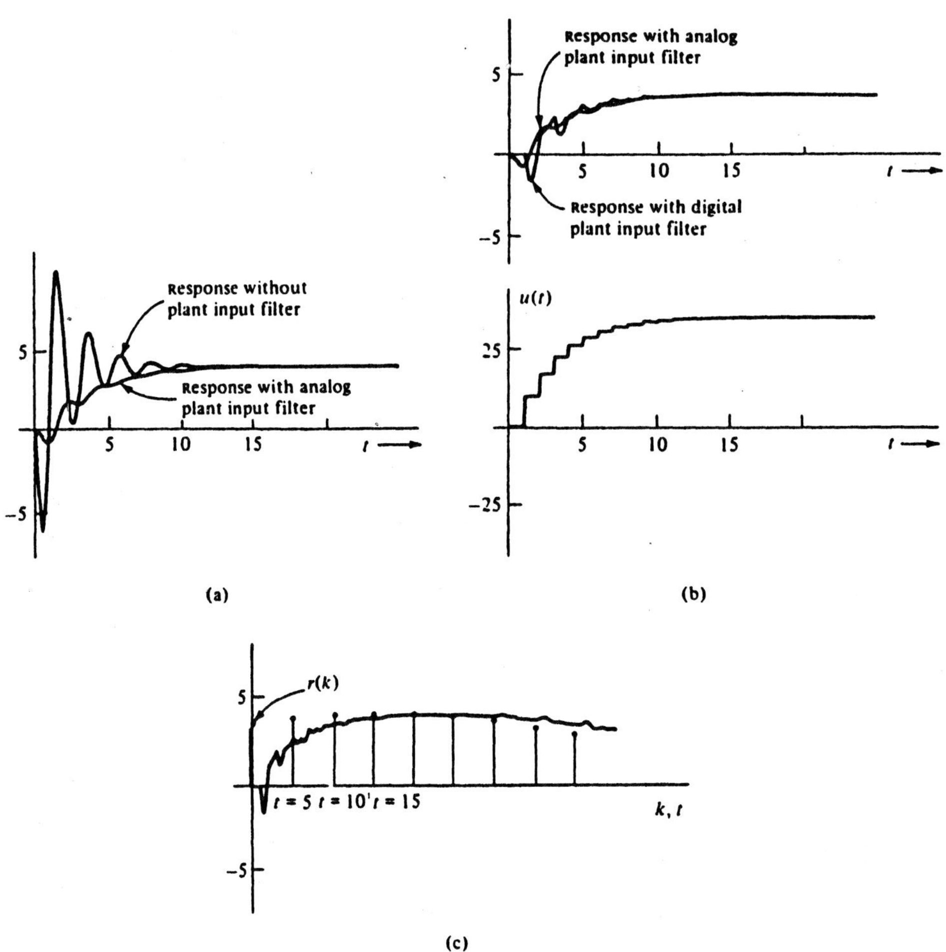

Figure 14.19 Improvement of between-sample plant response with high-speed digital input filtering. (a) Step response of the example system, with and without an analog input filter. (b) Step response with an analog input filter and with a digital input filter. (c) Tracking response of the system with a digital input filter.

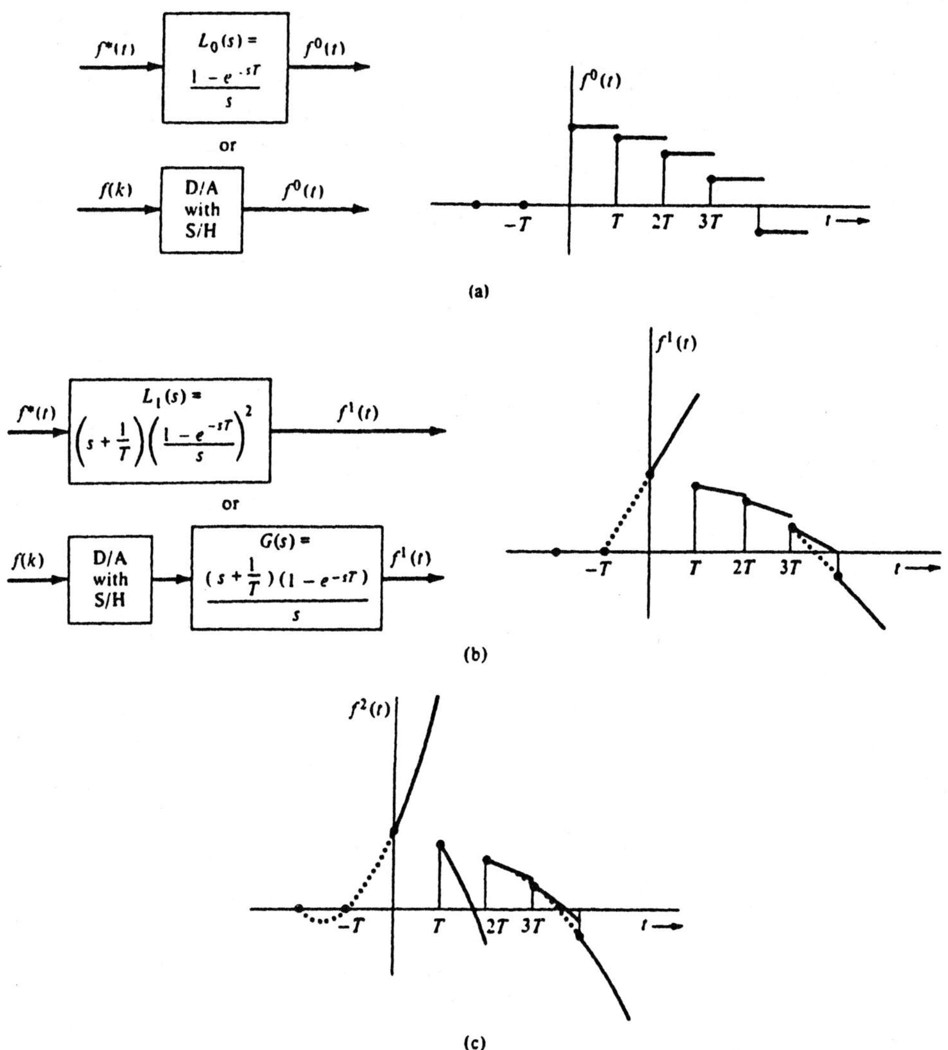

Figure 14.20 Typical response of holds. (a) Zero-order hold. (b) First-order hold. (c) Second-order hold.

References

[1] Santina, M.S., Stubberud, A.R., and Hostetter G.H., *Digital Control System Design,* 2nd ed., Saunders College Publishing, Philadelphia, 1994.

[2] DiStefano III, J.J., Stubberud, A.R., and Williams, I.J., *Feedback and Control Systems (Schaum's Outline),* 2nd ed., McGraw-Hill, New York, 1990.

[3] Kuo, B.C., *Digital Control Systems,* 2nd ed., Saunders Publishing, Philadelphia, 1992.

[4] Franklin, G.F., Powell, J.D., and Workman, M.L., *Digital Control of Dynamic Systems,* 2nd ed., Addison-Wesley, Reading, MA, 1990.

[5] Chen, C.T., *Linear System Theory and Design,* Saunders Publishing, Philadelphia, 1984.

[6] Åström, K.J. and Wittenmark, B., *Computer Controlled Systems,* Prentice Hall, Englewood Cliffs, NJ, 1987.

[7] Kailath, T., *Linear Systems,* Prentice Hall, Englewood Cliffs, NJ, 1980.

[8] Ogata, K., *Discrete-Time Control Systems,* Prentice Hall, Englewood Cliffs, NJ, 1987.

[9] Friedland, B., *Control System Design,* McGraw-Hill, New York, 1986.

[10] Wonham, W.M., *Linear Multivariable Control: A Geometric Approach,* 3rd ed., Springer, New York, 1985.

15

Quantization Effects

Mohammed S. Santina
The Aerospace Corporation, Los Angeles, CA

Allen R. Stubberud
University of California, Irvine, Irvine, CA

Gene H. Hostetter

15.1 Overview

The digital control system analysis and design methods that were presented in the previous chapters proceeded as if the controller signals and coefficients were of continuous amplitude. However, because digital controllers are implemented with finite word length registers and finite precision arithmetic, their signals and coefficients can attain only discrete values. Therefore, further analysis is needed to determine if the performance of the resulting digital controller in the presence of signal and coefficient quantization is acceptable.

In this chapter we discuss three error sources that may occur in digital processing of controllers. These error sources are (1) coefficient quantization, (2) quantization in A/D conversion, and (3) arithmetic operations. Limit cycles and deadbands are also discussed. Before discussing these errors, however, a brief review of fixed and floating point number arithmetic is presented.

15.2 Fixed-Point and Floating-Point Number Representations

There are many choices of arithmetic that can be used to implement digital controllers. The two most popular ones are *fixed-point* and *floating-point* binary arithmetic. Other nonstandard arithmetic such as *logarithmic* and *residue* representations [1] are also possibilities.

15.2.1 Fixed-Point Arithmetic

In general, an n-bit fixed-point binary number can be expressed as

$$\begin{aligned} N &= \sum_{j=-m}^{n-1} b_j 2^j = b_{n-1}2^{n-1} + b_{n-2}2^{n-2} + \cdots + b_1 2^1 \\ &\quad + b_0 2^0 + b_{-1}2^{-1} + b_{-2}2^{-2} + \cdots + b_{-m}2^{-m} \\ &= (b_{n-1}\cdots b_0 \bullet b_{-1}b_{-2}\cdots b_{-m})_2 \end{aligned} \tag{15.1}$$

where b_j is either zero or one.

The bit b_{n-1} is termed the *most significant bit* (MSB) and b_{-m} is termed the *least significant bit* (LSB). The *integer* portion of the number, $b_n b_{n-1} b_0$, is separated from the *fractional portion*, $b_{-1}b_{-2}\cdots b_{-m}$, by the *binary point* or *radix point*.

In the binary representation, each bit can be either a zero or a one. For example, the binary number 1101.101 has the decimal value

$$\begin{aligned} 1101.101 &= 1(2^3) + 1(2^2) + 0(2^1) + 1(2^0) \\ &\quad + 1(2^{-1}) + 0(2^{-2}) + 1(2^{-3}) \\ &= 13.625 \end{aligned}$$

In fixed-point arithmetic, numbers are always normalized as binary fractions (i.e., less than one) of the form $b_0 \bullet b_1 b_2 \cdots b_c$ where b_0 is the sign bit. The $C+1$ bit normalized number is stored in a register as shown in Figure 15.1 where the sign bit is separated from the C-bit number by a *fictitious* binary point. The binary point is fictitious because it does not occupy any bit location in the register. The *word length*, C, is defined as the

0-8493-8570-9/96/$0.00+$.50

number of bit locations in the register to the right of the binary point.

There are three commonly used methods for representing signed numbers:

1. signed-magnitude
2. two's complement
3. one's complement

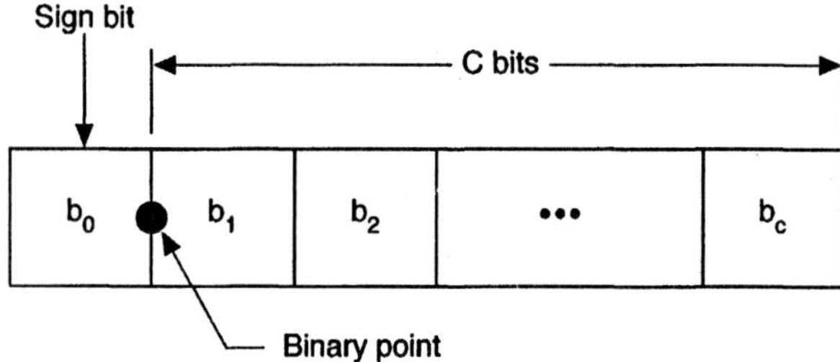

Figure 15.1 Normalized fixed-point numbers in a register.

Consider the $C + 1$ bit binary fraction $b_0 \bullet b_1 b_2 \cdots b_c$ where b_0 is the sign bit. In the *signed-magnitude* representation, the fractional number is positive if b_0 is zero, and it is negative if b_0 is one. For example, the decimal number 0.75 equals 0.11 in binary representation, and −0.75 equals 1.11.

In signed-magnitude representation, binary numbers can be converted to decimal numbers using the relation,

$$N = (-1)^{b_0} \sum_{i=1}^{C} b_i 2^{-i} \tag{15.2}$$

The *two's complement* representation of positive numbers is the same as signed-magnitude representation. The two's complement representations of negative numbers, however, are obtained by *complementing* (i.e., replacing every 1 with 0 and every 0 with 1) all of the bits of the positive number and adding one to the LSB of the complemented number. For example, the two's complement representation of the decimal number 0.75 is 0.11 and the two's complement representation of −0.75 is 1.01.

A decimal number can be recovered from its two's complement representation using the relationship,

$$N = -b_0 + \sum_{i=1}^{C} b_i 2^{-i} \tag{15.3}$$

The one's complement representation of fractional numbers is the same as the two's complement without the addition of one to the LSB. For example, the one's complement representation of 0.75 is 0.11 and the one's complement representation of −0.75 is 1.00. A decimal number can be recovered from its one's complement representation via the relationship,

$$N = b_0(2^{-c} - 1) + \sum_{i=1}^{C} b_i 2^{-i}. \tag{15.4}$$

The two's complement representation of binary numbers has several advantages over the signed-magnitude and the one's complement representations [2] and therefore is more popular.

In general, the sum of two normalized C-bit numbers using fixed-point arithmetic is a C-bit number while the product of two C-bit numbers is a $2C$-bit number. Hence, if the register word length is fixed to C bits, a *quantization* error will be introduced in multiplication but not in addition.[1] The product is quantized either by *rounding* or by *truncation*. For example, rounding the binary number 0.010101 to four bits after the binary point gives 0.0101 but rounding it to three bits yields 0.011. When a number is truncated, all of the bits to the right of its LSB are discarded. For example, truncating the number 0.010101 to three bits after the binary point gives 0.010.

15.2.2 Floating-Point Arithmetic

A major disadvantage of fixed-point arithmetic is the limited range of numbers that can be represented with a given word length. Another type of arithmetic which, for the same number of bits, has a much larger range of numbers is *floating-point arithmetic*. In general, a floating-point number can be expressed as

$$N = M \times 2^E \tag{15.5}$$

where M and E, both expressed in binary form, are termed the *mantissa* and the *exponent* of the number, respectively. In binary floating-point representation, numbers are always normalized by scaling M as a fraction whose decimal value lies in the range $0.5 \leq M < 1$.

Figure 15.2 shows a floating-point number stored in a register. The register is divided into the mantissa and the exponent. Both the mantissa and the exponent have fictitious binary points to separate the sign bits from the numbers. In floating-point arithmetic, negative mantissas and negative exponents are coded the same way as in fixed-point arithmetic using two's complement, signed-magnitude, or one's complement.

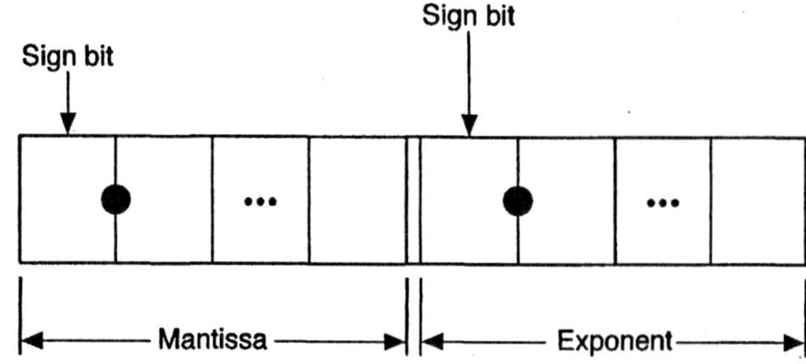

Figure 15.2 Normalized floating-point number in a register.

[1] When normalized signed numbers are added and the result is larger than one, then *overflow* occurs. Overflow does not occur in multiplication because the product of two normalized numbers is always less than one.

The product of two floating-point numbers is

$$(M_1 \times 2^{E_1})(M_2 \times 2^{E_2}) = (M_1 \times M_2)2^{(E1+E2)}$$

Thus, if the mantissa is limited to C bits, the product $M_1 \times M_2$ must be rounded or truncated to C bits. The sum of two floating-point numbers is performed by shifting the bits of the mantissa of the smaller number to the right and increasing its exponent until the two exponents are equal. Then the two mantissas are added, and, if necessary, normalized to satisfy Equation 15.5. The shifted mantissa may exceed its limited range and thus must be quantized. Hence, in floating-point arithmetic, quantization errors are introduced in both addition and multiplication. Roundoff or truncation errors will be introduced in the mantissa M but not in the exponent E, because the exponent, E, is always a positive or negative integer and integers have exact binary representations. Of course, if the number is too large or too small then over- or underflow can occur.

15.3 Truncation and Roundoff

Because of the finite word length of registers in digital computers, errors are always introduced when the numbers to be processed are quantized. These errors depend on (1) the way the numbers are represented (fixed- or floating-point arithmetic, signed-magnitude, two's or one's complement, etc.) and (2) how the numbers are quantized.

Consider the normalized binary number $b_0 \bullet b_1 b_2 \cdots b_c$ where b_0 is the sign bit, and $b_1 b_2 \cdots b_c$ is the binary code of a fixed-point number or the mantissa of a floating-point number. Denoting the number before quantization by x, the error introduced by quantization is

$$e_q = Q[x] - x$$

where $Q[x]$ is the quantized value of x. The range of quantization error depends on the type of arithmetic and the type of quantization used. Figure 15.3 shows the transfer characteristics of truncation and roundoff for signed-magnitude, two's complement, and one's complement representations.

For fixed-point arithmetic, the error caused by truncating a number to C bits is [3]

$$\begin{aligned} -2^{-c} < e_T \le 0, \quad x \ge 0 \\ 0 \le e_T < 2^{-c}, \quad x < 0 \end{aligned} \tag{15.6}$$

for the signed-magnitude and one's complement representations. For two's complement, the truncation error is

$$-2^{-c} < e_T \le 0 \tag{15.7}$$

for all x.

On the other hand, the error caused by rounding a number to C bits is

$$\frac{-2^{-c}}{2} \le e_R < \frac{2^{-c}}{2} \tag{15.8}$$

for signed-magnitude, one's complement and two's complement representations.

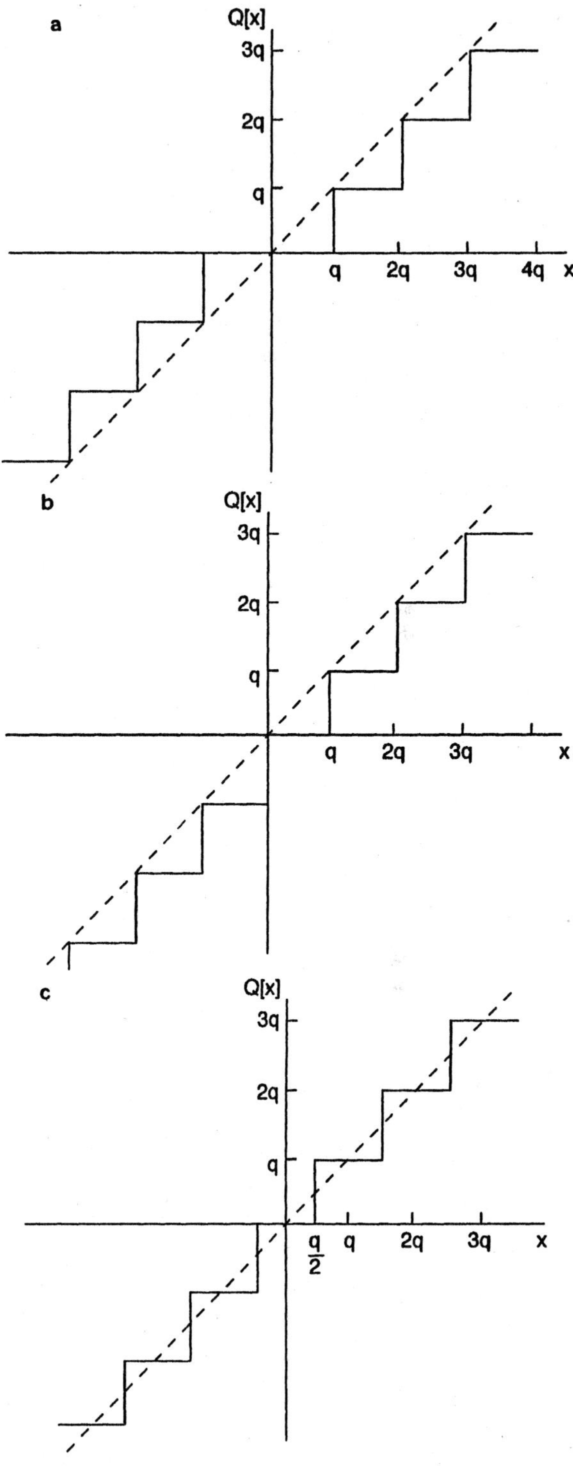

Figure 15.3 Transfer characteristics of truncation and rounding. (a) truncation with signed-magnitude and one's complement. (b) truncation with two's complement. (c) rounding with signed-magnitude, one's complement, and two's complement.

In fixed-point arithmetic, truncation or roundoff errors are independent of the magnitude of the original unquantized numbers. However, in floating-point arithmetic, these errors depend on the magnitude of the unquantized number. In floating point-arithmetic, roundoff and truncation errors occur only in the mantissa. Thus, if the mantissa is truncated to C bits the quantized number is

$$x_q = (1+\varepsilon)x$$

where ε is the relative error in x. In the case of truncation, for signed-magnitude and one's complement representations, the relative error in the value of the floating-point word is

$$-2.2^{-C} < \varepsilon \leq 0 \tag{15.9}$$

and for two's complement truncation, the error is

$$-2.2^{-C} < \varepsilon \leq 0, \quad x \geq 0, \tag{15.10}$$

$$0 \leq \varepsilon < 2.2^{-C}, \quad x < 0. \tag{15.11}$$

On the other hand, the roundoff error in the mantissa is of the form

$$-2^{-C} \leq \varepsilon \leq 2^{-C} \tag{15.12}$$

for all three types of representations. Figure 15.4 shows the probability density functions of fixed- and floating-point errors. As shown, the probability density function is *uniformly distributed* over the range of quantization. In the remainder of this article, and unless otherwise stated, we model fixed- and floating-point errors as stationary, white noise, random processes.

We now discuss the major sources of error caused by finite word length and then determine their effects on the behavior of digital controllers. These errors are coefficient quantization, quantization errors in A/D converters and quantization errors in arithmetic operations.

15.4 Effects of Coefficient Quantization

The digital control system design methods that were presented in the previous articles resulted in controllers whose coefficients are of arbitrary precision. However, because the controller is implemented with finite word length registers, each of its coefficients must be quantized. For example, consider the digital controller described by the transfer function

$$H(z) = \frac{z^3 + 1.584z^2 + 1.2769z + 0.5642}{z^4 + 2.689z^3 + 3.3774z^2 + 2.3823z + 0.6942} \tag{15.13}$$

which has poles located at $z_1 = 0.999$, $z_2 = 0.697$, and $z_{3,4} = 0.4965 \pm j0.8663$. If the binary form of the coefficients of this controller are truncated to three bits to the right of the binary point, then the quantized controller transfer function becomes

$$H_q(z) = \frac{z^3 + 1.5z^2 + 1.25z + 0.5}{z^4 + 2.625z^3 + 3.3749z^2 + 2.3748z + 0.6249} \tag{15.14}$$

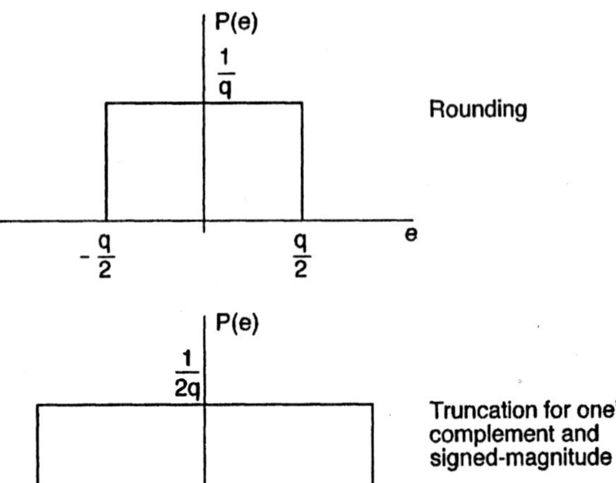

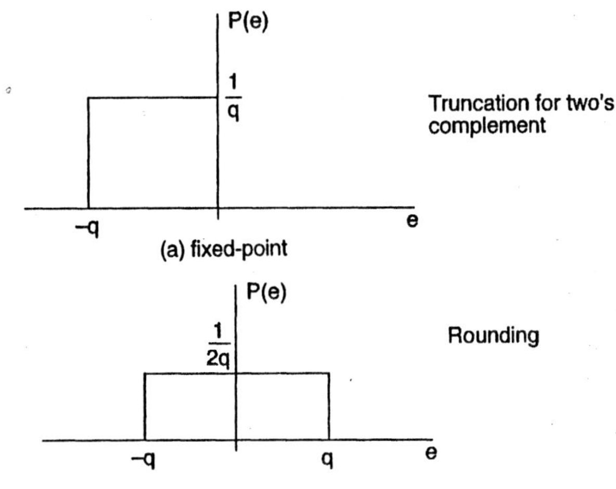

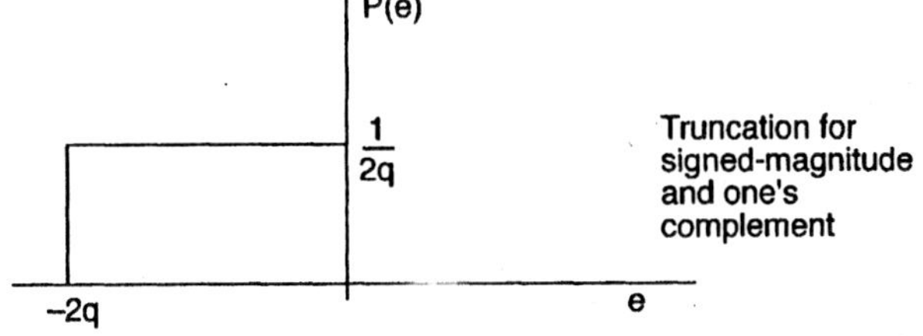

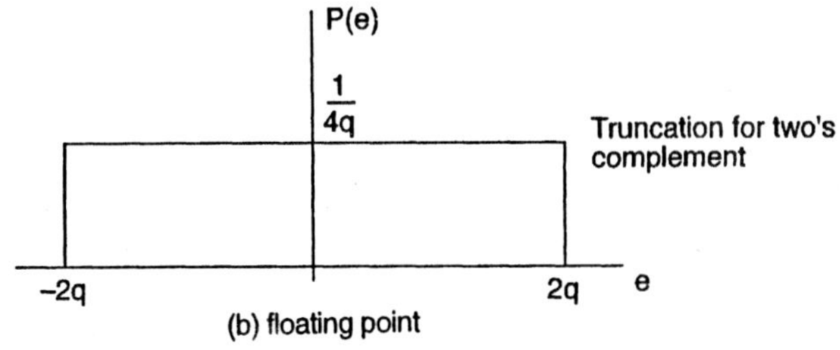

Figure 15.4 Probability density functions of fixed- and floating-point errors. (a) fixed-point errors. (b) floating point errors.

which has two poles outside the unit circle. Another example is quoted in reference [4] in which a stable fifth-order controller can become unstable even if it is realized with 18-bit arithmetic.

In general, consider the digital controller described by the transfer function

$$H(z) = \frac{\sum_{k=0}^{m} b_k z^{-k}}{1 - \sum_{k=1}^{n} a_k z^{-k}} \tag{15.15}$$

If the controller coefficients are quantized to C bits, then

$$\hat{a}_k = a_k + \delta_k$$

for fixed-point arithmetic or

$$\hat{a}_k = a_k(1 + \delta_k)$$

for floating-point arithmetic where δ_k is bounded in absolute value by 2^{-c}. Similarly,

$$\hat{b}_k = b_k + \eta_k$$

for fixed-point or

$$\hat{b}_k = b_k(1 + \eta_k)$$

for floating-point arithmetic.

In terms of the quantized coefficients, the controller transfer function becomes

$$H_q(z) = \frac{\sum_{k=0}^{m} \hat{b}_k z^{-k}}{1 - \sum_{k=1}^{n} \hat{a}_k z^{-k}} \tag{15.16}$$

One approach for analyzing the effects of coefficient quantization on system performance is to compare the response of the quantized controller with that of the ideal controller before quantization. One can also apply *sensitivity* analysis to determine the *variations* of the controller response to variations in its numerator and denominator coefficients. If the controller transfer function given by Equation 15.15 is rewritten in the form (assuming $b_o = 1$)

$$H(z) = \frac{\prod_{i=1}^{m}(1 + \beta_i z^{-1})}{\prod_{j=1}^{n}(1 - \alpha_j z^{-1})} = \frac{N(z^{-1})}{D(z^{-1})}$$

where α_i is the location of the ith pole of $H(z)$, and if the product of the quantized poles is expressed as

$$D_q(z^{-1}) = \prod_{j=1}^{n}(1 - \hat{\alpha}_j z^{-1})$$

where

$$\hat{\alpha}_j = \alpha_j + \Delta\alpha_j,$$

then [5]

$$\Delta\alpha_j = \sum_{k=1}^{n} \frac{\alpha_j^{n-k}}{\prod_{\substack{i=1 \\ i \neq j}}^{n} (\alpha_j - \alpha_i)} \Delta a_k \tag{15.17}$$

which relates the incremental changes of the jth pole of the controller transfer function to incremental changes in the a_k coefficient of the denominator polynomial of the controller transfer function. Recall that

$$\Delta a_k = \delta_k$$

for fixed-point arithmetic and

$$\Delta a_k = a_k \delta_k$$

for floating-point arithmetic. Similar results can also be obtained for the controller's zeros.

Equation 15.17 shows that, when the controller poles are close to each other, small changes in the coefficients a_k of the denominator polynomial cause large changes in the locations of the controller poles. As the order of the controller increases, the roots of its denominator become more sensitive to changes in the coefficients of the denominator polynomial.

To avoid the coefficient sensitivity problem, higher order controller transfer functions are decomposed into *cascaded* first- or second-order transfer functions as in Figure 15.5. When all the poles and zeros of the controller transfer function are real, the cascaded transfer functions are all of first order. Complex conjugate pairs of roots should be grouped into second-order subsystems to avoid complex number arithmetic operations.

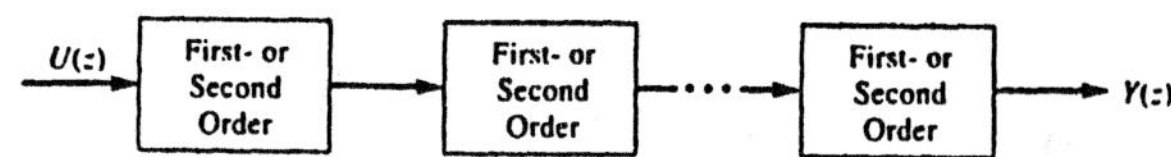

Figure 15.5 Cascade first- or second-order subsystems.

Another way to avoid the coefficient sensitivity problem is to use the *parallel* form as shown in Figure 15.6. The parallel form is obtained by decomposing the controller transfer function into first- or second-order subsystems using the method of partial fraction expansion. Using either form, that is cascade or parallel, each first-order subsystem can be realized using the structure shown in Figure 15.7(a) where

$$H(z) = \frac{1 + \beta_1 z^{-1}}{1 - \alpha_1 z^{-1}}$$

and each second-order subsystem may be realized as shown in Figure 15.7(b) where

$$H(z) = \frac{1 + \beta_1 z^{-1} + \beta_2 z^{-2}}{1 - \alpha_1 z^{-1} - \alpha_2 z^{-2}}$$

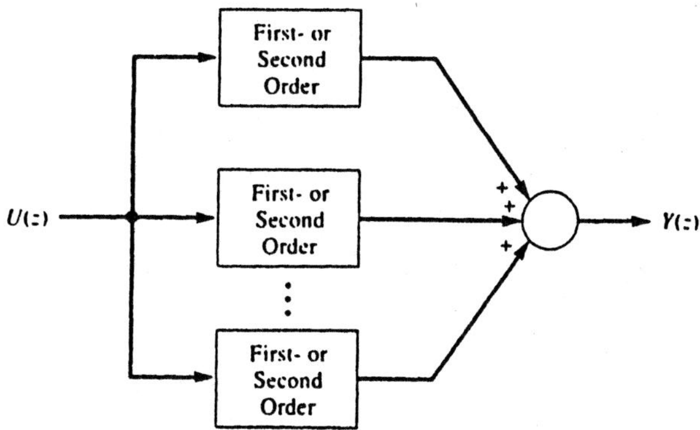

Figure 15.6 Parallel first- or second-order subsystems.

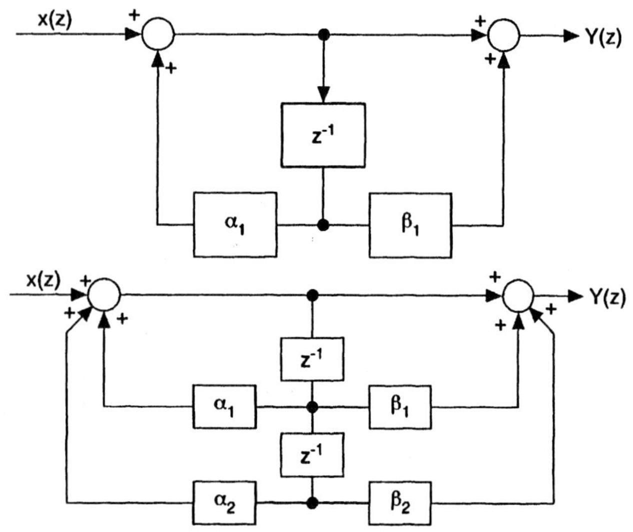

Figure 15.7 First and second-order realizations. (a) first-order realization. (b) second-order realization.

As a numerical example, consider again the controller transfer function given by Equation 15.13. Rewriting the transfer function in factored form yields

$$H(z) = \left[\frac{z+0.862}{z+0.999}\right]\left[\frac{1}{z+0.697}\right]\left[\frac{z^2+0.722z+0.6545}{z^2+0.993z+0.997}\right]$$

As in the previous example, if the binary representations of the coefficients of each factor are truncated to three bits to the right of the binary point, then the resulting quantized transfer function is

$$H_q(z) = \left[\frac{z+0.75}{z+0.875}\right]\left[\frac{1}{z+0.625}\right]\left[\frac{z^2+0.625z+0.625}{z^2+0.875z+0.875}\right]$$

which is stable and can be realized using first- and second-order subsystems. This controller is significantly different from the quantized controller given by Equation 15.14.

15.5 Quantization Effects in A/D Conversion

The second source of error that we shall discuss is quantization in analog-to-digital conversion. Conceptually, A/D conversion involves two steps: sampling and quantization. Sampling is the process of converting a continuous-time signal $x(t)$ to a discrete-time sequence $x(k)$, and quantization is the process of approximating each sample of the sequence with a digital code word, that is, each sample is rounded or truncated to fit into the finite-length register. Rounding approximates the sample by the nearest *quantization level* and truncating approximates the sample by the highest quantization level that is smaller than or equal to the sample value. Figure 15.8(a) shows a block diagram of an A/D converter in which the input signal $x(t)$ is sampled and then quantized to produce $x_q(k)$.

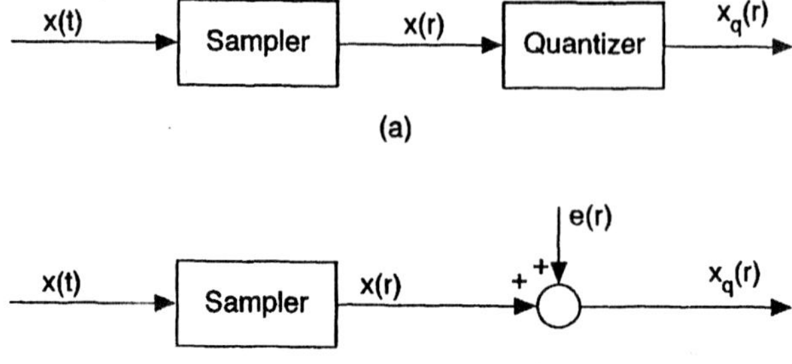

Figure 15.8 Block diagram of an A/D converter and its statistical model. (a) block diagram of an A/D converter. (b) statistical model of an A/D converter.

Let the word length of the A/D converter be C bits and let the number converted be of the form

$$x_q = b_1 2^{-1} + b_2 2^{-2} + b_3 2^{-3} + \cdots + b_c 2^{-c}$$

where $b_1, b_2, \ldots b_c$ are the binary code. The sign bit, however coded, will always be present. The largest x_q possible that is produced by the A/D converter is

$$\begin{aligned} x_q &= 2^{-1} + 2^{-2} + 2^{-3} + \cdots + 2^{-c} = \frac{1}{2}\sum_{i=0}^{C-1}\left(\frac{1}{2}\right)^i \\ &= 1 - 2^{-C} \end{aligned}$$

and the smallest nonzero x_q is 2^{-C}. The *dynamic range* of the converter is commonly defined as

$$DR = \frac{1-2^{-C}}{2^{-C}} = 2^C - 1.$$

Thus

$$C \geq \log_2(DR+1). \tag{15.18}$$

For roundoff, taking the error e as a random variable with a uniform probability density shown in Figure 15.4, then

$$E\{e\} = \int_{-q/2}^{q/2} xp(x)dx = 0 \tag{15.19}$$

where E denotes expected value, and

$$E\{e^2\} = \int_{-\infty}^{\infty} x^2 p(x)dx = \int_{-q/2}^{q/2} \frac{x^2}{q}dx = \frac{q^2}{12}.$$

In terms of the number of bits, C, the *variance* is

$$E\{e^2\} = \frac{2^{-2C}}{12} \tag{15.20}$$

For signed-magnitude and one's complement truncation, the error is uniformly distributed between $-q$ and q as shown in Figure 15.4. Then

$$E\{e\} = \int_{-\infty}^{\infty} xp(x)dx = 0 \tag{15.21}$$

and

$$E\{e^2\} = \int_{-\infty}^{\infty} x^2 p(x)dx = \int_{-q}^{q} \frac{x^2}{2q}dx = \frac{q^2}{3}.$$

In terms of the number of bits, C,

$$E\{e^2\} = \frac{2^{-2C}}{3}. \tag{15.22}$$

Comparing Equation 15.22 with Equation 15.20 for roundoff quantization gives

$$E\{e^2\} = \frac{q^2}{12} = \frac{2^{-2C}}{12} = \frac{2^{-2(C+1)}}{3}.$$

For two's complement truncation, the error is uniformly distributed as shown in Figure 15.4. Thus,

$$E\{e\} = \frac{-2^{-C}}{2} \tag{15.23}$$

and

$$E\{e^2\} = \frac{2^{-C}}{12} = \frac{2^{-2(C+1)}}{3} \tag{15.24}$$

As discussed in Section 15.6, the quantization error $e(k)$ can be viewed as an additive, stationary, white noise process, as shown in Figure 15.8(b), with mean and variance given by Equations 15.19 through 15.24 depending on whether the quantization is due to roundoff or truncation. Using superposition, the output of the digital controller can be decomposed into two parts, one due to the input $x(k)$ alone, and the other due to $e(k)$ which is assumed to be uncorrelated with $x(k)$.

15.5.1 Signal-to-Noise Ratio of an A/D Converter

Referring to Figure 15.8 again, if the input sequence $x(k)$ is modeled as a zero mean Gaussian probability density so that $3\sigma = 1$ (i.e., unity input level is a 3σ event), then

$$E\{x^2(k)\} = \sigma^2 = \left(\frac{1}{3}\right)^2 = \frac{1}{9}$$

Define the *signal-to-noise* ratio of an A/D converter as

$$\begin{aligned}(S/N)_{\text{dB}} &= 10\log_{10}\frac{\text{variance}\{\text{perfect signal}\}}{\text{variance}\{\text{error}\}} \\ &= 10\log_{10}\frac{\text{variance}\{x(k)\}}{\text{variance}\{e\}}\end{aligned}$$

For roundoff noise, the signal-to-noise ratio is

$$\begin{aligned}(S/N)_{\text{dB}} &= 10\log_{10}\frac{\frac{1}{9}}{\frac{2^{-2(C+1)}}{3}} \\ &= 10\log_{10}\left(\frac{1}{3}\right) - 10\log_{10}2^{-2(C+1)} \\ &= -4.77 + 6.02(C+1) \qquad (15.25)\end{aligned}$$

or

$$C = 0.166(S/N)_{\text{dB}} - 0.207$$

For truncation noise, $(C+1)$ is replaced by C in Equation 15.25. Hence,

$$(S/N)_{\text{dB}} = -4.77 + 6.02C \tag{15.26}$$

or

$$C = 0.166(S/N)_{\text{dB}} + 0.792$$

Equations 15.25 and 15.26 show that the signal-to-noise ratio increases 6 dB for every additional bit to the register. As a simple design procedure, one may choose the number of bits of the A/D converter to be the larger of the two values necessary for required dynamic range and required signal-to-noise ratio, that is,

$$C = \max\left\{\log_{10}(DR+1), 0.166(S/N)_{\text{dB}} - 0.207\right\}$$

for roundoff noise, or

$$C = \max\left\{\log_{10}(DR+1), 0.166(S/N)_{\text{dB}} + 0.792\right\}$$

for truncation noise.

The quantization error of an A/D converter is not a serious problem. In a 16-bit A/D converter, the maximum quantization error is $2^{-16} = 0.0015\%$ which is quite low compared with typical errors in analog sensors. This error, if taken to be "noise," gives a signal-to-noise ratio of $20\log_{10}(2^{-16}) = 96.3$ dB, which is much better than that of most high-fidelity audio systems. The designer must insure that enough bits are used to give the desired system accuracy.

15.6 Stochastic Analysis of Quantization Errors in Digital Processing

One approach for analyzing roundoff and truncation errors generated in digital processing of controllers is to derive deterministic upper bounds on the maximum errors that can possibly result from roundoff or truncation [6], [7]. In general, however, these bounds are *pessimistic* because the errors usually add up in the worst possible way.

Another popular approach for analyzing roundoff and truncation errors is to develop stochastic noise models of these errors first, and then determine their effects on system performance.

15.6.1 Fixed-Point Arithmetic

It was mentioned earlier that, in fixed-point arithmetic, quantization errors occur in multiplication and not in addition. Figure 15.9(a) shows a block diagram of a multiplier model in which two C-bit numbers are multiplied and then quantized to produce a C-bit number. An equivalent noise model of the multiplier, useful for analysis, is shown in Figure 15.9(b). The quantization error $e(k)$ is modeled as a stationary, additive, white noise sequence so that

$$\begin{aligned}E\{e(k)\} &= \mu_e \\ E\{e(i)e(j)\} &= \begin{cases} 0, & i \neq j \\ E\{e^2\}, & i = j \end{cases}\end{aligned}$$

where the mean and the variance of the error can be determined from the probability density function shown in Figure 15.4.

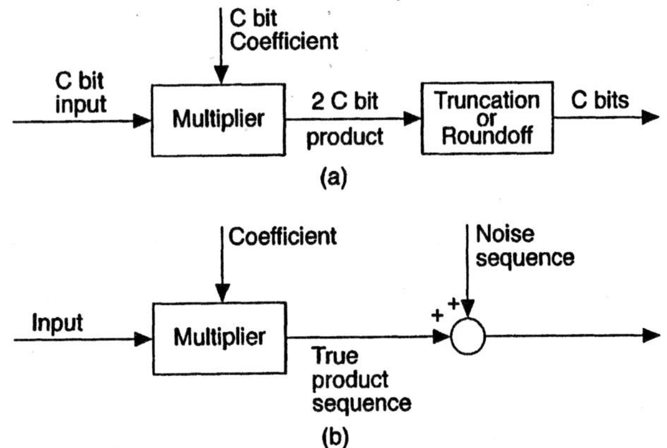

Figure 15.9 Model of multiplier. (a) multiplier model. (b) statistical model of multiplier.

Using superposition, the output of a digital controller due to the error $e(k)$ alone is given by the convolution solution

$$y_e(k) = \sum_{m=0}^{k} g(m)e(k-m)$$

where $g(m)$ is the unit pulse response of the system whose input is the error source and whose output is the digital controller output. The mean of the output is

$$\begin{aligned} E\{y_e(k)\} &= \sum_{m=0}^{k} g(m)E\{e(k-m)\} \\ &= \mu_e \sum_{m=0}^{k} g(m) \end{aligned} \quad (15.27)$$

and the variance of $y_e(k)$ is

$$\begin{aligned} E\{y_e^2(k)\} &= E\left\{\left[\sum_{m=0}^{k} g(m)e(k-m)\right] \cdot \left[\sum_{n=0}^{k} g(n)e(k-n)\right]\right\} \\ &= \sum_{m=0}^{k}\sum_{n=0}^{k} g(n)g(m)E\{e(k-m)e(k-n)\} \end{aligned}$$

But

$$E\{e(k-m)e(k-n)\} = E\{e^2\}\delta(m-n)$$

where δ is the unit pulse function. Then,

$$E\left\{y_e^2(k)\right\} = E\{e^2\}\textstyle\sum_{m=0}^{k} g^2(m)$$

Therefore as k approaches infinity, the mean and the variance of the output are

$$E\{y_e(k)\} = \mu_e \sum_{m=0}^{\infty} g(m) \quad (15.28)$$

and

$$E\left\{y_e^2(k)\right\} = E\{e^2\}\sum_{m=0}^{\infty} g^2(m) \quad (15.29)$$

respectively. Hence, the variance of the output equals the variance of the quantization noise times the noise power gain, NPG, where

$$(NPG) = \sum_{m=0}^{\infty} g^2(m) \quad (15.30)$$

Because digital controller transfer functions are realized using first- and second-order subsystems in parallel or cascade forms, the noise power gain of first- and second order-subsystems is now determined.

The transfer function of a first-order recursive subsystem is

$$G(z) = \frac{A(1+\beta_1 z^{-1})}{1-\alpha_1 z^{-1}} = k_0 + \frac{k_1}{1-\alpha_1 z^{-1}}$$

Hence,

$$g(m) = k_0\delta(m) + k_1\alpha_1^m, \quad m = 0, 1, 2, \ldots$$

and

$$\sum_{m=0}^{\infty} g(m) = k_0 + \frac{k_1}{1-\alpha_1}$$

Therefore, the mean of the output is

$$E\{y_e(k)\} = \mu_e\left[k_0 + \frac{k_1}{1-\alpha_1}\right] \quad (15.31)$$

Similarly,

$$g^2(m) = \begin{cases} (k_0+k_1)^2, & m = 0 \\ k_1^2\alpha_1^{2m}, & m = 1, 2, \ldots \end{cases}$$

and therefore the noise power gain becomes

$$(NPG) = k_0^2 + 2k_0k_1 + \frac{k_1^2}{1-\alpha_1^2} \quad (15.32)$$

On the other hand, the transfer function of a second-order recursive subsystem is

$$\begin{aligned} G(z) &= A\frac{1+\beta_1 z^{-1}+\beta_2 z^{-2}}{1-\alpha_1 z^{-1}-\alpha_2 z^{-2}} \\ &= k_0 + \frac{k_1}{1-r_1 z^{-1}} + \frac{k_2}{1-r_2 z^{-1}} \end{aligned} \quad (15.33)$$

Then,

$$g(m) = k_0\delta(m) + k_1 r_1^m + k_2 r_2^m, \quad m = 0, 1, 2, \ldots$$

and hence,

$$\sum_{m=0}^{\infty} g(m) = k_0 + \frac{k_1}{1-r_1} + \frac{k_2}{1-r_2} \quad (15.34)$$

Similarly,

$$g^2(m) = \begin{cases} (k_0 + k_1 + k_2)^2, & m = 0 \\ (k_1 r_1^m + k_2 r_2^m)^2, & m = 1, 2, \ldots \end{cases}$$

and the noise power gain becomes

$$(NPG) = k_0^2 + 2k_0k_1 + 2k_0k_2 + \frac{k_1^2}{1-r_1^2} + \frac{2k_1k_2}{1-r_1r_2} + \frac{k_2^2}{1-r_2^2} \quad (15.35)$$

which is real. For the special case

$$G(z) = \frac{1}{1 - \alpha_1 z^{-1} - \alpha_2 z^{-2}}$$

the (NPG), in terms of the polynomial coefficients, is

$$(NPG) = \frac{1-\alpha_2}{1+\alpha_2} \frac{1}{(1-\alpha_2)^2 - \alpha_1^2}$$

Equations 15.31 through 15.35 are used regularly in the stochastic analysis of quantization error.

Another method for calculating the noise power gain is to use contour integration as follows:

$$\begin{aligned}(NPG) &= \sum_{m=0}^{\infty} g^2(m) \\ &= \sum_{m=0}^{\infty} g(m) \frac{1}{2\pi j} \oint_C G(z) z^{m-1} dz\end{aligned}$$

where C is the contour of integration chosen as the unit circle $|z| = 1$. Thus,

$$(NPG) = \frac{1}{2\pi j} \oint_C G(z) z^{-1} \left(\sum_{m=0}^{\infty} g(m) z^m \right) dz$$

but

$$\textstyle\sum_{m=0}^{\infty} g(m) z^m = G(z^{-1})$$

Therefore,

$$(NPG) = \frac{1}{2\pi j} \oint_C G(z) G(z^{-1}) z^{-1} dz \quad (15.36)$$

which is the sum of residues of $G(z)G(z^{-1})z^{-1}$ within the unit circle. An efficient numerical method for solving Equation 15.36 is given in [8].

15.6.2 Quantization Noise Model of First Order Subsystems

Consider the first-order subsystem shown in Figure 15.10. Setting all the noise sources to zero, the transfer function of the ideal subsystem is

$$T(z) = \frac{Y(z)}{X(z)} = \frac{A(1 + \beta_1 z^{-1})}{1 - \alpha_1 z^{-1}}.$$

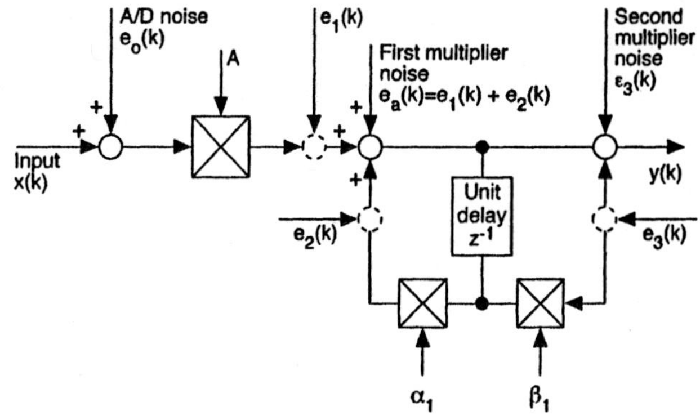

Figure 15.10 Fixed point quantization noise model of first-order subsystem.

The effect of the A/D converter on the output can be determined by setting all signals but e_0 to zero. The transfer function which relates e_0 to the filter output is

$$Y_0(z) = G_0(z) E_0(z)$$

where

$$G_0(z) = T(z) = k_0 + \frac{k_1}{1 - \alpha_1 z^{-1}}$$

Then, the mean of the output is given by Equation 15.31:

$$E\{y_0\} = \mu_0 \left[k_0 + \frac{k_1}{1 - \alpha_1} \right]$$

and the variance of the output is, using Equation 15.32,

$$\begin{aligned}E\left\{y_0^2\right\} &= \left[k_0^2 + 2k_0k_1 + \frac{k_1^2}{1-\alpha_1^2} \right] E\left\{e_0^2\right\} \\ &= (NPG)_0 E\left\{e_0^2\right\}\end{aligned}$$

The effect of quantization error due to multiplication on system output is determined as follows. Setting all signals but e_a to zero,

$$y_a(z) = G_a(z) E_a(z) = T(z) E_a(z)$$

Therefore,

$$E\{y_a(z)\} = \mu_a \left[k_0 + \frac{k_1}{1 - \alpha_1} \right]$$

and

$$E\left\{y_a^2\right\} = \left[k_0^2 + 2k_0k_1 + \frac{k_1^2}{1-\alpha_1^2} \right] E\left\{e_a^2\right\}$$

Assuming the multiplier errors e_1 and e_2 are uncorrelated, then

$$\text{variance}\{e_a = e_1 + e_2\} = E\left\{e_1^2\right\} + E\left\{e_2^2\right\}$$

Similarly, the second multiplier noise gain is calculated by setting all sources but e_3 to zero. The transfer function which relates e_3 to the output is

$$Y_3(z) = G_3(z) E_3(z)$$

where

$$G_3(z) = 1.$$

Then,

$$E\{y_3\} = E\{e_3\}$$

and

$$E\left\{y_3^2\right\} = E\left\{e_3^2\right\}$$

Also,

$$(NPG)_3 = 1.$$

Assuming the output noises y_0, y_a, and y_3 are uncorrelated, then

$$\text{variance}\{y_0 + y_a + y_3\} = (NPG)_0 E\left\{e_0^2\right\} + (NPG)_a\left[E\left\{e_1^2\right\} + E\left\{e_2^2\right\}\right] + (NPG)_3 E\left\{e_3^2\right\}$$

15.6.3 Quantization Noise Model of Second Order Subsystems

The previous analysis of first-order subsystems can easily be extended to second-order subsystems. The quantization noise model shown in figure 15.10 for first-order subsystems can easily be modified for second-order subsystems. Setting all the noise sources to zero, the ideal transfer function of a second-order subsystem which relates $X(z)$ to $Y(z)$ is

$$T(z) = \frac{Y(z)}{X(z)} = \frac{A(1 + \beta_1 z^{-1} + \beta_2 z^{-2})}{1 - \alpha_1 z^{-1} - \alpha_2 z^{-2}}.$$

Assuming that the output noises y_0, y_a, and y_b are uncorrelated, then

$$\text{variance}\{y_0 + y_a + y_b\} = (NPG)_0 E\left\{e_0^2\right\} + (NPG)_a\left[E\left\{e_1^2\right\} + E\left\{e_2^2\right\} + E\left\{e_3^2\right\}\right] + (NPG)_b\left[E\left\{e_4^2\right\} + E\left\{e_5^2\right\}\right]$$

where

$$(NPG)_0 = (NPG)_a$$

is determined from $T(z)$ and

$$(NPG)_b = 1.$$

15.6.4 Floating-Point Arithmetic

The analysis of quantization errors in floating-point digital controllers is more complicated than in fixed-point digital controllers [9]–[11]. It was mentioned earlier that in floating point arithmetic errors occur only in the mantissa. It was also mentioned that roundoff and truncation errors are introduced in both addition and multiplication. Let x_1 and x_2 be any two numbers before quantization. Quantizing the sum and the product of these two numbers gives

$$(x_1 + x_2)_q = (x_1 + x_2)(1 + \varepsilon_s) \quad (15.37)$$

$$(x_1 \cdot x_2)_q = (x_1 \cdot x_2)(1 + \varepsilon_p) \quad (15.38)$$

respectively, where the relative errors ε_s and ε_p, depending on the number representation, satisfy Equations 15.9 through 15.12. Each arithmetic operation introduces quantization error according to Equations 15.37 and 15.38. Detailed examples of roundoff and truncation errors accumulated in first- and second-order subsystems using floating-point arithmetic are given in [12].

15.7 Limit Cycle and Deadband Effects

When digital controllers are implemented with finite word length, *limit cycles*, or sustained oscillations, may appear at the controller output even in the absence of any applied input. Basically, there are two different kinds of limit cycles. One is due to roundoff in multiplication, termed the *deadband* effect, and the other is due to register overflow. Limit cycles exist in fixed-point digital controllers but can be ignored in floating-point controllers [9].

To illustrate the phenomenon of limit cycle due to roundoff, consider the first-order controller described by the difference equation

$$y(k) = ay(k-1) + x(k) \quad (15.39)$$

Let

$$x(k) = 0.9\delta(k), \quad a = 0.5, \quad y(-1) = 0.$$

If the controller equation is implemented with infinite word length registers, then

$$y(k) = 0.9(0.5)^k$$

As k approaches infinity, the steady-state value of the output $y(k)$ approaches zero. However, assuming that the controller equation is implemented with a 3-bit word length, then

$$y_q(k) = Q[0.5y_q(k-1)] + 0.75\delta(k)$$

Using decimal representation, the output can be calculated recursively as follows:

$$\begin{aligned}
y_q(0) &= Q[(0.5)(0)] + 0.75 = 0.75\\
y_q(1) &= Q[(0.5)(0.75)] = 0.375\\
y_q(2) &= Q[(0.5)(0.375)] = 0.25\\
y_q(3) &= Q[(0.5)(0.25)] = 0.125\\
y_q(4) &= Q[(0.5)(0.125)] = 0.125\\
&\vdots\\
y_q(k) &= Q[(0.5)(0.125)] = 0.125
\end{aligned}$$

Hence, as k approaches infinity, the steady-state value of $y_q(k)$ approaches 0.125 and not zero.

As another example, consider the system described by Equation 15.39. Let

$$x(k) = 0, \quad a = -0.5, \quad y(-1) = 0.75$$

Again, if the controller equation is implemented with infinite word length registers, then the output,

$$y(k) = 0.75(-0.5)^k$$

approaches zero as k approaches infinity. Assuming that the controller equation is implemented with a 3-bit word length, then

$$y_q(k) = Q[-0.5y_q(k-1)]$$

The output can be calculated recursively as follows:

$$\begin{aligned} y_q(0) &= Q[(-0.5)(0.75)] = Q[-0.375] = -0.375 \\ y_q(1) &= Q[(-0.5)(-0.375)] = 0.25 \\ y_q(2) &= Q[(-0.5)(0.25)] = -0.125 \\ y_q(3) &= Q[(-0.5)(-0.125)] = 0.125 \\ y_q(k) &= Q[(-0.5)(0.125)] = -0.125 \end{aligned}$$

and the output oscillates between 0.125 and −0.125 indefinitely.

An interesting example of limit cycle due to register overflow is given in [13]. Limit cycles due to roundoff and overflow are unwanted and their effect on control system performance should be minimized.

References

[1] Hanselmann, H., Implementation of Digital Controllers — A Survey, IFAC. 23(1), 7–32, 1987.

[2] Hwang, K., *Computer Arithmetic,* John Wiley & Sons, New York, 1979.

[3] Rabiner, L.R. and Gold, B., *Theory and Application of Digital Signal Processing,* Prentice Hall, Englewood Cliffs, NJ, 1975.

[4] Gold, B. and Rader, C.M., *Digital Processing of Signals,* McGraw-Hill, New York, 1969.

[5] Liu, B., Effect of Word Length on the Accuracy of Digital Filters—A Review, *IEEE Trans. Circuit Theory,* 18(6), 670–677, 1971.

[6] Slaughter, J.B., Quantization Errors in Digital Control Systems, *IEEE Trans. Automat. Contr.,* AC-9, 70–74, 1964.

[7] Franklin, G.F., Powell, J.D., and Workman, M.L., *Digital Control of Dynamic Systems,* 2nd ed., Addison-Wesley, Reading, MA, 1990.

[8] Åström, K.J, Jury, E.I., and Angiel, R.G., A Numerical Method for the Evaluation of Complex Integrals, *IEEE Trans. Automat. Contr.,* AC-15, 468–471, 1970.

[9] Sandberg, I.W., Floating-Point Roundoff Accumulation in Digital Filter Realization, *Bell Syst. Tech. J.,* 46, 1775–1791, 1967.

[10] Liu, B. and Kaneko, T., Error Analysis of Digital Filters Realized with Floating Point Arithmetic, *Proc. IEEE,* 57, 1735–1747, 1969.

[11] Aggarwal, J.K. and Kan, E.P., Error Analysis of Digital Filter Employing Floating-Point Arithmetic, IEEE Trans. CT., 18(6), 678–686, 1971.

[12] Oppenheim, A.V. and Schafer, R.W., *Digital Signal Processing,* Prentice Hall, Englewood Cliffs, NJ, 1975.

[13] Stubberud, P., *Digital Signal Processing,* Class Notes. University of Nevada, Las Vegas, 1990.

16

Sample-Rate Selection

Mohammed S. Santina
The Aerospace Corporation, Los Angeles, CA

Allen R. Stubberud
University of California, Irvine, Irvine, CA

Gene H. Hostetter

16.1 Introduction

In this chapter, the selection of sampling rate for a digital control system is briefly discussed. As the sampling rate is increased, the performance of the digital controller usually improves. Computer cost also increases because less time is available to process the controller equations. Reducing the sample rate for the sake of reducing cost, however, may degrade system performance or even cause instability. Additionally, for systems with analog-to-digital converters, higher sample rates require faster A/D conversion speed which may also be expensive.

Aside from cost, the selection of sampling rate for digital control systems depends on many factors, including smoothness of the time response, effects of disturbances and sensor noise, parameter variations and quantization. *The best sampling rate which can be chosen for a digital control system is the slowest rate that meets all performance requirements.* Before we discuss the selection of sampling rate, a statement of the sampling theorem is in order.

16.2 The Sampling Theorem

Sampling is the process of deriving a discrete-time sequence from a continuous-time function. Usually, but not always, the samples are evenly spaced in time. *Reconstruction* is the reverse; it is the formation of a continuous-time function from a sequence of samples. Many different continuous-time functions can have the same set of samples, so a reconstruction is not unique. Figure 16.1 shows two different continuous-time signals with the same samples, illustrating how, except in highly restricted circumstances, a sampled function is not uniquely determined by its samples. One important situation for which samples of a continuous-time function are unique occurs when the function is *bandlimited.* A signal $g(t)$ and its Fourier transform $G(\omega)$ are generally related by

$$\begin{aligned} G(\omega) &= \int_{-\infty}^{\infty} g(t)e^{-j\omega t}\,dt \\ g(t) &= \frac{1}{2\pi}\int_{-\infty}^{\infty} G(\omega)e^{j\omega t}\,d\omega \end{aligned} \tag{16.1}$$

This relationship is similar to Laplace transformation with $s =$

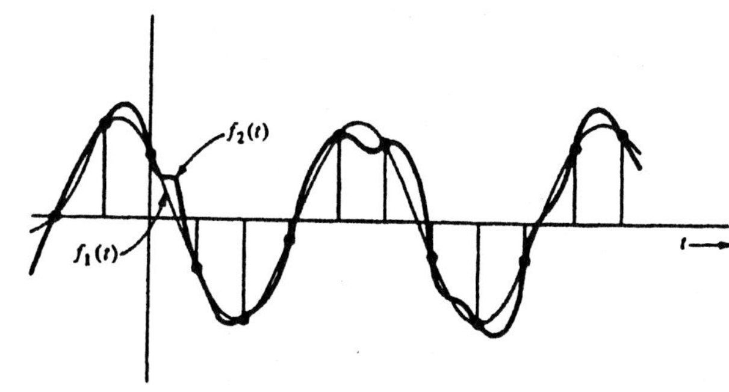

Figure 16.1 Two different continuous-time signals with the same samples.

$j\omega$, but the transform integral of Equation 16.1 extends over all time rather than from $t = 0^-$ on. The Fourier transform $G(\omega)$ is termed the *spectrum* of $g(t)$.

If a signal $g(t)$ is uniformly sampled with sampling period T to form the sequence

$$g(k) = g(t = kT)$$

then the corresponding impulse train that extends both ways in time

$$g^*(t) = \sum_{k=-\infty}^{\infty} g(kT)\delta(t - kT)$$

0-8493-8570-9/96/$0.00+$.50

is a continuous-time (hence it has a Fourier transform) signal that is equivalent to $g(kT)$ and has the Fourier transform

$$G^*(\omega) = \frac{1}{T}\sum_{n=-\infty}^{\infty} G(\omega - n\omega_s) \qquad (16.2)$$

where

$$\omega_s = 2\pi f_s = \frac{2\pi}{T}$$

To prove this result, the periodic function

$$s(t) = \sum_{k=0}^{\infty} \delta(t - kT)$$

is represented by an exponential Fourier series of the form

$$s(t) = \sum_{n=-\infty}^{\infty} d_n e^{(jn2\pi/T)t}$$

where

$$d_n = \frac{1}{T}\int_{-T/2}^{T/2} \sum_{k=-\infty}^{\infty} \delta(t - kT) e^{-(jn2\pi/T)t} dt = \frac{1}{T}$$

Hence,

$$s(t) = \frac{1}{T}\sum_{n=-\infty}^{\infty} e^{(jn2\pi/T)t}$$

Substituting this result in the impulse train,

$$g^*(t) = \sum_{k=-\infty}^{\infty} g(kT)\delta(t - kT)$$

gives

$$g^*(t) = \frac{1}{T} g(t) \sum_{n=-\infty}^{\infty} e^{(jn2\pi/T)t}$$

and taking the Fourier transform yields

$$G^*(\omega) = \frac{1}{T}\sum_{n=-\infty}^{\infty} \int_{-\infty}^{\infty} g(t) e^{(jn2\pi/T)t} e^{-j\omega t} dt$$

Therefore,

$$G^*(\omega) = \frac{1}{T}\sum_{n=-\infty}^{\infty} G\left(\omega - n\frac{2\pi}{T}\right) = \frac{1}{T}\sum_{n=-\infty}^{\infty} G(\omega - n\omega_s)$$

which completes the proof.

The function $G^*(\omega)$ in Equation 16.2 is periodic in ω, and each individual term in the series has the same form as the original $G(\omega)$, with the exception that the nth term is centered at

$$\omega = n\frac{2\pi}{T} \qquad n = \ldots, -2, -1, 0, 1, 2, \ldots$$

In general, then, if $G(\omega)$ is not limited to a finite frequency range, these terms overlap each other along the ω axis. A signal is bandlimited at (hertz) frequency f_B if

$$G(\omega) = 0 \quad \text{for} \quad |\omega| > 2\pi f_B = \omega_B$$

as shown in Figure 16.2(a). Equation 16.1 becomes

$$g(t) = \frac{1}{2\pi}\int_{-\omega_B}^{\omega_B} G(\omega) e^{j\omega t} d\omega$$

If the sampling frequency f_s is more than twice the bandlimit frequency f_B, the individual terms in Equation 16.2 do not overlap as shown in Figure 16.2(b), and $G(\omega)$ and thus $g(t)$ can be determined from $G^*(\omega)$, which in turn, is determined from the samples $g(k)$. Furthermore, if the sampling frequency f_s is exactly twice the bandlimit frequency f_B, the individual terms in Equation 16.2 do not overlap as shown in Figure 16.2(c).

In terms of the sampling period,

$$\omega_s = 2\omega_B$$

and

$$T = \frac{2\pi}{\omega_s}$$

Then,

$$T = \frac{\pi}{\omega_B} = \frac{1}{2f_B}$$

which relates the sampling period to the highest frequency f_B in the signal.

A statement of the sampling theorem is the following:

The uniform samples of a signal $g(t)$, that is bandlimited above (hertz) frequency f_B, are unique if, and only if, the sampling frequency is higher than $2f_B$.

In terms of the sampling period,

$$T < \frac{1}{2f_B} \qquad (16.3)$$

The frequency $2f_B$ is termed the *Nyquist frequency* for a bandlimited signal. As shown in Figure 16.2(d), if the sampling frequency does not exceed the Nyquist frequency, the individual terms in Equation 16.2 overlap, a phenomenon called *aliasing* (or *foldover*).

In digital signal processing applications, selection of the sampling period also depends on the reconstruction method used to recover the bandlimited signal from its samples [1]. Another statement of the sampling theorem related to signal reconstruction states that *when a bandlimited continuous-time signal is sampled at a rate higher than twice the bandlimit frequency, the samples can be used to reconstruct uniquely the original continuous-time signal.*

Åström and Wittenmark [2] suggest, by way of example, a criterion for the selection of the sample rate that depends on the magnitude of the error between the original signal and the reconstructed signal. The error decreases as the sampling rate is

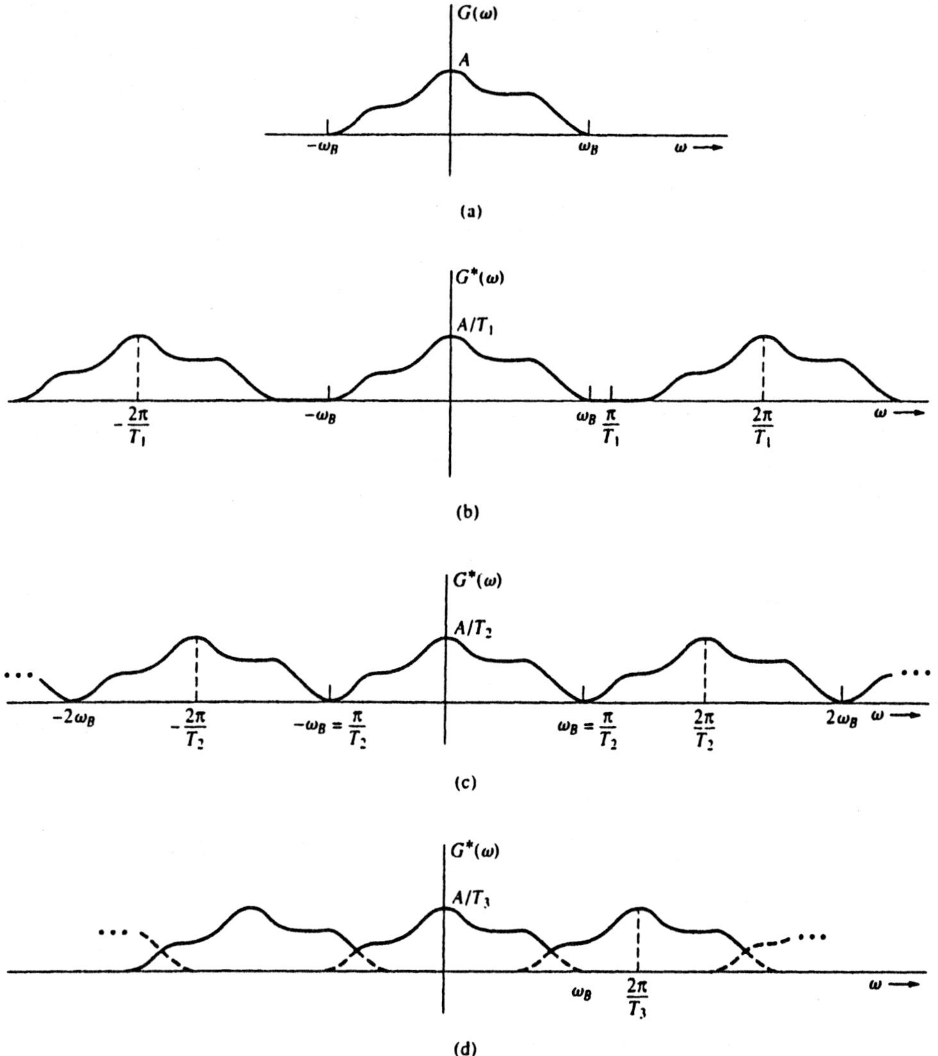

Figure 16.2 Frequency spectra of a signal sampled at various frequencies. (a) Frequency spectrum of an analog bandlimited signal $g(t)$. (b) Frequency spectrum of a sampled signal $g^*(t)$ with $f_{s1} > 2f_B(f_{s1} = 1/T_1)$. (c) Frequency spectrum of a sampled signal $g^*(t)$ with $f_{s2} = 2f_B(f_{s2} = 1/T_2)$. (d) Frequency spectrum of a sampled signal $g^*(t) < 2f_B(f_{s3} = 1/T_3)$.

increased considerably higher than the Nyquist rate. Depending on the hold device used for reconstruction, the number of samples required may be several hundreds per sampling period.

Although the sampling theorem is not applicable to most discrete-time control systems because the signals (e.g., steps and ramps) are not bandlimited and because good reconstruction requires long time delays, it does provide some guidance in selecting the sample rate and in deciding how best to filter sensor signals before sampling them.

16.3 Control System Response and the Sampling Period

The main objective of many digital control system designs is to select a controller so that the system-tracking output, as nearly as possible, tracks or "follows" the tracking command input. Perhaps, the first figure of merit that the designer usually selects is the closed-loop bandwidth, f_c (Hz), of the feedback system because f_c is related to the speed at which the feedback system should track the command input. Also, the bandwidth f_c is related to the amount of attenuation the feedback system must provide in the face of plant disturbances. It is then appropriate to relate the sampling rate to the bandwidth f_c, as suggested by the sampling theorem, because the bandwidth of the closed-loop system is related to the highest frequency of interest in the command input.

Consider the control system shown in Figure 13.3(a) in the article entitled "Discrete-Time Equivalents to Continuous-Time Systems" where the controller

$$G(s) = \frac{s + 10}{s(s + 4)}$$

has been designed so that the resulting feedback system has a 3 dB bandwidth $f_c = 0.45$ Hz. The step response of the digital control system using Euler's and Tustin's approximations for various sampling periods is shown in Figure 13.6. Raising the sample rate tends to decrease the amplitude of each step input change and thus reduces the amplitude of the undesirable between-sample response. As the sampling period is decreased from $T = 0.4$ sec. to $T = 0.1$ sec., or equivalently, the sampling rate is increased from 2.5 Hz (5.5 times f_c) to 10 Hz (22 times f_c), the step response of the feedback system using either approximation approaches the step response of the continuous-time system. However, as discussed in Chapter 13, Tustin's approximation usually gives better results than Euler's approximation for the same sampling period.

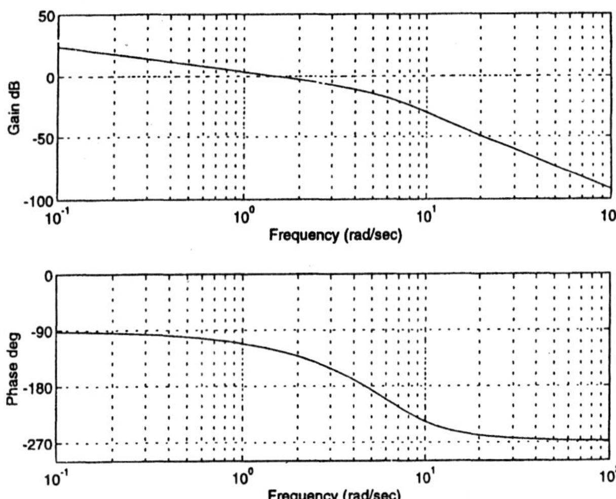

Figure 16.3 Frequency response of open-loop transfer function of the example system.

As a general rule of thumb, the sampling period should be chosen in the range

$$\frac{1}{30 f_c} < T < \frac{1}{5 f_c} \tag{16.4}$$

Of course, other design requirements may require even higher sample rates, but sampling rates less than 5 times f_c are not desirable and should be avoided if possible.

An interesting problem involving the sample rate selection is encountered in the control system design of flexible spacecraft. The spacecraft has a large number of bending modes of which the lowest bending mode frequency may be a fraction of 1 Hz and the highest frequency of interest may be 100 Hz or even higher. Typically, the closed-loop bandwidth of the spacecraft is an order of magnitude less than the lowest mode frequency, and as long as the controller does not excite any of the flexible modes, the sampling period may be selected solely based on the closed-loop bandwidth. Otherwise, these modes need to be attenuated[1] or controlled, and therefore, their frequencies will impact the sampling rate selection [3].

Another criterion for selecting the sampling period is based on the *rise time* of the feedback system so as to provide smoothness in the time response. It can easily be shown that the rise time (10% to 90%), T_r, of a first-order system of the form,

$$H(s) = \frac{1}{\tau s + 1}$$

is

$$T_r = 2.2\tau$$

The sampling period, in terms of the rise time, can be selected according to

$$0.095 T_r < T < 0.57 T_r$$

which is derived from Equation 16.4. Similarly, the rise time of the second-order system,

$$H(s) = \frac{\omega_n^2}{s^2 + 2\zeta\omega_n s + \omega_n^2}$$

is

$$T_r = \frac{\pi - \beta}{\omega_d}$$

where

$$\omega_d = \omega_n\sqrt{1-\zeta^2}$$

and

$$\beta = \sin^{-1}\sqrt{1-\zeta^2}$$

For a damping ratio $\zeta = 0.707$, the rise time is

$$T_r = \frac{3.33}{\omega_n}$$

Based on Equation 16.4, the sampling period is

$$0.06 T_r < T < 0.4 T_r. \tag{16.5}$$

Continuing with the previous example, according to Equation 16.5, the sampling period should be $T = 0.11$ sec. which agrees with the previous results.

In digital control systems, a time delay of up to a full sample period may be possible before the digital controller can respond to the next input command. Franklin et al. [3] suggest that the time delay be kept to about 10% of the rise time. Then, the sampling period should satisfy

$$T < \frac{0.05}{f_c}$$

Yet another criterion for selecting the sampling period, that depends on the frequency response of the continuous-time system, is given by Åström and Wittenmark [2]. The sampling rate is selected so that

$$0.15 < T\omega_0 < 0.5 \tag{16.6}$$

[1] When the modal parameters are well known, notch filters can be used to attenuate the modes, if necessary. In this case, the notch frequency of the filter attenuating the mode with the highest frequency may dictate the sampling rate of the digital controller.

where ω_0 is the gain crossover frequency of the continuous-time system in radians per second.

Continuing with the previous example, the open-loop transfer function is

$$G(s)H(s) = \frac{25(s+10)}{s(s^2+9s+40)(s+4)}$$

The frequency response of GH is shown in Figure 16.3 where the gain crossover frequency is $\omega_0 = 1.572$ rad/sec. According to Equation 16.6, the sampling period should be chosen in the range

$$0.095 < T < 0.32$$

which also agrees with the previous results.

16.4 Control System Response to External Disturbances

Plant disturbances are undesired, inaccessible plant inputs that the plant should *not* track. An example of disturbance is wind gusts buffeting a positioning system for a microwave antenna. Like initial conditions, the specific disturbance signals are normally unknown, although something is probably known about their character, their statistics, or both. As far as the selection of sampling rate is concerned, the most important plant disturbance to consider is random white noise because of its high frequency contents.

In general, when the controller is implemented digitally, it will perform less well than the analog controller in the face of white noise disturbance inputs. As the sampling rate is increased, the response of the digital controller usually approaches the response of the continuous-time controller.

Consider the continuous-time system described by

$$\dot{x} = \mathcal{A}x(t) + \mathcal{B}u(t) + Lw(t) \tag{16.7}$$

where $u(t)$ is the control input and $w(t)$ is a white noise process with covariance matrix

$$E\{w(t)w^{\dagger}(t+\tau)\} = Q\delta(\tau)$$

where E denotes expected value. If the control input is given by

$$u(t) = E_c x(t) \tag{16.8}$$

then

$$\begin{aligned}\dot{x} &= (\mathcal{A} + \mathcal{B}E_c)x(t) + Lw(t)\\ &= \mathcal{A}_c x(t) + Lw(t)\end{aligned} \tag{16.9}$$

Let the state covariance matrix be

$$P_c(t) = E\{x(t)x^{\dagger}(t)\}$$

It can be shown [4] that the steady state solution of the state covariance matrix P_c is

$$\mathcal{A}_c P_c + P_c\mathcal{A}_c^{\dagger} + LQL^{\dagger} = 0 \tag{16.10}$$

where P_c is a measure of the variation of the state vector about its mean. The solution of Equation 16.10 can be easily obtained using MATLAB (see LYAP.m) or some other computer-aided design tool.

When the controller is implemented digitally, however, the covariance of the discretized state vector will, in general, be higher than the covariance of the continuous-time state vector for identical disturbance inputs. This is to say that the amplitude of the state will be higher with the discrete controller than its continuous-time counterpart. When the continuous-time system described by Equation 16.7 is sampled with sampling period T, its discrete-time equivalent is (see Chapter 13).

$$x(k+1) = \Phi x(k) + Bu(k) + \omega(k) \tag{16.11}$$

where

$$\Phi = e^{\mathcal{A}T} = I + \mathcal{A}T + \frac{\mathcal{A}^2T^2}{2!} + \frac{\mathcal{A}^3T^3}{3!} + \cdots \tag{16.12}$$

and

$$B = \left(\int_0^T e^{\mathcal{A}\eta}d\eta\right)\mathcal{B} = \left[IT + \frac{\mathcal{A}T^2}{2!} + \frac{\mathcal{A}^2T^3}{3!} + \cdots\right] \tag{16.13}$$

The discrete-time noise is given by the integral

$$\omega(k) = \int_{kT}^{kT+T} \Phi(kT+T-\tau)Lw(\tau)d\tau$$

The covariance of the discrete-time noise is

$$\begin{aligned}Q_d &= E\{\omega(k)\omega^{\dagger}(k)\} = \int_{kT}^{kT+T}\int_{kT}^{kT+T} \Phi(kT+T-\tau)\cdot\\ &\quad LE\{w(\tau)w^{\dagger}(\lambda)\}L^{\dagger}\Phi^{\dagger}(kT+T-\lambda)d\lambda d\tau\\ &= \int_{kT}^{kT+T}\int_{kT}^{kT+T} \Phi(kT+T-\tau)\cdot\\ &\quad LQ\delta(\tau-\lambda)L^{\dagger}\Phi^{\dagger}(kT+T-\lambda)d\lambda d\tau\\ &= \int_{kT}^{kT+T} \Phi(kT+T-\tau)\cdot\\ &\quad LQL^{\dagger}\Phi^{\dagger}(kT+T-\tau)d\tau\end{aligned}$$

Let

$$\gamma = kT + T - \tau.$$

Then

$$Q_d = \int_0^T \Phi(\gamma)LQL^{\dagger}\Phi^{\dagger}(\gamma)d\gamma \tag{16.14}$$

Returning to Equation 16.11, the state feedback

$$u(k) = E_d x(k)$$

gives

$$\begin{aligned}x(k+1) &= (\Phi + BE_d)x(k) + \omega(k)\\ x(k+1) &= \Phi_c x(k) + \omega(k)\end{aligned} \tag{16.15}$$

Therefore, using Equation 16.15, the discrete-time state covariance matrix is

$$\begin{aligned} P_d(k+1) &= E\{x(k+1)x^\dagger(k+1)\} \\ &= E\{[\Phi_c x(k) + \omega(k)][\Phi_c x(k) + \omega(k)]^\dagger\} \\ &= \Phi_c P_d(k)\Phi_c^\dagger + Q_d \end{aligned}$$

where Q_d is given by Equation 16.14. Hence, the steady state covariance discrete-time matrix is

$$P_d = \Phi_c P_d \Phi_c^\dagger + Q_d \tag{16.16}$$

which can easily be solved using MATLAB (see DLYAP.m).

To illustrate the ideas involved in the selection of the sampling rate of a system driven by a white noise disturbance, consider the following simplified model for the roll attitude control of a spacecraft:

$$\begin{aligned} \dot{x}(t) &= \begin{bmatrix} 0 & 1 \\ 0 & 0 \end{bmatrix} x(t) + \begin{bmatrix} 0 \\ 1/J \end{bmatrix} u(t) + \begin{bmatrix} 0 \\ 1/J \end{bmatrix} w(t) \\ y(t) &= [1 \quad 0]x(t) \end{aligned}$$

where

x_1	=	roll attitude of the spacecraft in radians
x_2	=	roll rate in radians per second
u	=	control torque about the vehicle roll axis produced by the spacecraft actuators in foot-pounds
w	=	disturbance torque acting on the spacecraft in foot-pounds
J	=	moment of inertia of the vehicle about the roll axis at the vehicle center of mass in slug-feet squared.

For simplicity, we assume that $J = 1$. Suppose that it is desired to have both eigenvalues of the state feedback system at $s_{1,2} = -4.6$. Then the feedback gain vector is

$$e^\dagger = [-21.16 \quad -9.2]$$

If the noise covariance $Q = 1$, then the steady-state solution of the state covariance matrix given by Equation 16.10 is

$$Pc = \begin{bmatrix} 0.002568 & 0 \\ 0 & 0.0543476 \end{bmatrix}$$

and, therefore, the RMS of the spacecraft attitude,[2] x_1, is 0.0507, and the RMS of the spacecraft attitude rate, x_2, is 0.2231.

[2]Root-mean-square (RMS) of a random variable X is the square root of the mean square value (second moment) of X. If the random variable X has zero mean, then the RMS value and standard deviation of X are equal.

If the continuous-time model of the spacecraft is discretized with sampling period T, then, according to Equations 16.12, 16.13, and 16.14,

$$\begin{aligned} \Phi &= \begin{bmatrix} 1 & T \\ 0 & 1 \end{bmatrix} \\ b &= \begin{bmatrix} \frac{T^2}{2} \\ T \end{bmatrix} \end{aligned}$$

and

$$Q_d = \begin{bmatrix} \frac{T^3}{3} & \frac{T^2}{2} \\ \frac{T^2}{2} & T \end{bmatrix}$$

If the eigenvalues of the discrete-time system are located at $z_{1,2} = e^{-4.6T}$, then the feedback gain vector for the discrete-time system is

$$e^\dagger = \begin{bmatrix} \frac{2e^{-4.6T} - e^{-9.2T} - 1}{T^2} & \frac{2e^{-4.6T} + e^{-9.2T} - 3}{2T} \end{bmatrix}$$

Figure 16.4 shows the RMS values of the states x_1 and x_2 in terms of the sampling period generated with Equation 16.16. As the sampling period is increased, the RMS values of the states increase, and, as the sampling period is decreased, the RMS values decrease and eventually approach the RMS values of the continuous-time state variables calculated earlier. Examining the figure, an appropriate value of the sampling period is $T = 0.05$ seconds. The performance of the digital controller degrades as T is increased. This value of the sampling period also agrees with inequality Equation 16.4.

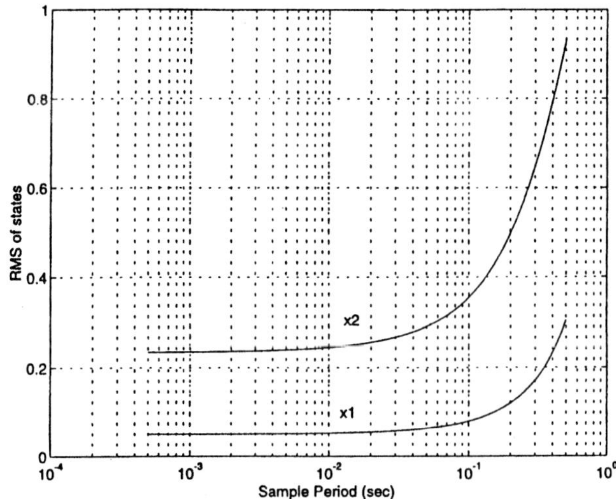

Figure 16.4 Response of digital control system as a function of a sampling period. As the sampling period is increased, the response degrades and, as the sampling period is decreased, response approaches continuous-time response.

16.5 Measurement Noise and Prefiltering

The sampling theorem is important to control system design because when A/D conversion is done on noisy signals with significant frequency components above half the sampling frequency, the high frequencies produce errors in the sampling *indistinguishable* from lower frequency errors. For this reason, low-pass filters, termed *prefilters,* or *antialiasing filters,* are used to reduce the high frequencies in sensor signals before their A/D conversion as in Figure 16.5.

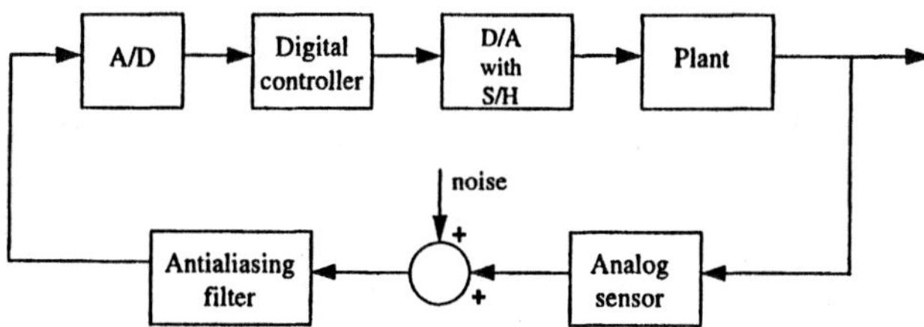

Figure 16.5 Antialiasing filters reduce high frequency noise in the sensor signal before A/D conversion.

For example, consider the noisy voltage signal, in Figure 16.6(a), composed of a 2 Hz sinusoidal signal of amplitude 2 volts and an 80 Hz sinusoidal noise of amplitude 0.2 volts. If this signal is sampled with $T = 1/45$ sec., then according to the sampling theorem, the sampled signal will be aliased as in Figure 16.6(b).

The first-order low-pass filter,

$$H(s) = \frac{50}{s+50},$$

selected with a corner frequency of about one-sixth the sampling frequency, will attenuate the 80 Hz noise as shown in Figure 16.6(c). Samples of the filtered analog signal are shown in Figure 16.6(d).

In feedback control systems, the phase lag of antialiasing filters may be significant enough to cause system instability. However, if the corner frequency of the filter is sufficiently higher than the control system bandwidth, the phase lag of the filter may not affect the performance of the system. On the other hand, if noise attenuation requires higher order filters or filters with corner frequencies close to the control system bandwidth, the filter should be treated as if it is part of the plant controlled by the discrete-time controller.

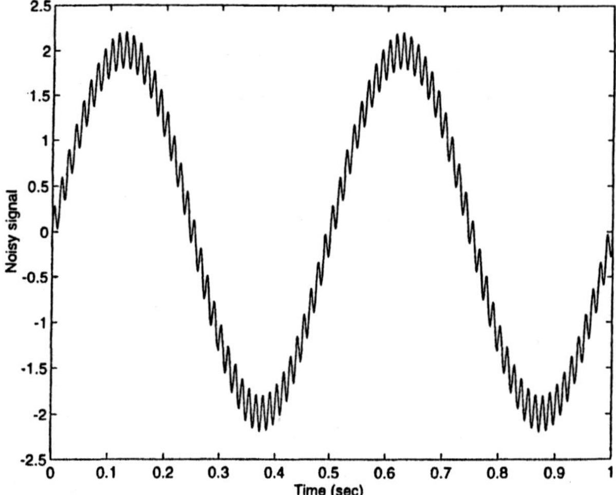

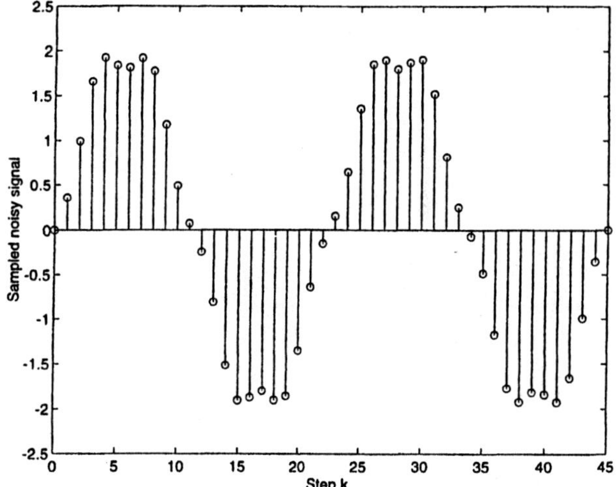

Figure 16.6 (a–d) Filtering of measurement noise using antialiasing filters. (a) Noisy signal; (b) aliased signal.

16.6 Effect of Sampling Rate on Quantization Error

Quantization errors in digital control systems are discussed in detail in the chapter entitled "Quantization Effects". In some applications, quantization errors cannot be ignored and their effect on system output depends on the sampling period.

As an example, consider the analog controller described by

$$H(s) = \frac{10^4}{s+1} \tag{16.17}$$

Its discrete-time equivalent, using the impulse invariant approximation, is

$$\begin{aligned} H(z) &= \frac{10^4 z}{z - e^{-T}} = \frac{10^4}{1 - e^{-T} z^{-1}} \\ &= k_0 + \frac{k_1}{1 - \alpha z^{-1}} \end{aligned}$$

where

$$k_0 = 0, \quad k_1 = 10^4, \quad \alpha = e^{-T}.$$

As discussed in the chapter entitled "Quantization Effects", the variance of the controller output equals the variance of the roundoff noise times the noise power gain. For the first order controller, the variance of the output is determined with Equations 15.20, 15.29, 15.30, and 15.32. Figure 16.7 shows the RMS of the controller output in terms of the sampling period and

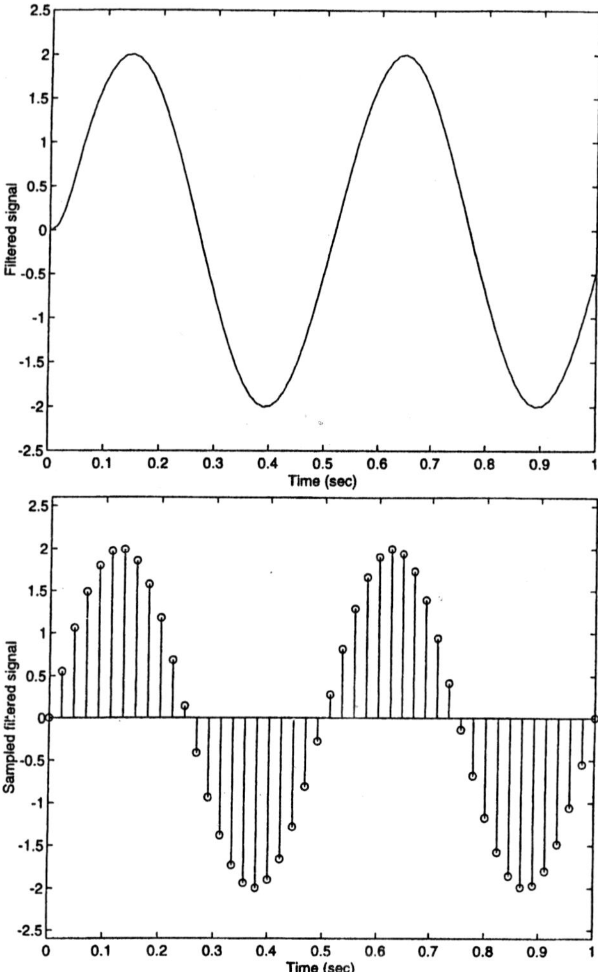

Figure 16.6 *(Continued)* Filtering of measurement noise using antialiasing filters. (c) filtered signal; (d) sampled filtered signal.

the word length C. As the sampling period is decreased, the RMS value of the output is increased. Also shown in the figure, increasing the word length C decreases the RMS value of the output. The RMS output does not necessarily increase as the sampling period is decreased. If the analog controller described by Equation 16.17 is discretized using Tustin's approximation, the discrete-time controller becomes

$$H(z) = \frac{10^4(z+1)}{\left(\frac{2}{T}+1\right)z + 1 - \frac{2}{T}}$$

Repeating the previous RMS analysis on this controller gives the results in Figure 16.8. Although the RMS of the controller output increases as T increases, the RMS value is less with Tustin's approximation than with the impulse-invariant approximation. Also shown in the figure, the RMS of the output increases as the word length is decreased.

It is through computer simulation of the plant and the controller that the best sample rate is achieved. It is good practice to investigate carefully the behavior of the controlled system for various sample rates when the arithmetic precision of the controller is reduced, when disturbances and noises are injected into the system at likely points, and when the plant model is changed in ways that might occur in practice.

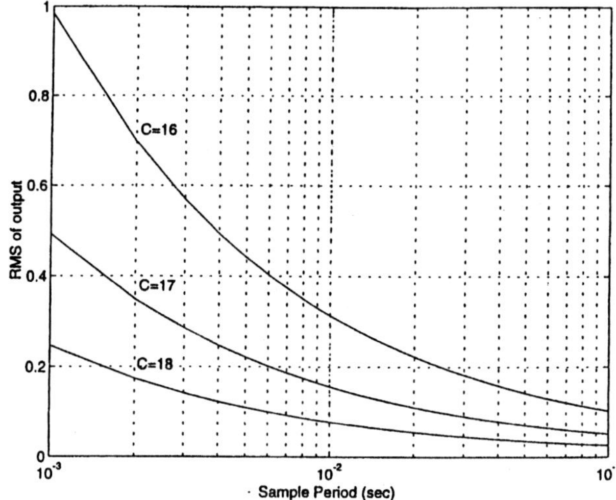

Figure 16.7 Response of controller using impulse invariant approximation as a function of sampling period for various word lengths. As the sampling period is increased, the error is decreased, and, as the word length is increased, the error is decreased.

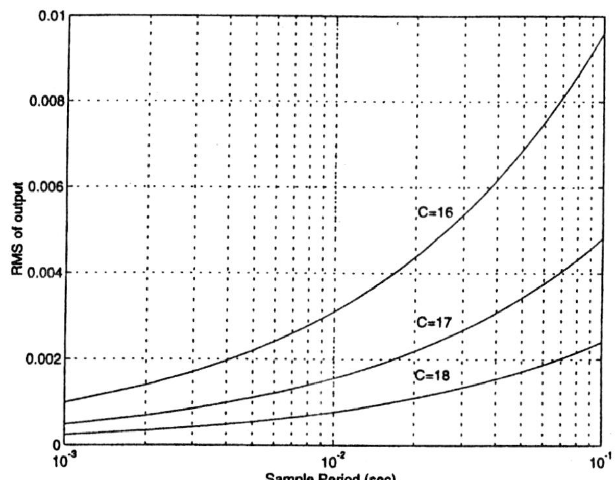

Figure 16.8 Response of controller using Tustin's approximation as a function of sampling period for various word lengths.

References

[1] Santina, M.S., Stubberud, A.R., and Hostetter, G.H., *Digital Control System Design*, 2nd ed., Saunders College Publishing, Philadelphia, 1994.

[2] Åström, K.J. and Wittenmark, B., *Computer Controlled Systems*, Prentice Hall, Englewood Cliffs, NJ, 1987.

[3] Franklin, G.F., Powell, J.D., and Workman, M.L., *Digital Control of Dynamic Systems,* 2nd ed., Addison-Wesley, Reading, MA, 1990.

[4] Bryson, A.E. and Ho, Y.C., *Applied Optimal Control,* Ginn, Waltham, MA, 1969.

17

Real-Time Software for Implementation of Feedback Control

David M. Auslander, John R. Ridgely, and Jason C. Jones
Mechanical Engineering Department, University of California at Berkeley

17.1 An Application Context

Digital computers have become the primary means of implementing feedback control algorithms. Real-time software is the medium in which these solutions are expressed. Computers are sequential devices, and, as such, can act only as sampled-data controllers, with time discretized. The fundamental characteristics of the real-time software, as distinct from "regular" software, are that the control algorithms must be run at their scheduled sample intervals (with some specified tolerance) and that associated software components, which interact with the sensors and actuators, can have critical time-window constraints.

The nature and difficulty of producing real-time software depends on the complexity and timing constraints of the problem. The tighter the time constraints with respect to the computer's basic computing speed limitations and the more things that need to be serviced simultaneously (at least "simultaneously" from the viewpoint of the control object), the more difficult it will be to complete the software design and implementation successfully.

17.2 The Software Hierarchy

Up to four distinct hierarchical levels can be present in software implementing feedback control. The extent to which all of these

0-8493-8570-9/96/$0.00+$.50

levels are used and the degree of interaction within and across levels determines the degree of complexity referred to above. The methods described here are appropriate for the types of demands placed on each of these levels by typical feedback control problems. Other problems, which might require more complex relationships within or across hierarchical levels, would require additional and/or different formalities for organizing the software design.

The four potential hierarchical levels are

1. Instrument/actuation activities: the lowest level of interaction with the instruments and actuators. Software at this level is usually short and might need to run frequently. It can be activated either by time or by an external signal. This level is also shared with peripheral controllers, such as disk, network, or printer controllers.
2. Feedback control algorithms: use the information from the instruments to compute actuation commands. Normally run on a specified time schedule.
3. Supervisory/sequence control: provide set point information to the feedback algorithms. Can also be used for corrections and adjustments to the feedback algorithms, such as adaptive control or set point scheduling.
4. Other: data logging, operator interface, communication with other computers, etc.

17.3 The Ground Rules

The following items describe the "ground rules" used to formulate the design methodology presented here.

17.3.1 Generic Solution

The methods and software developed here are designed for generic implementation. Although a number of proprietary means exist for implementing feedback control, for example, code translators that produce real-time code from a design, analysis, or simulation package, none of them is widely enough used to represent a *de facto* standard for feedback control implementation. For that reason, the methods here are intended for a broad variety of environments, without purchasing anything beyond basic computing tools.

17.3.2 Stand-Alone Feedback Control

The designs implemented are intended for stand-alone feedback control systems. When feedback control is a part of a larger software system, a variety of design constraints will be imposed based on the environment in which that system has been designed. The software methods described here will often be amenable to implantation in such environments, but, in some cases, may need substantial modification to be compatible.

17.3.3 Single Processor

The design formality used here extends only to applications that run in a single computer. Applications can be extended for multiprocessor implementation with *ad hoc* techniques for handling interprocessor communication.

17.4 Portability

Software portability, the ease with which software written in one environment can be used in another, is a major factor in overall development costs, time to market, maintainability, and upgradability. It has thus been a major focus in formulating feedback control implementation methodology.

17.4.1 Design Cycle

The design cycle for a control product typically goes through phases starting with simulation, followed by laboratory prototypes, preproduction prototypes, and finally, production systems. Each of these phases can use different computers and different development environments. Unless care is taken to assure software portability, large sections of code may need rewriting for each phase. In addition to the obvious cost consequences, there is a significant probability of introducing new bugs with each rewrite. To compound this problem, it is likely that software responsibility will rest with different people at each phase, placing a premium on consistent design methodology and documentation.

17.4.2 Life Cycle

Whereas current commercial life cycles for computers are two to three years, with overall capability approximately doubling in each new generation, the commercial lifetime of the equipment being controlled can be as long as twenty years. To keep a product up-to-date, it is often necessary to introduce new product models with minor changes in the physical design, but with new computational equipment greatly increasing the system's functionality, diagnostic abilities, communications, etc.

To accomplish this in a timely, cost-effective manner, it must be possible to build on existing software, as new models are introduced, and rapidly port old software to new platforms and operating environments.

17.4.3 Execution Shell

There are always parts of a control program that depend on the specific computational hardware and software being used. The *execution shell* isolates these environment-specific portions from the rest of the program. This part of the program will normally include interfaces to the:

- instrumentation/actuation hardware
- real-time operating system (if any)
- interrupt system

- clock
- operator interface

Concentrating this type of code in the execution shell insulates the control engineer from the need to do system-level programming, a critical part of portable software design.

17.4.4 Language

The general structure described here is independent of language, but *some* language must be chosen for the actual implementation. As of this writing, there are three top choices for languages to write portable control system implementation code: C, C++, and Ada.

Of these, C (currently the most widely used) is now an old design, which has been brought up-to-date with C++, a superset of C. Ada (a product of a committee of the US Department of Defense, USDOD) has facilities that make it more useful than C or C++ for highly complex projects. Ada's use is mandated for all appropriate USDOD projects.

C++ has been chosen for the samples in this chapter. The style of program writing that it encourages is much more conducive to writing code that can be maintained and modified more easily than C, and, because C++ is a superset of C, applications can be created by C programmers with only a recent introduction to C++. Ada, on the other hand, is not as widely used as C or C++, and the complexity needed to solve most feedback control problems does not warrant the use of Ada over C++. There are a large number of C++ instructional books currently available. The useful tutorial book by Lippman [3] introduces the C++ language without requiring previous background in C. The reference manual to C++ is written by Stroustrup [5], the inventor of the language.

17.5 Software Structure: Scan Mode

Successful control system design requires programming paradigms beyond the basic structure of algorithmic languages. In particular, the parallelism inherent in the control of physical systems and the notion of *duration,* are not concepts included within the syntax of standard languages.

17.5.1 Parallelism

Except for the simplest of control systems, several activities must be carried out simultaneously. A single computer, however, is a strictly sequential device, so that the parallelism viewed from the outside must be constructed by a rapid succession of sequential activities on the inside of the computer. The paradigm used for control system software must deal with the need for pseudosimultaneous execution of program components, and it must do that to meet the portability requirements. Multiple computers combined to do a single control job can exhibit true parallelism. In most practical cases, though, even when multiple computers are used, many of them must carry out more than one activity at a time.

A number of mechanisms, both commercial and *ad hoc,* exist for realizing parallelism in computing. The focus in this chapter will be on exploiting those mechanisms without compromising portability.

17.5.2 Nonblocking Code

The first step in constructing the paradigm for control system software design is to recognize the difference between *blocking* and *nonblocking* code. For a segment of code to be *nonblocking,* it must have a predictable computing time. The computing time need not be short, but it must be predictable.

A typical example of blocking code in many control programs is the "wait-for-something" statement based on a *while* loop,

```
while(-check-for-something-) ;
                          //Wait for an event
```

Because the event, in general, is asynchronous, there is no way to know when (or whether) the event will happen. The code is blocking because its execution time is not predictable. Likewise the *scanf* (or *cin*) statement is blocking because its execution time depends on when the user completes typing the requested input.

The methodology presented here is based on code that is completely nonblocking. Waiting for events, however, is fundamental to system control, so that a higher level of structure will be provided to accommodate such waits without the blocking code.

17.5.3 Scanned Code

The restriction to nonblocking code permits adopting a *scan* model for all software. In this model, all software elements are designed to be operated through repeated execution. It is the rapid repetition of repeats that gives the illusion of parallel execution even within a strictly sequential environment. By using the scan structure, parallelism can be maintained in computational environments, that do not normally support real-time multitasking, or across competitive and, therefore, usually incompatible real-time environments. In many respects, this software structure is an extension of the model used by programmable logic controllers (PLCs), which have successfully solved an important class of industrial control problems.

17.6 Control Software Design

Most important in assuring that software design will meet the engineering specifications of a project is a design structure that matches the problem reasonably well and allows for separating the solution into components matching the designers' interests and skills. In the feedback control software, the system engineering of the problem and its computational structure are partioned. This facilitates designing portable software, and match the skills for control system design to control engineer's knowledge.

17.6.1 System Engineering Structure: Tasks and States

A two-level structure characterizes the system engineering of the control job. *Tasks*, expressing the parallelism in the job, are a partition of the job into a set of semi-independent activities which, in general, are all active simultaneously. Tasks are internally characterized by *states*, indicating the particular action a task is carrying out.

The major creative engineering effort must consist of selecting and describing tasks and states. Once done, creating the actual computer code is relatively straightforward, because the formal definition of tasks and states modularizes the code, with code sections specifically connected to these design elements. This is very important for maintenance and upgrading where the code will need to be modified by people uninvolved in its original creation.

An additional benefit of the task/state paradigm is that the system operation is described in terms understandable to any engineers involved in the project. This opens the door to design review, sorely lacking in many software control projects.

Characterizing tasks as "semi-independent" recognizes that they are part of a system with a well-defined objective. Tasks will exchange data with other tasks and must synchronize their operations with other tasks. In all other aspects, they operate independently and asynchronously. They are asynchronous because their operation can be governed by activities outside of the computer which, in general, are not synchronous. Except for explicit synchronization, there is no *a priori* way to know how tasks will relate to each other computationally. This has important consequences for debugging and system reliability. In conventional software, erroneous situations can be repeatedly recreated until the cause of the problem is found. This is not possible in real-time systems because with asynchronous operation of tasks, with respect to each other and to external events, it may not be possible to reproduce an erroneous situation except in a statistical sense.

The concept of *state* in control theory means capturing information about the operation of a system in a set of variables. The state variable for a task also serves this purpose. Using the scan model for computing, the state provides the task with information indicating what action is required at each scan. It thus captures the scan history for the task. In this context states are represented as integer variables, with each task recognizing only a finite number of state variable values (i.e., states). The task is thus a *finite-state machine* and the whole program is a set of finite-state machines.

17.6.2 Computational Structure: Threads

The system engineering part of the design effort is founded on the operational description of the machine, but the computational structure is determined by detailed performance specifications. As long as the program is constructed from nonblocking code and the scan model for software structure, it is theoretically possible to accommodate *any* control job with a single, fast computer without special real-time hardware or software. However, computers are never as fast or inexpensive as desired. Fortunately, in most problems only a few tasks lack computational resources. In these cases, a computational structure is necessary to allow shifting resources from those tasks with excess to those that have need.

This is done by *threads*. Threads represent separate computing entities that can run asynchronously with respect to each other. They are activated by the *interrupt* mechanism of the computer. Resource shifting can be accomplished by putting selected groups of tasks into threads that are activated to receive more computational resources than would otherwise be the case.

In computing terminology, threads are distinguished as *lightweight* and *heavyweight* threads. Lightweight threads are executed asynchronously but share the same address space, whereas heavyweight threads do not share address spaces. In this chapter, the term *thread* will always refer to a lightweight thread.

Real-time environments are characterized by the types of thread structures implemented. By separating the system engineering from the computational structure, exactly the same application source code can be used in a variety of thread structures for the most effective implementation.

17.7 Design Formalism

Design formalisms are used in engineering to organize the design process, allow for effective communication, and to specify the documentation needed for modifying or analyzing the object designed. Software design has traditionally been a notable exception to this rule! Because real-time software adds the element of asynchronous operation, appropriate design formalism is even more important in designing control system software than conventional engineering software.

A major purpose for formalizing design is to establish a mechanism for modularizing the software. Modular software clearly connects sections of software with specific machine operations, keeps those sections short and readable, and clearly identifies interactions between software elements.

Given the state/scan model for control software, many tasks will have only a single action repeated on a fixed schedule. For those tasks, no further internal structure is needed. The task itself will have only one small or modest sized code module, and the task's description will relate very closely to that code.

17.7.1 Formality/Complexity

The degree of formality required is related to the complexity of the problem. Overly formalistic procedures can result in a rigid design environment absorbing too much overhead; too little formalism can create a chaotic environment and can fail to meet delivery and/or reliability commitments.

The stand-alone control system software described in this chapter falls into the range of low to moderate complexity. Some formalism is needed, but how it is applied will be left to the designer's discretion. In particular, the state-transition notation described below will be used in a manual construction mode to achieve maximum portability and flexibility. Because no specialized design software is required, the method can be used to design software for any target computer in any host environment. On the other hand, if more complex designs are attempted, a more organized approach is needed to control the design process.

The structure for control system software based on tasks and states has already been established. No further formal structure will be applied to tasks, but state transition logic will be used to characterize the relationships of states.

17.7.2 State-Transition Logic

A task can be in only one state at a time, characterized by the integer value of its state variable. *State-transition logic* describes how states change within a task. It is most commonly shown in diagramatic form, with circles for states and arrows for transitions between the states. These diagrams are widely used for sequential logic design, from which they can be directly converted to logic equations, and also to design software, for operating systems design ([1], [2], [4], [6]). A distinction in the usage described here, as well as in the more formal real-time usage referred to below, is that the software derived from the diagrams and the portable real-time implementation are tightly connected.

To make computer graphics easier, the transition logic diagram in Figure 17.1, for pulse-width modulation (PWM) generation, is drawn with rectangles for states and ovals for the transition conditions. In addition, the transition conditions are connected by dashed lines to the transition they describe. This avoids ambiguity if the transition description is close to more than one transition line. If none of the indicated transition conditions is true, no change of state takes place. Each state is identified by a name, and information about what the state does, if that is necessary.

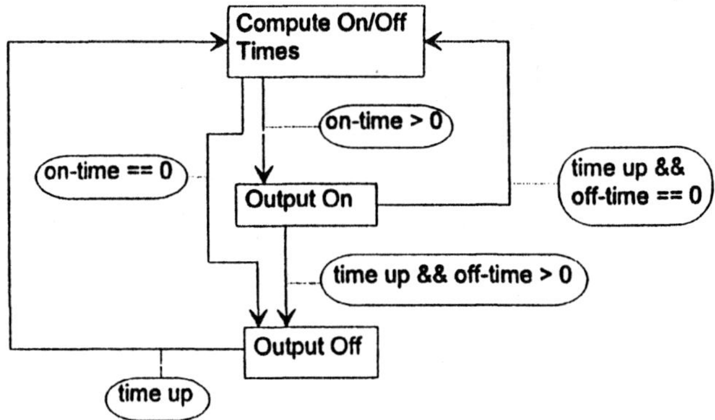

Figure 17.1 State transition logic for pulse-width modulation.

Software generated pulse-width modulation (PWM) can be used whenever the required frequency is not too high, for example, for running a heater. This logic diagram shows the task states needed to implement PWM, including the special cases of duty cycles of 0 or 1. This diagram translates directly into highly modular code, is specific enough so that it specifies in detail how the system should operate, and yet can be read by any engineer familiar with the application. Thus it is a primary document and a template for generating code in a relatively mechanical way.

PWM would normally be a low level task, forming the interface between a feedback control algorithm and an actuator. It would need to run often and would have a tight tolerance on its actual run times because errors in the run times will change the output power delivered by the actuator.

A higher level supervisory task, shown in Figure 17.2, would be used to generate the set point for a servo (mechanical position) controller or other types of controllers. The task would run less frequently and have more tolerant requirements for its actual run times.

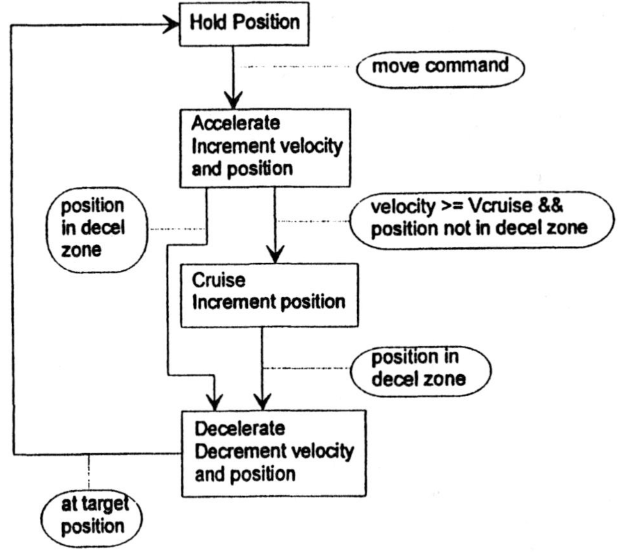

Figure 17.2 Motion profile generator.

The profile is based on a trapezoidal velocity, a constant acceleration section followed by a constant velocity section and, finally, a constant deceleration. The associated velocity and position are shown in Figure 17.3. In this case, the "path" is the motion of the motor under control; in the case of an XY motion, for example, the path velocity could be along a line in XY space with the X and Y motions derived from the path velocity and position.

17.7.3 Transition Logic Implementation

Implementation of the transition logic in code is up to the user, as is the maintenance of the transition logic diagrams and other design documentation. The execution shell will tend to the storage of state information and the scanning of the task code, but the user is responsible for all other aspects. This recommended design procedure is thus a semiformal method, and the user is responsibile for keeping the design documentation up-to-date. However, by avoiding specialized design tools, the method is much more widely applicable.

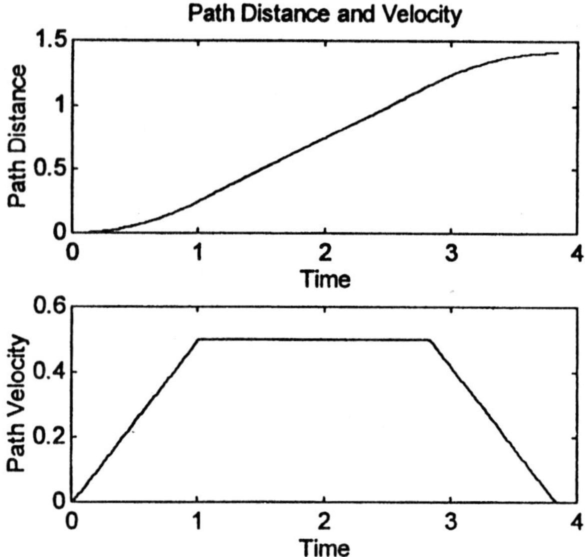

Figure 17.3 Trapezoidal velocity profile.

Transition logic code can be implemented directly with the *switch* statement of C++ (and C). Each *case* in the switch represents a state. The code associated with the state can be placed in-line if it is short, or as a separate function if it is longer. Within the code for the state, a decision is made as to which state should be executed next. The code is usually set up so that the default is to stay in the same state.

An example of implementating the PWM task is shown in Figure 17.4. In this implementation, it is assumed that the task is invoked at timed intervals, and that the task can control when it will next be invoked. Only the relevant transition logic code is shown here; variable definitions, etc., are not shown.

Each scan of this code executes only one state, so the scan will be extremely short. When `set_next_time()` is executed, the intention is that no more scans will take place until that time is reached. All code used is nonblocking as specified above. Any operations requiring waiting for some event to take place are coded at the transition logic level, rather than at the C++ code level. These logic elements tend to be the critical parts of control programs, so that placing them in the higher level of specification makes it easier to understand and critique a program.

This is not the only possible implementation. For example, the task could be written for an environment where the task is scanned continuously (see *continuous* tasks below). In that case, the transition conditions would include a test of time as well as velocity or position, and there would be no `set_next_time()` call. For the control system software implementation recommended here, a form roughly equivalent to the above code would be used.

17.7.4 Formal Implementation

The same basic methodology can be used for problems with considerably more complexity than the typical stand-alone control system. In that case, the transition logic would be formalized through a software design tool to capture the task and transition logic structure. In one particular version of such a tool the state structure is further defined by specification of *entry, action, test,* and *exit* functions. This provides for further modularization of the code, and a tool for the task and state-transition structure guarantees that the basic documentation is up-to-date. A code generator, part of the tool, emits C code that uses data structures to describe the system topology. An execution shell uses those data structures, combined with the user-written code for the state functions, to produce portable implementations of control software.

17.8 Scheduling

The computational technology needed to insure that all of the tasks meet their timing specifications is generally referred to as *scheduling.* The subsections below present various scheduling technologies in order of increasing complexity, and thus increasing cost and difficulty in development and utilization.

Because of the scan mode/nonblocking software paradigm, scheduling takes on a somewhat different flavor than it does in more general discussions. In particular, choosing a specific scheduling method is strictly a matter of performance. Any scheduling method could meet performance specifications if a fast enough computer were available. While this is a general statement of the portability principles given above, it is also an important design element because commitments to specific hardware and/or software execution environments can be delayed until sufficient software testing has been done to provide accurate performance data.

There are two general criteria for performance specification:

1. general rate of progress
2. executing within the assigned time slot

All tasks have a specification for (1). The job cannot be accomplished unless all tasks get enough computing resources to keep up with the demand for results. There is a large variation among tasks in the second performance specification. As noted above, some tasks have a very strict tolerance defining the execution time slot; others have none at all. The failure to meet time slot requirements is usually referred to as *latency* error, that is, the delay between the event that triggers a task and when the task is actually executed. The event is most often time, but can also be an external or internal signal.

17.8.1 Cooperative Scheduling

This is by far the simplest of all of the scheduling methods because it requires no special-purpose hardware or software. It will run on any platform that supports a C++ compiler and has a means of determining time (and even the time determination requirement can be relaxed in some cases). Cooperative scheduling utilizes a single computing thread encompassing the scheduler and all of the tasks. For the scheduler to gain access to the CPU, the tasks have to relinquish their control of the CPU voluntarily at timely intervals, thus the name "cooperative." In conventional program-

```
...
switch(state)
        {
        case 1: // Compute on/off times
                on_time = duty_cycle * period;
                off_time = period - on_time;

                if(on_time > 0)next_state = 2;  // Turn output on
                else next_state = 3;  // Turn output off directly
                break;  // Completes the scan for this state
        case 2: // Output on
                digital_out(bit_no,1);  // Turn on the relevant bit
                set_next_time(current_time + on_time);
                if(off_time > 0)next_state = 3;  // to output off
                else next_state = 1;  // Duty cycle is 1.0; no off time
                break;
        case 3: // Output off
                digital_out(bit_no,0);  // Turn off the relevant bit
                set_next_time(current_time + off_time);
                next_state = 1;  // Back to compute times
                break;
        } // End of switch
 ...
```

Figure 17.4 Implementation of the PWM task.

ming, the most difficult aspect of using cooperative scheduling is building the "relinquishment points" into the code. For control system code built with the scan/nonblocking rules, this is not an issue because control is given up after every scan.

The performance problems with cooperative scheduling are not general progress but latency. The cooperative scheduler is extremely efficient because it has no hardware-related overhead. If some tasks exceed general progress requirements and others are too slow, the relative allocation of computing resources can be adjusted by changing the number of scans each task gets. However, if the overall general progress requirements cannot be met, then either a faster processor must be found, the job must be redesigned for multiple processors, or the job requirements must be redefined.

The most efficient form of cooperative scheduler gives each task in turn some number of scans in round-robin fashion. The latency for any given task is thus the worst-case time it takes for a complete round-robin. This can easily be unacceptable for some low-level tasks. A *minimum-latency* cooperative scheduler can be designed. At the expense of efficiency, it checks high priority tasks between every scan. In this case, the maximum latency is reduced to the worst-case execution time for a single scan.

Cooperative scheduling is also a mechanism that allows for portability of the control code to a simulation environment. Because it will run on any platform with the appropriate compiler, all that needs to be done is replace the time-keeping mechanism with a computed (simulated) time and add a hook to the simulation of the control object.

When cooperative scheduling is used, all tasks occupy a single computing thread.

17.8.2 Interrupt Scheduling

When general progress requirements can be met, but latency specifications cannot, then some form of scheduling beyond cooperative must be used. All of these start with *interrupts*. In effect, the interrupt mechanism is a hardware-based scheduler. In its simplest usage, it can reduce the latency of the highest priority task from the worst-case single scan time in cooperative, minimum latency scheduling down to a few microseconds. It does this by operating in parallel with the CPU to monitor external electrical signals. When a specified signal changes, the interrupt controller signals the CPU to stop its current activity and change *context* to an associated interrupt task. Changing context requires saving internal CPU information and setting up to execute the interrupt task. When the interrupt task is done, the CPU resumes its former activity. Each interrupt occurrence activates a new computing thread.

Used by itself, the interrupt system can reduce the latency for a group of tasks that have very tight latency tolerances. The design rule for interrupt-based tasks is that they also have short execution times. This is important if there are several interrupt tasks, so that interference among them is minimal. An interrupt task will often prevent any other interrupt tasks from running until it is done. In other cases, an interrupt task will prevent interrupt tasks at equal or lower priority from running until it is done. Any priority system in this environment must be implemented as part of the interrupt hardware. The pulse-width modulation (PWM) task illustrated above would often be implemented as an interrupt task.

Many computer systems have fewer actual interrupt inputs than needed, but latency can still be reduced by "sharing" the existing interrupts. To do that, there must be software to determine which task to activate when the interrupt occurs. This scheme cannot match the low latencies achieved when each task is attached to its own interrupt, but there still can be a large improvement over cooperative scheduling.

17.8.3 Preemptive Scheduling

When tasks have latencies comfortably met by cooperative scheduling but too long (in execution time) for interrupt tasks, a preemptive scheduler must be used. The preemptive scheduler is itself activated by interrupts. It checks a set of tasks for priority, then runs the highest priority task that has requested computing resource. It then resets the interrupt system so further interrupts can occur. If, at the time of a later interrupt, a task of priority higher than the currently running task becomes ready to run, it will *preempt* the existing task, which will be suspended until the higher priority task completes. In the same manner, if an interrupt-based task is activated, it will take precedence over a task scheduled by the preemptive scheduler.

The preemptive scheduler is thus a software version of the interrupt controller. Because of the time needed for the scheduler itself to run, the latencies for the tasks it schedules are longer than for interrupt tasks. The priority structure, in this case, is software based.

In order for preemptive scheduling to work, interrupt-activated code must be allowed to be reentrant or to overlap in time. Each interrupt activates a new computing thread. If several interrupts overlap, there will be several computing threads active simultaneously.

The simplest form of preemptive scheduler, adequate for control problems, requires that, once a task is activated, it will run to completion before any other tasks at the same or lower priority level run. Other than suspension due to preemptive activity, the task cannot "suspend itself" until it completes (by executing a *return* statement).

17.8.4 Time Slice Scheduling

If several tasks with only weak latency requirements have scan execution times that vary greatly from one state to another, it may not be possible to balance the resource allocation by giving different numbers of scans to each. In the worst circumstances, it might be necessary to use some blocking code, or one task might be technically nonblocking (i.e., predictable), but have such a long execution time per scan that it might as well be blocking. A *time slicing* scheduler can balance the computing resource in these cases by allocating a given amount of computing time to each task in turn, rather than allocating scans as a cooperative scheduler does. This can be quite effective, but the time slicing scheduler itself is quite complex and thus better avoided, if possible. Time slice schedulers are usually part of more general scheduling systems that also include the ability of tasks to suspend themselves, for example, to wait for a resource to become available.

17.9 Task Type Preferences

Tasks that follow the state model (scan/nonblock) need no further differentiation, in theory, to solve all control software problems. Designing and building a code on this basis would, however, be bad practice. If a scheduling mode other than cooperative were required, changes to the code would be necessary for implementation, thereby destroying the important portability. For this reason, task types are defined for using the code in any scheduling environment needed to meet the performance specifications. Designation of a task type is an indication of *preference.* The actual implementation might use a simpler scheduling method if it can meet performance demands.

Tasks are broken down into two major categories, intermittent and continuous, and into several subcategories within the intermittent tasks.

17.9.1 Intermittent

Tasks that run in response to a specific event, complete their activity, and are then dormant until the next such event are categorized as *intermittent.* Most control system tasks fit into this category. Following the types of schedulers available, they are subcategorized as *unitary* or *preemptable.*

Unitary tasks can be scheduled as direct interrupt tasks. To simplify scheduling and make execution as fast as possible, each time a unitary task is run it gets only a single scan. Control is then returned to the interrupt system. This design is based on the assumption that most tasks of this type will be single-state tasks and will thus only need one scan.The PWM task shown above is an exception to this rule. The "compute on/off times" state is transient. Before returning from the interrupt, either the "output on" or "output off" state must be executed. Therefore, it sometimes uses a single scan when it is invoked, and sometimes uses two scans. To accommodate this, the repetition through the scans must be handled internally, a structural compromise, but one that improves the overall efficiency of these very high priority tasks.

No specific task priorities are associated with unitary tasks. Because any priority structure must exist in the interrupt controller, from the software perspective they are all considered to be of the same priority, the highest priority of any tasks in the project.

Preemptable tasks generally use more computing time than unitary tasks and are often more complex. For that reason, they are given as many scans as needed to complete their activity. When they are finished, they must signal the scanner that they are ready to become dormant. Software-based preemption is the characteristic of this class of tasks, so that each task must be assigned a priority.

The external control of scans is very important for preemptable tasks, and somewhat in contrast to the policy for unitary tasks. It is the scans that allow for effective cooperative scheduling. The external capture of state-transition information is necessary for an automatically produced audit trail of operation. The audit trail is a record of transitions that have taken place. It is particularly useful in verifying proper operation or in determining what each task has been doing just prior to a system malfunction. The exception for internal transitions in unitary tasks is based on assuming that many of them will not do any state transitions, and, when they do, they will be few and the total execution time will still be very short. State transitions in unitary tasks that occur across invocations are still tracked externally and are available for audit purposes.

17.9.2 Continuous

Continuous tasks have no clean beginning or ending points. They will absorb all the computing resource they can get. The scan/nonblock structure is absolutely essential for continuous tasks; otherwise they would never relinquish the CPU!

Continuous tasks represent the most primitive form of task. Given the state transition structure, continuous tasks are all that is needed to implement the "universal" solution referred to above. If all of the timing and event detection is done internally using transition logic, the same functionality can be produced as can be realized from unitary and preemptable tasks. However, that information will be entered in an *ad hoc* manner and so cannot be used to implement any schedulers other than cooperative.

17.9.3 Task Type Assignment

Control tasks are dominated by unitary and preemptable tasks. The choice between the two is based on the nature of the task and defines the type of scheduling used with the task to meet performance issues. Over the life of the software, it is likely that it will actually be executed in several different environments.

Unitary tasks are used for activities that are very short and high priority. Most low-level interactions with instruments and actuators fall into this category. Separate tasks are used when the data requires some processing as in pulse-width modulation, pulse timing, step-rate generation for a stepping motor, etc., or when the data is needed by several tasks and conflicts could arise from multiple access to the relevant I/O device. Use of analog-to-digital converters could fall into that category. Very little computation should be done in the unitary task, enough to insure the integrity of the data, but no more. Further processing is left to lower priority tasks.

Preemptable tasks are used for longer computations. This would include the control algorithms, supervisory tasks to coordinate several feedback loops, batch process sequencers, adaptation or identification algorithms, etc. Some tasks, such as a PID controller, could still be single state, but the amount of computation needed makes them better grouped with preemptable tasks rather than unitary tasks.

To meet the special needs of feedback control software, two types of tasks each for unitary and preemptable tasks are defined in the sample software. The general hardware interrupt task is defined as a unitary task and, as a special case, timer-interrupt tasks are defined. Because many computer systems have fewer hardware timers than needed, software is also included to simulate additional timers. In a parallel fashion, event-driven and sample-time preemptable tasks are defined. Event-driven tasks are completely general. They are activated from other tasks via a function call to "set" an event. Sample-time tasks are special cases of event-driven tasks. Sample-time and timer-interrupt tasks dominate control system software.

Continuous tasks are usually used for functions, such as the operator interface, data logging, communication with a host computer, generation of process statistics, and other such activities. They form the lowest priority group of activities and only receive CPU resource when preemptable and interrupt tasks are not active.

17.10 Intertask Communication

Tasks cooperate by exchanging information. Because the tasks can execute asynchronously (in separate threads) in some scheduling domains, the data exchange must be done to assure correct and timely information. The methods discussed here are valid for single computer systems; other methods for data exchange must be used for multicomputer systems.

17.10.1 Data Integrity

The primary danger in data exchange is that a mixture of old and new data will be used in calculations. For example, if a task is transferring a value to a variable in an ordinary C or C++ expression, an interrupt may occur during this transfer. If the interrupt occurs when the transfer is partially complete, the quantity represented will have a value consisting of some of the bits from its earlier value and some from the later value. For most C data types, that quantity will not necessarily lie anywhere between those two values (earlier, later) and might not even be a valid number.

At a higher level, if a task is carrying out a multiline calculation, an interrupt could occur somewhere in the middle. As a result of the interrupt, if a value that is used both above and below the point of interrupt is changed, the early part of the calculation will be carried out with the old value and the later part with the new value. In some cases, this could be benign, because no invalid data is used, but in other cases it could lead to erroneous results.

The most insidious factor associated with these errors is that, whether or not the error will occur is statistical, sometimes it happens, sometimes not. Because interrupts arise from sources whose timing is not synchronized with the CPU, the relative timing will never repeat. The "window" for such an error occurrence could be as small as a microsecond. Debugging in the conventional sense then becomes impossible. The usual strategy of repeating the program over and over again, each time watching how the error occurs and extracting different data, doesn't work here because the error occurrence has a probability of only one chance in several hundred thousand per second. Careful program design is the only antidote.

17.10.2 Mutual Exclusion

In single processor systems, the general method of guarding data is to identify critical data transactions and to protect them from interruption by globally disabling the computer's interrupt system. In effect this temporarily elevates the priority of the task, in which this action is taken, to the highest priority possible, because nothing can interrupt it. This is a drastic action, so that the time spent in this state must be minimized. Other less drastic ways can be used to protect data integrity, but they are more complex to program requiring substantially more overhead.

Assuring data integrity in information interchange has two components:

1. defining sets of variables in each task used solely for data exchange (exchange variables)
2. establishing regions of mutual exclusion in each task for protected exchange activities

The exchange variables can only be accessed under mutual exclusion conditions, which must be kept as short as possible. The best way to keep them short is to allow nothing but simple assignment statements in mutual exclusion zones. Under this method, the exchange variables themselves are never used in computations. This controls both low-level problems (mid-data-value interrupts) and high-level problems (ambiguous values) because the exchange values are changed only when data is being made available to other tasks.

17.10.3 Exchange Mechanisms

There are two direct mechanisms for implementing data exchange in C/C++. One is the use of global variables for the exchange variables; the other is to use local variables for the exchange variables and implement data exchange through function calls. Global variables (*statics* if task functions are in the same file, *externs* otherwise) are more efficient, but functions encapsulate data better and are more easily generalizable (for example, to multiprocessor systems). Unless computing efficiency is the critical design limitation, the function method is recommended. An example using the function call method follows.

Function Exchange Method: Assume that Task1 and Task2 are in the same file. Each of the tasks is defined as a C++ class in Figure 17.5.

The "exchange variables" are designated by the prefix `exch_`. They are defined within the *class* used for each task, and thus have scope only within that class.

A similar definition exists for Task2.

Within the Task1 function, the exchange now can be defined as in Figure 17.6.

The functions `_disable()` and `_enable()`, which act directly on the interrupt system, are compiler dependent. These definitions are from Microsoft Visual C++.

17.11 Prototype Platform

The sample programs are designed to run on an IBM-PC (or compatible) equipped for lab use. Hardware dependencies are isolated so that they can be easily modified for other target platforms. A complete software package implementing all of the principles in this chapter has been developed with an execution shell customized for use on the PC. It implements several different real-time and simulation modes. The software is available via anonymous ftp from *ftp.me.berkeley.edu/pub/dma/controls* (connect to *ftp.me.berkeley.edu* and then go to the directory */pub/dma/controls*). It is copyrighted software, distributed free of charge, with the restriction that it be used for non-commercial purposes only. It should be relatively easy to port the software to other platforms.

This environment is extremely convenient for prototyping, however, it is often not reliable enough for production control system applications.

Although space limitations preclude a detailed discussion here of implementation software, it provides a detailed example of the methods outlined here. The full source code is available for study or modification. A detailed study of an example control program is given below as a model of software usage and control program construction. Task structures and relevant details are given for several other sample problems.

17.11.1 Control Hardware

A laboratory I/O board is used for communicating with the physical system. In general, these boards provide analog-to-digital conversion, digital-to-analog conversion, digital input and digital output. Interface code is provided for the Data Translation DT10EZ and the Real-Time Devices ADA 2000. The ADA 2000 board has a single interrupt line; the DT10EZ has no interrupts, so the interrupt line from the parallel (printer) port is used.

17.11.2 Operating System

Most of the actual real-time implementation is done with MS/PCDOS. This operating system runs the processor in "real" 8086 emulation mode. In this mode, there is no memory protection at all, so the real-time operations, setting up interrupts, access to devices on the computer's bus, etc., are done by bypassing DOS and taking direct control of the interrupt and I/O systems. Although the 640Kb memory limitation of DOS is becoming outdated, it is normally adequate for real-time control software. Jobs within the complexity limits for the methods described here will normally fit in this memory space. The most common exceptions are jobs that save a lot of data.

In most cases, the control programs should be run in "pure" DOS rather than in a DOS session under Windows. Because Windows DOS sessions emulate much of the PC hardware, timing may not be accurate.

Simulation jobs can be run anywhere because they do not require any special hardware, just a C++ compiler.

```
...
        static
        class task1_class
              {
              private:
              float exch_a,exch_b;    // Exchange variables
              ...     // Other stuff
              public:
              float get_a(void)
                    {
                    float a;        // Local variable to hold copied data

                    _disable();
                    a = exch_a;
                    _enable();
                    return a;
                     }
              float get_b(void)
                    ...
              ...     // Other stuff
              }task1;
```

Figure 17.5 Task1 class.

```
        ...
        // Get information from Task2
        a = task2.get_a(); // The disable/enable is inside the 'get' function
        b = task2.get_b();
        ...     // Do some computation
        // Copy results to exchange variables
        _disable();
        exch_a = a;
        exch_b = b;
        _enable();
        ...
```

Figure 17.6 Task2 class.

17.11.3 Compiler

The sample software has been created with the Microsoft Visual C++ compiler (professional, v1.5) and the Borland C++ compiler (v4.5). Either compiler works for single-thread mode, but the Borland compiler is recommended for multi-thread operation. An attempt has been made to keep most of the code in generic C++, so it should be possible to port it to other compilers without too much trouble.

17.11.4 Timekeeping

Most of the events to which real-time programs respond are based on time, specific times, time elapsed since a particular event, etc. The PC platform does not have any free resources for timing, but there is just enough that can be "borrowed" from DOS to do the job. There is one clock chip accessible on the PC, with two of its three timers accessible. These are normally used by DOS, but not for essential functions. One of these can be attached to an interrupt and so can be used as the basis for running all of the timer-interrupt (unitary) and sample-time (preemptible) tasks. It can also provide the basis for keeping time, with a minimum granularity of between 500 and 1,000 microseconds. The other accessible clock cannot be connected to the interrupt system but can be used in a free-running mode. As long as it is sampled often enough (50 msec absolute maximum), it can be used for timing with about a 1 microsecond resolution. The actual timing accuracy is considerably less due to unpredictable latencies but is still much better than with the interrupt clock.

For slow processes (thermal, for example), the internal DOS timing can be used by making calls to `ftime()`. This, in fact, is a portable way to do timing, because most systems that support a C/C++ compiler support the *ftime* function. In DOS, however, the tick interval is about 50 msec, much too slow for most mechanical control problems.

For systems with very loose timing accuracy requirements, the most portable mode for timing is to run the system as if it were a simulation, but calibrate the "tick time" so that computed time in the program matches actual time. This is completely portable because no special facilities are required. It will actually solve a surprisingly large number of practical problems. Recalibration is needed whenever the program or the processor is changed, but the recalibration process is very easy.

17.11.5 Operator Interface

The operator interface is an extremely important part of any control system but far beyond the scope of this chapter to discuss adequately. The primary problem in prototype development is that nothing in C or C++ syntax is helpful in developing a suitable, portable operator interface package. Like every other part of the program, the operator interface must be nonblocking, but all of the basic console interaction in C or C++ is blocking (*scanf, cin,* etc.).

A simple, character-based interface is provided for use with the scheduling software. It is suitable for prototype use and is relatively easy to port. The versions with the sample programs support the DOS and Windows environments.

17.12 Program Structure: The Anatomy of a Control Program

A relatively simple control problem will illustrate the construction of a control program. It is a single-loop position control of an inertial object with a PID controller. To simplify the job, it is assumed that the position is measured directly (from an analog signal or from a wide enough decoder not requiring frequent scanning) and that the actuation signal is a voltage. Thus, no special measurement or actuation tasks are needed.

Four tasks are required:

1. feedback (PID) control, sample-time
2. supervisor, which gives set points to the feedback task, sample-time
3. operator interface, continuous
4. data logger, continuous

The supervisory task generates a set point profile for the move. The state-transition logic diagram is shown in Figure 17.2. The profile is trapezoidal, with constant acceleration, constant velocity, and constant deceleration sections.

Tasks are defined as C++ classes. The tasks communicate among themselves by making calls to public functions for data exchange, as discussed above. However, any task could call a function in any other task, so the order in which the parts of the class are defined is important. In implementing a small control job, it is convenient to put all of the tasks into a single file. For larger jobs, however, each task should have its own file, as should the 'main' section.

To meet these needs, the following program structure is recommended:

1. Define all of the classes for all of the tasks in a header file.
2. Put global references to each task (pointers) in the same header file.
3. The functions associated with each task can go in separate files or can be collected in fewer files.
4. The system setup information and the *UserMain()* function can go in the same or a separate file.
5. Use the *new* operator to instantiate all of the tasks in the same file as *UserMain()*.
6. Make sure the header file is *included* in all files.

The function *UserMain()* plays the role of a program *main* function. The actual main function definition will be in different places, depending on the environment to which the program is linked.

Because this is a relatively small job, only two files are used: *mass1.hpp* for the header and *mass1.cpp* for everything else.

17.12.1 Task Class Definition

Classes for the tasks are all derived from an internal class called *CTask*. This task contains all of the information needed to schedule and run tasks as well as a *virtual* function called *Run*. This function must have a counterpart in the derived class for the task, where it becomes the function that is executed each time the task is scanned. Other class-based functions are normally defined for data exchange and other additional computation.

The supervisory task is a good starting point. Its class definition is as in Figure 17.7.

17.12.2 Instantiating Tasks

Tasks represent instantiations of the task classes. In most cases, each class has only one associated task. However, as noted below, several tasks can be defined from the same class. All task references are made by pointers to the tasks. These are declared as global *(extern)* in the header file,

```
#ifdef CX_SIM_PROC
   extern CMassSim *mass_sim;
#endif
extern Mass1Control *Mass1;
extern CDataLogger *DataLogger;
extern CSupervisor *Supervisor;
extern COpInt *OpInt;
```

The pointers are defined in the file (or section of the file) containing *UserMain()*,

```
#ifdef CX_SIM_PROC
   CMassSim *mass_sim;
#endif
Mass1Control *Mass1;
CDataLogger *DataLogger;
CSupervisor *Supervisor;
COpInt *OpInt;
```

```
class CSupervisor : public CTask
  {
  private:
    float position_set;
    float vprof,xprof;  // Profile velocity and position
    float xinit,xtarget,dir;
    float dt;
    int newtarget;
    real_time t4,thold;      // Used to time the hold period
    float accel,vcruise;     // Accel (and decel), cruise velocity
    float exch_t,exch_v,exch_x,exch_target; // Exchange variables
  public:
   CSupervisor(real_time dtt,int priority);     // Constructor
   void Run(void);      // The 'run' method -- virtual from CTask
   void SetTarget(float xtrgt);
  };
```

Figure 17.7 Supervisory task class.

Then, within *UserMain()*, the actual tasks are instantiated with the memory allocation operator, *new*,

```
#ifdef CX_SIM_PROC
   mass_sim = new CMassSim;
#endif
Mass1 = new Mass1Control;
DataLogger = new CDataLogger;
Supervisor = new CSupervisor(0.02,5);
// Send sample time and priority
OpInt = new COpInt(0.2);
// Set the print interval.
```

Each of these variables has now been allocated memory, so that tasks can call functions in any of the other task classes. The simulation, *mass_sim*, is treated differently from the other tasks and will be discussed after the control tasks. The simulation material is only compiled with the program when it is compiled for simulation, as noted by the *#ifdef* sections.

17.12.3 Task Functions

The functions in the *public* section of *CSupervisor* are typical of the way tasks are defined. The constructor function for the *Supervisor* task (*CSupervisor* is the name of the class; *Supervisor* is the name of the task) sets initial values for variables and passes basic configuration information on to the parent *CTask* by a call to its constructor. This is the constructor for *Supervisor:*

```
CSupervisor::CSupervisor(real_time dtt,
  int priority) // Constructor
  : CTask("Supervisor", SAMPLE_TIME,
  priority,dtt)
  {

  dt = dtt;
  xinit = 0.0;
  xtarget = 1.2;
  dir = 1.0;
  accel = 0.5;
  vcruise = 0.6;
  thold = 0.1;
  // Hold for 100 ms after the
  // end of the profile
  t4 = 0.0;
  vprof = 0.0;
  xprof = xinit;
  newtarget = 0;
  State = 0;
  };
```

The *Run* function is where all the work is done. In *CSupervisor* it is based on a transition logic structure moving through the various stages of the profile, accelerate, cruise, decelerate, and hold. It will then wait for a new target position command, or time-out and stop the program if the command does not come. The virtualized *Run* function is called from the parent *CTask* every time the task is activated (in this case, every sample time). See Figure 17.8.

The *Idle()* at the end of the function is extremely important. It is the signal to the scheduler that this task needs no attention until the next time increment elapses. If this is left out of a *sample-time* or *event task*, the scheduler will continue to give it attention, at the expense of any lower priority tasks and all of the continuous tasks. Leaving out the *Idle()* statement or putting it in the wrong position is a common bug.

References to other tasks are made by calling *public* functions in those tasks. Both *Mass1* and *DataLogger* are referenced here. When functions from other tasks are called, they run at the priority level of the calling task, rather than the priority level of the task where they are defined.

The *SetTarget* function allows other tasks to change the target value, thus defining a new move. This would most likely be done from the operator interface task. Because the state structure is set up so that the check for a new target is made only in state 4,

```
void CSupervisor::Run(void)
  {
  float d_end,d_decel;
  real_time tt;   // Curent time

  tt = GetTimeNow();
  switch(State)
    {
    case 0:     // Initial call
     Mass1->SetSetpoint(xprof); // Send a new setpoint to the controller
     Mass1->SetStart(); // Send a start message to the controller
     if(xtarget > xinit)dir = 1.0; // Determine direction of the move
     else dir = -1.0;
     State = 1;
     break;

    case 1:     // Start profile -- Accelerate
     vprof += dir * accel * dt;// Integrate the velocity and position
     if(fabs(vprof) >= vcruise)
        {
        // This is the end of the acceleration section
        vprof = dir * vcruise;
        State = 2;      // Go on to next state
        }
     // Check whether
     // cruise should be skipped because decel should be started
     d_end = fabs(xprof - xtarget); // Absolute distance to end
     d_decel = vprof * vprof / (2.0 * accel); // Distance to decelerate
               // to stop at current velocity
     if(d_decel >= d_end)
        {
        // Yes, go straight to decel
        vprof -= dir * accel * dt;  // Start decel
        State = 3;
        }
     xprof += vprof * dt;
     break;

    case 2:     // Cruise -- constant velocity
     xprof += vprof * dt;
     d_end = fabs(xprof - xtarget);  // Absolute distance to end
     d_decel = vprof * vprof / (2.0 * accel); // Distance to decelerate
           // to stop at current velocity
     if(d_decel >= d_end)
       {
       // Yes, go to decel
       vprof -= dir * accel * dt;  // Start decel
       State = 3;
       }
     break;
```

Figure 17.8 Run function.

```
case 3:      // Deceleration
   d_end = fabs(xprof - xtarget); // Absolute distance to end
   vprof = dir * sqrt(2.0 * d_end * accel);// Velocity that will get
        // to stop in desired distance
   xprof += vprof *dt;
   if(fabs(xprof - xinit) >= fabs(xtarget - xinit))
    {
        // This is the end of the profile
    xprof = xtarget;
    vprof = 0.0;     // Stop

    t4 = GetTimeNow(); // Start a timer for the hold state
    State = 4;       // Go to HOLD state
    }
  break;

 case 4: // Hold final position until either a command for a
      // new target is sent, or time runs out
    // Check for new target
    DisableInterrupts();
    if(newtarget)
     {

     xinit = xtarget; // Start new profile where this one ended
     xtarget = exch_target;  // New target position
     newtarget = 0;
     vprof = 0.0;
     xprof = xinit;
     State = 0;       // Start the profile again
     break;
     }
     if((GetTimeNow() - t4) >= thold) // Check for timeout
      {
      // End the program
      TheMaster->Stop();
      }
     break;
    }
  DisableInterrupts();     // Copy data to exchange variables
  exch_t = tt;
  exch_v = vprof;
  exch_x = xprof;
  EnableInterrupts();
  Mass1->SetSet point(exch_x); // Send a new set point to the controller
  DataLogger->LogProfileVal(exch_t,exch_v,exch_x);
  Idle();
  };
```

Figure 17.8 Run function (continued).

a move will be completed before the next move is started. This also means that, if two new moves are sent while a previous move is still in progress, only the last will be recognized.

```
void CSupervisor::SetTarget(float trgt)
  {
  DisableInterrupts();
  exch_target = trgt;
  newtarget = 1;
  EnableInterrupts();
  }
```

17.12.4 A Generic Controller Class

Where several tasks of similar function will be used, the inheritance property of C++ can be used to great advantage. PID controllers are so common, that a class for PID controllers has been defined that can act as a parent class for any number of actual PID controllers. The class definition for this is in Figure 17.9.

The arguments to the *constructor* function have the data needed to customize a task to the scheduler. All of the variables are *protected* rather than *private* so that they can be referred to readily from the derived class. The major properties, distinguishing one actual derived PID control task from another, are where it gets its process data from and where it sends its actuation information. These are specified in the *GetProcVal* and *SetActVal* functions, listed here as *virtual*, and must be supplied in the derived class because the versions here are just dummies.

The *Run* function in the *PIDControl* class is not virtual so that the control calculation can be completely defined in the parent class (it is, of course, virtual in the higher level parent class, *Ctask)*. Its only connection to the derived class is in getting the process value and sending out the actuation value (Figure 17.10).

The use of the *waiting* state (state 0) prevents the controller from becoming active before appropriate initialization and setup work has been done by other tasks. No call to *Idle* is in the portion of state 0 that is doing the transition to state 1. This is done to prevent a delay between the start signal and the actual beginning of control.

Defining a class for an actual PID control is very simple, and has very little in it. Here is the definition for the position control task used in this sample problem:

```
class Mass1Control : public PIDControl
  {
  public:
    Mass1Control();      // Constructor
    void SetActVal(float val);
    // Set the actuation value
    float GetProcVal(void);
    // Get the process value --
  };
```

It has no private data at all, and only defines a constructor and the two virtual functions for getting process data and setting the actuation value. Its constructor is

```
Mass1Control::Mass1Control() :
  PIDControl("Mass1",10,0.02)
  // Call base class constructor also
 {
 kp = 1.5;
 // Initial controller gains
 ki = 0.0;
 kd = 2.0;
 set = 0.0;
 // Initial set point
 min = -1.0;
 max = 1.0;
 }
```

It first calls the constructor for the parent class to set the task name, priority, and sample time, and then sets the controller gains, limits and initial set point. These variables are *protected* in the parent class, and so are freely accessible from this derived class.

The *GetProcVal* and *SetActVal* functions are where much of the hardware-dependent code goes, at least as relates to the control system hardware. The versions shown here use *define* statements for the simulation code so that other sections can be easily added as the environment changes.

```
void Mass1Control::SetActVal(float val)
// Set the actuation value
  {
  #ifdef CX_SIM_PROC
  // The following code is for
  // simulation
  mass_sim->SetSimForce(val);
  // No mutual exclusion is needed for
  // simulation
  #endif
  }

float Mass1Control::GetProcVal(void)
  {
  float x,v;

  #ifdef CX_SIM_PROC
  // The following code is for
  // simulation
  mass_sim->GetSimVal(&x,&v);
  return(x);
  // Position is the controlled
  // variable
  #endif
  }
```

Other such generic definitions can be used to great advantage when multiple elements with similar function are in a control system. For example, the *CSupervisor* class could easily be generalized in a similar manner to allow for several simultaneous profile generating tasks.

```
class PIDControl : public CTask
 {
 protected:// This is just the generic part of a control task
    // so all variables are made accessible to the derived class
  float integ;
  float set,val;       // Set point and output (position) value
  float prev_set,prev_val;
  float kp,ki,kd,min,max; // Controller gains, limits
  real_time dt;
  float mc;        // Controller output
  int start_flag;      // Used for transition from initial state

  // Exchange variables, set in this class
  float exch_val,exch_mc;
  // Exchange variables obtained from other tasks
  float exch_set,exch_kp,exch_ki,exch_kd,exch_min,exch_max;
  int exch_newgains;// Use this as a flag to indicate that new gains
        // have been set
  int exch_start_flag;

 public:
  PIDControl(char *name,int priority,float dtt); // Constructor
  void Run (void);     // Run method
  float PIDCalc(void);     // Do the PID calculation
  void SetGains(float kpv,float kiv,float kdv,float minv,float maxv);
  void SetStart(void);     // Set the start flag to 1

  virtual void SetActVal(float val){}// Set the actuation value --
   // The real version of this must be supplied in the derived class
  void SetSetpoint(float sp);
  void GetData(float *pval,float *pmc,float *pset);

  virtual float GetProcVal(void){return 0.0;}  // Get the process value --
    //  The real version of this must be supplied in the derived class
};
```

Figure 17.9 PID controller class.

Because the remaining tasks are constructed in a similar manner and do not introduce any new elements, they will not be discussed in detail here.

17.12.5 The Simulation Function

Simulation is extremely important to control system development, but because it appears as "overhead," it is often neglected. In addition to whatever simulation has been done on the algorithm side, it is very important that the actual control system software be tested in simulation. The environment of simulation is so much more friendly than the actual control environment, that there is a large time saving for every bug or misplaced assumption found while simulating.

The simulation part of the software is treated as a separate category. In principle, it is handled similarly to the tasks, but is kept separate so that the task structure is not distorted by the simulation.

There are two ways to handle simulation. The method here is to define an internal simulation class, *CSim*, from which a class is derived to define the particular simulation. The simulation itself is then written in C++, with the simulation function called whenever time is updated. An alternative method is to provide an interface to a commercial simulation system, most of which have the ability to call C or C++ functions. To do this, the internal computing paradigm must be changed. The simulation system is in charge, so that the call to the simulation must also trigger the operation of the control system. This latter method is not implemented in the current software.

The simulation class defined for the sample position control problem is

```
void PIDControl::Run (void) // Task function
 {
 // This task has two states - a ``waiting'' state when it
 // first turns on and
 // a ``runnning'' state when other tasks have properly initialized
 // everything
 // The variable 'State' is inherited as a 'protected'
 // variable from the parent class, CTask.

switch(State)
 {
 case 0:      // Waiting for 'start' signal
  DisableInterrupts(); // copy relevant exchange
             // variables for this state
  start_flag = exch_start_flag;
  EnableInterrupts();
  if(start_flag)
   {
   State = 1;
   return;      // Set new state and return.
                // Next scan will go to 'run' state
   }
   else
    {
    // Stay in 'wait' state
    Idle(); // Indicate that the task can inactivate
            // until next sample time
    return;
    }
 case 1:      // Run the control algorithm

  {
  DisableInterrupts(); // copy relevant exchange variables
  if(exch_newgains) // Using this flag minimizes interrupt
                    // disabled time
   {
   kp = exch_kp;
   ki = exch_ki;
   kd = exch_kd;
   min = exch_min;
   max = exch_max;
   exch_newgains = 0;  //Turn off the flag
   }
  set = exch_set;
  EnableInterrupts();
```

Figure 17.10 Run function for the PIDControl class.

```
        val = GetProcVal(); // Get the process output
        mc = PIDCalc(); // Do the PID calculation
        SetActVal(mc);  // Send out the actuation value
        DisableInterrupts();
        // Set output exchange values
        exch_val = val;
        exch_set = set;
        exch_mc = mc;
        EnableInterrupts();
        Idle();      // Wait for next sample interval
        }
      }
    }
```

Figure 17.10 Run function for the PIDControl class (continued).

```
#ifdef CX_SIM_PROC
class CMassSim : public CSim
  {
  private:
    float dt;
    float x,v;  // position and velocity
    float m;    // mass
    float f;    // force
    FILE *file;
    public:
    CMassSim();
    void RunSimStep(real_time t,real_time dt);
    // The simulation function
    void GetSimVal(float *px,float *pv);
    void SetSimForce(float ff);
  };
#endif  // End of simulation section
```

Similarly to the tasks, the *RunSimStep* function is declared *virtual* in the base class and must be defined in the derived class. This is the function called when time is updated. For the position control problem, it contains a very simple simulation of an inertial object, based on the parameter values set in the constructor:

```
CMassSim::CMassSim() : CSim()
// Call base class constructor also
  {

  x = v = 0.0;
  m = 0.08;    // Mass value
  f = 0.0;     // Initial value of force
  file = fopen("mass_sim.dat","w");
  }

void CMassSim::RunSimStep(real_time t,
                          real_time dtt)
  {
  // Calculate one step
  // (Modified Euler method)
  dt = dtt;
  // Set internal value for step size
  v += f * dt / m;
  x += v * dt;
  fprintf(file,"%lg %g %g %g\n",t,v,x,f);
  }
```

The simulation output is also created directly from this function. Because the simulation is not present in other conditions, it is more appropriate to put this output here than to send it to the data logging task. The other functions in the class exchange data. Because the simulation is not used in a real-time environment, it is not necessary to maintain mutual exclusion on the simulation side of these exchanges.

17.12.6 Results

The graphs in Figure 17.11 show the results for the data sets given

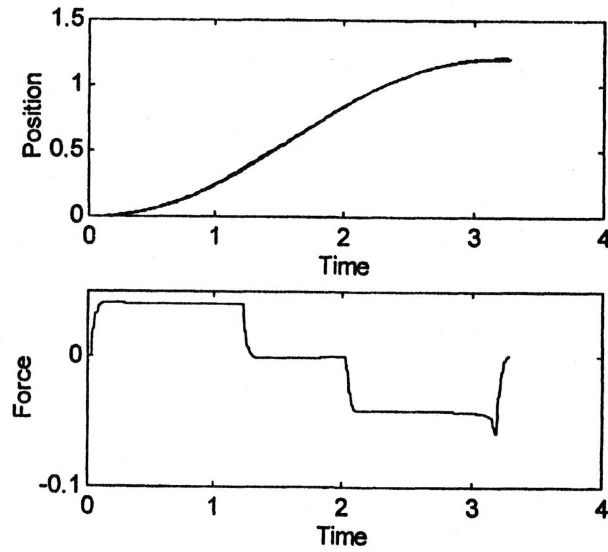

Figure 17.11 Profile position change, inertial system.

in the listings above. These results are for a simulation. The top graph shows the position of the mass (solid line) and the set point generated by the nearly overlaid profile (dashed line). The bottom graph shows the force applied to the mass to achieve these results.

These results were plotted from the output of the data logging task and the output generated directly by the simulation.

In addition to these, there is also a transition audit trail that is generated. The transition audit trail keeps track of the most recent transitions, which, in this case, includes all of the transitions that take place. The file is produced when the control is terminated:

```
CTL_EXEC State Transition Logic Trace File

Time     Task         From   To

0        OpInt        0      1
0.01     Supervisor   0      1
0.03     Mass1        0      1
0.2      OpInt        1      0
0.21     OpInt        0      1
0.41     OpInt        1      0
0.42     OpInt        0      1
0.62     OpInt        1      0
0.63     OpInt        0      1
0.83     OpInt        1      0
0.84     OpInt        0      1
1.04     OpInt        1      0
1.05     OpInt        0      1
1.23     Supervisor   1      2
1.25     OpInt        1      0
1.26     OpInt        0      1
1.46     OpInt        1      0
1.47     OpInt        0      1
1.68     OpInt        1      0
1.69     OpInt        0      1
1.9      OpInt        1      0
1.91     OpInt        0      1
2.01     Supervisor   2      3
2.11     OpInt        1      0
2.12     OpInt        0      1
2.32     OpInt        1      0
2.33     OpInt        0      1
2.53     OpInt        1      0
2.54     OpInt        0      1
2.74     OpInt        1      0
2.75     OpInt        0      1
2.96     OpInt        1      0
2.97     OpInt        0      1
3.15     Supervisor   3      4
3.18     OpInt        1      0
3.19     OpInt        0      1
```

This shows the *Supervisor* task turning on and then moving through states 1, 2, 3, and 4. Task *Mass1*, the control task, starts in its 'waiting' state, 0, and goes to the 'running' state, 1, in response to the command from the *Supervisor*. It also shows the transitions in the operator interface task going back and forth between a state that prints the ongoing results (state 0) and a state that waits (0.2 sec. in this case) for the next time to print its progress report. Because standard C/C++ input/output functions are used for this implementation, a progress check is about all that can be done in the operator interface because the input functions (*scanf* or *cin*) are blocking.

The audit trail is a fundamental debugging tool. It can be used to find out where a program went astray, whether critical timing constraints are being met, etc.

17.13 Sample Problems and Utilities

Sample problems are included with the source and library software needed to implement the methods described in this chapter. Each of the samples includes a simulation and a physical implementation. The physical implementation isolates the code for unique laboratory peripherals so that they can be modified. A set of utilities useful in general feedback control software development is also available. It includes a PID controller, a trapezoidal profiler, a quadrature decoder, a data logger, and an operator interface.

The sample set includes the following examples:

1. Position control (MASS1): the implementation of the control system described above.
2. Position control with an operator interface (MASS2): a modification of MASS1 to include an interactive operator interface. Other operational aspects have been changed as well to make it possible for the operator to make successive moves.
3. Control of an instrumented HO gage train (TRAIN): the train, on an oval track, is attached to a rotary encoder via a telescoping wand. The encoder pulses are slow enough for software decoding. The program includes a digital interrupt for the encoder and uses an event-type task to implement a cascade control.
4. Control of a hydraulic balance beam (BEAM): a beam, pivoted at its center has water tanks at each end and pumps connecting the tanks. The beam can be balanced by pumping water from one tank to the other. In its normal mode, the system is open-loop unstable. The system uses cascade control, feedback linearization, and nonlinear input compensation to derive the mass of water in each tank from a pressure transducer reading (which varies with angle).

Acknowledgement

This work was supported in part by the Synthesis Coalition, sponsored by the U.S. National Science Foundation.

References

[1] Auslander, D.M. and Sagues, P., *Microprocessors for Measurement and Control,* Osborne/McGraw-Hill, Berkeley, CA, 1981.

[2] Dornfeld, D.A., Auslander, D.M., and Sagues, P., Programming and Optimization of MultiMicroprocessor Controlled Manufacturing Processes, *Mech. Eng.,* 102(13), 34–41, 1980.

[3] Lippman, S. B., *C++ Primer,* 2nd Ed., Addison-Wesley, Reading, MA, 1991.

[4] Simms, M. J., Using State Machines as a Design and Coding Tool, *Hewlett-Packard J.,* 45(6), 1994.

[5] Stroustrup, B., *The C++ Programming Language,* 2nd Ed., Addison-Wesley, Reading, MA, 1991.

[6] Ward, P.T. and Mellor, S.J., *Structured Development for Real-Time Systems,* Prentice Hall, Englewood Cliffs, NJ, 1985.

18

Programmable Controllers

Gustaf Olsson
Dept. of Industrial Electrical Engineering and Automation, Lund Institute of Technology, Lund, Sweden

18.1 The Development of Programmable Controllers

Sequencing has traditionally been realized with relay techniques. Until the beginning of the 1970s, electromechanical relays and pneumatic couplings dominated the market. During the 1970s **programmable logical controllers** *(PLCs)* became more and more common, and today sequencing is normally implemented in software. Even if ladder diagrams are being phased out of many automation systems, they still are used to describe and document sequencing control implemented in software.

Programmable logical controllers *(PLCs)* are microcomputers developed to handle Boolean operations. A *PLC* produces *on/off* voltage outputs and can actuate elements such as electric motors, solenoids (and thus pneumatic and hydraulic valves), fans, heaters, and light switches. They are vital parts of industrial automation equipment found in all kinds of industries.

The *PLC* was initially developed by a group of engineers from General Motors in 1968, where the initial specification was formulated: it had to be easily programmed and reprogrammed, preferably in-plant, easily maintained and repaired, smaller than its relay equivalent, and cost-competitive with the solid-state and relay panels then in use. This provoked great interest from engineers of all disciplines using the *PLC* for industrial control. A microprocessor-based *PLC* was introduced in 1977 by Allen-Bradley Corporation in the USA, using an Intel 8080 microprocessor with circuitry to handle bit logic instructions at high speed.

The early *PLCs* were designed only for logic-based sequencing jobs (on/off signals). Today there are hundreds of different *PLC* models on the market. They differ in their memory size (from 256 bytes to several kilobytes) and I/O capacity (from a few lines to thousands). The difference also lies in the features they offer. The smallest *PLCs* serve just as relay replacers with added timer and counter capabilities. Many modern *PLCs* also accept proportional signals. They can perform simple arithmetic calculations and handle analog input and output signals and *PID* controllers. This is the reason why the letter *L* was dropped from *PLC*, but the term *PC* may cause confusion with personal computers so we keep the *L* here.

In 1994 the world market for *PLCs* was about $5 billion (according to Automation Research Corporation, ARC, USA). The largest share of the market is in Europe with some 45%. Japan has about 23% and USA 7%.

The logical decisions and calculations may be simple in detail, but the decision chains in large plants are very complex. This naturally raises the demand for structuring the problem and its implementation. Sequencing networks operate asynchronously, i.e., the execution is not directly controlled by a clock. The chain of execution may branch for different conditions, and concurrent operations are common. This makes it crucial to structure the programming and the programming languages. Grafcet, an

0-8493-8570-9/96/$0.00+$.50

international standard, is an important notation for describing binary sequences, including concurrent processes; it is used as a documentation tool and a programming language. Applications of function charts in industrial control problems are becoming more and more common. Graphical tools for programming and operator communication are also becoming a standard of many commercial systems. Furthermore, in a large automation system, communication between the computers becomes crucial for the whole operation.

In Section 18.2 hardware is described for binary sensors and actuators. Section 18.3 gives an elementary description of Boolean algebra. Ladder diagrams are still used to describe logical circuits and are described in Section 18.4. The structure of *PLCs* is outlined in Section 18.5, and their programming is described in Section 18.6. Their role in large automation systems is shown in Section 18.7, where communication is emphasized.

18.2 Binary Sensors and Actuators

In sequential control, measurements are of the *on/off* type and depend on binary sensors. In a typical process or manufacturing industry there are literally thousands of *on/off* conditions that have to be recorded. Binary sensors are used to detect the position of contacts, count discrete components in material flows, detect alarm limits of levels and pressures, and find end positions of manipulators.

18.2.1 Limit Switches

Limit switches have been used for decades to indicate positions. They consist of mechanically actuated electrical contacts. A contact opens or closes when some variable (position, level) has reached a certain value. There are hundreds of types of limit switches. Limit switches represent a crucial part of many control systems, and the system reliability depends to a great extent on them. Many process failures are due to limit switches. They are located 'where the action is' and are often subject to excessive mechanical forces or too large currents.

A normally open (n.o.) and a normally closed (n.c.) switch contact are shown in their normal and actuated positions in Figure 18.1. A switch can have two outputs, called change over or transfer contacts. In a circuit diagram it is common practice to draw each switch contact the way it appears with the system at rest.

The simplest type of sensor consists of a **single-pole, single-throw** (SPST) mechanical switch. Closing a mechanical switch causes a problem because it 'bounces' for a few milliseconds. When it is important to detect only the first closure, such as in a limit switch or in a 'panic button' the subsequent opening and closing bounces need not be monitored. When the opening of the switch must be detected after a closing, the switch must not be interrogated until after the switch 'settling time' has expired. A programmed delay is one means of overcoming the effects of switch bounce.

A **transfer contact** (sometimes called **single-pole double-throw**, SPDT) can be classified as either 'break-before-make' (BBM) or 'make-before-break' (MBB) (Figure 18.2). In a *BBM* contact, both contacts are open for a short moment during switching. In a *MBB* contact, there is a current in both contacts briefly during a switch.

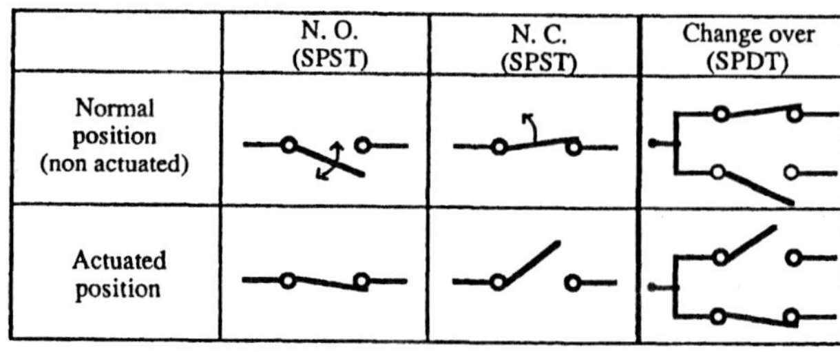

Figure 18.1 Limit switch symbols for different contact configurations.

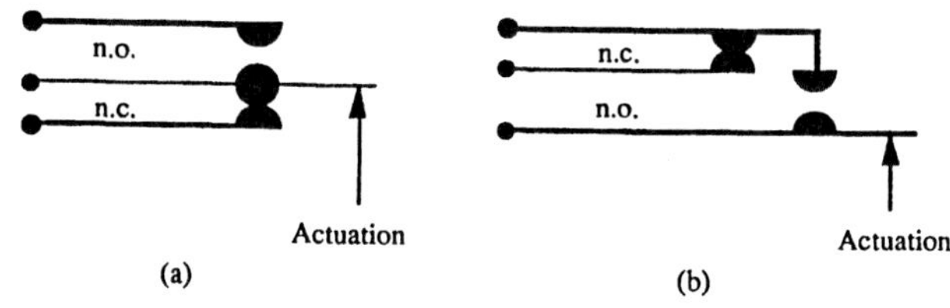

Figure 18.2 (a) Break-before-make (BBM). (b) Make-before-break (MBB) contact.

Contact debouncing in a *SPDT* switch can be produced in the hardware. When the grounded moving contact touches either input, the input is pulled low and the circuit is designed to latch the logic state corresponding to the first contact closure and to ignore the subsequent bounces.

18.2.2 Point Sensors

There are many kinds of measurement sensors that switch whenever a variable (level, pressure, temperature, or flow) reaches a certain point. Therefore, they are called **point sensors**. They are used as devices that actuate an alarm signal or shut down the process whenever some dangerous situation occurs. Consequently they have to be robust and reliable.

18.2.3 Digital Sensors

Digital measuring devices (**digital sensors**) generate discrete output signals, such as pulse trains or encoded data, that can be directly read by the processor. The sensor part of a digital measuring device is usually quite similar to that of their analog counterparts. There are digital sensors with microprocessors to perform numerical manipulations and conditioning locally, and to provide output signals in either digital or analog form. When the output of a digital sensor is a pulse signal, a counter is used to count the pulses or to count clock cycles over one pulse duration. The count is first represented as a digital word according to some code and then read by the computer.

18.2.4 Binary Actuators

In many situations, sufficient control of a system can be achieved if the actuator has only two states: one with electrical energy applied (*on*) and the other with no energy applied (*off*). In such a system no digital-to-analog converter is required and amplification can be performed by a simple switching device rather than by a linear amplifier.

Many types of actuators, such as magnetic valves controlling pneumatic or hydraulic cylinders, electromagnetic relays controlling electrical motors, and lamps, can receive digital signals from a controller. There are two main groups of binary actuators, monostable and bistable units. A **monostable** actuator has only one stable position and gives only one signal. A contactor for an electric motor is monostable. As long as a signal is sent to the contactor, the motor rotates, but, as soon as the signal is broken, the motor will stop.

A **bistable** unit remains in its given position until another signal arrives. In that sense the actuator is said to have a memory. For example, in order to move a cylinder, controlled by a bistable magnet valve, one signal is needed for the positive movement and another one for the negative movement.

18.2.5 Switches

The output lines from a computer output port can supply only small amounts of power. Typically a high-level output signal has a voltage between 2 and 5 V and a low-level output of less than 1 V. The current capacity depends on the connection of the load but is generally less than 20 mA, so that the output can switch power of less than 100 mW.

For higher power handling capability, it is more important to avoid a direct electrical connection between a computer output port and the switch. The switch may generate electrical noise which would affect the operation of the computer if there were a common electrical circuit between the computer and the switch. Also, if the switch fails, high voltage from the switch could affect the computer output port and damage the computer.

The most common electrically isolated switch in control applications has been the **electromechanical relay**. Low-power reed relays are available for many computer bus systems and can be used for isolated switching of signals. Relays for larger power ratings cannot be mounted on the computer board. A relay is a robust switch that can block both direct and alternating currents. Relays are available for a wide range of power, from reed relays used to switch millivolt signals to contactors for hundreds of kilowatts. Moreover, their function is well understood by maintenance personnel. Some of the disadvantages are that relays are relatively slow, switching in the order of milliseconds instead of microseconds. They suffer from contact bouncing problems which can generate electric noise, and, in turn, may influence the computer.

The switching of high power is easily done in solid-state switches which avoid many deficiencies of relays. A solid-state switch has a control input which is coupled optically or inductively to a solid-state power switching device. The control inputs to solid-state switches, designed to be driven directly from digital logic circuits, are quite easily adaptable to computer control.

18.3 Elementary Switching Theory

In this section we will describe elementary switching theory that is relevant for process control applications.

18.3.1 Notations

An electric switch or relay contact and a valve intended for logic circuits are both binary, designed to operate in the *on/off* mode. A transistor can also be used as a binary element operating only in on/off mode, either conducting or not conducting current.

A binary variable has the values 0 or 1. For a switch contact, relay contact, or a transistor (labelled X), the statement $X = 0$ means that the element is open (does not conduct current) and $X = 1$ means closed (conducts). For a push button or a limit switch, $X = 0$ means that the switch is not being actuated, and $X = 1$ indicates actuation.

Often a binary variable is represented as a voltage level. In *positive* logic, the higher level corresponds to logical 1 and the lower level to logical 0. In *TTL* (transistor-transistor logic), logical 0 is typically defined by levels between 0 and 0.8 V and logical 1 any voltage higher than 2 V. Similarly, in pneumatic systems $X = 0$ may mean that the line is exhausted to atmospheric pressure while $X = 1$ means a pressurized line.

18.3.2 Basic Logical Gates

Following is a brief recapitulation of Boolean algebra. The simplest logical operation is the negation (*NOT*), with one input and one output. If the input $U = 0$, then the output $Y = 1$ and vice versa. We denote negation of U by $\overline{U}$. The behavior of a switching circuit can be represented by **truth tables**, where the output value is given for all possible combinations of inputs. The symbol and the truth table for *NOT* is shown in Figure 18.3.

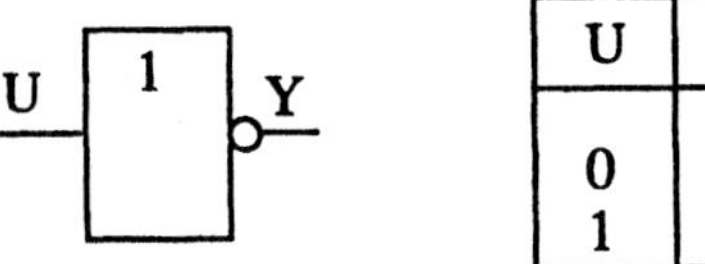

U	Y
0	1
1	0

Figure 18.3 The ISO symbol and truth table for *NOT*.

Two normally open switch contacts connected in series constitute an *AND* gate defined by **Boolean multiplication** as

$$Y = U1 \cdot U2.$$

$Y = 1$ only if both $U1$ and $U2$ are equal to 1, otherwise $Y = 0$ (Figure 18.4). The multiplication sign is often omitted, just as in ordinary algebra. An *AND* gate can have more than two inputs, because any number of switches can be connected in series. Adding a third switch results in $Y = U1 \cdot U2 \cdot U3$. We use the ISO (International Standards Organization) symbol for the gate.

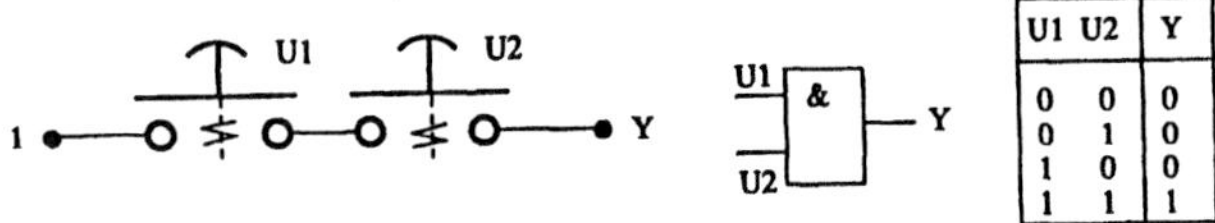

Figure 18.4 An *AND* gate, its ISO symbol, and truth table.

A common operation is a logical *AND* between two bytes in a process called **masking**. The first byte is the input register reference while the other byte is defined by the user to mask out bits of interest. The *AND* operation is made bit by bit of the two bytes (Figure 18.5). In other words, only where the mask byte contains 'ones' is the original bit of the reference byte copied to the output.

input register mask	11011000 01101101
output	01001000

Figure 18.5 Masking two bytes with an *AND* operation.

If two switches $U1$ and $U2$ are connected in parallel, the operation is a **Boolean addition** and the function is of the *OR* type. Here, $Y = 1$ if either $U1$ or $U2$ is actuated; otherwise $Y = 0$. The logic is denoted (Figure 18.6) by

$$Y = U1 + U2.$$

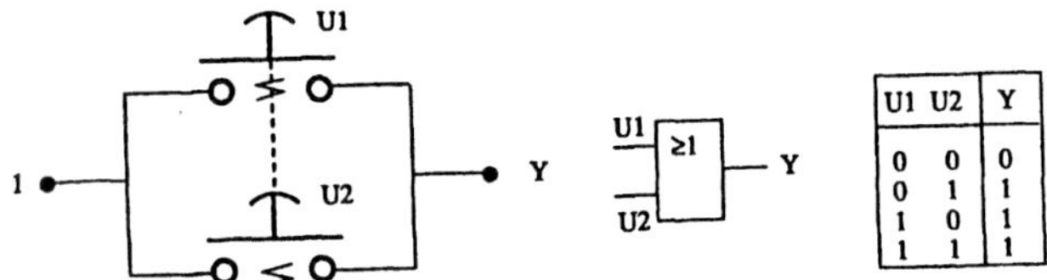

Figure 18.6 An *OR* gate, its ISO symbol, and truth table.

As for the *AND* gate, more switches can be added (in parallel), giving $Y = U1 + U2 + U3 \ldots$. The ≥ 1 designation inside the *OR* symbol means that gate output is 'high' if the number of 'high' input signals is equal to or greater than 1.

A logical *OR* between two bytes also makes a bit by bit logical operation (Figure 18.7). The *OR* operation can be used to set one or several bits unconditionally to 1.

There are some important theorems for one binary variable X, such as

$$\begin{aligned} X + X &= X, \\ X \cdot X &= X, \\ X + \overline{X} &= 1, \\ \text{and} \quad X \cdot \overline{X} &= 0. \end{aligned}$$

input register mask	11011000 01101101
output	11111101

Figure 18.7 Masking two bytes with an *OR* operation.

Similarly, for two variables we can formulate and easily verify

$$\begin{aligned} X + Y &= Y + X, \\ X \cdot Y &= Y \cdot X, \\ X + XY &= X, \\ X \cdot (X + Y) &= X, \\ (X + \overline{Y}) \cdot Y &= X \cdot Y, \\ X \cdot \overline{Y} + Y &= X + Y, \\ \text{and} \quad XY + \overline{Y} &= X + \overline{Y}. \end{aligned}$$

The *De Morgan* theorems are useful in manipulating Boolean expressions:

$$\begin{aligned} (\overline{X + Y + Z + \cdots}) &= \overline{X} \cdot \overline{Y} \cdot \overline{Z} \cdots, \\ \text{and} \quad (\overline{X \cdot Y \cdot Z \cdots}) &= \overline{X} + \overline{Y} + \overline{Z} \cdots \end{aligned}$$

The theorems can help in simplifying complex binary expressions, thus saving components for the actual implementation.

18.3.3 Additional Gates

Two normally closed gates in series may define a **NOR** gate, i.e., the system conducts if *neither* the first *nor* the second switch is actuated. According to De Morgan's theorem this can be expressed as

$$Y = \overline{U1} \cdot \overline{U2} = (\overline{U1 + U2}),$$

showing that the *NOR* gate can be constructed from the combination of a *NOT* and an *OR* gate (Figure 18.8). The circle at an input or output line of the symbol represents Boolean inversion.

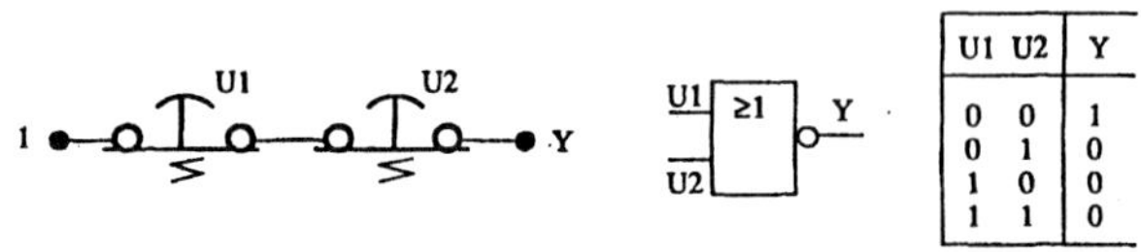

Figure 18.8 A *NOR* gate, its ISO symbol, and truth table.

A *NOR* gate is easily implemented electronically or pneumatically. Moreover any Boolean function can be obtained only from a *NOR* gate, which makes it a **universal gate.** For example, a *NOT* gate is a *NOR* gate with a single input. An *OR* gate can be obtained by connecting a *NOR* gate and a *NOT* gate in series. An *AND* gate, obtained by using two *NOT* gates and one *NOR* gate (Figure 18.9), is written as

$$Y = \overline{(\overline{U1}) + (\overline{U2})} = (\overline{\overline{U1}}) \cdot (\overline{\overline{U2}}) = U1 \cdot U2.$$

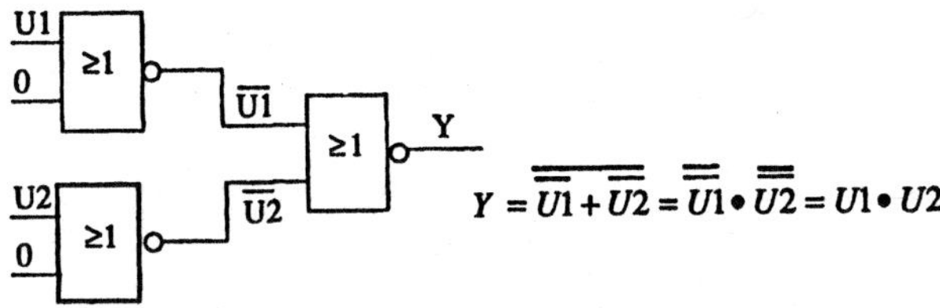

Figure 18.9 Three *NOR* gates acting as an *AND* gate (this is not the minimal realization of an *AND* gate).

A **NAND** gate is defined by

$$Y = \overline{(U1 \cdot U2)} = \overline{U1} + \overline{U2}.$$

The system does *not* conduct if both $U1$ *and* $U2$ are actuated, i.e., it conducts if either switch is not actuated. Like the *NOR* gate, the *NAND* gate is a universal gate (Figure 18.10).

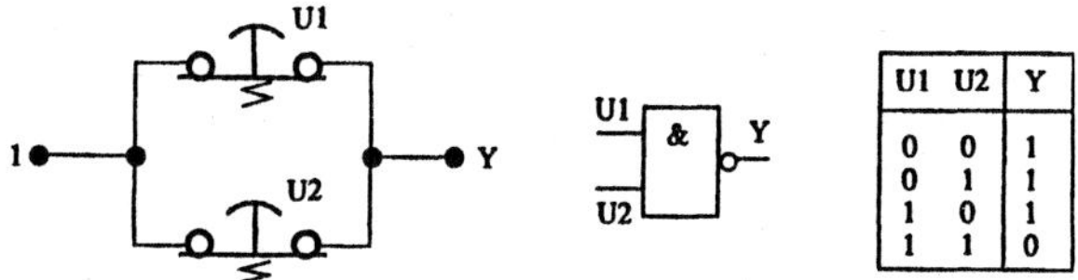

U1	U2	Y
0	0	1
0	1	1
1	0	1
1	1	0

Figure 18.10 A *NAND* gate, its ISO symbol, and its truth table.

The *NAND* and *NOR* operations are called **complete operations**, because all others can be derived from either of them. No other gate or operation has the same property.

A circuit with two switches, each having double contacts (one normally open and the other normally closed), is shown in Figure 18.11. This is an **exclusive OR (XOR)** circuit, and the output is defined by

$$Y = U1 \cdot \overline{U2} + \overline{U1} \cdot U2.$$

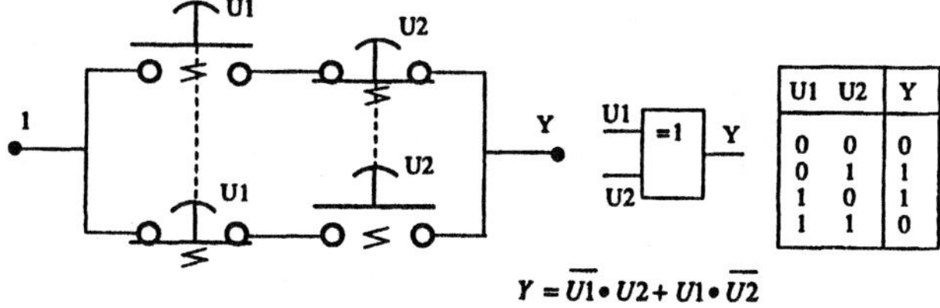

U1	U2	Y
0	0	0
0	1	1
1	0	1
1	1	0

Figure 18.11 An *exclusive-OR* gate, its ISO symbol, and its truth table.

The circuit conducts *only* if *either* $U1 = 1$ *or* $U2 = 1$, but $Y = 0$ if *both* $U1$ *and* $U2$ have the same sign (compare with the *OR* gate). For example, such a switch can be used to control the room light from two different switch locations $U1$ and $U2$. In digital computers *XOR* circuits are extremely important for binary addition.

An *exclusive OR (XOR)* between two bytes will copy the 1 in the input register only where the mask contains 0. Where the mask contains 1, the bits of the first operand are inverted. In other words, in the positions where the operands are equal, the result is 0 and, conversely, where the operands are not equal, the result is a 1 (Figure 18.12). This is often used to determine if and how a value of an input port has been changed between two readings.

input register mask	1 1 0 1 1 0 0 0 0 1 1 0 1 1 0 1
output	1 0 1 1 0 1 0 1

Figure 18.12 Masking two bytes with an *XOR* operation.

EXAMPLE 18.1: Simple combinatorial network

A simple example of a combinatorial circuit expressed in ISO symbols is shown in Figure 18.13. The logical expressions are:

$$\begin{aligned} y3 &= u1 \cdot \overline{u12} \\ y4 &= u2 \cdot y2 \\ y2 &= y4 + \overline{u1} \end{aligned}$$

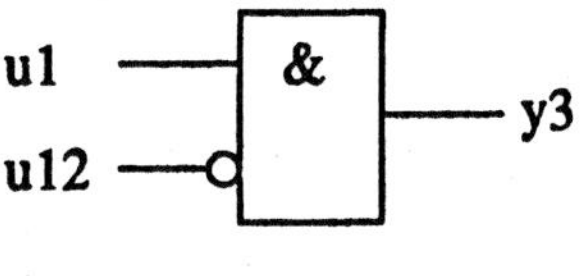

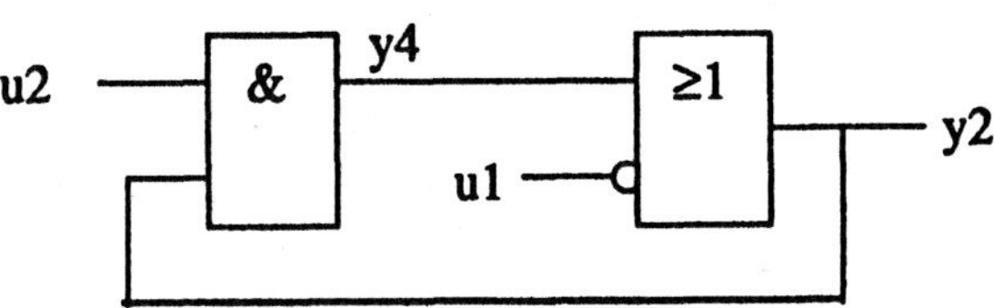

Figure 18.13 Simple combinatorial circuit.

The ISO organization that deals specifically with questions concerning electrotechnical and electronic standards is the IEC (International Electrotechnical Commission). Standards other than IEC are often used to symbolize switching elements. The IEC symbols are not universally accepted and in the USA there are at least three different sets of symbols. In Europe, the DIN standard is common. Three common standards are shown in Figure 18.14.

	DIN/IEC 113-7 DIN/IEC 617-12	old DIN 40700-14	USA
NOT	U 1 Y	U Y	U Y
OR	U1 U2 ≥1 Y	U1 U2 Y	U1 U2 Y
AND	U1 U2 & Y	U1 U2 Y	U1 U2 Y
NOR	U1 U2 ≥1 Y	U1 U2 Y	U1 U2 Y
NAND	U1 U2 & Y	U1 U2 Y	U1 U2 Y
XOR	U1 U2 =1 Y	U1 U2 Y	U1 U2 Y

Figure 18.14 Commonly used logical gate symbols.

In principle all switching networks can be tested by truth tables. Unfortunately the number of Boolean functions grows rapidly with the number of variables n, because the number of combinations becomes 2^n. It is outside the scope of this text to discuss different simplifications of Boolean functions. A method known as the Karnaugh map may be used if the number of variables is small. For systems with many variables (more than about 10), there are numerical methods to handle the switching network. The method by Quine-McCluskey may be the best known, and is described in standard textbooks on switching theory (e.g., [2]).

18.3.4 Flip-Flops

Hitherto we have described **combinatorial networks**, i.e., the gate output Y depends only on the *present* combination of input signals $\mathbf{U} = (U_1, U_2, \ldots)$, or

$$Y(t) = f[\mathbf{U}(t)].$$

Gates have *no memory* so the network is a *static* system. To introduce a memory function, we define **flip-flop** elements, whose output depends on the present state of the inputs and on the previous flip-flop state. The basic type of flip-flop is the **SR** flip-flop (Set-Reset). The two inputs S and R can be either 1 or 0. Both are, however, not permitted to be 1 or 0 at the same time. The output is called y and normally $\overline{y}$ also is an output. If $S = 1$ then $y = 1$ $(\overline{y} = 0)$ and the flip-flop becomes *set*. If S returns to 0, then the gate remembers that S had been 1 and keeps $y = 1$. If R becomes 1 (assuming that $S = 0$), the flip-flop is *reset*, and $y = 0$ $(\overline{y} = 1)$. Again R can return to 0 and y remains 0 until a new S signal appears. Let us call the states at consecutive moments y_n and y_{n+1}. Then the operation can be written as

$$y_{n+1} = \overline{R} \cdot (S + y_n).$$

An SR flip-flop can be illustrated by two logical elements (Figure 18.15).

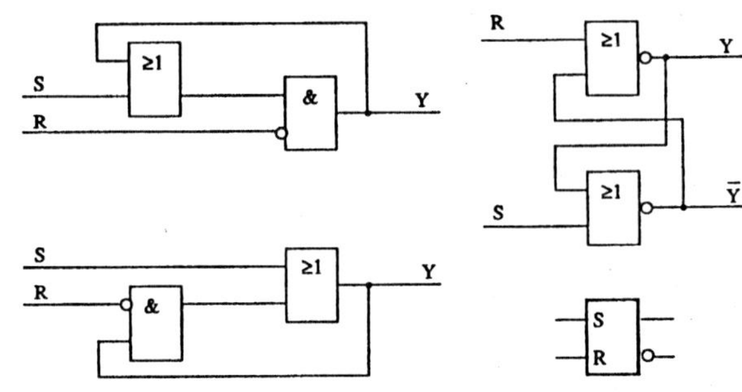

Figure 18.15 Three different illustrations of a flip-flop gate and its DIN/IEC symbol.

By adding two *AND* gates and a clock-pulse input to the flip-flop we obtain a delay (D) flip-flop or a **latch**. The delay flip-flop has two inputs, a data (D) and a clock pulse (CP) (Figure 18.16). Whenever a clock pulse appears, the output y accepts the D input value that existed before the appearance of the clock pulse. In other words, the D input is delayed by one clock pulse. The new state y_{n+1} is always independent of the old state.

By introducing a feedback from the output to the flip-flop input and a time delay in the circuit, we obtain a **trigger** or a **toggle** (T) flip-flop. The T flip-flop (Figure 18.16) is often used in counting and timing circuits as a 'frequency divider' or a 'scale-of-two' gate. It has only one input, T. Whenever an upgoing pulse appears in T, the output y flips to the other state.

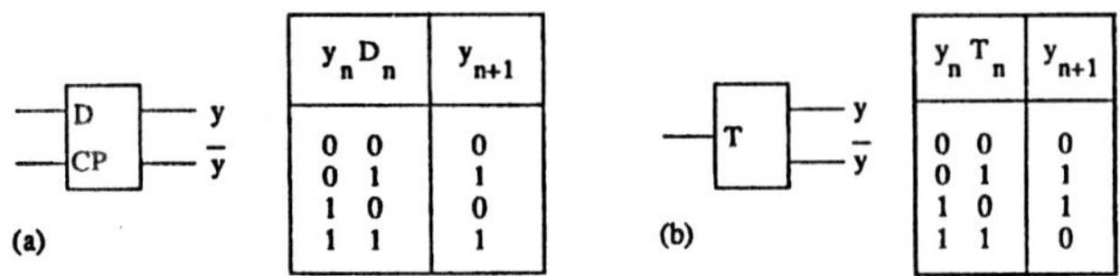

y_n	D_n	y_{n+1}
0	0	0
0	1	1
1	0	0
1	1	1

y_n	T_n	y_{n+1}
0	0	0
0	1	1
1	0	1
1	1	0

Figure 18.16 A delay (D) flip-flop and a trigger (T) flip-flop.

All three types of flip-flops can be realized in a **JK flip-flop**, with J being the set signal and K the reset signal. It frequently comes with a clock-pulse input. Depending on the input signals, the JK flip-flop can be a SR flip-flop, a latch or a trigger.

18.3.5 Realization of Switching

Switching functions for actuators are mostly made to switch *power* as opposed to *signals*. Here we focus on *signal switching* in an electronic environment. Electromechanical relays for signal switching have been mentioned. Electronic logic gates can be implemented with diodes that provide *AND* and *OR* gates. It is not suitable to cascade several gates in series, so they are not very useful.

A common way to implement the gates, however, is by transistor logic, because the signal then can be amplified back to normal levels. *TTL* logic has been dominant for a long time. Within the *TTL* family are several types with differing power consumption and speed.

Conventional *TTL* circuits have been largely replaced by *LS-TTL* (low-power Schottky *TTL*) elements. They contain so-called Schottky diodes to increase switching speed, and they use considerably less power than the older *TTL* type. Many *TTL* circuits today are replaced by *CMOS* (complementary metal-oxide semiconductor) based on *FETs* (field-effect transistors) rather than bipolar transistors. A *CMOS* circuit has about three orders of magnitude lower power consumption than corresponding *TTL* gates. The *CMOS* are also less sensitive to electrical noise and the supply voltage. Furthermore, *CMOS* circuits are slower in principle and are easily damaged by static electricity. A new generation of the *CMOS* circuits is the high speed *(HC) CMOS* logic.

Complex circuits can of course be manufactured as medium or large-scale integrated circuits. This is, however, not economically justifiable in very small quantities. By using so called **programmable logic devices** *(PLD)*, one can obtain semicustomized *ICs* quite inexpensively. *PLDs* are fuse-programmable chips mostly belonging to the *LS-TTL* family. The circuit contains a large array of programmable gates connected by microscopic fused links. These fuses can be selectively blown by using a special programming unit.

In the *PLD* family there are **programmable array logic** *(PAL)*, **field- programmable logic arrays** *(FPLA)* and **programmable read-only memory** *(PROM)*. In the *PAL* system there is a programmable *AND*-gate array connected to a fixed *OR*-gate array. In the FPLA system both the *AND* and the *OR* gates are programmable. In both *PAL* and *FPLA* there are chips available with *NOR, XOR*, or *D*- flip-flops so that a complete sequential system can be included in one chip, (Figure 18.17).

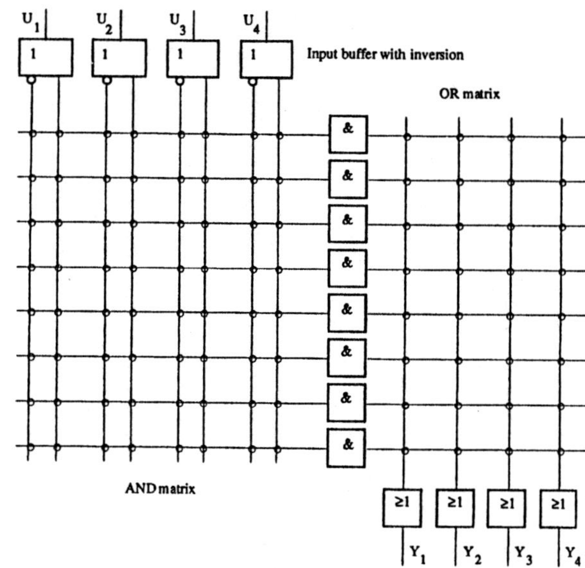

Figure 18.17 A functional circuit for PAL, FPLA and PROM circuits.

To make programming of *PLDs* less complex, several software packages are available for personal computers. They convert Boolean equations into data for feeding to the programming unit. Testing of the programmed chip is another attractive software feature.

18.4 Ladder Diagrams

Many switches are produced from solid-state gates, but electromechanical relays are still used in many applications. Statistics show that the share of electromechanical relays versus the total number of gates in use is decreasing. This does not mean that their importance is dwindling; relays remain, in fact, a necessary interface between the control electronics and the powered devices.

Relay circuits are usually drawn in the form of **ladder diagrams**. Even if the relays are replaced by solid-state switches or programmable logic, they are still quite popular for describing combinatorial circuits or sequencing networks. They are also a basis for writing programs for programmable controllers.

18.4.1 Basic Description

A ladder diagram reflects a conventional wiring diagram (Figure 18.18). A wiring diagram shows the physical arrangement of the various components (switches, relays, motors, etc.) and their interconnections, and is used by electricians to do the actual wiring of a control panel. The ladder diagrams are more schematic and show each branch of the control circuit on a separate horizontal row (the rungs of the ladder). They emphasize the function of each branch and the resulting sequence of operations. The base of the diagram shows two vertical lines, one connected to a voltage source and the other to ground.

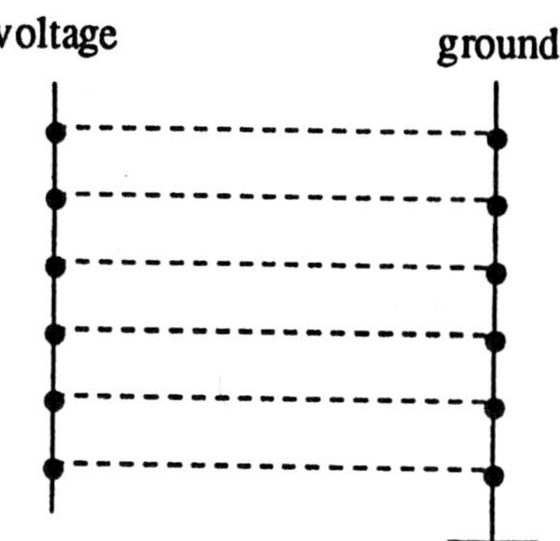

Figure 18.18 Framework of a ladder diagram.

Relay contacts are either normally open (n.o.) or normally closed (n.c.), where *normally* refers to the state in which the coil is not energized. Relays can implement elementary circuits such as *AND* and *OR* as well as *SR* flip-flops. The relay symbols are shown in Figure 18.19.

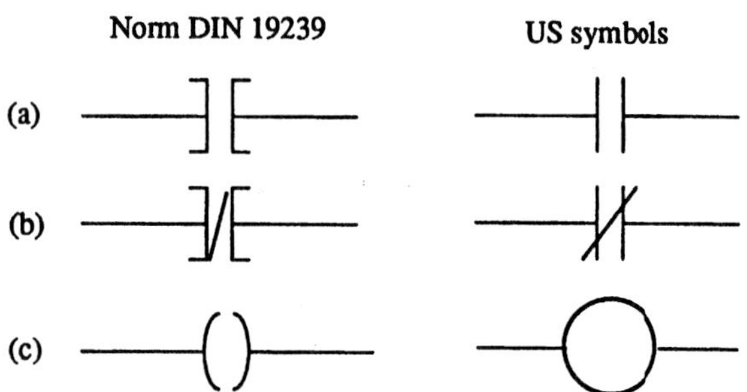

Figure 18.19 Relay symbols for n.o., n.c. contacts, and relay coil.

EXAMPLE 18.2: Combinatorial circuit

The combinatorial circuit of Figure 18.13 can be represented by a ladder diagram (Figure 18.20). All of the conditions have to be satisfied simultaneously. The series connection is a logical *AND* and the parallel connection a logical *OR*. The lower case characters (u, y) denote the switches and the capital symbols (Y; the ring symbol) denote the coil.

The relay contacts usually have negligible resistance, whether they are limit switches, pressure, or temperature switches. The output element (the ring) could be any resistive load (relay coil) or a lamp, motor, or any other electrical device that can be actuated. Each rung of the ladder diagram must contain at least one output element, otherwise a short circuit would occur.

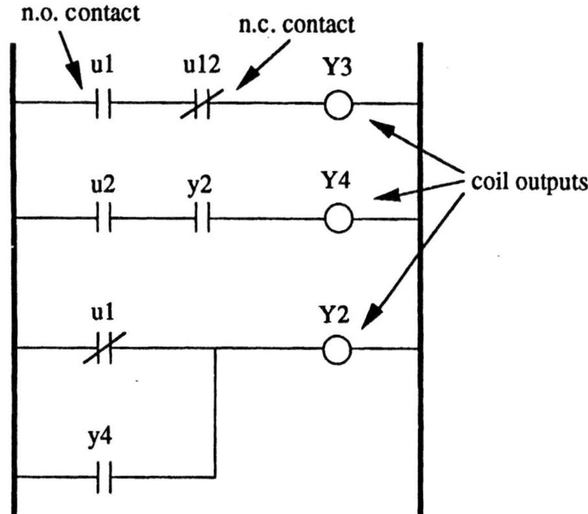

Figure 18.20 The combinatorial circuit in Figure 18.13 represented by a ladder diagram.

EXAMPLE 18.3: A flip-flop as a ladder diagram

A flip-flop (Figure 18.15) can also be described by a ladder diagram (Figure 18.21). When a set signal is given, the S relay conducts a current that reaches the relay coil Y. Note that the R is not touched. Energizing the relay coil closes the relay contact y in line 2. The **set** push button can now be released and current continues to flow to coil Y through the contact y, i.e., the flip-flop remains *set*. Thus, the y contact provides the 'memory' of the flip-flop. In industrial terminology, the relay is a **self-holding** or **latched** relay. At the moment the **reset** push button is pressed, the circuit to Y is broken and the flip-flop returns to its former **reset** state.

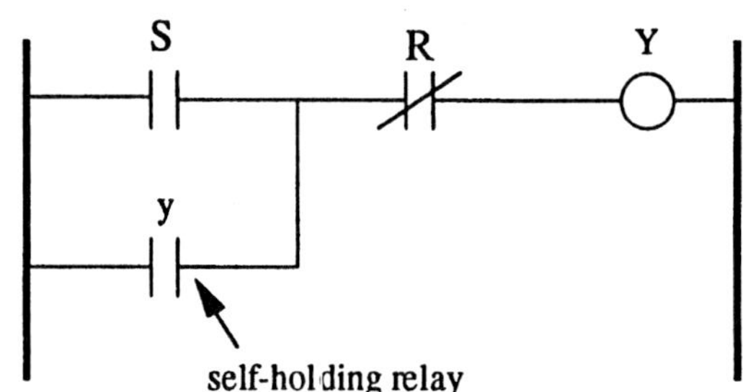

Figure 18.21 An SR flip-flop described by a ladder diagram.

18.4.2 Sequential Circuits

In a combinatorial network the outputs depend only on the momentary values of the inputs. In a sequence chart, however, the outputs also depend on earlier inputs. The sequence contains memory elements and must be described by states. Many sequence operations can be described by ladder diagrams and can be defined by a number of states, where each state is associated with a certain control action.

Only *one state at a time* can be active. Therefore a signal is needed to change from one state to another. The **reset** (R) signal in Figure 18.21 is such a signal. The sequence can be described as a series of SR flip-flops where each step is a rung on the ladder. When the signal is given, the next flip-flop is set. The structure of the sequence is shown as a ladder diagram in Figure 18.22. The execution jumps one step at a time and returns to step 1 after the last step.

Step 1 can be initiated with a **start** button. When running in an infinite loop, it can also be started from the *last step*. When the *last step* is active together with a new condition for the startup of *step 1*, then the *step 1* coil is activated, and the self-holding relay keeps it set. When the condition for *step 2* is satisfied, the relay *step 2* latches circuit 2 and at the same time guarantees that circuit 1 is broken. This is then continued in the same fashion. In order to insure a repetitive sequence, the last step has to be connected to *step 1* again.

This is an example of an **asynchronous** execution. In switching theory there are also **synchronous** charts, where the state changes are caused by a clock pulse. In industrial automation

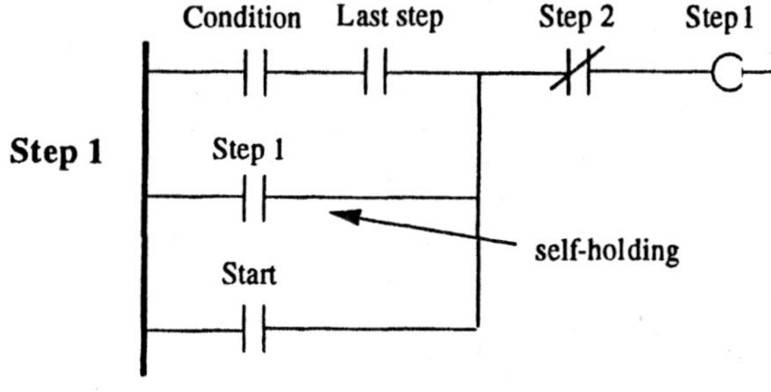

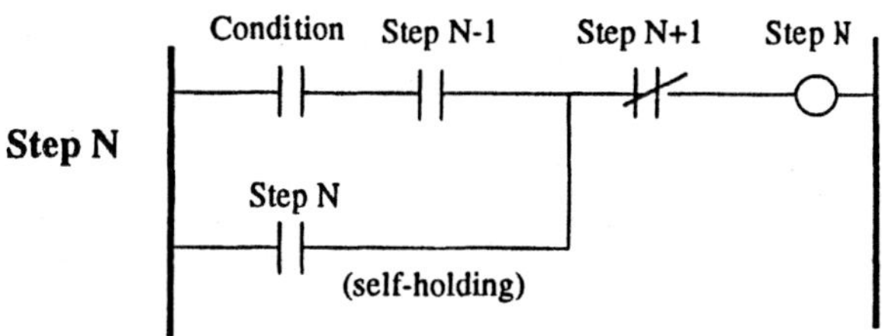

Figure 18.22 A sequence described by a ladder diagram.

applications, we mostly talk about asynchronous charts, because the state changes do depend not on clock pulses but on several conditions in different parts of the sequence. In other words, an asynchronous system is **event-based**, while a synchronous system is **time-based**. Moreover, we are dealing with design of asynchronous systems with **sustained input signals** rather than pulse inputs.

18.5 Programmable Controllers

Control equipment can be based on one of the following four principles:

1. relay systems
2. digital logic
3. computers
4. *PLC* systems

The best overall choice is a *PLC*. If there is a special requirement for speed or resistance to electric noise, then digital logic or relay systems may be required. For handling complex functions computers may be marginally superior, but *PLCs* today are built with increasingly complex features.

The overall comparison between different control equipment types can be based on

- price per function
- physical size
- operating speed
- electrical noise immunity
- installation (design, install, programming)
- capability for complicated operations
- ease of changing functions
- ease of maintenance

In smaller *PLCs* the functions are performed by individual printed circuit cards within a single compact unit. Larger *PLCs* are constructed on a modular basis with function modules slotted into the backplane connectors of the mounting rack. A programming unit is necessary to download control programs to the *PLC* memory.

18.5.1 Basic Structure

The basic operation of a *PLC* corresponds to a software-based equivalent of a relay panel. However, a *PLC* can also execute other operations, such as counting, delays and timers. Because a *PLC* can be programmed in easy-to-learn languages, it is naturally more flexible than any hardware relay system. A single *PLC* can replace hundreds of relays. *PLCs* are in fact more flexible than programmable logical devices but usually slower, so that *PLDs* and *PLCs* often coexist in industrial installations offering the best and most economical solutions.

PLCs can be programmed using both relay-type ladder diagrams (mostly in the United States) and logical gate symbols (mostly in Europe) but high level program languages are becoming more common.

Figure 18.23 shows the basic structure of a *PLC*. The inputs are read into the input memory register. This function is already included in the system software in the *PLC*. An input-output register is often not only a bit but a byte. Consequently one input instruction gives the status of 8 different input ports.

The instructions fetch the value from the input register and operate on only this or on several operands. The central processing unit *(CPU)* works towards a result register or accumulator (*A*). The result of an instruction is stored either in some intermediate register or directly in the output memory register that is written to the outputs. The output function is usually included in the system programs in a *PLC*. A typical commercial *PLC* is shown in Figure 18.24.

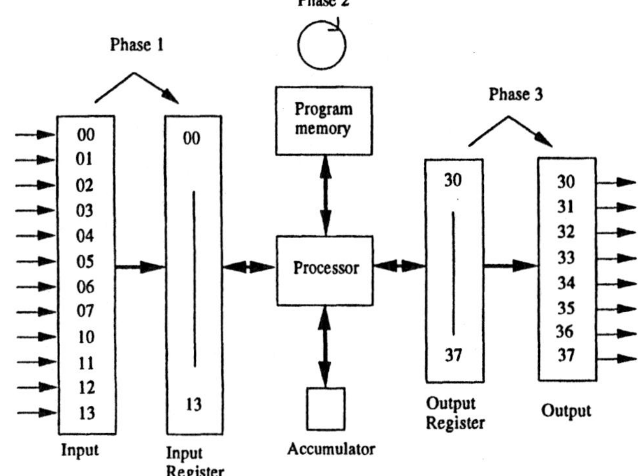

Figure 18.23 Basic structure of a *PLC*.

A *PLC* is specifically made to fit an industrial environment

Figure 18.24 A small *PLC* system, Micrologix, from Allen-Bradley. A simple *PLC* programmer is included (courtesy Allen-Bradley Company, Inc.).

where it is exposed to hostile conditions, such as heat, humidity, unreliable power, mechanical shocks, and vibrations. A *PLC* also comes with input-output modules for different voltages and can be easily interfaced to hundreds of input and output lines. *PLCs* have both hardware and software features that make them attractive as controllers of a wide range of industrial equipment. They are specially built computers with three functions, memory, processing, and input/output.

18.5.2 Basic Instructions and Execution

To make a *PLC* system for industrial automation, it has to work in real time. Consequently the controller has to act on external events very quickly, with a short response time. There are two principal ways to sense the external signals, by **polling** the input signals regularly or by using **interrupt** signals. The polling method's drawback is that some external event may be missed if the processor is not sufficiently fast. On the other hand such a system is simple to program.

A system with interrupts is more difficult to program but the risk of missing some external event is much smaller. The polling method is usually used in simpler automation systems while interrupts are used in more complex control systems.

'Programming' a *PLC* consists mainly of defining sequences. The input and output functions are already prepared. The instructions from a ladder diagram, a logical gate diagram, or Boolean expressions are translated into machine code. At the execution, the program memory is run through cyclically in an infinite loop. Every scan may take some 15–30 msec. in a small *PLC*, and the scanning time is approximately proportional to the memory size. In some *PLCs* the entire memory is always scanned even if the code is shorter. In other systems the execution stops at an **end** statement that concludes the code; thus the loop time can be shortened for short programs.

The response time of the *PLC* of course depends on the processing time of the code. While the instructions and the output executions are executed, the computer system cannot read any new input signals. Usually this is not a big problem, since most signals in industrial automation are quite slow or last for a relatively long time.

The ladder diagram can be considered as if every rung were executed at the same time. Thus it is not possible to visualize that the ladder diagram is executed sequentially on a row-by-row basis. The execution has to be very fast compared to the time scale of the process.

Four fundamental machine instructions can solve most sequencing problems:

ld, ldi A number from the computer input memory is loaded (LD) or inverted (LDI) before it is read into the accumulator (A).

and, ani An *AND* or *AND Inverse* instruction executes an *AND* logical operation between A and an input channel, and stores the result in A.

or, ori An *OR* or *OR Inverse* instruction executes an *OR* logical operation between A and an input channel, and stores the result in A.

out The instruction outputs A to the output memory register. The value remains in A, so that the same value can be sent to several output relays.

The logical operations may be performed on bits as well as on bytes.

EXAMPLE 18.4: Translation from a ladder diagram to machine code

The translation from ladder diagram to machine code is illustrated by Figure 18.25. The gate $y11$ gives a self-holding capability. Note that $y11$ is a memory cell, and $Y11$ is an output.

A logical sequence or ladder diagram is often branched. Then there is a need to store intermediate signals for later use. This can be done with special help relays, but in a *PLC* it is better to use two instructions **orb** (*OR* Block) or **anb** (*AND* Block). They use a memory stack area (last in, first out) in the *PLC* to store the output temporarily.

EXAMPLE 18.5: Using the block instruction and stack memory

The ladder diagram (Figure 18.26) can be coded with the following machine code:

ld x1 Channel 1 is read into the accumulator (A).

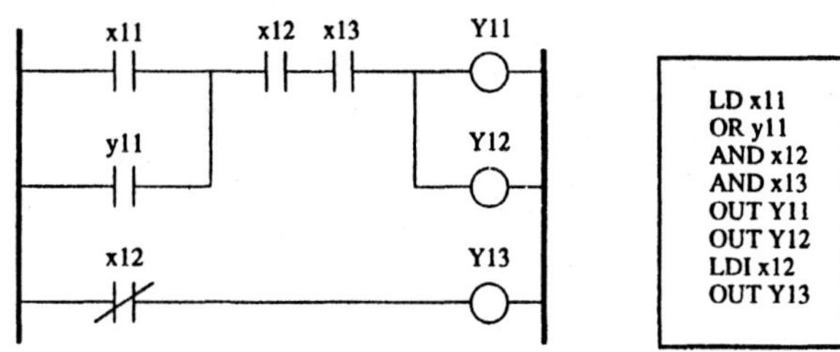

Figure 18.25 Translation of a ladder diagram into machine code.

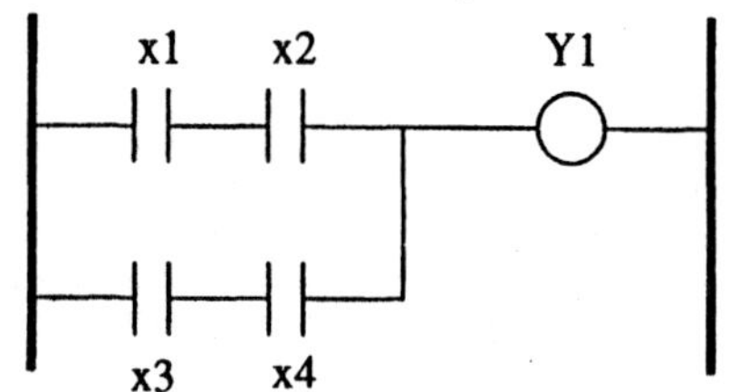

Figure 18.26 Example of using a stack memory.

and x2 The result of the *AND* operation is stored in *A*.

ld x3 The content of *A* is stored on the stack. Channel 3 is read into *A*.

and x4 The result of lines 3 and 4 is stored in *A*.

orb An *OR* operation between *A* and the stack. The result is stored in *A*. The last element of the stack is eliminated.

out Y1 Output of *A* on Channel 1.

EXAMPLE 18.6: Using the block instructions and the stack memory

The logical gates in Figure 18.27 are translated to machine code by using block instructions.

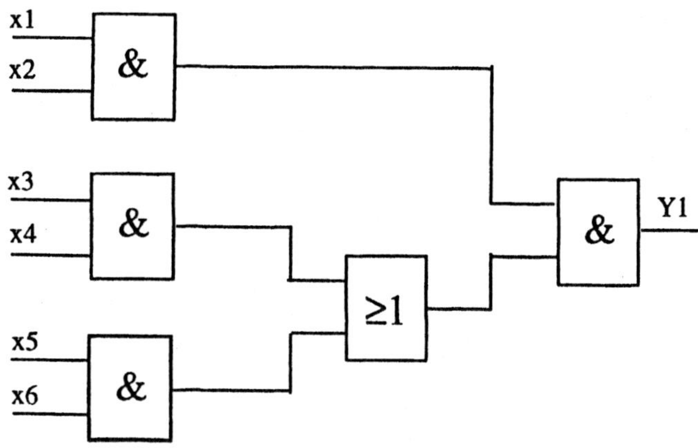

Figure 18.27 Example of a logical circuit.

The corresponding machine code is as follows:

ld x1 Load Channel 1.

and x2 The result is stored in *A*.

ld x3 The content of *A* is stored on the stack. Status of Channel 3 is loaded into *A*.

and x4 The result of lines 3 and 4 is stored in *A*.

ld x5 The content of *A* is stored on the stack. Status of Channel 5 is loaded into *A*.

and x6 The result of lines 5 and 6 is stored in *A*.

orb Operates on the last element in the stack (the result of lines 3 and 4) and the content of *A*. The result is stored in *A*. The last element of the stack is removed.

anb Operates on the last element in the stack (the result of lines 1 and 2) and the content of *A*. The result is stored in *A*. The last element of the stack is removed.

out Y1.

18.5.3 Additional PLC Instructions

For logical circuits there are also operations such as *XOR*, *NAND*, and *NOR* as described earlier. Modern *PLC* systems are supplied with instructions for alphanumerical or text handling and communication as well as composed functions such as timers, counters, memory, and pulses.

A pulse instruction *(PLS)* gives a short pulse, e.g., to reset a counter. A *PLC* may also contain delay gates or time channels so that a signal in an output register may be delayed for a certain time. Special counting channels can count numbers of pulses.

Different signals can be shaped, such as different combinations of delays, oscillators, rectangular pulses, ramp functions, shift registers, or flip-flops. As already mentioned, advanced *PLCs* also contain floating-point calculations as well as prepared functions for signal filtering and feedback control algorithms.

18.6 PLC Programming

A *PLC* is usually programmed via an external unit; this unit is unnecessary for the *PLC* operation and may be removed when the *PLC* is operating. Programming units range from small hand-held portable units, sometimes called 'manual programmers', to personal computers.

A manual *PLC* programmer looks like a large pocket calculator with a number of keys and a simple display (Figure 18.24 above). Each logic element of the ladder diagram is entered separately, one at a time, with series or parallel connections achieved by using keys for *AND*, *OR*, and *NOT*.

A more sophisticated programmer becoming increasingly common is a personal computer with a graphical display. The display typically shows several ladder diagram lines at a time and can also indicate the power flow within each line during operation to make debugging and testing simpler. Other units are programmed with logical gates instead of a ladder diagram. The program is entered by moving a cursor along the screen (using arrow keys or a mouse). When the cursor reaches the location where the next element is to be added, confirmation is given via additional keys.

An increasing number of *PLCs* are programmed in English-statement type languages, because of the increasing use of *PLCs* for analog control. With the combination of a Boolean language and several other types of instructions, the structuring of large programs soon becomes extremely difficult. Therefore, the demand for high level languages increases significantly as the *PLC* operations become complex.

A language standard being accepted in the *PLC* community is IEC 1131-3.

18.6.1 Specifying Industrial Sequences with Grafcet

The need to structure a sequential process problem is not apparent in small systems but becomes crucial very quickly. As a control system becomes more complex the need for better functional description increases. Each block must include more and more complex functions. This means, that logical expressions in terms of ladder diagrams or logical circuits are not sufficiently powerful to allow a structured description. For a more rational top-down analysis, the functional diagram **Grafcet** (GRAphe de Commande Etape-Transition) was developed by a French commission in the late 1970s and has been adopted as the French national standard. Since 1988 Grafcet has been specified in a European standard (IEC 848).

The Grafcet Diagram

Grafcet is a method developed for specifying industrial control sequences by a diagram. A similar method originated in Germany, called FUP (FUnction Plan). The basic ideas behind the two methods are the same and the differences are of minor importance.

As an example, we will illustrate the use of Grafcet for a batch process. A tank is to be filled with a liquid. When it is full, the liquid is heated until a certain temperature has been reached. After a specified time the tank is emptied, and the process starts all over again. The Grafcet diagram of the sequence is shown in Figure 18.28.

An indicator *Empty* signals that there is no liquid left. A *Start* signal together with the *Empty* indication initiates the sequence. In Step 2 the bottom valve is closed and the pump started. An indicator *Full* tells when to stop the pumping and causes a jump (called **transition**) to Step 3. The pump is switched off and a heater switched on. The heater remains on until the final temperature has been reached *(Temp)* and there is a jump to Step 4. The heater is switched off and a timer is started. When the waiting time has elapsed *(time_out)*, there is a transition to Step 5 and the outlet valve can be opened. The sequence then returns to *Start.*

The figure illustrates that the Grafcet consists of a column of numbered **blocks**, each one representing a **step**. The vertical lines joining adjacent blocks represent **transitions**. Each transition is associated with a logical condition called **receptivity**, defined by a Boolean expression written next to a short horizontal line crossing the transition line. If the receptivity is logical 1, the transition is executed and the system passes to the next step.

A Grafcet diagram basically describes two things according to specific rules:

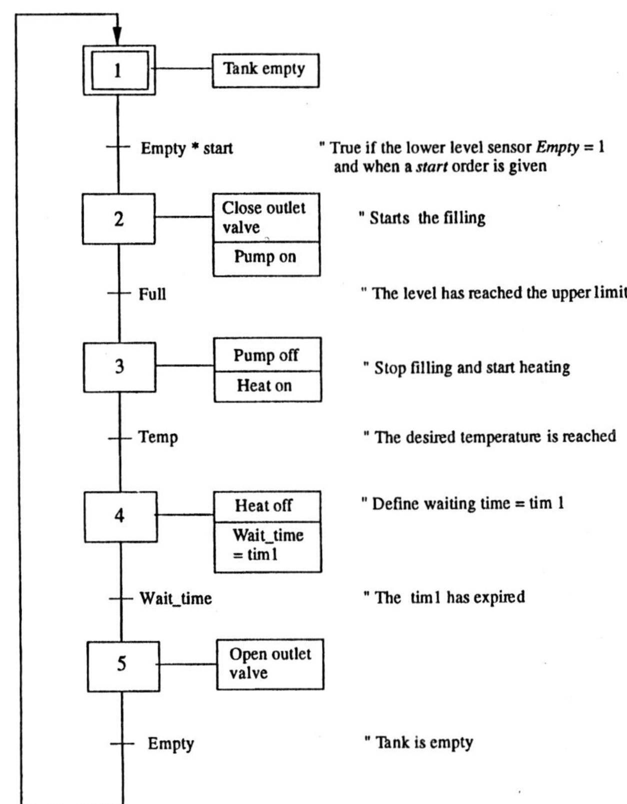

Figure 18.28 Grafcet diagram for the batch tank.

- which order to execute actions
- what to execute

The function diagram is split up into two parts. The part describing the order between the steps is called the **sequence part**. Graphically this is shown as the left part of Figure 18.28, including the five boxes. The sequence part does not describe the actions to execute. This is described by the **object part** of the diagram, consisting of the boxes to the right of the steps.

A *step* can be either active or passive, i.e., being executed or not. The *initial* step is the first step of the diagram, described by a box with a double frame. An **action** is a description of what is performed at a step. Every action has to be connected to a step and can be described either by a ladder, logical circuit, or Boolean algebra. When a step becomes active, its action is executed. However, a logical condition can be connected to the action so that it is not executed until the step is active and the logical condition is fulfilled. This feature is useful as a safety precaution.

Several actions can be connected to one step. They can be *outputs, timers,* or *counters* but can also be controller algorithms, filtering calculations, or routines for serial communication. A *transition* is an 'obstacle' between two steps and can originate only from an active step. Once a transition has taken place, the next step becomes active, and the previous one becomes inactive. The transition consists of a logical condition that has to be true to make the transition between two steps possible.

By combining the three building blocks *initial step, steps,* and *transitions,* it is possible to describe quite a large number of functions. The steps can be connected in

- simple sequences
- alternative parallel branches
- simultaneous parallel branches

In a **simple sequence** there is only one transition after a step, and after a transition there is only one step. No branching is made. In an **alternative parallel sequence** (Figure 18.29), there are two or more transitions after one step. This means that the flow of the diagram can take alternative paths. Typically this is an *if-then-else* condition useful to describe e.g., alarm situations.

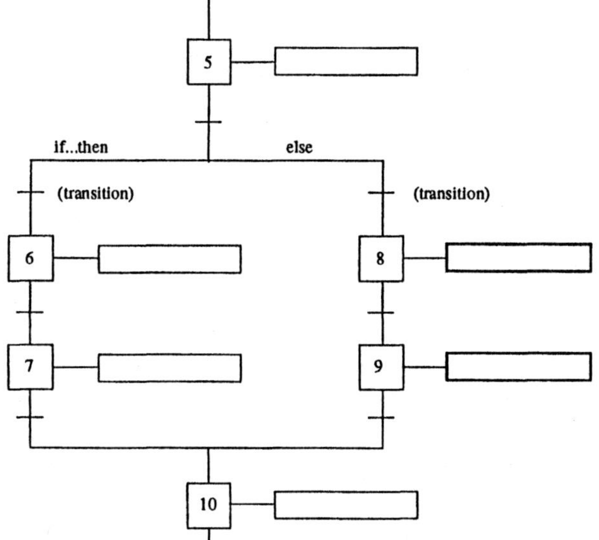

Figure 18.29 Grafcet for alternative parallel paths.

It is very important to insure that the transition condition located immediately before the alternative branching is consistent, so that the alternative branches are not allowed to start simultaneously. A branch of an alternative sequence always has to start and end with transition conditions.

In a **simultaneous parallel sequence** (Figure 18.30), there are two or more possible steps after a transition. Several steps can be active simultaneously, a concurrent (parallel) execution of several actions.

The double horizontal lines define the parallel processes. When the transition condition is true, both branches become active *simultaneously* and are executed *individually* and *concurrently.* A transition to the step below the lower parallel line cannot be executed until both concurrent processes are completed. In translating the Grafcet notation into a real-time program, synchronization tools, such as semaphores, have to be used. The branches become concurrent processes, that have to be synchronized so that racing conditions are avoided.

The three types of sequences can be mixed, but the mixing must be done correctly. For example, if two alternative branches are terminated with a parallel ending (two horizontal bars), then the sequence is locked, because the parallel end waits for both branches to finish, and the alternative start has started only one branch. Also, if simultaneous parallel branches are finished with an alternative ending (one horizontal bar), then there may be several active steps in the code, and it is not executed in a controlled manner.

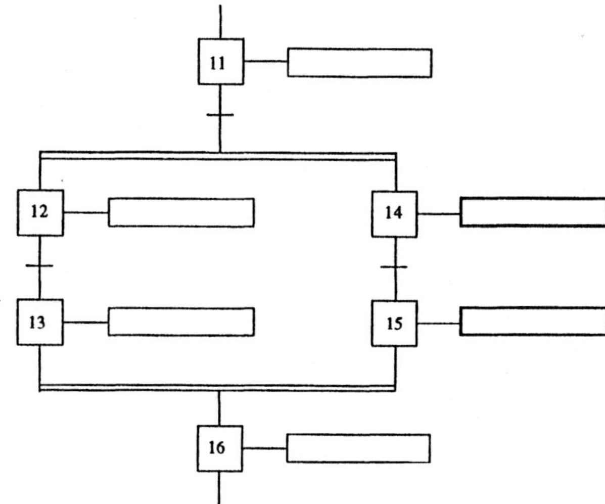

Figure 18.30 Grafcet for simultaneous parallel paths.

Computer Implementation of Grafcet

There are several inherent real-time features in Grafcet that must be observed for an implementation. The realization of real-time systems requires intensive effort with considerable investment in time and personnel. The translation of the Grafcet function diagram into computer code is not part of the definition and of course varies in the different systems. Obviously any implementation makes use of real-time programming tools.

Grafcet compilers are available for many different industrial control computers. Typically, the block programming and compilation are performed on a *PC.* After compilation the code is transferred to the *PLC* for execution. The *PC* is then removed in the real-time execution phase. More advanced *PLC* systems have Grafcet compilers built into the system.

The obvious advantage of Grafcet and similar types of abstract descriptions is their independence of specific hardware and their orientation to the task. Unfortunately high level languages such as Grafcet, do not yet enjoy the success they deserve. So many programmers start anew with programming in C or Assembler, but control tasks of the type we have seen are more easily solved with a functional block description.

As in any complex system description, the diagram or the code has to be structured suitably. A Grafcet implementation should allow dividing the system into smaller parts and the Grafcet diagram into several subgraphs. For example, each machine may have its own graph. Such hierarchical structuring is of fundamental importance in large systems. Of course Grafcet is also useful for less complex tasks. It is quite easy for the nonspecialist to understand the function compared to the function of a ladder diagram. By having a recognized standard for describing automation systems, the chances for reutilizing computer code are increased.

The translation of Grafcet to computer code depends on the specific *PLC*. Even if there is no automatic translator from Grafcet to machine code the functional diagram is very useful, because it allows the user to structure the problem. Many machine manufacturers today use Grafcet to describe the intended use and function of the machinery. Subsequent programming is much simpler if Grafcet can be used all the way. An implementation tool also may simulate the control code of a Grafcet diagram on the screen. During the simulation, the actual active state is shown.

18.6.2 Application of Function Charts in Industrial Control

The use of Grafcet for sequential programming is demonstrated for a manufacturing cell in a flexible manufacturing system. The cell consists of three NC machines (e.g., a drill, lathe, and mill), a robot for material handling, and a buffer storage (Figure 18.31).

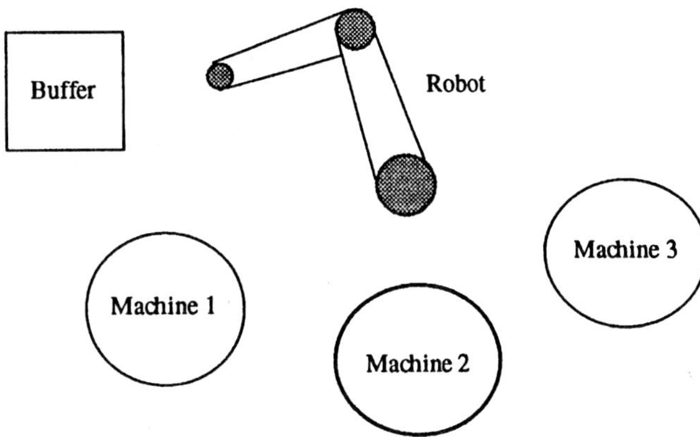

Figure 18.31 Layout of the manufacturing cell.

At the cell level we do not deal with the individual control loops of the machines or of the robot. They are handled by separate systems. The cell computer sends on/off commands to the machines and its main tasks are to control the individual sequencing of each machine (and the robot) and to synchronize the operations between the machines and the robot. The control task is a mixture of sequencing control and real-time synchronization. We will demonstrate how Grafcet expresses the operations. The implementation of the function chart is then left to the compiler.

The manufactured product has to be handled in the three machines in a predefined order (like a transfer line). The robot delivers new parts to each machine and moves them between the machines.

Synchronization of Tasks

The synchronization of the different machines is done by a **scheduler** graph with a structure indicated in Figure 18.32.

The scheduler communicates with each machine and with the robot and determines when they can start or when the robot can be used. It works like a scheduler in a real-time operating system, distributing the common resource , the robot, as efficiently as possible. The scheduler has to guarantee that the robot does not cause any deadlock. If the robot has picked up a finished part from a machine and has nowhere to place it, then the system will stop. Consequently the scheduler has to match demands from the machines with the available resources (robot and buffer capacity).

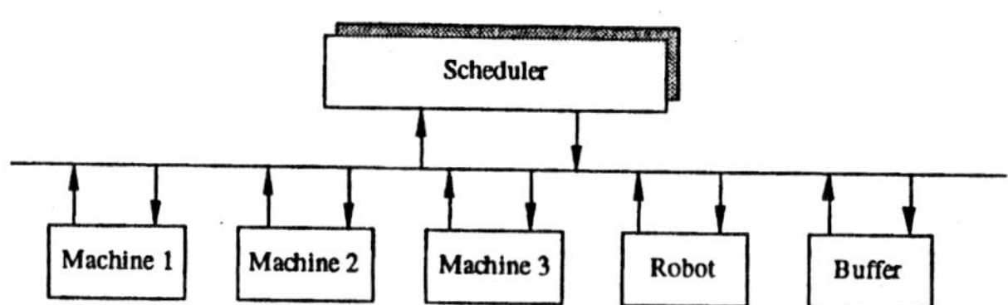

Figure 18.32 Logical structure of the machine cell.

The scheduler graph is described by a number of parallel branches, one for each machine, robot, and buffer. Because all of the devices are operating simultaneously, the scheduler has to handle all of them concurrently by sending and receiving synchronization signals of the type *start* and *ready*. When a machine gets a *start* command from the scheduler, it performs a task defined by its Grafcet diagram. When the machine has terminated, it sends a *ready* signal to the scheduler. Figure 18.32 shows that no machine communicates directly with the robot. Instead all the communication signals go via the scheduler. The signals are transition conditions in each Grafcet branch. By structuring the graph in this hierarchical way, new machines can be added to the cell without reprogramming any of the sequences of the other machines. The robot has to add the new operations to serve the new machines.

A good implementation of Grafcet supports a hierarchical structuring of the problem. The total operation can first be defined by a few complex operations, each one consisting of many steps. Then it is possible to go on to more and more detailed operations.

18.6.3 Analog Controllers

PLCs nowadays can handle not only binary signals, but analog-to-digital (A/D) and digital-to-analog (D/A) converters can be added to the *PLC* rack. The resolution, i.e., the number of bits to represent the analog signal, varies between the systems. The converted analog signal is placed in a digital register in the same way as a binary signal and is available for the standard *PLC* arithmetic and logical instructions.

In the event that plant signals do not correspond to any of the standard analog ranges, most manufacturers provide signal conditioning modules. Such a module provides buffering and scaling of plant signals to standard signals (typically 0 to 5 V or 0 to 10 V).

A *PLC* equipped with analog input channels may perform mathematical operations on the input values and pass the results directly to analog output modules to drive continuous actuators in the process directly. The sophistication of the control algorithms may vary, depending on the complexity of the *PLC*, but most systems today offer *PID* (proportional-integral-derivative) controller modules. The user has to tune the regulator parameters. To obtain sufficient capacity, many systems provide add-on *PID* function modules, containing input and output analog channels together with dedicated processors to carry out the necessary control calculations. This processor operates in parallel with the main *CPU*. When the main *CPU* requires status data from the *PID* module, it reads the relevant locations in the I/O memory where the *PID* processor places this information each time it completes a control cycle.

Many *PLC* systems also supply modules for digital filtering, for example, first-order smoothing (exponential filter) of input signals. The user gives the time constant of the filter as a parameter.

A *PLC* system may provide programming panels with a menu of questions and options relating to the setup of the control modules, such as gain parameters, integral and derivative times, filter time constants, sampling rate, and engineering unit conversions.

18.7 PLCs as Part of Automation Systems

The demands for communication in any process or manufacturing plant are steadily increasing. Any user today demands flexible and open communication, following some standard. Here we will just mention some of the crucial concepts essential for any nontrivial *PLC* installation.

Communication

A distributed system is more than simply connecting different units in a network. Certainly, the units in such a system can communicate, but the price is too much unnecessary communication, and the capacity of the systems cannot be fully used. Therefore, the architecture of the communication is essential. Reasons for installing network instead of point-to-point links are that

- all devices can access and share data and programs
- cabling for point-to-point becomes impractical and prohibitively expensive
- a network provides a flexible base for contributing communication architectures

To overcome the difficulties of dealing with a large number of incompatible standards, the International Organization for Standardization (ISO) has defined the **open systems interconnection** (*OSI*) scheme. *OSI* itself is not a standard, but offers a framework to identify and separate the different conceptual parts of the communication process. In practice, *OSI* does not indicate which voltage levels, transfer speeds, or protocols need to be used to achieve compatibility between systems. It says that there *has* to be compatibility of voltage levels, speed, and protocols as well as for a large number of other factors. The practical goal of *OSI* is optimal network interconnection, in which data can be transferred between different locations without wasting resources for conversion and creating related delays and errors.

PLC systems are an essential part of most industrial control systems. Below we will illustrate how they are connected at different levels of a plant network (Figure 18.33).

Fieldbus—Communication at the Sensor Level

There is a trend to replace conventional cables from sensors with a single digital connection. Thus, a single digital loop can replace a large number of 4 – 20 mA conductors. This has been implemented not only in manufacturing plants but also in aircraft and automobiles. Each sensor needs an interface to the bus, and standardization is necessary. This structure is known as **Fieldbus**. There is no single Fieldbus yet, but different solutions have been presented by the industry and by research institutions. In time, what has been proposed and is operating in the field will crystallize around one or more technologies that will become part of a more general Fieldbus.

When all communicating units located in a close work cell are connected to the same physical bus, there is no need for multiple end-to-end transfer checks as if the data were routed along international networks. To connect computers in the restricted environment of a factory, the data exchange definition of *OSI* layers 1 (physical layer) and 2 (data link layer) and an application protocol at the *OSI* level 7 are enough.

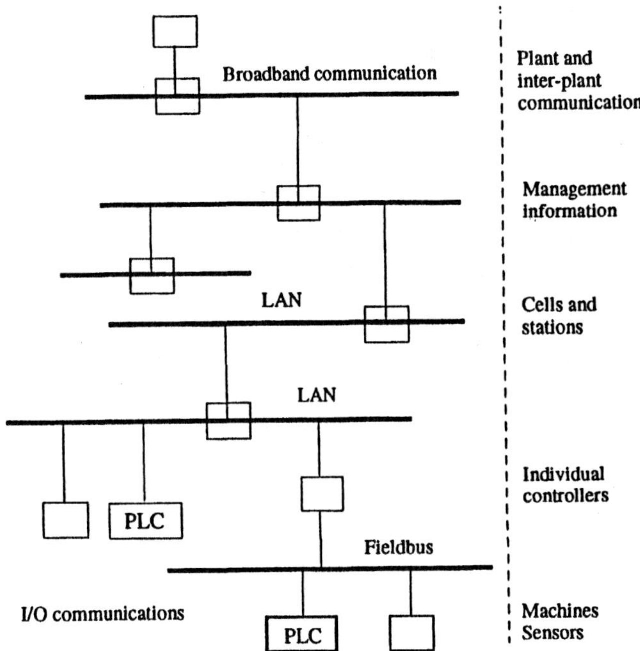

Figure 18.33 Structure of a plant network.

Fieldbuses open notable possibilities. A large share of the intelligence required for process control is moved out to the field. The maintenance of sensors becomes much easier because test and calibration operations can be remotely controlled and requiring less direct intervention by maintenance personnel. And as we have already pointed out, the quality of the collected data influences directly the quality of process control.

The International Electrotechnical Commission (IEC) is working on an international fieldbus standard. The standard should insure interconnectivity of different devices connected to the same physical medium. National projects have already started in different countries to define how the future standard will look. A final agreement has not been reached yet, but nobody wants to wait until a general standard is introduced. Some companies have already defined their products and are marketing them, and projects have been carried out in some countries to define national fieldbus standards. In the end, all experience and proposals may merge in a single, widely accepted standard, but different existing proposals may live in parallel.

Some examples of fieldbuses are FIP from France and PROFIBUS from Germany as well as the industrial Bitbus developed by Intel. There is a need for low cost complements. Therefore, buses like SDS (Smart Distributed System), based on the CAN (controller area network) bus (developed by Honeywell), ASI (actuator-sensor-interface) and Opus have been developed. They are more restricted than FIP or Profibus, and are meant to be a low end alternatives (but still compatible upwards) to the more advanced fieldbuses. Many semiconductor manufacturers, like Motorola, Intel, and NEC make circuits (e.g., for communication or single chip computers) with built-in CAN interfaces. Sensor and controller manufacturers form user groups for the different fieldbus concepts.

Local Area Networks (LANs)

To communicate between different *PLC* systems and computers within a plant there is a clear trend to use Ethernet as the medium. Ethernet is a widely used local area network *(LAN)* for both industrial and office applications. Jointly developed by the Xerox, Intel, and Digital Equipment, Ethernet was introduced in 1980. Ethernet follows the IEEE 802.3 specifications.

Ethernet has a bus topology with branch connections. Physically, Ethernet consists of a screened coaxial cable to which peripherals are connected with "taps". Ethernet does not have a network controlling unit. All devices decide independently when to access the medium. Consequently, because the line is entirely passive, there is no single-failure point on the network. Ethernet supports communication at different speeds, as the connected units do not need to decode messages not explicitly directed to them. The maximum data transfer rate is 10 Mbit/sec.

Ethernet's concept is flexible and open. There is little capital bound up in the medium, and the medium itself does not have active parts like servers or network control computers which could break down or act as bottlenecks to tie up communication capacity. Some companies offer complete Ethernet-based communication packages which may also implement higher layer services in the *OSI* hierarchy.

Plant Wide Networks

Two comprehensive concepts for information exchange in industrial processes are *MAP* (Manufacturing Automation Protocol) and *TOP* (Technical and Office Protocol). They are largely compatible with each other and are oriented to different aspects of industrial processes (production vs. administration). Both *MAP* and *TOP* are resource-intensive products supporting the interconnection of many devices in medium to large plants.

The **Manufacturing Automation Protocol** *(MAP)* is not a standard, an interface, or an electric cable. It is a comprehensive concept to realize interconnectivity between equipment on the plant floor and higher planning and control systems. But the realization of the conceptually simple principle to have different units communicate using common protocols has taken about thirty years and is far from complete. The principal goals of open communication are **interoperability** (all information should be understandable by the addressed units without conversion programs) and **interchangeability** (a device replaced with another of a different model or manufacturer should be able to operate without changes in the rest of the connected system).

MAP follows the *OSI* layering scheme. For every one of the *OSI* layers there is a defined standard as part of the *MAP* scheme. The standards at levels 1 to 6 are also used in applications other than *MAP;* the *MAP* specific part is the **Manufacturing Message Specification** *(MMS)*. *MMS* is a kind of language, a collection of abstract commands for remotely monitoring and controling industrial equipment. *MMS* defines the content of monitoring and control messages as well as the actions which should follow, the expectable reactions, acknowledging procedures, etc.

A *MAP* application must have a physical connection which follows the *LAN* Token Bus standard with Logical Link Control according to IEEE 802.2, must code data following ASN.1 (ISO 8824) and the Basic Encoding Rules of ISO 8825, and has to exchange MMS Messages (ISO 9506). Any other combination, even if technically feasible, is not consistent with the MAP scheme. For instance, a solution where Ethernet is used instead of Token Bus for the data link and physical connection is not a *MAP* application. However, in recent *MAP* applications, Ethernet has been accepted for the connections. Provided the net has sufficient transfer capacity, the real-time requirements can be satisfied.

In factory automation there are, generally speaking, three operational levels, as indicated in Figure 18.33: general management, process control or production line control, and field control. *MAP* supports the central levels of communication and coordinates the operations of multiple cells on a production line and of several lines at plant level. *MAP* is not apt for communication and control down to the sensor level. *MAP* is a very "heavy" product because of all the involved layers with the related protocols. *MAP* does not match the need for simple, fast, and cheap technology required at the lowest factory automation levels. Here the fieldbus is used. *MAP* remains the key concept for the practically realizing Computer-Integrated Manufacturing (CIM) applications.

Acknowledgments

The author has enjoyed numerous discussions on this topic with the Ph.D. students Gunnar Lindstedt, Lars Ericson, and Sven-Göran Bergh. Dr. Krister Mårtensson, Control Development AB has contributed with his deep insight. Special thanks to Dan Östman, Allen-Bradley AB, Sweden for his support.

Further Reading

An integrated view of computers for automation is presented in

[1] Olsson, G. and Piani, G., *Computer Systems in Automation and Control*, Prentice Hall Int., UK and USA, 1992.

In particular, more details on communication are presented. Switching theory is introduced in

[2] Lee, S.C., *Modern Switching Theory and Digital Design*, Prentice Hall, Englewood Cliffs, NJ, 1978.
[3] Fletcher, D.I., *An Engineering Approach to Digital Design*, Prentice Hall, Englewood Cliffs, NJ, 1980.

A good overview of sensors, actuators, and switching elements for both electric and pneumatic environments is contained in

[4] Pessen, D. W., *Industrial Automation: Circuit Design and Components*, John Wiley & Sons, New York, 1989.

Programmable logic devices are described in several articles in BYTE magazine (1987).

A lot of practical information on *PLCs*, their construction, use and applications is presented in

[5] Warnock, I. G., *Programmable Controllers Operation and Application*, Prentice Hall, Englewood Cliffs, NJ, 1988.

The manuals from the different *PLC* manufacturers provide full details of the facilities and programming methods for a given model.

There are two primary sources for Grafcet information:

[6] GRAFCET - A Function Chart for Sequential Processes, Publ. by ADEPA, 17, Rue Perier, B.P. No. 54, 92123 Montrouge Cedex, France, 1979.
[7] Manual 02.1987: Book 3 – Grafcet Language, Telemecanique Inc., 901 Baltimore Boulevard, Westminster, MD 21157, 1987.

FUP is defined by the German standard DIN 40719, and the European Grafcet standard is described in IEC Standard 848.

[8] Tanenbaum, A. S., *Computer Networks*, Prentice Hall, Englewood Cliffs, NJ, 1989.

This reference tells almost everything that is to be told about computer communication, at a very high level and yet not boring.

[9] Black, U. D., *Data Networks: Concepts, Theory and Practice*, Prentice Hall, Englewood Cliffs, NJ, 1989.

The reference above is a modern and comprehensive guide to several types of communication, not only related to data. Both Tanenbaum and Black treat everything from the physical line to network communication and follow quite strictly the *OSI* model.

The major ideas behind the *MAP* concept are described by

[10] Kaminski, M. A. Jr., Protocols for communicating in the factory, *IEEE Spectrum*, 56–62, April, 1986.

An overview of the proposals and expectations for a general Fieldbus standard is given in

[11] Wood, G. G., International Standards Emerging for Fieldbus, *Control Eng.*, 22–25, October, 1988.

A general introduction to PROFIBUS is given by

[12] Göddertz, J., PROFIBUS (in English), Klockner-Möller, Bonn, 1990.

Several articles on components, *PLCs* and market reviews appear regularly in journals such as *Control Engineering, Instrument & Control Systems, Machine and Design* and *Product Engineering.*

SECTION V

Analysis and Design Methods for Nonlinear Systems

19

Analysis Methods

Derek P. Atherton
School of Engineering, The University of Sussex

19.1 The Describing Function Method

19.1.1 Introduction

The describing function method, abbreviated as DF, was developed in several countries in the 1940s [4], to answer the question: "What are the necessary and sufficient conditions for the nonlinear feedback system of Figure 19.1 to be stable?" The problem still remains unanswered for a system with static nonlinearity, $n(x)$, and linear plant $G(s)$. All of the original investigators found limit cycles in control systems and observed that, in many instances with structures such as Figure 19.1, the wave form oscillation at the input to the nonlinearity was almost sinusoidal. If, for example, the nonlinearity in Figure 19.1 is an ideal relay, that is has an on-off characteristic, so that an odd symmetrical input wave form will produce a square wave at its output, the output of $G(s)$ will be almost sinusoidal when $G(s)$ is a low pass filter which attenuates the higher harmonics in the square wave much more than the fundamental. It was, therefore, proposed that the nonlinearity should be represented by its gain to a sinusoid and that the conditions for sustaining a sinusoidal limit cycle be evaluated to assess the stability of the feedback loop. Because of the nonlinearity, this gain in response to a sinusoid is a function of the amplitude of the sinusoid and is known as the describing function. Because describing function methods can be used other than for a single sinusoidal input, the technique is referred to as the single sinusoidal DF or sinusoidal DF.

19.1.2 The Sinusoidal Describing Function

For the reasons explained above, if we assume in Figure 19.1 that $x(t) = a\cos\theta$, where $\theta = \omega t$ and $n(x)$ is a symmetrical odd nonlinearity, then the output $y(t)$ will be given by the Fourier series,

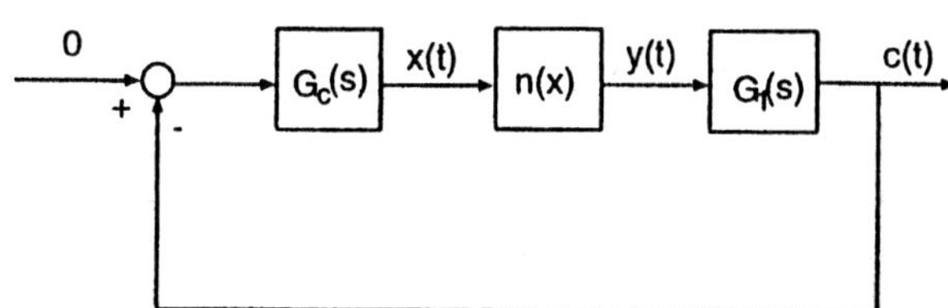

Figure 19.1 Block diagram of a nonlinear system.

$$y(\theta) = \sum_{n=0}^{\infty} a_n \cos n\theta + b_n \sin n\theta, \quad (19.1)$$

$$\text{where } a_0 = 0, \quad (19.2)$$

$$a_1 = (1/\pi)\int_0^{2\pi} y(\theta)\cos\theta d\theta, \quad (19.3)$$

and

$$b_1 = (1/\pi)\int_0^{2\pi} y(\theta)\sin\theta d\theta. \quad (19.4)$$

The fundamental output from the nonlinearity is $a_1\cos\theta + b_1\sin\theta$, so that the describing function, DF, defined as the fundamental output divided by the input amplitude, is complex and given by

$$N(a) = (a_1 - jb_1)/a \quad (19.5)$$

which may be written

$$N(a) = N_p(a) + jN_q(a) \quad (19.6)$$

where

$$N_p(a) = a_1/a \text{ and } N_q(a) = -b_1/a. \quad (19.7)$$

Alternatively, in polar coordinates,

$$N(a) = M(a)e^{j\Psi(a)} \quad (19.8)$$

where

0-8493-8570-9/96/$0.00+$.50

$$M(a) = (a_1^2 + b_1^2)^{1/2}/a$$

and

$$\Psi(a) = -\tan^{-1}(b_1/a_1). \quad (19.9)$$

If $n(x)$ is single valued, then $b_1 = 0$ and

$$a_1 = (4/\pi)\int_0^{\pi/2} y(\theta)\cos\theta d\theta \quad (19.10)$$

giving

$$N(a) = a_1/a = (4/a\pi)\int_0^{\pi/2} y(\theta)\cos\theta d\theta. \quad (19.11)$$

Although Equations 19.3 and 19.4 are an obvious approach to evaluating the fundamental output of a nonlinearity, they are indirect, because one must first determine the output wave form $y(\theta)$ from the known nonlinear characteristic and sinusoidal input wave form. This is avoided if the substitution $\theta = \cos^{-1}(x/a)$ is made. After some simple manipulations,

$$a_1 = (4/a)\int_0^a x n_p(x) p(x) dx \quad (19.12)$$

and

$$b_1 = (4/a\pi)\int_0^a n_q(x) dx. \quad (19.13)$$

The function $p(x)$ is the amplitude probability density function of the input sinusoidal signal given by

$$p(x) = (1/\pi)(a^2 - x^2)^{-1/2}. \quad (19.14)$$

The nonlinear characteristics $n_p(x)$ and $n_q(x)$, called the inphase and quadrature nonlinearities, are defined by

$$n_p(x) = [n_1(x) + n_2(x)]/2 \quad (19.15)$$

and

$$n_q(x) = [n_2(x) - n_1(x)]/2 \quad (19.16)$$

where $n_1(x)$ and $n_2(x)$ are the portions of a double-valued characteristic traversed by the input for $\dot{x} > 0$ and $\dot{x} < 0$, respectively. When the nonlinear characteristic is single-valued, $n_1(x) = n_2(x)$, so $n_p(x) = n(x)$ and $n_q(x) = 0$. Integrating Equation 19.12 by parts yields

$$a_1 = (4/\pi)n(0^+) + (4/a\pi)\int_0^a n'(x)(a^2 - x^2)^{1/2} dx \quad (19.17)$$

where $n'(x) = dn(x)/dx$ and $n(0^+) = \lim_{\varepsilon\to\infty} n(\varepsilon)$, a useful alternative expression for evaluating a_1.

An additional advantage of using Equations 19.12 and 19.13 is that they yield proofs of some properties of the DF for symmetrical odd nonlinearities. These include the following:

1. For a double-valued nonlinearity, the quadrature component $N_q(a)$ is proportional to the area of the nonlinearity loop, that is,

$$N_q(a) = -(1/a^2\pi) \text{ (area of nonlinearity loop)} \quad (19.18)$$

2. For two single-valued nonlinearities $n_\alpha(x)$ and $n_\beta(x)$, with $n_\alpha(x) < n_\beta(x)$ for all $0 < x < b$, $N_\alpha(a) < N_\beta(a)$ for input amplitudes less than b.
3. For a single-valued nonlinearity with $k_1 x < n(x) < k_2 x$ for all $0 < x < b$, $k_1 < N(a) < k_2$ for input amplitudes less than b. This is the sector property of the DF; a similar result can be obtained for a double-valued nonlinearity [6].

When the nonlinearity is single valued, from the properties of Fourier series, the DF, $N(a)$, may also be defined as:

1. the variable gain, K, having the same sinusoidal input as the nonlinearity, which minimizes the mean squared value of the error between the output from the nonlinearity and that from the variable gain, and
2. the covariance of the input sinusoid and the nonlinearity output divided by the variance of the input.

19.1.3 Evaluation of the Describing Function

To illustrate the evaluation of the DF several simple examples are considered.

Cubic Nonlinearity

For this nonlinearity $n(x) = x^3$ and using Equation 19.3,

$$\begin{aligned} a_1 &= (4/\pi)\int_0^{\pi/2} (a\cos\theta)^3 \cos\theta d\theta, \\ &= (4/\pi)a^3 \int_0^{\pi/2} \cos^4\theta d\theta, \\ &= (4/\pi)a^3 \int_0^{\pi/2} \frac{(1+\cos 2\theta)^2}{4} d\theta, \\ &= (4/\pi)a^3 \int_0^{\pi/2} \left(\frac{3}{8} + \frac{\cos 2\theta}{2} + \frac{\cos 4\theta}{8}\right) d\theta = 3a^3/4, \end{aligned}$$

giving $N(a) = 3a^2/4$.

Alternatively from Equation 19.12,

$$a_1 = (4/a)\int_0^a x^4 p(x) dx.$$

The integral $\mu_n = \int_{-\infty}^{\infty} x^n p(x) dx$ is known as the nth moment of the probability density function and, for the sinusoidal distribution with $p(x) = (1/\pi)(a^2 - x^2)^{-1/2}$, μ_n has the value

$$\mu_n = \begin{cases} 0, & \text{for } n \text{ odd, and} \\ a^n \frac{(n-1)}{n}\frac{(n-3)}{(n-2)}\cdots\frac{1}{2}, & \text{for } n \text{ even.} \end{cases} \quad (19.19)$$

Therefore,

$$\begin{aligned} a_1 &= (4/a)\frac{1}{2}.\frac{3}{4}.\frac{1}{2}a^4, \\ &= 3a^3/4 \text{ as before.} \end{aligned}$$

Saturation Nonlinearity

To calculate the DF, the input can alternatively be taken as $a \sin\theta$. For an ideal saturation characteristic, the nonlinearity output wave form $y(\theta)$ is as shown in Figure 19.2. Because of the symmetry of the nonlinearity, the fundamental of the output can be evaluated from the integral over a quarter period so that

$$N(a) = \frac{4}{a\pi}\int_0^{\pi/2} y(\theta)\sin\theta d\theta,$$

which, for $a > \delta$, gives

$$N(a) = \frac{4}{a\pi}\left[\int_0^{\alpha} ma\sin^2\theta d\theta + \int_{\alpha}^{\pi/2} m\delta\sin\theta d\theta\right]$$

where $\alpha = \sin^{-1}\delta/a$. Evaluation of the integrals gives

$$N(a) = (4m/\pi)\left[\frac{\alpha}{2} - \frac{\sin 2\alpha}{4} + \delta\cos\alpha\right]$$

which, on substituting for δ, gives the result

$$N(a) = (m/\pi)(2\alpha + \sin 2\alpha). \quad (19.20)$$

Because, for $a < \delta$, the characteristic is linear giving $N(a) = m$, the DF for ideal saturation is $mN_s(\delta/a)$ where

$$N_s(\delta/a) = \begin{cases} 1, & \text{for } a < \delta, \text{ and} \\ (1/\pi)[2\alpha + \sin 2\alpha], & \text{for } a > \delta, \end{cases} \quad (19.21)$$

where $\alpha = \sin^{-1}\delta/a$.

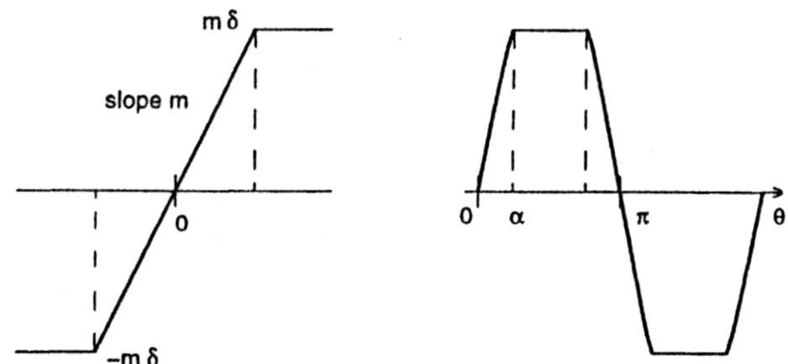

Figure 19.2 Saturation nonlinearity.

Alternatively, one can evaluate $N(a)$ from Equation 19.17, yielding

$$N(a) = a_1/a = (4/a^2\pi)\int_0^{\delta} m(a^2 - x^2)^{1/2}dx.$$

Using the substitution $x = a\sin\theta$,

$$N(a) = (4m/\pi)\int_0^{\alpha}\cos^2\theta d\theta = (m/\pi)(2\alpha + \sin 2\alpha)$$

as before.

Relay with Dead Zone and Hysteresis

The characteristic is shown in Figure 19.3 together with the corresponding input, assumed equal to $a\cos\theta$, and the corresponding output wave form. Using Equations 19.3 and 19.4

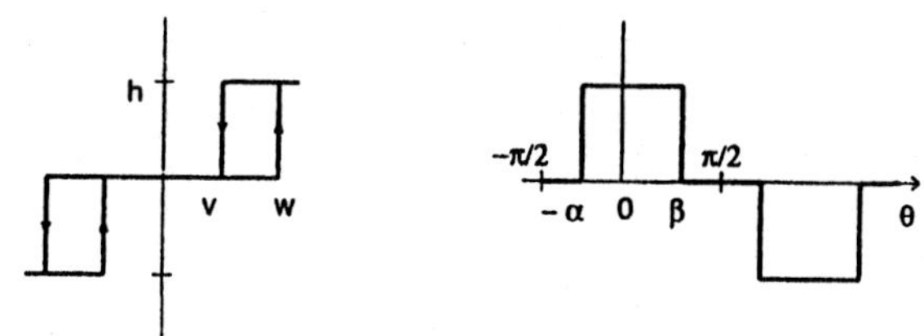

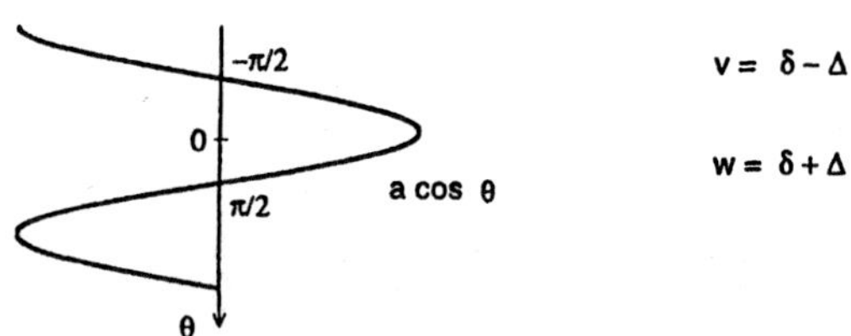

Figure 19.3 Relay with dead zone and hysteresis.

over the interval $-\pi/2$ to $\pi/2$ and assuming that the input amplitude a is greater than $\delta + \Delta$,

$$\begin{aligned} a_1 &= (2/\pi)\int_{-\alpha}^{\beta} h\cos\theta d\theta \\ &= (2h/\pi)(\sin\beta + \sin\alpha), \end{aligned}$$

where $\alpha = \cos^{-1}[(\delta - \Delta)/a]$ and $\beta = \cos^{-1}[(\delta + \Delta)/a]$, and

$$\begin{aligned} b_1 &= (2/\pi)\int_{-\alpha}^{\beta} h\sin\theta d\theta \\ &= (-2h/\pi)\left(\frac{(\delta+\Delta)}{a} - \frac{(\delta-\Delta)}{a}\right) = 4h\Delta/a\pi. \end{aligned}$$

Thus

$$\begin{aligned} N(a) &= \frac{2h}{a^2\pi}\left\{\left[a^2 - (\delta+\Delta)^2\right]^{1/2}\right. \\ &\quad \left.+\left[a^2 - (\delta-\Delta)^2\right]^{1/2}\right\} - \frac{j4h\Delta}{a^2\pi}. \end{aligned} \quad (19.22)$$

For the alternative approach, one must first obtain the in-phase and quadrature nonlinearities shown in Figure 19.4. Using Equations 19.12 and 19.13,

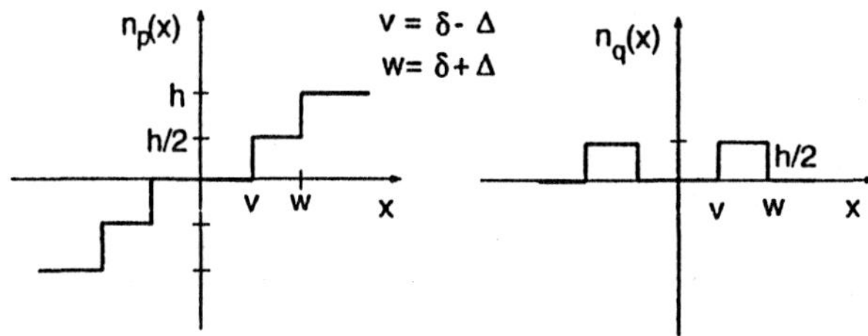

Figure 19.4 Functions $n_p(x)$ and $n_q(x)$ for the relay of Figure 19.3.

$$\begin{aligned} a_1 &= (4/a)\int_{\delta-\Delta}^{\delta+\Delta} x(h/2)p(x)dx + \int_{\delta+\Delta}^{a} xhp(x)dx, \\ &= \frac{2h}{a\pi}\left\{\left[a^2 - (\delta+\Delta)^2\right]^{1/2} + \left[a^2 - (\delta-\Delta)^2\right]^{1/2}\right\}, \end{aligned}$$

and

$$\begin{aligned} b_1 &= (4/a\pi)\int_{d-\Delta}^{\delta+\Delta}(h/2)dx = 4h\Delta/a\pi \\ &= (\text{Area of nonlinearity loop})/a\pi \end{aligned}$$

as before.

The DF of two nonlinearities in parallel equals the sum of their individual DFs, a result very useful for determining DFs, particularly of linear segmented characteristics with multiple break points. Several procedures [4] are available for approximating the DF of a given nonlinearity either by numerical integration or by evaluating the DF of an approximating nonlinear characteristic defined, for example, by a quantized characteristic, linear segmented characteristic, or Fourier series. Table 19.1 gives a list of DFs for some commonly used approximations of nonlinear elements. Several of the results are in terms of the DF for an ideal saturation characteristic of unit slope, $N_s(\delta/a)$, defined in Equation 19.21.

19.1.4 Limit Cycles and Stability

To investigate the possibility of limit cycles in the autonomous closed loop system of Figure 19.1, the input to the nonlinearity $n(x)$ is assumed to be a sinusoid so that it can be replaced by the amplitude-dependent DF gain $N(a)$. The open loop gain to a sinusoid is thus $N(a)G(j\omega)$ and, therefore, a limit cycle exists if

$$N(a)G(j\omega) = -1 \tag{19.23}$$

where $G(j\omega) = G_c(j\omega)G_1(j\omega)$. As in general, $G(j\omega)$ is a complex function of ω and $N(a)$ is a complex function of a, solving Equation 19.23 will yield both the frequency ω and amplitude a of a possible limit cycle.

Various procedures can be used to examine Equation 19.23; the choice is affected to some extent by the problem, for example, whether the nonlinearity is single- or double-valued or whether $G(j\omega)$ is available from a transfer function $G(s)$ or as measured frequency response data. Usually the functions $G(j\omega)$ and $N(a)$ are plotted separately on Bode or Nyquist diagrams, or Nichols charts. Alternatively, stability criteria (e.g., the Hurwitz-Routh) or root locus plots may be used with the characteristic equation

$$1 + N(a)G(s) = 0, \tag{19.24}$$

although the equation is appropriate only for $s \approx j\omega$.

Figure 19.5 illustrates the procedure on a Nyquist diagram, where the $G(j\omega)$ and $C(a) = -1/N(a)$ loci are plotted intersecting for $a = a_0$ and $\omega = \omega_0$. The DF method indicates therefore that the system has a limit cycle with the input sinusoid to the nonlinearity, x, equal to $a_0 \sin(\omega_0 t + \phi)$, where ϕ depends on the initial conditions. When the $G(j\omega)$ and $C(a)$ loci do not intersect, the DF method predicts that no limit cycle will exist if the Nyquist stability criterion is satisfied for $G(j\omega)$ with respect to any point on the $C(a)$ locus. Obviously, if the nonlinearity has unit gain for small inputs, the point $(-1, j0)$ will lie on $C(a)$ and may be used as the critical point, analogous to a linear system.

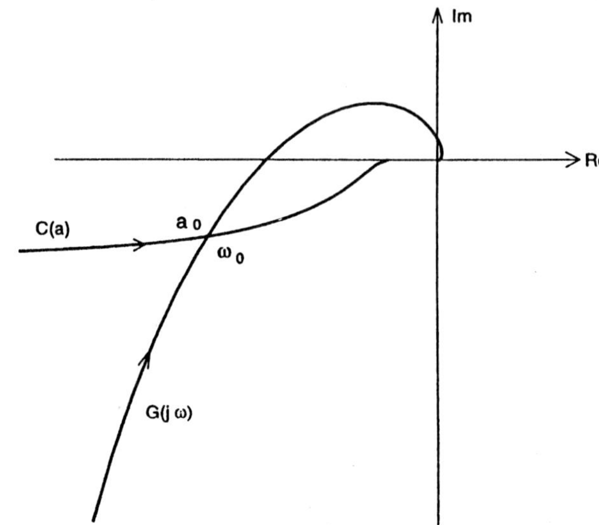

Figure 19.5 Nyquist plot showing solution for a limit cycle.

For a stable case, it is possible to use the gain and phase margin to judge the relative stability of the system. However, a gain and phase margin can be found for every amplitude a on the $C(a)$ locus, so it is usually appropriate to use the minimum values of the quantities [4]. When the nonlinear block includes dynamics so that its response is both amplitude and frequency dependent, that is $N(a, \omega)$, then a limit cycle will exist if

$$G(j\omega) = -1/N(a,\omega) = C(a,\omega). \tag{19.25}$$

To check for possible solutions of this equation, a family of $C(a, \omega)$ loci, usually as functions of a for fixed values of ω, is drawn on the Nyquist diagram.

When a solution to Equation 19.23 exists, an additional point of interest is whether the predicted limit cycle is stable. This is important if the control system is designed to have a limit cycle operation, as in an on-off temperature control system. It may also be important in other systems, because, in an unstable limit cycle condition, the signal amplitudes may not become bounded but continue to grow. Provided that only one possible limit cycle solution is predicted by the DF method, the stability of the limit cycle can be assessed by applying the Nyquist stability criterion to points on the $C(a)$ locus on both sides of the solution point. For this perturbation approach, if the stability criterion indicates instability (stability) for the point on $C(a)$ with $a < a_0$ and stability (instability) for the point on $C(a)$ with $a > a_0$, the limit cycle is stable (unstable).

When multiple solutions exist, the situation is more complicated and the criterion above is a necessary but not sufficient result for the stability of the limit cycle [5].

Normally in these cases, the stability of the limit cycle can be ascertained by examining the roots of the characteristic equation

$$1 + N_{i\gamma}(a)G(s) = 0 \tag{19.26}$$

where $N_{i\gamma}(a)$ is known as the incremental describing function (IDF). $N_{i\gamma}(a)$ for a single valued nonlinearity can be evaluated

TABLE 19.1 DFs of single-valued nonlinearities.

Nonlinearity	Condition	DF
General quantizer	$a < \delta_1$	$N_p = 0$
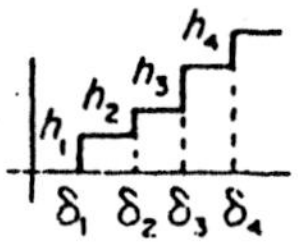	$\delta_{M+1} > a > \delta_M$	$N_p = (4/a^2\pi) \sum_{m=1}^{M} h_m(a^2 - \delta_m^2)^{1/2}$
Uniform quantizer $h_1 = h_2 = \cdots h$ $\delta_m = (2m-1)\delta/2$	$a < \delta$	$N_p = 0$
	$(2M+1)\delta > a > (2M-1)\delta$	$N_p = (4h/a^2\pi) \sum_{m=1}^{M} (a^2 - n^2\delta^2)^{1/2}$
	$n = (2m-1)/2$	
Relay with dead zone	$a < \delta$	$N_p = 0$
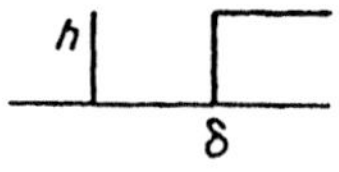	$a > \delta$	$N_p = 4h(a^2 - \delta^2)^{1/2}/a^2\pi$
Ideal relay 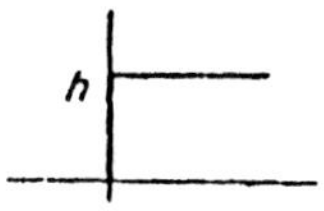		$N_p = 4h/a\pi$
Preload 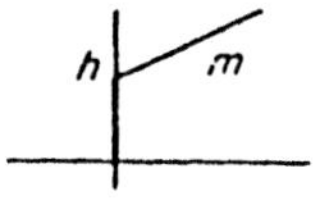		$N_p = (4h/a\pi) + m$
General piecewise linear	$a < \delta_1$	$N_p = (4h/a\pi) + m_1$
	$\delta_{M+1} > \alpha > \delta_M$	$N_p = (4h/a\pi) + m_{M+1} + \sum_{j=1}^{M} (m_j - m_{j+1})N_s(\delta_j/a)$
Ideal saturation 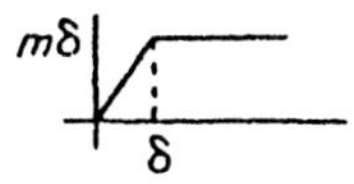		$N_p = mN_s(\delta/a)$
Dead zone 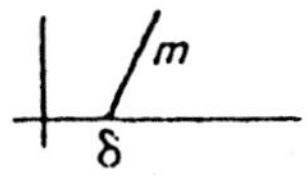		$N_p = m[1 - N_s(\delta/a)]$

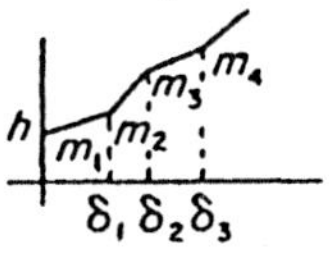

TABLE 19.2 DFs of single-valued nonlinearities. *cont.*

Nonlinearity	Condition	DF
Gain changing nonlinearity (m_1, m_2, δ)		$N_p = (m_1 - m_2)N_s(\delta/a) + m_2$
Saturation with dead zone (m, δ_1, δ_2)		$N_p = m[N_s(\delta_2/a) - N_s(\delta_1/a)]$
(m_1, m_2, δ_1, δ_2)		$N_p = -m_1 N_s(\delta_1/a) + (m_1 - m_2)N_s(\delta_2/a) + m_2$
(h, m, δ)	$a < \delta$	$N_p = 0$
	$a > \delta$	$N_p = 4h(a^2 - \delta^2)^{1/2}/a_2\pi + m - mN_s(\delta/a)$
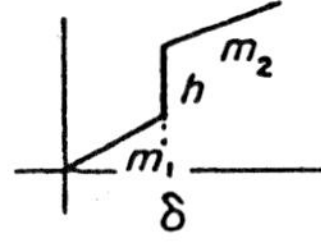	$a < \delta$	$N_p = m_1$
	$a > \delta$	$N_p = (m_1 - m_2)N_s(\delta/a) + m_2 + 4h(a^2 - \delta^2)^{1/2}/a^2\pi$
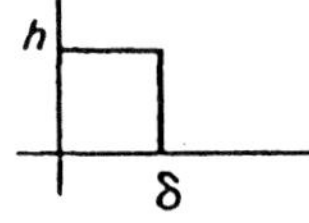	$a < \delta$	$N_p = 4h/a\pi$
	$a > \delta$	$N_p = 4h/[a - (a^2 - \delta^2)^{1/2}]/a^2\pi$
Limited field of view (m_1, $-m_2$, δ)		$N_p = (m_1 + m_2)N_s(\delta/a) - m_2 N_s[(m_1 + m_2)\delta/m_2 a]$
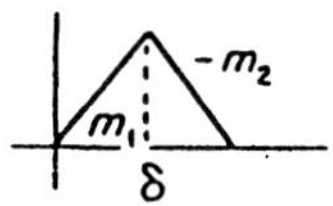	$a < \delta$	$N_p = m_1$
	$a > \delta$	$N_p = m_1 N_s(\delta/a) - 4m_1\delta(a^2 - \delta^2)^{1/2}/a^2\pi$
$y = x^m$	$m > -2$ Γ is the gamma function	$N_p = \dfrac{\Gamma(m+1)a^{m-1}}{2^{m-1}\Gamma[(3+m)/2]\Gamma[(1+m)/2]} = \dfrac{2}{\sqrt{\pi}} \dfrac{\Gamma[(m+2)/2]a^{m-1}}{\Gamma[(m+3)/2]}$

from

$$N_{i\gamma}(a) = \int_{-a}^{a} n'(x)p(x)dx \qquad (19.27)$$

where $n'(x)$ and $p(x)$ are as previously defined. $N_{i\gamma}(a)$ is related to $N(a)$ by the equation

$$N_{i\gamma}(a) = N(a) + (a/2)dN(a)/da. \qquad (19.28)$$

Thus, for example, for an ideal relay, making $\delta = \Delta = 0$ in Equation 19.22 gives $N(a) = 4h/a\pi$, also found directly from Equation 19.17, and, substituting this value in Equation 19.28, yields $N_{i\gamma}(a) = 2h/a\pi$. Some examples of feedback system analysis using the DF follow.

Autotuning in Process Control

In 1943 Ziegler and Nichols [18] suggested a technique for tuning the parameters of a PID controller. Their method was based on testing the plant in a closed loop with the PID controller in the proportional mode. The proportional gain was increased until the loop started to oscillate and then the value of gain and the oscillation frequency were measured. Formulae were given for setting the controller parameters based on the gain named the critical gain, K_c, and the frequency called the critical frequency, ω_c.

Assuming that the plant has a linear transfer function $G_1(s)$, then K_c is its gain margin and ω_c the frequency at which its phase shift is 180°. Performing this test in practice may prove difficult. If the plant has a linear transfer function and the gain is adjusted too quickly, a large amplitude oscillation may start to build up. In 1984 Astrom and Hagglund [2] suggested replacing the proportional control by a relay element to control the amplitude of the oscillation. Consider therefore the feedback loop of Figure 19.1 with $n(x)$ an ideal relay, $G_c(s) = 1$, and the plant with a transfer function $G_1(s) = 10/(s+1)^3$. The $C(a)$ locus, $-1/N(a) = -a\pi/4h$, and the Nyquist locus $G(j\omega)$ in Figure 19.6 intersect. The values of a and ω at the intersection

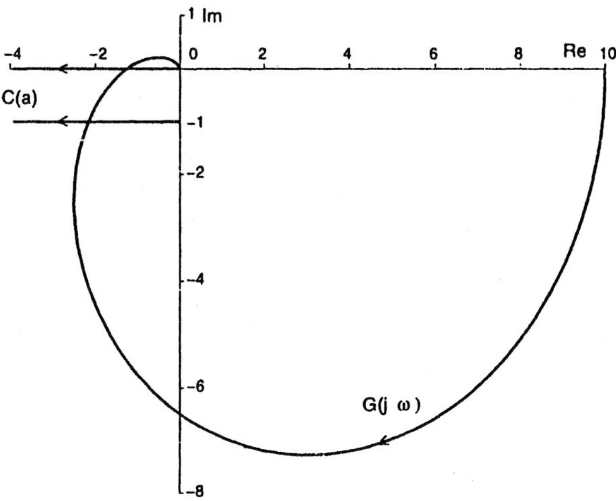

Figure 19.6 Nyquist plot for $10/(s+1)^3$ and $C(a)$ loci for $\Delta = 0$ and $4h/\pi$.

can be calculated from

$$-a\pi/4h = \frac{10}{(1+j\omega)^3} \qquad (19.29)$$

which can be written

$$\text{Arg}\left(\frac{10}{(1+j\omega)^3}\right) = 180°\text{, and} \qquad (19.30)$$

$$\frac{a\pi}{4h} = \frac{10}{(1+\omega^2)^{3/2}}. \qquad (19.31)$$

The solution for ω_c from Equation 19.30 is $\tan^{-1}\omega_c = 60°$, giving $\omega_c = \sqrt{3}$. Because the DF solution is approximate, the actual measured frequency of oscillation will differ from this value by an amount which will be smaller the closer the oscillation is to a sinusoid. The exact frequency of oscillation in this case will be 1.708 rads/sec in error by a relatively small amount. For a square wave input to the plant at this frequency, the plant output signal will be distorted by a small percentage. The distortion, d, is defined by

$$d = \left[\frac{\textit{M.S. value of signal} - \textit{M.S. value of fundamental harmonic}}{\textit{M.S. value of fundamental harmonic}}\right]^{1/2} \qquad (19.32)$$

Solving Equation 19.31 gives the amplitude of oscillation a as $5h/\pi$. The gain through the relay is $N(a)$ equal to the critical gain K_c. In the practical situation where a is measured, K_c equal to $4h/a\pi$, should be close to the known value of 0.8 for this transfer function.

If the relay has an hysteresis of Δ, then with $\delta = 0$ in Equation 19.22 gives

$$N(a) = \frac{4h(a^2-\Delta^2)^{1/2}}{a^2\pi} - j\frac{4h\Delta}{a^2\pi}$$

from which

$$C(a) = \frac{-1}{N(a)} = \frac{-\pi}{4h}\left[(a^2-\Delta^2)^{1/2} + j\Delta\right].$$

Thus on the Nyquist plot, $C(a)$ is a line parallel to the real axis at a distance $\pi\Delta/4h$ below it, as shown in Figure 19.6 for $\Delta = 1$ and $h = \pi/4$ giving $C(a) = -(a^2-1)^{1/2} - j$. If the same transfer function is used for the plant, then the limit cycle solution is given by

$$-(a^2-1)^{1/2} - j = \frac{10}{(1+j\omega)^3} \qquad (19.33)$$

where $\omega = 1.266$, which compares with an exact solution value of 1.254, and $a = 1.91$. For the oscillation with the ideal relay, Equation 19.26 with $N_{i\gamma}(a) = 2h/a\pi$ shows that the limit cycle is stable. This agrees with the perturbation approach which also shows that the limit cycle is stable when the relay has hysteresis.

Feedback Loop with a Relay with Dead Zone

For this example the feedback loop of Figure 19.1 is considered with $n(x)$ a relay with dead zone and $G(s) = 2/s(s+1)^2$. From Equation 19.22 with $\Delta = 0$, the DF for this relay, given by

$$N(a) = 4h(a^2-\delta^2)^{1/2}/a^2\pi \text{ for } a > \delta, \qquad (19.34)$$

is real because the nonlinearity is single valued. A graph of $N(a)$ against a is in Figure 19.7, and shows that $N(a)$ starts at zero, when $a = \delta$, increases to a maximum, with a value of $2h/\pi\delta$ at $a = \delta\sqrt{2}$, and then decreases toward zero for larger inputs. The $C(a)$ locus, shown in Figure 19.8, lies on the negative real axis starting at $-\infty$ and returning there after reaching a maximum value of $-\pi\delta/2h$. The given transfer function $G(j\omega)$ crosses the negative real axis, as shown in Figure 19.8, at a frequency of $\tan^{-1}\omega = 45°$, that, is $\omega = 1$ rad/sec and, therefore, cuts the $C(a)$ locus twice. The two possible limit cycle amplitudes at this frequency can be found by solving

$$\frac{a^2\pi}{4h(a^2-\delta^2)^{1/2}} = 1$$

which gives $a = 1.04$ and 3.86 for $\delta = 1$ and $h = \pi$. Using the perturbation method or the IDF criterion, the smallest amplitude limit cycle is unstable and the larger one is stable. If a condition similar to the lower amplitude limit cycle is excited in the system, an oscillation will build up and stabilize at the higher amplitude limit cycle.

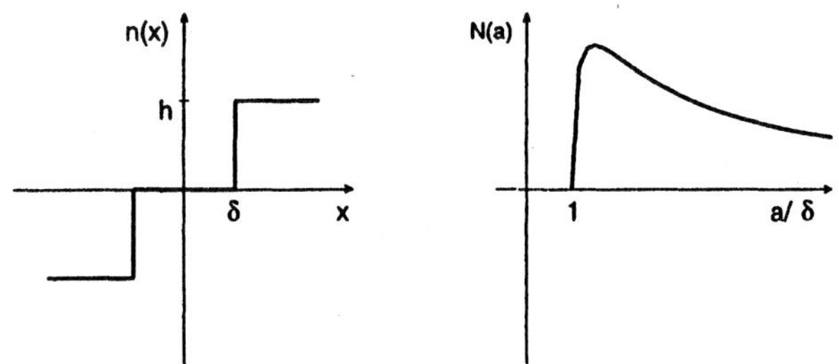

Figure 19.7 $N(a)$ for ideal relay with dead zone.

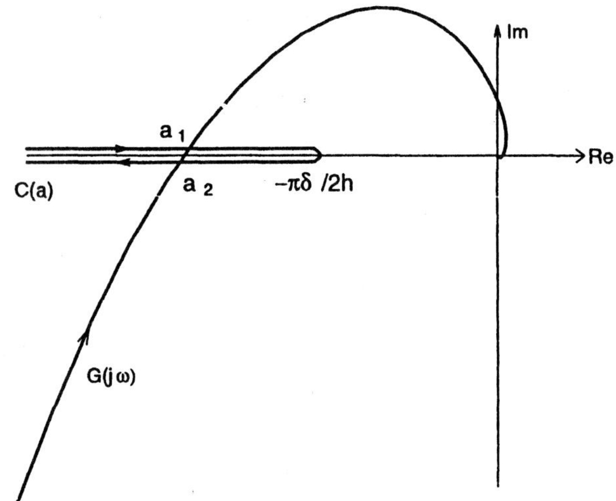

Figure 19.8 Two limit cycles: a_1, unstable; a_2, stable.

Other techniques show that the exact frequencies of the limit cycles for the smaller and larger amplitudes are 0.709 and 0.989, respectively. Although the transfer function is a good low pass filter, the frequency of the smallest amplitude limit cycle is not predicted accurately because the output from the relay, a wave form with narrow pulses, is highly distorted.

If the transfer function of $G(s)$ is $K/s(s+1)^2$, then no limit cycle will exist in the feedback loop, and it will be stable if

$$\left.\frac{K}{\omega(1+\omega^2)}\right|_{\omega=1} < \frac{\pi d}{2h},$$

that is, $K < \pi\delta/h$. If $\delta = 1$ and $h = \pi$, $K < 1$ which may be compared with the exact result for stability of $K < 0.96$.

Feedback Loop with a Polynomial Nonlinearity

In this example possibility of a limit cycle in a feedback loop with $n(x) = x - (x^3/6)$ and $G(s) = K(1-s)/s(s+1)$ is investigated. For the nonlinearity $N(a) = 1 - (a^2/8)$, the $C(a)$ locus starts at -1 on the Nyquist diagram. As a increases, the $C(a)$ locus moves along the negative real axis to $-\infty$ for $a = 2\sqrt{2}$. For a greater than $2\sqrt{2}$, the locus returns along the positive real axis from ∞ to the origin as a becomes large. For small signal levels $N(a) \approx 1$, an oscillation will start to build up, assuming the system is initially at rest with $x(t) = 0$, only if the feedback loop with $G(s)$ alone is unstable. The characteristic equation

$$s^2 + s + K - Ks = 0$$

must have a root with a positive real part, that is, $K > 1$. The phase of $G(j\omega) = 180°$, when $\omega = 1$ and the corresponding gain of $G(j\omega)$ is K. Thus the DF solution for the amplitude of the limit cycle is

$$|G(j\omega)|_{\omega=1} = \frac{1}{1-(a^2/8)}$$

resulting in

$$K = 8/(8-a^2),$$

and

$$a = 2\sqrt{2}[(K-1)/K]^{1/2}. \quad (19.35)$$

As K increases, the limit cycle becomes more distorted because of the shape of the nonlinearity. For example, if $K = 2.4$, the DF solution gives $\omega = 1$ and $a = 2.10$. If four harmonics are balanced [4], the limit cycle frequency is 0.719 and the amplitudes of the fundamental, third, fifth and seventh harmonics at the input to the nonlinearity are 2.515, 0.467, 0.161 and 0.065, respectively.

Because the DF approach is a method for evaluating limit cycles, it is sometimes argued that it cannot guarantee the stability of a feedback system, when instability is caused by an unbounded, not oscillatory, signal in the system. Fortunately another finding is helpful with this problem [17]. This states that, in the feedback system of Figure 19.1, if the symmetric odd nonlinearity $n(x)$ is such that $k_1x < n(x) < k_2x$, for $x > 0$, and $n(x)$ tends to k_3x for large x, where $k_1 < k_3 < k_2$, then the nonlinear system is either stable or possesses a limit cyle, provided that the linear system with gain K replacing N is stable for $k_1 < K < k_2$. For this situation, often true in practice, the nonexistence of a limit cycle indicates stability.

19.1.5 Stability and Accuracy

Because the DF method is an approximate procedure, it is desirable to judge its accuracy. Predicting that a system will be stable, when in practice it is not, may have unfortunate consequences. Many attempts have been made to solve this problem, but those obtained are difficult to apply or produce too conservative results [11].

The problem is illustrated by the system of Figure 19.1 with a symmetrical odd single-valued nonlinearity confined to a sector between lines of slope k_1 and k_2, that is, $k_1 x < n(x) < k_2 x$ for $x > 0$. For absolute stability, the circle criterion requires satisfying the Nyquist criterion for the locus $G(j\omega)$ for all points within a circle having its diameter on the negative real axis of the Nyquist diagram between the points $(-1/k_1, 0)$ and $(-1/k_2, 0)$, as shown in Figure 19.9. On the other hand, because the DF for this nonlinearity lies within the diameter of the circle, the DF method requires satisfying the Nyquist criterion for $G(j\omega)$ for all points on the circle diameter, if the autonomous system is to be stable.

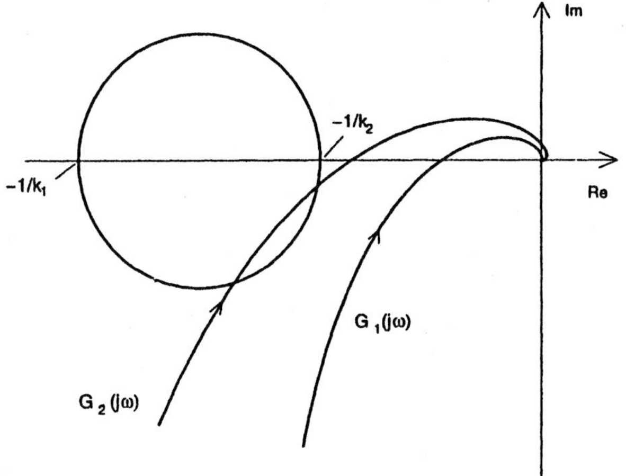

Figure 19.9 Circle criterion and stability.

Therefore, for a limit cycle in the system of Figure 19.1, errors in the DF method relate to its inability to predict a phase shift, which the fundamental harmonic may experience in passing through the nonlinearity, rather than an incorrect magnitude of the gain. When the input to a single-valued nonlinearity is a sinusoid together with some of its harmonics, the fundamental output is not necessarily in phase with the fundamental input, that is, the fundamental gain has a phase shift. The actual phase shift varies with the harmonic content of the input signal in a complex manner, because the phase shift depends on the amplitudes and phases of the individual input components.

From an engineering viewpoint one can judge the accuracy of DF results by estimating the distortion, d, in the input to the nonlinearity. This is straightforward when a limit-cycle solution is given by the DF method; the loop may be considered opened at the nonlinearity input, the sinusoidal signal corresponding to the DF solution can be applied to the nonlinearity, and the harmonic content of the signal fed back to the nonlinearity input can be calculated. Experience indicates that the percentage accuracy of the DF method in predicting the fundamental amplitude and frequency of the limit cycle is less than the percentage distortion in the fedback signal. In the previous section, the frequency of oscillation in the autotuning example, where the distortion was relatively small, was given more accurately than in the third example. Due to the relatively poor filtering of the plant in this example, the distortion in the fedback signal was much higher. As mentioned previously, the DF method may incorrectly predict stability. To investigate this problem, the procedure above can be used again, by taking, as the nonlinearity input, a sinusoid with amplitude and frequency corresponding to values of those parameters where the phase margin is small. If the calculated fedback distortion is high, say greater than 2% per degree of phase margin, the DF result should not be relied on.

The limit-cycle amplitude predicted by the DF is an approximation to the fundamental harmonic. The accuracy of this prediction cannot be assessed by using the peak value of the limit cycle to estimate an equivalent sinusoid. It is possible to estimate the limit cycle more accurately by balancing more harmonics, as mentioned earlier. Although this is difficult algebraically other than with loops whose nonlinearity is mathematically simply described, for example a cubic, software is available for this purpose [10]. The procedure involves solving sets of nonlinear algebraic equations but good starting guesses can usually be obtained for the magnitudes and phases of the other harmonic components from the wave form fedback to the nonlinearity, assuming its input is the DF solution. This procedure was used to balance four harmonics in obtaining a better solution for the distorted limit cycle in example 3 of the previous section.

19.1.6 Compensator Design

Although the design specifications for a control system are often in terms of step-response behavior, frequency domain design methods rely on the premise that the correlation between the frequency response and a step response yields a less oscillatory step response if the gain and phase margins are increased. Therefore the design of a suitable linear compensator for the system of Figure 19.1 using the DF method, is usually done by selecting for example a lead network to provide adequate gain and phase margins for all amplitudes. This approach may be used in example 2 of the previous section where a phase lead network could be added to stabilize the system, say for a gain of 1.5, for which it is unstable without compensation. Other approaches are the use of additional feedback signals or modification of the nonlinearity $n(x)$ directly or indirectly [4], [7].

When the plant is nonlinear, its frequency response also dependes on the input sinusoidal amplitude represented as $G(j\omega, a)$. In recent years several approaches [12], [14] use the DF method to design a nonlinear compensator for the plant, with the objective of closed-loop performance independent of the input amplitude.

19.1.7 Closed-Loop Frequency Response

When the closed-loop system of Figure 19.1 has a sinusoidal input $r(t) = R\sin(\omega t + \theta)$, it is possible to evaluate the closed-loop frequency response using the DF. If the feedback loop has no limit cycle when $r(t) = 0$ and, in addition, the sinusoidal input $r(t)$ does not induce a limit cycle, then, provided that $G_c(s)G_1(s)$ gives good filtering, $x(t)$, the nonlinearity input, almost equals the sinusoid $a\sin\omega t$. Balancing the components of frequency ω around the loop,

$$g_c R\sin(\omega t + \theta + \phi_c) - a g_1 g_c M(a) \sin[\omega t + \phi_1 + \phi_c + \Psi(a)] = a\sin\omega t \quad (19.36)$$

where $G_c(j\omega) = g_c e^{j\phi_c}$ and $G_1(j\omega) = g_1 e^{j\phi_1}$. In principle Equation 19.36, which can be written as two nonlinear algebraic equations, can be solved for the two unknowns a and θ and the fundamental output signal can then be found from

$$c(t) = aM(a)g_1\sin[\omega t + \Psi(a) + \phi_1] \quad (19.37)$$

to obtain the closed-loop frequency response for R and ω.

Various graphical procedures have been proposed for solving the two nonlinear algebraic equations resulting from Equation 19.36 [9], [13], [15]. If the system is lightly damped, the nonlinear equations may have more than one solution, indicating that the frequency response of the system has a jump resonance. This phenomenon of a nonlinear system has been studied by many authors, both theoretically and practically [8], [16].

Further Information

Many control engineering text books contain material on nonlinear systems where the describing function is discussed. The coverage, however, is usually restricted to the basic sinusoidal DF for determining limit cycles in feedback systems. The basic DF method, which is one of quasilinearisation, can be extended to cover other signals, such as random signals, and also to cover multiple input signals to nonlinearities and feedback system analysis. The two books with the most comprehensive coverage of this are Gelb and van der Velde [7] and Atherton [3]. More specialized books on nonlinear feedback systems usually cover the phase plane method, the subject of the next article, and the DF together with other topics such as absolute stability, exact linearization, etc.

References

[1] Andronov, A. A., Vitt, A. A., and Khaikin, S.E., *Theory of Oscillators,* Addison-Wesley, Reading, MA, 1966. (First edition published in Russia in 1937.)

[2] Astrom, K.J. and Haggland, T., *Automatic tuning of single regulators,* Proc IFAC Congress, Budapest, Vol 4, 267-272, 1984.

[3] Atherton, D.P., *Nonlinear Control Engineering, Describing Function Analysis and Design,* Van Nostrand Reinhold, London, 1975.

[4] Atherton, D.P., *Non Linear Control Engineering,* Student Ed., Van Nostrand Reinhold, New York, 1982.

[5] Choudhury, S.K., Atherton, D.P., Limit cycles in high order nonlinear systems, *Proc. Inst. Electr. Eng.,* 121, 717-24, 1974.

[6] Cook, P.A., Describing function for a sector nonlinearity, *Proc. Inst. Electr. Eng.,* 120, 143-44, 1973.

[7] Gelb, A. and van der Velde, W.E., *Multiple Input Describing Functions and Nonlinear Systems Design,* McGraw-Hill, New York, 1968.

[8] Lamba, S.S. and Kavanagh, R.J., The phenomenon of isolated jump resonance and its applications, *Proc. Inst. Electr. Eng.,* 118, 1047-50, 1971.

[9] Levinson, E., Some saturation phenomena in servomechanisms with emphasis on the techometer stabilised system, *Trans. Am. Inst. Electr. Eng.,* Part 2, 72, 1-9, 1953.

[10] McNamara, O.P. and Atherton, D.P., Limit cycle prediction in free structured nonlinear systems, *IFAC Congress,* Munich, Vol 8, 23-28, July 1987.

[11] Mees, A.I. and Bergen, A.R., Describing function revisited, *IEEE Trans. Autom. Control,* 20, 473-78, 1975.

[12] Nanka-Bruce, O. and Atherton, D.P., Design of nonlinear controllers for nonlinear plants, *IFAC Congress,* Tallinn, Volume 6, 75-80, 1990.

[13] Singh, T.P., Graphical method for finding the closed loop frequency response of nonlinear feedback control systems, *Proc. Inst. Electr. Eng.,* 112, 2167-70, 1965.

[14] Taylor, J.H. and Strobel, K.L., Applications of a nonlinear controller design approach based on the quasilinear system models, *Prof ACC,* San Diego, 817-824, 1984.

[15] West, J.C. and Douce, J.L., The frequency response of a certain class of nonlinear feedback systems, *Br. J. Appl. Phys.,* 5, 201-10, 1954.

[16] West, J.C., Jayawant, B.V., and Rea, D.P., Transition characteristics of the jump phenomenon in nonlinear resonant circuits, *Proc. Inst. Electr. Eng.,* 114, 381-92, 1967.

[17] Vogt, W.G. and George, J.H., On Aizerman's conjecture and boundedness, *IEEE Trans. Autom. Control,* 12, 338-39, 1967.

[18] Ziegler, J.G. and Nichols, N.B., Optimal setting for automatic controllers, *Trans. ASME,* 65, 433-444, 1943.

19.2 The Phase Plane Method

19.2.1 Introduction

The phase plane method was the first method used by control engineers for studying the effects of nonlinearity in feedback systems. The technique which can only be used for systems with second order models was examined and further developed for control engineering purposes for several major reasons,

1. The phase plane approach had been used for several studies of second order nonlinear differential equations arising in fields such as planetary motion, nonlinear mechanics and oscillations in vacuum tube circuits.
2. Many of the control systems of interest, such as servomechanisms, could be approximated by second order nonlinear differential equations.
3. The phase plane was particularly appropriate for dealing with nonlinearities with linear segmented characteristics which were good approximations for the nonlinear phenomena encountered in control systems.

The next section considers the basic aspects of the phase plane approach but later concentration is focused on control engineering applications where the nonlinear effects are approximated by linear segmented nonlinearities.

19.2.2 Background

Early analytical work [1], on second order models assumed the equations

$$\begin{aligned} \dot{x}_1 &= P(x_1, x_2) \\ \dot{x}_2 &= Q(x_1, x_2) \end{aligned} \tag{19.38}$$

for two first-order nonlinear differential equations. Equilibrium, or singular points, occur when

$$\dot{x}_1 = \dot{x}_2 = 0$$

and the slope of any solution curve, or trajectory, in the $x_1 - x_2$ state plane is

$$\frac{dx_2}{dx_1} = \frac{\dot{x}_2}{\dot{x}_1} = \frac{Q(x_1, x_2)}{P(x_1, x_2)} \tag{19.39}$$

A second order nonlinear differential equation representing a control system can be written

$$\ddot{x} + f(x, \dot{x}) = 0 \tag{19.40}$$

If this is rearranged as two first-order equations, choosing the phase variables as the state variables, that is $x_1 = x,\ x_2 = \dot{x}$, then Equation 19.40 can be written as

$$\dot{x}_1 = \dot{x}_2 \quad \dot{x}_2 = -f(x_1, x_2) \tag{19.41}$$

which is a special case of Equation 19.39. A variety of procedures has been proposed for sketching state [phase] plane trajectories for Equations 19.39 and 19.41. A complete plot showing trajectory motions throughout the entire state (phase) plane is known as a state (phase) portrait. Knowledge of these methods, despite the improvements in computation since they were originally proposed, can be particularly helpful for obtaining an appreciation of the system behavior. When simulation studies are undertaken, phase plane graphs are easily obtained and they are often more helpful for understanding the system behavior than displays of the variables x_1 and x_2 against time.

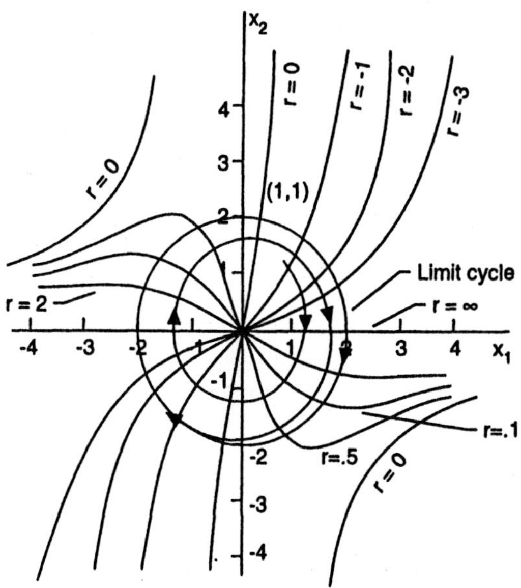

(a) μ = 0.2

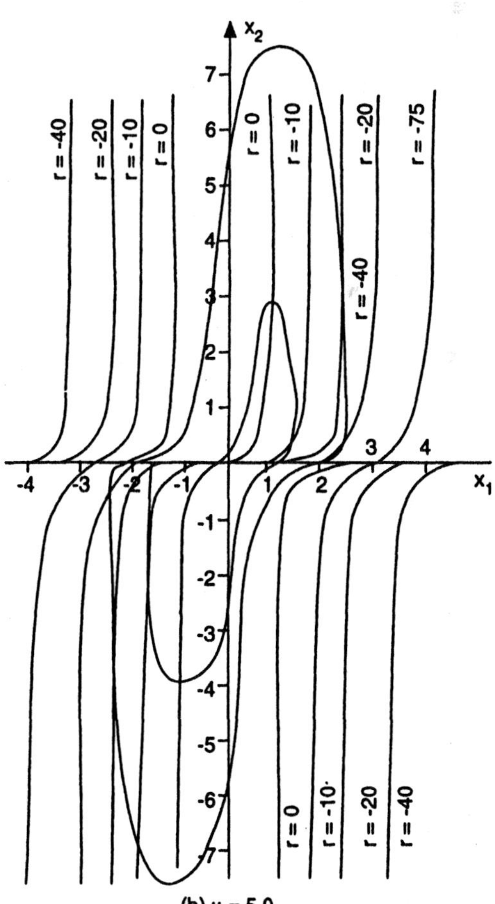

(b) μ = 5.0

Figure 19.10 Phase portraits of the Van der Pol equation for different values of μ.

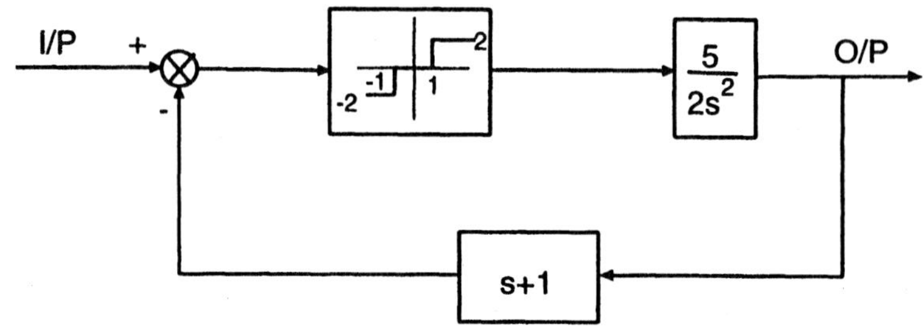

Figure 19.11 Relay system.

Many investigations using the phase plane technique were concerned with the possibility of limit cycles in the nonlinear differential equations. When a limit cycle exists, this results in a closed trajectory in the phase plane. Typical of such investigations was the work of Van der Pol, who considered the equation

$$\ddot{x} - \mu(1-x^2)\dot{x} + x = 0 \tag{19.42}$$

where μ is a positive constant. The phase plane form of this equation can be written as

$$\begin{aligned} \dot{x}_1 &= x_2 \\ \dot{x}_2 &= -f(x_1, x_2) = \mu(1-x_1^2)x_2 - x_1 \end{aligned} \tag{19.43}$$

The slope of a trajectory in the phase plane is

$$\frac{dx_2}{dx_1} = \frac{\dot{x}_2}{\dot{x}_1} = \frac{\mu(1-x_1^2)x_2 - x_1}{x_2} \tag{19.44}$$

and this is only singular (that is, at an equilibrium point), when the right hand side of Equation 19.44 is 0/0, that is $x_1 = x_2 = 0$.

The form of this singular point which is obtained from linearization of the equation at the origin depends upon μ, being an unstable focus for $\mu < 2$ and an unstable node for $\mu > 2$. All phase plane trajectories have a slope of r when they intersect the curve

$$rx_2 = \mu(1-x_1^2)x_2 - x_1 \tag{19.45}$$

One way of sketching phase plane behavior is to draw a set of curves given for various values of r by Equation 19.45 and marking the trajectory slope r on the curves. This procedure is known as the method of isoclines and has been used to obtain the limit cycles shown in Figure 19.10 for the Van der Pol equation with $\mu = 0.2$ and 4.

19.2.3 Piecewise Linear Characteristics

When the nonlinear elements occurring in a second order model can be approximated by linear segmented characteristics then the phase plane approach is usually easy to use because the nonlinearities divide the phase plane into various regions within which the motion may be described by different linear second-order equations [2]. The procedure is illustrated by the simple relay system in Figure 19.11.

The block diagram represents an "ideal" relay position control system with velocity feedback. The plant is a double integrator, ignoring viscous (linear) friction, hysteresis in the relay, or backlash in the gearing. If the system output is denoted by x_1 and its derivative by x_2, then the relay switches when $-x_1 - x_2 = \pm 1$; the equations of the dotted lines are marked switching lines on Figure 19.12.

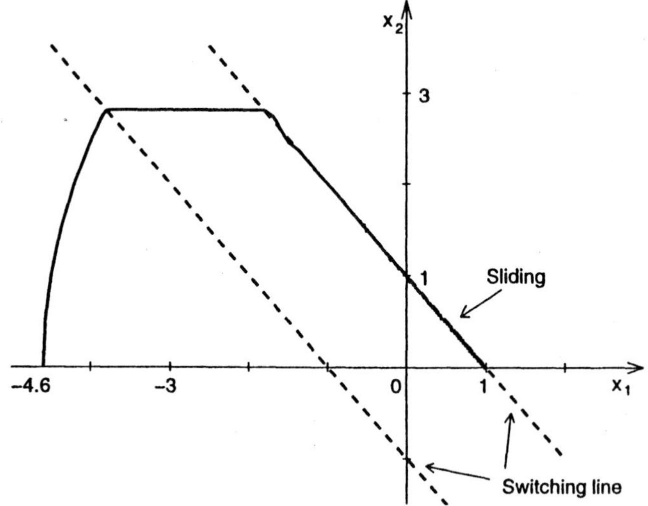

Figure 19.12 Phase plane for relay system.

Because the relay output provides constant values of ± 2 and 0 to the double integrator plant, if we denote the constant value by h, then the state equations for the motion are

$$\begin{aligned} \dot{x}_1 &= x_2 \\ \dot{x}_2 &= h \end{aligned} \tag{19.46}$$

which can be solved to give the phase plane equation

$$x_2^2 - x_{20}^2 = 2h(x_1 - x_{10}) \tag{19.47}$$

which is a parabola for h finite and the straight line $x_2 = x_{20}$ for $h = 0$, where x_{20} and x_{10} are the initial values of x_2 and x_1. Similarly, more complex equations can be derived for other second-order transfer functions. Using Equation 19.47 with the appropriate values of h for the three regions in the phase plane, the step response for an input of 4.6 units can be obtained as shown in Figure 19.12.

In the step response, when the trajectory meets the switching line $x_1 + x_2 = -1$ for the second time, trajectory motions at both sides of the line are directed towards it, resulting in a sliding motion down the switching line. Completing the phase portrait by drawing responses from other initial conditions shows that the autonomous system is stable and also that all responses will

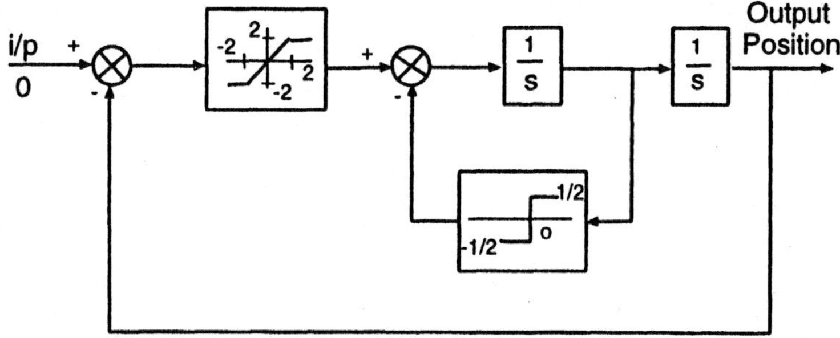

Figure 19.13 Block diagram of servomechanism.

finally slide down a switching line to equilibrium at $x_1 = \pm 1$.

An advantage of the phase plane method is that it can be used for systems with more than one nonlinearity and for those situations where parameters change as functions of the phase variables. For example, Figure 19.13 shows the block diagram of an approximate model of a servomechanism with nonlinear effects due to torque saturation and Coulomb friction.

The differential equation of motion in phase variable form is

$$\dot{x}_2 = f_s(-x_1) - (1/2)\,\mathrm{sgn}\,x_2 \qquad (19.48)$$

where f_s denotes the saturation nonlinearity and sgn the signum function, which is $+1$ for $x_2 > 0$ and -1 for $x_2 < 0$. There are six linear differential equations describing the motion in different regions of the phase plane. For x_2 positive, Equation 19.48 can be written

$$\ddot{x}_1 + f_s(x_1) + 1/2 = 0$$

so that for

(a) x_2+ve, $x_1 < -2$, we have $\dot{x}_1 = x_2, \dot{x}_2 = 3/2$, a parabola in the phase plane.

(b) x_2+ve $|x_1| < 2$, we have $\dot{x}_1 = x_2, \dot{x}_2 + x_1 + 1/2 = 0$, a circle in the phase plane.

(c) x_2+ve, $x_1 > 2$, we have $\dot{x}_1 = x_2, \dot{x}_2 = -5/2$, a parabola in the phase plane.

Similarly for x_2 negative,

(d) x_2−ve, $x_1 - 2$, we have $\dot{x}_1 = x_2, \dot{x}_2 = 5/2$, a parabola in the phase plane.

(e) x_2−ve, $|x_2| < 2$, we have $\dot{x}_1 = x_2, \dot{x}_2 + x_1 - 1/2 = 0$, a circle in the phase plane.

(f) x_2−ve,$x_1 > 2$, we have $\dot{x}_1 = x_2, \dot{x}_2 = -3/2$, a parabola in the phase plane.

Because all the phase plane trajectories are described by simple mathematical expressions, it is straightforward to calculate specific phase plane trajectories.

19.2.4 Discussion

The phase plane approach is useful for understanding the effects of nonlinearity in second order systems, particularly if it may be approximated by a linear segmented characteristic. Solutions for the trajectories with other nonlinear characteristics may not be possible analytically so that approximate sketching techniques were used in early work on nonlinear control. These approaches are described in many books, for example, [3],[4],[5],[6],[7], [8], [9], and [10]. Although the trajectories are now easily obtained with modern simulation techniques, knowledge of the topological aspects of the phase plane are still useful for interpreting the responses in different regions of the phase plane and appreciating the system behavior.

References

[1] Andronov, A.A., Vitt, A.A., and Khaikin, S.E., *Theory of Oscillators,* Addison-Wesley, Reading, MA, 1966. (First edition published in Russia in 1937.)

[2] Atherton, D.P., *Non Linear Control Engineering,* Student Ed., Van Nostrand Reinhold, New York, 1982.

[3] Blaquiere, A., *Nonlinear Systems Analysis,* Academic Press, New York, 1966.

[4] Cosgriff, R., *Nonlinear Control Systems,* McGraw-Hill, New York, 1958.

[5] Cunningham, W.J., *Introduction to Nonlinear Analysis,* McGraw-Hill, New York, 1958.

[6] Gibson, J.E., *Nonlinear Automatic Control,* McGraw-Hill, New York, 1963.

[7] Graham, D. and McRuer, D., *Analysis of Nonlinear Control Systems,* John Wiley & Sons, New York, 1961.

[8] Hayashi, C., *Nonlinear Oscillations in Physical Systems,* McGraw-Hill, New York, 1964.

[9] Thaler, G.J. and Pastel, M.P., *Analysis and Design of Nonlinear Feedback Control Systems,* McGraw-Hill, New York, 1962.

[10] West, J.C., *Analytical Techniques of Nonlinear Control Systems,* E.U.P., London, 1960.

20

Design Methods

R. H. Middleton
Department of Electrical and Computer Engineering, University of Newcastle, NSW, Australia

Stefan F. Graebe
PROFACTOR GmbH, Steyr, Austria

Anders Ahlén
Systems and Control Group, Department of Technology, Uppsala University, Uppsala, Sweden

Jeff S. Shamma
Center for Control and Systems Research, Department of Aerospace Engineering and Engineering Mechanics, The University of Texas at Austin, Austin, TX

20.1 Dealing with Actuator Saturation

R. H. Middleton, Department of Electrical and Computer Engineering, University of Newcastle, NSW, Australia

20.1.1 Description of Actuator Saturation

Essentially all plants have inputs (or manipulated variables) that are subject to hard limits on the range (or sometimes also rate) of variations that can be achieved. These limitations may be due to restrictions deliberately placed on actuators to avoid damage to a system and/or physical limitations on the actuators themselves. Regardless of the cause, limits that cannot be exceeded invariably exist. When the actuator has reached such a limit, the actuator is said to be "saturated" since no attempt to further increase the control input gives any variation in the actual control input. The simplest case of actuator saturation in a control system is to consider a system that is linear apart from an input saturation as depicted in Figure 20.1.

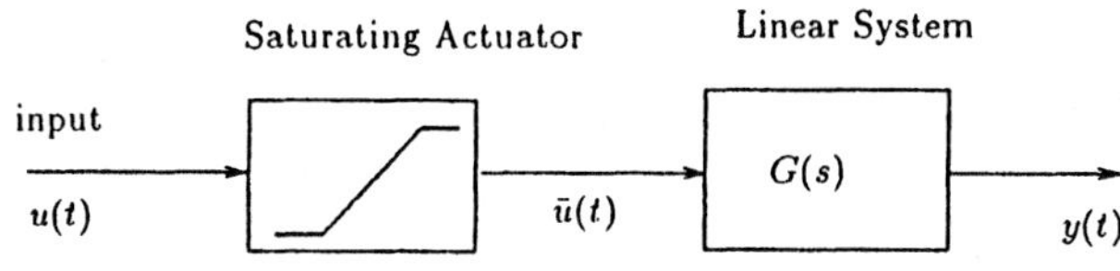

Figure 20.1 Linear plant model including actuator saturation.

Mathematically, the action of a saturating actuator can be described as:

$$\bar{u} = \left\{ \begin{array}{l} u_{max} : u \geq u_{max} \\ u : u_{min} < u < u_{max} \\ u_{min} : u_{min} \geq u \end{array} \right\}$$

Heuristically, it can be seen that once in saturation, the incremental (or small signal) gain of the actuator becomes zero. Alternatively, from a describing function viewpoint, a saturation is an example of a sector nonlinearity with a describing function as illustrated in Figure 20.2.

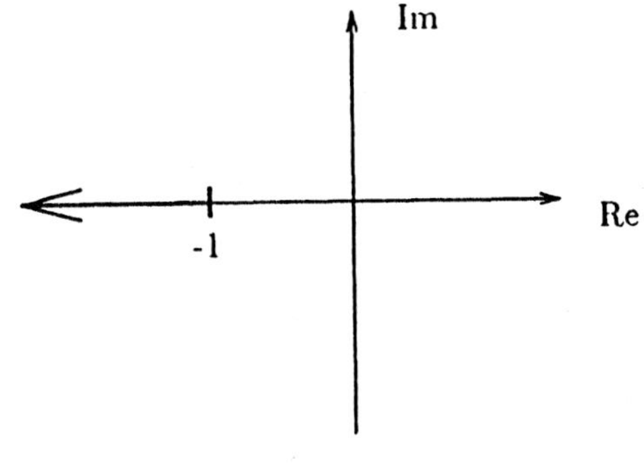

Figure 20.2 Describing function for a saturation.

This describing function gives exactly a range of gains starting at 1 and reducing to zero as amplitude increases. From both

0-8493-8570-9/96/$0.00+$.50

perspectives, actuator saturation can be seen to be equivalent to a nonlinear reduction in gain.

20.1.2 Effects of Actuator Saturation

The main possible effects of actuator saturation on a control system are poor performance and/or instability to large disturbances. These effects are seen as "large" disturbance effects since for "small" disturbances, actuator saturation may be averted, and a well-behaved linear response can occur. The following two examples illustrate the possible effects of actuator saturation.

EXAMPLE 20.1: Integral windup

In this example, we consider the control system depicted in Figure 20.3, with $u_{\min} = -1$ and $u_{\max} = 1$

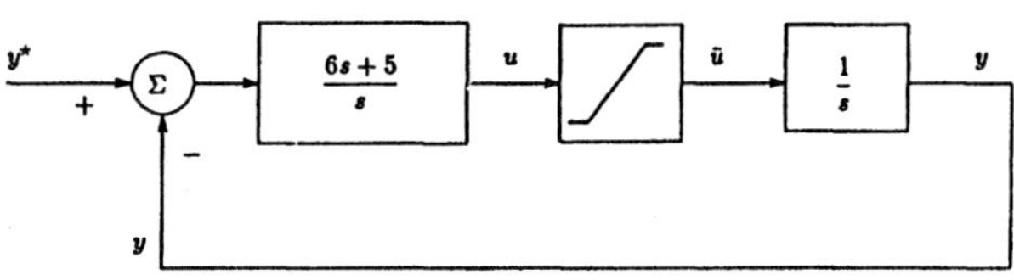

Figure 20.3 Example control system for saturating actuators.

The simulated response for this system with a step change in the set point of 0.4 is illustrated in Figure 20.4. Note that this step change corresponds to a "small" step where saturation is evident, but only to a small extent. In this case, the step response is well behaved, with only a small amount of overshoot occurring.

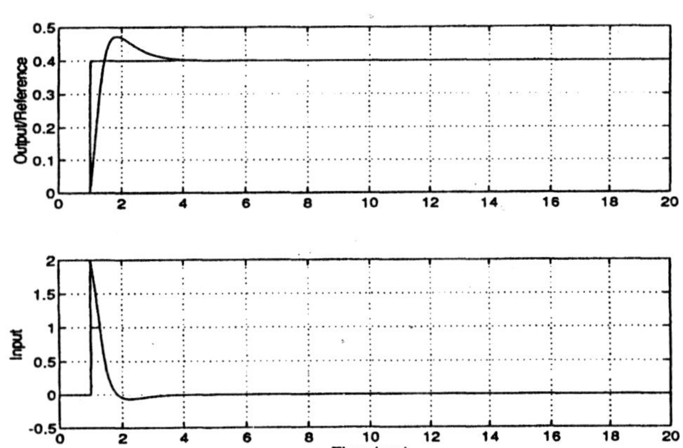

Figure 20.4 Response to a small step change for Example 20.1.

In contrast to this, Figure 20.5 shows simulation results for a step change of four units in the set point. In this case, note that the response is very poor with large overshoot and undershoot in the response. The input response shows why this is occurring, where the unsaturated input reaches very large values due to integral action in the controller. This phenomenon is termed integral (or reset[1]) windup.

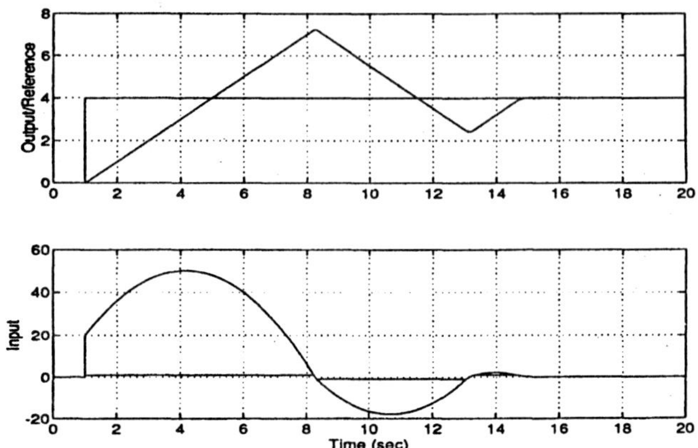

Figure 20.5 Response to a large step change.

An even more dramatic effect is shown in Figure 20.6 where an open loop system is strictly unstable[2].

EXAMPLE 20.2: Controller saturation for an unstable plant.

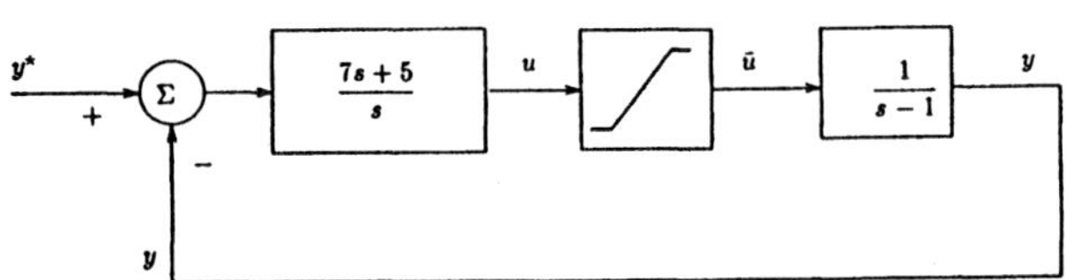

Figure 20.6 Open-loop unstable system with actuator saturation.

In this case (where again we take $u_{min} = -1$ and $u_{max} = 1$), a step change in the reference of 0.8 units causes a dramatic failure[3] of the control system as illustrated in Figure 20.7. This instability is caused solely by saturation of the actuator since, for small step changes, the control system is well behaved.

20.1.3 Reducing the Effects of Actuator Saturation

The effects of actuator saturation cannot always be completely avoided. However, there are ways of reducing some of the effects

[1]The term *reset* is commonly used in the process control industry for integral action.

[2]The strict sense is that the plant has an open-loop pole with positive real part.

[3]It has been reported (e.g., Stein, G.,[5]) that this type of failure was one of the factors that caused the Chernobyl nuclear disaster (in this case, a limit on the rate of change of the actuator exacerbated an already dangerous situation).

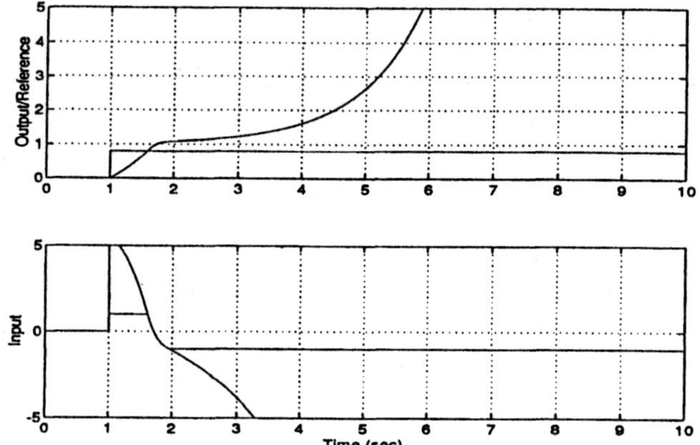

Figure 20.7 Actuator saturation causing instability in Example 20.2.

of actuator saturation, as indicated below.

1. *Where possible, avoid conditionally stable[4] control systems.* Conditionally stable feedback control systems are undesirable for several reasons. Included in these reasons is the effect of actuator saturations. Simple describing function arguments show that the combination of a conditionally stable control system and a saturating actuator give rise to limit cycle behavior. In most cases, this limit cycle behavior is unstable. Instability of such a limit cycle generally means that for slightly larger initial conditions or disturbances, compared with the limit cycle, the output diverges; and, conversely, for smaller initial conditions, stable convergence to a steady state occurs. This is clearly undesirable, but cannot always be avoided. Note that a controller for a plant can be designed that gives unconditional stability if and only if the plant has:

    a. No poles with positive real part

    b. No purely imaginary poles of repetition greater than 2.

    Therefore, a plant that is open loop strictly unstable can be only conditionally stabilized.

2. *Avoid applying unrealistic reference commands to a control system.* Note that in the examples presented previously, the reference commands were in many cases unrealistic. Take, for example, the situation shown in Figure 20.5. In this case, an instantaneous change of 4 units is being commanded. Clearly, however, because of the actuator saturation, the output, y, can never change by more than 1 unit per second. Therefore, we know that the commanded trajectory can never be achieved. A more sensible reference signal would be one that ramps up (at a rate of 1 unit per second) from 0 to 4 units. If this reference were applied instead of the step, a greatly reduced overshoot, etc. would be obtained. The implementation of this type of idea is often termed a *reference governor* or *reference conditioner* for the system. See, for example, [4] for more details.

3. *Utilize saturation feedback to implement the controller.* To implement saturation feedback we note that any linear controller of the form

$$U(s) = C(s)E(s) \tag{20.1}$$

can be rewritten as

$$U(s) = \frac{P(s)}{L(s)}E(s) \tag{20.2}$$

where $L(s) = s^n + l_{n-1}s^{k-1} + \ldots + l_o$ is a monic polynomial in s, and $P(s) = p_n s^n + p_{n-1}s^{n-1} + \ldots + p_o$ is a polynomial in s. Let the closed-loop poles be at $s = -\alpha_i; i = 1 \ldots N > n$. Then the controller can be implemented via saturation feedback as shown in Figure 20.8.

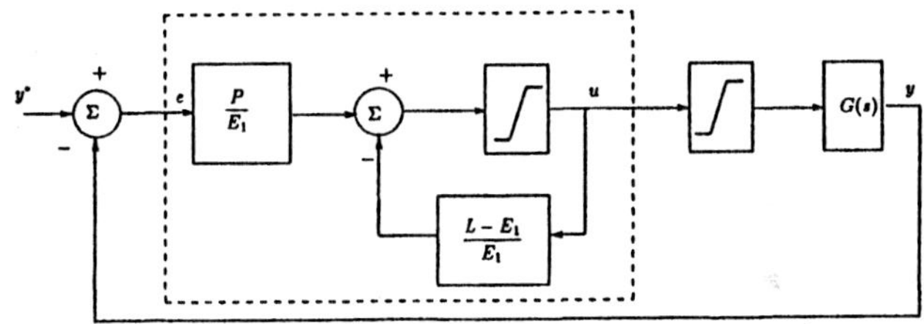

Figure 20.8 Controller implementation using saturation feedback.

In the above implementation $E_1(s)$ can, in principle, be any stable monic polynomial of degree n. The quality of the performance of this anti-integral windup scheme depends on the choice of $E_1(s)$. A simple choice that gives good results in most cases is

$$E_1(s) = (s + \alpha_{m_1})(s + \alpha_{m_2}) \ldots (s + \alpha_{m_n}) \tag{20.3}$$

where $\alpha_{m_1} \ldots \alpha_{m_n}$ are the n fastest closed-loop poles. Note that when the actuator is not saturated we have that $U(s) = \frac{P(s)}{E_1(s)}E(s) - \frac{L(s)-E_1(s)}{E_1(s)}U(s)$ and so $\frac{U(s)}{E(s)} = \frac{P(s)}{L(s)}$, which is precisely the desired linear transfer function. When the actuator does saturate, the fact that E_1 is stable improves the controller behavior. The following examples illustrate the advantages of this approach.

[4]A conditionally stable control system is one in which a reduction in the loop gain may cause instability.

EXAMPLE 20.3: Anti-integral windup (Example 1 revisited)

In this case $L(s) = s$; $P(s) = 6s + 5$; and the closed-loop poles are at $s = -1$ and $s = -5$. We therefore choose $E_1(s) = (s + 5)$ and obtain the control system structure illustrated in Figure 20.9.

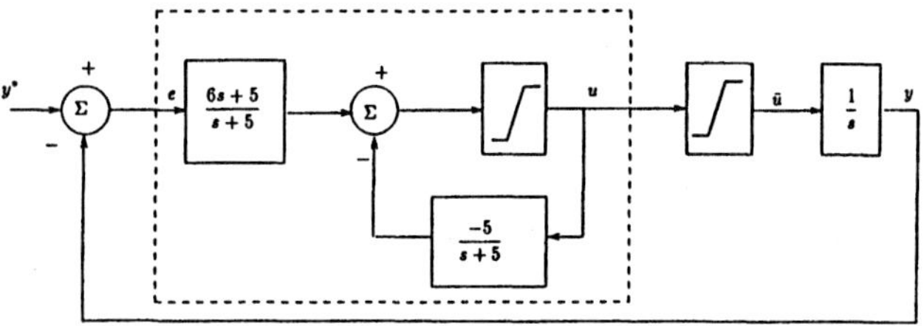

Figure 20.9 Control system for Example 20.3.

Figure 20.10 compares the performance of this revised arrangement with that of Figure 20.5, showing excellent performance in this case.

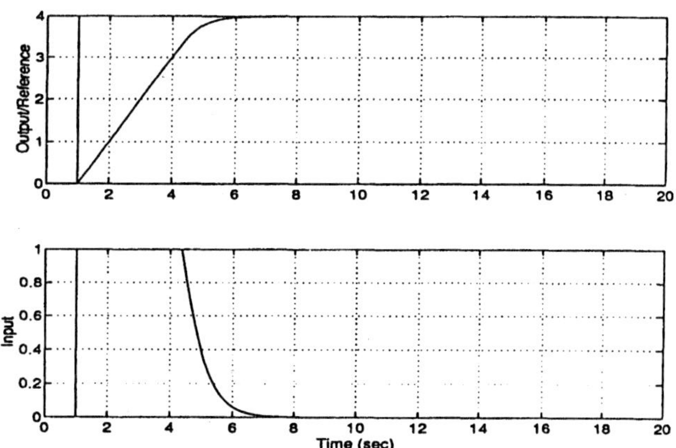

Figure 20.10 Response to a large step change for Example 20.3.

EXAMPLE 20.4: Improved control of unstable systems (Example 2 revisited)

Let us now consider Example 20.2 again. In this case, $P(s) = 7s + 5$; $L(s) = s$ and the closed-loop poles are again at $s = -1$ and $s = -5$. This suggests $E_1(s) = (s + 5)$ giving the following control system structure:

The comparative step response to Figure 20.7 is given for a slightly larger step (in this case, 1.0 units) in Figure 20.12. Note in this case that previously where instability arose, in this case very good response is obtained.

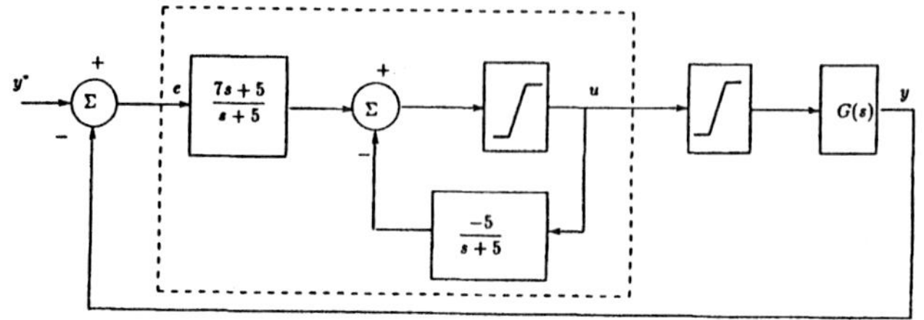

Figure 20.11 Controller structure for Example 20.4.

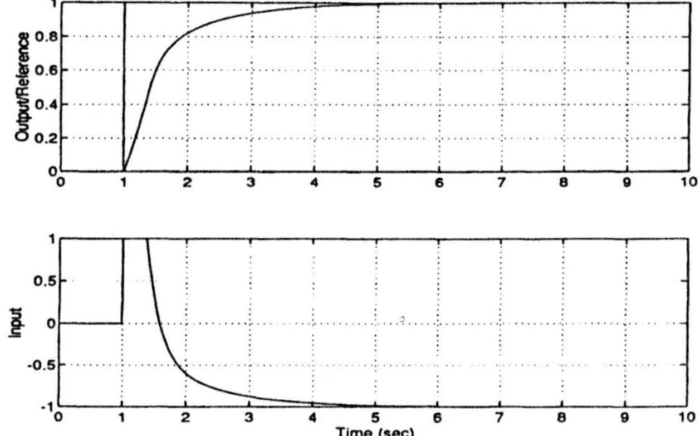

Figure 20.12 Response for Example 20.4.

References

[1] Astrom, K. and Wittenmark, B., *Computer Controlled Systems: Theory and Design,* 2nd ed., Prentice Hall, Englewood Cliffs, NJ, 1990.

[2] Braslavsky, J. and Middleton, R., On the stabilisation of linear unstable systems with control constraints, *Proc. IFAC World Congr.,* Sydney, Australia, 1993.

[3] Gilbert, E., Linear control systems with pointwise in time constraints: what do we do about them?, *Proc. Am. Control Conf.,* Chicago, 1992.

[4] Seron, S.G.M. and Goodwin, G., All stabilizing controllers, feedback linearization and anti-windup: a unified review, *Proc. Am. Control Conf.,* Baltimore, 1994.

[5] Stein, G., Bode lecture, 28th IEEE Conference on Decision and Control, Tampa, FL, 1989.

Further Reading

The idea of using saturation feedback to help prevent integral windup (and related phenomena) has been known for many years now. Astrom and Wittenmark [1] give a description of this and an interpretation in terms of observer design with nonlinear actuators.

More advanced, constrained optimization-based procedures are the subject of current research by many authors. Gilbert [3] gives an overview of this area (together with the problem of maintaining system states within desired constraints).

Another approach that may be useful where actuator saturation is caused by large changes in the reference signal (as opposed to disturbances or other effects) is that of a *reference governor* or *reference conditioner*. Seron and Goodwin [4] explore this and its relationship with the technique of *saturation feedback*. Also, as mentioned previously, it has long been known that strictly unstable systems with actuator constraints can never be globally stabilized; see, for example, [2] for a recent look at this problem.

20.2 Bumpless Transfer

Stefan F. Graebe, PROFACTOR GmbH, *Steyr, Austria*

Anders Ahlén, *Systems and Control Group, Department of Technology, Uppsala University, Uppsala, Sweden*

20.2.1 Introduction

Traditionally, the problem of bumpless transfer refers to the instantaneous switching between manual and automatic control of a process while retaining a smooth ("bumpless") control signal. As a simple example illustrating this issue, we consider a typical start-up procedure.

EXAMPLE 20.5:

Consider a system with open-loop dynamics

$$\begin{aligned} \dot{x}(t) &= -0.2x(t) + 0.2u(t), \qquad x(t_o) = x_o \\ y(t) &= x(t) \end{aligned} \tag{20.4}$$

where $u(t)$ is the control signal and $y(t)$ is a noise-free measurement of the state $x(t)$. With s denoting the Laplace transform complex variable, we also consider the proportional-integral (PI) controller

$$C(s) = 2.3\left(1 + \frac{1}{4.2s}\right),$$

digitally approximated as

$$\begin{aligned} X^I(t+\Delta) &= X^I(t) + \Delta 0.547e(t) \\ u_{\text{ctrl}}(t) &= X^I(t) + 2.3e(t) \ . \end{aligned} \tag{20.5}$$

In Equation 20.5, $X^I(t)$ denotes the integrator state of the PI controller, Δ is the sampling period, $e(t) \triangleq r(t) - y(t)$ is the control error, and $u_{\text{ctrl}}(t)$ is the control signal generated by the feedback controller. The reference signal is assumed to have a constant value of $r \equiv 4$. Then the following procedure, although simplified for this example, is typical of industrial start-up strategies.

The system is started from rest, $x_o = 0$, and the control is manually held at $u_{\text{man}}(t) = 4$ until, say at time t_s^-, the system output is close to the desired setpoint

$$y(t_s^-) \approx r = 4 \ .$$

At that point, control is switched to the PI controller for automatic operation. Figure 20.13 illustrates what happens if the above start-up procedure is applied blindly without bumpless transfer and the controller state is $X^I(t_s) = 0$ at switching time $t_s = 16$. With a manual control of $u_{\text{man}}(t) = 4$, $t \in [1, 16)$, the control error at switching time can be computed from Equation 20.4 as

$$e(t_s^-) = y(t_s^-) - r(t_s^-) = 0.2 \ .$$

Hence, from Equation 20.5, the automatic control at switching time for $X^I(t_s) = 0$ yields $u_{\text{ctrl}}(t_s^+) = 0.46$. As a result, there is a "bump" in the control signal, u, when switching from manual control, $u = u_{\text{man}}(t_s^-) = 4$, to automatic control, $u = u_{\text{ctrl}}(t_s^+) = 0.46$, and an unacceptable transient follows.

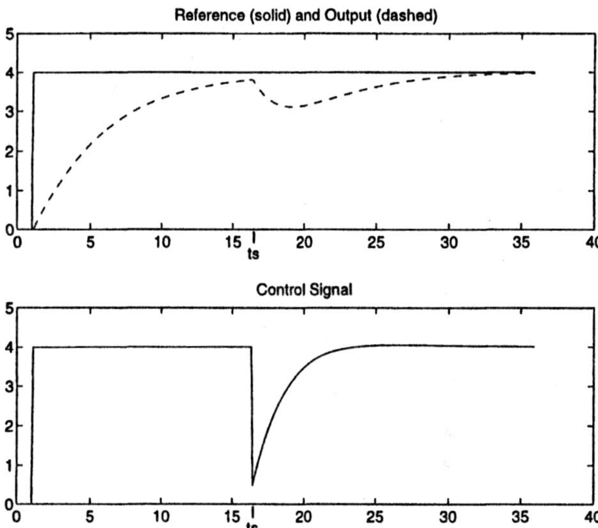

Figure 20.13 Without bumpless transfer mechanism, poor performance occurs at switching from manual to automatic control at time t_s. (Top) Reference signal, $r(t)$, (solid); output, $y(t)$, (dashed). (Bottom) Control signal, $u(t)$.

Avoiding transients after switching from manual to automatic control can be viewed as an initial condition problem on the output of the feedback controller. If the manual control just before switching is $u_{\text{man}}(t_s^-) = u_{\text{man}}^o$, then bumpless transfer requires that the automatic controller take that same value as initial condition on its output, u_{ctrl}, so that $u_{\text{ctrl}}(t_s^+) = u_{\text{man}}^o$. By mapping this condition to the controller states, bumpless transfer can be viewed as a problem of choosing the appropriate initial conditions on the controller states.

In the case of the PI controller in Equation 20.5, the initial condition on the controller state X^I that yields an arbitrary value u_{man}^o is trivially computed as

$$X^I(t_s^-) = u_{\text{man}}^o - 2.3e(t_s^-) \ . \tag{20.6}$$

By including Equation 20.6 as initial condition at the switching time on the PI controller in Equation 20.5, bumpless transfer is achieved as shown in Figure 20.14.

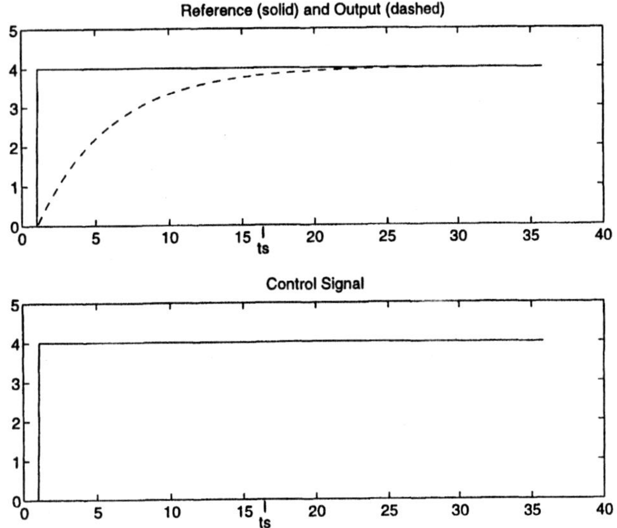

Figure 20.14 Bumpless transfer from manual to automatic control at switching time t_s. (Top) Reference signal, $r(t)$, (solid); output, $y(t)$, (dashed). (Bottom) Control signal, $u(t)$.

Taking a more general point of view, there are several practical situations that call for strategies that could be classified as bumpless transfer. We list these and their associated constraints in Section 2. In Section 3, we review a general framework in which bumpless transfer is considered to be a tracking control problem. Section 4 presents a number of other techniques and Section 5 provides a brief summary.

20.2.2 Applications of Bumpless Transfer

In this section, we present several scenarios that may all be interpreted as bumpless transfer problems. Since each of theses scenarios is associated with different constraints, they tend to favor different bumpless transfer techniques.

Switching Between Manual and Automatic Control

The ability to switch between manual and automatic control while retaining a smooth control signal is the traditional bumpless transfer problem. Its essence is described in Example 20.5, although the general case allows arbitrary controller complexity instead of being restricted to PI controllers. This is probably the simplest bumpless transfer scenario, as it tends to be associated with three favorable factors.

Firstly, the switching scheme is usually designed for a particular feedback loop. Thus, the controller and its structure are known and can be exploited. If it is a PI controller, for example, the simple strategy of Section 1 suffices. Other techniques take advantage of the particular structures of **observer-based controllers** [1], **internal model controllers (IMC)** [10], [3] or controllers implemented in incremental form [1].

Secondly, in contrast to the scenario of Section 2.4, manual to automatic switching schemes are usually implemented in the same process control computer as the controller itself. In that case, the exact controller state is available to the switching algorithm, which can manipulate it to achieve the state associated with a smooth transfer. The main challenge is therefore to compute the appropriate state for higher-order controllers.

Thirdly, switching between manual and automatic controllers usually occurs under fairly benign conditions specifically aimed at aiding a smooth transfer. Many strategies implemented in practice (see Section 4) are simple because they implicitly assume constant signals.

Filter and Controller Tuning

It is frequently desired to tune filter or controller parameters on-line and in response to experimental observations.

EXAMPLE 20.6:

Consider Figure 20.15, which shows the sinusoid

$$s(t) = 5sin(0.4\pi t) + 2$$

filtered by a filter F_1. Until the switching time, the filter is given by

$$F_1(s) = \frac{1}{0.5s + 1} \ .$$

Assume that, at time $t_s \approx 10$, it is desired to retune the time constant of the filter to obtain

$$F_2(s) = \frac{1}{2s + 1} \ .$$

Then, merely changing the filter time constant without adjusting the filter state results in the transient shown in Figure 20.15.

The scenario of Example 20.6 can be considered as a bumpless transfer problem between two dynamical systems, F_1 and F_2. Although these systems have the meaning of filters in Example 2, the same considerations apply, of course, to the retuning of controller parameters.

Assuming that the bumpless transfer algorithm is implemented in the same computer as the filter or controller to be tuned, this problem amounts to an appropriate adjustment of the state as discussed in the previous section. The main difference is now that one would like to commence with the tuning even during transients. Although the general techniques of Sections 3 and 4 could be applied to this case, a simpler scheme, sufficient for low-order filters and controllers, can be derived as follows.

Let the signal produced by the present filter or controller be denoted by u_1 and let the retuned filter or controller be implemented by the state-space model

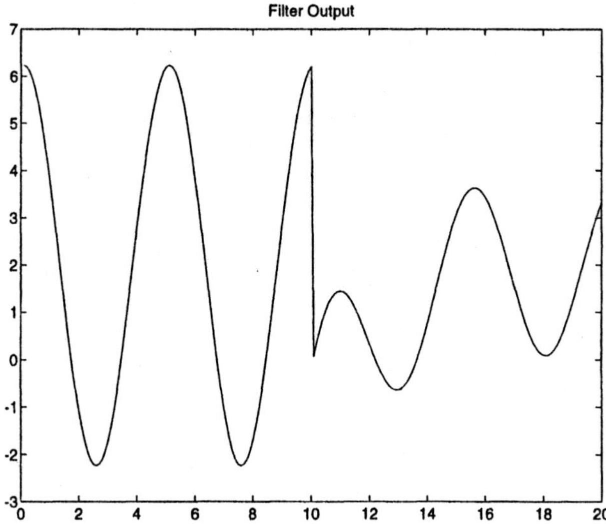

Figure 20.15 Transient produced by changing a filter time constant without adjusting the filter state.

$$\begin{aligned} \dot{x}(t) &= Ax(t) + Be(t) \\ u_2(t) &= Cx(t) + De(t) \; . \end{aligned} \tag{20.7}$$

Bumpless retuning at time t_s requires that the state of Equation 20.7 be such that

$$u_2(t_s) \approx u_1(t_s) \tag{20.8}$$

and

$$\left.\frac{d^k u_2(t)}{dt^k}\right|_{t=t_s} \approx \left.\frac{d^k u_1(t)}{dt^k}\right|_{t=t_s} \qquad k = 1, \ldots, n \tag{20.9}$$

for n as large as possible. This ensures that the retuning not only avoids discontinuous jumps, but also retains smooth derivatives. Substituting Equation 20.7 into Equation 20.8 yields, at time $t = t_s$,

$$\begin{aligned} u_1 &= Cx + De \\ \frac{du_1}{dt} &= CAx + CBe + D\frac{de}{dt} \\ \frac{d^2u_1}{dt^2} &= CA^2x + CABe + CB\frac{de}{dt} + D\frac{d^2e}{dt^2} \\ &\vdots \\ \frac{d^{n-1}u_1}{dt^{n-1}} &= CA^{n-1}x + CA^{n-2}Be + \ldots \\ &\quad + CB\frac{d^{n-2}e}{dt^{n-2}} + D\frac{d^{n-1}e}{dt^{n-1}} \; . \end{aligned}$$

Hence, assuming that the system in Equation 20.7 is observable and x has dimension n, the observatibility matrix

$$\mathcal{O} \triangleq \begin{pmatrix} C \\ CA \\ CA^2 \\ \vdots \\ CA^{n-1} \end{pmatrix}$$

is nonsingular and

$$\begin{aligned} x = \mathcal{O}^{-1} \Bigg\{ & \begin{pmatrix} u_1 \\ \frac{du_1}{dt} \\ \vdots \\ \frac{d^{n-1}u_1}{dt^{n-1}} \end{pmatrix} - \begin{pmatrix} D \\ CB \\ CAB \\ \vdots \\ CA^{n-2}B \end{pmatrix} e \\ & - \begin{pmatrix} 0 \\ D \\ CB \\ \vdots \\ CA^{n-3}B \end{pmatrix} \frac{de}{dt} - \begin{pmatrix} 0 \\ 0 \\ D \\ CB \\ \vdots \\ CA^{n-4}B \end{pmatrix} \frac{d^2e}{dt^2} - \ldots \\ & - \begin{pmatrix} 0 \\ 0 \\ \vdots \\ 0 \\ D \end{pmatrix} \frac{d^{n-1}e}{dt^{n-1}} \Bigg\} \end{aligned} \tag{20.10}$$

uniquely determines the state, x, that will match the $(n-1)$ first derivatives of the output of the retuned system to the corresponding derivatives of the output from the original system. Of course, the state cannot be computed directly from Equation 20.10, as this would require $(n-1)$ derivatives of u_1 and e, which could be noisy. A standard technique, however, is to approximate the required derivatives with band-pass filters as

$$\frac{d^k e}{dt^k} \approx \mathcal{L}^{-1}\left\{\frac{s^k}{e_m s^m + \ldots + e_1 s + 1} E(s)\right\}$$

where $E(s)$ denotes the Laplace transform of the control error, $\mathcal{L}^{-1}$ is the inverse Laplace transform, and $[e_m s^m + \ldots + e_1 s + 1]$, $m \geq n-1$, is an observer polynomial with roots and degree selected to suit the present noise level. Filters and controllers with on-line tuning interface can easily be augmented with Equation 20.10 to recompute a new state whenever the parameters are changed by a user.

Clearly, as the order n of Equation 20.7 increases, Equation 20.10 becomes increasingly noise sensitive. Therefore, it is not the approach we would most recommend, although its simplicity bears a certain attraction for low-order applications. Our primary reason for including it here is because it captures the essence of bumpless transfer as being the desire to compute the state of a dynamical system so its output will match another signal's value and derivatives. Indeed, Equation 20.10 can be interpreted as a simple observer that reconstructs the state by approximate differentiation of the output. As we will see in Section 4, this idea can be extended by considering more sophisticated observers with improved noise rejection properties.

Consider the setup of Example 20.6. Retuning the filter constant and adjusting the state according to Equation 20.10 yields the smooth performance shown in Figure 20.16.

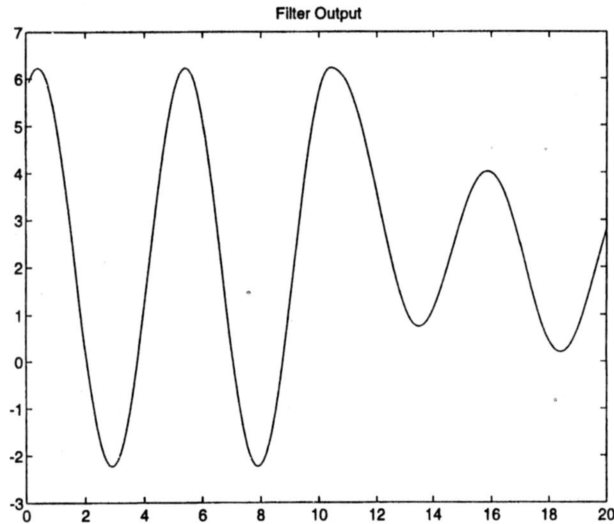

Figure 20.16 Smooth retuning of the filter constants by adjustment of the state according to Equation 20.10.

Scheduled and Adaptive Controllers

Scheduled controllers are controllers with time-varying parameters. These time variations are usually due to measured time-varying process parameters (such as a time delay varying with production speed) or due to local linearization in different operating ranges. If the time variations are occasional and the controller remains observable for all parameter settings, the principle of the previous section could be applied.

If, however, the controller order becomes large, the noise sensitivity of Equation 20.10 can become prohibitive. Furthermore, due to the inherent bandwidth limitations of evaluating the filtered derivatives, Equation 20.10 is not suitable for bumpless transfer if the parameters change significantly at every sampling interval, such as in certainty equivalence adaptive controllers. In that case, the techniques of Sections 3 and 4 are preferable.

Tentative Evaluation of New Controllers

This is a challenging, and only recently highlighted (see [4], [5]), bumpless transfer scenario. It is motivated by the need to test tentative controller designs safely and economically on critical processes during normal operation.

Consider, for example, a process operating in closed loop with an existing controller. Assume that the performance is mediocre, and that a number of new controller candidates have been designed and simulated. It is then desired to test these controllers, tentatively, on the plant to assess their respective performances. Frequently it is not possible or feasible to shut down the plant intermittently, and the alternative controllers therefore have to be brought on-line with a bumpless transfer mechanism during normal plant operation.

This is not a hypothetical situation; see [6] for a full-scale industrial example. Indeed, this scenario has considerable contemporary relevance, since economic pressures and ecological awareness require numerous existing plants to be retrofitted with high-performance advanced controllers during normal operation.

There are four primary factors that make the tentative evaluation phase particularly challenging for bumpless transfer: safety, economic feasibility, robustness and generality.

Safety. The actual performance of the new controllers is not reliably known, even if they have been simulated. In a worst case, one of them might drive the process unstable. It is then of overriding concern that bumpless transfer back to the original, stabilizing, controller is still possible. Due to this safety requirement, the technique should not rely on steady-state or constant signals, but be dynamic.

Economic feasibility. Since the achievable improvement due to the new controllers is not accurately known in advance, there tends to be a reluctance for costly modifications in hardware and/or software during the evaluation phase. Therefore, the bumpless transfer algorithm should be external to the existing controller and require only the commonly available signals of process input, output and reference. In particular, it should not require manipulating the states of the existing controller, as they may be analog. The technique should not only provide smooth transfer to the new controller, but also provide the transfer back to the existing controller. Thus, it should be bidirectional.

Robustness. Since the existing controller could very well be analog, it might be only approximately known. Even digital controllers are commonly implemented in programmable logic controllers (PLC) with randomly varying sampling rates that change the effective controller gains. Hence, the technique should be insensitive to inaccurately known controllers.

Generality. To be applicable as widely as possible, the bumpless transfer technique should not require the existing controller to have a particular order or structure, such as the so-called velocity form or such as constant feedback from a dynamical observer.

These objectives can be achieved by considering the tentative (also called the idle or latent) controller itself as a dynamic system and forcing its output to track the **active controller** by means of a tracking loop [4], [5]. This recasts bumpless transfer into an associated tracking problem to which systematic analysis and design theory may be applied.

20.2.3 Robust Bidirectional Transfer

In this section, we describe a general framework in which the problem of bumpless transfer is recast into a tracking problem. The solution is specifically aimed at the scenario described above. Beyond providing a practical solution to such cases, the framework also provides useful insights when analyzing other techniques described later in Section 20.2.4.

Consider Figure 20.17, where G denotes the transfer function of a single-input single-output (SISO) plant currently controlled by the active controller C_A. The bold lines in Figure 20.17 show the active closed loop

$$y = \frac{C_A G}{1 + C_A G} r \ . \qquad (20.11)$$

The regular lines make up an additional feedback configuration governed by

$$u_L = \frac{F_L T_L C_L}{1 + T_L C_L Q_L} u_A + \frac{C_L}{1 + T_L C_L Q_L}(r - y) \qquad (20.12)$$

which describes the **two-degree-of-freedom** tracking loop of Figure 20.18. Within this configuration, the **latent controller,** C_L, takes the role of a dynamical system whose output, u_L, is forced to track the active control signal, u_A, which is the reference signal to the tracking loop. Tracking is achieved by means of the tracking controller triplet (F_L, T_L, Q_L). Frequently, a **one-degree-of-freedom** tracking controller, in which $F_L = Q_L = 1$, is sufficient; we include the general case mainly for compatibility with other techniques, which we will mention in Section 20.2.4. Note that the plant control error, $r - y$, acts as an input disturbance in the tracking loop. Its effect can be eliminated by an appropriate choice of F_L, or it can be attenuated by designing (F_L, T_L, Q_L) for good input disturbance rejection.

While u_L is tracking u_A, the plant input can be switched bumplessly from the active controller to the latent controller (for graphical clarity, this switch is not shown in Figure 20.17). Simultaneously, the effect of the tracking loop is removed by opening the switch S_1 in Figure 20.17. Thus, C_L becomes the now-active controller regulating the plant, while the tracking loop (F_L, T_L, Q_L) is disconnected and never affects the plant control loop. Clearly, a second tracking loop (also not included in Figure 20.17 for clarity) can then be switched in to ensure that the previously active controller now becomes a latent controller in tracking mode.

The control problem associated with bumpless transfer, then, is the design of the triplet (F_L, T_L, Q_L) to guarantee a certain tracking bandwidth in spite of noise and controller uncertainty in C_L, which is the "plant" of the tracking loop.

Note that this strategy achieves the objectives set out in Section 2.4. Firstly, assume that a newly designed controller is temporarily activated for performance analysis (becoming C_A) and the existing controller is placed into tracking mode (becoming C_L).

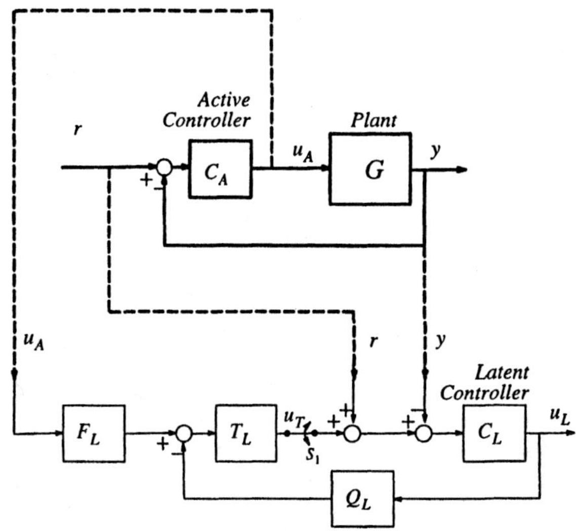

Figure 20.17 The unidirectional bumpless transfer diagram, in which the plant, G, is controlled by the active controller, C_A. The output, u_L, of the latent controller, C_L, is forced to track the active control signal, u_A, by means of the tracking controller, (F_L, T_L, Q_L). Any time u_L is tracking u_A, the plant input can be switched from the active controller to the latent controller and bumpless transfer is achieved. Simultaneously, the effect of the tracking loop (F_L, T_L, Q_L) is removed from the plant loop by opening switch S_1. Complementing the diagram with a second tracking circuit allows bidirectional transfer.

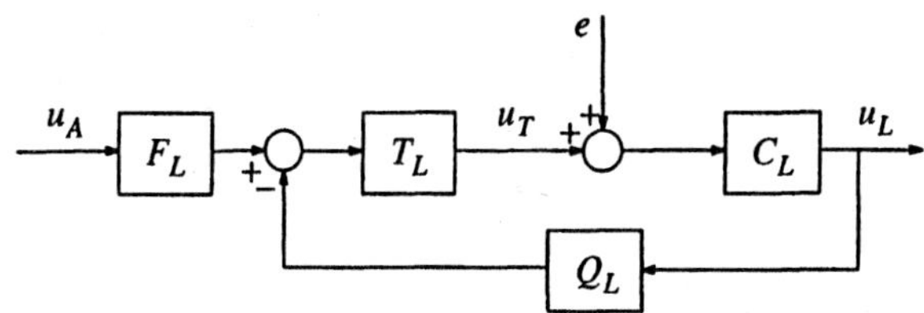

Figure 20.18 Tracking loop with latent control signal u_L tracking the active control signal u_A. The plant control error, $e = r - y$, acts as input disturbance to the latent controller C_L, here regarded as a "system" to be controlled.

Then, if the new controller inadvertently drives the plant unstable, the existing controller still retains stable tracking of the (unbounded) active control, u_L. Therefore, control can be bumplessly transferred back to the existing controller for immediate stabilization of the plant; see [4] for an example. This guarantees safety during the testing of newly designed controllers.

Secondly, the only signals accessed by the tracking loop are the plant reference, the plant output and the active control signals, all of which are commonly available. Furthermore, by adding a second tracking loop to the diagram of Figure 20.17, the scheme becomes bidirectional and does not require the existing controller to feature bumpless transfer facilities. Thirdly, a distinguishing feature compared to alternative techniques is that the tracking controllers are operating in closed loop for $Q_L \neq 0$. Thus, they can be designed to be insensitive to inaccurately known controllers. Fourthly and finally, the technique does not presuppose the plant controllers to have any particular structure. As long as one can conceive of a tracking controller for them, there is no requirement for them to be biproper, minimum phase, digital or linear.

Once bumpless transfer has been associated with a control problem in this way, tracking controllers can be created by considering the plant controllers as being, themselves, "plants" and designing regulators for them by any desired technique. For a discussion of tracking controller design and the associated robustness issues, see [5].

20.2.4 Further Bumpless Transfer Techniques

In this section we outline a number of further bumpless transfer techniques commonly referred to in the literature. Most of these techniques are presented in the context of anti-windup design. To be consistent with the literature we adopt this context here.

If a process exhibits actuator saturations that were neglected in a linear control design, the closed loop may suffer from unacceptable transients after saturation. This is due to the controller states becoming inconsistent with the saturated control signal and is known as *windup.*

Techniques designed to combat windup are known as *anti-windup.* Their aim is to ensure that the control signal never attempts to take a value beyond the saturation limits. In the sense that this requires the control output to track a signal (i.e., the saturation curve), the problem of anti-windup is structurally equivalent to the problem of bumpless transfer.

The perhaps most well-known techniques for anti-windup and bumpless transfer design are the **conditioning technique** of Hanus [7] and Hanus et al. [8] and the observer-based technique by Åström and Wittenmark [1]. The fundamental idea of the conditioning technique is to manipulate the reference signal such that the control signal under consideration is in agreement with the desired control signal, that is, the saturated signal in an anti-windup context and the active control signal in the bumpless transfer context.

The observer-based technique is built on the idea of feeding an observer with the desired (active) control signal and thereby obtaining controller states that are matched to the states of the active controller. Both the above-described techniques have recently been found to be special cases of a more general anti-windup and bumpless transfer structure presented by Rönnbäck [12] and Rönnbäck et al. [11]. We present the conditioning and observer-based techniques in this context next.

Consider the system

$$y = \frac{B}{A}v \tag{20.13}$$

and the controller

$$Fu = (F - PR)v + PTr - PSy \tag{20.14}$$

$$v = \text{sat}(u) \quad \triangleq \quad \begin{cases} u_{\min} & u \leq u_{\min} \\ u & u_{\min} \leq u \leq u_{\max} \\ u_{\max} & u \geq u_{\max}, \end{cases} \tag{20.15}$$

where v is a signal caused by actuator saturations, r is the reference signal, and P and F constitute additional design freedom to combating windup. See Figure 20.19.

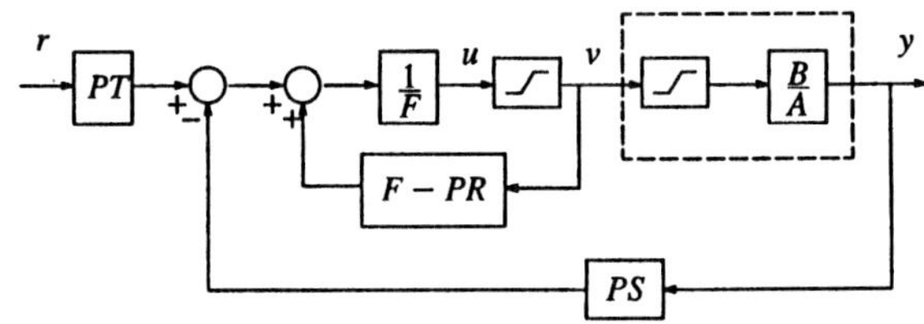

Figure 20.19 General anti-windup and bumpless transfer structure.

When $u(t)$ does not saturate, we obtain the nominal two-degree-of-freedom controller

$$Ru = Tr - Sy \tag{20.16}$$

whereas when $u(t)$ does saturate, feeding back from $v(t)$ prevents the controller states from winding up.

The observer-based technique of Åström and Wittenmark [1] is directly obtained by selecting the polynomials P and F above as $P = 1$ and $F = A_o$, where A_o is the characteristic polynomial of the observer. For a particular choice of observer polynomial, namely $A_o = T/t_0$, where t_0 represents the **high-frequency gain** of T/R in the nominal controller in Equation 20.16, we obtain the conditioning technique.

In the context of anti-windup design, bumpless transfer between an active control signal, u_A, and a latent control signal, u_L, is achieved by setting $u = u_L$, and $u_{\min} = u_{\max} = u_A$. Choosing bumpless transfer technique is thus a matter of choosing polynomials P and F in Equation 20.14. For details about the relations between anti-windup design and bumpless transfer tracking controller design, the reader is referred to Graebe and Ahlén [5]. More details about anti-windup design can be found in, e.g., [12], [11] as well as in another chapter of this book.

Another technique that is popularly used for both anti-windup and bumpless transfer is depicted in the diagram of Figure 20.20, where G is a plant being controlled by the active controller, C_A, and C_L is a latent alternative controller. If

$$K = K_0 \gg 1 \tag{20.17}$$

is a large constant [or a diagonal matrix in the multiple-input multiple-output (MIMO) case], then this technique is also known as *high gain conventional anti-windup and bumpless transfer.* In a slight variation, Uram [13] used the same configuration but proposed K to be designed as the high-gain integrator

$$K = \frac{K_0}{s}, \qquad K_0 \gg 1 \ . \tag{20.18}$$

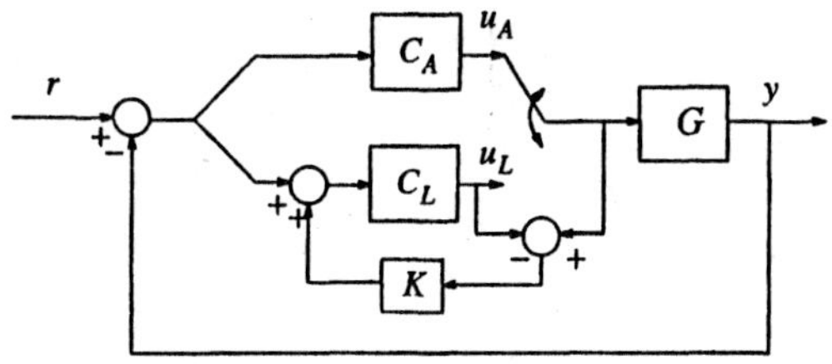

Figure 20.20 Conventional high-gain anti-windup and bumpless transfer strategy, with $K = K_0 \gg 1$ or $K = \frac{K_0}{s}$ with $K_0 \gg 1$.

Clearly, the configuration of Figure 20.20 is a special case of the general tracking controller approach of Figure 20.17 with the particular choices $F_L = 1$, $Q_L = 1$ and $T_L = K$, where K is given by either Equation 20.17 or Equation 20.18. One of the advantages of viewing bumpless transfer as a tracking problem is that we can immediately assess some of the implications of the choices in Equation 20.17 or Equation 20.18. The performance of these two schemes is determined by how well the latent controller C_L, viewed as a system, lends itself to wide bandwidth control by a simple proportional or integral controller such as in Equation 20.17 or Equation 20.18.

Campo et al. [2] and Kotare et al. [9] present a general framework that encompasses most of the known bumpless transfer and anti-windup schemes as special cases. This framework lends itself well for the analysis and comparison of given schemes. It is not as obvious, however, how to exploit the framework for synthesis. The fact that different design choices can indeed have a fairly strong impact on the achieved performance is nicely captured by Rönnbäck et al. [11] and Rönnbäck [12]. These authors focus on the problem of controller wind-up in the presence of actuator saturations. They gain interesting design insights by interpreting the difference between the controller output and the saturated plant input as a fictitious input disturbance. As discussed by Graebe and Ahlén [5], the proposal of Rönnbäck [12] is structurally equivalent to the tracking controller configuration of Section 20.2.3. It is rarely pointed out, however, that the design considerations for bumpless transfer and anti-windup can be rather different.

20.2.5 Summary

This section has discussed the problem of bumpless transfer, which is concerned with smooth switching between alternative dynamical systems. We have highlighted a number of situations in which this problem arises, including switching from manual to automatic control, on-line retuning of filter or controller parameters and tentative evaluation of new controllers during normal plant operation.

If it is possible to manipulate the states of the controller directly, there are several techniques to compute the value of the state vector that will give the desired output. If it is not possible to manipulate the states directly, such as in an analog controller, then the input to the controller can be used instead. Viewed in this way, bumpless transfer becomes a tracking problem, in which the inputs to the controller are manipulated so that its output will track an alternative control signal. Once bumpless transfer is recognized as a tracking problem, systematic design techniques can be applied to design appropriate tracking controllers. We have outlined several advantages with this approach and showed that some other known techniques can be interpreted within this setting.

20.2.6 Defining Terms

Active controller: A regulator controlling a plant at any given time. This term is used to distinguish the active controller from an alternative standby controller in a bumpless transfer context.

Conditioning technique: A technique in which the reference signal is manipulated in order to achieve additional control objectives. Typically, the reference signal is manipulated in order to avoid the control signal's taking a value larger than a known saturation limit.

High-frequency gain: The high-frequency for gain of a strictly proper transfer function is zero; of a biproper transfer function, it is the ratio of the leading coefficients of the numerator and denominator; and for an improper transfer function, it is infinity. Technically, the high-frequency gain of a transfer function H(s) is defined as $\lim_{s\to\infty} H(s)$.

Internal model controller (IMC): A controller parameterization in which the model becomes an explicit component of the controller. Specifically, $C = Q/(1 - Q\hat{G})$, where $\hat{G}$ is a model and Q is a stable and proper transfer function.

Latent controller: A standby controller that is not controlling the process, but that should be ready for a smooth takeover from an active controller in a bumpless transfer context.

Observer-based controller: A controller structure in which the control signal is generated from the states of an observer.

One-degree-of-freedom controller: A control structure in which the reference signal response is uniquely determined from the output disturbance response.

Two-degree-of-freedom controller: A control structure in which the reference signal response can be shaped, to a certain extent, independently of the disturbance response. This is usually achieved with a setpoint filter.

References

[1] Åström, K.J. and Wittenmark, B., *Computer Controlled Systems, Theory and Design*, Prentice Hall, Englewood Cliffs, NJ, 2nd ed., 1984.

[2] Campo, P.J., Morari, M., and Nett, C. N., Multivari-

able anti-windup and bumpless transfer: a general theory, *Proc. ACC '89*, 2, 1706–1711, 1989.

[3] Goodwin, G.C., Graebe, S. F., and Levine, W.S., Internal model control of linear systems with saturating actuators, *Proc. ECC '93*, Groningen, The Netherlands, 1993.

[4] Graebe, S. F. and Ahlén, A., Dynamic transfer among alternative controllers, *Prepr. 12th IFAC World Congr.*, Vol. 8, Sydney, Australia, 245–248, 1993.

[5] Graebe, S. F. and Ahlén, A., Dynamic transfer among alternative controllers and its relation to anti-windup controller design, *IEEE Trans. Control Syst. Technol.* To appear. Jan. 1996.

[6] Graebe, S. F., Goodwin, G.C., and Elsley, G., Rapid prototyping and implementation of control in continuous steel casting, Tech. Rep. EE9471, Dept. Electrical Eng., University of Newcastle, NSW 2308, Australia, 1994.

[7] Hanus, R., A new technique for preventing control windup, *Journal A*, 21, 15–20, 1980.

[8] Hanus, R., Kinnaert, M., and Henrotte, J-L., Conditioning technique, a general anti-windup and bumpless transfer method, *Automatica*, 23(6), 729–739, 1987.

[9] Kotare, M.V., Campo, P.J., Morari, M., and Nett, C.N., A unified framework for the study of anti-windup designs, CDS Tech. Rep. No. CIT/CDS 93-010, California Institute of Technology, Pasadena, 1993.

[10] Morari, M. and Zafiriou, E., *Robust Process Control*, Prentice Hall, Englewood Cliffs, NJ, 1989.

[11] Rönnbäck, S. R., Walgama, K. S., and Sternby, J., An extension to the generalized anti-windup compensator, in *Mathematics of the Analysis and Design of Process Control*, Borne, P., et al., Eds., Elsevier/North-Holland, Amsterdam, 1992.

[12] Rönnbäck, S., Linear control of systems with actuator constraints, Ph.D. thesis, Luleå University of Technology, Sweden, 1190, 1993.

[13] Uram, R., Bumpless transfer under digital control, *Control Eng.*, 18(3), 59–60, 1971.

20.3 Linearization and Gain-Scheduling

Jeff S. Shamma, Center for Control and Systems Research, Department of Aerospace Engineering and Engineering Mechanics, The University o' Texas at Austin, Austin, TX

20.3.1 Introduction

A classical dilemma in modeling physical systems is the trade-off between model accuracy and tractability. While sophisticated models might provide accurate descriptions of system behavior, the resulting analysis can be considerably more complicated. Simpler models, on the other hand, may be more amenable for analysis and derivation of insight, but might neglect important system behaviors. Indeed, the required fidelity of a model depends on the intended utility. For example, one may use a very simplified model for the sake of control design, but then use a sophisticated model to simulate the overall control system.

One instance where this dilemma manifests itself is the use of linear versus nonlinear models. Nonlinearities abound in most physical systems. Simple examples include saturations, rate limiters, deadzones, and backlash. Further examples include the inherently nonlinear behavior of systems such as robotic manipulators, aircraft, and chemical process plants. However, methods for analysis and control design are considerably more available for linear systems than nonlinear systems.

One approach is to directly address the nonlinear behavior of such systems, and nonlinear control design remains an topic of active research. An alternative is to linearize the system dynamics, i.e., to approximate the nonlinear model by a linear one. Some immediate drawbacks are that (1) the linear model can give only a local description of the system behavior and (2) some of the intricacies of the system behavior may be completely neglected by the linear approximation—even locally. In some cases, these consequences are tolerable, and one may then employ methods for linear systems.

One approach to address the local restriction of linearization-based controllers is to perform several linearization-based control designs at many operating conditions and then interpolate the local designs to yield an overall nonlinear controller. This procedure is known as *gain-scheduling*. It is an intuitively appealing but heuristic practice, which is used in a wide variety of control applications. It is especially prevalent in flight control systems.

This chapter presents a review of linearization and gain scheduling, exploring both the benefits and practical limitations of each.

20.3.2 Linearization

An Example

To illustrate the method of linearization, consider a single link coupled to a rotational inertia by a flexible shaft (Figure 20.21). The idea is to control the link through a torque on the rotational inertia. This physical system may be viewed as a very simplified model of a robotic manipulator with flexible joints.

The equations of motion are given by

$$\frac{d}{dt}\begin{pmatrix}\theta_1(t)\\ \theta_2(t)\\ \dot\theta_1(t)\\ \dot\theta_2(t)\end{pmatrix} = \begin{pmatrix}0\\ 0\\ \left(\left(mgL\sin(\theta_1(t))\right)\right)/J_1 - c\dot\theta_1(t)\left|\dot\theta_1(t)\right|\\ 0\end{pmatrix} + \begin{pmatrix}0 & 0 & 1 & 0\\ 0 & 0 & 0 & 1\\ -k/J_1 & k/J_1 & 0 & 0\\ k/J_2 & -k/J_2 & 0 & 0\end{pmatrix}\begin{pmatrix}\theta_1(t)\\ \theta_2(t)\\ \dot\theta_1(t)\\ \dot\theta_2(t)\end{pmatrix} + \begin{pmatrix}0\\ 0\\ 0\\ 1/J_2\end{pmatrix}T(t) \tag{20.19}$$

Here $\theta_1(t), \theta_2(t)$ are angles measured from vertical, $T(t)$ is the

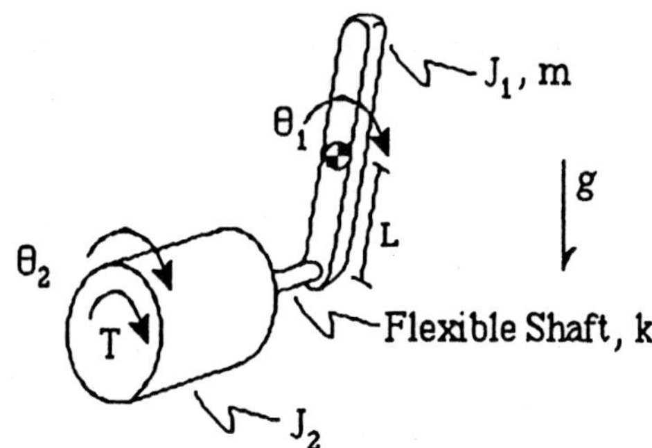

Figure 20.21 Rotational link.

torque input, k is a rotational spring constant, c is a nonlinear damping coefficient, J_1, J_2 are rotational inertias, L is the link length, and m is the link mass.

Now suppose the link is to be controlled in the vicinity of the upright stationary position. First-order approximations near this position lead to $\sin\theta_1 \simeq \theta_1$ and $c\dot{\theta}_1 \left|\dot{\theta}_1\right| \simeq 0$. The state equation (20.19) is then approximated by the equations,

$$\frac{d}{dt}\begin{pmatrix}\theta_1(t)\\ \theta_2(t)\\ \dot{\theta}_1(t)\\ \dot{\theta}_2(t)\end{pmatrix} \simeq \begin{pmatrix}0 & 0 & 1 & 0\\ 0 & 0 & 0 & 1\\ (mgL-k)/J_1 & k/J_1 & 0 & 0\\ k/J_2 & -k/J_2 & 0 & 0\end{pmatrix}\begin{pmatrix}\theta_1(t)\\ \theta_2(t)\\ \dot{\theta}_1(t)\\ \dot{\theta}_2(t)\end{pmatrix} + \begin{pmatrix}0\\ 0\\ 0\\ 1/J_2\end{pmatrix} T(t) \qquad (20.20)$$

The simplified dynamics are now in the general *linear* form

$$\dot{x}(t) = Ax(t) + Bu(t) \qquad (20.21)$$

Two consequences of the linearization are

- Global behavior, such as full angular rotations, are poorly approximated.
- The nonlinear damping, $c\dot{\theta}_1 \left|\dot{\theta}_1\right|$, is completely neglected, even locally.

Despite these limitations, an analysis or control design based on the linearization can still be of value for the nonlinear system, provided that the state vector and control inputs are close to the upright equilibrium.

Linearization of Functions

This section reviews some basic concepts from multivariable calculus. For a vector $x \in \mathcal{R}^n$, $|x|$ denotes the Euclidean norm, i.e.,

$$|x| = \left(\sum_{i=1}^{n} x_i^2\right)^{1/2} \qquad (20.22)$$

Let $f : \mathcal{R}^n \to \mathcal{R}^p$, i.e., f is a function which maps vectors in $\mathcal{R}^n$ to values in $\mathcal{R}^p$. In terms of individual components,

$$f(x) = \begin{pmatrix} f_1(x_1,\ldots,x_n)\\ \vdots \\ f_p(x_1,\ldots,x_n)\end{pmatrix} \qquad (20.23)$$

where the x_i are scalar components of the $\mathcal{R}^n$-vector x, and the f_i are scalar valued functions of $\mathcal{R}^n$.

The *Jacobian matrix* of f is denoted Df and is defined as the $p \times n$ matrix of partial derivatives

$$Df = \begin{pmatrix} \frac{\partial f_1}{\partial x_1} & \cdots & \frac{\partial f_1}{\partial x_n}\\ \vdots & \ddots & \vdots\\ \frac{\partial f_p}{\partial x_1} & \cdots & \frac{\partial f_p}{\partial x_n}\end{pmatrix} \qquad (20.24)$$

In case f is continuously differentiable at x_o, then the Jacobian matrix can be used to approximate f. A multivariable Taylor series expansion takes the form

$$f(x) = f(x_o) + Df(x_o)(x - x_o) + r(x) \qquad (20.25)$$

where the remainder, $r(x)$, represents higher-order terms which satisfy

$$\lim_{h\to 0} \frac{|r(x_o + h)|}{|h|} = 0 \qquad (20.26)$$

Now let $f : \mathcal{R}^n \times \mathcal{R}^m \to \mathcal{R}^p$, i.e., f is a function which maps a pair of vectors in $\mathcal{R}^n$ and $\mathcal{R}^m$, respectively, to values in $\mathcal{R}^p$. The notations $D_1 f$ and $D_2 f$ denote the Jacobian matrices with respect to the first variable and second variables, respectively. Thus, if

$$f(x,u) = \begin{pmatrix} f_1(x_1,\ldots,x_n,u_1,\ldots,u_m)\\ \vdots \\ f_p(x_1,\ldots,x_n,u_1,\ldots,u_m)\end{pmatrix} \qquad (20.27)$$

then $D_1 f$ denotes the $p \times n$ matrix

$$D_1 f = \begin{pmatrix} \frac{\partial f_1}{\partial x_1} & \cdots & \frac{\partial f_1}{\partial x_n}\\ \vdots & \ddots & \vdots\\ \frac{\partial f_p}{\partial x_1} & \cdots & \frac{\partial f_p}{\partial x_n}\end{pmatrix} \qquad (20.28)$$

and $D_2 f$ denotes the $p \times m$ matrix

$$D_2 f = \begin{pmatrix} \frac{\partial f_1}{\partial u_1} & \cdots & \frac{\partial f_1}{\partial u_m}\\ \vdots & \ddots & \vdots\\ \frac{\partial f_p}{\partial u_1} & \cdots & \frac{\partial f_p}{\partial u_m}\end{pmatrix} \qquad (20.29)$$

EXAMPLE 20.7: Rotational Link Jacobian Matrices

Consider again the rotational link example in Equation 20.19. Let x denote the state-vector, u denote the torque input, and $f(x,u)$ denote the right-hand side of Equation 20.19. The Jacobian matrices $D_1 f$ and $D_2 f$ are given by

$$D_1 f(x,u) = \begin{pmatrix} 0 & 0 & 1 & 0\\ 0 & 0 & 0 & 1\\ (mgL\cos\theta_1 - k)/J_1 & k/J_1 & 2c\left|\dot{\theta}_1\right| & 0\\ k/J_2 & -k/J_2 & 0 & 0\end{pmatrix},$$

$$D_2 f(x,u) = \begin{pmatrix}0\\ 0\\ 0\\ 1/J_2\end{pmatrix} \qquad (20.30)$$

Linearization About an Equilibrium

Approximation of System Dynamics A general form for nonlinear differential equations is

$$\dot{x}(t) = f(x(t), u(t)) \tag{20.31}$$

where $x(t)$ is the state vector, $u(t)$ is the input vector, and $f : \mathcal{R}^n \times \mathcal{R}^m \rightarrow \mathcal{R}^n$. The existence and uniqueness of solutions will be assumed.

The pair (x_o, u_o) is called an *equilibrium* if

$$0 = f(x_o, u_o) \tag{20.32}$$

The reasoning behind this terminology is that, starting from the initial condition $x(0) = x_o$ with a constant input $u(t) = u_o$, the solution *remains* at $x(t) = x_o$.

Assuming that f is continuously differentiable at (x_o, u_o), a multivariable Taylor series expansion yields

$$\begin{aligned} \dot{x}(t) &= f(x_o, u_o) + D_1 f(x_o, u_o)(x(t) - x_o) \\ &\quad + D_2 f(x_o, u_o)(u(t) - u_o) + r(x(t), u(t)) \end{aligned} \tag{20.33}$$

where the remainder, $r(x, u)$, satisfies

$$\lim_{(x,u)\rightarrow(x_o,u_o)} \frac{r(x,u)}{\sqrt{|x - x_o|^2 + |u - u_o|^2}} = 0 \tag{20.34}$$

Thus, the approximation is accurate up to first order. Define the deviation-from-equilibrium terms

$$\tilde{x}(t) = x(t) - x_o \tag{20.35}$$

$$\tilde{u}(t) = u(t) - u_o \tag{20.36}$$

Assuming that the equilibrium is *fixed*, i.e., $\frac{d}{dt}x_o = 0$, along with the condition $0 = f(x_o, u_o)$, leads to

$$\dot{\tilde{x}}(t) \simeq D_1 f(x_o, u_o)\tilde{x}(t) + D_2 f(x_o, u_o)\tilde{u}(t) \tag{20.37}$$

Equation 20.37 represents the *linearization* of the nonlinear dynamics (Equation 20.31) about the equilibrium point (x_o, u_o).

EXAMPLE 20.8: Rotational Link Linearization

Consider again the rotational link example of Equation 20.19. In addition to the upright equilibrium, there is a *family* of equilibrium conditions given by

$$x_o = \begin{pmatrix} q \\ (q - mgL\sin q)/k \\ 0 \\ 0 \end{pmatrix} \qquad u_o = -mgL\sin q \tag{20.38}$$

where q denotes the equilibrium angle for θ_1. When $q = 0$, Equation 20.38 yields the upright equilibrium.

For a fixed q, the linearized dynamics about the corresponding equilibrium point may be obtained by substituting the Jacobian matrices from Example 20.7 into Equation 20.37 to yield

$$\begin{aligned} \dot{\tilde{x}}(t) &= \begin{pmatrix} 0 & 0 & 1 & 0 \\ 0 & 0 & 0 & 1 \\ (mgL\cos q - k)/J_1 & k/J_1 & 0 & 0 \\ k/J_2 & -k/J_2 & 0 & 0 \end{pmatrix} \tilde{x}(t) \\ &\quad + \begin{pmatrix} 0 \\ 0 \\ 0 \\ 1/J_2 \end{pmatrix} \tilde{u}(t) \end{aligned} \tag{20.39}$$

As before, the nonlinear damping is completely neglected in the linearized equations.

Stability This section discusses how linearization of a nonlinear system may be used to analyze stability. First, some definitions regarding stability are reviewed.

Let x_o be an equilibrium for the unforced state equations:

$$\dot{x}(t) = f(x(t)) \tag{20.40}$$

i.e., $f(x_o) = 0$.

The equilibrium x_o is *stable* if for each $\varepsilon > 0$, there exists a $\delta(\varepsilon) > 0$ such that

$$|x(0) - x_o| < \delta(\varepsilon) \quad \Rightarrow \quad |x(t) - x_o| < \varepsilon, \quad \forall t \geq 0 \tag{20.41}$$

It is *asymptotically stable* if it is stable and for some δ^*

$$|x(0) - x_o| < \delta^* \quad \Rightarrow \quad |x(t) - x_o| \rightarrow 0, \text{ as } t \rightarrow \infty \tag{20.42}$$

It is *unstable* if it is not stable. Note that the conditions for stability pertain to a neighborhood of an equilibrium.

The following is a standard analysis result. Let $f : \mathcal{R}^n \rightarrow \mathcal{R}^n$ be continuously differentiable. The equilibrium x_o is asymptotically stable if all of the eigenvalues of $Df(x_o)$ have negative real parts. It is unstable if $Df(x_o)$ has an eigenvalue with a positive real part.

Since the eigenvalues of $Df(x_o)$ determine the stability of the linearized system

$$\dot{\tilde{x}}(t) = Df(x_o)\tilde{x}(t) \tag{20.43}$$

this result states that the linearization can provide *sufficient conditions* for stability of the nonlinear system in a neighborhood of an equilibrium. In case $Df(x_o)$ has purely imaginary eigenvalues, then nonlinear methods are *required* to assess stability of the nonlinear system.

EXAMPLE 20.9: Rotational Link Stability

Consider the *unforced* rotational link equations, i.e., torque $T(t) = 0$. In this case, the two equilibrium conditions are the upright position, $x_o = 0$, or the hanging position, $x_o = (\pi \;\; 0 \;\; 0 \;\; 0)^T$. Intuitively, the upright equilibrium is unstable, while the hanging equilibrium is stable.

Let $J_1, J_2, k, m, g, L, c = 1$. For the upright equilibrium, the linearized equations are

$$\dot{\tilde{x}}(t) = \begin{pmatrix} 0 & 0 & 1 & 0 \\ 0 & 0 & 0 & 1 \\ 0 & 1 & 0 & 0 \\ 1 & -1 & 0 & 0 \end{pmatrix} \tilde{x}(t) \tag{20.44}$$

which has eigenvalues of $(\pm 0.79,\ \pm 1.27j)$. Since one of the eigenvalues has a positive real part, the upright equilibrium of the original nonlinear system is unstable.

For the hanging equilibrium, the linearized equations are

$$\dot{\tilde{x}}(t) = \begin{pmatrix} 0 & 0 & 1 & 0 \\ 0 & 0 & 0 & 1 \\ -2 & 1 & 0 & 0 \\ 1 & -1 & 0 & 0 \end{pmatrix} \tilde{x}(t) \tag{20.45}$$

Note that $\tilde{x}(t)$ represents *different* quantities in the two linearizations, namely the deviation from the different corresponding equilibrium positions. The linearization now has eigenvalues of $(\pm 1.61j,\ \pm 0.61j)$. In this case, the linearization *does not* provide information regarding the stability of the nonlinear system. This is because the nonlinear damping term $c\dot{\theta}_1 \left|\dot{\theta}_1\right|$ is completely neglected. If this term were replaced by linear damping in (Equation 20.19), e.g., $c\dot{\theta}_1$, then the linearized dynamics become

$$\dot{\tilde{x}}(t) = \begin{pmatrix} 0 & 0 & 1 & 0 \\ 0 & 0 & 0 & 1 \\ -2 & 1 & -1 & 0 \\ 1 & -1 & 0 & 0 \end{pmatrix} \tilde{x}(t) \tag{20.46}$$

and the resulting eigenvalues are $(-0.35 \pm 1.5j,\ -0.15 \pm 0.63j)$. In this case, the linearized dynamics are asymptotically stable, which in turn implies that the nonlinear dynamics are asymptotically stable. In the case of the nonlinear damping, one may use alternate methods to show that the hanging equilibrium is indeed asymptotically stable. However, this could not be concluded from the linearization.

Stabilization Linearization of a nonlinear system also may be used to design stabilizing controllers. Let (x_o, u_o) be an equilibrium for the nonlinear equations (20.31). Let

$$y(t) = g(x(t)) \tag{20.47}$$

denote the outputs available for measurement. In case f and g are continuously differentiable, the linearized equations are

$$\dot{\tilde{x}}(t) = D_1(x_o, u_o)\tilde{x}(t) + D_2(x_o, u_o)\tilde{u}(t) \tag{20.48}$$

$$\tilde{y}(t) = Dg(x_o)\tilde{x}(t) \tag{20.49}$$

where $\tilde{x}(t) = x(t) - x_o$, $\tilde{u}(t) = u(t) - u_o$, and $\tilde{y}(t) = y(t) - g(x_o)$.

Using the results from the previous section, if the controller

$$\dot{z}(t) = Az(t) + B\tilde{y}(t) \tag{20.50}$$

$$\tilde{u}(t) = Cz(t) + D\tilde{y}(t) \tag{20.51}$$

asymptotically stabilizes the linearized system, then the control

$$u(t) = u_o + \tilde{u}(t) \tag{20.52}$$

asymptotically stabilizes the nonlinear system at the equilibrium (x_o, u_o).

Conversely, a linearized analysis under certain conditions can show that no controller with continuously differentiable dynamics is asymptotically stabilizing. More precisely, consider the controller

$$\dot{z}(t) = F(z(t), y(t)) \tag{20.53}$$

$$u(t) = G(z(t)) \tag{20.54}$$

where $(0, y_0)$ is an equilibrium and $u_o = G(0)$. Suppose $[D_1 f(x_o, u_o), D_2 f(x_o, u_o)]$ either is *not* a stabilizable pair or $[Dg(x_o), D_1 f(x_o, u_o)]$ is not a detectable pair. Then *no* continuously differentiable F and G lead to an asymptotically stablilizing controller.

EXAMPLE 20.10: Rotational Link Stabilization

Consider the rotational link equations with $J_1, J_2, k, m, g, L, c = 1$. The equilibrium position with the link at 45 degrees is given by

$$x_o = \begin{pmatrix} \pi/4 \\ \pi/4 - 1/\sqrt{2} \\ 0 \\ 0 \end{pmatrix}, \quad u_o = -1/\sqrt{2} \tag{20.55}$$

For this equilibrium, the linearized equations are

$$\dot{\tilde{x}}(t) = A\tilde{x}(t) + B\tilde{u}(t) \tag{20.56}$$

where

$$A = \begin{pmatrix} 0 & 0 & 1 & 0 \\ 0 & 0 & 0 & 1 \\ 1/\sqrt{2} - 1 & 1 & 0 & 0 \\ 1 & -1 & 0 & 0 \end{pmatrix} \tilde{x}, \quad B = \begin{pmatrix} 0 \\ 0 \\ 0 \\ 1 \end{pmatrix} \tilde{u} \tag{20.57}$$

One can show that the state feedback (which resembles proportional-derivative feedback)

$$\tilde{u}(t) = -K\tilde{x}(t) = -\begin{pmatrix} 2 & 4 & 4 & 2 \end{pmatrix} \tilde{x}(t) \tag{20.58}$$

is stabilizing. Therefore the control

$$u(t) = u_o - K(x(t) - x_o) \tag{20.59}$$

stabilizes the nonlinear system at the 45° equilibrium.

Now suppose only $\theta_1(t)$ is available for feedback, i.e.,

$$y(t) = Cx(t) = \begin{pmatrix} 1 & 0 & 0 & 0 \end{pmatrix} x(t) \tag{20.60}$$

Let $\tilde{y}(t) = y(t) - Cx_o$. A model-based controller which stabilizes the linearization is

$$\dot{z}(t) = Az(t) - BKz(t) + H(\tilde{y}(t) - Cz(t)) \tag{20.61}$$

$$\tilde{u}(t) = -Kz(t) \tag{20.62}$$

where

$$H = \begin{pmatrix} 1 \\ 2 \\ 2 \\ 1 \end{pmatrix} \tag{20.63}$$

Therefore, the control

$$\dot{z}(t) = Az(t) - BKz(t) + H(y(t) - y_o - Cz(t)) \tag{20.64}$$

$$u(t) = u_o - Kz(t) \tag{20.65}$$

stabilizes the nonlinear system.

Limitations of Linearization

This section presents several examples that illustrate some limitations in the utility of linearizations.

EXAMPLE 20.11: Hard Nonlinearities

Consider the system

$$\dot{x}(t) = Ax(t) + BN(u(t)) \tag{20.66}$$

This system represents linear dynamics where the input u first passes through a nonlinearity, N. Some common nonlinearities are saturation:

$$N(u) = \begin{cases} 1 & u \geq 1 \\ u & -1 \leq u \leq 1 \\ -1 & u \leq -1 \end{cases} \tag{20.67}$$

deadzone:

$$N(u) = \begin{cases} u - 1 & u \geq 1 \\ 0 & -1 \leq u \leq 1 \\ u + 1 & u \leq -1 \end{cases} \tag{20.68}$$

and relay:

$$N(u) = \begin{cases} 1 & u > 0 \\ -1 & u < 0 \end{cases} \tag{20.69}$$

Other nonlinearities are backlash and hysteresis. All of these nonlinearities do not lend themselves to linearization-based analysis. Even in the regions where the nonlinearities are differentiable, a linearization *completely removes* the intricacies that the nonlinearities cause.

EXAMPLE 20.12: Local Nature of Linearization

Consider the scalar equation

$$\dot{x}(t) = -\sin(x(t)) \tag{20.70}$$

The equilibrium $x_o = 0$ is asymptotically stable. However, the equilibrium $x_o = \pi$ is not.

EXAMPLE 20.13: Linearization not Asymptotically Stable

Consider the scalar equation

$$\dot{x}(t) = -x^3(t) \tag{20.71}$$

The equilibrium $x_o = 0$ is globally asymptotically stable, i.e., $x(t) \to 0$ as $t \to \infty$ for any initial condition. However, the linearization yields

$$\dot{\tilde{x}}(t) = 0 \tag{20.72}$$

which is inconclusive. A similar phenomenon was seen with the rotational link in the hanging equilibrium.

EXAMPLE 20.14: Linearization not Stabilizable

Consider the scalar equation

$$\dot{x}(t) = x(t) + x(t)u(t) \tag{20.73}$$

At the equilibrium $x_o, u_o = 0$, the linearization is

$$\dot{\tilde{x}}(t) = \tilde{x}(t) \tag{20.74}$$

This is not stabilizable, since no input term appears. However, the constant feedback $u(t) = -2$ in (20.73) yields

$$\dot{x}(t) = -x(t) \tag{20.75}$$

which is asymptotically stable.

EXAMPLE 20.15: Non-differentiable Feedback

Consider the second-order nonlinear system

$$\dot{x}_1(t) = u(t) \tag{20.76}$$

$$\dot{x}_2(t) = x_2(t) - x_1^3(t) \tag{20.77}$$

For the equilibrium $x_o, u_o = 0$, any stabilizing feedback law, $u = g(x)$, must satisfy $g(0) = 0$. Suppose that g is continuously differentiable. The linearization of the closed-loop system yields

$$\dot{\tilde{x}}(t) = \begin{pmatrix} \partial g/\partial x_1(0) & \partial g/\partial x_2(0) \\ 0 & 1 \end{pmatrix} \tilde{x}(t) \tag{20.78}$$

which is unstable. Therefore, no continuously differentiable feedback is stabilizing. However, one can show that the *non-differentiable* feedback

$$u(x) = -x_1 + x_2 + \frac{4}{3}x_2^{1/3} - x_1^3 \tag{20.79}$$

is stabilizing.

Linearization About a Trajectory

The previous sections addressed linearization about a single equilibrium point. Another situation in which linearization can be useful is where the nonlinear system is to follow a prescribed trajectory. Possible sources for this trajectory are repeated maneuvers of the nonlinear system or the outcome of some trajectory optimization (e.g., a robot following a specified optimal path or an aerospace vehicle executing an optimal flight path). The objective of the linearization is then to study the behavior of the system near the prescribed trajectory.

Let $x^*(t)$ and $u^*(t)$ satisfy the nonlinear differential equation (20.31). Let f be continuously differentiable. The objective is to examine the behavior of the nonlinear system near the *trajectory* $(x^*(t), u^*(t))$. Towards this end, let

$$\tilde{x}(t) = x(t) - x^*(t), \quad \tilde{u}(t) = u(t) - u^*(t) \tag{20.80}$$

Assuming that f is continuously differentiable, it may be approximated near the trajectory $(x^*(t), u^*(t))$ by

$$f(x(t), u(t)) \simeq f(x^*(t), u^*(t)) + D_1 f(x^*(t), u^*(t))\tilde{x}(t) + D_2 f(x^*(t), u^*(t))\tilde{u}(t) \tag{20.81}$$

Substituting this approximation into Equation 20.31 leads to

$$\dot{x}(t) \simeq f(x^*(t), y^*(t)) + D_1 f(x^*(t), u^*(t))\tilde{x}(t) + D_2 f(x^*(t), u^*(t))\tilde{u}(t) \tag{20.82}$$

Using $\dot{x}^*(t) = f(x^*(t), u^*(t))$ then leads to the linear *time-varying* dynamics:

$$\dot{\tilde{x}}(t) = D_1 f(x^*(t), u^*(t))\tilde{x} + D_2 f(x^*(t), u^*(t))\tilde{u}(t) \tag{20.83}$$

The time-varying nature of the linearization occurs even though the original nonlinear system is time-invariant.

As in the case of linearization about an equilibrium, the linearized dynamics may be used to infer properties of the nonlinear system when the state-trajectory is close to $x^*(t)$ and the input history is close to $u^*(t)$. Linearization along a trajectory does not restrict the nonlinear system to stay close to a single equilibrium point. The cost of this advantage is that the situation is more complicated, in that one must establish stability properties and/or design stabilizing controllers for linear—but time-varying—system dynamics. These issues are discussed in other articles.

EXAMPLE 20.16: Rotational Link Along a Trajectory

Consider the rotational link equations with $J_1, J_2, k, m, g, L, c = 1$. The nominal trajectory of interest is the link at a constant rate of rotation, i.e., $\theta_1^*(t) = \omega_o t$. Solving for the remaining states and torque leads to

$$\theta_2^*(t) = \omega_o + \omega_o^2 + \omega_t o - \sin\omega_o t \tag{20.84}$$
$$\dot{\theta}_1^*(t) = \omega_o \tag{20.85}$$
$$\dot{\theta}_2^*(t) = \omega_o - \omega_o \cos\omega_o t \tag{20.86}$$
$$T^*(t) = 2\omega_o + \omega_o^2 - \omega_o\cos\omega_o t - \sin\omega_o t \tag{20.87}$$

Linearizing along this trajectory yields

$$\dot{\tilde{x}}(t) = \begin{pmatrix} 0 & 0 & 1 & 0 \\ 0 & 0 & 0 & 1 \\ \cos\omega_o t - 1 & 1 & 2\omega_o & 0 \\ 1 & -1 & 0 & 0 \end{pmatrix} \tilde{x}(t) + \begin{pmatrix} 0 \\ 0 \\ 0 \\ 1 \end{pmatrix} \tilde{u}(t) \tag{20.88}$$

As expected, the linearized dynamics are time-varying (note the $\cos\omega_o t$ term).

20.3.3 Gain Scheduling

Motivation

A major drawback of linearization is that a control design based on the linearized dynamics need not exhibit good performance or even be stabilizing when operating away from the equilibrium. One possibility is to linearize along a trajectory which is not restricted to a local operating region. However, this trajectory must be known in advance in order to perform the control design. Such advance knowledge of the trajectory is often not available.

One approach to address the local restriction in linearization is a design procedure called *gain scheduling*. The main idea is to break the control design process into two steps. First, one designs local linear controllers based on linearizations of the nonlinear system at several different equilibria, usually called in this context operating conditions. Second, a global nonlinear controller for the nonlinear system is obtained by interpolating, or "scheduling," the local operating point designs.

One example is flight control. Here, the linearized systems correspond to the aircraft in a particular flight condition characterized by the atmospheric conditions, aircraft orientation, and aircraft velocity. The local linear controllers are adequate to control the aircraft near a particular operating condition. The global controller, formed by patching together local controllers, is needed to provide transitions between flight conditions.

While intuitively appealing, gain scheduling is an ad hoc practice guided by heuristic rules of thumb. Nevertheless, it does enjoy widespread usage in a variety of applications, such as aircraft control, missile autopilots, jet-engine control, and process control.

This section provides an outline of gain scheduling, its advantages, and limitations.

Gain-Scheduled Control Design

Nonlinear Systems Consider the nonlinear system

$$\dot{x}(t) = f(x(t), u(t)) \tag{20.89}$$
$$y(t) = g(x(t)) \tag{20.90}$$

where $y(t)$ denotes the measured output. Assume that there exists a parameterized family of equilibrium points (x_{eq}, u_{eq}), i.e.,

$$0 = f(x_{eq}(s), u_{eq}(s)) \tag{20.91}$$

where s takes its values in some specified operating region. The variable s, called the *scheduling variable*, will be measured upon operation of the control system and will be used to infer the equilibrium to which the system is near.

Now assuming that s is *fixed* leads to a family of linearizations

$$\dot{\tilde{x}}(t) = A(s)\tilde{x}(t) + B(s)\tilde{u}(t) \tag{20.92}$$
$$\tilde{y}(t) = C(s)\tilde{x}(t) \tag{20.93}$$

where

$$A(s) = D_1 f(x_{eq}(s), u_{eq}(s)),$$
$$B(s) = D_2 f(x_{eq}(s), u_{eq}(s)),$$

$$\begin{aligned} C(s) &= Dg(x_{eq}(s)), \quad (20.94) \\ \tilde{x}(t) &= x(t) - x_{eq}(s), \\ \tilde{u}(t) &= u(t) - u_{eq}(s), \\ \tilde{y}(t) &= y(t) - g(x_{eq}(s)). \quad (20.95) \end{aligned}$$

Under the appropriate stabilizability/detectability assumptions, one can design stabilizing linear controllers (using any of a variety of linear design methods). The result is an indexed collection of controllers,

$$\dot{z}(t) = \overline{A}(s)z(t) + \overline{B}(s)\tilde{y}(t) \quad (20.96)$$
$$\tilde{u}(t) = \overline{C}(s)z(t) + \overline{D}(s)\tilde{y}(t) \quad (20.97)$$

In practice, controllers usually are not designed at every value of s but rather at several operating points indexed by selected values $\{s_1, s_2, \ldots, s_N\}$. In between these points, the controller matrices are interpolated according to the scheduling variable, s.

Although the family of controllers is designed assuming that s is fixed, upon operation of the control system the controller matrices *vary* in time according to the evolution of s. The method of changing the controller matrices can either be smooth or discontinuous switching. In either case, the desired effect is to alleviate the restriction to the local operating region of any individual linearized design. Therefore, depending on the current region of operation (according to s), appropriate controller gains are employed. For example, Figure 20.22 shows a block diagram of gain scheduling for command following. The scheduling variable, s, can either be endogenous to the plant (e.g., a particular state-variable) or an exogenous signal (e.g., a function of some reference command r).

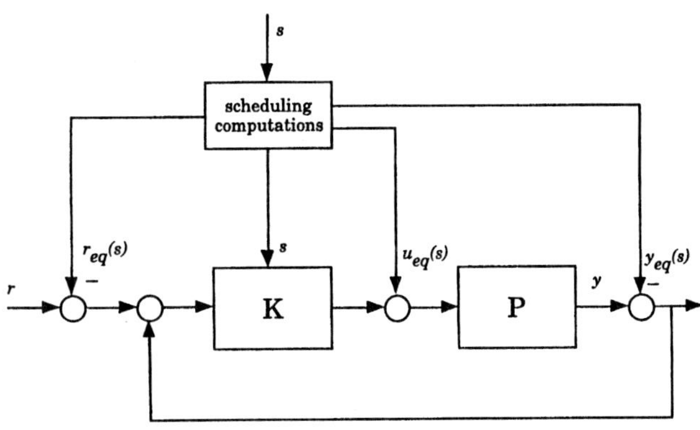

Figure 20.22 Gain-scheduled command following.

EXAMPLE 20.17: Gain-Scheduled Design for Simplified Rotational Link

Consider the rotational link example with the simplification that the rotational flexibility is now rigid. In this case, $\theta_1 = \theta_2$ and $J = J_1 + J_2$. Dropping the subscript on the angles, the equations simplify to

$$\frac{d}{dt}\begin{pmatrix} \theta(t) \\ \dot{\theta}(t) \end{pmatrix} = \begin{pmatrix} \theta(t) \\ mgL\sin(\theta(t))/J - c\dot{\theta}(t)\left|\dot{\theta}(t)\right| + T(t) \end{pmatrix} \quad (20.98)$$

which are in the form $\dot{x}(t) = f(x(t), u(t))$. Let $r(t)$ denote the reference command for $\theta(t)$. A family of equilibrium conditions is parameterized by

$$x_{eq}(s) = \begin{pmatrix} s \\ 0 \end{pmatrix}, \quad u_{eq}(s) = -mgL\sin s/J \quad (20.99)$$

For a fixed s, the linearization is

$$\dot{\tilde{x}}(t) = A(s)\tilde{x}(t) + B(s)\tilde{u}(t) \quad (20.100)$$

where

$$A(s) = \begin{pmatrix} 0 & 1 \\ mgL\cos s/J & 0 \end{pmatrix}, \quad B(s) = \begin{pmatrix} 0 \\ 1 \end{pmatrix} \quad (20.101)$$

Let $m, g, L, J = 1$. At any *fixed* equilibrium, the proportional-derivative feedback

$$\tilde{u}(t) = -\begin{pmatrix} \cos s + 2 & 2 \end{pmatrix}\tilde{x}(t) + 2\tilde{r}(t) \quad (20.102)$$

places the closed-loop poles at $-1 \pm j$ and has zero steady-state error to step commands, where $\tilde{r}(t) = r(t) - s$. The family of linearization-based controllers is then

$$u(t) = u_{eq}(s) - \begin{pmatrix} \cos s + 2 & 2 \end{pmatrix}(x(t) - x_{eq}(s)) + 2(r(t) - s) \quad (20.103)$$

At this point, the scheduling variable s will vary in time according to some scheduling variable. The decision now becomes how to schedule the gains and what to use as a scheduling variable. More precisely, the choices are $s(t) = \theta(t)$ vs. $s(t) = r(t)$ for the scheduling variable and smooth vs. switched scheduling. These options are described as follows:

Switched Scheduling on $\theta(t)$

In this case, the operating range is divided into several regions $\{R_1, \ldots, R_N\}$. Within each region is a representative equilibrium, say $\{\theta_1^*, \ldots, \theta_N^*\}$ and the scheduling variable varies according to

$$s(t) = \theta_i^* \quad \text{whenever } \theta(t) \in R_i \quad (20.104)$$

Smooth Scheduling on $\theta(t)$

Rather than switch between operating points, let $s(t) = \theta(t)$. A peculiar consequence of such scheduling is that the linearization of the overall closed-loop system *differs* from the original linearized plant and linear controller. This is due to terms that were constant, but now vary, such as $u_{eq}(s)$ and $\cos s$.

Switched Scheduling on $r(t)$

This scheduling is based on the *anticipation* that the angle $\theta(t)$ will follow the reference command. Similarly to switched scheduling on $\theta(t)$, set

$$s(t) = \theta_i^* \quad \text{whenever } r(t) \in R_i \quad (20.105)$$

Smooth Scheduling on $r(t)$

Again an inherent assumption is that the angle $\theta(t)$ does not lag the reference command dramatically. One possibility is $s(t) = r(t - T)$.

Note that the gain-scheduled design allows larger variations in $\theta(t)$ than would a design based on a single equilibrium. However, effects such as the nonlinear damping are still neglected. In case $\dot{\theta}(t)$ is large, the approximation accuracy of the fixed linearizations (upon which the gain-scheduled designs are based) suffers. Fast changes in the scheduling variable also increase the discrepancy between the resulting system dynamics and the design model linearization.

Linear Parameter Varying Systems A "linear parameter varying" (LPV) system is defined as a linear system whose coefficients depend on an exogenous time-varying parameter, e.g.,

$$\dot{x}(t) = A(\theta(t))x(t) + B(\theta(t))u(t) \tag{20.106}$$

$$y(t) = C(\theta(t))x(t) \tag{20.107}$$

The exogenous parameter, $\theta(t)$, is unknown *a priori*; however, it can be measured or estimated upon operation of the system. The reason for the special nomenclature is to distinguish LPV systems from linear time-varying systems for which the time variations are known beforehand (as in periodic systems). Typical *a priori* assumptions on $\theta(t)$ are bounds on its magnitude and rate of change.

LPV systems form a useful paradigm for the study of gain-scheduled control. Gain-scheduled control design traditionally starts with a family of linearizations of a nonlinear system indexed by a scheduling variable. This naturally leads to the LPV structure. An LPV structure also arises from simplifying assumptions on the internal structure of a nonlinear model. Rather than model the dynamic evolution of a particular variable, one can treat it as an exogenous independent parameter. For example, in flight control, the dynamic pressure is a dynamic function of the aircraft maneuvers. However, it is useful to model it as an independent time-varying variable.

EXAMPLE 20.18: Rotational Link as LPV

Recall that the rotational link model has a family of equilibrium points:

$$x_{eq}(s) = \begin{pmatrix} s \\ (s - mgL\sin s)/k \\ 0 \\ 0 \end{pmatrix} \qquad u_{eq}(s) = -mgL\sin s \tag{20.108}$$

Define the new state and input variables:

$$x_{new}(t) = \begin{pmatrix} \theta_1(t) \\ \theta_2(t) - \theta_{2,eq}(t) \\ \dot{\theta}_1(t) \\ \dot{\theta}_2(t) \end{pmatrix}$$

$$u_{new}(t) = u(t) - u_{eq}(\theta_1(t)) \tag{20.109}$$

Then, neglecting the nonlinear damping, the state dynamics can be written as:

$$\dot{x}_{new}(t) = \begin{pmatrix} 0 & 0 & 1 & 0 \\ 0 & 0 & mg\cos\theta_1(t)/K & 0 \\ 0 & k/J_2 & 0 & 0 \\ 0 & -k/J_2 & 0 & 0 \end{pmatrix} x_{new}(t) + \begin{pmatrix} 0 \\ 0 \\ 0 \\ 1 \end{pmatrix} u_{new}(t) \tag{20.110}$$

Note that the original state equations are *transformed* into a quasi-LPV form, with the "exogenous" parameter actually the angle $\theta_1(t)$. It is interesting to note that this quasi-LPV family *is not* the same family obtained by performing linearizations about equilibrium conditions.

Now suppose that the parameter in the LPV plant (Equation 20.106) satisfies the bounds $|\theta(t)| \leq 1$. A traditional gain-scheduled design approach is to assume that the parameter is *constant* and design a family parameter-dependent controller to achieve desired stability and performance specifications. This results in an LPV controller such as

$$\dot{z}(t) = \overline{A}(\theta(t))z(t) + \overline{B}(\theta(t))y(t) \tag{20.111}$$

$$u(t) = \overline{C}(\theta(t))z(t) + \overline{D}(\theta(t))y(t) \tag{20.112}$$

In practice, the LPV controller gains come from an interpolation of several designs throughout the parameter range of values. Upon operation of the control system, the controller gains are updated according to the parameter time variations.

Because the parameter is actually time varying, the gain-scheduled design can experience degradation of performance or even loss of stability. However, one can show that if the parameter variations are sufficiently *slow*, then the desired properties are maintained.

EXAMPLE 20.19: Time-Varying Oscillator

A classical example of parameter-varying instability from frozen parameter stability is the time-varying oscillator:

$$\dot{x}(t) = \begin{pmatrix} 0 & 1 \\ -(1 + \theta(t)/2) & -.2 \end{pmatrix} x(t) \tag{20.113}$$

These equations can be viewed as a mass-spring-damper system with time-varying spring stiffness. For *fixed* parameter values, $\theta(t) = \theta_o$ = a constant, the equilibrium $x_o = 0$ is asymptotically stable. However, for the parameter trajectory $\theta(t) = \cos 2t$, it becomes unstable. An intuitive explanation is that the stiffness variations are timed to pump energy into the oscillations.

Discussion

Conceptually, gain scheduling allows for greater operating regions than a design based on a single equilibrium. However, since the scheduling variable is no longer constant, the gain schedule introduces *time variations* in the overall control system.

Such time variations typically are not addressed in the original frozen-equilibrium design. One consequence is possible degradation in performance or even instability of the gain-scheduled system. Another consequence is that the state of the nonlinear system while in transition need not be close to any of the equilibrium points, and hence outside of the design regions of the linearized controllers. However, the effects of these phenomena are reduced in the case of slow transitions among the operating conditions. In the end, the quality of a gain-scheduled design is typically inferred from extensive computer simulations.

Despite its widespread popularity, gain scheduling has received relatively little theoretical attention. Some references are stated in the section for further information. However, the theoretical basis for gain scheduling can be summarized as follows. The overall design is based on a collection of *linearizations* at *fixed* equilibrium conditions. If these design models are reasonable representations of the system dynamics, then one can expect that the stability and performance properties of the linearized designs should carry over to the overall gain-scheduled design. If the nonlinearities dominate or if the transitions between operating conditions are fast, and if these phenomena are not recognized in the design process, then one should not expect that the gain-scheduled design will perform satisfactorily. This reasoning leads to the popular heuristic guideline for successful gain-scheduled designs to "schedule on a slow-variable which captures the nonlinearities."

Further Reading

References discussing linearization as well as other methods for nonlinear systems:

[1] Khalil, H.K., *Nonlinear Systems.* Macmillan, New York, 1992.

[2] Vidyasagar, M., *Nonlinear Systems Analysis,* 2nd ed. Prentice Hall, Englewood Cliffs, NJ, 1993.

For more specialized results when linearization methods are inconclusive (including a discussion of Example 20.15).

[3] Bacciotti, A., *Local Stabilizability of Nonlinear Control Systems,* World Scientific Publishing, Singapore, 1992.

References which give an overview of gain scheduling as well as some theoretical analyses:

[4] Rugh, W.J, Analytical framework for gain-scheduling. *IEEE Control Syst. Mag.*, 11(1), 79–84, 1991.

[5] Shamma, J.S. and Athans, M., Analysis of nonlinear gain-scheduled control systems. *IEEE Trans. Autom. Control.* 35(8), 898–907, 1990.

[6] Shamma, J.S. and Athans, M., Gain scheduling: Potential hazards and possible remedies. *IEEE Control Syst. Mag.*, 12(3), 101–107, 1992.

References discussing methods for linear parameter varying systems:

[7] Packard, A., Gain scheduling via linear fractional transformations. *Syst. Control Lett.*, 22, 79–92, 1994.

[8] Shahruz, S.M. and Behtash, S., Design of controllers for linear parameter-varying systems by the gain scheduling technique. *J. Math. Anal. Appl.*, 168(1), 125–217, 1992.

[9] Shamma, J.S. and Athans, M., Guaranteed properties of gain scheduled control of linear parameter varying plants. *Automatica*, 27(3), 559–564, 1991.

References presenting new approaches to gain scheduling:

[10] Kaminer, I., Pascoal, A.M., Khargonekar, P.P, and Coleman, E., A velocity algorithm for the implementation of gain-scheduled controllers, *Proceedings of the 1993 European Control Conference*, pp. 787–792, to appear in *Automatica*, 1993.

References with example applications of gain scheduling:

[11] Apkarian, P., Gahinet, P., and Biannic, J.-M., Self-scheduled H-infinity control of a missile via LMIs. *Proceedings of the 33rd IEEE Conference on Decision and Control.* pp. 3312–3317, 1994.

[12] Astrom, K.J. and Wittenmark, B., *Adaptive Control,* Addison-Wesley, New York, 1989, chapt. 9.

[13] Nichols, R.A., Reichert, R.T., and Rugh, W.J., Gain scheduling for H-infinity controllers: A flight control example, *IEEE Trans. Control Syst. Technol.*, 1(2), 69–79, 1993.

[14] Shamma, J.S. and Cloutier, J.R., Gain-scheduled missile autopilot design using linear parameter varying methods. *J. Guidance Control Dynam.*, 16(2), 256–263, 1993.

[15] Whatley, M.J. and Pott, D.C., Adaptive gain improves reactor control, *Hydrocarbon Processing*, May, pp. 75–78, 1984.

References discussing other linearization-based methods:

[16] Baumann, W.T. and Rugh, W.J., Feedback control of nonlinear systems by extended linearization, *IEEE Trans. Autom. Control,* 31(1), 40–46, 1986.

[17] Reboulet, C. and Champetier, C., A new method for linearizing nonlinear systems: The pseudolinearization. *Int. J. Control,* 40(4), 631–638, 1984.

[18] Lawrence, D.A. and Rugh, W.J., Input-output pseudolinearization for nonlinear systems, *IEEE Trans. Autom. Control,* 39(11), 2207–2218, 1994.

SECTION VI

Software for Control System Analysis and Design

21

Numerical and Computational Issues in Linear Control and System Theory

A.J. Laub
Department of Electrical and Computer Engineering, University of California, Santa Barbara, CA

R.V. Patel
Department of Electrical and Computer Engineering, Concordia University, Montreal, Quebec, Canada

P.M. Van Dooren
Department of Mathematical Engineering, Université Catholique de Louvain, Belgium

21.1 Introduction

This chapter provides a survey of various aspects of the numerical solution of selected problems in linear systems, control, and estimation theory. Space limitations preclude an exhaustive survey and extensive list of references. The interested reader is referred to [10] for sources of additional detailed information.

Many of the problems considered in this chapter arise in the study of the "standard" linear model

$$\dot{x}(t) = Ax(t) + Bu(t) \quad (21.1)$$

$$y(t) = Cx(t) + Du(t). \quad (21.2)$$

Here, $x(t)$ is an n-vector of states, $u(t)$ is an m-vector of controls or inputs, and $y(t)$ is a p-vector of outputs. The standard discrete-time analog of Equations 21.1 and 21.2 takes the form

$$x_{k+1} = Ax_k + Bu_k \quad (21.3)$$

$$y_k = Cx_k + Du_k. \quad (21.4)$$

Of course, considerably more elaborate models are also studied, including time-varying, stochastic, and nonlinear versions of the above, but these are not be discussed in this chapter. In fact, the above linear models are usually derived from linearizations of nonlinear models about selected nominal points.

The matrices considered here are, for the most part, assumed to have real coefficients and to be small (of order a few hundred or less) and dense, with no particular exploitable structure. Calculations for most problems in classical single-input, single-output control fall into this category. Large, sparse matrices or matrices with special, exploitable structures may involve significantly different concerns and methodologies than those to be discussed here.

The systems, control, and estimation literature is replete with *ad hoc* algorithms to solve the computational problems which arise in the various methodologies. Many of these algorithms work quite well on some problems (e.g., "small order" matrices) but encounter numerical difficulties, often severe, when "pushed" (e.g., on larger order matrices). The reason for this is that little or no attention has been paid to the way algorithms perform in "finite arithmetic," i.e., on a finite word length digital computer.

A simple example due to Moler and Van Loan [10](p. 649)[1] illustrates a typical pitfall. Suppose it is desired to compute the matrix e^A in single precision arithmetic on a computer which gives 6 decimal places of precision in the fractional part of floating-point numbers. Consider the case

$$A = \begin{bmatrix} -49 & 24 \\ -64 & 31 \end{bmatrix}$$

[1]The page number indicates the location of the appropriate reprint in [10].

0-8493-8570-9/96/$0.00+$.50

and suppose the computation is attempted with the Taylor series formula

$$e^A = \sum_{k=0}^{+\infty} \frac{1}{k!} A^k. \tag{21.5}$$

This is easily coded and it is determined that the first 60 terms in the series suffice for the computation, in the sense that terms for $k \geq 60$ of the order of 10^{-7} no longer add anything significant to the sum. The resulting answer is

$$\begin{bmatrix} -22.2588 & -1.43277 \\ -61.4993 & -3.47428 \end{bmatrix}.$$

Surprisingly, the true answer is (correctly rounded)

$$\begin{bmatrix} -0.735759 & 0.551819 \\ -1.47152 & 1.10364 \end{bmatrix}.$$

What happened here was that the intermediate terms in the series got very large before the factorial began to dominate. The 17th and 18th terms, for example, are of the order of 10^7 but of opposite signs so that the less significant parts of these numbers, while significant for the final answer, are "lost" because of the finiteness of the arithmetic.

For this particular example, various fixes and remedies are available. However, in more realistic examples one seldom has the luxury of having the "true answer" available so that it is not always easy simply to inspect or test a computed solution and determine that it is in error. Mathematical analysis (truncation of the series, in the example above) alone is simply not sufficient when a problem is analyzed or solved in finite arithmetic (truncation of the arithmetic). Clearly, a great deal of care must be taken.

The finiteness inherent in representing real or complex numbers as floating-point numbers on a digital computer manifests itself in two important ways: floating-point numbers have only finite precision and finite range. The degree of attention paid to these two considerations distinguishes many reliable algorithms from more unreliable counterparts.

The development in systems, control, and estimation theory, of stable, efficient, and reliable algorithms which respect the constraints of finite arithmetic began in the 1970s and is continuing. Much of the research in numerical analysis has been directly applicable, but there are many computational issues in control (e.g., the presence of hard or structural zeros) where numerical analysis does not provide a ready answer or guide. A symbiotic relationship has developed, especially between numerical linear algebra and linear system and control theory, which is sure to provide a continuing source of challenging research areas.

The abundance of numerically fragile algorithms is partly explained by the following observation:

> If an algorithm is amenable to "easy" hand calculation, it is probably a poor method if implemented in the finite floating-point arithmetic of a digital computer.

For example, when confronted with finding the eigenvalues of a 2×2 matrix, most people would find the characteristic polynomial and solve the resulting quadratic equation. But when extrapolated as a general method for computing eigenvalues and implemented on a digital computer, this is a very poor procedure for reasons such as roundoff and overflow/underflow. The preferred method now would generally be the double Francis QR algorithm (see [13] for details) but few would attempt that manually, even for very small order problems.

Many algorithms, now considered fairly reliable in the context of finite arithmetic, are not amenable to hand calculations (e.g., various classes of orthogonal similarities). This is a kind of converse to the observation quoted above. Especially in linear system and control theory, we have been too easily seduced by the ready availability of closed-form solutions and numerically naive methods to implement those solutions. For example, in solving the initial value problem

$$\dot{x}(t) = Ax(t); \quad x(0) = x_0 \tag{21.6}$$

it is not at all clear that one should explicitly compute the intermediate quantity e^{tA}. Rather, it is the vector $e^{tA}x_0$ that is desired, a quantity that may be computed by treating (21.6) as a system of (possibly stiff) differential equations and using an implicit method for numerically integrating the differential equation. But such techniques are definitely not attractive for hand computation.

Awareness of such numerical issues in the mathematics and engineering community has increased significantly in the last fifteen years or so. In fact, some of the background material well known to numerical analysts has already filtered down to undergraduate and graduate curricula in these disciplines. This awareness and education has affected system and control theory, especially linear system theory. A number of numerical analysts were attracted by the wealth of interesting numerical linear algebra problems in linear system theory. At the same time, several researchers in linear system theory turned to various methods and concepts from numerical linear algebra and attempted to modify them in developing reliable algorithms and software for specific problems in linear system theory. This cross-fertilization has been greatly enhanced by the widespread use of software packages and by recent developments in numerical linear algebra. This process has already begun to have a significant impact on the future directions and development of system and control theory, and on applications, as evident from the growth of computer-aided control system design (CACSD) as an intrinsic tool. Algorithms implemented as mathematical software are a critical "inner" component of a CACSD system.

In the remainder of this chapter, we survey some recent results and trends in this interdisciplinary research area. We emphasize numerical aspects of the problems/algorithms, which is why we also spend time discussing appropriate numerical tools and techniques. We discuss a number of control and filtering problems that are of widespread interest in control.

Before proceeding further we list here some notation to be used:

$\mathbb{F}^{n \times m}$	the set of all $n \times m$ matrices with coefficients in the field $\mathbb{F}$ ($\mathbb{F}$ is generally $\mathbb{R}$ or $\mathbb{C}$)
A^T	the transpose of $A \in \mathbb{R}^{n \times m}$
A^H	the complex-conjugate transpose of $A \in \mathbb{C}^{n \times m}$
A^+	the Moore-Penrose pseudoinverse of A
$\|A\|$	the spectral norm of A (i.e., the matrix norm subordinate to the Euclidean vector norm: $\|A\| = \max_{\|x\|_2=1} \|Ax\|_2$)
diag $(a_1, \cdots, a_n)$	the diagonal matrix $\begin{bmatrix} a_1 & & 0 \\ & \ddots & \\ 0 & & a_n \end{bmatrix}$
$\Lambda(A)$	the set of eigenvalues $\lambda_1, \cdots, \lambda_n$ (not necessarily distinct) of $A \in \mathbb{F}^{n \times n}$
$\lambda_i(A)$	the ith eigenvalue of A
$\Sigma(A)$	the set of singular values $\sigma_1, \cdots, \sigma_m$ (not necessarily distinct) of $A \in \mathbb{F}^{n \times m}$
$\sigma_i(A)$	the ith singular value of A.

Finally, let us define a particular number to which we make frequent reference following. The *machine epsilon* or *relative machine precision* is defined, roughly speaking, as the smallest positive number ϵ that, when added to 1 on our computing machine, gives a number greater than 1. In other words, any machine representable number δ less than ϵ gets " rounded off" when (floating-point) added to 1 to give exactly 1 again as the rounded sum. The number ϵ, of course, varies depending on the kind of computer being used and the precision of the computations (single precision, double precision, etc.). But the fact that such a positive number ϵ exists is entirely a consequence of finite word length.

21.2 Numerical Background

In this section we give a very brief discussion of two concepts fundamentally important in numerical analysis: *numerical stability* and *conditioning*. Although this material is standard in textbooks such as [6], it is presented here for completeness and because the two concepts are frequently confused in the systems, control, and estimation literature.

Suppose we have some mathematically defined problem represented by f which acts on data d belonging to some set of data $\mathcal{D}$, to produce a solution $f(d)$ in a solution set $\mathcal{S}$. These notions are kept deliberately vague for expository purposes. Given $d \in \mathcal{D}$ we desire to compute $f(d)$. Suppose d^* is some approximation to d. If $f(d^*)$ is "near" $f(d)$, the problem is said to be well-conditioned. If $f(d^*)$ may potentially differ greatly from $f(d)$ even when d^* is near d, the problem is said to be ill-conditioned. The concept of "near" can be made precise by introducing norms in the appropriate spaces. We can then define the condition of the problem f with respect to these norms as

$$\kappa[f(d)] = \lim_{\delta \to 0} \sup_{d_2(d,d^*)=\delta} \left[\frac{d_1(f(d), f(d^*))}{\delta} \right] \quad (21.7)$$

where d_i (. , .) are distance functions in the appropriate spaces. When $\kappa[f(d)]$ is infinite, the problem of determining $f(d)$ from d is *ill-posed* (as opposed to *well-posed*). When $\kappa[f(d)]$ is finite and *relatively large* (or *relatively small*), the problem is said to be *ill conditioned* (or *well conditioned*).

A simple example of an ill-conditioned problem is the following. Consider the $n \times n$ matrix

$$A = \begin{bmatrix} 0 & 1 & 0 & \cdot & \cdot & \cdot & 0 \\ \cdot & \cdot & \cdot & \cdot & \cdot & \cdot & \cdot \\ \cdot & & \cdot & \cdot & \cdot & \cdot & \cdot \\ \cdot & & & & \cdot & \cdot & 0 \\ \cdot & & & & & \cdot & 1 \\ 0 & \cdot & \cdot & \cdot & \cdot & \cdot & 0 \end{bmatrix}$$

with n eigenvalues at 0. Now consider a small perturbation of the data (the n^2 elements of A) consisting of adding the number 2^{-n} to the first element in the last (nth) row of A. This perturbed matrix then has n distinct eigenvalues $\lambda_1, \cdots, \lambda_n$ with $\lambda_k = 1/2 \exp(2k\pi j/n)$ where $j := \sqrt{-1}$. Thus, we see that this small perturbation in the data has been magnified by a factor on the order of 2^n resulting in a rather large perturbation in solving the problem of computing the eigenvalues of A. Further details and related examples can be found in [13].

Thus far we have not mentioned how the problem f above (computing the eigenvalues of A in the example) was to be solved. Conditioning is a function solely of the problem itself. To solve a problem numerically, we must implement some numerical procedure or algorithm which we denote by f^*. Thus, given d, $f^*(d)$ is the result of applying the algorithm to d (for simplicity, we assume d is "representable"; a more general definition can be given when some approximation d^{**} to d must be used). The algorithm f^* is said to be numerically (backward) stable if, for all $d \in \mathcal{D}$, there exists $d^* \in \mathcal{D}$ near d so that $f^*(d)$ is near $f(d^*)$ ($f(d^*)$ = the exact solution of a nearby problem). If the problem is well conditioned, then $f(d^*)$ is near $f(d)$ so that $f^*(d)$ is near $f(d)$ if f^* is numerically stable. In other words, f^* does not introduce any more sensitivity to perturbation than is inherent in the problem. Example 1 below further illuminates this definition of stability which, on a first reading, can seem somewhat confusing.

Of course, one cannot expect a stable algorithm to solve an ill-conditioned problem any more accurately than the data warrant but an unstable algorithm can produce poor solutions even to well-conditioned problems. Example 2, below, illustrates this phenomenon. There are thus two separate factors to consider in determining the accuracy of a computed solution $f^*(d)$. First, if the algorithm is stable, $f^*(d)$ is near $f(d^*)$, for some d^*, and second, if the problem is well-conditioned then, as above, $f(d^*)$ is near $f(d)$. Thus, $f^*(d)$ is near $f(d)$ and we have an "accurate" solution.

Rounding errors can cause unstable algorithms to give disastrous results. However, it would be virtually impossible to account for every rounding error made at every arithmetic operation in a complex series of calculations. This would constitute a *forward* error analysis. The concept of *backward* error analysis based on the definition of numerical stability given above provides a more practical alternative. To illustrate this, let us consider the singular value decomposition of an arbitrary $m \times n$ matrix A with coefficients in $\mathbb{R}$ or $\mathbb{C}$ [6] (see also part C of Section 21.3 of this chapter),

$$A = U\Sigma V^H . \tag{21.8}$$

Here U and V are $m \times m$ and $n \times n$ unitary matrices, respectively, and Σ is an $m \times n$ matrix of the form

$$\Sigma = \left[\begin{array}{c|c} \Sigma_r & 0 \\ \hline 0 & 0 \end{array}\right] ; \ \Sigma_r = \mathrm{diag}\{\sigma_1, \cdots, \sigma_r\} \tag{21.9}$$

with the *singular value* σ_i positive and satisfying $\sigma_1 \geq \sigma_2 \cdots \geq \sigma_r > 0$. The computation of this decomposition is, of course, subject to rounding errors. Denoting computed quantities by an overbar, for some *error matrix* E_A,

$$\overline{A} = A + E_A = \overline{U\Sigma V}^H . \tag{21.10}$$

The computed decomposition thus corresponds exactly to a *perturbed* matrix $\overline{A}$. When using the SVD algorithm available in the literature [6], this perturbation can be bounded by

$$\| E_A \| \leq \pi\epsilon \| A \| \tag{21.11}$$

where ϵ is the machine precision and π is some quantity depending on the dimensions m and n, but reasonably close to 1 (see also [10](p. 74)). Thus, the *backward error* E_A induced by this algorithm has roughly the same norm as the *input error* E_i resulting, for example, when reading the data A into the computer. Then, according to the definition of numerical stability given above, when a bound such as that in Equation 21.11 exists for the error induced by a numerical algorithm, the algorithm is said to be *backward stable* [13]. Notice that backward stability does not guarantee any bounds on the errors in the result $\overline{U}$, $\overline{\Sigma}$, and $\overline{V}$. In fact, this depends on how perturbations in the data (namely $E_A = \overline{A} - A$) affect the resulting decomposition (namely $E_U = \overline{U} - U$, $E_\Sigma = \overline{\Sigma} - \Sigma$, and $E_V = \overline{V} - V$). This is commonly measured by the condition $\kappa[f(A)]$.

Backward stability is a property of an algorithm and condition is associated with a problem and the specific data for that problem. The errors in the result depend on the stability of the algorithm used and the condition of the problem solved. A *good* algorithm should, therefore, be backward stable because the size of the errors in the result is then mainly due to the condition of the problem, not to the algorithm. An unstable algorithm, on the other hand, may yield a large error even when the problem is well-conditioned.

Bounds of the type Equation 21.11 are obtained by an error analysis of the algorithm used, and the condition of the problem is obtained by a sensitivity analysis, e.g., see [13].

We close this section with two simple examples to illustrate some of the concepts introduced.

EXAMPLE 21.1:

Let x and y be two floating-point computer numbers and let $fl(x * y)$ denote the result of multiplying them in floating-point computer arithmetic. In general, the product $x * y$ requires more precision to be represented exactly than was used to represent x or y. But for most computers

$$fl(x * y) = x * y(1 + \delta) \tag{21.12}$$

where $|\delta| < \epsilon$ (= relative machine precision). In other words, $fl(x * y)$ is $x * y$ correct to within a unit in the last place. Another way to write Equation 21.12 is as

$$fl(x * y) = x(1 + \delta)^{1/2} * y(1 + \delta)^{1/2} \tag{21.13}$$

where $|\delta| < \epsilon$. This can be interpreted as follows: the computed result $fl(x * y)$ is the exact product of the two slightly perturbed numbers $x(1 + \delta)^{1/2}$ and $y(1 + \delta)^{1/2}$. The slightly perturbed data (not unique) may not even be representable as floating-point numbers. The representation Equation 21.13 is simply a way of accounting for the roundoff incurred in the algorithm by an initial (small) perturbation in the data.

EXAMPLE 21.2:

Gaussian elimination with no pivoting for solving the linear system of equations

$$Ax = b \tag{21.14}$$

is known to be numerically unstable; see, for example, [6] and Section 21.3. The following data illustrates this phenomenon. Let

$$A = \begin{bmatrix} 0.0001 & 1.000 \\ 1.000 & -1.000 \end{bmatrix}, b = \begin{bmatrix} 1.000 \\ 0.000 \end{bmatrix}.$$

All computations are carried out in four-significant-figure decimal arithmetic. The "true answer" $x = A^{-1}b$ is

$$\begin{bmatrix} 0.9999 \\ 0.9999 \end{bmatrix}.$$

Using row 1 as the "pivot row" (i.e., subtracting 10,000 $\times$ row 1 from row 2) we arrive at the equivalent triangular system

$$\begin{bmatrix} 0.0001 & 1.000 \\ 0 & -1.000 \times 10^4 \end{bmatrix} \begin{bmatrix} x_1 \\ x_2 \end{bmatrix} = \begin{bmatrix} 1.000 \\ -1.000 \times 10^4 \end{bmatrix}.$$

The coefficient multiplying x_2 in the second equation should be $-10,001$, but because of roundoff, becomes $-10,000$. Thus, we compute $x_2 = 1.000$ (a good approximation), but back-substitution in the equation

$$0.0001x_1 = 1.000 - fl(1.000 * 1.000)$$

yields $x_1 = 0.000$. This extremely bad approximation to x_1 is the result of numerical instability. The problem, it can be shown, is quite well-conditioned.

21.3 Fundamental Problems in Numerical Linear Algebra

In this section we give a brief overview of some of the fundamental problems in numerical linear algebra which serve as building blocks or "tools" for the solution of problems in systems, control, and estimation.

A. Linear Algebraic Equations and Linear Least Squares Problems

Probably the most fundamental problem in numerical computing is the calculation of a vector x which satisfies the linear system

$$Ax = b \tag{21.15}$$

where $A \in \mathbb{R}^{n \times n}$ (or $\mathbb{C}^{n \times n}$) and has rank n. A great deal is now known about solving Equation 21.15 in finite arithmetic both for the general case and for a large number of special situations, e.g., see [6].

The most commonly used algorithm for solving Equation 21.15 with general A and small n (say $n \leq 200$) is Gaussian elimination with some sort of pivoting strategy, usually "partial pivoting." This amounts to factoring some permutation of the rows of A into the product of a unit lower triangular matrix L and an upper triangular matrix U. The algorithm is effectively stable, i.e., it can be proved that the computed solution is near the exact solution of the system

$$(A + E)x = b \tag{21.16}$$

with $|e_{ij}| \leq \phi(n)\,\gamma\,\beta\,\epsilon$ where $\phi(n)$ is a modest function of n depending on details of the arithmetic used, γ is a "growth factor" (which is a function of the pivoting strategy and is usually—but not always—small), β behaves essentially like $\|A\|$, and ϵ is the machine precision. In other words, except for moderately pathological situations, E is "small"—on the order of $\epsilon\,\|A\|$.

The following question then arises. If, because of rounding errors, we are effectively solving Equation 21.16 rather than Equation 21.15, what is the relationship between $(A + E)^{-1}b$ and $A^{-1}b$? To answer this question we need some elementary perturbation theory and this is where the notion of condition number arises. A condition number for Equation 21.15 is given by

$$\kappa(A) := \|A\|\;\|A^{-1}\|. \tag{21.17}$$

Simple perturbation results can show that perturbation in A and/or b can be magnified by as much as $\kappa(A)$ in the computed solution. Estimating $\kappa(A)$ (since, of course, A^{-1} is unknown) is thus a crucial aspect of assessing solutions of Equation 21.15 and the particular estimating procedure used is usually the principal difference between competing linear equation software packages. One of the more sophisticated and reliable condition estimators presently available is implemented in LINPACK [3] and its successor LAPACK [1]. LINPACK and LAPACK also feature many codes for solving Equation 21.14 in case A has certain special structures (e.g., banded, symmetric, or positive definite).

Another important class of linear algebra problems, and one for which codes are available in LINPACK and LAPACK, is the linear least squares problem

$$\min \|Ax - b\|_2 \tag{21.18}$$

where $A \in \mathbb{R}^{m \times n}$ and has rank k, with (in the simplest case) $k = n \leq m$, e.g., see [6]. The solution of Equation 21.18 can be written formally as $x = A^{+}b$. The method of choice is generally based upon the QR factorization of A (for simplicity, let $\text{rank}(A) = n$)

$$A = QR \tag{21.19}$$

where $R \in \mathbb{R}^{n \times n}$ is upper triangular and $Q \in \mathbb{R}^{m \times n}$ has orthonormal columns, i.e., $Q^T Q = I$. With special care and analysis, the case $k < n$ can also be handled similarly. The factorization is effected through a sequence of Householder transformations H_i applied to A. Each H_i is symmetric and orthogonal and of the form $I - 2uu^T/u^T u$ where $u \in \mathbb{R}^m$ is specially chosen so that zeros are introduced at appropriate places in A when it is premultiplied by H_i. After n such transformations,

$$H_n H_{n-1} \cdots H_1 A = \begin{bmatrix} R \\ 0 \end{bmatrix}$$

from which the factorization Equation 21.19 follows. Defining c and d by

$$\begin{bmatrix} c \\ d \end{bmatrix} := H_n H_{n-1} \cdots H_1 b$$

where $c \in \mathbb{R}^n$, it is easily shown that the least squares solution x of Equation 21.18 is given by the solution of the linear system of equations

$$Rx = c. \tag{21.20}$$

The above algorithm is numerically stable and, again, a well-developed perturbation theory exists from which condition numbers can be obtained, this time in terms of

$$\kappa(A) := \|A\|\;\|A^{+}\|.$$

Least squares perturbation theory is fairly straightforward when $\text{rank}(A) = n$, but is considerably more complicated when A is rank-deficient. The reason for this is that, although the inverse is a continuous function of the data (i.e., the inverse is a continuous function in a neighborhood of a nonsingular matrix), the pseudoinverse is discontinuous. For example, consider

$$A = \begin{bmatrix} 1 & 0 \\ 0 & 0 \end{bmatrix} = A^{+}$$

and perturbations

$$E_1 = \begin{bmatrix} 0 & 0 \\ \delta & 0 \end{bmatrix}$$

and

$$E_2 = \begin{bmatrix} 0 & 0 \\ 0 & \delta \end{bmatrix}$$

with δ small. Then

$$(A+E_1)^+ = \begin{bmatrix} \frac{1}{1+\delta^2} & \frac{\delta}{1+\delta^2} \\ 0 & 0 \end{bmatrix}$$

which is close to A^+ but

$$(A+E_2)^+ = \begin{bmatrix} 1 & 0 \\ 0 & \frac{1}{\delta} \end{bmatrix}$$

which gets arbitrarily far from A^+ as δ is decreased towards 0.

In lieu of Householder transformations, Givens transformations (elementary rotations or reflections) may also be used to solve the linear least squares problem [6]. Recently, Givens transformations have received considerable attention for solving linear least squares problems and systems of linear equations in a parallel computing environment. The capability of introducing zero elements selectively and the need for only local interprocessor communication make the technique ideal for "parallelization."

B. Eigenvalue and Generalized Eigenvalue Problems

In the algebraic eigenvalue/eigenvector problem for $A \in \mathbb{R}^{n\times n}$, one seeks nonzero solutions $x \in \mathbb{C}^n$ and $\lambda \in \mathbb{C}$ which satisfy

$$Ax = \lambda x. \tag{21.21}$$

The classic reference on the numerical aspects of this problem is Wilkinson [13]. A briefer textbook introduction is given in [6].

Quality mathematical software for eigenvalues and eigenvectors is available; the EISPACK [5], [11] collection of subroutines represents a pivotal point in the history of mathematical software. The successor to EISPACK (and LINPACK) is the recently released LAPACK [1] in which the algorithms and software have been restructured to provide high efficiency on vector processors, high performance workstations, and shared memory multiprocessors.

The most common algorithm now used to solve Equation 21.21 for general A is the QR algorithm of Francis [13]. A shifting procedure enhances convergence and the usual implementation is called the double-Francis-QR algorithm. Before the QR process is applied, A is initially reduced to upper Hessenberg form A_H ($a_{ij} = 0$ if $i - j \geq 2$). This is accomplished by a finite sequence of similarities of the Householder form discussed above. The QR process then yields a sequence of matrices orthogonally similar to A and converging (in some sense) to a so called quasi upper triangular matrix S also called the real Schur form (RSF) of A. The matrix S is block upper triangular with 1×1 diagonal blocks corresponding to real eigenvalues of A and 2×2 diagonal blocks corresponding to complex-conjugate pairs of eigenvalues. The quasi upper triangular form permits all arithmetic to be real rather than complex as would be necessary for convergence to an upper triangular matrix. The orthogonal transformations from both the Hessenberg reduction and the QR process may be accumulated in a single orthogonal transformation U so that

$$U^T A U = R \tag{21.22}$$

compactly represents the entire algorithm. An analogous process can be applied in the case of symmetric A, and considerable simplifications and specializations result.

Closely related to the QR algorithm is the QZ algorithm for the generalized eigenvalue problem

$$Ax = \lambda M x \tag{21.23}$$

where $A, M \in \mathbb{R}^{n\times n}$. Again, a Hessenberg-like reduction, followed by an iterative process, is implemented with orthogonal transformations to reduce Equation 21.23 to the form

$$QAZy = \lambda QMZy \tag{21.24}$$

where QAZ is quasi upper triangular and QMZ is upper triangular. For a review and references to results on stability, conditioning, and software related to Equation 21.23 and the QZ algorithm, see [6]. The generalized eigenvalue problem is both theoretically and numerically more difficult to handle than the ordinary eigenvalue problem, but it finds numerous applications in control and system theory [10](p. 109).

C. The Singular Value Decomposition and Some Applications

One of the basic and most important tools of modern numerical analysis, especially numerical linear algebra, is the singular value decomposition (SVD). Here we make a few comments about its properties and computation as well as its significance in various numerical problems.

Singular values and the singular value decomposition have a long history, especially in statistics, and more recently in numerical linear algebra. Even more recently the ideas are finding applications in the control and signal processing literature, although their use there has been overstated somewhat in certain applications. For a survey of the singular value decomposition, its history, numerical details, and some applications in systems and control theory, see [10](p. 74).

The fundamental result was stated in Section 21.2 (for the complex case). The result for the real case is similar and is stated below.

THEOREM 21.1 *Let $A \in \mathbb{R}^{m\times n}$ with rank(A) $= r$. Then there exist orthogonal matrices $U \in \mathbb{R}^{m\times m}$ and $V \in \mathbb{R}^{n\times n}$ so that*

$$A = U\Sigma V^T \tag{21.25}$$

where

$$\Sigma = \begin{bmatrix} \Sigma_r & 0 \\ 0 & 0 \end{bmatrix}$$

and $\Sigma_r = diag\{\sigma_1, \cdots, \sigma_r\}$ with $\sigma_1 \geq \cdots \geq \sigma_r > 0$.

The proof of Theorem 21.1 is straightforward and can be found, for example, in [6]. Geometrically, the theorem says that bases can be found (separately) in the domain and codomain spaces of a linear map with respect to which the matrix representation of the linear map is diagonal. The numbers $\sigma_1, \cdots, \sigma_r$ together with $\sigma_{r+1} = 0, \cdots, \sigma_n = 0$ are called the singular values of A, and they are the positive square roots of the eigenvalues of A^TA. The columns $\{u_k, k = 1. \cdots, m\}$ of U are called the left singular vectors of A (the orthonormal eigenvectors of AA^T), while the columns $\{v_k, k = 1, \cdots, n\}$ of V are called the right singular vectors of A (the orthonormal eigenvectors of A^TA). The matrix A can then also be written (as a dyadic expansion) in terms of the singular vectors as follows:

$$A = \sum_{k=1}^{r} \sigma_k u_k v_k^T.$$

The matrix A^T has m singular values, the positive square roots of the eigenvalues of AA^T. The r [= rank (A)] nonzero singular values of A and A^T are, of course, the same. The choice of A^TA rather than AA^T in the definition of singular values is arbitrary. Only the nonzero singular values are usually of any real interest and their number, given the SVD, is the rank of the matrix. Naturally, the question of how to distinguish nonzero from zero singular values in the presence of rounding error is a nontrivial task.

It is not generally advisable to compute the singular values of A by first finding the eigenvalues of A^TA, tempting as that is. Consider the following example, where μ is a real number with $|\mu| < \sqrt{\epsilon}$ (so that $fl(1 + \mu^2) = 1$ where $fl(\cdot)$ denotes floating-point computation). Let

$$A = \begin{bmatrix} 1 & 1 \\ \mu & 0 \\ 0 & \mu \end{bmatrix}.$$

Then

$$fl(A^TA) = \begin{bmatrix} 1 & 1 \\ 1 & 1 \end{bmatrix}.$$

So we compute $\hat{\sigma}_1 = \sqrt{2}$, $\hat{\sigma}_2 = 0$ leading to the (erroneous) conclusion that the rank of A is 1. Of course, if we could compute in infinite precision, we would find

$$A^TA = \begin{bmatrix} 1+\mu^2 & 1 \\ 1 & 1+\mu^2 \end{bmatrix}$$

with $\sigma_1 = \sqrt{2 + \mu^2}$, $\sigma_2 = |\mu|$ and thus rank $(A) = 2$. The point is that by working with A^TA we have unnecessarily introduced μ^2 into the computations. The above example illustrates a potential pitfall in attempting to form and solve the normal equations in a linear least squares problem and is at the heart of what makes square root filtering so attractive numerically. Very simplistically, square root filtering involves working directly on an "A-matrix," for example, updating it, as opposed to updating an "A^TA-matrix."

Square root filtering is usually implemented with the QR factorization (or some closely related algorithm) as described previously rather than SVD. The key thing to remember is that, in most current computing environments, the condition of the least-squares problem is squared unnecessarily in solving the normal equations. Moreover, critical information may be lost irrecoverably by simply forming A^TA.

Returning now to the SVD, two features of this matrix factorization make it attractive in finite arithmetic: first, it can be computed in a numerically stable way, and second, singular values are well-conditioned. Specifically, there is an efficient and numerically stable algorithm due to Golub and Reinsch [6] which works directly on A to give the SVD. This algorithm has two phases. In the first phase, it computes orthogonal matrices U_1 and V_1 so that $B = {U_1}^T A V_1$ is in bidiagonal form, i.e., only the elements on its diagonal and first superdiagonal are nonzero. In the second phase, the algorithm uses an iterative procedure to compute orthogonal matrices U_2 and V_2 so that ${U_2}^T B V_2$ is diagonal and nonnegative. The SVD defined in Equation 21.25 is then $\Sigma = U^T B V$, where $U = U_1 U_2$ and $V = V_1 V_2$. The computed U and V are orthogonal approximately to the working precision, and the computed singular values are the exact σ_i's for $A + E$ where $\|E\|/\|A\|$ is a modest multiple of ϵ. Fairly sophisticated implementations of this algorithm can be found in [3] and [5]. The well-conditioned nature of the singular values follows from the fact that if A is perturbed to $A + E$, then it can be proved that

$$|\sigma_i(A + E) - \sigma_i(A)| \leq \|E\|.$$

Thus, the singular values are computed with small absolute error although the relative error of sufficiently small singular values is not guaranteed to be small.

It is now acknowledged that the singular value decomposition is the most generally reliable method of determining rank numerically (see [10](p. 589) for a more elaborate discussion). However, it is considerably more expensive to compute than, for example, the QR factorization which, with column pivoting [3], can usually give equivalent information with less computation. Thus, while the SVD is a useful theoretical tool, its use for actual computations should be weighed carefully against other approaches.

Only recently has the problem of numerical determination of rank become well-understood. The essential idea is to try to determine a "gap" between "zero" and the "smallest nonzero singular value" of a matrix A. Since the computed values are exact for a matrix near A, it makes sense to consider the ranks of all matrices in some δ-ball (with respect to the spectral norm $\|\cdot\|$, say) around A. The choice of δ may also be based on measurement errors incurred in estimating the coefficients of A, or the coefficients may be uncertain because of rounding errors incurred in a previous computation. However, even with SVD, numerical determination of rank in finite arithmetic is a difficult problem.

That other methods of rank determination are potentially unreliable is demonstrated by the following example. Consider the Ostrowski matrix $A \in \mathbb{R}^{n \times n}$ whose diagonal elements are all -1, whose upper triangle elements are all $+1$, and whose lower triangle elements are all 0. This matrix is clearly of rank n, i.e., is invertible. It has a good "solid" upper triangular shape. All of its eigenvalues ($= -1$) are well away from zero. Its determinant $(-1)^n$ is definitely not close to zero. But this matrix is, in fact, very nearly singular and becomes more nearly so as n increases.

Note, for example, that

$$\begin{bmatrix} -1 & +1 & \cdots & & \cdots & +1 \\ 0 & \ddots & \ddots & & & \vdots \\ \vdots & \ddots & \ddots & \ddots & & \\ & & \ddots & \ddots & & \vdots \\ \vdots & & & \ddots & \ddots & +1 \\ 0 & \cdots & & \cdots & 0 & -1 \end{bmatrix} \begin{bmatrix} 1 \\ 2^{-1} \\ \vdots \\ 2^{-n+1} \end{bmatrix}$$

$$= \begin{bmatrix} -2^{-n+1} \\ -2^{-n+1} \\ \vdots \\ -2^{-n+1} \end{bmatrix} \rightarrow \begin{bmatrix} 0 \\ 0 \\ \vdots \\ 0 \end{bmatrix} (n \rightarrow +\infty).$$

Moreover, adding 2^{-n+1} to every element in the first column of A gives an exactly singular matrix. Arriving at such a matrix by, say Gaussian elimination, would give no hint as to the near singularity. However, it is easy to check that $\sigma_n(A)$ behaves as 2^{-n+1}. A corollary for control theory is that eigenvalues do not necessarily give a reliable measure of "stability margin." It is useful to note that in this example of an invertible matrix, the crucial quantity, $\sigma_n(A)$, which measures nearness to singularity, is simply $1/\|A^{-1}\|$, and the result is familiar from standard operator theory. There is nothing intrinsic about singular values in this example and, in fact, $\|A^{-1}\|$ might be more cheaply computed or estimated in other matrix norms.

Because rank determination, in the presence of rounding error, is a nontrivial problem, the same difficulties naturally arise in any problem equivalent to, or involving, rank determination, such as determining the independence of vectors, finding the dimensions of certain subspaces, etc. Such problems arise as basic calculations throughout systems, control, and estimation theory. Selected applications are discussed in more detail in [10](p. 74).

Finally, let us close this section with a brief example illustrating a totally inappropriate use of SVD. The rank condition

$$\text{rank } [B, AB, \cdots, A^{n-1}B] = n \qquad (21.26)$$

for the controllability of Equation 21.1 is too well-known. Suppose

$$A = \begin{bmatrix} 1 & \mu \\ 0 & 1 \end{bmatrix}, B = \begin{bmatrix} 1 \\ \mu \end{bmatrix}$$

with $|\mu| < \sqrt{\epsilon}$. Then

$$fl[B, AB] = \begin{bmatrix} 1 & 1 \\ \mu & \mu \end{bmatrix}$$

and now even applying SVD, the erroneous conclusion of uncontrollability is reached. Again the problem is in just forming AB; not even SVD can come to the rescue after that numerical *faux pas*.

21.4 Applications to Systems and Control

A reasonable approach to developing numerically reliable algorithms for computational problems in linear system theory would be to reformulate the problems as concatenations of subproblems for which numerically stable algorithms are available. Unfortunately, one cannot ensure that the stability of algorithms for the subproblems results in stability of the overall algorithm. This requires separate analysis which may rely on the sensitivity or condition of the subproblems. In the next section we show that delicate (i.e., badly-conditioned) subproblems should be avoided whenever possible; a few examples are given where a possibly badly-conditioned step is circumvented by carefully modifying or completing existing algorithms; see, e.g., [10](p. 109).

A second difficulty is the ill-posedness of some of the problems occurring in linear system theory. Two approaches can be adopted. One can develop an acceptable perturbation theory for such problems, using a concept such as *restricted condition* which is condition under perturbations for which a certain property holds, e.g., fixed rank [10](p. 109). One then looks for restricting assumptions that make the problem well-posed. Another approach is to delay any such *restricting choices* to the end and leave it to the user to decide which choice to make by looking at the results. The algorithm then provides quantitative measures that help the user make this choice; see, e.g., [10](p. 171, p. 529). By this approach one may avoid artificial restrictions of the first approach that sometimes do not respect the practical significance of the problem.

A third possible *pitfall* is that many users almost always prefer fast algorithms to slower ones. However, slower algorithms are often more reliable.

In the subsections that follow, we survey a representative selection of numerical linear algebra problems arising in linear systems, control, and estimation theory, which have been examined with some of the techniques described in the preceding sections. Many of these topics are discussed briefly in survey papers such as [7] and [12] and in considerably more detail in the papers included or referenced in [10]. Some of the scalar algorithms discussed here do not extend trivially to the matrix case. When they do, we mention only the matrix case. Moreover, we discuss only the numerical aspects here; for the system-theoretical background, we refer the reader to the control and systems literature.

21.4.1 Some Typical Techniques

Most of the reliable techniques in numerical linear algebra are based on the use of orthogonal transformations. Typical examples of this are the QR decomposition for least squares problems, the Schur decomposition for eigenvalue and generalized eigenvalue problems, and the singular value decomposition for rank determinations and generalized inverses. Orthogonal transformations also appear in most of the reliable linear algebra techniques for control theory. This is partly due to the direct application of existing linear algebra decompositions to problems in control. Obvious examples of this are the Schur approach for solving algebraic Riccati equations, both continuous- and discrete-time [10](p. 529, p. 562, p. 573), for solving Lyapunov equations [10](p. 430), and for performing pole placement [10](p. 415). New orthogonal decompositions have also been introduced that rely heavily on the same principles but were specifically developed for problems encountered in control. Orthogonal state-space transformations on a system $\{A, B, C\}$ result in a new state-space representation $\{U^H AU, U^H B, CU\}$ where U performs some kind of decomposition on the matrices A, B, and C. These special forms, termed "condensed forms," include

- the state Schur form [10](p. 415),
- the state Hessenberg form [10](p. 287),
- the observer Hessenberg form [10](p. 289, p. 392), and
- the controller Hessenberg form [10](p. 128, p. 357).

Staircase forms or block Hessenberg forms are other variants of these condensed forms that have proven useful in dealing with MIMO systems [10](p. 109, p. 186, p. 195).

There are two main reasons for using these orthogonal state-space transformations:

- The numerical sensitivity of the control problem being solved is not affected by these transformations because sensitivity is measured by norms or angles of certain spaces and these are unaltered by orthogonal transformations.
- Orthogonal transformations have minimum condition number, essential in proving bounded error propagation and establishing numerical stability of the algorithm that uses such transformations.

More details on this are given in the paper [10](p. 128) and in subsequent sections where some of these condensed forms are used for particular applications.

21.4.2 Transfer Functions, Poles, and Zeros

In this section, we discuss important structural properties of linear systems and the numerical techniques available for determining them. The transfer function $R(\lambda)$ of a linear system is given by a polynomial representation $V(\lambda)T^{-1}(\lambda)U(\lambda) + W(\lambda)$ or by a state-space model $C(\lambda I - A)^{-1}B + D$. The results in this subsection hold for both the discrete-time case (where λ stands for the shift operator z) and the continuous-time case (where λ stands for the differentiation operator D).

A. The Polynomial Approach

One is interested in a number of structural properties of the transfer function $R(\lambda)$ such as poles, transmission zeros, decoupling zeros, etc. In the scalar case, where $\{T(\lambda), U(\lambda), V(\lambda), W(\lambda)\}$ are scalar polynomials, all of this can be found with a greatest common divisor (GCD) extraction routine and a rootfinder, for which reliable methods exist. In the matrix case, the problem becomes much more complex and the basic method for GCD extraction, the Euclidean algorithm, becomes unstable (see [10](p. 109)). Moreover, other structural elements (null spaces, etc.) come into the picture, making the polynomial approach less attractive than the state-space approach [10](p. 109).

B. The State-Space Approach (see [10](p. 109), and references therein)

The structural properties of interest are poles and zeros of $R(\lambda)$, decoupling zeros, controllable and unobservable subspaces, supremal (A, B)-invariant and controllability subspaces, factorizability of $R(\lambda)$, left and right null spaces of $R(\lambda)$, etc. These concepts are fundamental in several design problems and have received considerable attention over the last few years; see, e.g., [10](pp. 74, 109, 174, 186, 529). In [10](p. 109), it is shown that all the concepts mentioned above can be considered generalized eigenstructure problems and that they can be computed via the Kronecker canonical form of the pencils

$$\begin{array}{cc} [\ \lambda I - A\ ] & [\ \lambda I - A \mid B\ ] \\ \left[\begin{array}{c} \lambda I - A \\ \hline -C \end{array}\right] & \left[\begin{array}{c|c} \lambda I - A & B \\ \hline -C & D \end{array}\right] \end{array} \tag{21.27}$$

or from other pencils derived from these. Backward stable software is also available for computing the Kronecker structure of an arbitrary pencil. A remaining problem here is that determining several of the structural properties listed above may be ill-posed in some cases in which one has to develop the notion of restricted condition (see [10](p. 109)). Sensitivity results in this area are still few but are slowly emerging. A completely different approach is to reformulate the problem as an approximation or optimization problem for which *quantitative measures* are derived, leaving the final choice to the user. Results in this vein are obtained for controllability, observability [10](pp. 171, 186, 195), (almost) (A, B)-invariant, and controllability subspaces.

21.4.3 Controllability and Other "Abilities"

The various "abilities," such as controllability, observability, reachability, reconstructibility, stabilizability, and detectability are basic to the study of linear control and system theory. These concepts can also be viewed in terms of decoupling zeros, controllable and unobservable subspaces, controllability subspaces, etc. mentioned in the previous section. Our remarks here are confined, but not limited, to the notion of controllability.

A large number of algebraic and dynamic characterizations of controllability have been given; see [7] for a sample. But every one of these has difficulties when implemented in finite arithmetic. For a survey of this topic and numerous examples, see [10](p. 186). Part of the difficulty in dealing with controllability numerically lies in the intimate relationship with the invariant subspace problem [10](p. 589). The controllable subspace associated with Equation 21.1 is the smallest A-invariant subspace (subspace spanned by eigenvectors or principal vectors) containing the range of B. Since A-invariant subspaces can be extremely sensitive to perturbation, it follows that, so too, is the controllable subspace. Similar remarks apply to the computation of the so-called controllability indices. The example discussed in the third paragraph of Section 21.2 dramatically illustrates these remarks. The matrix A has but one eigenvector (associated with 0) whereas the slightly perturbed A has n eigenvectors associated with the n distinct eigenvalues.

Recently, attempts have been made to provide numerically stable algorithms for the pole placement problem discussed in a later section. It suffices to mention here that the problem of pole placement by state feedback is closely related to controllability. Recent work on developing numerically stable algorithms for pole placement is based on the reduction of A to a Hessenberg form; see, e.g., [10](pp. 357, 371, 380). In the single-input case, a good approach is the controller Hessenberg form mentioned above where the state matrix A is upper Hessenberg and the input vector B is a multiple of $(1, 0, \cdots, 0)^T$. The pair (A, B) is then controllable if, and only if, all $(n-1)$ subdiagonal elements of A are nonzero. If a subdiagonal element is 0, the system is uncontrollable, and a basis for the uncontrollable subspace is easily constructed. The transfer function gain or first nonzero Markov parameter is also easily constructed from this "canonical form." In fact, the numerically more robust system Hessenberg form, playing an ever-increasing role in system theory is replacing the numerically more fragile special case of the companion or rational canonical or Luenberger canonical form.

A more important aspect of controllability is topological notions such as "near uncontrollability." But there are numerical difficulties here also, and we refer to Parts 3 and 4 of [10] for further details. Related to this is an interesting system-theoretic concept called "balancing" discussed in Moore's paper [10] (p. 171). The computation of "balancing transformations" is discussed in [10](p. 642).

There are at least two distinct notions of near-uncontrollability [7] in the parametric sense and in the energy sense. In the parametric sense, a controllable pair (A, B) is said to be near-uncontrollable if the parameters of (A, B) need be perturbed by only a relatively small amount for (A, B) to become uncontrollable. In the energy sense, a controllable pair is near uncontrollable if large amounts of control energy ($\int u^T u$) are required for a state transfer. The pair

$$A = \begin{pmatrix} 0 & 1 & & \cdots & 0 \\ \vdots & \ddots & \ddots & & \vdots \\ & & & & \\ \vdots & & & \ddots & 1 \\ 0 & \cdots & & \cdots & 0 \end{pmatrix}, \quad B = \begin{pmatrix} 0 \\ \vdots \\ 0 \\ 1 \end{pmatrix}$$

is very near uncontrollable in the energy sense but not as badly so in the parametric sense. Of course, both measures are coordinate dependent and "balancing" is one attempt to remove this coordinate bias. The pair (A, B) above is in "controllable canonical form." It is now known that matrices in this form (specifically, the A matrix in rational canonical form) almost always exhibit poor numerical behavior and are "close to" uncontrollable (unstable, etc.) as the size n increases. For details, see [10](p. 59).

21.4.4 Computation of Objects Arising in the Geometric Theory of Linear Multivariable Control

A great many numerical problems arise in the geometric approach to control of systems modeled as Equation 21.1 and 21.2. Some of these are discussed in the paper by Klema and Laub [10](p. 74). The power of the geometric approach derives in large part from the fact that it is independent of specific coordinate systems or matrix representations. Numerical issues are a separate concern.

A very thorough numerical treatment of numerical problems in linear system theory has been given by Van Dooren [10](p. 109). This work has applications for most calculations with linear state-space models. For example, one byproduct is an extremely reliable algorithm (similar to an orthogonal version of Silverman's structure algorithm) for the computation of multivariable system zeros [10](p. 271). This method involves a generalized eigenvalue problem (the Rosenbrock pencil), but the "infinite zeros" are first removed by deflating the given matrix pencil.

21.4.5 Frequency Response Calculations

Many properties of a linear system such as Equations 21.1 and 21.2 are known in terms of its frequency response matrix

$$G(j\omega) := C(j\omega I - A)^{-1}B + D; \quad (\omega \geq 0) \tag{21.28}$$

(or $G(e^{j\theta})$; $\theta \in [0, 2\pi]$ for Equations 21.3 and 21.4). In fact, various norms of the return difference matrix $I + G(j\omega)$ and related quantities have been investigated in control and system theory to providing robust linear systems with respect to stability, noise response, disturbance attenuation, sensitivity, etc.

Thus it is important to compute $G(j\omega)$ efficiently, given A, B, and C for a (possibly) large number of values of ω (for convenience we take D to be 0 because if it is nonzero it is trivial to add to G). An efficient and generally applicable algorithm for this problem is presented in [10](p. 287). Rather than solving the linear equation $(j\omega I - A)X = B$ with dense unstructured A, which would require $O(n^3)$ operations for each successive value of ω, the new method initially reduces A to upper Hessenberg form H. The orthogonal state-space coordinate transformations used to obtain the Hessenberg form of A are incorporated into B and C giving $\tilde{B}$ and $\tilde{C}$. As ω varies, the coefficient matrix in the linear equation $(j\omega I - H)X = \tilde{B}$ remains in upper Hessenberg form. The advantage is that X can now be found in $O(n^2)$ operations rather than $O(n^3)$ as before, a substantial saving. Moreover, the method is numerically very stable (via either LU or QR factorization) and has the advantage of being independent of the eigenstructure (possibly ill-conditioned) of A. Another efficient and reliable algorithm for frequency response computation [10](p. 289) uses the observer Hessenberg form mentioned in Section 21.4.1 together with a determinant identity and a property of the LU decomposition of a Hessenberg matrix.

The methods above can also be extended to state-space models in implicit form, i.e., where Equation 21.1 is replaced by

$$E\dot{x} = Ax + Bu. \tag{21.29}$$

Then Equation 21.28 is replaced with

$$G(j\omega) = C(j\omega E - A)^{-1}B + D, \tag{21.30}$$

and the initial triangular/Hessenberg reduction [6] can be employed again to reduce the problem to updating the diagonal of a Hessenberg matrix and consequently an $O(n^2)$ problem.

A recent advance for the frequency response evaluation problem is using matrix interpolation methods to achieve even greater computational efficiency.

21.4.6 Numerical Solution of Linear Ordinary Differential Equations and Matrix Exponentials

The "simulation" or numerical solution of linear systems of ordinary differential equations (ODEs) of the form,

$$\dot{x}(t) = Ax(t) + f(t), \quad x(0) = x_0, \tag{21.31}$$

is a standard problem that arises in finding the time response of a system in state-space form. However, there is still debate as to the most effective numerical algorithm, particularly when A is defective (i.e., when A is $n \times n$ and has fewer than n linearly independent eigenvectors) or nearly defective. The most common approach involves computing the matrix exponential e^{tA}, because the solution of Equation 21.31 can be written simply as

$$x(t) = e^{tA}x_0 + \int_0^t e^{(t-s)A} f(s)\, ds.$$

A delightful survey of computational techniques for matrix exponentials is given in [10](p. 649). Nineteen "dubious" ways are explored (there are many more ways not discussed) but no clearly superior algorithm is singled out. Methods based on Padé approximation or reduction of A to real Schur form are generally attractive while methods based on Taylor series or the characteristic polynomial of A are generally found unattractive. An interesting open problem is the design of a special algorithm for the matrix exponential when the matrix is known a priori to be stable ($\Lambda(A)$ in the left half of the complex plane).

The reason for the adjective "dubious" in the title of [10](p. 649) is that in many (maybe even most) circumstances, it is better to treat Equation 21.31 as a system of differential equations, typically stiff, and to apply various ODE techniques, specially tailored to the linear case. ODE techniques are preferred when A is large and sparse for, in general, e^{tA} is unmanageably large and dense.

21.4.7 Lyapunov, Sylvester, and Riccati Equations

Certain matrix equations arise naturally in linear control and system theory. Among those frequently encountered in the analysis and design of continuous-time systems are the Lyapunov equation

$$AX + XA^T + Q = 0, \tag{21.32}$$

and the Sylvester equation

$$AX + XF + Q = 0. \tag{21.33}$$

The appropriate discrete-time analogs are

$$AXA^T - X + Q = 0 \tag{21.34}$$

and

$$AXF - X + Q = 0. \tag{21.35}$$

Various hypotheses are posed for the coefficient matrices A, F, Q to ensure certain properties of the solution X.

The literature in control and system theory on these equations is voluminous, but most of it is *ad hoc*, at best, from a numerical point of view, with little attention to questions of numerical stability, conditioning, machine implementation, and the like.

For the Lyapunov equation the best overall algorithm in terms of efficiency, accuracy, reliability, availability, and ease of use is that of Bartels and Stewart [10](p. 430). The basic idea is to reduce A to quasi-upper-triangular form [or real Schur form (RSF)] and to perform a back substitution for the elements of X.

An attractive algorithm for solving Lyapunov equations has been proposed by Hammarling [10](p. 500). This algorithm is a variant of the Bartels-Stewart algorithm but instead solves directly for the Cholesky factor Y of X: $Y^TY = X$ and Y is upper triangular. Clearly, given Y, X is easily recovered if necessary. But in many applications, for example [10](p. 642), only the Cholesky factor is required.

For the Lyapunov equation, when A is stable, the solutions of the equations above are also equal to the reachability and observability Grammians $P_r(T)$ and $P_o(T)$, respectively, for $T = +\infty$ for the system $\{A, B, C\}$:

$$\begin{aligned} P_r(T) &= \int_0^T e^{tA}BB^Te^{tA^T}\,dt; \\ P_o(T) &= \int_0^T e^{tA^T}C^TCe^{tA}dt \\ P_r(T) &= \textstyle\sum_{k=0}^T A^kBB^T(A^T)^k; \\ P_o(T) &= \textstyle\sum_{k=0}^T (A^T)^kC^TCA^k. \end{aligned} \tag{21.36}$$

These can be used along with some additional transformations (see [10](pp. 171, 642)) to compute so-called *balanced* realizations $\{\tilde{A}, \tilde{B}, \tilde{C}\}$. For these realizations both P_o and P_r are equal and diagonal. These realizations have some nice sensitivity properties with respect to poles, zeros, truncation errors in digital filter implementations, etc. [10](p. 171). They are, therefore, recommended whenever the choice of a realization is left to the user. When A is not stable, one can still use the *finite range* Grammians Equation 21.36, for $T < +\infty$, for balancing [10](p. 171). A reliable method for computing integrals and sums of the type

Equation 21.36 can be found in [10](p. 681). It is also shown in [10](p. 171) that the reachable subspace and the unobservable subspace are the image and the kernel of $P_r(T)$ and $P_o(T)$, respectively. From these relationships, sensitivity properties of the spaces under perturbations of $P_r(T)$ and $P_o(T)$ can be derived.

For the Sylvester equation, the Bartels-Stewart algorithm reduces both A and F to real Schur form (RSF) and then a back substitution is done. It has been demonstrated in [10](p. 495) that some improvement in this procedure is possible by only reducing the larger of A and F to upper Hessenberg form. The stability of this method has been analyzed in [10](p. 495). Although only *weak stability* is obtained, this is satisfactory in most cases.

Algorithms are also available in the numerical linear algebra literature for the more general Sylvester equation

$$A_1 X {F_1}^T + A_2 X {F_2}^T + Q = 0$$

and its symmetric Lyapunov counterpart

$$AXF^T + FXA^T + Q = 0.$$

Questions remain about estimating the condition of Lyapunov and Sylvester equations efficiently and reliably in terms of the coefficient matrices. A deeper analysis of the Lyapunov and Sylvester equations is probably a prerequisite to at least a better understanding of condition of the Riccati equation for which, again, there is considerable theoretical literature but not as much known from a purely numerical point of view. The symmetric $n \times n$ algebraic Riccati equation takes the form

$$Q + AX + XA^T - XGX = 0 \qquad (21.37)$$

for continuous-time systems and

$$A^T XA - X - A^T XG_1(G_2 + G_1^T XG_1)^{-1} G_1^T XA + Q = 0 \qquad (21.38)$$

for discrete-time systems. These equations appear in several design/analysis problems, such as optimal control, optimal filtering, spectral factorization, e.g., see the papers in Part 7 of [10] and references therein. Again, appropriate assumptions are made on the coefficient matrices to guarantee the existence and/or uniqueness of certain kinds of solutions X. Nonsymmetric Riccati equations of the form

$$Q + A_1 X + XA_2 - XGX = 0 \qquad (21.39)$$

for the continuous-time case (along with an analog for the discrete-time case) are also studied and can be solved numerically by the techniques discussed below.

Several algorithms have been proposed based on different approaches. One of the more reliable general-purpose methods for solving Riccati equations is the Schur method [10](p. 529). For the case of Equation 21.37, for example, this method is based upon reducing the associated $2n \times 2n$ Hamiltonian matrix

$$\begin{pmatrix} A & -G \\ -Q & -A^T \end{pmatrix} \qquad (21.40)$$

to RSF. If the RSF is ordered so that its stable eigenvalues (there are exactly n of them under certain standard assumptions) are in the upper left corner, the corresponding first n vectors of the orthogonal matrix, which effects the reduction, forms a basis for the stable eigenspace from which the nonnegative definite solution X is then easily found.

Extensions to the basic Schur method have been made [10] (p. 562, p. 573) which were prompted by the following situations:

- G in Equation 21.37 is of the form $BR^{-1}B^T$ where R may be nearly singular, or G_2 in Equation 21.38 may be exactly or nearly singular.
- A in Equation 21.38 is singular (A^{-1} is required in the classical approach involving a symplectic matrix which plays a role analogous to Equation 21.40).

This resulted in the generalized eigenvalue approach requiring the computation of a basis for the deflating subspace corresponding to the stable generalized eigenvalues. For the solution of Equation 21.37, the generalized eigenvalue problem is given by

$$\lambda \begin{bmatrix} I & 0 & 0 \\ 0 & I & 0 \\ 0 & 0 & 0 \end{bmatrix} - \begin{bmatrix} A & 0 & B \\ -Q & -A^T & 0 \\ 0 & B^T & R \end{bmatrix}; \qquad (21.41)$$

for Equation 21.38, the corresponding problem is

$$\lambda \begin{bmatrix} I & 0 & 0 \\ 0 & A^T & 0 \\ 0 & {G_1}^T & 0 \end{bmatrix} - \begin{bmatrix} A & 0 & -G_1 \\ -Q & I & 0 \\ 0 & 0 & G_2 \end{bmatrix}. \qquad (21.42)$$

The extensions above can be generalized even further, as the following problem illustrates. Consider the optimal control problem

$$\min \frac{1}{2} \int_0^{+\infty} [x^T Qx + 2x^T Su + u^T Ru] dt \qquad (21.43)$$

subject to

$$E\dot{x} = Ax + Bu. \qquad (21.44)$$

The Riccati equation associated with Equations 21.43 and 21.44 then takes the form

$$\begin{aligned} &E^T XBR^{-1}B^T XE - (A - BR^{-1}S^T)^T XE \\ &- E^T X(A - BR^{-1}S^T) - Q + SR^{-1}S^T = 0 \end{aligned} \qquad (21.45)$$

or

$$\begin{aligned} &(E^T XB + S)R^{-1}(B^T XE + S^T) \\ &- A^T XE - E^T XA - Q = 0. \end{aligned} \qquad (21.46)$$

This so-called "generalized" Riccati equation can be solved by considering the associated matrix pencil

$$\begin{pmatrix} A & 0 & B \\ -Q & -A^T & -S \\ S^T & B^T & R \end{pmatrix} - \lambda \begin{pmatrix} E & 0 & 0 \\ 0 & E^T & 0 \\ 0 & 0 & 0 \end{pmatrix}. \tag{21.47}$$

Note that S in Equation 21.43 and E in Equation 21.44 are handled directly and no inverses appear. The presence of a nonsingular E in state-space models of the form Equation 21.44 adds no particular difficulty to the solution process and is numerically the preferred form if E is, for example, near singular or even sparse. Similar remarks apply to the frequency response problem in Equations 21.29 and 21.30 and, indeed, throughout all of linear control and system theory. The stability and conditioning of these approaches are discussed in [10](pp. 529, 573). Other methods, including Newton's method and iterative refinement, have been analyzed in, for example, [10](p. 517). Numerical algorithms for handling Equations 21.41, 21.42, and 21.47 and a large variety of related problems are described in [10](pp. 421, 573). A thorough survey of the Schur method, generalized eigenvalue/eigenvector extensions, and the underlying algebraic structure in terms of "Hamiltonian pencils" and "symplectic pencils" is included in [8], [2].

Schur techniques can also be applied to Riccati differential and difference equations and to nonsymmetric Riccati equations which arise, for example, in invariant imbedding methods for solving linear two-point boundary value problems.

As with the linear Lyapunov and Sylvester equations, only recently have satisfactory results been obtained concerning condition of Riccati equations, a topic of great interest independent of the solution method used, be it a Schur-type method or one of numerous alternatives.

A very interesting class of invariant-subspace-based algorithms for solving the Riccati equation and related problems uses the so-called matrix sign function. These methods, which are particularly attractive for very large order problems, are described in detail in [10](p. 486) and the references therein. These algorithms are based on Newton's method applied to a certain matrix equation. A new family of iterative algorithms of arbitrary order convergence has been developed in [10](p. 624). This family of algorithms can be parallelized easily and yields a viable method of solution for very high order Riccati equations.

21.4.8 Pole Assignment and Observer Design

Designing state or output feedback for a linear system, so that the resulting closed-loop system has a desired set of poles, can be considered an inverse eigenvalue problem. The state feedback pole assignment problem is as follows: Given a pair (A, B), one looks for a matrix F so that the eigenvalues of the matrix

$$A_F = A + BF$$

lie at specified locations or in specified regions. Many approaches have been developed for solving this problem. However, only recently has the emphasis shifted towards numerically reliable methods and consideration of the numerical sensitivity of the problem, e.g., see the papers in Part 6 of [10]. Special cases of the pole assignment problem arise in observer design [10](p. 407) and in deadbeat control for discrete-time systems (where $A + BF$ is required to be nilpotent) [10](p. 392). The numerically reliable methods for pole assignment are based on reducing A to either a real Schur form [10](p. 415), or to a Hessenberg or block Hessenberg (staircase) form [10](p. 357, p. 380). The latter may be regarded a numerically robust alternative to the controllable or Luenberger canonical form whose computation is known to be numerically unreliable [10](p. 59). For multi-input systems, the additional freedom available in the state-feedback matrix can be used for eigenvector assignment and sensitivity minimization for the closed-loop poles [10](p. 333). There the resulting matrix A_F is not computed directly but instead the matrices Λ and X of the decomposition

$$A_F = X \Lambda X^{-1}$$

are computed via an iterative technique. The iteration aims to minimize the sensitivity of the placed eigenvalues λ_i or to maximize the orthogonality of the eigenvectors x_i.

Pole assignment by output feedback is more difficult, theoretically as well as computationally. Consequently, there are few numerically reliable algorithms available [10](p. 371). Other recent work on pole assignment has been concerned with generalized state space or descriptor systems.

The problem of observer design for a given state-space system $\{A, B, C\}$ is finding matrices T, A_K, and K so that

$$TA_K - AT = KC \tag{21.48}$$

whereby the spectrum of A_K is specified. Because this is an underdetermined (and nonlinear) problem in the unknown parameters of T, A_K, and K, one typically sets $T = I$ and Equation 21.48 then becomes

$$A_K = A + KC$$

which is a transposed pole placement problem. In this case the above techniques of pole placement automatically apply here. In reduced order design, T is nonsquare and thus cannot be equated to the identity matrix. One can still solve Equation 21.48 via a recurrence relationship when assuming A_K in Schur form [10](p. 407).

21.4.9 Robust Control

In the last decade, there has been significant growth in the theory and techniques of robust control; see, e.g., [4] and the references therein. However, the area of robust control is still evolving and its numerical aspects have just begun to be addressed [9]. Consequently, it is premature to survey reliable numerical algorithms in the area. To suggest the flavor of the numerical and computational issues involved, in this section we consider a development in robust control that has attracted a great deal of attention, the so-called H_∞ approach. H_∞ and the related structured singular value approach have provided a powerful framework for synthesizing *robust* controllers for linear systems. The controllers are robust because they achieve desired system performance despite a significant amount of uncertainty in the system.

In this section, we denote by $\mathbb{R}(s)^{n\times m}$ the set of proper real rational matrices of dimension $n \times m$. The H_∞ norm of a stable matrix $G(s) \in \mathbb{R}(s)^{n\times m}$ is defined as

$$\|G(s)\|_\infty := \sup_{\omega\in\mathbb{R}} \sigma_{\max}[G(j\omega)] \tag{21.49}$$

where $\sigma_{\max}[\cdot]$ denotes the largest singular value of a (complex) matrix. Several iterative methods are available for computing this norm. In one approach, a relationship is established between the singular values of $G(jw)$ and the imaginary eigenvalues of a Hamiltonian matrix obtained from a state-space realization of $G(s)$. This result is then used to develop an efficient bisection algorithm for computing the H_∞ norm of $G(s)$.

To describe the basic H_∞ approach, consider a linear, time-invariant system described by the state-space equations

$$\begin{aligned} \dot{x}(t) &= Ax(t) + B_1 w(t) + B_2 u(t), \\ z(t) &= C_1 x(t) + D_{11} w(t) + D_{12} u(t), \quad (21.50) \\ \text{and} \quad y(t) &= C_2 x(t) + D_{21} w(t) + D_{22} u(t), \end{aligned}$$

where $x(t) \in \mathbb{R}^n$ denotes the state vector; $w(t) \in \mathbb{R}^{m_1}$ is the vector of disturbance inputs; $u(t) \in \mathbb{R}^{m_2}$ is the vector of control inputs, $z(t) \in \mathbb{R}^{p_1}$ is the vector of error signals, and $y(t) \in \mathbb{R}^{p_2}$ is the vector of measured variables. The transfer function relating the inputs $\begin{bmatrix} w \\ u \end{bmatrix}$ to the outputs $\begin{bmatrix} z \\ y \end{bmatrix}$ is

$$G(s) := \begin{bmatrix} G_{11}(s) & G_{12}(s) \\ G_{21}(s) & G_{22}(s) \end{bmatrix} \tag{21.51}$$

$$= \begin{bmatrix} D_{11} & D_{12} \\ D_{21} & D_{22} \end{bmatrix} + \begin{bmatrix} C_1 \\ C_2 \end{bmatrix} (sI - A)^{-1} \begin{bmatrix} B_1 & B_2 \end{bmatrix}. \tag{21.52}$$

Implementing a feedback controller defined by

$$u = K(s)y \tag{21.53}$$

where $K(s) \in \mathbb{R}(s)^{m_2\times p_2}$, we get the closed-loop transfer matrix $T_{zw}(s) \in \mathbb{R}(s)^{p_1\times m_1}$ from the disturbance w to the regulated output z

$$T_{zw} := G_{11} + G_{12}K(I - G_{22}K)^{-1}G_{21}. \tag{21.54}$$

Next, define the set $\mathcal{K}$ of all internally stabilizing feedback controllers for the system in Equation 21.50, i.e.,

$$\mathcal{K} := \{K(s) \in \mathbb{R}(s)^{m_2\times p_2} : T_{zw}(s) \text{ is internally stable}\}.$$

Now let $K(s) \in \mathcal{K}$, and define

$$\gamma := \|T_{zw}(s)\|_\infty. \tag{21.55}$$

Then the H_∞ control problem is to find a controller $K(s) \in \mathcal{K}$ that minimizes γ. The optimal value of γ is defined as

$$\gamma_{\text{opt}} := \inf_{K\in\mathcal{K}} \|T_{zw}(s)\|_\infty. \tag{21.56}$$

This problem was originally formulated in an input-output setting, and the early methods for computing γ_{opt} used either an iterative search involving spectral factorization and solving the resulting Nehari problem or computed the spectral norm of the associated Hankel plus Toeplitz operator. In a recent state-space formulation for computing γ_{opt}, promising from the viewpoint of numerical computation, the problem is formulated in terms of two algebraic Riccati equations which depend on a gain parameter γ. Then, under certain assumptions (see e.g. [9] for details), it can be shown that for a controller $K(s) \in \mathcal{K}$ to exist so that $\|T_{zw}\|_\infty < \gamma$, three conditions have to be satisfied, namely, stabilizing solutions exist for the two Riccati equations, and the spectral radius of the product of the solutions is bounded by γ^2. If these conditions are satisfied for a particular value of γ, the corresponding controller $K(s)$ can be obtained from the solutions of the Riccati equations. The optimal gain, γ_{opt}, is the infimum over all suboptimal values of γ such that the three conditions are satisfied.

The approach above immediately suggests a bisection-type algorithm for computing γ_{opt}. However, such an algorithm can be very slow in the neighborhood of the optimal value. To obtain speedup near the solution, a gradient approach is proposed in [9]. The behavior of the Riccati solution as a function of γ is used to derive an algorithm that couples a gradient method with bisection. It has been pointed out in [9] that the Riccati equation can become ill-conditioned as the optimal value of γ is approached. It is therefore recommended in [9] that, instead of computing the Riccati solutions explicitly, invariant subspaces of the associated Hamiltonian matrices should be used.

21.5 Mathematical Software

A. General Remarks

The previous sections have highlighted some topics from numerical linear algebra and their application to numerical problems arising in systems, control, and estimation theory. These problems represent only a very small subset of numerical problems of interest in these fields but, even for problems apparently "simple" from a mathematical viewpoint, the myriad of details which constitute a sophisticated implementation become so overwhelming that the only effective means of communicating an algorithm is through mathematical software. Mathematical or numerical software is an implementation on a computer of an algorithm for solving a mathematical problem. Ideally, such software would be reliable, portable, and unaffected by the machine or system environment.

The prototypical work on reliable, portable mathematical software for the standard eigenproblem began in 1968. EISPACK, Editions I and II ([5], [11]), were an outgrowth of that work. Subsequent efforts of interest to control engineers include LINPACK

[3] for linear equations and linear least squares problems, FUNPACK (Argonne) for certain function evaluations, MINPACK (Argonne) for certain optimization problems, and various ODE and PDE codes. High quality algorithms are published regularly in the *ACM Transactions on Mathematical Software*. Recently, LAPACK, the successor to LINPACK and EISPACK has been released [1]. This software has been designed to run efficiently on a wide range of machines, including vector processors, shared-memory multiprocessors, and high-performance workstations.

Technology to aid in developing mathematical software in Fortran has been assembled as a package called TOOLPACK. Mechanized code development offers other advantages with respect to modifications, updates, versions, and maintenance.

Inevitably, numerical algorithms are strengthened when their mathematical software is portable because they can be used widely. Furthermore, such software is markedly faster, by factors of 10 to 50, than earlier and less reliable codes.

Many other features besides portability, reliability, and efficiency characterize "good" mathematical software, for example,

- high standards of documentation and style so as to be easily understood and used,
- ease of use; ability of the user to interact with the algorithm,
- consistency/compatibility/modularity in the context of a larger package or more complex problem,
- error control, exception handling,
- robustness in unusual situations,
- graceful performance degradation as problem domain boundaries are approached,
- appropriate program size (a function of intended use, e.g., low accuracy, real-time applications),
- availability and maintenance,
- "tricks" such as underflow-/overflow-proofing, if necessary, and implementation of columnwise or rowwise linear algebra.

Clearly, the list can go on.

What becomes apparent from these considerations is that evaluating mathematical software is a challenging task. The quality of software is largely a function of its operational specifications. It must also reflect the numerical aspects of the algorithm being implemented. The language used and the compiler (e.g., optimizing or not) for that language have an enormous impact on quality, perceived and real, as does the underlying hardware and arithmetic. Different implementations of the same algorithm can have markedly different properties and behavior.

One of the most important and useful developments in mathematical software for most control engineers has been very high level systems such as Matlab, Xmath, Ctrl-C, etc. These systems spare the engineer the drudgery of working at a detailed level with languages such as Fortran and C, and they provide a large number of powerful computational "tools" (frequently through the availability of formal "toolboxes"). For many problems, the engineer must still have some knowledge of the algorithmic details embodied in such a system.

B. Mathematical Software in Control

Many aspects of systems, control, and estimation theory are at the stage from which one can start the research and design necessary to produce reliable, portable mathematical software. Certainly many of the underlying linear algebra tools (for example, in EISPACK, LINPACK, and LAPACK) are considered sufficiently reliable to be used as black, or at least gray, boxes by control engineers. Much of that theory and methodology can and has been carried over to control problems, but this applies only to a few basic control problems. Typical examples are Riccati equations, Lyapunov equations, and certain basic state-space transformations and operations. Much of the work in control, particularly design and synthesis, is simply not amenable to nice, "clean" algorithms. The ultimate software must be capable of enabling a dialogue between the computer and the control engineer, but with the latter probably still making the final engineering decisions.

21.6 Concluding Remarks

Several numerical issues and techniques from numerical linear algebra together with a number of important applications of these ideas have been outlined. A key question in these and other problems in systems, control, and estimation theory is what can be computed reliably and used in the presence of parameter uncertainty or structural constraints (e.g., certain "hard zeros") in the original model, and rounding errors in the calculations. However, because the ultimate goal is to solve real problems, reliable tools (mathematical software) and experience must be available to effect real solutions or strategies. The interdisciplinary effort during the last decade has significantly improved our understanding of the issues involved in reaching this goal and has resulted in some high quality control software based on numerically reliable and well-tested algorithms. This provides clear evidence of the fruitful symbiosis between numerical analysis and numerical problems from control. We expect this symbiotic relationship to flourish as control engineering realizes the full potential of the computing power becoming more widely available in multiprocessing systems and high-performance workstations. However, as in other applications areas, software continues to act as a constraint and a vehicle for progress. Unfortunately, high quality software is very expensive.

In this chapter we have focused only on dense numerical linear algebra problems in systems and control. Several related topics that have not been covered here are, for example, parallel algorithms, algorithms for sparse or structured matrices, optimization algorithms, ODE algorithms, algorithms for differential-algebraic systems, and approximation algorithms. These areas are well-established in their own right, but for control applications a lot of groundwork remains undone. The main reason we have confined ourselves to dense numerical linear algebra problems in systems and control is that, in our opinion, this area has reached a mature level where definitive statements and recommendations can be made about various algorithms and other developments. We hope that this has been reflected in this chapter.

References

[1] Anderson, E., Bai, Z., Bischof, C., Demmel, J., Dongarra, J., DuCroz, J., Greenbaum, A., Hammarling, S., McKenney, A., Ostrouchov, S., and Sorenson, D., *LAPACK Users' Guide*, SIAM, Philadelphia, PA, 1992.

[2] Bunse-Gerstner, A., Byers, R., and Mehrmann, V., Numerical methods for algebraic Riccati equations. In *Proc. Workshop on the Riccati Equation in Control, Systems, and Signals (Como, Italy)* Bittanti, S., Ed., Pitagora, Bologna, Italy, 1989, pp 107–116.

[3] Dongarra, J.J., Bunch, J.R., Moler, C.B., and Stewart, G.W., *LINPACK Users' Guide*, SIAM, Philadelphia, PA, 1979.

[4] Dorato, P. and Yedavalli, R.K., Eds., *Recent Advances in Robust Control*, Selected Reprint Series, IEEE, New York, 1990.

[5] Garbow, B.S., Boyle, J.M., Dongarra, J.J., and Moler, C.B., *Matrix Eigensystem Routines—EISPACK Guide Extension*, in *Lecture Notes in Computer Science*, Springer, New York, 1977, vol. 51.

[6] Golub, G.H. and Van Loan, C.F., *Matrix Computations*, 2nd ed., Johns Hopkins University Press, Baltimore, MD, 1989.

[7] Laub, A.J., Survey of computational methods in control theory, in *Electric Power Problems: The Mathematical Challenge*, Erisman, A.M., Neves, K.W., and Dwarakanath, M.H., Eds., SIAM, Philadelphia, PA, 1980, pp 231–260.

[8] Laub, A.J., Invariant subspace methods for the numerical solution of Riccati equations. In *The Riccati Equation* Bittanti, S., Laub, A.J., and Willems, J.C., Eds., Springer, Berlin, 1991, pp 163–196.

[9] Pandey, P. and Laub, A.J., Numerical issues in robust control design techniques, in *Control and Dynamic Systems – Advances in Theory and Applications: Digital and Numeric Techniques and Their Applications in Control Systems*, Leondes, C.T., Ed., Academic, San Diego, CA, 1993, vol. 55, pp 25–50.

[10] Patel, R.V., Laub, A.J., and Van Dooren, P.M., Eds., *Numerical Linear Algebra Techniques for Systems and Control*, Selected Reprint Series, IEEE Press, New York, 1994.

[11] Smith, B.T, Boyle, J.M., Dongarra, J.J., Garbow, B.S., Ikebe, Y., Klema, V.C., and Moler, C.B., *Matrix Eigensystem Routines – EISPACK Guide*, in *Lecture Notes in Computer Science*. Springer, New York, 1976, vol. 6.

[12] Van Dooren, P., Numerical aspects of system and control algorithms, *Journal A*, 30, 1989, pp 25–32.

[13] Wilkinson, J.H., *The Algebraic Eigenvalue Problem*, Oxford University Press, Oxford, England, 1965.

22

Software for Modeling and Simulating Control Systems

Martin Otter
Institute for Robotics and System Dynamics, German Aerospace Research Establishment Oberpfaffenhofen (DLR), Wessling, Germany

François E. Cellier
Department of Electrical and Computer Engineering, The University of Arizona, Tucson, AZ

22.1 Introduction

Software for the simulation of continuous-time systems was first standardized in 1967 [4]. The standard committee consisted largely of control engineers. Thus, one would expect that today's simulation languages for continuous system simulation should be particularly well suited for modeling and simulating control systems. This chapter answers the question of whether this expectation holds true or not.

There has always been a strong link between the control and simulation communities. On the one hand, simulation is an extremely important tool for every control engineer who is doing practical control system design in industry. For arbitrarily nonlinear plants, there is often no alternative to designing controllers by means of trial and error, using computer simulation. Thus, there is hardly any control engineer who wouldn't be using simulation, at least occasionally. On the other hand, although simulation can be (and has been) applied to virtually all fields of science and engineering (and some others as well), control engineers have always been among the most cherished of its customers — after all, they have paid for the butter on the bread of many a simulation software designer for years. Moreover, a good number of today's simulation researchers received their graduate education in control engineering.

There exist on the market many highly successful special-purpose simulation software tools, e.g., for the simulation of electronic circuitry, or for the simulation of multibody system dynamics, and there is (or at least used to be) a good reason for that. However, there is no market to speak of for special-purpose control system simulators, in spite of the fact that control is such an important application of simulation. The reason for this seeming discrepancy is that control systems contain not only a controller, but also a plant, which can be basically anything. Thus, a simulation tool that is able to simulate control systems must basically be able to simulate pretty much anything.

Hence a substantial portion of this chapter is devoted to a discussion of general-purpose simulation software. Yet, control systems do call for a number of special features in a simulation tool, and these features are pointed out explicitly.

This chapter is structured in three parts. In a first section, the special demands of control systems to a general-purpose simulation tool are outlined. In a second part, the chapter classifies the existing modeling and simulation tools and mentions a few of them explicitly. The chapter ends with a critical discussion of some of the shortcomings of the currently available simulation tools for modeling and simulating control systems.

This chapter is written with several different customer populations in mind. It should be useful reading for the average practical control engineer who needs to decide which simulation tool to acquire and, maybe even more importantly, what questions to ask when talking to a simulation vendor. It should, however, also be useful for simulation software vendors who wish to upgrade their tools to better satisfy the needs of an important subset of their customer base, namely the control engineers. It should finally appeal to the simulation research community by presenting a state-of-the-art picture of what has been accomplished in

0-8493-8570-9/96/$0.00+$.50

control system simulation so far, and where some of the still unresolved issues are that might be meaningful to address in the future.

22.2 Special Demands of Control Engineers for a Simulation Tool

This section discusses the special requirements of control engineers as far as simulation tools are concerned.

22.2.1 Block Diagram Editors

Block diagrams are the most prevalent modeling tool of the control engineer. Figure 22.1 shows a typical block diagram of a control loop around a single-input single-output (SISO) plant. Evidently, control engineers would like to describe their systems

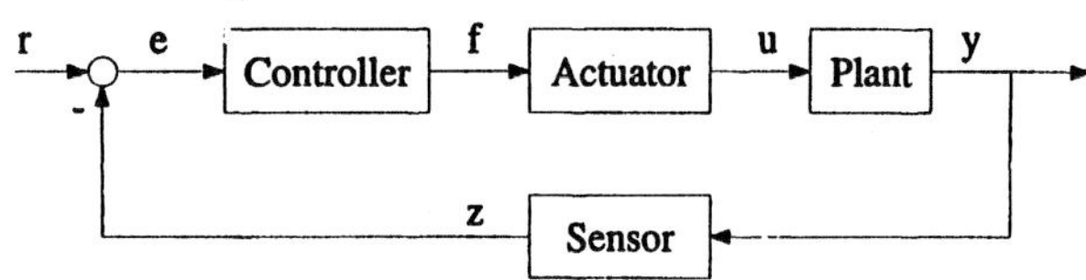

Figure 22.1 Typical control loop around a SISO plant.

in the simulation model in exactly the same fashion. After all, a "model" is simply an encoded form of the knowledge available about the system under study.

Why are block diagrams so essential to control engineers? Most control systems of interest are man-made. Thus, the control engineer has a say on how the signals inside the control system influence each other. In particular, control engineers have learned to design their control systems such that the behavior of each block is, for all practical purposes, independent of the block(s) that it feeds. This can be accomplished by placing voltage follower circuits in between neighboring blocks, as shown in Figure 22.2.

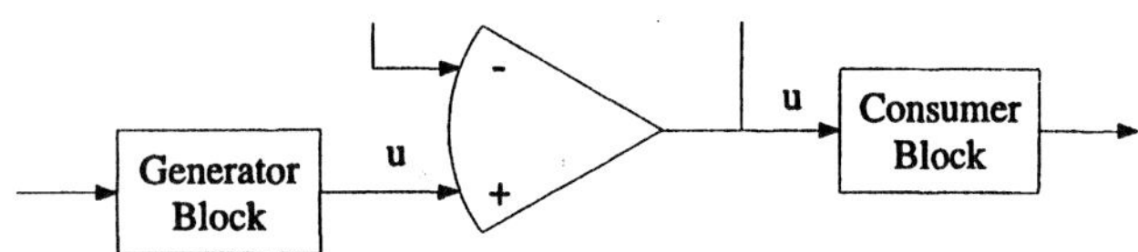

Figure 22.2 Voltage follower circuit for decoupling neighboring blocks.

In the block diagram, these voltage follower circuits are never actually shown. They are simply assumed to be present at all times. Control engineers do this because it simplifies the control system analysis, and thereby, indirectly, also the control system design. Furthermore, it is exactly the same mechanism that helps with decoupling the reaction of the control system to different control inputs. If you depress the gas pedal of your car, you want your car to speed up, and not make a left turn at the same time.

Evidently, it is possible to use block diagrams also to describe any other type of system, such as the simple electrical circuit shown in Figure 22.3. However, this is an abuse of the concept

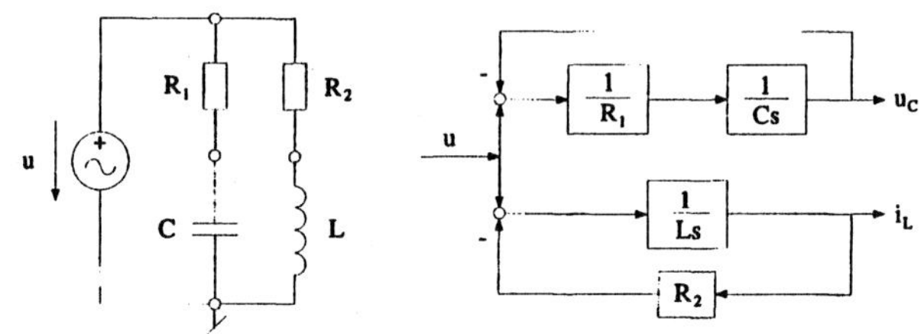

Figure 22.3 Electrical circuit described as block diagram.

of a block diagram. Signals, such as the electrical potential on, and the current through, a wire, that physically belong together and are inseparable from each other, get separated in the block diagram into two totally independent signals. Thereby, the block diagram loses all of its topological significance.

Chapter 6 discusses explicitly the use of block diagrams in control engineering. Because control engineers are so important to them, most simulation software vendors offer today a block-diagram editor as the graphical modeling tool. Unfortunately, block-diagram editors are not a panacea for all graphical modeling needs. Block-diagram editors are certainly not the right tool to describe, e.g., electrical circuits or multibody systems.

22.2.2 Hierarchical Modeling

Control systems are frequently built like an onion. One control loop encompasses another. For example, it is quite common that a local nonlinear control loop is built around a nonlinear robot arm, with the purpose of *linearizing* the behavior of the robot arm, such that, from the outside, the arm with its local controller looks like a linear system. This system then becomes a block in another control system at a hierarchically higher level. The purpose of that control layer may be to *decouple* the control inputs from each other, such that each control input drives precisely one link (the physical configuration may be different). This control system then turns into a series of blocks in a yet higher-level control configuration, in which each individual control input is controlled for performance.

Evidently, control engineers would like their block-diagram editors to behave in exactly the same fashion. One entire block diagram may become a single block at the next higher level. Most of the block-diagram editors currently on the market offer such a hierarchical decomposition facility.

22.2.3 Plant Modeling

One part of the control system to be simulated is the plant to be controlled. As was mentioned earlier, this plant can represent anything that is controllable. It can be a thermal power plant, or an aircraft, or an aquarium for tropical fish. In order to test his or her controller design, the control engineer should be able to simulate the plant with the control system around it.

As was mentioned before, block diagrams are hardly the right tool to model a physical plant. In Chapter 7, the problem of modeling physical systems is discussed in greater detail.

However, it should still be mentioned also that the controllers, after having been designed in an abstract fashion, need to be implemented using physical components. Although the control engineer can choose these components, they may still have some nonideal characteristics, the effect of which ought to be analyzed before the system is actually built. In this case, even the controller blocks become physical systems, and the same restrictions that were previously mentioned with respect to the physical plant to be controlled apply to them as well.

In summary, block diagrams are only useful to describe the higher levels of the control system architecture, but are rarely a good choice for describing the physical layer at the bottom of the hierarchy.

22.2.4 Coupling Models From Different Sources

Control systems are often interdisciplinary. A car manufacturer may want his control engineers to simulate the behavior of the engine before the new model is ever built. However, the engine contains the fuel delivery system, the electrical spark plugs, fans that blow air after being driven by some belts that are hooked to the mechanical subsystem, the thermal comportment of the engine, and vibrations produced by the interaction between the various components, to mention just a few of its facets.

Simulation is, in practice, mostly done to save money. If it takes more time to build a simulation model than to build the real system, simulation will hardly be a viable option, because time is money.

Car manufacturers don't build their entire product from scratch anymore. If you open the hood of your American-built car, you may encounter a Japanese engine, a German transmission, and a French fuel-injection system. Car manufacturers buy many of the components of their cars from other sources. More and more often, car manufacturers request that these components be delivered together with simulation models capturing their behavior, because it cannot be expected of the control engineers working for the car manufacturer that they first start modeling each of the second-source components separately. They would never have their simulation models ready by the time the new car model needs to be built. What cannot, however, be expected is that all these simulation models are delivered encoded in the same simulation language. The transmission may have been modeled in Adams, the electrical system in Spice, the fuel injection system in ACSL, etc.

The control engineers are at the top of the hierarchy. It is their job to ensure that all the components work properly together. Hence, it is important that a control engineer can bond together models encoded in different modeling languages in a single simulation environment. This is a very tough problem.

22.2.5 Linearization

One way that control engineers deal with control systems is to linearize the plant to be controlled or at least a part thereof. This then enables them to perform the controller design in a simplified fashion, since there exist analytical controller design strategies for linear systems, whereas the nonlinear control system design would have to be done by trial and error.

Control engineers want the linearization of the original model to be done in an automated fashion. Moreover, this has to happen inside the modeling environment, since the original nonlinear model needs to be interpreted in this process.

This feature is very important for control engineers. They want to be able to compare the behavior of the linearized model with that of the original nonlinear model before they go about designing the controller. Then, after the controller has been synthesized, they would like to simulate the control behavior of the controller when applied to the original nonlinear plant. Finally, they may want to use the linear control system design only as a first step on the way to determining an appropriate controller for the original nonlinear plant. The so-synthesized controller can be interpreted as an approximation of the ultimately used controller. It has the right structure, but the previously found parameter values are only approximations of the final parameter values. The parameters are then fine-tuned using the nonlinear plant model together with some parameter optimization algorithm.

Some of the currently available simulation environments, such as ACSL and SIMULINK, offer a limited model linearization capability. A linear model of the type:

$$\dot{\mathbf{x}} = \mathbf{A} \cdot \mathbf{x} + \mathbf{B} \cdot \mathbf{u} \tag{22.1}$$

is obtained from the original nonlinear model:

$$\dot{\mathbf{x}} = \mathbf{f}(\mathbf{x}, \mathbf{u}, t) \tag{22.2}$$

by approximating the two Jacobians:

$$\mathbf{A} = \frac{\partial \mathbf{f}}{\partial \mathbf{x}} \qquad \mathbf{B} = \frac{\partial \mathbf{f}}{\partial \mathbf{u}} \tag{22.3}$$

through numerical differences. The facility is limited in three ways: (1) There is no control over the quality of the numerical difference approximation, and thereby over the linearization. The problem can be arbitrarily poorly conditioned. A symbolic differentiation of the model to generate the Jacobians may be more suitable and is entirely feasible. (2) The approximation is necessarily local, i.e., limited to an operating point $< \mathbf{x_0}, \mathbf{u_0} >$. If, during simulation, the solution starts to deviate a lot from this operating point, the approximation may be meaningless. (3) The approximation makes the assumption that the state variables must be preserved. This assumption may be too strong. If a subsystem is represented by the state-space model:

$$\begin{aligned} \dot{\mathbf{x}} &= \mathbf{f}(\mathbf{x}, \mathbf{u}, t) \\ \mathbf{y} &= \mathbf{g}(\mathbf{x}, \mathbf{u}, t) \end{aligned} \tag{22.4}$$

all one may wish to preserve is the input-output behavior, but this behavior should be preserved over an entire trajectory or even set of trajectories. This can often be accomplished by a model of the type:

$$\dot{\mathbf{z}} \doteq \mathbf{A} \cdot \mathbf{z} + \mathbf{B} \cdot \mathbf{u} \tag{22.5}$$

$$\mathbf{y} = \mathbf{C} \cdot \mathbf{z} + \mathbf{D} \cdot \mathbf{u} \tag{22.6}$$

if only the length of the linear state vector $\mathbf{z}$ is chosen sufficiently larger than that of the original state vector $\mathbf{x}$ [21].

22.2.6 Parameter Identification

Contrary to the plant parameters that can be determined (at least in an approximate fashion) from physical considerations, controller parameters are technological parameters that can be freely chosen by the designer. Hence, a tool is needed to determine optimal controller parameter values in the sense of minimizing (or maximizing) a performance index.

Although some simulation environments offer special tools for parameter identification, they all proceed in a purely numerical fashion. The authors of this article are convinced that much can be done to improve both the convergence speed and the convergence range of the optimization by proceeding in a mixed symbolic and numerical fashion.

Let $\mathbf{p}$ be the vector of unknown parameters, and PI the performance index to be optimized. It is fairly straightforward to augment the model at compile time by a sensitivity model that computes the sensitivity vector $\partial PI/\partial \mathbf{p}$. If there are k parameters and n equations in the original model, the augmented model will have $n(k+1)$ equations.

The control engineer can then look at the magnitude of the sensitivity parameters as a function of time, and pick a subset of those (those with large magnitudes) for optimization. Let us assume the reduced set of parameters $\mathbf{pr}$ is of length $kr < k$. Optimizing $PI(\mathbf{pr})$ implies making $\partial PI/\partial \mathbf{pr} = 0$. The latter problem can be solved by Newton iteration:

$$\frac{\partial^2 PI_\ell}{\partial \mathbf{pr}_\ell^2} \cdot \delta_\ell = -\frac{\partial PI_\ell}{\partial \mathbf{pr}_\ell} \tag{22.7}$$

$$\mathbf{pr}_{\ell+1} = \mathbf{pr}_\ell + \delta_\ell \tag{22.8}$$

Each iteration implies solving the augmented set of the original equations and the equations partially differentiated with respect to design parameters. Even equations for the Hessian matrix (the second partial derivative) can be generated symbolically at compile time, if code is generated simultaneously that prevents these additional equations from being executed during each function evaluation.

22.2.7 Frequency Domain

Control engineers like to switch back and forth between the time domain and the frequency domain when they are dealing with linear (or linearized) systems. Most simulation systems offer the capability to enter blocks as transfer functions. The polynomial coefficients are used in a set of differential equations (using the controller-canonical form), thereby converting the transfer function back into the time domain.

Although this feature is useful, it doesn't provide the control engineer with true frequency analysis capabilities. Control engineers like to be able to find the bandwidth of a plant, or determine the loop gain of a feedback loop. Such operations are much more naturally performed in the frequency domain, and it seems useful, therefore, to have a tool that would transform a linear (or linearized) model into the frequency domain, together with frequency domain analysis tools operating on the so-transformed model.

22.2.8 Real-Time Applications

Control systems are often not fully automated, but represent a collaborative effort of human and automatic control. Complex systems (such as an aircraft or a power-generation system) cannot be controlled by human operators alone, because of the time-critical nature of the decisions that must be reached. Humans are not fast and not systematic enough for this purpose. Yet, safety considerations usually mandate at least some human override capability, and often, humans are in charge of the higher echelons of the control architecture, i.e., they are in control of those tasks that require more intelligence and insight, yet are less time-critical.

Simulation of such complex control systems should allow human operators to drive the simulation in just the same manner as they would drive the real system. This is useful for both system debugging as well as operator training. However, since humans cannot be time-scaled, it is then important to perform the entire simulation in real time.

Another real-time aspect of simulation is the need to download controller designs into the digital controllers that are used to control the actual plant once the design has been completed. It does not make sense to ask the control engineer to reimplement the final design manually in the actual controller, since this invariably leads to new bugs in the code. It is much better if the modeling environment offers a fully automated real-time code generation facility, generating real-time code in either C, Fortran, or Ada.

Finally, some simulators contain hardware in the loop. For example, flight simulators for pilot training are elaborate electromechanical devices by themselves. It is the purpose of these simulators to make the hardware components behave as closely as possible to those that would be encountered in the real system. This entails simulated scenery, simulated force feedback, possibly simulated vibrations, etc. Evidently, these simulations need to be performed in real time as well.

22.3 Overview of Modeling and Simulation Software

There currently exist hundreds of different simulation systems on the market. They come in all shades and prices, specialized for different application areas, for different computing platforms, and embracing different modeling paradigms. Many of them are simply competitors of each other. It does not serve too much purpose to try to survey all of them. A list of current products and vendors is published in [29].

Is such a multitude of products justified? Why are there many more simulation languages around than general-purpose programming languages? The answer is easy. The general-purpose programming market is much more competitive. Since millions of Fortran and C compilers are sold, these compilers are comparatively cheap. It is almost impossible for a newcomer to penetrate this market, because he or she would have to work under cost for too long a period to ever make it. Simulation software is sold in hundreds or thousands of copies, not millions. Thus, the software is comparatively more expensive, and individuals who sell ten copies may already make a modest profit. Yet, the bewildering diversification on the simulation software market is certainly not overly productive.

Rather than trying to be exhaustive, the authors decided to concentrate here on a few of the more widely used products, a discussion that, in addition, shall help explain the different philosophies embraced by these software tools. This serves the purpose of consolidating the classification of modeling and simulation paradigms that had already been attempted in the previous section of this chapter. A more elaborate discussion of modeling and simulation software in general (not focused on the modeling and simulation of control systems) can be found in [9].

22.3.1 Block Diagram Simulators

The natural description form of the higher echelons of a control architecture is block diagrams, i.e., a graphical representation of a system via input-output blocks (see also Chapter 6). As already mentioned, most of the major simulation software producers offer a block-diagram editor as a graphical front end to their simulation engines.

Three of the most important packages of this type currently on the market are briefly discussed. All of them allow the simulation of continuous-time (differential equation) and discrete-time (difference equation) blocks and mixtures thereof. This is of particular importance to control engineers, since it allows them to model and simulate sampled-data control systems. Some of the tools also support state events, but their numerical treatment is not always appropriate. Modeling is done graphically, and block diagrams can mostly be structured in a hierarchical fashion.

- **SIMULINK** from The MathWorks Inc. [22, 23]: An easy-to-use, point-and-click program. SIMULINK is an extension to MATLAB, the widely used program for interactive matrix manipulations and numerical computations in general. Of the three programs, SIMULINK offers the most intuitive user interface. MATLAB can be employed as a powerful pre- and postprocessing facility for simulation, allowing, e.g., parameter variation and optimization (although not employing the more advanced semisymbolic processing concepts that were discussed in the previous section of this chapter) as well as displaying the simulation results in a rich set of different formats. SIMULINK and MATLAB are available for a broad range of computing platforms and operating systems (PC/Windows, MacIntosh, Unix/X-Windows, VAX/VMS). SIMULINK supports the same philosophy that is used within MATLAB. By default, the equations of a SIMULINK model are preprocessed into an intermediate format, which is then interpreted. This has the advantage that the program is highly interactive, and simulations can run almost at once. It has recently become possible to alternatively compile built-in elements of SIMULINK into C, to be used in the simulation or in a real-time application. However, user-defined equations programmed in the powerful MATLAB language (as M-files) are still executed many times slower due to their being interpreted rather than compiled. SIMULINK enjoys a lot of popularity, especially in academia, where its highly intuitive and easily learnable user interface is particularly appreciated.
- **SystemBuild** from Integrated Systems Inc. [18]: Overall, SystemBuild offers more powerful features than SIMULINK. For example, it offers much better event specification and handling facilities. A separate editor for defining finite-state machines is available. Models can be described by differential-algebraic equations (DAEs), and even by overdetermined DAEs. The latter are needed if, e.g., general-purpose multibody system programs will be used within a block-diagram editor for the description of complex mechanical mechanisms such as vehicles.[1] The price to be paid for this flexibility and generality is a somewhat more involved user interface that is a little more difficult to master. For several years already, SystemBuild has offered the generation of real-time code in C, Fortran, and Ada. SystemBuild is an extension to Xmath (formerly $\mathrm{MATRIX_X}$, the main MATLAB competitor). Xmath is very similar to MATLAB, but supports more powerful data structures and a more intimate connection to X-Windows. This comes at a price, though. Xmath and SystemBuild are not available for PC/Windows or Macintosh computers. Due to their flexibility and the more advanced features offered by these tools, these products have a lot of appeal to industrial customers, whereas academic users may be more attracted to the ease of use and platform-independence offered by SIMULINK.

[1] Multibody programs that can be utilized within SystemBuild include SIMPACK [30] and DADS [32].

- **EASY-5** from Boeing:
  Available since 1981, EASY-5 is one of the oldest block-diagram editors on the market. It is designed for simulations of very large systems. The tool is somewhat less easy to use than either SIMULINK or SystemBuild. It uses fully compiled code from the beginning. After a block diagram has been built, code is generated for the model as a whole, compiled to machine code, and linked to the simulation run-time engine. This has the effect that the compilation of a block diagram into executable run-time code is rather slow; yet, the generated code executes generally faster than in the case of most other block diagram programs.

As already mentioned, block-diagram editors have the advantage that they are (usually) easy to master by even novice or occasional users, and this is the main reason for their great success. On the other hand, nearly all block-diagram editors on the market, including SIMULINK and SystemBuild, suffer from some severe drawbacks.

First, they don't offer a "true" component library concept in the sense used by a higher-level programming language. Especially, the user can store model components in a (so-called) "library" and retrieve the component by "dragging" it from the library to the model area, with the effect that the component is being *copied*. Consequently, every change in the library requires manual repetition of the copying process, which is error prone and tedious[2].

Second, it is often the case that differential equations have to be incorporated directly in textual form, because the direct usage of block-diagram components becomes tedious. In SIMULINK and SystemBuild, the only reasonable choice is to program such parts directly in C or Fortran, i.e., by using a modeling technique from the 1960s. In this respect, the general-purpose simulation languages, to be discussed in the next section of this chapter, offer much better support, because differential equations can be specified directly, using user-defined variable names rather than indices into an array. Furthermore, the equations can be provided in an arbitrary order, since the modeling compiler will sort them prior to generating code.

22.3.2 General-Purpose Simulation Languages

Block-diagram simulators became fashionable only after the recent proliferation of graphics workstations. Before that time, most general-purpose modeling and simulation was done using simulation languages that provided textual user interfaces similar to those offered by general-purpose programming languages. Due to the success of the aforementioned graphical simulation programs, most of these programs have meanwhile been enhanced by graphical front ends as well. However, the text-oriented origin of these programs often remains clearly visible through the new interface.

- **ACSL** from Mitchell & Gauthier Assoc. [24]:
  Available since 1975, ACSL has long been the unchallenged leader in the market of simulation languages. This situation changed in recent years due to the success of SIMULINK and SystemBuild. ACSL is a language based on the CSSL-standard [4]. An ACSL program is preprocessed to Fortran for platform independence. The resulting Fortran program is then compiled further to machine code. As a consequence, ACSL simulations always run efficiently, which is in contrast to the simulation code generated by most block-diagram simulators. User-defined Fortran, C, and Ada functions can be called from an ACSL model. ACSL can handle ODEs and DAEs, but no overdetermined DAEs. For a long time already, ACSL has supported state-event handling in a numerically reliable way (by means of the *schedule* statement), such that discontinuous elements can be handled. Recently, ACSL has been enhanced by a block-diagram front end, a post-processing package for visualization and animation, and a MATLAB-like numerical computation engine.
  A block in ACSLs block-diagram modeler can take any shape and the input/output points can be placed everywhere, contrary, e.g., to the much more restricted graphical appearance of SIMULINK models. Consequently, with ACSL it is easier to get a closer correspondence between reality and its graphical image. Unfortunately, ACSL is not (yet) truly modular. All variables stored in a block have *global* scope. This means that one has to be careful not to use the same variable name in different blocks. Furthermore, it is not possible to define a block once, and to use several copies of this block. As a result, it is not convenient to build up user-defined block libraries. ACSL is running on a wide variety of computing platforms ranging from PCs to supercomputers. Due to the 20 years of experience, ACSL is fairly robust, contains comparatively decent integration algorithms, and many small details that may help the simulation specialist in problematic situations. Although the ACSL vendors have lost a large percentage of their academic users to SIMULINK, ACSL is still fairly popular in industry.
- **Simnon** from SSPA Systems [10, 13]:
  Simnon was the first direct-executing, fully digital simulation system on the market. Designed originally as a university product, Simnon is a fairly small and easily manageable software system for the simulation of continuous-time and discrete-time systems. Simnon offered, from its conception, a mixture between a statement-oriented and a block-

[2]In a higher-level programming language, a change in a library function just requires repeating the linking process.

oriented user interface. Meanwhile, a graphical front end has been added as well. Simnon has been for years a low-cost alternative to ACSL, and enjoyed widespread acceptance, especially in academia. Due to its orientation, it suffered more than ACSL from the SIMULINK competition.

- **Desire** from G.A. & T.M. Korn [20]:
Desire is another direct-executing simulation language, designed to run on small computers at impressively high speed. It contains a built-in micro-compiler that generates machine code for Intel processors directly from the model specification. Since no detour is done through a high-level computer language, as is the case in most other compiled simulation languages, compilation and linking are nearly instantaneous. It is a powerful feature of the language that modeling and simulation constructs can be mixed. It is therefore easy to model and simulate systems with varying structure. For example, when simulating the ejector seat of an aircraft, several different models are simulated one after another. This is done by *chaining* several Desire models in sequence, which are compiled as needed and then run at once. Desire also offers fairly sophisticated high-speed matrix manipulation constructs, e.g., optimized for the formulation of neural network models. Desire is used both in academia and industry, and has found a strong market in real-time simulation of small- to medium-sized systems, and in digital instrumentation of measurement equipment.

22.3.3 Object-Oriented Modeling Languages

It was mentioned earlier that block-diagram languages are hardly the right choice for modeling physical systems. The reason is that the block-diagram languages, as well as their underlying general-purpose simulation languages, are assignment statement oriented, i.e., each equation has a natural *computational causality* associated with it. It is always clear, what the inputs of an equation are, and which is the output.

Unfortunately, physics doesn't know anything about computational causality. Simultaneous events are always acausal. Modeling an electrical resistor, it is not evident ahead of time, whether an equation of the type:

$$u = R \cdot i \tag{22.9}$$

will be needed, or one of the form:

$$i = \frac{u}{R} \tag{22.10}$$

It depends on the environment in which the resistor is embedded.

Consequently, the modeling tool should relax the artificial causality constraint that has been imposed on the model equations in the past. By doing so, a new class of modeling tools results. This concept has been coined the *object-oriented modeling* paradigm, since it provides the modeling language with a true one-to-one topological correspondence between the physical objects and their software counterparts inside the model. The details of this new modeling paradigm are discussed more thoroughly in Chapter 7 and are not repeated here.

- **Dymola** from Dynasim AB [11, 12]:
The idea of general object-oriented modeling, and the first modeling language implementing this new concept, Dymola, were created by Elmqvist as part of his Ph.D. dissertation [11]. Dymola then already offered a full topological description capability for physical systems, and demonstrated the impressive potential of this new modeling approach by means of an object-oriented model of a quite complex thermal power station. However, the demand for such general-purpose, large-scale system modeling tools had not arisen yet, and neither was the computer technology of the era ready for this type of tool. Consequently, Dymola remained for many years a university prototype with fairly limited circulation. The book *Continuous System Modeling* [8], which assigned a prominent role to object-oriented modeling and Dymola, reignited interest in this tool and, since the fall of 1992, Dymola has become a commercial product. Many new features have been added to Dymola since then, such as (even multiple) inheritance, a MATLAB-like matrix capability, a high-level, object-oriented event-handling concept able to deal correctly with multiple simultaneous events, handling of higher-index differential algebraic equations, to mention only a few. Dymola is a model compiler that symbolically manipulates the model equations and generates a simulation program in a variety of formats, including ACSL, Simnon, Desire, and SIMULINK (C-SimStruct). It also supports a simulator based on the DSblock format (discussed in the next subsection), called Dymosim. A graphical front end, called Dymodraw, has been developed. It is based on *object diagrams* rather than block diagrams. Models (objects) are represented by icons. Connections between icons are nondirectional, representing a physical connection between physical objects. An electrical circuit diagram is a typical example of an object diagram. Also available is a simulation animator, called Dymoview, for graphical representation of motions of two- and three-dimensional mechanical bodies.
- **Omola** from Lund Institute of Technology [1]–[3]:
Omola was created at the same department that had originally produced Dymola. At the time when Omola was conceptualized, the object-oriented programming paradigm had entered a phase of widespread proliferation, and researchers in Lund wanted to create a tool that made use of a terminology that would be closer to that currently employed in object-oriented programming software. Omola is still a

university prototype only. Its emphasis is primarily on language constructs, whereas Dymola's emphasis is predominantly on symbolic formula manipulation algorithms. Omola is designed for flexibility and generality, whereas Dymola is designed for high-speed compilation of large and complex industrial models into efficient simulation run-time code. Omola supports only its own simulator, called Omsim. In order to provide a user-friendly interface, Omola also offers an experimental object-diagram editor. Yet, Omola's object-diagram editor is considerably less powerful than Dymola's.

- **VHDL-A** a forthcoming IEEE standard [33, 34]: VHDL is an IEEE standard for hardware description languages. It provides a modeling language for describing digital circuitry. VHDL has been quite successful in the sense that nearly all simulators for logical devices on the market are based on this standard. This allows an easy exchange of models between different simulators. Even more importantly, libraries for logical devices have been developed by different companies and are being sold to customers, independently of the simulator in use. The VHDL standard is presently under revision for an analog extension, called VHDL-A [34], to include analog circuit elements. The main goal of VHDL-A is to define a product-independent language for mixed-level simulations of electrical circuits [33]. Different levels of abstractions and different physical phenomena will be describable within a single model. This development could be of interest to control engineers as well, since the VHDL-A definition is quite general. It includes assignment statement-based input/output blocks, as well as object-oriented (physical) model descriptions, and supports differential-algebraic equations. It may well be that VHDL-A becomes a standard not only for electronic circuit descriptions, but also for modeling other types of physical systems. In that case, this emerging development could gain central importance to the control community as well. However, the VHDL-A committee is currently focusing too much on language constructs without considering the implications of their decisions on efficient run-time code generation. The simulation of analog circuits (and other physical systems) is much more computation intensive than the simulation of digital circuitry. Thus, efficient run-time code is of the essence. A standard like VHDL-A would, however, solve many problems. First, model exchange between different simulation systems would become feasible. This is especially important for mixing domain-specific modeling systems with block-diagram simulators. Second, a new market for model component libraries would appear, because third-party vendors could develop and sell such libraries in a product-independent way. From a puristic point of view, VHDL-A is not truly object-oriented, because some features, such as inheritance, are missing. However, since VHDL-A contains the most important feature of object-oriented modeling systems, namely support for physical system modeling, it was discussed in this context.

In order to show the unique benefits of object-oriented modeling for control applications, as compared to the well-known but limited traditional modeling systems, additional issues are discussed in more detail in the following subsections.

Object Diagrams and Class Inheritance

The concept of object diagrams is well understood. The former "blocks" of the block diagrams are replaced by mnemonically shaped icons. Each icon represents an object. An icon can have an arbitrary number of pins (terminals) through which the object that the icon represents exchanges information with other objects. Objects can be hierarchically structured, i.e., when the user double-clicks on an icon ("opening" the icon), a new window may pop up showing another object diagram, the external connections of which correspond to the terminals of the icon that represents the object diagram. Connections are non-directional (they represent physical connections rather than information paths), and one connection can (and frequently does) represent more than one variable.

Figure 22.4 shows a typical object diagram as managed by the object-diagram editor of Dymola. Different object diagrams can use different modeling styles. The three windows to the right of Figure 22.4 show an electrical circuit diagram, a multibody system, and a block diagram (a special case of an object diagram). Another frequently used object-diagram representation would be a bond graph. Mechatronics systems use components from different domains, and hence it makes sense to use the modeling mechanism that is most natural to the individual domain, when modeling the different subsystems. Drive trains are attached to each joint of the robot (left part of the window *Mechanical*). A drive train class contains instances of the model classes *Control* and *Electrical*. The three windows in the second column from the left show different model libraries. Each model is represented by an icon that can be picked and dragged into the corresponding object-diagram window for composing models from components (in some cases hierarchical models by themselves) and their interconnections. Contrary to the case of block-diagram editors, these are true libraries, in the sense that changes in a library are reflected at once (after compilation) in the models using this library. This is due to the fact that the libraries contain model classes, i.e., definitions of generic model structures, rather than the model objects themselves, and dragging an icon into an object diagram only establishes a link to the desired model class rather than leading to an object instantiation being made at once.

One important aspect of object-oriented modeling has not been discussed yet. It concerns *class inheritance*. Resistors, capacitances, and inductors have in common that they are all one-

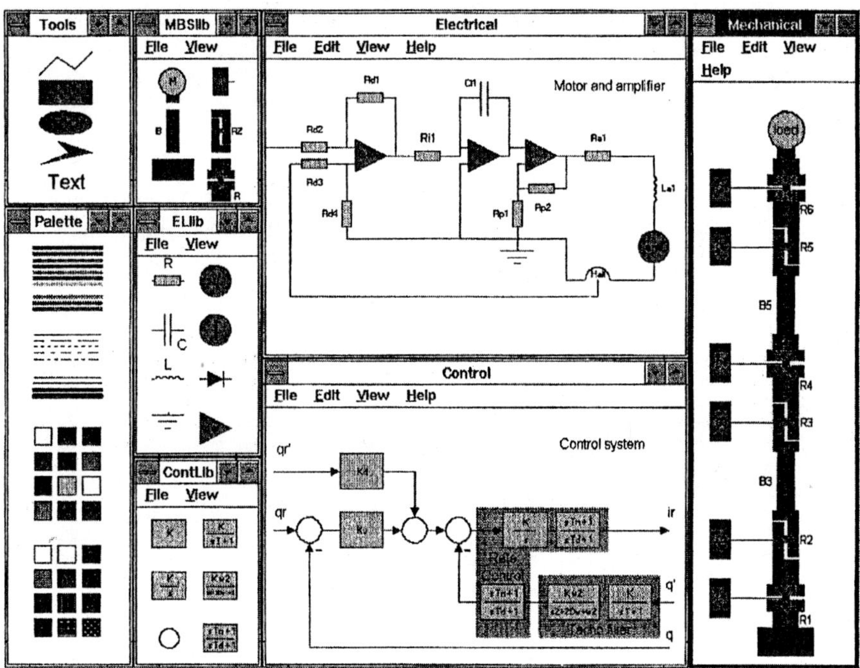

Figure 22.4 Object-oriented view of mechatronic model.

port elements. They all share the fact that they have two pins, each carrying a potential and a current, that the voltage drop across the one-port element can be computed as the difference between its two terminal potentials, and that current in equals current out.

It would be a pity if these common facts would have to be repeated for each model class. In terms of a Dymola notation, the generic superclass *OnePort* could be described as:

```
model class OnePort
  cut WireA(Va/i), WireB(Vb/ - i)
  local u
    u = Va - Vb
end
```

Resistors and *Capacitors* could then incorporate the properties of the superclass *OnePort* into their specific definitions through a mechanism of inheritance:

```
model class Resistor            model class Capacitor
  inherit OnePort                 inherit OnePort
  parameter R                     parameter C
    u = R * i                       C * der(u) = i
end                             end
```

The use of class inheritance enhances the robustness of the model, because the same code is never being manually copied and migrated to different places inside the code. Thereby, if the superclass is ever modified, the modification gets automatically migrated down into all individual model classes that inherit the superclass. Otherwise, it could happen that a user implements the modification as a local patch in one of the model classes only, being totally unaware of the fact that the same equations are also used inside other model classes.

The 3D-multibody system library supplied with Dymola makes extensive use of class inheritance in the definition of joints. *RevoluteJoint* and *PrismaticJoint* have in common that they both share the base class *OneDofJoint*. However, every *OneDofJoint* inherits the base class *Joint*.

Higher Index Models and Feedback Linearization

Higher index models are models with dependent storage elements. The simplest such model imaginable would be an electrical circuit with two capacitors in parallel or two inductors in series. Each capacitor or inductor is an energy storage element. However, the coupled models containing two parallel capacitors or two inductors in series still contain only one energy storage element, i.e., the coupled model is of first order only, and not of second order. Models of systems that contain algebraic equations which explicitly or implicitly relate state variables algebraically to each other, are called *higher index models*. To be more specific, the (perturbation) index of the DAE

$$\mathbf{f}\,(\dot{\mathbf{x}}(t), \mathbf{x}(t), \mathbf{w}(t), t) \;=\; \mathbf{0} \tag{22.11}$$

is the smallest number j such that after $j-1$ differentiations of Equation 22.11, $\dot{\mathbf{x}}$ and $\mathbf{w}$ can be uniquely determined as functions of $\mathbf{x}$ and t. Note that $\mathbf{w}$ are purely algebraic variables, whereas $\mathbf{x}$ are variables that appear differentiated within Equation 22.11. Currently available DAE solvers, such as DASSL [5, 28], are not designed to solve DAEs with an index greater than one without modifications in the code that depend on the model structure. The reasons for this property are beyond the scope of this article (see [15, 17] for details). Rather than modifying the DAE solvers such that they are able to deal with the higher index problems in a numerical fashion (which can be done, e.g., see [6]), it may make sense to preprocess the model symbolically in such a way that the (perturbation) index of the model is reduced to one. A very general and fast algorithm for this purpose was developed by Pantelides [27]. This algorithm constructs all the equations needed to express $\dot{\mathbf{x}}$ and $\mathbf{w}$ as functions of $\mathbf{x}$ and t by differentiating the necessary parts of Equation 22.11 sufficiently often. As a by-product, the algorithm determines in an automatic way the (structural) index of the DAE. The Pantelides algorithm has meanwhile been implemented in both Dymola and Omola.

Higher index modeling problems are closely related to *inverse models*, and in particular to *feedback linearization*, an important method for the control of nonlinear systems; e.g., see Chapter 57, or [19, 31]. Inverse models arise naturally in the following control problem: given a desired plant output, what is the plant input needed to make the real plant output behave as similarly as possible to the desired plant output? If only the plant dynamics model could be inverted, i.e., its outputs treated as inputs and its inputs as outputs, solving the control problem would be trivial. Of course, this cannot usually be done, because if the plant dynamics model is strictly proper (or in the nonlinear case: exhibits integral behavior), which is frequently the case, the inverse plant dynamics model is non-proper (exhibits differential behavior). This problem can be solved by introducing a reference model with a sufficient number of poles, such that the cascade model of the reference model and the inverse plant dynamics model is at least proper (does not exhibit differential behavior). This idea is illustrated in Figure 22.5.

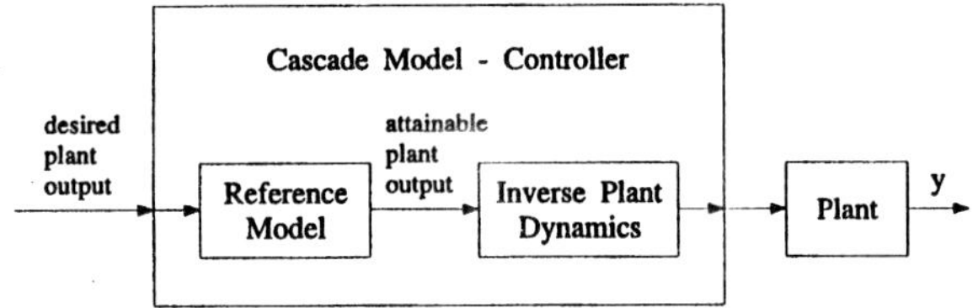

Figure 22.5 Control through inverse plant dynamics model.

Using the object-oriented modeling methodology, this approach to controller design can be implemented elegantly. The user would start out with the reference model and the plant dynamics model. The input of the reference model is then declared as external input, the output of the reference model is *connected* to the output of the plant dynamics model, and the input of the plant dynamics model is declared as external output. Object-oriented modeling systems, such as Dymola, are capable of generating either a DAE or an ODE model from such a description. However, the original set of equations resulting from connecting the submodels in such a fashion is invariably of higher index. The Pantelides algorithm is used to reduce the index down to one, leading to a DAE formulation containing algebraic loops but no dependent storage elements.

Inverse dynamic models can also be used for *input-output linearization*, a special case of *feedback linearization*. The main difference to the feedforward compensation discussed above consists in using the measured state of the system instead of reconstructing this state in a separate dynamic model. To be more specific, the output equation (22.13) of the state-space model

$$\dot{\mathbf{x}} = \mathbf{f}(\mathbf{x}) + \mathbf{B}(\mathbf{x}) \cdot \mathbf{u} \tag{22.12}$$

$$\mathbf{y} = \mathbf{g}(\mathbf{x}) \tag{22.13}$$

is differentiated sufficiently often, in order that the input $\mathbf{u}$ occurs in the differentiated output equations. Solving these equations for $\mathbf{u}$ allows the construction of a control law, such that the closed-loop system has purely linear dynamics. For details, see Chapter 57. By interpreting Equations 22.12 and 22.13 as a DAE (of the type of Equation 22.11), with $\mathbf{w} = \mathbf{u}$ and $\mathbf{y}$ as known functions of time, it can be noticed that the necessary differentiations to determine $\mathbf{u}$ and $\dot{\mathbf{x}}$ explicitly as functions of $\mathbf{x}$ correspond exactly to the differentiations needed to determine the index of the DAE. In other words, the Pantelides algorithm can be used to carry out this task, instead of forming the Lie brackets of Equation 22.13, as is usually done.

To summarize, inverse models and, in particular, input-output linearization compensators, can easily be constructed by object-oriented modeling tools such as Dymola and Omola. This practical approach was described in [25].

Discontinuity Handling and Events

Discontinuous models play an important role in control engineering. On the one hand, control engineers often employ discontinuous control actions, e.g., when they use *bang-bang control.* However, and possibly even more importantly, the actuators that transform the control signals to corresponding power signals often contain lots of nasty discontinuities. Typically, switching power converters may exhibit hundreds if not thousands of switching events within a single control response [16].

Proper discontinuity handling in simulation has been a difficult issue all along. The problem is that the numerical integration algorithms in use are incompatible with the notion of discontinuous functions. Event detection and handling mechanisms to adequately deal with discontinuous models have been described in [7]. However, many of the available modeling and simulation systems in use, such as SIMULINK, Simnon, and Desire, still don't offer appropriate event handling mechanisms. This is surprising, since discontinuous models are at the heart of a large percentage of engineering problems. Only ACSL, Dymosim, SystemBuild, and some other systems offer decent event handling capabilities.

Unfortunately, these basic event handling capabilities are still on such a low level that it is very difficult for a user to construct a valid model of a discontinuous system even in the simplest of cases. In order to justify this surprising statement, a control circuit for the heating of a room, as shown in Figure 22.5, is discussed. The heating process is described by a PT1 element.

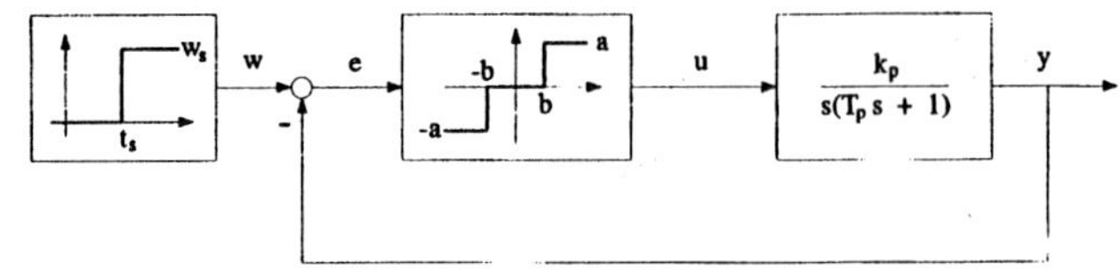

Figure 22.6 Simple control circuit with discontinuities.

The controller consists of a three point element together with an actuator with integral behavior, which is combined with the PT1 element in Figure 22.6. At a specific time instant t_s, the set point w jumps from zero to w_s. This system can be described by the following ACSL model:

```
program HeatControl
  initial
    constant kp = 1, Tp = 0.1, a = 1, b = 0.05, ws = 1,
             ts = 0.5, x10 = 0, x20 = 0
    integer mode

    ! initialize input and mode (valid for x10=0, x20=0)
      w    = 0
      mode = 0

    ! define time instant when w is jumping
      schedule setpoint .at. ts
  end

  dynamic
    derivative
      ! calculate model signals
        e = w - y
        u = mode * a
        x = integ(u, x10)
        y = kp*realpl(Tp, x, x20)

      ! define switching of 3-point controller
        schedule switch .xz. (e - b) .or. (e + b)
    end
    discrete setpoint
      w = ws
    end
    discrete switch
      if (e .lt. -b) then
        mode = -1
      elseif (e .gt. b) then
        mode = +1
      else
        mode = 0
      endif
    end
  end
end
```

During numerical integration, only the *derivative* section is executed. Since variable *mode* changes its value only when an event occurs, no discontinuity is present when integrating this section. An event occurs when $e - b$ or $e + b$ crosses zero in either direction, or when the integration reaches time t_s. However, the above code will not always work correctly. Let us analyze some problematic situations:

1. *Initialization*:
   Before the integration starts, the *initial*, *derivative*, and *discrete* sections are evaluated once in the order of appearance. However, this does not help with the proper initialization of the variable *mode*. From the block diagram of Figure 22.6, it is easy to deduce that for zero initial conditions ($x_{10} = 0$, $x_{20} = 0$) of the dynamic elements and zero input of w, the control error is zero, and therefore, *mode* has to be initialized with zero as well. However, when any of the initial conditions are non-zero, it is by no means obvious, how the variable *mode* must be initialized.
   A proper initialization can be done in the following way:

   ```
   initial
     ...
     ! initialize mode
       e = w - x20
       if (e .lt. -b) then
         mode = -1
       elseif (e .gt. b) then
         mode = +1
       else
         mode = 0
       endif
   end
   ```

   In other words, the plant dynamics must be analyzed in order to determine the correct initial value for variable *mode*. Usually this requires doubling of code of the *derivative* and *discrete* sections. This process becomes more involved as the plant grows in complexity, or when the plant itself contains discontinuous elements. Furthermore, it creates a serious barrier for modularization, because the correct initialization of a local element, such as the three point controller, requires global analysis of the model.
   It should be noted that, even with the above initialization scheme, the simulation will be incorrect if $x_{20} = b$ and $k_p \cdot x_{10} > x_{20}$. This is due to a subtle artifact of the crossing functions. If $x_{20} = b$, $mode = 0$ and the crossing function $e+b$ is identical to zero. If $k_p \cdot x_{10} > x_{20}$, y is growing, and therefore e decreases to a value smaller than $-b$ shortly after the integration starts. Since an event occurs only if a crossing function *crosses* zero, no event is scheduled. As a consequence, *mode* remains zero, although it should become -1 shortly after the integration starts. The initialization section will become even more involved if such situations are to be handled correctly.
2. *Simultaneous events*:
   What happens if a state event of the three point controller and the time event of the set point occur at the same time instant? In the above example, this situation can easily be provoked by simulating first with $w = 0$, determining the exact time instant at which a state event occurs and then use this time instant as initial value for t_s.
   When two events occur simultaneously, the corresponding discrete sections are executed in the order of appearance in the code. Obviously, this can lead to a wrong setting of variable *mode*. Assuming that at the time of the event, $w = 0$ and e crosses b in the negative direction, i.e., $e = b - \epsilon$. Due to the *discrete* section *switch*, the variable *mode* will be set to zero. However, when the integration starts again, $w = w_s > 0$ and $e > b$, i.e., *mode* should be 1. In other words, *mode* has the wrong value, independently of the ordering of the discrete sections! The correct treatment of such a situation requires merg-

ing the two discrete sections into one and doubling code from the derivative section. Again, this results in a serious barrier for modularization.

It should have become clear by now that separating the modeling code into *initial*, *derivative*, and *discrete* sections, as done in ACSL, Omola, VHDL-A and other systems, is not a good idea in the presence of state events. For the user it is nearly impossible to manually generate code that is correct in all circumstances.

In [14], a satisfactory solution to the problems mentioned above is proposed for *object-oriented* modeling systems, and the proposed solution has been implemented in Dymola. It is beyond the scope of this chapter to discuss all the details. In a nut shell, higher language elements are introduced that allow the selective activation of equations based on Boolean expressions *becoming* true. These *instantaneous equations* are treated in the same way as continuous equations. In particular, they are *sorted* together with all the other equations. The sorting process automatically guarantees that the code at the initial time and at event instants is executed in the correct sequence, so that the simultaneous event problem mentioned above can no longer occur. Furthermore, the model is iterated at the initial time and after event handling to find automatically the correct switching states to prevent the initialization problem explained above from ever occurring.

To summarize, the object-oriented modeling paradigm allows a satisfactory handling of discontinuous models. This is not the case with the traditional modeling systems in use today.

22.3.4 Coupling of Simulation Packages

In the last section, it was discussed that modeling languages could use some sort of standardization, in order to improve the capability of simulation users to exchange models and even entire model libraries among each other. VHDL-A was mentioned as one attempt at answering this demand.

However, it was also mentioned that more and more producers of technical equipment, such as car manufacturers, depend on second-source components *and second-source models thereof* in their system design. If every second-source provider could be forced to provide models encoded in a subset of VHDL-A, this might solve the problem. However, this will not happen for years to come. At least as an intermediate (and more easily achievable) solution, one could try to create a much lower-level standard, one for simulation run-time environments.

For efficient simulation, models have to be compiled into machine code. Portability issues suggest generation of code first in a high-level programming language, such as Fortran, C, or Ada, which is then compiled to machine code using available standard compilers. Therefore, it is natural to ask for a standardization of the *interfaces* of modeling and simulation environments at the programming language level. This allows generation of a program code from a modeling tool A, say a mechanical or electronic circuit modeling system, and use it as a component in another modeling tool B, say a block-diagram package. It is much easier to use a model at the level of a programming language with a defined interface, than writing a compiler to transform a VHDL-A model down to a programming language.

Some simulation researchers have recognized this need, and, in fact, several different low-level interface definitions are already in use:

- **DSblock** interface definition [26]:
  This was the first proposal for a neutral, product-independent low-level interface. It was originally specified in Fortran. The latest revision uses C as specification language, and supports the description of time-, state-, and step-event-driven ordinary differential equations in state-space form, as well as regular and overdetermined DAEs of indices 1 and 2. All signal variables are characterized by text strings that are supplied through the model interface. This allows an identification of signals by their names used in the higher-level modeling environment, and not simply by an array index. Presently, Dymola generates DSblock code as interface for its own simulator, Dymosim. Also, the general-purpose multibody program SIMPACK [30] can be optionally called as a DSblock.
- **SimStruct** from The MathWorks [23]:
  In the newest release of SIMULINK (Version 1.3), the interface to C-coded submodels is clearly defined, and has been named SimStruct. Furthermore, with the SIMULINK accelerator, and the SIMULINK C-Code generator, SIMULINK can generate a SimStruct model from a SIMULINK model consisting of any built-in SIMULINK elements and from SimStruct blocks (S-functions written in the MATLAB language cannot yet be compiled). A SimStruct block allows the description of input/output blocks in state-space form consisting of continuous- and discrete-time blocks, with multirate sampling of the discrete blocks[3]. However, neither DAEs nor state-events are supported. DAEs are needed in order to allow the incorporation of model code from domain-specific modeling tools like electric circuits or mechanical systems. State-events are needed in order to properly describe discontinuous modeling elements and variable structure systems.
- **User Code Block (UCB)** interface from Integrated Systems [18]:
  The UCB-interface used with SystemBuild allows the description of time- and state-event-dependent ordinary differential equations in state-space form, as well as regular and overdetermined DAEs of index 1. It is more general than the SimStruct interface. Some commercial multibody packages (e.g., SIMPACK [30], DADS [32]) already support this interface, i.e., can be used within SystemBuild as an

[3]Multirate sampling is a special case of time-events.

input/output block. Two serious drawbacks are still present in this definition. First, the dimensions of model blocks have to be defined in the SystemBuild environment. This means that model blocks from other modeling environments, such as mechanical and electrical systems, cannot be incorporated in a fully automated fashion, because the system dimensions depend on the specific model components. Contrarily, in the DSblock interface definition, the model dimensions are reported from the DSblock to the calling environment. Second, variables are identified by index in the SystemBuild environment. This restricts the practical use of the tool to models of low to medium complexity only.

22.4 Shortcomings of Current Simulation Software

As already discussed, a serious shortcoming of *most* simulation tools currently on the market is their inability to treat discontinuous models adequately. This is critical because most real-life engineering problems contain discontinuous components. Sure, a work-around for this type of problem consists in modeling the system in such a detail that no discontinuities are present any longer. This is done, e.g., in the standard electric circuit simulator SPICE. However, the simulation time increases then by a factor of 10 to 100, and this limits the size of systems that can be handled economically. Note that proper handling of discontinuous elements is *not* accomplished by just supplying language elements to handle state events, as is done, e.g., in ACSL or SystemBuild. The real problem has to do with determining the correct mode the discontinuous system is in at all times. Object-oriented modeling can provide an answer to this critical problem.

Block-diagram editors are, in the view of the authors of this chapter, a cul-de-sac. They *look* modern and attractive, because they employ modern graphical input/output technology. However, the underlying concept is unnecessarily and unjustifiably limited. Although it is trivial to offer block-diagram editing as a special case within a general-purpose object-diagram editor, the extension of block-diagram editors to object-diagram editors is far from trivial. It is easy to predict that block-diagram editors will be replaced by object-diagram editors in the future, in the same way as block-diagram editors have replaced the textual input of general-purpose simulation languages in the past. However, it may take several years before this will be accomplished. Most software vendors only react to pressure from their customers. It may still take a little while before enough simulation users tell their simulation software providers that object-diagram editors is what they need and want.

Although the introduction of block-diagram simulators has enhanced the user-friendliness of simulation environments a lot, there is still the problem with model component libraries. As previously explained, the "library" technique supported by block-diagram simulation systems, such as SIMULINK and SystemBuild, is only of limited use, because a modification in a component in a library cannot easily be incorporated into a model in which this component is being used. Again, the object-oriented modeling systems together with their object-diagram editors provide a much better and more satisfactory solution.

Beside the modeling and simulation issues discussed in some detail in this chapter, there exists the serious practical problem of organizing and documenting simulation experiments. To organize the storage of the results of many simulation runs, possibly performed by different people, and to keep all the information about the simulation runs necessary in order to reproduce these runs in the future, i.e., store the precise conditions under which the results have been produced, is a problem closely related to version control in general software development. At present, almost no support is provided for such tasks by available simulation systems.

22.5 Conclusions

The survey presented in this chapter is necessarily somewhat subjective. There exist several hundred simulation software packages currently on the market. It is evident that no single person can have a working knowledge of all these packages. Furthermore, software is a very dynamic field that is constantly enhanced and upgraded. The information provided here represents our knowledge as of July 1995. It may well be that some of our criticism will already be outdated by the time the reader lays his eyes on this handbook.

To summarize, the textual simulation languages of the past have already been largely replaced by block-diagram editors, since these programs are much easier to use. Most simulation programs entered through block-diagram editors still have problems with efficiency, because equations are interpreted rather than compiled into machine code. For larger systems, the right choice of simulation package is therefore still not easy. The future belongs definitely to the object-oriented modeling languages and their object-diagram editors, since these new techniques are much more powerful, reflect more closely the physical reality they try to capture, and contain the popular block-diagrams as a special case.

Acknowledgments

The authors would like to acknowledge the valuable discussions held with, and comments received from, Hilding Elmqvist and Ingrid Bausch-Gall.

References

[1] Andersson, M., *Omola — An Object–Oriented Language for Model Representation*, Licenciate thesis TFRT–3208, Department of Automatic Control, Lund Institute of Technology, Lund, Sweden, 1990.

[2] Andersson, M., Discrete event modelling and simulation in Omola, *Proc. IEEE Symp. Comp.–Aided Control Sys. Des.*, Napa, CA, pp. 262–268, 1992.

[3] Andersson, M., *Object-Oriented Modeling and Simulation of Hybrid Systems*, Ph.D. thesis TFRT–1043, Department of Automatic Control, Lund Institute of Technology, Lund, Sweden, 1994.

[4] Augustin, D.C., Fineberg, M.S., Johnson, B.B., Linebarger, R.N., Sansom, F.J., and Strauss, J.C., The SCi Continuous System Simulation Language (CSSL), *Simulation*, 9, 281–303, 1967.

[5] Brenan, K.E., Campbell, S.L., and Petzold, L.R., *Numerical Solution of Initial-Value Problems in Differential-Algebraic Equations*, Elsevier Science Publishers, New York, 1989.

[6] Bujakiewicz, P., *Maximum Weighted Matching for High Index Differential Algebraic Equations*, Ph.D. thesis, Technical University Delft, The Netherlands, 1984.

[7] Cellier, F.E., *Combined Continuous/Discrete System Simulation by Use of Digital Computers: Techniques and Tools*, Ph.D. dissertation, Diss ETH No 6483, ETH Zurich, Zurich, Switzerland, 1979.

[8] Cellier, F.E., *Continuous System Modeling*, Springer–Verlag, New York, 1991.

[9] Cellier, F.E., Integrated continuous–system modeling and simulation environments, in *CAD for Control Systems*, D.A. Linkens, Ed., Marcel Dekker, New York, 1993, 1.

[10] Elmqvist, H., *Simnon — An Interactive Simulation Program for Nonlinear Systems*, Report CODEN:LUTFD2/(TFRT–7502), Dept. of Automatic Control, Lund Inst. of Technology, Lund, Sweden, 1975.

[11] Elmqvist, H., *A Structured Model Language for Large Continuous Systems*, Ph.D. dissertation. Report CODEN:LUTFD2/(TFRT–1015), Department of Automatic Control, Lund Institute of Technology, Lund, Sweden, 1978.

[12] Elmqvist, H., *Dymola — User's Manual*, Dynasim AB, Research Park Ideon, Lund, Sweden, 1995.

[13] Elmqvist, H., Åström, K.J., Schönthal, T., and Wittenmark, B., *Simnon — User's Guide for MS–DOS Computers*, SSPA Systems, Gothenburg, Sweden, 1990.

[14] Elmqvist, H., Cellier, F.E., and Otter, M., Object-oriented modeling of hybrid systems. *Proc. ESS'93, European Simulation Symp.*, Delft, The Netherlands, pp. xxxi–xli, 1993.

[15] Gear, C.W., Differential–algebraic equation index transformations, *SIAM J. Sci. Stat. Comput.*, 9, 39–47, 1988.

[16] Glaser, J.S., Cellier, F.E., and Witulski, A.F., Object-oriented power system modeling using the Dymola modeling language, *Proc. Power Electronics Specialists Conf.*, Atlanta, GA, 2, 837–843, 1995.

[17] Hairer, E. and Wanner, G., *Solving Ordinary Differential Equations. II. Stiff and Differential–Algebraic Problems*, Springer–Verlag, Berlin, 1991.

[18] Integrated Systems Inc., *SystemBuild User's Guide*, Version 4.0, Santa Clara, CA, 1994.

[19] Isidori, A., *Nonlinear Control Systems: An Introduction*, Springer–Verlag, Berlin, 1989.

[20] Korn, G.A., *Interactive Dynamic–System Simulation*, McGraw–Hill, New York, 1989.

[21] Ljung, L., *System Identification*, Prentice Hall, Englewood Cliffs, NJ, 1987.

[22] Mathworks Inc., *SIMULINK — User's Manual*, South Natick, MA, 1992.

[23] Mathworks Inc., *SIMULINK — Release Notes Version 1.3*, South Natick, MA, 1994.

[24] Mitchell and Gauthier Assoc., *ACSL: Advanced Continuous Simulation Language — Reference Manual*, 10th ed., Mitchell & Gauthier Assoc., Concord, MA, 1991.

[25] Mugica, F. and Cellier, F.E., Automated synthesis of a fuzzy controller for cargo ship steering by means of qualitative simulation, *Proc. ESM'94, European Simulation MultiConference*, Barcelona, Spain, pp. 523–528, 1994.

[26] Otter, M., *DSblock: A Neutral Description of Dynamic Systems*, Version 3.2. Technical Report TR R81–92, DLR, Institute for Robotics and System Dynamics, Wessling, Germany, 1992. Newest version available via anonymous ftp from "rlg15.df.op.dlr.de" (129.247.181.65) in directory "pub/dsblock".

[27] Pantelides, C.C., The consistent initialization of differential–algebraic systems, *SIAM J. Sci. Stat. Comput.*, 9, 213–231, 1988.

[28] Petzold, L.R., A description of DASSL: A differential/algebraic system solver, *Proc. 10th IMACS World Congress*, Montreal, Canada, 1982.

[29] Rodrigues, J., Directory of simulation software. SCS — The Society For Computer Simulation, Vol. 5, ISBN 1-56555-064-1, 1994.

[30] Rulka, W., SIMPACK — a computer program for simulation of large–motion multibody systems, in *Multibody Systems Handbook*, ed. W. Schiehlen, Springer–Verlag, Berlin, 1990.

[31] Slotine, J.-J. E. and Li, W., *Applied Nonlinear Control*, Prentice Hall, Englewood Cliffs, NJ, 1991.

[32] Smith, R.C. and Haug, E.J., DADS — Dynamic Analysis and Design System, in *Multibody Systems Handbook*, ed. W. Schiehlen, Springer–Verlag, Berlin, 1990.

[33] Vachoux, A. and Nolan, K., Analog and mixed–level simulation with implications to VHDL. *Proc. NATO/ASI Fundamentals and Standards in Hardware Description Languages*, Kluwer Academic Publishers, Amsterdam, The Netherlands, 1993.

[34] Vachoux, A., VHDL-A archive site (IEEE DASC 1076.1 Working Group) on the Internet on machine "nestor.epfl.ch" under directories "pub/vhdl/standards/ieee/1076.1" to get to the read–only archive site, and "incoming/vhdl" to upload files, 1995.

23

Computer-Aided Control Systems Design

C. Magnus Rimvall
ABB Corporate Research and Development, Heidelberg, Germany

Christopher P. Jobling
Department of Electrical and Electronic Engineering, University of Wales, Swansea, Singleton Park, Wales, UK

23.1 Introduction

The use of computers in the design of control systems has a long and fairly distinguished history. It begins before the dawn of the modern information age with the analogue computing devices, which were used to create tables of ballistic data for artillery and anti-aircraft gunners and continues to the present day, in which modern desktop machines have computing power undreamed of when the classical and modern control theories were laid down in the middle years of the twentieth century.

Modern computer-aided control systems design (CACSD) has been made possible by the synthesis of several key developments in computing. The development and continued dominance of high-level procedural languages such as FORTRAN enabled the development and distribution of standard mathematical software. The emergence of fully interactive operating systems such as UNIX and its user "shells" influenced the development of CACSD packages, which have been constructed along similar lines. The ready availability and cheapness of raster-graphic displays has provided the on-screen display of data from control systems analysis, the creation of tools for modeling control systems using familiar block diagrams and have the potential to make order-of-magnitude improvements in the ease-of-use, ease-of-manipulation, and efficiency of the interaction between the control designer, his model, analysis tools, and end-product — software for embedded controllers. The driving force of all these developments is the seemingly continual increase in computing power year-on-year and the result has been to make computers accessible to large numbers of people while at the same time making them easier to use.

A control engineer often describes systems through the use of block diagrams. This is not only the traditional graphical representation of a control system, it is also an almost discipline-independent, and thus universally understandable, representation for dynamic systems. The diagrams may also constitute a complete documentation of the designed system. Block diagrams are self-documenting and, when appropriately annotated, may form complete and consistent specifications of control systems. It is therefore not surprising that a number of tools for modeling (control) systems through block diagrams have emerged on the market over the last 5 to 10 years.

In addition to serving as a documentation aid, the overall cost and cycle time for developing complex controllers is radically reduced if analysis/simulation code and/or real-time code is automatically generated from the block-diagrams. This eliminates time-consuming manual coding, and avoids the introduction of coding bugs.

In this chapter, we explore the state of the art in CACSD. We begin with a brief survey of the tools that have been developed over the years. We then focus on the matrix environments that provide the current standard and attempt to explain why they are important. We also examine modern block-diagram editors, simulation and code generation tools, and finally, speculate on the future.

23.2 A Brief History of CACSD

The term computer-aided control system design may be defined as:

"The use of digital computers as a primary tool

0-8493-8570-9/96/$0.00+$.50

during the modeling, identification, analysis and design phases of control engineering."

CACSD tools and packages typically provide well-integrated support for the analysis and design of linear plant and controllers, although many modern packages also provide support for the modeling, simulation, and linearization of nonlinear systems and some have the capability of implementing a control law in software.

Figure 23.1 (adapted and updated from Rimvall [8, 9]) illustrates the development of CACSD packages over the last four decades. In order to put events into proper context, other key influencing factors, chiefly hardware and software developments, are also shown. In this section we describe the background to the emergence of CACSD tools in more detail, starting with technological developments and then moving on to user interface aspects. The aim is to understand the current state of the art by examining the historical context in which these tools have been developed.

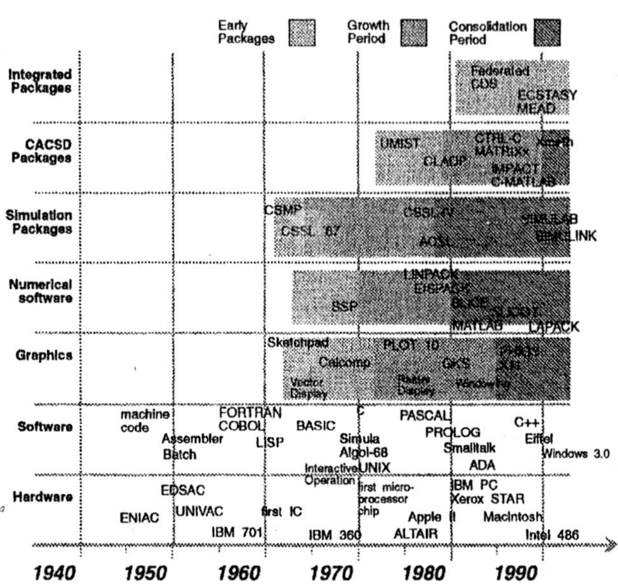

Figure 23.1 The historical development of interactive CACSD tools showing the availability of related hardware and software. Some actual products are included to indicate the state of the art.

23.2.1 Technological Developments

Computing Hardware

Since 1953, there has been a phenomenal growth in the capabilities and power of computing hardware. Observers estimate that the power of computing devices (in terms of both execution speed and memory availability) has doubled every second to third year, whereas the size and cost (per computational unit) of the hardware has halved at approximately the same rate.

In terms of CACSD, the chief effect of these developments has been to widen the range of applications for computing and at the same time to make computers, and therefore the applications, widely available to practitioners in all branches of the subject. For example control engineers, control theorists, and control implementors all benefit as described below.

- Desk-top machines with orders of magnitude more power than mainframe machines of two decades ago provide the means by which CACSD can be brought to the data analysis, model building, simulation, performance analysis and modification, control law synthesis, and documentation that is the day-to-day work of the control engineer.
- Without powerful computing hardware, many of the complex algorithms developed by control theorists for both analysis and implementation would otherwise be impractical.
- Embedded computer systems that implement controllers, smart actuators and smart sensors are routinely used to implement the control laws developed by control engineers and control theorists.

System Software

The development of system software, such as operating systems, programming languages and program execution environments, has been slower than that of hardware, but is nonetheless impressive. Less impressive is the steadily increasing cost of application software, estimated at about 80% of the total installation cost of a computing system, which developments in computer science have been largely unable to reduce. We are, in fact, in the midst of a software crisis, dating from about 1968, which is the result of ever-increasing improvements in hardware. Such improvements increase the possibilities for software, raise the expectations of users, and therefore raise the stakes in software production faster than improvements in software development technology can be made.

High-Level Languages

The invention of FORTRAN was a major breakthrough in engineering computing. A high-level language, FORTRAN and the *compilers* which convert it into machine code, allowed engineers to write programs in a language that was sufficiently close to mathematical notation so as to be quite natural. Since its invention, numerous other high-level languages have been created, although FORTRAN continues to dominate engineering "number-crunching." For the implementation of control algorithms, assembly languages are still popular, although high(er)-level languages like C, which is the predominant systems programming language, MODULA, and ADA are gaining acceptance in the marketplace.

Graphical Displays

Engineers are, in general, more comfortable with pictures than with text as a means of communicating their ideas. Hence, the wide availability of graphical displays is of prime importance to many areas of engineering computing. Indeed, the development of computer graphics has been the means by which certain control systems design techniques, such as multivariable control systems analysis, have been made practicable. Computer graphics have also been instrumental in providing improvements in human-machine interfaces such as schematic systems input and direct manipulation interfaces with windows, icons, pull-down menus, and pop-up dialogue boxes. Further improvements in user interfacing techniques such as *hypermedia* will continue to rely on developments in display technology.

For modern CACSD, the most significant development in display technology has been the development of cheap, high-resolution *raster graphics displays*, although, historically, great strides were made with less well known and more expensive *vector refresh* and *vector storage* display technology. The prime feature of raster-scan technology is that an area of the image may be made to appear to move on the screen by the application of simple logical operations. Raster graphics displays are, therefore, ideal for building direct manipulation graphics applications such as the block-diagram editors, discussed later. They are not so well suited to the direct display and manipulation of vector images, which are a key part of many engineering graphics applications. For example, it is difficult to move part of a vector image such as a bode-plot without destroying the rest of the picture or to display sloping lines that look smooth at low resolutions. However, the dominance of the technology has been a factor in ensuring that the deficiencies in the technology can be overcome by clever software.

Quality Numerical Software

Following the invention of FORTRAN there was a gradual development of useful general-purpose subroutines which could be archived into libraries, distributed and shared. This lead eventually to the development of standard subroutine libraries such as EISPACK [13], LINPACK [4] and LAPACK [1] (for solving eigenvalue problems and sets of linear equations) which have had a direct influence on the development of CACSD.

Simulation Languages

For many years before the predominance of digital computers, dynamic system behavior was simulated using analogue and hybrid computers. *Digital simulation* began to take over from analogue and hybrid simulation during the mid-1960s. Digital simulation programs can be used to model a wider range of nonlinear phenomena more reliably than analogue or hybrid computers, at the cost of losing real-time and introducing quantization problems. However, the disadvantages of the technology are more than outweighed by improvements in modeling possibilities and increases in productivity. Digital simulation has superseded analogue computation in all but a few specialized areas.

The first digital simulation systems were FORTRAN programs. Eventually, special purpose languages emerged which allowed statements written in a form close to state equation notation to be translated into FORTRAN, which enabled the engineer to concentrate on the problem description. In 1967, a standard language called CSSL (Continuous Systems Simulation Language) [2] was proposed by the U.S. Simulation Council and this forms the basis of most simulation languages in use today.

23.2.2 User Interfaces

Over the years, user interaction with computers has become progressively more direct. In the very early days, the user interface was another human being. These "operators" were gradually replaced by *operating systems* which provided communication first through the medium of punch-card and paper tape, then later by teletype machines, text-based visual display units, and, most recently, by windowed graphical user interfaces. Along with this change there has been a corresponding change in style for CACSD tools. Batch mode programs were collected into "packages" and provided with question and answer or menued interfaces. These in turn have been largely superseded by command-driven interfaces and direct-manipulation graphical user interfaces, currently used only for specialized tasks like block-diagram input, will have a wider role in future CACSD packages.

23.2.3 CACSD Packages of Note

As the supporting technology has developed, control engineers, mainly, it has to be said, working in academia, have been actively engaged in developing tools to support developments in control theory and in combining these tools into packages. Early pioneering work was carried out in Europe where the emphasis was on frequency response methods for multivariable control systems analysis and design. Some of the first CACSD packages were developed in the mid-1970s. In the U.S., control theory was concentrated in the time domain and made use of state-space models. Several packages of tools for state-space design were created and reached maturity in the late 1970s. These packages were usually written in FORTRAN and made use of a question-and-answer interface. Some of the better packages made use of standard numerical libraries like EISPACK and LINPACK, but many, it has to be said, made use of home-grown algorithms with sometimes dubious numerical properties.

One of the earliest standardization efforts was concerned with algorithms and there have been several attempts to create standard CACSD libraries. One of these, SLICOT [16], is still ongoing. But it has to be admitted that such efforts have had little success in the marketplace. The real break-through came with the development of the "matrix environments," which are discussed in the next section. Currently, although many research groups continue to develop specialist tools and packages in conventional languages like FORTRAN, most CACSD tool-makers now use these matrix environments as a high-level language for creating "toolboxes" of tools.

23.3 The State of the Art in CACSD

In this section we describe the matrix environments that have come to dominate CACSD, that is the analysis, synthesis, and design of linear controllers for linear plants. We then examine some of the requirements of CACSD that are less well served by the current generation of tools.

23.3.1 Consolidation in CACSD

As can be seen in Figure 23.1, the 1980s was a decade of consolidation during which CACSD technology matured. Menu driven and Q&A dialogues were superseded by command languages. The matrix environment has become the *de facto* standard for CACSD.

The reasons for this are due to the simplicity of the data structures and the interface model and the *flexibility* of the package. We illustrate these properties using MATLAB (MATrix LABoratory) [6], the original matrix environment. Originally designed as a teaching program for graduate students, giving interactive access to the linear algebra routines EISPACK and LINPACK, MATLAB was released into the public domain in around 1980.

In MATLAB, matrices and matrix operations are entered into the computer in the straightforward fashion illustrated in Figure 23.2. This elegant treatment of linear algebra readily appealed

```
>> a = [1 3 5
7 6 5; 0 0 5];

>> [vec, val] = eig(a)

vec =

   -0.7408   -0.3622   -0.1633
    0.6717   -0.9321   -0.8981
         0         0    0.4082


vec =
   -1.7202         0         0
         0    8.7202         0
         0         0    5.0000
```

Figure 23.2 Entering and manipulating matrices in MATLAB. In this example a matrix is defined and its eigenvectors and eigenvalues are determined.

to control scientists who realized that it was equally applicable to the solution of "modern control" problems based on linear state-space models (Figure 23.3).

```
>> A = [0,1,0;0,0,1;-2,-3,4];
>> B = [0, 0, 1];
>> C = [1, 0, 0];
>> poles = eig(A)

poles =
   -0.4142
    2.0000
    2.4142

>> stable = all(poles < 0)

stable =

     0
>>
```

Figure 23.3 Using state-space matrices. A simple stability test showing the power of the matrix functions built in to MATLAB. The Boolean function "all" returns the value TRUE (or 1) if all the elements of the argument are non-zero. The argument is itself a vector of Boolean values (that is those values of the vector of the poles of the **A** matrix that are negative). By treating matrices as "first-class objects," MATLAB provides many such opportunities for avoiding loops and other control structures required to do similar tasks in conventional languages.

However, powerful though the basic "matrix calculator" capabilities of MATLAB are, its real flexibility is due to its support of *macro files*. A macro file (M-file), in its simplest form, is just a collection of ordinary MATLAB commands which are stored in a file. When called, such a "script" of commands is executed just as if it had been typed by the user. MATLAB's real strength lies in its ability to use M-files to create new functions. Such a function is defined in Figure 23.4. Once defined in this way, the new function can be executed as if it was part of the language (Figure 23.5).

```
function qs = control(a, b)
% Returns the controllability
% matrix [b, ab, a^2b, ...]
% used as: qs = control(a, b)
[ma, na] = size(a); [mb, nb] = size(b);
if ma != na
        error('Non-square A matrix')
elseif ma != mb
        error('Unequal number of rows in A and B')
else
        qs = b; k = b;
        for i = 2:ma;
                k = a*k; qs = [qs, k];
        end
end
```

Figure 23.4 The extension of MATLAB by means of macro or M-files. Here is a routine for determining the controllability of a state-space model.

By creating a set of functions in this way, it is relatively easy to build up a "toolbox" of useful functions for a particular application domain. This is exactly what happened shortly after

```
>> qs=control(A,B)

qs =

     0     0     1
     0     1     4
     1     4    13

>>
```

Figure 23.5 Using a user-defined function as an extension to MATLAB.

the release of the original MATLAB. Entrepreneurs quickly realized that if they cleaned up the code, added control-oriented data types and functions and some graphics capability, MATLAB could be resold as a proprietary CACSD package. So, based mainly on the state-space methods in vogue in the U.S., several packages, such as MATRIXx and Ctrl-C, emerged and were a great success.

MATLAB itself underwent further development. It was rewritten in C for efficiency and enhanced portability and released as a commercial product in 1985. Like its competitors, the main market was initially the CACSD market, where, supported by two sets of toolbox extensions called the Control and Signal processing Toolboxes, MATLAB made rapid inroads into academia and industry. A recent development has been the provision of add-on graphical input of system models, in the form of block diagrams, support for "point-and-click" nonlinear simulation, and enhanced graphical functionality. At least one package, MATRIXx has evolved further by the addition of data structures and more sophisticated support for macro development. The result is the package X-Math described by Floyd et al. [5].

23.3.2 A Critique of Matrix Environments for CACSD

MATLAB and similar matrix environments are far from completely ideal. In 1987, Rimvall [8] gave the following requirements for a CACSD environment, which are largely still valid today.

- *Software packages must support the same entities used by human specialists in the field.*
- *The basic commands of an interactive environment must be fast yet flexible.*
- *CACSD packages should support an algorithmic interface.*
- *The transition from basic use to advanced use must be gradual.*
- *The system must be transparent.*
- *Small and large systems should be equally treated by the user interface.*
- *The system must be able to communicate with the outside world.*

Matrix environments do not meet all of these requirements. The following sections give a critical review of the state of the art.

Support of Control Entities

For a control engineer the entities of interest are

- numerical descriptions of systems (state-space models, transfer functions, etc.)
- symbolic elements for general system equations
- graphical elements for the definition of system topologies
- support of large-scale data management, e.g., in the form of a relational database
- support of small-scale data management, e.g., in the form of spreadsheets
- graphical displays of numerical computations, possibly together with graphical interaction for requirement specifications, etc.

MATLAB was developed by a numerical analyst for numerical analysts. Such people need, and MATLAB provides, only one data structure, the complex matrix. It is a credit to its flexibility that the package can be adapted to a control engineer's needs by the careful use of convention and toolbox extensions (Figure 23.6), but the price paid is increased complexity.

Take, as a simple example, single-input single-output control systems design. For each element in the system model, i.e., plant, controller, feedback network, the user has to look after four matrices for a state-space model or two polynomials for a transfer function. He cannot simply refer to the "transfer function G," but must refer instead to the numerator and the denominator polynomials (see Figure 23.7) that stand for G. These polynomials can, in turn, only be distinguished from row vectors by convention and context.

In MATRIXx, this problem was avoided by using packing techniques and a special data structure so that, for example the state-space model in Figure 23.3, would have been stored as shown in Figure 23.8 and additional data would be stored in the workspace of the program so that the **A**, **B**, **C**, **D** matrices could be extracted later when needed.

Such packing schemes are quite widely used by toolbox writers to overcome the limitations imposed by the two-dimensional matrix. One is usually advised, but not required, to manipulate such structures only through the packing and unpacking routines that usually accompany the toolbox code. For example, the packed state-space model might have a function **sstosys** to pack the data and **systoss** to unpack it into separate components, as shown in Figure 23.9. The advantage is that once packed, the state-space model G can be used in processing as if it was the single-system object which it represents. To see this, compare the code for simulation and analysis of a system given in Figure 23.10 (a) with the MATLAB Control System Toolbox code given in Figure 23.10 (b).

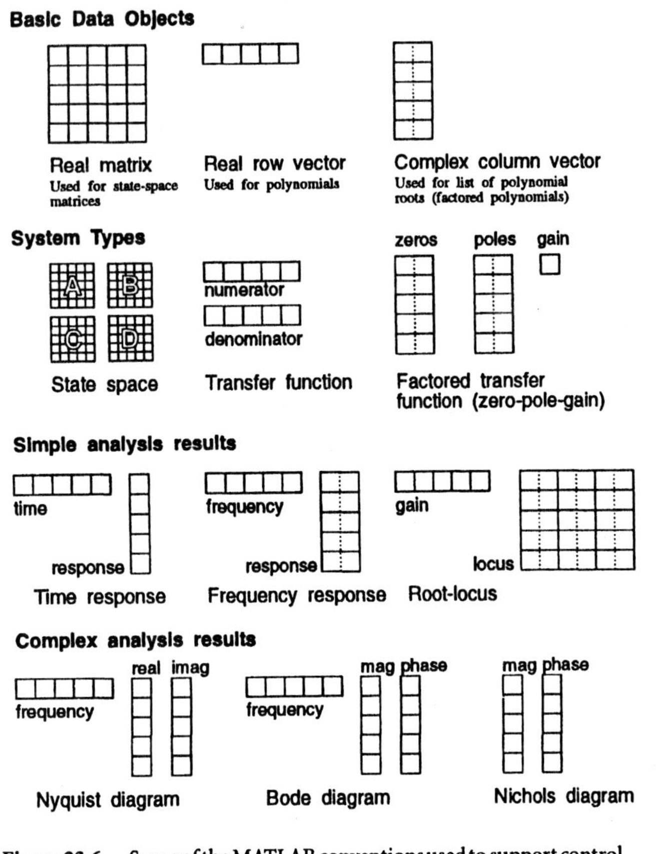

Figure 23.6 Some of the MATLAB conventions used to support control engineering data types.

```
>> % Plant: G(s) = 5/s(s^2 + 2s + 1)
>>  num_G = 5;
    den_G = conv([1 0],[1 2 1]);
>> % Controller: Gc(s) = 15(s + 1)/(s + 2)
>>   K_Gc = 15; Z_Gc = -1;
     P_Gc = -2;
>>   [num_Gc,den_Gc] =
     zp2tf(K_Gc,Z_Gc,P_Gc);
>> % Feedback: H(s) = 1/(s + 10)
>>   num_H = 1; den_H = [1 10];
```

Figure 23.7 Defining a control system in MATLAB.

```
>>  G = [ 0,  1, 0, 0
         0,  0, 0, 0
        -2, -3, 4, 1
         1,  0, 0, 0]
>> % size(A) = [3,3], size(B) = [3,1], ...
       size(C) = [1,3]
```

Figure 23.8 A packed "system matrix," additional values would have to be included to store the sizes of the relevant elements, but these are not shown for clarity.

```
>>  G = sstosys(A,B,C,D);
>>  [a,b,c,d] = systoss(G)

A =   1  0  0
      0  0  1
     -2 -3  4

B =   0
      0
      1

C =  1 0 0

D = 0
>>
```

Figure 23.9 Packing and unpacking system models.

```
>> Go = series(Gc,G)
>>  rlocus(Go)
```

a) Using packed data

```
>>  [Go_A,Go_B,Go_C,Go_D] = ...
       series(Gc_A,Gc_B,Gc_C,Gc_D,G_A,G_B,
       G_C,G_D)
>>  rlocus(Go_A,Go_B,Go_C,Go_D)
```

b) Using non-packed data, the MATLAB control systems toolbox

Figure 23.10 Use of a packed data structure to simplify interaction.

However, aside from the problem that packed data structures may be accidentally used as ordinary matrices, there is a more severe problem that results from a lack of standardization. There are now a number of toolboxes that are used in CACSD, and none of them takes a standard approach to packing data structures. Thus, the data structures used in the Multivariable Control Systems Toolbox are completely incompatible with those used in the Systems Identification Toolbox, which itself is incompatible with the standard Control Systems Toolbox. The consequence is that each toolbox must supply conversion tools and the situation is similar to the problems faced with integrating data from two different packages.

There is therefore an identified need for matrix environments to provide a wider range of, preferably user-definable, data types. These would be used in the same way as record data types are used in conventional programming systems and would be considerably safer to use, since the types expected and returned by functions could be specified in advance and the scope for misuse would be much reduced. In addition, the need to invent new types for each application would be somewhat reduced. This approach has been taken in the matrix environment X-Math and similar features are planned for a future release of MATLAB. Some of the other requirements listed above, such as graphical systems input, graphical display of results, and spreadsheet data manipulation are covered to a greater or lessor extent by the current generation of matrix environments. The others, namely symbolic data processing and database support, are not, but are considered to be outside the scope of this chapter.

Fast Yet Flexible Command Language

MATLAB clearly satisfies this criterion, as is evidenced by the natural interaction shown in Figure 23.3. For CACSD use, it is debatable whether the principle still holds, mainly because of the way that the package entities needed for control have to be constructed and managed by the user. Nonetheless, no one could complain that matrix environments are not flexible: the growing number of new control applications for them provides ample evidence for this.

Algorithmic Interface

The support of an algorithmic interface is simply a recognition of the fact that no package developer can anticipate the requirements of every user. So the package must be extensible by provision of user-defined macros and functions. MATLAB has these, and their provision is clearly important to the users of the package and developers of toolbox extensions. However, there is a limit to the software robustness of the mechanisms that MATLAB provides. MATLAB is an untyped language, all data structures used in extensions to MATLAB are implemented in terms of collections of matrices and vectors. It is therefore up to the programmer to develop conventions for using these data items such that the algorithms work properly. A strongly typed language, in which the user must specify the nature of each data object before it is used, is a much safer basis on which to provide extensions that are to be used by many other people.

Transition from Basic to Advanced Use

The user of a CACSD package is faced with two different types of complexity: the complexity of the user interface and the complexity of the underlying theory and algorithms. In both cases extra guidance is needed for novice users. Typically, the designers of CACSD packages do not wish to stand in the way of the expert users, so they provide direct access to the whole package and interfere in the use of the package as little as possible. This creates problems for novice or infrequent users of the package – novices because they are coming to the package without any knowledge of it, infrequent users because they have probably forgotten most of what they learned the last time they used the package.

In MATLAB, the user interface is deceptively simple. One can take a short tutorial and learn the basic concepts and underlying principles in perhaps one hour. But what happens when one is finished the tutorial and wants to do some actual work? The sheer number of commands in the system can be overwhelming. In basic MATLAB there are some 200 commands, add a few toolboxes and the number quickly increases. The only way to find out how to use a command is to know its name. If you don't know the name you can list all the commands available, but since each command name is limited to eight characters there is not necessarily going to be any relationship between command name and command function. Having found a command the next step is to learn how to use it. In a research prototype CACSD package called IMPACT, Rimvall and Bomholt [10] provided a latent question and answer mode feature which switches from normal command entry to step-by-step elicitation of parameters when requested by the user. Other ways of overcoming some of these difficulties [9] include providing a means of loading toolboxes only when they are needed, thereby reducing the instantaneous "name-space", and providing operator overloading so that the same named procedure can be used to operate on different data types. The latter facility is provided in X-Math [5] and enables, for example, the multiplication operator $*$ to mean matrix multiplication, series combination of systems, polynomial convolution, or time response evaluation, depending on the types of the operands.

Transparency

This is a fundamental principle of software engineering that simply means that there should be no hidden processing or reliance on side effects on which the package depends for its correct operation. Everything the package does and the package itself should at all times be in the complete control of the user.

Scalability

This simply means that account should always be taken of the limitations of numerical algorithms. The package should warn the user when limits are reached and the algorithms should thereafter "degrade gracefully." It is surprising how many applications have been programmed with artificial limits set on various arrays, which is fine so long as the user never presents the package with a problem that its designer never believed would ever be tackled (an inevitable event). Most matrix environments are limited only by the available memory.

23.3.3 "Open Systems"

The need to transfer data to other systems is simply a recognition that no one package can do all things equally well. In many applications, it makes sense to pass a processing task on to an expert. The ability of a package to be able to exchange data (both import and export) is the main feature of so-called open systems. At the very least it must be possible to save data in a form that can be retrieved by an external program. MATLAB and its cousins provide basic file transfer capabilities, but the ideal CACSD package would have some link to a much more convenient data-sharing mechanism, such as could be provided by a database.

23.3.4 Other Desirable Features

- *Form or menu driven input* is often more useful than a functional command driven interface for certain types of data entry. A good example is the plotting of results where the selection of options and parameters for axis scaling, tick marks, etc., are more conveniently specified by means of a dialogue box than by a series of function calls. Such a facility is provided in X-Math's graphics.

- *Graphical input* is useful for defining systems to be analyzed. Today, most of the major packages provide block diagram input, usually tied to nonlinear simulation. What is rarer is graphical input of more application-specific system representations, such as circuit diagrams.
- *Strong data typing*, as already discussed, is useful for toolbox developers since it provides a robust means of developing extra algorithms within the context of the CACSD package. On the other hand there is a fine balance between the needs of the algorithm developer and the algorithm implementor. The former is probably best served by a typeless environment in which it is easy and quick to try out new ideas (such an environment is often called a *rapid-prototyping environment*). The latter, who needs to ensure that the algorithms will work properly under all conditions, needs strong typing to ensure that this can be guaranteed. A similar dichotomy between inventors and implementors can be observed in software engineering.
- *Data persistence.* Unless explicitly saved, CACSD data is not maintained between sessions. Neither can data easily be shared between users. The evolution of models and results over time cannot be recorded. Hence CACSD packages need database support.
- *Matrix environments only support numerical computation.* It is often useful to be able to manipulate a symbolic representation of a control system. Delaying the replacement of symbolic parameters for numerical values for as long as possible can often yield great insight into such properties as stability, sensitivity, and robustness.

23.4 CACSD Block Diagram Tools

As we have discussed in the previous sections, the 1980s was an important decade for control engineering. Apart from new theories, better design methods, and more accurate numerical algorithms, this was the decade when powerful and easy-to-use interactive CACSD tools were put on the average control engineer's desk. Through the use of interactive and extendible programs, new methods and algorithms could be easily implemented and quickly brought to bear on real control engineering problems. Yet despite this tremendous improvement in the availability of good control design environments, the total cost and cycle time for a complex control design was still perceived by many groups and companies as being too high. One of the major remaining bottlenecks was the manual conversion of a control design into testable simulation code and, at a later stage, the conversion of the eventual design into the actual embedded real-time controller code.

A control engineer often describes a system through the use of block diagrams of different kinds. To bypass the bottleneck between theoretical design and actual real-time implementation, systems that took engineering block diagrams and automatically converted them into simulation and/or real-time code started to emerge in the middle of the 1980s. As an early example, in 1984 General Electric decided to develop a block-diagram-based tool with automatic code generation capabilities. This program allowed draftspersons to enter controls block diagrams and automatically convert the functionality of these diagrams into real-time code. Although it used limited graphics, this GE-Internal "Autocode" program successfully produced code at 50% of the cost of traditionally generated code, primarily due to error reduction of not hand coding. This reduction of costs provided the evidence that automatic translation of block diagrams is both feasible and desirable. However, due to advances in both computer graphics and code-generation techniques the first tool was obsolete by the late 1980s. In recent years, several commercial block-diagram-based tools have become available. These tools include System Build from Integrated Systems Incorporated, Model-C from Systems Control Technology, the PC-Based XAnalog from Xanalog, Simulab/Simulink from the Mathworks, and BEACON from General Electric. Some of these tools primarily serve as interfaces to analysis packages such as MATRIXx (System-Build), CTRL-C (Model-C), and MATLAB (Simulink). In some cases they can also be used to directly generate a computer language such as FORTRAN, ADA, or C. A summary of an early 1989 evaluation of the suitability of using System Build, CTRL-C, and Grumman's Protoblock for engine control is given in [14].

23.4.1 Basic Block Diagram System Representations

Some basic user requirements fulfilled by most modern block diagram-oriented CACSD packages are

1. A simple-to-use graphical user interface that can be used with little or no training. The graphical interface is usually based on the Macintosh, MS-Windows, and/or the X-Window System standard.
2. A set of rules for drawing controls-oriented diagrams, sometimes adhering to a standard diagram representations such as ICE-1331 or Petri Nets.
3. An object-based representation of the diagram entities and their graphical behavior. The underlying package must retain a semantic understanding of the diagram so that, for example, pertinent information such as signal types, dimensions, and ranges are propagated through the diagram, or connecting lines are retained when objects are moved.
4. Hierarchical structure, which allows individual blocks to reference either other block diagrams or external modules (e.g., precoded system primitives).
5. Efficient internal simulation capabilities and/or real-time code generation capabilities including optimization of execution speed and/or memory allocation,

As a consequence of the last two points, the block diagram tools must have an open architecture so that the modules created can be associated with external code in a modular fashion. There are two main reasons for this:

- When the block diagrams are used to simulate a physical system, the resulting models must frequently be interfaced with already existing submodels (e.g., from various FORTRAN libraries).
- When real-time controllers are implemented, the autogenerated code must be interfaced with operating system code and other "foreign" software.

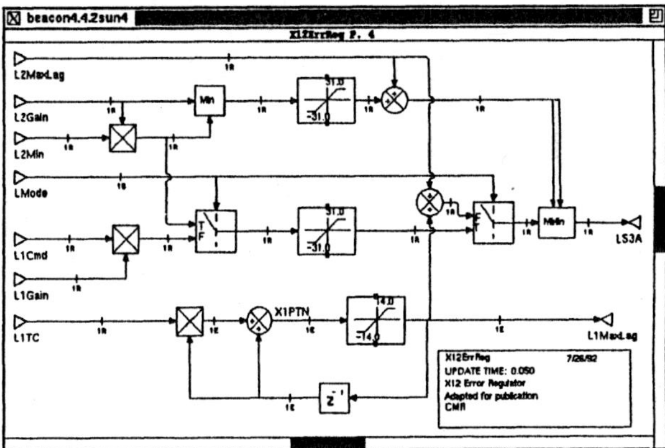

Figure 23.11 A signal flow diagram in the BEACON system.

All of today's block diagram CACSD tools use hierarchical *signal flow diagrams* as their main system representation. As illustrated in Figure 23.11, a signal flow diagram is a directed graph with the nodes representing standard arithmetic, dynamic and logic control blocks such as adders, delays, various filters, nonlinear blocks, and Boolean logic blocks. The connections between the blocks represent "signal" information which is transmitted from one block to another. The connections also indicate the order of execution of the various blocks. Signal flow diagrams are ideal for describing the dynamics of a system or controller.

Some CACSD packages also support some alternate system representations better suited for the logic and sequencing portion of a controller. Possible representations include *ladder-logic*, *dynamic truth-tables*, *flowcharts*, *Petri-nets*, or *state-transition diagrams*.

Figure 23.12 shows a typical control flow diagram or flowchart. The connections in this case represent the order of execution. The triangular blocks are decision blocks while the square blocks are variable assignment blocks written in a PASCAL-like language. Also shown are a multiway branch and a truth table. BEACON requires that the control flow diagrams produce structured code which equivalently means that a diagram can be implemented as sequence of if-then-else statements without go-to's.

Hierarchies greatly facilitate the drawing and organization of diagrams. They provide appropriate levels of abstraction so that individual diagrams can be understood without clutter from details. Hierarchies simplify individual diagrams, making the resulting code easier to test. One can build up a set of subdiagram libraries which can be linked into possibly several higher-level diagrams. Some block-diagram editors also allow the mixing of various diagram types in a hierarchical fashion (e.g., to call a low-level signal flow diagram implementing a control-law scheme from a decision-making flowchart diagram).

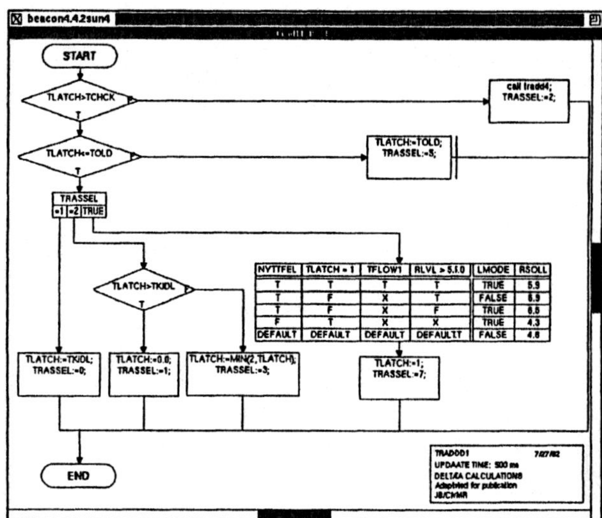

Figure 23.12 A BEACON control flow block diagram.

The graphical modeling environments cannot be viewed as replacements for the matrix environments described in the previous sections, as most of the block diagram environments have very limited analytical capabilities (usually only simulation and linearization). However, many of today's block diagram tools have been developed as companion packages by the same commercial vendors that also sell matrix environments. Through linearization, it thus becomes possible to transform a nonlinear block diagram to a linear representation, which can then be analyzed and used for design in the matrix environment. Unfortunately, such automatic transformations are only available between tools from the same vendor, cross-translations between arbitrary tools are not possible.

23.4.2 Architectures of Block-Diagram Systems

To illustrate typical features and capabilities of a block diagram-oriented simulation or code-generation package, examples will be drawn from BEACON, a CACSD environment developed at GE between 1989 and 1995. There are of course many other block diagram systems, but being commercial products, the essential features are difficult to describe in detail. That said, another system that is well documented and worthy of study is the BlockEdit tool which was part of ECSTASY, a CACSD package developed in the U.K. in the late 1980s [7]. BEACON has been in production use within GE since the first quarter of 1992. Through the use of BEACON, the company has been able to substantially reduce the overall cost and cycle time for developing complex controllers. The automatic generation of code not only eliminates the time-consuming manual coding, but also avoids the

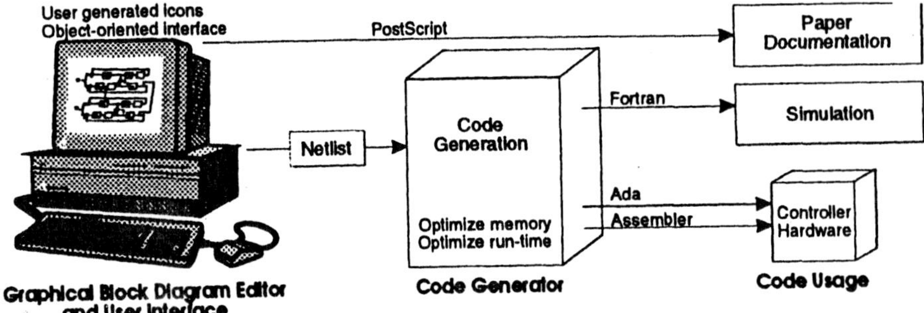

Figure 23.13 The BEACON architecture.

manual introduction of bugs into the code.

BEACON allows the user to graphically design a complete real-time controller as a series of hierarchical block diagrams. These diagrams can thereafter be automatically converted into a variety of computer languages for either control analysis, simulation, or real-time computer code, as illustrated in Figure 23.13.

As shown in this figure, the BEACON system consists of three major components:

- A graphical block-diagram editor with which the engineer designs the system to be simulated/coded [15]. Within this editor, the user may also create new graphical icons representing various numerical or logical blocks.
- A netlist generated from the diagram and containing a full description of that diagram. The netlist format is keyword-oriented, and it has a syntax resembling that of a higher-level language such as Pascal or Ada. To allow a variety of code generators and other uses such as the generation of I/O or termination lists or the automatic generation of test cases, all of the information except graphical location contained in the block diagram is written to the ASCII nestlist file.
- An automatic code generator which translates the block diagrams into simulation and/or real-time computer code [11]. For each block defined in the graphical editor, the code generator expects to find a Block Description Language (BDL) definition as described later.

The BEACON architecture is one of the most open and extendible in the industry, allowing for straightforward extensions to the capability of the system and easy interfacing to other systems. Therefore, the architecture of other block diagram environments are often variants of that of BEACON. Some of the most common differences found in other systems are

- *Built-in simulation capabilities.* Many of today's commercial systems have a nonlinear simulation engine directly built into the system, avoiding BEACON's explicit translation step. Simulation results may then also be directly displayed on or accessed from the original diagram (e.g., in the form of time histories). This allows the user to see immediately the effects of any changes made to the diagram. One drawback of this approach is that these non-compiled approaches all have some kind of threaded-code or interpretative model execution, leading to much slower simulations than explicitly compiled simulation models such as those coming out of BEACON. Some systems allow for either of the two approaches.
- *The avoidance of an explicit netlist.* Many systems have a monolithic architecture with no direct access to the information in a modeled system. This prevents users from directly interfacing the block-diagram editor to other tools or filters (as often performed on a quite ad hoc basis by the users within GE).
- *No code generation.* Some older systems have built-in simulation capabilities only, with no generation of real-time or explicit simulation code.

23.4.3 Open Architectures of Block-Diagram Editors

Flexible block diagrams have the capability of allowing users to develop or modify the graphical representation of symbols to meet the needs of various applications. In addition, it must be possible to add or modify the semantic meaning of the new or changed graphical symbols for simulation- or code-generation purposes.

The Editing of Block Diagram Symbols

In BEACON, all symbols were developed using a Symbol Editor as shown in Figures 23.14 and 23.15. This graphical editor is similar to most other object-oriented graphical editors, with the additional ability to describe diagram connectivity and the display of changing parameter values on the symbol itself. Each symbol is made up of a variety of separate objects (shapes) which are grouped together.

In Figure 23.14, we see a symbol editor session, with the edited switch symbol in the lower left window. The drawing primitives with its graphical shapes is the one in the middle. The large window to the right is an example of a block attributes window. In this case it is the connectivity definition attributes for the left edge of the switch block; these attributes are used to define the sides and vertices which allow inputs or outputs, the allowed number of connections, vector dimension, and types.

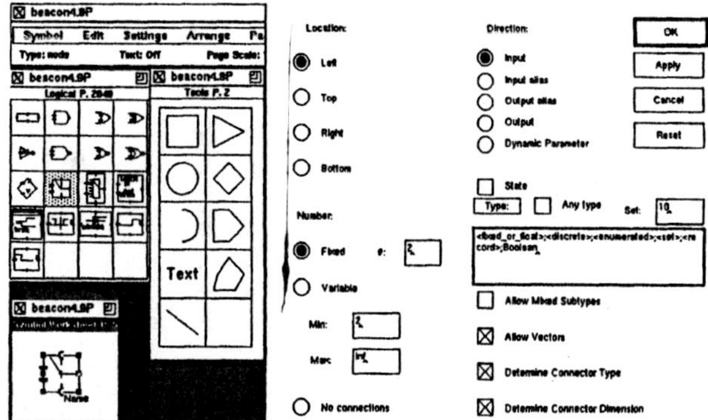

Figure 23.14 The BEACON symbol editor.

Associated with most BEACON block symbols is a parameter form. These forms are unique for each individual block, allowing the user to define the parameters of the block and the specific function. For example, the integrator allows specification of the type of integration to be implemented, as well as rate limits and initial conditions.

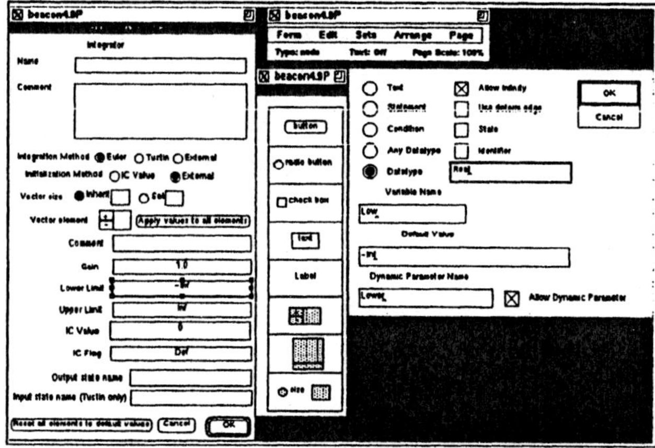

Figure 23.15 Examples of block parameter forms.

The forms are constructed during palette design using the forms editor shown in Figure 23.15. To the left of the screen we see the actual parameter form of the integrator block. In the middle we have the palette from which the primitive form elements may be picked. Each primitive forms object, such as text/value boxes and action buttons, have definable characteristics which will vary from element to element. To the right of Figure 23.15 we see the characteristics of the data-input box for the parameter "lower limit."

The Functional Description of Symbols

The BEACON code generator will process a netlist into FORTRAN, Ada, C, or 68000 code. It accomplishes this by merging the block ordering information, the connectivity information, and the block-specific parameter values found in the netlist with block-specific functional descriptions of each block type. These block descriptions are stored separately from the netlist. This process is illustrated in Figure 23.16.

Each block type supported by BEACON (e.g., adder, integrator, switch) will have a single block definition describing the functionality of the block. Whenever a new block symbol is added using the graphical symbol editor, a corresponding block definition file must be added to the system too. This block definition is written in "BEACON Block-Definition Language" (BDL), a special-purpose structured language which contains all the necessary elements for describing block connectivity, block parameters, algorithms, as well as implementational detail such as fixed-point scaling.

As illustrated in Figure 23.17, each BDL consists of a declarative section and a functional body section.

The declarative section serves four purposes:

- It defines the interfaces to the block, including inputs, outputs, and parameters. Corresponding instantiations of each interface can be found in the netlist.
- It declares variables to be generated in the resulting source code, such as states and temporary variables.
- It declares meta-variables to be used during the code-generation process, but not appearing in the final code.
- It contains checks to be performed on the netlist data, including checks on type, dimension, and scale. The completeness of the netlist is checked as each I/O or parameter declaration implies a corresponding netlist entry.

The body section serves two purposes:

- Some of the BDL instructions directly describe algorithm(s) to be implemented in the target code (*target-code instructions*). These statements are translated into equivalent instructions in the target language. As shown in Figure 23.17, all lines between `BODY_BEGIN` and `BODY_END` which are not preceded by ## are targetcode instructions (Figure 23.18).
- Other portions of the BDL instructions are used and processed by the code generator itself (*meta instructions*). Meta instructions include calls to error-checking routines, or if-/case-statements which have to be evaluated by the code generator to determine which target-code instructions are to be actually used (e.g., when several integration methods are available). All lines in the body which are preceded by ## are meta instructions (Figure 23.19).

Figure 23.20 shows a portion of the FORTRAN-77 code resulting from the lower branches of the diagram in Figure 23.11 (input `L1TC` to output `L1MaxLag`). The code resulting from a signal-flow diagram is a well-structured and yet locally optimized

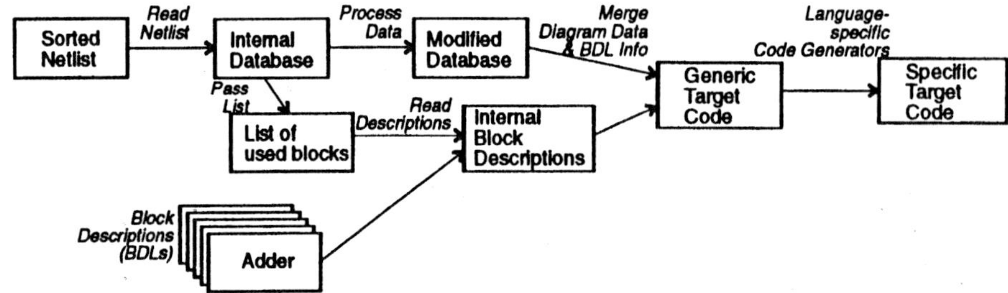

Figure 23.16 General principle of the workstation-based code generator.

```
BLOCK BEGIN Switch IS
  DEFINITION_BEGIN                                                      Declarations
      input(1,1..2)[<>]  : signal_in   %type(1,=)        -> Tin
                                       %vectordim(1,=)   -> Vin
                                       %scale(1,=)[:]    -> Bin[1..Vin]
                                       %toggle(:)        -> Sin(1..2);
      input(2,1)[<>]     : switch_in   %type             == boolean;
      output(<>)         : signal_out  %type             == Tin
                                       %vectordim        == Vin
                                       %scale[:]         == Bin[1..Vin]
   DEFINITION_END;

   BODY_BEGIN                                                           Functionality
      ## IF (Sin(1)) THEN
         IF (switch_in) THEN
            ## VECTORFOR i IN 1..Vin LOOP
               signal_out[i] := signal_in(1,1)[i];
            ## END LOOP;
         ELSE
            ## VECTORFOR i IN 1..Vin LOOP
               signal_out[i] := signal_in(1,2)[i];
            ## END LOOP;
         ENDIF;
      ## ELSEIF (Sin(2)) THEN
         IF (switch_in) THEN
            ## VECTORFOR i IN 1..Vin LOOP
               signal_out[i] := signal_in(1,2)[i];
            ## END LOOP;
         ELSE
            ## VECTORFOR i IN 1..Vin LOOP
               signal_out[i] := signal_in(1,1)[i];
            ## END LOOP;
         ENDIF;
      ## ELSE
         ## call ERROR("Neither of the two switch toggles are true");
      ## END IF;
   BODY_END;
BLOCK_END Switch;
```

Figure 23.17 Declarative and functional parts of a BDL.

```
IF (switch_in) THEN
   ...
END IF;
```

Figure 23.18 Example of a target-code instruction.

```
## VECTORFOR i IN 1..Vin LOOP
   ...
## END LOOP;
```

Figure 23.19 Example of a meta instruction.

implementation of the diagram. It has the following characteristics:

- Through processing the sorted netlist, each block on the diagram is individually mapped onto the target language. Named blocks are preceded by a comment stating that name (e.g., L1MAXLAG).
- Each connection on the diagram corresponds to a memory location in the code. To ensure readable code, each labeled connection is explicitly declared as a variable in the code (e.g., X1PTN). Unlabeled connections are mapped into reusable temporary variables or, in the case of assembler code, temporarily used registers. This ensures a locally optimized and yet fully readable code.
- States and other variables explicitly named on the diagram retain their name in the code. Unnamed states are automatically assigned unique names.
- Each numerical value is directly inserted into the code using the appropriate format of the target language and arithmetic type/precision used.

```
C **** DELAY_DELAY_34
C **** Out only
C
      IF ( ICFLAG ) THEN
         SS0001_XX_DELAY_34 = 0.0000E+00
         EETMP5 = 0.0000E+00
      ELSE
         EETMP5 = SS0001_XX_DELAY_34
      END IF
C
      EETMP6 = L1TC * EETMP5
      X1PTN = EETMP6 + EETMP5
      IF ( X1PTN .GT. 1.4000E+01 ) THEN
         EETMP5 = 1.4000E+01
      ELSE IF ( X1PTN .LT. -1.4000E+01) THEN
         EETMP5 = -1.4000E+01
      ELSE
         EETMP5 = X1PTN
      END IF
C
C **** SUBNET_OUTPUT L1MAXLAG
C
      L1MAXLAG = EETMP5
C
C **** DELAY _DELAY_34
C **** In only
C
      SS0001_XX_DELAY_34 = EETMP5
C
```

Figure 23.20 Portions of the FORTRAN-77 code generated from a signal flow diagram.

```
IF (TLATCH .GT. TCHCK) THEN
  IF (TLATCH .LE. TOLD) THEN
    IF (TRASSEL .EQ. 1) THEN
      TLATCH = TKIDL
      TRASSEL = 0
      RETURN
    ELSE
      IF (TRASSEL .EQ. 2) THEN
        IF (TLATCH .GT. TKIDL) THEN
          TLATCH = 0.0
          TRASSEL = 1
          RETURN
        ELSE
          TLATCH = MIN(2,TLATCH)
          TRASSEL = 3
          RETURN
        END IF
      ELSE
        LMODE = .FALSE.
        RSOLL = 4.8
        IF (NYTTFEL) THEN
          IF (TLATCH .EQ. 1) THEN
            IF (TFLOW1) THEN
              IF (RLVL .GT. 5.0) THEN
                LMODE = .TRUE.
                RSOLL = 5.9
              END IF
            END IF
          ELSE
            IF (RLVL .GT. 5.0) THEN
              LMODE = .FALSE.
              RSOLL = 6.9
            ELSE
              LMODE = .TRUE.
              RSOLL = 6.5
            END IF
          END IF
        ELSE
          IF (TLATCH .EQ. 1) THEN
            LMODE = .TRUE.
            RSOLL = 4.3
          END IF
        END IF
        TLATCH = 1
        TRASSEL = 7
        RETURN
      END IF
    ELSE
      TLATCH = TOLD
      TRASSEL = 5
      RETURN
    END IF
  ELSE
    CALL TRADD4
    TRASSEL = 2
    RETURN
  END IF
END IF
```

Figure 23.21 FORTRAN-77 code generated from a control flow diagram.

Code from Control Flow Diagrams

Control flow diagrams, such as the one shown in Figure 23.12, are processed in a similar manner to signal flow diagrams. The main difference is that, while a signal flow diagram uses a fixed set of blocks with well-defined semantics (the block interconnections and block parameters being the only variants between two blocks of the same type), the blocks in control flow diagrams may contain arbitrary expressions, assignment statements, and/or procedure calls (as shown in Figure 23.12). These BEACON language constructs must be translated into the primitives of each target language.

The BEACON graphical editor ensures that control flow diagrams are well structured, i.e., that the diagram can be mapped into structured code. Figure 23.21 shows some of the FORTRAN-77 code generated from the diagram in Figure 23.12. As can be seen from the example, the automatic translation of large truth-tables into complex structured code is particularly time saving.

23.5 Conclusions

In this chapter we have reviewed the tools available for the computer-aided design of control systems. The main features of the current state of the art are analysis tools built around a "matrix environment" and modeling, simulation, and code generation tools constructed around the block diagram representation. For the most part, control systems analysis and design is done from a textual interface and modeling, simulation, and code generation rely on a graphical user interface. There are links between the two "environments," usually provided by some form of linearization.

Future CACSD environments will have to give equal emphasis to "control data objects" as they now do for matrices. This is becoming urgent as the number of specialist toolboxes being written for MATLAB and similar packages increases. Only by having a set of commonly approved data types can the further development of incompatible data formats *within a single package* be prevented. Rimvall and Wette have defined an extended MATLAB-compatible command language to overcome such problems and the issues are discussed in [12].

As graphical user interfaces become more popular on computing devices, the possibilities for interactive manipulation of systems will have to be explored. We expect that graphical tools for control systems analysis and design will become commonplace over the next few years and may eventually replace textual interfaces for most users.

A final important area for development in CACSD will be driven by the need to embed control systems design into information systems for enterprise integration. To some extent this is already happening with the need for multidisciplinary teams of engineers to work on common problems. The computer-based support of such projects requires facilities for the development and exchange of models, the storage of design data, version control, configuration management, project management, and computer-supported cooperative work. It is likely that CACSD will have to develop into a much more open set of tools supported by databases, networks, and distributed computation. The implications of some of these developments are discussed in [3].

References

[1] Anderson, E., Bai, Z., Bischof, C., Demmel, J., Dongarra, J., DuCroz, J., Greenbaum, A., Hammarling, S., McKenney, A., and Soresen, D., LAPACK: A portable linear algebra library for supercomputers. Technical report, Argonne National Laboratory, 1989.

[2] Augustin, D., Strauss, J.C., Fineberg, M.S., Johnson, B.B., Linebarger, R.N., and Samson, F.J., The SCi continuous system simulation language (CSSL), *Simulation,* 9(6), 281–304, 1967.

[3] Barker, H.A., Chen, M., Grant, P.W., Jobling, C.P., and Townsend, P., Open architecture for computer-aided control engineering, *IEEE Control Syst. Mag.,* 12(3), 17–27, 1993.

[4] Dongarra, J.J., Bunch, J.R., Moler, C.B., and Stewart, G.W., LINPACK user's guide, *Lecture Notes in Computer Science,* 1979.

[5] Floyd, M.A., Dawes, P.J., and Milletti, U., X-Math: a new generation of object-oriented CACSD tools, in *Proceedings European Control Conference,* 3, 2232–2237, 1991, Grenoble, France.

[6] Moler, C., MATLAB — user's guide. Technical report, Department of Computer Science, University of New Mexico, Albuquerque, NM, 1980.

[7] Munro, N. and Jobling, C.P., ECSTASY: A control system CAD environment, in *CAD for Control Systems,* D.A. Linkens, Ed., Marcel Dekker, New York, 1994, 449.

[8] Rimvall, C.M., CACSD software and man-machine interfaces of modern control environments, *Trans. Inst. Meas. Control,* 9(2), 1987.

[9] Rimvall, C.M., Interactive environments for CACSD software, in *Preprints of 4th IFAC Symp. on Computer Aided Design in Control System CADCS '88,* Beijing, PRC, pages 17–26, Pergamon Press, New York, 1988.

[10] Rimvall, C.M. and Bomholt, L., A flexible man-machine interface for CACSD applications, in *Proc. 3rd IFAC Symp. on Computer Aided Design in Control and Engineering,* Pergamon Press, New York, 1985.

[11] Rimvall, C.M., Radecki, M., Komar, A., Wadhwa, A., Spang III, H.A., Knopf, R., and Idelchik, M., Automatic generation of real-time code using the BEACON CAE environment, in *Proc. 12th IFAC World Congress on Automatic Control,* 6, 99–104, 1993, Sidney, Australia.

[12] Rimvall, C.M. and Wette, M., Towards standards for CACE command syntax and graphical interface, in *Proc. 12th IFAC World Congress on Automatic Control,* 8, 87–390, 1993, Sydney, Australia.

[13] Smith, B.T., Boyle, J.M., Dongarra, J.J., Garbow, B.S., and Ikebe, Y., Matrix eigensystem routines — EISPACK guide extension, *Lecture Notes in Computer Science,* 51, 1977.

[14] Spang III, H.A., Rimvall, C.M., Sutherland, H.A., and Dixon, W., An evaluation of block diagram CAE tools, in *Proc. 11th IFAC World Congress on Automatic Control,* 9, 79–84, 1990, Tallinn, Estonia.

[15] Spang III, H.A., Wadhwa, A., Rimvall, C.M., Knopf, R., Radecki, M., and Idelchik, M., The BEACON block-diagram environment, in *Proc. 12th IFAC World Congress on Automatic Control,* 6, 105–110, 1993, Sydney, Australia.

[16] van den Boom, A., Brown, A., Dumortier, F., Geurts, A., Hammarling, S., Kool, R., Vanbegin, M., van Dooren, P., van Huffle, S., SLICOT: A subroutine library in control and system theory, in *Proc. 5th IFAC Symposium on Computer Aided Design in Control Systems — CADCS'91,* Swansea, U.K., 1991, 1–76.

Further Reading

Keeping up to date with developments in CACSD is not always easy but the Proceedings of the triennial IFAC symposium on Computer-Aided Design in Control Systems (CADCS) and the IEEE biennial workshop on CACSD are useful indicators of the latest trends. The proceedings of the last three of these meetings are given in the list of ideas for further reading given below. The other items give useful snapshots of the state of the art at various points in the last 10 years or so. In addition to these sources, the IEEE Control Systems Magazine regularly publishes articles on CACSD and is a good place to look for other information.

[1] Jamshidi, M. and Herget, C. J., Eds., Computer-Aided Control Systems Engineering, North-Holland, Amsterdam, 1985.

[2] CADCS '91, Proceedings of the 5th IFAC Symposium on Computer Aided Design in Control Systems, Swansea, U.K. Pergamon Press, Oxford, U.K., 1991.

[3] Jamshidi, M., Tarokh, M., and Shafai, B., Computer-aided analysis and design of control systems. Prentice Hall, Englewood Cliffs, NJ, 1991.

[4] CACSD '92, Proceedings of the IEEE Control Systems Society Symposium on CACSD, Napa, CA, 1992.

[5] Jamshidi, M. and Herget, C. J., Eds., Recent Advances in Computer-Aided Control Systems Engineering. Studies in Automation and Control. Elsevier, Amsterdam, 1992.

[6] CACSD '94, Proceedings of the IEEE/IFAC Joint Symposium on Computer-Aided Control System Design, Tucson, AZ. Pergamon Press, Oxford, U.K., 1994.

[7] Linkens, D. A., Ed., CAD for Control Systems, Marcel Dekker, New York, 1994.

PART B

ADVANCED METHODS OF CONTROL

SECTION VII

Analysis Methods for MIMO Linear Systems

24

Multivariable Poles, Zeros, and Pole-Zero Cancellations

Joel Douglas
Department of Electrical Engineering and Computer Science, Massachusetts Institute of Technology, Cambridge, MA

Michael Athans
Department of Electrical Engineering and Computer Science, Massachusetts Institute of Technology, Cambridge, MA

24.1 Introduction

In this chapter we will introduce the basic building blocks necessary to understand linear time-invariant, multivariable systems. We will examine solutions of linear systems in both the time domain and frequency domain. An important issue is our ability to change the system's response by applying different inputs. We will thus introduce the concept of controllability. Similarly, we will introduce the concept of observability to quantify how well we can determine what is happening internally in our model when we can observe only the outputs of the system. An important issue in controllability and observability is the role of zeros. We will define them for multivariable systems and show their role in these concepts. Throughout, we will introduce the linear algebra tools necessary for multivariable systems.

24.2 Unforced Linear Systems

24.2.1 Eigenvectors and Eigenvalues

Given a matrix $A \in \mathcal{R}^{n\times n}$, the eigenstructure of A is defined by n complex numbers λ_i. When the λ_i are all different, each λ_i has corresponding vectors $v_i \in C^n$ and $w_i \in C^n$ so that

$$Av_i = \lambda_i v_i; \quad w_i^H A = \lambda_i w_i^H; \quad i = 1...n \tag{24.1}$$

where w^H is the complex conjugate transpose of w. The complex numbers λ_i are called the eigenvalues of A. The vectors v_i are called the right eigenvectors, and the vectors w_i are called the left eigenvectors. Notice that any multiple of an eigenvector is also an eigenvector.

The left and right eigenvectors are mutually orthogonal, that is, they satisfy the property

$$w_i^H v_j = \delta_{ij} \stackrel{\Delta}{=} \begin{cases} 1 & \text{if } i = j \\ 0 & \text{if } i \neq j \end{cases} \tag{24.2}$$

We assume throughout that the eigenvalues are distinct; that is, they are all different. The case where eigenvalues repeat is much more complicated, both theoretically and computationally. The case of repeated eigenvalues is covered in Kailath [1].

One other formula useful for describing linear systems is the dyadic formula. This formula shows how the matrix A can be formed from its eigenvalues and eigenvectors. It is given by

$$A = \sum_{i=1}^{n} \lambda_i v_i w_i^H \tag{24.3}$$

24.2.2 The Matrix Exponential

The matrix exponential, e^A, is defined by

$$e^A = I + A + \frac{1}{2}A^2 + \frac{1}{6}A^3 + \ldots \tag{24.4}$$

$$= \sum_{k=0}^{\infty} \frac{1}{k!}A^k \tag{24.5}$$

The matrix exponential solves the following matrix differential equation

$$\dot{x}(t) = Ax(t), \quad x(0) = \xi \tag{24.6}$$

0-8493-8570-9/96/$0.00+$.50

The solution is

$$x(t) = e^{At}\xi \tag{24.7}$$

The matrix exponential can be calculated from the eigenstructure of the matrix A. If A has the eigenstructure as in Equation 24.1, then

$$e^{A} = \sum_{i=1}^{n} e^{\lambda_i} v_i w_i^H \quad \text{and} \quad e^{At} = \sum_{i=1}^{n} e^{\lambda_i t} v_i w_i^H \tag{24.8}$$

This can be seen by writing the matrix exponential using the infinite sum, substituting in the dyadic formula (Equation 24.3), and using Equation 24.2.

Taking the Laplace transform of Equation 24.6,

$$sx(s) - \xi = Ax(s) \tag{24.9}$$

where we have used the initial condition $x(0) = \xi$. Thus, the solution is

$$x(s) = (sI - A)^{-1}\xi \tag{24.10}$$

Therefore, $(sI - A)^{-1}$ is the Laplace transform of e^{At}.

24.2.3 Definition of Modes

The solution to the unforced system Equation 24.6 can be written in terms of the eigenstructure of A as

$$x(t) = e^{At}\xi = \sum_{i=1}^{n} e^{\lambda_i t} v_i (w_i^H \xi) \tag{24.11}$$

The ith mode is $e^{\lambda_i t} v_i$, defined by the direction of the right eigenvector v_i, and the exponential associated with the eigenvalue λ_i. $w_i^H \xi$ is a scalar, specifying the degree to which the initial condition ξ excites the ith mode.

Taking Laplace transforms,

$$x(s) = \sum_{i=1}^{n} \frac{1}{s - \lambda_i} v_i w_i^H \xi \tag{24.12}$$

From this equation,

$$x(s) = (sI - A)^{-1}\xi = \sum_{i=1}^{n} \frac{1}{s - \lambda_i} v_i w_i^H \xi \tag{24.13}$$

When the initial condition is equal to one of the right eigenvectors, only the mode associated with that eigenvector is excited. To see this, let $\xi = v_j$. Then, using Equation 24.2,

$$x(t) = \sum_{i=1}^{n} e^{\lambda_i t} v_i (w_i^H v_j) = e^{\lambda_j t} v_j \tag{24.14}$$

A mode is called stable if the dynamics of the modal response tend to zero asymptotically. This is, therefore, equivalent to

$$\text{Re}(\lambda_i) < 0 \tag{24.15}$$

A system is said to be stable if all of its modes are stable. Thus, for any initial condition, $x(t) \rightarrow 0$ if the system is stable.

24.2.4 Multivariable Poles

Consider the system

$$\begin{aligned} \dot{x}(t) &= Ax(t) + Bu(t) && (24.16) \\ y(t) &= Cx(t) + Du(t) && (24.17) \end{aligned}$$

The multivariable poles are defined as the eigenvalues of A. Thus, the system is stable if all of its poles are strictly in the left half of the s-plane. The poles are therefore the roots of the equation

$$\det(sI - A) = 0 \tag{24.18}$$

24.3 Forced Linear Time-Invariant Systems

24.3.1 Solution to Forced Systems

We consider the dynamical system

$$\begin{aligned} \dot{x}(t) &= Ax(t) + Bu(t), \quad x(0) = \xi && (24.19) \\ y(t) &= Cx(t) + Du(t) && (24.20) \end{aligned}$$

The solution $x(t)$ to this equation is

$$x(t) = e^{At}\xi + \int_0^t e^{A(t-\tau)} Bu(\tau) d\tau \tag{24.21}$$

This can be seen by differentiating the solution, and substituting in Equation 24.19. The output $y(t)$ is thus

$$\begin{aligned} y(t) &= Cx(t) + Du(t) = Ce^{At}\xi \\ &\quad + \int_0^t Ce^{A(t-\tau)} Bu(\tau) d\tau + Du(t) && (24.22) \end{aligned}$$

Using the eigenstructure of A, we can write

$$\begin{aligned} y(t) &= \sum_{i=1}^{n} e^{\lambda_i t} (Cv_i)(w_i^H \xi) && (24.23) \\ &\quad + \sum_{i=1}^{n} e^{\lambda_i t} (Cv_i)(w_i^H B) \int_0^t e^{-\lambda_i \tau} u(\tau) d\tau + Du(t) && (24.24) \end{aligned}$$

Applying the Laplace Transform to the system Equations 24.19 and 24.20,

$$y(s) = C(sI - A)^{-1} Bu(s) + Du(s) \tag{24.25}$$

Using the eigenstructure of A, we can substitute for $(sI - A)^{-1}$ to get the Laplace Transform equation

$$y(s) = \sum_{i=1}^{n} \frac{Cv_i w_i^H B}{s - \lambda_i} u(s) + Du(s) \tag{24.26}$$

The matrix $\frac{Cv_i w_i^H B}{(s-\lambda_i)}$ is called the residue matrix at the pole $s = \lambda_i$.

We can see that $w_i^H B$ is an indication of how much the ith mode is exited by the inputs, and Cv_i indicates how much the ith mode is observed in the outputs. This is the basis for the concepts of controllability and observability, respectively.

24.3.2 Controllability and Observability

Controllability is concerned with how much an input affects the states of a system. The definition is general enough to handle nonlinear systems and time-varying systems.

DEFINITION 24.1 The nonlinear time-invariant (LTI) system

$$\dot{x}(t) = f[x(t), u(t)], \quad x(0) = \xi, \tag{24.27}$$

is called completely controllable if, for any initial state ξ and any final state θ, we can find a piecewise, continuous bounded function $u(t)$ for $0 \leq t \leq T, T < \infty$, so that $x(T) = \theta$. A system which is not completely controllable is called uncontrollable.

Observability is defined similarly.

DEFINITION 24.2 The nonlinear time-invariant system

$$\dot{x}(t) = f[x(t), u(t)], \quad x(0) = \xi \tag{24.28}$$

$$y(t) = g[x(t), u(t)] \tag{24.29}$$

is observable if one can calculate the initial state ξ based upon measurements of the input $u(t)$ and output $y(t)$ for $0 \leq t \leq T$, $T < \infty$. A system which is not observable is called unobservable.

For LTI systems, there are simple tests to determine if a system is controllable and observable. The ith mode is uncontrollable if, and only if,

$$w_i^H B = 0 \tag{24.30}$$

Thus a mode is called uncontrollable if none of the inputs can excite the mode. The system is uncontrollable if it is uncontrollable from any mode. Thus, a system is controllable if

$$w_i^H B \neq 0 \quad i = 1, ..., n \tag{24.31}$$

Similarly, the ith mode is unobservable if, and only if,

$$Cv_i = 0 \tag{24.32}$$

Thus a mode is called unobservable if we can not see its effects in any of the outputs. The system is unobservable if it is unobservable from any mode. Thus, a system is unobservable if

$$Cv_i \neq 0, \quad i = 1, ..., n \tag{24.33}$$

24.3.3 Other Tests for Controllability and Observability

There is a simple algebraic test for controllability and observability. Let us define the controllability matrix M_c as

$$M_c = \begin{bmatrix} B & AB & A^2B & \dots & A^{n-1}B \end{bmatrix} \tag{24.34}$$

The system is controllable if, and only if, $rank(M_c) = n$. Note that if the rank of M_c is less than n, we do not know anything about the controllability of individual modes. We only know that at least one mode is uncontrollable.

There is a similar test to determine the observability of the system. We form the observability matrix M_o as

$$M_o = \begin{bmatrix} C \\ CA \\ CA^2 \\ \vdots \\ CA^{n-1} \end{bmatrix}. \tag{24.35}$$

The system is observable if, and only if, $rank(M_o) = n$. Again, this test provides no insight into the observability of the modes of the system. It only determines whether or not the whole system is observable.

We now have tests for controllability and observability. We will now relate the loss of controllability and observability to pole-zero cancellations. First, we need to define the concept of a zero for a multi-input multioutput system.

24.4 Multivariable Transmission Zeros

There are several ways in which zeros can be defined for multivariable systems. The one we will examine is based on a generalized eigenvalue problem and has an interesting physical interpretation. Another approach to defining and calculating the zeros of MIMO systems can be found in Kailath [1].

24.4.1 Definition of MIMO Transmission Zeros

To define the multi-input, multioutput (MIMO) transmission zeros, we will first assume that we have a system with the same number of inputs and outputs. This is referred to as a square system. We will later extend the definition to nonsquare systems. For square systems, we can represent the system in the time domain as

$$\dot{x}(t) = Ax(t) + Bu(t) \tag{24.36}$$

$$y(t) = Cx(t) + Du(t) \tag{24.37}$$

where $x(t) \in \mathcal{R}^n$, $u(t) \in \mathcal{R}^m$, and $y(t) \in \mathcal{R}^m$. We can also write the transfer function matrix as

$$G(s) = C(sI - A)^{-1}B + D \tag{24.38}$$

where $G(s) \in \mathcal{C}^{m \times m}$. Given this system, we have the following definition:

DEFINITION 24.3 The plant has a zero at the (complex) value z_k if vectors $\xi_k \in \mathcal{C}^n$ and $u_k \in \mathcal{C}^m$ exist which are not both zero, so that the solution to the equations

$$\dot{x}(t) = Ax(t) + Bu_k e^{z_k t}, \quad x(0) = \xi_k \tag{24.39}$$

$$y(t) = Cx(t) + Du(t) \tag{24.40}$$

has the property that

$$y(t) \equiv 0 \quad \forall t > 0 \tag{24.41}$$

This property of transmission zeros is sometimes called transmission blocking. When zeros repeat, this definition still holds but a more complicated transmission blocking property also holds [2].

As an example, let us show that this definition is consistent with the standard definition of zeros for single-input, single-output systems.

EXAMPLE 24.1:

Let us consider the following plant:

$$g(s) = \frac{s+1}{s(s-1)} \tag{24.42}$$

Then a state-space representation of this is

$$\dot{x}_1(t) = x_2(t) \tag{24.43}$$

$$\dot{x}_2(t) = x_2(t) + u(t) \tag{24.44}$$

$$y(t) = x_1(t) + x_2(t) \tag{24.45}$$

Let us define $u(t) = 2e^{-t}$, so that $u_k = 2$ and $z_k = -1$. Let us also define the vector ξ_k as

$$\xi_k = \begin{bmatrix} 1 \\ -1 \end{bmatrix} \tag{24.46}$$

Then

$$\dot{x_2}(t) = x_2(t) + 2e^{-t}, \quad x_2(0) = -1 \tag{24.47}$$

$$\Rightarrow x_2(t) = -e^t + e^t \int_0^t e^{-\tau} 2e^{-\tau} \partial\tau = -e^{-t} \tag{24.48}$$

$$\dot{x}_1(t) = x_2(t), \quad x_1(0) = 1 \tag{24.49}$$

$$\Rightarrow x_1(t) = 1 + e^{-t} - 1 = e^{-t} \tag{24.50}$$

Thus,

$$y(t) = x_1(t) + x_2(t) = 0 \tag{24.51}$$

So we have confirmed that $z = -1$ is a transmission zero of the system.

From the definition, we can see how transmission zeros got their name. If an input is applied at the frequency of the transmission zero in the correct direction (u_k), and the initial condition is in the correct direction ξ_k, then nothing is transmitted through the system.

24.4.2 Calculation of Transmission Zeros

To calculate the transmission zero, we can rewrite the definition of the transmission zero in matrix form as

$$\begin{bmatrix} z_k I - A & -B \\ -C & -D \end{bmatrix} \begin{bmatrix} \xi_k \\ u_k \end{bmatrix} = \begin{bmatrix} 0 \\ 0 \end{bmatrix} \tag{24.52}$$

This is in the form of a generalized eigenvalue problem, typically written as

$$G v_i = z_i M v_i \tag{24.53}$$

$$w_i^H G = z_i w_i^H M \tag{24.54}$$

where z_i is a generalized eigenvalue, with right and left generalized eigenvectors v_i and w_i, respectively. The generalized eigenvalues are the roots of the equation

$$\det(zM - G) = 0 \tag{24.55}$$

Note that if M is invertible, there are n generalized eigenvalues. Otherwise, there are less than n.

Equation 24.52 is a generalized eigenvalue problem with

$$G = \begin{bmatrix} A & B \\ C & D \end{bmatrix} \quad M = \begin{bmatrix} I & 0 \\ 0 & 0 \end{bmatrix} \tag{24.56}$$

Let us look at the implication of the generalized eigenvalue problem. Let z_k be a transmission zero. Then it must be true that

$$0 = \det \begin{bmatrix} z_k I - A & -B \\ -C & -D \end{bmatrix} \tag{24.57}$$

If there are no common poles and zeros, then $z_k I - A$ is invertible, and we can write

$$0 = \det(z_k I - A) \det[C(z_k I - A)^{-1} B + D] \tag{24.58}$$

$$= \det(z_k I - A) \det[G(z_k)] \tag{24.59}$$

Since we assume there are no common poles and zeros in the system, then $\det(z_k I - A) \neq 0$, and so it must be true that

$$\det(G(z_k)) = 0 \tag{24.60}$$

Thus, in the case that there are no common poles and zeros, the MIMO transmission zeros are the roots of Equation 24.60. To check for transmission zeros at the same frequencies as the poles, we must use the generalized eigenvalue problem.

Let us now give a multivariable example.

EXAMPLE 24.2:

Consider the system given by Equations 24.36 and 24.37, with the matrices given by

$$A = \begin{bmatrix} -2 & 0 & 0 \\ 0 & -3 & 0 \\ 0 & 0 & -4 \end{bmatrix} \quad B = \begin{bmatrix} -1 & 0 \\ -1 & 0 \\ 0 & 1 \end{bmatrix} \tag{24.61}$$

$$C = \begin{bmatrix} 1 & 0 & 0 \\ 0 & 1 & 1 \end{bmatrix} \quad D = \begin{bmatrix} 1 & 0 \\ 1 & 0 \end{bmatrix} \tag{24.62}$$

Solving the generalized eigenvalue problem Equation 24.52, we find that there are two zeros, given by

$$z_1 = -1 \quad z_2 = -3 \tag{24.63}$$

$$\xi_1 = \begin{bmatrix} 2 \\ 1 \\ 1 \end{bmatrix} \quad \xi_2 = \begin{bmatrix} 0 \\ 1 \\ -1 \end{bmatrix} \tag{24.64}$$

$$u_1 = \begin{bmatrix} -2 \\ 3 \end{bmatrix} \quad u_2 = \begin{bmatrix} 0 \\ -1 \end{bmatrix} \tag{24.65}$$

Let us also check the transfer function calculation. The transfer function matrix for this system is given by

$$G(s) = \begin{bmatrix} \frac{s+1}{s+2} & 0 \\ \frac{s+2}{s+3} & \frac{1}{s+4} \end{bmatrix} \tag{24.66}$$

To find transmission zeros at locations other than the poles of the system, we take the determinant and set it equal to zero to find

$$0 = \det(G(z)) = \frac{s+1}{(s+2)(s+4)} \tag{24.67}$$

Thus we find a transmission zero at $z = -1$. Notice that we have correctly found the transmission zero which is not at the pole frequency, but in this case we did not find the transmission zero at the same frequency as the pole at $s = -3$. It is also important to realize that, although one of the SISO transfer functions has a zero at $s = -2$, this is not a transmission zero of the MIMO system.

Also notice that this frequency domain method does not give us the directions of the zeros. We need to use the generalized eigenvalue problem to determine the directions.

24.4.3 Transmission Zeros for Nonsquare Systems

When we have a nonsquare system, we differentiate between right zeros and left zeros. In the following, we will assume we have a system with n states, m inputs, and p outputs.

DEFINITION 24.4 z_k is a right zero of the system,

$$\dot{x}(t) = Ax(t) + Bu(t) \tag{24.68}$$
$$y(t) = Cx(t) + Du(t) \tag{24.69}$$

if vectors $\xi_k \in C^n$ and $u_k \in C^m$ exist, both not zero, so that

$$\begin{bmatrix} z_k I - A & -B \\ -C & -D \end{bmatrix} \begin{bmatrix} \xi_k \\ u_k \end{bmatrix} = \begin{bmatrix} 0 \\ 0 \end{bmatrix} \tag{24.70}$$

DEFINITION 24.5 z_k is a left zero of the system,

$$\dot{x}(t) = Ax(t) + Bu(t) \tag{24.71}$$
$$y(t) = Cx(t) + Du(t) \tag{24.72}$$

if it is a right zero of the system,

$$\dot{x}(t) = A^T x(t) + C^T u(t) \tag{24.73}$$
$$y(t) = B^T x(t) + D^T u(t) \tag{24.74}$$

In other words, z_k is a left zero of our system if vectors $\eta_k \in C^n$ and $\gamma_k \in C^p$ exist, both not zero, so that

$$\begin{bmatrix} \eta_k & \gamma_k \end{bmatrix} \begin{bmatrix} z_k I - A & -B \\ -C & -D \end{bmatrix} = \begin{bmatrix} 0 & 0 \end{bmatrix} \tag{24.75}$$

For square systems, the left and right zeros coincide; any frequency which is a right zero is also a left zero.

24.5 Multivariable Pole-Zero Cancellations

Now that we have defined the zeros of a multivariable system, we are in a position to describe pole-zero cancellations and what they imply in terms of controllability and observability.

First, we give a SISO example, which shows that a pole-zero cancellation implies loss of controllability or observability.

EXAMPLE 24.3:

Consider the system given by the equations

$$\begin{bmatrix} \dot{x}_1(t) \\ \dot{x}_2(t) \end{bmatrix} = \begin{bmatrix} 0 & 1 \\ 1 & 0 \end{bmatrix} \begin{bmatrix} x_1(t) \\ x_2(t) \end{bmatrix} + \begin{bmatrix} 0 \\ 1 \end{bmatrix} u(t) \tag{24.76}$$
$$y(t) = \begin{bmatrix} 1 & -1 \end{bmatrix} \begin{bmatrix} x_1(t) \\ x_2(t) \end{bmatrix} \tag{24.77}$$

The transfer function for this system is given by

$$g(s) = \frac{-s+1}{s^2-1} = -\frac{s-1}{(s+1)(s-1)} \tag{24.78}$$

Thus, there is a pole-zero cancellation at $s = 1$. To check for loss of controllability and observability, we will perform an eigenvalue decomposition of the system. The eigenvalues and associated left and right eigenvalues are given by

$$\lambda_1 = -1 \quad v_1 = \begin{bmatrix} 1 \\ -1 \end{bmatrix} \quad w_1 = \begin{bmatrix} .5 & -.5 \end{bmatrix} \tag{24.79}$$

$$\lambda_2 = 1 \quad v_2 = \begin{bmatrix} 1 \\ 1 \end{bmatrix} \quad w_2 = \begin{bmatrix} .5 & .5 \end{bmatrix} \tag{24.80}$$

It is easy to verify that the dyadic decomposition Equation 24.3 holds. We now can use the modal tests for controllability and observability. We can see that

$$Cv_1 = 2 \quad w_1^T B = -.5 \tag{24.81}$$

This tells us that the first mode is both controllable and observable, as expected. For the second mode,

$$Cv_2 = 0 \quad w_2^T B = .5 \tag{24.82}$$

We see that the second mode is controllable, but unobservable. The conclusion is that this particular state-space realization has an unobservable mode, but is controllable.

Now let us examine what happens when we lose observability in more detail. We will assume we have the system described by Equations 24.36–24.38. Let us assume a pole with frequency λ_k and direction v_k, that is, A has an eigenvalue λ_k with associated eigenvalue v_k. Thus

$$(\lambda_k I - A)v_k = 0 \tag{24.83}$$

Let us also assume a (right) zero at frequency z_k with direction given by $\begin{bmatrix} \xi_k^T & u_k^T \end{bmatrix}^T$, that is,

$$(z_k I - A)\xi_k - Bu_k = 0 \tag{24.84}$$
$$C\xi_k + Du_k = 0 \tag{24.85}$$

If it is true that $\lambda_k = z_k$ and $\xi_k = \beta v_k$ with β any scalar, then

$$Bu_k = 0 \tag{24.86}$$

If we assume that the columns of B are linearly independent (which says we don't have redundant controls), then it must be true that $u_k = 0$. However, with $u_k = 0$,

$$Cv_k = 0 \Rightarrow \quad k^{th} \text{ mode is unobservable} \tag{24.87}$$

On the other hand, let us assume that the kth mode is unobservable. Then we know that

$$Cv_k = 0 \tag{24.88}$$

where v_k is the eigenvector associated with the kth mode. We will show that there must be a pole-zero cancellation. Let us choose $u_k = 0$. If λ_k is the eigenvalue associated with the kth mode, then we know

$$(\lambda_k I - A)v_k = 0 \Rightarrow (\lambda_k I - A)v_k - Bu_k = 0 \tag{24.89}$$

$$Cv_k + Du_k = 0 \tag{24.90}$$

Thus, λ_k is also a zero with direction

$$\begin{bmatrix} v_k \\ 0 \end{bmatrix} \tag{24.91}$$

We have now proven the following theorem:

THEOREM 24.1 *The kth mode, with eigenvalue λ_k and eigenvector v_k, is unobservable if, and only if, λ_k is a right transmission zero with direction given by Equation* 24.91.

Following the exact same steps, we show that loss of controllability implies a pole-zero cancellation. Using the same steps as before, we can prove the following theorem.

THEOREM 24.2 *Let the kth mode have eigenvalue λ_k and* left *eigenvector w_k. Then the kth mode is uncontrollable if, and only if, λ_k is a left transmission zero with direction given by*

$$\begin{bmatrix} w_k^T & 0 \end{bmatrix} \tag{24.92}$$

We conclude with an example.

EXAMPLE 24.4:

Consider the system Equations 24.19 and 24.20, with matrices

$$A = \begin{bmatrix} -2.5 & -.5 & .5 \\ .5 & -1.5 & .5 \\ 1 & 1 & -2 \end{bmatrix} \quad B = \begin{bmatrix} 1 \\ 1 \\ -2 \end{bmatrix} \tag{24.93}$$

$$C = \begin{bmatrix} 0 & 0 & 1 \\ 0 & 1 & 0 \end{bmatrix} \quad D = \begin{bmatrix} 0 \\ 0 \end{bmatrix} \tag{24.94}$$

A has the following eigenstructure:

$$\lambda_1 = -1 \quad v_1 = \begin{bmatrix} 0 \\ 1 \\ 1 \end{bmatrix} \quad w_1^T = \begin{bmatrix} \frac{1}{2} & \frac{1}{2} & \frac{1}{2} \end{bmatrix} \tag{24.95}$$

$$\lambda_2 = -2 \quad v_2 = \begin{bmatrix} 1 \\ -1 \\ 0 \end{bmatrix} \quad w_2^T = \begin{bmatrix} \frac{1}{2} & -\frac{1}{2} & \frac{1}{2} \end{bmatrix} \tag{24.96}$$

$$\lambda_3 = -3 \quad v_1 = \begin{bmatrix} 1 \\ 0 \\ -1 \end{bmatrix} \quad w_1^T = \begin{bmatrix} \frac{1}{2} & \frac{1}{2} & -\frac{1}{2} \end{bmatrix} . \tag{24.97}$$

Let us check for controllability and observability.

$$Cv_1 = \begin{bmatrix} 1 \\ 1 \end{bmatrix} \quad w_1^T B = 0 \tag{24.98}$$

$$Cv_2 = \begin{bmatrix} 0 \\ -1 \end{bmatrix} \quad w_2^T B = -1 \tag{24.99}$$

$$\text{and} \quad Cv_3 = \begin{bmatrix} -1 \\ 0 \end{bmatrix} \quad w_3^T B = 2. \tag{24.100}$$

The conclusion is, therefore, that all modes are observable and that the first mode is uncontrollable, but the second and third modes are controllable. It is easy to verify that $z = -1$ is a left zero of this system, with direction

$$\begin{bmatrix} \eta^T & \gamma^T \end{bmatrix} = \begin{bmatrix} \frac{1}{2} & \frac{1}{2} & \frac{1}{2} & 0 & 0 \end{bmatrix} . \tag{24.101}$$

Thus there is a pole-zero cancellation at $z = -1$, which makes the system uncontrollable.

References

[1] Kailath, T., *Linear Systems*, Prentice Hall, Englewood Cliffs, NJ, 1980.

[2] MacFarlane, A.G.J. and Karcanias, N., Poles and Zeros of Linear Multivariable Systems: A Survey of the Algebraic, Geometric, and Complex Variable Theory, *Int. J. Control*, 24, 33–74, 1976.

[3] Rosenbrock, H.H., *State-space and Multivariable Theory*, John Wiley & Sons, New York, 1973.

[4] Sain, M.K. and Schrader, C.B., The Role of Zeros in the Performance of Multiinput, Multioutput Feedback Systems, *IEEE Trans. Education*, 33(3), 1990.

[5] Schrader, C.B. and Sain, M.K., Research on System Zeros: A Survey, *Int. J. Control*, 50(4), 1989.

25

Fundamentals of Linear Time-Varying Systems

Edward W. Kamen
School of Electrical and Computer Engineering, Georgia Institute of Technology, Atlanta, GA

25.1 Introduction

Most systems arising in practice are time varying, that is, the values of the output response depend on when the input is applied. Time variation is a result of system parameters changing as a function of time, such as aerodynamic coefficients in high-speed aircraft, circuit parameters in electronic circuits, mechanical parameters in machinery, and diffusion coefficients in chemical processes. Time variation may also be a result of linearizing a nonlinear system about a family of operating points and/or about a time-varying operating point.

Time variation is often ignored in dealing with systems arising in practice. However, due to the desire to achieve better accuracy and quality in a wide range of applications, there is an increasing interest in including the effects of time variation when analyzing a system, or when designing observers and controllers. This requires that time variation be included in the model used to study a given system. In particular, one can consider parametric models such as an input/output differential equation with time-varying coefficients or a state model with time-varying coefficients. In this article, the focus is on the fundamentals of the state model for linear time-varying systems. Most of the material covered deals with the standard theory in existence since the 1960s. However, there are more recent approaches to linear time-varying systems that are not discussed here, but which the interested reader may want to consult. These include algebraic approaches [5], [6], operator-theoretic approaches [13], [14], [2], [12], and the parameter-variation (gain-scheduling) approach [11], [3]. Other references to more recent work are given in the section on further reading.

In this chapter, we first consider system analysis based on the state model including the construction of canonical forms and the study of the system properties of stability, controllability, and observability. Both the continuous-time and the discrete-time cases are considered. In the last part of the chapter, an example is given of the application to state observers and state feedback control, and examples are given on checking for stability.

25.2 Analysis of Continuous-Time Linear Time-Varying Systems

Consider a continuous-time system with single input $u(t)$ and single output $y(t)$, where $u(t)$ and $y(t)$ are real-valued functions of the continuous-time variable t. It is assumed that there is no initial energy in the system prior to the application of the input $u(t)$, that is, the system is initially **at rest** before the application of the input. Then if the system is causal, it can be modeled by the input/output relationship

$$y(t) = F(\{u(\tau) : -\infty < \tau \leq t\}, t) \tag{25.1}$$

where $y(t)$ is the output response resulting from $u(t)$ and F is a function that may be nonlinear. If the system is linear, then F is linear and Equation 25.1 becomes

$$y(t) = \int_{-\infty}^{t} h(t, \tau)u(\tau)d\tau \tag{25.2}$$

0-8493-8570-9/96/$0.00+$.50

where $h(t, \tau)$ is the **impulse response function**, that is, $h(t, \tau)$ is the response to the impulse $\delta(t - \tau)$ applied at time τ with no initial energy. It should be emphasized that $y(t)$ given by Equation 25.2 is the output response assuming that the system is at rest prior to the application of the input $u(t)$. Also, it is assumed that there are conditions on $h(t, \tau)$ and/or $u(t)$ which insure that the integral in Equation 25.2 exists.

The linear system given by Equation 25.2 is **time invariant** (or **constant**) if and only if

$$h(t + \gamma, \tau + \gamma) = h(t, \tau), \quad \text{for all real numbers} \quad t, \tau, \gamma. \tag{25.3}$$

Time invariance means that if $y(t)$ is the response to $u(t)$, then for any real number t_1, the shifted output $y(t - t_1)$ is the response to the shifted input $u(t - t_1)$. Setting $\gamma = -\tau$ in Equation 25.3 gives

$$h(t - \tau, 0) = h(t, \tau), \quad \text{for all real numbers} \quad t, \tau. \tag{25.4}$$

Hence, the system defined by Equation 25.2 is time invariant if and only if the impulse response function $h(t, \tau)$ is a function only of the difference $t - \tau$. In the time-invariant case, Equation 25.2 reduces to the convolution relationship

$$y(t) = h(t) * u(t) = \int_{-\infty}^{t} h(t - \tau)u(\tau)d\tau \tag{25.5}$$

where $h(t) = h(t, 0)$ is the impulse response (i.e., the response to the impulse $\delta(t)$ applied at time 0).

The linear system defined by Equation 25.2 is **finite-dimensional** or **lumped** if the input $u(t)$ and the output $y(t)$ are related by the nth-order differential equation

$$y^{(n)}(t) + \sum_{i=0}^{n-1} a_i(t)y^{(i)}(t) = \sum_{i=0}^{m} b_i(t)u^{(i)}(t) \tag{25.6}$$

where $y^{(i)}(t)$ is the ith derivative of $y(t)$, $u^{(i)}(t)$ is the ith derivative of $u(t)$, and $a_i(t)$ and $b_i(t)$ are real-valued functions of t. In Equation 25.6, it is assumed that $m \le n$. The linear system given by Equation 25.6 is time invariant if, and only if, all coefficients in Equation 25.6 are constants, that is, $a_i(t) = a_i$ and $b_i(t) = b_i$ for all i, where a_i and b_i are real constants.

25.2.1 State Model Realizations

A state model for the system given by Equation 25.6 can be constructed as follows. First, suppose that $m = 0$ so that Equation 25.6 becomes

$$y^{(n)}(t) + \sum_{i=0}^{n-1} a_i(t)y^{(i)}(t) = b_0(t)u(t). \tag{25.7}$$

Then defining the state variables

$$x_i(t) = y^{(i-1)}(t), \qquad i = 1, 2, \ldots, n, \tag{25.8}$$

the system defined by Equation 25.7 has the state model,

$$\dot{x}(t) = A(t)x(t) + B(t)u(t) \tag{25.9}$$

$$y(t) = Cx(t), \tag{25.10}$$

where the coefficient matrices $A(t)$, $B(t)$, and C are given by

$$A(t) = \begin{bmatrix} 0 & 1 & 0 & 0 & \cdots & 0 & 0 \\ 0 & 0 & 1 & 0 & \cdots & 0 & 0 \\ 0 & 0 & 0 & 1 & \cdots & 0 & 0 \\ \vdots & \vdots & \vdots & & & \vdots & \vdots \\ 0 & 0 & 0 & 0 & \cdots & 0 & 1 \\ -a_0(t) & -a_1(t) & -a_2(t) & -a_3(t) & \cdots & -a_{n-2}(t) & -a_{n-1}(t) \end{bmatrix}$$

$$B(t) = \begin{bmatrix} 0 \\ 0 \\ 0 \\ \vdots \\ 0 \\ b_0(t) \end{bmatrix} \tag{25.11}$$

$$C = \begin{bmatrix} 1 & 0 & 0 & \cdots & 0 & 0 \end{bmatrix} \tag{25.12}$$

and $x(t)$ is the n-dimensional state vector given by

$$x(t) = \begin{bmatrix} x_1(t) \\ x_2(t) \\ \vdots \\ x_{n-1}(t) \\ x_n(t) \end{bmatrix} \tag{25.13}$$

When $m \ge 1$ in Equation 25.6, the definition in Equation 25.8 of the state variables will not yield a state model of the form given in Equations 25.9 and 25.10. If $m < n$, a state model can be generated by first rewriting Equation 25.6 in the form

$$D^n y(t) + \sum_{i=0}^{n-1} D^i \left[\alpha_i(t)y(t)\right] = \sum_{i=0}^{m} D^i \left[\beta_i(t)u(t)\right] \tag{25.14}$$

where D is the derivative operator, the $\alpha_i(t)$ are functions of $a_i(t)$ and derivatives of $a_i(t)$, and the $\beta_i(t)$ are functions of $b_i(t)$ and derivatives of $b_i(t)$. If $a_i(t)$ are constants so that $a_i(t) = a_i$ for all t, then the $\alpha_i(t)$ are constants and $\alpha_i(t) = a_i$, $i = 0, 1, \ldots, n-1$. If the $b_i(t)$ are constants so that $b_i(t) = b_i$ for all t, then the $\beta_i(t)$ are constants and $\beta_i(t) = b_i$, $i = 0, 1, \ldots, m$. Let $\alpha(t) = [\alpha_0(t)\ \alpha_1(t) \cdots \alpha_{n-2}(t)\ \alpha_{n-1}(t)]^T$ denote the n-element column vector consisting of the time-varying coefficients appearing in the left-hand side of Equation 25.14 and let $a(t) = [a_0(t)\ a_1(t) \cdots a_{n-2}(t)a_{n-1}(t)]^T$ denote the vector of time-varying coefficients in the left-hand side of Equation 25.6. (Here superscript "T" denotes the transpose operation.) Then a can be expressed as a function of α, that is,

$$a = f(\alpha) \tag{25.15}$$

For example, when $n = 2$, Equation 25.15 is given by

$$a_0(t) = \alpha_0(t) + \dot{\alpha}_1(t) \quad (25.16)$$
$$a_1(t) = \alpha_1(t) \quad (25.17)$$

and when $n = 3$, Equation 25.15 is given by

$$a_0(t) = \alpha_0(t) + \dot{\alpha}_1(t) + \ddot{\alpha}_2(t) \quad (25.18)$$
$$a_1(t) = \alpha_1(t) + 2\dot{\alpha}_2(t) \quad (25.19)$$
$$a_2(t) = \alpha_2(t) \quad (25.20)$$

The function f relating a and α is invertible for any integer $n \geq 1$, and thus

$$\alpha = f^{-1}(a) \quad (25.21)$$

where f^{-1} is the inverse of f. For example, when $n = 2$, it follows directly from Equations 25.16 and 25.17 that f^{-1} is given by

$$\alpha_0(t) = a_0(t) - \dot{a}_1(t) \quad (25.22)$$
$$\alpha_1(t) = a_1(t) \quad (25.23)$$

Since f is invertible, there is a one-to-one and onto correspondence between the coefficients in the left-hand side of Equation 25.6 and the coefficients in the left-hand side of Equation 25.14. Similarly, there is a one-to-one and onto correspondence between the coefficients in the right-hand side of Equation 25.6 and the right-hand side of Equation 25.14.

Now given the system defined by Equation 25.6 written in the form of Equation 25.14, define the state variables

$$\begin{aligned}
x_n(t) &= y(t) \\
x_{n-1}(t) &= Dx_n(t) + \alpha_{n-1}(t)x_n(t) - \beta_{n-1}(t)u(t) \\
x_{n-2}(t) &= Dx_{n-1}(t) + \alpha_{n-2}(t)x_n(t) - \beta_{n-2}(t)u(t) \\
&\vdots \\
x_1(t) &= Dx_2(t) + \alpha_1(t)x_n(t) - \beta_1(t)u(t)
\end{aligned} \quad (25.24)$$

where $\beta_i(t) = 0$ for $i > m$. Then with the state vector $x(t)$ defined by Equation 25.13, the system defined by Equation 25.6 has the state model

$$\dot{x}(t) = A(t)x(t) + B(t)u(t) \quad (25.25)$$
$$y(t) = Cx(t) \quad (25.26)$$

where the coefficient matrices $A(t)$, $B(t)$, and C are given by

$$A(t) = \begin{bmatrix} 0 & 0 & 0 & \cdots & 0 & -\alpha_0(t) \\ 1 & 0 & 0 & \cdots & 0 & -\alpha_1(t) \\ 0 & 1 & 0 & \cdots & 0 & -\alpha_2(t) \\ \vdots & & & \cdots & & \vdots \\ 0 & 0 & 0 & \cdots & 0 & -\alpha_{n-2}(t) \\ 0 & 0 & 0 & \cdots & 1 & -\alpha_{n-1}(t) \end{bmatrix} \quad (25.27)$$

$$B(t) = \begin{bmatrix} \beta_0(t) \\ \beta_1(t) \\ \beta_2(t) \\ \vdots \\ \beta_{n-2}(t) \\ \beta_{n-1}(t) \end{bmatrix} \quad (25.28)$$

and

$$C = \begin{bmatrix} 0 & 0 & \cdots & 0 & 1 \end{bmatrix}$$

The state model with $A(t)$, $B(t)$, and C given by Equations 25.27 and 25.28 is said to be in **observer canonical form**, which is uniformly observable as discussed below. Note that for this particular state realization, the row vector C is constant (independent of t).

In addition to the state model defined by Equations 25.27 and 25.28, there are other possible state models for the system given by Equation 25.6. For example, another state model can be constructed in the case when the left-hand side of Equation 25.6 can be expressed in a factored operator form so that Equation 25.6 becomes

$$(D - p_1(t))(D - p_2(t))\cdots(D - p_n(t))y(t) = \sum_{i=0}^{m} b_i(t)u^{(i)}(t) \quad (25.29)$$

where again D is the derivative operator and the $p_i(t)$ are real-valued functions of the time variable t. For example, consider the $n = 2$ case for which

$$\begin{aligned}
(D - p_1(t))(D - p_2(t))y(t) &= (D - p_1(t))[\dot{y}(t) - p_2(t)y(t)] \\
&= \ddot{y}(t) - [p_1(t) + p_2(t)]\dot{y}(t) \\
&\quad + [p_1(t)p_2(t) - \dot{p}_2(t)]y(t)
\end{aligned} \quad (25.30)$$

With $n = 2$ and $m = 1$, the state variables may be defined by

$$x_1(t) = \dot{y}(t) - p_2(t)y(t) - b_1(t)u(t) \quad (25.31)$$
$$x_2(t) = y(t) \quad (25.32)$$

which results in the state model given by Equations 25.25 and 25.26 with

$$A(t) = \begin{bmatrix} p_1(t) & 0 \\ 1 & p_2(t) \end{bmatrix}$$
$$B(t) = \begin{bmatrix} b_0(t) - \dot{b}_1(t) + p_1(t)b_1(t) \\ b_1(t) \end{bmatrix} \quad (25.33)$$

and

$$C = [0 \quad 1] \quad (25.34)$$

In the general case given by Equation 25.29, there is a state model in the form of Equations 25.25 and 25.26 with $A(t)$ and C given by

$$A(t) = \begin{bmatrix} p_1(t) & 0 & 0 & 0 & \cdots & 0 & 0 \\ 1 & p_2(t) & 0 & 0 & \cdots & 0 & 0 \\ 0 & 1 & p_3(t) & 0 & \cdots & 0 & 0 \\ \vdots & & & & & \vdots & \vdots \\ 0 & 0 & 0 & 0 & \cdots & p_{n-1}(t) & 0 \\ 0 & 0 & 0 & 0 & \cdots & 1 & p_n(t) \end{bmatrix} \quad (25.35)$$

$$C = \begin{bmatrix} 0 & 0 & 0 & \cdots & 1 \end{bmatrix} \quad (25.36)$$

A very useful feature of the state model with $A(t)$ given by Equation 25.35 is the lower triangular form of $A(t)$. In particular, as discussed below, stability conditions can be specified in terms of the $p_i(t)$, which can be viewed as the "time-varying poles" of the system (see [15], [4]). Unfortunately, in the time-varying case the computation of the $p_i(t)$ requires that nonlinear differential equations must be solved, and as a result, generally there is no simple method for determining the $p_i(t)$ (when they exist). Nevertheless, the form given by Equation 25.35 can be determined in special cases, and thus is of interest in applications of the theory.

25.2.2 The State Model

For an m-input, p-output, linear n-dimensional, time-varying continuous-time system, the general form of the state model is

$$\dot{x}(t) = A(t)x(t) + B(t)u(t) \quad (25.37)$$

$$y(t) = C(t)x(t) + D(t)u(t) \quad (25.38)$$

where Equation 25.37 is the **state equation** and Equation 25.38 is the **output equation.** In Equations 25.37 and 25.38, $A(t)$ is the $n \times n$ **system matrix,** $B(t)$ is the $n \times m$ **input matrix,** $C(t)$ is the $p \times n$ **output matrix,** $D(t)$ is the $p \times m$ **direct feed matrix,** $u(t)$ is the m-dimensional input vector, $x(t)$ is the n-dimensional state vector, and $y(t)$ is the p-dimensional output vector. The term $D(t)u(t)$ in Equation 25.38 is of little significance in the theory, and thus $D(t)u(t)$ is usually omitted from Equation 25.38, which will be done here.

To solve Equation 25.37, first consider the homogeneous equation

$$\dot{x}(t) = A(t)x(t), \qquad t > t_0 \quad (25.39)$$

with the initial condition $x(t_0)$ at initial time t_0. For any $A(t)$ whose entries are piecewise continuous, it is known that for any initial condition $x(t_0)$, there is a unique continuous solution of Equation 25.39 given by

$$x(t) = \Phi(t, t_0)x(t_0), \qquad t > t_0 \quad (25.40)$$

where $\Phi(t, t_0)$ is a $n \times n$ matrix function of t and t_0, called the **state-transition matrix.** The state-transition matrix has the following fundamental properties:

$$\Phi(t, t) = I = n \times n \text{ identity matrix, for all } t \quad (25.41)$$

$$\Phi(t, \tau) = \Phi(t, t_1)\Phi(t_1, \tau), \text{ for all } t_1, t, \tau \quad (25.42)$$

$$\Phi^{-1}(t, \tau) = \Phi(\tau, t), \text{ for all } t, \tau \quad (25.43)$$

$$\frac{\partial}{\partial t}\Phi(t, \tau) = A(t)\Phi(t, \tau), \text{ for all } t, \tau \quad (25.44)$$

$$\frac{\partial}{\partial \tau}\Phi(t, \tau) = -\Phi(t, \tau)A(\tau), \text{ for all } t, \tau \quad (25.45)$$

$$\det \Phi(t, \tau) = \exp\left[\int_\tau^t \operatorname{tr}[A(\sigma)]d\sigma\right] \quad (25.46)$$

In Equation 25.46, "det" denotes the determinant and "tr" denotes the trace. Equation 25.42 is called the **composition property.** It follows from this property that $\Phi(t, \tau)$ can be written in the factored form

$$\Phi(t, \tau) = \Phi(t, 0)\Phi(0, \tau), \text{ for all } t, \tau \quad (25.47)$$

It follows from Equation 25.45 that the **adjoint equation**

$$\dot{\gamma}(t) = -A^T(t)\gamma(t) \quad (25.48)$$

has state-transition matrix equal to $\Phi^T(\tau, t)$, where again $\Phi(t, \tau)$ is the state-transition matrix for Equation 25.39.

If the system matrix $A(t)$ is constant over the interval $[t_1, t_2]$, that is, $A(t) = A$, for all $t \in [t_1, t_2]$, then the state-transition matrix is equal to the matrix exponential over $[t_1, t_2]$:

$$\Phi(t, \tau) = e^{A(t-\tau)} \text{ for all } t, \tau \in [t_1, t_2] \quad (25.49)$$

If $A(t)$ is time varying and $A(t)$ commutes with its integral over the interval $[t_1, t_2]$, that is,

$$A(t)\left[\int_\tau^t A(\sigma)d\sigma\right] = \left[\int_\tau^t A(\sigma)d\sigma\right]A(t), \quad \text{for all } t, \tau \in [t_1, t_2] \quad (25.50)$$

then $\Phi(t, \tau)$ is given by

$$\Phi(t, \tau) = \exp\left[\int_\tau^t A(\tau)d\tau\right], \text{ for all } t, \tau \in [t_1, t_2] \quad (25.51)$$

Note that the commutativity condition in Equation 25.50 is always satisfied in the time-invariant case. It is also always satisfied in the one-dimensional case ($n = 1$) because scalars commute. Thus $\Phi(t, \tau)$ is given by the exponential form in Equation 25.51 when $n = 1$. Unfortunately, the exponential form for $\Phi(t, \tau)$ does not hold for an arbitrary time-varying matrix $A(t)$ when $n > 1$, and, in general, there is no known closed-form expression for $\Phi(t, \tau)$ when $n > 1$. However, approximations to $\Phi(t, \tau)$ can be readily computed from $A(t)$ by numerical techniques, such as the method of successive approximations (e.g., see [8]). Approximations to $\Phi(t, \tau)$ can also be determined by discretizing the time variable as shown below.

Given the state transition matrix $\Phi(t, \tau)$, for any given initial state $x(t_0)$ and input $u(t)$ applied for $t \geq t_0$, the complete solution to Equation 25.37 is

$$x(t) = \Phi(t, t_0)x(t_0) + \int_{t_0}^t \Phi(t, \tau)B(\tau)u(\tau)d\tau, \ t > t_0 \quad (25.52)$$

Then, when $y(t) = C(t)x(t)$, the output response $y(t)$ is given by

$$\begin{aligned} y(t) &= C(t)\Phi(t, t_0)x(t_0) \\ &+ \int_{t_0}^{t} C(t)\Phi(t, \tau)B(\tau)u(\tau)d\tau, \ t > t_0 \end{aligned} \quad (25.53)$$

If the initial time t_0 is taken to be $-\infty$ and there is no initial energy at $t = -\infty$, Equation 25.53 becomes

$$y(t) = \int_{-\infty}^{t} C(t)\Phi(t, \tau)B(\tau)u(\tau)d\tau \quad (25.54)$$

Comparing Equation 25.54 with the m-input, p-output version of the input/output Equation 25.2 reveals that

$$H(t, \tau) = \begin{cases} C(t)\Phi(t, \tau)B(\tau) \text{ for } t \geq \tau \\ 0, \ t < \tau \end{cases} \quad (25.55)$$

where $H(t, \tau)$ is the $p \times m$ impulse response function matrix. Inserting Equation 25.47 into Equation 25.55 reveals that $H(t, \tau)$ can be expressed in the factored form,

$$H(t, \tau) = H_1(t)H_2(\tau), \ t \geq \tau \quad (25.56)$$

where

$$H_1(t) = C(t)\Phi(t, 0) \text{ and } H_2(\tau) = \Phi(0, \tau)B(\tau) \quad (25.57)$$

It turns out (see [8]) that a linear time-varying system with impulse response matrix $H(t, \tau)$ has a state realization given by Equations 25.37–25.38 with $D(t) = 0$ if, and only if, $H(t, \tau)$ can be expressed in the factored form given in Equation 25.56.

25.2.3 Change of State Variables

Suppose that the system under study has the n-dimensional state model

$$\begin{aligned} \dot{x}(t) &= A(t)x(t) + B(t)u(t) \quad (25.58) \\ y(t) &= C(t)x(t) \quad (25.59) \end{aligned}$$

In the following development, the system given by Equations 25.58–25.59 will be denoted by the triple $[A(t), B(t), C(t)]$.

Given any $n \times n$ invertible differentiable matrix $P(t)$, another state model can be generated by defining the new state vector $z(t) = P(t)x(t)$. The state variables for the new state model (i.e., the $z_i(t)$) are linear combinations with time-varying coefficients of the state variables of the given state model. The state equations for the new state model are given by

$$\begin{aligned} \dot{z}(t) &= \Big[P(t)A(t)P^{-1}(t) \\ &+ \dot{P}(t)P^{-1}(t)\Big]z(t) + P(t)B(t)u(t) \quad (25.60) \\ y(t) &= C(t)P^{-1}(t)z(t) \quad (25.61) \end{aligned}$$

This new model will be denoted by the triple $[\overline{A}(t), \overline{B}(t), \overline{C}(t)]$ where

$$\begin{aligned} \overline{A}(t) &= P(t)A(t)P^{-1}(t) + \dot{P}(t)P^{-1}(t) \quad (25.62) \\ \overline{B}(t) &= P(t)B(t) \quad (25.63) \\ \overline{C}(t) &= C(t)P^{-1}(t) \quad (25.64) \end{aligned}$$

The state models $[A(t), B(t), C(t)]$ and $[\overline{A}(t), \overline{B}(t), \overline{C}(t)]$ are said to be **algebraically equivalent**. The state-transition matrix for $[\overline{A}(t), \overline{B}(t), \overline{C}(t)]$ is given by

$$\overline{\Phi}(t, \tau) = P(t)\Phi(t, \tau)P^{-1}(\tau) \quad (25.65)$$

where $\Phi(t, \tau)$ is the state-transition matrix for $[A(t), B(t), C(t)]$. It is also possible to express the transformation matrix $P(t)$ in terms of the coefficient matrices of the state models as follows. Assuming that the entries of $B(t)$ are $n - 1$ times differentiable and the entries of $A(t)$ are $n - 2$ times differentiable, define the $n \times m$ matrices

$$\begin{aligned} K_0(t) &= B(t) \quad (25.66) \\ K_i(t) &= -A(t)K_{i-1}(t) + \dot{K}_{i-1}(t), \ i = 1, 2, \ldots, n-1 \quad (25.67) \end{aligned}$$

and let $K(t)$ denote the $n \times mn$ matrix whose ith block column is equal to $K_i(t)$, that is

$$K(t) = [K_0(t)K_1(t) \cdots K_{n-1}(t)] \quad (25.68)$$

Similarly, for the model $[\overline{A}(t), \overline{B}(t), \overline{C}(t)]$, define

$$\overline{K}(t) = \left[\overline{K}_0(t)\overline{K}_1(t) \cdots \overline{K}_{n-1}(t)\right] \quad (25.69)$$

where the $\overline{K}_i(t)$ are given by Equations 25.66 and 25.67 with $A(t)$ and $B(t)$ replaced by $\overline{A}(t)$ and $\overline{B}(t)$ respectively. Then

$$P(t)K(t) = \overline{K}(t) \quad (25.70)$$

Hence, if $K(t)$ has rank n for all t so that the $n \times n$ matrix $K(t)K^T(t)$ is invertible, Equation 25.70 can be solved for $P(t)$. This gives

$$P(t) = \overline{K}(t)K^T(t)\left[K(t)K^T(t)\right]^{-1} \quad (25.71)$$

As will be discussed below, the rank condition on $K(t)$ implies that the system $[A(t), B(t), C(t)]$ is uniformly controllable. It follows from Equation 25.70 that uniform controllability is preserved under a change of state variables.

It may also be possible to express $P^{-1}(t)$ in terms of $A(t)$, $C(t)$ and $\overline{A}(t)$, $\overline{C}(t)$ as follows. Assuming that the entries of $C(t)$ are $n - 1$ times differentiable and the entries of $A(t)$ are $n - 2$ times differentiable, define the $p \times n$ matrices

$$\begin{aligned} L_0(t) &= C(t) \quad (25.72) \\ L_i(t) &= L_{i-1}(t)A(t) + \dot{L}_{i-1}(t), \ i = 1, 2, \ldots, n-1 \quad (25.73) \end{aligned}$$

and let $L(t)$ denote the $pn \times n$ matrix whose ith block row is equal to $L_i(t)$, that is

$$L(t) = \begin{bmatrix} L_0(t) \\ L_1(t) \\ \vdots \\ L_{n-1}(t) \end{bmatrix} \quad (25.74)$$

Similarly, for the model $[\overline{A}(t), \overline{B}(t), \overline{C}(t)]$, define

$$\overline{L}(t) = \begin{bmatrix} \overline{L}_0(t) \\ \overline{L}_1(t) \\ \vdots \\ \overline{L}_{n-1}(t) \end{bmatrix} \tag{25.75}$$

where the $\overline{L}_i(t)$ are given by Equations 25.72 and 25.73 with $A(t)$ and $C(t)$ replaced by $\overline{A}(t)$ and $\overline{C}(t)$, respectively. Then

$$L(t)P^{-1}(t) = \overline{L}(t). \tag{25.76}$$

Hence, if $L(t)$ has rank n for all t so that the $n \times n$ matrix $L^T(t)L(t)$ is invertible, Equation 25.76 can be solved for $P^{-1}(t)$. This gives

$$P^{-1}(t) = \left[L^T(t)L(t)\right]^{-1} L^T(t)\overline{L}(t) \tag{25.77}$$

As will be discussed below, the rank condition on $L(t)$ implies that the system $[A(t), B(t), C(t)]$ is uniformly observable. It follows from Equation 25.76 that uniform observability is preserved under a change of state variables.

Given an n-dimensional state model $[A(t), B(t), C(t)]$ and any $n \times n$ constant matrix Γ, there is a transformation $P(t)$ such that $\overline{A}(t) = \Gamma$. To see this, first note that by Equations 25.43 and 25.47, $\Phi(t, \tau)$ can be written in the factored form

$$\Phi(t, \tau) = \Phi(t, 0)\Phi^{-1}(\tau, 0). \tag{25.78}$$

Then with $P(t) = \exp(\Gamma t)\Phi^{-1}(t, 0)$, using Equation 25.65 gives

$$\overline{\Phi}(t, \tau) = \left[e^{\Gamma t}\Phi^{-1}(t, 0)\right]\left[\Phi(t, 0)\Phi^{-1}(\tau, 0)\right] \times \left[\Phi(\tau, 0)e^{-\Gamma\tau}\right] \tag{25.79}$$

$$\overline{\Phi}(t, \tau) = e^{\Gamma(t-\tau)} \tag{25.80}$$

and thus, $\overline{A}(t) = \Gamma$ as claimed.

This result shows that via a change of state, $A(t)$ can be transformed to any desired constant matrix, such as a diagonal matrix. Unfortunately, system stability (defined below) is not necessarily preserved under a change of state, and thus equivalence of $A(t)$ to a constant matrix cannot be used in general to test for stability. To insure that stability is preserved, it is necessary to consider a stronger notion of equivalence called **topological equivalence**. This means that, in addition to being algebraically equivalent, the transformation $z(t) = P(t)x(t)$ has the properties

$$|\det P(t)| \geq c_0, \text{ for all } t, \tag{25.81}$$

$$\left|p_{ij}(t)\right| \leq c_1, \text{ for all } t \text{ and } i, j = 1, 2, \ldots, n, \tag{25.82}$$

where $p_{ij}(t)$ is the i, j entry of $P(t)$, and c_0 and c_1 are finite positive constants. The conditions given in Equations 25.81 and 25.82 are equivalent to requiring that $P(t)$ and its inverse $P^{-1}(t)$ be bounded matrix functions of t. A transformation $P(t)$ satisfying Equations 25.81 and 25.82 is called a **Lyapunov transformation**. As noted below, stability is preserved under a Lyapunov transformation.

25.2.4 Canonical Forms

Now suppose that the system with state model $[A(t), B(t), C(t)]$ has a single input ($m = 1$) so that $B(t)$ is an n-element column vector. Assuming that $B(t)$ and $A(t)$ can be differentiated an appropriate number of times, let $K(t)$ denote the $n \times n$ matrix

$$K(t) = [K_0(t)K_1(t)\cdots K_{n-1}(t)] \tag{25.83}$$

where the $K_i(t)$ are defined by Equations 25.66–25.67, and define

$$K_n(t) = -A(t)K_{n-1}(t) + \dot{K}_{n-1}(t). \tag{25.84}$$

Assuming that $K(t)$ is invertible for all t, we can define the n-element column vector as

$$\eta(t) = -K^{-1}(t)K_n(t) \tag{25.85}$$

The vector $\eta(t)$ is invariant under any change of state $z(t) = P(t)x(t)$. In other words, if Equations 25.83–25.85 are evaluated for the new state model $[\overline{A}(t), \overline{B}(t), \overline{C}(t)]$ resulting from the transformation $z(t) = P(t)x(t)$, Equation 25.85 will yield the same result for $\eta(t)$. In addition, if $A(t) = A$ and $B(t) = B$ where A and B are constant matrices, the vector η is constant and is given by

$$\eta = \left[a_0\ (-a_1)\cdots(-1)^i a_i\cdots(-1)^{n-1}a_{n-1}\right]^T \tag{25.86}$$

where the a_i are the coefficients of the characteristic polynomial of A, that is,

$$\det(sI - A) = s^n + \textstyle\sum_{i=0}^{n-1} a_i s^i. \tag{25.87}$$

Given the analogy with the time-invariant case, the vector $\eta(t)$ given by Equation 25.85 may be called a **characteristic vector.**

Now given $\eta(t)$ defined by Equation 25.85, let

$$f^{-1}(\eta) = [\Psi_0(t)\ -\ \Psi_1(t)\cdots (-1)^i\Psi_i(t)\cdots(-1)^{n-1}\Psi_{n-1}(t)]^T \tag{25.88}$$

where f^{-1} is the inverse of the function f defined by Equation 25.15. Then there is a transformation $P(t)$ which converts $[A(t), B(t), C(t)]$ into the **control canonical form** $[\overline{A}(t), \overline{B}, \overline{C}(t)]$ with $\overline{A}(t)$ and $\overline{B}$ given by

$$\overline{A}(t) = \begin{bmatrix} 0 & 1 & 0 & 0 & \cdots & 0 & 0 \\ 0 & 0 & 1 & 0 & \cdots & 0 & 0 \\ 0 & 0 & 0 & 1 & \cdots & 0 & 0 \\ \vdots & \vdots & \vdots & & & \vdots & \vdots \\ 0 & 0 & 0 & 0 & \cdots & 0 & 1 \\ -\Psi_0(t) & -\Psi_1(t) & -\Psi_2(t) & -\Psi_3(t) & \cdots & -\Psi_{n-2}(t) & -\Psi_{n-1}(t) \end{bmatrix} \tag{25.89}$$

$$\overline{B} = \begin{bmatrix} 0 \\ 0 \\ 0 \\ \vdots \\ 0 \\ 1 \end{bmatrix}$$

This form was first derived in [10]. The transformation $P(t)$ which yields the control canonical form can be determined using Equation 25.71.

Now consider a system with state model $[A(t), B(t), C(t)]$ having one output (and any number of inputs), so that $C(t)$ is a n-element row vector. Assuming that $A(t)$ and $C(t)$ can be differentiated an appropriate number of times, let $L(t)$ denote the $n \times n$ matrix

$$L(t) = \begin{bmatrix} L_0(t) \\ L_1(t) \\ \vdots \\ L_{n-1}(t) \end{bmatrix} \tag{25.90}$$

where the $L_i(t)$ are defined by Equations 25.72–25.73, and define

$$L_n(t) = L_{n-1}(t)A(t) + \dot{L}_{n-1}(t). \tag{25.91}$$

Assuming that $L(t)$ is invertible for all t, we can define the n-vector

$$\rho(t) = -\left[L^T(t)\right]^{-1} L_n^T(t) \tag{25.92}$$

As is the case for the vector $\eta(t)$ defined above, the n-vector $\eta(t)$ is invariant under any change of state $z(t) = P(t)x(t)$. In addition, if $A(t) = A$ and $B(t) = C$ where A and C are constant, the vector ρ is constant and is given by

$$\rho = [a_0\, a_1 \cdots a_i \cdots a_{n-1}]^T \tag{25.93}$$

where the a_i are the coefficients of the characteristic polynomial of A.

The vector $\rho(t)$ given by Equation 25.92 is another **characteristic vector** of the given time-varying system. Given $\rho(t)$, let

$$f^{-1}(\rho) = [\alpha_0(t)\, \alpha_1(t) \cdots \alpha_{n-1}(t)]^T \tag{25.94}$$

where again f^{-1} is the inverse of the function f defined by Equation 25.15. Then there is a transformation $P(t)$ which converts $[A(t), B(t), C(t)]$ into the observer canonical form $[\overline{A}(t), \overline{B}(t), \overline{C}]$ with $\overline{A}(t)$ given by Equation 25.27 and $\overline{C}$ given by Equation 25.28. The transformation $P(t)$ can be computed using Equation 25.77. In the section on applications it will be shown how the control and observer canonical forms can be used to design controllers and observers.

25.2.5 Stability

Given a system with n-dimensional state model $[A(t), B(t), C(t)]$, again consider the homogeneous equation

$$\dot{x}(t) = A(t)x(t), \; t > t_0 \tag{25.95}$$

with solution

$$x(t) = \Phi(t, t_0)x(t_0), \; t > t_0 \tag{25.96}$$

The system is said to be **asymptotically stable** if the solution $x(t)$ satisfies the condition $\|x(t)\| \to 0$ as $t \to \infty$ for any initial state $x(t_0)$ at initial time t_0. Here $\|x(t)\|$ denotes the **Euclidean norm** of the state $x(t)$ given by

$$\|x(t)\| = \sqrt{x_1^2(t) + x_2^2(t) + \cdots + x_n^2(t)} \tag{25.97}$$

where $x(t) = [x_1(t)\, x_2(t) \ldots x_n(t)]^T$. A system is asymptotically stable if, and only if,

$$\|\Phi(t, t_0)\| \to 0 \text{ as } t \to \infty \tag{25.98}$$

where $\|\Phi(t, t_0)\|$ is the matrix norm equal to the square root of the largest eigenvalue of $\Phi^T(t, t_0)\Phi(t, t_0)$.

A stronger notion of stability is **uniform exponential stability** which requires that positive constants c and λ exist so that

$$\|\Phi(t, t_0)\| \le ce^{-\lambda(t-t_0)}, \; t \ge t_0 \tag{25.99}$$

If Equation 25.99 is satisfied, it follows from Equation 25.96 that $x(t)$ satisfies the condition

$$\|x(t)\| \le ce^{-\lambda(t-t_0)}\|x(t_0)\|, \; t \ge t_0 \tag{25.100}$$

for any initial state $x(t_0)$.

Uniform exponential stability is also equivalent to requiring that positive constants T and γ exist with $\gamma < 1$ so that

$$\|\Phi(t + T, t)\| < \gamma, \text{ all } t \ge t_1 \text{ for some } t_1 \ge t_0 \tag{25.101}$$

Since $x(t+T) = \Phi(t+T, t)x(t)$, Equation 25.101 implies that

$$\|x(t + T)\| \le \gamma\|x(t)\|, \text{ for } t \ge t_1 \tag{25.102}$$

It also follows from Equation 25.99 that uniform exponential stability is preserved under a Lyapunov transformation $P(t)$. To see this, let $\overline{\Phi}(t, t_0)$ denote the state-transition matrix for $\dot{z}(t) = \overline{A}(t)z(t)$ where $\overline{A}(t) = P(t)A(t)P^{-1}(t) + \dot{P}(t)P^{-1}(t)$. Then using Equation 25.65 gives

$$\|\overline{\Phi}(t, t_0)\| \le \|P(t)\|\; \|\Phi(t, t_0)\|\; \|P^{-1}(t_0)\|. \tag{25.103}$$

But Equations 25.81 and 25.82 imply that

$$\|P(t)\| \le M_1 \text{ and } \|P^{-1}(\tau)\| \le M_2 \tag{25.104}$$

for some finite constants M_1 and M_2. Then inserting Equations 25.99 and 25.104 into Equation 25.103 yields

$$\|\overline{\Phi}(t, t_0)\| \le cM_1M_2e^{-\lambda(t-t_0)}, \; t \ge t_0, \tag{25.105}$$

which verifies that $\dot{z}(t) = \overline{A}(t)z(t)$ is uniformly exponentially stable.

A sufficient condition for uniform exponential stability is that a symmetric positive definite matrix $Q(t)$ exists with $c_1 I \leq Q(t) \leq c_2 I$ for some positive constants c_1, c_2, such that

$$A^T(t)Q(t) + Q(t)A(t) + \dot{Q}(t) \leq -c_3 I \quad (25.106)$$

for some positive constant c_3. Here $F \leq G$ means that F and G are symmetric matrices and $G - F$ is positive semidefinite (all eigenvalues are ≥ 0). This stability test is referred to as a Lyapunov criterion. For more details, see [8].

When $n = 1$ so that $A(t) = a(t)$ is 1×1, as noted above,

$$\Phi(t, t_0) = \exp\left[\int_{t_0}^{t} a(\tau)d\tau\right]. \quad (25.107)$$

It follows that the system is asymptotically stable if, and only if,

$$\int_{t_0}^{t} a(\sigma)d\sigma \to -\infty \text{ as } t \to \infty \quad (25.108)$$

and the system is uniformly exponentially stable if, and only if, a positive constant ν exists so that

$$\frac{1}{t}\int_{t_0}^{t} a(\sigma)d\sigma \leq -\nu, \text{ for } t > t_1, \text{ for some } t_1 \geq t_0. \quad (25.109)$$

When $n > 1$, if $A(t)$ commutes with its integral for $t > t_1$ for some $t_1 \geq t_0$ [see (25.50)], then

$$\Phi(t, t_1) = \exp\left[\int_{t_1}^{t} A(\sigma)d\sigma\right], \ t > t_1 \quad (25.110)$$

and a sufficient condition for uniform exponential stability is that the eigenvalues of

$$\frac{1}{t}\int_{t_2}^{t} A(\sigma)d\sigma \quad (25.111)$$

are bounded functions of t and have real parts $\leq -\lambda$ for $t > t_2$ for some $t_2 \geq t_1$, where λ is a positive constant. If $A(t)$ is upper or lower triangular with $p_1(t), p_2(t), \ldots, p_n(t)$ on the diagonal, then a sufficient condition for uniform exponential stability is that the off-diagonal entries of $A(t)$ be bounded and the scalar systems

$$\dot{x}_i(t) = p_i(t)x_i(t), \ t > t_0 \quad (25.112)$$

be uniformly exponentially stable for $i = 1, 2, \ldots, n$. Note that the matrix $A(t)$ given by Equation 25.35 is lower triangular.

Finally, if the entries of $B(t)$ and $C(t)$ are bounded, then uniform exponential stability implies that the system is **bounded-input, bounded-output (BIBO) stable**, that is, a bounded input $u(t)$ always results in a bounded output response $y(t)$.

25.2.6 Controllability and Observability

Given a system with the n-dimensional state model $[A(t), B(t), C(t)]$, it is now assumed that the entries of $A(t)$, $B(t)$, and $C(t)$ are at least continuous functions of t. The system is said to be **controllable** on the interval $[t_0, t_1]$, where $t_1 > t_0$, if for any states x_0 and x_1, a continuous input $u(t)$ exists that drives the system to the state $x(t_1) = x_1$ at time $t = t_1$ starting from the state $x(t_0) = x_0$ at time $t = t_0$.

Define the **controllability Gramian** which is the $n \times n$ matrix given by

$$W(t_0, t_1) = \int_{t_0}^{t_1} \Phi(t_0, t)B(t)B^T(t)\Phi^T(t_0, t)dt \quad (25.113)$$

The controllability Gramian $W(t_0, t_1)$ is symmetric positive semidefinite and is the solution to the matrix differential equation

$$\begin{aligned} \frac{d}{dt}W(t, t_1) &= A(t)W(t, t_1) + W(t, t_1)A^T(t) \\ &\quad - B(t)B^T(t), \quad (25.114) \\ W(t_1, t_1) &= 0 \end{aligned}$$

Then the system is controllable on $[t_0, t_1]$ if, and only if, $W(t_0, t_1)$ is invertible, in which case a continuous input $u(t)$ that drives the system from $x(t_0) = x_0$ to $x(t_1) = x_1$ is

$$\begin{aligned} u(t) &= -B^T(t)\Phi^T(t_0, t)W^{-1}(t_0, t_1)\left[x_0 - \Phi(t_0, t_1)x_1\right], \\ &\quad t_0 \leq t \leq t_1 \quad (25.115) \end{aligned}$$

When $A(t)$ is $n - 2$ times differentiable and $B(t)$ is $n - 1$ times differentiable, a sufficient condition for controllability is that the matrix $K(t) = [K_0(t)K_1(t)\ldots K_{n-1}(t)]$ defined by Equations 25.66–25.68 has rank n for at least one value of $t \in [t_0, t_1]$. This condition was first derived in [9]. As defined in this work, the system is said to be **uniformly controllable** on $[t_0, t_1]$ if the rank of $K(t)$ is equal to n for all $t \in [t_0, t_1]$.

Now suppose that the system input $u(t)$ is zero, so that the state model is given by

$$\dot{x}(t) = A(t)x(t) \quad (25.116)$$

$$y(t) = C(t)x(t) \quad (25.117)$$

From Equations 25.116–25.117, the output response $y(t)$ resulting from initial state $x(t_0)$ is

$$y(t) = C(t)\Phi(t, t_0)x(t_0), \ t > t_0 \quad (25.118)$$

Then the system is said to be observable on the interval $[t_0, t_1]$ if any initial state $x(t_0) = x_0$ can be determined from the output response $y(t)$ given by Equation 25.118 for $t \in [t_0, t_1]$. Define the **observability Gramian** which is the $n \times n$ matrix given by

$$M(t_0, t_1) = \int_{t_0}^{t_1} \Phi^T(t, t_0)C^T(t)C(t)\Phi(t, t_0)dt \quad (25.119)$$

The observability Gramian $W(t_0, t_1)$ is symmetric positive semidefinite and is the solution to the matrix differential equation

$$\begin{aligned} \frac{d}{dt}M(t, t_1) &= -A^T(t)M(t, t_1) - M(t, t_1)A(t) \\ &\quad - C^T(t)C(t), \quad (25.120) \\ M(t_1, t_1) &= 0 \end{aligned}$$

Then the system is **observable** on $[t_0, t_1]$ if, and only if, $M(t_0, t_1)$ is invertible, in which case the initial state $x(t_0)$ is given by

$$x_0 = M^{-1}(t_0, t_1)\int_{t_0}^{t_1} \Phi^T(t, t_0)C^T(t)y(t)dt \quad (25.121)$$

When $A(t)$ is $n-2$ times differentiable and $C(t)$ is $n-1$ times differentiable, a sufficient condition for observability is that the matrix $L(t)$ defined by Equations 25.72–25.74 has rank n for at least one value of $t \in [t_0, t_1]$. This condition was also first derived in [9]. The system is said to be **uniformly observable** on $[t_0, t_1]$ if the rank of $L(t)$ is equal to n for all $t \in [t_0, t_1]$.

Again given a system with state model $[A(t), B(t), C(t)]$, the **adjoint system** is the system with state model $[-A^T(t), C^T(t), B^T(t)]$. The system $[A(t), B(t), C(t)]$ is controllable (resp., observable) on an interval $[t_0, t_1]$ if, and only if, the adjoint system is observable (resp., controllable) on the interval $[t_0, t_1]$.

25.3 Discrete-Time Linear Time-Varying Systems

A discrete-time linear time-varying causal system with single input $u(k)$ and single output $y(k)$ can be modeled by the input/output relationship,

$$y(k) = \sum_{j=-\infty}^{k} h(k, j)u(j) \tag{25.122}$$

where k is an integer-valued variable (the discrete-time index) and $h(k, j)$ is the output response resulting from the unit pulse $\delta(k-j)$ (where $\delta(k-j) = 1$ for $k = j$ and $= 0$ for $k \neq j$) applied at time j with no initial energy in the system. The output $y(k)$ given by Equation 25.122 is the response resulting from the input $u(k)$ assuming that the system is at rest prior to the application of $u(k)$. It is assumed that $u(k)$ and/or $h(k, j)$ is constrained so that the summation in Equation 25.122 is well-defined. The system defined by Equation 25.122 is time invariant if, and only if, $h(k, j)$ is a function of only the difference $k - j$, in which case Equation 25.122 reduces to the convolution relationship,

$$y(k) = h(k) * u(k) = \sum_{j=-\infty}^{k} h(k-j)u(j) \tag{25.123}$$

where $h(k-j) = h(k-j, 0)$.

The system defined by Equation 25.122 is finite dimensional if the input $u(k)$ and the output $y(k)$ are related by the nth-order difference equation,

$$y(k+n) + \sum_{i=0}^{n-1} a_i(k)y(k+i) = \sum_{i=0}^{m} b_i(k)u(k+i) \tag{25.124}$$

where $m \leq n$ and the $a_i(k)$ and the $b_i(k)$ are real-valued functions of the discrete-time variable k. The system given by Equation 25.124 is time invariant if, and only if, all coefficients in Equation 25.124 are constants, that is, $a_i(k) = a_i$ and $b_i(k) = b_i$ for all i, where a_i and b_i are constants.

When $m < n$, the system defined by Equation 25.124 has the n-dimensional state model

$$x(k+1) = A(k)x(k) + B(k)u(k) \tag{25.125}$$

$$y(k) = Cx(k) \tag{25.126}$$

where

$$A(k) = \begin{bmatrix} 0 & 0 & 0 & \cdots & 0 & -a_0(k) \\ 1 & 0 & 0 & \cdots & 0 & -a_1(k-1) \\ 0 & 1 & 0 & \cdots & 0 & -a_2(k-2) \\ \vdots & & & \cdots & & \vdots \\ 0 & 0 & 0 & \cdots & 0 & -a_{n-2}(k-n+2) \\ 0 & 0 & 0 & \cdots & 1 & -a_{n-1}(k-n+1) \end{bmatrix} \tag{25.127}$$

$$B(k) = \begin{bmatrix} b_0(k) \\ b_1(k-1) \\ b_2(k-2) \\ \vdots \\ b_{n-2}(k-n+2) \\ b_{n-1}(k-n+1) \end{bmatrix} \tag{25.128}$$

and

$$C = \begin{bmatrix} 0 & 0 & \cdots & 0 & 1 \end{bmatrix}$$

where $b_i(k) = 0$ for $i > m$. This particular state model is referred to as the observer canonical form. As in the continuous-time case, there are other possible state realizations of Equation 25.124, but these will not be considered here. It is interesting to note that the entries of $A(k)$ and $B(k)$ in the observer canonical form are simply time shifts of the coefficients of the input/output difference Equation 25.124, whereas in the continuous-time case, this relationship is rather complicated.

25.3.1 State Model in the General Case

For an m-input p-output linear n-dimensional time-varying discrete-time system, the general form of the state model is

$$x(k+1) = A(k)x(k) + B(k)u(k) \tag{25.129}$$

$$y(k) = C(k)x(k) + D(k)u(k) \tag{25.130}$$

where $A(k)$ is $n \times n$, $B(k)$ is $n \times m$, $C(k)$ is $p \times n$ and $D(k)$ is $p \times m$. The state model given by Equations 25.129–25.130 may arise as a result of sampling a continuous-time system given by the state model

$$\dot{x}(t) = A(t)x(t) + B(t)u(t) \tag{25.131}$$

$$y(t) = C(t)x(t) + D(t)u(t). \tag{25.132}$$

If the sampling interval is equal to T, then setting $t = kT$ in Equation 25.132 yields an output equation of the form in Equation 25.130 where $C(k) = C(t)|_{t=kT}$ and $D(k) = D(t)|_{t=kT}$. To "discretize" Equation 25.131, first recall (see Equation 25.52) that the solution to Equation 25.131 is

$$x(t) = \Phi(t, t_0)x(t_0) + \int_{t_0}^{t} \Phi(t, \tau)B(\tau)u(\tau)d\tau, \ t > t_0 \tag{25.133}$$

Then setting $t = kT + T$ and $t_0 = kT$ in Equation 25.133 yields

$$x(kT+T) = \Phi(kT+T, kT)x(kT) + \int_{kT}^{kT+T} \Phi(kT+T, \tau)B(\tau)u(\tau)d\tau \quad (25.134)$$

The second term on the right-hand side of Equation 25.134 can be approximated by

$$\left[\int_{kT}^{kT+T} \Phi(kT+T, \tau)B(\tau)d\tau\right]u(kT)$$

and thus Equation 25.134 is in the form of Equation 25.129 with

$$A(k) = \Phi(kT+T, kT) \quad (25.135)$$

$$B(k) = \int_{kT}^{kT+T} \Phi(kT+T, \tau)B(\tau)d\tau \quad (25.136)$$

Note that the matrix $A(k)$ given by Equation 25.135 is always invertible since $\Phi(kT+T, kT)$ is always invertible (see the property given by Equation 25.43). As discussed below, this implies that discretzed or sampled data systems are "reversible."

From Equations 25.135–25.136 it is seen that the computation of $A(k)$ and $B(k)$ requires knowledge of the state-transition matrix $\Phi(t, \tau)$ for $t = kT + T$ and $\tau \in [kT, KT + T)$. If $A(t)$ in Equation 25.131 is a continuous function of t over each interval $[kT + T, kT)$ and the variation of $A(t)$ over each interval $[kT, kT + T)$ is sufficiently small, then $\Phi(kT + T, \tau)$ can be approximated by

$$\Phi(kT+T, \tau) = e^{A(kT)(kT+T-\tau)} \text{ for } \tau \in [kT, kT+T) \quad (25.137)$$

and hence $A(k)$ and $B(k)$ can be determined using

$$A(k) = e^{A(kT)T} \quad (25.138)$$

$$B(k) = \int_{kT}^{kT+T} e^{A(kT)(kT+T-\tau)}B(\tau)d\tau \quad (25.139)$$

Given the discrete-time system defined by Equation 25.129–25.130, the solution to Equation 25.129 is

$$x(k) = \Phi(k, k_0)x(k_0) + \sum_{j=k_0}^{k-1} \Phi(k, j+1)B(j)u(j), \quad k > k_0 \quad (25.140)$$

where the $n \times n$ state-transition matrix $\Phi(k, j)$ is given by

$$\Phi(k, k_0) = \begin{cases} \text{not defined for } k < k_0 \\ I, \quad k = k_0 \\ A(k-1)A(k-2)\cdots A(k_0), \quad k > k_0 \end{cases} \quad (25.141)$$

It follows directly from Equation 25.141 that $\Phi(k, k_0)$ is invertible for $k > k_0$ only if $A(k)$ is invertible for $k \geq k_0$. Thus, in general, the initial state $x(k_0)$ can not be determined from the relationship $x(k) = \Phi(k, k_0)x(k_0)$. In other words, a discrete-time system is not necessarily **reversible**, although any continuous-time system given by Equations 25.131–25.132 is reversible since $\Phi(t, t_0)$ is always invertible. However, as noted above, any sampled data system is reversible.

The state-transition matrix $\Phi(k, k_0)$ satisfies the composition property:

$$\Phi(k, k_0) = \Phi(k, k_1)\Phi(k_1, k_0) \text{ where } k_0 \leq k_1 \leq k \quad (25.142)$$

and in addition,

$$\Phi(k+1, k_0) = A(k)\Phi(k, k_0), \quad k \geq k_0 \quad (25.143)$$

If $A(k)$ is invertible for all k, $\Phi(k, k_0)$ can be written in the factored form

$$\Phi(k, k_0) = \Phi_1(k)\Phi_2(k_0), \quad k \geq k_0 \quad (25.144)$$

where

$$\Phi_1(k) = \begin{cases} A(k-1)A(k-2)\cdots A(0), \quad k \geq 1 \\ I, \quad k = 0 \\ A^{-1}(k-2)A^{-1}(k-3)\cdots A^{-1}(-1), \quad k < 0 \end{cases} \quad (25.145)$$

$$\Phi_2(k_0) = \begin{cases} A^{-1}(0)A^{-1}(1)\cdots A^{-1}(k_0-1), \quad k_0 > 0 \\ I, \quad k_0 = 0 \\ A(-1)A(-2)\cdots A(k_0), \quad k_0 < 0 \end{cases} \quad (25.146)$$

When $y(k) = C(k)x(k)$, the output response $y(k)$ is given by

$$y(k) = C(k)\Phi(k, k_0)x(k_0) + \sum_{j=k_0}^{k-1} C(k)\Phi(k, j+1)B(j)u(j), \quad k > k_0 \quad (25.147)$$

If the initial time k_0 is set equal to $-\infty$ and there is no initial energy at time $k = -\infty$, Equation 25.147 becomes

$$y(k) = \textstyle\sum_{j=-\infty}^{k-1} C(k)\Phi(k, j+1)B(j)u(j) \quad (25.148)$$

Comparing Equation 25.148 with the m-input, p-output version of the input/output Equation 25.122 reveals that

$$H(k, j) = \begin{cases} C(k)\Phi(k, j+1)B(j), & k > j \\ 0, & k \leq j \end{cases} \quad (25.149)$$

where $H(k, j)$ is the $p \times m$ unit-pulse response function matrix. Note that if $A(k)$ is invertible so that $\Phi(k, k_0)$ has the factorization given in Equation 25.144, then $H(k, j)$ can be expressed in the factored form

$$H(k, j) = [C(k)\Phi_1(k)][\Phi_2(j+1)B(j)], \quad \text{for } k > j. \quad (25.150)$$

As in the continuous-time case, this factorization is a fundamental property of unit-pulse response matrices $H(k, j)$ that are realizable by a state model (with invertible $A(k)$).

25.3.2 Change of State Variables and Canonical Forms

Given the discrete-time system defined by Equations 25.129–25.130 where $D(k) = 0$, in the following development the system will be denoted by the triple $[A(k), B(k), C(k)]$. For any $n \times n$ invertible matrix $P(k)$, another state model can be generated by defining the new state vector $z(k) = P(k)x(k)$. The new state model is given by

$$\begin{aligned} z(k+1) &= \overline{A}(k)z(k) + \overline{B}(k)u(k) && (25.151) \\ y(k) &= \overline{C}(k)z(k) && (25.152) \end{aligned}$$

where

$$\begin{aligned} &\overline{A}(k) = P(k+1)A(k)P^{-1}(k), \\ &\overline{B}(k) = P(k+1)B(k), \quad \overline{C}(k) = C(k)P^{-1}(k) \end{aligned} \qquad (25.153)$$

The state-transition matrix $\overline{\Phi}(k, k_0)$ for the new state model is given by

$$\overline{\Phi}(k, k_0) = P(k)\Phi(k, k_0)P^{-1}(k_0) \qquad (25.154)$$

where $\Phi(k, k_0)$ is the state-transition matrix for $[A(k), B(k), C(k)]$. The new state model, which will be denoted by $[\overline{A}(k), \overline{B}(k), \overline{C}(k)]$, and the given state model $[A(k), B(k), C(k)]$ are said to be algebraically equivalent.

Now define the $n \times m$ matrices

$$R_0(k) = B(k) \qquad (25.155)$$

$$R_i(k) = A(k)R_{i-1}(k-1), \qquad i = 1, 2, ..., n-1 \qquad (25.156)$$

and let $R(k)$ denote the $n \times nm$ matrix whose ith block column is equal to $R_i(k)$, that is,

$$R(k) = [R_0(k)R_1(k) \cdots R_{n-1}(k)] \qquad (25.157)$$

Similarly, for the model $[\overline{A}(k), \overline{B}(k), \overline{C}(k)]$ define

$$\overline{R}(k) = [\overline{R}_0(k)\overline{R}_1(k) \cdots \overline{R}_{n-1}(k)] \qquad (25.158)$$

where the $\overline{R}_i(k)$ are given by Equations 25.155–25.156 with $A(k)$ and $B(k)$ replaced by $\overline{A}(k)$ and $\overline{B}(k)$, respectively. Then the transformation $P(k)$ is given by

$$P(k+1)R(k) = \overline{R}(k) \qquad (25.159)$$

and thus, if $R(k)$ has rank n for all k so that $R(k)R^T(k)$ is invertible, Equation 25.159 can be solved for $P(k+1)$. This gives

$$P(k+1) = \overline{R}(k)R^T(k)[R(k)R^T(k)]^{-1} \qquad (25.160)$$

As discussed below, the rank condition on $R(k)$ implies that the system is uniformly n-step controllable. It follows from Equation 25.159 that uniform n-step controllability is preserved under a change of state variables.

Now suppose that the system with state model $[A(k), B(k), C(k)]$ has a single input ($m = 1$) so that $B(k)$ is an n-element column vector. Define

$$\begin{aligned} R_n(k) &= A(k)R_{n-1}(k-1) && (25.161) \\ \eta(k) &= -R^{-1}(k)R_n(k) && (25.162) \end{aligned}$$

where $R_{n-1}(k)$ is defined by Equation 25.156 and $R(k)$ is defined by Equation 25.157. The n-element column vector $\eta(k)$ is invariant under any change of state $z(k) = P(k)x(k)$, and in the time-invariant case, η is constant and is given by

$$\eta = [a_0 \, a_1 \ldots a_{n-1}]^T \qquad (25.163)$$

where the a_i are the coefficients of the characteristic polynomial of A given by Equation 25.87.

Given $\eta(k)$ defined by Equation 25.162, write $\eta(k)$ in the form

$$\eta(k) = [\eta_0(k) \, \eta_1(k) \ldots \eta_{n-1}(k)]. \qquad (25.164)$$

Then there is a transformation $P(k)$ which converts $[A(k), B(k), C(k)]$ into the control canonical form $[\overline{A}(k), \overline{B}, \overline{C}(k)]$, with $\overline{A}(k)$ and $\overline{B}$ given by

$$\overline{A}(k) = \begin{bmatrix} 0 & 1 & 0 & \cdots & 0 & 0 \\ 0 & 0 & 1 & \cdots & 0 & 0 \\ \vdots & \vdots & \vdots & \cdots & \vdots & \vdots \\ 0 & 0 & 0 & \cdots & 0 & 1 \\ -\eta_0(k) & -\eta_1(k+1) & -\eta_2(k+2) & \cdots & -\eta_{n-2}(k+n-2) & -\eta_{n-1}(k+n-1) \end{bmatrix} \qquad (25.165)$$

$$\overline{B} = \begin{bmatrix} 0 \\ 0 \\ \vdots \\ 0 \\ 1 \end{bmatrix}$$

The transformation $P(k)$ which yields the control canonical form can be determined using Equation 25.160.

Again consider the m-input, p-output case where the state model is given by $[A(k), B(k), C(k)]$. Define the $n \times p$ matrices

$$O_0(k) = C(k) \qquad (25.166)$$

$$O_i(k) = O_{i-1}(k+1)A(k), \qquad i = 1, 2, \ldots, n-1 \qquad (25.167)$$

and let $O(k)$ denote the $pn \times n$ matrix whose ith block row is equal to $O_i(k)$, that is,

$$O(k) = \begin{bmatrix} O_0(k) \\ O_1(k) \\ \vdots \\ O_{n-1}(k) \end{bmatrix} \qquad (25.168)$$

Given an algebraically equivalent state model $[\overline{A}(k), \overline{B}(k), \overline{C}(k)]$, define

$$\overline{O}(k) = \begin{bmatrix} \overline{O}_0(k) \\ \overline{O}_1(k) \\ \vdots \\ \overline{O}_{n-1}(k) \end{bmatrix} \tag{25.169}$$

where the $\overline{O}_i(k)$ are given by Equations 25.166–25.167 with $A(k)$ and $C(k)$ replaced by $\overline{A}(k)$ and $\overline{C}(k)$, respectively. Then

$$O(k)P^{-1}(k) = \overline{O}(k) \tag{25.170}$$

Then if $O(k)$ has rank n for all k, Equation 25.170 can be solved for $P^{-1}(k)$ which gives

$$P^{-1}(k) = [O^T(k)O(k)]^{-1}O^T(k)\overline{O}(k) \tag{25.171}$$

The rank condition on $O(k)$ is equivalent to uniform n-step observability of the system $[A(k), B(K), C(k)]$. It follows from Equation 25.170 that this property is preserved under a change of state.

Now suppose that the system with state model $[A(k), B(k), C(k)]$ has one output ($p = 1$) with any number of inputs. Define

$$O_n(k) = O_{n-1}(k+1)A(k) \tag{25.172}$$

$$\rho(k) = -[O^T(k)]^{-1}O_n^T(k) \tag{25.173}$$

where $O_{n-1}(k)$ is defined by Equation 25.167 and $O(k)$ is defined by Equation 25.168. The n-element column vector $\rho(k)$ is also invariant under a change of state and is given by Equation 25.163 in the time-invariant case.

Given $\rho(k)$ defined by Equation 25.173, write $\rho(k)$ in the form

$$\rho(k) = [a_0(k)\ a_1(k) \dots a_{n-1}(k)]^T \tag{25.174}$$

Then there is a transformation $P(k)$ that converts $[A(k), B(k), C(k)]$ into the observer canonical form $[\overline{A}(k), \overline{B}(k), \overline{C}]$ with $\overline{A}(k)$ given by Equation 25.127 and $\overline{C}$ given by Equation 25.128. The transformation $P(k)$ can be computed using Equation 25.171.

As in the continuous-time case, stability is not necessarily preserved under a transformation $P(k)$. However, as discussed below, stability is preserved if $P(k)$ is a Lyapunov transformation, that is, both $\|P(k)\|$ and $\|P^{-1}(k)\|$ are bounded functions of k, where $\| \quad \|$ denotes the matrix norm defined previously.

25.3.3 Stability

Given a discrete-time system with n-dimensional state model $[A(k), B(k), C(k)]$, consider the homogeneous equation

$$x(k+1) = A(k)x(k), \qquad k \geq k_0. \tag{25.175}$$

The solution is

$$x(k) = \Phi(k, k_0)x(k_0), \qquad k > k_0 \tag{25.176}$$

where $\Phi(k, k_0)$ is the state-transition matrix defined by Equation 25.141.

The system is said to be asymptotically stable if the solution $x(k)$ satisfies the condition $\|x(k)\| \to 0$ as $k \to \infty$ for any initial state $x(k_0)$ at the initial time k_0. This is equivalent to requiring that

$$\|\Phi(k, k_0)\| \to 0 \quad \text{as} \quad k \to \infty \tag{25.177}$$

The system is uniformly exponentially stable if positive constants c and ρ exist with $\rho < 1$ so that the solution $x(k)$ satisfies

$$\|x(k)\| \leq c\rho^{k-k_0}\|x(k_0)\|, \quad k > k_0 \tag{25.178}$$

for any initial state $x(k_0)$. This is equivalent to requiring that

$$\|\Phi(k, k_0)\| \leq c\rho^{k-k_0}, \quad k \geq k_0 \tag{25.179}$$

It follows from Equation 25.179 that uniform exponential stability is preserved under a Lyapunov transformation.

Uniform exponential stability is also equivalent to requiring that a positive integer q and a positive constant $\gamma < 1$ exist so that

$$\|\Phi(k+q, k)\| < \gamma, \quad \text{all} \quad k \geq k_1 \quad \text{for some} \quad k_1 \geq k_0. \tag{25.180}$$

Since $x(k+q) = \Phi(k+q, k)x(k)$, Equation 25.180 implies that

$$\|x(k+q)\| \leq \gamma\|x(k)\| \quad \text{for} \quad k \geq k_1. \tag{25.181}$$

Another necessary and sufficient condition for uniform exponential stability is that a symmetric positive definite matrix $Q(k)$ exists with $c_1 I \leq Q(k) \leq c_2 I$ for some positive constants c_1 and c_2 so that

$$A^T(k)Q(k+1)A(k) - Q(k) \leq -c_3 I \tag{25.182}$$

for some positive constant c_3.

25.3.4 Controllability and Observability

The system with state model $[A(k), B(k), C(k)]$ is said to be controllable on the interval $[k_0, k_1]$ with $k_1 > k_0$ if, for any states x_0 and x_1, an input $u(k)$ exists that drives the system to the state $x(k_1) = x_1$ at time $k = k_1$ starting from the state $x(k_0) = x_0$ at time $k = k_0$.

Define the $n \times nm$ **controllability** (or **reachability**) **matrix**

$$\begin{aligned} R(k_0, k_1) = {} & [B(k_1 - 1)\ \Phi(k_1, k_1 - 1)B(k_1 - 2) \\ & \Phi(k_1, k_1 - 2)B(k_1 - 3) \dots \\ & \Phi(k_1, k_0 + 1)B(k_0)] \end{aligned} \tag{25.183}$$

Then from Equation 25.140, the state $x(k_1)$ at time $k = k_1$ resulting from state $x(k_0)$ at time $k = k_0$ and the input sequence $u(k_0), u(k_0 + 1), \dots, u(k_1 - 1)$ is given by

$$x(k_1) = \Phi(k_1, k_0)x(k_0) + R(k_0, k_1)U(k_0, k_1) \tag{25.184}$$

where $U(k_0, k_1)$ is the mn-element column vector of inputs given by

$$U(k_0, k_1) = \left[u^T(k_1 - 1)\ u^T(k_1 - 2) \cdots u^T(k_0)\right]^T \tag{25.185}$$

Now for any states $x(k_0) = x_0$ and $x(k_1) = x_1$, from Equation 25.184 there is a sequence of inputs given by $U(k_0, k_1)$ that drives the system from x_0 to x_1 if, and only if, the matrix $R(k_0, k_1)$ has rank n. If this is the case, Equation 25.184 can be solved for $U(k_0, k_1)$, giving

$$\begin{aligned} U(k_0, k_1) &= R^T(k_0, k_1)\left[R(k_0, k_1)R^T(k_0, k_1)\right]^{-1} \\ &\quad [x_1 - \Phi(k_1, k_0)x_0] \end{aligned} \tag{25.186}$$

Hence rank $R(k_0, k_1) = n$ is a necessary and sufficient condition for controllability over the interval $[k_0, k_1]$.

Now set $k_0 = k - n + 1$ and $k_1 = k + 1$ in $R(k_0, k_1)$, which results in the matrix $R(k - n + 1, k + 1)$ which will be denoted by $R(k)$. The matrix $R(k)$ is the same as the matrix $R(k)$ defined by Equations 25.155–25.157. As previously noted, the system is said to be **uniformly n-step controllable** if rank $R(k) = n$ for all k.

Suppose that the system input $u(k)$ is zero, so that the state model is given by

$$x(k+1) = A(k)x(k), \tag{25.187}$$

$$y(k) = C(k)x(k). \tag{25.188}$$

From Equations 25.187–25.188, the output response $y(k)$ resulting from initial state $x(k_0)$ is given by

$$y(k) = C(k)\Phi(k, k_0)x(k_0), \qquad k \geq k_0. \tag{25.189}$$

Then the system is said to be observable on the interval $[k_0, k_1]$ if any initial state $x(k_0) = x_0$ can be determined from the output response $y(k)$ given by Equation 25.189 for $k = k_0, k_0 + 1, \ldots, k_1 - 1$. Using Equation 25.189,

$$\begin{bmatrix} y(k_0) \\ y(k_0+1) \\ \vdots \\ y(k_1-2) \\ y(k_1-1) \end{bmatrix} = \begin{bmatrix} C(k_0)x_0 \\ C(k_0+1)\Phi(k_0+1, k_0)x_0 \\ \vdots \\ C(k_1-2)\Phi(k_1-2, k_0)x_0 \\ C(k_1-1)\Phi(k_1-1, k_0)x_0 \end{bmatrix} \tag{25.190}$$

The right-hand side of Equation 25.190 can be written in the form $O(k_0, k_1)x_0$ where $O(k_0, k_1)$ is the $np \times n$ **observability matrix** defined by

$$O(k_0, k_1) = \begin{bmatrix} C(k_0) \\ C(k_0+1)\Phi(k_0+1, k_0) \\ \vdots \\ C(k_1-2)\Phi(k_1-2, k_0) \\ C(k_1-1)\Phi(k_1-1, k_0) \end{bmatrix} \tag{25.191}$$

Equation 25.190 can always be solved for x_0 if, and only if, rank $O(k_0, k_1) = n$, which is a necessary and sufficient condition for observability on $[k_0, k_1]$. If the rank condition holds, the solution of Equation 25.190 for x_0 is

$$x_0 = \left[O^T(k_0, k_1)O(k_0k_1)\right]^{-1} O^T(k_0, k_1)Y(k_0, k_1) \tag{25.192}$$

where

$$\begin{aligned} Y(k_0, k_1) &= \left[y^T(k_0)\, y^T(k_0+1) \cdots \right. \\ &\quad \left. y^T(k_1-2) y^T(k_1-1)\right]^T \end{aligned} \tag{25.193}$$

Setting $k_0 = k$ and $k_1 = k + n$ in $O(k_0, k_1)$ yields the matrix $O(k, k + n)$ which is equal to the matrix $O(k)$ defined by Equation 25.168. As noted above, if rank $O(k) = n$ for all k, the system is said to be uniformly n-step observable.

25.4 Applications and Examples

Consider the single-input single-output linear time-varying, continuous-time system given by the input/output differential equation

$$\ddot{y}(t) + (\cos t)\dot{y}(t) + (1 - \sin t)y(t) = \dot{u}(t) + u(t) \tag{25.194}$$

To determine the state model which is in observer canonical form, we need to write Equation 25.194 in the form

$$\ddot{y}(t) + D[\alpha_1(t)y(t)] + \alpha_0(t)y(t) = \dot{u}(t) + u(t) \tag{25.195}$$

In this case,

$$\alpha_1(t) = \cos t \quad \text{and} \quad \alpha_0(t) = 1 \tag{25.196}$$

Then from Equations 25.27–25.28, the observer canonical form of the state model is

$$\begin{aligned} \begin{bmatrix} \dot{x}_1(t) \\ \dot{x}_2(t) \end{bmatrix} &= \begin{bmatrix} 0 & -1 \\ 1 & -\cos t \end{bmatrix} \begin{bmatrix} x_1(t) \\ x_2(t) \end{bmatrix} \\ &\quad + \begin{bmatrix} 1 \\ 1 \end{bmatrix} u(t) \end{aligned} \tag{25.197}$$

$$y(t) = \begin{bmatrix} 0 & 1 \end{bmatrix} \begin{bmatrix} x_1(t) \\ x_2(t) \end{bmatrix} \tag{25.198}$$

The state variables $x_1(t)$ and $x_2(t)$ in this state model are given by

$$\begin{aligned} x_1(t) &= \dot{y}(t) + (\cos t)y(t) - u(t), \\ x_2(t) &= y(t) \end{aligned} \tag{25.199}$$

If the output $y(t)$ of the system can be differentiated, then $x_1(t)$ and $x_2(t)$ can be directly determined from the input $u(t)$ and the output $y(t)$ by using Equation 25.199. In practice, however differentiation of signals should be avoided, and thus directly determining $x_1(t)$ and $x_2(t)$ using Equation 25.199 is usually not viable. As discussed below, by using a state observer we can estimate the state variables without having to differentiate signals.

The general form of a state observer for a m-input, p-output, n-dimensional system with state model $[A(t), B(t), C(t)]$ is

$$\begin{aligned} \frac{d}{dt}\hat{x}(t) &= A(t)\hat{x}(t) + \Gamma(t)[y(t) - C(t)\hat{x}(t)] \\ &\quad + B(t)u(t) \end{aligned} \tag{25.200}$$

where $\Gamma(t)$ is the $n \times p$ observer gain matrix and $\hat{x}(t)$ is the estimate of $x(t)$. With the estimation error $e(t)$ defined by

$$e(t) = x(t) - \hat{x}(t) \tag{25.201}$$

the error is given by the differential equation

$$\dot{e}(t) = [A(t) - \Gamma(t)C(t)]\,e(t), \qquad t > t_0, \tag{25.202}$$

with initial error $e(t_0)$ at initial time t_0. The objective is to choose the gain matrix $\Gamma(t)$ so that, for any initial error $e(t_0)$, $\|e(t)\| \to 0$ as $t \to \infty$, with some desired rate of convergence. If there is one output ($p = 1$) and the state model is in observer canonical form, it is possible to "assign" the error dynamics (given by Equation 25.202) by selecting $\Gamma(t)$. To illustrate this, consider the state model given by Equations 25.197–25.198. In this case, the error expression Equation 25.202 is

$$\dot{e}(t) = \begin{bmatrix} 0 & -1 - \gamma_1(t) \\ 1 & -\cos t - \gamma_2(t) \end{bmatrix} e(t) \tag{25.203}$$

where $\Gamma(t) = [\gamma_1(t) \;\; \gamma_2(t)]^T$. From Equation 25.203, it is clear that by setting

$$\begin{aligned} \gamma_1(t) &= a_0 - 1, \\ \gamma_2(t) &= a_1 - \cos t \end{aligned} \tag{25.204}$$

where a_0 and a_1 are constants, the coefficient matrix on the right-hand side of Equation 25.203 is constant, and its eigenvalues can be assigned by choosing a_0 and a_1. Hence, any desired rate of convergence to zero can be achieved for the error $e(t)$.

The estimate $\hat{x}(t)$ of $x(t)$ can then be used to realize a **feedback control law** of the form

$$u(t) = -F(t)\hat{x}(t) \tag{25.205}$$

where $F(t)$ is the feedback gain matrix. The first step in pursuing this is to consider the extent to which the system can be controlled by state feedback of the form given in Equation 25.205 with $\hat{x}(t)$ replaced by $x(t)$; in other words, the true system state $x(t)$ is assumed to be available. In particular, we can ask whether or not there is a gain matrix $F(t)$ so that with $u(t) = -F(t)x(t)$, the resulting closed-loop system is uniformly exponentially stable with some desired rate of convergence to zero. For the system given by Equation 25.197, the state equation of the closed-loop system is

$$\begin{bmatrix} \dot{x}_1(t) \\ \dot{x}_2(t) \end{bmatrix} = \begin{bmatrix} -f_1(t) & -1 - f_2(t) \\ 1 - f_1(t) & -\cos t - f_2(t) \end{bmatrix} \begin{bmatrix} x_1(t) \\ x_2(t) \end{bmatrix} \tag{25.206}$$

where $F(t) = [f_1(t) f_2(t)]$. From Equation 25.206 it is not obvious whether there is an $F(t)$ which results in uniform exponential stability. This can be answered by attempting to transform the state model given by Equations 25.197–25.198 to control canonical form. The steps are as follows.

Evaluating the $K_i(t)$ given by Equations 25.66–25.67 gives

$$K_0(t) = B(t) = \begin{bmatrix} 1 \\ 1 \end{bmatrix}, \tag{25.207}$$

$$K_1(t) = -A(t)K_0(k) + \dot{K}_0(t) = \begin{bmatrix} 1 \\ -1 + \cos t \end{bmatrix} \tag{25.208}$$

Then

$$K(t) = [K_0(t) \;\; K_1(t)] = \begin{bmatrix} 1 & 1 \\ 1 & -1 + \cos t \end{bmatrix} \tag{25.209}$$

and

$$\det[K(t)] = -2 + \cos t \le -1 \text{ for all } t \tag{25.210}$$

where "det" denotes the determinant. Because $\det[K(t)] \neq 0$, $K(t)$ has rank 2 for all t, and thus the system given by Equations 25.197–25.198 is uniformly controllable and there is a control canonical form.

To determine the control canonical form, it is first necessary to compute the "characteristic vector" $\eta(t)$ given by Equation 25.85. In this case,

$$\begin{aligned} K_n(t) &= K_2(t) = -A(t)K_1(t) + \dot{K}_1(t) \\ &= \begin{bmatrix} -1 + \cos t \\ -1 + (\cos t)(-1 + \cos t) - \sin t \end{bmatrix} \end{aligned} \tag{25.211}$$

Then

$$\begin{aligned} \eta(t) &= -K^{-1}(t)K_2(t) \\ &= \frac{1}{2 - \cos t} \begin{bmatrix} 2 - \cos t + \sin t \\ -2\cos t + \cos^2 t - \sin t \end{bmatrix} \end{aligned} \tag{25.212}$$

Now let

$$[\psi_0(t) - \psi_1(t)]^T = f^{-1}(\eta) \tag{25.213}$$

where f^{-1} is defined by Equations 25.22–25.23, that is,

$$f^{-1}(\eta) = [\eta_0(t) - \dot{\eta}_1(t) \; \eta_1(t)]^T \tag{25.214}$$

Then

$$\begin{aligned} \psi_0(t) &= \eta_0(t) - \dot{\eta}_1(t) = \frac{2 - \cos t + \sin t}{2 - \cos t} \\ &\quad + \frac{d}{dt}\left[\frac{-2\cos t + \cos^2 t - \sin t}{2 - \cos t}\right] \end{aligned} \tag{25.215}$$

$$\begin{aligned} \psi_1(t) &= -\eta_1(t) \\ &= \frac{-1}{2 - \cos t}\left[-2\cos t + \cos^2 t - \sin t\right] \end{aligned} \tag{25.216}$$

and the control canonical form is given by

$$\begin{aligned} \begin{bmatrix} \dot{z}_1(t) \\ \dot{z}_2(t) \end{bmatrix} &= \begin{bmatrix} 0 & 1 \\ -\psi_0(t) & -\psi_1(t) \end{bmatrix} \begin{bmatrix} z_1(t) \\ z_2(t) \end{bmatrix} \\ &\quad + \begin{bmatrix} 0 \\ 1 \end{bmatrix} u(t) \end{aligned} \tag{25.217}$$

where $z(t) = P(t)x(t)$ with $P(t)$ given by

$$P(t) = \overline{K}(t)K^{-1}(t) \tag{25.218}$$

with

$$\overline{K}(t) = \begin{bmatrix} 0 & -1 \\ 1 & \psi_1(t) \end{bmatrix}$$
$$K^{-1}(t) = \frac{1}{-2+\cos t}\begin{bmatrix} -1+\cos t & -1 \\ -1 & 1 \end{bmatrix} \tag{25.219}$$

Inserting Equation 25.219 into Equation 25.218 yields

$$P(t) = \frac{1}{-2+\cos t}\begin{bmatrix} 1 & -1 \\ -1+\cos t - \psi_1(t) & -1+\psi_1(t) \end{bmatrix} \tag{25.220}$$

Since $\psi_1(t)$ given by Equation 25.216 is a bounded function of t, the transformation $P(t)$ is clearly bounded for all t. In addition,

$$P^{-1}(t) = K(t)[\overline{K}(t)]^{-1} = \begin{bmatrix} \psi_1(t)-1 & 1 \\ \psi_1(t)+1-\cos t & 1 \end{bmatrix} \tag{25.221}$$

which is also bounded. Hence $P(t)$ is a Lyapunov transformation.

Now consider the feedback control

$$u(t) = -\overline{F}(t)z(t) \tag{25.222}$$

where $\overline{F}(t) = [\overline{f}_1(t) \quad \overline{f}_2(t)]$. Inserting Equation 25.222 into Equation 25.217 results in the closed-loop equation

$$\begin{bmatrix} \dot{z}_1(t) \\ \dot{z}_2(t) \end{bmatrix} = \begin{bmatrix} 0 & 1 \\ -\psi_0(t) - \overline{f}_1(t) & -\psi_1(t) - \overline{f}_2(t) \end{bmatrix} \times \begin{bmatrix} z_1(t) \\ z_2(t) \end{bmatrix} \tag{25.223}$$

Then setting

$$\overline{f}_1(t) = -\psi_0(t) + b_0$$
$$\overline{f}_2(t) = -\psi_1(t) + b_1 \tag{25.224}$$

where b_0 and b_1 are constants, it is seen that the coefficient matrix on the right-hand side of Equation 25.223 is constant and thus the state-transition matrix is

$$\Phi_{cl}(t, t_0) = \exp\left[\overline{A}_{cl}(t - t_0)\right] \tag{25.225}$$

where

$$\overline{A}_{cl} = \begin{bmatrix} 0 & 1 \\ -b_0 & -b_1 \end{bmatrix} \tag{25.226}$$

and "cl" stands for "closed loop." Clearly, the eigenvalues of $\overline{A}_{cl}$ can be assigned by choosing b_0 and b_1,. Thus the transformed Equation 25.223 can be made uniformly exponentially stable with any desired rate of convergence to zero. This result then carries over to Equation 25.197 as follows.

First, because $z(t) = P(t)x(t)$, we can replace $z(t)$ in Equation 25.222 by $P(t)x(t)$ which results in the feedback control

$$u(t) = -\overline{F}(t)P(t)x(t) \tag{25.227}$$

Then using this control in Equation 25.197 gives the closed-loop equation

$$\dot{x}(t) = \left[A(t) - B\overline{F}(t)P(t)\right]x(t) \tag{25.228}$$

where

$$A(t) = \begin{bmatrix} 0 & -1 \\ 1 & \cos t \end{bmatrix} \quad \text{and} \quad B = \begin{bmatrix} 0 \\ 1 \end{bmatrix} \tag{25.229}$$

However, because $z(t) = P(t)x(t)$, the state transition matrix $\Phi_{cl}(t, t_0)$ for Equation 25.228 is given by

$$\Phi_{cl}(t, t_0) = P^{-1}(t)\exp\left[\overline{A}_{cl}(t - t_0)\right]P(t_0) \tag{25.230}$$

Then since both $P(t)$ and $P^{-1}(t)$ are bounded, it follows from Equation 25.230 that by assigning the eigenvalues of $\overline{A}_{cl}$, it is possible to have

$$\|\Phi_{cl}(t, t_0)\| \leq ce^{-\lambda(t-t_0)}, \quad \text{for } t > t_0, \tag{25.231}$$

for any desired $\lambda > 0$. This shows that via the state feedback control given by Equation 25.227, the resulting closed-loop system can be made uniformly exponentially stable with any desired rate λ of convergence to zero.

25.4.1 Exponential Systems

A system with n-dimensional state model $[A(t), B(t), C(t)]$ is said to be an **exponential system** if its state-transition matrix $\Phi(t, \tau)$ can be written in the matrix exponential form,

$$\Phi(t, \tau) = e^{\Gamma(t,\tau)}, \tag{25.232}$$

where $\Gamma(t, \tau)$ is a $n \times n$ matrix function of t and τ. The form given in Equation 25.232 is valid (at least locally, i.e., when t is close to τ) for a large class of time-varying systems. For a mathematical development of this, see [7].

As noted above, the exponential form in Equation 25.232 is valid for any system where $A(t)$ commutes with its integral (see Equation 25.50), in which case

$$\Gamma(t, \tau) = \int_\tau^t A(\sigma)d\sigma. \tag{25.233}$$

The class of systems for which $A(t)$ commutes with its integral is actually fairly large; in particular, this is the case for any $A(t)$ given by

$$A(t) = \sum_{i=1}^r f_i(t)A_i \tag{25.234}$$

where $f_i(t)$ are arbitrary real-valued functions of t and the A_i are arbitrary constant $n \times n$ matrices which satisfy the commutativity conditions

$$A_iA_j = A_jA_i, \quad \text{for all integers} \quad 1 \leq i,\ j \leq r \tag{25.235}$$

For example, suppose that

$$A(t) = \begin{bmatrix} f_1(t) & c_1f_2(t) \\ c_2f_2(t) & f_1(t) \end{bmatrix} \tag{25.236}$$

where $f_1(t)$ and $f_2(t)$ are arbitrary real-valued functions of t and c_1 and c_2 are arbitrary constants. Then

$$A(t) = f_1(t)A_1 + f_2(t)A_2 \quad (25.237)$$

where $A_1 = I$ and

$$A_2 = \begin{bmatrix} 0 & c_1 \\ c_2 & 0 \end{bmatrix} \quad (25.238)$$

Obviously, A_1 and A_2 commute, and thus $\Phi(t,\tau)$ is given by Equations 25.232 and 25.233. In this case, $\Phi(t,\tau)$ can be written in the form

$$\Phi(t,\tau) = \exp\left[\left(\int_\tau^t f_1(\sigma)d\sigma\right)I\right]\exp\left[\left(\int_\tau^t f_2(\sigma)d\sigma\right)A_2\right] \quad (25.239)$$

Given an n-dimensional system with exponential state-transition matrix $\Phi(t,\tau) = e^{\Gamma(t,\tau)}$, $\Phi(t,\tau)$ can be expressed in terms of scalar functions using the Laplace transform as in the time-invariant case. In particular, let

$$\Phi(t,\beta,\tau) = \text{inverse transform of } [sI - (1/\beta)\Gamma(\beta,\tau)]^{-1} \quad (25.240)$$

where $\Gamma(\beta,\tau) \doteq \Gamma(t,\tau)|_{t=\beta}$ and β is viewed as a parameter. Then

$$\Phi(t,\tau) = \Phi(t,\beta,\tau)|_{\beta=t} \quad (25.241)$$

For example, suppose that

$$A(t) = \begin{bmatrix} f_1(t) & f_2(t) \\ -f_2(t) & f_1(t) \end{bmatrix} \quad (25.242)$$

where $f_1(t)$ and $f_2(t)$ are arbitrary functions of t with the constraint that $f_2(t) \geq 0$ for all t. Then

$$\Gamma(t,\tau) = \int_\tau^t A(\sigma)d\sigma \quad (25.243)$$

and $\Phi(t,\beta,\tau) =$ inverse transform of

$$\begin{bmatrix} s-\gamma_1(\beta,\tau) & -\gamma_2(\beta,\tau) \\ \gamma_2(\beta\tau) & s-\gamma_1(\beta\tau) \end{bmatrix}^{-1} \quad (25.244)$$

where

$$\gamma_1(\beta,\tau) = (1/\beta)\int_\tau^\beta f_1(\sigma)d\sigma \quad (25.245)$$

$$\gamma_2(\beta,\tau) = (1/\beta)\int_\tau^\beta f_2(\sigma)d\sigma \quad (25.246)$$

Evaluating Equation 25.244 and using Equation 25.241 gives (for $t > 0$)

$$\Phi(t,\tau) = \begin{bmatrix} e^{\gamma_1(t,\tau)t}\cos[\gamma_2(t,\tau)t] & e^{\gamma_1(t,\tau)t}\sin[\gamma_2(t,\tau)t] \\ -e^{\gamma_1(t,\tau)t}\sin[\gamma_2(t,\tau)t] & e^{\gamma_1(t,\tau)t}\cos[\gamma_2(t,\tau)t] \end{bmatrix} \quad (25.247)$$

25.4.2 Stability

Again consider an n-dimensional exponential system $[A(t), B(t), C(t)]$ with state-transition matrix $\Phi(t,\tau) = e^{\Gamma(t,\tau)}$. A sufficient condition for the system to be uniformly exponentially stable is that the eigenvalues of the $n \times n$ matrix $(1/t)\Gamma(t,\tau)$ be bounded as functions of t and have real parts $\leq -\nu$ for all $t > \tau$ for some $\nu > 0$ and τ. For example, suppose that $A(t)$ is given by Equation 25.236 so that

$$(1/t)\Gamma(t,\tau) = \begin{bmatrix} \gamma_1(t,\tau) & c_1\gamma_2(t,\tau) \\ c_2\gamma_2(t,\tau) & \gamma_1(t,\tau) \end{bmatrix} \quad (25.248)$$

where $\gamma_1(t,\tau)$ and $\gamma_2(t,\tau)$ are given by Equations 25.245 and 25.246 with $\beta = t$. Then

$$\det[sI - (1/t)\Gamma(t,\tau)] = s^2 - 2\gamma_1(t,\tau)s + \gamma_1^2(t,\tau) - c_1c_2\gamma_2^2(t,\tau) \quad (25.249)$$

and the eigenvalues of $(1/t)\Gamma(t,\tau)$ have real parts $\leq -\nu$ for all $t > \tau$ for some $\nu > 0$ and τ if

$$\gamma_1(t,\tau) \leq \nu_1, \quad \text{for all} \quad t > \tau, \quad \text{for some} \quad \nu_1 < 0, \quad (25.250)$$

$$\gamma_1^2(t,\tau) - c_1c_2\gamma_2^2(t,\tau) \geq \nu_2 \text{ for all } t > \tau \text{ for some } \nu_2 > 0 \quad (25.251)$$

Therefore, if Equations 25.250–25.251 are satisfied and $\gamma_1(t,\tau)$ and $\gamma_2(t,\tau)$ are bounded functions of t, the system with $A(t)$ given by Equation 25.236 is uniformly exponentially stable.

For an n-dimensional system with state model $[A(t), B(t), C(t)]$, if $A(t)$ commutes with its integral, so that

$$(1/t)\Gamma(t,\tau) = (1/t)\int_\tau^t A(\sigma)d\sigma, \quad (25.252)$$

the above eigenvalue condition on $(1/t)\Gamma(t,\tau)$ is satisfied if the eigenvalues of $A(t)$ are bounded and are $\leq -\eta$ for all $t > \tau$, for some $\eta > 0$ and τ. Hence, a pointwise eigenvalue condition on $A(t)$ guarantees uniform exponential stability for systems where $A(t)$ commutes with its integral. Unfortunately this condition does not hold in general. For an example (taken from [8]), suppose that

$$A(t) = \begin{bmatrix} -1+\alpha(\cos^2 t) & 1-\alpha(\sin t)(\cos t) \\ -1-\alpha(\sin t)(\cos t) & -1+\alpha(\sin^2 t) \end{bmatrix} \quad (25.253)$$

where α is a real parameter. The eigenvalues of $A(t)$ are equal to

$$\frac{\alpha-2\pm\sqrt{\alpha^2-4}}{2} \quad (25.254)$$

which are strictly negative if $0 < \alpha < 2$. But

$$\Phi(t,0) = \begin{bmatrix} e^{(\alpha-1)t}(\cos t) & e^{-t}(\sin t) \\ -e^{(\alpha-1)t}(\sin t) & e^{-t}(\cos t) \end{bmatrix} \quad (25.255)$$

and thus the system is obviously not stable if $\alpha > 1$.

25.4.3 The Lyapunov Criterion

By using the Lyapunov criterion (see Equation 25.106), it is possible to derive sufficient conditions for uniform exponential stability without computing the state-transition matrix. For example, suppose that

$$A(t) = \begin{bmatrix} 0 & 1 \\ -1 & -a(t) \end{bmatrix} \tag{25.256}$$

where $a(t)$ is a real-valued function of t with $a(t) \geq c$ for all $t > t_1$, for some t_1 and some constant $c > 0$. Now in Equation 25.106, choose

$$Q(t) = \begin{bmatrix} a(t) + \frac{2}{a(t)} & 1 \\ 1 & \frac{2}{a(t)} \end{bmatrix} \tag{25.257}$$

Then, $c_1 I \leq Q(t) \leq c_2 I$, for all $t > t_1$ for some constants $c_1 > 0$ and $c_2 > 0$. Now

$$Q(t)A(t) + A^T(t)Q(t) + \dot{Q}(t) = \begin{bmatrix} -2 + \dot{a}(t) - \frac{\dot{a}(t)}{a^2(t)} & 0 \\ 0 & -1 - \frac{\dot{a}(t)}{a^2(t)} \end{bmatrix} \tag{25.258}$$

Hence if

$$-2 + \dot{a}(t) - \frac{\dot{a}(t)}{a^2(t)} \leq -c_3 \quad \text{for} \quad t > t_1 \quad \text{for some} \quad c_3 > 0 \tag{25.259}$$

and

$$-1 - \frac{\dot{a}(t)}{a^2(t)} \leq -c_4 \quad \text{for} \quad t > t_1 \quad \text{for some} \quad c_4 > 0, \tag{25.260}$$

the system is uniformly exponentially stable. For instance, if $a(t) = b - \cos t$, then Equations 25.259 and 25.260 are satisfied if $b > 2$, in which case the system is uniformly exponentially stable.

Now suppose that

$$A(t) = \begin{bmatrix} 0 & 1 \\ -a_1(t) & -a_2(t) \end{bmatrix} \tag{25.261}$$

As suggested on page 109 of [8], sufficient conditions for uniform exponential stability can be derived by taking

$$Q(t) = \begin{bmatrix} a_1(t) + a_2(t) + \frac{a_1(t)}{a_2(t)} & 1 \\ 1 & 1 + \frac{1}{a_2(t)} \end{bmatrix} \tag{25.262}$$

25.5 Defining Terms

State model: For linear time-varying systems, this is a mathematical representation of the system in terms of state equations of the form $\dot{x}(t) = A(t)x(t) + B(t)u(t)$, $y(t) = C(t)x(t) + D(t)u(t)$.

State-transition matrix: The matrix $\Phi(t, t_0)$ where $\Phi(t, t_0)x(t_0)$ is the state at time t starting with state $x(t_0)$ at time t_0 and with no input applied for $t \geq t_0$.

Exponential system: A system whose state-transition matrix $\Phi(t, \tau)$ can be written in the exponential form $e^{\Gamma(t,\tau)}$ for some $n \times n$ matrix function $\Gamma(t, \tau)$.

Reversible system: A system whose state-transition matrix is invertible.

Sampled data system: A discrete-time system generated by sampling the inputs and outputs of a continuous-time system.

Change of state: A transformation $z(t) = P(t)x(t)$ from the state vector $x(t)$ to the new state vector $z(t)$.

Algebraic equivalence: Refers to two state models of the same system related by a change of state.

Lyapunov transformation: A change of state $z(t) = P(t)x(t)$ where $P(t)$ and its inverse $P^{-1}(t)$ are both bounded functions of t.

Canonical form: A state model $[A(t), B(t), C(t)]$ with one or more of the coefficient matrices $A(t)$, $B(t)$, $C(t)$ in a special form.

Control canonical form: In the single-input case, a canonical form for $A(t)$ and $B(t)$ that facilitates the study of state feedback control.

Observer canonical form: In the single-output case, a canonical form for $A(t)$ and $C(t)$ that facilitates the design of a state observer.

Characteristic vector: A time-varying generalization corresponding to the vector of coefficients of the characteristic polynomial in the time-invariant case.

Asymptotic stability: Convergence to zero of the solution to $\dot{x}(t) = A(t)x(t)$ for any initial state $x(t_0)$.

Uniform exponential stability: Convergence to zero at an exponential rate of the solutions to $\dot{x}(t) = A(t)x(t)$.

Controllability: The existence of inputs that drive a system from any initial state to any desired state.

Observability: The ability to compute the initial state $x(t_0)$ from knowledge of the output response $y(t)$ for $t \geq t_0$.

State feedback control: A control signal of the form $u(t) = -F(t)x(t)$ where $F(t)$ is the gain matrix and $x(t)$ is the system state.

Observer: A system which provides an estimate $\hat{x}(t)$ of the state $x(t)$ of a system.

References

[1] Ball, J.A., Gohberg, I., and Kaashoek, M.A., Nevanlinna-Pick Interpolation for Time-Varying Input-Output Maps: The Discrete Case, *Operator Theory: Advances and Applications*, 56, 1–51, 1992.

[2] Ball, J.A., Gohberg, I., and Kaashoek, M.A., Nevanlinna-Pick Interpolation for Time-Varying Input-Output Maps: The Continuous Time Case, *Operator Theory: Advances and Applications*, 56, 52–89, 1992.

[3] Becker, G., Packard, A., Philbrick, D., and Balas, G., Control of Parametrically-Dependent Linear Systems: A Single Quadratic Lyapunov Approach, *Proc. Am. Control Conf.*, San Francisco, CA, 2795–2799, 1993.

[4] Kamen, E.W., The Poles and Zeros of a Linear Time-Varying System, *Linear Algebra and Its Applications*, 98, 263–289, 1988.

[5] Kamen, E.W. and Hafez, K.M., Algebraic Theory of Linear Time-Varying Systems, *SIAM J. Control and Optimiz.*, 17, 500–510, 1979.

[6] Kamen, E.W., Khargonekar, P.P., and Poolla, K.R., A Transfer Function Approach to Linear Time-Varying Discrete-Time Systems, *SIAM J. Control Optimiz.*, 23, 550–565, 1985.

[7] Magnus, W., On the Exponential Solution of Differential Equations for a Linear Operator, *Communications on Pure and Applied Mathematics*, VII, 649–673, 1954.

[8] Rugh, W.J., *Linear System Theory*, Second ed., Prentice Hall, Englewood Cliffs, NJ, 1996.

[9] Silverman, L.M. and Meadows, H.E., Controllability and Observability in Time-Variable Linear Systems, *SIAM J. Control Optimiz.*, 5, 64–73, 1967.

[10] Silverman, L.M., Transformation of Time-Variable Systems to Canonical (Phase-Variable) Form, *IEEE Trans. Automat. Control*, AC-11, 300, 1966.

[11] Shamma, J. and Athans, M., Guaranteed Properties of Gain Scheduled Control of Linear Parameter-Varying Plants, *Automatica*, 27, 559–564, 1991.

[12] Tadmor, G. and Verma, M., Factorization and the Nehari Theorem in Time-Varying Systems, *Math. Control, Signals, Syst.*, 5, 419–452, 1992.

[13] Zames, G. and Wang, L.Y., Local-Global Algebras for Slow H^{∞} Adaptation: Part I - Inversion and Stability, *IEEE Trans. Automat. Control*, 36, 130–142, 1991.

[14] Wang, L.Y. and Zames, G., Local-Global Algebras for Slow H^{∞} Adaptation: Part II - Optimization of Stable Plants, *IEEE Trans. Automat. Control*, 36, 143–151, 1991.

[15] Zhu, J.J., Well-Defined Series and Parallel D-Spectra for Linear Time-Varying Systems, *Proc. Am. Control Conf.*, Baltimore, MD, 1994, 734–738.

Further Reading

Two of the "classic texts" on linear time-varying systems are *Linear Time-Varying Systems* by H. D'Angelo (Allyn and Bacon, 1970) and *Finite Dimensional Linear Systems* by R. W. Brockett (Wiley, 1970). Another in-depth treatment of the time-varying case can be found in *Linear System Theory* by L. A. Zadeh and C. A. Desoer (McGraw-Hill, 1963). For more recent texts containing the theory of the time-varying case, see *Linear System Theory* by W. J. Rugh (Prentice Hall, 1996) and *Mathematical Control Theory* by E. D. Sontag (Springer-Verlag, 1990). For results on the adaptive control of time-varying systems, see *Linear Time-Varying Plants: Control and Adaptation* by K. S. Tsakalis and P. A. Ioannou (Prentice Hall, 1993). For a treatment focusing on the case of periodic coefficients, see *Analysis of Periodically Time-Varying Systems* by J. A. Richards (Springer-Verlag, 1983).

The theory of differential equations with time-varying coefficients is developed in *Ordinary Differential Equations* by R. K. Miller and A. N. Michel (Academic Press, 1982) and in *Stability of Differential Equations* by R. Bellman (McGraw-Hill, 1953). Textbooks focusing on the stability of systems including the time-varying case include *Stability of Motion* by W. Hahn (Springer-Verlag, 1967) and *Stability of Linear Systems* by C. J. Harris and J. F. Miles (Academic Press, 1980). Finally, for results on observers for time-varying systems, see *Observers for Linear Systems* by J. O'Reilly (Academic Press, 1983).

26

Geometric Theory of Linear Systems

Fumio Hamano

California State University, Long Beach

26.1 Introduction

In the late 1960s Basile and Marro [1] (and later Wonham and Morse, [10]) discovered that the behavior of time-invariant linear control systems could be seen as a manifestation of the subspaces similar to the invariant subspaces characterized by the system matrices. As a result, the system behavior could be predicted and the solvability of many control problems could be tested by examining the properties of such subspaces. In many instances one can understand essential issues intuitively in geometric terms. Moreover, thanks to good algorithms and software available in the literature (see [5]), the above subspaces can be generated and the properties can be readily examined by using personal computers. Thus, a large class of problems involving feedback control laws and observability of linear systems can be solved effectively by this geometric method, e.g., problems of disturbance localization, decoupling, unknown input observability and system inversion, observer design, regulation and tracking, robust control, etc. Comprehensive treatments of the basic theory and many applications, including the ones mentioned above, can be found in the excellent books by Basile and Marro [5] and Wonham [11]. The method is also useful in the analysis and design of decentralized control systems [6]. This chapter serves as an introduction to the subject. Extensive references can be found in the previously mentioned books. To prepare this chapter, Reference [5] has been used as the primary reference, and the majority of the proofs omitted in this chapter can be found in this reference.

Section 26.2 gives a review of elementary notions including invariant subspaces, reachability, controllability, observability, and detectability. It also provides convenient formulae for subspace calculations. Sections 26.3 through 26.7 describe the basic ingredients of the geometric theory (or approach). More specifically, Section 26.3 introduces the fundamental notions of $(A,\text{im}B)$-controlled and $(A,\text{ker}C)$-*conditioned invariants* (which are subspaces of the state space), and Section 26.4 provides some algebraic properties of these invariants. In Section 26.5 "largest" controlled and "smallest" conditioned invariants are given with respect to certain subspaces, and Section 26.6 discusses well-structured special classes of controlled and conditioned invariants. Section 26.7 analyzes the above invariants in relation to stabilization. Sections 26.8 through 26.10 describe applications to demonstrate the use of the basic tools developed in the previous sections. For this goal, the disturbance localization problem is chosen and it is discussed in three different situations with varying degrees of sophistication. The disturbance localization problems are chosen since the methods used to solve the problems can be used or extended to solve other more involved problems. It also has historical significance as one of the first problems for which the geometric method was used.

Notation: Capital letters A, B, etc. denote the matrices (or linear maps) with I and I_n reserved, respectively, for an identity matrix (of appropriate dimension) and an $n \times n$ identity matrix. Capital script letters such as $\mathcal{V}$, $\mathcal{W}$ represent vector spaces or subspaces. Small letters x, y, etc. are column vectors (or vectors in given vector spaces). Scalars are also denoted by small letters. The number 0 is used for a zero matrix, vector or scalar depending on the context. Notation ":=" means "(the left hand side, i.e., ":" side) is defined by (the right hand side, i.e., "="

0-8493-8570-9/96/$0.00+$.50

side)". Similarly for "=:" where the roles of the left and right hand sides are reversed. The image (or range) and the kernel (or null space) of M are respectively denoted by im$M$ and ker$M$. The expression $\mathcal{V}+\mathcal{W}$ represents the sum of two subspaces $\mathcal{V}$ and $\mathcal{W}$, i.e., $\mathcal{V}+\mathcal{W} := \{v+w : v \in \mathcal{V} \text{ and } w \in \mathcal{W}\}$. If $\mathcal{V}$ is a subspace of $\mathcal{W}$, we write $\mathcal{V} \subset \mathcal{W}$. If $\mathcal{V} \subset \mathcal{X}$, we use $A^{-1}\mathcal{V} := \{x \in \mathcal{X} : Ax \in \mathcal{V}\}$, i.e., the set of all $x \in \mathcal{X}$ satisfying $Ax \in \mathcal{V}$. Similarly, $A^{-k}\mathcal{V} := \{x \in \mathcal{X} : A^k x \in \mathcal{V}\}, k = 1, 2, \cdots$.

26.2 Review of Elementary Notions

In this section we will review invariant subspaces and some of their basic roles in the context of the linear systems.

DEFINITION 26.1 A subspace $\mathcal{V}$ of $\mathcal{X} := R^n$ is said to be *A-invariant* if

$$A\mathcal{V} \subset \mathcal{V}. \tag{26.1}$$

An A-invariant subspace plays the following obvious but important role for a free linear system. Consider the free linear system described by

$$\dot{x}(t) = Ax(t), x(0) = x_0 \tag{26.2}$$

where the column vectors $x(t) \in \mathcal{X} := R^n$ and $x_0 \in \mathcal{X}$ are, respectively, the state of the system at time $t \geq 0$ and the initial state, and A is an $n \times n$ real matrix. Now, suppose a subspace $\mathcal{V}$ is A-invariant. Clearly, if $x(t) \in \mathcal{V}$, then the rate of change $\dot{x}(t) \in \mathcal{V}$, which implies that the state remains in $\mathcal{V}$. More strongly, we have

LEMMA 26.1 Let $\mathcal{V} \subset \mathcal{X}$. For the free linear system (26.2), $x_0 \in \mathcal{V}$ implies $x(t) \in \mathcal{V}$ for all $t \geq 0$ if and only if $\mathcal{V}$ is A-invariant.

Let $\mathcal{V}$ be A-invariant, and introduce a new basis $\{e_1, \cdots, e_n\}$ such that

$$span\{e_1, \cdots, e_\nu\} = \mathcal{V}, \quad \nu \leq n, \tag{26.3}$$

and define a coordinate transformation by

$$x = [e_1 \ldots e_n] \begin{bmatrix} \tilde{x}_1 \\ \tilde{x}_2 \end{bmatrix}, x \in R^n, \ \tilde{x}_1 \in R^\nu, \ \tilde{x}_2 \in R^{n-\nu}. \tag{26.4}$$

Then, it is easy to see that, with respect to the new basis, the state equation (26.2) can be rewritten as

$$\begin{bmatrix} \dot{\tilde{x}}_1(t) \\ \dot{\tilde{x}}_2(t) \end{bmatrix} = \begin{bmatrix} \tilde{A}_{11} & \tilde{A}_{12} \\ 0 & \tilde{A}_{22} \end{bmatrix} \begin{bmatrix} \tilde{x}_1(t) \\ \tilde{x}_2(t) \end{bmatrix}, \tag{26.5}$$

$$\tilde{x}_1(0) = \tilde{x}_{10}, \ \tilde{x}_2(0) = \tilde{x}_{20}. \tag{26.6}$$

Clearly, if $\tilde{x}_{20} = 0$, then $\tilde{x}_2(t) = 0$ for all $t \geq 0$, i.e., $x_0 \in \mathcal{V}$ implies $x(t) \in \mathcal{V}$ for all $t \geq 0$ (which has been stated in Lemma 26.1).

Let $\mathcal{V}$ be an A-invariant subspace in $\mathcal{X}$. The *restriction* $A|\mathcal{V}$ of a linear map $A : \mathcal{X} \rightarrow \mathcal{X}$ (or $n \times n$ real matrix A) to a subspace $\mathcal{V}$ is a linear map from $\mathcal{V}$ to $\mathcal{V}$ mapping $v \mapsto Av$ for all $v \in \mathcal{V}$. For $x \in \mathcal{X}$, we write $x + \mathcal{V} := \{x + v : v \in \mathcal{V}\}$ called the *coset* of x modulo $\mathcal{V}$, which represents a hyper plane passing through a point x. The set of cosets modulo $\mathcal{V}$ is a vector space called the *factor space* (or *quotient space*) and it is denoted by $\mathcal{X}/\mathcal{V}$. An *induced map* $A|\mathcal{X}/\mathcal{V}$ is a linear map defined by $x + \mathcal{V} \mapsto Ax + \mathcal{V}, x \in \mathcal{X}$.

An A-invariant subspace $\mathcal{V}$ is said to be *internally stable* if $\tilde{A}_{11}$ in Equation 26.6 is stable (i.e., all the eigenvalues have negative real parts), or equivalently, if $A|\mathcal{V}$ is stable. Therefore, $x(t)$ converges to the zero state as $t \rightarrow \infty$ whenever $x_0 \in \mathcal{V}$ if and only if $\mathcal{V}$ is internally stable. Also, an A-invariant subspace $\mathcal{V}$ is said to be *externally stable* if $\tilde{A}_{22}$ is stable, i.e., if $A|\mathcal{X}/\mathcal{V}$ is stable. Clearly, $x_2(t)$ converges to zero as $t \rightarrow \infty$, i.e., $x(t)$ converges to $\mathcal{V}$ as $t \rightarrow \infty$ if and only if $\mathcal{V}$ is externally stable. Note that the eigenvalues of $\tilde{A}_{11}$ and $\tilde{A}_{22}$ do not depend on a particular choice of coordinates (as long as Equation 26.3 is satisfied).

Let us now consider a continuous time, time-invariant linear control system $\sum := [A, B, C]$ described by

$$\dot{x}(t) = Ax(t) + Bu(t), \ x(0) = x_0, \tag{26.7}$$

$$y(t) = Cx(t) \tag{26.8}$$

where the column vectors $x(t) \in \mathcal{X} := R^n, u(t) \in R^m$ and $y(t) \in R^p$ are, respectively, the state, input and output of the system at time $t \geq 0$, $x_0 \in R^n$ is the initial state, and A, B, C, and D are real matrices with consistent dimensions. We assume that $u(t)$ is piecewise continuous. We will also be interested in the closed loop system, namely, we apply a linear state feedback law of the form

$$u(t) = Fx(t), \tag{26.9}$$

then Equation 26.7 becomes

$$\dot{x}(t) = (A + BF)x(t), \ x(0) = x_0 \tag{26.10}$$

where F is an $m \times n$ real matrix.

Some invariant subspaces are associated with reachability (controllability) and observability. A state $\tilde{x}$ is said to be *reachable* (or *controllable*) if there is a control which drives the zero state to $\tilde{x}$ (or, respectively, $\tilde{x}$ to the zero state) in finite time, i.e., if there is $u(t), 0 \leq t \leq t_f$ such that $x(0) = 0$ (or, respectively, $\tilde{x}$) and $x(t_f) = \tilde{x}$ (or, respectively 0) for some $0 < t_f < \infty$. The set of reachable (or controllable) states forms a subspace and it is called the *reachable* (or, respectively, *controllable*) *subspace*, which will be denoted as $\mathcal{V}_{\text{reach}}$ (or, respectively, $\mathcal{V}_{contr}$). For an $n \times n$ matrix M and a subspace $\mathcal{I} \subset \mathcal{X}$, define

$$\mathcal{R}(M, \mathcal{I}) := \mathcal{I} + M\mathcal{I} + \cdots + M^{n-1}\mathcal{I}.$$

The reachable and controllable subspaces are characterized by the following:

THEOREM 26.1 *For the continuous-time system* $\sum := [A, B, C]$,

$$\begin{aligned} \mathcal{V}_{\text{reach}} &= \mathcal{R}(A, \text{im}B) = \text{im}[B \ AB \ldots A^{n-1} B] \\ &= \mathcal{V}_{contr}. \end{aligned} \tag{26.11}$$

REMARK 26.1 The subspace $\mathcal{V}_{reach} = \mathcal{R}(A, \text{im}B)$ is A-invariant. It is also $(A+BF)$-invariant for any $m \times n$ real matrix F.

The pair (A, B) or system $\sum := [A, B, C]$ is said to be *reachable* (or *controllable*) if $\mathcal{V}_{reach} = \mathcal{X}$ (or, respectively, $\mathcal{V}_{contr} = \mathcal{X}$). The set $\Lambda := \{\lambda_1, \cdots, \lambda_n\}$ of complex numbers is called a *symmetric* set if, whenever λ_i is not a real number, $\lambda_j = \lambda_i^*$ for some $j = 1, \cdots, n$ where λ_i^* is the complex conjugate of λ_i. Denote by $\sigma(A+BF)$ the spectrum (or the eigenvalues) of $A+BF$. Then, we have

THEOREM 26.2 *For any symmetric set $\Lambda := \{\lambda_1, \cdots, \lambda_n\}$ of complex numbers $\lambda_1, \cdots, \lambda_n$, there is an $m \times n$ real matrix F such that $\sigma(A+BF) = \Lambda$ if and only if the pair (A, B) is reachable (or controllable).*

PROOF 26.1 See [5] and [11].

REMARK 26.2 Let $\dim\mathcal{V}_{reach} = r \leq n$. For any symmetric set $\Lambda := \{\lambda_1, \cdots, \lambda_r\}$ of complex numbers $\lambda_1, \cdots, \lambda_r$, there is an $m \times n$ real matrix F such that $\sigma(A+BF|\mathcal{V}_{reach}) = \Lambda$. This can be seen by applying a coordinate transformation utilizing the A-invariance of $\mathcal{V}_{reach}$. (See Equation 26.5.)

The pair (A, B) is said to be *stabilizable* if there is a real matrix F such that the eigenvalues of $A+BF$ have negative real parts. We have (see [5])

COROLLARY 26.1 Pair (A, B) is stabilizable if and only if $\mathcal{V}_{reach}$ is externally stable.

The state $\tilde{x}$ of the system $\sum$ is said to be *unobservable* if it produces zero output when the input is not applied, i.e., if $x(0) = \tilde{x}$ and $u(t) = 0$ for all $t \geq 0$ implies $y(t) = 0$ for all $t \geq 0$. The set of unobservable states forms a subspace which is called the *unobservable subspace.* This will be denoted by $\mathcal{V}_{unobs}$.

THEOREM 26.3

$$\begin{aligned}\mathcal{V}_{unobs} &= ker\begin{bmatrix} C \\ CA \\ \vdots \\ CA^{n-1}\end{bmatrix} \\ &= KerC \cap A^{-1}kerC \cap \ldots \\ &\quad \cap A^{-(n-1)}kerC. \qquad (26.12)\end{aligned}$$

REMARK 26.3 The subspace $\mathcal{V}_{unobs}$ is A-invariant. It is also $(A+GC)$-invariant for any $n \times p$ real matrix G.

The pair (A, C) is said to be *observable* if $\mathcal{V}_{unobs} = 0$, and the pair (A, C) is said to be *detectable* if $A+GC$ is stable for some real matrix G. For observability and detectability, we have the following facts.

THEOREM 26.4 *For any symmetric set $\Lambda := \{\lambda_1, \cdots, \lambda_n\}$ of complex numbers $\lambda_1, \cdots, \lambda_n$, there is an $n \times p$ real matrix G such that $\sigma(A+GC) = \Lambda$ if and only if the pair (A, C) is observable.*

COROLLARY 26.2 Pair (A, C) is detectable if and only if $\mathcal{V}_{unobs}$ is internally stable.

The following formula is useful for subspace calculations.

LEMMA 26.2 Let $\mathcal{V}, \mathcal{V}_1, \mathcal{V}_2, \mathcal{V}_3 \subset \mathcal{X}$. Then, letting $\mathcal{V}^{\perp} := \{x \in \mathcal{X} : x'v = 0 \text{ for all } v \in \mathcal{V}\}$,

$$\begin{aligned}
&(\mathcal{V}^{\perp})^{\perp} = \mathcal{V}, \\
&(\mathcal{V}_1 + \mathcal{V}_2)^{\perp} = \mathcal{V}_1^{\perp} \cap \mathcal{V}2^{\perp}, \\
&(\mathcal{V}_1 \cap \mathcal{V}_2)^{\perp} = \mathcal{V}_1^{\perp} + \mathcal{V}_2^{\perp}, \\
&A(\mathcal{V}_1 + \mathcal{V}_2) = A\mathcal{V}_1 + A\mathcal{V}_2, \\
&A(\mathcal{V}_1 \cap \mathcal{V}_2) \subset A\mathcal{V}_1 \cap A\mathcal{V}_2, \\
&(A_1 + A_2)\mathcal{V} = A_1\mathcal{V} + A_2\mathcal{V} \text{ where } A_1 \text{ and } A_2 \text{ are} \\
&n \times n \text{ matrices}, \\
&(A\mathcal{V})^{\perp} = A'^{-1}\mathcal{V}^{\perp}, \\
&\mathcal{V}_1 + (\mathcal{V}_2 \cap \mathcal{V}_3) \subset (\mathcal{V}_1 + \mathcal{V}_2) \cap (\mathcal{V}_1 + \mathcal{V}_3), \\
&\mathcal{V}_1 \cap (\mathcal{V}_2 + \mathcal{V}_3) \supset (\mathcal{V}_1 \cap \mathcal{V}_2) + (\mathcal{V}_1 \cap \mathcal{V}_3), \\
&\mathcal{V}_1 \cap (\mathcal{V}_2 + \mathcal{V}_3) = \mathcal{V}_2 + \mathcal{V}_1 \cap \mathcal{V}_3 \text{ provided } \mathcal{V}_1 \supset \mathcal{V}_2.
\end{aligned}$$

26.3 (A,imB)-Controlled and (A,kerC)-Conditioned Invariant Subspaces and Duality

In this section we introduce important subspaces associated with system $\sum := [A, B, C]$ described by Equation 26.7 (or Equation 26.10) and Equation 26.8 with a state feedback law (Equation 26.9). According to Lemma 26.1, for a free linear system (Equation 26.2), an A-invariant subspace is a subspace having the property that any state trajectory starting in the subspace remains in it. However, for a linear system $\sum := [A, B, C]$ with input, A-invariance is not necessary in order for a subspace to have the above trajectory confinement property. In fact, let $\mathcal{V} \subset \mathcal{X}$. A state trajectory can be confined in $\mathcal{V}$ if and only if $\dot{x}(t) \in \mathcal{V}$, and to produce $\dot{x}(t) \in \mathcal{V}$ whenever $x(t) \in \mathcal{V}$, we need $\dot{x}(t) = Ax(t) + Bu(t) = v(t) \in \mathcal{V}$, i.e., $Ax(t) = v(t) - Bu(t)$ for some $u(t)$ and $v(t)$, which implies $A\mathcal{V} \subset \mathcal{V} + \text{im}B$. The converse also holds. Summarizing, we have

LEMMA 26.3 Consider the system described by Equation 26.7. For each initial state $x_0 \in \mathcal{V}$, there is an (admissible) input $u(t), t \geq 0$ such that the corresponding $x(t) \in \mathcal{V}$ for all $t \geq 0$ if and only if

$$A\mathcal{V} \subset \mathcal{V} + imB. \qquad (26.13)$$

Subspaces satisfying Equation 26.13 play a fundamental role in the geometric approach, and we will introduce

DEFINITION 26.2 A subspace $\mathcal{V}$ is said to be an $(A, \text{im}B)$-*controlled invariant* (*subspace*) (see [1] and [5]) or an (A, B)-*invariant subspace* (see [10] and [11]) if it is A-invariant modulo $\text{im}B$, i.e., if Equation 26.13 holds.

An important property of the above subspace is described by

THEOREM 26.5 *Let $\mathcal{V} \subset \mathcal{X}$. Then, there exists an $m \times n$ real matrix F such that*

$$(A + BF)\mathcal{V} \subset \mathcal{V} \tag{26.14}$$

if and only if $\mathcal{V}$ is an $(A, \text{im}B)$-controlled invariant.

REMARK 26.4 If the state feedback control law (Equation 26.9) is applied to the system $\sum := [A, B, C]$, the corresponding state equation is (Equation 26.10). Therefore, recalling Lemma 26.1, if $\mathcal{V}$ is an $(A, \text{im}B)$-controlled invariant, then there is an F for Equation 26.10 such that $x(t) \in \mathcal{V}$ for all $t \geq 0$ provided $x_0 \in \mathcal{V}$.

Another class of important subspaces is now introduced.

DEFINITION 26.3 A subspace $\mathcal{S}$ of $\mathcal{X}$ is said to be an $(A, \ker C)$-*conditioned invariant* (subspace) if

$$A(\mathcal{S} \cap \ker C) \subset \mathcal{S}. \tag{26.15}$$

There is a duality between controlled invariants and conditioned invariants in the following sense. By taking the orthogonal complements of the quantities on both sides of Equation 26.15, we see that Equation 26.15 is equivalent to $\{A(\mathcal{S} \cap \ker C)\}^{\perp} \supset \mathcal{S}^{\perp}$, which, in turn, is equivalent to $A'^{-1}(\mathcal{S}^{\perp} + \text{im}C') \supset \mathcal{S}^{\perp}$, i.e., $A'\mathcal{S}^{\perp} \subset \mathcal{S}^{\perp} + \text{im}C'$. Similarly, Equation 26.13 holds if and only if $A'(\mathcal{V}^{\perp} \cap \ker B') \subset \mathcal{V}^{\perp}$. Thus, we have

LEMMA 26.4 A subspace $\mathcal{S}$ is an $(A, \ker C)$-conditioned invariant if and only if $\mathcal{S}^{\perp}$ is an $(A', \text{im}C')$-controlled invariant. Also, a subspace $\mathcal{V}$ is an $(A, \text{im}B)$-controlled invariant if and only if $\mathcal{V}^{\perp}$ is an $(A', \ker C')$-conditioned invariant.

Due to Lemma 26.4, the previous theorem can be translated easily into the following property.

THEOREM 26.6 *Let $\mathcal{S} \subset \mathcal{X}$. Then, there exists an $n \times p$ real matrix G satisfying*

$$(A + GC)\mathcal{S} \subset \mathcal{S} \tag{26.16}$$

if and only if $\mathcal{S}$ is an $(A, \ker C)$-conditioned invariant.

A subspace may be both controlled and conditioned invariant. In such a case we have the following ([2], Section 5.1.1 of [5] and [6]):

LEMMA 26.5 There exists an $m \times p$ real matrix K such that

$$(A + BKC)\mathcal{V} \subset \mathcal{V} \tag{26.17}$$

if and only if $\mathcal{V}$ is both an $(A, \text{im}B)$-controlled invariant and an $(A, \ker C)$-conditioned invariant.

PROOF 26.2 [Only if part]: The controlled invariance and conditioned invariance follow trivially by Theorems 26.5 and 26.6, respectively.

[If part] : For $C = 0$, $\ker C = 0$ or $\mathcal{V} = 0$, the statement of the lemma trivially holds. So, we will prove the lemma assuming such trivialities do not occur. Since by assumption $\mathcal{V}$ is an $(A, \text{im}B)$-controlled invariant, there is an $m \times n$ real matrix $\bar{F}$ satisfying $(A + B\bar{F})\mathcal{V} \subset \mathcal{V}$. Let $\{v_1 \dots v_\mu\}$ be a basis of $\mathcal{V} \cap \ker C$. Complete the basis $\{v_1 \dots v_\mu \dots v_\upsilon \dots v_n\}$ of $\mathcal{X}$ in such a way that $\{v_1 \dots v_\mu \dots v_\upsilon\}$ is a basis of $\mathcal{V}$ where $1 \leq \mu \leq \upsilon \leq n$. Define an $m \times n$ real matrix $\widetilde{F}$ by

$$\widetilde{F}_{v_i} = \begin{cases} \bar{F} v_i, & i = \mu + 1, \dots, \upsilon \\ 0, & i = 1, \dots, \mu \\ \text{arbitrary}, & \text{otherwise} \end{cases} .$$

Choose K so that

$$\widetilde{F} = KC, \text{ i.e., } \widetilde{F}\left[v_{\mu+1} \dots v_\upsilon\right] = K\left[Cv_{\mu+1} \dots C_{v_\upsilon}\right].$$

Note that due to the particular choice of basis the columns of $\left[Cv_{\mu+1} \dots C_{v_\upsilon}\right]$ are linearly independent, and so the above K certainly exists.

26.4 Algebraic Properties of Controlled and Conditioned Invariants

An $(A, \text{im}B)$-controlled invariant has the following properties in addition to the ones discussed in the previous section. The proofs are omitted. They can be found in Chapter 4 of Reference [5].

LEMMA 26.6 If $\mathcal{V}_1$ and $\mathcal{V}_2$ are $(A, \text{im}B)$-controlled invariants, so is $\mathcal{V}_1 + \mathcal{V}_2$.

REMARK 26.5 But, the intersection of two $(A, \text{im}B)$-controlled invariants is not in general an $(A, \text{im}B)$-controlled invariant.

LEMMA 26.7 Let $\mathcal{V}_1$ and $\mathcal{V}_2$ be $(A, \text{im}B)$-controlled invariants. Then, there is an $m \times n$ real matrix F satisfying

$$(A + BF)\mathcal{V}_i \subset \mathcal{V}_i, \quad i = 1, 2 \tag{26.18}$$

if and only if $\mathcal{V}_1 \cap \mathcal{V}_2$ is an $(A, \text{im}B)$-controlled invariant.

By duality, we have

LEMMA 26.8 If $\mathcal{S}_1$ and $\mathcal{S}_2$ are $(A,\ker C)$-*conditioned invariants*, then so is $\mathcal{S}_1 \cap \mathcal{S}_2$.

REMARK 26.6 $\mathcal{S}_1 + \mathcal{S}_2$ is not necessarily an $(A,\ker C)$-conditioned invariant.

LEMMA 26.9 Let $\mathcal{S}_1$ and $\mathcal{S}_2$ be $(A,\ker C)$-conditioned invariants. Then, there is an $n \times p$ real matrix G satisfying

$$(A + GC)\mathcal{S}_i \subset \mathcal{S}_i, \; i = 1, 2 \tag{26.19}$$

if and only if $\mathcal{S}_1 + \mathcal{S}_2$ is a conditioned invariant.

26.5 Maximum Controlled and Minimum Conditioned Invariants

Let $\mathcal{K} \subset \mathcal{X}$, and consider the set of $(A,\text{im}B)$-controlled invariants contained in $\mathcal{K}$ (by subspace inclusion). Lemma 26.6 states that the set of $(A,\text{im}B)$-controlled invariants is closed under subspace addition. As a result, the set of $(A,\text{im}B)$-controlled invariants contained in $\mathcal{K}$ has a largest element or supremum. This element is a unique subspace that contains (by subspace inclusion) any other $(A,\text{im}B)$-controlled invariants contained in $\mathcal{K}$, and is called the *maximum* (or *supremum*) $(A,\text{im}B)$-*controlled invariant* contained in $\mathcal{K}$. This will be denoted in the sequel as $\mathcal{V}_{\max}(A,\text{im}B, \mathcal{K})$. Similarly, let $\mathcal{I} \subset \mathcal{X}$. Then, owing to Lemma 26.8, it can be shown that the set of $(A, \ker C)$-conditioned invariants containing $\mathcal{I}$ has a smallest element or infimum. This is a unique $(A, \ker C)$-conditioned invariant containing all other $(A, \ker C)$-conditioned invariants containing $\mathcal{I}$, and is called the *minimum* (or *infimum*) $(A, \ker C)$-*conditioned invariant* containing $\mathcal{I}$. This subspace will be denoted as $\mathcal{S}_{\min}(A,\ker C, \mathcal{I})$. The subspaces $\mathcal{V}_{\max}(A,\text{im}B, \mathcal{K})$ and $\mathcal{S}_{\min}(A,\ker C, \mathcal{I})$ are important because they can be computed in a finite number of steps (in at most n iterations) and because testing the solvability of control problems typically reduces to checking the conditions involving these subspaces. The geometric algorithms to compute $\mathcal{V}_{\max}(A,\text{im}B, \mathcal{K})$ and $\mathcal{S}_{\min}(A, B, \mathcal{I})$ are given below.

Algorithm to compute $\mathcal{V}_{\max}(A,\text{im}B, \mathcal{K})$:

$$\mathcal{V}_{\max}(A, \text{im}B, \mathcal{K}) = \mathcal{V}_{\dim \mathcal{K}} \tag{26.20a}$$

where

$$\mathcal{V}_0 := \mathcal{K} \tag{26.20b}$$

$$\mathcal{V}_i := \mathcal{K} \cap A^{-1}(\mathcal{V}_{i-1} + \text{im}B), \quad i = 1, \ldots, \dim \mathcal{K}. \tag{26.20c}$$

PROOF 26.3 See [5] or [11].

REMARK 26.7 The algorithm (Equations 26.20a to 26.20c) has the following properties:

1. $\mathcal{V}_1 \supset \mathcal{V}_2 \supset \ldots \supset \mathcal{V}_{\dim \mathcal{K}}$.
2. If $\mathcal{V}_\ell = \mathcal{V}_{\ell+1}$, then $\mathcal{V}_\ell = \mathcal{V}_{\ell+1} = \ldots = \mathcal{V}_{\dim \mathcal{K}}$.

REMARK 26.8 For an algorithm to compute F such that $(A + BF)\mathcal{V}_{\max}(A,\text{im}B, \mathcal{K}) \subset \mathcal{V}_{\max}(A,\text{im}B, \mathcal{K})$, see [5].

The following is the dual of the above algorithm (see [5]).

Algorithm to calculate $\mathcal{S}_{\min}(A,\ker C, \mathcal{I})$:

$$\mathcal{S}_{\min}(A, \ker C, \mathcal{I}) = \mathcal{S}_{n-\dim \mathcal{I}} \tag{26.21a}$$

where

$$\mathcal{S}_0 := \mathcal{I} \tag{26.21b}$$

$$\mathcal{S}_i := \mathcal{I} + A(\mathcal{S}_{i-1} \cap \ker C), \quad i = 1, \cdots, \text{n-}\dim \mathcal{I}. \tag{26.21c}$$

REMARK 26.9 The algorithm generates a monotonically nondecreasing sequence:

1. $\mathcal{S}_1 \subset \mathcal{S}_2 \subset \ldots \subset \mathcal{S}_{\text{n-}\dim \mathcal{I}}$,
2. If $\mathcal{S}_\ell = \mathcal{S}_{\ell+1}$, then $\mathcal{S}_\ell = \mathcal{S}_{\ell+1} = \ldots = \mathcal{S}_{\text{n-}\dim \mathcal{I}}$.

26.6 Self-Bounded (A, im B)-Controlled Invariant and (A, ker C)-Conditioned Invariant and Constrained Reachability and Observability

Let $\mathcal{V}_0$ be an $(A,\text{im}B)$-controlled invariant contained in $\mathcal{K}$, and consider all the possible state trajectories (with different controls) starting at x_0 in $\mathcal{V}_0$ and confined in $\mathcal{K}$. We know that there is at least one control for which the state trajectory remains in $\mathcal{V}_0$, but that there may be another control for which the state trajectory goes out of $\mathcal{V}_0$ while remaining in $\mathcal{K}$. However, some $(A,\text{im}B)$-controlled invariant contained in $\mathcal{K}$, say $\mathcal{V}$, has a stronger property that, for any initial state in $\mathcal{V}$, there is no control that drives the state (initially in $\mathcal{V}$) out of $\mathcal{V}$ while maintaining the state trajectory in $\mathcal{K}$, i.e., the state trajectory must go out of $\mathcal{K}$ if it ever goes out of $\mathcal{V} \subset \mathcal{K}$. Such an $(A,\text{im}B)$-controlled invariant contained in $\mathcal{K}$ is characterized by

$$\mathcal{V}_{\max}(A, \text{im}B, \mathcal{K}) \cap \text{im}B \subset \mathcal{V}, \tag{26.22}$$

and we have

DEFINITION 26.4 An $(A,\text{im}B)$-controlled invariant $\mathcal{V}$ contained in $\mathcal{K}$ is said to be *self-bounded* with respect to $\mathcal{K}$ if Equation 26.22 holds.

REMARK 26.10 The left-hand side of inclusion 26.22 represents the set of all possible influences of control on the state velocity at each instant of time that do not pull the state out of $\mathcal{K}$. Self-bounded $(A, \mathrm{im}B)$-controlled invariants have the following properties.

For each $\mathcal{K}$ we can choose a single state feedback control law which works for all the self-bounded $(A, \mathrm{im}B)$-controlled invariants with respect to $\mathcal{K}$. More precisely,

LEMMA 26.10 Let F be an $m \times n$ real matrix satisfying

$$(A + BF)\mathcal{V}_{\max}(A, \mathrm{im}B, \mathcal{K}) \subset \mathcal{V}_{\max}(A, \mathrm{im}B, \mathcal{K}). \quad (26.23)$$

Then, any self-bounded $(A, \mathrm{im}B)$-controlled invariant $\mathcal{V}$ with respect to $\mathcal{K}$ satisfies

$$(A + BF)\mathcal{V} \subset \mathcal{V}. \quad (26.24)$$

PROOF 26.4 See [5].

It can be shown that the set of self-bounded $(A, \mathrm{im}B)$-controlled invariants in $\mathcal{K}$ is closed under subspace intersection, i.e., if $\mathcal{V}_1$ and $\mathcal{V}_2$ are self-bounded $(A, \mathrm{im}B)$-controlled invariants with respect to $\mathcal{K}$, so is $\mathcal{V}_1 \cap \mathcal{V}_2$. Therefore, the above set has a minimum element which is called the *minimum self-bounded $(A, \mathrm{im}B)$-controlled invariant,* denoted in this chapter by $\mathcal{V}_{\mathrm{sb,min}}(A, \mathrm{im}B, \mathcal{K})$. The subspace $\mathcal{V}_{\mathrm{sb,min}}(A, \mathrm{im}B, \mathcal{K})$ is related to $\mathcal{V}_{\max}(A, \mathrm{im}B, \mathcal{K})$ and $\mathcal{S}_{\min}(A, \mathcal{K}, \mathrm{im}\,B)$ as follows

THEOREM 26.7

$$\mathcal{V}_{\mathrm{sb,min}}(A, imB, \mathcal{K}) = \mathcal{V}_{\max}(A, imB, \mathcal{K}) \cap \mathcal{S}_{\min}(A, \mathcal{K}, imB). \quad (26.25)$$

PROOF 26.5 See Section 4.1.2 of [5] and [7].

The minimum self-bounded $(A, \mathrm{im}B)$-controlled invariant is closely related to constrained reachability. The set of all states that can be reached from the zero state through state trajectories constrained in $\mathcal{K}$ is called the *reachable set* (or *reachable subspace,* since the set forms a subspace) *in* $\mathcal{K}$, or the *supremum* $(A, \mathrm{im}B)$*-reachability subspace contained in* $\mathcal{K}$. This set will be denoted by $\mathcal{V}_{\mathrm{reach}}(\mathcal{K})$. It is an $(A, \mathrm{im}B)$-controlled invariant satisfying the following properties.

LEMMA 26.11

$$\mathcal{V}_{\mathrm{reach}}(\mathcal{K}) \subset \mathcal{V}_{\max}(A, \mathrm{im}B, \mathcal{K}) \subset \mathcal{K}, \quad (26.26)$$

$$\mathcal{V}_{\mathrm{reach}}(\mathcal{V}_{\max}(A, \mathrm{im}B, \mathcal{K})) = \mathcal{V}_{\mathrm{reach}}(\mathcal{K}). \quad (26.27)$$

THEOREM 26.8 *Let F be a real matrix satisfying $(A + BF)\mathcal{V}_{\max} \subset \mathcal{V}_{\max}$ where $\mathcal{V}_{\max} := \mathcal{V}_{\max}(A, imB, \mathcal{K})$. Then,*

$$\mathcal{V}_{\mathrm{reach}}(\mathcal{K}) = \mathcal{V}_{sb,\min}(A, imB, \mathcal{K}) = \mathcal{R}(A + BF, imB \cap \mathcal{V}_{\max}). \quad (26.28)$$

REMARK 26.11 The last expression in Equation 26.28 represents the smallest $A + BF$ invariant subspace containing $\mathrm{im}B \cap \mathcal{V}_{\max}$.

Dual to the above results we have

DEFINITION 26.5 An $(A, \mathrm{ker}C)$-conditioned invariant $\mathcal{S}$ containing $\mathcal{I}$ is said to be *self-hidden with respect to $\mathcal{I}$* if

$$\mathcal{S} \subset \mathcal{S}_{\min}(A, \mathrm{ker}C, \mathcal{I}) + \mathrm{ker}C. \quad (26.29)$$

LEMMA 26.12 Let G be an $n \times p$ real matrix satisfying

$$(A + GC)\mathcal{S}_{\min}(A, \mathrm{ker}C, \mathcal{I}) \subset \mathcal{S}_{\min}(A, \mathrm{ker}C, \mathcal{I}). \quad (26.30)$$

Then, any $(A, \mathrm{ker}C)$-conditioned invariant $\mathcal{S}$ (containing $\mathcal{I}$) self-hidden with respect to $\mathcal{I}$ satisfies

$$(A + GC)\mathcal{S} \subset \mathcal{S}. \quad (26.31)$$

If $(A, \mathrm{ker}C)$-conditioned invariants $\mathcal{S}_1$ and $\mathcal{S}_2$ containing $\mathcal{I}$ are self-hidden with respect to $\mathcal{I}$, so is $\mathcal{S}_1 + \mathcal{S}_2$. Therefore, the above set has a maximum element which is called *the maximum self-hidden $(A, \mathrm{ker}C)$-conditioned invariant with respect to $\mathcal{I}$* denoted by $\mathcal{S}_{\mathrm{sh,max}}(A, \mathrm{ker}C, \mathcal{I})$. The subspace $\mathcal{S}_{\mathrm{sh,max}}(A, \mathrm{ker}C, \mathcal{I})$ is related to $\mathcal{V}_{\max}(A, \mathcal{I}, \mathrm{ker}C)$ and $\mathcal{S}_{\min}(A, \mathrm{ker}C, \mathcal{I})$ as follows

THEOREM 26.9

$$\mathcal{S}_{\mathrm{sh,max}}(A, kerC, \mathcal{I}) = \mathcal{S}_{\min}(A, kerC, \mathcal{I}) + \mathcal{V}_{\max}(A, \mathcal{I}, kerC). \quad (26.32)$$

Furthermore, we have

THEOREM 26.10

$$\begin{aligned} &\mathcal{S}_{\mathrm{sh,max}}(A, kerC, \mathcal{I}) = \\ &(kerC + \mathcal{S}_{\min}) \cap (A + GC)^{-1}(kerC + \mathcal{S}_{\min}) \\ &\cap \ldots \cap (A + GC)^{-(n-1)}(kerC + \mathcal{S}_{\min}) \end{aligned} \quad (26.33)$$

where $\mathcal{S}_{\min} := \mathcal{S}_{\min}(A, kerC, \mathcal{I})$ and the matrix G is such that $(A + GC)\mathcal{S}_{\min} \subset \mathcal{S}_{\min}$.

REMARK 26.12 The right-hand side of Equation 26.33 represents the largest $A + GC$ invariant subspace contained in $\mathrm{ker}C + \mathcal{S}_{\min}$.

REMARK 26.13 An interpretation of the right-hand side of Equation 26.33 exists in terms of observers. See section 4.1.3 for more details.

26.7 Internal, External Stabilizability

We now introduce the notions of stability associated with controlled and conditioned invariants.

DEFINITION 26.6 An $(A, \mathrm{im}B)$-controlled invariant $\mathcal{V}$ is said to be *internally stabilizable* if, for any initial state $x_0 \in \mathcal{V}$, there is a control $u(t)$ such that $x(t) \in \mathcal{V}$ for all $t \geq 0$ and $x(t)$ converges to the zero state as $t \to \infty$, or, alternatively (and, in fact, equivalently), if there exists an $m \times n$ real matrix F satisfying $(A + BF)\mathcal{V} \subset \mathcal{V}$ and $(A + BF)|\mathcal{V}$ is stable.

DEFINITION 26.7 An $(A, \mathrm{im}B)$-controlled invariant $\mathcal{V}$ is said to be *externally stabilizable* if, for any initial state $x_0 \in \mathcal{X}$, there is a control $u(t)$ such that $x(t)$ converges to $\mathcal{V}$ as $t \to \infty$, or, alternatively (and, in fact, equivalently), if there exists an $m \times n$ real matrix F satisfying $(A + BF)\mathcal{V} \subset \mathcal{V}$ and $(A + BF)|\mathcal{X}/\mathcal{V}$ is stable.

Internal and external stabilizabilities can be easily examined by applying appropriate coordinate transformations. For this define $n \times n$ and $m \times m$ nonsingular matrices $T = [T_1\ T_2\ T_3\ T_4]$ and $U = [U_1\ U_2]$, respectively, as follows. Let $\mathcal{V}$ be an $(A, \mathrm{im}B)$-controlled invariant. Choose T_1 and T_2 so that $\mathrm{im}T_1 = \mathcal{V}_{\mathrm{reach}}(\mathcal{V})$ and $\mathrm{im}[T_1\ T_2] = \mathcal{V}$. Noting that $\mathcal{V}_{\mathrm{reach}}(\mathcal{V}) = \mathcal{V} \cap \mathcal{S}_{\min}(A, \mathcal{V}, \mathrm{im}B)$, select T_3 satisfying $\mathrm{im}[T_1\ T_3] = \mathcal{S}_{\min}(A, \mathcal{V}, \mathrm{im}B)$. Also, choose U_1 so that $\mathrm{im}BU_1 = \mathcal{V} \cap \mathrm{im}B$. Then, we have

$$\widetilde{A} := T^{-1}AT = \begin{bmatrix} \widetilde{A}_{11} & \widetilde{A}_{12} & \widetilde{A}_{13} & \widetilde{A}_{14} \\ 0 & \widetilde{A}_{22} & \widetilde{A}_{23} & \widetilde{A}_{24} \\ \widetilde{A}_{31} & \widetilde{A}_{32} & \widetilde{A}_{33} & \widetilde{A}_{34} \\ 0 & 0 & \widetilde{A}_{43} & \widetilde{A}_{44} \end{bmatrix}, \tag{26.34}$$

$$\widetilde{B} := T^{-1}BU = \begin{bmatrix} \widetilde{B}_{11} & 0 \\ 0 & 0 \\ 0 & \widetilde{B}_{32} \\ 0 & 0 \end{bmatrix}. \tag{26.35}$$

Note that the zero matrix in the second block row of $\widetilde{A}$ is due to the following facts: $\mathcal{V}_{\mathrm{reach}}(\mathcal{V})$ is $(A + BF)$-invariant for F satisfying $(A+BF)\mathcal{V} \subset \mathcal{V}$ and the second block row of $\widetilde{B}$ is zero. The fourth block row of $\widetilde{A}$ has zero blocks since $\mathcal{V}$ is $(A + BF)$-invariant for some F, but the fourth block row of $\widetilde{B}$ is zero. Now, let F be such that $(A + BF)\mathcal{V} \subset \mathcal{V}$ holds. Then, noting that $\widetilde{A}_{31} + \widetilde{B}_{32}\widetilde{F}_{21} = 0$ and $\widetilde{A}_{32} + \widetilde{B}_{32}\widetilde{F}_{22} = 0$ by construction, we obtain

$$\widetilde{A} + \widetilde{B}\widetilde{F} := T^{-1}(A + BF)T = \begin{bmatrix} \widetilde{A}_{11} + \widetilde{B}_{11}\widetilde{F}_{11} & \widetilde{A}_{12} + \widetilde{B}_{11}\widetilde{F}_{12} & \widetilde{A}_{13} + \widetilde{B}_{11}\widetilde{F}_{13} & \widetilde{A}_{14} + \widetilde{B}_{11}\widetilde{F}_{14} \\ 0 & \widetilde{A}_{22} & \widetilde{A}_{23} & \widetilde{A}_{24} \\ 0 & 0 & \widetilde{A}_{33} + \widetilde{B}_{32}\widetilde{F}_{23} & \widetilde{A}_{34} + \widetilde{B}_{32}\widetilde{F}_{24} \\ 0 & 0 & \widetilde{A}_{43} & \widetilde{A}_{44} \end{bmatrix} \tag{26.36}$$

where

$$\widetilde{F} = \begin{bmatrix} \widetilde{F}_{11} & \widetilde{F}_{12} & \widetilde{F}_{13} & \widetilde{F}_{14} \\ \widetilde{F}_{21} & \widetilde{F}_{22} & \widetilde{F}_{23} & \widetilde{F}_{24} \end{bmatrix}.$$

Note that $\widetilde{A}_{22}$ cannot be altered by any linear state feedback satisfying $(A + BF)\mathcal{V} \subset \mathcal{V}$, i.e., the (2, 2)-block of $\widetilde{A} + \widetilde{B}\widetilde{F}$ remains $\widetilde{A}_{22}$ for any real $\widetilde{F}$ satisfying

$$(\widetilde{A} + \widetilde{B}\widetilde{F})\mathrm{im}\begin{bmatrix} I & 0 \\ 0 & I \\ 0 & 0 \\ 0 & 0 \end{bmatrix} \subset \mathrm{im}\begin{bmatrix} I & 0 \\ 0 & I \\ 0 & 0 \\ 0 & 0 \end{bmatrix} \tag{26.37}$$

where I's in the first block rows and I's in the second block rows are respectively $\dim\mathcal{V}_{\mathrm{reach}}(\mathcal{V})\times \dim\mathcal{V}_{\mathrm{reach}}(\mathcal{V})$ and $(\dim\mathcal{V}\text{-}\dim \mathcal{V}_{\mathrm{reach}}(\mathcal{V}))\times (\dim\mathcal{V}\text{-}\dim \mathcal{V}_{\mathrm{reach}}(\mathcal{V}))$ identity matrices and 0's are zero matrices with suitable dimensions. Thus,

LEMMA 26.13

$$\sigma((A + BF)|\mathcal{V}/\mathcal{V}_{\mathrm{reach}}(\mathcal{V})) = \sigma(\widetilde{A}_{22}) \tag{26.38}$$

(i.e., fixed) for all F satisfying $(A + BF)\mathcal{V} \subset \mathcal{V}$. Here, $(A + BF)|\mathcal{V}/\mathcal{V}_{\mathrm{reach}}(\mathcal{V})$ is the induced map of $A + BF$ restricted to $\mathcal{V}/\mathcal{V}_{\mathrm{reach}}(\mathcal{V})$.

Note also that by construction the pair $(\widetilde{A}_{11}, \widetilde{B}_{11})$ is reachable. Hence, by Theorem 26.2, we have

LEMMA 26.14 $\sigma((A + BF)|\mathcal{V}_{\mathrm{reach}}(\mathcal{V})) = \sigma(\widetilde{A}_{11} + \widetilde{B}_{11}\widetilde{F}_{11})$ can be freely assigned by an appropriate choice of F satisfying $(A + BF)\mathcal{V} \subset \mathcal{V}$ (or satisfying Equation 26.37).

The eigenvalues of $(A + BF)|\mathcal{V}/\mathcal{V}_{\mathrm{reach}}(\mathcal{V})$ are called the *internal unassignable eigenvalues* of $\mathcal{V}$. The internal unassignable eigenvalues of $\mathcal{V}_{\max}(A, \mathrm{im}B, \ker C)$ are called *invariant zeroes* of the system $\sum := [A, B, C]$ (or triple (A, B, C)), which are equal to the transmission zeroes of $C(sI - A)^{-1}B$ if (A, B) is reachable and (A, C) is observable. Table 26.1 shows how freely the eigenvalues can be assigned for $A + BF$ by choosing F satisfying $(A + BF)\mathcal{V} \subset \mathcal{V}$ given an $(A, \mathrm{im}B)$-controlled invariant $\mathcal{V}$, and Theorems 26.11 and 26.12 easily follow from the table.

THEOREM 26.11 *An $(A, \mathrm{im}B)$-controlled invariant $\mathcal{V}$ is internally stabilizable if and only if all its internal unassignable eigenvalues have negative real parts.*

THEOREM 26.12 *An $(A, \mathrm{im}B)$-controlled invariant $\mathcal{V}$ is externally stabilizable if and only if $\mathcal{V} + \mathcal{V}_{\mathrm{reach}}$ is externally stable.*

LEMMA 26.15 If the pair (A, B) is stabilizable, then all $(A, \mathrm{im}B)$-controlled invariants are externally stabilizable.

REMARK 26.14 The matrix F can be defined independently on $\mathcal{V}$ and on $\mathcal{X}/\mathcal{V}$.

Dual to the above internal and external stabilizabilities for a controlled invariant are, respectively, external and internal sta-

TABLE 26.1 Spectral assignability of $(A+BF)|\mathcal{W}$ given an $(A,\text{im}B)$-controlled invariant $\mathcal{V}^a$

$\mathcal{W}$	$\mathcal{X}/(\mathcal{V}+\mathcal{V}_{reach})$	$(\mathcal{V}+\mathcal{V}_{reach})/\mathcal{V}$	$\mathcal{V}/\mathcal{V}_{reach}$	$\mathcal{V}_{reach}$
assignability	fixed	free	fixed	free

[a]The table indicates that $\sigma((A+BF)|\mathcal{X}/(\mathcal{V}+\mathcal{V}_{reach}))$ is fixed for all F satisfying $(A+BF)\mathcal{V}\subset\mathcal{V}$, $\sigma((A+BF)|(\mathcal{V}+\mathcal{V}_{reach})/\mathcal{V})$ is freely assignable (up to a symmetric set) by choosing an appropriate F satisfying $(A+BF)\mathcal{V}\subset\mathcal{V}$, etc.

bilizabilities for a conditioned invariant which is defined as follows.

DEFINITION 26.8 An $(A,\ \text{ker}C)$-conditioned invariant $\mathcal{S}$ is said to be *externally stabilizable* if there exists an $n\times p$ real matrix G satisfying $(A+GC)\mathcal{S}\subset\mathcal{S}$ and $(A+GC)|\mathcal{X}/\mathcal{S}$ is stable.

DEFINITION 26.9 An $(A,\ \text{ker}C)$-conditioned invariant $\mathcal{S}$ is said to be *internally stabilizable* if there exists an $n\times p$ real matrix G satisfying $(A+GC)\mathcal{S}\subset\mathcal{S}$ and $(A+GC)|\mathcal{S}$ is stable.

How freely the eigenvalues can be chosen for $A+GC$ by means of G is given in Table 26.2 from which necessary and sufficient conditions of the stabilizabilities follow easily. (See [5] for details.)

26.8 Disturbance Localization (or Decoupling) Problem

One of the first problems to which the geometric notions were applied is the problem of disturbance localization (or disturbance decoupling). As we can see, the solution to the problem is remarkably simple in geometric terms. The solution and analysis of this problem can also be used to solve other more involved problems (e.g., model following, decoupling, and disturbance decoupling in decentralized systems).

We will be concerned with a time-invariant linear control system $\sum_d := [A,B,D,C]$ described by

$$\dot{x}(t) = Ax(t)+Bu(t)+Dw(t),\ x(0)=x_0, \quad (26.39)$$

$$y(t) = Cx(t) \quad (26.40)$$

where the column vectors $x(t)\in\mathcal{X}:=R^n$, $u(t)\in R^m$, $y(t)\in R^p$ and $w(t)\in R^{m_d}$ are the state, input, output and unknown "disturbance" of the system at time $t\geq 0$, $x_0\subset R^n$ is the initial state and the real matrices A, B, C, and D have consistent dimensions. We assume that $u(t)$ and $w(t)$ are piecewise continuous and that the disturbance $w(t)$ can neither be measured nor controlled. The above notation will be standard in the sequel. If we apply the linear state feedback control law

$$u(t) = Fx(t) \quad (26.41)$$

to the above system, we obtain the state equation

$$\dot{x}(t) = (A+BF)x(t)+Dw(t),\ x(0)=x_0 \quad (26.42)$$

where F is an $m\times n$ real matrix. Our problem is to choose the control law (26.41) so that the disturbance does not affect the output in the resulting closed loop system given by Equations 26.42 and 26.40, i.e., so that $y(t)=0$ for all $t\geq 0$ for any $w(t)$, $t\geq 0$ provided $x(0)=x_0=0$. By virtue of Theorem 26.1, note that at any time $t\geq 0$ all the possible states due to all the admissible $w(\tau)$, $0\leq\tau\leq t$ are characterized by

$$\mathcal{R}(A+BF,\text{im}D) := \text{im}D+(A+BF)\text{im}D+\cdots + (A+BF)^{n-1}\text{im}D. \quad (26.43)$$

Therefore, in algebraic terms, the above problem can be restated as

Disturbance Localization (or Disturbance Decoupling) Problem: Given $n\times n$, $n\times m$, $n\times m'$ and $p\times n$ real matrices A, B, D and C, find an $m\times n$ real matrix F satisfying

$$\mathcal{R}(A+BF,\text{im}D)\subset\text{ker}C. \quad (26.44)$$

It is not always possible to find F satisfying Equation 26.44. The following theorem gives the necessary and sufficient condition for the existence of such an F in terms of the given matrices. (See Section 4.2 in Reference [5] and Section 4.3 in Reference [11].)

THEOREM 26.13 *There exists an $m\times n$ real matrix F satisfying Equation 26.44 if and only if*

$$imD\subset\mathcal{V}_{max}(A,imB,kerC). \quad (26.45)$$

PROOF 26.6 [Only if part]: Trivially, $\text{im}D\subset\mathcal{R}(A+BF,\text{im}D)$. Since $\mathcal{R}(A+BF,\text{im}D)$ is an $(A+BF)$-invariant and, by assumption, is contained in $\text{ker}C$, we have $\mathcal{R}(A+BF,\text{im}D)\subset\mathcal{V}_{max}(A,\text{im}B,\text{ker}C)$. Therefore, Equation 26.45 holds.
[If part]: Let F be such that

$$(A+BF)\mathcal{V}_{max}(A,\text{im}B,\text{ker}C)\subset\mathcal{V}_{max}(A,\text{im}B,\text{ker}C). \quad (26.46)$$

The inclusion (26.45) implies $(A+BF)\,\text{im}D\subset\mathcal{V}_{max}(A,\text{im}B,\text{ker}C)$, $\ldots$, $(A+BF)^{n-1}\text{im}D\subset\mathcal{V}_{max}(A,\text{im}B,\text{ker}C)$. Therefore,

$$\mathcal{R}(A+BF,\text{im}D)\subset\mathcal{V}_{max}(A,\text{im}B,\text{ker}C)\subset\text{ker}C.$$

REMARK 26.15 Let $n^*:=\dim\mathcal{V}_{max}(A,\text{im}B,\text{ker}C)$, and let T be a real nonsingular matrix for which the first n^* columns form a basis of $\mathcal{V}_{max}(A,\text{im}B,\text{ker}C)$. Then, the inclusion (26.45)

TABLE 26.2 Spectral assignability of $(A + GC)|\mathcal{W}$ given an $(A, \ker C)$-cond. invariant $\mathcal{S}^a$

$\mathcal{W}$	$\mathcal{X}/\mathcal{S}_{\text{sh,max}}(A,\ker C, \mathcal{S})$	$\mathcal{S}_{\text{sh,max}}(A,\ker C, \mathcal{S})/\mathcal{S}$	$\mathcal{S}/\mathcal{S} \cap \mathcal{V}_{unobs}$	$\mathcal{S}$
assignability	free	fixed	free	fixed

[a] The table indicates that $\sigma((A + BF)|\mathcal{X}/\mathcal{S}_{\text{sh,max}}(A,\ker C, \mathcal{S}))$ is freely assignable (up to a symmetric set) by choosing an appropriate F satisfying $(A + GC)\mathcal{S} \subset \mathcal{S}$, $\sigma((A + BF)|\mathcal{S}_{\text{sh,max}}(A,\ker C, \mathcal{S})/\mathcal{S})$ is fixed for all F satisfying $(A + GC)\mathcal{S} \subset \mathcal{S}$, etc.

means that, with an $m \times n$ real matrix F satisfying $(A+BF)\mathcal{V}_{\max}(A,\text{im}B,\ker C) \subset \mathcal{V}_{\max}(A,\text{im}B,\ker C)$, the coordinate transformation

$x(t) = T\tilde{x}(t)$ transforms Equations 26.42 and 26.40 to

$$\begin{bmatrix} \dot{\tilde{x}}_1(t) \\ \dot{\tilde{x}}_2(t) \end{bmatrix} = \begin{bmatrix} \tilde{A}_{11} + \tilde{B}_1\tilde{F}_1 & \tilde{A}_{12} + \tilde{B}_1\tilde{F}_2 \\ 0 & \tilde{A}_{22} + \tilde{B}_2\tilde{F}_2 \end{bmatrix} \begin{bmatrix} \tilde{x}_1(t) \\ \tilde{x}_2(t) \end{bmatrix} + \begin{bmatrix} \tilde{D}_1 \\ 0 \end{bmatrix} w(t) \tag{26.47}$$

and

$$y(t) = \begin{bmatrix} 0 & \tilde{C}_2 \end{bmatrix} \begin{bmatrix} \tilde{x}_1(t) \\ \tilde{x}_2(t) \end{bmatrix} \tag{26.48}$$

where $\begin{bmatrix} \tilde{F}_1 & \tilde{F}_2 \end{bmatrix} = FT$.

26.9 Disturbance Localization with Stability

In the previous section, the disturbance localization problem without additional constraints was solved. In this section, the problem is examined with the important constraint of stability. For the system $\sum_d := [A, B, D, C]$ described by Equations 26.42 and 26.40, or for the matrices A, B, D, and C as before, we have the following:

Disturbance Localization Problem with Stability: Find (if possible) an $m \times n$ real matrix F such that (1) the inclusion (26.44) holds and (2) $A + BF$ is stable, i.e., the eigenvalues of $A + BF$ have negative real parts.

Trivially, for condition (2) to be true, it is necessary that the pair (A, B) is stabilizable. We have (See Section 4.2 in Reference [5])

THEOREM 26.14 *Let A, B, D, and C be as before, and assume that the pair (A, B) is stabilizable. Then, the disturbance localization problem with stability has a solution if and only if (1) inclusion (26.45) holds and (2) $\mathcal{V}_{\text{sb,min}}(A,imB+imD,kerC)$ is internally stabilizable.*

REMARK 26.16 An alternative condition can be found in [11]

We will first prove that the above two conditions are sufficient. This part of the proof leads to a constructive procedure to find a matrix F that solves the problem.

SUFFICIENCY PROOF OF THEOREM 26.14: Condition (2) means that there is a real matrix F such that $(A + BF)\mathcal{V}_{\text{sb,min}}(A,\text{im}B+\text{im}D, \ker C) \subset \mathcal{V}_{\text{sb,min}}(A,\text{im}B+ \text{im}D, \ker C)$ and $\mathcal{V}_{\text{sb,min}}(A, \text{im}B+ \text{im}D,\ker C)$ is internally stable. Since the pair (A, B) is stabilizable, $\mathcal{V}_{\text{sb,min}}(A,\text{im}B+\text{im}D,\ker C)$ is externally stabilizable by virtue of Lemma 26.15. Since the internal stability of $\mathcal{V}_{\text{sb,min}}(A, \text{im}B+ \text{im}D, \ker C)$ (with respect to $A + BF$) is determined by $F|\mathcal{V}_{\text{sb,min}}(A, \text{im}B+ \text{im}D, \ker C)$ and since the external stability depends only on $F|\mathcal{X}/\mathcal{V}_{\text{sb,min}}(A, \text{im}B+ \text{im}D, \ker C)$, the matrix F can be chosen so that the controlled invariant is both internally and externally stable, hence, $A + BF$ is stable. It remains to show that Equation 26.44 holds. Since $\mathcal{V}_{\text{sb,min}}(A, \text{im}B+ \text{im}D, \ker C)$ is self-bounded with respect to $kerC$, and since Equation 26.45 holds, we have $\mathcal{V}_{\text{sb,min}}(A, \text{im}B+ \text{im}D, \ker C) \supset (\text{im}B+\text{im}D) \cap \mathcal{V}_{\max}(A, \text{im}B+ \text{im}D, \ker C) \supset (\text{im}B+\text{im}D) \cap \mathcal{V}_{\max}(A, \text{im}B, \ker C) \supset \text{im}B \cap \mathcal{V}_{\max}(A, \text{im}B, \ker C)+ \text{im}D \supset \text{im}D$. This inclusion and the $(A+BF)$-invariance of $\mathcal{V}_{\text{sb,min}}(A, \text{im}B+\text{im}D, \ker C)$ imply $\mathcal{R}(A + BF,\text{im}D) \subset \mathcal{V}_{\text{sb,min}}(A, \text{im}B+ \text{im}D, \ker C) \subset \ker C$.

REMARK 26.17 In Remark 26.15 redefine $n^* :=\dim \mathcal{V}_{\text{sb,min}}(A,\text{im}B+\text{im}D,\ker C)$ and replace $\mathcal{V}_{\max}(A,\text{im}B,\ker C)$ by $\mathcal{V}_{\text{sb,min}}(A,\text{im}B+\text{im}D,\ker C)$. Then, $\sigma(A + BF|\, \mathcal{V}_{\text{sb,min}}(A, \text{im}B+\text{im}D, \ker C)) = \sigma\left(\tilde{A}_{11} + \tilde{B}_1\tilde{F}_1\right)$ and $\sigma(A + BF|\,\mathcal{X}/\mathcal{V}_{\text{sb,min}}(A,\text{im}B+ \text{im}D,\ker C)) = \sigma\left(\tilde{A}_{22} + \tilde{B}_2\tilde{F}_2\right)$. Therefore, we choose $\tilde{F}_1$ and $\tilde{F}_2$ so that $\tilde{A}_{11} + \tilde{B}_1\tilde{F}_1$ and $\tilde{A}_{22} + \tilde{B}_2\tilde{F}_2$ are stable and $\tilde{A}_{21} + \tilde{B}_2\tilde{F}_1 = 0$. Then, the desired F is given by $F = \begin{bmatrix} \tilde{F}_1 & \tilde{F}_2 \end{bmatrix} T^{-1}$.

To prove the necessity we will use the following:

LEMMA 26.16 If there exists an internally stabilizable $(A, \text{im}B)$-controlled invariant $\mathcal{V}$ satisfying $\mathcal{V} \subset \ker C$ and $\text{im}D \subset \mathcal{V}$, then $\mathcal{V}_{\text{sb,min}}(A,\text{im}B+\text{im}D,\ker C)$ is internally stabilizable, i.e., it is $(A+BF)$-invariant and $A+BF|\mathcal{V}_{\text{sb,min}}(A,\text{im}B+\text{im}D,\ker C)$ is stable for some real matrix F.

PROOF 26.7 See the proof of Lemma 4.2.1 in [5]. Also see [3] and [8].

NECESSITY PROOF OF THEOREM 26.14: Suppose that the problem has a solution, i.e., there is a real matrix F such that $A + BF$ is stable and $\mathcal{V} := \mathcal{R}(A + BF,\text{im}D) \subset \ker C$. Clearly, $\text{im}D \subset \mathcal{V}$. Also, $\mathcal{V}$ is an $(A + BF)$-invariant contained in $\ker C$. Hence, $\mathcal{V} \subset \mathcal{V}_{\max}(A,\text{im}B,\ker C)$. Therefore, condition (1) holds. To show condition (2), note that $\mathcal{V}$ is internally (and externally) stable with respect to $A + BF$, $\mathcal{V} \subset \ker C$ and $\text{im}D \subset \mathcal{V}$. Condition (2) follows by Lemma 26.16.

26.10 Disturbance Localization by Dynamic Compensator

In the previous section, we used the (static) state feedback control law $u(t) = Fx(t)$ to achieve disturbance rejection and stability. In this section, we use measurement output feedback with a dynamic compensator $\sum_c$ placed in the feedback loop to achieve the same objectives.

More specifically, we will be concerned with the system $\sum_{\rm dm} := [A, B, D, C, C_{meas}]$ described by

$$\begin{aligned} \dot{x}(t) &= Ax(t) + Bu(t) + Dw(t), \; x(0) = x_0 \\ y(t) &= Cx(t) \\ y_{meas}(t) &= C_{meas}x(t) \end{aligned} \tag{26.49}$$

where $y_{meas}(t) \in \mathcal{Y}_{meas} := R^{p_m}$ stands for a measurement output and C_{meas} is a $p_m \times n$ real matrix. Define the dynamic compensator $\sum_c := [A_c, B_c, C_c, K_c]$ by

$$\dot{x}_c(t) = A_c x_c(t) + B_c y_{meas}(t), \; x_c(0) = x_{c0}, \tag{26.50}$$

$$u(t) = C_c x_c(t) + K_c y_{meas}(t) \tag{26.51}$$

where $x_c(t), x_{c0} \in \mathcal{X}_c := R^{n_c}$, and A_c, B_c, C_c, and K_c are, respectively, $n_c \times n_c$, $n_c \times p_m$, $m \times n_c$, $m \times p_m$ real matrices. We wish to determine A_c, B_c, C_c, and K_c such that (1) $y(t) = 0$ for all $t \geq 0$ for any (admissible) $w(t)$, $t \geq 0$ provided $x_0 = 0$ and $x_{co} = 0$, and (2) $x(t)$ and $x_c(t)$ converges to zero as $t \to +\infty$ for all $x_0 \in \mathcal{X}$ and $x_{co} \in \mathcal{X}_c$ provided $w(t) = 0$ for all $t \geq 0$. For this, it is convenient to introduce the *extended state* defined by

$$\hat{x} := \begin{bmatrix} x \\ x_c \end{bmatrix} \in \hat{\mathcal{X}} := R^{n+n_c}$$

where $x \in \mathcal{X}$ and $x_c \in \mathcal{X}_c$. Then, the overall system (including $\sum_{dm}$ and $\sum_c$) can be rewritten as

$$\dot{\hat{x}}(t) = \hat{A}\hat{x}(t) + \hat{D}w(t), \; \hat{x}(0) = \hat{x}_0, \tag{26.52}$$

$$y(t) = \hat{C}\hat{x}(t) \tag{26.53}$$

where $\hat{x}_0 := [x_0' \; x_c']'$ and the matrices are defined by

$$\hat{A} := \begin{bmatrix} A + BK_c C_{meas} & BC_c \\ B_c C_{meas} & A_c \end{bmatrix}, \quad \hat{D} := \begin{bmatrix} D \\ 0 \end{bmatrix}, \; \hat{C} := [C \; 0]. \tag{26.54}$$

REMARK 26.18 Define

$$\begin{aligned} \hat{A}_0 &:= \begin{bmatrix} A & 0 \\ 0 & 0 \end{bmatrix}, \; \hat{B}_0 := \begin{bmatrix} B & 0 \\ 0 & I \end{bmatrix}, \\ \hat{C}_{meas} &:= \begin{bmatrix} C_{meas} & 0 \\ 0 & I \end{bmatrix}, \\ \hat{K}_c &:= \begin{bmatrix} K_c & C_c \\ B_c & A_c \end{bmatrix}. \end{aligned} \tag{26.55}$$

Then, it is easy to verify that

$$\hat{A} = \hat{A}_0 + \hat{B}_0 \hat{K}_c \hat{C}_{meas}. \tag{26.56}$$

(See Section 5.1 of [5] and [9].)

REMARK 26.19 An important special case of $\sum_c$ is a state estimate feedback through an observer. More specifically, let L_1 and L_2 satisfy

$$L_1 C_{meas} + L_2 = I_n, \tag{26.57}$$

and consider an asymptotic observer (or state estimator) described by

$$\begin{aligned} \dot{x}_c(t) &= Ax_c(t) + Bu(t) + G\{C_{meas}x_c(t) - y_{meas}(t)\}, \\ &\quad x_c(0) = x_{c0}, \end{aligned} \tag{26.58}$$

$$x_{est}(t) = L_2 x_c(t) + L_1 y_{meas}(t) \tag{26.59}$$

where $x_{est}(t) \in R^n$ is an "estimate" of $x(t)$ (under proper conditions) and G, L_1, and L_2 are real matrices with appropriate dimensions. If we use a state estimate feedback law

$$u(t) = Fx_{est}(t) \tag{26.60}$$

with an $m \times n$ real matrix F, then the overall system reduces to Equations 26.52 and 26.53 where $\hat{A}$, $\hat{D}$, and $\hat{C}$ are given, instead of Equation 26.54, by

$$\hat{A} := \begin{bmatrix} A + BFL_1 C_{meas} & BFL_2 \\ (BFL_1 - G)C_{meas} & A + BFL_2 + GC_{meas} \end{bmatrix}, \quad \hat{D} := \begin{bmatrix} D \\ 0 \end{bmatrix}, \; \hat{C} := [C \; 0]. \tag{26.61}$$

It should also be noted that, if we apply the coordinate transformation

$$\hat{x} = \begin{bmatrix} I_n & 0 \\ I_n & -I_n \end{bmatrix} \bar{x},$$

Equations 26.52 and 26.61 can be rewritten as

$$\dot{\bar{x}}(t) = \bar{A}\bar{x}(t) + \bar{D}w(t), \; \bar{x}(0) = \bar{x}_0 \tag{26.62}$$

where

$$\bar{A} := \begin{bmatrix} A + BF & -BFL_2 \\ 0 & A + GC_{meas} \end{bmatrix}, \; \bar{D} := \begin{bmatrix} D \\ D \end{bmatrix}. \tag{26.63}$$

(See Section 5.1.2 of [5] for further discussions.)

With the above notation, the problem can be restated in geometric terms as:

Disturbance Localization Problem with Stability by Dynamic Compensator: Find (if possible) a number $n_c (=\dim \mathcal{X}_c)$ and real matrices A_c, B_c, C_c, and K_c of dimensions $n_c \times n_c$, $n_c \times p_m$, $m \times n_c$, $m \times p_m$, respectively, such that

$$(1) \quad \mathcal{R}(\hat{A}, \mathrm{im}\hat{D}) \subset \ker\hat{C}, \tag{26.64}$$

and

$$(2) \quad \hat{A} \text{ is stable.} \tag{26.65}$$

REMARK 26.20 Noting that $\mathcal{R}(\hat{A}, \mathrm{im}\hat{D})$ is the minimum $\hat{A}$-invariant containing $\mathrm{im}\hat{D}$, it is easy to see that condition (1) is equivalent to the following condition (1)' there is an $\hat{A}$-invariant $\hat{\mathcal{V}}$ satisfying

$$\text{im}\hat{D} \subset \hat{\mathcal{V}} \subset \ker\hat{C}. \tag{26.66}$$

The conditions under which the above problem is solvable are given by the following theorems.

THEOREM 26.15 *Let (A, B) be stabilizable, and also let (A, C_{meas}) be detectable. Then, the disturbance localization problem with stability by dynamic compensator has a solution if and only if there exists an internally stabilizable (A, imB)-controlled invariant $\mathcal{V}$ and an externally stabilizable $(A, kerC_{meas})$-conditioned invariant $\mathcal{S}$ satisfying the condition*

$$imD \subset \mathcal{S} \subset \mathcal{V} \subset kerC. \tag{26.67}$$

PROOF 26.8 See Section 5.2 of [5] and [9].

The above condition is an existence condition and it is not convenient for testing. The following theorem (see Theorem 5.2-2 in [5]) gives constructive conditions and they can be readily tested.

THEOREM 26.16 *Assume that (A, B) is stabilizable and (A, C_{meas}) is detectable. The disturbance localization problem with stability by dynamic compensator has a solution if and only if the following conditions hold:*

(1) $\mathcal{S}_{\min}(A, kerC_{meas}, imD) \subset \mathcal{V}_{\max}(A, imB, kerC), \quad (26.68)$

(2) $\mathcal{S}_{\min}(A, kerC_{meas}, imD) + \mathcal{V}_{\max}(A, imD, kerC \cap kerC_{meas})$ *is externally stabilizable and*

(3) $\mathcal{V}_{sb,\min}(A, imB + imD, kerC) + \mathcal{V}_{\max}(A, imD, kerC \cap kerC_{meas})$ *is internally stabilizable.* (26.69)

PROOF OF NECESSITY: See Section 5.2 of [5] and [4].
PROOF OF SUFFICIENCY: Let

$$\begin{aligned} \mathcal{S} &:= \mathcal{S}_{\min}(A, \ker C_{meas}, \text{im}D) \\ &\quad + \mathcal{V}_{\max}(A, \text{im}D, \ker C \cap \ker C_{meas}), \quad (26.70) \\ \mathcal{V} &:= \mathcal{V}_{sb,\min}(A, \text{im}B + \text{im}D, \ker C) \\ &\quad + \mathcal{V}_{\max}(A, \text{im}D, \ker C \cap \ker C_{meas}). \quad (26.71) \end{aligned}$$

We will show that $\mathcal{S}$ and $\mathcal{V}$ satisfy the conditions of Theorem 26.15. By assumptions (2) and (3), $\mathcal{S}$ is externally stabilizable and $\mathcal{V}$ is internally stabilizable. Suppose condition (1) also holds. Then, clearly im$D \subset \mathcal{V}_{\max}(A,imB,kerC)$, which implies $\mathcal{V}_{\max}(A,$ im$B$+im$D$,ker$C) = \mathcal{V}_{\max}(A,imB$,ker$C)$ since trivially $\mathcal{V}_{\max}(A,$im$B$+im$D$,ker$C) \supset \mathcal{V}_{\max}(A,imB$,ker$C)$ and $A\mathcal{V}_{\max}(A,$im$B$+im$D$,ker$C) \subset \mathcal{V}_{\max}(A,$ im$B$+ im$D$, ker$C)$+ im$B$+ im$D = \mathcal{V}_{\max}(A,$ im$B$+ im$D$, ker$C)$+ im$B$. Also, (1) implies $\mathcal{S}_{\min}(A,$kerC_{mea},im$D) \subset$ ker$C$, from which it is straightforward to show $\mathcal{S}_{\min}(A,$kerC_{mea},im$D) \subset \mathcal{S}_{\min}(A,kerC$,im$B$+ im$D)$. By Theorem 26.7, it now follows that

$$\begin{aligned} &\mathcal{V}_{sb,\min}(A, \text{im}B + \text{im}D, \ker C) + \mathcal{S} \\ = \;& \mathcal{V}_{\max}(A, \text{im}B, \ker C) \cap \mathcal{S}_{\min}(A, \ker C, \text{im}B + \text{im}D) \\ & + \mathcal{S}_{\min}(A, \ker C_{mea}, \text{im}D) \\ & + \mathcal{V}_{\max}(A, \text{im}D, \ker C \cap \ker C_{meas}) \\ = \;& \{\mathcal{V}_{\max}(A, \text{im}B, \ker C) + \mathcal{S}_{\min}(A, \ker C_{mea}, \text{im}D)\} \\ & \cap \mathcal{S}_{\min}(A, \ker C, \text{im}B + \text{im}D) \\ & + \mathcal{V}_{\max}(A, \text{im}D, \ker C \cap \ker C_{meas}) \\ = \;& \mathcal{V}_{\max}(A, \text{im}B, \ker C) \cap \mathcal{S}_{\min}(A, \ker C, \text{im}B + \text{im}D) \\ & + \mathcal{V}_{\max}(A, \text{im}D, \ker C \cap \ker C_{meas}) \\ = \;& \mathcal{V}, \end{aligned} \tag{26.72}$$

which implies $\mathcal{S} \subset \mathcal{V}$. Clearly, im$D \subset \mathcal{S} \subset \mathcal{V} \subset$ ker$C$.
Procedure for finding A_C, B_C, C_C and K_C: Define $\mathcal{S}$ and $\mathcal{V}$ by Equations 26.70 and 26.71. Let $\mathcal{L}$ be a subspace of R^n satisfying

$$\mathcal{L} \oplus (\mathcal{S} \cap \ker C_{meas}) = \mathcal{S}. \tag{26.73}$$

Clearly,

$$\mathcal{L} \cap \ker C_{meas} = 0. \tag{26.74}$$

Now let $\mathcal{L}_{comp}$ be a subspace satisfying $\mathcal{L} \bigoplus \mathcal{L}_{comp} = R^n$ and $\mathcal{L}_{comp} \supset$ kerC_{meas}, and define the projection L_2 on $\mathcal{L}_{comp}$ along $\mathcal{L}$. Select L_1 such that

$$L_1 C_{meas} = I_n - L_2. \tag{26.75}$$

Such an L_1 exists since ker$(I_n - L_2) = \mathcal{L}_{comp} \supset$ kerC_{meas}. Choose real matrices F and G so that

$$(A + BF)\mathcal{V} \subset \mathcal{V}, \tag{26.76}$$

$$(A + GC_{meas})\mathcal{S} \subset \mathcal{S}, \tag{26.77}$$

and $A + BF$ and $A + GC_{meas}$ are stable. Then, use Equations 26.58 and 26.59 as a dynamic compensator, or equivalently, set $A_c := A + BFL_2 + GC_{meas}$, $B_c := BFL_1 - G$, $C_c := FL_2$ and $K_c := FL_1$.
From Equation 26.63, the overall system is clearly stable. It can also be shown that

$$\hat{\mathcal{V}} := \left\{ \begin{bmatrix} v \\ v - s \end{bmatrix} : v \in \mathcal{V}, \; s \in \mathcal{S} \right\}$$

is $\hat{A}$-invariant with im$\hat{D} \subset \hat{\mathcal{V}} \subset$ ker$\hat{C}$.

References

[1] Basile, G. and Marro, G., Controlled and conditioned invariant subspaces in linear system theory, *J. Optimiz. Th. Applic.*, 3(5), 305–315, 1969.

[2] Basile, G. and Marro, G., L'invariànza rispetto ai disturbi studiata nello spazio degli stati, *Rendiconti della LXX Riunione Annualle AEI*, paper 1-4-01, Rimini, Italy, 1969.

[3] Basile, G. and Marro, G., Self-bounded controlled invariant subspaces: a straight forward approach to constrained controllability, *J. Optimiz. Th. Applic.*, 38(1), 71–81, 1982.

[4] Basile, G., Marro, G., and Piazzi, A., Stability without eigenspaces in the geometric approach: some new results, in *Frequency Domain and State Space Methods for Linear Systems*, C. A. Byrnes and A. Lindquist, Eds., North Holland (Elsevier), Amsterdam, 1986, 441.

[5] Basile, G. and Marro, G., *Controlled and Conditioned Invariants in Linear System Theory*, Prentice Hall, Englewood Cliffs, NJ, 1992.

[6] Hamano, F. and Furuta, K., Localization of disturbances and output decomposition of linear multivariable systems, *Int. J. Control*, 22(4), 551–562, 1975.

[7] Morse, A.S., Structural invariants of linear multivariable systems, *SIAM J. Contr. Optimiz.*, 11(3), 446–465, 1973.

[8] Schumacher, J.M., On a conjecture of Basile and Marro, *J. Optimiz. Th. Applic.*, 41(2), 371–376, 1983.

[9] Willems, J. C. and Commault, C., Disturbance decoupling by measurement feedback with stability or pole placement, *SIAM J. Control Optimiz.*, 19(4), 490–504, 1981.

[10] Wonham, W. M. and Morse, A. S., Decoupling and pole assignment in linear multivariable systems: a geometric approach, *SIAM J. Control Optimiz.*, 8(1), 1–18, 1970.

[11] Wonham, W. M., *Linear Multivariable Control, A Geometric Approach*, 3rd ed., Springer-Verlag, New York, 1985.

27

Polynomial and Matrix Fraction Descriptions

David F. Delchamps
Cornell University, Ithaca, NY

27.1 Introduction

For control system design, it is useful to characterize multi-input, multi-output, time-invariant linear systems in terms of their transfer function matrices. The transfer function matrix of a real m-input, p-output continuous-time system is a $(p \times m)$ matrix-valued function $G(s)$, where s is the Laplace transform variable; the corresponding object in discrete time is a $(p \times m)$ matrix-valued function $G(z)$, where z is the z-transform variable. Things are particularly interesting when $G(s)$ or $G(z)$ is a **proper rational matrix** function of s or z, that is, when every entry in $G(s)$ or $G(z)$ is a ratio of two real polynomials in s or z whose denominator's degree is at least as large as its numerator's degree. In this case, the system has state space realizations of the form

$$\begin{aligned} \dot{x}(t) &= Ax(t) + Bu(t) \\ y(t) &= Cx(t) + Du(t) \end{aligned}$$

or

$$\begin{aligned} x(k+1) &= Ax(k) + Bu(k) \\ y(k) &= Cx(k) + Du(k)\,, \end{aligned}$$

where the state vector x takes values in $\mathbf{R}^n$. Any such realization defines a decomposition $G(s) = C(sI_n - A)^{-1}B + D$ or $G(z) = C(zI_n - A)^{-1}B + D$ for the system's transfer function matrix. A realization is minimal when the state vector dimension n is as small as it can be; the **MacMillan degree** $\delta_M(G(s))$ or $\delta_M(G(z))$ is the value of n in a minimal realization. A system's MacMillan degree is a natural candidate for the "order of the system."

It is not easy to construct minimal realizations for a multi-input, multi-output system with an arbitrary proper rational transfer function matrix; even determining such a system's MacMillan degree requires some effort. Circumstances are simpler for single-input, single-output (SISO) systems. Consider, for example, a real SISO continuous-time system with proper rational transfer function $g(s)$. We can express $g(s)$ as a ratio

$$g(s) = \frac{p(s)}{q(s)}\,, \tag{27.1}$$

where $q(s)$ is a polynomial of degree n, $p(s)$ is a polynomial of degree at most n, and $p(s)$ and $q(s)$ are coprime, which is to say that their only common factors are nonzero real numbers. The coprimeness of $p(s)$ and $q(s)$ endows the fractional representation Equation 27.1 with a kind of irreducibility; furthermore, Equation 27.1 is "minimal" in the sense that any other factorization $g(s) = \hat{p}(s)/\hat{q}(s)$ features a denominator polynomial $\hat{q}(s)$ whose degree is at least n.

The MacMillan degree $\delta_M(g(s))$ is precisely n, the degree of $q(s)$ in Equation 27.1. To see that $n \le \delta_M(g(s))$, suppose (A, B, C, D) are the matrices in a minimal realization for $g(s)$. Set $\hat{n} = \delta_M(g(s))$, so A is $(\hat{n} \times \hat{n})$. The matrix $(sI_{\hat{n}} - A)^{-1}$ has rational entries whose denominator polynomials have degrees at most $\hat{n}$. Multiplying on the left by $\hat{C}$ and on the right by $\hat{B}$ and finally adding $\hat{D}$ results in a rational function whose denominator degree in lowest terms is at most $\hat{n}$. This rational function, however, is $g(s)$, whose lowest-terms denominator degree is n, whence it follows that $n \le \hat{n}$.

To finish showing that $\delta_M(g(s)) = n$, it suffices to construct a realization whose A-matrix is $(n \times n)$. There are many ways to do this; here is one. Begin by setting $d = \lim_{|s|\to\infty} g(s)$; d is well-defined because $g(s)$ is proper. Then

$$g(s) = \frac{\tilde{p}(s)}{q(s)} + d\,,$$

where the degree of $\tilde{p}(s)$ is at most $n-1$. Cancel the coefficient of s^n from $q(s)$ and $\tilde{p}(s)$ so that

$$q(s) = s^n + q_1 s^{n-1} + \cdots + q_{n-1}s + q_n$$

and

$$\tilde{p}(s) = \gamma_1 s^{n-1} + \gamma_2 s^{n-2} + \cdots + \gamma_{n-1}s + \gamma_n\,.$$

0-8493-8570-9/96/$0.00+$.50

It is not hard to verify that $g(s) = C(sI_n - A)^{-1}B + D$ when

$$A = \begin{bmatrix} 0 & 1 & 0 & . & . & . & 0 \\ 0 & 0 & 1 & 0 & . & . & 0 \\ 0 & 0 & 0 & 1 & . & . & 0 \\ . & . & . & . & . & . & . \\ . & . & . & . & . & 1 & 0 \\ 0 & 0 & . & . & . & 0 & 1 \\ -q_n & -q_{n-1} & . & . & . & -q_2 & -q_1 \end{bmatrix},$$

$$B = \begin{bmatrix} 0 \\ 0 \\ . \\ . \\ . \\ 0 \\ 1 \end{bmatrix}, \tag{27.2}$$

and

$$C = \begin{bmatrix} \gamma_n & \gamma_{n-1} & . & . & . & \gamma_1 \end{bmatrix}, \quad D = d. \tag{27.3}$$

To summarize the foregoing discussion, if $g(s)$ is the transfer function of a real time-invariant continuous-time SISO linear system, and $g(s)$ is proper rational, then $g(s)$ possesses a fractional representation Equation 27.1 that

- is irreducible because $p(s)$ and $q(s)$ are coprime, or, equivalently,
- is minimal because the degree of $q(s)$ is as small as it can be in such a representation.

Moreover,

- the degree of $q(s)$ in any minimal and irreducible representation Equation 27.1 is equal to the MacMillan degree of $g(s)$.

Replacing s with z results in identical assertions about discrete-time SISO systems.

Our goal in what follows will be to generalize these results to cover continuous-time, multi-input, multi-output (MIMO) systems with $(p \times m)$ proper rational transfer function matrices $G(s)$. The development is purely algebraic and applies equally well to discrete-time MIMO systems with proper rational transfer function matrices $G(z)$; simply replace s with z throughout. In Section 27.2, we present MIMO versions of the fractional representation Equation 27.1, complete with appropriate matrix analogues of irreducibility and minimality. In Section 27.3, we relate the results of Section 27.2 to the MacMillan degree of $G(s)$. We obtain as a bonus a minimal realization of $G(s)$ that turns out to be the MIMO analogue of the realization in Equations 27.2–27.3. In Section 27.4 we prove a well-known formula for the MacMillan degree of $G(s)$ and discuss briefly some connections between the material in Section 27.2 and the theory of ARMA models for MIMO systems, with a nod toward some modern results about stable coprime factorization and robust control. The algebra we will be employing throughout, while tedious at times, is never terribly abstract, and much of it is interesting in its own right.

27.2 Polynomial Matrix Fraction Descriptions

We begin by establishing some basic terminology. A **real polynomial matrix** is a matrix each of whose entries is a polynomial in s with real coefficients. We often omit the adjective "real" but assume, unless stated otherwise, that all polynomial matrices are real. One can perform elementary operations on polynomial matrices such as addition and multiplication using the rules of polynomial algebra along with the standard formulas that apply to matrices of numbers. Likewise, one can compute the determinant $\det F(s)$ of a square polynomial matrix $F(s)$ by any of the usual formulas or procedures. All the familiar properties of the determinant hold for polynomial matrices, e.g., the product rule

$$\det[F_1(s)F_2(s)] = [\det F_1(s)][\det F_2(s)], \tag{27.4}$$

which applies when $F_1(s)$ and $F_2(s)$ are the same size.

There are two shades of invertibility for square polynomial matrices.

DEFINITION 27.1 A square polynomial matrix $F(s)$ is said to be *nonsingular* if, and only if, $\det F(s)$ is a nonzero polynomial and *unimodular* if, and only if, $\det F(s)$ is a nonzero real number.

Thus if $F(s)$ is **nonsingular**, we are free to compute $F^{-1}(s)$ using the adjugate-determinant formula for the inverse [12], namely,

$$F^{-1}(s) = \frac{1}{\det F(s)} \operatorname{adj} F(s).$$

The (i, j) entry of $\operatorname{adj} F(s)$ is $(-1)^{i+j}$ times the determinant of the matrix obtained from $F(s)$ by eliminating its jth row and ith column. $F^{-1}(s)$ is in general a rational matrix function of s. If $F(s)$ is not just nonsingular but also **unimodular**, then $F^{-1}(s)$ is actually a polynomial matrix; thus unimodular polynomial matrices are "polynomially invertible."

From Definition 27.1 along with the product rule Equation 27.4, the product of two nonsingular polynomial matrices is nonsingular and the product of two unimodular polynomial matrices is unimodular. Furthermore, we have the following "pointwise" characterization of unimodularity: a square polynomial matrix $F(s)$ is unimodular precisely when $F(s_o)$ is an invertible matrix of complex numbers for every complex number s_o. This statement follows from the fundamental theorem of algebra, which implies that if $\det F(s)$ is a polynomial of nonzero degree, then $\det F(s_o) = 0$ for at least one complex number s_o.

As an example, consider the two polynomial matrices

$$F_1(s) = \begin{bmatrix} s+2 & s+2 \\ s+1 & s+2 \end{bmatrix} \text{ and } F_2(s) = \begin{bmatrix} s+2 & s+4 \\ s & s+2 \end{bmatrix}.$$

$F_2(s)$ is unimodular and $F_1(s)$ is merely nonsingular; in fact,

$$F_1^{-1}(s) = \begin{bmatrix} 1 & -1 \\ -\frac{s+1}{s+2} & 1 \end{bmatrix} \text{ and}$$

$$F_2^{-1}(s) = \frac{1}{4}\begin{bmatrix} s+2 & -s-2 \\ -s & s+2 \end{bmatrix}.$$

We are ready to define the matrix generalization(s) of the fractional representation Equation 27.1.

DEFINITION 27.2 Let $G(s)$ be a real $(p \times m)$ proper rational matrix function of s.

- A *right matrix fraction description*, or *right MFD*, of $G(s)$ is a factorization of the form $G(s) = P(s)Q^{-1}(s)$, where $P(s)$ is a real $(p \times m)$ polynomial matrix and $Q(s)$ is a real $(m \times m)$ nonsingular polynomial matrix.
- A *left matrix fraction description*, or *left MFD*, of $G(s)$ is a factorization of the form $G(s) = Q_L^{-1}(s)P_L(s)$, where $P_L(s)$ is a real $(p \times m)$ polynomial matrix and $Q_L(s)$ is a real $(p \times p)$ nonsingular polynomial matrix.

It is easy to construct left and right matrix fraction descriptions for a given $(p \times m)$ proper rational matrix $G(s)$. For instance, let $q(s)$ be the lowest common denominator of the entries in $G(s)$. Set $P(s) = q(s)G(s)$; then $P(s)$ is a polynomial matrix. Setting $Q(s) = q(s)I_m$ makes $P(s)Q^{-1}(s)$ a right MFD for $G(s)$; setting $Q_L(s) = q(s)I_p$ makes $Q_L^{-1}(s)P(s)$ a left MFD of $G(s)$.

Next we introduce matrix versions of the notions of irreducibility and minimality associated with the fractional representation Equation 27.1. First consider minimality. Associated with each right (or left) MFD of $G(s)$ is the "denominator matrix" $Q(s)$ (or $Q_L(s)$); the degree of the nonzero polynomial det $Q(s)$ (or det $Q_L(s)$) plays a role analogous to that of the degree of the polynomial $q(s)$ in Equation 27.1.

DEFINITION 27.3 Let $G(s)$ be a real $(p \times m)$ proper rational matrix.

- A right MFD $P(s)Q^{-1}(s)$ of $G(s)$ is *minimal* if, and only if, the degree of det $\hat{Q}(s)$ in any other right MFD $\hat{P}(s)\hat{Q}^{-1}(s)$ of $G(s)$ is at least as large as the degree of det $Q(s)$.
- A left MFD $Q_L^{-1}(s)P_L(s)$ of $G(s)$ is *minimal* if, and only if, the degree of det $\hat{Q}_L(s)$ in any other left MFD $\hat{Q}_L^{-1}(s)\hat{P}_L(s)$ of $G(s)$ is at least as large as the degree of det $Q_L(s)$.

To formulate the matrix version of irreducibility, we require the following analogues of the coprimeness condition on the polynomials $p(s)$ and $q(s)$ appearing in Equation 27.1.

DEFINITION 27.4

- Two polynomial matrices $P(s)$ and $Q(s)$ possessing m columns are said to be *right coprime* if, and only if, the following condition holds: If $F(s)$ is an $(m \times m)$ polynomial matrix so that for some polynomial matrices $\hat{P}(s)$ and $\hat{Q}(s)$, $P(s) = \hat{P}(s)F(s)$ and $Q(s) = \hat{Q}(s)F(s)$, then $F(s)$ is unimodular.
- Two polynomial matrices $P_L(s)$ and $Q_L(s)$ possessing p rows are said to be *left coprime* if, and only if, the following condition holds: If $F(s)$ is a $(p \times p)$ polynomial matrix so that for some polynomial matrices $\hat{P}_L(s)$ and $\hat{Q}_L(s)$, $P(s) = F(s)\hat{P}_L(s)$ and $Q_L(s) = F(s)\hat{Q}_L(s)$, then $F(s)$ is unimodular.

Saying that two polynomial matrices are right (or left) coprime is the same as saying that they have no right (or left) common square polynomial matrix factors other than unimodular matrices. In this way, unimodular matrices play a role similar to that of nonzero real numbers in the definition of coprimeness for scalar polynomials. Any two unimodular matrices of the same size are trivially right and left coprime. Similarly, any $(r \times r)$ unimodular matrix is right coprime with any polynomial matrix that has r columns and left coprime with any polynomial matrix that has r rows. Equipped with Definition 27.4, we can explain what it means for a right or left MFD to be irreducible.

DEFINITION 27.5 *Let* $G(s)$ be a real $(p \times m)$ proper rational matrix.

- A right MFD $P(s)Q^{-1}(s)$ of $G(s)$ is *irreducible* if, and only if, $P(s)$ and $Q(s)$ are right coprime.
- A left MFD $Q_L^{-1}(s)P_L(s)$ of $G(s)$ is *irreducible* if, and only if, $P_L(s)$ and $Q_L(s)$ are left coprime.

It is time to begin reconciling Definition 27.3, Definition 27.5, and the scalar intuition surrounding irreducibility and minimality of the fractional representation Equation 27.1. The proof of the following central result will occupy most of the remainder of this section; the finishing touches appear after Lemmas 27.1 and 27.2 below.

THEOREM 27.1 *Let $G(s)$ be a real $(p \times m)$ proper rational matrix.*

- *A right MFD $P(s)Q^{-1}(s)$ of $G(s)$ is minimal if, and only if, it is irreducible. Furthermore, if $P(s)Q^{-1}(s)$ and $\hat{P}(s)\hat{Q}^{-1}(s)$ are two minimal right MFDs of $G(s)$, then an $(m \times m)$ unimodular matrix $V(s)$ exists so that $\hat{P}(s) = P(s)V(s)$ and $\hat{Q}(s) = Q(s)V(s)$.*
- *A left MFD $Q_L^{-1}(s)P_L(s)$ of $G(s)$ is minimal if, and only if, it is irreducible. Furthermore, if $Q_L^{-1}(s)P_L(s)$ and $\hat{Q}_L^{-1}(s)\hat{P}_L(s)$ are two minimal left MFDs of $G(s)$, then a $(p \times p)$ unimodular matrix $V(s)$ exists so that $\hat{P}_L(s) = V(s)P(s)$ and $\hat{Q}_L(s) = V(s)Q(s)$.*

Some parts of Theorem 27.1 are easier to prove than others. In particular, it is straightforward to show that a minimal right or left MFD is irreducible. To see this, suppose $G(s) = P(s)Q^{-1}(s)$ is a right MFD that is not irreducible. Since $P(s)$ and $Q(s)$ are not right coprime, factorizations $P(s) = \hat{P}(s)F(s)$ and $Q(s) = \hat{Q}(s)F(s)$ exist with $F(s)$ not unimodular. $F(s)$ and $\hat{Q}(s)$ are evidently nonsingular because $Q(s)$ is. Furthermore,

$$\begin{aligned} G(s) &= P(s)Q^{-1}(s) = \hat{P}(s)F(s)F^{-1}(s)\hat{Q}^{-1}(s) \\ &= \hat{P}(s)\hat{Q}^{-1}(s), \end{aligned}$$

which shows how to "reduce" the original MFD $P(s)Q^{-1}(s)$ by "canceling" the right common polynomial matrix factor $F(s)$ from $P(s)$ and $Q(s)$. Because $F(s)$ is not unimodular, det $F(s)$ is a polynomial of positive degree, and the degree of det $\hat{Q}(s)$ is therefore strictly less than the degree of det $Q(s)$. Consequently, the original right MFD $P(s)Q^{-1}(s)$ is not minimal. The argument demonstrating that minimality implies irreducibility for left MFDs is essentially identical.

Proving the converse parts of the assertions in Theorem 27.1 is substantially more difficult. The approach we adopt, based on the so-called Smith form for polynomial matrices, is by no means the only one. Rugh [10] follows a quite different route, as does Vidyasagar [13]. We have elected to center our discussion on the Smith form partly because the manipulations we will be performing are similar to the things one does to obtain LDU decompositions [12] for matrices of numbers via Gauss elimination.

We will be employing a famous result from algebra known as the *Euclidean algorithm*. It says that if $f(s)$ is a real polynomial and $g(s)$ is another real polynomial of lower degree, then unique real polynomials $\kappa(s)$ and $\rho(s)$ exist, where $\rho(s)$ has lower degree than $g(s)$, for which

$$f(s) = g(s)\kappa(s) + \rho(s).$$

It is customary to regard the zero polynomial as having degree $-\infty$ and nonzero constant polynomials as having degree zero. If $g(s)$ divides $f(s)$, then $\rho(s) = 0$; otherwise, one can interpret $\kappa(s)$ and $\rho(s)$ respectively as the quotient and remainder obtained from dividing $f(s)$ by $g(s)$.

The Smith form of a real $(k \times r)$ polynomial matrix $F(s)$ is a matrix obtained by performing elementary row and column operations on $F(s)$. The row and column operations are of three kinds:

- interchanging the ith and jth rows (or the ith and jth columns);
- replacing row i with the difference (row i) $- \kappa(s) \times$ (row j), where $\kappa(s)$ is a polynomial (or replacing column j with the difference (column j) $- \kappa(s) \times$ (column i)); and
- replacing row (or column) i with its multiple by a nonzero real number γ.

One can view each of these operations as the result of multiplying $F(s)$ on the left or the right by a unimodular matrix. To interchange the ith and jth rows of $F(s)$, multiply on the left by the permutation matrix Π^{ij}; Π^{ij} is a $(k \times k)$ identity matrix with the ith and jth rows interchanged. To replace row i with itself minus $\kappa(s)$ times row j, multiply on the left by the matrix $E^{ij}[-\kappa(s)]$, a $(k \times k)$ identity matrix except with $-\kappa(s)$ in the ij position. Because det $\Pi^{ij} = -1$ and det $E^{ij}[-\kappa(s)] = 1$, both matrices are unimodular. To multiply row i by γ, multiply $F(s)$ on the left by the $(k \times k)$ diagonal matrix with γ in the ith diagonal position and ones in all the others; this matrix is also unimodular, because its determinant is γ. The column operations listed above result from multiplying on the right by $(r \times r)$ unimodular matrices of the types just described.

The following result defines the Smith form of a polynomial matrix. It is a special case of a more general theorem [6, pp. 175–180].

THEOREM 27.2 **Smith Form** *Let $F(s)$ be a real $(k \times r)$ polynomial matrix with $k \geq r$. A $(k \times k)$ unimodular matrix $U(s)$ and an $(r \times r)$ unimodular matrix $R(s)$ exist so that $U(s)F(s)R(s)$ takes the form*

$$\Lambda(s) = \begin{bmatrix} d_1(s) & 0 & 0 & \cdot & \cdot & 0 \\ 0 & d_2(s) & 0 & \cdot & \cdot & 0 \\ 0 & 0 & d_3(s) & 0 & \cdot & 0 \\ \cdot & \cdot & \cdot & \cdot & \cdot & \cdot \\ \cdot & \cdot & \cdot & \cdot & \cdot & 0 \\ 0 & \cdot & \cdot & \cdot & 0 & d_r(s) \\ 0 & \cdot & \cdot & \cdot & \cdot & 0 \\ \cdot & \cdot & \cdot & \cdot & \cdot & \cdot \\ \cdot & \cdot & \cdot & \cdot & \cdot & \cdot \\ 0 & \cdot & \cdot & \cdot & \cdot & 0 \end{bmatrix} \tag{27.5}$$

Furthermore, each nonzero $d_j(s)$ is monic, and $d_j(s)$ divides $d_{j+1}(s)$ for all j, $1 \leq j \leq r-1$.

PROOF 27.1 Let $\delta_{\min}$ be the smallest of the degrees of the nonzero polynomials in $F(s)$. Now invoke the following procedure:

Reduction Step: Pick an entry in $F(s)$ whose degree is $\delta_{\min}$ and use row and column exchanges to bring it to the (1, 1) position; denote the resulting matrix by $G(s)$. For each i, $2 \leq i \leq k$, use the Euclidean algorithm to find $\kappa_i(s)$ so that $\rho_i(s) = [G(s)]_{i1} - \kappa_i(s)[G(s)]_{11}$ has strictly lower degree than $[G(s)]_{11}$, which has degree $\delta_{\min}$. Observe that $\rho_i(s) = 0$ if $[G(s)]_{11}$ divides $[G(s)]_{i1}$. Multiply $G(s)$ on the left by the sequence of matrices $E^{i1}[-\kappa_i(s)]$; this has the effect of replacing each $[G(s)]_{i1}$ with $\rho_i(s)$. Multiplying on the right by a similar sequence of E^{1j}-matrices replaces each $[G(s)]_{1j}$ with a polynomial whose degree is lower than $\delta_{\min}$.

The net result is a new matrix whose $\delta_{\min}$ is lower than the $\delta_{\min}$ of $F(s)$. Repeating the reduction step on this new matrix results in a matrix whose $\delta_{\min}$ is still lower. Because we cannot continue reducing $\delta_{\min}$ forever, iterating the reduction step on the successor matrices of $F(s)$ leads eventually to a matrix of the form

$$F_1(s) = \begin{bmatrix} d_1(s) & 0 & 0 & . & . & 0 \\ 0 & \pi & \pi & . & . & \pi \\ 0 & \pi & \pi & . & . & \pi \\ . & . & . & . & . & . \\ . & . & . & . & . & . \\ 0 & \pi & . & . & . & \pi \end{bmatrix}, \qquad (27.6)$$

where each of the πs is a polynomial.

Divisibility Check: If $d_1(s)$ does not divide all the πs in Equation 27.6, find a π that $d_1(s)$ does not divide and multiply $F_1(s)$ on the left by an E^{i1}-matrix so as to add the row containing the offending π to the first row of $F_1(s)$. Now repeat the reduction step on this new matrix. What results is a matrix of the form Equation 27.6 with a lower $\delta_{\min}$.

Repeat the divisibility check on this new matrix. The process terminates eventually in a matrix of the form Equation 27.6 whose $d_1(s)$ divides all the polynomials π. Note that this newest $F_1(s)$ can be written in the form $U_1(s)F(s)R_1(s)$ for some unimodular matrices $U_1(s)$ and $R_1(s)$. The next phase of the Smith form computation entails performing the reduction step and divisibility check on the $(k-1 \times r-1)$ matrix of πs in $F_1(s)$. What results is a matrix

$$F_2(s) = \begin{bmatrix} d_1(s) & 0 & 0 & . & . & 0 \\ 0 & d_2(s) & 0 & . & . & 0 \\ 0 & 0 & \pi & . & . & \pi \\ 0 & 0 & \pi & . & . & \pi \\ . & . & . & . & . & . \\ . & . & . & . & . & . \\ 0 & 0 & \pi & . & . & \pi \end{bmatrix}$$

in which $d_2(s)$ divides all of the πs. Moreover, $d_1(s)$ divides $d_2(s)$ because $d_2(s)$ is a polynomial linear combination of the πs in $F_1(s)$. Furthermore, there are unimodular matrices $U_2(s)$ and $R_2(s)$ so that $F_2(s) = U_2(s)F_1(s)R_2(s)$, whereby $F_2(s) = U_2(s)U_1(s)F(s)R_1(s)R_2(s)$.

Continuing in this fashion leads successively to $F_3(s), \ldots$, and finally $F_r(s)$, which has the same form as $\Lambda(s)$. To get $\Lambda(s)$, modify $F_r(s)$ by scaling all of the rows of $F_r(s)$ so that the nonzero $d_j(s)$-polynomials are monic. It is evident that there are unimodular matrices $U(s)$ and $R(s)$ so that $\Lambda(s) = U(s)F(s)R(s)$.

A few comments are in order. First, if $F(s)$ is a real $(k \times r)$ polynomial matrix with $k \leq r$, then applying Theorem 27.2 to $F^T(s)$ and transposing the result yields unimodular matrices $U(s)$ and $R(s)$ so that $U(s)F(s)R(s)$ takes the form

$$\Lambda(s) = \begin{bmatrix} d_1(s) & 0 & 0 & . & . & 0 & 0 & . & 0 \\ 0 & d_2(s) & 0 & . & . & . & . & . & 0 \\ 0 & 0 & d_3(s) & 0 & . & . & . & . & 0 \\ . & . & . & . & . & . & . & . & . \\ . & . & . & . & . & . & . & . & 0 \\ 0 & . & . & . & 0 & d_k(s) & 0 & . & 0 \end{bmatrix}; \qquad (27.7)$$

this is the Smith form of an $F(s)$ with more columns than rows.

Next, consider the polynomials $\{d_j(s)\}$ in Equation 27.5; these are called the *elementary divisors* of $F(s)$. It might not appear prima facie as if the proof of Theorem 27.2 specified them uniquely, but it does. In fact, we have the following explicit characterization.

FACT 27.1 *For each j, $1 \leq j \leq \min(k, r)$, the product $d_1(s)\cdots d_j(s)$ is the monic greatest common divisor of all of the $(j \times j)$ minors in $F(s)$.*

PROOF 27.2 The divisibility conditions on the $\{d_j(s)\}$ make it obvious that $d_1(s)\cdots d_j(s)$ is the monic greatest common divisor of the $(j \times j)$ minors in $\Lambda(s)$. Fact 27.1 follows immediately from the observation that the reduction procedure leading from $F(s)$ to $\Lambda(s)$ is such that for each i, the set of $(j \times j)$ minors of $F_i(s)$ has the same family of common divisors as the set of $(j \times j)$ minors of $F_{i+1}(s)$. (See also the proof of Corollary 27.1 below.) As a result, the monic greatest common divisor of the $(j \times j)$ minors does not change over the course of the procedure.

One final comment: the sequence of elementary divisors, in general, might start out with one or more 1s and terminate with one or more 0s. In fact, because every nonzero elementary divisor is a monic polynomial, each constant elementary divisor is either a 1 or a 0. If $F(s)$ is a square matrix, then det $F(s)$ is a nonzero real multiple of the product of its elementary divisors; this follows from the unimodularity of $U(s)$ and $R(s)$ in Theorem 27.2. Hence if $F(s)$ is nonsingular, none of its elementary divisors is zero. If $F(s)$ is unimodular, then all of its elementary divisors are equal to 1. In other words, the Smith form of a $(k \times k)$ unimodular matrix is simply the $(k \times k)$ identity matrix I_k.

The Smith form is the key to finishing the proof of Theorem 27.1. The following lemma, which offers useful alternative characterizations of coprimeness, is the first step in that direction.

LEMMA 27.1 Let $P(s)$ and $Q(s)$ be real polynomial matrices having respective sizes $(p \times m)$ and $(m \times m)$. The following three conditions are equivalent:

1. $P(s)$ and $Q(s)$ are right coprime.
2. There exist real polynomial matrices $X(s)$ and $Y(s)$ with respective sizes $(m \times m)$ and $(m \times p)$ so that $X(s)Q(s) + Y(s)P(s) = I_m$.
3. The $(m + p \times m)$ matrix

$$\begin{bmatrix} Q(s_o) \\ P(s_o) \end{bmatrix}$$

has full rank m for every complex number s_o.

PROOF 27.3 We show that $(1) \Longrightarrow (2) \Longrightarrow (3) \Longrightarrow (1)$. Set

$$F(s) = \begin{bmatrix} Q(s) \\ P(s) \end{bmatrix}$$

and, using the notation of Theorem 27.2, let $\Lambda(s) = U(s)F(s)R(s)$ be the Smith form of $F(s)$. Denote by $\Delta(s)$ the

$(m \times m)$ matrix comprising the first m rows of $\Lambda(s)$; the diagonal elements of $\Delta(s)$ are the elementary divisors of $F(s)$.

If (1) holds, we need $\Delta(s) = I_m$. To see this, partition $U^{-1}(s)$ as

$$U^{-1}(s) = W(s) = \begin{bmatrix} W_1(s) & W_2(s) \\ W_3(s) & W_4(s) \end{bmatrix};$$

then

$$F(s) = \begin{bmatrix} Q(s) \\ P(s) \end{bmatrix} = \begin{bmatrix} W_1(s) \\ W_3(s) \end{bmatrix} \Delta(s) R^{-1}(s),$$

so that $P(s)$ and $Q(s)$ have $\Delta(s)R^{-1}(s)$ is an $(m \times m)$ as a right common polynomial matrix factor. For (1) to hold, this last matrix must be unimodular, and that happens when $\Delta(s) = I_m$.

Hence, right coprimeness of $P(s)$ and $Q(s)$ implies that $\Delta(s) = I_m$. It follows immediately in this case that if we partition the first m rows of $U(s)$ as $\begin{bmatrix} U_1(s) & U_2(s) \end{bmatrix}$, then (2) holds with $X(s) = R(s)U_1(s)$ and $Y(s) = R(s)U_2(s)$. (3) is a straightforward consequence of (2) because, if $F(s_o)$ were rank-deficient for some s_o, then $X(s_o)Q(s_o) + Y(s_o)P(s_o)$ could not be I_m. Finally, (3) implies (1) because, if (3) holds and

$$F(s) = \begin{bmatrix} \hat{P}(s) \\ \hat{Q}(s) \end{bmatrix} \hat{R}(s)$$

is a factorization with $\hat{R}(s)$ $(m \times m)$ but not unimodular, then $\det \hat{R}(s_o) = 0$ for at least one complex number s_o, which contradicts (3).

Not surprisingly, Lemma 27.1 has the following left-handed analogue.

LEMMA 27.2 Let $P_L(s)$ and $Q_L(s)$ be real polynomial matrices with respective sizes $(p \times m)$ and $(p \times p)$. The following three conditions are equivalent:

1. $P_L(s)$ and $Q_L(s)$ are left coprime.
2. Real polynomial matrices $X(s)$ and $Y(s)$ exist with respective sizes $(p \times p)$ and $(m \times p)$ so that $Q_L(s)X(s) + P_L(s)Y(s) = I_p$.
3. The $(p \times p + m)$ matrix

$$\begin{bmatrix} Q_L(s_o) & P_L(s_o) \end{bmatrix}$$

has full rank p for every complex number s_o.

We are ready now to finish proving Theorem 27.1.

Proof of Theorem 27.1: We prove only the assertions about right MFDs. We have shown already that a minimal right MFD of $G(s)$ is irreducible. As for the converse, suppose $G(s) = P(s)Q^{-1}(s)$ is an irreducible right MFD and that $G(s) = \hat{P}(s)\hat{Q}^{-1}(s)$ is a minimal (hence irreducible) right MFD. By Lemma 27.1, there exist $X(s)$, $Y(s)$, $\hat{X}(s)$, and $\hat{Y}(s)$ of appropriate sizes satisfying

$$\begin{aligned} X(s)Q(s) + Y(s)P(s) &= I_m \\ \hat{X}(s)\hat{Q}(s) + \hat{Y}(s)\hat{P}(s) &= I_m . \end{aligned}$$

Invoke the fact that $P(s)Q^{-1}(s) = \hat{P}(s)\hat{Q}^{-1}(s)$; a straightforward manipulation yields

$$\begin{aligned} X(s)\hat{Q}(s) + Y(s)\hat{P}(s) &= Q^{-1}(s)\hat{Q}(s), \\ \hat{X}(s)Q(s) + \hat{Y}(s)P(s) &= \hat{Q}^{-1}(s)Q(s). \end{aligned}$$

The matrices on the right-hand sides are inverses of each other and are polynomial matrices; hence they must be unimodular. This implies, in particular, that the degrees of the determinants of $Q(s)$ and $\hat{Q}(s)$ are the same, and $P(s)Q^{-1}(s)$ is, therefore, also a minimal right MFD.

Finally, setting $V(s) = Q^{-1}(s)\hat{Q}(s)$ reveals that $\hat{P}(s) = P(s)V(s)$ and $\hat{Q}(s) = Q(s)V(s)$, proving that the two minimal right MFDs are related as in the theorem statement.

Theorem 27.1 establishes the equivalence of minimality and irreducibility for MFDs. It is worth remarking that its proof furnishes a means for reducing a nonminimal MFD to a minimal one. The argument supporting Lemma 27.1 provides the key. Let $G(s) = P(s)Q^{-1}(s)$ be a nonminimal right MFD. Employing the same notation as in the proof of Lemma 27.1,

$$\begin{aligned} F(s) &= \begin{bmatrix} Q(s) \\ P(s) \end{bmatrix} = \begin{bmatrix} W_1(s) \\ W_3(s) \end{bmatrix} \times \\ &\quad \Delta(s)R^{-1}(s) \stackrel{\text{def}}{=} \hat{F}(s)\Delta(s)R^{-1}(s). \end{aligned}$$

Setting $\hat{Q}(s) = W_1(s)$ and $\hat{P}(s) = W_2(s)$ makes $\hat{P}(s)\hat{Q}^{-1}(s)$ a right MFD of $G(s)$. By definition of $W(s)$ and $U(s)$,

$$U(s)\hat{F}(s) = \begin{bmatrix} I_m \\ 0 \end{bmatrix};$$

taken in conjunction with Theorem 27.2, this implies that $\hat{F}(s)$ has ones as its elementary divisors. As in the proof of Lemma 27.1, we conclude that $\hat{P}(s)\hat{Q}^{-1}(s)$ is an irreducible (hence minimal) right MFD.

Theorem 27.1 also makes an important assertion about the "denominator matrices" appearing in minimal MFDs. The theorem states that if $Q(s)$ and $\hat{Q}(s)$ are denominator matrices in two minimal right MFDs of a proper rational $G(s)$, then $Q(s)$ and $\hat{Q}(s)$ differ by a right unimodular matrix factor. It follows that the determinants of $Q(s)$ and $\hat{Q}(s)$ are identical up to a nonzero real multiple; in particular, $\det Q(s)$ and $\det \hat{Q}(s)$ have the same roots including multiplicities. The same is true about the determinants of the denominator matrices $Q_L(s)$ and $\hat{Q}_L(s)$ from two minimal left MFDs. It is clear that the poles of $G(s)$ must lie among the roots of $\det Q(s)$ and $\det Q_L(s)$. What is not so obvious is the fact, whose proof we postpone until the next section, that these last two polynomials are actually nonzero real multiples of each other.

27.3 Fractional Degree and MacMillan Degree

Conspicuously absent from Section 27.2 is any substantive discussion of relationships between right and left MFDs for a proper rational matrix $G(s)$. The most important connection enables us to close the circle of ideas encompassing the minimal MFDs of Section 27.2 and the irreducible fractional representation Equation 27.1 for a scalar rational function. A crucial feature of Equation 27.1 is that the degree of $q(s)$ is the same as the MacMillan degree of $g(s)$. Our principal goal in what follows is to prove a similar assertion about the MacMillan degree of $G(s)$ and the degrees of the determinants of the denominator matrices appearing in minimal right and left MFDs of $G(s)$.

Our first task is to demonstrate that, when $P(s)Q^{-1}(s)$ and $Q_L^{-1}(s)P_L(s)$ are right and left MFDs of a proper rational matrix $G(s)$, the degrees of the polynomials $\det Q(s)$ and $\det Q_L(s)$ are the same. To that end, we need the following technical lemma.

LEMMA 27.3 Suppose that the $(m+p \times m+p)$ polynomial matrix $W(s)$ is nonsingular and that the $(m \times m)$ submatrix $W_1(s)$ is also nonsingular, where

$$W(s) = \begin{bmatrix} W_1(s) & W_2(s) \\ W_3(s) & W_4(s) \end{bmatrix}.$$

Then,

1. $H(s) = W_4(s) - W_3(s)W_1^{-1}(s)W_2(s)$ is also nonsingular;
2. $\det W(s) = \det W_1(s) \det H(s)$; and
3. $W^{-1}(s)$ is given by

$$\begin{bmatrix} W_1^{-1}(s) + W_1^{-1}(s)W_2(s)H^{-1}(s)W_3(s)W_1^{-1}(s) & -W_1^{-1}(s)W_2(s)H^{-1}(s) \\ -H^{-1}(s)W_3(s)W_1^{-1}(s) & H^{-1}(s) \end{bmatrix}$$

PROOF 27.4 Statements (1) and (2) follow from the identity

$$\begin{bmatrix} I_m & 0 \\ -W_3(s)W_1^{-1}(s) & I_p \end{bmatrix} W(s) = \begin{bmatrix} W_1(s) & W_2(s) \\ 0 & H(s) \end{bmatrix},$$

because the matrix multiplying $W(s)$ has determinant 1. Multiplying the last equation on the left by

$$\begin{bmatrix} W_1^{-1}(s) & 0 \\ 0 & H^{-1}(s) \end{bmatrix} \begin{bmatrix} I_m & -W_2(s)H^{-1}(s) \\ 0 & I_p \end{bmatrix}$$

yields statement (3).

A direct consequence of Lemma 27.3 is the advertised relationship between determinants of the denominator matrices in right and left MFDs.

THEOREM 27.3 *Let $P(s)Q^{-1}(s)$ and $Q_L^{-1}(s)P_L(s)$, respectively, be minimal right and left MFDs of a real $(p \times m)$ proper rational matrix $G(s)$. Then* $\det Q_L(s)$ *is a nonzero real multiple of* $\det Q(s)$; *in particular, the two polynomials have the same degree and the same roots.*

PROOF 27.5 We learned in the proof of Lemma 27.1 that unimodular matrices $U(s)$ and $R(s)$ exist so that

$$U(s)\begin{bmatrix} Q(s) \\ P(s) \end{bmatrix} R(s) = \begin{bmatrix} I_m \\ 0 \end{bmatrix}.$$

Set

$$M(s) = \begin{bmatrix} R(s) & 0 \\ 0 & I_p \end{bmatrix} U(s)$$

and define $W(s) = M^{-1}(s)$; observe that $M(s)$ and $W(s)$ are unimodular. Partition $M(s)$ and $W(s)$ conformably as follows:

$$M(s) = \begin{bmatrix} M_1(s) & M_2(s) \\ M_3(s) & M_4(s) \end{bmatrix};$$

$$W(s) = \begin{bmatrix} W_1(s) & W_2(s) \\ W_3(s) & W_4(s) \end{bmatrix}.$$

It follows that $W_1(s) = Q(s)$ and $W_3(s) = P(s)$. In particular, $W_1(s)$ is nonsingular. By Lemma 27.3, $M_4(s)$ is nonsingular and equals $H^{-1}(s)$, where $H(s) = W_4(s) - W_3(s)W_1^{-1}(s)W_2(s)$. Because $M(s)W(s) = I_{m+p}$,

$$M_3(s)W_2(s) + M_4(s)W_4(s) = I_p;$$

hence $M_3(s)$ and $M_4(s)$ are left coprime by Lemma 27.1. Furthermore,

$$M_3(s)W_1(s) + M_4(s)W_3(s) = 0;$$

the fact that $G(s) = P(s)Q^{-1}(s) = W_3(s)W_1^{-1}(s)$ makes

$$G(s) = M_4^{-1}(s)[-M_3(s)]$$

a minimal left MFD of $G(s)$. By Theorem 27.1, $\det M_4(s)$ is a nonzero real multiple of the determinant of the matrix $Q_L(s)$ appearing in any minimal left MFD $Q_L^{-1}(s)P_L(s)$ of $G(s)$.

By item (2) in Lemma 27.3, $\det W(s) = \det W_1(s) \det H(s)$. Because $W_1(s) = Q(s)$ and $M_4(s) = H^{-1}(s)$,

$$\det Q(s) = \det W(s) \det M_4(s).$$

Unimodularity of $W(s)$ implies that $\det Q(s)$ and $\det M_4(s)$ differ by a nonzero real multiple.

Theorem 27.3 makes possible the following definition.

DEFINITION 27.6 The *fractional degree* $\delta_F[G(s)]$ of a real $(p \times m)$ proper rational matrix $G(s)$ is the degree of $\det Q(s)$ (or of $\det Q_L(s)$), where $Q(s)$ (or $Q_L(s)$) comes from a minimal right (or left) MFD $P(s)Q^{-1}(s)$ (or $Q_L^{-1}(s)P_L(s)$) of $G(s)$.

As promised, we will demonstrate below that the fractional degree of a proper rational matrix $G(s)$ is the same as its MacMillan degree $\delta_M[G(s)]$. Our approach will be first to show that $\delta_F \leq \delta_M$ and then to construct a state space realization for $G(s)$ with state vector dimension $\delta_F[G(s)]$. The existence of such a realization guarantees that $\delta_F \geq \delta_M$, from which it follows that the two degrees are equal.

Accordingly, let $G(s)$ be a given real $(p \times m)$ proper rational matrix. Let $D = \lim_{s\to\infty} G(s)$; set $Z(s) = G(s) - D$. If $P(s)Q^{-1}(s)$ is a minimal right MFD of $G(s)$, then $\tilde{P}(s)Q^{-1}(s)$ is a minimal right MFD for $Z(s)$, where $\tilde{P}(s) = P(s) - DQ(s)$. To see why $\tilde{P}(s)Q^{-1}(s)$ is minimal, note that the two matrices

$$F(s_o) = \begin{bmatrix} Q(s_o) \\ P(s_o) \end{bmatrix}, \quad \tilde{F}(s_o) = \begin{bmatrix} Q(s_o) \\ P(s_o) - DQ(s_o) \end{bmatrix}$$

have the same nullspace (and hence the same rank) for every complex number s_o. It follows from item 3 in Lemma 27.1 that $\tilde{P}(s)Q^{-1}(s)$ is irreducible and hence minimal. In addition, we can conclude that the fractional degrees of $G(s)$ and $Z(s)$ are the same.

Suppose that $n = \delta_M(G(s))$ and that (A, B, C, D) is a minimal realization of $G(s)$. The Popov-Belevitch-Hautus test for reachability [11] implies that the $(n \times n + m)$ matrix,

$$K(s_o) = \begin{bmatrix} (s_o I_n - A) & B \end{bmatrix}$$

has full rank n for every complex number s_o. By Lemma 27.1, $(sI_n - A)^{-1}B$ is an irreducible (hence minimal) left MFD of $K(s)$. Thus $K(s)$ has fractional degree n. Now,

$$Z(s) = C(sI_n - A)^{-1}B = CF(s);$$

it follows that the fractional degree of $Z(s)$ is at most equal to n. To see this, suppose $K(s) = P(s)Q^{-1}(s)$ is a minimal right MFD of $K(s)$; then $Z(s) = [CP(s)]Q^{-1}(s)$ is a right MFD of $Z(s)$, and this MFD need not be minimal. Hence $\delta_F(Z) \le \delta_F(K) = n$. The upshot is that the fractional degree of $G(s)$, which is the same as the fractional degree of $Z(s)$, is bounded from above by the MacMillan degree of $G(s)$. In other words,

$$\delta_F[G(s)] \le \delta_M[G(s)].$$

Proving the reverse inequality requires a bit more effort. Suppose we have a minimal right MFD $G(s) = P(s)Q^{-1}(s)$ and corresponding right MFD $Z(s) = \tilde{P}(s)Q^{-1}(s)$. The fractional degree of $G(s)$ is the same as the degree of det $Q(s)$. To show how the entries in $Q(s)$ combine to determine the degree of det $Q(s)$, we need the following definition.

DEFINITION 27.7 Let $Q(s)$ be a real nonsingular $(m \times m)$ polynomial matrix. The *jth column degree of $Q(s)$*, $\delta_j[Q(s)]$, is the highest of the degrees of the polynomials in the jth column of $Q(s)$, $1 \le j \le m$. The *high-order coefficient matrix of $Q(s)$*, Q_H, is the real $(m \times m)$ matrix whose (i, j) entry is the coefficient of $s^{\delta_j[Q(s)]}$ in $[Q(s)]_{ij}$.

The nonsingularity of $Q(s)$ guarantees that all of the column degrees are nonnegative integers. The classic expansion for the determinant [12, page 157] reveals det $Q(s)$ as the sum of m-fold products each of which contains precisely one element from each column of $Q(s)$. It follows that the degree of det $Q(s)$ is bounded from above by the sum of all the column degrees of $Q(s)$. In fact, if $\delta_1, \ldots, \delta_m$ are the column degrees of $Q(s)$, then the coefficient of the $s^{(\delta_1+\cdots+\delta_m)}$ in the expansion for det $Q(s)$ is exactly det Q_H, the determinant of the high-order coefficient matrix. In other words, det $Q(s)$ has degree *equal* to the sum of the column degrees when det $Q_H \neq 0$, which is the same as saying that the high-order coefficient matrix is invertible.

Our method for constructing for $G(s)$ a realization whose state dimension is $\delta_F[G(s)]$ hinges crucially on having a minimal right MFD $P(s)Q^{-1}(s)$ of $G(s)$ whose $Q(s)$ has an invertible high-order coefficient matrix. It turns out that such an MFD always exists. The idea is to start with an arbitrary minimal right MFD $\hat{P}(s)\hat{Q}^{-1}(s)$ and "operate on it" with a unimodular $V(s)$ via

$$P(s) = \hat{P}(s)V(s), \quad Q(s) = \hat{Q}(s)V(s)$$

to get another minimal right MFD $P(s)Q^{-1}(s)$ with an invertible Q_H. We construct $V(s)$ by looking at $\hat{Q}(s)$ only. Specifically, we prove the following assertion.

LEMMA 27.4 If $\hat{Q}(s)$ is a real nonsingular $(m \times m)$ polynomial matrix, a unimodular matrix $V(s)$ exists so that

- $Q(s) = \hat{Q}(s)V(s)$ has an invertible high-order coefficient matrix Q_H, and
- the column degrees $\{\delta_j(Q(s))\}$ are in decreasing order, *i.e.*,

$$\delta_1(Q(s)) \ge \delta_2(Q(s)) \cdots \ge \delta_m(Q(s)). \quad (27.8)$$

PROOF 27.6 If $\hat{Q}_H$ is already invertible, let $V(s)$ be a permutation matrix Π so that the columns of $\hat{Q}(s)\Pi$ are lined up in decreasing order of column degree. If $\hat{Q}_H$ is not invertible, after finding Π as above, choose a nonzero $w \in \mathbf{R}^m$ satisfying $\hat{Q}_H \Pi w = 0$. Assume without loss of generality that the first nonzero element in w is a 1 and occurs in the kth position. Let $E(s)$ be the $(m \times m)$ polynomial matrix all of whose columns except the kth are the same as those in the $(m \times m)$ identity matrix I_m; as for the the kth column, let $[E(s)]_{ik}$ be 0 when $i < k$ and $w_i s^{\delta_k - \delta_i}$ when $i > k$, where δ_j denotes the jth column degree of $\hat{Q}(s)\Pi$. $E(s)$ has determinant 1 and is therefore unimodular; furthermore, $\hat{Q}(s)\Pi E(s)$ has the same columns as $\hat{Q}(s)\Pi$ except for the kth column, and the choice of $E(s)$ guarantees that the kth column degree of $\hat{Q}(s)\Pi E(s)$ is lower than the kth column degree of $\hat{Q}(s)\Pi$.

The preceding paragraph describes a technique for taking a $\hat{Q}(s)$ with singular $\hat{Q}_H$ and finding a unimodular matrix $\Pi E(s)$ so that $\hat{Q}(s)\Pi E(s)$ has a set of column degrees whose sum is less than the sum of the column degrees of $\hat{Q}(s)$. If $Q(s) = \hat{Q}(s)\Pi E(s)$ still fails to have an invertible high-order coefficient matrix, we can repeat the column-permutation-and-reduction procedure on $\hat{Q}(s)\Pi E(s)$, and so on. This iteration must terminate after a finite number of steps because we cannot reduce the sum of the column degrees forever. When all is said and done, we will have a unimodular matrix $V(s)$ so that $Q(s) = \hat{Q}(s)V(s)$ has an invertible high-order coefficient matrix Q_H with columns arrayed in decreasing order of column degree, so that the column degrees of $Q(s)$ satisfy Equation 27.8.

Thus we can take an arbitrary minimal right MFD $G(s) = \hat{P}(s)\hat{Q}^{-1}(s)$ and form a new minimal right MFD $P(s)Q^{-1}(s)$ using $P(s) = \hat{P}(s)V(s)$ and $Q(s) = \hat{Q}(s)V(s)$; choosing $V(s)$ appropriately, using Lemma 27.4, makes Q_H invertible, ensuring in turn that

$$\delta_F[G(s)] = \text{degree}[\det Q(s)] = \delta_1[Q(s)] + \cdots + \delta_m[Q(s)].$$

Furthermore, we can assume that the column degrees of $Q(s)$ satisfy the ordering of Equation 27.8.

Our final step is to produce a realization (A, B, C, D) of $G(s)$ where A is $(\delta_F \times \delta_F)$. This will confirm that $\delta_F(G(s)) \geq \delta_M(G(s))$ and, hence, that the two degrees are equal. First define D and $Z(s) = G(s) - D$ as before, and let $Z(s) = \tilde{P}(s)Q^{-1}(s)$ be the corresponding minimal right MFD for $Z(s)$. Because

$$[\tilde{P}(s)]_{ij} = \sum_{k=1}^{m}[Z(s)]_{ik}[Q(s)]_{kj},$$

and because $Z(s)$ is strictly proper, the degree of $[\tilde{P}(s)]_{ij}$ is strictly less than δ_j for all i and j. In particular, if $\delta_j = 0$, then $[\tilde{P}(s)]_{ij} = 0$ for all i, $1 \leq i \leq p$ (as usual, the zero polynomial has degree $-\infty$).

Next define, for each $k > 0$, the polynomial k-vector $\mathbf{s}_k$ by

$$\mathbf{s}_k = \begin{bmatrix} 1 \\ s \\ s^2 \\ \cdot \\ \cdot \\ s^{k-1} \end{bmatrix}.$$

We now form a matrix $\mathbf{S}$ that has m columns and number of rows equal to δ_F, which is equal in turn to the sum of the column degrees $\{\delta_j\}$. The jth column of $\mathbf{S}$ has $\delta_1 + \cdots + \delta_{j-1}$ 0s at the top, the vector $\mathbf{s}_{\delta_j}$ in the next δ_j positions, and 0s in the remaining positions. For example, if $m = 3$ and $\delta_1 = 3$, $\delta_2 = 2$, and $\delta_3 = 0$, then

$$\mathbf{S} = \begin{bmatrix} 1 & 0 & 0 \\ s & 0 & 0 \\ s^2 & 0 & 0 \\ 0 & 1 & 0 \\ 0 & s & 0 \end{bmatrix}.$$

Our observation above concerning the degrees of the entries in $\tilde{P}(s)$ ensures that a real $(p \times \delta_F)$ matrix C exists so that $\tilde{P}(s) = C\mathbf{S}$. This C will be the C-matrix in our realization of $G(s)$. Since we want

$$\begin{aligned} Z(s) &= \tilde{P}(s)Q^{-1}(s) = C\mathbf{S}Q^{-1}(s)B \\ &= C(sI_{\delta_F} - A)^{-1}B, \end{aligned}$$

we will construct A and B so that

$$\mathbf{S}Q^{-1}(s) = (sI_{\delta_F} - A)^{-1}B,$$

or, equivalently,

$$s\mathbf{S} = A\mathbf{S} + BQ(s). \tag{27.9}$$

Recall first that Q_H, the high-order coefficient matrix of $Q(s)$, is invertible; by definition of Q_H and the column degrees $\{\delta_j\}$,

$$Q(s) = Q_H \begin{bmatrix} s^{\delta_1} & 0 & \cdot & \cdot & \cdot & 0 \\ 0 & s^{\delta_2} & 0 & \cdot & \cdot & \cdot \\ \cdot & 0 & \cdot & \cdot & \cdot & \cdot \\ \cdot & \cdot & \cdot & \cdot & 0 & \cdot \\ \cdot & \cdot & \cdot & 0 & s^{\delta_{m-1}} & 0 \\ \cdot & \cdot & \cdot & \cdot & 0 & s^{\delta_m} \end{bmatrix} + \tilde{Q}(s), \tag{27.10}$$

where $\tilde{Q}(s)$ satisfies the same constraints as $\tilde{P}(s)$ on the degrees of its entries. We may write $\tilde{Q}(s) = G\mathbf{S}$ for some real $(m \times \delta_F)$ matrix G. Denote by $\Sigma(s)$ the diagonal matrix on the right-hand side of Equation 27.10. Then A and B satisfy Equation 27.9 when

$$s\mathbf{S} = (A + BQ_HG)\mathbf{S} + BQ_H\Sigma(s). \tag{27.11}$$

Define B as follows: $B = \tilde{B}Q_H^{-1}$, where $\tilde{B}$ is the real $(\delta_F \times m)$ matrix whose jth column is zero when $\delta_j = 0$ and contains a single 1 at the $\delta_1 + \cdots + \delta_j$ position if $\delta_j \neq 0$. Observe that if $i = \delta_1 + \cdots + \delta_j$ for some j, then the ith row of BQ_HG is the same as the jth row of G; for all other values of i, the ith row of BQ_HG is zero.

Finally, define A. If $i = \delta_1 + \cdots + \delta_j$ for some j, then the ith row of A is the negative of the ith row of BQ_HG. For other values of i, the ith row of A contains a 1 in the $(i, i+1)$ position (just above the diagonal) and 0s elsewhere.

Verifying that A and B so defined satisfy the required relationship Equation 27.11 is straightforward. The important consequence of the construction is that $Z(s) = C(sI_{\delta_F} - A)^{-1}B$, so that (A, B, C, D) is a realization of $G(s)$ whose A-matrix has size $(\delta_F \times \delta_F)$. The following theorem summarizes the results of this section so far.

THEOREM 27.4 *If $G(s)$ is a real $(p \times m)$ proper rational matrix, then the MacMillan degree of $G(s)$ is the same as the fractional degree of $G(s)$. The procedure outlined above yields a minimal realization (A, B, C, D) of $G(s)$.*

The central role of Lemma 27.4 in the proof of Theorem 27.4 merits a closer look. In essence, Lemma 27.4 guarantees the existence of a minimal right MFD $G(s) = P(s)Q^{-1}(s)$ wherein the column degrees of $Q(s)$ sum to the degree of the determinant of $Q(s)$. The crucial enabling feature of $Q(s)$ is the invertibility of its high-order coefficient matrix Q_H. We will see presently that any minimal right MFD whose denominator matrix possesses this last property will have the same column degrees as $Q(s)$ up to a reordering. As a result, these special column degrees are a feature of certain right MFDs and of the transfer function matrix $G(s)$ itself.

DEFINITION 27.8 The *ordered column indices* $l_1[Q(s)], \ldots, l_m[Q(s)]$ of a real nonsingular $(m \times m)$ polynomial matrix $Q(s)$ are the column degrees of $Q(s)$ arranged in decreasing order.

THEOREM 27.5 *Let $Q(s)$ and $\hat{Q}(s)$ be real nonsingular $(m \times m)$ polynomial matrices appearing in minimal right MFDs $P(s)Q^{-1}(s)$ and $\hat{P}(s)\hat{Q}^{-1}(s)$ of a real $(p \times m)$ proper rational matrix $G(s)$. If the high-order coefficient matrices Q_H and $\hat{Q}_H$ are both invertible, then the ordered column indices of $Q(s)$ and $\hat{Q}(s)$ are identical.*

PROOF 27.7 Assume first that $Q(s)$ and $\hat{Q}(s)$ satisfy the second item in Lemma 27.4, i.e., have respective column degrees $\{\delta_j\}$ and $\{\hat{\delta}_j\}$ that are decreasing in j. Write Equation 27.10 along with a similar equation for $\hat{Q}(s)$ as follows:

$$\begin{aligned} Q(s) &= Q_H \Sigma(s) + \tilde{Q}(s) \\ \hat{Q}(s) &= \hat{Q}_H \hat{\Sigma}(s) + \tilde{\hat{Q}}(s) . \end{aligned}$$

By construction,

$$\begin{aligned} \lim_{|s|\to\infty} \tilde{Q}(s)\Sigma^{-1}(s) &\overset{\text{def}}{=} \lim_{|s|\to\infty} \Delta(s) = 0, \\ \lim_{|s|\to\infty} \tilde{\hat{Q}}(s)\hat{\Sigma}^{-1}(s) &\overset{\text{def}}{=} \lim_{|s|\to\infty} \hat{\Delta}(s) = 0 . \end{aligned}$$

Because $Q(s)$ and $\hat{Q}(s)$ both come from minimal right MFDs of $G(s)$, by Theorem 27.1 an $(m \times m)$ unimodular matrix $V(s)$ exists so that $\hat{Q}(s) = Q(s)V(s)$. Manipulation yields

$$\Sigma(s)U(s)\hat{\Sigma}^{-1}(s) = [I_m + \Delta(s)]^{-1} Q_H^{-1} \hat{Q}_H [I_m + \hat{\Delta}(s)] . \tag{27.12}$$

The right-hand side of Equation 27.12 approaches a constant limit as $|s| \to \infty$; note, in particular, that $I_m + \Delta(s)$ is nonsingular for $|s|$ large enough. Meanwhile, the (i, j) entry of the matrix on the left-hand side of Equation 27.12 is simply $s^{\delta_i - \hat{\delta}_j}[U(s)]_{ij}$. Hence we need $[U(s)]_{ij} = 0$ whenever $\delta_i > \hat{\delta}_j$. One by one we will show that $\delta_j \le \hat{\delta}_j$. If $\delta_1 > \hat{\delta}_1$, then, by the ordering on the $\hat{\delta}$s, $\delta_1 > \hat{\delta}_j$ for all j, and the entire first row of $U(s)$ must be zero, contradicting nonsingularity and, a fortiori, unimodularity of $U(s)$. Assume inductively that $\delta_j \le \hat{\delta}_j$ for $j < k$ but that $\delta_k > \hat{\delta}_k$. In this case, the orderings on the δs and $\hat{\delta}$s imply that $\delta_i > \hat{\delta}_k \ge \hat{\delta}_j$ for all $i \le k$ and all $j \ge k$; hence the entire upper right-hand $(k \times m - k)$ corner of $U(s)$ must be zero, which contradicts unimodularity of $U(s)$ once again.

Thus $\delta_j \le \hat{\delta}_j$ for all j, $1 \le j \le m$. It follows that $\delta_j = \hat{\delta}_j$ for every j because the sum of the δs and the sum of the $\hat{\delta}$s must both equal the fractional degree of $G(s)$, which is the common degree of the determinants of $Q(s)$ and $\hat{Q}(s)$. Our initial assumption that the columns of $Q(s)$ and $\hat{Q}(s)$ were arrayed in decreasing order of column degree means, in terms of Definition 27.8, that $Q(s)$ and $\hat{Q}(s)$ have the same ordered column indices. Finally, it is easy to eliminate this initial assumption; simply precede the argument with right multiplications by permutation matrices Π and $\hat{\Pi}$ that reorder the matrices' columns appropriately. In any event, the ordered column indices of $Q(s)$ and $\hat{Q}(s)$ are identical.

The principal consequence of Theorem 27.5 is that any two $Q(s)$-matrices with invertible Q_Hs appearing in minimal right MFDs of $G(s)$ have the same set of ordered column indices. These special ordered column indices are sometimes called the *Kronecker controllability indices* of $G(s)$. They have other names, as well; Forney [4], for example, calls them *invariant dynamical indices.* They relate to controllability because the realization (A, B, C, D) we constructed *en route* to Theorem 27.4 is precisely the MIMO analogue to the SISO realization given in Equations 27.2–27.3. The realizations are called *controllable canonical forms* [2], [7], [10]. Interested readers can verify that applying the realization procedure following Lemma 27.4 to a scalar irreducible fractional representation Equation 27.1 leads exactly to Equations 27.2–27.3.

It is worth making one final observation. Our proof of Theorem 27.4 relied on constructing a minimal realization of a proper rational transfer matrix $G(s)$ starting from a minimal right MFD $P(s)Q^{-1}(s)$ of $G(s)$. We could have worked instead with a minimal left MFD $G(s) = Q_L^{-1}(s)P_L(s)$, in which case we would have considered the *row degrees* and high-order coefficient matrix of $Q_L(s)$. Perhaps the simplest way to view this is to realize $G^T(s)$ by following the route we have already laid out beginning with the right MFD $P_L^T(s)(Q_L^T(s))^{-1}$ of $G^T(s)$ and subsequently transposing the end result.

27.4 Smith–MacMillan Form, ARMA Models, and Stable Coprime Factorization

The aim of this section is to tie up some loose ends and to point the reader toward some important modern control theoretic developments that rest heavily on the theory of polynomial MFDs described in the foregoing sections. First we discuss the so-called Smith–MacMillan form for proper rational matrices; many of the results detailed in Sections 27.2–27.3 have alternative derivations based on the Smith–MacMillan form. Next, we describe briefly the connection between MFDs and ARMA models for MIMO linear systems. We close with a quick introduction to stable coprime factorization and mention briefly its generalizations and applications in robust control theory.

The Smith–MacMillan form of a real $(p \times m)$ proper rational matrix $G(s)$ is basically a rational version of the Smith form of a polynomial matrix $F(s)$. It was introduced originally by MacMillan [9] and later exploited by Kalman in an important paper [8] that demonstrated correspondences between several notions of rational matrix degree. Given $G(s)$, begin by letting $q(s)$ be the monic lowest common denominator of its entries. Set $F(s) = q(s)G(s)$; by Theorem 27.2, we can find unimodular matrices $U(s)$ and $R(s)$ of respective sizes $(p \times p)$ and $(m \times m)$ so that $\Lambda(s) = U(s)F(s)R(s)$ has the form of Equation 27.5 or 27.7 depending on whether $p \ge m$ or $p \le m$, respectively.

Assuming temporarily that $p \geq m$, the matrix $U(s)G(s)R(s) = \frac{1}{q(s)}F(s) \stackrel{\text{def}}{=} \Lambda_{SM}(s)$ takes the form

$$\begin{bmatrix} \gamma_1(s)/\phi_1(s) & 0 & 0 & \cdot & \cdot & 0 \\ 0 & \gamma_2(s)/\phi_2(s) & 0 & \cdot & \cdot & 0 \\ 0 & 0 & \gamma_3(s)/\phi_3(s) & \cdot & 0 & 0 \\ \cdot & \cdot & \cdot & \cdot & \cdot & \cdot \\ \cdot & \cdot & \cdot & \cdot & \cdot & 0 \\ 0 & \cdot & \cdot & \cdot & 0 & \gamma_m(s)/\phi_m(s) \\ 0 & \cdot & \cdot & \cdot & \cdot & 0 \\ \cdot & \cdot & \cdot & \cdot & \cdot & \cdot \\ \cdot & \cdot & \cdot & \cdot & \cdot & \cdot \\ 0 & \cdot & \cdot & \cdot & \cdot & 0 \end{bmatrix}, \tag{27.13}$$

where $\gamma_k(s)/\phi_k(s)$ is the fraction $d_k(s)/q(s)$ expressed in lowest terms. If $d_k(s) = 0$, set $\phi_k(s) = 1$. The divisibility conditions on $\{d_k(s)\}$ guaranteed by Theorem 27.2 ensure that $\gamma_k(s)$ divides $\gamma_{k+1}(s)$ and $\phi_{k+1}(s)$ divides $\phi_k(s)$ for all k, $1 \leq k < m$.

Furthermore, if we set

$$P(s) = U^{-1}(s)\begin{bmatrix} \gamma_1(s) & 0 & 0 & \cdot & \cdot & 0 \\ 0 & \gamma_2(s) & 0 & \cdot & \cdot & 0 \\ 0 & 0 & \gamma_3(s) & 0 & \cdot & 0 \\ \cdot & \cdot & \cdot & \cdot & \cdot & \cdot \\ \cdot & \cdot & \cdot & \cdot & \cdot & 0 \\ 0 & \cdot & \cdot & \cdot & 0 & \gamma_m(s) \\ 0 & \cdot & \cdot & \cdot & \cdot & 0 \\ \cdot & \cdot & \cdot & \cdot & \cdot & \cdot \\ \cdot & \cdot & \cdot & \cdot & \cdot & \cdot \\ 0 & \cdot & \cdot & \cdot & \cdot & 0 \end{bmatrix} \stackrel{\text{def}}{=} U^{-1}(s)\Gamma(s)$$

and

$$Q(s) = R(s)\begin{bmatrix} \phi_1(s) & 0 & 0 & \cdot & \cdot & 0 \\ 0 & \phi_2(s) & 0 & \cdot & \cdot & 0 \\ 0 & 0 & \phi_3(s) & 0 & \cdot & 0 \\ \cdot & \cdot & \cdot & \cdot & \cdot & \cdot \\ \cdot & \cdot & \cdot & \cdot & \cdot & 0 \\ 0 & \cdot & \cdot & \cdot & 0 & \phi_m(s) \end{bmatrix} \stackrel{\text{def}}{=} R(s)\Phi(s),$$

then $P(s)Q^{-1}(s)$ is a right MFD of $G(s)$.

$P(s)Q^{-1}(s)$ is minimal because of the divisibility conditions on the $\{\gamma_j\}$ and $\{\phi_j\}$. The easiest way to see this is by checking that the $(m + p \times m)$ matrix

$$\begin{bmatrix} Q(s_o) \\ P(s_o) \end{bmatrix} = \begin{bmatrix} R(s_o) & 0 \\ 0 & U^{-1}(s_o) \end{bmatrix}\begin{bmatrix} \Phi(s_o) \\ \Gamma(s_o) \end{bmatrix}$$

has full rank m for every complex number s_o. The idea is that if, for example, $\gamma_k(s_o) = 0$ for some smallest value of k, then $\gamma_j(s_o) = 0$ for all $j \geq k$; hence $\phi_j(s_o) \neq 0$ for $j \geq k$ which, coupled with $\gamma_j(s_o) \neq 0$ for $j < k$, means that m of the $\{\gamma_j(s_o)\}$ and $\{\phi_j(s_o)\}$ are nonzero. Because $P(s)Q^{-1}(s)$ is a minimal right MFD of $G(s)$, it follows from Theorem 27.4 that the MacMillan degree of $G(s)$ is the degree of det $Q(s)$, which is the sum of the degrees of the polynomials $\{\phi_j(s)\}$.

The same sort of analysis works when $p \leq m$; in that case, $\Lambda_{SM}(s) = U(s)G(s)R(s)$ looks like a rational version of the matrix in Equation 27.7. In either case, $\Lambda_{SM}(s)$ is called the Smith–MacMillan form of $G(s)$. To summarize,

THEOREM 27.6 **(Smith–MacMillan Form)**

Let $G(s)$ be a real $(p \times m)$ proper rational matrix and let $m \wedge p$ be the minimum of m and p. Unimodular matrices $U(s)$ and $R(s)$ exist so that $\Lambda_{SM}(s) = U(s)G(s)R(s)$, where

- *$[\Lambda_{SM}(s)]_{ij} = 0$, except when $i = j$ and $[\Lambda_{SM}(s)]_{jj}(s) = \gamma_j(s)/\phi_j(s)$, $1 \leq j \leq m \wedge p$, where the $\{\gamma_j(s)/\phi_j(s)\}$ are ratios of coprime monic polynomials,*
- *$\gamma_j(s)$ divides $\gamma_{j+1}(s)$ and $\phi_{j+1}(s)$ divides $\phi_j(s)$, $1 \leq j < m \wedge p$, and*
- *the MacMillan degree of $G(s)$ is the sum of the degrees of the $\{\phi_j(s)\}$.*

An interesting consequence of Theorem 27.6 is the following characterization of MacMillan degree [9].

COROLLARY 27.1 The MacMillan degree of a real $(p \times m)$ proper rational matrix $G(s)$ is the degree of the lowest common denominator of all of the minor subdeterminants of $G(s)$.

PROOF 27.8 Assume $p \geq m$; the argument is similar when $p \leq m$. A glance at $\Lambda_{SM}(s)$ in Equation 27.13 and the divisibility conditions on the $\{\phi_j(s)\}$ in Theorem 27.6 reveals that for each k, $1 \leq k \leq m$, the product $\phi_1(s) \cdots \phi_k(s)$ is the monic lowest common denominator of all of the $(j \times j)$ minors in $\Lambda_{SM}(s)$ of order $j \leq k$. Hence the product of all of the $\phi_j(s)$ is the monic lowest common denominator of all of the minors of $\Lambda_{SM}(s)$. Now, $G(s)$ and $\Lambda_{SM}(s)$ are related via pre- and postmultiplication by unimodular matrices; hence the $(j \times j)$ minor determinants of each matrix are polynomial linear combinations of the $(j \times j)$ minors of the other matrix. It follows that any common denominator for one set of minors is a common denominator for the other set, and the proof is complete.

Observe that we could have generated many of the results in Sections 27.2 and 27.4 by appealing to the Smith–MacMillan form. A disadvantage of this approach is that Theorem 27.6 produces only right MFDs for $(p \times m)$ rational matrices when $p > m$ and only left MFDs when $p < m$.

We consider next some relationships between MFDs and some well-known time-domain representations of input-output linear systems. Let $G(s)$ be a real $(p \times m)$ proper rational matrix that is the transfer function of a real m-input, p-output, time-invariant linear system with input $u : \mathbf{R} \to \mathbf{R}^m$ and output $y : \mathbf{R} \to \mathbf{R}^p$. Let $\mathcal{L}\{y\}(s)$ and $\mathcal{L}\{u\}(s)$ be the Laplace transforms of y and u, so $\mathcal{L}\{y\}(s) = G(s)\mathcal{L}\{u\}(s)$. If $G(s) = Q_L^{-1}(s)P_L(s)$ is a left MFD of $G(s)$, then

$$Q_L(s)\mathcal{L}\{y\}(s) = P_L(s)\mathcal{L}\{u\}(s);$$

in the time domain, this last equation corresponds with the vector differential equation

$$Q_L(D)y(t) = P_L(D)u(t), \tag{27.14}$$

where D is the differential operator $\frac{d}{dt}$. Equation 27.14 is an *ARMA (autoregressive moving-average) representation* for the system's input-output relationship.

Similarly, if $G(s) = P(s)Q^{-1}(s)$ is a right MFD of $G(s)$, we can define $w : \mathbf{R} \to \mathbf{R}^m$ by means of the autoregressive (AR) differential equation

$$Q(D)w(t) = u(t) \tag{27.15}$$

and use w as the input to a moving-average (MA) specification of y, namely,

$$y(t) = P(D)w(t). \tag{27.16}$$

Because

$$\begin{aligned} \mathcal{L}\{y\}(s) &= P(s)\mathcal{L}\{w\}(s) \text{ and } \mathcal{L}\{w\}(s) = Q^{-1}(s)\mathcal{L}\{u\}(s), \\ \mathcal{L}\{y\}(s) &= P(s)Q^{-1}(s)\mathcal{L}\{u\}(s) = G(s)\mathcal{L}\{u\}(s), \end{aligned}$$

so Equations 27.15 and 27.16 together constitute another time-domain description of the input-output behavior of the system. Whereas Equation 27.14 gives an ARMA description for the system, Equations 27.15–27.16 split the input-output relation into autoregressive and moving-average parts. For a SISO system, any fractional representation of the form Equation 27.1 acts as a left and right "MFD," so that the two time-domain characterizations are identical.

We close by presenting a very brief introduction to some of the ideas underlying stable coprime factorization, which is the single most important off-shoot of the theory of MFDs for input-output systems. We call a rational matrix $H(s)$ *stable* if, and only if, the poles of the entries in $H(s)$ lie in the open left half-plane $\text{Re}\{s\} < 0$. As usual, let $G(s)$ be a real $(p \times m)$ proper rational matrix. It turns out to be possible to write $G(s)$ in the form $G(s) = H_1(s)H_2^{-1}(s)$, where

- $H_1(s)$ and $H_2(s)$ are stable proper rational matrices of respective sizes $(p \times m)$ and $(m \times m)$, and
- $H_1(s)$ and $H_2(s)$ are *right coprime over the ring of stable rational functions,* that is, the only common right stable $(m \times m)$ proper rational matrix factors of $H_1(s)$ and $H_2(s)$ have inverses that are also stable and proper.

Any such representation $G(s) = H_1(s)H_2^{-1}(s)$ is called a *stable right coprime factorization* of $G(s)$. One can define stable left coprime factorizations similarly.

It is not difficult to show that stable coprime factorizations exist. One approach, patterned after a technique due originally to Vidyasagar [14], goes as follows. Given $G(s)$, choose $\alpha > 0$ so that $-\alpha$ is not a pole of any entry in $G(s)$. Let $\sigma = 1/(s+\alpha)$, so that $s = (1 - \alpha\sigma)/\sigma$. Define $\tilde{G}(\sigma) = G((1 - \alpha\sigma)/\sigma)$. It follows that $\tilde{G}(\sigma)$ is a proper rational matrix function of σ. To see why it is proper, observe that the condition $\sigma \to \infty$ is the same as $s \to -\alpha$, and $-\alpha$ is not a pole of $G(s)$.

Now invoke the theory of Section 27.2 and find a minimal right MFD $\tilde{G}(\sigma) = P(\sigma)Q^{-1}(\sigma)$ of $\tilde{G}(\sigma)$. Finally, set $H_1(s) = P[1/(s+\alpha)]$ and $H_2(s) = Q[1/(s+\alpha)]$. Then

$$G(s) = \tilde{G}(\frac{1}{s+\alpha}) = H_1(s)H_2^{-1}(s).$$

Because $P(\sigma)$ and $Q(\sigma)$ are polynomial in σ, $H_1(s)$ and $H_2(s)$ are proper rational matrix functions of s. Moreover, all of the poles of $H_1(s)$ and $H_1(s)$ are at $-\alpha$, which means that $H_1(s)$ and $H_2(s)$ are stable. Furthermore, any stable $(m \times m)$ proper rational right common factor $H(s)$ of $H_1(s)$ and $H_2(s)$ defines, via $V(s) = H((1 - \alpha\sigma)/\sigma)$, a polynomial right common factor of $P(\sigma)$ and $Q(\sigma)$, which must have a polynomial inverse by minimality of $P(\sigma)Q^{-1}(\sigma)$. It follows that $H^{-1}(s) = V^{-1}[1/(s+\alpha)]$ is stable and proper, implying that $H_1(s)$ and $H_2(s)$ are right coprime over the ring of stable proper rational functions.

The principal application of stable coprime factorizations is to robust control system design. At the heart of such applications is the notion of the H^∞ *norm* of a stable proper rational matrix $H(s)$. Given such an $H(s)$, the H^∞ norm of $H(s)$ is the supremum over $\omega \in \mathbf{R}$ of the largest singular value of $H(i\omega)$. Given two possibly unstable $(p \times m)$ transfer function matrices $G_a(s)$ and $G_b(s)$, one can define the distance between $G_a(s)$ and $G_b(s)$ in terms of the H^∞ norm of the *stable* rational matrix

$$\begin{bmatrix} H_{a1}(s) \\ H_{a2}(s) \end{bmatrix} - \begin{bmatrix} H_{b1}(s) \\ H_{b2}(s) \end{bmatrix},$$

where $G_a(s) = H_{a1}(s)H_{a2}^{-1}(s)$ and $G_b(s) = H_{b1}(s)H_{b2}^{-1}(s)$ are stable coprime factorizations of $G_a(s)$ and $G_b(s)$.

Interested readers can consult [1], [3], [5], [13], and the references therein for a through development of the ideas underlying robust control system design and their dependence on the theory of stable coprime factorization. A by-product of Vidyasagar's approach [13] is a framework for understanding MFDs and stable coprime factorizations in terms of more general themes from abstract algebra, notably ring theory. This framework reveals that many of our results possess natural generalizations that apply in contexts broader than those considered here.

27.5 Defining Terms

proper rational matrix: A matrix whose entries are proper rational functions, i.e., ratios of polynomials each of whose numerator degrees is less than or equal to its denominator degree.

MacMillan degree: The dimension of the state in a minimal realization of a proper rational transfer function matrix.

real polynomial matrix: A matrix whose entries are polynomials with real coefficients.

nonsingular: A real square polynomial matrix is nonsingular if its determinant is a nonzero polynomial.

unimodular: A real square polynomial matrix is unimodular if its determinant is a nonzero real number.

References

[1] Boyd, S. P. and Barratt, C., *Linear Controller Design: Limits of Performance,* Prentice Hall, Englewood Cliffs, NJ, 1991.

[2] Delchamps, D. F., *State Space and Input-Output Linear Systems,* Springer, New York, 1988.

[3] Doyle, J. C., Francis, B.A., and Tannenbaum, A., *Feedback Control Theory,* MacMillan, New York, 1992.

[4] Forney, G. D., Minimal Bases of Rational Vector Spaces With Applications to Multivariable Linear Systems, *SIAM J. Control,* 13, 493–520, 1975.

[5] Francis, B., *A Course in H_∞ Control Theory,* Volume 88 In *Lecture Notes in Control and Information Sciences,* Springer, 1987.

[6] Jacobson, N., *Basic Algebra I,* W. H. Freeman, San Francisco, 1974.

[7] Kailath, T., *Linear Systems,* Prentice Hall, Englewood Cliffs, NJ, 1980.

[8] Kalman, R. E., Irreducible Realizations and the Degree of a Rational Matrix, *J. SIAM,* 13, 520–544, 1965.

[9] MacMillan, B., An Introduction to Formal Realizability Theory Parts I and II, *Bell Syst. Tech. J.,* 31, 217–279, 591-600, 1952.

[10] Rugh, W. J., *Linear System Theory,* Prentice Hall, Englewood Cliffs, NJ, 1993.

[11] Sontag, E., *Mathematical Control Theory,* Springer, New York, 1990.

[12] Strang, G., *Linear Algebra and Its Applications,* Academic, New York, 1976.

[13] Vidyasagar, M., *Control System Synthesis: A Factorization Approach,* MIT, Cambridge, MA, 1985.

[14] Vidyasagar, M., On the Use of Right-Coprime Factorizations in Distributed Feedback Systems Containing Unstable Subsystems, *IEEE Trans. Circuits Syst. CAS-25,* 916–921, 1978.

28

Robustness Analysis with Real Parametric Uncertainty

R. Tempo
CENS-CNR, Politecnico di Torino, Torino, Italy

F. Blanchini
Dipartimento di Matematica e Informatica, Università di Udine, Udine, Italy

28.1 Motivations and Preliminaries

In recent years, stability and performance of control systems affected by bounded perturbations have been studied in depth. The attention of researchers and control engineers concentrated on new robustness tools in the areas H_∞, Kharitonov (or real parametric uncertainty), L_1, Lyapunov, μ and quantitative feedback control (QFT). For further discussions on these topics and on the exposition of the main technical results, the reader may consult other chapters of this volume and [1].

One of the key features of this chapter is the concept of *robustness*. To explain, instead of a nominal system, we study a family of systems and we say that a certain property (e.g., stability or performance) is robustly satisfied if it is satisfied for all members of the family. In particular, we focus on linear, time-invariant, single-input, single-output systems affected by real parametric uncertainty. Stability of interval polynomials (that is, polynomials whose coefficients lie within given intervals) and the well-known *Theorem of Kharitonov* [20] are at the core of this research area. This theorem states that an interval polynomial has all its roots in the open left half-plane if and only if four specially constructed polynomials have roots in the open left half-plane. Subsequently, the *Edge Theorem* [10] studied the problem of affine dependence between the coefficients and more general regions than the open left half-plane. This result provides a computationally tractable solution proving that it suffices to check stability of the so-called one-dimensional exposed edges. Refer to Ackermann [4], Barmish, [5], Bhattacharyya, Chapellat and Keel [11] and Kogan [21] for a discussion of the extensive literature on this subject.

To explain more precisely robustness analysis with real parametric uncertainty, we consider a family of polynomials $p(s, q)$ of degree n whose real coefficients $a_i(q)$ are continuous functions of an ℓ-dimensional vector of real uncertain parameters q, each bounded in the interval $[q_i^-, q_i^+]$. More formally, we define

$$p(s,q) \doteq a_0(q) + a_1(q)s + a_2(q)s^2 + \ldots + a_n(q)s^n,$$

$$q \doteq (q_1, q_2, \ldots, q_\ell)$$

and the set

$$Q \doteq \{q : q_i^- \leq q_i \leq q_i^+, i = 1, 2, \ldots, \ell\}.$$

We assume that $p(s, q)$ is of degree n for all $q \in Q$; that is, we assume that $a_n(q) \neq 0$ for all $q \in Q$. Whenever the relations

0-8493-8570-9/96/$0.00+$.50

between the polynomial coefficients $a_i(q)$ and the vector q are specified, we study the root location of $p(s,q)$ for all $q \in Q$. Within this framework, the basic property we need to guarantee is robust stability. In particular, we say that $p(s,q)$ is robustly stable if $p(s,q)$ has roots in the open left half-plane for all $q \in Q$.

The real parametric approach can be also formulated for control systems. In this case, we deal with a family of plants denoted by $P(s,q)$. More precisely, we concentrate on robust stability or performance of a proper plant

$$P(s,q) \doteq \frac{N_P(s,q)}{D_P(s,q)}$$

where $N_P(s,q)$ and $D_P(s,q)$ are the numerator and denominator polynomials, respectively, whose real coefficients are continuous functions of q. We assume that $D_P(s,q)$ has invariant degree for all $q \in Q$. We also assume that there is no unstable pole-zero cancellation for all $q \in Q$; the reader may refer to Chockalingam and Dasgupta [12] for further discussions.

Robustness analysis is clearly of interest when the plant requires compensation. In practice, if the compensator is designed on the basis of the nominal plant, then robustness analysis can be performed by means of the closed-loop polynomial. That is, given a compensator transfer function

$$C(s) \doteq \frac{N_C(s)}{D_C(s)}$$

connected in a feedback loop with $P(s,q)$, we immediately write the closed-loop polynomial

$$p(s,q) = N_P(s,q)N_C(s) + D_P(s,q)D_C(s)$$

whose root location determines closed-loop stability.

To conclude this preliminary discussion, we remark that one of the main technical tools described here is the so-called value set (or template, in the QFT jargon [19]); see Barmish [5] for a detailed exposition of its properties. In particular, we show that if the polynomial or plant coefficients are affine functions of q, the value set can be easily constructed with two-dimensional (2-D) graphics. Consequently, robustness tests in the frequency domain can be readily performed. Finally, in this chapter, since our main goal is to introduce the basic concepts and tools available for robustness analysis with real parametric uncertainty, we do not provide formal proofs, but we make reference to the specific literature on the subject.

EXAMPLE 28.1: DC-electric motor with uncertain parameters.

For the sake of illustrative purposes, an example of a DC-electric motor is formulated and carried out throughout this chapter in various forms. Consider the system represented in Figure 28.1 of an armature-controlled DC-electric motor with independent excitation. The voltage-to-angle transfer function $P(s) = \Theta(s)/V(s)$ is given by

$$P(s) = \frac{K}{LJs^3 + (RJ + BL)s^2 + (K^2 + RB)s}$$

where L is the armature inductance, R the armature resistance, K the motor electromotive force-speed constant, J the moment of inertia and B the mechanical friction. Clearly, the values of some of these parameters may be uncertain. For example, the moment of inertia and the mechanical friction are functions of the load. Therefore, depending on the specific application, if the load is not fixed, the values of J and B are not precisely known. Similarly, the armature resistance R is a parameter that can be measured very accurately but is subject to temperature variations and the motor constant K is a function of the field magnetic flow, which may vary.

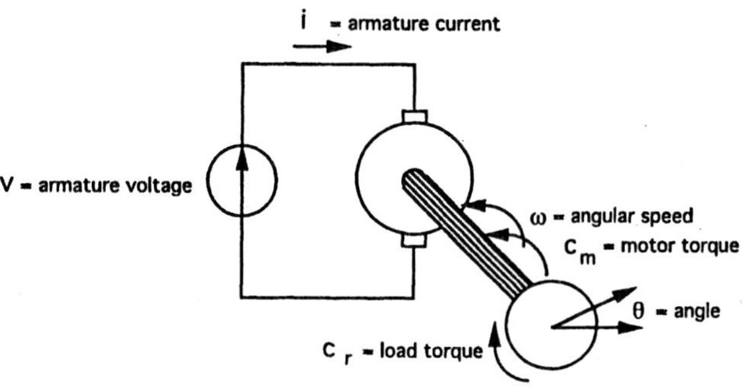

Figure 28.1

To summarize, it is reasonable to say that the motor parameters, or a subset of them, may be unknown but bounded within given intervals. More precisely, we can identify

$$q_1 = L;\ q_2 = R;\ q_3 = K;\ q_4 = J;\ q_5 = B$$

and specify a given interval $[q_i^-, q_i^+]$ for each q_i, $i = 1, 2, \ldots, 5$. Then, instead of $P(s)$, we write

$$P(s,q) = \frac{q_3}{q_1q_4s^3 + (q_2q_4 + q_1q_5)s^2 + (q_3^2 + q_2q_5)s}.$$

28.2 Description of the Uncertainty Structures

As discussed in Section 1, we consider a proper plant $P(s,q)$ whose coefficients are continuous functions of the uncertainty q, which is confined to the set Q. Depending on the specific problem under consideration, the coefficients of $N_P(s,q)$ and $D_P(s,q)$ may be linear or nonlinear functions of q. To explain more precisely, we consider the Example 28.1. Assume that the armature inductance L, the armature resistance R and the constant K are fixed while the moment of inertia J and the mechanical friction B are unknown. Then we take $q_1 = J$ and $q_2 = B$ as uncertain parameters; the resulting set Q is a two-dimensional rectangle. In this case, the plant coefficients are *affine*[1] functions

[1]An affine function $f : Q \to \mathbf{R}$ is the sum of a linear function and a constant. For example, $f(q) = 3q_1 + 2q_2 - 4$ is affine.

of q_1 and q_2

$$\begin{aligned} N_P(s,q) &= K; \\ D_P(s,q) &= Lq_1s^3 + (Rq_1 + Lq_2)s^2 + (K^2 + Rq_2)s \end{aligned}$$

and we say that the plant has an affine uncertainty structure. This situation arises in practice whenever, for example, the load conditions are not known. Other cases, however, may be quite different from the point of view of the uncertainty description. For example, if L, K and B are fixed and R and J are uncertain, we identify q_1 and q_2 with R and J, respectively. We observe that the denominator coefficient of s^2 contains the product of the two uncertain parameters q_1 and q_2

$$\begin{aligned} N_P(s,q) &= K; \\ D_P(s,q) &= Lq_2s^3 + (q_1q_2 + BL)s^2 + (K^2 + Bq_1)s. \end{aligned}$$

In this case, the plant coefficients are no longer affine functions of the uncertainties, but they are *multiaffine* functions[2] of q. This discussion can be further generalized. It is well known that the motor constant K is proportional to the magnetic flow. In an ideal machine, being excitation independent, such a flow is constant; but in a real machine, the armature reaction phenomenon causes magnetic saturation with the consequence that the constant K drops when the armature current exceeds a certain value. The final effect is that K is uncertain and we set $q_1 = K$. In turn, this implies that the plant coefficients are *polynomial* functions of the uncertainties. In addition, since q_1 enters both $N_P(s,q)$ and $D_P(s,q)$, we observe that there is coupling between numerator and denominator coefficients. In different situations, when this coupling is not present, we say that the numerator and denominator uncertainties are *independent.* An important class of independent uncertainties, in which all the coefficients of the numerator and denominator change independently within given intervals, is the so-called interval plant; e.g., see [6]. In other words, an interval plant is the ratio of two independent interval polynomials; recall that an interval polynomial

$$p(s,q) = q_0 + q_1s + q_2s^2 + \ldots + q_ns^n$$

has independent coefficients bounded in given intervals $q_i^- \leq q_i \leq q_i^+$ for $i = 0, 1, 2, \ldots, n$.

The choice of the uncertain parameters for a control system is a modeling problem, but robustness analysis is of increasing difficulty for more general uncertainty structures. In the following sections, we show that robustness analysis can be easily performed if the structure is affine, and we demonstrate that a "tight" approximate solution can be readily generated in the multiaffine case.

[2]A function $f : Q \to \mathbf{R}$ is said to be multiaffine if the following condition holds: If all components $q_1, \ldots, q_\ell$ except one are fixed, then f is affine. For example, $f(q) = 3q_1q_2q_3 - 6q_1q_3 + 4q_2q_3 + 2q_1 - 2q_2 + q_3 - 1$ is multiaffine.

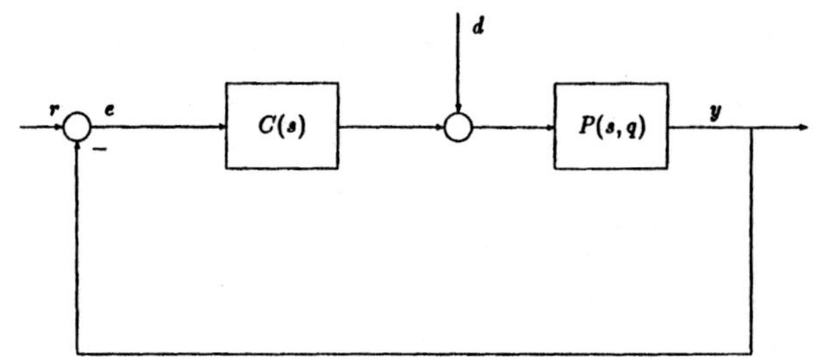

Figure 28.2

28.3 Uncertainty Structure Preservation with Feedback

In the previous sections, we described the classes of uncertainty structures entering into the open-loop plant. From the control system point of view, an important and closely related question arises. Namely, what are the conditions under which a certain uncertainty structure is preserved with feedback? To answer this question, we consider a plant $P(s,q)$ and a compensator $C(s)$ connected with the feedback structure shown in Figure 28.2.

Depending on the specific problem under consideration (for example, disturbance attenuation or tracking), we study sensitivity, complementary sensitivity and output-disturbance transfer functions (e.g., see [15])

$$S(s,q) \doteq \frac{1}{1+P(s,q)C(s)}; \quad T(s,q) \doteq \frac{P(s,q)C(s)}{1+P(s,q)C(s)};$$
$$R(s,q) \doteq \frac{P(s,q)}{1+P(s,q)C(s)}.$$

For example, it is immediate to show that the sensitivity function $S(s,q)$ takes the form

$$S(s,q) = \frac{D_P(s,q)D_C(s)}{N_P(s,q)N_C(s) + D_P(s,q)D_C(s)}.$$

If the uncertainty q enters affinely into the plant numerator and denominator, q also enters affinely into $S(s,q)$. We conclude that the affine structure is preserved with feedback. The same fact also holds for $T(s,q)$ and $R(s,q)$. Next, we consider the interval plant structure; recall that an interval plant has independent coefficients bounded in given intervals. It is easy to see that in general this structure is not preserved with compensation also for low-order controllers like proportional-integral (PI) or proportional-derivative (PD). Moreover, if the plant is affected by uncertainty entering independently into numerator and denominator coefficients, this decoupling is destroyed for all transfer functions $S(s,q)$, $T(s,q)$ and $R(s,q)$. Finally, it is important to notice that the multiaffine and polynomial uncertainty structures are preserved with feedback. Table 28.1 summarizes this discussion. In each entry of the first row of the table we specify the structure of the uncertain plant $P(s,q)$, and in the entry below, the corresponding structure of $S(s,q)$, $T(s,q)$ and $R(s,q)$.

TABLE 28.1 Uncertainty Structure with Feedback.

$P(s,q)$	Independent	Interval	Affine	Multiaffine	Polynomial
$S(s,q)$, $T(s,q)$, $R(s,q)$	Dependent	Affine	Affine	Multiaffine	Polynomial

28.4 Overbounding with Affine Uncertainty: The Issue of Conservatism

As briefly mentioned at the end of Section 3, the affine structure is very convenient for performing robustness analysis. However, in several real problems, the plant does not have this form; e.g., see [2]. In such cases, the nonlinear uncertainty structure can always be embedded into an affine structure by replacing the original family by a "larger" one. Even though this process has the advantage of handling much more general robustness problems, it has the obvious drawback of giving only an approximate but guaranteed solution. Clearly, the goodness of the approximation depends on the specific problem under consideration. To illustrate this simple overbounding methodology, we consider the DC-electric motor transfer function with two uncertain parameters and identify $q_1 = R$ and $q_2 = J$. As previously discussed, with this specific choice the plant has a multiaffine uncertainty structure. That is,

$$P(s,q) = \frac{K}{Lq_2 s^3 + (q_1 q_2 + BL)s^2 + (K^2 + Bq_1)s}.$$

To overbound $P(s,q)$ with an affine structure, we set $q_3 = q_1 q_2$. Given bounds $[q_1^-, q_1^+]$ and $[q_2^-, q_2^+]$ for q_1 and q_2, respectively, the range of variation $[q_3^-, q_3^+]$ for q_3 can be easily computed

$$\begin{aligned} q_3^- &= \min\{q_1^- q_2^-, q_1^- q_2^+, q_1^+ q_2^-, q_1^+ q_2^+\}; \\ q_3^+ &= \max\{q_1^- q_2^-, q_1^- q_2^+, q_1^+ q_2^-, q_1^+ q_2^+\}. \end{aligned}$$

Clearly, the new uncertain plant

$$P(s,\widetilde{q}) = \frac{K}{Lq_2 s^3 + (q_3 + BL)s^2 + (K^2 + Bq_1)s}$$

has three uncertain parameters $\widetilde{q} = (q_1, q_2, q_3)$ entering affinely into $P(s,\widetilde{q})$. Since $q_3 = q_1 q_2$, this new parameter is not independent and not all values of $[q_3^-, q_3^+]$ are physically realizable. However, since we assume that the coefficients q_i are independent, this technique leads to an affine overbounding of $P(s,q)$ with $P(s,\widetilde{q})$. We conclude that if a certain property is guaranteed for $P(s,\widetilde{q})$, then this same property is also guaranteed for $P(s,q)$. Unfortunately, the converse is not true. The control systems interpretation of this fact is immediate: Suppose that a certain compensator $C(s)$ does not stabilize $P(s,\widetilde{q})$. It may turn out that this same compensator does stabilize the family $P(s,q)$. Figure 28.3 illustrates the overbounding procedure for $q_1^- = 1.2, q_1^+ = 1.7, q_2^- = 1.7, q_2^+ = 2.2, q_3^- = 2.04$ and $q_3^+ = 3.74$.

To generalize this discussion, we restate the overbounding problem as follows: Given a plant $P(s,q)$ having nonlinear uncertainty structure and a set Q, find a new uncertain plant $P(s,\widetilde{q})$

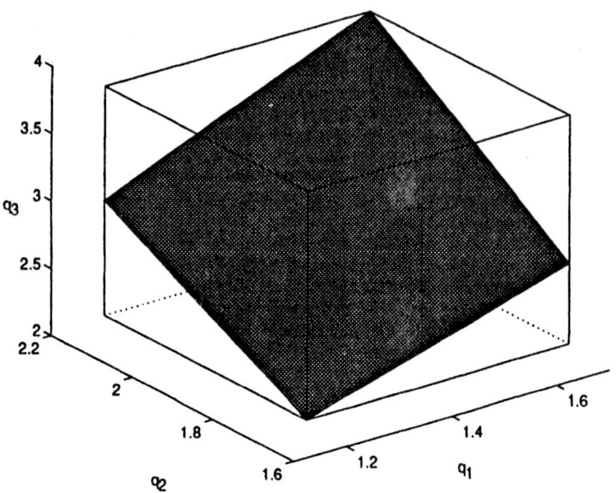

Figure 28.3

with affine uncertainty structure and a new set $\widetilde{Q}$. In general, since this process is not unique, there is no systematic procedure to construct an "optimal" overbounding. In practice, however, the control engineer may find via heuristic considerations a reasonably good method to perform it. The most natural way to obtain this bound may be to compute an interval overbounding for each coefficient of the numerator and denominator coefficients; that is, an interval plant overbounding. To illustrate, letting $a_i(q)$ and $b_i(q)$ denote respectively the numerator and denominator coefficients of $P(s,q)$, lower and upper bounds are given by

$$a_i^- = \min_{q \in Q} a_i(q); \quad a_i^+ = \max_{q \in Q} a_i(q)$$

and

$$b_i^- = \min_{q \in Q} b_i(q); \quad b_i^+ = \max_{q \in Q} b_i(q).$$

If $a_i(q)$ and $b_i(q)$ are affine or multiaffine functions and q lies in a rectangular set Q, these minimizations and maximizations can be easily performed. That is, if we denote by $q^1, q^2, \ldots, q^L \doteq q^{2^\ell}$ the vertices of Q, then

$$\begin{aligned} a_i^- &= \min_{q \in Q} a_i(q) = \min_{k=1,2,\ldots,L} a_i(q^k); \\ a_i^+ &= \max_{q \in Q} a_i(q) = \max_{k=1,2,\ldots,L} a_i(q^k) \end{aligned}$$

and

$$\begin{aligned} b_i^- &= \min_{q \in Q} b_i(q) = \min_{k=1,2,\ldots,L} b_i(q^k); \\ b_i^+ &= \max_{q \in Q} b_i(q) = \max_{k=1,2,\ldots,L} b_i(q^k). \end{aligned}$$

To conclude this section, we remark that for more general uncertainty structures than multiaffine, a tight interval plant overbounding may be difficult to construct.

28.5 Robustness Analysis for Affine Plants

In this section, we study robustness analysis of a plant $P(s,q)$ affected by affine uncertainty $q \in Q$. The approach taken here is an extension of the classical Nyquist criterion and requires the notion of *value set.*

For fixed frequency $s = j\omega$, we define the value set $P(j\omega, Q) \subset \mathbf{C}$ as

$$P(j\omega, Q) \doteq \{P(j\omega, q) : D_P(j\omega, q) \neq 0, q \in Q\}.$$

Roughly speaking, $P(j\omega, Q)$ is a set in the complex plane that graphically represents the uncertain plant. Without uncertainty, $P(j\omega)$ is a singleton and its plot for a range of frequencies is the Nyquist diagram. The nice feature is that this set is two-dimensional even if the number of uncertain parameters is large. Besides the issue of the value set construction, we now formally state a robustness criterion. This is an extension of the classical Nyquist criterion and holds for more general uncertainty structures than affine—continuity of the plant coefficients with respect to the uncertain parameters suffices. However, for more general classes of plants than affine, the construction of the value set is a hard problem.

28.5.1 Robustness Criterion for Uncertain Plants

The plant $P(s,q)$ is robustly stable for all $q \in Q$ if and only if the Nyquist stability criterion is satisfied for some $q \in Q$ and $-1 + j0 \notin P(j\omega, Q)$ for all $\omega \in \mathbf{R}$.

This criterion can be proved using continuity arguments; see [17]. To detect robustness, one should check if the Nyquist stability criterion holds for some $q \in Q$; without loss of generality, this check can be performed for the nominal plant. Secondly, it should be verified that the value set does not go through the point $-1 + j0$ for all $\omega \in \mathbf{R}$. In practice, however, one can discretize a bounded interval $\Omega \subset \mathbf{R}$ with a "sufficiently" high number of samples; continuity considerations guarantee that the intersampling is not a critical issue. Finally, by drawing the value set, the gain and phase margins can be graphically evaluated; similarly, the resonance peak of the closed-loop system can be computed using the well-known constant M-circles.

28.5.2 Value Set Construction for Affine Plants

In this section, we discuss the generation of the value set $P(j\omega, Q)$ in the case of affine plants. The reader familiar with Nyquist-type analysis and design is aware of the fact that a certain range of frequencies, generally close to the crossover frequencies, can be specified a priori. That is, a range $\Omega = [\omega^-, \omega^+]$ may be imposed by design specifications or estimated by performing a frequency analysis of the nominal system under consideration. In this section, we assume that $D_P(j\omega, q) \neq 0$ for all $q \in Q$ and $\omega \in \Omega$. We remark that if the frequency $\omega = 0$ lies in the interval Ω, this assumption is not satisfied for Type 1 or Type 2 systems; however, these systems can be easily handled with contour indentation techniques as in the classical Nyquist analysis. We also observe that the assumption that $P(s,q)$ does not have poles in the interval $[\omega^-, \omega^+]$ simply implies that $P(j\omega, Q)$ is bounded.

To proceed with the value set construction, we first need a preliminary definition. The one-dimensional *exposed edge* e^{ik} is a convex combination of the adjacent vertices[3] q^i and q^k of Q

$$e^{ik} \doteq \{\lambda q^i + (1-\lambda)q^k : \ \lambda \in [0, 1]\}.$$

Denote by E the set of all $q \in e^{ik}$ for some i, k. This set is the collection of all one-dimensional exposed edges of Q. Then, for fixed $\omega \in \Omega$, it can be shown (see [17]) that

$$\partial P(j\omega, Q) \subseteq P(j\omega, E) \doteq \{P(j\omega, q) : q \in E\}$$

where $\partial P(j\omega, Q)$ denotes the boundary of the value set and $P(j\omega, E)$ is the image in the complex plane of the exposed edges. This says that the construction of the value set requires computations involving only the one-dimensional exposed edges. The second important fact observed in [17] is that the image of the edge e^{ik} in the complex plane is an arc of a circle or a line segment. To explain this claim, in view of the affine dependence of both $N(s,q)$ and $D(s,q)$ vs. q, we write the uncertain plant corresponding to the edge e^{ik} in the form

$$\begin{aligned} P(s, e^{ik}) &= \frac{N_P(s, \lambda q^i + (1-\lambda)q^k)}{D_P(s, \lambda q^i + (1-\lambda)q^k)} \\ &= \frac{N_P(s, q^k) + \lambda N_P(s, (q^i - q^k))}{D_P(s, q^k) + \lambda N_P(s, (q^i - q^k))} \end{aligned}$$

for $\lambda \in [0, 1]$. For fixed $s = j\omega$, it follows that the mapping from the edge e^{ik} to the complex plane is bilinear. Then, it is immediate to conclude that the image of each edge is an arc of a circle or a line segment; the center and the radius of the circle and the extreme points of the segment can also be computed. Even though the number of one-dimensional exposed edges of the set Q is $\ell 2^{\ell-1}$, the set E is one-dimensional. Therefore, a fast computation of $P(j\omega, E)$ can be easily performed and the boundary of the value set $P(j\omega, Q)$ can be efficiently generated.

Finally, an important extension of this approach is robustness analysis of systems with time delay. That is, instead of $P(s,q)$, we consider

$$P_\tau(s, q) \doteq \frac{N(s,q)}{D(s,q)} e^{-\tau s}$$

where $\tau \geq 0$ is a delay. It is immediate to see that the value set of $P_\tau(s,q)$ at frequency $s = j\omega$ is given by the value set of the plant $\frac{N(s,q)}{D(s,q)}$ rotated with respect to the origin of the complex plane of an angle $\tau\omega$ in clockwise direction. Therefore, the robustness

[3]Two vertices are adjacent if they differ for only one component. For example, in Figure 3 the vertices $q^1 = (1.2, 2.2, 2.04)$ and $q^2 = (1.2, 2.2, 3.74)$ are adjacent.

criterion given in this section still applies; see Barmish and Shi [7] for further details.

EXAMPLE 28.2: The DC-electric motor revisited.

To illustrate the concepts discussed in this section, we revisit the DC-electric motor example (Example 28.1). We take two uncertain parameters

$$q_1 = J; \quad q_2 = B$$

where $q_1 \in [0.03, 0.15]$ and $q_2 \in [0.001, 0.03]$ with nominal values $J = 0.042\,\mathrm{kg\,m^2}$ and $B = 0.01625\,\mathrm{N\,m/rs}$. The remaining parameters take values $K = 0.9\,\mathrm{V/rs}$, $L = 0.025\,\mathrm{H}$ and $R = 5\Omega$. The voltage-to-angle uncertain plant is

$$P(s, q) = \frac{0.9}{0.025q_1 s^3 + (5q_1 + 0.025q_2)s^2 + (0.81 + 5q_2)s}.$$

To proceed with robustness analysis, we first estimate the critical range of frequencies obtaining $\Omega = [10, 100]$. We notice that the denominator of $P(s, q)$ is nonvanishing for all $q \in Q$ in this range. Then we study robust stability of the plant connected in a feedback loop with a proportional-integral-derivative (PID) compensator

$$C(s) = K_P + \frac{K_I}{s} + K_D s.$$

For closed-loop stability, we recall that the Nyquist criterion requires that the Nyquist plot of the open-loop system does not go through the point $-1 + j0$ and that it does encircle this point (in counterclockwise direction) a number of times equal to the number of unstable poles; e.g., see [18]. In this specific case, setting $K_P = 200$, $K_I = 5120$ and $K_D = 20$, we see that the closed-loop nominal system is stable with a phase margin $\phi \approx 63.7°$ and a crossover frequency $\omega_c \approx 78.8$ rad/s. As a result of the analysis carried out by sweeping the frequency, it turns out that the closed-loop system is not robustly stable, since at the frequency $\omega \approx 16$ rad/s the value set includes the point $-1 + j0$. Figure 28.4 shows the Nyquist plot of the nominal plant and the value set for 12 equispaced frequencies in the range [12, 34].

To robustly stabilize $P(s, q)$, we take $K_I = 2000$ and the same values of K_P and K_D as before. The reasons for choosing this value of K_I can be explained as follows: The compensator transfer function has phase zero at the frequency $\overline{\omega} = \sqrt{K_I/K_D}$. Thus, reducing K_I from 5120 to 2000 causes a drop of $\overline{\omega}$ from 16 rad/s to 10 rad/s. This implies that the phase lead effect begins at lower frequencies, "pushing" the value set out of the critical point $-1 + j0$. Since the nominal system is stable with a phase margin $\phi \approx 63.7°$ and a crossover frequency $\omega_c \approx 80.8$ rad/s, this new control system has nominal performance very close to the previous one. However, with this new PID compensator, the system becomes robustly stable. To see this, we generated the value sets for the same frequencies as before; see Figure 28.5. From this figure, we observe that $P(j\omega, q)$ does not include the point $-1 + j0$. We conclude that the closed-loop system is now robustly stable; the worst-case phase margin is $\phi \approx 57.9°$.

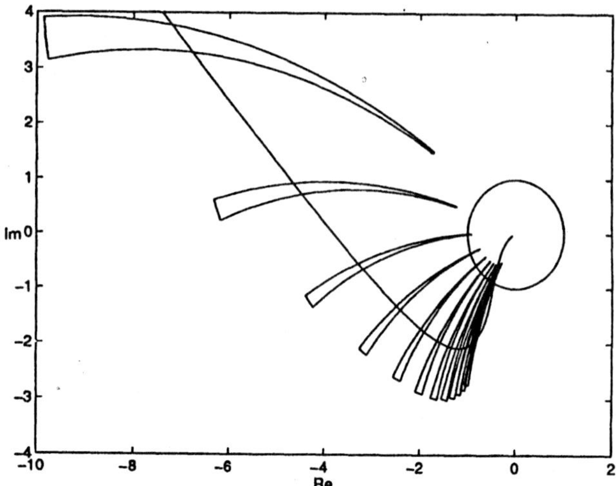

Figure 28.4

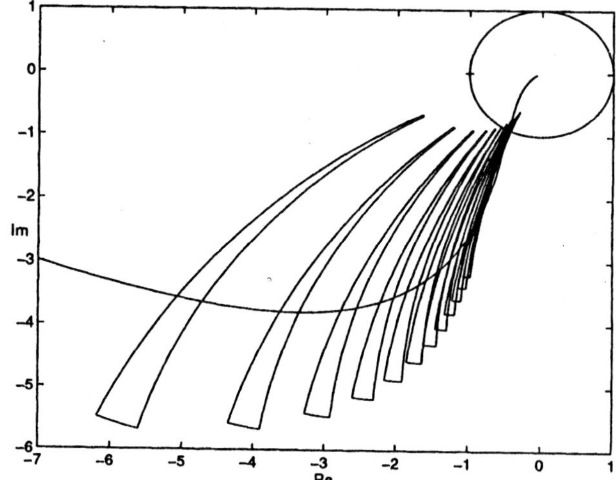

Figure 28.5

28.6 Robustness Analysis for Affine Polynomials

In this section, we study robustness analysis of the closed-loop polynomial

$$p(s, q) = N_P(s, q)N_C(s) + D_P(s, q)D_C(s)$$

when the coefficients are affine functions of q. The main goal is to provide an alternative criterion for polynomials instead of plants. With this approach, we do not need the nonvanishing condition about $D_P(s, q)$; furthermore, unstable pole-zero cancellations are not an issue. In this case, however, we lose the crucial insight given by the Nyquist plot. For fixed frequency $s = j\omega$, we define the value set $p(j\omega, Q) \subset \mathbf{C}$ as

$$p(j\omega, Q) \doteq \{p(j\omega, q) : q \in Q\}.$$

As in the plant case, $p(j\omega, Q)$ is a set in the complex plane that moves with frequency and which graphically represents the uncertain polynomial.

28.6.1 Zero Exclusion Condition for Uncertain Polynomials

The polynomial $p(s, q)$ is robustly stable for all $q \in Q$ if and only if $p(s, q)$ is stable for some $q \in Q$ and $0 \notin p(j\omega, Q)$ for all $\omega \in \mathbf{R}$.

The proof of this criterion requires elementary facts and, in particular, continuity of the roots of $p(s, q)$ vs. its coefficients; see [16]. Similarly to the discussion in Section 5, we notice that the criterion defined in this section is easily implementable—at least, whenever the value set can be efficiently generated. That is, given an affine polynomial family, we take any element in this family and we check its stability. This step is straightforward using the Routh table or any root-finding routine. Then we sweep the frequency ω over a selected range of critical frequencies $\Omega = [\omega^-, \omega^+]$. This interval can be estimated, for example, using some a priori information on the specific problem or by means of one of the bounds given by Marden [22]. If there is no intersection of $p(j\omega, Q)$ with the origin of the complex plane for all $\omega \in \Omega$, then $p(s, q)$ is robustly stable.

Remark: A very similar zero-exclusion condition can be stated for more general regions $\mathcal{D}$ than the open left half-plane. Meaningful examples of $\mathcal{D}$ regions are the open unit disk, a shifted left half-plane and a damping cone[4]. In this case, instead of sweeping the imaginary axis, we need to discretize the boundary of $\mathcal{D}$.

28.6.2 Value Set Construction for Affine Polynomials

In this section, we discuss the generation of the value set $p(j\omega, Q)$. Whenever the polynomial coefficients are affine functions of the uncertain parameters, the value set can be easily constructed. To this end, two key facts are very useful. First, for fixed frequency, we notice that $p(j\omega, Q)$ is a two-dimensional convex polygon. Secondly, letting $q^1, q^2, \ldots, q^L$ denote the vertices of Q as in Section 4, we notice that the vertices of the value set are a subset of the complex numbers $p(j\omega, q^1), p(j\omega, q^2), \ldots, p(j\omega, q^L)$. These two observations follow from the fact that, for fixed frequency, real and imaginary parts of $p(s, q)$ are both affine functions of q. Then the value set is a two-dimensional affine mapping of the set Q, and its vertices are generated by vertices of Q. Thus, for fixed ω, it follows that

$$p(j\omega, Q) = \text{conv}\,\{p(j\omega, q^1), p(j\omega, q^2), \ldots, p(j\omega, q^L)\}$$

where conv denotes the convex hull[5]. The use of this fact is immediate: For fixed frequency, one can generate the points $p(j\omega, q^1), p(j\omega, q^2), \ldots, p(j\omega, q^L)$ in the complex plane. The value set can be constructed by taking the convex hull of these points; this can be readily done with 2-D graphics. From the computational point of view, we notice that the number of edges of the polygon is, at most, 2ℓ at each frequency. This follows from the observation that any edge of the value set is the image of an exposed edge of Q. In addition, parallel edges of Q are mapped into parallel edges in the complex plane, and the edges of Q have only ℓ distinct directions. These facts can be used to efficiently compute $p(j\omega, Q)$. We now provide an example that illustrates the value set generation.

EXAMPLE 28.3: Value set generation.

Using the same data as in Example 28.2 and a PID controller with gains $K_P = 200$, $K_I = 5120$ and $K_D = 20$, we study the closed-loop polynomial

$$\begin{aligned} p(s, q) &= 0.025q_1 s^4 + s^3(5q_1 + 0.025q_2) \\ &\quad + s^2(5q_2 + 18.81) + 180s + 4608. \end{aligned}$$

Robustness analysis is then performed for 29 equispaced frequencies in the range [2, 30]. Figure 28.6 shows the polygonality of the value set and zero inclusion for $\omega \approx 16$ rad/s, which demonstrates instability. This conclusion coincides with that previously obtained in Example 28.2.

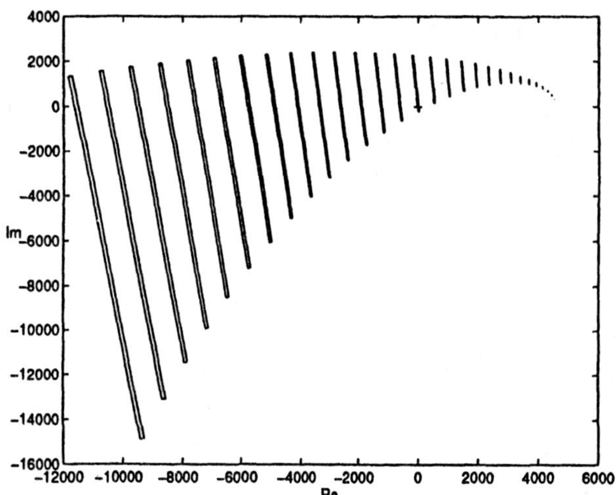

Figure 28.6

28.6.3 Interval Polynomials: Kharitonov's Theorem and Value Set Geometry

In the special case of interval polynomials, robustness analysis can be greatly facilitated via Kharitonov's Theorem [20]. To explain, we now recall this result. Given an interval polynomial

$$p(s, q) = q_0 + q_1 s + q_2 s^2 + \ldots + q_n s^n$$

[4]A damping cone is a subset of the complex plane defined as $\{s : Re(s) \leq -\alpha|Im(s)|\}$ for $\alpha > 0$.

[5]The convex hull conv S of a set S is the smallest convex set containing S.

of order n (that is, $q_n \neq 0$) and bounds $[q_i^-, q_i^+]$ for each coefficient q_i, form the following four polynomials:

$$\begin{aligned} p_1(s) &\doteq q_0^+ + q_1^+ s + q_2^- s^2 + q_3^- s^3 + q_4^+ s^4 + q_5^+ s^5 \\ &\quad + q_6^- s^6 + q_7^- s^7 + \ldots; \\ p_2(s) &\doteq q_0^- + q_1^- s + q_2^+ s^2 + q_3^+ s^3 + q_4^- s^4 + q_5^- s^5 \\ &\quad + q_6^+ s^6 + q_7^+ s^7 + \ldots; \\ p_3(s) &\doteq q_0^+ + q_1^- s + q_2^- s^2 + q_3^+ s^3 + q_4^+ s^4 + q_5^- s^5 \\ &\quad + q_6^- s^6 + q_7^+ s^7 + \ldots; \\ p_4(s) &\doteq q_0^- + q_1^+ s + q_2^+ s^2 + q_3^- s^3 + q_4^- s^4 + q_5^+ s^5 \\ &\quad + q_6^+ s^6 + q_7^- s^7 + \ldots . \end{aligned}$$

Then, $p(s,q)$ is stable for all $q \in Q$ if and only if the four Kharitonov polynomials $p_1(s)$, $p_2(s)$, $p_3(s)$ and $p_4(s)$ are stable. To provide a geometrical interpretation of this result, we notice that the value set for fixed frequency $s = j\omega$ is a rectangle with level edges parallel to real and imaginary axis. The four vertices of this set are the complex numbers $p_1(j\omega)$, $p_2(j\omega)$, $p_3(j\omega)$ and $p_4(j\omega)$; see [13]. If the four Kharitonov polynomials are stable, due to the classical Mikhailov's criterion [23] their phase is strictly increasing for ω increasing. In turn, this implies that the level rectangular value set moves in a counterclockwise direction around the origin of the complex plane. Next, we argue that the strictly increasing phase of the vertices and the parallelism of the four edges of the value set with real or imaginary axis guarantee that the origin does not lie on the boundary of the value set. By continuity, we conclude that the origin is outside the value set and zero-exclusion condition is satisfied; see [24].

28.6.4 From Robust Stability to Robust Performance

In this section, we point out the important fact that the polynomial approach discussed in this chapter can be also used for robust performance. To explain, we take an uncertain plant with affine uncertainty and we show how to compute the largest peak of the Bode plot magnitude for all $q \in Q$; that is, the worst-case H_∞ norm. Formally, for a stable, strictly proper plant $P(s,q)$, we define

$$\max_{q\in Q} \|P(s,q)\|_\infty \doteq \max_{q\in Q} \sup_{\omega} |P(j\omega,q)|.$$

Given a performance level $\delta > 0$, then

$$\max_{q\in Q} \|P(s,q)\|_\infty < \delta$$

if and only if

$$\left| \frac{N_P(j\omega,q)}{D_P(j\omega,q)} \right| < \delta$$

for all $\omega \geq 0$ and $q \in Q$. Since $P(j\omega,q) \to 0$ for $\omega \to \infty$, this is equivalent to checking whether the zero-exclusion condition

$$N_P(j\omega,q) - \delta D_P(j\omega,q) e^{j\phi} \neq 0$$

is satisfied for all $\omega \in \mathbf{R}$, $q \in Q$ and $\phi \in [0, 2\pi]$. In turn, this implies that the complex coefficients polynomial

$$p_\phi(s,q) = N_P(s,q) - \delta D_P(s,q) e^{j\phi}$$

has roots in the open left half-plane for all $q \in Q$ and $\phi \in [0, 2\pi]$. Clearly, for fixed $\phi \in [0, 2\pi]$, the criterion given in this section can be readily used; however, since $p_\phi(s,q)$ has complex coefficients it should be necessarily checked for all $\omega \in \mathbf{R}$, including negative frequencies.

28.6.5 Algebraic Criteria for Robust Stability

If the uncertain polynomial under consideration is affine, the well-known Edge Theorem applies [10]. This algebraic criterion is alternative to the frequency domain approach studied in this section. Roughly speaking, this result says that an affine polynomial family is robustly stable if and only if all the polynomials associated with the one-dimensional exposed edges of the set Q are stable. Even though this result is of algebraic nature, it can be explained by means of value set arguments. For affine polynomials and for fixed ω, we already observed in this section that the boundary of the value set $p(j\omega, Q)$ is the image of the one-dimensional exposed edges of Q. Thus, to guarantee zero-exclusion condition, we need to guarantee that all edge polynomials are nonvanishing for all $\omega \in \mathbf{R}$; otherwise, an instability occurs. We conclude that stability detection for affine polynomials requires the solution of a number of one-dimensional stability problems. Each of these problems can be stated as follows: Given polynomials $p_0(s)$ and $p_1(s)$ of order n and $m < n$, respectively, we need to study stability of

$$p(s,\lambda) = p_0(s) + \lambda p_1(s)$$

for all $\lambda \in [0, 1]$. A problem of great interest is to ascertain when robust stability of $p(s,\lambda)$ can be deduced from the stability of the extreme polynomials $p(s,0)$ and $p(s,1)$. This problem can be formulated in more general terms: To construct classes of uncertain polynomials for which stability of the vertex polynomials (or a subset of them) implies stability of the family. Clearly, the edges associated with an interval polynomial is one of such classes. Another important example is given by the edges of the closed-loop polynomial of a control system consisting of a first-order compensator and an interval plant; see [6]. Finally, see [25] for generalizations and for the concept of convex directions polynomials.

28.6.6 Further Extensions: The Spectral Set

In some cases it is of interest to generate the entire root location of a family of polynomials. This leads to the concept of *spectral set*; see [9]. Given a polynomial $p(s,q)$, we define the spectral set as

$$\sigma \doteq \{s \in \mathbf{C} : p(s,q) = 0 \text{ for some } q \in Q\}.$$

The construction of this set is quite easy for affine polynomials. Basically, the key idea can be explained as follows: For fixed

$s \in \mathbf{C}$, checking if s is a member of the spectral set can be accomplished by means of zero-exclusion condition. Next, it is easy to compute a bounded root confinement region $\overline{\sigma} \supseteq \sigma$; e.g., see [22]. Then the construction of the spectral set σ amounts to a two-dimensional gridding of $\overline{\sigma}$ and, for each grid point, checking zero exclusion.

The spectral set concept can be further extended to control systems consisting of a plant $P(s, q)$ with a feedback gain K_P that needs tuning. In this case, we deal with the so-called robust root locus [8]; that is, the generation of all the roots of the closed-loop polynomial when K_P ranges in a given interval. To illustrate, we consider the same data as in Example 28.2 and a feedback gain K_P, thus obtaining the closed-loop polynomial

$$p(s,q) = 0.025q_1 s^3 + (5q_1 + 0.025q_2)s^2 + (0.81 + 5q_2)s + 0.9K_P$$

where $q_1 \in [0.03, 0.15]$ and $q_2 \in [0.001, 0.03]$. In Figure 28.7 we show the spectral set for $K_P = 200$; however, the spectral set also includes a real root in the interval $[-200.58, -200.12]$, which is out of the plot.

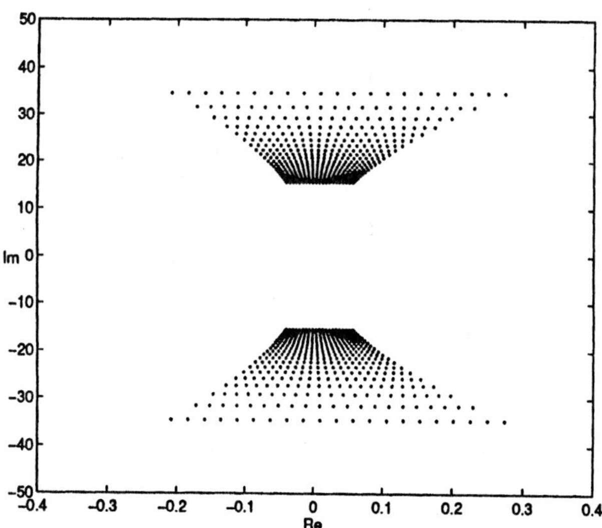

Figure 28.7

28.7 Multiaffine Uncertainty Structures

In this section, we discuss the generation of the value set for polynomials with more general uncertainty structures than affine. In particular, we study the case when the polynomial coefficients are multiaffine functions of the uncertain parameters. Besides the motivations provided in Section 3, we recall that this uncertainty structure is quite important for a number of reasons. For example, consider a linear state-space system of the form $\dot{x}(t) = A(q)x(t)$ where each entry of the matrix $A(q) \in \mathbf{R}^{m \times m}$ lies in a bounded interval. Then the characteristic polynomial required for stability considerations has a multiaffine uncertainty structure.

In the case of multiaffine uncertainty, the value set is generally not convex and its construction cannot be easily performed. However, we can easily generate a "tight" convex approximation of $p(j\omega, Q)$, this approximation being its convex hull conv $p(j\omega, Q)$. More precisely, the following fact, called the Mapping Theorem, holds: The convex hull of the value set conv $p(j\omega, Q)$ is given by the convex hull of the vertex polynomials $p(j\omega, q^1), p(j\omega, q^2), \ldots, p(j\omega, q^L)$. In other words, the parts of the boundary of $p(j\omega, Q)$ that are not line segments are always contained inside this convex hull. Clearly, if conv $p(j\omega, Q)$ is used instead of $p(j\omega, Q)$ for robustness analysis through zero exclusion, only a sufficient condition is obtained. That is, if the origin of the complex plane lies inside the convex hull, we do not know if is also inside the value set. We now formally state the Mapping Theorem [28].

28.7.1 The Mapping Theorem

THEOREM 28.1 *For fixed frequency $\omega \in \mathbf{R}$,*

$$conv\, p(j\omega, Q) = conv\{p(j\omega, q^1), p(j\omega, q^2), \ldots, p(j\omega, q^L)\}.$$

With regard to applicability and usefulness of this result, comments very similar to those made in Section 28.6 about the construction of the value set for affine polynomials can be stated. Figure 28.8 illustrates the Mapping Theorem for the polynomial

$$p(s,q) = s^3 + (q_2 + 4q_3 + 2q_1q_2)s^2 + (4q_2q_3 + q_1q_2q_4)s + q_3 + 2q_1q_3 - q_1q_2(q_4 - 0.5)$$

with four uncertain parameters q_1, q_2, q_3 and q_4 each bounded in the interval $[0, 1]$ and frequency $\omega = 1$ rad/s. The "true" value set shown in this figure is obtained via random generation of 10,000 samples uniformly distributed in the set Q.

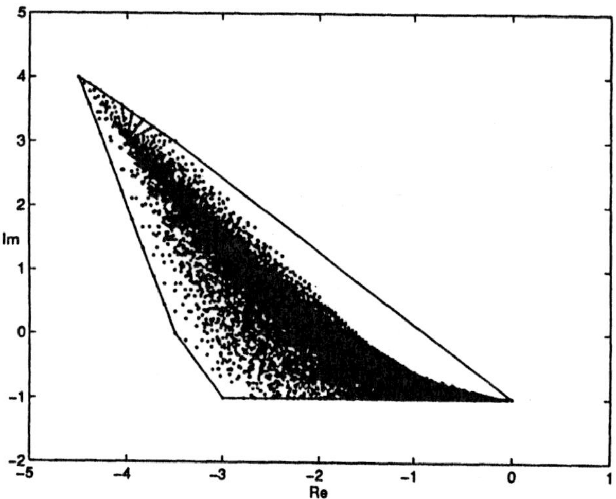

Figure 28.8

28.8 More General Uncertainty Structures

For other general uncertainty structures such as polynomial or nonlinear functions of q, there is no analytical tool that enables us to construct the value set. However, for systems with a few uncertain parameters, a brute force approach, as in the example of Section 7, can be taken by simply gridding the set Q with a sufficiently high number of points. With this procedure, one can easily obtain a "cloud" in the complex plane that looks like the value set. Obviously, this method is not practical for a large number of uncertain parameters. Hence, to facilitate robustness analysis for classes of nonlinear uncertainty structures, a number of different approaches have been proposed in the recent literature. Among the others, we recall the parameter space approach [3], the techniques developed for the computation of the multivariable gain margin [14] and [26], and the geometric programming methods [27].

28.9 Conclusion

In this chapter, we discussed some of the key ideas motivating the study of robustness of control systems affected by real parametric uncertainty. In addition, we provided two frequency domain criteria that can be easily implemented with 2-D graphics. An example of a DC-electric motor was carried out throughout the chapter to illustrate modeling of uncertainty and subsequent robustness analysis.

28.10 Acknowledgments

This work was supported by funds of CENS-CNR and by CNR Grant number 92 02887-CT07.

References

[1] Special issue on robust control, *Automatica*, 29, 1-252, 1993.

[2] Abate, M., Barmish, B.R., Murillo-Sanchez, C., and Tempo, R., Application of some new tools to robust stability analysis of spark ignition engines: a case study, *IEEE Trans. Control Syst. Technol.*, 2, 22–30, 1994.

[3] Ackermann, J.E., Parameter space design of robust control systems, *IEEE Trans. Autom. Control*, 25, 1058–1072, 1980.

[4] Ackermann, J., in cooperation with Bartlett, A., Kaesbauer, D., Sienel, W., and Steinhauser, R., *Robust Control, Systems with Uncertain Physical Parameters*, Springer-Verlag, London, 1993.

[5] Barmish, B.R., *New Tools for Robustness of Linear Systems*, Macmillan, New York, 1994.

[6] Barmish, B.R., Hollot, C.V., Kraus, F., and Tempo, R., Extreme point results for robust stabilization of interval plants with first order compensators, *IEEE Trans. Autom. Control*, 37, 707–714, 1992.

[7] Barmish, B.R. and Shi, Z., Robust stability of perturbed systems with time delays, *Automatica*, 25, 371–381, 1990.

[8] Barmish, B.R. and Tempo, R., The robust root locus, *Automatica*, 26, 283–292, 1990.

[9] Barmish, B.R. and Tempo, R., On the spectral set for a family of polynomials, *IEEE Trans. Autom. Control*, 36, 111–115, 1991.

[10] Bartlett, A.C., Hollot, C.V., and Huang, L., Root locations of an entire polytope of polynomials: it suffices to check the edges, *Math. Control, Signals Syst.*, 1, 61–71, 1988.

[11] Bhattacharyya, S.P., Chapellat, H., and Keel, L.H., *Robust Control: The Parametric Approach*, Prentice Hall, Englewood Cliffs, NJ, 1995.

[12] Chockalingam, G. and Dasgupta, S., Minimality, stabilizability and strong stabilizability of uncertain plants, *IEEE Trans. Autom. Control*, 38, 1651–1661, 1993.

[13] Dasgupta, S., Kharitonov's theorem revisited, *Syst. Control Lett.*, 11, 381–384, 1988.

[14] de Gaston, R.R.E. and Safonov, M.G., Exact calculation of the multiloop stability margin, *IEEE Trans. Autom. Control*, 33, 156–171, 1988.

[15] Doyle, J.C., Francis, B.A., and Tannenbaum, A.R., *Feedback Control Theory*, Macmillan, New York, 1992.

[16] Frazer, R.A. and Duncan, W.J., On the criteria for the stability of small motion, *Proc. of the R. Soc.*, A, 124, 642–654, 1929.

[17] Fu, M., Computing the frequency response of linear systems with parametric perturbations, *Syst. Control Lett.*, 15, 45–52, 1990.

[18] Horowitz, I.M., *Synthesis of Feedback Systems*, Academic Press, New York, 1963.

[19] Horowitz, I., Survey of quantitative feedback theory (QFT), *Int. J. Control*, 53, 255–291, 1991.

[20] Kharitonov, V.L., Asymptotic stability of an equilibrium position of a family of systems of linear differential equations, *Differential'nye Uraveniya*, 14, 1483–1485, 1978.

[21] Kogan, J., *Robust Stability and Convexity*, Lecture Notes in Control and Information Sciences, Springer-Verlag, 1995.

[22] Marden, M., *Geometry of Polynomials, Mathematical Surveys*, no. 3, American Mathematical Society, Providence, RI, 1966.

[23] Mikhailov, A.W., Methods of harmonic analysis in control theory, *Avtomatika i Telemekhanika*, 3, 27–81, 1938.

[24] Minnichelli, R.J., Anagnost, J.J., and Desoer, C.A., An elementary proof of Kharitonov's theorem with extensions, *IEEE Trans. Autom. Control*, 34, 995–998, 1989.

[25] Rantzer, A., Stability conditions for polytopes of polynomials, *IEEE Trans. Autom. Control*, 37, 79–89, 1992.

[26] Sideris, A. and Sanchez Pena, R.S., Fast calculation of the multivariable stability margin for real interrelated uncertain parameters, *IEEE Trans. Autom. Control*, 34, 1272–1276, 1989.

[27] Vicino, A., Tesi, A. and Milanese, M., Computation of nonconservative stability perturbation bounds for systems with nonlinearly correlated uncertainties, *IEEE Trans. Autom. Control*, 35, 835–841, 1990.

[28] Zadeh, L.A., and Desoer, C.A., *Linear System Theory*, McGraw-Hill, New York, 1963.

29

MIMO Frequency Response Analysis and the Singular Value Decomposition

Stephen D. Patek
Laboratory for Information and Decision Systems, Massachusetts Institute of Technology, Cambridge, MA

Michael Athans
Laboratory for Information and Decision Systems, Massachusetts Institute of Technology, Cambridge, MA

29.1 Modeling MIMO Linear Time-Invariant Systems in Terms of Transfer Function Matrices

Any multivariable linear time invariant (LTI) system is uniquely described by its impulse response matrix. The Laplace transform of this matrix function of time gives rise to the system's transfer function matrix (TFM). We assume that all systems in the sequel have TFM representations; frequency response analysis is always applied to TFM descriptions of systems. For the case of finite dimensional LTI systems, when we have a state-space description of the system, closed-form evaluation of the TFM is particularly easy. This is discussed in Example 29.1. More generally, *infinite*-dimensional LTI systems also have TFM representations, although closed-form evaluation of the TFM of these systems is less straightforward. Stability (in whatever sense) is *presumed*; without stability we cannot talk about the steady-state response of a system to sinusoidal inputs. For such systems, the TFM can be measured by means of sinusoidal inputs. We will not discuss the use of frequency-domain techniques in robustness analysis.

EXAMPLE 29.1: Finite-dimensional LTI systems

Suppose that we have a finite-dimensional LTI system G, with the following state equations

$$\begin{aligned} \dot{x}(t) &= Ax(t) + Bu(t), \\ y(t) &= Cx(t) + Du(t). \end{aligned}$$

Taking the Laplace transform of the state equation, we see that,

$$x(s) = (sI - A)^{-1}Bu(s)$$

where we have abused notation slightly with $x(s)$ and $u(s)$ representing the Laplace transforms of $x(t)$ and $u(t)$, respectively. Similarly,

$$y(s) = [C(sI - A)^{-1}B + D]u(s) \equiv G(s)u(s).$$

The matrix quantity $G(s)$ is the transfer function matrix associated with the system G.

As a matter of convention, we will refer to the input of the system generically as u. The more specific notation $u(t)$ and $u(s)$ will be used in reference to u represented as time-domain and frequency-domain (Laplace-transform) signals, respectively. A similar convention applies to the output, y. We will assume that u is a vector with m components and that the output y is a vector with p components. This makes the TFM a $p \times m$ matrix. To make this explicit,

$$G(s) = \begin{bmatrix} g_{11}(s) & \cdots & g_{1m}(s) \\ \vdots & & \vdots \\ g_{p1}(s) & \cdots & g_{pm}(s) \end{bmatrix} \tag{29.1}$$

$$u(s) = [u_1(s), \cdots, u_m(s)]^T \tag{29.2}$$

$$y(s) = [y_1(s), \cdots, y_p(s)]^T \tag{29.3}$$

Componentwise,

$$y_k(s) = \sum_{j=1}^{m} g_{kj}(s)u_j(s), \quad k = 1, \cdots, p. \tag{29.4}$$

0-8493-8570-9/96/$0.00+$.50

As Laplace transforms, $u(s)$, $y(s)$, and $G(s)$ are generally complex-valued quantities. In more formal mathematical notation, $u(s) \in C^m$, $y(s) \in C^p$, and $G(s) \in C^{p\times m}$.

29.2 Frequency Response for MIMO Plants

In discussing the frequency response of LTI systems, we focus our attention on systems which are *strictly* stable. This allows us to envision applying sinusoidal inputs to the system and measuring steady-state output signals which are appropriately scaled and phase-shifted sinusoids of the same frequency. Because we are dealing with MIMO systems, there are now additional factors which affect the nature of the frequency response, particularly the relative magnitude and phase of each of the components of the input vector u. These considerations will be discussed in detail below.

Suppose that we have in mind a complex exponential input, as below,

$$u(t) = \tilde{u}e^{j\omega t} \tag{29.5}$$

where $\tilde{u} = (\tilde{u}_1, \cdots, \tilde{u}_m)^T$ is a fixed (complex) vector in C^m. Note that by allowing $\tilde{u}$ to be complex, the individual components of $u(t)$ can have different phases relative to one another.

EXAMPLE 29.2:

Suppose that $m = 2$ and that $\tilde{u} = ((1+j1), (1-j1))^T$, then,

$$\begin{aligned} u(t) &= \begin{pmatrix} (1+j1) \\ (1-j1) \end{pmatrix} e^{j\omega t} \\ &= \begin{pmatrix} \sqrt{2}e^{j\pi/4} \\ \sqrt{2}e^{-j\pi/4} \end{pmatrix} e^{j\omega t} \\ &= \begin{pmatrix} \sqrt{2}e^{j(\omega t+\pi/4)} \\ \sqrt{2}e^{j(\omega t-\pi/4)} \end{pmatrix} \end{aligned}$$

Thus, the two components of $u(t)$ are phase shifted by $\pi/2$ radians (or 90 degrees).

Suppose that this input is applied to our stable LTI system G (of compatible dimension). We know from elementary linear systems theory that each component of the output of G can be expressed in terms of G's *frequency response* $G(j\omega)$. (We obtain the frequency response from $G(s)$, literally, by setting $s = j\omega$.) Thus, at steady state,

$$y_k(t) = \sum_{j=1}^{m} g_{kj}(j\omega)\tilde{u}_j e^{j\omega t}; \quad k = 1, \cdots, p. \tag{29.6}$$

We may now express the vector output $y(t)$ at steady state as follows:

$$y(t) = \tilde{y}e^{j\omega t}, \quad \tilde{y} = (\tilde{y}_1, \cdots, \tilde{y}_p)^T \in C^p, \tag{29.7}$$

where

$$\tilde{y}_k = \sum_{j=1}^{m} g_{kj}(j\omega)\tilde{u}_j, \quad k = 1, \cdots, p. \tag{29.8}$$

Putting all of this together,

$$\tilde{y} = G(j\omega)\tilde{u}. \tag{29.9}$$

Just as in the SISO case, the MIMO frequency response $G(j\omega)$ provides a convenient means of computing the output of an LTI system driven by a complex exponential input. *Analysis* of the frequency response, however, is now complicated by the fact that $G(j\omega)$ is a matrix quantity. A simple way is needed to characterize the "size" of the frequency response as a function of ω. The effect of $G(j\omega)$ on complex exponential input signals depends on the direction of $\tilde{u} \in C^m$, including the relative phase of the components. In fact, a whole range of "sizes" of $G(j\omega)$ exists, depending on the directional nature of $\tilde{u}$. Our characterization of size should thus provide both upper and lower bounds on the magnitude gain of the frequency response matrix. The mathematical tool we need here is the Singular Value Decomposition discussed briefly in the following section.

29.3 Mathematical Detour

We present here some mathematical definitions and basic results from linear algebra germane to our discussion of MIMO frequency response analysis. An excellent reference for this material can be found in [5].

29.3.1 Introduction to Complex Vectors and Complex Matrices

Given a complex-valued column vector $x \in C^n$, we may express x in terms of its real and imaginary components,

$$x = \begin{pmatrix} x_1 \\ \vdots \\ x_n \end{pmatrix} = \begin{pmatrix} a_1 + jb_1 \\ \vdots \\ a_n + jb_n \end{pmatrix} = a + jb$$

where a and b are both purely real-valued vectors in C^n. We define the row vector x^H as the complex-conjugate transpose of x, i.e.,

$$x^H = (x_1^*, \cdots, x_n^*) = a^T - jb^T.$$

The superscript asterix above denotes the complex-conjugate operation.

If we have two vectors x and y, both elements of C^n, then the *inner product of x and y* is given by the (complex) scalar $x^H y$. Two vectors x and y in C^n, which satisfy $x^H y = 0$, are said to be *orthogonal*.

The *Euclidean norm* of x, denoted by $\|x\|_2$, is given by the square root of the inner product of x with itself,

$$\|x\|_2 = \sqrt{x^H x} = \sqrt{a^T a + b^T b} = \sqrt{\sum_{i=1}^{n}(a_i{}^2 + b_i{}^2)}.$$

It is important to note that $\|\cdot\|_2$ is a real scalar function chosen arbitrarily to reflect the "size" of vectors in C^n. (It is true that, as a norm, $\|\cdot\|_2$ has to satisfy certain mathematical requirements, particularly positivity, scaling, and the triangle inequality. Aside from this, our definition of $\|\cdot\|_2$ is arbitrary.) Because all of the components of x are taken into account simultaneously, the value (and interpretation) of $\|x\|_2$ will depend on the *units* in which the components of x are expressed.

EXAMPLE 29.3:

Suppose that we are dealing with a high-power (AC) electronic device and that the state of the device is determined by a vector $x \in C^2$ made up of phased voltages at two distinct points in the circuitry. Suppose first that both quantities are expressed in terms of kilovolts (kV). For example,

$$x = (1 + j2, 2 - j3)^T \;\; kV$$

then,

$$\begin{aligned} \|x\|_2{}^2 &= [(1 - j2), (2 + j3)][(1 + j2), (2 - j3)]^T \\ &= (1 + 4) + (4 + 9) \\ &= 18 \end{aligned}$$

If, however, the first component is expressed in terms of Volts (V), then

$$\begin{aligned} \|x\|_2{}^2 &= [(1000 - j2000), (2 + j3)] \\ &\quad [(1000 + j2000), (2 - j3)]^T, \\ &= (10^6 + 4{\times}10^6) + (4 + 9) \\ &= 5000013, \end{aligned}$$

which is a far cry from what we had before! Note that this is not an entirely unreasonable example. In general, the components of x can consist of entirely different types of physical quantities, such as voltage, current, pressure, concentration, etc. The choice of units is arbitrary and will have an important impact on the "size" of x when measured in terms of the Euclidean norm.

A complex-valued matrix $M \in C^{p\times m}$ is a matrix whose individual components are complex valued. Since M is complex valued, we can express it as the sum of its real and imaginary matrix parts, i.e., $M = A + jB$, where A and B are both purely real-valued matrices in $C^{p\times m}$.

We define the *Hermitian of a complex matrix M* as the complex-conjugate transpose of M, that is, M^H is computed by taking the complex conjugate of the transpose of M. Mathematically,

$$M^H = A^T - jB^T \tag{29.10}$$

The following will play an important role in the next subsection.

Important Fact: Both $M^H M$ and MM^H have eigenvalues that are purely real valued and nonnegative. Moreover, their nonzero eigenvalues are identical even though $M^H M \neq MM^H$.

EXAMPLE 29.4:

Let

$$M = \begin{bmatrix} 1 & 1+j \\ -j & 2+j \end{bmatrix}$$

Then,

$$\begin{aligned} M^H &= \begin{bmatrix} 1 & j \\ 1-j & 2-j \end{bmatrix} \\ M^H M &= \begin{bmatrix} 2 & j3 \\ -j3 & 7 \end{bmatrix} \end{aligned}$$

and

$$MM^H = \begin{bmatrix} 3 & 3+j2 \\ 3-j2 & 6 \end{bmatrix}$$

Although $M^H M$ and MM^H are clearly not equal, a simple calculation easily reveals that both products have the same characteristic polynomial,

$$\det(\lambda I - M^H M) = \det(\lambda I - MM^H) = \lambda^2 - 9\lambda + 5$$

This implies that $M^H M$ and MM^H share the same eigenvalues.

A complex-valued matrix M is called *Hermitian* if $M = M^H$. A nonsingular, complex-valued matrix is called *unitary* if $M^{-1} = M^H$. Stated another way, a complex-valued matrix M is unitary if its column-vectors are mutually orthonormal.

The *spectral norm of a matrix* $M \in C^{p\times m}$, denoted $\|M\|_2$, is defined by

$$\|M\|_2 = \max_{\|x\|_2=1} \|Mx\|_2 \tag{29.11}$$

The best way to interpret this definition is to imagine the vector x rotating on the surface of the unit hypersphere in C^m, generating the vector Mx in C^p. The size of Mx, i.e., its Euclidean norm, will depend on the direction of x. For some direction of x, the vector Mx will attain its maximum value. This value is equal to the spectral norm of M.

29.3.2 The Singular Value Decomposition (SVD)

In this subsection we give a quick introduction to the singular value decomposition. This will be an essential tool in analyzing MIMO frequency response. For more details, the reader is referred to [5].

The Singular Values of a Matrix

Suppose that M is a $p\times m$ matrix, real or complex. Assume that the rank of M is k. We associate with M a total of k positive constants, denoted $\sigma_i(M)$, or simply σ_i, $i = 1, \cdots, k$. These are the *singular values of M*, computed as the positive square roots of the nonzero eigenvalues of either $M^H M$ or MM^H, that is,

$$\sigma_i(M) = \sqrt{\lambda_i(M^H M)} = \sqrt{\lambda_i(MM^H)} \;\; > 0, \quad i = 1, \cdots, k \tag{29.12}$$

where $\lambda_i(\cdot)$ is a shorthand notation for "the ith nonzero eigenvalue of". Note that the matrices $M^H M$ and $M M^H$ may have one or more zero valued eigenvalues in addition to the ones used to compute the singular values of M. It is common to index and rank the singular values as follows:

$$\sigma_1(M) \geq \sigma_2(M) \geq \cdots \geq \sigma_k(M) > 0 \tag{29.13}$$

The largest singular value of M, denoted $\sigma_{max}(M)$, is thus equal to $\sigma_i(M)$. Similarly, $\sigma_{min}(M) = \sigma_k(M)$.

While it is tricky, in general, to compute the eigenvalues of a matrix numerically, reliable and efficient techniques for computing singular values are available in commercial software packages, such as MATLAB.

The Singular Value Decomposition (SVD)

The SVD is analogous to matrix diagonalization. It allows one to write the matrix M in terms of its singular values and involves the definition of special directions in both the range and domain spaces of M.

To begin the definition of the SVD, we use the k nonzero singular values σ_i of M, computed above. First form a square matrix with the k singular values along the main diagonal. Next, add rows and columns of zeros until the resulting matrix Σ is $p \times m$. Thus,

$$\Sigma = \left[\begin{array}{cccc|c} \sigma_1 & 0 & \cdots & 0 & \\ 0 & \sigma_2 & \cdots & 0 & 0_{k\times(m-k)} \\ \cdots & \cdots & \cdots & \cdots & \\ 0 & 0 & \cdots & \sigma_k & \\ \hline & 0_{(p-k)\times k} & & & 0_{(p-k)\times(m-k)} \end{array}\right] \tag{29.14}$$

By convention, assume that

$$\sigma_1 \geq \sigma_2 \geq \cdots \geq \sigma_k$$

THEOREM 29.1 *(SVD) A $p \times p$ unitary matrix U (with $U = U^H$) and an $m \times m$ unitary matrix V (with $V = V^H$) exist so that*

$$M = U \Sigma V^H \qquad \Sigma = U^H M V \tag{29.15}$$

The p-dimensional column vectors u_i, $i = 1, \cdots, p$, of the unitary matrix U are called the *left singular vectors* of M. The m-dimensional column vectors v_i, $i = 1, \cdots, m$ of V are called the *right singular vectors* of M. Thus, we can visualize,

$$U = [u_1 \; u_2 \; \cdots \; u_p], \qquad V = [v_1 \; v_2 \; \cdots \; v_m]$$

Since U and V are each unitary,

$$\begin{aligned} u_i^H u_k &= \delta_{ik}, \quad i, k = 1, \cdots, p \\ v_i^H v_k &= \delta_{ik}, \quad i, k = 1, \cdots, m \end{aligned}$$

where δ_{ik} is the Kronecker delta. Because the left and right singular vectors are linearly independent, they can serve as basis vectors for C^p and C^m, respectively. Moreover, the left and right singular vectors associated with the (nonzero) singular values of M span the range and left-null spaces of the matrix M, respectively. Finally, a simple calculation shows that the left singular vectors of M are the normalized right eigenvectors of the $p \times p$ matrix MM^H. Similarly, the right singular vectors of M are the normalized left eigenvectors of the $p \times p$ matrix $M^H M$.

Some Properties of Singular Values

We list here some important properties of singular values. We leave the proofs to the reader. Some of the properties require that the matrix be square and nonsingular.

1. $\sigma_{max}(M) = \max_{\|x\|_2=1} \|Mx\|_2 = \|M\|_2 = \frac{1}{\sigma_{min}(M^{-1})}$.
2. $\sigma_{min}(M) = \min_{\|x\|_2=1} \|Mx\|_2 = \frac{1}{\|M^{-1}\|_2} = \frac{1}{\sigma_{max}(M^{-1})}$.
3. $\sigma_i(M) - 1 \leq \sigma_i(I + M) \leq \sigma_i(M) + 1, \quad i = 1, \cdots, k$.
4. $\sigma_i(\alpha M) = |\alpha| \sigma_i(M)$ for all $\alpha \in C, i = 1, \cdots, k$.
5. $\sigma_{max}(M_1 + M_2) \leq \sigma_{max}(M_1) + \sigma_{max}(M_2)$.
6. $\sigma_{max}(M_1 M_2) \leq \sigma_{max}(M_1) \cdot \sigma_{max}(M_2)$.

Property 1 indicates that maximum singular value $\sigma_{max}(M)$ is identical to the spectral norm of M. Thus, Properties 5 and 6 are restatements of the triangle inequality and submultiplicative property, respectively.

The SVD and Finite Dimensional Linear Transformations

We shall now present some geometric interpretations of the SVD result. Consider the linear transformation

$$y = Mu, \quad u \in C^m, y \in C^p. \tag{29.16}$$

Let M have the singular value decomposition discussed above, that is, $M = U \Sigma V^H$. It may help the reader to think of u as the input to a static system M with output y. From the SVD of M,

$$y = Mu = U \Sigma V^H u.$$

Suppose we choose u to be one of the right singular vectors, say v_i, of M. Let y_i denote the resulting "output" vector. Then,

$$y_i = M v_i = U \Sigma V^H v_i.$$

Because the right singular vectors of M are orthonormal,

$$V^H v_i = (0, \cdots, 0, 1, 0, \cdots, 0)^T,$$

where the ith component only takes on the value of 1. In view of the special structure of the matrix of singular values Σ,

$$\Sigma V^H v_i = (0, \cdots, 0, \sigma_i, 0, \cdots, 0)^T$$

where, again, only the ith component is potentially nonzero. Thus, finally,

$$y_i = U \Sigma V^H v_i = \sigma_i u_i. \tag{29.17}$$

Equation 29.17 interprets the unique relationship between singular values and singular vectors. In the context of M as a "static" system, when the input u is equal to a right singular vector v_i, the output direction is fixed by the corresponding left singular vector u_i. Keeping in mind that both u_i and v_i have unit magnitudes (in the Euclidean sense), the amplification (or attenuation) of the input is measured by the associated singular value σ_i. If we choose $u = v_i$, where $i > k$, then the corresponding output vector is zero because the matrix is not full rank and there are no more (nonzero) singular values.

Because Equation 29.17 holds for $i = 1, \ldots, k$, it is true in particular for the maximum and minimum singular values and associated singular vectors. By abuse of notation, we shall refer to these left and right singular vectors as *maximum and minimum singular vectors*, respectively, and use the subscripts "max" and "min" to distinguish them. Within the context of "static" systems, inputs along the *maximum* right singular vector generate the *largest* output along the direction of the *maximum* left singular vector. Similar comments apply to the case where inputs are in the direction of the minimum left singular vector.

EXAMPLE 29.5: The case where M is real and 2×2

Let us suppose that M is a real-valued, nonsingular matrix mapping $u \in R^2$ to $y = Mu \in R^2$. Let us suppose further that u rotates on the circumference of the unit circle. The image of u under the transformation of M will then trace an ellipse in the (output) plane, as illustrated in Figure 29.1.

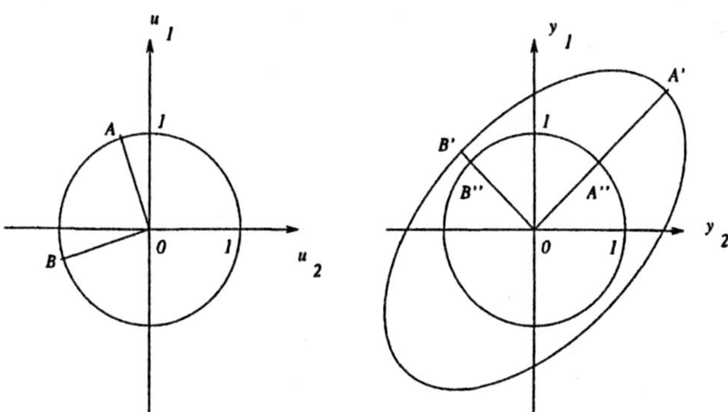

Figure 29.1 Visualization of SVD quantities.

Because M is a real, nonsingular, 2×2 matrix, the SVD analysis will give

$$\Sigma = \begin{bmatrix} \sigma_{max} & 0 \\ 0 & \sigma_{min} \end{bmatrix} \quad U = [u_{max} \; u_{min}] \quad V = [v_{max} \; v_{min}]$$

where Σ, U, and V are all real valued.

Suppose that when u equals the vector OA, the output y is the vector OA'. Suppose further that, $y = y_{max} \equiv \sigma_{max} u_{max}$. Thus, the maximum right singular vector v_{max} equals the (unit length) vector OA, and the maximum left singular vector u_{max} equals the (unit length) vector OA". Moreover, the maximum singular value, σ_{max} equals the length of the vector OA'.

Similarly, suppose that when u equals the vector OB, the output y is the vector OB'. Suppose further that, $y = y_{min} \equiv \sigma_{min} u_{min}$. Thus, the minimum right singular vector v_{min} equals the (unit length) vector OB, and the minimum left singular vector u_{min} equals the (unit length) vector OB". Moreover, the minimum singular value, σ_{min} equals the length of the vector OB'.

Notice in Figure 29.1 that the left singular vectors are normal to each other, as are the right singular vectors.

As the minimum singular value decreases, so does the semiminor axis of the ellipse. As this happens, the ellipse becomes more and more elongated. In the limit, as $\sigma_{min} \to 0$, the ellipse degenerates into a straight line segment, and the matrix M becomes singular. In this limiting case, there are directions in the output space that we cannot achieve.

If the matrix M were a 3×3 real nonsingular matrix, then we could draw a similar diagram, illustrating the unit sphere mapping into a three-dimensional ellipsoid. Unfortunately, diagrams for higher dimensional matrices are impossible. Similarly, diagrams for *complex* matrices (even 2×2 matrices) are impossible, because we need a plane to represent each complex number.

Using these geometric interpretations of SVD quantities, it is possible to be precise about the meaning of the "size" of a real or complex matrix. From an intuitive point of view, if we consider the "system" $y = Mu$, and if we restrict the input vector u to have unit length, then

1. The matrix M is "large" if $\|y\|_2 >> 1$, independent of the direction of the unit input vector u.
2. The matrix M is "small' if $\|y\|_2 << 1$, independent of the direction of the unit input vector u.

If we accept these definitions, then we can quantify size as follows:

1. The matrix M is "large" if its minimum singular value is large, i.e., $\sigma_{min}(M) >> 1$.
2. The matrix M is "small" if its maximum singular value is small, i.e., $\sigma_{max}(M) << 1$.

More Analytical Insights

Once we have computed an SVD for a matrix M, in $y = Mu$, then we can compute many other important quantities. In particular, suppose that M is $m \times m$ and nonsingular. It follows that M has m nonzero singular values. We saw earlier (in Equation 29.17) that

$$y_i = \sigma_i u_i$$

when $u = v_i$, $i = 1, \cdots, m$. Because the left singular vectors are orthonormal, they form a basis for the m-dimensional input space, so that we can write any (input) vector in C^m as a linear combination of the v_is. For example, let u be given as follows:

$$u = \gamma_1 v_1 + \gamma_2 v_2 + \cdots + \gamma_m v_m$$

where the γ_i are real or complex scalars. From the linearity of the transformation M,

$$y = Mu = \gamma_1 \sigma_1 u_1 + \gamma_2 \sigma_2 u_2 + \cdots + \gamma_m \sigma_m v_m.$$

Using the SVD, we can also gain insight on the inverse transformation $u = M^{-1} y$. From the SVD theorem, we know that $M = U\Sigma V^H$. Using the fact that U and V are unitary, $M^{-1} = V\Sigma^{-1}U^H$. Notice that

$$\Sigma^{-1} = \text{diag}\{\frac{1}{\sigma_1}, \frac{1}{\sigma_2}, \cdots, \frac{1}{\sigma_m}\}$$

Thus, if

$$y = \delta_1 u_1 + \delta_2 u_2 + \cdots + \delta_m u_m$$

then

$$u = M^{-1}y = \delta_1 \frac{1}{\sigma_1} v_1 + \delta_2 \frac{1}{\sigma_2} v_2 + \cdots + \delta_m \frac{1}{\sigma_m} v_m$$

This implies that the information in the SVD of M can be used to solve systems of linear equations without computing the inverse of M.

29.4 The SVD and MIMO Frequency Response Analysis

We now return to our discussion of MIMO frequency response with the full power of the SVD at our disposal. Once again, we shall focus our attention on the transfer function matrix (TFM) description of the strictly stable LTI system $G(s)$. As before, we will assume that G has m inputs and p outputs, making $G(s)$ a $p \times m$ matrix. In general we shall assume that $p \geq m$, so that, unless the rank k of $G(s)$ becomes less than m, the response of the system to a non-zero input is always non-zero.

Recall that if the input vector signal $u(t)$ is a complex exponential of the form $u(t) = \tilde{u}e^{j\omega t}$, with $\tilde{u}$ fixed in C^m, then at steady state, the output vector $y(t)$ will also be a complex exponential function, $y(t) = \tilde{y}e^{j\omega t}$, for some $\tilde{y} \in C^p$. Recall, also, that the complex vectors $\tilde{u}$ and $\tilde{y}$ are related by $G(s)$ evaluated at $s = j\omega$, that is,

$$\tilde{y} = G(j\omega)\tilde{u}.$$

It is important to note that $G(j\omega)$ is a complex matrix that changes with frequency ω. For any given fixed frequency, we can calculate the SVD of $G(j\omega)$:

$$G(j\omega) = U(j\omega)\Sigma(\omega)V^H(j\omega)$$

Note that, in general, all of the factors in the SVD of $G(j\omega)$ are explicitly dependent on omega:

1. The matrix $\Sigma(\omega)$ is a $p \times m$ matrix whose main diagonal is composed of the singular values of $G(j\omega)$,

$$\sigma_{max}(\omega) = \sigma_1(\omega), \sigma_2(\omega), \cdots, \sigma_{k_\omega}(\omega) = \sigma_{min}(\omega)$$

where k_ω is the rank of $G(j\omega)$.
2. The matrix $U(j\omega)$ is an $m \times m$ complex-valued matrix whose column vectors $\{u_j(j\omega)\}$ are the left singular vectors of $G(j\omega)$, and
3. The matrix $V(j\omega)$ is a $p \times p$ complex-valued matrix whose column vectors $\{v_j(j\omega)\}$ are the right singular vectors of $G(j\omega)$.

29.4.1 Singular Value Plots (SV Plots)

Once we calculate the maximum and minimum singular values of $G(j\omega)$ for a range of frequencies ω, we can plot them together on a Bode plot (decibels versus rad/sec in log-log scale). Figure 29.2 shows a hypothetical SV plot.

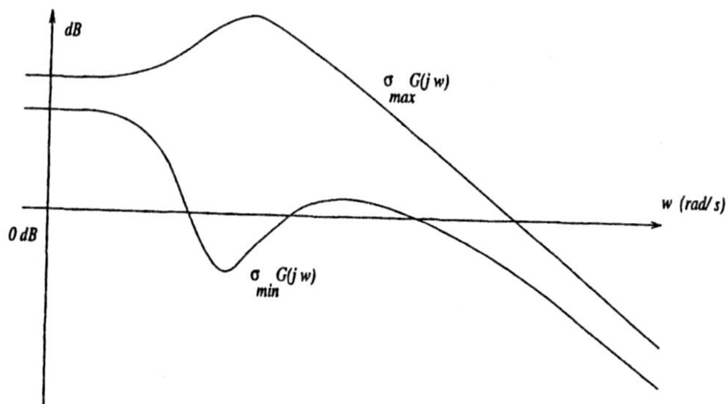

Figure 29.2 A hypothetical SV plot.

With the proper interpretation, the SV plot can provide valuable information about the properties of the MIMO system G. In particular, it quantifies the "gain-band" of the plant at each frequency, and shows how this changes with frequency. It is a natural generalization of the information contained in the classical Bode magnitude plot for SISO plants. One main difference here is that, in the multivariable case, this "gain-band" is described by two curves, not one.

It is crucial to interpret the information contained in the SV plot correctly. At each frequency ω we assume that the input is a *unit* complex exponential, $u(t) = \tilde{u}e^{j\omega t}$. Then, assuming that we have reached steady state, we know that the output is also a complex exponential with the same frequency, $y(t) = \tilde{y}e^{j\omega t}$, where $\tilde{y} = G(j\omega)\tilde{u}$. The magnitude $\|\tilde{y}\|_2$ of the output complex exponential thus depends on the direction of the input as well as on the frequency ω. Now, by looking at an SV plot, we can say that, at a given frequency:

1. The largest output size is $\|\tilde{y}\|_{2,max} = \sigma_{max}G(j\omega)$, for $\|\tilde{u}\|_2 = 1$,
2. The smallest output size is $\|\tilde{y}\|_{2,min} = \sigma_{min}G(j\omega)$, for $\|\tilde{u}\|_2 = 1$.

This allows us to discuss qualitatively the size of the plant gain as a function of frequency:

1. The plant has large gain at ω if $\sigma_{min}G(j\omega)) >> 1$,
2. The plant has small gain at ω if $\sigma_{max}G(j\omega)) << 1$.

29.4.2 Computing Directional Information

In addition to computing system gain as a function of frequency, we can also use the SVD to compute "directional information" about the system. In particular, we can compute the direction of maximum and minimum amplification of the unit, real-valued sinusoidal input. In the following, we present a step-by-step methodology for maximum amplification direction analysis. Minimum amplification direction analysis is completely analogous, and will therefore not be presented explicitly.

Maximum Amplification Direction Analysis

1. Select a specific frequency ω.
2. Compute the SVD of $G(j\omega)$, i.e., find $\Sigma(\omega)$, $U(j\omega)$, and $V(j\omega)$ such that $G(j\omega) = U(j\omega)\Sigma(\omega)V^H(j\omega)$ where U and V are unitary and Σ is the matrix of singular values.
3. In particular, find the maximum singular value $\sigma_{max}(\omega)$ of $G(j\omega)$.
4. Find the maximum right singular vector $v_{max}(\omega)$. This is the first column of the matrix $V(j\omega)$ found in the SVD. Note that $v_{max}(\omega)$ is a complex vector with m elements. Write the elements of $v_{max}(\omega)$ in polar form, i.e.,

$$[v_{max}(\omega)]_i = |a_i|e^{j\psi_i}, \quad i = 1, 2, \cdots, m.$$

Notice that a_i and ψ_i are really functions of ω; we suppress this frequency dependence for clarity.

5. Find the maximum left singular vector $u_{max}(\omega)$. This is the first column of the matrix $U(j\omega)$ found in the SVD. Note that $u_{max}(\omega)$ is a complex vector with p elements. Write the elements of $u_{max}(\omega)$ in polar form, i.e.,

$$[u_{max}(\omega)]_i = |b_i|e^{j\phi_i}, \quad i = 1, 2, \cdots, p.$$

Notice that b_i and ϕ_i are functions of ω; we suppress this frequency dependence for clarity.

6. We are now in a position to construct the real sinusoidal input signals that correspond to the direction of maximum amplification and to predict the output sinusoids that are expected at steady state. The input vector $u(t)$ is defined componentwise by

$$u_i(t) = |a_i| \sin(\omega t + \psi_i), \quad i = 1, 2, \cdots, m$$

where the parameters a_i and ψ_i are those determined above. Note that the amplitude and phase of each component sinusoid is distinct. We can utilize the implications of the SVD to predict the steady-state output sinusoids as

$$y_i(t) = \sigma_{max}(\omega)|b_i| \sin(\omega t + \phi_i), \quad i = 1, 2, \cdots, p.$$

Notice that all parameters needed to specify the output sinusoids are already available from the SVD.

When we talk about the "directions of maximum amplification", we mean input *sinusoids* of the form described above with very precise magnitude and phase relations to one another. The resulting output sinusoids also have very precise magnitude and phase relations, all as given in the SVD of $G(j\omega)$. Once again, a completely analogous approach can be taken to compute the minimum amplification direction associated with $G(j\omega)$.

It is important to remember that the columns of $U(j\omega)$ and $V(j\omega)$ are orthonormal. This means we can express *any* sinusoidal input vector as a linear combination of the right singular vectors of $G(j\omega)$ at a particular value of ω. The corresponding output sinusoidal vector will be a linear combination of the left singular vectors, after being scaled by the appropriate singular values.

Finally, because we measure system "size" in terms of the ratio of output Euclidean norm to input Euclidean norm, the "size" of the system is heavily dependent on the units of the input and output variables.

29.5 Frequency Response Analysis of MIMO Feedback Systems

In this section, we look at frequency domain-analysis for various control system configurations. We will pay particular attention to the classical unity-feedback configuration, where the variables to be controlled are used as feedback. Next we will look at a broader class of control system configurations relevant for some of the more modern controller design methodologies such as H_∞ and l_1 synthesis, as well as in robustness analysis and synthesis. MIMO frequency-domain analysis as discussed above will be pivotal throughout.

29.5.1 Classical Unity-Feedback Systems

Consider the unity-feedback system in the block diagram of Figure 29.3. Recall that the *loop* transfer function matrix is defined as

$$T(s) = G(s)K(s)$$

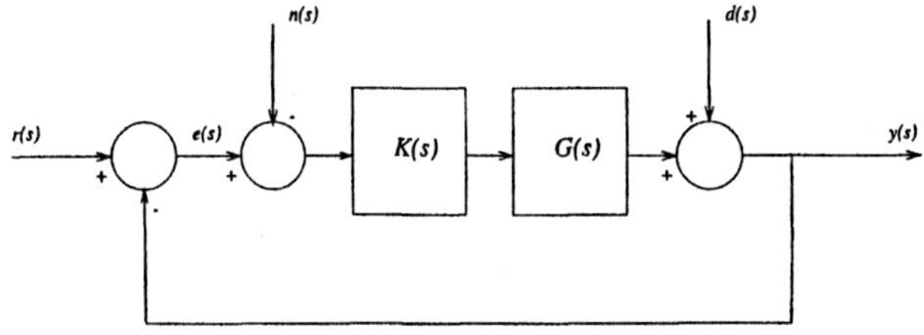

Figure 29.3 The unity feedback control system configuration.

The *sensitivity* $S(s)$ and *complementary-sensitivity* $C(s)$ transfer function matrices are, respectively, defined as

$$\begin{aligned} S(s) &= [I + T(s)]^{-1} \\ C(s) &= [I + T(s)]^{-1}G(s)K(s) = S(s)T(s) \end{aligned}$$

With these definitions,

$$e(s) = S(s)[r(s) - d(s)] + C(s)n(s) \tag{29.18}$$

The objective in control system design is to keep the error signal e "small". This means the *transfer* from the various disturbances to e must be small. Because it is always true that $S(s) + C(s) = I$, there is a trade-off involved. From Equation 29.18, we would like both $S(s)$ and $C(s)$ to be "small" for all s; but this is impossible because $S(s) + C(s) = I$. SVD analysis of the MIMO frequency response can be important in quantifying these issues.

Command Following

Suppose that the reference (command) signal $r(t)$ is sinusoidal, $r(t) = \tilde{r}e^{j\omega t}$. Then, as long as $d(s) = 0$ and $n(s) = 0$,

$$e(t) = \tilde{e}e^{j\omega t}$$

where $\tilde{e} = S(j\omega)\tilde{r}$. Thus,

$$\|\tilde{e}\|_2 \leq \sigma_{max}[S(j\omega)] \cdot \|\tilde{r}\|_2$$

Now suppose that $r(t)$ is the superposition of more than one sinusoid. Let Ω_r be the range of frequencies at which the input $r(t)$ has its energy. Then, in order to have good command following, we want

$$\sigma_{max}[S(j\omega)] << 1 \quad \forall\omega \in \Omega_r \tag{29.19}$$

Our objective now is to express this prescription for command following in terms of the loop-transfer function $T(s) = G(s)K(s)$. From our earlier discussion

$$\begin{aligned}\sigma_{max}[S(j\omega)] &= \sigma_{max}\{[I + T(j\omega)]^{-1}\} \\ &= \frac{1}{\sigma_{min}[I + T(j\omega)]}\end{aligned}$$

This implies that, for good command following, we must have $\sigma_{min}[I + T(j\omega)] >> 1$ for all $\omega \in \Omega_r$. However, as we saw earlier,

$$\sigma_{min}[I + T(j\omega)] \geq \sigma_{min}[T(j\omega)] - 1$$

so it is sufficient that, for all $\omega \in \Omega_r$, $\sigma_{min}[T(j\omega)] >> 1$ for good command following.

Disturbance Rejection

Suppose that the disturbance signal $d(t)$ is sinusoidal, $d(t) = \tilde{d}e^{j\omega t}$. Then, as long as $r(s) = 0$ and $n(s) = 0$,

$$e(t) = \tilde{e}e^{j\omega t}$$

where $\tilde{e} = -S(j\omega)\tilde{d}$. Thus,

$$\|\tilde{e}\|_2 \leq \sigma_{max}[S(j\omega)] \cdot \|\tilde{d}\|_2$$

Now suppose that $d(t)$ is the superposition of more than one sinusoid. Let Ω_d be the range of frequencies at which the input $d(t)$ has its energy. Then, just as with command following, for good disturbance rejection, we want

$$\sigma_{max}[S(j\omega)] << 1 \quad \forall\omega \in \Omega_d \tag{29.20}$$

Using the same argument given earlier, this prescription for disturbance rejection makes it sufficient that, for all $\omega \in \Omega_d$, $\sigma_{min}[T(j\omega)] >> 1$.

Relationships to C(s)

Here we wish to determine in a precise quantitative way the consequences of obtaining command following and disturbance rejection. As we shall see, a price is paid in constraints on the complementary-sensitivity function $C(s)$.

THEOREM 29.2 *Let $\Omega_p = \Omega_r \bigcup \Omega_d$. (Here "p" refers to "performance".) Consider δ so that $0 < \delta << 1$. If*

$$\sigma_{max}[S(j\omega)] \leq \delta << 1$$

for all $\omega \in \Omega_p$, then

$$1 << \frac{1-\delta}{\delta} \leq \sigma_{min}[T(j\omega)]$$

and

$$1 - \delta \leq \sigma_{min}[C(j\omega)] \leq \sigma_{max}[C(j\omega)] \leq 1 + \delta$$

for all $\omega \in \Omega_p$.

Thus, in obtaining a performance level of δ, it is necessary that all of the singular values of $C(j\omega)$ are within δ of 1. In fact, because $S(s) + C(s) = I$, we must have $C(j\omega) \approx I$. (We shall discuss below why this can be a problem.)

PROOF 29.1 We start by using the definition of $S(s)$,

$$\begin{aligned}\sigma_{max}[S(j\omega)] &= \sigma_{max}[(I + T(j\omega))^{-1}] \\ &= \frac{1}{\sigma_{min}[I + T(j\omega)]} \\ &\geq \frac{1}{1 + \sigma_{min}[T(j\omega)]}\end{aligned}$$

Using the hypothesis that $\sigma_{max}[S(j\omega)] \leq \delta$,

$$\frac{1}{1 + \sigma_{min}[T(j\omega)]} \leq \delta$$

which by solving for $\sigma_{min}[T(j\omega)]$ yields the first inequality. By cross-multiplying, we obtain the following useful expression:

$$\frac{1}{\sigma_{min}[T(j\omega)]} \leq \frac{\delta}{1-\delta} << 1 \tag{29.21}$$

Now consider the complementary-sensitivity function $C(s)$. $C(s) = [I + T(s)]^{-1}T(s)$. By taking the inverse of both sides, $C^{-1}(s) = T^{-1}(s)[I + T(s)] = I + T^{-1}(s)$. Thus

$$\frac{1}{\sigma_{min}[C(j\omega)]} = \sigma_{max}[C^{-1}(j\omega)] = \sigma_{max}[I + T^{-1}(j\omega)]$$

which implies

$$\begin{aligned}\frac{1}{\sigma_{min}[C(j\omega)]} &\leq 1 + \sigma_{max}[T^{-1}(j\omega)] \\ &= 1 + \frac{1}{\sigma_{min}[T(j\omega)]} \\ &\leq 1 + \frac{\delta}{1-\delta} = \frac{1}{1-\delta}\end{aligned}$$

(Notice that the second inequality follows from Equation 29.21.) Now,

$$\begin{aligned} 1-\delta &\le \sigma_{min}[C(j\omega)] \le \sigma_{max}[C(j\omega)] \\ &= \sigma_{max}[I - S(j\omega)] \le 1 + \sigma_{max}[S(j\omega)] \le 1+\delta \end{aligned}$$

which is the second desired inequality.

Measurement Noise Insensitivity: A Conflict!

Suppose that the measurement noise $n(t)$ is sinusoidal, $n(t) = \tilde{n}e^{j\omega t}$. Then, as long as $r(s) = 0$ and $d(s) = 0$,

$$e(t) = \tilde{e}e^{j\omega t}$$

where $\tilde{e} = C(j\omega)\tilde{d}$. Thus,

$$\|\tilde{e}\|_2 \le \sigma_{max}[C(j\omega)] \cdot \|\tilde{n}\|_2$$

Now suppose that $n(t)$ is the superposition of more than one sinusoid. Let Ω_n be the range of frequencies at which the input $n(t)$ has its energy. Then, in order to be insensitive to measurement noise, we want

$$\sigma_{max}[C(j\omega)] << 1 \quad \forall\omega \in \Omega_n \qquad (29.22)$$

THEOREM 29.3 *Let γ be such that $0 < \gamma << 1$. If*

$$\sigma_{max}[C(j\omega)] \le \gamma$$

for all $\omega \in \Omega_n$, then

$$\sigma_{min}[T(j\omega)] \le \sigma_{max}[T(j\omega)] \le \frac{\gamma}{1-\gamma} \approx \gamma << 1$$

and

$$1 \approx 1-\gamma \le \sigma_{min}[S(j\omega)] \le \sigma_{max}[S(j\omega)]$$

for all $\omega \in \Omega_n$.

Thus, if the complementary-sensitivity function $C(j\omega)$ has low gain on Ω_n, then so does the loop-transfer function $T(j\omega)$. This in turn implies that the sensitivity transfer function $S(j\omega)$ has nearly unity gain on Ω_n. In other words, wherever (in frequency) we are insensitive to measurement noise we are necessarily prone to poor command following *and* disturbance rejection. This is primarily a consequence of the fact that $C(s) + S(s) = I$.

PROOF 29.2 To prove the first relationship we use the fact that $C(s) = [I + T(s)]^{-1}T(s) = [T^{-1}(s) + I]^{-1}$ (proved using a few algebraic manipulations). This gives

$$\begin{aligned} \sigma_{max}[C(j\omega)] &= \frac{1}{\sigma_{min}[T^{-1}(j\omega) + I]} \\ &\ge \frac{1}{\sigma_{min}[T^{-1}(j\omega)] + 1} \\ &= \frac{\sigma_{max}[T(j\omega)]}{1 + \sigma_{max}[T(j\omega)]} \end{aligned}$$

Thus,

$$\begin{aligned} \sigma_{max}[T(j\omega)] &\le \sigma_{max}[C(j\omega)] \\ &\quad + \sigma_{max}[C(j\omega)]\sigma_{max}[T(j\omega)] \\ &\le \gamma + \gamma\sigma_{max}[T(j\omega)] \end{aligned}$$

which yields the desired inequality.

To prove the second relationship, observe that

$$\begin{aligned} \sigma_{max}[S(j\omega)] &\ge \sigma_{min}[S(j\omega)] \\ &= \sigma_{min}[(I + T(j\omega))^{-1}] \\ &= \frac{1}{\sigma_{max}[I + T(j\omega)]} \\ &\ge \frac{1}{\sigma_{max}[T(j\omega)] + 1} \\ &\ge 1-\gamma \end{aligned}$$

where the last inequality comes from the first relationship proved above.

Design Implications

Achievable control design specifications must have a wide separation (in frequency) between the sets $\Omega_p = \Omega_r \bigcup \Omega_d$ and Ω_n. We cannot obtain good command following and disturbance rejection when we have sensors that are noisy on Ω_p. Figure 29.4 illustrates a problem that is well-posed in terms of these constraints.

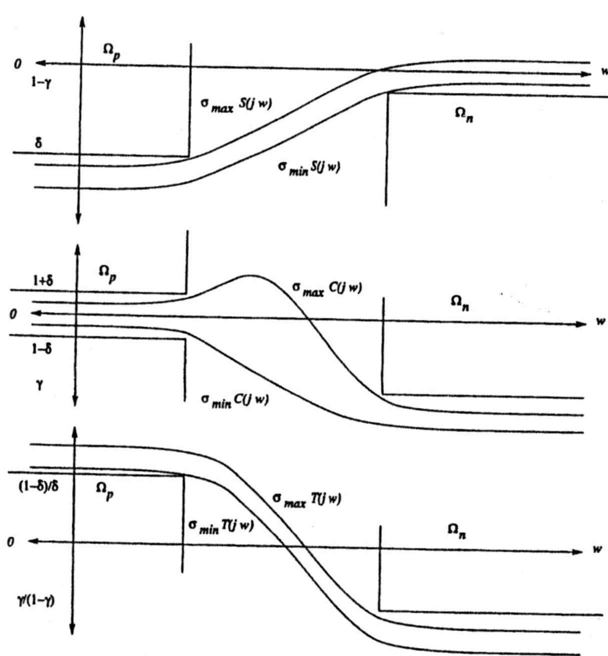

Figure 29.4 A well-posed problem: the singular value traces fall within the regions defined by Ω_p and δ (for performance) and Ω_n and γ (for noise insensitivity).

29.5.2 A More General Setting

Recent control design methodologies are convenient for more generalized problem descriptions. Consider the LTI system P shown in Figure 29.5, where u denotes the control vector input to the plant, y represents the measurement vector, w is a generalized disturbance vector, and z is a generalized performance vector. The assignment of physical variables to w and z here is arbitrary and is left to the control system analyst. One illustration is given in Example 29.6. We see that the transfer function matrix for this system can be partitioned as follows:

$$P(s) = \begin{bmatrix} P_{11}(s) & P_{12}(s) \\ P_{21}(s) & P_{22}(s) \end{bmatrix}$$

From this partition,

$$\begin{aligned} z(s) &= P_{11}(s)w(s) + P_{12}(s)u(s) \\ y(s) &= P_{21}(s)w(s) + P_{22}(s)u(s) \end{aligned}$$

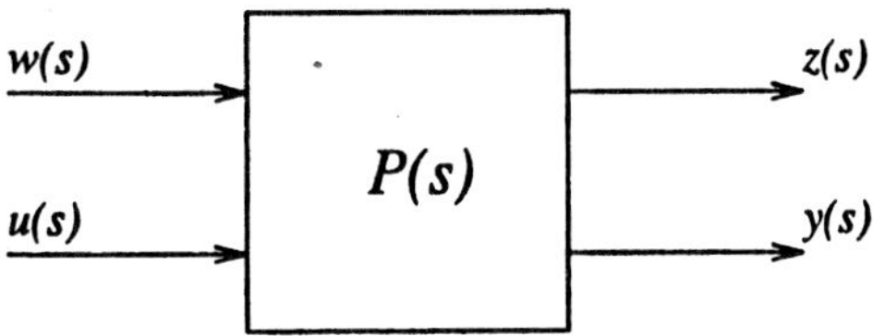

Figure 29.5 A generalized plant.

Let K be a feedback controller for the system so that $u(s) = K(s)y(s)$. This feedback interconnection is shown in Figure 29.6. It is simple to verify that

$$\begin{aligned} y(s) &= [I - P_{22}(s)K(s)]^{-1}P_{21}(s)w(s), \\ z(s) &= \{P_{11}(s) + P_{12}(s)K(s) \\ &\quad [I - P_{22}(s)K(s)]^{-1}P_{21}(s)\}w(s), \\ &\equiv F(P, K)(s)w(s). \end{aligned}$$

The closed-loop transfer function from $w(s)$ to $z(s)$, denoted $F(P, K)(s)$, is called the (lower) linear fractional transformation (of the plant P and K). This type of mathematical object is often very useful in describing feedback interconnections.

EXAMPLE 29.6: Relationships

In this example we show how the classical unity-feedback setup can be mapped into the general formulation of this section. Consider the block diagram shown in Figure 29.7. With four generalized disturbance inputs, a command input r, a sensor noise input n, a system-output disturbance d_o, and a system-input disturbance d_i. (These variables are lumped into the generalized disturbance vector w.) There are two generalized performance variables, control (effort) u and tracking error e. (These are lumped into the generalized performance vector z.) The variables u and y are the control inputs and sensor outputs of the generalized plant P.

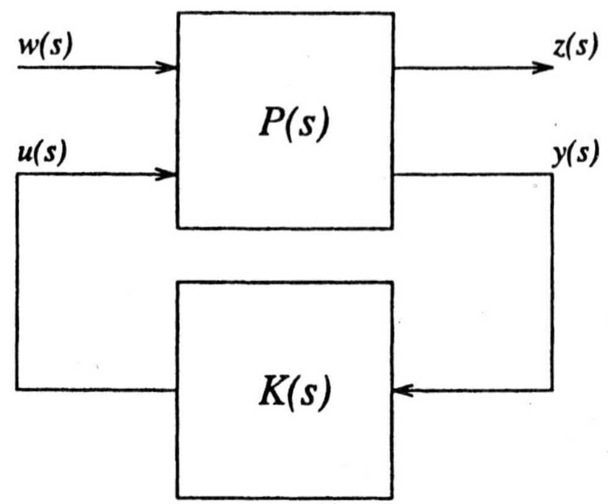

Figure 29.6 A feedback interconnection.

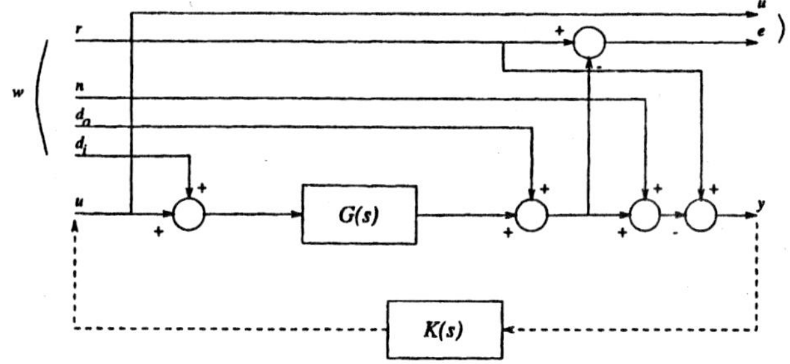

Figure 29.7 Unity feedback example.

Frequency Weights

In control design or analysis it is often necessary to incorporate extra information about the plant and its environment into the generalized description P. For example, we may know that system-input disturbances are always low-frequency in nature, and/or we may only care about noise rejection at certain high frequencies. This kind of information can be incorporated as "frequency weights" augmenting the generalized inputs and outputs of the plant P. Such weighting functions $W_d(s)$ and $W_p(s)$ are included in the block diagram of Figure 29.8. Examples are given below.

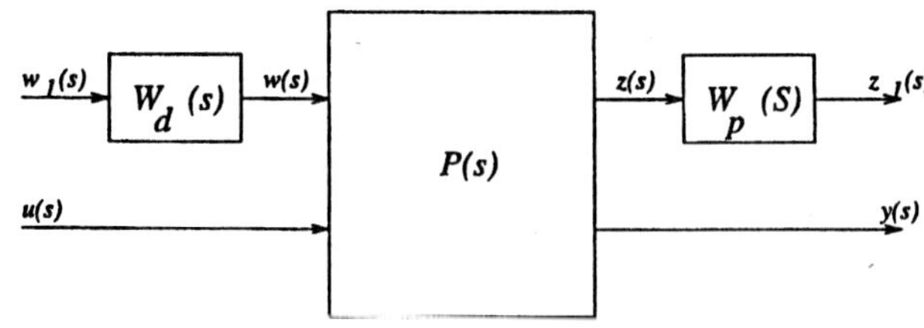

Figure 29.8 Weighting functions and the generalized plant.

With frequency-dependent weightings we can define a new generalized plant P_w. Referring to Figure 29.8 and using the block partition of P discussed earlier,

$$P_w(s) = \begin{bmatrix} W_p(s)P_{11}(s)W_d(s) & W_p(s)P_{12}(s) \\ P_{21}(s)W_d(s) & P_{22}(s) \end{bmatrix}.$$

With the weighting functions augmenting the plant description, P_w is ready (at least conceptually) for control design or analysis.

If state-space techniques are being employed, the *dynamics* (i.e., the state variables) for W_d and W_p must be included in the state-space representation of P_w. This can be accomplished by state augmentation.

Typically, weights W_d on the generalized disturbance vector emphasize the frequency regions of most significant disturbance strength. Here $w(s) = W_d(s)w_1(s)$. We choose W_d so that a complete description of the generalized disturbance is obtained with $\|w_1(j\omega)\| = 1$ for all ω. Thus, we may think of W_d as an active filter which emphasizes (or de-emphasizes) certain variables in specific frequency domains consistent with our understanding of the system's environment. Reference inputs and system disturbances are most often low frequency, so these parts of W_d are usually low-pass filters. On the other hand, certain noise inputs are notched (like 60 Hz electrical hum) or high frequency. These parts of $W_d(s)$ should be band-pass or high-pass filters, respectively.

Weights W_p on the generalized performance vector emphasize the *importance* of good performance in different frequency regions. For example, we may be very interested in good tracking at low frequencies (but not at high frequencies), so that the weighting on this part of $W_p(s)$ should be a low-pass filter. On the other hand, we may not want to use control energy at high frequencies, so we would choose this part of $W_p(s)$ as high-pass. The various components of $W_p(s)$ must be consistent with one another in gain. The relative weights on variables in z should make sense as a whole.

It is important to note that we must still operate within constraints to mutually achieve various types of performance, as was the case with the classical unity-feedback formulation. Because we are no longer constrained to that rigid type of feedback interconnection, general results are difficult to state. The basic idea, however, remains that there is an essential conflict between noise-rejection and command-following/disturbance-rejection. In general there must be a separation in frequency between the regions in which the respective types of performance are important.

Weighted Sensitivity

We consider here the case of only one generalized input variable, d, a system-output disturbance. We are also interested in only one performance variable y, the disturbed output of the plant. Specifically, we have in mind a weighting function W_p which reflects our specifications for y. The feedback configuration is shown in Figure 29.9.

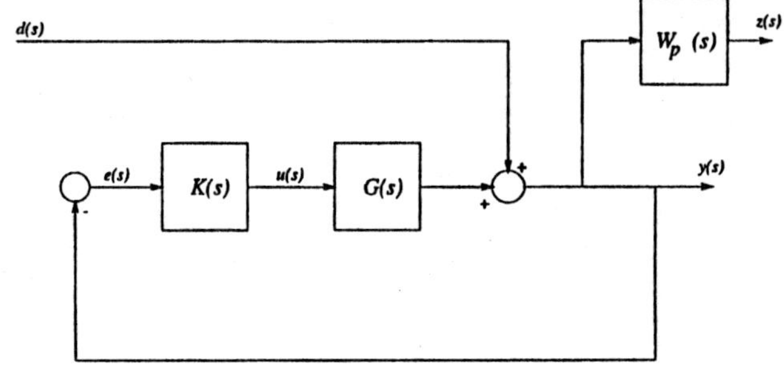

Figure 29.9 Feedback interconnection for weighted sensitivity.

It is not hard to see that $y(s) = S(s)d(s)$. Thus, $z(s) = W_p(s)S(s)d(s)$. We refer to $W_p(s)S(s)$ as the " weighted" sensitivity transfer function for the feedback system. If $\sigma_{max}[W_p(j\omega)S(j\omega)] < 1$ for all ω, then,

$$\begin{aligned}\sigma_{max}[S(j\omega)] &= \sigma_{max}[W_p^{-1}(j\omega)W_p(j\omega)S(j\omega)]\\ &\leq \sigma_{max}[W_p^{-1}(j\omega)]\sigma_{max}[W_p(j\omega)S(j\omega)]\\ &< \sigma_{max}[W_p^{-1}(j\omega)]\end{aligned}$$

To interpret this, when $\sigma_{max}[W_p(j\omega)S(j\omega)] < 1$ for all ω, then the largest singular value of the sensitivity transfer function is strictly less than the largest singular value of the inverse of the weighting function.

Weighted Complementary Sensitivity

We consider here the case of only one generalized input variable, n, a sensor noise input. We are also only interested in one performance variable y, the output of the plant. Once again, we have in mind a weighting function W_p which reflects our specifications for y. The feedback configuration is shown in Figure 29.10.

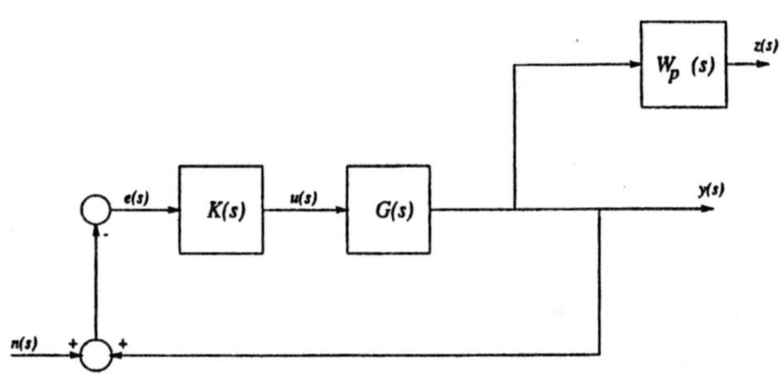

Figure 29.10 Feedback interconnection for weighted complementary sensitivity.

It is not hard to see that $y(s) = C(s)d(s)$. Thus, $z(s) = W_p(s)C(s)d(s)$. We refer to $W_p(s)C(s)$ as the " weighted" complementary-sensitivity transfer function for the feedback system. Notice that if $\sigma_{max}[W_p(j\omega)C(j\omega)] < 1$ for all ω then,

$$\begin{aligned}\sigma_{max}[C(j\omega)] &= \sigma_{max}[W_p^{-1}(j\omega)W_p(j\omega)C(j\omega)]\\ &\leq \sigma_{max}[W_p^{-1}(j\omega)]\sigma_{max}[W_p(j\omega)C(j\omega)]\\ &< \sigma_{max}[W_p^{-1}(j\omega)]\end{aligned}$$

To interpret this, when $\sigma_{max}[W_p(j\omega)C(j\omega)] < 1$ for all ω, then the largest singular value of the complementary-sensitivity transfer function is strictly less than the largest singular value of the inverse of the weighting function.

References

[1] Athans, M., Lecture Notes for Multivariable Control Systems I and II, Massachusetts Institute of Technol-

ogy, 1994. (This reference may not be generally available.)

[2] Freudenberg, J.S. and Looze, D.P., *Frequency Domain Properties of Scalar and Multivariable Feedback Systems,* Springer, Berlin, 1987.

[3] Kailath, T., *Linear Systems,* Prentice Hall, Englewood Cliffs, NJ, 1980.

[4] Maciejowski, J.M., *Multivariable Feedback Design,* Addison-Wesley, Wokingham, 1989.

[5] Strang, G., *Linear Algebra and its Applications,* Harcourt Brace Jovanovich, San Diego, 1988.

30

Stability Robustness to Unstructured Uncertainty for Linear Time Invariant Systems

Alan Chao
Laboratory for Information and Decision Systems, Massachusetts Institute of Technology, Cambridge, MA

Michael Athans
Laboratory for Information and Decision Systems, Massachusetts Institute of Technology, Cambridge, MA

30.1 Introduction

In designing feedback control systems, the stability of the resulting closed-loop system is a primary objective. Given a finite dimensional, linear time-invariant (FDLTI) model of the plant, $G(s)$, the stability of the nominal closed-loop system based on this model, Figure 30.1, can be guaranteed through proper design: in the Nyquist plane for single-input, single-output (SISO) systems or by using well-known design methodologies such as LQG and $\mathcal{H}_\infty$ for multi-input, multi-output (MIMO) systems. In any case, since the mathematical model is FDLTI, this nominal stability can be analyzed by explicitly calculating the closed-loop poles of the system. It is clear, however, that nominal stability is never enough since the model is never a true representation of the actual plant. That is, there are always modeling errors or uncertainty. As a result, the control engineer must ultimately ensure the stability of the actual closed-loop system, Figure 30.2. In other words, the designed controller, $K(s)$, must be robust to the model uncertainty. In this article, we address this topic of stability robustness. We present a methodology to analyze the stability of the actual closed-loop system under nominal stability to a given model and a certain representation of the uncertainty.

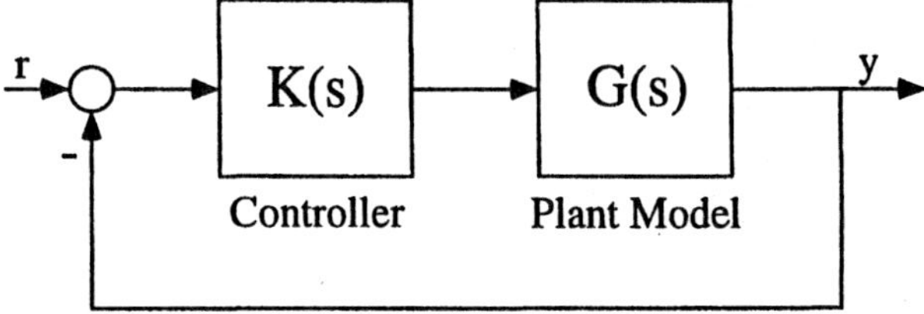

Figure 30.1 Block diagram of nominal feedback loop.

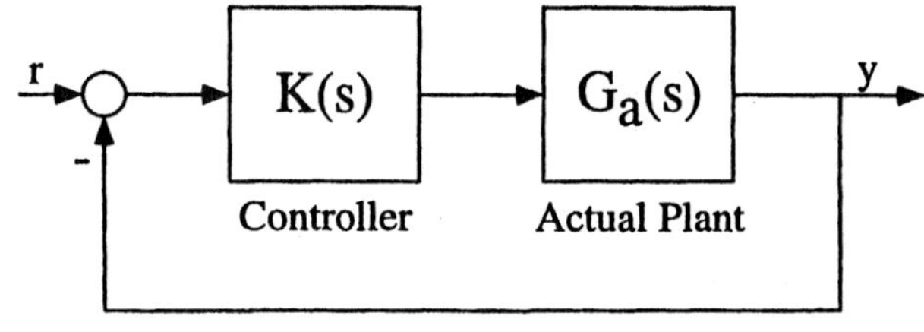

Figure 30.2 Block diagram of actual feedback loop.

The outline of the chapter is as follows. We first establish a representation of the uncertainty on which we will base our analysis. We then proceed to derive conditions that guarantee the stability of the actual closed-loop system. First, we concentrate on SISO systems where we use the familiar Nyquist stability criterion to derive the stability robustness condition. We then derive this same condition using the small gain theorem. The purpose of this is to provide a simple extension of our analysis to MIMO systems, which we present next. We then interpret the stability robustness conditions and examine their impact on attainable closed-loop performance, such as disturbance rejection and command following. Finally, we present a discussion

0-8493-8570-9/96/$0.00+$.50

on other possible representations of uncertainty and their respective stability robustness conditions. Examples are presented throughout the discussion.

30.2 Representation of Model Uncertainty

30.2.1 Sources of Uncertainty

Before we can analyze the stability of a closed-loop system under uncertainty, we must first understand the causes of uncertainty in the model so as to find a proper mathematical representation for it. Throughout the discussion we will assume that the actual plant, $G_a(s)$, is linear time-invariant (LTI) and that we have a nominal LTI model, $G(s)$. Although this assumption may seem unreasonable since actual physical systems are invariably nonlinear and since a cause of uncertainty is that we model them as LTI systems, we need this assumption to obtain simple, practical results. In practice, these results work remarkably well for a large class of engineering problems, because many systems are designed to be as close to linear time-invariant as possible.

The sources of modeling errors are both intentional and unintentional. Unintentional model errors arise from the underlying complexity of the physical process, the possible lack of laws for dynamic cause and effect relations, and the limited opportunity for physical experimentation. Simply put, many physical processes are so complex that approximations are inevitable in deriving a mathematical model. On the other hand, many modeling errors are intentionally induced. In the interest of reducing the complexity and cost of the control design process, the engineer will often neglect "fast" dynamics in an effort to reduce the order or the dimension of the state-space representation of the model. For example, one may neglect "fast" actuator and sensor dynamics, "fast" bending and/or torsional modes, and "small" time delays. In addition, the engineer will often use nominal values for the parameters of his model, such as time constants and damping ratios, even though he knows that the actual values will be different because of environmental and other external effects.

30.2.2 Types of Uncertainty

The resulting modeling errors can be separated into two types. The first type is known as parametric uncertainty. Parametric uncertainty refers to modeling errors, under the assumption that the actual plant is of the same order as the model, where the numerical values of the coefficients to the differential equation, which are related to the physical parameters of the system, between the actual plant and the model are different. The second type of uncertainty is known as unstructured uncertainty. In this case, the modeling errors refer to the difference in the dynamics between the finite dimensional model and the unknown and possibly infinite dimensional actual process.

In this chapter, we limit ourselves to addressing stability robustness with respect to unstructured uncertainty. We do this for two main reasons. First, we can capture the parametric errors in terms of the more general definition of unstructured uncertainty. Second and more importantly, unstructured uncertainty allows us to capture the effects of unmodeled dynamics. From our discussion above, we admit that we often purposely neglect "fast" dynamics in an effort to simplify the model. Furthermore, it can be argued that all physical processes are inherently distributed systems and that the modeling process acts to lump the dynamics of the physical process into a system that can be defined in a finite dimensional state-space. Therefore, unmodeled dynamics are always present and need to be accounted for in terms of stability robustness.

30.2.3 Multiplicative Representation of Unstructured Uncertainty

What we need now is a mathematical representation of unstructured uncertainty. The difficulty is that since the actual plant is never exactly known, we cannot hope to model the uncertainty to obtain this representation, for otherwise, it would not be uncertain. On the other hand, in practice, the engineer is never totally ignorant of the nature and magnitude of the modeling error. For example, from our arguments above, it is clear that unstructured uncertainty cannot be captured by a state-space representation, since the order of the actual plant is unknown, and thus we are forced to find a representation in terms of input-output relationships. In addition, if we choose to neglect "fast" dynamics, then we would expect that the magnitude of the uncertainty will be large at high frequencies in the frequency domain. In any case, the key here is to define a representation that employs the minimal information regarding the modeling errors and that is, in turn, sufficient to address stability robustness.

Set Membership Representation for Uncertainty

In this article we adopt a set membership representation of unstructured uncertainty. The idea is to define a bounded set of transfer function matrices, $\mathcal{G}$, which contains $G_a(s)$. Therefore, if $\mathcal{G}$ is properly defined such that we can show stability for all elements of $\mathcal{G}$, then we would have shown stability robustness. Towards this end, we define $\mathcal{G}$ as

$$\mathcal{G} = \{\tilde{G}(s) \mid \tilde{G}(s) = (I + w(s)\Delta(s))G(s), \ \|\Delta(j\omega)\|_{\mathcal{H}_\infty} \leq 1\} \tag{30.1}$$

where

1. $w(s)$ is a fixed, proper, and strictly stable scalar transfer function
2. $\Delta(s)$ is a strictly stable transfer function matrix (TFM)
3. No unstable or imaginary axis poles of $G(s)$ are cancelled in forming $\tilde{G}(s)$

This is known as the multiplicative representation for unstructured uncertainty. Since it is clear that the nominal model $G(s)$ is contained in $\mathcal{G}$, we view $\mathcal{G}$ as a set of TFMs perturbed from $G(s)$ that covers the actual plant. Our requirements that $w(s)\Delta(s)$ is strictly stable and that there are no unstable or imaginary axis pole-zero cancellations mean that the unstable and imaginary axis poles of our model and any $\tilde{G}(s) \in \mathcal{G}$ coincide. This assumes that the modeling effort is at least adequate enough to capture the unstable dynamics of $G_a(s)$.

In Equation 30.1, the term $w(s)\Delta(s)$ is known as the multiplicative error. We note that since the $\mathcal{H}_\infty$ norm of $\Delta(j\omega)$ varies between 0 and 1 and since the phase and direction of $\Delta(j\omega)$ are allowed to vary arbitrarily, the multiplicative error for any $\tilde{G}(s) \in \mathcal{G}$ is contained in a bounded hypersphere of radius $|w(j\omega)|$ at each frequency. Therefore, our representation of unstructured uncertainty is one in which we admit total lack of knowledge of the phase and direction of the actual plant with respect to the model, but that we have a bound on the magnitude of this multiplicative error. This magnitude-bound information is frequency dependent and is reflected in the fixed transfer function $w(s)$, which we will refer to as the weight. We note that we could have used a TFM $W(s)$ instead of a scalar $w(s)$ in our representation. However, in that case, the multiplicative error will reflect directional information of $W(s)$. Since we have presumed total lack of knowledge regarding directional information of the actual plant relative to the model, it is common in practice to choose $W(s)$ to be scalar.

SISO Interpretations for $\mathcal{G}$

To get a better feel for $\mathcal{G}$ and our representation for unstructured uncertainty, we first specialize to the SISO case in which our definition for $\mathcal{G}$ gives

$$\tilde{g}(s) = (1 + w(s)\Delta(s))g(s) \tag{30.2}$$

where

$$\|\Delta(j\omega)\|_{\mathcal{H}_\infty} = \sup_{\omega} |\Delta(j\omega)| \le 1 \tag{30.3}$$

Since the phase of $\Delta(j\omega)$ is allowed to vary arbitrarily and its magnitude varies from 0 to 1 at all frequencies, the set $\mathcal{G}$ is the set of transfer functions whose magnitude bode plot lies in an envelope surrounding the magnitude plot of $g(s)$, as shown in Figure 30.3. Therefore, the size of the unstructured uncertainty is represented by the size of this envelope. From the figure, the upper edge of the envelope corresponds to the plot of $(1 + |w(j\omega)|)|g(j\omega)|$ while the lower edge corresponds to the plot of $(1 - |w(j\omega)|)|g(j\omega)|$. Therefore, $|w(j\omega)|$ is seen as a frequency-dependent magnitude bound on the uncertainty. As mentioned beforehand, the size of the unstructured uncertainty typically increases with increasing frequency. Therefore, we would typically expect the size of the envelope containing $\mathcal{G}$ to increase with increasing frequency and also $|w(j\omega)|$ to increase with increasing frequency, as shown in Figure 30.3. Furthermore, we want to stress that since the phase of $\Delta(j\omega)$ is allowed to vary arbitrarily, the phase difference between any $\tilde{g}(j\omega) \in \mathcal{G}$ and $g(j\omega)$ can be arbitrarily large at any frequency.

For another interpretation, we can look at the multiplicative error $w(s)\Delta(s)$ in the SISO case. Solving for $w(s)\Delta(s)$ in Equation 30.2 gives

$$\frac{\tilde{g}(s) - g(s)}{g(s)} = w(s)\Delta(s) \tag{30.4}$$

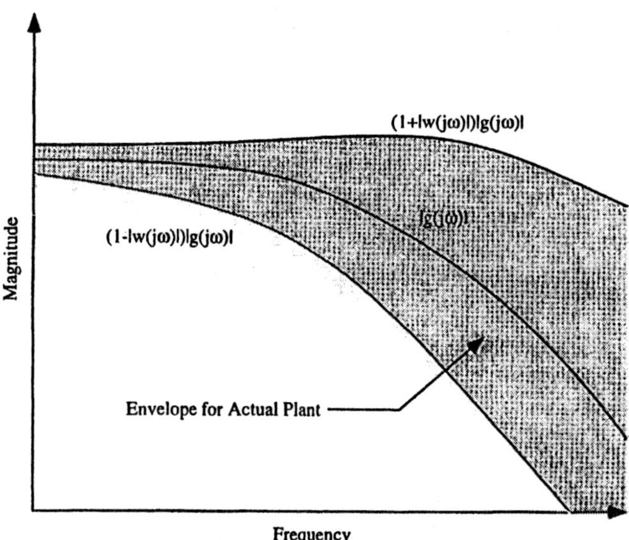

Figure 30.3 Bode plot interpretation of multiplicative uncertainty.

which shows that $w(s)\Delta(s)$ is the normalized error in the transfer function of the perturbed system with respect to the nominal model. Using Equation 30.3 and noting that everything is scalar, we take magnitudes on both sides of (30.4) for $s = j\omega$ to get

$$\begin{aligned} |\tilde{g}(j\omega) - g(j\omega)| &\le |w(j\omega)||\Delta(j\omega)||g(j\omega)| \\ &\le |w(j\omega)||g(j\omega)| \;\; \forall\, \omega \end{aligned} \tag{30.5}$$

As shown in Figure 30.4, for each ω, this inequality describes a closed disk in the complex plane of radius $|w(j\omega)||g(j\omega)|$ centered at $g(j\omega)$ which contains $\tilde{g}(j\omega)$. Since Equation 30.5 is valid for any $\tilde{g}(s) \in \mathcal{G}$, the set $\mathcal{G}$ is contained in that closed disk for each ω. In this interpretation, the unstructured uncertainty is represented by the closed disk, and therefore, we see that the direction and phase of the uncertainty is left arbitrary. However, we note that the radius of the closed disk does not necessarily increase with increasing frequency because it depends also on $|g(j\omega)|$, which typically decreases with increasing frequency at high frequencies due to roll-off.

Choosing $w(s)$

From our discussion, it is clear that our representation of the uncertainty only requires a nominal model, $G(s)$, and a scalar weight, $w(s)$, which reflects our knowledge on the magnitude bound of the uncertainty. The next logical question is how to choose $w(s)$ in the modeling process. From the definition in Equation 30.1, we know that we must choose $w(s)$ so that the actual plant is contained in $\mathcal{G}$. In the course of modeling, whether through experimentation, such as frequency response, and/or model reduction, we will arrive at a set of transfer function matrices. It is assumed that our modeling effort is thorough enough to adequately capture the actual process so that this set will cover the TFM of the actual plant. From this set we will choose a nominal model, $G(s)$. We assume that $G(s)$ is square, same number of inputs as outputs, and nonsingular along the $j\omega$-axis in the s-plane. With this assumption, we can calculate, at each frequency, the multiplicative error for each TFM $G_i(s)$

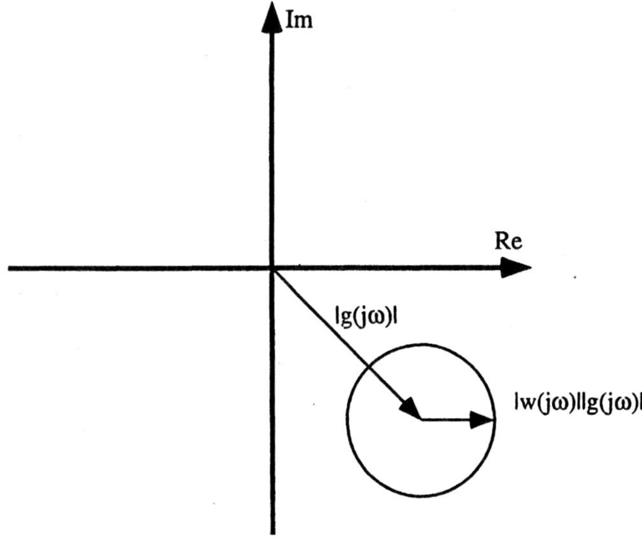

Figure 30.4 Interpretation of multiplicative uncertainty in the complex plane.

in our set using

$$w(j\omega)\Delta(j\omega) = G_i(j\omega)G^{-1}(j\omega) - I \qquad (30.6)$$

Taking maximum singular values on both sides of the above equation gives

$$\sigma_{\max}[w(j\omega)\Delta(j\omega)] = \sigma_{\max}[G_i(j\omega)G^{-1}(j\omega) - I] \qquad (30.7)$$

Since $\|\Delta(j\omega)\|_{\mathcal{H}_\infty} \leq 1$ and $w(j\omega)$ is a scalar, it is clear that we must choose

$$|w(j\omega)| \geq \sigma_{\max}[G_i(j\omega)G^{-1}(j\omega) - I] \;\; \forall\, \omega \geq 0 \qquad (30.8)$$

to include the TFM $G_i(s)$ in $\mathcal{G}$. Therefore, we must choose a stable, proper $w(s)$ such that

$$|w(j\omega)| \geq \max_i \sigma_{\max}[G_i(j\omega)G^{-1}(j\omega) - I] \;\; \forall\, \omega \geq 0 \qquad (30.9)$$

to ensure that we include all $G_i(s)$ and, thus, the actual plant in $\mathcal{G}$. This process is illustrated in the following example.

EXAMPLE 30.1: Integrator with Time Delay

Consider the set of SISO plants

$$g_\tau(s) = \frac{1}{s}\exp^{-s\tau}, \quad 0 \leq \tau \leq 0.2 \qquad (30.10)$$

which is the result of our modeling process on the actual plant. Since we cannot fully incorporate the delay in a state-space model, we choose to ignore it. As a result, our model of the plant is

$$g(s) = \frac{1}{s} \qquad (30.11)$$

To choose the appropriate $w(s)$ to cover all $g_\tau(s)$ in $\mathcal{G}$, we have to satisfy Equation 30.9, which in the SISO case is

$$|w(j\omega)| \geq \max_\tau \left|\frac{g_\tau(j\omega)}{g(j\omega)} - 1\right| \;\; \forall\, \omega \geq 0 \qquad (30.12)$$

Therefore, we need to choose $w(s)$ such that

$$|w(j\omega)| \geq \max_\tau \left|e^{-j\omega\tau} - 1\right| \;\; \forall\, \omega \geq 0 \qquad (30.13)$$

Using $e^{-j\omega\tau} = \cos(\omega\tau) - j\sin(\omega\tau)$ and a few trigonometry identities, the above inequality can be simplified as

$$|w(j\omega)| \geq \max_\tau \left|2\sin\left(\frac{\omega\tau}{2}\right)\right| \;\; \forall\, \omega \geq 0 \qquad (30.14)$$

A simple $w(s)$ that will satisfy Equation 30.14 is

$$w(s) = \frac{0.21s}{0.05s + 1} \qquad (30.15)$$

This is shown in Figure 30.5, where $|w(j\omega)|$ and $2\left|\sin(\frac{\omega\tau}{2})\right|$ are plotted together on a magnitude bode plot for $\tau = 0.2$, which is the worst-case value. We note that $w(s)$ is proper and strictly stable as required.

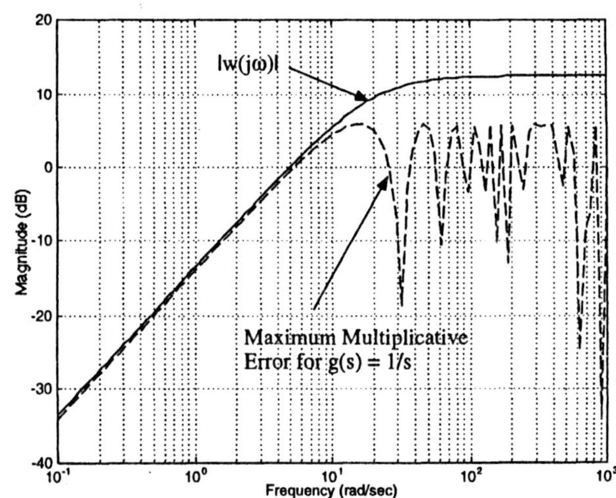

Figure 30.5 $w(s)$ for time delay uncertainty.

Now, let us suppose that we seek a better model by approximating the delay with a first-order Pade approximation

$$e^{-s\tau} \approx \frac{1 - \frac{s\tau}{2}}{1 + \frac{s\tau}{2}} \qquad (30.16)$$

Our model becomes

$$g(s) = \frac{1}{s}\left(\frac{1 - \frac{0.1s}{2}}{1 + \frac{0.1s}{2}}\right) \qquad (30.17)$$

where we approximate the delay at its midpoint value of 0.1 sec. To choose $w(s)$ in this case, we again need to satisfy Equation 30.12 which becomes

$$|w(j\omega)| \geq \max_\tau \left|-\frac{e^{-j\omega\tau}(j\omega + 20)}{(j\omega - 20)} - 1\right| \;\; \forall\, \omega \geq 0 \qquad (30.18)$$

A simple $w(s)$ that will satisfy Equation 30.18 is

$$w(s) = \frac{0.11s}{0.025s + 1} \tag{30.19}$$

This is shown in Figure 30.6. We again note that $w(s)$ is proper and strictly stable as required.

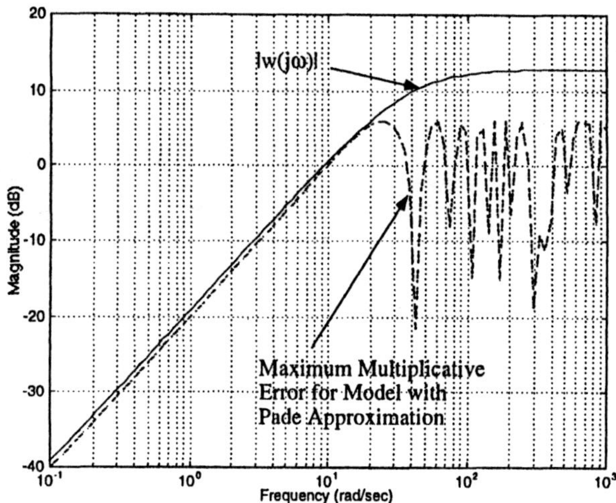

Figure 30.6 $w(s)$ for time delay uncertainty with Pade approximation.

Comparing the magnitudes of $w(s)$ for the two models in Figure 30.7, we note that, for the model with the Pade approximation, $|w(j\omega)|$ is less than that for the original model for $\omega \leq 100$ rad/sec. Since $|w(j\omega)|$ is the magnitude bound on the uncertainty, Figure 30.7 shows that the uncertainty for the model with the Pade approximation is smaller than that for the original model in this frequency range. Physically, this is because the model with the Pade approximation is a better model of the actual plant for $\omega \leq 100$ rad/sec. As a result, its uncertainty is smaller.

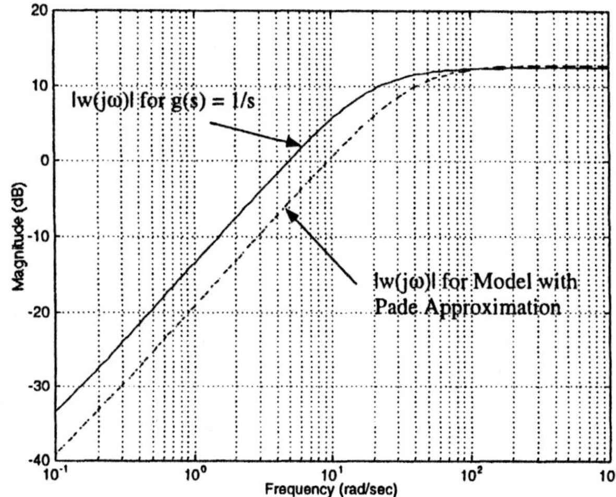

Figure 30.7 Comparing $w(s)$ for the two models.

We note that our choice of representing the uncertainty as being bounded in magnitude but arbitrary in phase and direction is, in general, an overbound on the set of TFMs that we obtained from the modeling process. That is, the set of TFMs from the modeling process may only be a small subset of $\mathcal{G}$. The benefit of using this uncertainty representation is that it allows for simple analysis of the stability robustness problem using minimal information regarding the modeling errors. However, the cost of such a representation is that the stability robustness results obtained may be conservative. This is because these results are obtained by showing stability with respect to the larger set $\mathcal{G}$ instead of the smaller set from our modeling process. Another way of looking at it is that the resulting set of stabilizing controllers for the larger set $\mathcal{G}$ will be smaller. As a result, the achievable performance may be worse. This conservatism will be discussed throughout as we develop our analysis for stability robustness, and the impact of stability robustness and this conservatism on closed-loop performance will be discussed in Section 30.4.

Reflecting Modeling Errors to the Input of the Plant

Finally, we note that in our representation of multiplicative unstructured uncertainty, we choose to lump the uncertainty at the plant output. In other words, we assume that the actual plant is of the form $(I + w(s)\Delta(s))G(s)$ where the uncertainty, $(I + w(s)\Delta(s))$ multiplying the model is at the plant output. To be more precise notationally, we should rewrite our definition for $\mathcal{G}$ as

$$\mathcal{G} = \{\tilde{G}(s) \mid \tilde{G}(s) = (I + w_o(s)\Delta(s))G(s), \|\Delta(j\omega)\|_{\mathcal{H}_\infty} \leq 1\} \tag{30.20}$$

where we have added the subscript "o" to $w(s)$ to distinguish it as the weight corresponding to uncertainty at the plant output. Alternatively, we can instead choose to lump the uncertainty at the plant input. In this case, we assume that the actual plant is of the form $G(s)(I+w_i(s)\Delta(s))$ where the uncertainty multiplying the model is at the plant input. As a result, the definition for $\mathcal{G}$ in this case is

$$\mathcal{G} = \{\tilde{G}(s) \mid \tilde{G}(s) = G(s)(I + w_i(s)\Delta(s)), \|\Delta(j\omega)\|_{\mathcal{H}_\infty} \leq 1\} \tag{30.21}$$

where $w_i(s)$ is the weight corresponding to the uncertainty at the plant input. Comparing the two representations reveals that, in general, the two sets defined by Equations 30.20 and 30.21 are not the same because matrix multiplication does not commute. Since $\Delta(s)$ is constrained similarly and we must choose the weight so that $\mathcal{G}$ covers the actual plant in both cases, we note that, in general, $w_o(s)$ is not the same as $w_i(s)$.

Of course, we do not physically lump the modeling errors to the plant input or output. Instead, in the course of modeling, we lump the model so that the modeling errors are reflected either to the input of the model or to the output. To be sure, modeling errors associated with the plant actuators and sensors, such as neglected actuator and sensor dynamics, are reflected more naturally to the model's input and output, respectively. However, modeling errors associated with internal plant dynamics, such as neglected flexible modes, are not naturally reflected to either the model's input or output. In this case, we have a choice as to where we wish to reflect these errors. As a final note, for the SISO case, the two representations are equivalent since scalar multiplication does commute. As a result, we can choose $w_o(s) = w_i(s) = w(s)$ so that it does not make a difference where we reflect our modeling errors.

30.3 Conditions for Stability Robustness

Having established a multiplicative representation for unstructured uncertainty, we proceed to use it to analyze the stability robustness problem. As mentioned beforehand, since the actual modeling error and, thus, the actual plant is never known, we cannot hope to simply evaluate the stability of the actual closed-loop system by using the Nyquist stability criterion or by calculating closed-loop poles. Instead, we must rely on our representation of the uncertainty to arrive at conditions or tests in the frequency domain that guarantee stability robustness. Throughout the discussion, we assume that we have a controller, $K(s)$, that gives us nominal stability for the feedback loop in Figure 30.1 and that we use this controller in the actual feedback loop, Figure 30.2. In addition, since we are interested in internal stability of the feedback loop, we assume throughout that for SISO systems, the transfer function $k(s)g(s)$ is stabilizable and detectable; that is, there are no unstable pole-zero cancellations in forming $k(s)g(s)$. For MIMO systems, the corresponding conditions are that both $K(s)G(s)$ and $G(s)K(s)$ are stabilizable and detectable.

30.3.1 Stability Robustness for SISO Systems

In this section we analyze the stability robustness of SISO feedback systems to multiplicative unstructured uncertainty. We first derive a sufficient condition for stability robustness using the familiar Nyquist stability criterion. We then derive the same condition using another method: the small gain theorem. The goal here is not only to show that one can arrive at the same answer but also to present the small gain approach to analyzing stability robustness, which can be easily extended to MIMO systems. Finally, we compare our notion of stability robustness to more traditional notions of robustness such as gain and phase margins.

Stability Robustness Using the Nyquist Stability Criterion

We begin the analysis of stability robustness to unstructured uncertainty for SISO systems with the Nyquist stability criterion. We recall from classical control theory that the Nyquist stability criterion is a graphical representation of the relationship between the number of unstable poles of an open-loop transfer function, $l(s)$, and the unstable zeros of the return difference transfer function, $1+l(s)$. Since the zeros of $1+l(s)$ correspond to the closed-loop poles of $l(s)$ under negative unity feedback, the Nyquist stability criterion is used to relate the number of unstable poles of $l(s)$ to the stability of the resulting closed-loop system. Specifically, the Nyquist stability criterion states that the corresponding closed-loop system is stable if and only if the number of positive clockwise encirclements (or negative counterclockwise encirclements) of the point $(-1,0)$ in the complex plane by the Nyquist plot of $l(s)$ is equal to $-P$, where P is the number of unstable poles of $l(s)$. Here, the Nyquist plot is simply the plot in the complex plane of $l(s)$ evaluated along the closed Nyquist contour D_r, which is defined in the usual way with counterclockwise indentations around the imaginary axis poles of $l(s)$ so that they are excluded from the interior of D_r. Notationally, we express the Nyquist stability criterion as

$$\aleph(-1, l(s), D_r) = -P \tag{30.22}$$

In our analysis we are interested in the stability of the two feedback loops given in Figures 30.1 and 30.2. For the nominal feedback loop, we define the nominal loop transfer function as

$$l(s) = g(s)k(s) \tag{30.23}$$

Since we assume that the nominal closed loop is stable, we have that

$$\aleph(-1, g(s)k(s), D_r) = -P \tag{30.24}$$

where P is the number of unstable poles of $g(s)k(s)$. For the actual feedback loop, we define the actual loop transfer function as

$$l_a(s) = g_a(s)k(s) = (1 + w(s)\Delta(s))g(s)k(s) \tag{30.25}$$

where the second equality holds for some $\|\Delta(j\omega)\|_{\mathcal{H}_\infty} \leq 1$ since we assumed that $g_a(s) \in \mathcal{G}$ in our representation of the unstructured uncertainty. In addition, since we assumed that the unstable poles of $g(s)$ and $g_a(s)$ coincide and since the same $k(s)$ appears in both loop transfer functions, the number of unstable poles of $g_a(s)k(s)$ and thus $l_a(s)$ is also P. Similarly, since we assumed that the imaginary axis poles of $g(s)$ and $g_a(s)$ coincide, the same Nyquist contour D_r can be used to evaluate the Nyquist plot of $l_a(s)$. Therefore, by the Nyquist stability criterion, the actual closed-loop system is stable if and only if

$$\begin{aligned}\aleph(-1, g_a(s)k(s), D_r) &= \aleph(-1, (1 + w(s)\Delta(s))g(s)k(s), \\ &\qquad D_r) \\ &= -P\end{aligned} \tag{30.26}$$

Since we do not know $g_a(s)$, we can never hope to use Equation 30.26 to evaluate or show the stability of the actual system. However, we note that Equation 30.26 implies that the actual closed-loop system is stable if and only if the number of counterclockwise encirclements of the critical point, $(-1,0)$, is the same for the Nyquist plot of $l_a(s)$ as that for $l(s)$. Therefore, if we can show that for the actual loop transfer function, we do not change the number of counterclockwise encirclements from that of the nominal loop transfer function, then we can guarantee that the actual closed-loop system is stable. The idea is to utilize our set membership representation of the uncertainty to ensure that it is impossible to change the number of encirclements of the critical point for any loop transfer function $\tilde{g}(s)k(s)$ with $\tilde{g}(s) \in \mathcal{G}$. This will guarantee that the actual closed-loop system is stable, since our representation is such that $g_a(s) \in \mathcal{G}$. This is the type of sufficient condition for stability robustness that we seek.

To obtain this condition, we need a relationship between the Nyquist plots of $l(s)$ and $l_a(s)$ using our representation of the unstructured uncertainty. To start, we separate the Nyquist plot into three parts corresponding to the following three parts of the Nyqust contour D_r:

1. The nonnegative $j\omega$ axis
2. The negative $j\omega$ axis
3. The part of D_r that encircles the right half s-plane where $|s| \to \infty$

For the first part, we note from Equations 30.23 and 30.25 that $l_a(s)$ can be expressed as

$$l_a(s) = (1 + w(s)\Delta(s))l(s) \tag{30.27}$$

for some $\|\Delta\|_{\mathcal{H}_\infty} \leq 1$. Therefore, the multiplicative uncertainty representation developed earlier holds equally for $l(s) = g(s)k(s)$ as for $g(s)$. Extending the SISO interpretation of this uncertainty representation in Figure 30.4 to Figure 30.8, we note that for any nonnegative ω, the Nyquist plot of $l_a(s)$ is a point contained in the disk of radius $|w(j\omega)||l(j\omega)|$ centered at the point $l(j\omega)$, which is the Nyquist plot of $l(s)$ at that ω. Figure 30.8, then, gives a relationship between the Nyquist plot of $l(s)$ and $l_a(s)$ at a particular frequency for the part of the Nyquist contour along the positive $j\omega$ axis. For the second part of the Nyquist contour, we note that the Nyquist plot is simply a mirror image, with respect to the real axis, of the Nyquist plot for the first part, and thus this relationship will be identical. For the third part of the Nyquist contour, we note that since the model $g(s)$ represents a physical system, it should be strictly proper, and therefore, $|g(s)| \to 0$ as $|s| \to \infty$. In addition, the same holds for $g_a(s)$ since it is a physical system. Since the controller $k(s)$ is proper, it follows that the both the Nyquist plots for $l(s)$ and $l_a(s)$ are at the origin for the part of the Nyquist contour that encircles the right half s-plane.

For stability robustness, we only need to consider the relationship between $l(s)$ and $l_a(s)$ for the first part of the Nyquist contour. This is because for the second part the relationship is identical and thus the conclusions for stability robustness will be identical. Finally, for the third part, since the Nyquist plots for both $l(s)$ and $l_a(s)$ are at the origin, they are identical and cannot impact stability robustness. As a result, we only need to consider the relationship between $l(s)$ and $l_a(s)$ that is presented in Figure 30.8. We illustrate this relationship along a typical Nyquist plot of the nominal $l(s)$, for $\omega \geq 0$, in Figure 30.9 where, for clarity, we only illustrate the uncertainty disk at a few frequencies. Here, we note that the radius of the uncertainty disk changes as a function of frequency since both $w(j\omega)$ and $l(j\omega)$ varies with frequency.

From Figure 30.9, we note that it is impossible for the Nyquist plot of $l_a(s)$ to change the number of encirclements of the critical point $(-1, 0)$ if the disks representing the uncertainty in $l(s)$ do

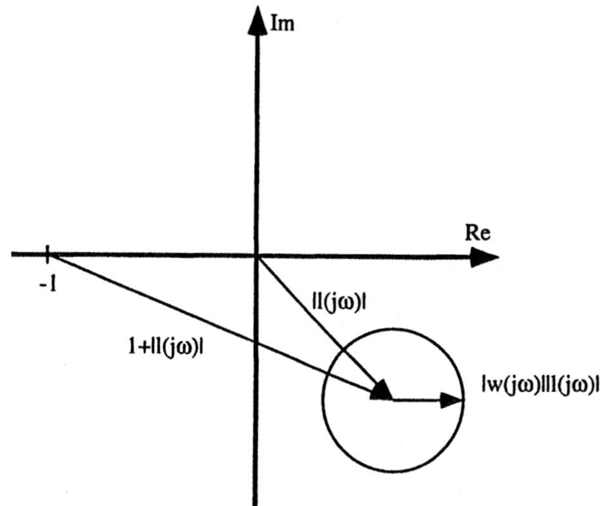

Figure 30.8 Multiplicative uncertainty in the Nyquist plot.

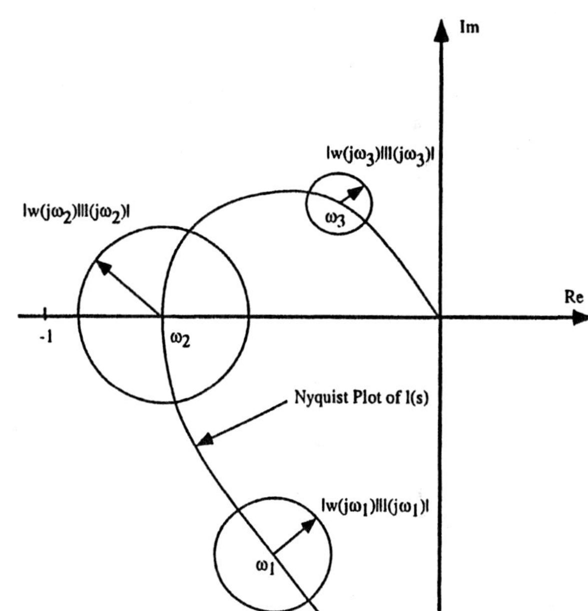

Figure 30.9 Stability robustness using the Nyquist stability criterion.

not intersect the critical point for all $\omega \geq 0$. To show this, let us prove the contrapositive. That is, if the Nyquist plot of $l_a(s)$ does change the number of encirclements of the critical point, then by continuity there exists a frequency ω^* where the line connecting $l(j\omega^*)$ and $l_a(j\omega^*)$ must intersect the critical point. Since the uncertainty disk at that frequency is convex, this uncertainty disk must also intersect the critical point. Graphically, then, we can guarantee stability robustness if the distance between the critical point and the Nyquist plot of $l(s)$ is strictly greater than the radius of the uncertainty disk for all $\omega \geq 0$. From Figure 30.8, the distance between the critical point and the Nyquist plot of the nominal loop transfer function, $l(s)$, is simply $|1 + l(j\omega)|$, which is the magnitude of the return difference. Therefore, a condition for stability robustness is

$$|w(j\omega)||l(j\omega)| < |1 + l(j\omega)| \;\; \forall\, \omega \geq 0 \tag{30.28}$$

or equivalently,

$$|w(j\omega)||g(j\omega)k(j\omega)| < |1 + g(j\omega)k(j\omega)| \;\; \forall\, \omega \geq 0 \quad (30.29)$$

We note that although the Nyquist stability criterion is both a necessary and sufficient condition for stability, the above condition for stability robustness is clearly only sufficient. That is, even if we violate this condition at a particular frequency or at a range of frequencies, the Nyquist plot of the actual system may not change the number of encirclements of the critical point and thus the actual closed-loop system may be stable. This is illustrated in Figure 30.10, where we note that if the Nyquist plot of $l_a(s)$ follows $l_{a2}(s)$, then the actual system is stable. The key here is that since we do not know $l_a(s)$, we cannot say whether or not the actual system is stable when the condition is violated. Therefore, to guarantee actual stability, we need to ensure a safe distance or margin, which may not be necessary, between the nominal Nyquist plot and the critical point. This margin is in terms of the uncertainty disk. We see that this conservatism stems from the fact that we admit total lack of knowledge concerning the phase of the actual plant, which led to the representation of the uncertainty as a disk in the complex plane.

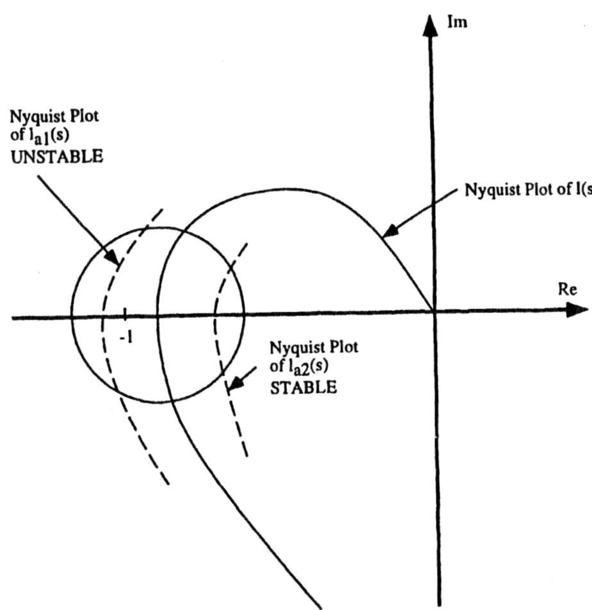

Figure 30.10 Sufficiency of the stability robustness condition.

Stability Robustness Using Small Gain Theorem

We now seek to derive an equivalent condition for stability robustness for SISO systems using the small gain theorem, see for example Dahleh and Diaz-Bobillo [1]. The goal here is to introduce another methodology whereby conditions for stability robustness can be derived. As we will see in the sequel, this methodology can be easily extended to MIMO systems. We begin with a statement of the small gain theorem, specialized to LTI systems, which addresses the stability of a closed-loop system in the standard feedback form given in Figure 30.11.

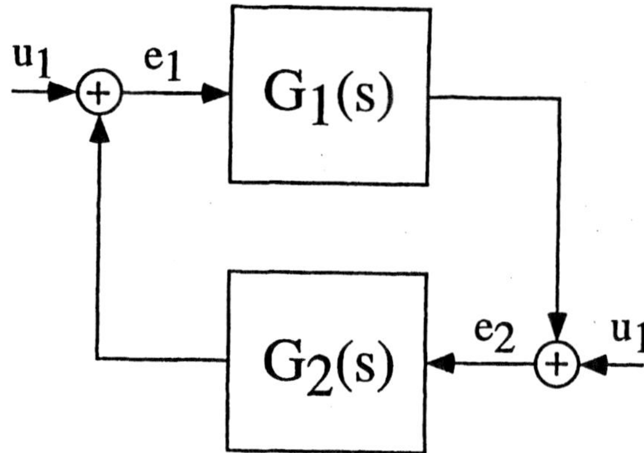

Figure 30.11 Standard feedback form.

THEOREM 30.1 **(Small Gain Theorem)** *Under the assumption that $G_1(s)$ and $G_2(s)$ are stable in the feedback system in Figure 30.11, the closed-loop transfer function matrix from (u_1, u_2) to (e_1, e_2) is stable if the small gain condition*

$$\|G_1(j\omega)\|_{\mathcal{H}_\infty} \|G_2(j\omega)\|_{\mathcal{H}_\infty} < 1 \quad (30.30)$$

is satisfied.

Proof: We first show that the sensitivity transfer function matrix, $S(s) = (I - G_1(s)G_2(s))^{-1}$ is stable. For this, we need to show that if the small gain condition Equation 30.30 is satisfied, then $S(s) = (I - G_1(s)G_2(s))^{-1}$ is analytic in the closed right-half s-plane. An equivalent statement is that the return difference, $D(s) = I - G_1(s)G_2(s)$ is nonsingular for all s in the closed right-half plane.

For arbitrary input u and for all complex s,

$$\begin{aligned} \|D(s)u\| &= \|(I - G_1(s)G_2(s))u\| \\ &= \|u - G_1(s)G_2(s)u\| \end{aligned} \quad (30.31)$$

where $\|\cdot\|$ represents the standard Euclidean norm. From the triangle inequality,

$$\|u - G_1(s)G_2(s)u\| \geq \|u\| - \|G_1(s)G_2(s)u\| \quad (30.32)$$

from the definition of the maximum singular value,

$$\|G_1(s)G_2(s)u\| \leq \sigma_{\max}[G_1(s)G_2(s)]\|u\| \quad (30.33)$$

and from the submultiplicative property of induced norms,

$$\sigma_{\max}[G_1(S)G_2(s)] \leq \sigma_{\max}[G_1(s)]\sigma_{\max}[G_2(s)] \quad (30.34)$$

Substituting Equations 30.32, 30.33, and 30.34 into Equation 30.31 gives for all complex s,

$$\begin{aligned} \|D(s)u\| &= \|u - G_1(s)G_2(s)u\| \\ &\geq \|u\| - \|G_1(s)G_2(s)u\| \\ &\geq \|u\| - \sigma_{\max}[G_1(s)G_2(s)]\|u\| \\ &\geq \|u\| - \sigma_{\max}[G_1(s)]\sigma_{\max}[G_2(s)]\|u\| \\ &= (1 - \sigma_{\max}[G_1(s)]\sigma_{\max}[G_2(s)])\|u\| \end{aligned} \quad (30.35)$$

Since G_1 and G_2 are stable, they are analytic in the closed right-half plane. Therefore, by the maximum modulus theorem [2],

$$\begin{aligned}\sigma_{\max}[G_1(s)] \le \sup_{\omega} \sigma_{\max}[G_1(j\omega)] &= \|G_1(j\omega)\|_{\mathcal{H}_\infty} \\ \sigma_{\max}[G_2(s)] \le \sup_{\omega} \sigma_{\max}[G_2(j\omega)] &= \|G_2(j\omega)\|_{\mathcal{H}_\infty}\end{aligned} \tag{30.36}$$

for all s in the closed right-half plane. Substituting Equation 30.36 into Equation 30.35 gives

$$\|D(s)u\| \ge (1 - \|G_1(j\omega)\|_{\mathcal{H}_\infty}\|G_2(j\omega)\|_{\mathcal{H}_\infty})\|u\| \tag{30.37}$$

for all s in the closed right-half plane. From the small gain condition, there exists an $\epsilon > 0$ such that

$$\|G_1(j\omega)\|_{\mathcal{H}_\infty}\|G_2(j\omega)\|_{\mathcal{H}_\infty} < 1 - \epsilon \tag{30.38}$$

Therefore, for all s in the closed right-half plane,

$$\|D(s)u\| \ge \epsilon\|u\| > 0 \tag{30.39}$$

for any arbitrary u, which implies that $D(s)$ is nonsingular in the closed right-half s-plane.

From a similar argument, we can show that $(I - G_2(s)G_1(s))$ is also stable. Therefore, the transfer function matrix relating (u_1, u_2) to (e_1, e_2), which is given by

$$\begin{bmatrix} (I - G_2(s)G_1(s))^{-1} & -(I - G_2(s)G_1(s))^{-1}G_2(s) \\ (I - G_1(s)G_2(s)^{-1}G_1(s) & (I - G_1(s)G_2(s))^{-1} \end{bmatrix} \tag{30.40}$$

is stable. □

We note that the small gain theorem is only a sufficient condition for stability. For example, in the SISO case, the small gain condition (Equation 30.30) can be expressed as

$$|g_1(j\omega)g_2(j\omega)| < 1 \;\; \forall\, \omega \ge 0 \tag{30.41}$$

which implies that the Nyquist plot of the loop transfer function $g_1(s)g_2(s)$ lies strictly inside the unit circle centered at the origin, as shown in Figure 30.12. This is sufficient for stability because $g_1(s)$ and $g_2(s)$ are stable; however, we note that it is clearly not necessary. In addition, we note that the small gain theorem applies equally well to MIMO systems, since the proof was actually done for the MIMO case. In fact, the general form of the small gain theorem applies to nonlinear, time-varying operators over any normed signal space. For a treatment of the general small gain theorem and its proof, the reader is referred to [3].

For stability robustness, we are, as before, interested in the stability of the two feedback loops given in Figures 30.1 and 30.2. From our representation of unstructured uncertainty, we can express the actual plant as

$$g_a(s) = (1 + w(s)\Delta(s))g(s) \tag{30.42}$$

for some $\|\Delta(j\omega)\|_{\mathcal{H}_\infty} \le 1$. Therefore, the actual feedback loop can also be represented by the block diagram in Figure 30.13.

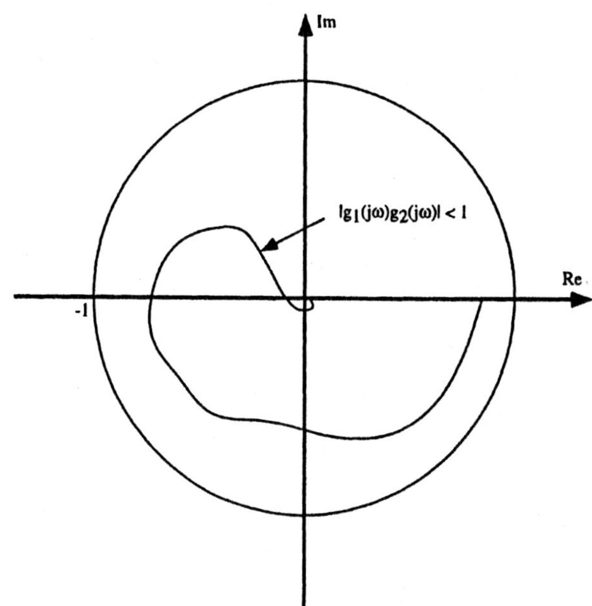

Figure 30.12 Sufficiency of SISO small gain theorem.

In the figure, we note that we choose, merely by convention, to reflect the unstructured uncertainty to the output of the plant instead of to the input since for the SISO case the two are equivalent. In addition, we note that Figure 30.13 can also be interpreted as being the feedback loop for all perturbed plants, $\tilde{g}(s)$, belonging to the set $\mathcal{G}$. The key for stability robustness, then, is to show stability for this feedback loop using the small gain theorem for all $\|\Delta(j\omega)\|_{\mathcal{H}_\infty} \le 1$ and, therefore, for all $\tilde{g}(s) \in \mathcal{G}$.

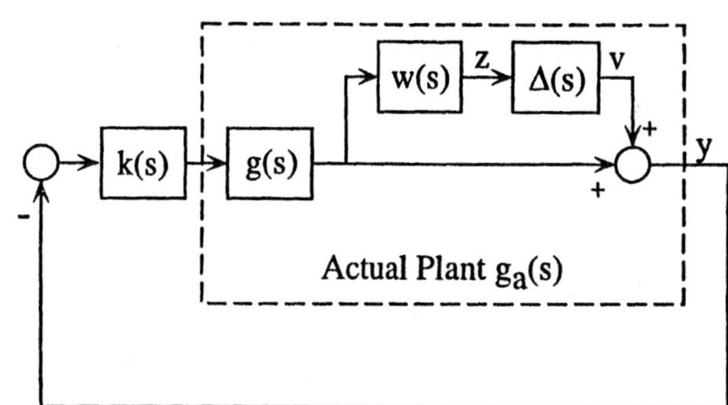

Figure 30.13 Actual feedback loop with uncertainty representation.

To apply the small gain theorem, we first reduce the feedback loop in Figure 30.13 to the standard feedback form in Figure 30.11. To do this, we isolate $\Delta(s)$ and calculate the transfer function from the output of Δ, v, to its input, z. From Figure 30.13,

$$\begin{aligned} z(s) &= -w(s)g(s)k(s)y(s) \\ y(s) &= v(s) - g(s)k(s)y(s) \end{aligned} \tag{30.43}$$

As a result, the transfer function seen by Δ is given by $m(s)$ where

$$m(s) = \frac{-w(s)g(s)k(s)}{1 + g(s)k(s)} \tag{30.44}$$

and the reduced block diagram is given in Figure 30.14. We note that $m(s)$ is simply the product of the complementary sensitivity transfer function for the nominal feedback loop and the weight $w(s)$. Since we assumed that the nominal closed loop is stable and $w(s)$ is stable, $m(s)$ is stable. Furthermore, since we also assumed that $\Delta(s)$ is stable, the assumption for the small gain theorem is satisfied for the closed-loop system in Figure 30.14. Applying the small gain theorem, this closed-loop system is stable if the small gain condition

$$\left\| \frac{-w(j\omega)g(j\omega)k(j\omega)}{1+g(j\omega)k(j\omega)} \right\|_{\mathcal{H}\infty} \|\Delta(j\omega)\| < 1 \tag{30.45}$$

is satisfied. Since $\|\Delta(j\omega)\|_{\mathcal{H}_\infty} \leq 1$, an equivalent condition for stability to all $\|\Delta(j\omega)\|_{\mathcal{H}_\infty} \leq 1$ is

$$\left\| \frac{w(j\omega)g(j\omega)k(j\omega)}{1+g(j\omega)k(j\omega)} \right\|_{\mathcal{H}_\infty} < 1 \tag{30.46}$$

which is a sufficient condition for the stability of the feedback loop in Figure 30.13 for all $\tilde{g}(s) \in \mathcal{G}$. Since $g_a(s) \in \mathcal{G}$, this is a sufficient condition for stability robustness.

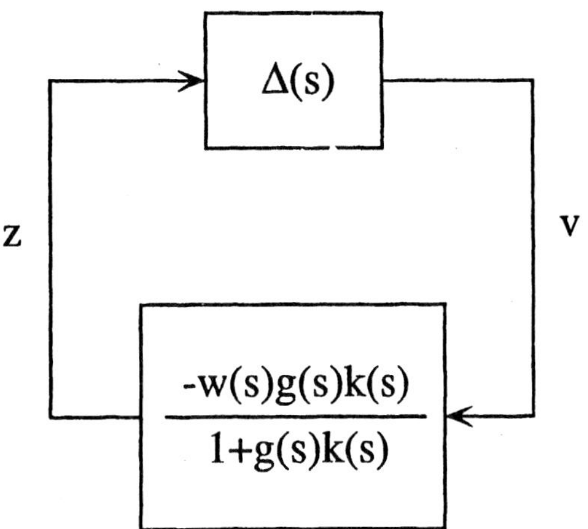

Figure 30.14 Actual feedback loop in standard form.

We now proceed to show that the stability robustness condition in Equation 30.46 is equivalent to that derived using the Nyquist stability criterion. From the definition of the $\mathcal{H}_\infty$ norm, an equivalent condition to Equation 30.46 is given by

$$\left| \frac{w(j\omega)g(j\omega)k(j\omega)}{1+g(j\omega)k(j\omega)} \right| < 1 \ \forall\, \omega \geq 0 \tag{30.47}$$

Since everything is scalar, this condition is equivalent to

$$|w(j\omega)||g(j\omega)k(j\omega)| < |1+g(j\omega)k(j\omega)| \ \forall\, \omega \geq 0 \tag{30.48}$$

which is exactly the condition for stability robustness derived using the Nyquist stability criterion. This equivalence is not due to the equivalence of the Nyquist stability criterion and the small gain theorem, since the former is a necessary and sufficient condition for stability while the latter is only sufficient. Rather, this equivalence is due to our particular approach in applying the small gain theorem and to our representation of the uncertainty. What this equivalence gives us is an alternative to the more familiar Nyquist stability criterion in analyzing the stability robustness problem. Unlike the Nyquist stability criterion, the small gain theorem applies equally well to MIMO systems. Therefore, our analysis is easily extended to the MIMO case, as we will do in Section 30.3.2.

Comparison to Gain and Phase Margins

In closing this section on stability robustness for SISO systems, we would like to compare our notion of robustness to more traditional notions such as gain and phase margins. As we recall, gain margin is the amount of additional gain that the SISO open-loop transfer function can withstand before the closed loop goes unstable, and phase margin is the amount of additional phase shift or pure delay that the loop transfer function can withstand before the closed loop goes unstable. To be sure, gain and phase margins are measures of robustness for SISO systems, but they are, in general, insufficient in guaranteeing stability in the face of dynamic uncertainty such as those due to unmodeled dynamics. This is because gain and phase margins only deal with uncertainty in terms of pure gain variations or pure phase variations but not a combination of both. That is, the open loop can exhibit large gain and phase margins and yet be close to instability as shown in the Nyquist plot in Figure 30.15. In the figure, we note that for frequencies between ω_1 and ω_2 the Nyquist plot is close to the critical point so that a combination of gain and phase variation along these frequencies such as that in the perturbed Nyquist plot will destabilize the closed loop. This combination of gain and phase variations can be the result of unmodeled dynamics. In such a case, gain and phase margins will give a false sense of stability robustness. In contrast, we could get a true sense of stability robustness by explicitly accounting for the dynamic uncertainty in terms of unstructured uncertainty.

In addition, gain and phase margins are largely SISO measures of stability robustness since they are inadequate in capturing the cross coupling between inputs and outputs of the dynamics of MIMO systems. For MIMO systems, we usually think of gain and phase margins as being independent gain and phase variations that are allowed at each input channel. These variations clearly cannot cover the combined gain, phase, and directional variations due to MIMO dynamic uncertainty. As a result, the utility of our notion of stability robustness over traditional gain and phase margin concepts becomes even more clear in the MIMO case.

30.3.2 Stability Robustness for MIMO Systems

In this section, we analyze the stability robustness of MIMO feedback systems to multiplicative unstructured uncertainty. As mentioned beforehand, we will use the small gain theorem since it offers a natural extension from the SISO case to the MIMO case. As shown in the SISO case, the general procedure for analyzing stability robustness using the small gain theorem is as follows:

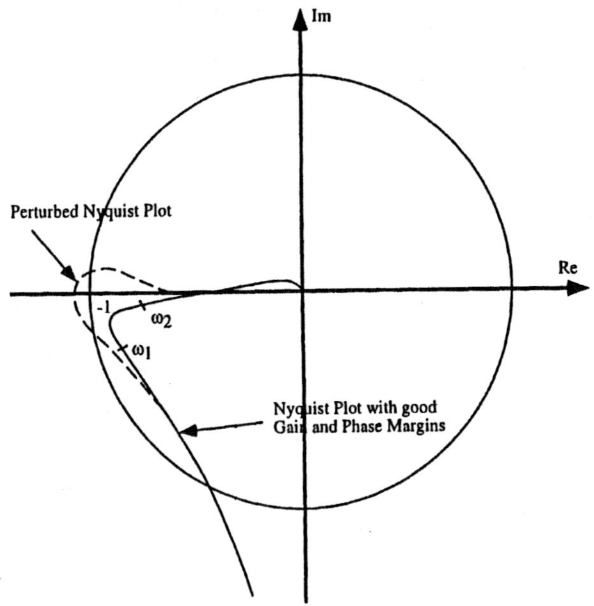

Figure 30.15 Insufficiency of gain and phase margins.

1. Start with the block diagram of the actual feedback loop with the actual plant represented by the nominal model perturbed by the uncertainty. Note that this block diagram also represents the feedback loop for all perturbed plants belonging to the set $\mathcal{G}$, which contains the actual plant.
2. Reduce the feedback loop to the standard feedback form by isolating the $\Delta(s)$ block and calculating the TFM from the output of Δ to the input of Δ. Denote this TFM as $M(s)$.
3. Apply the small gain theorem. In particular, since $\|\Delta(j\omega)\|_{\mathcal{H}_\infty} \leq 1$, the small gain theorem guarantees stability for the feedback loop for all perturbed plants in the set $\mathcal{G}$ and therefore guarantees robust stability if $\|M(j\omega)\|_{\mathcal{H}_\infty} < 1$.

We will follow this procedure in our analysis of MIMO stability robustness. For MIMO systems, as shown in Section 30.2.3, there is a difference between reflecting the modeling errors to the input and the output of the plant. As a consequence, we separate the two cases and derive a different stability robustness condition for each case. In the end, we will relate these two stability robustness tests and discuss their differences.

Uncertainty at Plant Output

We start with the case where the modeling errors are reflected to the plant output. In this case, the actual plant is of the form

$$G_a(s) = (I + w_o(s)\Delta(s))G(s) \tag{30.49}$$

for some $\|\Delta(j\omega)\|_{\mathcal{H}_\infty} \leq 1$. Following step 1 of the procedure, the block diagram of the actual feedback loop given this representation of the uncertainty is depicted in Figure 30.16(a). For step 2, we calculate the TFM from the output of Δ, v, to its input, z. From Figure 30.16(a),

$$\begin{aligned} z(s) &= -w_o(s)G(s)K(s)y(s) \\ y(s) &= v(s) - G(s)K(s)y(s) \end{aligned} \tag{30.50}$$

As a result, $M(s)$, the TFM seen by $\Delta(s)$, is

$$M(s) = -w_o(s)G(s)K(s)(I + G(s)K(s))^{-1} \tag{30.51}$$

and the reduced block diagram is given in Figure 30.16(b). We note that $M(s)$ is simply the product of the complementary sensitivity TFM, $C(s) = G(s)K(s)(I + G(s)K(s))^{-1}$, for the nominal feedback loop and the scalar weight $w_o(s)$. Since we assumed that the nominal closed loop and $w_o(s)$ are both stable, $M(s)$ is stable. Furthermore, since we also assumed that $\Delta(s)$ is stable, the assumption for the small gain theorem is satisfied for the closed loop system in Figure 30.16(b). For step 3, we apply the small gain theorem, which gives

$$\left\| w_o(j\omega)G(j\omega)K(j\omega)(I + G(j\omega)K(j\omega))^{-1} \right\|_{\mathcal{H}_\infty} < 1 \tag{30.52}$$

as a sufficient condition for stability robustness. Using the definition of the $\mathcal{H}_\infty$ norm, an equivalent condition for stability robustness is

$$\sigma_{\max}[w_o(j\omega)G(j\omega)K(j\omega)(I + G(j\omega)K(j\omega))^{-1}] < 1 \quad \forall\, \omega \geq 0 \tag{30.53}$$

Since $w_o(s)$ is scalar, another sufficient condition for stability robustness to multiplicative uncertainty at the plant output is

$$\sigma_{\max}[G(j\omega)K(j\omega)(I + G(j\omega)K(j\omega))^{-1}] < \frac{1}{|w_o(j\omega)|} \quad \forall\, \omega \geq 0 \tag{30.54}$$

Uncertainty at Plant Input

In the case where the modeling errors are reflected to the plant input, the actual plant is of the form

$$G_a(s) = G(s)(I + w_i(s)\Delta(s)) \tag{30.55}$$

for some $\|\Delta(j\omega)\|_{\mathcal{H}_\infty} \leq 1$. Following the procedure in Section 30.3.2, a sufficient condition for stability robustness is given by

$$\left\| w_i(j\omega)(I + K(j\omega)G(j\omega))^{-1}K(j\omega)G(j\omega) \right\|_{\mathcal{H}_\infty} < 1 \tag{30.56}$$

Using the definition of the $\mathcal{H}_\infty$ norm and the fact that $w_i(s)$ is scalar, an equivalent condition for stability robustness is

$$\sigma_{\max}[(I + K(j\omega)G(j\omega))^{-1}K(j\omega)G(j\omega)] < \frac{1}{|w_i(j\omega)|} \quad \forall\, \omega \geq 0 \tag{30.57}$$

Comparison between the sufficient conditions Equations 30.54 and 30.57 reveals that, although the two conditions both address the same stability robustness problem, they are indeed different.

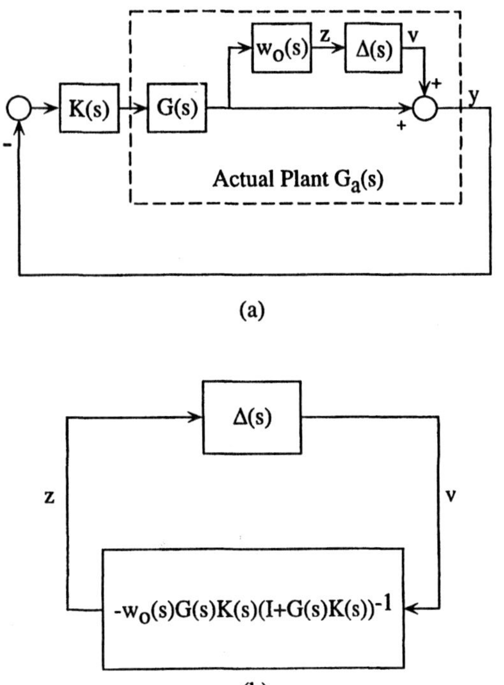

Figure 30.16 Stability robustness for multiplicative uncertainty at plant output.

When we reflect the modeling errors to the plant output, our stability robustness condition is based on the nominal complementary sensitivity TFM, $C(s)$. This TFM is the closed loop TFM formed from the loop TFM with the loop broken at the plant output. On the other hand, when we reflect the modeling errors to the plant input, our stability robustness condition is based on the nominal input complementary sensitivity TFM, $C_i(s)$. In general, $C_i(s)$ is different from $C(s)$. As mentioned in Section 30.2.3, this difference stems from the fact that, although the sets $\mathcal{G}$ for the two cases both contain the actual plant, they are different because matrix multiplication does not commute. As a result, the two different conditions can give different results. Since the two conditions are both only sufficient, only one of them needs to be satisfied in order to conclude stability robustness. The fact that we only need to satisfy one of the conditions to achieve stability robustness means that one condition is less conservative than the other. This is indeed true since the sets $\mathcal{G}$ are different, and thus, invariably, one is larger than the other, resulting in a more conservative measure of modeling error. Finally, we note that if $G(s)$ and $K(s)$ are both SISO, then both Equations 30.54 and 30.57 reduce to the sufficient condition derived for the SISO case.

30.4 Impact of Stability Robustness on Closed-Loop Performance

In the previous section, we have established, for both the SISO and MIMO cases, sufficient conditions for stability robustness under the multiplicative representation for unstructured uncertainty. To the control engineer, these conditions will have to be satisfied by design in order to ensure the stability of the actual closed-loop system. These conditions, then, become additional constraints on the design that will invariably impact the performance of the closed-loop system, such as command following and disturbance rejection. As shown in the chapter on MIMO frequency response analysis in this handbook, these performance specifications can be defined in the frequency domain in terms of the maximum singular value of the sensitivity TFM, $S(s) = (I + G(s)K(s))^{-1}$. In particular, we require $\sigma_{\max}[S(j\omega)]$ to be small in the frequency range of the command and/or disturbance signals. In this section, we seek an interpretation of our stability robustness conditions in the frequency domain in order to discuss its impact on command following and disturbance rejection.

We start with the SISO case and the stability robustness condition given in Equation 30.48. Since everything is scalar and positive, this condition is equivalent to

$$\left|\frac{g(j\omega)k(j\omega)}{1+g(j\omega)k(j\omega)}\right| < \frac{1}{|w(j\omega)|} \quad \forall\, \omega \geq 0 \tag{30.58}$$

where the left side is the magnitude of the closed-loop or complementary sensitivity transfer function, $c(j\omega)$, of the nominal design. This condition is illustrated in the frequency domain in Figure 30.17. Interpreting from the figure, the stability robustness condition states that the magnitude plot of the nominal closed-loop transfer function must lie strictly below the plot of the inverse of $|w(j\omega)|$. As is typically the case, the inverse of the magnitude of $w(j\omega)$ will be large at low frequencies and small at high frequencies, since modeling errors increase with increasing frequency. Specifically, the modeling errors become significant near and above the frequency ω_m defined by

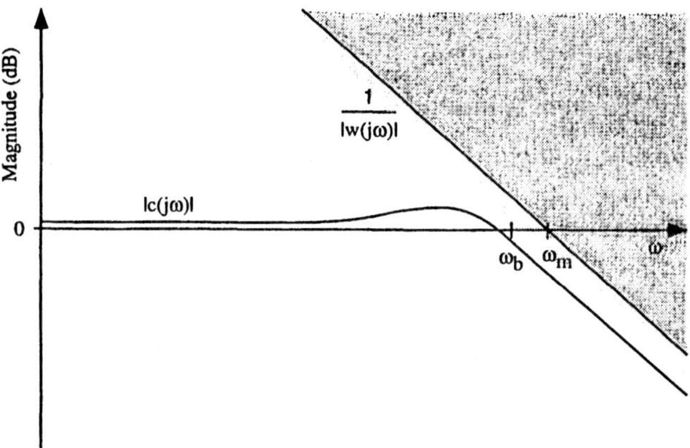

Figure 30.17 Interpretation of stability robustness for SISO systems.

$$|w(j\omega_m)| = 0 \text{ dB} \tag{30.59}$$

Therefore, the stability robustness condition limits the bandwidth of the nominal closed-loop design. That is, the bandwidth of the nominal closed-loop design, ω_b, is constrained to be less than ω_m, as shown in the figure. Indeed, as indicated in Figure 30.17, the presence of significant modeling errors at frequencies beyond ω_m forces the rapid roll-off of the designed closed-loop transfer function at high frequencies. Physically, this roll-off prevents energy at these frequencies from exciting the unmodeled dynamics and, therefore, prevents the possible loss of stability. In terms of closed-loop performance, we recall from the chapter on MIMO frequency response analysis that a necessary condition for $\sigma_{\max}[S(j\omega)]$ to be small at a certain frequency is that the singular values of the closed-loop or complementary sensitivity TFM, $C(s) = G(s)K(s)(I + G(s)K(s))^{-1}$, must be close to unity (0 dB) at that frequency. For SISO systems, this simply translates to requiring that the magnitude of the closed-loop transfer function be close to unity (0 dB). From Figure 30.17, it is clear that the stability robustness condition limits the range of frequencies over which we can expect to achieve good command following and/or output disturbance rejection to below ω_m. In other words, we should not expect good performance at frequencies where there are significant modeling errors.

For the MIMO case where the modeling errors are reflected to the plant output, the stability robustness condition given in Equation 30.54 can be illustrated in terms of singular value plots, as shown in Figure 30.18. This figure gives a similar interpretation concerning the impact of stability robustness on the nominal closed-loop performance as that for the SISO case. From the figure, the output bandwidth of the nominal closed-loop design, ω_{bo}, is constrained by the stability robustness condition to be less than ω_{mo}, where ω_{mo} is defined as the frequency at and beyond which the modeling error reflected to the plant output becomes significant. In other words, ω_{mo} is defined as

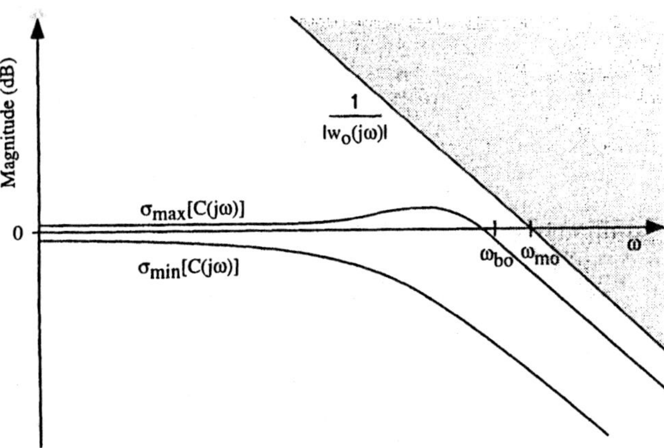

Figure 30.18 Interpretation of stability robustness for MIMO systems with uncertainty reflected to plant output.

$$|w_o(j\omega_{mo})| = 0 \text{ dB} \tag{30.60}$$

Therefore, as in the SISO case, the stability robustness condition limits the range of frequencies over which we can expect to achieve good command following and/or output disturbance rejection to below ω_{mo}. On the other hand, for the MIMO case where the modeling errors are reflected to the plant input, we do not get a similar interpretation. In this case, the stability robustness condition given in Equation 30.57 can be illustrated in terms of singular value plots, as shown in Figure 30.19. The difference lies in the fact that this condition depends on the input complementary sensitivity TFM, $C_i(s) = (I + K(s)G(s))^{-1}K(s)G(s)$, instead of $C(s)$. Under this stability robustness condition, the input bandwidth of the nominal closed-loop design, ω_{bi}, is constrained to be less than ω_{mi}, where ω_{mi} is defined as the frequency at and beyond which the modeling error reflected to the plant input becomes significant. Here, ω_{mi} is defined as

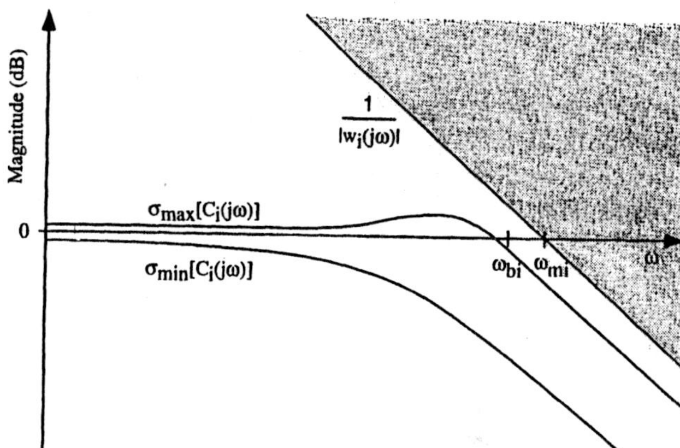

Figure 30.19 Interpretation of stability robustness for MIMO systems with uncertainty reflected to plant input.

$$|w_i(j\omega_{mi})| = 0 \text{ dB} \tag{30.61}$$

which may be different than ω_{mo} since $w_i(s)$ is, in general, not the same as $w_o(s)$. Since command following and output disturbance rejection depends on the sensitivity TFM, $S(s) = (I + G(s)K(s))^{-1}$, instead of the input sensitivity TFM, $S_i(s) = (I + K(s)G(s))^{-1}$, we cannot directly ascertain the impact of the stability robustness condition in Equation 30.57 on these performance measures. However, we can obtain insight on the impact of this stability robustness condition on input disturbance rejection because the TFM from the disturbance at the plant input to the plant output is equal to $G(s)(I + K(s)G(s))^{-1}$. Therefore, if the input sensitivity TFM is small then we have good input disturbance rejection. As before, a necessary condition for $\sigma_{\max}[S_i(j\omega)]$ to be small at a certain frequency is that the singular values of $C_i(j\omega)$ are close to unity (0 dB). From Figure 30.19, this stability robustness condition clearly limits the range of frequencies at which we can expect to achieve good input disturbance rejection to below ω_{mi}. This is because we must roll-off $C_i(j\omega)$ above ω_{mi} to satisfy the stability robustness condition.

To illustrate the concept of stability robustness to multiplicative unstructured uncertainty and its impact on closed-loop performance, we present the following example.

EXAMPLE 30.2: Integrator with Time Delay

Suppose that we wish to control the plant from Example 1 using a simple proportional controller. Our objective is to push the bandwidth of the nominal closed-loop system as high as possible. We use the simple integrator model

$$g(s) = \frac{1}{s} \tag{30.62}$$

for our design. In this case, the nominal loop transfer function is simply $g(s)k$ where k is the proportional gain of our controller, and the nominal closed loop transfer function is simply

$$c(s) = \frac{k}{s+k} \tag{30.63}$$

Since we require stability of the actual closed loop, we need to satisfy the stability robustness condition given in Equation 30.58 where, from Example 1, the weight $w(s)$ of the uncertainty for this model is

$$w(s) = \frac{0.21s}{0.05s+1} \tag{30.64}$$

As shown in Figure 30.20, the maximum value for k at which the stability robustness condition is satisfied is $k = 5.9$. With this value of k, the bandwidth achieved for the nominal system based on this model is 5.9 rad/sec.

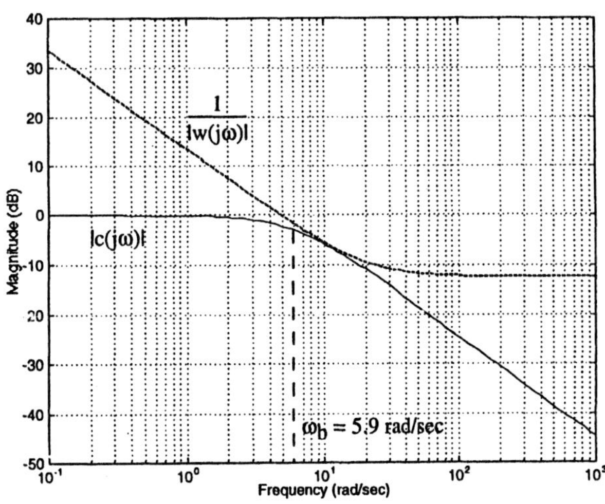

Figure 30.20 Stability robustness condition for 0.2 sec maximum delay uncertainty.

Now, consider the case where the time delay can vary up to 0.4 sec instead of 0.2 sec. In this case, we use the same model as above, but our uncertainty representation must change in order to cover this additional uncertainty. Specifically, we must choose a new weight $w(s)$. As in Example 1, we need to choose $w(s)$ such that

$$|w(j\omega)| \geq \max_{\tau} \left|2\sin\left(\frac{\omega\tau}{2}\right)\right| \quad \forall\, \omega \geq 0 \tag{30.65}$$

where τ now ranges from 0 to 0.4 sec. A simple $w(s)$ that will satisfy Equation 30.65 is

$$w(s) = \frac{0.41s}{0.1s+1} \tag{30.66}$$

This is shown in Figure 30.21, where $|w(j\omega)|$ and $2\left|\sin(\frac{\omega\tau}{2})\right|$ are plotted together on a magnitude Bode plot for $\tau = 0.4$, which is now the worst case value. We note that $w(s)$ is proper and strictly stable as required. Since we require stability of the actual closed

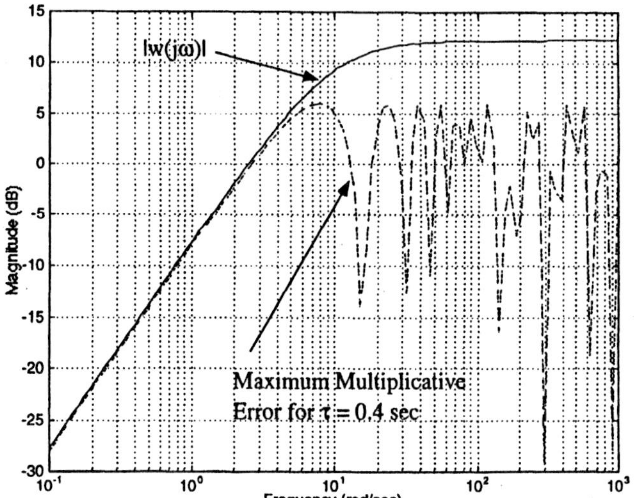

Figure 30.21 $w(s)$ for 0.4 sec maximum delay uncertainty.

loop, we again need to satisfy the stability robustness condition given in Equation 30.58. As shown in Figure 30.22, the maximum value for k at which the stability robustness condition is satisfied is now $k = 2.9$. With this value of k, the bandwidth achieved for the nominal system based on this model is 2.9 rad/sec.

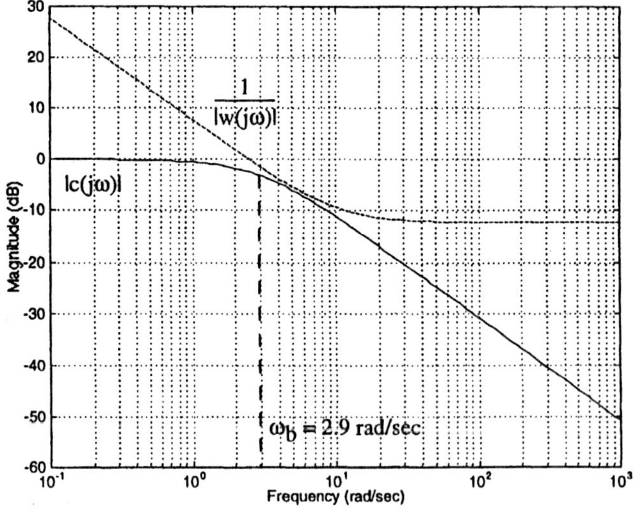

Figure 30.22 Stability robustness condition for 0.4 sec maximum delay uncertainty.

Comparing the two cases, it is clear that the uncertainty set for the second case is larger because the variation in the unknown time delay is larger. Physically, the larger time delay means larger unmodeled dynamics that we must cover in our uncertainty representation. This results in a larger magnitude for $w(s)$ at low frequencies, as shown in Figure 30.23. In particular, the frequency at which the modeling errors become significant, ω_m, is lower for the second case than the first, and therefore, further limiting the achievable bandwidth of the nominal closed loop. As a result, we are able to push the bandwidth higher for the first case. This results in better achievable performance, such as command following and output disturbance rejection, as shown by the sensitivity bode plots in Figure 30.24.

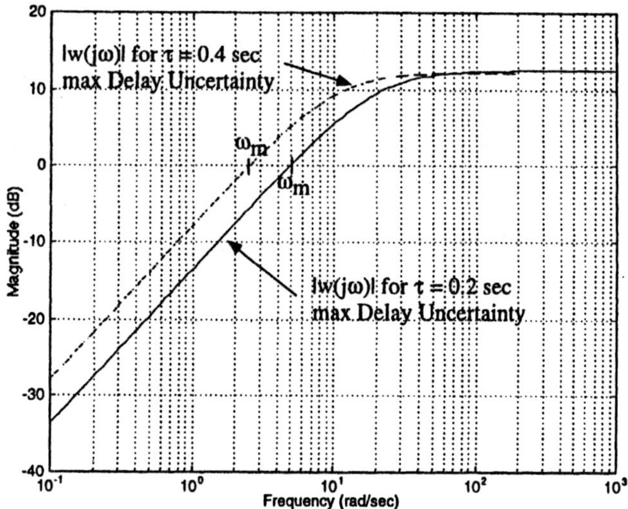

Figure 30.23 Comparing $|w(j\omega)|$.

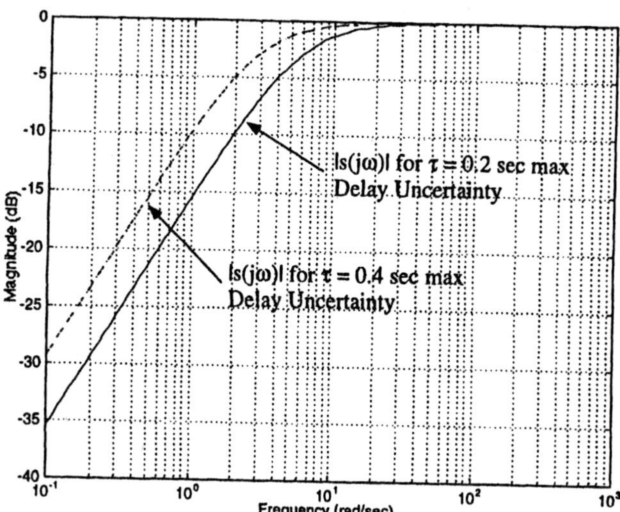

Figure 30.24 Comparing sensitivity magnitude plots.

In summary, the requirement of stability robustness has a definite impact on the achievable nominal performance of the closed loop. This impact is typically a limitation on the nominal closed-loop bandwidth which, in turn, constrains the frequency range at which good command following and output disturbance rejection is achievable. As is evident in the above example, there is a trade-off between stability robustness and performance. That is, a controller that is designed to be robust to a larger uncertainty set will result in poorer performance due to the fact that its bandwidth will be further limited. In addition, there is a connection between this trade-off and the conservatism of our stability robustness conditions. In particular, our representation of the uncertainty is one of overbounding the set of TFMs from our modeling process with a larger set. If we choose $w(s)$ where the overbounding is not tight, then this will further limit the amount of achievable performance that we can obtain for the closed loop. Therefore, we see that both the concept of unstructured uncertainty and the way we represent this uncertainty in the stability robustness problem have an impact on achievable nominal performance. As a result, to improve performance we may have to reduce the level of uncertainty by having a better model and/or by obtaining a better representation for the uncertainty. Finally, we wish to point out that only nominal performance or the closed-loop performance of the model without the uncertainty is discussed. The actual performance of the loop will be different. However, for low frequencies where we care about command following and output disturbance rejection, the modeling errors will be typically small. Therefore, the difference between actual performance and nominal performance at these frequencies will be small.

30.5 Other Representations for Model Uncertainty

In this section, we present some alternative set membership representations for unstructured uncertainty. For each representation, we will start with the definition of the set $\mathcal{G}$. In defining $\mathcal{G}$, we assume throughout that

1. The weight is a fixed, proper, and strictly stable scalar transfer function
2. $\Delta(s)$ is a strictly stable TFM
3. No unstable or imaginary axis poles of $G(s)$ are cancelled in forming any element $\tilde{G}(s)$ of the set $\mathcal{G}$

With this definition, the appropriate stability robustness condition is obtained using the procedure outlined in Section 30.3.2. Like those derived for the multiplicative uncertainty representation, these conditions are all conservative since they are derived using the small gain theorem. The focus here will be primarily on the presentation of results rather than detailed treatments of their derivation. MIMO systems are treated as SISO systems that are simply special cases. Comments on the usefulness of these representations and interpretations of their results are given as appropriate.

30.5.1 Additive Uncertainty

An alternative representation for unstructured uncertainty is additive uncertainty. In this case, the set $\mathcal{G}$ is defined as

$$\mathcal{G} = \{\tilde{G}(s) \mid \tilde{G}(s) = G(s) + w_a(s)\Delta(s), \|\Delta(j\omega)\|_{\mathcal{H}_\infty} \leq 1\} \tag{30.67}$$

Following the procedure in Section 30.3.2, the stability robustness condition for additive uncertainty is

$$\sigma_{\max}[(I + K(j\omega)G(j\omega))^{-1}K(j\omega)] < \frac{1}{|w_a(j\omega)|} \quad \forall\, \omega \geq 0 \tag{30.68}$$

From Equation 30.67, the additive error is defined as

$$E_a(s) = w_a(s)\Delta(s) = \tilde{G}(s) - G(s) \tag{30.69}$$

which is simply the difference between the perturbed plant and the model. As a result, the additive representation is a more natural representation for differences in the internal dynamics between the actual plant and the model. In particular, it is no longer necessary to reflect these modeling errors to the plant input or output. However, we note that the resulting stability robustness condition is explicitly dependent on the controller TFM, $K(s)$, and the loop TFM, $K(s)G(s)$, instead of simply on $K(s)G(s)$. This is because, unlike multiplicative uncertainty, the uncertainty representation here does not apply equally to the loop TFM as to the model, $G(s)$. The result is that there will be added complexity in designing a $K(s)$ that will satisfy the condition because shaping the TFM, $(I+K(j\omega)G(j\omega))^{-1}K(j\omega)$, to satisfy Equation 30.68 will require shaping both $K(j\omega)$ and $K(j\omega)G(j\omega)$. Due to this complication, the multiplicative uncertainty or another form of cascaded uncertainty representation, where the representation applies equally to $G(s)$ as to $K(s)G(s)$, is often used instead in practice.

30.5.2 Division Uncertainty

Another representation for unstructured uncertainty is the division uncertainty. Like the multiplicative uncertainty representation, the division uncertainty represents the modeling error in a cascade form, and therefore, the modeling error can be reflected either to the plant output or the plant input. In the case where the modeling errors are reflected to the plant output, the set $\mathcal{G}$ is defined as

$$\begin{aligned}\mathcal{G} &= \{\tilde{G}(s) \mid \tilde{G}(s) \\ &= (I + w_{do}(s)\Delta(s))^{-1}G(s), \|\Delta(j\omega)\|_{\mathcal{H}_\infty} \leq 1\}\end{aligned} \tag{30.70}$$

Following the procedure in Section 30.3.2, the stability robustness condition for division uncertainty at the plant output is given by

$$\sigma_{\max}[(I + G(j\omega)K(j\omega))^{-1}] < \frac{1}{|w_{do}(j\omega)|} \quad \forall\, \omega \geq 0 \tag{30.71}$$

Similarly, for the case where the modeling errors are reflected to the plant input, the set $\mathcal{G}$ is defined as

$$\begin{aligned}\mathcal{G} &= \{\tilde{G}(s) \mid \tilde{G}(s) \\ &= G(s)(I + w_{di}(s)\Delta(s))^{-1}, \|\Delta(j\omega)\|_{\mathcal{H}_\infty} \leq 1\}\end{aligned} \tag{30.72}$$

and the resulting stability robustness condition is given by

$$\sigma_{\max}[(I + K(j\omega)G(j\omega))^{-1}] < \frac{1}{|w_{di}(j\omega)|} \quad \forall\, \omega \geq 0 \tag{30.73}$$

As shown in Figure 30.25, the stability robustness condition for division uncertainty at the plant output requires the maximum singular value plot for the sensitivity TFM of the nominal system, $\sigma_{\max}[S(j\omega)]$, to lie strictly below the plot of the inverse of $|w_{do}(j\omega)|$. Similarly, for division uncertainty at the plant input, the stability robustness condition requires that the maximum singular value plot for the input sensitivity TFM of the nominal system, $\sigma_{\max}[S_i(j\omega)]$, must lie strictly below the plot of the inverse of $|w_{di}(j\omega)|$. As shown in the figure, the inverse of the magnitude of $w_{do}(s)$ is typically large at low frequencies and approaches unity (0 dB) at high frequencies where the modeling errors become significant. Since the stability robustness condition for division uncertainty at the plant output depends on the nominal sensitivity TFM, its impact on command following and output disturbance rejection is clear from Figure 30.25. In particular, if the modeling error becomes significant at a particular frequency, ω_{mo} such that the plot of the inverse of $|w_{do}(j\omega)|$ is close to 0 dB at that frequency and beyond, the control must be designed such that $\sigma_{\max}[S(j\omega)]$ is below this barrier. That is, $\sigma_{\max}[S(j\omega)]$ is not allowed to be much greater than 0 dB for $\omega \geq \omega_{mo}$. However, we know from the Bode integral theorem [4] that if we suppress the sensitivity at high frequencies then we cannot also suppress it at low frequencies. Therefore, we again see how modeling errors can place a limitation on the range of frequencies over which we can expect to achieve good command following and output disturbance rejection. The same can be said for input disturbance rejection with respect to the stability robustness condition for division uncertainty at the plant input.

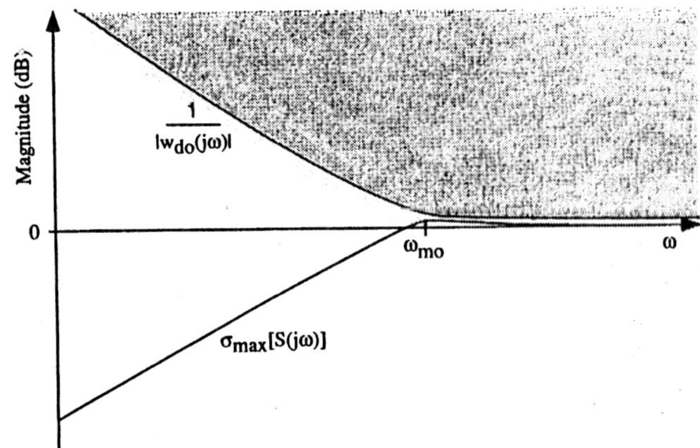

Figure 30.25 Interpretation of stability robustness for division uncertainty at plant output.

30.5.3 Representation for Parametric Uncertainty

Finally, we can also define the set $\mathcal{G}$ as

$$\begin{aligned} \mathcal{G} &= \{\tilde{G}(s) \mid \tilde{G}(s) \qquad (30.74) \\ &= (I + w_p(s)G(s)\Delta(s))^{-1}G(s), \ \|\Delta(j\omega)\|_{\mathcal{H}_\infty} \leq 1\} \end{aligned}$$

In this case, the stability robustness condition is given by

$$\sigma_{\max}[(I + G(j\omega)K(j\omega))^{-1}G(j\omega)] < \frac{1}{|w_p(j\omega)|} \quad \forall\, \omega \geq 0 \qquad (30.75)$$

As shown in the following example, we can use this representation to handle parametric uncertainty in the A matrix of the nominal model $G(s)$.

EXAMPLE 30.3: Parametric Uncertainty in the A Matrix

Consider the dynamics of the actual plant in state-space form

$$\begin{aligned} \dot{x} &= Ax + Bu \\ y &= Cx \end{aligned} \qquad (30.76)$$

where the parameters of the A matrix are uncertain but are known to be constant and contained in a certain interval. In particular, consider the case where the A matrix is given as

$$A = \bar{A} + \delta A \qquad (30.77)$$

where the elements of $\bar{A}$ contain the midpoint values for each corresponding element in A and each element of δA is known to exist in the interval

$$-1 \leq \delta A_{ij} \leq 1 \qquad (30.78)$$

We note that since

$$\|\delta A\| \leq n \qquad (30.79)$$

for all possible δA satisfying Equation 30.78 where n is the dimension of A, it is clear that the set of possible δA is contained in the set $\{w_p(s)\Delta(s) | \ \|\Delta(j\omega)\|_{\mathcal{H}_\infty} \leq 1, \, w_p(s) = n\}$. Therefore, we can express the A matrix as

$$A = \bar{A} + w_p(s)\Delta(s) \qquad (30.80)$$

for $w_p(s) = n$ and for some $\|\Delta(j\omega)\|_{\mathcal{H}_\infty} \leq 1$. In block diagram form, the state-space equations in Equation 30.76 can then be represented as in Figure 30.26, where the model $G(s)$ is taken to be $(sI - \bar{A})^{-1}$. We note that this is equivalent to the uncertainty representation given in Equation 30.74.

We note that the above representation for parametric uncertainty is, in general, very conservative. This is because the representation allows for complex perturbations of the matrix A since $\Delta(s)$ is complex. Since the parameters are real, the perturbed set of plants is much larger than the set that will actually be realized by the plant, which results in the conservatism.

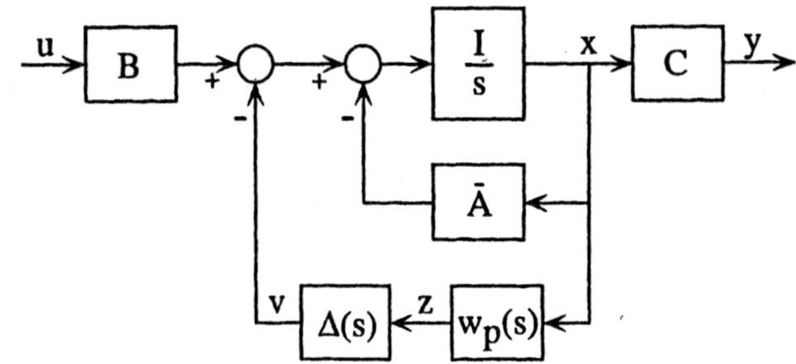

Figure 30.26 Representation for parametric uncertainty in the A matrix.

Notes

The majority of the material in this article is adopted from Doyle and Stein [5], Dahleh and Diaz-Bobillo [1], Maciejowski [6], and Doyle et al. [7]. Other excellent references include Green and Limebeer [8] and Morari and Zafiriou [9].

References

[1] Dahleh, M. A. and Diaz-Bobillo, I. J., *Control of Uncertain Systems: A Linear Programming Approach*, Englewood Cliffs, NJ: Prentice Hall, 1995.

[2] Hildebrand, F. B., *Advanced Calculus for Applications*, 2nd Ed., Englewood Cliffs, NJ: Prentice Hall, 1976.

[3] Desoer, C. A. and Vidyasagar, M., *Feedback Systems: Input-Output Properties*, New York: Academic Press, 1975.

[4] Freudenburg, J. S. and Looze, D. P., *Frequency Domain Properties of Scalar and Multivariable Feedback Systems*, Berlin: Springer-Verlag, 1987.

[5] Doyle, J.C. and Stein, G., "Multivariable Feedback Design: Concepts for a Classical/Modern Synthesis," *IEEE Trans. Autom. Control*, 26(1), 4–16, 1981.

[6] Maciejowski, J. M., *Multivariable Feedback Design*, Reading, MA: Addison-Wesley, 1989.

[7] Doyle J. C., Francis, B. A., and Tannenbaum, A. R., *Feedback Control Theory*, New York: Macmillan, 1992.

[8] Green, M. and Limebeer, D. J. N., *Linear Robust Control*, Englewood Cliffs, NJ: Prentice-Hall, 1995.

[9] Morari, M. and Zafiriou, E., *Robust Process Control*, Englewood Cliffs, NJ: Prentice Hall, 1989.

31

Tradeoffs and Limitations in Feedback Systems

Douglas P. Looze
Dept. Electrical and Computer Engineering, University of Massachusetts, Amherst, MA

James S. Freudenberg
Dept. Electrical Engineering & Computer Science, University of Michigan, Ann Arbor, MI

31.1 Introduction and Motivation

It is well known that feedback may be used in a control design to obtain desirable properties that are not achievable with open-loop control. Among these are the abilities to stabilize an unstable system, and to reduce the effects of plant disturbances and modeling errors upon the system output. On the other hand, use of feedback control can also have undesirable consequences: feedback may destabilize a system, introduce measurement noise, amplify the effects of disturbances and modeling errors, and generate large control signals. A satisfactory feedback design will, if possible, achieve the potential benefits of feedback without incurring excessive costs.

Unfortunately, feedback systems possess limitations that manifest themselves as design tradeoffs between the benefits and costs of feedback. For example, there exists a well-known tradeoff between the response of a feedback system to plant disturbances and to sensor noise. This tradeoff is an inherent consequence of feedback system topology; indeed, the effect of a plant disturbance cannot be attenuated without a measurement of its effect upon the system output. Other design tradeoffs are a consequence of system properties such as unstable poles, nonminimum phase zeros, time delays, and bandwidth limitations.

The study of feedback design limitations and tradeoffs dates back at least to the seminal work of Bode [1]. In his classic work, Bode stated the famous gain-phase relations and analyzed their impact upon the classical loop-shaping problem. He also derived the Bode sensitivity integral relation. The importance of this relation to feedback control design was emphasized by Horowitz [2]. Connections between the classical loop-shaping problem, including the gain-phase relations, and modern control techniques were developed by Doyle and Stein [3]. Further results pertaining to design limitations were derived by Freudenberg and Looze [4, 5], who studied inherent design limitations present when the system to be controlled has unstable poles, nonminimum phase zeros, time delays, and/or bandwidth constraints. In all these works, emphasis is placed upon frequency response properties and, in particular, the insight into feedback design that may be obtained from a Bode plot.

In this chapter, we describe several design tradeoffs that are present in feedback system design due to the structure of a feedback loop, bandwidth limitations, and plant properties, such as unstable poles, nonminimum phase zeros, and time delays. In Section 31.2 we show how the closed-loop transfer functions of a feedback system may be used to describe signal response, differential sensitivity, and stability robustness properties of the system. In Section 31.3 we describe design tradeoffs imposed by the topology of a feedback loop and the relation between feedback properties and open-loop gain and phase. Next, in Section 31.4, we describe design limitations imposed by the Bode gain-phase

0-8493-8570-9/96/$0.00+$.50

relations, Bode sensitivity integral, and the dual complementary sensitivity integral derived by Middleton and Goodwin [6]. These tradeoffs are present due to the fact that the closed-loop system is stable, linear, and time invariant. When the plant has nonminimum phase zeros, unstable poles, and/or time delays, additional design tradeoffs are present. As discussed in Section 31.5, these may be described using the Poisson integral. A brief summary is presented in Section 31.6.

31.2 Quantification of Performance for Feedback Systems

31.2.1 Overview

As noted in Section 31.1, feedback control can provide several advantages over open-loop control. These advantages include the ability to stabilize systems that are open-loop unstable; the ability to reduce the effects of unmeasureable disturbances on the system response; and the ability to reduce the effects of plant modeling errors and variations on the system response. The use of feedback can also detract from these properties: systems can be destabilized, and the effects of disturbances and plant uncertainty can be amplified.

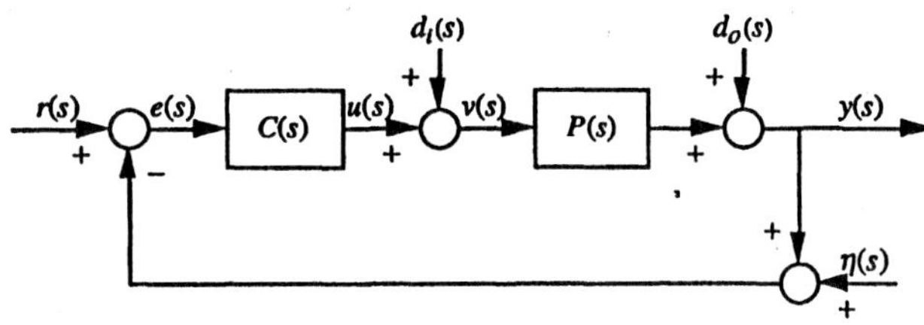

Figure 31.1 Linear, time-invariant feedback system.

The principal objective in any control problem is to select the input to the plant so that the plant output follows a desired, or commanded, signal while not using too much control effort[1]. The standard unity feedback architecture (see Figure 31.1) compares a measurement of the plant output with the command signal to generate the feedback error signal. The feedback error signal is then compensated to produce a control signal, which is applied to the plant. Given this feedback architecture, the control system analysis and design problem is to select the compensator transfer function to achieve the control objective.

The ability of the plant output to follow the command input is referred to as command following. To achieve command following, the closed-loop system must, at a minimum, be stable. As Figure 31.1 illustrates, exogenous signals can act at various points in the feedback loop to affect the response of the system. Signals that enter the loop between the control action and the output to be controlled are referred to as disturbance signals. The ability of the control system to eliminate the effects of the disturbances is referred to as disturbance rejection. Signals that enter the loop between the system output and the comparison of the measurement with the command input are referred to as measurement noise signals. Measurement noise only affects the output through the action of the compensator.

In addition to the uncertainty caused by the disturbance signals and measurement noise, the behavior of the plant as described by the model will differ from that of the true plant. A feedback system that is stable for the plant model is called nominally stable. A feedback loop that maintains stability when subject to variations in the plant model is robustly stable. Nominal performance is achieved when command following and disturbance rejection objectives are met for the plant model. Robust performance is achieved if the feedback loop maintains performance (command following or disturbance rejection) when subject to variations in the plant model.

The objective of this section is to provide a formal mechanism for quantifying the potential benefits and liabilities of feedback. In general, performance is quantified by defining a measure of size of the difference between the plant output and the command input. Input signal effort is quantified by defining a similar measure of the size of the control input. This chapter measures the size of signals in the frequency domain in terms of the maximum value of their spectrum. As will be seen, performance and robust stability can then be expressed in terms of the magnitude of certain closed-loop transfer functions.

31.2.2 Closed-Loop Transfer Functions

Consider the linear time-invariant feedback system shown in Figure 31.1. The transfer functions $P(s)$ and $C(s)$ are those of the plant and compensator, respectively. The signal $r(s)$ is a reference or command input, $y(s)$ is the system output, $d_o(s)$ is a disturbance at the plant output, $d_i(s)$ is a disturbance at the plant input, and $\eta(s)$ is sensor noise. Three transfer functions are essential in studying properties of this system. Define the open-loop transfer function

$$L(s) = P(s)C(s) \tag{31.1}$$

the sensitivity function

$$S(s) = \frac{1}{1 + L(s)} \tag{31.2}$$

and the complementary sensitivity function [7, 8]

$$T(s) = \frac{L(s)}{1 + L(s)} \tag{31.3}$$

31.2.3 Nominal Signal Response

The response of the system of Figure 31.1 to the exogenous inputs (i.e., the command input, the input disturbance, the output

[1] A precompensator can be used to shape the actual reference or command signal to help achieve the desired control objectives. Throughout this chapter, we assume that the command signal $r(s)$ is the result of any precompensation that is used.

disturbance, and the sensor noise) is governed by the closed-loop transfer functions (Equations 31.2 and 31.3). The output of the system is given by:

$$y(s) = S(s)\,[d_o(s) + P(s)d_i(s)] + T(s)[r(s) - \eta(s)] \tag{31.4}$$

The feedback loop error between the measured output and the command input is given by:

$$e(s) = S(s)\,[r(s) - d_o(s) - P(s)d_i(s) - \eta(s)] \tag{31.5}$$

The performance error between the output and the command input is given by:

$$\begin{aligned} e_p(s) &= r(s) - y(s) \\ &= S(s)\,[r(s) - d_o(s) - P(s)d_i(s)] - T(s)\eta(s) \end{aligned} \tag{31.6}$$

The control input generated by the compensator is given by:

$$u(s) = -T(s)d_i(s) + \frac{T(s)}{P(s)}[r(s) - \eta(s) - d_o(s)] \tag{31.7}$$

Because the principal objective for control system performance is to have the plant output track the command input without using too much control energy, it is desirable to have the performance error (Equation 31.6) and the plant input (Equation 31.7) be small. It is apparent from Equations 31.6 and 31.7 that each of these signals can be made small by making the sensitivity and complementary sensitivity transfer functions small. In particular, the performance error (Equation 31.6) can be made small at any frequency s by making the sensitivity function small relative to the command response and disturbances, and by making the complementary sensitivity small relative to the measurement noise. The plant input can also be kept within a desired range by bounding the magnitude of the complementary sensitivity function relative to the external signals. The presence of the plant transfer function in the denominator of the second term in Equation 31.7 implies that the magnitude of the control signal will increase as the plant magnitude decreases, unless there is a corresponding decrease in the magnitudes of either the complementary sensitivity function or the external command, measurement noise, and output disturbance signals. The presence of high-frequency measurement noise or disturbances together with a bound on the desired control magnitude thus imposes a bandwidth constraint on the closed-loop system.

31.2.4 Differential Sensitivity

Because the true plant behavior will be somewhat different from the behavior described by the plant model $P(s)$, it is important that performance be maintained for variations in the plant model that correspond to possible differences with reality. Although the precise characterization of robust performance is beyond the scope of this chapter (see [9]), incremental changes in the system response due to model variations can be characterized by the sensitivity function. Assume that the true plant model is given by:

$$\mathbf{P}(s) = P(s) + \Delta P(s) \tag{31.8}$$

Assuming that the disturbance and measurement noise signals are zero, the nominal closed-loop response (i.e., the response to the command reference for the plant model $P(s)$) is

$$y(s) = T(s)r(s) \tag{31.9}$$

The response for the feedback system using the true plant is

$$\mathbf{y}(s) = \mathbf{T}(s)r(s) \tag{31.10}$$

where

$$\mathbf{y}(s) = y(s) + \Delta y(s) \tag{31.11}$$

$$\mathbf{T}(s) = T(s) + \Delta T(s) \tag{31.12}$$

Thus, the variation in the command response is proportional to the variation in the true complementary sensitivity function.

Let $\mathbf{S}(s)$ denote the sensitivity function for the feedback system using the true plant transfer function. Then, it can be shown [10] that:

$$\frac{\Delta T(s)}{T(s)} = \mathbf{S}(s)\frac{\Delta P(s)}{P(s)} \tag{31.13}$$

Equation 31.13 shows that the true sensitivity function determines the relative deviation in the command response due to a relative deviation in the plant transfer function. If the true sensitivity function is small, the variation in command response will be small relative to the variation in the plant model.

Although Equation 31.13 provides insight into how the command response varies, it is not useful as a practical analysis and design tool because the relationship is expressed in terms of the unknown true plant transfer function. However, as the plant variation $\Delta P(s)$ becomes small, the true sensitivity function $\mathbf{S}(s)$ approaches the nominal sensitivity function $S(s)$. Thus, for $\Delta P(s) = dP(s)$, Equation 31.13 becomes

$$\frac{dT(s)}{T(s)} = S(s)\frac{dP(s)}{P(s)} \tag{31.14}$$

Equation 31.14 shows that the sensitivity function $S(s)$ governs the sensitivity of the system output to small variations in the plant transfer function. Hence, the variation in the command response will be small if the sensitivity function is small relative to the plant variation.

31.2.5 Robust Stability

Assuming the loop transfer function (Equation 31.1) has no pole-zero cancellations in the closed right-half plane (CRHP), the closed-loop system of Figure 31.1 will be stable if the sensitivity (Equation 31.2) and complementary sensitivity (Equation 31.3) transfer functions have no poles in the CRHP. A number of techniques, such as the Routh-Hurwitz and Nyquist criteria [11, 12], are available for evaluating whether the system of Figure 31.1 is stable for the assumed plant model $P(s)$.

The sensitivity and complementary sensitivity functions each characterize stability robustness of the system against particular classes of plant variations (e.g., see [13], Table 1). One of the most significant types of plant modeling error is the high-frequency uncertainty that is present in any model of a physical system. All finite-order transfer function models of physical systems deteriorate at high frequencies due to neglected dynamics (such as actuator and measurement lags, and flexible modes), the effects of approximating distributed systems with lumped models, and time delays. The result of these neglected physical phenomena is that the uncertainty in the gain of the true plant increases (eventually differing from the assumed plant model by more than 100%) and the phase of the true plant becomes completely uncorrelated with that of the model (i.e., there is $\pm 180^\circ$ uncertainty in the phase).

This type of uncertainty can be represented as a relative deviation from the nominal plant. The true plant transfer function is represented by a multiplicative error model:

$$\mathbf{P}(s) = P(s)(1 + \Delta(s)) \tag{31.15}$$

Given a true plant $\mathbf{P}(s)$, the multiplicative error $\Delta(s)$ is the relative plant error:

$$\Delta(s) = \frac{\mathbf{P}(s) - P(s)}{P(s)} \tag{31.16}$$

It will be assumed that the multiplicative error is stable. From Equation 31.16, it is apparent that this assumption will be satisfied if the number and location of any unstable poles of the true plant are the same as those of the plant model.

The characteristics of increasing gain and phase uncertainty can be represented by assuming that the multiplicative error is unknown except for an increasing upper bound on its magnitude. That is, it will be assumed that $\Delta(s)$ is any stable transfer function that satisfies:

$$|\Delta(j\omega)| < M(\omega) \tag{31.17}$$

where (typically) $M(\omega) \to \infty$ as $\omega \to \infty$.

If the only information available about the uncertainty is Equation 31.17, then $\Delta(s)$ is referred to as unstructured uncertainty. In that case, a necessary and sufficient condition for the feedback system to be stable for all plants described by Equations 31.15 to 31.17 is that the system be stable when $\Delta(s) = 0$ and that the complementary sensitivity function satisfy the bound [3]

$$|T(j\omega)| < \frac{1}{M(\omega)} \quad \forall\omega \tag{31.18}$$

Equation 31.18 demonstrates that the presence of high-frequency uncertainty forces the complementary sensitivity function to become small to ensure that the system is robustly stable. This in turn implies that an upper limit on bandwidth is imposed on the closed-loop system.

31.2.6 Summary of Feedback System Performance Specification

In the previous section we saw that the sensitivity and complementary sensitivity functions each characterize important properties of a feedback system. This fact motivates stating design specifications directly as frequency-dependent bounds upon the magnitudes of these functions. Hence, we typically require that

$$|S(j\omega)| \le M_S(\omega) \quad \forall\omega \tag{31.19}$$

and

$$|T(j\omega)| \le M_T(\omega) \quad \forall\omega \tag{31.20}$$

The bounds $M_S(\omega)$ and $M_T(\omega)$ will generally depend upon the size of the disturbance and noise signals, the level of plant uncertainty, and the extent to which the effect of these phenomena upon the system output must be diminished (see Equations 31.6, 31.7, 31.14, and 31.18).

It is interesting to note that the two design specifications represent two different aspects of control system design. The bound on the sensitivity function (Equation 31.19) typically represents the potential benefits of feedback, such as disturbance rejection, command following, and robust performance. The bound on the complementary sensitivity function represents desired limits on the cost of feedback: amplification of measurement noise, increased control signal requirements, and possible introduction of instability into the feedback loop.

31.3 Design Tradeoffs at a Single Frequency

31.3.1 Introduction

Often the requirements imposed by various design objectives are mutually incompatible. It is therefore important to understand when a given design specification is achievable and when it must be relaxed by making tradeoffs between conflicting design goals. The objective of this section is to explore the limitations that are imposed by the structure of the feedback loop (Figure 31.1).

The structural tradeoffs presented in this section express the limitation that there is only one degree of design freedom in the feedback loop. This degree of freedom can be exercised by specifying any one of the transfer functions associated with the feedback loop. Because the sensitivity and complementary sensitivity transfer functions are both uniquely determined by the loop transfer function, these transfer functions cannot be specified independently. Hence, the properties of the feedback loop are completely determined once any one of $L(s)$, $S(s)$, or $T(s)$ is specified, and these properties can be analyzed in terms of the chosen transfer function. In particular, the performance bounds (Equations 31.19 and 31.20) are not independent, and may lead to conflicts.

This section presents explicit relationships between the sensitivity function, the complementary sensitivity function, and the loop transfer function of the feedback system of Figure 31.1. These relationships can be used to explore whether design specifications in the form of Equations 31.19 and 31.20 are consistent, and if not, how the specifications might be modified to be achievable.

31.3.2 Relationship of Closed-Loop Transfer Functions and Design Specifications

One important design tradeoff may be quantified by noting that the sensitivity and complementary sensitivity functions satisfy the identity

$$S(j\omega) + T(j\omega) = 1 \tag{31.21}$$

The identity (Equation 31.21) is a structural identity in the sense that it is a consequence of the topology of the feedback loop.

From Equation 31.21 it follows that $|S(j\omega)|$ and $|T(j\omega)|$ cannot both be very small at the same frequency. Hence, at each frequency, there exists a tradeoff between those feedback properties, such as sensitivity reduction and disturbance response, that are quantified by $|S(j\omega)|$ and those properties, such as measurement noise response and robustness to high-frequency uncertainty, that are quantified by $|T(j\omega)|$.

In applications it often happens that levels of uncertainty and sensor noise become significant at high frequencies, while disturbance rejection and sensitivity reduction are generally desired over a lower frequency range. Hence the tradeoff imposed by Equation 31.21 is generally performed by requiring $M_S(\omega)$ to be small at low frequencies, $M_T(\omega)$ to be small at high frequencies, and neither $M_S(\omega)$ nor $M_T(\omega)$ to be excessively large at any frequency.

This situation is illustrated in Figure 31.2. The bound $M_S(\omega)$ is small at low frequencies, representing the frequencies over which disturbance rejection and command following are important. It increases with increasing frequency until it becomes greater than 1 at high frequencies. Conversely, the bound $M_T(\omega)$ begins somewhat greater than 1 at low frequencies, and rolls off (decreases in magnitude) at higher frequencies where the plant model error and measurement noise become significant. The sensitivity and complementary sensitivity functions shown in Figure 31.2 satisfy the design objectives (Equations 31.19 and 31.20).

The identity (Equation 31.21) can be used to derive conditions that the design objective functions $M_S(\omega)$ and $M_T(\omega)$ must satisfy. Taking the absolute value of the right hand side of Equation 31.21, using the triangle identity, and applying the bounds (Equations 31.19 and 31.20), the following inequality is obtained:

$$M_S(\omega) + M_T(\omega) \geq 1 \tag{31.22}$$

Inequality (Equation 31.22) reinforces the observations made previously: the design objectives cannot be specified to require both the sensitivity function and the complementary sensitivity function to be small at the same frequency.

The structural identity can also be used to explore limits on more detailed specifications of control system objectives. For example, suppose that at a given frequency ω the dominant objective that specifies the weighting on the complementary sensitivity function is that the control signal should remain bounded when subject to an output disturbance which is unknown but bounded:

$$|u(j\omega)| < M_u(\omega)$$

for all output disturbances that satisfy

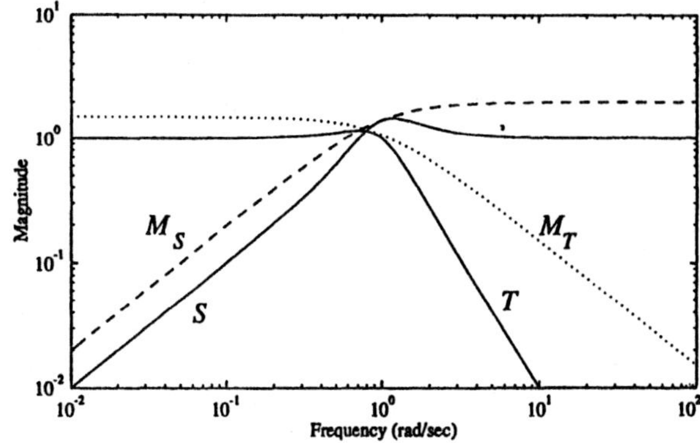

Figure 31.2 Typical sensitivity function, complementary sensitivity function, and specification bounds.

$$|d_o(j\omega)| \leq M_{do}(\omega) \tag{31.23}$$

Then, combining Equations 31.7 and 31.23, the design specification on the complementary sensitivity (Equation 31.20) becomes:

$$|T(j\omega)| \leq |P(j\omega)| \frac{M_u(\omega)}{M_{do}(j\omega)} \equiv M_T(\omega) \tag{31.24}$$

Substituting this value for the design specification bound on the complementary sensitivity into the inequality (Equation 31.22) yields:

$$M_S(\omega) + |P(j\omega)| \frac{M_u(\omega)}{M_{do}(\omega)} \geq 1 \tag{31.25}$$

Inequality (Equation 31.25) requires the design specification on the sensitivity function to increase as the plant magnitude decreases relative to the available control authority for a given disturbance. Thus, the desire to reject disturbances (reflected by smaller $M_S(\omega)$) is limited. The tradeoff between disturbance rejection and control amplitude is quantified by Equation 31.25.

31.3.3 Relation Between Open-Loop and Closed-Loop Specifications

Because specifying one of the transfer functions $S(s)$, $T(s)$, or $L(s)$ completely determines the others, any one of them can be used as a basis for analysis and design. Classical "loop-shaping" design methods proceeded by directly manipulating the loop transfer function $L(s)$ (using, for example, lead and lag filters) to alter the feedback properties of the system. As is well known, these methods are quite effective in coping with the type of design problems for which they were developed. One reason for the success of these methods is that, for a scalar system, open-loop gain and phase can be readily related to feedback properties. Indeed, the following relations are well known (e.g., [2, 3]) and can readily be deduced from Equations 31.2 and 31.3:

$$|L(j\omega)| \gg 1 \quad \Leftrightarrow \quad \begin{cases} |S(j\omega)| \ll 1 \\ \text{and} \\ T(j\omega) \approx 1 \end{cases} \tag{31.26}$$

and

$$|L(j\omega)| \ll 1 \quad \Leftrightarrow \quad \begin{cases} S(j\omega) \approx 1 \\ \text{and} \\ |T(j\omega)| \ll 1 \end{cases} \tag{31.27}$$

At frequencies for which open-loop gain is approximately unity, feedback properties depend critically upon the value of open-loop phase:

$$\left.\begin{array}{c}|L(j\omega)| \approx 1 \\ \text{and} \\ \angle L(j\omega) \approx \pm 180^\circ\end{array}\right\} \Leftrightarrow \left\{\begin{array}{c}|S(j\omega)| \gg 1 \\ \text{and} \\ |T(j\omega)| \gg 1\end{array}\right. \quad (31.28)$$

The approximations in Equations 31.26 to 31.28 yield the following rules of thumb useful in design. First, large loop gain yields small sensitivity and good disturbance rejection properties, although noise appears directly in the system output. Second, small loop gain is required for small noise response and for robustness against large multiplicative uncertainty. Finally, at frequencies near gain crossover ($|L(j\omega)| \approx 1$), the phase of the system must remain bounded sufficiently far away from $\pm180°$ to provide an adequate stability margin and to prevent amplifying disturbances and noise.

It is also possible to relate open-loop gain to the magnitude of the plant input. From Equations 31.26 and 31.2, it follows that

$$|L(j\omega)| \gg 1 \Leftrightarrow |S(j\omega)C(j\omega)| \approx |P^{-1}(j\omega)| \quad (31.29)$$

Hence, requiring loop gain to be large at frequencies for which the plant gain is small may lead to unacceptably large response of the plant input to noise and disturbances (see also Equation 31.25 and [2]).

Recall the discussion of the requirements imposed on the magnitudes of the sensitivity and complementary sensitivity by the performance objectives Equations 31.19 and 31.20; (see Figure 31.2). From that discussion and the approximations given by Equations 31.26 to 31.28, it follows [3] that corresponding specifications upon open-loop gain and phase might appear as in Figure 31.3. These specifications reflect the fact that loop gains must be large at low frequencies for disturbance rejection and sensitivity reduction and must be small at high frequencies to provide stability robustness. At intermediate frequencies the phase of the system must remain bounded away from $\pm180°$ to prevent excessively large values of $|S(j\omega)|$ and $|T(j\omega)|$. To obtain the benefits of feedback over as large a frequency range as possible, it is also desirable that ω_L be close to ω_H. Of course, gain and phase must also satisfy the encirclement condition dictated by Nyquist's stability criteria.

31.4 Constraints Imposed by Stability

31.4.1 Introduction

Implicit in our construction of design specifications such as those illustrated by Figures 31.2 and 31.3 is the assumption that the transfer functions can be prescribed independently in different frequency regions or that open-loop gain and phase can be independently manipulated. However, the requirement that the closed-loop system be stable imposes additional constraints on the transfer functions that relate the values of the functions in different frequency regions. These constraints will be referred to as *analytic constraints* because they result from the requirement that the transfer functions have certain *analytic* properties.

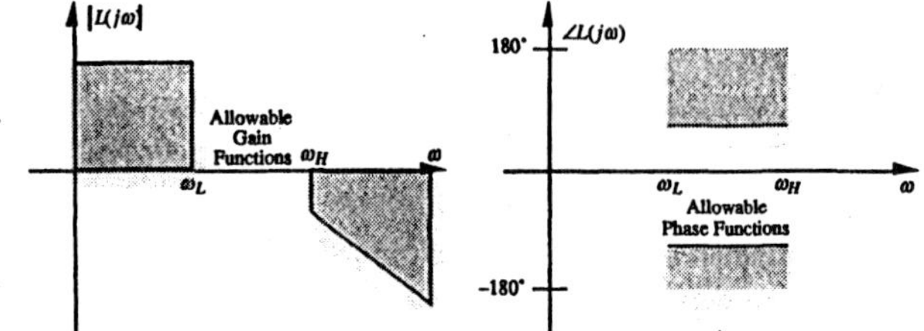

Figure 31.3 Gain and phase specifications.

As has been noted in Section 31.3, any of the transfer functions $S(s)$, $T(s)$, or $L(s)$ can be used to characterize the performance and properties of the feedback system. This section will present the constraints imposed by stability for each of these transfer functions.

31.4.2 Bode Gain-Phase Relation and Interpretations

We have shown in the previous section how classical control approaches view design specifications in terms of limits on the gain and phase of the open-loop transfer function. However, gain and phase are not mutually independent: the value of one is generally determined once the other is specified. There are many ways to state this relationship precisely; the one most useful for our purposes was derived by Bode [1]. This relation has been used by many authors (see [2, 4]) to analyze the implications that the gain-phase relation has upon feedback design.

THEOREM 31.1 *(Bode Gain-Phase Relation): Assume that $L(s)$ is a rational function with real coefficients and with no poles or zeros in the closed right-half plane. Then, at each frequency $s = j\omega$, the following integral relation must hold:*

$$\angle L(j\omega_0) - \angle L(0) = \frac{1}{\pi}\int_{-\infty}^{\infty} \frac{d\log|L|}{d\nu}\left[\log\coth\frac{|\nu|}{2}\right]d\nu \quad (31.30)$$

where

$$\nu = \log\left(\frac{\omega}{\omega_0}\right) \quad (31.31)$$

Theorem 2.5.1 states conditions ($L(s)$ rational, stable, and minimum phase) under which knowledge of open-loop gain along the $j\omega$-axis suffices to determine open-loop phase to within a factor of $\pm\pi$. Constraints analogous to Equation 31.30 hold for unstable or nonminimum phase systems. Hence, gain and phase cannot be manipulated independently in design.

Equation 31.30 shows that the phase of a transfer function is related to the slope (on a log-log scale) of the magnitude of the transfer function. The presence of the weighting function $\log\coth\frac{|\nu|}{2} = \log\left|\frac{\omega+\omega_0}{\omega-\omega_0}\right|$ shows that the dependence of $\angle L(j\omega_0)$ upon the rate of gain decrease at frequency ω diminishes rapidly as the distance between ω and ω_0 increases (see Figure 31.4). Hence this integral supports a rule of thumb stating that a $20N$ db/decade rate of gain decrease in the vicinity of frequency ω_0 implies that $\angle L(j\omega_0) \approx -90N°$. Most transfer functions are sufficiently well behaved that the value of phase at a frequency is largely determined by that of the gain over a decade-wide interval centered at the frequency of interest [11].

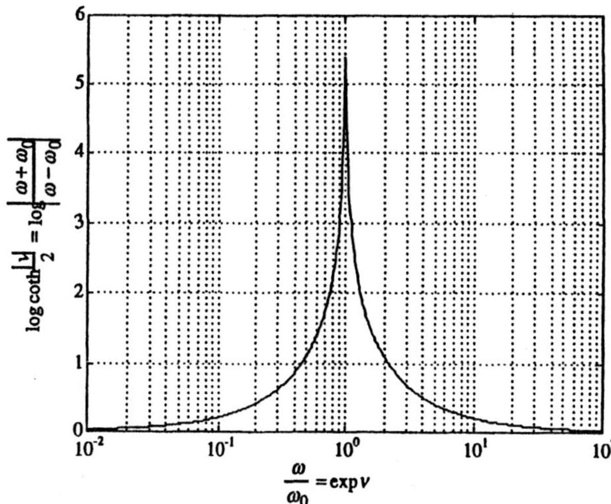

Figure 31.4 Weighting function in gain-phase integral.

The Bode gain-phase relation may be used to assess whether a design specification of the type shown in Figure 31.3 is achievable. Since a $20N$ db/decade rate of gain decrease in the vicinity of crossover implies that phase at crossover is roughly $-90N°$, it follows that the rate of gain decrease cannot be much greater than 20 db/decade if the Nyquist stability criterion is to be satisfied and if an acceptable phase margin is to be maintained. One implication of this fact is that the frequency ω_L in Figure 31.3 cannot be too close to the frequency ω_H. Hence, the frequency range over which loop gain can be large to obtain sensitivity reduction is limited by the need to ensure stability robustness against uncertainty at higher frequencies, and to maintain reasonable feedback properties near crossover. As discussed in [2] and [3], relaxing the assumption that $L(s)$ has no right-half plane poles or zeros does not lessen the severity of this tradeoff. Indeed, the tradeoff only becomes more difficult to accomplish. If one is willing to accept a system that is only conditionally stable, Horowitz [2] claims that larger values of low-frequency gain may be obtained.

31.4.3 The Bode Sensitivity Integral

The purpose of this section is to present and discuss the constraint imposed by stability on the sensitivity function. This constraint was first developed in the context of feedback systems in [1]. This integral quantifies a tradeoff between sensitivity reduction and sensitivity increase which must be performed whenever the open-loop transfer function has at least two more poles than zeros.

The magnitude of the sensitivity function of a scalar feedback system can be obtained easily using a Nyquist plot of $L(j\omega)$. Indeed, since $S(j\omega) = 1/[1 + L(j\omega)]$, the magnitude of the sensitivity function is just the reciprocal of the distance from the Nyquist plot to the critical point. In particular, sensitivity is less than one at frequencies for which $L(j\omega)$ is outside the unit circle centered at the critical point. Sensitivity is greater than one at frequencies for which $L(j\omega)$ is inside this unit circle.

To motivate existence of the integral constraint, consider the open-loop transfer function $L(s) = \frac{2}{(s+1)^2}$. As shown in Figure 31.5, there exists a frequency range over which the Nyquist plot of $L(j\omega)$ penetrates the unit circle and sensitivity is thus greater than one. In practice, the open-loop transfer function will generally have at least two more poles than zeros [2]. If $L(s)$ is stable then, using the gain-phase relation (Equation 31.30), it is straightforward to show that $L(j\omega)$ will asymptotically have phase lag at least $-180°$. Hence, there will always exist a frequency range over which sensitivity is greater than one. This behavior may be quantified using a classical theorem due to Bode [1], which was extended in [4] to allow unstable poles in the loop transfer function.

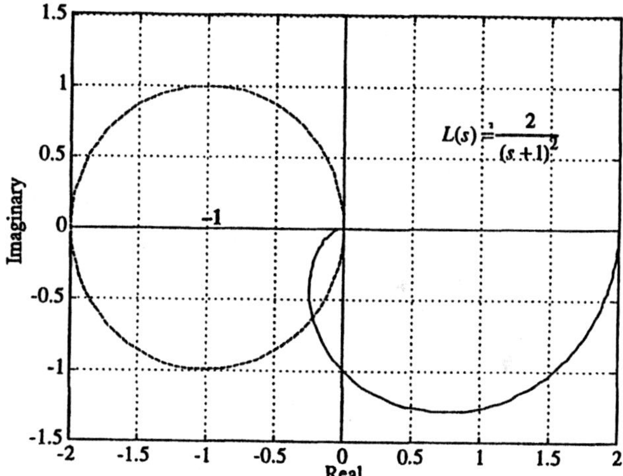

Figure 31.5 Effect of a two-pole rolloff upon the Nyquist plot.

THEOREM 31.2 (Bode Sensitivity Integral): Suppose that the open-loop transfer function $L(s)$ is rational and has right-half plane poles $\{p_1 : i = 1, \ldots, N_p\}$, with multiple poles included according to their multiplicity. If $L(s)$ has at least two more poles than zeros, and if the associated feedback system is stable, then the sensitivity function must satisfy[2]

$$\int_0^{\infty} \log |S(j\omega)| d\omega = \pi \sum_{i=1}^{N_p} Re[p_i] \tag{31.32}$$

[2]Throughout this chapter, the function log(*) will be used to denote the natural logarithm.

This theorem shows that a tradeoff exists between sensitivity properties in different frequency ranges. Indeed, for stable open-loop systems, the area of sensitivity reduction must equal the area of sensitivity increase on a plot of the logarithm of the sensitivity versus linear frequency (see Figure 31.6). In this respect, the benefits and costs of feedback are balanced exactly.

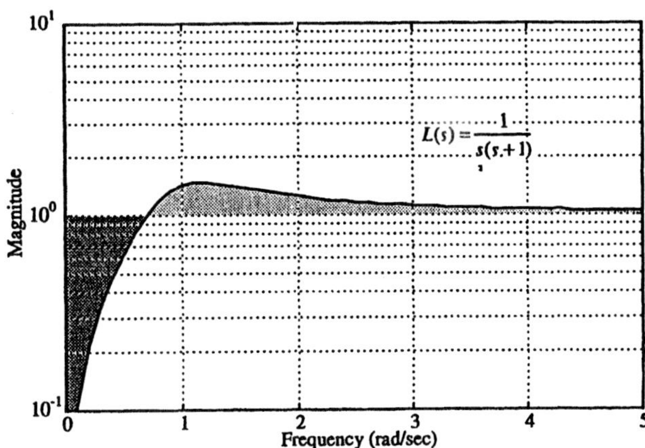

Figure 31.6 Areas of sensitivity reduction (dark gray) and sensitivity increase (light gray).

The extension of Bode's theorem to open-loop unstable systems shows that the area of sensitivity increase exceeds that of sensitivity reduction by an amount proportional to the distance from the unstable poles to the left-half plane. A little reflection reveals that this additional sensitivity increase is plausible for the following reason. When the system is open-loop unstable, then it is obviously necessary to use feedback to achieve closed-loop stability, as well as to obtain sensitivity reduction. One might expect that this additional benefit of feedback would be accompanied by a certain cost, and the integral (Equation 31.32) substantiates that hypothesis. Alternatively, we could interpret Equation 31.32 as implying that the area of sensitivity reduction must be less than that of sensitivity increase, thus indicating that a portion of the open-loop gain which could otherwise contribute to sensitivity reduction must instead be used to pull the unstable poles into the left-half plane.

By itself, the tradeoff quantified by Equation 31.32 does not impose a meaningful design limitation. Although it is true that requiring a large area of sensitivity reduction over a low-frequency interval implies that an equally large area of sensitivity increase must be present at higher frequencies, it does not follow that there must exist a peak in sensitivity which is bounded greater than one. It is possible to achieve an arbitrary large area of sensitivity increase by requiring $|S(j\omega)| = 1 + \delta, \forall \omega \in [\omega_1, \omega_2]$, where δ can be chosen arbitrarily small and the interval (ω_1, ω_2) is adjusted to be sufficiently large.

The analysis in the preceding paragraph ignores the effect of limitations upon system bandwidth that are always present in a practical design. For example, it is almost always necessary to decrease open-loop gain at high frequencies to maintain stability robustness against large modeling errors due to unmodeled dynamics. Small open-loop gain is also required to prevent sensor noise from appearing at the system output. Finally, requiring open-loop gain to be large at a frequency for which plant gain is small may lead to unacceptably large response of the plant input to noise and disturbances. Hence the natural bandwidth of the plant also imposes a limitation upon open-loop bandwidth.

One or more of the bandwidth constraints just cited is usually present in any practical design. It is reasonable, therefore, to assume that open-loop gain must satisfy a frequency-dependent bound of the form

$$|L(j\omega)| \leq \varepsilon \left(\frac{\omega_c}{\omega}\right)^{1+k} \quad \forall \omega \geq \omega_c \tag{31.33}$$

where $\varepsilon < 1/2$ and $k > 0$. This bound imposes a constraint upon the rate at which loop gain rolls off, as well as the frequency at which rolloff commences and the level of gain at that frequency.

When a bandwidth constraint such as Equation 31.33 is imposed, it is obviously not possible to require the sensitivity function to exceed one over an arbitrarily large frequency interval. When Equation 31.33 is satisfied, there is an upper bound on the area of sensitivity increase which can be present at frequencies greater than ω_c. The corresponding limitation imposed by the sensitivity integral (Equation 31.32) and the rolloff constraint (Equation 31.33) is expressed by the following result [5].

COROLLARY 31.1 Suppose, in addition to the assumptions of Theorem 31.2, that $L(s)$ satisfies the bandwidth constraint (Equation 31.33). Then the tail of the sensitivity integral must satisfy

$$\left| \int_{\omega_c}^{\infty} \log |S(j\omega)| d\omega \right| \leq \frac{3\varepsilon\omega_c}{2k} \tag{31.34}$$

The bound defined by Equation 31.34 implies that the sensitivity tradeoff imposed by the integral (Equation 31.32) must be accomplished primarily over a finite frequency interval. As a consequence, the amount by which $|S(j\omega)|$ must exceed one cannot be arbitrarily small. Suppose that the sensitivity function is required to satisfy the upper bound

$$|S(j\omega)| \leq \alpha < 1 \quad \forall \omega \leq \omega_l < \omega_c \tag{31.35}$$

If the bandwidth constraint (Equation 31.33) and the sensitivity bound (Equation 31.35) are both satisfied, then the integral constraint (Equation 31.32) may be manipulated to show [5] that

$$\sup_{\omega \in (\omega_l, \omega_c)} \log |S(j\omega)| \geq \frac{1}{\omega_c - \omega_l} \left\{ \pi \sum_{i=1}^{N_p} \operatorname{Re}[p_i] + \omega_l \cdot \log\left(\frac{1}{\alpha}\right) - \frac{3\varepsilon\omega_c}{2k} \right\} \tag{31.36}$$

The bound given in Equation 31.36 shows that increasing the area of low-frequency sensitivity reduction by requiring α to be very small or ω_l to be very close to ω_c, will necessarily cause a large peak in sensitivity at frequencies between ω_l and ω_c. Hence the integral constraint (Equation 31.32), together with the bandwidth constraint (Equation 31.33) imposes a tradeoff between sensitivity reduction and sensitivity increase, which must be accounted for in design.

It may be desirable to impose a bandwidth constraint such as Equation 31.33 directly on the complementary sensitivity function $T(s)$, since $T(s)$ directly expresses the feedback properties of sensor noise response and stability robustness, while $L(s)$ does so only indirectly. Analogous results to those stated in Corollary 31.1 can be obtained in this case.

31.4.4 Complementary Sensitivity Integral

The complementary sensitivity function is constrained by the stability requirement for the closed-loop system in a manner analogous to the sensitivity function.

THEOREM 31.3 *(Complementary Sensitivity Integral [6]): Suppose that the open-loop transfer function $L(s)$ is given by the product of a rational transfer function and a delay element:*

$$L(s) = \tilde{L}(s)e^{-s\tau} \tag{31.37}$$

$\tilde{L}(s)$ is assumed to be rational with right-half plane zeros $\{z_i : i = 1, \ldots, N_z\}$ (with multiple zeros included according to their multiplicity). If $L(s)$ has at least one pole at the origin (i.e., one integrator), and if the associated feedback system is stable, then the complementary sensitivity function must satisfy

$$\int_0^\infty \log|T(j\omega)| d\left(\frac{1}{\omega}\right) = \pi \sum_{i=1}^{N_z} Re\left[\frac{1}{z_i}\right] + \frac{\pi}{2}\tau - \frac{\pi}{2}K_v^{-1} \tag{31.38}$$

where K_v is the velocity constant of the system:

$$K_v = \lim_{s \to 0} sL(s) \tag{31.39}$$

The complementary sensitivity integral (Equation 31.38) has a similar interpretation to the Bode sensitivity integral. Recall that the complementary sensitivity function characterizes the response of the system to sensor noise and the robustness of the system to high-frequency model errors. Theorem 31.3 states that, if the loop transfer function is minimum phase (i.e., it has no right-half plane zeros, and no delay) and is a type II system, then the area of amplified sensor noise response must equal the area of attenuated sensor noise response. In the case of the complementary sensitivity function, the areas are computed with respect to an inverse frequency scale (see Figure 31.7). The presence of nonminimum phase zeros or delays worsens the tradeoff (i.e., increases the required area of noise amplification). For type I systems, the tradeoff is improved by the term involving the velocity constant on the right side of Equation 31.39.

As for the sensitivity integral, the complementary sensitivity integral does not imply that the peak in the complementary sensitivity transfer function must be large. It is possible to accommodate the required increase in the magnitude of the complementary sensitivity function by allowing it to be only slightly greater

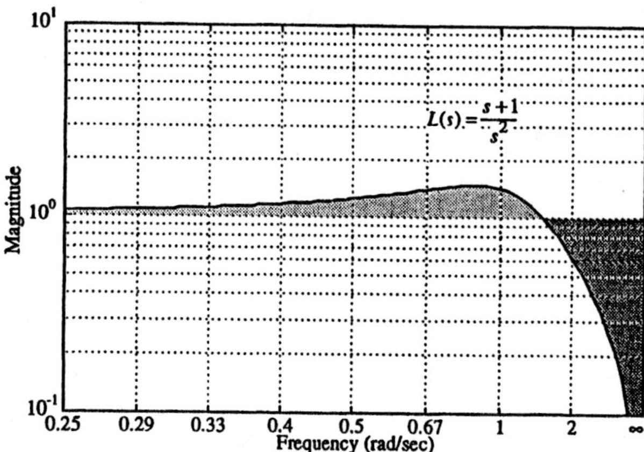

Figure 31.7 Areas of complementary sensitivity increase (light gray) and complementary sensitivity decrease (dark gray).

than one over a large interval of low frequencies (since the area is computed with respect to inverse frequency). However, when combined with tracking requirements imposed at low frequencies (analogous to the rolloff requirement of Equation 31.33), the integral (Equation 31.38) can be used to develop a lower bound on the peak of the complementary sensitivity function.

Assume that the open-loop transfer function satisfies

$$|L(j\omega)| \geq \delta\left(\frac{\omega_p}{\omega}\right)^{1+k} \quad \forall \omega \leq \omega_p \tag{31.40}$$

where $\delta > 2$ and $k > 0$. This bound imposes a constraint upon the tracking performance of the system.

When a performance constraint such as Equation 31.40 is imposed, it is obviously not possible to require the complementary sensitivity function to exceed one over an arbitrarily large inverse frequency interval. When Equation 31.40 is satisfied, there is an upper bound on the area of complementary sensitivity increase which can be present at frequencies less than ω_p. The corresponding limitation imposed by the complementary sensitivity integral (Equation 31.38) and the rolloff constraint (Equation 31.40) is expressed by the following result.

COROLLARY 31.2 Suppose, in addition to the assumptions of Theorem 31.3, that $L(s)$ satisfies the performance constraint (Equation 31.40). Then the low-frequency tail of the complementary sensitivity integral must satisfy

$$\left|\int_0^{\omega_p} \log|T(j\omega)| d\omega\right| \leq \frac{3}{2k\delta\omega_p} \tag{31.41}$$

The bound given by Equation 31.41 implies that the complementary sensitivity tradeoff imposed by the integral (Equation 31.38) must be accomplished primarily over a finite inverse frequency interval. As a consequence, the amount by which $|T(j\omega)|$ must exceed one cannot be arbitrarily small. Suppose that the complementary sensitivity function is required to satisfy the upper bound

$$|T(j\omega)| \leq \alpha < 1 \quad \forall \omega \geq \omega_h > \omega_p \tag{31.42}$$

If the performance constraint (Equation 31.40) and the complementary sensitivity bound (Equation 31.42) are both satisfied, then the integral constraint (Equation 31.38) may be manipulated to show that

$$\sup_{\omega\in(\omega_p,\omega_h)} \log|T(j\omega)| \geq \frac{1}{\frac{1}{\omega_p}-\frac{1}{\omega_h}}$$
$$\left\{\pi\sum_{i=1}^{N_z} \text{Re}\left[\frac{1}{z_i}\right]+\frac{1}{\omega_h}\cdot\log\left(\frac{1}{\alpha}\right)-\frac{3}{2k\delta\omega_p}\right\} \tag{31.43}$$

The bound given by Equation 31.43 shows that increasing the area of high-frequency complementary sensitivity reduction by requiring α to be very small or ω_h to be very close to ω_p will necessarily cause a large peak in sensitivity at frequencies between ω_p and ω_h. Hence, the integral constraint (Equation 31.38), together with the performance constraint (Equation 31.40), imposes a tradeoff between complementary sensitivity reduction and complementary sensitivity increase, which must be accounted for in design.

31.5 Limitations Imposed by Right Half-Plane Poles and Zeros

31.5.1 Introduction

As discussed in Section 31.2, design specifications are often stated in terms of frequency-dependent bounds on the magnitude of closed-loop transfer functions. It has long been known that control system design is more difficult for nonminimum phase or unstable systems. The sensitivity and complementary sensitivity integrals presented in Section 31.4 indicated that nonminimum phase zeros and unstable poles could worsen the individual design tradeoffs. In fact, right half-plane poles and zeros impose additional constraints upon the control system design. This section examines these limitations in detail.

31.5.2 Limitations for Nonminimum Phase Systems

Suppose that the plant possesses zeros in the open right half-plane. Then the internal stability requirement dictates that these zeros also appear, with at least the same multiplicity, in the open-loop transfer function $L(s) = P(s)C(s)$. Let the set of all open right half-plane zeros of $L(s)$ (including any present in the compensator) be denoted by

$$\mathbf{Z} \equiv \{z_i : i = 1,\ldots,N_z\} \tag{31.44}$$

Defining the Blaschke product (all-pass filter)

$$B_z(s) \equiv \prod_{i=1}^{N_z} \frac{z_i-s}{z_i+s} \tag{31.45}$$

we can factor the open-loop transfer function into the form

$$L(s) = L_m(s)\prod_{i=1}^{N_z} \frac{z_i-s}{z_i+s} \tag{31.46}$$

where $L_m(s)$ has no zeros in the open right-half plane. Note that

$$|L(j\omega)| = |L_m(j\omega)| \quad \forall\omega \tag{31.47}$$

and

$$\angle\frac{z_i-j\omega}{z_i+j\omega} \to -180^\circ \quad \text{as} \quad \omega\to\infty \tag{31.48}$$

These facts show that open right half-plane zeros contribute additional phase lag without changing the gain of the system (hence the term "nonminimum phase zero"). The effect that this additional lag has upon feedback properties can best be illustrated using a simple example.

Consider the nonminimum phase plant $P(s) = \frac{1}{s+1}\frac{1-s}{1+s}$ and its minimum phase counterpart $P_m(s) = \frac{1}{s+1}$. Figure 31.8 shows that the additional phase lag contributed by the zero at $s = 1$ causes the Nyquist plot to penetrate the unit circle and the sensitivity to be larger than one. Experiments with various compensation schemes reveal that using large loop gain over some frequency range to obtain small sensitivity in that range tends to cause sensitivity to be large at other frequencies.

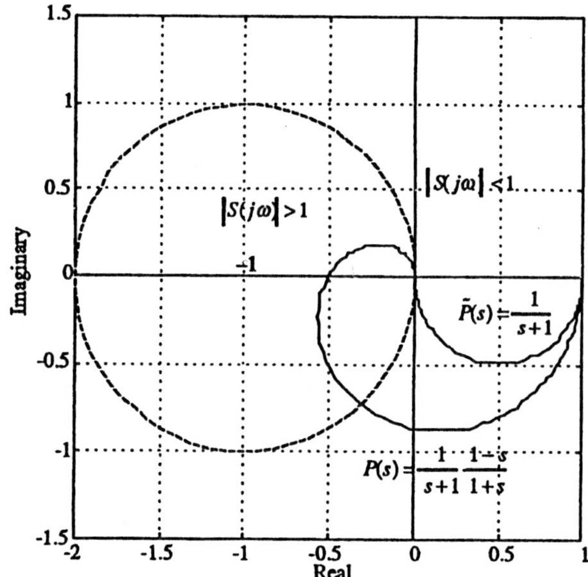

Figure 31.8 Additional phase lag contributed by a nonminimum phase zero.

Assume that the open-loop transfer function can be factored as

$$L(s) = L_0(s)B_z(s)B_p^{-1}(s)e^{-s\tau} \tag{31.49}$$

where $\tau \geq 0$ represents a possible time delay, $L_0(s)$ is a proper rational function with no poles or zeros in the open right plane, and $B_z(s)$ is the Blaschke product (Equation 31.45) containing the open right half-plane zeros of the plant plus those of the compensator. The Blaschke product

$$B_p(s) = \prod_{i=1}^{N_p} \frac{p_i - s}{p_i + s} \tag{31.50}$$

contains all the poles of both plant and compensator in the open right half-plane, again counted according to multiplicity. We emphasize once more that internal stability requirements dictate that all right-half plane poles and zeros of the plant must appear with at least the same multiplicity in $L(s)$ and hence cannot be canceled by right-half plane zeros and poles of the compensator.

One constraint that right half-plane zeros impose upon the sensitivity function is immediately obvious from the definition $S(j\omega) = 1/[1 + L(j\omega)]$. Suppose that $L(s)$ has a zero at $s = z$. It follows that

$$S(z) = 1 \tag{31.51}$$

Poles of $L(s)$ also constrain the sensitivity function. If $L(s)$ has a pole at $s = p$, then

$$S(p) = 0 \tag{31.52}$$

From Equations 31.51 and 31.52, it is clear that if the plant (and thus $L(s)$) has zeros or poles at points of the open right-half plane, then the value of the sensitivity function is constrained at those points. Naturally, the value of sensitivity along the $j\omega$-axis, where the design specifications are imposed and the conjectured tradeoff must take place, is of more concern.

THEOREM 31.4 *Suppose that the open-loop transfer function $L(s)$ has a zero, $z = x + jy$, with $x = 0$. Assume that the associated feedback system is stable. Then the sensitivity function must satisfy*

$$\int_0^\infty \log|S(j\omega)| W(z,\omega) d\omega = \pi \log\left|B_p^{-1}(z)\right| \tag{31.53}$$

where $W(z, \omega)$ is a weighting function. For a real zero, $z = x$,

$$W(x,\omega) = \frac{2x}{x^2 + \omega^2} \tag{31.54}$$

and, for a complex zero, $z = x + jy$,

$$W(z,\omega) = \frac{x}{x^2 + (y-\omega)^2} + \frac{x}{x^2 + (y+\omega)^2} \tag{31.55}$$

A number of remarks about this theorem are in order. First, as discussed in [4], the integral relations are valid even if $S(s)$ has zeros (or poles) on the $j\omega$-axis. Second, a zero z with multiplicity $m > 1$ imposes additional interpolation constraints on the first $m - 1$ derivatives of $\log S(s)$ evaluated at the zero. These interpolation constraints also have equivalent statements as integral relations [4].

We now show that Equation 31.53 imposes a sensitivity tradeoff. To see this, note first that the weighting function satisfies $W(z, \omega) > 0, \forall\omega$, and that the Blaschke product satisfies $\log\left|B_p^{-1}(z)\right| \geq 0$. Using these facts, it follows easily from Equation 31.53 that requiring sensitivity reduction ($\log|S(j\omega)| < 0$) over some frequency range implies that there must be sensitivity amplification ($\log|S(j\omega)| > 0$) at other frequencies. Hence, if the plant is nonminimum phase, one cannot use feedback to obtain the benefits of sensitivity reduction over one frequency range unless one is willing to pay the attendant price in terms of increased sensitivity elsewhere.

Theorem 31.4 verifies the conjecture that a sensitivity tradeoff is present whenever a system is nonminimum phase. Recall it was also conjectured that the severity of the tradeoff is a function of the phase lag contributed by the zero at frequencies for which sensitivity reduction is desired. This conjecture can be verified by using the form of the weighting function $W(z, \omega)$ (Equations 31.54 and 31.55).

Consider first the case of a real zero $z = x$. Equation 31.46 shows that, as a function of frequency, the additional phase lag contributed by this zero is

$$\theta(x,\omega) \equiv \angle\frac{x - j\omega}{x + j\omega} \tag{31.56}$$

Noting that

$$\frac{d\theta(x,\omega)}{d\omega} = \frac{-2x}{x^2 + \omega^2} \tag{31.57}$$

it follows that the weighting function in Equation 31.54 satisfies

$$W(x,\omega) = -\frac{d\theta(x,\omega)}{d\omega} \tag{31.58}$$

Hence, the weighting function appearing in the sensitivity constraint is equal to (minus) the rate at which the phase lag due to the zero increases with frequency.

One can use the weighting function (Equation 31.58) to compute the weighted length of a frequency interval. Note that sensitivity reduction is typically required over a low-frequency interval $\Omega \equiv [0, \omega_1]$ and that the weighted length of such an interval equals

$$\begin{aligned} W(x,\Omega) &\equiv \int_0^{\omega_1} W(x,\omega) d\omega \\ &= -\theta(x,\omega_1) \end{aligned} \tag{31.59}$$

Hence, the weighted length of the interval is equal to (minus) the phase lag contributed by the zero at the upper endpoint of the interval. It follows that, as $\omega_1 \to \infty$, the weighted length of the $j\omega$-axis equals π.

For a complex zero, the weighting function (Equation 31.55) is equal to $(-1/2)$ the sum of the additional phase lag contributed by the zero and by its complex conjugate:

$$W(z,\omega) = -\frac{1}{2}\left[\frac{d\theta(z,\omega)}{d\omega} + \frac{d\theta(\bar{z},\omega)}{d\omega}\right] \tag{31.60}$$

Hence, the weighted length of the frequency interval $\Omega \equiv [0, \omega_1]$ is

$$W(z,\omega) = -\frac{1}{2}\left[\theta(z,\omega_1) + \theta(\bar{z},\omega_1)\right] \tag{31.61}$$

As we have already remarked, the integral constraint (Equation 31.53) implies that a tradeoff exists between sensitivity reduction and sensitivity increase in different frequency ranges. An interesting interpretation of this tradeoff is available using the weighting function. Suppose first that $L(s)$ has no poles in the open right-half plane. Then the integral constraint is

$$\int_0^\infty \log|S(j\omega)| W(x,\omega) d\omega = 0 \tag{31.62}$$

Equation 31.62 states that the weighted area of sensitivity increase must equal the weighted area of sensitivity reduction. Since the weighted length of the $j\omega$-axis is finite, it follows that the amount by which sensitivity must exceed one at higher frequencies cannot be made arbitrarily small.

If the open-loop system has poles in the open right-half plane, then the weighted area of sensitivity increase must exceed that of sensitivity reduction. In particular,

$$\log\left|B_p^{-1}(z)\right| = \sum_{i=1}^{N_p} \log\left|\frac{\bar{p}_i + z}{p_i - z}\right| \tag{31.63}$$

The right side of Equation 31.63 is always greater than zero, and becomes large whenever the zero z approaches the value of one of the unstable poles p_i. It follows (unsurprisingly) that systems with approximate pole-zero cancellations in the open right-half plane can have especially bad sensitivity properties.

We can use the integral constraint (Equation 31.53) to obtain some simple lower bounds on the size of the peak in sensitivity accompanying a given level of sensitivity reduction over a low-frequency interval. Bounds of this type were first discussed by Francis and Zames ([14], Theorem 3). The results presented here will show how the relative location of the zero to the interval of sensitivity reduction influences the size of the peak in sensitivity outside that interval.

Suppose that the sensitivity function is required to satisfy the upper bound

$$|S(s)| \leq \alpha < 1 \quad \forall \omega \in \Omega \tag{31.64}$$

where $\Omega = [0, \omega_1]$ is a low-frequency interval of interest. Define the infinity norm of the sensitivity function:

$$\|S\|_\infty \equiv \sup_{\omega \geq 0} |S(j\omega)| \tag{31.65}$$

Assuming that the upper bound (Equation 31.64) is satisfied, the integral (Equation 31.53) can be used to compute a lower bound on $\|S\|_\infty$ for each nonminimum phase zero of $L(s)$.

COROLLARY 31.3 Suppose that the conditions in Theorem 31.3 are satisfied and that the sensitivity function is bounded as in Equation 31.64. Then the following lower bound must be satisfied at each nonminimum phase zero of $L(s)$:

$$\|S\|_\infty \geq \left(\frac{1}{\alpha}\right)^{\frac{W(z,\Omega)}{\pi - W(z,\Omega)}} \left|B_p^{-1}(z)\right|^{\frac{\pi}{\pi - W(z,\Omega)}} \tag{31.66}$$

The bound defined by Equation 31.66 shows that if the sensitivity is required to be very small over the interval $(0, \omega_1)$ then there necessarily exists a large peak in sensitivity outside this interval. Furthermore, the smallest possible size of this peak will become larger if the open-loop system has unstable poles near any zero.

The size of the sensitivity peak also depends upon the location of the interval $(0, \omega_1)$ relative to the zero. Assume for simplicity that the system is open-loop unstable and the zero is real. Then

$$\|S\|_\infty \geq \left(\frac{1}{\alpha}\right)^{\frac{W(z,\Omega)}{\pi - W(z,\Omega)}} \tag{31.67}$$

Recall that the weighted length of the interval $\Omega = [0, \omega_1]$ is just equal to (minus) the phase lag contributed by the zero at the upper endpoint of that interval. Since the zero eventually contributes 180° phase lag, it follows that as $\omega_1 \to \infty$, $W(x, \Omega) \to \pi$. Thus, the exponent in Equation 31.67 becomes unbounded and, since $\alpha < 1$, so does the peak in sensitivity. To summarize, requiring sensitivity to be small throughout a frequency range extending into the region where the nonminimum phase zero contributes a significant amount of phase lag implies that there will necessarily exist a large peak in sensitivity at higher frequencies. On the other hand, if the zero is located so that it contributes only a negligible amount of phase lag at frequencies for which sensitivity reduction is desired, then it does not impose a serious limitation upon sensitivity properties of the system. Analogous results hold, with appropriate modifications, for a complex zero.

Suppose now that the open-loop system has poles in the open right-half plane. It is interesting to note that, in this case, the bound (Equation 31.66) implies the existence of a peak in sensitivity even if no sensitivity reduction is present!

Recall next the approximation given by Equation 31.26 which shows that small sensitivity can be obtained only by requiring open-loop gain to be large. It is easy to show that $|S(j\omega)| \leq \alpha < 1$ implies that $|L(j\omega)| \geq /\alpha - 1$. Inequality (Equation 31.67) implies that, to prevent poor feedback properties, open-loop gain should not be large over a frequency interval extending into the region for which a nonminimum phase zero contributes significant phase lag. This observation substantiates a classical design rule of thumb: loop gain must be rolled off before the phase lag contributed by the zero becomes significant.

However, if one is willing and able to adopt some nonstandard design strategies (such as having multiple gain crossover frequencies) then [15] it is possible to manipulate the design tradeoff imposed by a nonminimum phase zero to obtain some benefits of large loop gain at higher frequencies. One drawback of these strategies is that loop gain must be small, and hence the benefits of feedback must be lost, over an intermediate frequency range.

31.5.3 Limitations for Unstable Systems

We show in this section that unstable poles impose constraints upon the complementary sensitivity function which, loosely speaking, are dual to those imposed upon the sensitivity function by nonminimum phase zeros. That such constraints exist might be conjectured from the existence of the interpolation constraint (Equation 31.52) and the algebraic identity (Equation 31.21). Together, these equations show that if $L(s)$ has a pole $s = p$, then the complementary sensitivity function satisfies

$$T(p) = 1 \tag{31.68}$$

Furthermore, if $L(s)$ has a zero at $s = z$, then

$$T(z) = 0 \tag{31.69}$$

The previous results for the sensitivity function, together with the fact that $T(s)$ is constrained to equal one at open right half-plane poles of $L(s)$, suggests that similar constraints might exist for $|T(j\omega)|$ due to the presence of such poles. It is also possible to motivate the presence of the integral constraint on $|T(j\omega)|$ using an argument based upon the inverse Nyquist plot [16] and the fact that $|T(j\omega)| > 1$ whenever $L^{-1}(j\omega)$ is inside the unit circle centered at the critical point.

As in Section 31.5.2, it is assumed that $L(s)$ has the form of Equation 31.49. The following theorem states the integral constraint on the complementary sensitivity function due to unstable poles.

THEOREM 31.5 *Suppose that the open-loop transfer function has a pole, $p = x + jy$, with $x > 0$. Assume that the associated feedback system is stable. Then the complementary sensitivity function must satisfy*

$$\int_0^{\infty} \log|T(j\omega)| W(p, \omega) d\omega = \pi \log \left| B_z^{-1}(p) \right| + \pi x \tau \tag{31.70}$$

where $W(p, \omega)$ is a weighting function. For a real pole, $p = x$,

$$W(x, \omega) = \frac{2x}{x^2 + \omega^2} \tag{31.71}$$

and, for a complex pole, $p = x + jy$

$$W(p, \omega) = \frac{x}{x^2 + (y - \omega)^2} + \frac{x}{x^2 + (y + \omega)^2} \tag{31.72}$$

Remarks analogous to those following Theorem 31.4 apply to this result also. The integral relations are valid even if $T(s)$ has zeros on the $jw - axis$, and there are additional constraints on the derivative of $\log T(s)$ at poles with multiplicity greater than one.

The integral (Equation 31.70) shows that there exists a tradeoff between sensor noise response properties in different frequency ranges whenever the system is open-loop unstable. Since $|T(j\omega)|$ is the reciprocal of the stability margin against multiplicative uncertainty, it follows that a tradeoff between stability robustness properties in different frequency ranges also exists. Using analysis methods similar to those in the preceding section, one can derive a lower bound on the peak in the complementary sensitivity function present whenever $|T(j\omega)|$ is required to be small over some frequency interval. One difference is that $|T(j\omega)|$ is generally required to be small over a high, rather than a low, frequency range.

It is interesting that time delays worsen the tradeoff upon sensor noise reduction imposed by unstable poles. This is plausible for the following reason. Use of feedback around an open-loop unstable system is necessary to achieve stability. Time delays, as well as nonminimum phase zeros, impede the processing of information around a feedback loop. Hence, it is reasonable to expect that design tradeoffs due to unstable poles are exacerbated when time delays or nonminimum phase zeros are present. This interpretation is substantiated by the fact that the term due to the time delay in Equation 31.70 is proportional to the product of the length of the time delay and the distance from the unstable pole to the left half-plane.

31.5.4 Summary

Nonminimum phase or unstable systems impose additional tradeoffs for control system design. Nonminimum phase zeros limit the frequency range over which control system performance can be achieved, while unstable poles require active control over certain frequency ranges and reduce the overall performance that can be achieved. Quantitative expressions of these tradeoffs are given by the integral constraints of Theorems 31.4 and 31.5. These constraints can be used together with bounds on the desired performance to compute approximations that provide useful insight into the design tradeoffs.

31.6 Summary and Conclusions

In this chapter, we have discussed design limitations and tradeoffs present in feedback design problems. It is important that working control engineers have a firm knowledge of these tradeoffs. On the one hand, such knowledge provides insight into how a feedback design problem should be approached, and into when a control design is achieving a reasonable compromise between conflicting design goals. On the other hand, if a reasonable compromise cannot be achieved for a particular system, then knowledge of design tradeoffs may be used to modify the plant, to improve sensors and actuators, and to develop better models so that a tractable design problem is obtained.

References

[1] Bode, H.W., *Network Analysis and Feedback Amplifier Design*, Van Nostrand, Princeton, NJ, 1945.

[2] Horowitz, I.M., *Synthesis of Feedback Systems*, Academic Press, New York, 1963.

[3] Doyle, J.C. and Stein, G., Multivariable feedback design: concepts for a classical/modern synthesis, *IEEE Trans. Autom. Control*, 26, 1981.

[4] Freudenberg, J.S. and Looze, D.P., Right half plane poles and zeros, and design tradeoffs in feedback systems, *IEEE Trans. Autom. Control*, 30, 1985.

[5] Freudenberg, J.S. and Looze, D.P., *Frequency Domain Properties of Scalar and Multivariable Feedback Systems*, Springer-Verlag, Berlin, 1988.

[6] Middleton, R.H. and Goodwin, G.C., *Digital Control and Estimation: A Unified Approach*, Prentice Hall, Englewood Cliffs, NJ, 1990.

[7] Kwakernaak, H., Robustness optimization of linear feedback systems, *Proc. 22nd IEEE Conference on Decision and Control,* San Antonio, TX, 1983.

[8] Kwakernaak, H., Minimax frequency domain performance and robustness optimization of linear feedback systems, *IEEE Trans. Autom. Control,* 30, 1985.

[9] Doyle, J.C., Francis, B.A., and Tannenbaum, A.R., *Feedback Control Theory,* Macmillan, New York, 1992.

[10] Cruz, J.B. and Perkins, W.R., A new approach to the sensitivity problem in multivariable feedback design, *IEEE Trans. Autom. Control,* 9, 1964.

[11] Franklin, G.F., Powell, J.D., and Emami-Naeini, A., *Feedback Control of Dynamic Systems,* Addison-Wesley, Reading, MA, 1991.

[12] Dorf, R.C., *Modern Control Systems,* Addison-Wesley, Reading, MA, 1992.

[13] Doyle, J.C., Wall, J.E., Jr., and Stein, G., Performance and robustness analysis for structured uncertainty, *Proc. 21st IEEE Conference on Decision and Control,* Orlando, FL, 1982.

[14] Francis, B.A. and Zames, G., On optimal sensitivity theory for SISO feedback systems, *IEEE Trans. Autom. Control,* 29, 1984.

[15] Horowitz, I.M. and Liao, Y.-K., Limitations of non-minimium phase feedback systems, *Int. J. Control,* 40(5), 1984.

[16] Rosenbrock, H.H., *Computer-Aided Control System Design,* Academic Press, London, 1974.

32

Modeling Deterministic Uncertainty

Jörg Raisch
Institut für Systemdynamik und Regelungstechnik, Universität Stuttgart, Stuttgart, FR Germany

Bruce A. Francis
Department of Electrical and Computer Engineering, University of Toronto, Toronto, Ontario, Canada

32.1 Introduction

At first glance, the notion of deterministic uncertainty seems to be a contradiction in terms; after all, the word "deterministic" is often used to signify the absence of any form of uncertainty. We will explain later on that, properly interpreted, this choice of words does indeed make sense. For the moment, we concentrate on the notion of uncertainty.

Uncertainty in the control context comes in two basic versions: uncertain signals and uncertainty in the way the plant maps input signals into output signals ("plant uncertainty").

Most processes we wish to control are subject to influences from their environment: some of them known (measured disturbances, reference inputs); others, uncertain signals (unmeasured disturbances, noise corrupting measurements). All these signals are labelled "external," because they originate in the "outside world[1]."

A plant model, by definition, is a simplified representation of reality and is usually geared towards a specific purpose. Models that are meant to be used for feedback controller design tend to be especially crude. This is because (1) most popular design techniques can handle only very restricted classes of models and (2) in a feedback configuration, one can potentially get away with more inaccurate models than in applications that are based on a pure feedforward structure.

Meaningful analysis and design, however, are possible only if signal and plant uncertainty are, in some sense, "limited." In other words, we have to assume that there exists some, however incomplete, knowledge about signal and plant uncertainty; we need an *uncertainty model*. Such an uncertainty model defines an admissible *set of plant models*, $\mathcal{G}$, and an admissible *set of uncertain external input signals*, $\mathcal{W}$ (Figure 32.1).

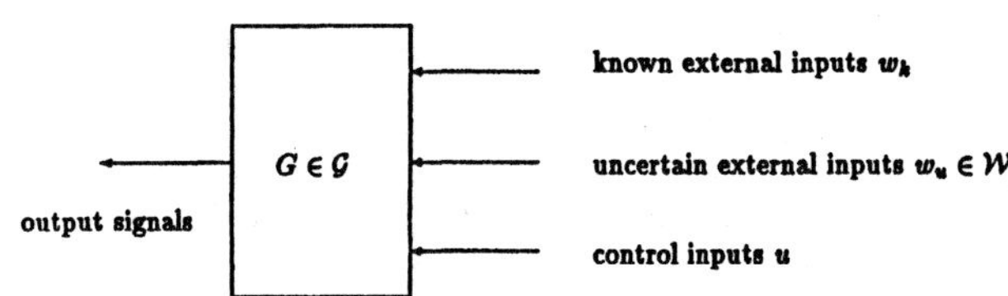

Figure 32.1 Signal and plant uncertainty models.

The adjective in "deterministic uncertainty model" points to the fact that we do not attempt to assign probabilities (or probability densities) to the elements of the sets $\mathcal{G}$ and $\mathcal{W}$; every element is considered to be as "good" as any other one. Based on this, one can ask the key (robustness) questions in control systems analysis: Is closed-loop stability guaranteed for every plant model in $\mathcal{G}$? Do desired closed-loop performance properties hold for every external input in $\mathcal{W}$ and every $G \in \mathcal{G}$?

One has to keep in mind, however, that no class of mathematical models is able to describe every detail of reality; i.e., the physical plant cannot be an element in $\mathcal{G}$. Robustness of a desired closed-loop property with respect to any specific plant uncertainty model does not therefore guarantee that the *real* control system will also have this property. If the uncertainty model is chosen in a sensible way (that's what this chapter is about), it will, however, increase the likelihood for the real system to "function properly."

We will first discuss signal uncertainty. Then we will summarize the most common plant uncertainty models. Finally, we will briefly mention the topic of model validation; i.e., whether a given set of experimental data is compatible with given uncertainty models $\mathcal{G}$ and $\mathcal{W}$. We will work in a continuous-time

[1] Control inputs, on the other hand, are generated within the control loop, and are called "internal input signals."

0-8493-8570-9/96/$0.00+$.50

framework, and all signals will be real and (except for some of the examples) vector valued.

32.2 Characterization of Uncertain Signals

Formulating a signal uncertainty model almost always involves a trade-off between conflicting principles. One wants the model to be "tight", i.e., to contain only signals that make physical sense. Tightening uncertainty, however, typically implies imposing additional restrictions on the model; it gets unwieldy and more difficult to use for analysis and design purposes.

The following widely used uncertainty models are on the extreme (simple) end of the spectrum:

$$\mathcal{W}_2(c) := \left\{ w_u(t) \;\middle|\; \|w_u\|_2 := \left(\int_{-\infty}^{\infty} w_u(t)^T w_u(t)\, dt\right)^{\frac{1}{2}} \leq c \right\}, \tag{32.1}$$

i.e., the admissible input set consists of all signals with $\mathcal{L}_2$-norm (energy) less than or equal to a given constant c;

$$\mathcal{W}_\infty(c) := \left\{ w_u(t) \;\middle|\; \|w_u\|_\infty := \sup_t \max_i |w_{u_i}(t)| \leq c \right\}, \tag{32.2}$$

i.e., the admissible input set consists of all signals with $\mathcal{L}_\infty$-norm (maximum magnitude) less than or equal to a given constant c.

If necessary, these models can be refined by introducing suitable weights or filters. In this case, admissible input signals are

$$\bar{w}_u(t) := \int_{-\infty}^{t} W(t-\tau) w_u(\tau)\, d\tau, \tag{32.3}$$

where $w_u(t)$ ranges over the sets defined by Equation 32.1 or Equation 32.2. Even such a modified uncertainty description remains pretty crude. Whether it is adequate depends on the control problem at hand. If the answer turns out to be "no," one has to cut back the "size" of the admissible signal classes by bringing in additional a priori information. This is illustrated in Example 32.1:

EXAMPLE 32.1:

Suppose we want to control room temperature. Clearly, outdoor temperature, T_o, is a disturbance signal for our control problem. Assume that, for one reason or another, we cannot measure T_o. In Toronto, T_o can go up to +30°C in summer and down to −30°C in winter. In this case, a simple uncertainty model is given by

$$T_o(t) \in \mathcal{W}_\infty(30^\circ) = \{w_u(t) \mid \sup_t |w_u(t)| \leq 30^\circ \text{ C}\}.$$

Clearly, this set contains many signals that do not make physical sense; it admits, for example, temperatures of +30° during a winter night and −30° at noon in summer. A tighter uncertainty set is obtained if we write $T_o(t) = T_a(t) + T_\Delta(t)$, where $T_a(t)$ represents the average seasonal and daily variation of temperature, and $T_\Delta(t)$ is a deviation term. $T_a(t)$ could, for example, be modeled as the output of an autonomous dynamic system (a so-called *exosystem*) with pairs of poles at $s = \pm j(365\text{ days})^{-1}$ (accounting for seasonal variations) and $s = \pm j(24\text{ hours})^{-1}$ (accounting for daily variations). For the deviation term, we can now assume a much stricter $\mathcal{L}_\infty$-norm bound, e.g., 10°. Furthermore, it makes sense to restrict the maximal rate of change of $T_\Delta(t)$ to, say, 5° per hour. This is achieved if we define the admissible set of deviations via

$$T_\Delta(t) = \int_{-\infty}^{t} W(t-\tau) w_u(\tau)\, d\tau,$$

where the weighting function $W(t)$ is given by

$$W(t) = \begin{cases} 5\frac{1}{\text{hours}} e^{-\frac{t}{2\text{hours}}} & \text{for } t \geq 0 \\ 0 & \text{for } t < 0 \end{cases}$$

and $w_u(t)$ lives in the set $\mathcal{W}_\infty(1^\circ)$.

32.3 Characterization of Uncertain Plant Models

It is convenient to characterize the plant family $\mathcal{G}$ by specifying a nominal plant model together with a family of perturbations, denoted $\mathcal{D}$, away from the nominal. As for the set of admissible input signals, we have two requirements for $\mathcal{D}$:

1. $\mathcal{D}$ should be *tight* (i.e., contain only perturbations that somehow "mirror" the difference between our nominal model and the real world). If we include perturbations that make no physical sense, we will end up with a controller that tries to accommodate "too many" models and might therefore be "too conservative."
2. $\mathcal{D}$ should be easy to use for analysis and design purposes.

Again, both requirements rarely go together. By imposing *structure* on a perturbation model, we can, in general, capture uncertainty more precisely; *unstructured* perturbation models, on the other hand, are in general easier to use in design procedures.

For lack of space, we can present only the most common plant uncertainty models, the ones that are widely used for controller design and synthesis purposes. It is not surprising that, in a sense, this purpose reflects back onto the model itself. Some models we will be dealing with involve restrictive assumptions that can be justified only through a utilitarian argument; without these assumptions it would be a lot harder (or even impossible) to give robustness tests. We will look at unstructured perturbation models first. Such models are typically specified by giving, at every frequency, an upper bound on the maximum singular value of the transfer function matrix, this being a matrix generalization of the magnitude of a complex number. Then we will deal with a specific class of structured perturbations that can be graphically characterized by Nyquist arrays. Finally, we will discuss a fairly general class of structured perturbations that includes the others.

32.3.1 Unstructured Plant Uncertainty Models

In what follows, $G(s)$ denotes the nominal plant model; $G_r(s)$, the transfer function matrix of a perturbed plant model; and $\Delta(s)$, the transfer function matrix of a perturbation. The dimensions of $G(s)$ are $p \times q$. It is assumed that the elements of $G(s)$, $G_r(s)$, $\Delta(s)$ are proper (numerator degree less than or equal to denominator degree) real-rational transfer functions. The number of *unstable* poles of G and G_r, that is, those in the closed right half-plane, are denoted m_G and m_{G_r}.

Additive Modeling Perturbations

Additive perturbations are defined by (see Figure 32.2)

$$G_r(s) := G(s) + \Delta_A(s). \tag{32.4}$$

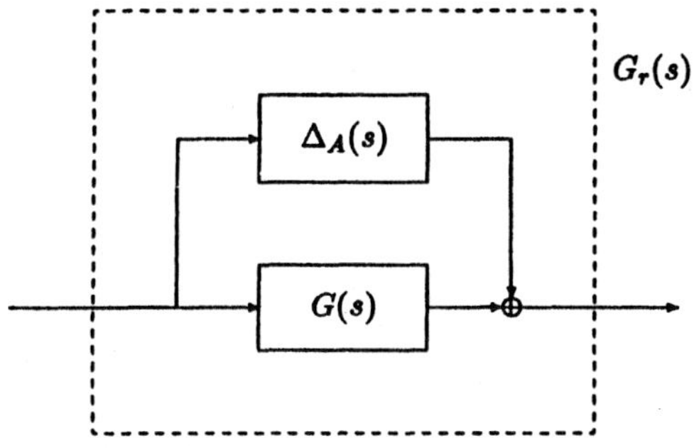

Figure 32.2 Additive model perturbation.

The class of unstructured additive perturbations we will be looking at is given by

$$\mathcal{D}_A := \{\Delta_A \mid \bar{\sigma}[\Delta_A(j\omega)] < l_A(\omega); \quad m_{G_r} = m_G\}. \tag{32.5}$$

Thus, the set of admissible Δ_A is characterized by two assumptions: 1. For each frequency ω, we know a (finite) upper bound $l_A(\omega)$ for the size of Δ_A (in the sense of the maximal singular value) 2. Δ_A cannot change the number of unstable poles of the model.

Thus l_A is an *envelope function* on the size of the perturbation. The regularity of the function l_A is not too important; it can be assumed to be, for example, piecewise continuous. The second assumption seems pretty artificial, but is needed if we want to give simple robustness tests[2]. It trivially holds for all stable perturbation matrices. Obviously, $l_A(\omega)$ being finite for every ω implies that Δ_A does not have any poles on the imaginary axis. A typical perturbation bound $l_A(\omega)$ is shown in Figure 32.3. The DC gain is usually known precisely [$\Delta_A(0) = 0$]; at high frequencies, we often have $G_r \to 0$, $G \to 0$ (G_r and G are strictly proper) and therefore $\Delta_A \to 0$.

[2]A stability robustness test was first reported by Cruz, Freudenberg and Looze [2].

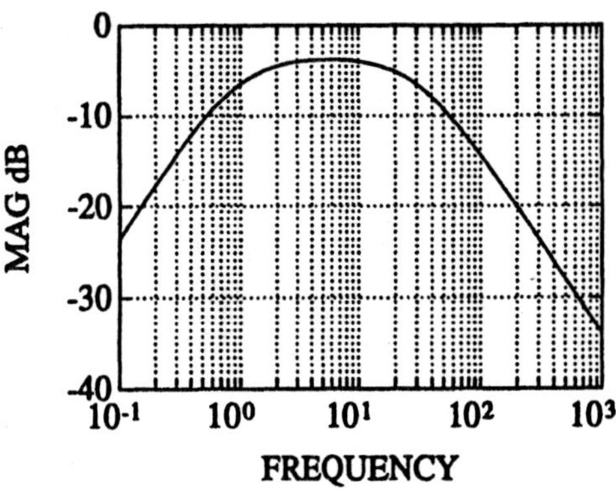

Figure 32.3 Typical bound for additive modeling perturbation.

Example 32.2 illustrates how an additive perturbation model typically can be obtained from frequency-response data.

EXAMPLE 32.2:

For simplicity, assume the plant is single-input/single-output (SISO). Suppose that the plant is stable and its transfer function is arrived at by means of frequency-response experiments. Magnitude and phase are measured at a number of frequencies, ω_i, $i = 1, \ldots, M$, and this experiment is repeated several, say N, times. Let the magnitude-phase measurement for frequency ω_i and experiment k be denoted (G_{ik}, ϕ_{ik}). Based on these data, select nominal magnitude-phase pairs (G_i, ϕ_i) for each frequency ω_i, and fit a nominal transfer function $G(s)$ to these data. Then fit a weighting function $l_A(s)$ so that

$$\left|G_{ik}e^{j\phi_{ik}} - G_i e^{j\phi_i}\right| < |l_A(j\omega_i)|,$$

$$i = 1, \ldots, M; \; k = 1, \ldots, N.$$

Example 32.3 shows how to "cover" a parameter-uncertainty model by an additive perturbation model.

EXAMPLE 32.3:

Consider the plant model

$$\frac{k}{s+1}, \quad 5 < k < 10.$$

Thus, the gain k is uncertain. Let us take the midpoint for the nominal plant:

$$G(s) = \frac{7.5}{s+1} \quad .$$

Then the envelope function is determined via

$$\left|\frac{k}{j\omega + 1} - G(j\omega)\right| < l_A(\omega), \quad 5 < k < 10,$$

that is, $l_A(\omega) = \left|\frac{2.5}{j\omega+1}\right|$.

Multiplicative Perturbations

Multiplicative (or proportional) perturbations at the plant output are defined by the following relation between G and G_r (see Figure 32.4):

$$G_r(s) := (I_p + \Delta_M(s))G(s) \quad . \tag{32.6}$$

Such proportional perturbations are invariant with respect to

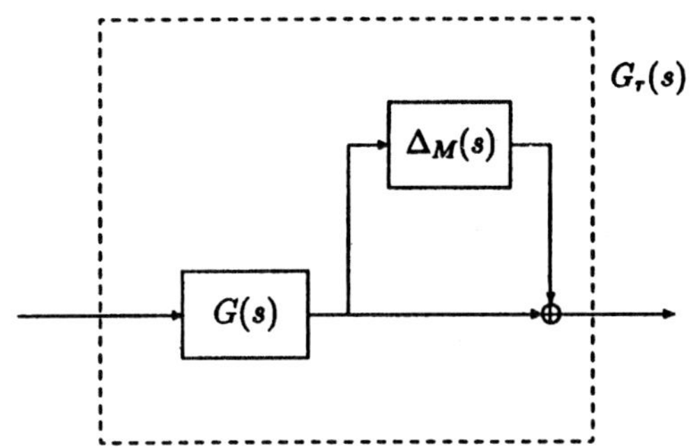

Figure 32.4 Multiplicative (proportional) model perturbation.

multiplication (from the right) by a known transfer function matrix (e.g., a compensator).

We consider the following class of (proportional perturbation) models:

$$\begin{aligned} \mathcal{D}_M &:= \mathcal{D}_{M1} \cup \mathcal{D}_{M2}, & (32.7)\\ \mathcal{D}_{M1} &:= \{\Delta_M \mid \bar{\sigma}[\Delta_M(j\omega)] < l_M(\omega); \\ &\qquad \Delta_M \text{ stable}\}, & (32.8)\\ \mathcal{D}_{M2} &:= \{\Delta_M \mid \bar{\sigma}[\Delta_M(j\omega)] < l_M(\omega); \\ &\qquad m_{G_r} = m_G\}. & (32.9)\end{aligned}$$

Hence, admissible perturbations either are stable or do not change the number of unstable poles of the plant model. Again, in both cases, we assume that we know an upper bound for the perturbation frequency response $\Delta_M(j\omega)$. A typical perturbation bound l_M is shown in Figure 32.5: exact knowledge of DC gain implies $\Delta_M(0) = 0$; neglecting high-order dynamics often causes Δ_M to be greater than 1 at high frequency.

REMARK 32.1 The perturbation class $\mathcal{D}_A$ is of a simpler form than Equations 32.7 to 32.9, because an additive stable perturbation transfer function matrix does not affect the number of unstable poles. As Example 32.4 shows, this is not always true for multiplicative perturbations: a stable proportional perturbation *can* cancel unstable model poles.

EXAMPLE 32.4:

$$G(s) = \frac{-4}{s-1}$$

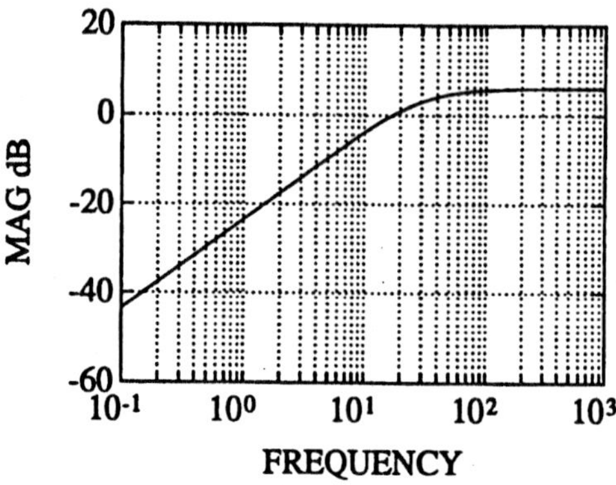

Figure 32.5 Typical bound for multiplicative model perturbation.

$$\begin{aligned} \Delta_M(s) &= \frac{-2}{s+1}\\ G_r(s) &= [1+\Delta_M(s)]G(s)\\ &= \frac{-4}{s+1}. \end{aligned}$$

Stability robustness results for this class of perturbations have been given by Doyle and Stein [6].

Coprime Factor Perturbations

For this uncertainty description, we need the concept of a left coprime factorization over $\mathcal{RH}_\infty$ [3] [14]. Here we give only a very brief introduction to this subject. Details can be found elsewhere in this book. Consider a finite-dimensional linear time-invariant SISO system. Clearly, its transfer function, $G(s)$, can be represented by a pair of polynomials with real coefficients— $N_P(s)$, a numerator, and $M_P(s)$, a denominator polynomial: $G(s) = \frac{N_P(s)}{M_P(s)}$. Furthermore, $N_P(s)$ and $M_P(s)$ can always be chosen to be coprime, meaning that all common divisors of $N_P(s)$, $M_P(s)$ are invertible in the set of polynomials (i.e., are real constants). Thus, there is no pole-zero cancellation when forming $\frac{N_P(s)}{M_P(s)}$.

For several reasons, it proves to be an advantage to use a straightforward generalization and to replace polynomials by proper stable transfer functions (i.e., elements from the set $\mathcal{RH}_\infty$). We write $G(s) = \frac{N(s)}{M(s)}$, where $N(s)$, $M(s) \in \mathcal{RH}_\infty$ can always be chosen to be coprime. This is called a *coprime factorization over* $\mathcal{RH}_\infty$. Coprimeness in the $\mathcal{RH}_\infty$ context means that all common divisors are invertible in $\mathcal{RH}_\infty$ (i.e., stable, minimum-phase transfer functions with equal numerator and denominator degrees). Hence, when forming $\frac{N(s)}{M(s)}$, no cancellation of poles and zeros in the closed right half-plane can occur. Obviously, such a factorization is non-unique. However, it can be made unique up to sign by requiring $|N(j\omega)|^2 + |M(j\omega)|^2 = 1$; that is, the 1×2 matrix $[N(s)M(s)]$ is allpass. This is called a *normalized* coprime factorization [15].

These concepts can be easily extended to the multivariable

[3] $\mathcal{RH}_\infty$ denotes the set of proper stable real-rational transfer functions.

case. However, as matrix multiplication is not commutative, we have to distinguish between *left* and *right coprime factorizations*: A pair $\tilde{N}(s)$, $\tilde{M}(s)$ [a pair $N(s)$, $M(s)$] of stable transfer function matrices with appropriate dimensions is called a left (right) coprime factorization of $G(s)$, if $G(s) = \tilde{M}^{-1}(s)\tilde{N}(s)$ [if $G(s) = N(s)M^{-1}(s)$] and all common left divisors of $\tilde{N}(s)$ and $\tilde{M}(s)$ [all common right divisors of $N(s)$ and $M(s)$] are invertible in $\mathcal{RH}_\infty$.

Now we are in a position to define coprime factor perturbations. These are additive stable perturbation terms in both numerator and denominator of a left (or right) coprime factorization (see Figure 32.6):

$$\begin{aligned} G &= \tilde{M}^{-1}\tilde{N}; \qquad \tilde{M},\ \tilde{N}\ldots\text{left coprime} \qquad (32.10)\\ G_r &= \underbrace{(\tilde{M}+\Delta_M)^{-1}}_{\tilde{M}_r^{-1}}\underbrace{(\tilde{N}+\Delta_N)}_{\tilde{N}_r}; \qquad \tilde{M}_r,\ \tilde{N}_r\ldots\text{left coprime.} \end{aligned}$$

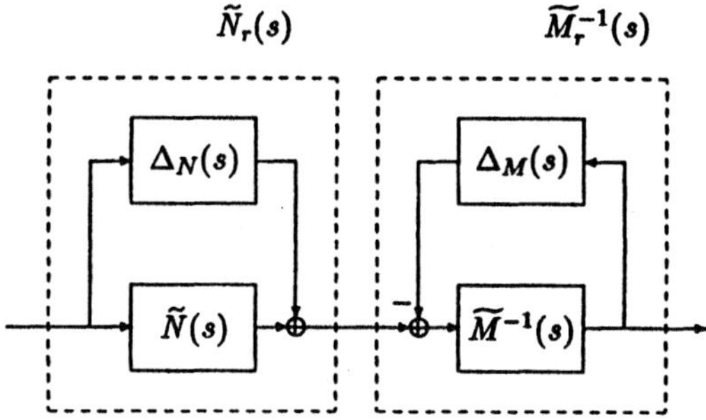

Figure 32.6 Coprime factor perturbation.

Specifically, we consider the following class of unstructured coprime factor perturbations:

$$\mathcal{D}_{MN} := \{[\Delta_M \quad \Delta_N] \,|\, \bar{\sigma}[\Delta_M(j\omega)\ \Delta_N(j\omega)] < l_{MN}(\omega)\ ;\quad \Delta_M,\ \Delta_N \text{ stable}\}. \qquad (32.11)$$

A typical perturbation bound $l_{MN}(\omega)$ may look as shown in Figure 32.3. Restricting the class $\mathcal{D}_{MN}$ to stable transfer function matrices does not imply any loss of generality, as both $\tilde{M}$, $\tilde{N}$ (denominator, numerator of the nominal model) and $\tilde{M}+\Delta_M$, $\tilde{N}+\Delta_N$ (denominator, numerator of any perturbed model) are stable by definition. Note that we do not have any restrictive assumptions regarding the number of unstable poles of admissible models $G_r(s)$. Perturbations from the set $\mathcal{D}_{MN}$ can both increase and decrease the number of unstable model poles.

EXAMPLE 32.5:

$$G(s) = \frac{1}{s+\varepsilon} = \underbrace{\left[\frac{s+\varepsilon}{s+1}\right]^{-1}}_{\tilde{M}(s)^{-1}}\underbrace{\left[\frac{1}{s+1}\right]}_{\tilde{N}(s)}$$

$$\begin{aligned} \Delta_N &= 0\\ \Delta_M(s) &= \frac{-2\varepsilon}{s+1}\\ G_r(s) &= \frac{1}{s-\varepsilon} = \underbrace{\left[\frac{s-\varepsilon}{s+1}\right]^{-1}}_{\tilde{M}_r(s)^{-1}}\underbrace{\left[\frac{1}{s+1}\right]}_{\tilde{N}_r(s)}. \end{aligned}$$

Coprime factor perturbations are especially useful for describing uncertainty in flexible structures. They allow covering a family of transfer function matrices with slightly damped uncertain pole pairs by a (relatively speaking) small perturbation set.

EXAMPLE 32.6:

Consider a nominal transfer function

$$G(s) = \frac{10}{(s+0.05-5j)(s+0.05+5j)}.$$

Suppose a perturbed model has a slightly different pair of poles:

$$G_r(s) = \frac{10}{(s+0.05-6j)(s+0.05+6j)}.$$

We first determine the magnitude of the smallest additive perturbation which, centered around $G(s)$, covers $G_r(s)$ and scale it (at each frequency) by $1/|G(j\omega)|$ [this is, of course, equivalent to computing the magnitude of the multiplicative perturbation connecting $G(s)$ and $G_r(s)$]. The result is shown in the left part of Figure 32.7.

We now compute normalized coprime factorizations $\{N(s), M(s)\}$ and $\{N_r(s), M_r(s)\}$ for $G(s)$ and $G_r(s)$ and plot the size (maximal singular value) of the perturbation matrix $[N_r(j\omega) - N(j\omega)\ M_r(j\omega) - M(j\omega)]$ over frequency. This gives a *relative (scaled)* measure of perturbation size, as both $[N(s)\ M(s)]$ and $[N_r(s)\ M_r(s)]$ are allpass. It is shown in the right half of Figure 32.7.

Clearly, there is an order of magnitude difference between the peak values of both perturbations.

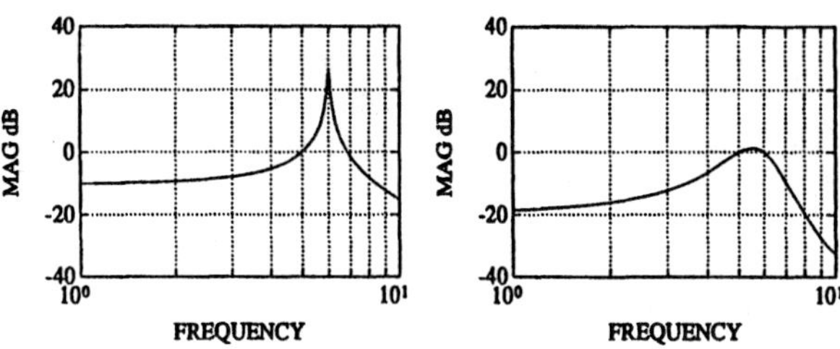

Figure 32.7 Perturbation sizes (normalized coprime factorization case shown on the right).

Normalized coprime factor perturbations and their use in H_∞-design methods have been extensively investigated by McFarlane and Glover [11].

Generalized Unstructured Perturbations

Additive, multiplicative, and coprime factor uncertainty can be conveniently represented in a unified way via *generalized unstructured perturbations* [11]. For this purpose, we introduce a "new" (proper) transfer function matrix

$$\Theta(s) = \begin{bmatrix} \Theta_{11}(s) & \Theta_{12}(s) \\ \Theta_{21}(s) & \Theta_{22}(s) \end{bmatrix}. \tag{32.12}$$

Its partitioning implies a partitioning of its input and output vector. The lower parts of both vectors represent the "usual" plant input and output signals. The upper part of the output signal, z_Δ, is fed back into the upper part of the input signal, w_Δ, via a perturbation $\Delta(s)$ (Figure 32.8).

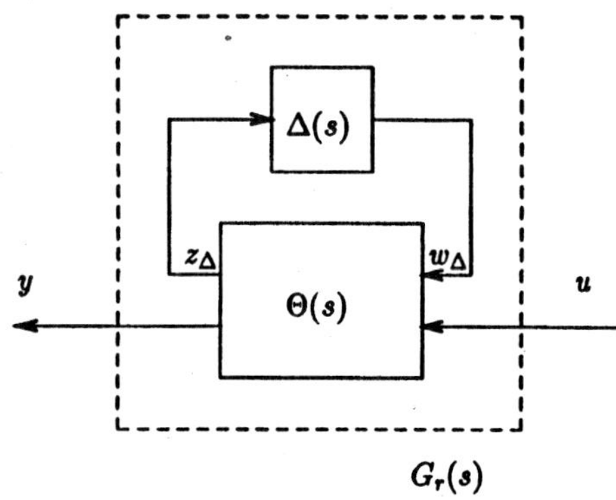

Figure 32.8 Generalized unstructured perturbation.

Denote the number of rows and columns of Δ by m_Δ and l_Δ, respectively. Then the transfer function matrix G_r of the perturbed model is given as an *upper linear fractional transformation* of Δ with respect to Θ:

$$G_r = \Theta_{22} + \Theta_{21}\Delta(I_{l_\Delta} - \Theta_{11}\Delta)^{-1}\Theta_{12} \tag{32.13}$$

$$= \Theta_{22} + \Theta_{21}(I_{m_\Delta} - \Delta\Theta_{11})^{-1}\Delta\Theta_{12}. \tag{32.14}$$

For $\Delta = 0$ (zero perturbation), we expect to recover the nominal plant model as transfer function matrix between plant input and output vector. Hence, we must have $\Theta_{22} = G$. For Equations 32.13 and 32.14 to make sense, we also have to assume that

1. $[I_{l_\Delta} - \Theta_{11}(\infty)\Delta(\infty)]$ (or, equivalently, $[I_{m_\Delta} - \Delta(\infty)\Theta_{11}(\infty)]$) is nonsingular (i.e., G_r is uniquely defined and proper) for all admissible perturbations.
2. The transfer function matrix $\Theta(s)$ can be stabilized from the plant input and detected from the plant output.

Specifically, we consider the following class of perturbations:

$$\mathcal{D} := \mathcal{D}_1 \cup \mathcal{D}_2, \tag{32.15}$$

$$\mathcal{D}_1 := \{\Delta \mid \bar{\sigma}[\Delta(j\omega)] < l(\omega),\ \Delta \text{ stable}\}, \tag{32.16}$$

$$\mathcal{D}_2 := \left\{\Delta \mid \bar{\sigma}[\Delta(j\omega)] < l(\omega),\ m_{G_r} = m_G\right\}. \tag{32.17}$$

We admit all bounded (on $s = j\omega$) perturbation transfer function matrices that either are stable or do not change the number of unstable poles of the plant model.

We have not specified yet how to choose Θ if we want to translate additive, multiplicative, and coprime factor uncertainty into this general framework. By substituting into Equation 32.13, it is easy to check that

- for additive perturbations,

$$\Theta = \begin{bmatrix} 0 & I_q \\ I_p & G \end{bmatrix}, \quad \Delta = \Delta_A \tag{32.18}$$

- for multiplicative perturbations,

$$\Theta = \begin{bmatrix} 0 & G \\ I_p & G \end{bmatrix}, \quad \Delta = \Delta_M \tag{32.19}$$

- for coprime factor perturbations,

$$\Theta = \begin{bmatrix} \begin{bmatrix} -\tilde{M}^{-1} \\ 0 \end{bmatrix} & \begin{bmatrix} -G \\ I_q \end{bmatrix} \\ \tilde{M}^{-1} & G \end{bmatrix},$$
$$\Delta = \begin{bmatrix} \Delta_M & \Delta_N \end{bmatrix}.$$

For additive uncertainty, $\mathcal{D}_1 \subset \mathcal{D}_2$ (stable perturbations do not change the number of unstable model poles). Hence, in this case, $\mathcal{D}$ and $\mathcal{D}_A$ are the same. In the case of coprime factor uncertainty, all perturbation terms are stable by definition. We can therefore write $\mathcal{D} = \mathcal{D}_{MN}$ without restricting generality.

This framework is especially useful for theoretical investigations. Instead of proving, say, robust stability for the three sets of perturbations $\mathcal{D}_A$, $\mathcal{D}_M$, and $\mathcal{D}_{MN}$, it suffices to establish this result for the class $\mathcal{D}$ of generalized perturbations.

Which Unstructured Perturbation Model?

All of the above perturbation models are reasonably simple and easy to use in design procedures. Hence, the decision on which model to choose hinges primarily on their potential to cover a given set $\mathcal{G}$ without including too many "physically impossible" plant models. In other words, we want the least conservative perturbation set that "does the job." Whether this is a set of additive, multiplicative or coprime factor perturbations depends on the problem at hand. A useful rule of thumb is: Whenever one deals with low-damped mechanical systems, it is a good idea to try coprime factor uncertainty first (compare example 32.6). For stable non-oscillatory processes, one might try an additive perturbation model first, it is more intuitive, and, if the nominal model is stable, excludes any unstable $G_r(s)$.

32.3.2 Structured Plant Uncertainty

Uncertain Parameters

Physical (or theoretical), as opposed to experimental, model building usually results in a state model, and model uncertainty is often in the form of parameter uncertainty: the (real-valued) parameters of the state model have physical meaning and are assumed to lie within given intervals. When converting to a transfer function matrix, however, one usually gets a fairly complicated set of admissible parameters. This is due to the fact that, in general, each parameter of the resulting transfer function matrix depends on several parameters of the underlying state model. Only in exceptionally simple cases (as in Example 32.7) can we expect the set of admissible parameters to form a parallelepiped.

EXAMPLE 32.7:

Consider the spring-damper system in Figure 32.9.

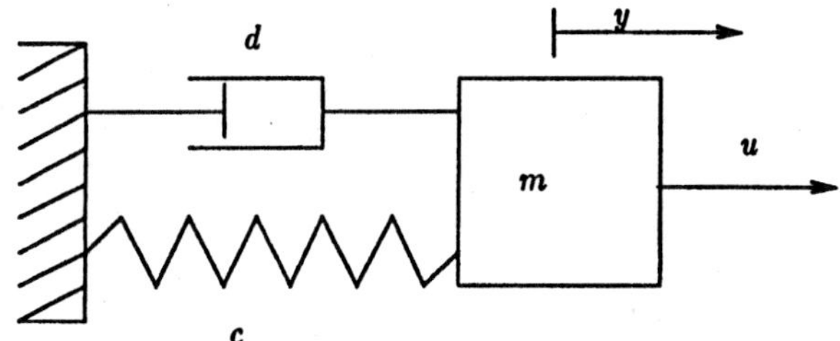

Figure 32.9 Spring-damper system.

Force is denoted by u, position by y. Newton's law gives the transfer function

$$G(s) = \frac{1}{ms^2 + ds + c} \quad .$$

In most cases, one cannot expect to know the precise values of mass, damping and spring constant. Hence, it makes sense to define a class of models by stating admissible parameter intervals:

$$\begin{aligned}
\Delta_m &\in [m_u - m,\ m_o - m] := \mathcal{D}_m \\
\Delta_d &\in [d_u - d,\ d_o - d] := \mathcal{D}_d \\
\Delta_c &\in [c_u - c,\ c_o - c] := \mathcal{D}_c \\
\mathcal{G}_{mdc} &:= \left\{ \frac{1}{(m+\Delta_m)s^2 + (d+\Delta_d)s + (c+\Delta_c)} \;\middle|\; \Delta_m \in \mathcal{D}_m,\ \Delta_d \in \mathcal{D}_d,\ \Delta_c \in \mathcal{D}_c \right\}.
\end{aligned}$$

Independent Additive Perturbations in the Elements of a Transfer Function Matrix

Another straightforward (and often very useful) structured uncertainty description is the following. Consider an additive perturbation matrix $\Delta_A(s) = G_r(s) - G(s)$ and give frequency-dependent bounds $l_{ik}(\omega)$ for the magnitude of each of its elements:

$$\underbrace{\begin{bmatrix} |\Delta_{A_{11}}(j\omega)| & \dots & |\Delta_{A_{1q}}(j\omega)| \\ \vdots & \ddots & \vdots \\ |\Delta_{A_{p1}}(j\omega)| & \dots & |\Delta_{A_{pq}}(j\omega)| \end{bmatrix}}_{:=|\Delta_A|_e} \leq_e \underbrace{\begin{bmatrix} l_{11}(\omega) & \dots & l_{1q}(\omega) \\ \vdots & \ddots & \vdots \\ l_{p1}(\omega) & \dots & l_{pq}(\omega) \end{bmatrix}}_{:=L(\omega)} \tag{32.20}$$

($\leq_e$ denotes "elementwise less or equal"; $|\Delta_A|_e$ is a matrix with entries $|\Delta_{A_{ik}}|$). Again, we make the additional assumption that no admissible perturbation shall change the number of unstable poles. Hence, we get the following classes of perturbations and plant models:

$$\mathcal{D}_{Ae} := \{\Delta_A \mid |\Delta_A(j\omega)|_e \leq_e L(\omega); \; m_{G_r} = m_G\} \tag{32.21}$$

$$\mathcal{G}_{Ae} := \{G + \Delta_A \mid \Delta_A \in \mathcal{D}_{Ae}\}. \tag{32.22}$$

Such a set of models can be easily represented in a graphical way (Figure 32.10). Consider first the Nyquist plots of each element $g_{ik}(j\omega)$ of the nominal model ("Nyquist array"). For each frequency, draw a circle with radius $l_{ik}(\omega)$ around $g_{ik}(j\omega)$. This gives a set of bands covering the nominal Nyquist plots. Then a transfer function matrix G_r is a member of the model set of Equation 32.22 if and only if the Nyquist plot of each element $g_{r_{ik}}(j\omega)$ is contained in the appropriate band, and G_r and G have the same number of unstable poles.

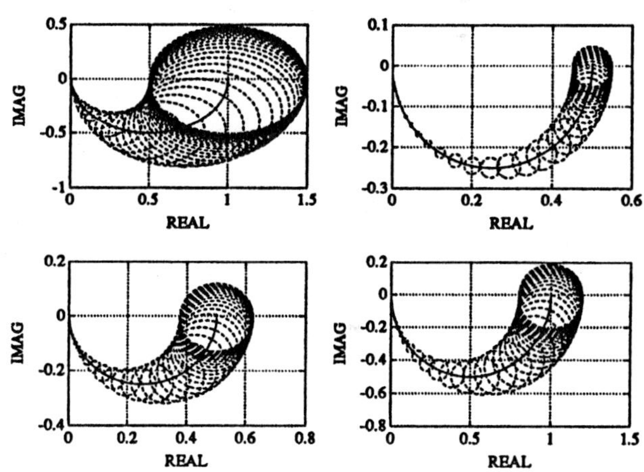

Figure 32.10 Graphical representation of the model class $\mathcal{G}_{Ae}$.

The usefulness of this uncertainty model for purposes of controller design stems mainly from the fact that it fits "naturally" into the framework of Nyquist array methods. Stability robustness results can be found in [12] and [8].

Generalized Structured Plant Uncertainty

A more general structured perturbation model has become very popular recently, as the so-called μ-*theory* [3], [4] provides analysis and synthesis tools to deal with such uncertainty sets. As a motivation for this general perturbation model, consider the following example from Maciejowski [10]:

EXAMPLE 32.8:

Let $G(s)$ be a 2×2 plant model with additive unstructured uncertainty $\Delta_A(s)$ and independent multiplicative perturbations $\tilde{\delta}_1(s), \tilde{\delta}_2(s)$ in each input channel (Figure 32.11).

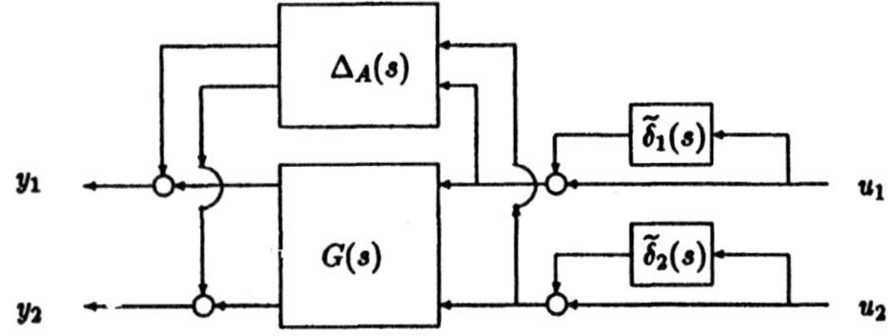

Figure 32.11 Example for a model with independent perturbations.

Assume that $\Delta_A(s)$, $\tilde{\delta}_1(s)$, and $\tilde{\delta}_2(s)$ are stable, proper and bounded by

$$\begin{aligned} |\tilde{\delta}_1(j\omega)| &< l_1(\omega), \\ |\tilde{\delta}_2(j\omega)| &< l_2(\omega), \\ \overline{\sigma}[\Delta_A(j\omega)] &< l_3(\omega). \end{aligned}$$

It is easy to see that we could subsume all three uncertainty terms in a single (additive) perturbation matrix $\overline{\Delta}_A(s)$:

$$\begin{aligned} G_r &= (G+\Delta_A)\left(I_2 + \begin{bmatrix} \tilde{\delta}_1 & 0 \\ 0 & \tilde{\delta}_2 \end{bmatrix}\right) \\ &= G + \underbrace{\left(\Delta_A \begin{bmatrix} 1+\tilde{\delta}_1 & 0 \\ 0 & 1+\tilde{\delta}_2 \end{bmatrix} + G\begin{bmatrix} \tilde{\delta}_1 & 0 \\ 0 & \tilde{\delta}_2 \end{bmatrix}\right)}_{:=\overline{\Delta}_A}. \end{aligned}$$

Using the properties of singular values, we can derive both structured and unstructured bounds for the overall perturbation $\overline{\Delta}_A$:

$$\begin{aligned} \overline{\sigma}[\overline{\Delta}_A(j\omega)] &\leq \overline{\sigma}[\Delta_A(j\omega)]\,\overline{\sigma}\begin{bmatrix} 1+\tilde{\delta}_1(j\omega) & 0 \\ 0 & 1+\tilde{\delta}_2(j\omega) \end{bmatrix} \\ &\quad + \overline{\sigma}[G(j\omega)]\,\overline{\sigma}\begin{bmatrix} \tilde{\delta}_1(j\omega) & 0 \\ 0 & \tilde{\delta}_2(j\omega) \end{bmatrix} \\ &< l_3(\omega)\ (1+\max(l_1(\omega), l_2(\omega))) \\ &\quad + \overline{\sigma}[G(j\omega)]\max(l_1, l_2) \end{aligned}$$

or

$$\left|\overline{\Delta}_{Aik}(j\omega)\right| < l_3(\omega)(1+l_k(\omega)) + |g_{ik}(j\omega)| l_k(\omega).$$

In both cases, the perturbation bounds will probably be very *conservative*; i.e., the resulting uncertainty class will cover more than the perturbations $\Delta_A(s)$, $\tilde{\delta}_1(s)$ and $\tilde{\delta}_2(s)$ we started off with. It is therefore a good idea to preserve perturbation structure when combining different "sources" of model uncertainty. This can be accomplished in the following way. As in the section on generalized unstructured perturbations, we define a "new" transfer function matrix $\Theta(s)$ that, apart from the plant input and output, contains another pair of input and output vectors, $\tilde{w}_\Delta$ and $\tilde{z}_\Delta$. $\tilde{z}_\Delta$ is fed back into $\tilde{w}_\Delta$ via a *blockdiagonal* perturbation matrix. Graphically, this corresponds to "pulling out" all perturbations from Figure 32.11 and rearranging them in a blockdiagonal structure (Figure 32.12).

Note that Figure 32.12 looks like Figure 32.8, the only difference being the blockdiagonal structure of the perturbation matrix in the feedback loop, which mirrors the structure of the underlying perturbation model.

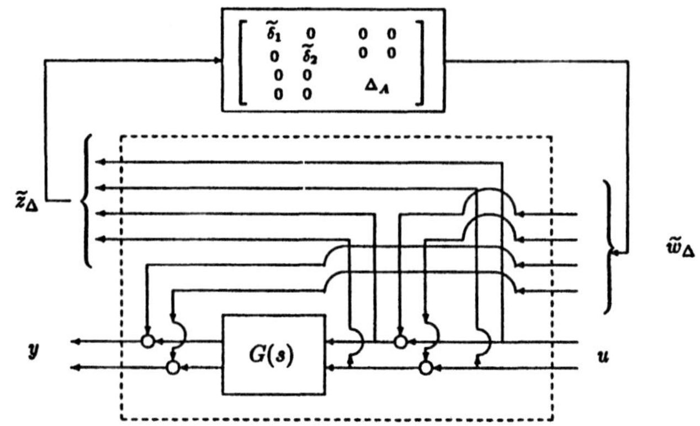

Figure 32.12 Generalized structured perturbation model.

In the general case, one proceeds in exactly the same way. The class of admissible models is represented by an upper linear fractional transformation of a blockdiagonal perturbation transfer function matrix $\Delta_s(s)$ with respect to a suitably defined transfer function matrix $\Theta(s)$:

$$G_r = \mathcal{F}_U(\Theta, \Delta_s). \tag{32.23}$$

Structure and dimension of Δ_s depend, of course, on the number of independent perturbation terms and their dimensions. Without loss of generality, we can assume that scalar uncertainty is always listed first in Δ_s. In this general framework, we can also restrict Δ_s to be stable. If unstable perturbations have to be considered, we can always circumvent this problem by introducing a coprime factorization with stable perturbations. Often, notation is simplified by normalizing the size of the perturbation blocks. This can be easily done if each perturbation bound $l_i(\omega)$ can be written as the magnitude of a frequency response $w_i(j\omega)$. In this case, we just have to multiply Θ_{11} and Θ_{12} from the left (or Θ_{11} and Θ_{21} from the right) by a suitably dimensioned diagonal matrix containing the $w_i(s)$. As only magnitude is important, we can always choose the transfer functions $w_i(s)$ to be stable and minimum phase. Hence, we get the following class of perturbations:

$$\begin{aligned} \mathcal{D}_s &:= \{\Delta_s \mid \Delta_s \\ &= \begin{bmatrix} \delta_1 I_{l_1} & 0 & \dots & & & 0 \\ 0 & \ddots & & & & \vdots \\ \vdots & & \delta_k I_{l_k} & & & \\ & & & \Delta_{k+1} & & \\ & & & & \ddots & 0 \\ 0 & \dots & & & 0 & \Delta_v \end{bmatrix}, \\ &\quad \Delta_s \text{ stable},\ \overline{\sigma}[\Delta_s(j\omega)] < 1\}. \end{aligned} \tag{32.24}$$

EXAMPLE 32.9:

Let us look at Example 32.8 again. Suppose $w_1(s)$, $w_2(s)$, and $w_3(s)$ are stable minimum-phase transfer functions with

$$|w_1(j\omega)| = l_1(\omega),$$

$$|w_2(j\omega)| = l_2(\omega),$$
$$|w_3(j\omega)| = l_3(\omega).$$

Then we get

$$\Theta(s) = \begin{bmatrix} \begin{bmatrix} 0 & 0 \\ w_3 I_2 & 0 \end{bmatrix} & \begin{bmatrix} \operatorname{diag}\{w_1, w_2\} \\ w_3 I_2 \end{bmatrix} \\ [G \;\; I_2] & G \end{bmatrix}$$

$$\Delta_s = \begin{bmatrix} \delta_1 & 0 & 0 & 0 \\ 0 & \delta_2 & 0 & 0 \\ 0 & 0 & & \\ 0 & 0 & & \Delta_3 \end{bmatrix}$$
$$= \begin{bmatrix} \frac{\tilde{\delta}_1}{w_1} & 0 & 0 & 0 \\ 0 & \frac{\tilde{\delta}_2}{w_2} & 0 & 0 \\ 0 & 0 & & \\ 0 & 0 & & \frac{1}{w_3}\Delta_A \end{bmatrix}$$

and, as expected,

$$G_r = \mathcal{F}_U(\Theta, \Delta_s)$$
$$= G + \left(\Delta_A \begin{bmatrix} 1+\tilde{\delta}_1 & 0 \\ 0 & 1+\tilde{\delta}_2 \end{bmatrix} + G \begin{bmatrix} \tilde{\delta}_1 & 0 \\ 0 & \tilde{\delta}_2 \end{bmatrix} \right).$$

REMARK 32.2 All structured perturbation models in this section can be written as in Equations 32.23 and 32.24. However, when "translating" parameter perturbations into this framework, one has to take into account that $\mathcal{D}_s$ admits complex perturbations whereas parameters (and therefore parameter perturbations) in a state or transfer function model are real.

32.4 Model Validation

Model validation is understood to be the procedure of establishing whether a set of experimental data is compatible with given signal and plant uncertainty models, $\mathcal{W}$ and $\mathcal{G}$. It is *not* the (futile) attempt to show that an uncertainty model can *always* explain the true plant's input/output behavior; future experiments might well provide data that are inconsistent with $\mathcal{W}$ and $\mathcal{G}$.

Model validation is a rapidly expanding area, and, for lack of space, we can hope only to give a flavor of the subject by looking at a comparatively simple version of the problem. We assume that

1. $\mathcal{W}$ is a singleton (i.e., there is no signal uncertainty).
2. Plant uncertainty is in the form of stable generalized unstructured perturbation [i.e., $G_r(s) = \mathcal{F}_U[\Theta(s), \Delta(s)]$, where $\Delta(s) \in \mathcal{D}_l$; see Section 32.3], and $\Theta_{21}(s)$ is invertible.
3. Experimental data are given in the form of M frequency response measurements $G_r(j\omega_1), \ldots, G_r(j\omega_M)$.

The plant uncertainty model of item 2 contains the cases of coprime factor perturbations and of stable additive and multiplicative perturbations.

With the invertibility condition for Θ_{21} in force, we can rewrite the perturbed plant model as follows:

$$G_r = \mathcal{F}_U(\Theta, \Delta) \quad (32.25)$$
$$= \Theta_{22} + \Theta_{21}(I_{m_\Delta} - \Delta\Theta_{11})^{-1}\Delta\Theta_{12} \quad (32.26)$$
$$= (\Delta\Gamma_{12} + \Gamma_{22})^{-1}(\Delta\Gamma_{11} + \Gamma_{21}), \quad (32.27)$$

where

$$\begin{bmatrix} \Gamma_{11} & \Gamma_{12} \\ \Gamma_{21} & \Gamma_{22} \end{bmatrix} := \begin{bmatrix} \Theta_{12} - \Theta_{11}\Theta_{21}^{-1}\Theta_{22} & -\Theta_{11}\Theta_{21}^{-1} \\ \Theta_{21}^{-1}\Theta_{22} & \Theta_{21}^{-1} \end{bmatrix}. \quad (32.28)$$

This can be easily checked by substituting Equation 32.28 into Equation 32.27. Multiplying Equation 32.27 by $(\Delta\Gamma_{12} + \Gamma_{22})$ from the left gives

$$\Delta \underbrace{(\Gamma_{12}G_r - \Gamma_{11})}_{:=W} = \underbrace{\Gamma_{21} - \Gamma_{22}G_r}_{:=U}. \quad (32.29)$$

Consistency of experimental data $G_r(j\omega_1), \ldots, G_r(j\omega_M)$ and uncertainty model is equivalent to the existence of a transfer function matrix $\Delta(s) \in \mathcal{D}_l$ that solves Equation 32.29 for $\omega_1, \ldots, \omega_M$. Clearly, a necessary condition for this is that the system of linear equations over $\mathbb{C}$

$$\Delta_i W(j\omega_i) = U(j\omega_i) \quad (32.30)$$

must have a solution Δ_i with $\bar{\sigma}[\Delta_i] < l(\omega_i)$ for each $i \in \{1, \ldots, M\}$. Using interpolation theory, it has been shown by Boulet and Francis [1] that this condition is also sufficient[4]: if suitable Δ_i exist, one can always find a stable transfer function matrix $\Delta(s)$ such that $\Delta(j\omega_i) = \Delta_i$, $i = 1, \ldots, M$, and $\bar{\sigma}[\Delta(j\omega)] < l(\omega)$ for all $\omega \in \mathbb{R}$.

References

[1] Boulet, B. and Francis, B.A., Consistency of experimental frequency-response data with coprime factor plant models, Systems Control Group Report 9402, Department of Electrical and Computer Engineering, University of Toronto, 1994.

[2] Cruz, J.B., Freudenberg, J.S., and Looze, D.P., A relationship between sensitivity and stability of multivariable feedback systems, *IEEE Trans. Autom. Control*, 26, 66–74, 1981.

[3] Doyle, J.C., Analysis of feedback systems with structured uncertainties, *Proc. IEE Part D*, 129, 242–250, 1982.

[4] Doyle, J.C., Structured uncertainty in control system design, in Proc. 24th IEEE Conf. Decision and Control, Ft. Lauderdale, FL, 1985, 260–265.

[5] Doyle, J.C., Francis, B.A., and Tannenbaum, A.R., *Feedback Control Theory*, Macmillan, New York, 1992.

[4]Their proof is for coprime factor perturbations. It carries over to the slightly more general case considered here.

[6] Doyle, J.C. and Stein, G., Multivariable feedback design: concepts for a classical/modern synthesis, *IEEE Trans. Autom. Control,* 2b, 4–16, 1981.
[7] *IEEE Trans. Autom. Control,* Special issue on system identification for robust control design, July 1992.
[8] Lunze, J., Robustness tests for feedback control systems using multidimensional uncertainty bounds, *Syst. Control Lett.,* 4, 85–89, 1984.
[9] Lunze, J., *Robust Multivariable Control,* Prentice Hall, New York, 1989.
[10] Maciejowski, J.M., *Multivariable Feedback Design,* Addison-Wesley, Reading, MA, 1989.
[11] McFarlane, D.C. and Glover, K., *Robust Controller Design Using Normalized Coprime Factor Plant Descriptions,* Springer-Verlag, Berlin, 1989.
[12] Owens, D.H. and Chotai, A., On eigenvalues, eigenvectors and singular values in robust stability analysis, *Int. J. Control,* 40, 285–296, 1984.
[13] Raisch, J., *Mehrgrößenregelung im Frequenzbereich,* R. Oldenbourg-Verlag, Munich, 1994.
[14] Vidyasagar, M., *Control System Synthesis – A Factorization Approach,* MIT Press, MA, 1985.
[15] Vidyasagar, M., Normalized coprime factorizations for non-strictly proper systems, *IEEE Trans. Autom. Control,* 33, 300–301, 1988.

Further Reading

Any book on robust control contains information on plant model uncertainty. We especially recommend Vidyasagar [14], McFarlane and Glover [11], and Lunze [9]. Parts of this contribution are based on Raisch [13]. See also Doyle, Francis and Tannenbaum [5].

Several papers related to model validation can be found in a special issue of the *IEEE Trans. Autom. Control* dealing with system identification for robust control design [7].

33

The Use of Multivariate Statistics in Process Control

Michael J. Piovoso
DuPont Central Science & Engineering, Wilmington, DE

Karlene A. Kosanovich
Department of Chemical Engineering, University of South Carolina, Columbia, SC

33.1 Introduction

Advancements in automation and distributed control systems make possible the collection of large quantities of data. But without the adequate analytical tools, it is not possible to interpret the data. Every modern industrial site believes that its data bank is a gold mine of information if only the *important* and relevant information could be extracted painlessly and quickly. Timely interpretation of data would improve quality and safety, reduce waste, and improve business profits allowing for undetected sensor failures, uncalibrated and misplaced sensors, lack of integrity of the data historian and of data compression techniques for storing the data, and transcription errors. In the face of these problems data analysis methods may appear to be inadequate. Meanwhile, the data bank grows without appearing to garner any useful information. Without accurate and timely measurements, feedback control of the process to reach some specified objective is very difficult. For example, in the chemical industry, composition measurements are usually not made on-line; rather they are sampled and analyzed off-line. The delay between sampling and the results is usually on the order of hours. Thus, timely information about composition is unavailable to take remedial control action.

In a typical chemical process, it is common to sample and store hundreds of process variable measurements. These data can be characterized as being, noisy and collinear. In addition, there are instances when measurements are not in the data set, and also when the values of variables are grossly erroneous. To handle these data requires tools capable of handling the redundancy[1], noise, and missing information.

Good design of experiments and a priori knowledge of the process would allow the use of standard techniques such as Multiple Linear Regression (MLR) to develop predictive models. In practice, carrying out experiments on an operating process is unlikely because of production requirements and financial loss. In the majority of situations, only historical data are available. MLR works best when the independent variables are noise free and uncorrelated, which is unrealistic for real process data.

This chapter discusses the use of more appropriate multivariate statistical techniques to analyze process data (historical) and to develop predictive models in support of process monitoring and control. In particular, the multivariate methods of Partial Least Squares or Projection to Latent Structures (PLS), Principal Component Analysis/Principal Component Regression (PCA/PCR) will be presented starting with the theoretical development and followed by examples to illustrate concepts and implementation in real industrial processes.

PCA is a method for modeling a set of data assembled in a matrix X, where the rows are the sampled process variables at a fixed sampling time, and a column is an uniformly sampled variable. PCA produces a mapping of the data set X onto a reduced subspace defined by the span of a chosen subset of eigenvectors of the variance-covariance matrix of the X data. This set of eigenvectors or directions in the X space is referred to as the PCA loadings. This technique has the advantage of allowing the development of

[1]Although there are hundreds of measurements, there are not hundreds of different events occurring.

0-8493-8570-9/96/$0.00+$.50

a linear model which produces an orthogonal set of pseudomeasurements containing the significant variations of the X data. The first pseudomeasurement or principal component explains the greatest amount of variation and the second the next largest amount after removal of the first effect, and so on. These pseudomeasurements, the inner products of the true measurements with the loadings, are called the scores. The entire sets of scores and loadings define the process data, and the loadings are the statistical process model. A subset of the first few scores provides information in a lower dimensional space of the behavior of the process during the period in which the measurements were made. This set of scores and the PCA loadings can be used to determine if the present process operation has changed its behavior relative to the data that was used to define the scores and loadings [8]. In addition this score space has properties which make it attractive for doing multivariate statistical process control [2].

Although PCA is suitable for process monitoring, when there is a specific control objective, PCA is not the appropriate tool. For example, if a critical measurement is not readily available, such as when a laboratory analysis is required, then PLS or PCR are tools to consider. PLS, like PCA, provides a model for the X space of data. However, it is not the same model as the PCA model. The PLS X-space model is a rotated version of the PCA model. The rotation is defined so that the scores of the measurements provide the maximum information about the quantity to predict called the Y-data. An example in which PLS might be an appropriate tool is the control of the nonmeasured composition of the output of a distillation column [4]. Traditionally, this is accomplished by selecting a tray temperature which best correlates with the actual composition and holding that temperature at a prescribed value. Generally, there are many tray temperatures available; PLS might be used to model composition using the redundant information contained in multiple tray temperatures. Now, the control objective is not to hold one tray temperature constant, but rather to allow all the tray temperatures to move as needed to maintain the desired composition set point.

PCR is an extension of PCA to the modeling of some Y data from the X data. The approach to defining this relationship is accomplished in two steps. The first is to perform PCA on the X data and then to regress the scores onto the Y data. Unlike PLS, PCR establishes its loadings independent of the Y data set.

33.2 Multivariate Statistics

In this section, we review the theoretical foundations of the multivariate statistical methods providing only information that will allow the reader to grasp the important concepts. A thorough review of PCA can be found in the article by Wold et al. [5], and that of PLS and PCR by Martens and Næs [3].

33.2.1 Principal Component Analysis

PCA involves several steps. First, the data are mean centered, and often normalized by the standard deviation. Mean centering implies that the average value for each variable is subtracted from the corresponding measurement. Scaling is necessary to avoid problems associated with having some measurements with large values and others with small ones. For example, pressure may be measured in the thousands of Pascals while temperature may be in units of hundreds of degrees Celsius. Scaling puts all the numbers on the same magnitude by multiplying the mean centered data by an appropriate constant, usually the inverse of the standard deviation.

From this normalized data, the variance-covariance matrix is generated by the relationship $X^T X$, where X is the normalized data matrix. The $X^T X$ matrix is positive semidefinite[2] and it defines the directions in the X space where most of the variability occurs. These directions constitute the eigenvectors of $X^T X$. The eigenvalues are related to the amount of variability explained by each eigenvector.

PCA compresses the information in the X matrix, $X = (x_1^T, x_2^T, \cdots, x_K^T)^T$, into a set of pseudovariables $T = (t_1, t_2, \cdots, t_A)$. The row vector x_k represents process measurement at time k, and the column vector t_a represents the time history of the projection of all the measurements onto the ath eigenvector.

PCA is illustrated in Figure 33.1. In this example, two measurements are being made on a given process. Observe that the data are not linearly independent because, as measurement one increases, measurement two does so as well. In general, the measurements define a K dimensional hyperspace, where K is the rank of $X^T X$. The first eigenvector is the direction in which the data exhibit the greatest variability, illustrated by the vector, p_1. The second eigenvector will be orthogonal to the first (in the direction of greatest variability of the residual, $(X - t_1 p_1^T)$), denoted by p_2. Some eigenvectors may define directions of extraneous information, for example, p_2, which is needed only to explain the noise in the data. If only one eigenvector, p_1, is used to represent X, then a smoothed reconstruction of the X data is possible.

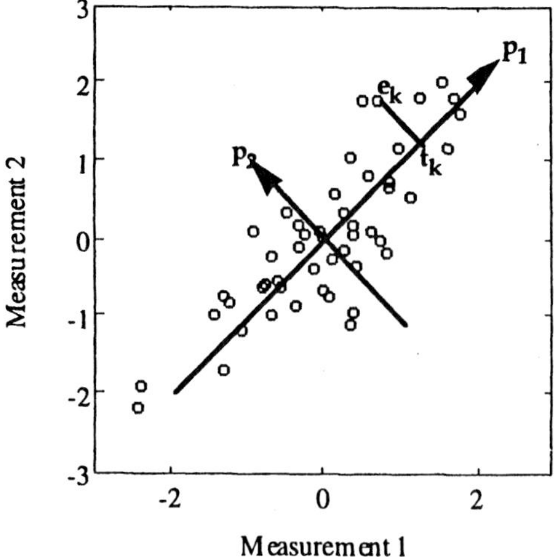

Figure 33.1 PCA illustration.

[2] eigenvalues are all nonnegative

Consider measurement vector at the kth time interval, x_k. This data point has a projection onto the first eigenvector. The distance from the origin along the vector p_1 to the projection point is the associated score. The coordinates of this projection point on the line spanned by the first eigenvector represent a reconstruction of the data from a single eigenvector. The projection error, e_k, is a point in the subspace orthogonal to that containing the reconstructed data point.

Generalizing, PCA can be decomposed as follows. Let $P = (p_1, p_2, \cdots, p_A)$, where p_j is the jth eigenvector of the covariance matrix $X^T X$, and

$$\tau_a = t_a^T t_a. \tag{33.1}$$

The scalar, τ_a, defines the amount of variability in the normalized measurement data explained by the ath eigenvector. Furthermore, the matrix P has the property[3] that

$$P^T P = I \tag{33.2}$$

where I is the identity matrix. The relationship between **P** and τ_a is

$$X^T X p_a = p_a \tau_a. \tag{33.3}$$

If all of the eigenvectors have been extracted, then X can be reconstructed perfectly from

$$X = T P^T \tag{33.4}$$

If only $A < K$ eigenvectors are used, then X is only approximately recovered. In this case,

$$X = T P^T + E_x \tag{33.5}$$

and E_x are the reconstruction errors. Figure 33.1 shows the projection of a measurement onto an eigenvector producing a score, and the component of the orthogonal error.

33.2.2 Principal Component Regression

PCR is an extension of PCA applied to the modeling of Y data from the X or measurement data. For example, if X is composed of temperatures and pressures, Y may be the set of compositions resulting from thermodynamic considerations. The approach to defining this relationship is accomplished in two steps. The first is to perform a PCA on the X data which yields a set of scores for each measurement vector, that is, if x_j is the jth vector of the K measurements at a time j, t_j is the corresponding j^{th} vector of A scores. Given the matrix of scores T, the Y data are regressed on the matrix of scores by

$$Y = Tq + E_y \tag{33.6}$$

Using the orthogonality of the matrix of eigenvectors, P and Equation 33.4, T is related to data matrix X by

$$T = XP \tag{33.7}$$

[3] orthonormal

Substituting this in Equation 33.6 gives,

$$Y = XPQ + E_y \tag{33.8}$$

or

$$B = PQ \tag{33.9}$$

where B is the principal component regression coefficient of X onto Y.

33.2.3 Partial Least Squares

PLS is a term for a family of philosophically and technically related mathematical techniques originally proposed by Herman Wold's fundamental concepts of iterative fitting of bilinear models in several blocks of variables [6], [7]. The original applications were to concrete data-analytical problems in economics and social sciences. This new technique was developed to address the shortcomings of standard methods when dealing with a modest number of observations, highly collinear variables, and data with noise in both the X and Y data sets. Standard techniques such as multiple linear regression had parameter identification and convergence problems.

PLS, sometimes called Projections onto Latent Structures, is similar to PCR. Both decompose the X data into a smaller score space T. They differ in how they relate the scores to the Y data. In PCR, the scores from the PCA decomposition of the X data are regressed onto the Y data. By contrast, in PLS, both the Y data and the X data are decomposed into scores and loadings. The orthogonal sets (T, U) which result are generated to maximize the covariance between the scores for the X data and those for the Y data. This is attractive, particularly where not all of the major sources of variability in X are correlated with the variability in the Y data. PLS attempts to find a different set of orthogonal representations for the X data to give better predictions of the Y data. Thus, a given number A of orthogonal vectors will yield a *poorer* representation of the X data, while the scores for this same set of vectors will yield a *better* prediction of Y than would be possible with PCR.

Mathematically, X is decomposed into a model for the first principal component as

$$X = t_1 p_1^T + E_{x,1}. \tag{33.10}$$

The Y data is similarly decomposed as

$$Y = u_1 q_1^T + F_{y,1}. \tag{33.11}$$

Having found t_1, p_1, u_1, and q_1, the procedure is repeated for the residual matrices $E_{x,1}$ and $F_{y,1}$ to find t_2, p_2, u_2 and q_2. This continues until the residuals contain no useful information. The t's and u's are the scores, and the p's and q's are the loadings of the X and Y data, respectively. Each t_i and u_i are related by linear regression to form

$$u_i = b_i t_i + r_i. \tag{33.12}$$

Excellent treatments with more detail are found in Geladi and Kowalski [1] and Martens and Næs [3].

The implementation of the PLS algorithm is slightly different for a single y variable as opposed to the multiple y variables. For the former, PLS is a single pass solution; for the latter, it is iterative. The approach for the single y variable is to introduce an additional set of loading *weights* W which permits an easier interpretation of the results than otherwise possible. The first step in this modeling problem, like that of PCR, is to mean center and scale the X data so that one variable does not overwhelm another by its size. Also, the y variable needs to be mean centered. The maximum number of loadings, $Amax$, to investigate must be specified. The following steps need to be performed for each factor a from $1 \leq a \leq Amax$. Here, $X(0)$ is the mean-corrected and scaled version of X, and $y(0)$ is the mean-corrected version of the single y-variable.

1. Find $\hat{w}_a$ that maximizes $\hat{w}_a^T X_{a-1}^T y_{a-1}$ subject to $\hat{w}_a^T \hat{w}_a = 1$.
2. Find the scores t_a as the projection of X_{a-1} on $\hat{w}_a$. Thus
$$t_a = X_{a-1}\hat{w}_a. \quad (33.13)$$
3. Regress X_{a-1} on t_a to find the loadings p_a via
$$p_a = X_{a-1}^T t_a/(t_a^T t_a). \quad (33.14)$$
4. Regress t_a on y_{a-1} to find
$$q_a = y_{a-1}^T t_a/(t_a^T t_a). \quad (33.15)$$
5. Subtract $t_a\, p_a^T$ from X_{a-1} and call this X_a.
6. Subtract $t_a\, q_a^T$ from y_{a-1}, and call this y_a.
7. Increment a by 1 and, if $a \neq Amax + 1$, go to step 1.

No iteration is required for finding loadings and scores when there is a single y variable.

The above algorithm needs to be slightly modified and iterated for multiple y's. The term y_{a-1} in step 1 above is replaced by a temporary u_a. The quantity u_a is iterated until convergence. Initially, u_a is chosen as one of the columns of the Y data. The algorithm for this consists of these steps:

1. $\hat{w}_a = u_a^T X_{a-1}/(u_a^T u_a)$ (regression of X onto u).
2. Normalize $\hat{w}_a$ to unit length.
3. $t_a = X_{a-1}\hat{w}_a/(\hat{w}_a^T \hat{w}_a)$ (compute scores).
4. $q_a = t_a^T Y_{a-1}/(t_a^T t_a)$ (regression of columns of Y onto t).
5. $u_a = Y_{a-1}^T q_a/(q_a^T q_a)$ (upgrade estimate of u).
6. If u_a has not converged go to 1, otherwise
7. $p_a = X_{a-1}^T t_a/(t_a^T t_a)$ (compute the x-loadings).
8. Calculate residuals $X_a = X_{a-1} - t_a p_a^T$ and $Y_a = Y_{a-1} - t_a q_a^T$
9. Increment a by 1 and, if $a \neq Amax+ 1$, set u_a to a column of Y_{a-1} and go to 1.

The use of simultaneous information in the X and Y *data* does have some disadvantages for the case of multiple y's. In order to have orthogonal sets of score vectors (Q and T), two sets of *loadings* or basis vectors (generally termed W and P) are needed for the X data. W is referred to as the *weights* and P the *loadings*. The W is orthonormal, but the P is not. Without the two, the score vectors are not orthogonal. One then loses computational advantages of having orthogonal scores (note item 8 above). If the scores are correlated, then the simultaneous regression of Y and X on all the scores T is necessary. Also, a lack of orthogonality leads to greater variance in the regression results. The need to remove one vector at a time is of overwhelming importance; hence, using the two sets of loading vectors is preferable to the simultaneous regression.

PLS, like any data modeling paradigm, may underfit or overfit the data. By underfitting, not enough loadings are used, and the model fails to capture some of the information. By overfitting, too many loadings are used, and the model tends to fit some of the noise. Both cases produce suboptimal models. Thus, it becomes of paramount importance to validate the model to avoid these problems. Although there are many ways of doing this, we will discuss only the cross-validation method. The reader is referred to Martens and Næs [3] for more detail of this and other validation methods. In cross-validation, the data, both X and corresponding Y, are segregated into groups, typically 4 – 10. Using all but one of the groups, a new PLS model is generated as the number of loadings vary from 1 to $Amax$. Each of these models is used to predict the Y data in the group withheld. The prediction error sum of squares (PRESS) is computed for each model. This procedure is repeated until each group is withheld once and only once. Then the overall PRESS is generated for a given number of loadings, a, by summing the prediction errors for all withheld data. A plot of the PRESS vs. loading number will typically reach a minimum and then start to increase again. The value corresponding to the minimum PRESS is taken as the number of loadings needed. Fewer than this number tend to underfit the data and more begin to overfit it.

33.2.4 Example

An example demonstrates how PCR and PLS compare with multiple linear regression (MLR). Consider the following X data.

$$X = \begin{bmatrix} 1.0 & 0.0 & -1.9985 \\ -1.0 & 1.0 & 3.4944 \\ 0.0 & -1.0 & -1.5034 \\ 1.0 & 1.0 & -0.4958 \end{bmatrix} \quad (33.16)$$

This X data can be viewed as three process measurements used to predict a single y variable. There are four observations of $y = [-3.2475\ 5.2389\ -2.0067\ -1.2417]^T$. The correct model relating the X data to the observation y is $y = -0.58x_1 + 1.333x_3$ where x_1 and x_3 are the first and third columns of the X data respectively.

For these data, the model identified by MLR is $y = 0.75x_1 - x_2 + 1.5x_3$. The discrepancy between the true result and this estimate is due to the poor conditioning of the X data. Column 3 is highly correlated with columns 1 and 2. In fact, $x_3 \approx -2x_1 +$

$1.5x_2$. This collinearity makes the MLR solution highly sensitive to outliers and noise in the data. To illustrate this, consider the solution when 0.001 is added to each element of y. The new solution is given by $y = 0.0703x_1 - 0.3827x_2 + 1.5889x_3$. Subtracting 0.001 from each element of y produces a solution $y = 1.5793x_1 - 1.6173x_2 + 2.4111x_3$.

The problem is not only related to noise in the *Y* data. Suppose that we add zero–mean, normally–distributed, noise with variance 10^{-4} to each element of the *X* matrix. The MLR solution now becomes $y = -5.5716x_1 + 3.7403x_2 - 1.1518x_3$.

Notice the wide changes in the model with small changes in the data. Models that exhibit high sensitivity to small errors are not very robust. They will usually yield bogus results in a predictive mode. This is a serious drawback of MLR when applied to data that is highly collinear. On the other hand, the method works well when the *X* data are orthogonal as in experimental designs.

Now, compare the results of MLR with those obtained using PCR. The PCA model requires two eigenvectors or loadings, and the resulting PCR model for the same set of data is $y = -0.6288x_1 + 0.0364x_2 + 1.3093x_3$. This is much closer to the true model than was found with MLR. Furthermore, the PCR solution is less sensitive to small deviations in either the *Y* data or *X* data. The model is essentially unchanged if 0.001 is added or subtracted from the original *Y* data. The same is true when a the small noise signal (10^{-4} variance) is added to the *X* data.

PLS works equally well. A two-loading model provides a similar solution to PCR, $y = -0.62831x_1 + 0.03697x_2 + 1.3093x_3$. Like PCR, the PLS solution is less sensitive to small errors in the *Y* and *X* data than the MLR method. Clearly, PLS and PCR have a considerable advantage over MLR when there are few observations of correlated *X* data.

33.3 Areas of Applications

Multivariate statistical methods have many applications in the process industry. This section will deal with two application areas: data analysis/estimation and inferential measurement for control. Data analysis can be used to develop a model of the expected behavior of the process using PCA/PLS techniques. In theory, deviations from the model are an indication that something is amiss. Inferential measurements are used in cases where the quantity to be controlled is measured by infrequent laboratory analysis. A statistical model can be used to estimate its value for control purposes. A method for controlling in the score space is developed. Examples are presented to demonstrate these concepts.

33.3.1 Data Analysis

Chemical processes typically have automated methods for collecting large amounts of data. These data may be a potential gold mine of information, but because tools are not in place to separate pertinent information quickly from irrelevant noise. This potential is not realized. Multivariate statistical methods may possibly aid in separating the significant information from the noise.

Although many measurements are made, there are only a few physical phenomena occurring. Thus, many of the variables are highly correlated. Trying to learn about the process by looking at the data in a univariate fashion, only confuses the issue. Multivariate methods, such as PLS and PCA, often capture the essence of the information in a lower dimensional space defined by two or three primary loadings or latent vectors. Observing these data and understanding the significance of any clusters in this lower dimensional space generally leads to insights about the process.

A process can be viewed as an instrument which can be calibrated with multivariate statistical tools [8]. Process data provide information about the process. If this instrument is calibrated correctly, that is, it provides the same measures for all periods in which the outputs or the quality variables are on target, then it can provide the kind of information that process engineers and operators need. If the process is drifting away from the quality targets, then this should be detected easily on-line and adjusted appropriately to prevent a poorer quality product and to reduce losses and downtime. To achieve this goal, reference data are gathered when the process is producing top grade product. Process variability may be captured in a PCA model providing a *fingerprint* of the process and setting the standard by which process operation is judged. New data are compared to this multivariate model to determine if they are consistent with the normal operation.

Process Monitoring and Detection

Real time monitoring addresses the classification of new process data relative to the reference data. Careful selection of the reference process data (normal operating conditions) must be used to develop the calibration model. Detection, on the other hand, judges the appropriateness of the model. Both real time monitoring and detection will be discussed in the context of the score space.

The Mahalanobis distance, h_i, is a measure of the extent to which the scores for new data, x_i can be classified as belonging to the set of scores used to generate the calibration model [9]. Thus, h_i measures the goodness of fit *within the model space.* For a new data vector x_i of size $(1 \times m)$, h (scalar) is calculated as

$$h_i = (x_i - \mu)^T S^{-1} (x_i - \mu) \tag{33.17}$$

where μ is the centroid of the calibration model, and S is the variance-covariance matrix of the calibration set X. Because t is the projection of data, x, onto a reduced space defined by the first A eigenvectors, and because the mean of the score vectors is zero due to mean centering of the data, h_i can be computed in the score space defined by the principal components as

$$h_i = \frac{1}{m} + \sum_{a=1}^{A} \frac{t_{ia}^2}{t_a' t_a} \tag{33.18}$$

where m is the number of process measurements used in the model generation, t_{ia} is the ath score for the data vector, x_i, and t_a is the vector of scores corresponding to the ath principal component for the data used to generate the model. Following Shah and Gemperline [9], if h_i is beyond the 90th percentile of the calibration set for A principal components, then the scores for x_i are considered outside the model space; otherwise the new data vector is similar to the reference data.

Process Description

An example is presented to illustrate the application of process monitoring and detection to a continuous chemical process [8]. In the first stage of that process, several chemical reactions occur that produce a viscous polymer product; in the second stage, the polymer is treated mechanically to prepare it for the third and final stage. Critical properties, such as viscosity and density, if altered significantly, will affect the final product resulting in a loss of revenue and operability problems for the customer. Moreover, it is difficult to determine which stage is responsible for the quality degradation. A lack of on-line sensors to measure continuously the critical properties makes it impossible to relate any specific changes to a particular stage. Indeed, property changes are detected by laboratory measurements with delays of 8 hours or more. Because the analytical results represent past information, the current state of operations may not reflect the process state.

Infrequent measurements make the control of product quality difficult. At best, the operators have learned a set of heuristics, that, if adhered to, usually produces a good product. However, unforeseen disturbances and undetected equipment degradations may occur which will also affect the product. There are periods of operations when the final process step produces a degraded product in spite of near-perfect upstream operations.

Since stage two operates in support of stage three and to compensate for property errors in the first stage, this work will focus on process monitoring and detection of stage two. A simplified description of stage two is as follows. The feed from the reaction stage is combined with a solvent in a series of mixers operated at carefully controlled speeds and temperatures. A sample of the mixture is taken for lab analysis at the exit from the final mixer. From there, the process fluid is sent to a blender whose level and speed are controlled to impart certain properties. The fluid is then filtered to remove undissolved particulates before it is sent to the final stage of the process. A significant amount of mechanical energy is necessary to move the viscous fluid through a complex transport network of pipes, pumps, and filters. The life span of the equipment is unpredictable. Even the same type of equipment placed in service at the same time may need replacing at widely differing times. Problems of incipient failure, pluggage, and unscheduled downtimes are an accepted, albeit undesirable, part of the operations. To prolong continuous operations, pumps and filters are installed in pairs so that the load on one can be temporarily increased while the other is being serviced. Tight control of the process fluid is desirable because the equipment settings in the final stage of the process are preset to receive a uniform process fluid. As such, small deviations in the fluid properties may result in machine failure and nonsaleable product.

Frequent equipment maintenance is not the only source of control problems. Abrupt changes also occur due to throughput demands in the final stage of the process. For example, if there is a decrease in demand due to downstream equipment failure, or a sudden increase due to the addition of new or reserviced equipment, the second stage must reduce or increase production as quickly as possible. It is more dramatic when throughput has to be reduced because the process fluid properties will change if not treated immediately. Because situations are frequent and unscheduled, the second stage of this process changes significantly and never quite reaches equilibrium. Clearly, throughput has dominant effect on the variability in the sensor values and process performance.

Model Development

A calibration model to monitor and improve second stage process operations is developed because it is here that critical properties are imparted to the process fluid. Intuitively, if the second stage can be monitored to anticipate shifts in normal process operation or to detect equipment failure, then corrective action can be taken to minimize these effects on the final product. One of the limitations of this concept is that disturbances, that may affect the final product, will not manifest themselves in the variables from which the model is created. The converse is also true, that disturbances in the monitored variables may not affect the final product. However, faced with few choices, a calibration model is a rational approach to monitor and detect unusual process behavior for improved process understanding.

Because throughput has an effect on all of the measurements, a PCA analysis would be overwhelmed by this effect, and it would obscure other information about the state of the process. By first eliminating the throughput effect using PLS analysis, an examination of the residuals using PCA reveals other sources of variations critical to process operations.

Routine process data are collected over a period of several months. Cross-validation detects and removes outliers, and only data corresponding to *normal* process operations, that is, when top grade product is made, are used in the model development. Two calibration models are developed; both are reduced order models that capture dominant directions of data variability. The PLS model shows that two loadings are needed to explain approximately 60% of the variance in the measurements. A third loading does not significantly change the total explained variance, and 100% of the variance in the throughput variables are explained. PCA analysis on the residuals shows that five principal components explain 90% of the residual variability. Additional components provide no added statistical significance.

Figure 33.2 shows the observations of the residual data plotted in the score space of principal components 1 and 2. The observations are spread over a wide region and include a wide range of rate settings. In a related work, Piovoso et al. [12] show how filtering the data with a nonlinear, finite impulse, median hybrid filter can magnify differences in the operations and reveal structure in the data. By examining the loadings of the PCA, one may be able to relate the principal components to some physical process phenomena. This is particularly true for the early components because they explain most of the variability.

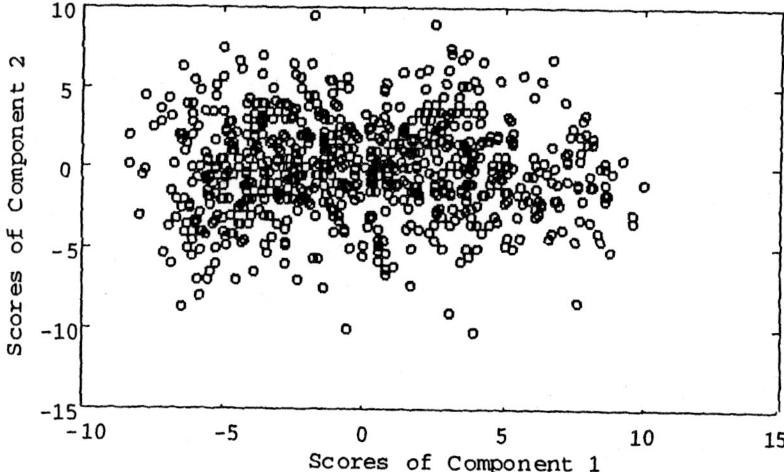

Figure 33.2 Scores of component 1 vs. scores of component 2, PCA model.

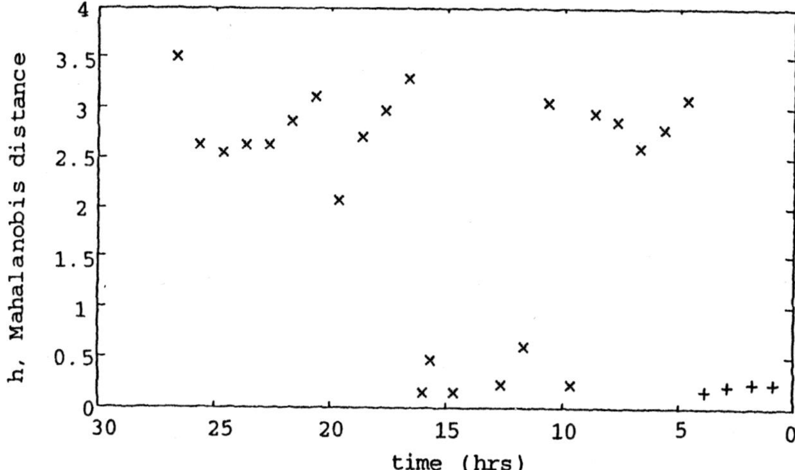

Figure 33.3 Mahalanobis distance for 30 hours of on-line operation. + : most recent 4 hours; x: most distant history.

On-Line Monitoring and Detection

The PLS and PCA models are used on-line to monitor and to detect statistically significant deviations in process operations. The system configuration is given in Piovoso et al. [8]. The Mahalanobis distance is calculated for each new data vector. The Mahalanobis criterion gives a measure of the location of new data within the model subspace. If h_i is within the model space, then no alert is sent to the operators. Figure 33.3 illustrates a plot of the normalized Mahalanobis distance for a period of 30 hours with the most recent time at the right side of the plot. The normalization is done with respect to the largest Mahalanobis distance in the calibration model. The operations from the fifth to the eleventh hour in the past indicate that the process variables fell outside the region of normal plant behavior. The reason for the unusual behavior for that period was an unexpected pump failure. There is indication of this failure at the eleventh hour and actual failure occurred at the ninth hour. During the most distant period, the process variables are again outside of the desired region because there was a filter pluggage due to large particulates in the process fluid.

Along with the Mahalanobis distance plots, diagrams such as that shown in Figure 33.4 are used by operating personnel to monitor the process operations easily. The o denote the scores for the calibration model, and the + denote the most recent period of operation. The arrows are used to indicate the process history. Clearly, there is a time period in which the process operation can be judged dissimilar to the calibration set. Either an unusual control action or a disturbance shifted the process state from the desired region as evidenced by the two operating points in the upper left hand side. Eventually, appropriate control actions are applied and the process returns within the region defined by the calibration set.

One current limitation is the lack of good representative data. A design of experiments is necessary to map the entire range of operations, as is discussed for the binary distillation column example presented in Section 33.3.4. If only a portion of the operating space is known, it is possible that the correlations between the inputs/outputs under closed-loop operations may be incorrect if the process moves to a regime that is not a part of the reference set used to develop the model.

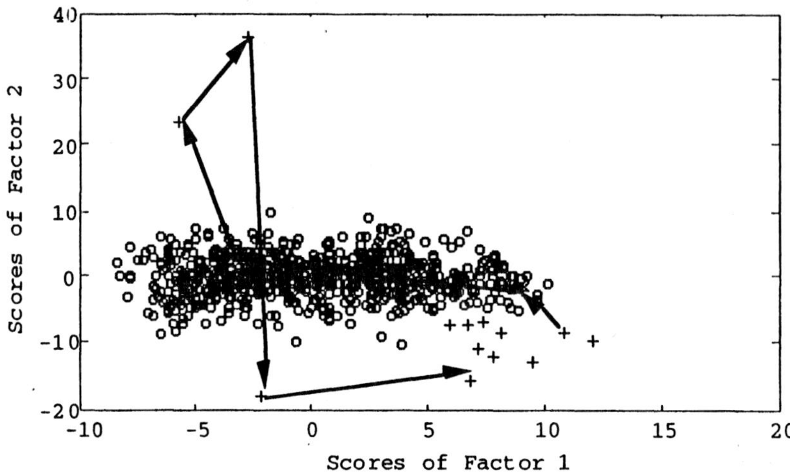

Figure 33.4 Score plot. + : most recent operation, o: desired operating region.

33.3.2 Batch Processes

Batch and semibatch processes play an important role in the chemical industry due to their low volume, high value products. Examples include reactors, crystallizers, injection molding processes, and the manufacture of polymers. Batch processes are characterized by a prescribed processing of materials for a finite duration. Successful operation means tracking a prescribed recipe with a high degree of reproducibility from batch to batch. Temperature and pressure profiles are implemented with servo-controllers, and precise sequencing operations are produced with tools such as programmable logic controllers.

The main characteristics of batch processes, flexibility, finite duration, and nonlinear behavior, are associated with both their success and their incompatibility with the usual techniques for monitoring and control. However, disturbances and the absence of on-line quality measurements often affect the reproducibility of batch processes. Nomikos and MacGregor [13] propose Statistical Process Control (SPC) schemes for batch processes, based on Multiway PCA (MPCA), that use on–line measurements directly to recognize systematically and scientifically significant deviations from normal process operating behavior. Analogous to the prior example, an empirical model, based on the MPCA analysis of data obtained when the process is operating well, is used to characterize normal process behavior. The evolution of future batches is then monitored by comparing them against this MPCA model using the statistical control limits developed from the reference database. Kosanovich et al. [14] discuss the application to an industrial batch process.

MPCA Method

MPCA is an extension of PCA to handle data in three–dimensional arrays. The three dimensions arise from batch trajectories that consist of batch runs, variables, and sample times. These data are organized into an array X of dimension $(I \times J \times K)$ where I is the number of batches, J is the number of variables, and K is the number of samples over the duration of the batch. MPCA is equivalent to performing ordinary PCA on a two–dimensional matrix formed by unfolding X so that each of its vertical slices contains the observed variables for all batches at a given time. In this approach, MPCA explains the variation of variables about their mean trajectories. MPCA decomposes X into a summation of the product of t–score vectors and p–loading matrices, plus a residual matrix (E_x) that is minimized in a least-squares sense,

$$X = \sum_{r=1}^{R} t_r \bigotimes P_r + E_x$$

and R is the number of principal components used in the analysis[4].

This decomposition summarizes and compresses the data, with respect to both variables and time, into low dimensional score spaces. These spaces represent the major variability over the batches at all points in time. Each p-loading matrix summarizes the major time variations of the variables about their average trajectories over all the batches. By doing this, MPCA utilizes the magnitude of the deviation of each variable from its mean trajectory and the correlations among them. The appropriate number of principal components may be found by cross-validation.

To analyze the performance of a set of batch runs, an MPCA analysis can be performed on all of the batches, and the scores for each batch can be plotted in the space of the principal components. All batches exhibiting similar time histories will have scores which cluster in the same region of the principal component space. Batches that exhibit deviations from normal behavior will have scores falling outside the main cluster, but batches with similar behavior will cluster in the same region.

Process Description

The chemical process from which data are taken is a single batch polymer reactor [14]. The critical properties that must be controlled for the final product are related to the extent of reaction (e.g., molecular weight distribution). The product's critical

[4] $\bigotimes$ is the tensor product

properties are determined by off-line chemical measurements, and not every batch has its critical properties measured. The property measurement results are available 12 hours or more after the completion of each batch. These results cannot be used in a timely fashion to compensate for poor product quality. Furthermore, it is often difficult to establish the root cause of property deviation when a bad batch is manufactured.

The total time for the batch cycle is less than 2 hours. For this analysis, we can consider the two dominant phenomena, vaporization and polymerization, that produce the polymer. During the first part of the batch cycle, the solvent is vaporized and removed from the reactor; this takes approximately one hour. In the latter part, polymerization occurs to attain a desired molecular weight distribution. The finished product is then expelled from the vessel under pressure to complete the cycle. The total batch time is monitored carefully, and under the continuously supplied external heat source is the main control knob. There is also statistical control of the vessel temperature. The known sources of variability from batch to batch within a product type are variations in the heat content of the heat source, various levels of impurities in the ingredients, and residual polymer buildup over the operating life of a reactor.

Analysis and Results

Data for 50 batches made in the same reactor, for the same batch recipe, are collected at one-minute intervals. The database variables contain information about the state of the reactor (temperatures, pressures) and the state of the external heat source. This is not unusual as temperatures and pressures reflect the progress of the reaction in the vessel. Initial analysis indicates that changes in the level and quality of the external heat source result in clustering within the score space. A score plot of principal components 1 and 2 is shown in Figure 33.5. To eliminate that effect, only those batches with the same setting are studied. The data are further segregated into two groups reflecting vaporization and polymerization. In the interest of space, only analysis of the vaporization stage is discussed.

Applying MPCA to the vaporization stage data reveals that the first direction of variability is related to the reactor temperature rise and the second, to the quality of the heat supplied by the heat source. These two principal components explain approximately 55% of the total variance. Heat effects dominate because boiling is the primary event, and the boiling rate and subsequent temperature rise depend on the heat transfer rate from the heat source to the reactor content. What is learned, however, is that, for a constant heat input and similar reactor initial conditions, the rate at which the reactor temperature rises differs from batch to batch. This implies that some batches may boil more quickly than others and that maintaining the prescribed boiling time may vaporize more than just the solvent, thereby affecting the final polymer composition. A control strategy forcing the batch to follow a prescribed reactor center temperature profile rather than a timed sequence should reduce the variability in the temperature rise from batch to batch. Additional work by Kosanovich and Schnelle [17] on the same process supports this recommendation.

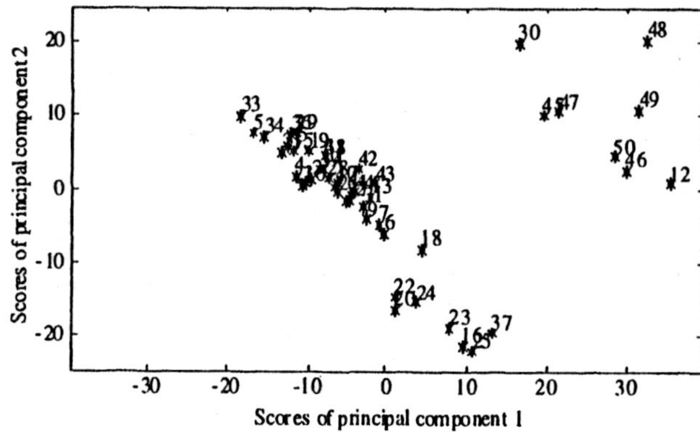

Figure 33.5 Score plot of the first two principal components for 50 batches using stage 1 data.

33.3.3 Inferential Control

Multivariate statistical methods can be used to develop nonparametric models for feedback control. This is useful particularly in cases where instruments are not available to provide on-line measurements of the quantity to be controlled. One example is composition measurement. Models can be used to predict the information needed, and first-principles models are best suited for this. However, such models are time consuming to develop and are generally not available. Multivariate statistical models are an alternative [15]. One can develop a nonparametric model to predict the quantity to be controlled and use this as the measurement in a closed-loop configuration [2]. Alternatively, closed-loop control can be formulated and implemented within the reduced space defined by a PCA model [4], [16].

PCA/PCR Controller Design

A controller design is proposed based on a PCA model. Within the chemical industry, processes involve many stages, all influencing the final product properties. Maintaining these properties at their specifications requires good control of all stages. Processes of this type have a large number of exogenous variables and a few manipulated ones per process stage. The exogenous variables indicate the process state, and the manipulated variables control indirectly unmeasured quantities. This approach provides no automatic mechanism to adjust controller set points when disturbances occur. Careful monitoring of these variables will allow correlation of good and bad product properties with the exogenous variables' variations. To the extent that quality data can be obtained, a relationship between the exogenous and the quality variables can define an acceptable region of operation. When the operation is inside this region, the process is functioning as it did when good product properties were observed. If the process is deviating from the desired process region, suitable control action based on the relationship between the exogenous and the manipulated variables will return the process within the desired process region.

More precisely, we can develop a PCA model to represent the desired process region in the score space, and then design a controller in the score space that maintains operation within this region. The control moves in the score space are then mapped to the real variable space and implemented in the process. In this fashion, the process is kept within the desired region provided that the PCA model has correctly established the relationships between the exogenous variables and the manipulated ones. This proposed control formulation is analogous to modal control. A high purity binary distillation column operated at atmospheric pressure is selected as an example to explain the development of the controller. The control objective is to maintain the distillate product purity, x_d at 99.5%; the tray temperatures are the exogenous variables, X_{ex}, and reflux rate is the manipulated variable, x_{mp}.

Let X be composed of two types of variables X_{ex} and X_{mp}. For various operating conditions,

$$X = [X_{ex}|X_{mp}] \tag{33.19}$$

The development of a PCR model yields,

$$[X_{ex}|X_{mp}] = TP^T + E_x \tag{33.20}$$

and

$$x_d = Tq^T + f_y \tag{33.21}$$

The equivalent controller set point in the score space is determined from $x_{d,sp}$ by

$$t_{sp} = x_{d,sp}(q^T)^\dagger \tag{33.22}$$

where $(q^T)^\dagger$ is the pseudoinverse of q^T, a (A x 1) vector. The score vector, t, can be computed from the projection of x onto the matrix of eigenvectors P

$$t = xP. \tag{33.23}$$

Define $\Delta t = t_{sp} - t$ in the score space as the error between the desired score set point and the scores associated with the vector, x, at a sample time. Conversely, the error in the score space can be reconstructed as an error in the X space by

$$\Delta x = \Delta t P^T \tag{33.24}$$

The temperature variables cannot be manipulated arbitrarily; only the reflux rate can be changed to drive the process in a direction that produces a new x vector so that $t \rightarrow t_{sp}$. In the X space, this implies that the required changes in the reflux rate ought to drive the temperatures toward the values that produce $x_{d,sp}$ (t_{sp}). From the score space perspective, the manipulated variable changes must be determined so as to generate changes in the exogenous variables consistent with remaining in the desirable part of the score space. To achieve this, the relationship between the exogenous and manipulated variables in the score space must be defined.

Consider the partition of the matrix of eigenvectors P as

$$P^T = [P_{ex}|P_{mp}] \tag{33.25}$$

P_{ex} is an $(A \times r)$ matrix, where r is the number of exogenous variables, and P_{mp} is an $(A \times (m - r))$ matrix, where $(m - r)$ is the number of manipulated variables. The relationship between the exogenous and the manipulated variables can be found from the relationship between P_{ex} and P_{mp}, as

$$P_{mp} = P_{ex}\Lambda \tag{33.26}$$

where Λ is $(r \times (m-r))$ matrix of coefficients that defines the relationship between the manipulated and the exogenous variables in the score space.

The justification for the linear relationship, Equation 33.26, lies in the fact that the eigenvectors define a hyperplane within the X space. Because the data are mean centered, the hyperplane goes through the origin and the eigenvectors, p_j^T, are unit vectors that lie on the hyperplane. Hence, p_j^T defines a point on this hyperplane. Solving the above equation for Λ yields,

$$\Lambda = P_{ex}^\dagger P_{mp} \tag{33.27}$$

where $P_{ex}^\dagger$ is the pseudoinverse of P_{ex}.

Implementation

The PCA/PCR controller would function in the following way. Given $x_{d,sp}$ the corresponding t_{sp} can be determined using Equation 33.22. Similarly, given x, t is found from Equation 33.23. The difference between t and t_{sp} represents the desired change in the scores, Δt, which can be used to generate a corresponding Δx from Equation 33.24. Using the Λ matrix from Equation 33.27, the change in the reflux rate that would drive future changes in tray temperatures closer to zero, so that $x_d \rightarrow x_{d,sp}$, is found.

Because this is a steady-state model, the resulting controller moves may be exceptionally large, violating constraints in the X space. Thus, only a fraction of the change can be implemented. With no knowledge of the process dynamics built into the model, this fraction becomes a tuning parameter. If the control interval is long compared to the plant dynamics, all of the computed changes could be implemented. On the other hand, if the control interval is short compared to the time constants of the plant, the fully implemented calculated control move might be too large because the plant will not be able to respond fast enough. Figure 33.6 illustrates this scheme.

33.3.4 Binary Distillation Column

The distillation column is by far the most commonly studied process in the chemical engineering literature. In this work, we attempt to estimate and control the distillate composition of a binary column at a 99.5% purity. For our purposes, the following assumptions are made: constant molal overflow (CMO), 100% stage efficiency, and constant column pressure. A simple, linear liquid hydraulic relationship between the liquid leaving the nth tray and the holdup on the nth tray is assumed [10]. There are a total of 20 trays, the condenser is a total condenser, and the reboiler is modeled as another tray in the column. The vapor mole fraction is obtained by a bubble point calculation, and we assume ideal vapor phase and Raoult's law. Two proportional-plus-integral (PI) controllers are used at both ends of the column

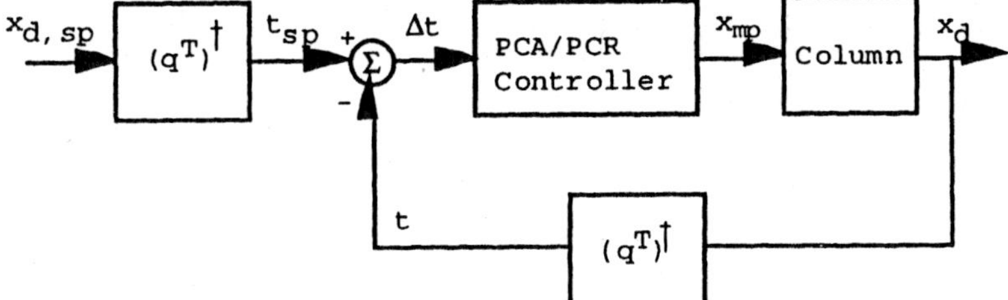

Figure 33.6 Block diagram of a PCA/PCR controller.

to maintain the inventory in the reflux drum and the level in the bottom of the column. Two additional PI controllers are used to control the distillate and bottoms compositions in an L/V configuration. Nominal conditions for the column are: reflux ratio of 2 and distillate/feed ratio of 0.5.

Composition Estimation

It is typical to control the end point compositions using the temperatures on a selected tray, in the rectifying section, and in the stripping section. In most situations, this is effective; however, there are cases which are not pathological, showing that the relationship between a single temperature and product composition cannot account for changes in feed composition and the product changes at the other end of the column. In most industrial columns, temperature measurements are available at more than one location; it would seem prudent to make use of all observed system data to infer the end point compositions. In this way dependence on a single tray temperature is eliminated and the use of process information maximized. Mejdell and Skogestad [11] proposed the use of such a composition estimator using PCR or PLS to estimate end point compositions in a binary and a multicomponent distillation column. Their results indicate improved estimates of the product compositions and robustness to measurement noise. Their work demonstrates the use of static estimators for dynamic control when no composition measurements are available.

The data sets are the steady-state temperature and distillate profiles obtained from a four-factor, five-level designed experiment where feed rate, feed composition, vapor and reflux rates are varied. Variations on the order of ± 20% changes in feed composition and flow rates are contained in the data. In addition, the data are collected with an uncorrelated, uniformly distributed noise of $\pm 0.2^o$C on the temperature measurements. A PLS model with three loading vectors using only eight tray temperatures is sufficient based upon cross-validation statistics, to explain 98% of the total variance in the distillate composition. This PLS model based on tray temperatures is used only to provide an estimate of the distillate composition for feedback control using a PI controller (see Figure 33.7 (left)).

It is conceivable to develop a composition estimator based on temperatures and reflux (see Figure 33.7 (right)). In the PLS model, reflux is not scaled as are the temperatures because it is not the same type of measurement [2]. Only mean centering is applied to the reflux data, and the data are obtained as described previously. As before, a PLS model with three loadings explains 97% of the total variance in the composition.

Both models predict the distillate composition adequately. The temperature–reflux prediction model is not as accurate as the temperature only estimator which is consistent with the findings of Mejdell and Skogestad [11]. For this reason, they did not consider the contribution of manipulated variable measurements to the development of the composition estimator. Feedback control, using either of these two estimation schemes for distillate composition set point and input load changes, results in satisfactory prediction of composition and good closed-loop performance.

PCA/PCR Estimation and Control

To apply the PCA/PCR controller formulation to control the column requires that a PCA model be developed from data composed of temperatures and reflux information. The scores of that model are then regressed on the distillate composition information (PCR). By cross-validation, it is determined that three principal components are needed to explain 96% of the total variability in the distillate composition. The controller is implemented as discussed in Section 33.3.3. This provides an integral only controller. The response of the system to both a set point change and a load disturbance is more aggressive compared to the temperature only/PI and temperature-reflux/PI cases. The Ziegler-Nichols tuning methodology is used to set the PI controller parameter values. Figures 33.8 and 33.9 summarize the system's response in both the PCA/PCR controller and the temperature-reflux/PI controller studies. The PCA/PCR controller case produces less noise in the manipulated variable than does the temperature-reflux/PI controller case.

An appropriate fraction of the control move can be implemented in situations where the controller action is too aggressive. For the distillation column we would trade off speed of response with noise in the control moves and in the estimate of the distillate composition. In some situations, if the portion of the control move implemented is too large, the closed-loop system may become unstable. Conversely, if it is too small, the response might be too sluggish. Constraints on the size of the manipulated variable or on the rate of change of the manipulated variable can be incorporated in a straightforward fashion for any of the models discussed here.

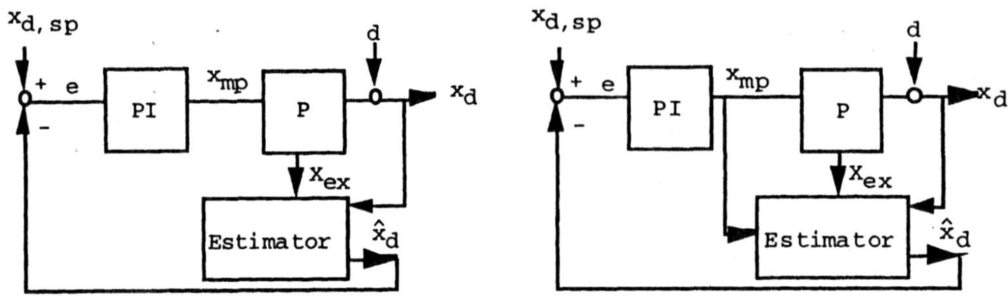

Figure 33.7 Feedback control structure with temperature only (left) and temperature-reflux (right).

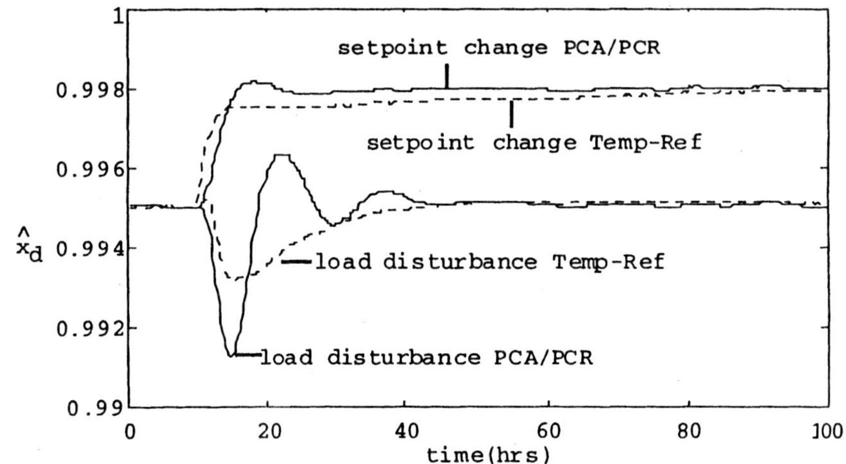

Figure 33.8 Estimated distillate composition, PCA/PCR controller scheme.

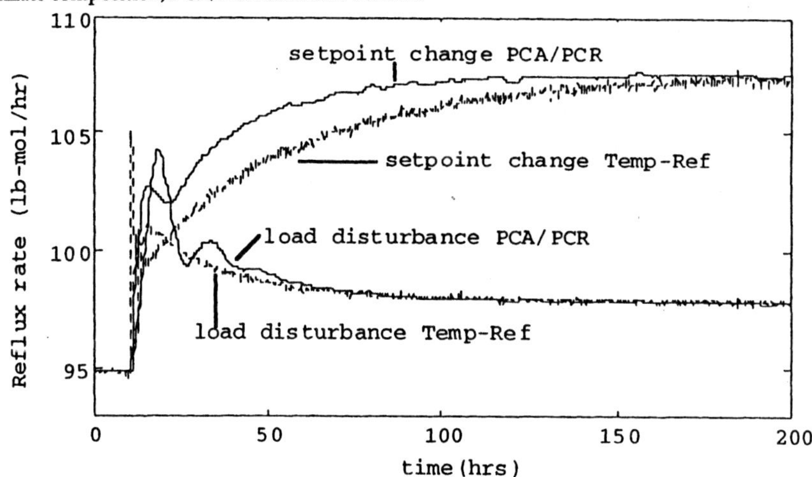

Figure 33.9 Reflux rate, PCA/PCR controller scheme.

33.4 Summary

Data are gathered in many chemical processes at very high rates. Unfortunately, much of that data is analyzed infrequently unless a major operating problem occurs. This chapter presents multivariate statistical techniques to reduce the vast array of numbers into a smaller, more meaningful set, to deal with noise, collinearity, and to provide a basis for improved control.

Three such techniques are presented: Principal Component Analysis, Principal Component Regression, and Partial Least Squares. A generic example is provided to highlight the utility of these techniques and to compare it to the more traditional multiple linear regression. A real industrial example illustrates the combined use of PLS and PCA for monitoring and detection of abnormal events. Data are first analyzed to determine normal variability in the process. Subsequently, models are developed which compactly define that variability. The information in the model is then used to classify the current process state as to whether it is a member of the class of normal variations. If it is not, information (not root cause) why the data are not of the expected form is made available to the operator to allow for corrective action. Additional discussion on the use of the Mahalanobis distance is provided as a discriminant statistic.

An extension to PCA, Multiway Principal Component Analysis is developed for batch processes. This is necessary to handle the three-dimensional nature of batch data. Its usage is demonstrated on an industrial batch polymer reactor to provide new insights, to corroborate existing process knowledge, and to propose meaningful controls improvement. The analysis indicates clearly that a division of the data into sets corresponding to two chemical phenomena provide clarity of information within the data and allow for interpretation based on process understanding. Doing so leads to the identification of the principal directions of variability associated with each phenomenon and suggests where to improve the existing control strategy.

A novel controller design using PCA and PCR is presented and demonstrated on a high purity distillation column. Its development is based on producing manipulated variable moves that are a function of the exogenous variables that produce a set of scores within the score space of the PCA model. Such a controller belongs to the class of modal controllers. The results on the distillation column, when the entire control action is implemented, show that the controller action is aggressive with less noisy estimates and faster settling times. Implementing a fraction of the controller action gives comparable results when compared to the temperature-reflux/PI controller case.

References

[1] Geladi, P. and Kowalski, B., Partial least-squares regression: a tutorial, *Analytica Chim. Acta,* 185:1–17, 1986.

[2] Kresta, J.V. and Macgregor, J. F., Multivariate statistical monitoring of process operating performance, *Can. J. Chem. Eng.,* 69:35–47, 1991.

[3] Martens, H. and Næs, T., *Multivariate Calibration,* John Wiley & Sons, New York, 1989.

[4] Piovoso, M. J. and Kosanovich, K. A., Applications of multivariate statistical methods to process monitoring and controller design, *Int. J. Control,* 59(3):743–765, 1994.

[5] Wold, S., Esbensen, K., and Geladi, P., Principal component analysis, *Chemo. Intel. Lab. Sys.,* 2:37–52, 1987.

[6] Wold, H., Soft Modelling. The Basic Design and Some Extensions, in *Systems Under Indirect Observations,* K. J″oreskog and H. Wold, eds., Elsevier Science, North Holland, Amsterdam, 1982.

[7] Wold, S., Ruhe, A., Wold, H., and Dunn, W., The collinearity problem in linear regression. The partial least squares (PLS) approach to generalized inverses, *SIAM J. Sci. Stat. Comp.,* 5:735–743, 1984.

[8] Piovoso, M. J., Kosanovich, K. A., and Yuk, J. P., Process data chemometrics, *IEEE Trans. Instrum. Meas.,* 41(2):262–268, 1992.

[9] Shah, N. K. and Gemperline, P. J., Combination of the mahalanobis distance and residual variance pattern recognition techniques for classification of near-infrared reflection spectra, *Am. Chem. Soc.,* 62:465–470, 1990.

[10] Luyben, W. L., *Process Modeling, Simulation and Control for Chemical Engineers,* McGraw Hill, New York, 1990.

[11] Mejdell, T. and Skogestad, S., Estimation of distillation composition from multiple temperature measurements using partial least squares regression, *Ind. Eng. Chem. Res.,* 30:2543–2555, 1991.

[12] Piovoso, M. J., Kosanovich, K. A., and Pearson, R. K., Monitoring process performance in real-time, *Proc. Am. Control Conf.,* Chicago, IL, 3:2359–2363, 1992.

[13] Nomikos, P. and MacGregor, J. F., Monitoring of batch processes using multi-way PCA, *AIChE J.,* 40:1361-1375, 1994.

[14] Kosanovich, K. A., Piovoso, M. J., Dahl, K. S., MacGregor, J. F., and Nomikos, P., Multi-way PCA applied to an industrial batch process, *Proc. Am. Control Conf.,* Baltimore, MD, 2:1294–1298, 1994.

[15] Ljung, L. 1989. *System Identification, Theory for the User,* Prentice Hall, Englewood Cliffs, NJ, 1989.

[16] Kasper, M. H. and Ray, W. H., Chemometric methods for process monitoring and high performance controller design, *AIChE J.,* 38:1593–1608, 1992.

[17] Kosanovich, K. A. and Schnelle, P. D., Improved regulation of an industrial batch reactor, *Spring AIChE Conf.,* Houston, TX, 1995.

SECTION VIII

Kalman Filter and Observers

34

Linear Systems and White Noise

William S. Levine
Department of Electrical Engineering, University of Maryland, College Park, MD

34.1 Introduction

A linear system with white noise added to the input and, often but not always, white noise added to the output is the most common model for randomness in control systems. It is the basis for Kalman filtering and the linear quadratic Gaussian (LQG) or H_2 optimal regulator. This chapter presents the basic facts about linear systems and white noise and the intuition and ideas underlying them. Results are emphasized, not mathematically rigorous proofs. It is assumed that the reader is familiar with the elementary aspects of probability at, for example, the level of Leon-Garcia [1].

There are many reasons why people so commonly use a linear system driven by white noise as a model of randomness despite the fact that no system is truly linear and no noise is truly white. One of the reasons is that such a model is both elementary and tractable. The calculations are easy. The results can be understood without a deep knowledge of the mathematics of stochastic processes.

A second reason is that a large class of stochastic processes can be represented as the output of a linear system driven by white noise. The precise result can be found later in this article. The generality and tractability of this model can be quite dangerous because they often lead people to use it inappropriately. Some of the limitations of the model will also be discussed.

Scalar discrete time stochastic processes are described first, because this is the simplest case. This is followed by a discussion of the ways single-input single-output (SISO) discrete-time linear systems operate on scalar discrete-time stochastic processes. Vector discrete-time stochastic processes and multiple-input multiple-output (MIMO) linear systems, a notationally more difficult but conceptually identical situation, are then briefly covered.

The second half of this chapter describes continuous time stochastic processes and linear systems in the same order as was used for discrete time.

34.2 Discrete Time

The only difference between an n-dimensional vector random variable and a scalar discrete-time stochastic process over n time steps is in how they are interpreted. Mathematically, they are identical. As an aid to understanding discrete time stochastic processes, this equivalence will be emphasized in the following.

34.2.1 Basics

The usual precise mathematical definition of a random variable [2,3] is unnecessary here. It is sufficient to define an n-dimensional random variable by means of its probability density function.

DEFINITION 34.1 An n-dimensional random variable, denoted $\underline{\mathbf{x}} = [\mathbf{x}_1\ \mathbf{x}_2 \ldots \mathbf{x}_n]'$, takes values $\underline{x} \epsilon R^n$ according to a probability that can be determined from the probability density function (pdf) $p_{\underline{\mathbf{x}}}(\underline{x})$. Unfortunately, the notation needed to describe stochastic processes precisely is very complicated. Thus, it is important to recognize that bold letters always denote random variables while the values that may be taken by a random vari-

0-8493-8570-9/96/$0.00+$.50

able are denoted by standard letters. For example, **x** is a random variable that may take values x. Underlined lower case letters will always denote vectors or discrete-time stochastic processes. Underlined capital letters will denote matrices. A list of all the notation used in this chapter appears at the end of the chapter. The reader is assumed to know the basic properties of pdfs.

DEFINITION 34.2 A scalar discrete-time stochastic process over n time steps, $\mathbf{x}(1), \mathbf{x}(2), \ldots, \mathbf{x}(n)$, denoted $\underline{\mathbf{x}}$, takes values $x(k) \epsilon R$, where R denotes the real numbers and $k = 1, 2, \ldots, n$, according to a probability that can be determined from the pdf $p_{\underline{\mathbf{x}}}(\underline{x})$.

The equivalence is obvious once $\mathbf{x}(1), \mathbf{x}(2), \ldots, \mathbf{x}(n)$ is written as a vector $[\mathbf{x}(1)\, \mathbf{x}(2) \ldots \mathbf{x}(n)]' = [\mathbf{x}_1\, \mathbf{x}_2 \ldots \mathbf{x}_n]'$ where $'$ denotes the transpose. The equivalence is emphasized by using the same notation for both. The context will make it clear which is meant whenever it matters.

It is important to be able to specify the relationships among a collection of random variables. A simple special case is that they are completely unrelated.

DEFINITION 34.3 The n-vector random variable (equivalently, discrete-time stochastic process) $\underline{\mathbf{x}}$, is composed of n independent random variables $\mathbf{x}_1, \mathbf{x}_2, \ldots, \mathbf{x}_n$ if and only if their joint pdf has the form

$$p_{\underline{\mathbf{x}}}(\underline{x}) = \prod_{k=1}^{n} p_{\mathbf{x}_k}(x_k) = p_{\mathbf{x}_1}(x_1) p_{\mathbf{x}_2}(x_2) \ldots p_{\mathbf{x}_n}(x_n) \quad (34.1)$$

Independence is an extreme case. More typically, and more interestingly, the individual elements of a vector random variable or a scalar discrete-time stochastic process are related. One way to characterize these relationships is by means of the conditional pdf, $p_{\underline{\mathbf{x}}_1|\underline{\mathbf{x}}_2}(\underline{x}_1|\underline{\mathbf{x}}_2)$. When $p_{\underline{\mathbf{x}}_2}(\underline{x}_2) \neq 0$ the conditional pdf is given by

$$p_{\underline{\mathbf{x}}_1|\underline{\mathbf{x}}_2}(\underline{x}_1|\underline{\mathbf{x}}_2) = \frac{p_{\underline{\mathbf{x}}}(\underline{x})}{p_{\underline{\mathbf{x}}_2}(\underline{x}_2)} \quad (34.2)$$

where $\underline{\mathbf{x}}_1$ is the $m -$ vector $[\mathbf{x}_1\, \mathbf{x}_2 \ldots \mathbf{x}_m]'$; $\underline{\mathbf{x}}_2$ is the $(n-m) -$ vector $[\mathbf{x}_{m+1}\, \mathbf{x}_{m+2} \ldots \mathbf{x}_n]'$; $\underline{\mathbf{x}}$ is the $n -$ vector $[\mathbf{x}_1\, \mathbf{x}_2 \ldots \mathbf{x}_n]'$; and the $\underline{x}_i$s have the same dimensions as the corresponding $\underline{\mathbf{x}}_i$s. It is often possible to avoid working directly with pdfs, especially in the study of linear systems. Instead, one uses expectations.

34.2.2 Expectation

DEFINITION 34.4 The expected value, expectation, or mean, of a scalar random variable, **x**, is denoted by $E(\mathbf{x})$ or m and given by

$$m \triangleq E(\mathbf{x}) \triangleq \int_{-\infty}^{\infty} x p_{\mathbf{x}}(x) dx \quad (34.3)$$

Applying Definition 34.4 to the scalar random variable $\mathbf{x}(k)$, the kth element of the scalar discrete-time stochastic process, $\underline{\mathbf{x}}$, gives

$$m(k) \triangleq E(\mathbf{x}(k)) = \int_{-\infty}^{\infty} \ldots \int_{-\infty}^{\infty} x(k) p_{\underline{\mathbf{x}}}(\underline{x}) dx_1 dx_2 \ldots dx_n \quad (34.4)$$

Note that the integrations over x_ℓ, $\ell \neq k$, simply give the necessary marginal density $p_{\mathbf{x}(k)}(x(k))$.

The computation in Equation 34.4 can be repeated for all $k = 1, 2, \ldots, n$ and the result organized as an n-vector

$$E(\underline{\mathbf{x}}) = [E(\mathbf{x}(1))\, E(\mathbf{x}(2)) \ldots E(\mathbf{x}(n))]' \quad (34.5)$$

One can also take the expectation with respect to functions of $\underline{\mathbf{x}}$. Two of these are particularly important.

DEFINITION 34.5 The covariance of the scalar random variables, $\mathbf{x}_k$ and $\mathbf{x}_\ell$, is denoted by $r_{k\ell}$, and given by

$$r_{k\ell} \triangleq E((\mathbf{x}_k - m(k))(\mathbf{x}_\ell - m(\ell))) = \int_{-\infty}^{\infty} \int_{-\infty}^{\infty} (x_k - m(k))(x_\ell - m(\ell)) p_{\mathbf{x}_k \mathbf{x}_\ell}(x_k, x_\ell) dx_k dx_\ell \quad (34.6)$$

Definition 34.5 can be applied to each pair $\mathbf{x}(k)\mathbf{x}(\ell)$, $k, \ell = 1, 2, \ldots, n$ of elements of the discrete-time scalar stochastic process $\underline{\mathbf{x}}$. The result is a collection of n^2 elements $r_{k\ell}$, $k, \ell = 1, 2 \ldots n$. It is conventional to emphasize the time dependence by defining

$$r(k, \ell) \triangleq r_{k\ell} \quad (34.7)$$

It should be obvious from Equation 34.6 that

$$r(k, \ell) = r(k, \ell) \text{ for all } k, \ell \quad (34.8)$$

and that (because $p_{\underline{\mathbf{x}}}(\underline{x}) \geq 0$ for all $\underline{x}$)

$$r(k, k) \geq 0 \text{ for all } k \quad (34.9)$$

In the context of discrete-time stochastic processes $r(k, \ell)$ is known as the autocovariance function of the stochastic process $\underline{\mathbf{x}}$. It can be helpful, in trying to understand the properties of the autocovariance function, to write it as a matrix.

$$\underline{R} = \begin{bmatrix} r(1,1) & r(1,2) & \ldots & r(1,n) \\ r(2,1) & r(2,2) & & \\ \vdots & & \ddots & \vdots \\ r(n,1) & \ldots & & r(n,n) \end{bmatrix} \quad (34.10)$$

In the context of n-vector random variables the matrix $\underline{R}$ is known as the covariance matrix. The autocovariance will be discussed further subsequently. Another important expectation will be defined first.

DEFINITION 34.6 The characteristic function of a random variable, **x**, is denoted by $f_{\mathbf{x}}(w)$ and given by

$$f_{\mathbf{x}}(w) \triangleq E(e^{jw\mathbf{x}}) = \int_{-\infty}^{\infty} e^{jwx} p_{\mathbf{x}}(x) dx \quad (34.11)$$

where

$$j \stackrel{\Delta}{=} \sqrt{-1}$$

Note that $f_{\mathbf{x}}(w)$ is a deterministic function of w, not a random variable. Note also that $f_{\mathbf{x}}(-w)$ is the Fourier transform of $p_{\mathbf{x}}(x)$ and is thus equivalent to $p_{\mathbf{x}}(x)$ in the same way that Fourier transform pairs are usually equivalent; there is a unique correspondence between a function and its transform.

The generalization to the case of n-vector random variables or discrete time stochastic processes over n time steps is as follows.

DEFINITION 34.7 The characteristic function of an n-vector random variable (or discrete time stochastic process), $\underline{\mathbf{x}}$, is denoted $f_{\underline{\mathbf{x}}}(\underline{w})$ and given by

$$f_{\underline{\mathbf{x}}}(\underline{w}) = E(e^{j\sum_{k=1}^{n}\mathbf{x}_k w_k}) = E(e^{j\underline{\mathbf{x}}'\underline{w}}) \tag{34.12}$$

The characteristic function is particularly useful for studying the effect of linear mappings on a stochastic process. As an example, consider the operation of an $m \times n$ real matrix $\underline{L}$ on the n-vector random variable $\underline{\mathbf{x}}$.

Let $\underline{\mathbf{y}} = \underline{L}\underline{\mathbf{x}}$

$$f_{\underline{\mathbf{y}}}(\underline{w}) = E(e^{j\underline{w}'\underline{\mathbf{y}}}) = E(e^{j\underline{w}'\underline{L}\underline{\mathbf{x}}}) = E(e^{j(\underline{L}'\underline{w})'\underline{\mathbf{x}}}) = f_{\underline{\mathbf{x}}}(\underline{L}'\underline{w}) \tag{34.13}$$

34.2.3 Example—Discrete-Time Gaussian Stochastic Processes

DEFINITION 34.8 An n-vector random variable (or discrete-time stochastic process), $\underline{\mathbf{x}}$, is a Gaussian (normal) random variable (discrete-time stochastic process) if and only if it has the n-dimensional Gaussian (normal) pdf

$$p_{\underline{\mathbf{x}}}(\underline{x}) = \frac{1}{(2\pi)^{n/2}(det\underline{R})^{1/2}} e^{-\frac{1}{2}(\underline{x}-\underline{m})'\underline{R}^{-1}(\underline{x}-\underline{m})} \tag{34.14}$$

where

$$\begin{aligned}\underline{m} &= E(\underline{\mathbf{x}})\\ \underline{R} &= E((\underline{\mathbf{x}}-\underline{m})(\underline{\mathbf{x}}-\underline{m})'),\\ &\quad \text{the covariance matrix of } \underline{\mathbf{x}}\end{aligned}$$

The definition implies that $\underline{R}$ is symmetric ($\underline{R} = \underline{R}'$, see Equations 34.8 and 34.10) and that $\underline{R}$ must be positive definite ($\underline{y}'\underline{R}\underline{y} > 0$ for all n-vectors $\underline{y} \neq 0$ and often indicated by $\underline{R} > 0$).

It is easy to demonstrate that the characteristic function of the n-vector Gaussian random variable, $\underline{\mathbf{x}}$, with mean $\underline{m}$ and covariance matrix $\underline{R}$ is

$$f_{\underline{\mathbf{x}}}(\underline{w}) = e^{j\underline{w}'\underline{m} - \frac{1}{2}\underline{w}'\underline{R}\underline{w}} \tag{34.15}$$

Any n-vector random variable or stochastic process that has a characteristic function in the form of Equation 34.15 is Gaussian. In fact, some authors [2] define a scalar Gaussian random variable by Equation 34.15 with m and R scalars, rather than by Equation 34.14. The reason is that Equation 34.15 is well defined when $R = 0$, whereas Equation 34.14 blows up. A scalar Gaussian random variable $\mathbf{x}$ with variance $R = 0$ and mean m makes perfectly good sense. It is the deterministic equality $\mathbf{x} = m$.

One other special case of the n-vector Gaussian random variable is particularly important. When $r_{k\ell} = 0$ for all $k \neq \ell$, $\underline{R}$ is a diagonal matrix. The pdf becomes

$$\begin{aligned}p_{\underline{\mathbf{x}}}(\underline{x}) &= \frac{1}{(2\pi)^{n/2} r_{11}^{1/2} r_{22}^{1/2} \ldots r_{nn}^{1/2}}\\ &\quad e^{-\frac{1}{2}\left(\frac{(x_1-m_1)^2}{r_{11}} + \frac{(x_2-m_2)^2}{r_{22}} + \ldots + \frac{(x_n-m_n)^2}{r_{nn}}\right)}\\ &= \left(\frac{1}{\sqrt{2\pi} r_{11}^{1/2}} e^{-\frac{1}{2}\frac{(x_1-m_1)^2}{r_{11}}}\right)\\ &\quad \left(\frac{1}{\sqrt{2\pi} r_{22}^{1/2}} e^{-\frac{1}{2}\frac{(x_2-m_2)^2}{r_{22}}}\right)\ldots\\ &\quad \left(\frac{1}{\sqrt{2\pi} r_{nn}^{1/2}} e^{\frac{1}{2}\frac{(x_n-m_n)^2}{r_{nn}}}\right)\\ &= p_{\mathbf{x}_1}(x_1)p_{\mathbf{x}_2}(x_2)\ldots p_{\mathbf{x}_n}(x_n)\end{aligned} \tag{34.16}$$

In other words, the $\mathbf{x}_k$ are independent random variables. It is an important property of Gaussian random variables that they are independent if and only if their covariance matrix is diagonal, as has just been proven.

Lastly, the characteristic function will be used to prove that if $\underline{\mathbf{y}} = \underline{L}\underline{\mathbf{x}}$ where $\underline{L}$ is an $m \times n$ real matrix and $\underline{\mathbf{x}}$ is an n-vector Gaussian random variable with mean $\underline{m}$ and covariance $\underline{R}$, then $\underline{\mathbf{y}}$ is an m-vector Gaussian random vector with mean $\underline{L}\underline{m}$ and covariance $\underline{L}\underline{R}\underline{L}'$. The proof is as follows:

$$\begin{aligned}\underline{\mathbf{y}} &= \underline{L}\underline{\mathbf{x}}\\ f_{\underline{\mathbf{y}}}(\underline{w}) &= f_{\underline{\mathbf{x}}}(\underline{L}'\underline{w})\end{aligned}$$

by Equation 34.15

$$\begin{aligned}&= e^{j(\underline{L}'\underline{w})'\underline{m} - \frac{1}{2}(\underline{L}'\underline{w})'\underline{R}(\underline{L}\underline{w})}\\ &= e^{j\underline{w}'\underline{L}\underline{m} - \frac{1}{2}\underline{w}'\underline{L}\underline{R}\underline{L}'\underline{w}}\\ &= e^{j\underline{w}'(\underline{L}\underline{m}) - \frac{1}{2}\underline{w}'(\underline{L}\underline{R}\underline{L}')\underline{w}}\end{aligned} \tag{34.17}$$

Finally, the uniqueness of Fourier transforms (characteristic functions), and the fact that Equation 34.17 is the characteristic function of an m-vector Gaussian random variable with mean $\underline{L}\underline{m}$ and covariance matrix $\underline{L}\underline{R}\underline{L}'$, complete the proof.

34.2.4 Stationarity, Ergodicity, and White Noise

Several important properties of a discrete-time stochastic process are, strictly speaking, properly defined only for processes for which $k = \ldots -2, -1, 0, 1, 2, \ldots$ (that is, $-\infty < k < \infty$).

Let

$$\underline{\mathbf{x}} = [\ldots \mathbf{x}_{-2}\ \mathbf{x}_{-1}\ \mathbf{x}_0\ \mathbf{x}_1\ \mathbf{x}_2 \ldots]' \tag{34.18}$$

denote either the scalar discrete-time stochastic process on the interval $-\infty < k < \infty$ or the equivalent infinite-dimensional vector random variable. It is difficult to visualize and write explicitly the pdf for an infinite-dimensional random variable, but that is not necessary. The pdfs for all possible finite-dimensional subsets of the elements of $\underline{\mathbf{x}}$ completely characterize the pdf of $\underline{\mathbf{x}}$. Similarly, $E(\underline{\mathbf{x}}) = \underline{m}$ is computable term by term from the $\mathbf{x}_i$ taken one at a time and $E((\underline{\mathbf{x}} - \underline{m})(\underline{\mathbf{x}} - \underline{m})') = \underline{R}$ is computable from all pairs $\mathbf{x}_i, \mathbf{x}_j$ taken two at a time.

EXAMPLE 34.1:

Let $\underline{\mathbf{x}}$ be a Gaussian scalar discrete-time stochastic process with mean $E(\mathbf{x}_k) = m_k = 0$ for all k, $-\infty < k < \infty$ and covariance

$$E(\mathbf{x}_k \mathbf{x}_\ell) = r(k, \ell) = \begin{cases} 1 & k = \ell \\ 1/2 & |k - \ell| = 1 \\ 0 & \text{otherwise} \end{cases}$$

Note that this completely describes the pdf of $\underline{\mathbf{x}}$ even though $\underline{\mathbf{x}}$ is infinite-dimensional.

This apparatus makes it possible to define two forms of time-invariance for discrete-time stochastic processes.

DEFINITION 34.9 A scalar discrete-time stochastic process $\underline{\mathbf{x}}$, defined on $-\infty < k < \infty$ is stationary if and only if

$$p_{\mathbf{x}_{k_1}, \mathbf{x}_{k_2}, \ldots, \mathbf{x}_{k_n}}(x_{k_1}, x_{k_2}, \ldots, x_{k_n}) = \qquad (34.19)$$
$$p_{\mathbf{x}_{k_1+k_p}, \mathbf{x}_{k_2+k_p}, \ldots, \mathbf{x}_{k_n+k_p}}(x_{k_1+k_p}, x_{k_2+k_p}, \ldots, x_{k_n+k_p})$$

for all possible choices $-\infty < k_\ell, k_p < \infty$, $\ell = 1, 2, \ldots, n$ and all finite n.

The region of definition of stochastic processes defined on $(-\infty, \infty)$ will be emphasized by using the notation $\mathbf{x}_{(-\infty,\infty)}$ to denote such processes.

It can be difficult to verify Definition 34.9. There is a weaker and easier to use form of stationarity. It requires a preliminary definition.

DEFINITION 34.10 A scalar discrete-time stochastic process, $\underline{\mathbf{x}}$, is a second-order process if, and only if, $E(\mathbf{x}_k^2) < \infty$ for all k, $-\infty < k < \infty$.

DEFINITION 34.11 A second-order scalar discrete-time stochastic process, $\mathbf{x}_{(-\infty,\infty)}$ is wide-sense stationary if, and only if, its mean $\underline{m}$ and autocovariance $r(k, \ell)$ satisfy

$$m(k) = m, \text{ a constant for all } k, -\infty < k < \infty \qquad (34.20)$$
$$r(k, \ell) = r(k + i, \ell + i) \text{ for all } k, \ell, i, -\infty < k, \ell, i < \infty \qquad (34.21)$$

It is customary to define

$$r(k) \triangleq r(k + \ell, \ell) \text{ for all } k, \ell, -\infty < k, \ell < \infty \qquad (34.22)$$

It is obvious that a stationary discrete-time stochastic process is also wide-sense stationary because the invariance of the pdfs implies invariance of the expectations. Because the pdf of an n-vector Gaussian random variable is completely defined by its mean, $\underline{m}$, and covariance, $\underline{R}$, a scalar wide-sense Gaussian stationary discrete-time stochastic process is also stationary.

The apparatus needed to define discrete-time white noise is now in place.

DEFINITION 34.12 A scalar second-order discrete-time stochastic process, $\mathbf{x}_{(-\infty,\infty)}$, is a *white-noise* process if and only if

$$m(k) = E(\mathbf{x}_k) = 0 \quad \text{for all } k, -\infty < k < \infty \qquad (34.23)$$
$$r(k) = E(\mathbf{x}_{\ell+k}\mathbf{x}_\ell) = r\delta(k) \quad \text{for all } -\infty < k, \ell < \infty \qquad (34.24)$$

where

$$r \geq 0$$

and

$$\delta(k) = \begin{cases} 1 & k = 0 \\ 0 & \text{otherwise} \end{cases} \qquad (34.25)$$

If, in addition, $\mathbf{x}_k$ is Gaussian for all k, the process is Gaussian white noise. An explanation of the term "white noise" requires an explanation of transforms of stochastic processes. This will be forthcoming shortly. First, the question of estimating the pdf will be introduced.

In the real world of engineering, someone has to determine the pdf of a given random variable or stochastic process. The typical situation is that one has some observations of the process and some prior knowledge about the process. For example, the physics often indicates that a discrete-time stochastic process, $\underline{\mathbf{x}}$, $-\infty < k < \infty$, is wide-sense stationary—at least as an adequate approximation to reality. If one is content with second-order properties of the process, this reduces the problem to determining m and $r(k)$, $-\infty < k < \infty$, from observations of the process. The discussion here will be limited to this important, but greatly simplified, version of the general problem.

In order to have a precisely specified problem with a well-defined solution assume that the data available is one observation (sample) of the stochastic process over the complete time interval $-\infty < k < \infty$. This is certainly impossible for individuals having a finite life span but the idealized mathematical result based on this assumption clarifies the practical situation.

Define

$$\mathbf{m}_\ell = \frac{1}{2\ell} \sum_{k=-\ell}^{\ell} \mathbf{x}_k \qquad (34.26)$$

$$\mathbf{r}_\ell(k) = \frac{1}{2\ell} \sum_{i=-\ell}^{\ell} \mathbf{x}_{i+k}\mathbf{x}_i \qquad (34.27)$$

Note that $\mathbf{m}_\ell$ and $\mathbf{r}_\ell(k)$ are denoted by bold letters in Equations 34.26 and 34.27. This indicates that they are random variables, unlike the expectations m and $r(k)$, which are deterministic quantities. It can then be proved that

$$\lim_{\ell \to \infty} E[(\mathbf{m}_\ell - m)^2] = 0 \qquad (34.28)$$

provided

$$\lim_{|k|\to\infty} r(k) = 0 \tag{34.29}$$

A similar result holds for $\mathbf{r}_\ell(k)$ (see Wong [2], pp. 77-80).

In order to apply this result to a real problem one must know, *a priori*, that Equation 34.29 holds. Again, this is often known from the physics.

The convergence result in Equation 34.28 is fairly weak. It would be preferable to prove that $\mathbf{m}_\ell$ converges to m almost surely (with probability 1). It is possible to prove this stronger result if the process is ergodic as well as stationary. This subject is both complicated and technical. See [2], [4] for engineering-oriented discussions and [3] for a more mathematical introduction.

34.2.5 Transforms

In principle, one can apply any transform that is useful in the analysis of discrete-time signals to discrete-time stochastic processes. The two obvious candidates are the Z-transform and the discrete Fourier transform [5]. There is a slight theoretical complication. To see this, consider the discrete Fourier transform of the discrete-time stochastic process $\mathbf{x}_{(-\infty,\infty)}$

$$\mathbf{x}_f(\Omega) \triangleq \sum_{k=-\infty}^{\infty} \mathbf{x}(k)e^{-j\Omega k} \tag{34.30}$$

Notice that $\mathbf{x}_f(\Omega)$ is an infinite-dimensional random vector for each fixed value of Ω (it is a function of the random variables $\mathbf{x}(k)$ so it is a random variable) and $\mathbf{x}_f(\Omega)$ is defined for all Ω, $-\infty < \Omega < \infty$, not just integer values of Ω. In other words $\mathbf{x}_f(\Omega)$ is a stochastic process in the continuous variable Ω. As is usual with the discrete Fourier transform, $\mathbf{x}_f(\Omega)$ is periodic in Ω with period 2π.

A very important use of transforms in the study of discrete-time stochastic processes is the spectral density function.

DEFINITION 34.13 Let $\mathbf{x}_{(-\infty,\infty)}$ be a wide-sense stationary scalar discrete-time stochastic process with mean m and autocovariance function $r(k)$. Assume $\sum_{k=-\infty}^{\infty} |r(k)| < \infty$. Then the discrete Fourier transform of the autocovariance function $r(k)$,

$$s(\Omega) \triangleq \sum_{k=-\infty}^{\infty} r(k)e^{-j\Omega k} \tag{34.31}$$

is well defined and known as the *spectral density function* of $\mathbf{x}_{(-\infty,\infty)}$.

Notice, as the notation emphasizes, no part of Equation 34.31 is random. Both the spectral density and the autocovariance are deterministic descriptors of the stochastic process, $\mathbf{x}_{(-\infty,\infty)}$.

The inverse of Equation 34.31 is the usual inverse of the discrete Fourier transform. That is,

$$r(k) = \frac{1}{2\pi}\int_{2\pi} s(\Omega)e^{j\Omega k}d\Omega \tag{34.32}$$

where $\int_{2\pi}$ means the integral is taken over any interval of duration 2π.

When $m = 0$

$$r(0) = E(\mathbf{x}_k^2) = \int_{2\pi} s(\Omega)\frac{d\Omega}{2\pi} \tag{34.33}$$

If $\mathbf{x}_{(-\infty,\infty)}$ is a voltage, current, or velocity, then $r(0)$ can be interpreted as average power, at least to within a constant of proportionality. With this interpretation of $r(0)$, $s(\Omega_0)$ (where Ω_0 is any fixed frequency) must be the average power per unit frequency in $\mathbf{x}_{(-\infty,\infty)}$ at the frequency Ω_0 [2]. This is why it is called the spectral density.

EXAMPLE 34.2:

Suppose $\mathbf{x}_{(-\infty,\infty)}$ is a white noise process (see Definition 34.12) with $r(k) = \delta(k)$. The spectral density function for this process is

$$s(\Omega) = \sum_{k=-\infty}^{\infty} \delta(k)e^{-j\Omega k} = 1 \tag{34.34}$$

This is why it is called "white noise." Like white light, all frequencies are equally present in a white noise process. A reasonable conjecture is that it is called "noise" because, in the early days of radio and telephone, such a stochastic process was heard as a recognizable and unwanted sound.

34.2.6 Single-Input Single-Output Discrete-Time Linear Systems

It is convenient, both pedagogically and notationally, to begin with single-input single-output (SISO) discrete-time linear systems described by their impulse response, $h(k, \ell)$. The notation is shown in Figure 34.1.

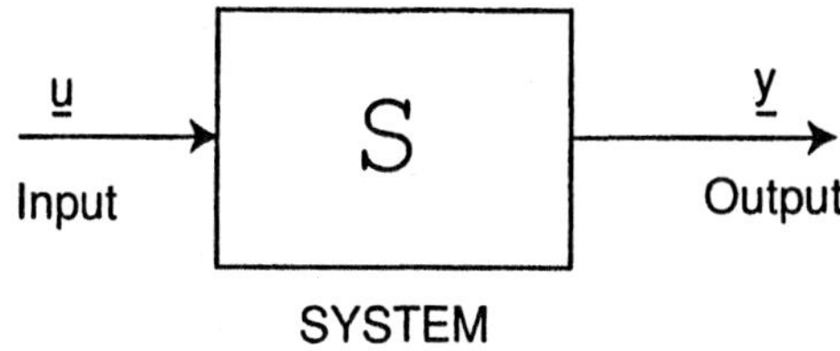

Figure 34.1 A symbolic representation of a discrete-time linear system, S, with input $\underline{u} = [\ldots u_{-2}\ u_{-1}\ u_0\ u_1 \ldots]'$ and output $\underline{y} = [\ldots y_{-1}\ y_0\ y_1 \ldots]'$.

DEFINITION 34.14 The impulse response of a SISO discrete-time linear system is denoted by $h(k, \ell)$, where $h(k, \ell)$ is the value of the output at instant k, when the input is a unit impulse at instant ℓ.

The response of a linear system S to an arbitrary input,

$u_{(-\infty,\infty)}$, is then given by a convolution sum

$$y(k) = \sum_{\ell=-\infty}^{\infty} h(k,\ell)u(\ell); \quad -\infty < k < \infty \tag{34.35}$$

The crucial point is that S is a linear map, as is easily proved from Equation 34.35. Linear maps take Gaussian stochastic processes (random variables) into Gaussian stochastic processes. A special case of this, when S can be written as an $n \times m$ matrix, was proved earlier (Equation 34.17).

EXAMPLE 34.3:

Consider a SISO discrete-time linear system, S, with input, $\mathbf{u}_{(-\infty,\infty)}$, a discrete-time Gaussian stochastic process having mean $m_u(k)$ and autocovariance $r_u(k,\ell)$. Knowing that the system is linear and that the input is Gaussian, the output, $\mathbf{y}_{(-\infty,\infty)}$, must be a Gaussian stochastic process; what are its mean and autocovariance? They can be calculated element by element:

$$\begin{aligned}
m_y(k) &= E(\mathbf{y}(k)) = E\left(\sum_{\ell=-\infty}^{\infty} h(k,\ell)\mathbf{u}(\ell)\right) \\
&= \sum_{\ell=-\infty}^{\infty} h(k,\ell)E(\mathbf{u}(\ell)) \\
&= \sum_{\ell=-\infty}^{\infty} h(k,\ell)m_u(\ell); \quad -\infty < k < \infty \quad (34.36) \\
r_y(k,\ell) &= E\left((\mathbf{y}(k)-m_u(k))(\mathbf{y}(\ell)-m_u(\ell))\right) \\
&= E\left(\left(\sum_{i=-\infty}^{\infty} h(k,i)(\mathbf{u}(i)-m_u(i))\right)\right. \\
&\quad \left.\left(\sum_{j=-\infty}^{\infty} h(\ell,j)(\mathbf{u}(j)-m_u(j))\right)\right) \\
&= \sum_{i=-\infty}^{\infty}\sum_{j=-\infty}^{\infty} h(k,i)h(\ell,j)E \\
&\quad ((\mathbf{u}(i)-m_u(i))\,(\mathbf{u}(j)-m_u(j))) \\
&= \sum_{i=-\infty}^{\infty}\sum_{j=-\infty}^{\infty} h(k,i)h(\ell,j)r_u(i,j) \quad (34.37)
\end{aligned}$$

Of course, not every stochastic process is Gaussian. However, careful review of the previous example shows that Equations 34.36 and 34.37 are valid formulas for the mean and autocovariance of $\mathbf{y}_{(-\infty,\infty)}$ whenever $\mathbf{u}_{(-\infty,\infty)}$ is a second-order scalar discrete-time stochastic process with mean $m_u(k)$ and autocovariance $r_u(i,j)$. This is a very important point. The second-order properties of a stochastic process are easily computed. The calculations are even simpler if the input process is wide-sense stationary and the linear system is time-invariant.

EXAMPLE 34.4:

Consider a SISO discrete-time linear time-invariant (LTI) systems, S, with input, $\mathbf{u}_{(-\infty,\infty)}$, a wide-sense stationary second-order discrete-time stochastic process having mean m_u and autocovariance $r_u(k)$. Denote the impulse response of S by $h(k)$, where $h(k) \triangleq h(k+\ell,\ell)$.

$$\begin{aligned}
m_y(k) &= E(\mathbf{y}(k)) = E(\sum_{\ell=-\infty}^{\infty} h(k-\ell)\mathbf{u}(\ell)) \\
&= \sum_{\ell=-\infty}^{\infty} h(k-\ell)E(\mathbf{u}(\ell)) \\
&= \sum_{\hat{\ell}=-\infty}^{\infty} h(\hat{\ell})E(\mathbf{u}(k-\hat{\ell})) \\
&= \alpha m_u \quad (34.38)
\end{aligned}$$

where

$$\alpha = \sum_{\hat{\ell}=-\infty}^{\infty} h(\hat{\ell})$$

Similarly, using Equation 34.37,

$$\begin{aligned}
r_y(k,\ell) &= \sum_{i=-\infty}^{\infty}\sum_{j=-\infty}^{\infty} h(k-i)h(\ell-j)r_u(i-j) \\
&= \sum_{\hat{\ell}=-\infty}^{\infty}\sum_{\hat{k}=-\infty}^{\infty} h(\hat{\ell})h(\hat{k})r_u(k-\ell-\hat{\ell}+\hat{k})
\end{aligned} \tag{34.39}$$

Notice that the convolution sum in Equation 34.39 depends only on the difference $k-\ell$. Thus, Equations 34.38 and 34.39 prove that $m_y(k) = m_y$, a constant, and that $r_y(k,\ell) = r_y(k-\ell) = r_y(\bar{k})$ where $\bar{k} \triangleq k-\ell$. This proves that $\mathbf{y}_{(-\infty,\infty)}$ is also wide-sense stationary.

Although the computations in Equation 34.39 still appear to be difficult, they clearly involve deterministic convolution. It is well known that the Fourier transform can be used to simplify such calculations. The impulse responses of many discrete-time LTI systems, $h(k)$, $-\infty < k < \infty$, have discrete Fourier transforms, $h_f(\Omega)$, $0 \le \Omega < 2\pi$. The exceptions include unstable systems. Assume $h_f(\Omega)$ exists in the preceding example and that $\mathbf{u}_{(-\infty,\infty)}$ is wide-sense stationary with mean m_u and spectral density function $s(\Omega)$. From the definition of the discrete Fourier transform

$$h_f(\Omega) = \sum_{k=-\infty}^{\infty} h(k)e^{-j\Omega k} \tag{34.40}$$

it is evident that Equation 34.38 becomes

$$m_y = h_f(0)m_u \tag{34.41}$$

Using the inversion formula (Equation 34.32) in Equation 34.39 gives

$$r_y(k) = \sum_{\ell=-\infty}^{\infty}\sum_{i=-\infty}^{\infty} h(\ell)h(i)$$

$$\left(\frac{1}{2\pi}\int_{2\pi} s_u(\Omega)e^{j\Omega(k-\ell+i)}d\Omega\right)$$
$$= \int_{2\pi}\sum_{i=-\infty}^{\infty} h(i)s_u(\Omega)e^{j\Omega(k+i)}\frac{d\Omega}{2\pi}$$
$$\left(\sum_{\ell=-\infty}^{\infty} h(\ell)e^{-j\Omega\ell}\right)\frac{d\Omega}{2\pi}$$
$$= \int_{2\pi}\left(\sum_{i=-\infty}^{\infty} h(i)e^{j\Omega k}\right)h_f(\Omega)s_u(\Omega)e^{j\Omega k}\frac{d\Omega}{2\pi}$$
$$= \int_{2\pi} h_f(-\Omega)h_f(\Omega)s_u(\Omega)e^{j\Omega k}\frac{d\Omega}{2\pi}$$
$$= \int_{2\pi} |h_f(\Omega)|^2 s_u(\Omega)e^{j\Omega k}\frac{d\Omega}{2\pi} \quad (34.42)$$

By the uniqueness of Fourier transforms, Equation 34.42 implies

$$s_y(\Omega) = |h_f(\Omega)|^2 s_u(\Omega) \quad (34.43)$$

SISO linear systems in state-space form generally have an n-vector state. When either the initial state is random or the input is a stochastic process, this state is a vector stochastic process. The ideas and notation for such processes are described in the following section. This is followed by a description of linear systems in state-space form.

34.2.7 Vector Discrete-Time Stochastic Processes and LTI Systems

The vector case involves much more complicated notation but no new concepts.

DEFINITION 34.15 An m-vector discrete-time stochastic process over n time steps, denoted $\underline{\mathbf{X}} \triangleq \{[\mathbf{x}_1(k)\,\mathbf{x}_2(k)\ldots\mathbf{x}_m(k)]';\ k = 1, 2, \ldots n\}$, takes values $\underline{X} = \{[x_1(h)x_2(k)\ldots x_m(k)]';\ k = 1, 2, \ldots, n\}$ according to a probability that can be determined from the pdf $p_{\underline{\mathbf{X}}}(\underline{X})$.

It can be helpful to visualize such a process as an $m \times n$ matrix

$$\underline{\mathbf{X}} = \begin{bmatrix} \mathbf{x}_1(1) & \mathbf{x}_1(2) & \ldots & \mathbf{x}_1(n) \\ \mathbf{x}_2(1) & \mathbf{x}_2(2) & \ldots & \mathbf{x}_2(n) \\ \vdots & \vdots & \ddots & \vdots \\ \mathbf{x}_m(1) & \mathbf{x}_m(2) & \ldots & \mathbf{x}_m(n) \end{bmatrix} \quad (34.44)$$

The notation is that bold capital underlined letters denote vector stochastic processes.

All of the scalar results apply, with obvious modifications, to the vector case. For example, $E(\mathbf{X})$ can be computed element by element from Definition 34.4.

$$E(x_\ell(k)) = \int_{-\infty}^{\infty} x p_{\mathbf{x}_\ell(k)}(x)dx \quad (34.45)$$

for all $\ell = 1, 2, \ldots, m$, $k = 1, 2, \ldots, n$. (See Equation 34.4.)

Then, the nm results of Equation 34.45 can be organized as an m-vector over n time steps, $\underline{m}(k) = [m_1(k)\, m_2(k)\ldots m_m(k)]'$, where

$$m_\ell(k) = E(\mathbf{x}_\ell(k)) \quad (34.46)$$

Similarly, the autocovariance of $\underline{\mathbf{X}}$ in Equation 34.44 can be computed element by element using Definition 34.5 and Equation 34.7. The results are conventionally organized as an autocovariance matrix at each pair of times. That is

$$\underline{R}(k,\ell) = \begin{bmatrix} r_{11}(k,\ell) & r_{12}(k,\ell) & \ldots & r_{1m}(k,\ell) \\ r_{21}(k,\ell) & r_{22}(l,\ell) & \ldots & r_{2m}(k,\ell) \\ \vdots & \vdots & \ddots & \vdots \\ r_{m1}(k,\ell) & r_{m2}(k,\ell) & \ldots & r_{mm}(k,\ell) \end{bmatrix} \quad (34.47)$$

where

$$r_{ij}(k,\ell) \triangleq E((x_i(k) - m_i(k))(x_j(\ell) - m_j(\ell)))$$

EXAMPLE 34.5: Discrete-time m-vector white noise:

Any white noise must be wide-sense stationary and have zero mean. Thus, letting $\underline{\xi}(k)$ denote an m-vector white noise at the instant k,

$$\underline{m}_\xi(k) = E(\underline{\xi}(k)) = \underline{0} \text{ for all } k,\ -\infty < k < \infty \quad (34.48)$$

$$\underline{R}_\xi(k,\ell) = E(\underline{\xi}(k)\underline{\xi}'(\ell)) = \underline{0} \text{ for all } k \neq \ell,\ -\infty < k, \ell < \infty$$

because white noise must have autocovariance equal to zero except when $k = \ell$. When $k = \ell$, the autocovariance can be any positive semidefinite matrix. Thus,

$$\underline{R}_\xi(k,\ell) = \underline{R}_\xi\delta(k-\ell) \quad (34.49)$$

where $\underline{R}_\xi$ can be any positive semidefinite matrix ($\underline{y}'\underline{R}_\xi\underline{y} \geq 0$ for all m-vectors $\underline{y}$).

Finally, as is customary with stationary processes

$$\underline{R}_\xi(k) \triangleq \underline{R}_\xi(k+\ell,\ell) = \underline{R}_\xi\delta(k) \quad (34.50)$$

Reading vector versions of all the previous results would be awfully tedious. Thus, the vector versions will not be written out here. Instead, an example of the way linear systems operate on vector discrete-time stochastic processes will be presented.

The previous discussion of linear systems dealt only with impulse responses and, under the added assumption of time-invariance, Fourier transforms. Initial conditions were assumed to be zero. The following example includes nonzero initial conditions and a state-space description of the linear system. The system is assumed to be LTI. The time-varying case is not harder, but the notation is messy.

EXAMPLE 34.6:

Let $\underline{\mathbf{x}}_0$ be an n-dimensional second-order random variable with mean $\underline{m}_{x_0}$ and covariance $\underline{R}_{x_0}$. Let $\underline{\xi}$ and $\underline{\theta}$ be, respectively, m-dimensional and p-dimensional second-order discrete-time stochastic processes on $0 \leq k \leq k_f$. Let $E(\underline{\xi}) = \underline{0}$, $E(\underline{\theta}) =$

$\underline{0}$, $E(\underline{\xi}(k)\underline{\xi}'(\ell)) = \Xi\delta(k-\ell)$, $E(\underline{\theta}(k)\underline{\theta}'(\ell)) = \Theta\delta(k-\ell)$, and $E(\underline{\xi}(k)\underline{\theta}'(\ell)) = \underline{0}$ for all $0 \leq k, \ell \leq k_f$. Strictly speaking, Ξ and Θ are not white noise processes because they are defined only on a finite interval. However, in the context of state-space analysis it is standard to call them white noise processes. Finally, assume $E((\underline{x}_0 - \underline{m}_{x_0})\underline{\xi}'(k)) = 0$ and $E((\underline{x}_0 - \underline{m}_{x_0})\underline{\theta}'(k)) = 0$ for all $0 \leq k \leq k_f$.

Suppose now that the n-vector random variable $\underline{x}(k+1)$ and the p-vector random variable $\underline{y}(t)$ are defined recursively for $k = 0, 1, 2, \ldots, k_f$ by

$$\underline{x}(k+1) = \underline{A}\underline{x}(k) + \underline{L}\underline{\xi}(k); \quad \underline{x}(0) = \underline{x}_0 \tag{34.51}$$

$$\underline{y}(k) = \underline{C}\underline{x}(k) + \underline{\theta}(k) \tag{34.52}$$

where $\underline{A}$ is a deterministic $n \times n$ matrix, $\underline{L}$ is a deterministic $n \times m$ matrix, and $\underline{C}$ is a deterministic $p \times n$ matrix. What can be said about the stochastic processes $\underline{X} = \{\underline{x}(k);\ k = 0, 1, 2, \ldots, k_f + 1\}$ and $\underline{Y} = \{\underline{y}(k);\ k = 0, 1, 2, \ldots, k_f\}$?

The means and autocovariances of $\underline{X}$ and $\underline{Y}$ are easily computed. For example,

$$\begin{aligned} E(\underline{x}(k+1)) &= E(\underline{A}\underline{x}(k) + \underline{L}\underline{\xi}(k)) \\ &= \underline{A}E(\underline{x}(k)) + \underline{L}E(\underline{\xi}(k)) \\ &= \underline{A}E(\underline{x}(k)) \end{aligned} \tag{34.53}$$

Similarly,

$$E(\underline{y}(k)) = \underline{C}E(\underline{x}(k)) \tag{34.54}$$

Note that Equation 34.53 is a simple deterministic recursion for $E(\underline{x}(k+1))$ starting from $E(\underline{x}(0)) = \underline{m}_{x_0}$. Thus,

$$E(\underline{x}(k)) = \underline{m}_x(k) = \underline{A}^k \underline{m}_{x_0} \tag{34.55}$$

$$E(\underline{y}(k)) = \underline{C}\underline{A}^k \underline{m}_{x_0} \tag{34.56}$$

Similarly, recursive equations for the autocovariances of $\underline{X}$ and $\underline{Y}$ can be derived. Because the expressions for $\underline{Y}$ are just deterministic algebraic transformations of those for $\underline{X}$, only those for $\underline{X}$ are given here.

$$\begin{aligned} \underline{R}_x(k+1, k+1) &\triangleq E((\underline{x}(k+1) - \underline{m}_x(k+1)) \\ &\quad (\underline{x}(k+1) - \underline{m}_x(k+1))') \\ &= E((\underline{A}(\underline{x}(k) - \underline{m}_x(k)) + \underline{L}\underline{\xi}(k)) \\ &\quad (\underline{A}(\underline{x}(k) - \underline{m}_x(k)) + \underline{L}\underline{\xi}(k))') \\ &= \underline{A}E((\underline{x}(k) - \underline{m}_x(k)) \\ &\quad (\underline{x}(k) - \underline{m}(k))')\underline{A}' + \underline{L}\Xi\underline{L}' \\ &= \underline{A}\underline{R}_x(k,k)\underline{A}' + \underline{L}\Xi\underline{L}' \end{aligned} \tag{34.57}$$

$$\begin{aligned} \underline{R}_x(k+1, k) &\triangleq E((\underline{x}(k+1) \\ &\quad - \underline{m}_x(k+1))(\underline{x}(k) - \underline{m}_x(k))') \\ &= \underline{A}\underline{R}_x(k,k) \end{aligned} \tag{34.58}$$

Note that Equations 34.57 and 34.58 use the assumption that $E((\underline{x}(k) - \underline{m}_x(k))\underline{\xi}(\ell)) = 0$ for all $0 \leq k, \ell \leq k_f$. Furthermore, the recursion in Equation 34.57 begins with $\underline{R}_x(0,0) = \underline{R}_{x_0}$.

34.3 Continuous Time

Continuous-time stochastic processes are technically much more complicated than discrete-time stochastic processes. Fortunately, most of the complications can be avoided by the restriction to second-order processes and linear systems. This is what will be done in this article because linear systems and second-order processes are by far the most common analytical basis for the design of control systems involving randomness.

A good understanding of the discrete-time case is helpful because the continuous-time results are often analogous to those in discrete time.

34.3.1 Basics

It is very difficult to give a definition of continuous-time stochastic process that is both elementary and mathematically correct. For discrete-time stochastic processes every question involving probability can be answered in terms of the finite-dimensional pdfs. For continuous-time stochastic processes the finite-dimensional pdfs are not sufficient to answer every question involving probability. See Wong [2], pp. 59-62 for a readable discussion of the difficulty.

In the interest of simplicity and because the questions of interest here are all answerable in terms of the finite-dimensional pdfs, the working definition of a continuous-time stochastic process will be as follows.

DEFINITION 34.16 A scalar continuous-time stochastic process, denoted by $\mathbf{x}_{[t_s,t_f]}$, takes values $x(t) \in R$ for all real t, $t_s \leq t \leq t_f$, according to a probability that can be determined from the family of finite-dimensional pdfs. $p_{\mathbf{x}(t_1),\mathbf{x}(t_2),\ldots,\mathbf{x}(t_n)}(x(t_1), x(t_2), \ldots, x(t_n))$ for all finite collections of t_k, t_k real, $t_s \leq t_k \leq t_f$ $k = 1, 2, \ldots, n$.

The essential feature of this definition is the idea of a signal that is a function of the continuous time, t, and also random. The definition itself will not be used directly. Its main purpose is to allow the definitions of expectation and second-order process.

34.3.2 Expectations

As for discrete-time stochastic processes, the expectation, expected value, or mean of a scalar continuous time stochastic process follows directly from Definitions 34.16 and 34.4.

$$m(t) \triangleq E(\mathbf{x}(t)) = \int_{-\infty}^{\infty} x p_{\mathbf{x}(t)}(x)dx \tag{34.59}$$

Note that $m(t)$ is defined where $\mathbf{x}_{[t_s,t_f]}$ is, that is, on the interval $t_s \leq t \leq t_f$.

The covariance function is defined and computed for continuous-time stochastic processes in exactly the same way as for discrete-time stochastic processes. See Equations 34.6 and 34.7.

DEFINITION 34.17 The covariance function of a scalar continuous time stochastic process $\mathbf{x}_{[t_s,t_f]}$ is denoted $r(t,\tau)$ and is given by

$$\begin{aligned} r(t,\tau) &= E((\mathbf{x}(t)-m(t))(\mathbf{x}(\tau)-m(\tau)) \quad (34.60)\\ &= \int_{-\infty}^{\infty}\int_{-\infty}^{\infty}(x-m(t))(y-m(\tau)) \\ &\quad p_{\mathbf{x}(t),\mathbf{x}(\tau)}(x,y)dxdy \end{aligned}$$

As in the discrete-time case, it should be obvious from Equation 34.60 that

$$r(t,\tau) = r(\tau,t) \text{ for all } t,\ \tau,\ t_s \le t, \tau \le t_f \quad (34.61)$$

and

$$r(t,t) \ge 0 \text{ for all } t,\ t_s \le t \le t_s \quad (34.62)$$

Because $r(t,\tau)$ is a function of two variables and not a matrix it is necessary to extend the idea of nonnegative definiteness to such functions in order to define and demonstrate this aspect of the "shape" of the covariance function. The idea behind the following definition is to form every possible symmetric matrix from time samples of $r(t,\tau)$ and then to require that all those matrices be positive semidefinite in the usual sense.

DEFINITION 34.18 A real-valued function $g(t,\tau)$; $t,\tau \in R, t_s \le t,\tau \le t_f$; is positive semidefinite if, for every finite collection $t_1, t_2, \ldots, t_n$; $t_s \le t_i \le t_f$ for $i = 1, 2, \ldots, n$ and every real n-vector $\underline{\alpha} = [\alpha_1,\ \alpha_2, \ldots, \alpha_n]'$

$$\sum_{k-1}^{n}\sum_{\ell-1}^{n}\alpha_k\alpha_\ell g(t_k,t_\ell) \ge 0 \quad (34.63)$$

The function $g(t,\tau)$ is positive definite if strict inequality holds in Equation 34.63 whenever $t_1, t_2, \ldots, t_n$ are distinct and $\underline{\alpha} \ne \underline{0}$.

It is easy to prove that an autocovariance function, $r(t,\tau)$, must be positive semidefinite. First,

$$E\left(\left(\sum_{k=1}^{n}\alpha_k(\mathbf{x}(t_k)-m(t_k))^2\right)\right) \ge 0 \quad (34.64)$$

because the expectation of a perfect square must be ≥ 0.

Then,

$$\begin{aligned} 0 &\le E((\sum_{k=1}^{n}\alpha_k(\mathbf{x}(t_k)-m(t_k))^2) \\ &= E\left(\sum_{k=1}^{n}\sum_{\ell=1}^{n}\alpha_k\alpha_\ell(\mathbf{x}(t_k)-m(t_k))(\mathbf{x}(t_\ell)-m(t_\ell))\right) \\ &= \sum_{k=1}^{n}\sum_{\ell=1}^{n}\alpha_k\alpha_\ell E((\mathbf{x}(t_k)-m(t_k))(\mathbf{x}(t_\ell)-m(t_\ell))) \\ &= \sum_{k=1}^{n}\sum_{\ell=1}^{n}\alpha_k\alpha_\ell r(t_k,t_\ell). \end{aligned}$$

The definition of the characteristic function in the case of a continuous-time stochastic process is obvious from Equations 34.11, 34.59, and 34.60. Because it is generally a function of time as well as w it is not as useful in the continuous-time case. For this reason it is not given here.

34.3.3 Example—Continuous-Time Gaussian Stochastic Processes

DEFINITION 34.19 A scalar continuous-time stochastic process $\mathbf{x}_{[t_s,t_f]}$ is a Gaussian process if the collection of random variables $\{\mathbf{x}(t_1), \mathbf{x}(t_2), \ldots, \mathbf{x}(t_n)\}$ is an n-vector Gaussian random variable for every finite set $\{t_1, t_2, \ldots, t_n;\ t_s \le t_i \le t_f, i = 1, 2, \ldots, n\}$.

It follows from Definition 34.8 in Section 34.2.3 that a Gaussian process is completely specified by its mean

$$m(t) = E(\mathbf{x}(t)) \quad (34.65)$$

and autocovariance

$$r(t,\tau) = E((\mathbf{x}(t)-m(t))(\mathbf{x}(\tau)-m(\tau))) \quad (34.66)$$

EXAMPLE 34.7: Wiener process (also known as Brownian motion):

Let $\mathbf{x}_{[0,\infty)}$ be a Gaussian process with

$$m(t) = 0$$

$$r(t,\tau) = \min(t,\tau) \quad (34.67)$$

Notice that the interval of definition is infinite and open at the right. This is a very minor expansion of Definition 34.16.

The Wiener process plays a fundamental role in the theory of stochastic differential equations (see Chapter 60) so it is worthwhile to derive one of its properties. Consider any ordered set of times $t_0 < t_1 < t_2 < \ldots < t_n$.

Define

$$\mathbf{y}_k = \mathbf{x}_{t_k} - \mathbf{x}_{t_{k-1}} \quad k = 1, 2, \ldots, n$$

Then $\mathbf{y}_k$ is an increment in the process $\mathbf{x}_{[0,\infty)}$. Consider the collection of increments

$$\underline{\mathbf{y}} = \{\mathbf{y}_k, k = 1, 2, \ldots, n\}$$

Because $\mathbf{x}_{[0,\infty)}$ is a Wiener process each of the $\mathbf{x}_{t_k}$ is a Gaussian random variable (see Definition 34.19). Therefore $\mathbf{y}_k$, the difference of Gaussian random variables, is also a Gaussian random variable for every k and $\underline{\mathbf{y}}$ is an n-vector Gaussian random variable. Therefore, $\underline{\mathbf{y}}$ is completely characterized by its mean and covariance. The mean is zero. The covariance is diagonal, as can be seen from the following calculation.

Consider, for $j \le k-1$ (the means are zero)

$$\begin{aligned} E(\mathbf{y}_k\mathbf{y}_j) &= E((\mathbf{x}_{t_k}-\mathbf{x}_{t_{k-1}})(\mathbf{x}_{t_j}-\mathbf{x}_{t_{j-1}})) \\ &= E(\mathbf{x}_{t_k}\mathbf{x}_{t_j}) - E(\mathbf{x}_{t_{k-1}}\mathbf{x}_{t_j}) \\ &\quad - E(\mathbf{x}_{t_k}\mathbf{x}_{t_{j-1}}) + E(\mathbf{x}_{t_{k-1}}\mathbf{x}_{t_{j-1}}) \\ &= t_j - t_j - t_{j-1} + t_{j-1} = 0 \end{aligned}$$

A similar calculation for $j \ge k+1$ completes the demonstration that the covariance of $\mathbf{y}$ is diagonal. Because $\mathbf{y}$ is Gaussian, the fact that its covariance is diagonal proves that the $\underline{\mathbf{y}}_k$ $k = 1, 2, \ldots, n$ are independent. This is a very important property of the Wiener process. It is usually described thus: the Wiener process has independent increments.

34.3.4 Stationarity, Ergodicity, and White Noise

The concepts of stationarity and wide-sense stationarity for continuous-time stochastic processes are virtually identical to those for discrete-time stochastic processes. All that is needed is to define the continuous-time stochastic process on the infinite interval

$$\mathbf{x}_{(-\infty,\infty)} \triangleq \mathbf{x}(t) \text{ for all } t, \quad -\infty < t < \infty$$

and adjust the notation in Definitions 34.9 and 34.11. Only the latter is included here so as to emphasize the focus on second-order processes.

DEFINITION 34.20 A scalar continuous-time stochastic process, $\mathbf{x}_{(-\infty,\infty)}$, is a second-order process if and only if $E(x^2(t)) < \infty$ for all t, $-\infty < t < \infty$.

DEFINITION 34.21 A second-order, continuous-time stochastic process $\mathbf{x}_{(-\infty,\infty)}$ is wide-sense stationary if and only if its mean $m(t)$ and covariance function $r(t,\tau)$ satisfy

$$m(t) = m, \text{ a constant for all } t, -\infty < t < \infty \quad (34.68)$$

$$r(t,\tau) = r(t+\sigma, \tau+\sigma) \text{ for all } t, \tau, \sigma, -\infty < t, \tau, \sigma < \infty \quad (34.69)$$

It is customary to define

$$r(t) = r(t+\tau, \tau) \text{ for all } t, \tau, -\infty < t < \infty \quad (34.70)$$

The apparatus necessary to define continuous-time white noise is now in place.

DEFINITION 34.22 A scalar second-order continuous-time stochastic process, $\mathbf{x}_{(-\infty,\infty)}$ is a white noise process if and only if

$$m(t) = E(\mathbf{x}(t)) = 0 \text{ for all } t, -\infty < t < \infty \quad (34.71)$$

$$r(t) = r\delta(t), \qquad r > 0, \quad -\infty < t < \infty \quad (34.72)$$

As in discrete time, the process is Gaussian white noise if, in addition, $\mathbf{x}(t)$ is Gaussian for all t.

As in discrete time, it is often necessary to determine the properties of a stochastic process from a combination of prior knowledge and observations of one sample function. In order for this to be possible it is necessary that the process be stationary and that, in some form, averages over time of $\mathbf{x}(t)$, $-\infty < t < \infty$, converge to expectations. Mathematically, this can be expressed as follows. Given a stationary continuous-time stochastic process $\mathbf{x}_{(-\infty<t<\infty)}$, one wants

$$\lim_{t\to\infty} \frac{1}{2t}\int_{-t}^{t} f(\mathbf{x}(t))dt = E(f(\mathbf{x}(0))) \quad (34.73)$$

to hold for any reasonable function $f(\cdot)$. This is true, with some minor technical conditions for processes that are stationary and ergodic. See [4] for an introduction to this important but difficult subject.

34.3.5 Transforms

Again, in parallel to the discrete-time case, any transform that is useful in the analysis of continuous-time signals can be applied to continuous-time stochastic processes. The two obvious candidates are the Laplace and Fourier transforms [5]. The most important application is the completely deterministic Fourier transform of the autocovariance of a wide-sense stationary continuous-time stochastic process.

DEFINITION 34.23 Let $\mathbf{x}_{(-\infty,\infty)}$ be a wide-sense stationary scalar continuous-time stochastic process with mean m and autocovariance $r(t)$. Assume $\int_{-\infty}^{\infty} |r(t)|dt < \infty$. Then the Fourier transform of the autocovariance function $r(t)$,

$$s(w) = \int_{-\infty}^{\infty} r(t)e^{-jwt}dt \quad (34.74)$$

is well defined and known as the spectral density of $\mathbf{x}_{(-\infty,\infty)}$.

The inverse of Equation 34.74 is the usual inverse of the Fourier transform.

$$r(t) = \frac{1}{2\pi}\int_{-\infty}^{\infty} s(w)e^{jwt}dw \quad (34.75)$$

When $m = 0$,

$$r(0) = E(\mathbf{x}^2(t)) = \int_{-\infty}^{\infty} s(w)\frac{dw}{2\pi} \quad (34.76)$$

If $\mathbf{x}_{(-\infty,\infty)}$ is a voltage, current, or velocity, then $r(0)$ can be interpreted as average power, at least to within a constant of proportionality. Then $s(w_0)$, where w_0 is any fixed frequency, must be the average power per unit frequency in $\mathbf{x}_{(-\infty,\infty)}$ at the frequency w_0 [2]. This is why $s(w)$ is called the spectral density.

EXAMPLE 34.8:

Suppose $\mathbf{x}_{(-\infty,\infty)}$ is a white noise process (see Definition 34.22) with $r = 1$. The spectral density for this process is

$$s(w) = \int_{-\infty}^{\infty} \delta(t)e^{-jwt}dt = 1$$

Because all frequencies are equally present, as in white light, a process having $s(w) = 1$ is called a white noise process.

34.3.6 SISO Continuous-Time Linear Systems

It is convenient to begin with linear systems described by their impulse response, $h(t, \tau)$. The notation is shown in Figure 34.2.

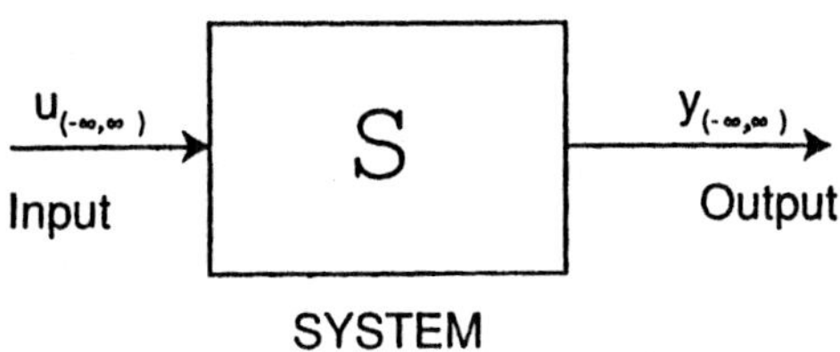

Figure 34.2 Representation of a system with input $u_{(-\infty,\infty)}$ and output $y_{(-\infty,\infty)}$.

DEFINITION 34.24 The impulse response of a SISO continuous-time system is denoted by $h(t, \tau)$, where $h(t, \tau)$ is the value of the output at instant t when the input is a unit impulse, $\delta(t - \tau)$, at instant τ.

The response of linear system, S, to an arbitrary input, $u_{(-\infty,\infty)}$ is given by

$$y(t) = \int_{-\infty}^{\infty} h(t, \tau)u(\tau)d\tau \qquad -\infty < t < \infty \tag{34.77}$$

Because the second-order properties of $\mathbf{y}_{(-\infty,\infty)}$ are completely described by its mean and autocovariance, it is worthwhile to know how to compute them from $h(t, \tau)$ and the second-order properties of $\mathbf{u}_{(-\infty,\infty)}$. The trick is to compute them at each time, t. That is

$$m_y(t) \triangleq E(\mathbf{y}(t)) = E(\int_{-\infty}^{\infty} h(t, \tau)\mathbf{u}(\tau)d\tau)$$

By the linearity of integration

$$= \int_{-\infty}^{\infty} E(h(t, \tau)\mathbf{u}(\tau))d\tau$$

Because $h(t, \tau)$ is deterministic,

$$= \int_{-\infty}^{\infty} h(t, \tau)E(\mathbf{u}(\tau))d\tau$$

That is,

$$m_y(t) = \int_{-\infty}^{\infty} h(t, \tau)m_u(\tau)d\tau \tag{34.78}$$

where

$$m_u(t) \triangleq E(\mathbf{u}(t))$$

Similarly,

$$\begin{aligned}
r_y(t, \tau) &= E((\mathbf{y}(t) - m_y(t))(\mathbf{y}(\tau) - m_y(\tau))) \\
&= E(\int_{-\infty}^{\infty} h(t, \sigma_1)(\mathbf{u}(\sigma_1) - m_u(\sigma_1))d\sigma_1 \\
&\quad \int_{-\infty}^{\infty} h(\tau, \sigma_2)(\mathbf{u}(\sigma_2) - m_u(\sigma_2))d\sigma_2) \\
&= E(\int_{-\infty}^{\infty} h(t, \sigma_1)h(\tau, \sigma_2)(\mathbf{u}(\sigma_1) - m_u(\sigma_1)) \\
&\quad (\mathbf{u}(\sigma_2) - m_u(\sigma_2))d\sigma_1 d\sigma_2) \\
&= \int_{-\infty}^{\infty}\int_{-\infty}^{\infty} h(t, \sigma_1)h(\tau, \sigma_2) \\
&\quad E((\mathbf{u}(\sigma_1) - m_u(\sigma_1))\mathbf{u}(\sigma_2) - m_u(\sigma_2)))d\sigma_1 d\sigma_2 \\
&= \int_{-\infty}^{\infty}\int_{-\infty}^{\infty} h(t, \sigma_1, \sigma_2)r_u(\sigma_1, \sigma_2)d\sigma_1 d\sigma_2
\end{aligned} \tag{34.79}$$

where

$$r_u(t, \tau) \triangleq E((\mathbf{u}(t) - m_u(t))(\mathbf{u}(\tau) - m_u(\tau)))$$

As in the discrete-time case, the Fourier transform can be used to simplify the calculations when the linear system is also time-invariant and $\mathbf{u}_{(-\infty,\infty)}$ is wide-sense stationary. Denote the impulse response by $h(t)$, where $h(t) = h(t + \tau, \tau)$. Let $m_u = E(\mathbf{u}(t))$, and $r_u(t) = E((\mathbf{u}(t + \tau) - m_u)(\mathbf{u}(\tau) - m_u))$. It is then easy to show, by paralleling the argument leading to Equation 34.41, that

$$m_y = h_f(0)m_u \tag{34.80}$$

where

$$h_f(w) = \int_{-\infty}^{\infty} h(t)e^{-jwt}dt, \text{ the Fourier transform of } h(t).$$

Similarly,

$$r_y(t) = \int_{-\infty}^{\infty} |h_f(w)|^2 s_u(w)e^{jwt}\frac{dw}{2\pi} \tag{34.81}$$

where $s_u(w)$ denotes the spectral density of $\mathbf{u}_{(-\infty,\infty)}$ (see Equation 34.74). This follows from virtually the same argument as Equation 34.42. By the uniqueness of Fourier transforms, Equation 34.81 implies

$$s_y(w) = |h_f(w)|^2 s_u(w) \tag{34.82}$$

It is necessary to define the notation for n-vector continuous-time stochastic processes before describing continuous-time systems in state-space form. Both of these are done in the following section.

34.3.7 Vector Continuous-Time Stochastic Process and LTI Systems

The vector case involves no new concepts but requires a more complicated notation. Because only second-order stochastic processes are discussed here all that is needed is a notation for the process, its mean, and its autocovariance.

An n-vector continuous-time stochastic process, $\underline{\mathbf{x}}_{[t_2,t_2]}$, is an m-vector random variable, $\underline{\mathbf{x}}(t) = [\mathbf{x}_1(t)\mathbf{x}_2(t)\ldots\mathbf{x}_m(t)]'$ at each instant of time t, $t_1 \le t \le t_2$. As throughout this chapter, vectors are denoted by lower-case underlined letters. Random variables and stochastic processes are bold letters.

The mean of a vector stochastic process is defined at each instant of time; the autocovariance is defined for paired times. Thus, when $\underline{\mathbf{x}}(t)$ is an n-vector random variable defined for all t, $t_1 \le t \le t_2$, (that is, $\mathbf{x}_{[t_1,t_2]}$ is an n-vector stochastic process)

$$\underline{m}_x(t) = E(\underline{\mathbf{x}}(t)) \quad t_1 \le t \le t_2 \tag{34.83}$$

will denote its mean and

$$\begin{aligned} \underline{R}_x(t,\tau) &= E((\underline{\mathbf{x}}(t) - \underline{m}_x(t))(\underline{\mathbf{x}}(\tau) - \underline{m}_x(\tau))') \\ & \quad t_1 \le t, \tau \le t_2 \end{aligned} \tag{34.84}$$

will denote its autocovariance. Note the transpose in Equation 34.84 and the capital $\underline{R}_x$. This emphasizes that the autocovariance is an $n \times n$ matrix for each pair t and τ. As in the discrete-time case, Equations 34.83 and 34.84 are evaluated element by element. Thus,

$$m_{x_i}(t) = E(\mathbf{x}_i(t)) \qquad i = 1, 2, \ldots, m \tag{34.85}$$

$$r_{x_{ij}}(t,\tau) = E((\mathbf{x}_i(t) - m_{x_i}(t))(\mathbf{x}_j(\tau) - m_{\mathbf{x}_j}(\tau))) \tag{34.86}$$

EXAMPLE 34.9: Continuous-time m-vector white noise:

Any white noise must be wide-sense stationary and have zero mean. Letting $\underline{\xi}(t)$ denote an m-vector continuous-time white noise at the instant t,

$$\underline{m}_\xi(t) = E(\underline{\xi}(t)) = \underline{0} \text{ for all } t, -\infty < t < \infty \tag{34.87}$$

$$\begin{aligned} \underline{R}_\xi(t,\tau) &= E(\underline{\xi}(t)\underline{\xi}'(\tau)) = \underline{0} \\ & \text{for all } t \ne \tau \; -\infty < t, \tau < \infty \end{aligned} \tag{34.88}$$

because the autocovariance of white noise must be equal to 0 except when $t = \tau$. At $t = \tau$, the autocovariance reduces to a covariance matrix (denoted by $\underline{\Xi}$, which must be positive semidefinite because it is a covariance matrix) times a unit impulse. A rigorous derivation of this is hard but the analogy to Equation 34.49 should be convincing. Thus, for all $-\infty < t, \tau < \infty$

$$\underline{R}_\xi(t,\tau) = \underline{\Xi}\delta(t-\tau) \tag{34.89}$$

where $\underline{\Xi}$ is any positive semidefinite matrix.

Now that vector continuous-time stochastic processes and vector white noise have been introduced it is possible to develop the state-space description of linear continuous-time systems driven by continuous-time stochastic processes. For convenience, only the case of LTI systems is described.

EXAMPLE 34.10: LTI system and white noise:

Let $\underline{\mathbf{x}}_0$ be an n-dimensional second-order random variable with mean $\underline{m}_{x_0}$ and covariance $\underline{R}_{x_0}$. Let $\underline{\xi}_{[0,\infty)}$ and $\underline{\theta}_{[0,\infty)}$ be q- and p-dimensional wide-sense stationary second-order continuous-time stochastic processes satisfying, for all $t \in [0,\infty)$

$$E(\underline{\xi}(t)) = \underline{0} \text{ and } E(\underline{\theta}(t)) = \underline{0} \tag{34.90}$$

$$\underline{R}_\xi(t) = E(\underline{\xi}(t+\tau)\underline{\xi}'(\tau)) = \underline{\Xi}\delta(t) \tag{34.91}$$

$$\underline{R}_\theta(t) = E(\underline{\theta}(t+\tau)\underline{\theta}'(\tau)) = \underline{\Theta}\delta(t) \tag{34.92}$$

where $\underline{\Xi}$ and $\underline{\Theta}$ are symmetric positive semidefinite matrices of appropriate dimension.

Strictly speaking $\underline{\xi}_{[0,\infty)}$ and $\underline{\theta}_{[0,\infty)}$ are not white noise processes because they are defined only in $[0,\infty)$. However, in the context of state-space analysis it is standard to call them white noise processes. Obviously, what matters is not the name, but Equations 34.90 to 34.92.

For convenience, assume that

$$E((\underline{\mathbf{x}}_0 - \underline{m}_{x_0})\underline{\xi}'(t)) = \underline{0} \tag{34.93}$$

$$E((\underline{\mathbf{x}}_0 - \underline{m}_{x_0})\underline{\theta}'(t)) = \underline{0} \tag{34.94}$$

and

$$E(\underline{\xi}(t)\underline{\theta}'(\tau)) = \underline{0} \text{ for all } t, \tau \in [0,\infty) \tag{34.95}$$

The assumptions in Equations 34.93 to 34.95 remove some terms from the subsequent expressions, thereby saving space and the effort to compute them, but are otherwise inessential.

Define the n- and p-vector random variables $\underline{\mathbf{x}}(t)$ and $\underline{\mathbf{y}}(t)$ by

$$\frac{d\underline{\mathbf{x}}}{dt}(t) = \underline{A}\underline{\mathbf{x}}(t) + \underline{B}\underline{u}(t) + \underline{L}\underline{\xi}(t); \quad \underline{\mathbf{x}}(0) = \underline{\mathbf{x}}_0 \tag{34.96}$$

$$\underline{\mathbf{y}}(t) = \underline{C}\underline{\mathbf{x}}(t) + \underline{\theta}(t) \tag{34.97}$$

where $\underline{u}(t)$ is a deterministic m-vector, the control, and all the matrices are real, deterministic, and have the appropriate dimensions (that is, $\underline{A}$ is $n \times n$, $\underline{B}$ is $n \times m$, $\underline{C}$ is $p \times n$, and $\underline{L}$ is $n \times q$). The structure of this system is shown in Figure 34.3.

It follows from Equations 34.96 and 34.97 that $\underline{\mathbf{x}}_{[0,\infty)}$ and $\underline{\mathbf{y}}_{[0,\infty)}$ are vector continuous-time stochastic processes. Assuming that $\underline{u}(t)$ is bounded for all t guarantees that $\underline{\mathbf{x}}_{[0,t_f)}$ and $\underline{\mathbf{y}}_{[0,t_f)}$ are second-order for all finite t_f. What are their second-order statistics?

To compute the mean of $\underline{\mathbf{x}}(t)$ take the expected value of both sides of Equation 34.96. That is,

$$E(\dot{\underline{\mathbf{x}}}(t)) = E(\underline{A}\underline{\mathbf{x}}(t) + \underline{B}\underline{u}(t) + \underline{L}\underline{\xi}(t))$$

Interchanging the order of expectation and d/dt (both are linear so their order can be reversed) and using linearity on the right-hand side of the equation gives

$$\frac{d}{dt}E(\underline{\mathbf{x}}(t)) = \underline{A}E(\underline{\mathbf{x}}(t)) + \underline{B}\underline{u}(t); \quad E(\underline{\mathbf{x}}(0)) = \underline{m}_{x_0} \tag{34.98}$$

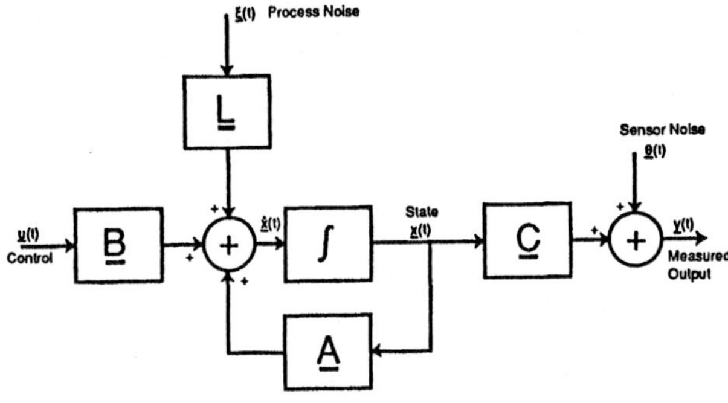

Figure 34.3 A block diagram of the LTI system described by Equations 34.96 and 34.97.

or

$$\dot{\underline{m}}_x(t) = \underline{A}\underline{m}_x(t) + Bu(t); \quad \underline{m}_x(0) = \underline{m}_{x0}$$

This is just a deterministic ordinary differential equation in state-space form for $E(\underline{\mathbf{x}}(t))$. Once $E(\underline{\mathbf{x}}(t))$ has been computed, taking expectation of both sides of Equation 34.97 gives

$$E(\underline{\mathbf{y}}(t)) = \underline{C}E(\underline{\mathbf{x}}(t)) \tag{34.99}$$

This is a deterministic algebraic equation.

A derivation of the autocovariance of $\underline{\mathbf{x}}_{[0,t_f)}$ is beyond the scope of this chapter. A derivation can be found in ([6], pp. 111-113).

The result is that

$$\underline{R}_x(t,t) = E((\underline{\mathbf{x}}(t) - \underline{m}_x(t))(\underline{\mathbf{x}}(t) - \underline{m}_x(t))')$$

satisfies the differential equation

$$\dot{\underline{R}}_x(t,t) = \underline{A}\underline{R}_x(t,t) + \underline{R}_x(t,t)\underline{A}' + \underline{L}\underline{\Xi}\underline{L}' \tag{34.100}$$

with initial condition $\underline{R}_x(0,0) = \underline{R}_{x_0}$ and $\underline{R}_x(t,\tau)$, with $t \geq \tau$, satisfies the differential equation

$$\frac{\partial \underline{R}_x(t,\tau)}{\partial t} = \underline{A}\underline{R}_x(t,\tau) \tag{34.101}$$

while $\underline{R}_x(\tau,t)$, again with $t \geq \tau$, satisfies

$$\frac{\partial R_x(\tau,t)}{\partial t} = \underline{R}_x(\tau,t)\underline{A}' \tag{34.102}$$

The initial condition for both Equations 34.101 and 34.102 is $\underline{R}_x(\tau,\tau)$, computed from Equation 34.100.

34.4 Conclusions

The second-order properties of many zero mean stochastic processes can be reproduced by a LTI system with white noise input. The precise result for scalar continuous-time systems is the following [4]. Let $s(w)$ be a real non-negative absolutely integrable function such that

$$\int_{-\infty}^{\infty} \frac{|\ln s(w)|}{1+w^2} dw < \infty \tag{34.103}$$

Then there exists a square integrable function $h_f(w)$, $-\infty < w < \infty$, such that

$$|h_f(w)|^2 = s(w) \tag{34.104a}$$

$$h(t) = \int_{-\infty}^{\infty} h_f(w)e^{jwt}\frac{dw}{2\pi} = 0 \quad \text{for } t < 0 \tag{34.104b}$$

$$h_f(\sigma + jw) = \int_{-\infty}^{\infty} e^{j(\sigma+jw)t}h(t)dt \neq 0, \quad w > 0 \tag{34.104c}$$

Because the spectral density of white noise is one for all w, Equation 34.104 means that the spectral density, $s(w)$, is identical to the spectral density of the output of a LTI system with impulse response $h(t)$ driven by a white noise input. Equation 34.104 means that $h(t)$ is causal. Equation 34.104 means that the transfer function $h_f(w)$ is minimum phase. Similar results hold in discrete time and for vector signals [7], [8].

The restriction to second-order processes with zero mean is unimportant because the mean is deterministic. Thus, the only significant limitation on the use of a linear system driven by white noise as a model for second-order stochastic processes is Equation 34.103. The major limitation of the model is that the second-order statistics are not always a good description of a stochastic process. The second-order statistics are most useful when the process is approximately Gaussian. Processes that are very different from Gaussian, especially in the vicinity of the mean, are not usually well modeled by their second-order statistics.

34.5 Notation

Lower case letters such as x, y, r denote real, deterministic scalars. The same letters, underlined, denote vectors.

Lower case letters i, j, k, ℓ, m, n denote integers (j is also used to indicate $\sqrt{-1}$ and m for mean—the meaning should be evident from the context).

Upper case, underlined, capital letters such as $\underline{A}$, $\underline{B}$, $\underline{C}$, $\underline{L}$, $\underline{\Theta}$, $\underline{\Xi}$ denote real deterministics matrices.

Lower case, bold letters such as $\mathbf{x}$, $\mathbf{y}$ denote scalar random variables. The same letters, underlined, denote vector random variables (or discrete-time stochastic processes).

Upper case, underlined bold letters; such as $\underline{\mathbf{X}}$, $\underline{\mathbf{Y}}$, $\underline{\boldsymbol{\Theta}}$, $\underline{\boldsymbol{\Xi}}$ denote vector discrete-time stochastic processes.

Lower case, bold letters with interval subscripts such as $\mathbf{x}_{(-\infty,\infty)}$ denote stochastic processes.

References

[1] Leon-Garcia, A., *Probability and Random Processes for Electrical Engineering*, Addison-Wesley, Reading, MA, 1989.

[2] Wong, E., *Introduction to Random Processes*, Springer-Verlag, Amsterdam, 1983.

[3] Breiman, L., *Probability*, SIAM (republication of a 1968 book published by Addison-Wesley), 1992.

[4] Wong, E., *Stochastic Processes in Information and Dynamical Systems*, McGraw-Hill, New York, 1971.

[5] Oppenheim, A.V. and Willsky, A.S., with I.T. Young, *Signals and Systems*, Prentice Hall, Englewood Cliffs, NJ, 1983.

[6] Davis, M.H.A., *Linear Estimation and Stochastic Control*, Chapman and Hall, 1977.

[7] Astrom, K.J., *Introduction to Stochastic Control Theory*, Academic Press, New York, 1970.

[8] Caines, P.E., *Linear Stochastic Systems*, John Wiley & Sons, New York, 1988.

35

Kalman Filtering

Michael Athans
Massachusetts Institute of Technology, Cambridge, MA

35.1 Introduction

The purpose of this chapter is to provide an overview of Kalman filtering concepts. We cover only a small portion of the material associated with Kalman filters. Our choice of material is primarily motivated by the material needed in the design and analysis of multivariable feedback control systems for linear time-invariant (LTI) systems and, more specifically, by the design philosophy commonly called the linear quadratic Gaussian (LQG) method. Thus, from a technical perspective, we cover, without proofs, the so-called steady-state, constant-gain LTI Kalman filters.

More details regarding different formulations and solutions of Kalman filtering problems (continuous-time linear time-varying, discrete-time linear time-varying, and time-invariant) can be found in several classic and standard textbooks on the subject of estimation theory. Please see Further Reading at the end of this chapter.

The steady-state constant-gain Kalman filter is an algorithm that is used to estimate the state variables of a continuous-time LTI system, subject to stochastic disturbances, on the basis of noisy measurements of certain output variables. Thus, the Kalman filtering algorithm combines the information regarding the plant dynamics, the probabilistic information regarding the stochastic disturbances that influence the plant state variables, as well as that regarding the measurement noise that corrupts the sensor measurements, and the deterministic controls.

Credit for "inventing" this algorithm is usually given to Dr. Rudolph E. Kalman, who presented the key ideas in the late 1950s and early 1960s. Although the Kalman filter is an alternative representation of the Wiener filter [1], Kalman's contribution was to tie the state estimation problem to the state-space models, and the (then new) concepts of controllability and observability [2] to [5]. Thousands of papers have been written about the Kalman filter and its numerous applications to navigation, tracking, and estimator and controller design in defense and industrial applications.

35.2 Problem Definition

In this section we present the definition of the basic stochastic estimation problem for which the Kalman filter (KF) yields an "optimal" solution. First we present the plant state dynamics. We deal with a finite-dimensional LTI system whose state vector $\underline{x}(t)$, $\underline{x}(t) \in R^n$ (a real n-vector), obeys the stochastic differential equation

$$\dot{\underline{x}}(t) = \underline{A}\underline{x}(t) + \underline{B}\underline{u}(t) + \underline{L}\underline{\xi}(t) \tag{35.1}$$

where $\underline{u}(t)$, $\underline{u}(t) \in R^m$ is the deterministic control vector (assumed known), $\underline{\xi}(t)$, $\underline{\xi}(t) \in R^q$ is a vector-valued stochastic process, often called the *process noise*, that acts as a disturbance to the plant dynamics. Vectors are underlined, lower case letters and matrices are underlined, upper case letters. The process noise $\underline{\xi}(t)$ is assumed to have certain statistical properties, corresponding to a stationary (time-invariant) continuous-time *white* Gaussian noise with zero mean, i.e.,

$$E\{\underline{\xi}(t)\} = 0, \text{ for all } t \tag{35.2}$$

0-8493-8570-9/96/$0.00+$.50

and its covariance matrix is defined by

$$\text{cov}\,[\underline{\xi}(t);\underline{\xi}(\tau)] = E\{\underline{\xi}(t)\underline{\xi}'(\tau)\} = \underline{\Xi}\delta(t-\tau) \tag{35.3}$$

with $\delta(t-\tau)$ being the Dirac delta function (impulse at $t=\tau$). The matrix $\underline{\Xi}$ is called the intensity matrix of $\underline{\xi}(t)$ and it is a symmetric positive definite matrix, i.e.,

$$\underline{\Xi} = \underline{\Xi}' > 0 \tag{35.4}$$

REMARK 35.1 Continuous time white noise does not exist in nature; it is the limit of a broadband stochastic process. In the frequency domain, continuous-time white noise corresponds to a stochastic process with constant power spectral density as a function of frequency. This implies that continuous-time white noise has constant power at all frequencies, and therefore has infinite energy! White noise is completely unpredictable, as can be seen from Equation 35.3 because it is uncorrelated for any $t \neq \tau$, while it has finite variance and standard deviation at $t=\tau$. This is obviously an approximation to reality. As with the Dirac delta function, $\delta(t)$, white noise creates some subtle mathematical issues, but is nonetheless extremely useful in engineering.

REMARK 35.2 The state $\underline{x}(t)$, the solution of Equation 35.1, is a well-defined physical stochastic process, a so-called colored Gaussian random process, and it has finite energy. Its power spectral density rolls off at high frequencies.

Next, we turn our attention to the measurement equation. We assume that our sensors cannot directly measure all of the physical state variables, the components of the vector $\underline{x}(t)$ of the plant given in Equation 35.1. Rather, in the classical Kalman filter formulation we assume that we can measure only certain output variables (linear combinations of the state variables) in the presence of additive continuous-time white noise.

The mathematical model of the measurement process is as follows:

$$\underline{y}(t) = \underline{C}\underline{x}(t) + \underline{\theta}(t) \tag{35.5}$$

The vector $\underline{y}(t) \in R^p$ represents the sensor measurement. The measurement or sensor noise $\underline{\theta}(t) \in R^p$ is assumed to be a continuous-time white Gaussian random process, independent of $\underline{\xi}(t)$, with zero mean, i.e.,

$$E\{\underline{\theta}(t)\} = 0, \text{ for all } t \tag{35.6}$$

and covariance matrix

$$\text{cov}\,[\underline{\theta}(t);\underline{\theta}(\tau)] = E\{\underline{\theta}(t)\underline{\theta}'(\tau)\} = \underline{\Theta}\delta(t-\tau) \tag{35.7}$$

where the sensor noise intensity matrix is symmetric and positive definite, i.e.,

$$\underline{\Theta} = \underline{\Theta}' > 0 \tag{35.8}$$

Figure 35.1 shows a visualization, in block diagram form, of Equations 35.1 and 35.5.

35.2.1 The State Estimation Problem

Imagine that we have been observing the control $\underline{u}(\tau)$ and the output $\underline{y}(\tau)$ over the infinite past up to the present time t. Let

$$U(t) = \{\underline{u}(\tau);\ -\infty < \tau \leq t\} \tag{35.9}$$
$$Y(t) = \{\underline{y}(\tau);\ -\infty < \tau \leq t\} \tag{35.10}$$

denote the past histories of the control and output, respectively.

The state estimation problem is as follows: given $U(t)$ and $Y(t)$, find a vector $\hat{\underline{x}}(t)$, at time t, which is an "optimal" estimate of the present state $\underline{x}(t)$ of the system defined by Equation 35.1.

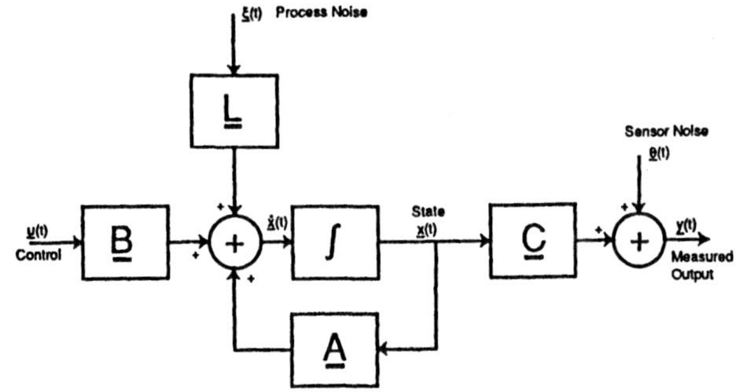

Figure 35.1 A stochastic linear dynamic system.

Under the stated assumptions regarding the Gaussian nature of $\underline{\xi}(t)$ and $\underline{\theta}(t)$ the "optimal" state estimate is the same for an extremely large class of optimality criteria [6]. This generally optimal estimate is the conditional mean of the state, i.e.,

$$\hat{\underline{x}}(t) \triangleq E\{\underline{x}(t)|U(t), Y(t)\} \tag{35.11}$$

One can relax the Gaussian assumption and define the optimality of the state estimate $\hat{\underline{x}}(t)$ in different ways. One popular way is to demand that $\hat{\underline{x}}(t)$ be generated by a *linear* transformation on the past "data" $U(t)$ and $Y(t)$, such that the state estimation error $\tilde{\underline{x}}(t)$

$$\tilde{\underline{x}} \triangleq \underline{x}(t) - \hat{\underline{x}}(t) \tag{35.12}$$

has zero mean, i.e.,

$$E\{\tilde{\underline{x}}(t)\} = \underline{0} \tag{35.13}$$

and the cost functional

$$J = E\left\{\sum_{i=1}^{n} \tilde{x}_i^2(t)\right\} = E\left\{\tilde{\underline{x}}'(t)\tilde{\underline{x}}(t)\right\} = tr[E\{\tilde{\underline{x}}(t)\tilde{\underline{x}}'(t)\}] \tag{35.14}$$

is minimized.

The cost functional J has the physical interpretation that it is the sum of the error variances $E\{\tilde{x}_i^2(t)\}$ for each state variable. If we let $\underline{\Sigma}$ denote the covariance matrix (stationary) of the state estimation error

$$\underline{\Sigma} \triangleq E\{\tilde{\underline{x}}(t)\tilde{\underline{x}}'(t)\} \tag{35.15}$$

then the cost, J, of Equation 35.14 can also be written as

$$J = tr[\underline{\Sigma}] \tag{35.16}$$

Bottom Line: We need an algorithm that translates the signals we can observe, $\underline{u}(t)$ and $\underline{y}(t)$, into a state estimate $\hat{\underline{x}}(t)$, such that the state estimation error $\tilde{\underline{x}}$ is "small" in some well-defined sense. The KF is the algorithm that does just that!

35.3 Summary of the Kalman Filter Equations

35.3.1 Additional Assumptions

In this section we summarize the on-line and off-line equations that define the Kalman filter. Before we do that we make two additional "mild" assumptions

$$[\underline{A}, \underline{L}] \text{ is stabilizable (or controllable)} \tag{35.17}$$

$$[\underline{A}, \underline{C}] \text{ is detectable (or observable)} \tag{35.18}$$

$[\underline{A}, \underline{L}]$ is controllable, means that the process noise $\underline{\xi}(t)$ excites all modes of the system defined by Equation 35.1; $[\underline{A}, \underline{C}]$ is observable means that the "noiseless" output $\underline{y}(t) = \underline{C}\underline{x}(t)$ contains information about all state variables. If $[\underline{A}, L]$ is stabilizable, the modes of the system that are not excited by $\underline{\xi}(t)$ are asymptotically stable; if $[A, C]$ is detectable, the unobserved modes are asymptotically stable.

35.3.2 The Kalman Filter Dynamics

The function of the KF is to generate in real time the state estimate $\hat{\underline{x}}(t)$ of the state $\underline{x}(t)$. The KF is actually an LTI dynamic system, of identical order (n) to the plant Equation 35.1, and is driven by (1) the deterministic control input $\underline{u}(t)$, and (2) the measured output vector $\underline{y}(t)$. The Kalman filter dynamics are given as follows:

$$\frac{d\hat{\underline{x}}(t)}{dt} = \underline{A}\hat{\underline{x}}(t) + \underline{B}\underline{u}(t) + \underline{H}[\underline{y}(t) - \underline{C}\hat{\underline{x}}(t)] \tag{35.19}$$

A block diagram visualization of Equation 35.19 is shown in Figure 35.2. Note that in Equation 35.19 all variables have been defined previously, except for the KF gain matrix $\underline{H}$, whose calculation is carried out off-line and is discussed shortly.

The filter gain matrix $\underline{H}$ multiplies the so-called *residual* or *innovations* vector

$$\underline{r}(t) \overset{\Delta}{=} \underline{y}(t) - \underline{C}\hat{\underline{x}}(t) \tag{35.20}$$

and updates the time rate of change, $d\hat{\underline{x}}(t)/dt$, of the state estimate $\hat{\underline{x}}(t)$. The residual $\underline{r}(t)$ is like an "error" between the *measured* output $\underline{y}(t)$, and the *predicted* output $\underline{C}\hat{\underline{x}}(t)$.

REMARK 35.3 From an intuitive point of view the KF, defined by Equation 35.19 and illustrated in Figure 35.2, can be thought as a model-based observer or state reconstructor. The reader should carefully compare the structures depicted in Figures 35.1 and 35.2. The plant/sensor properties, reflected by the matrices $\underline{A}$, $\underline{B}$, and $\underline{C}$, are duplicated in the KF[1]. The state estimate $\hat{\underline{x}}(t)$ is continuously updated by the actual sensor measurements, through the formation of the residual $\underline{r}(t)$ and the "closing" of the loop with the filter gain matrix $\underline{H}$.

The KF dynamics of Equation 35.19 can also be written in the form

$$\frac{d\hat{\underline{x}}(t)}{dt} = [\underline{A} - \underline{H}\underline{C}]\hat{\underline{x}}(t) + \underline{B}\underline{u}(t) + \underline{H}\underline{y}(t) \tag{35.21}$$

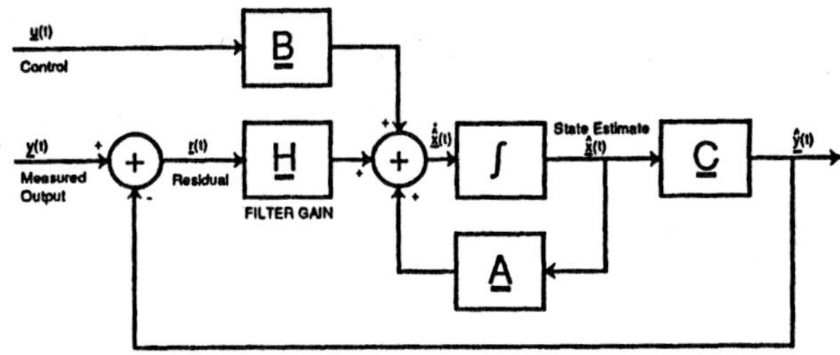

Figure 35.2 The structure of the Kalman Filter. The control, $\underline{u}(t)$, and measured output, $\underline{y}(t)$, are those associated with the stochastic system of Figure 35.1. The filter gains matrix $\underline{H}$ is computed in a special way.

From the structure of Equation 35.21 we can immediately see that the stability of the KF is governed by the matrix $\underline{A} - \underline{H}\underline{C}$. At this point of our development we remark that the assumption in Equation 35.18, i.e., the detectability of $[\underline{A}, \underline{C}]$, guarantees the existence of at least one filter gain matrix $\underline{H}$ such that the KF is stable, i.e.,

$$Re\lambda_i[\underline{A} - \underline{H}\underline{C}] < 0; \quad i = 1, 2, \ldots, n \tag{35.22}$$

where λ_i is the ith eigenvalue of $[\underline{A} - \underline{H}\underline{C}]$.

35.3.3 Properties of Model-Based Observers

We have remarked that the KF gain matrix $\underline{H}$ is calculated in a very special way. However, it is extremely useful to examine the structure of Equations 35.19 or 35.21 and Figure 35.2 with a filter gain matrix $\underline{H}$ that is arbitrary except for the requirement that Equation 35.22 holds. Thus, for the development that follows in this subsection think of $\underline{H}$ as being a fixed matrix.

As before let $\tilde{\underline{x}}(t)$ denote the state estimation error vector

$$\tilde{\underline{x}}(t) \overset{\Delta}{=} \underline{x}(t) - \hat{\underline{x}}(t) \tag{35.23}$$

It follows that

$$\frac{d\tilde{\underline{x}}(t)}{dt} = \frac{d\underline{x}(t)}{dt} - \frac{d\hat{\underline{x}}(t)}{dt} \tag{35.24}$$

[1] No signal corresponding to $\underline{L}\underline{\xi}(t)$ shows up in Figure 35.2. This is because we assumed that $\underline{\xi}(t)$ had zero-mean and was completely unpredictable. Thus, the best estimate for $\underline{\xi}(t)$ given data up to time t is 0.

Next, we substitute Equations 35.1, 35.5, and 35.21 into Equation 35.24 and use Equation 35.23 as appropriate. After some easy algebraic manipulations we obtain the following stochastic vector differential equation for the state estimation error $\underline{\tilde{x}}(t)$:

$$\frac{d\underline{\tilde{x}}(t)}{dt} = [\underline{A} - \underline{HC}]\underline{\tilde{x}}(t) + \underline{L\xi}(t) - \underline{H\theta}(t) \quad (35.25)$$

Note that, in view of Equation 35.22, the estimation error dynamic system is stable. Also note that the deterministic signal $\underline{Bu}(t)$ does not appear in the error equation (35.25).

Under our assumptions that the system is stable and was started at the indefinite past ($t_0 \rightarrow -\infty$), it is easy to verify that

$$E\{\underline{\tilde{x}}(t)\} = \underline{0} \quad (35.26)$$

This implies that any stable model-based estimator of the form shown in Figure 35.2, with any filter gain matrix $\underline{H}$, gives us *unbiased* (that is, $E(\underline{\hat{x}}(t)) = E(\underline{x}(t))$ estimates.

Using next elementary facts from stochastic linear system theory one can calculate the error covariance matrix $\underline{\Sigma}$ of the state estimation error $\underline{\tilde{x}}(t)$

$$\underline{\Sigma} \triangleq \text{cov}[\underline{\tilde{x}}(t); \underline{\tilde{x}}(t)] = E\{\underline{\tilde{x}}(t)\underline{\tilde{x}}'(t)\} \quad (35.27)$$

The matrix $\underline{\Sigma}$ is the solution of the so-called *Lyapunov matrix equation* (linear in $\underline{\Sigma}$)

$$[\underline{A} - \underline{HC}]\,\underline{\Sigma} + \underline{\Sigma}(\underline{A} - \underline{HC})' + \underline{L\Xi L}' + \underline{H\Theta H}' = \underline{0} \quad (35.28)$$

with

$$\underline{\Sigma} = \underline{\Sigma}' \geq \underline{0} \quad (35.29)$$

Thus, for any given filter gain matrix $\underline{H}$ we can calculate[2] the associated error covariance matrix $\underline{\Sigma}$ from Equation 35.28. Recalling Equation 35.16, we can evaluate, for a given $\underline{H}$, the quality of the estimator by calculating

$$J = tr[\underline{\Sigma}] \quad (35.30)$$

The specific way that the KF gain is calculated is by solving a constrained static deterministic optimization problem. Minimize Equation 35.30 with respect to the elements h_{ij} of the matrix $\underline{H}$ subject to the algebraic constraints given in Equations 35.28 and 35.29.

35.3.4 The Kalman Filter Gain and Associated Filter Algebraic Riccati Equation (FARE)

We now summarize the off-line calculations that define fully the Kalman filter (Equation 35.19 or 35.21).

The KF gain matrix $\underline{H}$ is computed by

$$\underline{H} = \underline{\Sigma C}'\underline{\Theta}^{-1} \quad (35.31)$$

[2]The MATLAB and MATRIX-X software packages can solve Lyapunov equations.

where $\underline{\Sigma}$ is the unique, symmetric, and at least positive semidefinite solution matrix of the so-called filter algebraic Riccati equation (FARE)

$$\underline{0} = \underline{A\Sigma} + \underline{\Sigma A}' + \underline{L\Xi L}' - \underline{\Sigma C}'\underline{\Theta}^{-1}\underline{C\Sigma} \quad (35.32)$$

with

$$\underline{\Sigma} = \underline{\Sigma}' \geq \underline{0} \quad (35.33)$$

REMARK 35.4 The formula for the KF gain can be obtained by setting

$$\frac{\partial}{\partial h_{ij}} tr[\underline{\Sigma}] = 0 \quad (35.34)$$

where $\underline{\Sigma}$ is given by Equation 35.28. The result is Equation 35.31. Substituting Equation 35.31 into Equation 35.28 one deduces the FARE (Equation 35.32).

35.3.5 Duality Between the KF and LQ Problems

The mathematical problems associated with the solution of the LQ and KF are dual. This duality was recognized by R.E. Kalman as early as 1960 [2].

The duality can be used to deduce several properties of the KF simply by "dualizing" the results of the LQ problem. A summary of the KF properties is given in Section 35.4.

35.4 Kalman Filter Properties

35.4.1 Introduction

In this section we summarize the key properties of the Kalman filter. These properties are the "dual" of those for the LQ controller.

35.4.2 Guaranteed Stability

Recall that the KF algorithm is

$$\frac{d\underline{\hat{x}}(t)}{dt} = [\underline{A} - \underline{HC}]\underline{\hat{x}}(t) + \underline{Bu}(t) + \underline{Hy}(t) \quad (35.35)$$

Then, under the assumptions of Section 35.3, the matrix $[\underline{A} - \underline{HC}]$ is strictly stable, i.e.,

$$Re\lambda_i[\underline{A} - \underline{HC}] < 0; \qquad i = 1, 2, \ldots, n \quad (35.36)$$

35.4.3 Frequency Domain Equality

One can readily derive a frequency domain equality for the KF. In the development that follows, let

$$\underline{\Xi} = \underline{I} \quad (35.37)$$

Let us make the following definitions: let $\underline{G}_{KF}(s)$ denote the KF loop-transfer matrix

$$\underline{G}_{KF}(s) \triangleq \underline{C}(s\underline{I} - \underline{A})^{-1}\underline{H} \quad (35.38)$$

$$\underline{G}_{KF}^{H}(s) \triangleq \underline{H}'(-s\underline{I} - \underline{A}')^{-1}\underline{C}' \quad (35.39)$$

where $[A]^H$ denotes the complex conjugate of the transpose of an arbitrary complex matrix A. Let $\underline{G}_{FOL}(s)$ denote the filter open-loop transfer matrix (from $\underline{\xi}(t)$ to $\underline{y}(t)$)

$$\underline{G}_{FOL}(s) \triangleq \underline{C}(s\underline{I} - \underline{A})^{-1}\underline{L} \quad (35.40)$$

$$\underline{G}_{FOL}^{H}(s) \triangleq \underline{L}'(-s\underline{I} - \underline{A}')^{-1}\underline{C}' \quad (35.41)$$

Then the following equality holds

$$[\underline{I} + \underline{G}_{KF}(s)]\underline{\Theta}[\underline{I} + \underline{G}_{KF}(s)]^H = \underline{\Theta} + \underline{G}_{FOL}(s)\underline{G}_{FOL}^{H}(s) \quad (35.42)$$

If

$$\underline{\Theta} = \mu \underline{I} \qquad \mu > 0 \quad (35.43)$$

then Equation 35.42 reduces to

$$[\underline{I} + \underline{G}_{KF}(s)][\underline{I} + \underline{G}_{KF}(s)]^H = \underline{I} + \frac{1}{\mu}\underline{G}_{FOL}(s)\underline{G}_{FOL}^{H}(s) \quad (35.44)$$

35.4.4 Guaranteed Robust Properties

The KF enjoys the same type of robustness properties as the LQ regulator. The following properties are valid if

$$\underline{\Theta} = \text{ diagonal matrix} \quad (35.45)$$

From the frequency domain equality (Equation 35.42) we deduce the inequality

$$[\underline{I} + \underline{G}_{KF}(s)][\underline{I} + \underline{G}_{KF}(s)]^H \geq \underline{I} \quad (35.46)$$

From the definition of singular values we then deduce that

$$\sigma_{\min}[\underline{I} + \underline{G}_{KF}(s)] \geq 1 \quad \text{or} \quad \sigma_{\max}[\underline{I} + \underline{G}_{KF}(s)]^{-1} \leq 1 \quad (35.47)$$

$$\sigma_{\min}[\underline{I} + \underline{G}_{KF}^{-1}(s)] \geq \tfrac{1}{2} \quad \text{or} \quad \sigma_{\max}[\underline{I} + \underline{G}_{KF}(s)]^{-1}\underline{G}_{KF} \leq 2 \quad (35.48)$$

35.5 The Accurate Measurement Kalman Filter

We summarize the properties of the Kalman Filter (KF) problem when the intensity of the sensor noise approaches zero. In a mathematical sense this is the "dual" of the so-called "cheap-control" LQR problem. The results are fundamental to the loop transfer recovery (LTR) method applied at the plant input.

35.5.1 Problem Definition

Consider as before, the stochastic LTI system

$$\dot{\underline{x}}(t) = \underline{A}\underline{x}(t) + \underline{L}\underline{\xi}(t) \quad (35.49)$$

$$\underline{y}(t) = \underline{C}\underline{x}(t) + \underline{\theta}(t) \quad (35.50)$$

We assume that the process noise $\underline{\xi}(t)$ is white, zero-mean, and with unit intensity, i.e.,

$$E\{\underline{\xi}(t)\underline{\xi}'(\tau) = \underline{I}\delta(t-\tau) \quad (35.51)$$

We also assume that the measurement noise $\underline{\theta}(t)$ is white, zero-mean, and with intensity indexed by μ, i.e.,

$$E\{\underline{\theta}(t)\underline{\theta}'(\tau) = \mu\underline{I}\delta(t-\tau) \quad (35.52)$$

DEFINITION 35.1 The accurate measurement KF problem is defined by the limiting case

$$\mu \to 0 \quad (35.53)$$

corresponding to essentially noiseless measurements.

Under the assumptions that $[\underline{A}, \underline{L}]$ is stabilizable and that $[\underline{A}, \underline{C}]$ is detectable we know that the KF is a stable system and generates the state estimates $\hat{\underline{x}}(t)$ by

$$\frac{d\hat{\underline{x}}(t)}{dt} = [\underline{A} - \underline{H}_\mu\underline{C}]\hat{\underline{x}}(t) + \underline{H}_\mu\underline{y}(t) \quad (35.54)$$

where we use the subscript μ to stress the dependence of the KF gain matrix $\underline{H}_\mu$ upon the parameter μ.

We recall that $\underline{H}_\mu$ is computed by

$$\underline{H}_\mu = \frac{1}{\mu}\underline{\Sigma}_\mu\underline{C}' \quad (35.55)$$

where the error covariance matrix $\underline{\Sigma}_\mu$, also dependent upon μ, is calculated by the solution of the FARE:

$$\underline{0} = \underline{A}\underline{\Sigma}_\mu + \underline{\Sigma}_\mu\underline{A}' + \underline{L}\underline{L}' - \frac{1}{\mu}\underline{\Sigma}_\mu\underline{C}'\underline{C}\underline{\Sigma}_\mu \quad (35.56)$$

We seek insight about the limiting behavior of both $\underline{\Sigma}_\mu$ and $\underline{H}_\mu$ as $\mu \to 0$.

35.5.2 The Main Result

In this section we summarize the main result in terms of a theorem.

THEOREM 35.1 *Suppose that the transfer function matrix from the white noise $\underline{\xi}(t)$ to the output $\underline{y}(t)$ for the system defined by Equations 35.49 and 35.50, i.e., the transfer function matrix*

$$\underline{W}(s) \triangleq \underline{C}(s\underline{I} - \underline{A})^{-1}\underline{L} \quad (35.57)$$

is minimum phase. Then,

$$\lim_{\mu\to 0}\underline{\Sigma}_\mu = \underline{0} \quad (35.58)$$

and

$$\lim_{\mu\to 0}\sqrt{\mu}\underline{H}_\mu = \underline{L}\underline{W}; \quad \underline{W}'\underline{W} = \underline{I} \quad (35.59)$$

PROOF 35.1 This is theorem 4.14 in Kwakernaak and Sivan [7], pp. 370-371.

REMARK 35.5 It can be shown that the requirement that $\underline{W}(s)$, given by Equation 35.57, be minimum phase is both a necessary and sufficient condition for the limiting properties given by Equations 35.58 and 35.59.

REMARK 35.6 The implication of Equation 35.58 is that, in the case of exact measurements upon a minimum phase plant, the KF yields exact state estimates, since the error covariance matrix is zero. This assumes that the KF has been operating upon the data for a sufficiently long time so that initial transient errors have died out.

REMARK 35.7 For a non-minimum phase plant

$$\lim_{\mu \to 0} \underline{\Sigma}_{\mu} \neq \underline{0} \tag{35.60}$$

Hence, perfect state estimation is impossible for non-minimum phase plants.

REMARK 35.8 The limiting behavior (with $\underline{L} = \underline{B}$) of the Kalman Filter gain

$$\lim_{\mu \to 0} \sqrt{\mu} \underline{H}_{\mu} = \underline{B}\,\underline{W}; \quad \underline{W}'\underline{W} = \underline{I} \tag{35.61}$$

is the precise dual of the limiting behavior of the LQ control gain

$$\lim_{\rho \to 0} \sqrt{\rho} \underline{G}_{\rho} = \underline{W}\,\underline{C}; \quad \underline{W}'\underline{W} = \underline{I} \tag{35.62}$$

for the minimum phase plant

$$\underline{G}(s) = \underline{C}(s\underline{I} - \underline{A})^{-1}\underline{B} \tag{35.63}$$

The relation (Equation 35.61) has been used by Doyle and Stein [8] to apply the LTR method at the plant input, while Equation 35.62 has been used by Kwakernaak [9] to apply the LTR method at the plant output (see also Kwakernaak and Sivan [7], pp. 419-427).

References

[1] Wiener, N., *Extrapolation, Interpolation and Smoothing of Stationary Time Series*, MIT Press and John Wiley & Sons, New York, 1950 (reprinted from a publication restricted for security reasons in 1942).

[2] Kalman, R.E., A new approach to linear filtering and prediction problems, *ASME J. Basic Eng., Ser. D*, 82, 34–45, 1960.

[3] Kalman, R.E., New methods and results in linear prediction and filtering theory, *Proc. Symp. Engineering Applications of Random Function Theory and Probability*, John Wiley & Sons, New York, 1961.

[4] Kalman, R.E. and Bucy, R.S., New results in linear filtering and prediction theory, *ASME J. Basic Eng., Ser. D*, 83, 95–108, 1961.

[5] Kalman, R.E., New methods in Wiener filtering, *Proc. First Symp. Engineering Applications of Random Function Theory and Probability*, John Wiley & Sons, New York, 1963, chap. 9.

[6] Van Trees, H.L., *Detection Estimation, and Modulation Theory*, J. Wiley & Sons, New York, 1968.

[7] Kwakernaak, H. and Sivan, R., *Linear Optimal Control Systems*, John Wiley & Sons, New York, 1972.

[8] Doyle, J.C. and Stein, G., Multivariable feedback design: concepts for a classical/modern synthesis, *IEEE Trans. Autom. Control*, 1981.

[9] Kwakernaak, H., Optimal low sensitivity linear feedback systems, *Automatica*, 5, 279–286, 1969.

Further Reading

There is a vast literature on Kalman filtering and its applications. A reasonable starting point is the recent textbook *Kalman Filtering: Theory and Practice* by M.S. Grewal and A.P. Andrews, Prentice Hall, Englewood Cliffs, NJ, 1993.

The books by Gelb et al., Jazwinski, and Maybeck have been standard references for those interested primarily in applications for some time.

[1] Gelb, A., Kasper, J.F., Jr., Nash, R.A., Jr., Price, C.F., and Sutherland, A.A., Jr., *Applied Optimal Estimation*, MIT Press, Cambridge, MA, 1974.

[2] Jazwinski, A.H., *Stochastic Processes and Filtering*, Academic Press, New York, 1970.

[3] Maybeck, P.S., *Stochastic Models, Estimation, and Control*, Vol. 1, Academic Press, San Diego, 1979.

[4] Maybeck, P.S., *Stochastic Models, Estimation, and Control*, Vol. 2, Academic Press, San Diego, 1982.

The reprint volume edited by H. Sorenson, *Kalman Filtering: Theory and Application*, IEEE Press, New York, 1985 contains reprints of many of the early theoretical papers on Kalman filtering, as well as a collection of applications.

A very readable introduction to the theory is

[5] Davis, M.H.A., *Linear Estimation and Stochastic Control*, Halsted Press, New York, 1977.

36

Riccati Equations and their Solution

Vladimír Kučera
Institute of Information Theory and Automation, Prague, Academy of Sciences of the Czech Republic

36.1 Introduction

An ordinary differential equation of the form

$$\dot{x}(t) + f(t)x(t) - b(t)x^2(t) + c(t) = 0 \qquad (36.1)$$

is known as a *Riccati equation*, deriving its name from Jacopo Francesco, Count Riccati (1676–1754)[1], who studied a particular case of this equation from 1719 to 1724.

For several reasons a differential equation of the form of Equation 36.1, and generalizations thereof comprise a highly significant class of nonlinear ordinary differential equations. First, they are intimately related to ordinary linear homogeneous differential equations of the second order. Second, the solutions of Equation 36.1 possess a very particular structure in that the general solution is a fractional linear function in the constant of integration. In applications, Riccati differential equations appear in the classical problems of the calculus of variations and in the associated disciplines of optimal control and filtering.

The *matrix* Riccati differential equation refers to the equation

$$\begin{aligned}&\dot{X}(t) + X(t)A(t) - D(t)X(t)\\ &\quad - X(t)B(t)X(t) + C(t) = 0\end{aligned} \qquad (36.2)$$

defined on the vector space of real $m \times n$ matrices. Here, A, B, C, D are real matrix functions of the appropriate dimensions. Of particular interest are the matrix Riccati equations that arise in optimal control and filtering problems and which enjoy certain symmetry properties. This chapter is concerned with these symmetric matrix Riccati differential equations and concentrates on the following four major topics:

- Basic properties of the solutions
- Existence and properties of constant solutions
- Asymptotic behavior of the solutions
- Methods for the numerical solution of the Riccati equations.

36.2 Optimal Control and Filtering: Motivation

The following problems of optimal control and filtering are of great engineering importance and serve to motivate our study of the Riccati equations.

A linear-quadratic *optimal control* problem consists of the following. Given a linear system

$$\begin{aligned}\dot{x}(t) &= Fx(t) + Gu(t), \quad x(t_0) = c\\ y(t) &= Hx(t)\end{aligned} \qquad (36.3)$$

where x is the n-vector state, u is the q-vector control input, y is the p-vector of regulated variables, and F, G, H are constant real matrices of the appropriate dimensions. One seeks to determine a control input function u over some fixed time interval $[t_1, t_2]$ such that a given quadratic cost functional of the form

$$\begin{aligned}\eta(t_1, t_2, T) &= \int_{t_1}^{t_2} [y'(t)y(t) + u'(t)u(t)]dt\\ &\quad + x'(t_2)Tx(t_2),\end{aligned} \qquad (36.4)$$

with T a constant real symmetric ($T = T'$) and non-negative definite ($T \geq 0$) matrix, is afforded a minimum in the class of all solutions of Equation 36.3.

0-8493-8570-9/96/$0.00+$.50

A unique optimal control exists for all finite $t_2 - t_1 > 0$ and has the form

$$u(t) = -G'P(t, t_2, T)x(t)$$

where $P(t, t_2, T)$ is the solution of the matrix Riccati differential equation

$$-\dot{P}(t) = P(t)F + F'P(t) - P(t)GG'P(t) + H'H \quad (36.5)$$

subject to the terminal condition

$$P(t_2) = T.$$

The optimal control is a linear state feedback, which gives rise to the closed-loop system

$$\dot{x}(t) = [F - GG'P(t, t_2, T)]x(t)$$

and yields the minimum cost

$$\eta^*(t_1, t_2, T) = c'P(t_1, t_2, T)c. \quad (36.6)$$

A Gaussian *optimal filtering* problem consists in the following. Given the p-vector random process z modeled by the equations

$$\begin{aligned} \dot{x}(t) &= Fx(t) + Gv(t) \\ z(t) &= Hx(t) + w(t) \end{aligned} \quad (36.7)$$

where x is the n-vector state and v, w are independent Gaussian white random processes (respectively, q-vector and p-vector) with zero means and identity covariance matrices. The matrices F, G, H are constant real ones of the appropriate dimensions.

Given known values of z over some fixed time interval $[t_1, t_2]$ and assuming that $x(t_1)$ is a Gaussian random vector, independent of v and w, with zero mean and covariance matrix S, one seeks to determine an estimate $\hat{x}(t_2)$ of $x(t_2)$ such that the variance

$$\sigma(S, t_1, t_2) = Ef'[x(t_2) - \hat{x}(t_2)][x(t_2) - \hat{x}(t_2)]'f \quad (36.8)$$

of the error encountered in estimating any real-valued linear function f of $x(t_2)$ is minimized.

A unique optimal estimate exists for all finite $t_2 - t_1 > 0$ and is generated by a linear system of the form

$$\begin{aligned} \dot{\hat{x}}(t) &= F\hat{x}(t) + Q(S, t_1, t)H'e(t), \quad \hat{x}(t_0) = 0 \\ e(t) &= z(t) - H\hat{x}(t) \end{aligned}$$

where $Q(S, t_1, t)$ is the solution of the matrix Riccati differential equation

$$\dot{Q}(t) = Q(t)F' + FQ(t) - Q(t)H'HQ(t) + GG' \quad (36.9)$$

subject to the initial condition

$$Q(t_1) = S.$$

The minimum error variance is given by

$$\sigma^*(S, t_1, t_2) = f'Q(S, t_1, t_2)f. \quad (36.10)$$

Equations 36.5 and 36.9 are special cases of the matrix Riccati differential Equation 36.2 in that A, B, C and D are constant real $n \times n$ matrices such that

$$B = B', \quad C = C', \quad D = -A'.$$

Therefore, symmetric solutions $X(t)$ are obtained in the optimal control and filtering problems.

We observe that the control Equation 36.5 is solved *backward* in time while the filtering Equation 36.9 is solved *forward* in time. We also observe that the two equations are *dual* to each other in the sense that

$$P(t, t_2, T) = Q(S, t_1, t)$$

on replacing F, G, H, T and $t_2 - t$ in Equation 36.5 respectively by F', H', G', S and $t - t_1$ or, vice versa, on replacing F, G, H, S and $t - t_1$ in Equation 36.9 respectively, by F', H', G', T and $t_2 - t$. This makes it possible to dispense with both cases by considering only one prototype equation.

36.3 Riccati Differential Equation

This section is concerned with basic properties of the prototype matrix Riccati differential equation

$$\dot{X}(t) + X(t)A + A'X(t) - X(t)BX(t) + C = 0 \quad (36.11)$$

where A, B, C are constant real $n \times n$ matrices with B and C symmetric and non-negative definite,

$$B = B', \quad B \geq 0 \text{ and } C = C', \quad C \geq 0. \quad (36.12)$$

By definition, a *solution* of Equation 36.11 is a real $n \times n$ matrix function $X(t)$ that is absolutely continuous and satisfies Equation 36.11 for t on an interval on the real line R.

Generally, solutions of Riccati differential equations exist only locally. There is a phenomenon called finite escape time: the equation

$$\dot{x}(t) = x^2(t) + 1$$

has a solution $x(t) = \tan t$ on the interval $(-\frac{\pi}{2}, 0]$ that cannot be extended to include the point $t = -\frac{\pi}{2}$. However, Equation 36.11 with the sign-definite coefficients as shown in Equation 36.12 does have global solutions.

Let $X(t, t_2, T)$ denote the solution of Equation 36.11 that passes through a constant real $n \times n$ matrix T at time t_2. We shall assume that

$$T = T' \text{ and } T \geq 0. \quad (36.13)$$

Then the solution exists on every finite subinterval of R, is symmetric, non-negative definite and enjoys certain monotone properties.

THEOREM 36.1 *Under the assumptions of Equations* 36.12 *and* 36.13, *Equation* 36.11 *has a unique solution* $X(t, t_2, T)$ *satisfying*

$$X(t, t_2, T) = X'(t, t_2, T), \quad X(t, t_2, T) \geq 0$$

for every T and every finite t, t_2 such that $t \leq t_2$.

This can most easily be seen by associating Equation 36.11 with the optimal control problem described in Equations 36.3 to 36.6. Indeed, using Equation 36.12, one can write $B = GG'$ and $C = H'H$ for some real matrices G and H. The quadratic cost functional η of Equation 36.4 exists and is non-negative for every T satisfying Equation 36.13 and for every finite $t_2 - t$. Using Equation 36.6, the quadratic form $c'X(t, t_2, T)c$ can be interpreted as a particular value of η for every real vector c.

A further consequence of Equations 36.4 and 36.6 follows.

THEOREM 36.2 *For every finite t_1, t_2 and τ_1, τ_2 such that $t_1 \leq \tau_1 \leq \tau_2 \leq t_2$,*

$$X(t_1, \tau_1, 0) \leq X(t_1, \tau_2, 0)$$
$$X(\tau_2, t_2, 0) \leq X(\tau_1, t_2, 0)$$

and for every $T_1 \leq T_2$,

$$X(t_1, t_2, T_1) \leq X(t_1, t_2, T_2).$$

Thus, the solution of Equation 36.11 passing through $T = 0$ does not decrease as the length of the interval increases, and the solution passing through a larger T dominates that passing through a smaller T.

The Riccati Equation 36.11 is related in a very particular manner with linear Hamiltonian systems of differential equations.

THEOREM 36.3 *Let*

$$\Phi(t, t_2) = \begin{bmatrix} \Phi_{11} & \Phi_{12} \\ \Phi_{21} & \Phi_{22} \end{bmatrix}$$

be the fundamental matrix solution of the linear Hamiltonian matrix differential system

$$\begin{bmatrix} \dot{U}(t) \\ \dot{V}(t) \end{bmatrix} = \begin{bmatrix} A & -B \\ -C & -A' \end{bmatrix} \begin{bmatrix} U(t) \\ V(t) \end{bmatrix}$$

that satisfies the transversality condition

$$V(t_2) = TU(t_2).$$

If the matrix $\Phi_{11} + \Phi_{12}T$ is nonsingular on an interval $[t, t_2]$, then

$$X(t, t_2, T) = (\Phi_{21} + \Phi_{22}T)(\Phi_{11} + \Phi_{12}T)^{-1} \quad (36.14)$$

is a solution of the Riccati Equation 36.11.

Thus, if $V(t_2) = TU(t_2)$, then $V(t) = X(t, t_2, T)U(t)$ and the formula of Equation 36.14 follows.

Let us illustrate on a simple example. The Riccati equation

$$\dot{x}(t) = x^2(t) - 1, \quad x(0) = T$$

satisfies the hypotheses of Equations 36.12 and 36.13. The associated linear Hamiltonian system of equations reads

$$\begin{bmatrix} \dot{u}(t) \\ \dot{v}(t) \end{bmatrix} = \begin{bmatrix} 0 & -1 \\ -1 & 0 \end{bmatrix} \begin{bmatrix} u(t) \\ v(t) \end{bmatrix}$$

and has the solution

$$\begin{bmatrix} u(t) \\ v(t) \end{bmatrix} = \begin{bmatrix} \cosh t & -\sinh t \\ -\sinh t & \cosh t \end{bmatrix} \begin{bmatrix} u(0) \\ v(0) \end{bmatrix}$$

where $v(0) = Tu(0)$. Then the Riccati equation has the solution

$$x(t, 0, T) = \frac{-\sinh t + T \cosh t}{\cosh t - T \sinh t}$$

for all $t \leq 0$. The monotone properties of the solution are best seen in Figure 36.1.

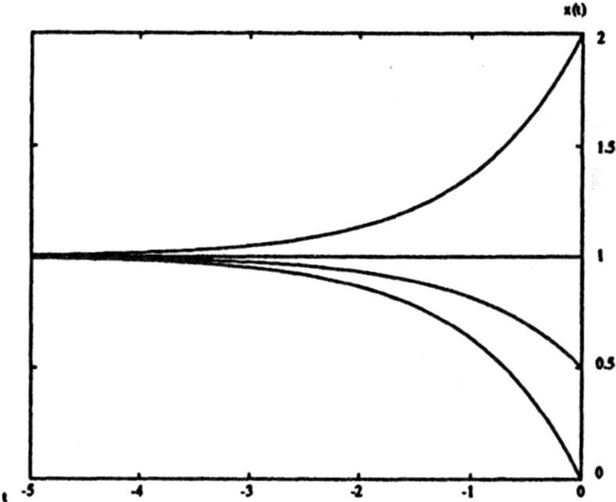

Figure 36.1 Graph of solutions.

36.4 Riccati Algebraic Equation

The constant solutions of Equation 36.11 are just the solutions of the quadratic equation

$$XA + A'X - XBX + C = 0, \quad (36.15)$$

called the *algebraic Riccati equation*. This equation can have real $n \times n$ matrix solutions X that are symmetric or nonsymmetric, sign definite or indefinite, and the set of solutions can be either finite or infinite. These solutions will be studied under the standing assumption of Equation 36.12, namely

$$B = B', \quad B \geq 0 \text{ and } C = C', \quad C \geq 0.$$

36.4.1 General Solutions

The solution set of Equation 36.15 corresponds to a certain class of n-dimensional invariant subspaces of the associated $2n \times 2n$ matrix

$$H = \begin{bmatrix} A & -B \\ -C & -A' \end{bmatrix}. \quad (36.16)$$

This matrix has the *Hamiltonian* property

$$\begin{bmatrix} 0 & I \\ -I & 0 \end{bmatrix} H = -H' \begin{bmatrix} 0 & I \\ -I & 0 \end{bmatrix}.$$

It follows that H is similar to $-H'$, and therefore the spectrum of H is symmetrical with respect to the imaginary axis.

Now suppose that X is a solution of Equation 36.15. Then

$$H \begin{bmatrix} I \\ X \end{bmatrix} = \begin{bmatrix} I \\ X \end{bmatrix} (A - BX).$$

Denote $J = U^{-1}(A - BX)U$ the Jordan form of $A - BX$ and put $V = XU$. Then

$$H \begin{bmatrix} U \\ V \end{bmatrix} = \begin{bmatrix} U \\ V \end{bmatrix} J$$

which shows that the columns of

$$\begin{bmatrix} U \\ V \end{bmatrix}$$

are Jordan chains for H, i. e., sets of vectors $x_1, x_2, ..., x_r$ such that $x_1 \neq 0$ and for some eigenvalue λ of H

$$Hx_1 = \lambda x_1$$
$$Hx_j = \lambda x_j + x_{j+1}, \quad j = 2, 3, ..., r.$$

In particular, x_1 is an eigenvector of H. Thus, we have the following result.

THEOREM 36.4 *Equation 36.15 has a solution X if and only if there is a set of vectors $x_1, x_2, ...x_n$ forming a set of Jordan chains for H, and if*

$$x_i = \begin{bmatrix} u_i \\ v_i \end{bmatrix}$$

where u_i is an n-vector, then $u_1, u_2, ..., u_n$ are linearly independent.

Furthermore, if

$$U = [u_1 \ ... \ u_n], \quad V = [v_1 \ ... \ v_n]$$

every solution of Equation 36.15 has the form $X = VU^{-1}$ for some set of Jordan chains $x_1, x_2, ..., x_n$ for H.

To illustrate, consider the scalar equation

$$X^2 + pX + q = 0$$

where p, q are real numbers and $q \leq 0$. The Hamiltonian matrix

$$H = \begin{bmatrix} -\frac{p}{2} & -1 \\ q & \frac{p}{2} \end{bmatrix}$$

has eigenvalues λ and $-\lambda$, where

$$\lambda^2 = (\frac{p}{2})^2 - q.$$

If $\lambda \neq 0$ there are two eigenvectors of H, namely

$$x_1 = \begin{bmatrix} 1 \\ -\frac{p}{2} + \lambda \end{bmatrix}, \quad x_2 = \begin{bmatrix} 1 \\ -\frac{p}{2} - \lambda \end{bmatrix},$$

which correspond to the solutions

$$X_1 = -\frac{p}{2} + \lambda, \quad X_2 = -\frac{p}{2} - \lambda.$$

If $\lambda = 0$ there exists one Jordan chain,

$$x_1 = \begin{bmatrix} 1 \\ -\frac{p}{2} \end{bmatrix}, \quad x_2 = \begin{bmatrix} 0 \\ 1 \end{bmatrix},$$

which yields the unique solution

$$X_1 = -\frac{p}{2}.$$

Theorem 36.4 suggests that, generically, the number of solutions of Equation 36.15 to be expected will not exceed the binomial coefficient $\binom{2n}{n}$, the number of ways in which the vectors $x_1, x_2, ..., x_n$ can be chosen from a basis of $2n$ eigenvectors for H. The solution set is infinite if there is a continuous family of Jordan chains. To illustrate this point consider equation 36.15 with

$$A = \begin{bmatrix} 0 & 0 \\ 0 & 0 \end{bmatrix}, \quad B = \begin{bmatrix} 1 & 0 \\ 0 & 1 \end{bmatrix}, \quad C = \begin{bmatrix} 0 & 0 \\ 0 & 0 \end{bmatrix}.$$

The Hamiltonian matrix

$$H = \begin{bmatrix} 0 & 0 & -1 & 0 \\ 0 & 0 & 0 & -1 \\ 0 & 0 & 0 & 0 \\ 0 & 0 & 0 & 0 \end{bmatrix}$$

has the eigenvalue 0, associated with two Jordan chains

$$x_1 = \begin{bmatrix} a \\ b \\ 0 \\ 0 \end{bmatrix}, \quad x_2 = \begin{bmatrix} c \\ d \\ -a \\ -b \end{bmatrix},$$

and

$$x_3 = \begin{bmatrix} c \\ d \\ 0 \\ 0 \end{bmatrix}, \quad x_4 = \begin{bmatrix} a \\ b \\ -c \\ -d \end{bmatrix}$$

where a, b and c, d are real numbers such that $ad - bc = 1$. The solution set of Equation 36.15 consists of the matrix

$$X_{13} = \begin{bmatrix} 0 & 0 \\ 0 & 0 \end{bmatrix}$$

and two continuous families of matrices

$$X_{12}(a, b) = \begin{bmatrix} ab & -a^2 \\ b^2 & -ab \end{bmatrix}$$

and

$$X_{34}(c, d) = \begin{bmatrix} -cd & c^2 \\ -d^2 & cd \end{bmatrix}.$$

Having in mind the applications in optimal control and filtering, we shall be concerned with the solutions of Equation 36.15 that are symmetric and non-negative definite.

36.4.2 Symmetric Solutions

In view of Theorem 36.4, each solution X of Equation 36.15 gives rise to a factorization of the characteristic polynomial χ_H of H as

$$\chi_H(s) = (-1)^n q(s) q_1(s)$$

where $q = \chi_{A-BX}$. If the solution is *symmetric*, $X = X'$, then $q_1(s) = q(-s)$. This follows from

$$\begin{bmatrix} I & 0 \\ X & I \end{bmatrix}^{-1} \begin{bmatrix} A & -B \\ -C & -A' \end{bmatrix} \begin{bmatrix} I & 0 \\ X & I \end{bmatrix} = \begin{bmatrix} A - BX & -B \\ 0 & -(A - BX)' \end{bmatrix}.$$

There are two symmetric solutions that are of particular importance. They correspond to a factorization

$$\chi_H(s) = (-1)^n q(s) q(-s)$$

in which q has all its roots with nonpositive real part; it follows that $q(-s)$ has all its roots with non-negative real part. We shall designate these solutions X_+ and X_-.

One of the basic results concerns the existence of these particular solutions. To state the result, we recall some terminology. A pair of real $n \times n$ matrices (A, B) is said to be *controllable* (respectively, *stabilizable*) if the $n \times 2n$ matrix $[\lambda I - A \quad B]$ has linearly independent rows for every complex λ (respectively, for every complex λ such that $Re\,\lambda \geq 0$). The numbers λ for which $[\lambda I - A \quad B]$ loses rank are the eigenvalues of A that are not controllable (respectively, stabilizable) from B. A pair of real $n \times n$ matrices (A, C) is said to be *observable* (respectively, *detectable*) if the $2n \times n$ matrix $\begin{bmatrix} \lambda I - A \\ C \end{bmatrix}$ has linearly independent columns for every complex λ (respectively, for every complex λ such that $Re\,\lambda \geq 0$). The numbers λ for which $\begin{bmatrix} \lambda I - A \\ C \end{bmatrix}$ loses rank are the eigenvalues of A that are not observable (respectively, detectable) in C. Finally we let dim $\mathcal{V}$ denote the dimension of a linear space $\mathcal{V}$ and Im M, Ker M the image and the kernel of a matrix M, respectively.

THEOREM 36.5 *There exists a unique symmetric solution X_+ of Equation 36.15 such that all eigenvalues of $A - BX_+$ have nonpositive real part if and only if (A, B) is stabilizable.*

THEOREM 36.6 *There exists a unique symmetric solution X_- of Equation 36.15 such that all eigenvalues of $A - BX_-$ have non-negative real part if and only if $(-A, B)$ is stabilizable.*

We observe that both (A, B) and $(-A, B)$ are stabilizable if and only if (A, B) is controllable. It follows that both solutions X_+ and X_- exist if and only if (A, B) is controllable.

For two real symmetric matrices X_1 and X_2, the notation $X_1 \geq X_2$ means that $X_1 - X_2$ is non-negative definite. Since $A - BX_+$ has no eigenvalues with positive real part, neither has $X_+ - X$. Hence $X_+ - X \geq 0$. Similarly one can show that $X - X_- \geq 0$, thus introducing a partial order among the set of symmetric solutions of Equation 36.15.

THEOREM 36.7 *Suppose that X_+ and X_- exist. If X is any symmetric solution of Equation 36.15, then*

$$X_+ \geq X \geq X_-.$$

That is why X_+ and X_- are called the *extreme* solutions of Equation 36.15; X_+ is the maximal symmetric solution while X_- is the minimal symmetric solution. The set of all symmetric solutions of Equation 36.15 can be related to a certain subset of the set of invariant subspaces of the matrix $A - BX_+$ or the matrix $A - BX_-$. Denote $\mathcal{V}_0$ and $\mathcal{V}_+$ the invariant subspaces of $A - BX_+$ that correspond respectively to the pure imaginary eigenvalues and to the eigenvalues having negative real part. Denote $\mathcal{W}_0$ and $\mathcal{W}_-$ the invariant subspaces of $A - BX_-$ that correspond respectively to the pure imaginary eigenvalues and to the eigenvalues having positive real part. Then it can be shown that $\mathcal{V}_0 = \mathcal{W}_0$ is the kernel of $X_+ - X_-$ and the symmetric solution set corresponds to the set of all invariant subspaces of $A - BX_+$ contained in $\mathcal{V}_+$ or, equivalently, to the set of all invariant subspaces of $A - BX_-$ contained in $\mathcal{W}_-$.

THEOREM 36.8 *Suppose that X_+ and X_- exist. Let X_1, X_2 be symmetric solutions of Equation 36.15 corresponding to the invariant subspaces $\mathcal{V}_1$, $\mathcal{V}_2$ of $\mathcal{V}_+$ (or $\mathcal{W}_1$, $\mathcal{W}_2$ of $\mathcal{W}_-$). Then $X_1 \geq X_2$ if and only if $\mathcal{V}_1 \supset \mathcal{V}_2$ (or if and only if $\mathcal{W}_1 \subset \mathcal{W}_2$).*

This means that the symmetric solution set of Equation 36.15 is a complete *lattice* with respect to the usual ordering of symmetric matrices. The maximal solution X_+ corresponds to the invariant subspace $\mathcal{V}_+$ of $A - BX_+$ or to the invariant subspace $\mathcal{W} = 0$ of $A - BX_-$, whereas the minimal solution X_- corresponds to the invariant subspace $\mathcal{V} = 0$ of $A - BX_+$ or to the invariant subspace $\mathcal{W}_-$ of $A - BX_-$.

This result allows one to count the distinct symmetric solutions of Equation 36.15 in some cases. Thus, let α be the number of distinct eigenvalues of $A - BX_+$ having negative real part and let $m_1, m_2, ..., m_\alpha$ be the multiplicities of these eigenvalues. Owing to the symmetries in H, the matrix $A - BX_-$ exhibits the same structure of eigenvalues with positive real part.

THEOREM 36.9 *Suppose that X_+ and X_- exist. Then the symmetric solution set of Equation 36.15 has finite cardinality if and only if $A - BX_+$ is cyclic on $\mathcal{V}_+$ (or if and only if $A - BX_-$ is cyclic on $\mathcal{W}_-$). In this case, the set contains exactly $(m_1 + 1)...(m_\alpha + 1)$ solutions.*

Simple examples are most illustrative. Consider Equation 36.15 with

$$A = \begin{bmatrix} 0 & 0 \\ 0 & 1 \end{bmatrix}, \quad B = \begin{bmatrix} 1 & 0 \\ 0 & 1 \end{bmatrix}, \quad C = \begin{bmatrix} 1 & 0 \\ 0 & 3 \end{bmatrix}$$

and determine the lattice of symmetric solutions. We have

$$\chi_H(s) = s^4 - 5s^2 + 4$$

and the following eigenvectors of H

$$x_1 = \begin{bmatrix} 1 \\ 0 \\ -1 \\ 0 \end{bmatrix}, \quad x_2 = \begin{bmatrix} 1 \\ 0 \\ 1 \\ 0 \end{bmatrix},$$

$$x_3 = \begin{bmatrix} 0 \\ 1 \\ 0 \\ -1 \end{bmatrix}, \quad x_4 = \begin{bmatrix} 0 \\ 1 \\ 0 \\ 3 \end{bmatrix}$$

are associated with the eigenvalues $1, -1, 2, -2$, respectively. Hence, the pair of solutions

$$X_+ = \begin{bmatrix} 1 & 0 \\ 0 & 3 \end{bmatrix}, \quad X_- = \begin{bmatrix} -1 & 0 \\ 0 & -1 \end{bmatrix}$$

corresponds to the factorization

$$\chi_H(s) = (s^2 - 3s + 2)(s^2 + 3s + 2)$$

and the solutions

$$X_{2,3} = \begin{bmatrix} 1 & 0 \\ 0 & -1 \end{bmatrix}, \quad X_{14} = \begin{bmatrix} -1 & 0 \\ 0 & 3 \end{bmatrix}$$

correspond to the factorization

$$\chi_H(s) = (s^2 - s - 2)(s^2 + s - 2).$$

There are four subspaces invariant under the matrices

$$A - BX_+ = \begin{bmatrix} -1 & 0 \\ 0 & -2 \end{bmatrix}, \quad A - BX_- = \begin{bmatrix} 1 & 0 \\ 0 & 2 \end{bmatrix}$$

each corresponding to one of the four solutions above. The partial ordering

$$X_+ \geq X_{2,3} \geq X_-, \quad X_+ \geq X_{1,4} \geq X_-$$

defines the lattice visualized in Figure 36.2.

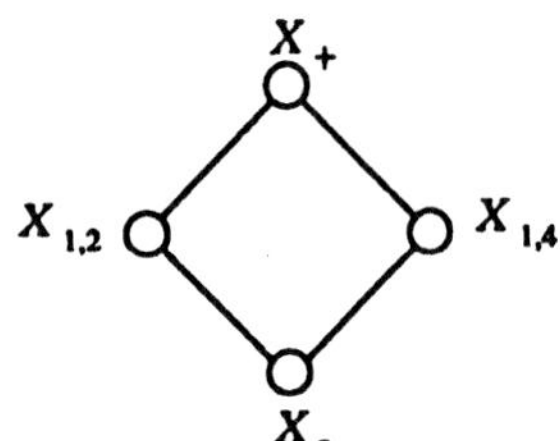

Figure 36.2 Lattice of solutions.

As another example, we consider Equation 36.15 where

$$A = \begin{bmatrix} 0 & 0 \\ 0 & 0 \end{bmatrix}, \quad B = \begin{bmatrix} 1 & 0 \\ 0 & 1 \end{bmatrix}, \quad C = \begin{bmatrix} 1 & 0 \\ 0 & 1 \end{bmatrix}$$

and classify the symmetric solution set.

We have

$$\chi_H(s) = (s - 1)^2(s + 1)^2$$

and a choice of eigenvectors corresponding to the eigenvalues $1, -1$ of H is

$$x_1 = \begin{bmatrix} 1 \\ 0 \\ -1 \\ 0 \end{bmatrix}, \quad x_2 = \begin{bmatrix} 0 \\ 1 \\ 0 \\ -1 \end{bmatrix},$$

$$x_3 = \begin{bmatrix} 1 \\ 0 \\ 1 \\ 0 \end{bmatrix}, \quad x_4 = \begin{bmatrix} 0 \\ 1 \\ 0 \\ 1 \end{bmatrix}.$$

Hence

$$X_+ = \begin{bmatrix} 1 & 0 \\ 0 & 1 \end{bmatrix}, \quad X_- = \begin{bmatrix} -1 & 0 \\ 0 & -1 \end{bmatrix}$$

are the extreme solutions.

We calculate

$$A - BX_+ = \begin{bmatrix} -1 & 0 \\ 0 & -1 \end{bmatrix}, \quad A - BX_- = \begin{bmatrix} 1 & 0 \\ 0 & 1 \end{bmatrix}$$

and observe that the set of subspaces invariant under $A - BX_+$ or $A - BX_-$ (other than the zero and the whole space, which correspond to X_+ and X_-) is the family of 1-dimensional subspaces parameterized by their azimuth angle ϑ. These correspond to the solutions

$$X_\vartheta = \begin{bmatrix} \cos\vartheta & \sin\vartheta \\ \sin\vartheta & -\cos\vartheta \end{bmatrix}.$$

Therefore, the solution set consists of X_+, X_- and the continuous family of solutions X_ϑ. It is a complete lattice and $X_+ \geq X_\vartheta \geq X_-$ for every ϑ.

36.4.3 Definite Solutions

Under the standing assumption (36.12), namely

$$B = B', \quad B \geq 0, \quad \text{and } C = C', \quad C \geq 0$$

one can prove that $X_+ \geq 0$ and $X_- \leq 0$. The existence of X_+, however, excludes the existence of X_- and vice versa, unless (A, B) is controllable.

If X_+ does exist, any other solution $X \geq 0$ of Equation 36.15 corresponds to a subspace $\mathcal{W}$ of $\mathcal{W}_-$ that is invariant under $A - BX$. From Equation 36.15,

$$X(A - BX) + (A - BX)'X = -XBX - C.$$

The restriction of $A - BX$ to $\mathcal{W}$ has eigenvalues with positive real part. Since $-XBX - C \leq 0$, it follows from Lyapunov theory that X restricted to $\mathcal{W}$ is nonpositive definite, hence zero. We conclude that the solutions $X \geq 0$ of Equation 36.15 correspond to those subspaces $\mathcal{W}$ of $\mathcal{W}_-$ that are invariant under A and contained in Ker C.

The set of symmetric non-negative definite solutions of Equation 36.15 is a sublattice of the lattice of all symmetric solutions. Clearly X_+ is the largest solution and it corresponds to the invariant subspace $\mathcal{W} = 0$ of A. The smallest non-negative definite solution will be denoted by X_* and it corresponds to $\mathcal{W}_*$, the largest invariant subspace of A contained in Ker C and associated with eigenvalues having positive real part.

The non-negative definite solution set of Equation 36.15 has finite cardinality if and only if A is cyclic on $\mathcal{W}_*$. In this case, the set contains exactly $(p_1+1)\ldots(p_\rho+1)$ solutions, where ρ is the number of distinct eigenvalues of A associated with $\mathcal{W}_*$ and $p_1, p_2, \ldots, p_\rho$ are the multiplicities of these eigenvalues.

Analogous results hold for the set of symmetric solutions of Equation 36.15 that are nonpositive definite. In particular, if X_- exists, then any other solution $X \leq 0$ of Equation 36.15 corresponds to a subspace $\mathcal{V}$ of $\mathcal{V}_+$ that is invariant under A and contained in Ker C. Clearly X_- is the smallest solution and it corresponds to the invariant subspace $\mathcal{V} = 0$ of A. The largest nonpositive definite solution is denoted by $X_\times$ and it corresponds to $\mathcal{W}_\times$, the largest invariant subspace of A contained in Ker C and associated with eigenvalues having negative real part.

Let us illustrate on a simple example. Consider Equation 36.15 where

$$A = \begin{bmatrix} 1 & 1 \\ 0 & 1 \end{bmatrix}, \quad B = \begin{bmatrix} 0 & 0 \\ 0 & 1 \end{bmatrix}, \quad C = \begin{bmatrix} 0 & 0 \\ 0 & 0 \end{bmatrix}$$

and classify the two sign-definite solution sets. We have

$$X_+ = \begin{bmatrix} 8 & 4 \\ 4 & 4 \end{bmatrix}, \quad X_- = \begin{bmatrix} 0 & 0 \\ 0 & 0 \end{bmatrix}.$$

The matrix A has one eigenvalue with positive real part, namely 1, and a basis for $\mathcal{W}_*$ is

$$x_1 = \begin{bmatrix} 1 \\ 0 \end{bmatrix}, \quad x_2 = \begin{bmatrix} 0 \\ -1 \end{bmatrix}.$$

Thus, there are three invariant subspaces of $\mathcal{W}_*$ corresponding to the three non-negative definite solutions of Equation 36.15

$$X_+ = \begin{bmatrix} 8 & 4 \\ 4 & 4 \end{bmatrix}, \quad X_1 = \begin{bmatrix} 0 & 0 \\ 0 & 2 \end{bmatrix}, \quad X_* = \begin{bmatrix} 0 & 0 \\ 0 & 0 \end{bmatrix}.$$

These solutions make a lattice and

$$X_+ \geq X_1 \geq X_*.$$

The matrix A has no eigenvalues with negative real part. Therefore $\mathcal{V}_* = 0$ and X_- is the only nonpositive definite solution of Equation 36.15.

Another example for Equation 36.15 is provided by

$$A = \begin{bmatrix} 0 & 0 \\ 0 & 0 \end{bmatrix}, \quad B = \begin{bmatrix} 1 & 0 \\ 0 & 0 \end{bmatrix}, \quad C = \begin{bmatrix} 0 & 0 \\ 0 & 0 \end{bmatrix}.$$

It is seen that neither (A, B) nor $(-A, B)$ is stabilizable; hence, neither X_+ nor X_- exists. The symmetric solution set consists of one continuous family of solutions

$$X_\alpha = \begin{bmatrix} 0 & 0 \\ 0 & \alpha \end{bmatrix}$$

for any real α. Therefore, both sign-definite solution sets are infinite; the non-negative solution set is unbounded from above while the nonpositive solution set is unbounded from below.

36.5 Limiting Behavior of Solutions

The length of the time interval $t_2 - t_1$ in the optimal control and filtering problems is rather artificial. For this reason, an infinite time interval is often considered. This brings in the question of the limiting behavior of the solution $X(t, t_2, T)$ for the Riccati differential Equation 36.11.

In applications to optimal control, it is customary to fix t and let t_2 approach $+\infty$. Since the coefficient matrices of Equation 36.11 are constant, the same result is obtained if t_2 is held fixed and t approaches $-\infty$. The limiting behavior of $X(t, t_2, T)$ strongly depends on the terminal matrix $T \geq 0$. For a suitable choice of T, the solution of Equation 36.11 may converge to a constant matrix $X \geq 0$, a solution of Equation 36.15. For some matrices T, however, the solution of Equation 36.11 may fail to converge to a constant matrix, but it may converge to a periodic matrix function.

THEOREM 36.10 *Let (A, B) be stabilizable. If t and T are held fixed and $t_2 \to \infty$, then the solution $X(t, t_2, T)$ of Equation 36.11 is bounded on the interval $[t, \infty)$.*

This result can be proved by associating an optimal control problem with Equation 36.11. Then stabilizability of (A, B) implies the existence of a stabilizing (not necessarily optimal) control. The consequent cost functional Equation 36.4 is finite and dominates the optimal one.

If (A, B) is stabilizable, then X_+ exists and each real symmetric non-negative definite solution X of Equation 36.15 corresponds to a subset $\mathcal{W}$ of $\mathcal{W}_*$, the set of A-invariant subspaces contained in Ker C and associated with eigenvalues having positive real part. The convergence of the solution $X(t, t_2, T)$ of Equation 36.11 to X depends on the properties of the image of $\mathcal{W}_*$ under T.

For simplicity, it is assumed that the eigenvalues $\lambda_1, \lambda_2, \ldots, \lambda_p$ of A associated with $\mathcal{W}_*$ are simple and, except for pairs of complex conjugate eigenvalues, have different real parts. Let the corresponding eigenvectors be ordered according to decreasing real parts of the eigenvalue

$$v_1, v_2, \ldots, v_p,$$

and denote $\mathcal{W}_i$ the A-invariant subspace of $\mathcal{W}_*$ spanned by $v_1, v_2, \ldots, v_i$.

THEOREM 36.11 *Let (A, B) be stabilizable and the subspaces $\mathcal{W}_i$ of $\mathcal{W}_*$ satisfy the above assumptions. Then, for all fixed t and a given terminal condition $T \geq 0$, the solution $X(t, t_2, T)$ of Equation 36.11 converges to a constant solution of Equation 36.15 as $t_2 \to \infty$ if and only if the subspace $\mathcal{W}_{k+1}$ corresponding to any pair λ_k, λ_{k+1} of complex conjugate eigenvalues is such that $\dim T\mathcal{W}_{k+1}$ equals either $\dim T\mathcal{W}_{k-1}$ or $\dim T\mathcal{W}_{k-1} + 2$.*

Here is a simple example. Let

$$A = \begin{bmatrix} 1 & -1 \\ 1 & 1 \end{bmatrix}, \quad B = \begin{bmatrix} 1 & 0 \\ 0 & 1 \end{bmatrix}, \quad C = \begin{bmatrix} 0 & 0 \\ 0 & 0 \end{bmatrix}.$$

The pair (A, B) is stabilizable and A has two eigenvalues $1 + j$ and $1 - j$. The corresponding eigenvectors

$$v_1 = \begin{bmatrix} j \\ 1 \end{bmatrix}, \quad v_2 = \begin{bmatrix} -j \\ 1 \end{bmatrix}$$

span $\mathcal{W}_*$. Now consider the terminal condition

$$T = \begin{bmatrix} 1 & 0 \\ 0 & 0 \end{bmatrix}.$$

Then

$$T\mathcal{W}_0 = 0, \quad T\mathcal{W}_2 = \text{Im}\begin{bmatrix} 1 \\ 0 \end{bmatrix}.$$

Theorem 36.11 shows that $X(t, t_2, T)$ does not converge to a constant matrix; in fact,

$$X(t, t_2, T) = \frac{1}{1 + e^{2(t-t_2)}} \begin{bmatrix} 2\cos^2(t - t_2) & -\sin 2(t - t_2) \\ -\sin 2(t - t_2) & 2\sin^2(t - t_2) \end{bmatrix}$$

tends to a periodic solution if $t_2 \to \infty$. On the other hand, if we select

$$T_0 = \begin{bmatrix} 0 & 0 \\ 0 & 0 \end{bmatrix}$$

we have

$$T_0\mathcal{W}_0 = 0, \quad T_0\mathcal{W}_2 = 0$$

and $X(t, t_2, T_0)$ does converge. Also, if we take

$$T_1 = \begin{bmatrix} 1 & 0 \\ 0 & 1 \end{bmatrix}$$

we have

$$T_1\mathcal{W}_0 = 0, \quad T_1\mathcal{W}_2 = R^2$$

and $X(t, t_2, T_1)$ converges as well.

If the solution $X(t, t_2, T)$ of Equation 36.11 converges to a constant matrix X_T as $t_2 \to \infty$, then X_T is a real symmetric nonnegative definite solution of Equation 36.15. Which solution is attained for a particular terminal condition?

THEOREM 36.12 *Let (A, B) be stabilizable. Let*

$$X_T = \lim_{t_2 \to \infty} X(t, t_2, T)$$

for a fixed $T \geq 0$. Then $X_T \geq 0$ is the solution of Equation 36.15 corresponding to the subspace $\mathcal{W}_T$ of $\mathcal{W}_*$, defined as the span of the real vectors v_i such that $T\mathcal{W}_i = T\mathcal{W}_{i-1}$ and of the complex conjugate pairs v_k, v_{k+1} such that $T\mathcal{W}_{k+1} = T\mathcal{W}_{k-1}$.

The cases of special interest are the extreme solutions X_+ and X_*. The solution $X(t, t_2, T)$ of Equation 36.11 tends to X_+ if and only if the intersection of $\mathcal{W}_*$ with Ker T is zero, and to X_* if and only if $\mathcal{W}_*$ is contained in Ker T.

This is best illustrated on the previous example, where

$$A = \begin{bmatrix} 1 & -1 \\ 1 & 1 \end{bmatrix}, \quad B = \begin{bmatrix} 1 & 0 \\ 0 & 1 \end{bmatrix}, \quad C = \begin{bmatrix} 0 & 0 \\ 0 & 0 \end{bmatrix}$$

and $\mathcal{W}_* = R^2$. Then $X(t, t_2, T)$ converges to X_+ if and only if T is positive definite; for instance, the identity matrix T yields the solution

$$X(t, t_2, I) = \frac{2}{1 + e^{2(t-t_2)}} \begin{bmatrix} 1 & 0 \\ 0 & 1 \end{bmatrix}$$

which tends to

$$X_+ = \begin{bmatrix} 2 & 0 \\ 0 & 2 \end{bmatrix}.$$

On the other hand, $X(t, t_2, T)$ converges to X_* if and only if $T = 0$; then

$$X(t, t_2, 0) = 0$$

and $X_* = 0$ is a fixed point of Equation 36.11.

36.6 Optimal Control and Filtering: Application

The problems of optimal control and filtering introduced in Section 2 are related to the matrix Riccati differential Equations 36.5 and 36.9, respectively. These problems are defined over a finite horizon $t_2 - t_1$. We now apply the convergence properties of the solutions to study the two optimal problems in case the horizon gets large.

To fix ideas, we concentrate on the optimal control problem. The results can easily be interpreted in the filtering context owing to the duality between Equations 36.5 and 36.9.

We recall that the finite horizon optimal control problem is that of minimizing the cost functional of Equation 36.4,

$$\eta(t_2) = \int_{t_1}^{t_2} [\, y'(t)y(t) + u'(t)u(t)\,]dt + x'(t_2)Tx(t_2)$$

along the solutions of Equation 36.3,

$$\dot{x}(t) = Fx(t) + Gu(t)$$
$$y(t) = Hx(t).$$

The optimal control has the form

$$u_0(t) = -G'X(t, t_2, T)x(t)$$

where $X(t, t_2, T)$ is the solution of Equation 36.11,

$$\dot{X}(t) + X(t)A + A'X(t) - X(t)BX(t) + C = 0$$

subject to the terminal condition $X(t_2) = T$, and where

$$A = F, \quad B = GG', \quad C = H'H.$$

The optimal control can be implemented as a state feedback and the resulting closed-loop system is

$$\begin{aligned} \dot{x}(t) &= [\, F - GG'X(t, t_2, T)\,]x(t) \\ &= [\, A - BX(t, t_2, T)\,]x(t). \end{aligned}$$

Hence, the relevance of the matrix $A - BX$, which plays a key role in the theory of the Riccati equation.

The *infinite horizon* optimal control problem then amounts to finding

$$\eta_* = \inf_{u(t)} \lim_{t_2 \to \infty} \eta(t_2) \qquad (36.17)$$

and the corresponding optimal control $u_*(t)$, $t \geq t_1$ achieving this minimum cost.

The *receding horizon* optimal control problem is that of finding

$$\eta_{**} = \lim_{t_2 \to \infty} \inf_{u(t)} \eta(t_2) \qquad (36.18)$$

and the limiting behavior $u_{**}(t)$, $t \geq t_1$ of the optimal control $u_o(t)$.

The question is whether η_* is equal to η_{**} and whether u_* coincides with u_{**}. If so, the optimal control for the infinite horizon can be approximated by the optimal control of the finite horizon problem for a sufficiently large time interval.

It turns out that these two control problems have different solutions corresponding to different solutions of the matrix Riccati algebraic Equation 36.15,

$$XA + A'X - XBX + C = 0.$$

THEOREM 36.13 *Let (A, B) be stabilizable. Then the infinite horizon optimal control problem of Equation 36.17 has a solution*

$$\eta_* = x'(t_1)X_o x(t_1), \quad u_*(t) = -G'X_o x(t)$$

where $X_o \geq 0$ is the solution of Equation 36.15 corresponding to $\mathcal{W}_o$, the largest A-invariant subspace contained in $\mathcal{W}_* \cap \operatorname{Ker} T$.

THEOREM 36.14 *Let (A, B) be stabilizable. Then the receding horizon optimal control problem of Equation 36.18 has a solution if and only if the criterion of Theorem 36.11 is satisfied and, in this case,*

$$\eta_{**} = x'(t_1)X_T x(t_1), \quad u_{**}(t) = -G'X_T x(t)$$

where $X_T \geq 0$ is the solution of Equation 36.15 corresponding to $\mathcal{W}_T$ and defined in Theorem 36.12.

The equivalence result follows.

THEOREM 36.15 *The solution of the infinite horizon optimal control problem is exactly the limiting case of the receding horizon optimal control problem if and only if the subspace $\mathcal{W}_* \cap \operatorname{Ker} T$ is invariant under A.*

A simple example illustrates these points. Consider the finite horizon problem defined by

$$\dot{x}_1(t) = 2x_1(t) + u_1(t)$$
$$\dot{x}_2(t) = x_2(t) + u_2(t)$$

and

$$\eta(t_2) = [\,x_1(t_2) + x_2(t_2)\,]^2 + \int_{t_1}^{t_2} [\,u_1^2(\tau) + u_2^2(\tau)\,]d\tau,$$

which corresponds to the data

$$A = \begin{bmatrix} 2 & 0 \\ 0 & 1 \end{bmatrix}, \quad B = \begin{bmatrix} 1 & 0 \\ 0 & 1 \end{bmatrix}, \quad C = \begin{bmatrix} 0 & 0 \\ 0 & 0 \end{bmatrix}$$

and

$$T = \begin{bmatrix} 1 & 1 \\ 1 & 1 \end{bmatrix}.$$

Clearly $\mathcal{W}_* = R^2$ and the subspace

$$\mathcal{W}_* \cap \operatorname{Ker} T = \operatorname{Im} \begin{bmatrix} 1 \\ -1 \end{bmatrix}$$

is not invariant under A. Hence, the infinite and receding horizon problems are not equivalent.

The lattice of symmetric non-negative definite solutions of Equation 36.11 has the four elements

$$X_+ = \begin{bmatrix} 4 & 0 \\ 0 & 2 \end{bmatrix}, \quad X_1 = \begin{bmatrix} 0 & 0 \\ 0 & 2 \end{bmatrix},$$
$$X_2 = \begin{bmatrix} 4 & 0 \\ 0 & 0 \end{bmatrix}, \quad X_* = \begin{bmatrix} 0 & 0 \\ 0 & 0 \end{bmatrix}$$

depicted in Figure 36.3.

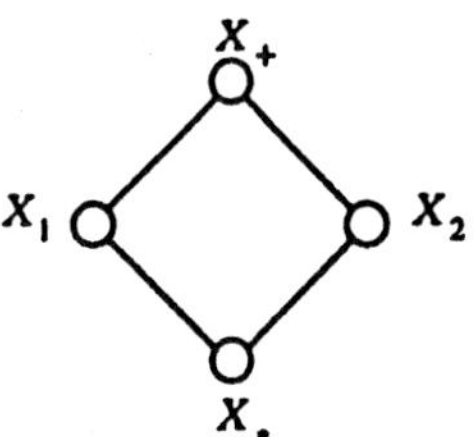

Figure 36.3 Lattice of solutions.

Since the largest A-invariant subspace of $\mathcal{W}_* \cap \operatorname{Ker} T$ is zero, the optimal solution X_o of Equation 36.11 is the maximal element X_+. The infinite horizon optimal control reads

$$u_{1*}(t) = -4x_1(t)$$
$$u_{2*}(t) = -2x_2(t)$$

and affords the minimum cost

$$\eta_* = 4x_1^2(t_1) + 2x_2^2(t_1).$$

Now the eigenvectors of A spanning $\mathcal{W}_*$ are

$$v_1 = \begin{bmatrix} 1 \\ 0 \end{bmatrix}, \quad v_2 = \begin{bmatrix} 0 \\ 1 \end{bmatrix}$$

and their T-images

$$Tv_1 = \begin{bmatrix} 1 \\ 1 \end{bmatrix}, \quad Tv_2 = \begin{bmatrix} 1 \\ 1 \end{bmatrix}$$

are linearly dependent. Hence, $\mathcal{W}_T$ is spanned by v_2 only,

$$\mathcal{W}_T = \text{Im}\begin{bmatrix} 0 \\ 1 \end{bmatrix}$$

and the optimal limiting solution X_T of Equation 36.11 equals X_2. The receding horizon optimal control reads

$$u_{1**}(t) = -4x_1(t)$$
$$u_{2**}(t) = 0$$

and affords the minimum cost

$$\eta_{**}(t) = 4x_1^2(t_1).$$

The optimal control problems with large horizon are practically relevant if the optimal closed-loop system

$$\dot{x}(t) = (A - BX)x(t)$$

is stable. A real symmetric non-negative definite solution X of Equation 36.15 is said to be *stabilizing* if the eigenvalues of $A - BX$ all have negative real part. It is clear that the stabilizing solution, if it exists, is the maximal solution X_+. Thus, the existence of a stabilizing solution depends on $A - BX_+$ having eigenvalues with only negative real part.

THEOREM 36.16 *Equation* 36.15 *has a stabilizing solution if and only if* (A, B) *is stabilizable and the Hamiltonian matrix* H *of Equation* 36.16 *has no pure imaginary eigenvalue.*

The optimal controls over large horizons have a certain stabilizing effect. Indeed, if $X \geq 0$ is a solution of Equation 36.15 that corresponds to an A-invariant subspace $\mathcal{W}$ of $\mathcal{W}_*$, then the control $u(t) = -G'Xx(t)$ leaves unstable in $A - BX$ just the eigenvalues of A associated with $\mathcal{W}$; all the remaining eigenvalues of A with positive real part are stabilized. Of course, the pure imaginary eigenvalues of A, if any, cannot be stabilized; they remain intact in $A - BX$ for any solution X of Equation 36.15.

In particular, the infinite horizon optimal control problem leaves unstable the eigenvalues of A associated with $\mathcal{W}_o$, which are those not detectable either in C or in T, plus the pure imaginary eigenvalues. It follows that the infinite horizon optimal control results in a stable system if and only if X_o is the stabilizing solution of Equation 36.15. This is the case if and only if the hypotheses of Theorem 36.16 hold and $\mathcal{W}_o$, the largest A-invariant subspace contained in $\mathcal{W}_* \cap \operatorname{Ker} T$, is zero. Equivalently, this corresponds to the pair

$$\left(\begin{bmatrix} C \\ T \end{bmatrix}, A\right)$$

being detectable.

The allocation of the closed-loop eigenvalues for the receding horizon optimal control problem is different, however. This control leaves unstable all eigenvalues of A associated with $\mathcal{W}_T$, where W_T is a subspace of $\mathcal{W}_*$ defined in Theorem 36.12. Therefore, the number of stabilized eigenvalues may be lower, equal to the dimension of $T\mathcal{W}_*$, whenever $\operatorname{Ker} T$ is not invariant under A. It follows that the receding horizon optimal control results in a stable system if and only if X_T is the stabilizing solution of Equation 36.15. This is the case if and only if the hypotheses of Theorem 36.16 hold and $\mathcal{W}_T$ is zero. Equivalently, this corresponds to $\mathcal{W}_* \cap \operatorname{Ker} T = 0$. Note that this case occurs in particular if $T \geq X_+$.

It further follows that under the standard assumption, namely that

$$(A, B) \text{ stabilizable}$$
$$(A, C) \text{ detectable},$$

both infinite and receding horizon control problems have solutions; these solutions are equivalent for any terminal condition T; and the resulting optimal system is stable.

36.7 Numerical Solution

The matrix Riccati *differential* Equation 36.11 admits an analytic solution only in rare cases. A numerical integration is needed and the Runge-Kutta methods can be applied.

A number of techniques are available for the solution of the matrix Riccati *algebraic* Equation 36.15. These include invariant subspace methods and the matrix sign function iteration. We briefly outline these methods here with an eye on the calculation of the stabilizing solution to Equation 36.15.

36.7.1 Invariant Subspace Method

In view of Theorem 36.4, any solution X of Equation 36.15 can be computed from a Jordan form reduction of the associated $2n \times 2n$ Hamiltonian matrix

$$H = \begin{bmatrix} A & -B \\ -C & -A' \end{bmatrix}.$$

Specifically, compute a matrix of eigenvectors V to perform the following reduction

$$V^{-1}HV = \begin{bmatrix} -J & 0 \\ 0 & J \end{bmatrix}$$

where $-J$ is composed of Jordan blocks corresponding to eigenvalues with negative real part only. If the stabilizing solution X exists, then H has no eigenvalues on the imaginary axis and J is indeed $n \times n$. Writing

$$V = \begin{bmatrix} V_{11} & V_{12} \\ V_{21} & V_{22} \end{bmatrix}$$

where each V_{ij} is $n \times n$, the solution sought is found by solving a system of linear equations,

$$X = V_{21}V_{11}^{-1}.$$

However, there are numerical difficulties with this approach when H has multiple or near-multiple eigenvalues. To ameliorate these difficulties, a method has been proposed in which a nonsingular matrix V of eigenvectors is replaced by an orthogonal matrix U of Schur vectors so that

$$U'HU = \begin{bmatrix} S_{11} & S_{12} \\ 0 & S_{22} \end{bmatrix}$$

where now S_{11} is a quasi-upper triangular matrix with eigenvalues having negative real part and S_{22} is a quasi-upper triangular matrix with eigenvalues having positive real part. When

$$U = \begin{bmatrix} U_{11} & U_{12} \\ U_{21} & U_{22} \end{bmatrix}$$

then we observe that

$$\begin{bmatrix} V_{11} \\ V_{21} \end{bmatrix}, \begin{bmatrix} U_{11} \\ U_{21} \end{bmatrix}$$

span the same invariant subspace and X can again be computed from

$$X = U_{21}U_{11}^{-1}.$$

36.7.2 Matrix Sign Function Iteration

Let M be a real $n \times n$ matrix with no pure imaginary eigenvalues. Let M have a Jordan decomposition $M = V\ J\ V^{-1}$ and let $\lambda_1, \lambda_2, ..., \lambda_n$ be the diagonal entries of J (the eigenvalues of M repeated according to their multiplicities). Then the *matrix sign function* of M is given by

$$\operatorname{sgn} M = V \begin{bmatrix} \operatorname{sgn} \operatorname{Re}\lambda_1 & & \\ & \ddots & \\ & & \operatorname{sgn} \operatorname{Re}\lambda_n \end{bmatrix} V^{-1}.$$

It follows that the matrix $Z = \operatorname{sgn} M$ is diagonalizable with eigenvalues ± 1 and $Z^2 = I$. The key observation is that the image of $Z + I$ is the M-invariant subspace of R^n corresponding to the eigenvalues of M with negative real part.

This property clearly provides the link to Riccati equations, and what we need is a reliable computation of the matrix sign. Let $Z_0 = M$ be an $n \times n$ matrix whose sign is desired. For $k = 0, 1, ...$ perform the iteration

$$Z_{k+1} = \frac{1}{2c}(Z_k + c^2 Z_k^{-1})$$

where $c = |\det Z_k|^{1/n}$. Then

$$\lim_{k\to\infty} Z_k = Z = \operatorname{sgn} M.$$

The constant c is chosen to enhance convergence of this iterative process. If $c = 1$, the iteration amounts to Newton's method for solving the equation

$$Z^2 - I = 0.$$

Naturally, it can be shown that the iteration is ultimately quadratically convergent.

Thus, to obtain the stabilizing solution X of Equation 36.15, provided it exists, we compute $Z = \operatorname{sgn} H$, where H is the Hamiltonian matrix of Equation 36.16. The existence of X guarantees that H has no eigenvalues on the imaginary axis.

Writing

$$Z = \begin{bmatrix} Z_{11} & Z_{12} \\ Z_{21} & Z_{22} \end{bmatrix}$$

where each Z_{ij} is $n \times n$, the solution sought is found by solving a system of linear equations

$$\begin{bmatrix} Z_{12} \\ Z_{22} + I \end{bmatrix} X = - \begin{bmatrix} Z_{11} + I \\ Z_{21} \end{bmatrix}.$$

36.7.3 Concluding Remarks

We have discussed two numerical methods for obtaining the stabilizing solution of the matrix Riccati algebraic Equation 36.15. They are both based on the intimate connection between the Riccati equation solutions and invariant subspaces of the associated Hamiltonian matrix. The method based on Schur vectors is a direct one while the method based on the matrix sign function is iterative.

The Schur method is now considered one of the more reliable for Riccati equations and has the virtues of being simultaneously efficient and numerically robust. It is particularly suitable for Riccati equations with relatively small dense coefficient matrices, say, of order a few hundred or less. The matrix sign function method is based on the Newton iteration and features global convergence, with ultimately quadratic order. Iteration formulas can be chosen to be of arbitrary order convergence in exchange for, naturally, an increased computational burden. The effect of this increased computation can, however, be ameliorated by parallelization.

The two methods are not limited to computing the stabilizing solution only. The matrix sign iteration can also be used to calculate X_-, the antistabilizing solution of Equation 36.15, by considering the matrix $\operatorname{sgn} H - I$ instead of $\operatorname{sgn} H + I$. The Schur approach can be used to calculate any, not necessarily symmetric, solution of Equation 36.15 by ordering the eigenvalues on the diagonal of S accordingly.

References

Historical documents:

[1] Riccati, J. F., Animadversationes in aequationes differentiales secundi gradus, *Acta Eruditorum Lipsiae*, 8, 67–73, 1724.

[2] Boyer, C.B., *The History of Mathematics*, Wiley, New York, 1974.

Tutorial textbooks:

[3] Reid, W. T., *Riccati Differential Equations*, Academic Press, New York, 1972.

[4] Bittanti, S., Laub, A. J., and Willems, J. C., Eds., *The Riccati Equation*, Springer-Verlag, Berlin, 1991.

Survey paper:

[5] Kučera, V., A review of the matrix Riccati equation, *Kybernetika*, 9, 42–61, 1973.

Original sources on optimal control and filtering:

[6] Kalman, R. E., Contributions to the theory of optimal control, *Bol. Soc. Mat. Mexicana*, 5, 102–119, 1960.

[7] Kalman, R. E. and Bucy, R. S., New results in linear filtering and prediction theory, *J. Basic Eng. (ASME Trans.)*, 83D, 95–108, 1961.

Original sources on the algebraic equation:

[8] Willems, J. C., Least squares stationary optimal control and the algebraic Riccati equation, *IEEE Trans. Autom. Control*, 16, 612–634, 1971.

[9] Kučera, V., A contribution to matrix quadratic equations, *IEEE Trans. Autom. Control*, 17, 344–347, 1972.

Original sources on the limiting behavior:

[10] Callier, F. M. and Willems, J. L., Criterion for the convergence of the solution of the Riccati differential equation, *IEEE Trans. Autom. Control*, 26, 1232–1242, 1981.

[11] Willems, J. L. and Callier, F. M., Large finite horizon and infinite horizon LQ-optimal control problems, *Optimal Control Appl. Methods*, 4, 31–45, 1983.

Original sources on the numerical methods:

[12] Laub, A. J., A Schur method for solving algebraic Riccati equations, *IEEE Trans. Autom. Control*, 24, 913–921, 1979.

[13] Roberts, J. D., Linear model reduction and solution of the algebraic Riccati equation by use of the sign function, *Int. J. Control*, 32, 677–687, 1980.

37

Observers

Bernard Friedland
Department of Electrical and Computer Engineering, New Jersey Institute of Technology, Newark, NJ

37.1 Introduction

An observer for a dynamic system $\mathcal{S}(x, y, u)$ with state x, output y, and input u is another dynamic system $\hat{\mathcal{S}}(\hat{x}, y, u)$ having the property that the state $\hat{x}$ of the observer $\hat{\mathcal{S}}$ converges to the state x of the process $\mathcal{S}$, independent of the input u or the state x.

Among the various applications for observers, perhaps the most important is for the implementation of closed-loop control algorithms designed by state-space methods. The control algorithm is designed in two parts: a "full-state feedback" part based on the assumption that all the state variables can be measured; and an observer to estimate the state of the process based upon the observed output. The concept of separating the feedback control design into these two parts is known as the *separation principle,* which has rigorous validity in linear systems and in a limited class of nonlinear systems. Even when its validity cannot be rigorously established, the separation principle is often a practical solution to many design problems.

The concept of an observer for a dynamic process was introduced in 1966 by Luenberger [1]. The generic "Luenberger observer," however, appeared several years after the Kalman filter, which is in fact an important special case of a Luenberger observer—an observer optimized for the noise present in the observations and in the input to the process.

37.2 Linear Full-Order Observers

37.2.1 Continuous-Time Systems

Consider a linear, continuous-time dynamic system

$$\dot{x} = Ax + Bu \tag{37.1}$$

$$y = Cx \tag{37.2}$$

The more generic output

$$y = Cx + Du$$

can be treated by defining a modified output

$$\bar{y} = y - Du$$

and working with $\bar{y}$ instead of y. The direct coupling Du from the input to the output is absent in most physical plants.

A full-order observer for the linear process defined by Equations 37.1 and 37.2 has the generic form

$$\dot{\hat{x}} = \hat{A}\hat{x} + Ky + Hu \tag{37.3}$$

where the dimension of state $\hat{x}$ of the observer is equal to the dimension of process state x.

The matrices $\hat{A}$, K, and H appearing in Equation 37.3 must be chosen to conform with the required property of an observer: that the observer state must converge to the process state independent of the state x and the input u. To determine these matrices, let

$$e := x - \hat{x} \tag{37.4}$$

be the estimation error. From Equations 37.1, 37.2, and 37.3

$$\begin{aligned} \dot{e} &= Ax + Bu - \hat{A}(x - e) - GCx - Hu \\ &= \hat{A}e + (-\hat{A} + A - KC)x + (B - H)u \end{aligned} \tag{37.5}$$

From Equation 37.5 it is seen that for the error to converge to zero, independent of x and u, the following conditions must be satisfied:

$$\hat{A} = A - KC \tag{37.6}$$

$$H = B \tag{37.7}$$

0-8493-8570-9/96/$0.00+$.50

When these conditions are satisfied, the estimation error is governed by

$$\dot{e} = \hat{A}e \tag{37.8}$$

which converges to zero if $\hat{A}$ is a "stability matrix." When $\hat{A}$ is constant, this means that its eigenvalues must lie in the (open) left half-plane.

Since the matrices A, B, and C are defined by the plant, the only freedom in the design of the observer is in the selection of the gain matrix K.

To emphasize the role of the observer gain matrix, and accounting for requirements of Equations 37.6 and 37.7, the observer can be written

$$\dot{\hat{x}} = A\hat{x} + Bu + K(y - C\hat{x}) \tag{37.9}$$

A block diagram representation of Equation 37.9, as given in Figure 37.1, aids in the interpretation of the observer. Note that the observer comprises a model of the process with an added input:

$$K(y - C\hat{x}) = Kr$$

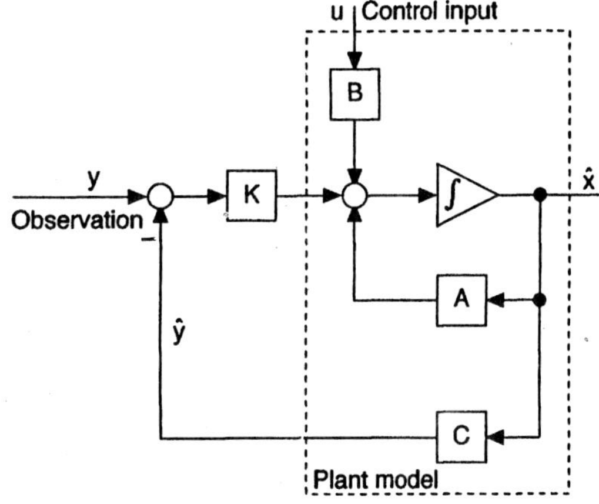

Figure 37.1 Full-order observer for linear process.

The quantity

$$r := y - C\hat{x} = y - \hat{y} \tag{37.10}$$

often called the *residual*, is the difference between the actual observation y and the "synthesized" observation

$$\hat{y} = C\hat{x}$$

produced by the observer. The observer can be viewed as a feedback system designed to drive the residual to zero: as the residual is driven to zero, the input to Equation 37.9 due to the residual vanishes and the state of Equation 37.9 looks like the state of the original process.

The fundamental problem in the design of an observer is the determination of the observer gain matrix K such that the closed-loop observer matrix

$$\hat{A} = A - KC \tag{37.11}$$

is a stability matrix.

There is considerable flexibility in the selection of the observer gain matrix. Two methods are standard: optimization and pole placement.

Optimization

Since the observer given by Equation 37.9 has the structure of a Kalman filter (see Chapter 35), its gain matrix can be chosen as a Kalman filter gain matrix, i.e.,

$$K = PC'R^{-1} \tag{37.12}$$

where P is the covariance matrix of the estimation error and satisfies the matrix Riccati equation

$$\dot{P} = AP + PA' - PC'R^{-1}CP + Q \tag{37.13}$$

where R is a positive definite matrix and Q is a positive semidefinite matrix.

In most applications the steady-state covariance matrix is used in Equation 37.12. This matrix is given by setting $\dot{P}$ in Equation 37.13 to zero. The resulting equation is known as the *algebraic Riccati equation.* Algorithms to solve the algebraic Riccati equation are included in popular control system software packages such as MATLAB, MATRIX-X, CONTROL-C.

In order for the gain matrix given by Equations 37.12 and 37.13 to be genuinely optimum, the process noise and the observation noise must be white with the matrices Q and R being their spectral densities. It is rarely possible to determine these spectral density matrices in practical applications. Hence, the matrices Q and R are best treated as design parameters that can be varied to achieve overall system design objectives.

If the observer is to be used as a state estimator in a closed-loop control system, an appropriate form for the matrix Q is

$$Q = q^2BB' \tag{37.14}$$

As has been shown by Doyle and Stein [2], as $q \to \infty$, this observer tends to "recover" the stability margins assured by a full-state feedback control law obtained by quadratic optimization.

Pole-Placement

An alternative to solving the algebraic Riccati equation to obtain the observer gain matrix is to select K to place the poles of the observer, i.e., the eigenvalues of $\hat{A}$ in Equation 37.11.

When there is a single observation, K is a column vector with exactly as many elements as eigenvalues of $\hat{A}$. Hence, specification of the eigenvalues of $\hat{A}$ uniquely determines the gain matrix K. A number of algorithms can be used to determine the gain matrix, some of which are incorporated into the popular control system design software packages. Some of the algorithms have been found to be numerically ill conditioned, so caution should be exercised in using the results.

This author has found the Bass-Gura [3] formula effective in most applications. This formula gives the gain matrix as

$$K = (OW)'^{-1}(\hat{a} - a) \tag{37.15}$$

where

$$a = [\ a_1 \quad a_2 \quad \cdots \quad a_n\ ]' \tag{37.16}$$

is the vector formed from the coefficients of the characteristic polynomial of the process matrix A

$$|sI - A| = s^n + a_1 s^{n-1} + \cdots + a_{n-1}s + a_n \tag{37.17}$$

and $\hat{a}$ is the vector formed from the coefficients of the desired characteristic polynomial

$$|sI - \hat{A}| = s^n + \hat{a}_1 s^{n-1} + \cdots + \hat{a}_{n-1}s + \hat{a}_n \tag{37.18}$$

The other matrices in Equation 37.15 are given by

$$O = [\ C' \quad A'C' \quad \cdots \quad A'^{n-1}C'\ ] \tag{37.19}$$

which is the *observability matrix* of the process, and

$$W = \begin{bmatrix} 1 & a_1 & \cdots & a_n \\ 0 & 1 & \cdots & a_{n-1} \\ \vdots & \vdots & \vdots & \vdots \\ 0 & 0 & \cdots & 1 \end{bmatrix} \tag{37.20}$$

The determinant of W is 1, so it is not singular. If the observability matrix O is not singular, the inverse matrix required in Equation 37.15 exists. Hence, the gain matrix K can be found that places the observer poles at arbitrary locations if (and only if) the process for which an observer is sought is observable. Numerical problems occur, however, when the observability matrix is nearly singular. Other numerical problems can arise in determination of the characteristic polynomial $|sI - A|$ for high-order systems and in the determination of $sI - \hat{A}$ when the individual poles, and not the characteristic polynomial, are specified. In such instances, it may be necessary to use an algorithm designed to handle difficult numerical calculations.

When two or more quantities are observed, there are more elements in the gain matrix than eigenvalues of $\hat{A}$, so specification of the eigenvalues of $\hat{A}$ does not uniquely specify the gain matrix K. In addition to placing the eigenvalues, more of the "eigenstructure" of $\hat{A}$ can be specified. This method of selecting the gain matrix is fraught with difficulty, however, and the use of the algebraic Riccati equation is usually preferable.

37.3 Linear Reduced-Order Observers

The observer described in the previous section has the same order as the plant, irrespective of the number of independent observations. A reduced-order observer of order $n - m$, where n is the dimension of the state vector and m is the number of observations, can also be specified. When the number of observations is comparable to the dimension of the state vector, the reduced-order observer may represent a considerable simplification.

The description of the reduced-order observer is simplified if the state vector can be partitioned into two substates:

$$x = \begin{bmatrix} x_1 \\ \cdots \\ x_2 \end{bmatrix} \tag{37.21}$$

such that

$$x_1 = y = Cx \tag{37.22}$$

is the observation vector (of dimension m) and x_2 (of dimension $n - m$) comprises the components of the state vector that cannot be measured directly.

In terms of x_1 and x_2 the plant dynamics are written

$$\dot{x}_1 = A_{11}x_1 + A_{12}x_2 + B_1 u \tag{37.23}$$

$$\dot{x}_2 = A_{21}x_1 + A_{22}x_2 + B_2 u \tag{37.24}$$

Since x_1 is directly measured, no observer is required for that substate, i.e.,

$$\hat{x}_1 = x_1 = y \tag{37.25}$$

For the remaining substate, we define the reduced-order observer by

$$\hat{x}_2 = Ky + z \tag{37.26}$$

where z is the state of a system of order $n - m$:

$$\dot{z} = \hat{A}z + Ly + Hu \tag{37.27}$$

A block diagram representation of the reduced-order observer is given in Figure 37.2a.

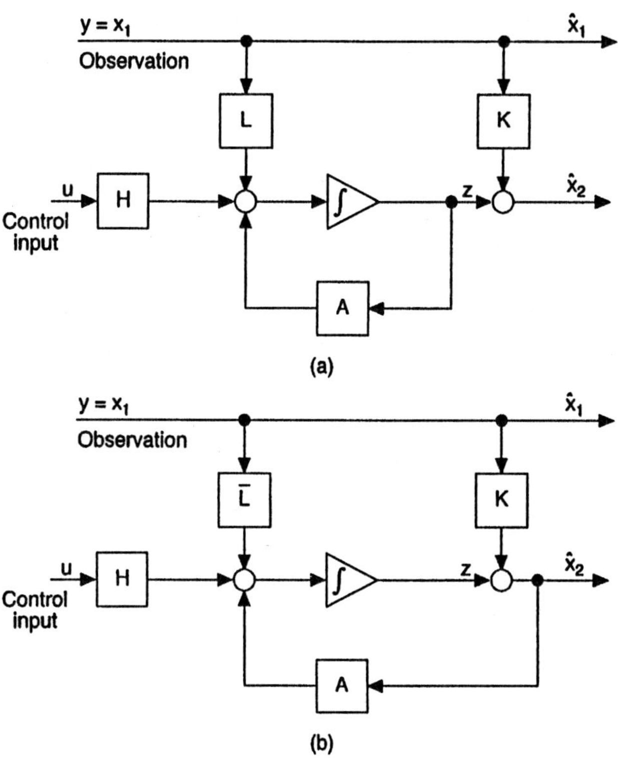

Figure 37.2 Reduced-order observer for linear process. (a) Feedback from z; (b) feedback from $\hat{x}_2$.

The matrices $\hat{A}$, L, H, and K are chosen, as in the case of the full-order observer, to ensure that the error in the estimation of the state converges to zero, independent of x, y, and u.

Since there is no error in estimation of x_1, i.e.,

$$e_1 = x_1 - \hat{x}_1 = 0 \tag{37.28}$$

by virtue of Equation 37.25, it is necessary to ensure only the convergence of

$$e_2 = x_2 - \hat{x}_2 \tag{37.29}$$

to zero.

From Equations 37.24 to 37.27

$$\dot{e}_2 = (A_{21} - KA_{11} + \hat{A}K - L)x_1 + (A_{22} - KA_{12} - \hat{A})x_2 + \hat{A}e_2 + (B_2 - KB_1 - H)u \tag{37.30}$$

As in the case of the full-order observer, to make the coefficients of x_1, x_2, and u vanish it is necessary that the matrices in Equations 37.25 and 37.27 satisfy

$$\hat{A} = A_{22} - KA_{12} \tag{37.31}$$
$$L = A_{21} - KA_{11} + \hat{A}K \tag{37.32}$$
$$H = B_2 - KB_1 \tag{37.33}$$

Two of these conditions (Equations 37.31 and 37.33) are analogous to Equations 37.6 and 37.7 for the full-order observer; Equation 37.32 is a new requirement for the additional matrix L that is required by the reduced-order observer.

When these conditions are satisfied, the error in estimation of x_2 is given by

$$e_2 = \hat{A}e_2$$

Hence, the gain matrix K must be chosen such that the eigenvalues of $\hat{A} = A_{22} - KA_{12}$ lie in the (open) left half-plane; A_{22} and A_{12} in the reduced-order observer take the roles of A and C in the full-order observer; once the gain matrix K is chosen, there is no further freedom in the choice of L and H.

The specific form of the new matrix L in Equation 37.32 suggests another option for implementation of the dynamics of the reduced-order observer, namely

$$\dot{z} = \hat{A}\hat{x}_2 + \bar{L}y + Hu \tag{37.34}$$

where

$$\bar{L} = A_{21} - KA_{11} \tag{37.35}$$

A block diagram representation of this option is given in Figure 37.2b.

The selection of the gain matrix K of the reduced-order observer may be accomplished by any of the methods that can be used to select the gains of the full-order observer. In particular, pole placement, using any convenient algorithm, is feasible. Or the gain matrix can be obtained as the solution of a reduced-order Kalman filtering problem. For this purpose, one can use Equations 37.12 and 37.13, with A and C therein replaced by A_{22} and A_{12} of the reduced-order problem.

A more rigorous solution, taking into account the cross-correlation between the observation noise and the process noise [4], is available. Suppose the dynamic process is governed by

$$\dot{x}_1 = A_{11}x_1 + A_{12}x_2 + B_1u + F_1v \tag{37.36}$$
$$\dot{x}_2 = A_{21}x_1 + A_{22}x_2 + B_2u + F_2v \tag{37.37}$$

with the observation being noise free:

$$y = x_1 \tag{37.38}$$

In this case, the gain matrix is given by

$$K = (PA_{12}' + F_2QF_1')R^{-1} \tag{37.39}$$

where

$$R = F_1QF_1'$$

and P is the covariance matrix of the estimation error e_2, as given by

$$\dot{P} = \tilde{A}P + P\tilde{A}' - PA_{12}R^{-1}A_{12}'P + \tilde{Q} \tag{37.40}$$

where

$$\tilde{A} = A_{22} - F_2QF_1'R^{-1}A_{12} \tag{37.41}$$
$$\tilde{Q} = F_2QF_2' - F_2QF_1'R^{-1}F_1QF_2' \tag{37.42}$$

Note that Equation 37.40 becomes homogeneous when

$$\tilde{Q} = 0 \tag{37.43}$$

One of the solutions of Equation 37.40 could be

$$P = 0 \tag{37.44}$$

which would imply that the error in estimating x_2 converges to zero! We can't expect to achieve anything better than this. Unfortunately, $P = 0$ is not the only possible solution to Equation 37.43. To test whether it is, it is necessary to check whether the resulting observer dynamics matrix

$$\hat{A} = A_{22} - F_2F_1^{-1}A_{12} \tag{37.45}$$

is a stability matrix. If not, Equation 37.43 is not the correct solution to Equation 37.40.

The eigenvalues of the "zero steady-state variance" observer dynamics matrix Equation 37.45 have an interesting interpretation: as shown in [4], these eigenvalues are the transmission zeros of the plant with respect to the noise input to the process. Hence, the variance of the estimation error converges to zero if the plant is "minimum phase" with respect to the noise input.

For purposes of robustness, as discussed in Section 4, suggest that the noise distribution matrix F include a term proportional to the control distribution matrix B, i.e.,

$$F = \bar{F} + q^2BB'$$

In this case, the zero-variance observer gain would satisfy

$$H = B_2 - KB_1 = 0 \tag{37.46}$$

as $q \to \infty$.

If Equation 37.46 is satisfied the observer poles are located at the transmission zeros of the plant. Thus, in order to use the gain given by Equation 37.46, it is necessary for the plant to be minimum-phase with respect to the input. Rynaski [5] has defined observers meeting this requirement as *robust observers*.

37.4 Discrete-Time Systems

Observers for discrete-time systems can be defined in a manner analogous to continuous-time systems.

Consider a discrete-time linear system

$$x_{n+1} = \Phi x_n + \Gamma u_n \tag{37.47}$$

with observations defined by

$$y_n = Cx_n \tag{37.48}$$

A full-order observer for Equation 37.47 is a dynamic system of the same order as the process whose state is to be estimated, excited by the inputs and outputs of that process and having the property that the estimation error (i.e., the difference between the state x_n of the process and the state $\hat{x}_n$ of the observer) converges to zero as $n \to \infty$, independent of the state of the process or its inputs and outputs.

Let the observer be defined by the general linear difference equation

$$\hat{x}_{n+1} = \hat{\Phi}\hat{x}_n + Ky_n + Hu_n \tag{37.49}$$

The goal is to find conditions on the matrices Φ, K, and H such that the requirements stated above are met. To find these conditions subtract Equation 37.49 from Equation 37.47

$$x_{n+1} - \hat{x}_{n+1} = \Phi x_n + \Gamma u_n - \hat{\Phi}\hat{x}_n - Ky_n - Hu_n \tag{37.50}$$

Letting

$$e_n = x_n - \hat{x}_n$$

and using Equation 37.48 we obtain from Equation 37.50

$$e_{n+1} = \Phi e_n + (\Phi - KC - \hat{\Phi})x_n + (\Gamma - H)u_n \tag{37.51}$$

Thus, in order to meet the requirements stated above, the transition matrix $\hat{\Phi}$ of the observer must be stable (i.e., the eigenvalues of Φ must lie within the unit circle) and, moreover,

$$\begin{aligned} \hat{\Phi} &= \Phi - KC & (37.52)\\ H &= \Gamma & (37.53) \end{aligned}$$

By virtue of these relations the observer can be expressed as

$$\hat{x}_{n+1} = \Phi x_n + \Gamma u_n + K(y_n - C\hat{x}_n) \tag{37.54}$$

It is seen from Equation 37.52 that the observer has the same dynamics as the underlying process, except that it has an additional input

$$K(y_n - C\hat{x}_n)$$

i.e., a gain matrix K multiplying the *residual*

$$r_n = y_n - C\hat{x}_n$$

As in the continuous-time case, the observer can be interpreted as a feedback system, the role of which is that of driving residual r_n to zero.

The observer design thus reduces to the selection of the gain matrix K that makes the eigenvalues of $\hat{\Phi} = \Phi - KC$ lie at suitable locations within the unit circle.

If the discrete-time system is observable, the eigenvalues of $\Phi_c = \Phi - KC$ can be put anywhere. For a single-output plant, the Bass-Gura formula or other well-known algorithm can be used. For both single and multiple output processes, the observer gain matrix can be selected to make the observer a Kalman filter (i.e., a minimum variance estimator).

The gain matrix of the discrete-time Kalman filter is given by

$$K = \Phi PC'(PCP' + R)^{-1} \tag{37.55}$$

where P is the covariance matrix of the estimation error, given (in the steady state) by the *discrete-time algebraic Riccati equation*

$$P = \Phi[P - PC'(CPC' + R)^{-1}CP]\Phi' + Q \tag{37.56}$$

The matrices Q and R are the covariance matrices of the excitation noise and the observation noise, respectively. As in the case of continuous-time processes, it is rarely possible to determine these matrices with any degree of accuracy. Hence, these matrices can be regarded as design parameters that can be adjusted by the user to provide desirable observer characteristics.

37.5 The Separation Principle

The predominant use of an observer is to estimate the state for purposes of feedback control. In particular, in a linear system with a control designed on the assumption of full-state feedback

$$u = -Gx \tag{37.57}$$

when the state x is not directly measured, the state $\hat{x}$ of the observer is used in place of the actual state x in Equation 37.57. Thus, the control is implemented using

$$u = -G\hat{x} \tag{37.58}$$

where

$$\hat{x} = x - e \tag{37.59}$$

Hence, when an observer is used, the closed-loop dynamics are given in part by

$$\begin{aligned} \dot{x} &= Ax - BG(x - e) \\ &= (A - BG)x + BGe & (37.60) \end{aligned}$$

This equation, together with the equation for the propagation of the error, define the complete dynamics of the closed-loop system.

When a full-order observer is used

$$\dot{e} = \hat{A}e = (A - KC)e \tag{37.61}$$

Thus, the complete closed-loop dynamics are

$$\begin{bmatrix} \dot{x} \\ \dot{e} \end{bmatrix} = \begin{bmatrix} A - BG & BG \\ 0 & A - KC \end{bmatrix} \begin{bmatrix} x \\ e \end{bmatrix} \tag{37.62}$$

The closed-loop dynamics are governed by the upper triangular matrix

$$\mathbf{A} = \begin{bmatrix} A - BG & BG \\ 0 & A - KC \end{bmatrix} \quad (37.63)$$

the eigenvalues of which are given by

$$|sI - \mathbf{A}| = |sI - A + BG||sI - A + KC| = 0 \quad (37.64)$$

i.e., the closed-loop eigenvalues are the eigenvalues of $A - BG$, the full-state feedback system; and the eigenvalues of $A - KC$, the dynamics matrix of the observer. This is a statement of the well-known *separation principle*, which permits one to design the observer and the full-state feedback control independently, with the assurance that the poles of the closed-loop dynamic system will be the poles selected for the full-state feedback system and those selected for the observer.

When a reduced-order observer is used, it is readily established that the closed-loop dynamics are given by

$$\begin{bmatrix} \dot{x} \\ \dot{e}_2 \end{bmatrix} = \begin{bmatrix} A - BG & BG_2 \\ 0 & A_{22} - KA_{12} \end{bmatrix} \begin{bmatrix} x \\ e_2 \end{bmatrix} \quad (37.65)$$

and hence that the eigenvalues of the closed-loop system are given by

$$|sI - A + BG||sI - A_{22} + KA_{12}| = 0 \quad (37.66)$$

Thus, the separation principle also holds when a reduced-order observer is used.

It is important to recognize, however, that the separation principle applies only when the model of the process used in the observer agrees exactly with the actual dynamics of the physical process. It is not possible to meet this requirement in practice and, hence, the separation principle is an approximation at best. To assess the effect of a model discrepancy on the closed-loop dynamics, consider the following possibilities:

Case 1. Error in dynamics matrix

$$A = \bar{A} + \delta A$$

Case 2. Error in control distribution matrix

$$B = \bar{B} + \delta B$$

Case 3. Error in observation matrix

$$C = \bar{C} + \delta C$$

Using the "metastate"

$$\mathbf{x} = \begin{bmatrix} x \\ e \end{bmatrix}$$

it is readily determined [6] that the characteristic polynomial of the complete, closed-loop system for cases 1 and 3 is given by

$$|sI - \mathbf{A}| = \begin{vmatrix} sI - A_c & -BG \\ \delta A + K\delta C & sI - \hat{A} \end{vmatrix} \quad (37.67)$$

where

$$A_c = \bar{A} - BG, \quad \hat{A} = \bar{A} - K\bar{C}$$

Similarly, using the metastate

$$\mathbf{x} = \begin{bmatrix} \hat{x} \\ e \end{bmatrix}$$

it is found that the characteristic polynomial for case 2 is given by

$$|sI - \mathbf{A}| = \begin{vmatrix} sI - \hat{A} & -\delta BG \\ KC & sI - A_c \end{vmatrix} \quad (37.68)$$

where

$$A_c = A - \bar{B}G, \quad \hat{A} = A - KC$$

To assess the effect of perturbations of the dynamics matrices on the characteristic polynomial, the following determinantal identity can be used:

$$\begin{vmatrix} \mathcal{A} & \mathcal{B} \\ \mathcal{C} & \mathcal{D} \end{vmatrix} = |\mathcal{D}||\mathcal{A} - \mathcal{C}\mathcal{D}^{-1}\mathcal{B}| \quad (37.69)$$

Apply Equation 37.69 to Equation 37.68 to obtain

$$\begin{aligned} |sI - \mathbf{A}| &= |sI - A_c||sI - \hat{A} + \delta BG(sI - A_c)^{-1}KC| \\ &= |sI - A_c||sI - \hat{A}||I + \delta BG(sI - A_c)^{-1} \\ &\quad KC(sI - \hat{A})^{-1}| \end{aligned} \quad (37.70)$$

upon use of

$$|\mathcal{A}\mathcal{B}| = |\mathcal{A}||\mathcal{B}|$$

The separation principle would continue to hold if the coefficient of δB in Equation 37.70 were to vanish. It does vanish if observer matrix K satisfies the *Doyle-Stein condition* [2]

$$K(I + C\Phi K)^{-1} = B(C\Phi B)^{-1} \quad (37.71)$$

where

$$\Phi = (sI - A)^{-1}$$

is the plant resolvent.

To verify this, note that

$$\begin{aligned} (sI - \hat{A})^{-1} &= (sI - A + KC)^{-1} = (\Phi^{-1} + KC)^{-1} \\ &= \Phi - \Phi K(I + C\Phi K)^{-1}C\Phi \end{aligned} \quad (37.72)$$

When the Doyle-Stein condition (Equation 37.71) holds, Equation 37.72 becomes

$$(sI - \hat{A})^{-1} = \Phi - \Phi B(C\Phi B)^{-1}C\Phi$$

and so

$$C(sI - \hat{A})^{-1} = C\Phi - C\Phi B(C\Phi B)^{-1}C\Phi = 0$$

which ensures that the coefficient of δB in Equation 37.70 vanishes and, hence, that the separation principle applies.

Regarding the Doyle-Stein condition the following remarks are in order:

- The Doyle-Stein condition can rarely be satisfied exactly. But, as shown [2], it can be satisfied approximately by making the observer a Kalman filter with a noise matrix of the form given by Equation 37.14.
- The Doyle-Stein condition is not the only way the coefficient of δB can vanish. However, the Doyle-Stein condition ensures other robustness properties.
- An analogous condition for δA and δC can be specified.

In carrying out a similar analysis for a reduced-order observer it is found that the characteristic polynomial for the closed-loop control system, when a reduced-order observer is used and the actual control distribution matrix $B = \bar{B} + \delta B$ differs from the nominal (design) value $\bar{B}$, is given by

$$|sI - A| = \begin{vmatrix} sI - F + \Delta G_2 & \Delta G \\ -BG_2 & sI - A_c \end{vmatrix} \tag{37.73}$$

where

$$\Delta = K\delta B_1 - B_2 \tag{37.74}$$

It is seen that the characteristic polynomial of the closed-loop system reduces to that of Equation 37.66 when

$$\Delta = 0 \tag{37.75}$$

It is noted that Equation 37.75 can hold in a single-input system in which the loop gain is the only variable parameter. In this case

$$\delta B_1 = \rho B_1, \quad \delta B_2 = \rho B_2 \tag{37.76}$$

and thus

$$\Delta = \rho(KB_1 - B_2) = -\rho H$$

Hence, if the observer is designed with

$$H = B_2 - KB_1 = 0$$

the separation principle holds for arbitrary changes in the loop gain.

If Equation 37.75 cannot be satisfied, then, as shown in [7], condition analogous to the Doyle-Stein condition can be derived from Equation 37.73 in the case of a scalar control input (Equation 37.76):

$$[I - K(I + A_{12}\Phi_{22}K)^{-1}A_{12}\Phi_{22}](B_2 - KB_1) = 0 \tag{37.77}$$

where

$$\Phi_{22} = (sI - A_{22})^{-1}$$

37.6 Nonlinear Observers

The concept of an observer carries over to nonlinear systems. However, for a nonlinear system, the structure of the observer is not nearly as obvious as it is for a linear system. The design of observers for nonlinear systems has been addressed by several authors, such as Thau [8] and Kou et. al. [9].

An observer for a plant, consisting of a dynamic system

$$\dot{x} = f(x, u) \tag{37.78}$$

with observations given by

$$y = g(x, u) \tag{37.79}$$

is another dynamic system, the state of which is denoted by $\hat{x}$, excited by the output y of the plant, having the property that the error

$$e = x - \hat{x} \tag{37.80}$$

converges to zero in the steady state.

One way of obtaining an observer is to imitate the procedure used in a linear system, namely to construct a model of the original system (Equation 37.1) and force it with the "residual":

$$r = y - \hat{y} = y - g(\hat{y}, u) \tag{37.81}$$

The equation of the observer thus becomes

$$\dot{\hat{x}} = f(\hat{x}, u) + \kappa(y - g(\hat{x}, u)) \tag{37.82}$$

where $\kappa(\ )$ is a suitably chosen nonlinear function. (How to choose this function will be discussed later.) A block diagram representation of a general nonlinear observer is shown in Figure 37.3.

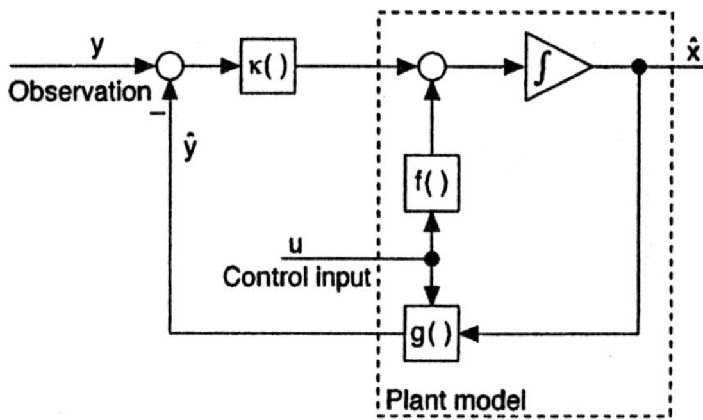

Figure 37.3 Structure of nonlinear observer.

The differential equation for the error e can be used to study its behavior. This equation is given by

$$\begin{aligned} \dot{e} &= \dot{x} - \dot{\hat{x}} \\ &= f(x, u) - f(\hat{x}, u) - \kappa(g(x, u) - g(\hat{x}, u)) \\ &= f(x, u) - f(x - e, u) \\ &\quad + \kappa(g(x - e, u) - g(x, u)) \end{aligned} \tag{37.83}$$

Suppose that by the proper choice of $\kappa(\ )$ the error Equation 37.83 can be made asymptotically stable, so that an equilibrium state is reached for which

$$\dot{e} = 0$$

Then, in equilibrium, Equation 37.83 becomes

$$0 = f(x, u) - f(x - e, u) + \kappa(g(x - e, u) - g(x, u)) \tag{37.84}$$

Since the right-hand side of Equation 37.84 becomes zero when $e = 0$, *independent* of x and u, it is apparent that $e = 0$ is an equilibrium state of Equation 37.83. This implies that if $\kappa(\,)$ can be chosen to achieve asymptotic stability, the estimation error e converges to zero.

It is very important to appreciate that the right-hand side of Equation 37.84 becomes zero independent of x and u only when the nonlinear functions $f(\cdot,\cdot)$ and $g(\cdot,\cdot)$ used in the observer are exactly the same as in Equations 37.78 and 37.79, which define the plant dynamics and observations, respectively. Any discrepancy between the corresponding functions generally prevents the right-hand side of Equation 37.84 from vanishing and hence leads to a steady-state estimation error. Since the mathematical model of a physical process is always an approximation, in practice the steady-state estimation error generally does not go to zero. But, by careful modeling, it is usually possible to minimize the discrepancies between the f and g functions of the true plant and the model used in the observer. This usually keeps the steady-state estimation error acceptably small.

For the same reason that the model of the plant and the observation that is used in the observer must be accurate, it is important that the control u that goes into the plant is the very same control used in the observer. If the control to the plant is subject to saturation, for example, then the nonlinear function that models the saturation must be included in the observer. Failure to observe this precaution can cause difficulties.

Including control saturation in the observer is particularly important as a means for avoiding the phenomenon known as *integrator windup*: the compensator, which has a pole at the origin, provides integral action. Imagine the transfer function of the compensator represented by an integrator in parallel with a second-order subsystem. The control signal to the integrator is oblivious to the fact that the input to the plant has saturated and hence keeps the integrator "winding up"; the error signal changes sign when the desired output reaches the set point, but the control signal does not drop from its maximum value. When the saturation is included in the observer, on the other hand, the control signal drops from its maximum value even before the error changes sign, thus correctly taking the dynamics (i.e., the lag) of the process into account.

The function $\kappa(\,)$ in the observer must be selected to ensure asymptotic stability of the origin ($e = 0$ in Equation 37.84). By the theorem of Lyapunov's first method (see Chapter "Lyapunov Stability"), the origin is asymptotically stable if the Jacobian matrix of the dynamics, evaluated at the equilibrium state, corresponds to an asymptotically stable linear system. For the dynamics of the error Equation 37.83 the Jacobian matrix with respect to the error e evaluated at $e = 0$ is given by

$$A_c(x) = (\partial f/\partial x) - K(\partial g/\partial x) \tag{37.85}$$

This is the nonlinear equivalent of the closed-loop observer equation of a linear system

$$A_c = A - KC$$

where A and C are the plant dynamics and observation matrices, respectively. The problem of selecting the gain matrix for a nonlinear observer is analogous to that of a linear observer, but somewhat more complicated by the presence of the nonlinearities that make the Jacobian matrices in Equation 37.85 dependent on the state x of the plant. Nevertheless, the techniques used for selecting the gain for a linear observer can typically be adapted for a nonlinear observer. Pole placement is one method; another is to make the observer an extended Kalman filter which, as explained later, entails on-line computation of the gains via the linearized variance equation.

It should be noted that the observer closed-loop dynamics matrix depends on the actual state of the system and hence is time varying. The stability of the observer thus cannot be rigorously determined by the locations of the eigenvalues of A_C.

The choice of $\kappa(\,)$ may be aided through the use of Lyapunov's second method. Using this method, Thau [8] considered a "mildly nonlinear" process

$$\dot{x} = f(x) = Ax + \mu\phi(x) \tag{37.86}$$

where μ is a small parameter, with linear observations

$$y = Cx$$

For this case, κ can be simply a gain matrix chosen to stabilize the linear portion of the system

$$\kappa(r) = Kr$$

where K is chosen to stabilize

$$\hat{A} = A - KC$$

This choice of K ensures asymptotic stability of the observer if $\phi(\,)$ satisfies a Lipschitz condition

$$||\phi(u) - \phi(v)|| \leq k||u - v||$$

and when

$$P\hat{A} + \hat{A}'P = -Q \leq -c_0 I$$

In this case, asymptotic stability of the observer

$$\dot{\hat{x}} = A\hat{x} + \mu\phi(\hat{x}) + K(y - C\hat{x})$$

is assured for

$$\mu < \frac{c_0}{2k||P||}$$

This analysis was substantially extended by Kou et al. [9].

Suggestions for the choice of the nonlinear function $\kappa(\,)$ have appeared in the technical literature. One suggestion [10], for example is to use

$$\kappa(r) = \Psi(\hat{x})^{-1}Kr$$

where K is a constant, possibly diagonal, matrix and $\Psi(x)$ is the Jacobian matrix defined by

$$\Psi(x) = \frac{\partial\Phi(x)}{\partial x}$$

with

$$\Phi(x) = \begin{bmatrix} L_f g(x) \\ L_f^2 g(x) \\ \vdots \\ L_f^n g(x) \end{bmatrix}$$

in which $L_f^n g(x)$ is the nth Lie derivative of $g(x)$ with respect to $f(x)$. (See Chapter "Differential Geometry and Lie Algebraic Methods".) This approach can possibly be viewed as a nonlinear generalization of the Bass-Gura formula given previously, since the matrix Ψ is akin to the observability matrix O defined earlier.

37.6.1 Using Zero-Crossing or Quantized Observations

The ability of an observer to utilize nonlinear observations is not limited to observations that exhibit only moderate nonlinearity; even highly nonlinear observations can be accommodated. Perhaps the most nonlinear observation is that of a zero-crossing detector, in which

$$y = \text{sgn}(x) = \begin{cases} 1, & x > 0 \\ -1, & x < 0 \end{cases}$$

This is the extreme special case of a quantizer, in which the output is quantized to only two levels.

Suppose, for example, that the only observation is of the zero-crossing of x_1. The observer for this process is then given by

$$\dot{\hat{x}} = f(\hat{x}, u) - k[y - \text{sgn}(\hat{x}_1)] \quad (37.87)$$

as illustrated in Figure 37.4.

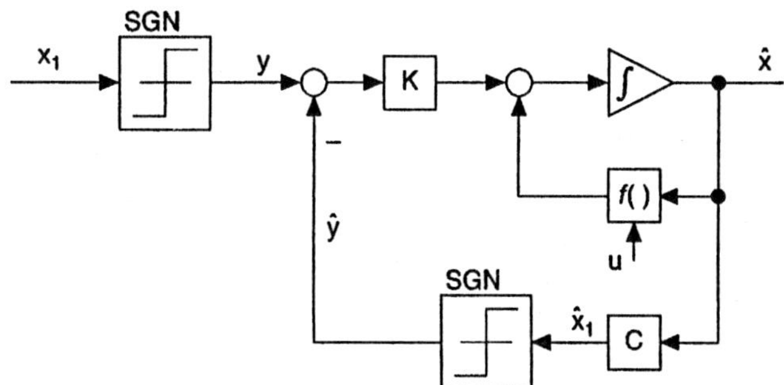

Figure 37.4 Observer using zero-crossing data.

Provided that a gain k, in this case a scalar parameter, can be found that stabilizes the observer at $e = 0$, the estimation error will be reduced to zero. The partial derivative of the nonlinear function with respect to the observation does not exist in this case because the observation is discontinuous with respect to the state. The stability of the observer cannot be established by linearizing about the origin. You have to use some other method, such as Lyapunov's second method, or determine the appropriate range of k by simulation.

Some insight into how the observer operates can be gained by considering that both y and $\text{sgn}(x_1)$ are signals that take on the values of ± 1; their difference, which is the residual that appears in Equation 37.87, is either 0 or ± 2. Suppose the observer is working well; most of the time y and $\text{sgn}(x_1)$ will have the same sign and the residual will be zero. The residual will be nonzero for the short time interval in which y and $\text{sgn}(x_1)$ have different signs. The residual will thus consist of a train of narrow pulses, each of height ± 2 and of width proportional to the phase difference between y and $\text{sgn}(x_1)$, as shown in Figure 37.5. The effect of each pulse is to nudge the state of the observer to agree with the state of the plant.

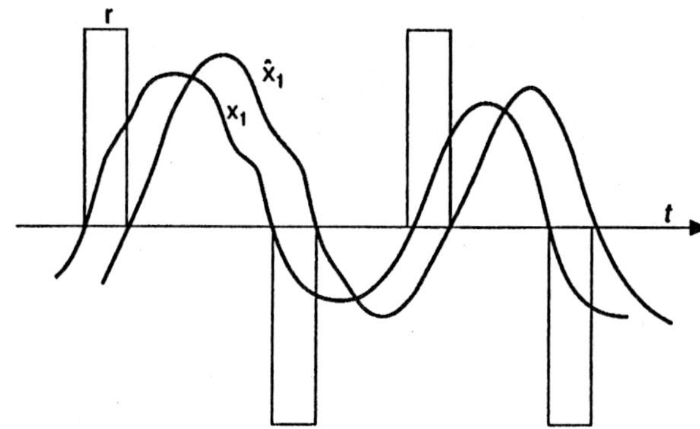

Figure 37.5 Residual for zero-crossing observations consists of pulses.

The same idea extends to quantized observations, since a quantizer can be regarded as a multiple-level crossing detector.

Of course, if there are other observations in addition to the zero-crossing observation, they can be combined with the latter and, with an appropriate choice of gains, can provide enhanced performance.

37.6.2 Reduced-Order Observers

Nonlinear reduced-order observers can be developed by the same methods one uses for linear, reduced-order observers.

Suppose that the state is partitioned into two substates as given by Equation 37.21 with the observation given by

$$y = g(x_1, u)$$

Provided that this expression can be solved for x_1 as a function of y and u

$$x_1 = \psi(y, u) = \bar{y}$$

we can use $\bar{y}$ as the observation. The completely general case in which y contains more state variables than its dimension can probably be handled in a manner similar to that used for linear systems, as discussed by Friedland [6]. The derivation is very tortuous in linear systems and is likely to be even more so in nonlinear systems and is, hence, omitted.

Corresponding to the partitioning of the state vector x as in Equation 37.21, the dynamic equations are written:

$$\dot{x}_1 = f_1(x_1, x_2, u) \quad (37.88)$$

$$\dot{x}_2 = f_2(x_1, x_2, u) \quad (37.89)$$

The nonlinear reduced-order observer is assumed to have the same structure as the corresponding linear observer. For the estimate of the substate x_1 we use the observation itself

$$\hat{x}_1 = y \quad (37.90)$$

while the substate x_2 is estimated using an observer of the form

$$\hat{x}_2 = Ky + z \quad (37.91)$$

where z is the state of a dynamic system of the same order as the dimension of the subvector x_2 and is given by

$$\dot{z} = \phi(y, \hat{x}_2, u) \quad (37.92)$$

A block diagram representation of the observer having the structure of Equations 37.90 to 37.92 is given in Figure 37.6.

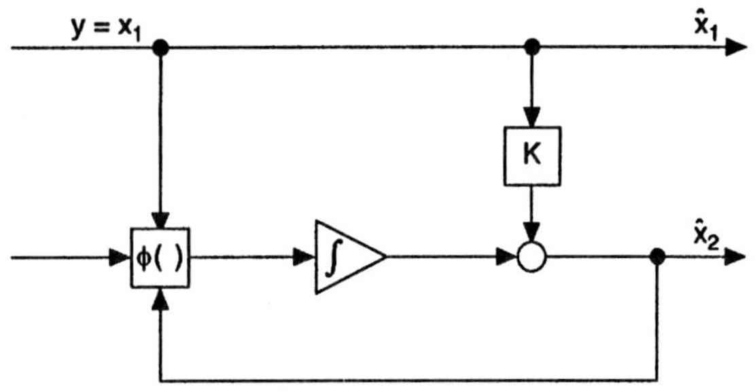

Figure 37.6 Reduced-order nonlinear observer.

The object of the observer design is the determination of the gain matrix K and the nonlinear function ϕ. As for the full-order observer, these are to be selected such that:

- The steady-state error in estimating x_2 converges to zero, independent of x and u. (The error in estimating x_1 is already zero when $\hat{x}_1 = y$.)
- The observer is asymptotically stable.

As in the case of the full-order observer, we proceed by deriving the differential equation for the estimation error

$$e = x_2 - \hat{x}_2 \quad (37.93)$$

Using Equations 37.89, 37.91, and 37.88, we get

$$\begin{aligned} \dot{e} &= \dot{x}_2 - \dot{\hat{x}}_2 \\ &= f_2(y, x_2, u) - Kf_1(y, x_2, u) - \phi(y, x_2 - e, u) \end{aligned} \quad (37.94)$$

In order for the right-hand side of Equation 37.94 to vanish when $e = 0$, it is necessary that the function $\phi(\cdot, \cdot, \cdot)$ satisfy

$$\phi(y, x_2, u) = f_2(y, x_2, u) - Kf_1(y, x_2, u) \quad (37.95)$$

for all values of y, x_2, and u.

To achieve asymptotic stability, the linearized system

$$\dot{e} = A(x_2)e \quad (37.96)$$

with

$$A(x_2) = (\partial\phi/\partial x_2) = (\partial f_2/\partial x_2) - K(\partial f_1/\partial x_2) \quad (37.97)$$

must be asymptotically stable.

As in the case of the full-order observer, there are several techniques for selecting an appropriate gain matrix.

The reduced-order nonlinear observer can be further generalized by replacing the linear function Ky in (Equation 37.91) by a nonlinear function $\kappa(y)$.

37.6.3 Extended Separation Principle

When an observer having the structure described above is used to estimate the state of a linear system, and the estimate is used in place of the actual state, the poles of the closed-loop system comprise the poles of the observer and the poles that would be present if full-state feedback were implemented. This is the separation principle of linear systems. (But remember that this result holds only when the model of the plant used in implementing the observer is a faithful model of the physical plant.)

The separation principle of linear systems can be extended to nonlinear systems. Consider, in particular, the nonlinear system

$$\dot{x} = f(x, u) \quad (37.98)$$

for which a control law

$$u = \gamma(x) \quad (37.99)$$

has been designed. Use of Equation 37.99 in Equation 37.98 gives the closed-loop dynamics

$$\dot{x} = f(x, \gamma(x)) = F(x) \quad (37.100)$$

Assume that the closed-loop dynamics of the system with full-state feedback, as represented by $F(x)$, has been designed—by whatever method might be appropriate—to achieve satisfactory behavior. How will the process behave when the state $\hat{x}$ of an observer is used in place of the true process state x [i.e., when $u = \gamma(\hat{x})$] ? As earlier, let

$$\hat{x} = x - e$$

where e is the error in estimating the state. Then Equation 37.99 becomes

$$u = \gamma(\hat{x}) = \gamma(x - e)$$

Then Equation 37.100 becomes

$$\dot{x} = f(x, \gamma(x - e)) \quad (37.101)$$

which together with Equation 37.83, which now becomes

$$\begin{aligned} \dot{e} &= f(x, \gamma(\hat{x})) - f(x - e, \gamma(\hat{x})) \\ &\quad + K[g(x - e, \gamma(\hat{x})) - g(x, \gamma(\hat{x}))] \end{aligned} \quad (37.102)$$

define the closed-loop dynamics.

There is not much that can be done with Equations 37.101 and 37.102 when f, g, and γ are general functions. But suppose these functions are sufficiently smooth to permit the use of Taylor's theorem, i.e.,

$$f(x,\gamma(x-e)) = f(x,\gamma(\hat{x})) - (\partial f/\partial\gamma)(\partial\gamma/\partial e)e + O(e^2) \quad (37.103)$$

$$f(x-e,\gamma(\hat{x})) = f(x,\gamma(\hat{x})) - (\partial f/\partial x)e + O(e^2) \quad (37.104)$$

$$g(x-e,\gamma(\hat{x})) = g(x,\gamma(\hat{x})) - (\partial g/\partial x)e + O(e^2) \quad (37.105)$$

where $O(e^2)$ represents terms that go to zero as $\|e\|^2$. Then Equations 37.101 and 37.102 become

$$\dot{x} = F(x) - (\partial f/\partial\gamma)(\partial\gamma/\partial e)e + O(e^2) \quad (37.106)$$

$$\dot{e} = [\partial f/\partial x + K(\partial g/\partial x)]e + O(e^2) \quad (37.107)$$

With the terms of $O(e^2)$ omitted, a block diagram representation of Equations 37.106 and 37.107 is shown in Figure 37.7. Note that the equation for the error has no input from the state estimation. The error thus converges asymptotically to zero as in the linear case. Since the error is the input to the full-state feedback control system, if the latter is asymptotically stable, the effect of the estimation error vanishes.

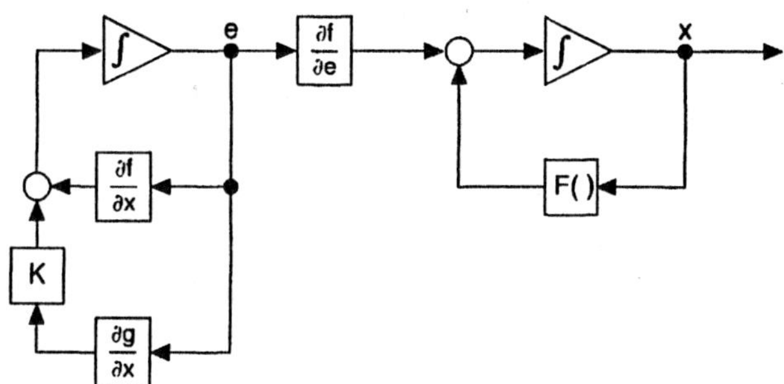

Figure 37.7 Illustration of extended separation principle.

37.6.4 Extended Kalman Filter

Sometimes the nonlinear function $\kappa(\,)$ can simply be a constant gain. Often, however, there is no obvious method of choosing this gain and a more systematic method is required. It is often appropriate to make the observer an "extended Kalman filter," i.e., to calculate the gain matrix on-line from the solution of the variance equation of the Kalman filter.

Few of the many applications for which Kalman filters have been used have met the linearity requirements of the theory. Nevertheless, the theory has been successfully, if not rigorously, applied. This is done by using a nonlinear observer of the form of Equation 37.84, but with the gain matrix K therein being computed, along with the state estimate, using Equations 37.12 and 37.13. The matrices A and C in these equations are the Jacobian matrices of the dynamics and observations

$$A = [\partial f/\partial x]_{x=\hat{x}} \quad (37.108)$$

$$C = [\partial g/\partial x]_{x=\hat{x}} \quad (37.109)$$

for the nonlinear process defined by

$$\dot{x} = f(x,u) \quad (37.110)$$

$$y = g(x,u) \quad (37.111)$$

In the linear case, the error covariance matrix, and through it the gain matrix K, do not depend on the estimated state. In principle, these matrices can be computed before the filter is implemented and stored in the filter's memory. In the nonlinear case, however, the matrices A and C that are used in computing P and K depend on the state estimate. Hence, in the extended Kalman filter, the observer and the Kalman filter gain matrix computation are coupled. This means that the equations for both the variance equation and the observer must be implemented on-line as shown schematically in Figure 37.8.

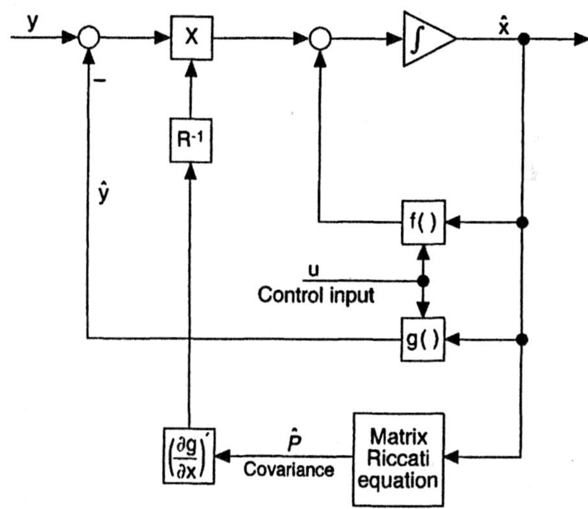

Figure 37.8 Schematic of extended Kalman filter, showing coupling between state estimation and covariance computation.

The requirement for on-line computation of the extended Kalman filter gain matrix can be a computational burden. Even considering that the covariance equation P is symmetric, there still are $k(k-1)/2$ scalar differential equations in Equation 37.13 that must be integrated numerically in addition to the k scalar observer equations for a kth-order dynamic process. In a tenth-order process, for example, a total of 55 equations must be integrated. It does not take a process of much higher order to overwhelm even a supercomputer. Moreover, the matrix Riccati equation (Equation 37.13) for P is notorious for being poorly behaved. Unless special measures are taken, the numerical solution to Equation 37.13 is likely to lose its positive definite character as the theory requires. If this happens, the resulting state estimate $\hat{x}$ will probably be useless.

Fortunately, it is rarely necessary to be a stickler for accuracy in the implementation of Equation 37.13. In the first place, the entire theory of the extended Kalman filter is only approximate. Moreover, the spectral density matrices Q and R that appear in Equation 37.13 are hardly ever known to be better than an order of magnitude. Hence, any computational method that gives a reasonable approximation to P and K is usually acceptable. Some of the approximations that have been considered include the following:

- Regard P as being piecewise constant and compute it relatively infrequently, using the discrete-time version of the Kalman filter.
- Simulate the observer and Equation 37.13 off-line; examine the results and use appropriate approximations. It may be possible, for example, to approximate some of the gains by constants. The effect of the approximations must be evaluated by further simulation.
- Use a simpler model to represent the process in Equation 37.13 than is used as the process model in the observer; but be careful not to use an overly simple model in the implementation of the process dynamics.

Acknowledgment

The author is grateful to Dr. Sergey Levkov of the Institute of Cybernetics in Kiev, Ukraine, for reviewing the manuscript and bringing to his attention some new references on nonlinear observers.

References

[1] Luenberger, D., Observers for multivariable systems, *IEEE Trans. Autom. Control,* 11, 190–197, 1966.

[2] Doyle, J.C. and Stein, G., Robustness with Observers, *IEEE Trans. Autom. Control,* 24, 607–611, 1979.

[3] Bass, R.W. and Gura, I., High-order system design via state-space considerations, Proc. Jt. Autom. Control Conf., Troy, NY, 311–318, June 1965.

[4] Friedland, B., On the properties of reduced-order Kalman filters, *IEEE Trans. Autom. Control,* 34, (3), 321–324, 1989.

[5] Rynaski, E.G., Flight control synthesis using robust observers, Proc. AIAA Guidance Control Conf., San Diego, CA, 825–831, September 1982.

[6] Friedland, B., *Control System Design: An Introduction to State-Space Methods,* McGraw-Hill, New York, 1986.

[7] Madiwale, A.N. and Williams, D.E., Some extensions of loop transfer recovery, Proc. Am. Control Conf., Boston, MA, 790–795, June 1985.

[8] Thau, F.E., Observing the state of nonlinear dynamic systems, *Int. J. Control,* 17, 471-479, 1973.

[9] Kou, S.R., Elliot, D.L., and Tarn, T.J., Exponential observers for nonlinear dynamic systems, *Inf. Control,* 29, (3), 204–216.

[10] Ciccarella, G., DallaMora, M., and Germani, A., A Luenberger-like observer for nonlinear systems, *Int. J. Control,* 57, (3), 537–556, 1993.

SECTION IX

Design Methods for MIMO LTI Systems

38

Eigenstructure Assignment

Kenneth M. Sobel
Department of Electrical Engineering, The City College of New York, New York, NY

Eliezer Y. Shapiro
HR Textron, Valencia, CA

Albert N. Andry, Jr.
Teledyne Electronic Devices, Marina del Rey, CA

38.1 Introduction

Eigenstructure assignment is a useful tool[1] that allows the designer to satisfy damping, settling time, and **mode decoupling** specifications directly by choosing eigenvalues and eigenvectors. Andry et al. [1] have applied eigenstructure assignment to designing a **stability augmentation system** for the lateral dynamics of the L-1011 aircraft. Both constant gain output feedback and **gain suppression** designs are proposed. Sobel and Shapiro [22] used eigenstructure assignment to design dynamic compensators for the L-1011 aircraft. First- and second-order compensators were proposed for the case in which sideslip angle could not be measured. Later, Sobel et al. [25] proposed a systematic method for choosing the elements of the feedback gain matrix which can be suppressed to zero with minimal effect on the eigenvalue and eigenvector assignment. A design of an eigenstructure assignment gain-suppression flight controller is shown for F-18 HARV aircraft.

Specialized task tailored modes for highly maneuverable fighter aircraft have been designed by using eigenstructure assignment. Sobel and Shapiro [21] designed an eigenstructure assignment pitch pointing and vertical translation controller for the AFTI F-16 aircraft. Pilot command tracking was achieved by using a special case of O'Brien and Broussard's [14] command generator tracker. Sobel and Shapiro [20] designed a yaw pointing and lateral translation controller for the linearized lateral dynamics of the Flight Propulsion Control Coupling (FPCC) aircraft. This conceptual control-configured vehicle has a vertical canard in addition to the more conventional control surfaces.

The chapter is organized as follows. Section 38.2 describes eigenstructure assignment using constant gain output feedback. Section 38.3 extends eigenstructure assignment to allow the designer to suppress chosen elements of the feedback gain matrix to zero. A systematic method for choosing the entries to be suppressed is discussed. Section 38.4 describes eigenstructure assignment using dynamic compensation. Each section includes an application of eigenstructure assignment to the design of a stability augmentation system for the linearized lateral dynamics of the F-18 High Angle of Attack Research Vehicle (HARV). Finally, in Section 38.5 we present a robust sampled data, eigenstructure assignment control law design for the yaw pointing/lateral translation maneuver of the FPCC aircraft.

38.2 Eigenstructure Assignment Using Output Feedback

Consider a system modeled by the linear time-invariant matrix differential equation described by

$$\dot{x} = Ax + Bu \tag{38.1}$$

[1]Reprinted from International Journal of Control, 59(1), 13–37, 1994. (With permission.)

0-8493-8570-9/96/$0.00+$.50

$$y = Cx \tag{38.2}$$

where x is the state vector ($n \times 1$), u the control vector ($m \times 1$), and y the output vector ($r \times 1$). It is assumed that the m inputs and the r outputs are independent. Also, as is usually the case in aircraft problems, it is assumed that $m < r < n$. If there are no exogenous inputs such as pilot commands, the feedback control vector u equals a matrix times the output vector y :

$$u = -Fy \tag{38.3}$$

The feedback problem can be stated as follows: Given a set of desired eigenvalues, (λ_i^d), $i = 1, 2, \ldots, r$ and a corresponding set of desired eigenvectors, (v_i^d), $i = 1, 2, \ldots, r$, find the real $m \times r$ matrix F such that the eigenvalues of $A - BFC$ contain (λ_i^d) as a subset, and the corresponding eigenvectors of $A - BFC$ are close to the respective members of the set (v_i^d).

Srinathkumar [26] has shown that, if (A, B) is a controllable pair, then the feedback gain matrix F will exactly assign r eigenvalues. It will also assign the corresponding eigenvectors, provided that v_i^d is chosen to be in the subspace spanned by the columns of $(\lambda_i I - A)^{-1}B$ for $i = 1, 2, \ldots, r$. This subspace is of dimension m, which is the number of independent control variables. In general, a chosen or desired eigenvector v_i^d will not reside in the prescribed subspace and, hence, cannot be achieved. Instead, a "best possible" choice for an achievable eigenvector is made. Andry et al. [1] showed that the best possible eigenvector is the projection of v_i^d onto the subspace spanned by the columns of $(\lambda_i I - A)^{-1}B$. An alternative representation, described by Kautsky et al. [7], showed that the subspace in which the eigenvector v_i must reside is also given by the null space of $U_1^T(\lambda_i I - A)$. The matrix U_1 is obtained from the singular value decomposition of B, given by

$$B = [U_0, U_1]\begin{bmatrix} \Sigma W^T \\ 0 \end{bmatrix} \tag{38.4}$$

The method of Kautsky et al. [7] for computing the subspaces is the preferred method for numerical computation.

Finally, the complete controllability assumption may be removed by using results derived by Liebst and Garrard [9] and by Liebst et al. [10]. These results allow the designer to alter eigenvectors which correspond to uncontrollable eigenvalues.

In many practical situations, complete specification of v_i^d is neither required or known, but rather the designer is interested only in certain elements of the eigenvector. Thus, assume that v_i^d has the following structure:

$$v_i^d = \left[v_{i1}, x, x, x, x, v_{ij}, x, x, v_{in}\right]^T$$

where v_{ij} are designer specified components and x is an unspecified component. Define, as shown by Andry et al. [1], a reordering operation $\{\,\}^{R_i}$ so that

$$\left\{v_i^d\right\}^{R_i} = \begin{bmatrix} \ell_i \\ d_i \end{bmatrix} \tag{38.5}$$

where ℓ_i is a vector of specified components of v_i^d and d_i is a vector of unspecified components of v_i^d. The rows of the matrix $(\lambda_i I - A)^{-1}B$ are also reordered to conform with the reordered components of v_i^d. Thus,

$$\left\{(\lambda_i I - A)^{-1}B\right\}^{R_i} = \begin{bmatrix} \tilde{L}_i \\ D_i \end{bmatrix}. \tag{38.6}$$

Then, as shown by Andry et al. [1], the achievable eigenvector v_i^a is given by

$$v_i^a = (\lambda_i I - A)^{-1}B\tilde{L}_i^{\dagger}\ell_i \tag{38.7}$$

where $(\cdot)^{\dagger}$ denotes the appropriate pseudoinverse of $(\cdot)$.

The output feedback gain matrix using eigenstructure assignment [1] is described by

$$F = -(Z - A_1V)(CV)^{-1} \tag{38.8}$$

where A_1 is the first m rows of the matrix A in Equation 38.1, V is the matrix whose columns are the r achievable eigenvectors, Z is a matrix whose columns are $\lambda_i z_i$ where the ith eigenvector v_i is partitioned as $v_i = \begin{bmatrix} z_i \\ w_i \end{bmatrix}$ with z_i an $m \times 1$ vector, and C is the output matrix in Equation 38.2. The result of Equation 38.8 assumes that the system described by Equations 38.1 and 38.2 has been transformed into a system in which the control distribution matrix B is a lead block identity matrix.

An alternative representation for the feedback gain matrix F as developed by Sobel et al. [24] is

$$F = -V_b\Sigma_b^{-1}U_{b0}^T(V\Lambda - AV)V_r\Sigma_r^{-1}U_{r0}^T \tag{38.9}$$

where the singular value decompositions of the matrices B and CV are given by

$$B = [U_{b0}U_{b1}]\begin{bmatrix} \Sigma_b W^T \\ 0 \end{bmatrix} \tag{38.10}$$

$$CV = [U_{r0}U_{r1}]\begin{bmatrix} \Sigma_r W_r^T \\ 0 \end{bmatrix} \tag{38.11}$$

and where Λ is an $r \times r$ diagonal matrix with entries λ_i, $i = 1, 2, \ldots, r$. The method described by Equation 38.9 is the preferred method for numerical computation.

We conclude the discussion with a comment about the closed-loop system stability. Unfortunately, it is not yet possible to insure that stable open-loop eigenvalues do not move into the right half of the complex plane when an eigenstructure assignment output feedback controller is utilized. This is still an open area for further research. However, for aircraft flight control systems, the closed-loop stability requirement is neither necessary nor sufficient. For example, some modes, such as the dutch roll mode, are required to meet minimum frequency and damping specifications, as described in MIL-F-8785C [13]. For these modes, stability alone is not sufficient. Other modes, such as the spiral mode, may be unstable provided that the time to double amplitude is sufficiently large. For these modes, stability may not be necessary.

38.2.1 F-18 Harv Linearized Lateral Dynamics Design Example

Consider the lateral directional dynamics of the F-18 HARV aircraft linearized at a Mach number of 0.38, an altitude of 5000 ft, and an angle of attack of five degrees. The aerodynamic model is augmented with first order actuators and a yaw rate washout filter. The eight state variables are aileron deflection δ_a, stabilator deflection δ_s, rudder deflection δ_r, sideslip angle β, roll rate p, yaw rate r, bank angle ϕ, and washout filter state x_8. The three control variables are aileron command δ_{ac}, stabilator command δ_{sc}, and rudder command δ_{rc}. The four measurements are r_{wo}, p, β, ϕ, where r_{wo} is the washed out yaw rate. All quantities are in the body axis frame of reference with units of degrees or degrees per second. The state space matrices A, B, and C that completely describe the model are shown in the Appendix.

An output feedback gain matrix is now computed by using eigenstructure assignment. The desired dutch roll eigenvalues are chosen with a damping ratio of 0.707 and a natural frequency in the vicinity of 3 rad/sec. The roll subsidence and spiral modes are chosen to be merged into a complex mode, as suggested by [1]. The desired eigenvalues are

dutch roll mode:

$$\lambda_{dr}^d = -2 \pm j2$$

roll mode:

$$\lambda_{roll}^d = -3 \pm j2$$

The desired eigenvectors are chosen to keep the quantity $|\phi/\beta|$ small. Therefore, the desired dutch roll eigenvectors will have zero entries in the rows corresponding to bank angle and roll rate. The desired roll mode eigenvectors will have zero entries in the rows corresponding to yaw rate, sideslip, and x_8 (which is filtered yaw rate). The desired eigenvectors are

$$v_{dr}^d = \begin{bmatrix} x \\ x \\ x \\ 1 \\ 0 \\ x \\ 0 \\ x \end{bmatrix} \pm j \begin{bmatrix} x \\ x \\ x \\ x \\ 0 \\ 1 \\ 0 \\ x \end{bmatrix} \qquad v_{roll}^d = \begin{bmatrix} x \\ x \\ x \\ 0 \\ 1 \\ 0 \\ x \\ 0 \end{bmatrix} \pm j \begin{bmatrix} x \\ x \\ x \\ 0 \\ x \\ 0 \\ 1 \\ 0 \end{bmatrix} \begin{matrix} \delta_a \\ \delta_s \\ \delta_r \\ \beta \\ p \\ r \\ \phi \\ x_8 \end{matrix}$$

The achievable eigenvectors are computed by taking the orthogonal projection onto the null space of $U_1^T(\lambda_i I - A)$. However, care must be taken when computing the pseudoinverse of $\tilde{L}_i$ because this matrix is ill-conditioned. Press et al. [17] suggest computing the pseudoinverse by using a singular value decomposition in which the singular values, that are significantly smaller than the largest singular value, are treated as zero. The achievable eigenvectors in this paper were computed by using MATLAB® function PINV with TOL = 0.01. The achievable eigenvectors are shown in Table 38.1 where the underlined numbers indicate the small couplings between p, ϕ, and the dutch roll mode and between β, r, x_8, and the roll mode. Hence, the ratio $|\phi/\beta|$ can be expected to be small.

The feedback gain matrix is computed by using Equation 38.9 and is shown in Table 38.2. From the open-loop state responses to a one degree initial sideslip, shown in Figure 38.1, we conclude that the aircraft is poorly damped with strong coupling between the dutch roll mode and roll mode. The closed-loop state response is shown in Figure 38.2. Observe that the maximum absolute values of the bank angle and roll rate are 0.0532 degrees and 0.2819°/sec., respectively.

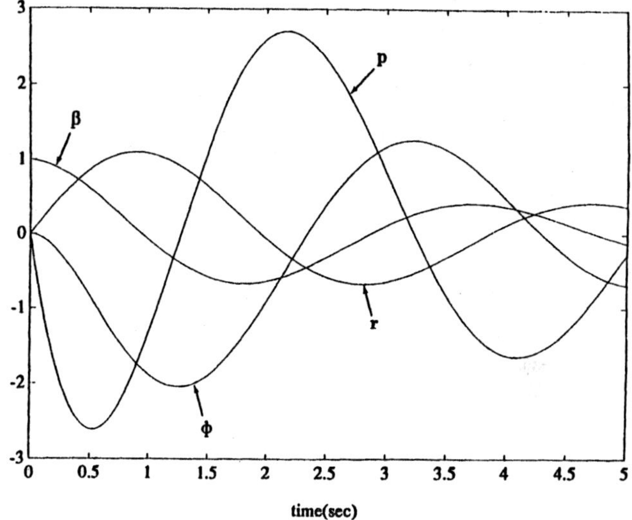

Figure 38.1 F-18 open-loop state responses.

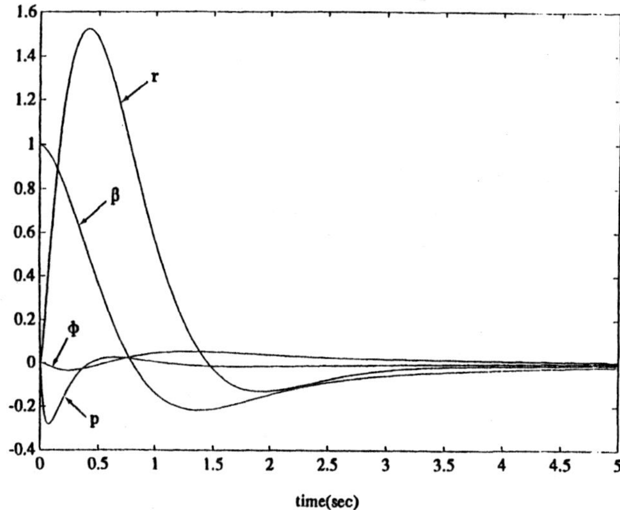

Figure 38.2 F-18 closed-loop state responses for output feedback.

Finally, we consider the multivariable gain and phase margins for our design. Suppose that the modeling errors may be described by the matrix L given by

$$L = Diag\left(\ell_1 e^{j\phi_1}, \ell_2 e^{j\phi_2}, \ldots, \ell_m e^{j\phi_m}\right).$$

Then, as shown by Lehtomaki [8], multivariable gain and phase

TABLE 38.1 Achievable Eigenvectors for Constant Gain Output Feedback

Dutch roll mode			Roll mode			
1.1781		0.2929	0.3212		0.2785	δ_a
−2.7860		1.9039	0.3259		0.2058	δ_s
10.1406		−4.1580	−0.1617		0.0693	δ_r
0.9777	$\pm j$	1.7260	−0.0558	$\pm j$	0.0791	β
−0.0559		−0.0094	0.9983		−4.9603	p
5.7061		0.9932	−0.0016		−0.0246	r
−0.0915		−0.1302	−0.9938		0.9916	ϕ
−0.5258		−1.0322	−0.0022		0.0032	x_8

TABLE 38.2 Constant Gain Output Feedback Control Law

Feedback gain matrix, degree/degree				Gain and phase margins	$\max\|\phi\|$
r_{wo}	p	β	ϕ	(at inputs δ_{ac}, δ_{sc}, δ_{rc})	$\max\|p\|$
−0.1704	0.1380	0.0277	0.4092	[−5.50 dB, 18.65 dB]	0.0532°
0.7164	0.1075	−1.7252	0.4867	±52.41°	0.2819°/sec.
−2.2741	0.0173	4.4961	−0.3877		

margins at the inputs may be defined.

Let $\sigma_{min}[I + FG(s)] > \alpha$ where the plant transfer matrix is given by $G(s) = C(sI - A)^{-1}B$. Then, the upward gain margin is at least as large as $1/(1-\gamma)$ and the gain reduction margin is at least as small as $1/(1+\gamma)$. The phase margins are at least $\pm SIN^{-1}(\gamma/2)$. The multivariable gain and phase margins for the constant gain output feedback design are shown in Table 38.2. We conclude that the margins are acceptable, especially because these margins are considered conservative.

38.3 Eigenstructure Assignment Using Constrained Output Feedback

The feedback gain matrix given by Equation 38.8 feeds back every output to every input. We now consider the problem of constraining certain elements of F to be zero. By suppressing certain gains to zero, the designer reduces controller complexity and increases reliability.

Using the development of [1], define

$$\Omega = CV \tag{38.12}$$

$$\Psi = Z - A_1 V \tag{38.13}$$

Then the expression for the feedback gain matrix F is given by (see Equation 38.8)

$$F = -\Psi\Omega^{-1}. \tag{38.14}$$

By using the Kronecker product and the lead block identity structure of the matrix B, each row of the feedback gain matrix can be computed independently of all the other rows [1]. Let ψ_i be the ith row of the matrix Ψ. Then the solution for f_i, which is the ith row of the feedback gain matrix F, is given by

$$f_i = -\psi_i\Omega^{-1}. \tag{38.15}$$

If f_{ij} is chosen to be constrained to zero, then Andry et al. [1] show that f_{ij} should be deleted from f_i and that the jth row of Ω should be deleted. Let $\tilde{\Omega}$ be the matrix Ω with its jth row deleted and $\tilde{f}_i$, be the row vector f_i with its jth entry deleted. Then, by using a pseudoinverse, the solution for $\tilde{f}_i$, whose entries are the remaining active gains in the ith row of the matrix F, is given by

$$f_i = -\psi_i\tilde{\Omega}^{\dagger} \tag{38.16}$$

where $(\cdot)^{\dagger}$ denotes the appropriate pseudoinverse of $(\cdot)$. If more than one gain in a row of F is to be set to zero, the $\tilde{f}_i$ and $\tilde{\Omega}$ must be appropriately modified.

38.3.1 Eigenvalue/Eigenvector Derivative Method for Choosing Gain-Suppression Structure

Calvo-Ramon [3] has proposed a method for choosing a priori which gains should be set to zero based on the sensitivities of the eigenvalues to changes in the feedback gains. The first-order sensitivity of the hth eigenvalue to changes in the ijth entry of the matrix F is denoted by $\partial\lambda_h/\partial f_{ij}$. The expected shift in the eigenvalue λ_h when constraining feedback gain f_{ij} to zero is given by

$$S^h_{ij} = (f_{ij})\frac{\partial\lambda_h}{\partial f_{ij}}. \tag{38.17}$$

Next, combine all the eigenvalue shifts that are related to the same feedback gain f_{ij} to form a decision matrix $D^{\lambda} = \{d_{ij}\}$, $D^{\lambda}\varepsilon R^{m\times r}$, wherc

$$d^{\lambda}_{ij} = \frac{1}{n}\left[\sum_{h=1}^{n}\left(\bar{S}^h_{ij}\right)\left(S^h_{ij}\right)\right]^{1/2} \tag{38.18}$$

and $\bar{(\cdot)}$ denotes the complex conjugate of $(\cdot)$.

The decision matrix D^{λ} is used to determine which feedback gains f_{ij} should be set to zero. If d^{λ}_{ij} is "small", then setting f_{ij} to zero will have a small effect on the closed-loop eigenvalues. Conversely, if d^{λ}_{ij} is "large", then setting f_{ij} to zero will have a significant effect on the closed-loop eigenvalues. The control system designer must determine which d^{λ}_{ij} are "small" and which are "large" for a particular problem. In this regard, it is assumed that the states, inputs, and outputs are scaled so that these variables are expressed in the same or equivalent units.

The decision matrix D^λ was used by Calvo-Ramon [3] to design a constrained output feedback controller using eigenstructure assignment. However, the sensitivities of the eigenvectors with respect to the gains were not considered when deciding which feedback gains should be set to zero. Recall that the eigenvalues determine transient response characteristics such as overshoot and settling time, whereas the eigenvectors determine mode decoupling. This mode decoupling is related, for example, to the $|\phi/\beta|$ ratio in an aircraft lateral dynamics problem. This ratio must be small, as specified in [13], which implies that the closed-loop aircraft should exhibit a significant degree of decoupling between the dutch roll mode and the roll mode. The approach of Calvo-Ramon [3] may yield a constrained controller with acceptable overshoot and settling time, but the mode decoupling may be unacceptable. Thus, consideration of both eigenvalue and eigenvector sensitivities is important when choosing which feedback gains should be constrained to zero.

Sobel et al. [25] extended the results of Calvo-Ramon [3] to include both eigenvalue and eigenvector sensitivities to the feedback gains. The eigenvector derivatives are used to compute the expected shift in eigenvector ν_h when constraining feedback gain f_{ij} to be zero. This expected shift in eigenvector ν_h is given by

$$\bar{s}_{ij}^{h} = (f_{ij})(\partial \nu_h / \partial f_{ij}) \tag{38.19}$$

Then, all the eigenvector shifts related to the same feedback gain f_{ij} are combined to form an eigenvector decision matrix D^ν,

$$D^\nu = \frac{1}{n}\left[\sum_{h=1}^{n}\left(\bar{s}_{ij}^{h}\right)^{*}\left(\bar{s}_{ij}^{h}\right)\right]^{1/2} \tag{38.20}$$

where $(\cdot)^*$ denotes the complex-conjugate transpose of $(\cdot)$. The gains that should be set to zero are determined by first eliminating those f_{ij} corresponding to entries of D^λ that are considered to be small. Then, those entries of D^ν corresponding to those f_{ij} that were chosen to be set to zero based on D^λ are reviewed. In this way, the designer can determine whether some of the f_{ij} that may be set to zero based on eigenvalue considerations should not be constrained based on eigenvector considerations.

38.3.2 F-18 Harv Linearized Lateral Dynamics Design Example

We return to the example which was first considered in Section 38.2.1 but now we seek a constrained-output feedback controller. The eigenvalue decision matrix D^λ and the eigenvector decision matrix D^ν are shown in Table 38.3. The entries of D^λ considered large are underlined in Table 38.3. Observe that only 7 of the 12 feedback gains are needed when only eigenvalue sensitivities are considered. The constrained-output feedback gain matrix based on using only the information available from the eigenvalue decision matrix D^λ is shown in Table 38.4. The state responses to a one degree initial sideslip are shown in Figure 38.3. Observe the significantly increased coupling between sideslip and bank angle as compared to the unconstrained design of Section 38.2.1. The maximum absolute values of the bank angle and roll rate are now 0.5102 degrees and 1.7881°/sec. compared with 0.0532 degrees and 0.2819°/sec. obtained with the unconstrained feedback gain matrix. The increased coupling is due to ignoring the eigenvector sensitivities and illustrates the importance of the eigenvectors in achieving adequate mode decoupling.

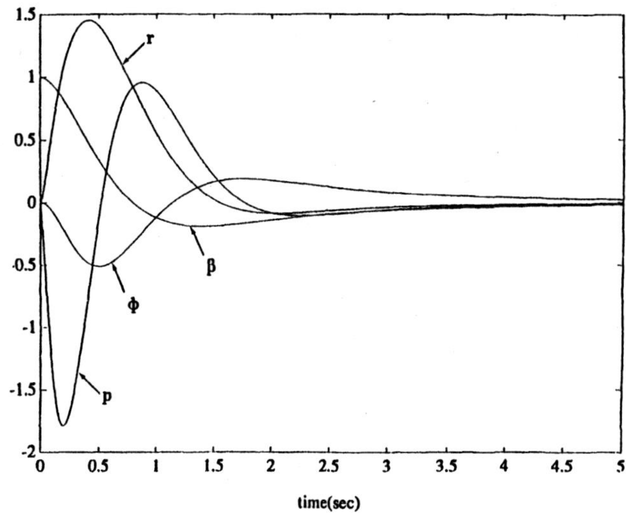

Figure 38.3 F-18 closed-loop state responses for constrained-output feedback (using only D^λ).

Next, consider the entries of D^ν that correspond to those f_{ij} that were chosen to be set to zero based on the eigenvalue decision matrix. The two largest d_{ij}^{ν} that belong to this class are d_{11}^{ν} and d_{23}^{ν}. A new constrained-output feedback gain matrix is computed in which f_{11} and f_{23} are not set to zero. Nine gains now need to be unconstrained when using both eigenvalue and eigenvector information. The new feedback gain matrix is given in Table 38.4 and the state responses for a one degree initial sideslip are shown in Figure 38.4. Observe that these time responses are almost identical to the responses in Figure 38.2, which were obtained by using all 12 feedback gains. Thus, a simpler controller is obtained with a negligible change in the aircraft time responses. Finally, the multivariable gain and phase margins are shown in Table 38.4. The values of these margins are considered acceptable because singular value based multivariable stability margin computation is conservative.

38.4 Eigenstructure Assignment Using Dynamic Compensation

We now generalize the eigenstructure assignment flight control design methodology to include the design of low-order dynamic compensators of any given order ℓ, $0 \leq \ell \leq n - r$. Recall that n and r are the dimensions of the aircraft state and output vectors, respectively. Consider the linear time-invariant aircraft

TABLE 38.3 Eigenvalue and Eigenvector Decision Matrices

Eigenvalue decision D^{λ}				Eigenvalue decision matrix D^{ν}				
r_{wo}	p	β	ϕ	r_{wo}	p	β	ϕ	
0.0710	$\underline{0.5117}$	0.0019	$\underline{0.2379}$	$\underline{0.2379}$	0.4675	0.0107	0.1835	δ_a
$\underline{0.2286}$	$\underline{0.3150}$	0.0843	$\underline{0.3362}$	0.7927	0.2882	$\underline{0.5270}$	0.1693	δ_s
$\underline{0.6754}$	0.0162	$\underline{0.3347}$	0.0482	0.7780	0.0259	0.3194	0.0871	δ_r

TABLE 38.4 Comparison of Constant Gain Control Laws

Feedback gain matrix degree/degree				Gain and phase margins (at inputs $\delta_{ac}, \delta_{sc}, \delta_{rc}$)	max $\lvert\phi\rvert$ / max $\lvert p\rvert$
r_{wo}	p	β	ϕ		
Unconstrained					
−0.1704	0.1380	0.0277	0.4092	[−5.50 dB, 18.65 dB]	0.0532°
0.7164	0.1075	−1.7252	0.4867	±52.41°	0.2819°/sec.
−2.2741	0.0173	4.4961	−0.3877		
Constrained D^{λ} only					
0.0	0.1926	0.0	0.6417	[−5.38 dB, 16.90 dB]	0.5102°
0.3371	0.2010	0.0	0.7554	±50.75°	1.7881°/sec.
−2.2648	0.0	4.4891	0.0		
Constrained D^{λ} and D^{ν}					
−0.1643	0.1365	0.0	0.4049	[−5.32 dB, 16.22 dB]	0.0505°
0.7164	0.1075	−1.7252	0.4867	±50.01°	0.2662°/sec.
−2.2648	0.0	4.4891	0.0		

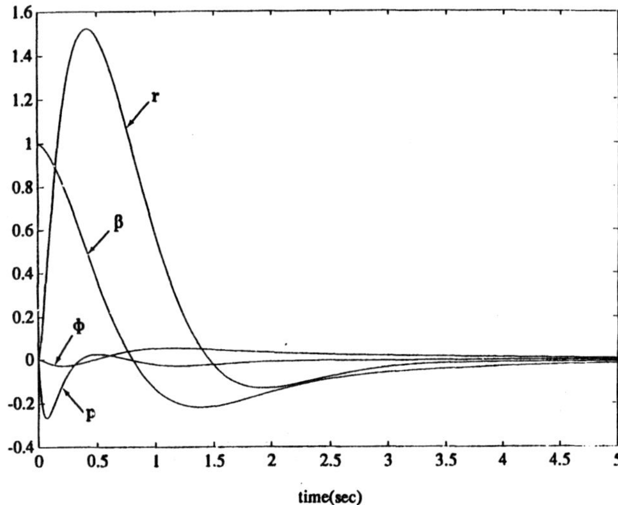

Figure 38.4 F-18 closed-loop state responses for constrained-output feedback (using both D^{λ} and D^{ν}).

described by Equations 38.1 and 38.2 with a linear time-invariant dynamic controller specified by

$$\dot{z}(t) = Dz(t) + Ey(t), \tag{38.21}$$

$$u(t) = F(z) + Gy(t), \tag{38.22}$$

where the controller state vector $z(t)$ is of dimension ℓ, $0 \leq \ell \leq n - r$.

It is convenient to model the aircraft and compensator by the composite system originally proposed by Johnson and Athans [6]. Thus, define

$$\dot{\bar{x}} = \bar{A}\bar{x} + \bar{B}\bar{u} \tag{38.23}$$

$$\bar{y} = \bar{C}\bar{x} \tag{38.24}$$

$$\bar{u} = \bar{F}\bar{y} \tag{38.25}$$

where

$$\bar{x} = \begin{bmatrix} x \\ \hline z \end{bmatrix}, \bar{A} = \begin{bmatrix} A|0 \\ \hline 0|0 \end{bmatrix}, \bar{B} = \begin{bmatrix} B|0 \\ \hline 0|I \end{bmatrix},$$

$$\bar{C} = \begin{bmatrix} C|0 \\ \hline 0|I \end{bmatrix}, \bar{F} = \begin{bmatrix} G|F \\ \hline E|D \end{bmatrix}$$

Furthermore, the eigenvectors of the composite system may be described by

$$\nu_i = \begin{bmatrix} \nu_i(x) \\ \hline \nu_i(z) \end{bmatrix} \tag{38.26}$$

where $\nu_i(x)$ is the ith subeigenvector corresponding to the aircraft and $\nu_i(z)$ the ith subeigenvector corresponding to the compensator. Once the compensator dimension is chosen, the problem is solved as previously shown for the constant gain case.

The dynamic compensator design problem may be stated as follows. Given a set of desired aircraft eigenvalues $\{\lambda_i^d\}$, $i = 1, 2, \ldots, r + \ell$ and a corresponding set of desired aircraft subeigenvectors $\nu_i^d(x)$, $i = 1, 2, \ldots, r + \ell$, find real matrices $D(\ell \times \ell)$, $E(\ell \times r)$, $F(m \times \ell)$, and $G(m \times r)$ so that the eigenvalues of $A + BFC$ contain $\{\lambda_i^d\}$ as a subset and corresponding subeigenvectors $\{\nu_i(x)\}$ are close to the respective members of the set $\{\nu_i^d(x)\}$.

38.4.1 F-18 Harv Linearized Lateral Dynamics Design Example

We again return to the example first considered in Section 38.2.1. If $y = [r_{wo}, p, \beta, \phi]^T$, as was the case in Sections 38.2.1 and 38.3.1, then the designer can specify both the dutch roll mode and roll mode eigenvalues. The designer might also specify three entries in the real and imaginary parts of the dutch roll eigenvectors and four entries in the real and imaginary parts of the roll mode eigenvectors as was done in Section 38.2.1. Then achievable eigenvectors are computed.

Now suppose that the measurement is given by $y = [r_{wo}, p, \phi]^T$, but the designer is still required to assign both the dutch roll and roll mode eigenvalues. Using the results of Section 38.4, we might utilize a first order dynamic compensator with state z_1. The composite system has state vector $\bar{x} = [\delta_a, \delta_s, \delta_r, \beta, p, r, \phi, x_8, z_1]^T$ and the measurement vector given by $\bar{y} = [r_{wo}, p, \phi, z_1]^T$. Thus, as before, the designer might choose to specify three or four entries of the real and imaginary parts of $v_i(x)$, i =1,2,3,4, which are the entries of the dutch roll and roll mode eigenvectors corresponding to the original aircraft state variables. Again, achievable eigenvectors will be computed.

We might ask whether the eigenvalue/eigenvector specifications are identical for both of the proposed problems. Certainly, the eigenvalue specifications and the corresponding $v_i^d(x)$ subeigenvectors may be identical. However, the $v_i^d(z)$ subeigenvector specifications must now be properly chosen. Otherwise, the modal matrix for the composite system may become numerically singular.

Finally, we remark that if the first-order compensator does not perform acceptably, then the designer might try a higher order compensator. In the case of a second-order compensator, the composite system has state vector $\bar{x} = [\delta_a, \delta_s, \delta_r, \beta, p, r, \phi, x_8, z_1, z_2]^T$ and the measurement vector is given by $\bar{y} = [r_{wo}, p, \phi, z_1, z_2]^T$. In this case, the designer can also specify one of the compensator eigenvalues and some entries of its corresponding eigenvector.

Now consider the case when only the washed out yaw rate, roll rate, and bank angle are measured. We form the composite system described by Equations 38.23–38.25 by appending a first-order compensator to the aircraft dynamics. We specify that the roll mode and dutch roll mode eigenvalues are the same as in the constant gain feedback problem in Section 38.2.1. The compensator pole is not specified, but it is chosen by the eigenstructure assignment algorithm to obtain eigenvalue and eigenvector assignment for the aircraft modes. Furthermore, because we have three sensors plus one compensator state, the eigenstructure assignment algorithm will allow us to specify four closed-loop eigenvalues. We specify that the desired aircraft subeigenvectors $v_i^d(x)$ are the same as the desired eigenvectors in the constant gain output feedback problem. The desired compensator subeigenvectors are chosen so that the dutch roll and roll modes participate in the compensator state solution. The control law is described by

$$\begin{bmatrix} \delta_a \\ \delta_s \\ \delta_r \end{bmatrix} = \begin{bmatrix} 0.1708 & -0.1383 & -0.4079 \\ -0.2723 & -0.3328 & -0.2408 \\ 1.0897 & 0.5836 & -1.5548 \end{bmatrix} \begin{bmatrix} r_{wo} \\ p \\ \phi \end{bmatrix} + \begin{bmatrix} -0.0040 \\ -1.4112 \\ 3.7731 \end{bmatrix} z_1, \quad (38.27)$$

$$\dot{z}_1 = -2.4784 z_1 + 0.5906 r_{wo} + 0.9817 p + 3.0796\phi. \quad (38.28)$$

The desired eigenvalues, achievable eigenvalues, and desired eigenvectors are shown in Table 38.5. The separation principle does not apply to the dynamic compensator described by Equations 38.21 and 38.22. Thus, the composite system has eigenvalues at -1.7646 and -0.7549 which are due to the compensator and the yaw rate washout filter, respectively. The time responses of the aircraft states are shown in Figure 38.5 from which we observe that the responses with the dynamic compensator are slower than the responses with constant gain output feedback. However, the dynamic compensator controller is implemented without the need for a sideslip sensor. In addition, the ratio $|\phi/\beta| \approx 0.8$ as compared to $|\phi/\beta| \approx 4$ for the open-loop aircraft.

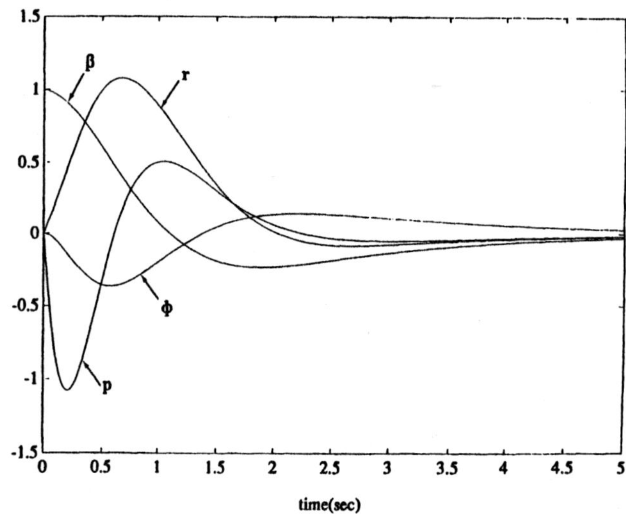

Figure 38.5 F-18 closed-loop state responses for first-order dynamic compensator.

The multivariable stability margins are computed using $\sigma_{\min}[I + KG(s)]$ where $G(s) = C(sI - A)^{-1}B$ and $K = -[F(sI - D)^{-1}E + G]$. The multivariable gain margins are $GM\varepsilon[-2.47$ dB, 3.46 dB] and the multivariable phase margin is ± 18.89 degrees. If the designer considers these margins inadequate, then an optimization may be used to compute an eigenstructure assignment controller with a constraint on the minimum of the smallest singular value of the return difference matrix. In the next section, we present a robust sampled data eigenstructure controller for the yaw pointing/lateral translation maneuver of the FPCC aircraft.

38.5 Robust Sampled Data Eigenstructure Assignment

Sobel and Shapiro [20] used eigenstructure assignment to design a continuous-time controller for the yaw pointing/lateral

TABLE 38.5 Eigenvalues and Eigenvectors for Dynamic Compensator

Desired eigenvalues	Achievable eigenvalues	Desired eigenvectors: Dutch roll mode Re	Dutch roll mode Im	Roll mode Re	Roll mode Im	
		x	x	x	x	δ_a
	$\lambda_{dr} = -2 \pm j2$	x	x	x	x	δ_s
	$\lambda_{roll} = -3 \pm j2$	x	x	x	x	δ_r
$\lambda_{dr} = -2 \pm j2$	$\lambda_{act} = -30.0000$	x	1	0	0	β
$\lambda_{roll} = -3 \pm j2$	$\lambda_{act} = -27.6993$	0	0	x	1	p
	$\lambda_{act} = -25.2634$	1	x	0	0	r
	$\lambda_{filt} = -0.7549$	0	0	1	x	ϕ
	$\lambda_{comp} = -1.7646$	x	x	0	0	x_8
		1	x	1	x	z_1

translation maneuver of the Flight Propulsior Control Coupling (FPCC) aircraft. This conceptual control-configured aircraft has a vertical canard and is difficult to control because the control distribution matrix has a minimum singular value of 0.0546. The design of Sobel and Shapiro [20] is characterized by perfect decoupling, but the minimum of the smallest singular value of the return difference matrix at the aircraft inputs was only 0.18.

Sobel and Lallman [24] proposed a pseudocontrol strategy for reducing the dimension of the control space by using the singular value decomposition. The FPCC yaw pointing/lateral translation design of Sobel and Lallman [24] yields a minimum of the smallest singular value of the return difference matrix at the aircraft inputs of 0.9835, but the lateral translation transient response has significant coupling to the heading angle.

Sobel and Shapiro [23] have proposed an extended pseudocontrol strategy. Piou and Sobel [15] extended eigenstructure assignment to linear time-invariant plants which are represented by Middleton and Goodwin's [12] unified delta model which is valid both for continuous-time and sampled data operation of the plant. Piou et al. [16] have extended Yedavalli's [27] Lyapunov approach for stability robustness of a linear time-invariant system to the unified delta system. In this section, we design a robust, sampled data, extended pseudocontrol, eigenstructure assignment flight control law for the yaw pointing/lateral translation maneuver of the FPCC aircraft. The main goal of this section is to describe a design methodology which incorporates robustness into the eigenstructure assignment method. However, a sampled data design is used for illustration.

38.5.1 Problem Formulation

Consider a nominal linear time-invariant system described by (A, B, C). The corresponding sampled data system is described by (A_δ, B_δ, C) and the unified delta model is described by (A_ρ, B_ρ, C). Suppose that the nominal delta system is subject to linear time-invariant uncertainties in the entries of A_ρ, B_ρ described by dA_ρ and dB_ρ, respectively. Then, the delta system with uncertainty is given by $(A_\rho + dA_\rho, B_\rho + dB_\rho, C)$. Here $dA_\rho = dA$, $dB_\rho = dB$ in continuous time and $dA_\rho = dA_\delta, dB_\rho = dB_\delta$ in discrete time. Furthermore, suppose that bounds are available on the maximum absolute values of the elements of dA and dB so that $\{dA : dA^+ \leq A_{\max}\}$ and $\{dB : dB^+ \leq B_{\max}\}$ and where "$\leq$" is applied element by element to matrices.

Consider the constant gain output feedback control law described by $u(t) = F_\rho y(t)$ where $F_\rho = F$ in continuous time and $F_\rho = F_\delta$ in discrete time. Then, the nominal-closed loop unified delta system is given by $\rho x(t) = A_{\rho c} x(t)$ where $A_{\rho c} = A + BFC$ in continuous time and $A_\delta + B_\delta F_\delta C$ in discrete time. The uncertain closed-loop unified delta system is given by $\rho x(t) = A_{\rho c} x(t) + dA_{\rho c} x(t)$ where $dA_{\rho c} = dA + dB(FC)$ in continuous time and $dA_\delta + dB_\delta(F_\delta C)$ in discrete time. The reader is referred to Middleton and Goodwin [12] for a more detailed description.

38.5.2 Pseudocontrol and Robustness Results

The purpose of a pseudocontrol is to reduce the dimension of the control space. This reduction is needed for systems whose control distribution matrix B has a minimum singular value that is very small. After the eigenstructure assignment design is complete, the controller is mapped back into the original control space. Consider the singular value decomposition of the matrix B_ρ given by

$$B_\rho = [U_1 U_2 U_0] \begin{bmatrix} \Sigma_1 & & \\ & \Sigma_2 & \\ & & 0 \end{bmatrix} \begin{bmatrix} V_1^T \\ V_2^T \\ V_0^T \end{bmatrix} \tag{38.29}$$

where $\Sigma_1 = \text{diag}\,[\sigma_1, \ldots, \sigma_a]$ and $\Sigma_2 = \text{diag}\,[\sigma_{a+1}, \ldots, \sigma_b]$ and where $\sigma_b \leq \sigma_{b-1} \leq \ldots \leq \sigma_{a+1} \leq \varepsilon$ with ε small.

LEMMA 38.1 Let the system with the pseudo-control $\tilde{u}(t)$ be described by

$$\rho x(t) = A\rho x(t) + \tilde{B}_\rho \tilde{u}(t), \tag{38.30}$$

$$y(t) = Cx(t), \tag{38.31}$$

where

$$\tilde{B}_\rho = U_1 + U_2[\alpha_1, \alpha_2]. \tag{38.32}$$

We design a feedback pseudocontrol for the system described by Equations 38.30–38.32. Then, the true control $u(t)$ for the system described by (A_ρ, B_ρ, C) is given by

$$u(t) = \left[V_1 \Sigma_1^{-1} + V_2 \Sigma_2^{-1} \alpha\right] \tilde{u}(t). \tag{38.33}$$

Furthermore, when $\alpha = [0, 0]^T$, the control law $u(t)$ given by Equation 38.33 reduces to the control law given by Equation 20 in Sobel and Lallman [24].

THEOREM 38.1 *The system matrix $A_{\rho c} + dA_{\rho c}$ is stable if*

$$\sigma_{\max}\left(E_{2\max}^T P_\rho^+ E_{1\max}\right)_s < 1 \qquad (38.34)$$

where

$$E_{1\max} = A_{\rho\max} + B_{\rho\max}(F_\rho C)^+ \quad and$$
$$E_{2\max} = \left\{I_n + \Delta\left[A_\rho + B_\rho(F_\rho C)\right]\right\}^+ + (\Delta/2)E_{1\max}$$

and where P_ρ satisfies the Lyapunov equation given by

$$A_{\rho c}^T P_\rho + P_\rho A_{\rho c} + \Delta A_{\rho c} P_\rho A_{\rho c}^T = -2I_n$$

and where P_ρ^+ is the matrix formed by the modulus of the entries of the matrix P_ρ and $(\cdot)_s$ denotes the symmetric part of a matrix.

38.5.3 FPCC Yaw Pointing/Lateral Translation Controller Design Using Robust, Sampled Data, Eigenstructure Assignment

We consider the FPCC aircraft linearized lateral dynamics which is described by Sobel and Lallman [24]. The state space matrices A and B are shown in the appendix. The state variables are sideslip angle β, bank angle ϕ, roll rate p, and lateral directional flight path angle ($\gamma = \Psi + \beta$), where Ψ is the heading angle. The control variables are rudder δ_r, ailerons δ_a, and vertical canard δ_c. The angles and surface deflections are in degrees, and the angular rates are in degrees per second. The five measurements are $\beta, \phi, p, r, \gamma$.

First, we design an eigenstructure assignment control law by using an orthogonal projection. The delta state-space matrices A_δ and B_δ are computed by using the MATLAB® Delta Toolbox. The sampling period Δ is chosen to be 0.02 seconds for illustrative purposes. The desired dutch roll, roll mode, and flight path mode eigenvalues are achieved exactly because five measurements are available for feedback. The achievable eigenvectors are computed by using the orthogonal projection of the ith desired eigenvector onto the subspace which is spanned by the columns of $(\gamma_i I - A_\delta)^{-1}B_\delta$. The closed-loop delta eigenvalues $\gamma_i, i = 1, \ldots, n$, and the feedback gain matrix F_δ are shown in Table 38.6. The desired closed-loop eigenvectors are shown in Table 38.7.

The orthogonal projection solution is characterized by excellent decoupling with the minimum of the smallest singular value of $(I + FG)$ equal to 0.18. Here the transfer function matrix of the delta plant is given by $G([e^{j\omega\Delta} - 1]/\Delta)$ where $0 < \omega < \pi/T$. Furthermore, the Lyapunov robust stability condition of Equation 38.34 is not satisfied.

To improve the minimum singular value of $(I + FG)$, we design a controller by using an orthogonal projection with the pseudocontrol of Sobel and Lallman [24]. This pseudocontrol mapping is given by Equation 38.33 with $\alpha = [0, 0]$. This design is characterized by a lateral translation response with significant coupling between γ and Ψ with the minimum of the smallest singular value of $(I + FG)$ equal to 0.9835. The Lyapunov sufficient robust stability condition of Equation 38.34 is not satisfied. We note that the yaw pointing responses exhibit excellent decoupling for both of the orthogonal projection designs.

Next, to obtain a robust design with excellent decoupling, we design a robust pseudocontrol law by using the design method proposed by Piou et al. [16]. This new design method minimizes an objective function which weights the heading angle due to a lateral flight path angle command and the lateral flight path angle due to a heading command. Constraints are placed on the time constants of the dutch roll, roll, and flight path modes, the damping ratios of the dutch roll and roll modes, and the new sufficient condition for robust stability. Mathematically, the objective function to be minimized is given by

$$J = \sum_{k=1}^{100}\left[(1-\alpha)\left(\Psi_k^2\right)_{\gamma c} + \alpha\left(\gamma_k^2\right)_{\Psi c}\right]. \qquad (38.35)$$

The upper limit on the index k is chosen to include the time interval $k\Delta\varepsilon[0, 2]$ during which most of the transient response occurs. Of course, computation of Equation 38.35 requires that two linear simulations be performed during each function evaluation of the optimization. The constraints for continuous time and the corresponding constraints for discrete time are shown in Table 38.8 where ζ is the damping ratio. For illustrative purposes we have chosen $A_{\max} = 0.085A^+$ *and* $B_{\max} = 0$. After many trials, we found that a good value for the weight a in Equation 38.35 is $a = 0.0075$.

The parameter vector contains the quantities which may be varied by the optimization. This seventeen dimensional vector includes Re γ_{dr}, Im γ_{dr}, Re γ_{roll}, Im γ_{roll}, γ_{fp}, Re $z_1(1)$, Re $z_1(2)$, Im $z_1(1)$, Im $z_1(2)$, Re $z_3(1)$, Re $z_3(2)$, Im $z_3(1)$, Im $z_3(2)$, $z_5(1)$, $z_5(2)$, and the two-dimensional pseudocontrol vector α of Equation 38.32. Here, the two-dimensional complex vectors z_i contain the free eigenvector parameters, that is, the ith eigenvector v_i may be written as

$$v_i = L_i z_i \qquad (38.36)$$

where the columns of $L_i = (\gamma_i I - A_\delta)^{-1}\tilde{B}_\delta$ are a basis for the subspace in which the ith eigenvector must reside. Thus, the free parameters are the vectors z_i rather than the eigenvectors v_i. The vectors z_i are two dimensional because the optimization is performed in the two-dimensional pseudocontrol space.

The optimization uses subroutine *constr* from the MATLAB® Optimization Toolbox and subroutine *delsim* from the MATLAB® Delta Toolbox. The optimization is initialized with the orthogonal projection pseudocontrol design which yields an initial value of 20.6 for the objective function of Equation 38.35 and a value of 1.66 for the right hand-side (RHS) of the robustness condition of Equation 38.34. Unfortunately, the minimum of the smallest singular value of $(I + FG)$ is only 0.2607 which is less than desired.

To achieve a design with excellent time responses, Lyapunov robustness, and an acceptable minimum of the smallest singular

TABLE 38.6 Comparison of FPCC Designs ($\Delta = 0.02$ sec.)

Closed-loop eigenvalues			Feedback gain matrix					
			β	ϕ	p	r	γ	
Orthogonal projection design	γ_{dr}	$= -1.9990 \pm j1.921$	1.4688	−0.2866	−0.0022	0.3799	−1.5332	δ_r
	γ_{roll}	$= -2.95 \pm j1.883$	0.5652	−2.2990	−0.9044	−0.6691	−0.8890	δ_a
	γ_{fp}	$= 0.4975$	9.2964	−1.2143	−0.514	−1.4219	−15.57	δ_c
			β	ϕ	p	r	γ	
Orthogonal projection design with pseudocontrol	γ_{dr}	$= -1.999 \pm j1.921$	−0.4929	−0.0332	0.0076	0.6883	2.7584	δ_r
	γ_{roll}	$= -2.95 \pm j1.883$	0.9517	−2.353	−0.9093	−0.7565	−2.5147	δ_a
	γ_{fp}	$= -0.4975$	0.1203	−0.0530	−0.0245	−0.1535	−0.6023	δ_c
			β	ϕ	p	r	γ	
Robust pseudocontrol design	$\gamma_{1,2}$	$= -3.37 \pm j1.56$	−0.2833	−0.0340	−0.0062	0.2377	0.5336	δ_r
	$\gamma_{3,4}$	$= -3.16 \pm j1.473$	1.5501	−2.4760	−1.0370	−0.9812	−1.7788	δ_a
	γ_{fp}	$= -0.4310$	5.6151	−0.0012	−0.1486	−4.6301	−10.285	δ_c
			β	ϕ	p	r	γ	
Robust pseudocontrol design with singular value constraint	$\gamma_{1,2}$	$= -2.55 \pm j1.19$	−0.1658	−0.1967	−0.0808	0.4895	0.8616	δ_r
	$\gamma_{3,4}$	$= -2.60 \pm j1.22$	0.8390	−1.5479	−0.7663	−0.8565	−1.2679	δ_a
	γ_{fp}	$= -0.3248$	1.1525	−0.8794	−0.4764	−1.9712	−3.2578	δ_c

Eigenvalues are computed by using feedback gains with significant digits to machine precision.

TABLE 38.7 FPCC Desired Closed-Loop Eigenvectors

Dutch roll mode	Roll mode	Flight path mode	
$\begin{bmatrix} x \\ 0 \\ 0 \\ 1 \\ 0 \end{bmatrix} \pm j \begin{bmatrix} 1 \\ 0 \\ 0 \\ x \\ 0 \end{bmatrix}$	$\begin{bmatrix} 0 \\ x \\ 1 \\ 0 \\ 0 \end{bmatrix} \pm j \begin{bmatrix} 0 \\ 1 \\ x \\ 0 \\ 0 \end{bmatrix}$	$\begin{bmatrix} x \\ 0 \\ 0 \\ 0 \\ 1 \end{bmatrix}$	$\begin{matrix} \beta \\ \phi \\ p \\ r \\ \gamma \end{matrix}$

value of $(I + FG)$, we repeat the optimization with the additional constraint that $\sigma_{\min}(I + FG) \geq 0.55$. Once again we initialize the optimization at the orthogonal projection pseudocontrol design. The optimization yields an optimal objective function of 0.0556 and a value of 0.999 for the RHS of the robustness condition. The time responses for yaw pointing and lateral translation are shown in Figures 38.6 and 38.7, respectively. The lateral translation response is deemed excellent even though it has some small increase in coupling compared to the design without the additional singular value constraint. Thus, we have obtained a controller which simultaneously achieves excellent time responses, Lyapunov robustness, and an acceptable minimum of the smallest singular value of the return difference matrix at the aircraft inputs.

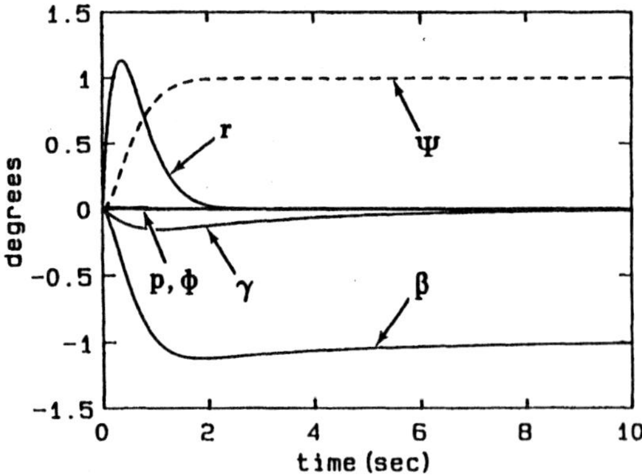

Figure 38.6 FPCC yaw pointing; robust pseudocontrol with singular value constraint.

38.6 Defining terms

stability augmentation system: A feedback control designed to modify the inherent aerodynamic stability of the airframe.

gain suppression: A feedback control in which one or more entries of the gain matrix are constrained to be zero.

specialized task tailored modes: A feedback control system designed for a specific maneuver such as bomb-

TABLE 38.8 Constraints for the FPCC Designs

Continuous time	Discrete time
For complex eigenvalues:	
$\text{Re}\lambda \in [-4, -1.5]$	$\|1+\Delta\gamma\| \in [e^{-4\Delta}, e^{-1.5\Delta}]$
$\zeta \in [0.4, 0.9]$	$\|1+\Delta\gamma\| \in \left[\exp\left(\frac{-0.9\phi}{[1-(0.9)^2]^{1/2}}\right), \exp\left(\frac{-0.4\phi}{[1-(0.4)^2]^{1/2}}\right),\right]$ where $\phi = \arg(1+\Delta\gamma)$
For the real eigenvalue:	
$\lambda \in [-1, -0.05]$	$\|1+\Delta\gamma\| \in [e^{-\Delta}, e^{-0.05\Delta}]$
For Lyapunov robustness:	
	$\sigma_{\max}\left(E^T_{2\max} P^+_\rho E_{1\max}\right) < 0.999$
For multivariable stability margins (final design only):	
	$\min_{\omega} \sigma_{\min} (I + FG) \geq 0.55;\ 0 < \omega < \pi/T$

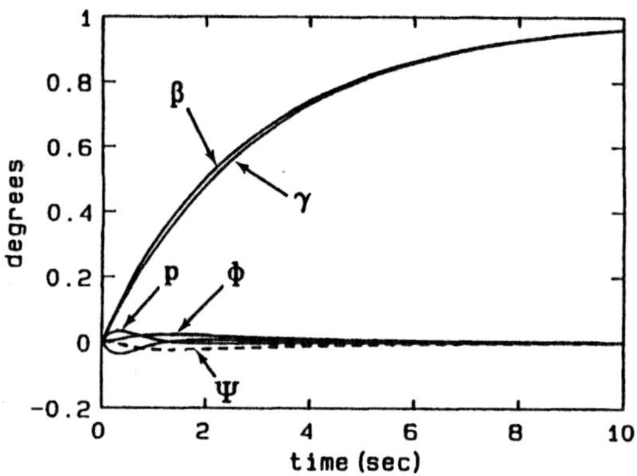

Figure 38.7 FPCC lateral translation; robust pseudocontrol with singular value constraint.

ing, strafing, air-to-air combat, etc.

mode decoupling: The elimination of interactions between modes, for example, in an aircraft mode decoupling is important so that a sideslip angle disturbance will not cause rolling motion.

eigenstructure assignment: A feedback control designed to modify both the system's eigenvalues and eigenvectors.

References

[1] Andry, A.N., Shapiro, E.Y., and Chung, J.C., Eigenstructure assignment for linear systems, *IEEE Trans. on Aerospace Electron. Syst.*, 19, 711–729, 1983.

[2] Burrows, S.P., Patton, R.J., and Szymanski, J.E., Robust eigenstructure assignment with a control design package, *IEEE Control Syst. Mag.*, 9, 29–32, 1989.

[3] Calvo-Ramon, J.R., Eigenstructure assignment by output feedback and residue analysis, *IEEE Trans. Autom. Control*, 31, 247–249, 1986.

[4] Doyle, J.C. and Stein, G., Multivariable feedback design: concepts for a classical/modern synthesis, *IEEE Trans. Autom. Control*, 26, 4–16, 1981.

[5] Gavito, V.F. and Collins, D.J., Application of eigenstructure assignment to design of robust decoupling controllers in mimo systems, *Proc. AIAA Guidance, Navigation, and Control Conf.*, 1986, AIAA Paper No. 86-2246, 828–834.

[6] Johnson, T.L. and Athans, M., On the design of optimal constrained dynamic compensation for linear constant systems, *IEEE Trans. Autom. Control*, 15, 658–660, 1970.

[7] Kautsky, J., Nichols, N.K., and Van Dooren, P., Robust pole assignment in linear state feedback, *Int. J. Control*, 41, 1129–1155, 1985.

[8] Lehtomaki, N.A., Practical robustness measures in multivariable control system analysis, Ph.D. Thesis, Massachusetts Institute of Technology, Cambridge, MA, 1981.

[9] Liebst, B.S. and Garrard, W.L., Application of Eigenspace Techniques to Design of Aircraft Control Systems, *Proc. Am. Control Conf.*, 1985, 475–480.

[10] Liebst, B.S., Garrard, W.L., and Adams, W.M., Design of an Active Flutter Suppression System, *J. Guidance, Control, Dyn.*, 9, 64–71, 1986.

[11] Lu, J., Chiang, H.D. and Thorp, J.S., Eigenstructure assignment by decentralized feedback control, *IEEE Trans. Auto. Control*, 38(4), 587–594, 1993.

[12] Middleton, R.H. and Goodwin, *Digital Control and Estimation: A Unified Approach*, Prentice Hall, Englewood Cliffs, NJ, 1990.

[13] Military Specifications-Flying Qualities of Piloted Airplanes, 1980, MIL-F-8785C.

[14] O'Brien, M.J. and Broussard, J.R., Feedforward control to track the output of a forced model, *Proc. 17th IEEE Conf. Decision Control*, San Diego, 1978, 1149–1155.

[15] Piou, J.E. and Sobel, K.M., Robust sampled data eigenstructure assignment using the delta operator, *Proc. Guidance, Navigation, and Control Conf.*, Hilton Head, SC, 1992.

[16] Piou, J.E., Sobel, K.M., and Shapiro, E.Y., Robust Lyapunov constrained sampled data eigenstructure assignment using the delta operator with application to flight control design, *Proc. First IEEE Conf. Control Appl.*, Dayton, Ohio, 1992, 1000–1005.

[17] Press, W.H., Flannery, B.P., Teukolsky, S.A., and Vetterling, W.T., *Numerical Recipes: The Art of Scientific Computing*, Cambridge University Press, New York, 1987.

[18] Schulz, M.J. and Inman, D.J., Eigenstructure assignment and controller optimization for mechanical systems, *IEEE Trans. Control Syst. Tech.*, 2(2), 88–100, 1994.

[19] Sobel, K.M., Banda, S.S., and Yeh, H.-H., Robust control for linear systems with structured state space uncertainty, *Int. J. Control*, 50, 1991–2004, 1989.

[20] Sobel, K.M. and Shapiro, E.Y., Application of eigensystem assignment to lateral translation and yaw pointing flight control, *Proc. 23rd IEEE Conf. Decision Control*, Las Vegas, 1984, 1423–1428.

[21] Sobel, K.M. and Shapiro, E.Y., A design methodology for pitch pointing flight control systems, *J. Guidance, Control, and Dyn.*, 8, 181–187, 1985.

[22] Sobel, K.M. and Shapiro, E.Y., Application of eigenstructure assignment to flight control design: some extensions, *J. Guidance, Control, Dyn.*, 10, 73–81, 1987.

[23] Sobel, K.M. and Shapiro, E.Y., An extension to the pseudocontrol strategy with application to an eigenstructure assignment yaw pointing/lateral translation control law, *Proc. 30th IEEE Conf. Decision Control*, Brighton, U.K., 1991, 515–516.

[24] Sobel, K.M. and Lallman, F.J., Eigenstructure assignment for the control of highly augmented aircraft, *J. Guidance, Control, and Dyn.*, 12, 318–324, 1989.

[25] Sobel, K.M., Yu, W., and Lallman, F.J., Eigenstructure assignment with gain suppression using eigenvalue and eigenvector derivatives, *J. Guidance, Control, Dyn.*, 13, 1008–1013, 1990.

[26] Srinathkumar, S., Eigenvalue/Eigenvector Assignment Using Output Feedback, *IEEE Trans. Autom. Control*, AC-23(1), 79–81, 1978.

[27] Yedavalli, R.K., Perturbation bounds for robust stability in linear state space models, *Int. J. Control*, 42, 1507–1517, 1985.

[28] Yu, W., Piou, J.E., and Sobel, K.M., Robust eigenstructure assignment for the extended medium range air to air missile, *Automatica*, 29(4), 889–898, 1993.

Further Reading

Kautsky et al. [7] suggested that eigenstructure assignment can be used to obtain a design with eigenvalues, which are least sensitive to parameter variation, by reducing one of several sensitivity measures. Among these measures are the quadratic norm condition number of the closed-loop modal matrix and the sum of the squares of the quadratic norms of the left eigenvectors. Burrows et al. [2] have proposed a stability augmentation system for a well-behaved light aircraft by using eigenstructure assignment with an optimization which minimizes the condition number of the closed-loop modal matrix. Such an approach may sometimes produce an acceptable controller.

In contrast to eigensensitivity, Doyle and Stein [4] have characterized the stability robustness of a multi-input, multioutput system by the minimum of the smallest singular value of the return difference matrix at the plant input or output. Gavito and Collins [5] have proposed an eigenstructure assignment design in which a constraint is placed on the minimum of the smallest singular value of the return difference matrix at the inputs of both an L-1011 aircraft and a CH-47 helicopter.

Other applications of eigenstructure assignment include air-to-air missiles [28], mechanical systems [18], and power systems [11].

APPENDIX

Data for the F-18 HARV Lateral Directional Dynamics at M = 0.38, H = 5000 ft., and $\alpha = 5°$

$$A = \begin{bmatrix} -30.0000 & 0 & 0 & 0 & 0 & 0 & 0 & 0 \\ 0 & -30.0000 & 0 & 0 & 0 & 0 & 0 & 0 \\ 0 & 0 & -30.0000 & 0 & 0 & 0 & 0 & 0 \\ -0.0070 & -0.0140 & 0.0412 & -0.1727 & 0.0873 & -0.9946 & 0.0760 & 0 \\ 15.3225 & 12.0601 & 2.2022 & -11.0723 & -2.1912 & 0.7096 & 0 & 0 \\ -0.3264 & 0.2041 & -1.3524 & 2.1137 & -0.0086 & -0.1399 & 0 & 0 \\ 0 & 0 & 0 & 0 & 1.0000 & 0.0875 & 0 & 0 \\ 0 & 0 & 0 & 0 & 0 & 0.5000 & 0 & -0.5 \end{bmatrix}$$

$$B = \begin{bmatrix} 30.0000 & 0 & 0 \\ 0 & 30.0000 & 0 \\ 0 & 0 & 30.0000 \\ 0 & 0 & 0 \\ 0 & 0 & 0 \\ 0 & 0 & 0 \\ 0 & 0 & 0 \\ 0 & 0 & 0 \end{bmatrix}$$

$$C = \begin{bmatrix} 0 & 0 & 0 & 0 & 0 & 1 & 0 & -1 \\ 0 & 0 & 0 & 0 & 1 & 0 & 0 & 0 \\ 0 & 0 & 0 & 1 & 0 & 0 & 0 & 0 \\ 0 & 0 & 0 & 0 & 0 & 0 & 1 & 0 \end{bmatrix}$$

Data for the FPCC Lateral Directional Dynamics

$$A = \begin{bmatrix} -0.340 & 0.0517 & 0.001 & -0.997 & 0 \\ 0 & 0 & 1 & 0 & 0 \\ -2.69 & 0 & -1.15 & 0.738 & 0 \\ 5.91 & 0 & 0.138 & -0.506 & 0 \\ -0.340 & 0.0517 & 0.001 & 0.0031 & 0 \end{bmatrix}$$

$$B = \begin{bmatrix} 0.0755 & 0 & 0.0246 \\ 0 & 0 & 0 \\ 4.48 & 5.22 & -0.742 \\ -5.03 & 0.0998 & 0.984 \\ 0.0755 & 0 & 0.0246 \end{bmatrix}$$

39

Linear Quadratic Regulator Control

Leonard Lublin
Space Engineering Research Center, Massachusetts Institute of Technology, Cambridge, MA

Michael Athans
Space Engineering Research Center, Massachusetts Institute of Technology, Cambridge, MA

39.1 Introduction

The linear quadratic regulator problem, commonly abbreviated as LQR, plays a key role in many control design methods. Not only is LQR a powerful design method, but in many respects it is also the mother of many current, systematic control design procedures for linear multiple-input, multiple output (MIMO) systems. Both the linear quadratic Gaussian, LQG or $\mathcal{H}_2$, and $\mathcal{H}_\infty$ controller design procedures have a usage and philosophy that are similar to the LQR methodology. As such, studying the proper usage and philosophy of LQR controllers is an excellent way to begin building an understanding of even more powerful design procedures.

From the time of its conception in the 1960s, the LQR problem has been the subject of volumes of research. Yet, as will be detailed, the LQR is nothing more than the solution to a convex, least squares optimization problem that has some very attractive properties. Namely the optimal controller automatically ensures a stable closed-loop system, achieves guaranteed levels of stability robustness, and is simple to compute. To provide a control systems engineer with the knowledge needed to take advantage of these attractive properties, this chapter is written in a tutorial fashion and from a user's point of view, without the complicated theoretical derivations of the results. A more in-depth reference that contains the theory for much of the material presented here is the text by Anderson and Moore [1], while the text by Kwakernaak and Sivan [2] is considered to be the classic text on the subject.

In addition to the standard LQR results, robustness properties and useful variations of the LQR are also discussed. In particular, we describe how sensitivity weighted LQR (SWLQR) can be used to make high-gain LQR controllers robust to parametric modeling errors, as well as how to include frequency domain specifications in the LQR framework through the use of frequency weighted cost functionals. Since this chapter can be viewed as an introduction to optimal control for MIMO, linear, time-invariant systems, we briefly discuss the full state feedback $\mathcal{H}_\infty$ controller and its relation to a mini-max quadratic optimal control problem for completeness.

39.2 The Time-Invariant LQR Problem

To begin, we simply state the LQR problem, its solution, and the assumptions used in obtaining the solution. The rest of this chapter is devoted to discussing the properties of the controller, how the mathematical optimization problem relates to the physics of control systems, and useful variations of the standard LQR.

THEOREM 39.1 *[Steady State LQR] Given the system dynamics*

$$\dot{x}(t) = Ax(t) + Bu(t)\,; \quad x(t=0) = x_0 \tag{39.1}$$

with $x(t) \in \mathbb{R}^n$ and $u(t) \in \mathbb{R}^m$ along with a linear combination of states to keep small

$$z(t) = Cx(t) \tag{39.2}$$

with $z(t) \in \mathbb{R}^p$. We define a quadratic cost functional

$$J = \int_0^\infty \left[z^T(t)z(t) + u^T(t)Ru(t)\right] dt \tag{39.3}$$

in which the size of the states of interest, $z(t)$, is weighted relative to the amount of control action in $u(t)$ through the weighting matrix

0-8493-8570-9/96/$0.00+$.50

R.

If the following assumptions hold:

1. *The entire state vector $x(t)$ is available for feedback.*
2. *$[A \quad B]$ is stabilizable and $[A \quad C]$ is detectable.*
3. *$R = R^T > 0$*

Then

1. *The linear quadratic controller is the unique, optimal, full state feedback control law*

$$u(t) = -Kx(t) \qquad with \qquad K = R^{-1}B^T S \tag{39.4}$$

that minimizes the cost, J, subject to the dynamic constraints imposed by the open-loop dynamics in Equation 39.1.

2. *S is the unique, symmetric, positive semidefinite solution to the algebraic Riccati equation*

$$SA + A^T S + C^T C - SBR^{-1}B^T S = 0 \tag{39.5}$$

3. *The closed-loop dynamics arrived at by substituting Equation 39.4 into Equation 39.1*

$$\dot{x}(t) = [A - BK]x(t) \tag{39.6}$$

are guaranteed to be asymptotically stable.

4. *The minimum value of the cost J in Equation 39.3 is $J = x_0^T S x_0$.*

While the theorem states the LQR result, it is still necessary to discuss how to take advantage of it. Before doing so, note that the control gains, K from Equation 39.4, can be readily computed for large-order systems using standard software packages such as MATLAB and Matrix$_X$. Hence, the remainder of this chapter is really about how the values of the design variables used in the optimization problem influence the behavior of the controllers. Here the design variables are z, the states to keep small, and R, the control weighting matrix. We begin the discussion by describing the physical motivation behind the optimization problem.

39.2.1 Physical Motivation for the LQR

The LQR problem statement and cost can be motivated in the following manner. Suppose that the system Equation 39.1 is initially excited, and that the net result of this excitation is reflected in the initial state vector x_0. This initial condition can be regarded as an undesirable deviation from the equilibrium position of the system, $x(t) = 0$. Given these deviations, the objective of the control can essentially be viewed as selecting a control vector $u(t)$ that regulates the state vector $x(t)$ back to its equilibrium position of $x(t) = 0$ as quickly as possible.

If the system Equation 39.1 is controllable, then it is possible to drive $x(t)$ to zero in an arbitrarily short period of time. This would require very large control signals which, from an engineering point of view, are unacceptable. Large control signals will saturate actuators and if implemented in a feedback form will require high-bandwidth designs that may excite unmodeled dynamics. Hence, it is clear that there must be a balance between the desire to regulate perturbations in the state to equilibrium and the size of the control signals needed to do so.

Minimizing the quadratic cost functional from Equation 39.3 is one way to quantify our desire to achieve this balance. Realize that the quadratic nature of both terms in the cost

$$u^T(t)Ru(t) > 0 \qquad \text{for} \qquad u(t) \neq 0 \tag{39.7}$$

$$z^T(t)z(t) = x^T(t)C^T Cx(t) \geq 0 \qquad \text{for} \qquad x(t) \neq 0 \tag{39.8}$$

ensures that they will be non-negative for all t. Further since the goal of the control law is to make the value of the cost as small as possible, larger values of the terms (39.7) and (39.8) are penalized more heavily than smaller ones. More specifically, the term (39.7) represents a penalty that helps the designer keep the magnitude of $u(t)$ "small". Hence the matrix R, which is often called the control weighting matrix, is the designer's tool which influences how "small" $u(t)$ will be. Selecting large values of R leads to small values of $u(t)$, which is also evident from the control gain K given in Equation 39.4.

The other term, Equation 39.8, generates a penalty in the cost when the states that are to be kept small, $z(t)$, are different from their desired equilibrium value of zero. The selection of which states to keep small, that is the choice of C in Equation 39.2, is the means by which the control system designer communicates to the mathematics the relative importance of individual state variable deviations. That is, which errors are bothersome and to what degree they are so.

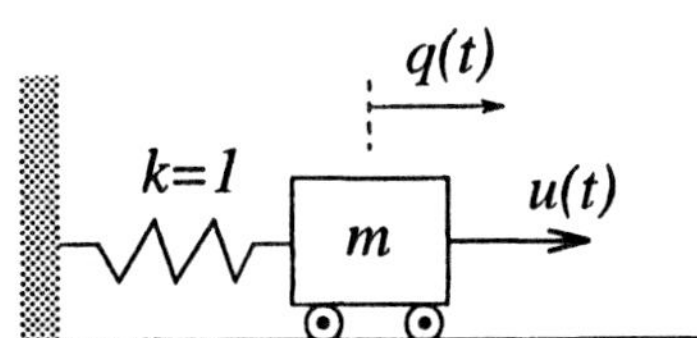

Figure 39.1 Single degree of freedom oscillator with mass m, spring stiffness $k = 1$, and control force input $u(t)$.

EXAMPLE 39.1:

Consider the single degree of freedom, undamped, harmonic oscillator shown in Figure 39.1. Using Newton's laws, the equations of motion are

$$m\ddot{q}(t) + kq(t) = u(t) \tag{39.9}$$

Letting $x_1(t) = q(t)$ denote the position of the mass and $x_2(t) = \dot{q}(t)$ its velocity, the dynamics (Equation 39.9) can be expressed in the state space as

$$\dot{x}_1(t) \quad = \quad x_2(t) \tag{39.10}$$

$$\dot{x}_2(t) = -\omega^2 x_1(t) + \omega^2 u(t)$$

or

$$\dot{x}(t) = Ax(t) + Bu(t)$$

with

$$x(t) = \begin{bmatrix} x_1(t) \\ x_2(t) \end{bmatrix} \quad A = \begin{bmatrix} 0 & 1 \\ -\omega^2 & 0 \end{bmatrix}$$

$$B = \begin{bmatrix} 0 \\ \omega^2 \end{bmatrix}$$

where $\omega^2 = 1/m$ is the square of the natural frequency of the system with $k = 1$.

In the absence of control, initial conditions will produce a persistent sinusoidal motion of the mass at a frequency of ω rad/sec. As such, we seek a control law that regulates the position of the mass to its equilibrium value of $q(t) = 0$. Thus, we define the states of interest, z, as $x_1(t)$.

$$z(t) = Cx(t) \quad \text{with} \quad C = \begin{bmatrix} 1 & 0 \end{bmatrix} \tag{39.11}$$

Since the control is scalar, the cost (Equation 39.3) takes the form

$$J = \int_0^\infty \left[z^2(t) + \rho u^2(t)\right] \mathrm{d}t \quad \text{with} \quad \rho > 0$$

where ρ is the control weighting parameter.

The optimal LQR control law takes the form $u(t) = -Kx(t)$ where the LQ gain K is a row vector $K = \begin{bmatrix} k_1 & k_2 \end{bmatrix}$. To determine the value of the gain, we must solve the algebraic Riccati equation (39.5). Recalling that we want a symmetric solution to the Riccati equation, letting

$$S = \begin{bmatrix} S_{11} & S_{12} \\ S_{12} & S_{22} \end{bmatrix} \tag{39.12}$$

and using the values of A, B, and C given above, the Riccati equation (39.5) becomes

$$\begin{aligned} 0 = & \begin{bmatrix} S_{11} & S_{12} \\ S_{12} & S_{22} \end{bmatrix} \begin{bmatrix} 0 & 1 \\ -\omega^2 & 0 \end{bmatrix} \\ & + \begin{bmatrix} 0 & -\omega^2 \\ 1 & 0 \end{bmatrix} \begin{bmatrix} S_{11} & S_{12} \\ S_{12} & S_{22} \end{bmatrix} + \begin{bmatrix} 1 \\ 0 \end{bmatrix} \begin{bmatrix} 1 & 0 \end{bmatrix} \\ & - \frac{1}{\rho} \begin{bmatrix} S_{11} & S_{12} \\ S_{12} & S_{22} \end{bmatrix} \begin{bmatrix} 0 \\ \omega^2 \end{bmatrix} \begin{bmatrix} 0 & \omega^2 \end{bmatrix} \begin{bmatrix} S_{11} & S_{12} \\ S_{12} & S_{22} \end{bmatrix} \end{aligned}$$

Carrying out the matrix multiplications leads to the following three equations in the three unknown S_{ij} from Equation 39.12.

$$0 = 2\omega^2 S_{12} - 1 + \frac{1}{\rho}\omega^4 S_{12}^2 \tag{39.13}$$

$$0 = -S_{11} + \omega^2 S_{22} + \frac{1}{\rho}\omega^4 S_{12} S_{22} \tag{39.14}$$

$$0 = -2S_{12} + \frac{1}{\rho}\omega^4 S_{22}^2 \tag{39.15}$$

Solving Equation 39.13 for S_{12} and simplifying yields

$$S_{12} = \frac{-\rho \pm \rho\sqrt{1 + 1/\rho}}{\omega^2}$$

Both the positive and negative choices for S_{12} are valid solutions of the Riccati equation. While we are only interested in the positive semidefinite solution, we still need more information to resolve which choice of S_{12} leads to the unique choice of $S \geq 0$. Rewriting Equation 39.15 as

$$2S_{12} = \frac{1}{\rho}\omega^4 S_{22}^2 \tag{39.16}$$

indicates that S_{12} must be positive to satisfy the equality of Equation 39.16, since the right-hand side of the equation must always be positive. Equation 39.16 indicates that there will also be a $\pm$ sign ambiguity in selecting the appropriate S_{22}. To resolve the ambiguity we use Sylvester's Test, which says that for $S \geq 0$ both

$$S_{11} \geq 0 \quad \text{and} \quad S_{11}S_{22} - S_{12}^2 \geq 0 \tag{39.17}$$

Solving Equation 39.15 and 39.13 using the relations in Equation 39.17, which clearly show that $S_{22} > 0$, gives the remaining elements of S

$$\begin{aligned} S_{11} &= \frac{\rho}{\omega}\sqrt{2(1 + 1/\rho)(\sqrt{1 + 1/\rho} - 1)} \\ S_{12} &= \frac{-\rho + \rho\sqrt{1 + 1/\rho}}{\omega^2} \\ S_{22} &= \frac{\rho}{\omega^3}\sqrt{2(\sqrt{1 + 1/\rho} - 1)} \end{aligned}$$

The final step in computing the controller is to evaluate the control gains K from Equation 39.4. Doing so gives

$$k_1 = \sqrt{1 + 1/\rho} - 1 \qquad k_2 = \frac{1}{\omega}\sqrt{2(\sqrt{1 + 1/\rho} - 1)}$$

Given the control gains as a function of the control weighting ρ it is useful to examine the locus of the closed-loop poles for the system as ρ varies over $0 < \rho < \infty$. Evaluating the eigenvalues of the closed-loop dynamics from Equation 39.6 leads to a pair of complex conjugate closed-loop poles

$$\begin{aligned} \lambda_{1,2} = & -\frac{\omega}{2}\sqrt{2(\sqrt{1 + 1/\rho} - 1)} \\ & \pm \jmath\frac{\omega}{2}\sqrt{2(\sqrt{1 + 1/\rho} + 1)}. \end{aligned}$$

A plot of the poles as ρ varies over $0 < \rho < \infty$ is shown in Figure 39.1. Notice that for large values of ρ, the poles are near their open-loop values at $\pm \jmath\omega$, and as ρ decreases the poles move farther out into the left half-plane. This is consistent with how ρ influences the cost. Large values of ρ place a heavy penalty on the control and lead to low gains with slow transients, while small values of ρ tell the mathematics that large control gains with fast transients are acceptable.

39.2.2 Designing LQR Controllers

While the above section presents all the formulas needed to start designing MIMO LQR controllers, the point of this section is to inform the potential user of the limitations of the methodology.

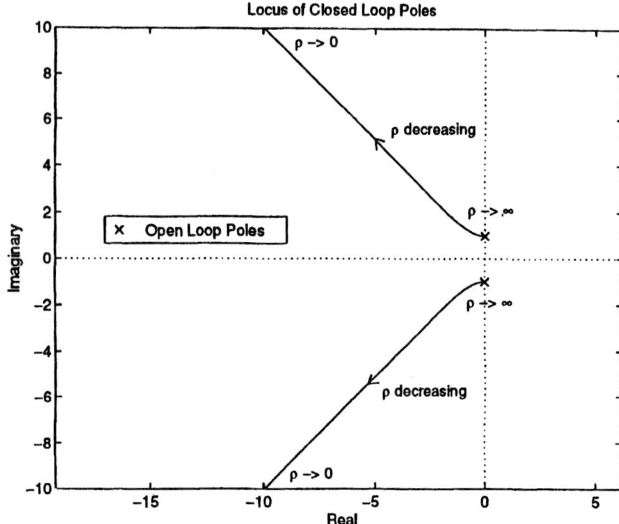

Figure 39.2 Locus of closed-loop pole locations for the single degree of freedom oscillator with $\omega = 1$ as ρ varies over $0 < \rho < \infty$.

The most restrictive aspects of LQR controllers is that they are full-state feedback controllers. This means that every state that appears in the model of the physical system Equation 39.1 must be measured by a sensor. In fact, the notation of $z(t) = Cx(t)$ for the linear combination of states that are to be regulated to zero is deliberate. We do not call $z(t)$ the outputs of the system because all the states $x(t)$ must be measured in real time to implement the control law Equation 39.4.

Full-state feedback is appropriate and can be applied to systems whose dynamics are described by a finite set of differential equations and whose states can readily be measured. An aircraft in steady, level flight is an example of a system whose entire state can be measured with sensors. In fact, LQR control has been used to design flight control systems for modern aircraft.

On the other hand, full-state feedback is typically not appropriate for flexible systems. The dynamics of flexible systems are described by partial differential equations that often require very high-order, if not infinite-dimensional, state-space models. As such, it is not feasible to measure all the states of systems that possess flexible dynamics. Returning to the aircraft example, one could not use a LQR controller to regulate the aerodynamically induced vibrations of the aircraft's wing. The number of sensors that would be needed to measure the state of the vibrating wing prohibit this.

Having discussed the implications of full-state feedback, the next restrictive aspect of LQR to discuss is the gap between what the LQR controller achieves and the desired control system performance. Recall that the LQR is the control that minimizes the quadratic cost of Equation 39.3 subject to the constraints imposed by the system dynamics. This optimization problem and the resulting optimal controller have very little to do with more meaningful control system specifications like levels of disturbance rejection, overshoot in tracking, and stability margins. This gap must always be kept in mind when using LQR to design feedback controllers. The fact that LQR controllers are in some sense optimal is of no consequence if they do not meet the performance goals. Further since control system specifications are not given in terms of minimizing quadratic costs, it becomes the job of the designer to use the LQR tool wisely. To do so it helps to adopt a means-to-an-end design philosophy where the LQR is viewed as a tool, or the means, used to achieve the desired control system performance, or the end.

One last issue to point out is that LQR controller design is an iterative process even though the methodology systematically produces optimal, stabilizing controllers. Since the LQR formulation does not directly allow one to achieve standard control system specifications, trial and error iteration over the values of the weights in the cost is necessary to arrive at satisfactory controllers. Typically LQR designs are carried out by choosing values for the design weights, synthesizing the control law, evaluating how well the control law achieves the desired robustness and performance, and iterating through this process until a satisfactory controller is found. In the sequel, various properties and weight selection tools will be presented that provide good physical and mathematical guidance for selecting values for the LQR design variables. Yet these are only guides, and as such they will not eliminate the iterative nature of LQR controller design.

39.3 Properties of LQR

The LQR has several very important properties, which we summarize below. It is important to stress that the properties of LQR designs hinge upon the fact that full-state feedback is used and the specific way that the control gain matrix K is computed from the solution of the Riccati equation.

39.3.1 Robustness Properties

To visualize the robustness properties of LQR controllers, it is necessary to consider the loop transfer function matrix that results when implementing the control law of Equation 39.4. The LQR loop transfer function matrix, denoted by $G_{LQ}(s)$, induced by the control scheme of Equation 39.4 is given by

$$G_{LQ}(s) = K(sI - A)^{-1}B$$

and the closed-loop dynamics of Equation 39.6 using this representation are shown in Figure 39.3. An interesting fact is that $G_{LQ}(s)$ is always square and minimum phase. Note that as a consequence of this feedback architecture any unstructured modeling errors must be reflected to the inputs of the LQR loop, location 1 in Figure 39.3.

Under the *assumption that the control weight matrix* $R = R^T > 0$ *is diagonal,* the LQR loop transfer matrix is guaranteed to satisfy both of the inequalities

$$\begin{aligned} \sigma_{\min}\left[I + G_{LQ}(j\omega)\right] &\geq 1 \quad \forall\omega \\ \sigma_{\min}\left[I + G_{LQ}(j\omega)^{-1}\right] &\geq \tfrac{1}{2} \quad \forall\omega \end{aligned} \tag{39.18}$$

where $\sigma_{\min}$ denotes the minimum singular value. Since the multivariable robustness properties of any design depend on the

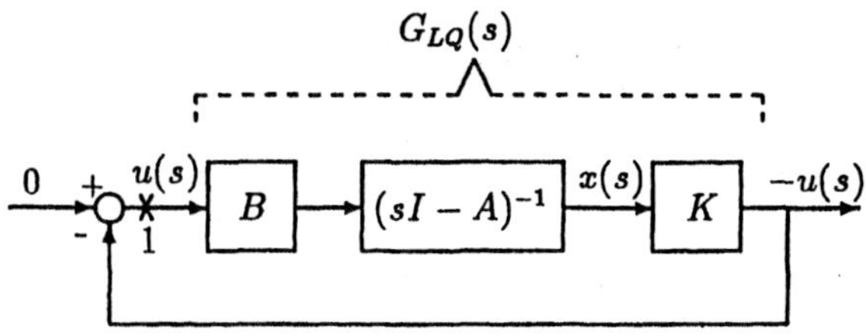

Figure 39.3 The LQR loop transfer matrix, $G_{LQ}(s) = K(sI - A)^{-1}B$.

size of $\sigma_{\min}[I + G(j\omega)]$ and $\sigma_{\min}\left[I + G(j\omega)^{-1}\right]$ the following guaranteed multivariable gain and phase margins are inherent to LQR controllers as a result of Equation 39.18.

LQR Stability Robustness Properties

1. Upward gain margin is infinite
2. Downward gain margin is at least 1/2
3. Phase margin is at least $\pm 60°$

These gain and phase margins can occur independently and simultaneously in all m control channels. To visualize this, consider Figure 39.4, where the $f_i(\cdot)$ can be viewed as perturbations to the individual inputs, $u_i = f_i(\cdot)\mu_i$. As a result of the gain margin properties (1) and (2), the LQR system shown in Figure 39.4 is guaranteed to be stable for any set of scalar gains β_i with $f_i = \beta_i$ where the β_i lie anywhere in the range $1/2 < \beta_i < \infty$. The phase margin property (3) ensures the LQR system shown in Figure 39.4 is guaranteed to be stable for any set of scalar phases ϕ_i with $f_i = e^{j\phi_i}$ where the ϕ_i can lie anywhere in the range $-60° < \phi_i < +60°$.

These inherent robustness properties of LQR designs are useful in many applications. To further appreciate what they mean, consider a single input system with a single variable we wish to keep small

$$\dot{x}(t) = Ax(t) + bu(t) \qquad \text{with} \qquad z(t) = c^T x(t)$$

and let the control weight be a positive scalar, $R = \rho > 0$. Then the resulting LQR loop transfer function is scalar, and its robustness properties can be visualized by plotting the Nyquist diagram of $G_{LQ}(j\omega)$ for all ω. Figure 39.5 contains a Nyquist plot for a scalar $G_{LQ}(j\omega)$ that illustrates why the LQR obtains good gain and phase margins. Essentially, inequality Equation 39.18 guarantees that the Nyquist plot will not penetrate the unit circle centered at the critical point at (-1,0).

39.3.2 Asymptotic Properties of LQR Controllers

It is clear that the closed-loop poles of the LQR design will depend upon the values of the design parameters $z(t)$ and R. The exact numerical values of the closed-loop poles can only be determined using a digital computer, since we have to solve the Riccati equation (39.5). However, it is possible to qualitatively predict the asymptotic behavior of the closed-loop poles for the LQR as the size of the control gain is varied without solving the

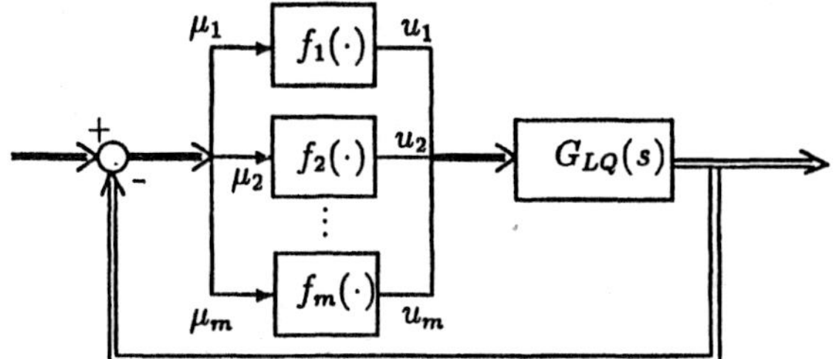

Figure 39.4 LQR loop used to visualize the guaranteed robustness properties.

Riccati equation, evaluating the control gains, and computing the closed-loop poles.

Assumptions for Asymptotic LQR Properties

1. The number of variables to keep small is equal to the number of controls. That is $\dim[z(t)] = \dim[u(t)] = m$.
2. The control weight is chosen such that $R = \rho\tilde{R}$ where ρ is a positive scalar and $\tilde{R} = \tilde{R}^T > 0$.
3. $G_z(s)$ is defined to be the square transfer function matrix between the variables we wish to keep small and the controls with the loop open, $z(s) = G_z(s)u(s)$ where $G_z(s) = C(sI - A)^{-1}B$, and q is the number of transmission zeros of $G_z(s)$.

Adjusting ρ directly influences the feedback control gain. When $\rho \to \infty$, we speak of the LQR controller as having low gain since under Assumption (2)

$$u(t) = -\frac{1}{\rho}\tilde{R}^{-1}B^T Sx(t) \to 0$$

as $\rho \to \infty$. Likewise, when $\rho \to 0$ we speak of the high gain controller since $u(t)$ will clearly become large. The asymptotic properties of the LQR closed-loop poles under the stated assumptions are as follows.

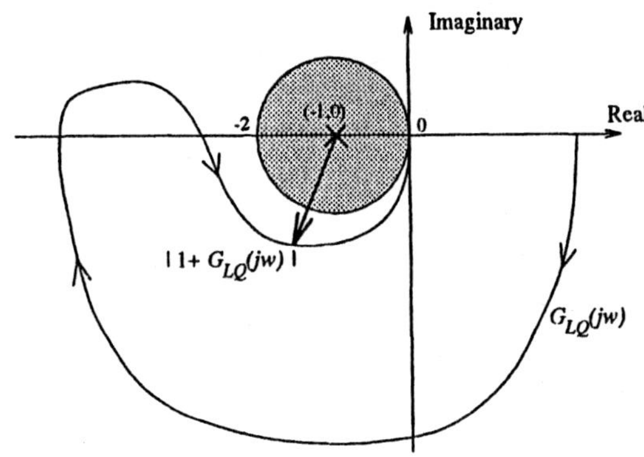

Figure 39.5 A typical Nyquist plot for a single-input, single-output LQR controller.

LQR Asymptotic Properties

1. **Low Gain:** As $\rho \to \infty$, the closed-loop poles start at
   a. The stable open-loop poles of $G_z(s)$
   b. The mirror image, about the $j\omega$-axis, of any unstable open-loop poles of $G_z(s)$
   c. If any open-loop poles of $G_z(s)$ lie exactly on the $j\omega$-axis, the closed-loop poles start just to the left of them
2. **High Gain:** As $\rho \to 0$, the closed-loop poles will
   a. Go to cancel any minimum phase zeros of $G_z(s)$
   b. Go to the mirror image, about the $j\omega$-axis, of any non-minimum phase zeros of $G_z(s)$
   c. The rest will go off to infinity along stable Butterworth patterns

Realize that these rules are not sufficient for constructing the entire closed-loop pole root locus. However, they do provide good insight into the asymptotic behavior of LQR controllers. In particular, the second rule highlights how the choice of $z(t)$, which defines the zeros of $G_z(s)$, influences the closed-loop behavior of LQR controllers.

Extensions of these results exist for the case where Assumption (1) is not satisfied, that is when $\dim[z(t)] \neq \dim[u(t)]$. When $\dim[z(t)] \neq \dim[u(t)]$, the low gain asymptotic result will still hold, which is to be expected. The difficulty for non-square $G_z(s)$ arises in the high-gain case as $\rho \to 0$ because it is not a simple manner to evaluate the zeros that the closed-loop poles will go toward. For details of how to evaluate the zero locations that the closed-loop poles will go toward in the cheap control limit, one can refer to [3] and [2, *problem 3.14*].

39.4 LQR Gain Selection Tools

In this section we present some variations on the standard LQR problem given in Theorem 39.1. It is best to view the following variations on LQR as tools a designer can use to arrive at controllers that meet a set of desired control system specifications. While the tools are quite useful, the choice of which tool to use is very problem specific. Hence, we also present some physical motivation to highlight when the various tools might be useful. Realize that these tools do not remove the iterative nature of LQR control design. In fact, these tools are only helpful when used iteratively and when used with an understanding of the underlying physics of the design problem and the limitations of the design model. The variations on LQR presented here are by no means a complete list.

39.4.1 Cross-Weighted Costs

Consider the linear time-invariant system with dynamics described by the state equation in Equation 39.1 and the following set of variables we wish to keep small

$$z(t) = Cx(t) + Du(t) \tag{39.19}$$

Substituting Equation 39.19 into the quadratic cost from Equation 39.3 produces the cost function

$$J = \int_0^\infty \Big\{x^T(t)C^TCx(t) + 2x^T(t)C^TDu(t) + u^T(t)\left[R + D^TD\right]u(t)\Big\}\, dt \tag{39.20}$$

which contains a cross penalty on the state and control, $2x^T(t)C^T Du(t)$. The solution to the LQR controller that minimizes the cost of Equation 39.20 will be presented here, but first we give some motivation for such a cost and choice of performance variables.

Cross-weighted cost functions and performance variables with control feed-through terms are common. The control feed-through term, $Du(t)$ in Equation 39.19, arises physically when the variables we wish to keep small are derivatives of the variables that appear in the state vector of the open-loop dynamics Equation 39.1. To see this, consider the single degree of freedom oscillator from Example 39.1 where the acceleration of the mass is the variable we wish to keep small, $z(t) = \ddot{q}(t)$. Using the definitions from the example and Equations 39.11, this can be expressed as

$$\begin{aligned} z(t) &= \ddot{q}(t) = \dot{x}_2(t) \\ &= -\omega^2 x_1(t) + \omega^2 u(t) \\ &= [-\omega^2 \quad 0]x(t) + \omega^2 u(t) \end{aligned}$$

Clearly, penalizing acceleration requires the control feed-through term in the definition of the variables we wish to keep small. Control feed-through terms also arise when penalizing states of systems whose models contain neglected or unknown, higher-order modes. Such models often contain control feed-through terms to account for the low-frequency components of the dynamics of the modes that are not present in the design model. Further, as will be seen below, cross-coupled costs such as Equation 39.20 arise when frequency-dependent weighting terms are used to synthesize LQR controllers.

THEOREM 39.2 *[Cross-Weighted LQR] Consider the system dynamics*

$$\dot{x}(t) = Ax(t) + Bu(t) \qquad x(t=0) = x_0 \tag{39.21}$$

with $x(t) \in \mathbb{R}^n$ and $u(t) \in \mathbb{R}^m$ and the quadratic cost with a cross penalty on state and control

$$J = \int_0^\infty \Big[x^T(t)R_{xx}x(t) + 2x^T(t)R_{xu}u(t) + u^T(t)R_{uu}u(t)\Big]\, dt \tag{39.22}$$

in which the size of the states is weighted relative to the amount of *control action in $u(t)$ through the state weighting matrix R_{xx}, the control weighting matrix R_{uu}, and the cross weighting matrix R_{xu}.*

If the following assumptions hold:

1. *The entire state vector $x(t)$ is available for feedback*
2. $R_{xx} = R_{xx}^T \geq 0$, $R_{uu} = R_{uu}^T > 0$, *and* $\begin{bmatrix} R_{xx} & R_{xu} \\ R_{xu}^T & R_{uu} \end{bmatrix} \geq 0$
3. $[A \quad B]$ *is stabilizable and* $[A \quad R_{xx}^{1/2}]$ *is detectable*[1]

Then

1. *The linear quadratic controller is the unique, optimal, full-state feedback control law*
$$\begin{aligned} u(t) &= -Kx(t) \quad \text{with} \\ K &= R_{uu}^{-1}\left(R_{xu}^T + B^T S\right) \end{aligned} \tag{39.23}$$
that minimizes the cost, J of Equation 39.22, subject to the dynamic constraints imposed by the open-loop dynamics in Equation 39.21.
2. *Defining $A_r = \left(A - BR_{uu}^{-1}R_{xu}^T\right)$, S is the unique, symmetric, positive semi-definite solution to the algebraic Riccati equation*
$$\begin{aligned} SA_r \;+\; A_r^T S + \left(R_{xx} - R_{xu}R_{uu}^{-1}R_{xu}^T\right) \\ - SBR_{uu}^{-1}B^T S = 0 \end{aligned} \tag{39.24}$$
3. *The closed-loop dynamics arrived at by substituting Equation 39.23 into Equation 39.21 are guaranteed to be asymptotically stable*

Note that when the cost is defined by Equation 39.3 and the variables to be kept small are defined by Equation 39.19, $R_{xx} = C^T C$, $R_{xu} = C^T D$, and $R_{uu} = \left[R + D^T D\right]$ as can be seen from Equation 39.20. Also, if $R_{xu} = 0$ this theorem reduces to the standard LQR result presented in Theorem 39.1.

The assumptions used in Theorem 39.2 are essentially those used in the standard LQR result. One difference is that here we require $R_{xx} = R_{xx}^T \geq 0$, while in Theorem 39.1 this was automatically taken care of by our definition of the state cost as $x^T(t)C^T Cx(t)$ because $C^T C \geq 0$. Unfortunately, the guaranteed robustness properties presented in the previous section are not necessarily true for the LQR controllers that minimize the cost (Equation 39.22).

39.4.2 The Stochastic LQR Problem

In the stochastic LQR problem, we allow the inclusion of a white noise process in the state dynamics. Namely, the state vector $x(t)$ now satisfies the stochastic differential equation

$$\dot{x}(t) = Ax(t) + Bu(t) + L\xi(t) \tag{39.25}$$

where $\xi(t)$ is a vector valued, zero-mean, white noise process. We still assume that all the state variables can be measured exactly and used for feedback as necessary. Physically, the term $L\xi(t)$ in Equation 39.25 should be viewed as the impact of a persistent set of disturbances on a system that corrupt the desired equilibrium value of the state, $x(t) = 0$. The L matrix describes how the disturbances impact the system, and the stochastic process $\xi(t)$ captures the dynamics of how the disturbances evolve with time. In particular, since the disturbances are a white noise process, they are completely unpredictable from moment to moment.

As with the standard LQR problem, we would like to find the control law that minimizes some quadratic cost functional which captures a trade off between state cost and control cost. However, the cost functional for the standard LQR problem, Equation 39.3, cannot be used here as a result of the stochastic nature of the states. The value of the cost functional would be infinity because of the continuing excitation from $\xi(t)$. To accommodate the stochastic nature of the state, we seek the optimal control that minimizes the expected value, or mean, of the standard LQR cost functional

$$J = \mathrm{E}\left\{\lim_{\tau\to\infty}\frac{1}{\tau}\int_0^{\tau}\left[z^T(t)z(t) + u^T(t)Ru(t)\right]dt\right\} \tag{39.26}$$

where E is the expected value operator. The normalization by the integration time τ is necessary to ensure that the cost functional is finite. Surprisingly, the optimal control for this stochastic version of the LQR problem is identical to that of the corresponding deterministic LQR problem. In particular, Theorem 39.1 also describes the unique optimal full-state feedback control law that minimizes the cost Equation 39.26 subject to the constraints Equation 39.25. This is a result of the fact that $\xi(t)$ is a white noise process. Since the disturbances are completely unpredictable, the optimal thing to do is ignore them. As a consequence of all this, the properties of the deterministic LQR problem from Theorem 39.1 discussed in Section 39.3 are also true for this stochastic LQR problem.

Unfortunately, these results do not make it any easier to design LQR controllers to specifically reject disturbances that directly impact the state dynamics. This is because physical disturbances are typically not white in nature, and the stochastic LQR result only applies for white noise disturbances. However, these results do play a key role in synthesizing multivariable controllers when some of the states cannot be measured and fed back for control.

39.4.3 Sensitivity Weighted LQR (SWLQR)

Synthesizing LQR controllers requires an accurate model of the system dynamics. Not only are the control gains, K, a function of the system matrices A and B, but in the high-gain limit we know that the closed-loop poles will go to cancel the zeros of $G_z(s)$. If the models upon which the control design is based are inaccurate, the closed-loop system at a minimum will not achieve the predicted performance and will, in the worst case, be unstable.

---

[1] $R_{xx}^{1/2}$ is the matrix square root of R_{xx} which always exists for positive semidefinite matrices. For the cross-coupled cost in Equation 39.20, $R_{xx} = C^T C$ and $R_{xx}^{1/2} = C$.

This is particularly relevant for systems with lightly damped poles and zeros near the $j\omega$ axis. Hence, it is important to address the issue of stability robustness, since designers often must work with models that do not capture the dynamical behavior of their systems exactly.

Modeling errors can be classified into two main groups: unmodeled dynamics and parametric errors. Unmodeled dynamics are typically the result of unmodeled non-linear behavior and neglected high-frequency dynamics which lie beyond the control bandwidth of interest. How one can design LQR controllers that are robust to unmodeled dynamics will be discussed in the following section on frequency weighted LQR. Parametric uncertainty arises when there are errors in the values of the system parameters used to form the model. Sensitivity weighted LQR, SWLQR, is a variation on the standard LQR problem that can be used to increase the stability robustness of LQR controllers to parametric modeling errors. Before presenting the SWLQR results, we will discuss parametric modeling errors.

Parametric Modeling Errors

When parametric errors exist, the model will be able to capture the system's nominal dynamics, but the model cannot capture the exact behavior. Parametric errors occur often because it is difficult to know the exact properties of the physical devices that make up a system and because the physical properties may vary with time or the environment in which they are placed. For example, consider the single degree of freedom oscillator from Example 39.1 as the model of a car tire. While we might have some knowledge of the tire's stiffness, it is unrealistic to expect to know its exact value, as the tire is a complex physical device whose properties change with wear and temperature. Hence, the value of the spring stiffness k, which models the tire's elasticity, will be parametrically uncertain.

To deal with any sort of modeling error, one must formulate a model of the uncertainty in the nominal design model. Parametrically uncertain variables may appear directly as the elements of, or be highly nonlinear functions of, the elements of the A and B matrices. When the latter is the case, one often reduces the uncertainty model to parametric uncertainty in more fundamental system quantities such as the frequency and/or damping of the system's poles and zeros. In either case one must decide which elements of the model, be they pole frequencies or elements of the A matrix, are uncertain and by how much they are uncertain.

This digression on parametric modeling errors and the presentation of the SWLQR results that follows is vital. To achieve high levels of performance, which means using high-gain control, LQR requires an accurate plant model. If a model contains parametric modeling errors and high-gain LQR control ($\rho \rightarrow 0$) is used, then the closed-loop poles will go to cancel zeros of $G_z(s)$ which will be in different locations than the true zeros. Such errors often lead to instability, and this is why we worry about parametric modeling errors.

SWLQR

Designing feedback controllers that are robust to parametric uncertainty is a very difficult problem. Research on this problem is ongoing, and a great deal of research has gone into this problem in the past, see [4] for a survey on this topic. Here we present only one of the many methods, SWLQR, that can be used to design LQR controllers that are robust to parametric modeling errors. While the treatment is by no means complete, it should give one a sense of how to deal with parametric modeling errors when designing LQR controllers.

The philosophy behind SWLQR is that it aims, as the title implies, to desensitize the LQR controller to parametric uncertainty. The advantage of this philosophy is that it produces a technique that is nothing more than a special way to choose the weighting matrices of the cross-weighted LQR problem presented in Section 39.4.1. A shortcoming of the method is that it does not result in a controller that is guaranteed to be robust to the parametric uncertainty in the model.

In presenting the SWLQR results, the vector of uncertain parameters to which you wish to desensitize a standard LQR controller will be denoted by α, and α_i will denote the ith element of α. To desensitize LQR controllers to the uncertain variables in α, a quadratic penalty on the sensitivity of the state to each of the uncertain variables α_i,

$$\frac{\partial x^T(t)}{\partial \alpha_i} R_{\alpha\alpha_i} \frac{\partial x(t)}{\partial \alpha_i} \tag{39.27}$$

is added to the cross-weighted LQR cost functional (Equation 39.22). In this expression $\frac{\partial x(t)}{\partial \alpha_i}$ is known as the sensitivity state of the system and $R_{\alpha\alpha_i}$ is the sensitivity state weighting matrix, which must be positive semidefinite. Quadratic penalties for each α_i (Equation 39.27) are included in the LQR cost functional to tell the mathematics that we care not only about deviations in the state from its equilibrium position but that we also care about how sensitive the value of the state is to the known uncertain parameters of the system. The larger the value of $R_{\alpha\alpha_i}$, the more we tell the cost that we are uncertain about the influence of parameter α_i in our model on the trajectory of the state. In turn, the larger the uncertainty in the influence of parameter α_i on the state trajectory, the more robust the resulting SWLQR controller will be to the uncertain parameter α_i.

Making certain assumptions about the dynamics of the sensitivity states leads to the SWLQR result [7, 6]. It is not necessary to understand or discuss these assumptions to design SWLQR controllers, and we thus simply present the SWLQR result.

The SWLQR Result *Given a stable open-loop system (Equation 39.21), let α denote the vector of uncertain variables to which you want to desensitize a LQR controller and define the sensitivity state weighting matrices $R_{\alpha\alpha_i}$ such that*

$$R_{\alpha\alpha_i} = R_{\alpha\alpha_i}^T \geq 0 \qquad \text{where} \qquad R_{\alpha\alpha_i} \in \mathbb{R}^{n\times n}$$

Then the SWLQR controller results from synthesizing the cross-coupled cost LQR controller from Theorem 39.2 with the following set of design weights

$$R_{xx} = \tilde{R}_{xx} + \sum_{i=1}^{n_\alpha} \frac{\partial A^T}{\partial \alpha_i} A^{-T} R_{\alpha\alpha_i} A^{-1} \frac{\partial A}{\partial \alpha_i} \quad (39.28)$$

$$R_{xu} = \tilde{R}_{xu} + \sum_{i=1}^{n_\alpha} \frac{\partial A^T}{\partial \alpha_i} A^{-T} R_{\alpha\alpha_i} A^{-1} \frac{\partial B}{\partial \alpha_i} \quad (39.29)$$

$$R_{uu} = \tilde{R}_{uu} + \sum_{i=1}^{n_\alpha} \frac{\partial B^T}{\partial \alpha_i} A^{-T} R_{\alpha\alpha_i} A^{-1} \frac{\partial B}{\partial \alpha_i} \quad (39.30)$$

In these expressions n_α denotes the number of uncertain parameters in α, and $\tilde{R}_{xx}$, $\tilde{R}_{xu}$, and $\tilde{R}_{uu}$ are the nominal values of the weights from a standard, unrobust LQR design. The matrix derivatives $\frac{\partial A}{\partial \alpha_i}$ and $\frac{\partial B}{\partial \alpha_i}$ are taken term by term. For example, $\frac{\partial A}{\partial \alpha_i}$ is a matrix of the same size as A whose elements are the partial derivatives with respect to α_i of the corresponding elements in the A matrix. For

$$A = \begin{bmatrix} A_{11} & A_{12} & \cdots & A_{1n} \\ A_{21} & & & \\ \vdots & \ddots & & \\ A_{n1} & & & A_{nn} \end{bmatrix}$$

$$\frac{\partial A}{\partial \alpha_i} = \begin{bmatrix} \frac{\partial A_{11}}{\partial \alpha_i} & \frac{\partial A_{12}}{\partial \alpha_i} & \cdots & \frac{\partial A_{1n}}{\partial \alpha_i} \\ \frac{\partial A_{21}}{\partial \alpha_i} & & & \\ \vdots & \ddots & & \\ \frac{\partial A_{n1}}{\partial \alpha_i} & & & \frac{\partial A_{nn}}{\partial \alpha_i} \end{bmatrix}$$

While it might not be obvious that the modified cost expressions (Equations 39.28 to 39.30) lead to controllers that are more robust to the uncertain parameters in α, extensive empirical and experimental results verify that they do [7]. To see this mathematically, consider the state cost, $x^T(t)R_{xx}x(t)$ from Equation 39.28, when the variables we wish to keep small are defined by $z(t) = Cx(t)$,

$$\begin{aligned} x^T(t)R_{xx}x(t) &= x^T(t)C^TCx(t) \\ &+ x^T(t)\frac{\partial A^T}{\partial \alpha_1} A^{-T} R_{\alpha\alpha_1} A^{-1} \frac{\partial A}{\partial \alpha_1} x(t) \\ &+ \cdots + x^T(t)\frac{\partial A^T}{\partial \alpha_{n_\alpha}} A^{-T} R_{\alpha\alpha_{n_\alpha}} \\ &A^{-1} \frac{\partial A}{\partial \alpha_{n_\alpha}} x(t) \end{aligned}$$

This expression can also be expressed as $z^T(t)z(t)$ with $z(t) = \bar{C}x(t)$ where $\bar{C}^T = [\ C^T \quad C_{\alpha_1}^T \quad \cdots \quad C_{\alpha_{n_\alpha}}^T\ ]$ and C_{α_i} is the minimal order matrix square root of

$$\frac{\partial A^T}{\partial \alpha_i} A^{-T} R_{\alpha\alpha_i} A^{-1} \frac{\partial A}{\partial \alpha_i}.$$

From this we see that the SWLQR cost modifications essentially add new variables to the vector of variables we wish to keep small, $z(t)$. In addition to the standard variables we wish to keep small for performance defined by $Cx(t)$, we tell the cost to keep the variables $C_{\alpha_i}x(t)$ small, and these are related to the sensitivity of the state to parametrically uncertain variables. Furthermore with this new $z(t)$ vector, the zeros of $G_z(s)$ will change. As a result, in the high-gain limit as $\rho \to 0$ the closed-loop poles will not go to cancel the uncertain zeros defined by the original vector of variables we wish to keep small, $z(t) = Cx(t)$.

Selecting the values of the state sensitivity weights $R_{\alpha\alpha_i}$ is, as to be expected, an iterative process. Typically $R_{\alpha\alpha_i}$ is chosen to be either $\beta_i I$ or $\beta_i \tilde{R}_{xx}$ where β_i is a positive scalar constant. Both these choices simplify the selection of $R_{\alpha\alpha_i}$ by reducing the choice to a scalar variable. When using $R_{\alpha\alpha_i} = \beta_i \tilde{R}_{xx}$, one tends to directly trade off performance, reflected by the nominal state cost $x^T(t)\tilde{R}_{xx}x(t)$ with robustness to the parametric error now captured by the state cost

$$\beta_i x^T(t)\frac{\partial A^T}{\partial \alpha_i} A^{-T} \tilde{R}_{xx} A^{-1} \frac{\partial A}{\partial \alpha_i} x(t)$$

On the other hand, selecting $R_{\alpha\alpha_i} = \beta_i I$ is a way of telling the cost that you care about both performance and robustness since the various terms in the state cost are not directly related as when $R_{\alpha\alpha_i} = \beta_i \tilde{R}_{xx}$.

EXAMPLE 39.2:

Consider the mass, spring, dashpot system shown in Figure 39.6. The four masses are coupled together by dampers with identical coefficients, $c_i = .05\ \forall i$, and springs with identical stiffnesses, $k_i = 1\ \forall i$. In this system, the mass m_3 is uncertain, but known to lie within the interval $.25 \le m_3 \le 1.75$ while the values of m_1, m_2, and m_4 are all known to be 1. A unit intensity, zero-mean, white noise disturbance force $\xi(t)$ acts on m_4, and the control $u(t)$ is a force exerted on m_2. The performance variable of interest is the position of the tip mass m_4, so $z(t) = q_4(t)$. The design goals are to synthesize a controller which is robust to the uncertainty in m_3 and that rejects the effect of the disturbance source on the position of m_4. A combination of SWLQR and the stochastic LQR results from Section 39.4.2 will be used to meet these design goals.

Using Newton's laws, the dynamics of this system can be expressed as

$$\mathcal{M}\ddot{q}(t) + \mathcal{C}\dot{q}(t) + \mathcal{K}q(t) = T_1u(t) + T_2\xi(t) \quad (39.31)$$

where $q^T(t) = [q_1(t)\ \ q_2(t)\ \ q_3(t)\ \ q_4(t)]^T$,

$$\mathcal{M} = \begin{bmatrix} m_1 & 0 & 0 & 0 \\ 0 & m_2 & 0 & 0 \\ 0 & 0 & m_3 & 0 \\ 0 & 0 & 0 & m_4 \end{bmatrix}$$

$$T_1 = \begin{bmatrix} 0 \\ 1 \\ 0 \\ 0 \end{bmatrix} \quad T_2 = \begin{bmatrix} 0 \\ 0 \\ 0 \\ 1 \end{bmatrix}$$

$$\mathcal{K} = \begin{bmatrix} k_1 + k_2 & -k_2 & 0 & 0 \\ -k_2 & k_2 + k_3 & -k_3 & 0 \\ 0 & -k_3 & k_3 + k_4 & -k_4 \\ 0 & 0 & -k_4 & k_4 \end{bmatrix}$$

$$\mathcal{C} = \begin{bmatrix} c_1+c_2 & -c_2 & 0 & 0 \\ -c_2 & c_2+c_3 & -c_3 & 0 \\ 0 & -c_3 & c_3+c_4 & -c_4 \\ 0 & 0 & -c_4 & c_4 \end{bmatrix}$$

Then by defining the state vector as

$$x(t) = \begin{bmatrix} q(t) \\ \dot{q}(t) \end{bmatrix} \tag{39.32}$$

the dynamics from Equation 39.31 can be represented in the state space as

$$\dot{x}(t) = Ax(t) + Bu(t) + L\xi(t)$$

with

$$A = \begin{bmatrix} 0 & I \\ -\mathcal{M}^{-1}\mathcal{K} & -\mathcal{M}^{-1}\mathcal{C} \end{bmatrix}$$

$$B = \begin{bmatrix} 0 \\ \mathcal{M}^{-1}T_1 \end{bmatrix} \quad L = \begin{bmatrix} 0 \\ \mathcal{M}^{-1}T_2 \end{bmatrix}$$

With the states defined by Equation 39.32, the performance variables we wish to keep small are

$$z(t) = Cx(t) \text{ where } C = [0\ 0\ 0\ 1\ 0\ 0\ 0\ 0] \tag{39.33}$$

The first step in designing a SWLQR controller is to determine satisfactory values for the nominal design weights $\tilde{R}_{xx}$, $\tilde{R}_{xu}$, and $\tilde{R}_{uu}$. This can be done by ignoring the uncertainty and designing a LQR controller that achieves the desired system performance. Given the single-input, single-output (SISO) system with the desire to control the position of m_4, we find a nominal LQR controller by minimizing the stochastic LQR cost functional

$$J = \mathrm{E}\left\{\lim_{\tau\to\infty}\frac{1}{\tau}\int_0^{\tau}\left[z^T(t)z(t) + \rho u(t)^2\right]\mathrm{d}t\right\} \qquad \rho > 0 \tag{39.34}$$

Substituting $z(t)$ from Equation 39.33 into the cost brings the integrand into the form of that in the cross-weighted cost from Equation 39.22 with $\tilde{R}_{xx} = C^TC$, $\tilde{R}_{xu} = 0$, and $\tilde{R}_{uu} = \rho$. Since m_3 is uncertain, we let it take on its median value of $m_3 = 1$ to synthesize actual controllers. After some iteration over the value of ρ, we decide that $\rho = .01$ leads to a nominal LQR controller that achieves good performance.

Figure 39.7 shows a set of transient responses for the mass spring system to an impulse applied to the disturbance source of the system. Responses for various values of the uncertain mass m_3 are shown, to illustrate how well the controllers behave in the presence of the parametric uncertainty. Notice that the LQR controller designed assuming $m_3 = 1$ is not stable over the entire range of possible values for m_3 and that even when the LQR controller is stable for values of $m_3 \neq 1$ the performance degrades. A better way to view the performance robustness of this controller to the uncertain mass in the presence of a white noise disturbance source is shown in Figure 39.8. This figure compares the root mean square (RMS) value of the tip mass position, $q_4(t)$, normalized by its open-loop RMS value as a function of the value of the uncertain mass m_3. Such plots are often called cost "buckets," for obvious reasons. When the closed-loop system is unstable, the cost blows up so that the wider the cost "bucket" the more robust the system will be to the parametric uncertainty. The cost bucket of Figure 39.8 clearly shows that the LQR controller is not robust to the parametric modeling error in m_3, since the cost "bucket" completely lies within the bounds on m_3.

Now to robustify this nominal LQR controller, we use the SWLQR modified costs from Equations 39.28 to 39.30. Here $\alpha = m_3$, $\frac{\partial B}{\partial m_3} = 0$ and

$$\frac{\partial A}{\partial m_3} = \begin{bmatrix} 0 & 0 \\ -\frac{\partial(\mathcal{M}^{-1})}{\partial m_3}\mathcal{K} & -\frac{\partial(\mathcal{M}^{-1})}{\partial m_3}\mathcal{C} \end{bmatrix}$$

$$\text{where} \quad \frac{\partial \mathcal{M}^{-1}}{\partial m_3} = \begin{bmatrix} 0 & 0 & 0 & 0 \\ 0 & 0 & 0 & 0 \\ 0 & 0 & -m_3^{-2} & 0 \\ 0 & 0 & 0 & 0 \end{bmatrix}$$

Using $R_{\alpha\alpha} = \beta I$ and $m_3 = 1$, we found that $\beta = .01$ led to a design that was significantly more robust than the LQR design. This can be seen in Figures 39.7 and 39.8 that compare the transient response and cost "buckets" of the SWLQR controller to the LQR controller. Both figures demonstrate that the SWLQR controller is stable over the entire range of possible values for m_3. As to be expected though, we do sacrifice some nominal performance to achieve the stability robustness.

39.4.4 LQR with Frequency Weighted Cost Functionals

As discussed in Section 39.2, there is often a gap between what a LQR controller achieves and the desired control system performance. This gap can essentially be viewed as the discrepancy between the time domain optimization method, LQR, and the control system specifications. Often, both control system performance and robustness specifications can be expressed and analyzed in the frequency domain, but the time domain nature of the LQR problem makes it difficult to take advantage of these attributes. Using frequency weighted cost functionals to synthesize LQR controllers leads to a useful and practical variation on the standard LQR problem that narrows the gap between the time and frequency domains.

Recall that the quadratic cost functional is our means of communicating to the mathematics the desired control system performance. We've discussed the physical interpretation of the cost in the time domain, and now to understand it in the frequency domain we apply Parseval's Theorem to it.

THEOREM 39.3 *[Parseval's] For a vector valued signal $h(t)$ defined on $-\infty < t < +\infty$ with*

$$\int_{-\infty}^{\infty} h^T(t)h(t)\mathrm{d}t < \infty$$

if we denote the Fourier transform of $h(t)$ by $h(\jmath\omega)$ then

$$\int_{-\infty}^{\infty} h^T(t)h(t)\mathrm{d}t = \frac{1}{2\pi}\int_{-\infty}^{\infty} h^*(\jmath\omega)h(\jmath\omega)\mathrm{d}\omega$$

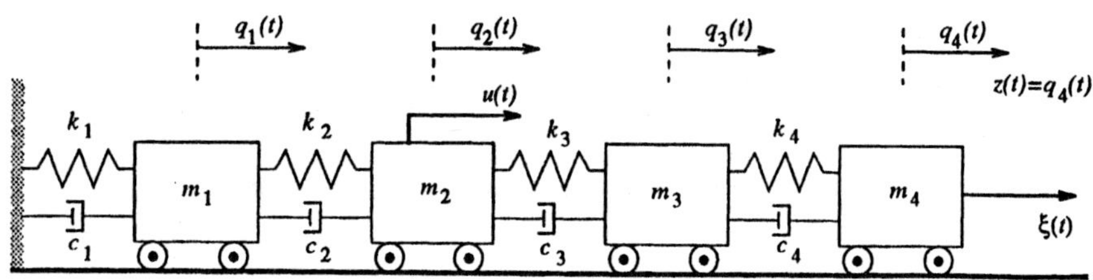

Figure 39.6 Mass, spring, dashpot system from Example 39.2.

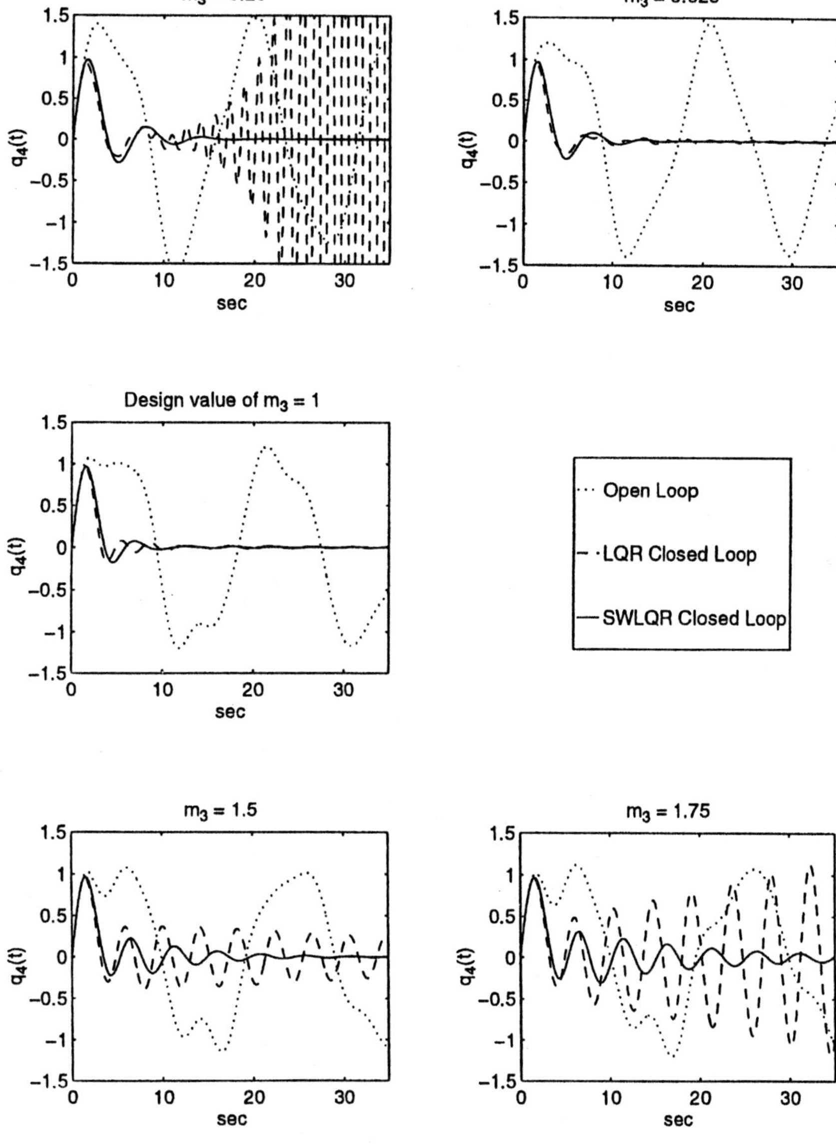

Figure 39.7 Transient response of the 4 mass, spring, dashpot system to an impulse applied at the disturbance source $\xi(t)$ for various values of m_3. The open-loop transients are shown with dots, the LQR transients are shown with a dashed line, and the SWLQR transients are shown with a solid line.

where $h^*(j\omega) = h^T(-j\omega)$ is the complex conjugate transpose of $h(j\omega)$.

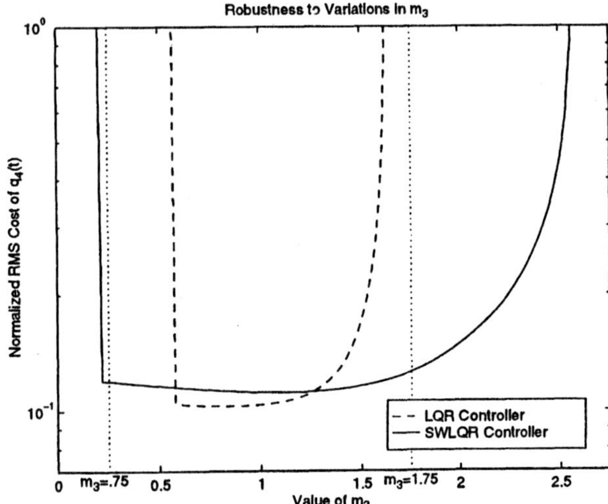

Figure 39.8 Comparison of cost "buckets" between the LQR and SWLQR controllers.

Applying this theorem to the quadratic cost functional from Equation 39.3 with $R = \rho I$ and performance variables $z(t) = Cx(t)$ produces the cost

$$J = \frac{1}{2\pi}\int_{-\infty}^{\infty}\left[z^*(j\omega)z(j\omega) + \rho u^*(j\omega)u(j\omega)\right]d\omega$$

Notice that the time domain weight, ρ, remains constant in the frequency domain. This illustrates that the standard LQR cost functional tells the mathematics to weight the control equally at all frequencies.

To influence the desired control system performance on a frequency by frequency basis, we employ the frequency weighted cost functional

$$\begin{aligned} J &= \frac{1}{2\pi}\int_{-\infty}^{\infty}\left[z^*(j\omega)W_1^*(j\omega)W_1(j\omega)z(j\omega)\right. \\ &\quad \left. + u^*(j\omega)W_2^*(j\omega)W_2(j\omega)u(j\omega)\right]d\omega \end{aligned} \quad (39.35)$$

in which both $W_1(j\omega)$ and $W_2(j\omega)$ are user-specified frequency weights [8]. Looking at the cost (Equation 39.35) for a SISO system in which the design weights will be scalar functions produces

$$J = \frac{1}{2\pi}\int_{-\infty}^{\infty}\left[|W_1(j\omega)|^2|z(j\omega)|^2 + |W_2(j\omega)|^2|u(j\omega)|^2\right]d\omega$$

and helps to illustrate the benefit of using frequency weighted cost functionals. The frequency weights allow us to place distinct penalties on the state and control cost at various frequencies, which is not possible to do when constant weights are used. Their main advantage is that they facilitate the incorporation of known and desired control system behavior into the synthesis process. This is why frequency weighting plays a key role in modern control design [9]. If one knows over what frequency range it is important to achieve performance and where the control energy must be small, this knowledge can be reflected into the frequency weights. Since we seek to minimize the quadratic cost (Equation 39.35), large terms in the integrand incur greater penalties than small terms and more effort is exerted to make them small. For example, if there is a rigorous bandwidth constraint set by a region of high-frequency unmodeled dynamics and if the control weight $W_2(j\omega)$ is chosen to have a large magnitude over this region then the resulting controller would not exert substantial energy in the region of unmodeled dynamics. This in turn would limit the controller's bandwidth. Similar arguments can be used to select $W_1(j\omega)$ to tell the controller synthesis at what frequencies the size of the performance variables need to be small.

To synthesize LQR controllers that minimize the frequency weighted cost functional (Equation 39.35), it is necessary to transform the cost functional back into the time domain. Doing so requires state space realization of the frequency weights $W_1(j\omega)$ and $W_2(j\omega)$. Hence we let

$$\begin{aligned} z_1(s) &= W_1(s)z(s) \text{ with} \\ W_1(s) &= C_1(sI - A_1)^{-1}B_1 + D_1 \quad (39.36) \\ z_2(s) &= W_2(s)u(s) \text{ with} \\ W_2(s) &= C_2(sI - A_2)^{-1}B_2 + D_2 \quad (39.37) \end{aligned}$$

Using these definitions and Parseval's Theorem, the cost (Equation 39.35) becomes

$$\begin{aligned} J &= \frac{1}{2\pi}\int_{-\infty}^{\infty}\left[z_1^*(j\omega)z_1(j\omega) + z_2^*(j\omega)z_2(j\omega)\right]d\omega \\ &= \int_{-\infty}^{\infty}\left[z_1^T(t)z_1(t) + z_2^T(t)z_2(t)\right]dt \quad (39.38) \end{aligned}$$

This cost can be brought into the form of an LQR problem that we know how to solve by augmenting the dynamics of the weights (Equations 39.36 and 39.37) to the open-loop dynamics (Equations 39.1 and 39.2). Using $x_1(t)$ and $x_2(t)$ to denote, respectively, the states of $W_1(s)$ and $W_2(s)$ to carry out the augmentation produces the system

$$\begin{aligned} \dot{\mathcal{X}}(t) &= \mathcal{A}\mathcal{X}(t) + \mathcal{B}u(t) \quad (39.39) \\ \mathcal{Z}(t) &= \mathcal{C}\mathcal{X}(t) + \mathcal{D}u(t) \end{aligned}$$

with

$$\mathcal{A} = \begin{bmatrix} A & 0 & 0 \\ B_1C & A_1 & 0 \\ 0 & 0 & A_2 \end{bmatrix} \quad \mathcal{B} = \begin{bmatrix} B \\ 0 \\ B_2 \end{bmatrix}$$

$$\mathcal{C} = \begin{bmatrix} D_1C & C_1 & 0 \\ 0 & 0 & C_2 \end{bmatrix} \quad \mathcal{D} = \begin{bmatrix} 0 \\ D_2 \end{bmatrix}$$

where $\mathcal{X}^T(t) = [x^T(t) \quad x_1^T(t) \quad x_2^T(t)]$ and $\mathcal{Z}^T(t) = [z_1^T(t) \; z_2^T(t)]$ contains the outputs of $W_1(s)$ and $W_2(s)$. Notice that

$$\begin{aligned} \mathcal{Z}^T(t)\mathcal{Z}(t) &= z_1^T(t)z_1(t) + z_2^T(t)z_2(t) \\ &= \mathcal{X}^T(t)\mathcal{C}^T\mathcal{C}\mathcal{X}(t) + 2\mathcal{X}^T(t)\mathcal{C}^T\mathcal{D}u(t) \\ &\quad + u^T(t)\mathcal{D}^T\mathcal{D}u(t) \end{aligned}$$

and thus the frequency weighted cost (Equation 39.38) becomes

$$J = \int_{-\infty}^{\infty} \Big[\mathcal{X}^T(t)\mathcal{C}^T\mathcal{C}\mathcal{X}(t) + 2\mathcal{X}^T(t)\mathcal{C}^T\mathcal{D}u(t) + u^T(t)\mathcal{D}^T\mathcal{D}u(t)\Big]\,dt$$

Realize that minimizing this cost subject to the dynamic constraints (Equation 39.39) is not only equivalent to minimizing the frequency weighted cost functional (Equation 39.35), but it is also a simple manner to carry out the optimization by using the cross-weighted LQR results from Theorem 39.2 with

$$R_{xx} = \mathcal{C}^T\mathcal{C} \quad R_{xu} = \mathcal{C}^T\mathcal{D} \quad R_{uu} = \mathcal{D}^T\mathcal{D} \tag{39.40}$$

Applying Theorem 39.2 produces the control law

$$u(t) = -K\mathcal{X}(t) = -[K_x \;\; K_1 \;\; K_2]\begin{bmatrix} x(t) \\ x_1(t) \\ x_2(t) \end{bmatrix} \tag{39.41}$$

where $K = R_{uu}^{-1}\left(R_{xu}^T + \mathcal{B}^T S\right)$, S is the solution to the Riccati equation

$$S\left(\mathcal{A} - \mathcal{B}R_{uu}^{-1}R_{xu}^T\right) + \left(\mathcal{A}^T - R_{xu}R_{uu}^{-1}\mathcal{B}^T\right)S + \left(R_{xx} - R_{xu}R_{uu}^{-1}R_{xu}^T\right) - S\mathcal{B}R_{uu}^{-1}\mathcal{B}^T S = 0,$$

and the control weights are defined by Equation 39.40.

Clearly the augmented state vector $\mathcal{X}(t)$ must be fed back to implement the control law (Equation 39.41). Since $x_1(t)$ and $x_2(t)$ are the fictitious states of the weights $W_1(s)$ and $W_2(s)$, it is necessary to deliberately create them in real time to apply the feedback control law (Equation 39.41). That is, using frequency weighted cost functionals to synthesize LQR controllers leads to dynamic compensators. The control law is no longer a set of static gains multiplying the measurement of the state vector $x(t)$. It is a combination of the dynamics of the weighting matrices and the static gains K. A block diagram showing the feedback architecture for the dynamic LQR controller arrived at by minimizing the frequency weighted cost functional of Equation 39.35 is shown in Figure 39.9. Realize that having to implement a dynamic compensator is the price we pay for using frequency weights to incorporate known frequency domain information into the LQR controller synthesis. Though since the order of the compensator is only equal to the combined order of the design weights, this fact should not be troublesome.

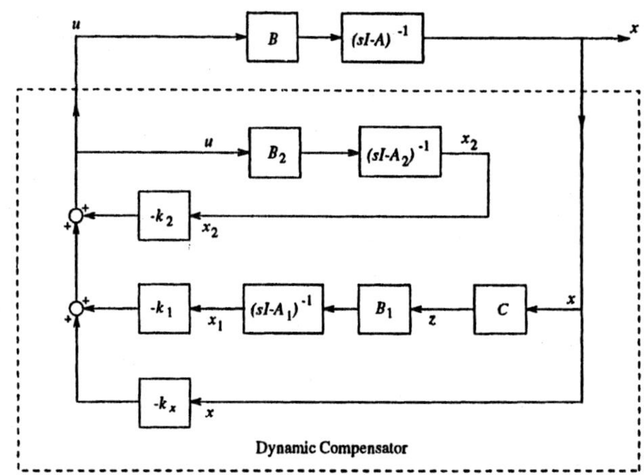

Figure 39.9 Block diagram for the LQR controller synthesized using the frequency weighted cost from Equation 39.35.

Selecting the Frequency Weights

Although we do not present the results here, it is possible to use any combination of frequency weights and constant weights when synthesizing LQR controllers with frequency weighted costs. For example, one could use a constant state penalty, $W_1(s) = I$, with a frequency weighted control. In any case, to satisfy the underlying LQR assumptions one must use stable design weights and ensure that $D_2^T D_2 > 0$ (so that R_{uu}^{-1} exists).

When synthesizing LQR controllers with frequency weighted cost functionals, the iteration over the values of the design weights involves selecting the dynamics of multivariable systems. Although it is not necessary, choosing the weighting matrices to be scalar functions multiplying the identity matrix,

$$W_1(s) = w_1(s)I \qquad \text{and} \qquad W_2(s) = w_2(s)I$$

simplifies the process of selecting useful weights. Then the process of selecting the weights reduces to selecting the transfer functions $w_1(s)$ and $w_2(s)$ so that their magnitudes have the desired effect on the cost.

To arrive at a useful set of weights, one must use one's knowledge of the physics of the system and the desired control system performance. Keeping this in mind, simply select the magnitude of $w_2(s)$ to be large relative to $|w_2(0)|$ over the frequency range where you want the control energy to be small and choose $w_1(s)$ to have a large magnitude relative to $|w_1(0)|$ over the frequency regions where you want the performance variables to be small. It is important to note that here large is relative to the values of the DC gains of the weights, $|w_1(0)|$ and $|w_2(0)|$. The DC gains specify the nominal penalties on the state and control cost in a manner similar to the constant weights of a standard LQR problem. As such, when choosing the frequency weights, it is beneficial to break the process up into two steps. The first step is to choose an appropriate DC gain to influence the overall characteristics of the controller, and the second step is to choose the dynamics so that the magnitudes of the weights reflect the relative importance of the variables over the various frequency ranges.

When choosing the frequency weights it is vital to keep in mind that the controls will have to be large over the frequency ranges where the performance variables are to be small. Likewise, if the control energy is specified to be small over a frequency range, it will not be possible to make the performance variables small there. Even though we could tell the cost to make both the performance variables and the control signals small in the same frequency range through the choice of the frequency weights, doing so will most likely result in a meaningless controller since we are asking the mathematical optimization problem to defy the underlying physics.

39.5 Mini-Max and $\mathcal{H}_\infty$ Full-State Feedback Control

Consider the problem of rejecting the effect of the disturbances, $d(t)$, on the performance variables of interest, $z(t)$, for the system

$$\begin{aligned} \dot{x}(t) &= Ax(t) + Bu(t) + Ld(t) \\ z(t) &= Cx(t) \end{aligned} \tag{39.42}$$

LQR control is not well suited to handle this problem because the optimal control that minimizes the quadratic cost (Equation 39.3) subject to the dynamic constraints (Equation 39.42) wants to know the future values of the disturbances, which is not realistic. The stochastic version of the LQR problem is also inappropriate unless $d(t)$ is white noise, which is rarely the case. To deal optimally with the disturbances using a full-state feedback controller, it is necessary to adopt a different philosophy than that of the LQR. Rather than treating the disturbances as known or white noise signals, they are assumed to behave in a "worst case" fashion. Treating the disturbances in this way leads to the so called $\mathcal{H}_\infty$ full-state feedback controller. $\mathcal{H}_\infty$ controllers have become as popular as LQR controllers in recent years as a result of their own attractive properties [10]. We introduce $\mathcal{H}_\infty$ controllers here using the quadratic cost functional optimization point of view.

The "worst case" philosophy for dealing with the disturbances arises by including them in the quadratic cost functional with their own weight, γ, much in the same way that the controls are included in the LQR cost functional. That is we seek to optimize the quadratic cost

$$\begin{aligned} J(u,d) &= \frac{1}{2}\int_0^\infty \Big[z^T(t)z(t) + \rho u^T(t)u(t) \\ &\quad - \gamma^2 d^T(t)d(t)\Big]\,dt \\ &\quad \gamma, \rho > 0 \end{aligned} \tag{39.43}$$

subject to the dynamic constraints of Equation 39.42. In this optimization problem both the controls and disturbances are the unknown quantities that the cost is optimized over. Note that since the disturbances enter the cost functional as a negative quadratic, they will seek to maximize $J(u, d)$. At the same time the controls seek to minimize $J(u, d)$ since they enter the cost functional as a positive quadratic. Hence, by using the cost (Equation 39.43), we are playing a mini-max differential game in which nature tries to maximize the cost through the choice of the disturbances, and we as control system designers seek to minimize the cost through the choice of the control $u(t)$. This mini-max optimization problem can be compactly stated as

$$\min_u \max_d J(u,d). \tag{39.44}$$

Since nature is allowed to pick the disturbances $d(t)$ which maximize the cost, this optimization problem deals with the disturbances by producing a control law that is capable of rejecting specific worst case disturbances.

The solution to the mini-max optimization problem (Equation 39.44) is not guaranteed to exist for all values of γ. When the solution does exist, it produces a full-state feedback control law similar in structure to the LQR controller. The solution to the mini-max differential game is summarized in the following theorem.

THEOREM 39.4 *[Mini-Max Differential Game] Given the system dynamics*

$$\dot{x}(t) = Ax(t) + Bu(t) + Ld(t) \tag{39.45}$$

with $x(t) \in \mathbb{R}^n$, $u(t) \in \mathbb{R}^m$, and $d(t) \in \mathbb{R}^q$ along with the performance variables we wish to keep small $z(t) = Cx(t)$ with $z(t) \in \mathbb{R}^p$, we define the mini-max quadratic cost functional

$$J(u,d) = \frac{1}{2}\int_0^\infty \Big[z^T(t)z(t) + \rho u^T(t)u(t) - \gamma^2 d^T(t)d(t)\Big]\,dt$$

$$\gamma, \rho > 0$$

in which ρ and γ are user-specified design variables that weight the relative influence of the controls and disturbances. Under the following assumptions

1. *The entire state vector $x(t)$ is available for feedback*
2. *$d(t)$ is a deterministic, bounded energy signal with $\int_0^\infty d^T(t)d(t)dt < \infty$*
3. *Both $[A \quad B]$ and $[A \quad L]$ are stabilizable, and $[A \quad C]$ is detectable*

If the optimum value of the cost $J(u, d)$ constrained by the system dynamics (Equation 39.45) exists, it is a unique saddle point of $J(u, d)$ where

1. *The optimal mini-max control law is*

$$u(t) = -Kx(t) \quad \text{with} \quad K = \frac{1}{\rho}B^T S \tag{39.46}$$

2. *The optimal, worst case, disturbance is*

$$d(t) = \frac{1}{\gamma^2}L^T Sx(t)$$

3. *S is the unique, symmetric, positive semidefinite solution of the matrix Riccati equation*

$$SA + A^T S + C^T C - S\left(\frac{1}{\rho}BB^T - \frac{1}{\gamma^2}LL^T\right)S = 0 \tag{39.47}$$

4. *The closed-loop dynamics for Equation 39.45 using Equation 39.46*

$$\dot{x}(t) = (A - BK)x(t) + Ld(t) \tag{39.48}$$

are guaranteed to be asymptotically stable.

If the solution to this optimization problem exists, it produces a stabilizing full-state feedback controller with the same structure as the LQR controller but with a different scheme for evaluating the feedback gains. Since the L matrix of Equation 39.45 appears in the Riccati equation (39.47), the mini-max control law directly incorporates the information of how the disturbances impact the system dynamics. The facts that the min-max controller guarantees a stable closed-loop and takes into consideration the nature of the disturbances make it an attractive alternative to LQR for synthesizing controllers.

39.5.1 Synthesizing Mini-Max Controllers

Mini-max controllers are not guaranteed to exist for arbitrary values of the design weight γ in the quadratic cost functional $J(u, d)$. Since γ influences the size of the $-\frac{1}{\gamma^2}LL^T$ term in the Riccati equation (39.47), there will exist values of $\gamma > 0$ for which there is either no solution to the Riccati equation or for which S will not be positive semidefinite. It turns out that there is a minimum value of γ, $\gamma_{\min}$, for which the mini-max optimization problem has a solution. Hence, useful values of γ will lie in the interval $\gamma_{\min} \leq \gamma < \infty$. Note that as $\gamma \to \infty$, the Riccati equation (39.47) becomes identical to the LQR one from Equation 39.5, and we recover the LQR controller. Likewise when $\gamma = \gamma_{\min}$, we have another special case which is known as the full-state feedback $\mathcal{H}_\infty$ controller[2]. For any other value of γ in $\gamma_{\min} \leq \gamma < \infty$ we still have an admissible stabilizing mini-max controller.

As with LQR, synthesizing mini-max controllers requires solving an algebraic Riccati equation. However, the presence of the $-\frac{1}{\gamma^2}LL^T$ term in the mini-max Riccati equation (39.47) makes the process of computing an $S = S^T \geq 0$ that satisfies Equation 39.47 more complicated than finding an $S = S^T \geq 0$ that satisfies the LQR Riccati equation (39.5). The reasons for this are directly related to the issue of whether or not a solution to the mini-max optimization problem exists.

To understand how one computes mini-max controllers, it is necessary to understand how current algebraic Riccati equation solvers work. While there is a rich theory for the topic, we summarize the key results in the following theorem.

THEOREM 39.5 *[The Algebraic Riccati Equation] The Riccati equation*

$$A^T S + SA + SVS - Q = 0 \tag{39.49}$$

is solved by carrying out a spectral factorization[3] of its associate Hamiltonian matrix

$$H = \begin{bmatrix} A & V \\ Q & -A^T \end{bmatrix}$$

If $V = -BB^T$, $Q = -C^T C$, $[A \quad B]$ is stabilizable, and $[A \quad C]$ is detectable then the Riccati equation solvers produce the unique, symmetric, positive semidefinite solution, $S = S^T \geq 0$, to the Riccati equation (39.49). Otherwise, as long as H has no $j\omega$-axis eigenvalues, the spectral factorization can be performed and a solution, S, which satisfies Equation 39.49 is produced.

[2]See Chapter 40 for details.

[3]Spectral factorizations are essentially eigenvalue decompositions.

The Hamiltonian matrix for the mini-max Riccati equation, H_γ, is

$$H_\gamma = \begin{bmatrix} A & \frac{1}{\gamma^2}LL^T - \frac{1}{\rho}BB^T \\ -C^T C & -A^T \end{bmatrix} \tag{39.50}$$

The sign indefinite nature of the $1/\gamma^2 LL^T - 1/\rho BB^T$ term in H_γ makes it quite difficult to numerically test whether or not a solution to the mini-max optimization problem exists. Thus, a constructive algorithm based on the existing algebraic Riccati equation solvers is used to synthesize mini-max controllers.

THEOREM 39.6 *[Algorithm for Computing Mini-Max Controllers] Pick a value of γ and check to see if H_γ from Equation 39.50 has any $j\omega$-axis eigenvalues. If it does, increase γ and start over. If it does not, use an algebraic Riccati equation solver to produce a solution S to Equation 39.47. Test if $S \geq 0$. If it is not, increase gamma and start over. If $S \geq 0$, check to see if the closed-loop dynamics from Equation 39.48 are stable. If they are not, increase gamma and start over. If they are, you have constructed a mini-max controller, since you've found a $S = S^T \geq 0$ that satisfies the Riccati equation (39.47).*

Theoretically, the final step of the algorithm, which requires checking the closed-loop stability is not necessary. However, we highly recommend it since the numerical stability of the solution to the Riccati equations for values of γ near $\gamma_{\min}$ can be questionable.

From the algorithm for computing mini-max controllers, it can be seen that evaluating $\mathcal{H}_\infty$ controllers for a fixed value of ρ will require a trial-and-error search over γ. To compute $\gamma_{\min}$ it is best to use a bisection search over γ in the algorithm to find the smallest value of γ for which $S = S^T \geq 0$ and the closed-loop system (Equation 39.48) is stable. Current control system design packages such as MATLAB and Matrix$_X$ employ such algorithms for computing the $\mathcal{H}_\infty$ controller that in turn determines $\gamma_{\min}$.

While the mini-max and $\mathcal{H}_\infty$ controllers are quite distinct from LQR controllers all the advice given for designing LQR controllers applies to mini-max controllers as well. Namely, it is necessary to iterate over the values of the design weights and independently check the robustness and performance characteristics for each design when synthesizing mini-max and $\mathcal{H}_\infty$ controllers. It can be shown through manipulating the Riccati equation (39.47) that the robustness properties of LQR controllers from Section 39.3 apply to $\mathcal{H}_\infty$ and mini-max controllers as well. Furthermore, modifications of the powerful sensitivity and frequency weighted LQR design tools from Section 39.4 do exist and can be used to incorporate stability robustness and known design specifications into the controller synthesis of mini-max and $\mathcal{H}_\infty$ controllers.

References

[1] Anderson, B.D.O. and Moore, J.B., *Optimal Control: Linear Quadratic Methods,* Prentice Hall, Englewood Cliffs, NJ, 1990.

[2] Kwakernaak, H. and Sivan, R., *Linear Optimal Control Systems,* John Wiley & Sons, New York, 1972.

[3] Emami-Naeini, A. and Rock, S.M., On asymptotic behavior of non-square linear optimal regulators, in *Proc. 23rd Conf. Decision and Control,* Las Vegas, NV, Dec. 1984, pp.1762–1763.

[4] Weinmann, A., *Uncertain Models and Robust Control,* Springer-Verlag, New York, 1991.

[5] Grocott, S.C.O., Comparision of Control Techniques for Robust Performance on Uncertain Structural Systems, Master's thesis, Massachusetts Institute of Technology, 1994. MIT SERC report No. 2-94.

[6] Sesak, J.R., Sensitivity Constrained Linear Optimal Control Analysis and Synthesis, Ph.D. thesis, University of Wisconsin, 1974.

[7] Grocott, S.C.O., How, J.P., and Miller, D.W., A comparison of robust control techniques for uncertain structural systems, *Proc. AIAA Guidance Navigation and Control Conference,* Scottsdale, AZ, Aug. 1994, pp. 261–271.

[8] Gupta, N., Frequency-shaped cost functionals: extension of linear-quadratic-Gaussian methods, *J. Guidance Control Dynam.,* 3(6), 529–535, 1980.

[9] Doyle, J., Francis, B., and Tannenbaum, A., *Feedback Control Theory,* Macmillan, New York, 1992.

[10] Kwakernaak, H., Robust control and $\mathcal{H}_\infty$ optimization-tutorial paper, *Automatica,* 29(2), 255–273, 1993.

40

$\mathcal{H}_2$ (LQG) and $\mathcal{H}_\infty$ Control

Leonard Lublin, Simon Grocott, and Michael Athans
Space Engineering Research Center, Massachusetts Institute of Technology, Cambridge, MA

40.1 Introduction

The fundamentals of output feedback $\mathcal{H}_2$ (linear quadratic Gaussian or LQG) and $\mathcal{H}_\infty$ controllers, which are the primary synthesis tools available for linear time-invariant systems, are presented in an analogous and tutorial fashion without rigorous mathematics. Since $\mathcal{H}_2$ and $\mathcal{H}_\infty$ syntheses are carried out in the modern control design paradigm, a review of the paradigm is presented, along with the definitions of the $\mathcal{H}_2$ and $\mathcal{H}_\infty$ norms and the methods used to compute them. The state-space formulae for the optimal controllers, under less restrictive assumptions than are usually found in the literature, are provided in an analogous fashion to emphasize the similarities between them. Rather than emphasizing the derivation of the controllers, we elaborate on the physical interpretation of the results and how one uses frequency weights to design $\mathcal{H}_\infty$ and $\mathcal{H}_2$ controllers. Finally, a simple disturbance rejection design for the longitudinal motion of an aircraft is provided to illustrate the similarities and differences between $\mathcal{H}_\infty$ and $\mathcal{H}_2$ controller synthesis.

40.2 The Modern Paradigm

$\mathcal{H}_2$ and $\mathcal{H}_\infty$ syntheses are carried out in the modern control paradigm. In this paradigm both performance and robustness specifications can be incorporated in a common framework along with the controller synthesis. In the modern paradigm, all of the information about a system is cast into the generalized block diagram shown in Figure 40.1 [1, 2, 3]. The generalized plant, P, which is assumed to be linear and time-invariant throughout this article contains all the information a designer would like to incorporate into the synthesis of the controller, K. System dynamics, models of the uncertainty in the system's dynamics, frequency weights to influence the controller synthesis, actuator dynamics, sensor dynamics, and implementation hardware dynamics from

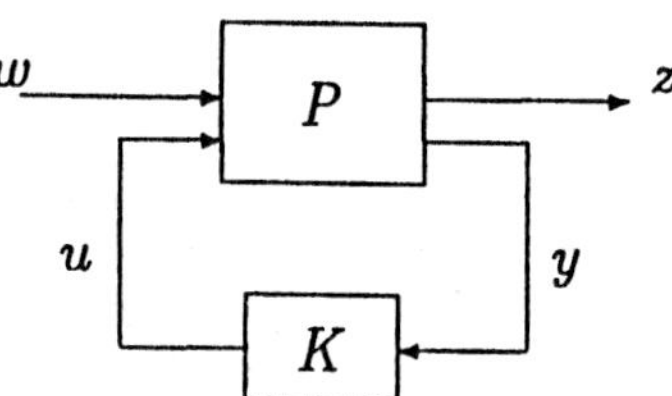

Figure 40.1 Generalized block diagram of the modern paradigm.

amplifiers, and analog-to-digital and digital-to-analog converters are all included in P. The inputs and outputs of P are, in general, vector valued signals. The sensor measurements that are used by the feedback controller are denoted y, and the inputs generated by the controller are denoted u. The components of w are all the exogenous inputs to the system. Typically these consist of disturbances, sensor noise, reference commands, and fictitious signals that drive frequency weights and models of the uncertainty in the dynamics of the system. The components of z are all the variables we wish to control. These include the performance variables of interest, tracking errors between reference signals and plant outputs, and the actuator signals which cannot be arbitrarily large and fast.

The general control problem in this framework is to synthesize a controller that will keep the size of the performance variables, z, small in the presence of the exogenous signals, w. For a classical disturbance rejection problem, z would contain the performance variables we wish to keep small in the presence of the disturbances contained in w that would tend to drive z away from zero. Hence, the disturbance rejection performance would depend on the "size" of the closed-loop transfer function from w to z, which we shall denote as $T_{zw}(s)$. This is also true for a command following control problem in which z would contain the tracking error that we would like to keep small in the presence of the commands in w that drive the tracking error away from zero.

0-8493-8570-9/96/$0.00+$.50

Clearly then, the "size" of $T_{zw}(s)$ influences the effect that the exogenous signals in w have on z. Thus, in this framework, we seek controllers that minimize the "size" of the closed-loop transfer function $T_{zw}(s)$. Given that $T_{zw}(s)$ is a transfer function matrix, it is necessary to use appropriate norms to quantify its size. The two most common and physically meaningful norms that are used to classify the "size" of $T_{zw}(s)$ are the $\mathcal{H}_2$ and $\mathcal{H}_\infty$ norms. As such, we seek controllers that minimize either the $\mathcal{H}_2$ or $\mathcal{H}_\infty$ norm of $T_{zw}(s)$ in the modern control paradigm.

40.2.1 System Norms

Here we define and discuss the $\mathcal{H}_2$ and $\mathcal{H}_\infty$ norms of the linear, time-invariant, stable system with transfer function matrix

$$G(s) = C(sI - A)^{-1}B$$

This notation is meant to be general, and the reader should not think of $G(s)$ as only the actuator to sensor transfer function of a system. Realize that $G(s)$ is a system and thus requires an appropriate norm to classify its size. By a norm, we mean a positive, scalar number that is a measure of the size of $G(s)$ over all points in the complex s-plane. This is quite different from, for example, the maximum singular value of a matrix, $\sigma_{\max}[A]$, which is a norm that classifies the size of the matrix A.

The $\mathcal{H}_2$ Norm

DEFINITION 40.1 [$\mathcal{H}_2$ Norm] The $\mathcal{H}_2$ norm of $G(s)$, denoted $\|G\|_2$, is defined as

$$\begin{aligned} \|G\|_2 &= \left(\frac{1}{2\pi}\int_{-\infty}^{\infty} \text{trace}\left[G(j\omega)G^*(j\omega)\right] d\omega\right)^{\frac{1}{2}} \\ &= \left(\frac{1}{2\pi}\int_{-\infty}^{\infty} \sum_{i=1}^{r} \sigma_i^2[G(j\omega)]d\omega\right)^{\frac{1}{2}} \end{aligned}$$

where σ_i denotes the ith singular value, $G^*(j\omega)$ is the complex conjugate transpose of $G(j\omega)$, and r is the rank of $G(j\omega)$.

The $\mathcal{H}_2$ norm has an attractive, physically meaningful interpretation. If we consider $G(s)$ to be the transfer function matrix of a system driven by independent, zero mean, unit intensity white noise, u, then the sum of the variances of the outputs y is exactly the square of the $\mathcal{H}_2$ norm of $G(s)$. That is

$$\mathrm{E}\left[y^T(t)y(t)\right] = \|G(s)\|_2^2 \tag{40.1}$$

The $\mathcal{H}_2$ norm of $G(s)$ thus gives a precise measure of the "power" or signal strength of the output of a system driven with unit intensity white noise. Note that in the scalar case $\sqrt{\mathrm{E}[y^T(t)y(t)]}$ is the RMS or root mean squared value for $y(t)$ so the $\mathcal{H}_2$ norm specifies the RMS value of $y(t)$. A well-known fact for stochastic systems is that the mean squared value of the outputs can be computed by solving the appropriate Lyapunov equation [4]. As such, a state space procedure for computing the $\mathcal{H}_2$ norm of $G(s)$ is as follows [2].

Computing the $\mathcal{H}_2$ Norm If L_c denotes the controllability Gramian of (A, B) and L_o the observability Gramian of (A, C), then

$$AL_c + L_cA^T + BB^T = 0 \qquad A^TL_o + L_oA + C^TC = 0$$

and

$$\|G\|_2 = \left[\text{trace}(CL_cC^T)\right]^{\frac{1}{2}} = \left[\text{trace}(B^TL_oB)\right]^{\frac{1}{2}}$$

Note that this procedure for computing the $\mathcal{H}_2$ norm involves the solution of linear Lyapunov equations and can be done without iteration.

The $\mathcal{H}_\infty$ Norm

DEFINITION 40.2 [$\mathcal{H}_\infty$ Norm] The $\mathcal{H}_\infty$ norm of $G(s)$, denoted $\|G\|_\infty$, is defined as

$$\|G\|_\infty = \sup_\omega \sigma_{\max}[G(j\omega)]$$

In this definition "sup" denotes the supremum or least upper bound of the function $\sigma_{\max}[G(j\omega)]$, and thus the $\mathcal{H}_\infty$ norm of $G(s)$ is nothing more than the maximum value of $\sigma_{\max}[G(j\omega)]$ over all frequencies ω. The supremum must be used in the definition since, strictly speaking, the maximum of $\sigma_{\max}[G(j\omega)]$ may not exist even though $\sigma_{\max}[G(j\omega)]$ is bounded from above.

$\mathcal{H}_\infty$ norms also have a physically meaningful interpretation when considering the system $y(s) = G(s)u(s)$. Recall that when the system is driven with a unit magnitude sinusoidal input at a specific frequency, $\sigma_{\max}[G(j\omega)]$ is the largest possible output size for the corresponding sinusoidal output. Thus, the $\mathcal{H}_\infty$ norm is the largest possible amplification over all frequencies of a unit sinusoidal input. That is, it classifies the greatest increase in energy that can occur between the input and output of a given system. A state space procedure for calculating the $\mathcal{H}_\infty$ norm is as follows.

Computing the $\mathcal{H}_\infty$ Norm *Let $\|G\|_\infty = \gamma_{\min}$. For the transfer function $G(s) = C(sI - A)^{-1}B$ with A stable and $\gamma > 0$, $\|G\|_\infty < \gamma$ if and only if the Hamiltonian matrix*

$$H = \begin{bmatrix} A & \frac{1}{\gamma^2}BB^T \\ -C^TC & -A^T \end{bmatrix}$$

has no eigenvalues on the $j\omega$-axis. This fact lets us compute a bound, γ, on $\|G\|_\infty$ such that $\|G\|_\infty < \gamma$. So to find $\gamma_{\min}$, select a $\gamma > 0$ and test if H has eigenvalues on the $j\omega$-axis. If it does, increase γ. If it does not, decrease γ and recompute the eigenvalues of H. Continue until $\gamma_{\min}$ is calculated to within the desired tolerance.

The iterative computation of the $\mathcal{H}_\infty$ norm, which can be carried out efficiently using a bisection search over γ, is to be expected given that by definition we must search for the largest value of $\sigma_{\max}[G(j\omega)]$ over all frequencies.

Note, the $\mathcal{H}_2$ norm is not an induced norm, whereas the $\mathcal{H}_\infty$ norm is. Thus, the $\mathcal{H}_2$ norm does not obey the submultiplicative property of induced norms. That is, the $\mathcal{H}_\infty$ norm satisfies

$$\|G_1G_2\|_\infty \leq \|G_1\|_\infty\|G_2\|_\infty$$

but the $\mathcal{H}_2$ norm does not have the analogous property. This fact makes synthesizing controllers that minimize $\|T_{zw}(s)\|_\infty$ attractive when one is interested in directly shaping loops to satisfy norm bounded robustness tests[1]. On the other hand, given the aforementioned properties of the $\mathcal{H}_2$ norm, synthesizing controllers that minimize $\|T_{zw}(s)\|_2$ is attractive when the disturbances, w, are stochastic in nature. In fact, $\mathcal{H}_2$ controllers are nothing more than linear quadratic Gaussian (LQG) controllers so the vast amount of insight into the well-understood LQG problem can be readily applied to $\mathcal{H}_2$ synthesis.

EXAMPLE 40.1:

In this example the $\mathcal{H}_2$ and $\mathcal{H}_\infty$ norms are calculated for the simple four-spring, four-mass system shown in Figure 39.6. The equations of motion for this system can be found in Example 2 in Chapter 39. The system has force inputs on the second and fourth masses along with two sensors that provide a measure of the displacement of these masses. The singular values of the transfer function from the inputs to outputs, which we denote by $G(s)$, are shown in Figure 40.3. The $\mathcal{H}_\infty$ norm of the system is equal to the peak of $\sigma_1 = 260.4$, and the $\mathcal{H}_2$ norm of the system is equal to the square root of the sum of the areas under the square of each of the singular values, 14.5. Note that when considering the $\mathcal{H}_2$ norm, observing the log log plot of the transfer function can be very deceiving, since the integral is of σ_i, not $\log(\sigma_i)$, over ω, not $\log \omega$.

As pointed out in the example, the differences between the $\mathcal{H}_\infty$ and $\mathcal{H}_2$ norms for a system $G(s)$ are best viewed in the frequency domain from a plot of the singular values of $G(\jmath\omega)$. Specifically, the $\mathcal{H}_\infty$ norm is the peak value of $\sigma_{\max}[G(\jmath\omega)]$ while the $\mathcal{H}_2$ norm is related to the area underneath the singular values of $G(\jmath\omega)$. For a more in-depth treatment of these norms the reader is referred to [1, 2, 5, 6].

40.3 Output Feedback $\mathcal{H}_\infty$ and $\mathcal{H}_2$ Controllers

Given that all the information a designer would like to include in the controller synthesis is incorporated into the system P, the synthesis of $\mathcal{H}_2$ and $\mathcal{H}_\infty$ controllers is quite straightforward. In this respect all of the design effort is focused on defining P. Below, we discuss how to define P using frequency weights to meet typical control system specifications. Here we simply present the formulas for the controllers.

All the formulas will be based on the following state-space realization of P,

$$P := \left[\begin{array}{c|cc} A & B_1 & B_2 \\ \hline C_1 & D_{11} & D_{12} \\ C_2 & D_{21} & D_{22} \end{array}\right]$$

This notation is a shorthand representation for the system of equations

$$\begin{aligned} \dot{x}(t) &= Ax(t) + B_1w(t) + B_2u(t) && (40.2)\\ z(t) &= C_1x(t) + D_{11}w(t) + D_{12}u(t) && (40.3)\\ y(t) &= C_2x(t) + D_{21}w(t) + D_{22}u(t) && (40.4) \end{aligned}$$

Additionally, the following assumptions concerning the allowable values for the elements of P are made.

Assumptions on P

1. $D_{11} = 0$ (A.1)
2. $[A \quad B_2]$ is stabilizable (A.2)
3. $[A \quad C_2]$ is detectable (A.3)
4. $V = \begin{bmatrix} B_1 \\ D_{21} \end{bmatrix}\begin{bmatrix} B_1^T & D_{21}^T \end{bmatrix} := \begin{bmatrix} V_{xx} & V_{xy} \\ V_{xy}^T & V_{yy} \end{bmatrix} \geq 0$ with $V_{yy} > 0$ (A.4)
5. $R = \begin{bmatrix} C_1^T \\ D_{12}^T \end{bmatrix}\begin{bmatrix} C_1 & D_{12} \end{bmatrix} := \begin{bmatrix} R_{xx} & R_{xu} \\ R_{xu}^T & R_{uu} \end{bmatrix} \geq 0$ with $R_{uu} > 0$ (A.5)

Assumption 1 ensures that none of the disturbances feed through to the performance variables which is necessary for $\mathcal{H}_2$ synthesis but may be removed for $\mathcal{H}_\infty$ synthesis (see [7] for details.) Assumptions 2 and 3 are needed to guarantee the existence of a stabilizing controller while the remaining assumptions are needed to guarantee the existence of positive semidefinite solutions to the Riccati equations associated with the optimal controllers.

THEOREM 40.1 *[$\mathcal{H}_2$ Output Feedback] Assuming that $w(t)$ is a unit intensity white noise signal, $E[w(t)w^T(\tau)] = I\delta(t-\tau)$, the unique, stabilizing, optimal controller which minimizes the $\mathcal{H}_2$ norm of $T_{zw}(s)$ is*

$$K_2 := \left[\begin{array}{c|c} A + B_2F_2 + L_2C_2 + L_2D_{22}F_2 & -L_2 \\ \hline F_2 & 0 \end{array}\right] \quad (40.5)$$

where

$$\begin{aligned} F_2 &= -R_{uu}^{-1}\left(R_{xu}^T + B_2^TX_2\right) \\ L_2 &= -\left(Y_2C_2^T + V_{xy}\right)V_{yy}^{-1} && (40.6) \end{aligned}$$

and X_2 and Y_2 are the unique, positive semidefinite solutions to the following Riccati equations

$$\begin{aligned} 0 &= X_2A_r + A_r^TX_2 + R_{xx} - R_{xu}R_{uu}^{-1}R_{xu}^T \\ &\quad - X_2B_2R_{uu}^{-1}B_2^TX_2 && (40.7)\\ 0 &= A_eY_2 + Y_2A_e^T + V_{xx} - V_{xy}V_{yy}^{-1}V_{xy}^T \\ &\quad - Y_2C_2^TV_{yy}^{-1}C_2Y_2 && (40.8) \end{aligned}$$

---

[1] See Chapter 42 for a detailed exposition of this concept.

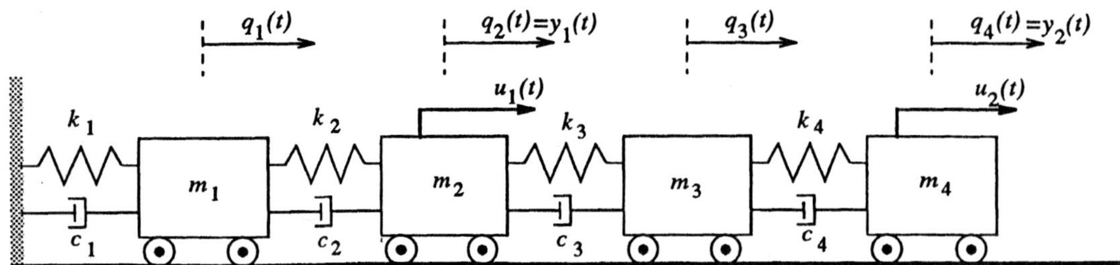

Figure 40.2 Mass, spring, dashpot system from Example 40.1. For the example $k_i = m_i = 1\ \forall i$, and $c_i = .05\ \forall i$.

where

$$A_r = \left(A - B_2 R_{uu}^{-1} R_{xu}^T\right) \quad \text{and}$$
$$A_e = \left(A - V_{xy} V_{yy}^{-1} C_2\right)$$

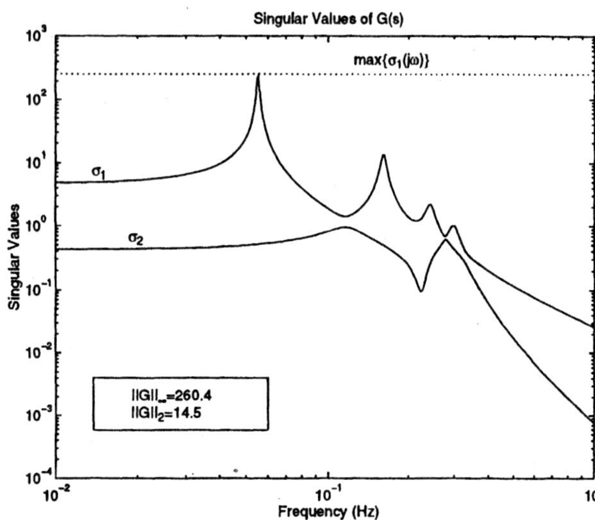

Figure 40.3 Singular values of the transfer function between the inputs and outputs of the mass, spring system shown in Figure 40.2.

THEOREM 40.2 *[$\mathcal{H}_\infty$ Output Feedback [8]] Assuming that $w(t)$ is a bounded $\mathcal{L}_2$ signal, $\int_{-\infty}^{\infty} w^T(t)w(t)dt < \infty$, a stabilizing controller which satisfies $\|T_{zw}(j\omega)\|_\infty < \gamma$ is*

$$K_\infty := \left[\begin{array}{c|c} A & -Z_\infty L_\infty \\ \hline F_\infty & 0 \end{array}\right] \tag{40.9}$$

where

$$A_\infty = A + (B_1 + L_\infty D_{21}) W_\infty + B_2 F_\infty + Z_\infty L_\infty C_2 + Z_\infty L_\infty D_{22} F_\infty$$

where

$$F_\infty = -R_{uu}^{-1}\left(R_{xu}^T + B_2^T X_\infty\right) \quad W_\infty = \frac{1}{\gamma^2} B_1^T X_\infty$$
$$L_\infty = -\left(Y_\infty C_2^T + V_{xy}\right) V_{yy}^{-1} \quad Z_\infty = \left(I - \frac{1}{\gamma^2} Y_\infty X_\infty\right)^{-1}$$

and X_∞ and Y_∞ are the solutions to the following Riccati equations

$$0 = X_\infty A_r + A_r^T X_\infty + R_{xx} - R_{xu} R_{uu}^{-1} R_{xu}^T - X_\infty \left(B_2 R_{uu}^{-1} B_2^T - \frac{1}{\gamma^2} B_1 B_1^T\right) X_\infty \tag{40.10}$$

$$0 = A_e Y_\infty + Y_\infty A_e^T + V_{xx} - V_{xy} V_{yy}^{-1} V_{xy}^T - Y_\infty \left(C_2^T V_{yy}^{-1} C_2 - \frac{1}{\gamma^2} C_1^T C_1\right) Y_\infty \tag{40.11}$$

that satisfy the following conditions

1. *$X_\infty \geq 0$*
2. *The Hamiltonian matrix for Equation 40.10,*

$$\begin{bmatrix} A - B_2 R_{uu}^{-1} R_{xu}^T & -B_2 R_{uu}^{-1} B_2^T + \frac{1}{\gamma^2} B_1 B_1^T \\ -R_{xx} + R_{xu} R_{uu}^{-1} R_{xu}^T & -\left(A - B_2 R_{uu}^{-1} R_{xu}^T\right)^T \end{bmatrix}$$

   has no $j\omega$-axis eigenvalues, or equivalently $A + B_1 W_\infty + B_2 F_\infty$ is stable
3. *$Y_\infty \geq 0$*
4. *The Hamiltonian matrix for Equation 40.11,*

$$\begin{bmatrix} \left(A - V_{xy} V_{yy}^{-1} C_2\right)^T & -C_2^T V_{yy}^{-1} C_2 + \frac{1}{\gamma^2} C_1^T C_1 \\ -V_{xx} + V_{xy} V_{yy}^{-1} V_{xy}^T & -A + V_{xy} V_{yy}^{-1} C_2 \end{bmatrix}$$

   has no $j\omega$-axis eigenvalues, or equivalently $A + L_\infty C_2 + \frac{1}{\gamma^2} Y_\infty C_1^T C_1$ is stable
5. *$\rho(Y_\infty X_\infty) < \gamma^2$, where $\rho(\cdot) = \max_i |\lambda_i(\cdot)|$ is the spectral radius*

The (sub)optimal central $\mathcal{H}_\infty$ controller which minimizes $\|T_{zw}\|_\infty$ to within the desired tolerance is K_∞ with γ equal to the smallest value of $\gamma > 0$ that satisfies conditions 1 to 5.

Unlike the $\mathcal{H}_2$ controller, the $\mathcal{H}_\infty$ controller presented here is not truly optimal. Since there is no closed-form, state-space solution to the problem of minimizing the infinity norm of a multiple-input, multiple-output (MIMO) transfer function matrix $T_{zw}(s)$, the connections between the mini-max optimization problem

$$\inf_u \sup_w \int_0^\infty \left[z^T(t)z(t) - \gamma^2 w^T(t)w(t)\right] dt \tag{40.12}$$

and $\mathcal{H}_\infty$ optimization are used to arrive at the constructive approach for synthesizing suboptimal $\mathcal{H}_\infty$ controllers given in Theorem 40.2. In fact, satisfying the conditions 1 to 5 of Theorem 40.2 is analogous to finding a saddle point of the optimization problem (Equation 40.12), and the search for $\gamma_{\min}$ is analogous to finding the global minimum over all the possible saddle points. As such, any value of $\gamma > \gamma_{\min}$ will also satisfy conditions 1 to 5 of Theorem 40.2, and thus produce a stabilizing controller. Such controllers are neither $\mathcal{H}_2$ nor $\mathcal{H}_\infty$ optimal. Since in the limit as $\gamma \to \infty$ the equations from Theorem 40.2 reduce to the equations for the $\mathcal{H}_2$ optimal controller, controllers with values of γ between $\gamma_{\min}$ and infinity provide a trade off between $\mathcal{H}_\infty$ and $\mathcal{H}_2$ performance. Along these lines, it is also worth noting that there is a rich theory for mixed $\mathcal{H}_2/\mathcal{H}_\infty$ controllers that minimize the $\mathcal{H}_2$ norm of $T_{zw}(s)$ subject to additional $\mathcal{H}_\infty$ constraints. See [9, 10, 11] for details.

The value of $w(t)$ that maximizes the cost in Equation 40.12 is known as the worst case disturbance, as it seeks to maximize the detrimental effect the disturbances have on the system. In this regard, $\mathcal{H}_\infty$ controllers provide optimal disturbance rejection to worst case disturbance, whereas the $\mathcal{H}_2$ controllers provide optimal disturbance rejection to stochastic disturbances.

Both $\mathcal{H}_2$ and $\mathcal{H}_\infty$ controllers are observer-based compensators [2], which can be seen from their block diagrams, shown in Figures 40.4 and 40.5. The regulator gains F_2 and F_∞ arise from synthesizing the full-state feedback controller, which minimizes the appropriate size of $z^T(t)z(t)$ constrained by the system dynamics Equation 40.2. Then the control law is formed by applying these regulator gains to an estimate of the states $x(t)$. The states, $x(t)$, are estimated using the noisy measurements of $y(t)$ from Equation 40.4, and L_2 and $Z_\infty L_\infty$ are the corresponding filter gains of the estimators.

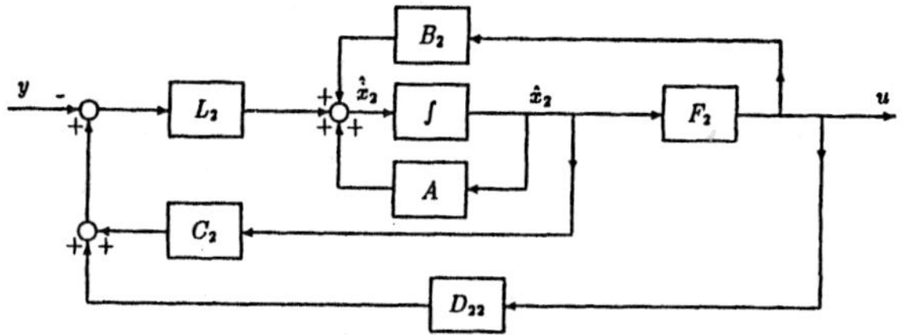

Figure 40.4 Block diagram of K_2 from Equation 40.9. Note, the Kalman Filter estimate of the states $x(t)$ from Equation 40.2, $\hat{x}_2(t)$, are the states of K_2.

In particular, F_2 is the full-state feedback LQR gain that minimizes the quadratic cost

$$J_{\mathrm{LQ}} = \mathrm{E}\left\{\lim_{\tau\to\infty}\frac{1}{\tau}\int_0^\tau\left[z^T(t)z(t)\right]dt\right\}$$

constrained by the dynamics of Equation 40.2, and L_2 is the Kalman filter gain from estimating the states x based on the measurements $y(t)$. Under the assumption that $z(t)$ is an ergodic process[2]

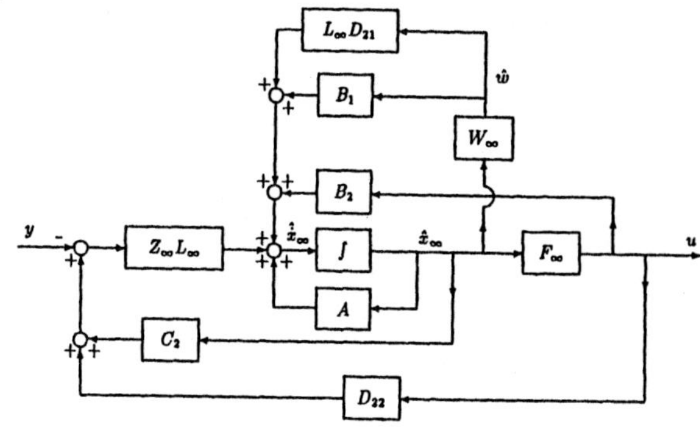

Figure 40.5 Block diagram of K_∞ from Equation 40.9. Note, the $\mathcal{H}_\infty$ optimal estimate of the states $x(t)$ from Equation 40.2, $\hat{x}_\infty(t)$, are the states of K_∞, and $\hat{w}(t)$ is an estimate of the worst case disturbance.

$$J_{\mathrm{LQ}} = \lim_{\tau\to\infty}\frac{1}{\tau}\int_0^\tau z^T(t)z(t)dt = \mathrm{E}\left[z^T(t)z(t)\right] = \|T_{zw}\|_2^2 \tag{40.13}$$

and this is exactly why $\mathcal{H}_2$ synthesis is nothing more than LQG control.

Analogously, F_∞ is the full-state feedback $\mathcal{H}_\infty$ control gain that results from optimizing the mini-max cost of Equation 40.12, and W_∞ is the full-state feedback gain that produces the worst case disturbance which maximizes the cost of Equation 40.12[3]. Unlike the Kalman filter in the $\mathcal{H}_2$ controller, the $\mathcal{H}_\infty$ optimal estimator must estimate the states of P in the presence of the worst case disturbance which is evident from the block diagram of K_∞ shown in Figure 40.5 [12]. This is why the filter gain of the $\mathcal{H}_\infty$ optimal estimator, $Z_\infty L_\infty$, is coupled to the regulator portion of the problem through X_∞ from Equation 40.10.

Since the $\mathcal{H}_2$ controller is an LQG controller, the closed-loop poles of $T_{zw}(s)$ separate into the closed-loop poles of the regulator, $\mathrm{eig}(A - B_2F_2)$, and estimator, $\mathrm{eig}(A - L_2C_2)$. A consequence of this separation property is that the $\mathcal{H}_2$ Riccati equations (Equations 40.7 and 40.8) can be solved directly without iteration. Since the worst case disturbance must be taken into consideration when synthesizing the $\mathcal{H}_\infty$ optimal estimator, the regulator and estimator problems in the $\mathcal{H}_\infty$ synthesis are coupled. Thus, the $\mathcal{H}_\infty$ controller does not have a separation structure that is analogous to that of the $\mathcal{H}_2$ controller. In addition, the $\mathcal{H}_\infty$ Riccati equations (Equations 40.10 and 40.11) are further coupled through the γ parameter, and we must iterate over the value of γ to find solutions of the $\mathcal{H}_\infty$ Riccati equations that satisfy conditions 1 to 5 of Theorem 40.2.

Note that in the literature the following set of additional, simplifying assumptions on the values of the elements of P are often made to arrive at less complicated sets of equations for the optimal $\mathcal{H}_\infty$ and $\mathcal{H}_2$ controllers [6, 13, 14].

[2]Assuming $z(t)$ is ergodic implies that its mean can be computed from the time average of a measurement of $z(t)$ as $t \to \infty$ [4].

[3]See the section on $\mathcal{H}_\infty$ Full State Feedback in Chapter 39 for details.

Additional Assumptions on P

1. $D_{22} = 0$	(No control feed-through term)
2. $C_1^T D_{12} = 0$	(No cross penalty on control and state)
3. $B_1 D_{21}^T = 0$	(Uncorrelated process and sensor noise)
4. $D_{12}^T D_{12} = I$	(Unity penalty on every control)
5. $D_{21}^T D_{21} = I$	(Unit intensity sensor noise on every measurement)

40.4 Designing $\mathcal{H}_2$ and $\mathcal{H}_\infty$ Controllers

The results presented in the previous section are powerful because they provide the control system designer with a systematic means of designing controllers for systems whose entire state cannot be fed back. In order to take full advantage of these powerful tools, it is up to the designer to communicate to the optimization problems the desired control system performance and robustness. In the modern paradigm, this is done through the choice of the system matrix P. Since systems and their associated desired performance are diverse, there is no systematic procedure for definirg P. However, by exploiting the rich mathematics of the $\mathcal{H}_2$ and $\mathcal{H}_\infty$ optimization problems along with their physical interpretations, it is possible to formulate guidelines for selecting appropriate systems P for a wide variety of problems.

Regardless of the synthesis employed, P will contain both the system model and the design weights used to communicate to the optimization the desired control system performance. Any linear interconnection of design weights and model can be selected so long as Assumptions A.1 to A.5 are satisfied. To satisfy the assumption that $R_{uu} > 0$, all of the control signals must appear explicitly in z. This is to be expected, since we cannot allow the synthesis to produce arbitrarily large control signals. Similarly, to ensure $V_{yy} > 0$, every measurement y must be corrupted by some sensor noise so as to avoid singular estimation problems.

Frequency-dependent weighting matrices are often included in P, since they provide greater freedom in telling the synthesis the desired control system performance. The synthesis of $\mathcal{H}_2$ and $\mathcal{H}_\infty$ controllers with frequency weights is just as straightforward as classical LQG synthesis with constant weights. Once the interconnection of the model with the defined performance variables, disturbances, and weights is specified, it is just a simple matter of state augmentation and block diagram manipulation to realize the state space form of P in Equations 40.2 to 40.4. Then given a state space representation of P, the formulas for the optimal controllers found in Theorems 40.1 and 40.2 can be used. The ability to use any admissible system interconnection with any combination of frequency weights is a direct consequence of the fact that we build an estimator for the entire state of P into the controllers. As such, the dynamics of any frequency weights will be reflected in the compensator whose order will be the same as that of P.

In either the $\mathcal{H}_2$ or $\mathcal{H}_\infty$ framework, arriving at a satisfactory design will involve iteration over the values of the frequency weights. Thus, it is vital to have an in-depth understanding of the dynamics of the system when choosing the system interconnection and the values for the design variables.

40.4.1 $\mathcal{H}_2$ Design

Given that the $\mathcal{H}_2$ optimal controller is an LQG controller, it is useful to adopt a stochastic framework and use the insights afforded by the well-known LQG problem when selecting P for $\mathcal{H}_2$ synthesis, see [15, 13]. In this respect, $w(t)$ must contain both the process and sensor noises, while $z(t)$ must contain linear combinations of both the states and controls. Furthermore, the system P must be comprised of the system model and all the design weights such as the noise intensities and the state and control weighting matrices. For example, Figure 40.6 illustrates a possible system interconnection for the classical LQG problem of minimizing a weighted sum of state and control penalties given a system whose dynamics

$$\begin{aligned} \dot{x}(t) &= Ax(t) + Bu(t) + Ld(t) \\ y(t) &= Cx(t) + v(t) \end{aligned} \tag{40.14}$$

are driven by the uncorrelated stochastic disturbances, $d(t)$, and sensor noise, $v(t)$.

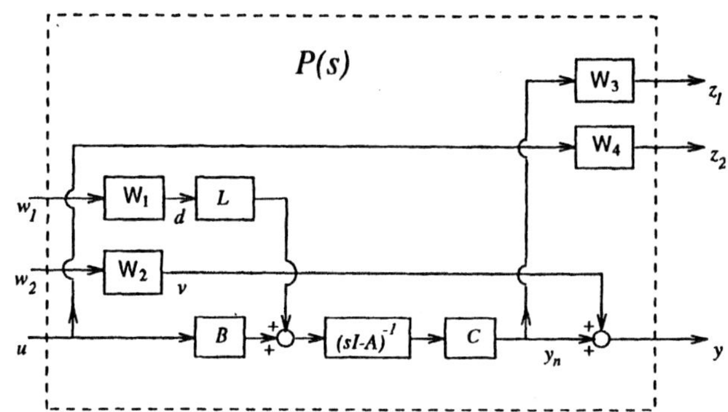

Figure 40.6 Block diagram interconnection for a typical $P(s)$.

In the interconnection of Figure 40.6, W_i are weighting matrices, or design variables, that the designer selects. For the classical LQG problem, all the W_i are constant matrices. Since $w^T(t) = [w_1^T(t) \quad w_2^T(t)]$ must be a unit intensity, white noise process, W_1 and W_2 are the matrix square roots of the intensity matrices for the process and sensor noises d and v such that

$$\mathrm{E}\left\{\begin{bmatrix} d(t) \\ v(t) \end{bmatrix}\begin{bmatrix} d^T(\tau) & v^T(\tau) \end{bmatrix}\right\}$$

$$= \begin{bmatrix} \mathsf{W}_1\mathsf{W}_1^T & 0 \\ 0 & \mathsf{W}_2\mathsf{W}_2^T \end{bmatrix}\delta(t-\tau)$$

As for the performance variable weights, W_3 is a weight on the outputs that produces a particular state weighting, and W_4 is the matrix square root of the control weighting matrix. These define the cost J_{LQ} from Equation 40.13 to be

$$J_{LQ} = \mathrm{E}\left\{\lim_{\tau\to\infty}\frac{1}{\tau}\int_0^\tau\left[x^T(t)C^T\mathsf{W}_3^T\mathsf{W}_3Cx(t) + u^T(t)\mathsf{W}_4^T\mathsf{W}_4u(t)\right]dt\right\}$$

A drawback of classical LQG synthesis is that the weighting matrices are constant and thus limit our ability to place distinct penalties on the disturbances and performance variables at various frequencies. When synthesizing $\mathcal{H}_2$ controllers the weights W_i can, in general, be functions of frequency. Since performance and robustness specifications are readily visualized in the frequency domain, using frequency weights provides much more freedom in telling the optimization problem the desired control system behavior.

When choosing the values of the frequency weights, one should use the fact that $\mathcal{H}_2$ synthesis is equivalent to LQG synthesis. Any frequency weights that appear on the performance variables can be chosen using the insights afforded by the LQR problem with frequency weighted cost functionals as a result of Equation 40.13 [4]. In brief, one uses Parseval's Theorem to arrive at a frequency domain representation of J_{LQ} from Equation 40.13. For the system interconnection shown in Figure 40.6 with scalar frequency weights $\mathsf{W}_3(s) = \mathsf{w}_3(s)I$ and $\mathsf{W}_4(s) = \mathsf{w}_4(s)I$

$$J_{LQ} \approx \frac{1}{2\pi}\int_{-\infty}^{\infty}\left[|\mathsf{w}_3(j\omega)|^2 y_n^*(j\omega)y_n(j\omega) + |\mathsf{w}_4(j\omega)|^2 u^*(j\omega)u(j\omega)\right]d\omega$$

From this expression of the $\mathcal{H}_2$ cost, it is clear that the weights should be chosen to have a large magnitude over the frequencies where we want the outputs, y_n, and controls, u, to be small.

Frequency weights that appear on the disturbance signals should be viewed as shaping filters that specify the spectral content of the process and sensor noises. The values of the weights can then be chosen to capture the true spectral content of the disturbances, as $w(t)$ must be unit intensity white noise to apply Theorem 40.2, or they can be chosen to influence the controller to produce some desired behavior. For example, in the system shown in Figure 40.6 if $\mathsf{W}_1(s) = \mathsf{w}_1(s)I$, then the control will work hard to reject the disturbances, d, over the frequencies where $|\mathsf{w}_1(j\omega)|$ is large. Likewise, if $\mathsf{W}_2(s) = \mathsf{w}_2(s)I$ and $|\mathsf{w}_2(j\omega)|$ is large over a particular frequency range, then the controller will not exert much effort there because we are telling the synthesis that the sensor measurements are very noisy there.

40.4.2 $\mathcal{H}_\infty$ Design

In the $\mathcal{H}_\infty$ framework it is possible to use loop shaping, see [1], to achieve performance and robustness specifications that can be expressed in the frequency domain. This is due to the fact that

$$\|T_{zw}\|_\infty < \gamma \Rightarrow \|(T_{zw})_{ij}\|_\infty < \gamma \quad \forall i, j \tag{40.15}$$

where $(T_{zw})_{ij}$ denotes the closed-loop transfer function matrix between exogenous disturbance w_j and performance variable z_i. To take advantage of Equation 40.15, it is necessary to define P so that the closed-loop transfer function matrices we wish to shape appear directly in $T_{zw}(s)$ and are multiplied by frequency-dependent design weights.

These concepts are best illustrated through an example. Consider the system Equation 40.14, which can be represented in the frequency domain as

$$\begin{aligned} y(s) &= G_1(s)d(s) + G_2(s)u(s) + v(s) \\ G_1(s) &= C(sI - A)^{-1}L \\ G_2(s) &= C(sI - A)^{-1}B \end{aligned}$$

where the disturbances are now considered to be unknown but bounded $\mathcal{L}_2$ signals. Suppose that we are interested in designing a controller that rejects the effect of the disturbances $d(t)$ on the outputs $y(t)$ and that is robust to an unstructured additive error in the input to output system model. Then it is necessary to independently shape the closed-loop transfer function between d and y, $S(s)G_1(s)$, and the closed-loop transfer function $K(s)S(s)$. In particular, we require $S(s)G_1(s)$ to have a desirable shape, and we need to satisfy the standard additive error stability robustness test

$$\sigma_{\max}[K(j\omega)S(j\omega)] < \frac{1}{|e_a(j\omega)|}$$

where $S(s) = [I - G_2(s)K(s)]^{-1}$ and $e_a(s)$ is a transfer function whose magnitude bounds the additive error [5]. If $\mathsf{W}_1 = I$ and $\mathsf{W}_2 = I$, then the system interconnection shown in Figure 40.6 is suitable for designing $\mathcal{H}_\infty$ controllers that achieve the loop-shaping objectives, because

$$T_{zw}(s) = \begin{bmatrix} \mathsf{W}_3S(s)G_1(s) & \mathsf{W}_3C(s) \\ \mathsf{W}_4K(s)S(s)G_1(s) & \mathsf{W}_4K(s)S(s) \end{bmatrix} \tag{40.16}$$

where $C(s) = S(s)G_2(s)K(s)$.

Notice that both of the loops of interest, $S(s)G_1(s)$ and $K(s)S(s)$, appear directly in Equation 40.16 multiplied by the design weights. By selecting scalar frequency-dependent weights, $\mathsf{W}_3 = \mathsf{w}_3(s)I$ and $\mathsf{W}_4 = \mathsf{w}_4(s)I$, an $\mathcal{H}_\infty$ controller that achieves a specific value of γ ensures that

$$\sigma_{\max}[S(j\omega)G_1(j\omega)] < \frac{\gamma}{|\mathsf{w}_3(j\omega)|} \quad \forall\omega \tag{40.17}$$

$$\sigma_{\max}[K(j\omega)S(j\omega)] < \frac{\gamma}{|\mathsf{w}_4(j\omega)|} \quad \forall\omega \tag{40.18}$$

as a result of Equation 40.15. Similar bounds will also hold for the other $(T_{zw})_{ij}$ in Equation 40.16. To take advantage of Equations 40.17 and 40.18, set $\gamma = 1$ and select the values of $\mathsf{w}_3(s)$ and $\mathsf{w}_4(s)$ to provide desirable bounds on $S(s)G_1(s)$ and $K(s)S(s)$. For example, let $\mathsf{w}_4(s) = e_a(s)$. Then if the $\mathcal{H}_\infty$ controller based on these values of the weights achieves $\|T_{zw}\|_\infty \approx 1$, the desired

[4]Section 4.4 in Chapter 39 has a detailed exposition of this.

[5]See Chapter 30.

loops will in fact be shaped to satisfy Equation 40.17 and 40.18. This is how one should choose the values of the design variables to shape the loops of interest in an $\mathcal{H}_\infty$ design.

In using this method of weight selection there are a few issues the designer must keep in mind. First of all, realize that the bounds implied by Equation 40.15 and exemplified by Equation 40.17 are not necessarily tight over all frequencies. As a result it helps to graphically inspect all the constraints implicit in the choice of $T_{zw}(s)$ as one iterates through the values of the design variables. More importantly, simply assuming $\gamma = 1$ when the values of the weights are chosen does not ensure an $\mathcal{H}_\infty$ controller that achieves $\|T_{zw}\|_\infty \approx 1$. In fact, when $\|T_{zw}\|_\infty \gg 1$ it is a strong indication that the values of the design variables impose unrealistic constraints on the system's dynamics. One cannot choose $\mathrm{w}_3(s)$ and $\mathrm{w}_4(s)$ arbitrarily. They must complement each other. Another reason why the design variables cannot be chosen arbitrarily involves the fact that $\| (T_{zw})_{ij} \|_\infty < \gamma \;\; \forall i, j$. Not only will $\mathrm{w}_3(s)$ shape the weighted sensitivity transfer function $S(s)G_1(s)$, it will also shape $C(s)$. Since $S(s) + C(s) = I$, there will clearly be restrictions on the choice of $\mathrm{w}_3(s)$. While loops such as $C(s)$ may not be of primary interest, they will influence the overall performance of the controller and should be kept in mind when selecting the values of the weights.

The choice of P in Figure 40.6 with $\mathrm{W}_1 = \mathrm{W}_2 = I$ could also have been made using structured singular value concepts [6]. In this context, the performance variables z_2 and disturbances w_2 can be viewed as the inputs and outputs to an unknown but norm bounded unstructured uncertainty that captures the additive error in the input to output model. Likewise, the performance variables z_1 and disturbances w_1 can be viewed as the inputs and outputs to a fictitious, unknown, norm bounded unstructured uncertainty that captures the desire to reject the disturbances d at the outputs y_n. Then selecting the values of the design weights is akin to scaling the system in the same way that the D-scales, used in the D-K iteration, scale the system.

40.4.3 Additional Notes for Selecting Design Weights

To ensure that Assumptions A.1 to A.5 are satisfied once the dynamics of the frequency weights are augmented to the system model, it is necessary to use proper, stable, minimum phase weights. For example, in the system shown in Figure 40.6, $\mathrm{W}_4(s)$ must contain an output feedthrough term to ensure $R_{uu} > 0$.

An important issue to be aware of when using frequency weights is that it is possible to define a set of weights with repetitive information. For example, in the system of Figure 40.6 with $\mathrm{W}_2(s) = \mathrm{w}_2(s)I$ and $\mathrm{W}_4(s) = \mathrm{w}_4(s)I$, specifying the magnitudes of $\mathrm{w}_2(s)$ and $\mathrm{w}_4(s)$ to be large over the same frequency region tells both optimization problems the same information, make the controls small there. Not only is such information redundant, it is also undesirable, since the order for the compensator is equal to the order of P.

[6]See Chapter 42 for more details.

40.5 Aircraft Design Example

To illustrate more clearly how one uses frequency weights to design $\mathcal{H}_2$ and $\mathcal{H}_\infty$ controllers, we shall discuss the design of a wind gust disturbance rejection controller for a linearized model of an F-8 aircraft. As you shall see, the modern paradigm allows us to incorporate frequency domain performance and robustness specifications naturally and directly into the controller synthesis.

The F-8 is an "old-fashioned" aircraft that has been used by NASA as part of their digital fly-by-wire research program. Assuming that the aircraft is flying at a constant altitude in equilibrium flight allows us to linearize the nonlinear equations of motion. In doing so, the longitudinal dynamics decouple from the lateral dynamics. The variables needed to characterize the longitudinal motion, which are defined in the schematic drawing of the F-8 shown in Figure 40.7, are the horizontal velocity, $v(t)$, pitch angle, $\theta(t)$, pitch rate, $q(t) = \dot{\theta}(t)$, angle of attack, $\alpha(t)$, and flight path angle, $\beta(t) = \theta(t) - \alpha(t)$. To control the longitudinal motion, elevators, $\delta_e(t)$, and flaperons, $\delta_f(t)$, which are just like the elevators except that they move in the same direction, were used. While the thrust also influences the longitudinal motion of the aircraft, it is considered to be constant in our designs. The measurements are the pitch and flight path angles, $y^T(t) = [\theta(t) \quad \beta(t)]$

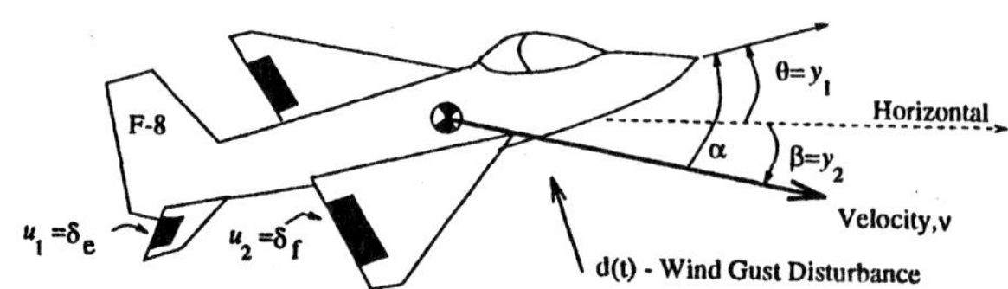

Figure 40.7 Definition of variables for the longitudinal dynamics of the F-8.

The effect of wind gust disturbances, which primarily corrupt the angle of attack, is modeled as the output of a shaping filter driven with unit intensity white noise, $d(t)$. Using the state vector

$$x^T(t) = [\theta(t) \quad \beta(t) \quad q(t) \quad v(t) \quad x_d(t)]$$

in which $x_d(t)$ is the state of the first-order shaping filter of the wind gust disturbance model, the linearized, longitudinal equations of the F-8 aircraft are

$$\begin{aligned} \dot{x}(t) &= Ax(t) + Bu(t) + Ld(t) \\ y(t) &= Cx(t) + v(t) \end{aligned} \tag{40.19}$$

with $u^T(t) = [\delta_e(t) \quad \delta_f(t)]$ and

$$A = \begin{bmatrix} 0.0 & 0.0 & 1.0 & 0.0 & 0.0 \\ 1.50 & -1.50 & 0.0 & 0.0057 & 1.50 \\ -12.0 & 12.0 & -0.60 & -0.0344 & -12.0 \\ -0.8520 & 0.290 & 0.0 & -0.0140 & -0.290 \\ 0.0 & 0.0 & 0.0 & 0.0 & -0.730 \end{bmatrix}$$

$$B = \begin{bmatrix} 0.0 & 0.0 \\ 0.160 & 0.80 \\ -19.0 & -3.0 \\ -0.0115 & -0.0087 \\ 0.0 & 0.0 \end{bmatrix}$$

$$L^T = \begin{bmatrix} 0.0 & 0.0 & 0.0 & 0.0 & 1.1459 \end{bmatrix}$$

$$C = \begin{bmatrix} 1 & 0 & 0 & 0 & 0 \\ 0 & 1 & 0 & 0 & 0 \end{bmatrix}$$

The units for the angles and control signals are in degrees while the velocity has units of ft/s. The outputs are modeled as a nominal signal with additive white noise $v(t)$ that has an intensity of $\mu = 0.01$ deg^2/s, $\mathrm{E}\left\{v^T(t)v(\tau)\right\} = \mu I\delta(t-\tau)$, to capture the limited accuracy of the sensors.

The objective is to design controllers that reduce the effect of the wind disturbance on the system. Specifically we would like the magnitude of each output to be less than 0.25 degrees up to 1.0 rad/sec as the aircraft passes through wind gusts. In addition, we require the control system to be robust to an unstructured multiplicative error reflected to the output of the plant whose magnitude is bounded by the function

$$e_m(s) = 0.1s^2$$

This multiplicative error captures the unmodeled dynamics associated with the flexibility of the aircraft's airframe. It will essentially constrain the bandwidth of the design to prevent these unmodeled modes from being excited.

Both of the design specifications can be represented in the frequency domain. To meet the performance specification, we require the closed-loop transfer function from d to y, $S(s)G_1(s)$, to satisfy

$$\sigma_{\max}[S(j\omega)G_1(j\omega)] < .25 \quad \text{for} \quad 0 < \omega \leq 1.0 \text{ rad/sec} \tag{40.20}$$

where

$$\begin{aligned} S(s) &= [I - G_2(s)K(s)]^{-1} \\ G_2(s) &= C(sI-A)^{-1}B \\ G_1(s) &= C(sI-A)^{-1}L \end{aligned}$$

To ensure stability robustness to the multiplicative error we require that

$$\sigma_{\max}[C(j\omega)] < \frac{1}{|e_m(j\omega)|} \quad \forall\omega \tag{40.21}$$

where $C(s) = S(s)G_2(s)K(s)$.

Given this representation of the design goals, we shall synthesize $\mathcal{H}_2$ and $\mathcal{H}_\infty$ controllers that shape the closed-loop transfer functions $S(s)G_1(s)$ and $C(s)$ to meet these constraints. Using the system interconnection for P shown in Figure 40.8 makes good mathematical and physical sense for this problem. Mathematically, $P(s)$ shown in Figure 40.8 leads to the following closed-loop transfer function matrix,

$$T_{zw}(s) = \begin{bmatrix} \mathrm{w}_1(s)S(s)G_1(s) & \sqrt{\mu}\mathrm{w}_2(s)C(s) \\ \rho \mathrm{w}_1(s)K(s)S(s)G_1(s) & \rho\sqrt{\mu}\mathrm{w}_2(s)K(s)S(s) \end{bmatrix} \tag{40.22}$$

Notice that the loops of interest, $S(s)G_1(s)$ and $C(s)$, appear directly in Equation 40.23 and are directly influenced by the scalar frequency weights $\mathrm{w}_1(s)$ and $\mathrm{w}_2(s)$. Realize that the coloring filter dynamics for the wind gust disturbance are already included in the system dynamics Equation (40.19), so that $\mathrm{w}_1(s)$ should not be viewed as a shaping filter for d. Rather $\mathrm{w}_1(s)$ is a design variable that overemphasizes the frequency range in which the impact of d is most vital, and it is chosen to reflect in the optimization problem our desire to appropriately shape $S(s)G_1(s)$. The scalar constant weight ρ, which is a penalty on the control that must be included in the synthesis to satisfy Assumption A.5, was allowed to vary, whereas μ was held fixed to capture the limited sensor accuracy. $P(s)$ also makes good sense in terms of the physics of the design objectives. It includes the effects of both the process and sensor noises, and its performance variables, z, contain the outputs we wish to keep small in the presence of the disturbances.

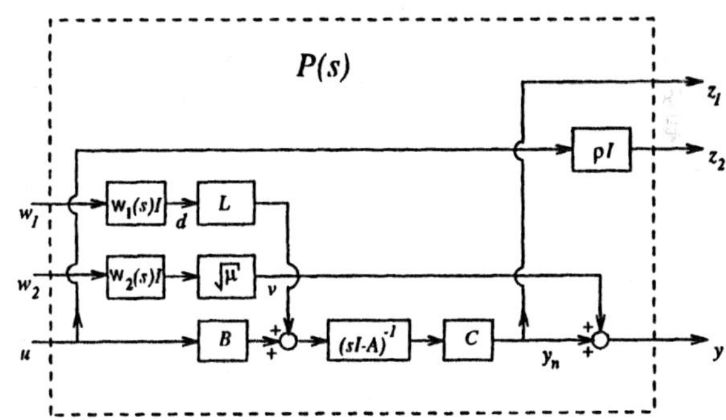

Figure 40.8 Generalized system $P(s)$ used in the F-8 designs.

To illustrate that there is a strong connection between the physical, stochastic motivation used to select the values of the weights in the $\mathcal{H}_2$ framework and the more mathematical norm bound motivation used in the $\mathcal{H}_\infty$ framework, we compare the results of an $\mathcal{H}_\infty$ and an $\mathcal{H}_2$ design that both use the same values of the weights. The weights were chosen as described in the previous section. After some iteration we found that with

$$\rho = 0.01, \quad \mathrm{w}_1(s) = \frac{0.1(s+100)}{(s+1.25)}, \quad \text{and}$$

$$\mathrm{w}_2(s) = \frac{5000(s+3.5)}{3.5(s+1000)}$$

both the $\mathcal{H}_\infty$ and $\mathcal{H}_2$ designs met the desired performance and robustness specifications. Note that as a result of using these frequency weights, the controllers had eight states.

Figures 40.9 and 40.10 show that the loops of interest have in fact been shaped in accordance with the design goals. As seen in Figure 40.9, which compares the open and closed-loop disturbance to output transfer functions, both designs meet the performance goal from Equation 40.20. In this figure, $1/|\mathrm{w}_1(j\omega)|$ is also shown to illustrate how the value of $\mathrm{w}_1(s)$ was chosen. From an $\mathcal{H}_2$ perspective, the $|\mathrm{w}_1(j\omega)|$ is large over $0 < \omega \leq 1.0$ rad/sec and small elsewhere to tell the synthesis that the intensity of the disturbance is large where we desire good disturbance rejection.

In the context of the $\mathcal{H}_\infty$ design, we assumed that $\gamma = 1$ and selected $1/|w_1(j\omega)|$ to appropriately bound $\sigma_{max}[S(j\omega)G_1(j\omega)]$ in accordance with the fact that

$$\sigma_{max}[S(j\omega)G_1(j\omega)] < \frac{\gamma}{|w_1(j\omega)|} \quad \forall\omega \tag{40.23}$$

for any $\mathcal{H}_\infty$ design which achieves $\|T_{zw}\|_\infty < \gamma$. For the values of the weights given above, $\|T_{zw}\|_\infty = 0.95$ for the $\mathcal{H}_\infty$ design, which ensures that Equation 40.23 was satisfied. A difference in the performance achieved by both designs is expected since the same values of the weights are used to minimize different measures of the size of $T_{zw}(s)$.

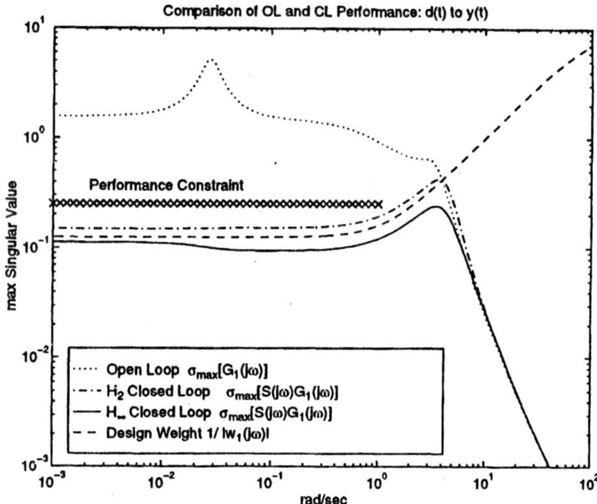

Figure 40.9 Comparison of the open- and closed-loop disturbance to output transfer functions for the $\mathcal{H}_\infty$ and $\mathcal{H}_2$ designs. The frequency weight $w_1(s)$ used to shape $S(s)G_1(s)$ is also shown.

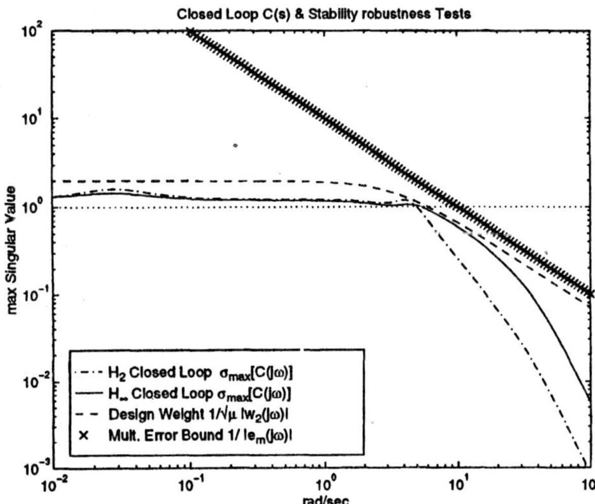

Figure 40.10 Comparison of $\sigma_{max}[C(j\omega)]$ for the $\mathcal{H}_\infty$ and $\mathcal{H}_2$ designs. The bound on the multiplicative error, $1/|e_m(j\omega)|$, and the frequency weight $w_2(s)$ used to shape $C(s)$ are also shown.

Figure 40.10, which compares $\sigma_{max}[C(j\omega)]$ for the designs, illustrates that the stability robustness criterion from Equation 40.21 is also satisfied. Again, $1/\sqrt{\mu}|w_2(j\omega)|$ is shown to illustrate the manner in which the value of $w_2(s)$ was chosen. As seen in the figure, $w_2(s)$ forces the $\mathcal{H}_\infty$ design to roll off below the stability robustness bound, $1/|e_m(j\omega)|$, in accordance with the fact that

$$\sigma_{max}[C(j\omega)] < \frac{\gamma}{\sqrt{\mu}|w_2(j\omega)|} \quad \forall\omega. \tag{40.24}$$

Since $|w_2(j\omega)|$ is large beyond 5.0 rad/sec, it tells the $\mathcal{H}_2$ synthesis that the sensor noise is large there, which in turn limits the control energy beyond 5.0 rad/sec.

The value of ρ also played an important role in the designs. In the $\mathcal{H}_2$ design, adjusting ρ directly influenced the amount of control effort used, just as a control weight would in LQG synthesis. For the $\mathcal{H}_\infty$ design, ρ minimized the constraints that the values of $w_1(s)$ and $w_2(s)$ placed on the closed-loop transfer functions $K(s)S(s)G_1(s)$ and $K(s)S(s)$ in Equation 40.23. Since these loops are not of primary interest, choosing a small value of ρ ensured that $w_1(s)$ and $w_2(s)$ would not overly constrain these since, for example,

$$\sigma_{max}[K(j\omega)S(j\omega)G_1(j\omega)] < \frac{\gamma}{\rho|w_1(j\omega)|} \quad \forall\omega$$

for any $\mathcal{H}_\infty$ design.

The similarities in the achieved loop shapes are not coincidental. In fact, the dynamics of the $\mathcal{H}_\infty$ and $\mathcal{H}_2$ controllers presented here are quite similar. There is a clear reason why the similarity exists even though the optimization problems used are distinct. Once all of the desired control system performance is incorporated into $P(s)$ via the design variables, the task of minimizing the $\mathcal{H}_2$ norm of $T_{zw}(s)$ becomes nearly identical to the task of minimizing the $\mathcal{H}_\infty$ norm of $T_{zw}(s)$. This can be seen in Figure 40.11, which compares the values of $\sigma_{max}[P_{zw}(j\omega)]$ and $\sigma_{max}[T_{zw}(j\omega)]$ for the two designs. Here $P_{zw}(s)$ denotes the open-loop transfer function matrix between w and z of $P(s)$. As such, $\sigma_{max}[P_{zw}(j\omega)]$ is an indication of the nominal cost that the controllers seek to minimize. Specifically, to minimize the $\mathcal{H}_\infty$ norm of $T_{zw}(s)$ the peak in $\sigma_{max}[P_{zw}(j\omega)]$ must be flattened out so that it looks like a low pass filter. Then the DC gain of the filter must be reduced to further minimize the $\mathcal{H}_\infty$ norm of $T_{zw}(s)$. This is also the case for minimizing the $\mathcal{H}_2$ norm of $T_{zw}(s)$ which is dominated by the area under the spike in $\sigma_{max}[P_{zw}(j\omega)]$ (recall that the area is evaluated linearly and that we are using a log log plot). While the optimization problems are distinct, the manner in which the cost is minimized is similar.

Figure 40.11 also provides a clear indication of how the optimization problems differ. Notice that the $\mathcal{H}_2$ design rolls off faster than the $\mathcal{H}_\infty$ design. This is because the $\mathcal{H}_2$ design minimizes energy, or the area under $\sigma_{max}[T_{zw}(j\omega)]$, at the expense of its peak value, whereas the $\mathcal{H}_\infty$ design seeks to minimize the peak of $\sigma_{max}[P_{zw}(j\omega)]$ at the expense of allowing there to be more energy at higher frequencies.

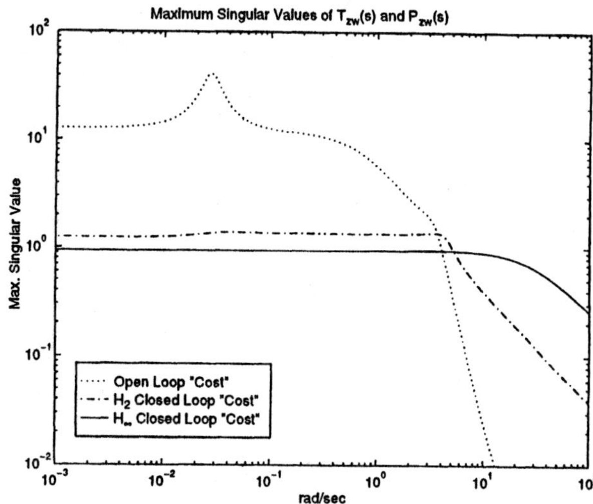

Figure 40.11 Comparison of the open-loop "cost", $\sigma_{\max}[P_{zw}(j\omega)]$, and the closed-loop "cost", $\sigma_{\max}[T_{zw}(j\omega)]$, for the $\mathcal{H}_\infty$ and $\mathcal{H}_2$ controllers.

It would be improper to draw conclusions about which synthesis approach is better based on these designs, especially since the same values for the weights in $P(s)$ were used. Rather, our intent has been to illustrate the connections between the $\mathcal{H}_\infty$ and $\mathcal{H}_2$ frameworks and how one can go about synthesizing $\mathcal{H}_\infty$ and $\mathcal{H}_2$ controllers to meet frequency domain design specifications. We should also note that the methodology used in this example has been applied to and experimentally verified on a much more complex system [16].

References

[1] Doyle, J., Francis, B., and Tannenbaum, A., *Feedback Control Theory*, Macmillan, New York, 1992.

[2] Boyd, S. P. and Barrat, C. H., *Linear Controler Design: Limits of Performance*, Prentice Hall, Englewood Cliffs, NJ, 1991.

[3] Zhou, K. Doyle, J., and Glover, K., *Robust and Optimal Control*, Prentice Hall, Englewood Cliffs, NJ, 1995.

[4] Papoulis, A., *Probability, Random Variables, and Stochastic Processes*, 3rd ed., McGraw Hill, New York, 1991.

[5] Francis, B. A., *A Course in $\mathcal{H}_\infty$ Control Theory*, Springer-Verlag, Berlin, 1987.

[6] Doyle,J., Glover,K., Khargonekar, P., and Francis, B., State space solutions to standard $\mathcal{H}_2/\mathcal{H}_\infty$ control problems, *IEEE Trans. Autom. Control*, 34(8), 831–847, 1989.

[7] Maciejowski, J., *Multivariable Feedback Design*, Addison-Wesley, Wokingham, England, 1989.

[8] Glover, K. and Doyle, J., State-space formulae for all stabilizing controllers that satisfy an $\mathcal{H}_\infty$-norm bound and relations to risk sensitivity, *Syst. Control Lett.*, 11, 167–172, 1988.

[9] Bernstein, D.S. and Haddad, W.M., LQG control with an $\mathcal{H}_\infty$ performance bound: a Riccati equation approach, *IEEE Trans. Autom. Control*, 34(3), 293–305, 1989.

[10] Doyle, J., Zhou, K., Glover, K., and Bodenheimer, B., Mixed $\mathcal{H}_2$ and $\mathcal{H}_\infty$ performance objectives, II. Optimal control, *IEEE Trans. Autom. Control*, 39(8), 1575–1587, 1994.

[11] Scherer, C.W., Multiobjective $\mathcal{H}_2/\mathcal{H}_\infty$ control, *IEEE Trans. Autom. Control*, 40(6), 1054–1062, 1995.

[12] Nagpal, K. and Khargonekar, P., Filtering and smoothing in a $\mathcal{H}_\infty$ setting, *IEEE Trans. Autom. Control*, 36(2), 152–166, 1991.

[13] Kwakernaak, H. and Sivan, R., *Linear Optimal Systems*, John Wiley & Sons, New York, 1972.

[14] Kwakernaak, H., Robust control and $\mathcal{H}_\infty$ optimization—tutorial paper, *Automatica*, 29(2), 255–273, 1993.

[15] Anderson, B.D.O. and Moore, J.B., *Optimal Control: Linear Quadratic Methods*, Prentice Hall, Englewood Cliffs, NJ, 1990.

[16] Lublin, L. and Athans, M., An experimental comparison of $\mathcal{H}_2$ and $\mathcal{H}_\infty$ designs for an interferometer testbed, in *Lecture Notes in Control and Information Sciences: Feedback Control, Nonlinear Systems, and Complexity*, Francis, B. and Tannenbaum, A., Eds., Springer-Verlag, Amsterdam, 1995, 150–172.

41

ℓ_1 Robust Control: Theory, Computation and Design

Munther A. Dahleh
Lab. for Information and Decision Systems, M.I.T., Cambridge, MA

41.1 Introduction[1]

Feedback controllers are designed to achieve certain performance specifications in the presence of both plant and signal uncertainty. Typical controller design formulations consider quadratic cost functions on the errors primarily for mathematical convenience. However, practical situations may dictate other kinds of measures. In particular, the peak values of error signals are often more appropriate for describing desired performance. This can be a consequence of uniform tracking problems, saturation constraints, rate limits, or simply a disturbance rejection problem. In addition, disturbance and noise are in general bounded and persistent because they continue to act on the system as long as the system is in operation. Such signals are better described by giving information about both the signals' frequency content and time-domain bounds on peak values.

The above class of problems motivated a formulation that involves the Peak-to-Peak gain of a system, which is mathematically given by the ℓ_1 norm of the pulse response. This formulation was first reported in [13], and the problem was completely solved in several subsequent articles [5, 6, 7, 8, 11]. The extension of the theory to incorporate plant uncertainty was reported in [3, 4, 9, 10]. An extensive coverage of this theory with detailed references can be found in [2].

The need for developing this problem was further intensified by the failure of the frequency-domain techniques to address time-domain specifications. For instance, attempting to achieve an overshoot constraint using $\mathcal{H}_\infty$ or $\mathcal{H}_2$ by appropriately adjusting the weighting matrices can be a very frustrating experience. On the other hand, solutions to such problems will no longer be in closed form due to the complexity of the performance objectives. Exact or approximate solutions will be obtained from solving equivalent yet simpler optimization problems. The derivation of such simpler problems and the computational properties are essential components of this research direction.

41.1.1 A Design Tool

The motivation behind research in ℓ_1 theory is developing a design tool for MIMO uncertain systems. A powerful design tool in general should have three ingredients:

1. **Practicality:** The ability to translate a large set of performance specifications into conditions or constraints readily acceptable by the design tool. It is evident that not all design specifications can be immediately translated into mathematical conditions. However, the mathematical formulation should well approximate these objectives.
2. **Computability:** It is in general straightforward to formulate controller design problems as constrained optimization problems. What is not so straightforward is formulating problems that can be solved efficiently and with acceptable complexity.

[1] Research Supported by AFOSR under grant AFOSR-91-0368, by NSF under grant 9157306-ECS, and by Draper Laboratory under grant DL-H-467128.

0-8493-8570-9/96/$0.00+$.50

3. **Flexibility:** The ability to alter a design to achieve additional performance specifications with small marginal cost.
   It is evident that practicality and computability are conflicting ingredients. Computational complexity grows as a function of several parameters, which include the dimension of the system, the uncertainty description, and the performance specifications. Flexibility makes it possible to design a controller in stages, i.e., by altering a nominally good design to achieve additional specifications.

41.1.2 Motivation

We give some reasons behind the development of such a design tool, by quoting from the book: *Control of Uncertain System: A Linear Programming Approach.*

1. **Complex Systems:** Many of today's systems, ranging from space structures to high purity distillation columns, are quite complex. The complexity comes from the very high order of the system as well as the large number of inputs and outputs. Modeling such systems accurately is a difficult task and may not be possible. A powerful methodology that deals systematically with multiple inputs and outputs and with various classes of structured uncertainty is essential.
2. **High Performance Requirement:** Systems are built to perform specific jobs with high accuracy. Robots are already used to perform accurate jobs such as placing components on integrated circuit boards. Aircraft are built with high maneuverability and are designed specifically for such tasks. Classical SISO design techniques cannot accommodate these problems resulting in designs that are conservative and perform poorly.
3. **Limits of Performance:** In complex systems, it is time-consuming to establish, by trial and error, whether a system can meet certain performance objectives (even without uncertainty). Thus, it is necessary to develop systematic methods to quantify the fundamental limitations of systems and to highlight the trade-offs of a given design.
4. **A Systematic Design Process:** It is inevitable that designing a controller for a system will involve iterations. Unless this procedure is made systematic, the design process can become very cumbersome and slow. The design procedure should not only exhibit a controller. It should also provide the designer with indicators for the next iteration, by showing which of the constraints are the limiting ones and which part of the uncertainty is causing the most difficulty. Note also that a general procedure should be able to accommodate a variety of constraints, both in the time and in the frequency domain.
5. **Computable Methods:** It is quite straightforward to formulate important control problems, but it is not so easy to formulate solvable problems that can provide computable methods for analysis and synthesis. Much of the current research invokes high level mathematics to provide simple computable results. The computability of a methodology is the test of its success. By computability we do not mean necessarily closed form solutions. Problems that give rise to such solutions have limited applicability. However, computability means that we can synthesize controllers via computable algorithms and simultaneously obtain qualitative information about the solution.
6. **Technological Advancement:** Many aspects of technological development will affect the design of control systems. The availability of "cheaper" sensors and actuators that are well-modeled allows for designing control systems with a large number of inputs and outputs. The availability of fast microprocessors, as well as memory, makes it possible to implement more complex feedback systems with high order. The limiting factor in controller implementation is no longer the order of the controller. Instead, it is the computational power and the memory availability.
7. **Available Methods:** The available design techniques have concentrated on frequency-domain performance specifications by considering errors in terms of energy rather than peak values. These methods (such as $\mathcal{H}_2$ and $\mathcal{H}_\infty$) have elegant solutions. However, this elegance is lost if additional time-domain specifications (e.g., overshoot, undershoot, settling time ..) are added. This created a need for a time-domain based computational methodology that can accommodate additional frequency-domain constraints. This design tool aims at achieving this objective.

In the sequel we will summarize the ℓ_1 design tool by discussing the three ingredients, *practicality, computability,* and *flexibility.*

41.2 Practicality

41.2.1 Examples

To motivate the formulation of the ℓ_1 problem, we begin with two examples.

The first is the control design for an *Earth Observing System (EOS).* EOS is a spacecraft that orbits the earth and points in a specific location. It carries on its platform various sensory instruments, with the objective of collecting data from earth. An example of such an instrument is an array of cameras intended to provide images of various points and landscapes on earth. The spacecraft is subjected to various kinds of disturbances: external pressures, noise generated from the instruments on board, and the spacecraft itself. The objective of the control design is to point the spacecraft accurately in a specific direction, otherwise known as attitude control.

The second example is the control design of an active suspension of an automobile. A simplified one-dimensional problem is shown in Figure 41.1. The objective of the controller design

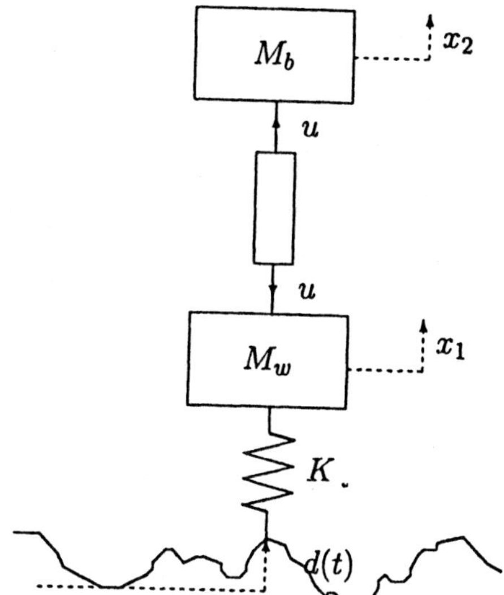

Figure 41.1 Active suspension.

is to maximize ride comfort, while simultaneously maintaining handling (road holding ability) in the presence of bad road conditions.

These examples have several common features:

1. The objectives in both problems are to keep the maximum deviations of signals from set points bounded by some prescribed value to attain uniform tracking or disturbance rejection. In the EOS example, performance is measured in terms of maximum deviations of the attitude angles from a set point. In the active suspension problem, performance is measured in terms of the maximum force acting on the system and the maximum deviation of the distance of the wheel from the ground.
In mathematical terms, both performance specifications are stated in terms of *peak* values of signals, i.e.,
$$\|w\|_\infty = \max_{i=1,\ldots n} \sup_t |w(t)|.$$
This is known as the ℓ_∞-norm of the signal (or its peak value). For the active suspension problem, the performance can be stated as
$$\left\| \begin{pmatrix} x_1 - d \\ u \end{pmatrix} \right\|_\infty \leq \gamma,$$
for some performance bound γ. It is straightforward to incorporate additional scalings and weights into the performance.
2. The disturbances in both examples are unknown, persistent in time, but, bounded in magnitude (peak value). A good model for the disturbance in both cases is given by
$$d = Ww, \qquad \|w\|_\infty \leq 1,$$
where W is a linear time-invariant filter that gives information about the frequency content of the disturbance. This model of disturbance accommodates persistent disturbances that are not necessarily periodic. It does not assume that the signal dies out asymptotically.
3. In both problems, saturation constraints are quite important and play a role in limiting the performance. In the active suspension problem, the saturation constraint is given by the maximum deflection of the hydraulic actuator, i.e.,
$$\|x_1 - x_2\|_\infty \leq \gamma_{sat}.$$
These constraints combined with the performance objectives have to be satisfied for all disturbances w such that $\|w\|_\infty \leq 1$.
4. Both examples are difficult to model precisely, and thus the control strategies have to accommodate unmodelled dynamics. In this article, we will not discuss in detail the robust performance problems. We refer the reader to [2] for details.

It is evident from the above discussion that peak values of signals are natural quantities in stating design specifications.

41.2.2 Formulation

Figure 41.2 shows a general setup for posing performance specifications. The variables as defined in the figure are

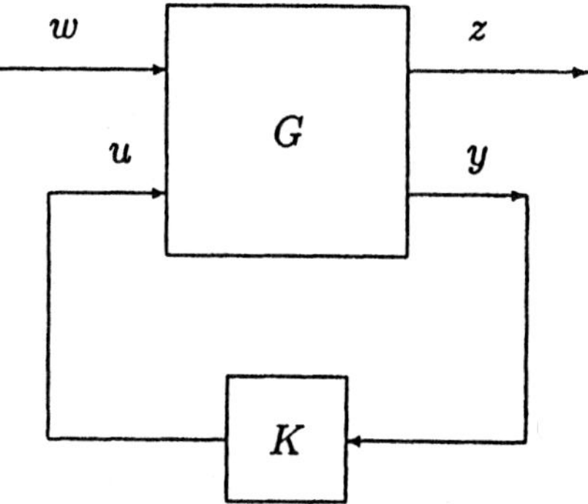

Figure 41.2 General setup.

$u =$ control inputs
$y =$ measured outputs

$$w = \text{exogenous Inputs} = \begin{cases} \text{fixed commands} \\ \text{unknown commands} \\ \text{disturbances} \\ \text{noise} \\ \vdots \end{cases}$$

$$z = \text{regulated outputs} = \begin{cases} \text{tracking Errors} \\ \text{control Inputs} \\ \text{measured Outputs} \\ \text{states} \\ \vdots \end{cases}$$

The operator G is a 2×2 block matrix mapping the inputs w and u to the outputs z and y:

$$\begin{bmatrix} z \\ y \end{bmatrix} = \begin{bmatrix} G_{11} & G_{12} \\ G_{21} & G_{22} \end{bmatrix} \begin{bmatrix} w \\ u \end{bmatrix}.$$

The actual process or *plant* is the submatrix G_{22}. Both the exogenous inputs and the regulated outputs are auxiliary signals that need not be part of the closed loop system. The feedback controller is denoted by K.

From our discussion above, the set of exogenous inputs consists of unknown, persistent, but bounded, disturbances,

$$\mathcal{D} = \{w \in \ell_\infty \mid \|w\|_\infty \le 1\}.$$

The performance measure (combined with constraints) is stated as

$$\|z\|_\infty \le \gamma, \ \forall w \in \mathcal{D}.$$

If Φ is the linear time-invariant system mapping w to z, then

$$\|z\|_\infty \le \gamma, \text{ for all } w \in \mathcal{D} \Longleftrightarrow \|\Phi\|_1 \le \gamma,$$

where

$$\|\Phi\|_1 = \max_{1 \le i \le n_z} \sum_{j=1}^{n_w} \sum_{k=0}^{\infty} |\phi_{ij}(k)|.$$

The latter is the expression for the ℓ_1 norm of the system. In conclusion, the ℓ_1 norm of a system is the *peak-to-peak* gain of the system and can directly describe time-domain performance specifications. The nominal performance problem can be stated as

$$\inf_{K \text{ stabilizing}} (\sup_w \|\Phi\|_1).$$

41.2.3 Comparison to $\mathcal{H}_\infty$

Suppose the exogenous inputs are such that $\|w\|_2 \le 1$ but are otherwise arbitrary ($\|w\|_2$ is the energy contained in the signal). If the objective is to minimize the energy of the regulated output, then the nominal performance problem is defined as

$$\inf_{K \text{ stabilizing}} (\sup_w \|\Phi w\|_2) = \inf_{K \text{ stabilizing}} \sup_\theta \sigma_{max}[\hat{\Phi}(e^{i\theta})].$$

Both of these norm minimization problems fall under the same paradigm of minimax optimality. Minimizing the $\mathcal{H}_\infty$ norm results in attenuating the energy of the regulated signal but may still result in signals that have large amplitudes. Minimizing the ℓ_1 norm results in attenuating the amplitude of the regulated output, and overbounds the maximum energy due to bounded energy inputs because

$$\|\hat{\Phi}\|_{\mathcal{H}_\infty} \le \sqrt{m}\|\Phi\|_1,$$

where m is the number of rows in Φ. On the other hand, the following inequality holds:

$$\|\Phi\|_1 \le 2(N+1)\sqrt{n}\|\hat{\Phi}\|_{\mathcal{H}_\infty},$$

where N is the McMillan degree of the system, and n is the number of columns of Φ. The latter bound is the tightest possible bound (i.e., equality holds for certain classes of systems) and it shows that the gap between these measures can be large if the order of the system is high.

41.2.4 Prototypes

The following prototypes have been discussed in [2]. These are representative problems quite common in applications. We will use these prototypes to illustrate the significance of the ℓ_1 design methodology.

Disturbance Rejection

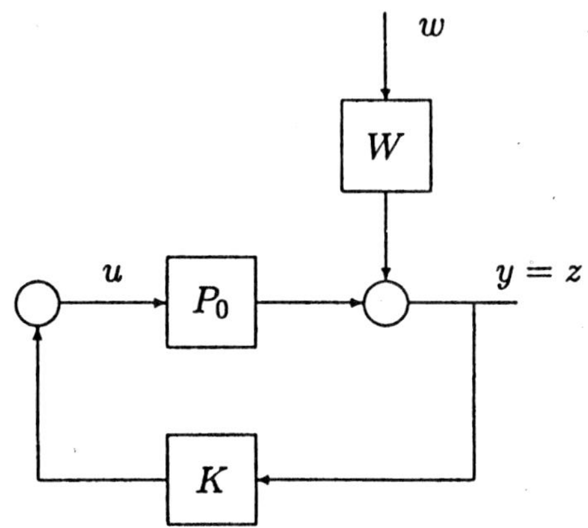

Figure 41.3 A disturbance rejection problem.

In the context of ℓ_∞ signals, the disturbance rejection problem is defined as follows: Find a feedback controller that minimizes the maximum amplitude of the regulated output over all possible disturbances of bounded magnitude. The two-input two-output system shown in Figure 41.3 depicts the particular case where the disturbance enters the system at the plant output. Its mathematical representation is given by

$$\begin{aligned} z &= P_0 u + W w, \\ y &= P_0 u + W w. \end{aligned}$$

The disturbance rejection problem provides a general enough structure to represent a broad class of interesting control problems.

Command Following with Saturation

The command following problem, equivalent to a disturbance rejection problem, is shown in Figure 41.4. We will show how to pose this problem in the presence of saturation nonlinearities at the input of the plant, as an ℓ_1-optimal control problem. Define the function

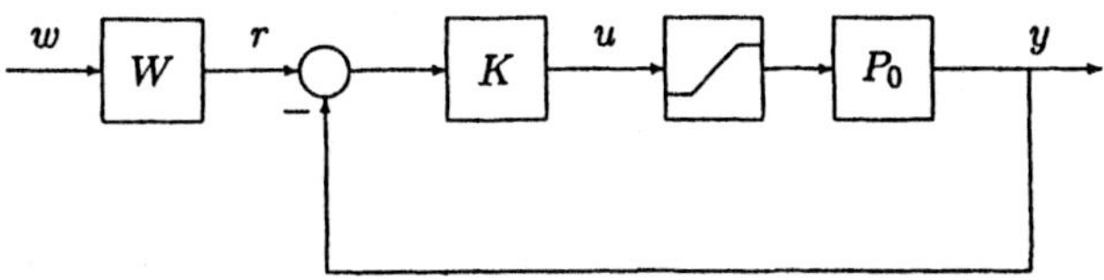

Figure 41.4 Command following with input saturation.

$$\text{Sat}(u) = \begin{cases} u & |u| \leq U_{\max} \\ U_{\max} sgn(u) & |u| \geq U_{\max} \end{cases}.$$

Let the plant be described as

$$Pu = P_0 \text{Sat}(u)$$

where P_0 is LTI. Let the commands be modeled as

$$r = Ww \text{ where } \|w\|_\infty \leq 1.$$

The objective is to find a controller K so that y follows r uniformly in time. Keeping in mind the saturation function, and in order to stay in the linear region of operation, the allowable control inputs must have $\|u\|_\infty \leq U_{\max}$. Let γ be the (tracking) performance level desired, and define

$$z = \begin{bmatrix} (y - r)/\gamma \\ u/U_{\max} \end{bmatrix}$$

with

$$y = P_0 u.$$

The problem is equivalent to finding a controller so that

$$\sup_w \|z\|_\infty < 1,$$

which is an ℓ_1-optimal control problem.

The above closed loop system will remain stable even if the input saturates, as long as it does so infrequently. The solution to the above problem will determine the limits of performance when the system is required to operate in the linear region. Also, the stability for such a system will mean the local stability of the nonlinear system.

Saturation and Rate Limits

In the previous example, actuator limitations may require that the rate of change of the control input be bounded. This is captured in the condition

$$\left| \frac{u(k) - u(k-1)}{T_s} \right| \leq U_{\text{der}}$$

where T_s is the sampling period. Let

$$W_{\text{der}} = \frac{1 - \lambda}{T_s U_{\text{der}}}.$$

This condition can be easily incorporated in the objective function by defining z as

$$z = \begin{bmatrix} (y - r)/\gamma \\ u/U_{\max} \\ W_{\text{der}} u \end{bmatrix}.$$

The result is a standard ℓ_1-optimal control problem.

41.2.5 Stability and Performance Robustness

The power of any design methodology is in its ability to accommodate plant uncertainty. The ℓ_1 norm gives a good measure of performance. Because it is a gain over a class of signals, it will also provide a good measure for robustness. This is a consequence of the small gain theorem which is stated below [2].

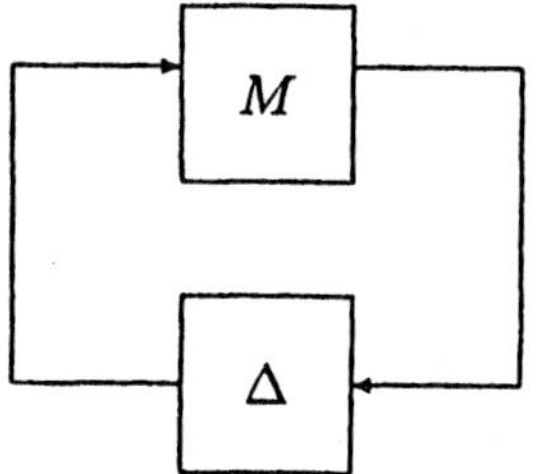

Figure 41.5 Stability robustness problem.

Let M be a linear time-invariant system and Δ be a strictly proper ℓ_∞-stable perturbation. The closed-loop system shown in Figure 41.5 is ℓ_∞-stable for all Δ with $\|\Delta\|_{\ell_\infty-\text{ind}} \leq 1$ if and only if $\|M\|_1 < 1$.

The above result indicates that the ℓ_1 norm is the exact measure of robustness when the perturbations are known to be BIBO stable, bounded gain, and possibly nonlinear or time-varying. The result can be adapted to derive stability robustness conditions for a variety of plant uncertainty descriptions. We describe one such situation below.

Unstructured Multiplicative Perturbations

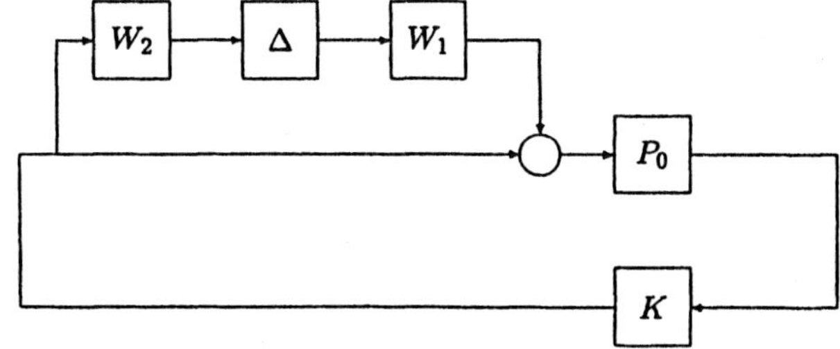

Figure 41.6 Multiplicative perturbations.

Consider the case where the system has input uncertainty in a multiplicative form as in Figure 41.6, i.e., let

$$\Omega = \{P \mid P = P_o(I + W_1 \Delta W_2) \text{ and } ||\Delta||_{\ell_\infty-\text{ind}} \leq 1\}.$$

If a controller is designed to stabilize P_0, under what conditions will it stabilize every system in the set Ω? By simple manipulations of the closed-loop system, the problem is equivalent to the stability robustness of the feedback system in Figure 41.5, with $M = W_2(I - KP_o)^{-1}KP_oW_1$. In general this manipulation is done in a systematic way: Cut the loop at the inputs and outputs of Δ, and then calculate the map from the output of Δ, w, to the input of Δ, z. A sufficient condition for robust stability is then given by $\|M\|_1 < 1$. The resulting two-input two-output description is given by

$$\begin{aligned} y &= P_0 u + P_0 W_1 w, \\ z &= W_2 u. \end{aligned}$$

This is a standard ℓ_1 minimization problem.

Structured Uncertainty

In many applications, uncertainty is described in a structured fashion where independent perturbations are introduced in various locations of the closed loop system. It turns out that one can derive an exact necessary and sufficient condition in terms of a scaled ℓ_1 norm of the system to guarantee stability robustness in the presence of such structured perturbations. It can also be shown that the problem of achieving robust performance (where performance is measured in terms of the ℓ_1 norm) is equivalent to robustly stabilizing a plant with structured uncertainty. In this article, we will not discuss this problem. Instead, we refer the reader to [2] for more details.

41.3 Computability

Since it is quite hard, in general, to obtain closed form solutions to general optimization problems, we need to be precise about the meaning of a "solution". A closed form solution has two important features: the first is the ability to compute the optimal solution through efficient algorithms, and the second is to provide a qualitative insight into the properties of the optimal solution. A numerical solution should offer both of these ingredients. In particular, it should provide

1. the exact solution whenever it is possible
2. upper and lower approximations of the objective function when exact solutions cannot be obtained and a methodology for synthesizing suboptimal controllers
3. qualitative information about the controller, e.g., the order of the controller

Solutions based on general algorithms even for convex optimization problems offer only approximate upper bounds on the solution. To obtain more information, one needs to invoke duality theory. Duality theory provides a simple reformulation of the optimization problem from which lower bounds on the objective function and, possibly, exact solutions can be found.

41.3.1 Summary of Results

To minimize the ℓ_1 norm, first the Youla parameterization of all stabilizing controllers is invoked. The resulting optimization problem can be stated as an infinite-dimensional linear program in the free parameter. Two cases occur:

1. The infinite dimensional LP is exactly equivalent to a finite-dimensional LP. This happens if the dimension of the control input is at least as large as the dimension of the regulated variables and the dimension of the measured output is at least as large as the exogenous inputs. This means that the controller has a lot of degrees of freedom.
2. If any of the above conditions is violated, then the problem is inherently infinite-dimensional. However, duality theory can be used to provide approximate solutions to this problem with guaranteed performance levels.

The details of the computations for both of the above cases can be found in [2]. The most successful algorithm for computing solutions for the second case is not based on straightforward approximation, but rather on embedding the problem in another that falls under the first case. This procedure generates approximate solutions with converging upper and lower bounds, and also provides information about the order of the actual optimal controller. Other emerging techniques are based on dynamic programming and viability theory and can be found in [1, 12]

41.4 Flexibility

Flexibility is the ability to use the design tool to alter a given nominal design so that additional specifications are met with minimal expense. Examples of additional specifications include fixed input constraints and frequency-domain constraints. The computational cost of alteration should be much less than the incorporation of the specification directly in the problem. In addition, it is desirable to maintain the qualitative information of the original solution.

Since the general synthesis problem is equivalent to an infinite dimensional LP, many additional specifications (not directly addressed by ℓ_1 norms) can be incorporated as additional linear constraints. Frequency-domain constraints can be well-approximated by linear constraints. Below we consider an example of adding fixed input constraints to the ℓ_1 problem.

41.4.1 ℓ_1 Performance Objective with Fixed Input Constraints

Consider the case where the specifications given are such that the control signal resulting from a step input must be constrained

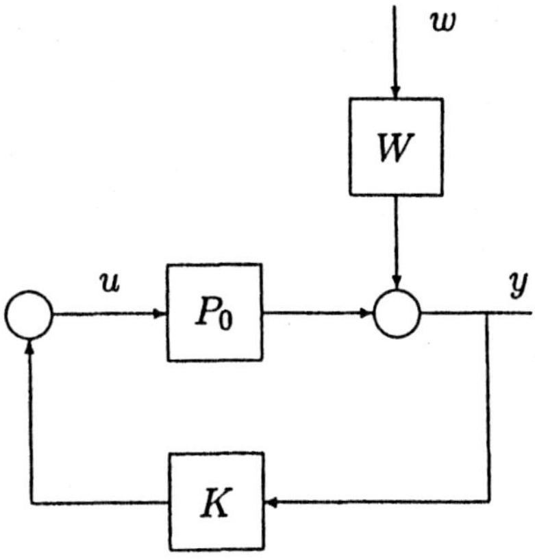

Figure 41.7 A disturbance rejection problem.

uniformly in time (e.g., to avoid actuator saturation). We want to bound the controller response to a step input and at the same time minimize the ℓ_1 norm of the transfer functions from the disturbance to both the control signal and plant output. In such a case we augment the basic ℓ_1 problem in the following way:

$$\inf_{K \text{ stab.}} \left\| \begin{array}{c} K(I+PK)^{-1}W \\ (I+PK)^{-1}W \end{array} \right\|_1$$

subject to

$$\|K(I+PK)^{-1}w_f\|_\infty \leq U_{max}$$

where w_f is a unit step input disturbance and U_{max} is the specified bound. The above modification results in adding infinitely many constraints on the sequence $K(I+PK)^{-1}$ (i.e., convolution of a unit step with $K(I+PK)^{-1}$). However, since the peak is typically achieved in early samples, only a finite number of constraints must be included (the rest being inactive). This is a particular case of nondecaying template constraints which arise frequently in control system design.

Trade-Offs in Design

We take these specifications a step further by asking the following questions: What are the trade-offs in the design? How does the bound on the control signal step response affect the overall performance? And, how does it affect the structure of the optimal solution?

These questions can be readily answered with the ℓ_1 machinery. It amounts to solving a family of mixed ℓ_1 problems parameterized in U_{max}. Solutions for a range of values of U_{max} are presented in Figures 41.8 and 41.9 by showing the performance degradation and the controller order growth as U_{max} decreases. The numerical values are based on a model for the X29 aircraft (for details, see [2]). The following conclusions can be drawn from this analysis:

1. The results present the trade-offs clearly. It is possible to reduce U_{max} by 50% without losing too much performance in terms of the ℓ_1 norm of the system. This implies that the controller can be altered to satisfy stricter conditions on the step response without losing the nominal performance. The curve in Figure 41.8 also shows the smallest possible achievable U_{max}.
2. The trade-offs in the order of the controller are valuable. The trade-off curve in Figure 41.9 shows that, by adding two additional states, U_{max} can be reduced to about 50% of its unconstrained value.
3. To compute such solutions, the unconstrained problem is solved first. The performance for a step input is then checked, and constraints are added only at the time instants where the peak value of the input u is larger than U_{max}. This is a simpler problem than incorporating the infinite-horizon constraints at all the time instants of the step response.

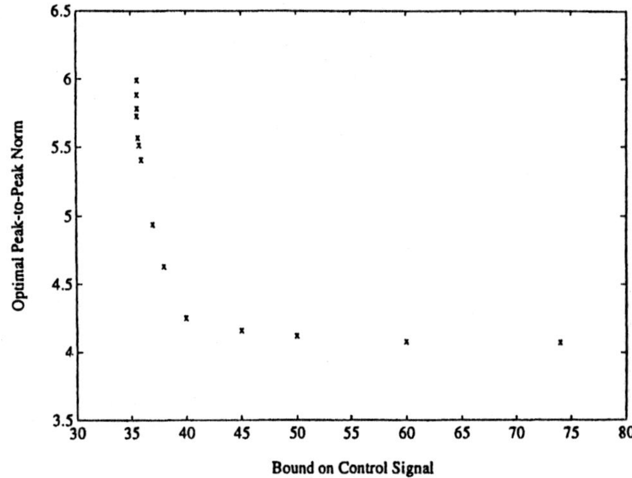

Figure 41.8 X29: Trade-offs in performance vs. control signal bound, U_{max}.

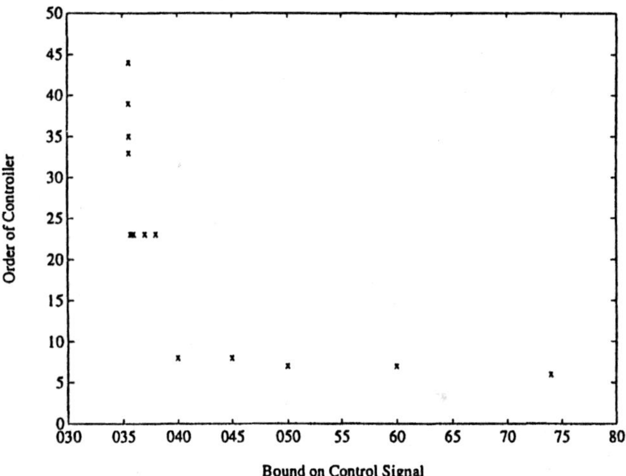

Figure 41.9 X29: Trade-offs in controller order vs. control signal bound, U_{max}.

Finally, such constraints are hard to deal with by selecting weights and solving an ℓ_1 problem or an $\mathcal{H}_\infty$ problem. The advantage that the ℓ_1 problem has over $\mathcal{H}_\infty$ is that such constraints can be incorporated in the problem, as described earlier, and then solved using the same solution techniques.

41.5 Conclusions

In this chapter, we gave an overview of the ℓ_1 theory for robust control. The presentation was not detailed. However, it was intended to serve as an introduction to a more detailed account of the theory that can be found in the book, *Control of Uncertain Systems: A Linear Programming Approach*, and references therein.

We highlighted three ingredients of the ℓ_1 design tool, practicality, computability and flexibility. These properties allow for implementing a computer-aided-design environment based on ℓ_1 nominal designs, in which the designer has the flexibility to incorporate frequency-domain and fixed-input constraints, without losing the qualitative information about the structure of the controllers obtained from the nominal designs. Such an environment has proven very powerful in designing controllers for real applications.

References

[1] Barabanov, A. and Sokolov, A., The geometrical approach to l_1 optimal control, *Proc. 33rd IEEE Conf. Decision Control,* 1994.

[2] Dahleh, M.A. and Diaz-Bobillo, I., *Control of Uncertain Systems: A Linear Programming Approach,* Prentice Hall, Englewood Cliffs, NJ, 1995.

[3] Dahleh, M.A. and Khammash, M.H., Controller design for plants with structured uncertainty, *Automatica,* 29(1), 1993.

[4] Dahleh, M.A. and Ohta, Y., A necessary and sufficient condition for robust BIBO stability, *Syst. Contr. Lett.,* 11, 1988.

[5] Dahleh, M.A. and Pearson, J.B., $\mathcal{L}_1$ optimal feedback compensators for continuous time systems, *IEEE Trans. Automat. Control,* 32, October 1987.

[6] Dahleh, M.A. and Pearson, J.B., ℓ_1 optimal feedback controllers for mimo discrete-time systems, *IEEE Trans. Automat. Control,* April 1987.

[7] Dahleh, M.A. and Pearson, J.B., Optimal rejection of persistent disturbances, robust stability and mixed sensitivity minimization, *IEEE Trans. Automat. Control,* 33, August 1988.

[8] Diaz-Bobillo, I.J. and Dahleh, M.A., Minimization of the maximum peak-to-peak gain: The general multiblock problem, *IEEE Trans. Automat. Control,* October 1993.

[9] Khammash, M. and Pearson, J.B., Performance robustness of discrete-time systems with structured uncertainty, *IEEE Trans. Automat. Control,* 36, 1991.

[10] Khammash, M. and Pearson, J.B., Robust disturbance rejection in ℓ_1-optimal control systems, *Syst. Control Lett.,* 14, 1990.

[11] McDonald, J.S. and Pearson, J.B., ℓ_1-optimal control of multivariable systems with output norm constraints, *Automatica,* 27, 1991.

[12] Shamma, J.S., Nonlinear state feedback for ℓ_1 optimal control, *Syst. Control Lett.,* to appear.

[13] Vidyasagar, M., Optimal rejection of persistent bounded disturbances, *IEEE Trans. A-C,* 31(6), 1986.

42

The Structured Singular Value (μ) Framework

Gary J. Balas
Aerospace Engineering and Mechanics, University of Minnesota, Minnesota, MN

Andy Packard
Mechanical Engineering, University of California, Berkeley, CA

42.1 Introduction

This chapter gives a brief overview of the structured singular value (μ). The μ-based methods discussed are useful for analyzing the performance and robustness properties of linear feedback systems. Computational software for μ-based analysis and synthesis is available in commercial software products [2], [4]. The interested reader might also consult the tutorials in references [14] and [8], and application-oriented papers, such as [6], [13], and [1].

42.2 Shortcomings of Simple Robustness Analysis

Many of the theoretical robustness results for single-input, single-output (SISO) systems show that if a single-loop system has good robust stability characteristics, and good nominal performance characteristics, then, necessarily, it has reasonably good robust performance characteristics. Unfortunately, this "fact" is not, in general, true for multiloop systems. Also, for multiloop systems, checking the robustness via individual loop-at-a-time calculations can be misleading, because the interactions between the deviations are not accounted for in such an analysis. In this chapter, we illustrate these difficulties with examples and introduce the structured singular value as an analytical tool for uncertain, multivariable systems.

The first example concerns control of the angular velocity of a satellite spinning about one of its principal axes. Its mathematical origins are due to Doyle, and are alluded to in [14]. The closed-loop multivariable (MIMO) system is shown in Figure 42.1.

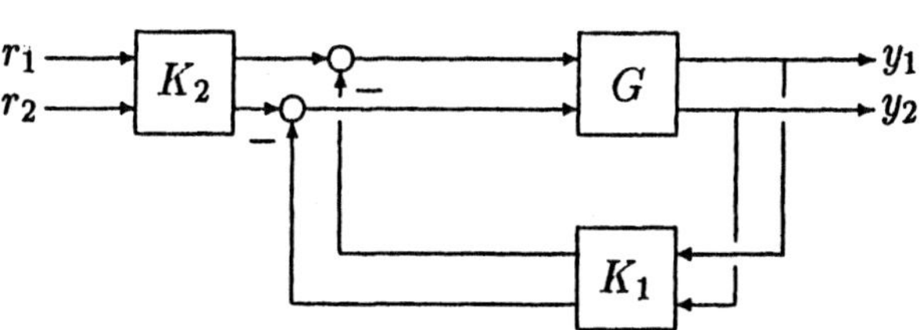

Figure 42.1 Nominal multiloop feedback system.

Set $a := 10$, and define

$$
\begin{aligned}
G &:= \frac{1}{s^2+a^2}\begin{bmatrix} s-a^2 & a(s+1) \\ -a(s+1) & s-a^2 \end{bmatrix}, \\
K_1 &:= \begin{bmatrix} 1 & 0 \\ 0 & 1 \end{bmatrix}, \\
K_2 &:= \frac{1}{1+a^2}\begin{bmatrix} 1 & -a \\ a & 1 \end{bmatrix}.
\end{aligned}
$$

0-8493-8570-9/96/$0.00+$.50

A minimal, state-space realization for the plant G is

$$G = \left[\begin{array}{c|c} A_G & B_G \\ \hline C_G & D_G \end{array}\right] = \left[\begin{array}{cc|cc} 0 & a & 1 & 0 \\ -a & 0 & 0 & 1 \\ \hline 1 & a & 0 & 0 \\ -a & 1 & 0 & 0 \end{array}\right] \tag{42.1}$$

In order to assess the robustness margins to perturbations in the input channels into the plant, consider the four-input, four-output system, denoted by M, in Figure 42.2. The lines from r_1 and r_2 which run above K_2 and G are included to define the tracking error, e_1 and e_2 explicitly.

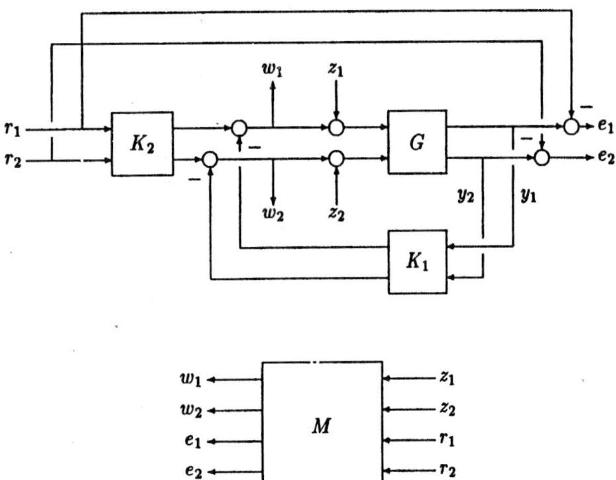

Figure 42.2 Closed-Loop system with uncertainty model.

Some important transfer functions are

$$M_{ry} = \frac{1}{s+1} I_2 \quad , \quad M_{w_1,z_1} = M_{w_2,z_2} = -\frac{1}{s+1}.$$

These imply that the nominal closed-loop system has decoupled command response, with a bandwidth of 1 rad/sec, the crossover frequency in the first feedback loop is 1 rad/sec, with phase margin of 90°, the gain margin in the first channel is infinite, the crossover frequency in the second loop is 1 rad/sec, with phase margin of 90°, and the gain margin in the second channel is infinite. These suggest that the performance of the closed-loop system is excellent and that it is quite robust to perturbations in each input channel. Yet, consider a 5% variation in each channel at the input to the plant. Referring to Figure 42.3, let $\delta_1 = 0.05$, and $\delta_2 = -0.05$. The output response $y(t)$ to a unit-step reference input in channel 1 is shown in Figure 42.4 (along with the nominal responses). Note that the ideal behavior of the nominal system has degraded sharply despite the seemingly innocuous perturbations and excellent gain/phase margins in the closed-loop system. In fact, for a slightly larger perturbation, $\delta_1 = 0.11, \delta_2 = -0.11$, the closed-loop system is actually unstable. Why do these small perturbations cause such a significant degradation in performance? To answer this, calculate the 4×4 transfer matrix M represented in Figure 42.2, giving

$$\begin{bmatrix} w_1 \\ w_2 \\ e_1 \\ e_2 \end{bmatrix} = \begin{bmatrix} -\frac{1}{s+1} & -\frac{a}{s+1} & \frac{s-a^2}{s+1} & -a \\ \frac{a}{s+1} & -\frac{1}{s+1} & a & \frac{s-a^2}{s+1} \\ \frac{1}{s+1} & \frac{a}{s+1} & -\frac{s}{s+1} & 0 \\ -\frac{a}{s+1} & \frac{1}{s+1} & 0 & -\frac{s}{s+1} \end{bmatrix} \begin{bmatrix} z_1 \\ z_2 \\ r_1 \\ r_2 \end{bmatrix} =: M \begin{bmatrix} z_1 \\ z_2 \\ r_1 \\ r_2 \end{bmatrix}$$

Note that the earlier calculations about the closed-loop system yielded information only about the (1, 1), (2, 2), and (3 : 4, 3 : 4) entries. The notation (3 : 4, 3 : 4) denotes the 2×2 matrix formed by rows 3 to 4 and columns 3 to 4 of M. In particular, these entries are all small, in some sense. The neglected entries, (1, 2), (2, 1), (1 : 2, 3 : 4), (3 : 4, 1 : 2) are all quite large, because $a = 10$. It is these large off-diagonal entries, and the manner in which they enter, that causes the extreme sensitivity of the closed-loop system's performance to the perturbations δ_1 and δ_2. For instance, with $\delta_2 \equiv 0$, the perturbation δ_1 can only cause instability by making the transfer function $\left(1 - M_{w1,z1}\delta_1\right)^{-1}$ unstable. Similarly, with $\delta_1 \equiv 0$, the perturbation δ_2 can only cause instability by making the transfer function $\left(1 - M_{w2,z2}\delta_2\right)^{-1}$ unstable. Because both $M_{w1,z1}$ and $M_{w2,z2}$ are "small", this requires large perturbations, and the single-loop gain/phase margins reported earlier are accurate. However, acting together, the perturbations can cause instability by making

$$\left[\begin{bmatrix} 1 & 0 \\ 0 & 1 \end{bmatrix} - \begin{bmatrix} -\frac{1}{s+1} & -\frac{a}{s+1} \\ \frac{a}{s+1} & -\frac{1}{s+1} \end{bmatrix} \begin{bmatrix} \delta_1 & 0 \\ 0 & \delta_2 \end{bmatrix}\right]^{-1}$$

unstable. The denominator of this multivariable transfer function is

$$s^2 + (2 + \delta_1 + \delta_2)\, s + \left[1 + \delta_1 + \delta_2 + (a^2 + 1)\delta_1\delta_2\right].$$

By choosing $\delta_1 = \frac{1}{\sqrt{a^2+1}} \approx 0.1$, and $\delta_2 = -\delta_1$, the characteristic equation has a root at $s = 0$, indicating marginal stability. For slightly larger perturbations, a root moves into the right half-plane. The simultaneous nature of the perturbations has resulted in a much smaller destabilizing perturbation than predicted by the gain/phase margin calculations.

In terms of robust stability, the loop-at-a-time gain/phase margins only depended on the scalar transfer functions $M_{w1,z1}$ and $M_{w2,z2}$, but the robust stability properties of the closed-loop system to simultaneous perturbations actually depend on the 2×2 transfer function matrix $M_{w,z}$. Similarly, assessing the robust performance characteristics of the closed-loop system involves additional transfer functions ignored in the simple-minded analysis. Consider the perturbed closed-loop system in Figure 42.3. In terms of the transfer function matrix M, the perturbed $r \rightarrow e$ transfer function can be drawn as shown in Figure 42.5. Partition the transfer function matrix M into four 2×2 blocks, as

$$M = \left[\begin{array}{c|c} M_{11} & M_{12} \\ \hline M_{21} & M_{22} \end{array}\right].$$

Then the perturbed closed-loop transfer function from r to e can be written as

$$e = \left[M_{22} + M_{21}\Delta\left(I - M_{11}\Delta\right)^{-1} M_{12}\right] d$$

where Δ is the structured matrix of perturbations, $\Delta = \text{diag}\left[\delta_1, \delta_2\right]$. Our initial information about the closed-loop system consisted of the diagonal entries of M_{11} and the entire matrix M_{22}. We have seen that the large off-diagonal entries of M_{11} created destabilizing interactions between the two perturbations. In the robust performance problem, there are additional relevant transfer functions, M_{21} and M_{12}, which were not analyzed in the loop-at-a-time robustness tests, or in the $r \to y$ nominal command response, though it is clear that these transfer functions may play a pivotal role in the robust performance characteristics of the closed-loop system.

Hence, by calculating the two single loop-at-a-time robustness tests and a nominal performance test, 10 of the 16 elements of the relevant 4×4 transfer function matrix are ignored. Any test which accounts for simultaneous perturbations along with the subsequent degradation of performance must be performed on the whole matrix. The point of this example is to show that there are some interesting issues in multivariable system analysis and standard SISO ideas that cannot be made into useful analytical tools simply by applying loop-at-a-time analysis. The structured singular value (μ), introduced in the next section, is a linear algebra tool useful for these types of MIMO system analyses.

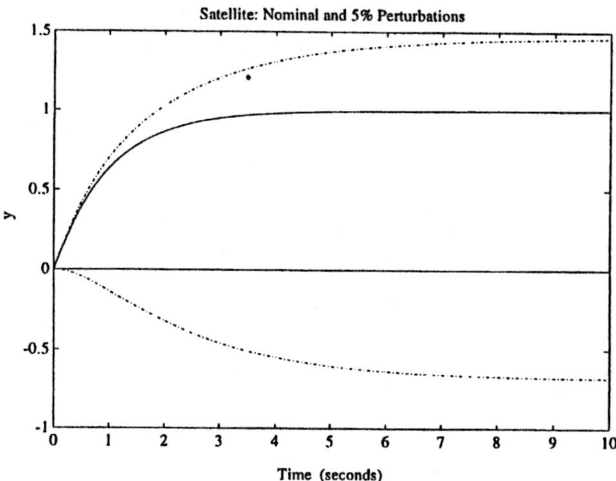

Figure 42.4 Satellite: Nominal (solid) and 5% perturbation responses (dashed).

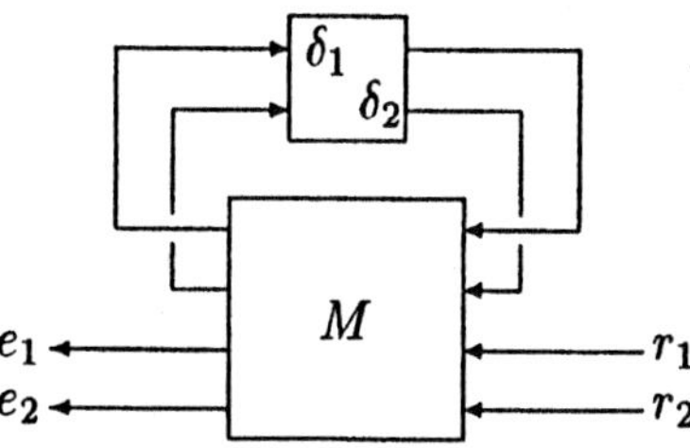

Figure 42.5 Perturbed system.

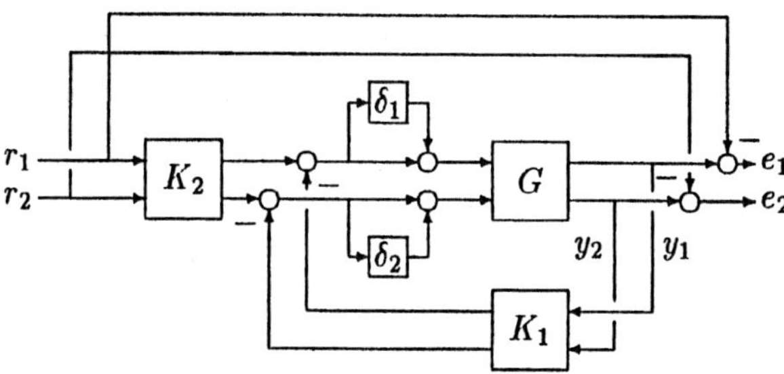

Figure 42.3 Satellite: Closed-Loop system with uncertain elements, δ_1 and δ_2.

42.3 Complex Structured Singular Value

This section is devoted to defining the structured singular value, a matrix function denoted by $\mu(\cdot)$ [5]. The notation we use is standard from linear algebra and systems theory. **R** denotes the set of real numbers, **C** denotes the set of complex numbers, $|\cdot|$ is the absolute value of elements in **R** or **C**, $\mathbf{R}^n$ is the set of real n vectors, $\mathbf{C}^n$ is the set of complex n vectors, $\|v\|$ is the Euclidean norm for $v \in \mathbf{C}^n$, $\mathbf{R}^{n\times m}$ is the set of $n \times m$ real matrices, $\mathbf{C}^{n\times m}$ is the set of $n \times m$ complex matrices, I_n is the $n \times n$ identity matrix, and $0_{n\times m}$ is an entirely zero matrix. For $M \in \mathbf{C}^{n\times m}$, M^T is the transpose of M, M^* is the complex-conjugate transpose of M, and $\bar{\sigma}(M)$ is the maximum singular value of M. For $M \in \mathbf{C}^{n\times n}$, $\lambda_i(M)$ is an eigenvalue of M, $\rho(M) := \max_i |\lambda_i(M)|$ is the **spectral radius of** M, and $\rho_R(M)$ is the real spectral radius of M, $\rho_R(M) := \max\{|\lambda| : \lambda \in \mathbf{R}, \det(\lambda I - M) = 0\}$, with $\rho_R(M) := 0$ if M has no real eigenvalues. If $M \in \mathbf{C}^{n\times n}$ satisfies $M = M^*$, then $M > 0$ denotes that M is positive definite, and $M^{\frac{1}{2}}$ means the unique positive definite Hermitian square root. For $M = M^*$, then $\lambda_{\max}(M)$ denotes the most positive eigenvalue.

We consider matrices $M \in \mathbf{C}^{n\times n}$. In the definition of $\mu(M)$, there is an underlying structure Δ, (a prescribed set of block diagonal matrices) on which everything following depends. This structure may be defined differently for each problem depending on the uncertainty and performance objectives. Defining the structure involves specifying three things: the total number of blocks, the type of each block, and their dimensions.

42.3.1 Purely Complex μ

Two types of blocks—*repeated scalar* and *full* blocks are considered. Two nonnegative integers, S and F, denote the number of *repeated scalar* blocks and the number of *full* blocks, respectively. To track the block dimensions, we introduce positive integers $r_1, \ldots, r_S$; $m_1, \ldots, m_F$. The ith repeated scalar block is $r_i \times r_i$, while the jth full block is $m_j \times m_j$. With those integers given, define $\Delta \subset \mathbf{C}^{n\times n}$ as

$$\mathbf{\Delta} := \left\{ \text{diag}\left[\delta_1 I_{r_1}, \ldots, \delta_S I_{r_S}, \Delta_{S+1}, \ldots, \Delta_{S+F}\right] : \delta_i \in \mathbf{C}, \Delta_{S+j} \in \mathbf{C}^{m_j \times m_j}, 1 \le i \le S, 1 \le j \le F \right\}. \tag{42.2}$$

For consistency among all the dimensions, $\sum_{i=1}^{S} r_i + \sum_{j=1}^{F} m_j$ must equal n. Often, we will need norm bounded subsets of $\mathbf{\Delta}$, and we introduce the notation $\mathbf{B_\Delta} := \{\Delta \in \mathbf{\Delta} : \bar{\sigma}(\Delta) \le 1\}$. Note that in Equation 42.2 all of the repeated scalar blocks appear first, followed by the full blocks. This is done to simplify the notation and can easily be relaxed. The full blocks are also assumed to be square, but again, this is only to simplify notation.

DEFINITION 42.1 For $M \in \mathbf{C}^{n \times n}$, $\mu_{\mathbf{\Delta}}(M)$ is defined

$$\mu_{\mathbf{\Delta}}(M) := \frac{1}{\min\{\bar{\sigma}(\Delta) : \Delta \in \mathbf{\Delta}, \det(I - M\Delta) = 0\}} \tag{42.3}$$

unless no $\Delta \in \mathbf{\Delta}$ makes $I - M\Delta$ singular, in which case $\mu_{\mathbf{\Delta}}(M) := 0$.

It is instructive to consider a "feedback" interpretation of $\mu_{\mathbf{\Delta}}(M)$ at this point. Let $M \in \mathbf{C}^{n \times n}$ be given, and consider the loop shown in Figure 42.6. This picture represents the loop equations $u = Mv$, $v = \Delta u$. As long as $I - M\Delta$ is nonsingular, the only solutions of u, v to the loop equations are $u = v = 0$. However, if $I - M\Delta$ is singular, then there are infinitely many solutions to the equations, and the norms $\|u\|$, $\|v\|$ of the solutions can be arbitrarily large. Motivated by connections with stability of systems, we call this constant matrix feedback system "unstable". Likewise, the term "stable" will describe the situation when the only solutions are identically zero. In this context then, $\mu_{\mathbf{\Delta}}(M)$ provides a measure of the smallest structured Δ that causes "instability" of the constant matrix feedback loop in Figure 42.6. The norm of this "destabilizing" Δ is exactly $1/\mu_{\mathbf{\Delta}}(M)$.

Consider $M \in \mathbf{C}^{5 \times 4}$,

$$M := \begin{bmatrix} 0.100 + 0.070i & -0.154 + 0.162i & 0 - 0.560i & 0 + 42.000i \\ 0 - 0.273i & -0.300 - 0.280i & 2.860 + 0.546i & -26.000 + 72.800i \\ 0.100 + 0.175i & 0.077 - 0.108i & -0.400 + 0.210i & 5.000 + 3.500i \\ 0 + 0.002i & -0.004 - 0.002i & 0.006 + 0.011i & 0.200 + 0.420i \\ 0.024 + 0.028i & 0 + 0.027i & -0.066 + 0.042i & 0 + 0.700i \end{bmatrix} \tag{42.4}$$

to show the dependence of $\mu_{\mathbf{\Delta}}(M)$ on the set $\mathbf{\Delta}$. Two different block structures compatible (in the sense of dimensions) with M are

$$\mathbf{\Delta}_1 = \left\{\text{diag}\left[\delta_1, \delta_2, \Delta_3\right] : \delta_1, \delta_2 \in \mathbf{C}, \Delta_3 \in \mathbf{C}^{2\times 3}\right\}, \text{ and}$$
$$\mathbf{\Delta}_2 = \left\{\text{diag}\left[\delta_1, \delta_2, \delta_3, \Delta_4\right] : \delta_i \in \mathbf{C}, \Delta_4 \in \mathbf{C}^{1\times 2}\right\}.$$

The definition yields $\mu_{\mathbf{\Delta}_1}(M) \approx 8.32$, while $\mu_{\mathbf{\Delta}_2}(M) \approx 2.42$.

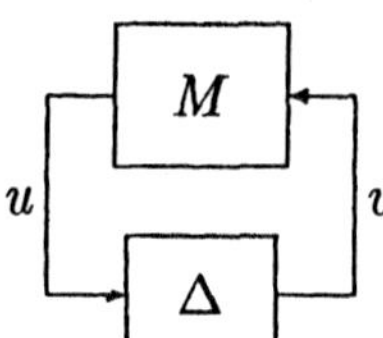

Figure 42.6 $M - \Delta$ interconnection.

An alternative expression for $\mu_{\mathbf{\Delta}}(M)$ follows easily from the definition.

LEMMA 42.1 $\mu_{\mathbf{\Delta}}(M) = \max_{\Delta \in \mathbf{B_\Delta}} \rho(\Delta M)$

Continuity of the function $\mu : \mathbf{C}^{n\times n} \to \mathbf{R}$ is apparent from this lemma. In general, though, the function $\mu : \mathbf{C}^{n\times n} \to \mathbf{R}$ is not a norm, because it doesn't satisfy the triangle inequality. However, for any $\alpha \in \mathbf{C}$, $\mu(\alpha M) = |\alpha| \mu(M)$, so it is related to how "big" the matrix is.

We can relate $\mu_{\mathbf{\Delta}}(M)$ to familiar linear algebra quantities when $\mathbf{\Delta}$ is one of two extreme sets:

- If $\mathbf{\Delta} = \{\delta I : \delta \in \mathbf{C}\}$ ($S = 1, F = 0, r_1 = n$), then $\mu_{\mathbf{\Delta}}(M) = \rho(M)$.

  Proof: The only Δ's in $\mathbf{\Delta}$ which satisfy the $\det(I - M\Delta) = 0$ constraint are reciprocals of nonzero eigenvalues of M. The smallest one of these is associated with the largest (magnitude) eigenvalue, so, $\mu_{\mathbf{\Delta}}(M) = \rho(M)$.

- If $\mathbf{\Delta} = \mathbf{C}^{n\times n}$ ($S = 0, F = 1, m_1 = n$), then $\mu_{\mathbf{\Delta}}(M) = \bar{\sigma}(M)$.

  Proof: If $\bar{\sigma}(\Delta) < \frac{1}{\bar{\sigma}(M)}$, then $\bar{\sigma}(M\Delta) < 1$, so $I - M\Delta$ is nonsingular. Applying Equation 42.3 implies $\mu_{\mathbf{\Delta}}(M) \le \bar{\sigma}(M)$. On the other hand, let u and v be unit vectors satisfying $Mv = \bar{\sigma}(M)u$, and define $\Delta := \frac{1}{\bar{\sigma}(M)} vu^*$. Then $\bar{\sigma}(\Delta) = 1/\bar{\sigma}(M)$ and $I - M\Delta$ is obviously singular. Hence, $\mu_{\mathbf{\Delta}}(M) \ge \bar{\sigma}(M)$.

Obviously, for a general $\mathbf{\Delta}$ as in Equation 42.2, $\{\delta I_n : \delta \in \mathbf{C}\}$ must be included in $\mathbf{\Delta} \subset \mathbf{C}^{n\times n}$. Hence directly from the definition of μ, and the two special cases above, we conclude that

$$\rho(M) \le \mu_{\mathbf{\Delta}}(M) \le \bar{\sigma}(M). \tag{42.5}$$

These bounds by themselves may provide little information about the value of μ, because the gap between ρ and $\bar{\sigma}$ can be large. To see this, consider the matrix

$$M = \begin{bmatrix} 0 & \alpha \\ 0 & 0 \end{bmatrix}$$

and two different block structures, $\mathbf{\Delta}_1 := \{\delta I_2 : \delta \in \mathbf{C}\}$ and $\mathbf{\Delta}_2 := \{\Delta \in \mathbf{C}^{2\times 2}\}$. μ with respect to $\mathbf{\Delta}_1$, which corresponds to $\rho(M)$, is 0, independent of α. μ with respect to $\mathbf{\Delta}_2$, which corresponds to $\bar{\sigma}(M)$, is $|\alpha|$.

The bounds on μ can be refined with transformations on M that do not affect $\mu_\Delta(M)$, but do affect ρ and $\bar{\sigma}$. To do this, define two subsets, $\mathbf{Q}_\Delta$ and $\mathbf{D}_\Delta$ of $\mathbf{C}^{n\times n}$,

$$\mathbf{Q}_\Delta = \{Q \in \mathbf{\Delta} : Q^* Q = I_n\} \tag{42.6}$$

$$\mathbf{D}_\Delta = \left\{ \begin{array}{c} \operatorname{diag}\left[D_1, \ldots, D_S, d_{S+1} I_{m_1}, \ldots, d_{S+F} I_{m_F}\right] : \\ D_i \in \mathbf{C}^{r_i \times r_i}, D_i = D_i^* > 0, d_{S+j} \in \mathbf{R}, d_{S+j} > 0 \end{array} \right\} \tag{42.7}$$

For any $\Delta \in \mathbf{\Delta}$, $Q \in \mathbf{Q}_\Delta$, and $D \in \mathbf{D}_\Delta$,

$$\begin{aligned} Q^* \in \mathbf{Q}_\Delta, \quad Q\Delta \in \mathbf{\Delta}, \quad \Delta Q \in \mathbf{\Delta} \\ \bar{\sigma}(Q\Delta) = \bar{\sigma}(\Delta Q) = \bar{\sigma}(\Delta) \quad (42.8) \\ D^{\frac{1}{2}} \Delta = \Delta D^{\frac{1}{2}} \quad (42.9) \end{aligned}$$

THEOREM 42.1 *For all $Q \in \mathbf{Q}_\Delta$ and $D \in \mathbf{D}_\Delta$*

$$\mu_\Delta(MQ) = \mu_\Delta(QM) = \mu_\Delta(M) = \mu_\Delta\left(D^{\frac{1}{2}} M D^{-\frac{1}{2}}\right) \tag{42.10}$$

PROOF 42.1 For all $D \in \mathbf{D}_\Delta$ and $\Delta \in \mathbf{\Delta}$,

$$\begin{aligned} \det(I - M\Delta) &= \det\left(I - MD^{-\frac{1}{2}} D^{\frac{1}{2}} \Delta\right) \\ &= \det\left(I - MD^{-\frac{1}{2}} \Delta D^{\frac{1}{2}}\right) \\ &= \det\left(D^{-\frac{1}{2}} D^{\frac{1}{2}} - D^{-\frac{1}{2}} M D^{-\frac{1}{2}} \Delta\right) \\ &= \det\left(I - D^{\frac{1}{2}} M D^{-\frac{1}{2}} \Delta\right) \end{aligned}$$

because D commutes with Δ. Therefore $\mu_\Delta(M) = \mu_\Delta\left(D^{\frac{1}{2}} M D^{-\frac{1}{2}}\right)$. Also, for each $Q \in \mathbf{Q}_\Delta$, $\det(I - M\Delta) = 0$ if and only if $\det\left(I - MQQ^*\Delta\right) = 0$. Because $Q^*\Delta = \Delta$ **and** $\bar{\sigma}\left(Q^*\Delta\right) = \bar{\sigma}(\Delta)$, $\mu_\Delta(MQ) = \mu_\Delta(M)$ as desired. The argument for QM is the same.

Therefore, the bounds in Equation 42.5 can be tightened to

$$\begin{aligned} \max_{Q \in \mathbf{Q}} \rho(QM) &\leq \max_{\Delta \in \mathbf{B}_\Delta} \rho(\Delta M) \\ &= \mu_\Delta(M) \leq \inf_{D \in \mathbf{D}} \bar{\sigma}\left(D^{\frac{1}{2}} M D^{-\frac{1}{2}}\right) \end{aligned} \tag{42.11}$$

where the equality comes from Lemma 42.1.

The lower bound, $\max_{Q \in \mathbf{Q}} \rho(QM)$, is always an equality [5]. Unfortunately, the quantity $\rho(QM)$ can have multiple local maxima which are not global. Thus local search cannot be guaranteed to obtain μ, but can only yield a lower bound. The upper bound can be reformulated as a convex optimization problem, so that the global minimum can, in principle, be found. Unfortunately, the upper bound is not always equal to μ. For block structures $\mathbf{\Delta}$ satisfying $2S + F \leq 3$, the upper bound is always equal to $\mu_\Delta(M)$, and for block structures with $2S + F > 3$, matrices exist for which μ is less than the infimum [5], [9].

Convexity properties make the upper bound computationally attractive. The simplest convexity property is given in the following theorem, which shows that the function $\bar{\sigma}\left(D^{\frac{1}{2}} M D^{-\frac{1}{2}}\right)$ has convex level sets.

THEOREM 42.2 *Let $M \in \mathbf{C}^{n\times n}$ be given, along with a scaling set $\mathbf{D}_\Delta$, and $\beta > 0$. Then the set $\left\{D \in \mathbf{D}_\Delta : \bar{\sigma}\left(D^{\frac{1}{2}} M D^{-\frac{1}{2}}\right) < \beta\right\}$ is convex.*

PROOF 42.2 The following chain of equivalences comprises the proof:

$$\begin{aligned} & \bar{\sigma}\left(D^{\frac{1}{2}} M D^{-\frac{1}{2}}\right) < \beta \\ \Leftrightarrow\ & \lambda_{\max}\left(D^{-\frac{1}{2}} M^* D^{\frac{1}{2}} D^{\frac{1}{2}} M D^{-\frac{1}{2}}\right) < \beta^2 \\ \Leftrightarrow\ & D^{-\frac{1}{2}} M^* D^{\frac{1}{2}} D^{\frac{1}{2}} M D^{-\frac{1}{2}} - \beta^2 I < 0 \\ \Leftrightarrow\ & M^* D M - \beta^2 D < 0 \end{aligned} \tag{42.12}$$

The latter is clearly a convex condition in D because it is linear.

The final condition in Equation 42.12 is called a *Linear Matrix Inequality* (LMI) in the variable D. In [3], a large number of control synthesis and analysis problems are cast as solutions of LMI's.

42.3.2 Mixed μ: Real and Complex Uncertainty

Until this point, this section has dealt with complex-valued perturbation sets. In specific instances, it may be more natural to describe modeling errors with real perturbations, for instance, when the real coefficients of a linear differential equation are uncertain. Although these perturbations can be treated simply as complex, proceeding with a complex μ analysis, the results may be conservative. Hence, theory and algorithms to test for robustness and performance degradation with mixed (real blocks and complex blocks) perturbation have been developed.

Definition 42.1 of μ can be used for more general sets $\mathbf{\Delta}$, such as those containing real and complex blocks. There are 3 types of blocks, *repeated real scalar*, *repeated complex scalar*, and *complex full* blocks. As before in Section 42.3, S and F, denote the number of *repeated, complex scalar* blocks and the number of *complex full* blocks, respectively. V denotes the number of *repeated, real scalar* blocks. The block dimensions of the real block are denoted by the positive integers $t_1, \ldots, t_V$. With these integers given, and r_i and m_j as defined in Section 42.3, define $\mathbf{\Delta}$ as

$$\begin{aligned} \mathbf{\Delta} = \Big\{ & \operatorname{diag}\left[\delta_1^r I_{t_1}, \ldots, \delta_V^r I_{t_V}, \delta_{V+1}^c I_{r_1}, \ldots,\right. \\ & \left.\delta_{V+S}^c I_{r_S}, \Delta_{V+S+1}, \ldots, \Delta_{V+S+F}\right] : \\ & \delta_k^r \in \mathbf{R}, \delta_{V+i}^c \in \mathbf{C}, \Delta_{V+S+j} \in \mathbf{C}^{m_j \times m_j}, \\ & 1 \leq k \leq V, 1 \leq i \leq S, 1 \leq j \leq F \Big\}. \end{aligned} \tag{42.13}$$

For consistency among all the dimensions, $\sum_{k=1}^{V} t_k + \sum_{i=1}^{S} r_i + \sum_{j=1}^{F} m_j$ must equal $= n$.

The mixed μ function inherits many of the properties of the purely complex μ function, [5] and [7]. However, in some aspects such as continuity, the mixed μ problem can be fundamentally different from the complex μ problem. It is now well-known that real μ problems can be discontinuous in the problem data. Beside adding computational difficulties to the problem, the utility of real μ is doubtful as a robustness measure in such cases, the system model is always a mathematical abstraction from the real world, computed to finite precision. It has been shown that, for many **mixed** μ problems, μ is continuous. The main idea is that, *if a mixed μ problem has complex uncertainty blocks that "count", then the function is continuous.* From an engineering viewpoint, "count" implies that the complex uncertainty affects the value of μ. This is the usual case because one is usually interested in robust performance problems (which therefore contain at least one complex block – see Section 42.5.2), or robust stability problems with some unmodeled dynamics, which are naturally covered with complex uncertainty. Purely real problems can be made continuous by adding a small amount of complex uncertainty to each real uncertainty (see [1] for an example). Consequently, a small amount of phase uncertainty is added to the gain uncertainty.

The theory for bounding (both lower and upper) mixed real/complex bounds is much more complicated to describe than the bounding theory for complex μ. The lower bound for the mixed case is a real eigenvalue maximization problem. Techniques to solve approximately for a mixed μ lower bound using power algorithms have been derived, and are similar to those used for a lower bound for complex μ [15]. The mixed μ upper bound takes the form of a more complicated version of the same problem, involving an additional "G scaling matrix" which scales only the real uncertainty blocks. This minimization, involving an LMI expression similar to Equation 42.12, is computed using convex optimization techniques similar to those for the purely complex upper bound. See references [7] and [15] for more computational details.

42.3.3 Frequency Domain, Robust Stability tests with μ

The best-known use of μ as a robustness analysis tool is in the frequency domain [5], [14]. Suppose $\hat{G}(s)$ is a stable, multi-input, multioutput transfer function of a linear system. For clarity, assume $\hat{G}$ has n_z inputs and n_w outputs. Let Δ be a block structure, as in Equation 42.2, and assume that the dimensions are such that $\Delta \subset \mathbf{C}^{n_z \times n_w}$. We want to consider feedback perturbations to $\hat{G}$, themselves dynamical systems, with the block-diagonal structure of the set Δ, in Figure 42.7.

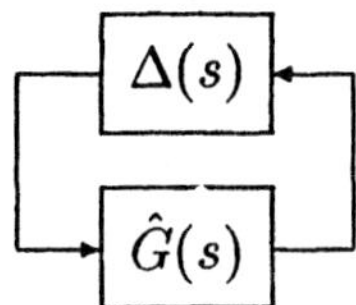

Figure 42.7 $\hat{G} - \Delta$ feedback loop block diagram.

Let $\mathbf{S}$ denote the set of real-rational, proper, stable, transfer matrices (of appropriate dimensions, which should be clear from context). Associated with any block structure Δ, let $\mathbf{S}_\Delta$ denote the set of all block diagonal, stable rational transfer functions, with diagonal block structure as in Δ.

$$\mathbf{S}_\Delta := \left\{ \Delta \in \mathbf{S} : \Delta(s_o) \in \Delta \text{ for all } s_o \in \bar{\mathbf{C}}_+ \right\}.$$

THEOREM 42.3 *Let $\beta > 0$. The loop in Figure 42.7 is well-posed and internally stable for all $\Delta \in \mathbf{S}_\Delta$ with $\|\Delta\|_\infty < \frac{1}{\beta}$, where $\|\Delta(s)\|_\infty := \max_{\omega \in \mathcal{R}} \bar{\sigma}\left(\Delta(j\omega)\right)$ if, and only if,*

$$\|G\|_\Delta := \sup_{\omega \in \mathbf{R}} \mu_\Delta\left(\hat{G}(j\omega)\right) \leq \beta$$

In summary, the peak value on the μ plot of the frequency response of the known linear system that the perturbation "sees" determines the size of perturbations against which the loop is robustly stable.

REMARK 42.1 If the peak occurs at $\omega = 0$, there are systems G where the theorem statement needs to be modified to be correct. In particular, it may be impossible to do what the theorem statement implies, that is, construct a real-rational perturbation $\Delta \in \mathbf{S}_\Delta$ with $\|\Delta\|_\infty = \left(\|G\|_\Delta\right)^{-1}$ and $(I - G\Delta)^{-1}$ unstable. Rather, for any $\epsilon > 0$, there will be a real-rational perturbation $\Delta \in \mathbf{S}_\Delta$ with $\|\Delta\|_\infty = \left(\|G\|_\Delta\right)^{-1} + \epsilon$ and $(I - G\Delta)^{-1}$ unstable. These facts can be ascertained from results in [9], [10].

The overall implication of this modification can be viewed in two opposite ways. If the theorem is used for actual robustness analysis, the original viewpoint that $\left(\|G\|_\Delta\right)^{-1}$ is the size of the smallest real-rational perturbation causing instability is certainly the "right" mental model to use, because the small correction that may be needed is arbitrarily small, and hence of little engineering relevance. On the other hand, if Theorem 42.3 is being used to prove another theorem, then one needs to be very careful.

In the next section, the linear algebra results which extend μ from a robust stability tool to a robust performance tool are covered. These linear algebra results are then applied to give a frequency-domain robust performance test in Section 42.5.2.

42.4 Linear Fractional Transformations and μ

The use of μ in control theory depends to a great extent on its intimate relationship with a class of general linear feedback loops called **Linear Fractional Transformations** (LFTs) [11]. This section explores this relationship with some simple theorems that can be obtained almost immediately from the definition of μ. To introduce these, consider a complex matrix M partitioned as

$$M = \begin{bmatrix} M_{11} & M_{12} \\ M_{21} & M_{22} \end{bmatrix} \tag{42.14}$$

and suppose that there is a defined block structure $\mathbf{\Delta_1}$ compatible in size with M_{11} (for any $\Delta_1 \in \mathbf{\Delta_1}$, $M_{11}\Delta_1$ is square). For $\Delta_1 \in \mathbf{\Delta_1}$, consider the loop equations

$$z = M_{11}w + M_{12}d; \quad e = M_{21}w + M_{22}d; \quad w = \Delta_1 z \tag{42.15}$$

which correspond to the block diagram in Figure 42.8 (note the similarity to Figure 42.5 in Section 42.2).

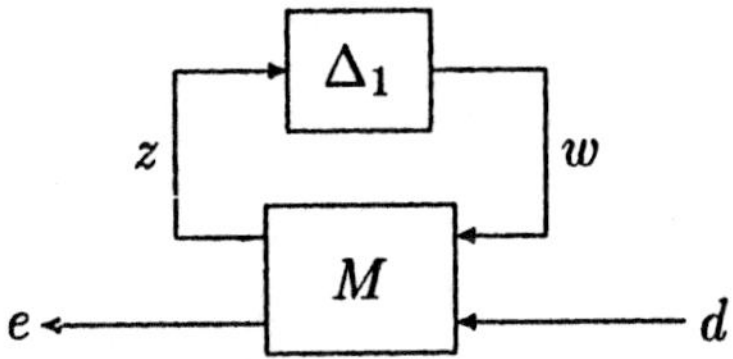

Figure 42.8 Linear fractional transformation.

The set of Equations 42.15 is called **well posed** if, for any vector d, unique vectors w, z, and e exist satisfying the loop equations. The set of equations is well-posed if, and only if, the inverse of $I - M_{11}\Delta_1$ exists. If not, then depending on d and M, there is either no solution to the loop equations, or there are an infinite number of solutions. When the inverse does exist, the vectors e and d must satisfy $e = F_u(M, \Delta_1)d$, where

$$F_u(M, \Delta_1) := M_{22} + M_{21}\Delta_1(I - M_{11}\Delta_1)^{-1}M_{12}. \tag{42.16}$$

$F_u(M, \Delta_1)$ is called a Linear Fractional Transformation on M by Δ_1 and, in a feedback diagram, appears as in Figure 42.8. $F_u(M, \Delta_1)$ denotes that the "upper" loop of M is closed by Δ_1. An analogous formula describes $F_l(M, \Delta_2)$ which is the resulting matrix obtained by closing the "lower" loop of M with a matrix $\Delta_2 \in \mathbf{\Delta_2}$.

In this formulation, the matrix M_{22} is assumed to be something nominal, and $\Delta_1 \in \mathbf{B}_{\mathbf{\Delta_1}}$ is viewed as a norm-bounded perturbation from an allowable perturbation class, $\mathbf{\Delta_1}$. The matrices M_{12}, M_{21}, and M_{22} and the formula $F_u(M, \Delta_1)$ reflect prior knowledge showing how the unknown perturbation affects the nominal map, M_{22}. This type of uncertainty, called *linear fractional*, is natural for many control problems and encompasses many other special cases considered by researchers.

42.4.1 Well-Posedness and Performance for Constant LFT's

Let M be a complex matrix partitioned as

$$M = \begin{bmatrix} M_{11} & M_{12} \\ M_{21} & M_{22} \end{bmatrix} \tag{42.17}$$

and suppose that there are two defined block structures, $\mathbf{\Delta_1}$ and $\mathbf{\Delta_2}$, compatible in size with M_{11} and M_{22} respectively. Define a third structure $\mathbf{\Delta}$ as

$$\mathbf{\Delta} = \left\{ \begin{bmatrix} \Delta_1 & 0 \\ 0 & \Delta_2 \end{bmatrix} : \Delta_1 \in \mathbf{\Delta_1}, \Delta_2 \in \mathbf{\Delta_2} \right\}. \tag{42.18}$$

Now there are three structures for which we may compute μ. The notation we use to keep track of this is as follows: $\mu_1(\cdot)$ is with respect to $\mathbf{\Delta_1}$, $\mu_2(\cdot)$ is with respect to $\mathbf{\Delta_2}$, : $\mu_{\mathbf{\Delta}}(\cdot)$ is with respect to $\mathbf{\Delta}$. In view of this, $\mu_1(M_{11})$, $\mu_2(M_{22})$ and $\mu_{\mathbf{\Delta}}(M)$ all make sense, though, for instance, $\mu_1(M)$ does not. For notation, let $\mathbf{B}_1 := \{\Delta_1 \in \mathbf{\Delta_1} : \bar{\sigma}(\Delta_1) \leq 1\}$.

Clearly, the linear fractional transformation $F_u(M, \Delta_1)$ is well-posed for all $\Delta_1 \in \mathbf{B}_{\mathbf{\Delta_1}}$ if, and only if, $\mu_1(M_{11}) < 1$. As the "perturbation" Δ_1 deviates from zero, the matrix $F_u(M, \Delta_1)$ deviates from M_{22}. The range of values for $\mu_2(F_u(M, \Delta_1))$ is intimately related to $\mu_{\mathbf{\Delta}}(M)$, as follows:

THEOREM 42.4 **Main Loop Theorem:** *The following are equivalent:*

$$\mu_{\mathbf{\Delta}}(M) < 1 \quad \Longleftrightarrow \quad \begin{cases} \mu_1(M_{11}) < 1, \text{ and} \\ \max\limits_{\Delta_1 \in \mathbf{B}_1} \mu_2(F_u(M, \Delta_1)) < 1 \end{cases}.$$

PROOF 42.3 The proof of this theorem is based on the definition of μ and Schur formulae for determinants of block partitioned matrices as in [9]. The Main Loop Theorem forms the basis for all uses of μ in linear system robustness analysis, whether from a state-space, frequency-domain, or Lyapunov approach.

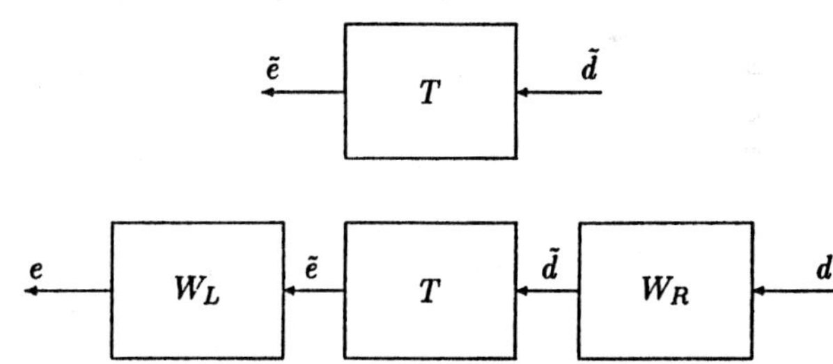

Figure 42.9 Unweighted and weighted MIMO systems.

42.5 Robust Performance Tests using μ and Main Loop Theorem

Often, stability is not the only property of a closed-loop system that must be robust to perturbations. Typically there are exogenous disturbances acting on the system (wind gusts, sensor noise) which result in tracking and regulation errors. Under perturbation, the effect of these disturbances on error signals can greatly increase. In most cases, long before the onset of instability, the closed-loop performance will be unacceptably degraded. Hence the need for a "robust performance" test to indicate the worst-case level of performance degradation for a given level of perturbations.

42.5.1 Characterization of Performance in μ setting

Within the structured singular value setting, the most natural (mathematical) way to characterize acceptable performance is in terms of MIMO $\|\cdot\|_\infty$ ($\mathcal{H}_\infty$) norms, discussed in detail in other Chapters (29 and 40) of this Handbook. In this section, we quickly review the $\mathcal{H}_\infty$ norm, and interpretations.

Suppose T is a MIMO stable linear system, with transfer function matrix $T(s)$. For a given driving signal $\tilde{d}(t)$, define $\tilde{e}$ as the output, as in the left-hand diagram of Figure 42.9.

Assume that the dimensions of T are $n_e \times n_d$. Let $\beta > 0$ be defined as

$$\beta := \|T\|_\infty := \max_{\omega\in\mathbf{R}} \bar{\sigma}\left[T(j\omega)\right]. \quad (42.19)$$

A time-domain, sinusoidal, steady-state interpretation of this quantity is as follows:

Fact: For any frequency $\bar{\omega} \in \mathbf{R}$, any vector of amplitudes $a \in \mathbf{R}_{n_d}$, and any vector of phases $\phi \in \mathbf{R}^{n_d}$, with $\|a\|_2 \leq 1$, define a time signal

$$\tilde{d}(t) = \begin{bmatrix} a_1 \sin(\bar{\omega}t + \phi_1) \\ \vdots \\ a_{n_d} \sin\left(\bar{\omega}t + \phi_{n_d}\right) \end{bmatrix}.$$

Apply this input to the system T, resulting in a steady-state response $\tilde{e}_{ss}$ of the form

$$\tilde{e}_{ss}(t) = \begin{bmatrix} b_1 \sin(\bar{\omega}t + \psi_1) \\ \vdots \\ b_{n_e} \sin\left(\bar{\omega}t + \psi_{n_e}\right) \end{bmatrix}.$$

The vector $b \in \mathbf{R}^{n_e}$ will satisfy $\|b\|_2 \leq \beta$. Moreover, β, as defined in Equation 42.19, is the smallest number for which this fact is true for every $\|a\|_2 \leq 1$, $\bar{\omega}$, and ϕ.

Multivariable performance objectives are represented by a single, MIMO $\|\cdot\|_\infty$ objective on a closed-loop transfer function. Because many objectives are being lumped into one matrix and the associated cost is the norm of the matrix, it is important to use frequency-dependent weighting functions, so that different requirements can be meaningfully combined into a single cost function.

In the weighting function selection, diagonal weights are most easily interpreted. Consider the right-hand diagram of Figure 42.9. Assume that W_L and W_R are diagonal, stable transfer function matrices, with diagonal entries denoted L_i and R_i. Bounds on the quantity $\|W_L T W_R\|_\infty$ will imply bounds about the sinusoidal steady-state behavior of the signals $\tilde{d}$ and $\tilde{e}(= T\tilde{d})$. Specifically, for sinusoidal signal $\tilde{d}$, the steady-state relationship between $\tilde{e}(= T\tilde{d})$, $\tilde{d}$ and $\|W_L T W_R\|_\infty$ follows. The steady-state solution $\tilde{e}_{ss}$, denoted as

$$\tilde{e}_{ss}(t) = \begin{bmatrix} \tilde{e}_1 \sin(\bar{\omega}t + \psi_1) \\ \vdots \\ \tilde{e}_{n_e} \sin\left(\bar{\omega}t + \psi_{n_d}\right) \end{bmatrix},$$

satisfies $\sum_{i=1}^{n_e} |L_i(j\bar{\omega})\tilde{e}_i|^2 \leq 1$ for all sinusoidal input signals $\tilde{d}$ of the form,

$$\tilde{d}(t) = \begin{bmatrix} \tilde{d}_1 \sin(\bar{\omega}t + \phi_i) \\ \vdots \\ \tilde{d}_{n_d} \sin\left(\bar{\omega}t + \phi_{n_d}\right) \end{bmatrix},$$

satisfying

$$\sum_{i=1}^{n_d} \frac{\left|\tilde{d}_i\right|^2}{|R_i(j\bar{\omega})|^2} \leq 1$$

if, and only if, $\|W_L T W_R\|_\infty \leq 1$.

42.5.2 Frequency-Domain Robust Performance Tests

Recall from Section 42.3.1, that if Δ is a single full complex block, then the function μ_Δ is simply the maximum singular value function. We can use this fact, along with the Main Loop Theorem (Theorem 42.4) and the $\mathcal{H}_\infty$ notion of performance, to obtain the central robust performance theorem for perturbed transfer functions.

Assume G is a stable linear system, with real-rational, proper transfer function $\hat{G}$. The dimension of G is $n_z + n_d$ inputs and $n_w + n_e$ outputs. Partition G in the obvious manner, so that G_{11} has n_z inputs and n_w outputs, and so on. Let $\Delta \subset \mathbf{C}^{n_w \times n_z}$ be a block structure, as in Equation 42.2. For $\Delta \in \mathbf{S}_\Delta$, consider the behavior of the perturbed system in Figure 42.10.

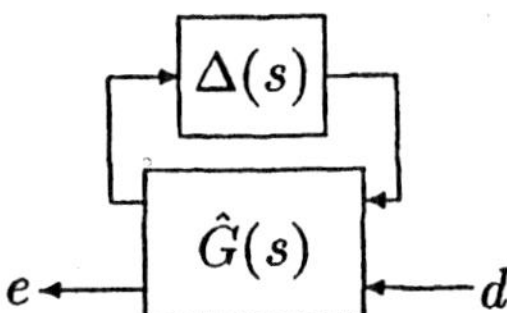

Figure 42.10 Robust performance LFT.

Define an augmented block structure

$$\Delta_P := \left\{ \begin{bmatrix} \Delta & 0 \\ 0 & \Delta_F \end{bmatrix} : \Delta \in \mathbf{\Delta}, \Delta_F \in \mathbf{C}^{n_d \times n_e} \right\}.$$

Δ_F corresponds to the Δ_2 block of the Main Loop Theorem. It is used to compute bounds on $\bar{\sigma}(\cdot)$ of the perturbed transfer function $F_u\left(\hat{G}, \Delta\right)$ as Δ takes on values in $\mathbf{S}_\Delta$.

THEOREM 42.5 *Let $\beta > 0$. For all $\Delta \in \mathbf{S}_\Delta$ with $\|\Delta\|_\infty < \frac{1}{\beta}$, the perturbed system in Figure 42.10 is well-posed, internally stable, and $\left\|F_u\left(\hat{G}, \Delta\right)\right\|_\infty \leq \beta$ if, and only if,*

$$\|G\|_{\Delta_P} := \sup_{\omega\in\mathbf{R}} \mu_{\Delta_P}\left(\hat{G}(j\omega)\right) \leq \beta.$$

See [14] for details of the proof. The robust performance theorem provides a test to determine if the performance of the system

$F_u\left(\hat{G}, \Delta\right)$ remains acceptable for all possible norm-bounded perturbations.

42.5.3 Robust Performance Example

It is instructive to carry out these steps on a simple example. Here, we analyze the robust stability of a simple single-loop feedback regulation system with two uncertainties. The plant is a lightly-damped, nominal two-state system with uncertainty in the (2, 1) entry of the A matrix (the frequency-squared coefficient) and unmodeled dynamics (in the form of multiplicative uncertainty) at the control input. The overall block diagram of the uncertain closed-loop system is shown in Figure 42.11.

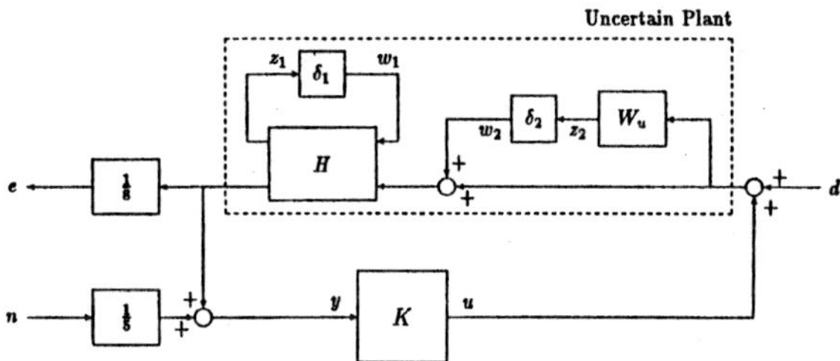

Figure 42.11 Robust Stability/Performance Example.

The two-state system with uncertainty in the A matrix is represented as an upper linear fractional transformation about a two-input, two-output, two-state system H, whose realization is

$$A = \begin{bmatrix} 0 & 1 \\ -16 & -0.16 \end{bmatrix}, \quad B = \begin{bmatrix} 0 & 0 \\ 1 & 1 \end{bmatrix},$$
$$C = \begin{bmatrix} 6.4 & 0 \\ 16 & 0 \end{bmatrix}, \quad \text{and} \quad D = \begin{bmatrix} 0 & 0 \\ 0 & 0 \end{bmatrix}$$

The resulting second order system takes the form

$$F_u(H, \delta_1) = \frac{16}{s^2 + 0.16s + 16(1 + 0.4\delta_1)}$$

If we assume that δ_1 is unknown, but satisfies $|\delta_1| \le 1$, then we interpret the second-order system to have 40% uncertainty in the denominator entry of the natural frequency-squared coefficient.

The plant is also assumed to have unmodeled dynamics at the input. This could arise from an unmodeled, or unreliable, actuator, for instance. The uncertainty is assumed to be about 20% at low frequency, rising to 100% at 6.5 radians/second. We model it using the multiplicative uncertainty model, using a first-order weight, $W_u = \frac{6.5s + 8}{s + 42}$. In the block diagram, this is represented with the simple linear fractional transformation involving δ_2.

The closed-loop performance objective is $\|T\|_\infty \le 1$, where T is the transfer function from input disturbance d and sensor noise n to the output error e,

$$e = T\begin{bmatrix} d \\ n \end{bmatrix} = [T_1 \; T_2]\begin{bmatrix} d \\ n \end{bmatrix}$$

Note that the closed-loop T is a function of both δ_1 and δ_2. The scalar blocks which weight the error and the noise are used to normalize the two transfer functions that make up T. Finally, for comparison, the open-loop system has $\|T_1\|_\infty \approx 6$, and $\|T_2\|_\infty = 0$.

For this example, the controller is chosen as

$$K = \frac{-12.56s^2 + 17.32s + 67.28}{s^3 + 20.37s^2 + 136.74s + 179.46}.$$

Finally, $G(s)$ in Figure 42.12 denotes the closed-loop transfer function matrix from Figure 42.11. The dimensions of G are two states, four inputs and three outputs.

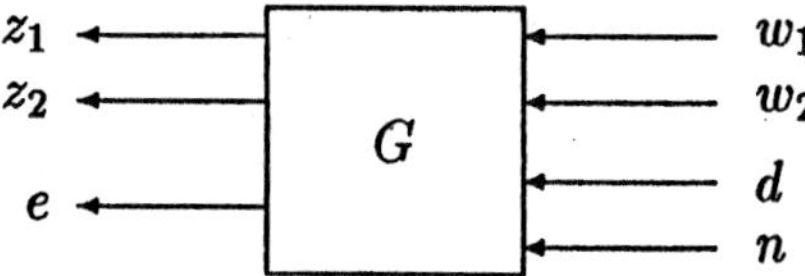

Figure 42.12 Closed-loop interconnection.

In terms of G, we have

$$T = F_u\left(G, \begin{bmatrix} \delta_1 & 0 \\ 0 & \delta_2 \end{bmatrix}\right)$$

Hence, using Theorems 42.3 and 42.5, the robust stability and robust performance of the closed-loop system can be ascertained by appropriate structured singular value calculations on G (or particular subblocks of G). In the next section, we analyze the robust stability and robust performance of the closed-loop system for a variety of assumptions on the uncertain elements, δ_1 and δ_2.

Analysis

For notational purposes, partition $G(s)$ into

$$G = \begin{bmatrix} G_{11} & G_{12} \\ G_{21} & G_{22} \end{bmatrix} \tag{42.20}$$

where $G_{11}(s)$ is 2×2, and $G_{22}(s)$ is 2×1. The first two inputs and outputs of G are associated with the perturbation structure and the third and fourth inputs and third output correspond to an exogenous multivariable disturbance signal, and associated error. For robust stability calculations we are only interested in the submatrix G_{11}, and for robust performance calculations the entire matrix G.

Robust stability calculations are performed with respect to two different block structures:

$$\Delta_1 := \left\{\text{diag}\,[\delta_1, \delta_2] : \delta_1, \delta_2 \in \mathbf{C}\right\},$$
$$\Delta_2 := \left\{\text{diag}\,[\delta_1, \delta_2] : \delta_1 \in \mathbf{R}, \delta_2 \in \mathbf{C}\right\}$$

For robust performance calculations, a 2×1 full block is appended to Δ_i for the performance calculation, yielding $\Delta_P \subset$

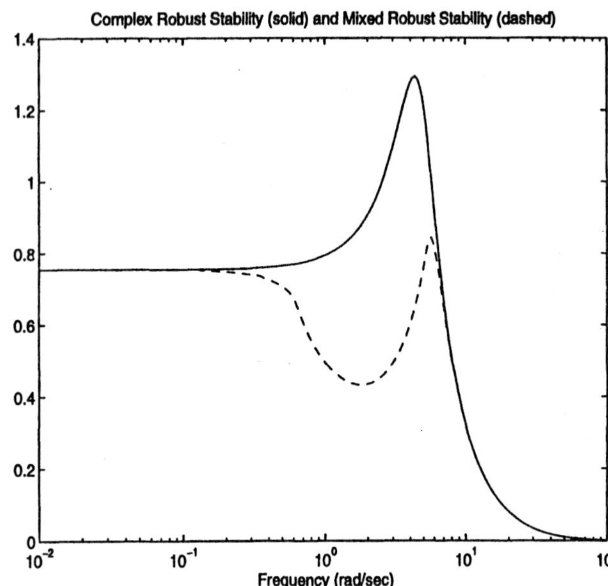

Figure 42.13 Complex robust stability (solid) and Mixed robust stability (dashed) plots.

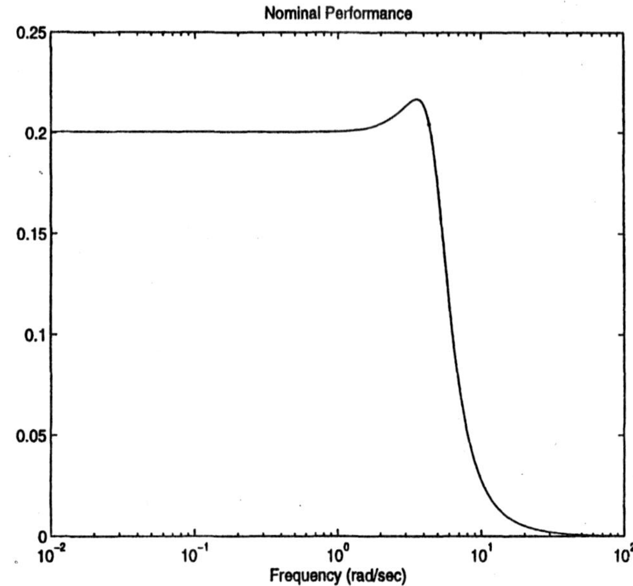

Figure 42.14 Nominal performance plot.

$\mathbf{C}^{4\times 3}$. The two block structures used to evaluate robust performance are:

$$\Delta_{P_1} := \left\{\text{diag}\,[\delta_1, \delta_2, \Delta_F] : \delta_1, \delta_2 \in \mathbf{C}, \Delta_F \in \mathbf{C}^{2\times 1}\right\},$$
$$\Delta_{P_2} := \left\{\text{diag}\,[\delta_1, \delta_2, \Delta_F] : \delta_1 \in \mathbf{R}, \delta_2 \in \mathbf{C}, \Delta_F \in \mathbf{C}^{2\times 1}\right\}$$

All the upper and lower bounds μ calculations are performed using the *μ Analysis and Synthesis Toolbox* [2].

Robust Stability

The robustness of the closed-loop system with respect to linear, complex time-invariant structured perturbations, Δ_1, is a μ test on $G_{11}(j\omega)$. The complex, robust stability bounds from the μ calculation are shown in Figure 42.13 (Note that the upper and lower bounds are identical.). The peak μ value is about 1.29, hence for any $\Delta(s) \in \mathbf{S}_{\Delta_1}$ stability is preserved as long as $\|\Delta(s)\|_\infty < \frac{1}{1.29}$, and there is a perturbation $\Delta_{\text{dest}}(s)$, of the correct structure, with $\|\Delta_{\text{dest}}\|_\infty = \frac{1}{1.29}$ that does cause instability.

The nominal performance of this system is defined by the $\mathcal{H}_\infty$ norm of the transfer function G_{22} is $\|G_{22}\|_\infty = 0.22$. The maximum singular value of G_{22} is plotted across frequency in Figure 42.14. The robustness and performance measures were originally scaled to be less than 1 when they were achieved. Therefore, the system *is not* robustly stabilized with respect to linear, time-invariant structured complex perturbations of size 1, but it achieves the performance objective on the nominal system.

Recall that the first uncertainty, δ_1, corresponds to uncertainty in the $A(2, 1)$ coefficient and the second uncertainty, δ_2, corresponds to input multiplicative modeling error. The $A(2, 1)$ coefficient uncertainty can be treated as a real uncertainty. This would imply that the magnitude $A(2, 1)$ varies between 9.6 and 22.4. In the initial robust stability analysis, both of these uncertainties were modeled as complex perturbations, a potentially more conservative representation of the uncertainty. Let us reanalyze the system with respect to Δ_2 where δ_1 is treated as a real perturbation.

We can analyze the robust stability of the system with respect to one of the uncertainties being real and the other uncertainty complex. This is shown in the mixed robust stability plot shown in Figure 42.13. Notice that when the $A(2, 1)$ uncertainty is treated as a real perturbation, and the input multiplicative uncertainty is complex, the mixed robust stability μ value is reduced from 1.29 to 0.84. Hence the system *is* robustly stabilized with respect to real uncertainty in the A(2,1) coefficient and complex input multiplicative uncertainty. In this example, it is very conservative to treat the variation in the coefficient, $A(2, 1)$, as a complex uncertainty.

Robust Performance

The closed-loop system under perturbations becomes $F_u(G, \Delta)$. To analyze the degradation of performance due to the uncertainty, we use Theorem 42.5, and the augmented block structure Δ_{P_1}. The plot in Figure 42.15 of $\mu_{\Delta_{P_1}}(G(j\omega))$ is shown. The peak is approximately at 1.41. Applying Theorem 42.5 implies that for any structured $\Delta(s) \in \mathbf{S}_{\Delta_{\mathbf{P1}}}$ with $\|\Delta(s)\|_\infty < \frac{1}{1.41}$, the perturbed loop remains stable **and**, the $\|\cdot\|_\infty$ norm of $F_u(G, \Delta)$ is guaranteed to be ≤ 1.41. Also, the converse of the theorem shows that there is a perturbation Δ, whose $\|\cdot\|_\infty$ is arbitrarily close to $\frac{1}{1.41}$ that causes $\left\|F_u\left(\Delta, \hat{G}\right)\right\|_\infty > 1.41$. Therefore robust performance was not achieved.

Figure 42.15 also shows the results of a mixed μ analysis on $G(j\omega)$ with respect to Δ_{P_2}. The peak value of μ is 0.99. This implies that for a real perturbation δ_1 and a finite dimensional, linear

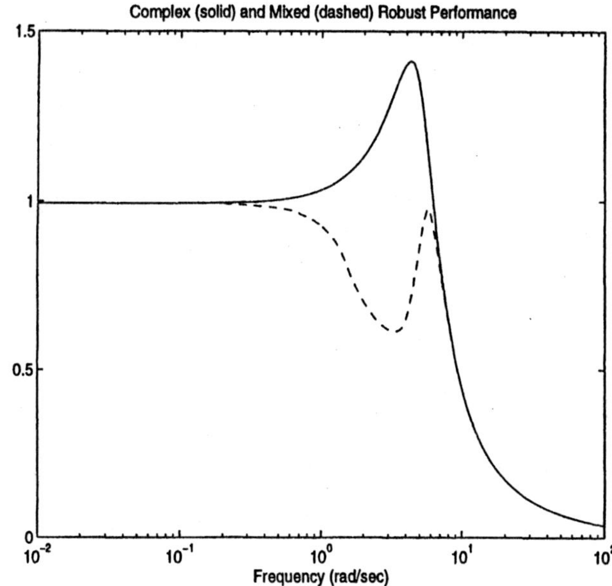

Figure 42.15 Complex (solid) and Mixed (dashed) robust performance plots.

time-invariant complex perturbation $\delta_2(s)$, stability is preserved and the performance objective achieved. Therefore the robust performance objective is achieved when the frequency-squared coefficient is treated as a real perturbation and the input multiplicative uncertainty is treated as a complex perturbation.

42.6 Spinning Satellite: Robust Performance Analysis with μ

Consider the 4×4 transfer function M shown (along with its with internal structure) in Figure 42.2. The perturbations δ_1 and δ_2 enter as shown in Figure 42.3. The appropriate block structure for the robust performance assessment of this example is $\{\text{diag}(\delta_1, \delta_2, \Delta_F) : \delta_i \in \mathbf{C}, \Delta_F \in \mathbf{C}^{2\times 2}\}$. This implies that there is independent uncertainty in each of the actuators but that the rest of the model is accurate. Computing the structured singular value of $\mu_\Delta(M)$ across frequency yields a peak of about 11. This implies that a diagonal perturbation $\mathbf{diag}\,[\delta_1, \delta_2]$ of size 1/11 exists so that the perturbed reference-to-error transfer function has a singular value peak of approximately 11. This μ analysis clearly detects the poor robust performance characteristics of the closed-loop system. In the next section, we turn our attention to design techniques which use the structured singular value as a design objective.

42.7 Control Design via μ Synthesis

Consider the standard linear fractional description of the control problem shown in Figure 42.16. The P block represents the open-loop interconnection and contains all of the known elements including the nominal plant model, uncertainty structure, and performance and uncertainty weighting functions. The Δ_{pert} block represents the structured set of norm-bounded uncertainty being considered and K represents the controller. Δ_{pert} parameterizes all of the assumed model uncertainty in the problem. Three groups of inputs enter P, perturbations z, disturbances d, and controls u, and three groups of outputs are generated, perturbations w, errors e, and measurements y. The set of systems to be controlled is described by the LFT

$$\left\{F_u\left(\Delta_{pert}, P\right) : \Delta_{pert} \in \mathbf{S}_{\Delta_{\text{pert}}}\right\}.$$

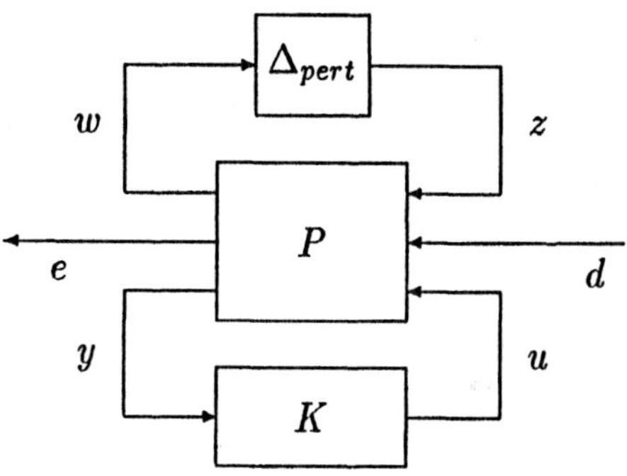

Figure 42.16 Linear fractional transformation description of control problem.

The design objective is to find a stabilizing controller K, so that, for all $\Delta_{pert} \in \mathbf{S}_{\Delta_{\text{pert}}}$, $\|\Delta_{pert}\|_\infty \leq 1$, the closed-loop system is stable and satisfies

$$\|F_u\left[F_l(P, K), \Delta_{pert}\right]\|_\infty \leq 1.$$

The performance objective involves a robust performance test on the linear fractional transformation $F_l(P, K)$. To assess the robust performance of the closed-loop system, define an augmented perturbation structure, Δ,

$$\Delta = \left\{\begin{bmatrix} \Delta_{pert} & 0 \\ 0 & \Delta_F \end{bmatrix} : \Delta_{pert} \in \mathbf{\Delta}_{pert}, \ \Delta_F \in \mathbf{C}^{n_d \times n_e}\right\}.$$

The goal of μ synthesis is to minimize overall stabilizing controllers K, the peak value of $\mu_\Delta(\cdot)$ of the closed-loop transfer function $F_l(P, K)$. More formally,

$$\min_{\substack{K \\ \text{stabilizing}}} \ \max_\omega \ \mu_\Delta[F_l(P, K)(j\omega)] \tag{42.21}$$

For tractability of the μ synthesis problem, $\mu_\Delta[\cdot]$ is replaced by the upper bound for μ, $\bar{\sigma}\left[D(\cdot)D^{-1}\right]$. The scaling matrix D is a member of the appropriate set of scaling matrices $\mathbf{D}$ for the perturbation set Δ. One can reformulate this optimization problem as follows:

$$\min_{\substack{K \\ \text{stabilizing}}} \ \max_\omega \ \min_{D_\omega \in \mathbf{D}} \ \bar{\sigma}\left[D_\omega F_l(P, K)(j\omega)D_\omega^{-1}\right]. \tag{42.22}$$

Here, the D minimization is an approximation to the $\mu_\Delta[F_l(P,K)(j\omega)]$. D_ω is chosen from the set of scalings, **D**, independently at every ω. Hence,

$$\min_{\substack{K\\ \text{stabilizing}}} \min_{\substack{D_{(\cdot)}\\ D_\omega\in\mathbf{D}}} \max_\omega \bar{\sigma}\left[D_\omega F_l(P,K)(j\omega)D_\omega^{-1}\right]. \tag{42.23}$$

The expression $\max_\omega \bar{\sigma}[\cdot]$ corresponds to $\|[\cdot]\|_\infty$, leaving

$$\min_{\substack{K\\ \text{stabilizing}}} \min_{\substack{D_{(\cdot)}\\ D_\omega\in\mathbf{D}}} \left\|\left[D.F_l(P,K)(j\cdot)D_\cdot^{-1}\right]\right\|_\infty. \tag{42.24}$$

Assume, for simplicity, that the uncertainty block Δ_{pert} has only full blocks. Then the set $\mathbf{D}_\Delta$ is of the form

$$\mathbf{D} = \left\{\text{diag}\left[d_1 I, d_2 I, \ldots, d_{F-1}I, I\right] : d_i > 0\right\}. \tag{42.25}$$

For any complex matrix M, the elements of $\mathbf{D}_\Delta$, which were originally defined as real and positive, can take on any nonzero complex values and without changing the value of the upper bound, $\inf_{D\in\mathcal{D}} \bar{\sigma}\left(DMD^{-1}\right)$. Hence, we can restrict the scaling matrix to be a real-rational, stable, minimum-phase transfer function, $\hat{D}(s)$. The optimization is now

$$\min_{\substack{K\\ \text{stabilizing}}} \min_{\substack{\hat{D}(s)\in\mathbf{D}\\ \text{stable, min-phase}}} \|\hat{D}F_l(P,K)\hat{D}^{-1}\|_\infty. \tag{42.26}$$

This approximation to μ synthesis, is currently "solved" by an iterative approach, referred to as "$D - K$ iteration."

To solve Equation 42.26, first consider holding $\hat{D}(s)$ fixed. Given a stable, minimum-phase, real-rational $\hat{D}(s)$, solve the optimization $\min_{\substack{K\\ \text{stabilizing}}} \|\hat{D}F_l(P,K)\hat{D}^{-1}\|_\infty$. This equation is an $\mathcal{H}_\infty$ optimization control problem. The solution to the $\mathcal{H}_\infty$ problem is well-known, consisting of solving algebraic Riccati equations in terms of the state-space system.

Now suppose that a stabilizing controller, $K(s)$, is given, we then solve the following minimization corresponding to the upper bound for μ.

$$\min_{D_\omega\in\mathbf{D}} \bar{\sigma}\left[D_\omega F_l(P,K)(j\omega)D_\omega^{-1}\right]$$

This minimization is done over the real, positive D_ω from the set $\mathbf{D}_\Delta$ defined in Equation 42.25. Recall that the addition of phase to each d_i does not affect the value of $\bar{\sigma}\left[D_\omega F_l(P,K)(j\omega)D_\omega^{-1}\right]$. Hence, each discrete function, d_i, of frequency is fit (in magnitude) by a proper, stable, minimum-phase transfer function, $\hat{d}_{R_i}(s)$. These are collected together in a diagonal transfer function matrix $\hat{D}(s)$,

$$\hat{D}(s) = \text{diag}\left[\hat{d}_{R_1}(s)I, \hat{d}_{R_2}(s)I, \ldots \hat{d}_{R_{F-1}}(s)I, I\right],$$

and absorbed into the original open-loop generalized plant P. Iterating on these two steps comprises the current approach to $D - K$ iteration.

There are several problems with the $D - K$ iteration control design procedure. The first is that we have approximated $\mu_\Delta(\cdot)$ by its upper bound. This is not serious because the value of μ and its upper bound are often close. The most serious problem, that the $D - K$ iteration does not always converge to a global, or even, local minimum, [14] is a more severe limitation of the design procedure. However, in practice the $D - K$ iteration control design technique has been successfully applied to many engineering problems such as vibration suppression for flexible structures, flight control, chemical process control problems, and acoustic reverberation suppression in enclosures.

42.8 F-14 Lateral-Directional Control System Design

Consider the design of a lateral-directional axis controller for the F-14 aircraft during powered approach to landing. The linearized F-14 model is found at an angle-of-attack (α) of 10.5 *degrees* and airspeed of 140 *knots*. The problem is posed as a robust performance problem with multiplicative plant uncertainty at the plant input and minimization of weighted-output transfer functions as the performance criterion. A diagram for the closed-loop system, which includes the feedback structure of the plant and controller and elements associated with the uncertainty models and performance objectives, is shown in Figure 42.17.

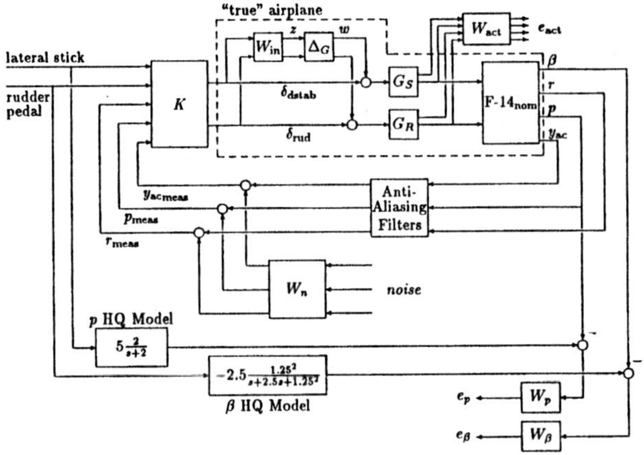

Figure 42.17 F-14 control block diagram.

The overall performance objective is to have the "true" airplane, represented by the dashed box in Figure 42.17, respond effectively to the pilot's lateral stick and rudder pedal inputs. The performance objective includes

1. Decoupled response of the lateral stick, δ_{lstk}, to roll rate, p, and rudder pedals, δ_{rudp}, to side-slip angle, β. The lateral stick and rudder pedals have a maximum deflection of ± 1 *inch*. Therefore they are represented as unweighted signals in Figure 42.17.
2. The aircraft handling quality (HQ) response from the lateral stick to roll rate should be a first-order system, $5(2)/(s+2)\frac{degrees/sec.}{inch}$. The aircraft handling

quality response from the rudder pedals to side-slip angle should be $-2.5\frac{1.25^2}{s^2+2.5s+1.25^2}\frac{degrees/sec.}{inch}$.

3. The stabilizer actuators have $\pm 20°$ and $\pm 90°/sec.$ deflection and deflection rate limits. The rudder actuators have $\pm 30°$ and $\pm 125°/sec$ deflection and deflection rate limits.
4. The three measurement signals, roll rate, yaw rate and lateral acceleration, are passed through second-order antialiasing filters prior to being fed to the controller. The natural frequency and damping values for the yaw rate and lateral acceleration filters are 12.5Hz and 0.5, respectively and 4.1Hz and 0.7 for the roll rate filter. The antialiasing filters have unity gain at DC (see Figure 42.17). These signals are also corrupted by noise.

The performance objectives are accounted for in this framework via minimizing weight transfer function norms. Weighting functions serve two purposes in the $\mathcal{H}_\infty$ and μ framework: they allow the direct comparison of different performance objectives with the same norm and they allow incorporating frequency information into the analysis. The F-14 performance weighting functions include:

1. Limits on the actuator deflection magnitude and rates are included via the W_{act} weight. W_{act} is a 4×4 constant, diagonal scaling matrix described by $W_{act} = \text{diag}(1/90, 1/20, 1/125, 1/30)$. These weights correspond to the stabilizer and rudder deflection rate and deflection limits.
2. W_n is a 3×3 diagonal, frequency varying weight used to model the magnitude of the sensor noise. $W_n = \text{diag}(0.025, 0.025, 0.0125\frac{s+1}{s+100})$ which corresponds to the noise levels in the roll rate, yaw rate and lateral acceleration channels.
3. The desired δ_{lstk}-to-p and δ_{rudp}-to-β responses of the aircraft are formulated as a model matching problem in the μ framework. The difference between the ideal response of the transfer functions, δ_{lstk} filtered through the roll rate HQ model and δ_{rudp} filtered through the side-slip angle HQ model, and the aircraft response, p and β, is used to generate an error that is to be minimized. The W_p transfer function, see Figure 42.17, weights the difference between the idealized roll rate response and the actual aircraft response, p.

$$W_p = \frac{0.05s^4 + 2.90s^3 + 105.93s^2 + 6.17s + 0.16}{s^4 + 9.19s^3 + 30.80s^2 + 18.33s + 3.95}.$$

The magnitude of W_p emphasizes the frequency range between 0.06 and 30 rad/sec. The desired performance frequency range is limited due to a right half-plane zero in the model at 0.002 rad/sec., therefore, accurate tracking of sinusoids below 0.002 rad/sec. isn't required. Between 0.06 and 30 rad/sec., a roll rate tracking error of less than 5% is desired. The performance weight on the β tracking error, W_β, is just $2 \times W_p$. This also corresponds to a 5% tracking error objective.

All the weighted performance objectives are scaled for an $\mathcal{H}_\infty$ less than 1 when they are achieved. The performance of the closed-loop system is evaluated by calculating the maximum singular value of the weighted transfer functions from the disturbance and command inputs to the error outputs, as in Figure 42.18.

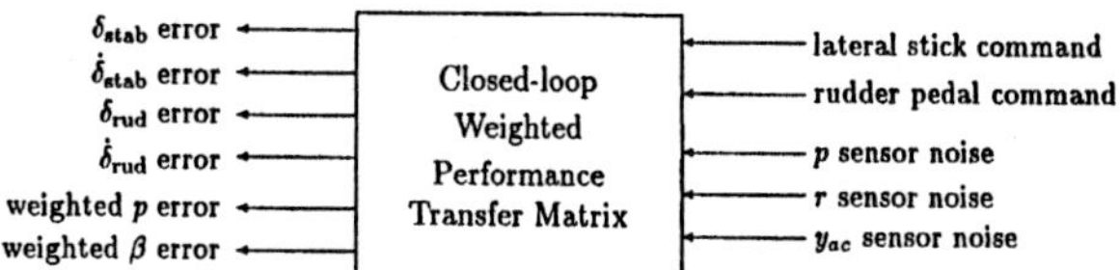

Figure 42.18 F-14 weighted performance objectives transfer matrix.

42.8.1 Nominal Model and Uncertainty Models

The pilot has the ability to command the lateral directional response of the aircraft with the lateral stick (δ_{lstk}) and rudder pedals (δ_{rped}). The aircraft has two control inputs, differential stabilizer deflection (δ_{dstab}, $degrees$) and rudder deflection (δ_{rud}, $degrees$), three measured outputs, roll rate ($p, degs/sec$), yaw rate (r, $degs/sec$) and lateral acceleration (y_{ac}, g's), and a calculated output side-slip angle (β). Note that β is not a measured variable but is used as a performance measure. The lateral directional F-14 model, F-14$_{nom}$, has four states, lateral velocity (v), yaw rate (r), roll rate (p) and roll angle (ϕ). These variables are related by the state-space equations

$$\begin{bmatrix} \dot{v} \\ \dot{r} \\ \dot{p} \\ \dot{\phi} \\ \beta \\ p \\ r \\ y_{ac} \end{bmatrix} = \left[\begin{array}{c|c} A & B \\ \hline C & D \end{array}\right] \begin{bmatrix} v \\ r \\ p \\ \phi \\ \delta_{dstab} \\ \delta_{drud} \end{bmatrix}$$

$$A = \begin{bmatrix} -1.16e-1 & -2.27e+2 & 4.30e+1 & 3.16e+1 \\ 2.65e-3 & -2.59e-1 & -1.45e-1 & 0.00e+0 \\ -2.11e-2 & 6.70e-1 & -1.36e+0 & 0.00e+0 \\ 0.00e+0 & 1.85e-1 & 1.00e+0 & 0.00e+0 \end{bmatrix},$$

$$B = \begin{bmatrix} 6.22e-02 & 1.01e-1 \\ -5.25e-03 & -1.12e-2 \\ -4.67e-02 & 3.64e-3 \\ 0.00e+00 & 0.00e+0 \end{bmatrix},$$

$$C = \begin{bmatrix} 2.47e-1 & 0.00e+0 & 0.00e+0 & 0.00e+0 \\ 0.00e+0 & 0.00e+0 & 5.73e+1 & 0.00e+0 \\ 0.00e+0 & 5.73e+1 & 0.00e+0 & 0.00e+0 \\ -2.83e-3 & -7.88e-3 & 5.11e-2 & 0.00e+0 \end{bmatrix},$$

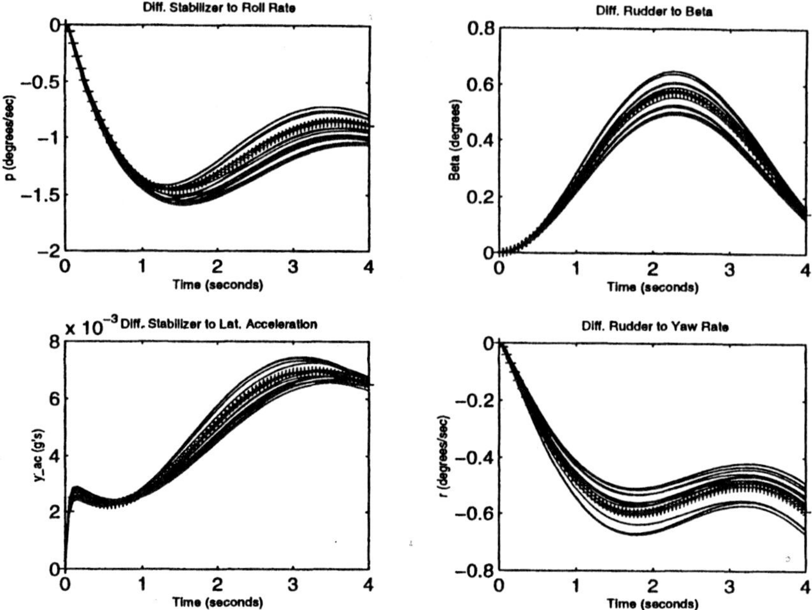

Figure 42.19 Unit-step responses of the nominal model (+) and 15 perturbed models from $\mathcal{G}$.

$$D = \begin{bmatrix} 0.00e+00 & 0.00e+0 \\ 0.00e+00 & 0.00e+0 \\ 0.00e+00 & 0.00e+0 \\ 2.89e-03 & 2.27e-3 \end{bmatrix},$$

The dashed box represents the "true" airplane, corresponding to a set of F-14 plant models define by $\mathcal{G}$. Inside the box is the nominal model of the airplane dynamics, F-14$_{\text{nom}}$, models of the actuators, G_S and G_R, and two elements, W_{in} and Δ_G, which parameterize the uncertainty in the model. This type of uncertainty is called multiplicative plant input uncertainty. The transfer function W_{in} is assumed known and reflects the amount of uncertainty in the model. The transfer function Δ_G is assumed stable and unknown, except for the norm condition, $\|\Delta_G\|_\infty \leq 1$.

A "first principles" set of uncertainties in the aircraft model would include

1. Uncertainty in the stabilizers and the rudder actuators. The electrical signals that command deflections in these surfaces must be converted to actual mechanical deflections by the electronics and hydraulics of the actuators. Unlike the models, this is not done perfectly in the actual system.
2. Uncertainty in the forces and moments generated on the aircraft, due to specific deflections of the stabilizers and rudder. As a first approximation, this arises from the uncertainties in the aerodynamic coefficients, which vary with flight conditions, as well as uncertainty in the exact geometry of the airplane.
3. Uncertainty in the linear and angular accelerations produced by the aerodynamically generated forces and moments. This arises from the uncertainty in the various inertial parameters of the airplane, in addition to neglected dynamics, such as fuel slosh and airframe flexibility.
4. Other forms of uncertainty that are less well understood.

In this example, we choose not to model the uncertainty in this detailed manner but rather to lump all of these effects together into one, complex full-block, multiplicative uncertainty at the input of the rigid body aircraft nominal model.

The stabilizer and rudder actuators, G_S and G_R, are modeled as first order transfer functions, $25/(s+25)$. Given the actuator and aircraft nominal models (denoted by $G_{\text{nom}}(s)$), we also specify a stable, 2×2 transfer function matrix $W_{\text{in}}(s)$ called the uncertainty weight. These transfer matrices parameterize an entire set of plants, $\mathcal{G}$, which must be suitably controlled by the robust controller K.

$$\mathcal{G} := \{G_{\text{nom}}(I + \Delta_G W_{\text{del}}) : \Delta_G \text{ stable}, \|\Delta_G\|_\infty \leq 1\}.$$

All of the uncertainty in modeling the airplane is captured in the normalized, unknown transfer function Δ_G. The unknown transfer function $\Delta_G(s)$ is used to parameterize the potential differences between the nominal model $G_{\text{nom}}(s)$, and the actual behavior of the real airplane, denoted by $\mathcal{G}$.

In this example, the uncertainty weight W_{in} is of the form, $W_{\text{in}}(s) := \text{diag}(w_1(s), w_2(s))I_2$, for particular scalar valued functions $w_1(s)$ and $w_2(s)$. The $w_1(s)$ weight associated with the differential stabilizer input is selected to be $w_1(s) = \frac{2(s+4)}{s+160}$. The $w_2(s)$ weight associated with the differential rudder input is selected to be $w_2(s) = \frac{1.5(s+20)}{s+200}$. Hence the set of plants that are represented by this uncertainty weight

$$\mathcal{G} := \left\{ \text{F-14}_{\text{nom}} \begin{bmatrix} \frac{25}{s+25} & 0 \\ 0 & \frac{25}{s+25} \end{bmatrix} \left(I_2 + \begin{bmatrix} \frac{2(s+4)}{s+100} & 0 \\ 0 & \frac{1.5(s+20)}{s+200} \end{bmatrix} \Delta_G(s) \right) : \Delta_G(s) \text{ stable}, \|\Delta_G\|_\infty \leq 1 \right\}$$

Note that the weighting functions are used to normalize the size of the unknown perturbation Δ_G. At any frequency ω, the value of $|w_1(j\omega)|$ and $|w_2(j\omega)|$ can be interpreted as the percentage of uncertainty in the model at that frequency. The dependence of the uncertainty weight on frequency indicates that the level of uncertainty in the airplane's behavior depends on frequency.

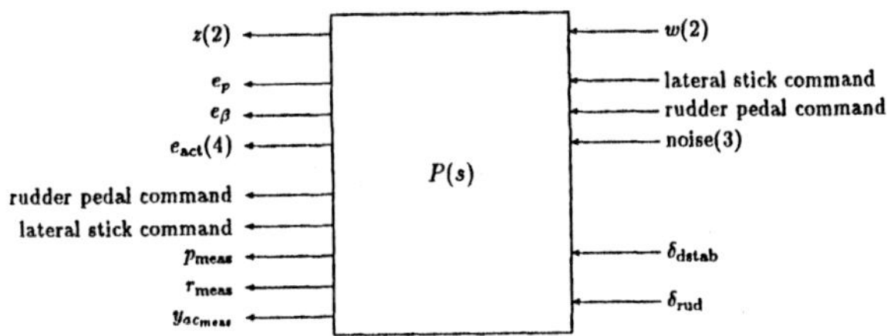

Figure 42.20 F-14 generalized plant.

The particular uncertainty weights chosen imply that, in the differential stabilizer channel at low frequency, there is potentially a 5% modeling error, and at a frequency of 93 rad/sec., the uncertainty in channel 1 can be as much as 100%, and can get larger at higher frequencies. The rudder channel has more uncertainty at low frequency, up to 15% modeling error, and at a frequency of 177 rad/sec., the uncertainty is at 100%. To illustrate the variety of plants represented by the set $\mathcal{G}$, some step responses of different systems from $\mathcal{G}$ are shown in Figure 42.19.

The control design objective is a stabilizing controller K so that for all stable perturbations $\Delta_G(s)$, with $\|\Delta_G\|_\infty \leq 1$, the perturbed closed-loop system remains stable, and the perturbed weighted performance transfer functions has an $\mathcal{H}_\infty$ norm less than 1 for all such perturbations. These mathematical objectives fit exactly into the structured singular value framework.

42.8.2 Controller Design

The control design block diagram shown in Figure 42.17 is redrawn as $P(s)$, shown in Figure 42.20. $P(s)$, the 25-state, six-input, six-output open-loop transfer matrix, corresponds to the P in the linear fractional block diagram in Figure 42.16.

The first step in the $D-K$ iteration control design procedure is to design an $\mathcal{H}_\infty$ (sub)optimal controller for the open-loop interconnection, P. In terms of the $D-K$ iteration, this amounts to holding the d variable fixed (at 1) and minimizing the $\|\cdot\|_\infty$ norm of $F_l(P, K)$ over the controller variable K. The resulting controller is labeled K_1.

The second step in the $D-K$ iteration involves solving a μ analysis problem corresponding to the closed-loop system, $F_l(P, K_1)$. This calculation produces a frequency dependent scaling variable d_ω, the (1,1) entry in the scaling matrix. In a general problem (with more than two blocks), there would be several d variables, and the overall matrix is referred to as "the D-scales." The varying variables in the D-scales are fit (in magnitude) with proper, stable, minimum-phase rational functions and absorbed into the generalized plant for additional iterations. These scalings are used to "trick" the $\mathcal{H}_\infty$ minimization to concentrate more on minimizing μ rather than $\bar{\sigma}$ across frequency. For the first iteration in this example, the d scale data is fit with a first-order transfer function.

The new generalized plant used in the second iteration has 29 states, four more states than the original 25-state generalized plant, P. These extra states are due to the D-scale data being fitted with a rational function and absorbed into the generalized plant for the next iteration. Four $D-K$ iterations are performed until μ reaches a value of 1.02. Information about the $D-K$ iterations is shown in Table 42.1. All the analysis and synthesis results were obtained with *The μ Analysis and Synthesis Toolbox, Version 2.0* [2].

TABLE 42.1 F-14 $D-K$ iteration information.

Iteration number	1	2	3	4
Total D-scale order	0	4	4	4
Controller order	25	29	29	29
$\mathcal{H}_\infty$ norm achieved	1.562	1.079	1.025	1.017
Peak μ value	1.443	1.079	1.025	1.017

Analysis of the Controllers

The robust performance properties of the controllers can be analyzed using μ analysis. Robust performance is achieved if, and only if, for every frequency, $\mu_\Delta(F_l(P, K)(j\omega))$ of the closed-loop frequency response is less than 1. Plots of μ of the closed-loop system with K_1 and K_4 implemented are shown in Figure 42.21.

The controlled system with K_1 implemented **does not** achieve **robust performance.** This conclusion follows from the μ plot, which peaks to a value of 1.44, at a frequency of 7 rad/sec.. Because μ is 1.44, there is a perturbation matrix Δ_G, so that $\|\Delta_G\|_\infty = \frac{1}{1.44}$, and the perturbed weighted performance transfer functions gets "large." After four $D-K$ iterations the peak robust performance μ value is reduced to 1.02 (Figure 42.21), thereby, nearly achieving all of our robust performance objectives.

Illustrating the robustness of the closed-loop system in the time domain, time responses of the ideal model, the nominal closed-loop system and the "worst-case" closed-loop system from $\mathcal{G}$ (using perturbations of size 1) are shown in Figure 42.22. Controller K_4 is implemented in the closed-loop simulations. A 1-inch lateral stick command is given at 9 sec., held at 1 inch until 12 sec., and then returns to zero. The rudder is commanded at 1 second with a positive 1 inch pedal deflection and held at 1 inch until 4 seconds. At 4 seconds a -1-inch pedal deflection is commanded, held to 7 seconds, and then returned to zero. One

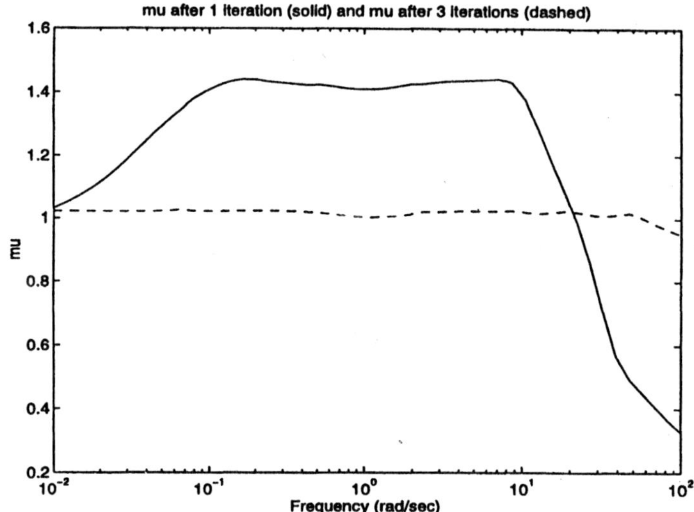

Figure 42.21 F-14 robust performance μ plots with K_1 and K_4 implemented.

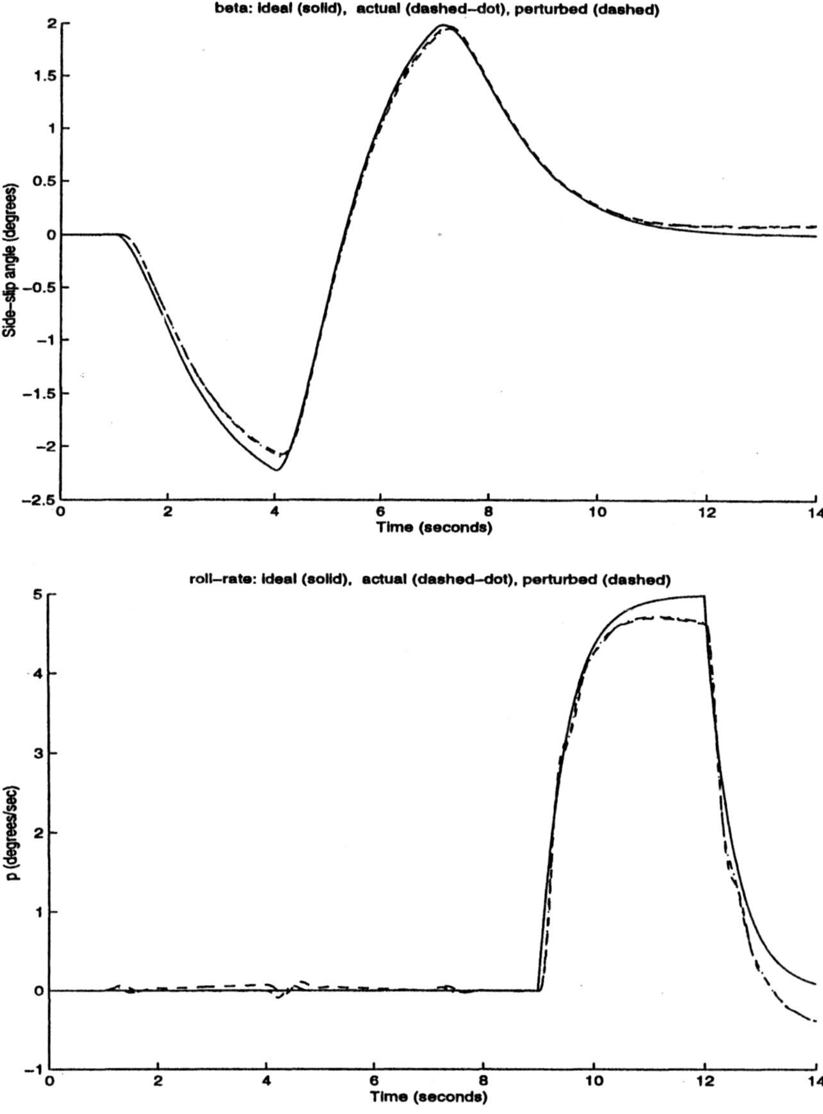

Figure 42.22 Time response plots of the F-14 lateral directional control system.

can see from the time responses that the closed-loop response is nearly identical for the nominal closed-loop system and the "worst-case" closed-loop system. The ideal time response for β and p are plot for reference.

42.9 Conclusion

This chapter outlined the usefulness of the structured singular value (μ) analysis and synthesis techniques in designing and analyzing multiloop feedback control systems. Through examples, we have shown some pitfalls with simple-minded analytical techniques, and illustrated the usefulness of the analytic framework provided by the structured singular value. We outlined an approach to robust controller synthesis, the $D - K$ iteration. As an example, these techniques were applied to the design of a lateral directional control system for the F-14 aircraft.

References

[1] Balas, G.J. and Doyle, J.C., Control of lightly damped, flexible modes in the controller crossover region, *AIAA J. Guidance, Dyn. Control*, 17(2), 370–377, 1994.

[2] Balas, G.J., Doyle, J.C., Glover, K., Packard, A., and Smith, R., The μ Analysis and Synthesis Toolbox, MathWorks, Natick, MA, 1991.

[3] Boyd, S., El Ghaoui, L., Feron, E., Balakrishnan, V., *Linear Matrix Inequalities in System and Control Theory*, SIAM Studies in Applied Mathematics, SIAM, Philadelphia, 1994.

[4] Chiang, R. and Safonov, M., Robust Control Toolbox, MathWorks, Natick, MA, 1988.

[5] Doyle, J.C., Analysis of feedback systems with structured uncertainties, *IEEE Proc.*, 129 Part D(6), 242–250, 1982.

[6] Doyle, J.C., Lenz, K., and Packard, A., Design Examples using μ synthesis: Space Shuttle Lateral Axis FCS during reentry, *IEEE CDC*, 1986, 2218–2223; also *Modelling, Robustness and Sensitivity Reduction in Control Systems*, NATO ASI Series, Curtain, R.F., Ed., Springer, 1987, vol. F34, 128–154.

[7] Fan, M., Tits, A., and Doyle, J.C., Robustness in the presence of joint parametric uncertainty and unmodeled dynamics, *IEEE Trans. Automat. Control*, 36, 25–38, 1991.

[8] Packard, A., Doyle, J.C., and Balas, G.J., Linear, multivariable robust control with a μ perspective, *ASME J. Dyn., Meas. Control, Special Ed. on Control*, 115(2b), 426–438, 1993.

[9] Packard, A. and Doyle, J.C., The complex structured singular value, *Automatica*, 29(1), 71–109, 1993.

[10] Packard, A. and Pandey, P., Continuity properties of the real/complex structured singular value, *IEEE Trans. Automat. Control*, 38(3), 415–428, 1993.

[11] Redheffer, On a certain linear fractional transformation, *J. Math. Phys.*, 39, 269–286, 1960.

[12] Safonov, M.G., Stability margins of diagonally perturbed multivariable feedback systems, *Proc. IEEE*, 129 Part D(6), 251–256, 1982.

[13] Skogestad, S., Morari, M., and Doyle, J.C., Robust control of ill-conditioned plants: high-purity distillation, *IEEE Trans. Automat. Control*, 33(12), 1092–1105, 1988.

[14] Stein, G. and Doyle, J.C., Beyond singular values and loopshapes, *AIAA J. Guidance and Control*, 14(1), 5–16, 1991.

[15] Young, P.M., Newlin, M.P., and Doyle, J.C., *μ analysis with real parametric uncertainty*, In *Proc. 30th IEEE Conf. Decision and Control*, Hawaii, 1991, 1251–1256.

43

Algebraic Design Methods

Vladimír Kučera
Institute of Information Theory and Automation, Prague, Academy of Sciences of the Czech Republic

43.1 Introduction

One of the features of modern control theory is the growing presence of algebra. Algebraic formalism offers several useful tools for control system design, including the so-called "factorization" approach.

This approach is based on the input-output properties of linear systems. The central idea is that of "factoring" the transfer matrix of a (not necessarily stable) system as the "ratio" of two *stable* transfer matrices. This is a natural step for the linear systems whose transfer matrices are rational, i.e., for the lumped-parameter systems. Under certain conditions, however, this approach is productive also for the distributed-parameter systems.

The starting point of the factorization approach is to obtain a simple parameterization of *all* controllers that stabilize a given plant. One could then, in principle, choose the best controller for various applications. The key point here is that the parameter appears in the closed-loop system transfer matrix in a linear manner, thus making it easier to meet additional design specifications.

The actual design of control systems is an engineering task that cannot be reduced to algebra. Design contains many additional aspects that have to be taken into account: sensor placement, computational constraints, actuator constraints, redundancy, performance robustness, among many others. There is a need for an understanding of the control process, a feeling for what kinds of performance objectives are unrealistic, or even dangerous, to ask for. The algebraic approach to be presented, nevertheless, is an elegant and useful tool for the mathematical part of the controller design.

43.2 Systems and Signals

The fundamentals of the factorization approach will be explained for linear systems with *rational* transfer functions whose input u and output y are scalar quantities. We suppose that u and y live in a space of functions mapping a time set into a value set. The time set is a subset of real numbers bounded on the left, say R_+ (the non-negative reals) in the case of continuous-time systems and Z_+ (the non-negative integers) for discrete-time systems. The value set is taken to be the set of real numbers R.

Let the input and output spaces of a continuous-time system be the spaces of locally (Lebesgue) integrable functions f from R_+ into R and define a p-norm

$$||f||_{L_p} = \left[\int_0^\infty |f(t)|^p \mathrm{d}t\right]^{1/p} \quad \text{if } i = 1 \le p < \infty$$

$$||f||_{L_\infty} = \operatorname*{ess\,sup}_{t\ge 0} |f(t)| \quad \text{if } p = \infty .$$

The corresponding normed space is denoted by L_p.

Systems having the desirable property of preserving these functional spaces are called *stable*. More precisely, a system is said to be L_p stable if any input $u \in L_p$ gives rise to an output $y \in L_p$. The systems that are L_∞ stable are also termed to be bounded-input bounded-output (BIBO) stable.

0-8493-8570-9/96/$0.00+$.50

The transfer function of a continuous-time system is the Laplace transform of its impulse response $g(t)$,

$$G(s) = \int_0^\infty g(t)e^{-st}\mathrm{dt}.$$

It is well known that a system with a rational transfer function $G(s)$ is BIBO stable if and only if $G(s)$ is proper and Hurwitz stable, i.e., bounded at infinity with all poles having negative real parts.

In the study of discrete-time systems, we let the input and output spaces be the spaces of infinite sequences $f = (f_0, f_1, ...)$ mapping Z_+ into R and define a p-norm as follows:

$$||f||_{l_p} = \left[\sum_{i=0}^{\infty} |f_i|^p\right]^{1/p} \quad \text{if } 1 \le p < \infty$$

$$||f||_{l_\infty} = \sup_{i\ge 0} |f_i| \quad \text{if } p = \infty .$$

A discrete-time system is said to be l_p stable if it transforms any input $u \in l_p$ to an output $y \in l_p$. The systems that are l_∞ stable are also known as BIBO stable systems.

The transfer function of a discrete-time system is defined as the z-transform of its unit pulse response $(h_0, h_1, ...)$,

$$H(z) = \sum_{i=0}^{\infty} h_i z^{-i} ,$$

and it is always proper. A system with a proper rational transfer function $H(z)$ is BIBO stable if and only if $H(z)$ is Schur stable, i.e., its poles all have modulus less than one.

Of particular interest are discrete-time systems that are finite-input finite-output (FIFO) stable. Such a system transforms finite input sequences into finite output sequences, its unit pulse response is finite, and its transfer function $H(z)$ has no poles outside the origin $z = 0$, i.e., $H(z)$ is a polynomial in z^{-1}.

43.3 Fractional Descriptions

Consider a rational function $G(s)$. By definition, it can be expressed as the ratio

$$G(s) = \frac{B(s)}{A(s)}$$

of two qualified rational functions A and B.

A well-known example is the *polynomial* description, in which case A and B are coprime polynomials, i.e., polynomials having no roots in common.

Another example is to take for A and B two coprime, *proper* and *Hurwitz-stable* rational functions. When $G(s)$ is, say,

$$G(s) = \frac{s+1}{s^2+1}$$

then one can take

$$A(s) = \frac{s^2+1}{(s+\lambda)^2}, \quad B(s) = \frac{s+1}{(s+\lambda)^2}$$

where $\lambda > 0$ is a real number. We recall that two proper and Hurwitz-stable rational functions are coprime if they have no infinite nor unstable zeros in common. Therefore, in the example above, the denominator of A and B can be any strictly Hurwitz polynomial of degree exactly two; if its degree is lower, then A would not be proper and if it is higher, then A and B would have a common zero at infinity. The set of proper and Hurwitz-stable rational functions is denoted by $R_H(s)$.

The proper rational functions $H(z)$ arising in discrete-time systems can be treated in a similar manner. One can write

$$H(z) = \frac{B(z)}{A(z)}$$

where A and B are coprime, *Schur-stable* (hence proper) rational functions. The coprimeness means having no unstable zeros (i.e., in the closed disc $|z| \ge 1$) in common. For example, if

$$H(z) = \frac{1}{z-1}$$

then one can take

$$A(z) = \frac{z-1}{z-\lambda}, \quad B(z) = \frac{1}{z-\lambda}$$

for any real number λ such that $|\lambda| < 1$. The set of Schur-stable rational functions is denoted by $R_S(z)$.

The particular choice of $\lambda = 0$ in the example above leads to

$$A(z) = \frac{z-1}{z} = 1 - z^{-1}, \quad B(z) = \frac{1}{z} = z^{-1} .$$

In this case A and B are in fact polynomials in z^{-1}.

43.4 Feedback Systems

To control a system means to alter its dynamics so that a desired behavior is obtained. This can be done by feedback. A typical feedback system consists of two subsystems, S_1 and S_2, connected as shown in Figure 43.1.

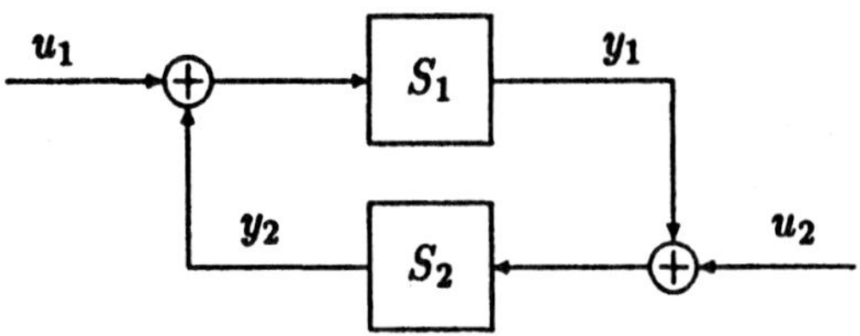

Figure 43.1 Feedback system.

In most applications, it is desirable that the feedback system be BIBO stable in the sense that whenever the exegenous inputs u_1 and u_2 are bounded in magnitude, so too are the output signals y_1 and y_2.

In order to study this property, we express the transfer functions of S_1 and S_2 as ratios of proper stable rational functions and seek for conditions under which the transfer function of the feedback system is proper and stable.

To fix ideas, consider continuous-time systems and write

$$S_1 = \frac{B(s)}{A(s)}, \quad S_2 = -\frac{Y(s)}{X(s)}$$

where A, B and X, Y are two couples of coprime rational functions from $R_H(s)$. The transfer matrix of the feedback system

$$\begin{bmatrix} y_1 \\ y_2 \end{bmatrix} = \begin{bmatrix} \frac{S_1}{1-S_1S_2} & \frac{S_1S_2}{1-S_1S_2} \\ \frac{S_1S_2}{1-S_1S_2} & \frac{S_2}{1-S_1S_2} \end{bmatrix} \begin{bmatrix} u_1 \\ u_2 \end{bmatrix}$$

is then given by

$$\begin{bmatrix} y_1 \\ y_2 \end{bmatrix} = \frac{1}{AX+BY} \begin{bmatrix} BX & -BY \\ -BY & -AY \end{bmatrix} \begin{bmatrix} u_1 \\ u_2 \end{bmatrix}.$$

We observe that the numerator matrix has all its elements in $R_H(s)$ and that no infinite or unstable zeros of the denominator can be absorbed in all these elements. We therefore conclude that the transfer functions belong to $R_H(s)$ if and only if the inverse of $AX + BY$ is in $R_H(s)$.

We illustrate with the example where S_1 is a differentiator and S_2 is an invertor such that

$$S_1(s) = s, \quad S_2(s) = -1.$$

We take

$$A(s) = \frac{1}{s+\lambda}, \quad B(s) = \frac{s}{s+\lambda}$$

for any real $\lambda > 0$ and

$$X(s) = 1, \quad Y(s) = 1.$$

Then

$$(AX+BY)^{-1}(s) = \frac{s+\lambda}{s+1}$$

resides in $R_H(s)$ and hence the feedback system is BIBO stable.

The above analysis applies also to discrete-time systems; the set $R_H(s)$ is just replaced by $R_S(z)$. However, we note that any closed loop around a discrete-time system involves some information delay, no matter how small. Indeed, a control action applied to S_1 cannot affect the measurement from which it was calculated in S_2. Therefore, either $S_1(z)$ or $S_2(z)$ must be strictly proper; we shall assume that it is $S_1(z)$ that has this property.

To illustrate the analysis of discrete-time systems, consider a summator S_1 and an amplifier S_2,

$$S_1(z) = \frac{1}{z-1}, \quad S_2(z) = -k.$$

Taking

$$A(z) = \frac{z-1}{z-\lambda}, \quad B(z) = \frac{1}{z-\lambda}$$

for any real λ in magnitude less than 1 and

$$X(z) = 1, \quad Y(z) = k$$

one obtains

$$(AX+BY)^{-1}(z) = \frac{z-\lambda}{z-(1-k)}.$$

Therefore, the closed-loop system is BIBO stable if and only if $|1-k| < 1$.

To summarize, the fractional representation used should be matched with the goal of the analysis. The denominators A, X and the numerators B, Y should be taken from the set of stable transfer functions, either $R_H(s)$ or $R_S(z)$, depending on the type of the stability studied. This choice makes the analysis more transparent and leads to a simple algebraic condition: the inverse of $AX + BY$ is stable. Any other type of stability can be handled in the same way provided one can identify the set of the transfer functions that these stable systems will have.

43.5 Parameterization of Stabilizing Controllers

The design of feedback control systems consists of the following: given one subsystem, say S_1, we seek to determine the other subsystem, S_2, so that the resulting feedback system shown in Figure 43.1 meets the design specifications. We call S_1 the *plant* and S_2 the *controller*. Our focus is first on achieving BIBO stability. Any controller S_2 that BIBO stabilizes the plant S_1 is called a *stabilizing* controller for this plant.

Suppose S_1 is a continuous-time plant that gives rise to the transfer function

$$S_1(s) = \frac{B(s)}{A(s)}$$

for some coprime elements A and B of $R_H(s)$. It follows from the foregoing analysis that a stabilizing controller exists and that all controllers that stabilize the given plant are generated by all solution pairs X, Y with $X \neq 0$ of the Bézout equation

$$AX + BY = 1$$

over $R_H(s)$. There is no loss of generality in setting $AX + BY$ to the identity rather than to any rational function whose inverse is in $R_H(s)$: this inverse is absorbed by X and Y and therefore cancels in forming

$$S_2(s) = -\frac{Y(s)}{X(s)}.$$

The solution set of the equation $AX + BY = 1$ with A and B coprime in $R_H(s)$ can be parameterized as

$$X = X' + BW, \quad Y = Y' - AW$$

where X', Y' is a particular solution of the equation and W is a free parameter, which is an arbitrary function in $R_H(s)$.

The parameterization of the family of all stabilizing controllers S_2 for the plant S_1 now falls out almost routinely:

$$S_2(s) = -\frac{Y'(s) - A(s)W(s)}{X'(s) + B(s)W(s)}$$

where the parameter W varies over $R_H(s)$ while satisfying $X' + BW \neq 0$.

In order to determine the set of all controllers S_2 that stabilize the plant S_1, one needs to do two things: (1) express $S_1(s)$ as a ratio of two coprime elements from $R_H(s)$ and (2) find a particular solution in $R_H(s)$ of a Bézout equation, which is equivalent to finding one stabilizing controller for S_1. Once these two steps are completed, the formula above provides a parameterization of the set of all stabilizing controllers for S_1. The condition $X' + BW \neq 0$ is not very restrictive, as $X' + BW$ can identically vanish for at most one choice of W.

As an example, we shall stabilize an integrator plant S_1. Its transfer function can be expressed as

$$S_1(s) = \frac{\frac{1}{s+1}}{\frac{s}{s+1}}$$

where $s+1$ is an arbitrarily chosen Hurwitz polynomial of degree one. Suppose that using some design procedure we have found a stabilizing controller for S_1, namely

$$S_2(s) = -1.$$

This corresponds to a particular solution $X' = 1,\ Y' = 1$ of the Bézout equation

$$\frac{s}{s+1}X + \frac{1}{s+1}Y = 1.$$

The solution set in $R_H(s)$ of this equation is

$$X(s) = 1 + \frac{1}{s+1}W(s), \quad Y(s) = 1 - \frac{s}{s+1}W(s).$$

Hence, all controllers S_2 that BIBO stabilize S_1 have the transfer function

$$S_2(s) = -\frac{1 - \frac{s}{s+1}W(s)}{1 + \frac{1}{s+1}W(s)}$$

where W is any function in $R_H(s)$.

It is clear that the result is independent of the particular fraction taken to represent S_1. Indeed, if $s+1$ is replaced by another Hurwitz polynomial $s+\lambda$ in the above example, one obtains

$$S_2(s) = -\frac{\lambda - \frac{s}{s+\lambda}W'(s)}{1 + \frac{1}{s+\lambda}W'(s)},$$

which is the same set when

$$W'(s) = \left(\frac{s+\lambda}{s+1}\right)^2 W(s) + \frac{s+\lambda}{s+1}(\lambda - 1).$$

43.6 Parameterization of Closed-Loop Transfer Functions

The utility of the fractional approach derives not merely from the fact that it provides a parameterization of all controllers that stabilize a given plant in terms of a free parameter W, but also from the simple manner in which this parameter enters the resulting (stable) closed-loop transfer matrix.

In fact,

$$\begin{bmatrix} y_1 \\ y_2 \end{bmatrix} = \begin{bmatrix} B(X' + BW) & -B(Y' - AW) \\ -B(Y' - AW) & -A(Y' - AW) \end{bmatrix} \begin{bmatrix} u_1 \\ u_2 \end{bmatrix}$$

and we observe that all four transfer functions are *linear* in the free parameter W.

This result serves to parameterize the performance specifications, and it is the starting point for the selection of the best controller for the application at hand. The search for S_2 is thus replaced by a search for W. The crucial point is that the resulting selection/optimization problem is linear in W while it is nonlinear in S_2.

43.7 Optimal Performance

The performance specifications often involve a norm minimization.

Let us consider the problem of *disturbance attenuation*. We are given, say, a continuous-time plant S_1 having two inputs: the control input u and an unmeasurable disturbance d (see Figure 43.2). The objective is to determine a BIBO stabilizing controller S_2 for the plant S_1 such that the effect of d on the plant output y is minimized in some sense.

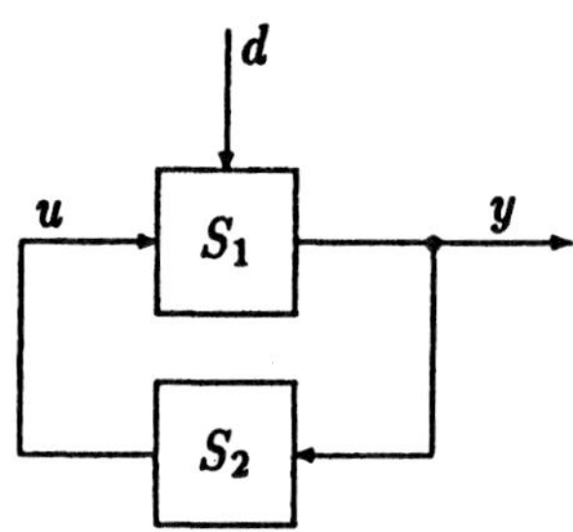

Figure 43.2 Disturbance attenuation.

We describe the plant by two transfer functions

$$S_{1u}(s) = \frac{B(s)}{A(s)}, \quad S_{1d}(s) = \frac{C(s)}{A(s)}$$

where A, B and C is a triple of coprime functions from $R_H(s)$. The set of stabilizing controllers for S_1 is given by the transfer function

$$S_2(s) = -\frac{Y'(s) - A'(s)W(s)}{X'(s) + B'(s)W(s)}$$

where A', B' is a *coprime* fraction over $R_H(s)$ for S_{1u},

$$\frac{B(s)}{A(s)} = \frac{B'(s)}{A'(s)}$$

and X', Y' is a particular solution over $R_H(s)$ of the equation

$$A'X + B'Y = 1$$

such that $X' + B'W \neq 0$.

The transfer function, $G(s)$, between d and y in a stable feedback system is

$$G = \frac{S_{1d}}{1 - S_{1u}S_2} = C(X' + B'W)$$

and it is linear in the proper and Hurwitz-stable rational parameter W.

Now suppose that the disturbance d is any function from L_∞, i.e., any essentially bounded real function on R_+. Then

$$||y||_{L_\infty} \leq ||G||_1 \; ||d||_{L_\infty}$$

where

$$||G||_1 = \int_0^\infty |g(t)| \, \mathrm{dt}$$

and $g(t)$ is the impulse response corresponding to $G(s)$. The parameter W can be used to minimize the norm $||G||_1$ and hence the maximum output amplitude.

If d is a stationary white noise, the steady-state output variance equals

$$Ey^2 = ||G||_2^2 \; Ed^2,$$

where

$$||G||_2^2 = \int_0^\infty |g(t)|^2 \, \mathrm{dt} = \frac{1}{2\pi j} \oint G(-s)G(s)\mathrm{d}s.$$

The last integral is a contour integral up the imaginary axis and then around an infinite semicircle in the left half-plane. Again, W can be selected so as to minimize the norm $||G||_2$, thus minimizing the steady-state output variance.

Finally, suppose that d is any function from L_2, i.e., any finite-energy real function on R_+. Then one obtains

$$||y||_{L_2} \leq ||G||_\infty \; ||d||_{L_2}$$

where

$$||G||_\infty = \sup_{Re\, s > 0} |G(s)|.$$

Therefore, choosing W to make the norm $||G||_\infty$ minimal, one minimizes the maximum output energy.

The above system norms provide several examples showing how the effect of the disturbance on the plant output can be measured. The optimal attenuation is achieved by minimizing these norms.

Minimizing the 1-norm involves a linear program while minimizing the ∞-norm requires a search. The 2-norm minimization has a closed-form solution, which will be now described.

We recall that

$$G(s) = P(s) + Q(s)W(s)$$

where $P = CX'$ and $Q = CB'$. The norm $||G||_2$ is finite if and only if G is strictly proper and has no poles on the imaginary axis, so we assume that Q has no zeros on the imaginary axis. We factorize

$$Q = Q_{ap} \, Q_{mp}$$

where Q_{ap} satisfies $Q_{ap}(-s)Q_{ap}(s) = 1$ (the so-called all-pass function) and Q_{mp} is such that Q_{mp}^{-1} is in $R_H(s)$ (the so-called minimum-phase function); this factorization is unique up to the sign. Let Q_{ap}^* denote the function $Q_{ap}^*(s) = Q_{ap}(-s)$. Then

$$\begin{aligned} ||G||_2^2 &= ||P + QW||_2^2 \\ &= ||Q_{ap}^* P + Q_{mp} W||_2^2 . \end{aligned}$$

Decompose $Q_{ap}^* P$ as

$$Q_{ap}^* P = (Q_{ap}^* P)_{st} + (Q_{ap}^* P)_{un}$$

where $(Q_{ap}^* P)_{st}$ is in $R_H(s)$ and $(Q_{ap}^* P)_{un}$ is unstable but strictly proper; this decomposition is unique. Then

$$||G||_2^2 = ||(Q_{ap}^* P)_{un}||_2^2 + ||(Q_{ap}^* P)_{st} + Q_{mp} W||_2^2 .$$

Since the first term is independent of W,

$$\min_W ||G||_2 = ||(Q_{ap}^* P)_{un}||_2$$

and this minimum is attained by

$$W = -\frac{(Q_{ap}^* P)_{st}}{Q_{mp}} .$$

Here is an illustrative example. The plant is given by

$$S_{1u}(s) = \frac{s-2}{s+1}, \quad S_{1d}(s) = 1$$

and we seek to find a stabilizing controller S_2 such that

$$G(s) = \frac{S_{1d}(s)}{1 - S_{1u}(s)S_2(s)}$$

has minimum 2-norm.

We write

$$A(s) = 1, \quad B(s) = \frac{s-2}{s+1}, \quad C(s) = 1$$

and find all stabilizing controllers first. Since the plant is already stable, these are given by

$$S_2(s) = -\frac{-W(s)}{1 + \frac{s-2}{s+1} W(s)}$$

where W is a free parameter in $R_H(s)$.

Then

$$G(s) = 1 + \frac{s-2}{s+1} W(s),$$

so that

$$P(s) = 1, \quad Q(s) = \frac{s-2}{s+1} .$$

Clearly

$$Q_{ap}(s) = \frac{s-2}{s+2}, \quad Q_{mp}(s) = \frac{s+2}{s+1}$$

and

$$Q_{ap}^*(s)P(s) = \frac{s+2}{s-2} = 1 + \frac{4}{s-2}.$$

Therefore,

$$||G||_2^2 = ||\frac{4}{s-2}||_2^2 + ||1 + \frac{s+2}{s+1}W||_2^2$$

so that the least norm

$$\min_W ||G||_2 = ||\frac{4}{s-2}||_2 = ||\frac{4}{s+2}||_2 = 2$$

is attained by

$$W(s) = -\frac{s+1}{s+2}.$$

43.8 Robust Stabilization

The actual plant can differ from its nominal model. We suppose that a nominal plant description is available together with a description of the plant uncertainty. The objective is to design a controller that stabilizes all plants lying within the specified domain of uncertainty. Such a controller is said to *robustly* stabilize the family of plants.

The plant uncertainty can be modeled conveniently in terms of its fractional description. To fix ideas, we shall consider discrete-time plants factorized over $R_S(z)$ and endow $R_S(z)$ with the ∞-norm: for any function $H(z)$ from $R_S(z)$,

$$||H||_\infty = \sup_{|z|>1} |H(z)|.$$

For any two such functions, $H_1(z)$ and $H_2(z)$, we define

$$\begin{aligned} ||[H_1 \;\; H_2]||_\infty &= ||\begin{bmatrix} H_1 \\ H_2 \end{bmatrix}||_\infty \\ &= \sup_{|z|>1} (|H_1(z)|^2 + |H_2(z)|^2)^{1/2}. \end{aligned}$$

Let S_{10} be a nominal plant giving rise to a strictly proper transfer function

$$S_{10}(z) = \frac{B(z)}{A(z)}$$

where A and B are coprime functions from $R_S(z)$. We denote $S_1(A, B, \mu)$ the family of plants having strictly proper transfer functions

$$S_1(z) = \frac{B(z) + \Delta B(z)}{A(z) + \Delta A(z)}$$

where ΔA and ΔB are functions from $R_S(z)$ such that

$$||[\Delta A \;\; \Delta B]||_\infty < \mu$$

for some non-negative real number μ.

Now, let S_2 be a BIBO stabilizing controller for S_{10}. Therefore,

$$S_2 = -\frac{Y' - AW}{X' + BW}$$

where $AX' + BY' = 1$ and W is an element of $R_S(z)$. Then S_2 will BIBO stabilize all plants from $S_1(A, B, \mu)$ if and only if the inverse of

$$\begin{aligned} &(A + \Delta A)(X' + BW) + (B + \Delta B)(Y' - AW) \\ &= 1 + [\Delta A \;\; \Delta B]\begin{bmatrix} X' + BW \\ Y' - AW \end{bmatrix} \end{aligned}$$

is in $R_S(z)$. This is the case whenever

$$||[\Delta A \;\; \Delta B]\begin{bmatrix} X' + BW \\ Y' - AW \end{bmatrix}||_\infty < 1$$

so we have the following condition of robust stability:

$$\mu||\begin{bmatrix} X' + BW \\ Y' - AW \end{bmatrix}||_\infty \le 1.$$

The best controller that robustly stabilizes the plant corresponds to the parameter W that minimizes the ∞-norm above. This requires a search; closed-form solutions exist only in special cases. One such case is presented next.

Suppose the nominal model

$$S_{10}(z) = \frac{1}{z-1}$$

has resulted from

$$S_1(z) = \frac{z+\delta}{(z-1)(z-\varepsilon)}$$

by neglecting the second-order dynamics, where $\delta \ge 0$ and $0 \le \varepsilon < 1$. Rearranging,

$$S_1(z) = \frac{\frac{1}{z} + \frac{1}{z}\frac{\delta+\varepsilon}{z-\varepsilon}}{\frac{z-1}{z}}$$

and one identifies

$$\Delta A = 0, \quad \Delta B = \frac{1}{z}\frac{\delta+\varepsilon}{z-\varepsilon}.$$

Hence,

$$||[\Delta A \;\; \Delta B]||_\infty = \frac{\delta+\varepsilon}{1-\varepsilon}$$

and the true plant belongs to the family

$$S_1\left(\frac{z-1}{z}, \frac{1}{z}, \frac{\delta+\varepsilon}{1-\varepsilon}\right).$$

All controllers that BIBO stabilize the nominal plant S_{10} are given by

$$S_2(z) = -\frac{1 - \frac{z-1}{z}W(z)}{1 + \frac{1}{z}W(z)}$$

where W is a free parameter in $R_S(z)$. Which controller yields the best stability margin against δ and ε? The one that minimizes the ∞-norm in

$$\frac{\delta+\varepsilon}{1-\varepsilon}||\begin{bmatrix} 1 + \frac{1}{z}W \\ 1 - \frac{z-1}{z}W \end{bmatrix}||_\infty < 1.$$

Suppose we wish to obtain a controller of McMillan degree zero, $S_2(z) = -K$. Then

$$W(z) = (1-K)\frac{z}{z-(1-K)}$$

and $|1-K| < 1$. The norm

$$\left\| \begin{bmatrix} 1+\frac{1}{z}W \\ 1-\frac{z-1}{z}W \end{bmatrix} \right\|_\infty = \sqrt{(1+K^2)}\left\|\frac{z}{z-(1-K)}\right\|_\infty$$

attains the least value of $\sqrt{2}$ by $K = 1$, which corresponds to $W(z) = 0$. It follows that the controller

$$S_2(z) = -1$$

stabilizes all plants $S_1(z)$ for which

$$\frac{\delta+\varepsilon}{1-\varepsilon} < \frac{1}{\sqrt{2}} .$$

43.9 Robust Performance

The performance specifications often result in divisibility conditions. A typical example is the problem of *reference tracking.*

Suppose we are given a discrete-time plant S_1, with transfer function

$$S_1(z) = \frac{B(z)}{A(z)}$$

in coprime fractional form over $R_S(z)$, together with a reference r whose z-transform is of the form

$$r = \frac{E(z)}{D(z)}$$

where only D is specified. We recall that $S_1(z)$ is strictly proper. The objective is to design a BIBO stabilizing controller S_2 such that the plant output y asymptotically tracks the reference r (see Figure 43.3). The controller can operate on both r (feedforward)

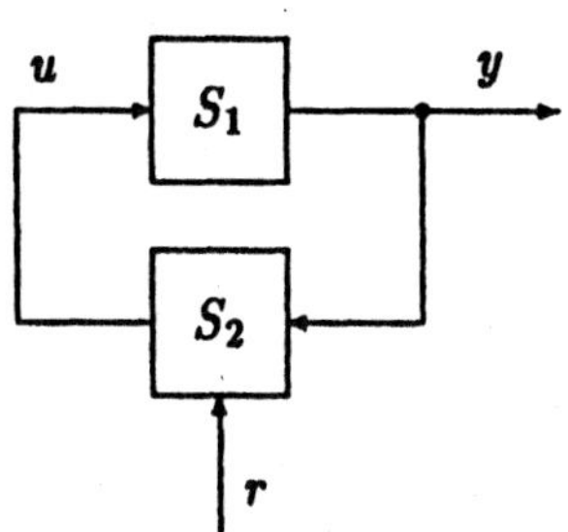

Figure 43.3 Reference tracking.

and y (feedback), so it is described by two transfer functions

$$S_{2y}(z) = -\frac{Y(z)}{X(z)}, \quad S_{2r}(z) = \frac{Z(z)}{X(z)}$$

where X, Y and Z are from $R_S(z)$.

The requirement of tracking imposes that the tracking error

$$e = r - y = (1 - \frac{BZ}{AX+BY})\frac{E}{D}$$

belong to $R_S(z)$. Since $AX + BY$ has inverse in $R_S(z)$ for every stabilizing controller and E is unspecified, D must divide $1-BZ$ in $R_S(z)$. Hence, there must exist a function V in $R_S(z)$ such that $1 - BZ = DV$. Therefore, S_2 exists if and only if B and D are coprime in $R_S(z)$, and the two controller transfer functions evolve from solving the two Bézout equations

$$\begin{aligned} AX + BY &= 1 \\ DV + BZ &= 1 \end{aligned}$$

where the function V serves to express the tracking error as

$$e = VE.$$

The reference tracking is said to be *robust* if the specifications are met even as the plant is slightly perturbed. We call S_{10} the nominal plant and $S_1(A, B, \mu)$ the neighborhood of S_{10} defined by

$$S_1 = \frac{B+\Delta B}{A+\Delta A}$$

where ΔA and ΔB are functions of $R_S(z)$ such that

$$\|[\,\Delta A \quad \Delta B\,]\|_\infty < \mu$$

for some non-negative real number μ. We recall that all $S_1(z)$ are strictly proper.

Now $A+\Delta A$ and $B+\Delta B$ are not specified, but $(A+\Delta A)X + (B+\Delta B)Y$ still has inverse in $R_S(z)$; call it U. We have

$$e = \left(\frac{A+\Delta A}{U}X + \frac{B+\Delta B}{U}(Y-Z)\right)\frac{E}{D}.$$

Hence, for robust tracking, D must divide both X and $Y - Z$ in $R_S(z)$. But, it is sufficient that D divides X; this condition already implies the other one as can be seen on subtracting the two Bézout equations above.

We illustrate on a discrete-time plant S_1 given by

$$S_1(z) = \frac{1}{z-2}$$

whose output is to track every sinusoidal sequence of the form

$$r = \frac{az+b}{z^2-z+1}$$

where a, b are unspecified real numbers. Taking

$$A(z) = \frac{z-2}{z}, \quad B(z) = \frac{1}{z}, \quad D(z) = \frac{z^2-z+1}{z^2}$$

and solving the pair of Bézout equations

$$\begin{aligned} \frac{z-2}{z}X(z) + \frac{1}{z}Y(z) &= 1 \\ \frac{z-z+1}{z^2}V(z) + \frac{1}{z}Z(z) &= 1 \end{aligned}$$

yields the tracking controllers in parametric form

$$S_{2y}(z) = -\frac{2 - \frac{z-2}{z} W_1(z)}{1 + \frac{1}{z} W_1(z)}$$

$$S_{2r}(z) = \frac{\frac{z-1}{z} - \frac{z^2-z+1}{z^2} W_2(z)}{1 + \frac{1}{z} W_1(z)}$$

for any elements W_1, W_2 of $R_S(z)$. The resulting error is

$$e = [\, 1 + \frac{1}{z} W_2(z) \,] \frac{az+b}{z^2}.$$

Not all of these controllers, however, achieve a robust tracking of the reference. The divisibility condition is fulfilled if and only if W_1 is restricted to

$$W_1(z) = -\frac{z-1}{z} + \frac{z^2-z+1}{z^2} W(z)$$

where W is free in $R_S(z)$.

It is to be noted that the requirement of asymptotic tracking leaves enough degrees of freedom to meet additional design specifications.

43.10 Finite Impulse Response

Transients in discrete-time systems can settle in finite time. Systems having the property that any input sequence with a finite number of nonzero elements produces an output sequence with a finite number of nonzero elements have been called FIFO stable. We recall that a system with proper rational transfer function $H(z)$ is FIFO stable if and only if $H(z)$ is a polynomial in z^{-1}.

Let us consider the feedback system shown in Figure 43.1 and focus on achieving FIFO stability. To this end we write the transfer function of the plant as

$$S_1(z) = \frac{B(z)}{A(z)}$$

where this time A and B are coprime polynomials in z^{-1}. We recall that the plant incorporates the necessary delay so that $S_1(z)$ is strictly proper. Repeating the arguments used to design a BIBO stable system, we conclude that all controllers S_2 that FIFO stabilize the plant S_1 have the transfer function

$$S_2(z) = -\frac{Y(z)}{X(z)}$$

where X, Y is the solution class of the polynomial Bézout equation

$$AX + BY = 1.$$

In particular, if X' and Y' define any FIFO stabilizing controller for S_1, the set of all such controllers can be parameterized as

$$S_2(z) = -\frac{Y' - AW}{X' + BW}$$

where $W(z)$ is a free polynomial in z^{-1}.

It is a noteworthy fact that the parametric expressions for the sets of BIBO stable and FIFO stable controllers are the same; the only difference is that the free parameter of FIFO stabilizing controllers is permitted to range over only the smaller set of polynomials in z^{-1}, whereas in BIBO stabilizing controllers it is permitted to range over the larger set of Schur-stable rational functions in z. Indeed, FIFO stability is more restrictive than BIBO stability.

The design options offered by FIFO stability are remarkable. The parameter W can be selected so as to minimize the McMillan degree of S_2, or to achieve the shortest impulse response of the closed-loop system. Various norm minimizations can also be performed.

A well-known example is the deadbeat controller. We consider a double-summator plant with transfer function

$$S_1(z) = \frac{1}{(z-1)^2}$$

and interpret the exogenous inputs u_1 and u_2 as accounting for the effect of the initial conditions of S_1 and S_2. The requirement of FIFO stability is then equivalent to achieving finite responses y_1 and y_2 for all initial conditions. Since in this case

$$A(z) = (1 - z^{-1})^2, \quad B(z) = z^{-2}$$

and the Bézout equation

$$(1 - z^{-1})^2 X(z) + z^{-2} Y(z) = 1$$

has a particular solution

$$X'(z) = 1 + z^{-1}, \quad Y'(z) = 3 - 2z^{-1}$$

we obtain all deadbeat (or FIFO stabilizing) controllers as

$$S_2(z) = -\frac{3 - 2z^{-1} - (1 - z^{-1})^2 W(z)}{1 + 2z^{-1} + z^{-2} W(z)}.$$

The deadbeat controller of least McMillan degree ($= 1$) is obtained for $W(z) = 0$. The choice $W(z) = -3$ leads to a deadbeat controller that rejects step disturbances u_1 (hence, persistent) at the plant output y_1 in finite time. And when u_1 is a stationary white noise, then $W(z) = 2.5$ minimizes the steady-state variance of y_1 among all deadbeat controllers of McMillan degree 2.

43.11 Multivariable Systems

Up until now we have considered only single-input single-output (SISO) plants and controllers. In the case of multiple inputs and/or outputs, the input-output properties of linear systems are represented by a *matrix* of transfer functions. The additional intricacies introduced by these systems stem mainly from the fact that the matrix multiplication is not commutative.

Consider a rational transfer matrix $G(s)$ whose dimensions are, say, $m \times n$. Then it is always possible to factorize G as follows:

$$\begin{aligned} G(s) &= B_R(s)A_R^{-1}(s) \\ &= A_L^{-1}(s)B_L(s) \end{aligned}$$

where the factors B_R, A_R and A_L, B_L are respectively $m \times n$, $n \times n$ and $m \times m$, $m \times n$ matrices of qualified rational functions, say from $R_H(s)$, such that

$$A_R,\ B_R \text{ are right coprime}$$

$$A_L,\ B_L \text{ are left coprime}\,.$$

These "matrix fractions" are unique except for the possibility of multiplying the "numerator" and the "denominator" matrices by a matrix whose determinant has inverse in $R_H(s)$. That is, if $G(s)$ can also be expressed as

$$\begin{aligned} G(s) &= B_R'(s)A_R'^{-1}(s) \\ &= A_L'^{-1}(s)B_L'(s) \end{aligned}$$

where the factors are matrices of functions from $R_H(s)$ such that

$$A_R',\ B_R' \text{ are right coprime}$$

$$A_L',\ B_L' \text{ are left coprime}\,,$$

then

$$A_R'(s) = A_R(s)U_R(s), \quad B_R'(s) = B_R(s)U_R(s)$$

$$A_L'(s) = U_L(s)A_L(s), \quad B_L'(s) = U_L(s)B_L(s)$$

for some matrices U_R and U_L over $R_H(s)$ whose determinants have stable inverses in $R_H(s)$.

Analogous results hold for discrete-time systems. To illustrate, consider the transfer matrix

$$G(z) = \begin{bmatrix} \frac{1}{z-1} & \frac{2-z}{z^2-z} \\ \frac{1}{z-1} & \frac{1}{z-1} \end{bmatrix}$$

and determine its left and right coprime factorizations over $R_S(z)$. One obtains, for instance,

$$\begin{aligned} G(z) &= \begin{bmatrix} \frac{1}{z} & \frac{2-z}{z^2-\lambda z} \\ 0 & \frac{1}{z-\lambda} \end{bmatrix} \begin{bmatrix} 1 & 0 \\ -1 & \frac{z-1}{z-\lambda} \end{bmatrix}^{-1} \\ &= \begin{bmatrix} 1 & -1 \\ 0 & \frac{z-1}{z-\mu} \end{bmatrix}^{-1} \begin{bmatrix} 0 & -\frac{2}{z} \\ \frac{1}{z-\mu} & \frac{1}{z-\mu} \end{bmatrix} \end{aligned}$$

for any real λ and μ with modulus less than one.

Let us now consider the feedback system shown in Figure 43.1 where S_1 and S_2 are multivariable systems and analyze its BIBO stability. We therefore factorize the two transfer matrices over $R_H(s)$,

$$S_1(s) = B_R(s)A_R^{-1}(s) = A_L^{-1}(s)B_L(s),$$

$$S_2(s) = -X_L^{-1}(s)Y_L(s) = -Y_R(s)X_R^{-1}(s)$$

where the two pairs A_R, B_R and X_R, Y_R are right coprime while the two pairs A_L, B_L and X_L, Y_L are left coprime. The transfer matrix of the feedback system

$$\begin{bmatrix} y_1 \\ y_2 \end{bmatrix} = \begin{bmatrix} S_1(I-S_2S_1)^{-1} & S_1(I-S_2S_1)^{-1}S_2 \\ S_2(I-S_1S_2)^{-1}S_1 & S_2(I-S_1S_2)^{-1} \end{bmatrix} \begin{bmatrix} u_1 \\ u_2 \end{bmatrix}$$

then reads

$$\begin{bmatrix} y_1 \\ y_2 \end{bmatrix} = \begin{bmatrix} B_R(X_LA_R+Y_LB_R)^{-1}X_L & -B_R(X_LA_R+Y_LB_R)^{-1}Y_L \\ I - A_R(X_LA_R+Y_LB_R)^{-1}X_L & -A_R(X_LA_R+Y_LB_R)^{-1}Y_L \end{bmatrix} \begin{bmatrix} u_1 \\ u_2 \end{bmatrix}$$

or alternatively

$$\begin{bmatrix} y_1 \\ y_2 \end{bmatrix} = \begin{bmatrix} X_R(A_LX_R+B_LY_R)^{-1}B_L & X_R(A_LX_R+B_LY_R)^{-1}A_L - I \\ -Y_R(A_LX_R+B_LY_R)^{-1}B_L & -Y_R(A_LX_R+B_LY_R)^{-1}A_L \end{bmatrix} \begin{bmatrix} u_1 \\ u_2 \end{bmatrix}.$$

The feedback system is BIBO stable if and only if this transfer matrix has entries in $R_H(s)$. We therefore conclude that the feedback system is BIBO stable if and only if the common denominator $X_LA_R + Y_LB_R$, or alternatively $A_LX_R + B_LY_R$, has inverse with entries in $R_H(s)$.

A parameterization of all controllers S_2 that BIBO stabilize the plant S_1 is now at hand. Given left and right coprime factorizations over $R_H(s)$ of the plant transfer matrix

$$S_1 = B_RA_R^{-1} = A_L^{-1}B_L\,,$$

we select matrices X_L', Y_L' and X_R', Y_R' with entries in $R_H(s)$ such that

$$X_L'A_R + Y_L'B_R = I, \quad A_LX_R' + B_LY_R' = I\,.$$

Then the family of all stabilizing controllers has the transfer matrix

$$\begin{aligned} S_2 &= -(X_L' + W_LB_L)^{-1}(Y_L' - W_LA_L) \\ &= -(Y_R' - A_RW_R)(X_R' + B_RW_R)^{-1} \end{aligned}$$

where W_L is a matrix parameter whose entries vary over $R_H(s)$ such that $X'_L + W_L B_L$ is nonsingular, and W_R is a matrix parameter whose entries vary over $R_H(s)$ such that $X'_R + B_R W_R$ is nonsingular.

As an example, determine all BIBO stabilizing controllers for the discrete-time plant considered earlier, with the transfer matrix

$$S_1(z) = \begin{bmatrix} \frac{1}{z-1} & \frac{2-z}{z^2-z} \\ \frac{1}{z-1} & \frac{1}{z-1} \end{bmatrix}.$$

The left and right coprime factors over $R_S(z)$ can be taken as

$$A_R(z) = \begin{bmatrix} 1 & 0 \\ -1 & \frac{z-1}{z} \end{bmatrix}, \quad B_R(z) = \begin{bmatrix} \frac{1}{z} & \frac{2-z}{z^2} \\ 0 & \frac{1}{z} \end{bmatrix}$$

and

$$A_L(z) = \begin{bmatrix} 1 & -1 \\ 0 & \frac{z-1}{z} \end{bmatrix}, \quad B_L(z) = \begin{bmatrix} 0 & -\frac{2}{z} \\ \frac{1}{z} & \frac{1}{z} \end{bmatrix}.$$

The Bézout equations

$$X'_L A_R + Y'_L B_R = I, \quad A_L X'_R + B_L Y'_R = I$$

have particular solutions

$$X'_L(s) = \begin{bmatrix} 1 & 0 \\ 1 & 1 \end{bmatrix}, \quad Y'_L(s) \begin{bmatrix} 0 & 0 \\ 0 & 1 \end{bmatrix}$$

and

$$X'_R(s) = \begin{bmatrix} 1 & 1 \\ 0 & 1 \end{bmatrix}, \quad Y'_R(s) \begin{bmatrix} 0 & 1 \\ 0 & 0 \end{bmatrix}.$$

The set of stabilizing controllers is given by

$$\begin{aligned} S_2(s) &= -\left(\begin{bmatrix} 1 & 0 \\ 1 & 1 \end{bmatrix} + W_L \begin{bmatrix} 0 & -2z^{-1} \\ z^{-1} & z^{-1} \end{bmatrix}\right)^{-1} \\ &\quad \left(\begin{bmatrix} 0 & 0 \\ 0 & 1 \end{bmatrix} - W_L \begin{bmatrix} 1 & -1 \\ 0 & 1-z^{-1} \end{bmatrix}\right) \end{aligned}$$

where W_L varies over $R_S(z)$, or by

$$\begin{aligned} S_2(s) &= -\left(\begin{bmatrix} 0 & 1 \\ 0 & 0 \end{bmatrix} - \begin{bmatrix} 1 & 0 \\ -1 & 1-z^{-1} \end{bmatrix} W_R\right) \\ &\quad \left(\begin{bmatrix} 1 & 1 \\ 0 & 1 \end{bmatrix} + \begin{bmatrix} z^{-1} & -z^{-1}+2z^{-2} \\ 0 & z^{-1} \end{bmatrix} W_R\right)^{-1} \end{aligned}$$

where W_R varies over $R_S(z)$ as well.

It is clear that the two parameterizations of S_2 are equivalent. To each controller S_2 there is a unique parameter W_L such that $S_2 = -(X'_L + W_L B_L)^{-1}(Y'_L - W_L A_L)$ as well as a unique parameter W_R such that $S_2 = -(Y'_R - A_R W_R)(X'_R + B_R W_R)^{-1}$, and these two are related by

$$W_R - W_L = X'_L Y'_R - Y'_L X'_R.$$

It is easy to see that the transfer matrix of the closed-loop system is *linear* in the free parameter W_L or W_R. Indeed,

$$\begin{bmatrix} y_1 \\ y_2 \end{bmatrix} = \begin{bmatrix} B_R(X'_L + W_L B_L) & -B_R(Y'_L - W_L A_L) \\ I - A_R(X'_L + W_L B_L) & -A_R(Y'_L - W_L A_L) \end{bmatrix} \begin{bmatrix} u_1 \\ u_2 \end{bmatrix}$$

or alternatively

$$\begin{bmatrix} y_1 \\ y_2 \end{bmatrix} = \begin{bmatrix} (X'_R + B_R W_R) B_L & (X'_R + B_R W_R) A_L - I \\ -(Y'_R - A_R W_R) B_L & -(Y'_R - A_R W_R) A_L \end{bmatrix} \begin{bmatrix} u_1 \\ u_2 \end{bmatrix}.$$

Thus, control synthesis problems beyond stabilization can be handled by determining the parameters W_L or W_R as described for SISO systems.

Let us consider the disturbance attenuation problem for the discrete-time plant

$$S_{1u}(z) = \begin{bmatrix} \frac{1}{z-1} & \frac{2-z}{z^2-z} \\ \frac{1}{z-1} & \frac{1}{z-1} \end{bmatrix}, \quad S_{1d}(z) = \begin{bmatrix} \frac{1}{z-1} \\ \frac{1}{z-1} \end{bmatrix}$$

where the disturbance d is assumed to be an arbitrary l_∞ sequence. We seek to find a BIBO stabilizing controller that minimizes the maximum amplitude of the plant output y.

We write

$$\begin{aligned} S_{1u}(z) &= A_L^{-1}(z) B_L(z) = B_R(z) A_R^{-1}(z) \\ S_{1d}(z) &= A_L^{-1}(z) C_L(z) \end{aligned}$$

where

$$A_L(z) = \begin{bmatrix} 1 & -1 \\ 0 & 1-z^{-1} \end{bmatrix}, \quad B_L(z) = \begin{bmatrix} 1 & -2z^{-1} \\ z^{-1} & z^{-1} \end{bmatrix},$$

$$C_L(z) = \begin{bmatrix} 0 \\ z^{-1} \end{bmatrix}$$

$$A_R(z) = \begin{bmatrix} 1 & 0 \\ -1 & 1-z^{-1} \end{bmatrix},$$

$$B_R(z) = \begin{bmatrix} z^{-1} & -z^{-1}+2z^{-1} \\ 0 & z^{-1} \end{bmatrix}.$$

The set of BIBO stabilizing controllers has been found to have the transfer matrix

$$S_2(z) = -\left(\begin{bmatrix} 0 & 1 \\ 0 & 0 \end{bmatrix} - \begin{bmatrix} 1 & 0 \\ -1 & 1-z^{-1} \end{bmatrix} W_R\right)$$
$$\left(\begin{bmatrix} 1 & 1 \\ 0 & 1 \end{bmatrix} + \begin{bmatrix} z^{-1} & -z^{-1}+2z^{-2} \\ 0 & z^{-1} \end{bmatrix} W_R\right)^{-1}$$

where W_R varies over $R_S(z)$.

The disturbance-output transfer matrix equals

$$\begin{aligned} G(z) &= (I - S_{1u}S_2)^{-1}S_{1d} \\ &= (X_R' + B_R W_R)^{-1} C_L . \end{aligned}$$

When

$$W_R = \begin{bmatrix} W_{11} & W_{12} \\ W_{21} & W_{22} \end{bmatrix}$$

one obtains the expression

$$G(z) = \begin{bmatrix} z^{-1} + z^{-2}W_{12} - (z^{-2} - 2z^{-3})W_{22} \\ z^{-1} + z^{-2}W_{22} \end{bmatrix}.$$

The 1-norm of an $m \times n$ matrix $G(z)$ with entries

$$G_{ij}(z) = \sum_{k=0}^{\infty} g_{ij,k}\, z^{-k}$$

is defined by

$$||G||_1 = \max_{i=1,\ldots,m} \sum_{j=1}^{n} ||G_{ij}||_1$$

where

$$||G_{ij}||_1 = \sum_{k=0}^{n} |g_{ij,k}| .$$

In our case

$$\begin{aligned} ||G||_1 &= \max\Big(||z^{-1} + z^{-2}W_{12} - (z^{-2} - 2z^{-3}) \\ &\quad W_{22}||_1,\ ||z^{-1} + z^{-2}W_{22}||_1\Big) \end{aligned}$$

and it is clear by inspection that $||G||_1$ attains its minimum for $W_{12}(z) = 0$, $W_{22}(z) = 0$ and

$$\min_{W_R} ||G||_1 = \max(1, 1) = 1 .$$

The corresponding optimal BIBO stabilizing controllers are

$$S_2(z) = -\begin{bmatrix} -W_{11} & 1 \\ W_{11} - (1 - z^{-1})W_{21} & 0 \end{bmatrix}$$
$$\begin{bmatrix} 1 + z^{-1}W_{11} - (z^{-1} - 2z^{-2})W_{22} & 1 \\ z^{-1}W_{21} & 1 \end{bmatrix}^{-1}$$

for any functions W_{11} and W_{21} in $R_S(z)$. The one of least McMillan degree reads

$$S_2(z) = \begin{bmatrix} 0 & -1 \\ 0 & 0 \end{bmatrix}.$$

43.12 Extensions

The factorization approach presented here for linear time-invariant systems with rational transfer matrices can be generalized to extend the scope of the theory to include distributed-parameter systems, time-varying systems, and even nonlinear systems.

The transfer matrices of distributed-parameter systems are no longer rational and coprime factorizations cannot be assumed a priori to exist. The coefficients of time-varying systems are functions of time, and the operations of multiplication and differentiation do not commute. In nonlinear systems, transfer matrices are replaced by input-output maps. Suitable factorizations of these maps may not exist and, if they do, they are not commutative in general.

For many systems of physical and engineering interest, these difficulties can be circumvented and the algebraic factorization approach carries over with suitable modifications.

References

Tutorial textbooks:

[1] Desoer, C. A. and Vidyasagar, M., *Feedback Systems: Input-Output Properties,* Academic Press, New York, 1975.

[2] Kučera, V., *Discrete Linear Control: The Polynomial Equation Approach,* Wiley, Chichester, 1979.

[3] Vidyasagar, M., *Control System Synthesis: A Factorization Approach,* MIT Press, Cambridge, MA, 1985.

[4] Doyle, J. C., Francis, B. A., and Tannenbaum, A. R., *Feedback Control Theory,* Macmillan, New York, 1992.

Survey paper:

[5] Kučera, V., Diophantine equations in control — A survey, *Automatica,* 29, 1361–1375, 1993.

Original sources on the parameterization of all stabilizing controllers:

[6] Kučera, V., Stability of discrete linear feedback systems, Paper 44.1, *Prepr. 6th IFAC World Congr.,* Vol. 1, Boston, 1975.

[7] Youla, D. C., Jabr, H. A., and Bongiorno, J. J., Modern Wiener–Hopf design of optimal controllers. II. The multivariable case, *IEEE Trans. Autom. Control,* 21, 319–338, 1976.

[8] Desoer, C. A., Liu, R. W., Murray, J., and Saeks, R., Feedback system design: the fractional representation approach to analysis and synthesis, *IEEE Trans. Autom. Control,* 25, 399–412, 1980.

Original sources on norm minimization:

[9] Francis, B. A., On the Wiener–Hopf approach to optimal feedback design, *Syst. Control Lett.,* 2, 197–201, 1982.

[10] Chang, B. C. and Pearson, J. B., Optimal disturbance reduction in linear multivariable systems, *IEEE Trans. Autom. Control,* 29, 880–888, 1984.

[11] Dahleh, M. A. and Pearson, J. B., l^1-optimal feedback controllers for MIMO discrete-time systems, *IEEE Trans. Autom. Control,* 32, 314–322, 1987.

Original sources on robust stabilization:

[12] Vidyasagar, M. and Kimura, H., Robust controllers for uncertain linear multivariable systems, *Automatica,* 22, 85–94, 1986.

[13] Dahleh, M. A., BIBO stability robustness in the presence of coprime factor perturbations, *IEEE Trans. Autom. Control,* 37, 352–355, 1992.

Original source on FIFO stability and related designs:

[14] Kučera, V. and Kraus, F. J., FIFO stable control systems, *Automatica,* 31, 605–609, 1995.

44

Quantitative Feedback Theory (QFT) Technique

Constantine H. Houpis
Air Force Institute of Technology, Wright-Patterson AFB, OH

44.1 Introduction

44.1.1 Quantitative Feedback Theory

(QFT)[1] is a very powerful design technique for the achievement of assigned performance tolerances over specified ranges of structured plant parameter uncertainties without and with control effector failures [9]. It is a frequency domain design technique utilizing the Nichols chart (NC) to achieve a desired robust design over the specified region of plant parameter uncertainty. This chapter presents an introduction to QFT analog and discrete design techniques for both multiple-input single-output (MISO) [1] [5] [13] and multiple-input multiple-output (MIMO) [3] [4] [6] [7] [10] [11] [12] control systems. QFT computer-aided design (CAD) packages are readily available to expedite the design process. The purposes of this chapter are (1) to provide a basic understanding of QFT; (2) to provide the minimum amount of mathematics necessary to achieve this understanding; (3) to discuss the basic design steps; and (4) to present a practical example.

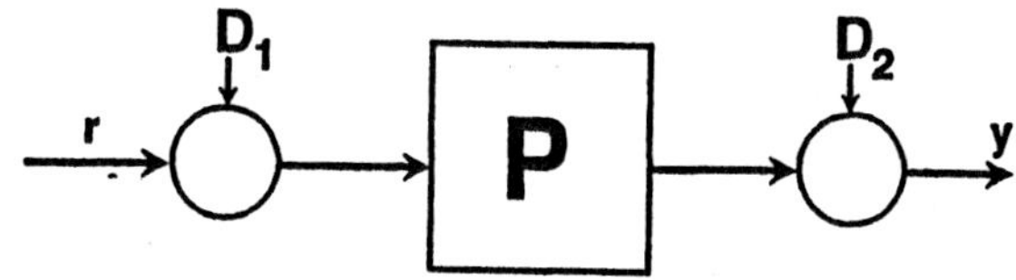

Figure 44.1 An open-loop system (basic plant).

[1]The original version of this material was first published by the Advisory Group for Aerospace Research and Development, North Atlantic Treaty Organization (AGARD/NATO) in Lecture Series LS-191 "on Linear Dynamics and Chaos" in June 1993.

0-8493-8570-9/96/$0.00+$.50

44.1.2 Why Feedback?

For the answer to the question of "Why do you need QFT?" consider the following system. The plant P responds to the input $r(t)$ with the output $y(t)$ in the face of disturbances $d_1(t)$ and $d_2(t)$ (see Figure 44.1). If it is desired to achieve a specified system transfer function $T(s)[= Y(s)/R(s)]$, then it is necessary to insert a prefilter whose transfer function is $T(s)/P(s)$, as shown in Figure 44.2. This compensated system produces the desired output as long as the plant does not change and there are no disturbances. This type of system is sensitive to changes in the plant (or uncertainty in the plant), and the disturbances are reflected directly into the output. Thus, it is necessary to feed back the information in the output in order to reduce the output sensitivity to parameter variation and to attenuate the effect of disturbances on the plant output.

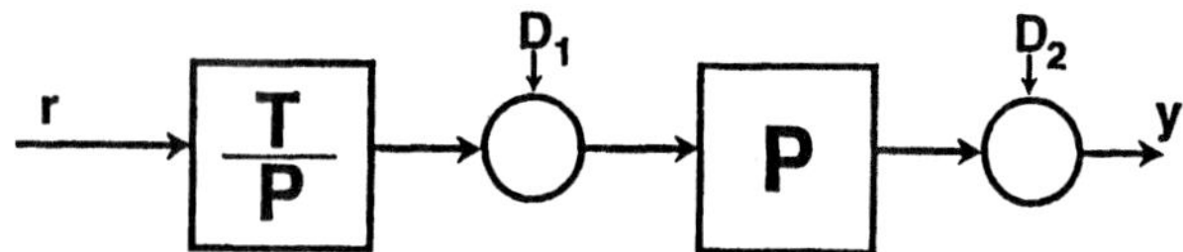

Figure 44.2 A compensated open-loop system.

In designing a feedback control system, it is desired to utilize a technique that:

- Addresses all known plant variations up front
- Incorporates information on the desired output tolerances
- Maintains reasonably low loop gain (reduce the "cost of feedback")

This last item is important in order to avoid the problems associated with high loop gains such as sensor noise amplification, saturation, and high-frequency uncertainties.

44.1.3 What Can QFT Do?

Assume that the characteristics of a plant that is to be controlled over a specified region of operation vary; that is, a plant with structured parameter uncertainty. This plant parameter uncertainty may be described by the Bode plots of Figure 44.3. This figure represents the range of variation of plant magnitude (dB) and phase over a specified frequency range. The bounds of this variation, for this example, can be described by six linear time-invariant (LTI) plant transfer functions. By the application of QFT for a MISO control system containing this plant, a single compensator and a prefilter may be designed to achieve a specified robust design.

44.1.4 Benefits of QFT

The benefits of QFT may be summarized as follows:

- The result is a robust design that is insensitive to plant variation

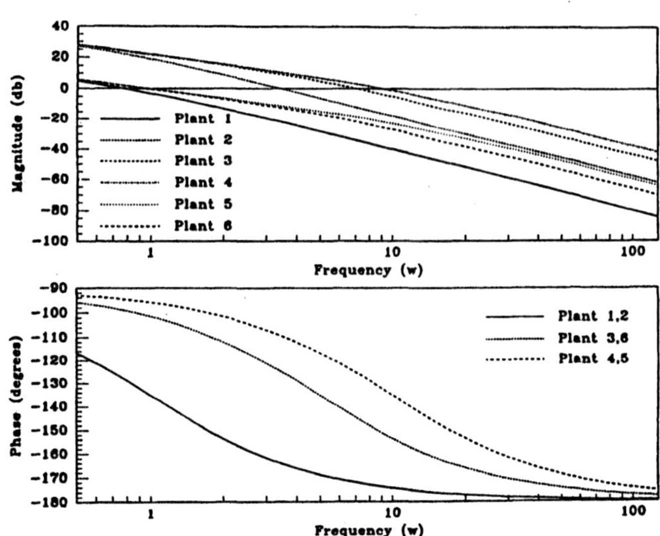

Figure 44.3 The Bode plots of six LTI plants that represent the range of the plant's parameter uncertainty.

- There is one design for the full envelope (no need to verify plants inside templates)
- Any design limitations are apparent up front
- In comparison to other multivariable design techniques there is less development time for a full envelope design
- One can determine what specifications are achievable early in the design process
- One can redesign quickly for changes in the specifications
- The structure of compensator (controller) is determined up front

44.2 The MISO Analog Control System [1]

44.2.1 Introduction

The mathematical proof that an $m \times m$ feedback control system can be represented by m^2 equivalent MISO feedback control systems is given in Section 44.4.2. A 3×3 MIMO control system can be represented by the m^2 MISO equivalent loops shown in Figure 44.4. Thus, this and the next section present an introduction to the QFT technique by considering only a MISO feedback control system.

44.2.2 MISO System

The overview of the MISO QFT design technique is presented in terms of the minimum-phase (m.p.) LTI MISO system of Figure 44.5. The control ratios for tracking ($D = 0$) and for disturbance rejection ($R = 0$) are, respectively,

$$T_R = \frac{F(s)G(s)P(s)}{1+G(s)P(s)} = \frac{F(s)L(s)}{1+L(s)} \tag{44.1}$$

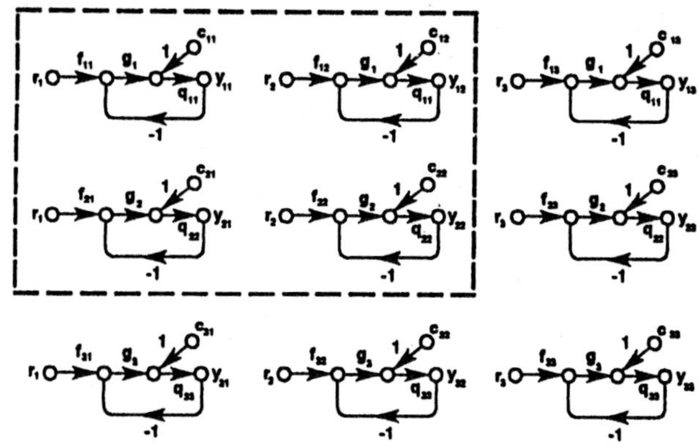

Figure 44.4 m^2 MISO equivalent of a $3x3$ MIMO feedback control system.

$$T_D = \frac{P(s)}{1+G(s)P(s)} = \frac{P(s)}{1+L(s)} \tag{44.2}$$

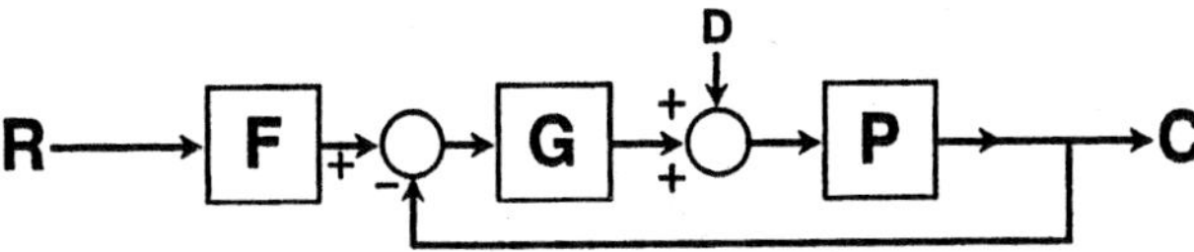

Figure 44.5 A MISO plant.

The design objective is to design the prefilter $F(s)$ and the compensator $G(s)$ so the specified robust design is achieved for the given region of plant parameter uncertainty. The design procedure to accomplish this objective is as follows:

Step 1: Synthesize the desired tracking model.

Step 2: Synthesize the desired disturbance model.

Step 3: Specify the J LTI plant models that define the boundary of the region of plant parameter uncertainty.

Step 4: Obtain plant templates, at specified frequencies, that pictorially describe the region of plant parameter uncertainty on the NC.

Step 5: Select the nominal plant transfer function $P_o(s)$.

Step 6: Determine the stability contour (U-contour) on the NC.

Steps 7-9: Determine the disturbance, tracking, and optimal bounds on the NC.

Step 10: Synthesize the nominal loop transmission function $L_o(s) = G(s)P_o(s)$ that satisfies all the bounds and the stability contour.

Step 11: Based upon Steps 1 to 10 synthesize the prefilter $F(s)$.

Step 12: Simulate the system in order to obtain the time response data for each of the J plants.

The following sections illustrate this design procedure.

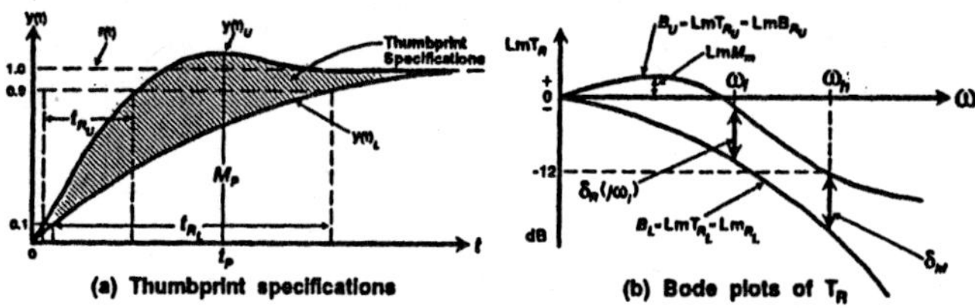

Figure 44.6 Desired response characteristic: (a) thumbprint specifications; (b) Bode plots of T_R.

44.2.3 Synthesize Tracking Models

The tracking thumbprint specifications, based upon satisfying some or all of the step-forcing function figures of merit for underdamped (M_p, t_p, t_s, t_r, K_m) and overdamped (t_s, t_r, K_m) responses, respectively, for a simple-second order system, are depicted in Figure 44.6a. The Bode plots corresponding to the time responses $y(t)_U$ (Equation 44.3) and $y(t)_L$ (Equation 44.4) in Fig. 44.6b represent the upper bound B_U and lower bound B_L, respectively, of the thumbprint specifications; i.e., an acceptable response $y(t)$ must lie between these bounds. Note that for $m.p.$ plants, only the tolerance on $|T_R(j\omega_i)|$ need be satisfied for a satisfactory design. For nonminimum-phase ($n.m.p.$) plants, tolerances on $\angle T_R(j\omega_i)$ must also be specified and satisfied in the design process [4] [5]. It is desirable to synthesize the tracking control ratios

$$T_{R_U} = \frac{(\omega_n^2/a)(s+a)}{s^2+2\zeta\omega_n s+\omega_n^2} \tag{44.3}$$

$$T_{R_L} = \frac{K}{(s-\sigma_1)(s-\sigma_2)(s-\sigma_3)} \tag{44.4}$$

corresponding to the upper and lower bounds T_{R_U} and T_{R_L}, respectively, so that $\delta_R(j\omega_i) = B_U - B_L$ increases as ω_i increases above the $0-dB$ crossing frequency of T_{R_U}. This characteristic of δ_R simplifies the process of synthesizing $L_o(s) = G(s)P_o(s)$. This synthesis process requires the determination of the tracking bounds $B_R(j\omega_i)$ that are obtained based upon $\delta_R(j\omega_i)$. The achievement of the desired performance specification is based upon the frequency bandwidth (BW), $0 < \omega \leq \omega_h$, which is determined by the intersection of the -12-dB line and the B_U curve in Figure 44.6b.

44.2.4 Disturbance Model

The simplest disturbance control ratio model specification is $|T_D(j\omega)| = |Y(j\omega)/D(j\omega)| \leq \alpha_p$ a constant [the maximum magnitude of the output based upon a unit-step disturbance input (d_1 of Figure 44.1)]. Thus, the frequency domain disturbance specification is log magnitude (Lm) $T_D(j\omega) \leq Lm\ \alpha_p$ over the desired specified BW $0 \leq \omega \leq \omega_h$ as defined in Figure 44.6b. Thus, the disturbance specification is represented by only an upper bound on the NC over the specified BW.

44.2.5 J LTI Plant Models

The simple plant

$$P_j(s) = \frac{Ka}{s(s+a)} \quad (44.5)$$

where $K \epsilon \{1, 10\}$ and $a \epsilon \{1, 10\}$, is used to illustrate the MISO QFT design procedure. The region of plant parameter uncertainty is illustrated by Figure 44.7. This region of uncertainty may be described by J LTI plants, where $j = 1, 2, \ldots J$. These plants lie on the boundary of this region of uncertainty; that is, the boundary points 1 to 6 are utilized to obtain six LTI plant models that adequately define the region of plant parameter uncertainty.

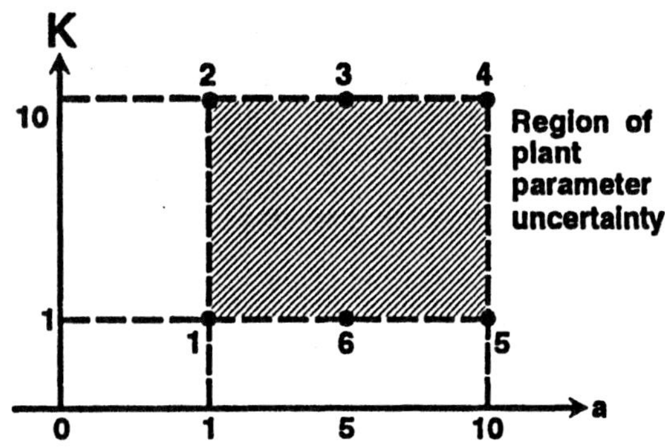

Figure 44.7 Region of plant uncertainty characterizing Equation 44.5.

44.2.6 Plant Templates of $P_j(s)$, $\Im P(j\,\omega_i)$

With $L = GP$, Equation 44.1 yields

$$LmT_R = LmF - Lm\left[\frac{L}{1+L}\right] \quad (44.6)$$

The change in T_R due to the uncertainty in P, since F is LTI, is

$$\Delta(LmT_R) = LmT_R - LmF = Lm\left[\frac{L}{1+L}\right] \quad (44.7)$$

By the proper design of $L_o = GP_o$ and F, this change in T_R is restricted so that the actual value of Lm T_R always lies between B_U and B_L of Figure 44.6b. The first step in synthesizing an L_o is to make NC templates that characterize the variation of the plant uncertainty (see Figure 44.8), as described by $j = 1, 2, \ldots, J$ plant transfer functions, for various values of ω_i over a specified frequency range. The boundary of the plant template can be obtained by mapping the boundary of the plant parameter uncertainty region, Lm $P_j(j\omega_i)$ vs $\angle P_j(j\omega_i)$, as shown on the NC in Figure 44.8. A curve is drawn through the points 1, 2, 3, 4, 5, and 6 where the shaded area is labeled $\Im P(j1)$, which can be represented by a plastic template. Templates for other values of ω_i are obtained in a similar manner. A characteristic of these templates is that, starting from a "low value" of ω_i, the templates widen (angular width becomes larger) for increasing values of ω_i, then, as ω_i takes on larger values and approaches infinity, they become narrower and eventually approach a straight line of height V dB (see Equation 44.9).

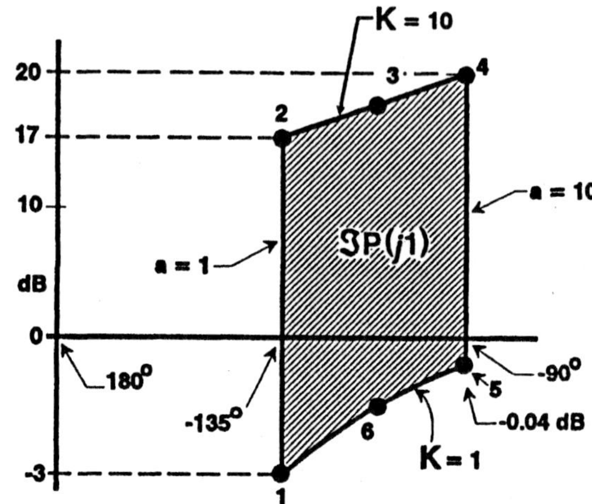

Figure 44.8 The template $\Im P(j1)$ characterizing Equation 44.5.

44.2.7 Nominal Plant

While any plant case can be chosen, select whenever possible a plant whose NC template point is always at the lower left corner for all frequencies for which the templates are obtained.

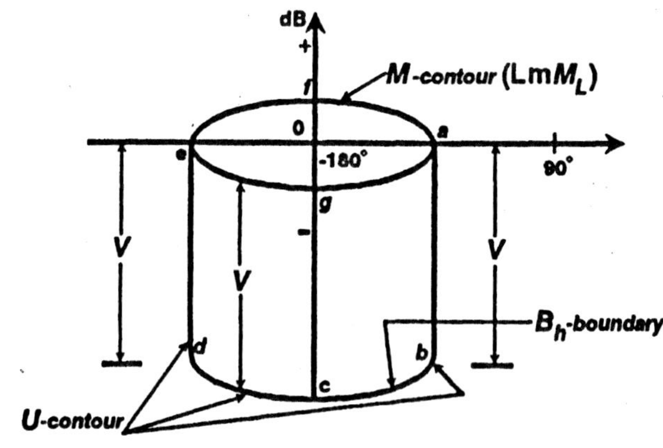

Figure 44.9 U-contour construction.

44.2.8 U-contour (Stability Bound)

The specifications on system performance in the frequency domain (see Figure 44.6b) identify a minimum damping ratio ζ for the dominant roots of the closed-loop system, which becomes a bound on the value $M_p \approx M_m$. On the NC this bound on $M_m = M_L$ (see Figures 44.6b and 44.9) establishes a region that must not be penetrated by the templates and the loop transmission function $L(j\omega)$ for all ω. The boundary of this region is referred to as the universal high-frequency boundary (UHFB) or stability bound—the U-contour, because this becomes the dominating constraint on $L(j\omega)$. Therefore, in Figure 44.9 the top portion, efa, of the M_L contour becomes part of the U-contour. For a large problem class, as $\omega \to \infty$, the limiting value of the plant transfer function approaches

$$\lim_{\omega\to\infty}[P(j\omega)] = \frac{K}{\omega^{\lambda}} \tag{44.8}$$

where λ represents the excess of poles over zeros of $P(s)$. The plant template for this problem class approaches a vertical line of length equal to

$$\begin{aligned}\Delta \lim_{(\omega\to\infty)} [Lm P_{\max} - Lm P_{\min}] = \\ Lm K_{\max} - Lm K_{\min} = VdB\end{aligned} \tag{44.9}$$

If the nominal plant is chosen at $K = K_{\min}$, then measuring VdB down from the bottom portion *age* of the constraint M_L gives the *bcd* portion of the U-contour *abcdefa* of Figure 44.9. The remaining portions, *ab* and *de*, of the stability contour are determined during the process of determining the tracking bounds.

44.2.9 Optimal Bounds $B_o(j\omega_i)$ on $L_o(j\omega_i)$

The determination of the tracking $B_R(j\omega_i)$ and the disturbance $B_D(j\omega_i)$ bounds are required in order to yield the optimal bounds $B_o(j\omega_i)$ on $L_o(j\omega_i)$.

Tracking Bounds

The solution for $B_R(j\omega_i)$ requires that the condition (actual)$\Delta T_R(j\omega_i) \leq \delta_R(j\omega_i)$ dB (see Figure 44.6b) must be satisfied. Thus, it is necessary to determine the resulting constraint, or bound $B_R(j\omega_i)$, on $L(j\omega_i)$. The procedure is to pick a nominal plant $P_o(s)$ and to derive tracking bounds on the NC, at specified values of frequency, by use of templates or a CAD package. That is, along a phase angle grid line on the NC, move the nominal point on the template $\Im P(j\omega_i)$ up or down, without rotating the template, until it is tangent to two M−contours whose difference in M values is essentially equal to δ_R. When this condition has been achieved, the location of the nominal point on the template becomes a point on the tracking bound $B_R(j\omega_i)$ on the NC. This procedure is repeated on sufficient angle grid lines on the NC to provide sufficient points to draw $B_R(j\omega_i)$ and for all values of frequency for which templates have been obtained. In general, the templates are moved from right to left starting from a phase angle grid line to the right of the M_L contour. When the templates become tangent to the M_L contour, the nominal point on the templates yields points on the *ab* and *de* portions of the stability contour. For m.p. systems, the condition $\Delta T_R(j\omega_i) \leq \delta_R(j\omega_i)$ requires that the synthesized loop transmission must satisfy the requirement that Lm $L(j\omega_i)$ is on or above the corresponding tracking bound Lm $B_R(j\omega_i)$.

Disturbance Bounds

The general procedure for determining disturbance bounds for the MISO control system of Figure 44.5 is outlined as follows, but more details are given in [1]. From Equation 44.2 the following equation is obtained:

$$T_D = \frac{P_o}{\frac{P_o}{P}+L_o} = \frac{P_o}{W} \tag{44.10}$$

where $W = (P_o/P) + L_o$. From Equation 44.10, setting Lm $T_D = \delta_D =$ Lm α_p, the following relationship is obtained:

$$Lm W = Lm P_o - \delta_D \tag{44.11}$$

For each value of frequency for which the NC templates are obtained, the magnitude of $|W(j\omega_i)|$ is obtained from Equation 44.11. This magnitude, in conjunction with the equation $W(j\omega_i) = [P_o(j\omega_i)/P(j\omega_i)]$, is utilized to obtain a graphical solution for $B_D(j\omega_i)$ as shown in Figure 44.10. Note that in this figure the template is plotted in rectangular or polar coordinates.

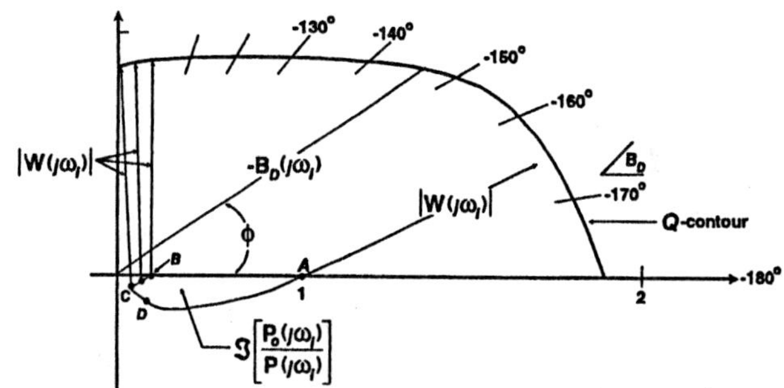

Figure 44.10 Graphical evaluation of $B_D(j\omega_i)$.

Optimal Bounds

For the case shown in Figure 44.11, $B_o(j\omega_i)$ is composed of those portions of each respective bound $B_R(j\omega_i)$ and $B_D(j\omega_i)$ that have the largest dB values. The synthesized $L_o(j\omega_i)$ must lie on or just above the bound $B_o(j\omega_i)$ of Figure 44.11.

44.2.10 Synthesizing (or Loop-Shaping) $L_o(s)$ and $F(s)$

The shaping of $L_o(j\omega)$ is shown by the dashed curve in Figure 44.11. A point such as Lm $L_o(j2)$ must be on or above $B_o(j2)$. Further, in order to satisfy the specifications, $L_o(j\omega)$ cannot violate the U−contour. In this example, a reasonable $L_o(j\omega)$ closely follows the U−contour up to $\omega = 40 rad/s$ and must stay below it above $\omega = 40$, as shown in Figure 44.11. It also must be at least a Type 1 $L_o(s)$ transfer function (one or more poles at the origin) for tracking a step-forcing function with zero steady-state error [1]. Synthesizing a rational function $L_o(s)$ that satisfies the above specification involves building up the function

$$\begin{aligned} L_o(j\omega) &= L_{ok}(j\omega) \\ &= P_o(j\omega)\prod_{k=0}^{w}[K_k G_k(j\omega)] \end{aligned} \tag{44.12}$$

where for $k = 0$, $G_0 = 1\angle 0°$ and $K = \prod_{k=0}^{w} K_k$. In order to minimize the order of the compensator, a good starting point

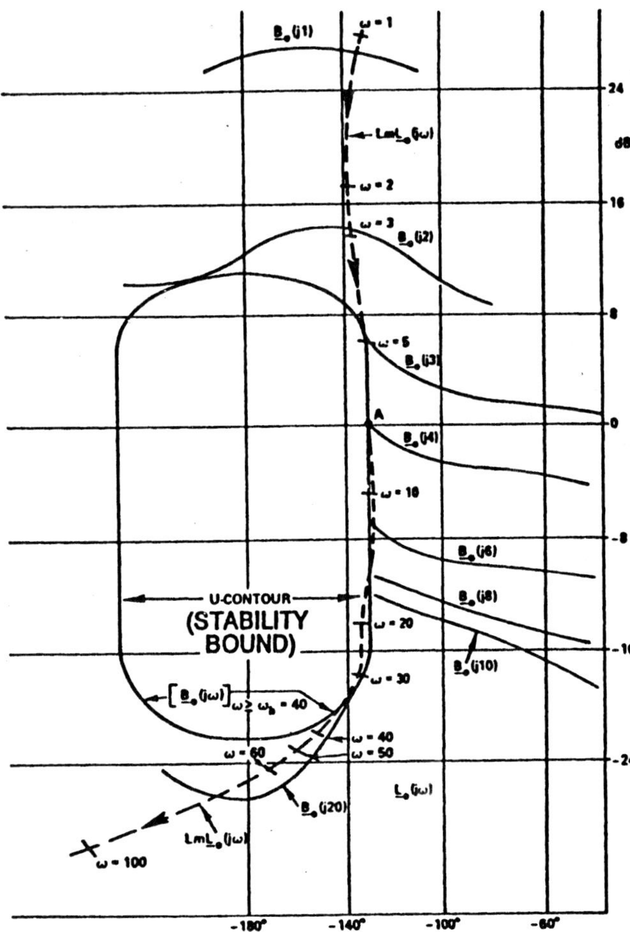

Figure 44.11 Bounds $B_o(j\omega_i)$ and loop-shaping.

for "building up" the loop transmission function is to assume initially that $L_{o0}(j\omega) = P_o(j\omega)$ as indicated in Equation 44.13. $L_o(j\omega)$ is built up term-by-term or by a CAD loop-shaping routine [8], in order (1) that the point $L_o(j\omega_i)$ lies on or above the corresponding optimal bound $B_o(j\omega_i)$, (2) that it passes close to the trough of the low frequency bounds, for achieving minimal gain, and (3) to stay just outside the U−contour in the NC of Figure 44.11. The design of a proper $L_o(s)$ guarantees only that the variation in $|T_R(j\omega_i)|$ is less than or equal to that allowed; i.e., $\delta_R(j\omega_i)$. The purpose of the prefilter $F(s)$ is to position Lm $[T(j\omega)]$ within the frequency domain specifications; i.e., that it always lies between B_U and B_L (see Figure 44.6b) for all J plants. The method for determining $F(s)$ is discussed in the next section. Once a satisfactory $L_o(s)$ is achieved, then the compensator is given by $G(s) = L_o(s)/P_o(s)$. Note that for this example $L_o(j\omega)$ slightly intersects the U−contour at frequencies above ω_h. Because of the inherent overdesign feature of the QFT technique, as a first trial design no effort is made to fine-tune the synthesis of $L_o(s)$. If the simulation results are not satisfactory, then a fine tuning of the design can be made. The available CAD packages simplify and expedite this fine tuning.

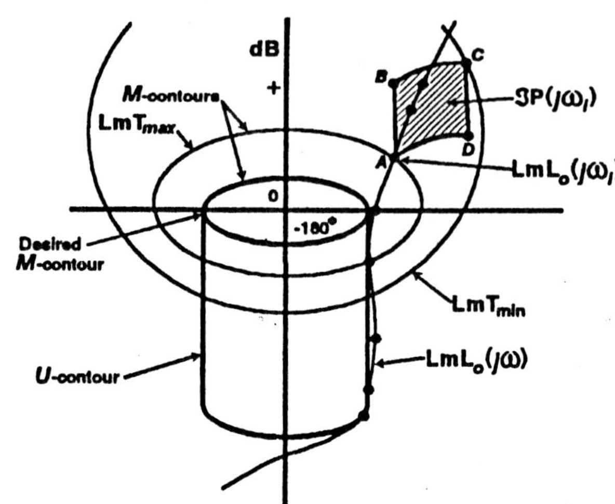

Figure 44.12 Prefilter determination.

44.2.11 Prefilter Design

[1, 2, 4, 5]

Design of a proper $L_o(s)$ guarantees only that the variation in $|T_R(j\omega)|$ is less than or equal to that allowed; i.e., $[LmT_R(j\omega)] \leq \delta_R(j\omega)$ The purpose of the prefilter $F(s)$ is to position

$$LmT(j\omega) = Lm\frac{L(j\omega)}{1+L(j\omega)} \tag{44.13}$$

within the frequency domain specifications. A method for determining the bounds on $F(s)$ is as follows:

Step 1. Place the nominal point of the ω_i plant template on the $L_o(j\omega_i)$ point on the $L_o(j\omega)$ curve on the NC (see Figure 44.12).

Step 2. Traversing the template, determine the M-contours that yield the maximum Lm $T_{\max}$ and the minimum Lm $T_{\min}$ values of Equation 44.13.

Step 3. Based upon obtaining sufficient data points, by repeating Step 2 within the desired frequency bandwidth for various values of ω_i, and in conjunction with the data used to obtain Figure 44.6b the plots of Figure 44.13 are obtained.

Step 4. Utilizing Figure 44.13, the straight-line Bode technique, and the condition

$$\lim_{s\to 0} F(s) = 1 \tag{44.14}$$

for a step-forcing function, an $F(s)$ is synthesized that lies within the upper and lower plots in Figure 44.13.

44.2.12 Simulation

The "goodness" of the synthesized $L_o(s)$ and $F(s)$ is determined by simulating the QFT-designed control system for all J plants. MISO QFT CAD packages, as discussed in Section 44.2.4, are available to expedite this simulation phase of the complete design process.

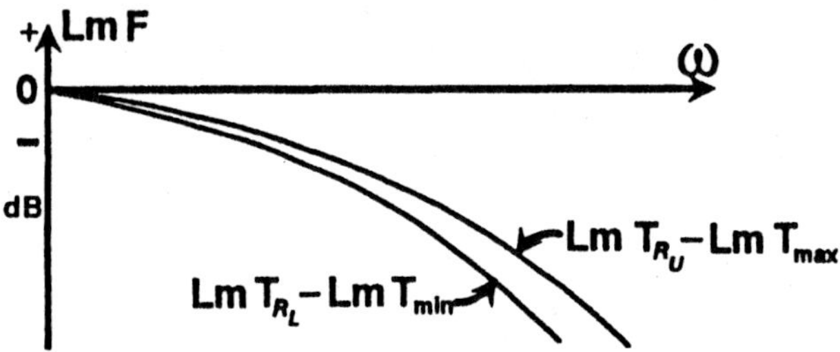

Figure 44.13 Frequency bounds on $F(s)$.

44.2.13 MISO QFT CAD Packages

The first usable MISO QFT CAD package was developed in 1986 for the analog design and in 1991 for the discrete design at the Air Force Institute of Technology (AFIT). These CAD packages have been a catalyst in assisting the newcomer to QFT to understand the fundamentals of this powerful design technique. The QFT CAD package illustrated in Appendix B can be used for both MISO and MIMO control system designs.

MISO QFT CAD

The flowchart of the MISO QFT CAD options in the AFIT package called TOTAL-P.C is shown in Appendix A. Those desiring a copy of this package can contact Professor C. H. Houpis, AFIT/ENG, Wright-Patterson AFB, OH 45433. This package has been designed as an educational tool. The QFT CAD package of Appendix B can also be used for the design of a MISO control system.

MISO QFT PC CAD

Dr. Yossi Chait, University of Massachusetts, and Dr. Oded Yaniv, Tel-Aviv University, Israel, have developed a MISO QFT PC CAD package for both analog and discrete system design available in MATLAB.

44.3 The MISO Discrete Control System

[13]

44.3.1 Introduction

The bilinear transformation, z−domain to the w'−domain and vice-versa, is utilized in order to accomplish the QFT design for both MISO and MIMO sampled-data (discrete) control system design in the w'−domain. This transformation enables the MISO QFT analog design technique to be readily used, with minor exceptions, to perform the QFT design for the controller $G(w')$. If the w'−domain simulations satisfy the desired performance specification, then by use of the bilinear transformation the z−domain controller $G(z)$ is obtained. With this z−domain controller, a discrete-time domain simulation is obtained to verify the "goodness" of the design. The QFT technique requires the determination of the minimum sampling frequency $(\omega_s)_{\min}$ BW that is needed for a satisfactory design [13, 14]. The larger the plant uncertainty and the narrower the system performance tolerances, the larger must be the value of $(\omega_s)_{\min}$. *Henceforth, the prime is omitted from w'; whenever the symbol w is used it is to be interpreted as w'.*

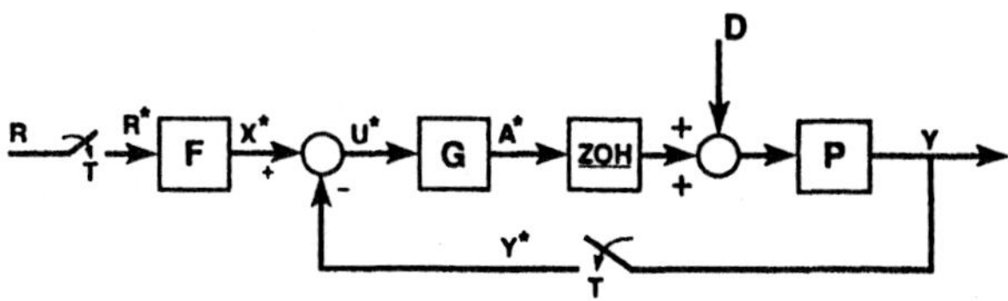

Figure 44.14 A MISO sampled-data control system.

44.3.2 The MISO Sampled-Data Control System

Figure 44.14 represents the MISO discrete control system, having plant uncertainty, that is to be designed by the QFT technique. The equations that describe this system are as follows:

$$\begin{aligned} P_z(z) &= G_{zo}P(z) = (1-z^{-1})Z\left[\frac{P(s)}{s}\right] \\ &= (1-z^{-1})P_e(z) \qquad (44.15) \\ L(z) &= G_{zo}P(z)G_1(z), \quad P_e \equiv \frac{P(s)}{s}, \\ P_e(z) &= Z\left[\frac{P(s)}{s}\right] = Z[P_e] \qquad (44.16) \\ D(s) &= \frac{1}{s}\ P_e(s) = P(s)D(s) \\ P_e(z) &= Z[P(s)D(s)] = PD(z) \qquad (44.17) \\ T_R &= \frac{F(z)L(z)}{1+L(z)} \quad Y_D = \frac{PD(z)}{1+L(z)} \qquad (44.18) \\ Y(z) &= \left[\frac{L(z)F(z)}{1+L(z)}\right]R(z) + \frac{PD(z)}{1+L(z)} \\ &= Y_R(z) + Y_D(z) \\ &= T_R(z)R(z) + Y_D(z) \qquad (44.19) \end{aligned}$$

44.3.3 w-Domain

The pertinent $s-$, $z-$, and $w-$plane relationships are as follows:

$$\begin{aligned} \alpha^2 &= \left(\frac{\sigma T}{2}\right)^2 < 2, \qquad \frac{\omega T}{2} \le 0.297 \qquad (44.20) \\ s &= \sigma + j\omega, \quad (a) \\ w &= u + jv = \left(\frac{2}{T}\right)\left[\frac{z-1}{z+1}\right] \quad (b) \qquad (44.21) \\ z &= \frac{Tw+2}{-Tw+2}, \\ v &= \frac{2}{T}\tan\left(\frac{\omega T}{2}\right) = \left(\frac{\pi}{\omega_s}\right)\tan\left(\frac{\omega\pi}{\omega_s}\right) \qquad (44.22) \\ \omega_s &= 2\pi/T, \quad z = \epsilon^{\sigma T}\angle\omega T = |z|\angle\omega T \qquad (44.23) \end{aligned}$$

44.3.4 Assumptions

For this chapter, the following assumptions are made:

- Minimum-phase (m.p.) stable plants
- The analog design models, Equations 44.3 and 44.4, yield the desired time response characteristics for the discrete-time system.
- The sampling time T is small enough so that over the BW, $0 < \omega < \omega_h$, Equation 44.23 is valid permitting the approximation $s \approx w$ and, in turn,

$$T_R(w) \approx [T_R(s)]_{s=w} \qquad (44.24)$$

Both the upper and lower bound w−domain tracking models are obtained in this manner. The disturbance specification is the same as for the analog case.

44.3.5 Nonminimum Phase $L_o(w)$

It is important to note that in the w domain any practical $L(w)$ is n.m.p. containing a zero at $2/T$ (the sampling zero). This result is due to the fact that any practical $L(z)$ has an excess of at least one pole over zeros. Thus, the design technique *for a stable uncertain plant* is modified [14] to incorporate the allpass filter (apf)

$$A(w) = \frac{w-\frac{2}{T}}{w+\frac{2}{T}} = -A'(w) = -\left[\frac{\frac{2}{T}-w}{\frac{2}{T}+w}\right] \qquad (44.25)$$

as follows. Let the nominal loop transmission be defined as

$$L_o \equiv -L_{mo}(w)A(w) = L_{mo}(w)A'(w) \qquad (44.26)$$

From Equation 44.26 it is seen that

$$\angle L_{mo}(jv) = \angle L_o(jv) - \angle A'(jv) \qquad (44.27)$$

where

$$-\angle A'(jv) = 2\tan^{-1}\frac{vT}{2} > 0 \qquad (44.28)$$

An analysis of Equations 44.26 to 44.28 reveals that the bounds $B_o'(jv_i)$ on $L_o(jv)$ become the bounds $B_{mo}(jv_i)$ on $L_{mo}(jv)$ by shifting, over the desired BW, $B_o'(jv_i)$ positively (to the right on the NC) by the angle $\angle A'(jv_i)$, as shown in Figure 44.15. The U−contour (B_h') must also be shifted to the right by the same amount, at the specified frequencies v_i, to obtain the shifted U−contour $B_h(jv_i)$. The contour B_h' is shifted to the right until it reaches the vertical line $\angle L_{mo_1}(jv_K) = 0°$. The value of v_K, which is a function of ω_s and the phase margin angle γ as shown in Figure 44.15, [13] is given by

$$2\tan^{-1}\left(\frac{v_k T}{2}\right) = 180° - \gamma \qquad (44.29)$$

It should be mentioned that loop-shaping or synthesizing $L_o(w)$ can be done directly without the use of an apf.

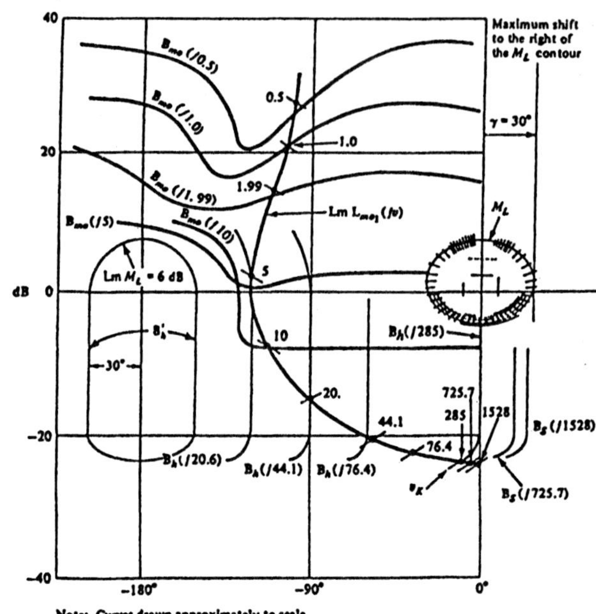

Figure 44.15 The shifted bounds on the NC.

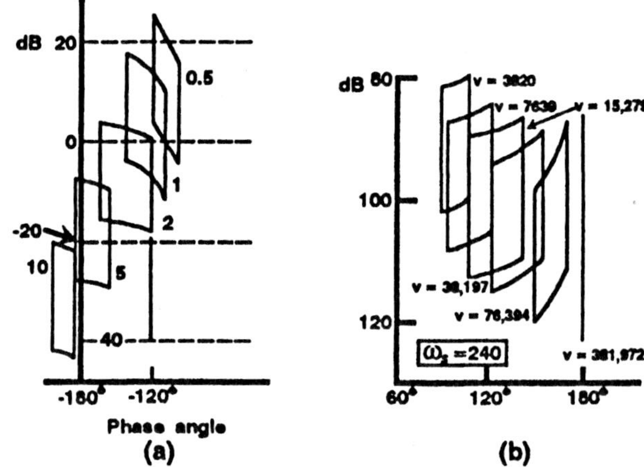

Figure 44.16 w−domain plant templates.

44.3.6 Plant Templates $\Im$ P (jv_i)

The plant templates in the w−domain have the same characteristic as those for the analog case (see Section 44.2.4) for the frequency range $0 < \omega_i \leq \omega_s/2$ as shown in Figure 44.16a. In the frequency range $\omega_s/2 < \omega_i < \infty$, the w−domain templates widen once again, then eventually approach a vertical line as shown in Figure 44.16b.

44.3.7 Synthesizing $L_{mo}(w)$

The frequency spectrum can be divided into four general regions for the purpose of synthesizing (loop-shaping) an $L_{mo}(w)$ that satisfies the desired system performance specifications for the plant having plant parameter uncertainty. These four regions are

Region 1. For the frequency range where Equation 44.20 is essentially satisfied, use the analog templates; i.e., $\Im P(j\omega_i) \approx \Im P(jv_i)$. The w−domain tracking, disturbance, and optimal bounds and the U–contour are essentially the same as those for the analog system. The templates are used to obtain these bounds on the NC in the same manner as for the analog system.

Region 2. For the frequency range $v_{0.25} < v_i \leq v_h$, where $\omega_i \leq 0.25\omega_s$, use the w–domain templates. These templates are used to obtain all three types of bounds, in the same manner as for the analog system, in this region; and the corresponding $B_h'(jv_i)$ contours are also obtained. Depending on the value of T, v_h may be less than $v_{0.25}$.

Region 3. For the frequency range $v_h < v_i \leq v_K$, for the specified value of ω_s, only the B_h' contours are plotted.

Region 4. For the frequency range $v_h > v_K$, use the w–domain templates. Since the templates $\Im P_e(jv_i)$ broaden out again for $v_i > v_K$, as shown in Figure 44.16, it is necessary to obtain the more stringent (stability) bounds B_S shown in Figure 44.17. The templates are used only to determine the stability bounds B_S.

The synthesis (or loop-shaping) of $L_{mo}(w)$ involves the synthesis of the following function:

$$L_{mo}(jv) = P_{eo}(jv) \prod_{k=0}^{w} [K_k G_k(jv)] \qquad (44.30)$$

where the nominal plant $P_{eo}(w)$ is the plant from the J plants that has the smallest dB value and the largest (most negative) phase lag characteristic. The final synthesized $L_{mo}(w)$ function must be one that satisfies the following conditions:

1. In Regions 1 and 2 the point on the NC that represents the dB value and phase angle of $L_{mo}(jv_i)$ must be such that it lies on or above the corresponding $B_{mo}(jv_i)$ bound (see Figure 44.15).
2. The values of Equation 44.30 for the frequency range of region 3 must lie to the right of or just below the corresponding B_h' contour (see Figure 44.15).
3. The value of Equation 44.30 for the frequency range of region 4 must lie below the B_S contour for negative phase angles on the NC (see condition 4 next).
4. In utilizing the bilinear transformation of Equation 44.21, the w–domain transfer functions are all equal order over equal order.
5. The Nyquist stability criterion dictates that the $L_{mo}(jv)$ plot is on the "right side" or the "bottom right side" of the $B_h(jv_i)$ contours for the frequency range of $0 \leq v_i \leq v_K$. It has been shown that [3]
   a. $L_{mo}(jv)$ must reach the *right-hand bottom* of $B_h(jv_K)$, (i.e., approximately point K in Figure 44.17) at a value of $v \leq v_K$.
   b. $\angle L_{mo}(jv_K) < 0°$ in order that there exists a practical L_{mo} that satisfies the bounds $B(jv)$ and provides the required stability.
6. For the situation where one or more of the J LTI plants that represent the uncertain plant parameter characteristics represent unstable plants and one of these unstable plants is selected as the nominal plant, *then the apf to be used in the QFT design must include all right half-plane (RHP) zeros of P_{zo}*. This situation is not discussed in this chapter. **Note:** For experienced QFT control system designers, $L_o(v)$ can be synthesized without the use of apf. This approach also is not covered in this chapter.

The synthesized $L_{mo}(w)$ obtained following the guidelines of this section is shown in Figure 44.17.

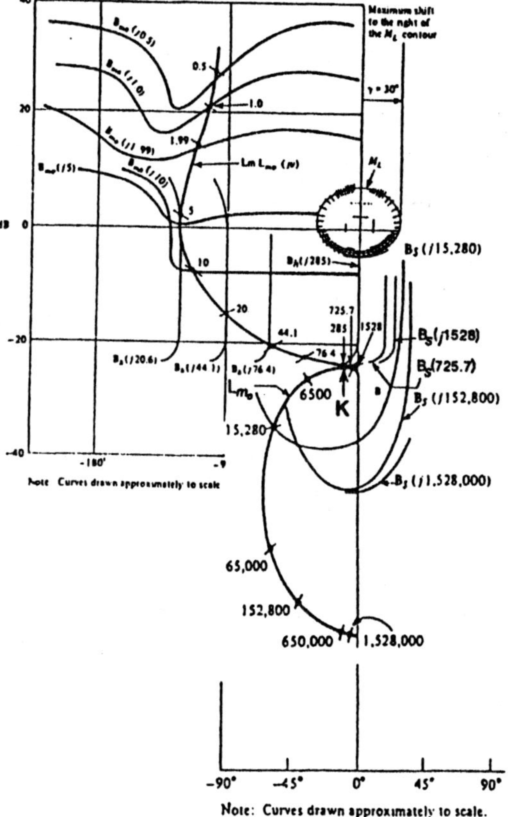

Figure 44.17 A satisfactory design: $L_{mo}(jv)$ at $\omega_s = 240$.

44.3.8 Prefilter Design

The procedure for synthesizing $F(w)$ is the same as for the analog case (see Section 44.2.11) over the frequency range $0 < v_i \leq v_h$. In order to satisfy condition 4 of Section 44.3.7, a nondominating zero or zeros ("far left" in the w–plane) are inserted so that the final synthesized $F(w)$ is equal order over equal order.

44.3.9 w-Domain Simulation

The "goodness" of the synthesized $L_{mo}(w)$ [or $L_o(w)$] and $F(w)$ is determined by first simulating the QFT w–domain designed control system for all J plants in the $w-$domain (an "analog" time domain simulation). See Section 44.2.13 for MISO QFT CAD packages that expedite this simulation.

44.3.10 z-Domain

The two tests of the "goodness" of the $w-$domain QFT-designed system is a discrete-time domain simulation of the system shown in Figure 44.14. To accomplish this simulation, the $w-$domain transfer functions $G(w)$ and $F(w)$ are transformed to the $z-$domain by use of the bilinear transformation of Equation 44.21. This transformation is utilized since the degree of the numerator and denominator polynomials of these functions are equal and the controller and prefilter do not contain a zero-order-hold device.

Comparison of the Controller's w- and z-Domain Bode Plots

Depending on the value of the sampling time T, warping may be sufficient to alter the loop-shaping characteristics of the controller when it is transformed from the $w-$domain into the $z-$domain. For the warping effect to be minimal, the Bode plots (magnitude and angle) of the $w-$ and $z-$domain controllers must essentially lie on top of one another within the frequency range $0 < \omega \leq [(2/3)(\omega_s/2)]$. If the warping is negligible, then a discrete-time simulation can proceed. If not, a smaller value of sampling time needs to be selected.

Accuracy

The CAD packages that are available to the designer determines the degree of accuracy of the calculations and simulations. The smaller the value of T, the greater the degree of accuracy that is required to be maintained. The accuracy can be enhanced by implementing $G(z)$ and $F(z)$ as a set of g and f cascaded transfer functions, respectively; that is,

$$\begin{aligned} G(z) &= G_1(z)G_2(z)\cdots G_g(z), \\ F(z) &= F_1(z)F_2(z)\cdots F_f(z) \end{aligned} \tag{44.31}$$

Analysis of Characteristic Equation $Q_j(z)$

Depending on the value of T and the plant parameter uncertainty, the pole-zero configuration in the vicinity of the $-1+j0$ point in the $z-$plane for one or more of the J LTI plants can result in an unstable discrete-time response. Thus, before proceeding with a discrete-time domain simulation, an analysis of the characteristic equation $Q_j(z)$ for all J LTI plants must be made. If an unstable system exists, an analysis of $Q_j(z)$ and the corresponding root locus may reveal that a slight relocation of one or more controller poles in the vicinity of the $-1+j0$ point toward the origin may ensure a stable system for all J plants without essentially affecting the desired loop-shaping characteristic of $G(z)$.

Simulation and CAD Packages

With the "design checks" of Sections 44.3.10 through 44.3.10 satisfied, then a discrete-time simulation is performed to verify that the desired performance specifications have been achieved. In order to enhance the MISO QFT discrete control system design procedure that is presented in this chapter, the CAD flow chart of Section 44.2.13 is shown in Appendix B.

44.4 MIMO Systems

[2, 12, 16, 17]

44.4.1 Introduction

Figure 44.18 represents an $m \times m$ MIMO closed-loop system in which F, G, and P are each $m \times m$ matrices, and $\mathcal{P} = \{P\}$ is a set of J matrices due to plant parameter uncertainty. There are m^2 closed-loop system transfer functions (transmissions) t_{ij} contained within its system transmission matrix (i.e., $T = \{t_{ij}\}$) relating the outputs y_i to the inputs r_j (e.g., $y_i = t_{ij}r_j$). These relationships hold for both the $s-$ and $w-$domain analysis of a MIMO system. In a quantitative problem statement there are tolerance bounds on each t_{ij}, giving a set of m^2 acceptable regions τ_{ij} that are to be specified in the design; thus, $t_{ij} \epsilon \tau_{ij}$ and $\mathfrak{T} = \{\tau_{ij}\}$. From Figure 44.18 the system control ratio relating r to y is:

$$T = [I + PG]^{-1}PGF \tag{44.32}$$

The t_{ij} expressions derived from this expression are very complex and not suitable for analysis. The QFT design procedure systematizes and simplifies the manner of achieving a satisfactory system design for the entire range of plant uncertainty. In order to readily apply the QFT technique, another mathematical system description is presented in the next section. The material presented in this chapter pertains to both the $s-$ and $w-$domain analysis of MIMO systems.

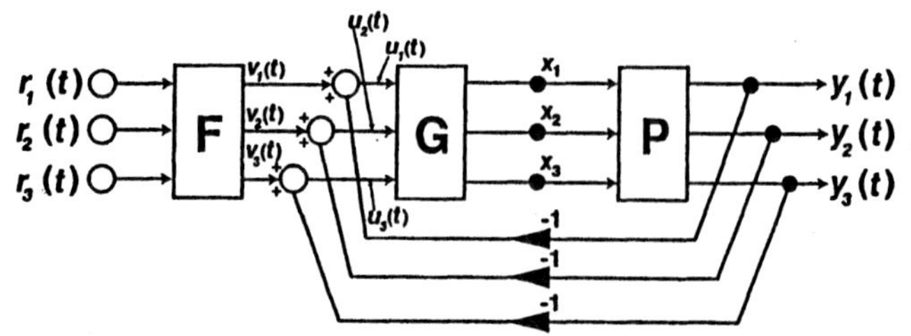

Figure 44.18 A 3 × 3 MIMO feedback control system.

44.4.2 Derivation of m^2 MISO System Equivalents

The G, F, P and P^{-1} matrices are defined as follows:

$$G = \begin{bmatrix} g_1 & 0 & \cdots & 0 \\ 0 & g_2 & \cdots & 0 \\ \vdots & \vdots & \vdots & \vdots \\ 0 & 0 & \cdots & g_m \end{bmatrix}$$

$$F = \begin{bmatrix} f_{11} & f_{12} & \cdots & f_{1m} \\ f_{21} & f_{22} & \cdots & f_{2m} \\ \vdots & \vdots & \vdots & \vdots \\ f_{m1} & f_{m2} & \cdots & f_{mm} \end{bmatrix} \quad (44.33)$$

$$P = \begin{bmatrix} p_{11} & p_{12} & \cdots & p_{1m} \\ p_{21} & p_{22} & \cdots & p_{2m} \\ \vdots & \vdots & \vdots & \vdots \\ p_{m1} & p_{m2} & \cdots & p_{mm} \end{bmatrix}$$

$$P^{-1} = \begin{bmatrix} p_{11}^* & p_{12}^* & \cdots & p_{1m}^* \\ p_{21}^* & p_{22}^* & \cdots & p_{2m}^* \\ \vdots & \vdots & \vdots & \vdots \\ p_{m1}^* & p_{m2}^* & \cdots & p_{mm}^* \end{bmatrix} \quad (44.34)$$

Although only a diagonal G matrix is considered, the use of a nondiagonal G matrix may allow the designer more design flexibility [2]. The m^2 effective plant transfer functions are based upon defining:

$$q_{ij} \equiv \frac{1}{p_{ij}^*} = \frac{\det P}{adj P_{ij}} \quad (44.35)$$

There is a requirement that det.P be m.p. The Q matrix is then formed as:

$$Q = \begin{bmatrix} q_{11} & q_{12} & \cdots & q_{1m} \\ q_{21} & q_{22} & \cdots & q_{2m} \\ \vdots & \vdots & \vdots & \vdots \\ q_{m1} & q_{m2} & \cdots & q_{mm} \end{bmatrix} = \begin{bmatrix} \frac{1}{p_{11}^*} & \frac{1}{p_{12}^*} & \cdots & \frac{1}{p_{1m}^*} \\ \frac{1}{p_{21}^*} & \frac{1}{p_{22}^*} & \cdots & \frac{1}{p_{2m}^*} \\ \vdots & \vdots & \vdots & \vdots \\ \frac{1}{p_{m1}^*} & \frac{1}{p_{m2}^*} & \cdots & \frac{1}{p_{mm}^*} \end{bmatrix} \quad (44.36)$$

The matrix P^{-1} is partitioned to the form:

$$P^{-1} = [P_{ij}^*] = \left[\frac{1}{q_{ij}}\right] = \Lambda + B \quad (44.37)$$

where Λ is the diagonal part of P^{-1} and B is the balance of P^{-1}; thus $\lambda_{ij} = 1/q_{ii} = p_{ii}^*$, $b_{ii} = 0$, and $b_{ij} = 1/q_{ij} = p_{ij}^*$ for $i \neq j$. Premultiplying Equation 44.32 by $[I + PG]$ yields

$$[I + PG]T = PGF \rightarrow [P^{-1} + G]T = GF \quad (44.38)$$

where P is nonsingular. Using Equation 44.37 with G diagonal, Equation 44.38 can be rearranged to the form:

$$T = [\Lambda + G]^{-1}[GF - BT] \quad (44.39)$$

Equation 44.39 is used to define the desired fixed-point mapping where each of the m^2 matrix elements on the right-hand side of this equation can be interpreted mathematically as representing a MISO system. Proof of the fact that design of each MISO system yields a satisfactory MIMO design is based on the Schauder fixed-point theorem [7]. This theorem is described, based upon a *unit impulse input,* by defining a mapping

$$Y(T_i) = [\Lambda + G]^{-1}[GF - BT_i] \equiv T_j \quad (44.40)$$

where each member of T is from the acceptable set $\Im$. If this mapping has a fixed point [i.e., $T \epsilon \Im$ such that $Y(T_i) = T_j$], then this T is a solution of Equation 44.39. The y_{11} output obtained from Equation 44.40, for the 3×3 case, is given by:

$$y_{11} = \frac{q_{11}}{1+g_1 q_{11}}\left[g_1 f_{11} - \left(\frac{t_{21}}{q_{12}} + \frac{t_{31}}{q_{13}}\right)\right] \quad (44.41)$$

Based upon the derivation of all the y_{ij} expressions from Equation 44.40 yields the four effective MISO loops (in the boxed area) in Figure 44.4, resulting from a 2×2 system and the nine effective MISO loops resulting from a 3×3 system [4]. The control ratios for the desired tracking inputs r_i by the corresponding outputs y_i for each feedback loop of Equation 44.40 have the form

$$y_{ii} = w_{ij}(v_{ij} + c_{ij}) \quad (44.42)$$

where $w_{ii} = q_{ii}/(1 + g_i q_{ii})$ and $v_{ij} = g_i f_{ij}$. The interaction between the loops has the form

$$c_{ij} = -\sum_{k \neq i}\left[\frac{t_{kj}}{q_{ik}}\right] \quad k = 1, 2, 3, \ldots, m \quad (44.43)$$

and appears as a "cross-coupling effect" input in each of the feedback loops. *Thus, Equation 44.42 represents the control ratio of the ith MISO system.* The transfer function $w_{ii}v_{ij}$ relates the "desired" ith output to the jth input r_j, and the transfer function $w_{ii}c_{ij}$ relates the ith output to the jth cross-coupling effect input c_{ij}. The outputs given in Equation 44.42 can thus be expressed as

$$y_{ij} = (y_{ij})_{r_i} + (y_{ij})_{c_{ij}} = y_{r_i} + y_{c_{ij}} \quad (44.44)$$

or, based on a *unit impulse input,*

$$t_{ij} = t_{r_{ij}} + t_{c_{ij}} \quad (44.45)$$

where

$$t_{r_{ij}} = y_{r_i} = w_{ii}v_{ij} \quad t_{c_{ij}} = y_{c_{ij}} = w_{ii}c_{ij} \quad (44.46)$$

and where now the upper bound, in the low-frequency range $(0 < \omega \leq \omega_h)$, is expressed as b_{ij}'. Thus,

$$\tau_{c_{ij}} = b_{ij} - b_{ij}' \quad (44.47)$$

represents the maximum portion of b_{ij} allocated toward the cross-coupling effect rejection, and b_{ij}' represents the upper bound for the tracking portion, respectively, of t_{ij}. For each MISO system there is a cross-coupling effect input that is a function of all the other loop outputs. The object of the design is to have each loop track its desired input while minimizing the outputs due to the disturbance (cross-coupling effects) inputs.

In each of the nine structures of Figure 44.4 it is necessary that the control ratio t_{ij} must be a member of the acceptable $t_{ij} \epsilon \tau_{ij}$. All the g_i and f_{ij} must be chosen to ensure that this condition is satisfied, thus constituting nine MISO design problems. If all of these MISO problems are solved, there exists a fixed point; then y_{ij} on the left-hand side of Equation 44.40 may be replaced by t_{ij}, and all the elements of T on the right-hand side by t_{kj}. This means that there exist nine t_{ij} and t_{kj}, each in its acceptable set, which is a solution to Figure 44.18. If each element is 1:1, then this solution must be unique. A more formal and detailed treatment is given in [7].

44.4.3 Tracking and Cross-Coupling Specifications

The presentation for the remaining portion of this chapter is based upon not only a diagonal G matrix but also a diagonal F matrix. Thus, in Figure 44.4 the t_{ij} terms, for $i \neq j$, represent disturbance responses due to the cross-coupling effect whereas the t_{ij} terms, for $i = j$ (see Equation 44.45) is composed of a desired tracking term t_r and of an unwanted cross-coupling term t_c. Therefore, the desired tracking specifications for the diagonal MISO systems of Figure 44.4 contain an upper-bound and a lower bound as shown in Figure 44.6. The disturbance specification for all MISO loops is given by only an upper bound. These performance specifications are shown in Figure 44.19 for a 2×2 (in the boxed area) and for a 3×3 MIMO feedback control system.

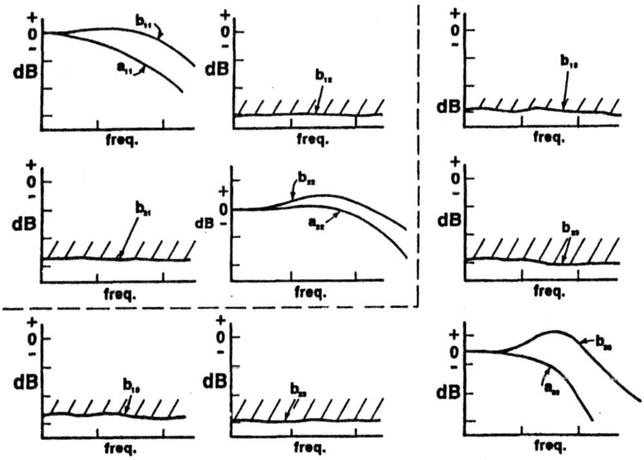

Figure 44.19 Tracking and cross-coupling specifications for a $2X2$ (in boxed area) and for a $3X3$ MIMO system.

Tracking Specifications

Based upon the analysis of Equations 44.45 to 44.47, the specifications for the t_{ii} responses shown in Figure 44.19 need to be modified as shown in Figure 44.20. As shown in this figure, a portion of $\delta_R(j\omega_i)$ (see Figure 44.6) has been allocated [2, 7] for the disturbance (cross-coupling effect) specification. Thus, based upon this modification and given an uncertain plant $\mathcal{P} = \{P_j\}(j = 1, 2, \ldots, J)$ and the BW ω_h above which output sensitivity is ignored, it is desired to synthesize G and F such that for all $P \epsilon \mathcal{P}$

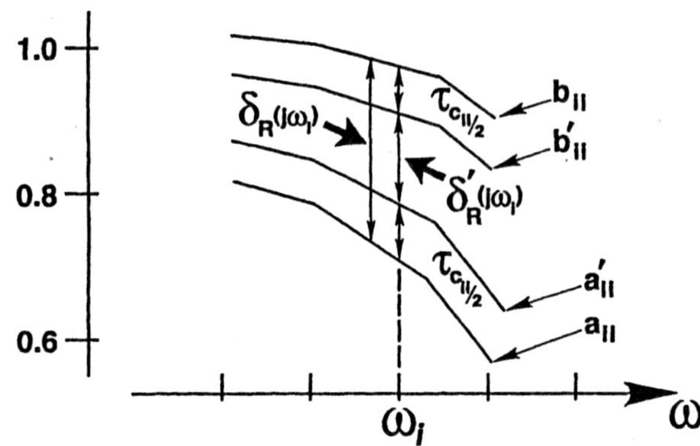

Figure 44.20 Allocation for tracking and cross-coupling specifications for the t_{ii} responses.

$$a'_{ii} \leq |t_{ii}(j\omega)| \leq b'_{ii} \quad \text{for} \quad \omega \leq \omega_h \tag{44.48}$$

A finite ω_h is recommended because, in strictly proper systems, feedback is not effective in the high-frequency range.

Disturbance Specification (Cross-Coupling Effect)

Based upon the previous discussion the disturbance specification, an upper bound, is expressed as

$$\left|t_{c_{ik}}\right| \leq \left|b_{ij}\right| \tag{44.49}$$

Thus, the synthesis of G must satisfy both Equations 44.48 and 44.49.

44.4.4 Determination of Tracking, Cross-Coupling, and Optimal Bounds

The remaining portion of the MIMO QFT approach is confined to a 2×2 system. The reader can refer to the references for higher-order systems ($m > 2$). From Equation 44.39 the following equations are obtained:

$$t_{11} = \frac{L_1 f_{11} + c_{11} q_{11}}{1 + L_1}$$

where

$$c_{11} = -\frac{t_{21}}{q_{12}}, \quad L_1 = q_{11} g_1 \tag{44.50}$$

$$t_{12} = \frac{c_{12} q_{11}}{1 + L_1}$$

where

$$c_{12} = -\frac{t_{22}}{q_{12}}, \quad f_{12} = 0 \tag{44.51}$$

$$t_{21} = \frac{c_{21} q_{22}}{1 + L_2}$$

where

$$c_{21} = -\frac{t_{11}}{q_{21}}, \quad L_2 = q_{22} g_2, \quad f_{21} = 0 \tag{44.52}$$

$$t_{22} = \frac{L_2 f_{22} + c_{22} q_{22}}{1 + L_2}$$

where

$$c_{22} = -\frac{t_{12}}{q_{21}} \quad (44.53)$$

Equations 44.51 and 44.52 correspond to the MISO systems for the first row of loops in Figure 44.4, and Equations 44.53 and 44.53 correspond to the MISO loops for the second row.

Tracking Bounds

The tracking bounds for the ii MISO system is determined in the same manner as for the MISO system of PART II (see Section 44.2.9). By use of the templates for the ii loop plant, the value of $\delta_R(j\omega_i)'$, shown in Figure 44.20, is used to satisfy the constraint of Equation 44.48.

Cross-Coupling Bounds

From Equations 44.51 and 44.53, considering only the first row of MISO loops in Figure 44.4, the following cross-coupling transfer functions are obtained (see Figures 44.20 and 44.19, respectively):

$$|t_{c_{11}}| = \left|\frac{c_{11}q_{11}}{1+L_1}\right| \leq \tau_{c_{ij}} \equiv \tau_{c_{11}} \quad \cdots \quad (44.54)$$

$$|t_{c_{12}}| = \left|\frac{c_{12}q_{11}}{1+L_1}\right| \leq b_{ij} \equiv b_{12} \quad \cdots \quad (44.55)$$

Substituting for c_{11} and c_{12} into Equations 44.54 and 44.55, respectively, and replacing t_{21} and t_{22} by their respective upper bound values b_{21} and b_{22}, and rearranging these equations yields:

$$\left|\frac{1}{1+L_1}\right| =\leq \left|\frac{q_{12}}{q_{11}}\right| \frac{\tau_{c_{11}}}{b_{21}} = M_{m_{11}} \quad (44.56)$$

$$\left|\frac{1}{1+L_1}\right| \leq \left|\frac{q_{12}}{q_{11}}\right| \frac{b_{12}}{b_{22}} = M_{m_{12}} \quad (44.57)$$

Substituting into these equations $L_1 = 1/l_1$ yields:

$$\left|\frac{l_1}{1+l_1}\right| \leq M_{m_{11}} \quad \cdots \quad \left|\frac{l_1}{1+l_1}\right| \leq M_{m_{12}} \quad (44.58)$$

By analyzing these equations for each of the J plants over the desired BW, the maximum value M_m that each of these equations can have, for each value of ω_i (or v_i), is readily determined by use of a CAD package. Thus, since $L_1 = 1/l_1$, the reciprocals of these values yield the value of the corresponding M−contours or cross-coupling bounds, for $\omega = \omega_i$ (or v_i), on the NC.

Optimal Bounds

The points on the optimal bound for a given value of frequency and for a given row of MISO loops of Figure 44.4 are determined by selecting the largest dB value, for a given NC phase angle, from all the tracking and cross-coupling bounds for these loops at this frequency. The MIMO QFT CAD package is designed to perform this determination of the optimal bounds.

44.4.5 QFT Methods of Designing MIMO Systems

There are two methods of achieving a QFT MIMO design. Method 1 involves synthesizing the loop transmission function L_i and the prefilter f_{ii} independent of the previous synthesized loop transmission functions and prefilters. Method 2 substitutes the synthesized g_i and f_{ii} of the first (or prior) MISO loop(s) that is (are) designed into the equations that describe the remaining loops to be designed. For Method 2, it is necessary to make the decision as to the order that the L_i functions are to be synthesized. Generally, the loops are chosen on the basis of the phase margin frequency ω_ϕ requirements. That is, the loop having the smallest value of ω_ϕ is chosen as the first loop to be designed; the loop having the next smallest value of ω_ϕ is selected as the second loop to be designed; etc. This is an important requirement for Method 2.

Method 1

This method involves overdesign (worst-case scenario), i.e., in getting the M_m values of Equations 44.56 and 44.57, for the 2 × 2 case, the maximum magnitude that q_{12} and the minimum magnitude that q_{11} can have, for each value of ω_i, over the entire J LTI plants are utilized. This method requires that the diagonal dominance condition [2, 7] be satisfied. When this condition is not satisfied, then Method 2 needs to be utilized.

Method 2

Once the order in which the loops are to be designed is designated accordingly (loop 1, loop 2, etc.), then the compensator g_1 and the prefilter f_{11} are synthsized. These are now known LTI functions, which are utilized to define the loop 2 effective plant transfer function. That is, substitute Equation 44.51 into Equation 44.53, then rearrange the result to obtain a new expression for t_{12} in terms of g_1 and f_{11} as follows:

$$t_{21} = -\left[\frac{\frac{f_{11}L_1q_{22_e}}{q_{21}(1+L_1)}}{1+g_2q_{22_e}}\right] \quad (44.59)$$

where the effective loop 2 transfer function is

$$q_{22_e} = \frac{q_{22}(1+L_1)}{(1+L_1-\gamma_{12}} \quad \text{where } \gamma_{12} = \frac{q_{11}q_{22}}{q_{12}q_{21}} \quad (44.60)$$

Repeating a similar procedure, the expression for t_{22} is

$$t_{22} = \frac{f_{22}g_2q_{22_e}}{1+g_2q_{22_e}} \quad (44.61)$$

Remember that a diagonal prefilter matrix has been specified. Note that Equations 44.59 to 44.61 involve the known f_{11} and g_1, which reduces the overdesign of loop 2.

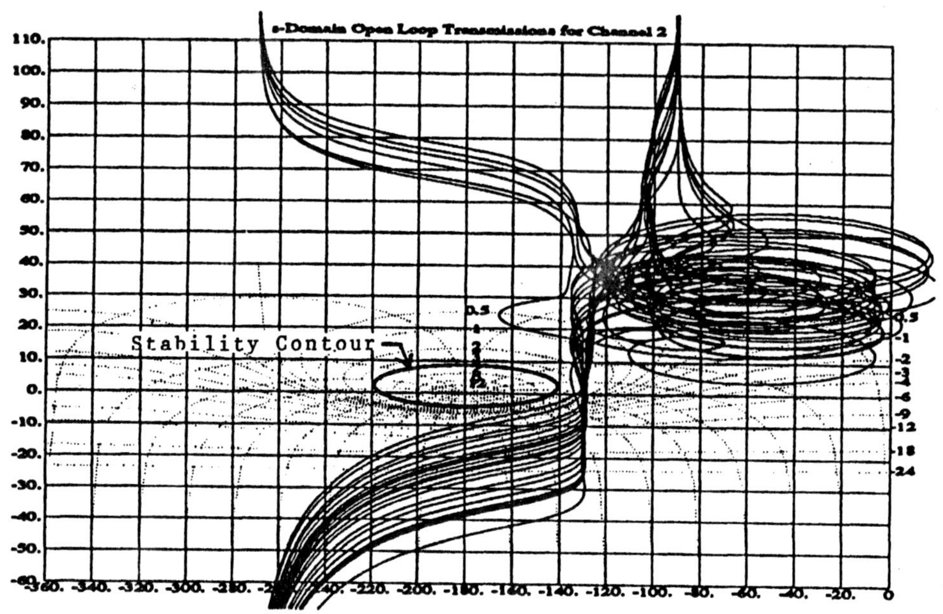

Figure 44.21 Open-loop transmissions on NC.

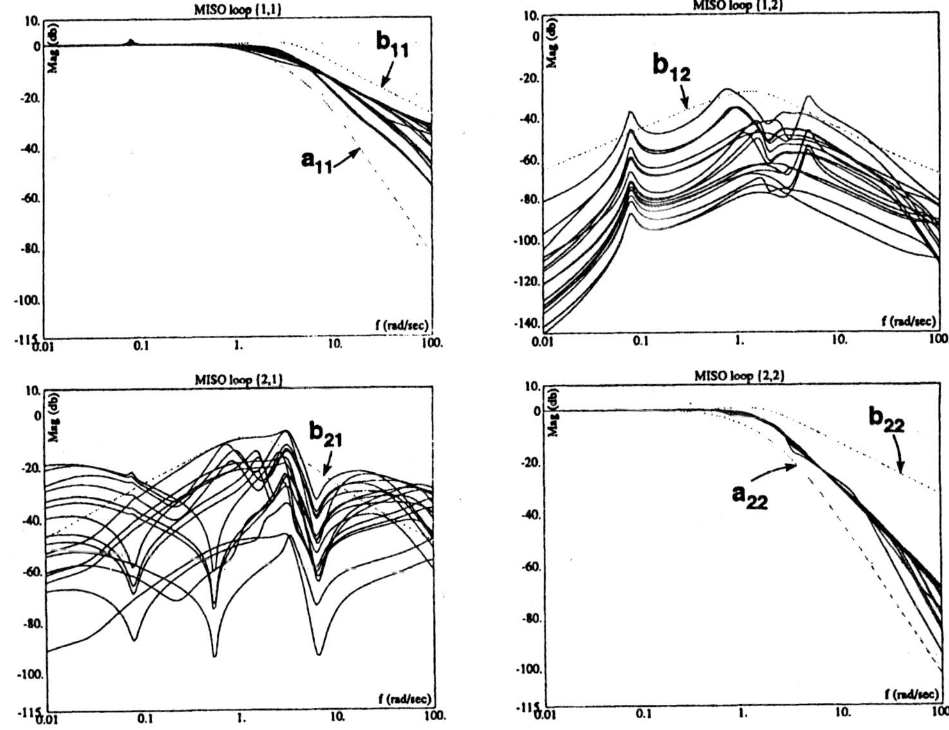

Figure 44.22 Closed-loop transmissions for an analog design system.

44.4.6 Synthesizing the Loop Transmission and Prefilter Functions

Once the optimal bound has been determined for each L_i loop, then the synthesis procedures for determining the loop transmission and prefilter functions are the same as for the MISO analog and discrete systems as discussed in Sections 44.2 and 44.3, respectively.

44.4.7 Overview of the MIMO QFT CAD Package

[15]

The MIMO QFT CAD package, implemented using Mathematica, is capable of carrying an analog or a discrete MISO or MIMO QFT design from problem setup through the design process to a frequency domain analysis of the compensated system. For analog control problems, the design process is performed in the s−plane. The design process is performed in either the w−plane or the s−plane using the pseudo-continuous-time (PCT) representation of the sampled-data system. A flowchart of the CAD package is given in Appendix B.

44.5 QFT Application

A MIMO QFT example, [18] a 2 x 2 analog flight control system (see the boxed-in loops of Figure 44.4), is presented to illustrate the power of this design technique. Also, this example illustrates the increased accuracy and efficiency achieved by the MIMO QFT CAD package [15] and the straightforward method for designing an analog MIMO control system. The specifications require a robust analog design for an aircraft that provides stability and meets time domain performance requirements for the specified four flight conditions (Table 44.1) and the six aircraft failure modes (Table 44.2). Table 44.3 lists the resulting set of 24 plant cases that incorporate these flight conditions and failure modes. For stability, a 45° phase margin is required for each of the two feedback loops. Frequency domain performance specifications, when met, result in the desired closed-loop system performance in the time domain. The frequency domain specifications are shown as dashed lines on the Bode plots of Figure 44.22.

TABLE 44.1 Flight Conditions.

	Aircraft Parameters	
Flight Condition	Mach	Altitude
1	0.2	30
2	0.6	30,000
3	0.9	20,000
4	1.6	30,000

The specifications, the plant models [18] for the 24 cases, and the weighting matrix are entered into the QFT CAD package. The automated features accessed through the designer interface of the CAD package result in synthesizing the compensators $g_1(s)$ and $g_2(s)$ in the manner described in Section 44.2.10. That is, the nominal loop transmission functions $L_{1_0}(s) = g_1(s)q_{11_0}(s)$ and $L_{2_0}(s) = g_2(s)q_{22_0}(s)$ are synthesized (or shaped) so that they satisfy their respective stability bounds and their respective optimal bounds $B_{1_0}(j\omega_i)$ and $B_{2_0}(j\omega_i)$. Note that q_{11_0} and q_{22_0} are the nominal plant transfer functions. The first step in a validation check is to plot the loop transmission functions $L_{2_\iota}(s)$, where $\iota = 1, \ldots, 24$, for all 24 cases on the NC. This is accomplished by a CAD routine, as shown in Figure 44.21 for the purpose of a stability check (m.p. plants). As is seen, none of the cases violate the M_L stability contour (the dark ellipse). In this design, when synthesizing $L_{2_0}(s)$ a trade-off exists between performance and bandwidth. In this example, the designer chooses to accept the consequences of violating the disturbance bound for $\omega = 2rad/s$. With $L_{1_0}(s)$ and $L_{2_0}(s)$ synthesized, the automated features of the CAD package expedite the design of the prefilters $f_{11}(s)$ and $f_{22}(s)$.

TABLE 44.2 Failure Modes.

Failure Mode	Failure Condition
1	Healthy aircraft
2	One horizontal tail fails
3	One flaperon fails
4	One horizontal tail and one flaperon fail, same side
5	One horizontal tail and one flaperon fail, opposite side
6	Both flaperons fail

TABLE 44.3 Plant Models.

	Flight Condition			
Failure Mode	1	2	3	4
1	#1	#7	#13	#19
2	#2	#8	#14	#20
3	#3	#9	#15	#21
4	#4	#10	#16	#22
5	#5	#11	#17	#23
6	#6	#12	#18	#24

For the second step in the design validation process, the 2×2 array of Bode plots shown in Figure 44.22 is generated, showing on each plot the 24 possible closed-loop transmissions from an input to an output of the completed system. The consequence of violating the channel 2 disturbance bound for $\omega = 2rad/s$ is seen where the closed-loop transmissions violate b_{21}, denoted by dashed line, beginning at $\omega = 2rad/s$. Violation of performance bounds during loop-shaping may result in violation of the performance specifications for the closed-loop system.

As seen in Figure 44.22, a robust design has been achieved for this 2×2 MIMO analog flight control system. The time domain results, although not drawn, meet all specifications.

References

[1] D'Azzo, J.J., and Houpis, C.H., *Linear Control System Analysis and Design*, 4th Ed., McGraw-Hill, New York, 1995.

[2] Houpis, C. H., Quantitative Feedback Theory (QFT) For the Engineer: A Paradigm for the Design of Control Systems for Uncertain Nonlinear Plants, WL-TR-95-3061, AF Wright Aeronautical Laboratory, Wright-Patterson AFB, OH, 1987. (Available from National Technical Information Service, 5285 Port Royal Road, Springfield, VA 22151, document number AD-A297574.)

[3] Horowitz, I. M. and Sidi, M., Synthesis of feedback systems with large plant ignorance for prescribed time domain tolerances, *Int. J. Control,* 16, 287–309, 1972.

[4] Horowitz, I. M. and Loecher, C., Design of a 3 × 3 multivariable feedback system with large plant uncertainty, *Int. J. Control,* 33, 677–699, 1981.

[5] Horowitz, I. M., Optimum loop transfer function in single-loop minimum phase feedback systems, *Int. J. Control,* 22, 97–113, 1973.

[6] Horowitz, I. M., Synthesis of feedback systems with non-linear time uncertain plants to satisfy quantitative performance specifications, *IEEE Proc.,* 64, 123–130, 1976.

[7] Horowitz, I. M., Quantitative synthesis of uncertain multiple-input multiple-output feedback systems, *Int. J. Control,* 30, 81–106, 1979.

[8] Thompson, D. F. and Nwokah, O.D.I., Optimal loop synthesis in quantitative feedback theory, Proc. Am. Control Conf., San Diego, CA, 1990, 626–631.

[9] Houpis, C. H. and Chandler, P.R., Eds., Quantitative Feedback Theory Symposium Proceedings, WL-TR-92-3063, Wright Laboratories, Wright-Patterson AFB, OH, 1992.

[10] Keating, M. S., Pachter, M., and Houpis, C.H., Damaged aircraft control system design using QFT, *Proc. Nat. Aerosp. Electron. Conf.* (NAECON), Vol. 1, Ohio, May 1994, 621–628.

[11] Reynolds, O.R., Pachter, M., and Houpis, C.H., Design of a subsonic flight control system for the Vista F-16 using quantitative feedback theory, *Proc. Am. Control Conf.,* 1994, 350–354.

[12] Trosen, D. W., Pachter, M., and Houpis, C.H., Formation flight control automation, *Proc. Am. Inst. Aeronaut. Astronaut.* (AIAA) Conf., Scottsdale, AZ, 1994, 1379–1404.

[13] Houpis, C. H. and Lamont, G., *Digital Control Systems: Theory, Hardware, Software,* 2nd ed., McGraw-Hill, NY, 1992.

[14] Horowitz, I.M. and Liao, Y.K., Quantitative feedback design for sampled-data system, *Int. J. Control,* 44, 665–675, 1986.

[15] Sating, R.R., Horowitz, I.M., and Houpis, C.H., Development of a MIMO QFT CAD Package (Version 2), paper presented at the 1993 Am. Control Conf., Air Force Institute of Technolgy, Wright-Patterson AFB, Ohio.

[16] Boyum, K. E., Pachter, M., and Houpis, C.H., High angle of attack velocity rolls, *13th IFAC Symp. Autom. Control Aerosp.,* 51–57, Palo Alto, CA, September 1994.

[17] Schneider, D. L., QFT Digital Flight Control Design as Applied to the AFTI/F-16, M.S. thesis, AFIT/GE/ENG/86D-4, School of Engineering, Air Force Institute of Technolgy, Ohio, 1986.

[18] Arnold, P. B., Horowitz, I.M., and Houpis, C.H., YF-16CCV Flight Control System Reconfiguration Design Using Quantitative Feedback Theory, *Proc. Nat. Aerosp. Electron. Conf.* (NAECON), Vol. 1, 1985, 578–585.

44.6 Appendix A

CAD flowchart for MISO QFT design.

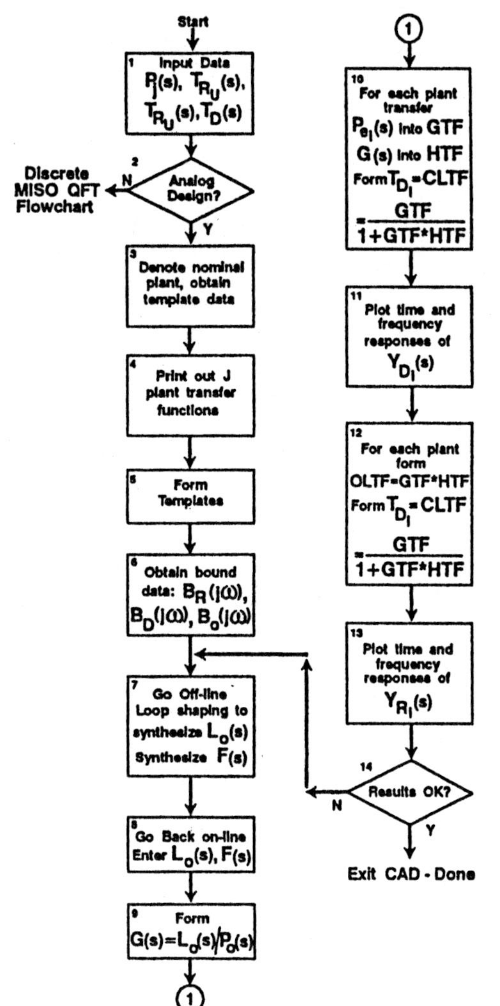

44.7 Appendix B

MIMO QFT flowchart for analog and discrete control systems [15].

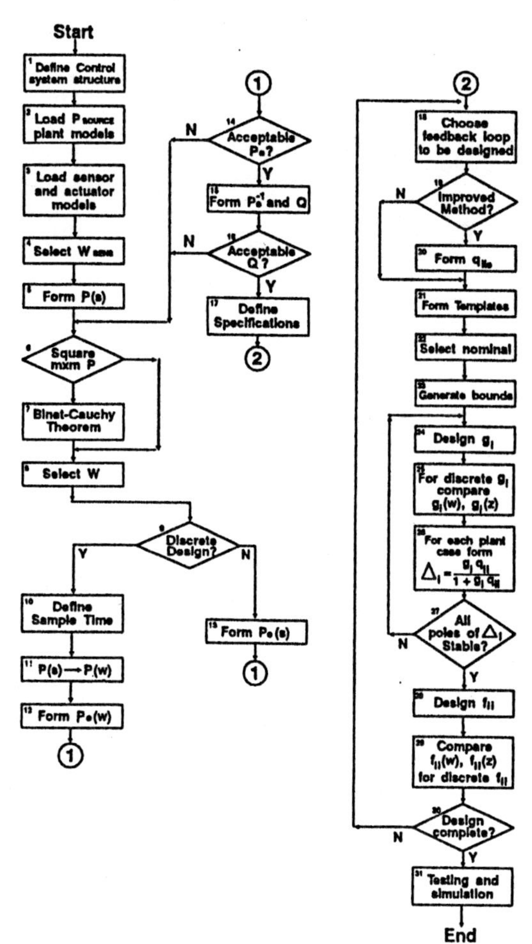

45

The Inverse Nyquist Array and Characteristic Locus Design Methods

Neil Munro
UMIST, Manchester, England

John M. Edmunds
UMIST, Manchester, England

45.1 Introduction

During the decade 1970–1980, the well established frequency-domain approach of Nyquist was extended [9], [19] to deal with multi-input, multi-output systems, where significant interaction is present. Rosenbrock created his Inverse (and Direct) Nyquist Array design methods and MacFarlane created his Characteristic Locus design method. In these approaches, the uncontrolled system is described by an $m \times m$ transfer-function matrix $G(s)$, and a forward-path controller matrix $K(s)$ and prefilter $P(s)$ are to be designed so that, when the feedback loops are closed, as in Figure 45.1, the resulting system

$$H(s) = [I + G(s)K(s)]^{-1}G(s)K(s)P(s), \tag{45.1}$$

is stable, has satisfactory disturbance rejection properties, and good tracking properties.

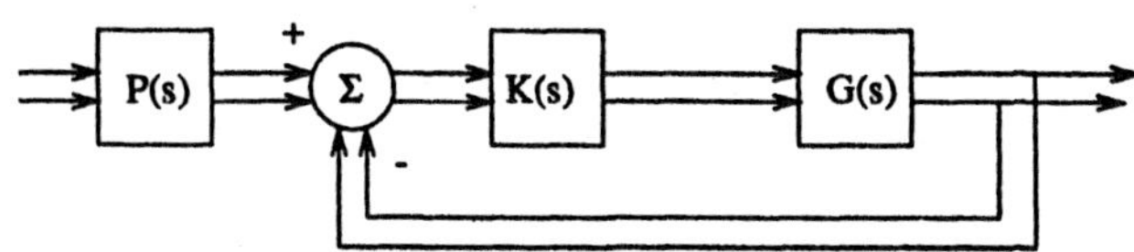

Figure 45.1 A multivariable feedback system.

In these two approaches, the desired compensator $K(s)$ is achieved by two different processes, as described later in this article. However, since the original development of these tools, other researchers have examined the problems of choosing the best scaling for the system's input and output variables [5], [12], [13], [14] and the systematic decomposition of the original system into simpler design problems [3], where possible. Current research is now focused on the establishment of robust versions of these techniques for systems with uncertain parameters [2], [16].

45.2 The Design Techniques

45.2.1 Design Preliminaries

The initial stages of the frequency-response design methods are to decide the purpose of the control, normalize the input and output units, and then split the design into smaller problems, if possible. The required compensation can be calculated for each of the resulting problems by using either the Nyquist Array or the Characteristic Locus design approaches.

Design Aims. If possible these aims should be quantified to enable the evaluation of the success or failure of the design. Multivariable frequency-response design aims often include the following:

1. To ensure a suitable input-output response; in particular, to ensure that the output tracks the reference input in the required manner.
2. To stabilize unstable systems.
3. To make the system insensitive to noise disturbances.
4. To make the system insensitive to small parameter variations.
5. To insure robustness for larger parameter variations.
6. To allow independent control of the different inputs and outputs by removing interaction.

Input and Output Scaling. For multivariable systems with different types of outputs, the choice of units for the input

0-8493-8570-9/96/$0.00+$.50

and output variables have significant effects. Although the choice of output units does not affect the system, it has an obvious effect on the numbers representing the interactions [11]. Because the multivariable design methods tend to balance these interactions, it is important to choose the right output units. The best choice will depend on the proposed use of the model. On some occasions, there is sufficient specialized knowledge about the system and the use for the model to permit this choice. However, there are occasions when this knowledge is not available, in which case a scaling algorithm has to be used to balance the units.

One algorithm for this scaling attempts to give a unity sum for the magnitudes of the elements in each row and column of the transfer function matrix $G(s)$. This insures that the response to each input has a similar magnitude, and that each of the outputs is stimulated by a similar amount. Define the row and column dominance ratios as the absolute value of the diagonal element divided by the sum of the absolute values of the elements in the corresponding row or column. After the scaling algorithm has been applied, the row and column dominance ratios (see Equations 45.2 and 45.3) for the loops will be equal to the diagonal elements of the matrix and so will be balanced. Moreover, because the algorithm sets all the row and column sums to unity, any reordering of the inputs and outputs will lead to new dominance ratios, which are equal to the new diagonal elements. Hence, the row and column dominance will be balanced for any pairing of inputs and outputs. This algorithm will give the same scaling factors independently of the order of the inputs and outputs. In contrast, the Perron-Frobenius eigenvector approach [12] will insure that all the row dominance ratios are equal and that all the column dominance ratios are equal, but it will produce different scalings for different orders of inputs and outputs, because it aims at just maximizing the minimum dominance ratio for a particular order.

Problem Decomposition. It is advantageous to split large problems into several smaller ones if this can be done, because the complexity of control increases considerably as the number of loops increases. There are two main ways of reducing the system to smaller problems. Some loops are orders of magnitude faster than others, in which case they should be closed first, because the multivariable design methods tend to give similar bandwidths on all loops. Some subsystems may be almost independent of others, in which case they should be controlled separately. A test for independence is whether the system has block diagonal dominance [3]. First the inputs and outputs are rearranged to make the system block diagonal dominant with as many almost independent subsystems as possible. Then, if the dominance ratio is about 10:1 for the matrix of gains of the subsystems, the separate subsystems are considered independently. When testing for this interaction, it is those frequencies near the desired bandwidth that are significant, because, at low frequencies, the designed high loop gains will remove interaction problems, and, at very high frequencies, nothing much will come through the system.

For large systems it is not practical to try all the $m!$ possible output orders. One suitable reordering algorithm first moves the most dominant element to the diagonal of the matrix fixing the first input-output pair, and then successively moves the most dominant remaining element to the diagonal. Having chosen the pairing of the inputs and outputs, there now remain $m!$ possible orders for these pairs. To make the matrix as block diagonally dominant as possible, an order is chosen which puts related loops near each other. One approach is to use the most dominant loop as the first loop, and then, successively add the loops, which are most highly coupled, to the previous loops. If it is possible to get diagonal dominance just by scaling and reordering, this will lead to a more reliable control scheme, because each loop will depend on just one sensor and just one actuator. In particular, if it is possible to get block diagonal dominance with good dominance margins by scaling and reordering, then decentralized control can be used.

45.2.2 The Inverse Nyquist Array Design Method

Rosenbrock [19] has shown that the stability and performance of a multivariable system, which is diagonal dominant, can be inferred directly from the stability and performance of each resulting loop. The frequency-domain design method proposed consists of determining a multivariable precompensator matrix K_p, which is as simple as possible (preferably a matrix of gain terms) so that the resulting inverse system $\hat{Q}(s) = [G(s)K_p]^{-1}$ is almost decoupled or 'diagonal dominant'. When this condition has been achieved, a diagonal matrix $K_d(s)$ can be used to implement single-loop compensators to meet the overall design specifications.

A rational $m \times m$ matrix $\hat{Q}(s)$ is row diagonal dominant on some closed contour D in the complex plane if

$$|\hat{q}_{ii}(s)| > \sum_{\substack{j=1\\ j\neq i}}^{m} |\hat{q}_{ij}(s)| \tag{45.2}$$

for $i = 1, \ldots, m$ and all s on D. Column diagonal dominance is defined similarly. Here, D is the usual Nyquist contour, i.e., the $j\omega$-axis followed by a semicircle of infinite radius.

The dominance of a rational matrix $\hat{Q}(s)$ can be determined by a simple graphical construction. Let $\hat{q}_{ii}(s)$ map D into $\hat{\Gamma}_i$ as in Figure 45.2. This will look like an inverse Nyquist plot, but does not represent anything directly measurable on the physical system. For each s on D, draw a circle of radius

$$d_i(s) = \sum_{\substack{j=1\\ j\neq i}}^{m} |\hat{q}_{ij}(s)| \tag{45.3}$$

centered on the appropriate point of $\hat{q}_{ii}(s)$, as in Figure 45.2. Do the same for the other diagonal elements of $\hat{Q}(s)$. If each of the bands so produced excludes the origin, for $i = 1, \ldots, m$, then $\hat{Q}(s)$ is row dominant on D. A similar test for column dominance can be defined.

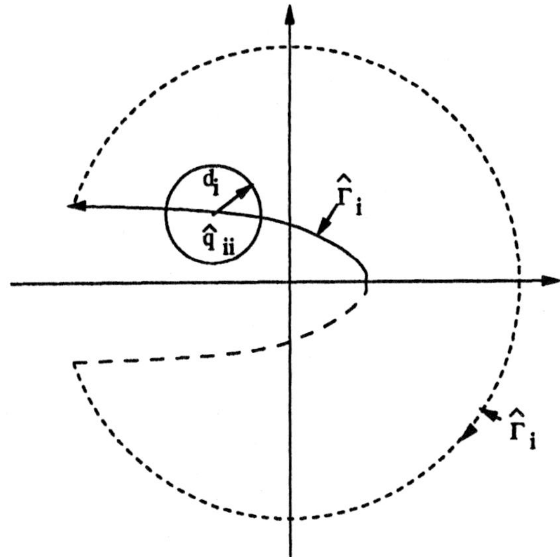

Figure 45.2 Complex plane mapping.

If for each diagonal element $\hat{q}_{ii}(s)$, the band swept out by its circles does not touch the segment of the negative real axis between the origin and the point $-f_i$, where f_i represents the feedback gain chosen for loop i, then the generalized form of the inverse Nyquist stability criterion, defined below, is satisfied.

THEOREM 45.1 *If $\hat{Q}$ and $\hat{H}$, where $\hat{H} = H^{-1} = \hat{Q} + I$, are dominant on the Nyquist D contour, then, because we are considering inverse polar plots, stability is determined using*

$$\hat{N}_H - \hat{N}_Q = -p_o + p_c \tag{45.4}$$

where $\hat{N}_H$ is the total number of clockwise encirclements of the critical points $-f_i$ by the $\hat{q}_{ii}(s)$, $\hat{N}_Q$ is the total number of clockwise encirclements of the origin by the $\hat{q}_{ii}(s)$, as s traverses the Nyquist D contour clockwise, p_o is the number of right half plane zeros of the open-loop system characteristic polynomial, and p_c is the number of right half plane zeros of the closed-loop system characteristic polynomial, where $p_c = 0$ for closed-loop stability.

In general, the $\hat{q}_{ii}(s)$ do not represent anything directly measurable on the system. However, using a theorem due to Ostrowski [17],

$$h_i^{-1}(s) = h_{ii}^{-1}(s) - f_i \tag{45.5}$$

is contained within the band swept out by the circles centered on $\hat{q}_{ii}(s)$, and this remains true for all intermediate values of gain f_j in each other loop j between zero and f_j. Note that $h_{ii}^{-1}(s)$ is the inverse transfer function seen between input i and output i with all loops closed. The transfer function $h_i(s)$ is that seen in the ith loop when this is open, but the other loops are closed. It is this transfer function for which we must design a single-loop controller for the ith loop. The theorems above tell us that as the gain in each other loop changes, $h_{ii}^{-1}(s)$ will also change but always remains inside a certain band. The band in which $h_i^{-1}(s)$ lies can be further narrowed. If $\hat{Q}$ and $\hat{H}$ are dominant, and if

$$\phi_i(s) = \max_{\substack{j \\ j \neq i}} \frac{d_j(s)}{|f_j + \hat{q}_{jj}(s)|}, \tag{45.6}$$

then $h_i^{-1}(s)$ lies within a band based on $\hat{q}_{ii}(s)$ and defined by circles of radius

$$r_i(s) = \phi_i(s) d_i(s) \tag{45.7}$$

Thus, once the closed-loop system gains have been chosen so that stability is achieved in terms of the larger Gershgorin bands, then a measure of the gain margin for each loop can be determined by drawing the smaller Ostrowski bands, using the 'shrinking factors' $\phi_i(s)$ defined by Equation 45.6, with circles of radius r_i. These narrower bands also reduce the region of uncertainty as to the actual location of the inverse transfer function $h_{ii}^{-1}(s)$ for each loop.

Using the ideas developed above, the design is carried out using the inverse transfer function matrix. We are trying to determine an inverse precompensator $\hat{K}_p(s)$ so that $\hat{Q}(s) = \hat{K}_p(s)\hat{G}(s)$ is diagonal dominant. One method of determining $\hat{K}_p(s)$ is to build up the required matrix out of elementary row operations using a graphical display of all of the elements of $\hat{Q}(s)$ as a guide. This approach has proven successful in practice and has, in most cases considered to date, resulted in $K_p(s)$ being a simple matrix of real constants which can be readily realized. Another useful approach is to choose $\hat{K}_p = G(0)$, if $|G(0)|$ is nonsingular. Here again $\hat{K}_p(s)$ is a matrix of real constants which simply diagonalizes the plant at zero frequency.

A further approach, perhaps more systematic than those mentioned above, is to determine $\hat{K}_p$ as the 'best', in a least-mean-squares sense, wholly real matrix which most nearly diagonalizes the system $\hat{Q}$ at some frequency $s = j\omega$ [7], [19]. This choice of $\hat{K}_p$, can be considered as the best matrix of real constants which makes the sum of the moduli of the off-diagonal elements in each row of $\hat{Q}$ as small as possible compared with the modulus of the diagonal element at some frequency $s = j\omega$. We may choose a different frequency ω for each row of $\hat{K}$. It is also possible to pseudodiagonalize each row of $\hat{Q}$ at a weighted sum of frequencies and the formulation of this latter problem again results in an eigenvalue/eigenvector problem [19]. However, although this form of pseudodiagonalization frequently produces useful results, the constraint that the control vector $\hat{K}_j$ should have unit norm does not prevent the diagonal term $\hat{q}_{jj}$ from becoming very small, although the row is diagonal dominant, or from vanishing altogether. So, if instead of this constraint, we substitute the alternative constraint that $|\hat{q}_{jj}(j\omega)| = 1$, then a similar analysis leads to a generalized eigenvalue problem, which must be solved using the appropriate numerical method.

Since the introduction of the Nyquist Array design methods in the early 1970s, several further ways of defining diagonal dominance have been the subject of other research. These are summarized briefly below, along with the original definition:

(i) row/column dominance $\{d_R, d_C\}$	Rosenbrock, 1974.
(ii) generalized dominance $\{d_G\}$	Limebeer, 1982.
(iii) M-matrix based dominance $\{d_M\}$	Araki and Nwokah, 1975
(iv) fundamental dominance $\{d_F\}$	Yeung and Bryant, 1992.
(v) L-dominance $\{d_L\}$	Yeung and Bryant, 1992.

The size of the dominance circles produced by each of these criteria can be ordered as follows:

$$d_L < \{d_M = d_F\} < d_G < \min\{d_R, d_C\}.$$

From a design point of view, the generalized, the fundamental and the M-matrix based dominance conditions are most useful as analysis tools. The original definition of dominance, although yielding the most conservative dominance test, nevertheless provides the benefit of independent loop design. The L-dominance criterion is based on the use of a sequential loop design approach, and requires an LU decomposition of the return-difference matrix of the system. However, many industrial multivariable control problems have now been solved using this design method with Rosenbrock's definition of diagonal dominance. A later section presents an example where the required multivariable precompensator is determined using only elementary row operations.

45.2.3 The Characteristic Locus Method

The characteristic gains are the eigenvalues $g_1, \ldots, g_m$ of the frequency response matrix $G(s)$. Characteristic loci are multivariable extensions of Nyquist diagrams formed by plotting the characteristic gains for frequencies around the Nyquist D contour. As with the single-input Nyquist diagrams, the resulting loci separate the images of the right- and left-half frequency planes and so can be used for predicting the number of closed-loop unstable poles for different feedback gains. For linear systems, the loci give an exact stability criterion. A multivariable system will be closed-loop stable with a feedback gain I/g, where I is the unit matrix of order m, if the number of encirclements of the critical point, g, in an anticlockwise direction, by the characteristic loci $g_i(s)$, as s travels round the Nyquist D contour in a clockwise direction, is equal to the number of open-loop unstable poles p_0.

A characteristic loci plot for a compensated two-input/two-output chemical reactor with two unstable poles, showing stability for negative feedback gains greater than 0.09, is shown in Figure 45.3. The compensated state-space matrices are

$$A = \begin{bmatrix} 1.4000 & -0.2077 & 6.7150 & -5.6760 \\ -0.5814 & -4.2900 & 0 & 0.6750 \\ 1.0670 & 4.2730 & -6.6540 & 5.8930 \\ 0.0480 & 4.2730 & 1.3430 & -2.1040 \end{bmatrix} \quad (45.8)$$

$$B = \begin{bmatrix} 0 & 0 \\ -0.0969 & 30.3952 \\ 30.8515 & 6.2564 \\ -0.0194 & 6.0801 \end{bmatrix} \quad (45.9)$$

$$C = \begin{bmatrix} 1 & 0 & 1 & -1 \\ 0 & 1 & 0 & 0 \end{bmatrix} \quad (45.10)$$

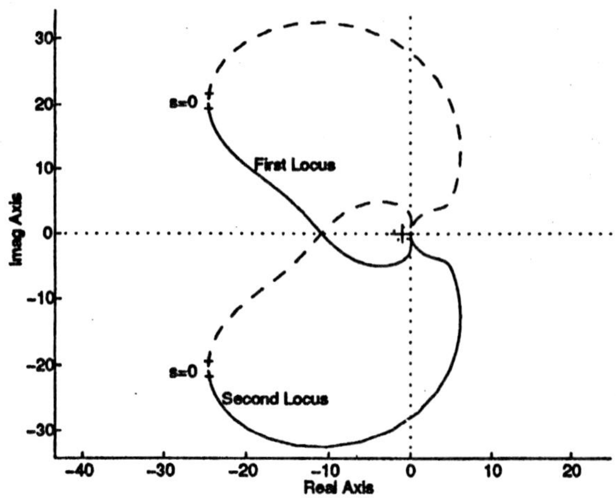

Figure 45.3 Compensated characteristic loci for a two-input/two-output chemical reactor.

A characteristic locus design starts by compensating the high frequencies to remove high frequency interaction, proceeds to the intermediate frequencies, and finishes with the low frequencies. By considering the compensators in this order, they can be designed more independently of each other. The method is based on the ways that interaction can be reduced in a multivariable system. In absolute terms, the frequencies, which are considered to be high, depend on the type of system being controlled, because some types of systems, such as turbines, act a lot faster than others, such as distillation columns. So, high frequencies are defined as those greater than about half the required closed-loop cut-off frequency w_c. Similarly, low frequencies are defined as those below about $0.1w_c$. The frequencies between the low and high frequencies will be called intermediate frequencies.

By definition, at high frequencies the gain of the system is not large. The only way that the interaction can be made small is by making the system approximately diagonal. If the relative gains of the various elements of the transfer function change at high frequencies, a similar speed of response will be required on each of the multivariable outputs in order to maintain the diagonal dominance of the system over the full range of high frequencies. Several algorithms have been suggested to attempt to decrease the high frequency interaction. Some of these require a computer for the calculations but some can be done by hand [4]. The original algorithm proposed for removing the high frequency interaction was called the ALIGN algorithm [10], [9]. This is a way of calculating a real matrix, K_h, which is approximately the inverse of a complex matrix, and is applied at the desired bandwidth frequency. The elements of K_h are chosen to minimize the sum of the squares of the elements of an error matrix E where $J = G(jw_c)K_h + E$. Here, J is a diagonal matrix with all of its diagonal elements having unit modulus.

The required dynamic compensation $k_i(s)$, for each of the characteristic loci g_i of a system with high frequency compensation, is chosen by considering each locus as a single-input, single-output system and by using the Bode or Nyquist plots to design the loop phase and gain functions, with the restriction that each of these compensators $k_i(s)$ approaches unity at high frequencies. This requirement insures that the high frequency compensation previously applied is not disturbed. Once the individual required compensators have been designed, they cannot be implemented directly because the individual characteristic loci do not correspond to particular input-output pairs. The approximate commutative controller [10] approximates the eigenvector structure to allow the compensation to be applied to the correct loci. The resulting compensator is $K_m(s) = A\Lambda_k(s)B$, where the ALIGN algorithm is used to find real approximations A of the eigenvector matrix, and B of the inverse eigenvector matrix. $\Lambda_k(s)$ is the diagonal matrix of the compensators $k_i(s)$. This compensator works best if the required compensation for the locus is not too different. Integral action can also be added at this stage. More than one approximate commutative controller may be required, if the intermediate frequency compensation extends over a large frequency range.

Integral action can be introduced if there is a requirement for not having steady-state errors. It removes steady state-errors by putting an infinite gain in the feedback loop at zero frequency. The integral action may be introduced when using an approximate commutative controller to do the intermediate frequency compensation, if a lag compensator was required. It can also be introduced in the form, $K_I(s) = aK_I/s + I$, where the matrix K_I balances the low frequency gains, and the 'a' is a tuning coefficient to control the frequency range over which the integral action is to extend. The complete control scheme is then $[K_h A\Lambda_k(s)BK_I]u(s)$.

45.3 Applications

The system to be considered is an automotive gas turbine, shown schematically in Figure 45.4. The two system outputs to be controlled are the gas generator speed (y_1) and the interturbine temperature (y_2), indicated by a thermocouple. The two inputs to the system are fuel pump excitation (u_1) in mAmps and nozzle actuator excitation (u_2) in volts. A locally linearized model of the system, obtained at 80% gas generator speed (and 85% power turbine speed), is given by the transfer function matrix,

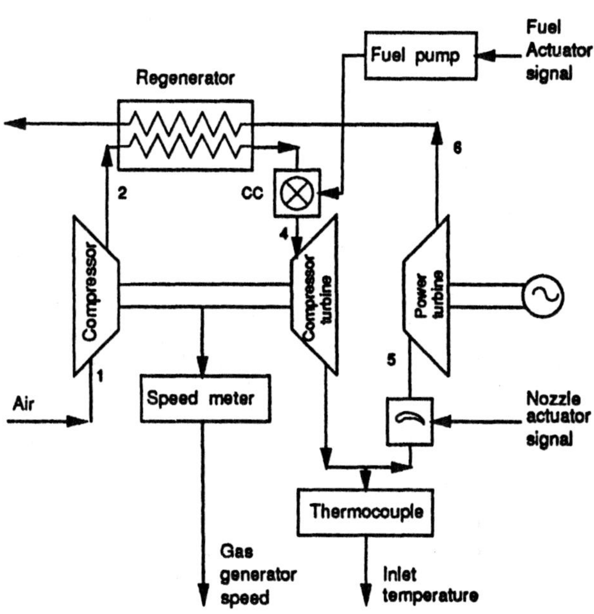

Figure 45.4 Schematic of an automotive gas turbine.

$$G(s) = \begin{bmatrix} \frac{.806s+.264}{s^2+1.15s+.202} & \frac{-(15.0s+1.42)}{s^3+12.8s^2+13.6s+2.36} \\ \frac{1.95s^2+2.12s+.490}{s^3+9.15s^2+9.39s+1.62} & \frac{7.14s^2+25.8s+9.35}{s^4+20.8s^3+116.4s^2+111.6s+18.8} \end{bmatrix} \tag{45.11}$$

This transfer function matrix $G(s)$ includes the characteristics of the fuel system and nozzle actuators and the thermocouple used to measure the interturbine temperature. The time constant associated with the speed transducer indicating the gas generator speed is very fast with respect to the dynamics of the gas generator and has been ignored. Further details of the nonlinear model of the gas turbine are given by Winterbone, Munro and Lourtie [21].

The Inverse Nyquist Array Approach. Figure 45.5 shows the inverse Nyquist array for this $G(s)$, i.e., $\hat{G}(j\omega) = G^{-1}(j\omega)$, for ω varying from 0 to 25 radians/second. This diagram depicts, in inverse polar form, the frequency behavior of the system without any form of controller. An inspection of Figure 45.5 reveals that the uncompensated system is not 'diagonal dominant'.

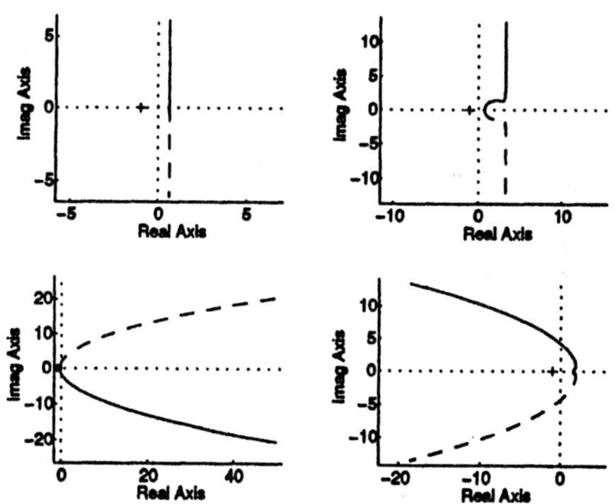

Figure 45.5 Inverse Nyquist array of the uncompensated system with Gershgorin bands superimposed.

However, the compensated system $\hat{Q}(s) = \hat{K}(s)\hat{G}(s)$ is diagonal dominant for

$$K(s) = \begin{bmatrix} 0.361 & 0.450 \\ -1.130 & 1.00 \end{bmatrix} \tag{45.12}$$

i.e.,

$$\hat{K}(s) = \begin{bmatrix} 1.15 & -0.5175 \\ 1.30 & 0.415 \end{bmatrix} \tag{45.13}$$

where $\hat{K}$ is built up as the following series of elementary row operations:

$$\begin{aligned} \text{Row 1} &= \text{Row 1} - 0.45\,\text{Row 2} \\ \text{Row 2} &= \text{Row 2} + 1.30\,\text{Row 1} \\ \text{Row 1} &= 1.15\,\text{Row 1} \end{aligned}$$

The inverse Nyquist array for $\hat{Q}(j\omega) = \hat{K}\hat{G}(j\omega)$ is shown in Figure 45.6, for the same frequency range, with the column

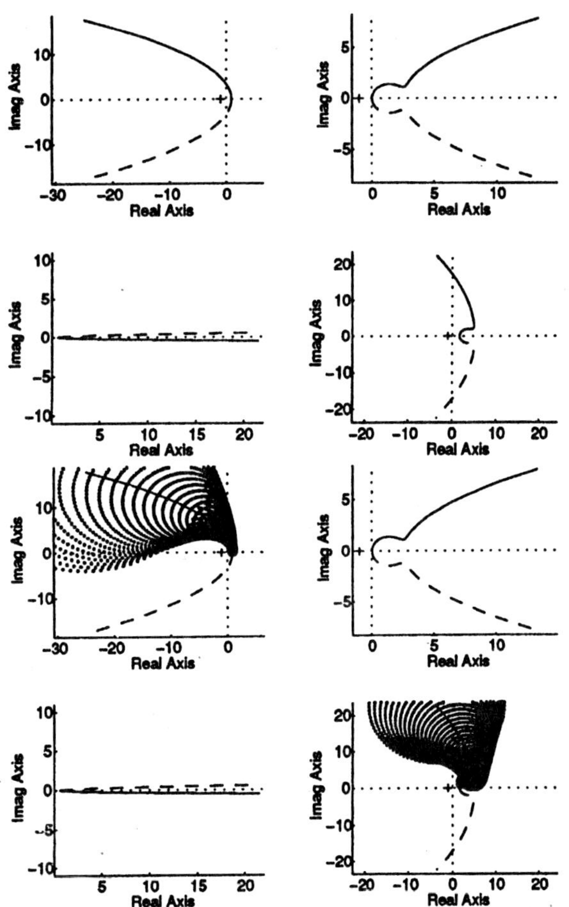

Figure 45.6 Inverse Nyquist array of a compensated system with Gershgorin bands superimposed.

dominance circles superimposed on the diagonal elements. Since the Gershgorin bands do not include the origins, the compensated system $\hat{Q}(s)$ is (column) diagonal dominant. Further, the multivariable system shown is now sufficiently decoupled to allow applying classical control compensation techniques to each of the two prime control loops, as required. For feedback gains $f_1 = 5$, $f_2 = 5$, the closed-loop system is Nyquist stable with respect to the Gershgorin bands, which in fact contain the elements $q_{ii}^{-1}(s)$.

However, in general, $\hat{q}_{ii} \neq q_{ii}^{-1}$. Now, using the theorem due to Ostrowski, the size of the Gershgorin bands can be reduced by a shrinking factor $\phi(s)$, defined by Equation 45.6. The inverse Nyquist array of $\hat{Q}$ with the Ostrowski bands is shown in Figure 45.7 for $f_1 = 5$ and $f_2 = 5$.

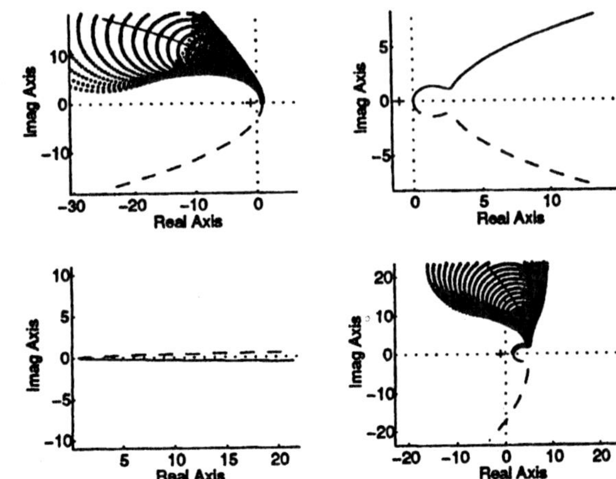

Figure 45.7 Ostrowski bands superimposed on Figure 45.6.

The time domain behavior of the resulting closed-loop system was examined and was found considerably better than that of the open-loop system. Integral action was introduced into both feedback loops to remove steady-state error, and the final compensator was set to

$$K^* = \operatorname{diag} 5(s+2)/s,\ 10(s+2)/sK. \tag{45.14}$$

The open-loop system step responses, shown for a small step change in fuel pump excitation in Figure 45.8, clearly show the

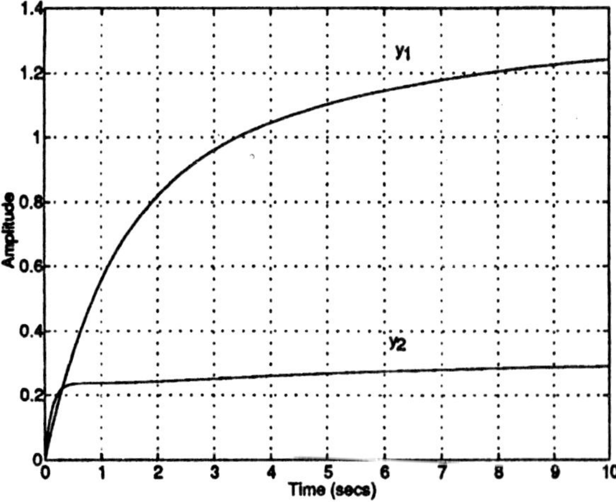

Figure 45.8 Time response of the open-loop system to a step change in u_1.

interaction that exists between the interturbine temperature output and fuel input in the uncontrolled system. Figure 45.9 shows the response of the two system outputs for a small step change in nozzle actuator excitation, and again the interaction between the gas generator speed output and the nozzle area input are also clearly shown.

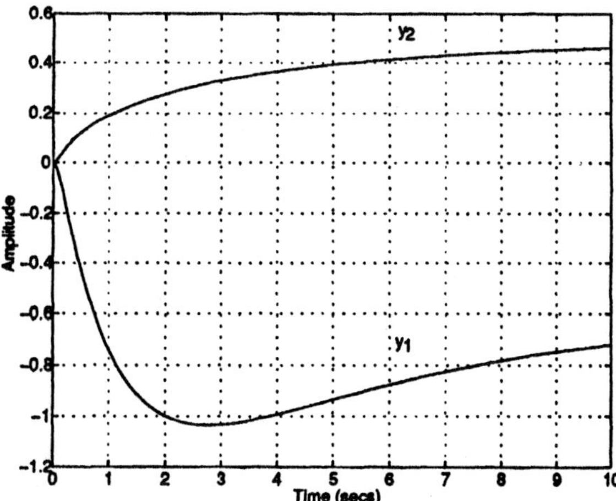

Figure 45.9 Time response of the open-loop system to a step change in u_2.

The response of gas generator speed (y_1) and inter turbine (thermocouple indicated) temperature (y_2) for a step change in the speed reference input to the closed-loop system is shown in Figure 10. The requested change in gas generator speed has been

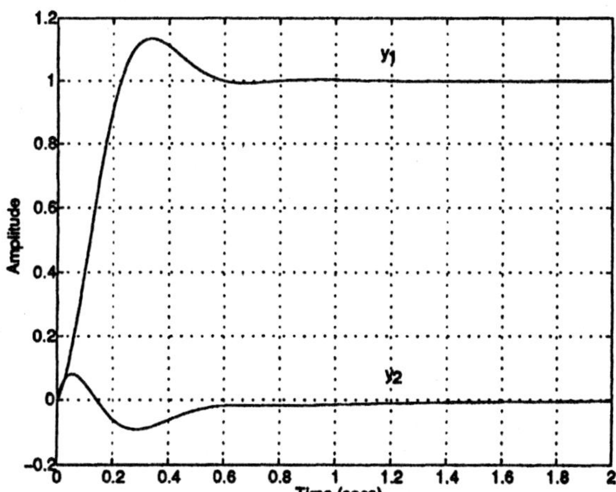

Figure 45.10 Time response of the closed-loop system to a step in r_1.

achieved with only a small transient variation in interturbine temperature. The gas generator response is second order with a small overshoot. This correlates well with the response predicted by element $\hat{q}_{11}(j\omega)$. Also, because the requested change in this reference input is relatively small, no saturation effects were observed on either of the actuator signals. The responses of the two closed-loop system outputs for a step change in the interturbine temperature reference input are shown in Figure 45.11. These results are again acceptable, because the temperature response characteristic is faster and the interaction (i.e., gas-generator speed variation) is reduced considerably compared with the open-loop uncompensated case. The correlation is again good between element $\hat{q}_{22}(j\omega)$ in the time and frequency domains. From the observations made above, the responses of the controlled system are satisfactory and the interaction observed in the uncontrolled

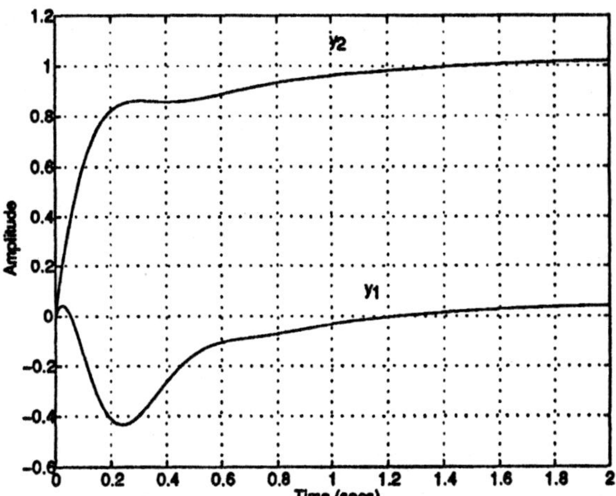

Figure 45.11 Time response of the closed-loop system to a step in r_2.

system has been greatly reduced. The multivariable controller required to obtain these results consists of a proportional action precompensator K_p and two single-loop proportional-plus-integral action controllers. However, this controller does not contain any limits on fuel input rate or surge limitations with respect to the compressor. Nevertheless, responses obtained from the full nonlinear model were very similar to those obtained from the linear model for small input perturbations.

45.4 The Characteristic Locus Approach

The aim of this example is to illustrate the Characteristic Locus design method, using an automotive gas turbine. Because the gains have already been balanced, the characteristic loci are drawn to start the design process (Figure 45.12). The design is done here with a System Toolbox [6], written at UMIST. The characteristic loci do not have to start on the real axis because they are plots of the eigenvalues of the frequency response matrix. This frequency response will be real at zero frequency but the eigenvalues of a real matrix do not have to be real. The frequency dependent nature of the response can be seen better from the Bode plots of the characteristic gains, the $g_i(j\omega)$.

The ALIGN algorithm calculates a real constant matrix K_h to make the product $G(s)K_h$ as nearly diagonal as possible with the resulting diagonal elements having approximately unit modulus. The required bandwidth is approximately 10 rad/sec so that the ALIGN algorithm is applied to $g(j12)$. This will also improve diagonal dominance at this frequency in order to minimize interaction. This compensator

$$K_h = \begin{pmatrix} 1.6112 & 6.1459 \\ -11.1189 & 4.2511 \end{pmatrix} \tag{45.15}$$

is now added as a precompensator and the resulting Bode plots of the characteristic gains are drawn (Figure 45.13).

The characteristic gains have approximately unit magnitude at the frequency used for the ALIGN algorithm, hence this will become the closed-loop bandwidth. In order to minimize the

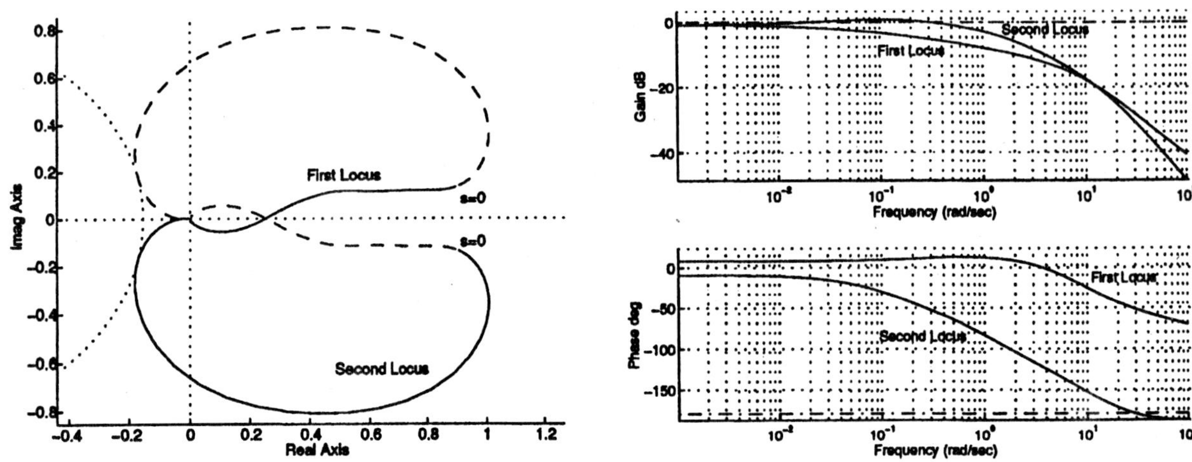

Figure 45.12 Nyquist and Bode plots of the uncompensated characteristic gains.

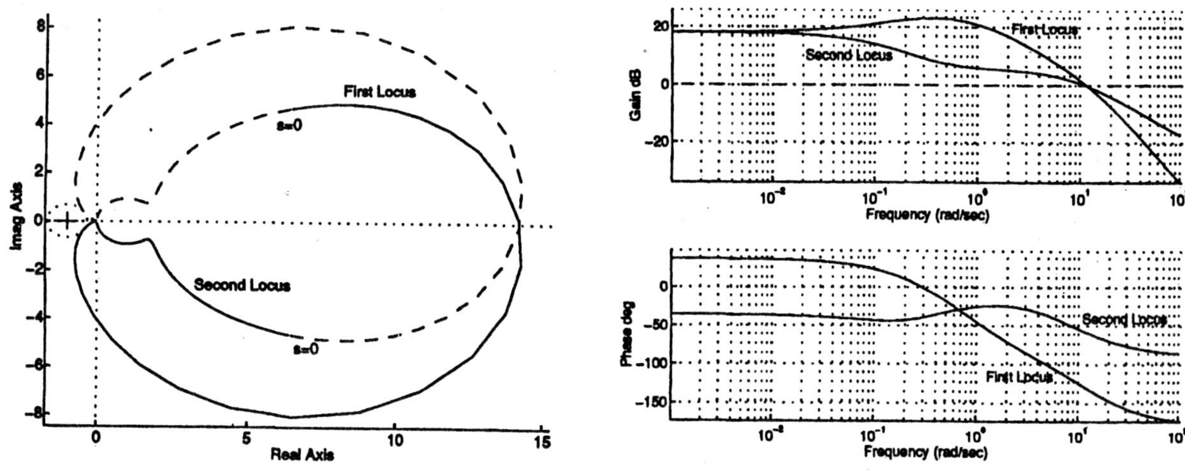

Figure 45.13 Nyquist and Bode plots of the characteristic gains with K_h added.

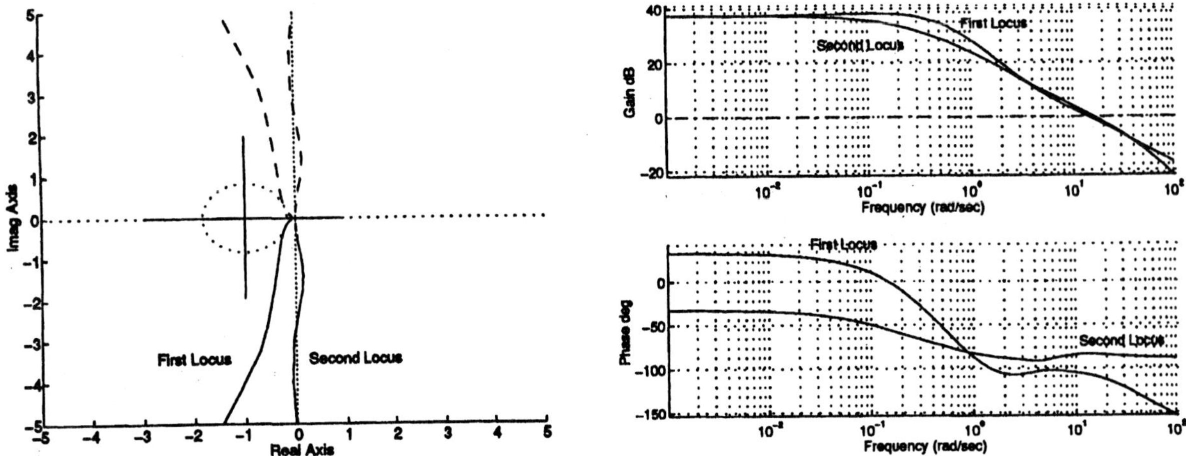

Figure 45.14 Nyquist and Bode plots of characteristic Loci with dynamic compensation.

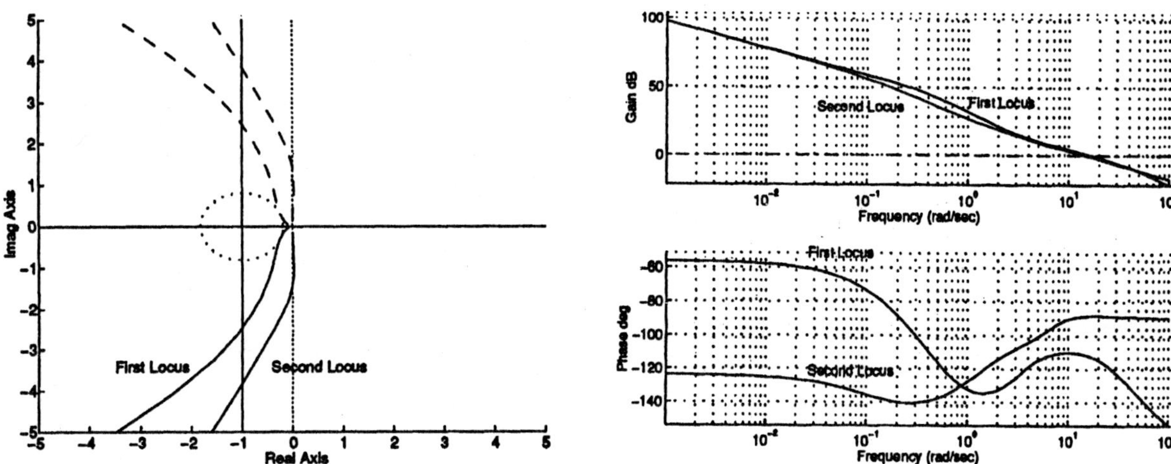

Figure 45.15 Nyquist and Bode plots of Characteristic Loci with integral compensation.

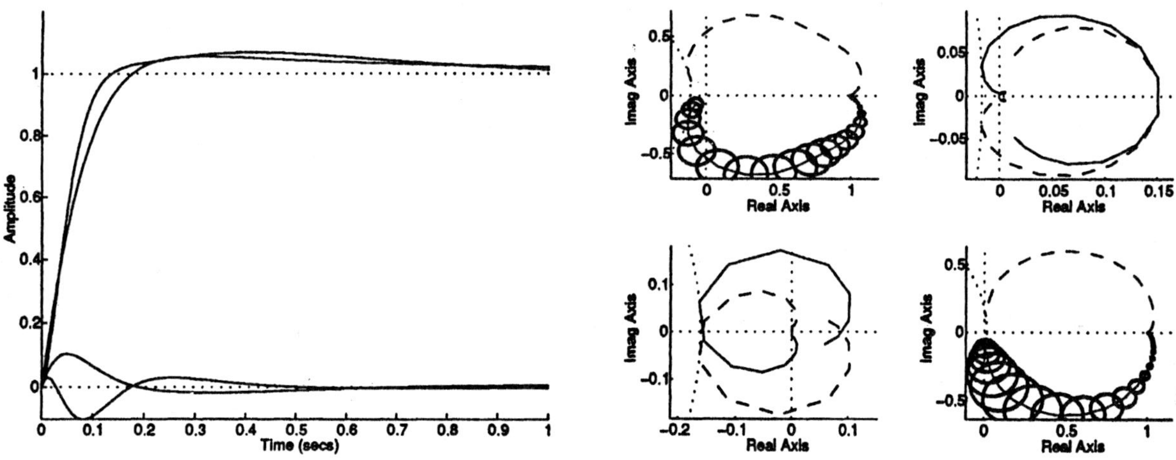

Figure 45.16 Closed-loop step responses and closed-loop Direct Nyquist Array of H(s).

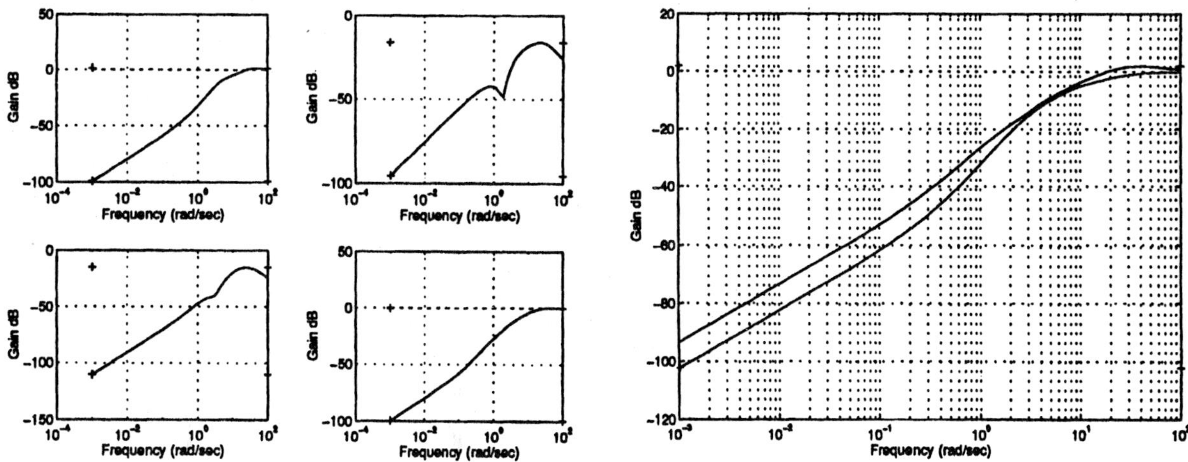

Figure 45.17 Bode plots of $S(s)$ and its singular values.

sensitivity of the system to parameter variations and noise, we need to increase the feedback loop gains as fast as possible for decreasing frequencies, while maintaining phase margins of more than 60° to insure good stability margins. The first locus needs a small increase of phase margin so that the compensation for this loop is a lag-lead compensator. The second locus has a large phase margin, so that a simple lag is used for this locus. The resulting dynamic compensation is

$$K_m(s) = \begin{bmatrix} \frac{5.5(s+5)(s+5)}{(s+0.5)(s+50)} & 0 \\ 0 & \frac{(s+9)}{(s+0.5)} \end{bmatrix} \quad (45.16)$$

This has to be applied to the system using an approximate commutative controller calculated at a frequency of 3 radians/sec, because we are compensating the characteristic loci which are not particularly associated with either input. The frequency of 3 radians/sec was chosen because the biggest differences in the control action occur at about this frequency. The resulting compensated system is $G(s)K_h A\Lambda_k(s)B$. Because $G(s)K_h = U * \Lambda * U^{-1}$, A should approximate U, and B equals its inverse.

$$A = \begin{pmatrix} -0.9881 & -0.0520 \\ 0.2026 & -1.0093 \end{pmatrix} \quad (45.17)$$

$$B = \begin{pmatrix} 1.0014 & 0.0516 \\ -0.2010 & -0.9804 \end{pmatrix} \quad (45.18)$$

Next, the characteristic loci of this compensated system are calculated in order to check stability and to see the required low frequency compensation.

In this case, the characteristic loci show no encirclements, so that the system will be stable. There are also good gain and phase margins which can be seen better using Bode plots of the characteristic gains. From these plots, we decide to increase the low frequency gain for more benefit from the feedback. The resulting dynamic compensation chosen is

$$K_l m(s) = \begin{pmatrix} \frac{5.5(s+1)(s+5)(s+5)}{s(s+0.5)(s+50)} & 0 \\ 0 & \frac{(s+1)(s+9)}{s(s+0.5)} \end{pmatrix} \quad (45.19)$$

This is applied to the system using the same approximate commutative controller calculated at a frequency of 3 rad/sec. The compensated characteristic loci are displayed in order to check that stability has been maintained and that the compensation has affected the loci in the required manner (Figure 45.15).

The characteristic loci still show no encirclements, so the system will be stable. The Bode plots show that the bandwidth is still approximately 10 rad/sec, with phase margins of about 60–80° and good gain margins. The integral action has given high gains at low frequencies, removing low frequency interaction.

The bandwidth of the closed loop response is about 10 rad/sec. The off-diagonal plots (Figure 45.16) indicate that the peak interaction is about 0.1. The interaction for the first reference input corresponds to element 2,1 in the closed-loop time response array. The 15% interaction in the frequency response results in just less than 10% interaction on the time response. The response of y_1 corresponds to element 1,1 in the closed-loop Nyquist array. The slight resonance leads to the overshoot. The bandwidth of about 12 rad/sec leads to the rise time of just over 1/10 sec. The interaction for the second reference input corresponds to element 1,2 in the closed-loop time response array. The response of y_2 corresponds to element 2,2 in the closed-loop Nyquist array.

The diagonal elements of the closed-loop Nyquist array H(s) (Figure 45.16) show good stability margins for changes in individual loop gains, because the diagonal elements show that large extra feedback gains can be introduced without causing instability. The closed-loop Gershgorin bands indicate the stability for simultaneous changes in sensor gains.

The noise rejection properties of the control system are evaluated by using the plots of the sensitivity matrix $S(s) = (I + L(s))^{-1} = (I - H(s))$, where L is the loop gain (Figure 45.17). Next look at the singular values of this closed-loop array to find the robustness for general perturbations. The closed-loop noise is multiplied by $S(s)$, and so the noise is rejected at those frequencies where the sensitivity is small.

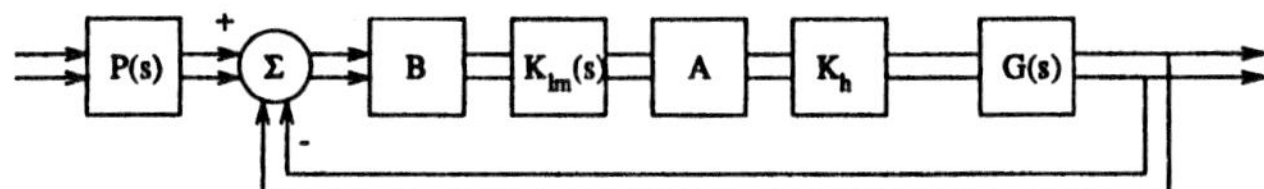

Figure 45.18 Final control configuration.

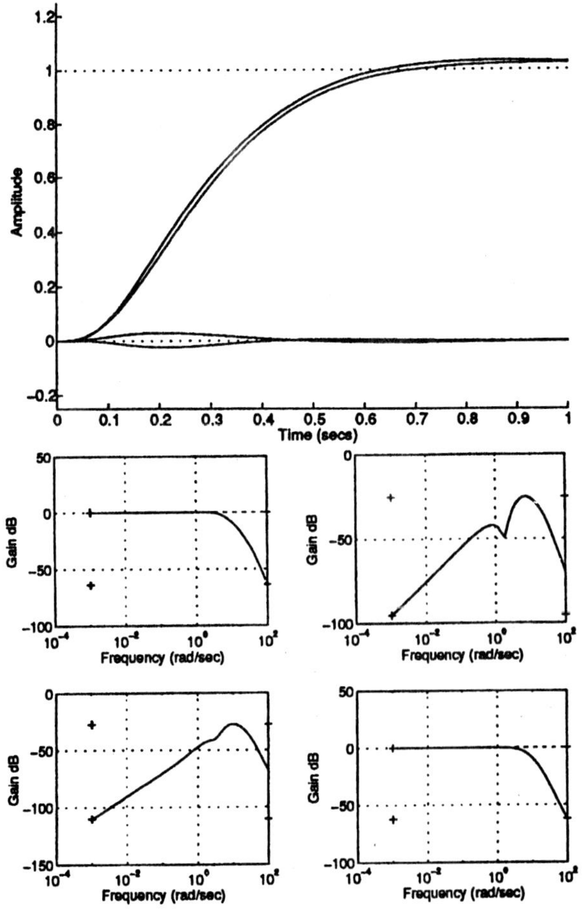

Figure 45.19 Final closed-loop step responses and direct magnitude array.

There is some noise amplification at frequencies between 10 and 100 rad/sec. The remaining interaction, shown by the off-diagonal elements is at frequencies near the bandwidth. This interaction of 10% can be decreased, at the expense of slowing down the response, by putting in a low pass filter $P(s) = 64/(s+8)(s+8)$ on each reference input. This compensator makes the sensitivity to system gain changes become $[I + L(s)]^{-1}P(s)$. The interaction has been considerably reduced to 2–3% at the expense of slowing down the system.

References

[1] Araki, M. and Nwokah, O.I., Bounds for closed-loop transfer functions, *IEEE Trans. Automat. Control,* Oct., 666–670, 1975.

[2] Amin. R. and Edmunds, J.M., Confidence limits of identified frequency responses of multivariable systems, *Int. J. Control,* 48, 1988.

[3] Bennett, W.H. and Baras, J.S., *Conference on Sciences and Systems,* John Hopkins University, Baltimore, MD, 1979.

[4] Edmunds, J.M. and Kouvaritakis, B., Extensions of the frame alignment technique and their use in the characteristic locus design method, *Int. J. Control,* 29, 787, 1979.

[5] Edmunds, J.M., Input Output Scaling for Multivariable Systems and Large Scale Systems, *Control Systems Centre Report No. 760,* UMIST, Manchester, England, 1992.

[6] Edmunds, J.M., The System Toolbox, in *MATLAB Toolboxes and Applications for Control,* A.J. Chipperfield and P.J. Fleming, Eds., IEE Control Series, No. 48, Peter Perigrinus Ltd., Stevenage, UK, 1993.

[7] Hawkins, D.J., Pseudodiagonalization and the inverse Nyquist array method, *Proc. IEE,* 119, 337–342, 1972.

[8] Limebeer, D.J.N., The application of generalized diagonal dominance to linear system stability theory, *Int. J. Control,* 36(2), 185–212, 1982.

[9] MacFarlane, A.G.J., *Complex Variable Methods for Linear Multivariable Feedback Systems,* Taylor and Francis, London, 1980.

[10] MacFarlane, A.G.J. and Kouvaritakis, B., A design technique for linear multivariable feedback systems, *Int. J. Control,* 25, 837–874, 1977.

[11] Maciejowski, J.M., *Multivariable Feedback Design,* Addison-Wesley, 1989.

[12] Mees, A.I., Achieving Diagonal Dominance, *Syst. Control Lett.,* 1, 155–158, 1981.

[13] Munro, N., Recent extensions to the inverse Nyquist array method, *Proc. 24th IEEE Conf. Decision Control,* Miami, FL, 1852–57, 1985.

[14] Munro, N., Computer-Aided Design I: The Inverse Nyquist Array (INA) Design Method, in *Multivariable Control for Industrial Applications,* J. O'Reilly, Ed., 1987, 211.

[15] Munro, N., Computer-Aided Design II: Applications of the INA Method, in *Multivariable Control for Industrial Applications,* J. O'Reilly, Ed., 1987, 229.

[16] Munro, N. and Kontogiannis, E., The Robust Direct Nyquist Array, *Control Systems Centre Report No. 794,* UMIST, Manchester, England, 1994.

[17] Ostrowski, A.M., Note on bounds for determinants with dominant principal diagonal, *Proc. Am. Math. Soc.,* 3, 26–30, 1952.

[18] Patel, R.V. and Munro, N., *Multivariable System Theory and Design,* Pergamon Press, London, 1982.

[19] Rosenbrock, H.H.R., *Computer Aided Control System Design,* Academic Press, London, 1974.

[20] Yeung, L-F. and Bryant, G.F., New dominance concepts for multivariable control systems design, *Int. J. Control,* 55(4), 969–980, 1992.

[21] Winterbone, D.E., Munro, N., and Lourtie, P.M.G., Design of a Multivariable Controller for an Automotive Gas-Turbine, *ASME Gas-Turbine Conference,* Washington, Paper No 73-GT-14, 1973.

Further Reading

A fuller description of the inverse Nyquist array design method and the Characteristic Locus design method with further application examples can be found in *Multivariable System Theory and Design* [1982] by R. V. Patel and N. Munro, *Multivariable Control for Industrial Applications* [1987] edited by J. O'Reilly, and *Multivariable Feedback Design* [1989] by J. M. Maciejowski.

46

Robust Servomechanism Problem

Edward J. Davison
Department of Electrical & Computer Engineering, University of Toronto, Toronto, Ontario, Canada

46.1 Introduction

The so-called *servomechanism problem* is one of the most basic problems to occur in the field of automatic control, and it arises in almost all application problems of the aerospace and process industries. In the servomechanism problem, it is desired to design a controller for a plant (or "system") so that the outputs of the plant are independent, as much as possible, from disturbances which may affect the system (i.e., *regulation* occurs), and also such that the outputs asymptotically track any specified reference input signals applied to the system (i.e., *tracking* occurs), subject to the requirements of maintaining closed-loop system stability.

This chapter examines some aspects of controller synthesis for the multivariable servomechanism problem when the plant to be controlled is subject to uncertainty. In this case, a controller is to be designed so that desired regulation and tracking takes place in spite of the fact that the plant dynamics or/and parameters may vary by arbitrary, large amounts, subject only to the condition that the resultant closed-loop perturbed system remains stable. This problem is called the *robust servomechanism problem* (RSP).

0-8493-8570-9/96/$0.00+$.50

46.2 Preliminary Results

46.2.1 Plant Model

The plant to be controlled is assumed to be described by the following linear time-invariant (LTI) model:

$$\begin{aligned} \dot{x} &= Ax + Bu + E\omega \\ y &= Cx + Du + F\omega \\ y_m &= C_m x + D_m u + F_m \omega \\ e &= y - y_{\text{ref}} \end{aligned} \tag{46.1}$$

where $x \in \mathbb{R}^n$ is the state, $u \in \mathbb{R}^m$ are the inputs that can be manipulated, $y \in \mathbb{R}^r$ are the outputs that are to be regulated and $y_m \in \mathbb{R}^{r_m}$ are the outputs which can be measured. Here $\omega \in \mathbb{R}^{\Omega}$ correspond to the disturbances in the system, which in general cannot necessarily be measured, and $e \in \mathbb{R}^r$ is the error in the system, which is the difference between the output y and the reference input signal y_{ref}, in which it is desired that the outputs y should track.

46.2.2 Class of Tracking/Disturbance Signals

It is assumed that the disturbances ω arise from the following class of systems:

$$\dot{\eta}_1 = \mathcal{A}_1\eta_1, \quad \omega = \mathcal{C}_1\eta_1; \quad \eta_1 \in \mathbb{R}^{n_1} \tag{46.2}$$

and that the reference input signals y_{ref} arise from the following class of systems:

$$\dot{\eta}_2 = \mathcal{A}_2\eta_2, \quad \rho = \mathcal{C}_2\eta_2, \quad y_{ref} = G\rho; \quad \eta_2 \in \mathbb{R}^{n_2} \quad (46.3)$$

It is assumed for nontriviality that $sp(\mathcal{A}_1) \subset \mathbb{C}^+$, $sp(\mathcal{A}_2) \subset \mathbb{C}^+$, where $sp(\cdot)$ denotes the eigenvalues of $(\cdot)$ and $\mathbb{C}^+$ denotes the closed right complex half-plane. It is also assumed with no loss of generality that $(\mathcal{C}_1, \mathcal{A}_1)$, $(\mathcal{C}_2, \mathcal{A}_2)$ are observable and that $\text{rank}\begin{pmatrix} E \\ F \end{pmatrix} = \text{rank}\, \mathcal{C}_1 = \dim(\omega)$, $\text{rank}\, G = \text{rank}\, \mathcal{C}_2 = \dim(\rho)$.

This class of signals is quite broad and includes most classes of signals that occur in application problems, e.g., constant, polynomial, sinusoidal, polynomial-sinusoidal, etc.

The following definitions will be used in the development to follow.

DEFINITION 46.1 Given the systems represented by Equations 46.2, and 46.3, let $\{\lambda_1, \lambda_2, ..., \lambda_p\}$ be the zeros of the least common multiple of the nominal polynomial of $\mathcal{A}_1$ and minimal polynomial of $\mathcal{A}_2$ (multiplicities repeated), and call

$$\Lambda := \{\lambda_1, \lambda_2, ..., \lambda_p\} \quad (46.4)$$

the *disturbances/tracking poles* of Equations 46.2 and 46.3.

DEFINITION 46.2 Given the model represented by Equation 46.1, consider the system

$$\begin{aligned} \dot{x} &= Ax + Bu; \quad u \in R^m, \; y \in R^r, \; x \in R^n \\ y &= Cx + Du \end{aligned}$$

Then $\lambda \in \mathbb{C}$ is said to be a *transmission zero* (TZ) [3] of (C, A, B, D) if

$$\text{rank}\begin{bmatrix} A - \lambda I & B \\ C & D \end{bmatrix} < n + \min(r, m)$$

In particular, the transmission zeros are the zeros (multiplicities included) of the greatest common divisor of all $[n + \min(r, m)] \times [n + \min(r, m)]$ minors of $\begin{bmatrix} A - \lambda I & B \\ C & D \end{bmatrix}$.

DEFINITION 46.3 Given the system (C, A, B, D), assume that one or more of the transmission zeros of (C, A, B, D) are contained in the closed right complex half-plane; then (C, A, B, D) is said to be a *nonminimum-phase* system. If (C, A, B, D) is not nonminimum phase, then it is said to be a *minimum-phase* system.

46.2.3 Robust Servomechanism Problem

The *robust servomechanism problem* (RSP) for Equation 46.1 consists in finding an LTI controller that has inputs y_m, y_{ref} and outputs u for the plant so that:

1. The resultant closed-loop system is asymptotically stable.
2. Asymptotic tracking occurs; that is,

$$\lim_{t\to\infty} e(t) = 0, \quad \forall x(0) \in \mathbb{R}^n, \quad \forall \eta_1(0) \in \mathbb{R}^{n_1}, \; \forall \eta_2(0) \in \mathbb{R}^{n_2}$$

   for all controller initial conditions.
3. Condition 2 holds for any arbitrary perturbations in the plant model Equation 46.1 (e.g., plant parameters or plant dynamics, including changes in model order) that do not cause the resultant closed-loop system to become unstable.

In this problem statement, there is no requirement made regarding the transient behavior of the closed-loop system; thus, the following problem statement is now made.

Perfect RSP

Given the plant represented by Equation 46.1, it is desired to find a controller such that:

1. It solves the RSP for the class of disturbances/tracking signals given by Equations 46.2 and 46.3.
2. The controller gives *perfect error regulation* when applied to the nominal plant model Equation 46.1, i.e., given $x(0)$, $z_1(0)$, $z_2(0)$, located on the unit sphere, with $\eta(0) = 0$, where $\eta(0)$ is the initial condition of the servo-compensator (see Equation 46.9), then $\forall\epsilon > 0$, there exists a controller (parameterized by ϵ) that satisfies property 1 and has the property that $\int_0^\infty e'(\tau)e(\tau)d\tau < \epsilon$, with no unbounded peaking occurring in the response of e; i.e., there exists a constant ρ, independent of ϵ, such that $\sup_{t\geq 0} |e(t)| < \rho$.

Thus, in the perfect RSP, arbitrarily good transient error, with no unbounded peaking in the error response of the system, can be obtained for any initial condition of the plant and for any disturbance/tracking signals that belong to Equations 46.2 and 46.3.

46.3 Main Results

The following results are obtained concerning the existence of a solution to the RSP [6], [5].

THEOREM 46.1 *There exists a solution to the RSP for Equation 46.1, if and only if the following conditions are all satisfied:*

1. *(C_m, A, B) is stabilizable and detectable.*
2. *$m \geq r$.*
3. *The transmission zeros of (C, A, B, D) exclude the disturbance/tracking poles λ_i, $i = 1, 2, ..., p$.*
4. *$y \subset y_m$; i.e., the outputs y are measurable.*

REMARK 46.1 The conditions 2 and 3 are equivalent to the condition:

$$\text{rank}\begin{bmatrix} A-\lambda_i I & B \\ C & D \end{bmatrix} = n+r, \quad i=1,2,\ldots,p \qquad (46.5)$$

The following existence results are obtained concerning the existence of a solution to the perfect RSP [12]:

THEOREM 46.2 *There exists a solution to the perfect RSP for Equation 46.1, if and only if the following conditions are all satisfied:*

1. *(C_m, A) is detectable.*
2. *$m \geq r$.*
3. *(C, A, B, D) is minimum phase.*
4. *$y \subset y_m$.*

REMARK 46.2 If $m = r$, the above conditions simplify to just conditions 3 and 4.

The following definitions of a stabilizing compensator and servo-compensator are required in the development to follow.

DEFINITION 46.4 Given the stabilizable, detectable system (C_m, A, B, D) obtained from Equation 46.1, an LTI *stabilizing compensator*

$$\begin{aligned} \dot{\xi} &= \Lambda_1\xi + \Lambda_2 y_m \\ u &= K_1\xi + K_2 y_m \end{aligned} \qquad (46.6)$$

is defined to be a controller that asymptotically stabilizes the resultant closed-loop system, such that "desired" transient behavior occurs.

This compensator is not a unique device and may be designed by using a number of different techniques.

DEFINITION 46.5 Given the disturbance/reference input poles $\lambda_i, i = 1, 2, \ldots, p$, the matrix $\mathcal{C} \in R^{p\times p}$ and vector $\gamma \in R^p$ are defined by

$$\mathcal{C} := \begin{bmatrix} 0 & 1 & 0 & \ldots & 0 \\ 0 & 0 & 1 & \ldots & 0 \\ \vdots & \vdots & \vdots & & \vdots \\ -\delta_1 & -\delta_2 & -\delta_3 & \ldots & -\delta_p \end{bmatrix}, \quad \gamma := \begin{bmatrix} 0 \\ 0 \\ \vdots \\ 0 \\ 1 \end{bmatrix} \qquad (46.7)$$

where the coefficients $\delta_i, i = 1, 2, \ldots, p$ are given by the coefficients of the polynomial $\prod_{i=1}^{p}(\lambda - \lambda_i)$; i.e.,

$$\lambda^p + \delta_p\lambda^{p-1} + \ldots + \delta_2\lambda + \delta_1 := \prod_{i=1}^{p}(\lambda - \lambda_i) \qquad (46.8)$$

The following compensator, called a *servo-compensator*, is of fundamental importance in the design of controllers to solve the RSP [5].

DEFINITION 46.6 Consider the class of disturbance/tracking signals given by Equations 46.2 and 46.3, and consider the system of Equation 46.1; then a servo-compensator for Equation 46.1 is a controller with input $e \in R^r$ and output $\eta \in R^{rp}$ given by

$$\dot{\eta} = \mathcal{C}^*\eta + \mathcal{B}^*e \qquad (46.9)$$

where

$$\mathcal{C}^* := \text{block diag}(\underbrace{\mathcal{C}, \mathcal{C}, \ldots, \mathcal{C}}_{r}) \qquad (46.10)$$

$$\mathcal{B}^* := \text{block diag}(\underbrace{\gamma, \gamma, \ldots, \gamma}_{r}) \qquad (46.11)$$

where $\mathcal{C}, \gamma$ are given by Equation 46.7.

The servo-compensator is unique within the class of coordinate transformations and nonsingular input transformations.

Given the servo-compensator of Equation 46.9, now let $\mathcal{D} \in R^{r\times rp}$ be defined by

$$\mathcal{D} := \text{block diag}(\underbrace{\delta, \delta, \ldots, \delta}_{r}) \qquad (46.12)$$

where $\delta \in R^{1\times p}$ is given by:

$$\delta := (1\ 0\ 0 \ldots 0) \qquad (46.13)$$

The servo-compensator has the following properties:

LEMMA 46.1 [12] Given the plant represented by Equation 46.1, assume that the existence conditions of Theorem 46.1 all hold; then

1. The system

$$\left\{\begin{pmatrix} C_m & 0 \\ 0 & \mathcal{D} \end{pmatrix}, \begin{pmatrix} A & 0 \\ \mathcal{B}^*C & \mathcal{C}^* \end{pmatrix}, \begin{pmatrix} B \\ \mathcal{B}D \end{pmatrix}\right\}$$

is stabilizable and detectable and has centralized fixed modes [8] (i.e., those modes of the system that are not both simultaneously controllable and observable) equal to the centralized fixed modes of $(C_m, A,$ $B, D_m)$.

2. The transmission zeros of

$$\left\{(0\ \ \mathcal{D}), \begin{pmatrix} A & 0 \\ \mathcal{B}^*C & \mathcal{C}^* \end{pmatrix}, \begin{pmatrix} B \\ \mathcal{B}D \end{pmatrix}\right\}$$

are equal to the transmission zeros of (C, A, B, D).

46.3.1 Robust Servomechanism Controller

Consider the system of Equation 46.1, and assume that the existence conditions of Theorem 46.1 hold; then any LTI controller that solves the RSP for Equation 46.1 consists of the following structure [6] (see Figure 46.1):

$$u = \xi + K\eta \tag{46.14}$$

where $\eta \in R^{rp}$ is the output of the servo-compensator (Equation 46.9), and ξ is the output of a stabilizing compensator $\mathcal{S}$ with inputs y_m, y_{ref}, η, u, where $\mathcal{S}$, K are found to stabilize and give "desired behavior" to the following stabilizable and detectable system:

$$\begin{aligned} \begin{pmatrix} \dot{x} \\ \dot{\eta} \end{pmatrix} &= \begin{bmatrix} A & 0 \\ \mathcal{B}^*C & \mathcal{C}^* \end{bmatrix} \begin{pmatrix} x \\ \eta \end{pmatrix} + \begin{pmatrix} B \\ \mathcal{B}^*D \end{pmatrix} y_{ref} \\ \tilde{y}_m &= \begin{bmatrix} C_m & 0 \\ 0 & I \\ 0 & 0 \end{bmatrix} \begin{pmatrix} x \\ \eta \end{pmatrix} + \begin{pmatrix} D_m \\ 0 \\ I \end{pmatrix} y_{ref} \end{aligned} \tag{46.15}$$

where, from Lemma 46.1, the centralized fixed modes (if any) of

$$\left\{ \begin{pmatrix} C_m & 0 \\ 0 & I \end{pmatrix}, \begin{bmatrix} A & 0 \\ \mathcal{B}^*C & \mathcal{C}^* \end{bmatrix}, \begin{pmatrix} B \\ \mathcal{B}^*D \end{pmatrix} \right\}$$

are equal to the centralized fixed modes of (C_m, A, B); i.e., there always exists a coordinate transformation and nonsingular input transformation, by which any controller that solves the RSP for Equation 46.1 can be described by Equation 46.14. It is to be noted that this controller always has order $\geq rp$.

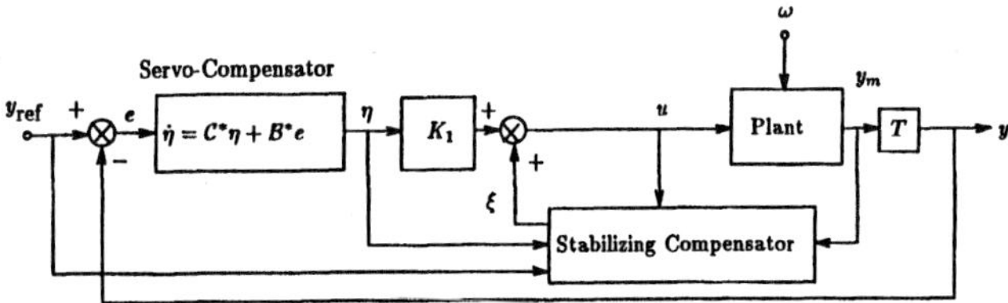

Figure 46.1 General controller to solve robust servomechanism problem.

Properties of Robust Servomechanism Controller

Some properties of the robust servomechanism controller (RSC) represented by Equation 46.14 are as follows [5]:

1. In the above controller, it is required to know only the disturbance/reference input poles $\{\lambda_1, \lambda_2, ..., \lambda_p\}$; i.e., it is not necessary to know E, F of Equation 46.1 nor $\mathcal{A}_1$, $\mathcal{A}_2$, $\mathcal{C}_1$, $\mathcal{C}_2$, G of Equations 46.2 and 46.3.
2. A controller exists generically [2] for almost all plants described by Equation 46.1, provided that (a) $m \geq r$, and (b) the outputs y can be measured; if either condition (a) or (b) fails to hold, there is no solution to the RSP.

46.3.2 Various Classes of Stabilizing Compensators

Various special classes of stabilizing compensators $\mathcal{S}$ that can be used in the RSC (Equation 46.14) are as follows. It is assumed that the existence conditions of Theorem 46.1 are all satisfied in order to implement these proposed controllers.

Multivariable Three-Term Controller

(See Figure 46.2). In order to use this controller, it is assumed that:

1. The plant of Equation 46.1 is open loop asymptotically stable.
2. The disturbance/tracking poles $\{\lambda_1, \lambda_2, ..., \lambda_p\}$ are of the polynomial-sinusoidal type; i.e., it is assumed that $\text{Re}(\lambda_i) = 0, i = 1, 2, ..., p$.

If these assumptions hold, then the following generalized three-term controller solves the RSP [9]:

$$\begin{aligned} u &= (K_0 + K_1 s)e + K_2\eta \\ \dot{\eta} &= \mathcal{C}^*\eta + \mathcal{B}^*e \end{aligned} \tag{46.16}$$

An algorithm is given [7], that shows that a stabilizing K_2, with $K_0 = 0$, $K_1 = 0$, can always be found for this controller.

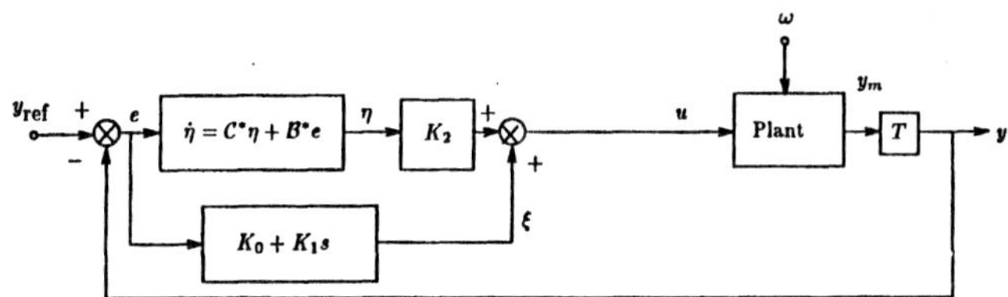

Figure 46.2 Generalized three term controller to solve the robust servomechanism problem.

46.3.3 Complementary Controller

(See Figure 46.3) In order to use this controller, it is assumed that the plant of Equation 46.1 is open loop asymptotically stable. If this assumption holds, then the following controller, called a complementary controller, will solve the RSP [5]:

$$\begin{aligned} u &= K_0\hat{x} + K_1\eta \\ \dot{\eta} &= \mathcal{C}^*\eta + \mathcal{B}^*e \\ \dot{\hat{x}} &= A\hat{x} + Bu \end{aligned} \tag{46.17}$$

where (K_0, K_1) are found to stabilize the stabilizable system

$$\dot{\tilde{x}} = \begin{bmatrix} A & 0 \\ \mathcal{B}^*C & \mathcal{C}^* \end{bmatrix} \tilde{x} + \begin{pmatrix} B \\ \mathcal{B}^*D \end{pmatrix} u \tag{46.18}$$

using state feedback; i.e., $u = (K_0 \ K_1)\tilde{x}$.

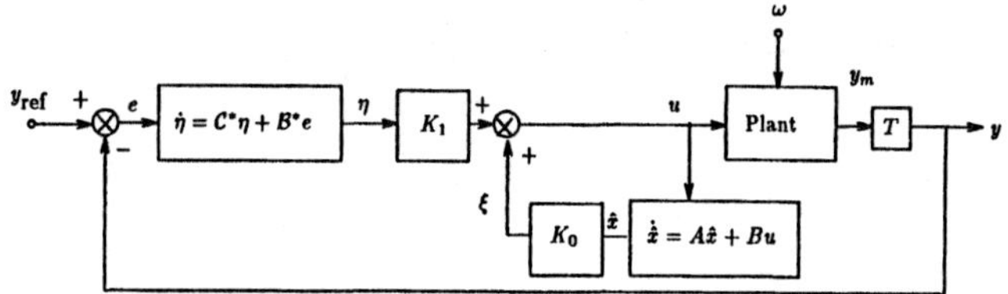

Figure 46.3 Complementary controller to solve the robust servomechanism problem.

46.3.4 Observer-Based Stabilizing Controller

(See Figure 46.4) No additional assumptions are required in order to implement this controller. The controller is given as follows [9]:

$$\begin{aligned} u &= K_0\hat{x} + K_1\eta \\ \dot{\eta} &= C^*\eta + \mathcal{B}^*e \\ \dot{\hat{x}} &= \{A + (B - \Lambda D_m)K_0 - \Lambda C_m\}\hat{x} \\ &\quad + (B - \Lambda D_m)K_1\eta + \Lambda y_m \end{aligned} \tag{46.19}$$

where (K_0, K_1) are found to stabilize the system of Equation 46.18, and Λ is an observer gain matrix found to stabilize the system matrix $(A - \Lambda C_m)$ where (C_m, A) is detectable. (A reduced-order observer could also be used to replace the full-order observer in Equation 46.19.)

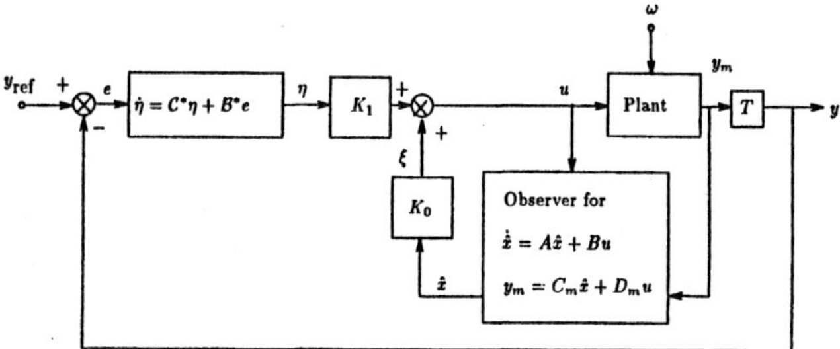

Figure 46.4 Observer-based stabilizing compensator to solve the robust servomechanism problem.

46.4 Applications and Example Calculations

We initially demonstrate the theory by considering the design of a controller for a nonminimum-phase plant (Example 46.1), and a minimum-phase plant (Example 46.2), and thence conclude with a case study on the control of a nontrivial distillation column system.

EXAMPLE 46.1: Nonminimum-phase system

Consider the following system

$$\begin{aligned} \dot{x} &= \begin{pmatrix} 2 & -1 \\ 0 & 0 \end{pmatrix} x + \begin{pmatrix} 1 \\ 1 \end{pmatrix} u \\ y &= (1 \;\; 0)x + \omega; \qquad y_m = y \end{aligned} \tag{46.20}$$

which is open loop unstable and nonminimum phase (with a transmission zero at 1), and in which it is desired to design a controller to solve the RSP for the case of constant disturbances and constant reference input signals. In this case, the disturbance/tracking poles = {0}, and it can be directly verified that the existence conditions for a solution to the problem are satisfied from Theorem 46.1. In the controller design, it is initially assumed that the control input should not be "excessively large."

Controller Development

On applying the servo-compensator of Equation 46.9 for constant disturbances/reference input signals to Equation 46.20, the following system is obtained:

$$\begin{aligned} \begin{pmatrix} \dot{x} \\ \dot{\eta} \end{pmatrix} &= \begin{pmatrix} 2 & -1 & 0 \\ 0 & 0 & 0 \\ 1 & 0 & 0 \end{pmatrix} \begin{pmatrix} x \\ \eta \end{pmatrix} + \begin{pmatrix} 1 \\ 1 \\ 0 \end{pmatrix} \omega \\ &\quad + \begin{pmatrix} 0 \\ 0 \\ -1 \end{pmatrix} y_{ref} \\ y &= (1 \;\; 0 \;\; 0)\begin{pmatrix} x \\ \eta \end{pmatrix} + 1\omega \end{aligned} \tag{46.21}$$

and, on minimizing the performance index [12] for Equation 46.21

$$J_\epsilon = \int_0^\infty (e'e + \epsilon\dot{u}'\dot{u})d\tau, \qquad \epsilon = 1 \tag{46.22}$$

the following controller is obtained:

$$u = (k_1 \;\; k_2)\hat{x} + k\int_0^t (y - y_{ref}) \tag{46.23}$$

where

$$k_1 = -13.77, \quad k_2 = 8.721, \quad k = 1.000$$

and where $\hat{x}$ is the output of an observer for Equation 46.20. On using a reduced-order observer with observer pole $= -1$, the following controller is thence obtained.

Robust Servomechanism Controller

$$\begin{aligned} u &= -22.498y + \eta + 8.7214\sigma \\ \dot{\eta} &= y - y_{ref} \\ \dot{\sigma} &= 16.443\sigma + 2\eta - 41.996y \end{aligned} \tag{46.24}$$

which, when applied to Equation 46.20, gives the following closed-loop poles:

$$\begin{pmatrix} -0.552 \pm j0.456 \\ -1.00 \\ -1.95 \end{pmatrix}$$

Properties of Controller

On applying the controller of Equation 46.24 to the plant of Equation 46.20, the following closed-loop system is obtained:

$$\begin{pmatrix} \dot{x} \\ \dot{\eta} \\ \dot{\sigma} \end{pmatrix} = \begin{bmatrix} -20.498 & -1 & 1 & 8.7214 \\ -22.498 & 0 & 1 & 8.7214 \\ 1 & 0 & 0 & 0 \\ -41.996 & 0 & 2 & 16.443 \end{bmatrix} \begin{pmatrix} x \\ \eta \\ \sigma \end{pmatrix} + \begin{pmatrix} 0 \\ 0 \\ -1 \\ 0 \end{pmatrix} y_{\text{ref}} + \begin{pmatrix} -22.49 \\ -22.49 \\ 1 \\ -41.99 \end{pmatrix} \omega$$

$$y = [1\ 0\ 0\ 0] \begin{pmatrix} x \\ \eta \\ \sigma \end{pmatrix} + \omega$$

$$u = [-22.498\ 0\ 1\ 8.7214] \begin{pmatrix} x \\ \eta \\ \sigma \end{pmatrix} + (-22.498)\omega$$

which gives the Bode magnitude response between e, u and input y_{ref} in Figure 46.5, e, u and input ω in Figure 46.6, and unit-step responses in y_{ref} in Figure 46.7 and in ω in Figure 46.8, with zero initial conditions. It is seen that satisfactory tracking/regulation occurs using the controller.

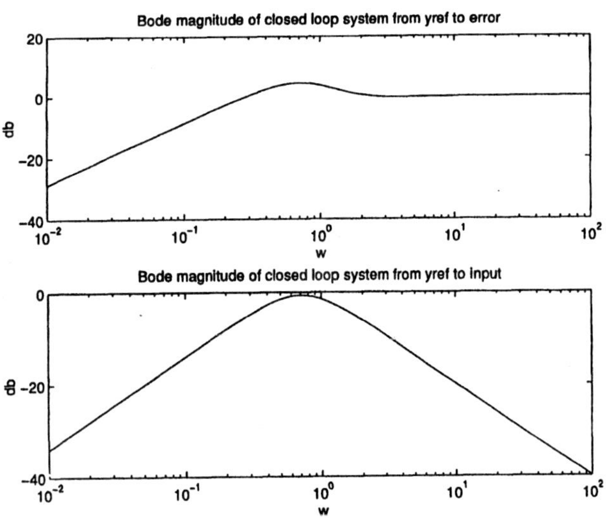

Figure 46.5 Bode plot magnitude of closed-loop system using robust servomechanism controller from reference input signal: (a) to error; (b) to control input.

Response of Closed-Loop System to Unbounded Reference Input/Disturbance Signals

According to Theorem 3 in [12], the RSC of Equation 46.14 has the property of not only achieving asymptotic tracking/regulation for the class of constant reference input/disturbance signals, but also bringing about *exact* asymptotic tracking/

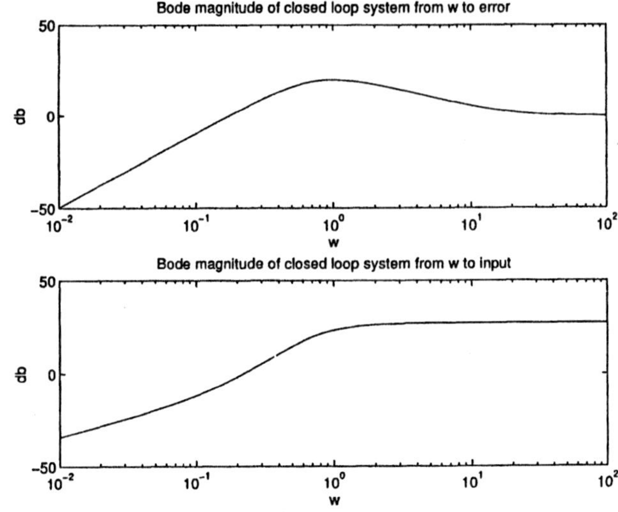

Figure 46.6 Bode plot magnitude of closed-loop system using robust servomechanism controller from disturbance signal: (a) to error; (b) to control input.

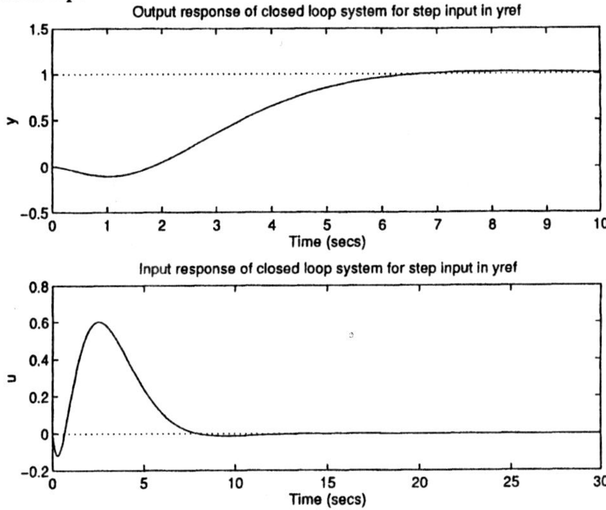

Figure 46.7 Response of closed-loop system using robust servomechanism controller for unit step in reference input signal: (a) output; (b) input.

regulation for unbounded signals that have the property that:

$$\begin{aligned} \lim_{t\to\infty} \dot{y}_{\text{ref}}(t) &= 0 \\ \lim_{t\to\infty} \dot{\omega}(t) &= 0 \end{aligned} \tag{46.25}$$

To illustrate this result, Figure 46.9 gives the response of the closed-loop system for the case when the tracking input signal is given by

$$y_{\text{ref}}(t) = t^{1/4}, \quad t \geq 0 \tag{46.26}$$

with zero initial conditions. It is seen that exact asymptotic tracking indeed does take place using this controller.

Optimum Response of Nonminimum-Phase System

The RSC of Equation 46.24 was obtained so that the magnitude of the control input signal is constrained by letting $\epsilon = 1$

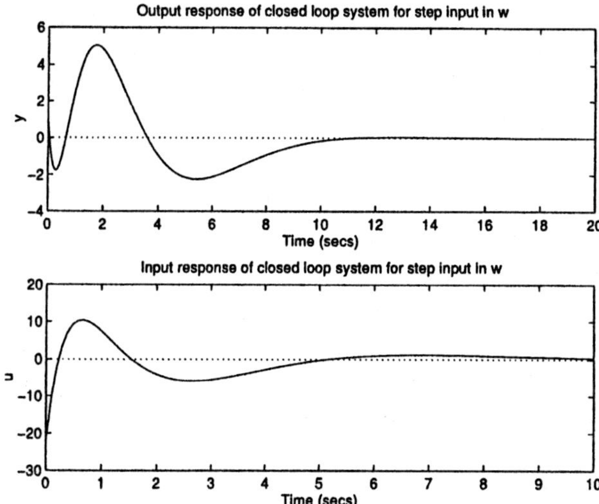

Figure 46.8 Response of closed-loop system using robust servomechanism controller for unit step in disturbance input signal: (a) output; (b) input.

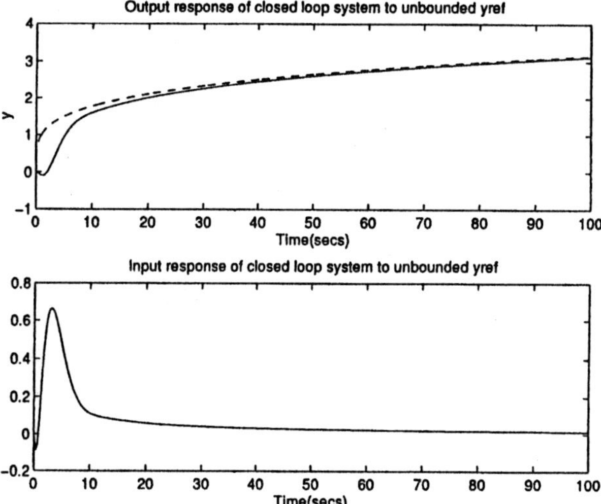

Figure 46.9 Response of closed-loop system using robust servomechanism controller for unbounded reference input signal $= t^{1/4}$: (a) output; (b) input.

in the performance index of Equation 46.22. It is of interest to determine what type of transient performance can be obtained for the system if the controller is designed without any regard for the magnitude of the control input signal; i.e., by letting $\epsilon = 10^{-8}$, say, in the performance index (Equation 46.22). According to Theorem 46.2, "perfect control" cannot be achieved for the system; in particular, if the plant's and controller's initial conditions are equal to zero and $\omega = 0$, then the limiting performance that can be achieved is given by [13]

$$\lim_{\epsilon \to 0} J_\epsilon = 2 \sum_{i=1}^{l} \frac{1}{\lambda_i^*} y_{\text{ref}}^2 \tag{46.27}$$

where $\{\lambda_i^*, \quad i = 1, 2, ..., l\}$ are the nonminimum TZ of the system, and J_ϵ is given by Equation 46.22. In this case, $l = 1$, $\lambda_1^* = 1$, so that the limiting performance index is given by

$\lim_{\epsilon \to 0} J_\epsilon = 2y_{\text{ref}}^2$.

The following optimal controller is now obtained on putting $\epsilon = 10^{-8}$ in the performance index of Equation 46.22:

$$\begin{aligned} u &= (-2.043 \times 10^4 \;\; 2.029 \times 10^4)x + 10^4 \eta \\ \dot{\eta} &= y - y_{\text{ref}} \end{aligned} \tag{46.28}$$

This controller results in the following closed-loop system poles:

$$\begin{pmatrix} -70.7 \pm j70.7 \\ -1.00 \end{pmatrix}$$

with the following optimal performance index, for $x(0) = 0$, $\omega = 0$:

$$J = 2.014 y_{\text{ref}}^2 \tag{46.29}$$

which is "close" to the optimal limiting performance index of $2y_{\text{ref}}^2$.

A response of the closed-loop system using the controller of Equation 46.28 for a unit step in y_{ref}, with zero initial conditions for the plant, is given in Figure 46.10. It is seen that the response of the closed-loop system is only about twice as fast as that obtained using the controller of Equation 46.24, in spite of the fact that the control signal is now some 40 times larger than that obtained using Equation 46.24, which confirms the fact that nonminimum-phase systems are fundamentally "difficult to control."

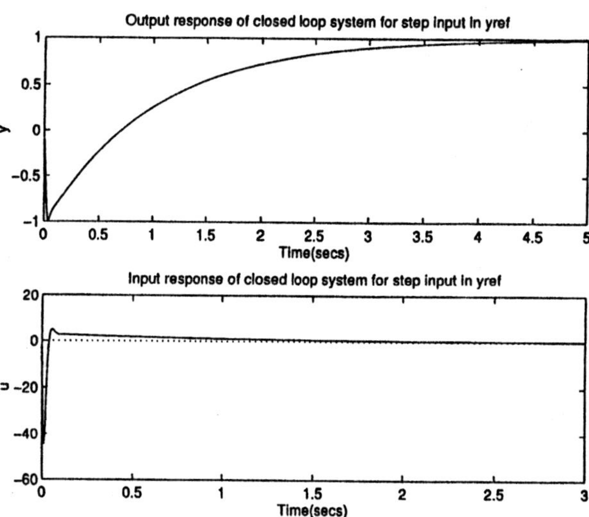

Figure 46.10 Response of closed-loop system using robust servomechanism controller with high gain for unit step in reference input signal: (a) output; (b) input.

High-Gain Servomechanism Control

The same example as considered in the previous sections will now be considered, except that a high-gain servomechanism controller (HGSC) [12], which is simpler than an RSC, will now be applied and compared with the RSC of Equation 46.24.

Plant Model

$$\begin{aligned}\dot{x} &= \begin{pmatrix} 2 & -1 \\ 0 & 0 \end{pmatrix} x + \begin{pmatrix} 1 \\ 1 \end{pmatrix} u \\ y &= (1 \quad 0)x + \omega\end{aligned} \tag{46.30}$$

Controller Development

Given the cheap control performance index

$$\tilde{J}_\epsilon = \int_0^\infty (y'y + \epsilon u'u) d\tau \tag{46.31}$$

let $\epsilon = 1$; in this case, the optimal control that minimizes $\tilde{J}_\epsilon$ for the system of Equation 46.30 with $\omega = 0$ is given by

$$u = kx, \quad k = (-10.292 \quad 5.6458) \tag{46.32}$$

which results in the closed-loop system of Equations 46.30 and 46.32 having poles given by $(-0.457, -2.19)$.

On letting $\mathcal{K} := -[c(A + Bk)^{-1}B]^{-1} = -1$, the HGSC is now given by [12]:

$$u = \mathcal{K}(y_{\text{ref}} - y) + (k + \mathcal{K}c)\hat{x} \tag{46.33}$$

where $\hat{x}$ is the output of an observer for Equation 46.30. On choosing a reduced-order observer with observer gain $= -1$, and simplifying, the following controller is finally obtained:

High-Gain Servomechanism Controller

$$\begin{aligned} u &= -y_{\text{ref}} - 15.94y - 5.646\xi \\ \dot{\xi} &= 10.29\xi + 28.87y + 2y_{\text{ref}} \end{aligned} \tag{46.34}$$

Properties of HGSC

On applying Equation 46.34 to Equation 46.30, the following closed-loop system is obtained:

$$\begin{aligned}
\begin{pmatrix} \dot{x} \\ \dot{\xi} \end{pmatrix} &= \begin{bmatrix} -13.94 & -1 & -5.646 \\ -15.94 & 0 & -5.646 \\ 28.87 & 0 & 10.29 \end{bmatrix} \begin{pmatrix} x \\ \xi \end{pmatrix} \\
&\quad + \begin{pmatrix} -1 \\ -1 \\ 2 \end{pmatrix} y_{\text{ref}} + \begin{pmatrix} 0 \\ 0 \\ 28.87 \end{pmatrix} \omega \\
e &= [1 \quad 0 \quad 0] \begin{pmatrix} x \\ \xi \end{pmatrix} - 1y_{\text{ref}} + 1\omega \\
y &= [1 \quad 0 \quad 0] \begin{pmatrix} x \\ \xi \end{pmatrix} + 1\omega \\
u &= [-15.94 \quad 0 \quad -5.646] \begin{pmatrix} x \\ \xi \end{pmatrix} \\
&\quad - 1y_{\text{ref}} - 15.94\omega
\end{aligned}$$

which has closed-loop poles given by $(-0.457, -1.00, -2.19)$. The Bode magnitude response of this system with respect to output e, u and input y_{ref} is given in Figure 46.11. The unit-step function response of this system for an increase in y_{ref} with zero initial conditions is given in Figure 46.12. It is seen that satisfactory tracking occurs using this controller.

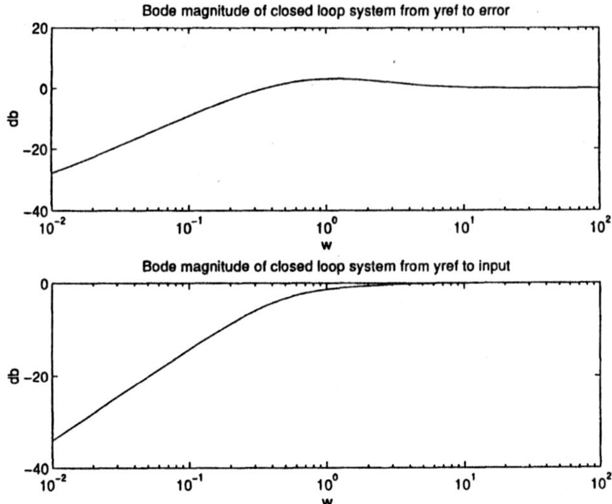

Figure 46.11 Bode plot magnitude of closed-loop system using high-gain servomechanism controller from reference input signal: (a) to error; (b) to control input.

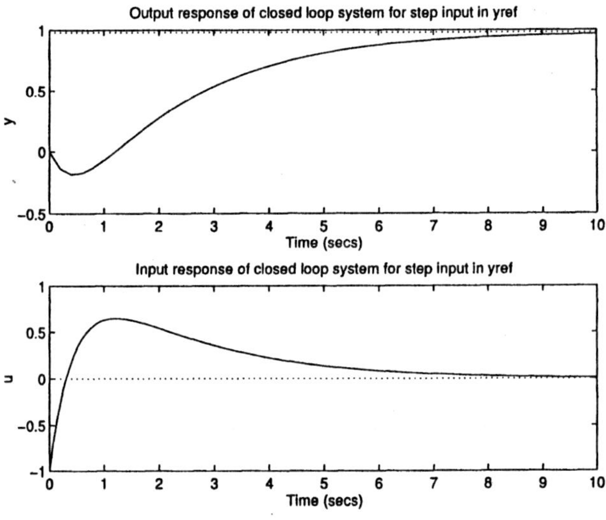

Figure 46.12 Response of closed-loop system using high-gain servomechanism controller for unit step in reference input signal: (a) output; (b) input.

Comparison of Robustness Properties of HGSC to RSP

Although the HGSC of Equation 46.34 is "simpler" than the RSC of Equation 46.24, the outstanding advantage of the RSP is that it is robust; i.e., it provides *exact* asymptotic tracking/regulation for any perturbations of the plant model, that do not destabilize the perturbed closed-loop system. As an example of this behavior, consider the HGSC (Equation 46.34) and RSC (Equation 46.24) controlling the following (slightly) perturbed model of Equation 46.20:

Perturbed Model

$$\begin{aligned} \dot{x} &= \begin{pmatrix} 2 & -1 \\ 0 & -0.1 \end{pmatrix} x + \begin{pmatrix} 1 \\ 1 \end{pmatrix} u \\ y &= (1 \;\; 0)x + \omega \end{aligned} \qquad (46.35)$$

In this case, the resultant perturbed closed-loop system remains stable for each of the controllers, but the HGSC no longer provides tracking; e.g., $\lim_{t\to\infty} y(t) = 0.3$ when $y_{ref} = 1$. In contrast, it may be verified that the RSC (Equation 46.24) still provides exact tracking when applied to Equation 46.35.

EXAMPLE 46.2: Minimum-phase system

Consider the following system, which is a model of a pressurized head box in paper manufacturing:

$$\begin{aligned} \dot{x} &= \begin{pmatrix} 0.395 & 0.01145 \\ -0.011 & 0 \end{pmatrix} x + \begin{pmatrix} 0.03362 & 1.038 \\ 0.000966 & 0 \end{pmatrix} u \\ y &= \begin{pmatrix} 1 & 0 \\ 0 & 1 \end{pmatrix} x; \quad y_m = y \end{aligned} \qquad (46.36)$$

which has open-loop eigenvalues $= (-3.19 \times 10^{-4}, -3.95 \times 10^{-1})$. In this case, the plant represented by Equation 46.36 has no transmission zeros, which implies that it is minimum phase, and so there exists a solution to the RSP for the class of constant disturbances/reference input signals such that perfect control occurs (see Theorem 46.2).

The following development illustrates how "perfect control" in a minimum-phase system results in *arbitrarily good approximate error regulation* (AGAER) [12] occurring; i.e., although the servocompensator is designed to give exact asymptotic error regulation for constant disturbances/reference input signals, AGAER will occur for other classes of disturbances/reference input signals, e.g., for the class of sinusoidal signals $y_{ref} = \bar{y}\sin(\omega t)$.

The following perfect controller is now obtained:

$$u = K_0 y + K \int_0^t (y - y_{ref}) d\tau \qquad (46.37)$$

which, on minimizing the performance index

$$J_\epsilon = \int_0^\infty (e'e + \epsilon \dot{u}'\dot{u}) d\tau \qquad (46.38)$$

gives

$$\begin{aligned} K_0 &= \begin{pmatrix} -4.33 & --4549 \\ -138.3 & 158.7 \end{pmatrix}, \\ K &= \begin{pmatrix} -371.9 & -10^4 \\ -10^4 & 371.9 \end{pmatrix} \\ &\text{for } \epsilon = 10^{-8} \\ K_0 &= \begin{pmatrix} -13.8 & -1.439 \times 10^4 \\ -438.2 & 477.3 \end{pmatrix}, \\ K &= \begin{pmatrix} -338.8 & -9.994 \times 10^4 \\ -9.994 \times 10^4 & 3388 \end{pmatrix} \\ &\text{for } \epsilon = 10^{-10} \end{aligned} \qquad (46.39)$$

It may now be demonstrated that AGAER occurs; e.g., when the reference input signal $y_{ref} = \bar{y}\sin(\omega t)$, $\omega = 0.1$ is applied, we obtain the following *error coefficient* [12]:

$$\begin{aligned} {}_0K_t^0(j\omega) &= \begin{pmatrix} 1.4 \times 10^{-3} & 1.0 \times 10^{-4} \\ 1.0 \times 10^{-4} & 4.6 \times 10^{-2} \end{pmatrix} \\ &\quad \text{for } \epsilon = 10^{-8} \\ &= \begin{pmatrix} 4.4 \times 10^{-4} & 1.0 \times 10^{-5} \\ 1.0 \times 10^{-5} & 1.4 \times 10^{-2} \end{pmatrix} \\ &\quad \text{for } \epsilon = 10^{-10} \end{aligned} \qquad (46.40)$$

which implies that excellent (but approximate) tracking occurs as $\epsilon \to 0$ for the tracking signal $y_{ref} = \bar{y}\sin(\omega t)$, $\omega = 0.1$.

Stability Robustness Concerns

The previous example studies have ignored "stability robust concerns"; i.e., the recognition that any plant model is really only an approximation to a model of the actual physical system, and hence that a given physical system may become unstable under feedback if the model used is sufficiently inaccurate. The following shows that for some classes of systems, a system may be relatively insensitive to plant perturbations, whereas for other classes of systems, a system may be highly sensitive to plant perturbations. This implies that any controller that is designed to solve the RSP must *always* take into account stability robustness considerations for the particular problem being studied.

Consider the head box example (Equation 46.36) and the controller

$$u = K_0 y + K \int_0^t (y - y_{ref}) d\tau \qquad (46.41)$$

which has been designed for the class of constant disturbances/reference input signals. It is now desired to determine the optimal controller gain matrices K_0, K that minimize the performance index

$$J_\epsilon = \int_0^\infty (e'e + \epsilon \dot{u}'\dot{u}) d\tau \qquad \epsilon = 10^{-8} \qquad (46.42)$$

with and without a *gain margin* (GM) constraint [10] imposed on the system. This constraint can be imposed by using the computer-aided design (CAD) approach of Davison and Ferguson [9]. Let Γ denote the optimal cost matrix of Equation 46.42, and define $J_{opt} = \text{trace}(\Gamma)$. The following results are now obtained (see Table 46.1)

In this case, the optimal controller that minimizes J_ϵ (Equation 46.42), subject to a fairly demanding GM constraint of (0.2,2) is only about "two times slower" than the optimal controller, which does not take gain margin into account; i.e., the system is relatively insensitive to plant perturbations.

Consider now, however, the following system:

$$\begin{aligned} \dot{x} &= \begin{pmatrix} -1 & 0 \\ 0 & -2 \end{pmatrix} x + \begin{pmatrix} 7 & 8 \\ 12 & 14 \end{pmatrix} u \\ y &= \begin{pmatrix} 7 & -8 \\ -6 & 7 \end{pmatrix} x; \quad y_m = y \end{aligned} \qquad (46.43)$$

TABLE 46.1 Optimal Controller Parameters Obtained.

GM Constraint	J_{opt}	(K_0, K)	Closed-Loop Poles
None	0.469	$\begin{bmatrix} -4.33 & -4550 & -371.9 & -9993 \\ -138 & 158.7 & -9993 & 371.9 \end{bmatrix}$	$\begin{pmatrix} -2.20 \pm j2.20 \\ -72.1 \pm j72.1 \end{pmatrix}$
(0.2,2)	0.700	$\begin{bmatrix} 2330 & 245 & 6700 & 281 \\ -504 & -820 & -1960 & -1580 \end{bmatrix}$	$\begin{pmatrix} -1.43 \pm j1.51 \\ -5.31 \\ -437 \end{pmatrix}$

TABLE 46.2 Optimal Controller Parameters Obtained.

GM Constraint	J_{opt}	(K_0, K)	Closed-Loop Poles
None	0.100	$\begin{bmatrix} -58.1 & -15.2 & -326 & -9995 \\ 220 & -185 & 9995 & -326 \end{bmatrix}$	$\begin{pmatrix} -10.1 \pm j10.1 \\ -700 \pm j700 \end{pmatrix}$
(0.9,1.1)	24.4	$\begin{bmatrix} -0.600 & -0.962 & -0.636 & 0.109 \\ 0.740 & 0.759 & 0.620 & -0.380 \end{bmatrix}$	$\begin{pmatrix} -0.0538 \\ -0.350 \\ -0.897 \pm j4.21 \end{pmatrix}$

and the controller of Equation 46.41. It is desired now to find optimal controller parameters K_0, K so as to minimize the performance index J_ϵ (Equation 46.42) with and without a modest GM of (0.9,1.1) applied. The following results are obtained from Davison and Copeland [10] (see Table 46.2).

In this case, it can be seen that a very modest demand that the controller should have a GM of only 10% has produced a dramatic effect in terms of the controller. The controller obtained with the 10% GM constraint being imposed has a performance index that is some 10^5 times "worse" than the case when no GM constraint is imposed; i.e., the system is extremely sensitive to plant perturbations. Thus, this example emphasizes the need to always apply some type of stability robustness constraint when solving "high-performance controller" problems.

More Complex Tracking/Disturbance Signal Requirements

The following example illustrates the utilization of more complex servo-compensator construction. In this case, the head box problem modeled by Equation 46.36 is to be controlled for the class of disturbance/reference input signals that have the structure:

$$y_{\text{ref}} = \bar{y}_1 \sin(\omega_1 t) + \bar{y}_2 \sin(\omega_2 t) + \bar{y}_3 \sin(\omega_3 t) \qquad (46.44)$$

where $\omega_1 = \pi$, $\omega_2 = 3\pi$, $\omega_3 = 5\pi$, and $\bar{y}_1, \bar{y}_2, \bar{y}_3$ are arbitrary real two-dimensional vectors.

From Theorem 46.1, there exists a solution to the problem, and the servo-compensator is now given from Equation 46.9 by:

$$\dot{\eta} = \text{block diag}\left[\begin{pmatrix} 0 & I \\ -\omega_1^2 I & 0 \end{pmatrix}, \begin{pmatrix} 0 & I \\ -\omega_2^2 I & 0 \end{pmatrix}, \begin{pmatrix} 0 & I \\ -\omega_3^2 I & 0 \end{pmatrix}\right]\eta + \begin{bmatrix} 0 \\ I \\ 0 \\ I \\ 0 \\ I \end{bmatrix}(y - y_{\text{ref}}) \qquad (46.45)$$

and the following controller is now obtained:

$$u = K_0 y + (K_1 \; K_2 \; K_3)\eta \qquad (46.46)$$

In this case, the following performance index

$$\bar{J}_\epsilon = \int_0^\infty \left\{(x' \; \eta')Q\begin{pmatrix} x \\ \eta \end{pmatrix} + \epsilon u'u\right\} d\tau \qquad (46.47)$$

where

$$\begin{aligned} Q &:= (0 \; I \; 0 \; I \; 0 \; I \; 0)'(0 \; I \; 0 \; I \; 0 \; I \; 0) \\ \epsilon &= 10^{-8} \end{aligned} \qquad (46.48)$$

is to be minimized in order to determine the optimal controller parameters (K_0, K_1, K_2, K_3). As before, let Γ denote the optimal cost matrix of Equation 46.47, corresponding to a minimization of $\bar{J}_\epsilon$, and define $\bar{J}_{opt} = \text{trace}(\Gamma)$. The following results are now obtained:

$$\bar{J}_{opt} = 36.3$$

and the optimal K_0, K_1, K_2, K_3 are given in Table 46.3.

The eigenvalues of the resultant closed-loop system using this controller are given in Table 46.4.

In this case, the response of the resultant closed-loop system for a triangular tracking signal of period 4 seconds is given in Figure 46.13. It is seen that the tracking performance of the system is excellent; it is to be noted that the dominant harmonics of this triangular wave are given by π, 3π, 5π rad/s, and thus the servo-compensator is approximating the tracking of a triangular wave, in this case, by tracking/regulating the main harmonic terms of the periodic signal.

46.4.1 CAD Approach

In designing controllers for actual physical systems, it is often important to impose on the controller construction constraints that are related to the specific system being considered. Such constraints can be incorporated directly using a CAD approach (e.g., see [9]). The following example illustrates the type of results that may be obtained.

TABLE 46.3 Optimal Value of $K0$, $K1$, $K2$, $K3$ Obtained.

K0=	−1.8047e+00	−2.1315e+03		
	−5.8028e+01	8.0400e+01		
K1=	−3.3062e+02	8.0559e+03	−4.0912e+01	−1.8825e+03
	−9.7821e+03	−2.4767e+02	−6.5109e+02	8.0833e+01
K2=	−2.6297e+02	9.7588e+03	−2.1678e+01	−2.2892e+02
	−8.2362e+03	−3.2042e+02	−6.0072e+02	9.6246e+00
K3=	−1.8240e+02	9.9091e+03	−1.7523e+01	−8.3041e+01
	−5.7704e+03	−3.2592e+02	−5.1951e+02	3.3729e+00

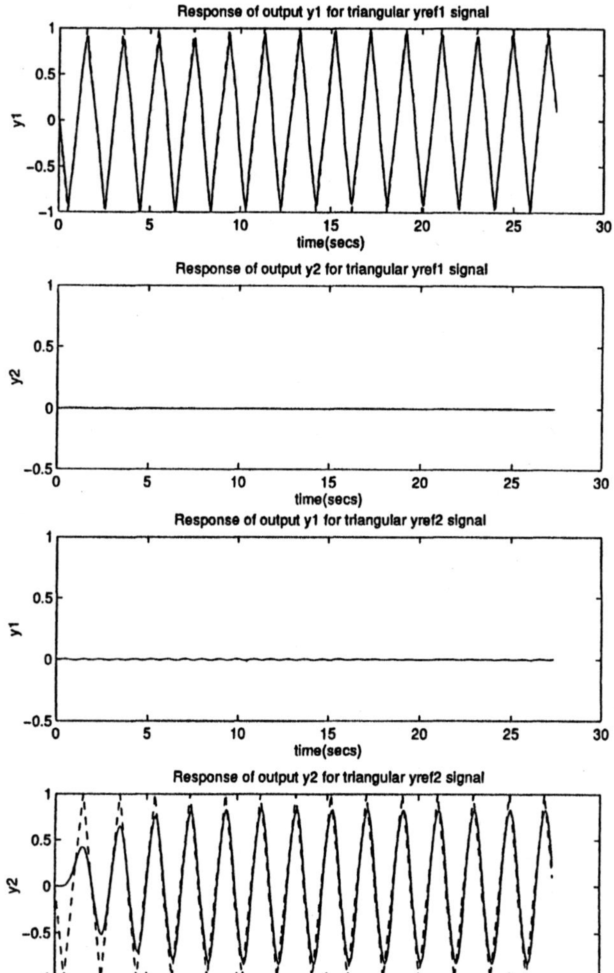

Figure 46.13 Response of closed-loop system for head box example using robust servomechanism controller for triangular reference input signal.

TABLE 46.4 Closed-Loop Eigenvalues.

−1.5016e+01	+2.8300e+01i
−1.5016e+01	−2.8300e+01i
−3.0344e+01	
−1.2005e−01	+1.3596e+01i
−1.2005e−01	−1.3596e+01i
−1.9565e−02	+1.5708e+01i
−1.9565e−02	−1.5708e+01i
−5.4357e−02	+9.4249e+00i
−5.4357e−02	−9.4249e+00i
−3.5863e−02	+6.7460e+00i
−3.5863e−02	−6.7460e+00i
−1.0298e+00	
−4.4094e−01	+3.2555e+00i
−4.4094e−01	−3.2555e+00i

Consider the head box problem modeled by Equation 46.36, controlled by the controller

$$u = K_0 y + K \int_0^t (y - y_{\text{ref}}) d\tau \tag{46.49}$$

and assume that it is desired to determine the controller gain matrices K_0, K_1, so as to minimize the performance index:

$$J_\epsilon = \int_0^\infty (e'e + \epsilon \dot{u}'\dot{u}) d\tau, \quad \epsilon = 10^{-8} \tag{46.50}$$

such that all elements of K_0, K satisfy the constraint:

$$|k_{ij}| \leq 100$$

This constraint could arise, for example, in terms of attempting to regulate the control signal magnitude level for a system. The following results are obtained in this case [9] (see Table 46.5).

Consider now the following model, which approximately describes the behavior of a DC motor:

$$\begin{aligned} \dot{x} &= \begin{bmatrix} -7.535 \times 10^{-2} & 5.163 \\ -209.4 & -198.1 \end{bmatrix} x \\ &+ \begin{bmatrix} 0 \\ 188.7 \end{bmatrix} u + \begin{bmatrix} -4.651 \\ 0 \end{bmatrix} \omega \\ y &= [1 \quad 0] x; \quad y_m = x \end{aligned} \tag{46.51}$$

and consider the following controller:

$$u = K_0 y_m + K \int_0^t (y - y_{\text{ref}}) d\tau \tag{46.52}$$

where (K_0, K) are to be obtained so as to minimize the performance index

$$J_\epsilon = \int_0^\infty (e'e + \dot{u}'\dot{u}) d\tau, \quad \epsilon = 10^{-8} \tag{46.53}$$

TABLE 46.5 Results Obtained.

Constraint	J_{opt}	(K_0, K)	Closed-Loop Poles
None	0.516	$\begin{bmatrix} -4.33 & -4550 & -372 & -9993 \\ -138 & 159 & -9993 & 372 \end{bmatrix}$	$\begin{pmatrix} -2.20 \pm j2.20 \\ -72.1 \pm j72.1 \end{pmatrix}$
$\lvert k_{ij}\rvert \leq 100$	22.2	$\begin{bmatrix} -70 & -100 & -100 & -100 \\ -2.6 & 78 & -29 & 22 \end{bmatrix}$	$\begin{pmatrix} -0.157 \pm j0.330 \\ -2.61 \pm j5.62 \end{pmatrix}$

TABLE 46.6 Results Obtained.

Constraint	J_{opt}	(K_0, K)	Closed-Loop Poles
None	9.93_{10}^{-3}	$[-98.1 \; -1.49 \; -10000]$	$\begin{pmatrix} -116 \pm j161 \\ -249 \end{pmatrix}$
$\zeta \geq 1$	1.00_{10}^{-2}	$[-109 \; -1.62 \; -9480]$	$\begin{pmatrix} -153 \pm j153 \\ -197 \end{pmatrix}$

subject to the constraint that the damping factor of the closed-loop system should have the property that $\zeta \geq 1$, in order to prevent an excessively oscillatory response, say. The following results are obtained [9] (see Table 46.6).

Discrete Systems

The previous results have considered a continuous system. For discrete-time systems, equivalent conditions for the existence of a solution to the RSP and the necessary controller structure can be obtained (e.g., see [4]). The following example illustrates this point.

Consider the following discrete system:

$$\begin{aligned} x_{k+1} &= \begin{pmatrix} 0.9512 & 0 \\ 0 & 0.09048 \end{pmatrix} x_k \\ &+ \begin{pmatrix} 4.877 & 4.877 \\ -1.1895 & 3.369 \end{pmatrix} u_k \\ y_k &= \begin{pmatrix} 1 & 0 \\ 0 & 1 \end{pmatrix} x_k; \quad y^m = y \end{aligned} \tag{46.54}$$

in which it is desired to solve the RSP for the class of constant disturbances/constant tracking signals. In this case, the above system is controllable and observable and has no transmission zeros, so that there exists a solution to the RSP; the servomechanism controller becomes in this case:

$$\begin{aligned} u_k &= K_0 y_k + K\eta_k \\ \eta_{k+1} &= \eta_k + y_k - y_{\text{ref}} \end{aligned} \tag{46.55}$$

and, on minimizing the performance index

$$J_\epsilon^* = \sum_{k=1}^{\infty}[e_k' e_k + \epsilon(u_{k+1} - u_k)'(u_{k+1} - u_k)]; \quad \epsilon = 10^{-8} \tag{46.56}$$

the following controller gain matrices are obtained:

$$\begin{aligned} &J_{opt}^* = 6.00 \\ &(K_0, K)_{opt} = \\ &\begin{bmatrix} -0.296 & 0.239 & -0.152 & 0.219 \\ -0.104 & -0.239 & -0.0535 & -0.219 \end{bmatrix} \end{aligned} \tag{46.57}$$

The closed-loop poles obtained by applying the above controller to the plant modeled by Equation 46.54 are, in this case, given by $-8.4 \times 10^{-10} \pm j1.6 \times 10^{-5}, 7.2 \times 10^{-11} \pm j8.1 \times 10^{-6}$); i.e., a "dead-beat" closed-loop system time response is obtained.

46.4.2 Case Study Problem – Distillation Column

The following model of a binary distillation column with pressure variation is considered:

$$\begin{aligned} \dot{x} &= Ax + Bu + E\omega \\ y &= Cx \end{aligned} \tag{46.58}$$

where (C, A, B, E) are given in Table 46.7. Here y_1 is the composition of the more volatile component in the bottom of the column, y_2 is the composition of the more volatile component in the top of the column, and y_3 is the pressure in the column; ω_1 is the input feed disturbance in the column; and u_1 the reheater input, u_2 the condensor input, and u_3 the reflux in the system.

Eigenvalues and Transmission Zeros of Distillation Column

The open-loop eigenvalues and transmission zeros of the distillation column are given in Table 46.8, which implies that the system is minimum phase.

It is desired now to find a controller that solves the RSP problem for this system for the case of constant disturbances and constant reference input signals. In this case, the existence conditions of Theorem 46.1 hold, so that a solution to the problem exists; in particular, there exists a solution to the "perfect control robust servomechanism" problem (see Theorem 46.2) for the system.

Perfect Robust Controller

The following controller is obtained from Zhang and Davison [14], and it can be shown to produce "perfect control" (i.e., the transient error in the system can be made arbitrarily small) in the system as $\epsilon \to 0$:

$$u = \frac{1}{\epsilon}\frac{(s+1)^2}{(\epsilon^2 s + 1)^2}\frac{\Theta}{s}(y - y_{\text{ref}}) \tag{46.59}$$

where

TABLE 46.7 Data Matrices for Distillation Column Model.

A=

	x1	x2	x3	x4	x5
x1	−0.01400	0.00430	0	0	0
x2	0.00950	−0.01380	0.00460	0	0
x3	0	0.00950	−0.01410	0.00630	0
x4	0	0	0.00950	−0.01580	0.01100
x5	0	0	0	0.00950	−0.03120
x6	0	0	0	0	0.02020
x7	0	0	0	0	0
x8	0	0	0	0	0
x9	0	0	0	0	0
x10	0	0	0	0	0
x11	0.02550	0	0	0	0

	x6	x7	x8	x9	x10
x1	0	0	0	0	0
x2	0	0	0	0	0
x3	0	0	0	0	0
x4	0	0	0	0	0
x5	0.01500	0	0	0	0
x6	−0.03520	0.02200	0	0	0
x7	0.02020	−0.04220	0.02800	0	0
x8	0	0.02020	−0.04820	0.03700	0
x9	0	0	0.02020	−0.05720	0.04200
x10	0	0	0	0.02020	−0.04830
x11	0	0	0	0	0.02550

	x11
x1	0
x2	0.00050
x3	0.00020
x4	0
x5	0
x6	0
x7	0
x8	0.00020
x9	0.00050
x10	0.00050
x11	−0.01850

B=

	u1	u2	u3
x1	0	0	0
x2	5.00000e−06	−4.00000e−05	0.00250
x3	2.00000e−06	−2.00000e−05	0.00500
x4	1.00000e−06	−1.00000e−05	0.00500
x5	0	0	0.00500
x6	0	0	0.00500
x7	−5.00000e−06	1.00000e−05	0.00500
x8	−1.00000e−05	3.00000e−05	0.00500
x9	−4.00000e−05	5.00000e−06	0.00250
x10	−2.00000e−05	2.00000e−06	0.00250
x11	0.00046	0.00046	0

TABLE 46.7 Data Matrices for Distillation Column Model (cont.).

C=						
		x1	x2	x3	x4	x5
	y1	0	0	0	0	0
	y2	1.00000	0	0	0	0
	y3	0	0	0	0	0
		x6	x7	x8	x9	x10
	y1	0	0	0	0	1.00000
	y2	0	0	0	0	0
	y3	0	0	0	0	0
		x11				
	y1	0				
	y2	0				
	y3	1.00000				
E=						
		w1				
	x1	0				
	x2	0				
	x3	0				
	x4	0				
	x5	0.01				
	x6	0				
	x7	0				
	x8	0				
	x9	0				
	x10	0				
	x11	0				

$$\Theta := \begin{bmatrix} 1.7599 \times 10^0 & -3.4710 \times 10^6 & -1.0869 \times 10^3 \\ -1.7599 \times 10^0 & 3.4710 \times 10^6 & -1.0870 \times 10^3 \\ -3.9998 \times 10^2 & -3.0545 \times 10^4 & -7.8258 \times 10^0 \end{bmatrix} \tag{46.60}$$

Properties of Closed-Loop System

In order to determine the potential "speed of response" that may be obtained by the controller modeled by Equation 46.59, the following closed-loop eigenvalues of the system are obtained with $\epsilon = 0.1$ (see Table 46.9).

Using the controller of Equation 46.59 with $\epsilon = 0.1$, the response of Figure 46.14 is then obtained for the case of a unit-step increase in $y_{\text{ref}} = (1\ 0\ 0)'$, $(0\ 1\ 0)'$, $(0\ 0\ 1)'$, respectively, with zero initial conditions. It is seen that "perfect control" indeed does take place; i.e., all transients have died down in less than 1 second, the system displays "low interaction," and no "excessive peaking occurs." In real life, however, this controller would not be used because the control inputs are excessively large (see Figure 46.14).

The following decentralized controller, obtained by the procedure given in [11], would be quite realistic to implement, however, since the control input signals do not "peak" now as they did in the previous controller.

46.4.3 Decentralized Robust Controller

$$\begin{aligned} u &= \mathcal{K}_1 y + \mathcal{K}_2 \eta + \mathcal{K}_3 \rho \\ \dot{\eta} &= y - y_{\text{ref}} \\ \dot{\rho} &= \mathcal{K}_5 \rho + \mathcal{K}_4 y \end{aligned} \tag{46.61}$$

where $\mathcal{K}_1, \mathcal{K}_2, \mathcal{K}_3, \mathcal{K}_4, \mathcal{K}_5$ are given in Table 46.10.

Properties of Closed-Loop System

The following closed-loop eigenvalues are obtained by applying the controller of Equation 46.61 to Equation 46.58 (see Table 46.11).

Using the controller of Equation 46.61, the response of Figure 46.15 is then obtained for the case of a unit step in $y_{\text{ref}} = (1\ 0\ 0)'$, $(0\ 1\ 0)'$, $(0\ 0\ 1)'$, respectively, and for the case of a unit step in the disturbance term ω with zero initial conditions. It is seen that excellent tracking/regulation takes place; i.e., the output responses obtained show little interaction effects with no peaking occurring, and the control input signals are now quite realistic to implement. In addition, the controller has the additional advantage of being decentralized; i.e., the controller is particularly simple to implement. The time response of the closed-loop system, however, is now much slower compared to the case when perfect control is applied.

46.5 Concluding Remarks

In this overview on the RSP, the emphasis has been placed on the control of LTI continuous systems, and existence conditions and

TABLE 46.8 Properties of Distillation Column Model.

Open-Loop Eigenvalues	Transmission Zeros
−9.6031e−02	−9.1024e−02
−7.0083e−02	−6.6830e−02
−5.0545e−02	−5.0627e−02
−1.2152e−04	−2.9083e−02
−3.2355e−03	−2.2078e−02
−3.3900e−02	−9.5168e−03
−7.7594e−03	−6.4089e−03
−1.9887e−02	
−1.8154e−02	
−2.4587e−02	
−1.4196e−02	

TABLE 46.9 Closed-Loop Eigenvalues.

−9.9102e+01	+3.1595e+02i
−9.9102e+01	−3.1595e+02i
−9.9097e+01	+3.1594e+02i
−9.9097e+01	−3.1594e+02i
−1.2776e+02	
−5.9778e+01	
−1.0030e+01	
−1.6633e+00	
−7.8470e−01	
−9.1245e−01	+2.8213e−01i
−9.1245e−01	−2.8213e−01i
−9.1207e−01	+2.8314e−01i
−9.1207e−01	−2.8314e−01i
−9.1027e−02	
−6.6832e−02	
−6.4090e−03	
−9.5169e−03	
−2.9083e−02	
−5.0626e−02	
−2.2078e−02	

TABLE 46.10 Decentralized Controller Gains Obtained.

K1 =	1.7409e+05	0	0
	0	1.9140e+04	0
	0	0	−7.7460e+03
K2 =	4.5284e+02	0	0
	0	1.2143e+01	0
	0	0	−3.7682e+00
K3 =	5.7841e+05	0	0
	0	3.0466e+05	0
	0	0	−1.0636e+07
K4 =	1.0000e+00	0	0
	0	1.0000e+00	0
	0	0	1.0000e+00
K5 =	−1.1024e+05	0	0
	0	−8.1558e+07	0
	0	0	−1.3506e+04

TABLE 46.11 Closed-Loop Eigenvalues.

−8.1558e+07	
−1.3506e+04	
−1.1023e+03	
−1.7595e+00	+4.1365e+01i
−1.7595e+00	−4.1365e+01i
−9.5681e−03	+7.5529e−02i
−9.5681e−03	−7.5529e−02i
−8.9273e−02	
−6.5720e−02	
−5.0556e−02	
−2.9058e−02	
−2.2081e−02	
−9.5034e−03	
−6.4076e−03	
−2.5926e−03	
−6.2808e−04	
−4.4157e−04	

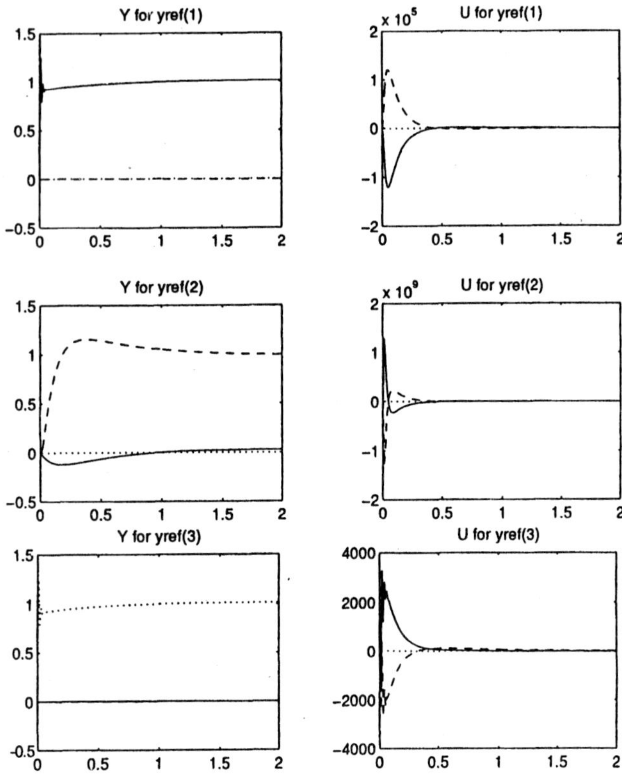

Figure 46.14 Response of closed-loop system for distillation column example using perfect robust servomechanism controller for unit step in y_{ref} given by $y_{ref}(1) = (1\,0\,0)'$, $y_{ref}(2) = (0\,1\,0)'$, $y_{ref}(3) = (0\,0\,1)'$.

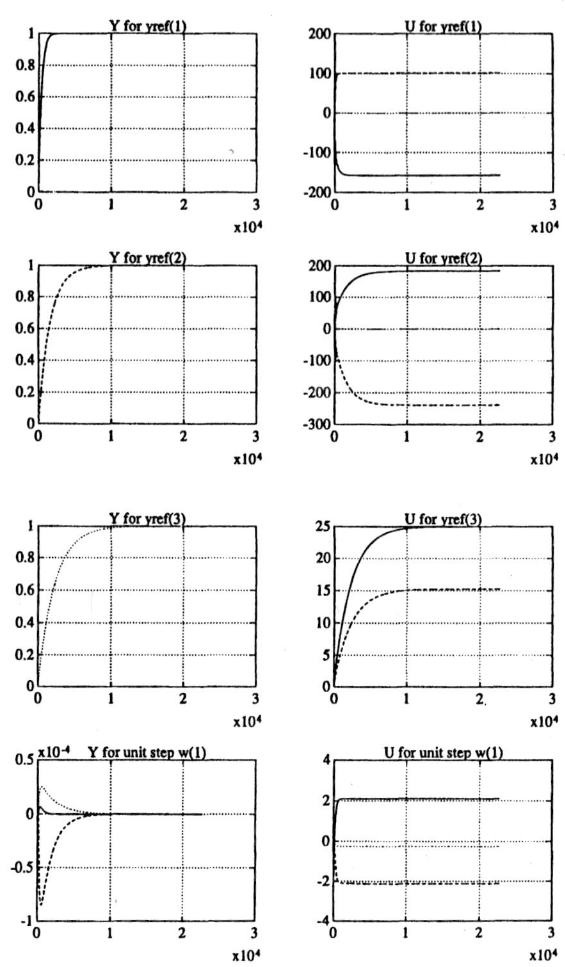

Figure 46.15 Response of closed-loop system for distillation column example using decentralized robust servomechanism controller for unit-step in y_{ref} given by $y_{ref}(1) = (1\,0\,0)'$, $y_{ref}(2) = (0\,1\,0)'$, $y_{ref}(3) = (0\,0\,1)'$ and for a unit-step disturbance.

corresponding required controller construction to solve the RSP have been reviewed. To demonstrate the principles involved, various simple nonminimum- and minimum-phase examples were initially considered, and then a case study of a nontrivial system example was studied.

46.6 Defining Terms

Arbitrarily good approximate error regulation (AGAER): The property of a closed-loop system that permits arbitrarily good regulation to occur for arbitrary disturbance/tracking signals of a specified class.

Centralized fixed modes: Those modes of an LTI system that are not both simultaneously controllable and observable.

Decentralized control: Refers to a controller in which the information flow between the inputs and outputs is constrained to be block diagonal.

Error coefficient: The steady-state error coefficient matrix associated with a closed-loop system for a given class of disturbance or reference input signals.

Gain margin (GM): Given a stable closed-loop system, the GM (θ, β) refers to the largest perturbation of gain in the system's transfer function matrix that may occur before instability occurs.

High-gain servomechanism controller (HGSC): A controller that gives perfect tracking for continuous minimum-phase systems.

Minimum phase: A system whose transmission zeros are all contained in the open left complex half-plane.

Nonminimum phase: A system that is not minimum phase.

Perfect control: The ability of a controller to provide arbitrarily good transient response in the system.

Perfect robust controller: A controller that solves the RSP such that perfect control occurs.

Robust servomechanism problem (RSP): The problem of finding a controller to solve the servomechanism problem that has the property of providing exact asymptotic error regulation, independent of any perturbations in the plant that do not destabilize the system.

Servomechanism problem: The problem of finding a controller to provide asymptotic error regulation and tracking for a system, subject to a specified class of disturbances and tracking signals.

Servo-compensator: A compensator that is used in the construction of a controller to solve the RSP.

Stabilizing compensator: A controller that stabilizes a system.

Transmission zero: A generalization of the notion of a zero of a single-input/single-output system to multivariable systems.

References

[1] Davison, E.J., The feedforward control of linear time invariant multivariable systems, *Automatica,* 9(5), 561–573, 1973.

[2] Davison, E.J. and Wang, S.H., Properties of linear time invariant multivariable systems subject to arbitrary output and state feedback, *IEEE Trans. Autom. Control,* 18, 24–32, 1973.

[3] Davison, E.J. and Wang, S.H., Properties and calculation of transmission zeros of linear multivariable time-invariant systems, *Automatica,* 10, 643–658, 1974.

[4] Goldenberg, A. and Davison, E.J., The feedforward and robust control of a general servomechanism problem with time lag, *8th Annu. Princeton Conf. Inf. Sci. Syst.,* 80–84, 1974.

[5] Davison, E.J. and Goldenberg, A., The robust control of a general servomechanism problem: the servo compensator, *Automatica,* 11, 461–471, 1975.

[6] Davison, E.J., The robust control of a servomechanism problem for linear time-invariant multivariable systems, *IEEE Trans. Autom. Control,* 21, 25–34, 1976.

[7] Davison, E.J., Multivariable tuning regulators: The feedforward and robust control of a general servomechanism problem, *IEEE Trans. Autom. Control,* 21, 35–47, 1976.

[8] Davison, E.J., The robust decentralized control of a general servomechanism problem, *IEEE Trans. Autom. Control,* 21, 14–24, 1976.

[9] Davison, E.J. and Ferguson, I.J., The design of controllers for the multivariable robust servomechanism problem using parameter optimization methods, *IEEE Trans. Autom. Control,* 26, 93–110, 1981.

[10] Davison, E.J. and Copeland, B., Gain margin and time lag tolerance constraints applied to the stabilization problem and robust servomechanism problem, *IEEE Trans. Autom. Control,* 30, 229–239, 1985.

[11] Davison, E.J. and Chang, T., Decentralized controller design using parameter optimization methods, *Control: Theor. Adv. Technol.* 2, 131–154, 1986.

[12] Davison, E.J. and Scherzinger, B., Perfect control of the robust servomechanism problem, *IEEE Trans. Autom. Control,* 32(8), 689–702, 1987.

[13] Qiu, Li and Davison, E.J., Performance limitations of non-minimum phase systems in the servomechanism problem, *Automatica,* 29(2), 337–349, 1993.

[14] Zhang, H. and Davison, E.J., A uniform high gain compensator for multivariable systems, *1994 IEEE Control Decision Conf.,* 892–897, 1994.

Further Reading

There are a number of important issues that have not yet been considered in this chapter. For example, when disturbances are measurable, so-called *feedforward control* [1], [7] can be highly effective in minimizing the effects of disturbances in the servomechanism problem.

In many classes of problems, the controller must often be constrained to be decentralized, e.g., in process control systems, power system problems, transportation system problems. A treatment of the so-called *decentralized robust servomechanism problem,* which arises in this case, is given in [8] and [11].

The effect of transportation delay in a system is often of critical importance in the design of controllers to solve the servomechanism problem. A treatment of systems that have *time lag* is given in [4].

Finally, it is often the case that no mathematical model is actually available to describe the plant that is to be controlled. In this case, if the plant is open-loop asymptotically stable, so-called *tuning regulator theory* [7] can be applied to obtain existence conditions and to design a controller, which can then be applied to the plant to solve the servomechanism problem.

The above treatment of the servomechanism problem has been carried out in a state-space setting; alternative treatments may be found using other settings such as geometric methods, frequency-domain methods, polynomial matrix representation methods, coprime matrix factorization methods, etc.

47

Numerical Optimization-Based Design

V. Balakrishnan
Purdue University

A. L. Tits
University of Maryland

47.1 Introduction

The fundamental control system design problem is the following: *Given a system that does not perform satisfactorily, devise a strategy to improve its performance.* Often, quantitative measures are available to evaluate the performance of the system. **Numerical optimization-based design**, or optimization-based design, for short, is design with the goal of optimizing these measures of performance by iterative optimization methods implemented on a computer.

Optimization-based design has a relatively short history compared to other areas of control, because the factors that have led it to evolve into a widely used design procedure are of recent vintage. The first is the exploding growth in computer power, the second, recent breakthroughs in optimization theory and algorithms, especially **convex optimization**, and the third, recent advances in numerical linear algebra.

The optimization-based design methods that we will consider in this chapter must be contrasted against classical optimal control methods. Classical optimal control methods rely on theoretical techniques (variational arguments or dynamic programming, for example) to derive the optimal control strategy (or conditions that characterize it). This is a very powerful approach because "exact" strategies can be obtained and limits of performance can be derived. However, this approach applies only to very restricted classes of models (such as lumped linear time-invariant (LTI) systems, for instance) and the class of performance measures that can be handled is very limited. Also, many design constraints cannot be handled by this approach. The methods we consider in this chapter, on the other hand, do not yield analytical solutions; instead the design problem is posed as a numerical optimization problem, typically either minimization (or maximization) or min-max (or max-min). This optimization problem is then solved numerically on a computer. This approach can tackle a much wider variety of problems compared with the analytical approach, usually at the cost of increased computation. Still, in some cases (especially with convex optimization methods), the computation required for optimization-based design is comparable to that required for the evaluation of related analytical solutions. For this reason, optimization-based design methods often compete quite favorably with related analytical design methods.

Our intent in this chapter is to introduce the reader to several important control system analysis and design methods that rely on numerical optimization. Whenever possible, we have kept tedious details to the minimum, because these obscure the ideas underlying the methods; the reader interested in details or technical conditions can turn to the list of references. We will also not cover several important optimization-based methods discussed elsewhere in the handbook; see Section VII. We will only be considering continuous-time systems in this chapter; most design methods that we discuss can be extended to discrete-time systems in a straightforward manner.

0-8493-8570-9/96/$0.00+$.50

47.2 A Framework for Controller Design

The framework for discussion is shown in Figure 47.1. P is the model of the plant, that is, of the system to be controlled. K is the controller that implements the control strategy for improving the system performance. y is the signal that the controller has access to, and u is the output of the controller that drives the plant. w and z represent inputs and outputs of interest, so that the map from w to z contains all the input-output maps of interest. We let n_w, n_u, n_z and n_y denote the sizes (i.e., number of components) of w, u, z and y respectively.

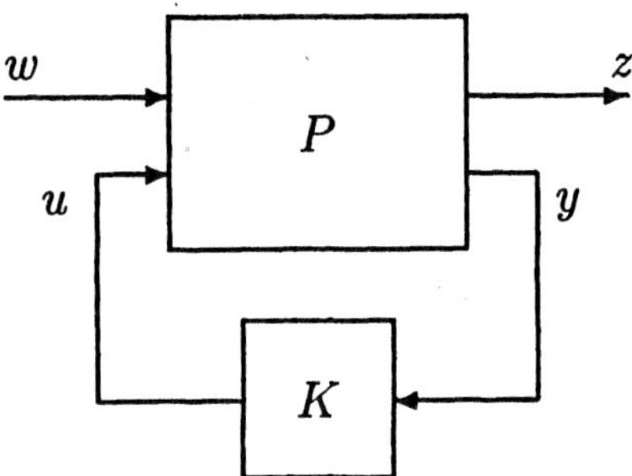

Figure 47.1 A standard controller design framework.

The choice of an approach for tackling a control design problem within this framework must consider the following three factors:[1] *The plant model P:* Depending on the particular modeling, system identification, and validation procedures, the plant model P may or may not be linear and may or may not be time-invariant; it may also be unspecified except for the constraint that it lie within a given set Π ("uncertainty modeling").
The class of controllers K: The controller could be static or dynamic, linear or nonlinear, time-invariant or not; it could be a state-feedback (i.e., the measured output y is the state of P) or an output-feedback controller; nonlinear controllers could be restricted to gain-scheduled controllers or to linear controllers augmented with output saturation; the order of the controller or its internal structure also could be specified or restricted.
The performance specifications: Perhaps the most important of these requires that the nominal closed-loop system be stable. Typical performance specifications, beyond stability, require some desirable property for the map from w to z. Often, it is required that K minimize some norm of the map from w to z, subject to constraints on other norms, as well as other constraints on the behavior of the closed-loop system (e.g., the decay rate of trajectories, the response to some specific reference inputs such as steps, etc.). When the plant is not completely specified and is known only to lie in some set Π, these performance specifications (in particular, closed-loop stability) may be required for every $P \in \Pi$ (this is called "worst-case design"), or they may be required "on the average" (this means that with the controller K in place, the "average behavior" of the closed-loop system is satisfactory).

In the simplest cases, numerical optimization can be bypassed altogether. A typical example is the linear quadratic Gaussian (LQG) problem, which applies to LTI models with a quadratic performance index; see Chapters 35, 39, and 40 for details. In most cases, however, numerical optimization is called for. These cases constitute the focus of this chapter. Sections 47.3 and 47.4 deal with a class of problems that enjoy certain convexity properties, making them, in a sense that we will describe, easier to solve. In Section 47.3, we consider the simplest paradigm, that of designing LTI controllers for a plant which is assumed to be LTI and known exactly. For a wide variety of performance specifications, this problem can be solved "easily", using convex optimization. Then, in Section 47.4, we turn to the harder problem of worst-case design of LTI controllers when the plant is known to belong to a certain class. When the controller is restricted to be static linear state-feedback, the design problem can be solved via convex optimization in a number of cases. The class of problems that can be tackled by convex optimization is large enough to be of definite interest. However, in practical situations, it is often necessary to deal with models or impose restrictions that rule out convex optimization. In such cases, one can resort to local optimization methods. While it is not guaranteed that the global optimum will be achieved (at least with reasonable computation), the class of problems that can be tackled is very large; in particular, there is little restriction on the possible plant models, and the main restriction on the class of controllers is that it must be parameterized by a finite number of real parameters, as discussed in Section 47.5. When the plant model is highly nonlinear and the controller is allowed to be time-varying, the solution of an open-loop optimal control problem is often a key stepping stone. Numerical methods for the solution of such problems are considered in Section 47.6. Finally, in Section 47.7, we discuss multiobjective problems and trade-off exploration.

47.3 LTI Controllers for LTI Systems: Design Based on the Youla Parameter

For the special case when the plant is modeled as a finite-dimensional LTI system, and LTI controllers are sought, the **Youla parameterization** of the set of all achievable stable closed-loop maps can be combined with convex optimization to design optimal (or, more often, suboptimal) controllers.

Consider the controller design framework in Figure 47.1, where the plant P has a transfer function

$$P(s) = \begin{bmatrix} P_{zw}(s) & P_{zu}(s) \\ P_{yw}(s) & P_{yu}(s) \end{bmatrix}, \tag{47.1}$$

where P_{zw} is the open-loop (i.e., with the controller removed from the loop) transfer function of the plant from w to z, P_{zu}

[1] These factors are in general not "God-given", and a trial and error approach is generally required.

that from u to z, P_{yw} that from w to y, and P_{yu} that from u to y. Then, with $K(s)$ denoting the transfer function of the LTI controller, the transfer function from w to z is

$$H_{cl}(s) = P_{zw}(s) + P_{zu}(s)K(s)\left(I - P_{yu}(s)K(s)\right)^{-1} P_{yw}(s). \tag{47.2}$$

H_{cl} is called the "closed-loop" transfer function from w to z. Note that z can include components of the control input u, so that specifications on the control input, such as bounds on the control effort, can be handled. The set of achievable, stable closed-loop maps from w to z is given by

$$\mathcal{H}_{cl} = \left\{H_{cl} : H_{cl} \text{ satisfies (47.2), } K \text{ stabilizes the system}\right\}. \tag{47.3}$$

The set of the controllers K that stabilize the system is in general not a convex set. Thus optimizing over $\mathcal{H}_{cl}$ using the description (47.3), with K as the (infinite-dimensional) optimization variable, is a very hard numerical problem. However, (see [1] for details) the set $\mathcal{H}_{cl}$ can be also written as

$$\mathcal{H}_{cl} = \{H_{cl} : H_{cl}(s) = T_1(s) + T_2(s)Q(s)T_3(s),\ Q \text{ is stable}\}, \tag{47.4}$$

where T_1, T_2 and T_3 are fixed, stable transfer matrices that can be computed from the data characterizing P. Moreover, given Q, the corresponding controller K can be immediately computed as $K = \mathcal{K}(Q)$(where $\mathcal{K}(\cdot)$ is a certain rational function of Q). The most important observation about this reparametrization of $\mathcal{H}_{cl}$ is that it is *affine* in the infinite-dimensional parameter Q; it is therefore a convex parameterization of the set of achievable stable closed-loop maps from w to z. (The parameter Q is also referred to as the Youla parameter.) This fact has an important ramification—it is possible now to use convex optimization techniques to find an optimal parameter Q_{opt}, and therefore an optimal controller $\mathcal{K}(Q_{opt})$.

EXAMPLE 47.1:

We demonstrate the affine reparametrization of $\mathcal{H}_{cl}$ with a simple example. Let

$$P(s) = \begin{bmatrix} P_{zw}(s) & P_{zu}(s) \\ P_{yw}(s) & P_{yu}(s) \end{bmatrix} = \begin{bmatrix} \frac{1}{s+2} & 1 \\ 1 & \frac{1}{s+2} \end{bmatrix}. \tag{47.5}$$

Then,

$$\mathcal{H}_{cl} = \left\{H_{cl} : H_{cl} = \frac{1}{s+2} + K(s)\left(1 - \frac{1}{s+2}K(s)\right)^{-1}, \right. \\ \left. K \text{ stabilizes system}\right\}. \tag{47.6}$$

The affine parameterization is

$$\mathcal{H}_{cl} = \left\{H_{cl} : H_{cl} = \frac{1}{s+2} + Q(s),\ Q \text{ is stable}\right\}. \tag{47.7}$$

Moreover, given the Youla parameter Q, the corresponding controller is given by

$$K(s) = \left(1 + \frac{Q(s)}{s+2}\right)^{-1} Q(s). \tag{47.8}$$

The reparametrization in this example is particularly simple—$T_1 = P_{zw}$, $T_2 = P_{zu}$ and $T_3 = P_{yw}$—because the open-loop transfer function from u to y (i.e., P_{yu}) is stable to start with. In cases when P_{yu} is unstable, the parameterization is more complicated; see for example [1].

The general procedure for designing controllers using the Youla parameterization proceeds as follows. Let ϕ_0, ϕ_1, ..., ϕ_m be (not necessarily differentiable) convex functionals on the closed-loop map that represent performance measures. These performance measures may be norms (typically $\mathbf{H}_2$ or $\mathbf{H}_\infty$ norms), certain time-domain quantities (step response overshoot, steady-state errors), etc.

Then the problem,

$$\begin{array}{ll} \text{minimize, w.r.t. } Q: & \phi_0(T_1 + T_2QT_3) \\ \text{subject to:} & \phi_1(T_1 + T_2QT_3) \le \alpha_1 \\ & \vdots \\ & \phi_m(T_1 + T_2QT_3) \le \alpha_m, \end{array} \tag{47.9}$$

is a convex optimization problem (with an infinite-dimensional optimization variable Q), because it has the form "Minimize a convex function subject to convex constraints". This problem is to minimize a measure of performance of the closed-loop system subject to other performance constraints.

In practice, problem (47.9) is solved by searching for Q over a finite-dimensional subspace. Typically, Q is restricted to lie in the set,

$$\{Q : Q = \beta_1 Q_1 + \cdots + \beta_n Q_n\}, \tag{47.10}$$

where $Q_1, \ldots, Q_n$ are stable, fixed, transfer matrices, and scalars $\beta_1, \ldots, \beta_n$ are the optimization variables. This enables us to solve problem (47.9) "approximately" by solving the following problem with a finite number of scalar optimization variables:

$$\begin{array}{ll} \text{minimize, w.r.t. } \beta_1, \ldots, \beta_n: & \phi_0(T_1 + T_2Q(\beta)T_3) \\ \text{subject to:} & \phi_1(T_1 + T_2Q(\beta)T_3) \le \alpha_1 \\ & \vdots \\ & \phi_m(T_1 + T_2Q(\beta)T_3) \le \alpha_m \end{array} \tag{47.11}$$

where $Q(\beta) = \beta_1 Q_1 + \cdots + \beta_n Q_n$.

Approximating the infinite-dimensional parameter Q by a finite-dimensional quantity is referred to as a "Ritz approximation". Evidently, the transfer matrices Q_i and their number, n, should be so chosen that the optimal parameter Q can be approximated with sufficient accuracy.

EXAMPLE 47.2:

With the same plant as in (47.5), take ϕ_0 as the $\mathbf{H}_\infty$ norm (i.e., peak value of the transfer function magnitude) of H_{cl}, ϕ_1 as the steady-state magnitude of z for a unit-step input at w (this is just the transfer function magnitude at DC), and $\alpha_1 = 0.1$. The design problem may thus be viewed as the minimization of the closed-loop $\mathbf{H}_\infty$ norm, subject to a DC disturbance rejection constraint.

Let us approximate Q by $Q(\beta) = \beta_1 + \beta_2 \frac{1}{s+1}$. Thus,

$$\phi_0(T_1 + T_2 Q T_3) = \left\| \frac{1}{s+2} + \beta_1 + \beta_2 \frac{1}{s+1} \right\|_\infty,$$
$$\phi_1(T_1 + T_2 Q T_3) = |0.5 + \beta_1 + \beta_2|. \quad (47.12)$$

Then, the optimization problem (47.11) becomes

$$\begin{array}{ll} \text{minimize, w.r.t. } \beta_1, \beta_2: & \left\| \frac{1}{s+2} + \beta_1 + \beta_2 \frac{1}{s+1} \right\|_\infty \\ \text{subject to:} & |0.5 + \beta_1 + \beta_2| \leq 0.1. \end{array} \quad (47.13)$$

The most important observation concerning our reduction of the LTI controller design to problem (47.11) is that it is a convex optimization problem with a finite number of optimization variables. Convexity has several important implications:

- Every stationary point of the optimization problem (47.11) is also a global minimizer.
- The problem can be solved in polynomial-time.
- We can immediately write necessary and sufficient optimality conditions.
- There is a well-developed duality theory.

From a practical standpoint, there are effective and powerful algorithms for the solution of problems such as (47.11), that is, algorithms that rapidly compute the global optimum, with non-heuristic stopping criteria. These algorithms range from simple descent-type or quasi-Newton methods for smooth problems to sophisticated cutting-plane or interior-point methods for non-smooth problems. A comprehensive literature is available on algorithms for convex programming; see for example, [11] and [17]; see also [1].

47.4 LTI Controllers for Uncertain Systems: LMI Synthesis

We now outline one convex optimization-based approach applicable when the plant P is not known exactly, but is known only to belong to a set Π of a certain type. This approach is controller design based on **Linear Matrix Inequalities** (LMIs).

47.4.1 Optimization Over Linear Matrix Inequalities

A linear matrix inequality is a matrix inequality of the form

$$F(\zeta) \triangleq F_0 + \sum_{i=1}^{m} \zeta_i F_i > 0, \quad (47.14)$$

where $\zeta \in \mathbf{R}^m$ is the variable, and $F_i = F_i^T \in \mathbf{R}^{n \times n}, i = 0, \ldots, m$ are given. The inequality symbol in (47.14) means that $F(\zeta)$ is positive-definite, i.e., $u^T F(\zeta) u > 0$ for all nonzero $u \in \mathbf{R}^n$. The set $\{\zeta \mid F(\zeta) > 0\}$ is convex. (We have used strict inequality mostly as a convenience; inequalities of the form $F(\zeta) \geq 0$ are also readily handled.)

Multiple LMIs

$$F_1(\zeta) > 0, \ldots, F_n(\zeta) > 0 \quad (47.15)$$

can be expressed as the single LMI

$$\begin{bmatrix} F_1(\zeta) & 0 & \ldots & 0 \\ 0 & F_2(\zeta) & \ldots & 0 \\ \vdots & \vdots & \ddots & \vdots \\ 0 & 0 & \ldots & F_n(\zeta) \end{bmatrix} > 0. \quad (47.16)$$

Therefore we will not distinguish between a set of LMIs and a single LMI, i.e., "the LMI $F_1(\zeta) > 0, \ldots, F_n(\zeta) > 0$" will mean "the LMI $\mathbf{diag}\,(F_1(\zeta), \ldots, F_n(\zeta)) > 0$". When the matrices F_i are diagonal, the LMI $F(\zeta) > 0$ is just a set of linear inequalities.

For many problems, the variables are matrices, e.g.,

$$A^T P + PA < 0 \quad (47.17)$$

where $A \in \mathbf{R}^{n \times n}$ is given and $P = P^T$ is the variable. In this case we will not write out the LMI explicitly[2] in the form $F(\zeta) > 0$, but instead make clear which matrices are the variables. The phrase "the LMI $A^T P + PA < 0$ in P" means that the matrix P is the variable.

LMI feasibility problems. Given an LMI $F(\zeta) > 0$, the corresponding LMI Problem (LMIP) is to find ζ^{feas} such that $F(\zeta^{\text{feas}}) > 0$ or determine that the LMI is infeasible. This is a convex feasibility problem. We will say "solving the LMI $F(\zeta) > 0$" to mean solving the corresponding LMIP.

Eigenvalue problems. The eigenvalue problem (EVP) is to minimize[3] the maximum eigenvalue of a matrix, subject to an LMI, which is equivalently expressed as

$$\begin{array}{ll} \text{minimize, w.r.t. } \zeta \text{ and } \lambda: & \lambda \\ \text{subject to:} & \lambda I - A(\zeta) > 0, \; B(\zeta) > 0. \end{array} \quad (47.18)$$

Here, A and B are symmetric matrices that depend affinely on the optimization variable ζ. This is a convex optimization problem.

Another LMI-based optimization problem that arises frequently in optimization-based design is the generalized eigenvalue problem or GEVP; see [3] for details.

47.4.2 Analysis and Design of Uncertain Control Systems Using LMIs

We next describe the set of plants for which we will synthesize controllers using LMI-based optimization. The set Π is described by the following state equations:

$$\begin{aligned} \dot{x} &= A(t)x + B_u(t)u + B_w(t)w, \quad x(0) = x_0, \\ z &= C_z(t)x + D_{zu}(t)u + D_{zw}(t)w, \end{aligned} \quad (47.19)$$

[2]This can be done as follows: Let $P_1, \ldots, P_m$ be a basis for symmetric $n \times n$ matrices ($m = n(n+1)/2$). Then take $F_0 = 0$ and $F_i = -A^T P_i - P_i A$.

[3]Technically, we seek the infimum, rather than the minimum (and corresponding parameter values in the closure of the constraint set).

where the matrices in (47.19) are unknown except for the fact that they satisfy

$$\begin{bmatrix} A(t) & B_u(t) & B_w(t) \\ C_z(t) & D_{zu}(t) & D_{zw}(t) \end{bmatrix} \in \Omega \qquad \text{for all } t \geq 0, \tag{47.20}$$

where $\Omega \subseteq \mathbf{R}^{(n+n_z)\times(n+n_u+n_w)}$ is a convex set of a certain type. In some applications we can have one or more of the integers n_u, n_w, and n_z equal to zero, which means that the corresponding variable is not used. For example, when $n_u = n_w = n_z = 0$, the set Π is described by $\{\dot{x} = A(t)x \mid A(t) \in \Omega\}$.

With the appropriate choice of Ω, a number of common control system models can be described in the framework of (47.20): LTI systems; polytopic systems, norm-bound systems, structured norm-bound systems; systems with parametric perturbations; systems with structured, bounded LTI perturbations, etc. For details, see [3] . For purposes of illustration, we now consider the polytopic system model (PS) in detail.

Polytopic system models arise when the uncertain plant is modeled as a linear time-varying system with state-space matrices that lie in a polytope. Thus, Ω is a polytope described as the convex hull of its vertices

$$\mathbf{Co}\left\{\begin{bmatrix} A_1 & B_{u,1} & B_{w,1} \\ C_{z,1} & D_{zu,1} & D_{zw,1} \end{bmatrix}, \ldots, \begin{bmatrix} A_L & B_{u,L} & B_{w,L} \\ C_{z,L} & D_{zu,L} & D_{zw,L} \end{bmatrix}\right\}, \tag{47.21}$$

where the matrices in (47.21) are given, and **Co** stands for the convex hull, defined by

$$\mathbf{Co}\{G_1, \ldots, G_L\} \triangleq \left\{G : G = \sum_{i=1}^{L} \lambda_i G_i,\ \lambda_i \geq 0,\ \sum_{i=1}^{L} \lambda_i = 1\right\}. \tag{47.22}$$

EXAMPLE 47.3:

Consider the linear time-varying system with two parameters $q_1(t)$ and $q_2(t)$:

$$\begin{aligned} \frac{dx}{dt} &= \begin{bmatrix} 0 & 1 \\ -q_1(t) & -q_2(t) \end{bmatrix} x(t) + \begin{bmatrix} 0 \\ 1 \end{bmatrix} u(t); \\ y(t) &= [1\ \ 0]\, x(t), \end{aligned} \tag{47.23}$$

with $q_1(t) \in [-1, 1]$ and $q_2(t) \in [-2, 2]$ for all $t \geq 0$. The corresponding polytopic system model is

$$\frac{dx}{dt} = A(t)x(t) + \begin{bmatrix} 0 \\ 1 \end{bmatrix} u(t), \quad y(t) = [1\ \ 0]x(t), \tag{47.24}$$

where

$$A(t) \in \mathbf{Co}\left\{\begin{bmatrix} 0 & 1 \\ -1 & -2 \end{bmatrix}, \begin{bmatrix} 0 & 1 \\ 1 & -2 \end{bmatrix}, \begin{bmatrix} 0 & 1 \\ -1 & 2 \end{bmatrix}, \begin{bmatrix} 0 & 1 \\ 1 & 2 \end{bmatrix}\right\}. \tag{47.25}$$

Suppose we have a physical plant that is well-modeled as a (possibly time-varying) linear system. We collect many sets of input-output measurements, obtained at different times, under different operating conditions, or from different instances of the system (e.g., different units from a manufacturing run). It is important that we have data sets from enough plants or plant conditions to characterize the plant variation. For each data set we estimate a time-invariant linear system model of the plant. To simplify the problem, we will assume that the state in this model is accessible, so that the different models refer to the same physical state vector. These models should be fairly close, but of course not exactly the same. Then, the original plant can be modeled as a PS with the vertices given by the estimated linear system models. In other words, we model the plant as a time-varying linear system, with system matrices that can jump around among any of the models we estimated. It seems reasonable to conjecture that a controller that works well with this PS is likely to work well when connected to the original plant.

A number of problems in the analysis and design for the aforementioned system models can be reformulated as one of the LMI optimization problems, namely LMIP or EVP. We will demonstrate this reformulation for PS, for a sample set of simple problems; we will also list a few more problems towards the end of this section. For a more comprehensive list, we refer the reader to [3].

Stability of Polytopic Systems

First, consider the problem of stability of the polytopic system PS, i.e., whether all the trajectories of the system

$$\frac{dx}{dt} = A(t)x(t), \quad A(t) \in \mathbf{Co}\{A_1, \ldots, A_L\} \tag{47.26}$$

converge to zero as $t \longrightarrow \infty$. A sufficient condition for this is the existence of a quadratic positive function $V(z) = z^T P z$ such that $dV(x(t))/dt < 0$ for any trajectory of (47.26). Since

$$\frac{d}{dt} V(x(t)) = x(t)^T \left(A(t)^T P + PA(t)\right) x(t), \tag{47.27}$$

a sufficient condition for stability is the existence of a P satisfying the condition

$$P > 0, \qquad A(t)^T P + PA(t) < 0,\ A(t) \in \mathbf{Co}\{A_1, \ldots, A_L\}. \tag{47.28}$$

If such a P exists, the PS (47.26) is "quadratically stable".

Condition (47.28) is equivalent to

$$P > 0, \quad A_i^T P + PA_i < 0, \quad i = 1, \ldots, L, \tag{47.29}$$

which is an LMI in P. V is sometimes called a "simultaneous quadratic Lyapunov function" because it proves stability of each of $A_1, \ldots, A_L$. Thus, determining quadratic stability is an LMIP.

Stabilizing State-Feedback Synthesis for Polytopic Systems

Consider the system (47.26) with state-feedback,

$$\frac{dx}{dt} = A(t)x(t) + B_u(t)u(t), \quad u(t) = Kx(t), \tag{47.30}$$

where

$$[A(t)\ B_u(t)] \in \mathbf{Co}\left\{[A_1\ B_{u,1}], \ldots, [A_L\ B_{u,L}]\right\}. \quad (47.31)$$

Our objective is to design the matrix K such that (47.30) is quadratically stable. This is the "quadratic stabilizability" problem.

System (47.30) is quadratically stable for some state-feedback K if P and K exist so that

$$P > 0, \quad (A_i+B_iK)^T P+P(A_i+B_iK) < 0, \quad i = 1, \ldots, L. \quad (47.32)$$

This matrix inequality is *not* jointly convex in P and K. However, with the bijective transformation $Y \triangleq P^{-1}$, $W \triangleq KP^{-1}$, we may rewrite it as

$$\begin{aligned} Y > 0, \quad & (A_i + B_iWY^{-1})^T Y^{-1} \\ & + Y^{-1}(A_i + B_iWY^{-1}) < 0 \\ & i = 1, \ldots, L. \end{aligned} \quad (47.33)$$

Multiplying the second inequality on the left and right by Y (such a congruence preserves the inequality) yields an LMI in Y and W,

$$Y > 0, \quad YA_i^T + W^T B_i^T + A_iY + B_iW < 0, \quad i = 1, \ldots, L. \quad (47.34)$$

If this LMIP in Y and W has a solution, then the Lyapunov function $V(z) = z^T Y^{-1} z$ proves the quadratic stability of the closed-loop system with state-feedback $u(t) = WY^{-1}x(t)$.

In other words, we can synthesize a linear state-feedback for the PS (47.26) by solving an LMIP.

More General Stabilizing Feedback Synthesis for PS

We have shown that synthesizing state feedback that renders a PS quadratically stable is equivalent to solving an LMIP. However, in general, the entire state of the PS may not be available for feedback. What is typically available is some linear combination of the states, so that

$$\frac{dx}{dt} = A(t)x(t) + B_u(t)u(t), \quad y(t) = C_y x(t) \quad (47.35)$$

where once again,

$$[A(t)\ B_u(t)] \in \mathbf{Co}\left\{[A_1\ B_{u,1}], \ldots, [A_L\ B_{u,L}]\right\}. \quad (47.36)$$

Our objective is to design the *output-feedback* matrix K so that with $u(t) = Ky(t)$, the PS (47.30) is quadratically stable.

Though this problem appears to be only a minor variation of the state-feedback synthesis problem, it turns out to be *much harder*. Here is why: Using the same steps as with state-feedback synthesis, we conclude that system (47.35) is quadratically stable for some output-feedback K if $P > 0$ and K exist so that

$$(A_i + B_iKC)^T P + P(A_i + B_iKC) < 0, \quad i = 1, \ldots, L. \quad (47.37)$$

This matrix inequality is *not* jointly convex in P and K; moreover, unlike the state-feedback case, there are no known procedures (change of variables, for instance) that convert this problem to a convex feasibility problem. We are, therefore, left with solving a nonconvex feasibility problem.

A simple heuristic iterative procedure, along the lines of the D-K iteration of μ-synthesis (see Chapter 42), can be employed to tackle this nonconvex optimization problem. Basically, the nonconvex matrix inequality 47.37 is feasible if, and only if, the minimum value of the optimization problem

$$\begin{array}{ll} \text{minimize, w.r.t. } P, K \text{ and } \gamma: & \gamma \\ \text{subject to:} & (A_i + B_iKC)^T P \\ & + P(A_i + B_iKC) < \gamma I, \\ & i = 1, \ldots, L \end{array} \quad (47.38)$$

is negative. Now, problem (47.38) is an EVP in P and γ, for fixed K, and is an EVP in K and γ for fixed P. Therefore, one can iterate between solving EVPs with respect to P and γ, and K and γ, respectively, to solve the inequality (47.37). Details can be found in [5].

Other Problems from Control

A number of other control problems can be solved using LMI-based optimization; see [3] for details. The list includes computation of ellipsoidal bounds on the state, computation of bounds on the decay-rate for state trajectories, stability margin bounds, ellipsoidal bounds on the set of states reachable with various input constraints, bounds on the output energy given a certain initial state condition, bounds on the RMS value of the output for white noise inputs, bounds on the RMS gain from given inputs to outputs, etc. Moreover, feedback design to optimize all these bounds can be tackled as well. Other problems on the list are synthesis of gain-scheduled state feedback, some linear controller design problems using the Youla parameter and Ritz approximations, synthesis of multipliers for stability and performance analysis of systems with structured perturbations.

47.4.3 Solving LMI Problems

We have already observed that the LMI-based optimization problems described are convex optimization problems with a finite number of variables. Thus the comments about convex optimization problems at the end of Section 47.3 apply to LMIP and EVP. Furthermore, a number of specialized interior-point algorithms that exploit the special structure of LMI problems have been recently developed, for example [17]; software packages implementing these algorithms are available as well (LMI Lab [7] and LMITOOL [6]).

47.5 More General Optimization for Controller Design

The design approaches outlined in Sections 47.3 and 47.4 are efficient and mathematically elegant. However, this is achieved at the cost of reduced flexibility in various respects: (1) Specifications without certain convexity properties cannot be handled gracefully, (2) Constraints on the controller structure (e.g., controller order) cannot be dealt with, and (3) The approaches are restricted to LTI models, or to cases when the plant sets Π are restricted to certain specific classes. In this section, we briefly survey approaches applicable to some of these more general situations.

47.5.1 Branch and Bound Methods

Often, the controller design problem can be posed as one of scalar parameter selection to optimize some performance objective, where each parameter is to be chosen from some specified interval (thus the parameters are required to lie in a "rectangle"). Such a situation arises, for instance, when one chooses the coefficients of the transfer function of the controller.

Suppose that it is easy to compute upper and lower bounds for the optimum value of the performance objective over any given rectangle of values for the parameters, with the property that the difference between the upper and lower bounds uniformly converges to zero when the rectangle shrinks to a point. This is the case, for example, with common bounds on quantities such as the decay-rate of trajectories or $\mathbf{H}_2$ and $\mathbf{H}_\infty$ norms of transfer functions of interest [2]. Then **branch and bound algorithms** can be used to find the globally optimal parameter values and the corresponding value of the performance objective. The branch and bound scheme proceeds as follows. First, upper and lower bounds are computed over the original parameter rectangle. These bounds are further refined by breaking up the parameter rectangle into subrectangles ("branching") to derive bounds for the global optimum over the original rectangle ("bounding"). The branching is based on some heuristic rules. As they progress, branch and bound algorithms maintain upper and lower bounds for the global optimum; thus termination at any time yields guaranteed bounds for the optimum.

For an account of the solution of some parameter problems from control via branch and bound methods, see [2].

47.5.2 Local Methods

Suppose that we are still dealing with an LTI model and LTI controller, with no prespecified controller structure, but that some constraints are not readily expressible by convex or quasi-convex functions. The stability requirement can still be handled implicitly via the Youla parameterization. If all other specifications are "smooth enough", such problems can be tackled by "classical" **nonlinear programming** techniques. Under mild assumptions, such techniques will construct a local minimizer (maximizer), which might be global. Often, such problems will include "functional" constraints, i.e., constraints of the type

$$\phi(\zeta, \alpha) \leq 0 \quad \forall \alpha \in \Omega, \tag{47.39}$$

where ζ is the vector of optimization variables (for instance, the vector of components β_i as in (47.11)), α is an index (e.g., time or frequency), and Ω is a compact set. A functional constraint of this type arises, for instance, when the frequency response of the controller must satisfy $\|K(j\omega)\| \leq 1$ for $\omega \in \mathbf{R}$ (where $\|\cdot\|$ is some suitable norm). Such a constraint would be useful in limiting the size of the actuator signal u.

Problems with functional constraints of the form (47.39) are called "semi-infinite"; see, e.g., [10] or [19]. Even if all specifications are convex (or quasi-convex), classical techniques may constitute a valid alternative to modern convex optimization techniques for such problems.

A more drastic departure from the approach discussed in the previous section is necessary when the controller structure is constrained, e.g., when the order of the controller is prescribed. Such restrictions cannot be expressed simply in terms of the Q parameter. In such cases, it is much more natural to use a controller parameterization directly linked to the structure, e.g., coefficients of transfer functions numerators and denominators, location of zeros and poles, etc. However, stability of the closed-loop system is then no longer automatically insured, but must be enforced explicitly. A similar situation arises when the model P or the class of controller under consideration is not LTI.

The simplest way to impose closed-loop stability in such cases is to use its definition directly. For most engineering problems, this is tantamount to the condition that bounded inputs give rise to bounded outputs, provided, of course, that the inputs and outputs are chosen wisely (in particular, any internal instability of the system should be reflected in its input-output behavior). In practice, it makes sense to require that for bounded inputs, the outputs be less than some specific values at all times or, in practice, on a long enough time interval. If both the plant and the controller are LTI, however, closed-loop stability can also be enforced by invoking stability criteria such as those of Routh-Hurwitz, Nyquist, or Lyapunov. For instance, suppose that ψ denotes the vector of controller parameters and $A(\psi)$ denotes the state dynamics matrix of the closed-loop system. Then, as discussed in Section 47.3, stability is equivalent to the existence of a symmetric matrix P satisfying

$$P > 0, \quad A(\psi)^T P + PA(\psi) < 0, \tag{47.40}$$

or equivalently,

$$\langle u, Pu\rangle > 0 \quad \forall u \in \{u \in \mathbf{R}^n : \|u\|_2 = 1\} \text{ and} \tag{47.41}$$

$$\langle u, (A(\psi)^T P + PA(\psi))u\rangle > 0 \quad \forall u \in \{u \in \mathbf{R}^n : \|u\|_2 = 1\}. \tag{47.42}$$

The latter constraints are of the form (47.39) (semi-infinite problem) with the components of ζ being those of P and ψ, and where the role of α is now played by u. This and other schemes are discussed in [15].

In Section 47.4 it was stressed that LMIs make a powerful tool for the design of *robust* control systems when the models $P \in \Pi$ are linear and certain other conditions are met. In other cases when the model is LTI, tools, such as $\mathbf{H}_\infty$ synthesis (see Chapters 39 and 40), μ-synthesis (see Chapter 42) and ℓ_1-synthesis (see Chapter 41) can be used. Constraints on the spectral norm (largest singular value; see Chapters 3 and 29), on the structured singular value (or, rather, on its upper bound), or on the ℓ_1 structured norm of selected transfer functions (e.g., sensitivity and complementary sensitivity) can also be included together with the other constraints in an optimization problem to be solved by general nonlinear programming methods (see [15] for a semi-infinite formulation of spectral norm constraints). However, the resulting problem is, in general, nonsmooth and may require sophisticated techniques for its solution (see, e.g., [12]) (but classical techniques often perform satisfactorily). In situations where the nominal model is not LTI, however, it is typically necessary to resort to heuristics, e.g., replace the model set Π with a finite set and split each specification into a set of specifications, one for each model in this finite set.

Finally, note that the general nonlinear programming approach is not limited to the case of LTI models, but can be also used in the general case of nonlinear, time-varying models.

47.5.3 A General Approach for LTI Models

The ultimate approach, in the case of LTI models, may be a combination of convex optimization techniques with the Youla parameterization and local techniques. The following scheme can be contemplated. First, ignore any restriction on controller structure and other nonconvex constraints (or approximate them with convex constraints). Globally solve the simplified problem using efficient convex optimization techniques. Next, use controller order reduction techniques to obtain a suboptimal controller of required order. Finally, use this suboptimal controller as an "initial guess" for the full optimization problem to be solved by local techniques.

47.6 Open-Loop Optimal Control

An entire arsenal of techniques have been developed for problems of **open-loop optimal control** of very general systems of the following type. Given the dynamics

$$\dot{x}(t) = f(x(t), u(t)), \quad t \in [0, T], \tag{47.43}$$

where $T > 0$ is given, and given an initial state

$$x(0) = x_0, \tag{47.44}$$

determine a function $u(\cdot)$ over the interval $[0, T]$ so that a certain objective function depending on $u(\cdot)$ and on the corresponding state trajectory $x_u(\cdot)$ achieves its minimum (maximum) value, possibly subject to constraints on the control and state trajectories.

One may wonder, however, how useful such an open-loop control would be in practice. Indeed, the state is bound to be affected by external disturbances, so that any precomputed control signal is unlikely to be close to optimal even after a short time. There are, however, important classes of applications, within the framework of Section 47.2, for which knowledge of an optimal (or suboptimal) open-loop control can be of great interest. We consider two such classes.

First suppose that the system to be controlled is highly nonlinear and that the effect of external disturbances is reasonably small compared to desired variations of the state (in particular, when the time horizon is relatively short). A typical example is that of a rendezvous between two spacecraft or that of a moon landing. The idea would then be as follows. First, compute an open-loop optimal (or suboptimal) control $u_0(\cdot)$ based on the nonlinear model, and compute the corresponding state trajectory $x_0(\cdot)$. Next, linearize the system dynamics around $u_0(\cdot)$ and $x_0(\cdot)$. Finally, design a controller for the linearized system, using, e.g., LMI techniques. The resulting controller is thus closed-loop and nonlinear (or affine-linear).

Alternatively, suppose that the system to be controlled has a long time horizon but evolves relatively slowly, as in a continuously operating chemical process. A promising technique for such systems is known as "model predictive control" or sometimes "receding horizon control" (see Chapters 10.7 and 10.8). The idea is, at discrete intervals of time t_i, to minimize the objective function of interest (possibly subject to constraints) for the time interval $[t_i, t_i + T]$ with the actual current state $x(t_i)$ as initial state, where T is a fixed quantity larger than $t_{i+1} - t_i$. Then the computed optimal control is applied until time t_{i+1}, at which point a new optimal control is computed. (It is assumed that the CPU time necessary to compute the optimal control is small compared to $t_{i+1} - t_i$.) This control scheme can be thought of as intermediate between open-loop and closed-loop control schemes.

47.7 Multiple Objectives—Trade-off Exploration

In the previous sections, we have considered problems of minimizing (or maximizing) one objective function possibly subject to constraints. Some cost functions considered were the $\mathbf{H}_\infty$ (or $\mathbf{H}_2$) gain from w to z or some measure of the stability margin. (We also have encountered pure feasibility problems not involving any objective functions.)

In practical design situations, however, it is generally the case that several, often conflicting, objective functions are to be jointly minimized/maximized, i.e., a (possibly constrained) **multiobjective** (or **multi-criterion) optimization problem** is to be solved (see for example [13]). Singling out one of the specifications as an objective function, as we did in Section 47.3, for example, is often artificial. In a strict sense, several objectives can seldom be jointly minimized (when one of them achieves its minimum, the others typically do not). Suppose ζ is the vector of optimization variables and $\phi_1, \ldots, \phi_n$ are the objective functions to be minimized. For simplicity of exposition, suppose that there are no constraints. To give a precise meaning to a multiobjective optimization problem, a "utility function" must be defined that encapsulates ϕ_1 through ϕ_n. Two such functions are commonly used: weighted sum and weighted maximum, i.e.,

$$\phi_S(\zeta) = \sum w_i \phi_i(\zeta), \tag{47.45}$$

where $w_i > 0, i = 1, \ldots, n$, and

$$\phi_M(\zeta) = \max \frac{\phi_i(\zeta)}{c_i}, \tag{47.46}$$

where $c_i > 0$, $i = 1, \ldots, n$. It may also be appropriate to replace each ϕ_i in (47.45) and (47.46) with $\phi_i(\zeta) - \hat{\phi}_i$, where $\hat{\phi}_i$ is a "goal" or "good value[4]" associated with ϕ_i. If (47.45) is used, a standard "smooth" single objective problem is obtained. In contrast, ϕ_M is generally nondifferentiable (it has "corners" at values of ζ where the "max" is achieved by more than one function). This nonsmoothness is benign, however, and can be removed by means of a simple trick: include an additional, artificial scalar optimization variable ζ^0, and solve

$$\text{minimize } \zeta^0 \text{ with respect to } (\zeta, \zeta^0) \tag{47.47}$$

$$\text{subject to } \frac{\phi_i(\zeta)}{c_i} \leq \zeta^0 \quad i = 1, \ldots, n, \tag{47.48}$$

which is a smooth constrained problem. Note however that the transformation just outlined hides the structure of the minimax problem. For this reason, numerical methods specifically designed for the latter are generally preferable (e.g., [9],[19], and [21]).

Whether utility function (47.45) or utility function (47.46) is selected, appropriate values of the weights are seldom obvious. Rather, it is typical that, after a trial solution has been obtained, the designer would need to modify the relative emphases on the various objective functions. This is the process of trade-off exploration, best performed in a congenial graphical environment in which the graphical output conveys information to the designer on the current design, on which specifications are competing, and on how much improvement can be expected with respect to selected specifications if others are relaxed. Moreover, the graphical input tools must allow the designer to convey his/her inclinations to the optimization process. One example of such an environment is described in [18].

47.8 Defining Terms

Branch and bound algorithm: An exhaustive search method for finding the maximum (or minimum) of a function over a rectangular region of parameters. The algorithm proceeds by systematically breaking up the parameter region into sub-rectangles ("branching"), computing bounds on the optimal objective over these sub-rectangles ("bounding"), and thereby computing bounds on the optimal objective over the original rectangle.

Convex optimization: The solution of an optimization problem requiring the minimization of a convex function subject to convex constraints on the optimization variables.

[4] c_i can also be thought of as the difference $\phi_i^b - \phi_i^g$ between a "bad" value to be avoided and a "good value" to aspire to; see [18].

Linear Matrix Inequality: A matrix inequality of the form

$$F(\zeta) \triangleq F_0 + \sum_{i=1}^{m} \zeta_i F_i > 0, \tag{47.49}$$

where $\zeta \in \mathbf{R}^m$ is the variable, and $F_i = F_i^T \in \mathbf{R}^{n \times n}$, $i = 0, \ldots, m$ are given. The inequality symbol in (47.49) means that $F(\zeta)$ is positive-definite, i.e., $u^T F(\zeta) u > 0$ for all nonzero $u \in \mathbf{R}^n$.

Multiobjective or multicriterion optimization problem: An optimization problem where conflicting objective functions are to be jointly minimized/maximized, subject to constraints.

Nonlinear Programming: The solution of an optimization problem requiring the minimization (or maximization) of a nonlinear objective subject to nonlinear constraints on the optimization variables. The term "nonlinear" helps to contrast the methods with Linear Programming methods.

Numerical optimization-based design: Design with the goal of optimizing measures of performance, using iterative optimization methods implemented on a computer.

Open-loop optimal control: For the system

$$\dot{x}(t) = f(x(t), u(t)), \quad t \in [0, T], \tag{47.50}$$

where $T > 0$ is given, and given an initial state

$$x(0) = x_0, \tag{47.51}$$

the problem of determining a function $u(\cdot)$ over the interval $[0, T]$ such that a certain objective function depending on $u(\cdot)$ and on the corresponding state trajectory $x_u(\cdot)$ achieves its minimum (maximum) value, possibly subject to constraints on the control and state trajectories.

Robustness: The ability of a control system (a closed-loop system consisting of a plant and controller) to continue performing satisfactorily despite variations in the plant dynamics.

Youla Parameterization: An affine parameterization of the set of achievable stable closed-loop transfer functions for a control system with a linear time-invariant plant and a linear time-invariant controller.

References

[1] Boyd, S. and Barratt, C., *Linear Controller Design: Limits of Performance,* Prentice Hall, Englewood Cliffs, NJ, 1991.

[2] Balakrishnan, V. and Boyd, S., Global optimization in control system analysis and design, in *Control and Dynamic Systems: Advances in Theory and Applications,* C. T. Leondes, Ed., Academic Press, New York, 1992, vol. 53, 1–55.

[3] Boyd, S., El Ghaoui, L., Feron, E., and Balakrishnan, V., *Linear Matrix Inequalities in System and Control Theory*, in *Studies in Applied Mathematics*, SIAM, Philadelphia, PA, 1994, vol 15.

[4] Dunn, J., Gradient-related constrained minimization algorithms in function spaces: Convergence properties and computational implications, in *Large Scale Optimization: State of the Art*, W. W. Hager, D. W. Hearn and P. M. Pardalos, Eds., Kluwer Academic, Dordrecht, 1993.

[5] El Ghaoui, L. and Balakrishnan, V., Synthesis of fixed-structure controllers via numerical optimization, in *Proc. IEEE Conf. on Decision and Control*, Orlando, FL, 1994, 2678–2683.

[6] El Ghaoui, L., Delebecque, F., and Nikoukhah, R., *LMITOOL: A User-friendly Interface for LMI Optimization*. ENSTA/INRIA, Paris, 1995.

[7] Gahinet, P. and Nemirovskii, A., *LMI Lab: A Package for Manipulating and Solving LMIs*. INRIA, Paris, 1993.

[8] Gill, P.E., Murray, W., and Wright, M., *Practical Optimization*. Academic, London, 1981.

[9] Han, S.P., Variable metric methods for minimizing a class of nondifferentiable functions. *Mathematical Programming*, 20, 1–13, 1981.

[10] Hettich, R. and Kortanek, K.O., Semi-infinite programming: Theory, methods, and applications. *SIAM Review*, 35(3), 380–429, 1993.

[11] Hiriart-Urruty, J-B. and Lemaréchal, C., *Convex Analysis and Minimization Algorithms I & II*, in *Grundlehren der mathematischen Wissenschaften*, Springer, Berlin, 1993, vols. 305 and 306.

[12] Lemaréchal, C., Nondifferentiable optimization, in *Optimization, Handbooks in Operations Research and Management Science*, Nemhauser, G., Rinnooy-Kan, A., and Todd, M., Eds., North Holland, 1989.

[13] Maciejowski, J.M., *Multivariable Feedback Design*. Addison-Wesley, Wokingham, England, 1989.

[14] Mayne, D.Q. and Polak, E., First-order strong variation algorithms for optimal control, *J. Optimiz. Theory Appl.*, 16(3/4), 277–301, 1975.

[15] Mayne, D.Q. and Polak, E., Optimization based design and control of constrained dynamic systems, in *Proc. 12th IFAC World Congress on Automatic Control*, Sydney, Australia, 1993, 129–138.

[16] Moré, J.J. and Wright, S.J., *Optimization Software Guide*, in *Frontiers in Applied Mathematics*, SIAM, Philadelphia, PA, 1993, vol. 14.

[17] Nesterov, Y. and Nemirovskii, A., *Interior-point polynomial methods in convex programming*, in *Studies in Applied Mathematics*, SIAM, Philadelphia, PA, 1994, vol. 13.

[18] Nye, W.T. and Tits, A.L., An application-oriented, optimization-based methodology for interactive design of engineering systems, *Int. J. Control*, 43(6), 1693–1721, 1986.

[19] Polak, E., *Computational Methods in Optimization: A Unified Approach*, 2nd ed., Academic Press, New York, 1996.

[20] Polak, E. and Mayne, D.Q., First-order strong variation algorithms for optimal control problems with terminal inequality constraints, *J. Optimiz. Theory Appl.*, 16(3/4), 303–325, 1975.

[21] Zhou, J.L. and Tits, A.L., An SQP algorithm for finely discretized continuous minimax problems and other minimax problems with many objective functions, *SIAM J. on Optimization*, 6(2), 1996.

Further Reading

The details of control system design using a combination of Youla parameterization and convex optimization can be found in the book *Linear Controller Design: Limits of Performance* by Boyd and Barratt. This book also discusses the framework for controller design presented in Section 47.2.

Control system analysis and design using Linear Matrix Inequalities is discussed in great detail in the book *Linear Matrix Inequalities in System and Control Theory*, by Boyd et al. This book also contains an extensive bibliography.

Local nonlinear programming methods are discussed, e.g., in *Practical Optimization* by Gill et al.. The book *Optimization Software Guide*, by Moré and Wright, has useful pointers to software. The survey paper "Optimization based design and control of constrained dynamic systems" by Mayne and Polak gives a quick idea of the versatility of local optimization methods, in particular semi-infinite optimization methods, in their application to controller design.

A major chapter of the book *Computational Methods in Optimization: A Unified Approach*, by Polak, is devoted to algorithms for open-loop optimal control. The focus is on the extension to functions spaces of algorithms initially developed for finite dimensional optimization problems. The review paper by Dunn also addresses this issue. Algorithms that make use of optimal control theory, namely of Pontryagin's Maximum Principle, are discussed, e.g., in two 1975 papers by Mayne and Polak.

48

Optimal Control

F. L. Lewis
Automation and Robotics Research Institute, The University of Texas at Arlington, Ft. Worth, TX

48.1 Introduction to Optimal Control Design

Many systems occurring naturally in fields such as biology and sociology use feedback control to achieve homeostasis, or equilibrium conducive to existence. Because the bounds within which life can continue are small (e.g., temperature changes of a few degrees can eliminate populations) and the *resources available* are limited, it is remarkable yet not expected that most of these feedback control systems have evolved into *optimal* systems where performance objectives are achieved efficiently with a minimum of control effort. Since naturally occurring systems are optimal, it makes sense to design man-made controllers from the view point of optimality.

48.1.1 The Philosophy of Classical Control

Classical control theory, developed in the mid 1900s, imparts a great deal of engineering insight. It was best developed for linear systems. Since computers were not available to solve complex design equations, the design algorithms are heuristic in terms of Bode plots, Nyquist plots, the root locus, and other single-input/single-output (SISO) graphical techniques that offer intuition and rely on the design engineer's expertise. Since most design was in the frequency domain, robustness to unknown disturbances, modeling errors, and noise was automatically built in.

Complex modern systems have multiple inputs and outputs. Examples include aircraft, satellites, and automobile engines, which, though nonlinear, can often be linearized about a desired operating point or trajectory. In such applications, classical design relies on successive loop closures based on *one-loop-at-a-time SISO design*. Unfortunately, using this approach, neither stability nor robustness of the overall system can be guaranteed, since one loop closure can destroy what has been gained in the design of previous loops.

48.1.2 The Philosophy of Optimal Control Design

About 1960, modern optimal control theory began developing for complex multivariable systems. It developed coincidentally with the space age (Sputnik was launched in 1957), the computer age, and the age of robotics. Optimal control is a branch of modern control theory that deals with designing controls for dynamical systems by minimizing a *performance index* that depends on the system variables. The performance index might include, for instance, a measure of operating error, a measure of control "effort", or any other characteristic important to the user of the control system. Under some mild assumptions, making the per-

0-8493-8570-9/96/$0.00+$.50

formance index small also guarantees that the system variables will be small, thus insuring *closed-loop stability*.

Classical design is concerned with directly selecting the feedback gains K in the inner *real-time control loops*. On the other hand, modern control design offers standard algorithms for implementing an *outer design loop* that automatically selects the inner loop feedback gains in such a fashion that *closed-loop stability and performance for MIMO systems is guaranteed*. In contrast to classical one-loop-at-a-time design, in modern control, *all of the feedback loops are closed simultaneously* by computing the feedback gains by solving standard *matrix design equations*. Special purpose software (e.g., MATLAB [10], $MATRIX_X$ [9]) is commercially available to solve these equations, so that control design for complex systems is straightforward with a personal computer (PC).

In this chapter, optimal control design is discussed for deterministic systems with a well-known mathematical model. The major emphasis is on linear systems in the state-space form. It will be assumed that *full state-variable feedback* is available; in the event that only partial information is available on the system states, the chapter on "Output Feedback" or the references should be consulted as well. The discussion in this chapter will center around continuous-time systems, with a follow-up discussion on discrete-time systems. Several design techniques will be covered, including the regulator problem, the tracker problem, minimum-time control, and polynomial design. Robustness of the LQR will be discussed using multivariable frequency-domain design techniques, which allow one to draw close connections with classical control theory. This is a condensed version of presentations available in [7], [8], and [11] where derivations and computer software appear. Other references appear at the end of the chapter.

48.2 Optimal Control of Continuous-Time Systems

In this section nonlinear control design will be covered for general nonlinear systems. Then, the linear quadratic regulator (LQR) will be developed for linear systems. The LQR is a cornerstone of modern control theory design.

48.2.1 The General Continuous-Time Optimal Control Problem

A state-variable model for a nonlinear time-varying dynamical system is given by Equation 48.1 in Table 48.1, where $x(t) \in \mathcal{R}^n$ is the vector of internal states and $u(t) \in \mathcal{R}^m$ is the vector of control inputs. This is the plant to be controlled. A broad range of performance objectives may be achieved by selecting the control $u(t)$ to minimize a *performance index* (PI) or a *cost* given by Equation 48.2 with t_0 the initial time and T the final time of interest. The *final-state weighting function* $\phi(x(T), T)$ and weighting function $L(x, u, t)$ are selected depending on the performance objectives.

TABLE 48.1 Continuous Nonlinear Optimal Controller.

System model:

$$\dot{x} = f(x, u, t), \qquad t \geq t_0, \quad t_0 \ fixed. \tag{48.1}$$

Performance index:

$$J(t_0) = \phi(x(T), T) + \int_{t_0}^{T} L(x, u, t)\, dt. \tag{48.2}$$

Final state constraint:

$$\psi(x(T), T) = 0. \tag{48.3}$$

Optimal Controller:

Hamiltonian:

$$H(x, u, t) = L(x, u, t) + \lambda^T f(x, u, t). \tag{48.4}$$

State equation:

$$\dot{x} = \frac{\partial H}{\partial \lambda} = f, \qquad t \geq t_0. \tag{48.5}$$

Costate equation:

$$-\dot{\lambda} = \frac{\partial H}{\partial x} = \frac{\partial f^T}{\partial x}\lambda + \frac{\partial L}{\partial x}, \qquad t \leq T. \tag{48.6}$$

Stationarity condition:

$$0 = \frac{\partial H}{\partial u} = \frac{\partial L}{\partial u} + \frac{\partial f^T}{\partial u}\lambda. \tag{48.7}$$

Boundary conditions:

$$x(t_0) \text{ given, initial condition,} \tag{48.8}$$

$$\begin{aligned} &(\phi_x + \psi_x^T \nu - \lambda)^T \mid_T dx(T) \\ &+ (\phi_t + \psi_t^T \nu + H) \mid_T dT = 0, \\ &\text{final condition.} \end{aligned} \tag{48.9}$$

The *optimal control problem* is to determine a control input $u(t)$ for the system that minimizes the PI and also insures that the *final state constraint* (Equation 48.3) is satisfied for a given function $\psi \in \mathcal{R}^p$. The roles of the final weighting function ϕ and the final constraint ψ should not be confused. The former is a function one would like to minimize, such as the final energy $x^T(T)S(T)x(T)$, with $S(T)$ a specified weighting matrix. On the other hand, $\psi(x(T), T)$ must be exactly equal to zero. A sample problem might be to find the control input $u(t)$ that drives a satellite, with dynamics described by Equation 48.1, from a given initial position $x(t_0)$ to an specified orbit, described by Equation 48.3, while minimizing the expended energy, as described by the PI (Equation 48.2).

Solution of the Nonlinear Optimal Control Problem

To solve the optimal control problem, Lagrange multipliers are used to adjoin the constraints (Equation 48.1) and (Equation 48.3) to the performance index (Equation 48.2). Since the system Equation 48.1 is an equality constraint which must hold at each time, an associated multiplier $\lambda(t) \in \mathcal{R}^n$ is required that is a function of time. Thus, the *Hamiltonian function* is defined as (Equation 48.4). Using the theory of Lagrange multipliers and the calculus of variations, the solution to the optimal control problem given in Table 48.1 is determined. It is assumed that the initial time t_0 and the initial state $x(t_0)$ are both known and fixed. In the boundary conditions, partial derivatives are denoted by subscripts (e.g., ψ_x represents $\frac{\partial \psi}{\partial x}$).

The equations in the table may be used as *design equations* for determining the control $u(t)$ that minimizes the PI. They are *necessary conditions* for the solution of the nonlinear optimal control problem. Any control $u(t)$ that results in a minimum value of the PI, when it is applied to the system, must satisfy the equations given there. Conditions under which these equations are sufficient as well are addressed in [3], [7].

The structure of the equations is worth discussing. According to the table, the Lagrange multiplier $\lambda(t)$ is a dynamical variable that satisfies its own dynamical Equation 48.6; it is called the *costate.* The optimal control $u(t)$ is then generally determined in terms of $x(t)$ and $\lambda(t)$ by using the *stationarity condition* (Equation 48.7) (so named because this is the condition that guarantees a minimum or stationary point with respect to changes in $u(t)$). The value of $\lambda(t)$ is usually of no concern ultimately, but it is an *intermediate variable* which must be determined to solve for the optimal control $u(t)$, that minimizes the PI $J(t_0)$ while insuring that constraints (Equation 48.1) and (Equation 48.3) are satisfied. The appearance of intermediate variables, that are required to solve for the variables of interest, is typical of optimal control design.

The dynamical state and costate equations, along with the control specified by the stationarity condition, are called the *Hamiltonian system.* These equations may be used to derive Lagrange's and Hamilton's equations of motion in physics (see Example 48.1). The costate equation and stationarity condition are called *Euler's equations.* In the time-invariant case, f and L are not explicit functions of t, so that neither is H. In this situation

$$\dot{H} = 0. \tag{48.10}$$

Thus for time-invariant systems and cost functions, the Hamiltonian is a *constant* on the optimal trajectory. This is a general statement of the principle of conservation of energy.

Two-Point Boundary-Value Problems

The solution for the optimal control $u(t)$ in Table 48.1 depends on solving two coupled differential equations, the state Equation 48.5 and the costate Equation 48.6, each of which is of order n. These two dynamical equations comprise the Hamiltonian system once the stationarity condition has been used to eliminate $u(t)$. The costate equation develops *backward in time* (by defining a backward time variable $\tau = T - t, d\tau = -dt$), with the final condition $\lambda(T)$ determined by Equation 48.9. The boundary conditions consist of the initial conditions on the state

$$n \text{ conditions:} \qquad x(t_0) \ \textit{given} \tag{48.11}$$

and the final conditions on the costate

$$p \text{ conditions:} \qquad \psi(x(T), T) = 0 \tag{48.12}$$

$$n - p \text{ conditions:} \qquad (\phi_x + \psi_x^T \nu - \lambda)^T \mid_T \, dx(T) = 0, \tag{48.13}$$

where it has been assumed for simplicity that the final time T is specified and hence fixed, so that $dT = 0$ in condition (Equation 48.9).

Since n boundary conditions are specified at the initial time t_0 and n conditions are specified at the final time T, this is a *two-point boundary-value problem.* There are many methods available for solving such problems, including the *shooting point method* and the *unit solution method*; good software is available for this purpose. The solution of the optimal control problem for nonlinear systems is often difficult, though, for some nonlinear plants, the design equations can be explicitly solved for the optimal control $u(t)$, yielding a great deal of insight. This includes the *Thrust Angle Programming* and *Intercept and Rendezvous* problems. In the special case that the plant is linear and the PI is quadratic, a solution is available, given subsequently.

EXAMPLE 48.1: Hamilton's principle

Both Lagrange's and Hamilton's equations of motion may be derived from Table 48.1.

a. Lagrange's equations of motion.

Define the generalized coordinate state vector q and the 'control input' as the generalized velocities $u = \dot{q}$. Define the Lagrangian $L(q, u) \equiv T(q, u) - U(q)$ as the difference between the kinetic and potential energies. Then the 'plant' (Equation 48.1) is

$$\dot{q} = u \equiv f(q, u)$$

where the function $f(\cdot)$ is given by the physics of the problem.

To find the trajectories of the motion, Hamilton's principle says we may minimize the performance index

$$J = \int_0^T L(q, u) \ dt,$$

so that the Hamiltonian (48.4) is $H = L + \lambda^T u$. According to Table 48.1, for a minimum

$$\begin{aligned} -\dot{\lambda} &= \frac{\partial H}{\partial q} = \frac{\partial L}{\partial q} \\ 0 &= \frac{\partial H}{\partial u} = \frac{\partial L}{\partial u} + \lambda. \end{aligned}$$

Combining these equations yields Lagrange's equations of motion,

$$\frac{\partial L}{\partial q} - \frac{d}{dt}\frac{\partial L}{\partial \dot{q}} = 0.$$

b. Hamilton's equations of motion.

Defining the generalized momentum vector by $\lambda = -\partial L/\partial \dot{q}$, the equations of motion may be expressed in Hamilton's form as

$$\dot{q} = \frac{\partial H}{\partial \lambda},$$

and

$$-\dot{\lambda} = \frac{\partial H}{\partial q}.$$

48.2.2 Continuous-Time Linear Quadratic Regulator

The nonlinear optimal control design equations in Table 48.1 are not easy to solve, and there is no design algorithm for doing so. In this section the design of optimal controllers for linear systems with quadratic performance indices will be discussed, the so-called *linear quadratic regulator (LQR) problem*. The LQR is a cornerstone of modern optimal control design consisting of *explicit matrix design equations easily solved on a digital computer*. It has a wide range of relevance, because many systems are linear to begin with, and many nonlinear systems may be considered linear when operating near an equilibrium point. The LQR solution is given as a *closed-loop feedback control*.

Consider the multivariable linear system (Equation 48.14) in Table 48.2 with state $x \in \mathcal{R}^n$ and control input $u \in \mathcal{R}^m$. The plant matrices may be time-varying (e.g., A(t), B(t)), though for notational convenience this dependence will not be shown explicitly.

Choose the control that minimizes the quadratic PI (Equation 48.15). The *control weighting R, state weighting Q*, and *final state weighting $S(T)$* are symmetric matrices of design parameters chosen by the designer depending on the control objectives. For instance, if the elements of $S(T)$ are selected larger, then the control will force the final state $x(T)$ to be smaller to keep the PI small. Weight matrices Q and $S(T)$ are assumed positive-semidefinite ($Q \geq 0$, $S(T) \geq 0$). Thus Q and $S(T)$ have nonnegative eigenvalues so that $x^T Qx$ and $x^T(T)S(T)x(T)$ are nonnegative for all $x(t)$. Likewise, it will be assumed that R is positive definite ($R > 0$), that is, R has positive eigenvalues so that $u^T Ru > 0$ for all $u(t)$. In this case, J is always bounded below by zero, so that a sensible minimization problem results. Since the squares of the states and control inputs occur in Equation 48.15, the PI is a form of generalized energy (consider the case when some of the state components are velocities, or currents and voltages) and minimizing it will keep the states and controls small.

Using the equations in Table 48.1, the solution to the LQR optimal control problem may be derived, as given in Table 48.2. The state and costate equations are

$$\dot{x} = Ax + Bu \tag{48.20}$$

and

$$-\dot{\lambda} = Qx + A^T\lambda, \tag{48.21}$$

where the negative sign indicates that the costate equation must be solved backward in time. The stationarity condition gives the control in terms of the costate as $u(t) = -R^{-1}B^T\lambda(t)$. To find the optimal control, the two-point boundary-value problem associated with the state and costate dynamics is analytically solved using the *sweep method* [3] where it is assumed that $\lambda(t) = S(t)x(t)$ for some unknown auxiliary matrix $S(t)$. The optimal control is given in terms of this intermediate matrix as follows. First, it can be determined that the auxiliary matrix $S(t)$ satisfies the bilinear *matrix Riccati equation* (48.16). In terms of the Riccati solution S(t), the optimal control is given by $u(t) = -R^{-1}B^TS(t)x(t)$. Thus, defining the *optimal feedback gain* by Equation 48.17, one may write the optimal control as the *state-feedback control law* (Equation 48.18).

A block diagram of the LQR is shown in Figure 48.1. It is a *feedback control system* with time-varying feedback gains $K(t)$, and a formal *design outer loop*. Even if the system (A, B) is time-invariant, the optimal control $u(t)$ is a *time-varying state feedback*. This is why the optimal LQ controller may not be determined using classical frequency-domain techniques. If the system model (Equation 48.14) is not an exact description of the plant, the LQR still performs well if it is fairly close. In fact, it will be seen in Section 48.2.4 that the LQR has important *guaranteed robustness properties*.

LQR design. To design the optimal LQR, the design engineer first selects the design parameter weight matrices Q, R, $S(T)$. Then, the Riccati equation is solved for the auxiliary matrix function $S(t)$, which is used to compute $K(t)$. The Riccati

TABLE 48.2 Continuous-Time Linear Quadratic Regulator.

System model:

$$\dot{x} = Ax + Bu, \quad t \geq t_0, \quad x(t_0) = x_0 \ given. \tag{48.14}$$

Performance index:

$$J(t_0) = \frac{1}{2}x^T(T)S(T)x(T) + \frac{1}{2}\int_{t_0}^{T}(x^TQx + u^TRu)\,dt, \tag{48.15}$$

with

$$S(T) \geq 0, \quad Q \geq 0, \quad R > 0.$$

Optimal feedback control:

Riccati equation:

$$-\dot{S} = A^TS + SA - SBR^{-1}B^TS + Q, \quad t \leq T, \quad S(T) \ given. \tag{48.16}$$

Optimal feedback gain:

$$K = R^{-1}B^TS. \tag{48.17}$$

Time-varying feedback:

$$u = -K(t)x. \tag{48.18}$$

Optimal cost:

$$J(t_0) = \frac{1}{2}x_0^TS(t_0)x_0. \tag{48.19}$$

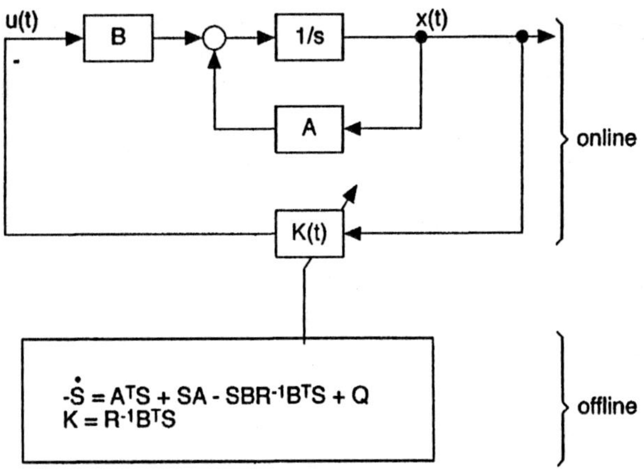

Figure 48.1 Linear quadratic regulator.

equation is solved *backward in time* using, as a final condition, the value of the weighting matrix $S(T)$ selected for the PI. This equation may be solved *off-line* for $S(t)$, and the optimal feedback gain $K(t)$ computed and stored. Next, a *computer simulation* is generally performed to verify the closed-loop performance. If the performance is not suitable, new design matrices Q, R, $S(T)$ are selected and the entire procedure is repeated. With commercially available software, the entire process is fast and convenient. Finally, during the implementation or control run, the states are measured and the feedback control $u(t) = -K(t)x$ applied to the plant.

Thus, by the sweep method, the LQR problem has been decomposed into two stages: off-line computation of the optimal gains using a backward differential equation, followed by the actual control of the plant using feedback. Such *hierarchical control schemes*, consisting of an inner linear feedback loop whose gain is computed by an outer quadratic design equation, are typical of modern control schemes. Optimal control is fundamentally a *noncausal design algorithm* requiring future information about the plant and performance objectives.

The LQR design procedure is in stark contrast to classical control design, where the gain matrix K is selected directly. In modern optimal control design, some parameter matrices Q, R, and $S(T)$ are selected by the engineer. Then, the feedback gain K is automatically given by *matrix design equations*. This has the significant advantages of allowing *all the control loops in a multiloop system to be closed simultaneously, while guaranteeing closed-loop stability.*

The optimal cost of using this controller is given in terms of the initial state by Equation 48.19. The initial state of the plant is known. Therefore, this expression allows computation of the optimal cost *before* the control is actually applied to the plant, or even before the optimal gain $K(t)$ is computed and it is simulated on a computer. If the cost is too high, the engineer can select different weighting matrices Q, R, and $S(T)$ in the performance index and try another design. This *preview feature* is typical of optimal control design using state feedback.

EXAMPLE 48.2: LQR for armature-controlled DC motor

The system equations of an armature-controlled DC motor are

$$\begin{aligned} \dot{i} &= -ai - k'\omega + bu \\ \dot{\omega} &= -\alpha\omega + ki \end{aligned} \qquad (48.22)$$

with $i(t)$ the armature current, $\omega(t)$ the motor speed, control input $u(t)$ the armature voltage, $1/a$ the electrical time constant, $1/\alpha$ the mechanical time constant, and the remaining variables other motor parameters.

Defining the state as $x = [i \quad \omega]^T$,

$$\dot{x} = \begin{bmatrix} -a & -k' \\ k & -\alpha \end{bmatrix} x + \begin{bmatrix} b \\ 0 \end{bmatrix} u \equiv Ax + Bu. \qquad (48.23)$$

It is required to determine $u(t)$ to minimize the PI

$$\begin{aligned} J &= \frac{1}{2}x^T(T)\begin{bmatrix} s_i & 0 \\ o & s_\omega \end{bmatrix} x(T) \\ &+ \frac{1}{2}\int_0^T \left[x^T \begin{bmatrix} q_i & 0 \\ 0 & q_\omega \end{bmatrix} x + ru^2\right] dt \end{aligned} \qquad (48.24)$$

with s_i, s_ω the final state weights, q_i, q_ω the (intermediate) state weights, and r the control weight. These are design parameters that may be adjusted or *tuned* using computer simulations to yield suitable closed-loop behavior, as will be seen in this example. The minimization of J corresponds to the regulation objective of driving the motor to a speed of zero from any initial speed, while keeping the control energy small. If the model represents a linearization of a nonlinear motor about a set point, the control will regulate the motor speed to that set point.

Since the Riccati equation solution $S(t)$ is symmetric, one may assume that

$$S = \begin{bmatrix} s_1 & s_2 \\ s_2 & s_3 \end{bmatrix} \qquad (48.25)$$

where the scalars $s_i(t)$ are to be determined. Substituting A, B from the state equation and $S(T)$, Q, R from the PI in the Riccati equation in Table 48.2 yields the three nonlinear scalar coupled differential equations

$$\begin{aligned} -\dot{s_1} &= -2as_1 + 2ks_2 - \beta s_1^2 + q_i, \\ -\dot{s_2} &= -(a+\alpha)s_2 - k's_1 + ks_3 - \beta s_1 s_2, \\ -\dot{s_3} &= -2\alpha s_3 - 2k's_2 - \beta s_2^2 + q_\omega, \end{aligned} \qquad (48.26)$$

where $\beta \equiv b^2/r$.

Writing the feedback gain as $K(t) = [k_i \quad k_\omega]$, the table shows that $K = R^{-1}B^T S$, so that

$$k_i = \frac{b}{r}s_1, \qquad k_\omega = \frac{b}{r}s_2. \qquad (48.27)$$

Then, the optimal control is given by the time-varying feedback $u = -k_i i - k_\omega \omega$.

Although the equations for $s_i(t)$ are difficult to solve analytically, it is easy to use computer software to solve the Riccati equation. Using such software, the optimal state trajectories and control voltage are plotted for $r = 1$ and several values of $q = q_i = q_\omega$. The results are displayed in Figure 48.2. Note that the states go to zero more quickly as q increases, while the controls become larger. Final weights of $s_i = s_\omega = 0$ were used. Based on the simulation results, suitable values for the PI

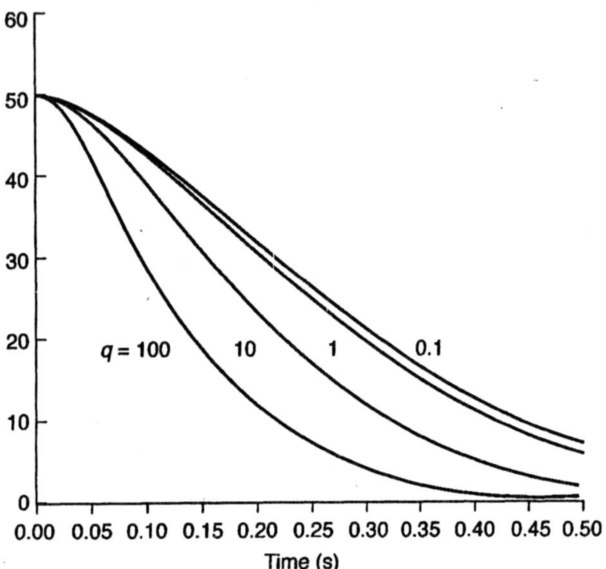

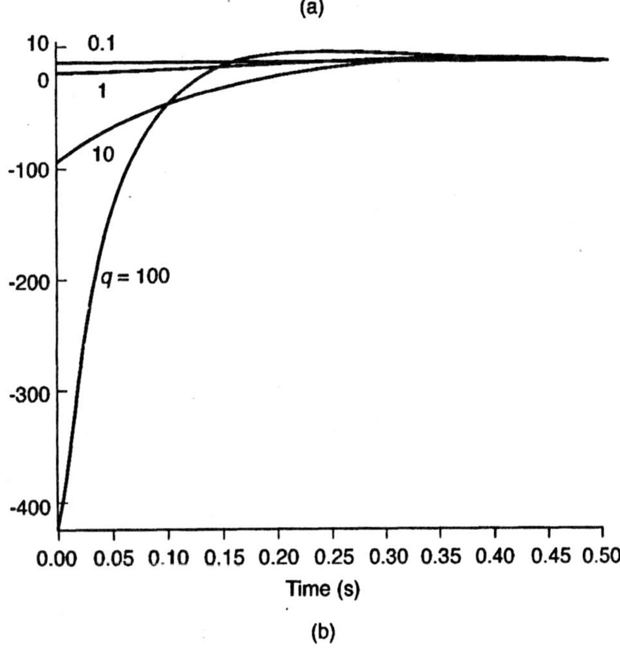

Figure 48.2 Results of DC motor simulation. (a) Motor speed. (b) Optimal control voltage.

weights can be selected. Then, the associated K(t) may be stored in memory and applied to the actual motor during the control implementation run.

48.2.3 Steady-State and Suboptimal Control

Even for time-invariant plants, the optimal LQ control is a *time-varying* state-variable feedback. Such feedbacks are inconvenient to implement, because they require the storage in computer memory of time-varying gains. An alternative control scheme will now be given in which the time-varying optimal gain $K(t)$ is replaced by its constant steady-state (e.g., $t \rightarrow \infty$) value. In most practical applications, this use of the steady-state feedback gain is adequate.

The results of this section are important. Showing how to find *multi-loop feedback gains for multi-input systems that are guaranteed to stabilize the closed-loop system.* The gains are determined simply by solving a *matrix design equation* using computer routines available in standard software packages such as MATLAB, and $MATRIX_X$. This goes far beyond what can be achieved with classical design techniques, which revolve around one-loop-at-a-time procedures offering no stability guarantees.

Steady-State Control—Guaranteed Stability of the LQR

Suppose the plant to be controlled has the linear description

$$\dot{x} = Ax + Bu, \tag{48.28}$$

with $x \in \mathcal{R}^n$ and control input $u \in \mathcal{R}^m$. For this section it will be necessary to assume that the plant is time-invariant.

Now, the control should be selected to minimize the quadratic PI

$$J(t_0) = \frac{1}{2}\int_0^\infty (x^TQx + u^TRu)\,dt, \tag{48.29}$$

with $Q \geq 0$ and $R > 0$. Since the integration interval is infinite, this is called an *infinite horizon* performance index; the performance objectives are referred to an infinite control interval $[0, \infty)$.

The control law of Table 48.2 still applies; however, because the control horizon is infinite, the Riccati equation (RE) may reach a steady-state solution where $\dot{S} = 0$. In this case, the RE may be replaced by the *algebraic Riccati equation (ARE)*

$$0 = A^TS + SA - SBR^{-1}B^TS + Q. \tag{48.30}$$

This is a symmetric matrix quadratic equation. $S(T)$ no longer appears in this steady-state formulation.

The ARE can have multiple solutions. However, if certain mild assumptions on the system and PI matrices hold, then there is a single *positive definite* solution S_∞, namely, the limiting solution to the time-varying RE for any $S(T)$. Then, the optimal infinite-horizon gain is the *constant* matrix given by

$$K_\infty = R^{-1}B^TS_\infty. \tag{48.31}$$

Thus, the optimal steady-state control is the *constant state-variable feedback*

$$u(t) = -K_\infty x(t). \tag{48.32}$$

Moreover, the optimal cost is given in terms of the initial state by

$$J = \frac{1}{2}x^T(0)S_\infty x(0). \tag{48.33}$$

Under the influence of the steady-state control the closed-loop plant has the *time-invariant* dynamics

$$\dot{x} = (A - BK_{\infty})x \equiv A_c x. \tag{48.34}$$

The advantages of this simplified control, that uses a constant feedback, are clear. The next theorem is vital to modern control theory, and shows that the steady-state LQR is *guaranteed to stabilize the system*, even if it has multiple inputs, as long as the plant satisfies some basic properties. The system (Equation 48.28) is said to be *stabilizable* if the control input $u(t)$ can be selected to stabilize all the modes in the closed-loop. This is a weaker property than *reachability*, which requires that there is a $u(t)$ that drives any given initial state to any desired final state. These are both controllability properties of the plant. The system is said to be *observable* through the output $y = Cx$ if measurements of only $y(t)$ can be used to reconstruct the entire initial state $x(0)$.

Stabilizability and observability are basic *open-loop properties* that hold for any well-behaved system. There are simple tests for these properties. In fact, if n is equal to the number of states, the system is reachable if, and only if, the matrix

$$U = [B \;\; AB \;\; A^2B \; \dots \; A^{n-1}B] \tag{48.35}$$

has rank n. This implies stabilizability as well. The system is observable if, and only if, the matrix

$$V = \begin{bmatrix} C \\ CA \\ \vdots \\ CA^{n-1} \end{bmatrix} \tag{48.36}$$

has rank n. Standard software packages such as MATLAB and $MATRIX_X$ offer routines to compute these block matrices.

THEOREM 48.1 **Stability of closed-loop system:**

Let C be any square root of Q so that $Q = C^TC$. Suppose (C, A) is observable and (A, B) is stabilizable. Then,

1) There is a unique symmetric positive-definite limiting solution S_{∞} to the Riccati equation (48.16) independent of the choice of $S(T)$. Furthermore, S_{∞} is the unique positive-definite solution of the ARE, and

2) The closed-loop plant A_c is asymptotically stable.

This result means that, as long as the system and PI satisfy certain basic controllability and observability requirements, *the steady-state LQ regulator will yield gains that stabilize the system.* Considering the difficulty encountered by classical control techniques in stabilizing multi-input systems, this is a remarkable property. Exactly as in classical control theory, the theorem predicts the *closed-loop* stability properties of the system in terms of *open-loop* system properties that are easily tested using matrix rank techniques.

The observability of $(\sqrt{Q}, A)$ is needed for the stability result. This property means roughly that all the plant modes should be weighted in the PI, which imposes a design requirement on the engineer as he selects the weighting matrix Q. If the control can be selected so that J is bounded, which the controllability property insures, then the integrand in (Equation 48.29) vanishes with time. If, as also required by the theorem, all states are observable in the PI, this will in turn insure that $x(t)$ vanishes with time, thereby guaranteeing closed-loop stability. It is not technically required that $(\sqrt{Q}, A)$ be observable to guarantee a stable closed-loop system. All that is required is the weaker condition of *detectability*, which corresponds to the observability of the unstable modes of A. However, detectability is more difficult to test for than observability.

The closed-loop poles will depend on the selection of the design matrices Q and R; however, the poles will always be stable as long as the engineer selects $R > 0$ and $Q \geq 0$ with $(\sqrt{Q}, A)$ observable. Thus, the elements of Q and R may be varied during an interactive computer-aided design procedure to obtain suitable closed-loop performance. The optimal gain K is found for given values of Q and R, and the closed-loop time responses are found by simulation. If these responses are unsuitable, new values for Q and R are selected and the design is repeated. Given good software to solve for K, this procedure is quite convenient. Such software is available, for instance, in MATLAB and $MATRIX_X$.

Suboptimal Control — Constant Feedback Gains

Even if the control interval $[0, T]$ is not infinite, the engineer may decide to use the steady-state gain K_{∞} instead of the optimal time-varying gain $K(t)$ given in Table 48.2. The theorems guarantee stability of the closed-loop system using the steady-state LQR. On a finite interval $[0, T]$, the constant gain K_{∞} is *suboptimal*, but the convenience gained by not having to implement a time-varying gain can more than compensate for the loss of optimality. Moreover, as T becomes large, the optimal gain $K(t)$ tends to K_{∞} so that the decision to use the steady-state gain makes more and more sense. In addition to the ease of implementation of constant feedback gains, this suboptimal controller has other important advantages: (1) *it guarantees stability even for complex multi-loop systems*, and (2) there are efficient numerical routines available for the solution of the ARE (e.g., MATLAB and $MATRIX_X$).

EXAMPLE 48.3: Inverted pendulum

Figure 48.3 shows a rod attached to a cart through a pivot. A force $u(t)$ is applied to the cart through a motor attached to an axle. The control objective is to use $u(t)$ to balance the pendulum upright while simultaneously keeping the horizontal movement $p(t)$ of the cart small. This is known as the *inverted pendulum problem.*

The state is $x = [\theta \;\; \dot{\theta} \;\; p \;\; \dot{p}]^T$. A force/moment balance approach or a Lagrangian approach may be used to obtain the dynamics. Assuming that $M = 5kg, m = 0.5kg, L = 1m$, we thus obtain the state equations by linearizing the dynamics about $\theta = \dot{\theta} = p = \dot{p} = 0$:

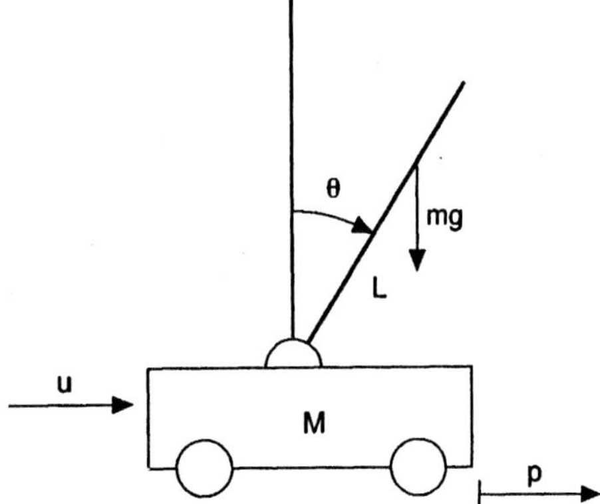

Figure 48.3 Inverted pendulum on a cart.

$$\dot{x} = \begin{bmatrix} 0 & 1 & 0 & 0 \\ 10.78 & 0 & 0 & 0 \\ 0 & 0 & 0 & 1 \\ -0.98 & 0 & 0 & 0 \end{bmatrix} x + \begin{bmatrix} 0 \\ -0.2 \\ 0 \\ 0.2 \end{bmatrix} u = Ax + Bu. \tag{48.37}$$

The open-loop poles are at $s = 0, 0, \pm 3.283$, so that, with no control input, the rod will clearly fall over due to the unstable pole at $s = 3.283$.

It is desired to select K in

$$u = -Kx = -(k_\theta \theta + k_{\dot{\theta}} \dot{\theta} + k_p p + k_{\dot{p}} \dot{p}) \tag{48.38}$$

to regulate the state x to zero. For this purpose, select the PI (Equation 48.29); then, K is determined by using the matrix design Equations 48.30, 48.31. Values of $R = 1$ and $Q = diag\{100, 100, 10, 10\}$ were selected. The motivation for choosing this Q was to place heavy emphasis on keeping the angle $\theta(t)$ small; the cart position control does not matter if the rod falls over. Using MATLAB subroutines from the Control System Toolbox, the optimal gain was easily found to be $K = [-156.16 \quad -49.21 \quad -3.16 \quad -8.72]$, which yields closed-loop poles at $s = -0.60 \pm j0.45, -2.48, -4.41$.

Using MATLAB routines, the closed-loop response was *simulated* and plotted. The angle $\theta(t)$ and position $p(t)$ in response to an initial condition offset of $\theta(0) = 0.1 rad \approx 6°$, $p(0) = 0.1m$ are shown in Figure 48.4a. The required control force u(t) is shown in Figure 48.4b. These plots are quite interesting and bear discussion. Due to the initial offset of 6° in angle, a large control must be applied immediately to push the cart under the rod to catch it so it does not fall. Subsequent smaller control motions begin to move the cart slowly back to the desired horizontal position of $p = 0$ while balancing the rod. Thus, the fast closed-loop poles correspond to the rod motion. The slow complex pole pair is associated with the cart position. This *two-time scale behavior was induced* by the widely disparate weightings selected in the design matrix Q.

In fact, the value for Q was selected by performing *several design iterations* with different Q until computer simulation finally showed a good time response. Such design iterations, coupled with computer simulation, are common in modern control and are very easy using software like MATLAB.

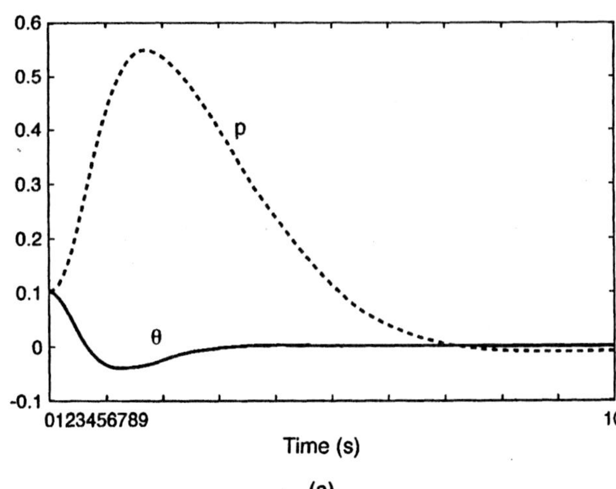

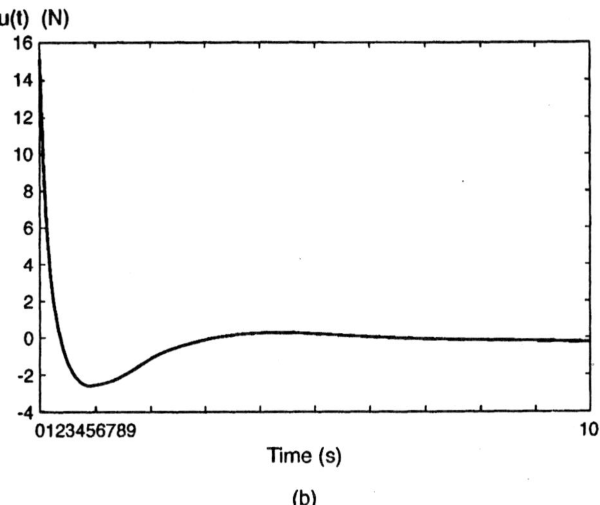

Figure 48.4 Inverted pendulum response. (a) Angle $\theta(t)$ and position $p(t)$. (b) Control input $u(t)$.

Eigenstructure LQR Design

In most computer software routines for solving for the LQR gains, the ARE solution is determined from the eigenstructure of the Hamiltonian matrix H in the *Hamiltonian system*

$$\begin{bmatrix} \dot{x} \\ \dot{\lambda} \end{bmatrix} = \begin{bmatrix} A & -BR^{-1}B^T \\ -Q & -A^T \end{bmatrix} \begin{bmatrix} x \\ \lambda \end{bmatrix} \equiv H \begin{bmatrix} x \\ \lambda \end{bmatrix} \tag{48.39}$$

which consists of the state Equation 48.20 and the costate Equation 48.21, with $u(t)$ replaced by $-R^{-1}B^T\lambda(t)$.

The Hamiltonian matrix H enjoys the special property of having n stable poles and n unstable poles (their images in the $j\omega$ axis), where n is the dimension of the state vector x. If (A, B) is stabilizable, and $(\sqrt{Q}, A)$ is observable, so that the conditions of Theorem 48.1 hold, then the stable eigenvalues of H are also the poles of the optimal closed-loop system

$$\dot{x} = (A - BK_\infty)x \equiv A_c x, \tag{48.40}$$

with K_∞ the steady-state feedback gain. This provides an alternative proof of the stability of the optimal closed-loop system.

Select the eigenvectors of the stable eigenvalues of H, and partition them as $[X_i^T \quad \Lambda_i^T]^T$. Let X be an $n \times n$ matrix whose columns are X_i, and Λ be an $n \times n$ matrix whose columns are Λ_i. Then the solution to the ARE is given in terms of the eigenstructure of H by

$$S_\infty = \Lambda X^{-1}, \tag{48.41}$$

and the steady-state gain is given by

$$K_\infty = R^{-1} B^T \Lambda X^{-1}. \tag{48.42}$$

(If the eigenvalues are complex, then, in the definitions of X and Λ, it is necessary to use the real and imaginary parts of the associated vectors X_i and Λ_i instead of the complex conjugate vectors themselves.)

48.2.4 Frequency-Domain Results and Robustness of the LQR

It is possible to discuss the LQR from the view point of the frequency domain. Classical frequency-domain results are given in terms of scalar SISO systems and involve notions like the loop gain, return difference, sensitivity, and so on. Such ideas are extended in modern optimal design using multivariable transfer functions and the notions of the *singular value* and the *multivariable Bode plot.* For more details, see Chapter 29 or the references. An important property of any closed-loop system is *robustness* to uncertainties, including modeling errors, disturbances, and noise. The LQR has some important robustness properties that are detailed here.

LQR Frequency-Domain Relationships

Suppose that the plant is time-invariant and, in Figure 48.1, K is the constant optimal LQ state-feedback gain determined using the LQR ARE as in Subsection 48.2.3. Define the plant transfer function as $G(s) \equiv (sI - A)^{-1}B$. Then, the *loop gain* referred to the input is

$$KG(s) = K(sI - A)^{-1}B = KG(s) \tag{48.43}$$

and the *closed-loop return difference* is $[I + K(sI - A)^{-1}B] = I + KG(s)$.

Optimal return difference relationship. Two key results are the following. The *optimal characteristic polynomial relationship* is

$$\Delta_c(s) = |I + K(sI - A)^{-1}B|\Delta(s) \tag{48.44}$$

where the open-loop characteristic polynomial is $\Delta(s) = |sI - A|$ and the closed-loop characteristic polynomial is $\Delta_c(s) = |sI - (A - BK)|$. The *optimal return difference relationship* is

$$\begin{aligned}[I + K(-sI - A)^{-1}B]^T R\, [I + K(sI - A)^{-1}B] \\ = R + B^T(-sI - A)^{-T} Q(sI - A)^{-1}B.\end{aligned} \tag{48.45}$$

These are extremely important because, exactly as in classical control theory, they express *closed-loop properties* in terms of *open-loop properties* that can be computed before the optimal controller is designed. They allow, for instance, the development of the *Chang-Letov design approach for LQR* which is an extension of root locus design to MIMO systems.

Optimal singular value relationships. Select the control weighting matrix as $R = \rho I$, with ρ a positive design parameter. Denoting the i-th singular value of a matrix M as $\sigma_i(M)$, Equation 48.45 yields the *optimal singular value relationship* of the LQR

$$\sigma_i[I + KG(j\omega)] = \left[1 + \frac{1}{\rho}\sigma_i^2[H(j\omega)]\right]^{\frac{1}{2}} \tag{48.46}$$

with

$$H(s) \equiv C(sI - A)^{-1}B \tag{48.47}$$

and matrix C defined by $Q = C^T C$. This is important because the right-hand side is known in terms of *open-loop* quantities before the optimal feedback gain is found by solution of the ARE, while the left-hand side is the closed-loop return difference.

According to this relationship, for all ω, the minimum singular value, denoted $\underline{\sigma}$, satisfies the *LQ optimal singular value constraint*

$$\underline{\sigma}[I + KG(j\omega)] \geq 1. \tag{48.48}$$

Thus, the LQ regulator always results in a *decreased sensitivity.*

Guaranteed Robustness of the Linear-Quadratic Regulator

The linear-quadratic regulator using full state feedback has many useful properties, including guaranteed closed-loop stability and ease of design by solving matrix design equations. It will now be shown that the steady-state LQR has certain *guaranteed robustness properties* that make it even more useful. These conclusions may be discovered using the *multivariable Nyquist criterion,* which shall be referred to the polar plot of the return difference $I + KG(s)$, where the origin is the critical point. A typical polar plot of $\underline{\sigma}[I + KG(j\omega)]$ is shown in Figure 48.5, where the optimal singular value constraint appears as the condition that *all the singular values remain outside the unit disc.*

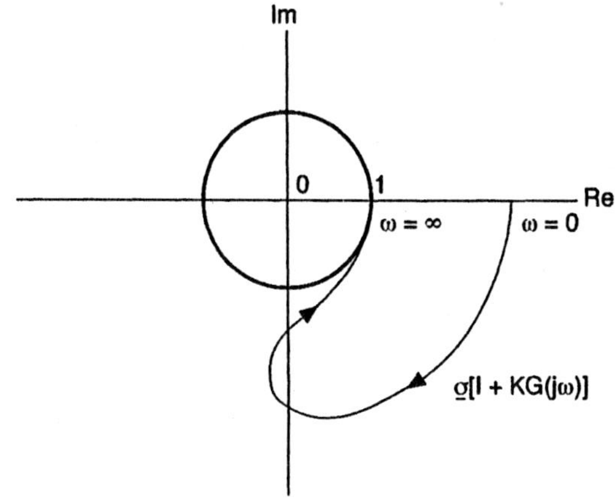

Figure 48.5 Typical polar plot for optimal LQ return difference.

Guaranteed stability. The multivariable Nyquist criterion says that (as long as the open-loop system $G(s)$ is stable) the closed-loop system is stable if none of the singular value plots of $I + KG(j\omega)$ encircle the origin in the figure. Due to the optimal singular value constraint, no encirclements are possible. This constitutes a proof of the *guaranteed stability* of the LQR discussed in Subsection 48.2.3.

Gain margin. Multiplying the optimal feedback K by any positive scalar gain $k > 1$ results in a loop gain of $kKG(s)$, which has a minimum singular value plot identical to the one in Figure 48.5 except that it is scaled outward; that is, the $\omega \to 0$ limit (i.e., the DC gain) will be larger, but the $\omega \to \infty$ limit will still be 1. Thus, the closed-loop system will still be stable. In classical terms, the LQ regulator with full state feedback has *an infinite gain margin.*

Phase margin. For multivariable systems the *phase margin* may be defined as the angle marked 'PM' in Figure 48.6. As in the classical case, it is the angle through which the polar plot of $\underline{\sigma}[I + KG(j\omega)]$ must be rotated (about the point 1) clockwise to make the plot go through the critical point. By combining

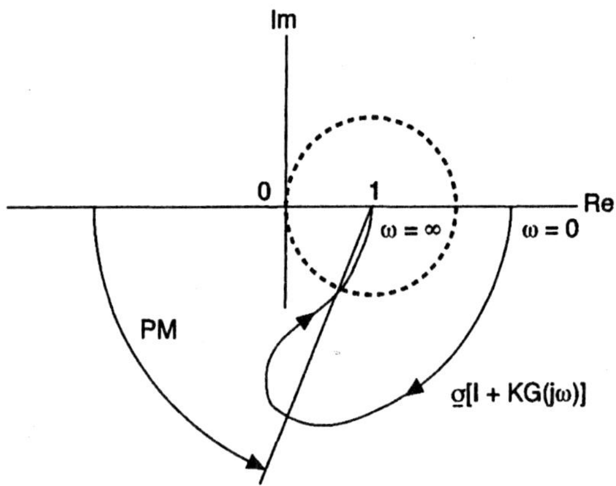

Figure 48.6 Definition of multivariable phase margin.

Figure 48.5 with Figure 48.6, it can be determined that, due to the LQ singular value constraint, the plot of $\underline{\sigma}[I + KG(j\omega)]$ must be rotated through at least 60° to make it pass through the origin. The LQR with full state feedback thus has a *guaranteed phase margin of at least* 60°. This means that a phase shift of up to 60° may be introduced in any of the m paths in Figure 48.1, or in all paths simultaneously, as long as the paths are not coupled to each other in the process. (Here, m is the number of control inputs, i.e., the number of control loops.)

This phase margin is excessive; it is higher than that normally required in classical control system design. This overdesign means that, in other performance aspects, the LQ regulator may have some deficiencies. One of these is that, at the *crossover frequency* (loop gain= 1), the slope of the *multivariable Bode plot* is -20 dB/decade, a relatively slow attenuation rate. (By allowing a Q weighting matrix in the PI that is not positive semidefinite, it is possible to obtain better LQ designs that have higher roll off rates at high frequencies.)

Stability with multiplicative uncertainty. It can be shown that Equation 48.48 implies that

$$\underline{\sigma}[I + (KG(j\omega))^{-1}] \geq \frac{1}{2}. \tag{48.49}$$

This corresponds to the fact that the LQR with state feedback remains stable for all *multiplicative uncertainties* in the plant transfer function which satisfies $m(\omega) < \frac{1}{2}$.

48.3 The Tracker Problem

The function of the LQ regulator is to hold the states near zero, that is, to guarantee closed-loop stability. Another fundamental design problem in systems engineering is to control a system so that a specified output follows a given nonzero *reference trajectory* $r(t)$. An example is controlling an aircraft to follow a desired step input command (e.g., change in altitude). This is called the *tracking or servodesign* problem. For this purpose, the regulator control law must be modified. The fundamental issue here is that for optimal tracking some additional *feedforward terms* must be added to the control input besides the basic LQR feedback loop that gives closed-loop stability.

Consider the linear system given by Equation 48.51, with $z(t)$ in Equation 48.52 a *performance output* that should track the given reference input $r(t)$. In contrast to classical control, it is easy to include here the case where both $z(t)$ and $r(t)$ are vectors, that is, the case of multivariable tracking. The control input is given by

$$u = -Kx + v, \tag{48.50}$$

where $v(t)$ is a feedforward signal required for good tracking performance. The feedback gain K and feedforward signal $v(t)$ are determined so as to keep the *tracking error* (Equation 48.53) small.

The solution to the tracking problem is significantly more involved than the regulator problem. It is now discussed from three points of view. In this section, full state-variable feedback is assumed. In actual applications, only *output feedback* is allowed. In that case refer to Chapter 5.1 or to [8], [11], where these techniques are extended.

48.3.1 Optimal LQ Tracker

The optimal LQ tracker can be derived along the same lines as the Hamiltonian approach in Section 48.2.2. The result is given in Table 48.3. The feedback structure is basically the same as the LQ regulator in Table 48.2, but now the PI contains the tracking error, not the state, because e(t) is to be small in this application. The presence of $e(t)$ in the PI has the effect of adding a feedforward term $v(t)$ to the control signal. The tracker gain K and signal $v(t)$ determined with the LQ optimal approach are given in the table. The feedback gain K is found in the same manner as in the LQR case. The structure of the LQ optimal tracker is given in Figure 48.7.

The feedforward signal is computed using the dynamical system (Equation 48.57), which is called the *adjoint system*; like the

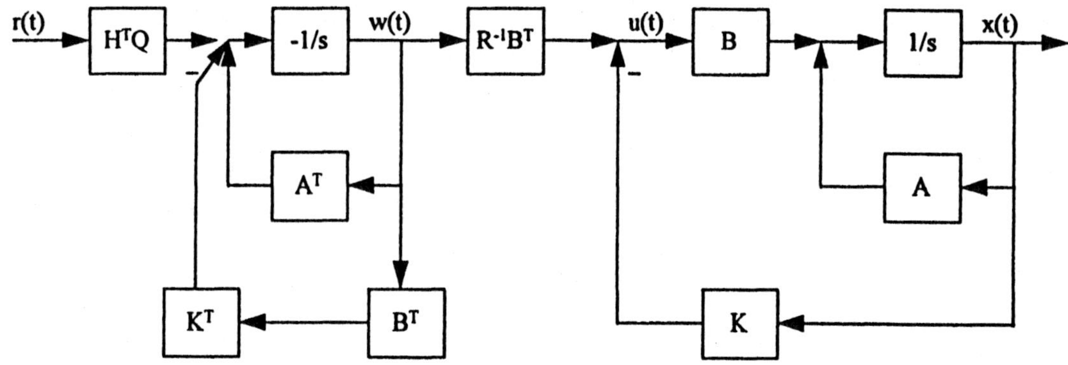

Figure 48.7 Optimal LQ Tracker.

TABLE 48.3 Optimal Continuous-Time Linear Quadratic Tracker.

System model:

$$\dot{x} = Ax + Bu, \quad t \geq t_0, \; x(t_0) = x_0 \; given. \tag{48.51}$$

Performance output and tracking error:

$$z = Hx, \tag{48.52}$$
$$e = r - z. \tag{48.53}$$

Performance index:

$$J(t_0) = \frac{1}{2}e^T(T)Pe(T) + \frac{1}{2}\int_{t_0}^{T}(e^TQe + u^TRu)\,dt, \tag{48.54}$$

with

$$P \geq 0, \quad Q \geq 0, \quad R > 0.$$

Optimal tracking controller:

Riccati equation:

$$-\dot{S} = A^TS + SA - SBR^{-1}B^TS + H^TQH, \quad t \leq T, \; S(T) = H^TPH. \tag{48.55}$$

Optimal feedback gain:

$$K = R^{-1}B^TS. \tag{48.56}$$

Feedforward system:

$$-\dot{w} = (A - BK)^Tw + H^TQr, \quad t \leq T, \; w(T) = H^TPr(T). \tag{48.57}$$

Feedback plus feedforward control:

$$u(t) = -K(t)x(t) + R^{-1}B^Tw(t). \tag{48.58}$$

Riccati equation, it is integrated *backward* in time. Thus, the Riccati equation and the adjoint system must be integrated off-line before the control run. In fact, the optimal LQ tracker is *noncausal*, because future values of the reference input $r(t)$ are needed to compute $w(t)$. The ramifications of this noncausal nature of the optimal tracker are illustrated in Figure 48.8, which shows the optimal tracker response for a scalar system using control weighting $R = 1$ and different values of the error weighting Q. The system output begins to change *before* the reference $r(t)$ does, so that the system anticipates the changes in $r(t)$. This *anticipatory behavior* is an important feature of the optimal tracker.

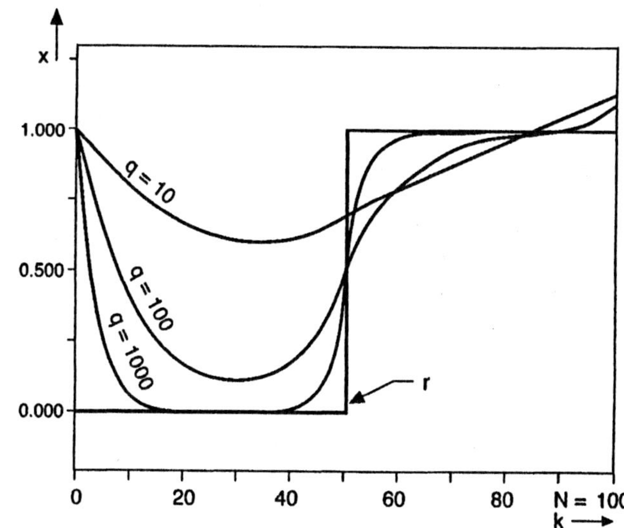

Figure 48.8 Anticipatory response of the optimal LQ tracker.

The noncausal nature of the optimal LQ tracker means that it cannot be implemented in practice when $r(t)$ is not predetermined. Therefore, two suboptimal strategies are now outlined that yield implementable tracking systems.

48.3.2 Conversion of an LQR to an LQ Tracker

As an alternative to the optimal tracking solution just presented, a causal tracker can be obtained as follows. First, the LQ regulator is designed using Table 48.2. Then, it is converted to an LQ tracker by adding a feedforward term. In the case where the reference signal $r(t)$ is a constant (i.e., step function) with magnitude r_0, the tracking control with state feedback is given by

$$u = -Kx + H_c^{-1}(0)r_0, \tag{48.59}$$

where the closed-loop transfer function is

$$H_c(s) = H(sI - (A - BK))^{-1}B \tag{48.60}$$

and $H_c(0)$ is the *DC gain* of the closed-loop system $(A - BK)$. The control gain K is found using the LQR design equations in Table 48.2. The structure of this suboptimal tracker is shown in Figure 48.9. Unfortunately, if the DC gain is not well known this tracker structure does not perform well, that is, this tracker is not *robust*.

48.3.3 A Practical Suboptimal Tracker

This subsection shows how to design a suboptimal tracker that works well for practical applications and is robust to uncertainties and disturbances. The key is in the use of engineering design insight and common sense to formulate the problem. One uses a *unity feedback gain outer loop*, which has proven effective in classical control approaches. This technique also relies on converting an LQR to a tracker, but differs from the work in the previous subsection.

Problem Formulation A general class of systems is described by the equations

$$\begin{aligned} \dot{x} &= Ax + Bu + Er \\ z &= Hx \end{aligned} \tag{48.61}$$

which can contain both the plant plus some desirable *compensator dynamics*. The control input is allowed to have the form

$$u = -Kx - KFr, \tag{48.62}$$

which consists of state feedback plus a feedforward term of a special composition. Placing the control into the system yields the closed-loop system

$$\dot{x} = (A - BK)x + (E - BKF)r \equiv A_c x + B_c r. \tag{48.63}$$

Matrices E and F are chosen to have a structure that is sensible from a design point of view. Specifically, it is very desirable to incorporate a unity-gain outer tracking loop in the controller, as shown in Example 48.4.

Deviation system and LQR design step Assume that the reference input is a unit step of magnitude r_0. Then, the *steady-state* system is

$$0 = A_c\bar{x} + B_c r_0,$$

where overbars denote steady-state values, so that the steady-state value of the state is $\bar{x} = -A_c^{-1}B_c r_0$. Though the reference input is assumed constant for design purposes, this is to allow good closed-loop rise time and overshoot qualities. Then, the designed controller works for *any reference input $r(t)$, even though time-varying.*

Define the *deviations*

$$\begin{aligned} \tilde{x} = x - \bar{x}, \quad \tilde{z} = z - \bar{z}, \\ \tilde{u} = u - \bar{u}, \quad \tilde{e} = e - \bar{e}. \end{aligned} \tag{48.64}$$

Then, the deviations satisfy the dynamics of the *deviation system*

$$\dot{\tilde{x}} = A\tilde{x} + B\tilde{u} \tag{48.65}$$

$$\tilde{z} = H\tilde{x} \tag{48.66}$$

$$\tilde{u} = -K\tilde{x}. \tag{48.67}$$

Because $e = r - z$, the tracking error deviation is $\tilde{e} = -\tilde{z}$. To induce tracking behavior, define the performance index

$$J(t_0) = \frac{1}{2}\int_0^\infty (\tilde{x}^T Q\tilde{x} + \tilde{u}^T R\tilde{u})\, dt, \tag{48.68}$$

which makes the entire deviation state, and therefore $\tilde{e}$, small.

Tracker design The tracking problem may now be solved as follows. First, solve the LQ regulator problem for the deviation system using Table 48.2. Then, the tracking control input is given by Equation 48.62. This tracker has a much different structure than the DC-gain-based tracker in Figure 48.9. The next example shows that a sensible choice for matrices E and F based on classical control notions gives a robust tracker with a unity gain outer loop. Then, a sensible choice for the PI design matrices Q and R gives good control gains and *guaranteed stability*, even for complex multiloop tracking systems.

Note that $e = \tilde{e} + \bar{e}$, where $\bar{e}$ is the steady-state value of the tracking error. Because this technique only guarantees that $\tilde{e}$ is small, special steps must be taken to guarantee that $\bar{e}$ is also small. One way to do this is to include *integrators in all the feedforward loops*, as in the next example. As an alternative, a term involving $\bar{e}$ can be added to the PI (48.68). This gives more involved design equations, which are nevertheless still easily solved by digital computer. The details are in [8]. Finally, although the gain determined in this fashion is optimal for the deviation system, it is not optimal for the tracking problem in terms of the original dynamics (48.61). In practical applications, however, it is suitable provided that the design matrices are sensibly selected.

EXAMPLE 48.4: Aircraft pitch-rate control system

This example illustrates the tracker design procedure just presented. Good tracker system design relies on a sensible selection of the structure matrices E and F, and good feedback gains rely on a sensible selection of the design weighting matrices Q, R. *Compensator dynamics* can be accounted for using this procedure. Because this is an LQ-based approach, a reasonable formulation of the problem should result in *guaranteed closed-loop stability*. This is an important feature of modern control design techniques, and is in complete contrast to classical techniques where stability in multi-loop systems can be difficult to achieve.

a. Aircraft and control system dynamics

In a pitch-rate control system, the control input is elevator actuator voltage $u(t)$ and r is a reference step input corresponding

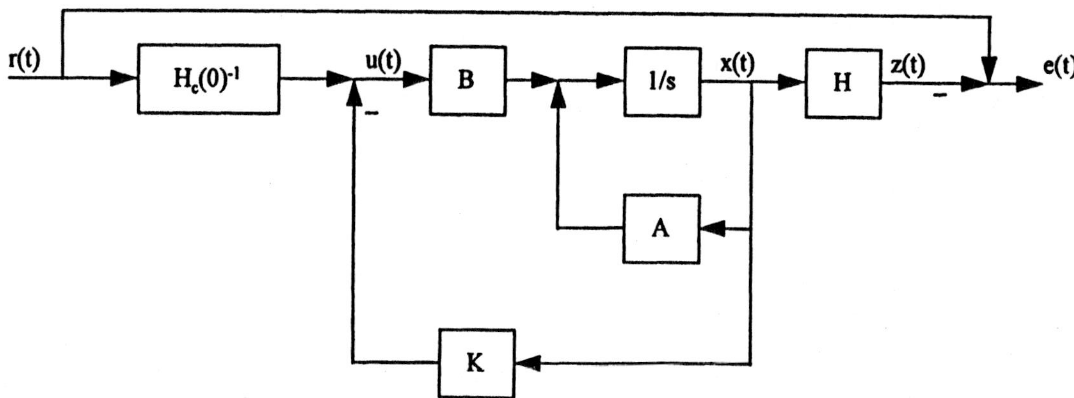

Figure 48.9 Tracker based on DC gain.

to the desired pitch command. The performance output $z(t)$ is the pitch rate q. To ensure zero steady-state error, an integrator is added in the feedforward channel; this corresponds to *compensator dynamics*, and is easily dealt with in this approach. The integrator output is ϵ. It is assumed here that all states are available as measurements for feedback purposes; in practice, the output-feedback design technique in [11] will be required.

The design is based on a short period approximation to the F-16 dynamics linearized about a nominal flight condition of 502 ft/s, 0 ft altitude, level flight, with the center of gravity at 0.35. The basic aircraft states of interest are q and angle of attack α. An additional state is introduced by the elevator actuator, whose deflection is δ_e. The states of the plant plus compensator are $x = [\alpha \quad q \quad \delta_e \quad \epsilon]^T$ and the system dynamics are described by Equation 48.61 with

$$A = \begin{bmatrix} -1.01887 & 0.90506 & -0.00215 & 0 \\ 0.82225 & -1.07741 & -0.17555 & 0 \\ 0 & 0 & -20.2 & 0 \\ 0 & -57.2958 & 0 & 0 \end{bmatrix}, \quad (48.69)$$

$$B = \begin{bmatrix} 0 \\ 0 \\ 20.2 \\ 0 \end{bmatrix}, E = \begin{bmatrix} 0 \\ 0 \\ 0 \\ 1 \end{bmatrix}, \quad (48.70)$$

and

$$H = \begin{bmatrix} 0 & 57.2958 & 0 & 0 \end{bmatrix} \quad (48.71)$$

The factor of 57.2958 is added to convert angles from radians to degrees. The last line of the state equation using this A and E matrix describes the integrator, $\dot{\epsilon} = -57.2958q + r$.

b. Control design

Select the control input $u(t)$ to yield good closed-loop response to a step input at r, which corresponds to a single-input/multi-output tracker design problem. Since the integrator makes the system Type I, the steady-state error $\tilde{e}$ is equal to zero and $e(t) = \tilde{e}(t)$. Thus, the design method just described is appropriate.

The control input is

$$\begin{aligned} u &= -Kx = -\begin{bmatrix} k_\alpha & k_q & k_{\delta_e} & k_I \end{bmatrix} x \\ &= -k_\alpha \alpha - k_q q - k_{\delta_e}\delta_e - k_I \epsilon. \end{aligned} \quad (48.72)$$

Therefore, referring to Equation 48.62 it is evident that $F = 0$; however, including the integrator output as a state variable in the dynamics (1) adds the feedforward path required for tracking behavior, that is, element k_I of the feedback matrix K is actually a 'feedforward' gain.

To determine the gain matrix K, select the PI (Equation 48.68), and try weighting matrices $R = 1$, $Q = diag\{1, 10, 1, 1\}$. Now use the LQR routine from the MATLAB Control Systems Toolbox to determine the optimal gain $K = [-0.046 \quad -1.072 \quad 0 \quad 3.381]$. Using MATLAB routines, the corresponding closed-loop poles are $s = -8.67 \pm j9.72, -9.85, -4.07, -1.04$. The resulting step response is shown in Figure 48.10, which displays good performance.

48.4 Minimum-Time and Constrained-Input Design

An important class of control problems is concerned with achieving the performance objectives in *minimum time*. A suitable performance index for these problems is

$$J = \int_{t_0}^{T} 1\, dt = T - t_0. \quad (48.73)$$

Several sorts of minimum-time problems are now discussed.

48.4.1 Nonlinear Minimum-Time Problems

Suppose the objective is to drive the system

$$\dot{x} = f(x, u) \quad (48.74)$$

from a given initial state $x(t_0) \in \mathcal{R}^n$ to a specified final state $x(T)$ in minimum time. Then, from Table 48.1 the Hamiltonian is

$$H = 1 + \lambda^T f \quad (48.75)$$

and the Euler equations are the costate equation

$$-\dot{\lambda} = \frac{\partial f^T}{\partial x}\lambda \quad (48.76)$$

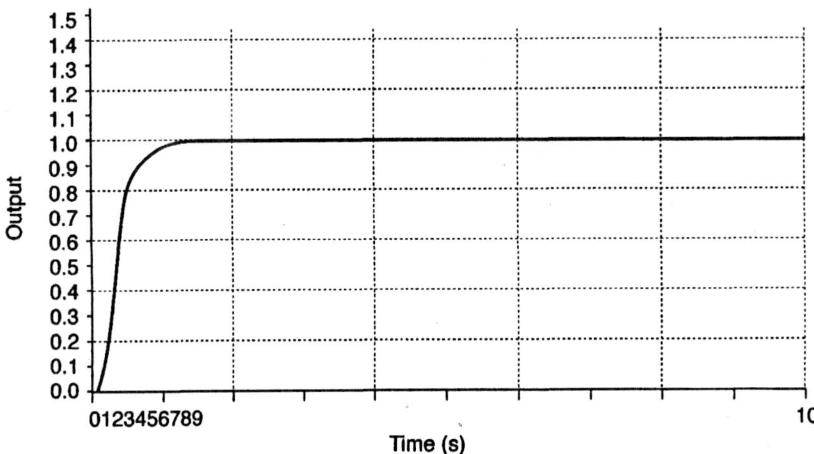

Figure 48.10 Pitch-rate step response.

plus the stationarity condition

$$0 = \frac{\partial f^T}{\partial u}\lambda. \tag{48.77}$$

Since the final state is fixed (so that $dx(T) = 0$) but the final time is free, the final condition in Table 48.1 says that

$$0 = H(T) = 1 + \lambda^T(T)f[x(T), u(T)]. \tag{48.78}$$

If $f(x, u)$ is not an explicit function of time, then according to the conservation principle (Equation 48.10), $H(t)$ is zero for all time.

The stationarity condition (Equation 48.77) may often be used to solve for $u(t)$ in terms of $\lambda(t)$. Then, $u(t)$ may be eliminated in the state and costate equations to obtain the Hamiltonian system. To solve this, we require n initial conditions ($x(t_0)$ given) and n final conditions ($x(T)$ specified). However, the final time T is now unknown. The function of Equation 48.78 is to provide one more equation so that T can be solved for. Several nonlinear design problems can be explicitly solved, yielding great insight into the minimum-time control structure. Examples include *Zermelo's Problem* and the *Brachistochrone Problem.*

48.4.2 Linear Quadratic Minimum-Time Design

The general solution procedure given in the previous subsection for the nonlinear minimum-time problem is difficult to apply. Moreover, a reasonable solution may not exist. A general class of practical problems is covered by the case where it is required to find an optimal control for the linear system

$$\dot{x} = Ax + Bu \tag{48.79}$$

that minimizes the performance index

$$J = \frac{1}{2}x^T(T)S_T x(T) + \frac{1}{2}\int_{t_0}^{T}(1 + x^TQx + u^TRu)\,dt \tag{48.80}$$

with $S_T \geq 0$, $Q \geq 0$, $R > 0$, and the final time T free. There is no constraint on the final state; thus, the control objective is to make the final state sufficiently small. Due to the term $\frac{1}{2}(T - t_0)$ arising from the integral, this must be accomplished in a short time period. This is a general sort of PI that allows for a trade-off between the minimum-time objective and a desire to keep the states and the controls small. Thus, if the engineer selects smaller Q and R, the term $\frac{1}{2}(T - t_0)$ in the PI dominates, and the control tries to make the transit time smaller. This is called the *linear quadratic (LQ) minimum-time problem.*

The solution for this problem is the same as in Table 48.2. The control is a linear time-varying state feedback given by

$$u = -K(t)x \tag{48.81}$$

with optimal gain

$$K = R^{-1}B^TS \tag{48.82}$$

and $S(t)$ the solution determined by integrating the Riccati equation backward from time T. Unfortunately, there is a problem in that the final time T is unknown.

To determine the value of T that minimizes the PI, an *extra condition* is needed, given by Equation 48.78, which yields

$$x^T(t_0)\dot{S}x(t_0) = 1, \tag{48.83}$$

with $x(t_0)$ the specified initial condition of the plant. The solution procedure for the LQ minimum-time problem is to integrate the Riccati equation

$$-\dot{S} = A^TS + SA + Q - SBR^{-1}B^TS \tag{48.84}$$

backward from some time τ using $S(\tau) = S_T$ as the final condition. At each time t, the left-hand side of Equation 48.83 is computed using the known initial state and $\dot{S}(t)$. Then, the minimum interval $(T - t_0)$ is equal to $(\tau - t)$ where t is the time for which Equation 48.83 first holds. This specifies the minimum final time T, and then allows the computation of the optimal feedback gain $K(t)$ on the interval $[t_0, T]$.

The Riccati derivative $\dot{S}$ is used to determine the optimal time interval, while S is used to determine the optimal feedback gain $K(t)$.

More details on this control scheme may be found in [12]. It is important to note that condition (Equation 48.83) may never hold. Then, the optimal solution is $T - t_0 = 0$, that is, the PI is minimized by using *no control.* Roughly speaking, if $x(t_0)$ and/or Q and $S(T)$ are selected large enough, then it makes sense to apply a nonzero control $u(t)$ to make $x(t)$ decrease. On the other hand, if Q and $S(T)$ are selected too small for the given initial state $x(t_0)$, then it is not worthwhile to apply any control to decrease $x(t)$, because a nonzero control and a nonzero time interval will increase the PI.

48.4.3 Constrained-Input Design and Bang-Bang Control

Up to this point minimum-time control has been presented based on the conditions of Table 48.1, which were derived using the calculus of variations. Under some smoothness assumptions on $f(x, u, t)$ and $L(x, u, t)$, the resulting controls are also smooth. Here, a fundamentally different sort of control strategy will be presented.

If the linear system

$$\dot{x} = Ax + Bu \tag{48.85}$$

with $x \in \mathcal{R}^n$, $u \in \mathcal{R}^m$ is prescribed, there are problems with using the pure minimum-time PI,

$$J(t_0) = \int_{t_0}^{T} 1\, dt, \tag{48.86}$$

where T is free. The way to minimize the time is to use infinite control energy! Since this optimal strategy is not acceptable, it is necessary to find a way to reformulate the minimum-time problem for linear systems.

Therefore, the control input now must satisfy the *magnitude constraint*

$$\|u(t)\| \leq 1 \tag{48.87}$$

for all $t \in [t_0, T]$. This constraint means that *each component* of the m-vector $u(t)$ must be no greater than 1. Thus, the control is constrained to an *admissible region* (in fact, a hypercube) of $\mathcal{R}^m$. If the constraints on the components of $u(t)$ have a value different from 1, then one may appropriately scale the corresponding columns of the B matrix to obtain the constraints in the form of Equation 48.87. A requirement like Equation 48.87 arises in many problems where the control magnitude is limited by physical considerations; for instance, the thrust of a rocket certainly has a maximum possible value, as has the armature voltage of a DC motor.

Referring to Table 48.1, the optimal control problem posed here is to find a control $u(t)$ that drives a given $x(t_0)$ to a final state $x(T)$ satisfying the final state constraint, minimizes the PI, and satisfies Equation 48.87 at all times. Intuitively, to minimize the time, the optimal control strategy appears to be to apply maximum effort (i.e., plus or minus 1) over the entire time interval. This idea will now be formalized. When a control component takes on a value at the boundary of its admissible region (i.e., ± 1), it is said to be *saturated.* Pontryagin and co-workers have shown that, in the case of constrained control, Table 48.1 still applies if the stationarity condition is replaced by the more general condition, known as *Pontryagin's Minimum Principle,*

$$H(x^*, u^*, \lambda^*, t) \leq H(x^*, u, \lambda^*, t), \text{ all admissible u.} \tag{48.88}$$

This is an extremely powerful result which can be employed to derive the following solution to the linear constrained-input minimum-time problem.

Define the *signum function* for scalar w as

$$sgn(w) = \begin{cases} 1, & w > 0 \\ indeterminate, & w = 0 \\ -1, & w < 0. \end{cases} \tag{48.89}$$

If w is a vector, define $v = sgn(w)$ as $v_i = sgn(w_i)$ for each i, where v_i, w_i are the components of v and w. Then, in terms of the costate, the optimal control is given by

$$u^*(t) = -sgn[B^T \lambda(t)]. \tag{48.90}$$

This may be interpreted as follows. For each column b_i of B, if $\lambda^T(t)b_i$ is positive, we should select $u_i(t) = -1$ to get the largest possible negative value of $\lambda^T(t)b_i u_i(t)$. On the other hand, if $\lambda^T(t)b_i$ is negative, we should select $u_i(t)$ as its maximum admissible value of 1 to make $\lambda^T(t)b_i u_i(t)$ as negative as possible. If $\lambda^T(t)b_i$ is zero at a single point t in time, then $u_i(t)$ can be assigned any value at that time, because then $\lambda^T(t)b_i u_i(t)$ is zero for all values of $u(t)$.

The quantity $B^T\lambda(t)$ is called the *switching function.* A sample switching function and the optimal control it determines are shown in Figure 48.11. When the switching function changes sign, the control switches from one of its extreme values to another. The control in the figure switches four times. The optimal linear minimum-time control is always saturated since it switches back and forth between its extreme values, so it is called *bang-bang control.* In some problems, a component $b_i^T\lambda(t)$ of the switching function $B^T\lambda(t)$ can be zero over a finite time interval. If this happens, component $u_i(t)$ of the optimal control is not well-defined by Equation 48.90. This is called a *singular condition.* If this does not occur, the time-optimal problem is called *normal.*

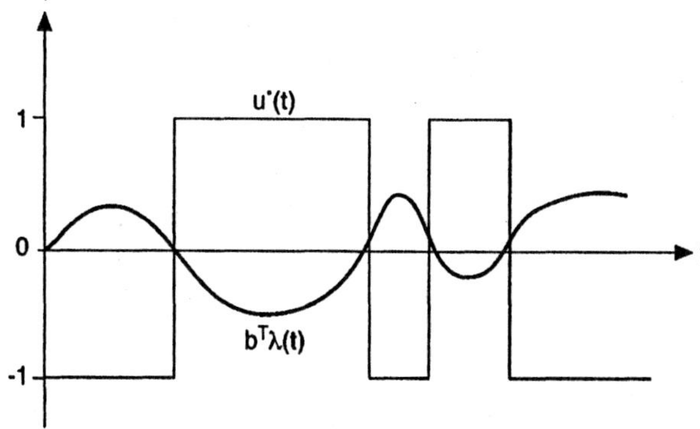

Figure 48.11 Sample switching function and associated optimal control.

The time-invariant plant (Equation 48.85) is reachable if, and only if, the reachability matrix (Equation 48.35) has full rank n. If b_i is the ith column of $B \in \mathcal{R}^{n\times m}$, then the plant is *normal* if

$$U_i = [b_i \; Ab_i \; \dots \; A^{n-1}b_i] \tag{48.91}$$

has full rank n for each $i = 1, 2, \dots, m$, that is, if the plant is reachable by each separate component u_i of $u \in \mathcal{R}^m$. Normality of the plant and normality of the minimum-time control problem are equivalent. Let the plant be normal (and hence reachable), and suppose it is desired to drive a given $x(t_0)$ to a desired fixed final state $x(T)$ in minimum time with a control satisfying Equation 48.87. Then, the following results have been achieved for time-invariant plants by Pontryagin and co-workers:

1. *If the desired final state $x(T)$ is equal to zero, then a minimum-time control exists if the plant has no poles with positive real parts (i.e., no poles in the open right half plane).*
2. *For any fixed $x(T)$, if a solution to the minimum-time problem exists, then it is unique.*
3. *Finally, if the n plant poles are all real and if the minimum-time control exists, then each component $u_i(t)$ of the time-optimal control can switch at most $n-1$ times.*

In both its computation and its final appearance, bang-bang control is fundamentally different from the smooth controls seen previously. The minimum principle leads to the expression (Equation 48.90) for $u^*(t)$, but it is difficult to solve explicitly for the optimal control. Instead, this condition specifies several different control laws, and it is necessary to select which among these is the optimal control. Thus, the minimum principle keeps one from having to examine all possible control laws for optimality, giving a small subset of potentially optimal controls to be investigated. In many cases, it is still possible to express $u^*(t)$ as a state-feedback control law.

EXAMPLE 48.5: Bang-Bang Control

Any system obeying Newton's laws for point-mass motion is described by

$$\begin{aligned} \dot{y} &= v, \\ \dot{v} &= u, \end{aligned} \tag{48.92}$$

with $y(t)$ the position, $v(t)$ the velocity, and $u(t)$ the input acceleration. The state is $x = [y \;\; v]^T$.

Let the acceleration input u be constrained in magnitude by

$$|u(t)| \le 1. \tag{48.93}$$

The control objective is to bring the state from any initial point (y_0, v_0) to the origin in the minimum time T. The final state must be fixed at

$$\psi(x(T), T) = \begin{bmatrix} y(T) \\ v(T) \end{bmatrix} = 0. \tag{48.94}$$

Using (48.88) the minimum-time control takes on only values of $u = \pm 1$. Moreover, there is at most one control switching because the maximum number of switchings is $n-1$ when the plant poles are all real. The *phase plane* is a coordinate system whose axes are the state variables. Phase plane plots of the state trajectories of (1) for $u = 1$ and for $u = -1$ are parabolas in the phase plane as shown in Figure 48.12. These parabolas represent minimum-time trajectories. The arrows indicate the direction of increasing time. For example, if the initial state (y_0, v_0) is as shown in Figure 48.12, then, under the influence of the control $u = -1$, the state will develop downward along the parabola, eventually passing through the point $(y = 0, v = -2)$. On the other hand, if a control of $u = 1$ is applied, the state will move upward and to the right.

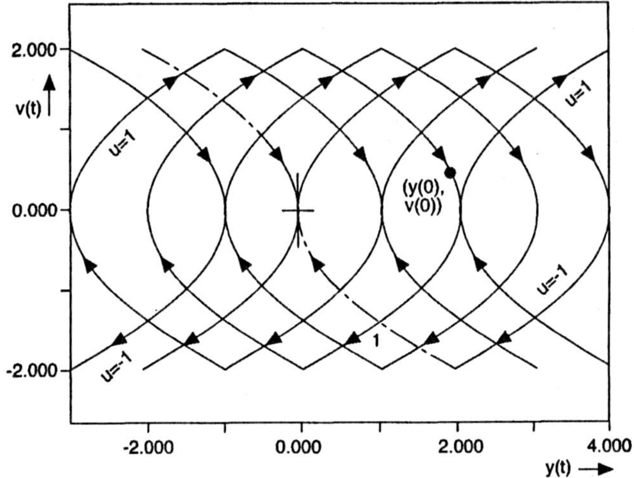

Figure 48.12 Phase plane trajectories for $u = 1$ and $u = -1$.

It will now be argued that this figure represents a *state-feedback control law*, which brings any state to the origin in minimum time. Suppose the initial state is as shown in Figure 48.12. Then the only way to arrive at the origin, while satisfying the Pontryagin conditions, is to apply $u = -1$ to move the state along a parabola to the dashed curve. At this point (labeled "a"), the control is switched to $u = 1$ to drive the state into the origin. Hence, the resulting seemingly roundabout trajectory is in fact a minimum-time path to the origin. The dashed curve is known as the *switching curve*. For initial states on this curve, a control of $u = 1$ (if $v_0 < 0$) or $u = -1$ (if $v_0 > 0$) for the entire control interval will bring the state to zero. For initial states off this curve, the state must first be driven onto the switching curve, and then the control must be switched to its other extreme value to bring the final state to zero. The switching curve is described by the equation $y = -\frac{1}{2}v\,|v|$.

Simply put, for initial states above the switching curve, the optimal control is $u = -1$, followed by $u = 1$, with the switching occurring when $y(t) = \frac{1}{2}v^2(t)$. For initial states below the switching curve, the optimal control is $u = 1$, followed by $u = -1$, with the switching occurring when $y(t) = -\frac{1}{2}v^2(t)$. Because the control at each time t is completely determined by the state (i.e., by the phase plane location), Figure 48.12 yields a feedback control law. This feedback law, represented graphically in the figure, can be stated as

$$u = \begin{cases} -1 & \text{if } y > -\frac{1}{2}v\,|v| \\ & \text{or if } y = -\frac{1}{2}v\,|v| \text{ and } y < 0 \\ 1 & \text{if } y < -\frac{1}{2}v\,|v| \\ & \text{or if } y = -\frac{1}{2}v\,|v| \text{ and } y > 0. \end{cases} \quad (48.95)$$

which makes it clear that the minimum-time control is indeed a state-feedback.

48.5 Optimal Control of Discrete-Time Systems

The discussion so far has applied to continuous-time (analog) systems. The discussion of the LQR problem for discrete-time systems

$$x_{k+1} = Ax_k + Bu_k, \quad (48.96)$$

is identical in form, though more complicated in its details. The problem is to select the state-feedback matrix K in

$$u_k = -Kx_k \quad (48.97)$$

to minimize a performance index specified by the design engineer.

48.5.1 Discrete-Time LQR

In general the optimal discrete-time linear quadratic regulator is a time-varying matrix gain sequence K_k. However, the practically useful solution is the optimal steady-state feedback gain obtained by using the infinite horizon PI (Equation 48.99). The design equations for the discrete-time LQR are given in Table 48.4. The discrete-time LQR equations are more complicated than the continuous-time equivalents; however, commercially available software (e.g., MATLAB) makes this irrelevant to the control designer. In practice, discrete-time design is as straightforward as continuous-time design.

All the results discussed for continuous-time systems in Section 48.2 have their discrete-time counterparts (see the references). Thus, as long as (A, B) is stabilizable and $(A, \sqrt{Q})$ observable, the discrete LQR has guaranteed properties of stability and robustness. Discrete versions of the tracker design problem are also given in the references.

48.5.2 Digital Control of Continuous-Time Systems

Using the discrete-time LQR design equations in Table 48.4, optimal digital controllers may be designed for continuous-time systems. In fact, standard techniques are available for determining a discrete-time description (Equation 48.98) (see Chapter 17) given the continuous-time dynamics $\dot{x} = Ax + Bu$ and a specified sampling period T. Then, the table allows the design of digital controllers, because the feedback gain Equation 48.102 is expressed in discrete-time, meaning that it can be directly programmed on a microprocessor or digital signal processor (DSP) and applied every T seconds to the plant. The next example shows some of the issues involved in digital control design, including selection of the sampling period and discretization of the plant.

TABLE 48.4 Discrete-Time Linear Quadratic Regulator.

System model:

$$x_{k+1} = Ax_k + Bu_k, \quad x_0 \ given. \quad (48.98)$$

Performance index:

$$J = \frac{1}{2}\sum_0^\infty (x_k^T Q x_k + u_k^T R u_k) \quad (48.99)$$

with

$$Q \geq 0, \quad R > 0.$$

Optimal feedback control:

Discrete-time algebraic Riccati equation:

$$0 = S - A^T SA + A^T SB(B^T SB + R)^{-1} B^T SA - Q. \quad (48.100)$$

Optimal feedback gain:

$$K = (B^T SB + R)^{-1} B^T SA. \quad (48.101)$$

Feedback control:

$$u_k = -Kx_k. \quad (48.102)$$

Optimal cost:

$$J = \frac{1}{2} x_0^T S x_0. \quad (48.103)$$

EXAMPLE 48.6: Digital inverted pendulum controller

In Example 48.3 a continuous-time controller was designed for an inverted pendulum on a cart; it is now desired to design a digital controller.

a. Discrete inverted pendulum dynamics

The continuous-time inverted pendulum dynamics are given in Example 48.3. Standard techniques for system discretization are covered in the chapter on Digital Control. The time histories and closed-loop poles in Example 48.3 reveal that a sampling period of $T = 0.1$ sec is very small compared to the speed of the plant response (e.g., about 1/10 of the smallest plant time constant). Therefore, this sampling period is selected.

Using the MATLAB Control System Toolbox to compute the zero-order-hold/step-invariant sampled dynamics yields the system

$$x_{k+1} = \begin{bmatrix} 1.054386 & 0.101806 & 0 & 0 \\ 1.097473 & 1.054386 & 0 & 0 \\ -0.004944 & -0.000164 & 1 & 0.1 \\ -0.099770 & -0.004944 & 0 & 1 \end{bmatrix} x_k$$

$$+\begin{bmatrix} -0.001009 \\ -0.020361 \\ 0.001001 \\ 0.020033 \end{bmatrix} u_k = Ax_k + Bu_k \quad (48.104)$$

where the state is $x_k = [\theta_k \;\; \dot{\theta}_k \;\; p_k \;\; \dot{p}_k]^T$.

The continuous system has poles at $s = 0, 0, 3.28, -3.28$. The discrete system has poles at $z = 1, 1, 1.3886, 0.7201$ which corresponds to the sampling transformation $z = e^{sT}$.

b. Digital controller design

To determine stabilizing control gains in

$$u_k = -Kx_k = -(k_\theta \theta_k + k_{\dot{\theta}} \dot{\theta}_k + k_p p_k + k_{\dot{p}} \dot{p}_k), \quad (48.105)$$

we may use the discrete-time LQR in Table 48.4. Note that this is a *multiloop design problem*, yet the LQR approach easily deals with it. Trying weighting matrices of $R = 1$, $Q = diag\{10, 10, 1, 1\}$ and using the discrete ARE solver in MATLAB yields the gains $K = [-1.294 \quad -10.02 \quad 3.648 \quad 16.94]^T$ and corresponding closed-loop poles at $z = 0.37, 0.72, 0.82 \pm j.029$.

A simulation is easily performed to obtain the closed-loop response shown in Figure 48.13. It is very instructive to compare this with the response obtained in Example 48.3. The advantage of discrete-time design is that the control input (2) may be computed every $T = 0.1\ s$ on a microprocessor and applied to the plant for real time control. The continuous-time feedback law needs to be applied using analog techniques or a very high sampling rate.

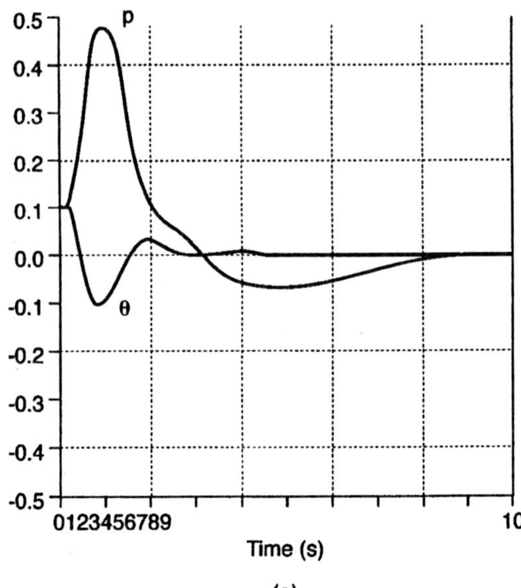

(a)

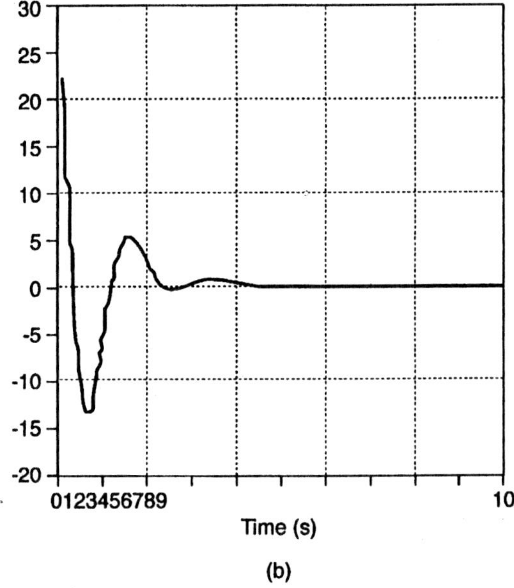

(b)

Figure 48.13 Response of inverted pendulum digital controller. (a) Rod angle $\theta(t)$ and cart position $p(t)$. (b) Control input $u(t)$.

48.6 Optimal LQ Design for Polynomial Systems

The discussion thus far has focused on the state-space formulation. A dynamical system may be equally well described in transfer function or *polynomial form* as

$$A(z^{-1})y_k = z^{-d}B(z^{-1})u_k, \quad (48.106)$$

with y_k the output and u_k the control input. The *system delay* is denoted d. This is a discrete-time formulation with z^{-1} denoting the unit delay. For simplicity we discuss the single-input/single-output case; these notions may be extended to multivariable polynomial systems using the *matrix fraction descriptions* of the plant. The denominator polynomial

$$A(z^{-1}) = 1 + a_1 z^{-1} + a_2 z^{-2} + \ldots + a_n z^{-n} \quad (48.107)$$

has roots specifying the system poles, and the numerator

$$B(z^{-1}) = b_0 + b_1 z^{-1} + \ldots + b_m z^{-m} \quad (48.108)$$

has roots at the system zeros.

In contrast to the PI selected for state-space systems, which is a sum of squares, for polynomial systems, it is more convenient to select the *square of sums* PI

$$J_k = \left[\sum_{i=0}^{n_P} p_i y_{k+d-i} - \sum_{i=0}^{n_Q} q_i y_{k-i}\right]^2 + \left[\sum_{i=0}^{n_R} r_i u_{k-i}\right]^2. \quad (48.109)$$

The constants p_i, q_i, r_i are weighting coefficients (design parameters) selected by the engineer and w_k is a reference or command signal. Defining the *weighting polynomials*

$$\begin{aligned} P(z^{-1}) &= 1 + p_1 z^{-1} + \ldots + p_{n_P} z^{-n_P}, \\ Q(z^{-1}) &= q_0 + q_1 z^{-1} + \ldots + q_{n_Q} z^{-n_Q}, \\ R(z^{-1}) &= r_0 + r_1 z^{-1} + \ldots + r_{n_R} z^{-n_R}, \end{aligned} \quad (48.110)$$

the PI may be written in the streamlined form

$$J_k = (Py_{k+d} - Qw_k)^2 + (Ru_k)^2. \quad (48.111)$$

This is a very general sort of PI. For instance, the tracking problem may be solved if we select $P = Q = 1$, $R = r_0$, for then

$$J_k = (y_{k+d} - w_k)^2 + (r_0 u_k)^2 \quad (48.112)$$

and a delayed version of the output y_k tries to follow a reference input w_k. The system delay d is explicitly accounted for. Thus, the polynomial tracker is very easy to compute and implement. In fact, it is *causal*, in contrast to the state-space LQR tracker where a noncausal feedforward signal was needed.

As another example, the regulator problem results if the weights are selected as $P = 1$, $Q = 0$, $R = r_0$, for then

$$J_k = (y_{k+d})^2 + (r_0 u_k)^2 \tag{48.113}$$

and the control tries to hold the output at zero without using too much energy.

The optimal control u_k that minimizes the PI is straightforward to determine. In the minimum-phase core (e.g., all roots of $B(z^{-1})$ stable), one solves the *Diophantine equation*

$$1 = AF + z^{-d}G \tag{48.114}$$

for the intermediate polynomials $F(z^{-1})$ and $G(z^{-1})$. Well-known routines are available for this. In fact, one may simply divide $A(z^{-1})$ into 1 until the remainder has a multiplier of z^{-d}. Then the quotient is $F(z^{-1})$ and the remainder yields $G(z^{-1})$. In terms of the Diophantine equation solution, the optimal control sequence is then given by the equation

$$(PBF + \frac{r_0}{b_0}R)u_k = -PGy_k + Qw_k. \tag{48.115}$$

This is nothing but a *difference equation* that gives the current control u_k in terms of y_k, w_k, and previous values of the control; it is easily implemented using a digital computer or microprocessor. Figure 48.14 shows the structure of the optimal LQ polynomial controller. Because it has a feedback and a feedforward component, it is called a *two-degrees-of-freedom regulator*. Such a controller can influence the closed-loop poles *as well as zeros*. Note that this controller actually requires full state feedback because the complete state is given by $y_k, y_{k-1}, \ldots, y_{k-n}, u_{k-d}, u_{k-d-1}, \ldots, u_{k-d-m}$.

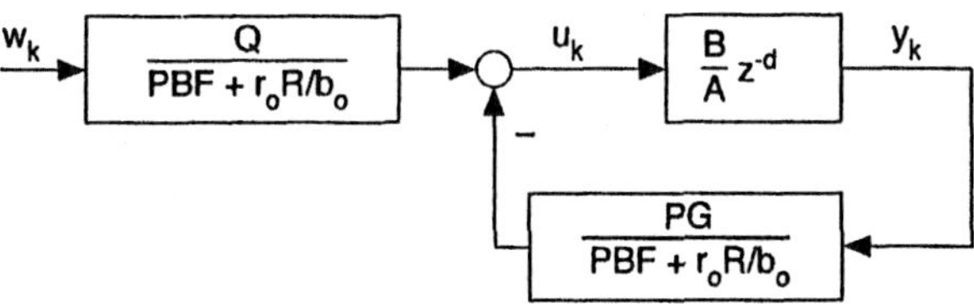

Figure 48.14 Optimal polynomial LQ regulator drawn as a two-degrees-of-freedom regulator.

Some fundamental points in polynomial LQ design, as contrasted to state-space design, are (1) the PI is a square of sums, (2) the role of the Riccati equation in state-space design is played by the Diophantine equation in polynomial design, and (3) the optimal tracker problem is easy to solve and implement since it is *causal.*

EXAMPLE 48.7: Polynomial LQ tracker

It is desired for the plant

$$y_k - 2y_{k-1} + \frac{3}{4}y_{k-2} = u_{k-1} - \frac{1}{2}u_{k-2} \tag{48.116}$$

to follow a given reference signal w_k using a fairly smooth control signal u_k. The control delay is $d = 1$. To accomplish the design, select the PI

$$J_k = (y_{k+1} - w_k)^2 + r^2(u_k - u_{k-1})^2. \tag{48.117}$$

This PI is motivated by (48.112), but the *first difference* of the control is weighted to keep u_k smooth, as per the specifications. The scalar r is a *design parameter* used to tune the closed-loop performance at the end of the design (e.g., for suitable damping ratio, overshoot, etc.).

Inspecting the plant and PI, the polynomials defined in the discussion are

$$\begin{aligned} A(z^{-1}) &= 1 - 2z^{-1} + 0.75z^{-2}, & B(z^{-1}) &= 1 - 0.5z^{-1}, \\ P(z^{-1}) &= Q(z^{-1}) = 1, & R(z^{-1}) &= r(1 - z^{-1}). \end{aligned} \tag{48.118}$$

To find the required tracking controller, the Diophantine equation is easily solved (simply perform long division of $A(z^{-1})$ into 1 to obtain the quotient $F(z^{-1})$ and remainder $z^{-1}G(z^{-1})$), resulting in the intermediate quantities

$$\begin{aligned} F(z^{-1}) &= 1, \\ G(z^{-1}) &= 2 - 0.75z^{-1}. \end{aligned} \tag{48.119}$$

According to Equation 48.115, therefore, the control is given by

$$[(1 + r^2) - z^{-1}(0.5 + r^2)]u_k = -(2 - 0.75z^{-1})y_k + w_k. \tag{48.120}$$

The variable r is a *design parameter* that can be varied by the engineer as he performs computer simulations of the closed-loop system (1), (5). Then, based on the simulations, the best value of r is selected and the resulting controller is applied to the actual plant. Selecting, for instance, $r = \frac{1}{2}$, yields the difference equation

$$1.25u_k = 0.75u_{k-1} - 2y_k + 0.75y_{k-1} + w_k, \tag{48.121}$$

which is easily solved for the current control input u_k in terms of u_{k-1}, current and previous values of y_k, and the current w_k. The controller is of the form shown in Figure 48.14.

References

[1] Anderson, B.D.O, and Moore, J.B, *Optimal Control*, Prentice Hall, Englewood Cliffs, NJ, 1990.

[2] Athans, M, and Falb, P., *Optimal Control*, McGraw-Hill, New York, 1966.

[3] Bryson, A.E., Jr. and Ho, Y.-C., *Applied Optimal Control*, Hemisphere, New York, 1975.

[4] Grimble, M.J. and Johnson, M.A., *Optimal Control and Stochastic Estimation: Theory and Applications*, John Wiley & Sons, New York, 1988, Vol. 1.

[5] Kirk, D.E., *Optimal Control Theory*, Prentice Hall, Englewood Cliffs, NJ, 1970.

[6] Kwakernaak, H. and Sivan, R., *Linear Optimal Control Systems*, John Wiley & Sons, New York, 1972.

[7] Lewis, F.L., *Optimal Control*, John Wiley & Sons, New York, 1986.

[8] Lewis, F.L., *Applied Optimal Control and Estimation*, Prentice Hall, Englewood Ciiffs, NJ, 1992.

[9] $MATRIX_X$, Integrated Systems, Inc., 2500 Mission College Blvd., Santa Clara, CA 95054, 1989.

[10] Moler, C., Little, J., and Bangert, S., *PC-Matlab*, The Mathworks, Inc., 20 North Main St., Suite 250, Sherborn, MA 01770, 1987.

[11] Stevens, B.L., and Lewis, F.L., *Aircraft Modelling, Dynamics, and Control*, John Wiley & Sons, New York, 1992.

[12] Verriest, E.I. and Lewis, F.L., "On the linear quadratic minimum-time problem," *IEEE Trans. Automat. Control*, 859–863, 1991.

Further Reading

Further information may be obtained in the references, and in Chapters 17 and 29.

49

Decentralized Control

M. E. Sezer
Bilkent University, Ankara, Turkey

D. D. Šiljak
Santa Clara University, Santa Clara, CA

49.1 Introduction

The complexity and high performance requirements of present–day industrial processes place increasing demands on control technology. The orthodox concept of driving a large system by a central computer has become unattractive for either economic or reliability reasons. New emerging notions are subsystems, interconnections, distributed computing, parallel processing, and information constraints, to mention a few. In complex systems, where databases are developed around the plants with distributed sources of data, a need for fast control action in response to local inputs and perturbations dictates the use of distributed (that is, decentralized) information and control structures.

The accumulated experience in controlling complex industrial processes suggests three basic reasons for using decentralized control structures:

1. dimensionality,
2. information structure constraints, and
3. uncertainty.

Because the amount of computation required to analyze and control a system of large dimension grows faster than its size, it is beneficial to decompose the system into subsystems, and design controls for each subsystem independently based on the local subsystem dynamics and its interconnections. In this way, special structural features of a system can be used to devise feasible and efficient decentralized strategies for solving large control problems previously impractical to solve by "one–shot" centralized methods.

A restriction on what and where the information is delivered in a system is a standard feature of interconnected systems. For example, the standard automatic generation control in power systems is decentralized because of the cost of excessive information requirements imposed by a centralized control strategy over distant geographic areas. The structural constraints on information make the centralized methods for control and estimation design difficult to apply, even to systems with small dimensions.

It is a common assumption that neither the internal nor the external nature of complex systems can be known precisely in deterministic or stochastic terms. The essential uncertainty resides in the interconnections between different parts of the system (subsystems). The local characteristics of each individual subsystem can be satisfactorily modeled in most practical situations. Decentralized control strategies are inherently robust with respect to a wide variety of structured and unstructured perturbations in the interconnections. The strategies can be made reliable to both interconnection and controller failures involving individual subsystems.

In decentralized control design, it is customary to use a wide variety of disparate methods and techniques that originated in system and control theory. Graph–theoretic methods have been devised to identify the special structural features of the system, which may help us cope with dimensionality problems and formulate a suitable decentralized control strategy. The concept of vector Liapunov functions, each component of which determines the stability of a part of the system where others do not, is a powerful method for the stability analysis of large interconnected systems. Stochastic modeling and decentralized control

0-8493-8570-9/96/$0.00+$.50

have been used in a broad range of situations, involving LQG design, Kalman filtering, Markov processes, and stability analysis and design. Robustness considerations of decentralized control have been carried out since the early stages of its evolution, often preceding a similar development in the centralized control theory. Especially popular have been the adaptive decentralized schemes because of their flexibility and ability to cope efficiently with perturbations in both the interactions and the subsystems of a large system.

The objective of this chapter is to introduce the concept and methods of decentralized control. Due to a large number of results and techniques available, only the basic theory and practice of decentralized control will be reviewed. At the end of the chapter is a discussion of the larger background listing the books and survey papers on the subject. References related to more sophisticated treatment of decentralized control and the relevant applications are also discussed.

49.2 The Decentralized Control Problem

To introduce the decentralized control problem, consider two inverted penduli coupled by a spring as shown in Figure 49.1. The control objective is to keep the penduli in the upright position by applying feedback control via the inputs u_1 and u_2. The linearized equations of motion in the vicinity of $\theta_1 = \theta_2 = 0$ are

$$\begin{aligned} m\ell^2\ddot{\theta}_1 &= mg\ell\theta_1 - ka^2(\theta_1 - \theta_2) + u_1, \\ m\ell^2\ddot{\theta}_2 &= mg\ell\theta_2 - ka^2(\theta_2 - \theta_1) + u_2. \end{aligned} \tag{49.1}$$

By choosing the state vector $x = (\theta_1, \dot{\theta}_1, \theta_2, \dot{\theta}_2)^T$ and the input vector $u = (u_1, u_2)^T$, the state space representation of the system is

$$\mathbf{S}:\ \dot{x} = \left[\begin{array}{cc|cc} 0 & 1 & 0 & 0 \\ \frac{g}{\ell} - \frac{ka^2}{m\ell^2} & 0 & \frac{ka^2}{m\ell^2} & 0 \\ \hline 0 & 0 & 0 & 1 \\ \frac{ka^2}{m\ell^2} & 0 & \frac{g}{\ell} - \frac{ka^2}{m\ell^2} & 0 \end{array}\right] x + \left[\begin{array}{c|c} 0 & 0 \\ \frac{1}{m\ell^2} & 0 \\ \hline 0 & 0 \\ 0 & \frac{1}{m\ell^2} \end{array}\right] u\,. \tag{49.2}$$

The fundamental restriction in choosing the feedback laws to control the system $\mathbf{S}$ is that each input u_1 and u_2 can depend only on the local states $x_1 = (\theta_1, \dot{\theta}_1)^T$ and $x_2 = (\theta_2, \dot{\theta}_2)^T$ of the corresponding penduli, that is, $u_1 = u_1(x_1)$ and $u_2 = u_2(x_2)$. This restriction is called the *decentralized information structure constraint.*

Since the system $\mathbf{S}$ is linear, a natural choice is the linear control laws

$$u_1 = k_1^T x_1\,,\quad u_2 = k_2^T x_2 \tag{49.3}$$

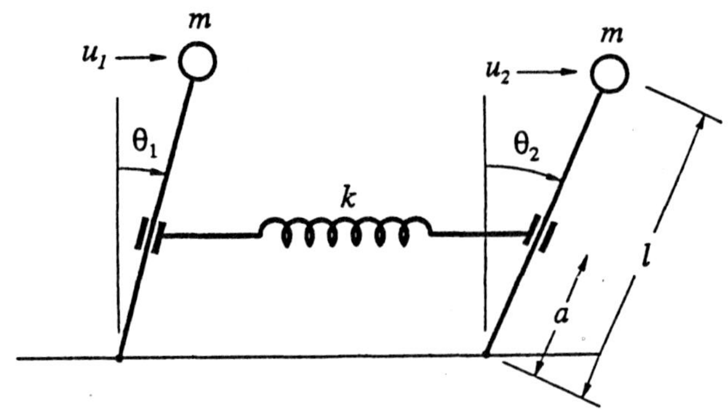

Figure 49.1 Inverted penduli.

where the feedback gain vectors $k_1 = (k_{11}, k_{12})^T$ and $k_2 = (k_{21}, k_{22})^T$ should be selected to *stabilize* the system $\mathbf{S}$, that is, hold the penduli in the upright position.

In control design, it is fruitful to recognize the structure of the system $\mathbf{S}$ as an interconnection

$$\begin{aligned} \mathbf{S}:\ \dot{x}_1 &= \begin{bmatrix} 0 & 1 \\ \alpha & 0 \end{bmatrix} x_1 + \begin{bmatrix} 0 \\ \beta \end{bmatrix} u_1 \\ &\quad + e\begin{bmatrix} 0 & 0 \\ -\gamma & 0 \end{bmatrix} x_1 + e\begin{bmatrix} 0 & 0 \\ \gamma & 0 \end{bmatrix} x_2, \\ \dot{x}_2 &= \begin{bmatrix} 0 & 1 \\ \alpha & 0 \end{bmatrix} x_2 + \begin{bmatrix} 0 \\ \beta \end{bmatrix} u_1 \\ &\quad + e\begin{bmatrix} 0 & 0 \\ \gamma & 0 \end{bmatrix} x_1 + e\begin{bmatrix} 0 & 0 \\ -\gamma & 0 \end{bmatrix} x_2, \end{aligned} \tag{49.4}$$

of two subsystems

$$\begin{aligned} \mathbf{S}_1:\ \dot{x}_1 &= \begin{bmatrix} 0 & 1 \\ \alpha & 0 \end{bmatrix} x_1 + \begin{bmatrix} 0 \\ \beta \end{bmatrix} u_1, \\ \mathbf{S}_2:\ \dot{x}_2 &= \begin{bmatrix} 0 & 1 \\ \alpha & 0 \end{bmatrix} x_2 + \begin{bmatrix} 0 \\ \beta \end{bmatrix} u_2, \end{aligned} \tag{49.5}$$

where $\alpha = g/\ell$, $\beta = 1/m\ell^2$, $\gamma = \bar{a}^2 k/m\ell^2$, and $e = (a/\bar{a})^2$. One reason is that, in designing control for interconnected systems, the designer has to account for essential *uncertainty* in the interconnections among the subsystems. Though models of the subsystems are commonly available with sufficient accuracy, the shape and size of the interconnections cannot be predicted satisfactorily either for modeling or operational reasons. In the example, the interconnection parameter $e = a/\bar{a}$ is the uncertain height of the spring which is normalized by its nominal value $\bar{a}$.

An equally important reason for decomposition is present when controlling large dynamic systems. In complex systems with many variables, most of the variables are *weakly coupled,* if coupled at all, and the behavior of the overall system is dominated by strongly connected variables. Considerable conceptual and numerical simplification can be gained by controlling the strongly coupled variables with decentralized control.

49.3 Plant and Feedback Structures

Consider a linear constant system

$$\begin{aligned} \mathbf{S}\colon \dot{x} &= Ax + Bu, \\ y &= Cx, \end{aligned} \tag{49.6}$$

as an interconnected system

$$\begin{aligned} \mathbf{S}\colon \dot{x}_i &= A_i x_i + B_i u_i + \sum_{j \in \mathcal{N}} (A_{ij} x_j + B_{ij} u_j), \\ y_i &= C_i x_i + \sum_{j \in \mathcal{N}} C_{ij} x_j, \quad i \in \mathcal{N}, \end{aligned} \tag{49.7}$$

which is composed of N subsystems

$$\begin{aligned} \mathbf{S}_i\colon \dot{x}_i &= A_i x_i + B_i u_i, \\ y_i &= C_i x_i, \quad i \in \mathcal{N}, \end{aligned} \tag{49.8}$$

where $x_i(t) \in \mathbb{R}^{n_i}$, $u_i(t) \in \mathbb{R}^{m_i}$, $y_i(t) \in \mathbb{R}^{\ell_i}$ are the state, input, and output of the subsystem $\mathbf{S}_i$ at a fixed time $t \in \mathbb{R}$. All matrices have proper dimensions, and $\mathcal{N} = \{1, 2, \ldots, N\}$. At present we are interested in *disjoint* decompositions, that is,

$$\begin{aligned} x &= (x_1^T, x_2^T, \ldots, x_N^T)^T, \\ u &= (u_1^T, u_2^T, \ldots, u_N^T)^T, \\ y &= (y_1^T, y_2^T, \ldots, y_N^T)^T, \end{aligned} \tag{49.9}$$

and where $x(t) \in \mathbb{R}^n$, $u(t) \in \mathbb{R}^m$, and $y(t) \in \mathbb{R}^\ell$ are the state, input, and output of the overall system $\mathbf{S}$, so that

$$\begin{aligned} \mathbb{R}^n &= \mathbb{R}^{n_1} \times \mathbb{R}^{n_2} \times \ldots \times \mathbb{R}^{n_N}, \\ \mathbb{R}^m &= \mathbb{R}^{m_1} \times \mathbb{R}^{m_2} \times \ldots \times \mathbb{R}^{m_N}, \\ \mathbb{R}^\ell &= \mathbb{R}^{\ell_1} \times \mathbb{R}^{\ell_2} \times \ldots \times \mathbb{R}^{\ell_N}. \end{aligned} \tag{49.10}$$

A compact description of the interconnected system $\mathbf{S}$ is

$$\begin{aligned} \mathbf{S}\colon \dot{x} &= A_D x + B_D u + A_C x + B_C u \\ y &= C_D x + C_C x, \end{aligned} \tag{49.11}$$

where

$$\begin{aligned} A_D &= \text{diag}\{A_1, A_2, \ldots, A_N\}, \\ B_D &= \text{diag}\{B_1, B_2, \ldots, B_N\}, \\ C_D &= \text{diag}\{C_1, C_2, \ldots, C_N\}, \end{aligned} \tag{49.12}$$

and the coupling block matrices are

$$A_C = (A_{ij}), \quad B_C = (B_{ij}), \quad C_C = (C_{ij}). \tag{49.13}$$

The collection of N decoupled subsystems is described by

$$\begin{aligned} \mathbf{S}_D\colon \dot{x} &= A_D x + B_D u \\ y &= C_D x, \end{aligned} \tag{49.14}$$

obtained from (49.11) by setting the coupling matrices to zero.

Important special classes of interconnected systems are input ($B_C = 0$) and output ($C_C = 0$) decentralized systems, where inputs and outputs are not shared among the subsystems. Input–output decentralized systems are described as

$$\begin{aligned} \mathbf{S}\colon \dot{x} &= A_D x + B_D u + A_C x \\ y &= C_D x, \end{aligned} \tag{49.15}$$

where both B_C and C_C are zero. This structural feature helps to a great extent when decentralized controllers and estimators are designed for large plants.

A *static decentralized state feedback*,

$$u = -K_D x, \tag{49.16}$$

is characterized by a block–diagonal gain matrix,

$$K_D = \text{diag}\{K_1, K_2, \ldots, K_N\}, \tag{49.17}$$

which implies that each subsystem $\mathbf{S}_i$ has its individual control law,

$$u_i = -K_i x_i, \quad i \in \mathcal{N}, \tag{49.18}$$

with a constant gain matrix K_i. The control law u of (49.16), which is equivalent to the totality of subsystem control laws (49.18), obeys the decentralized information structure constraint requiring that each subsystem $\mathbf{S}_i$ is controlled on the basis of its locally available state x_i. The closed–loop system is described as

$$\hat{\mathbf{S}}\colon \dot{x} = (A_D - B_D K_D C_D)x + A_C x. \tag{49.19}$$

When *dynamic output feedback* is used under decentralized constraints, then controllers of the following type are considered:

$$\begin{aligned} \mathbf{C}_i\colon \dot{z}_i &= F_i z_i + G_i y_i, \\ u_i &= -H_i z_i - K_i y_i, \quad i \in \mathcal{N}, \end{aligned} \tag{49.20}$$

which can be written in a compact form as a single decentralized controller defined as

$$\begin{aligned} \mathbf{C}_D\colon \dot{z} &= F_D z + G_D y, \\ u &= -H_D z - K_D y, \end{aligned} \tag{49.21}$$

where

$$\begin{aligned} z &= (z_1^T, z_2^T, \ldots, z_N^T)^T, \; y = (y_1^T, y_2^T, \ldots, y_N^T)^T, \\ u &= (u_1^T, u_2^T, \ldots, u_N^T)^T, \end{aligned} \tag{49.22}$$

are the state $z \in \mathbb{R}^r$, input $y \in \mathbb{R}^\ell$, and output $u \in \mathbb{R}^m$ of the controller $\mathbf{C}_D$. By combining the system $\mathbf{S}$ and the decentralized dynamic controller $\mathbf{C}_D$, we get the composite closed–loop system as

$$\mathbf{S}\&\mathbf{C}_D\colon \begin{bmatrix} \dot{x} \\ \dot{z} \end{bmatrix} = \tag{49.23}$$

$$\begin{bmatrix} A_D - B_D K_D C_D + A_C & -B_D H_D \\ G_D C_D & F_D \end{bmatrix} \begin{bmatrix} x \\ z \end{bmatrix}.$$

49.4 Decentralized Stabilization

The fundamental problem in decentralized control theory and practice is choosing individual subsystem inputs to stabilize the overall interconnected system. In the previous section, the plant structures have been described, where the plant, inputs and outputs are all decomposed with each local controller responsible for the corresponding subsystem. While this is the most common situation in practice, it is by no means all inclusive. It is often advantageous, and sometime necessary, to decentralize the inputs and outputs without decomposing the plant. This is the situation that we consider first.

49.4.1 Decentralized Inputs and Outputs

Suppose that only the inputs and outputs, but not states, of system **S** in (49.6) are partitioned as in (49.9), and **S** is described as

$$\mathbf{S}:\ \dot{x} = Ax + \sum_{i\in\mathcal{N}} \tilde{B}_i u_i, \qquad y_i = \tilde{C}_i x\,,\ i \in \mathcal{N}. \tag{49.24}$$

Then, the controllers $\mathbf{C}_i$ of (49.20) still operate on local measurements y_i to generate local controls u_i, but now they are collectively responsible for the whole system. In this case,

$$\mathbf{S\&C}_D:\ \begin{bmatrix} \dot{x} \\ \dot{z} \end{bmatrix} = \begin{bmatrix} A - BK_DC & -BH_D \\ G_DC & F_D \end{bmatrix} \begin{bmatrix} x \\ z \end{bmatrix}. \tag{49.25}$$

It is well–known that without the decentralization constraint on the controller, the closed–loop system of (49.25) can be stabilized if, and only if, the uncontrollable or unobservable modes of the open–loop system **S** are stable; or equivalently, the set of (centralized) fixed modes of **S**, which is defined as

$$\Lambda_C = \bigcap_K \sigma(A - BKC) \tag{49.26}$$

is included in the open left half plane, where $\sigma(.)$ denotes the set of eigenvalues of the indicated matrix. This basic result has been extended in [34] to decentralized control of **S**, where it was shown that the closed–loop system (49.25) can be made stable with suitable choice of the decentralized controllers $\mathbf{C}_i$ if, and only if, the set of decentralized fixed modes

$$\Lambda_D = \bigcap_{K_D} \sigma(A - BK_DC) = \bigcap_{K_1,\ldots,K_N} \sigma(A - \sum_{i\in\mathcal{N}} \tilde{B}_i K_i \tilde{C}_i) \tag{49.27}$$

is included in the open left half plane.

The result of [34] has been followed by extensive research on the following topics:

- state–space and frequency domain characterization of decentralized fixed modes,
- development of various techniques for designing decentralized controllers (*e.g.*, using static output feedback in all but one channel, distributing the control effort among channels, sequential stabilization, etc.)
- generalization of the concept of decentralized fixed modes to arbitrary feedback structure constraints,
- formulation of the concept of structurally fixed modes, and their algebraic and graph–theoretical characterization.

A useful and simple characterization of decentralized fixed modes was provided in [1]. For any subset $\mathcal{I} = \{i_1, \ldots, i_P\}$ of the index set $\mathcal{N}$, let $\mathcal{I}^C = \{j_1, \ldots, j_{N-P}\}$ denote the complement of $\mathcal{I}$ in $\mathcal{N}$, and define

$$\tilde{B}_{\mathcal{I}} = [\tilde{B}_{i_1}, \tilde{B}_{i_2}, \ldots, \tilde{B}_{i_P}]\,, \qquad \tilde{C}_{\mathcal{I}^C} = \begin{bmatrix} \tilde{C}_{j_1} \\ \tilde{C}_{j_2} \\ \vdots \\ \tilde{C}_{j_{N-P}} \end{bmatrix}. \tag{49.28}$$

Then a complex number $\lambda \in \mathbb{C}$ is a decentralized fixed mode of **S** if, and only if,

$$\operatorname{rank} \begin{bmatrix} A - \lambda I & \tilde{B}_{\mathcal{I}} \\ \tilde{C}_{\mathcal{I}^C} & 0 \end{bmatrix} < n \tag{49.29}$$

for some $\mathcal{I} \subset \mathcal{N}$. This result relates decentralized fixed modes to transmission zeros of the systems $(A, \tilde{B}_{\mathcal{I}}, \tilde{C}_{\mathcal{I}^C})$, called the complementary subsystems. Thus, appearance of a fixed mode corresponds to a special pole–zero cancellation, which can not be removed by constant decentralized feedback. However, under mild conditions, such fixed modes can be eliminated by time–varying decentralized feedback.

The characterization of decentralized fixed modes above prompts a generalization of the concept to arbitrary feedback structures. Let $\bar{K} = (\bar{k}_{ij})$ be an $m \times l$ binary matrix such that $\bar{k}_{ij} = 1$ if, and only if, a feedback link from output y_j to input u_i is allowed. Thus $\bar{K}$ specifies a constraint on the feedback structure, a special case of which is decentralized feedback. In this case, permissible controllers have the structure

$$\mathbf{C}_{\bar{K}}:\ \dot{z}_i = F_i z_i + \sum_{j\in\mathcal{J}_i} g_{ij} y_j, \qquad u_i = -h_i^T z_i - \sum_{j\in\mathcal{J}_i} k_{ij} y_i \tag{49.30}$$

where $\mathcal{J}_i = \{j: \bar{k}_{ij} = 1\}$.

Let K denote any feedback matrix conforming to the structure of $\bar{K}$, that is, one with $k_{ij} = 0$ whenever $\bar{k}_{ij} = 0$. Then, the set

$$\Lambda_{\bar{K}} = \bigcap_K \sigma(A - BKC) \tag{49.31}$$

can conveniently be defined as the set of fixed modes with respect to the decentralized feedback structure constraint specified by $\bar{K}$. Then the closed–loop system consisting of **S** and the constrained controller $\mathbf{C}_{\bar{K}}$ can be stabilized if, and only if, $\Lambda_{\bar{K}}$ is included in the open left half-plane. Finally, it remains to characterize $\Lambda_{\bar{K}}$ as in (49.29). This, however, is quite automatic; consider the index sets $\mathcal{I} \subset \mathcal{M} = \{1, 2, \ldots, M\}$ and replace $\mathcal{I}^C$ by $\mathcal{J} = \cup_{i\in\mathcal{I}^C} \mathcal{J}_i$, where now $\mathcal{I}^C$ refers to the complement of $\mathcal{I}$ in $\mathcal{M}$.

49.4.2 Structural Analysis

Structural analysis of large scale systems via graph–theoretic concepts and methods offers an appealing alternative to quantitative analysis which often faces difficulties due to high dimensionality and lack of exact knowledge of system parameters. Equipped with the powerful tools of graph theory, structural analysis provides valuable information concerning certain qualitative properties of the system under study by practical tests and algorithms [30].

One of the earliest problems of structural analysis is the graph–theoretic formulation of controllability [20]. Consider an uncontrollable pair (A, B). Loss of controllability is either due to a perfect matching of system parameters or due to an insufficient number of nonzero parameters, indicating a lack of sufficient linkage among system variables. In the latter case, the pair (A, B) is structurally uncontrollable in the sense that all pairs having the same structure as (A, B) are uncontrollable. Since the structure of (A, B) can be described by a directed graph (as explained below for a more general case), structural controllability can be checked by graph–theoretic means. Indeed, (A, B) is structurally controllable if, and only if, the system graph is input reachable (that is, each state variable is affected directly or indirectly by at least one input variable), and contains no dilations (that is, no subset of state variables exists whose number exceeds the total number of all state and input variables directly affecting these variables). These two conditions are equivalent to the spanning of the system graph by a minimal subgraph, called a cactus, which has a special structure.

The idea of treating controllability in a structural framework has led to formulation and graph–theoretic characterization of *structurally fixed modes* under constrained feedback [26]. Let $\mathbf{D} = (\mathcal{V}, \mathcal{E})$ be a directed graph associated with the system **S** of (49.6), where $\mathcal{V} = \mathcal{U} \cup \mathcal{X} \cup \mathcal{Y}$ is a set of vertices corresponding to inputs, states, and outputs of **S**, and $\mathcal{E}$ is a set of directed edges corresponding to nonzero parameters of the system matrices A, B, and C. To every nonzero a_{ij}, there corresponds an edge from vertex x_j to vertex x_i, to every nonzero b_{ij}, an edge from u_j to x_i, and to every nonzero c_{ij}, one from x_j to y_i. Given a feedback pattern $\bar{K}$ and adding to **D** a feedback edge from y_j to u_i for every $\bar{k}_{ij} = 1$, one gets a digraph $\mathbf{D}_{\bar{K}} = (\mathcal{V}, \mathcal{E} \cup \mathcal{E}_{\bar{K}})$ completely describing the structure of both the system **S** and the feedback constraint specified by $\bar{K}$.

Two systems are said to be structurally equivalent if they have the same system graphs. A system **S** is said to have structurally fixed modes with respect to a given $\bar{K}$ if every system structurally equivalent to **S** has fixed modes with respect to $\bar{K}$. Having structurally fixed modes is a common property of a class of systems described by the same system graph; if a system has no structurally fixed modes, then either it has no fixed modes, or if it does, arbitrarily small perturbations of system parameters can eliminate the fixed modes. As a result, if a system has no structurally fixed modes with respect to $\bar{K}$, then generically it can be stabilized by a constrained controller of the form defined in (49.30).

It was shown in [26] that a system **S** has no structurally fixed modes with respect to a feedback pattern $\bar{K}$ if, and only if

1. all state vertices of $\mathbf{D}_{\bar{K}}$ are covered by vertex disjoint cycles, and
2. no strong component of $\mathbf{D}_{\bar{K}}$ contains only state vertices, where a strong component is a maximal subgraph whose vertices are reachable from each other.

This simple graph–theoretic criterion has been used in an algorithmic way in problems such as choosing a minimum number of feedback links (or, if each feedback link is associated with a cost, choosing the cheapest feedback pattern) that avoid structurally fixed modes. As an example, consider a system with a system graph as in Figure 49.2. Let the costs of setting up feedback links (dotted lines) from each output to each input be given by a matrix

$$\begin{bmatrix} 6 & 2 \\ 3 & 7 \end{bmatrix}.$$

It can easily be verified that any feedback pattern of the form

$$\begin{bmatrix} 1 & * \\ * & * \end{bmatrix} \quad \text{or} \quad \begin{bmatrix} * & 1 \\ 1 & * \end{bmatrix},$$

where $*$ stands for either a 0 or a 1, avoids structurally fixed modes. Clearly, the feedback patterns which contain the least number of links and which cost the least are, respectively,

$$\bar{K}_1 = \begin{bmatrix} 1 & 0 \\ 0 & 0 \end{bmatrix} \quad \text{or} \quad \bar{K}_2 = \begin{bmatrix} 0 & 1 \\ 1 & 0 \end{bmatrix}.$$

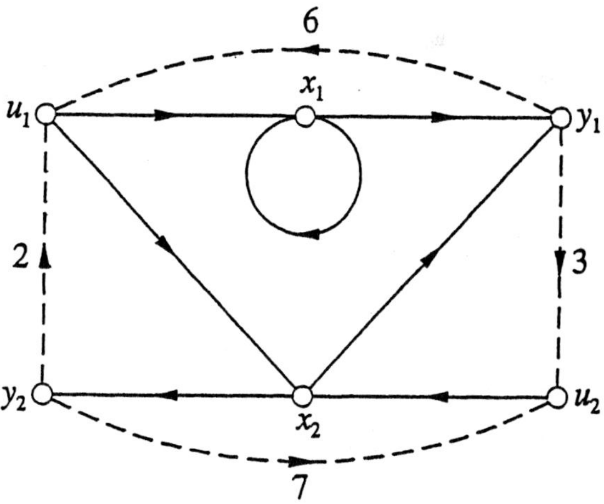

Figure 49.2 System graph.

49.4.3 Decentrally Stabilizable Structures

Consider an interconnected system

$$\begin{aligned} \mathbf{S}: \dot{x}_i &= A_i x_i + B_i(u_i + \sum_{j \in \mathcal{N}} D_{ij} x_j) \\ y_i &= x_i \quad i \in \mathcal{N} \end{aligned} \tag{49.32}$$

which is a special case of the system **S** in (49.7) in that $A_{ij} = B_i D_{ij}$, $B_{ij} = 0$, $C_i = I$, and $C_{ij} = 0$. Assuming that the decoupled subsystems described by the pairs (A_i, B_i) are controllable, it is easy to verify that **S** has no decentralized fixed modes. Thus **S** can be stabilized using a decentralized dynamic feedback controller of the form (49.21). However, because the subsystem outputs are the states, there should be no need to use dynamic controllers.

Choose the decentralized constant state feedbacks in (49.18) to place the subsystem poles at $-\mu_{il}\rho$, $i \in \mathcal{N}$, $l = 1, 2, \ldots, n_i$, where $-\mu_{il}$ are distinct negative real numbers, and ρ is a parameter. Then a suitable change of coordinate frame transforms the closed-loop system of (49.19) into the form

$$\hat{\mathbf{S}}\colon\ \dot{x} = (-\rho M + \hat{A}_C)x\ , \tag{49.33}$$

where $M = \mathrm{diag}\{M_1, M_2, \ldots, M_N\}$, with $M_i = \mathrm{diag}\{\mu_1, \mu_2, \ldots, \mu_{in_i}\}$, and $\hat{A}_C$ is independent of the parameter ρ. Clearly, $\hat{\mathbf{S}}$ is stable for a sufficiently large ρ.

The success of this high–gain decentralized stabilization technique results from the special structure of the interconnections among the subsystems. The interconnections from other subsystems affect a particular subsystem in the same way its local input does. This makes it possible to neutralize potentially destabilizing effects of the interconnections by a local state feedback and provide a high degree of stability to the decoupled subsystems. This special interconnection structure is termed the "matching conditions" [18].

Decentralized stabilizability of interconnected systems satisfying the matching conditions has motivated research in characterizing other decentrally stabilizable interconnection structures. Below, another such interconnection structure is described, where single–input subsystems are considered for convenience.

Let the interconnected system be described as

$$\mathbf{S}\colon\ \dot{x}_i = A_i x_i + b_i u_i + \sum_{j\in\mathcal{N}} A_{ij} x_j\ ,\quad i \in \mathcal{N} \tag{49.34}$$

where, without loss of generality, the subsystem pairs (A_i, b_i) are assumed to be in controllable canonical form. For each interconnection matrix A_{ij}, define an integer m_{ij} as

$$m_{ij} = \begin{cases} \max\{q - p : a_{pq}^{ij} \neq 0\}, & A_{ij} \neq 0, \\ -n, & A_{ij} = 0, \end{cases} \tag{49.35}$$

Thus, m_{ij} is the distance between the main diagonal and a line parallel to the main diagonal which borders all nonzero elements of A_{ij}.

For an index set $\mathcal{I} \subset \mathcal{N}$, let $\mathcal{I}_P$ denote any permutation of $\mathcal{I}$. Then, the system **S** in (49.34) is stabilizable by decentralized constant state feedback if

$$\sum_{\substack{i\in\mathcal{I}\\ j\in\mathcal{I}_P}} (m_{ij} - 1) < 0 \tag{49.36}$$

for all $\mathcal{I}$ and all permutations $\mathcal{I}_P$ [14], [30]. In the case of matching interconnections, $m_{ij} = n_j - n_i$, so that (49.36) guarantees decentralized stabilizability even when the elements of the interconnection matrices A_{ij} are bounded nonlinear, time–varying functions of the state variables. Therefore, the condition (49.36) and, thus, the matching conditions, are indeed structural conditions.

49.4.4 Vector Liapunov Functions

A general way to establish the stability of nonlinear interconnected systems is to apply the Matrosov–Bellman concept of vector Liapunov functions [17]. The concept has been developed to provide an efficient method of checking the stability of linear interconnected systems controlled by decentralized feedback [30]. First, each subsystem is stabilized using local state or output feedback. Then, for each stable closed–loop (but decoupled) subsystem, a Liapunov function is chosen using standard methods. These functions are stacked to form a vector of functions, which can then be used to form a single scalar Liapunov function for the overall system. The function establishes stability if we show positivity of the leading principal minors of a constant aggregate matrix whose dimension equals the number of subsystems.

Consider the linear interconnected system of (49.7),

$$\begin{aligned}\mathbf{S}\colon\ \dot{x}_i &= A_i x_i + B_i u_i \\ &\quad + \sum_{j\in\mathcal{N}} e_{ij} A_{ij} x_j\ ,\quad i \in \mathcal{N},\end{aligned} \tag{49.37}$$

where the output y_i is not included and $B_{ij} = 0$. We inserted the elements of $e_{ij} \in [0, 1]$ of the $N \times N$ interconnection matrix $E = (e_{ij})$ to capture the presence of uncertainty in coupling between the subsystems

$$\mathbf{S}_i\colon\ \dot{x}_i = A_i x_i + B_i u_i\ , \tag{49.38}$$

as illustrated by the example of the two penduli above.

We assume that each pair (A_i, B_i) is controllable and assign the eigenvalues $-\sigma_1^i \pm j\omega_1^i, \ldots, -\sigma_{p_i}^i \pm j\omega_{p_i}^i, \ldots, -\sigma_{2p_i+1}^i, \ldots, -\sigma_{n_i}^i$ to each closed–loop subsystem

$$\hat{\mathbf{S}}_i\colon\ \dot{x}_i = (A_i - B_i K_i)x_i \tag{49.39}$$

by applying decentralized feedback

$$u_i = -K_i x_i\ . \tag{49.40}$$

Using a nonsingular transformation,

$$x_i = T_i \tilde{x}_i, \tag{49.41}$$

we can obtain the closed–loop subsystems as

$$\tilde{\mathbf{S}}_i\colon\ \dot{\tilde{x}}_i = \Lambda_i \tilde{x}_i\ , \tag{49.42}$$

where the matrix $\Lambda_i = T_i^{-1}(A_i - B_i K_i)T_i$ has the diagonal form

$$\begin{aligned}\Lambda_i &= \mathrm{diag}\left\{\begin{bmatrix} -\sigma_1^i & \omega_1^i \\ -\omega_1^i & -\sigma_1^i \end{bmatrix}, \ldots, \begin{bmatrix} -\sigma_{p_i}^i & \omega_{p_i}^i \\ -\omega_{p_i}^i & -\sigma_{p_i}^i \end{bmatrix},\right. \\ &\qquad \left. -\sigma_{2p_i+1}, \ldots, -\sigma_{n_i}\right\}\ .\end{aligned} \tag{49.43}$$

For each transformed subsystem, there exists a suitable Liapunov function v: $\mathbb{R}^{n_i} \to \mathbb{R}_+$ of the form

$$v_i(\tilde{x}_i) = (\tilde{x}_i H_i \tilde{x}_i)^{\frac{1}{2}}, \tag{49.44}$$

where $H_i = I_i$ is the solution of the Liapunov matrix equation

$$\Lambda_i H_i + H_i \Lambda_i = -G_i \tag{49.45}$$

for $G_i = \text{diag}\{\sigma_1^i, \sigma_1^i, \ldots, \sigma_{p_i}^i, \sigma_{2p_i+1}^i, \ldots, \sigma_{n_i}^i\}$.

To determine the stability of the overall interconnected closed–loop system

$$\tilde{\mathbf{S}}\text{: } \dot{\tilde{x}}_i = \Lambda_i \tilde{x}_i + \sum_{j \in \mathcal{N}} e_{ij} \Delta_{ij} \tilde{x}_j \tag{49.46}$$

from the stability of the decoupled closed–loop subsystems $\tilde{\mathbf{S}}_i$, we consider subsystem functions v_i as components of a *vector Liapunov function* $v = (v_1, v_2, \ldots, v_N)^T$, and form a candidate Liapunov function V: $\mathbb{R}^n \to \mathbb{R}_+$ for the overall system $\tilde{\mathbf{S}}$ as

$$V(\tilde{x}) = \sum_{i \in \mathcal{N}} d_i v_i(\tilde{x}_i)\,, \tag{49.47}$$

where the existence of positive numbers d_i for stability of $\tilde{\mathbf{S}}$ has yet to be established, and $\Delta_{ij} = T_i^{-1} A_{ij} T_j$.

Taking the total time derivative of $V(\tilde{x})$ with respect to $\tilde{\mathbf{S}}$, after lengthy but straightforward computations [30],

$$\dot{V}(\tilde{x}) \le -d^T \bar{W} z\,, \tag{49.48}$$

with $d = (d_1, d_2, \ldots, d_N)^T$, $z = (\|\tilde{x}_1\|, \|\tilde{x}_2\|, \ldots, \|\tilde{x}_N\|)^T$, and $\bar{W} = (\bar{w}_{ij})$ is the $N \times N$ aggregate matrix defined as

$$\bar{w}_{ij} = \begin{cases} \frac{1}{2}\sigma_m^i - \bar{e}_{ii}\lambda_M^{1/2}(\Delta_{ii}^T \Delta_{ii}), & i = j \\ -\bar{e}_{ij}\lambda_M^{1/2}(\Delta_{ij}^T \Delta_{ij}), & i \ne j \end{cases} \tag{49.49}$$

where σ_m^i is the minimal value of all σ_k^i, and $\lambda_M(\cdot)$ is the maximal eigenvalue of the indicated matrix.

The elements $\bar{e}_{ij}$ of the fundamental interconnection matrix $\bar{E} = (\bar{e}_{ij})$ are binary numbers defined as

$$\bar{e}_{ij} = \begin{cases} 1, & \mathbf{S}_j \text{ acts on } \mathbf{S}_i \\ 0, & \mathbf{S}_j \text{ does not act on } \mathbf{S}_i\,. \end{cases} \tag{49.50}$$

In this way, the binary matrix describes the basic interconnection structure of the system **S**. In the case of two penduli,

$$\bar{E} = \begin{bmatrix} 1 & 1 \\ 1 & 1 \end{bmatrix}. \tag{49.51}$$

It has been shown in [30] that stability of $-\bar{W}$ (all eigenvalues of $-\bar{W}$ have negative real parts) implies stability of the closed–loop system $\tilde{\mathbf{S}}$ and, hence, $\hat{\mathbf{S}}$. To explain this fact, we note first that $w_{ii} > 0$, $w_{ij} \le 0$ $(i \ne j)$, which makes $\bar{W}$ an M–matrix (e.g., [30]) if, and only if, there exists a positive vector d ($d_i > 0$, $i \in \mathcal{N}$), so that the vector

$$c^T = d^T \bar{W} \tag{49.52}$$

is a positive vector as well. Positivity of c and d imply $V(\tilde{x}) > 0$ and $\dot{V}(\tilde{x}) < 0$ and, therefore, stability of $\hat{\mathbf{S}}$ by the standard Liapunov argument. Finally, the M–matrix property of $\bar{W}$ is equivalent to stability of $-\bar{W}$.

Several comments are in order. First, we note that the M–matrix property of $\bar{W}$ can be tested by a simple determinantal condition

$$\begin{vmatrix} \bar{w}_{11} & \bar{w}_{12} & \ldots & \bar{w}_{1k} \\ \bar{w}_{21} & \bar{w}_{22} & \ldots & \bar{w}_{2k} \\ \ldots & \ldots & \ldots & \ldots \\ \bar{w}_{k1} & \bar{w}_{k2} & \ldots & \bar{w}_{kk} \end{vmatrix} > 0\,, \quad k \in \mathcal{N}\,. \tag{49.53}$$

Another important feature of the concept of vector Liapunov functions is the *robustness* information about decentrally stabilized interconnected system $\hat{\mathbf{S}}$. The determinantal condition (49.53) is equivalent to the quasidominant diagonal property of $\bar{W}$,

$$\bar{w}_{ii} > d_i^{-1} \sum_{j \ne i}^{N} d_j |\bar{w}_{ij}|\,, \quad i \in \mathcal{N}\,. \tag{49.54}$$

where the d_i's are positive numbers. From (49.54), it is obvious that, if $\bar{W}$ is an M–matrix, so is W for any $E \le \bar{E}$, where the inequality is taken element by element; the system $\hat{\mathbf{S}}$ is *connectively stable* [30]. When a system is connectively stabilized by decentralized feedback, stability is robust and can tolerate variations in coupling among the subsystems. When the two penduli are stabilized for any given position $\bar{a}$ of the spring, including the entire length ℓ of the penduli, the penduli are stable for any position $a \le \bar{a}$. In other words, if the penduli are stabilized for the fundamental interconnection matrix $\bar{E}$ of (51), they are stabilized for any interconnection matrix

$$E = \begin{bmatrix} e & e \\ e & e \end{bmatrix}, \tag{49.55}$$

whenever $e \in [0, 1]$.

Finally, the decentrally stabilized system can tolerate nonlinearities in the interconnections among the subsystems. The nonlinear interconnections need not be known since only their size is required to be limited. Once the closed–loop system $\hat{\mathbf{S}}$ is shown to be stable, it follows [30] that a nonlinear time–varying version

$$\hat{\mathbf{S}}_N\text{: } \dot{\tilde{x}}_i = (A_i - B_i K_i)\tilde{x}_i + h_i(t, \tilde{x}), \quad i \in \mathcal{N} \tag{49.56}$$

of $\hat{\mathbf{S}}$ is connectively stable, provided the conical constraints

$$\|h_i(t, \tilde{x})\| \le \sum_{j=1}^{N} \bar{e}_{ij} \xi_{ij} \|\tilde{x}_j\|\,, \quad i \in \mathcal{N} \tag{49.57}$$

on interconnection functions h_i: $\mathbb{R} \times \mathbb{R}^n \to \mathbb{R}^{n_i}$ hold, where the nonnegative numbers ξ_{ij} do not exceed $\lambda_M^{1/2}(\Delta_{ij}^T \Delta_{ij})$. This robustness result is useful in practice because, typically, interconnections are poorly known, or they are changing during operation of the controlled system.

49.5 Optimization

There is no general method for designing optimal decentralized controls for interconnected systems, even if they are linear and time invariant. For this reason, standard design practice is to optimize each decoupled subsystem using Linear Quadratic (LQ) control laws. Then, suboptimality of the interconnected closed-loop system, which is driven by the union of the locally optimal LQ control laws, is determined with respect to the sum of the quadratic costs chosen for the subsystems. The suboptimal decentralized control design is attractive because, under relatively mild conditions, suboptimality implies stability. Furthermore, the degree of suboptimality can serve as a measure of robustness with respect to a wide spectrum of uncertainties residing in both the subsystems and their interactions.

Consider again the interconnected system

$$\mathbf{S}\colon \dot{x}_i = A_i x_i + B_i u_i + \sum_{i \in \mathcal{N}} A_{ij} x_j\,, \quad i \in \mathcal{N} \qquad (49.58)$$

in the compact form

$$\mathbf{S}\colon \dot{x} = A_D x + B_D u + A_C x\,. \qquad (49.59)$$

We assume that the subsystems

$$\mathbf{S}_i\colon \dot{x}_i = A_i x_i + B_i u_i \qquad (49.60)$$

or, equivalently, their union

$$\mathbf{S}\colon \dot{x} = A_D x + B_D u\,, \qquad (49.61)$$

is controllable, that is, all pairs (A_i, B_i) are controllable.

With $\mathbf{S}_D$ we associate a quadratic cost

$$J_D(x_0, u) = \int_0^\infty (x^T Q_D x + u^T R_D u)dt\,, \qquad (49.62)$$

where $Q_D = \text{diag}\{Q_1, Q_2, \ldots, Q_N\}$ is a symmetric nonnegative definite matrix, $R_D = \text{diag}\{R_1, R_2, \ldots, R_N\}$ is a symmetric positive definite matrix, and the pair $(A_D, Q_D^{1/2})$ is observable. The cost J_D can be considered as a sum of subsystem costs

$$J_i(x_{i0}, u_i) = \int_0^\infty (x_i^T Q_i x_i + u_i^T R u_i)dt\,. \qquad (49.63)$$

In order to satisfy the decentralized constraints on the control law, we solve the standard LQ optimal control problem $(\mathbf{S}_D, J_D)$ to get

$$u_D^{\odot} = -K_D x\,, \qquad (49.64)$$

where $K_D = \text{diag}\{K_1, K_2, \ldots, K_N\}$ is given as

$$K_D = R_D^{-1} B_D^T P_D\,,$$

and $P_D = \text{diag}\{P_1, P_2, \ldots, P_N\}$ is the unique symmetric positive definite solution of the algebraic Riccati equation

$$A_D^T P_D + P_D A_D - P_D B_D R_D^{-1} B_D^T P_D + Q_D = 0\,. \qquad (49.65)$$

The control $u_D^{\odot}$, when applied to $\mathbf{S}_D$, results in the closed-loop system

$$\hat{\mathbf{S}}_D^{\odot}\colon \dot{x} = (A_D - B_D K_D)x\,, \qquad (49.66)$$

which is optimal and produces the optimal cost

$$J_D^{\odot}(x_0) = x_0^T P_D x_0\,. \qquad (49.67)$$

The important fact about the locally optimal control $u_D^{\odot}$ is that it is decentralized. Each component

$$u_i^{\odot} = -K_i x_i \qquad (49.68)$$

of $u_D^{\odot}$ uses only the local state x_i. Generally, the proposed control strategy is not globally optimal, but we can proceed to determine if the cost $J_D^{\oplus}(x_0)$ corresponding to the closed-loop interconnected system

$$\hat{\mathbf{S}}^{\oplus}\colon \dot{x} = (A_D - B_D K_D + A_C)x \qquad (49.69)$$

is finite. If it is, then $\mathbf{S}^{\oplus}$ is suboptimal and a positive number μ exists such that

$$J_D^{\oplus}(x_0) \le \mu^{-1} J_D^{\odot}(x_0) \qquad (49.70)$$

for all $x_0 \in \mathbb{R}^n$. The number μ is called the degree of suboptimality of $u_D^{\odot}$.

We can determine the index μ by first computing the performance index

$$J_D^{\oplus}(x_0) \;=\; x_0^T H x_0\,, \qquad (49.71)$$

where

$$H \;=\; \int_0^\infty \exp(\hat{A}^T t) G_D \exp(\hat{A} t) dt,$$
$$G_D \;=\; Q_D + P_D B_D R_D^{-1} B_D^T P_D, \qquad (49.72)$$

and the closed-loop matrix is

$$\hat{A} \;=\; A_D - B_D K_D + A_C\,. \qquad (49.73)$$

It is important to note that $u_D^{\odot}$ is suboptimal if, and only if, the symmetric matrix H exists. The existence of H is guaranteed by the stability of $\hat{\mathbf{S}}$, in which case we can compute H as the unique solution of the Liapunov matrix equation

$$\hat{A}^T H + H \hat{A} = -G_D\,. \qquad (49.74)$$

The degree of suboptimality, which is the largest we can obtain in this context, is given as

$$\mu^* = \lambda_M^{1/2}(H P_D^{-1})\,. \qquad (49.75)$$

Details of this development, as well as the broad scope of suboptimality, were described in [30], where special attention was devoted to the robustness implications of suboptimality. First, we can explicitly characterize suboptimality in terms the of interconnection matrix A_C. The system $\hat{\mathbf{S}}^{\oplus}$ is suboptimal with degree μ if the matrix

$$F(\mu) \;=\; A_C^T P_D + P_D A_C - (1-\mu)$$
$$(Q_D + P_D B_D R_D^{-1} B_D^T P_D) \qquad (49.76)$$

is nonpositive definite. This is a sufficient condition for suboptimality, but one that implies stability if the pair $\{A_D + A_C, Q_D^{1/2}\}$ is detectable.

Another important aspect of nonpositivity of $F(\mu)$ is that it implies stability even if each control $u_i^{\ominus}$ is replaced by a nonlinearity $\phi_i(u_i^{\ominus})$, which is contained in a sector, or by a linear time–invariant dynamic element. Furthermore, if the subsystems are single–input systems, then each subsystem feedback loop has infinite gain margin, at least $\pm\cos^{-1}(1-\frac{1}{2}\mu)$ phase margin, and at least $50\mu\%$ gain reduction tolerance. These are the standard robustness characteristics of an optimal LQ control law, which are modified by the degree of suboptimality. It is interesting to note that the optimal robustness characteristics can be recovered by solving the inverse problem of optimal decentralized control. The matching conditions are one of the conditions that guarantee the solution of the problem.

The concept of suboptimality extends to the case of *overlapping subsystems*, when subsystems share common parts, and control is required to conform with the *overlapping information structure constraints*. By expanding the underlying state space, the subsystems become disjoint and decentralized control can be designed for the expanded system by standard techniques. Finally, the control laws obtained are contracted for implementation in the original system. This expansion–contraction framework is known as the Inclusion Principle. For a comprehensive presentation of the Principle, see [30].

49.6 Adaptive Decentralized Control

As mentioned in the section on decentrally stabilizable structures, many large scale interconnected systems with a good interconnection structure can be stabilized by a high–gain type decentralized control. How high the gain should be depends on how strong the interconnections are. If a bound on the interconnections is known, then stability can be guaranteed by a fixed high–gain controller. However, if such a bound is not available, then one has to use an adaptive controller which adjusts the gain to a value needed for overall stability.

Consider an interconnected system consisting of single–input subsystems

$$\mathbf{S}\colon\ \dot{x}_i(t) = A_i x_i(t) + b_i[u_i(t) + h_i(t, x(t))]\,, \quad i \in \mathcal{N} \tag{49.77}$$

where, without loss of generality, the pairs (A_i, b_i) are assumed to be in controllable canonical form, and the nonlinear matching interconnections $h_i\colon \mathbb{R}\times\mathbb{R}^n \to \mathbb{R}$ are assumed to satisfy

$$|h_i(t,x)| \le \sum_{j\in\mathcal{N}} \alpha_{ij}\|x_j\| \tag{49.78}$$

for some *unknown* constants $\alpha_{ij} \ge 0$. Let a decentralized state feedback

$$u_i(t) = -\rho(t)k_i^T R_i(\rho(t))\,, \quad i \in \mathcal{N} \tag{49.79}$$

be applied to S, where $R_i(\rho) = \operatorname{diag}\{\rho^{n_i-1}, \ldots, \rho, 1\}$, with $\rho(t)$ being a time–varying gain, and k_i^T are such that the matrices $\hat{A}_i = A_i - b_i k_i^T$ have distinct eigenvalues λ_{il}, $i \in \mathcal{N}$, $l = 1, 2, \ldots, n_i$. Let T_i denote the modal matrices of $\hat{A}_i$, i.e., $T_i\hat{A}_iT_i^{-1} = M_i = \operatorname{diag}\{\lambda_{i1}, \lambda_{i2}, \ldots, \lambda_{in_i}\}$. Then a time-varying coordinate transformation $z_i(t) = T_iR_i(\rho(t))x_i(t)$ transforms the closed–loop system $\hat{\mathbf{S}}$ into

$$\hat{\mathbf{S}}\colon\ \dot{z}_i(t) = \rho(t)M_iz_i(t) + g_i(t, z(t))\,, \quad i \in \mathcal{N}, \tag{49.80}$$

where, provided $0 \le \dot{\rho}(t) \le 1 \le \rho(t)$,

$$\|g_i(t,z)\| \le \sum_{j\in\mathcal{N}} \beta_{ij}\|z_j\| \tag{49.81}$$

for some *unknown* constants $\beta_{ij} \ge 0$. From (49.80) and (49.81) it follows that there exists a $\rho^* > 0$ so that $\hat{\mathbf{S}}$ is stable for all $\rho(t)$ satisfying $0 \le \dot{\rho}(t) \le 1 \le \rho^* \le \rho(t)$, as can be shown by the vector Liapunov approach. However, the crucial point is that ρ^* depends on the unknown bounds β_{ij}. Fortunately, the difficulty can be overcome by increasing $\rho(t)$ adaptively until it is high enough to guarantee stability of $\hat{\mathbf{S}}$. A simple adaptation rule that serves the purpose is

$$\dot{\rho}(t) = \min\{1, \gamma\|x(t)\|\} \tag{49.82}$$

where $\gamma > 0$ is arbitrary. Although the control law is decentralized, $\rho(t)$ is adjusted based on complete state information.

The same idea can also be used in constructing adaptive decentralized dynamic output feedback controllers for various classes of large scale systems with structured nonlinear, time–varying interconnections. A typical example is a system described by

$$\begin{aligned}\mathbf{S}\colon\ \dot{x}_i(t) &= A_ix_i(t) + b_iu_i(t) + h_i(t, x(t))\,,\\ y_i(t) &= c_i^Tx_i(t)\,, \quad i \in \mathcal{N}\end{aligned} \tag{49.83}$$

where

1. the decoupled subsystems described by the triples (A_i, b_i, c_i^T) are controllable and observable,
2. the transfer functions $G_i(s) = c_i^T(sI - A_i)^{-1}b_i$ of the decoupled systems are minimum phase, have *known* relative degree q_i and *known* high frequency gain $\kappa_i = \lim_{s\to\infty} s^{q_i}G_i(s)$, and
3. the nonlinear interconnections $h_i\colon \mathbb{R}\times\mathbb{R}^n \to \mathbb{R}^{n_i}$ are of the form $h_i(t,x) = b_if_i(t,x) + g_i(t,y)$ where $f_i\colon \mathbb{R}\times\mathbb{R}^n \to \mathbb{R}$ and $g_i\colon \mathbb{R}\times\mathbb{R}^n \to \mathbb{R}^{n_i}$ satisfy

$$\begin{aligned}|f_i(t,x)| &\le \sum_{j\in\mathcal{N}} \alpha_{ij}^f\|x_j\|\\ \|g_i(t,x)\| &\le \sum_{j\in\mathcal{N}} \alpha_{ij}^g\|y_j\|\end{aligned} \tag{49.84}$$

for some *unknown* constants α_{ij}^f, α_{ij}^g where $x(t) = [x_1^T(t), x_2^T(t), \ldots x_N^T(t)]^T$ and $y(t) = [y_1(t), y_2^T(t), \ldots y_N(t)]^T$ are the state and the output of the overall system.

Finally, suitable adaptive decentralized control schemes can be developed by forcing an interconnected system of the form

(49.83) to track a decoupled stable linear reference model described as

$$\begin{aligned} \mathbf{S}_M\colon \dot{x}_{Mi}(t) &= A_{Mi}x_{Mi}(t) + b_{Mi}r_i(t), \\ y_{Mi}(t) &= c_{Mi}^T x_{Mi}(t)\,, \quad i \in \mathcal{N}, \end{aligned} \tag{49.85}$$

under reasonable assumptions on $\mathbf{S}$ and $\mathbf{S}_M$.

49.7 Discrete and Sampled–Data Systems

Most of the results concerning the stability and stabilization of continuous–time interconnected systems can be carried over to the discrete case with suitable modifications. Yet, there is a distinct approach to the stability analysis of discrete systems, which is to translate the problem in to that of a continuous system for which abundant results are available. For an idea of this approach, consider a system

$$\mathbf{S}_{SD}\colon x(t+1) = (A_0 + \sum_{k \in \mathcal{K}} p_k A_k)x(t) \tag{49.86}$$

where A_0 is a stable matrix additively perturbed by $p_k A_k$, $k \in \mathcal{K} = \{1, 2, \ldots, K\}$ with p_k standing for one of K perturbation parameters. The purpose is to find the largest region in the parameter space within which $\mathbf{S}_{SD}$ remains stable. By choosing a Liapunov function $v(x) = x^T P x$, where P is the positive definite solution of the discrete Liapunov equation,

$$A_0^T P A_0 - P = -I, \tag{49.87}$$

it can be shown that $\mathbf{S}_{SD}$ is stable, provided $I - W(p)$ is positive definite, where

$$\begin{aligned} W(p) &= \sum_{k \in \mathcal{K}} p_k (A_k^T P A_0 + A_0^T P A_k) \\ &\quad + \sum_{k,l \in \mathcal{K}} p_k p_l A_k^T P A_l. \end{aligned} \tag{49.88}$$

Since the perturbation parameters appear nonlinearly in $W(p)$, characterization of a stability region in the parameter space is not easy. However, $I - W(p)$ is positive definite if the continuous system

$$\dot{\xi}(t) = (-I + \sum_{k \in \mathcal{K}} p_k E_k)\xi(t) \tag{49.89}$$

is stable, where

$$E_k = \begin{bmatrix} 0 & P^{1/2}A_k \\ E_k^T P^{1/2} & A_k^T P A_0 + A_0^T P A_k \end{bmatrix}. \tag{49.90}$$

An analysis of the stability of the perturbed continuous system in (49.89) provides a sufficient condition for the stability of the discrete system in (49.86). This idea can be generalized to the stability analysis of discrete interconnected systems by treating the interconnections as perturbations to nominal stable decoupled subsystems.

A major difference between discrete and continuous systems is that characterizing decentrally stabilizable interconnections for discrete systems is not as easy as for continuous systems. For example, there is no discrete counterpart to the matching conditions. On the other hand, most existing control schemes for continuous systems seem applicable to sampled–data systems provided the sampling rate is sufficiently high. To illustrate this observation, consider the decentralized control of an interconnected system,

$$\begin{aligned} \mathbf{S}\colon \dot{x}_i(t) &= A_i x_i(t) + b_i[u_i(t) \\ &\quad + \sum_{j \in \mathcal{N}} d_{ij}^T x_j(t)]\,, \quad i \in \mathcal{N}, \end{aligned} \tag{49.91}$$

using sampled–data feedback of the form

$$\begin{aligned} u_i(t) &= -k_i^T(t - t_m)x_i(t_m), \\ &\quad t_m \le t < t_{m+1}\,, \end{aligned} \tag{49.92}$$

where t_m are the sampling instants, and $k_i(t)$ are time–varying local feedback gains. With $T_m = t_{m+1} - t_m$ denoting the mth sampling period, it can be shown that the choice of

$$k_i^T(t) = [\delta^{n_i}(t) \ldots \delta'(t)\, \delta(t)] \tag{49.93}$$

or similar feedback gains having impulsive behavior, stabilize $\mathbf{S}$ provided T_m are sufficiently small. How small the sampling periods should be requires knowledge of the bounds on the interconnections. If these bounds are not available, then a simple centralized adaptation scheme, such as

$$T_{m+1}^{-1} = T_m^{-1} + \sum_{j \in \mathcal{N}} \gamma_i \|x_i(t_{m-m_i})\|, \tag{49.94}$$

with $\gamma_i > 0$, decreases T_m to the value needed for stability. Clearly, this is a high–gain stabilization scheme coupled with fast sampling, owing its success to the matching structure of the interconnections [36]. Similar adaptive sampled–data control schemes are available for more general classes of interconnected systems.

49.8 Graph–Theoretic Decompositions

Decomposition of large scale systems and their associated problems is often desirable for computational reasons. In such cases, decentralization or any other structural constraints on the controllers, estimators, or the design process itself, is preferred rather than necessary. Depending on the particular problem in hand, one may be interested in obtaining Lower Block Triangular (LBT) decompositions, input and/or output reachable acyclic decompositions, ϵ–decompositions, overlapping decompositions, etc. [30]. In all of these decomposition schemes, the problem is to find a suitable partitioning and reordering of the input, state, or output variables so that the resulting decomposed system has some desirable structural properties. As expected, the system graph plays the key role, with graph–theory providing the tools.

49.8.1 LBT Decompositions

LBT decompositions are used to reorder the states of system S in (49.6), so that the subsystems have a hierarchical interconnection pattern as

$$\begin{aligned} \mathbf{S}\colon \dot{x}_i &= \sum_{j=1}^{i} A_{ij}x_j + B_i u\,, \quad i \in \mathcal{N}, \\ y &= \sum_{i\in\mathcal{N}} C_i x_i. \end{aligned} \tag{49.95}$$

Such a decomposition corresponds to transforming the A matrix into a Lower Block–Triangular form by symmetric row and column permutations (hence the name LBT decomposition). In terms of system graph, LBT decomposition is the almost trivial problem of identifying the strong components of the truncated digraph $\mathbf{D}_x = (\mathcal{X}, \mathcal{E}_x)$, where $\mathcal{E}_x \subset \mathcal{E}$ contains only the edges connecting state vertices.

LBT decompositions offer computational simplification in the standard state feedback or observer design problems. For example, the problem of designing a state feedback

$$u = -Kx = -\sum_{i\in\mathcal{N}} K_i x_i \tag{49.96}$$

for arbitrary pole placement, can be reduced to computation of the individual blocks K_i of K in a recursive scheme involving the subsystems only.

49.8.2 Acyclic IO Reachable Decompositions

In acyclic Input–Output (IO) reachable decompositions, the purpose is to decompose S into the form

$$\begin{aligned} \mathbf{S}\colon \dot{x}_i &= \sum_{j=1}^{i} A_{ij}x_j + \sum_{j=1}^{i} B_{ij}u_j, \\ y_i &= \sum_{j=1}^{i} C_{ij}x_j\,, \quad i \in \mathcal{N}. \end{aligned} \tag{49.97}$$

That is, in addition to the A matrix, the B and C matrices must have LBT structure. In addition to the desired structure of the system matrices, it is also necessary that the decoupled subsystems represented by (A_{ii}, B_{ii}, C_{ii}) are at least structurally controllable and observable, and that none is further decomposable.

Because the LBT structure is concerned with the reachability properties of the system, both this structure and input and/or output reachability requirements for the subsystems, which are necessary for structural controllability and/or observability, can be taken care of by a suitable decomposition scheme based on binary operations on the reachability matrix of the system digraph. The requirement that the subsystems be dilation free, which is the second condition for structural controllability and/or observability, is of a different nature, however, and should be checked separately after the input–output reachability decomposition has been obtained.

When outputs are of no concern, it is easy to identify all possible acyclic, irreducible, input reachable decompositions of a given system. If some of the resulting decoupled subsystems turn out to contain dilations (destroying structural controllability), then they can suitably be combined with one or more subsystems at a higher level of hierarchy to eliminate the dilations without destroying the LBT structure. Provided that the overall system is structurally controllable, this process eventually gives an acyclic, irreducible decomposition in which all subsystems are structurally controllable. Of course, dual statements are valid for acyclic output reachable decompositions.

Once an acyclic decomposition into controllable subsystems is obtained, many design problems can be decomposed accordingly. An obvious example is the state feedback structure in (49.96). A more complicated problem is the suboptimal state feedback design discussed in the section on optimization. For the system in (49.97), the test matrix $F(\mu)$, with the inclusion of the input coupling terms B_{ij}, becomes

$$F(M_D) = F_D(M_D) + F_C(M_D) + F_C^T(M_D)\,, \tag{49.98}$$

where $M_D = \mathrm{diag}\{\mu_1, \mu_2, \ldots, \mu_N\}$, allowing different μ_i's for $\mathbf{S}_i$'s, $F_D(M_D) = [(1-\mu_i^{-1})(Q_i + K_i^T R_i K_i)]$, and $F_C(M_D) = [F_{ij}(\mu_i)]$ with

$$F_{ij}(\mu_i) = \begin{cases} \mu_i^{-1} P_i(A_{ij} - B_{ij}K_j)\,, & i > j \\ 0, & i \le j. \end{cases} \tag{49.99}$$

From the structure of $F(M_D)$ it is clear that the choice $\mu_i = \epsilon^{N+1-i}$, $i \in \mathcal{N}$, results in a negative definite $F(M_D)$ for sufficiently small ϵ. This guarantees existence of a suboptimal state feedback control law with the degree of suboptimality $\mu = \epsilon^N$. In practice, it is possible to achieve a much better μ by a careful choice of the weight matrices Q_i and R_i.

In a similar way, acyclic, structurally observable decompositions can be used to design suboptimal state estimators, which are discussed below in the context of sequential optimization for acyclic IO decompositions.

To illustrate the use of acyclic IO decompositions in a standard LQG optimization problem, it suffices to consider decomposition of a discrete–time system into only two subsystems as

$$\begin{aligned} \mathbf{S}_1\colon x_1(t+1) &= A_{11}x_1(t) + B_{11}u_1(t) + w_1(t), \\ y_1(t) &= C_{11}x_1(t) + v_1(t), \\ \mathbf{S}_2\colon x_2(t+1) &= A_{21}x_1(t) + A_{22}x_2(t) \\ &\quad + B_{21}u_1(t) + B_{22}u_2(t) + w_2(t), \\ y_2(t) &= C_{21}x_1(t) + C_{22}x_2(t) + v_2(t), \end{aligned} \tag{49.100}$$

with the usual assumptions on the input and measurement noises ω_i and v_i, $i = 1, 2$. Let each subsystem be associated with a performance criterion

$$\begin{aligned} \mathcal{E}J_i &= \mathcal{E}\left\{ \lim_{T\to\infty} T^{-1} \sum_{t=0}^{T-1} \left[x_i^T(t) Q_i x_i(t) \right.\right. \\ &\quad \left.\left. + u_i^T(t) R_i u_i(t) \right] \right\}, \quad i = 1, 2 \end{aligned} \tag{49.101}$$

where $\mathcal{E}$ denotes expectation.

The sequential optimization procedure consists of minimizing $\mathcal{E}J_1$ and $\mathcal{E}J_2$ subject to the dynamic equations for the systems $\mathbf{S}_1$ and $(\mathbf{S}_1, \mathbf{S}_2)$, respectively. The first problem has the standard solution $u_1^*(t) = -K_1\hat{x}_1(t)$, where K_1 is the optimal control gain found from the solution of the associated Riccati equation, and $\hat{x}_1(t)$ is the best estimate of $x_1(t)$ given the output information $\mathcal{Y}_1^{t-1} = \{y_1(0), \ldots, y_1(t-1)\}$. The estimate $\hat{x}_1(t)$ is generated by the Kalman filter

$$\begin{aligned}\hat{x}_1(t+1) &= A_{11}\hat{x}_1(t) + B_{11}u_1^*(t) \\ &\quad + L_1[y_1(t) - c_{11}\hat{x}_1(t)] \qquad (49.102)\end{aligned}$$

where L_1 is the steady–state estimator gain. With the control u_1^* applied to $\mathbf{S}_1$, the overall system becomes

$$\mathbf{S}: \begin{bmatrix} \hat{x}_1(t+1) \\ x_1(t+1) \\ x_2(t+1) \end{bmatrix} = \qquad (49.103)$$

$$\begin{bmatrix} A_{11} - B_{11}K_1 - L_1C_{11} & L_1C_{11} & 0 \\ -B_{11}K_1 & A_{11} & 0 \\ -B_{21}K_1 & A_{21} & A_{22} \end{bmatrix} \begin{bmatrix} \hat{x}_1(t) \\ x_1(t) \\ x_2(t) \end{bmatrix} + \begin{bmatrix} 0 \\ 0 \\ B_{22} \end{bmatrix} u_2(t) + \begin{bmatrix} L_1v_1(t) \\ w_1(t) \\ w_2(t) \end{bmatrix}$$

which preserves the LBT structure of the original system. Assuming that both $\mathcal{Y}_1^{t-1}$ and $\mathcal{Y}_2^{t-1} = \{y_2(0), \ldots, y_2(t-1)\}$ are available for constructing the control u_2^* (which is consistent with the idea of sequential optimization), the problem reduces to minimization of $\mathcal{E}J_2$ subject to (103). An analysis of the standard solution procedure reveals that the optimal control law can be expressed as

$$u_2^*(t) = -K_{21}\hat{x}_1(t) - K_{22}\hat{\xi}(t) \qquad (49.104)$$

where $K = [K_{21}\ K_{22}]$ is the optimal control gain, and $\hat{\xi}(t)$ is the optimal estimate of $x(t) = [x_1^T(t)\ x_2^T(t)]^T$, given $\mathcal{Y}_1^{t-1}$ and $\mathcal{Y}_2^{t-1}$. Furthermore, the $2n_1 + n_2$–dimensional Riccati equation, from which K is constructed, can be decomposed into an n_2–dimensional Riccati equation involving the parameters of the second isolated subsystem and a Liapunov equation corresponding to an $n_2 \times 2n_1$ dimensional matrix. This results in considerably simplifying the solution of the optimal control gain. However, the Kalman filter for $\hat{\xi}(t)$ still requires the solution of an $(n_1 + n_2)$–dimensional Riccati equation.

Other sequential optimization schemes based on various information structure constraints can be analyzed similarly; for details, see [30].

49.8.3 Nested Epsilon Decompositions

Epsilon decomposition of a square matrix M is concerned with transforming M by symmetric row and column permutations into a form

$$P^TMP = M_D + \epsilon M_C \qquad (49.105)$$

where M_D is block diagonal, and ϵ is a prescribed small number [27]. The problem is equivalent to identifying the connected components of a subgraph $\mathbf{D}^\epsilon$ of the digraph $\mathbf{D}$ associated with M, which is obtained by deleting all edges of $\mathbf{D}$ corresponding to those elements of M with magnitude smaller than ϵ. All of the vertices of a connected component of $\mathbf{D}^{\epsilon_1}$ appear in the same connected component of $\mathbf{D}^{\epsilon_2}$ for any $\epsilon_2 < e_1$. Thus one can identify a number of distinct values $\epsilon_1 > \epsilon_2 > \ldots > \epsilon_K$ such that

$$\begin{aligned}P^TMP &= (\ldots((M_0 + \epsilon_1M_1) + \epsilon_2M_2) \\ &\quad + \ldots + \epsilon_KM_K), \qquad (49.106)\end{aligned}$$

which is a *nested epsilon decomposition* of M as illustrated in Figure 49.3.

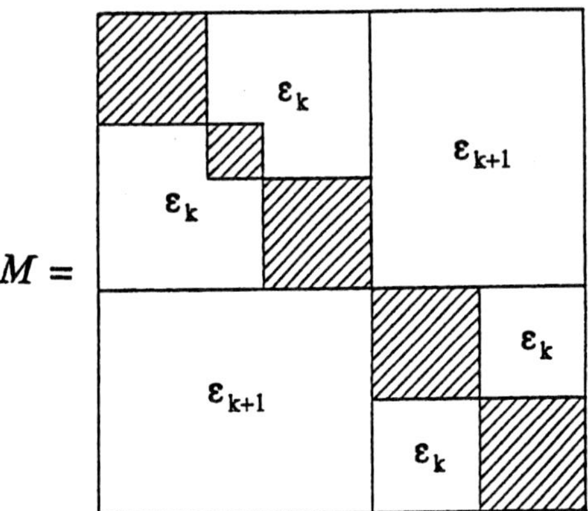

Figure 49.3 Nested epsilon decompositions.

As seen from the figure, a large ϵ results in a finer decomposition than a small ϵ does. Thus the choice of ϵ provides a compromise between the size and the number of components and the strength of the interconnections among them. A nice property of nested epsilon decompositions is that once the decomposition corresponding to some ϵ_k is obtained, the decomposition corresponding to ϵ_{k+1} can be found by working with a smaller digraph obtained by condensing $\mathbf{D}^{\epsilon_k}$ with respect to its components.

An immediate application of the nested epsilon decompositions is the stability analysis of a large scale system via vector Liapunov functions, where the matrix M is identified with the matrix A of the system in (6). Provided the subsystems resulting from the decomposition are stable, the stability of the overall system can easily be established by means of the aggregate matrix W in (49), whose off–diagonal elements are of the order of ϵ.

The nested epsilon decomposition algorithm can also be applied with some modifications to decompose a system with inputs as

$$\begin{aligned}\dot{x}_i &= A_{ii}x_i + B_{ii}u_i + \epsilon\sum_{j\neq i}^{N}(A_{ij}x_j + B_{ij}u_j), \\ &\quad i \in \mathcal{N}. \qquad (49.107)\end{aligned}$$

If each decoupled subsystem identified by a pair (A_{ii}, B_{ii}) is stabilized by a local state feedback of the form $u_i = -K_ix_i$, $i \in \mathcal{N}$, with the local gains not excessively high, then the closed–loop system preserves the weak–coupling property of the open–loop system, providing an easy way to stabilize the overall system. The same idea can also be employed in designing decentralized estimators [30] based on a suitable epsilon decomposition of the pair (A, C).

49.8.4 Overlapping Decompositions

Consider a system

$$\tilde{\mathbf{S}}: \dot{\tilde{x}}(t) = \tilde{A}\tilde{x}(t) \tag{49.108}$$

with an $\tilde{n}$–dimensional state vector $\tilde{x}$. Let columns of the matrix $V \in \mathbb{R}^{\tilde{n}\times n}$ form a basis for an n–dimensional A–invariant subspace of $\mathbb{R}^{\tilde{n}}$, and let A be the restriction of $\tilde{A}$ to $\mathrm{Im}V \simeq \mathbb{R}^n$, that is, $\tilde{A}V = VA$. Then the smaller order system

$$\mathbf{S}: \dot{x}(t) = Ax(t) \tag{49.109}$$

is called a restriction of $\tilde{\mathbf{S}}$. Conversely, starting with the system $\mathbf{S}$, one can obtain an expansion $\tilde{\mathbf{S}}$ of $\mathbf{S}$ by defining $\tilde{A} = VAV^L + M$, where V^L is any left inverse of V, and M is any complementary matrix satisfying $MV = 0$. The very definition of a restriction implies that $\mathbf{S}$ is stable if $\tilde{\mathbf{S}}$ is.

In many problems associated with large scale systems, it may be desirable to expand a system $\mathbf{S}$ to a larger dimensional one which possess some nice structural properties. The increase in dimensionality of the problem may very well be offset by the nice structure of the expansion. As an example, consider a system $\mathbf{S}$ with

$$A = \left[\begin{array}{cc|c} A_{11} & A_{12} & \epsilon A_{13} \\ \hline \epsilon A_{21} & \multicolumn{1}{|c|}{A_{22}} & \epsilon A_{23} \\ \hline \epsilon A_{31} & \multicolumn{1}{|c}{A_{32}} & A_{33} \end{array}\right] \tag{49.110}$$

where ϵ is a small parameter. Letting

$$V = \begin{bmatrix} I_1 & 0 & 0 \\ 0 & I_2 & 0 \\ 0 & I_2 & 0 \\ 0 & 0 & I_3 \end{bmatrix} \tag{49.111}$$

where I_k denotes an identity matrix of order n_k, one obtains an expansion $\tilde{\mathbf{S}}$ with

$$\tilde{A} = \left[\begin{array}{cc|cc} A_{11} & A_{12} & 0 & \epsilon A_{13} \\ \epsilon A_{21} & A_{22} & 0 & \epsilon A_{23} \\ \hline \epsilon A_{21} & 0 & A_{22} & \epsilon A_{23} \\ \epsilon A_{31} & 0 & A_{32} & A_{33} \end{array}\right]. \tag{49.112}$$

Since $\tilde{\mathbf{S}}$ has an obvious decomposition into two weakly coupled subsystems, one can take advantage of this structural property in stability analysis, which is not available for the original system $\mathbf{S}$.

One can easily notice from the structure of V in (114) that the expansion $\tilde{\mathbf{S}}$ of $\mathbf{S}$ is obtained simply by repeating the equation for the middle part x_2 of the state vector $x = [x_1^T\ x_2^T\ x_3^T]^T$. In some sense, x_2 is treated as common to two overlapping components $\tilde{x}_1 = [x_1^T\ x_2^T]^T$ and $\tilde{x}_2 = [x_2^T\ x_3^T]^T$ of x. Thus the partitioning of the A matrix in (113) is termed the *overlapping decomposition.*

Although the expansion matrix V can be any matrix with full column rank, if it is restricted to contain one and only one unity element in each row (which corresponds, as in the case above, to repeating some of the state equations in the expanded domain), then one can develop a suitable graph–theoretic algorithm to find the smallest expansion which has a disjoint decomposition (into decoupled or ϵ–coupled components) with the property that no component is further decomposable.

The idea of overlapping decompositions via expansions can be extended to systems with inputs. A system

$$\tilde{\mathbf{S}}: \dot{\tilde{x}}(t) = \tilde{A}\tilde{x}(t) + \tilde{B}u(t) \tag{49.113}$$

is said to be an expansion of

$$\mathbf{S}: \dot{x}(t) = Ax(t) + Bu(t) \tag{49.114}$$

if $\tilde{B} = VB$ in addition to $\tilde{A}V = VA$. Consider the optimal control problems of minimizing the performance criteria

$$\begin{aligned} J &= \int_0^\infty [x^T(t)Qx(t) + u^T(t)Ru(t)]dt \\ \tilde{J} &= \int_0^\infty [\tilde{x}^T(t)\tilde{Q}\tilde{x}(t) + u^T(t)Ru(t)]dt \end{aligned} \tag{49.115}$$

associated with $\mathbf{S}$ and $\tilde{\mathbf{S}}$. The optimal solutions are

$$u(t) = -Kx(t), \text{ and } u(t) = -\tilde{K}\tilde{x}(t), \tag{49.116}$$

respectively, resulting in closed–loop systems

$$\begin{aligned} \hat{\mathbf{S}}: \dot{x}(t) &= (A - BK)x(t), \\ \hat{\tilde{\mathbf{S}}}: \dot{\tilde{x}}(t) &= (\tilde{A} - \tilde{B}\tilde{K})\tilde{x}(t). \end{aligned} \tag{49.117}$$

Thus, $\hat{\mathbf{S}}$ is a restriction of $\hat{\tilde{\mathbf{S}}}$ if $(\tilde{A} - \tilde{B}\tilde{K})V = V(A - BK)$, or equivalently, if $\tilde{K} = KV$. The last condition is satisfied if $\tilde{Q}$ and Q are related as $Q = V^T\tilde{Q}V$, in which case the optimal cost matrices are also related as $P = V^T\tilde{P}V$. This analysis shows that, if the cost matrices $\tilde{Q}$ and R of the expanded system are chosen to be block diagonal with diagonal blocks associated with the decoupled expanded subsystems, then its optimal (in case of complete decoupling) or suboptimal (in case of weak decoupling) solution can be contracted back to an optimal or suboptimal solution of the original system with respect to a suitably chosen performance criterion.

References

[1] Anderson, B. D. O. and Clements, D. J., Algebraic characterization of fixed modes in decentralized control, *Automatica*, 17, 703–712, 1981.

[2] Brusin, V. A. and Ugrinovskaya, E. Ya., Decentralized adaptive control with a reference model, *Avtomatika i Telemekhanika*, 10, 29–36, 1992.

[3] Chae, S. and Bien, Z., Techniques for decentralized control for interconnected systems, in *Control and Dynamic Systems*, C. T. Leondes, Ed., Academic Press, Boston, 41, 273–315, 1991.

[4] Chen, Y. H., Decentralized robust control design for large–scale uncertain systems: The uncertainty is time–varying, *J. Franklin Institute*, 329, 25–36, 1992.

[5] Chen, Y. H. and Han, M. C., Decentralized control design for interconnected uncertain systems, in *Control and Dynamic Systems*, C. T. Leondes, Ed., Academic Press, Orlando, FL, 56, 219–266, 1993.

[6] Cheng, C. F., Wang, W. J., and Lin, Y. P., Decentralized robust control of decomposed uncertain interconnected systems, *Trans. ASME*, 115, 592–599, 1993.

[7] Cho, Y. J. and Bien, Z., Reliable control via an additive redundant controller, *Int. J. Control*, 50, 385–398, 1989.

[8] Date, R. A. and Chow, J. H., A parametrization approach to optimal H_2 and H_∞ decentralized control problems, *Automatica*, 29, 457–463, 1993.

[9] Datta, A., Performance improvement in decentralized adaptive control: A modified model reference scheme, *IEEE Trans. Automatic Control*, 38, 1717–1722, 1993.

[10] Gajić, Z. and Shen, X., *Parallel Algorithms for Optimal Control of Large Scale Linear Systems*, Springer–Verlag, Berlin, Germany, 1993.

[11] Geromel, J. C., Bernussou, J., and Peres, P. L. D., Decentralized control through parameter space optimization, *Automatica*, 30, 1565–1578, 1994.

[12] Gündes, A. N. and Kabuli, M. G., Reliable decentralized control, *Proc. Am. Control Conf.*, Baltimore, MD, pp. 3359–3363, 1994.

[13] Iftar, A., Decentralized estimation and control with overlapping input, state, and output decomposition, *Automatica*, 29, 511-516, 1993.

[14] Ikeda, M., Decentralized Control of Large Scale Systems, in *Three Decades of Mathematical System Theory*, H. Nijmeiyer and J. M. Schumacher, Eds., Springer–Verlag, New York, 1989, 219–242.

[15] Jamshidi, M., *Large–Scale Systems. Modeling and Control*, North–Holland, New York, 1983.

[16] Lakshmikantham, V. and Liu, X. Z., *Stability Analysis In Terms of Two Measures*, World Scientific, Singapore, 1993.

[17] Lakshmikantham, V., Matrosov, V. M., and Sivasundaram, S., *Vector Liapunov Functions and Stability Analysis of Nonlinear Systems*, Kluwer, The Netherlands, 1991.

[18] Leitmann, G., One approach to the control of uncertain systems, *ASME J. Dyn. Syst., Meas., and Control*, 115, 373–380, 1993.

[19] Leondes, C. T., Ed., *Control and Dynamic Systems*, Vols. 22-24, Decentralized/Distributed Control and Dynamic Systems, Academic Press, Orlando, FL, 1985.

[20] Lin, C. T., Structural controllability, *IEEE Trans. Auto. Control*, AC–19, 201–208, 1974.

[21] Lyon, J., Note on decentralized adaptive controller design, *IEEE Trans. Auto. Control*, 40, 89–91, 1995.

[22] Michel, A. N., On the status of stability of interconnected systems, *IEEE Trans. Circuits Syst.*, CAS–30, 326–340, 1983.

[23] Mills, J. K., Stability of robotic manipulators during transition to and from compliant motion, *Automatica*, 26, 861–874, 1990.

[24] Sandell, N. R., Jr., Varaiya, P., Athans, M., and Safonov, M. G., Survey of decentralized control methods for large scale systems, *IEEE Trans. Auto. Control*, AC–23, 108–128, 1978.

[25] Savastuk, S. V. and Šiljak, D. D., Optimal decentralized control, *Proc. Am. Control Conf.*, Baltimore, MD, pp. 3369–3373, 1994.

[26] Sezer, M.E. and Šiljak, D. D., Structurally fixed modes, *Syst. Control Lett.*, 1, 60–64, 1981.

[27] Sezer, M. E. and Šiljak, D. D., Nested ϵ–decomposition and clustering of complex systems, *Automatica*, 22, 321–331, 1986.

[28] Shi, L. and Singh, S. K., Decentralized adaptive controller design for large–scale systems with higher order interconnections, *IEEE Trans. Auto. Control*, 37, 1106–1118, 1992.

[29] Šiljak, D. D., *Large–Scale Dynamic Systems: Stability and Structure*, North–Holland, New York, 1978.

[30] Šiljak, D. D., *Decentralized Control of Complex Systems*, Academic Press, Cambridge, MA, 1991.

[31] Tamura, H. and Yoshikawa, T., *Large–Scale Systems Control and Decision Making*, Marcel Dekker, New York, 1990.

[32] Ünyelioğlu, K. A. and Özgüler, A. B., Reliable decentralized stabilization of feed–forward and feedback interconnected systems, *IEEE Trans. Auto. Control*, 37, 1119–1132, 1992.

[33] Voronov, A. A., Present state and problems of stability theory, *Automatika i Telemekhanika*, 5, 5–28, 1982.

[34] Wang, S. H. and Davison, E. J., On the stabilization of decentralized control systems, *IEEE Trans. Auto. Control*, AC–18, 473–478, 1973.

[35] Wu, W. and Lin, C., Optimal reliable control system design for steam generators in pressurized water reactors, *Nuclear Technology*, 106, 216–224, 1994.

[36] Yu, R., Ocali, O., and Sezer, M. E., Adaptive robust sampled–data control of a class of systems under structured perturbations, *IEEE Trans. Auto. Control*, 38, 1707–1713, 1993.

Further Reading

There is a number of survey papers on decentralized control and large scale systems [3], [14], [24]. The books on the subject are [10], [15], [19], [29], [31]. For a comprehensive treatment of decentralized control theory, methods, and applications, with a large number of references, see [30].

For further information on vector Liapunov functions and stability analysis of large scale interconnected systems, see the survey papers [22], [33], and books [16], [17].

Adaptive decentralized control has been of widespread recent interest, see [2], [9], [21], [23], [28], [30], [36].

Robustness of decentralized control to both structured and unstructured perturbations has been one of the central issues in the control of large scale systems. For the background of robustness issues in control, which are relevant to decentralized control, see [18], [30]. For new and interesting results on the subject, see [4], [5], [6], [8].

There is a number of papers devoted to design of decentralized control via parameter space optimization, which rely on powerful convex optimization methods. For recent results and references, see [11].

Overlapping decentralized control and the Inclusion Principle are surveyed in [30]. Useful extensions were presented in [13]. The concept of overlapping is basic to reliable control under controller failures using multiple decentralized controllers [30]. For more information about this area, see [7], [12], [32], [35].

In a recent development [25], it has been shown how optimal decentralized control of large scale interconnected systems can be obtained in the classical optimization framework of Lagrange. Both sufficient and necessary conditions for optimality are derived in the context of Hamilton-Jacobi equations and Pontryagin's maximum principle.

50

Decoupling

Trevor Williams
Department of Aerospace Engineering and Engineering Mechanics, University of Cincinnati, Cincinnati, OH

Panos J. Antsaklis
Department of Electrical Engineering, University of Notre Dame, Notre Dame, IN

50.1 Introduction

Multi-input/multi-output systems are usually difficult for human operators to control directly, since changing any one input generally affects many, if not all, outputs of the system. As an example, consider the vertical landing of a vertical take off and landing jet or of a lunar landing rocket. Moving to a desired landing point to the side of the current position requires tilting the thrust vector to the side; but this reduces the vertical thrust component, which was balancing the weight of the craft. The aircraft therefore starts to descend, which is not desired. To move to the side at a constant height thus requires smooth, simultaneous use of both attitude control and throttle. It would be simpler for the pilot if a single control existed to do this maneuver; hence the interest in control methods that make the original system behave in a way that is easier to control manually. One example of such technique is when a compensator is sought that makes the compensated system diagonally dominant. If this can be achieved, it is then possible to regard the system as, to first order, a set of independent single-input/single-output systems, which is far easier to control than the original plant. Another approach is that of decoupling, where the system transfer matrix is made to be exactly diagonal. Each output variable is therefore affected by only one input signal, and each input/output pair can then be controlled by an easier-to-design single-input/single-output controller or manually by a human operator.

This chapter studies the problem of making the transfer function matrix of the system diagonal using feedback control and, in particular, state feedback, state feedback with dynamic precompensation, and constant output feedback control laws. This problem is referred to as the *dynamical decoupling* problem, as it leads to a compensated system where the input actions are decoupled; it is also called a noninteracting control problem for similar reasons. Stability is an important issue and it also is examined here. Conditions for decoupling with stability and algorithms to determine such control laws are described. The problems of block diagonal or triangular decoupling are also addressed. They are of interest when full diagonal decoupling using a particular form of feedback control, typically state feedback, is not possible. Note that the approach taken in this chapter follows the development in [14]. Static decoupling is also briefly discussed; references for approximate diagonal decoupling are provided in "Further Reading."

50.1.1 Diagonal Decoupling

Diagonal decoupling of a system with equal numbers of inputs and outputs is the most straightforward type of problem in the field of noninteracting control. The goal is to apply some form of control law to the system so as to make the i-th output of the closed-loop system independent of all but the i-th closed-loop input signal. Each output can then be controlled by a dedicated simpler single-input/single-output controller, or by a human operator. The main questions to be answered when investigating diagonal decoupling of a given system are

- Can it be decoupled at all?
- If so, what form of controller is required to achieve this?

Three classes of controllers that are customarily considered are

1. Constant output feedback $u = Hy + Gr$, where the output y of the system is simply multiplied by a constant gain matrix H and this is fed back as the control signal u, with r the new external input to the system and G a constant gain matrix
2. Linear state feedback $u = Fx + Gr$, where the control signal consists of a constant matrix F multiplying the internal state variable vector x of the system
3. State feedback plus precompensation, where a feedforward dynamic control system is added to the state feedback controller.

0-8493-8570-9/96/$0.00+$.50

Note that the compensator in class 3 corresponds to dynamic output feedback, where the input and output signal vectors r and y are multiplied by dynamic transfer function gain matrices rather than constant ones.

The problem of diagonally decoupling a square system was the first decoupling question to be studied, and it can be answered in a fairly straightforward fashion. First of all, diagonal decoupling by state feedback plus precompensation, or by dynamic output feedback, amounts to finding a transfer matrix that, when the open-loop transfer matrix is multiplied by it, produces a diagonal closed-loop transfer matrix. This problem is therefore closely related to the problem of finding an inverse for the open-loop plant. As a result of this, any square plant that has a full rank transfer matrix can be diagonally decoupled by this type of control. This result was proved by Rekasius [10]. A system that does not satisfy this condition does not have linearly independent outputs, so it follows that it is impossible to decouple by any form of controller. It is of great practical interest to establish whether a given plant can actually be decoupled by a simpler type of controller than this. Falb and Wolovich [3] established the necessary and sufficient condition under which diagonal decoupling by state feedback alone is possible. This condition, which can be easily tested from either a state-space or a transfer matrix model of the plant, can be expressed as follows.

A square system can be diagonalized by state feedback alone if and only if the constant matrix $B^\star$ is nonsingular, where this matrix is defined below first from the state-space and then from the transfer matrix description of the system.

State-space representation. Let the given system be $\dot{x} = Ax + Bu$, $y = Cx + Du$ in the continuous-time case, or $x(k+1) = Ax(k) + Bu(k)$, $y(k) = Cx(k) + Du(k)$ in the discrete-time case; let A, B, C, D be $n \times n, n \times p, p \times n, p \times p$ real matrices, respectively; and assume for simplicity that the system is controllable and observable. Then the $p \times p$ matrix $B^\star$ is constructed as follows: If the i-th row of the direct feedthrough matrix D is nonzero, this becomes the i-th row of the constant matrix $B^\star$. Otherwise, find the lowest integer, f_i, for which the i-th row of $CA^{f_i-1}B$ is nonzero. This then becomes the i-th row of the constant matrix $B^\star$.

Transfer matrix representation. Let $T(s)$, with s the Laplace transform variable, be the $p \times p$ transfer function matrix of the continuous-time system; or $T(z)$, with z the Z-transform variable, be the transfer function matrix of the discrete-time system. Let $D(s)$ [or $D(z)$] be the diagonal matrix $D(s) = diag(s^{f_i})$ where the nonnegative integers $\{f_i\}$ are so that all rows of $\lim_{s\to\infty} D(s)T(s)$ are constant and nonzero. This limit matrix is $B^\star$; that is,

$$\lim_{s\to\infty} D(s)T(s) = B^\star \tag{50.1}$$

The integers $\{f_i\}$ are known as the decoupling indices of the system. They can be determined from either the state-space or the transfer function descriptions as described above; note that $f_i = 0$ corresponds to the i-th row of D being nonzero. In either case, of course, the resulting matrix $B^\star$ is the same. It should be noted that systems will generically satisfy the decoupling condition; that is, if all entries of the A, B, C (and D) matrices are chosen at random, the resulting $B^\star$ will have full rank. Diagonal decoupling by state feedback is therefore likely to be feasible for a wide variety of systems.

EXAMPLE 50.1:

$$A = \begin{pmatrix} 0 & 1 & 0 \\ 0 & 0 & 1 \\ -6 & -11 & -6 \end{pmatrix}, \quad B = \begin{pmatrix} 0 & 0 \\ 0 & 0 \\ -1 & 2 \end{pmatrix},$$
$$C = \begin{pmatrix} 3 & 6 & 1 \\ 2 & 0 & 0 \end{pmatrix}, \quad D = \begin{pmatrix} 0 & 0 \\ 0 & 0 \end{pmatrix}$$

This gives $f_1 = 1$, $f_2 = 3$, and $B^\star = \begin{pmatrix} -1 & 2 \\ -2 & 4 \end{pmatrix}$. This matrix is clearly singular; therefore, the system cannot be decoupled by state feedback.

EXAMPLE 50.2:

$$T(s) = \begin{pmatrix} \frac{1}{s} & \frac{2}{s+1} \\ \frac{4}{s+3} & \frac{8s}{s+4} \end{pmatrix}$$

This gives $B^\star = \begin{pmatrix} 1 & 2 \\ 0 & 8 \end{pmatrix}$, with decoupling indices $f_1 = 1$, $f_2 = 0$. This system can therefore be diagonally decoupled by state feedback.

EXAMPLE 50.3:

$$T(s) = \begin{pmatrix} \frac{1}{s} & \frac{2}{s+1} \\ \frac{4}{s+3} & \frac{8}{s+4} \end{pmatrix}$$

The same as previously, but with the (2, 2) entry divided by s. We now obtain $B^\star = \begin{pmatrix} 1 & 2 \\ 4 & 8 \end{pmatrix}$, with decoupling indices $f_1 = 1$, $f_2 = 1$. $B^\star$ is now singular, so this system cannot be diagonally decoupled by state feedback alone.

50.1.2 Diagonal Decoupling with Internal Stability

A question of great practical interest is whether the closed-loop system that is obtained after decoupling can be made stable. It can be shown constructively (for instance, by use of the algorithm given below) that all of the poles that are evident from the diagonal closed-loop transfer matrix can be assigned any desired values. The question therefore becomes: Can the closed-loop system be made internally stable, where there are no "hidden" cancellations between unstable poles and zeros? Such unstable modes are particularly dangerous in practice, as they will not be revealed by an examination of the transfer matrix. However, the hidden unstable state behavior they represent will very likely cause problems, such as burnout of internal electronic components of the system. It was shown by Gilbert [5] that a given plant may indeed have hidden fixed modes when it is diagonally decoupled by state feedback, with or without precompensation. Wolovich and Falb [15] then showed that these modes are the same for both cases; furthermore, they are a subset of the transmission zeros of the plant. In fact, they are those transmission zeros z_i that do not make any of the rows of the transfer matrix $T(s)$ equal to zero when evaluating $T(z_i)$; they are called *diagonal coupling zeros*. Thus, any plant with square, full-rank transfer matrix for which all the diagonal coupling zeros are in the left half-plane can be diagonally decoupled with internal stability by state feedback plus precompensation; or by state feedback alone if $B^\star$ is nonsingular. Therefore, there will never be any problems with internal stability when decoupling a minimum-phase system, as all of its transmission zeros are in the left half-plane.

An algorithm to diagonally decouple a system, when $B^\star$ has full rank, using state feedback is now presented. This algorithm is based on a procedure to obtain a stable inverse of a system that is described below. This procedure is applied to the system $D(s)T(s) = \hat{T}(s)$, where $D(s) = diag(s^{f_i})$ as in Equation 50.1, that can be shown to have a state-space realization $\{A, B, \hat{C}, \hat{D}\}$. In fact $\hat{D} = B^\star$, which is assumed to have full rank p; and this implies that a proper right inverse of the system $\hat{T}(s)$ exists. Here it is assumed that the system has the same number of inputs and outputs, and this simplifies the selection of F, G as in this case they are unique; see the algorithm for the inverse below for the nonsquare case. In particular, if the state feedback $u = Fx + Gr$ with

$$F = -(B^\star)^{-1}\hat{C}, \quad G = (B^\star)^{-1} \tag{50.2}$$

is applied to the system $\dot{x} = Ax + Bu$, $y = \hat{C}x + B^\star u$, then it can be shown that $\hat{T}_{F,G}(s) = D(s)T_{F,G}(s) = I_p$. This implies that if the state feedback $u = Fx + Gr$ with F, G as in Equation 50.2 is applied to the given system $\dot{x} = Ax + Bu$, $y = Cx + Du$ with transfer matrix $T(s)$, then

$$T_{F,G}(s) = D^{-1}(s) \tag{50.3}$$

which is diagonal with entries s^{-f_i}. Note that here the state feedback matrix F assigns all the n closed-loop eigenvalues at the locations of the n zeros of $\hat{T}(s)$; that is, at the zeros of $T(s)$ and of $D(s)$. The closed-loop eigenvectors are also appropriately assigned so the eigenvalues cancel all the zeros to give $D(s)T_{F,G}(s) = I_p$. This explains the control mechanism at work here and also makes quite apparent the changes necessary to ensure internal stability. Simply instead of $D(s)$ use $\hat{D}(s) = diag\{p_i(s)\}$ with $p_i(s)$ stable polynomials of degree s^{f_i}; that is, $p_i(s) = s^{f_i}$ + lower-degree terms. Then it can be shown that $\lim_{s\to\infty} \hat{D}(s)T(s) = B^\star$ and that $\{A, B, \tilde{C}, B^\star\}$ is a realization of $\hat{D}(s)T(s) = \tilde{T}(s)$. State feedback with

$$F = -(B^\star)^{-1}\tilde{C}, \quad G = (B^\star)^{-1} \tag{50.4}$$

gives

$$T_{F,G}(s) = \hat{D}^{-1}(s) = diag\{p_i^{-1}(s)\} \tag{50.5}$$

which is stable. Note that in this case the closed-loop eigenvalues are at the assumed stable zeros of $T(s)$ and at the selected stable zeros of the polynomials $p_i(s)$, $i = 1, .., p$.

EXAMPLE 50.4:

Let $T(s) = \begin{pmatrix} \frac{s+1}{s^2} & 0 \\ \frac{1}{s(s-1)} & \frac{-1}{s-1} \end{pmatrix}$.

Here

$$\lim_{s\to\infty} D(s)T(s) = \lim_{s\to\infty} diag(s, s)T(s) = \begin{pmatrix} 1 & 0 \\ 0 & -1 \end{pmatrix} = B^\star.$$

Since $B^\star$ has full rank, the system can be decoupled using state feedback $u = Fx + Gr$. The system has one transmission zero at -1 and there are no diagonal coupling zeros, so it can be decoupled with internal stability. Let $\hat{D}(s) = \begin{pmatrix} s+1 & 0 \\ 0 & s+2 \end{pmatrix}$. A minimal (controllable and observable) realization of $\tilde{T}(s) = \hat{D}(s)T(s)$ is $\{A, B, \tilde{C}, B^\star\}$ where

$$A = \begin{pmatrix} 0 & 1 & 0 \\ 0 & 0 & 0 \\ -1 & 0 & 1 \end{pmatrix}, \quad B = \begin{pmatrix} 0 & 0 \\ 1 & 0 \\ 0 & 1 \end{pmatrix},$$

$$\tilde{C} = \begin{pmatrix} 1 & 2 & 0 \\ 3 & 1 & -3 \end{pmatrix}.$$

In view now of Equations 50.4 and 50.5, for

$$F = -(B^\star)^{-1}\tilde{C} = \begin{pmatrix} -1 & -2 & 0 \\ 3 & 1 & -3 \end{pmatrix}$$

and

$$G = (B^\star)^{-1} = \begin{pmatrix} 1 & 0 \\ 0 & -1 \end{pmatrix},$$

$$T_{F,G}(s) = \hat{D}(s)^{-1} = \begin{pmatrix} \frac{1}{s+1} & 0 \\ 0 & \frac{1}{s+2} \end{pmatrix}.$$

The closed-loop eigenvalues are in this case located at the transmission zero of the plant at -1 and at the selected locations -1 and -2, the poles of $\hat{D}(s)^{-1}$. Note that it is not necessary to cancel the transmission zero at -1 in order to decouple the system since it is not a coupling zero; it could appear as a zero in the decoupled system instead. To illustrate this, consider Example 50.5 where $T(s)$ is the same except that the zero is now unstable at +1.

EXAMPLE 50.5:

Let $T(s) = \begin{pmatrix} \frac{s-1}{s^2} & 0 \\ \frac{1}{s(s-1)} & \frac{-1}{s-1} \end{pmatrix}$ where again

$$\lim_{s\to\infty} D(s)T(s) = \lim_{s\to\infty} diag(s, s)T(s) = \begin{pmatrix} 1 & 0 \\ 0 & -1 \end{pmatrix} = B^\star.$$

Since $B^\star$ has full rank, the system can be decoupled using state feedback. Since there are no diagonal coupling zeros, the system can be decoupled with internal stability. Write $T(s) = \begin{pmatrix} s-1 & 0 \\ 0 & 1 \end{pmatrix} T_N(s)$ and apply the algorithm to diagonally decouple $T_N(s)$. Now $D_N(s) = \begin{pmatrix} s^2 & 0 \\ 0 & s \end{pmatrix}$ and take

$$\hat{D}_N(s) = \begin{pmatrix} (s+2)(s+3) & 0 \\ 0 & s+1 \end{pmatrix}.$$

A minimal (controllable and observable) realization of $\tilde{T}_N(s) = \hat{D}_N(s)T_N(s)$ is $\{A, B, \tilde{C}_N, B_N^\star\}$ where

$$A = \begin{pmatrix} 0 & 1 & 0 \\ 0 & 0 & 0 \\ -1 & 0 & 1 \end{pmatrix}, \quad B = \begin{pmatrix} 0 & 0 \\ 1 & 0 \\ 0 & 1 \end{pmatrix},$$

$$\tilde{C}_N = \begin{pmatrix} 6 & 5 & 0; \\ 2 & 1 & -2 \end{pmatrix}$$

and

$$B_N^\star = \begin{pmatrix} 1 & 0 \\ 0 & -1 \end{pmatrix} = B^\star.$$

In view now of Equations 50.4 and 50.5, for $F = -(B^\star)^{-1}\tilde{C}_N = \begin{pmatrix} -6 & -5 & 0 \\ 2 & 1 & -2 \end{pmatrix}$ and

$$\begin{aligned} G &= (B_N^\star)^{-1} = \begin{pmatrix} 1 & 0 \\ 0 & -1 \end{pmatrix}, \\ (T_N)_{F,G}(s) &= \hat{D}_N(s)^{-1} = \begin{pmatrix} \frac{1}{(s+2)(s+3)} & 0 \\ 0 & \frac{1}{s+1} \end{pmatrix}. \end{aligned}$$

If now this state feeback is applied to the minimal realization $\{A, B, C\}$ of $T(s)$—note that A, B are the same as above—then

$$T_{F,G}(s) = \begin{pmatrix} s-1 & 0 \\ 0 & 1 \end{pmatrix} \hat{D}_N^{-1} = \begin{pmatrix} \frac{s-1}{(s+2)(s+3)} & 0 \\ 0 & \frac{1}{s+1} \end{pmatrix}.$$

Note that the unstable noncoupling transmission zero at +1 appears on the diagonal of the compensated system; the closed-loop eigenvalues are at the arbitrarily chosen stable locations -1,-2 and -3.

Algorithm to Obtain a Proper Stable Right Inverse Using State Feedback

Let $\dot{x} = Ax + Bu$, $y = Cx + Du$ with A, B, C, D $n \times n$, $n \times m$, $p \times n$, $p \times m$ real matrices, respectively, and assume that the system is controllable and observable. Let $T(s)$ be its transfer function matrix. It is known that there exists a proper right inverse $T_R(s)$, such that $T(s)T_R(s) = I_p$, if and only if $rank D = p$. If, in addition, all the zeros of $T(s)$ (that is, the transmission zeros of the system) are stable, then a stable right inverse of order n can be constructed with $k(< n)$ of its poles equal to the k stable zeros of $T(s)$ with the remaining $n-k$ poles arbitrarily assignable. This can be accomplished as follows:

Let $T_{eq} = F[sI - (A+BF)]^{-1}BG + G$ where F, G are $n \times m$, $m \times p$, respectively, and note that

$$\begin{aligned} T(s)T_{eq}(s) &= [C(sI-A)^{-1}B + D] \\ &\quad [F[sI-(A+BF)]^{-1}BG + G] \\ &= (C+DF)[sI-(A+BF)]^{-1}BG + DG \\ &= T_{F,G}(s) \end{aligned} \qquad (50.6)$$

which is the transfer matrix one obtains when the state feedback control law $u = Fx + Gr$ is applied to the given system. Note that the second line of Equation 50.6 results from application of a well-known formula for the matrix inverse. If now F, G are such that

$$C + DF = 0, \quad DG = I_p \qquad (50.7)$$

then $T_{F,G}(s) = I_p$ and T_{eq} is a proper right inverse $T_R(s)$. The additional freedom in the choice of F when $p < m$ is now used to derive a stable inverse; when $p = m$, F, G are uniquely determined from $F = -D^{-1}C$, $G = D^{-1}$.

If the nonsingular $m \times m$ matrix M is such that $DM = (I_p \;\; 0)$, then $C + DF = C + DMM^{-1}F = C + (I_p \;\; 0)\begin{pmatrix} \hat{F}_1 \\ \hat{F}_2 \end{pmatrix} = 0$ from which $F = M\begin{pmatrix} -C \\ \hat{F}_2 \end{pmatrix}$ with $\hat{F}_2$ arbitrary. Also, from $DG = DMM^{-1}G = I_p$, $G = M\begin{pmatrix} I_p \\ \hat{G}_2 \end{pmatrix}$ with $\hat{G}_2$ arbitrary. The eigenvalues of $A + BF = A + BM\begin{pmatrix} -C \\ \hat{F}_2 \end{pmatrix} = A + (\hat{B}_1 \;\; \hat{B}_2)\begin{pmatrix} -C \\ \hat{F}_2 \end{pmatrix} = A - \hat{B}_1C + \hat{B}_2\hat{F}_2$ are the poles of $T_R(s)$. It can be shown that the uncontrollable eigenvalues of $(A - \hat{B}_1C \;\; \hat{B}_2)$ are exactly the (k) zeros of the system; they cannot be changed via $\hat{F}_2$. The remaining $n-k$ controllable eigenvalues can be arbitrarily assigned using $\hat{F}_2$. In summary, the steps to derive a stable proper inverse are

Step 1. Find nonsingular $m \times m$ matrix M such that $DM = (I_p \;\; 0)$.

Step 2. Calculate $(\hat{B}_1 \;\; \hat{B}_2) = BM$, and $A - \hat{B}_1C$.

Step 3. Find $\hat{F}_2$ that assigns the controllable eigenvalues of $(A - \hat{B}_1C \;\; \hat{B}_2)$ to the desired locations. The remaining uncontrollable eigenvalues are the stable zeros of the system.

Step 4.

$$\left\{ A + BM\begin{pmatrix} -C \\ \hat{F}_2 \end{pmatrix}, BM\begin{pmatrix} I_p \\ \hat{G}_2 \end{pmatrix}, M\begin{pmatrix} -C \\ \hat{F}_2 \end{pmatrix}, M\begin{pmatrix} I_p \\ \hat{G}_2 \end{pmatrix} \right\} \qquad (50.8)$$

where $\hat{G}_2$, a $(m-p) \times p$ arbitrary real matrix, is a stable right inverse.

$T_{eq}(s)$ above is the open-loop equivalent to the state feedback control law. In view of Equations 50.6 and 50.7 the above algorithm selects F, G in a state feedback control law $u = Fx + Gr$

so that the closed-loop transfer matrix $T_{F,G}(s) = I_p$ and the closed-loop system is internally stable; that is, all the n eigenvalues of $A + BF$ are stable. Note that when $p = m$, then F, G are uniquely given by $F = -D^{-1}C$, $G = D^{-1}$; the eigenvalues of $A+BF$ are then the n zeros of the system. In this development of stable inverses via state feedback, the approach in [1] was taken; see also [12] and [7].

In order to implement decoupling by state feedback in practice, it is often necessary to estimate the internal state variables by means of an observer. Certain plants can be decoupled by *constant output feedback*, avoiding the need for an observer. The necessary and sufficient conditions under which this is possible were proved by Wolovich [18]: it is that $B^\star$ not only be nonsingular, but also that the modified inverse transfer matrix $B^\star T^{-1}(s)$ have only constant off-diagonal elements. This appears to be a very stringent condition, so diagonal decoupling by means of constant output feedback is not likely to be possible for any but a relatively small class of plants. This is in clear contrast with the state feedback case, as mentioned previously. If diagonal decoupling by output feedback is possible, any gain matrix H that achieves it must give all off-diagonal entries of $B^\star H$ equal to those of $B^\star T^{-1}(s)$. It can therefore be seen that any constant matrix of the form $(B^\star)^{-1}Z$ can be added to H, where Z is an arbitrary diagonal matrix, and still give a gain matrix that satisfies the required condition. There is thus a small amount of controller design freedom available, which can be used, for instance, to assign closed-loop poles to some extent. However, it does not appear possible to quantify this pole-placement freedom in any straightforward manner.

50.1.3 Block Decoupling

If diagonal decoupling by linear state feedback is not possible, an alternative to applying precompensation may still exist. It may be possible to use state feedback, or perhaps even output feedback, to reduce the system to a set of smaller subsystems that are independent; that is, decoupled. Controlling each of these small systems can then be performed in isolation from all the others, thus reducing the original plant control problem to several simpler ones. This is the idea behind block decoupling, where the goal is to transform the plant transfer matrix to one that is block diagonal rather than strictly diagonal. For square plants, each of these k diagonal blocks will also be square: the i-th will be taken to have p_i inputs and p_i outputs, with $\sum p_i = p$.

One question associated with block decoupling can be answered immediately: namely, any plant with nonsingular transfer matrix can be block decoupled by linear state feedback plus precompensation. This follows from the fact that any such system can be diagonally decoupled by this form of compensation and so is trivially of any desired block diagonal form. The two types of compensation that have to be addressed here are therefore state feedback and constant output feedback.

If we are interested in block decoupling a given system by state feedback, this implies that it was not fully diagonalizable by state feedback. Hence, the matrix $B^\star$ must have been singular. As the inverse of this matrix played a significant role in the development of diagonal decoupling compensators, it seems likely that overcoming this singularity may lead toward designing block decoupling compensators for systems that cannot be diagonalized by state feedback. An equivalent way of stating that $B^\star$ is singular is to note that, although all rows of $\lim_{s\to\infty} D(s)T(s)$ are certainly finite and nonzero, some of these rows must have been linearly dependent on the preceding ones. Suppose the i-th row is one such. It is then possible to add multiples of rows $1, \ldots, i-1$ to row i in order to zero out the i-th row in $B^\star$; that is, to make what had been the leading coefficient vector of this row of $D(s)T(s)$ zero. If the new leading term in this row is now of order s^{-k}, multiplying the row by s^k yields a new finite and nonzero limit as s goes to infinity. If this row vector is independent of the preceding ones, we have now increased the rank of the modified $B^\star$-like matrix; if not, the same process can be repeated until successful. This basic procedure leads to the following definition, which has proved to be very useful for studying block decoupling problems.

The interactor $X_T(s)$ of $T(s)$ is the unique polynomial matrix of the form $X_T(s) = H(s)\Delta(s)$, where $\Delta(s) = diag(s^{\hat{f}_i})$ and $H(s)$ is a lower triangular polynomial matrix with 1s on the diagonal and the nonzero off-diagonal elements divisible by s, for which

$$\lim_{s\to\infty} X_T(s)T(s) = K_T \tag{50.9}$$

is finite and full rank. The interactor can be found from the transfer matrix of the system [16]; from a state-space representation [4]; or from a polynomial matrix fraction description for it [13]. The basic procedure can be illustrated by applying it to two examples discussed previously.

EXAMPLE 50.6:

$$T(s) = \begin{pmatrix} \frac{1}{s} & \frac{2}{s+1} \\ \frac{4}{s+3} & \frac{8s}{s+4} \end{pmatrix}$$

. This gives $B^\star = \begin{pmatrix} 1 & 2 \\ 0 & 8 \end{pmatrix}$, with decoupling indices $f_1 = 1$, $f_2 = 0$. $B^\star$ is nonsingular, so it satisfies the definition of the desired matrix K_T. Thus, $K_T = B^\star = \begin{pmatrix} 1 & 2 \\ 0 & 8 \end{pmatrix}$ here, and $X_T(s) = diag(s^{f_1}, s^{f_2}) = \begin{pmatrix} s & 0 \\ 0 & 1 \end{pmatrix}$.

EXAMPLE 50.7:

$$T(s) = \begin{pmatrix} \frac{1}{s} & \frac{2}{s+1} \\ \frac{4}{s+3} & \frac{8}{s+4} \end{pmatrix}$$

$B^\star = \begin{pmatrix} 1 & 2 \\ 4 & 8 \end{pmatrix}$, which is singular, with decoupling indices $f_1 = 1$, $f_2 = 1$. Subtracting 4 times row 1 of $diag(s^{f_i})T(s)$ from row 2 eliminates the linearly dependent leading coefficient vector. The resulting lower-degree polynomial row vector can then be multiplied by s, so as to again obtain a finite limit as s goes to infinity. We then have

$$\hat{T}_1(s) = \begin{pmatrix} 1 & 0 \\ 0 & s \end{pmatrix}\begin{pmatrix} 1 & 0 \\ -4 & 1 \end{pmatrix}\begin{pmatrix} s & 0 \\ 0 & s \end{pmatrix}$$

$$T(s) = \begin{pmatrix} 1 & \frac{2s}{s+1} \\ \frac{-12s}{s+3} & \frac{-24s^2}{(s+1)(s+4)} \end{pmatrix}.$$

Unfortunately, $\hat{T}_1(s)$ has limit as s goes to infinity of

$$\begin{pmatrix} 1 & 2 \\ -12 & -24 \end{pmatrix},$$

which is still singular. We therefore have to repeat the procedure, this time adding 12 times row 1 to row 2 to eliminate the leading coefficients and multiplying the resulting row by s to give it a finite limit. We then obtain $\begin{pmatrix} 1 & 0 \\ 0 & s \end{pmatrix}\begin{pmatrix} 1 & 0 \\ 12 & 1 \end{pmatrix}\hat{T}_1(s) = \begin{pmatrix} 1 & \frac{2s}{s+1} \\ \frac{36s}{s+3} & \frac{96s^2}{(s+1)(s+4)} \end{pmatrix}$, which has limit as s goes to infinity of $\begin{pmatrix} 1 & 2 \\ 36 & 96 \end{pmatrix}$. This is clearly nonsingular, so $K_T = \begin{pmatrix} 1 & 2 \\ 36 & 96 \end{pmatrix}$ for this system. The interactor is then

$$\begin{aligned} X_T(s) &= \begin{pmatrix} 1 & 0 \\ 0 & s \end{pmatrix}\begin{pmatrix} 1 & 0 \\ 12 & 1 \end{pmatrix}\begin{pmatrix} 1 & 0 \\ 0 & s \end{pmatrix} \\ &\quad \begin{pmatrix} 1 & 0 \\ -4 & 1 \end{pmatrix}\begin{pmatrix} s & 0 \\ 0 & s \end{pmatrix} \\ &= \begin{pmatrix} s & 0 \\ -4s^3 + 12s^2 & s^3 \end{pmatrix} \\ &= \begin{pmatrix} 1 & 0 \\ -4s^2 + 12s & 1 \end{pmatrix}\begin{pmatrix} s & 0 \\ 0 & s^3 \end{pmatrix} \end{aligned}$$

which is of the desired form $H(s)\Delta(s)$.

It can be seen that, if $B^\star$ is nonsingular, no additional row operations are needed to modify it to give the nonsingular K_T. Thus, in this case $B^\star = K_T$ and $D(s) = X_T(s)$. But we already know that diagonal decoupling by state feedback is possible in this case; that is, diagonalization by state feedback is possible if and only if the interactor of the system is diagonal. This suggests the following general result.

A square system can be block decoupled by state feedback if and only if its interactor is of this same block diagonal structure.

A proof of this result is based on the fact that state feedback matrices F, G can always be found that make the closed-loop transfer matrix equal to the inverse of its interactor; see the algorithms discussed previously and [6], [2]. Thus, if this matrix is block diagonal, so is the closed-loop transfer matrix. The state feedback that achieves this form can be found in an analogous manner to the state feedback matrices determined above that diagonally decouple the system.

Note that the structure algorithm of Silverman [11] is quite closely related to the interactor. This method determines a polynomial matrix $X(s)$ such that $\lim_{s\to\infty} X(s)T(s)$ is finite and nonsingular; however, $X(s)$ is not of any particular structure, unlike the interactor. This makes $X_T(s)$ better suited to obtaining clear block decoupling results.

Another question that generalizes naturally from the diagonal case is that of stability. The only fixed modes when diagonalizing were the diagonal coupling zeros, which were all zeros of the original plant that were not also zeros of any of the rows of the plant transfer matrix. In the case of block decoupling, the only fixed poles are the block coupling zeros, which are all zeros of the plant that are not also zeros of one of the $(p_i \times m)$ row blocks of $T(s)$. As in the diagonal case, these zeros must be cancelled by closed-loop poles in the decoupled transfer matrix, so creating unobservable modes; all other poles can be assigned arbitrarily.

Finally, it may be possible to achieve block decoupling by the simpler constant output feedback compensation. It can be shown that the interactor also allows a simple test for this question. In fact, block decoupling by constant output feedback is possible if and only if the interactor of the system is block diagonal and the modified inverse system $K_T T^{-1}(s)$ has only constant entries outside the diagonal blocks. The output feedback gain matrix H that achieves block decoupling is such that $K_T H$ is equal to the constant term in $K_T T^{-1}(s)$ outside the diagonal blocks. This is very similar to the diagonal decoupling result. As there, a certain degree of flexibility exists in the design of H, due to the fact that the diagonal blocks of $K_T T^{-1}(s)$ are essentially arbitrary; this can be used to provide a small amount of pole assignment flexibility when decoupling.

50.1.4 Decoupling Nonsquare Systems

The previous development has been primarily for plants with equal numbers of inputs and outputs. Plants that are not square present additional complications when studying decoupling. For instance, if there are more outputs than inputs, it is clearly impossible to assign a separate input to control each output individually; diagonal decoupling in its standard form is therefore not feasible. Similarly, decoupling the system into several independent square subsystems is also impossible. On the other hand, plants with more inputs than outputs present the opposite difficulty: there are now more input variables than are required to control each output individually.

Fortunately, the classical decoupling problem can be generalized in a straightforward fashion to cover nonsquare plants as well as square ones. In view of the preceding remarks, it is clear that systems with more outputs than inputs ($p > m$) must be analyzed separately from those with more inputs than outputs ($p < m$). The former case leads to decoupling results that are barely more complicated than those for the square case; the additional design freedom available in the latter case means that conditions that were necessary and sufficient for $p = m$ become only sufficient for $p < m$.

Taking the case of more inputs than outputs ($p < m$), the following results can be shown to hold for diagonal decoupling. First, any such plant that is right-invertible (that is, for which the transfer matrix is of full rank, p) can be decoupled by state feedback plus precompensation; this follows from the close connections between this type of decoupling control law and finding a right inverse of the system. If we restrict ourselves to state feedback, two sufficient conditions for diagonal decoupling can be stated. First, the plant can be diagonalized by state feedback if its matrix $B^\star$ is of full row rank, p. This is extremely easy to test, but can be somewhat conservative. A tighter sufficient condition is as follows: The plant can be diagonalized by state feedback if a constant $(m \times p)$ matrix G can be found for which the $B^\star$ matrix of the square-modified transfer matrix $T(s)G$ is nonsingular.

It may be thought that these two sufficient conditions are identical. To see that they are not, consider the following simple example: $T(s) = \begin{pmatrix} \frac{1}{s} & \frac{1}{s^2} & \frac{1}{s} \\ \frac{s+2}{s^2} & \frac{1}{s^2} & \frac{1}{s} \end{pmatrix}$ has $B^\star = \begin{pmatrix} 1 & 0 & 1 \\ 1 & 0 & 1 \end{pmatrix}$, which has only rank 1. The first sufficient condition for diagonal decoupling is therefore violated. However, post-multiplying $T(s)$ by the matrix

$$G = \begin{pmatrix} 1 & 0 \\ 0 & 1 \\ -1 & 0 \end{pmatrix}$$

gives

$$T(s)G = \frac{1}{s^2}\begin{pmatrix} 0 & 1 \\ 2 & 1 \end{pmatrix},$$

which clearly has nonsingular $B^\star$ matrix of $\begin{pmatrix} 0 & 1 \\ 2 & 1 \end{pmatrix}$. The role of the G matrix is basically to cancel those higher-power terms in s in $T(s)$ that give rise to linearly dependent rows in $B^\star$; in the example, the first column of G, $(1\ 0\ -1)^T$, can be seen to be orthogonal to the repeated row vector $[1\ 0\ 1]$ in the original $B^\star$. Lower-power terms in $T(s)$ then become the leading terms, so their coefficients contribute to the new $B^\star$; these terms may well be independent of the first ones. An algorithm that goes into the details of constructing such a G, if it exists, for any right-invertible $T(s)$ is given by Williams [13].

Very similar results apply to the problem of block decoupling a system with more inputs than outputs ($p < m$) by means of state feedback. The more conservative sufficient condition states that the plant can be block decoupled if its interactor matrix has the desired block diagonal structure. This can then be tightened somewhat by proving that the plant $T(s)$ can be block decoupled if there exists some $(m \times p)$ constant matrix G that has interactor of the desired block diagonal form. Furthermore, the algorithm described previously for block decoupling of square plants can be applied equally in this case, either to $T(s)$ or $T(s)G$. The only distinction of significance between the square case and $p < m$ is that the algorithm was proved to use decoupling precompensation of lowest possible order in the square case; for nonsquare plants, minimality of this order cannot be proven.

In the case of plants with more outputs than inputs ($p > m$), the main complication is in modifying the definition of a "decoupled" closed-loop structure. Once this is done, the actual technical results are rather straightforward. As already noted, it is no longer possible to assign a single input to each individual output, as is required in the classical diagonal decoupling problem. The closest analog to this problem is one where the closed-loop system is decoupled into a set of m independent single-input/multi-output subsystems; each closed-loop control input influences a set of outputs, but does not affect any of the others. Similarly, it is not possible to assign equal numbers of independent inputs and outputs to each decoupled subsystem, as holds for square block decoupling. What we must do instead is to define decoupled subsystems that generally have more outputs than inputs; that is, they are of dimensions $p_i \times m_i$, where $p_i \leq m_i$; of course, $\sum p_i = p \leq \sum m_i = m$.

It can be shown that a very simple rank condition on the plant transfer matrix determines whether or not these decoupling problems have a solution. The simplest question to answer is whether the desired decoupled structure is achievable by means of a combination of state feedback and precompensation. The test is as follows:

Take the p_i rows of the plant transfer function corresponding to the outputs that are to be assigned to the i-th decoupled subsystem. If this $p_i \times m$ transfer matrix has rank m_i, and this holds for each i, then the plant can be decoupled into $p_i \times m_i$ subsystems by means of state feedback plus precompensation.

The significance of this result is easiest to see for the special case where $m_i = 1$ for each i, the closest analog to diagonal decoupling for systems with $p > m$. If decoupling is to be possible, we must have that each $p_i \times m$ transfer matrix of the i-th subsystem is of rank 1. This implies that the rows of this transfer matrix are all polynomial multiples of some common factor row vector. In other words, the p_i outputs of this subsystem are all made up of combinations of derivatives of a single "underlying" output variable. Similarly, decoupling into $p_i \times m_i$ subsystems is possible if and only if the p_i outputs making up the i-th subsystem are actually made up of some combinations of m_i "underlying" output variables.

In practice, applying these rank conditions to the plant transfer matrix dictates what block dimensions are possible as closed-loop block decoupled structure. They also show which outputs must be taken as members of the same decoupled subsystem. For instance, if we wish to achieve $p_i \times 1$ decoupling and two rows of the plant transfer matrix are linearly dependent, the corresponding outputs must clearly be placed in the same subsystem.

But this approach also has one further very important implication. Consider a system that satisfies these submatrix rank conditions. If we take the m_i "underlying" output variables for each of the r subsystems, write down the corresponding $m_i \times m$ transfer matrix, and then concatenate these, we obtain a new $m \times m$ transfer matrix, denoted by $T_m(s)$. It can then be shown (see [13]) that a controller will decouple $T(s)$ into $p_i \times m_i$ blocks if and only if it also decouples $T_m(s)$ into square $m_i \times m_i$ blocks. We can therefore take all of the decoupling results derived previously for square plants and use them to solve the problem of decoupling systems with more outputs than inputs. In particular, $T(s)$ can be decoupled into $p_i \times m_i$ blocks by state feedback if and only if it satisfies the submatrix rank conditions and the interactor matrix of $T_m(s)$ is $m_i \times m_i$ block diagonal. Also, it can be shown that $T_m(s)$ has precisely the same zeros as $T(s)$. The two systems therefore clearly also have the same coupling zeros, so the fixed poles when decoupling $T(s)$ are the same as the fixed poles when decoupling $T_m(s)$. Finally, decoupling by means of output feedback can also be studied by applying the existing results for square systems to the associated $T_m(s)$.

EXAMPLE 50.8:

The state-space model

$$A = \begin{pmatrix} 0 & 0 & 1 \\ 1 & 0 & 1 \\ 1 & 1 & 0 \end{pmatrix},$$

$$B = \begin{pmatrix} 1 & 0 \\ 1 & 0 \\ 1 & 1 \end{pmatrix}, \quad C = I_3$$

has transfer matrix

$$T(s) = \frac{1}{(s+1)(s^2-s-1)} \begin{pmatrix} s(s+1) & s \\ (s+1)^2 & s+1 \\ (s+1)^2 & s^2 \end{pmatrix}.$$

Clearly, the first two rows are linearly dependent, so this system can be decoupled into the block diagonal form $\begin{pmatrix} \star & 0 \\ \star & 0 \\ 0 & \star \end{pmatrix}$ by state feedback plus precompensation. In fact, the associated invertible transfer function for this system is

$$T_m(s) = \frac{1}{(s+1)(s^2-s-1)} \begin{pmatrix} s+1 & 1 \\ (s+1)^2 & s^2 \end{pmatrix},$$

which has interactor $\begin{pmatrix} s^2 & 0 \\ 0 & s \end{pmatrix}$ diagonal [with $K_T = \begin{pmatrix} 1 & 0 \\ 1 & 1 \end{pmatrix}$]. Thus, block decoupling is actually possible for this system using state feedback alone.

As a final point on general block decoupling, note that this problem can also be studied using the geometric approach; see [19]. This state-space technique is based on considering the supremal (A, B)-invariant subspaces contained in the kernels of the various subsystems formed by deleting the outputs corresponding to each desired block in turn. The ranges of these subspaces determine whether decoupling is possible by state feedback. If it is not, the related "efficient extension" approach allows a precompensator of relatively low order to be found that will produce the desired block diagonal structure. This approach is somewhat involved, and the interested reader is referred to Morse and Wonham [8] for further details.

50.1.5 Triangular Decoupling

There is a form of "partially decoupled" system that can be of particular value for certain plants. This is the triangularized form, where all entries of the closed-loop transfer matrix above its leading diagonal are made zero. The first closed-loop output, y_1, is therefore affected only by the first input r_1; the second, y_2, is influenced only by inputs r_1 and r_2; etc. This type of transfer matrix can be used in the following sequential control scheme. First, input r_1 is adjusted until output y_1 is as desired, and the control is then frozen. Output variable y_2 is then affected only by r_2 and the fixed r_1, so r_2 can be adjusted until this output is also as desired. The third input, r_3, can then be used to set output y_3, etc. This scheme can be seen to be less powerful than diagonal decoupling, as the outputs must be adjusted sequentially rather than fully independently. However, it has one strong point in its favor: *any right-invertible plant can be triangularized by state feedback alone*, regardless of whether additional precompensation is required to make it diagonally decoupled. Proof of this follows directly from the fact that there always exists some state feedback gains F, G for which $T_{F,G}(s) = X_T^{-1}(s)$, and the interactor is, by definition, lower triangular. Of course, similar results apply for generalized rather than standard interactors also. Therefore, it can be shown, as originally proved by Morse and Wonham [9], that all closed-loop poles of the triangularly decoupled system can be arbitrarily assigned.

Finally, it may also be possible to triangularize a system by means of the simpler constant output feedback. If the original plant is square and strictly proper ($D = 0$), it can be shown that this is possible if and only if all entries of the modified inverse transfer matrix $K_T T^{-1}(s)$ that lie above the leading diagonal are constant. This is quite a simple condition to test and is very similar to the test for diagonal decoupling by output feedback. The required gain matrix H is given from the fact that the upper triangular part of $K_T H$ is precisely the upper triangular constant part of $K_T T^{-1}(s)$. It can be noted that there is therefore some non-uniqueness in the choice of the gain H: in particular, we can add a term of the form $K_T^{-1} Z$ to H, where Z is any lower triangular constant matrix, and still get a suitable output gain matrix. If it is possible to triangularize a given system by output feedback, there is consequently some freedom to assign closed-loop poles also. However, it is difficult to quantify this freedom in any concrete way.

50.1.6 Static Decoupling

Static decoupling, as opposed to dynamic decoupling already described, is much easier to achieve. A system is statically decoupled if a step change in the (static) steady-state level of the i-th input is reflected by a change in the steady-state level of the i-th output and only that output. To derive the conditions for static decoupling, assume that the system is described by a $p \times p$ transfer matrix $T(s)$ that is bounded-input/bounded-output stable; that is, all of its poles are in the open left half of the s-plane and none is on the imaginary axis. Note that stability is necessary for the steady-state values of the outputs to be well defined. Assume now that the p inputs are step functions described by $u_i(s) = \frac{k_i}{s}$, $i = 1, .., p$. The steady-state value of the output vector y, y_{ss}, can then be found using the final value theorem, as follows:

$$y_{ss} = \lim_{s\to\infty} y(t) = \lim_{s\to 0} sT(s)\frac{1}{s}\begin{pmatrix} k_1 \\ k_2 \\ \vdots \\ k_p \end{pmatrix} = T(0)\begin{pmatrix} k_1 \\ k_2 \\ \vdots \\ k_p \end{pmatrix} \tag{50.10}$$

It is now clear that $T(s)$ is statically decoupled if and only if $T(0)$ is a diagonal nonsingular matrix; that is, all the off-diagonal entries of $T(s)$ must be divisible by s, while the entries on the diagonal should not be divisible by s. It can be shown easily that a system described by a $p \times p$ transfer matrix $T(s)$ that is bounded-input/bounded-output stable can be statically decoupled, via $u = Gr$, if and only if

$$rank T(0) = p \tag{50.11}$$

that is, if and only if there is no transmission zero at $s = 0$. Note that this condition, if a controllable and observable state-space description is given, is

$$rank\begin{pmatrix} A & B \\ C & D \end{pmatrix} = n + p \tag{50.12}$$

If this is the case, any feedforward constant gain G, in $u = Gr$, such that $T(0)G$ is a diagonal and nonsingular matrix will statically decouple the system. To illustrate, consider the following example:

EXAMPLE 50.9:

$$T(s) = \begin{pmatrix} \frac{s+2}{s+1} & \frac{2}{s+3} \\ \frac{s(s+1)}{(s+3)^2} & \frac{1}{s+1} \end{pmatrix}$$

Here $T(0) = \begin{pmatrix} 2 & \frac{2}{3} \\ 0 & 1 \end{pmatrix}$, which has full rank; therefore, it can be statically decoupled. Let $T(0)G = \begin{pmatrix} 2 & 0 \\ 0 & 1 \end{pmatrix}$; then $G = \begin{pmatrix} 1 & \frac{-1}{3} \\ 0 & 1 \end{pmatrix}$. Note that

$$T(s)G = \begin{pmatrix} \frac{s+2}{s+1} & \frac{-s(s-1)}{3(s+1)(s+3)} \\ \frac{s(s+1)}{(s+3)^2} & \frac{-s^3+s^2+17s+27}{3(s+1)(s+3)^2} \end{pmatrix}$$

where all the off-diagonal entries of $T(s)$ are divisible by s, while the entries on the diagonal are not divisible by s. If now the input $\frac{1}{s}\begin{pmatrix} k_1 \\ k_2 \end{pmatrix}$ is applied to $T(s)G$, the steady-state output is $T(0)G\begin{pmatrix} k_1 \\ k_2 \end{pmatrix} = \begin{pmatrix} 2k_1 \\ k_2 \end{pmatrix}$.

50.2 Defining Terms

Decoupling: Separating the system into a number of independent subsystems.

Non-interacting control: The control inputs and the outputs can be partitioned into disjoint subsets; each subset of outputs is controlled by only one subset of inputs, and each subset of inputs affects only one subset of outputs. From an input/output viewpoint, the system is split into independent subsystems; it is called decoupled.

References

[1] Antsaklis, P.J., Stable proper n-th order inverses, *IEEE Trans. Autom. Control*, 23, 1104–1106, 1978.

[2] Antsaklis, P.J., Maximal order reduction and supremal (A,B)-invariant and controllability subspaces, *IEEE Trans. Automat. Control*, 25, 44–49, 1980.

[3] Falb, P.L. and Wolovich, W.A., Decoupling in the design and synthesis of multivariable control systems, *IEEE Trans. Autom. Control*, 12, 651–659, 1967.

[4] Furuta, K. and Kamiyama, S., State feedback and inverse system, *Intern. J. Control*, 25(2), 229–241, 1977.

[5] Gilbert, E.G., The decoupling of multivariable systems by state feedback, *SIAM J. Control*, 50–63, 1969.

[6] Kamiyama, S. and Furuta, K., Integral invertibility of linear time-invariant systems, *Intern. J. Control*, 25(3), 403–412, 1977.

[7] Moore, B.C. and Silverman, L.M., A new characterization of feedforward, delay-free inverses, *IEEE Trans. Inf. Theory*, 19, 126–129, 1973.

[8] Morse, A.S. and Wonham, W.M., Decoupling and pole assignment by dynamic compensation, *SIAM J. Control*, 317–337, 1970.

[9] Morse, A.S. and Wonham, W.M., Triangular decoupling of linear multivariable systems, *IEEE Trans. Autom. Control*, 447–449, 1970.

[10] Rekasius, Z.V., Decoupling of multivariable systems by means of state variable feedback, *Proc. 3rd Allerton Conf.*, 439–448, 1965.

[11] Silverman, L.M., Decoupling with state feedback and precompensation, *IEEE Trans. Autom. Control*, 15, 487–489, 1970.

[12] Silverman, L.M. and Payne, H.J., Input-output structure of linear systems with application to the decoupling problem, *SIAM J. Control*, 9, 188–233, 1971.

[13] Williams, T., Inverse and Decoupling Problems in Linear Systems, Ph.D. thesis, Imperial College, London, 1981.

[14] Williams, T. and Antsaklis, P.J., A unifying approach to the decoupling of linear multivariable systems, *Intern. J. Control*, 44(1), 181–201, 1986.

[15] Wolovich, W.A. and Falb, P.L., On the structure of multivariable systems, *SIAM J. Control*, 437–451, 1969.

[16] Wolovich, W.A. and Falb, P.L., Invariants and canonical forms under dynamic compensation, *SIAM J. Control*, 996–1008, 1976.

[17] Wolovich, W.A., *Linear Multivariable Systems*, Springer-Verlag, New York, 1974.

[18] Wolovich, W.A., Output feedback decoupling, *IEEE Trans. Autom. Control*, 148–149, 1975.

[19] Wonham, W.M., *Linear Multivariable Control: A Geometric Approach*, Springer-Verlag, New York, 1979.

Further Reading

Making the system diagonally dominant is a powerful design approach. Details on how to achieve diagonal dominance using Rosenbrock's Inverse Nyquist Array method can be found in Rosenbrock, H.H. 1974. *Computer-Aided Control System Design*, Academic Press, New York.

A good introduction to the geometric approach and to the decoupling problem using that approach can be found in Wohnam, W.M. 1985. *Linear Multivariable Control: A Geometric Approach*, Springer-Verlag, New York. The problem of disturbance decoupling or disturbance rejection, where a disturbance in the state equations must become unobservable from the output, is also studied there using the geometric approach.

A geometric approach has also been used to study noninteracting control in nonlinear systems; see, for example, Battilotti, S. 1994. *Noninteracting Control with Stability for Nonlinear Systems*, Springer-Verlag, New York.

For the decoupling of singular systems see, for example Paraskevopoulos, P.N. and Koumboulis, F.N. 1992. The decoupling of generalized state-space systems with state feedback, *IEEE Trans. Autom. Control*, pp. 148–152, vol.37.

The following journals report advances in all areas of decoupling including diagonal, block and triangular decoupling: *IEEE Transactions on Automatic Control*, *International Journal of Control* and *Automatica*.

51

Predictive Control

A.W. Pike, M.J. Grimble, M.A. Johnson, A.W. Ordys, and S. Shakoor
Industrial Control Centre, University of Strathclyde, Glasgow, Scotland, U.K.

51.1 Introduction

Control design and implementation uses methods which are either model-free or else are model-based. The autotune method of Haggland and Åström [22] is one example of a model-free method which has found wide industrial application. However, for more complex industrial systems, the use of a system model for fully understanding process behavior is essential. In this demanding design context, the control design method used is termed Model-Based Control (MBC). The class of such techniques is very wide and encompasses almost all of the modern approaches to control design.

Model-Based Predictive Control (MBPC) utilizes the available system model to incorporate the predicted future behavior of the process into the controller design procedure. This method of control design and implementation usually comprises

1. a process model; often a linear discrete system model obtained experimentally,
2. a predictor equation; this is run forward for a fixed number of time steps to predict the likely process behavior,
3. a known future reference trajectory, and
4. a cost function; this is usually quadratic and costs future process output errors (w.r.t the known reference) and controls.

Optimizing the cost function subject to future process outputs and controls leads to an explicit expression for the controls. However, one further capability of the method which has created extensive industrial interest in MBPC is that it can easily incorporate process operational constraints into the method. Constraint handling often provides the major economic return on the control of complex processes in the petroleum and chemical industries.

Model-based predictive control algorithms are versatile and robust in applications, outperforming minimum variance and PID in challenging control situations [10]. Systems which are difficult to control are open-loop unstable, have a nonminimum phase response, have dead-time, or are multivariable or highly nonlinear. In such cases conventional PID loops (which give perfectly good control performance in the large majority of industrial applications) provide poor control [33].

During the last decade, predictive control has been successful in industrial applications. Today, there are about 300 commissioned applications of predictive control [33] ranging from the petrochemical, steel, and glass industries to robot manipulators, aviation, and tracking systems. In the petrochemical industry, for example, the changing demands of the market often require the rapid production of different grades of high quality oil. Integral to industrial production will be the use of advanced controls which can 'adapt' to a wide range of conditions [19], hence optimizing the process. One requirement for economic operation is limiting the variances of end product quality. It is also necessary to operate machinery within physical constraints and health and safety regulations. These multiple operational constraints are handled by MBPC in a straightforward way, a major practical advantage of MBPC which other techniques, such as Linear Quadratic Gaussian (LQG) optimal control, do not possess in their basic form [18]. Garcia et al, [19] claimed that MBPC methods are currently the only methods to handle constraints in a systematic manner. Typical payback periods, from the use of

0-8493-8570-9/96/$0.00+$.50

Predictive Control schemes are less than a year. One industrial supplier cited a petroleum application with "savings of about one million dollars per year" [14]. Thus the attraction of MBPC for private industry is immediately seen because of potential financial gain coupled with improved environmental performance.

Over the years, several MBPC methods have evolved but the techniques have a common underlying framework [36]. The main differences between methods are the ways by which future predictions are calculated, namely, the type of model used and how the cost function is defined.

Model-based predictive control methods were first applied by Richalet et al. [31] via the development of the IDentification and COMmand algorithm (IDCOM). In this method, the model contains only moving average terms (only zeros, no poles, in the transfer function) hence restricting its use. This algorithm, was soon improved by the development of the Dynamic Matrix Control (DMC) algorithm of Cutler and Ramaker [15]. Following this there was a proliferation of reported predictive control methods. Some of the more well-known are

MAC: Model Algorithmic Control [31]
EHAC: Extended Horizon Adaptive Control [37]
EPSAC: Extended Predictive Self Adaptive Control [16]
GPC: Generalized Predictive Control [7, 8]
PFC: Performance Functional Control [32]
APCS: Adaptive Predictive Control System [26]

Some of these algorithms have been developed as commercial products, DMC™, IDCOM™, and Connoisseur™ being typical examples.

51.2 A Review of MBPC Methods

The origin of the MBPC technique can be traced historically from the class of Minimum Variance (MV) algorithms. The development of discrete-time, minimum-variance control began with the algorithm of Åström [1]. This controller minimizes the variance of the output signal from the system at the kth step in the future, where k is the estimated plant dead time. Denoting the current time by t, then the task is to choose the control u_t so that the variance of the output y_{t+k} is minimized:

$$u_t = \text{Min}\left(E\left\{{y_{t+k}}^2|t\right\}\right) \tag{51.1}$$

where $E\{.|t\}$ denotes the conditional *expectation operator*.

The next stage in the development of minimum-variance algorithms is the work of Hastings-James [23], which was employed by Clarke and Gawthrop [9] in their self-tuning algorithm. The innovation added a control cost to the performance index:

$$J_t = E\left\{\left({y_{t+k}}^2 + \lambda \cdot {u_t}^2\right)\right\}. \tag{51.2}$$

The resulting Generalized Minimum Variance (GMV) controller is more robust in many applications (especially for *non-minimum phase* plants). Subsequently, most *minimum-variance* regulators include a costing on control.

The quadratic cost function J_t of Equation 51.2 contains only one output term k steps into the future. Model-based predictive control algorithms extend this idea by using a cost function which includes output and control values over a range or horizon of future steps. Thus a typical MBPC cost function would be

$$J_t = E\left\{\left(\sum_{j=0}^{N}\left[y_{t+j+1}^2 + \lambda_j U_{t+j}^2\right]\right)|t\right\} \tag{51.3}$$

where t is the current time step, N is the number of time steps in the cost horizon, and λ_j is a control weighting.

The characteristics of different MBPC techniques often have their origin in the format or type of process model used. Thus it is necessary to understand the potential model types before discussing different methods.

51.2.1 Process Models for MBPC

Most real processes exhibit highly nonlinear dynamics. However, many predictive controllers are designed on the basis of a linear model. For a nonlinear process, a linear model can only be a locally valid approximation for the dynamics about the specified operating point. Thus it is necessary to update the linear model automatically as the operating point changes. In MBPC, the control is a function of the model parameters. If the model is continuously updated on-line, resulting in corresponding controller updates, then the control is termed *adaptive*. Many of the model-based predictive controllers are designed to be adaptive.

The model must capture the process dynamics with sufficient accuracy to insure good control performance. Because there will be some modeling error, the controller should be designed to be robust to a certain degree of uncertainty in the model parameters. Furthermore the model structure must have the potential to calculate predictions of future outputs. Two main approaches exist for obtaining a suitable linear model:

1. model identification procedures using process input/output data, known as *'black box modeling'*, and

2. mathematical modeling based on fundamental physical laws. The model, which is usually nonlinear, may then be linearized at different operating points.

The type of linear model derived depends on the application.

Finite-Impulse or Step-Response Models

For stable processes Finite-Impulse Response (FIR) or Finite-Step Response (FSR) models may be obtained. Typical plants that are open-loop stable, possessing relatively slow and simple dynamics, are often encountered in the petrochemical and chemical process industries. The generic FIR/FSR model is described by

$$y(k) = \sum_{j=0}^{n_r-1} \phi_j u(k-j-1) \tag{51.4}$$

where ϕ_j is the jth impulse-response or step-response coefficient, respectively, n_r is the truncation order of the model, $u(k-j-1)$ is the process input $j+1$ steps into the past, and $y(k)$ is the process output at the current kth step. For this type of model the implicit assumption is that ϕ_j becomes insignificant for $j > n_r$. However, only for strictly stable models does ϕ_j tend asymptotically to zero as j increases. Hence this form is unsuitable for unstable processes which would require an infinite number of ϕ_j parameters. The model is attractive because it is easily obtained by applying step-response tests to an industrial plant. This type of model is used in the Dynamic Matrix Control and Model Algorithmic Controller algorithms.

Transfer Function Models

A more general class of models is the transfer function model, of which Equation 51.4 is a special case. Both stable and unstable processes may be represented by

$$y(k) = \frac{q^{-d}B(q^{-1})}{A(q^{-1})}u(k-1). \tag{51.5}$$

If a model of the process disturbances is included then Equation 51.5 becomes

$$y(k) = \frac{q^{-d}B(q^{-1})}{A(q^{-1})}u(k-1) + \frac{C(q^{-1})}{D(q^{-1})}\zeta(k). \tag{51.6}$$

A, B, C, D are polynomials in the backward shift operator q^{-1} (that is, $q^{-1}x(k) = x(k-1)$):

$$\begin{aligned}
A(q^{-1}) &= 1 + a_1 q^{-1} + a_2 q^{-2} + \cdots + a_{n_A} q^{-n_A} \\
B(q^{-1}) &= b_0 + b_1 q^{-1} + b_2 q^{-2} + \cdots + b_{n_B} q^{-n_B} \\
C(q^{-1}) &= 1 + c_1 q^{-1} + c_2 q^{-2} + \cdots + c_{n_C} q^{-n_C} \\
D(q^{-1}) &= 1 + d_1 q^{-1} + d_2 q^{-2} + \cdots + d_{n_D} q^{-n_D}
\end{aligned}$$

where d is the process pure time delay in sampling intervals, and $\zeta(k)$ is a Gaussian *white noise* sequence with zero mean.

Note that if D is selected as $A(q^{-1})\Delta$ where $\Delta = (1 - q^{-1})$, this gives an integrated moving average for the process disturbances, which is appropriate for modeling *Wiener processes.* Equation 51.6 then describes a Controlled Auto Regressive and Integrated Moving Average (CARIMA) model. One natural advantage of the CARIMA formulation is that integral control action normally results (which eliminates *steady state offsets*), with the controller expressed in incremental form. It is important to obtain good estimates of the process dynamic order n_A and the time delay d for good control performance. The $C(q^{-1})$ polynomial can be selected to fit the actual disturbance model or can be treated as a design weighting to provide greater robustness to unmodeled process dynamics [10].

51.2.2 The Methods of Model Based Predictive Control

In this section some of the more well-known MBPC algorithms will be briefly discussed, highlighting their distinguishing features and their comparative advantages and disadvantages.

The Dynamic Matrix Controller (DMC)

The DMC algorithm uses an FSR model (as described in Section 51.2.1. The process is optimized through the following stochastic version of the cost function:

$$J_{DMC} = E\left\{\left(\sum_{j=N_1}^{N_2}[\hat{y}(t+j|t) - r(t+j)]^2 + \sum_{j=1}^{N_u}[\lambda\Delta u(t+j-1)]^2\right) | t\right\} \tag{51.7}$$

where $\hat{y}(t+j|t)$ is the predicted output j steps into the future based upon information available at time t, $r(t+j)$ is the reference signal j steps into the future, $\Delta u(t) = (1-q^{-1})u(t) = u(t) - u(t-1)$, and the DMC tuning parameters are N_1 minimum costing horizon, N_2 maximum costing horizon, N_u control horizon, and λ control weighting factor.

Each of the above parameters has a specific role in tuning of DMC algorithm. These tuning parameters do not exclusively belong to DMC, but are a characteristic feature of MBPC algorithms in general. Using the accumulated experience of applying predictive algorithms, a number of engineering rules have been identified to obtain appropriate values of the parameters for good performance in different applications [8]. For example, nothing is gained by costing future errors in Equation 51.7 that cannot be influenced by future control actions. Thus it is sensible to select N_1 equal to the plant dead time and N_2 no greater than the settling time of the plant. In fact, the recommended default value for N_2 is the plant rise time. The role of λ is to penalize excessive incremental control actions. The larger the value of λ, the more sluggish the control will become. The facility exists to set the control horizon N_u at less than the maximum costing horizon N_2. The integer N_u specifies the degrees of freedom in selecting future controls, so that, after N_u future sampling intervals, the control increments are assumed to be zero, giving a constant control signal. A basic rule for selecting N_u in GPC algorithms is to set it at least equal to the number of unstable poles in the plant. For stable systems, setting $N_u = 1$ usually gives acceptable control performance. Note also that the parameter set chosen has a direct influence on the size of matrices required to compute the optimal control, and thus on the amount of computation involved.

DMC Advantages

- Very easy to implement the model–simple calculations
- Attractive for use by industrial personnel without extensive training
- No assumption about the order of the process is required

Disadvantage

- Open-loop unstable processes cannot be modeled or controlled.

DMC can be made adaptive by including a recursive identification algorithm, which estimates an impulse-response model. The step response can subsequently be derived [17].

The DMC algorithm has been developed into a very successful commercial package with more than 200 applications, mainly in the petrochemical industry.

The Model Algorithmic Controller (MAC)

The MAC algorithm is similar to the DMC approach, but possesses fewer tuning parameters:

1. An FIR model is used
2. N_u is fixed equal to N_2
3. N_1 is fixed at 1

The tuning parameters are thus N_2 and λ. The cost function for this method is

$$J_{MAC} = E\left\{\left(\sum_{i=1}^{N_2}\left(e^T e + \lambda \Delta u^T u\right)\right) | t\right\} \tag{51.8}$$

where the superscript 'T' denotes the matrix transpose, and

$$e = r(t+i) - \hat{y}(t+i|t) - H\Delta u(t+i-1) \tag{51.9}$$

where

$$H = \begin{bmatrix} h(1) & 0 & \cdots & \cdots & 0 \\ h(2) & h(1) & 0 & \cdots & 0 \\ \vdots & & \ddots & \ddots & \vdots \\ \vdots & & & \ddots & 0 \\ h(N_2) & \cdots & \cdots & & h(1) \end{bmatrix}. \tag{51.10}$$

The H matrix contains the plant impulse response coefficients, where $w, u, \hat{y} \in R^{N_2}$ and $H \in R^{N_2 X N_2}$.

Jacques Richalet was a pioneer of the application of MBPC techniques through the development of the IDCOM and subsequently the MAC controllers. More recently, a more advanced controller, PFC, has evolved from the original concepts. The PFC algorithm is available in the MATLAB environment as an MBPC toolbox. As a result, the PFC toolbox can be used in conjunction with the extensive MATLAB libraries.

Reported applications [13] include defense, chemical batch reactors, automobile, steel industry, robots, and river dam control.

The Generalized Predictive Controller (GPC)

The output predictions of the GPC controller are based upon using a CARIMA model (as described in Section 51.2.1). The GPC algorithm (as with all algorithms using transfer function models), can easily be implemented in an adaptive mode by using an on-line estimation algorithm such as recursive least squares. This algorithm will be illustrated in more detail subsequently. The cost function used in the GPC algorithm is

$$J_{GPC} = E\left\{\left(\sum_{i=N_1}^{N_2}[\hat{y}(t+i|t) - r(t+i)]^2 + \sum_{i=1}^{N_u}\lambda\Delta u^2(t+i-1)\right)|t\right\} \tag{51.11}$$

GPC Advantages

- The GPC is normally able to stabilize and control open-loop unstable processes through judicious choice of the tuning parameters N_1, N_2, λ, and N_u.
- The GPC approach is related to the properties of LQ control [4].
- The theoretical basis of the GPC algorithm has been developed. It has been shown by Clarke & Mohtadi [10], that, for limiting cases of parameter choices, the GPC algorithm is stable and also that certain well-known controllers (such as mean level and state deadbeat control) are inherent in the GPC structure.

Disadvantages

- There are no guaranteed stability properties for GPC except under special conditions.
- In general, all basic MBPC algorithms suffer from this drawback.
- Care must be taken in tuning parameter choices. If $N_u = N_2$ and the control weighting λ is set to zero, then GPC reduces to a Minimum Variance controller [20], which is known to be unstable on nonminimum phase processes.

The Extended Self-Adaptive Controller (EPSAC)

This method uses a CARIMA model, similar to that of the GPC algorithm, for prediction purposes. The optimal control law is found by minimizing

$$J_{EPSAC} = E\left\{\sum_{i=1}^{N_2}\rho(i)\left[\hat{y}(t+i|t) - P(q^{-1})r(t+i)\right]^2 |t\right\} \tag{51.12}$$

where $P(q^{-1})$ is a design polynomial and $\rho(i)$ an exponential weighting factor. Three factors are worth noting :

- The prefilter, $P(q^{-1})$, can be used as a predesign parameter to affect the disturbance rejection properties.
- Open-loop unstable plants can be controlled by appropriately selecting $P(q^{-1})$, N_2 and $\rho(i)$, however, the tuning is more involved.
- The absence of the control signal from the cost function implies that undesired large control signal variations cannot be suppressed.

The Extended-Horizon Adaptive Controller (EHAC)

The EHAC algorithm assumes an ARMA model of the form,

$$A(q^{-1})y(t) = B(q^{-1})u(t-d), \tag{51.13}$$

where d is time delay. Using this model, the optimal controls are found by minimizing the cost function,

$$J_{EHAC} = E\{[y(t+N_2) - r(t+N_2)]^2 |t\} \tag{51.14}$$

subject to

$$E\{y(t+N_2)-r(t+N_2)\}=0. \qquad (51.15)$$

The EHAC approach is to compute a sequence of inputs $[u(t), u(t+1), \ldots, u(t+N_2-d)]$, subject to the constraint Equation 51.15, at every instant. The solution to Equation 51.15 is *not* unique unless $N_2 = d$, [17] resulting in a number of different ways to finding the control sequence. Possible approaches include assuming that the control is constant over an interval, so that,

$$u(t)=u(t+1)=\cdots=u(t+N_2-d) \qquad (51.16)$$

or choosing some strategy where the control effort is minimized:

$$J=\sum_{i=0}^{N_2-d} u^2(t+i). \qquad (51.17)$$

EHAC Disadvantages

- Because only one tuning parameter is involved, a compromise between closed-loop performance and stability must be made.
- Finding the optimal control law is more involved when compared to the other MBPC methods.

51.3 A Tutorial Example: Unconstrained GPC

As has already been noted, there are many different formulations of predictive control. For tutorial purposes it will be sufficient to develop one of these. The Generalized Predictive Controller (GPC) [8], is among the most popular and possesses the main characteristics of predictive controllers. It is sufficiently 'general' so that it may be applied to a wide class of plants, including those that are nonminimum phase and/or open-loop unstable.

In the original paper [8], the GPC algorithm was developed in the polynomial domain. This approach will be adopted here, although the algorithm can also be developed in state space [29].

51.3.1 Derivation of the Optimal Predictor

The GPC algorithm uses a transfer function model of the form,

$$A(q^{-1})y(t)=B(q^{-1})u(t-1)+\frac{C(q^{-1})}{\Delta}\zeta(t), \qquad (51.18)$$

(all symbols having their usual meaning, and are as defined in Equation 51.6). Rearranging gives

$$A(q^{-1})\Delta y(t)=B(q^{-1})\Delta u(t-1)+C(q^{-1})\zeta(t). \qquad (51.19)$$

Now introduce the following *Diophantine equation,*

$$E_j(q^{-1})A(q^{-1})\Delta+q^{-j}F_j(q^{-1})=C(q^{-1}). \qquad (51.20)$$

The polynomials $E_j(q^{-1})$ and $F_j(q^{-1})$ are uniquely defined in Equation 51.20 and are of order $j-1$ and n_A, respectively. The solution can be found by a computationally efficient recursive algorithm [20]. Note from Eqtuation 51.20 that $E_j(q^{-1})$ represents the first j terms in the plant response to an impulse disturbance. To simplify the notation, the arguments q^{-1} will be omitted in the following analysis, and the algebra will be simplified by assuming $C(q^{-1})=1$. Equation 51.20 enables Equation 51.19 to be split into past and future terms, facilitating the derivation of the optimal j step ahead predictor. To derive the predictor, first multiply Equation 51.19 by E_jq^j to obtain

$$E_jA\Delta y(t+j)=E_jB\Delta u(t-1+j)+E_j\zeta(t+j). \qquad (51.21)$$

Substituting Equation 51.20 into Equation 51.21 yields

$$y(t+j)=E_jB\Delta u(t-1+j)+F_jy(t)+E_j\zeta(t+j). \qquad (51.22)$$

Thus the optimal predictor is obtained from Equation 51.22 by noting that the expected value of the future noise signal component $E_j\zeta(t+j)$ is zero by the definition of ζ. Thus the optimal predictor follows as

$$\hat{y}(t+j|t)=G_j\Delta u(t+j-1)+F_jy(t) \qquad (51.23)$$

where

$$G_j=EB=\frac{B\left[1-q^{-j}F_j\right]}{\Delta A}. \qquad (51.24)$$

From Equation 51.24 the first j terms of the G_j polynomial are the first j terms of the plant step response.

51.3.2 Derivation of the Optimal Control

By letting $f(t+j)$ denote the signals *that are known at time* $t+j$, then the predictor Equation 51.23 can be reformulated as follows:

$$\begin{aligned}
y(t+j) &= G_j\Delta u(t+j-1)+F_jy(t),\\
&= \left(g_{j0}+g_{j1}q^{-1}+\cdots+g_{jj}q^{-j}+\cdots\right.\\
&\quad \left.+g_{jn_{B+j-1}}q^{-(n_{B+j-1})}\right)\Delta u(t+j-1)\\
&\quad + F_jy(t),\\
&= \left(g_{j0}+g_{j1}q^{-1}+\cdots+g_{jj-1}q^{-j+1}\right)\\
&\quad \Delta u(t+j-1)\\
&\quad \text{(future, `forced' response component)}\\
&\quad +\left(g_{jj}q^{-j}+\cdots+g_{jn_{B+j-1}}q^{-(n_{B+j-1})}\right)\\
&\quad \Delta u(t+j-1)+F_jy(t),\\
&\quad \text{(past, `free' response component)}\\
&= G_j^*\Delta u(t+j-1)+f(t+j). \qquad (51.25)
\end{aligned}$$

Thus the predicted future output contains two components. The first component is the free response due to all past influences acting on the system. The second component is the forced response due to the future control actions. Stacking the output predictions for $j=1$ to N_p, results in the matrix equation

$$\hat{y}_{t,N_p}=\Gamma_{N_p}\Delta u_{t,N_p}+f_{t,N_p} \qquad (51.26)$$

where

$$\hat{y}_{t,N_p} = [\hat{y}(t+1)\cdots\hat{y}(t+N_p)]^T, \quad (51.27)$$
$$\Delta u_{t,N_p} = [\Delta u(t)\cdots\Delta u(t+N_p-1)]^T, \quad (51.28)$$
$$f_{t,N_p} = [f(t+1)\cdots f(t+N_p)]^T, \quad (51.29)$$

and

$$\Gamma_{N_p} = \begin{bmatrix} g_0 & 0 & \cdots & 0 \\ g_1 & g_0 & \cdots & 0 \\ \vdots & & \ddots & \\ g_{N_p-1} & g_{N_p-2} & \cdots & g_0 \end{bmatrix}. \quad (51.30)$$

The Γ_{N_p} matrix contains the step-response coefficients of dimensions $N_p x N_p$, where N_p is the prediction horizon. If the control horizon N_u was selected so that $N_u < N_p$, and the minimum and the maximum costing horizons were selected as $N_1 > 1$ and $N_2 = N_p$, then the matrix reduces to a $(N_p - N_1 + 1)xN_u$ matrix, thus reducing the computational requirements of the GPC algorithm.

Recalling the GPC cost function,

$$J_{GPC} = \left\{ \left(\sum_{i=N_1}^{N_2} [y(t+i|t) - r(t+i)]^2 + \sum_{i=1}^{N_u} \lambda \Delta u^2(t+i-1) \right) |t \right\}. \quad (51.31)$$

Denote the *costing range*, extending from the lower bound N_1 to the upper bound N_2, by N. To simplify the development, assume in the following that $N_1 = 1$, $N_2 = N_p$, and $N_u = N_p$. Hence substituting in Equation 51.31 for the future output, using the optimal predictor Equation 51.26, yields

$$J_t = \left(\Gamma_N \Delta u_{t,N} + f_{t,N} - r_{t,N}\right)^T \left(\Gamma_N \Delta u_{t,N} + f_{t,N} - r_{t,N}\right) + \lambda \left(\Delta u_{t,N}\right)^T \left(\Delta u_{t,N}\right). \quad (51.32)$$

The optimal control is then obtained by setting the partial derivative of Equation 51.32 w.r.t $\Delta u_{t,N}$ to zero. After some straightforward algebra,

$$\Delta u_{t,N} = \left(\Gamma_N{}^T \Gamma_N + \lambda I\right)^{-1} \Gamma_N{}^T \left(r_{t,N} - f_{t,N}\right). \quad (51.33)$$

From Equation 51.33, if the predicted free response exactly matches the future reference, then the implemented control will remain constant. Although the solution of Equation 51.33 gives the control N_p steps into the future, only the first element of $\Delta u_{t,N}$, Δu_t, is actually implemented. $\Delta u_{t,N}$ represents the best future control policy for the minimizing Equation 51.31, *based upon currently available information.* At the next sampling instant, new information becomes available which can be utilized to recalculate $\Delta u_{t,N}$ for that instant. This strategy is known as *Receding Horizon Control.* The receding horizon strategy is a common feature of most MBPC algorithms and is illustrated in Figure 51.1.

51.3.3 An Illustrative Example

An application of the unconstrained GPC algorithm, illustrating the effect of adjusting the control horizon N_u, is presented.

The plant considered is nonminimum phase with system matrices corresponding to the general form Equation 51.18 as follows:

$$A\left(q^{-1}\right) = \left(1-q^{-1}\right)\left(1-0.9q^{-1}\right),$$
$$B\left(q^{-1}\right) = 0.1q^{-1}\left(1-2q^{-1}\right), \quad \text{and} \quad (51.34)$$
$$C\left(q^{-1}\right) = 1.$$

The disturbance ζ is white noise, zero mean with unity variance. The GPC controller parameters were set as follows: $N_p = 10$ (prediction horizon), $N_1 = 1$ (minimum cost horizon), $N_2 = 10$ (maximum cost horizon), $\lambda = 0$ (control weighting), $N_u = 1, 2,$ and 3 (control horizon).

The controlled plant was simulated for 100 seconds, with a unity step reference change at $t = 20$ sec. and at $t = 80$ sec. The control and output signal mean values are shown in Figure 51.2, and the variances in Figure 51.3.

From the figures increasing N_u, which is equivalent to increasing the degrees of freedom in the control, leads to

1. improved reference tracking in the mean,
2. reduced output signal variance,
3. increased peak mean control signal, and
4. increased control signal variance.

The control horizon N_u, may be increased further, and stable control results for $N_u < N_2$. However when $N_u = N_2$, with the control weighting set to zero, the GPC controller becomes equivalent to minimum variance control [20], which is known to be unstable for a nonminimum phase plant, because the roots of the B polynomial are also roots of the closed-loop characteristic equation under minimum variance control [2]. Thus any process zeros existing outside the unit circle, become unstable closed-loop poles.

51.4 Constraint Handling

Operational constraints arise when process variables must lie within given limits. In industrial plant operation, when driving the process toward the most profitable operating conditions, constraints become active at the optimum point [30]. MBPC is increasingly popular as a modern advanced control methodology because it is the only known technique that incorporates constraints naturally and systematically.

Constraints on control increments and control absolute values are termed *hard constraints,* because they are associated with fundamental physical limitations of the actuator devices. For example, in marine systems, fin stabilizers [21] have a maximum operating angle of $\theta \leq 23°$ and a maximum rate of travel of $\theta_{rate} \leq \pm 20°/\text{sec}$. By contrast, constraints with regulated outputs are termed *soft constraints* because regulated outputs may

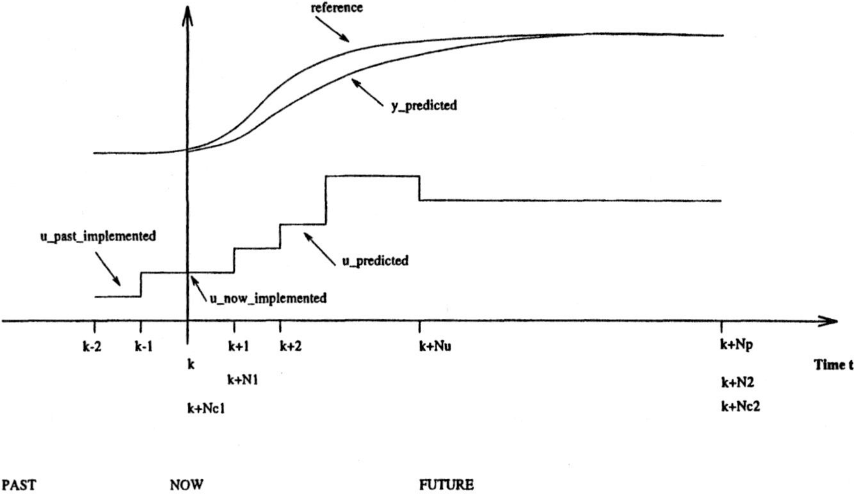

Key : Nu = control horizon, after which predicted controls are static, ie $\Delta u(t+k) = 0$ for all t+k>Nu.
N = costing range, extending from a lower bound t+N1 to an upper bound t+N2. Over this range input and output terms are included in the cost function.
Nc = constraint range, extending from a lower bound t+Nc1 to an upper bound t+Nc2. Over this range constraints on inputs and outputs are taken account of.
Np = prediction horizon up to which input and output signals are predicted.

Figure 51.1 The receding horizon concept.

exceed the constraints imposed upon them. Even if an output constraint is critical from a safety viewpoint, it is still soft because, under certain operating conditions, the constraint can be violated. For example, a process temperature output would have to satisfy $T_{\min} \leq T^{\circ C} \leq T_{\max}$ to prevent overstressing the installation.

In general, unlike the unconstrained problem, no analytic solution exists for the constrained optimal control problem. (However, in certain special cases an analytic solution can be found. If only terminal output constraints exist [11], or if the active constraint set at the optimum point is known [24], then Lagrange multiplier methods may be applied. If it is known that the constraints are active only for a finite time into the future, then an infinite horizon predictive controller can be formulated [27].)

The most common approach to solution is applying an iterative Quadratic Programming (QP) method. Popular techniques include the active set method, gradient projection [25], and the ellipsoid method [34]. Camacho [5] transforms the standard QP problem into an equivalent Linear Complementary problem (LCP). He demonstrates that the amount of computation can be significantly reduced for LCP compared to QP, crucial when the real application possesses fast dynamics.

Motivated by the desire to avoid using the computationally intensive mathematical programming techniques, Bemporad and Mosca [3] introduce a method to shape the reference signal $r_{t,N}$ so as to avoid constraint violations and to achieve minimum settling time.

It is possible for a constrained MBPC problem to be ill-posed. This means that no feasible solution exists satisfying all specified operational constraints. Scokaert and Clarke [35] provide a full discussion of the deep interelationship that exists between feasibility and stability in the closed loop. Infeasibility is often a precursor to instability. In real applications temporary infeasibility can arise as a natural consequence of driving toward optimal operating conditions. Thus the issue of reformulating control objectives in the presence of infeasibilities is critical. To quote Clarke [12] on this point,"...interesting problems now lie in the 'best' resolution of *infeasibilities* and in the *deliberate* and programmed choice of constraints for enhanced performance."

Finally to illustrate how the constrained MBPC problem is formulated, consider the GPC algorithm discussed previously. Using the cost function Equation 51.32, the QP problem becomes

$$\min_{\Delta u_{t,N_u}} \Big\{ J_t = \left(\Gamma_N \Delta u_{t,N_u} + f_{t,N} - r_{t,N}\right)^T \left(\Gamma_N \Delta u_{t,N_u} + f_{t,N} - r_{t,N}\right) + \lambda \left(\Delta u_{t,N_u}\right)^T \left(\Delta u_{t,N_u}\right) \Big\}$$

subject to

$$\begin{aligned}
&\Delta u^l \leq \Delta u_{t+k} \leq \Delta u^u, \forall k : N_{c1} \leq t+k \leq N_{c2} \\
&\qquad \text{(incremental control constraint)} \\
&u^l \leq u_{t+k} \leq u^u, \forall k : N_{c1} \leq t+k \leq N_{c2} \\
&\qquad \text{(absolute control constraint)} \\
&y^l \leq \hat{y}_{t+k} \leq y^u, \forall k : N_{c1} \leq t+k \leq N_{c2} \\
&\qquad \text{(predictor output inequality constraint)}
\end{aligned} \tag{51.35}$$

The default setting for N_{c1}, the constraint range lower bound, is 0. The default setting for N_{c2}, the constraint range upper bound, is the settling time of the plant.

51.5 Conclusions

For many control applications, there is no point in applying modern MBPC methods if the conventional PID controller performs acceptably. However, there are challenging practical control problems associated with multivariable, constrained nonlinear

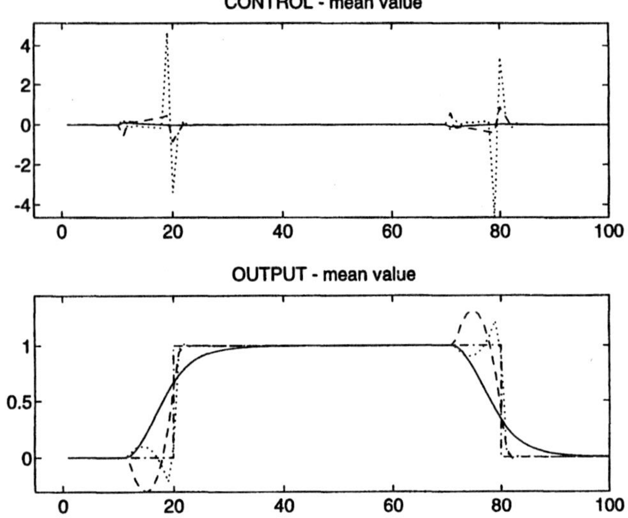

Figure 51.2 Signal mean values for the GPC example, $N_u = 1$: solid line, $N_u = 2$: dashed line, $N_u = 3$: dotted line.

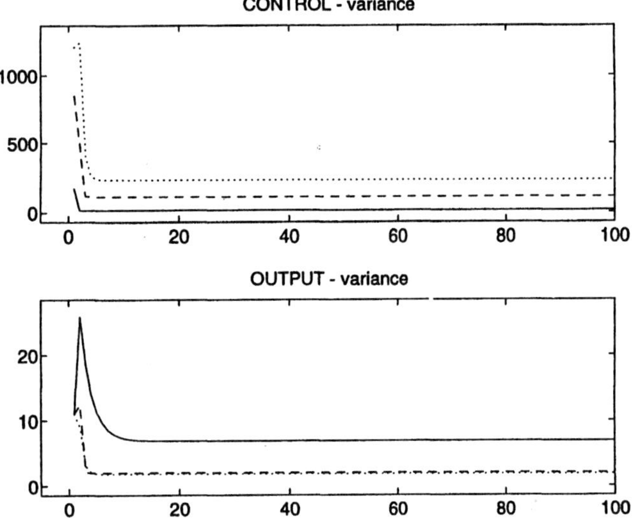

Figure 51.3 Signal variances for the GPC example, $N_u = 1$: solid line, $N_u = 2$: dashed line, $N_u = 3$: dotted line.

plants, where performance optimization is important. In such cases, MBPC is potentially the only realistic control solution.

The primary advantages of MBPC methods are

1. relatively easy to tune
2. able to handle constraints
3. can be extended to nonlinear processes (with appropriate modifications)
4. open methodology in designing predictive controllers within the common MBPC framework
5. applicable to a wide class of processes including those with unknown time delay (dead time), nonminimum phase, and/or open-loop instability

The primary disadvantages are

1. An accurate process model is required.
2. MBPC methods assume that future set points are known. This is often the case in the petrochemical industries and for many batch operations.
3. Theoretical guaranteed properties for MBPC methods is still a developing subject.

51.6 Defining Terms

CARIMA model: The origin of this term is easily seen if Equation 51.6 is rewritten in a different format. For $D(q^{-1} = A(q^{-1})\Delta$, Equation 51.6 may be written as

$$\begin{aligned} y(k) &= -a_1 y(k-1) - a_2 y(k-2) - \cdots \\ &\quad - a_{n_A} y(k - n_A) \\ &\quad \text{(output Auto Regressive component)} \\ &\quad + b_0 u(k-d-1) + b_1 u(k-d-2) + \cdots \\ &\quad + b_{n_B} u(k-d-n_B-1) \\ &\quad \text{(Control Moving Average component)} \\ &\quad + \frac{\zeta(k) + c_1\zeta(k-1) + c_2\zeta(k-2) + \cdots}{\Delta} \\ &\quad \frac{+ c_{n_C}\zeta(k-n_C)}{\Delta} \\ &\quad \text{(disturbance Integrated Moving} \\ &\quad \text{Average component).} \end{aligned}$$

The capitalized letters produce the acronym CARIMA.

Diophantine equation: Given in general form as the linear equation $AX + BY = C$ where A, B, C are known polynomials. The problem is to determine the X, Y polynomials that satisfy the above. An example of a Diophantine equation, when A, B, C are scalars, is given by $7x + 2y = 5$, and the problem is to determine the x and y integers that satisfy this equation. Here the integer solutions are given by $(1 + 2n, -1 - 7n)$.

Expectation operator: If $g(\zeta)$ is a function of a random variable ζ described by the probability density $\phi(\zeta)$, the expectation operator of g is defined as

$$E\{g(\zeta)\} = \int_{-\infty}^{\infty} (g(\zeta)\phi(\zeta))d\zeta.$$

The conditional expectation will be obtained if the probability density function $\phi(\zeta)$ is replaced by conditional probability $\phi(\zeta|\eta)$, where η is another random variable,

$$E\{g(\zeta|\eta)\} = \int_{-\infty}^{\infty} (g(\zeta)\phi(\zeta|\eta))d\zeta.$$

The conditional probability can be obtained from Bayes equation,

$$\phi(\zeta|\eta) = \frac{\phi(\zeta, \eta)}{\phi(\eta)}.$$

Linear Quadratic Gaussian (LQG): A dynamic optimization problem described by a linear model of the system (either continuous-time or discrete-time) with additive disturbances which are stochastic processes with Gaussian probability density functions. The performance index to be minimized is a finite or infinite horizon sum involving quadratic terms on the state of the system and/or on the control. The Gaussian probability density function of the random vector ζ of dimension n is given by the expression,

$$\phi(\zeta) = \frac{1}{\sqrt{(2\pi)^n \det(P)}} \exp\left[-\frac{1}{2}(\zeta - \bar{\zeta})^T P^{-1}(\zeta - \bar{\zeta}\right],$$

where $\bar{\zeta}$ is the mean value and P is the covariance matrix.

Minimum phase: The system is minimum phase if the open-loop transfer function has only left half-plane zeros (in the continuous-time case). In the discrete-time case this corresponds to all open-loop zeros of $B(z^{-1})$ lying inside the unit circle. Practically, minimum-phase behaviour of the system means that the step response of the system starts in a "right direction," the right-hand side limit of the time derivative of the system step response has the same sign as the steady-state gain.

Nonminimum phase: Descriptive of a plant transfer function containing zeros outside the unit circle. The characteristic step response of such plant is inverse, e.g., initially going negative, then positive until steady state is reached.

Steady-state offset: The difference between the set point value of the output and the actual steady-state value of the output is called steady-state offset. It is desirable to reduce the steady-state offset (to zero, if possible). It is a well known property that systems with integral action have zero steady state-offset.

White noise: A stochastic process $v(t)$ is called white noise if its realizations in time are uncorrelated, i.e., the autocorrelation function : $E\{v(t)v(t+\delta)\}$ has nonzero value only for $\delta = 0$. Note that this implies $E\{v(t)\} = 0$ for all t. The process is called Gaussian white noise if, in addition to the above assumption, at each time instant the probability density function is Gaussian.

Wiener process: A Wiener process is defined as the time integral of Gaussian white noise: $w(t) = \int_0^t e(s)ds$. This is a model for a random walk, where successive increments of w, dw, are random and statistically independent with respect to one another.

References

[1] Åström K.J., Computer control of a paper machine — an application of linear stochastic control theory. *IBM J. Res. Dev.*, 11, July 1967.

[2] Åström K.J. and Wittenmark, B., *Computer Controlled Systems — Theory and Design*, 2nd ed., Prentice Hall International, 1990.

[3] Bemporad A. and Mosca, E., Constraint fulfilment in feedback control via predictive reference management, *Proc. 3rd IEEE CCA*, Glasgow, pp.1909–1914, 1994.

[4] Bitmead, R.R., Gevers, M., and Wertz, V., *Adaptive Optimal Control*, Prentice Hall International, 1990.

[5] Camacho, E.F., Constrained Generalised predictive Control, *IEEE Trans. Auto. Control*, 38(2), 327, 1993.

[6] Camacho, E.F. and Bordons, C., *Model Predictive Control in the Process Industries*, Springer-Verlag, 1995.

[7] Clarke, D.W., Mohtadi, C., and Tuffs, P.S., Generalised predictive control — Part 1. The basic algorithm, *Automatica*, 23(2), 137, 1987.

[8] Clarke, D.W., Mohtadi, C., and Tuffs, P.S., Generalised predictive control — Part 2. Extensions and Interpretations, *Automatica*, 23(2), 149, 1987.

[9] Clarke D.W. and Gawthrop, P.J., Self-tuning controller, *Proc. IEEE*, 122(9), 929, 1975.

[10] Clarke, D.W. and Mohtadi, C., Properties of Generalised Predictive Control, *Automatica*, (6), 859, 1989.

[11] Clarke, D.W. and Scattolini, R., Constrained receding-horizon predictive control, *IEEE Proc. – D*, 138(4), 1991.

[12] Clarke, D.W., Ed., *Advances in Model-Based Predictive Control*, Oxford University Press, 1994.

[13] Compas, J.M., Decarreau, P., Lanquetin, G., Estival, J.L., Fulget, N., Martin, R., and Richalet, J., Industrial applications of Predictive Functional Control to rolling mill, fast robot, river dam, *Proc. 3rd IEEE conf. Control Appl.*, Glasgow, U.K., 1994.

[14] Connoisseur™ — Information pack, Predictive Control Ltd, Royal Mews, Gadbrook Park, Northwich, Cheshire CW9 7UD.

[15] Cutler, C.R. and Raemaker, B.L., Dynamic matrix control—A computer control algorithm, *Proc. JACC*, San Francisco, CA, 1980.

[16] De Keyser, R.M.C. and van Cauwenberghe, A.R., Extended prediction self-adaptive control, *Proc. 7th IFAC Symp. Ident. Syst. Parameter Est.*, York , UK, pp 1255–1260, 1985.

[17] De Keyser, R.M.C., Van de Velde, P.H.G.A., and Dumortier, F.A.G., A comparative study of self-adaptive long range predictive control methods, *Automatica*, 24(2), 149, 1988.

[18] ESPRIT CIM, Europe workshop on industrial applications of model based predictive control, Workshop proceedings, Cambridge, U.K., September 1992.

[19] Garcia, C.E, Prett, D.M., and Morari, M., Model predictive control: Theory and Practice—A survey, *Automatica*, (3), 335, 1989.

[20] Grimble, M.J., Generalised predictive control: An introduction to the advantages and limitations, *Int. J. Syst. Sci.*, 23, 85, 1992.

[21] Grimble, M.J., *Robust Industrial Control: Optimal Design Approach for Polynomial Systems*, Prentice Hall, Hemel Hempstead, 1994.

[22] Hagglund T. and Åström, K.J., U.S. Patent 4,549,123, 1985.

[23] Hastings-James, R., *A linear stochastic controller for regulation of systems with pure time-delay*, University of Cambridge Department of Engineering, Report No. CN/70/3, 1970.

[24] Heise, S.A. and Maciejowski, J.M., Stability of constrained MBPC using an internal model control structure, in *Advances in Model Based Predictive Control*, D. Clarke, Ed., Oxford University Press, 1994.

[25] Luenberger, D.G., *Linear and Nonlinear Programming*, Addison Wesley Inc., 1984.

[26] Martin-Sánchez, J., A new solution to adaptive control, *Proc. IEEE*, 16/64(8), 1976.

[27] Morari, M., Model predictive control: multivariable control technique of choice in the 1990's, in *Advances in Model Based Predictive Control*, D. Clarke, Ed., Oxford University Press, 1994.

[28] Morari, M. and Lee, J.H., Model predictive control: the good, the bad and the ugly, Arkun, Y. and Ray, W.H., Eds., *Proc. 4th Int. Conf. Chem. Proc. Control*, South Padre Island, Texas, 1991.

[29] Ordys, A.W. and Clarke, D.W., A state-space description for GPC controllers, *Int. J. Syst. Sci.*, 23(2), 1993.

[30] Prett D.M. and Garcia, C.E., *Fundamental Process Control*, Butterworths, 1988.

[31] Richalet, J., Rault, A., Testaud, L., and Papon, J., Model predictive heuristic control: Applications to industrial processes, *Automatica*, 14, 413, 1978.

[32] Richalet, J., Abu el Ata-Doss, S., Arber, Ch., Kuntze, H.B., Jacubasch, A., and Schill, W., Predictive functional control. Application to fast and accurate robots, *Proc. 10th IFAC World Cong.*, Munich, F.R.G., 1987.

[33] Richalet, J., Industrial applications of model based predictive control, *Automatica*, 29(5), 1251, 1993.

[34] Schijver, A., *Theory of Linear and Integer Programming*, John Wiley & Sons, New York, 1986.

[35] Scokaert, P.O.M and Clarke, D.W., Stability and feasibility in constrained predictive control, in *Advances in Model Based Predictive Control*, D.W. Clarke, Ed., Oxford University Press, 1994.

[36] Soeterboek, R., *Predictive Control, a Unified Approach*, Prentice Hall International, 1992.

[37] Ydstie, B.E., Extended horizon adaptive control, *Proc. 9th IFAC World Cong.*, Budapest, Hungary, 1984.

Further Reading

Two introductory textbooks giving an overview of the predictive control field and its application are *Model Predictive Control in the Process Industry* by E.F. Camacho and C. Bordons, Springer Verlag, 1995, ISBN 3-540-19924-1 and *Predictive Control - A Unified Approach* by R. Soeterboek, Prentice Hall International, 1992, ISBN 0-13-678350-3.

For more recent trends and active research issues in predictive control, the reader is referred to *Advances in Model-Based Predictive Control*, Editor, D.W. Clarke, Oxford University Press, 1994, ISBN 0-19-856292-6.

For more information about DMC, contact MathWorks Inc., 24 PrimePark Way, Natick, MA 01760-1500, USA; about IDCOM, PFC, contact 7 Bd du Marechal Juin, B.P. 52, 91371 Verrieres-le-buisson, FRANCE; and about SCAP, (APCS), contact SCAP Europa, S.A, Paseo de la Castellana, 173, 28046 Madrid, SPAIN.

SECTION X

Adaptive Control

52

Automatic Tuning of PID Controllers

Tore Hägglund
Department of Automatic Control, Lund Institute of Technology, Lund, Sweden

Karl J. Åström
Department of Automatic Control, Lund Institute of Technology, Lund, Sweden

52.1 Introduction

Methods for automatic tuning of PID controllers have been one of the results of the active research on adaptive control. A result of this development is that the design of PID controllers is going through a very interesting phase. Practically all PID controllers that are designed now have at least some features for automatic tuning. Automatic tuning has also made it possible to generate automatically gain schedules. Many controllers also have adaptation of feedback and feedforward gains.

The most important component of the adaptive controllers and automatic tuning procedures is the design method. The next section presents some of the most common design methods for PID controllers. These design methods are divided into three categories: (1) future based techniques, (2) analytical methods, (3) and methods that are based on optimization.

Section 52.3 treats adaptive techniques. An overview of different uses of these techniques is first presented, followed by a more detailed treatment of automatic tuning, gain scheduling, and adaptive control. Section 52.4 gives an overview of how the adaptive techniques have been used in commercial controllers. References are at the end of the chapter.

52.2 Design Methods

To obtain rational methods for designing controllers it is necessary to deal with specifications and models. In the classical Ziegler-Nichols methods, the process dynamics are characterized by two parameters, a gain and a time. Another approach is used in the analytical design methods, where the controller parameters are obtained from the specifications and the process transfer function by a direct calculation. Optimization methods allow for compromise between several different criteria. These approaches are discussed here.

52.2.1 Specifications

When solving a control problem it is necessary to understand the primary goal of control. Two common control objectives are to follow the setpoint and to reject disturbances. It is also important to have an assessment of the major limitations, which can be system dynamics, nonlinearities, disturbances, or process uncertainty. Typical specifications on a control system may include attenuation of load disturbances, setpoint following, robustness to model uncertainty, and lack of sensitivity to measurement noise.

Attenuation of Load Disturbances:

Attenuation of load disturbances is of primary concern for process control. The disturbances may enter the system in many different ways, but it is often assumed that they enter at the process input. Let e be the error caused by a unit step load disturbance at the process input. Typical quantities used to characterize the error are: maximum error, time to reach maximum, settling time, decay ratio, and the integrated absolute error, which is defined by

$$IAE = \int_0^\infty |e(t)|dt \tag{52.1}$$

Setpoint Following:

Setpoint following is often less important than load disturbance attenuation for process control applications. Setpoint changes are often only made when the production rate is altered. Furthermore, the response to setpoint changes can be improved by feeding the setpoint through ramping functions or by adjust-

0-8493-8570-9/96/$0.00+$.50

ing the setpoint weightings. Specifications on setpoint following may include requirements on rise time, settling time, decay ratio, overshoot, and steady-state offset for step changes in setpoint.

Robustness to Model Uncertainty

It is important that the controller parameters are chosen in such a way that the closed-loop system is not too sensitive to changes in process dynamics. There are many ways to specify the sensitivity. Many different criteria are conveniently expressed in terms of the Nyquist plot of the loop transfer function, and its distance to the critical point -1. Maximum sensitivity M_s is a good measure since $1/M_s$ is the shortest distance between the Nyquist plot and the critical point. Gain and phase margins are other related measures.

Sensitivity to Measurement Noise:

Care should always be taken to reduce measurement noise by appropriate filtering, since it will be fed into the system through the feedback. It will generate control actions and control errors. Measurement noise is typically of high frequency. The high-frequency gain of a PID controller is

$$K_{hf} = K(1 + N)$$

where K is the controller gain and N is the derivative gain limitation factor. See Chapter 10. Notice that $N = 0$ corresponds to PI control. Multiplication of the measurement noise by K_{hf} gives the fluctuations in the control signal that are caused by the measurement noise. Also notice that there may be a significant difference in K_{hf} for PI and PID control. It is typically an order of magnitude larger for a PID controller, since the gain normally is higher for a PID controller than for a PI controller, and N is typically around 10.

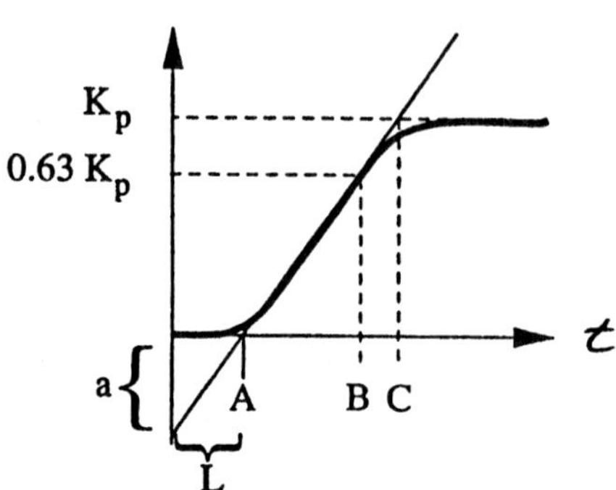

Figure 52.1 Determining a first-order plus dead-time model from a step response. Time constant T can be obtained either as the distance AB or the distance AC.

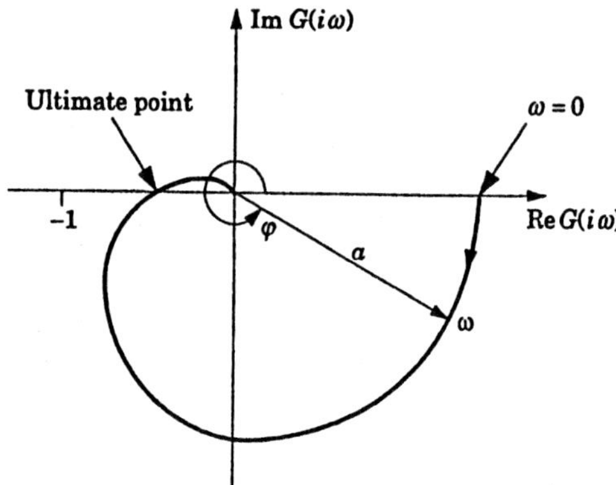

Figure 52.2 Static gain K_p, ultimate gain K_u, and ultimate period T_u defined in the Nyquist diagram. Static gain K_p is the point on the Nyquist plot at $\omega = 0$. Ultimate gain K_u is -1 divided by the ultimate point. Ultimate period T_u is 2π divided by the frequency corresponding to the ultimate point.

52.2.2 Feature-Based Techniques

The simplest design methods are based on a few features of the process dynamics that are easy to obtain experimentally. Typical time-domain features are static gain K_p, dominant time constant T, and dominant dead time L. They can all be determined from a step response of the process, see Figure 52.1. Static gain K_p, dominant time constant T, and dominant dead time L can be used to obtain an approximate first order plus dead-time model for the process as given in Equation 52.2.

$$G_p(s) = \frac{K_p}{1 + sT} e^{-sL} \tag{52.2}$$

Typical frequency-domain features are static gain K_p, ultimate gain K_u, and ultimate period T_u. They are defined in Figure 52.2.

Ziegler-Nichols Methods

In 1942, Ziegler and Nichols presented two design methods for PID controllers, one time-domain and one frequency-domain method [10]. The methods are based on determination of process dynamics in terms of only two parameters, a gain and a time. The controller parameters are then expressed in terms of these parameters by simple formulas. In both methods, the design specification of quarter amplitude decay ratio was used. The decay ratio is the ratio between two consecutive maxima of the control error after a step change in setpoint or load.

The time-domain method is based on a registration of the open-loop step response of the process. Ziegler and Nichols have given PID parameters directly as functions of a and L, defined in Figure 52.1. These are given in Table 52.1.

TABLE 52.1 Controller parameters obtained from the Ziegler-Nichols step response method.

Controller	K	T_i	T_d
P	$1/a$		
PI	$0.9/a$	$3L$	
PID	$1.2/a$	$2L$	$L/2$

The second method presented by Ziegler and Nichols is based on the frequency response of the process. They have given simple formulas for the parameters of the controller in terms of ultimate gain K_u and the ultimate period T_u. These parameters can be determined in the following way. Connect a controller to the process, set the parameters so that control action is proportional, i.e., $T_i = \infty$ and $T_d = 0$. Increase the gain slowly until the process starts to oscillate. The gain when this occurs is K_u and the period of the oscillation is T_u. The parameters can also be determined approximately by relay feedback as is discussed in Section 52.3. The controller parameters are given in Table 52.2.

TABLE 52.2 Controller parameters obtained from the Ziegler-Nichols frequency response method.

Controller	K	T_i	T_d
P	$0.5K_u$		
PI	$0.4K_u$	$0.8T_u$	
PID	$0.6K_u$	$0.5T_u$	$0.12T_u$

Modifications of the Ziegler-Nichols Methods

The Ziegler-Nichols methods do not give satisfactory control. The reason is that they give closed-loop systems with very poor damping. The design criterion "quarter amplitude decay ratio" corresponds to a relative damping of $\zeta \approx 0.2$ which is much too small for most applications. The maximum sensitivity is also much too large, which means that the closed-loop systems obtained are too sensitive to parameter variations.

The Ziegler-Nichols methods do, however, have the advantage of being very easy to use. Many efforts have therefore been made to obtain tuning methods that retain the simplicity of the Ziegler-Nichols methods but give improved robustness.

Chien, Hrones, and Reswick modified the coefficients in the Ziegler-Nichols methods [3]. In this way, they obtained tuning rules that give better damping. They developed different tuning rules for setpoint changes and load disturbances. Using the two-degrees of freedom structure of the PID controller, there is, however, normally no need to compromise between these two disturbances. If the tuning is performed to give good responses to load disturbances, the setpoint changes can be treated by, e.g., setpoint weighting (see Chapter 10). Table 52.3 gives the controller parameters based on load disturbances.

TABLE 52.3 Controller parameters obtained from the Chien, Hrones and Reswick load disturbance response method.

Overshoot	0%			20%		
Controller	K	T_i	T_d	K	T_i	T_d
P	$0.3/a$			$0.7/a$		
PI	$0.6/a$	$4L$		$0.7/a$	$2.3L$	
PID	$0.95/a$	$2.4L$	$0.42L$	$1.2/a$	$2L$	$0.42L$

Kappa-Tau Tuning

Significantly better tuning rules can be obtained if the process dynamics are described in terms of three parameters instead of two. An early step in this direction was made by Cohen and Coon, who assumed that the process was given by Equation 52.2, which has three parameters [4]. Their design did, however, also give very sensitive systems.

The Kappa-Tau method (see [1]) is a recent method that was developed for automatic tuning. It is given in two versions, one that is based on a step response experiment, and one that is based on the frequency response of the process. In both methods, maximum sensitivity M_s is used as a design variable. The Kappa-Tau method was obtained from extensive simulation investigations of the dominant pole design method applied on typical process control models [1].

In the step response method, the process is characterized by static gain K_p, gain a, and apparent deadtime L (see Figure 52.1). The controller parameters are given as functions of normalized dead time τ, which is defined as

$$\tau = \frac{L}{L+T} \tag{52.3}$$

In the frequency domain method, the process is characterized by static gain K_p, ultimate gain K_u, and ultimate period T_u. The controller parameters are given as functions of gain ratio κ, which is defined as

$$\kappa = \frac{1}{K_p K_u} \tag{52.4}$$

The Kappa-Tau method gives both PI and PID controller parameters. Figure 52.3 shows the PID controller parameters for the frequency response method.

The solid lines in the figure correspond to functions having the form

$$f(\tau) = a_0 e^{a_1 \tau + a_2 \tau^2}$$

that are fitted to the data. The function parameters are given in Table 52.4.

52.2.3 Analytical Methods

If the process can be described well by a simple model, the controller parameters can be obtained by a direct calculation. This approach is treated in this section.

Pole Placement

If the process is described by a low-order transfer function, a complete pole-placement design can be performed. Consider for example the process described by the second-order model

$$G(s) = \frac{K_p}{(1+sT_1)(1+sT_2)} \tag{52.5}$$

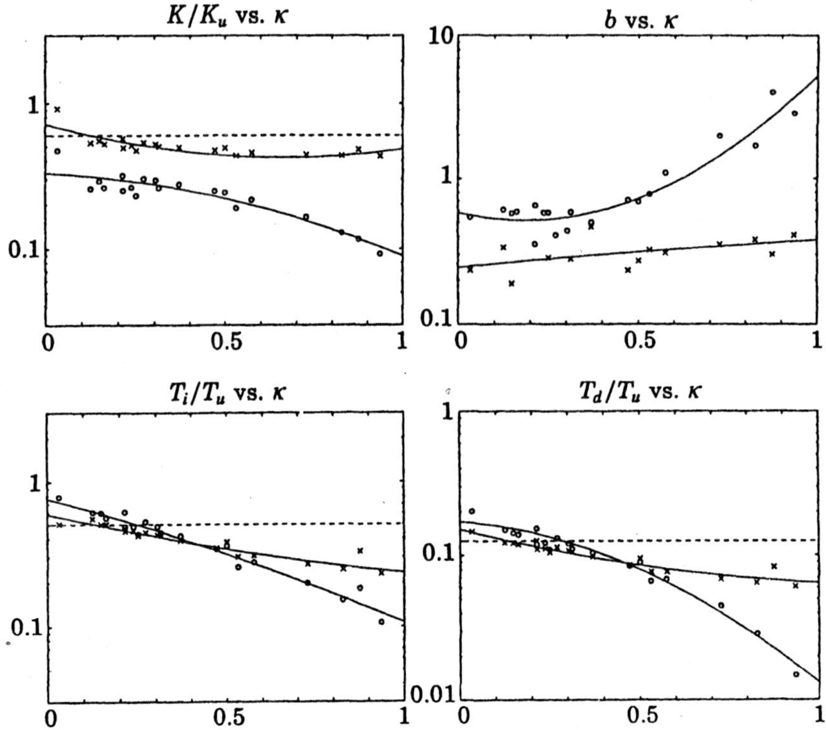

Figure 52.3 Tuning diagrams for PID control based on K_u, T_u and κ. Controller parameters are obtained by applying dominant pole design with $M_s = 1.4$, marked with o, and $M_s = 2$ marked with x, to processes in a test batch. The dashed lines correspond to the Ziegler-Nichols tuning rule.

TABLE 52.4 Tuning formula for PID control based on the Kappa-Tau method. The table gives parameters of functions of the form $f(\tau) = a_0 exp(a_1\tau + a_2\tau^2)$ for the normalized controller parameters.

	$M_s = 1.4$			$M_s = 2.0$		
	a_0	a_1	a_2	a_0	a_1	a_2
K/K_u	0.33	-0.31	-1.0	0.72	-1.6	1.2
T_i/T_u	0.76	-1.6	-0.36	0.59	-1.3	0.38
T_d/T_u	0.17	-0.46	-2.1	0.15	-1.4	0.56
b	0.58	-1.3	3.5	0.25	0.56	-0.12

This model has three parameters. By using a PID controller, which also has three parameters, it is possible to arbitrarily place the three poles of the closed-loop system. The transfer function of the PID controller in parallel form can be written as

$$G_c(s) = \frac{K(1 + sT_i + s^2T_iT_d)}{sT_i}$$

The characteristic equation of the closed-loop system becomes

$$s^3 + s^2\left(\frac{1}{T_1} + \frac{1}{T_2} + \frac{K_pKT_d}{T_1T_2}\right) + s\left(\frac{1}{T_1T_2} + \frac{K_pK}{T_1T_2}\right) + \frac{K_pK}{T_iT_1T_2} = 0$$

A suitable closed-loop characteristic equation of a third-order system is

$$(s + \alpha\omega)(s^2 + 2\zeta\omega s + \omega^2) = 0$$

which contains two dominant poles with relative damping (ζ) and frequency (ω), and a real pole located at $-\alpha\omega$. Identifying the coefficients in these two characteristic equations determines the PID parameters K, T_i and T_d. The solution is

$$\begin{aligned} K &= \frac{T_1T_2\omega^2(1 + 2\zeta\alpha) - 1}{K_p} \\ T_i &= \frac{T_1T_2\omega^2(1 + 2\zeta\alpha) - 1}{T_1T_2\alpha\omega^3} \\ T_d &= \frac{T_1T_2\omega(\alpha + 2\zeta) - T_1 - T_2}{T_1T_2(1 + 2\zeta\alpha)\omega^2 - 1} \end{aligned}$$

If the model is given by Equation 52.5, any value of ζ and ω can be chosen. There are, however, restrictions on reasonable choices of ω. Criteria like IAE will be smaller the larger the bandwidth is. Pure PI control is obtained for

$$\omega_c = \frac{T_1 + T_2}{T_1T_2(\alpha + 2\zeta)}$$

The choice of ω may be critical. The derivative time, T_d, is negative for $\omega < \omega_c$. The frequency ω_c thus gives a lower bound to the bandwidth. The gain increases rapidly with ω. The upper bound to the bandwidth is given by the validity of the simplified model. With this approach, it still remains to give a suitable value of ω.

λ-Tuning

Let G_p and G_c be the transfer functions of the process and the controller. The closed-loop transfer function obtained with error feedback is then

$$G_0 = \frac{G_p G_c}{1 + G_p G_c}$$

Solving this equation for G_c gives

$$G_c = \frac{1}{G_p} \cdot \frac{G_0}{1 - G_0} \tag{52.6}$$

If the closed-loop transfer function G_0 is specified and G_p is known, it is thus easy to compute G_c.

The method, called λ-tuning, was developed for processes with long dead time L [5]. Consider a process with the transfer function

$$G_p = \frac{K_p}{1 + sT} e^{-sL} \tag{52.7}$$

Assume that the desired closed-loop transfer function is specified as

$$G_0 = \frac{e^{-sL}}{1 + s\lambda T} \tag{52.8}$$

where λ is a tuning parameter. The time constants of the open- and closed-loop systems are the same when $\lambda = 1$. The closed-loop system responds faster than the open-loop system if $\lambda < 1$. It is slower when $\lambda > 1$.

It follows from Equation 52.6 that the controller transfer function becomes

$$G_c = \frac{1 + sT}{K_p(1 + \lambda sT - e^{-sL})}$$

When $L = 0$ this becomes a PI controller with gain $K = 1/(\lambda K_p)$ and integral time $T_i = T$. The sensitivity function obtained with λ-tuning is given by

$$S(s) = 1 - \frac{e^{-sL}}{1 + s\lambda T} = \frac{1 + s\lambda T - e^{-sL}}{1 + s\lambda T}$$

The maximum sensitivity M_s is always less than 2 if the model is correct. With unmodeled dynamics, the sensitivity may be larger. The parameter λ should be small to give a low IAE, but a small value of λ increases the sensitivity.

IMC

The internal model principle is a general method for design of control systems that can be applied to PID control. A block diagram of such a system is shown in Figure 52.4. It is assumed that all disturbances acting on the process are reduced to an equivalent disturbance d at the process output. In the figure, G_m denotes a model of the process, $G_m^\dagger$ is an approximate inverse of G_m, and G_f is a low-pass filter. The name internal model controller derives from the fact that the controller contains a model of the process. This model is connected in parallel with the process.

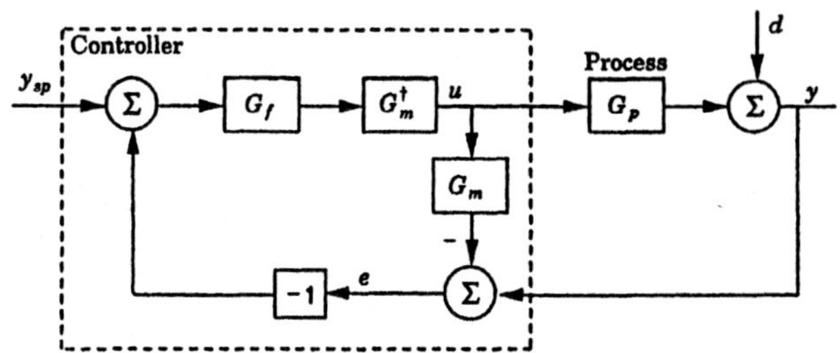

Figure 52.4 Block diagram of a closed-loop system with a controller based on the internal model principle.

If the model matches the process, i.e., $G_m = G_p$, the signal e is equal to the disturbance d for all control signals u. If $G_f = 1$ and $G_m^\dagger$ is an exact inverse of the process, then the disturbance d will be canceled perfectly. The filter G_f is introduced to obtain a system that is less sensitive to modeling errors, and to ensure that the system $G_f G_m^\dagger$ is realizable. A common choice is $G_f(s) = 1/(1 + sT_f)$, where T_f is a design parameter.

The controller obtained by the internal model principle can be represented as an ordinary series controller with the transfer function

$$G_c = \frac{G_f G_m^\dagger}{1 - G_f G_m^\dagger G_m} \tag{52.9}$$

From this expression it follows that controllers of this type cancel process poles and zeros. The controller is normally of high order. Using simple models it is, however, possible to obtain PI or PID controllers. To see this consider a process with the transfer function

$$G_p(s) = \frac{K_p}{1 + sT} e^{-sL}$$

An approximate inverse, where no attempt is made to find an inverse of the time delay, is given by

$$G_m^\dagger(s) = \frac{1 + sT}{K_p}$$

Choosing the filter

$$G_f(s) = \frac{1}{1 + sT_f}$$

and approximating the time delay by

$$e^{-sL} \approx 1 - sL$$

Equation 52.9 now gives

$$G_c(s) = \frac{1 + sT}{K_p s(L + T_f)}$$

which is a PI controller. If the time delay is approximated instead by a first-order Padé approximation

$$e^{-sL} \approx \frac{1 - sL/2}{1 + sL/2}$$

Equation 52.9 gives instead the PID controller

$$G_c(s) = \frac{(1+sL/2)(1+sT)}{K_p s(L+T_f+sT_fL/2)} \approx \frac{(1+sL/2)(1+sT)}{K_p s(L+T_f)}$$

An interesting feature of the internal model controller is that robustness is considered explicitly in the design. Robustness can be adjusted by selecting the filter G_f properly. A trade-off between performance and robustness can be made by using the filter constant as a design parameter.

The IMC method can be designed to give excellent responses to setpoint changes. Since the design method inherently implies that poles and zeros of the plant are canceled, the response to load disturbances may be poor if the canceled poles are slow in comparison with the dominant poles. This is discussed in the next section.

52.2.4 Optimization-Based Methods

A third category of design methods is based on optimization techniques.

Direct Criteria Optimization

Optimization is a powerful tool for design of controllers. The method is conceptually simple. A controller structure with a few parameters is specified. Specifications are expressed as inequalities of functions of the parameters. The specification that is most important is chosen as the function to optimize. The method is well suited for PID controllers where the controller structure and the parameterization are given. There are several pitfalls when using optimization. Care must be exercised when formulating criteria and constraints; otherwise, a criterion will indeed be optimal, but the controller may still be unsuitable because of a neglected constraint. Another difficulty is that the loss function may have many local minima. A third is that the computations required may be excessive. Numerical problems may also arise. Nevertheless, optimization is a good tool that has successfully been used to design PID controllers.

Popular optimization criteria are the integrated absolute error (IAE), the integrated time absolute error (ITAE), and the integrated square error (ISE). They are mostly done for the first-order plus dead-time model as given in Equation 52.2. Many tables that provide controller parameters based on optimization have been published [7].

Modulus Optimum (BO) and Symmetrical Optimum (SO)

Modulus Optimum (BO) and Symmetrical Optimum (SO) are two design methods that are based on optimization. The acronyms BO and SO are derived from the German words Betrags Optimum and Symmetrische Optimum. These methods are based on the idea of finding a controller that makes the frequency response from setpoint to plant output as close to one as possible for low frequencies. If $G(s)$ is the transfer function from the setpoint to the output, the controller is determined in such a way that $G(0) = 1$ and that $d^n|G(i\omega)|/d\omega^n = 0$ at $\omega = 0$ for as many n as possible.

If the closed-loop system is given by

$$G(s) = \frac{\omega_0^2}{s^2+\sqrt{2}\omega_0 s+\omega_0^2}$$

the first three derivatives of $|G(i\omega)|$ will vanish at the origin. If the transfer function G in the example is obtained by error feedback of a system with the loop transfer function G_{BO}, the loop transfer function is

$$G_{\mathrm{BO}}(s) = \frac{G(s)}{1-G(s)} = \frac{\omega_0^2}{s(s+\sqrt{2}\omega_0)}$$

which is the desired loop transfer function for the method called modulus optimum.

If the closed-loop transfer function is given by

$$G(s) = \frac{\omega_0^3}{(s+\omega_0)(s^2+\omega_0 s+\omega_0^2)} \tag{52.10}$$

the first five derivatives of $|G(i\omega)|$ will vanish at the origin. A system with this closed-loop transfer function can be obtained with a system having error feedback and the loop transfer function

$$G_\ell(s) = \frac{\omega_0^3}{s(s^2+2\omega_0 s+2\omega_0^2)}$$

The closed-loop transfer function (52.10) can also be obtained from other loop transfer functions if a two-degree of freedom controller is used. For example, if a process with the transfer function

$$G_p(s) = \frac{\omega_0^2}{s(s+2\omega_0)}$$

is controlled by a PI controller having parameters $K = 2$, $T_i = 2/\omega_0$ and $b = 0$, the loop transfer function becomes

$$G_{\mathrm{SO}} = \frac{\omega_0^2(2s+\omega_0)}{s^2(s+2\omega_0)} \tag{52.11}$$

The symmetric optimum aims at obtaining the loop transfer function given by Equation 52.11. Notice that the Bode diagram of this transfer function is symmetrical around the frequency $\omega = \omega_0$. This is the motivation for the name symmetrical optimum.

The methods BO and SO can be called loop-shaping methods since both methods try to obtain a specific loop transfer function. The design methods can be described as follows. It is first established which of the transfer functions, G_{BO} or G_{SO}, is most appropriate. The transfer function of the controller $G_c(s)$ is then chosen so that $G_\ell(s) = G_c(s)G_p(s)$, where G_ℓ is the chosen loop transfer function.

52.2.5 Cancellation of Process Poles

The PID controller has two zeros. Many design methods choose these zeros so that they cancel one or two of the dominant process poles. This often results in a simple design as in IMC or SO optimization. The response to set-point changes is good, but the methods will often result in poor responses to load disturbances. An exception to this is in the case of large dead time, in which case the settling time is already quite long relative to the time constant being canceled.

Figure 52.5 illustrates the problem. A process with a time constant $T = 10$ is controlled with a PI controller with integral time $T_i = 10$ and a suitable controller gain. The response to setpoint changes is good, but the load disturbance response is very sluggish. The figure also shows a retuned controller, where pole cancellation is avoided.

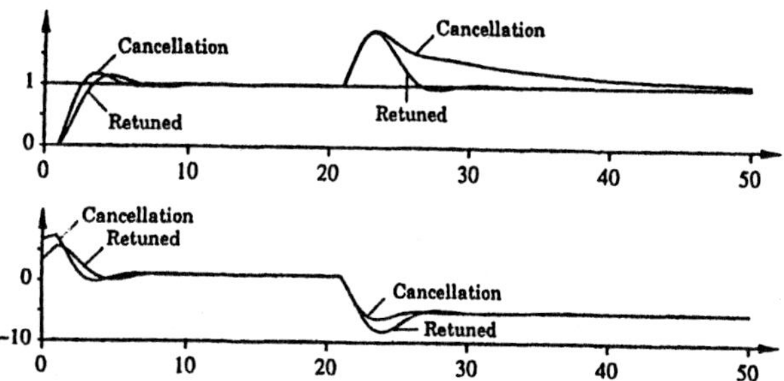

Figure 52.5 Simulation of a closed-loop system obtained by pole cancellation. The process transfer function is $G(s) = e^{-s}/(10s + 1)$, and the controller parameters are $K = 6.67$ and $T_i = 10$. The upper diagram shows setpoint $y_{sp} = 1$ and process output y, and the lower diagram shows control signal u. The figure also shows the responses to a retuned controller with $K = 6.67$, $T_i = 3$ and $b = 0.5$.

52.3 Adaptive Techniques

This section gives an overview of adaptive techniques. It starts with a discussion of uses of the different techniques, followed by a more detailed presentation of automatic tuning, gain scheduling, and adaptive control. The section ends with a presentation of how the adaptive techniques have been used in industrial controllers.

52.3.1 Use of the Adaptive Techniques

The word *adaptive techniques* is used to cover auto-tuning, gain scheduling and adaptation. Although research on adaptive techniques has almost exclusively focused on adaptation, experience has shown that auto-tuning and gain scheduling have much wider industrial applicability. Figure 52.6 illustrates the appropriate use of the different techniques.

Controller performance is the first issue to consider. If requirements are modest, a controller with constant parameters and conservative tuning can be used. Other solutions should be considered when higher performance is required.

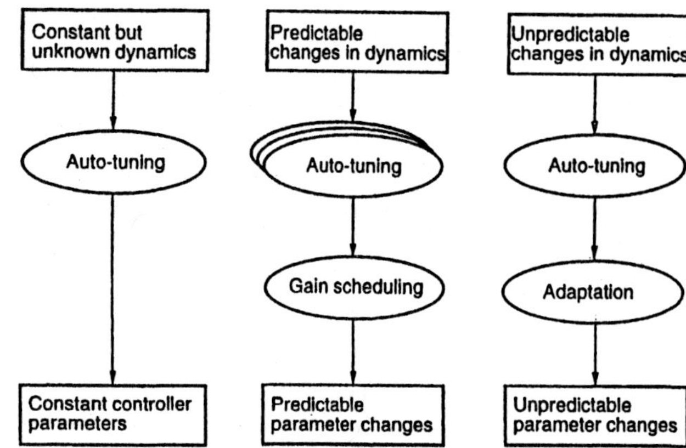

Figure 52.6 When to use different adaptive techniques.

If the process dynamics are constant, a controller with constant parameters should be used. The parameters of the controller can be obtained by auto-tuning.

If the process dynamics or the character of the disturbances are changing it is useful to compensate for these changes by changing the controller. If the variations can be predicted from measured signals, gain scheduling should be used since it is simpler and gives superior and more robust performance than continuous adaptation. Typical examples are variations caused by nonlinearities in the control loop. Auto-tuning can be used to build up the gain schedules automatically.

There are also cases where the variations in process dynamics are not predictable. Typical examples are changes due to unmeasurable variations in raw material, wear, fouling etc. These variations cannot be handled by gain scheduling but must be dealt with by adaptation. An auto-tuning procedure is often used to initialize the adaptive controller. It is then sometimes called pre-tuning or initial tuning.

52.3.2 Automatic Tuning

Automatic tuning (or auto-tuning) is a method where a controller is tuned automatically on demand from the user. Typically the user will either push a button or send a command to the controller. Automatic tuning is sometimes called tuning on demand or one-shot tuning.

Automatic tuning can also be performed by external devices which are connected to the control loop during the tuning phase. Since these devices are supposed to work together with controllers from different manufacturers, they must be provided with quite a lot of information about the controller structure and parameterization in order to provide appropriate controller parameters. Such information includes controller structure (series or parallel form), sampling rate, filter time constants, and units of the different controller parameters (gain or proportional band, minutes or seconds, time or repeats/time) (see Chapter 10).

The automatic tuning procedures can be divided into methods that are based on step response experiments, and methods based on frequency response experiments.

Step Response Methods

Most methods for automatic tuning of PID controllers are based on step response experiments. When the operator wishes to tune the controller, an open-loop step response experiment is performed. A process model is then obtained from the step response, and controller parameters are determined. This is usually done using simple look-up tables as in the Ziegler-Nichols method.

The greatest difficulty in carrying out the tuning automatically is in selecting the amplitude of the step. The user naturally wants the disturbance to be as small as possible so that the process is not disturbed more than necessary. On the other hand, it is easier to determine the process model if the disturbance is large. The result of this dilemma is usually that the operator himself has to decide how large the step in the control signal should be.

Controllers with automatic tuning which are based on this technique have become very common in the last few years. This is especially true of temperature controllers.

The Relay Autotuner

Frequency-domain characteristics of the process can be determined from experiments with relay feedback in the following way. When the controller is to be tuned, a relay with hysteresis is introduced in the loop, and the PID controller is temporarily disconnected, see Figure 52.7. For large classes of processes, re-

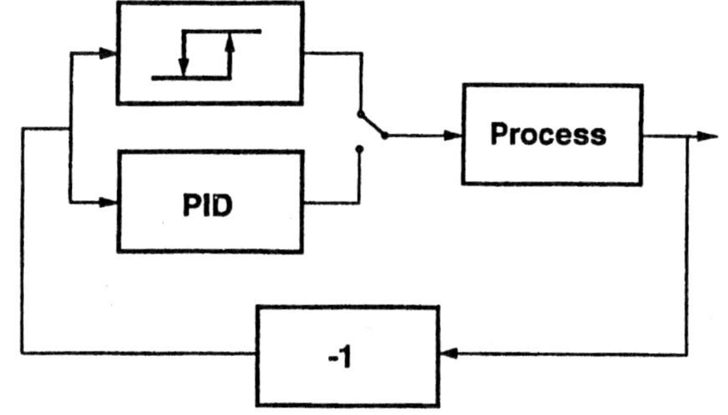

Figure 52.7 The relay autotuner. In the tuning mode the process is connected to relay feedback.

lay feedback gives an oscillation with period close to the ultimate frequency ω_u. The gain of the transfer function at this frequency is also easy to obtain from amplitude measurements. Describing function analysis can be used to determine the process characteristics. The describing function of a relay with hysteresis is

$$N(a) = \frac{4d}{\pi a}\left(\sqrt{1-\left(\frac{\epsilon}{a}\right)^2} - i\,\frac{\epsilon}{a}\right)$$

where d is the relay amplitude, ϵ the relay hysteresis and a is the amplitude of the input signal. The negative inverse of this describing function is a straight line parallel to the real axis, see Figure 52.8. The oscillation corresponds to the point where the negative inverse describing function crosses the Nyquist curve of

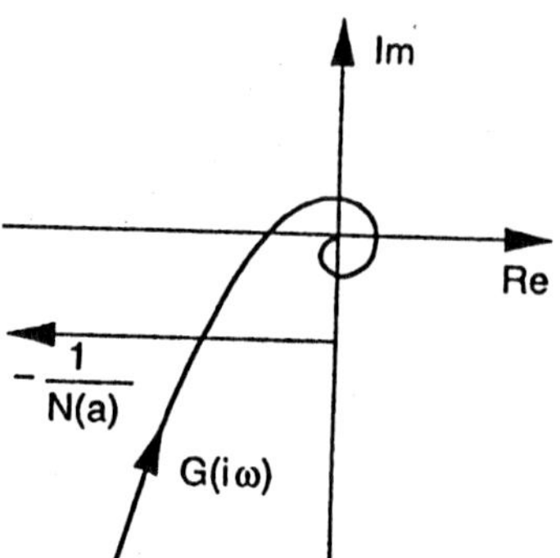

Figure 52.8 The negative inverse describing function of a relay with hysteresis $-1/N(a)$ and a Nyquist curve $G(i\omega)$.

the process, i.e., where

$$G(i\omega) = -\frac{1}{N(a)}$$

Since $N(a)$ is known, $G(i\omega)$ can be determined from the amplitude a and the frequency ω of the oscillation.

52.3.3 Gain Scheduling

By gain scheduling we mean a system where controller parameters are changed depending on measured operating conditions. The scheduling variable can, for instance, be the measurement signal, controller output or an external signal. For historical reasons the word gain scheduling is used even if other parameters like derivative time or integral time are changed. Gain scheduling is a very effective way of controlling systems whose dynamics change with the operating conditions.

The notion of gain scheduling was originally used for flight control systems, but it is being used increasingly in process control. It is, in fact, a standard ingredient in some single-loop PID controllers. For process control applications, significant improvements can be obtained by using just a few sets of controller parameters.

52.3.4 Adaptive Control

An adaptive controller is a controller whose parameters are continuously adjusted to accommodate changes in process dynamics and disturbances. Parameters can be adjusted directly or indirectly via estimation of process parameters. There is a large number of both direct and indirect methods available. Adaptation can be applied both to feedback and feedforward control parameters. Adaptive controls can be described conveniently in terms of the methods used for modeling and control design.

Model-Based Methods

All indirect systems can be represented by the block diagram in Figure 52.9. There is a parameter estimator that determines the parameters of the model based on observations of process inputs and outputs. There is also a design block that

TABLE 52.5 Industrial adaptive process controllers.

Manufacturer	Controller	Automatic tuning	Gain scheduling	Adaptive feedback	Adaptive feedforward
Bailey Controls	CLC04	Step	Yes	Model based	–
Control Techniques	Expert controller	Ramps	–	Model based	–
Fisher Controls	DPR900	Relay	Yes	–	–
	DPR910	Relay	Yes	Model based	Model based
Foxboro	Exact	Step	–	Rule based	–
Fuji	CC-S:PNA 3	Steps	Yes	–	–
Hartmann & Braun	Protronic P	Step	–	–	–
	Digitric P	Step	–	–	–
Honeywell	UDC 6000	Step	Yes	Rule based	–
Alfa Laval Automation	ECA40	Relay	Yes	–	–
	ECA400	Relay	Yes	Model based	Model based
Siemens	SIPART DR22	Step	Yes	–	–
Toshiba	TOSDIC-215D	PRBS	Yes	Model based	–
	EC300	PRBS	Yes	Model based	–
Turnbull Control Systems	TCS 6355	Steps	–	Model based	–
Yokogawa	SLPC-171,271	Step	Yes	Rule based	–
	SLPC-181,281	Step	Yes	Model based	–

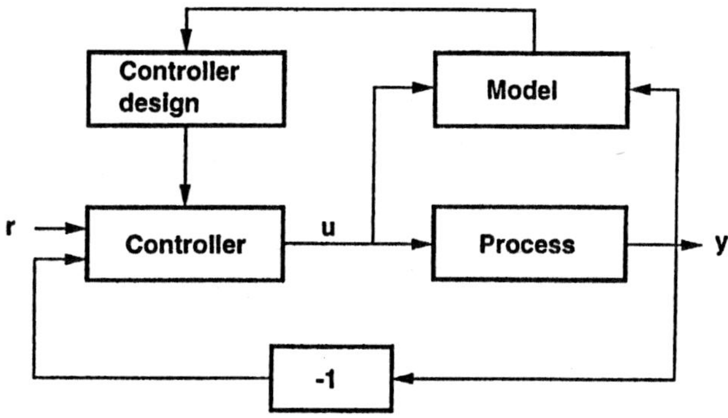

Figure 52.9 Block diagram of indirect systems.

computes controller parameters from the model parameters. The parameters can either be estimated recursively or batch-wise.

Rule-Based Methods

In the direct methods, the key issues are to find suitable features that characterize relevant properties of the closed-loop system and appropriate ways of changing the controller parameters so that the desired properties are obtained.

The majority of the PID controllers in industry are tuned manually by instrument engineers. The tuning is done based on past experience and heuristics. By observing the pattern of the closed-loop response to a set-point change, the instrument engineers use heuristics to directly adjust the controller parameters. The heuristics have been captured in tuning charts that show the responses of the system for different parameter values. A considerable insight into controller tuning can be developed by studying such charts and performing simulations. The heuristic rules have also been captured in knowledge bases in the form of crisp or fuzzy rules. Rules of this type are used in several commercial adaptive controllers. Most products will wait for set-point changes or major load disturbances. When these occur, properties like damping, overshoot, period of oscillations and static gains are estimated. Based on these properties, rules for changing the controller parameters to meet desired specifications are executed.

52.3.5 Adaptive Feed Forward

Feedforward control deserves special mention. It is a very powerful method for dealing with measurable disturbances. Use of feedforward control requires, however, good models of process dynamics. It is difficult to tune feedforward control loops automatically on demand, since the operator often cannot manipulate the disturbance used for the feedforward control. To tune the feedforward controller it is therefore necessary to wait for an appropriate disturbance. Adaptation is therefore particularly useful for the feedforward controller.

52.4 Some Commercial Products

Commercial PID controllers with adaptive techniques have been available since the beginning of the eighties.

There is a distinction between *temperature controllers* and *process controllers*. Temperature controllers are primarily designed for temperature control, whereas process controllers are supposed to work for a wide range of control loops in the process industry such as flow, pressure, level, and pH control loops. Automatic tuning and adaptation are easier to implement in temperature controllers, since most temperature control loops have several common features. This is the main reason why automatic tuning has been introduced more rapidly in these controllers.

The process controllers can be separate hardware boxes for single loops, or distributed control systems (DCS) where many loops are handled by one system.

Since the processes that are controlled with process controllers may have large differences in their dynamics, tuning and adaptation becomes more difficult compared to the pure temperature control loops. In Table 52.5, a collection of process controllers is presented together with information about their adaptive techniques.

Automatic tuning is the most common adaptive technique in the industrial products. The usefulness of this technique is also obvious from Figure 52.6, where it is shown that the auto-tuning procedures are used not only to tune the controller, but also to obtain a comfortable operation of the other adaptive techniques. Most auto-tuning procedures are based on step response analysis.

Gain scheduling is often not available in the controllers. This is surprising, since gain scheduling is found to be more useful than continuous adaptation in most situations. Furthermore, the technique is much simpler to implement than automatic tuning or adaptation.

It is interesting to see that many industrial adaptive controllers are rule based instead of model based. The research on adaptive control at universities has been almost exclusively focused on model-based adaptive control.

One of the earliest rule-based adaptive controllers is the Foxboro EXACT. It was released in 1984. In this controller, the user specifies a maximum damping and a maximum overshoot. Whenever the control loop is subjected to a larger load disturbance or setpoint change, the response is investigated and the controller parameters are adjusted according to certain rules to meet the specifications.

Adaptive feedforward control is seldom provided in the industrial controllers. This is surprising, since adaptive feedforward control is known to be of great value. Furthermore, it is easier to develop robust adaptive feedforward control than adaptive feedback control.

Alfa Laval Automation's ECA400 and Fisher Controls DPR910 are identical. This controller has automatic tuning, gain scheduling, adaptive feedback, and adaptive feedforward. The automatic tuning procedure is based on relay feedback. The automatic tuning procedure is also used to build the gain schedule automatically, and to initiate the adaptive feedback and feedforward. In this way, there is no need for the user to supply any information about the process dynamics. All adaptive features can be used automatically.

References

[1] Åström, K.J. and Hägglund, T., *PID Control — Theory, Design and Tuning,* Instrument Society of America, Research Triangle Park, NC, 2nd ed., 1995.

[2] Åström, K.J., Hägglund, T., Hang, C.C., and Ho, W.K., Automatic tuning and adaptation for PID controllers — A survey, *Control Eng. Prac.,* 1(4), 699, 1993.

[3] Chien, K.L., Hrones, J.A., and Reswick, J.B., On the automatic control of generalized passive systems, *Trans. ASME,* 74, 175, 1952.

[4] Cohen, G.H. and Coon, G.A., Theoretical consideration of retarded control, *Trans. ASME,* 75, 827, 1953.

[5] Dahlin, E.B., Designing and tuning digital controllers, *Instr. Control Syst.,* 42, 77, 1968.

[6] Hägglund, T. and Åström, K.J, Industrial adaptive controllers based on frequency response techniques, *Automatica,* 27, 599, 1991.

[7] Kaya, A. and Scheib, T.J., Tuning of PID controls of different structures, *Control Eng.,* July, 62, 1988.

[8] Kraus, T.W. and Myron, T.J., Self-tuning PID controller uses pattern recognition approach, *Control Eng.,* June, 106, 1984.

[9] Morari, M. and Zafiriou, E., *Robust Process Control,* Prentice Hall, Englewood Cliffs, New Jersey, 1989.

[10] Ziegler, J.G. and Nichols, N.B., Optimum settings for automatic controllers, *Trans. ASME,* 64, 759, 1942.

53

Self-Tuning Control

David W. Clarke
Department of Engineering Science, Parks Road, Oxford, UK

53.1 Introduction

Most control theory is concerned with the design of feedback systems for time-invariant plants with *known* mathematical models, e.g., in the form of *given* transfer functions. The assumption of constant, known models is not valid for many modern technical or nontechnical systems, such as:

Robotics. Inertias as seen by the drive motors vary with the end-effector position and the load mass so that the dynamic model varies with the robot's attitude.

Chemical reactors. Transfer functions vary according to the mix of reagents and catalyst in the vessel and change as the reaction progresses.

People. The gain of the function {injected anaesthetic $\rightarrow$ unconsciousness} depends on the patient's metabolism.

So how can controllers be designed to cope with these initial model uncertainties and time variations in dynamics? One approach is to design a *robust* fixed controller that ensures stability for all possible plant dynamics, but this is often at the expense of "detuned" behavior. Instead we can embed algorithms inside computers that "learn from experience" and *self-tune* the controllers so as to improve closed-loop performance. Often this learning process builds up a mathematical model based on experimental input/output data; this operation is known as *system identification* or *parameter estimation.* The model could be a complete transfer function or simply the gain and phase of the plant at a given input frequency. In the design of "learning" controllers there are two main themes:

- For controlling systems that have *unknown but constant* dynamics, using single-shot methods when the controller is first commissioned: If the target control law is the "industry-standard" PID (proportional-integral-derivative, or three-term) form, the approach is often called *autotuning*. If the controller is more complex, for example, when the plant has significant dead time for which PID control is not "tight", a full process model is estimated using system identification methods, and an analytic design procedure uses the model to *self-tune* the coefficients of a fixed control law.
- For *time-varying* systems: Here the identification algorithm operates all the time to "update" the model, and the coefficients of the control law are then automatically adjusted, as shown in Figure 53.1. This scheme must be "alert" so as to track *variations* in the plant's transfer function. This is the full *adaptive* control approach.

0-8493-8570-9/96/$0.00+$.50

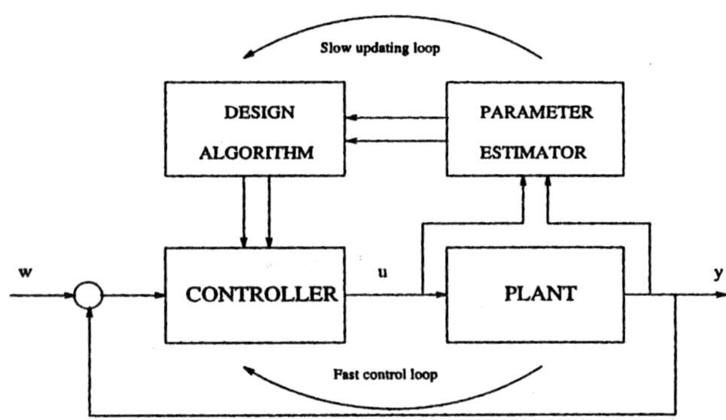

Figure 53.1 Structure of a self-tuning controller.

Figure 53.1 shows the basic structure of a self-tuning controller: the parameter estimator builds up the plant model from input/output data $\{u(t), y(t)\}$; a control design algorithm takes the estimated model parameters $\hat{\theta}$ and determines the "best" set of controller parameter θ_c (such as gains and time constants); and a controller applies these gains in a feedback loop. Such self-tuning designs have two time scales: a fast inner control loop and a slow outer "updating" loop.

53.1.1 Examples of Unknown and Time-Varying Systems

An annealing furnace thermally soaks metal over some prescribed temperature/time profile to attain desired material properties. The equation describing the furnace temperature T as a function of heat input u is

$$mc_p \frac{dT}{dt} = u - \beta T,$$

where m is the mass of the "burden," which differs from load to load, and β gives the heat loss from Newton's law of cooling. Hence the time-constant of the system, mc_p/β, varies with the burden, so that a fixed controller will lead to transients that differ accordingly. For example, a controller tuned for a large load would heat up a small mass too quickly. A self-tuned controller that deduces m (or an equivalent parameter) during warm-up provides more consistent results.

A hydraulically powered shaking table is used to test models of buildings to evaluate their earthquake resistance. It is required that the spectral density of the shaking should correspond to that of typical earthquakes at the design site. However, the model has mass and other dynamics that affect the behavior of the table (just as an inertial load added to a position control affects the loop dynamics). A dynamically complex (i.e., more than just PID) self-tuning controller is required to ensure that in closed loop the spectral density of the shaking is accurate. A similar problem occurs when testing suspension systems for racing cars using hydraulic actuators, where the objective is to replay the surface deviations of a given racing track in order to "tune" the suspension for best performance.

An exothermic chemical reactor generates heat $Q = kxe^{\alpha T}$, which varies exponentially with temperature T and depends on the proportion $x(t)$ of reagents left in the mixture. A linear model for small excursions around some nominal operating point T_0 can be deduced, so that the transfer function between u (control) and ΔT (deviations) is

$$G(s) = \frac{1}{sc + \beta - \gamma\alpha},$$

where $\gamma = kx \exp(\alpha T_0)$. Consider the behavior during a batch with $x(0) = 1$ initially and where the objective is to end up with $x = 0$ as quickly as possible. First T is small and $G(s)$ is stable. Heat is then added via u so the temperature and hence γ increase. This might cause the pole of G to become unstable. As the reagent "strength" $x(t)$ decreases during the reaction, the plant becomes stable again. In practice, many reactors are used for producing a range of chemicals for which variations *between* batches are highly significant. In such cases, a self-tuned controller that adjusts parameters according to the individual batch might suffice. Even better performance is possible if an adaptive controller is used, adjusting its parameters as the reaction progresses.

A materials testing machine has dynamics that depend on the stiffness of the specimen under test. If it is testing a manufactured item such as a bump-stop for a car suspension, the stiffness changes radically during the cycle. With a fixed controller, the response is either "sluggish" during one phase and acceptable in the other, or it can be successively well tuned initially and then oscillatory. The output of a "specimen stiffness estimator" can be used to "gain schedule" a PI (proportional-integrator) regulator to get good response over both parts of the cycle. This is full adaptive control, as rapid variations in parameters are seen.

The above examples indicate the range of possible controller tuning problems: PI settings for a simple fixed plant; a more general controller for a plant with rich dynamics; regulating a slowly changing plant; a rapidly time-varying plant for which the dynamics depend on some changing but measurable parameter.

Successful applications of self-tuning control have been for those cases where engineering knowledge leads to a simple model of the underlying dynamics for which bounds on parameter variation can be deduced. Early hopes for an effective "general-purpose" self-tuner have not been realized in practice. In particular, this is because of the requirement for "persistency of excitation" in the plant's input/output data, which cannot in general be guaranteed. Hence, in the following we concentrate on the fundamental ideas; the section "References" contains more details of, for example, multivariable self-tuning designs.

53.2 Some Simple Methods

One of the most basic problems in self-tuning is to find a "good" dynamic model of a plant. Suppose we take the classical first-order system $G(s) = K/(1 + sT)$, where K, T are unknown numbers that characterize its behavior. The simplest test is to inject a step of amplitude U and inspect the response and its derivative:

$$y(t) = KU(1 - e^{-t/T}) \;\rightarrow\; \dot{y}(t) = \frac{KU}{T} e^{-t/T}. \tag{53.1}$$

We note two things: (1) $y(t) \to KU$ as $t \to \infty$, and (2) the tangent at $t = 0$ has slope KU/T and meets the line $y = KU$ when $t = T$. Hence the "final value" gives K, and the meet of the tangent at the origin with the final value gives T. Now many real plant responses can be *approximated* by a first-order system together with a dead time, giving:

$$G(s) \approx e^{-sT_d} \frac{K}{1+sT}. \tag{53.2}$$

The response of this model is simply the above first-order response shifted by the dead time T_d, so that T_d is found just by inspecting when the output starts to move after the input step U is injected. To give a sense of the accuracy of this approximation, Figure 53.2 shows the step response of $1/(s+1)(1+0.1s)^2$, having a pole at $s = -1$ and two "fast" poles at $s = -10$. It is seen that the initial dynamics look closely like those of a dead-time of about 0.2 seconds.

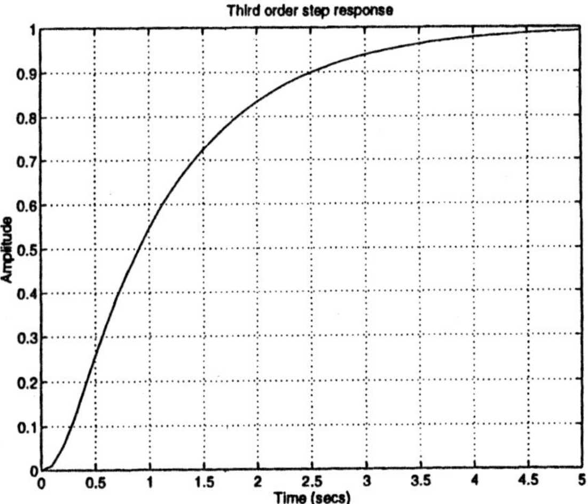

Figure 53.2 A third-order system with a dominant pole.

How can we use the above simple idea for finding K, T, T_d? What we can look for is the point of inflection of the output curve and extrapolate, as shown in Figure 53.3.

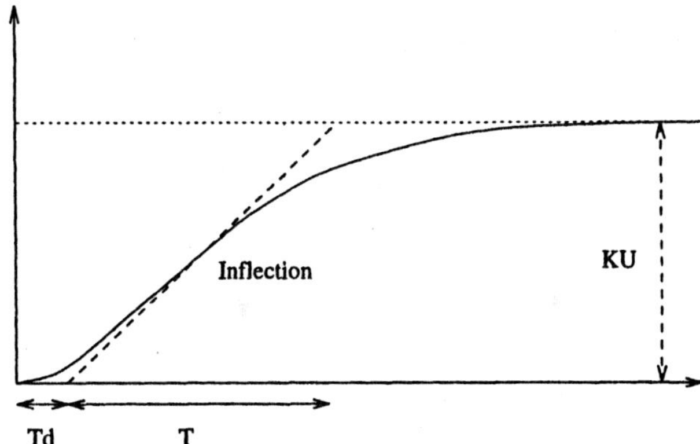

Figure 53.3 Finding K, T, T_d from a step-test.

This is easily done by inspection, but such a procedure is actually quite hard to do in a computer. Even more tricky is to derive a method that is reliable with real data where the output response is corrupted by noise. We might be able to get "reasonable" values of the model parameters "by eye", but a computer algorithm will be completely confused. What is necessary is a method that considers the *overall* response without getting locked into local features. Hence, looking for points of inflection is not the answer. Revert to the original first-order system, and write it as a differential equation:

$$T\frac{dy}{dt} + y = Ku(t).$$

Integrating with respect to t from $t = t1$ to $t = t2$ gives:

$$T[y(t2) - y(t1)] + \int_{t1}^{t2} y\,dt = K\int_{t1}^{t2} u\,dt. \tag{53.3}$$

Given a series of output samples $y(0), y(h), y(2h), \ldots, y(ih), \ldots$, where h is the sample interval, we can *approximate* the integrals in Equation 53.3 by, say, using the rectangular approximation:

$$\int_{jh}^{kh} y\,dt \approx h \sum_{i=j}^{i=k-1} y(ih),$$

leading to:

$$Ta_1 + a_2 = Kb_1, \tag{53.4}$$

where a_1, a_2, b_1 are available numerically. By repeating the procedure over another time period, a second such equation results, allowing the unknowns K, T to be deduced. We note that the use of the integrated equation to a certain extent "smoothes out" the noise.

Suppose (as with the annealing furnace) it is desired that the closed-loop transfer function is to be first-order with unit *dc* gain and a *fixed* time constant T_c *irrespective* of the values of K, T. With a PI controller of the form $C(s) = K_c(1 + 1/sT_i)$, the choices $T_i = \hat{T}$ and $K_c = T_i/(\hat{K}T_c)$ are appropriate. This completes the self-tuning design.

53.2.1 A Plant with an Unknown Time-Varying Gain

Consider now controlling an unknown-gain plant using discrete-time methods. At the sample instants, the output measurement $y(t)$ is made via an analog-to-digital converter, and the control $u(t)$ is calculated and applied via a digital-to-analog converter. We treat the general self-tuning design problem in two stages:

1. The design of a controller assuming *known* transfer function parameters
2. The estimation of the plant's dynamic parameters from the input/output data sequences

For discrete-time systems we will use a mixed Z-transform/shift-operator notation, where the z is considered to be the forward-shift operator: $zx(t) \to x(t+1)$. The current sample of a variable x is $x(t)$, the previous sample is $x(t-1)$, and so on. In particular, Δ is the backward-difference operator $\Delta = 1 - z^{-1}$, giving $\Delta x(t) \rightarrow x(t) - x(t-1)$.

A plant with known gain K has a sampled-data model $y(t+1) = Ku(t)$, as the control asserted at sample t affects the output measurement one sample later, so given the set point $w(t)$ the "best" *open-loop* control for attaining this value one sample later is clearly $u(t) = w(t)/K$. But we generally want to have *closed-loop* control so we compute an error signal:

$$e(t) = w(t) - y(t)$$

or:

$$e(t) = w(t) - Ku(t-1),$$

in the usual way. But the "best" control is such that $w(t) = Ku(t)$, so replace $w(t)$ by $Ku(t)$ and rearrange to get the feedback control:

$$u(t) = u(t-1) + e(t)/K,$$

or:

$$\Delta u(t) = u(t) - u(t-1) = e(t)/K,$$

i.e.,

$$u(t) = \frac{1}{K\Delta} e(t), \tag{53.5}$$

where $e(t)$ is the system error. Hence, the controller has the transfer function $1/K\Delta$: an integrator. We can modify the controller gain by a factor μ to give $\mu/K\Delta$ for which the "gain" $\mu = 1$ gives "ideal" single-step response to a change in w. This is the first step in a self-tuning design.

Now we design an estimator for the unknown plant gain parameter K. The basic idea is that if the *output = Gain*input*, then the *Gain estimate = output/input.* The way we will do it, however, will generalize to larger problems. The idea is to use the model to "predict the present" and to compare the prediction with the actual measured plant response.

A *prediction model* gives a forecast $\hat{y}(t+1|t)$ of the future output $y(t+1)$ depending on available input/output data and values of the model parameters. In our case, the forecast of the *current* output depends on the existing value of the gain estimate $\hat{K}(t-1)$ calculated at the previous sample:

$$\hat{y}(t|t-1) = \hat{K}(t-1)u(t-1),$$

and we define the *prediction error* $\epsilon(t)$ to be

$$\epsilon(t) = y(t) - \hat{y}(t|t-1). \tag{53.6}$$

The estimator updates at each t its "best guess" of plant gain. An "open-loop" best estimator would give $\hat{K}(t) = y(t)/u(t-1)$; but, as with control design, we prefer a *closed-loop* algorithm of the form:

$$\hat{K}(\text{new}) = \hat{K}(\text{old}) + f(\text{data}).\epsilon(t).$$

Hence we write, as for the controller design:

$$\begin{aligned} \hat{K}(t) &= y(t)/u(t-1) \\ &= [\hat{K}(t-1)u(t-1) + \epsilon(t)]/u(t-1), \end{aligned}$$

noting that

TABLE 53.1 The Signals in the Simple Self-Tuner.

t	w	y	e	K	$\hat{K}$	$\hat{y}$	ϵ	u
0	0	0	0	2	1	0	0	0
1	4	0	4	2	1	0	0	4
2	4	8	-4	2	2	4	4	2
3	4	4	0	2	2	4	0	2
4	4	2	2	1	1	4	-2	4
5	4	4	0	1	1	4	0	4

$$\begin{aligned} y(t) &= \hat{y}(t) + \epsilon(t) \\ &= \hat{K}(t-1) + \frac{\lambda}{u(t-1)}\epsilon(t), \end{aligned} \tag{53.7}$$

where the "estimator gain" λ is 1 for "optimal" performance. The sequence of operations is first to estimate $\epsilon(t)$ by comparing the old model's prediction $\hat{y}(t)$ of the newly acquired output data $y(t)$, then to update the model parameters depending on the chosen value of "adaptive gain" λ.

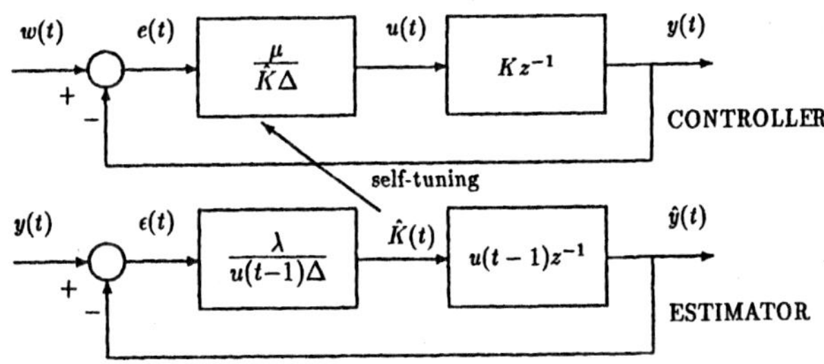

Figure 53.4 The controller and estimator as feedback loops.

To get the full self-tuner we just couple the estimator of plant gain of Equation 53.7 to the feedback controller of Equation 53.5 by passing $\hat{K}$ from one to the other. Note the resemblance between controller and estimator as shown in Figure 53.4: both are integrators. The use of feedback in control is for good performance *despite* "bad" models and disturbances; similarly, with the estimator we can get good results despite noise added to y. Just as with control where loop gain μ is reduced to improve stability, we reduce λ to improve the estimator's robustness. With "optimal" values of μ and λ, the algorithm takes just two successive sample times to get perfect control: step 1 gives an accurate $\hat{K}$ and step 2 attains the set-point. As an example, consider a plant that at $t = 0$ has $K = 2$ and our model $\hat{K}(0) = 1$. At $t = 1$ the set point w is made 4, and at $t = 3$ the plant gain changes to 1. Iterating the equations for the first five samples gives Table 53.1.

53.3 Plant Models

A mathematical model of a plant is required in order to design a controller. The design procedure needs to be analytic so that a unique control comes from a given model and performance criterion. In most cases, we consider a model as a *predictor* of the behavior of the system, e.g., its output response $y(t+i)$ to inputs $u(t-j)$. The model will have a set of parameters θ; *identification* will provide values for these, which can be fed into the control design. It is to be hoped that the model can in fact represent the "real" behavior of the plant. (Aside: A distillation column might be described by four hundred simultaneous nonlinear differential equations—far too complex for our needs. Fortunately, it is often well represented by a second-order transfer function with dead time.)

We adopt a *model structure* $\mathcal{M}(\theta)$ that is *parameterized* by the set θ; for example, we could choose:

$$\mathcal{M}(\theta) = \frac{b_0}{a_1 s + a_0}, \text{ i.e. } \theta = \{a_0, a_1, b_0\},$$

but this would not be sensible as an infinite number of values θ would give the same transfer function. Clearly it is better to divide through by a_1 and assume:

$$\mathcal{M}(\theta) = \frac{b_0}{s + a_0}, \text{ i.e., } \theta = \{a_0, b_0\},$$

with one fewer parameter to estimate and a unique answer obtainable.

The choice of model structure is therefore important in the design of general-purpose system identification or self-tuning schemes:

- Can $\mathcal{M}$ represent a general class of plants (e.g., unstable, lightly damped, high order, ...) so that the self-tuner does not need to be redesigned for each case?
- Does $\mathcal{M}$ use a *minimal* set of parameters so that the "true" input/output behavior is given by unique values of θ?
- Is the structure of $\mathcal{M}$ such that estimation is simple? So that it is *robust* to bad assumptions (e.g., about the number of poles and zeros in the real plant)?

We shall concentrate on dynamic models in which the underlying plant is continuous time and assumed to be in the locally linearized form:

$$y(t) = \frac{B(s)}{A(s)} u(t - T_d) + d(t), \tag{53.8}$$

where T_d is the dead time and $d(t)$ is an unmeasured disturbance. The inclusion of a dead time is for two reasons: (1) many industrial processes involve mass transport with corresponding delays, but also (2) plants with complex dynamics (e.g., many poles) can be *approximated* by a low-order model with dead time. The disturbance reflects reality: it corresponds to any factor that affects the output not included in the $G(s)$ such as measurement noise, "unmodeled dynamics", nonlinear effects, real disturbances (such as load torques) acting on the system. It will be found that estimation is a *compromise* between "alertness" to changes in the plant's $G(s)$ and susceptibility of the estimator to noise (fast estimators adapt rapidly to new output data, which sadly could be noise rather than real signals).

As most identification and self-tuning algorithms are implemented digitally, we shall assume *discrete-time* models obtained from $G(s)$ via sampling. These models can come in many forms, though all are usable as predictors. The simplest general-purpose form is the *pulse response* or weighting sequence. Consider injecting a unit pulse (i.e., of height 1 and over the sample interval h) into the plant and sampling the output to get a sequence $\{h_i\}$. *For a general input* $u(t)$, the sampled output $y(t)$ is then given by the *convolution sum*:

$$y(t) = \sum_{i=1}^{\infty} h_i u(t - i). \tag{53.9}$$

The only assumptions here are superposition and open-loop stability: the arbitrary linear plant has a parameter set $\theta = \{h_i\}$ being the sampled points on the pulse response.

Theoretically there are an *infinite* number of parameters in this model, but in practice we *truncate* after some point N, assuming $[h_i = 0, i > N]$ and where h_0 is zero, as the plant cannot respond instantaneously. If there is a dead time of k samples between a control and its initial effect on the output, then $[h_i = 0, \text{ for } i = 1 \cdots k]$ also, so that the model handles dead time simply by having leading coefficients of zero.

One problem with FIR models is that they require a very large number of parameters accurately to represent "stiff" dynamic systems (i.e., with fast and slow modes in the same plant), or even a simple lightly damped pole pair. The sample interval h must be less than the smallest time constant of interest, and the model "length" must be such that Nh exceeds the plant settling time, which is typically five times the largest time constant. Hence, even with only a 1:10 range of time constants, at least 50 parameters may be necessary. Nevertheless, the associated computations are very simple (easily embedded in VLSI), and FIRs are commonly used in signal processing and some process control designs. It sometimes proves to be useful to consider the input to a plant as a series of *increments* $\Delta u(t) = u(t) - u(t-1)$ (i.e., steps or control "moves" as in a stepper-motor) and the response (by superposition) to be

$$y(t) = s_1 \Delta u(t-1) + s_2 \Delta u(t-2) + \cdots + s_i \Delta u(t-i) + \cdots,$$

where s_i is the ith point on the plant's *unit-step response*. It is easy to show that the predictor that gives the next plant output is

$$y(t+1) = y(t) + s_1 \Delta u(t) + \sum_{i=1}^{N} h_i \Delta u(t - i). \tag{53.10}$$

The predicted output $y(t+1)$ is the sum of three components: the current output $y(t)$, the *forced response* due to the current control "move" $\Delta u(t)$, plus the *free response* due to previous control moves.

53.3.1 Transfer-Function or Difference-Equation Models

By far the most popular model in self-tuning control is the DARMA (deterministic autoregressive and moving average) or general difference-equation form:

$$\begin{aligned} y(t) + a_1 y(t-1) + a_2 y(t-2) + \cdots + a_{na} y(t - na) = \\ b_1 u(t-1) + b_2 u(t-2) + \cdots + b_{nb} u(t - nb). \end{aligned} \tag{53.11}$$

Defining $A(z^{-1})$ and $B(z^{-1})$ to be polynomials in the backward-shift operator:

$$A(z^{-1}) = 1 + a_1 z^{-1} + \cdots + a_{na} z^{-na} \quad (53.12)$$

$$B(z^{-1}) = b_1 z^{-1} + \cdots + b_{nb} z^{-nb}, \quad (53.13)$$

this can be written

$$A(z^{-1})y(t) = B(z^{-1})u(t),$$

i.e., a transfer function

$$G(z^{-1}) = \frac{B(z^{-1})}{A(z^{-1})}. \quad (53.14)$$

The values of the parameters are obtained by taking Z-transforms of $G(s)$ (+ZOH) as previously shown. Note that:

- There is a unique correspondence [via the mapping $z = \exp(sh)$] between the continuous- and discrete-time poles; the degree na is the same as that of $G(s)$. A pole $s = -\alpha$ in the left half-plane (LHP) in s (stable) maps into $z = \exp(-\alpha h)$; i.e., within the unit circle.
- There is *no* simple mapping of zeros. Indeed, even if $G(s)$ has all its zeros in the stability region (LHP), this does *not* mean that $G(z^{-1})$ will be likewise (in unit circle). It is, in fact, extremely common for so-called *nonminimum-phase* discrete-time models to appear (zeros outside unit circle), e.g., when there is *fractional dead time* or when controlling a high-order plant with a small sample interval h.

53.3.2 Incorporating Disturbances

In continuous-time random processes it is useful to define "white noise"—a signal with a constant spectral power at all frequencies. In discrete time, the corresponding signal is a sequence of random independent (uncorrelated) variables with zero mean and common variance σ^2 [hence called a $(0, \sigma^2)$ uncorrelated random sequence (URS)]. Here we do not have an "infinite variance" signal, but instead something easily produced by a random signal generator such as RAND in Matlab. Such a signal has the following properties:

$$\mathcal{E}e(t) = 0; \;\; \mathcal{E}e^2(t) = \sigma_e^2; \;\; \mathcal{E}e(i)e(j) = 0, \text{ for } i \neq j, \quad (53.15)$$

where $\mathcal{E}$ denotes the expectation operation. As in continuous time, a *general* stationary discrete-time random process is modeled by *white noise passing through a transfer function* in z.

Suppose the controlled part of a plant is $G = B_1/A_1$ and the output y is affected by additive disturbances $C_1/D_1 e(t)$. Then we have:

$$y(t) = \frac{B_1}{A_1}u(t) + \frac{C_1}{D_1}e(t),$$

which, when multiplied up, gives:

$$A_1 D_1 y(t) = B_1 D_1 u(t) + C_1 A_1 e(t).$$

Hence, we make the common overall plant assumption of the CARMA (controlled autoregressive and moving average) model:

$$A(z^{-1})y(t) = B(z^{-1})u(t) + C(z^{-1})e(t), \quad (53.16)$$

where $e(t)$ is a URS sequence of independent $(0, \sigma^2)$ random variables. The corresponding difference equation is then generated using the polynomials A, B, C.

However, the CARMA form is insufficient to characterize *offsets* for which $u = 0$ is not accompanied by $y = 0$. A model could include *deviations* from the mean levels u_0, y_0, say, but in control loops these mean levels would not, in general, be constant or known. Indeed, it is often found that disturbances are "steps"; for example, when passengers enter an elevator, there are steps in the load torque acting on the motor: these correspond to shifts in mean levels. One way to deal with this problem is to add an offset parameter d to the output. In a full model used in self-tuning, the value of d would need to be estimated along with the other (dynamic) parameters in the A, B, C polynomials.

In practice, the disturbance is likely to be a combination of factors such that no consistent values of the parameters $C(z^{-1})$ can be estimated. But it is known from the internal model principle that *disturbance elimination is best achieved by having an internal model of the disturbance in the control law.* Inspection provides the following "theorem": *the ubiquitous nature of PID regulators in industry implies that an integrator is the internal model of most practical disturbances.* Hence, the best *general* assumption is of a CARIMA (integrated) model in which random disturbances are integrated:

$$A(z^{-1})y(t) = B(z^{-1})u(t) + C(z^{-1})x(t), \quad (53.17)$$

where $x(t)$ is of the form a/Δ for deterministic step disturbances, or $\Delta x(t) = e(t)$ [with $e(t)$ a URS] for random disturbances. Incorporating into the model gives an *incremental* form:

$$A(z^{-1})\Delta y(t) = B(z^{-1})\Delta u(t) + C(z^{-1})e(t), \quad (53.18)$$

or

$$\begin{aligned} y(t) = \; & y(t-1) - a_1 \Delta y(t-1) - \cdots a_{na} \Delta y(t-na) \\ & + b_1 \Delta u(t-1) + \cdots b_{nb} \Delta u(t-nb) + e(t) \\ & + c_1 e(t-1) + \cdots + c_{nc} e(t-nc). \end{aligned} \quad (53.19)$$

Note that the model deals with *increments* of the input/output data such as $\Delta y(t-1) = y(t-1) - y(t-2)$ and hence no *dc* term is involved as Δ.constant = 0. It is found that injecting a *pulse* into x (or e) gives a step change in y, and injecting a "white" random sequence or URS gives "Brownian motion" for which y "drifts" (Brownian motion is quite a good model of stock exchange prices).

53.4 Recursive Prediction Error (RPE) Estimators

The job of an estimator is to provide values for the model parameters based on fitting the model's responses to the measured plant input/output data. A *recursive* estimator *updates* the estimates at each sample instant based on the newly available information. One important point about recursive estimators is that the computational load must *not* increase with time. Note that potentially the amount of data is always increasing with more available from each sample. *Prediction error* methods consider the model to be a forecaster $\hat{y}(t|t-1)$ of the actual outcome $y(t)$, the difference or residual $\epsilon = y - \hat{y}$ being used to correct the estimates. Now define the following n vectors:

1. $\theta = [\theta_1, \ldots, \theta_n]'$ are the n unknown plant parameters.
2. $\hat{\theta}(t) = [\hat{\theta}_1(t), \ldots, \hat{\theta}_n(t)]'$ are the corresponding estimates at time t.
3. $\mathbf{x}(t) = [x_1(t), \ldots, x_n(t)]'$ are *known* data associated with the parameters.

A *prediction model* generates a forecast $\hat{y}(t)$ depending on $\hat{\theta}(t-1)$ and $\mathbf{x}(t)$:

$$\hat{y}(t) == \hat{y}(t|t-1) = f(\hat{\theta}(t-1), \mathbf{x}(t)). \tag{53.20}$$

The models we will consider are all in *linear-in-the-parameters* (LITP) form with:

$$\hat{y}(t) = \sum_{i=1}^{n} \hat{\theta}_i(t-1)x_i(t), \tag{53.21}$$

and where the measured plant output is assumed to satisfy the LITP equation:

$$y(t) = \sum_{i=1}^{n} \theta_i x_i(t) + e(t), \tag{53.22}$$

with $e(t)$ corresponding to "noise" or "disturbances" and assumed to be independent of the data $\mathbf{x}(t)$. The scalar *prediction error* is defined to be

$$\epsilon(t) = y(t) - \hat{y}(t) = y(t) - f(\hat{\theta}(t-1), \mathbf{x}(t)), \tag{53.23}$$

where $y(t)$ is the new output measurement. The general structure is shown in Figure 53.5.

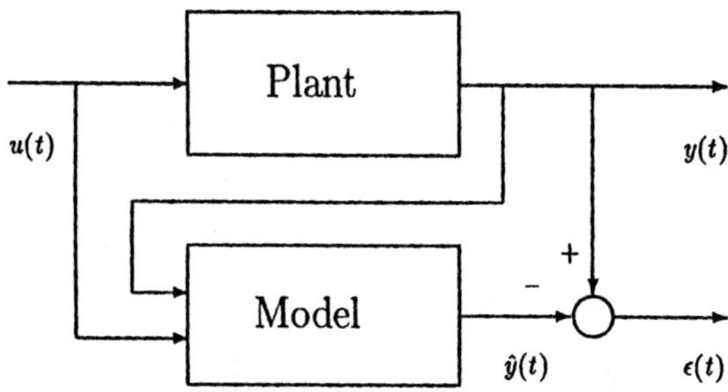

Figure 53.5 General RPE structure.

Typical models that come under the general description of Equation 53.21 are:

Pulse response

$$\begin{aligned} y(t) &= \sum_{i=1}^{n} h_i u(t-i) \\ \text{where: } \theta &= [h_1, h_2, \ldots, h_n]' \\ \text{and: } \mathbf{x}(t) &= [u(t-1), u(t-2), \ldots, \\ &\quad u(t-n)]'. \end{aligned}$$

In this model, the parameter vector are points on the pulse response, and the data vector contains n past values of the control signal.

DARMA

$$\begin{aligned} y(t) &= -a_1 y(t-1) - \ldots + b_1 u(t-1) \\ &\quad + \ldots + b_{nb} u(t-nb) \\ \text{where: } \theta &= [a_1, \ldots a_{na}, b_1, \ldots, b_{nb}]' \\ \text{and: } \mathbf{x}(t) &= [-y(t-1), \ldots, -y(t-na), \\ &\quad u(t-1), \ldots, u(t-nb)]'. \end{aligned}$$

In this model, the parameters are the difference-equation constants, and the data vector contains past values of inputs *and* outputs.

Note that the LITP equations can be written concisely as:

$$\hat{y}(t) = \mathbf{x}'(t)\hat{\theta}(t-1),$$

and equivalently

$$y(t) = \mathbf{x}'(t)\theta + e(t),$$

so that if the *parameter error vector* $\tilde{\theta}(t)$ is defined to be $\theta - \hat{\theta}(t)$, then:

$$\epsilon(t) = y(t) - \hat{y}(t) = \mathbf{x}'(t)\tilde{\theta}(t-1) + e(t). \tag{53.24}$$

This important equation shows that the prediction error, has two components: the model error and the unknown disturbance. Our use of this result depends on whether we want to identify a time-invariant system (in which case we average over a lot of data to eliminate the effects of noise) or whether we want to track a time-varying plant (which is possible only if the signal to noise ratio is large and the variations are not too fast). We can associate with the prediction error a cost function J such as $J(t) = 0.5\epsilon^2(t)$. If there is no noise, this cost depends only on $\tilde{\theta}$ and so we can imagine a set of equal-cost contours in the n-space of the parameters $\hat{\theta}$ with the minimum value at the "true" parameters. If there is noise, these contours will correspond to average or expected values.

An RPE algorithm updates the estimates using:

$$\hat{\theta}(t) = \hat{\theta}(t-1) + a(t)\mathbf{M}(t)\mathbf{x}(t)\epsilon(t), \tag{53.25}$$

where:

- $a(t)$ is a scalar "gain factor", giving the step length.

- $\mathbf{x}(t)\epsilon(t)$ is along the gradient vector $-\nabla J$ of the cost-function surface, pointing down the slope of local steepest descent.
- $\mathbf{M}(t)$ is a rotation matrix that modifies the parameter update direction away from the steepest descent route.

We can reduce the parameter error using a "large" value of the gain $a(t)$, but this increases the effect of the disturbance $e(t)$. The compromise between rapid model error reduction and insensitivity to noise is a fundamental design issue: in practice, we want rapid initial convergence, followed by good noise immunity for steady-state tracking. In view of its long history of success in many applications, the most common estimator used in self-tuning is based on variants of the *least-squares* (LS) method, for which the current estimate $\hat{\theta}(t)$ minimizes the "loss function":

$$J(t) = (\hat{\theta}(t) - \hat{\theta}(0))'\mathbf{S}(0)(\hat{\theta}(t) - \hat{\theta}(0)) + \sum_{i=1}^{t}(y(i) - \mathbf{x}'(i)\hat{\theta}(t))^2 \qquad (53.26)$$

It can be shown (e.g., Isermann et al. [8]) that $J(t)$ is minimized by the recursions:

$$\mathbf{S}(t) = \mathbf{S}(t-1) + \mathbf{x}(t)\mathbf{x}'(t) \qquad (53.27)$$

$$\hat{\theta}(t) = \hat{\theta}(t-1) + \mathbf{S}(t)^{-1}\mathbf{x}(t)\epsilon(t), \qquad (53.28)$$

with $\hat{\theta}(0)$ being the initial "guess" of the parameters and $\mathbf{S}(0)$ an assertion about their likely accuracy (i.e., small entries in $\mathbf{S}$ implies low accuracy). It can be shown that:

$$\mathbf{S}(t)\hat{\theta}(t) = \mathbf{S}(0)\hat{\theta}(0) + [\mathbf{S}(t) - \mathbf{S}(0)]\theta, \qquad (53.29)$$

so that as $\mathbf{S}(t)$ increases, the effect of the initial assumptions declines, and a solution is possible only if $\mathbf{S}$ has full rank: a "persistency of excitation" condition.

In practice, we prefer to propagate $\mathbf{P}(t) = \mathbf{S}^{-1}(t)$ to avoid the inversion in Equation 53.28; using $\mathbf{P}$, the important recursive least-squares (RLS) equations become:

Kalman gain vector:

$$\mathbf{k}(t) = \frac{\mathbf{P}(t-1)\mathbf{x}(t)}{1 + \mathbf{x}'(t)\mathbf{P}(t-1)\mathbf{x}(t)} \qquad (53.30)$$

parameter update:

$$\hat{\theta}(t) = \hat{\theta}(t-1) + \mathbf{k}(t)\epsilon(t) \qquad (53.31)$$

covariance update:

$$\mathbf{P}(t) = [\mathbf{I} - \mathbf{k}(t)\mathbf{x}'(t)]\mathbf{P}(t-1) \qquad (53.32)$$

The algorithm needs to be initialized. Note that $\mathbf{P}$ is proportional to the "errors in the parameters," as $\mathbf{P} = \mathbf{S}^{-1}$, so that at $t = 0$ where nothing (or little) is known, the appropriate choice is $\gamma\mathbf{I}$ where γ is large. It can then be shown that the estimates obtained after n samples using RLS are the same as with "off-line" LS. For smaller γ, though, slower parameter movement is seen, for that choice indicates the chosen $\hat{\theta}(0)$ are fairly accurate. In this way it is possible to start recursive estimation with a reasonable "guessed" model rather than simply $(0, 0, 0\ldots)'$.

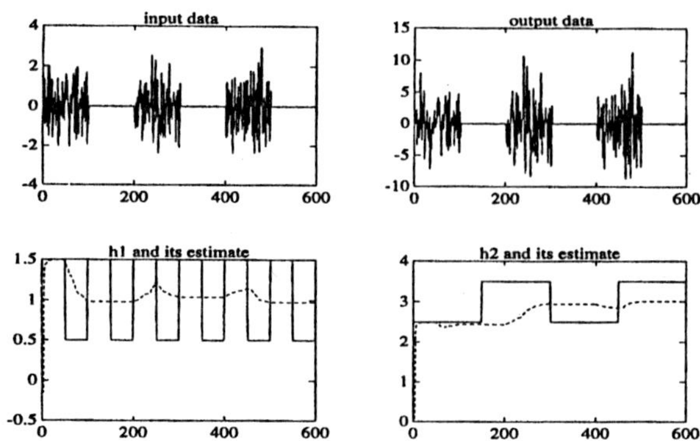

Figure 53.6 RLS with no forgetting.

What happens for large t? Assuming that there are good persistently exciting data, then $\mathbf{S} = \sum_0^t \mathbf{x}\mathbf{x}'$ increases all the time and hence $||\mathbf{k}|| \to 0$. This means that asymptotically RLS loses its "alertness" (though by that time there should be convergence to the "true" values despite noise). This is a natural consequence of the built-in assumptions: a fixed parameter plant where accurate estimates are wanted. Hence, the method is possibly suitable for self-tuning but not for adaptive control. Figure 53.6 shows RLS in action with data from a FIR model in which the parameters $\{h_1(t), h_2(t)\}$ each change by square waves. The input is white noise in "bursts" so it is not exciting during quiescent periods. The estimates do not move at all when there is zero excitation and change at an increasingly slow rate even when the data are "rich." Hence, RLS concentrates on accuracy (assuming fixed plant parameters) and not on tracking parameter changes.

53.4.1 Forgetting Factors

The cost function in LS weights all residuals $\epsilon(t-j)$ equally, no matter how far back in the past the data were acquired. If it is expected that the model dynamics varies with time, then recent data are more significant than old data. Thus, the idea is to "forget" so that the effect of data on the estimates decays in time. It is convenient to consider a simple single-parameter continuous-time estimator to explore the ideas. As a preliminary, consider the integral:

$$I = \int_0^t e^{-\alpha(t-\lambda)} f(\lambda)\, d\lambda. \qquad (53.33)$$

Then, from first principles:

$$\frac{dI}{dt} = f(t) - \alpha I. \qquad (53.34)$$

A fixed forgetting factor exponentially weights past data at a rate α, so that using the continuous-time exponentially weighted version of Equation 53.26, the estimate $\hat{\theta}(t)$ minimizes:

$$J(t) = e^{-\alpha t}s(0)(\hat{\theta}(t) - \hat{\theta}(0))^2 + \int_0^t e^{-\alpha(t-\lambda)}(y(\lambda) - \hat{\theta}(t)x(\lambda))^2\, d\lambda, \qquad (53.35)$$

giving, by setting the partial differential with respect to $\hat{\theta}$ to zero:

$$s(t)\hat{\theta}(t) = e^{-\alpha t}s(0)\hat{\theta}(0) + \int_0^t e^{-\alpha(t-\lambda)}y(\lambda)x(\lambda)\,d\lambda \quad (53.36)$$

where:

$$s(t) = e^{-\alpha t}s(0) + \int_0^t e^{-\alpha(t-\lambda)}x^2(\lambda)\,d\lambda. \quad (53.37)$$

Differentiating with respect to t and using the preliminary result gives the recursive equations of continuous-time RLS for a single parameter:

$$\frac{ds}{dt} + \alpha s = x^2 \quad (53.38)$$

$$\frac{d\hat{\theta}}{dt} + \frac{x^2}{s}\hat{\theta} = \frac{xy}{s}, \quad (53.39)$$

with initial conditions $\{s(0), \hat{\theta}(0)\}$. The Equation 53.39 shows that, for constant excitation $x = X$, $s(t)$ rises exponentially to X^2/α; for zero "forgetting" $\alpha = 0$, $s(t) \to \infty$ if the data are persistently exciting. In practice, the convention is to use the current prediction error $\epsilon = y - \hat{\theta}x$, so substituting we get as the update equation:

$$\frac{d\hat{\theta}}{dt} = \frac{x}{s}\epsilon. \quad (53.40)$$

It is instructive to find how the modeling error $\tilde{\theta} = \theta - \hat{\theta}$ changes with time. Substituting for y using $y = \theta x + e$ and assuming that θ varies in time we get:

$$\frac{d\tilde{\theta}}{dt} + \frac{x^2}{s}\tilde{\theta} = \frac{d\theta}{dt} - \frac{x}{s}e.$$

The left-hand side tries to drive $\tilde{\theta}$ to zero at a rate depending on x^2/s while the right-hand drives it away from zero. In RLS it is conventional to use the inverse $p = 1/s$ in the calculations. Substituting in the update for s we get:

$$\frac{dp}{dt} - \alpha p = -p^2x^2,$$

or:

$$\frac{dp}{dt} + (x^2p - \alpha)p = 0. \quad (53.41)$$

The behavior of forgetting factors such as α is clearly shown in this formulation: the equation has an unstable mode when $x^2p < \alpha$. If there is no excitation $[x(t) = 0]$, $p(t)$ increases exponentially at a rate determined by α. If x is constant p increases until $p = \alpha/x^2$ and then becomes constant. Hence, clearly one method to avoid "blow up" is to add a small quantity x_0^2 to x^2 in the updating equation for p.

53.4.2 Variable Forgetting Factors

The loss function J can be written in terms of ϵ:

$$J(t) = J_0(t) + \int_0^t e^{-\alpha(t-\lambda)}\epsilon(\lambda)^2\,d\lambda.$$

Suppose now that estimation is perfect (i.e., $\hat{\theta} = \theta$) for all t so that $\epsilon(t) = e(t)$ always and the variance of the noise is σ_0^2. Then the expected value of the irreducible or "ideal loss" is given by:

$$\mathcal{E}\{J_{opt}\} = \int_0^t e^{-\alpha(t-\lambda)}\sigma_0^2\,d\lambda = \frac{\sigma_0^2}{\alpha}(1 - \exp^{-\alpha t}).$$

For large t, this approaches the value $\sigma_0^2/\alpha = T_c\sigma_0^2$, say, where T_c corresponds to the "asymptotic averaging time"; i.e., the period of past time that contains the data "most influential" in providing the estimate.

Consider the differential equation for J, which is:

$$\frac{dJ}{dt} + \alpha J = \epsilon^2(t).$$

We would expect the average value for ϵ^2 to be greater than σ_0^2, given that $\epsilon(t) = x\tilde{\theta}(t) + e(t)$. In particular, we expect that $\epsilon^2 \gg \sigma_0^2$ when the estimate is far from the true value of θ. One way to proceed is to assert that the value J (or $1/J$) contains the "information" about the parameter and that this should be constant in the steady state. This implies that a good value for α is obtained by making $\alpha J = \epsilon^2 = \alpha J_{opt}$, giving:

$$\alpha = \frac{1}{T_c}\left(\frac{\epsilon(t)}{\sigma_0}\right)^2,$$

and the update equation for s now reads:

$$\frac{ds}{dt} + \frac{1}{T_c}\left(\frac{\epsilon(t)}{\sigma_0}\right)^2 s = x^2. \quad (53.42)$$

Hence, adaptation is faster with a prediction error that is large compared with the assumed SD of the underlying noise.

53.4.3 Forgetting with Multiparameter Models

For discrete-time estimation we write $\beta = \exp(-\alpha h)$, where h is the sample interval (small $\alpha \to \beta \approx 1 - \alpha h$,) and then the RLS equations become:

$$\mathbf{k}(t) = \frac{\mathbf{P}(t-1)\mathbf{x}}{\beta + \mathbf{x}'\mathbf{P}(t-1)\mathbf{x}} \quad (53.43)$$

$$\hat{\theta}(t) = \hat{\theta}(t-1) + \mathbf{k}(t)\epsilon(t) \quad (53.44)$$

$$\mathbf{P}(t) = [\mathbf{I} - \mathbf{k}(t)\mathbf{x}'(t)]\mathbf{P}(t-1)/\beta \quad (53.45)$$

A useful measure of the amount of data effectively contributing to the current estimate is the *asymptotic sample length* (ASL) or time constant, given by $1/(1 - \beta)$. A rapidly varying system might need an ASL of 20 samples ($\beta = 0.95$) to more than 1000 ($\beta = 0.999$). Figure 53.7 repeats the experiment of Figure 53.6 but using $\beta = 0.95$. Again, when the input is not exciting, the parameters freeze (and $\mathbf{P}$ "blows up"), but otherwise track changes rapidly. Moral: Use forgetting, but then make sure the input perturbs the plant.

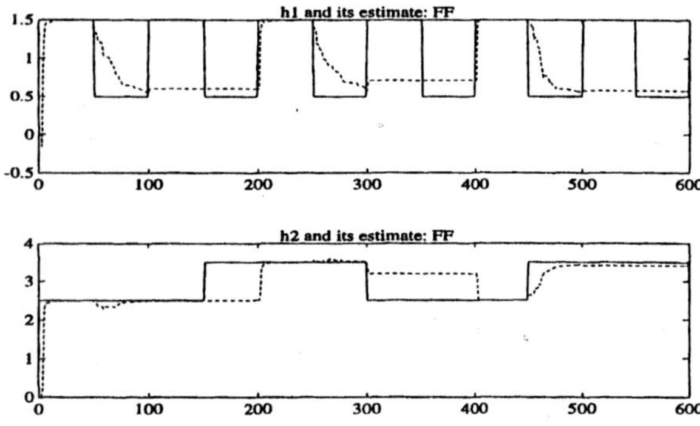

Figure 53.7 RLS with forgetting.

53.5 Predictive Models

Industrial processes have characteristics that make advanced control useful:

Disturbances: Fluctuations in the raw materials and the operating environment; sensor noise.

Dead-time: The effect of the current control is not seen in the measured response for a number of samples, because of material transport times.

Typical examples of processes like this are:

Steel rolling: Controls are via screws on the stands; gauge responses are after the end of the mill; X-ray gauge measurements are noisy.

Papermaking: The controls are on the head box at entry to the Fourdrinier wire; the basis weight of the dried paper is measured after the heating rolls.

Strip annealing: Inductive heating causes temperature changes, later measured by a pyrometer with large random fluctuations in its signal.

Distillation: The multiple lags (several hundred) arising from the thermal behavior of each tray appear like dead time; composition is measured by a chromatograph with sampling delays; ambient thermal variations induce disturbances.

Delays cause phase lag, which means that PID control gains must be reduced. Predictive control can give "perfect" control provided the delay is known; even better performance is obtainable if the disturbance process can be predicted also. Consider the problem of predicting the plant output with its two components:

Deterministic effects: Old inputs and outputs give initial conditions, from which the "free response" can be determined via the model. The "forced response" is the additional effect due to current and future controls.

Disturbances: Old URS values $e(t-i)$ can be reconstructed from the model and known data. The free response can then be computed; as nothing is known about future white noise, the best approach is simply to assume its mean value of 0.

Consider an *autoregressive* (AR) process driven by white noise giving the measured output $y(t)$:

$$y(t) = 0.9y(t-1) + e(t), \text{ or } y(t) = \frac{1}{(1-0.9z^{-1})}e(t).$$

By expanding as a power series, the model can be written in pulse form:

$$\begin{aligned} y(t) &= e(t) + 0.9e(t-1) + 0.9^2e(t-1) \\ &\quad + 0.9^3e(t-2) + \cdots. \end{aligned}$$

Consider its value two steps into the future:

$$\begin{aligned} y(t+2) &= e(t+2) + 0.9e(t+1) + 0.9^2e(t) \\ &\quad + 0.9^3e(t-1) + \cdots. \end{aligned}$$

As $e(t+1)$, $e(t+2)$ are not known at time t, the best prediction of $y(t+2)$ is

$$\begin{aligned} \hat{y}(t+2|t) &= 0.9^2e(t) + 0.9^3e(t-1) + \cdots \\ &= \frac{0.9^2}{(1-0.9z^{-1})}e(t). \end{aligned}$$

But the noise model gives $e(t) = (1-0.9z^{-1})y(t)$, giving the simple result that $\hat{y}(t+2|t) = 0.9^2y(t)$ (the free response of the noise model). The error in the prediction is

$$\tilde{y}(t+2|t) = e(t+2) + 0.9e(t+1).$$

To find the variance or mean square of this signal, simply square and take averages, taking $\mathcal{E}e(t+i)^2 = \sigma^2$; $\mathcal{E}e(t+i)e(t+j) = 0$, and the following should be noted:

1. The prediction error (p.e.) is *independent* of the "known" data [in this case, $e(t), e(t-1), \cdots$]; i.e., the maximum possible information has been extracted.
2. The variance of the p.e. is $(1+0.9^2)\sigma^2$, where σ^2 is the variance of the URS e.
3. The p.e. variance increases with the prediction *horizon*; here, two-steps ahead.
4. "Sluggish" disturbances (pole near 1) are predicted with more accuracy than rapidly moving disturbances.

To generalize this example, consider a noise model in *moving-average* (MA) form:

$$y(t) = N(z^{-1})e(t),$$

so that:

$$\begin{aligned} y(t+k) &= (1 + n_1z^{-1} + \cdots + n_{k-1}z^{-k+1})e(t+k) \\ &\quad + (n_kz^{-k} + n_{k+1}z^{-k-1} + \cdots)e(t+k) \\ &= N_k^*(z^{-1})e(t+k) + N_k(z^{-1})e(t). \qquad (53.46) \end{aligned}$$

The disturbance is split into "future" and "past" components, and the prediction uses known data:

$$\hat{y}(t+k|t) = N_k(z^{-1})e(t) = n_ke(t) + n_{k+1}e(t-1) + \cdots.$$

How does this procedure work with a transfer function (ARMA) structure:

$$y(t) = C(z^{-1})/A(z^{-1})e(t)?$$

Performing long division by A and stopping after k terms gives:

$$\frac{C(z^{-1})}{A(z^{-1})} = E(z^{-1}) + z^{-k}\frac{F(z^{-1})}{A(z^{-1})} = N_k^* + z^{-k}N_k.$$

In fact, instead of doing long division we multiply each side by A to get:

$$C(z^{-1}) = E(z^{-1})A(z^{-1}) + z^{-k}F(z^{-1}). \quad (53.47)$$

This key equation is known as a *Diophantine* (or *Bezoutian*) identity from which E and F can be obtained for given A, C, k by equating powers of z^{-1}. For example, consider:

$$(1 - 0.9z^{-1})y(t) = (1 + 0.7z^{-1})e(t) \rightarrow A = (1 - 0.9z^{-1});$$
$$C = (1 + 0.7z^{-1}).$$

Hence, the Diophantine identity of Equation 53.47 for $k = 2$ becomes:

$$(1 + 0.7z^{-1}) = E(z^{-1})(1 - 0.9z^{-1}) + z^{-2}F(z^{-1}),$$

or

$$(1 + 0.7z^{-1}) = (e_0 + e_1z^{-1})(1 - 0.9z^{-1}) + z^{-2}f_0. \quad (53.48)$$

Equating coefficients of increasing powers of z^{-1} in Equation 53.48:

$$z^0 : 1 = e_0;\ z^{-1} : 0.7 = -0.9 + e_1;$$
$$z^{-2} : 0 = -0.9e_1 + f_0.$$

Hence we have:

$$e_0 = 1;\ e_1 = 1.6;$$
$$f_0 = 1.44, \text{ i.e., } E(z^{-1}) = 1 + 1.6z^{-1}$$
$$\text{and } F(z^{-1}) = 1.44, \quad (53.49)$$

and so the two-step-ahead prediction becomes:

$$\hat{y}(t+2|t) = N_k(z^{-1})e(t) = \frac{F}{A}e(t) = \frac{1.44}{(1 - 0.9z^{-1})}e(t).$$

But we can reconstruct $e(t)$ from the measured value of $y(t)$ and the inverted model:

$$e(t) = \frac{A(z^{-1})}{C(z^{-1})}y(t),$$

giving the predictor:

$$\hat{y}(t+2|t) = \frac{1.44}{1 + 0.7z^{-1}}y(t). \quad (53.50)$$

The prediction error $\tilde{y}$ is given by $E(z^{-1})e(t+2)$, or:

$$\tilde{y}(t+2|t) = e(t+2) + 1.6e(t+1),$$

which has a variance of $(1 + 1.6^2)\sigma^2 = 3.56\sigma^2$. It is interesting to note that the actual variance of $y(t+2)$ is $14.47\sigma^2$, meaning that our predictor "explains" roughly three fourths of the output variance.

53.6 Minimum-Variance (MV) Control

A growing requirement in manufacturing is guaranteed and quantified quality, as measured, for example, by the proportion of a product lying outside some prespecified limit. In continuous processes, such as papermaking, it is important that the output (at worst) exceeds some lower quality limit (e.g., thickness of paper). To ensure that this is so, the *average* thickness must be set greater than the minimum by an amount dependent on the variance of the controlled output. Hence, if this variance is minimized, the manufacturer can reduce the average, as shown in Figure 53.8.

[Aside: the worst manufacturers sometimes make the best product. If they have a large *spread* in quality, they have to test everything and reject out-of-spec items; i.e., those below the lower statistical limit (LSL) of the figure. Hence, the majority of sales is at a higher quality than really needed. It is best to be "just good enough" = low spread of quality, and hence be profitable.]

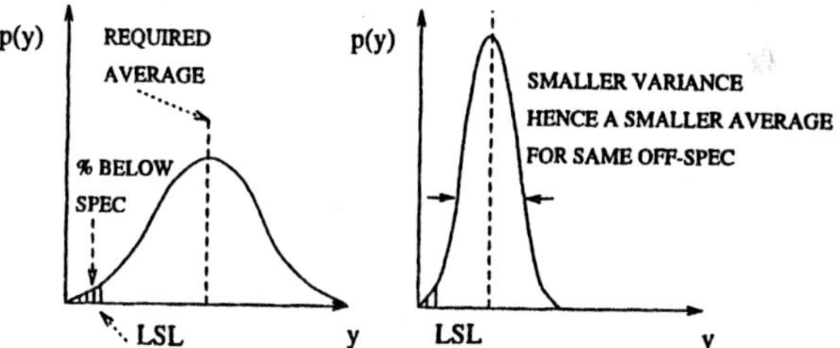

Figure 53.8 Using minimum-variance control.

Consider the plant with dead time k samples and with model:

$$A(z^{-1})y(t) = B(z^{-1})u(t-k) + C(z^{-1})e(t), \quad (53.51)$$

or, dividing:

$$y(t+k) = \frac{B(z^{-1})}{A(z^{-1})}u(t) + \frac{C(z^{-1})}{A(z^{-1})}e(t+k). \quad (53.52)$$

The second right-hand side term is the effect of the disturbances on the output, which can be predicted using the ideas of the previous section; the first term is the effect of the control (which by assumption can affect the output only after k samples). The idea of MV control, in essence, is to choose the control $u(t)$ that will counteract the *predicted* disturbance at time $t + k$. The development first solves the Diophantine identity Equation 53.47 to provide $E(z^{-1})$ and multiplies:

$$A(z^{-1})y(t+k) = B(z^{-1})u(t) + C(z^{-1})e(t+k)$$

(each side) by $E(z^{-1})$ to give:

$$EAy(t+k) = EBu(t) + ECe(t+k)$$

or, as $EA = C - z^{-k}F$ from Equation 53.47:

$$Cy(t+k) - Fy(t) = Gu(t) + ECe(t+k),$$

where $G(z^{-1}) = E(z^{-1})B(z^{-1})$. Hence, we have the equation:

$$y(t+k) = \frac{Fy(t) + Gu(t)}{C} + Ee(t+k). \quad (53.53)$$

The first term on the right-hand side is the k-step-ahead predictor and the second is the prediction error. Hence, we can write the key prediction equations:

k step prediction:

$$\hat{y}(t+k|t) = \frac{F(z^{-1})y(t) + G(z^{-1})}{C(z^{-1})}u(t) \quad (53.54)$$

k step error:

$$\tilde{y}(t+k|t) = E(z^{-1})e(t+k). \quad (53.55)$$

As an example take the plant:

$$(1 - 0.9z^{-1})y(t) = 0.5u(t-2) + (1 + 0.7z^{-1})e(t),$$

where we have already solved the Diophantine identity of Equation 53.47 giving Equation 53.49 and so:

$$y(t+2) = \frac{1.44y(t) + (0.5 + 0.8z^{-1})u(t)}{(1 + 0.7z^{-1})} + (1+1.6z^{-1})e(t+2).$$

The MV control is then easy to determine: Simply choose $u(t)$ so that the first right-hand side term becomes 0; all that remains on the controlled output is the prediction error, which cannot be minimized further as it is comprised of only future white noise components $e(t+1)$ and $e(t+2)$. The feedback control law for our example is then:

$$1.44y(t) + (0.5 + 0.8z^{-1})u(t) = 0,$$

or:

$$u(t) = -1.6u(t-1) - 2.88y(t).$$

In the general case, the control becomes:

$$u(t) = -\frac{F}{G}y(t) = -\frac{F(z^{-1})}{B(z^{-1})E(z^{-1})}y(t); \quad (53.56)$$

this controller *cancels* the zeros of the plant transfer function. In closed loop, the characteristic equation is

$$1 + \frac{F}{BE}z^{-k}BA = 0,$$

or:

$$B(EA + z^{-k}F) = 0,$$

so that, using the Diophantine identity of Equation 53.47, this reduces to:

$$B(z^{-1})C(z^{-1}) = 0. \quad (53.57)$$

The closed-loop modes are defined by those of C (which, in fact, are stable) and of $B(z^{-1})$. There is, therefore, a potential instability problem with MV control in cases where B has roots outside the unit-circle stability region (so-called *nonminimum-phase* zeros) as these appear as unstable poles of the closed loop; such nonminimum-phase zeros occur much more frequently in discrete systems than in continuous-time control.

53.7 Minimum-Variance Self-Tuning

All the machinery is now available for self-tuning: connect a parameter estimator to an MV controller by solving the Diophantine identity of Equation 53.47. Note that MV controller design requires knowledge of k, A, B, *and* C. In difference equation terms the CARMA plant model is

$$\begin{aligned} y(t) &= -a_1y(t-1) - \cdots + b_0u(t-k) + \\ &\cdots + e(t) + c_1e(t-1) + \cdots, \end{aligned}$$

but the standard estimators can estimate only A, B; the driving noise $e(t)$ is not measurable and, hence, cannot be placed into the $\mathbf{x}$-vector to estimate C. There are methods (such as extended least squares) for estimating C, but these tend to be unreliable in practice. However, it transpires we can obtain self-tuned MV (giving a *self-tuning regulator*) by using a standard LS estimator without needing knowledge of C (in effect, assuming $C = 1$)!

There is potentially a further problem: the effect of feedback control on the parameter estimates. Suppose, for example, that a plant:

$$y(t) = ay(t-1) + bu(t-1) + e(t) \quad (53.58)$$

has a simple proportional controller (with zero set point):

$$u(t) = -\alpha y(t),$$

or:

$$\alpha y(t-1) + u(t-1) = 0. \quad (53.59)$$

Then adding a fraction μ of Equation 53.59 to Equation 53.58 gives:

$$y(t) = (a + \mu\alpha)y(t-1) + (b + \mu)u(t-1) + e(t).$$

If we now use an estimator based on the two-parameter model:

$$y(t) = \theta_1 y(t-1) + \theta_2 u(t-1) + e(t) = \mathbf{x}'\theta + e,$$

then $\hat{\theta}_1 = (a + \mu\alpha)$, $\hat{\theta}_2 = (b + \mu)$ will be obtained, where μ is arbitrary. Hence, if we use LS estimation in a closed-loop mode, the estimated $\hat{\theta}$ does not converge to a unique point but to a line where the estimated parameters can wander up and down in unison. This is a problem of using closed-loop data with only internal signals such as $e(t)$ stimulating the loop; to get an consistent estimate we must do one of the following:

1. Use externally generated test signals, such as step changes in set point.
2. Have a controller that is more complex (higher order) than the plant.
3. Have a time-varying controller.

This third solution is appropriate for self-tuning, though it is still best to make the data "rich" by exciting the plant with external signals (e.g., a PRBS added to the set point).

How is it that we can use LS? The key idea is not to go *estimate* $\rightarrow$ *design* $\rightarrow$ *controller* (giving what is called the *indirect*

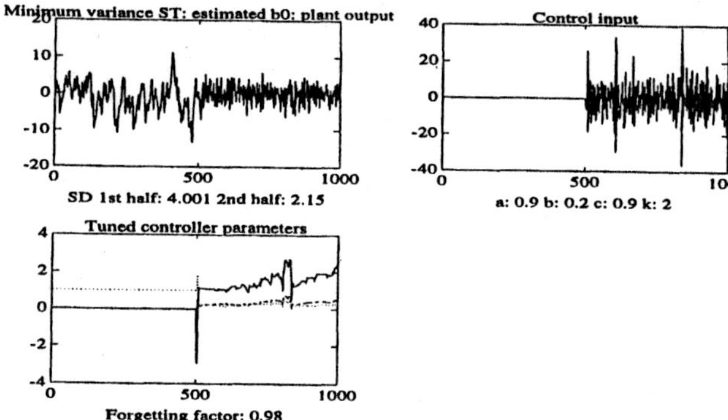

Figure 53.9 MV self-tuning, all parameters estimated.

approach), but instead to proceed *estimate* → *controller* (the *direct* approach). What are estimated are the *controller* (in fact, the k-step-ahead predictor) rather than the *plant* parameters. How this is done is seen below.

Recall the prediction Equation 53.54 when multiplied up by $C(z^{-1})$:

$$(1 + c_1 z^{-1} + c_2 z^{-2} + \cdots)\hat{y}(t+k|t) = Fy(t) + Gu(t),$$

giving at time t:

$$\hat{y}(t|t-k) = Fy(t-k) + Gu(t-k) - \sum c_i z^{-i}\hat{y}(t|t-k).$$

But the point about MV control is that it makes the prediction zero by correct choice of u. Hence all the terms in the sum on the right-hand side are set to zero by *previous* controls, so that from:

$$y(t) = \hat{y}(t|t-k) + \tilde{y}(t|t-k), \qquad (53.60)$$

we have:

$$y(t) = F(z^{-1})y(t-k) + G(z^{-1})u(t-k) + E(z^{-1})e(t). \qquad (53.61)$$

This is the crucial equation: it obeys the LS rules of an LITP model with:

$$\begin{aligned} \mathbf{x}(t) &= [\,y(t-k), y(t-k-1), \cdots, \\ &\qquad u(t-k), u(t-k-1), \cdots]' \qquad (53.62) \\ \theta &= [\,f_0, f_1, \cdots, g_0, g_1, \cdots]', \qquad (53.63) \end{aligned}$$

and, most importantly, the data $\mathbf{x}(t)$ are independent of the error term as the data are from $t-k$ backwards, whereas $E(z^{-1})e(t)$ finishes at $e_{k-1}e(t-k+1)$. Hence, LS leads directly to the required $\hat{F}$ and $\hat{G}$ parameters, so we get a self-tuner with feedback law:

$$\hat{F}(z^{-1})y(t) + \hat{G}(z^{-1})u(t) = 0. \qquad (53.64)$$

The procedure, then, is as follows:

1. Assemble old data into the $\mathbf{x}$-vector as in Equation 53.63.
2. Use RLS to get $\hat{\theta} = \hat{F}, \hat{G}$.
3. Use the estimated parameters in the feedback law of Equation 53.64.

Of course, the above is simply a *plausibility* argument; in fact, the algorithm can by lengthy algebra be shown to converge to give the required control signals; i.e., satisfying the *self-tuning* property. The speed of convergence is found to depend on the roots of $C(z^{-1})$. As an example, consider the first-order system:

$$(1 - 0.9z^{-1})y(t) = 0.2u(t-2) + (1 + 0.9z^{-1})e(t),$$

which has the two-step-ahead prediction equation:

$$(1 + 0.9z^{-1})\hat{y}(t+2|t) = 1.62y(t) + 0.2(1 + 1.8z^{-1})u(t),$$

for which the MV controller with *known* parameters is

$$1.62y(t) + 0.2u(t) + 0.36u(t-1) = 0.$$

The corresponding model to estimate in self-tuning is

$$y(t) = f_0 y(t-2) + g_0 u(t-2) + g_1 u(t-3) + \epsilon(t).$$

The estimator for a self-tuner will have data and parameter vectors:

$$\begin{aligned} \mathbf{x} &= [y(t-2), u(t-2), u(t-3)]' \\ \theta &= [f_0, g_0, g_1]' \;\rightarrow\; [1.62, 0.2, 0.36]'. \end{aligned}$$

The system was simulated for 1000 samples, the first 500 being "open loop." At $t = 500$ the self-tuner was switched on, giving the results seen in Figure 53.9. Observe how the variance has been reduced by STMV and how the estimated parameters "wander about."

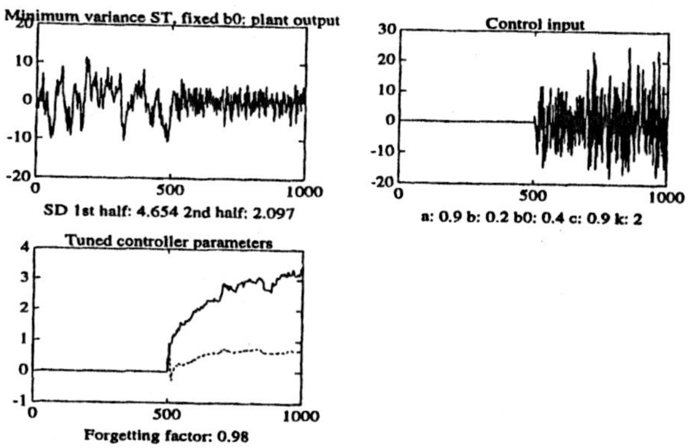

Figure 53.10 Simulation of a simple self-tuner with a fixed parameter $\bar{g}_0$.

As discussed above, the estimates of the parameters are not unique, as the control Equation 53.64 can be multiplied by an arbitrary factor μ without affecting $u(t)$. In principle, this is not a problem, but to avoid excessively large or small estimates we can "fix a parameter" to a guessed value and estimate the others. Typically, the fixed parameter is the value of g_0: the multiplier of the current control $u(t)$, whose nominal value is b_0. Suppose $\bar{g}_0$ is the choice. In our example, it means that the model becomes:

$$y(t) - \bar{g}_0 u(t-2) = y_1(t) = f_0 y(t-2) + g_1 u(t-3) + \epsilon(t).$$

Then the model to use in RLS has data and parameter vectors:

$$\begin{aligned} \mathbf{x}(t) &= [y(t-k), y(t-k-1), \cdots, \\ &\quad u(t-k-1), \cdots]' \qquad (53.65) \\ &= [y(t-2), u(t-3)]' \qquad (53.66) \end{aligned}$$

in our example;

$$\theta = [f_0, f_1, \cdots, g_1, \cdots]', \qquad (53.67)$$

and with "output" $y_1(t) = y(t) - \bar{g}_0 u(t-2)$. The control to use is like Equation 53.64 but is based on the *chosen* fixed $\bar{g}_0$ and the remaining estimates:

$$u(t) = -[\sum \hat{f}_i y(t-i) + \sum \hat{g}_i u(t-i)]/\bar{g}_0.$$

A self-tuner based on this idea is coded in Matlab as:

```
% M-file for the simple first-order A-W
% minimum-variance self-tuner.

nt = 1000; na = 1; nb = k-1; np = na+nb;
th=zeros(nt,np); P = 100*eye(np);
II = eye(np);

for i=k+1:nt,
 e(i) = rand;
 y(i) = a*y(i-1) + b*u(i-k) + e(i) + c*e(i-1);
                                      % plant
  if i > nt/2,
   x  = [y(i-k) u(i-2*k+1:i-k-1)];
                                      % data vector
   ep = y(i)-g0*u(i-k) - x*th(i-1,:)';
                                      % pred error
   kk = x*P/(beta+x*P*x');            % RLS
   th(i,:) = th(i-1,:) + kk*ep;  % ..
   P=(II - kk'*x)*P/beta;             % update
   u(i) = - (th(i,1)*y(i) +
           u(i-nb:i-1)*th(i,2:np)')/g0;
  end;
end;
```

Does it matter if the wrong value of $\bar{g}_0$ is chosen? No, provided that:

$$\frac{1}{2} < \frac{\bar{g}_0}{b_0} < \infty. \qquad (53.68)$$

This means that a "large" value of the fixed parameter is safe.

Figure 53.10 and Figure 53.11 show the behavior of a self-tuner with the parameter $\bar{g}_0$ set to 0.4 and then 2. Both are perfectly well behaved, though with slower convergence for $\bar{g}_0 = 2$. Note that in the original example with no "fixing" there is a point where the parameter estimates "jump." It is found that without fixing and with no external signals a self-tuner tends to "burst" like this every now and then.

Generalized minimum-variance (GMV) control was developed to overcome the problems of (1) MV's instability when B is nonminimum-phase and (2) the large control variance produced by MV, particularly when using "fast sampling." An *auxiliary* output $\phi(t)$ is defined:

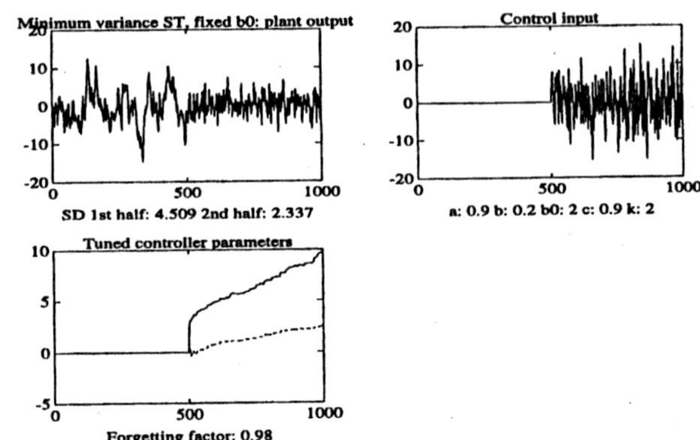

Figure 53.11 Simulation of STMV with a larger fixed parameter.

$$\phi(t) = P(z^{-1})y(t) - R(z^{-1})w(t-k) + Q(z^{-1})u(t-k), \qquad (53.69)$$

where P, Q, R are *design polynomials* whose choice gives a range of possible closed-loop objectives; see Harris and Billings [7] or Wellstead and Zarrop [11] for more details. GMV self-tuning simply uses the same approach as developed for MV:

1. Estimate polynomials $\hat{F}, \hat{G}, \hat{H}$ in the predictor model:

$$\begin{aligned} \phi(t) &= \hat{F}(z^{-1})y(t-k) + \hat{G}(z^{-1})u(t-k) \\ &\quad + \hat{H}(z^{-1})w(t-k) + \epsilon(t). \qquad (53.70) \end{aligned}$$

2. Using the estimates from Equation 53.70, compute the control:

$$\hat{G}(z^{-1})u(t) = -[\hat{F}(z^{-1})y(t) + \hat{H}(z^{-1})w(t)]. \qquad (53.71)$$

A simple case of GMV is when $P = R = 1$ and $Q = \lambda$ for which it can be shown that the cost:

$$J_{GMV} = \mathcal{E}[(y(t+k) - w(t))^2 + \frac{\lambda}{b_0}u^2(t)|t] \qquad (53.72)$$

is minimized, where the expectation is conditional on data available up to time t. It can further be shown that the characteristic polynomial now has a factor:

$$B(z^{-1}) + \frac{\lambda}{b_0}A(z^{-1}),$$

so that λ can be considered a root-locus parameter; for plants with a stable A polynomial, a large enough value of λ ensures loop stability even for nonminimum-phase plants. Unfortunately for self-tuning applications, a prior "good" value of λ needs to be known; nevertheless, λ can be used to trade off output variance against control activity.

53.8 Pole-Placement (PP) Self-Tuning

The problem with MV regulators if that they are unstable if the plant has a zero outside the unit circle, as the controller attempts to *cancel* the zero by having an unstable pole. Hence, there is a mode in the control signal that grows without limit. Therefore, alternative strategies such as GMV must be used for more complex plants; one such appeals to classical control design.

Recall the root-locus approach: it is a method of *analysis* showing how the poles of the closed-loop transfer function vary as some parameter, such as the controller gain, is changed. Design considers some pole-zero structure of a controller and then fixes the gain to give "nice" closed-loop pole positions, such as large ω_n for a desired value of ζ such as $1/\sqrt{2}$. An alternative procedure that is often used in discrete-time control is *first to choose* the desired pole positions and then back-calculate the appropriate controller.

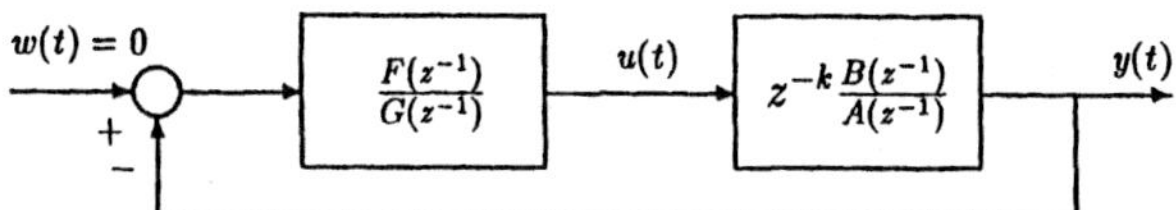

Figure 53.12 Feedback controller for pole placement.

Suppose that the feedback controller of Figure 53.12 is:

$$F(z^{-1})y(t) + G(z^{-1})u(t) = 0, \text{ where } g_0 = 1, \tag{53.73}$$

which gives the control action as:

$$\begin{aligned} u(t) &= -g_1 u(t-1) - g_2 u(t-1) - \cdots \\ &\quad - f_0 y(t) - f_1 y(t-1) - \cdots. \end{aligned}$$

If the plant has the CARMA model:

$$A(z^{-1})y(t) = z^{-k}B(z^{-1})u(t) + C(z^{-1})e(t),$$

where the dead time k is explicitly factored from the zero polynomial, the closed loop is given by:

$$\begin{aligned} [A(z^{-1})G(z^{-1}) + z^{-k}B(z^{-1})F(z^{-1})]y(t) &= \\ C(z^{-1})G(z^{-1})e(t). \end{aligned} \tag{53.74}$$

Suppose that the polynomials F and G are obtained by solving the Diophantine identity:

$$AG + z^{-k}BF = CT(z^{-1}), \tag{53.75}$$

where $T(z^{-1})$ is a polynomial *chosen by the designer* of the form:

$$T(z^{-1}) = 1 + t_1 z^{-1} + t_2 z^{-2} + \cdots = \prod_i (1 - \alpha_i z^{-1}),$$

with the α_i corresponding to the *desired closed-loop pole positions*. (For hand calculation, we compare powers of z^{-1} as before; in a computer, we use Euclid's algorithm.) By substituting Equation 53.75 into Equation 53.74, the closed loop is given by:

$$C(z^{-1})T(z^{-1})y(t) = C(z^{-1})G(z^{-1})e(t),$$

or:

$$y(t) = \frac{G(z^{-1})}{T(z^{-1})}e(t), \text{ and } u(t) = -\frac{F(z^{-1})}{T(z^{-1})}e(t). \tag{53.76}$$

Hence, the closed-loop poles are given by the user-chosen polynomial T; as the B polynomial is *not* cancelled by this law, there is no longer any unstable mode in the control signal even if the plant is nonminimum-phase. As an example, consider again the system:

$$(1 - 0.9z^{-1})y(t) = 0.5u(t-2) + (1 + 0.7z^{-1})e(t),$$

and choose $T(z^{-1}) = 1 - 0.5z^{-1}$, giving a relatively fast closed-loop pole at $\alpha = 0.5$. Then the Diophantine identity of Equation 53.75 becomes:

$$\begin{aligned} (1 - 0.9z^{-1})(1 + g_1 z^{-1} + \cdots) + z^{-2}0.5(f_0 + f_1 z^{-1}\cdots) \\ = (1 + 0.7z^{-1})(1 - 0.5z^{-1}). \end{aligned}$$

Comparing coefficients of increasing powers of z^{-1} gives $g_1 = 1.1$ and $f_0 = 1.28$, all other coefficients being zero. Hence, the control law is given by

$$u(t) = -[1.1u(t-1) + 1.28y(t)],$$

and in closed loop we have from Equation 53.76:

$$y(t) = \frac{1 + 1.1z^{-1}}{1 - 0.5z^{-1}}e(t),$$

and

$$u(t) = -\frac{1.28}{1 - 0.5z^{-1}}e(t).$$

It can be shown that the output variance due to this control law is $5.6\sigma^2$, roughly double the MV result: *you don't get something for nothing*.

It is interesting to see that pole assignment (for this regulator case) can also be self-tuned using RLS *without* knowing or estimating the polynomial C. To see why this is so, let $\mathcal{A}(z^{-1})$ and $\mathcal{B}(z^{-1})$ be the solutions to a new Diophantine identity *that does not include the* C *polynomial* (compare with Equation 53.75):

$$\mathcal{A}G(z^{-1}) + z^{-k}\mathcal{B}F(z^{-1}) = T(z^{-1}). \tag{53.77}$$

The polynomials F and G provided as "input" to this identity are those obtained by using the previous identity of Equation 53.75 with the "true" plant polynomials A, B, C. Let us take our example, giving the new Diophantine identity of Equation 53.77:

$$\mathcal{A}(1 + 1.1z^{-1}) + z^{-2}\mathcal{B}1.28 = (1 - 0.5z^{-1}),$$

giving $\mathcal{A} = 1 - 1.6z^{-1}, \mathcal{B} = 1.6 \times 1.1/1.28 = 1.375$.

Consider operating the plant in closed loop using the "correct" controller $-F/G$ as computed by Equation 53.75; then the sequence:

$$\mathcal{A}(z^{-1})y(t) - z^{-k}\mathcal{B}(z^{-1})u(t) = \frac{\mathcal{A}G + z^{-k}\mathcal{B}F}{T(z^{-1})}e(t) = e(t), \tag{53.78}$$

using the closed-loop behavior from Equation 53.76 and then the second Diophantine identity of Equation 53.77. This result gives the key idea: when in closed loop with the right PP controller, there exists a relationship between the input and output signals given by the new polynomials $\mathcal{A}$ and $\mathcal{B}$ such that the error term is just the white driving noise.

Equation 53.78 shows that RLS can be used to get unbiased estimates of $\mathcal{A}$ and $\mathcal{B}$, as the right-hand side noise term is uncorrelated. To estimate the parameters, simply choose the data vector to be

$$\mathbf{x}(t) = [-y(t-1), -y(t-2), \cdots; u(t-1), u(t-2), \cdots]'$$

and with model output $y(t)$. Hence, if the data sequences follow those of the pole-placed closed loop of Equation 53.76, then LS simply gives $\hat{\theta} \to [\mathcal{A}, \mathcal{B}]$. If these estimates are placed as *input* to the *second* Diophantine identity of Equation 53.77, we can use the identity to recompute the controller F and G polynomials; i.e., the procedure of going $[A, B, C, T] \to [F, G]$ via 53.75 $\to$ $[\mathcal{A}, \mathcal{B}]$ via RLS $\to$ $[F, G]$ via 53.77 is self-consistent.

Hence, the self-tuned version of PP goes through the following steps:

1. Use RLS to obtain estimates $\hat{\mathcal{A}}$ and $\hat{\mathcal{B}}$ from the model $\mathcal{A}y(t) = z^{-k}\mathcal{B}u(t) + e(t)$ (i.e., no C estimated).
2. Resolve the Diophantine identity of Equation 53.77 at each sample for user-chosen T (using as input the estimates $\hat{\mathcal{A}}$ and $\hat{\mathcal{B}}$) to obtain $\hat{F}$ and $\hat{G}$, by equating powers of z^{-1} as usual.
3. Assert the control $\hat{F}y(t) + \hat{G}u(t) = 0$, or $u(t) = -[\hat{F}/\hat{G}]y(t)$.

As with self-tuned MV, the above argument is simply to give plausibility to the approach; in practice, the procedure converges provided that there is a solution to the Diophantine identity of Equation 53.77 (which is not possible if there are common roots in the estimated model polynomials).

53.9 Long-Range Predictive Control

For practical applications, an adaptive controller must be *robust* against the prior assumptions made about the plant to be controlled. For example, we must choose sometimes arbitrary values for the degrees of the estimated polynomials and have no assurance that, for all t, our model is of neither too high nor too low an order compared with the true plant model within the bandwidth of our closed loop. The estimates may be affected over periods of time by disturbances not fully captured by our assumptions about $C(z^{-1})$. There may be occasions when there are common factors in the estimated TFs. The dead time of the plant may vary so that the k assumed in k-step-ahead prediction for MV control is not correct (so that the "true" value of g_0 is zero). Fractional dead time and fast sampling might cause the model to become nonminimum phase. Hence:

Minimum-variance. We might get instability by assuming too small a value of k, or if the plant is nonminimum phase.

Pole placement. There is no solution to the Diophantine identity if there are common factors in the estimated $\mathcal{A}$, $\mathcal{B}$ polynomials.

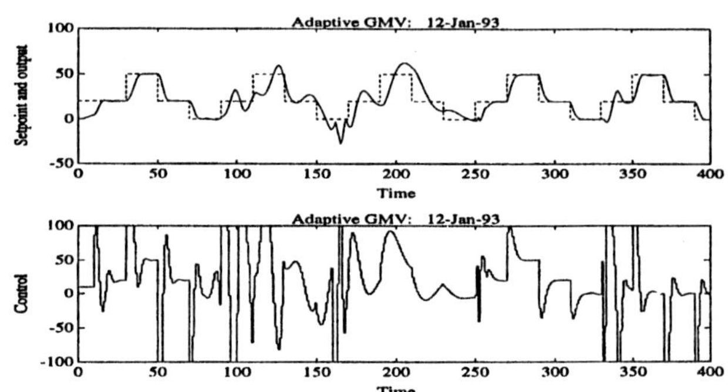

Figure 53.13 Simulation of the adaptive MV control of a dynamically time-varying plant.

Results of controlling a time-varying plant with an adaptive MV algorithm with a forgetting factor of 0.98 are shown in Figure 53.13. The plant changes its order n and delay k at various stages during the run. When the actual plant delay exceeds the assumed delay of the algorithm, MV goes unstable. Similarly, it is found that with adaptive PP, the control goes unstable if the assumed model order is greater than the actual model order. With a fixed PID control the initial results are good, but instability sets in when the plant changes its dynamics. In practice, good engineering design using prior knowledge and proper choice of sample interval (e.g., at most 0.1 times the open loop rise time) can give good results with MV and PP. On the other hand, you can be unlucky.

Long-range predictive control (LRPC) is a more modern approach that overcomes many of the above problems and has many extra features, which makes it useful in applications. The basic idea is to predict the future output as the sum of a *free response* (based on past known data) and a *forced response* depending on *current and future control* actions, as shown in Figure 53.14. Consider the noise-free incremental model:

$$A(z^{-1})\Delta y(t) = B(z^{-1})\Delta u(t-1), \tag{53.79}$$

or:

$$y(t) = y(t-1) - a_1\Delta y(t-1) - \cdots + b_1\Delta u(t-1) + \cdots. \tag{53.80}$$

Consider using this model to give the prediction $p(t+1)$ of the free output response based on maintaining the control signal equal to the previous value $u(t-1)$:

$$p(t+1) = y(t) - a_1\Delta y(t) - \cdots + b_2\Delta u(t-1) + \cdots,$$

as by assumption $\Delta u(t) = 0$. Continuing the iteration further:

$$p(t+2) = p(t+1) - a_1\Delta p(t+1) - \cdots + b_3\Delta u(t-1) + \cdots,$$

where the term $\Delta p(t+1) = p(t+1) - y(t)$, the difference between the prediction and the available data $y(t)$. After some stage all the $\Delta u(t-1)$ terms drop out as the polynomial B has been exhausted, leaving the iterations:

$$p(t+i) = p(t+i-1) - a_1 \Delta p(t+i-1) - a_2 \Delta p(t+i-2) - \cdots . \tag{53.81}$$

Hence, the predictions can be expressed verbally as "*iterate the plant equations forward in time, assuming current and future control increments (moves) are zero, and using existing old Δu, Δy to initialize the data.*"

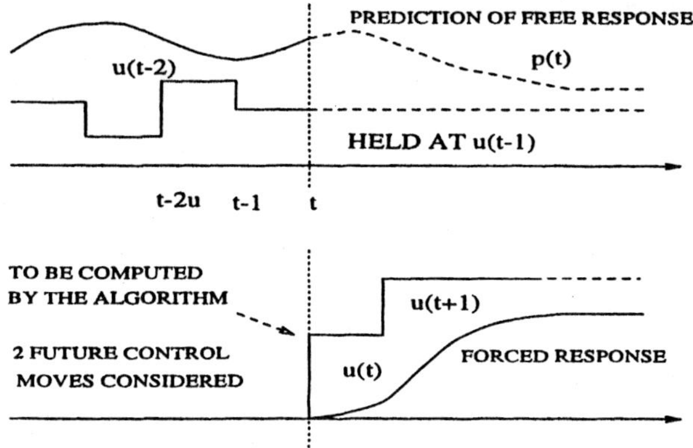

Figure 53.14 Long-range predictive control.

Now consider the forced component of response, with the control input being a set of moves $\Delta u(t), \Delta u(t+1), \cdots$, which are to be determined by the algorithm. This is simply a series of "steps" so that the output is just the superposition of a set of step responses $\{s_i\}$, giving the total response as:

$$\begin{aligned} y(t+1) &= s_1 \Delta u(t) + p(t+1) \\ y(t+2) &= s_1 \Delta u(t+1) + s_2 \Delta u(t) + p(t+2) \\ &\cdots \\ y(t+j) &= s_1 \Delta u(t+j-1) + s_2 \Delta u(t+j-2) + \\ &\cdots + s_j \Delta u(t) + p(t+j) \\ &\cdots \end{aligned}$$

Hence, we can collect N such equations into vector-matrix form:

$$\mathbf{y} = \mathbf{G}\Delta\mathbf{u} + \mathbf{p}, \tag{53.82}$$

where:

$$\begin{aligned} \mathbf{y} &= [y(t+1), y(t+2), \cdots, y(t+N)]' \\ \Delta\mathbf{u} &= [\Delta u(t), \Delta u(t+1), \cdots, \Delta u(t+N-1)]' \\ \mathbf{p} &= [p(t+1), p(t+2), \cdots, p(t+N)]' \cdot \end{aligned}$$

and:

$$\mathbf{G} = \begin{bmatrix} s_1 & 0 & \cdots & 0 \\ s_2 & s_1 & \cdots & 0 \\ \cdots & & & \\ s_N & s_{N-1} & \cdots & s_1 \end{bmatrix}.$$

Suppose that we had a "future set point" sequence $\{w(t+1) \cdots w(t+N)\}$ available at time t. In robotics, this could be the required future trajectory, whereas in process control we would normally assume the future set point to equal the current value. Either way, we can collect the sequence into a vector and hence define a vector of "future system errors":

$$\mathbf{e} = [w(t+1) - y(t+1), w(t+2) - y(t+2), \cdots, w(t+N) - y(t+N)]',$$

giving:

$$\mathbf{e} = \mathbf{w} - (\mathbf{G}\Delta u + \mathbf{p}).$$

The only unknowns in this set of equations are the future controls, so we minimize $S = \sum e(t+j)^2$ over the predictions by the set of controls:

$$\Delta\mathbf{u} = (\mathbf{G}'\mathbf{G})^{-1}\mathbf{G}'(\mathbf{w} - \mathbf{p}). \tag{53.83}$$

This gives a set of "best" future control actions; we then use a *receding-horizon* strategy, which simply asserts the first of this sequence and repeats the whole calculation at each sample instant. Note, however, that $\mathbf{G}$ is an $N \times N$ matrix and we get a solution only if it is invertible, giving:

$$\Delta\mathbf{u} = \mathbf{G}^{-1}(\mathbf{w} - \mathbf{p}). \tag{53.84}$$

This solution is exact such that the control sequence would drive *all* the future system errors to zero. Good in theory; bad in practice as it would require excessive control signals. Moreover, what happens if the plant delay is, say, 2 so that $s_1 = 0$? The result is failure, as we cannot then invert $\mathbf{G}$.

How can we derive more equations than unknowns to let LS "smooth" out our future errors? We can make *assumptions* about what controls will be exerted in the future. Consider controlling a simple Type 0 plant. We could inject a large initial signal to get it moving and then maintain a constant control of sufficient size to get it to the final set point. This would mean that at the initial time only two control increments are considered. Hence, we take:

$$\Delta\mathbf{u} = [\Delta u(t), \Delta u(t+1), 0, 0, \cdots, 0]',$$

so that now we have fewer unknowns than equations. In general we can allow only NU control increments to be nonzero (called the "control horizon"; see Figure 53.14) and will define the "future control increment vector" to be

$$\Delta\mathbf{u} = [\Delta u(t), \Delta u(t+1), \ldots, \Delta u(t+NU-1)]'.$$

This means that our set of equations for $\mathbf{e}$ involve a nonsquare matrix $\mathbf{G}$.

One special case is where only *one* control increment is considered at time t; i.e., $NU = 1$. In essence we are asking, "What step change in control will minimize the sum-of-squares of the future system errors?" Suppose we make the prediction horizon N very large. Then if there were any steady-state error, the corresponding sum-of-squares would be large compared with errors accruing during the transient. The outcome is that the control step will be just the right size to make the steady-state error zero, and there will be just the same dynamics in closed loop as in open loop. This simple approach is called *mean-level control* (see Figure 53.15). Given that $NU = 1$, the matrix inversion is simple, as $\mathbf{G}'\mathbf{G}$ is just $\sum_1^N s_i^2$: a scalar.

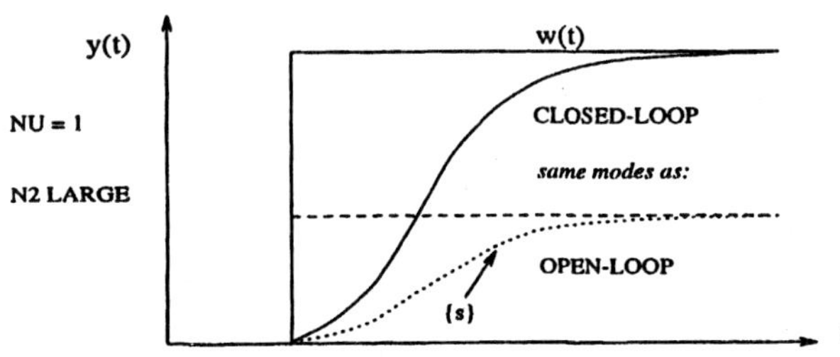

Figure 53.15 Mean-level control.

Consider the case where N is not large. Then the initial transient errors become increasingly important in comparison with the steady-state error so that the initial control is made larger to reduce them; i.e., we obtain a faster response as N reduces.

53.10 The Generalized Predictive Control (GPC) Cost Function

We can include other ideas into the LRPC algorithm. For example, if it is known that the plant has a dead time, then it is clear that the current control $u(t)$ *cannot affect* the future errors until the dead time is cleared. Hence, there is no point in putting $e(t+1)\cdots e(t+k-1)$ into the LS cost function. We might also want to have a mechanism for "trading" the cost of control (e.g., control variance) against output performance (i.e., error variance). Combining all these we get the generalized predictive control (GPC) cost-function:

$$J_{GPC}(N1, N2, NU, \lambda) = \sum_{i=N1}^{N2} e(t+i)^2 + \lambda \sum_{i=1}^{NU} \Delta u(t+i-1)^2,$$

where:

- $N1$ is the *lower costing horizon*
- $N2$ is the *upper costing horizon*
- $N1 \to N2$ is the *costing range*
- NU is the *control horizon;*, i.e., the "degrees of freedom" in the control
- λ is the *control weighting*

subject to the condition that assumed future control-increment sequence is zero after the control horizon.

Using GPC we have the solution

$$\Delta\mathbf{u} = (\mathbf{G}'\mathbf{G} + \lambda\mathbf{I})^{-1}\mathbf{G}'(\mathbf{w} - \mathbf{p}), \qquad (53.85)$$

where the matrix $\mathbf{G}$ is now $N2 - N1 + 1$ by NU:

$$\mathbf{G} = \begin{bmatrix} s_{N1} & s_{N1-1} & \cdots & 0 \\ s_{N1+1} & s_{N1} & s_{N1-1} & \cdots \\ \vdots & & & \\ s_{N2} & s_{N2-1} & \cdots & s_{N2-NU+1} \end{bmatrix}.$$

There are many possible combinations of the four "design parameters" ($N1$, $N2$, NU, λ) but in practice, two main choices are made. The first is mean-level control as above: $[k, \text{large}, 1, 0]$, where "large" means about 10. The other is based on "dead beat" control where the idea is to attain the set point as rapidly as possible such that the error *and all its derivatives* become zero simultaneously. This can be shown to be achieved by the following choices of horizon:

$$N1 = n,\ N2 \geq 2n - 1,\ NU = n \text{ and } \lambda = 0,$$

where n is the largest power of z^{-1} found in the model (including the extra because of the $\Delta = 1 - z^{-1}$). Suppose we choose $N2 = 2n - 1$. Then $\mathbf{G}$ is $n \times n$, and, provided it is invertible the GPC solution gives $J = 0$ (as the equations are solved exactly) and hence $e(t+n)\cdots e(t+2n-1)$ are all 0. This means that the control is such as to make n successive future system errors zero, a prerequisite for state deadbeat control.

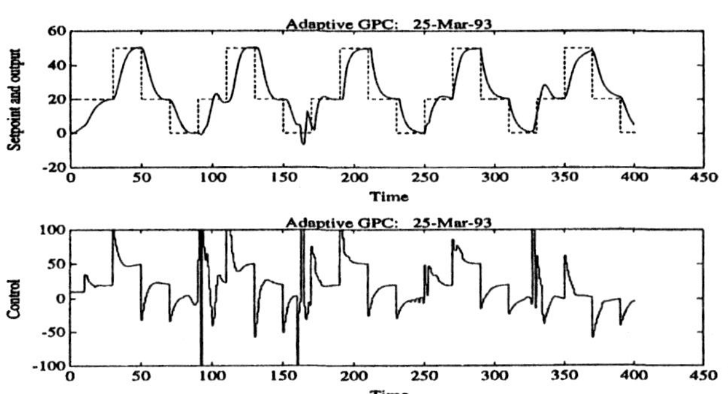

Figure 53.16 Adaptive GPC control of a time-varying plant.

It is not just the flexibility of GPC (see for example Soeterboek [10]) that gives it the power. It can handle realistic control problems that cannot be treated by other designs. Suppose that we know the control signal to be *constrained* (as with torque saturation in motors) or that some internal variable (such as temperature in a catalytic cracker) must not exceed some limit. In principle, we can use the prediction idea to test if any of our variables is likely to hit constraints and, hence, modify the control signal suitably. By running a plant nearer to constraints, we enhance quality and profitability, so LRPC is increasingly popular in industry.

By connecting an RLS estimator to GPC on the plant as in Figure 53.13, using a prediction horizon of 10 and a control horizon of 1 (approximating to mean-level control) we obtain the results of Figure 53.16: much better!

The problem with "simple" GPC as presented above is the lack of stability proofs, except for some limiting cases such as mean-level control; with practical applications, mean-level control has been used most often and this has not been found to cause difficulties. However, in order to get stability *guarantees,* we can use infinite-horizon LQ (e.g., Bitmead et al. [2]) or adopt *terminal constraints* where the objective now is to minimize a cost function *subject* to the output $y(t+j)$ exactly matching the set point over some future constraint range (see Clarke [3], and Mosca [9], for details).

53.11 Robustness of Self-Tuning Controllers

Applications of self-tuning control have to take into account the following practical problems:

- Disturbances acting on a process are likely to be nonstationary and to have inconsistent behavior, e.g., during plant start-up.
- Certain patterns of disturbance can lead to the estimation of a poor plant model.
- The dynamic order of the process is likely to be significantly greater than that assumed by the self-tuning control design.
- Actuator nonlinearities (e.g., stiction) give unrepresentative small-signal behavior.

Some of these difficulties (in particular, the problem of *unmodeled dynamics*) have been treated in some detail; others (e.g., nonlinear actuation) require careful attention to engineering detail. In general, a "good" self-tuner requires (1) a robust estimator; (2) a robust control design; and (3) "jacketing" software that takes into account practical process features.

For improving the estimator, it is possible to add *normalization* and a *dead zone.* Normalization simply takes the regressor vector and produces a factor $m(t-1) = \max\{m, ||\mathbf{x}(t-1)||\}$ and divides the prediction error Equation 53.23 by $m(t-1)$. A dead zone takes into account that the "true" prediction is of the form:

$$\hat{y}(t) = \sum_{1}^{n} \hat{\theta}_i^m x_m(t-i) + \sum_{1}^{n} \hat{\theta}_i^u x_u(t-i),$$

where x_m is the "modeled" data input and x_u, the "unmodeled" input. If, after normalization, the prediction error is smaller than a certain amount, then this is deemed to be due to the (assumed small) effect of unmodeled dynamics and the estimation is temporarily frozen. While there is a good theoretical background for this approach (see, for example, Mosca [9]), there is not much practical experience in how to choose an appropriate value for the dead zone parameter.

Perhaps the most important advance for ensuring robustness is the reconsideration of the CARIMA model in the form:

$$A(z^{-1})y(t) = B(z^{-1})u(t) + \frac{T(z^{-1})}{\Delta}e(t), \qquad (53.86)$$

where $T(z^{-1})$ is now considered to be an *assigned design* rather than an estimated polynomial for enhancing estimation and control robustness. Multiplying up in Equation 53.86 gives:

$$A(z^{-1})y^f(t) = B(z^{-1})u^f(t) + e(t), \qquad (53.87)$$

where the signals are given by $y^f = \Delta y/T$, $u^f = \Delta u/T$; i.e., they are band-pass filtered by Δ/T. The effect of Δ is to remove *dc*-offset or constant-load disturbances; the effect of $1/T$ is to filter out high-frequency effects so that the estimation concentrates on low-frequency behavior.

The above assumption about the disturbance also affects predictive controller design, as optimal predictions take into account the assumed model for the disturbance. For example, it can be shown that a predictive controller in closed loop satisfies:

$$\alpha T(z^{-1})w(t) = R(z^{-1})\Delta u(t) + S(z^{-1})y(t), \qquad (53.88)$$

where the polynomials R, S satisfy the Diophantine identity:

$$R(z^{-1})A(z^{-1})\Delta + B(z^{-1})S(z^{-1}) = P_c(z^{-1})T(z^{-1}), \qquad (53.89)$$

where P_c gives the closed-loop poles (e.g., for mean-level control $P_c = A$). Then the closed loop is stable for *fixed* estimated polynomials $\hat{A}$, $\hat{B}$ provided that for all frequencies up to Nyquist:

$$\left|\frac{B}{A} - \frac{\hat{B}}{\hat{A}}\right| < \left|\frac{P_c}{\hat{A}}\right| \cdot \left|\frac{T}{S}\right|, \qquad (53.90)$$

where S is deduced from Equation 53.89 using the estimated parameters. Hence, the right-hand side of Equation 53.90 is known from the estimated model, and in particular, T can be chosen to ensure the bound is satisfied, particularly at high frequencies where the undermodeling problem arises; see [3] for more details. It has been found by "benchmark" studies that good choice of T is highly significant. It is possible to consider *different* designs of T for the estimator and controller, but in practice it is convenient and near optimal to use the same polynomial. In general, the choice of $T = \hat{A}(1-\gamma z^{-1})^m$, where γ is in the neighborhood of a dominant plant pole, is fairly effective.

References

[1] Åström, K.J. and Wittenmark, B., *Adaptive Control,* Addison-Wesley, Reading, MA, 1989.

[2] Bitmead, R.R., Gevers, M., and Wertz, V., *Adaptive Optimal Control,* Prentice Hall, Englewood Cliffs, NJ, 1990.

[3] Clarke, D.W., Ed., *Advances in Model-Based Predictive Control,* Oxford University Press, UK, 1994.

[4] Gawthrop, P.J., *Continuous-Time Self-Tuning Control,* Research Studies Press, Letchworth, UK, 1987.

[5] Goodwin, G.C. and Sin, K.S., *Adaptive Filtering, Prediction and Control,* Prentice Hall, Englewood Cliffs, NJ, 1984.

[6] Hang, C.C., Lee, T.H., and Ho, W.K., *Adaptive Control,* Instrument Society of America, NCa, 1993.

[7] Harris, C.J. and Billings, S.A., Eds., *Self-Tuning and Adaptive Control: Theory and Applications,* Peter Perigrinus Ltd., Stevenage, UK, 1981.

[8] Isermann, R., Lachmann, K.-H., and Drago, D., *Adaptive Control Systems*, Prentice Hall, Englewood Cliffs, NJ, 1992.

[9] Mosca, E., *Optimal Predictive and Adaptive Control*, Prentice Hall, Englewood Cliffs, NJ, 1995.

[10] Soeterboek, R., *Predictive Control: A Unified Approach*, Prentice Hall, Englewood Cliffs, NJ, 1992.

[11] Wellstead, P.E. and Zarrop, M.B., *Self-Tuning Systems*, Wiley, New York, 1991.

54

Model Reference Adaptive Control

Petros Ioannou

University of Southern California, EE-Systems, MC-2562, Los Angeles, CA

54.1 Introduction

Research in adaptive control has a long history of intense activity involving debates about the precise definition of adaptive control, examples of instabilities, stability and robustness proofs, and applications. Starting in the early 1950s, the design of autopilots for high-performance aircraft motivated an intense research activity in adaptive control. High-performance aircraft undergo drastic changes in their dynamics when they fly from one operating point to another. These changes cannot be handled by constant gain feedback control. A sophisticated controller, such as an adaptive controller, that would be able to learn and accommodate changes in the aircraft dynamics was needed. Model reference adaptive control (MRAC) was suggested by Whitaker et al. [9] to solve the autopilot control problem. The sensitivity method and the MIT rule [18] were used to design the adjustment or adaptive laws for estimating the unknown parameters for the various proposed MRAC schemes.

The work on adaptive flight control was characterized by "a lot of enthusiasm, bad hardware and nonexisting theory" [1]. The lack of stability proofs and the lack of understanding of the properties of the proposed adaptive control schemes, coupled with a disaster in a flight test [8] caused the interest in adaptive control in the late 1950s and early 1960s to diminish.

The 1960s became the most important period for the development of control theory and adaptive control in particular. State-space techniques and stability theory based on Lyapunov were introduced. Developments in dynamic programming [11], dual control [13], and stochastic control in general and in system identification and parameter estimation [10] played a crucial role in the reformulation and redesign of adaptive control. By 1966 Parks [6] and others found a way of redesigning the MIT rule-based adaptive laws used in the MRAC schemes of the 1950s by applying the Lyapunov design approach. Their work, even though applicable to a special class of linear, time-invariant (LTI) plants, set the stage for further rigorous stability proofs in MRAC for more general classes of plant models.

The advances in stability theory and the progress in control theory in the 1960s improved the understanding of adaptive control and contributed to a strong renewed interest in the field in the 1970s. On the other hand, the simultaneous development and progress in computers and electronics that made the implementation of complex controllers, such as the adaptive ones, feasible, contributed to an increased interest in applications of adaptive control. The 1970s witnessed several breakthrough results in the design of adaptive control. MRAC schemes using the Lyapunov design approach were developed and analyzed and the concepts of positivity and hyperstability were used to develop a wide class of MRAC schemes with well-established stability properties [4, 5, 12, 18]. At the same time, parallel efforts for discrete-time plants in a deterministic and stochastic environment produced several classes of adaptive control schemes with rigorous stability proofs [14]. The excitement of the 1970s and the development of a wide class of adaptive control schemes with well-established stability properties was accompanied by a number of successful applications [15].

The successes of the 1970s, however, were soon followed by controversies over the practicality of adaptive control. As early as 1979 it was pointed out that the MRAC schemes of the 1970s could easily go unstable in the presence of small disturbances [12]. The nonrobust behavior of adaptive control became very controversial in the early 1980s when more examples of instabilities were published, demonstrating lack of robustness in the presence of unmodeled dynamics and or bounded disturbances

0-8493-8570-9/96/$0.00+$.50

[7, 16]. These examples stimulated many researchers, whose objective was to understand the mechanisms of instabilities and find ways to counteract them. By the mid 1980s, a number of new redesigns and modifications were proposed and analyzed, leading to a body of work known as robust adaptive control. An adaptive controller is defined to be robust if it guarantees signal boundedness in the presence of "reasonable" classes of unmodeled dynamics and bounded disturbances as well as performance error bounds that are of the order of the modeling error. The work on robust adaptive control continued throughout the 1980s and involved the understanding of the various robustness modifications and their unification under a more general framework [2, 12].

The solution of the robustness problem in adaptive control led to the solution of the long-standing problem of controlling a linear plant whose parameters are unknown and changing with time. By the end of the 1980s several breakthrough results were published in the area of adaptive control and in particular MRAC for linear time-varying plants [19].

The focus of adaptive control research in the late 1980s and early 1990s was on performance properties and on extending the results of the 1980s to certain classes of nonlinear plants with unknown parameters. These efforts led to new classes of MRAC-type schemes motivated from nonlinear system theory [3], as well as to MRAC schemes with improved transient and steady-state performance [2].

Adaptive control has been traditionally divided into two classes, the MRAC-type schemes and adaptive pole placement control (APPC) schemes. In MRAC both the poles and zeros of the plant are changed so that the closed-loop plant has the same input-output properties as those of a given reference model. In APPC only the poles of the plant are changed. In this chapter we concentrate on MRAC for continuous-time plants that attracted considerable interest in the literature of adaptive control. For information on APPC and discrete-time adaptive control, the reader is referred to [17] and [14].

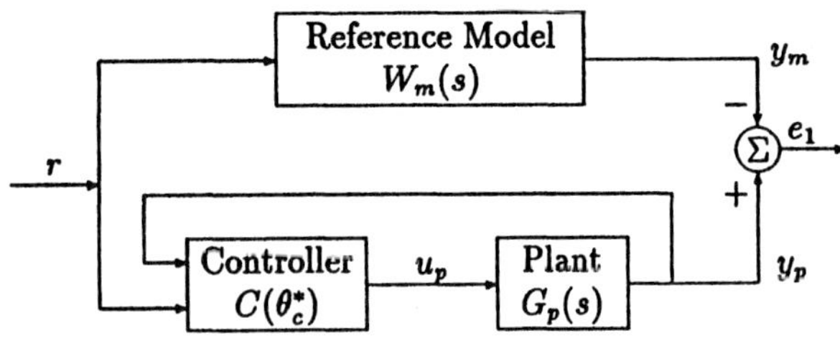

Figure 54.1 The diagram shows the basic structure of model reference control (MRC).

54.2 MRAC Schemes

Model reference adaptive control (MRAC) is derived from the model-following problem or model reference control (MRC) problem. In MRC, a good understanding of the plant and the performance requirements it has to meet allows the designer to come up with a model, referred to as the reference model, that describes the desired input-output properties of the closed-loop plant. The objective of MRC is to find the feedback control law that changes the structure and dynamics of the plant so that its input-output properties are exactly the same as those of the reference model. The structure of an MRC scheme for a LTI, single-input, single-output (SISO) plant is shown in Figure 54.1. The transfer function $W_m(s)$ of the reference model is designed so that, for a given reference input signal $r(t)$, the output $y_m(t)$ of the reference model represents the desired response the plant output $y_p(t)$ has to follow. The feedback controller, denoted by $C(\theta_c^*)$, is designed so that all signals are bounded and the closed-loop plant transfer function from r to y_p is equal to $W_m(s)$. This transfer function matching guarantees that for any given reference input $r(t)$, the tracking error $e_1(t)$, which represents the deviation of the plant output y_p from the desired trajectory y_m, converges to zero with time. The transfer function matching is achieved by cancelling the zeros of the plant transfer function $G_p(s)$ and replacing them with those of $W_m(s)$ through the use of the feedback controller $C(\theta_c^*)$. The cancellation of the plant zeros puts a restriction on the plant to be minimum phase, i.e., have stable zeros. If any plant zero is unstable, its cancellation may easily lead to unbounded signals.

The design of $C(\theta_c^*)$ requires knowledge of the coefficients of the plant transfer function $G_p(s)$. If θ^* is a vector containing all the coefficients of $G_p(s) = G_p(s, \theta^*)$, then the controller parameter vector θ_c^* may be computed by solving an algebraic equation of the form

$$\theta_c^* = F(\theta^*) \tag{54.1}$$

When θ^* is unknown the MRC scheme of Figure 54.1 cannot be implemented, since θ_c^* cannot be calculated using Equation 54.1 and is therefore unknown. One way of dealing with the unknown parameter case is to use the certainty equivalence approach [14, 17]. In this context, the certainty equivalence approach is to replace the unknown θ_c^* in the control law with its estimate $\theta_c(t)$ obtained using the direct or the indirect approach. The resulting control schemes are known as MRAC and can be classified as indirect MRAC, shown in Figure 54.2, and direct MRAC, shown in Figure 54.3. In indirect MRAC the controller parameter vector θ_c is calculated at each time using the estimate of the plant parameter vector θ^* and the mapping defined by Equation 54.1. In direct MRAC the vector θ_c is adjusted directly without any intermediate calculations that involve estimates of θ^*. In this case the plant transfer function $G_p(s, \theta^*)$ is parameterized with respect to θ_c^* to obtain $G_p(s, \theta_c^*)$, whose form is used to estimate θ_c^* directly.

Different choices of on-line parameter estimators lead to further classifications of MRAC.

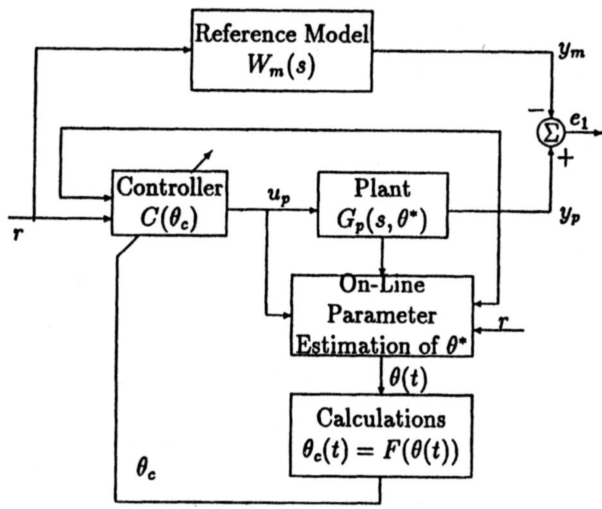

Figure 54.2 The diagram shows the basic structure of indirect MRAC.

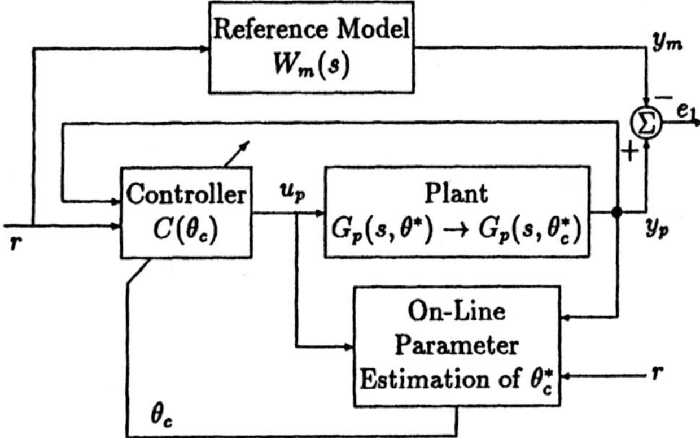

Figure 54.3 The diagram shows the basic structure of direct MRAC.

54.2.1 Model Reference Control

Consider the SISO, LTI plant described by the vector differential equation

$$\begin{aligned} \dot{x}_p &= A_p x_p + B_p u_p, \quad x_p(0) = x_0 \\ y_p &= C_p^T x_p \end{aligned} \tag{54.2}$$

where $x_p \in R^n$; $y_p, u_p \in R^1$ and A_p, B_p, C_p have the appropriate dimensions. The transfer function of the plant is given by

$$y_p = G_p(s) u_p = k_p \frac{Z_p(s)}{R_p(s)} u_p \tag{54.3}$$

where $Z_p(s)$, $R_p(s)$ are monic polynomials and k_p is a constant referred to as the high-frequency gain. The reference model, selected by the designer to describe the desired characteristics of the closed-loop system, is given by

$$y_m = W_m(s) r = k_m \frac{Z_m(s)}{R_m(s)} r \tag{54.4}$$

where $Z_m(s)$, $R_m(s)$ are monic polynomials of degree q_m, p_m, respectively, and k_m is a constant.

The MRC objective is to determine the plant input u_p so that all signals are bounded and the plant output y_p tracks the reference model output y_m as closely as possible for any given reference input $r(t)$ that is bounded and continuous. We refer to the problem of finding the desired u_p to meet the control objective as the MRC problem.

In order to meet the MRC objective with a control law that uses signals that are available for measurement, we assume that the plant and reference models satisfy the following assumptions:

Plant assumptions:

P1. $Z_p(s)$ is a monic Hurwitz polynomial of degree m_p, i.e., $Z_p(s)$ is a monic polynomial of degree m_p that has all roots in the open left half s-plane

P2. An upper bound n of the degree n_p of $R_p(s)$

P3. the relative degree $n^* = n_p - m_p$ of $G_p(s)$ and

P4. the sign of the high-frequency gain k_p

are known.

Reference model assumptions:

M1. $Z_m(s)$, $R_m(s)$ are monic Hurwitz polynomials of degree q_m, p_m, respectively, where $p_m \leq n$

M2. The relative degree $n_m^* = p_m - q_m$ of $W_m(s)$ is the same as that of $G_p(s)$, i.e., $n_m^* = n^*$

In addition to assumptions P1 to P4 and M1, M2, let us also assume that the plant parameters, i.e., the coefficients of $G_p(s)$, are known exactly. Since the plant is LTI and known, the design of the MRC scheme is achieved using linear system theory.

The MRC objective is met if u_p is chosen so that the closed-loop transfer function from r to y_p has stable poles and is equal to $W_m(s)$, the transfer function of the reference model. Such transfer function matching guarantees that for any reference input signal $r(t)$ the plant output y_p converges to y_m exponentially fast.

We consider the feedback control law

$$u_p = \theta_1^{*T} \omega_1 + \theta_2^{*T} \omega_2 + \theta_3^* y_p + c_0^* r = \theta_c^{*T} \omega \tag{54.5}$$

where $\theta_c^* = [\theta_1^{*T}, \theta_2^{*T}, \theta_3^*, c_0^*]^T$, $\omega = [\omega_1^T, \omega_2^T, y_p, r]^T$, $\alpha(s) = [s^{n-2}, s^{n-3}, \cdots, s, 1]^T$ $\omega_1 = \frac{\alpha(s)}{\Lambda(s)} u_p$, $\omega_2 = \frac{\alpha(s)}{\Lambda(s)} y_p$, $c_0^*, \theta_3^* \in R^1$; $\theta_1^*, \theta_2^* \in R^{n-1}$ are constant parameters to be designed and $\Lambda(s)$ is an arbitrary monic Hurwitz polynomial of degree $n-1$ that contains $Z_m(s)$ as a factor, i.e., $\Lambda(s) = \Lambda_0(s) Z_m(s)$, which implies that $\Lambda_0(s)$ is monic, Hurwitz, and of degree $n_0 = n - 1 - q_m$. The controller parameter vector $\theta_c^* \in R^{2n}$ is to be chosen so that the transfer function from r to y_p i.e., $y_p = G_c(s) r$ given by

$$G_c(s) = \frac{c_0^* k_p Z_p \Lambda^2}{\Lambda[(\Lambda - \theta_1^{*T}\alpha) R_p - k_p Z_p (\theta_2^{*T}\alpha + \theta_3^* \Lambda)]} \tag{54.6}$$

is stable and is equal to $W_m(s) = k_m \frac{Z_m}{R_m}$ for all s.

Since the degree of the denominator of $G_c(s)$ is n_p+2n-2 and that of $R_m(s)$ is $p_m \leq$ n, for the matching equation

$$\frac{c_0^* k_p Z_p \Lambda^2}{\Lambda[(\Lambda - \theta_1^{*T}\alpha) R_p - k_p Z_p (\theta_2^{*T}\alpha + \theta_3^* \Lambda)]} = k_m \frac{Z_m}{R_m} \tag{54.7}$$

to hold, an additional $n_p + 2n - 2 - p_m$ zero-pole cancellations must occur in $G_c(s)$. Now since $Z_p(s)$ is Hurwitz by assumption, and $\Lambda(s) = \Lambda_0(s)Z_m(s)$ is designed to be Hurwitz, it follows that all the zeros of $G_c(s)$ are stable and therefore any zero-pole cancellation can only occur in C^-, the open left half of the complex plane. Choosing

$$c_0^* = \frac{k_m}{k_p} \tag{54.8}$$

and using $\Lambda(s) = \Lambda_0(s)Z_m(s)$ the matching Equation 54.7 becomes

$$(\Lambda - \theta_1^{*T}\alpha)R_p - k_p Z_p(\theta_2^{*T}\alpha + \theta_3^*\Lambda) = Z_p\Lambda_0 R_m \tag{54.9}$$

Dividing both sides of Equation 54.9 by $R_p(s)$, we obtain

$$\Lambda - \theta_1^{*T}\alpha - k_p\frac{Z_p}{R_p}(\theta_2^{*T}\alpha + \theta_3^*\Lambda) = Z_p(Q + k_p\frac{\Delta^*}{R_p})$$

where $Q(s)$ (of degree $n-1-m_p$) is the quotient and $k_p\Delta^*$ (of degree at most $n_p - 1$) is the remainder of $\Lambda_0 R_m/R_p$, respectively. Then the solution for $\theta_i^*, i = 1, 2, 3$ can be found by inspection, i.e.,

$$\theta_1^{*T}\alpha(s) = \Lambda(s) - Z_p(s)Q(s)$$

$$\theta_2^{*T}\alpha(s) + \theta_3^*\Lambda(s) = \frac{Q(s)R_p(s) - \Lambda_0(s)R_m(s)}{k_p} \tag{54.10}$$

where the equality in the second equation is obtained by substituting for $\Delta^*(s)$ using the identity

$$\frac{\Lambda_0 R_m}{R_p} = Q + \frac{k_p\Delta^*}{R_p}$$

The parameters $\theta_i^*, i = 1, 2, 3$ can now be obtained directly by equating the coefficients of the powers of s on both sides of Equation 54.10. Equation 54.10 indicates that, in general, the controller parameters $\theta_i^*, i = 1, 2, 3$ are nonlinear functions of the coefficients of the plant polynomials $Z_p(s)$, $R_p(s)$ due to the dependence of $Q(s)$ on the coefficients of $R_p(s)$. When $n = n_p$ and $n^* = 1$, however, $Q(s) = 1$ and the θ_i^*s are linear functions of the coefficients of $Z_p(s)$, $R_p(s)$.

LEMMA 54.1

(i) Let the degrees of R_p, Z_p, Λ, Λ_0 and R_m be as specified in Equation 54.5. Then the solution $\overline{\theta}_c^* = [\theta_1^{*T}, \theta_2^{*T}, \theta_3^{*T}]$ of Equation 54.9 or Equation 54.10 always exists.

(ii) In addition, if R_p, Z_p are coprime and $n = n_p$, then the solution $\overline{\theta}_c^*$ is unique.

The proof is based on the solution of certain Diophantine equations and is given in [17].

It can be shown that the control law (Equation 54.5) with θ_c^* calculated from Equations 54.8 and 54.10 guarantees that the closed-loop plant is stable and the tracking error $e_1 = y_p - y_m$ converges exponentially to zero for any given bounded reference input r.

54.2.2 Direct MRAC

A direct MRAC scheme is formed by combining the control law (Equation 54.5), with θ_c^* replaced by its estimate $\theta_c(t)$ at time t, i.e.,

$$u_p = \theta_c^T(t)\omega \tag{54.11}$$

with an adaptive law that generates $\theta_c(t)$ at each time t.

The estimate $\theta_c(t)$ of θ_c^* is generated by first obtaining an appropriate parameterization of the plant in terms of θ_c^* and then using parameter estimation techniques to form the adaptive law for $\theta_c(t)$. Such a parameterization is developed by using the plant and matching equations to obtain

$$e_1 = W_m(s)\rho^*(u_p - \theta_c^{*T}\omega) \tag{54.12}$$

where $e_1 = y_p - y_m$, and $\rho^* = \frac{1}{c_0^*}$. Using Equation 54.12 a wide class of adaptive laws may be developed to estimate θ_c^*, ρ^*. The adaptive laws may be split into two major classes; those with unnormalized signals and those with normalized signals leading to direct MRAC without normalization and direct MRAC with normalization.

Direct MRAC Without Normalization

The derivation and complexity of the MRAC scheme depends on the relative degree n^* of the plant. For $n^* = 1$ we choose the transfer function $W_m(s)$ of the reference model to be strictly positive real (SPR) [17, 18]. Substituting for $u_p = \theta_c^T(t)\omega$ in Equation 54.12 we obtain the error equation

$$e_1 = W_m(s)\rho^*\tilde{\theta}_c^T\omega \tag{54.13}$$

where $\tilde{\theta}_c = \theta_c - \theta_c^*$ is the parameter error. Since $W_m(s)$ is SPR we can use the SPR-Lyapunov design approach [17] to generate the adaptive law

$$\dot{\theta}_c = -\Gamma e_1 \omega sgn(\rho^*) \tag{54.14}$$

where

$$\Gamma = \Gamma^T > 0$$

which, together with Equation 54.11 forms the MRAC scheme. The adaptive law (Equation 54.14) is chosen so that a certain positive definite function V of the error states of Equation 54.13 and Equation 54.14 is a Lyapunov function with the property

$$\frac{dV}{dt} \leq 0$$

which implies uniform stability and with additional arguments, $e_1(t) \to 0$ as $t \to \infty$.

When $n^* = 2$, $W_m(s)$ cannot be designed to be SPR because of assumption M2 and the fact that a transfer function of relative degree 2 cannot be SPR. In this case we rewrite Equation 54.12 as

$$e_1 = W_m(s)(s+p)\rho^*[u_f - \theta_c^{*T}\omega_f]$$

where $u_f = \frac{1}{s+p}u_p$, $\omega_f = \frac{1}{s+p}\omega$, and $p > 0$ is chosen so that $\overline{W}_m(s) = W_m(s)(s+p)$ is SPR. If we choose $u_f = \theta_c^T\omega_f$ then

$$e_1 = \overline{W}_m(s)\rho^*(\tilde{\theta}_c^T\omega_f)$$

which has the same form as Equation 54.13 and leads to the adaptive law

$$\dot{\theta}_c = -\Gamma e_1\omega_f sgn(\rho^*) \tag{54.15}$$

The adaptive law (Equation 54.15) is generated by using $u_f = \theta_c^T\omega_f$. Since $u_f = \frac{1}{s+p}u_p$ the control input u_p has to be chosen so that $u_f = \theta_c^T\omega_f$. We have

$$u_p = (s+p)\theta_c^T\omega_f = \theta_c^T\omega + \dot{\theta}_c^T\omega_f$$

where the second equality is obtained by treating s as a differential operator. Since $\dot{\theta}_c$ is given by Equation 54.15, the MRAC scheme when $n^* = 2$ is given by

$$\begin{aligned} u_p &= \theta_c^T\omega + \dot{\theta}_c^T\omega_f \\ \dot{\theta}_c &= -\Gamma e_1\omega_f sgn(\rho^*) \end{aligned} \tag{54.16}$$

When $n^* = 3$ we can use the same procedure as in the case of $n^* = 2$ to obtain the error equation

$$e_1 = W_m(s)(s+p_1)(s+p_2)\rho^*[u_f - \theta_c^{*T}\phi]$$

where

$$u_f = \frac{1}{(s+p_1)(s+p_2)}u_p, \quad \phi = \frac{1}{(s+p_1)(s+p_2)}\omega$$

and $p_1, p_2 > 0$ are chosen so that $\overline{W}_m(s) = W_m(s)(s+p_1)(s+p_2)$ is SPR. In this case we cannot choose $u_f = \theta_c^T\phi$ since such a choice will have to include second derivatives of θ_c in the expression for u_p that are not available for measurement. We go around this difficulty by using the nonlinear tools "backstepping" and "nonlinear damping" as explained in [3, 17] to obtain the MRAC scheme

$$\begin{aligned} u_p &= \theta_c^T\omega + \dot{\theta}_c^T\phi_1 - (s+p_2)\alpha_0(\phi^T\Gamma\phi)^2 r_0 \\ \dot{r}_0 &= -[p_1 + \alpha_0(\phi^T\Gamma\phi)^2]r_0 + \phi^T\Gamma\phi\epsilon_1 \mathrm{sgn}(\rho^*) \\ \dot{\theta}_c &= -\Gamma\epsilon_1\phi\mathrm{sgn}(\rho^*), \quad \dot{\rho} = \gamma\epsilon_1 r_0 \\ \phi_1 &= \frac{1}{s+p_2}\omega, \quad \epsilon_1 = e_1 - \frac{1}{s+q_0}\rho r_0 \end{aligned} \tag{54.17}$$

where $\alpha_0 > 0$ is a design constant and $\overline{W}_m(s) = \frac{1}{s+q_0}$. For $n^* > 3$ the procedure is the same, but it leads to much more complex MRAC schemes.

The above MRAC schemes guarantee that all signals are bounded and the tracking error converges to zero. If r is sufficiently rich then the MRAC scheme for $n^* = 1, 2$ guarantees exponential convergence of the parameter error $\tilde{\theta}_c$ and that the tracking error e_1 goes to zero. For $n^* \geq 3$ the convergence of $\tilde{\theta}_c$ to zero is asymptotic [17].

Direct MRAC with Normalized Adaptive Laws

This class of MRAC schemes dominated the literature of adaptive control due to the simplicity of their design as well as their robustness properties in the presence of modeling errors. The adaptive laws of these schemes are driven by a normalized error signal that "slows" down adaptation and improves robustness with respect to plant uncertainties. For this reason they are referred to as normalized adaptive laws. The MRAC law $u_p = \theta_c^T\omega$ in Equation 54.11 remains unchanged and the parametric model (Equation 54.12) is used to generate the adaptive law for θ_c. The parametric model (Equation 54.12) may be rewritten in various other forms, giving rise to a wide class of adaptive laws. For example, we can rewrite Equation 54.12 as

$$z = \theta_c^{*T}\phi_p \tag{54.18}$$

where $z = W_m(s)u_p$, $\phi_p = [W_m(s)\omega_1^T, W_m(s)\omega_2^T, W_m(s)y_p, y_p]^T$ or

$$e_1 = \rho^*(u_f - \theta_c^{*T}\phi) \tag{54.19}$$

where $u_f = W_m(s)u_p$, $\phi = W_m(s)\omega$. Equation 54.18 is obtained by first rewriting Equation 54.12 as $y_p - y_m = \rho^*[z - \theta_c^{*T}\phi_p - c_o^* y_m + c_o^* y_p]$ and then using the identity $\rho^* c_o^* = 1$.

Using Equation 54.12 and the SPR-Lyapunov approach [17], we have the adaptive law

$$\dot{\theta}_c = -\Gamma\epsilon\phi\mathrm{sgn}(\rho^*) \quad \dot{\rho} = \gamma\epsilon\xi \tag{54.20}$$

where

$$\begin{aligned} \epsilon &= e_1 - \hat{e}_1 - W_m L(\epsilon n_s^2) \\ \hat{e}_1 &= W_m(s)L(s)[\rho(u_f - \theta_c^T\phi)] \\ \xi &= u_f - \theta_c^T\phi, \quad \phi = L^{-1}(s)\omega \\ u_f &= L^{-1}(s)u_p, \quad n_s^2 = \phi^T\phi + u_f^2 \end{aligned}$$

$L(s)$ is chosen so that $W_m L$ is SPR and proper, and $L^{-1}(s)$ is proper and stable.

Using Equation 54.19 and the gradient method [17] we have

$$\dot{\theta}_c = -\Gamma\epsilon\phi sgn(\rho^*) \quad \dot{\rho} = \gamma\epsilon\xi \tag{54.21}$$

where ε, ϕ, ξ are the same as in Equation 54.20 with $L^{-1}(s) = W_m(s)$.

Using Equation 54.18 we can generate a wide class of adaptive laws using the gradient method with different cost functions as well as least squares [17]. The gradient algorithm is given by

$$\dot{\theta}_c = Pr[\Gamma\epsilon\phi_p] \tag{54.22}$$

and the least squares is given by

$$\begin{aligned} \dot{\theta} &= Pr[P\epsilon\phi_p] \\ \dot{P} &= \overline{Pr}[-\frac{P\phi_p\phi_p^T P}{m^2}] \end{aligned} \tag{54.23}$$

where $\epsilon = \frac{z-\hat{z}}{m^2}$, $\hat{z} = \theta_c^T\phi_p$, $m^2 = 1 + \phi_p^T\phi_p$ and $Pr[.]$ is the projection operator that constrains $c_0(t)$, the estimate of c_0^*, to satisfy $|c_0(t)| \geq c_m$, $\forall t \geq 0$, where $c_m > 0$ is a lower bound for $|c_0^*|$. The $\overline{Pr}(.)$ operator sets $\dot{P} = 0$, when $|c_0(t)| = c_m$ and $\dot{c}_0 < 0$. The projection is used to guarantee that $\frac{1}{c_0(t)}$ is bounded for all $t \geq 0$, a property that is used in the stability analysis of the MRAC scheme (Equation 54.11) with θ_c generated by Equation 54.22 or Equation 54.23. For the implementation of projection we require the knowledge of c_m, a lower bound for $|c_0^*|$ and the sign of c_0^*.

The control law (Equation 54.11) with any one of the adaptive laws (Equation 54.20, 54.21, 54.22, or 54.23) forms a direct MRAC scheme. As shown in [17] these schemes guarantee signal boundedness and convergence of the tracking error to zero. If, in addition, the reference input r(t) is sufficiently rich of order 2n then both the parameter and tracking errors converge to zero. The rate of convergence in the case of Equation 54.11 with Equation 54.22 or 54.23 is exponential, whereas for the case of the MRAC scheme (Equations 54.11 and 54.20, or 54.11 and 54.21), the convergence is asymptotic.

54.2.3 Indirect MRAC

In indirect MRAC the controller parameter vector θ_c(t) in the control law (Equation 54.11) is calculated at each time t using the estimates of k_p and of the coefficients of $Z_p(s)$, $R_p(s)$ that are generated using an adaptive law. The calculation of $\theta_c(t)$ is achieved by using the mapping defined by the matching Equations 54.8 and 54.10.

As in the direct MRAC case the adaptive laws for the estimated coefficients of $Z_p(s)$, $R_p(s)$ could be normalized or unnormalized. We concentrate on the normalized adaptive laws and refer the reader to [17] for results using unnormalized adaptive laws.

The adaptive law for estimating k_p and the coefficients of $Z_p(s)$, $R_p(s)$ is generated using the parametric plant model

$$z = \theta_p^{*T}\phi \qquad (54.24)$$

where

$$z = \frac{s^n}{\Lambda_p(s)}y_p, \quad \phi = [\frac{\alpha_{n-1}^T(s)}{\Lambda_p(s)}u_p, \frac{-\alpha_{n-1}^T(s)}{\Lambda_p(s)}y_p]^T$$

$\theta_p^* = [0, \cdots, 0, b_m, p_1^T, p_2^T]^T \in R^{2n}$, $p_1 = [b_{m-1}, \cdots, b_0]^T$ and $p_2 = [a_{n-1}, \cdots, a_0]^T$ are the coefficient vectors of $k_p[Z_p(s) - s^m]$, $R_p(s) - s^n$, respectively, $\Lambda_p(s)$ is an nth order monic Hurwitz polynomial, $a_{n-1}(s) = [s^{n-1}, \cdots, s, 1]^T$ and $b_m = k_p$.

Using Equation 54.24 the estimate $\theta_p(t)$ of θ_p^* may be generated using adaptive laws that are based on the gradient or the least squares methods. The controller parameter vector $\theta_c(t) = [\theta_1^T(t), \theta_2^T(t), \theta_3(t), c_0(t)]^T$ is calculated from $\theta_p(t) = [0, ...0, \hat{k}_p, \hat{p}_1^T, \hat{p}_2^T]^T$, the estimate of θ_p^* at each time t, as follows:

$$\begin{aligned}
c_0(t) &= \frac{k_m}{\hat{k}_p} \\
\theta_1^T\alpha_{n-2}(s) &= \Lambda(s) - \hat{Z}_p(s,t) \bullet \hat{Q}(s,t) \\
\theta_2^T\alpha_{n-2}(s) + \theta_3\Lambda(s) &= \frac{1}{\hat{k}_p}[\hat{Q}(s,t) \bullet \hat{R}_p(s,t) \\
&\quad - \Lambda_0(s)R_m(s)] \qquad (54.25)
\end{aligned}$$

where

$$\begin{aligned}
\hat{Q}(s,t) &= \text{quotient of } \frac{\Lambda_0(s)R_m(s)}{\hat{R}_p(s,t)}, \\
\hat{Z}_p(s,t) &= \hat{k}_p s^m + \hat{p}_1^T\alpha_{m-1}(s) \\
\hat{R}_p(s,t) &= s^n + \hat{p}_2^T\alpha_{n-1}(s), \ \alpha_i(s) = [s^i, s^{i-1}, \cdots, s, 1]^T
\end{aligned}$$

and $A \bullet B$ denotes pointwise in time multiplication. From Equation 54.25 it is clear that the adaptive law for $\hat{k}_p$ has to be modified using projection so that $|\hat{k}_p(t)| \geq k_m \geq 0$, where $k_m > 0$ is a lower bound for k_p. As an example of an adaptive law consider the gradient algorithm

$$\begin{aligned}
\dot{\theta}_p &= Pr[\Gamma\epsilon\phi], \qquad (54.26) \\
\epsilon &= \frac{z-\hat{z}}{m^2}, \quad \hat{z} = \theta_p^T\phi, \quad m^2 = 1 + \phi^T\phi
\end{aligned}$$

where $Pr[\bullet]$ is the projection operator that guarantees $|\hat{k}_p| \geq k_m > 0$ for all t$\geq$ 0. The projection operator requires the knowledge of the sign of k_p and the lower bound k_m of k_p. The indirect MRAC scheme (Equations 54.11, 54.25, and 54.26) guarantees that $\theta_c(t)$ given by Equation 54.25 exists and is bounded for any bounded estimate θ_p, all signals in the closed-loop plant are bounded and the tracking error converges to zero with time. If, in addition, the reference signal r is sufficiently rich of order 2n then the parameter and tracking errors converge to zero exponentially fast [17].

54.2.4 Robust MRAC

The MRAC schemes presented above are designed for the plant model (Equation 54.2) that is free of disturbances and unmodeled dynamics. In the presence of disturbances and/or unmodeled dynamics the above schemes may be driven unstable, as shown by several examples in [16]. These schemes can be made robust by modifying the adaptive laws using leakage, dead-zone, projections and their by-products [17]. For the MRAC schemes without normalization these modifications guarantee the existence of a region of attraction in which all signals are bounded and the tracking error converges to a smaller residual set. For the MRAC schemes with normalization the region of attraction becomes the whole space provided a special normalizing signal is used to bound from above all the modeling error terms that are required to be small in the low-frequency range.

As an example, let us modify the direct MRAC scheme (Equations 54.11 and 54.20) for robustness using a leakage type of modification known as σ-modification. We have

$$\begin{aligned}
u_p &= \theta_c^T\omega \\
\dot{\theta}_c &= -\Gamma\epsilon\phi\text{sgn}(\rho^*) - \sigma\Gamma\theta_c, \quad \dot{\rho} = -\gamma\epsilon\xi - \sigma_2\gamma\rho \\
n_s^2 &= \phi^T\phi + u_f^2 + m_s \\
\dot{m}_s &= -\delta_0 m_s + u_p^2 + y_p^2, \quad m_s(0) = 0
\end{aligned}$$

where $\sigma_1, \sigma_2 > 0$ are small positive constants, m_s is the dynamic normalizing signal, and $\delta_0 > 0$ is chosen so that the m_s bounds from above any modeling error term in the plant. The rest of the signals are as defined in Equation 54.20. If the above robust MRAC scheme is applied to the plant

$$y_p = G_p(s)(1 + \Delta_m(s))u_p$$

where $\Delta_m(s)$ is a multiplicative plant uncertainty with the property that $\Delta_m(s - \delta_0/2)$ has stable poles, then for small $\Delta_\infty \stackrel{\Delta}{=} \|W(s-\delta_0/2)\Delta_m(s-\delta_0/2)\|_\infty$, $\Delta_2 \stackrel{\Delta}{=} \|W(s-\delta_0/2)\Delta_m(s-\delta_0/2)\|_2$ where $W(s-\delta_0/2)$ is an arbitrary stable transfer function with stable $W^{-1}(s-\delta_0/2)$ and $W(s)\Delta_m(s)$ is strictly proper, we have signal boundedness for any finite initial condition. Furthermore, the tracking error has a mean square value of the order of Δ_∞, Δ_2. The details of the design and analysis of robust MRAC schemes are given in [17].

54.3 Examples

In this section, we present several examples that illustrate the design and analysis of the MRAC schemes described in the previous sections.

54.3.1 Scalar Example: Adaptive Regulation

Consider the following scalar plant:

$$\dot{x} = ax + u, \quad x(0) = x_0 \tag{54.27}$$

where a is a constant but unknown. The control objective is to determine a bounded function $u = f(t, x)$ such that the state $x(t)$ is bounded and converges to zero as $t \to \infty$ for any given initial condition x_0. Let $-a_m$ be the desired closed-loop pole where $a_m > 0$ is chosen by the designer. In this case the reference model is

$$\dot{x}_m = -a_m x_m, \quad x_m(0) = x_0 \tag{54.28}$$

Control law: If the plant parameter a is known the control law

$$u = -k^* x \tag{54.29}$$

with $k^* = a + a_m$ could be used to meet the control objective, i.e., with Equation 54.29, the closed-loop plant is

$$\dot{x} = -a_m x, \quad x(0) = x_0$$

whose equilibrium $x_e = 0$ is exponentially stable in the large.

Since a is unknown, k^* cannot be calculated and therefore Equation 54.29 cannot be implemented. A possible procedure to follow in the unknown parameter case is to use the same control law as given in Equation 54.29, but with k^* replaced by its estimate $k(t)$, i.e., we use

$$u = -k(t)x \tag{54.30}$$

and search for an adaptive law to update $k(t)$ continuously with time.

Adaptive law: The adaptive law for generating $k(t)$ is developed by viewing the problem as an on-line identification problem for k^*. This is accomplished by first obtaining an appropriate parameterization for the plant (Equation 54.27) in terms of the unknown k^*, as follows.

We add and subtract the desired control input $-k^*x$ in the plant Equation 54.27 to obtain

$$\dot{x} = ax - k^*x + k^*x + u$$

Since $a - k^* = -a_m$ we have

$$\dot{x} = -a_m x + k^* x + u$$

or

$$x = \frac{1}{s + a_m}(u + k^* x) \tag{54.31}$$

Equation 54.31 is a parameterization of the plant Equation 54.27 in terms of the unknown controller parameter k^*. Since x, u are measured and $a_m > 0$ is known, many adaptive laws may be generated using Equation 54.31 as shown in [17].

Substituting for the control $u = -k(t)x$ in Equation 54.31, we obtain the error equation that relates the parameter error $\tilde{k} = k - k^*$ with the estimation error $\epsilon_1 = x$, i.e.,

$$\dot{\epsilon}_1 = -a_m \epsilon_1 - \tilde{k}x, \quad \epsilon_1 = x \tag{54.32}$$

Due to Equation 54.31 the estimation error ϵ_1, which is defined as the error that reflects the parameter error $\tilde{k}$, is equal to the regulation error x. The error Equation 54.32 is in a convenient form for choosing an appropriate Lyapunov function to design the adaptive law for $k(t)$. We assume that the adaptive law is of the form

$$\dot{\tilde{k}} = \dot{k} = f_1(\epsilon_1, x, u) \tag{54.33}$$

where f_1 is some function to be selected, and propose

$$V(\epsilon_1, \tilde{k}) = \frac{\epsilon_1^2}{2} + \frac{\tilde{k}^2}{2\gamma}$$

for some $\gamma > 0$ as a potential Lyapunov function for the system defined by Equations 54.32 and 54.33. The time derivative of V along the trajectory of this system is given by

$$\dot{V} = -a_m \epsilon_1^2 - \tilde{k}\epsilon_1 x + \frac{\tilde{k} f_1}{\gamma}$$

Choosing $f_1 = \gamma \epsilon_1 x$, i.e.,

$$\dot{k} = \gamma \epsilon_1 x = \gamma x^2, \qquad k(0) = k_0 \tag{54.34}$$

we have

$$\dot{V} = -a_m \epsilon_1^2 \leq 0$$

Analysis: Since V is a positive definite function and $\dot{V} \leq 0$, we have $V \in \mathcal{L}_\infty$, which implies that $\epsilon_1, \tilde{k} \in \mathcal{L}_\infty$. Since $\epsilon_1 = x$, we also have that $x \in \mathcal{L}_\infty$ and therefore all signals in the closed-loop plant are bounded. Furthermore, $\epsilon_1 = x \in \mathcal{L}_2$, and $\dot{\epsilon}_1 = \dot{x} \in \mathcal{L}_\infty$ which imply that $\epsilon_1(t) = x(t) \to 0$ as $t \to \infty$. From $x(t) \to 0$ and the boundedness of k, we establish that $\dot{k}(t) \to 0$, $u(t) \to 0$ as $t \to \infty$.

We have shown that the combination of the control law (Equation 54.30) with the adaptive law (Equation 54.34) meets the control objective, in the sense that it forces the plant state to converge to zero while guaranteeing signal boundedness.

It is worth mentioning that we cannot establish that $k(t)$ converges to k^*, i.e., that the pole of the closed-loop plant converges to that of the reference model given by $-a_m$. The lack of parameter convergence is less crucial in adaptive control than in parameter identification, since in most cases the control objective can be achieved without requiring the parameters to converge to their true values. The simplicity of this scalar example, however, allows us to solve for $\varepsilon_1 = x$ explicitly, and study the properties of $k(t)$, $x(t)$ as they evolve with time. We can verify that

$$\begin{aligned} \epsilon_1(t) &= \frac{2ce^{-ct}}{c+k_0-a+(c-k_0+a)e^{-2ct}}\epsilon_1(0), \quad \epsilon_1 = x \\ k(t) &= \alpha + \frac{c[(c+k_0-\alpha)e^{2ct}-(c-k_0+\alpha)]}{(c+k_0-\alpha)e^{2ct}+(c-k_0+\alpha)} \end{aligned} \tag{54.35}$$

where $c^2 = \gamma x_0^2 + (k_0 - a)^2$, satisfy the differential Equations 54.32 and 54.34 of the closed-loop plant. Equation 54.35 can be used to investigate the effects of initial conditions and adaptive gain γ on the transient and asymptotic behavior of $x(t)$, $k(t)$. We have $\lim_{t\to\infty} k(t) = a + c$, if $c > 0$, and $\lim_{t\to\infty} k(t) = a - c$ if $c < 0$, i.e.,

$$\lim_{t\to\infty} k(t) = k_\infty = a + [\gamma x_0^2 + (k_0 - a)^2]^{1/2}$$

Therefore for $x_0 \neq 0$, $k(t)$ converges to a stabilizing gain whose value depends on γ and the initial condition x_0, k_0. It is clear from Equation 54.35 that the value of k_∞ is independent of whether k_0 is a destabilizing gain, i.e., $0 < k_0 < a$, or a stabilizing one, i.e., $k_0 > a$, as long as $(k_0 - a)^2$ is the same. The use of different k_0, however, will affect the transient behavior as it is obvious from Equation 54.35. In the limit as $t \to \infty$, the closed-loop pole converges to $-(k_\infty - a)$ which may be different from $-a_m$. Since the control objective is to achieve signal boundedness and regulation of the state $x(t)$ to zero, the convergence of $k(t)$ to k^* is not crucial.

54.3.2 Scalar Example: Adaptive Tracking

Consider the following first-order plant:

$$\dot{x} = ax + bu \tag{54.36}$$

where a, b are unknown parameters but the sign of b is known. The control objective is to choose an appropriate control law u such that all signals in the closed-loop plant are bounded and x tracks the state x_m of the reference model given by

$$x_m = \frac{b_m}{s + a_m} r$$

for any bounded piecewise continuous signal $r(t)$, where $a_m > 0$, b_m are known and $x_m(t)$, $r(t)$ are measured at each time t. It is assumed that a_m, b_m and r are chosen so that x_m represents the desired state response of the plant.

Control law: In order for x to track x_m for any reference input signal r, the control law should be chosen so that the closed-loop plant transfer function from the input r to the output x is equal to that of the reference model. We propose the control law

$$u = -k^*x + l^*r \tag{54.37}$$

where k^*, l^* are calculated so that

$$\frac{x(s)}{r(s)} = \frac{bl^*}{s - a + bk^*} = \frac{b_m}{s + a_m} = \frac{x_m(s)}{r(s)} \tag{54.38}$$

Equation 54.38 is satisfied if we choose

$$l^* = \frac{b_m}{b} \quad k^* = \frac{a_m + a}{b} \tag{54.39}$$

provided of course that $b \neq 0$, i.e., the plant is controllable. The control law (Equations 54.37 and 54.39) guarantees that the transfer function of the closed-loop plant, i.e., $\frac{x(s)}{r(s)}$ is equal to that of the reference model. Such a transfer function matching guarantees that $x(t) = x_m(t), \forall t \geq 0$ when $x(0) = x_m(0)$ or $|x(t) - x_m(t)| \to 0$ exponentially fast when $x(0) \neq x_m(0)$, for any bounded reference signal $r(t)$.

When the plant parameters a, b are unknown, Equation 54.37 cannot be implemented. Therefore, instead of Equation 54.37, we propose the control law

$$u = -k(t)x + l(t)r \tag{54.40}$$

where $k(t)$, $l(t)$ is the estimate of k^*, l^*, respectively, at time t, and search for an adaptive law to generate $k(t)$, $l(t)$ on-line.

Adaptive law: As before, we can view the problem as an on-line identification problem of the unknown constants k^*, l^*. We start with the plant equation, which we express in terms of k^*, l^* by adding and subtracting the desired input term $-bk^*x + bl^*r$ to obtain

$$x = \frac{b_m}{s + a_m} r + \frac{b}{s + a_m}(k^*x - l^*r + u) \tag{54.41}$$

Since $x_m = \frac{b_m}{s+a_m} r$ is a known bounded signal, we express Equation 54.41 in terms of the tracking error defined as $\epsilon_1 = x - x_m$, i.e.,

$$\epsilon_1 = \frac{b}{s + a_m}(k^*x - l^*r + u) \tag{54.42}$$

Substituting for $u = -k(t)x + l(t)r$ in Equation 54.42 and defining the parameter errors $\tilde{k} \triangleq k - k^*$, $\tilde{l} \triangleq l - l^*$, we have

$$\begin{aligned} \dot{\epsilon}_1 &= -a_m\epsilon_1 + b(-\tilde{k}x + \tilde{l}r) \\ \epsilon_1 &= x - x_m \end{aligned} \tag{54.43}$$

The development of the differential Equation 54.43 relating the estimation error with the parameter error is a significant step in deriving the adaptive laws for updating k(t), l(t). We assume that the structure of the adaptive law is given by

$$\dot{k} = f_1(\epsilon_1, x, r, u) \quad \dot{l} = f_2(\epsilon_1, x, r, u) \tag{54.44}$$

where the functions f_1, f_2 are to be designed.

Consider the function

$$V(\epsilon_1, \tilde{k}, \tilde{l}) = \frac{\epsilon_1^2}{2} + \frac{\tilde{k}^2}{2\gamma_1}|b| + \frac{\tilde{l}^2}{2\gamma_2}|b|$$

where $\gamma_1, \gamma_2 > 0$, as a Lyapunov candidate for the system (Equations 54.43 and 54.44). The time derivative $\dot{V}$ along any trajectory of the system is given by

$$\dot{V} = -a_m\epsilon_1^2 - b\tilde{k}\epsilon_1 x + b\tilde{l}\epsilon_1 r + \frac{|b|\tilde{k}}{\gamma_1}f_1 + \frac{|b|\tilde{l}}{\gamma_2}f_2 \quad (54.45)$$

Since $|b| = b\text{sgn}(b)$, the indefinite terms in Equation 54.43 disappear if we choose $f_1 = \gamma_1\epsilon_1 x\text{sgn}(b)$, $f_2 = -\gamma_2\epsilon_1 r\text{sgn}(b)$. Therefore, for the adaptive law

$$\dot{k} = \gamma_1\epsilon_1 x\text{sgn}(b), \quad \dot{l} = -\gamma_2\epsilon_1 r\text{sgn}(b) \quad (54.46)$$

we have

$$\dot{V} = -a_m\epsilon_1^2$$

Analysis: Treating $x_m(t), r(t)$ in Equation 54.43 as bounded arbitrary functions of time, it follows that V is a Lyapunov function for the third-order differential equation (54.43 and 54.46) and the equilibrium $\epsilon_{1e} = 0$, $\tilde{k}_e = 0$, $\tilde{l}_e = 0$ is uniformly stable. Furthermore, $\epsilon_1, \tilde{k}, \tilde{l} \in \mathcal{L}_\infty$ and $\epsilon_1 \in \mathcal{L}_2$. Since $\epsilon_1 = x - x_m$, $x_m \in \mathcal{L}_\infty$, we also have $x \in \mathcal{L}_\infty$ and $u \in \mathcal{L}_\infty$ and therefore all signals in the closed-loop plant are bounded. Now from Equation 54.43 we have $\dot{\epsilon}_1 \in \mathcal{L}_\infty$, which together with $\epsilon_1 \in \mathcal{L}_2$, implies that $\epsilon_1(t) \to 0$, as $t \to \infty$. We have established that the control law (Equation 54.40), together with the adaptive law (Equation 54.46) guarantees boundedness for all signals in the closed-loop system. In addition, the plant state $x(t)$ tracks the state of the reference model x_m asymptotically with time for any reference input signal r which is bounded and piecewise continuous. These results do not imply that $k(t) \to k^*$, $l(t) \to l^*$ as $t \to \infty$ i.e., that the transfer function of the closed-loop plant approaches that of the reference model as $t \to \infty$. In order to achieve such a result, the reference input r has to be sufficiently rich of order 2. A sufficiently rich input is one that excites all the modes of the system [14, 17]. For example $r(t) = \sin\omega t$ for some $\omega \neq 0$ is sufficiently rich of order 2 and guarantees the exponential convergence of $x(t)$ to $x_m(t)$ and of $k(t), l(t)$ to k^*, l^*, respectively. In general, a sufficiently rich reference input $r(t)$ is not desirable in cases where the control objective involves tracking of signals that are not rich in frequencies.

54.3.3 Example: Direct MRAC without Normalization (n* = 1)

Let us consider the second-order plant

$$y_p = \frac{k_p(s+b_0)}{s^2 + a_1 s + a_0}u_p$$

where $k_p > 0, b_0 > 0, k_p, b_0, a_1, a_0$ are unknown constants. The desired performance of the plant is specified by the reference model

$$y_m = \frac{1}{s+1}r$$

Using the results of Section 54.2.2 the control law is designed as

$$\begin{aligned}\dot{\omega}_1 &= -2\omega_1 + u_p, \quad \omega_1(0) = 0\\ \dot{\omega}_2 &= -2\omega_2 + y_p, \quad \omega_2(0) = 0\\ u_p &= \theta_1\omega_1 + \theta_2\omega_2 + \theta_3 y_p + c_0 r\end{aligned}$$

by choosing $\Lambda(s) = s+2$ in the general control law. The adaptive law is given by

$$\dot{\theta}_c = -\Gamma e_1\omega, \quad \theta_c(0) = \theta_0$$

where $e_1 = y_p - y_m$, $\theta_c = [\theta_1, \theta_2, \theta_3, c_0]^T$, $\omega = [\omega_1, \omega_2, y_p, r]^T$ and $\Gamma = \Gamma^T$ is any positive definite matrix.

Analysis: From Equation 54.12 we have that the tracking error e_1 satisfies

$$e_1 = \frac{1}{s+1}\rho^*\tilde{\theta}_c^T\omega$$

where $\rho^* = k_p$, $\tilde{\theta}_c = \theta_c - \theta_c^*$, i.e., $\dot{e}_1 = -e_1 + k_p\tilde{\theta}_c^T\omega$.

We choose the positive definite function

$$V = \frac{e_1^2}{2} + k_p\frac{\tilde{\theta}_c^T\Gamma^{-1}\tilde{\theta}_c}{2}$$

then

$$\dot{V} = -e_1^2 + k_p\tilde{\theta}_c^T e_1\omega - k_p\tilde{\theta}_c^T e_1\omega = -e_1^2 \leq 0$$

Therefore, e_1, θ_c are bounded, i.e., $e_1, \theta_c \in \mathcal{L}_\infty$ and e_1 is square integrable, i.e., $e_1 \in \mathcal{L}_2$. Since $y_m, e_1 \in \mathcal{L}_\infty$, we have $y_p \in \mathcal{L}_\infty$ and therefore $\omega_2 \in \mathcal{L}_\infty$. Now

$$\omega_1 = \frac{1}{s+2}u_p = \frac{(s^2 + a_1 s + a_0)}{(s+2)k_p(s+b_0)}y_p$$

Since $b_0 > 0$, i.e., the plant is minimum phase and $y_p \in \mathcal{L}_\infty$, we have $\omega_1 \in \mathcal{L}_\infty$. Hence, $\omega \in \mathcal{L}_\infty$, which implies that $u_p \in \mathcal{L}_\infty$. Since $e_1, \tilde{\theta}_c^T\omega \in \mathcal{L}_\infty$, we have $\dot{e}_1 \in \mathcal{L}_\infty$, which together with $e_1 \in \mathcal{L}_2$, implies that $e_1(t) \to 0$ as $t \to \infty$.

For parameter convergence, we choose r to be sufficiently rich of order 4. As an example, we select $r(t) = A_1\sin\omega_1 t + A_2\sin\omega_2 t$ for some nonzero constants $A_1, A_2, \omega_1, \omega_2$ with $\omega_1 \neq \omega_2$.

54.3.4 Example: Direct MRAC without Normalization (n* = 2)

Let us consider the second-order plant

$$y_p = \frac{k_p}{s^2 + a_1 s + a_0}u_p$$

where $k_p > 0$, a_1, a_0 are constants. The reference model is chosen as

$$y_m = \frac{5}{(s+5)^2}r$$

Using the results of Section 54.2.2 the control law is chosen as

$$\begin{aligned}\dot{\omega}_1 &= -2\omega_1 + u_p, \quad \dot{\omega}_2 = -2\omega_2 + y_p\\ \dot{\phi} &= -\phi + \omega\\ u_p &= \theta_c^T\omega - \phi^T\Gamma\phi e_1\end{aligned}$$

where $\omega = [\omega_1, \omega_2, y_p, r]^T$, $e_1 = y_p - y_m$, $p = 1$, $\Lambda(s) = s + 2$ and $\frac{5(s+1)}{(s+5)^2}$ is SPR. The adaptive law is given by

$$\dot{\theta}_c = -\Gamma e_1 \phi$$

where $\Gamma = \Gamma^T > 0$ is arbitrary and $\theta_c = [\theta_1, \theta_2, \theta_3, c_0]^T$.

Analysis: From Equation 54.12 by substituting for u_p we have that

$$e_1 = W_m(s) k_p (\tilde{\theta}_c^T \omega + \dot{\theta}_c^T \omega)$$

or

$$e_1 = W_m(s)(s+1) k_p \tilde{\theta}_c^T \phi \tag{54.47}$$

Since $W_m(s)(s+1)$ is SPR and $k_p > 0$ we can establish using the Lyapunov-like function

$$V = \frac{e^T P e}{2} + \frac{\tilde{\theta}_c^T \Gamma^{-1} \tilde{\theta}_c}{2} k_p$$

where e is the state of a state-space representation of Equation 54.47 and $P = P^T > 0$ satisfies the Lefschetz-Kalman-Yakubovich lemma [17] that

$$\dot{V} \leq -c e_1^2$$

for some constant $c > 0$. This implies that $e_1, \tilde{\theta}_c \in \mathcal{L}_\infty$ and $e_1 \in \mathcal{L}_2$. Proceeding as in the case of $n^* = 1$ we can establish that all signals are bounded and $e_1(t) \to 0$ as $t \to \infty$. For parameter convergence, the input r is chosen as $r(t) = A_1 \sin \omega_1 t + A_2 \sin \omega_2 t$ for some $A_1, A_2 \neq 0, \omega_1 \neq \omega_2$.

54.3.5 Example: Direct MRAC with Normalization

In contrast to the direct MRAC schemes without normalization the complexity of the design and analysis of direct MRAC schemes with normalization does not change with the relative degree of the plant. We demonstrate the design and analysis of MRAC with normalization using the first-order plant

$$\dot{x} = ax + bu$$

where a, b are unknown and $b > 0$. The closed-loop plant is required to be stable and the state x is required to track the state x_m of the reference model

$$x_m = \frac{b_m}{s + a_m} r$$

for any given bounded reference input signal r. If a, b were known the control law

$$u = -k^* x + l^* r \tag{54.48}$$

with $k^* = \frac{a_m + a}{b}$, $l^* = \frac{b_m}{b}$ could be used to meet the control objective exactly. Since a, b are unknown, we replace Equation 54.48 with

$$u = -k(t) x + l(t) r \tag{54.49}$$

where $k(t), l(t)$ are the on-line estimates of k^*, l^*, respectively. We design the adaptive laws for updating $k(t), l(t)$ by first developing appropriate parametric models for k^*, l^*. We can show that the tracking error e_1 satisfies

$$e_1 = \frac{b}{s + a_m}[u - (-k^* x + l^* r)]$$

which may be written in the form of Equation 54.19 in Section 54.2.2, i.e.,

$$e_1 = b(u_f - \theta_c^{*T} \phi)$$

where $\theta_c^* = [k^*, l^*]^T$, $\phi = W_m(s)[-x, r]^T$, $u_f = \frac{1}{s + a_m} u$. Using the gradient method and the fact that $b > 0$ we have

$$\dot{\theta}_c = -\Gamma \epsilon \phi, \quad \dot{\hat{b}} = \gamma \epsilon \xi \tag{54.50}$$

where θ_c, $\hat{b}$ are the estimates of θ_c^*, b, respectively,

$$\epsilon = \frac{e_1 - \hat{e}_1}{m^2}, \quad \hat{e}_1 = \hat{b}[u_f - \theta_c^T \phi]$$

$$\xi = u_f - \theta_c^T \phi, \quad m^2 = 1 + \phi^T \phi + u_f{}^2$$

The stability analysis of the MRAC examples is accomplished as follows. First we show that Equation 54.50 guarantees that $\theta_c, \hat{b}, \epsilon, \epsilon m \in \mathcal{L}_\infty$ and $\epsilon, \epsilon m, \dot{\theta}_c, \dot{\hat{b}} \in \mathcal{L}_2$ independent of the choice of u and the boundedness of ϕ, u, e_1. These properties are then used to establish the boundedness of all signals in the control loop and the convergence of the tracking error e_1 to zero. The details of the analysis are given in [17].

54.3.6 Example: Indirect MRAC

Consider the following third-order plant:

$$y_p = \frac{1}{s^2(s + a)} u_p \tag{54.51}$$

where a is the only unknown parameter. The output y_p is required to track the output of y_m of the reference model

$$y_m = \frac{1}{(s + 2)^3} r$$

The control law is given by

$$\begin{aligned} u_p \quad = \quad & \theta_{11} \frac{s}{(s + \lambda_1)^2} u_p + \theta_{12} \frac{1}{(s + \lambda_1)^2} u_p \\ & + \theta_{21} \frac{s}{(s + \lambda_1)^2} y_p + \theta_{22} \frac{1}{(s + \lambda_1)^2} y_p + \theta_3 y_p + c_0 r \end{aligned}$$

where $\theta_c = [\theta_{11}, \theta_{12}, \theta_{21}, \theta_{22}, c_0]^T \in R^6$. In direct MRAC, θ_c is generated by a sixth-order adaptive law. In indirect MRAC, θ_c is calculated from the adaptive law as follows. Using the results of Section 54.2.2 the estimate $\hat{a}$ of the only unknown plant parameter a is given by

$$\dot{\hat{a}} = \gamma_a \phi_a \epsilon$$

$$\begin{aligned}\epsilon &= \frac{z-\hat{z}}{1+\phi^T\phi}, \quad \hat{z}=\theta_p^T\phi, \quad z=y_p+\lambda_p^T\phi_2 \\ \phi &= [\phi_1^T,\phi_2^T]^T, \quad \phi_1=\frac{[s^2,s,1]^T}{(s+\lambda_1)^3}u_p, \\ \phi_2 &= -\frac{[s^2,s,1]^T}{(s+\lambda_1)^3}y_p\end{aligned}$$

where $\theta_p = [0,0,1,\hat{a},0,0]^T$, $\Lambda(s)$ is chosen as $\Lambda(s) = (s+\lambda_1)^3, \lambda_p = [3\lambda_1, 3\lambda_1^2, 3\lambda_1^3]^T$, $\phi_a = [0,0,0,1,0,0]\phi = -\frac{s^2}{(s+\lambda_1)^3}y_p$ and $\gamma_a > 0$ is a constant. The controller parameter vector is calculated as $c_0 = 1$,

$$\theta_1^T[s,1]^T = (s+\lambda_1)^2 - \hat{Q}(s,t)$$

$$\theta_2^T[s,1]^T + \theta_3(s+\lambda_1)^2 = \hat{Q}(s,t)\bullet[s^3+\hat{a}s^2] - (s+\lambda_1)^2(s+2)^3$$

where $\hat{Q}(s,t)$ is the quotient of $\frac{(s+\lambda_1)^2(s+2)^3}{s^3+\hat{a}s^2}$.

The example demonstrates that for the plant Equation 54.50, the indirect scheme requires a first-order adaptive law, whereas the direct scheme requires a sixth-order one.

References

[1] Åström, K.J., Theory and applications of adaptive control—a survey, *Automatica*, 19(5), 471–486, 1983.

[2] Ioannou, P.A. and Datta, A., Robust adaptive control: a unified approach, *Proc. IEEE*, 79(12), 1735–1768, 1991.

[3] Kanellakopoulos, I., Kokotovic, P.V., and Morse, A.S., Systematic design of adaptive controllers for feedback linearizable systems, *IEEE Trans. Autom. Control*, 36, 1241–1253, 1991.

[4] Morse, A.S., Global stability of parameter adaptive control systems, *IEEE Trans. Autom. Control*, 25, 433–439, 1980.

[5] Narendra, K.S., Lin, Y.H., and Valavani, L.S., Stable adaptive controller design, II. Proof of stability, *IEEE Trans. Autom. Control*, 25(3), 440–448, 1980.

[6] Parks, P.C., Lyapunov redesign of model reference adaptive control systems, *IEEE Trans. Autom. Control*, 11, 362–367, 1966.

[7] Rohrs, C.E., Valavani, L., Athans, M., and Stein, G., Robustness of continuous-time adaptive control algorithms in the presence of unmodeled dynamics, *IEEE Trans. Autom. Control*, 30(9), 881–889, 1985.

[8] Taylor, L.W. and Adkins, E.J., Adaptive control and the X-15, *Proc. Princeton University Conference on Aircraft Flying Qualities*, Princeton University, 1965.

[9] Whitaker, H.P., Yamron, J., and Kezer, A., Design of Model Reference Adaptive Control Systems for Aircraft, Report R-164, Instrumentation Laboratory, Massachusetts Institute of Technology, Cambridge, 1958.

[10] Åström, K.J. and Eykhoff, P., System identification - a survey, *Automatica*, 20(1), 123, 1971.

[11] Bellman, R.E., *Dynamic Control Processes—A Guided Tour*, Princeton University Press, Princeton, NJ, 1961.

[12] Egardt, B., *Stability of Adaptive Controllers*, Lecture Notes in Control and Information Sciences, Vol. 20, Springer-Verlag, Berlin, 1979.

[13] Fel'dbaum, A.A., *Optimal Control of Systems*, Academic Press, New York, 1965.

[14] Goodwin, G.C, and Sin, K.C., *Adaptive Filtering Prediction and Control*, Prentice Hall, Englewood Cliffs, NJ, 1984.

[15] Harris, C.J. and Billings, S.A., Eds., *Self-Tuning and Adaptive Control: Theory and Applications*, Peter Peregrinus, London, 1981.

[16] Ioannou, P.A. and Kokotovic, P.V., *Adaptive Systems with Reduced Models*, Lecture Notes in Control and Information Sciences, Vol. 47, Springer-Verlag, New York, 1983.

[17] Ioannou, P.A. and Sun, J., *Robust Adaptive Control*, Prentice Hall, Englewood Cliffs, NJ, 1996.

[18] Landau, I.D., *Adaptive Control: The Model Reference Approach*, Marcel Dekker, New York, 1979.

[19] Tsakalis, K.S. and Ioannou, P.A., *Linear Time Varying Systems: Control and Adaptation*, Prentice Hall, Englewood Cliffs, NJ, 1993.

Further Reading

Further details on MRAC for continuous-time plants can be found in the following textbooks:

[1] *Robust Adaptive Control*, by P. Ioannou and J. Sun, Prentice Hall, Englewood Cliffs, NJ, 1996.

[2] *Stable Adaptive Systems*, by K.S. Narendra and A. M. Annaswamy, Prentice Hall, Englewood Cliffs, NJ, 1989.

[3] *Adaptive Control: Stability, Convergence and Robustness*, by S. Sastry and M. Bodson, Prentice Hall, Englewood Cliffs, NJ, 1989.

[4] *Adaptive Control: The Model Reference Approach*, by I. D. Landau, Marcel Dekker, New York, 1979.

[5] *Adaptive Control*, by K. J. Åström and B.Wittenmark, Addison-Wesley, Reading, MA, 1989.

More information on the design and analysis of MRAC for linear, time-varying plants can be found in the following monograph:

[6] *Linear Time-Varying Systems: Control and Adaptation*, by K. Tsakalis and P. Ioannou, Prentice Hall, Englewood Cliffs, NJ, 1993.

More information on the design and analysis of robust MRAC can be found in the books 1, 2, 3, and 6 given above.

Details on MRAC for discrete-time plants can be found in the book 4 by Landau given above and in:

[7] *Adaptive Filtering Prediction and Control* by G. Goodwin and K. Sin, Prentice Hall, Englewood Cliffs, NJ, 1984.

For applications of MRAC the reader is referred to the following books.

[8] *Self-Tuning and Adaptive Control: Theory and Applications,* Edited by C. J. Harris and S. A. Billings, Peter Peregrinus, London, 1981.

[9] *Adaptive and Learning Systems: Theory and Applications,* Edited by K. S. Narendra, Plenum Press, New York, 1986.

SECTION XI

Analysis and Design of Nonlinear Systems

55

Analysis and Design of Nonlinear Systems

V. Jurdjevic
Department of Mathematics, University of Toronto, Ontario, Canada

Hassan K. Khalil
Michigan State University

Françoise Lamnabhi–Lagarrigue
Laboratoire des Signaux et Systèmes CNRS, Supélec, Gif-sur-Yvette, France

55.1 The Lie Bracket and Control

V. Jurdjevic, Department of Mathematics, University of Toronto, Ontario, Canada

55.1.1 Introduction

Time-dependent events $a(t)$ and $b(t)$ are said to commute if the occurrence of $a(t)$ during a time interval T followed by the occurrence of $b(t)$ during an interval S leads to an outcome that does not change when the order of events is reversed. Denoting by $b(S)a(T)$ the occurrence of a followed by b, then $a(T)b(S)$ denotes the reversed order, and a and b commute if $a(T)b(S) = b(S)a(T)$. Otherwise, a and b are said to be noncommutative events.

Most events do not commute, as is evident from common experience. For instance, filling the swimming pool with water and then diving into the pool results in a state different from one in which the order is reversed. Driving an automobile relies on a more subtle use of noncommutativity (based on nonholonomy). Any automobile driver knows that the rotations of the steering wheel do not commute with either forward or backward motions of the automobile. The ability to park the automobile in a tight spot, a most demanding challenge for an inexperienced driver, is only possible because of the noncommuting nature of these events. An experienced driver knows that it is possible to execute a parallel displacement of an automobile in any space which is large enough to allow some forward and backward movement, although the number of maneuvers may be so large that it might very well be advisable to look for another parking spot with more space to spare.

Control of many dynamic systems, like the control of an automobile, consists of a time-sequential application of noncommuting events. Knowing which event to apply at a given time is a basic requirement of successful control. The recognition of noncommutativity as a fundamental issue of control is a starting point for geometric control; this chapter describes the main mathematical tools required for its understanding.

55.1.2 Notations and Basic Assumptions

The subsequent discussion is confined to control systems described by a differential equation $\frac{dx}{dt} = F(x, u)$ in which the control functions $u(t) = (u_1(t), \ldots, u_m(t))$ take values in a fixed set U in R^m. Most of the theory presented in this chapter is extracted through F and its derivatives, and for that reason it is expedient to assume that the state variable $x(t)$ belongs to an analytic manifold M and that $F(x, u)$ is an analytic vector field for each u in U. The reader not familiar with these notions may assume at the beginning that M is a finite dimensional vector space and that for each $u \in U$, $F(x, u)$ can be represented by its Taylor series at each point x in M. The extensions to arbitrary manifolds will be defined as needed.

Throughout this chapter $\mathcal{A}(x, T)$ will denote the set of states reachable from an initial state x in exactly T units of time. Then, $\mathcal{A}(x, \leq T) = \bigcup_{0\leq t\leq T} \mathcal{A}(x, t)$, and $\mathcal{A}(x) = \bigcup_{t\geq 0} \mathcal{A}(x, t)$. As will be explained later, a differential control system may also be viewed as a family of vector fields $\mathcal{F}$, in which case the reachable sets will

0-8493-8570-9/96/$0.00+$.50

be subscripted $\mathcal{A}_{\mathcal{F}}(x, T)$, $\mathcal{A}_{\mathcal{F}}(x, \leq T)$, and $\mathcal{A}_{\mathcal{F}}(x)$ in order to emphasize their dependence on a given system $\mathcal{F}$. These notations will be needed particularly when discussing several systems simultaneously.

55.1.3 Vector Fields

Geometric control theory begins with a distinctive view of a differential equation, initiated by H. Poincaré in the latter part of the 19th century, that the solutions of

$$\frac{dx}{dt} = F(x(t)) \qquad x(t) \text{ in } M \qquad (55.1)$$

can be analyzed in simple mathematical terms without ever having to solve the differential equation. The corresponding theory is based on a single assumption that for each initial state x_0 differential Equation 55.1 admits a unique solution through that state defined for all times t. Assuming that this is the case, let $x(x_0, t)$ denote the solution of Equation 55.1 for which $x(x_0, 0) = x_0$. The mapping $(x_0, t) \to x(x_0, t)$ is called the flow, or dynamic system induced by the vector field F. Its essential properties are

1. $x(x_0, 0) = x_0$ for each x_0 in M
2. $x(x_0, t+s) = x(x(x_0, t), s)) = x(x(x_0, s), t)$ for all x_0 in M and s, t in R
3. $\frac{\partial}{\partial t}x(x_0, t) = F(x(x_0, t))$

At each instant of time, t, the flow of F induces a transformation on M which will be denoted by $\exp tF$: $\exp tF$ maps each initial point x_0 onto $x(x_0, t)$. It follows from (1) and (2) that $\exp 0F$ = Identity and that $\exp(t+s)F = (\exp tF)(\exp sF) = (\exp sF)(\exp tF)$ for any s and t, with $(\exp sF)(\exp tF)$ denoting the composition of mappings. Since $\exp 0F$ = Identity, it follows that $(\exp tF)^{-1} = \exp -tF$, and therefore, the mappings $\{\exp tF : t \in R\}$ form a commutative group. This group is called the one-parameter group of transformations induced by F.

EXAMPLE 55.1:

If F is a linear vector field, i.e., $F(x) = Ax$ for some linear mapping on M then $x(x_0, t) = e^{tA}x_0$ with $e^{tA} = \sum_{k=0}^{\infty} \frac{t^k}{k!}A^k$. Thus, in this case $\exp tF$ is a linear transformation on M equal to the exponential of a linear mapping.

EXAMPLE 55.2:

If $F(x) = b$ is a constant vector field, then $x(x_0, t) = x_0 + tb$, and therefore each transformation $\exp tF$ is a translation in the direction of b.

Vector fields, no matter how nonlinear, admit linear interpretations on an infinite dimensional vector space in which $\exp tF$ becomes the exponential of a linear mapping. In this interpretation vector fields act linearly on functions on M as follows: let

$$(\exp tF)(f)(x_0) = f(x(x_0, t))$$

for any real-valued function, f. Then Ff is the function on M defined by

$$Ff = \frac{d}{dt}(\exp tF)(f)|_{t=0}$$

Ff is called the derivation of f along the vector field F.

Ff admits a simple description in any system of coordinates on M. Assuming that M is a linear vector space, then any basis $a_1, \ldots, a_n$ in M induces coordinates $x_1, \ldots, x_n$ on M by the formula $x = \sum_{i=1}^{n} x_i a_i$. Then $F(x) = \sum_{i=1}^{n} F_i(x_1, \ldots, x_n)a_i$ and Equation 55.1 is written as a system of differential equations in R^n

$$\frac{dx_i}{dt} = F_i(x_1, \ldots, x_n) \qquad i = 1, \ldots, n. \qquad (55.2)$$

Any real-valued function on M becomes a function on R^n by the correspondence $f(x) = f(x_1, \ldots x_n)$. Then,

$$\frac{d}{dt}f(x_1(t), \ldots, x_n(t))|_{t=0} = \sum_{i=1}^{n} \frac{\partial f}{\partial x_i}(x_1, \ldots, x_n)F_i(x_1, \ldots, x_n)$$

and therefore

$$(Ff)(x_1, \ldots, x_n) = \sum_{i=1}^{n} \frac{\partial f}{\partial x_1}F_i(x_1, \ldots, x_n)$$

It follows that $(Ff)(x)$ is equal to the directional derivative of f at x in the direction of $F(x)$. In particular, when $f = x_i$, then $Fx_i = F_i, i = 1, \ldots, n$. Evidently F acts linearly on functions (as a directional derivative) and therefore satisfies further properties:

$Ff = 0$ for constant functions f, and $F(fg) = f(Fg) + g(Ff)$ with respect to the products of functions. (fg denotes the product of f and g, i.e., $(fg)(x) = f(x)g(x)$ and $f(Fg)$ is the product of f with Fg. Note that $f(Fg) \neq (Ff)g$.)

As a mapping on the space of functions $\exp tF$ satisfies

$$\begin{aligned} \frac{d}{dt}\exp tF &= F(\exp tF) = (\exp tF)F \\ \text{and} \quad \frac{d^n}{dt^n}\exp tF &= F^n \exp tF \end{aligned} \qquad (55.3)$$

with each product denoting the composition of linear mappings. Analytic vector fields can be represented by their Taylor series and therefore $(\exp tF)(f) = \sum_{k=0}^{\infty} \frac{t^k}{k!}F^k(f)$ for any analytic function f. The reader can easily verify that in each system of coordinates, $F^2 f = F(Ff) = F\left(\sum_{i=1}^{n} \frac{\partial f}{\partial x_i}F_i\right) = \sum_{i,j=1}^{n}\left(\frac{\partial^2 f}{\partial x_i \partial x_j}F_i F_j + \frac{\partial f}{\partial x_i}\frac{\partial F_i}{\partial x_j}\right)$. Then $F^3 f$ is the directional derivative of $F^2 f$ in the direction F and so on for each derivate $F^n f$.

55.1.4 The Lie Bracket

With these notions and notations at our disposal, let us return to the commutativity issue raised in the introduction, and consider commutativity properties of $\exp tF$ and $\exp sG$ corresponding to vector fields F and G. This question may be further motivated through the following control theoretic context:

Consider a control system

$$\frac{dx}{dt} = u(t)F(x(t)) + (1 - u(t))G(x(t)))$$

with switching control u which can only take values 0 or 1 (on-off controls). During any time interval that the control is turned on the system follows F and during the time intervals that the control is switched off the system follows G.

Let $u_1(t)$ and $u_2(t)$ denote the following control functions defined in an interval $[0, T + S]$:

$$u_1(t) = \begin{cases} 1, & t \in [0, T) \\ 0, & t \in [T, T + S] \end{cases}$$

$$u_2(t) = \begin{cases} 0, & t \in [0, S) \\ 1, & t \in [S, T + S] \end{cases}$$

The corresponding trajectories $x_1(t)$ and $x_2(t)$, both initiating from the same initial point x_0, are given by:

$$x_1(t) = (\exp tF)(x_0), \quad t \in [0, T)$$

$$x_1(t) = (\exp(t - T)G)(\exp tTF)(x_0), \quad t \in [T, T + S]$$

and

$$x_2(t) = (\exp tG)(x_0), \quad t \in [0, S)$$

$$x_2(t) = (\exp(t - S)F)(\exp SG)(x_0) \text{ for } t \in [S, S + T]$$

At the terminal time $t = T + S$, $x_1(t) = (\exp SG)(\exp TF)(x_0)$, and $x_2(t) = (\exp TF)(\exp SG)(x_0)$.

Assuming that the control actions of $u_1(t)$ and $u_2(t)$ lead to the same terminal state independently of the initial point x_0 and of the switching times S and T, then $(\exp SG)(\exp TF) = (\exp TF)(\exp SG)$. This equality remains unaltered when the domain is extended to the space of functions and therefore, $\frac{\partial}{\partial t}\frac{\partial}{\partial s}(\exp sG)(\exp tF)(f) = \frac{\partial}{\partial s}\frac{\partial}{\partial t}(\exp tF)(\exp sG)(f)$ for any function f.

Taking advantage of formulas (Equation 55.3)

$$\frac{\partial}{\partial t}\frac{\partial}{\partial s}(\exp sG)(\exp tF)f$$
$$= \frac{\partial}{\partial t}G(\exp sG)(\exp tF)(f)$$
$$= G(\exp sG)(F\exp tF)f, \text{ and}$$
$$\frac{\partial}{\partial s}\frac{\partial}{\partial t}(\exp tF)(\exp sG)(f) = F(\exp tF)(G\exp sG)f$$

Evaluating these derivatives at $t = s = 0$ gives

$$G(Ff) = F(Gf)$$

DEFINITION 55.1 Let F and G be any vector fields on M. Then $[F, G](f) = G(Ff) - F(Gf)$ for any real-valued function f on M. $[F, G]$ is called the Lie bracket of F with G.

It follows that $[F, G]$ is a vector field for any vector fields F and G. Note that $[F, G] = -[G, F]$. The ith coordinate of the Lie bracket $[F, G]$ is given by $[F, G](x_i)$. Therefore,

$$[F, G](x_i) = G(F_i) - F(G_i) = \sum_{j=1}^{n} \frac{\partial F_i}{\partial x_j}G_j - \frac{\partial G_i}{\partial x_j}F_j \tag{55.4}$$

The calculations above show that if $(\exp sG)(\exp tF) = (\exp tF)(\exp sG)$ for all s and t then $[F, G] = 0$. Somewhat remarkably, the converse is also true; i.e., if $[F, G] = 0$, then $(\exp tF)(\exp sG) = (\exp sG)(\exp tF)$.

EXAMPLE 55.3:

If $F(x) = b$ and $G(x) = c$ are any constant vector fields, then their coordinates are constant functions, and therefore $[F, G] = 0$ (as can be seen from Equation 55.4). Then, $(\exp tF)(x) = x + tb$ and $(\exp sG)(x) = x + sc$, and therefore $(\exp tF)(\exp sG)(x) = (x + sc) + tb = (x + tb) + sc = (\exp sG)(\exp tF)(x)$, confirming that the flows commute.

EXAMPLE 55.4:

Let $F(x) = Ax$ be a linear vector field and $G(x) = b$ a constant vector field. In terms of any linear coordinates $x_1, \ldots, x_n$, $F_i(x_1, \ldots, x_n) = \sum_{j=1}^{n} A_{ij} \cdot x_j$. (A_{ij}) is the matrix of A relative to this basis. Then

$$[F, G](x_i) = \sum_{j=1}^{n} \frac{\partial F_i}{\partial x_j}G_j - \frac{\partial G_i}{\partial x_j}F_j = \sum_{j=1}^{n} A_{ij}b_j - O \cdot F_j = \sum_{j=1}^{n} A_{ij}b_j$$

Thus $[G, F]$ is a constant vector field equal to Ab. $[G, F] = 0$ if and only if $b \in \ker A$.

For instance, the rotation $A = \begin{pmatrix} 0 & -1 & 0 \\ 1 & 0 & 0 \\ 0 & 0 & 0 \end{pmatrix}$ and $b = \begin{pmatrix} 0 \\ 0 \\ 1 \end{pmatrix}$ commute, while A does not commute with $b = \begin{pmatrix} 1 \\ 0 \\ 0 \end{pmatrix}$ (Figure 55.1).

EXAMPLE 55.5:

Let $F(x) = Ax$ and $G(x) = Bx$ be linear vector fields on M. Then $[F, G]$ is a linear vector field given by $[F, G](x) = (AB - BA)(x)$.

For example, if $A = \begin{pmatrix} 1 & 0 \\ 0 & -1 \end{pmatrix}$ and $B = \begin{pmatrix} 0 & 1 \\ 1 & 0 \end{pmatrix}$ are the matrices corresponding to F and G (relative to a linear system of coordinates) then $C = 2\begin{pmatrix} 0 & 1 \\ -1 & 0 \end{pmatrix}$ is the matrix that corresponds to $[F, G]$. The flows of these fields are shown in Figure 55.2.

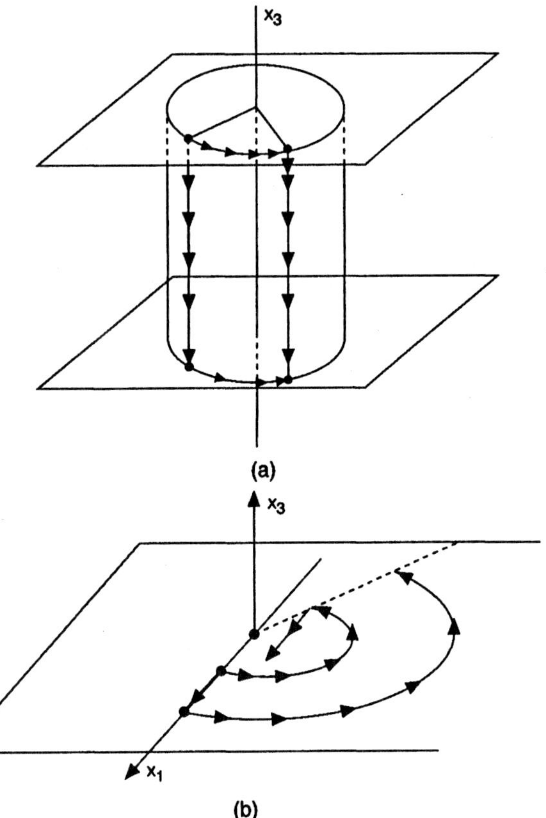

Figure 55.1 Commutativity: a. Commuting case, b. Noncommuting case.

Return now to the general discussion, and consider the expression

$$(\exp -tG)(\exp -tF)(\text{expt}G)(\text{expt}F) \qquad (55.5)$$

As shown earlier, this expression reduces to the identity when $[F, G] = 0$. Begin with a simple case with F and G linear vector fields defined by linear mappings A and B. Then $\text{expt}F = e^{tA} = \sum_{k=0}^{\infty} \frac{t^k}{k!} A^k$, and $\text{expt}G = \sum_{k=0}^{\infty} \frac{t^k}{k!} B^k$. We shall be interested only in the terms up to and including the second-order terms in t. Then,

$$e^{tB}e^{tA} = I + t(A+B) + t^2BA + \frac{t^2}{2!}(A^2+B^2)$$

and

$$e^{-tB}e^{-tA} = I - t(A+B) + t^2BA + \frac{t^2}{2!}(A^2+B^2)$$

Therefore,

$$\begin{aligned} & e^{-tB}e^{-tA}e^{tB}e^{tA} \\ = \; & (I - t(A+B) + t^2BA + \frac{t^2}{2!}(A^2+B^2))(I + t(A+B) \\ & + t^2BA + \frac{t^2}{2!}(A^2+B^2)) \\ = \; & I + t(A+B) + t^2BA + \frac{t^2}{2}(A^2+B^2) - t(A+B) \\ & - t^2(A+B)^2 + t^2BA + \frac{t^2}{2}(A^2+B^2) \end{aligned}$$

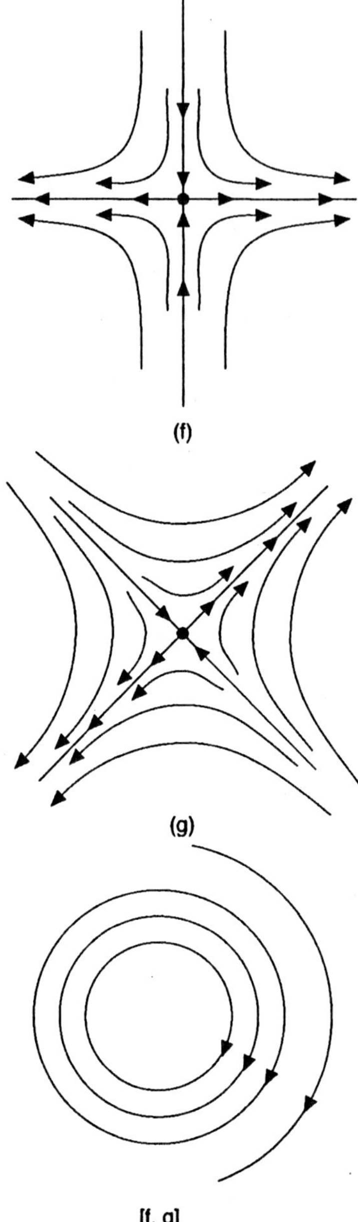

Figure 55.2 Lie brackets of vector fields.

$$= I - t^2(AB + BA) + 2t^2BA = I + t^2(BA - AB)$$

These calculations show that up to second-order terms in t,

$$(\exp -tG)(\exp -tF)(\text{expt}G)(\text{expt}F) = I + t^2[G, F]$$

This result extends to arbitrary vector fields, as can be demonstrated through the formulas $\text{expt}F = \sum_{k=0}^{\infty} \frac{t^k}{k!} F^k(f)$ and $(\text{expt}G)(f) = \sum_{k=0}^{\infty} \frac{t^k}{k!} G^k(f)$ described earlier. These arguments show that the proof for the general flows is the same as it is for the linear flows.

The asymptotic formula obtained above shows that the curve $\sigma(t) = (\exp -\sqrt{t}G)(\exp -\sqrt{t}F)(\text{expt}\sqrt{t}G)(\text{expt}\sqrt{t}F)(x_0)$ satisfies $\sigma(0) = x_0$ and $\frac{d\sigma}{dt}(0) = [G, F](x_0)$.

The preceding developments may be stated in a control theoretic context, as follows.

THEOREM 55.1 *Suppose that $\frac{dx}{dt} = F(x, u)$ is any system in M with the control functions taking place in a set $U \subset R^m$. Suppose that $F_1(x) = F(x, u_1)$ and $F_2(x) = F(x, u_2)$ for some choices u_1 and u_2 of control values. Assume that there exist control values u_3 and u_4 in U such that $-F_1(x) = F(x, u_3)$ and $-G(x) = F(x, u_4)$ for all x. Then there is a curve $\sigma(t)$ contained in the reachable set $A(x, \leq \epsilon)$ for any $\epsilon > 0$ such that $\frac{d\sigma}{dt}(0) = [F_1, F_2](x)$. That is, the system can move infinitesimally in the direction of the Lie bracket.*

EXAMPLE 55.6:

Let $A = \begin{pmatrix} 0 & 1 \\ 1 & 0 \end{pmatrix}$ and $B = \begin{pmatrix} 1 & 0 \\ 0 & -1 \end{pmatrix}$ define the bilinear system $\frac{dx}{dt} = (1-u)Ax + uBx$ in $M = R^2$, with $0 \leq u(t) \leq 1$. Then, $u = 0$ follows $F(x) = Ax$ and $u = 1$ follows $G(x) = Bx$. The Lie bracket $[F, G]$ is a linear field given by the rotation matrix $2\begin{pmatrix} 0 & 1 \\ -1 & 0 \end{pmatrix}$. Since

$$\frac{dx_1}{dt} = ux_1 + (1-u)x_2$$

and

$$\begin{aligned} \frac{dx_2}{dt} &= -x_2u + x_1(1-u), \\ \frac{d}{dt}x_1(t)x_2(t) &= (ux_1 + (1-u)x_2)x_2 + (-x_2u \\ &\quad + x_1(1-u))x_1 \\ &= (1-u)(x_1^2 + x_2^2) \geq 0. \end{aligned}$$

Thus, $x_1(t)x_2(t)$ is increasing for any admissible control function. The system cannot move infinitesimally in the direction of $[F, G]$ because neither $-F$ nor $-G$ can be traced by the admissible controls (Figure 55.3).

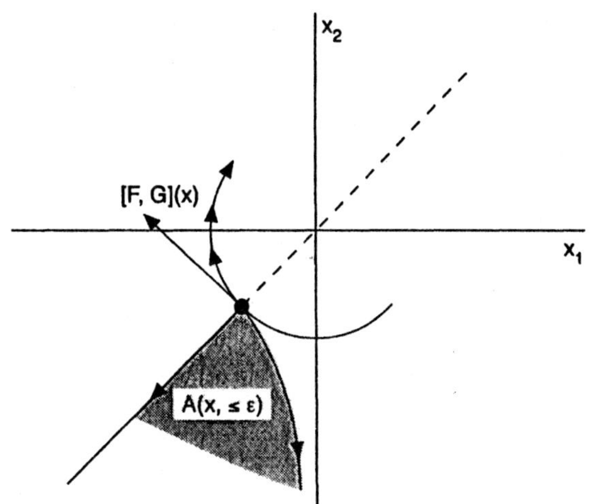

Figure 55.3 Noncontrollability of the Lie bracket.

55.1.5 The Lie Algebras

From a geometric point of view, a control system is a family of vector fields parameterized by controls. Each control value determines a vector field, and the corresponding trajectory is a solution curve of this field. As the control switches to a new value so does the vector field and the trajectory begins to follow the direction of the new field. From this view a trajectory generated by a piecewise constant control is a continuous curve in M having discontinuous derivatives with the breaks in the derivative corresponding to the changes in vector fields caused by the switches in the control. Such curves are conveniently called continuous broken trajectories. Between any consecutive breaks, the trajectory is a solution curve of a vector field in the family.

From this perspective, a linear control system $\frac{dx}{dt} = Ax + Bu$, $u \in U$ is a collection of affine vector fields $\mathcal{F}$ of the form $F_u(x) = Ax + Bu$ with $u \in U$. An affine vector field F is any vector field in M which satisfies $F(\sum \lambda_i x_i) = \sum \lambda_i F(x_i)$ for any affine combination $\sum \lambda_i x_i$ with $\sum \lambda_i = 1$. It can be shown that any affine field is the sum of a linear vector field and a constant vector field. The exponential map $\text{expt}F$ of any affine vector field $F(x) = Ax + b$ is given by $(\text{expt}F)(x) = e^{At}(x + \int_0^t e^{-As}b\,ds)$. For instance, if $b \in \ker A$, then $e^{-As}b = b$, and $(\text{expt}F)(x) = e^{At}x + bt$.

Vector fields form a vector space under pointwise addition and multiplication by scalars. The solution curves of F and λF, λ a real number, differ only by reparameterization of time; i.e., if $\frac{dx}{dt}(t) = F(x(t))$ then $\frac{d}{dt}x(\lambda t) = \lambda F(x(\lambda t))$, and consequently $y(t) = x(\lambda t)$ is a solution curve of λF.

For sums of vector fields the situation is different. Thus, $F(x) = \begin{pmatrix} 0 & 1 \\ 0 & 0 \end{pmatrix}\begin{pmatrix} x_1 \\ x_2 \end{pmatrix}$, $G(x) = \begin{pmatrix} 0 & 0 \\ 1 & 0 \end{pmatrix}\begin{pmatrix} x_1 \\ x_2 \end{pmatrix}$ add to $H(x) = \begin{pmatrix} 0 & 1 \\ 1 & 0 \end{pmatrix}\begin{pmatrix} x_1 \\ x_2 \end{pmatrix}$.

The phase portraits of these flows are shown in Figure 55.4.

It is known that every solution curve of the sum $F + G$ can be approximated by a curve that oscillates between the solution curves of F and the solution curves of G, as shown in Figure 55.5.

The vector space structure, together with the Lie bracket operation, turns the set of all vector fields into an algebra called the Lie algebra.

Any set of vector fields $\mathcal{F}$ generates the smallest sub-algebra that contains $\mathcal{F}$. We will use Lie($\mathcal{F}$) to denote the Lie algebra generated by a family of vector fields $\mathcal{F}$.

EXAMPLE 55.7:

Let $\mathcal{F}$ consist of two vector fields F and G in R^2 given by $Ff = x_1^2 \frac{\partial f}{\partial x_1}$ and $Gf = x_1^2 \frac{\partial f}{\partial x_2}$. Then, $[F, G]f = -2x_1^3 \frac{\partial f}{\partial x_2}$. Furthermore, $[F, [F, G]] = 6x_1^4 \frac{\partial}{\partial x_2}$. It then follows by induction that each vector field $x_1^n \frac{\partial f}{\partial x_2}$ is contained in Lie($\mathcal{F}$), and therefore Lie($\mathcal{F}$) is an infinite dimensional algebra of vector fields.

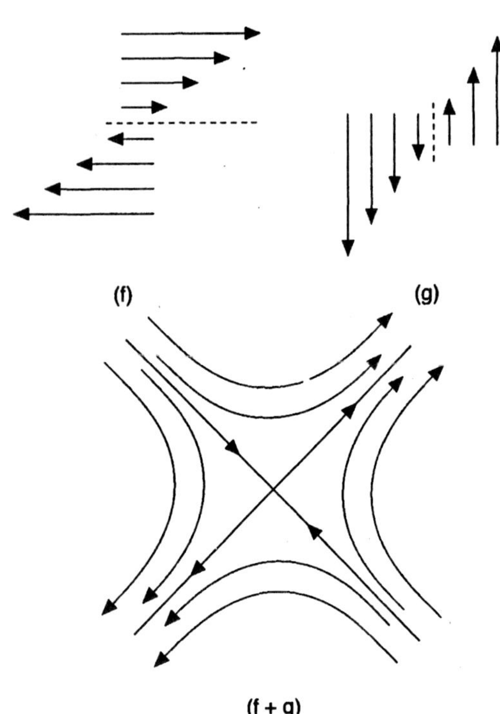

Figure 55.4 Sums of vector fields.

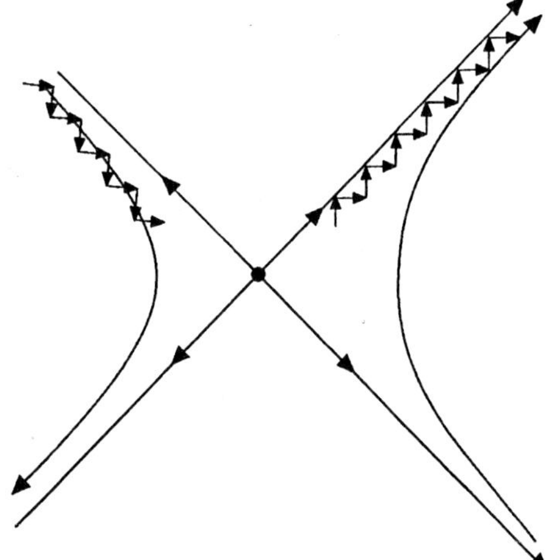

Figure 55.5 Chattering along the sum of vector fields.

EXAMPLE 55.8:

Consider now the family of affine vector fields $\mathcal{F} = \{F_u(x) = Ax + Bu : u \in U\}$ defined by a linear system with a constraint set U. When the linear span of points in U spans R^m, Lie($\mathcal{F}$) coincides with the Lie algebra generated by $F_0(x) = Ax$ and the constant vector fields with values in the range of B.

As we have already shown in Example 2, $[F_0, G] = Ab$ for any constant vector field $G = b$. Let $\mathcal{B}$ denote the range space of B. Then, it follows by the preceding observation that the space of constant vector fields in $\mathcal{B} + A\mathcal{B}$ is contained in Lie($\mathcal{F}$). Denote by $\mathcal{L}_k$ the vector space of all constant vector fields with values in $\mathcal{B} + A\mathcal{B} + \cdots + A^{k-1}\mathcal{B}$. It follows that $\mathcal{L}_{k+1} = [F_0, \mathcal{L}_k] + \mathcal{L}_k$ and therefore each space $\mathcal{L}_k$ is contained in Lie($\mathcal{F}$). Since each power A^{n+k} is a linear combination of $\{A^k : 0 \le k \le n-1\}$, as a consequence of the Cayley-Hamilton theorem, it follows that Lie($\mathcal{F}$) is equal to the linear span of F_0 and $\mathcal{L}_n$, and is therefore a finite dimensional Lie algebra. The space $\mathcal{L}_n$ is usually described as the range space of the controllability matrix

$$(B\,AB\ldots A^{n-1}B)$$

An arbitrary family of vector fields $\mathcal{F}$ in Lie($\mathcal{F}$), when evaluated at each point x, defines a linear space of tangent vectors at x. For instance, in Example 7, each vector in Lie($\mathcal{F}$) evaluated at $x_1 = 0$ is equal to zero. Therefore, each element in Lie($\mathcal{F}$) is equal to zero at such points, and consequently the corresponding space of tangent vectors is zero-dimensional. At any other point of the plane G and F are linearly independent, and their span is two-dimensional.

The Lie algebra in Example 8 evaluated at the origin is equal to the range space of the controllability matrix. The Lie algebra evaluated at any other point is the vector sum of Ax and the range of the controllability matrix.

We shall use $\text{Lie}_x(\mathcal{F})$ to denote the vector space of all tangent vectors $F(x)$ with F in Lie($\mathcal{F}$). Then $\dim \text{Lie}_x(\mathcal{F})$ will denote the dimension of this vector space.

The following theorem is of fundamental importance for geometric control theory.

THEOREM 55.2 *Let $\mathcal{F}$ be any family of vector fields, and let $A_{\mathcal{F}}(x, \le T)$ denote its reachable set from x in T units of time (in accordance with the conventions outlined earlier). Then* $\dim \text{Lie}_x(\mathcal{F}) = \dim M$ *is a necessary and sufficient condition that $\mathcal{A}_{\mathcal{F}}(x, \le T)$ has a non-empty interior in M. Furthermore, when* $\dim \text{Lie}_x(M) = \dim M$, *then*

$$cl\mathcal{A}_{\mathcal{F}}(x, \le T) \subseteq \text{int}\,\mathcal{A}_{\mathcal{F}}(x, \le T+\epsilon) \subseteq \mathcal{A}_{\mathcal{F}}(x, \le T+\epsilon)$$

for any $T > 0$ and any $\epsilon > 0$.

In the preceding notation, cl(A) denotes the topological closure of a set A, which means that cl(A) consists of A along with all points of M which are limit points of elements in A. The interior of any set A, denoted by int (A), consists of all points a in A contained in an open ball in M centered at a which is entirely contained in A.

Typically, the initial point x belongs to the boundary of the reachable set $A(x, \le T)$, as in Example 6. x is said to be small-time controllable by $\mathcal{F}$ whenever x belongs to the interior of $\mathcal{A}_{\mathcal{F}}(x, \le T)$ for any $T > 0$. The following example shows that x can be small-time locally controllable even for families of two vector fields neither of which vanishes at x.

EXAMPLE 55.9:

Let $\mathcal{F}$ be the family consisting of linear vector fields F and G in R^2 described by $A = \begin{pmatrix} 0 & 1 \\ 1 & 0 \end{pmatrix}$ and $B = \begin{pmatrix} -1 & 0 \\ 0 & -2 \end{pmatrix}$. The solution curves of F are hyperbolas $x_2^2 - x_1^2 = \text{const}$ while the integral curves of G are parabolas $x_2 = cx_1^2$. The curves are tangent to each other along the lines $x_1 = \pm\sqrt{2}x_2$. Any initial point x along such a line is in the interior of $\mathcal{A}_{\mathcal{F}}(x, \leq T)$, as Figure 55.6 shows.

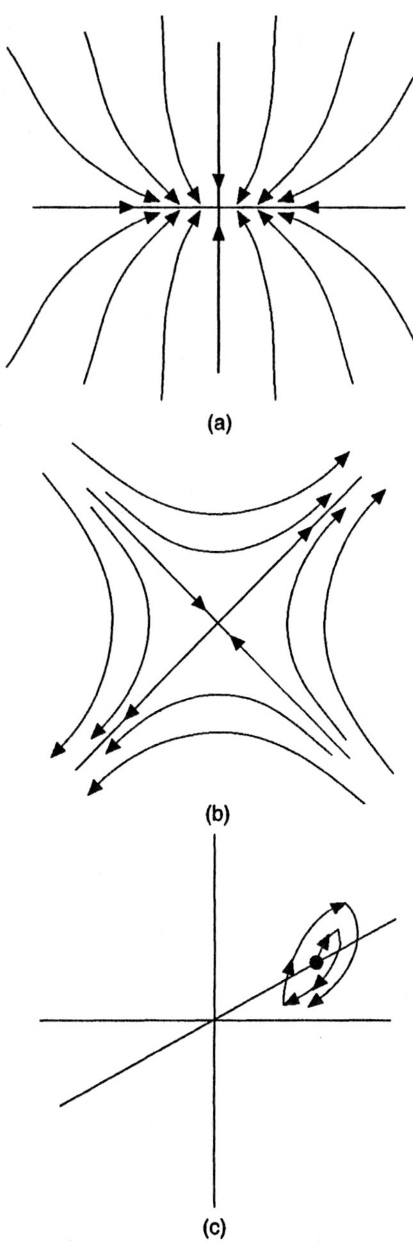

Figure 55.6 Small-time local controllability: a. Solution curves of G; b. Solution curves of F; c. Closed cycles.

For linear systems, any trajectory $x(t)$ which originates at $x_0 = 0$ is of the form $x(t) = e^{At}\int_0^t e^{-As}Bu(s)\,ds$, and is therefore necessarily contained in the range space of the controllability matrix, as can be seen from the expression

$$e^{A(t-s)}Bu(s) = \sum_{k=0}^{\infty} \frac{(t-s)^k}{k!} A^k Bu(s)\,ds$$

confirming the necessity of the Lie algebraic criterion stated in Theorem 55.2.

55.1.6 The Lie Saturate

We now shift to the invariance properties of control systems using the closure of the reachable sets as the basic criterion for invariance. Let $\mathcal{F}_1$ and $\mathcal{F}_2$ be any families of vector fields. Then $\mathcal{F}_1$ is said to be strongly equivalent to $\mathcal{F}_2$ if

1. $\text{Lie}_x(\mathcal{F}_1) = \text{Lie}_x(\mathcal{F}_2)$ for all x in M
2. $\text{cl}\mathcal{A}_{\mathcal{F}_1}(x, \leq T) = \text{cl}\mathcal{A}_{\mathcal{F}_2}(x, \leq T)$ for all $T > 0$ and all x in M

$\mathcal{F}_1$ and $\mathcal{F}_2$ are said to be equivalent if (2) is replaced by $\text{cl}\mathcal{A}_{\mathcal{F}_1}(x) = \text{cl}\mathcal{A}_{\mathcal{F}_2}(x)$.

DEFINITION 55.2 The (strong) Lie saturate of a given control system is the largest family of vector fields (strongly) equivalent to $\mathcal{F}$. The (strong) Lie saturate will be denoted by $\mathcal{LS}_s(\mathcal{F})$, and $\mathcal{LS}(\mathcal{F})$ will denote the Lie saturate of $\mathcal{F}$.

DEFINITION 55.3 A control system $\mathcal{F}$ is said to be strongly controllable if $\mathcal{A}_{\mathcal{F}}(x, \leq T) = M$ for each x in M and each $T > 0$. It is said to be controllable if $\mathcal{A}_{\mathcal{F}}(x) = M$ for each x in M.

THEOREM 55.3 *(i) A control system $\mathcal{F}$ is strongly controllable if and only if the strong Lie saturate of $\mathcal{F}$ is equal to Lie($\mathcal{F}$).*
(ii) $\mathcal{F}$ is controllable if and only if the Lie saturate of $\mathcal{F}$ is equal to Lie($\mathcal{F}$).

The above theorem is called the Lie saturate criterion of controllability. As elegant as the criterion seems, its practical value rests on the constructive means of calculating the Lie saturate. The next theorems describe permissible system enlargements, prolongations, which respect the reachable sets and may be used to generate a procedure for calculating the Lie saturate. We say that $\mathcal{F}_2$ is a prolongation of $\mathcal{F}_1$ if $\mathcal{F}_1 \subset \mathcal{F}_2$ and if $\mathcal{F}_1$ and $\mathcal{F}_2$ have the same Lie saturates.

We begin by noting that the convex hull of any family of vector fields is contained in the strong Lie saturate of the family. Recall that the convex hull of $\mathcal{F}$ consists of all convex combinations $\sum_{i=1}^m \lambda_i F_i$, with $\sum_{i=1}^m \lambda_i = 1$, $\lambda_i \geq 0$, and each F_i in $\mathcal{F}$. It is known that

$$\left(\exp\left(t\sum \lambda_i F_i\right)\right)(x) \in \text{cl}\mathcal{A}_{\mathcal{F}}(x, t)$$

for each $t > 0$, because any trajectory of a convex combination can be approximated by a trajectory of $\mathcal{F}$. The approximation is achieved by switching sufficiently fast along the trajectories of $F_1, \ldots, F_m$ around the trajectory of the convex sum (as illustrated in Figure 55.5). It may happen that the terminal point of the trajectory of the convex sum is not reachable by the original system, as the example below shows.

EXAMPLE 55.10:

The point $x = 1, y = 0$ cannot be reached in time $t = 1$ from the origin by the trajectories of

$$\begin{aligned} \frac{dx}{dt} &= -y^2 + 1 \\ \frac{dy}{dt} &= u \\ \text{with} \quad u(t) &= \pm 1 \end{aligned}$$

The convex hull of $U = \{-1, 1\}$ is the closed interval $-1 \leq u \leq 1$. Therefore, $u(t) = 0$ is in the convex hull of $U = \{-1, 1\}$. The corresponding trajectory which originates at 0 is given by $x(t) = t$, $y(t) = 0$ and reaches $x = 1$, $y = 0$ at $t = 1$.

Having taken the closure of the reachable sets as the criterion for equivalence, it becomes natural to pass to topologically closed families of vector fields. The choice of topology for the space of vector fields is not particularly important. In this context it is convenient to topologize the space of all vector fields by the C^∞ topology on compact subsets of M. Rather than going into the mathematical details of this topology, let us illustrate the use with an example.

Suppose that $X_\lambda(x) = \lambda(Ax + \frac{1}{\lambda}b)$ is a family of affine vector fields parameterized by λ. For each $\lambda \neq 0$, $(\text{expt}X_\lambda)x = e^{t\lambda A}x + \int_0^t e^{\lambda(t-s)A}b\,ds$, $\lim_{\lambda\to 0}(\text{expt}X_\lambda)(x) = x + bt$ because $\lim_{\lambda\to 0} e^{t\lambda A} = I$ uniformly in t. Thus the limiting curve $x + bt$ is equal to $(\text{expt}X_0)(x)$ with $X_0 = \lim_{\lambda\to 0} X_\lambda$.

It can be shown in general that if a sequence of vector fields converges to a vector field F then each curve $\sigma_n(t) = (\text{expt}F_n)(x_0)$ converges uniformly in t to $\sigma(t) = (\text{expt}F)(x_0)$. Therefore, each family $\mathcal{F}$ may be prolonged to its topological closure.

In addition to the convexification and the topological closure, there is another means of prolonging a given family of vector fields based on reparameterizations of trajectories.

Note that $y(t) = x(\lambda t)$ remains in the reachable set $\mathcal{A}_\mathcal{F}(x_0, \leq T)$ for any trajectory $x(t)$ of $\mathcal{F}$ for which $x(0) = x_0$ provided that $0 \leq \lambda \leq 1$. $y(t) \in \mathcal{A}_\mathcal{F}(x_0)$ for any $\lambda \geq 0$. Thus, $\lambda F \in \mathcal{LS}_s(\mathcal{F})$ for any $0 \leq \lambda \leq 1$ and any F in $\mathcal{LS}_s(\mathcal{F})$. It will be useful for further references to assemble these prolongations into a theorem.

THEOREM 55.4 *(i) The Lie saturate of any system is a closed convex cone, i.e., $\sum_{i=1}^m \lambda_i F_i \in \mathcal{LS}(\mathcal{F})$ for any vector fields $F_1, \ldots, F_m$ in $\mathcal{LS}(\mathcal{F})$ and any numbers $\lambda_1 \geq 0, \ldots, \lambda_m \geq 0$.*

(ii) The strong Lie saturate of any family of vector fields is a closed convex body, i.e., $\sum_{i=1}^m \lambda_i F_i \in \mathcal{LS}_s(\mathcal{F})$ for any elements $F_1, \ldots, F_m$ in $\mathcal{LS}_s(\mathcal{F})$ and any non-negative numbers $\lambda_1, \ldots, \lambda_n$ such that $\sum_{i=1}^m \lambda_i \leq 1$.

We now describe another operation which may be used to prolong the system without altering its reachable sets. This operation is called the normalization of the system.

An invertible map $\Phi : M \to M$ is called a strong normalizer for $\mathcal{F}$ if $\Phi(\mathcal{A}_\mathcal{F}(\Phi^{-1}(x), \leq T)) \subset \text{cl}\mathcal{A}_\mathcal{F}(x, \leq T)$ for all x in M and $T > 0$. Φ is called a normalizer for $\mathcal{F}$ if $\Phi\mathcal{A}_\mathcal{F}(\Phi^{-1}(x)) \subset \text{cl}\mathcal{A}_\mathcal{F}(x)$. It may be also said that Φ is a strong normalizer if both $\Phi(x)$ and $\Phi^{-1}(x)$ are contained in $\text{cl}\mathcal{A}_\mathcal{F}(x, \leq T)$ and that Φ is a normalizer if both $\Phi(x)$ and $\Phi^{-1}(x)$ belong to $\text{cl}\mathcal{A}_\mathcal{F}(x)$. In this notation $\Phi(\mathcal{A}_\mathcal{F}(\Phi^{-1}(x), \leq T))$ is equal to the set of points $\Phi(y)$ with y belonging to $\mathcal{A}_\mathcal{F}(\Phi^{-1}(x), \leq T)$. If Φ is any invertible transformation, and if F is any vector field then $(\Phi)(\text{expt}F)\Phi^{-1}$ is a one-parameter group of transformations and is itself generated by a vector field. That is, there is a vector field G such that

$$(\text{expt}G) = \Phi(\text{expt}F)\Phi^{-1}$$

It can be shown that $G = (d\Phi)F(\Phi^{-1})$ where $d\Phi$ denotes the derivative of Φ. We shall use $\Phi_\#(F)$ to denote the vector field $(d\Phi)(F\Phi^{-1})$.

EXAMPLE 55.11:

(i) Let Φ be a transformation $\Phi(x) = x + b$, and F a linear vector field $F(x) = Ax$. Then,

$$\Phi\text{expt}F\Phi^{-1}(x) = e^{At}(x - b) + b$$

Therefore, $\frac{d}{dt}e^{At}(x-b)+b|_{t=0} = A(x-b) = Ax - Ab$. Thus, $\Phi_\# F$ is an affine vector field.
(ii) If Φ is a linear transformation, then $d\Phi$ is also linear, and therefore, $\Phi_\# F = \Phi A \Phi^{-1}$, i.e., $\Phi_\# F$ is a linear vector field for any linear field F.

THEOREM 55.5 *(i) If Φ is a strong normalizer for a family of vector fields F then,*

$$\Phi_\#(\mathcal{LS}_s(\mathcal{F})) \cap \text{Lie}(\mathcal{F}) \subset \mathcal{LS}_s(\mathcal{F})$$

(ii) If Φ is a normalizer for $\mathcal{F}$, then

$$\Phi_\#(\mathcal{LS}(\mathcal{F})) \cap \text{Lie}(\mathcal{F}) \subset \mathcal{LS}(\mathcal{F})$$

55.1.7 Applications to Controllability

The geometric ideas that led to the Lie saturate criterion of controllability provide a beautiful proof of controllability of linear systems, demonstrating at the same time that linearity plays an inessential role. This proof goes as follows.

We use the induction on k to show that each controllability space $\mathcal{L}_k = \mathcal{B} + A\mathcal{B} + \cdots + A^{k-1}\mathcal{B}$ defined in Example 8 is contained in the strong Lie saturate of the system.

Let $\mathcal{F}$ denote the family of affine vector fields $F_u(x) = Ax + Bu$ defined by the linear system $\frac{dx}{dt} = Ax + Bu$. For each real number λ, $0 \leq \lambda \leq 1$ and each F_u in $\mathcal{F}$

$$F_{\lambda,u}(x) = \lambda(Ax + \frac{1}{\lambda}Bu)$$

belongs to $\mathcal{LS}_s(\mathcal{F})$ by Theorem 55.4 (ii). Its limit as $\lambda \to 0$ also belongs to $\mathcal{LS}_s(\mathcal{F})$ since the latter is closed. It follows that $\lim_{\lambda\to 0} F_{\lambda,u} = Bu$ and therefore $\mathcal{L}_1 = \mathcal{B}$ is contained in $\mathcal{LS}_s(\mathcal{F})$.

Now assume that $\mathcal{L}_{k-1} \subset \mathcal{LS}_s(\mathcal{F})$. Let b be any element of $\mathcal{L}_{k-1}$ and let α be any real number. The constant vector field $F_\alpha = \alpha b$ is in $\mathcal{L}_{k-1}$ for each α. Let $\Phi_\alpha = \exp F_\alpha$. Then $(\Phi_\alpha)^{-1} = \exp F_{-\alpha}$ and therefore both $\Phi_\alpha(x)$ and $\Phi_\alpha^{-1}(x)$ remain in $\text{cl}\mathcal{A}_\mathcal{F}(x, \leq T)$ for any $x \in M$ and $T > 0$. Therefore, Φ_α is a strong normalizer for $\mathcal{F}$. According to Theorem 55.5, $(\Phi_\alpha)_\#(Fu) \subset \mathcal{LS}_s(\mathcal{F})$ provided that $(\Phi_\alpha)_\#(F_u) \in \text{Lie}(\mathcal{F})$. Then $((\Phi_\alpha)_\#(F_0))(x) = \alpha\Phi_\alpha A\Phi_{-\alpha}(x) = A(x - \alpha b)$ because the derivative map of a translation is equal to the identity map. Thus, $(\Phi_\alpha)_\#(Fu)$ belongs to Lie($\mathcal{F}$). An analogous argument used in the first step of the induction procedure applied to the limit of $\lambda(\Phi_{\frac{\alpha}{\lambda}})_\#(F_0)$ as λ tends to 0 shows that the constant vector field $-\alpha Ab$ is contained in $\mathcal{LS}_s(\mathcal{F})$ for each real number α. But then $\mathcal{L}_k \subset \mathcal{LS}_s(\mathcal{F})$ because the convex hull of two vector spaces is the vector space spanned by their sum, i.e., $\mathcal{L}_k = \mathcal{L}_{k-1} + A\mathcal{L}_{k-1}$. Therefore, each $\mathcal{L}_k$ is in $\mathcal{LS}_s(\mathcal{F})$.

When $\mathcal{L}_{n-1} = M$, the space of all constant vector fields is in the strong Lie saturate and hence $\text{cl}(\mathcal{A}_\mathcal{F}(x, \leq T)) = M$ for each $x \in M$ and $T > 0$. But then it follows from Theorem 55.2 that $M = \text{cl}\mathcal{A}_\mathcal{F}(x, \leq T) \subset \text{int}\,\mathcal{A}_\mathcal{F}(x, \leq T+\epsilon) \subset \mathcal{A}_\mathcal{F}(x, \leq T+\epsilon)$. Therefore, the system is strongly controllable.

The inductive procedure can also be described pictorially as follows:

Step 1: Prolong the original system to its closed convex body $\mathcal{F}_1$. Geometrically $\mathcal{F}(x)$ is the translate of $\mathcal{B}$ to Ax. For each u, $Ax + \lambda Bu$ is the line through Ax parallel to Bu, as shown in Figure 55.7a. $\mathcal{F}_1(x)$ is the union of all translates of $\mathcal{B}$ to points λAx, $0 \leq \lambda \leq 1$, as shown in Figure 55.7b.

Step 2: $\mathcal{F}_1$ contains the vector space $\mathcal{B}$ as its edge. Conjugate the original family $\mathcal{F}$ by $\mathcal{B}$ to obtain a prolonged family $\mathcal{F}_2$ given by $\frac{dx}{dt} = Ax + Bu + ABv$ with both u and v as controls. $\mathcal{F}_2(x)$ is the translate of $\mathcal{B} + A\mathcal{B}$ to Ax, while the convex body $\mathcal{F}_3$ generated by $\mathcal{F}_2$ at each point x is the union of all translates of $\mathcal{B} + A\mathcal{B}$ to λAx as λ ranges in the interval [0, 1]. Figure 55.8 illustrates their differences.

Step 3: Conjugate the original family by the edge $\mathcal{B} + A\mathcal{B}$ of $\mathcal{F}_3$. The prolonged family is given by $\frac{dx}{dt} = Ax + Bu + ABv + A^2Bw$.

A repetition of these steps embodied in the induction scheme leads to the saturated system from which the controllability properties are evident.

We now illustrate the importance of the Lie saturate by considering controllability of linear systems with bounded controls. Strong controllability is not possible when the constraint set U is compact, because each set $\mathcal{A}_\mathcal{F}(x, \leq T)$ is compact. It is also known that controllability is not possible whenever the drift vector field Ax has an eigenvalue with non-zero real parts. We will now use the geometric framework provided by the Lie saturate criterion to obtain affirmative controllability results when the real part of the spectrum of A is zero. For simplicity the proof will be given for a particular case only when all the eigenvalues of A are zero, i.e., when A is nilpotent.

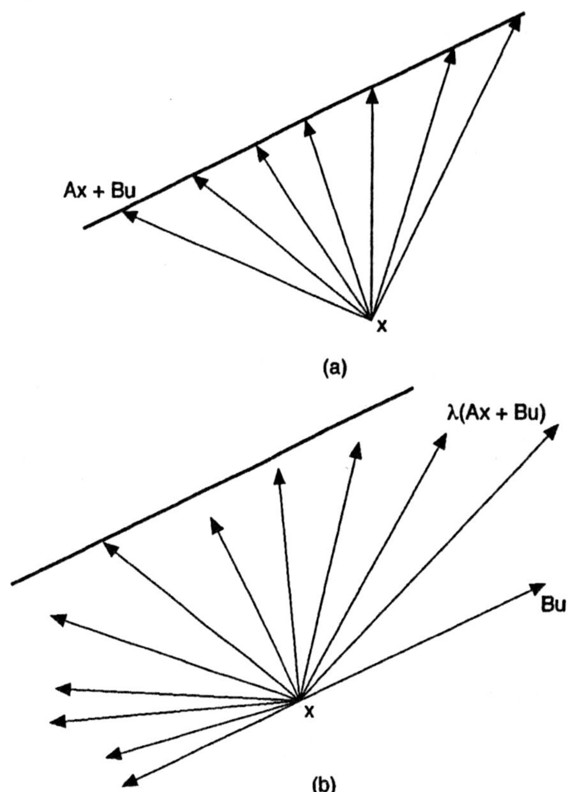

Figure 55.7 Illustration for step 1: a. $\mathcal{F}(x)$, b. $\mathcal{F}_1(x)$.

THEOREM 55.6 *Suppose that U is a compact neighborhood of the origin in R^m, and suppose further that the linear drift is nilpotent, i.e., suppose that there is a positive integer p such that $A^p \neq 0$ but $A^{p+1} = 0$. Then $\frac{dx}{dt} = Ax + Bu$, $u \in U$ is controllable, provided that the rank of $(B\ AB\ \cdots\ A^{n-1}B)$ is equal to* dim M.

Proof: There is no loss of generality in assuming that U is the cube $|u_i| \leq \epsilon$ $i = 1, \ldots, m$. Then the reachable set $\mathcal{A}(0)$ is a convex neighborhood of the origin in M. Any trajectory $x(t)$ which originates at $x(0) = 0$ is of the form

$$x(t) = \int_0^t e^{A(t-s)}Bu(s)\,ds = \sum_{k=0}^{p} \frac{A^kB}{k!}\int_0^t (t-s)^k u(s)\,ds$$

For any $u \in R^m$, and any real number λ, there exists $T > 0$ such that $\frac{|\lambda u_i|}{T^p} < \epsilon$ for all $i = 1, \ldots, m$. Let $u(T) = \frac{\lambda u}{T^p}$. The corresponding response $x(T)$ is equal to

$$\lambda\left(\frac{Bu}{T^p} + \frac{ABu}{T^{p-1}} + \cdots + \frac{A^{p-1}Bu}{p!T} + \frac{A^pBu}{(p+1)!}\right),$$

and therefore, $\lim_{T\to\infty} x(T) = \frac{\lambda A^pBu}{(p+1)!}$. Therefore, the line through A^pBu is contained in the closure of $\mathcal{A}(0)$. The convex hull of these lines as u ranges over R^m is equal to the vector space $A^p\mathcal{B}$.

Take now $u(T) = \frac{\lambda u}{T^{p-1}}$. The corresponding trajectory $x(T)$ is given by $\lambda(\frac{Bu}{T^{p-1}} + \cdots + \frac{A^{p-1}Bu}{p!} + \frac{TA^pBu}{(p+1)!})$. Then, $\frac{1}{2}(x(T) - \frac{\lambda TA^pBu}{(p+1)!}) \in \text{cl}\mathcal{A}(0)$, since the latter is convex. But then $\lim_{T\to\infty}\frac{1}{2}(x(T) - \frac{\lambda TA^pBu}{(p+1)!}) = \frac{\lambda A^{p-1}Bu}{p!}$. A repetition of

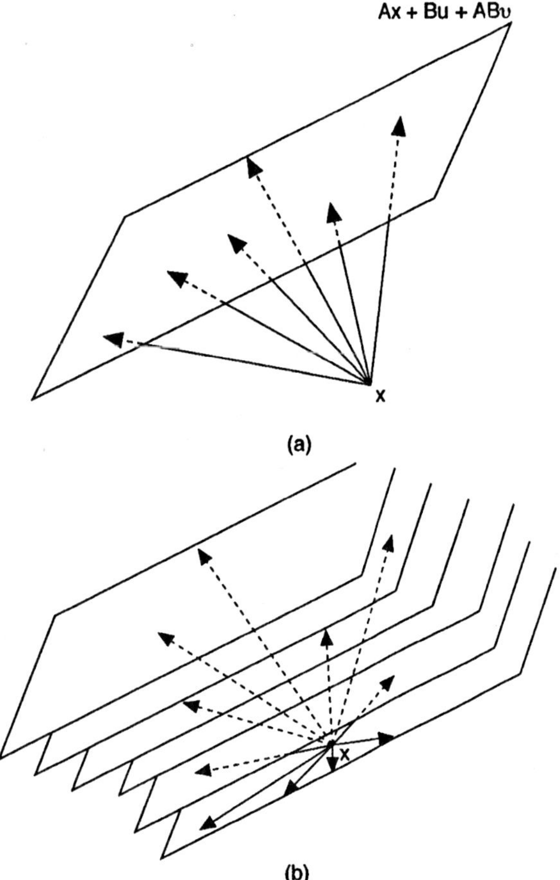

Figure 55.8 Illustration for step 2: a. $\mathcal{F}_2(x)$ b. $\mathcal{F}_3(x)$.

the previous argument shows that the sum of $A^{p-1}\mathcal{B}$ and $A^p\mathcal{B}$ is contained in cl$\mathcal{A}(0)$. Further repetitions of the same argument show that cl$\mathcal{A}(0) = \mathcal{B} + A\mathcal{B} + \cdots + A^p\mathcal{B}$. The latter is equal to M by the rank assumption.

Since $-A$ is also nilpotent, the above proof is applicable to the time reversed system $\frac{dx}{dt} = -Ax - Bu$, with $u \in U$, to show that its reachable set from the origin is the entire space M. Therefore, any initial point x_0 can be steered to the origin in some finite time T_1 using the time-reversed system. But then the origin can be steered to any terminal state x_1 as a consequence of the fact proved above that $\mathcal{A}(0) = M$. Thus, $\mathcal{A}(x_0) = M$ for any x_0 in M and our proof is finished.

REMARK 55.1 We have implicitly used the Lie saturate criterion to conclude that $\mathcal{A}(x) = M$ for all $x \in M$ whenever cl$\mathcal{A}(x) = M$ for all x in M.

55.1.8 Rotations

The group of rotations in R^3 is a natural state space for many mechanical control problems, because the kinematics of a rigid body can be described by the movements of an orthonormal frame fixed on the body relative to an orthonormal frame fixed in the ambient space. Recall that the rotation group consists of all linear transformations R which leave the Euclidean metric $\langle\, ,\, \rangle$ in R^3 invariant. A Euclidean metric in R^3 is any positive definite scalar product. So if e_1, e_2, e_3 is any orthonormal basis in R^3 and if $x = \sum_{i=1}^3 x_i e_i$, and $y = \sum_{i=1}^3 y_i e_i$ then $\langle x, y\rangle = \sum_{i=1}^3 x_i y_i$. R is a rotation if $\langle Rx, Ry\rangle = \langle x, y\rangle$ for all x and y in R^3.

Denoting by a_1, a_2, a_3 an orthonormal frame fixed on the body, then any motion of the body is monitored by the rotation through which the moving frame a_1, a_2, a_3 undergoes relative to the fixed frame e_1, e_2, e_3. This rotation, when expressed relative to the basis e_1, e_2, e_3, becomes a 3×3 matrix whose columns consist of the coordinates of a_1, a_2, a_3 relative to the fixed basis e_1, e_2, e_3.

The group of all such matrices whose determinant is equal to 1 is called the special orthogonal group and is denoted by $SO_3(R)$. $SO_3(R)$ is a three-dimensional manifold, which, together with its group structure, accounts for a rich geometric base, which needs to be properly understood as a prerequisite for effective control of mechanical systems.

Let us first outline the manifold structure of $SO_3(R)$. To begin with, the tangent space of $SO_3(R)$ at any point R_0 consists of all tangent vectors $\frac{d}{d\epsilon}R(\epsilon)|_{\epsilon=0}$ for curves $R(\epsilon)$ in $SO_3(R)$ which satisfy $R(0) = R_0$. The tangent space at the group identity I plays a special role and is called the Lie algebra of $SO_3(R)$. It consists of all matrices A for which $e^{A\epsilon} \in SO_3(R)$. Each such matrix A is antisymmetric because the rotations satisfy $R^{-1} = R^*$ with R^* equal to the transpose, and $e^{-A\epsilon} = (e^{A\epsilon})^* = e^{A^*\epsilon}$. Consequently, $A^* = -A$.

The space of 3×3 antisymmetric matrices is a three-dimensional vector space and is denoted by $so_3(R)$. Each rotation, consisting of orthonormal column vectors, is defined by six orthonormality relations in a nine-dimensional group of all 3×3 matrices. Therefore, $SO_3(R)$ is a three-dimensional manifold, and consequently, each tangent space is three-dimensional. But then, the tangent space at I is equal to $so_3(R)$.

Consider now the tangent space at an arbitrary point R_0. For any antisymmetric matrix A each of the curves $R_1(\epsilon) = R_0 e^{A\epsilon}$ and $R_2(\epsilon) = e^{A\epsilon}R_0$ is a curve in $SO_3(R)$ which passes through R_0 at $\epsilon = 0$. Therefore, both $\frac{dR_1}{d\epsilon}(0) = R_0 A$ and $\frac{dR_2}{d\epsilon}(0) = AR_0$ are tangent vectors at R_0. These vectors are different from each other because of noncommutativity of R_0 with A. The first vector is called the left-translation of A by R_0, and the second is called the right-translation of A by R_0. It follows that the tangent space at R_0 can be described by either left- or right-translations of $so_3(R)$.

Denote by A_1, A_2, A_3 the standard basis of $so_3(R)$,

$$A_1 = \begin{pmatrix} 0 & 0 & 0 \\ 0 & 0 & -1 \\ 0 & 1 & 0 \end{pmatrix}$$
$$A_2 = \begin{pmatrix} 0 & 0 & 1 \\ 0 & 0 & 0 \\ -1 & 0 & 0 \end{pmatrix}$$
$$A_3 = \begin{pmatrix} 0 & -1 & 0 \\ 1 & 0 & 0 \\ 0 & 0 & 0 \end{pmatrix}$$

Since $A_i e_i = 0$, it follows that each $e^{A_i\epsilon}$ is a rotation about the axis containing e_i, $i = 1, 2, 3$. For any antisymmetric matrix $A = \begin{pmatrix} 0 & -a_2 & a_2 \\ a_2 & 0 & -a_1 \\ -a_2 & a_1 & 0 \end{pmatrix}$ we will use $\hat{A}$ to denote the column vector $\begin{pmatrix} a_1 \\ a_2 \\ a_3 \end{pmatrix}$. $\hat{A}$ is the coordinate vector of A relative to the standard basis.

Any antisymmetric matrix A induces vector fields on $SO_3(R)$. The first vector field is given by $F_l(R) = RA$, and the second is $F_r(R) = AR$. F_l is called the left-invariant vector field induced by A because its tangent vector at R is a left translation by R of its tangent vector at the group identity. Similar explanations apply to right-invariant vector fields F_r. We will use $\overrightarrow{A_l}$ to denote the left-invariant vector field whose tangent at I is equal to A, i.e., $\overrightarrow{A_l}(R) = RA$. Similarly, $\overrightarrow{A_r}$ denotes the right-invariant field $\overrightarrow{A_r}(R) = AR$.

Then $(\overrightarrow{A_1})_l$, $(\overrightarrow{A_2})_l$ and $(\overrightarrow{A_3})_l$ is a basis of left-invariant vector fields which span each tangent space and $(\overrightarrow{A_1})_r$, $(\overrightarrow{A_2})_r$ and $(\overrightarrow{A_3})_r$ is a basis of right-invariant vector fields with the same property.

Any differentiable curve $R(t)$ in $SO_3(R)$ defines a curve of tangent vectors $\frac{dR}{dt}$ at $R(t)$, which can be expressed by either right or left basis. Let $\frac{dR}{dt} = \sum_{i=1}^{3} \omega_i(t)(\overrightarrow{A_i})_r(R) = \sum_{i=1}^{3} \Omega_i(t)(\overrightarrow{A_i})_l(R(t))$ denote the corresponding coordinates of $\frac{dR}{dt}$. Vectors $\omega(t) = \begin{pmatrix} \omega_1(t) \\ \omega_2(t) \\ \omega_3(t) \end{pmatrix}$, and $\Omega(t) = \begin{pmatrix} \Omega_1(t) \\ \Omega_2(t) \\ \Omega_3(t) \end{pmatrix}$ are called the angular velocities of $R(t)$. In analogy with the kinematics of a rigid body, the first angular velocity is called the (absolute) angular velocity, while the second is called the body angular velocity. The above differential equations can be rewritten as

$$\begin{pmatrix} 0 & -\omega_3(t) & \omega_2(t) \\ \omega_3(t) & 0 & -\omega_1(t) \\ -\omega_2(t) & \omega_1(t) & 0 \end{pmatrix} R(t)$$

$$= \frac{dR(t)}{dt} = R(t) \begin{pmatrix} 0 & -\Omega_3(t) & \Omega_2(t) \\ \Omega_3(t) & 0 & -\Omega_1(t) \\ -\Omega_2(t) & \Omega_1(t) & 0 \end{pmatrix}$$

It can be shown that $\Omega(t) = R^{-1}(t)\omega(t)$.

Any differentiable curve $R(t)$ whose angular velocity is constant is a solution curve of an invariant vector field. If $\omega(t)$ is constant, then $\frac{dR}{dt} = AR$ with $\hat{A} = \omega$, and if $\Omega(t)$ is constant then $\frac{dR}{dt} = RA$ with $\hat{A} = \Omega$. In the first case, $R(t) = e^{At}R_0$ while in the second case $R(t) = R_0 e^{At}$.

It can be shown that the Lie bracket of a right (respectively, left) invariant vector fields is a right (respectively, left) invariant vector field, with $[\overrightarrow{A_l}, \overrightarrow{B_l}](R) = R(BA - AB)$ and $[\overrightarrow{A_r}, \overrightarrow{B_r}](R) = (AB - BA)R$.

It is easy to verify that the commutator $AB - BA$ can also be expressed in terms of the cross-product of $\hat{A}$ and $\hat{B}$ in R^3 as follows: let $[A, B]_r = AB - BA$ and $[A, B]_l = BA - AB$. Then, $\widehat{[A, B]}_r = \hat{A} \times \hat{B}$, while $\widehat{[A, B]}_l = \hat{B} \times \hat{A}$.

Except for the cross-product correspondence, all of these concepts extend to the rotation group $SO_n(R)$ of R^n, and its $\frac{n(n-1)}{2}$-dimensional Lie algebra $so_n(R)$ of $n \times n$ antisymmetric matrices.

55.1.9 Controllability in $so_n(R)$

A unit sphere which rolls on a horizontal plane without slipping and without spinning along the axis perpendicular to the point of contact, can be described by the following equations:

$$\frac{dx_1}{dt} = u_1(t)$$

$$\frac{dx_2}{dt} = u_2(t)$$

$$\frac{dR(t)}{dt} = \begin{pmatrix} 0 & 0 & u_1 \\ 0 & 0 & u_2 \\ -uo & -u_2 & 0 \end{pmatrix} R(t)$$

$x_1(t)$ and $x_2(t)$ are the coordinates of the center of the sphere ($x_3 = 1$), and $R(t)$ is the orientation of the sphere relative to an absolute frame e_1, e_2, e_3. The angular velocity $\omega(t) = \begin{pmatrix} -u_2 \\ u_1 \\ 0 \end{pmatrix}$ of the sphere is always orthogonal to the velocity of its center.

The rotational kinematics of the sphere may be viewed as a left-invariant control system on $SO_3(R)$ with two controls u_1 and u_2. This control system has no drift, and therefore, according to a well-known theorem of geometric control theory, the system is strongly controllable whenever the Lie algebra generated by the controlling vector fields is equal to the Lie algebra of the group (in this case $SO_3(R)$). It follows that the controlling vector fields are $F_1(R) = A_2R$ and $F_2(R) = -A_1R$ corresponding to $u_1 = 1$, $u_2 = 0$ and $u_2 = 0$, $u_1 = 1$. The rotational part is strongly controllable since $[F_1, F_2](R) = A_3R$. It can also be shown that the overall system in $R^3 \times SO_3(R)$ is strongly controllable because the Lie algebra generated by the controlling vector fields is equal to $R^2 \times so_3(R)$.

There is a simple argument showing that any states in $R^2 \times SO_3(R)$ can be transferred to each other by two switches in controls. Note first that for any angular velocity $\hat{A}$ the corresponding rotation e^A is the rotation about $\hat{A}$ through the angle $\|\hat{A}\|$. Figure 55.9 shows that any rotation can be achieved by one switch in controls (two angular velocities ω_1 and ω_2).

The proof begins with the observation that each unit circle in the e_1, e_2 plane centered at the origin has a line ω in common with the circle in the a_1, a_2 plane also centered at the origin. ω is in the plane $\omega_3 = 0$ as shown in the picture. The first move consists of rotating about ω so that a_3 coincides with $-e_1$. Then rotate through π radians along the midpoint of the arc between a_2 and e_2. These two moves rotate any frame a_1, a_2, a_3 into the standard frame. The remaining moves are used to roll for the position of the point of contact along a line segment whose length is an integral multiple of 2π. Such moves do not alter the orientation of the ball. Any two points in the plane can be joined along the sides of an isosceles triangle with equal sides equal to $2\pi m$, as shown in Figure 55.10.

The reader may note the similarity of this argument with the one used to show that any rotation in R^3 may be achieved by the

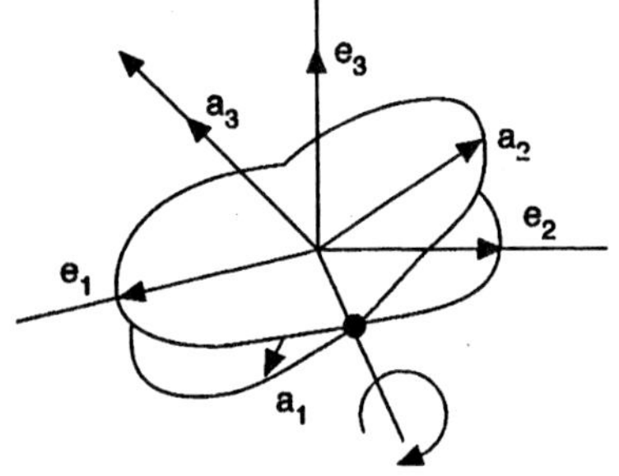

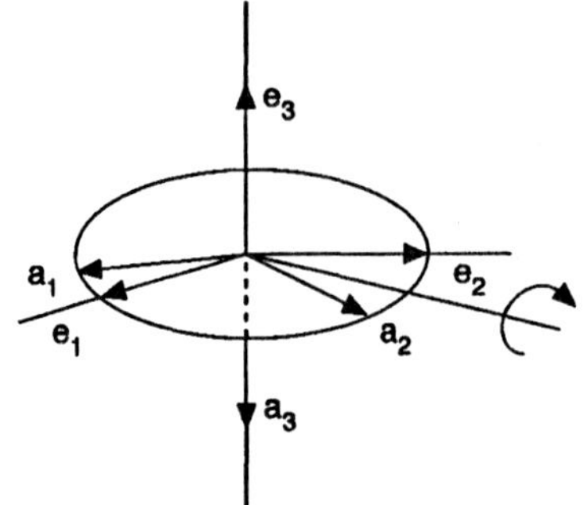

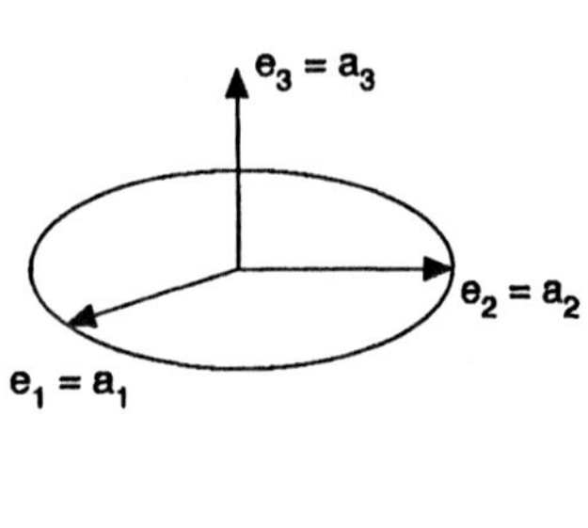

Figure 55.9 Rotational kinematics.

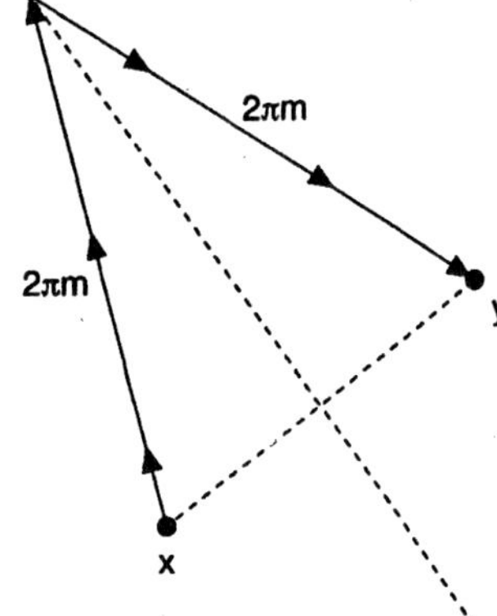

Figure 55.10 Translational kinematics.

rotations through the Euler angles ϕ, θ, and ψ.

This exposition ends with a controllability theorem whose proof also relies on the Lie saturate.

THEOREM 55.7 *Suppose that $\mathcal{F}$ is any family of right (or left) invariant vector fields on $SO_n(R)$ (or any other compact Lie group G). Then $\mathcal{F}$ is controllable if and only if Lie($\mathcal{F}$), evaluated at I, is equal to the Lie algebra of $SO_n(R)$ (or G).*

The proof consists in showing that $-\mathcal{F}$ is contained in the Lie saturate of $\mathcal{F}$. Therefore, the vector span of $\mathcal{F}$ is contained in $\mathcal{L}\ S(\mathcal{F})$ by the convexity property of $\mathcal{L}S(\mathcal{F})$. But then the Lie algebra of $\mathcal{F}$ is contained in the Lie saturate and, hence, must be equal to it. The controllability result then follows from the Lie saturate criterion. So, the proof will be complete once we showed that $(\exp -tF)(R) \in \text{cl}\mathcal{A}_{\mathcal{F}}(R)$ for any $t > 0$ and any $F \in \mathcal{F}$. Let $F(R) = AR$. Then, $(\exp -tF)(R) = e^{-tA}R$, and therefore $(\exp -tF)(R)$ belongs to $\text{cl}\mathcal{A}_{\mathcal{F}}(R)$ if and only if e^{-tA} belongs to the closure of the reachable set from the group identity.

$S0_n(R)$ is a compact group and therefore there exists a sequence of times t_n tending to ∞ such that $\lim e^{t_nA}$ exists. Let $\lim_{t\to\infty} e^{t_nA} = R_0$. Then, $R_0^{-1} = \lim_{n\to\infty} e^{-At_n}$. If necessary, choose a subsequence so that $t_{n+1} - t_n$ also tends to ∞. Then, $I = R_0^{-1}R_0 = (\lim_{n\to\infty} e^{-t_nA})(\lim e^{t_{n+1}A}) = \lim_{n\to\infty} e^{(t_{n+1}-t_n)A}$.

The preceding argument shows that e^{tA} comes arbitrarily close to the identity for large values of time. Then,

$$e^{-tA} = e^{-tA}(\lim_{n\to\infty} e^{(t_{n+1}-t_n)A})$$

$$= \lim_{n\to\infty} e^{((t_{n+1}-t_n)-t)A}$$

Since $t_{n+1} - t_n \to \infty$, $(t_{n+1} - t_n) - t > 0$ for large values of n and therefore $e^{-tA} \in \text{cl}\mathcal{A}_{\mathcal{F}}(I)$. The proof is now finished.

Theorem 55.7 might be used to show that the orientation of a rigid body may be controlled by any number of gyros situated on the body as long as the Lie algebra generated by their angular velocities has full rank.

Further Reading

The proofs of all theorems quoted in this paper can be found in the forthcoming book titled *Geometric Control Theory* by V. Jurdjevic, (to appear in Studies in Advanced Mathematics, Cambridge University Press.) The material for this article is taken out of the first part of the book dealing with the reachable sets of Lie determined systems (which includes analytic systems).

The reader may also find some of this material in the following publications:

[1] Jurdjevic, V. and Kupka, I.A., Polynomial control system, *Math. Ann.*, 361–368, 1985.

[2] Jurdjevic, V. and Sussmann, H.J., Control systems on Lie groups, *J. Diff. Eqs.*, 12, 313–329, 1972.

[3] Sussmann, H.J. and Jurdjevic, V., Controllability of non-linear systems, *J. Diff. Eqs.*, 12, 95–116, 1972.

Convexification of control systems is also known as the relaxation of controls in the early literature on control. See for instance,

[4] Hermes, H. and LaSalle, J.P., *Functional Analysis and Time Optimal Control*, Academic Press, New York, 1969.

[5] Warga, T., *Optimal Control of Differential and Functional Equations*, Academic Press, New York, 1972.

For other applications of Lie theory to control systems the reader may consult Geometric Methods in Systems Theory, Proceedings of the NATO Advanced Study Series, Editors: R. Brockett and D. Q. Mayne, Reidel Publishing, 1973.

55.2 Two-Time-Scale and Averaging Methods

Hassan K. Khalil, *Michigan State University*

55.2.1 Introduction

In this chapter we present the asymptotic methods of averaging and singular perturbation. Suppose we are given the state equation $\dot{x} = f(t, x, \epsilon)$, where ϵ is a "small" positive parameter, and, under certain conditions, the equation has an exact solution $x(t, \epsilon)$. Equations of this type are encountered in many applications. The goal of an asymptotic method is to obtain an approximate solution $\tilde{x}(t, \epsilon)$ such that the approximation error $x(t, \epsilon) - \tilde{x}(t, \epsilon)$ is small, in some norm, for small ϵ and the approximate solution $\tilde{x}(t, \epsilon)$ is expressed in terms of equations simpler than the original equation. The practical significance of asymptotic methods is in revealing underlying multiple-time-scale structures inherent in many practical problems. Quite often the solution of the state equation exhibits the phenomenon that some variables move in time faster than other variables, leading to the classification of variables as "slow" and "fast." Both the averaging method and the singular perturbation method deal with the interaction of slow and fast variables.

55.2.2 Asymptotic Methods

We start by a brief description of the classical perturbation method that seeks an approximate solution as a finite Taylor expansion of the exact solution. Then, we introduce the averaging method in its simplest form, which is sometimes called "periodic averaging" since the right-hand side function is periodic in time. Finally, we introduce the singular perturbation model and give its two-time-scale properties.

The Perturbation Method

Consider the system

$$\dot{x} = f(t, x, \epsilon), \quad x(t_0) = \eta \tag{55.6}$$

where f is sufficiently smooth in its arguments in the domain of interest, and ϵ is a positive parameter. The solution of Equation 55.6 depends on the parameter ϵ, a point that we shall emphasize by writing the solution as $x(t, \epsilon)$. The goal of the perturbation method is to exploit the "smallness" of the perturbation parameter ϵ to construct approximate solutions that will be valid for sufficiently small ϵ. The simplest approximation results by setting $\epsilon = 0$ in equation 55.6 to obtain the nominal or unperturbed problem:

$$\dot{x} = f(t, x, 0), \quad x(t_0) = \eta \tag{55.7}$$

Suppose this problem has a unique solution $x_0(t)$ defined on $[t_0, t_1]$. By continuous dependence of the solutions of differential equations on parameters, we know that, for sufficiently small ϵ, the system of Equation 55.6 has a unique solution $x(t, \epsilon)$ defined on $[t_0, t_1]$. Moreover,

$$\|x(t, \epsilon) - x_0(t)\| \leq k\epsilon, \quad \forall\, \epsilon < \epsilon_1, \ \forall\, t \in [t_0, t_1]$$

for some $k > 0$ and $\epsilon_1 > 0$. In this case, we say that the error is of order $O(\epsilon)$ and write $x(t, \epsilon) - x_0(t) = O(\epsilon)$. This order-of-magnitude notation will be used frequently. It is defined as follows.

DEFINITION 55.4 $\delta_1(\epsilon) = O(\delta_2(\epsilon))$ if there exist positive constants k and c such that

$$|\delta_1(\epsilon)| \leq k|\delta_2(\epsilon)|, \quad \forall\, |\epsilon| < c$$

Higher-order approximations for the solution of Equation 55.6 can be obtained in a straightforward manner. We construct a finite Taylor series

$$x(t, \epsilon) = x_0(t) + \sum_{k=1}^{N-1} x_k(t)\epsilon^k + \epsilon^N x_R(t, \epsilon) \tag{55.8}$$

Substitution of Equation 55.8 in Equation 55.6 yields

$$\sum_{k=0}^{N-1} \dot{x}_k(t)\epsilon^k + \epsilon^N \dot{x}_R(t, \epsilon) = f(t, x(t, \epsilon), \epsilon) \stackrel{\text{def}}{=} h(t, \epsilon) \tag{55.9}$$

where the coefficients of the Taylor series of $h(t, \epsilon)$ are functions of the coefficients of the Taylor series of $x(t, \epsilon)$. Since the equation holds for all sufficiently small ϵ, it must hold as an identity in ϵ. Hence, coefficients of like powers of ϵ must be equal. Matching those coefficients we can derive the equations that must be satisfied by x_0, x_1, and so on. The zeroth-order term $h_0(t)$ is given by $h_0(t) = f(t, x_0(t), 0)$. Hence, matching coefficients of ϵ^0 in Equation 55.9, we determine that $x_0(t)$ satisfies

$$\dot{x}_0 = f(t, x_0, 0), \quad x_0(t_0) = \eta$$

which, not surprisingly, is the unperturbed problem of Equation 55.7. The first-order term $h_1(t)$ is given by

$$h_1(t) = \frac{\partial f}{\partial x}(t, x_0(t), 0)\, x_1(t) + \frac{\partial f}{\partial \epsilon}(t, x_0(t), 0)$$

Matching coefficients of ϵ in Equation 55.9 we find that $x_1(t)$ satisfies

$$\dot{x}_1 = A(t)x_1 + g_1(t, x_0(t)), \quad x_1(t_0) = 0 \tag{55.10}$$

where

$$A(t) = \frac{\partial f}{\partial x}(t, x_0(t), 0), \quad g_1(t, x_0(t)) = \frac{\partial f}{\partial \epsilon}(t, x_0(t), 0)$$

This linear equation has a unique solution defined on $[t_0, t_1]$. This process can be continued to derive the equations satisfied by x_2, x_3, and so on. By straightforward error analysis, it can be established that

$$x(t, \epsilon) - \sum_{k=0}^{N-1} x_k(t)\epsilon^k = O(\epsilon^N) \tag{55.11}$$

The Taylor series approximation is effective when

- The unperturbed state Equation 55.7 is considerably simpler than the ϵ-dependent state Equation 55.6.
- ϵ is reasonably small that an "acceptable" approximation can be achieved with a few terms in the series.

The $O(\epsilon^N)$ error bound in Equation 55.11 is valid only on finite [order $O(1)$] time intervals $[t_0, t_1]$. It does not hold on intervals like $[t_0, T/\epsilon]$ nor on the infinite time interval $[t_0, \infty)$. The reason is that the constant k in the bound $k\epsilon^N$ depends on t_1 in such a way that it grows unbounded as t_1 increases. The error bound in Equation 55.11 can be extended to the infinite time interval $[t_0, \infty)$ if some additional conditions are added to ensure stability of the solution of the nominal system of Equation 55.7. In particular, suppose that Equation 55.7 has an **exponentially stable equilibrium point** $x = p^*$, then Equation 55.11 holds on the infinite time interval $[t_0, \infty)$, provided η is sufficiently close to p^*. We recall that *a solution $\bar{x}(t)$ of a state equation is exponentially stable if for $x(0)$ sufficiently close to $\bar{x}(0)$, the inequality*

$$\|x(t) - \bar{x}(t)\| \leq k\|x(0) - \bar{x}(0)\| \exp(-\gamma t), \quad \forall t \geq 0$$

is satisfied with some positive constants k and γ. This definition applies whether the $\bar{x}$ is an equilibrium point or a **periodic solution.** For **autonomous systems**, an equilibrium point is exponentially stable if the **linearization** of the system at this point results in a **Hurwitz matrix.**

Averaging

The averaging method applies to a system of the form

$$\dot{x} = \epsilon f(t, x, \epsilon) \tag{55.12}$$

where ϵ is a small positive parameter and $f(t, x, \epsilon)$ is periodic in t with period T; that is,

$$f(t, x, \epsilon) = f(t + T, x, \epsilon)$$

for all x and ϵ. We assume that f is sufficiently smooth in its arguments over the domain of interest. The method approximates the solution of Equation 55.12 by the solution of the autonomous **averaged system**

$$\dot{x} = \epsilon f_{av}(x) \tag{55.13}$$

where

$$f_{av}(x) = \frac{1}{T}\int_0^T f(\tau, x, 0)\, d\tau \tag{55.14}$$

The intuition behind this approximation can be seen as follows. The right-hand side of Equation 55.12 is multiplied by a positive constant ϵ. When ϵ is small, the solution x varies "slowly" with t relative to the periodic fluctuation of $f(t, x, \epsilon)$. It is intuitively clear that if the response of a system is much slower than the excitation, then such response will be determined predominantly by the average of the excitation. This intuition has its roots in linear system theory where we know that if the bandwidth of the system is much smaller than the bandwidth of the input, then the system acts as a low-pass filter that rejects the high-frequency component of the input. If the solution of Equation 55.12 is determined predominantly by the average of the right-hand side fluctuation, then it is reasonable, in order to get a first-order approximation with an order $O(\epsilon)$, that the function be replaced by its average.

The basic problem in the averaging method is to determine in what sense the behavior of the autonomous system of Equation 55.13 approximates the behavior of the more complicated nonautonomous system of Equation 55.12. We shall address this problem by showing, via a change of variables, that the nonautonomous system (Equation 55.12) can be represented as a perturbation of the autonomous system (Equation 55.13). Define

$$u(t, x) = \int_0^t [f(t, x, 0) - f_{av}(x)]\, d\tau \tag{55.15}$$

Since $f(t, x, 0) - f_{av}(x)$ is periodic in t of period T and has zero mean, the function $u(t, x)$ is periodic in t of period T. It can be also shown that $\partial u/\partial t$ and $\partial u/\partial x$ are periodic in t of period T. The change of variables

$$x = y + \epsilon u(t, y) \tag{55.16}$$

transforms Equation 55.12 into the form

$$\dot{y} = \epsilon f_{av}(y) + \epsilon^2 q(t, y, \epsilon) \tag{55.17}$$

where $q(t, y, \epsilon)$ is periodic in t of period T. This equation is a perturbation of the averaged system of Equation 55.13. It can be represented as a standard perturbation problem by changing the time variable from t to $s = \epsilon t$. In the s time scale, the equation is given by

$$\frac{dy}{ds} = f_{av}(y) + \epsilon q\left(\frac{s}{\epsilon}, y, \epsilon\right) \tag{55.18}$$

where $q(s/\epsilon, y, \epsilon)$ is periodic in s of period ϵT.

The problem has now been reduced to the perturbation problem we studied in the previous section. Therefore, if, for a given initial state $x(0) = \eta$, the averaged system

$$\frac{dy}{ds} = f_{av}(y), \quad y(0) = \eta$$

has a unique solution $\bar{y}(s)$ defined on $[0, b]$, then for sufficiently small ϵ the perturbed system of Equation 55.18 will have a unique solution defined for all $s \in [0, b]$ and the two solutions will be $O(\epsilon)$ close. Since $t = s/\epsilon$ and $x - y = O(\epsilon)$, by Equation 55.16, *the solution of the averaged system of Equation 55.13 provides an $O(\epsilon)$ approximation for the solution of Equation 55.12 over the time interval $[0, b/\epsilon]$ in the t time scale. If the averaged system of Equation 55.13 has an exponentially stable equilibrium point $x = p^*$, then there is a domain $\|x - p^*\| < \rho$ such that for all initial states within that domain, the $O(\epsilon)$ approximation will be valid for all $s \geq 0$, that is, for all $t \geq 0$.*

Investigation of Equation 55.18 reveals another interesting relationship between the nonautonomous Equation 55.12 and the averaged Equation 55.13. *If Equation 55.13 has an exponentially stable equilibrium point $x = p^*$, then Equation 55.12 has a unique exponentially stable periodic solution of period T in an $O(\epsilon)$ neighborhood of p^*.*

The averaging method can be extended to systems where the right-hand side of Equation 55.12 is not periodic in t, if an average of $f(t, x, 0)$ can be defined by the limit

$$f_{av}(x) = \lim_{T\to\infty} \frac{1}{T} \int_t^{t+T} f(\tau, x, 0)\, d\tau$$

A simple example is the case when $f(t, x, \epsilon) = f_1(t, x, \epsilon) + f_2(t, x, \epsilon)$, where f_1 is periodic in t while f_2 decays to zero as t tends to infinity, uniformly in (x, ϵ).

Singular Perturbation

While the perturbation method applies to state equations that depend smoothly on a small parameter ϵ, in this section we face a more difficult perturbation problem characterized by discontinuous dependence of system properties on the perturbation parameter ϵ. We shall study the singular perturbation model

$$\dot{x} = f(t, x, z, \epsilon) \tag{55.19}$$

$$\epsilon\dot{z} = g(t, x, z, \epsilon) \tag{55.20}$$

where setting $\epsilon = 0$ causes a fundamental and abrupt change in the dynamic properties of the system, as the differential equation $\epsilon\dot{z} = g$ degenerates into the algebraic or transcendental equation $0 = g(t, x, z, 0)$ The essence of the theory described in this section is that the discontinuity of solutions caused by singular perturbations can be avoided if analyzed in separate time scales. This multi-time-scale approach is a fundamental characteristic of the singular perturbation method.

Consider the singularly perturbed system of Equations 55.19) and 55.20 where $x \in R^n$ and $z \in R^m$. We assume that the functions f and g are sufficiently smooth in the domain of interest. When we set $\epsilon = 0$ in Equation 55.20, the dimension of the state equation reduces from $n + m$ to n because the differential Equation 55.20 degenerates into the equation

$$0 = g(t, x, z, 0) \tag{55.21}$$

We shall say that the model of Equations 55.19 and 55.20 is in **standard form** if and only if Equation 55.21 has $k \geq 1$ isolated real roots

$$z = h_i(t, x), \quad i = 1, 2, \ldots, k \tag{55.22}$$

for each (t, x) in the domain of interest. This assumption ensures that a well-defined n-dimensional **reduced model** will correspond to each root of Equation 55.21. To obtain the ith reduced model, we substitute Equation 55.22 into Equation 55.19, at $\epsilon = 0$, to obtain

$$\dot{x} = f(t, x, h(t, x), 0) \tag{55.23}$$

where we have dropped the subscript i from h. It is usually clear from the context which root of Equation 55.22 is being used. This model is called quasi-steady-state model because z, whose velocity $\dot{z} = g/\epsilon$ can be large when ϵ is small and $g \neq 0$, may rapidly converge to a root of Equation 55.21 which is the equilibrium of Equation 55.20. The model of Equation 55.23 is also known as the slow model or the reduced model.

Singular perturbations cause a two-time-scale behavior characterized by the presence of slow and fast transients in the system's response. Loosely speaking, the slow response is approximated by the reduced model of Equation 55.23, while the discrepancy between the response of the reduced model (Equation 55.23) and that of the full model of Equations 55.19 and 55.20 is the fast transient. To see this point, let us consider the problem of solving the state equation

$$\dot{x} = f(t, x, z, \epsilon), \quad x(t_0) = \xi \tag{55.24}$$

$$\epsilon\dot{z} = g(t, x, z, \epsilon), \quad z(t_0) = \eta \tag{55.25}$$

Let $x(t, \epsilon)$ and $z(t, \epsilon)$ denote the solution of the full problem of Equations 55.24 and 55.25. When we define the corresponding problem for the reduced model of Equation 55.23, we can specify only n initial conditions since the model is nth order. Naturally we retain the initial state for x, to obtain the reduced problem

$$\dot{x} = f(t, x, h(t, x), 0), \quad x(t_0) = \xi \tag{55.26}$$

Denote the solution of Equation 55.26 by $\bar{x}(t)$. Since the variable z has been excluded from the reduced model and substituted by its "quasi-steady-state" $h(t, x)$, the only information we can obtain about z by solving Equation 55.26 is to compute $\bar{z}(t) \stackrel{\text{def}}{=} h(t, \bar{x}(t))$, which describes the quasi-steady-state behavior of z when $x = \bar{x}$. By contrast to the original variable z, starting at t_0 from a prescribed η, the quasi-steady-state $\bar{z}$ is not free to start from a prescribed value, and there may be a large discrepancy between its initial value $\bar{z}(t_0) = h(t_0, \xi)$ and the prescribed initial state η. Thus $\bar{z}(t)$ cannot be a uniform approximation of $z(t, \epsilon)$. The best we can expect is that the estimate $z(t, \epsilon) - \bar{z}(t) = O(\epsilon)$ will hold on an interval excluding t_0, that is, for $t \in [t_b, t_1]$ where $t_b > t_0$. On the other hand, it is reasonable to expect the estimate $x(t, \epsilon) - \bar{x}(t) = O(\epsilon)$ to hold uniformly for all $t \in [t_0, t_1]$ since $x(t_0, \epsilon) = \bar{x}(t_0)$.

If the error $z(t, \epsilon) - \bar{z}(t)$ is indeed $O(\epsilon)$ over $[t_b, t_1]$, then it must be true that during the initial ("boundary-layer") interval $[t_0, t_b]$ the variable z approaches $\bar{z}$. Let us remember that the speed of z can be large since $\dot{z} = g/\epsilon$. In fact, having set $\epsilon = 0$ in Equation 55.20, we have made the transient of z instantaneous whenever $g \neq 0$. To analyze the behavior of z in the boundary layer, we set $y = z - h(t, x)$, to shift the quasi-steady-state of z to the origin, and change the time scale from t to $\tau = (t - t_0)/\epsilon$. The new time variable τ is "stretched"; that is, if ϵ tends to zero, τ tends to infinity even for finite t only slightly larger than t_0 by a fixed (independent of ϵ) difference. In the τ time scale, y satisfies the equation

$$\frac{dy}{d\tau} = g(t, x, y + h(t, x), \epsilon) - \epsilon\frac{\partial h}{\partial t} - \epsilon\frac{\partial h}{\partial x} f(t, x, y + h(t, x), \epsilon), \quad y(0) = \eta - h(t_0, \xi) \tag{55.27}$$

The variables t and x in the foregoing equation will be slowly varying since, in the τ time scale, they are given by

$$t = t_0 + \epsilon\tau, \quad x = x(t_0 + \epsilon\tau, \epsilon)$$

Setting $\epsilon = 0$ freezes these variables at their initial values and reduces Equation 55.27 to the autonomous system

$$\frac{dy}{d\tau} = g(t_0, \xi, y+h(t_0, \xi), 0), \quad y(0) = \eta(0) - h(t_0, \xi) \quad (55.28)$$

which has equilibrium at $y = 0$. The frozen parameters (t_0, ξ_0) in Equation 55.28 depend on the given initial time and initial state for the problem under consideration. In our investigation of the stability of the origin of Equation 55.28 we should allow the frozen parameters to take any values in the region of the slowly varying parameters (t, x). We rewrite Equation 55.28 as

$$\frac{dy}{d\tau} = g(t, x, y + h(t, x), 0) \quad (55.29)$$

where (t, x) are treated as fixed parameters. We shall refer to Equation 55.29 as the **boundary-layer model** or the boundary-layer system. Sometimes we shall refer also to Equation 55.28 as the boundary-layer model. This should cause no confusion since Equation 55.28 is an evaluation of Equation 55.29 for a given initial time and initial state. The model of Equation 55.29 is more suitable when we study stability properties of the boundary-layer system. The crucial stability property we need for the boundary-layer system is exponential stability of its origin, uniformly in the frozen parameters. The following definition states this property precisely.

DEFINITION 55.5 The equilibrium $y = 0$ of the boundary-layer system of Equation 55.29 is exponentially stable uniformly in (t, x) if there exist positive constants k, γ, and ρ_0 such that the solutions of Equation 55.27 satisfy

$$\|y(\tau)\| \leq k\|y(0)\| \exp(-\gamma\tau), \ \forall \ \|y(0)\| < \rho_0, \ \forall \ \tau \geq 0 \quad (55.30)$$

for all (t, x) in the domain of interest.

We note that if the Jacobian matrix $[\partial g/\partial y]$ satisfies the eigenvalue condition

$$Re\left[\lambda\left\{\frac{\partial g}{\partial y}(t, x, h(t, x), 0)\right\}\right] \leq -c < 0 \quad (55.31)$$

for all (t, x) in the domain of interest, then there exist constants k, γ, and ρ_0 for which inequality of Equation 55.30 is satisfied.

Under the boundary-layer stability condition, the fundamental result of singular perturbation, known as Tikhonov's theorem, states that *if the reduced problem of Equation 55.23 has a unique solution $\bar{x}(t)$, defined on $[t_0, t_1]$, then for sufficiently small ϵ, the full problem of Equations 55.24 and 55.25 has a unique solution $(x(t, \epsilon), z(t, \epsilon))$ defined on $[t_0, t_1]$, and*

$$x(t, \epsilon) - \bar{x}(t) = O(\epsilon) \quad (55.32)$$

$$z(t, \epsilon) - h(t, \bar{x}(t)) - \hat{y}(t/\epsilon) = O(\epsilon) \quad (55.33)$$

hold uniformly for $t \in [t_0, t_1]$, where $\hat{y}(\tau)$ is the solution of the boundary-layer model of Equation 55.28. Moreover, given any $t_b > t_0$,

$$z(t, \epsilon) - h(t, \bar{x}(t)) = O(\epsilon) \quad (55.34)$$

holds uniformly for $t \in [t_b, t_1]$.

55.2.3 Examples

We give four examples to illustrate the averaging and singular perturbation methods. Example 55.12 illustrates the application of the averaging method to a first-order system. Example 55.13 shows how the averaging method can be used to detect the existence of limit cycles in weakly nonlinear second-order systems. Example 55.14 is a physical example that shows how singularly perturbed models arise in physical situations. Example 55.15 illustrates the application of the singular perturbation method.

EXAMPLE 55.12:

Consider the scalar system

$$\dot{x} = \epsilon(x \sin^2 t - 0.5x^2) = \epsilon f(t, x)$$

The function $f(t, x)$ is periodic in t of period π. The averaged function $f_{av}(x)$ is given by

$$f_{av}(x) = \frac{1}{\pi}\int_0^{\pi} (x \sin^2 t - 0.5x^2)\, dt = 0.5(x - x^2)$$

The averaged system

$$\dot{x} = 0.5\epsilon(x - x^2)$$

has two equilibrium points at $x = 0$ and $x = 1$. The Jacobian df_{av}/dx evaluated at these equilibria is given by

$$\left.\frac{df_{av}}{dx}\right|_{x=0} = (0.5 - x)|_{x=0} = 0.5$$

$$\left.\frac{df_{av}}{dx}\right|_{x=1} = (0.5 - x)|_{x=1} = -0.5$$

The averaged system has an exponentially stable equilibrium point at $x = 1$. Thus, for sufficiently small ϵ, the system has an exponentially stable periodic solution of period π in an $O(\epsilon)$ neighborhood of $x = 1$. Moreover, for initial states sufficiently near $x = 1$, solving the averaged system with the same initial state as the original system yields the approximation

$$x(t, \epsilon) - x_{av}(t, \epsilon) = O(\epsilon), \ \forall \, t \geq 0$$

Figure 55.11 shows the solutions of the exact system and the averaged system for the initial state $x(0) = 0.5$ and $\epsilon = 0.2$.

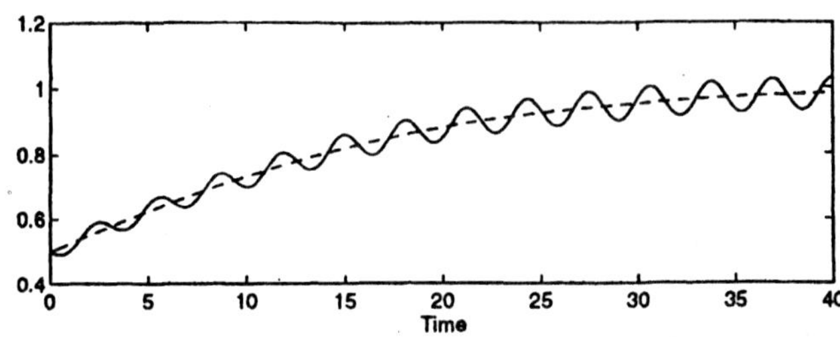

Figure 55.11 The exact (solid) and averaged (dashed) solutions of Example 55.12 with $\epsilon = 0.2$.

The figure illustrates clearly how the solution of the averaged system averages the exact solution.

EXAMPLE 55.13:

Consider the Van der Pol equation

$$\begin{aligned} \dot{x}_1 &= x_2 \\ \dot{x}_2 &= -x_1 + \epsilon x_2(1 - x_1^2) \end{aligned} \tag{55.35}$$

Representing the system in the polar coordinates

$$x_1 = r \sin\phi, \quad x_2 = r\cos\phi$$

it can be shown that

$$\dot{r} = \epsilon r \cos^2\phi \, (1 - r^2 \sin^2\phi) \tag{55.36}$$

$$\dot{\phi} = 1 - \epsilon \sin\phi\cos\phi \, (1 - r^2\sin^2\phi) \tag{55.37}$$

Divide Equation 55.36 by Equation 55.37 to obtain

$$\frac{dr}{d\phi} = \frac{\epsilon r \cos^2\phi \, (1 - r^2\sin^2\phi)}{1 - \epsilon \sin\phi\cos\phi \, (1 - r^2\sin^2\phi)} \stackrel{\text{def}}{=} \epsilon f(\phi, r, \epsilon) \tag{55.38}$$

If we view ϕ as an independent variable, then Equation 55.38 takes the form of Equation 55.12, where $f(\phi, r, \epsilon)$ is periodic in ϕ of period 2π. The averaged system is given by

$$\frac{dr}{d\phi} = \epsilon \left(\tfrac{1}{2} r - \tfrac{1}{8} r^3\right) \tag{55.39}$$

It has three equilibrium points at $r = 0$, $r = 2$, and $r = -2$. Since by definition $r \geq 0$, the negative root is rejected. We check stability of the equilibria via linearization. The Jacobian matrix is given by

$$\frac{df_{av}}{dr} = \tfrac{1}{2} - \tfrac{3}{8} r^2$$

and

$$\left.\frac{df_{av}}{dr}\right|_{r=0} = \tfrac{1}{2} > 0; \quad \left.\frac{df_{av}}{dr}\right|_{r=2} = -1 < 0$$

Thus, the equilibrium point $r = 2$ is exponentially stable. Therefore, for sufficiently small ϵ, Equation 55.38 has an exponentially stable periodic solution $r = R(\phi, \epsilon)$ in an $O(\epsilon)$ neighborhood of $r = 2$. In other words, the Van der Pol equation has a **stable limit cycle** in an $O(\epsilon)$ neighborhood of $r = 2$. Going back to the original system of Equation 55.35, it can be shown that, for sufficiently small ϵ, Equation 55.35 has a periodic solution of periodic $T = 2\pi + O(\epsilon)$.

EXAMPLE 55.14:

An armature-controlled DC motor can be modeled by the second-order state equation

$$\begin{aligned} J\frac{d\omega}{dt} &= ki \\ L\frac{di}{dt} &= -k\omega - Ri + u \end{aligned}$$

where i, u, R, and L are the armature current, voltage, resistance, and inductance, respectively; J is the moment of inertia; ω is the angular speed; and ki and $k\omega$ are, respectively, the torque and the back e.m.f. (electromotive force) developed with constant excitation flux ϕ. The first state equation is a mechanical torque equation, and the second one is an equation for the electric transient in the armature circuit. Typically L is "small" and can play the role of our parameter ϵ. This means that, with $\omega = x$ and $i = z$, the motor's model is in the standard form of Equations 55.19 and 55.20 whenever $R \neq 0$. Neglecting L, we solve

$$0 = -k\omega - Ri + u$$

to obtain

$$i = \frac{u - k\omega}{R}$$

which is the only root, and substitute it in the torque equation. The resulting model

$$J\dot{\omega} = -\frac{k^2}{R}\omega + \frac{k}{R}u$$

is the commonly used first-order model of the DC motor. In formulating perturbation models it is preferable to choose the perturbation parameter ϵ as a dimensionless ratio of two physical parameters. To that end, let us define the dimensionless variables

$$\omega_r = \frac{\omega}{\Omega}; \quad i_r = \frac{iR}{k\Omega}; \quad u_r = \frac{u}{k\Omega}$$

and rewrite the state equation as

$$\begin{aligned} T_m \frac{d\omega_r}{dt} &= i_r \\ T_e \frac{di_r}{dt} &= -\omega_r - i_r + u_r \end{aligned}$$

where $T_m = JR/k^2$ is the mechanical time constant and $T_e = L/R$ is the electrical time constant. Since $T_m \gg T_e$, we let T_m be the time unit; that is, we introduce the dimensionless time variable $t_r = t/T_m$ and rewrite the state equation as

$$\begin{aligned} \frac{d\omega_r}{dt_r} &= i_r \\ \frac{T_e}{T_m}\frac{di_r}{dt_r} &= -\omega_r - i_r + u_r \end{aligned}$$

This scaling has brought the model into the standard form with a physically meaningful dimensionless parameter

$$\epsilon = \frac{T_e}{T_m} = \frac{Lk^2}{JR^2}$$

EXAMPLE 55.15:

Consider the singular perturbation problem

$$\begin{aligned} \dot{x} &= x^2(1+t)/z, & x(0) &= 1 \\ \epsilon\dot{z} &= -[z + (1+t)x]\, z\, [z - (1+t)], & z(0) &= \eta \end{aligned}$$

Setting $\epsilon = 0$ yields the equation

$$0 = -[z + (1+t)x]\, z\, [z - (1+t)]$$

which has three isolated roots

$$z = -(1+t)x; \quad z = 0; \quad z = 1+t$$

in the region $\{t \geq 0 \text{ and } x \geq k > 0\}$. Consider first the root $z = -(1+t)x$. The boundary-layer model of Equation 55.29 is

$$\frac{dy}{d\tau} = -y[y - (1+t)x][y - (1+t)x - (1+t)]$$

By sketching the right-hand side function, it can be shown that the origin is asymptotically stable with $y < (1+t)x$ as its region of attraction. Using the Lyapunov function $V(y) = \frac{1}{2}y^2$, it can be verified that the boundary-layer stability condition is satisfied for $|y(0)| \leq \rho < (1+t)x$. The reduced problem

$$\dot{x} = -x, \quad x(0) = 1$$

has the unique solution $\bar{x}(t) = \exp(-t)$ for all $t \geq 0$. The boundary-layer problem with $t = 0$ and $x = 1$ is

$$\frac{dy}{d\tau} = -y(y-1)(y-2), \quad y(0) = \eta + 1$$

and has a unique exponentially decaying solution $\hat{y}(\tau)$ for $\eta + 1 \leq 1 - a$, that is, for $\eta \leq -a < 0$, where $a > 0$ can be arbitrarily small. Consider next the root $z = 0$. The boundary-layer model of Equation 55.29 is

$$\frac{dy}{d\tau} = -[y + (1+t)x]\, y\, [y - (1+t)]$$

By sketching the right-hand side function, it can be shown that the origin is unstable. Hence, Tikhonov's theorem does not apply to this case. Finally, the boundary-layer model for the root $z = 1+t$ is

$$\frac{dy}{d\tau} = -[y + (1+t) + (1+t)x][y + (1+t)]y$$

Similar to the first case, it can be shown that in this case also the origin is exponentially stable uniformly in (t, x). The reduced problem

$$\dot{x} = x^2, \quad x(0) = 1$$

has the unique solution $\bar{x}(t) = 1/(1-t)$ for all $t \in [0, 1)$. Notice that $\bar{x}(t)$ has a finite escape time at $t = 1$. However, Tikhonov's theorem still holds for $t \in [0, t_1]$ with $t_1 < 1$. The boundary-layer problem, with $t = 0$ and $x = 1$,

$$\frac{dy}{d\tau} = -(y+2)(y+1)y, \quad y(0) = \eta - 1$$

has a unique exponentially decaying solution $\hat{y}(\tau)$ for $\eta_0 > a > 0$. Among the three roots of Equation 55.21, only two roots, $h = -(1+t)x$ and $h = 1+t$, give rise to valid reduced models. Tikhonov's theorem applies to the root $h = -(1+t)x$ if $\eta < 0$ and to the root $h = 1+t$ if $\eta > 0$. Figure 55.12 shows simulation results at $\epsilon = 0.3$. It shows the exact and approximate solutions of x and z for $\eta = -2$. The z solution clearly exhibits a two-time-scale behavior. It starts with a fast transient from η to the reduced solution $\bar{z}(t)$. After the decay of this transient, it remains close to $\bar{z}(t)$. The approximate solution composed as $\bar{z}(t) + \hat{y}(t/\epsilon)$ gives an $O(\epsilon)$ approximation for all t, while the reduced solution $\bar{z}(t)$ gives an $O(\epsilon)$ approximation after the boundary-layer period. As for the slow variable x, the reduced solution $\bar{x}(t)$ gives an $O(\epsilon)$ approximation for all t.

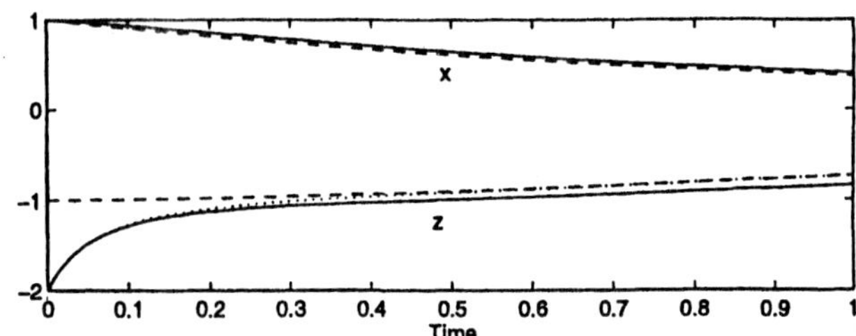

Figure 55.12 The exact (solid), reduced (dashed), and approximate (dotted) solutions of Example 55.14 with $\epsilon = 0.3$.

55.2.4 Defining Terms

Autonomous system: A state equation where the right-hand side function is independent of the time variable.

Averaged system: An autonomous system obtained by averaging the fast periodic right-hand side function of a nonautonomous system.

Boundary-layer model: A reduced-order model that describes the motion of the fast variables of a singularly perturbed system in a fast time scale where the slow variables are treated as constant parameters.

Equilibrium point: A constant solution x^* of $\dot{x} = f(t, x)$. For an autonomous system $\dot{x} = f(x)$, equilibrium points are the real roots of the equation $0 = f(x)$.

Exponentially stable solution: A solution (e.g., equilibrium point or periodic solution) is exponentially stable if other solutions in its neighborhood converge to it faster than an exponentially decaying function.

Hurwitz matrix: A square real matrix is Hurwitz if all its eigenvalues have negative real parts.

Linearization: Approximation of the nonlinear state equation in the vicinity of a nominal solution by a linear state equation, obtained by dropping second- and higher-order terms of the Taylor expansion (about the nominal solution) of the right-hand side function.

Periodic solution (orbit): A periodic solution $x(t)$ of $\dot{x} = f(t, x)$ satisfies the condition $x(t+T) = x(t)$, for all $t \geq 0$ for some positive constant T. The image of a periodic solution in the state space is a closed orbit.

Reduced model: A reduced-order model that describes the motion of the slow variables of a singularly perturbed system. The model is obtained by setting $\epsilon = 0$ and eliminating the fast variables.

Stable limit cycle: An isolated periodic orbit such that all trajectories in its neighborhood asymptotically converge to it.

Standard singularly perturbed model: A singularly perturbed model where upon setting $\epsilon = 0$, the degenerate equation has one or more isolated roots.

Reference

[1] Khalil, H.K., *Nonlinear Systems,* Macmillan, New York, 1992.

Further Reading

Our presentation of the asymptotic methods is based on the textbook by H.K. Khalil (see [1]). For further information on this topic, the reader is referred to Chapters 7 and 8 of Khalil's book. Chapter 7 covers the perturbation method and averaging, and Chapter 8 covers the singular perturbation method. Proofs of the results stated here are given in the book.

For a broader view of the use of singular perturbation methods in systems and control, the reader may consult Kokotovic, P.V., Khalil, H.K., and O'Reilly. 1986. *Singular Perturbations in Systems and Control,* Academic Press, New York. Kokotovic, P.V. and Khalil, H.K., Eds. 1986. *Singular Perturbations in Systems and Control,* IEEE Press. The latter book contains two survey papers that list over 500 references.

For a broader view of the averaging method, the reader may consult Sanders, J.A. and Verhulst, F. 1985. *Averaging Methods in Nonlinear Dynamical Systems,* Springer-Verlag.

A chapter on the use of averaging in adaptive control can be found in Sastry, S. and Bodson, M. 1989. *Adaptive Control,* Prentice Hall, Englewood Cliffs, NJ.

55.3 Volterra and Fliess Series Expansions for Nonlinear Systems

Françoise Lamnabhi-Lagarrigue,
Laboratoire des Signaux et Systèmes CNRS, Supélec, Gif-sur-Yvette, France

55.3.1 Motivation

Some Simple Examples

Consider the linear system

$$\begin{cases} \dot{x}(t) &= Ax(t) + Bu(t),\ x \in \mathbf{R}^n,\ u \in \mathbf{R},\ x(0) = x_0 \\ y(t) &= Cx(t) \end{cases}$$

Its solution may be written in the form

$$y(t) = Ce^{At}x_0 + \int_0^t Ce^{A(t-\tau)}Bu(\tau)d\tau. \tag{55.40}$$

On the other hand, the scalar time-varying linear system

$$\begin{cases} \dot{x}(t) &= a(t)x + b(t)u(t), \quad x, y, u \in \mathbf{R}, \quad x(0) = x_0 \\ y(t) &= c(t)x(t) \end{cases},$$

has a solution

$$\begin{aligned} y(t) &= c(t)e^{(\int_0^t a(\tau)d\tau)}x_0 \\ &\quad + \int_0^t c(t)e^{(\int_\tau^t a(\sigma)d\sigma)}b(\tau)u(\tau)d\tau. \end{aligned} \tag{55.41}$$

Now consider the system

$$\begin{cases} \dot{x}(t) &= (Dx(t) + B)u(t),\ x \in \mathbf{R}^2,\ y, u \in \mathbf{R}, \\ & \quad x(0) = (0,0)^T \\ y(t) &= Cx(t) \end{cases}$$

where

$$D = \begin{bmatrix} 0 & 0 \\ 1 & 0 \end{bmatrix},\ B = \begin{bmatrix} 1 \\ 0 \end{bmatrix},\ \text{and } C = \begin{bmatrix} 0 & 1 \end{bmatrix}.$$

The solution of this system may be written in the form

$$x(t) = C\int_0^t e^{(D\int_\tau^t u(\sigma)d\sigma)}Bu(\tau)d\tau.$$

Since $D^2 = 0$, the series definition of the exponential gives

$$e^{(D\int_\tau^t u(\sigma)d\sigma)} = \begin{bmatrix} 1 & 0 \\ \int_\tau^t u(\sigma)d\sigma & 1 \end{bmatrix}$$

and therefore

$$y(t) = \int_0^t \int_\tau^t u(\sigma)u(\tau)d\sigma d\tau$$

or

$$y(t) = \int_0^t \int_0^t \mu(\sigma - \tau)u(\sigma)u(\tau)d\sigma d\tau \tag{55.42}$$

where μ is the step function $\mu(t) = \begin{cases} 1 & t \geq 0 \\ 0 & t < 0 \end{cases}$.

Before introducing various types of expansions for the response of nonlinear control systems, let us summarize some classical results for the solution of linear differential equations.

Linear Differential Equations

Let us consider the linear time-varying differential equation

$$\dot{x}(t) = \sum_{i=1}^{m} \alpha_i(t) A_i x(t), \; x \in \mathbf{R}^n, \; x(0) = x_0 \tag{55.43}$$

where for $i = 1, \ldots, m$, $\alpha_i : \mathbf{R} \to \mathbf{R}$ are locally Lebesgue integrable functions and A_i are constant $n \times n$ matrices. We may also write

$$x(t) = x_0 + \sum_{i=1}^{m} \int_0^t \alpha_i(\sigma) A_i x(\sigma) d\sigma$$

From the classical Peano–Baker scheme, there exists a series solution of Equation 55.43 of the form [1]

$$\begin{aligned} x(t) &= x_0 + \textstyle\sum_{i=1}^{m} \left(\int_0^t \alpha_i(\sigma_1) d\sigma_1 \right) A_i x_0 \\ &+ \textstyle\sum_{i,j=1}^{m} \left(\int_0^t \int_0^{\sigma_1} \alpha_i(\sigma_1) \alpha_j(\sigma_2) d\sigma_1 d\sigma_2 \right) A_i A_j x_0 + \ldots \\ &+ \textstyle\sum_{i_1,\ldots,i_k=1}^{m} \left(\int_0^t \int_0^{\sigma_1} \cdots \int_0^{\sigma_{k-1}} \alpha_{i_1}(\sigma_1) \ldots \right. \\ &\quad \left. \alpha_{i_k}(\sigma_k) d\sigma_1 \ldots d\sigma_k \right) A_{i_1} \ldots A_{i_k} x_0 \\ &+ \ldots . \end{aligned} \tag{55.44}$$

This series expansion was used in quantum electrodynamics [11].

Under certain unspecified conditions of convergence, the solution of Equation 55.43 may also be written [37]

$$x(t) = e^{\Omega(t)} x_0 \tag{55.45}$$

with

$$\begin{aligned} \Omega(t) &= \textstyle\sum_{i=1}^{m} \left(\int_0^t \alpha_i(\sigma_1) d\sigma_1 \right) A_i + \frac{1}{2} \sum_{i,j=1}^{m} \\ &\quad \left(\int_0^t \int_0^{\sigma_1} \alpha_i(\sigma_1) \alpha_j(\sigma_2) d\sigma_1 d\sigma_2 \right) [A_i, A_j] \\ &+ \tfrac{1}{4} \textstyle\sum_{i,j,k=1}^{m} \left(\int_0^t \int_0^{\sigma_1} \int_0^{\sigma_2} \alpha_i(\sigma_1) \alpha_j(\sigma_2) \alpha_k(\sigma_3) \right. \\ &\quad \left. d\sigma_1 d\sigma_2 d\sigma_3 \right) \\ &\quad [A_i, [A_j, A_k]] \\ &+ \tfrac{1}{12} \textstyle\sum_{i,j,k=1}^{m} \left(\int_0^t \int_0^{\sigma_1} \int_0^{\sigma_2} \alpha_i(\sigma_1) \alpha_j(\sigma_2) \alpha_k(\sigma_3) \right. \\ &\quad \left. d\sigma_1 d\sigma_2 d\sigma_3 \right) \\ &\quad [[A_i, A_j], A_k] \\ &+ \ldots, \end{aligned} \tag{55.46}$$

where the commutator product or *Lie-product* is defined by

$$[A_i, A_j] = A_i A_j - A_i A_j.$$

Indeed, for instance, the first terms of the expansion 55.45 are given by

$$\begin{aligned} &\textstyle\sum_{i=1}^{m} \left(\int_0^t \alpha_i(\sigma_1) d\sigma_1 \right) A_i + \frac{1}{2} \sum_{i,j=1}^{m} \\ &\left(\int_0^t \int_0^{\sigma_1} \alpha_i(\sigma_1) \alpha_j(\sigma_2) d\sigma_1 d\sigma_2 \right) [A_i, A_j] \\ &+ \tfrac{1}{2!} \left(\textstyle\sum_{i=1}^{m} \left(\int_0^t \alpha_i(\sigma_1) d\sigma_1 \right) A_i \right)^2 \end{aligned}$$

Using an integration by parts, this leads to

$$\sum_{i=1}^{m} \left(\int_0^t \alpha_i(\sigma_1) d\sigma_1 \right) A_i + \sum_{i,j=1}^{m} \left(\int_0^t \int_0^{\sigma_1} \alpha_i(\sigma_1) \alpha_j(\sigma_2) d\sigma_1 d\sigma_2 \right) A_i A_j$$

which corresponds to the first two terms of the expansion 55.44.

55.3.2 Functional Expansions for Nonlinear Control Systems

The general problem we consider here is how to generalize Equations 55.40, 55.41, 55.42, 55.44 or 55.45 to nonlinear control systems of the form

$$\frac{dx}{dt} = f_0(x(t)) + \sum_{i=1}^{m} u_i(t) f_i(x(t)) \tag{55.47}$$

where $f_0, f_1, \ldots, f_m$ are C^∞ vector fields on a n-dimensional manifold M, x takes values in M and $u_i : \mathbf{R}^+ \to \mathbf{R}$, $i = 1, \ldots, m$ are piecewise continuous. In a local coordinate chart, $x = (x_1, \ldots, x_n)^T$, Equation 55.47 can be written

$$\dot{x}^k(t) = f_0^k(x(t)) + \sum_{i=1}^{m} u_i(t) f_i^k(x(t)), \quad 1 \le k \le n \tag{55.48}$$

where the functions $f_i^k : \mathbf{R}^n \to \mathbf{R}$ are C^∞. Let $h \in C^\infty(M)$. Then

$$\frac{d}{dt} h(x) = dh(x)\dot{x} = L_{f_0} h(x(t)) + \sum_{i=1}^{m} u_i(t) L_{f_i} h(x(t)),$$

where $L_{f_i} h$ is the *Lie derivative of* h along the vector field f_i, $i = 0, \ldots n$,

$$L_{f_i} h = \sum_{j=1}^{n} f_i^j(x) \frac{\partial}{\partial x_j} h(x).$$

Thus,

$$h(x(t)) = h(x(0)) + \sum_{i=0}^{m} \int_0^t u_i(\sigma) L_{f_i} h(x(\sigma)) d\sigma$$

where $u_0(t) = 1$, $t \ge 0$. Also,

$$L_{f_i} h(x(t)) = L_{f_i} h(x(0)) + \sum_{j=0}^{m} \int_0^t u_j(\sigma) L_{f_j} L_{f_i} h(x(\sigma)) d\sigma,$$

so that

$$\begin{aligned} h(x(t)) &= h(x(0)) + \textstyle\sum_{i=0}^{m} \left(\int_0^t u_i(\sigma) d\sigma \right) L_{f_i} h(x(0)) \\ &+ \textstyle\sum_{i,j=0}^{m} \int_0^t \int_0^{\sigma_2} u_i(\sigma_2) u_j(\sigma_1) L_{f_j} L_{f_i} h(x(\sigma_1)) \\ &\quad d\sigma_1 d\sigma_2. \end{aligned}$$

Iterating this procedure yields

$$\begin{aligned} h(x(t)) &= h(x(0)) + \textstyle\sum_{i=0}^{m} \\ &\quad \left(\int_0^t u_i(\sigma) d\sigma \right) L_{f_i} h(x(0)) \\ &+ \textstyle\sum_{\nu \ge 2}^{N} \sum_{j_1,\ldots,j_\nu=0}^{m} \\ &\quad \left(\int_0^t \int_0^{\sigma_\nu} \cdots \int_0^{\sigma_2} u_{j_\nu}(\sigma_\nu) \ldots u_{j_2}(\sigma_2) u_{j_1}(\sigma_1) \right. \\ &\quad \left. d\sigma_1 d\sigma_2 \ldots d\sigma_\nu \right) \\ &\quad \times L_{f_{j_\nu}} \cdots L_{f_{j_2}} L_{f_{j_1}} h(x(0)) \\ &+ \textstyle\sum_{j_1,\ldots,j_{N+1}=0}^{m} \\ &\quad \int_0^t \int_0^{\sigma_{N+1}} \cdots \int_0^{\sigma_2} u_{j_{N+1}}(\sigma_{N+1}) \ldots u_{j_2}(\sigma_2) u_{j_1}(\sigma_1) \\ &\quad \times L_{f_{j_{N+1}}} \cdots L_{f_{j_2}} L_{f_{j_1}} h(x(\sigma_1))(\sigma_1) d\sigma_1 \\ &\quad d\sigma_2 \ldots d\sigma_{N+1} \end{aligned} \tag{55.49}$$

It is not difficult to see that the remainder [45]

$$R_N = \sum_{j_1,\ldots,j_{N+1}=0}^{m} \int_0^t \int_0^{\sigma_{N+1}} \cdots \int_0^{\sigma_2} u_{j_{N+1}}(\sigma_{N+1}) \ldots u_{j_2}(\sigma_2) u_{j_1}(\sigma_1) \times L_{f_{j_\nu}} \ldots L_{f_{j_2}} L_{f_{j_1}} h(x(\sigma_1))(\sigma_1) d\sigma_1 d\sigma_2 \ldots d\sigma_{N+1}$$

is such that

$$\|R_N\| \leq \frac{A_t^{N+1} t^{N+1}}{(N+1)!} C_N,$$

where C_N is such that

$$|L_{f_{j_{N+1}}} \ldots L_{f_{j_2}} L_{f_{j_1}} h(x)| \leq C_N$$

on some compact set and

$$A_t = \sup(1, \max_{0 \leq \tau \leq t, 1 \leq i \leq n} |u_i(\tau)|).$$

If the vector fields f_i are analytic, and the function h is also analytic, then the previous result can be strengthened. One can actually prove [12], [45] that the series

$$h(x(0)) + \sum_{i=0}^{m} \left(\int_0^t u_i(\sigma) d\sigma \right) L_{f_i} h(x(0)) + \sum_{\nu \geq 2}^{N} \sum_{j_1,\ldots,j_\nu=0}^{m} \left(\int_0^t \int_0^{\sigma_\nu} \cdots \int_0^{\sigma_2} u_{j_\nu}(\sigma_\nu) \ldots u_{j_1}(\sigma_1) d\sigma_1 \ldots d\sigma_\nu \right) \times L_{f_{j_\nu}} \ldots L_{f_{j_2}} L_{f_{j_1}} h(x(0)) \tag{55.50}$$

converges to $h(x(t))$. Indeed in this case, there exists a constant $C > 0$ such that

$$|L_{f_{j_\nu}} \ldots L_{f_{j_2}} L_{f_{j_1}} h(x)| \leq (\nu)! C^\nu.$$

for all $\nu \geq 1$.

Volterra Series

In the following, we consider scalar input, scalar output nonlinear systems on $\mathbf{R}^n$ called *linear-analytic*,

$$\begin{cases} \dot{x}(t) &= f(x(t)) + u(t) g(x(t)), \quad x(0) = x_0 \\ y(t) &= h(x(t)) \end{cases} \tag{55.51}$$

We always assume that f, g, and h are analytic functions in x, in some neighborhood of the free response (when $u(t) = 0$, $\forall t \geq 0$). Analyticity is important but the restriction to scalar inputs and outputs can be easily removed. We say that a linear–analytic system admits a Volterra series representation if there exist locally bounded, piecewise continuous functions

$$w_n : \mathbf{R}^n \to \mathbf{R}, \quad n = 0, 1, 2, \ldots,$$

such that for each $T > 0$ there exists $\epsilon(T) > 0$ with the property that for all piecewise continuous functions $u(.)$ with $| u(T) | \leq \epsilon$ on $[0, T]$ we have

$$y(t) = w_0(t) + \sum_{n=1}^{\infty} \int_0^t \cdots \int_0^t w_n(t, \sigma_1, \ldots, \sigma_n) u(\sigma_1) \ldots u(\sigma_n) d\sigma_1 \ldots d\sigma_n \tag{55.52}$$

with the series converging absolutely and uniformly on $[0, T]$.

Bilinear Systems

It is not difficult to show that the following class of nonlinear systems called *bilinear systems*

$$\begin{cases} \dot{x}(t) &= [A(t) + u(t) B(t)] x(t), \ x(0) = x_0 \\ y(t) &= c(t) x(t) \end{cases} \tag{55.53}$$

admits a Volterra series representation.

Let Φ_A denote the transition matrix for $A(t)$. Make the change of variable $z(t) = \Phi_A(0, t) x(t)$ in order to eliminate A. This gives

$$\begin{cases} \dot{z}(t) &= u(t) \tilde{B}(t) z(t), \\ y(t) &= \tilde{c}(t) z(t) \end{cases}$$

where $\tilde{B}(t) = \Phi_A(0, t) B(t) \Phi_A(t, 0)$ and $\tilde{c}(t) = c(t) \Phi_A(0, t)$. Applying the Peano-Baker formula (Equation 55.44) construction, we have

$$z(t) = \left(I + \int_0^t u(\sigma_1) \tilde{B}(\sigma_1) d\sigma_1 + \int_0^t \int_0^{\sigma_2} u(\sigma_2) \tilde{B}(\sigma_2) u(\sigma_1) \tilde{B}(\sigma_1) d\sigma_1 d\sigma_2 + \ldots \right) z(0)$$

Thus, the Volterra kernels for $y(t)$ are given in triangular form by

$$w_n(t, \sigma_1, \sigma_2, \ldots, \sigma_n) = \begin{cases} c(t) \Phi_A(t, \sigma_n) B(\sigma_n) \Phi_A(\sigma_n, \sigma_{n-1}) B(\sigma_{n-1}) \ldots \ldots B(\sigma_1) \Phi_A(\sigma_1, 0) x(0) \\ 0 \quad \text{if} \quad \sigma_{i+p} < \sigma_p, \quad i, p = 1, 2, 3, \ldots \end{cases}$$

For A, B, and u bounded this series converges uniformly on any compact interval.

The existence and the computation of the Volterra series for more general nonlinear systems is less straightforward. Several authors [25], [36], [6], [15] gave the main results at about the same time.

For the existence of the Volterra series, let us recall for instance Brockett's result: Suppose that

$$f(.,.) : \mathbf{R} \times \mathbf{R}^n \to \mathbf{R}^n \text{ and } g(.,.) : \mathbf{R} \times \mathbf{R}^n \to \mathbf{R}^n$$

are continuous with respect to their first argument and analytic with respect to their second. Given any interval $[0, T]$ such that the solution of

$$\dot{x}(t) = f(t, x(t)), \quad x(0) = 0,$$

exists on $[0, T]$, there exists an $\epsilon > 0$ and a Volterra series for

$$\dot{x}(t) = f(t, x(t)) + u(t) g(t, x(t)), \ x(0) = 0, \tag{55.54}$$

with the Volterra series converging uniformly on $[0, T]$ to the solution of Equation 55.54 provided $|u(t)| < \epsilon$.

Although the computation of the Volterra kernels is given in the previous referenced papers, their expressions may also be obtained from the Fliess algebraic framework [13] summarized in the next section.

Fliess Series

Let us recall some definitions and results from the Fliess algebraic approach [12]. Let $u_1(t), u_2(t), \ldots, u_m(t)$ be some piecewise continuous inputs and $Z = \{z_0, z_1, \ldots, z_m\}$ be a finite set called the alphabet. We denote by Z^* the set of words generated by Z. The algebraic approach introduced by Fliess may be sketched as follows. Let us consider the letter z_0 as an operator which codes the integration with respect to time and the letter $z_i, i = 1, \ldots, m$, as an operator which codes the integration with respect to time after multiplying by the input $u_i(t)$. In this way, any word $w \in Z^*$ gives rise to an iterated integral, denoted by $\mathbf{I}^t\{w\}$, which can be defined recursively as follows:

$$\begin{aligned} \mathbf{I}^t\{\emptyset\} &= 1 \\ \mathbf{I}^t\{w\} &= \begin{cases} \int_0^t d\tau \mathbf{I}^\tau\{v\} & \text{if } w = z_0 v \\ \int_0^t u_i(\tau) d\tau \mathbf{I}^\tau\{v\} & \text{if } w = z_1 v \end{cases}, \quad v \in Z^*. \end{aligned} \tag{55.55}$$

Using the previous formalism and an iterative scheme like Equation 55.49, the solution $y(t)$ of the nonlinear control system

$$\begin{cases} \dot{x}(t) &= f(x(t)) + \sum_{i=1}^m u_i(t) g_i(x(t)), \ x(0) = x_0 \\ y(t) &= h(x(t)) \end{cases} \tag{55.56}$$

may be written [12]

$$\begin{aligned} y(t) \quad = \quad & h(x_0) + \sum_{\nu \geq 0} \sum_{j_0, j_1, \ldots, j_\nu = 0}^{m} L_{f_{j_\nu}} \cdots \\ & \ldots L_{f_{j_2}} L_{f_{j_1}} h(x_0) \mathbf{I}^t\{z_{j_0} z_{j_1} \ldots z_{j_\nu}\} \end{aligned} \tag{55.57}$$

with the series converging uniformly for small t and small $|u_i(\tau)|$, $0 \leq \tau \leq t$, $1 \leq i \leq m$. This functional expansion is called the *Fliess fundamental formula* or *Fliess expansion* of the solution. To this expansion can be also associated [12] an absoluting converging power series for small t and small $|u_i(\tau)|$, $0 \leq \tau \leq t$, $1 \leq i \leq m$, called the *Fliess generating power series* or *Fliess series* denoted by $\mathbf{g}$ of the following form

$$\begin{aligned} \mathbf{g} \quad = \quad & h(x_0) + \sum_{\nu \geq 0} \sum_{j_0, j_1, \ldots, j_\nu = 0}^{m} L_{f_{j_\nu}} \cdots \\ & \ldots L_{f_{j_2}} L_{f_{j_1}} h(x_0) z_{j_0} z_{j_1} \ldots z_{j_\nu}. \end{aligned} \tag{55.58}$$

This algebraic setting allows us to generalize to the nonlinear domain the Heaviside calculus for linear systems. This will appear clearly in the next section devoted to the effective computation of the Volterra series.

A lot of work uses this formalism, see for instance the work on bilinear realizability [44], some analytic aspects and local realizability of generating power series, on realization and input–output relations [49], or works establishing links with other algebras [10], [17].

Links Between Volterra and Fliess Series

The following result [6], [36], [31], [13] gives the expression of the Volterra kernels of the response of the nonlinear control system (55.51) in terms of the vector fields and the output function defining the system,

$$\begin{aligned} y(t) = \\ w_0(t) \quad + \quad & \sum_{n=1}^{\infty} \int_0^t \int_0^{\tau_2} \cdots \int_0^{\tau_n} w_n(t, \tau_n, \cdots, \tau_1) u(\tau_n) \cdots \\ & u(\tau_1) d\tau_n \cdots d\tau_1, \end{aligned} \tag{55.59}$$

where the kernels are analytic functions of the form

$$\begin{aligned} w_0(t) \quad & = \textstyle\sum_{\nu \geq 0} L_f^\nu h(x_0) \frac{t^\nu}{\nu!} = e^{t L_f} h(x_0), \\ w_1(t, \tau_1) \quad & = \textstyle\sum_{\nu_0, \nu_1 \geq 0} L_f^{\nu_0} L_g L_f^{\nu_1} h(x_0) \frac{(t-\tau_1)^{\nu_1} \tau_1^{\nu_0}}{\nu_1! \nu_0!} \\ & = e^{\tau_1 L_f} L_g e^{(t-\tau_1) L_f} h(x_0), \\ & \vdots \\ w_n(t, \tau_n, & \\ \tau_{n-1}, \cdots, \tau_1) \quad & = \textstyle\sum_{\nu_0, \nu_1, \cdots, \nu_n \geq 0} L_f^{\nu_0} L_g L_f^{\nu_1} \cdots L_g L_f^{\nu_n} h(x_0) \\ & \quad \frac{(t-\tau_n)^{\nu_n} \cdots \tau_1^{\nu_0}}{\nu_n! \cdots \nu_0!} \\ & = e^{\tau_1 L_f} L_g e^{(\tau_2 - \tau_1) L_f} \ldots L_g e^{(t-\tau_n) L_f} h(x_0) \end{aligned} \tag{55.60}$$

In order to show this, let us use the fundamental formula (55.57). The zero order kernel is the free response of the system. Indeed, from Equation 55.57 we have

$$\begin{aligned} w_0(t) \quad = \quad & h(x_0) + \sum_{\nu \geq 0} \sum_{j_0, \cdots, j_\nu = 0} L_{f_{j_\nu}} \cdots \\ & \ldots L_{f_{j_2}} L_{f_{j_1}} h(x_0) \int_0^t d\xi_{j_\nu} d\xi_{j_{\nu-1}} \cdots d\xi_{j_0}, \end{aligned}$$

which can also be written as

$$y(t) = \sum_{l \geq 0} L_f^l h(x_0) \frac{t^l}{l!},$$

or using a formal notation,

$$y(t) = e^{t L_f} h(x_0).$$

This formula is nothing other than the classical formula given in [16].

For the computation of the first order kernel, let us consider the terms of Equation 55.57 which contain only one contribution of the input u; therefore,

$$\begin{aligned} \int_0^t w_1(t, \tau_1) u(\tau_1) d\tau_1 \quad = \quad & \sum_{\nu_0, \nu_1 \geq 0} L_f^{\nu_0} L_g L_f^{\nu_1} h(x_0) \\ & \int_0^t \underbrace{d\xi_0 \cdots d\xi_0}_{\nu_1\text{- times}} d\xi_1 \underbrace{d\xi_0 \cdots d\xi_0}_{\nu_0\text{- times}}. \end{aligned}$$

But the iterated integral inside can be proven to be equal to

$$\int_0^t \frac{(t - \tau_1)^{\nu_1} \tau_1^{\nu_0}}{\nu_1! \nu_0!} u(\tau_1) d\tau_1.$$

So, the first order kernel may be written as

$$w_1(t,\tau_1) = \sum_{\nu_0,\nu_1\geq 0} L_f^{\nu_0} L_g L_f^{\nu_1} h(x_0) \frac{(t-\tau_1)^{\nu_1}\tau_1^{\nu_0}}{\nu_1!\nu_0!} = e^{\tau_1 L_f} L_g e^{(t-\tau_1)L_f} h(x_0).$$

For the computation of the second order kernel, let us regroup the terms of Equation 55.57, which contain exactly two contributions of the input u; therefore,

$$\int_0^t \int_0^{\tau_2} w_2(t,\tau_1,\tau_2)u(\tau_1)u(\tau_2)d\tau_1 d\tau_2 = \sum_{\nu_0,\nu_1,\nu_2\geq 0} L_f^{\nu_0} L_g L_f^{\nu_1} L_g L_f^{\nu_2} h(x_0) \int_0^t \int_0^{\tau_2} \underbrace{d\xi_0\cdots d\xi_0}_{\nu_2\text{- times}} d\xi_1 \underbrace{d\xi_0\cdots d\xi_0}_{\nu_1\text{- times}} d\xi_1 \underbrace{d\xi_0\cdots d\xi_0}_{\nu_0\text{- times}}.$$

The iterated integral inside this expression can be proven to be equal to

$$\int_0^t \int_0^{\tau_2} \frac{(t-\tau_2)^{\nu_2}(\tau_2-\tau_1)^{\nu_1}\tau_1^{\nu_0}}{\nu_2!\nu_1!\nu_0!} u(\tau_1)u(\tau_2)d\tau_1 d\tau_2.$$

Thus, the second order kernel may be written as

$$w_2(t,\tau_1,\tau_2) = \sum_{\nu_0,\nu_1,\nu_2\geq 0} L_f^{\nu_0} L_g L_f^{\nu_1} L_g L_f^{\nu_2} h(x_0) \frac{(t-\tau_2)^{\nu_2}(\tau_2-\tau_1)^{\nu_1}\tau_1^{\nu_0}}{\nu_2!\nu_1!\nu_0!} = e^{\tau_2 L_f} L_g e^{(\tau_1-\tau_2)L_f} L_g e^{(t-\tau_1)L_f} h(x_0).$$

The higher order is obtained in the same way.

Using the Campbell-Baker-Hausdorff formula

$$e^{\sigma L_f} L_g e^{-\sigma L_f} h(x_0) = \sum_{i=1}^{\infty} \frac{\sigma^i}{i!} ad^i_{L_f} L_g,$$

the expressions for the kernels (55.60) may be written,

$$\begin{aligned}
w_0(t) &= e^{tL_f} h(x_0),\\
w_1(t,\tau_1) &= e^{\tau_1 L_f} L_g e^{(t-\tau_1)L_f} h(x_0)\\
&= \textstyle\sum_{i=1}^{\infty} \frac{\tau_1^i}{i!} ad^i_{L_f} L_g e^{tL_f} h(x_0)\\
w_2(t,\tau_n,\tau_{n-1},\cdots,\tau_1) &= e^{\tau_1 L_f} L_g e^{(\tau_2-\tau_1)L_f} L_g e^{(t-\tau_2)L_f} h(x_0)\\
&= \textstyle\sum_{i,j=1}^{\infty} \frac{\tau_1^i}{i!}\frac{\tau_2^j}{j!} ad^i_{L_f} L_g ad^j_{L_f} L_g e^{tL_f} h(x_0)\\
&\vdots
\end{aligned} \tag{55.61}$$

These kernel expressions lead to techniques which may, for example, be used in singular optimal control problems [32]. This will be sketched in the next section.

55.3.3 Effective Computation of Volterra Kernels

Example

Let us consider the system [39],

$$\ddot{y}(t) + (\omega^2 + u(t))y(t) = 0,\ t \geq 0,\ y(0) = 0,\ \dot{y}(0) = 1,$$

After two integrations, we obtain

$$y(t) + \omega^2 \int_0^t \int_0^\tau y(\sigma)d\sigma d\tau + \int_0^t \int_0^\tau u(\sigma)y(\sigma)d\sigma d\tau - t = 0$$

The associated algebraic equation for the Fliess series (see Section 55.3.2) $\mathbf{g}$ is given by

$$(1+\omega^2 z_0^2)\mathbf{g} + z_0 z_1 \mathbf{g} - z_0 = 0.$$

In order to solve this equation, let us use the following iterative scheme

$$\mathbf{g} = \mathbf{g}_0 + \mathbf{g}_1 + \mathbf{g}_2 + \cdots + \mathbf{g}_i + \cdots$$

where $\mathbf{g}_i$ contains all the terms of the solution $\mathbf{g}$ having exactly i occurrences in the variable z_1,

$$\begin{aligned}
\mathbf{g}_0 &= (1+\omega^2 z_0^2)^{-1} z_0,\\
\mathbf{g}_1 &= -(1+\omega^2 z_0^2)^{-1} z_0 z_1 \mathbf{g}_0\\
&= -(1+\omega^2 z_0^2)^{-1} z_0 z_1 (1+\omega^2 z_0^2)^{-1} z_0,\\
\mathbf{g}_2 &= -(1+\omega^2 z_0^2)^{-1} z_0 z_1 \mathbf{g}_1\\
&= (1+\omega^2 z_0^2)^{-1} z_0 z_1 (1+\omega^2 z_0^2)^{-1} z_0 z_1 (1+\omega^2 z_0^2)^{-1} z_0,\\
&\vdots
\end{aligned}$$

Each $\mathbf{g}_i$, $i \geq 0$ is a (rational) generating power series of functionals y_i, $i \geq 0$ which represents the i-th order term of the Volterra series associated with the solution $y(t)$. Let us now compute $y_i(t)$ associated with $\mathbf{g}_i$, $i \geq 0$. First, note that

$$(1-az_0)^{-1} = \sum_{n\geq 0} a^n z_0^n, \qquad a \in \mathbf{C},$$

represents in the algebraic domain, the function e^{-at}. Indeed,

$$\mathbf{I}^t(z_0^n) = \frac{t^n}{n!}.$$

Consider now

$$\mathbf{g}_0 = -\frac{1}{2j\omega}(1+j\omega z_0)^{-1} + \frac{1}{2j\omega}(1-j\omega z_0)^{-1}.$$

Hence,

$$y_0(t) = w_0(t) = -\frac{1}{2j\omega}e^{-j\omega t} + \frac{1}{2j\omega}e^{j\omega t} = \frac{1}{\omega}\sin(wt).$$

The power series

$$\mathbf{g}_1 = -(1+\omega^2 z_0^2)^{-1} z_0 z_1 (1+\omega^2 z_0^2)^{-1} z_0,$$

after decomposing into partial fractions the term on the right and on the left of z_1,

$$[\tfrac{1}{2j\omega}(1+j\omega z_0)^{-1} - \tfrac{1}{2j\omega}(1-j\omega z_0)^{-1}]$$
$$z_1[-\tfrac{1}{2j\omega}(1+j\omega z_0)^{-1} + \tfrac{1}{2j\omega}(1-j\omega z_0)^{-1}]$$

or

$$\begin{aligned} g_1 &= \tfrac{1}{4\omega^2}\left[(1+j\omega z_0)^{-1}z_1(1+j\omega z_0)^{-1}\right. \\ &\quad -(1+j\omega z_0)^{-1}z_1(1-j\omega z_0)^{-1} \\ &\quad -(1-j\omega z_0)^{-1}z_1(1+j\omega z_0)^{-1} \\ &\quad \left.+(1-j\omega z_0)^{-1}z_1(1-j\omega z_0)^{-1}\right] \end{aligned}$$

In order to obtain the equivalent expression in the time domain, we need the following result [31], [27].

The rational power series

$$(1-a_0z_0)^{-p_0}z_1(1-a_1z_0)^{-p_1}z_1\cdots z_1(1-a_lz_0)^{-p_l}, \quad (55.62)$$

where $a_0, a_1, \cdots, a_l \in \mathbf{C}$, $p_0, p_1, \cdots, p_l \in \mathbf{N}$, in the symbolic representation of

$$\int_0^t\int_0^{\tau_l}\cdots\int_0^{\tau_2} f_{a_0}^{p_0}(t-\tau_l)\cdots f_{a_{l-1}}^{p_{l-1}}(\tau_2-\tau_1)f_{a_l}^{p_l}(\tau_1) \, u(\tau_l)\cdots u(\tau_1)d\tau_l\cdots d\tau_1,$$

where $f_a^p(t)$ denotes the exponential polynomial

$$\left(\sum_{j=0}^{p-1}\frac{\binom{j}{p-1}}{j!}a^jt^j\right)e^{at}.$$

For the previous example

$$\begin{aligned} y_1(t) &= \int_0^t \frac{1}{2j\omega}e^{-j\omega(t-\tau)}u(\tau)[\frac{-1}{2j\omega}e^{-j\omega\tau}+\frac{1}{2j\omega}e^{j\omega\tau}]d\tau \\ &- \int_0^t \frac{1}{2j\omega}e^{j\omega(t-\tau)}u(\tau)[\frac{-1}{2j\omega}e^{-j\omega\tau}+\frac{1}{2j\omega}e^{j\omega\tau}]d\tau. \end{aligned}$$

Therefore,

$$y_1(t) = \int_0^t w_1(t,\tau)u(\tau)d\tau,$$

with $w_1(t,\tau) = -\frac{1}{\omega^2}\sin[\omega(t-\tau)]\sin\omega t$.

The higher-order kernels can be computed in the same way after decomposing into partial fraction each rational power series.

Noncommutative Padé-Type Approximants

Assume that the functions $f_i^k : \mathbf{R}^n \to \mathbf{R}$ of Equation 55.48 are C^ω, with

$$\begin{aligned} f_i^k(x_1,\ldots,x^n) &= \textstyle\sum_{j_1,\ldots,j_n\geq 0} a_{j_1,\ldots,j_n}^{k,i}(x_1)^{j_1}\ldots(x_n)^{j_n} \\ h(x_1,\ldots,x^n) &= \textstyle\sum_{j_1,\ldots,j_n\geq 0} h_{j_1,\ldots,j_n}(x_1)^{j_1}\ldots(x_n)^{j_n}. \end{aligned}$$

Let γ denote an equilibrium point of the system (55.48) and let

$$x_{j_1,\ldots,j_n}^{<p>}$$

denote the monomial or new state

$$(x_1)^{j_1}\ldots(x_n)^{j_n}, \qquad j_1+\cdots j_n \leq p.$$

Then the Brockett bilinear system

$$\left\{\begin{aligned} \dot{x}_{j_1,\ldots,j_n}^{<p>} &= \textstyle\sum_{k=1}^n j_k\left(\sum_{i=0}^m u_i \right.\\ &\quad \left.\textstyle\sum_{i_1,\ldots,i_n} a_{j_1,\ldots,j_n}^{k,i} x_{j_1+i_1,\ldots,j_k+i_k-1,\ldots,j_n+i_n}^{<p>}\right), \\ y^{<p>} &= \textstyle\sum_{i_1,\ldots,i_n} h_{j_1,\ldots,j_n} x_{j_1,\ldots,j_n}^{<p>} \end{aligned}\right. \quad (55.63)$$

with initial conditions

$$x_{j_1,\ldots,j_n}^{<p>}(0) = (\gamma_1)^{j_1}\ldots(\gamma_n)^{j_n}$$

where

$$x_{j_1,\ldots,-1,\ldots,j_n}^{<p>} = 0$$

for all $j_1,\ldots,j_n \geq 0$ and

$$x_{j_1,\ldots,j_n}^{<p>} = 0$$

if $j_1+\ldots+j_n > p$, has the same Volterra series up to order p as the Volterra series of the nonlinear system (55.48).

This system may be interpreted in the algebraic context by defining the generating power series $\mathbf{g}_{j_1,\ldots,j_n}^{<p>}$ associated with $x_{j_1,\ldots,j_n}^{<p>}$ and $\mathbf{g}^{<p>}$ associated with $y^{<p>}$,

$$\left\{\begin{aligned} \mathbf{g}_{j_1,\ldots,j_n}^{<p>} &= \textstyle\sum_{k=1}^n j_k\left(\sum_{i=0}^m z_i \right.\\ &\quad \left.\textstyle\sum_{i_1,\ldots,i_n} a_{j_1,\ldots,j_n}^{k,i} \mathbf{g}_{j_1+i_1,\ldots,j_k+i_k-1,\ldots,j_n+i_n}^{<p>}\right), \\ \mathbf{g}^{<p>} &= \textstyle\sum_{i_1,\ldots,i_n} h_{j_1,\ldots,j_n} \mathbf{g}_{j_1,\ldots,j_n}^{<p>} \end{aligned}\right. \quad (55.64)$$

The rational power series $\mathbf{g}^{<p>}$ may be seen as a *noncommutative Padé-type approximant* for the forced differential system (55.48) which generalizes the notion of Padé-type approximant obtained in [4]. Using algebraic computing, these approximants may be derived explicitly [38]. These algebraic tools for the first time enable one to derive the Volterra kernels.

Other techniques have recently been introduced in order to compute approximate solutions to the response of nonlinear control systems. The method in [34] is based on the combinatorial notion of L-species. Links between this combinatorial method and the Fliess algebraic setting have been studied in [33] and [35]. Another approach using automata representations [41] is proposed in [23].

The Volterra series (55.52) terminating with the term involving the pth kernel is called a Volterra series *of length p*. In the following we will summarize some properties of an input-output map having a finite Volterra series. The main question is how to characterize a state space representation (55.51) that admits a finite Volterra series representation [8]. This study lead in particular to the introduction of a large class of approximating systems, having a solvable but not necessarily nilpotent Lie algebra [9] (see also [24]).

55.3.4 Approximation Abilities of Volterra Series

Analysis of Responses of Systems

In this part we show how to compute the response of nonlinear systems to typical inputs. We assume here $m = 1$. This method, based on the use of the formal representation of the Volterra kernels (55.64), is also easily implementable on a computer using formal languages [38]. These algebraic tools allow us to derive exponential polynomial expressions depending explicitly on time for the truncated Volterra series associated with the response [31], [27] and therefore lead to a finer analysis than pure numerical results.

To continue our use of algebraic tools, let us introduce the Laplace–Borel transform associated with a given analytic function input

$$u(t) = \sum_{n \geq 0} a_n \frac{t^n}{n!}.$$

Its Laplace–Borel transform is

$$g_u = \sum_{n \geq 0} a_n z_0^n.$$

For example, the Borel transformation of

$$\cos \omega t = \frac{1}{2} e^{j\omega t} + \frac{1}{2} e^{-j\omega t}.$$

is given by

$$g_u = \frac{1}{2}(1 - j\omega t z_0)^{-1} + \frac{1}{2}(1 + j\omega t z_0)^{-1} = (1 + \omega^2 z_0^2)^{-1}.$$

Before seeing the algebraic computation itself in order to compute the first terms of the response to typical inputs, let us introduce a new operation on formal power series, *the shuffle product.*

Given two formal power series,

$$g_1 = \sum_{w \in Z^*} (g_1, w) w \quad \text{and} \quad g_2 = \sum_{w \in Z^*} (g_2, w) w.$$

The *shuffle product* of two formal power series g_1 and g_2 is given by

$$g_1 \diamond g_2 = \sum_{w_1, w_2 \in Z^*} (g_1, w_1)(g_2, w_2) w_1 \diamond w_2,$$

where shuffle product of two words is defined as follows,

- $1 \diamond 1 = 1$
- $\forall z \in Z, \qquad 1 \diamond z = z \diamond 1 = z$
- $\forall z, z' \in Z, \qquad \forall w, w' \in Z^*$
  $zw \diamond z'w' = z[w \diamond z'w'] + z'[zw \diamond w']$.

This operation consists of shuffling all the letters of the two words by keeping the order of the letters in the two words. For instance,

$$z_0 z_1 \diamond z_1 z_0 = 2 z_0 z_1^2 z_0 + z_0 z_1 z_0 z_1 + z_1 z_0 z_1 z_0 + z_1 z_0^2 z_1.$$

It has been shown that the Laplace–Borel transform of Equation 55.62, for a given input $u(t)$ with the Laplace-Borel transform g_u, is obtained by substituting from the right, each variable z_1 by the operator $z_0[g_u \diamond \cdot]$.

Therefore, in order to apply this result, we need to know how to compute shuffle product of algebraic expressions of the form

$$\begin{aligned} g_n &= (1 + a_0 z_0)^{-1} z_{i_1} (1 + a_1 z_0)^{-1} z_{i_2} \cdots \\ &\quad (1 + a_{n-1} z_0)^{-1} z_{i_n} (1 + a_n z_0)^{-1} \end{aligned}$$

where $i_1, i_2, \cdots, i_n \in \{0, 1\}$. This computation is very simple as it amounts to adding some singularities. For instance

$$(1 + a z_0)^{-1} \diamond (1 + b z_0)^{-1} = (1 + (a + b) z_0)^{-1}.$$

Consider two generating power series of the form (55.3.4),

$$\begin{aligned} g_p &= (1 + a_0 z_0)^{-1} z_{i_1} (1 + a_1 z_0)^{-1} z_{i_2} \cdots \\ &\quad (1 + a_{p-1} z_0)^{-1} z_{i_p} (1 + a_p z_0)^{-1} \end{aligned}$$

and

$$\begin{aligned} g_q &= (1 + b_0 z_0)^{-1} z_{j_1} (1 + b_1 z_0)^{-1} z_{j_2} \cdots \\ &\quad (1 + b_{q-1} z_0)^{-1} z_{j_q} (1 + b_q z_0)^{-1} \end{aligned}$$

where p and $q \in \mathbf{N}$, the indices $i_1, i_2, \cdots, i_p \in \{0, 1\}$, $j_1, j_2, \cdots, j_q \in \{0, 1\}$ and $a_i, b_j \in \mathbf{C}$. The shuffle product of these expressions is given by induction on the length

$$\begin{aligned} g_p \diamond g_q &= g_p \diamond g_{q-1} z_{j_q} (1 + (a_p + b_q) z_0)^{-1} + \\ &\quad g_{p-1} \diamond g_q z_{i_p} (1 + (a_p + b_q) z_0)^{-1}. \end{aligned}$$

See [30] for case-study examples and some other rules for computing directly the stationary response to harmonic inputs or the response of a Dirac function and see [14] for the algebraic computation of the response to white noise inputs. This previous effective computation of the rational power series **g** and of the response to typical entries has been applied to the analysis of nonlinear electronics circuits [2] and to the study of laser semiconductors [18].

Optimality

Volterra series expansions have been used in order to study control variations for the output of nonlinear systems combined with some multiple integral identities [10]. This analysis [32], [42], [43], demonstrates links between the classical Hamiltonian formalism and the Lie algebra associated with the nonlinear control problem. To be more precise, let us consider the control system

$$\Sigma \begin{cases} \dot{x}(t) = f(x(t), u(t)), \\ x(0) = x_0 \end{cases}, \qquad (55.65)$$

where $f : \mathbf{R}^n \times \mathbf{R}^m \to \mathbf{R}^n$ is a smooth mapping. Let $h : \mathbf{R}^n \to \mathbf{R}$ be a smooth function, let

$$\gamma(t, x_0) = e^{tL_f} x_0$$

be the free response of the system and let $x(t, x_0, u)$ be the solution relative to the control u, u being an integrable function taking values in some given bounded open set $U \in \mathbf{R}^m$. An example of control problem is the following: find necessary conditions such that

$$h(\gamma(T)) = \min_u h(x(T, x_0, u)). \qquad (55.66)$$

Let $w_0 : [0, T] \times \mathbf{R}^n \to \mathbf{R}$ be defined as follows,

$$w_0(t, x) = h(e^{(T-t)L_f} x).$$

It is easy to see that the map $\lambda : [0, T] \to (\mathbf{R})^*$ given by

$$\lambda(t) = \frac{\partial w_0}{\partial x}(t, \gamma(t))$$

is the solution of the adjoint equation

$$-\dot{\lambda}(t) = \lambda(t)\frac{\partial f(x,0)}{\partial x}(t,\gamma(t))$$

and $\lambda(T) = dh(\gamma(t))$.

A first order necessary condition is provided by the application of the Maximum Principle: If $\gamma(t)$ satisfies (55.66) for $t \in [0,T]$, then

$$ad_{f_0} f_i w_0(t,\gamma(t)) = 0 \quad \text{for} \quad t \in [0,T] \tag{55.67}$$

and the matrix

$$\Big(\Big(\; f_{ij} w_0(t,\gamma(t)) \; \Big)\Big) \tag{55.68}$$

is a non-negative matrix for $t \in [0,T]$ where

$$f_0(x) = f(0,x), \quad f_i(x) = \frac{\partial f}{\partial u_i}(x,0),$$

and

$$f_{ij}(x) = \frac{\partial^2 f}{\partial u_i \partial u_j}(x,0), i, j = 1,\dots,m.$$

The reference trajectory γ is said *extremal* if it satisfies Equation 55.67 and is said *singular* if it is extremal and if all the terms in the matrix (55.68) vanish. If γ is singular and it satisfies Equation 55.66, then it can be shown for instance (see [32]) that if there exists $s \geq 1$ such that for $t \in [0,T]$ and $i, j = 1,\dots,m$,

$$[ad_{f_0}^{k+1} f_i, ad_{f_0}^{k} f_j] w_0(t,\gamma(t)) = 0 \quad \text{for} \quad k = 0, 1, s-1,$$

then

$$[ad_{f_0}^{k_1} f_i, ad_{f_0}^{k_2} f_j] \quad w_0 \quad (t,\gamma(t)) = 0 \quad \text{for} \quad k_1, k_2 \geq 0 \quad \text{with} \quad k_1 + k_2 = 0,\dots,2s$$

and the matrix

$$\Big(\Big([ad_{f_0}^{s+1} f_i, ad_{f_0}^{s} f_j] w_0(t,\gamma(t))\Big)\Big)$$

is a symmetric non-negative matrix for $t \in [0,T]$.

As a dual problem, sufficient conditions for local controllability have been derived using Volterra series expansions (see for instance [3]).

Search for Limit Cycles and Bifurcation Analysis

The Hopf bifurcation theorem deals with the appearance and growth of a limit cycle as a parameter is varied in a nonlinear system. Several authors have given a rigorous proof using various mathematical tools, series expansions, central manifold theorem, harmonic balance, Floquet theory, or Lyapunov methods. Using Volterra series [47] did provide a conceptual simplification of the Hopf proof. In many problems, the calculations involved are simplified leading to practical advantages as well as theoretical ones.

55.3.5 Other Approximations: Application to Motion Planning

Let us consider a control system

$$\dot{x} = \sum_{i=1}^{m} u_i(t) f_i(x). \tag{55.69}$$

The dynamical exact motion planning problem is the following: given two state vectors p and q, find an input function $u(t) = (u_1(t), u_2(t), \dots, u_m(t))$ that drives exactly the state vector from p to q. In order to solve this problem, several expansions for the solutions which are intimately linked with the Fliess series are used.

When the vector fields f_i, $i = 1,\dots,m$ are real analytic and complete and such that the generated control Lie algebra is everywhere of full rank and nilpotent, then in [46], [26], the authors described a complete solution of the previous control problem using P. Hall basis.

Let $A(z_1, z_2, \dots, z_m)$ denote the algebra of noncommutative polynomials in $(z_1, z_2, \dots, z_m)$ and let $L(z_1, z_2, \dots, z_m)$ denote the Lie subalgebra of $A(z_1, z_2, \dots, z_m)$ generated by $z_1, z_2, \dots, z_m$ with the Lie bracket defined by $[P,Q] = PQ - QP$. The elements of $L(z_1, z_2, \dots, z_m)$ are known as *Lie polynomials* in $z_1, z_2, \dots, z_m$. Let $\mathcal{F}_m$ be the set of *formal Lie monomials* in $z_1, z_2, \dots, z_m$. A *P. Hall* basis of $L(z_1, z_2, \dots, z_m)$ is a totally ordered subset $(\mathcal{B}, <)$ of $\mathcal{F}_m$ such that,

The z_i belong to $\mathcal{B}$.

If $A, B \in \mathcal{B}$, and degree$(A)<$ degree(B), then $A<B$.

If $P \in \mathcal{F}_m$ and P is not one of the z_i, then $P \in \mathcal{B}$ if and only if $P = [A,B]$ with $A, B \in \mathcal{B}$, $A<B$, and either $B = z_i$ for some i or $B = [C,D]$ with $C, D \in \mathcal{B}$, $C \leq A$.

If $L_k(z_1, z_2, \dots, z_m)$ denote the nilpotent version of $L(z_1, z_2, \dots, z_m)$ obtained by killing all the monomials of degree $k+1$, it is not difficult to see that $\{M \in \hat{\mathcal{B}} : degree(\mathrm{M}) \leq k\}$ is a basis of $L_k(z_1, z_2, \dots, z_m)$, where $\hat{\mathcal{B}}$ is the set of all elements of $L(z_1, z_2, \dots, z_m)$ obtained by actually evaluating the Lie brackets.

Now, let $z_1, \dots, z_m, z_{m+1}, \dots z_r$ a P. Hall basis of $L_k(z_1, z_2, \dots, z_m)$ and let E_f be the evaluation map that assigns to each P the vector field obtained by plugging in the $f_i, i = 1, \dots m$, for the corresponding z_i. We assume that the vector fields $f_{m+1}, \dots, f_r$ are given by $f_j = E_f(z_j)$ for $j = m+1, \dots r$.

An expansion for Equation 55.69, and consequently a solution to the exact motion planning problem, is then obtained [26] from the solution of

$$\dot{S}(t) = S(t)\left(u_1(t)z_1 + u_2(t)z_2 + \dots + u_r(t)z_r\right), \; S(0) = 1,$$

written as a product

$$S(t) = e^{h_r(t)z_r} e^{h_{r-1}(t)z_{r-1}} \dots e^{h_2(t)z_2} e^{h_1(t)z_1}.$$

For example, in the case $k = 3$, $m = 2$, we may choose $z_3 = [z_1, z_2]$, $z_4 = [z_1, [z_1, z_2]]$ and $z_5 = [z_2, [z_1, z_2]]$. For this choice, the functions h_j, $j = 1, \ldots, r$ are computed by solving

$$\begin{aligned} \dot{h}_1 &= u_1 \\ \dot{h}_2 &= u_2 \\ \dot{h}_3 &= h_1 u_2 + u_3 \\ \dot{h}_4 &= \tfrac{1}{2} h_1^2 u_2 + h_1 u_3 + u_4 \dot{h}_5 = h_2 u_3 + h_1 h_2 u_2 \end{aligned}$$

with $h_j(0) = 0, \quad j = 1, \ldots, 5.$

In order to take into account systems with drift, for the system (55.69) where $u_1(t)$ may be identically equal to 1, a different basis, called the Lyndon basis, has been used in [21]. Without going into the details, for the previous case, it is shown that the solution is given by the following exponential product expansion

$$x(t) = e^{\xi_1(t) f_1} e^{\xi_2(t) f_2} e^{\xi_3(t) f_3} e^{\xi_4(t) f_4} e^{\xi_5(t) f_5} . Id(x)|_{x(0)}$$

with here $z_3 = [z_2, z_1]$, $z_4 = [z_2, [z_1, z_1]]$, $z_5 = [z_2, [z_2, z_1]]$, $f_j = E_f(z_j)$ for $j = 3, \ldots, 5$ and

$$\begin{aligned} \xi_1 &= \int_0^t u_1(\tau) d\tau \\ \xi_2 &= \int_0^t u_2(\tau) d\tau \\ \xi_3 &= \int_0^t \xi_2 d\xi_1 \\ \xi_4 &= \int_0^t \xi_3 d\xi_1 \\ \xi_5 &= \tfrac{1}{2} \int_0^t \xi_2^2 d\xi_1 . \end{aligned}$$

References

[1] d'Alessandro, P., Isidori, A., and Ruberti, A., Realizations and structure theory of bilinear dynamical systems, *SIAM J. Control*, 12, 517–535, 1974.

[2] Baccar, S., Lamnabhi-Lagarrigue, F., and Salembier, G., Utilisation du calcul formel pour la modélisation et la simulation des circuits électroniques faiblement non linéaires, *Annales des Télécommunications*, 46, 282–288, 1991.

[3] Bianchini, R.M. and Stefani, G., A high order maximum principle and controllability, 1991.

[4] Brezinski, C., *Padé-type approximants and general orthogonal polynomials*, INSM, 50, Birkhaüser, 1980.

[5] Brockett, R.W., Nonlinear systems and differential geometry, *Proceedings IEEE*, 64, 61–72, 1976.

[6] Brockett, R.W., Volterra series and geometric control theory, *Automatica*, 12, 167–176, 1976; Brockett, R.W. and Gilbert, E.G., An addendum to Volterra series and geometric control theory, *Automatica*, 12, 635, 1976.

[7] Chen, K.T., Integration of paths, geometric invariants and a generalized Baker-Hausdorff formula, *Ann.Math.*, 65, 163–178, 1957.

[8] Crouch, P.E., Dynamical realizations of finite Volterra series, *SIAM J. Control and Optimization*, 19, 177–202, 1981.

[9] Crouch, P.E., Solvable approximations to control systems, *SIAM J. Control and Optimization*, 22(1), 40–54, 1984.

[10] Crouch, P.E. and Lamnabhi-Lagarrigue, F., Algebraic and multiple integral identities, *Acta Applicanda Mathematicae*, 15, 235–274, 1989.

[11] Feynman, R.P., An operator calculus having applications in quantum electrodynamics, *Physical Review*, 84, 108–128, 1951.

[12] Fliess, M., Fonctionnelles causales non linéaires et indéterminées non commutatives, *Bull. Soc. Math. France*, 109, 3–40, 1981.

[13] Fliess, M., Lamnabhi, M., and Lamnabhi-Lagarrigue, F., Algebraic approach to nonlinear functional expansions, *IEEE CS*, 30, 550–570, 1983.

[14] Fliess, M., and Lamnabhi-Lagarrigue, F., Application of a new functional expansion to the cubic anharmonic oscillator, *J. Math. Physics*, 23, 495–502, 1982.

[15] Gilbert, E.G., Functional expansions for the response of nonlinear differential systems, *IEEE AC*, 22, 909-921, 1977.

[16] Gröbner, W., *Die Lie-Reihen und ihre Anwendungen* (2e édition), VEB Deutscher Verlag der Wissenschaften, Berlin, 1967.

[17] Grunenfelder, L., Algebraic aspects of control systems and realizations, *J. Algebra*, 165, 446–464, 1994.

[18] Hassine, L., Toffano, Z., Lamnabhi-Lagarrigue, F. Destrez, A., and Birocheau, C., Volterra functional series expansions for semiconductor lasers under modulation, *IEEE J. Quantum Electron.*, 30, 918–928, 1994; Hassine, L., Toffano, Z., Lamnabhi-Lagarrigue, F., Joindot, I., and Destrez, A., Volterra functional series for noise in semiconductor lasers, *IEEE J. Quantum Electron.*, 30, 2534–2546, 1994.

[19] Hoang Ngoc Minh, V. and Jacob, G., Evaluation transform and its implementation in MACSYMA, in *New Trends in Systems Theory*, G. Conte, A.M. Perdon, B. Wyman, Eds., Progress in Systems and Control Theory, Birkhäuser, Boston, 1991.

[20] Jacob, G., Algebraic methods and computer algebra for nonlinear systems' study, *Proc. IMACS Symposium MCTS*, 1991.

[21] Jacob, G., Motion planning by piecewise constant or polynomial inputs, *Proc. NOLCOS'92*, 1992.

[22] Jacob, G. and Lamnabhi-Lagarrigue, F., *Algebraic computing in control*, Lect. Notes Contr. Inform. Sc., 165, Springer-Verlag, 1991.

[23] Hespel, C. and Jacob, G., Truncated bilinear approximants: Carleman, finite Volterra, Padé-type, geometric and structural automata, in *Algebraic Computing in Control*, Lect. Note Contr. Inform. Sc., G. Jacob and F. Lamnabhi-Lagarrigue, Eds., Springer-Verlag, 165, 1991.

[24] Jakubczyk, B. and Kaskosz, B., Realizability of Volterra series with constant kernels, *Nonlinear Analysis, Theory, Methods and Applications*, 5, 167–183, 1981.

[25] Krener, A.J., Local approximations of control systems, *J. Differential Equations*, 19, 125–133, 1975.

[26] Lafferriere, G. and Sussmann, H.J., Motion planning for controllable systems without drift: a preliminary report, *Report SYCON-90-04*, Rutgers Center for Systems and Control, June 1990.

[27] Lamnabhi, M., A new symbolic response of nonlinear systems, *Systems and Control Letters*, 2, 154–162, 1982.

[28] Lamnabhi, M., Functional analysis of nonlinear circuits: a generating power series approach, *IEE Proceedings*, 133, 375–384, 1986.

[29] Lamnabhi–Lagarrigue, F., *Séries de Volterra et commande optimale singulière*, Thèse d'état, Université Paris–Sud, Orsay, 1985.

[30] Lamnabhi–Lagarrigue, F., *Analyse des systèmes non linéaires*, Ed. Hermes, 1994.

[31] Lamnabhi–Lagarrigue F. and Lamnabhi, M., Détermination algébrique des noyaux de Volterra associés à certains systèmes non linéaires. *Ricerche di Automatica*, 10, 17–26, 1979.

[32] Lamnabhi-Lagarrigue, F. and Stefani, G., Singular optimal control problems: on the necessary conditions for optimality, *SIAM J. Control Optim.*, 28, 823–840, 1990.

[33] Lamnabhi–Lagarrigue, F., Leroux, P., and Viennot, X.G., Combinatorial approximations of Volterra series by bilinear systems, in *Analyse des Systèmes Contrôlés*, B. Bonnard, B. Bride, J.P. Gautier, and I. Kupka, Eds., Progress in Systems and Control Theory, Birkhäuser, 1991, 304–315.

[34] Leroux, P. and Viennot, X.G., A combinatorial approach to nonlinear functional expansions, *Theoritical Computer Science*, 79, 179–183, 1991.

[35] Leroux, P., Martin A., and Viennot, X.G., Computing iterated derivatives along trajectories of nonlinear systems, NOLCOS'92, M. Fliess, Ed., Bordeaux, 1992.

[36] Lesiak, C. and Krener, A.J., The existence and uniqueness of Volterra series for nonlinear systems, *IEEE AC*, 23, 1090–1095, 1978.

[37] Magnus, W., On the exponential solution of differential equations for a linear operator, *Comm. Pure Appl. Math.*, 7, 649–673, 1954.

[38] Martin, A., Calcul d'approximations de la solution d'un système non linéaire utilisant le logiciel SCRATCHPAD, in *Algebraic Computing in Control*, Lect. Note Contr. Inform. Sc., G. Jacob and F. Lamnabhi-Lagarrigue, Eds., Springer-Verlag, 165, 1991.

[39] Rugh, W.J., *Nonlinear System Theory*, The John Hopkins University Press, Baltimore, MD, 1981.

[40] Schetzen, I.W., *The Volterra and Wiener Theories of Nonlinear Systems*, John Wiley & Sons, New York, 1980.

[41] Schutzenberger, M.P., On the definition of a family of automata, *Information and Control*, 4, 245–270, 1961.

[42] Stefani, G., Volterra approximations, *NOLCOS'90*, Nantes, 1990.

[43] Stefani, G. and Zezza, P., A new type of sufficient optimality conditions for a nonlinear constrained optimal control problem, *NOLCOS'92*, Bordeaux, 1992.

[44] Sontag, E.D., Bilinear realizability is equivalent to existence of a singular affine differential input/output equation, *Syst. Control Lett.*, 11, 181–187, 1988.

[45] Sussmann, H.J., Lie brackets and local controllability: a sufficient condition for scalar-input systems, *SIAM J. Control Optimiz.*, 21, 686–713, 1983.

[46] Sussmann, H.J., Local controllability and motion planning for some classes of systems with drift, *Proc. 30th IEEE CDC*, 1110–1113, 1991.

[47] Tang, Y.–S., Mees, A.I., and Chua, L.O., Hopf bifurcation via Volterra series, *IEEE AC*, 28(1), 42–53, 1983.

[48] Volterra, V., *Theory of Functionals* (translated from Spanish), Blackie, London, 1930 (reprinted by Dover, New York, 1959).

[49] Wang, Y. and Sontag, E.D., Generating series and nonlinear systems: Analytic aspects, local realizability, and input–output representations, *Forum Math.*, 4, 299–322, 1992; Algebraic differential equations and rational control systems, *SIAM J. Control Optimiz.*,

[50] Wei, J. and Norman, E., On global representations of the solutions of linear differential equations as a product of exponentials, *Proc. Am. Math. Soc.*, 15, 327–334, 1964.

[51] Wiener, N., *Nonlinear Problems in Random Theory*, John Wiley & Sons, New York, 1958.

56

Stability

Hassan K. Khalil
Michigan State University

A.R. Teel
Department of Electrical Engineering, University of Minnesota

T.T. Georgiou
Department of Electrical Engineering, University of Minnesota

L. Praly
Centre Automatique et Systèmes, École Des Mines de Paris

E. Sontag
Department of Mathematics, Rutgers University

56.1 Lyapunov Stability

Hassan K. Khalil, Michigan State University

56.1.1 Introduction

Stability theory plays a central role in systems theory and engineering. Generally speaking, stability theory deals with the behavior of a system over a long time period. There are several ways to characterize stability. For example, we may characterize stability from an input-output viewpoint by requiring the output of a system to be "well behaved" in some sense whenever the input is well behaved. Alternatively, we may characterize stability by studying the asymptotic behavior of the state of the system near steady-state solutions, like equilibrium points or periodic orbits.

In this chapter we deal with stability of equilibrium points, usually characterized in the sense of Lyapunov. An equilibrium point is stable if all solutions starting at nearby points stay nearby; otherwise it is unstable. It is asymptotically stable if all solutions starting at nearby points not only stay nearby, but also tend to the equilibrium point as time approaches infinity. These notions are made precise later on.

We introduce Lyapunov's method for determining the stability of equilibrium points. Lyapunov laid the foundation of this method over a century ago, but of course the method as we use it today is the result of intensive research efforts by many engineers and applied mathematicians.

There are attractive features of Lyapunov's method which, include a solid theoretical foundation, the ability to conclude stability without knowledge of the solution (no extensive simulation effort), and an analytical framework that makes it possible to study the effect of model perturbations and to design feedback control. Its main drawback is the need to search for an auxiliary function that satisfies certain conditions. The existence of such function cannot necessarily be asserted from a particular given model.

56.1.2 Stability of Equilibrium Points

We consider a nonlinear system represented by the state model

$$\dot{x} = f(x) \tag{56.1}$$

where the components of the n-dimensional vector $f(x)$ are **locally Lipschitz functions** of x, defined for all x in a domain $D \subset R^n$. A function $f(x)$ is locally Lipschitz at a point x_0 if it satisfies the **Lipschitz condition** $\|f(x) - f(y)\| \leq L\|x - y\|$ for all x, y in some neighborhood of x_0, where L is a positive constant and $\|\cdot\|$ denotes the Euclidean norm of an n-dimensional vector; that is, $\|x\| = \sqrt{x_1^2 + x_2^2 + \cdots + x_n^2}$. The Lipschitz condition guarantees that Equation 56.1 has a unique solution that satisfies the initial condition $x(0) = x_0$. Suppose $\bar{x} \in D$ is an **equilibrium point** of Equation 56.1; that is, $f(\bar{x}) = 0$. Whenever the state of the system starts at $\bar{x}$, it will remain at $\bar{x}$ for all future time. Our goal is to characterize and study the stability of $\bar{x}$. For convenience, we take $\bar{x} = 0$. There is no loss of generality in doing this because any equilibrium point $\bar{x}$ can be shifted to

0-8493-8570-9/96/$0.00+$.50

the origin via the change of variables $y = x - \bar{x}$. Therefore, we shall always assume that $f(0) = 0$, and study stability of the origin $x = 0$.

The equilibrium point $x = 0$ of Equation 56.1 is **stable** if, for each $\epsilon > 0$, there is $\delta = \delta(\epsilon) > 0$ such that $\|x(0)\| < \delta$ implies that $\|x(t)\| < \epsilon$ for all $t \geq 0$. It is said to be **asymptotically stable** if it is stable and δ can be chosen such that $\|x(0)\| < \delta$ implies that $x(t)$ approaches the origin as t tends to infinity. When the origin is asymptotically stable, the **region of attraction** (also called region of asymptotic stability, domain of attraction, or basin) is defined as the set of all points x such that the solution of Equation 56.1 that starts from x at time $t = 0$ approaches the origin as t tends to ∞. When the region of attraction is the whole space R^n, we say that the origin is **globally asymptotically stable.**

The ϵ-δ requirement for stability takes a challenge-answer form. To demonstrate that the origin is stable, then, for any value of ϵ that a challenger may care to designate, we must produce a value of δ, possibly dependent on ϵ, such that a trajectory starting in a δ neighborhood of the origin will never leave the ϵ neighborhood.

Lyapunov Method

In 1892, Lyapunov introduced a method to determine the stability properties of an equilibrium point without solving the state equation. Let $V(x)$ be a continuously differentiable scalar function defined in a domain $D \subset R^n$ that contains the origin. A function $V(x)$ is said to be **positive definite** if $V(0) = 0$ and $V(x) > 0$ for $x \neq 0$. It is said to be **positive semidefinite** if $V(x) \geq 0$ for all x. A function $V(x)$ is said to be **negative definite** or **negative semidefinite** if $-V(x)$ is positive definite or positive semidefinite, respectively. The derivative of V along the trajectories of Equation 56.1 is given by $\dot{V}(x) = \sum_{i=1}^{n} \frac{\partial V}{\partial x_i}\dot{x}_i = \frac{\partial V}{\partial x} f(x)$, where $\frac{\partial V}{\partial x}$ is a row vector whose ith component is $\frac{\partial V}{\partial x_i}$.

Lyapunov's stability theorem states that *the origin is stable if there is a continuously differentiable positive definite function $V(x)$ so that $\dot{V}(x)$ is negative semidefinite, and it is asymptotically stable if $\dot{V}(x)$ is negative definite.* A function $V(x)$ satisfying the conditions for stability is called a **Lyapunov function**. The surface $V(x) = c$, for some $c > 0$, is called a **Lyapunov surface** or a level surface. Using Lyapunov surfaces, Figure 56.1 makes the theorem intuitively clear.

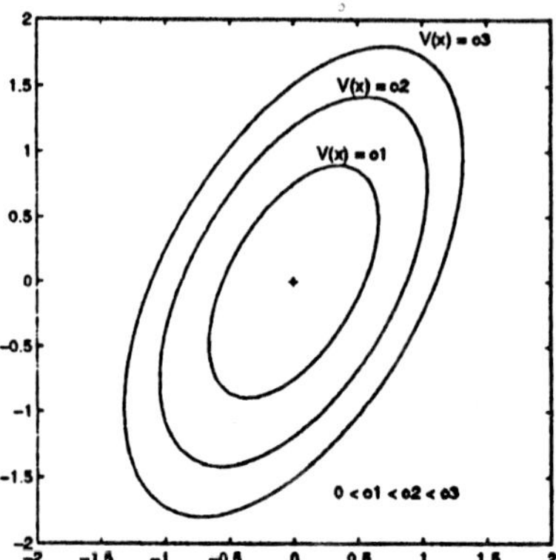

Figure 56.1 Level surfaces of a Lyapunov function.

It shows Lyapunov surfaces for decreasing constants $c3 > c2 > c1 > 0$. The condition $\dot{V} \leq 0$ implies that when a trajectory crosses a Lyapunov surface $V(x) = c$, it moves inside the set $\Omega_c = \{V(x) \leq c\}$ and can never come out again, since $\dot{V} \leq 0$ on the boundary $V(x) = c$. When $\dot{V} < 0$, the trajectory moves from one Lyapunov surface to an inner Lyapunov surface with a smaller c. As c decreases, the Lyapunov surface $V(x) = c$ shrinks to the origin, showing that the trajectory approaches the origin as time progresses. If we know only that $\dot{V} \leq 0$, we cannot be sure that the trajectory will approach the origin, but we can conclude that the origin is stable since the trajectory can be contained inside any ϵ neighborhood of the origin by requiring the initial state $x(0)$ to lie inside a Lyapunov surface contained in that neighborhood.

When $\dot{V}(x)$ is only negative semidefinite, we may still be able to conclude asymptotic stability of the origin if we can show that no solution can stay forever in the set $\{\dot{V}(x) = 0\}$, other than the trivial solution $x(t) \equiv 0$. Under this condition, $V(x(t))$ must decrease toward 0, and consequently $x(t) \to 0$, as $t \to \infty$. This extension of the basic theorem is usually referred to as *the invariance principle.*

Lyapunov functions can be used to estimate the region of attraction of an asymptotically stable origin, that is, to find sets contained in the region of attraction. If there is a Lyapunov function that satisfies the conditions of asymptotic stability over a domain D, and if $\Omega_c = \{V(x) \leq c\}$ is bounded and contained in D, then every trajectory starting in Ω_c remains in Ω_c and approaches the origin as $t \to \infty$. Thus, Ω_c is an estimate of the region of attraction. If the Lyapunov function $V(x)$ is radially unbounded (that is, $\|x\| \to \infty$ implies that $V(x) \to \infty$), then any point $x \in R^n$ can be included in the interior of a bounded set Ω_c. Therefore, *the origin is globally asymptotically stable if there is a continuously differentiable, radially unbounded function $V(x)$ such that for all $x \in R^n$, $V(x)$ is positive definite and $\dot{V}(x)$ is either negative definite or negative semidefinite, but no solution can stay forever in the set $\{\dot{V}(x) = 0\}$ other than the trivial solution $x(t) \equiv 0$.*

Lyapunov's method is a very powerful tool for studying the stability of equilibrium points. However, there are two drawbacks of the method that we should be aware of. First, there is no systematic method for finding a Lyapunov function for a given system. Second, the conditions of the method are only sufficient; they are not necessary. Failure of a Lyapunov function candidate to satisfy the conditions for stability or asymptotic stability does not mean that the origin is not stable or asymptotically stable.

Linear Systems

The linear time-invariant system

$$\dot{x} = Ax \tag{56.2}$$

has an equilibrium point at the origin. The equilibrium point is isolated if and only if $\det(A) \neq 0$. Stability properties of the origin can be characterized by the locations of the eigenvalues of the matrix A. Recall from linear system theory that the solution of Equation 56.2 for a given initial state $x(0)$ is given by $x(t) = \exp(At)x(0)$ and that for any matrix A there is a nonsingular matrix P (possibly complex) that transforms A into its Jordan form; that is,

$$P^{-1}AP = J = \text{block diag}[J_1, J_2, \ldots, J_r]$$

where J_i is the Jordan block associated with the eigenvalue λ_i of A. Therefore,

$$\exp(At) = P\exp(Jt)P^{-1} = \sum_{i=1}^{r}\sum_{k=1}^{m_i} t^{k-1}\exp(\lambda_i t)R_{ik} \tag{56.3}$$

where m_i is the order of the Jordan block associated with the eigenvalue λ_i. If one of the eigenvalues of A is in the open right complex half-plane, the corresponding exponential term $\exp(\lambda_i t)$ in Equation 56.3 will grow unbounded as $t \to \infty$. Therefore, to guarantee stability we must restrict the eigenvalues to be in the closed left complex half-plane. But those eigenvalues on the imaginary axis (if any) could give rise to unbounded terms if the order of the associated Jordan block is higher than one, due to the term t^{k-1} in Equation 56.3. Therefore, we must restrict eigenvalues on the imaginary axis to have Jordan blocks of order one. For asymptotic stability of the origin, $\exp(At)$ must approach 0 as $t \to \infty$. From Equation 56.3, this is the case if and only if $Re\lambda_i < 0$, $\forall\, i$.

When all eigenvalues of A satisfy $Re\lambda_i < 0$, A is called a **Hurwitz matrix** or a stability matrix . The origin of Equation 56.2 is asymptotically stable if and only if A is Hurwitz. Asymptotic stability of the origin can be also investigated using Lyapunov's method. Consider a quadratic Lyapunov function candidate $V(x) = x^T Px$ where P is a real symmetric positive definite matrix. The derivative of V along the trajectories of the linear system of Equation 56.2 is given by

$$\dot{V}(x) = x^T P\dot{x} + \dot{x}^T Px = x^T(PA + A^T P)x = -x^T Qx$$

where Q is a symmetric matrix defined by

$$PA + A^T P = -Q \tag{56.4}$$

If Q is positive definite, we can conclude that the origin is asymptotically stable, that is, $Re\lambda_i < 0$, for all eigenvalues of A. Suppose we start by choosing Q as a real symmetric positive definite matrix, and then solve Equation 56.4 for P. If Equation 56.4 has a positive definite solution, then again we can conclude that the origin is asymptotically stable. Equation 56.4 is called the **Lyapunov equation.** It turns out that *A is Hurwitz if and only if for any given positive definite symmetric matrix Q there exists a positive definite symmetric matrix P that satisfies the Lyapunov Equation 56.4. Moreover, if A is Hurwitz, then P is the unique solution of Equation 56.4.*

Equation 56.4 is a linear algebraic equation that can be solved by rearranging it in the form $Mx = y$ where x and y are defined by stacking the elements of P and Q in vectors. Almost all commercial software programs for control systems include commands for solving the Lyapunov equation.

Linearization

Consider the nonlinear system (Equation 56.4) and suppose that $f(x)$ is continuously differentiable for all $x \in D \subset R^n$. The Jacobian matrix $\frac{\partial f}{\partial x}$ is an $n \times n$ matrix whose (i, j) element is $\frac{\partial f_i}{\partial x_j}$. Let A be the Jacobian matrix evaluated at the origin $x = 0$. By the mean value theorem, it can be shown that $f(x) = Ax + g(x)$, where $\frac{\|g(x)\|}{\|x\|} \to 0$ as $\|x\| \to 0$. Suppose A is Hurwitz and Let P be the solution of the Lyapunov Equation 56.4 for some positive definite Q. Taking $V(x) = x^T Px$, it can be shown the $\dot{V}(x)$ is negative definite in the neighborhood of the origin. Hence, *the origin is asymptotically stable if all the eigenvalues of A have negative real parts.* Using some advanced results of Lyapunov stability, it can be also shown that *the origin is unstable if one (or more) of the eigenvalues of A has a positive real part.* This provides us with a simple procedure for determining stability of the origin of a nonlinear system by calculating the eigenvalues of its linearization about the origin. Note, however, that linearization fails when $Re\lambda_i \le 0$ for all i, with $Re\lambda_i = 0$ for some i.

Nonautonomous Systems

Equation 56.1 is called autonomous because f does not depend on t. The more general system

$$\dot{x} = f(t, x) \tag{56.5}$$

is called nonautonomous. Lyapunov's method applies to nonautonomous systems. In this case, we may allow the Lyapunov function candidate V to depend on t. Let $V(t, x)$ be a continuously differentiable function defined for all $t \ge 0$ and all $x \in D$. The derivative of V along the trajectories of Equation 56.5 is given by $\dot{V}(t, x) = \frac{\partial V}{\partial t} + \frac{\partial V}{\partial x} f(t, x)$. If there are positive definite functions $W_1(x)$, $W_2(x)$ and $W_3(x)$ such that

$$W_1(x) \le V(t, x) \le W_2(x) \tag{56.6}$$

$$\dot{V}(t, x) \le -W_3(x) \tag{56.7}$$

for all $t \ge 0$ and all $x \in D$, then the origin is uniformly asymptotically stable, where "uniformly" indicates that the ϵ–δ definition of stability and the convergence of $x(t)$ to zero are independent of the initial time t_0. Such uniformity annotation is not needed with autonomous systems since the solution of an autonomous state equation starting at time t_0 depends only on the difference $t - t_0$, which is not the case for nonautonomous systems. If inequalities Equations 56.6 and 56.7 hold globally and $W_1(x)$ is radially unbounded, then the origin is globally uniformly asymptotically stable.

56.1.3 Examples and Applications

Examples

EXAMPLE 56.1:

A simple pendulum moving in a vertical plane can be modeled by the second-order differential equation

$$ml\ddot{\theta} = -mg\sin\theta - kl\dot{\theta} \tag{56.8}$$

where l is the length of the rod, m is the mass of the bob, θ is the angle subtended by the rod and the vertical line through the pivot point, g is the acceleration due to gravity, and k is a coefficient of friction. Taking $x_1 = \theta$ and $x_2 = \dot{\theta}$ as the state variables, we obtain the state equation

$$\begin{aligned}\dot{x}_1 &= x_2\\ \dot{x}_2 &= -a\sin x_1 - bx_2\end{aligned} \tag{56.9}$$

where $a = g/l > 0$ and $b = k/m \geq 0$. The case $b = 0$ is an idealized frictionless pendulum. To find the equilibrium points, we set $\dot{x}_1 = \dot{x}_2 = 0$ and solve for x_1 and x_2. The first equation gives $x_2 = 0$ and the second one gives $\sin x_1 = 0$. Thus, the equilibrium points are located at $(n\pi, 0)$, for $n = 0, \pm1, \pm2, \ldots$. From the physical description of the pendulum it is clear that the pendulum has only two equilibrium positions corresponding to the equilibrium points $(0, 0)$ and $(\pi, 0)$. Other equilibrium points are repetitions of these two positions, which correspond to the number of full swings the pendulum would make before it rests at one of the two equilibrium positions. Let us use Lyapunov's method to study the stability of the equilibrium point at the origin. As a Lyapunov function candidate, we use the energy of the pendulum, which is defined as the sum of its potential and kinetic energies, namely,

$$\begin{aligned}V(x) &= \int_0^{x_1} a\sin y\,dy + \tfrac{1}{2}x_2^2\\ &= a(1-\cos x_1) + \tfrac{1}{2}x_2^2\end{aligned} \tag{56.10}$$

The reference of the potential energy is chosen such that $V(0) = 0$. The function $V(x)$ is positive definite over the domain $-2\pi < x_1 < 2\pi$. The derivative of $V(x)$ along the trajectories of the system is given by

$$\dot{V}(x) = a\dot{x}_1\sin x_1 + x_2\dot{x}_2 = -\,bx_2^2 \tag{56.11}$$

When friction is neglected ($b = 0$), $\dot{V}(x) = 0$ and we can conclude that the origin is stable. Moreover, $V(x)$ is constant during the motion of the system. Since $V(x) = c$ forms a closed contour around $x = 0$ for small $c > 0$, we see that the trajectory will be confined to one such contour and will not approach the origin. Hence, the origin is not asymptotically stable. On the other hand, in the case with friction ($b > 0$), $\dot{V}(x) = -bx_2^2 \leq 0$ is negative semidefinite and we can conclude that the origin is stable. Notice that $\dot{V}(x)$ is only negative semidefinite and not negative definite because $\dot{V}(x) = 0$ for $x_2 = 0$ irrespective of the value of x_1. Therefore, we cannot conclude this asymptotic stability using Lyapunov's stability theorem. Here comes the role of the invariance principle. Consider the set $\{\dot{V}(x) = 0\} = \{x_2 = 0\}$. Suppose that a solution of the state equation stays forever in this set. Then

$$x_2(t) \equiv 0 \Rightarrow \dot{x}_2(t) \equiv 0 \Rightarrow \sin x_1 = 0$$

Hence, on the segment $-\pi < x_1 < \pi$ of the $x_2 = 0$ line, the system can maintain the $\dot{V}(x) = 0$ condition only at the origin $x = 0$. Noting that the solution is confined to a set $\{V(x) \leq c\}$ and for sufficiently small c, $\{V(x) \leq c\} \subset \{-\pi < x_1 < \pi\}$, we conclude that no solution can stay forever in the set $\{V(x) \leq c\} \cap \{x_2 = 0\}$ other than the trivial solution $x(t) \equiv 0$. Hence, the origin is asymptotically stable. We can also estimate the region of attraction by the set $\{V(x) \leq c\}$ where $c < \min_{\{x_1 = \pm\pi\}} V(x) = 2a$ is chosen such that $V(x) = c$ is a closed contour contained in the strip $\{-\pi < x_1 < \pi\}$.

The pendulum equation has two equilibrium points at $(0,\ 0)$ and $(\pi,\ 0)$. Let us investigate stability of both points using linearization. The Jacobian matrix is given by

$$\frac{\partial f}{\partial x} = \begin{bmatrix} \frac{\partial f_1}{\partial x_1} & \frac{\partial f_1}{\partial x_2}\\ \frac{\partial f_2}{\partial x_1} & \frac{\partial f_2}{\partial x_2}\end{bmatrix} = \begin{bmatrix} 0 & 1\\ -a\cos x_1 & -b\end{bmatrix}$$

To determine stability of the origin we evaluate the Jacobian matrix at $x = 0$.

$$A = \left.\frac{\partial f}{\partial x}\right|_{x=0} = \begin{bmatrix} 0 & 1\\ -a & -b\end{bmatrix}$$

The eigenvalues of A are $\lambda_{1,2} = -\frac{b}{2} \pm \frac{1}{2}\sqrt{b^2 - 4a}$. For all positive values of a and b, the eigenvalues satisfy $Re\lambda_i < 0$. Hence, the equilibrium point at the origin is asymptotically stable. In the absence of friction ($k = 0$), both eigenvalues are on the imaginary axis. In this case we cannot determine stability of the origin through linearization. We have seen before that in this case the origin is a stable equilibrium point, as determined by an energy Lyapunov function. To determine stability of the equilibrium point at $(\pi,\ 0)$, we evaluate the Jacobian matrix at this point. This is equivalent to performing a change of variables $z_1 = x_1 - \pi$, $z_2 = x_2$ to shift the equilibrium to the origin, and then evaluating the Jacobian $[\partial f/\partial z]$ at $z = 0$.

$$\tilde{A} = \left.\frac{\partial f}{\partial x}\right|_{x_1=\pi, x_2=0} = \begin{bmatrix} 0 & 1\\ a & -b\end{bmatrix}$$

The eigenvalues of $\tilde{A}$ are $\lambda_{1,2} = -\frac{b}{2} \pm \frac{1}{2}\sqrt{b^2 + 4a}$. For all $a > 0$ and $b \geq 0$, there is one eigenvalue in the open right half-plane. Hence, the equilibrium point at $(\pi,\ 0)$ is unstable.

EXAMPLE 56.2:

Consider the system

$$\begin{aligned} \dot{x}_1 &= x_2 \\ \dot{x}_2 &= -g_1(x_1) - g_2(x_2) \end{aligned} \tag{56.12}$$

where $g_1(\cdot)$ and $g_2(\cdot)$ are locally Lipschitz and satisfy

$$g_i(0) = 0, \quad yg_i(y) > 0, \quad \forall\, y \neq 0, \quad i = 1, 2$$

and $\int_0^y g_1(z)\, dz \to \infty$, as $|y| \to \infty$. The system has an isolated equilibrium point at the origin. The equation of this system can be viewed as a generalized pendulum equation with $g_2(x_2)$ as the friction term. Therefore, a Lyapunov function candidate may be taken as the energy-like function $V(x) = \int_0^{x_1} g_1(y)\, dy + \frac{1}{2}x_2^2$, which is positive definite in R^2 and radially unbounded. The derivative of $V(x)$ along the trajectories of the system is given by

$$\dot{V}(x) = g_1(x_1)x_2 + x_2[-g_1(x_1) - g_2(x_2)] = -x_2 g_2(x_2) \leq 0$$

Thus, $\dot{V}(x)$ is negative semidefinite. Note that $\dot{V}(x) = 0$ implies $x_2 g_2(x_2) = 0$, which implies $x_2 = 0$. Therefore, the only solution that can stay forever in the set $\{x \in R^2 \mid x_2 = 0\}$ is the trivial solution $x(t) \equiv 0$. Thus, the origin is globally asymptotically stable.

EXAMPLE 56.3:

The second-order system

$$\begin{aligned} \dot{x}_1 &= -x_2 \\ \dot{x}_2 &= x_1 + (x_1^2 - 1)x_2 \end{aligned}$$

has a unique equilibrium point at the origin. Linearization at the origin yields the matrix

$$A = \left.\frac{\partial f}{\partial x}\right|_{x=0} = \begin{bmatrix} 0 & -1 \\ 1 & -1 \end{bmatrix}$$

which is Hurwitz. Hence, the origin is asymptotically stable. A Lyapunov function for the system can be found by solving the Lyapunov equation $PA + A^T P = -I$ for P. The unique solution is the positive definite matrix

$$P = \begin{bmatrix} 1.5 & -0.5 \\ -0.5 & 1 \end{bmatrix}$$

The quadratic function $V(x) = x^T P x$ is a Lyapunov function for the system in a certain neighborhood of the origin. Let us use this function to estimate the region of attraction. The function $V(x)$ is positive definite for all x. We need to determine a domain D about the origin where $\dot{V}(x)$ is negative definite and a set $\Omega_c \subset D$, which is bounded. We are interested in the largest set Ω_c that we can determine, that is, the largest value for the constant c, because Ω_c will be our estimate of the region of attraction. The derivative of $V(x)$ along the trajectories of the system is given by

$$\dot{V}(x) = -(x_1^2 + x_2^2) - x_1^2 x_2(x_1 - 2x_2)$$

The right-hand side of $\dot{V}(x)$ is written as the sum of two terms. The first term, $-\|x\|^2$, is the contribution of the linear part Ax, while the second term is the contribution of the nonlinear term $g(x) = f(x) - Ax$. Using the inequalities $|x_1 - 2x_2| \leq \sqrt{5}\|x\|$ and $|x_1 x_2| \leq \frac{1}{2}\|x\|^2$, we see that $\dot{V}(x)$ satisfies the inequality $\dot{V}(x) \leq -\|x\|^2 + (\sqrt{5}/2)\|x\|^4$. Hence $\dot{V}(x)$ is negative definite in the region $\{\|x\| < r\}$, where $r^2 = 2/\sqrt{5}$. We would like to choose a positive constant c such that $\{V(x) \leq c\}$ is a subset of this region. Since $x^T P x \geq \lambda_{min}(P)\|x\|^2$, we can choose $c < \lambda_{min}(P) r^2$. Using $\lambda_{min}(P) \geq 0.69$, we choose $c = 0.615 < 0.69(2/\sqrt{5}) = 0.617$. The set $\{V(x) \leq 0.615\}$ is an estimate of the region of attraction.

EXAMPLE 56.4:

Consider the nonautonomous system

$$\begin{aligned} \dot{x}_1 &= -x_1 - g(t)x_2 \\ \dot{x}_2 &= x_1 - x_2 \end{aligned}$$

where $g(t)$ is continuously differentiable and satisfies $0 \leq g(t) \leq k$ and $\dot{g} \leq g(t)$ for all $t \geq 0$. The system has an equilibrium point at the origin. Consider a Lyapunov function candidate $V(t, x) = x_1^2 + [1 + g(t)]x_2$. The function V satisfies the inequalities

$$x_1^2 + x_2^2 \leq V(t, x) \leq x_1^2 + (1 + k)x_2^2$$

The derivative of V along the trajectories of the system is given by

$$\dot{V} = -2x_1^2 + 2x_1 x_2 - [2 + 2g(t) - \dot{g}(t)]x_2^2$$

Using the bound on $\dot{g}(t)$, we have $2 + 2g(t) - \dot{g}(t) \geq 2 + 2g(t) - g(t) \geq 2$. Therefore,

$$\dot{V} \leq -2x_1^2 + 2x_1 x_2 - 2x_2^2 = -x^T \begin{bmatrix} 2 & -1 \\ -1 & 2 \end{bmatrix} x = -x^T Q x$$

The matrix Q is positive definite. Hence, the origin is uniformly asymptotically stable. Since all inequalities are satisfied globally and $x_1^2 + x_2^2$ is radially unbounded, the origin is globally uniformly asymptotically stable.

Feedback Stabilization

Consider the nonlinear system

$$\begin{aligned} \dot{x} &= f(x, u) \\ y &= h(x) \end{aligned}$$

where x is an n-dimensional state, u is an m-dimensional control input, and y is a p-dimensional measured output. Suppose f and h are continuously differentiable in the domain of interest, and $f(0, 0) = 0$ and $h(0) = 0$ so that the origin is an open-loop equilibrium point. Suppose we want to design an output feedback control to stabilize the origin; that is, to make the origin an asymptotically stable equilibrium point of the closed-loop system. We can pursue this design problem via linearization.

The linearization of the system about the point $(x = 0,\ u = 0)$ is given by

$$\begin{aligned} \dot{x} &= Ax + Bu \\ y &= Cx \end{aligned}$$

where

$$A = \left.\frac{\partial f}{\partial x}\right|_{x=0,u=0}, \quad B = \left.\frac{\partial f}{\partial u}\right|_{x=0,u=0}, \quad C = \left.\frac{\partial h}{\partial x}\right|_{x=0}$$

Assuming that (A, B) is stabilizable and (A, C) is detectable (that is, uncontrollable and unobservable eigenvalues, if any, have negative real parts), we can design a dynamic output feedback controller

$$\begin{aligned} \dot{z} &= Fz + Gy \\ u &= Hz + Ky \end{aligned}$$

where z is a q-dimensional vector, such that the closed-loop matrix

$$\mathcal{A} = \begin{bmatrix} A + BKC & BH \\ GC & F \end{bmatrix}$$

is Hurwitz. When the feedback control of Equation 56.13 is applied to the nonlinear system of Equation 56.13 it results in a system of order $(n + q)$, whose linearization at the origin is the matrix $\mathcal{A}$. Hence, the origin of the closed-loop system is asymptotically stable. By solving the Lyapunov equation $\mathcal{P}\mathcal{A} + \mathcal{A}^T\mathcal{P} = -\mathcal{Q}$ for some positive definite matrix $\mathcal{Q}$, we can use $V = \mathcal{X}^T\mathcal{P}\mathcal{X}$, where $\mathcal{X} = [x \ \ z]^T$, as a Lyapunov function for the closed-loop system, and we can estimate the region of attraction of the origin, as illustrated in Example 56.3.

56.1.4 Defining Terms

Asymptotically stable equilibrium: A stable equilibrium point with the additional feature that all trajectories starting at nearby points approach the equilibrium point as time approaches infinity.

Equilibrium point: A constant solution x^* of $\dot{x} = f(t, x)$. For an autonomous system $\dot{x} = f(x)$, equilibrium points are the real roots of the equation $0 = f(x)$.

Globally asymptotically stable equilibrium: A stable equilibrium point with the additional feature that all trajectories approach the equilibrium point as time approaches infinity.

Hurwitz matrix: A square real matrix is Hurwitz if all its eigenvalues have negative real parts.

Linearization: Approximation of the nonlinear state equation in the vicinity of an equilibrium point by a linear state equation, obtained by dropping second- and higher-order terms of the Taylor expansion (about the equilibrium point) of the right-hand side function.

Lipschitz condition: A condition imposed on a function $f(x)$ to ensure that it has a finite slope. For a vector-valued function, it takes the form $\|f(x) - f(y)\| \leq L\|x - y\|$ for some positive constant L.

Locally Lipschitz function: A function $f(x)$ is locally Lipschitz at a point if it satisfies the Lipschitz condition in the neighborhood of that point.

Lyapunov equation: A linear algebraic matrix equation of the form $PA + A^TP = -Q$, where A and Q are real square matrices and Q is symmetric and positive definite. The equation has a (unique) positive definite matrix solution P if and only if A is Hurwitz.

Lyapunov function: A scalar positive definite function of the state whose derivative along the trajectories of the system is negative semidefinite.

Lyapunov surface: A set of the form $V(x) = c$ where $V(x)$ is a Lyapunov function and c is a positive constant.

Negative (semi-) definite function: A scalar function of a vector argument $V(x)$ is negative (semi-) definite if it vanishes at the origin and is negative (nonpositive) for all points in the neighborhood of the origin, excluding the origin.

Negative (semi-) definite matrix: A square symmetric real matrix P is negative (semi-) definite if the quadratic form $V(x) = x^TPx$ is a negative (semi-) definite function.

Positive (semi-) definite function: A scalar function of a vector argument $V(x)$ is positive (semi-) definite if it vanishes at the origin and is positive (nonnegative) for all points in the neighborhood of the origin, excluding the origin.

Positive (semi-) definite matrix: A square symmetric real matrix P is positive (semi-) definite if the quadratic form $V(x) = x^TPx$ is a positive (semi-) definite function.

Region of attraction: The set of all points with the property that the trajectories starting at these points asymptotically approach the origin.

Stable equilibrium: An equilibrium point where all solutions starting at nearby points stay nearby.

Reference

[1] Khalil, H.K., *Nonlinear Systems*, Macmillan, New York, 1992.

Further Reading

Our presentation of Lyapunov stability is based on the textbook by H.K. Khalil (see [1]). For further information on Lyapunov stability, the reader is referred to Chapters 3, 4, and 5 of Khalil's book. Chapters 3 and 4 give a detailed treatment of the material covered in this chapter. They also address important topics, like the use of the center manifold theorem when linearization fails, and the use of Lyapunov's method when there is uncertainty in the state model. Chapter 5 illustrates the application of Lyapunov's method to control problems like the design of robust and adaptive control.

Other engineering textbooks where Lyapunov stability is emphasized include Vidyasagar, M. 1993. *Nonlinear Systems Analysis*, 2nd ed., Slotine, J.-J. and Li, W. 1991. *Applied Nonlinear Control*, Atherton, D.P. 1981. *Stability of Nonlinear Systems*, Hsu, J.C. and Meyer, A.U. 1968. *Modern Control Principles and Application*.

For a deeper look into the theoretical foundation of Lyapunov stability, there are excellent references, which include Rouche, N., Habets, P., and Laloy, M. 1977. *Stability Theory by Lyapunov's Direct Method*, Hahn, W. 1967. *Stability of Motion*, Krasovskii, N.N. 1963. *Stability of Motion*.

Periodic journals on control theory and systems often include articles where Lyapunov's method is used in system analysis or control design. Examples are the *IEEE Transactions on Automatic Control* and the IFAC journal *Automatica*.

56.2 Input-Output Stability

A.R. Teel, Department of Electrical Engineering, University of Minnesota

T.T. Georgiou, Department of Electrical Engineering, University of Minnesota

L. Praly, Centre Automatique et Systèmes, École Des Mines de Paris

E. Sontag, Department of Mathematics, Rutgers University

56.2.1 Introduction

A common task for an engineer is to design a system that reacts to stimuli in some specific and desirable way. One way to characterize appropriate behavior is through the formalism of input-output stability. In this setting a notion of well-behaved input and output signals is made precise and the question is posed: do well-behaved stimuli (inputs) produce well-behaved responses (outputs)?

General input-output stability analysis has its roots in the development of the electronic feedback amplifier of H.S. Black in 1927 and the subsequent development of classical feedback design tools for linear systems by H. Nyquist and H.W. Bode in the 1930s and 1940s, all at Bell Telephone Laboratories. These latter tools focused on determining input-output stability of linear feedback systems from the characteristics of the feedback components. Generalizations to nonlinear systems were made by several researchers in the late 1950s and early 1960s. The most notable contributions were those of G. Zames, then at M.I.T., I.W. Sandberg at Bell Telephone Laboratories, and V.M. Popov. Indeed, much of this chapter is based on the foundational ideas found in [5], [7] and [10], with additional insights drawn from [6]. A thorough understanding of nonlinear systems from an input-output point of view is still an area of ongoing and intensive research.

The strength of input-output stability theory is that it provides a method for anticipating the qualitative behavior of a feedback system with only rough information about the feedback components. This, in turn, leads to notions of robustness of feedback stability and motivates many of the recent developments in modern control theory.

56.2.2 Systems and Stability

Throughout our discussion of input-output stability, a **signal** is a "reasonable" (e.g., piecewise continuous) function defined on a finite or semi-infinite time interval, i.e., an interval of the form $[0, T)$ where T is either a strictly positive real number or infinity. In general, a signal is vector-valued; its components typically represent actuator and sensor values. A **dynamical system** is an object which produces an output signal for each input signal.

To discuss stability of dynamical systems, we introduce the concept of a **norm function**, denoted $||\cdot||$, which captures the "size" of signals defined on the semi-infinite time interval. The significant properties of a norm function are that 1) the norm of a signal is zero if the signal is identically zero, and is a strictly positive number otherwise, 2) scaling a signal results in a corresponding scaling of the norm, and 3) the triangle inequality holds, i.e., $||u_1 + u_2|| \leq ||u_1|| + ||u_2||$. Examples of norm functions are the p-**norms**. For any positive real number $p \geq 1$, the p-norm is defined by

$$||u||_p := \left(\int_0^\infty |u(t)|^p\right)^{\frac{1}{p}} \tag{56.13}$$

where $|\cdot|$ represents the standard Euclidean norm, i.e., $|u| = \sqrt{\sum_{i=1}^{n} u_i^2}$. For $p = \infty$, we define

$$||u||_\infty := \sup_{t \geq 0} |u(t)| . \tag{56.14}$$

The ∞-norm is useful when amplitude constraints are imposed on a problem, and the 2-norm is of more interest in the context of energy constraints. The norm of a signal may very well be infinite. We will typically be interested in measuring signals which may only be defined on finite time intervals or measuring truncated versions of signals. To that end, given a signal u defined on $[0, T)$ and a strictly positive real number τ, we use u_τ to denote the **truncated signal** generated by extending u onto $[0, \infty)$ by defining $u(t) = 0$ for $t \geq T$, if necessary, and then truncating, i.e., u_τ is equal to the (extended) signal on the interval $[0, \tau]$ and is equal to zero on the interval (τ, ∞).

Informally, a system is **stable** in the input-output sense if small input signals produce correspondingly small output signals. To make this concept precise, we need a way to quantify the dependence of the norm of the output on the norm of the input applied to the system. To that end, we define a **gain function** as a function from the nonnegative real numbers to the nonnegative real numbers which is continuous, nondecreasing, and zero when its argument is zero. For notational convenience we will say that the "value" of a gain function at ∞ is ∞. A dynamical system is stable (with respect to the norm $||\cdot||$) if there is a gain function γ which gives a bound on the norm of truncated output signals as a function of the norm of truncated input signals, i.e.,

$$||y_\tau|| \le \gamma(||u_\tau||), \qquad \text{for all } \tau . \tag{56.15}$$

In the very special case when the gain function is linear, i.e., there is at most an amplification by a constant factor, the dynamical system is **finite gain stable.** The notions of finite gain stability and closely related variants are central to much of classical input-output stability theory, but in recent years much progress has been made in understanding the role of more general (nonlinear) gains in system analysis.

The focus of this chapter will be on the stability analysis of interconnected dynamical systems as described in Figure 56.2. The composite system in Figure 56.2 will be called a **well-defined interconnection** if it is a dynamical system with $\begin{pmatrix} d_1 \\ d_2 \end{pmatrix}$ as input and $\begin{pmatrix} y_1 \\ y_2 \end{pmatrix}$ as output, i.e., given an arbitrary input signal $\begin{pmatrix} d_1 \\ d_2 \end{pmatrix}$, a signal $\begin{pmatrix} y_i \\ y_2 \end{pmatrix}$ exists so that, for the dynamical system Σ_1, the input $d_1 + y_2$ produces the output y_1 and, for the dynamical system Σ_2, the input $d_2 + y_1$ produces the output y_2. To see that not every interconnection is well-defined, consider the case where both Σ_1 and Σ_2 are the identity mappings. In this case, the only input signals $\begin{pmatrix} d_1 \\ d_2 \end{pmatrix}$ for which an output $\begin{pmatrix} y_1 \\ y_2 \end{pmatrix}$ can be found are those for which $d_1 + d_2 = 0$. The dynamical systems which make up a well-defined interconnection will be called its **feedback components.**

For stability of well-defined interconnections, it is not necessary for either of the feedback components to be stable nor is it sufficient for both of the feedback components to be stable. On the other hand, necessary and sufficient conditions for stability of a well-defined interconnection can be expressed in terms of the set of all possible input-output pairs for each feedback component. To be explicit, following are some definitions. For a given dynamical system Σ with input signals u and output signals y, the set of its ordered input-output pairs (u, y) is referred to as the **graph** of the dynamical system and is denoted $\mathcal{G}_\Sigma$. When the input and output are exchanged in the ordered pair, i.e., (y, u), the set is referred to as the **inverse graph** of the system and is denoted $\mathcal{G}^I_\Sigma$. Note that, for the system in Figure 56.2, the inverse graph of Σ_2 and the graph of Σ_1 lie in the same Cartesian product space called the **ambient space.** We will use as norm on the ambient space the sum of the norms of the coordinates.

The basic observation regarding input-output stability for a well-defined interconnection says, in informal terms, that if a signal in the inverse graph of Σ_2 is near any signal in the graph of Σ_1 then it must be small. To formalize this notion, we need the concept of the **distance** to the graph of Σ_1 from signals x in the ambient space. This (truncated) distance is defined by

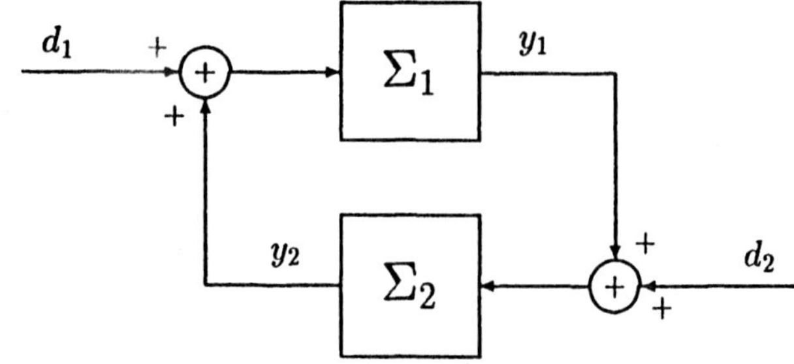

Figure 56.2 Standard feedback configuration.

$$d_\tau(x, \mathcal{G}_{\Sigma_1}) := \inf_{z \in \mathcal{G}_{\Sigma_1}} ||(x - z)_\tau|| . \tag{56.16}$$

THEOREM 56.1 **Graph separation theorem:** *A well-defined interconnection is stable if, and only if, a gain function γ exists which gives a bound on the norm of truncated signals in the inverse graph of Σ_2 as a function of the truncated distance from the signals to the graph of Σ_1, i.e.,*

$$x \in \mathcal{G}^I_{\Sigma_2} \implies ||x_\tau|| \le \gamma\left(\; d_\tau\left(x, \mathcal{G}_{\Sigma_1}\right) \; \right), \quad \textit{for all } \tau . \tag{56.17}$$

In the special case where γ is a linear function, the well-defined interconnection is finite gain stable.

The idea behind this observation can be understood by considering the signals that arise in the closed loop which belong to the inverse graph of Σ_2, i.e., the signals $(y_2, y_1 + d_2)$. (Stability with these signals taken as output is equivalent to stability with the original outputs.) Notice that, for the system in Figure 56.2, signals in the graph of Σ_1 have the form $(y_2 + d_1, y_1)$. Consequently, signals $x \in \mathcal{G}^I_{\Sigma_2}$ and $z \in \mathcal{G}_{\Sigma_1}$, which satisfy the feedback equations, also satisfy

$$(x - z)_\tau = (d_1, -d_2)_\tau \tag{56.18}$$

and

$$||(x - z)_\tau|| = ||(d_1, d_2)_\tau|| \tag{56.19}$$

for truncations within the interval of definition. If there are signals x in the inverse graph of Σ_2 with large truncated norm but small truncated distance to the graph of Σ_1, i.e., there exists some $z \in \mathcal{G}_{\Sigma_1}$ and $\tau > 0$ such that $||(x - z)_\tau||$ is small, then we can choose (d_1, d_2) to satisfy Equation 56.18 giving, according to Equation 56.19, a small input which produces a large output. This contradicts our definition of stability. Conversely, if there is no z which is close to x, then only large inputs can produce large x signals and thus the system is stable.

The distance observation presented above is the unifying idea behind the input-output stability criteria applied in practice. However, the observation is rarely applied directly because of the difficulties involved in exactly characterizing the graph of a dynamical system and measuring distances. Instead, various simpler conditions have been developed which constrain the graphs of the feedback components to guarantee that the graph of Σ_1

and the inverse graph of Σ_2 are sufficiently separated. There are many such sufficient conditions, and, in the remainder of this chapter, we will describe a few of them.

56.2.3 Practical Conditions and Examples

The Classical Small Gain Theorem

One of the most commonly used sufficient conditions for graph separation constrains the graphs of the feedback components by assuming that each feedback component is finite gain stable. Then, the appropriate graphs will be separated if the product of the coefficients of the linear gain functions is sufficiently small. For this reason, the result based on this type of constraint has come to be known as the small gain theorem.

THEOREM 56.2 **Small gain theorem:** *If each feedback component is finite gain stable and the product of the gains (the coefficients of the linear gain functions) is less than one, then the well-defined interconnection is finite gain stable.*

Figure 56.3 provides the intuition for the result. If we were to draw an analogy between a dynamical system and a static map whose graph is a set of points in the plane, the graph of Σ_1 would be constrained to the darkly shaded conic region by the finite gain stability assumption. Likewise, the inverse graph of Σ_2 would be constrained to the lightly shaded region. The fact that the product of the gains is less than one guarantees the positive aperture between the two regions and, in turn, that the graphs are separated sufficiently.

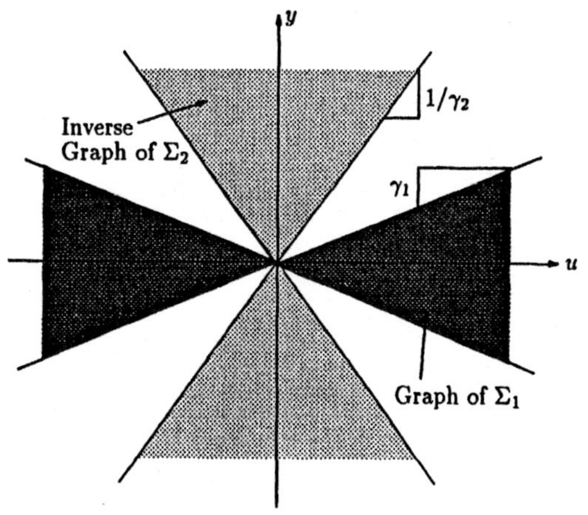

Figure 56.3 Classical small gain theorem.

To apply the small gain theorem, we need a way to verify that the feedback components are finite gain stable (with respect to a particular norm) and to determine their gains. In particular, any linear dynamical system that can be represented with a real, rational transfer function $G(s)$ is finite gain stable in any of the p-norms if, and only if, all of the poles of the transfer function have negative real parts. A popular norm to work with is the 2-norm. It is associated with the energy of a signal. For a single-input, single-output (SISO) finite gain stable system modeled by a real, rational transfer function $G(s)$, the smallest possible coefficient for the stability gain function with respect to the 2-norm, is given by

$$\bar{\gamma} := \sup_{\omega} |G(j\omega)| . \tag{56.20}$$

For multi-input, multioutput systems, the magnitude in Equation 56.20 is replaced by the maximum singular value. In either case, this can be established using **Parseval's theorem.** For SISO systems, the quantity in Equation 56.20 can be obtained from a quick examination of the Bode plot or Nyquist plot for the transfer function. If the Nyquist plot of a stable SISO transfer function lies inside a circle of radius $\bar{\gamma}$ centered at the origin, then the coefficient of the 2-norm gain function for the system is less than or equal to $\bar{\gamma}$.

More generally, consider a dynamical system that can be represented by a finite dimensional ordinary differential equation with zero initial state:

$$\begin{aligned} \dot{x} &= f(x,u) \quad , \qquad x(0)=0 \\ \text{and } y &= h(x,u) \quad . \end{aligned} \tag{56.21}$$

Suppose that f has globally bounded partial derivatives and that positive real numbers ℓ_1 and ℓ_2 exist so that

$$|h(x,u)| \leq \ell_1|x| + \ell_2|u| . \tag{56.22}$$

Under these conditions, if the trajectories of the unforced system with nonzero initial conditions,

$$\dot{x} = f(x,0) \quad , \qquad x(0) = x_\circ , \tag{56.23}$$

satisfy

$$|x(t)| \leq k \exp(-\lambda t)|x_\circ|, \tag{56.24}$$

for some positive real number k and λ and any $x_\circ \in \mathbb{R}^n$, then the system (Equation 56.21) is finite gain stable in any of the p-norms. This can be established using Lyapunov function arguments that apply to the system (Equation 56.23). The details can be found in the textbooks on nonlinear systems mentioned later in Section 56.10.

EXAMPLE 56.5:

Consider a nonlinear control system modeled by an ordinary differential equation with state $x \in \mathbb{R}^n$, input $v \in \mathbb{R}^m$ and disturbance $d_1 \in \mathbb{R}^m$:

$$\dot{x} = f(x, v + d_1) . \tag{56.25}$$

Suppose that f has globally bounded partial derivatives and that a control $v = \alpha(x)$ can be found, also with a globally bounded partial derivative, so that the trajectories of the system

$$\dot{x} = f(x, \alpha(x)) \quad , \qquad x(0) = x_\circ \tag{56.26}$$

satisfy the bound

$$|x(t)| \leq k \exp(-\lambda t)|x_\circ| \tag{56.27}$$

for some positive real numbers k and λ and for all $x_\circ \in \mathbb{R}^n$. As mentioned above, for any function h satisfying the type of bound in Equation 56.22, this implies that the system

$$\begin{aligned} \dot{x} &= f(x, \alpha(x) + d_1) \quad , \qquad x(0) = 0 \\ \text{and } y &= h(x, d_1) \end{aligned} \tag{56.28}$$

has finite 2-norm gain from input d_1 to output y. We consider the output

$$y := \dot{\alpha} = \frac{\partial \alpha}{\partial x} f(x, \alpha(x) + d_1) \tag{56.29}$$

which satisfies the type of bound in Equation 56.22 because α and f both have globally bounded partial derivatives.

We will show, using the small gain theorem, that disturbances d_1 with finite 2-norm continue to produce outputs y with finite 2-norm even when the actual input v to the process is generated by the following fast dynamic version of the commanded input $\alpha(x)$:

$$\begin{aligned} \epsilon \dot{z} &= Az + B(\alpha(x)) + d_2 \, , \; z(0) = -A^{-1}B\alpha(x(0)) \\ v &= Cz \, . \end{aligned} \tag{56.30}$$

Here, ϵ is a small positive parameter, the eigenvalues of A all have strictly negative real part (thus A is invertible), and $-CA^{-1}B = I$. This system may represent unmodeled actuator dynamics.

To see the stability result, we will consider the composite system in the coordinates x and $\zeta = z + A^{-1}B\alpha(x)$. Using the notation from Figure 56.2,

$$\Sigma_1 : \quad \begin{aligned} \dot{x} &= f(x, \alpha(x) + u_1) \, , \; x(0) = 0 \\ y_1 &= A^{-1}B\dot{\alpha}(x), \end{aligned} \tag{56.31}$$

and

$$\Sigma_2 : \quad \begin{aligned} \dot{\zeta} &= \epsilon^{-1}A\zeta + u_2 \, , \; \zeta(0) = 0 \\ y_2 &= C\zeta, \end{aligned} \tag{56.32}$$

with the interconnection conditions

$$u_1 = y_2 + d_1, \text{ and } u_2 = y_1 + \epsilon^{-1}d_2 \, . \tag{56.33}$$

Of course, if the system is finite gain stable with the inputs d_1 and $\epsilon^{-1}d_2$, then it is also finite gain stable with the inputs d_1 and d_2. We have already discussed that the system Σ_1 in Equation 56.31 has finite 2-norm gain, say γ_1. Now consider the system Σ_2 in Equation 56.32. It can be represented with the transfer function

$$\begin{aligned} G(s) &= C(sI - \epsilon^{-1}A)^{-1}, \\ &= \epsilon C(\epsilon sI - A)^{-1}, \\ &=: \epsilon \bar{G}(\epsilon s) \, . \end{aligned} \tag{56.34}$$

Identifying $\bar{G}(s) = C(sI - A)^{-1}$, we see that, if

$$\gamma_2 := \sup_\omega \sigma(\bar{G}(j\omega)), \tag{56.35}$$

then

$$\sup_\omega \sigma(G(j\omega)) = \epsilon \gamma_2 \, . \tag{56.36}$$

We conclude from the small gain theorem that, if $\epsilon < \frac{1}{\gamma_1\gamma_2}$, then the composite system (Equations 56.31–56.33), with inputs d_1 and d_2 and outputs $y_1 = A^{-1}B\dot{\alpha}(x)$ and $y_2 = C\zeta$, is finite gain stable.

The Classical Passivity Theorem

Another very popular condition used to guarantee graph separation is given in the *passivity theorem*. For the most straightforward passivity result, the number of input channels must equal the number of output channels for each feedback component. We then identify the relative location of the graphs of the feedback components using a condition involving the integral of the product of the input and the output signals. This operation is known as the **inner product**, denoted $\langle \cdot, \cdot \rangle$. In particular, for two signals u and y of the same dimension defined on the semi-infinite interval,

$$\langle u, y \rangle := \int_0^\infty u^T(t)y(t)dt \, . \tag{56.37}$$

Note that $\langle u, y \rangle = \langle y, u \rangle$ and $\langle u, u \rangle = ||u||_2^2$. A dynamical system is **passive** if, for each input-output pair (u, y) and each $\tau > 0$, $\langle u_\tau, y_\tau \rangle \geq 0$. The terminology used here comes from the special case where the input and output are a voltage and a current, respectively, and the energy absorbed by the dynamical system, which is the inner product of the input and output, is nonnegative.

Again by analogy to a static map whose graph lies in the plane, passivity of a dynamical system can be viewed as the condition that the graph is constrained to the darkly shaded region in Figure 56.4, i.e., the first and third quadrants of the plane. This graph and the inverse graph of a second system would be separated if, for example, the inverse graph of the second system were constrained to the lightly shaded region in Figure 56.4, i.e., the second and fourth quadrants but bounded away from the horizontal and vertical axes by an increasing and unbounded distance. But, this is the same as asking that the graph of the second system followed by the scaling '-1', i.e., all pairs $(u, -y)$, be constrained to the first and third quadrants, again bounded away from the axes by an increasing and unbounded distance, as in Figure 56.5a. For classical passivity theorems, this region is given a linear boundary as in Figure 56.5b. Notice that, for points $(u_\circ, y_\circ)$ in the plane, if $u_\circ \cdot y_\circ \geq \epsilon(u_\circ^2 + y_\circ^2)$ then $(u_\circ, y_\circ)$ is in the first or third quadrant, and $(\epsilon)^{-1}|u_\circ| \geq |y_\circ| \geq \epsilon|u_\circ|$ as in Figure 56.5b. This leads to the following stronger version of passivity. A dynamical system is **input and output strictly passive** if a strictly positive real number ϵ exists so that, for each input-output pair (u, y) and each $\tau > 0$, $\langle u_\tau, y_\tau \rangle \geq \epsilon \left(||u_\tau||_2^2 + ||y_\tau||_2^2 \right)$.

There are intermediate versions of passivity which are also useful. These correspond to asking for an increasing and unbounded distance from either the horizontal axis or the vertical axis but not both. For example, a dynamical system is **input strictly passive** if a strictly positive real number ϵ exists so that, for each input-output pair (u, y) and each $\tau > 0$, $\langle u_\tau, y_\tau \rangle \geq \epsilon ||u_\tau||_2^2$. Similarly, a dynamical system is **output strictly passive** if a strictly

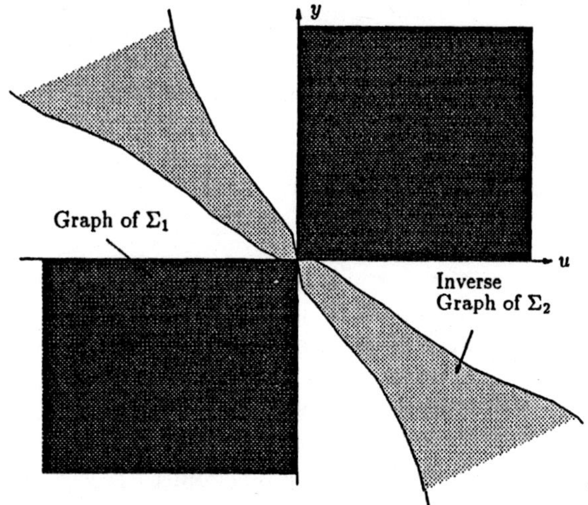

Figure 56.4 General passivity-based interconnection.

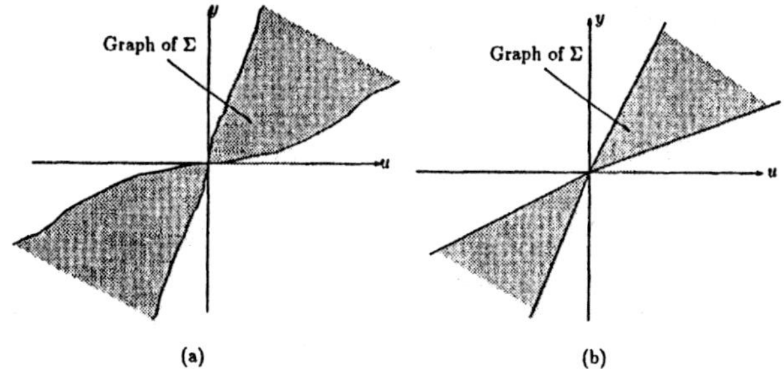

Figure 56.5 Different notions of input and output strict passivity.

positive real number ϵ exists so that, for each input-output pair (u, y) and each $\tau > 0$, $\langle u_\tau, y_\tau \rangle \geq \epsilon \|y_\tau\|_2^2$. It is worth noting that input and output strict passivity is equivalent to input strict passive plus finite gain stability. This can be shown with standard manipulations of the inner product. Also, the reader is warned that all three types of strict passivity mentioned above are frequently called "strict passivity" in the literature.

Again by thinking of a graph of a system as a set of points in the plane, output strict passivity is the condition that the graph is constrained to the darkly shaded region in Figure 56.6, i.e., the first and third quadrants with an increasing and unbounded distance from the vertical axis. To complement such a graph, consider a second dynamical system which, when followed by the scaling '-1', is also output strictly passive. Such a system has a graph (without the '-1' scaling) constrained to the second and fourth quadrants with an increasing and unbounded distance from the vertical axis. In other words, its inverse graph is constrained to the lightly shaded region of Figure 56.6, i.e., to the second and fourth quadrants but with an increasing and unbounded distance from the *horizontal* axis. The conclusions that we can then draw, using the graph separation theorem, are summarized in the following passivity theorem.

THEOREM 56.3 **Passivity theorem:** *If one dynamical system and the other dynamical system followed by the scaling '-1' are*

- *both input strictly passive, OR*
- *both output strictly passive, OR*
- *respectively, passive* and *input and output strictly passive,*

then the well-defined interconnection is finite gain stable in the 2-norm.

To apply this theorem, we need a way to verify that the (possibly scaled) feedback components are appropriately passive. For stable SISO systems with real, rational transfer function $G(s)$, it again follows from Parseval's theorem that, if

$$\text{Re } G(j\omega) \geq 0,$$

for all real values of ω, then the system is passive. If the quantity $\text{Re } G(j\omega)$ is positive and uniformly bounded away from zero for all real values of ω, then the linear system is input and output strictly passive. Similarly, if $\epsilon > 0$ exists so that, for all real values of ω,

$$\text{Re } G(j\omega - \epsilon) \geq 0, \tag{56.38}$$

then the linear system is output strictly passive. So, for SISO systems modeled with real, rational transfer functions, passivity and the various forms of strict passivity can again be easily checked by means of a graphical approach such as a Nyquist plot.

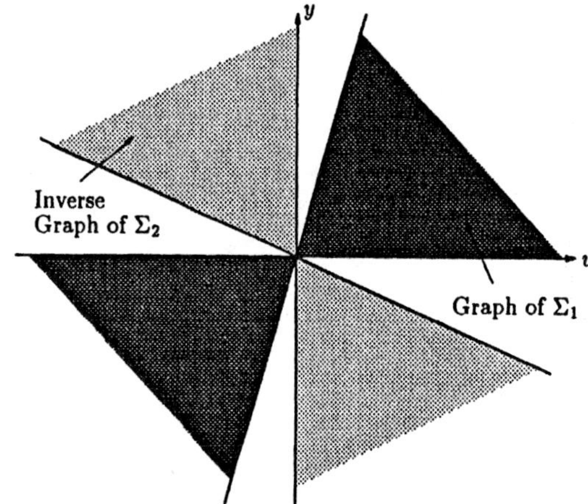

Figure 56.6 Interconnection of output strictly passive systems.

More generally, for any dynamical system that can be modeled with a smooth, finite dimensional ordinary differential equation,

$$\begin{aligned} \dot{x} &= f(x) + g(x)u \quad , \qquad x(0) = 0 \\ y &= h(x) \, , \end{aligned} \tag{56.39}$$

if a strictly positive real number ϵ exists and a nonnegative function $V : \mathbb{R}^n \to \mathbb{R}_{\geq 0}$ with $V(0) = 0$ satisfying

$$\frac{\partial V}{\partial x}(x) f(x) \leq -\epsilon h^T(x) h(x), \tag{56.40}$$

and

$$\frac{\partial V}{\partial x}(x) g(x) = h^T(x), \tag{56.41}$$

then the system is output strictly passive. With $\epsilon = 0$, the system is passive. Both of these results are established by integrating $\dot{V}$ over the semi-infinite interval.

EXAMPLE 56.6:

(This example is prompted by the work in Berghuis and Nijmeijer, *Syst. Control Lett.*, 1993, 21, 289–295.) Consider a "completely controlled dissipative Euler-Lagrange" system with generalized "forces" F, generalized coordinates q, uniformly positive definite "inertia" matrix $I(q)$, Rayleigh dissipation function $R(\dot{q})$ and, say positive, potential $V(q)$ starting from the position q_d. Let the dynamics of the system be given by the Euler-Lagrange-Rayleigh equations,

$$\overbrace{\frac{\partial L}{\partial \dot{q}}(q,\dot{q})}^{\cdot} = \frac{\partial L}{\partial q}(q,\dot{q}) + F^{\mathrm{T}} - \frac{\partial R}{\partial \dot{q}}(\dot{q}),$$
$$q(0) = q_d\,, \quad \dot{q}(0) = 0, \tag{56.42}$$

where L is the Lagrangian

$$L(q,\dot{q}) = \tfrac{1}{2}\dot{q}^{\mathrm{T}} I(q)\,\dot{q} - V(q)\,. \tag{56.43}$$

Along the solution of Equation 56.42,

$$\begin{aligned}
\dot{L} &= \frac{\partial L}{\partial \dot{q}}\ddot{q} + \frac{\partial L}{\partial q}\dot{q} \\
&= \frac{\partial L}{\partial \dot{q}}\ddot{q} + \left[\overbrace{\frac{\partial L}{\partial \dot{q}}}^{\cdot} - F^{\mathrm{T}} + \frac{\partial R}{\partial \dot{q}}\right]\dot{q}, \qquad (56.44)\\
&= \overbrace{\left(\frac{\partial L}{\partial \dot{q}}\dot{q}\right)}^{\cdot} - \left[F^{\mathrm{T}} - \frac{\partial R}{\partial \dot{q}}\right]\dot{q} \\
&= \overbrace{\left(\dot{q}^{\mathrm{T}} I(q)\,\dot{q}\right)}^{\cdot} - \left[F^{\mathrm{T}} - \frac{\partial R}{\partial \dot{q}}\right]\dot{q}, \qquad (56.45)\\
&= 2\,\overbrace{L + V}^{\cdot} - \left[F^{\mathrm{T}} - \frac{\partial R}{\partial \dot{q}}\right]\dot{q} \\
&= -2\,\dot{V} + \left[F^{\mathrm{T}} - \frac{\partial R}{\partial \dot{q}}\right]\dot{q}. \qquad (56.46)
\end{aligned}$$

We will suppose that $\epsilon > 0$ exists so that

$$\frac{\partial R}{\partial \dot{q}}(\dot{q})\,\dot{q} \geq \epsilon\,|\dot{q}|^2\,. \tag{56.47}$$

Now let V_d be a function so that the modified potential

$$V_m = V + V_d \tag{56.48}$$

has a global minimum at $q = q_d$, and let the generalized "force" be

$$F = -\frac{\partial V_d}{\partial q}(q) + F_m\,. \tag{56.49}$$

We can see that the system (Equation 56.42) combined with (Equation 56.49), having input F_m and output $\dot{q}$, is output strictly passive by integrating the derivative of the defined Hamiltonian,

$$H = \tfrac{1}{2}\dot{q}^{\mathrm{T}} I(q)\,\dot{q} + V_m(q) = L + 2\,V + V_d\,. \tag{56.50}$$

Indeed the derivative is

$$\dot{H} = \left[F_m^{\mathrm{T}} - \frac{\partial R}{\partial \dot{q}}\right]\dot{q} \tag{56.51}$$

and, integrating, for each τ

$$\left\langle \left[F_{m_\tau} - \frac{\partial R}{\partial \dot{q}}(\dot{q}_\tau)^{\mathrm{T}}\right], \dot{q}_\tau \right\rangle = H(\tau) - H(0)\,. \tag{56.52}$$

Since $H \geq 0$, $H(0) = 0$ and Equation 56.47 holds

$$\langle F_{m\tau}, \dot{q}_\tau \rangle \geq \epsilon \|\dot{q}_\tau\|_2^2\,. \tag{56.53}$$

Using the notation from Figure 56.2, let Σ_1 be the system (Equations 56.42 and 56.49) with input F_m and output $\dot{q}$. Let Σ_2 be any system that, when followed by the scaling '-1', is output strictly passive. Then, according to the passivity theorem, the composite feedback system as given in Figure 56.2 is finite gain stable using the 2-norm. One possibility for Σ_2 is minus the identity mapping. However, there is interest in choosing Σ_2 followed by the scaling '-1' as a linear, output strictly passive compensator which, in addition, has no direct feed-through term. The reason is that if, d_2 in Figure 56.2 is identically zero, we can implement Σ_2 with measurement only of q and without $\dot{q}$. In general,

$$G(s)\dot{q} = G(s)s\left(\frac{1}{s}\dot{q}\right) = G(s)s(q - q_d), \tag{56.54}$$

and the system $G(s)s$ is implementable if $G(s)$ has no direct feed-through terms. To design an output strictly passive linear system without direct feed-through, let A be a matrix having all eigenvalues with strictly negative real parts so that, by a well-known result in linear systems theory, a positive definite matrix P exists satisfying

$$A^T P + PA = -I\,. \tag{56.55}$$

Then, for any B matrix of appropriate dimensions, the system modeled by the transfer function,

$$G(s) = -B^T P(sI - A)^{-1}B\,, \tag{56.56}$$

followed by the scaling '-1', is output strictly passive. To see this, consider a state-space realization

$$\begin{aligned}
\dot{x} &= Ax + Bu \quad x(0) = 0 \\
y &= B^T P x,
\end{aligned} \tag{56.57}$$

and note that

$$\begin{aligned}
\overbrace{x^{\mathrm{T}} P x}^{\cdot} &= -x^{\mathrm{T}}x + 2x^{\mathrm{T}} P B u \qquad (56.58)\\
&= -x^{\mathrm{T}}x + 2y^{\mathrm{T}}u\,. \qquad (56.59)
\end{aligned}$$

But, with Equation 56.57, for some strictly positive real number c,

$$2c\,y^{\mathrm{T}} y \leq x^{\mathrm{T}} x\,. \tag{56.60}$$

So, integrating Equation 56.59 and with P positive definite, for all τ,

$$\langle y_\tau, u_\tau \rangle \geq c\|y_\tau\|_2^2\,. \tag{56.61}$$

As a point of interest, one could verify that

$$G(s)s = -B^T PA(sI - A)^{-1}B - B^T PB\,. \tag{56.62}$$

Simple Nonlinear Separation Theorems

In this section we illustrate how allowing regions with nonlinear boundaries in the small gain and passivity contexts may be useful. First we need a class of functions to describe nonlinear boundaries. A **proper separation function** is a function from the nonnegative real numbers to the nonnegative real numbers which is continuous, zero at zero, strictly increasing and unbounded. The main difference between a gain function and a proper separation function is that the latter is invertible, and the inverse is another proper separation function.

Nonlinear passivity We will briefly discuss a definition of nonlinear input and output strict passivity. To our knowledge, this idea has not been used much in the literature. The notion replaces the linear boundaries in the input and output strict passivity definition by nonlinear boundaries as in Figure 56.5a. A dynamical system is **nonlinearly input and output strictly passive** if a proper separation function ρ exists so that, for each input-output pair (u, y) and each $\tau > 0$, $\langle u_\tau, y_\tau \rangle \geq \|u_\tau\|_2 \rho(\|u_\tau\|_2) + \|y_\tau\|_2 \rho(\|y_\tau\|_2)$. (Note that in the classical definition of strict passivity, $\rho(\zeta) = \epsilon\zeta$ for all $\zeta \geq 0$.)

THEOREM 56.4 **Nonlinear passivity theorem:** *If one dynamical system is passive and the other dynamical system followed by the scaling '-1' is nonlinearly input and output strictly passive, then the well-defined interconnection is stable using the 2-norm.*

EXAMPLE 56.7:

Let Σ_1 be a single integrator system,

$$\begin{aligned} \dot{x}_1 &= u_1 \qquad x_1(0) = 0 \\ y_1 &= x_1 \,. \end{aligned} \tag{56.63}$$

This system is passive because

$$\begin{aligned} 0 \leq \tfrac{1}{2} x_1(\tau)^2 &= \int_0^\tau \frac{d}{dt} \tfrac{1}{2} x(t)^2 dt \\ &= \int_0^\tau y_1(t) u_1(t) dt \\ &= \langle y_{1_\tau}, u_{1_\tau} \rangle \,. \end{aligned} \tag{56.64}$$

Let Σ_2 be a system which scales the instantaneous value of the input according to the energy of the input:

$$\begin{aligned} \dot{x}_2 &= u_2^2 \qquad x_2(0) = 0 \\ y_2 &= -u_2 \left(\frac{1}{1 + |x_2|^{0.25}} \right) . \end{aligned} \tag{56.65}$$

This system followed by the scaling '-1' is nonlinearly strictly passive. To see this, first note that

$$x_2(t) = \|u_{2_t}\|_2^2 \tag{56.66}$$

which is a nondecreasing function of t. So,

$$\begin{aligned} \langle -y_{2_\tau}, u_{2_\tau} \rangle &= \int_0^\tau u_2^2(t) \left(\frac{1}{1 + |x_2(t)|^{0.25}} \right) dt, \\ &\geq \left(\frac{1}{1 + |x_2(\tau)|^{0.25}} \right) \int_0^\tau u_2^2(t) dt, \\ &= \left(\frac{1}{1 + \|u_{2_\tau}\|_2^{0.5}} \right) \|u_{2_\tau}\|_2^2 \,. \end{aligned} \tag{56.67}$$

Now we can define

$$\rho(\zeta) := \frac{0.5\zeta}{1 + \zeta^{0.5}}, \tag{56.68}$$

which is a proper separation function, so that

$$\langle -y_{2_\tau}, u_{2_\tau} \rangle \geq 2\rho(\|u_{2_\tau}\|_2) \|u_{2_\tau}\|_2 \,. \tag{56.69}$$

Finally, note that

$$\|y_{2_\tau}\|_2^2 = \int_0^\tau u_2^2(t) \frac{1}{\left(1 + x_2^{0.25}(t)\right)^2} dt \leq \|u_{2_\tau}\|_2^2, \tag{56.70}$$

so that

$$\langle -y_{2_\tau}, u_{2_\tau} \rangle \geq \rho(\|u_{2_\tau}\|_2) \|u_{2_\tau}\|_2 + \rho(\|y_{2_\tau}\|_2) \|y_{2_\tau}\|_2 \,. \tag{56.71}$$

The conclusion that we can then draw from the nonlinear passivity theorem is that the interconnection of these two systems:

$$\begin{aligned} \dot{x}_1 &= -(x_1 + d_2) \left(\frac{1}{1 + |x_2|^{0.25}} \right) + d_1, \\ \dot{x}_2 &= (x_1 + d_2)^2, \\ y_1 &= x_1, \\ \text{and } y_2 &= -(x_1 + d_2) \left(\frac{1}{1 + |x_2|^{0.25}} \right) \end{aligned} \tag{56.72}$$

is stable when measuring input (d_1, d_2) and output (y_1, y_2) using the 2-norm.

Nonlinear small gain Just as with passivity, the idea behind the small gain theorem does not require the use of linear boundaries. Consider a well-defined interconnection where each feedback component is stable but not necessarily finite gain stable. Let γ_1 be a stability gain function for Σ_1 and let γ_2 be a stability gain function for Σ_2. Then the graph separation condition will be satisfied if the distance between the curves $(\zeta, \gamma_1(\zeta))$ and $(\gamma_2(\xi), \xi)$ grows without bound as in Figure 56.7. This is equivalent to asking whether it is possible to add to the curve $(\zeta, \gamma_1(\zeta))$ in the vertical direction and to the curve $(\gamma_2(\xi), \xi)$ in the horizontal direction, by an increasing and unbounded amount, to obtain new curves $(\zeta, \gamma_1(\zeta) + \rho(\zeta))$ and $(\gamma_2(\xi) + \rho(\xi), \xi)$ where ρ is a proper separation function, so that the modified first curve is never above the modified second curve. If this is possible, we will say that the composition of the functions γ_1 and γ_2 is a **strict contraction**. To say that a curve $(\zeta, \tilde{\gamma}_1(\zeta))$ is never above a second curve $(\tilde{\gamma}_2(\xi), \xi)$ is equivalent to saying that $\tilde{\gamma}_1(\tilde{\gamma}_2(\zeta)) \leq \zeta$ or $\tilde{\gamma}_2(\tilde{\gamma}_1(\zeta)) \leq \zeta$ for all $\zeta \geq 0$. (Equivalently, we will write $\tilde{\gamma}_1 \circ \tilde{\gamma}_2 \leq \mathrm{Id}$ or $\tilde{\gamma}_2 \circ \tilde{\gamma}_1 \leq \mathrm{Id}$.) So, requiring that the composition of γ_1 and γ_2 is a strict contraction is equivalent to requiring that a strictly proper separation function ρ exists so that $(\gamma_1+\rho)\circ(\gamma_2+\rho) \leq \mathrm{Id}$ (equivalently $(\gamma_2+\rho)\circ(\gamma_1+\rho) \leq \mathrm{Id}$). This condition was made precise in [3]. (See also [2].) Note that it is not enough to add to just one curve because it is possible for the vertical or horizontal distance to grow without bound while the total distance remains bounded. Finally, note that, if the gain functions are linear, the condition is the same as the condition that the product of the gains is less than one.

THEOREM 56.5 **Nonlinear small gain theorem:** *If each feedback component is stable (with gain functions γ_1 and γ_2) and the composition of the gains is a strict contraction, then the well-defined interconnection is stable.*

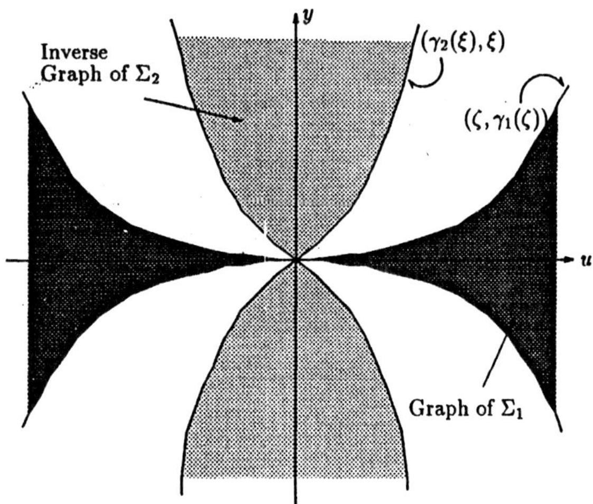

Figure 56.7 Nonlinear small gain theorem.

To apply the nonlinear small gain theorem, we need a way to verify that the feedback components are stable. To date, the most common setting for using the nonlinear small gain theorem is when measuring the input and output using the ∞-norm. For a nonlinear system which can be represented by a smooth, ordinary differential equation,

$$\begin{aligned} \dot{x} &= f(x,u) \quad , \qquad x(0)=0, \\ \text{and } y &= h(x,u), \end{aligned} \tag{56.73}$$

where $h(0,0)=0$, the system is stable (with respect to the ∞-norm) if there exist a positive definite and radially unbounded function $V : \mathbb{R}^n \to \mathbb{R}_{\geq} 0$, a proper separation function ψ, and a gain function $\tilde{\gamma}$ so that

$$\frac{\partial V}{\partial x} f(x,u) \leq -\psi(|x|) + \tilde{\gamma}(|u|) \,. \tag{56.74}$$

Since V is positive definite and radially unbounded, additional proper separation functions $\underline{\alpha}$ and $\bar{\alpha}$ exist so that

$$\underline{\alpha}(|x|) \leq V(x) \leq \bar{\alpha}(|x|) \,. \tag{56.75}$$

Also, because h is continuous and zero at zero, gain functions ϕ_x and ϕ_u exist so that

$$|h(x,u)| \leq \phi_x(|x|) + \phi_u(|u|) \,. \tag{56.76}$$

Given all of these functions, a stability gain function can be computed as

$$\gamma = \phi_x \circ \underline{\alpha}^{-1} \circ \bar{\alpha} \circ \psi^{-1} \circ \tilde{\gamma} + \phi_u \,. \tag{56.77}$$

For more details, the reader is directed to [8].

EXAMPLE 56.8:

Consider the composite system,

$$\begin{aligned} \dot{x} &= Ax + B\mathrm{sat}(z+d_1) \;, & x(0) &= 0 \\ \dot{z} &= -z + \epsilon(\exp(|x|+d_2)-1) \;, & z(0) &= 0, \end{aligned} \tag{56.78}$$

where $x \in \mathbb{R}^n$, $z \in \mathbb{R}$, the eigenvalues of A all have strictly negative real part, ϵ is a small parameter, and $\mathrm{sat}(s) = \mathrm{sgn}(s)\min\{|s|, 1\}$. This composite system is a well-defined interconnection of the subsystems

$$\Sigma_1 : \begin{aligned} \dot{x} &= Ax + B\mathrm{sat}(u_1) \,, \; x(0)=0 \\ y_1 &= |x| \end{aligned} \tag{56.79}$$

and

$$\Sigma_2 : \begin{aligned} \dot{z} &= -z + \epsilon\left(\exp(u_2)-1\right) \,, \; z(0)=0 \\ y_2 &= z \,. \end{aligned} \tag{56.80}$$

A gain function for the Σ_1 system is the product of the ∞-gain for the linear system

$$\begin{aligned} \dot{x} &= Ax + Bu \,, \; x(0)=0 \\ y &= x \,, \end{aligned} \tag{56.81}$$

which we will call $\bar{\gamma}_1$, with the function $\mathrm{sat}(s)$, i.e., for the system Σ_1,

$$\|y\|_\infty \leq \bar{\gamma}_1 \mathrm{sat}(\|u_1\|_\infty) \,. \tag{56.82}$$

For the system Σ_2,

$$\|z\|_\infty \leq |\epsilon| \left(\exp\left(\|u_2\|_\infty\right) - 1\right) \,. \tag{56.83}$$

The distance between the curves $(\zeta, \bar{\gamma}_1 \mathrm{sat}(\zeta))$ and $(|\epsilon|\left(\exp(\xi) - 1\right), \xi)$ must grow without bound. Graphically, one can see that a necessary and sufficient condition for this is that

$$|\epsilon| < \frac{1}{\exp(\bar{\gamma}_1) - 1} \,. \tag{56.84}$$

General Conic Regions

There are many different ways to partition the ambient space to establish the graph separation condition in Equation 56.17. So far we have looked at only two very specific sufficient conditions, the small gain theorem and the passivity theorem. The general idea in these theorems is to constrain signals in the graph of Σ_1 within some conic region, and signals in the inverse graph of Σ_2 outside of this conic region. Conic regions more general than those used for the small gain and passivity theorems can be generated by using operators on the input-output pairs of the feedback components.

Let **C** and **R** be operators on truncated ordered pairs in the ambient space, and let γ be a gain function. We say that the graph of Σ_1 is **inside** CONE(**C**, **R**, γ) if, for each $(u, y) =: z$ belonging to the graph of Σ_1,

$$||\mathbf{C}(z_\tau)|| \le \gamma(||\mathbf{R}(z_\tau)||), \quad \text{for all } \tau\,. \tag{56.85}$$

On the other hand, we say that the inverse graph of Σ_2 is **strictly outside** CONE(**C**, **R**, γ) if a proper separation function ρ exists so that, for each $(y, u) =: x$ belonging to the inverse graph of Σ_2,

$$||\mathbf{C}(x_\tau)|| \ge \gamma \circ (\mathrm{Id} + \rho)(||\mathbf{R}(x_\tau)||) + \rho(||x_\tau||), \quad \text{for all } \tau. \tag{56.86}$$

We will only consider the case where the maps **C** and **R** are **incrementally stable**, i.e., a gain function $\bar{\gamma}$ exists so that, for each x_1 and x_2 in the ambient space and all τ,

$$\begin{aligned} ||\mathbf{C}(x_{1_\tau}) - \mathbf{C}(x_{2_\tau})|| &\le \bar{\gamma}(||x_{1_\tau} - x_{2_\tau}||) \\ ||\mathbf{R}(x_{1_\tau}) - \mathbf{R}(x_{2_\tau})|| &\le \bar{\gamma}(||x_{1_\tau} - x_{2_\tau}||)\,. \end{aligned} \tag{56.87}$$

In this case, the following result holds.

THEOREM 56.6 **Nonlinear conic sector theorem:** *If the graph of Σ_1 is inside* CONE(**C**, **R**, γ) *and the inverse graph of Σ_2 is strictly outside* CONE(**C**, **R**, γ), *then the well-defined interconnection is stable.*

When γ and ρ are linear functions, the well-defined interconnection is finite gain stable.

The small gain and passivity theorems we have discussed can be interpreted in the framework of the nonlinear conic sector theorem. For example, for the nonlinear small gain theorem, the operator **C** is a projection onto the second coordinate in the ambient space, and **R** is a projection onto the first coordinate; γ is the gain function γ_1, and the small gain condition guarantees that the inverse graph of Σ_2 is strictly outside of the cone specified by this **C**, **R** and γ.

In the remaining subsections, we will discuss other useful choices for the operators **C** and **R**.

The classical conic sector (circle) theorem For linear SISO systems connected to memoryless nonlinearities, there is an additional classical result, known as the circle theorem, which follows from the nonlinear conic sector theorem using the 2-norm and taking

$$\begin{aligned} \mathbf{C}(u, y) &= y + cu \\ \mathbf{R}(u, y) &= ru \quad r \ge 0 \\ \gamma(\zeta) &= \zeta\,. \end{aligned} \tag{56.88}$$

Suppose ϕ is a memoryless nonlinearity which satisfies

$$|\phi(u, t) + cu| \le |ru| \quad \text{for all } t, u\,. \tag{56.89}$$

Graphically, the constraint on ϕ is shown in Figure 56.8. (In the case shown, $c > r > 0$.) We will use the notation SECTOR$[-(c + r), -(c - r)]$ for the memoryless nonlinearity. It is also clear that the graph of ϕ lies in the CONE(**C**, **R**, γ) with **C**, **R**, γ defined in Equation 56.88. For a linear, time invariant, finite dimensional SISO system, whether its inverse graph is strictly outside of this cone can be determined by examining the Nyquist plot of its transfer function. The condition on the Nyquist plot is expressed relative to a disk $\mathcal{D}_{c,r}$ in the complex plane centered on the real axis passing through the points on the real axis with real parts $-1/(c + r)$ and $-1/(c - r)$ as shown in Figure 56.9.

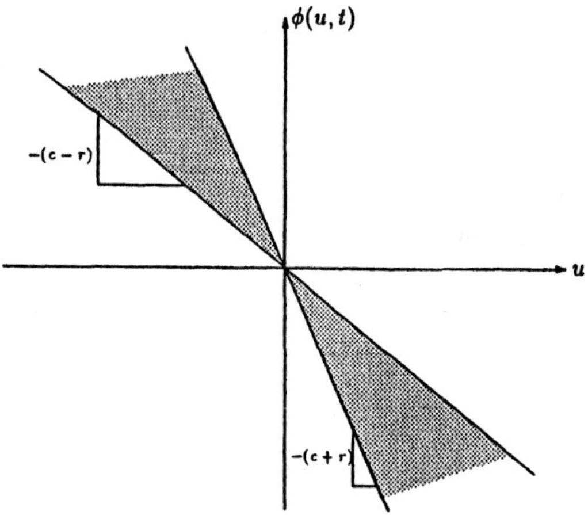

Figure 56.8 Instantaneous sector.

THEOREM 56.7 **Circle theorem:** *Let $r \ge 0$, and consider a well-defined interconnection of a memoryless nonlinearity belonging to* SECTOR$[-(c + r), -(c - r)]$ *with a SISO system having a real, rational transfer function $G(s)$. If*

- *$r > c$, $G(s)$ is stable and the Nyquist plot of $G(s)$ lies in the interior of the disc $\mathcal{D}_{c,r}$, or*
- *$r = c$, $G(s)$ is stable and the Nyquist plot of $G(s)$ is bounded away and to the right of the vertical line passing through the real axis at the value $-1/(c + r)$, or*
- *$r < c$, the Nyquist plot of $G(s)$ (with Nyquist path indented into the right-half plane) is outside of and*

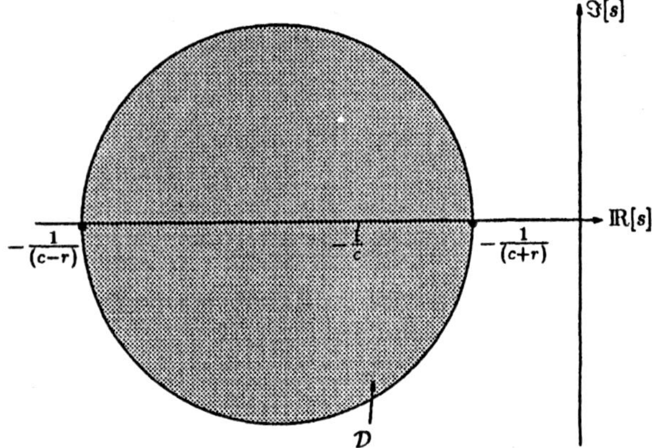

Figure 56.9 A disc in the complex plane.

> *bounded away from the disc $\mathcal{D}_{c,r}$, and the number of times the plot encircles this disc in the counterclockwise direction is equal to the number of poles of $G(s)$ with strictly positive real parts,*

then the interconnection is finite gain stable.

Case 1 is similar to the small gain theorem, and case 2 is similar to the passivity theorem. We will now explain case 3 in more detail. Let $n(s)$ and $d(s)$ represent, respectively, the numerator and denominator polynomials of $G(s)$. Since the point $(-1/c, 0)$ is inside the disc $\mathcal{D}_{c,r}$, it follows, from the assumption of the theorem together with the well-known Nyquist stability condition, that all of the roots of the polynomial $d(s)+cn(s)$ have negative real parts. Then $y = G(s)u = N(s)D(s)^{-1}u$ where

$$\begin{aligned} D(s) &:= \frac{d(s)}{d(s)+cn(s)}, \\ \text{and } N(s) &:= \frac{n(s)}{d(s)+cn(s)}, \end{aligned} \tag{56.90}$$

and, by taking $z = D(s)^{-1}u$, we can describe all of the possible input-output pairs as

$$(u, y) = (\ D(s)z, \quad N(s)z\ ) \ . \tag{56.91}$$

Notice that $D(s) + cN(s) = 1$, so that

$$||u+cy||_2 = ||z||_2 \ . \tag{56.92}$$

To put a lower bound on this expression in terms of $||u||_2$ and $||y||_2$, to show that the graph is strictly outside of the cone defined in Equation 56.88, we will need the 2-norm gains for systems modeled by the transfer functions $N(s)$ and $D(s)$. We will use the symbols γ_N and γ_D for these gains. The condition of the circle theorem guarantees that $\gamma_N < r^{-1}$. To see this, note that

$$N(s) = \frac{G(s)}{1+cG(s)} \tag{56.93}$$

implying

$$\gamma_N := \sup_{\omega\in\mathbb{R}} \left| \frac{G(j\omega)}{1+cG(j\omega)} \right| \ . \tag{56.94}$$

But

$$\begin{aligned} &|1+c\ G(j\omega)|^2 - r^2|G(j\omega)|^2 \\ &= (c\ \mathrm{Re}\{G(j\omega)\}+1)^2 + c^2\ \mathrm{Im}^2\{G(j\omega)\} \\ &\quad - r^2\ \mathrm{Re}^2\{G(j\omega)\} - r^2\ \mathrm{Im}^2\{G(j\omega)\}, \\ &= (c^2-r^2)\left(\mathrm{Re}\{G(j\omega)\} + \frac{c}{c^2-r^2}\right)^2 \\ &\quad + (c^2-r^2)\ \mathrm{Im}^2\{G(j\omega)\} - \frac{r^2}{c^2-r^2} \ . \end{aligned} \tag{56.95}$$

Setting the latter expression to zero defines the boundary of the disc $\mathcal{D}_{c,r}$. Since the expression is positive outside of this disc, it follows that $\gamma_N < r^{-1}$.

Returning to the calculation initiated in Equation 56.92, note that $\gamma_N < r^{-1}$ implies that a strictly positive real number ϵ exists so that

$$(1-\epsilon\gamma_D)\gamma_N^{-1} \geq r + 2\epsilon \ . \tag{56.96}$$

So,

$$\begin{aligned} ||u+cy||_2 &= ||z||_2 = (1-\epsilon\gamma_D)||z||_2 + \epsilon\gamma_D||z||_2, \\ &\geq (1-\epsilon\gamma_D)\gamma_N^{-1}||y||_2 + \epsilon||u||_2, \\ &\geq (r+\epsilon)||y||_2 + \epsilon(||u||_2 + ||y||_2) \ . \end{aligned} \tag{56.97}$$

We conclude that the inverse graph of the linear system is strictly outside of the CONE(**C**, **R**, γ) as defined in Equation 56.88.

Note, incidentally, that $N(s)$ is the closed loop transfer function from d_1 to y_1 for the special case where the memoryless nonlinearity satisfies $\phi(u) = -cu$. This suggests another way of determining stability: first make a preliminary loop transformation with the feedback $-cu$, changing the original linear system into the system with transfer function $N(s)$ and changing the nonlinearity into a new nonlinearity $\tilde{\phi}$ satisfying $|\tilde{\phi}(u,t)| \leq r|u|$. Then apply the classical small gain theorem to the resulting feedback system.

EXAMPLE 56.9:

Let

$$G(s) = \frac{175}{(s-1)(s+4)^2} \ . \tag{56.98}$$

The Nyquist plot of $G(s)$ is shown in Figure 56.10. Because G(s) has one pole with positive real part, only the third condition of the circle theorem can apply. A disc centered at -8.1 on the real axis and with radius 2.2 can be placed inside the left loop of the Nyquist plot. Such a disc corresponds to the values $c = 0.293$ and $r = 0.079$. Because the Nyquist plot encircles this disc once in the counterclockwise direction, it follows that the standard feedback connection with the feedback components $G(s)$ and a memoryless nonlinearity constrained to the SECTOR[-0.372,-0.214] is stable using the 2-norm.

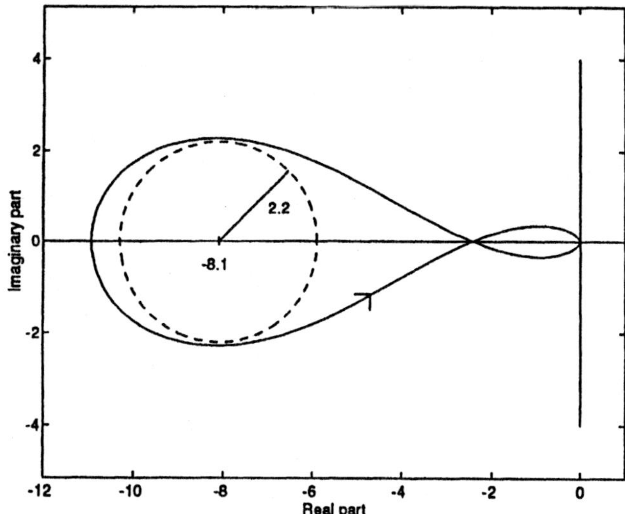

Figure 56.10 The Nyquist plot for $G(s)$ in Example 56.9

Coprime fractions Typical input-output stability results based on stable coprime fractions are corollaries of the conic sector theorem. For example, suppose both Σ_1 and Σ_2 are modeled by transfer functions $G_1(s)$ and $G_2(s)$. Moreover, assume stable (in any p-norm) transfer functions N_1, D_1, $\tilde{N}_1$, $\tilde{D}_1$, N_2 and D_2 exist so that D_1, D_2 and $\tilde{D}_1$ are invertible, and

$$\begin{aligned} G_1 &= N_1 D_1^{-1} = \tilde{D}_1^{-1} \tilde{N}_1 \\ G_2 &= N_2 D_2^{-1} \\ \mathrm{Id} &= \tilde{D}_1 D_2 - \tilde{N}_1 N_2 \, . \end{aligned} \tag{56.99}$$

Let $\mathbf{C}(u, y) = \tilde{D}_1(s)y - \tilde{N}_1(s)u$, which is incrementally stable in any p-norm, let $\mathbf{R}(u, y) = 0$, and let $\gamma \equiv 0$. Then, the graph of Σ_1 is inside and the inverse graph of Σ_2 is strictly outside $\text{CONE}(\mathbf{C}, \mathbf{R}, \gamma)$ and thus the feedback loop is finite gain stable in any p-norm.

To verify these claims about the properties of the graphs, first recognize that the graph of Σ_i can be represented as

$$\mathcal{G}_{\Sigma_i} = \begin{pmatrix} D_i(s)z \, , & N_i(s)z \end{pmatrix} \tag{56.100}$$

where z represents any reasonable signal. Then, for signals in the graph of Σ_1,

$$\begin{aligned} \mathbf{C}\left(D_1(s)z_\tau, N_1(s)z_\tau\right) &= \tilde{D}_1(s)N_1(s)z_\tau \\ &\quad - \tilde{N}_1(s)D_1(s)z_\tau \equiv 0 \, . \end{aligned} \tag{56.101}$$

Conversely, for signals in the inverse graph of Σ_2,

$$\begin{aligned} &\left\| \mathbf{C}\begin{pmatrix} N_2(s)z_\tau \, , D_2(s)z_\tau \end{pmatrix} \right\| \\ &= \left\| \tilde{D}_1(s)D_2(s)z_\tau - \tilde{N}_1(s)N_2(s)z_\tau \right\| \\ &= \|z_\tau\| \\ &\geq \epsilon \left\| \begin{pmatrix} N_2(s)z_\tau \, , D_2(s)z_\tau \end{pmatrix} \right\| \end{aligned} \tag{56.102}$$

for some strictly positive real number ϵ. The last inequality follows from the fact that D_2 and N_2 are finite gain stable.

EXAMPLE 56.10:

(This example is drawn from the work in Potvin, M-J., Jeswiet, J., and Piedboeuf, J.-C., *Trans. NAMRI/SME 1994, XXII*, pp 373–377.) Let Σ_1 represent the fractional Voigt–Kelvin model for the relation between stress and strain in structures displaying plasticity. For suitable values of Young's modulus, damping magnitude, and order of derivative for the strain, the transfer function of Σ_1 is

$$g_1(s) = \frac{1}{1 + \sqrt{s}} .$$

Integral feedback control, $g_2(s) = -\frac{1}{s}$, may be used for asymptotic tracking. Here

$$\begin{aligned} N_1(s) &= \frac{1}{s+1}, & D_1(s) &= \frac{1+\sqrt{s}}{s+1}, \\ N_2(s) &= -\frac{s+1}{1+s(1+\sqrt{s})}, & D_2(s) &= \frac{s(s+1)}{1+s(1+\sqrt{s})} . \end{aligned} \tag{56.103}$$

It can be shown that these fractions are stable linear operators, and thereby incrementally stable in the 2-norm. (This fact is equivalent to proving nominal stability and can be shown using Nyquist theory.) Moreover, it is easy to see that $D_1 D_2 - N_1 N_2 = 1$ so that the feedback loop is stable and finite gain stable.

Robustness of stability and the gap metric It is clear from the original graph separation theorem that, if a well-defined interconnection is stable, i.e., the appropriate graphs are separated in distance, then modifications of the feedback components will not destroy stability if the modified graphs are close to the original graphs.

Given two systems Σ_1 and Σ, define $\vec{\delta}(\Sigma_1, \Sigma) = \alpha$ if α is the smallest number for which

$$x \in \mathcal{G}_\Sigma, \implies d_\tau(x, \mathcal{G}_{\Sigma_1}) \leq \alpha \|x\|_\tau \quad \text{for all } \tau \, .$$

The quantity $\vec{\delta}(\cdot, \cdot)$ is called the "directed gap" between the two systems and characterizes basic neighborhoods where stability as well as closed-loop properties are preserved under small perturbations from the nominal system Σ_1 to a nearby system Σ.

More specifically, if the interconnection of (Σ_1, Σ_2) is finite gain stable, we define the gain $\beta_{\Sigma_1, \Sigma_2}$ as the smallest real number so that

$$\left\| \begin{pmatrix} d_1 + y_2 \\ y_1 \end{pmatrix} \right\|_\tau \leq \beta_{\Sigma_1, \Sigma_2} \left\| \begin{pmatrix} d_1 \\ d_2 \end{pmatrix} \right\|_\tau, \quad \text{for all } \tau \, .$$

If Σ is such that

$$\vec{\delta}(\Sigma_1, \Sigma)\beta_{\Sigma_1, \Sigma_2} < 1 \, ,$$

then the interconnection of (Σ, Σ_2) is also finite gain stable.

As a special case, let Σ, Σ_1, Σ_2 represent linear systems acting on finite energy signals. Further, assume that stable transfer functions N, D exist where D is invertible, $G_1 = ND^{-1}$, and N and

D are normalized so that $D^T(-s)D(s) + N^T(-s)N(s) = \mathrm{Id}$. Then, the class of systems in a ball with radius $\gamma \geq 0$, measured in the directed gap, is given by CONE(**C**, **R**, γ), where $\mathbf{R} = \mathrm{Id}$ and

$$\mathbf{C} = \mathrm{Id} - \begin{pmatrix} D(s) \\ N(s) \end{pmatrix} \Pi_+(D^T(-s), N^T(-s))$$

where Π_+ designates the truncation of the Laplace transform of finite energy signals to the part with poles in the left half plane. At the same time, if $\beta_{\Sigma_1,\Sigma_2} < 1/\gamma$, then it can be shown that Σ_2 is strictly outside the cone CONE(**C**, **R**, γ) and, therefore, stability of the interconnection of Σ with Σ_2 is guaranteed for any Σ inside CONE(**C**, **R**, γ).

Given Σ and Σ_1, the computation of the directed gap reduces to a standard $\mathcal{H}_\infty$-optimization problem (see [1]). Also, given Σ_1, the computation of a controller Σ_2, which stabilizes a maximal cone around Σ_1, reduces to a standard $\mathcal{H}_\infty$-optimization problem ([1]) and forms the basis of the $\mathcal{H}_\infty$-loop shaping procedure for linear systems introduced in [4].

A second key result which prompted introducing the gap metric is the claim that the behavior of the feedback interconnection of Σ and Σ_2 is "similar" to that of the interconnection of Σ_1 and Σ_2 if, and only if, the distance between Σ and Σ_1, measured using the gap metric, is small (i.e., Σ lies within a "small aperture" cone around Σ_1). The "gap" function is defined as

$$\delta(\Sigma_1, \Sigma) = \max\{\vec{\delta}(\Sigma_1, \Sigma), \vec{\delta}(\Sigma, \Sigma_1)\}$$

to "symmetrize" the distance function $\vec{\delta}(\cdot, \cdot)$ with respect to the order of the arguments. Then, the above claim can be stated more precisely as follows: for each $\epsilon > 0$, a $\zeta(\epsilon) > 0$ exists so that

$$\delta(\Sigma_1, \Sigma) < \zeta(\epsilon) \Longrightarrow \|x - x_1\|_\tau < \epsilon \|d\|_\tau$$

where $d = \begin{pmatrix} d_1 \\ d_2 \end{pmatrix}$ is an arbitrary signal in the ambient space and x (resp. x_1) represents the response $\begin{pmatrix} d_1 + y_2 \\ y_1 \end{pmatrix}$ of the feedback interconnection of (Σ, Σ_2) (resp. (Σ_1, Σ_2)). Conversely, if $\|x - x_1\|_\tau < \epsilon\|d\|_\tau$ for all d and τ, then $\delta(\Sigma_1, \Sigma) \leq \epsilon$.

Defining Terms

Ambient space: the Cartesian product space containing the inverse graph of Σ_2 and the graph of Σ_1.

Distance (from a signal to a set): measured using a norm function; the infimum, over all signals in the set, of the norm of the difference between the signal and a signal in the set; see equation (56.16); used to characterize necessary and sufficient conditions for input-output stability; see Section 56.2.2.

Dynamical system: an object which produces an output signal for each input signal.

Feedback components: the dynamical systems which make up a well-defined interconnection.

Finite gain stable system: a dynamical system is finite gain stable if a nonnegative constant exists so that, for each input-output pair, the norm of the output is bounded by the norm of the input times the constant.

Gain function: a function from the nonnegative real numbers to the nonnegative real numbers which is continuous, nondecreasing and zero when its argument is zero; used to characterize stability; see Section 56.2.2; some form of the symbol γ is usually used.

Graph (of a dynamical system): the set of ordered input-output pairs (u, y).

Inner product: defined for signals of the same dimension defined on the semi-infinite interval; the integral from zero to infinity of the component-wise product of the two signals.

Inside (or strictly outside) CONE(**C**, **R**, γ): used to characterize the graph or inverse graph of a system; determined by whether or not signals in the graph or inverse graph satisfy certain inequalities involving the operators **C** and **R** and the gain function γ; see Equations 56.85 and 56.86; used in the conic sector theorem; see Section 56.8.

Inverse graph (of a dynamical system): the set of ordered output-input pairs (y, u).

Norm function ($\|\cdot\|$): used to measure the size of signals defined on the semi-infinite interval; examples are the p-norms $p \in [1, \infty]$ (see Equations 56.13 and 56.14).

Parseval's theorem: used to make connections between properties of graphs for SISO systems modeled with real, rational transfer functions and frequency domain characteristics of their transfer functions; Parseval's theorem relates the inner product of signals to their Fourier transforms if they exist. For example, it states that, if two scalar signals u and y, assumed to be zero for negative values of time, have Fourier transforms $\hat{u}(j\omega)$ and $\hat{y}(j\omega)$ then

$$\langle u, y \rangle = \frac{1}{2\pi} \int_{-\infty}^{\infty} \hat{y}^*(j\omega)\hat{u}(j\omega)d\omega \, .$$

Passive: terminology resulting from electrical network theory; a dynamical system is passive if the inner product of each input-output pair is nonnegative.

Proper separation function: a function from the nonnegative real numbers to the nonnegative real numbers which is continuous, zero at zero, strictly increasing and unbounded; such functions are invertible on the nonnegative real numbers; used to characterize nonlinear separation theorems; see Section 56.6; some form of the symbol ρ is usually used.

Semi-infinite interval: the time interval $[0, \infty)$.

Signal: a "reasonable" vector-valued function defined on a finite or semi-infinite time interval; by "reasonable" we mean piecewise continuous or measurable.

SISO systems: an abbreviation for single input, single output systems.

Stable system: a dynamical system is stable if a gain function exists so that, for each input-output pair, the norm of the output is bounded by the gain function evaluated at the norm of the input.

Strict contraction: the composition of two gain functions γ_1 and γ_2 is a strict contraction if a proper separation function ρ exists so that $(\gamma_1 + \rho) \circ (\gamma_2 + \rho) \leq \text{Id}$, where $\text{Id}(\zeta) = \zeta$ and $\tilde{\gamma}_1 \circ \tilde{\gamma}_2(\zeta) = \tilde{\gamma}_1(\tilde{\gamma}_2(\zeta))$. Graphically, this is the equivalent to the curve $(\zeta, \gamma_1(\zeta) + \rho(\zeta))$ never being above the curve $(\gamma_2(\xi) + \rho(\xi), \xi)$. This concept is used to state the nonlinear small gain theorem. See Section 56.7.

Strictly passive: We have used various notions of strictly passive including input-, output-, input and output-, and nonlinear input and output-strictly passive. All notions strengthen the requirement that the inner product of the input-output pairs be positive by requiring a positive lower bound that depends on the 2-norm of the input and/or output. See Section 56.5.

Truncated signal: A signal defined on the semi-infinite interval which is derived from another signal (not necessarily defined on the semi-infinite interval) by first appending zeros to extend the signal onto the semi-infinite interval and then keeping the first part of the signal and setting the rest of the signal to zero. Used to measure the size of finite portions of signals.

Well-defined interconnection: An interconnection of two dynamical systems in the configuration of Figure 56.2 which results in another dynamical system, i.e., one in which an output signal is produced for each input signal.

References

[1] Georgiou, T.T. and Smith, M.C., Optimal robustness in the gap metric, *IEEE Trans. Auto. Control*, 35, 673–686, 1990.

[2] Jiang, Z.P., Teel, A.R., and Praly, L., Small-gain theorem for ISS systems and applications, *Math. Control, Sign., Syst.*, 7(2), 95–120, 1995.

[3] Mareels, I.M.Y. and Hill, D.J., Monotone stability of nonlinear feedback systems, *J. Math. Syst., Est. Control*, 2(3), 275–291, 1992.

[4] McFarlane, D.C. and Glover, K., *Robust controller design using normalized coprime factor plant descriptions*, Lecture Notes in Control and Information Sciences, Springer-Verlag, vol. 138, 1989.

[5] Popov, V.M., Absolute stability of nonlinear systems of automatic control, *Auto. Remote Control*, 22, 857–875, 1961.

[6] Safonov, M., *Stability and robustness of multivariable feedback systems*, The MIT Press, Cambridge, MA, 1980.

[7] Sandberg, I.W., On the $\mathcal{L}_2$-boundedness of solutions of nonlinear functional equations, *Bell Sys. Tech. J.*, 43, 1581–1599, 1964.

[8] Sontag, E., Smooth stabilization implies coprime factorization, *IEEE Trans. Auto. Control*, 34, 435–443, 1989.

[9] Zames, G., On the input-output stability of time-varying nonlinear feedback systems, Part I: Conditions using concepts of loop gain, conicity, and positivity, *IEEE Trans. Auto. Control*, 11, 228–238, 1969.

[10] Zames, G., On the input-output stability of time-varying nonlinear feedback systems, Part II: Conditions involving circles in the frequency plane and sector nonlinearities, *IEEE Trans. Auto. Control*, 11, 465–476, 1966.

Further Reading

As mentioned at the outset, the material presented in this chapter is based on the results in [6], [7], [9], and [10]. In the latter, a more general feedback interconnection structure is considered where nonzero initial conditions can also be consider as inputs.

Other excellent references on input-output stability include *The Analysis of Feedback Systems*, 1971, by J.C. Willems and *Feedback Systems: Input-Output Properties*, 1975, by C. Desoer and M. Vidyasagar. A nice text addressing the factorization method in linear systems control design is *Control Systems Synthesis: A Factorization Approach*, 1985, by M. Vidyasagar. A treatment of input-output stability for linear, infinite dimensional systems can be found in Chapter 6 of *Nonlinear Systems Analysis*, 1993, by M. Vidyasagar. That chapter also discusses many of the connections between input-output stability and state-space (Lyapunov) stability. Another excellent reference is *Nonlinear Systems*, 1992, by H. Khalil.

There are results similar to the circle theorem that we have not discussed. They go under the heading of "multiplier" results and apply to feedback loops with a linear element and a memoryless, nonlinear element with extra restrictions such as time invariance and constrained slope. Special cases are the well-known Popov and off-axis circle criterion. Some of these results can be recovered using the general conic sector theorem although we have not taken the time to do this. Other results, like the Popov criterion, impose extra smoothness conditions on the external inputs which are not found in the standard problem. References for these problems are *Hyperstability of Control Systems*, 1973, V.M. Popov, the English translation of a book originally published in 1966, and *Frequency Domain Criteria for Absolute Stability*, 1973, by K.S. Narendra and J.H. Taylor.

Another topic closely related to these multiplier results is the structured small gain theorem for linear systems which lends to much of the μ-synthesis control design methodology. This is described, for example, in *μ-Analysis and Synthesis Toolbox*, 1991, by G. Balas, J. Doyle, K. Glover, A. Packard and R. Smith.

There are many advanced topics concerning input-output stability that we have not addressed. These include the study of small-signal stability, well-posedness of feedback loops, and control design based on input-output stability principles. Many articles on these topics frequently appear in control and systems theory journals such as *IEEE Transactions on Automatic Control, Automatica, International Journal of Control, Systems and Control Letters, Mathematics of Control, Signals, and Systems*, to name a few.

57

Design Methods

Alberto Isidori
Maria Domenica Di Benedetto
Christopher I. Byrnes
Randy A. Freeman
Petar V. Kokotović
R. A. DeCarlo
S. H. Żak
S. V. Drakunov
Eyad H. Abed
Hua O. Wang
Alberto Tesi
J. Baillieul
B. Lehman
Miroslav Krstić
Kevin M. Passino
Stephen Yurkovich
Jay A. Farrell

57.1 Feedback Linearization of Nonlinear Systems

Alberto Isidori, Dipartimento di Informatica e Sistemistica, Università di Roma "La Sapienza", Rome, and Department of Systems Sciences and Mathematics, Washington University, St.Louis, MO

Maria Domenica Di Benedetto, Dipartimento di Ingegneria Elettrica, Università de L'Aquila, Monteluco di Roio (L'Aquila)

0-8493-8570-9/96/$0.00+$.50

57.1.1 The Problem of Feedback Linearization

A basic problem in control theory is how to use feedback in order to modify the original internal dynamics of a controlled plant so as to achieve some prescribed behavior. In particular, feedback may be used for the purpose of imposing, on the associated closed-loop system, the (unforced) behavior of some prescribed *autonomous linear system*. When the plant is modeled as a linear time-invariant system, this is known as the problem of pole placement, while, in the more general case of a nonlinear model, this is known as the problem of *feedback linearization* (see [1], [2], [3], [4]).

The purpose of this chapter is to present some of the basic features of the theory of feedback linearization.

Consider a *plant* modeled by nonlinear differential equations of the form

$$\dot{x} = f(x) + g(x)u \quad (57.1)$$
$$y = h(x) \quad (57.2)$$

having *internal state* $x = (x_1, x_2, \cdots, x_n) \in \mathbb{R}^n$, *control input* $u \in \mathbb{R}$ and *measured output* $y \in \mathbb{R}$. The functions

$$f(x) = \begin{pmatrix} f_1(x_1, x_2, \cdots, x_n) \\ f_2(x_1, x_2, \cdots, x_n) \\ \cdots \\ f_n(x_1, x_2, \cdots, x_n) \end{pmatrix},$$
$$g(x) = \begin{pmatrix} g_1(x_1, x_2, \cdots, x_n) \\ g_2(x_1, x_2, \cdots, x_n) \\ \cdots \\ g_n(x_1, x_2, \cdots, x_n) \end{pmatrix},$$
$$h(x) = h(x_1, x_2, \cdots, x_n)$$

are nonlinear functions of their arguments that are assumed to be differentiable a sufficient number of times.

Changes in the description and in the behavior of this system will be investigated under two types of transformations: (1) changes of coordinates in the state space and (2) static state feedback control laws, i.e., *memoryless* state feedback laws.

In the case of a *linear* system,

$$\dot{x} = Ax + Bu \quad (57.3)$$
$$y = Cx \quad (57.4)$$

a static state feedback control law takes the form

$$u = Fx + Gv, \quad (57.5)$$

in which v represents a new control input and F and G are matrices of appropriate dimensions. Moreover, only linear changes of coordinates are usually considered. This corresponds to the substitution of the original state vector x with a new vector z related to x by a transformation of the form

$$z = Tx$$

where T is a nonsingular matrix. Accordingly, the original description of the system of Equations 57.3 and 57.4 is replaced by a new description

$$\dot{z} = \tilde{A}z + \tilde{B}u$$
$$y = \tilde{C}z$$

in which

$$\tilde{A} = TAT^{-1}, \ \tilde{B} = TB, \ \tilde{C} = CT^{-1}.$$

In the case of a nonlinear system, a static state feedback control law is a control law of the form

$$u = \alpha(x) + \beta(x)v, \quad (57.6)$$

where v represents a new control input and $\beta(x)$ is assumed to be nonzero for all x. Moreover, *nonlinear* changes of coordinates are considered, i.e., transformations of the form

$$z = \Phi(x) \quad (57.7)$$

where z is the new state vector and $\Phi(x)$ represents a (n-vector-valued) function of n variables,

$$\Phi(x) = \begin{pmatrix} \phi_1(x) \\ \phi_2(x) \\ \cdots \\ \phi_n(x) \end{pmatrix} = \begin{pmatrix} \phi_1(x_1, x_2, \cdots, x_n) \\ \phi_2(x_1, x_2, \cdots, x_n) \\ \cdots \\ \phi_n(x_1, x_2, \cdots, x_n) \end{pmatrix}$$

with the following properties:

1. $\Phi(x)$ is invertible; i.e., there exists a function $\Phi^{-1}(z)$ such that

$$\Phi^{-1}(\Phi(x)) = x, \qquad \Phi(\Phi^{-1}(z)) = z$$

for all $x \in \mathbb{R}^n$ and all $z \in \mathbb{R}^n$.

2. $\Phi(x)$ and $\Phi^{-1}(z)$ are both smooth mappings; i.e., continuous partial derivatives of any order exist for both mappings.

A transformation of this type is called a *global diffeomorphism.* The first property is needed to guarantee the invertibility of the transformation to yield the original state vector as

$$x = \Phi^{-1}(z)$$

while the second one guarantees that the description of the system in the new coordinates is still a smooth one.

Sometimes a transformation possessing both these properties and defined for all x is hard to find and the properties in question are difficult to check. Thus, in most cases, transformations defined only in a neighborhood of a given point are of interest. A transformation of this type is called a *local diffeomorphism.* To check whether a given transformation is a local diffeomorphism, the following result is very useful.

PROPOSITION 57.1 Suppose $\Phi(x)$ is a smooth function defined on some subset U of $\mathbb{R}^n$. Suppose the Jacobian matrix

$$\frac{\partial \Phi}{\partial x} = \begin{pmatrix} \frac{\partial \phi_1}{\partial x_1} & \frac{\partial \phi_1}{\partial x_2} & \cdots & \frac{\partial \phi_1}{\partial x_n} \\ \frac{\partial \phi_2}{\partial x_1} & \frac{\partial \phi_2}{\partial x_2} & \cdots & \frac{\partial \phi_2}{\partial x_n} \\ \cdot & \cdot & \cdots & \cdot \\ \frac{\partial \phi_n}{\partial x_1} & \frac{\partial \phi_n}{\partial x_2} & \cdots & \frac{\partial \phi_n}{\partial x_n} \end{pmatrix}$$

is nonsingular at a point $x = x^0$. Then, for some suitable open subset U^0 of U, containing x^0, $\Phi(x)$ defines a local diffeomorphism between U^0 and its image $\Phi(U^0)$.

The effect of a change of coordinates on the description of a nonlinear system can be analyzed as follows. Set

$$z(t) = \Phi(x(t))$$

and differentiate both sides with respect to time to yield

$$\dot{z}(t) = \frac{dz}{dt} = \frac{\partial \Phi}{\partial x}\frac{dx}{dt} = \frac{\partial \Phi}{\partial x}(f(x(t)) + g(x(t))).$$

Then, expressing $x(t)$ as $\Phi^{-1}(z(t))$, one obtains

$$\dot{z}(t) = \tilde{f}(z(t)) + \tilde{g}(z(t))u(t)$$
$$y(t) = \tilde{h}(z(t))$$

$$\text{where}\quad \begin{aligned}\tilde{f}(z) &= \left(\frac{\partial \Phi}{\partial x} f(x)\right)_{x=\Phi^{-1}(z)}\\ \tilde{g}(z) &= \left(\frac{\partial \Phi}{\partial x} g(x)\right)_{x=\Phi^{-1}(z)}\\ \tilde{h}(z) &= (h(x))_{x=\Phi^{-1}(z)}\,.\end{aligned}$$

The latter are the formulas relating the new description of the system to the original one.

Given the nonlinear system of Equation 57.1, the problem of *feedback linearization* consists of finding, if possible, a change of coordinates of the form of Equation 57.7 and a static state feedback of the form of Equation 57.6 such that the composed dynamics of Equations 57.1 to 57.6, namely the system

$$\dot{x} = f(x) + g(x)\alpha(x) + g(x)\beta(x)v, \tag{57.8}$$

expressed in the new coordinates z, is the linear and controllable system

$$\begin{aligned}\dot{z}_1 &= z_2\\ \dot{z}_2 &= z_3\\ &\cdots\\ \dot{z}_{n-1} &= z_n\\ \dot{z}_n &= v.\end{aligned}$$

57.1.2 Normal Forms of Single-Input Single-Output Systems

Single-input single-output nonlinear systems can be locally given, by means of a suitable change of coordinates in the state space, a "normal form" of special interest, in which several important properties can be put in evidence and which is useful in solving the problem of feedback linearization. In this section, methods for obtaining this normal form are presented.

Given a real-valued function of $x = (x_1, \cdots, x_n)$

$$\lambda(x) = \lambda(x_1, \cdots, x_n)$$

and a (n-vector)-valued function of x

$$f(x) = \begin{pmatrix} f_1(x_1, \cdots, x_n)\\ \cdots \\ f_n(x_1, \cdots, x_n)\end{pmatrix},$$

we define a new real-valued function of x, denoted $L_f\lambda(x)$, in the following way

$$L_f\lambda(x) = L_f\lambda(x_1, \cdots, x_n) = \sum_{i=1}^{n} \frac{\partial \lambda}{\partial x_i} f_i(x_1, \cdots, x_n).$$

Setting

$$\frac{\partial \lambda}{\partial x} = \left(\frac{\partial \lambda}{\partial x_1} \quad \cdots \quad \frac{\partial \lambda}{\partial x_n}\right)$$

the function $L_f\lambda(x)$ can be expressed in the simple form

$$L_f\lambda(x) = \frac{\partial \lambda}{\partial x} f(x).$$

The new function $L_f\lambda(x)$ thus defined is sometimes called the derivative of $\lambda(x)$ along $f(x)$. Repeated use of this operation is possible. Thus, for instance, by differentiating $\lambda(x)$ first along $f(x)$ and then along $g(x)$, we may construct the function

$$L_g L_f \lambda(x) = \frac{\partial L_f \lambda}{\partial x} g(x),$$

or, differentiating k times $\lambda(x)$ along $f(x)$, we may construct a function recursively defined as

$$L_f^k \lambda(x) = \frac{\partial L_f^{k-1}\lambda}{\partial x} f(x).$$

With the help of this operation, we introduce the notion of *relative degree* of a system.

DEFINITION 57.1 The single-input single-output nonlinear system

$$\begin{aligned}\dot{x} &= f(x) + g(x)u\\ y &= h(x)\end{aligned}$$

has relative degree r at x^0 if:

1. $L_g L_f^k h(x) = 0$ for all x in a neighborhood of x^0 and all $k < r - 1$.
2. $L_g L_f^{r-1} h(x^0) \neq 0$.

EXAMPLE 57.1:

Consider the system

$$\dot{x} = \begin{pmatrix} 0\\ x_1^2 + \sin x_2\\ -x_2\end{pmatrix} + \begin{pmatrix}\exp(x_2)\\ 1\\ 0\end{pmatrix} u \tag{57.9}$$

$$y = h(x) = x_3\,. \tag{57.10}$$

For this system we have

$$\begin{aligned}\frac{\partial h}{\partial x} &= (0 \quad 0 \quad 1), & L_g h(x) = 0,\quad L_f h(x) = -x_2\\ \frac{\partial (L_f h)}{\partial x} &= (0 \quad -1 \quad 0), & L_g L_f h(x) = -1\,.\end{aligned}$$

Thus, the system has relative degree 2 at any point x°. However, if the output function were, for instance

$$y = h(x) = x_2$$

then $L_g h(x) = 1$ and the system would have relative degree 1 at any point x°.

The notion of relative degree has the following interesting interpretation. Suppose the system at some time t^0 is in the state $x(t^0) = x^0$. Calculate the value of $y(t)$, the output of the system, and of its derivatives with respect to time, $y^{(k)}(t)$, for $k = 1, 2, \cdots$, at $t = t^0$, to obtain

$$y(t^0) = h(x(t^0)) = h(t^0)$$

$$\begin{aligned}y^{(1)}(t) &= \frac{\partial h}{\partial x}\frac{dx}{dt} = \frac{\partial h}{\partial x}(f(x(t)) + g(x(t))u(t))\\ &= L_f h(x(t)) + L_g h(x(t))u(t).\end{aligned}$$

If the relative degree r is larger than 1, for all t such that $x(t)$ is near x^0, i.e., for all t near $t = t^0$, we have $L_g h(x(t)) = 0$ and therefore

$$y^{(1)}(t) = L_f h(x(t)).$$

This yields

$$\begin{aligned} y^{(2)}(t) &= \frac{\partial L_f h}{\partial x}\frac{dx}{dt} = \frac{\partial L_f h}{\partial x}(f(x(t)) + g(x(t))u(t)) \\ &= L_f^2 h(x(t)) + L_g L_f h(x(t))u(t) \end{aligned}$$

Again, if the relative degree is larger than 2, for all t near $t = t^0$ we have $L_g L_f h(x(t)) = 0$ and

$$y^{(2)}(t) = L_f^2 h(x(t)).$$

Continuing in this way, we get

$$y^{(k)}(t) = L_f^k h(x(t))$$

for all $k < r$ and all t near $t = t^0$, and

$$y^{(r)}(t^0) = L_f^r h(x^0) + L_g L_f^{r-1} h(x^0) u(t^0).$$

Thus, r is exactly the number of times $y(t)$ has to be differentiated at $t = t^0$ for $u(t^0)$ to appear explicitly.

The above calculations suggest that the functions $h(x)$, $L_f h(x)$, $\cdots$, $L_f^{r-1} h(x)$ have a special importance. As a matter of fact, it is possible to show that they can be used in order to define, at least partially, a local coordinate transformation near x^0, where x^0 is a point such that $L_g L_f^{r-1} h(x^0) \neq 0$. This is formally expressed in the following statement.

PROPOSITION 57.2 Let the system of Equations 57.1 and 57.2 be given and let r be its relative degree at $x = x^0$. Then $r \leq n$. Set

$$\begin{aligned} \phi_1(x) &= h(x) \\ \phi_2(x) &= L_f h(x) \\ &\cdots \\ \phi_r(x) &= L_f^{r-1} h(x). \end{aligned}$$

If r is strictly less than n, it is always possible to find $n - r$ additional functions $\phi_{r+1}(x), \cdots, \phi_n(x)$ such that the mapping

$$\Phi(x) = \begin{pmatrix} \phi_1(x) \\ \cdots \\ \phi_n(x) \end{pmatrix}$$

has a Jacobian matrix that is nonsingular at x^0 and therefore qualifies as a local coordinates transformation in a neighborhood of x^0. The value at x^0 of these additional functions can be fixed arbitrarily. Moreover, it is always possible to choose $\phi_{r+1}(x), \cdots, \phi_n(x)$ in such a way that

$$L_g \phi_i(x) = 0$$

for all $r + 1 \leq i \leq n$ and all x around x^0.

The description of the system in the new coordinates $z_i = \phi_i(x)$, $1 \leq i \leq n$ can be derived very easily. From the previous calculations, we obtain for $z_1, \cdots, z_r$

$$\frac{dz_1}{dt} = \frac{\partial \phi_1}{\partial x}\frac{dx}{dt} = \frac{\partial h}{\partial x}\frac{dx}{dt} = L_f h(x(t)) = \phi_2(x(t)) = z_2(t)$$

$$\cdots$$

$$\begin{aligned} \frac{dz_{r-1}}{dt} &= \frac{\partial \phi_{r-1}}{\partial x}\frac{dx}{dt} = \frac{\partial L_f^{r-2} h}{\partial x}\frac{dx}{dt} \\ &= L_f^{r-1} h(x(t)) = \phi_r(x(t)) = z_r(t). \end{aligned}$$

For z_r we obtain

$$\frac{dz_r}{dt} = L_f^r h(x(t)) + L_g L_f^{r-1} h(x(t)) u(t).$$

On the right-hand side of this equation we must now replace $x(t)$ by its expression as a function of $z(t)$, i.e., $x(t) = \Phi^{-1}(z(t))$. Thus, set

$$\begin{aligned} a(z) &= L_g L_f^{r-1} h(\Phi^{-1}(z)) \\ b(z) &= L_f^r h(\Phi^{-1}(z)), \end{aligned}$$

to obtain

$$\frac{dz_r}{dt} = b(z(t)) + a(z(t))u(t).$$

As far as the remaining coordinates are concerned, we cannot expect any special structure for the corresponding equations. If $\phi_{r+1}(x), \cdots, \phi_n(x)$ have been chosen so that $L_g \phi_i(x) = 0$, then

$$\begin{aligned} \frac{dz_i}{dt} &= \frac{\partial \phi_i}{\partial x}(f(x(t)) + g(x(t))u(t)) \\ &= L_f \phi_i(x(t)) + L_g \phi_i(x(t))u(t) \\ &= L_f \phi_i(t). \end{aligned}$$

Setting

$$q_i(z) = L_f \phi_i(\Phi^{-1}(z))$$

for $r + 1 \leq i \leq n$, the latter can be rewritten as

$$\frac{dz_i}{dt} = q_i(z(t)).$$

Thus, in summary, the state space description of the system in the new coordinates is as follows

$$\begin{aligned} \dot{z}_1 &= z_2 \\ \dot{z}_2 &= z_3 \\ &\cdots \\ \dot{z}_{r-1} &= z_r \\ \dot{z}_r &= b(z) + a(z)u \\ \dot{z}_{r+1} &= q_{r+1}(z) \\ &\cdots \\ \dot{z}_n &= q_n(z). \end{aligned}$$

In addition to these equations, one has to specify how the output of the system is related to the new state variables. Since $y = h(x)$, it is immediately seen that

$$y = z_1$$

The equations thus defined are said to be in *normal form*. Note that at point $z^0 = \Phi(x^0)$, $a(z^0) \neq 0$ by definition. Thus, the coefficient $a(z)$ is nonzero for all z in a neighborhood of z^0.

EXAMPLE 57.2:

Consider the system of Equations 57.9 and 57.10). In order to find the normal form, we set

$$\begin{aligned} z_1 &= \phi_1(x) = h(x) = x_3 \\ z_2 &= \phi_2(x) = L_f h(x) = -x_2 . \end{aligned}$$

We now have to find a function $\phi_3(x)$ that completes the coordinate transformation and is such that $L_g\phi_3(x) = 0$, i.e.,

$$\frac{\partial \phi_3}{\partial x} g(x) = \frac{\partial \phi_3}{\partial x_1} \exp(x_2) + \frac{\partial \phi_3}{\partial x_2} = 0 .$$

The function

$$\phi_3(x) = 1 + x_1 - \exp(x_2)$$

satisfies the equation above. The transformation $z = \Phi(x)$ defined by the functions $\phi_1(x)$, $\phi_2(x)$, and $\phi_3(x)$ has a Jacobian matrix

$$\frac{\partial \Phi}{\partial x} = \begin{pmatrix} 0 & 0 & 1 \\ 0 & -1 & 0 \\ 1 & -\exp(x_2) & 0 \end{pmatrix}$$

which is nonsingular for all x, and $\Phi(0) = 0$. Hence, $z = \Phi(x)$ defines a global change of coordinates. The inverse transformation is given by

$$\begin{aligned} x_1 &= z_3 - 1 + \exp(-z_2) \\ x_2 &= -z_2 \\ x_3 &= z_1 . \end{aligned}$$

In the new coordinates the system is described by

$$\begin{aligned} \dot{z}_1 &= z_2 \\ \dot{z}_2 &= (1 - z_3 - \exp(-z_2))^2 + \sin z_2 - u \\ \dot{z}_3 &= \exp(-z_2)(\sin z_2 - (z_3 - 1 + \exp(-z_2))^2) . \end{aligned}$$

These equations describe the system in normal form and are globally valid because the coordinate transformation we considered is global.

57.1.3 Conditions for Exact Linearization via Feedback

In this section, conditions and constructive procedures are given for a single-input single-output nonlinear system to be transformed into a linear and controllable system via change of coordinates in the state space and static state feedback.

The discussion is based on the normal form developed in the previous section. Consider a nonlinear system having at some point $x = x^0$ relative degree equal to the dimension of the state space, i.e., $r = n$. In this case, the change of coordinates required to construct the normal form is given by the function $h(x)$ and its first $n - 1$ derivatives along $f(x)$, i.e.,

$$\Phi(x) = \begin{pmatrix} \phi_1(x) \\ \phi_2(x) \\ \cdots \\ \phi_n(x) \end{pmatrix} = \begin{pmatrix} h(x) \\ L_f h(x) \\ \cdots \\ L_f^{n-1} h(x) \end{pmatrix} . \qquad (57.11)$$

No additional functions are needed to complete the transformation. In the new coordinates

$$z_i = \phi_i(x) = L_f^{i-1}(x) \qquad 1 \le i \le n$$

the system is described by equations of the form:

$$\begin{aligned} \dot{z}_1 &= z_2 \\ \dot{z}_2 &= z_3 \\ &\cdots \\ \dot{z}_{n-1} &= z_n \\ \dot{z}_n &= b(z) + a(z)u \end{aligned}$$

where $z = (z_1, z_2, \cdots, z_n)$. Recall that at the point $z^0 = \Phi(x^0)$, and hence at all z in a neighborhood of z^0, the function $a(z)$ is nonzero.

Choose now the following state feedback control law

$$u = \frac{1}{a(z)}(-b(z) + v), \qquad (57.12)$$

which indeed exists and is well defined in a neighborhood of z^0. The resulting closed-loop system is governed by the equations

$$\begin{aligned} \dot{z}_1 &= z_2 && (57.13) \\ \dot{z}_2 &= z_3 && (57.14) \\ &\cdots && (57.15) \\ \dot{z}_{n-1} &= z_n && (57.16) \\ \dot{z}_n &= v, && (57.17) \end{aligned}$$

i.e., it is linear and controllable. Thus we conclude that any nonlinear system with relative degree n at some point x^0 can be transformed into a system that is linear and controllable by means of (1) a local change of coordinates, and (2) a local static state feedback.

The two transformations used in order to obtain the linear form can be interchanged: one can first apply a feedback and then change the coordinates in the state space, without altering the result. The feedback needed to achieve this purpose is exactly the feedback of Equation 57.12, but now expressed in the x coordinates as

$$u = \frac{1}{a(\Phi(x))}(-b(\Phi(x)) + v). \qquad (57.18)$$

Comparing this with the expressions for $a(z)$ and $b(z)$ given in the previous section, one immediately realizes that this feedback, expressed in terms of the functions $f(x)$, $g(x)$, $h(x)$, which characterize the original system, has the form

$$u = \frac{1}{L_g L_f^{n-1}(x)}(-L_f^n(x) + v)$$

An easy calculation shows that the feedback of Equation 57.18, together with the same change of coordinates used so far (Equation 57.11), exactly yields the same linear and controllable system.

If x^0 is an equilibrium point for the original nonlinear system, i.e., if $f(x^0) = 0$, and if also $h(x^0) = 0$, then

$$\phi_1(x^0) = h(x^0) = 0$$

and

$$\phi_i(x^0) = \frac{\partial L_f^{i-1} h}{\partial x} f(x^0) = 0$$

for all $2 \leq i \leq n$, so that $z^0 = \Phi(x^0) = 0$. Note that a condition like $h(x^0) = 0$ can always be satisfied by means of a suitable translation of the origin of the output space. Thus, we conclude that, if x^0 is an equilibrium point for the original system, and this system has relative degree n at x^0, there is a feedback control law (defined in a neighborhood of x^0) and a coordinate transformation (also defined in a neighborhood of x^0) that change the system into a linear and controllable one, defined in a neighborhood of the origin.

New feedback controls can be imposed on the linear system thus obtained; for example,

$$v = Kz$$

where

$$K = (k_1 \quad k_2 \quad \cdots \quad k_n)$$

can be chosen to meet some given control specifications, e.g., to assign a specific set of eigenvalues or to satisfy an optimality criterion. Recalling the expression of the z_is as functions of x, the feedback in question can be rewritten as

$$v = k_1 h(x) + k_2 L_f h(x) + \cdots + k_n L_f^{n-1} h(x)$$

i.e., in the form of a nonlinear feedback from the state x of the original description of the system.

Up to this point of the presentation, the existence of an "output" function $h(x)$ relative to which the system of Equations 57.1 and 57.2 has relative degree exactly equal to n (at x^0) has been key in making it possible to transform the system into a linear and controllable one. Now, if such a function $h(x)$ is not available beforehand, either because the actual output of the system does not satisfy the conditions required to have relative degree n or simply because no specific output is defined for the given system, the question arises whether it is possible to find an appropriate $h(x)$ that allows output linearization. This question is answered in the remaining part of this section.

Clearly, the problem consists of finding a function, $h(x)$, satisfying the conditions

$$L_g h(x) = L_g L_f h(x) = \cdots = L_g L_f^{n-2} h(x) = 0$$

for all x near x^0 and

$$L_g L_f^{n-2} h(x^0) \neq 0.$$

We shall see that these conditions can be transformed into a partial differential equation for $h(x)$, for which conditions for existence of solutions as well as constructive integration procedures are well known. In order to express this, we need to introduce another type of differential operation. Given two (n-vector)-valued functions of $x = (x_1, \cdots, x_n)$, $f(x)$ and $g(x)$, we define a new (n-vector)-valued function of x, denoted $[f, g](x)$, in the following way

$$[f, g](x) = \frac{\partial g}{\partial x} f(x) - \frac{\partial f}{\partial x} g(x)$$

where $\frac{\partial g}{\partial x}$ and $\frac{\partial f}{\partial x}$ are the Jacobian matrices of $g(x)$ and $f(x)$, respectively. The new function thus defined is called the *Lie product* or *Lie bracket* of $f(x)$ and $g(x)$. The Lie product can be used repeatedly. Whenever a function $g(x)$ is "Lie-multiplied" several times by a function $f(x)$, the following notation is frequently used

$$\begin{aligned} ad_f g(x) &= [f, g](x) \\ ad_f^2 g(x) &= [f, [f, g]](x) \\ &\cdots \\ ad_f^k g(x) &= [f, ad_f^{k-1} g](x). \end{aligned}$$

We shall see now that the conditions a function $h(x)$ must obey in order to be eligible as "output" of a system with relative degree n can be re-expressed in a form involving the gradient of $h(x)$ and a certain number of the repeated Lie products of $f(x)$ and $g(x)$. For, note that, since

$$L_{[f,g]} h(x) = L_f L_g h(x) - L_g L_f h(x),$$

if $L_g h(x) = 0$, the two conditions $L_{[f,g]} h(x) = 0$ and $L_g L_f h(x) = 0$ are equivalent. Using this property repeatedly, one can conclude that a system has relative degree n at x^0 if and only if

$$L_g h(x) = L_{ad_f g} h(x) = \cdots = L_{ad_f^{n-2} g} h(x) = 0$$

for all x near x^0, and

$$L_{ad_f^{n-1} g} h(x^0) \neq 0.$$

Keeping in mind the definition of derivative of $h(x)$ along a given (n-vector)-valued function, the first set of conditions can be rewritten in the following form

$$\frac{\partial h}{\partial x} (g(x) \quad ad_f g(x) \quad \cdots \quad ad_f^{n-2} g(x)) = 0. \tag{57.19}$$

This partial differential equation for $h(x)$ has important properties. Indeed, if a function $h(x)$ exists such that

$$L_g h(x) = L_{ad_f g} h(x) = \cdots = L_{ad_f^{n-2} g} h(x) = 0$$

for all x near x^0, and

$$L_{ad_f^{n-1} g} h(x^0) \neq 0,$$

then necessarily the n vectors

$$g(x^0) \quad ad_f g(x^0) \quad \cdots \quad ad_f^{n-2} g(x^0) \quad ad_f^{n-1} g(x^0)$$

must be linearly independent. So, in particular, the matrix

$$(g(x) \quad ad_f g(x) \quad \cdots \quad ad_f^{n-2} g(x))$$

has rank $n-1$. The conditions for the existence of solutions to a partial differential equation of the form of Equation 57.19 where the matrix

$$(g(x) \quad ad_f g(x) \quad \cdots \quad ad_f^{n-2} g(x))$$

has full rank are given by the well-known *Frobenius' theorem.*

THEOREM 57.1 *Consider a partial differential equation of the form*

$$\frac{\partial h}{\partial x}(X_1(x) \quad X_2(x) \quad \cdots \quad X_k(x)) = 0,$$

in which $X_1(x), \cdots, X_k(x)$ are (n-vector)-valued functions of x. Suppose the matrix

$$(X_1(x) \quad X_2(x) \quad \cdots \quad X_k(x))$$

has rank k at the point $x = x^0$. There exist $n-k$ real-valued functions of x, say $h_1(x), \cdots, h_{n-k}(x)$, defined in a neighborhood of x^0, that are solutions of the given partial differential equation, and are such that the Jacobian matrix

$$\begin{pmatrix} \frac{\partial h_1}{\partial x} \\ \cdots \\ \frac{\partial h_{n-k}}{\partial x} \end{pmatrix}$$

has rank $n-k$ at $x = x^0$ if and only if, for each pair of integers (i, j), $1 \le i, j \le k$, the matrix

$$(X_1(x) \quad X_2(x) \quad \cdots \quad X_k(x) \quad [X_i, X_j](x))$$

has rank k for all x in a neighborhood of x^0.

REMARK 57.1 A set of k (n-vector)-valued functions $\{X_1(x), \cdots, X_k(x)\}$, such that the matrix:

$$(X_1(x) \quad X_2(x) \quad \cdots \quad X_k(x))$$

has rank k at the point $x = x^0$, is said to be *involutive* near x^0 if, for each pair of integers (i, j), $1 \le i, j \le k$, the matrix

$$(X_1(x) \quad X_2(x) \quad \cdots \quad X_k(x) \quad [X_i, X_j](x))$$

still has rank k for all x in a neighborhood of x^0. Using this terminology, the necessary and sufficient condition indicated in the previous theorem can be simply referred to as the *involutivity* of the set $\{X_1(x), \cdots, X_k(x)\}$.

The arguments developed thus far can be summarized formally as follows.

PROPOSITION 57.3 Consider a system:

$$\dot{x} = f(x) + g(x)u$$

There exists an "output" function $h(x)$ for which the system has relative degree n at a point x^0 if and only if the following conditions are satisfied:

1. The matrix

$$(g(x^0) \quad ad_f g(x^0) \quad \cdots \quad ad_f^{n-2} g(x^0) \quad ad_f^{n-1} g(x^0))$$

has rank n.

2. The set $\{g(x), ad_f g(x), \cdots, ad_f^{n-2} g(x)\}$ is involutive near x^0.

In view of the results illustrated at the beginning of the section, it is now possible to conclude that conditions 1 and 2 listed in this statement are necessary and sufficient conditions for the existence of a state feedback and of a change of coordinates transforming, at least locally around the point x^0, a given nonlinear system of the form

$$\dot{x} = f(x) + g(x)u$$

into a linear and controllable one.

REMARK 57.2 For a nonlinear system whose state space has dimension $n = 2$, condition 2 is always satisfied since $[g, g](x) = 0$. Hence, by the above result, any nonlinear system whose state space has dimension $n = 2$ can be transformed into a linear system, via state feedback and change of coordinates, around a point x^0 if and only if the matrix

$$(g(x^0) \quad ad_f g(x^0))$$

has rank 2. If this is the case, the vector $g(x^0)$ is nonzero and it is always possible to find a function $h(x) = h(x_1, x_2)$, defined locally around x^0, such that

$$\frac{\partial h}{\partial x} g(x) = \frac{\partial h}{\partial x_1} g_1(x_1, x_2) + \frac{\partial h}{\partial x_2} g_2(x_1, x_2) = 0.$$

If a nonlinear system of the form of Equations 57.1 and 57.2 having relative degree strictly less than n meets requirements 1 and 2 of the previous proposition, there exists a different "output" function, say $k(x)$, with respect to which the system has relative degree exactly n. Starting from this new function, it is possible to construct a feedback $u = \alpha(x) + \beta(x)v$ and a change of coordinates $z = \Phi(x)$, that transform the system

$$\dot{x} = f(x) + g(x)u$$

into a linear and controllable one. However, in general, the real output of the system expressed in the new coordinates

$$y = h(\Phi^{-1}(z))$$

is still a nonlinear function of the state z. Then the question arises whether there exist a feedback and a change of coordinates transforming the entire description of the system, output function included, into a linear and controllable one. The appropriate conditions should include the previous ones with some additional constraints arising from the need to linearize the output map. For the sake of completeness, a possible way of stating these conditions is given hereafter.

PROPOSITION 57.4 Let the system of Equations 57.1 and 57.2 be given and let r be its relative degree at $x = x^0$. There exist a static state feedback and a change of coordinates, defined locally around x^0, so that the system is transformed into a linear and controllable one

$$\begin{aligned} \dot{x} &= Ax + Bu \\ y &= Cx \end{aligned}$$

if and only if the following conditions are satisfied:

1. The matrix

$$\begin{pmatrix} g(x^0) & ad_f g(x^0) & \cdots & ad_f^{n-2} g(x^0) & ad_f^{n-1} g(x^0) \end{pmatrix}$$

has rank n.

2. The (n-vector)-valued functions defined as

$$\tilde{f}(x) = f(x) - \frac{L_f^r h(x)}{L_g L_f^{r-1} h(x)} g(x) \quad \tilde{g}(x) = \frac{1}{L_g L_f^{r-1} h(x)} g(x)$$

are such that

$$[ad_{\tilde{f}}^i \tilde{g}, ad_{\tilde{f}}^j \tilde{g}] = 0$$

for all pairs (i, j) such that $0 \le i, j \le n$.

EXAMPLE 57.3:

Consider the system of Equations 57.9 and 57.10. In order to see if this system can be transformed into a linear and controllable system via static state feedback and change of coordinates, we have to check conditions 1 and 2 of Proposition 57.3. We first compute $ad_f g(x)$ and $ad_f^2 g(x)$:

$$ad_f g(x) = \begin{pmatrix} \exp(x_2)(x_1^2 + \sin x_2) \\ -2x_1 \exp(x_2) - \cos x_2 \\ 1 \end{pmatrix}$$

$$ad_f^2 g(x) =$$

$$\begin{pmatrix} \exp(x_2)(x_1^2 + \sin x_2)(x_1^2 + \sin x_2 + \cos x_2) \\ x_1^2(\sin x_2 - 4x_1 \exp(x_2)) + 1 - 4x_1 \exp(x_2) \sin x_2 + 2x_1 \exp(x_2) \cos x_2 \\ -2x_1 \exp(x_2) - \cos x_2 \end{pmatrix}.$$

The matrix

$$\begin{pmatrix} g(x) & ad_f g(x) & ad_f^2 g(x) \end{pmatrix}$$

has rank 3 at all points x where its determinant, an analytic function of x, is different from zero. Hence, condition 1 is satisfied almost everywhere. Note that at point $x = 0$ the matrix

$$\begin{pmatrix} g(x) & ad_f g(x) & ad_f^2 g(x) \end{pmatrix}$$

has rank 2, and this shows that condition 1 is not satisfied at the origin.

The product $[g, ad_f g](x)$ has the form

$$[g, ad_f g](x) = \begin{pmatrix} 4x_1 \exp(2x_2) + \exp(x_2)(x_1^2 + \sin x_2 + 2\cos x_2) \\ \sin x_2 - 2\exp(2x_2) - 2x_1 \exp(x_2) \\ 0 \end{pmatrix}.$$

Then one can see that the matrix

$$\begin{pmatrix} g(x) & ad_f g(x) & [g, ad_f g](x) \end{pmatrix}$$

has rank 2 at all points x for which its determinant

$$\exp(x_2)(2\exp(2x_2) + 6x_1 \exp(x_2) + 2\cos x_2 + x_1^2)$$

is zero. This set of points has measure zero. Hence, condition 2 is not satisfied at any point x of the state space.

In summary, the system of Equations 57.9 and 57.10 satisfies condition 1 almost everywhere but does not satisfy condition 2. Hence, it is not locally feedback linearizable.

EXAMPLE 57.4:

Consider the system

$$\begin{aligned} \dot{x} &= \begin{pmatrix} x_3 - x_2 \\ 0 \\ x_3 + x_1^2 \end{pmatrix} + \begin{pmatrix} 0 \\ \exp(x_1) \\ \exp(x_1) \end{pmatrix} u \\ y &= x_2 . \end{aligned}$$

This system has relative degree 1 at all x since $L_g h(x) = \exp(x_1)$. It is easily checked that conditions 1 and 2 of Proposition 57.3 are satisfied. Hence, there exists a function $h(x)$ for which the system has relative degree 3. This function has to satisfy

$$\frac{\partial h}{\partial x} \begin{pmatrix} g(x) & ad_f g(x) \end{pmatrix} = 0.$$

A solution to this equation is given by

$$h(x) = x_1 .$$

The system can be transformed into a linear and controllable one by means of the static state feedback

$$u = \frac{-L_f^3 h(x) + v}{L_g L_f^2 h(x)} = \\ = \frac{2x_1x_2 - 2x_1x_3 - x_3 - x_1^2 + v}{\exp(x_1)}$$

and the coordinates transformation

$$\begin{aligned} z_1 &= h(x) = x_1 \\ z_2 &= L_f h(x) = x_3 - x_2 \\ z_3 &= L_f^2 h(x) = x_3 + x_1^2 . \end{aligned}$$

The original output of the system $y = x_2$ is a nonlinear function of z:

$$y = -z_2 + z_3 - z_1^2.$$

To determine whether the entire system, output function included, can be transformed into a linear and controllable one, condition 2 of Proposition 57.4 should be checked. Since $L_f h(x) = 0$, $\widetilde{f}(x) = f(x)$ and $\widetilde{g}(x) = \begin{pmatrix} 0 \\ 1 \\ 1 \end{pmatrix}$. Easy calculations yield

$$[ad_{\widetilde{f}}\widetilde{g}] = \begin{pmatrix} 0 \\ 0 \\ -1 \end{pmatrix}$$

$$[ad_{\widetilde{f}}^2\widetilde{g}] = \begin{pmatrix} 1 \\ 0 \\ 1 \end{pmatrix}$$

$$[ad_{\widetilde{f}}^3\widetilde{g}] = \begin{pmatrix} -1 \\ 0 \\ -2x_1 - 1 \end{pmatrix}$$

One can check that $[ad_{\widetilde{f}}^2\widetilde{g}, ad_{\widetilde{f}}^3\widetilde{g}] \neq 0$. Hence, condition 2 of Proposition 57.4 is not satisfied. Therefore, the system with its output cannot be transformed into a linear and controllable one.

References

[1] Jakubczyk, B. and Respondek, W., On linearization of control systems, *Bull. Acad. Polonaise Sci. Ser. Sci. Math.*, 28, 517–522, 1980.

[2] Su, R., On the linear equivalents of nonlinear systems, *Syst. Control Lett.*, 2, 48–52, 1982.

[3] Isidori, A., Krener, A.J., Gori-Giorgi, C., and Monaco, S., Nonlinear decoupling via feedback: a differential geometric approach, *IEEE Trans. Autom. Control*, 26, 331–345, 1981.

[4] Hunt, L.R., Su, R., and Meyer, G., Design for multi-input systems, in *Differential Geometric Control Theory*, Brockett, R.W., Millman, R.S., and Sussmann, H.J., Eds., Birkhauser, 1983, 268–298.

[5] Isidori, A., *Nonlinear Control Systems*, 2nd Ed., Springer-Verlag, 1989.

57.2 Nonlinear Zero Dynamics

Alberto Isidori, Dipartimento di Informatica e Sistemistica, Università di Roma "La Sapienza", Rome, and Department of Systems Sciences and Mathematics, Washington University, St.Louis, MO

Christopher I. Byrnes, Department of Systems Sciences and Mathematics, Washington University, St.Louis, MO

57.2.1 Input-Output Feedback Linearization

Consider a nonlinear single-input single-output system, described by equations of the form

$$\begin{aligned} \dot{x} &= f(x) + g(x)u \\ y &= h(x) \end{aligned} \tag{57.20}$$

and suppose $x = 0$ is an equilibrium of the vector field $f(x)$; i.e., $f(0) = 0$, and $h(0) = 0$. Assume also that this system has relative degree $r < n$ at $x = 0$. Then there is a neighborhood U of $x = 0$ in $\mathbb{R}^n$ and a local change of coordinates $z = \Phi(x)$ defined on U [and satisfying $\Phi(0) = 0$] such that, in the new coordinates, the system is described by equations of the form (see Chapter 57.1 for details)

$$\begin{aligned} \dot{z}_1 &= z_2 \\ \dot{z}_2 &= z_3 \\ &\cdots \\ \dot{z}_{r-1} &= z_r \\ \dot{z}_r &= v \\ \dot{z}_{r+1} &= b(z) + a(z)u \\ \dot{z}_{r+1} &= q_{r+1}(z) \\ &\cdots \\ \dot{z}_n &= q_n(z) \\ y &= z_1 . \end{aligned} \tag{57.21}$$

Equation 57.21, which describes the system in the new coordinates, can be more conveniently represented as follows. Set

$$\xi = \begin{pmatrix} z_1 \\ z_2 \\ \cdots \\ z_r \end{pmatrix} \qquad \eta = \begin{pmatrix} z_{r+1} \\ z_{r+2} \\ \cdots \\ z_n \end{pmatrix},$$

and recall that, in particular,

$$\xi = \begin{pmatrix} h(x) \\ L_f h(x) \\ \cdots \\ L_f^{r-1} h(x) \end{pmatrix}.$$

Moreover, define

$$A = \begin{pmatrix} 0 & 1 & 0 & \cdots & 0 & 0 \\ 0 & 0 & 1 & \cdots & 0 & 0 \\ \cdot & \cdot & \cdot & \cdots & \cdot & \cdot \\ 0 & 0 & 0 & \cdots & 0 & 1 \\ 0 & 0 & 0 & \cdots & 0 & 0 \end{pmatrix}, \qquad B = \begin{pmatrix} 0 \\ 0 \\ \cdot \\ 0 \\ 1 \end{pmatrix},$$

$$C = (1 \quad 0 \quad 0 \quad \cdots \quad 0 \quad 0)$$

and set

$$q(z) = \begin{pmatrix} q_{r+1}(z) \\ q_{r+2}(z) \\ \cdots \\ q_n(z) \end{pmatrix}.$$

Then, Equation 57.21 reduces to equations of the form

$$\begin{aligned}\dot{\xi} &= A\xi + B(b(\xi,\eta) + a(\xi,\eta)u)\\ \dot{\eta} &= q(\xi,\eta)\\ y &= C\xi\ .\end{aligned} \tag{57.22}$$

Suppose now the input u to the system of Equation 57.22 is chosen as

$$u = \frac{1}{a(\xi,\eta)}(-b(\xi,\eta) + v)\ . \tag{57.23}$$

This feedback law yields a closed-loop system that is described by equations of the form

$$\begin{aligned}\dot{\xi} &= A\xi + Bv\\ \dot{\eta} &= q(\xi,\eta)\ .\end{aligned} \tag{57.24}$$

This system clearly appears decomposed into a *linear subsystem*, of dimension r, which is the only one responsible for the input-output behavior, and a possibly nonlinear subsystem, of dimension $n-r$, whose behavior does not affect the output. In other words, this feedback law has changed the original system so as to obtain a new system whose input-output behavior coincides with that of a linear (controllable and observable) system of dimension r having transfer function

$$H(s) = \frac{1}{s^r}\ .$$

REMARK 57.3 To interpret the role played by the feedback law of Equation 57.23, it is instructive to examine the effect produced by a feedback of this kind on a *linear system*. In this case, the system of Equations 57.22 is modeled by equations of the form

$$\begin{aligned}\dot{\xi} &= A\xi + B(R\xi + S\eta + Ku)\\ \dot{\eta} &= P\xi + Q\eta\\ y &= C\xi\ .\end{aligned}$$

in which R and S are row vectors, of suitable dimensions, of real numbers; K is a nonzero real number; and P and Q are matrices, of suitable dimensions, of real numbers. The feedback of Equation 57.23 is a feedback of the form

$$u = -\frac{R}{K}\xi - \frac{S}{K}\eta + \frac{1}{K}v\ . \tag{57.25}$$

A feedback of this type indeed modifies the eigenvalues of the system on which it is imposed. Since, from the previous analysis, it is known that the transfer function of the resulting closed-loop system has no zeros and r poles at $s = 0$, it can be concluded that the effect of the feedback of Equation 57.25 is such as to place r eigenvalues at $s = 0$ and the remaining $n-r$ eigenvalues exactly where the $n-r$ zeros of the transfer function of the open-loop system are located. The corresponding closed-loop system, having $n-r$ eigenvalues coinciding with its $n-r$ zeros, is unobservable and its minimal realization has a transfer function that has no zeros and r poles at $s = 0$.

57.2.2 The Zero Dynamics

In this section we discuss an important concept that in many instances plays a role exactly similar to that of the "zeros" of the transfer function in a linear system.

Given a single-input single-output system, having relative degree $r < n$ at $x = 0$ represented by equations of the form of Equation 57.22, consider the following problem, which is sometimes called the *Problem of Zeroing the Output*. Find, if any, pairs consisting of an initial state x° and of an input function $u^\circ(\cdot)$, defined for all t in a neighborhood of $t = 0$, such that the corresponding output $y(t)$ of the system is identically zero for all t in a neighborhood of $t = 0$. Of course, the interest is to find *all* such pairs (x°, u°) and not simply the trivial pair $x^\circ = 0$, $u^\circ = 0$ (corresponding to the situation in which the system is initially at rest and no input is applied).

Recalling that, in the normal form of Equation 57.22

$$y(t) = \xi_1(t)\ ,$$

we observe that the constraint $y(t) = 0$ for all t implies

$$\xi_1(t) = \xi_2(t) = \ldots = \xi_r(t) = 0\ ,$$

that is, $\xi(t) = 0$ for all t. In other words, if the output of the system is identically zero, its state necessarily respects the constraint $\xi(t) = 0$ for all t. In addition, the input $u(t)$ must necessarily be the unique solution of the equation

$$0 = b(0,\eta(t)) + a(0,\eta(t))u(t)$$

[recall that $a(0,\eta(t)) \neq 0$ if $\eta(t)$ is close to 0]. As far as the variable $\eta(t)$ is concerned, it is clear that, $\xi(t)$ being identically zero, its behavior is governed by the differential equation

$$\dot{\eta}(t) = q(0,\eta(t))\ . \tag{57.26}$$

From this analysis it is possible to conclude the following. In order to have the output $y(t)$ of the system identically zero, necessarily the initial state must be such that $\xi(0) = 0$, whereas $\eta(0) = \eta^\circ$ can be chosen arbitrarily. According to the value of η°, the input must be set equal to the following function

$$u(t) = -\frac{b(0,\eta(t))}{a(0,\eta(t))}$$

where $\eta(t)$ denotes the solution of the differential equation

$$\dot{\eta}(t) = q(0,\eta(t))$$

with initial condition $\eta(0) = \eta^\circ$. Note also that for each set of initial conditions $(\xi,\eta) = (0,\eta^\circ)$ the input thus defined is the *unique* input capable of keeping $y(t)$ identically zero.

The dynamics of Equation 57.26 correspond to the dynamics describing the "internal" behavior of the system when input and initial conditions have been chosen in such a way as to constrain the output to remain identically zero. These dynamics, which are rather important in many instances, are called the *zero dynamics* of the system.

The previous analysis interprets the trajectories of the $(n-r)$-dimensional system

$$\dot{\eta} = q(0, \eta) \tag{57.27}$$

as "open-loop" trajectories of the system, when the latter is *forced* (by appropriate choice of input and initial condition) to constrain the output to be identically zero. However, the trajectories of Equation 57.27 can also be interpreted as *autonomous* trajectories of an appropriate "closed-loop system." In fact, consider again a system in the normal form of Equation 57.22 and suppose the feedback control law of Equation 57.23 is imposed, under which the input-output behavior becomes identical with that of a linear system. The corresponding closed-loop system thus obtained is described by Equations 57.24. If the linear subsystem is initially at rest and no input is applied, then $y(t) = 0$ for all values of t, and the corresponding internal dynamics of the whole (closed-loop) system are exactly those of Equation 57.27, namely, the zero dynamics of the open-loop system.

REMARK 57.4 In a linear system, the dynamics of Equation 57.27 are determined by the *zeros* of the transfer function of the system itself. In fact, consider a linear system having relative degree r and let

$$H(s) = K\frac{b_0 + b_1 s + \cdots + b_{n-r-1}s^{n-r-1} + s^{n-r}}{a_0 + a_1 s + \cdots + a_{n-1}s^{n-1} + s^n}$$

denote its transfer function. Suppose the numerator and denominator polynomials are relatively prime and consider a minimal realization of $H(s)$

$$\begin{aligned}\dot{x} &= Ax + Bu\\ y &= Cx\end{aligned}$$

with

$$A = \begin{pmatrix} 0 & 1 & 0 & \cdots & 0\\ 0 & 0 & 1 & \cdots & 0\\ \cdot & \cdot & \cdot & \cdots & \cdot\\ 0 & 0 & 0 & \cdots & 1\\ -a_0 & -a_1 & -a_2 & \cdots & -a_{n-1}\end{pmatrix} \quad B = \begin{pmatrix} 0\\ 0\\ \cdots\\ 0\\ K\end{pmatrix}$$

$$C = (b_0 \quad b_1 \quad \cdots \quad b_{n-r-1} \quad 1 \quad 0 \quad \cdots \quad 0);$$

The realization in question can easily be reduced to the form of Equation 57.21. For the ξ coordinates one has to take

$$\begin{aligned}\xi_1 &= Cx = b_0x_1 + b_1x_2 + \cdots + b_{n-r-1}x_{n-r} + x_{n-r+1}\\ \xi_2 &= CAx = b_0x_2 + b_1x_3 + \cdots\\ &\quad + b_{n-r-1}x_{n-r+1} + x_{n-r+2}\\ &\cdots\\ \xi_r &= CA^{r-1}x = b_0x_r + b_1x_{r+1} + \cdots\\ &\quad + b_{n-r-1}x_{n-1} + x_n.\end{aligned}$$

while for the η coordinates it is possible to choose

$$\begin{aligned}\eta_1 &= x_1\\ \eta_2 &= x_2\\ &\cdots\\ \eta_{n-r} &= x_{n-r}.\end{aligned}$$

In the new coordinates we obtain equations in normal form, which, because of the linearity of the system, have the following structure

$$\begin{aligned}\dot{\xi} &= A\xi + B(R\xi + S\eta + Ku)\\ \dot{\eta} &= P\xi + Q\eta\end{aligned}$$

where R and S are row vectors and P and Q are matrices of suitable dimensions. The zero dynamics of this system, according to our previous definition, are those of

$$\dot{\eta} = Q\eta .$$

The particular choice of the last $n-r$ new coordinates (i.e., of the elements of η) entails a particularly simple structure for the matrices P and Q. As a matter of fact, it is easily checked that

$$P = \begin{pmatrix} 0\\ 0\\ \cdot\\ 0\\ 1\end{pmatrix}, \quad Q = \begin{pmatrix} 0 & 1 & 0 & \cdots & 0\\ 0 & 0 & 1 & \cdots & 0\\ \cdot & \cdot & \cdot & \cdots & \cdot\\ 0 & 0 & 0 & \cdots & 1\\ -b_0 & -b_1 & -b_2 & \cdots & -b_{n-r-1}\end{pmatrix}.$$

From the particular form of this matrix, it is clear that the eigenvalues of Q coincide with the zeros of the numerator polynomial of $H(s)$, i.e., with the zeros of the transfer function. Thus, it is concluded that in a linear system the zero dynamics are linear dynamics with eigenvalues coinciding with the zeros of the transfer function of the system.

These arguments also show that the linear approximation, at $\eta = 0$, of the zero dynamics of a system coincides with the zero dynamics of the linear approximation of the system at $x = 0$. In order to see this, consider for $f(x)$, $g(x)$ and $h(x)$ expansions of the form

$$\begin{aligned}f(x) &= Ax + f_2(x)\\ g(x) &= B + g_1(x)\\ h(x) &= Cx + h_2(x)\end{aligned}$$

where

$$A = \left[\frac{\partial f}{\partial x}\right]_{x=0}, \qquad B = g(0), \qquad C = \left[\frac{\partial h}{\partial x}\right]_{x=0}.$$

An easy calculation shows, by induction, that

$$L_f^k h(x) = CA^k x + d_k(x)$$

where $d_k(x)$ is a function such that

$$\left[\frac{\partial d_k}{\partial x}\right]_{x=0} = 0 .$$

From this, one deduces that

$$\begin{aligned}CA^kB &= L_gL_f^kh(0) = 0 \qquad \text{for all } k < r-1\\ CA^{r-1}B &= L_gL_f^{r-1}h(0) \neq 0\end{aligned}$$

i.e., that the relative degree of the linear approximation of the system at $x = 0$ is exactly r.

From this fact, it is concluded that taking the linear approximation of equations in normal form, based on expansions of the form

$$\begin{aligned} b(\xi,\eta) &= R\xi + S\eta + b_2(\xi,\eta) \\ a(\xi,\eta) &= K + a_1(\xi,\eta) \\ q(\xi,\eta) &= P\xi + Q\eta + q_2(\xi,\eta) \end{aligned}$$

yields a linear system in normal form. Thus, the Jacobian matrix

$$Q = \left[\frac{\partial q}{\partial \eta}\right]_{(\xi,\eta)=0}$$

which describes the linear approximation at $\eta = 0$ of the zero dynamics of the original nonlinear system, has eigenvalues that coincide with the zeros of the transfer function of the linear approximation of the system at $x = 0$.

57.2.3 Local Stabilization of Nonlinear Minimum-Phase Systems

In analogy with the case of linear systems, which are traditionally said to be "minimum phase" when all their transmission zeros have negative real part, nonlinear systems (of the form of Equation 57.20) whose zero dynamics (Equation 57.27) have a locally (globally) asymptotically stable equilibrium at $z = 0$ are also called locally (globally) *minimum-phase* systems. As in the case of linear systems, minimum-phase nonlinear systems can be asymptotically stabilized via state feedback. We discuss first the case of local stabilization.

Consider again a system in normal form of Equation 57.22 and impose a feedback of the form

$$u = \frac{1}{a(\xi,\eta)}(-b(\xi,\eta) - c_0\xi_1 - c_1\xi_2 - \cdots - c_{r-1}\xi_r) \quad (57.28)$$

where $c_0, c_1, \cdots, c_{r-1}$ are real numbers.

This choice of feedback yields a closed-loop system of the form

$$\begin{aligned} \dot{\xi} &= (A+BK)\xi \\ \dot{\eta} &= q(\xi,\eta) \end{aligned} \quad (57.29)$$

with

$$A + BK = \begin{pmatrix} 0 & 1 & 0 & \cdots & 0 \\ 0 & 0 & 1 & \cdots & 0 \\ \cdot & \cdot & \cdot & \cdots & \cdot \\ 0 & 0 & 0 & \cdots & 1 \\ -c_0 & -c_1 & -c_2 & \cdots & -c_{r-1} \end{pmatrix}.$$

In particular, the matrix $A + BK$ has a characteristic polynomial

$$p(s) = c_0 + c_1 s + \cdots + c_{r-1}s^{r-1} + s^r.$$

From this form of the equations describing the closed-loop system we deduce the following interesting property.

PROPOSITION 57.5 Suppose the equilibrium $\eta = 0$ of the zero dynamics of the system is locally asymptotically stable and all the roots of the polynomial $p(s)$ have negative real part. Then the feedback law of Equation 57.28 locally asymptotically stabilizes the equilibrium $(\xi,\eta) = (0,0)$.

This is a consequence of the fact that the closed-loop system has a triangular form. According to a well-known property of systems in triangular form, since by assumption the subsystem

$$\dot{\eta} = q(0,\eta)$$

has a locally asymptotically stable equilibrium at $\eta = 0$ and the subsystem

$$\dot{\xi} = (A+BK)\xi$$

has a (globally) asymptotically stable equilibrium at $\xi = 0$, the equilibrium $(\xi,\eta) = (0,0)$ of the entire system is locally asymptotically stable.

Note that the matrix

$$Q = \left[\frac{\partial q(\xi,\eta)}{\partial \eta}\right]_{(\xi,\eta)=(0,0)}$$

characterizes the linear approximation of the zero dynamics at $\eta = 0$. If this matrix had all its eigenvalues in the left complex half-plane, then the result stated in Proposition 57.5 would have been a trivial consequence of the Principle of Stability in the First Approximation, because the linear approximation of Equation 57.29 has the form

$$\begin{pmatrix} \dot{\xi} \\ \dot{\eta} \end{pmatrix} = \begin{pmatrix} A & 0 \\ \star & Q \end{pmatrix} \begin{pmatrix} \xi \\ \eta \end{pmatrix}.$$

However, Proposition 57.5 establishes a stronger result, because it relies only upon the assumption that $\eta = 0$ is *simply* an asymptotically stable equilibrium of the zero dynamics of the system, and this (as is well known) does not necessarily require, for a nonlinear dynamics, asymptotic stability of the linear approximation (i.e., all eigenvalues of Q having negative real part). In other words, the result in question may also hold in the presence of some eigenvalue of Q with zero real part.

In order to design the stabilizing control law there is no need to know explicitly the expression of the system in normal form, but only to know *the fact* that the system has a zero dynamics with a locally asymptotically stable equilibrium at $\eta = 0$. Recalling how the ξ coordinates and the functions $a(\xi,\eta)$ and $b(\xi,\eta)$ are related to the original description of the system, it is easily seen that, in the original coordinates, the stabilizing control law assumes the form

$$\begin{aligned} u &= \frac{1}{L_g L_f^{r-1}h(x)}(-L_f^r h(x) - c_0 h(x) - c_1 L_f h(x) - \cdots \\ &\quad - c_{r-1}L_f^{r-1}h(x)) \end{aligned}$$

which is particularly interesting because expressed in terms of quantities that can be immediately calculated from the original data.

If an output function is not defined, the zero dynamics are not defined as well. However, it may happen that one is able to *design* a suitable dummy output whose associated zero dynamics have an asymptotically stable equilibrium. In this case, a control law of the form discussed before will guarantee asymptotic stability. This procedure is illustrated in the following simple example, taken from [5].

EXAMPLE 57.5:

Consider the system

$$\begin{aligned} \dot{x}_1 &= x_1^2 x_2^3 \\ \dot{x}_2 &= x_2 + u \end{aligned}$$

whose linear approximation at $x = 0$ has an uncontrollable mode corresponding to the eigenvalue $\lambda = 0$. Suppose one is able to find a function $\gamma(x_1)$ such that

$$\dot{x}_1 = x_1^2 [\gamma(x_1)]^3$$

is asymptotically stable at $x_1 = 0$. Then, setting

$$y = h(x) = \gamma(x_1) - x_2$$

a system with an asymptotically stable zero dynamics is obtained. As a matter of fact, we know that the zero dynamics are those induced by the constraint $y(t) = 0$ for all t. This, in the present case, requires that the x_1 and x_2 respect the constraint

$$\gamma(x_1) - x_2 = 0 .$$

Thus, the zero dynamics evolve exactly according to

$$\dot{x}_1 = x_1^2 [\gamma(x_1)]^3$$

and the system can be locally stabilized by means of the procedure discussed above. A suitable choice of $\gamma(x_1)$ will be, e.g.,

$$\gamma(x_1) = -x_1 .$$

Accordingly, a locally stabilizing feedback is the one given by

$$\alpha(x) = \frac{1}{L_g h(x)}\big(-L_f h(x) - c h(x)\big) = -c x_1 - (1+c) x_2 - x_1^2 x_2^3$$

with $c > 0$.

57.2.4 Global Stabilization of Nonlinear Minimum-Phase Systems

In this section we consider a special class of nonlinear system that can be *globally* asymptotically stabilized via state feedback. The systems in question are systems that can be transformed, by means of a globally defined change of coordinates and/or feedback, into a system having this special normal form

$$\begin{aligned} \dot{z} &= f_0(z, \xi_1) \\ \dot{\xi}_1 &= \xi_2 \\ &\cdots \\ \dot{\xi}_{r-1} &= \xi_r \\ \dot{\xi}_r &= u . \end{aligned} \tag{57.30}$$

REMARK 57.5 Note that a system in the normal form of Equation 57.21, considered in the previous sections, can indeed be changed, via feedback, into a system of the form

$$\begin{aligned} \dot{\xi}_1 &= \xi_2 \\ &\cdots \\ \dot{\xi}_{r-1} &= \xi_r \\ \dot{\xi}_r &= u \\ \dot{\eta} &= q(\xi, \eta) . \end{aligned} \tag{57.31}$$

Moreover, if the normal form of Equation 57.21 is globally defined, so also is the feedback yielding the (globally defined) normal form of Equation 57.31. The form of Equation 57.30 is a special case of Equation 57.31, the one in which the function $q(\xi, \eta)$ depends only on the component ξ_1 of the vector ξ. In Equation 57.30, for consistency with the notations more frequently used in the literature on global stabilization, the vector z replaces the vector η of Equation 57.31 and the places of z and ξ are interchanged.

In order to describe how systems of the form of Equation 57.30 can be globally stabilized, we begin with the analysis of the (very simple) case in which $r = 1$. For convenience of the reader, we recall that a smooth function $V : \mathbb{R}^n \to \mathbb{R}$ is said to be *positive definite* if $V(0) = 0$ and $V(x) > 0$ for $x \neq 0$, and *proper* if, for any $a \in \mathbb{R}$, the set $V^{-1}([0, a]) = \{x \in \mathbb{R}^n : 0 \leq V(x) \leq a\}$ is compact.

Consider a system described by equations of the form

$$\begin{aligned} \dot{z} &= f(z, \xi) \\ \dot{\xi} &= u \end{aligned} \tag{57.32}$$

in which $(z, \xi) \in \mathbb{R}^n \times \mathbb{R}$, and $f(0, 0) = 0$. Suppose the *subsystem*

$$\dot{z} = f(z, 0)$$

has a globally asymptotically stable equilibrium at $z = 0$. Then, in view of a converse Lyapunov theorem, there exists a smooth positive definite and proper function $V(z)$ such that $\frac{\partial V}{\partial z} f(z, 0)$ is negative for each nonzero z. Using this property, it is easy to show that the system of Equation 57.32 can be globally asymptotically stabilized. In fact, observe that the function $f(z, \xi)$ can be put in the form

$$f(z, \xi) = f(z, 0) + p(z, \xi)\xi \tag{57.33}$$

where $p(z, \xi)$ is a smooth function. For it suffices to observe that the difference

$$\bar{f}(z, \xi) = f(z, \xi) - f(z, 0)$$

is a smooth function vanishing at $\xi = 0$, and to express $\bar{f}(z, \xi)$ as

$$\bar{f}(z, \xi) = \int_0^1 \frac{\partial \bar{f}(z, s\xi)}{\partial s} ds = \int_0^1 \left[\frac{\partial \bar{f}(z, \zeta)}{\partial \zeta}\right]_{\zeta = s\xi} \xi ds .$$

Now consider the positive definite and proper function

$$W(z, \xi) = V(z) + \frac{1}{2}\xi^2 , \tag{57.34}$$

and observe that

$$\begin{pmatrix} \frac{\partial W}{\partial z} & \frac{\partial W}{\partial \xi} \end{pmatrix} \begin{pmatrix} f(z,\xi) \\ u \end{pmatrix} = \frac{\partial V}{\partial z} f(z,0) + \frac{\partial V}{\partial z} p(z,\xi)\xi + \xi u \,.$$

Choosing

$$u = u(z,\xi) = -\xi - \frac{\partial V}{\partial z} p(z,\xi) \qquad (57.35)$$

yields

$$\begin{pmatrix} \frac{\partial W}{\partial z} & \frac{\partial W}{\partial \xi} \end{pmatrix} \begin{pmatrix} f(z,\xi) \\ u(z,\xi) \end{pmatrix} < 0$$

for all nonzero (z, ξ). By the direct Lyapunov theorem, it is concluded that the system

$$\begin{aligned} \dot{z} &= f(z,\xi) \\ \dot{\xi} &= u(z,\xi) \end{aligned}$$

has a globally asymptotically stable equilibrium at $(z, \xi) = (0, 0)$.

In other words, it has been shown that, if $\dot{z} = f(z, 0)$ has a globally asymptotically stable equilibrium at $z = 0$, then the equilibrium $(z, \xi) = (0, 0)$ of the system of Equation 57.32 can be rendered globally asymptotically stable by means of a smooth feedback law $u = u(z, \xi)$.

The result thus proven can be easily extended by showing that, for the purpose of stabilizing the equilibrium $(z, \xi) = (0, 0)$ of Equation 57.32, it suffices to assume that the equilibrium $z = 0$ of

$$\dot{z} = f(z,\xi)$$

is *stabilizable* by means of a smooth law $\xi = v^\star(z)$.

LEMMA 57.1 Consider a system described by equations of the form of Equation 57.32. Suppose there exists a smooth real-valued function

$$\xi = v^\star(z)\,,$$

with $v^\star(0) = 0$, and a smooth real-valued function $V(z)$, which is positive definite and proper, such that

$$\frac{\partial V}{\partial z} f(z, v^\star(z)) < 0$$

for all nonzero z. Then, there exists a smooth static feedback law $u = u(z, \xi)$ with $u(0, 0) = 0$, and a smooth real-valued function $W(z, \xi)$, which is positive definite and proper, such that

$$\begin{pmatrix} \frac{\partial W}{\partial z} & \frac{\partial W}{\partial \xi} \end{pmatrix} \begin{pmatrix} f(z,\xi) \\ u(z,\xi) \end{pmatrix} < 0$$

for all nonzero (z, ξ).

In fact, it suffices to consider the (globally defined) change of variables

$$y = \xi - v^\star(z)\,,$$

which transforms Equation 57.32 into

$$\begin{aligned} \dot{z} &= f(z, v^\star(z) + y) \\ \dot{y} &= -\frac{\partial v^\star}{\partial z} f(z, v^\star(z) + y) + u\,, \end{aligned} \qquad (57.36)$$

and observe that the feedback law

$$u = \frac{\partial v^\star}{\partial z} f(z, v^\star(z) + y) + u'$$

changes the latter into a system satisfying the hypotheses that are at the basis of the previous construction.

Using repeatedly the property indicated in Lemma 57.1, it is straightforward to derive the following stabilization result about a system in the form of Equation 57.30.

THEOREM 57.2 *Consider a system of the form of Equation 57.30. Suppose there exists a smooth real-valued function*

$$\xi_1 = v^\star(z)\,,$$

with $v^\star(0) = 0$, and a smooth real-valued function $V(z)$, which is positive definite and proper, such that

$$\frac{\partial V}{\partial z} f_0(z, v^\star(z)) < 0$$

for all nonzero z. Then, there exists a smooth static feedback law

$$u = u(z, \xi_1, \ldots, \xi_r)$$

with $u(0, 0, \ldots, 0) = 0$, which globally asymptotically stabilizes the equilibrium $(z, \xi_1, \ldots, \xi_r) = (0, 0, \ldots, 0)$ of the corresponding closed-loop system.

Of course, a special case in which the result of Theorem 57.2 holds is when $v^\star(z) = 0$, i.e., when $\dot{z} = f_0(z, 0)$ has a globally asymptotically stable equilibrium at $z = 0$. This is the case of a system whose *zero dynamics* have a globally asymptotically stable equilibrium at $z = 0$, i.e., the case of a globally minimum-phase system.

The stabilization procedure outlined above is illustrated in the following example, taken from [1].

EXAMPLE 57.6:

Consider the problem of globally asymptotically stabilizing the equilibrium $(x_1, x_2, x_3) = (0, 0, 0)$ of the nonlinear system

$$\begin{aligned} \dot{x}_1 &= x_2^3 \\ \dot{x}_2 &= x_3^3 \\ \dot{x}_3 &= u\,. \end{aligned} \qquad (57.37)$$

To this end, observe that a "dummy output" of the form

$$y = x_3 - v^\star(x_1, x_2)$$

yields a system having relative degree $r = 1$ at each $x \in \mathbb{R}^3$ and two-dimensional zero dynamics. The latter, i.e., the dynamics obtained by imposing on Equation 57.37 the constraint $y = 0$, are those of the autonomous system

$$\begin{aligned} \dot{x}_1 &= x_2^3 \\ \dot{x}_2 &= (v^\star(x_1, x_2))^3\,. \end{aligned} \qquad (57.38)$$

From the discussion above we know that, if it is possible to find a function $v^\star(x_1, x_2)$ that globally asymptotically stabilizes the equilibrium $(x_1, x_2) = (0, 0)$ of Equation 57.38, then there exists an input $u(x_1, x_2, x_3)$ that globally asymptotically stabilizes the equilibrium $(x_1, x_2, x_3) = (0, 0, 0)$ of Equation 57.37. It is easy to check that the function

$$v^\star(x_1, x_2) = -x_1 \exp(x_1 x_2)$$

accomplishes this task. In fact, consider the system

$$\begin{aligned} \dot{x}_1 &= x_2^3 \\ \dot{x}_2 &= -x_1^3 \exp(3x_1 x_2) , \end{aligned} \tag{57.39}$$

and choose a candidate Lyapunov function

$$V(x_1, x_2) = x_1^4 + x_2^4 ,$$

which yields

$$\dot{V} = 4(x_1 x_2)^3 (1 - \exp(3x_1 x_2)) .$$

This function is nonpositive for all (x_1, x_2) and zero only at $x_1 = 0$ or $x_2 = 0$. Since no nontrivial trajectory of Equation 57.39 is contained in the set

$$M = \{(x_1, x_2) : \dot{V} = 0\} ,$$

by Lasalle's invariance principle it is concluded that the equilibrium $(x_1, x_2) = (0, 0)$ of Equation 57.39 is globally asymptotically stable.

In order to obtain the input function that globally stabilizes the equilibrium $(x_1, x_2, x_3) = (0, 0, 0)$ of Equation 57.37, it is necessary to use the construction indicated in the proof of Lemma 57.1. In fact, consider the change of variables

$$y = x_3 - v^\star(x_1, x_2)$$

which transforms Equation 57.37 into

$$\begin{aligned} \dot{x}_1 &= x_2^3 \\ \dot{x}_2 &= (y + v^\star(x_1, x_2))^3 \\ \dot{y} &= u - \frac{\partial v^\star}{\partial x_1} x_2^3 - \frac{\partial v^\star}{\partial x_2} (y + v^\star(x_1, x_2))^3 . \end{aligned} \tag{57.40}$$

Choosing a preliminary feedback

$$u = \frac{\partial v^\star}{\partial x_1} x_2^3 + \frac{\partial v^\star}{\partial x_2} (y + v^\star(x_1, x_2))^3 + u'$$

yields

$$\begin{aligned} \dot{x}_1 &= x_2^3 \\ \dot{x}_2 &= (y + v^\star(x_1, x_2))^3 \\ \dot{y} &= u' . \end{aligned} \tag{57.41}$$

which has exactly the form of Equation 57.32, namely,

$$\begin{aligned} \dot{z} &= f(z, \xi) \\ \dot{\xi} &= u' , \end{aligned}$$

with

$$\begin{aligned} z &= \begin{pmatrix} x_1 \\ x_2 \end{pmatrix} \\ f(z, \xi) &= \begin{pmatrix} x_2^3 \\ (v^\star(x_1, x_2))^3 \end{pmatrix} + \\ & \quad \xi \begin{pmatrix} 0 \\ 3(v^\star(x_1, x_2))^2 + 3v^\star(x_1, x_2)\xi + \xi^2 \end{pmatrix} \end{aligned}$$

and $\dot{z} = f(z, 0)$ has a globally asymptotically stable equilibrium at $z = 0$. As a consequence, this system can be globally asymptotically stabilized by means of a feedback law $u' = u'(z, \xi)$ of the form of Equation 57.35.

References

[1] Byrnes, C.I. and Isidori, A., New results and examples in nonlinear feedback stabilization, *Syst. Control Lett.*, 12, 437–442, 1989.

[2] Tsinias, J., Sufficient Lyapunov-like conditions for stabilization, *Math. Control Signals Syst.*, 2, 424–440, 1989.

[3] Byrnes, C.I. and Isidori, A., Asymptotic stabilization of minimum phase nonlinear systems, *IEEE Trans. Autom. Control*, 36, 1122–1137, 1991.

[4] Byrnes, C.I. and Isidori, A., On the attitude stabilization of a rigid spacecraft, *Automatica*, 27, 87–96, 1991.

[5] Isidori, A., *Nonlinear Control Systems*, 2nd. ed., Springer-Verlag, 1989.

57.3 Nonlinear Output Regulation

Alberto Isidori, Dipartimento di Informatica e Sistemistica, Università di Roma "La Sapienza", Rome and Department of Systems Sciences and Mathematics, Washington University, St.Louis, MO

57.3.1 The Problem of Output Regulation

A classical problem in control theory is to impose, via feedback, a prescribed steady-state response to every external command in a given family. This may include, for instance, the problem of having the output $y(\cdot)$ of a controlled plant asymptotically track any prescribed reference signal $y_{\text{ref}}(\cdot)$ in a certain class of functions of time, as well as the problem of having $y(\cdot)$ asymptotically reject any undesired disturbance $w(\cdot)$ in a certain class of disturbances. In both cases, the issue is to force the so-called *tracking error*, i.e., the difference between the reference output and the actual output, to be a function of time

$$e(t) = y_{\text{ref}}(t) - y(t)$$

that decays to zero as time tends to infinity, for every reference output and every undesired disturbance ranging over prespecified families of functions.

The problem in question can be characterized as follows. Consider a nonlinear system modeled by equations of the form

$$\begin{aligned} \dot{x} &= f(x, w, u) \\ e &= h(x, w)\,. \end{aligned} \tag{57.42}$$

The first equation of Equation 57.42 describes the dynamics of a *plant*, whose *state* x is defined in a neighborhood U of the origin in $\mathbb{R}^n$, with *control input* $u \in \mathbb{R}^m$ and subject to a set of *exogenous input* variables $w \in \mathbb{R}^r$, which includes *disturbances* (to be rejected) and/or *references* (to be tracked). The second equation defines an *error* variable $e \in \mathbb{R}^m$, which is expressed as a function of the state x and of the exogenous input w.

For the sake of mathematical simplicity, and also because in this way a large number of relevant practical situations can be covered, it is assumed that the family of the exogenous inputs $w(\cdot)$ that affect the plant, and for which asymptotic decay of the error is to be achieved, is the family of all functions of time that are solutions of a (possibly nonlinear) homogeneous differential equation

$$\dot{w} = s(w) \tag{57.43}$$

with initial condition $w(0)$ ranging on some neighborhood W of the origin of $\mathbb{R}^r$. This system, which is viewed as a mathematical model of a "generator" of all possible exogenous input functions, is called the *exosystem*.

It is assumed that $f(x, w, u), h(x, w), s(w)$ are smooth functions. Moreover, it is also assumed that $f(0, 0, 0) = 0, s(0) = 0$, $h(0, 0) = 0$. Thus, for $u = 0$, the composite system of Equations 57.42 and 57.43 has an equilibrium state $(x, w) = (0, 0)$ yielding zero error.

The control action to Equation 57.42 is to be provided by a *feedback controller* that processes the information received from the plant in order to generate the appropriate control input. The structure of the controller usually depends on the amount of information available for feedback. The most favorable situation, from the point of view of feedback design, occurs when the set of measured variables includes all the components of the state x of the plant and of the exogenous input w. In this case, it is said that the controller is provided with *full information* and can be constructed as a *memoryless* system, whose output u is a function of the states x and w of the plant and of the exosystem, respectively,

$$u = \alpha(x, w)\,. \tag{57.44}$$

The interconnection of Equation 57.42 and Equation 57.44 yields a closed-loop system described by the equations

$$\begin{aligned} \dot{x} &= f(x, w, \alpha(x, w)) \\ \dot{w} &= s(w)\,. \end{aligned} \tag{57.45}$$

In particular, it is assumed that $\alpha(0, 0) = 0$, so that the closed-loop system of Equation 57.45 has an equilibrium at $(x, w) = (0, 0)$.

A more realistic, and rather common, situation is the one in which only the components of the error e are available for measurement. In this case, it is said that the controller is provided with *error feedback*, and it can be useful to synthesize the control signal by means of a *dynamical* nonlinear system, modeled by equations of the form

$$\begin{aligned} \dot{\xi} &= \eta(\xi, e) \\ u &= \theta(\xi)\,, \end{aligned} \tag{57.46}$$

with internal state ξ defined in a neighborhood Ξ of the origin in $\mathbb{R}^\nu$. The interconnection of Equation 57.42 and Equation 57.46 yields, in this case, a closed-loop system characterized by the equations

$$\begin{aligned} \dot{x} &= f(x, w, \theta(\xi)) \\ \dot{\xi} &= \eta(\xi, h(x, w)) \\ \dot{w} &= s(w)\,. \end{aligned} \tag{57.47}$$

Again, it is assumed that $\eta(0, 0) = 0$ and $\theta(0) = 0$, so that the triplet $(x, \xi, w) = (0, 0, 0)$ is an equilibrium of the closed-loop system of Equation 57.47.

The purpose of the control is to obtain a closed-loop system in which, for every exogenous input $w(\cdot)$ (in the prescribed family) and every initial state (in some neighborhood of the origin), the output $e(\cdot)$ decays to zero as time tends to infinity. When this is the case, the closed-loop system is said to have the *property of output regulation*.

The property of output regulation is particularly meaningful when the exogenous inputs are "persistent" in time, as it is in the case of any periodic (and bounded) function. In these cases, in fact, the system may exhibit a "steady-state response" that is itself a persistent function of time, and whose characteristics depend entirely on the specific input imposed on the system and not on the state in which the system was at the initial time. To ensure that the exogenous inputs generated by an exosystem of the form of Equation 57.43 are bounded, it suffices to assume that the point $w = 0$ is a stable equilibrium (in the ordinary sense of Lyapunov) of the vector field $s(w)$ and to choose the initial condition at time $t = 0$ in some appropriate neighborhood $W^\circ \subset W$ of the origin. To guarantee that the inputs are persistent in time (that is, to exclude the possibility that some input decays to zero as time tends to infinity), it is convenient to assume that every point w of W° is Poisson stable.

We recall that a point w° is said to be *Poisson stable* [1] if the flow $\Phi_t^s(w^\circ)$ of the vector field $s(w)$ is defined for all $t \in \mathbb{R}$ and, for each neighborhood U° of w° and for each real number $T > 0$, there exists a time $t_1 > T$ such that $\Phi_{t_1}^s(w^\circ) \in U^\circ$, and a time $t_2 < -T$ such that $\Phi_{t_2}^s(w^\circ) \in U^\circ$. In other words, a point w° is Poisson stable if the trajectory $w(t)$ that originates in w° passes arbitrarily close to w° for arbitrarily large times, in both the forward and backward directions. Thus, it is clear that if every point w° of W° is Poisson stable, no trajectory of Equation 57.43 can decay to zero as time tends to infinity.

In what follows, we assume that the vector field $s(w)$ has the two properties indicated above; namely, that the point $w = 0$ is

[1] See, e.g., [9] for the definitions of *vector field* and *flow* of a vector field.

a stable equilibrium (in the ordinary sense) and there exists an open neighborhood of the point $w = 0$ in which every point is Poisson stable. For convenience, these two properties together will be referred to as the property of *neutral stability.*

REMARK 57.6 Note that the hypothesis of neutral stability implies that the matrix

$$S = \left[\frac{\partial s}{\partial w}\right]_{w=0}$$

which characterizes the linear approximation of the vector field $s(w)$ at $w = 0$, has all its eigenvalues on the imaginary axis.

If the exosystem is *neutrally stable* and the closed-loop system

$$\dot{x} = f(x, 0, \alpha(x, 0)) \tag{57.48}$$

is *asymptotically stable in the first approximation,* then the response of the composite system of Equation 57.45 from any initial state $[x(0), w(0)]$ in a suitable neighborhood of the origin $(0, 0)$ converges, as t tends to ∞, to a well-defined *steady-state response,* which is independent of $x(0)$ and depends only on $w(0)$. If this steady-state response is such that the associated tracking error is identically zero, then the closed-loop system of Equation 57.45 has the required property of output regulation. This motivates the following definition.

Full Information Output Regulation Problem

Given a nonlinear system of the form of Equation 57.42 and a neutrally stable exosystem modeled by Equation 57.43, find, if possible, a mapping $\alpha(x, w)$ such that

- $(S)_{FI}$ the equilibrium $x = 0$ of Equation 57.48 is asymptotically stable in the first approximation.
- $(R)_{FI}$ there exists a neighborhood $V \subset U \times W$ of $(0, 0)$ such that, for each initial condition $[x(0), w(0)] \in V$, the solution of Equation 57.45 satisfies

$$\lim_{t\to\infty} h(x(t), w(t)) = 0\,.$$

Again, if the exosystem is *neutrally stable* and the closed-loop system

$$\begin{aligned} \dot{x} &= f(x, 0, \theta(\xi)) \\ \dot{\xi} &= \eta(\xi, h(x, 0)) \end{aligned} \tag{57.49}$$

is *asymptotically stable in the first approximation,* then the response of the composite system of Equation 57.47 from any initial state $[x(0), \xi(0), w(0)]$ in a suitable neighborhood of the origin $(0, 0, 0)$ converges, as t tends to ∞, to a well-defined *steady-state response,* which is independent of $[x(0), \xi(0)]$ and depends only on $w(0)$. If this steady-state response is such that the associated tracking error is identically zero, then the closed-loop system of Equation 57.47 has the required property of output regulation.

Error Feedback Output Regulation Problem

Given a nonlinear system of the form of Equation 57.42 and a neutrally stable exosystem modeled by Equation 57.43, find, if possible, an integer ν and two mappings $\theta(\xi)$ and $\eta(\xi, e)$, such that

- $(S)_{EF}$ the equilibrium $(x, \xi) = (0, 0)$ of Equation 57.49 is asymptotically stable in the first approximation.
- $(R)_{EF}$ there exists a neighborhood $V \subset U \times \Xi \times W$ of $(0, 0, 0)$ such that, for each initial condition $[x(0), \xi(0), w(0)] \in V$, the solution of Equation 57.47 satisfies

$$\lim_{t\to\infty} h(x(t), w(t)) = 0\,.$$

Since one of the specifications in the problem of output regulation is that of achieving stability in the first approximation, it is clear that the properties of *stabilizability* and *detectability* of the linear approximation of the controlled plant at the equilibrium $(x, w, u) = (0, 0, 0)$ play a determinant role in the solution of this problem. For notational convenience, observe that the closed-loop system of Equation 57.45 can be written in the form

$$\begin{aligned} \dot{x} &= (A + BK)x + (P + BL)w + \phi(x, w) \\ \dot{w} &= Sw + \psi(w) \end{aligned}$$

where $\phi(x, w)$ and $\psi(w)$ vanish at the origin with their first-order derivatives, and A, B, P, K, L, S are matrices defined by

$$\begin{aligned} A &= \left[\frac{\partial f}{\partial x}\right]_{(0,0,0)} \quad & B &= \left[\frac{\partial f}{\partial u}\right]_{(0,0,0)} \quad & P &= \left[\frac{\partial f}{\partial w}\right]_{(0,0,0)} \\ K &= \left[\frac{\partial \alpha}{\partial x}\right]_{(0,0)} \quad & S &= \left[\frac{\partial s}{\partial w}\right]_{(0)} \quad & L &= \left[\frac{\partial \alpha}{\partial w}\right]_{(0,0)}. \end{aligned} \tag{57.50}$$

On the other hand, the closed-loop system of Equation 57.47 can be written in the form

$$\begin{aligned} \dot{x} &= Ax + BH\xi + Pw + \phi(x, \xi, w) \\ \dot{\xi} &= F\xi + GCx + GQw + \chi(x, \xi, w) \\ \dot{w} &= Sw + \psi(w) \end{aligned}$$

where $\phi(x, \xi, w)$, $\chi(x, \xi, w)$ and $\psi(w)$ vanish at the origin with their first-order derivatives, and C, Q, F, H, G are matrices defined by

$$\begin{aligned} C &= \left[\frac{\partial h}{\partial x}\right]_{(0,0)} \quad & Q &= \left[\frac{\partial h}{\partial w}\right]_{(0,0)} & & \\ F &= \left[\frac{\partial \eta}{\partial \xi}\right]_{(0,0)} \quad & G &= \left[\frac{\partial \eta}{\partial e}\right]_{(0,0)} \quad & H &= \left[\frac{\partial \theta}{\partial \xi}\right]_{(0)}. \end{aligned} \tag{57.51}$$

Using this notation, it is immediately realized that the requirement $(S)_{FI}$ is the requirement that all the eigenvalues of the Jacobian matrix of (Equation 57.48) at $x = 0$,

$$J = A + BK$$

have negative real part, whereas $(S)_{EF}$ is the requirement that all the eigenvalues of the Jacobian matrix of Equation 57.49 at $(x, \xi) = (0, 0)$,

$$J = \begin{pmatrix} A & BH \\ GC & F \end{pmatrix}$$

have negative real part.

From the theory of linear systems, it is then easy to conclude that:

1. $(S)_{FI}$ can be achieved *only if* the pair of matrices (A, B) is *stabilizable* [i.e., there exists K such that all the eigenvalues of $(A + BK)$ have negative real part].
2. $(S)_{EF}$ can be achieved *only if* the pair of matrices (A, B) is stabilizable and the pair of matrices (C, A) is *detectable* [i.e., there exists G such that all the eigenvalues of $(A + GC)$ have negative real part].

These properties of the linear approximation of the plant of Equation 57.42 at $(x, w, u) = (0, 0, 0)$ are indeed necessary conditions for the solvability of a problem of output regulation.

57.3.2 Output Regulation in the Case of Full Information

In this section, we show how the problem of output regulation via full information can be solved.

THEOREM 57.3 *The full information output regulation problem (Section 57.3.1) is solvable if and only if the pair (A, B) is stabilizable and there exist mappings $x = \pi(w)$ and $u = c(w)$, with $\pi(0) = 0$ and $c(0) = 0$, both defined in a neighborhood $W^\circ \subset W$ of the origin in $\mathbb{R}^r$, satisfying the conditions*

$$\begin{aligned} \frac{\partial \pi}{\partial w} s(w) &= f(\pi(w), w, c(w)) \\ 0 &= h(\pi(w), w) \end{aligned} \qquad (57.52)$$

for all $w \in W^\circ$.

To see that the conditions of Theorem 57.3 are sufficient, observe that, by hypothesis, there exists a matrix K such that $(A + BK)$ has eigenvalues with negative real part. Suppose the conditions of Equation 57.52 are satisfied for some $\pi(w)$ and $c(w)$, and define a feedback law in the following way

$$\alpha(x, w) = c(w) + K(x - \pi(w)).$$

It is not difficult to see that this is a solution of the full information output regulation problem. In fact, this choice clearly satisfies the requirement $(S)_{FI}$ because $\alpha(x, 0) = Kx$. Moreover, by construction, the function $\alpha(x, w)$ is such that

$$\frac{\partial \pi}{\partial w} s(w) = f(\pi(w), w, \alpha(\pi(w), w)) .$$

This identity shows that the graph of the mapping $x = \pi(w)$ is an *invariant* (and locally exponentially attractive) *manifold* for Equation 57.45. In particular, there exist real numbers $M > 0$ and $a > 0$ such that

$$\|x(t) - \pi(w(t))\| \le Me^{-at}\|x(0) - \pi(w(0))\|$$

for all $t \ge 0$ and for every sufficiently small $[x(0), w(0)]$. Thus, since the error map $e = h(x, w)$ is zero on the graph of the mapping $x = \pi(w)$, the condition $(R)_{FI}$ is satisfied.

REMARK 57.7 The result expressed by Theorem 57.3 can be interpreted as follows. Observe that, if the identities of Equation 57.52 hold, the graph of the mapping $x = \pi(w)$ is an invariant manifold for the composite system

$$\begin{aligned} \dot{x} &= f(x, w, c(w)) \\ \dot{w} &= s(w) \end{aligned} \qquad (57.53)$$

and the error map $e = h(x, w)$ is zero at each point of this manifold. From this interpretation, it is easy to observe that, for any initial (namely, at time $t = 0$) state w^* of the exosystem, i.e., for any exogenous input

$$w^*(t) = \Phi_t^s(w^*) ,$$

if the plant is in the initial state $x^* = \pi(w^*)$ and the input is equal to

$$u^*(t) = c(w^*(t))$$

then $e(t) = 0$ for all $t \ge 0$. In other words, the *control input* generated by the autonomous system

$$\begin{aligned} \dot{w} &= s(w) \\ u &= c(w) \end{aligned} \qquad (57.54)$$

is precisely the control required to impose, for any *exogenous input*, a response producing an identically zero error, provided that the initial condition of the plant is appropriately set [namely, at $x^* = \pi(w^*)$].

The question of whether such a response is actually the steady-state response [that is, whether the error converges to zero as time tends to infinity when the initial condition of the plant is other than $x^* = \pi(w^*)$] depends indeed on the asymptotic properties of the equilibrium $x = 0$ of $f(x, 0, 0)$. If the equilibrium at $x = 0$ is not stable in the first approximation, then in order to achieve the required steady-state response, the control law must also include a stabilizing component, as in the case of the control law $\alpha(x, w)$ indicated above. Under this control law, the composite system

$$\begin{aligned} \dot{x} &= f(x, w, c(w) + K(x - \pi(w))) \\ \dot{w} &= s(w) \end{aligned}$$

still has an invariant manifold of the form $x = \pi(w)$, but the latter is now locally exponentially attractive. In this configuration, from any initial condition in a neighborhood of the origin, the response of the *closed-loop* system

$$\dot{x} = f(x, w, c(w) + K(x - \pi(w)))$$

to any exogenous input $w^*(\cdot)$ converges towards the response of the *open-loop system*

$$\dot{x} = f(x, w, u)$$

produced by the same exogenous input $w^*(\cdot)$, by the control input $u^*(\cdot) = c(w^*(\cdot))$, with initial condition $x^* = \pi(w^*)$.

REMARK 57.8 If the controlled plant is a linear system, the conditions of Equation 57.52 reduce to linear matrix equations. In this case, the system in question can be written in the form

$$\begin{aligned} \dot{x} &= Ax + Pw + Bu \\ \dot{w} &= Sx \\ e &= Cx + Qw , \end{aligned}$$

and, if the mappings $x = \pi(w)$ and $u = c(w)$ are put in the form

$$\begin{aligned} \pi(w) &= \Pi w + \tilde{\pi}(w) \\ c(w) &= \Gamma w + \tilde{c}(w) , \end{aligned}$$

with

$$\Pi = \left[\frac{\partial \pi}{\partial w}\right]_{w=0} \qquad \Gamma = \left[\frac{\partial c}{\partial w}\right]_{w=0} ,$$

the conditions of Equation 57.52 have a solution if and only if the linear matrix equations

$$\begin{aligned} \Pi S &= A\Pi + P + B \\ 0 &= C\Pi + Q \end{aligned}$$

are solved by some Π and Γ. Note that, if this is the case, the mappings $\pi(w)$ and $c(w)$ that solve Equation 57.52 are actually linear mappings [i.e., $\pi(w) = \Pi w$ and $c(w) = \Gamma w$].

We describe now how the existence conditions of Equation 57.52 can be tested in the particular case in which $m = 1$ (one-dimensional control input and one-dimensional error) and the conditions of Equation 57.42 assume the form

$$\begin{aligned} \dot{x} &= f(x) + g(x)u \\ e &= h(x) + p(w) , \end{aligned} \tag{57.55}$$

which corresponds to the case of a single-input single-output system whose output is required to track any reference trajectory produced by

$$\begin{aligned} \dot{w} &= s(w) \\ y_{\text{ref}} &= -p(w) . \end{aligned}$$

We also assume that the triplet $\{f(x), g(x), h(x)\}$ has relative degree r at $x = 0$ so that coordinate transformation to a normal form is possible.[2] In the new coordinates, the system in question assumes the form

$$\begin{aligned} \dot{z}_1 &= z_2 \\ &\cdots \\ \dot{z}_{r-1} &= z_r \\ \dot{z}_r &= b(\xi, \eta) + a(\xi, \eta)u \\ \dot{\eta} &= q(\xi, \eta) \\ e &= z_1 + p(w) . \end{aligned}$$

In order to check whether the conditions of Equation 57.52 can be solved, it is convenient to set

$$\pi(w) = \text{col}(k(w), \lambda(w))$$

with

$$k(w) = \text{col}(k_1(w), \ldots, k_r(w)) .$$

In this case, the equations in question reduce to

$$\begin{aligned} \frac{\partial k_1(x)}{\partial w} s(w) &= k_2(w) \\ &\cdots \\ \frac{\partial k_{r-1}(x)}{\partial w} s(w) &= k_r(w) \\ \frac{\partial k_r(x)}{\partial w} s(w) &= b(k(w), \lambda(w)) + a(k(w), \lambda(w))c(w) \\ \frac{\partial \lambda(x)}{\partial w} s(w) &= q(k(w), \lambda(w)) \\ 0 &= k_1(w) + p(w) . \end{aligned}$$

The last one of these, together with the first $r - 1$, yields immediately

$$k_i(w) = -L_s^{i-1} p(w) \tag{57.56}$$

for all $1 \le i \le r$. The r-th equation can be solved by

$$c(w) = \frac{L_s k_r(w) - b(k(w), \lambda(w))}{a(k(w), \lambda(w))} \tag{57.57}$$

and, therefore, we can conclude that the solvability of the conditions of Equation 57.52 is, in this case, equivalent to the solvability of

$$\frac{\partial \lambda}{\partial w} s(w) = q(k(w), \lambda(w)) \tag{57.58}$$

for some mapping $\eta = \lambda(w)$.

We can therefore conclude that the full information output regulation problem is solvable if and only if the pair (A, B) is stabilizable and Equation 57.58 can be solved by some mapping $\lambda(w)$, with $\lambda(0) = 0$.

An application of these results is illustrated in the following example, taken from [9].

EXAMPLE 57.7:

Consider the system already in normal form

$$\begin{aligned} \dot{x}_1 &= x_2 \\ \dot{x}_2 &= u \\ \dot{\eta} &= \eta + x_1 + x_2^2 \\ y &= x_1 \end{aligned}$$

[2] See, e.g., Chapter 57.1 for the definitions of *relative degree* and *normal form*.

and suppose it is desired to asymptotically track any reference output of the form

$$y_{\text{ref}}(t) = M\sin(at+\phi)$$

where a is a fixed (positive) number, and M, ϕ arbitrary parameters.

In this case, any desired reference output can be imagined as the output of an exosystem defined by

$$\begin{aligned} s(w) &= \begin{pmatrix} aw_2 \\ -aw_1 \end{pmatrix} \\ p(w) &= -w_1 \end{aligned}$$

and therefore we could try to solve the problem via the theory developed in this section, i.e., posing a full information output regulation problem.

Following the procedure illustrated above, one has to set

$$\begin{aligned} k_1(w) &= -p(w) = w_1 \\ k_2(w) &= L_s k_1(w) = aw_2 \end{aligned}$$

and then search for a solution $\lambda(w_1, w_2)$ of the partial differential Equation 57.58, i.e.,

$$\frac{\partial\lambda}{\partial w_1}aw_2 - \frac{\partial\lambda}{\partial w_2}aw_1 = \lambda(w_1,w_2) + w_1 + (aw_2)^2 .$$

An elementary calculation shows that this equation can be solved by a complete polynomial of second degree, i.e.,

$$\lambda(w_1,w_2) = a_1w_1 + a_2w_2 + a_{11}w_1^2 + a_{12}w_1w_2 + a_{22}w_2^2 .$$

Once $\lambda(w_1, w_2)$ has been calculated, from the previous theory it follows that the mapping

$$\pi(w) = \begin{pmatrix} k_1(w) \\ k_2(w) \\ \lambda(w_1,w_2) \end{pmatrix} = \begin{pmatrix} w_1 \\ aw_2 \\ \lambda(w_1,w_2) \end{pmatrix}$$

and the function

$$c(w) = L_s k_2(w) = -a^2 w_1$$

are solutions of the conditions of Equation 57.52. In particular, a solution of the regulator problem is provided by

$$\alpha(x,w) = c(w) + K(x - \pi(w))$$

in which $K = (c_1\ c_2\ c_3)$ is any matrix that places the eigenvalues of

$$\begin{pmatrix} 0 & 1 & 0 \\ 0 & 0 & 0 \\ 1 & 0 & 1 \end{pmatrix} + \begin{pmatrix} 0 \\ 1 \\ 0 \end{pmatrix} K$$

in the left complex half-plane.

As expected, the difference

$$\begin{pmatrix} \xi_1 \\ \xi_2 \\ \xi_3 \end{pmatrix} = \begin{pmatrix} x_1 \\ x_2 \\ \eta \end{pmatrix} - \pi(w) = \begin{pmatrix} x_1 - w_1 \\ x_2 - aw_2 \\ \eta - \lambda(w_1,w_2) \end{pmatrix}$$

is asymptotically decaying to zero, and so is the error $e(t)$, which in this case is exactly equal to x_1. In fact, the variables ξ_1, ξ_2, ξ_3 satisfy

$$\begin{pmatrix} \dot\xi_1 \\ \dot\xi_2 \\ \dot\xi_3 \end{pmatrix} = \begin{pmatrix} 0 & 1 & 0 \\ c_1 & c_2 & c_3 \\ 1 & 0 & 1 \end{pmatrix}\begin{pmatrix} \xi_1 \\ \xi_2 \\ \xi_3 \end{pmatrix} + \begin{pmatrix} 0 \\ 0 \\ \xi_2^2 + 2a\xi_2 w_2(t) \end{pmatrix} .$$

57.3.3 Output Regulation in the Case of Error Feedback

In this section, it will be shown that the existence of a solution of the problem of output regulation in the case of error feedback depends, among other things, on a particular property of the autonomous system of Equation 57.54, which, as we have seen, may be thought of as a generator of those input functions that produce responses yielding zero error. In order to describe this property, we need first to introduce the notion of *immersion* of a system into another system (see [2]).

Consider a pair of smooth autonomous systems with outputs

$$\dot x = f(x), \quad y = h(x)$$

and

$$\dot{\tilde x} = \tilde f(\tilde x), \quad y = \tilde h(\tilde x)$$

defined on two different state spaces, X and $\widetilde X$, but having the same output space $Y = \mathbb{R}^m$. Assume, as usual, $f(0) = 0$, $h(0) = 0$ and $\tilde f(0) = 0$, $\tilde h(0) = 0$ and let the two systems in question be denoted, for convenience, by $\{X, f, h\}$ and $\{\widetilde X, \tilde f, \tilde h\}$, respectively.

System $\{X, f, h\}$ is said to be *immersed* into system $\{\widetilde X, \tilde f, \tilde h\}$ if there exists a (continuosly differentiable) mapping $\tau : X \to \widetilde X$, with $k \geq 1$, satisfying $\tau(0) = 0$ and such that

$$\begin{aligned} \frac{\partial\tau}{\partial x}f(x) &= \tilde f(\tau(x)) \\ h(x) &= \tilde h(\tau(x)) \end{aligned} \tag{57.59}$$

for all $x \in X$.

It is easy to realize that the two conditions indicated in this definition express nothing else than the property that any output response generated by $\{X, f, h\}$ is also an output response of $\{\widetilde X, \tilde f, \tilde h\}$. In fact, the first condition implies that the flows $\Phi_t^f(x)$ and $\Phi_t^{\tilde f}(\tilde x)$ of the two vector fields f and $\tilde f$ (which are τ-related), satisfy

$$\tau(\Phi_t^f(x)) = \Phi_t^{\tilde f}(\tau(x))$$

for all $x \in X$ and all $t \geq 0$, from which the second condition yields

$$h(\Phi_t^f(x)) = \tilde h(\tau(\Phi_t^f(x))) = \tilde h(\Phi_t^{\tilde f}(\tau(x))),$$

for all $x \in X$ and all $t \geq 0$, thus showing that the output response produced by $\{X, f, h\}$, when its initial state is any $x \in X$, is a response that can also be produced by $\{\widetilde X, \tilde f, \tilde h\}$, if the latter is set in the initial state $\tau(x) \in \widetilde X$.

The notion of immersion is relevant because sometimes $\{\widetilde{X}, \tilde{f}, \tilde{h}\}$ may have some special property that $\{X, f, h\}$ does not have. For example, any linear system can always be immersed into an *observable* linear system, and a similar thing occurs, under appropriate hypotheses, also in the case of a nonlinear system. Or, for instance, one may wish to have a *nonlinear* system immersed into a *linear* system, if possible. The notion of immersion is important for the solution of the problem of output regulation in the case of error feedback because the possibility of having the autonomous system of Equation 57.54 immersed into a system with special properties is actually a necessary and sufficient condition for the existence of such a solution.

In fact, the following result holds.

THEOREM 57.4 *The error feedback output regulation problem (Section 57.3.1) is solvable if and only if there exist mappings $x = \pi(w)$ and $u = c(w)$, with $\pi(0) = 0$ and $c(0) = 0$, both defined in a neighborhood $W^\circ \subset W$ of the origin in $\mathbb{R}^r$, satisfying the conditions*

$$\begin{aligned} \frac{\partial \pi}{\partial w} s(w) &= f(\pi(w), w, c(w)) \\ 0 &= h(\pi(w), w), \end{aligned} \tag{57.60}$$

for all $w \in W^\circ$ and such that the autonomous system with output $\{W^\circ, s, c\}$ is immersed into a system

$$\begin{aligned} \dot{\xi} &= \varphi(\xi) \\ u &= \gamma(\xi), \end{aligned}$$

defined on a neighborhood Ξ° of the origin in $\mathbb{R}^\nu$, in which $\varphi(0) = 0$ and $\gamma(0) = 0$, and the two matrices

$$\Phi = \left[\frac{\partial \varphi}{\partial \xi}\right]_{\xi=0}, \qquad \Gamma = \left[\frac{\partial \gamma}{\partial \xi}\right]_{\xi=0} \tag{57.61}$$

are such that the pair

$$\begin{pmatrix} A & 0 \\ NC & \Phi \end{pmatrix}, \quad \begin{pmatrix} B \\ 0 \end{pmatrix} \tag{57.62}$$

is stabilizable for some choice of the matrix N, and the pair

$$\begin{pmatrix} C & 0 \end{pmatrix}, \quad \begin{pmatrix} A & B\Gamma \\ 0 & \Phi \end{pmatrix} \tag{57.63}$$

is detectable.

To see that the conditions indicated in Theorem 57.4 are sufficient, choose N so that Equation 57.62 is stabilizable. Then, observe that, as a consequence of the hypotheses on Equations 57.62 and 57.63, the triplet

$$\begin{pmatrix} A & B\Gamma \\ NC & \Phi \end{pmatrix}, \quad \begin{pmatrix} B \\ 0 \end{pmatrix}, \quad \begin{pmatrix} C & 0 \end{pmatrix}$$

is stabilizable and detectable. Choose K, L, M so that all the eigenvalues of

$$\begin{pmatrix} \begin{pmatrix} A & B\Gamma \\ NC & \Phi \\ L(C & 0) \end{pmatrix} & \begin{pmatrix} B \\ 0 \end{pmatrix} M \\ & K \end{pmatrix}$$

have negative real part.

Now, consider the controller

$$\begin{aligned} \dot{\xi}_0 &= K\xi_0 + Le \\ \dot{\xi}_1 &= \varphi(\xi_1) + Ne \\ u &= M\xi_0 + \gamma(\xi_1). \end{aligned} \tag{57.64}$$

It is easy to see that the controller thus defined solves the problem of output regulation. In fact, it is immediately seen that all the eigenvalues of the Jacobian matrix of the vector field

$$F(x, \xi_0, \xi_1) = \begin{pmatrix} f(x, 0, M\xi_0 + \gamma(\xi_1)) \\ K\xi_0 + Lh(x, 0) \\ \varphi(\xi_1) + Nh(x, 0) \end{pmatrix}$$

at $(x, \xi_0, \xi_1) = (0, 0, 0)$, which has the form

$$\begin{pmatrix} A & BM & B\Gamma \\ LC & K & 0 \\ NC & 0 & \Phi \end{pmatrix},$$

have negative real part. Moreover, by hypothesis, there exist mappings $x = \pi(w)$, $u = c(w)$ and $\xi_1 = \tau(w)$ such that the conditions of Equation 57.60 hold and

$$\frac{\partial \tau}{\partial w} s(w) = \varphi(\tau(w)), \qquad c(w) = \gamma(\tau(w)).$$

This shows that the graph of the mapping

$$\begin{pmatrix} \xi_0 \\ \xi_1 \end{pmatrix} = \begin{pmatrix} 0 \\ \tau(w) \end{pmatrix}$$

is an invariant (and locally exponentially attractive) manifold for the corresponding closed-loop system of Equation 57.47 and this shows, as in the case of full information, that the condition $(R)_{EF}$ is satisfied.

REMARK 57.9 The controller of Equation 57.64 consists of the *parallel connection* of two subsystems: the subcontroller

$$\begin{aligned} \dot{\xi}_1 &= \varphi(\xi_1) + Ne \\ u &= \gamma(\xi_1), \end{aligned} \tag{57.65}$$

and the subcontroller

$$\begin{aligned} \dot{\xi}_0 &= K\xi_0 + Le \\ u &= M\xi_0. \end{aligned} \tag{57.66}$$

The role of the second subcontroller, which is a linear system, is nothing else than that of stabilizing in the first approximation the interconnection

$$\begin{aligned} \dot{x} &= f(x, w, \gamma(\xi_1) + u) \\ \dot{\xi}_1 &= \varphi(\xi_1) + Nh(x, w) \\ e &= h(x, w), \end{aligned}$$

that is, the interconnection of the controlled plant and the first subcontroller. The role of the first subcontroller, on the other hand, is that of producing an input that generates the desired steady-state response. As a matter of fact, the identities

$$\begin{aligned} \frac{\partial \pi}{\partial w} s(w) &= f(\pi(w), w, \gamma(\tau(w))) \\ \frac{\partial \tau}{\partial w} s(w) &= \varphi(\tau(w)) \end{aligned}$$

(which hold by construction) show that the submanifold

$$M_c = \{(x, \xi_0, \xi_1, w) : x = \pi(w), \xi_0 = 0, \xi_1 = \tau(w)\}$$

is an invariant manifold of the composite system

$$\begin{aligned} \dot{x} &= f(x, w, \gamma(\xi_1) + M\xi_0) \\ \dot{\xi}_0 &= K\xi_0 + Lh(x, w) \\ \dot{\xi}_1 &= \varphi(\xi_1) + Nh(x, w) \\ \dot{w} &= s(w) \end{aligned}$$

i.e., of the closed-loop system driven by the exosystem, and on this manifold the error map $e = h(x, w)$ is zero.

The role of the subcontroller of Equation 57.65 is that of producing, for each initial condition in M_c, an input that keeps the trajectory of this composite system evolving on M_c (and thereby producing a response for which the error is zero). For this reason, the subcontroller of Equation 57.65 is sometimes referred to as an *internal model* of the generator of exogenous inputs. The role of the subcontroller of Equation 57.66 is that of rendering M_c locally exponentially attractive so that every motion starting in a sufficiently small neighborhood of the equilibrium $(x, \xi_0, \xi_1, w) = (0, 0, 0, 0)$ exponentially converges towards the desired steady-state response.

The statement of Theorem 57.4 essentially says that the problem of output regulation in the case of error feedback is solvable if and only if it is possible to find a mapping $c(w)$ that renders the identities of Equation 57.60 satisfied for some $\pi(w)$ and, moreover, is such that the autonomous system with output

$$\begin{aligned} \dot{w} &= s(w) \\ u &= c(w) \end{aligned} \tag{57.67}$$

satisfies a certain number of special conditions, which are expressed as properties of the linear approximation of an "auxiliary" system in which the latter is requested to be immersed.

It is important to observe that the condition that the pair of Equations 57.62 is stabilizable implies the condition that the pair (A, B) is stabilizable; and, similarly, the condition that the pair of Equations 57.63 is detectable implies the condition that the pair (C, A) is detectable. Thus, the conditions of Theorem 57.4 include, as expected, the trivial necessary conditions mentioned earlier for the fulfillment of $(S)_{EF}$. Moreover, the condition that the pair of Equations 57.63 is detectable also implies the condition that the pair (Γ, Φ) is detectable. Therefore, a necessary condition for the solution of the problem of output regulation in the case of error feedback is that, for some $c(w)$ satisfying Equation 57.60, the autonomous system with outputs of Equation 57.67 is immersed into a system whose linear approximation at the equilibrium $\xi = 0$ is detectable. This can always be achieved if Equation 57.67 is a linear system, but it may not be possible in general.

If Equation 57.67 is immersed into a *linear* detectable system, then the previous construction shows that the error feedback output regulation problem is solvable by a *linear controller*. Fortunately, there are simple conditions to test whether a given nonlinear system can be immersed into a linear system, and the use of these conditions may yield interesting and powerful corollaries of Theorem 57.4.

COROLLARY 57.1 The error feedback output regulation problem is solvable by means of a *linear* controller if the pair (A, B) is stabilizable, the pair (C, A) is detectable, there exist mappings $x = \pi(w)$ and $u = c(w)$, with $\pi(0) = 0$ and $c(0) = 0$, both defined in a neighborhood $W^\circ \subset W$ of the origin, satisfying the conditions (of Equation 57.60) and such that, for some set of q real numbers $a_0, a_1, \ldots, a_{q-1}$,

$$L_s^q c(w) = a_0 c(w) + a_1 L_s c(w) + \cdots + a_{q-1} L_s^{q-1} c(w)\,, \tag{57.68}$$

and, moreover, the matrix

$$\begin{pmatrix} A - \lambda I & B \\ C & 0 \end{pmatrix} \tag{57.69}$$

is nonsingular for every λ, which is a root of the polynomial

$$p(\lambda) = a_0 + a_1\lambda + \ldots + a_{q-1}\lambda^{q-1} - \lambda^q\,,$$

having nonnegative real part.

PROOF 57.1 The condition given by Equation 57.68 implies that $\{W^\circ, s, c\}$ is immersed into a linear observable system. In particular, it is very easy to check that $\{W^\circ, s, c\}$ is immersed into the linear system

$$\begin{aligned} \dot{\xi} &= \Phi\xi \\ u &= \Gamma\xi \end{aligned}$$

in which

$$\begin{aligned} \Phi &= \operatorname{diag}(\widetilde{\Phi}, \ldots, \widetilde{\Phi}) \\ \Gamma &= \operatorname{diag}(\widetilde{\Gamma}, \ldots, \widetilde{\Gamma}) \end{aligned}$$

and

$$\widetilde{\Phi} = \begin{pmatrix} 0 & 1 & 0 & \cdots & 0 \\ 0 & 0 & 1 & \cdots & 0 \\ \cdot & \cdot & \cdot & \cdots & \cdot \\ 0 & 0 & 0 & \cdots & 1 \\ a_0 & a_1 & a_2 & \cdots & a_{q-1} \end{pmatrix},$$

$$\widetilde{\Gamma} = \begin{pmatrix} 1 & 0 & 0 & \cdots & 0 \end{pmatrix}.$$

Note that the minimal polynomial of Φ is equal to $p(\lambda)$. After having chosen a matrix N such that the pair (Φ, N) is stabilizable, for instance

$$N = \operatorname{diag}(\widetilde{N}, \ldots, \widetilde{N})$$

with

$$\widetilde{N} = \operatorname{col}(0, 0, \ldots, 0, 1)\,,$$

using a standard result in linear system theory, one can conclude that the remaining conditions of Theorem 57.4 hold because the matrix of Equation 57.69 is nonsingular for every λ that is an eigenvalue of Φ.

57.3.4 Structurally Stable Regulation

Suppose now that the mathematical model of the controlled plant depends on a vector $\mu \in \mathbb{R}^p$ of parameters, which are assumed to be fixed, but whose actual value is not known

$$\begin{aligned} \dot{x} &= f(x, w, u, \mu) \\ e &= h(x, w, \mu). \end{aligned} \tag{57.70}$$

Without loss of generality, we suppose $\mu = 0$ to be the nominal value of the parameter μ and, for consistency with the analysis developed earlier, we assume $f(x, w, u, \mu)$ and $h(x, w, \mu)$ to be smooth functions of their arguments. Moreover, it is assumed also that $f(0, 0, 0, \mu) = 0$ and $h(0, 0, \mu) = 0$ for each value of μ.

The structurally stable output regulation problem is to find a controller, described by equations of the form of Equation 57.46, that solves the error feedback output regulation problem for each value of μ in some neighborhood $\mathcal{P}$ of $\mu = 0$ in $\mathbb{R}^p$.

The solution of the problem in question is easily provided by the results illustrated in the previous section. In fact, it suffices to look at w and μ as if they were components of an "augmented" exogenous input

$$w^a = \begin{pmatrix} w \\ \mu \end{pmatrix},$$

which is generated by the "augmented" exosystem

$$\dot{w}^a = s^a(w^a) = \begin{pmatrix} s(w) \\ 0 \end{pmatrix},$$

and regard the family of plants modeled by Equation 57.70 as a single plant of the form of Equation 57.42, modeled by equations of the form

$$\begin{aligned} \dot{x} &= f^a(x, w^a, u) \\ e &= h^a(x, w^a). \end{aligned}$$

It is easy to realize that a controller that solves the problem of output regulation for the plant thus defined also solves the problem of structurally stable output regulation for the family of Equation 57.70. In fact, by construction, this controller will stabilize in the first approximation the equilibrium $(x, \xi) = (0, 0)$ of

$$\begin{aligned} \dot{x} &= f^a(x, 0, \theta(\xi)) \\ \dot{\xi} &= \eta(\xi, h^a(x, 0)), \end{aligned}$$

that is, the equilibrium $(x, \xi) = (0, 0)$ of

$$\begin{aligned} \dot{x} &= f(x, 0, \theta(\xi), \mu) \\ \dot{\xi} &= \eta(\xi, h(x, 0, \mu)), \end{aligned}$$

for $\mu = 0$. Since the property of stability in the first approximation is not destroyed by small parameter variations, the controller in question stabilizes any plant of the family of Equation 57.70, so long as μ stays in some open neighborhood $\mathcal{P}$ of the origin in the parameter space. Moreover, this controller will be such that $\lim_{t\to\infty} e(t) = 0$ for every $[x(0), \xi(0), w^a(0)]$ in a neighborhood of the origin. Since

$$w^a(0) = \begin{pmatrix} w(0) \\ \mu \end{pmatrix},$$

the controller in question trivially yields the required property of output regulation for any plant of the family of Equation 57.70, so long as μ stays on some open neighborhood $\mathcal{P}$ of the origin in the parameter space.

The conditions provided in Theorem 57.4 or in Corollary 57.1 can be easily translated into necessary and sufficient conditions for the existence of solutions of the problem of structurally stable regulation. For instance, as far as the latter is concerned, the following result holds. Set

$$A(\mu) = \left[\frac{\partial f}{\partial x}\right]_{(0,0,0,\mu)}, \quad B(\mu) = \left[\frac{\partial f}{\partial u}\right]_{(0,0,0,\mu)},$$
$$C(\mu) = \left[\frac{\partial h}{\partial x}\right]_{(0,0,\mu)}$$

and also observe that, because of the special form of the vector field $s^a(w^a)$,

$$\frac{\partial \pi^a}{\partial w^a} s^a(w^a) = \frac{\partial \pi^a(w, \mu)}{\partial w} s(w),$$

and the derivative of any function $\lambda(w, \mu)$ along $s^a(w^a)$ reduces to

$$L_{s^a}\lambda(w, \mu) = \frac{\partial \lambda(w, \mu)}{\partial w} s(w).$$

For convenience, the latter will be indicated simply as

$$L_s\lambda(w, \mu).$$

COROLLARY 57.2 The structurally stable output regulation problem is solvable by means of a *linear* controller if the pair $[A(0), B(0)]$ is stabilizable, the pair $[C(0), A(0)]$ is detectable, there exist mappings $x = \pi^a(w, \mu)$ and $u = c^a(w, \mu)$, with $\pi^a(0, \mu) = 0$ and $c^a(0, \mu) = 0$, both defined in a neighborhood $W^\circ \times \mathcal{P} \subset W \times \mathbb{R}^p$ of the origin, satisfying the conditions

$$\begin{aligned} \frac{\partial \pi^a(w, \mu)}{\partial w} s(w) &= f(\pi^a(w, \mu), w, c^a(w, \mu), \mu) \\ 0 &= h(\pi^a(w, \mu), w, \mu) \end{aligned} \tag{57.71}$$

and such that, for some set of q real numbers $a_0, a_1, \ldots, a_{q-1}$,

$$\begin{aligned} L_s^q c^a(w, \mu) &= a_0 c^a(w, \mu) + a_1 L_s c^a(w, \mu) + \cdots \\ &\quad + a_{q-1} L_s^{q-1} c^a(w, \mu), \end{aligned} \tag{57.72}$$

for all $(w, \mu) \in W^\circ \times \mathcal{P}$, and, moreover, the matrix

$$\begin{pmatrix} A(0) - \lambda I & B(0) \\ C(0) & 0 \end{pmatrix} \tag{57.73}$$

is nonsingular for every λ that is a root of the polynomial

$$p(\lambda) = a_0 + a_1\lambda + \ldots + a_{q-1}\lambda^{q-1} - \lambda^q$$

having nonnegative real part.

It is interesting to observe that Corollary 57.2 contains, as particular cases, a number of results about structurally stable regulation of linear and nonlinear systems. In a linear system, the solutions $\pi^a(w, \mu)$ and $c^a(w, \mu)$ of the regulator modeled by Equation 57.71 are linear in w and, therefore, for a linear exosystem

$$\dot{w} = Sw,$$

the condition of Equation 57.72 is indeed satisfied. The roots of $p(\lambda)$ coincide with those of S (possibly with different multiplicity) and thus the matrix of Equation 57.73 is required to be nonsingular for every eigenvalue λ of S. Note also that this condition guarantees the existence of solutions on the regulator modeled by Equation 57.71 for each value of μ.

In the case of nonlinear systems and constant exogenous inputs, the condition of Equation 57.72 is trivially satisfied by $q = 1$ and $a_0 = 0$.

Finally, another case in which the condition of Equation 57.72 is satisfied is that of a nonlinear system in which the solutions $\pi^a(w, \mu)$ and $c^a(w, \mu)$ of the regulator modeled by Equation 57.71 are such that $c^a(w, \mu)$ is a *polynomial* in w whose degree does not exceed a fixed number independent of μ, and the exosystem is any arbitrary linear system.

References

[1] Davison, E.J., The robust control of a servomechanism problem, *IEEE Trans. Autom. Control*, 21, 25–34, 1976.

[2] Fliess, M., Finite dimensional observation spaces for nonlinear systems, in *Feedback Control of Linear and Nonlinear Systems*, Hinrichsen, D. and Isidori, A., Eds., Springer-Verlag, 1982, 73–77.

[3] Francis, B.A., The linear multivariable regulator problem, *SIAM J. Control Optimiz.*, 14, 486–505, 1976.

[4] Francis, B.A. and Wonham, W.M., The internal model principle of control theory, *Automatica*, 12, 457–465, 1977.

[5] Hepburn, J.S.A. and Wonham, W.M., Error feedback and internal model on differentiable manifolds, *IEEE Trans. Autom. Control*, 29, 397–403, 1984.

[6] Hepburn, J.S.A and Wonham, W.M., Structurally stable nonlinear regulation with step inputs, *Math. Syst. Theory*, 17, 319–333, 1984.

[7] Huang, J. and Lin, C.F., On a robust nonlinear servomechanism problem, in Proc. 30th IEEE Conf. Decision Control, Brighton, England, December 1991, 2529–2530.

[8] Huang, J. and Rugh, W.J., On the nonlinear multivariable servomechanism problem, *Automatica*, 26, 963–972, 1990.

[9] Isidori, A., *Nonlinear Control Systems*, 2nd. ed., Springer-Verlag, 1989.

[10] Isidori, A. and Byrnes, C.I., Output regulation of nonlinear systems, *IEEE Trans. Autom. Control*, 35, 131–140, 1990.

57.4 Lyapunov Design

Randy A. Freeman, University of California, Santa Barbara

Petar V. Kokotović, University of California, Santa Barbara

57.4.1 Introduction

Lyapunov functions represent the primary tool for the stability analysis of nonlinear systems. They verify the stability of a given trajectory, and they also provide an estimate of its region of attraction. The purpose of this text is to illustrate the utility of Lyapunov functions in the *synthesis* of nonlinear control systems. We will focus on recursive state-feedback design methods which guarantee robust stability for systems with uncertain nonlinearities. Lyapunov design is used in many other contexts, such as dynamic feedback, output feedback, gain assignment, estimation, and adaptive control, but such topics are beyond the scope of this chapter.

Given a state-space model of a plant, the Lyapunov design strategy is conceptually straightforward and consists of two main steps:

1. Construct a candidate Lyapunov function V for the closed-loop system.
2. Construct a controller which renders its derivative $\dot{V}$ negative for all admissible uncertainties.

Such a controller design guarantees, by standard Lyapunov theorems, the robust stability of the closed-loop system. The difficulty lies in the first step, because only carefully constructed Lyapunov functions can lead to success in the second step. In other words, for an arbitrary Lyapunov function candidate V, it is likely that no controller can render $\dot{V}$ negative in the entire region of interest. Those select candidates which do lead to success in the second step are called *control Lyapunov functions.* Our first design step should, therefore, be to construct a control Lyapunov function for the given system; this will then insure the existence of controllers in the second design step.

In Section 57.4.2 we review the *Lyapunov redesign* method, in which a Lyapunov function is known for the *nominal system* (the system without uncertainties) and is used as the control Lyapunov function for the uncertain system. We will see that this method is essentially limited to systems whose uncertainties satisfy a restrictive *matching condition.* In Section 57.4.3 we show how such limitations can be avoided by taking the uncertainty into account while building the control Lyapunov function. We then present a recursive robust control design procedure in Section 57.4.4 for a class of uncertain nonlinear systems. Flexibilities in this recursive design are discussed in Section 57.4.6.

57.4.2 Lyapunov Redesign

A standard method for achieving robustness to state-space uncertainty is *Lyapunov redesign;* see [12]. In this method, one begins with a Lyapunov function for a nominal closed-loop system and then uses this Lyapunov function to construct a controller which guarantees robustness to given uncertainties. To illustrate this method, we consider the system,

$$\dot{x} = F(x) + G(x)u + \Delta(x,t), \tag{57.74}$$

where F and G are known functions comprising the *nominal system* and Δ is an uncertain function known only to lie within some bounds. For example, we may know a function $\rho(x)$ so that $|\ \Delta(x,t)\ | \leq \rho(x)$. A more general uncertainty Δ would also depend on the control variable u, but for simplicity we do not consider such uncertainty here. We assume that the nominal system is stabilizable, that is, that some state feedback $u_{nom}(x)$ exists so that the nominal closed-loop system,

$$\dot{x} = F(x) + G(x)u_{nom}(x), \tag{57.75}$$

has a globally asymptotically stable equilibrium at $x = 0$. We also assume knowledge of a Lyapunov function V for this system so that

$$\nabla V(x)\,[F(x) + G(x)u_{nom}(x)] < 0 \tag{57.76}$$

whenever $x \neq 0$. Our task is to design an additional robustifying feedback $u_{rob}(x)$ so that the composite feedback $u = u_{nom} + u_{rob}$ robustly stabilizes the system (Equation 57.74), that is, guarantees stability for every admissible uncertainty Δ. It suffices that the derivative of V along closed-loop trajectories is negative for all such uncertainties. We compute this derivative as follows:

$$\begin{aligned}\dot{V} \;=\; & \nabla V(x)\,[F(x) + G(x)u_{nom}(x)] \\ & + \nabla V(x)\,[G(x)u_{rob}(x) + \Delta(x,t)]\end{aligned} \tag{57.77}$$

Can we make this derivative negative by some choice of $u_{rob}(x)$? Recall from Equation 57.76 that the first of the two terms in Equation 57.77 is negative; it remains to examine the second of these terms. For those values of x for which the coefficient $\nabla V(x) \cdot G(x)$ of the control $u_{rob}(x)$ is nonzero, we can always choose the value of $u_{rob}(x)$ large enough to overcome any finite bound on the uncertainty Δ and thus make the second term in Equation 57.77 negative. The only problems occur on the set where $\nabla V(x) \cdot G(x) = 0$, because on this set

$$\dot{V} = \nabla V(x) \cdot F(x) + \nabla V(x) \cdot \Delta(x,t) \tag{57.78}$$

regardless of our choice for the control. Thus to guarantee the negativity of $\dot{V}$, the uncertainty Δ must satisfy

$$\nabla V(x) \cdot F(x) + \nabla V(x) \cdot \Delta(x,t) \leq 0 \tag{57.79}$$

at all points where $\nabla V(x) \cdot G(x) = 0$. This inequality constraint on the uncertainty Δ is necessary for the Lyapunov redesign method to succeed. Unfortunately, there are two undesirable aspects of this necessary condition. First, the allowable size of the uncertainty Δ is dictated by F and V and can thus be severely restricted. Second, this inequality (Equation 57.79) cannot be checked a priori on the system (Equation 57.74) because it depends on the choice for V.

These considerations lead to the following question. Are there structural conditions that can be imposed on the uncertainty Δ so that the necessary condition (Equation 57.79) is automatically satisfied? One such structural condition is obvious. If we require that the uncertainty Δ is of the form,

$$\Delta(x,t) = G(x) \cdot \bar{\Delta}(x,t), \tag{57.80}$$

for some uncertain function $\bar{\Delta}$, then clearly $\nabla V \cdot \Delta = 0$ at all points where $\nabla V \cdot G = 0$, and thus the necessary condition (Equation 57.79) is satisfied. In the literature, Equation 57.80 is called the *matching condition* because it allows the system (Equation 57.74) to be written

$$\dot{x} = F(x) + G(x)[u + \bar{\Delta}(x,t)] \tag{57.81}$$

where now the uncertainty $\bar{\Delta}$ is *matched* with the control u, that is, it enters the system through the same channel as the control [2], [4], [12].

There are many methods available for the design of $u_{rob}(x)$ when the matching condition (Equation 57.80) is satisfied. For example, if the uncertainty $\bar{\Delta}$ is such that $|\ \bar{\Delta}(x,t)\ | \leq \bar{\rho}(x)$ for some known function $\bar{\rho}$, then the control,

$$u_{rob}(x) = -\bar{\rho}(x)\frac{(\nabla V(x) \cdot G(x))^T}{|\ \nabla V(x) \cdot G(x)\ |}, \tag{57.82}$$

yields

$$\begin{aligned}\dot{V} \leq\; & \nabla V(x)\,[F(x) + G(x)u_{nom}(x)] \\ & + |\ \nabla V(x) \cdot G(x)\ |\,\big[-\bar{\rho}(x) + |\ \bar{\Delta}(x,t)\ |\big]\end{aligned} \tag{57.83}$$

The first term in Equation 57.83 is negative from the nominal design (Equation 57.76), and the second term is also negative because we know that $|\ \bar{\Delta}(x,t)\ | \leq \bar{\rho}(x)$. The composite control $u = u_{nom} + u_{rob}$ thus guarantees stability and robustness to the uncertainty $\bar{\Delta}$. This controller (Equation 57.82), proposed, for example, by [8], is likely to be discontinuous at points where $\nabla V(x) \cdot G(x) = 0$. Indeed, in the scalar input case, Equation 57.82 becomes

$$u_{rob}(x) = -\bar{\rho}(x)\mathrm{sgn}\,(\nabla V(x) \cdot G(x)) \tag{57.84}$$

which is discontinuous unless $\bar{\rho}(x) = 0$ whenever $\nabla V(x) \cdot G(x) = 0$. Corless and Leitmann [4] introduced a continuous approximation to this controller which guarantees convergence, not to the point $x = 0$, but to an arbitrarily small prescribed neighborhood of this point. We will return to this continuity issue in Section 57.4.5.

We have seen that, because of the necessary condition (Equation 57.79), the Lyapunov redesign method is essentially limited to systems whose uncertainties satisfy the restrictive matching condition. In the next sections, we will take a different look at Equation 57.79 and obtain much weaker structural conditions on the uncertainty, which still allow a systematic robust controller design.

57.4.3 Beyond Lyapunov Redesign

In the previous section, we have seen that, if a Lyapunov function V is to guarantee robustness to an uncertainty Δ, then the inequality

$$\nabla V(x) \cdot F(x) + \nabla V(x) \cdot \Delta(x,t) \leq 0 \qquad (57.85)$$

must be satisfied at all points where $\nabla V(x) \cdot G(x) = 0$. In the Lyapunov redesign method, this inequality was viewed as a constraint on the uncertainty Δ. Now let us instead view this inequality as a constraint on the Lyapunov function V. This new look at Equation 57.85 will lead us beyond Lyapunov redesign: our construction of V will be based on Equation 57.85 rather than on the nominal system. In other words, we will take the uncertainty Δ into account during the construction of V itself.

To illustrate our departure from Lyapunov redesign, consider the second-order, single-input uncertain system,

$$\dot{x}_1 = x_2 + \Delta_1(x,t) \qquad (57.86)$$

$$\dot{x}_2 = u + \Delta_2(x,t) \qquad (57.87)$$

where Δ_1 and Δ_2 are uncertain functions which satisfy some known bounds. Let us try Lyapunov redesign. The first step would be to find a state feedback $u_{nom}(x)$ so that the nominal closed-loop system,

$$\dot{x}_1 = x_2 \qquad (57.88)$$

$$\dot{x}_2 = u_{nom}(x) \qquad (57.89)$$

has a globally asymptotically stable equilibrium at $x = 0$. Because the nominal system is linear, this step can be accomplished with a linear control law $u_{nom}(x) = Kx$, and we can obtain a quadratic Lyapunov function $V(x) = x^T Px$ for the stable nominal closed-loop system. In this case, the necessary condition Equation 57.85 becomes

$$2x^T P \begin{bmatrix} x_2 + \Delta_1(x,t) \\ \Delta_2(x,t) \end{bmatrix} \leq 0 \qquad (57.90)$$

at all points where $2x^T P[0 \quad 1]^T = 0$, that is, where $x_2 = -cx_1$ for some constant $c > 0$. We substitute $x_2 = -cx_1$ in Equation 57.90, and, after some algebra, we obtain

$$x_1 \Delta_1(x_1, -cx_1, t) \leq cx_1^2 \qquad (57.91)$$

for all $x_1, t \in R$. Now suppose our knowledge of the uncertainty $\Delta_1(x_1, -cx_1, t)$ consists of a bound $\rho_1(x_1)$ so that $|\Delta_1(x_1, -cx_1, t)| \leq \rho_1(x_1)$. Then Equation 57.91 implies that $\rho_1(x_1) \leq c \mid x_1 \mid$, that is, that the uncertainty Δ_1 is restricted to exhibit only *linear* growth in x_1 at a rate determined by the constant c. In other words, if the uncertainty Δ_1 does not satisfy this c-linear growth, then this particular Lyapunov redesign fails. This was to be expected because the uncertainty Δ_1 does not satisfy the matching condition.

The above Lyapunov redesign failed because it was based on the linear nominal system which suggested a *quadratic* Lyapunov function V. Let us now ignore the nominal system and base our search for V directly on the inequality (Equation 57.85). Let $\mu(x_1)$ be a smooth function so that $\mu(0) = 0$, and consider the Lyapunov function

$$V(x) = x_1^2 + [x_2 - \mu(x_1)]^2 . \qquad (57.92)$$

This function V is smooth, positive definite, and radially unbounded and thus qualifies as a candidate Lyapunov function for our system (Equations 57.86–57.87). We will justify this choice for V in the next section; our goal here is to illustrate how we can use our freedom in the choice for the function μ to derive a necessary condition on the uncertainty Δ_1 which is much less restrictive than Equation 57.91.

For V in Equation 57.92, $\nabla V(x) \cdot G(x) = 0$ if, and only if, $x_2 = \mu(x_1)$, so that the necessary condition Equation 57.85 becomes

$$x_1 \mu(x_1) + x_1 \Delta_1(x_1, \mu(x_1), t) \leq 0 \qquad (57.93)$$

for all $x_1, t \in R$. Because we have left the choice for μ open, this inequality can be viewed as a constraint on the choice of V (through μ) rather than a constraint on the uncertainty Δ_1. We need only impose a structural condition on Δ_1 which guarantees the *existence* of a suitable function μ. An example of such a condition would be the knowledge of a bound $\rho_1(x_1)$ so that $|\Delta_1(x_1, x_2, t)| \leq \rho_1(x_1)$; then Equation 57.93 becomes

$$x_1 \mu(x_1) + \mid x_1 \mid \rho_1(x_1) \leq 0 \qquad (57.94)$$

for all $x_1 \in R$. It is then clear that we can satisfy Equation 57.94 by choosing, for example,

$$\mu(x_1) = -x_1 - \rho_1(x_1)\mathrm{sgn}(x_1). \qquad (57.95)$$

A technical detail is that this μ is not smooth at $x_1 = 0$ unless $\rho_1(0) = 0$, which means V in Equation 57.92 may not strictly qualify as a Lyapunov function. As we will show in Section 57.4.5, however, smooth approximations always exist that will end up guaranteeing convergence to a neighborhood of $x = 0$ in the final design. What is important is that this design succeeds for any function $\rho_1(x_1)$, regardless of its growth. Thus the c-linear growth condition on Δ_1 which appeared in the above Lyapunov redesign through Equation 57.91 is gone; this new design allows arbitrary growth (in x_1) of the uncertainty Δ_1.

We have not yet specified the controller design; rather, we have shown how the limitations of Lyapunov redesign can be overcome through a reinterpretation of the necessary condition (Equation 57.85) as a constraint on the choice of V. Let us now return to the controller design problem and motivate our choice of V in Equation 57.92.

57.4.4 Recursive Lyapunov Design

Let us consider again the system (Equations 57.86–57.87):

$$\dot{x}_1 = x_2 + \Delta_1(x,t) \quad (57.96)$$

$$\dot{x}_2 = u + \Delta_2(x,t) \quad (57.97)$$

We assume knowledge of two bounding functions $\rho_1(x_1)$ and $\rho_2(x)$ so that all admissible uncertainties are characterized by the inequalities,

$$|\Delta_1(x,t)| \leq \rho_1(x_1) \quad (57.98)$$

$$|\Delta_2(x,t)| \leq \rho_2(x) \quad (57.99)$$

for all $x \in R^2$ and all $t \in R$. Note that the bound ρ_1 on the uncertainty Δ_1 is allowed to depend only on the state x_1; this is the structural condition suggested in the previous section and will be characterized more completely below. We will take a recursive approach to the design of a robust controller for this system. This approach is based on the *integrator backstepping* technique developed by [11] for the adaptive control of nonlinear systems. The first step in this approach is to consider the scalar system,

$$\dot{x}_1 = \bar{u} + \Delta_1(x_1, \bar{u}, t), \quad (57.100)$$

which we obtain by treating the state variable x_2 in Equation 57.96 as a control variable $\bar{u}$. This new system (Equation 57.100) is only conceptual; its relationship to the actual system (Equations 57.96–57.97) will be explored later. Let us next design a robust controller $\bar{u} = \mu(x_1)$ for this conceptual system. By construction, this new system satisfies the matching condition, and so we may use the Lyapunov redesign method to construct the feedback $\bar{u} = \mu(x_1)$. The nominal system is simply $\dot{x}_1 = \bar{u}$ which can be stabilized by a nominal feedback $\bar{u}_{nom} = -x_1$. A suitable Lyapunov function for the nominal closed-loop system $\dot{x}_1 = -x_1$ would be $V_1(x_1) = x_1^2$. We then choose $\bar{u} = \bar{u}_{nom} + \bar{u}_{rob}$, where $\bar{u}_{rob}$ is given, for example, by Equation 57.82 with $\bar{\rho} = \rho_1$. The resulting feedback function for $\bar{u}$ is

$$\mu(x_1) = -x_1 - \rho_1(x_1)\mathrm{sgn}(x_1). \quad (57.101)$$

If we now apply the feedback $\bar{u} = \mu(x_1)$ to the conceptual system (Equation 57.100), we achieve $\dot{V}_1 \leq -2x_1^2$ and thus guarantee stability for every admissible uncertainty Δ_1. Let us assume for now that this function μ is (sufficiently) smooth; we will return to the question of smoothness in Section 57.4.5.

The idea behind the backstepping approach is to use the conceptual controller (Equation 57.101) in constructing a control Lyapunov function V for the actual system (Equations 57.96–57.97). If we treat the Lyapunov function as a penalty function, it is reasonable to include in V a term penalizing the difference between the state x_2 and the conceptual control $\bar{u}$ designed for the conceptual system (Equation 57.100):

$$V(x) = V_1(x_1) + [x_2 - \mu(x_1)]^2. \quad (57.102)$$

We have seen in the previous section that this choice satisfies the necessary condition (Equation 57.85) and is thus a good candidate for our system (Equations 57.96–57.97). It is no coincidence that we arrived at the same expression for μ in Equations 57.95 and 57.101 both from the viewpoint of the necessary condition (Equation 57.85) and from the viewpoint of the conceptual system (Equation 57.100).

Let us now verify that the choice of Equation 57.102 for V leads to a robust controller design for the system (Equations 57.96–57.97). Computing $\dot{V}$ we obtain

$$\begin{aligned} \dot{V} = {} & 2x_1[x_2 + \Delta_1(x,t)] + 2[x_2 - \mu(x_1)] \\ & \left[u + \Delta_2(x,t) - \mu'(x_1)[x_2 + \Delta_1(x,t)]\right] \end{aligned} \quad (57.103)$$

where $\mu' := d\mu/dx$. Rearranging terms and using Equation 57.101,

$$\begin{aligned} \dot{V} \leq {} & -2x_1^2 + 2[x_2 - \mu(x_1)] \\ & \left[x_1 + u + \Delta_2(x,t) - \mu'(x_1)[x_2 + \Delta_1(x,t)]\right] \end{aligned} \quad (57.104)$$

The effect of backstepping is that the uncertainties enter the expression (Equation 57.104) with the same coefficient as the control variable u; in other words, the uncertainties effectively satisfy the matching condition. As a result, we can again apply the Lyapunov redesign technique; the feedback control $u = u_{nom} + u_{rob}$ with a nominal control,

$$u_{nom}(x) = -[x_2 - \mu(x_1)] - x_1 + \mu'(x_1)x_2, \quad (57.105)$$

yields

$$\begin{aligned} \dot{V} \leq {} & -2x_1^2 - 2[x_2 - \mu(x_1)]^2 + 2[x_2 - \mu(x_1)] \\ & \left[u_{rob} + \Delta_2(x,t) - \mu'(x_1)\Delta_1(x,t)\right]. \end{aligned} \quad (57.106)$$

We may complete the design by choosing u_{rob} as in Equation 57.82:

$$u_{rob}(x) = -\bar{\rho}(x)\mathrm{sgn}[x_2 - \mu(x_1)] \quad (57.107)$$

where $\bar{\rho}$ is some function satisfying $|\Delta_2(x,t) - \mu'(x_1)\Delta_1(x,t)| \leq \bar{\rho}(x)$. This yields

$$\dot{V} \leq -2x_1^2 - 2[x_2 - \mu(x_1)]^2 \quad (57.108)$$

for all admissible uncertainties Δ_1 and Δ_2, and thus robust stability is guaranteed.

EXAMPLE 57.8:

The above second-order design applies to the following system:

$$\dot{x}_1 = x_2 + \Delta_1(x,t) \quad (57.109)$$

$$\dot{x}_2 = u \quad (57.110)$$

where we let the uncertainty Δ_1 be any function satisfying $|\Delta_1(x,t)| \leq |x_1|^3$. In this case, the function μ in Equation 57.101 would be $\mu(x_1) = -x_1 - x_1^3$ which is smooth as required. The nominal control u_{nom} is given by Equation 57.105, which, for this example, becomes

$$u_{nom}(x) = -\left[x_2 + x_1 + x_1^3\right] - x_1 - (1 + 3x_1^2)x_2 \quad (57.111)$$

Adding the robustifying term (Equation 57.107), we obtain the following robust controller:

$$\begin{aligned} u(x) &= -\left[x_2 + x_1 + x_1^3\right] - x_1 - \left(1 + 3x_1^2\right)x_2 \\ &\quad - \left(| x_1 |^3 + 3 | x_1 |^5\right) \operatorname{sgn}\left(x_2 + x_1 + x_1^3\right). \end{aligned} \quad (57.112)$$

This robust controller is not continuous at points where $x_2 + x_1 + x_1^3 = 0$; an alternate smooth design will be proposed in Section 57.4.5.

The above controller design for the system (Equations 57.96–57.97) is a two-step design. In the first step, we considered the scalar system (Equation 57.100) and designed a controller $\bar{u} = \mu(x_1)$ which guaranteed robust stability. In the second step, we used the Lyapunov function (Equation 57.102) to construct a controller (Equations 57.105 + 57.107) for the actual system (Equations 57.96–57.97). This step-by-step design can be repeated to obtain controllers for higher order systems. For example, suppose that instead of the system (Equations 57.96–57.97), we have the system

$$\begin{aligned} \dot{x}_1 &= x_2 + \Delta_1(x, t) && (57.113) \\ \dot{x}_2 &= z + \Delta_2(x, t) \\ \dot{z} &= v + \Delta_3(x, z, t) && (57.114) \end{aligned}$$

where Δ_1 and Δ_2 are as in Equations 57.98–57.99, Δ_3 is a new uncertainty, and v is the control variable. We can use the controller $u(x) := u_{nom}(x) + u_{rob}(x)$ designed above in Equations 57.105 + 57.107 to obtain the following Lyapunov function W for our new system:

$$W(x, z) = V(x) + [z - u(x)]^2 \quad (57.115)$$

Here V is the Lyapunov function (Equation 57.102) used above, and we have simply added a term which penalizes the difference between the state variable z and the controller $u(x)$ designed above for the old system (Equations 57.96– 57.97). If $u(x)$ is smooth, then Equation 57.115 qualifies as a candidate Lyapunov function for our new system; see Section 57.4.5 below for details on choosing a smooth function $u(x)$. We can now construct a controller $v = v(x, z)$ for our new system in the same manner as the construction of $\mu(x_1)$ and $u(x)$ above.

We have thus obtained a systematic method for constructing Lyapunov functions and robust controllers for systems of the form,

$$\begin{aligned} \dot{x}_1 &= x_2 + \Delta_1(x, t), && (57.116) \\ \dot{x}_2 &= x_3 + \Delta_2(x, t), \\ &\vdots \\ \dot{x}_n &= u + \Delta_n(x, t). && (57.117) \end{aligned}$$

The Lyapunov function will be of the form

$$V(x) = x_1^2 + \sum_{i=1}^{n-1} [x_{i+1} - \mu_i(x_1, \ldots, x_i)]^2 \quad (57.118)$$

where the functions μ_i are constructed according to the recursive procedure described above. For this approach to succeed, it is sufficient that the uncertainties Δ_i satisfy the following *strict feedback condition*:

$$| \Delta_i(x, t) | \leq \rho_i(x_1, \ldots, x_i) \quad (57.119)$$

for known functions ρ_i. The restriction here is that the ith bound ρ_i can depend only on the first i states $(x_1, ..., x_i)$.

This recursive Lyapunov design technique applies to uncertain systems more general than Equations 57.116–57.117. Multi-input versions are possible, and the strict feedback condition (Equation 57.119) can be relaxed to allow the bound ρ_i to depend also on the state x_{i+1}. In particular, the bound ρ_n on the last uncertainty Δ_n can also depend on the control variable u. More details can be found in [5], [14], [16], [17].

57.4.5 Smooth Control Laws

The control law μ_i designed in the ith step of the recursive design becomes part of the Lyapunov function (Equation 57.118) in the next step. It is imperative, therefore, that each such function μ_i be differentiable. To illustrate how smooth functions can be obtained at each step, let us return to the second-order design in Section 57.4.4. The first step was to design a robust controller $\bar{u} = \mu(x_1)$ for the conceptual system,

$$\dot{x}_1 = \bar{u} + \Delta_1(x_1, \bar{u}, t) \quad (57.120)$$

with $| \Delta_1 | \leq \rho_1(x_1)$. In general, when $\rho_1(0) \neq 0$, we cannot choose μ as in Equation 57.101 because of the discontinuity at $x_1 = 0$. One alternative is to approximate Equation 57.101 by smooth function as follows:

$$\mu(x_1) = -x_1 - \rho_1(x_1)\frac{x_1}{| x_1 | + \delta(x_1)} \quad (57.121)$$

where $\delta(x_1)$ is a smooth, strictly positive function. This choice for μ is once differentiable, and it reduces to the discontinuous function (Equation 57.101) when $\delta \equiv 0$. We compute the derivative of $V_1(x_1) = x_1^2$ for the system (Equation 57.120) with the smooth feedback (Equation 57.121):

$$\begin{aligned} \dot{V} &\leq -2x_1^2 - 2\rho_1(x_1)\frac{x_1^2}{| x_1 | + \delta(x_1)} \\ &\quad + 2\rho_1(x_1) | x_1 | \\ &\leq -2x_1^2 + 2\rho_1(x_1)\frac{\delta(x_1) | x_1 |}{| x_1 | + \delta(x_1)} \\ &\leq -2x_1^2 + 2\rho_1(x_1)\delta(x_1) \end{aligned} \quad (57.122)$$

If we now choose $\delta(x_1)$ so that $\rho_1\delta$ is small, we see that $\dot{V}_1 < 0$ except in a small neighborhood of $x_1 = 0$.

In the second design step, we will choose u_{nom} as before in Equation 57.105, but instead of Equation 57.106 we obtain the Lyapunov derivative,

$$\begin{aligned}\dot{V} &\leq -2x_1^2 + 2\rho_1(x_1)\delta(x_1) - 2\left[x_2 - \mu(x_1)\right]^2 \\ &\quad + 2\left[x_2 - \mu(x_1)\right] \\ &\quad \left[u_{rob} + \Delta_2(x,t) - \mu'(x_1)\Delta_1(x,t)\right] \quad (57.123)\end{aligned}$$

where the extra term $2\rho_1\delta$ comes from Equation 57.122 and is a result of our smooth choice for $\mu(x_1)$. Our remaining task is to construct the robustifying term u_{rob}. Using the bound $|\ \Delta_2(x,t) - \mu'(x_1)\Delta_1(x,t)\ | \leq \bar{\rho}(x)$ as before, we obtain

$$\begin{aligned}\dot{V} &\leq -2x_1^2 + 2\rho_1(x_1)\delta(x_1) - 2\left[x_2 - \mu(x_1)\right]^2 \\ &\quad + 2\left[x_2 - \mu(x_1)\right]u_{rob} + 2\left|x_2 - \mu(x_1)\right|\bar{\rho}(x) \\ &\quad (57.124)\end{aligned}$$

We cannot choose u_{rob} as before in Equation 57.107 because it is not continuous at points where $x_2 = \mu(x_1)$. We could choose a smooth approximation to Equation 57.107, as we did above for the function μ, but, to illustrate an alternative approach, we will instead make use of Young's inequality,

$$2ab \leq \frac{1}{\varepsilon}a^2 + \varepsilon b^2 \quad (57.125)$$

which holds for any $a, b \in R$ and any $\varepsilon > 0$. Applying this inequality to the last term in Equation 57.124, we obtain

$$\begin{aligned}\dot{V} &\leq -2x_1^2 + 2\rho_1(x_1)\delta(x_1) \\ &\quad - \left[2 - \frac{1}{\varepsilon(x)}\right]\left[x_2 - \mu(x_1)\right]^2 \\ &\quad + 2\left[x_2 - \mu(x_1)\right]u_{rob} + \varepsilon(x)\left[\bar{\rho}(x)\right]^2 (57.126)\end{aligned}$$

where $\varepsilon(x)$ is a smooth, strictly positive function to be chosen below. Thus

$$u_{rob}(x) = -\frac{1}{2\varepsilon(x)}\left[x_2 - \mu(x_1)\right] \quad (57.127)$$

is smooth and yields

$$\begin{aligned}\dot{V} &\leq -2x_1^2 - 2\left[x_2 - \mu(x_1)\right]^2 \\ &\quad + 2\rho_1(x_1)\delta(x_1) + \varepsilon(x)\left[\bar{\rho}(x)\right]^2. \quad (57.128)\end{aligned}$$

It is always possible to choose $\delta(x_1)$ and $\varepsilon(x)$ so that the right-hand side of Equation 57.128 is negative, except possibly in a neighborhood of $x = 0$. Thus we have gained smoothness in the control law but lost exact convergence of the state to zero.

57.4.6 Design Flexibilities

The degrees of freedom in the recursive Lyapunov design procedure outlined above are numerous and allow for the careful shaping of the closed-loop performance. However, this procedure is new, and guidelines for exploiting design flexibilities are only beginning to appear. Our purpose in this section is to point out several of these degrees of freedom and discuss the consequences of various design choices.

We have already seen that the choices for the functions μ_i in the Lyapunov function (Equation 57.118) are by no means unique, nor is it the final choice for the control law. For example, when choosing a robustifying term u_{rob} in some step of the design, should we choose a smooth approximation to Equation 57.82 as in Equation 57.121, or should we use Young's inequality and choose Equation 57.127? Also, how should we choose the nominal term u_{nom}? Is it always good to cancel nonlinearities and apply linearlike feedback as in Equation 57.105, and if so, which gain should we use? Such questions are difficult in general, but there are some guidelines that apply in many situations. For example, consider the task of robustly stabilizing the point $x = 0$ of the simple scalar system,

$$\dot{x} = -x^3 + u + \Delta(x,t) \quad (57.129)$$

where Δ is an uncertain function with a known bound $|\ \Delta(x,t)\ | \leq |\ x\ |$. Because the matching condition is satisfied, we can apply the Lyapunov redesign method and choose $u = u_{nom} + u_{rob}$. One choice for the nominal control would be $u_{nom}(x) = x^3 - x$, which, together with a robustifying term as in Equation 57.82 yields a control law

$$u(x) = x^3 - 2x \quad (57.130)$$

This control law is valid from the viewpoint of Lyapunov redesign, and it indeed guarantees robustness to the uncertainty Δ. In such a choice, however, large positive feedback x^3 is used to cancel the nonlinearity $-x^3$ in Equation 57.129. This is absurd because the nonlinearity $-x^3$ is *beneficial* for stabilization, and positive feedback x^3 will lead to unnecessarily large control effort and will cause robustness problems more severe than those caused by the uncertainty Δ. Clearly, a much more reasonable choice is $u(x) = -2x$, but how can we identify better choices in a more general setting? One option would be to choose the control to minimize some meaningful cost functional. For example, the control

$$u(x) = x^3 - x - x\sqrt{x^4 - 2x^2 + 2} \quad (57.131)$$

minimizes the worst-case cost functional

$$J = \sup_{|\Delta| \leq |x|} \int_0^\infty \left[x^2 + u^2\right] dt \quad (57.132)$$

for this system Equation 57.129. The two control laws (Equations 57.130 and 57.131) are shown in Figure 57.1. We see that the optimal control (Equation 57.131) recognizes the benefit of the nonlinearity $-x^3$ in Equation 57.129 and accordingly produces little control effort for large x. Moreover, this optimal control is never positive feedback. Unfortunately, the computation of the optimal control (Equation 57.131) requires the solution of a Hamilton–Jacobi–Isaacs partial differential equation, and will be difficult and expensive for all but the simplest problems.

As a compromise between the benefits of optimality and its computational burden, we might consider the *inverse* optimal

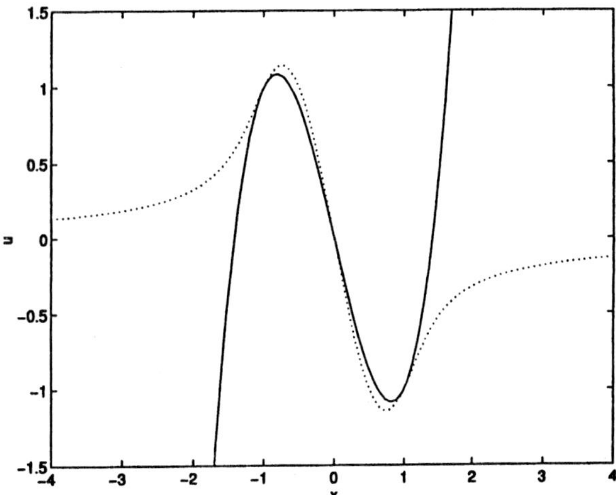

Figure 57.1 Comparison of the control laws Equation 57.130 (solid) and Equation 57.131 (dotted).

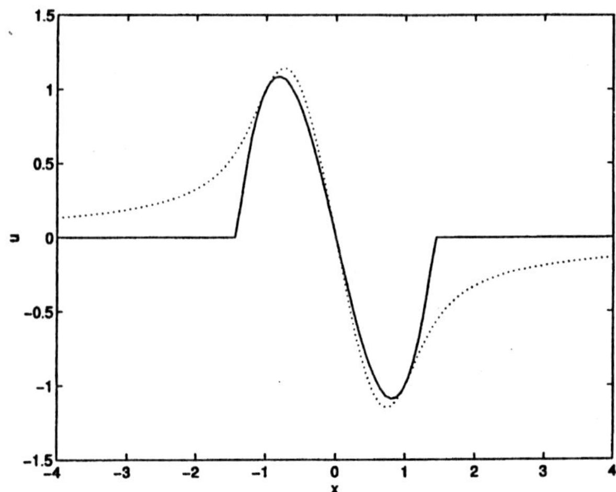

Figure 57.2 A comparison of the control laws Equation 57.135 (solid) and Equation 57.131 (dotted).

control problem, summarized for example by [7]. In this approach, we start with a Lyapunov function as in the Lyapunov redesign method above. We then show that this Lyapunov function is in fact the value function for some meaningful optimal stabilization problem, and we use this information to compute the corresponding optimal control. Freeman and Kokotović [6] have shown that the pointwise solutions of static minimization problems of the form,

$$\text{minimize} \quad u^T Su, \quad S = S^T > 0, \tag{57.133}$$

$$\text{subject to} \quad \sup_{\Delta} \left[\dot{V}(x, u, \Delta) + \sigma(x)\right] \leq 0, \tag{57.134}$$

yield optimal controllers (in this inverse sense) for meaningful cost functionals, where V is a given control Lyapunov function and σ is a positive function whose choice represents a degree of freedom. When the system is jointly affine in the control u and the uncertainty Δ, this optimization problem (Equations 57.133–57.134) is a quadratic program with linear constraints and thus has an *explicit* solution $u(x)$. For example, the solution to Equations 57.133–57.134 for the system (Equation 57.129) with $V = x^2$ and $\sigma = 2x^2$ yields the control law,

$$u(x) = \begin{cases} x^3 - 2x & \text{when} \quad x^2 \leq 2, \\ 0 & \text{when} \quad x^2 \geq 2. \end{cases} \tag{57.135}$$

As shown in Figure 57.2, this control (Equation 57.135) is qualitatively the same as the optimal control (Equation 57.131); both recognize the benefit of the nonlinearity $-x^3$ and neither one is ever positive feedback. The advantage of the inverse optimal approach is that the controller computation is simple once a control Lyapunov function is known.

We thus have some guidelines for choosing the intermediate control laws μ_i at each step of the design, namely, we can avoid the wasteful cancellations of beneficial nonlinearities. However, these guidelines, when used at early steps of the recursive design, have not yet been proved beneficial for the final design.

We have shown that the choices for the functions μ_i in Equation 57.118 represent important degrees of freedom in the recursive Lyapunov design procedure. Perhaps even more important is the choice of the form of V itself. Recall that, given a Lyapunov function V_i and a control μ_i at the ith design step, we constructed the new Lyapunov function V_{i+1} as follows:

$$\begin{aligned} V_{i+1}(x_1, \ldots, x_{i+1}) &= V_i(x_1, \ldots, x_i) \\ &+ [x_{i+1} - \mu_i(x_1, \ldots, x_i)]^2 . \end{aligned} \tag{57.136}$$

This choice for V_{i+1} is not the only choice that will lead to a successful design, and we are thus confronted with another degree of freedom in the design procedure. For example, instead of Equation 57.136, we can choose

$$\begin{aligned} V_{i+1}(x_1, \ldots, x_{i+1}) &= \kappa\left[V_i(x_1, \ldots, x_i)\right] \\ &+ [x_{i+1} - \mu_i(x_1, \ldots, x_i)]^2 . \end{aligned} \tag{57.137}$$

where $\kappa : R_+ \rightarrow R_+$ is any smooth, positive-definite, unbounded function whose derivative is everywhere strictly positive. This function κ represents a nonlinear weighting on the term V_i and can have a large effect on the control laws obtainable in future steps.

The last degree of freedom we wish to discuss involves the quadraticlike term in Equations 57.136 and 57.137. Praly et al. [15] have shown that Equation 57.137 can be replaced by the more general expression

$$\begin{aligned} V_{i+1}(x_1, \ldots, x_{i+1}) &= \kappa\left[V_i(x_1, \ldots, x_i)\right] \\ &+ \int_{\mu_i(x_1,\ldots,x_i)}^{x_{i+1}} \phi(x_1, \ldots, x_i, s)ds \end{aligned} \tag{57.138}$$

for a suitable choice of the function ϕ. Indeed, Equation 57.137 is a special case of Equation 57.138 for $\phi = 2[s - \mu_i(x_1, \ldots, x_i)]$. This degree of freedom in the choice of ϕ can be significant; for example, it allowed extensions of the recursive design to the nonsmooth case by [15]. It can also be used to reduce the unnecessarily large control gains often caused by the quadratic term in Equation 57.136. To illustrate this last point, let us return to the second-order example (Equations 57.109–57.110) given by

$$\dot{x}_1 = x_2 + \Delta_1(x, t) \tag{57.139}$$

$$\dot{x}_2 = u \tag{57.140}$$

where the uncertainty Δ_1 is any function satisfying $|\Delta_1(x,t)| \leq |x_1|^3$. Recall that using the Lyapunov function,

$$V(x) = x_1^2 + \left[x_2 + x_1 + x_1^3\right]^2 \qquad (57.141)$$

we designed the following robust controller (Equation 57.112) for this system:

$$u(x) = -\left[x_2 + x_1 + x_1^3\right] - x_1 - \left(1 + 3x_1^2\right)x_2 - \left(|x_1|^3 + 3|x_1|^5\right)\operatorname{sgn}\left(x_2 + x_1 + x_1^3\right). \qquad (57.142)$$

This controller is not continuous at points where $x_2+x_1+x_1^3 = 0$. In other words, the local controller gain $\partial u/\partial x_2$ is *infinite* at such points. Such infinite gain will cause high-amplitude chattering along the manifold M described by $x_2 + x_1 + x_1^3 = 0$. As a result of such chattering, this control law may not be implementable because of the unreasonable demands on the actuator. However, as was shown in Section 57.4.5, we can use this same Lyapunov function (Equation 57.141) to design a *smooth* robust controller $\bar{u}$ for our system. Will such a smooth controller eliminate the chattering caused by the discontinuity in Equation 57.142? Surprisingly, the answer is no. One can show that the local controller gain $\partial\bar{u}/\partial x_2$, although finite because of the smoothness of $\bar{u}$, grows like x_1^6 along the manifold M. Thus for large signals, this local gain is extremely large and can cause chattering as if it were infinite. Figure 57.3 shows the Lyapunov function V in Equation 57.141, plotted as a function of the two variables, $z_1 := x_1$ and $z_2 := x_2 + x_1 + x_1^3$. A smooth control law $\bar{u}$ designed using this V is shown in Figure 57.4, again plotted as a function of z_1 and z_2. Note that the x_1^6 growth of the local gain of $\bar{u}$ along the manifold $z_2 = 0$ is clearly visible in this figure. One might conclude that such high gain is unavoidable for this particular system. This conclusion is false, however, because the x_1^6 growth of the local gain is an artifact of the *quadratic* form of the Lyapunov function (Equation 57.141) and has nothing to do with robust stability requirements for this system. Freeman and Kokotović [6] have shown how to choose the function ϕ in Equation 57.138 to reduce greatly the growth of the local gain. For this example (Equations 57.139–57.140), they achieved a growth of x_1^2 as opposed to the x_1^6 growth caused by the quadratic V in Equation 57.141. Their new, *flattened* Lyapunov function is shown in Figure 57.5, and the corresponding control law is shown in Figure 57.6. The x_1^2 versus x_1^6 growth of the local gain is evident when comparing Figures 57.6 and 57.4. The control signals generated from a particular initial condition are compared in Figure 57.7. These controls produce virtually the same state trajectories, but the chattering caused by the old control law has been eliminated by the new one.

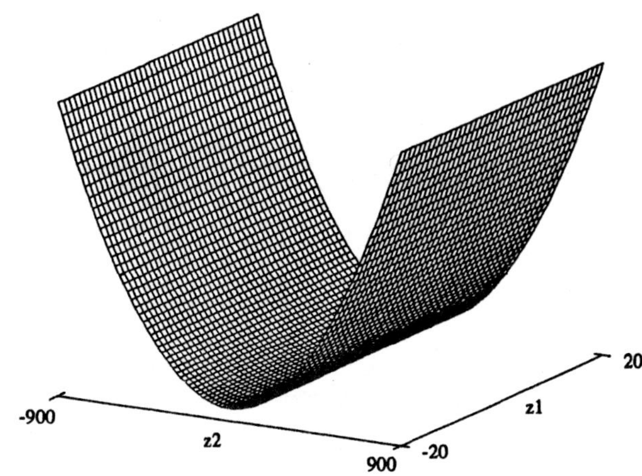

Figure 57.3 Quadraticlike Lyapunov function (Equation 57.141). (From Freeman, R. A. and Kokotović, P. V., *Automatica*, 29(6), 1425–1437, 1993. With permission.)

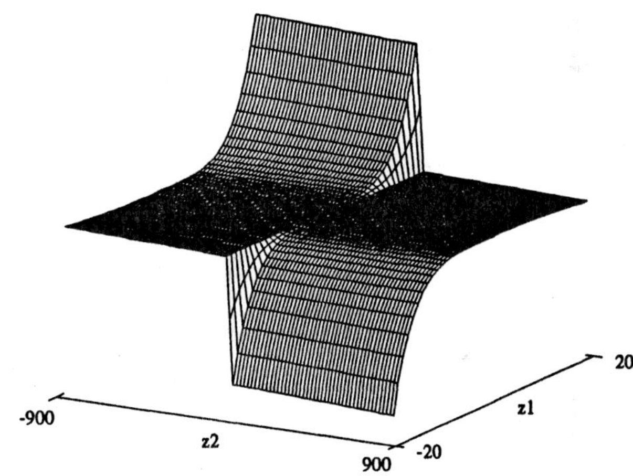

Figure 57.4 Controller from quadratic-like Lyapunov function (Equation 57.141). (From Freeman, R. A. and Kokotović, P. V., *Automatica* 29(6), 1425–1437, 1993. With permission.)

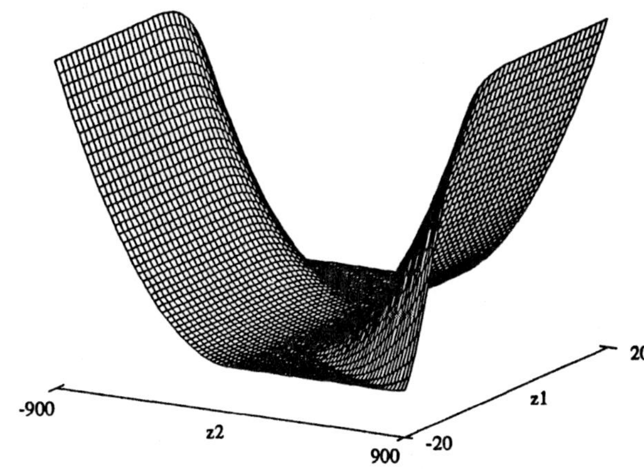

Figure 57.5 New flattened Lyapunov function. (From Freeman, R. A. and Kokotović, P. V., *Automatica*, 29(6), 1425–1437, 1993. With permission.)

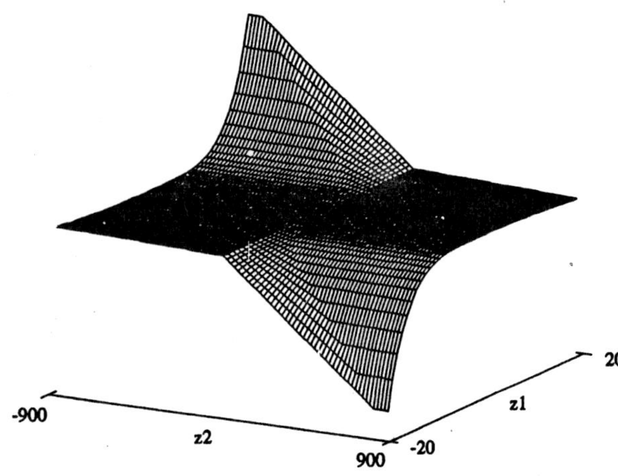

Figure 57.6 Controller from flattened Lyapunov function. (From Freeman, R. A. and Kokotović, P. V., *Automatica,* 29(6), 1425–1437, 1993. With permission.)

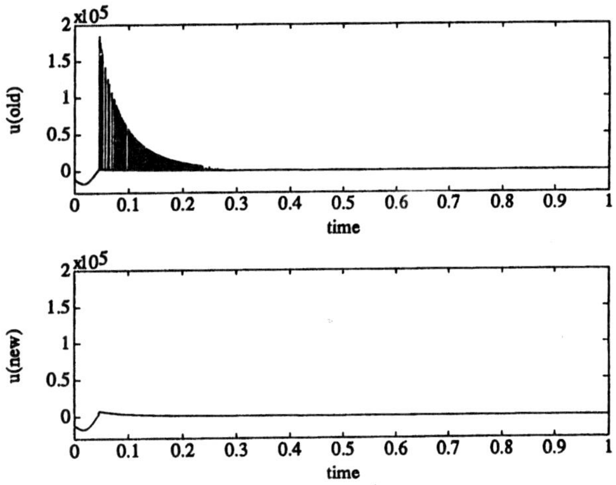

Figure 57.7 Comparison of control signals. (From Freeman, R. A. and Kokotović, P. V., *Automatica,* 29(6), 1425–1437, 1993. With permission.)

References

[1] Artstein, Z., Stabilization with relaxed controls, *Nonlinear Anal.* 7(11), 1163–1173, 1983.

[2] Barmish, B. R., Corless, M. J., and Leitmann, G., A new class of stabilizing controllers for uncertain dynamical systems, *SIAM J. Cont. Optimiz.,* 21, 246–255, 1983.

[3] Corless, M. J., Robust stability analysis and controller design with quadratic Lyapunov functions, in *Variable Structure and Lyapunov Control,* Zinober, A., Ed., Springer, Berlin, 1993.

[4] Corless, M. J. and Leitmann, G., Continuous state feedback guaranteeing uniform ultimate boundedness for uncertain dynamic systems, *IEEE Trans. Automat. Control,* 26(5), 1139–1144, 1981.

[5] Freeman, R. A. and Kokotović, P. V., Design of 'softer' robust nonlinear control laws, *Automatica* 29(6), 1425–1437, 1993.

[6] Freeman, R. A. and Kokotović, P. V., Inverse optimality in robust stabilization, *SIAM J. Control Optimiz.,* to appear revised November 1994.

[7] Glad, S. T., Robustness of nonlinear state feedback — A survey, *Automatica,* 23(4), 425–435, 1987.

[8] Gutman, S., Uncertain dynamical systems — Lyapunov min-max approach, *IEEE Trans. Automat. Control,* 24, 437–443, 1979.

[9] Jacobson, D. H., *Extensions of Linear-Quadratic Control, Optimization and Matrix Theory,* Academic, London, 1977.

[10] Jurdjevic, V. and Quinn, J. P., Controllability and stability, *J. Diff. Eq.,* 28, 381–389, 1978.

[11] Kanellakopoulos, I., Kokotović, P. V., and Morse, A. S., Systematic design of adaptive controllers for feedback linearizable systems, *IEEE Trans. Automat. Control,* 36(11), 1241–1253, 1991.

[12] Khalil, H. K., *Nonlinear Systems,* Macmillan, New York, 1992.

[13] Krasovsky, A. A., A new solution to the problem of a control system analytical design, *Automatica,* 7, 45–50, 1971.

[14] Marino, R. and Tomei, P., Robust stabilization of feedback linearizable time-varying uncertain nonlinear systems, *Automatica,* 29(1), 181–189, 1993.

[15] Praly, L., d'Andréa Novel, B., and Coron, J.-M., Lyapunov design of stabilizing controllers for cascaded systems, *IEEE Trans. Automat. Control,* 36(10), 1177–1181, 1991.

[16] Qu, Z., Robust control of nonlinear uncertain systems under generalized matching conditions, *Automatica,* 29(4), 985–998, 1993.

[17] Slotine, J. J. E. and Hedrick, K., Robust input-output feedback linearization, *Int. J. Control,* 57, 1133–1139, 1993.

[18] Sontag, E. D., A Lyapunov-like characterization of asymptotic controllability, *SIAM J. Control Optimiz.,* 21(3), 462–471, 1983.

Further Reading

Contributions to the development of the Lyapunov design methodology for systems with no uncertainties include [1], [9], [10], [13], [18]. A good introduction to Lyapunov redesign can be found in Chapter 5.5 of [12]. Corless [3] has recently surveyed various methods for the design of robust controllers for nonlinear systems using *quadratic* Lyapunov functions. Details of the recursive design presented in Section 57.4.4 can be found in [6], [14], [16]. The state-space techniques discussed in this chapter can be combined with nonlinear input/output techniques to obtain more advanced designs (see Chapter 56.2). Finally, when the uncertain nonlinearities are given by *constant* parameters multiplying *known* nonlinearities, adaptive control techniques apply (see Chapter 57.8).

57.5 Variable Structure, Sliding-Mode Controller Design

R. A. DeCarlo, School of Electrical and Computer Engineering, Purdue University, West Lafayette, IN
S. H. Żak, School of Electrical and Computer Engineering, Purdue University, West Lafayette, IN
S. V. Drakunov, Department of Electrical Engineering, Tulane University, New Orleans, LA

57.5.1 Introduction and Background

This chapter investigates Variable Structure Control (VSC) as a high-speed switched feedback control resulting in a sliding mode. For example, the gains in each feedback path switch between two values according to a rule that depends on the value of the state at each instant. The purpose of the switching control law is to drive the nonlinear plant's state trajectory onto a prespecified (user-chosen) surface in the state space and to maintain the plant's state trajectory on this surface for all subsequent time. This surface is called a *switching surface.* When the plant state trajectory is "above" the surface, a feedback path has one gain and a different gain if the trajectory drops "below" the surface. This surface defines the rule for proper switching. The surface is also called a *sliding surface* (sliding manifold) because, ideally speaking, once intercepted, the switched control maintains the plant's state trajectory on the surface for all subsequent time and the plant's state trajectory then slides along this surface. The plant dynamics restricted to this surface represent the controlled system's behavior. The first critical phase of a VSC design is to define properly a switching surface so that the plant, restricted to the surface, has desired dynamics, such as stability to the origin, tracking, regulation, etc.

In summary, a VSC control design breaks down into two phases. The first phase is to design or choose a switching surface so that the plant state restricted to the surface has desired dynamics. The second phase is to design a switched control that will drive the plant state to the switching surface and maintain it on the surface upon interception. A Lyapunov approach is used to characterize this second design phase. Here a generalized Lyapunov function, that characterizes the motion of the state trajectory to the surface, is defined in terms of the surface. For each chosen switched control structure, one chooses the "gains" so that the derivative of this Lyapunov function is negative definite, thus guaranteeing motion of the state trajectory to the surface.

As an introductory example, consider the first order system $\dot{x}(t) = u(x,t)$ with control

$$u(x,t) = -\text{sgn}(x) = \begin{cases} -1 & \text{if } x > 0 \\ +1 & \text{if } x < 0 \end{cases}$$

Hence, the system with control satisfies $\dot{x} = -\text{sgn}(x)$ with trajectories plotted in Figure 57.8(a). Here the control $u(x,t)$ switches, changing its value between ± 1 around the surface $\sigma(x,t) = x = 0$. Hence for any initial condition x_0, a finite time t_1 exists for which $x(t) = 0$ for all $t \geq t_1$. Now suppose $\dot{x}(t) = u(x,t) + v(t)$ where again $u(x,t) = -\text{sgn}(x)$ and $v(t)$ is a bounded disturbance for which $\sup_t |v(t)| < 1$. As before, the control $u(x,t)$ switches its value between ± 1 around the surface $\sigma(x,t) = x = 0$. It follows that if $x(t) > 0$, then $\dot{x}(t) = -\text{sgn}[x(t)] + v(t) < 0$ forcing motion toward the line $\sigma(x,t) = x = 0$, and if $x(t) < 0$, then $\dot{x}(t) = -\text{sgn}[x(t)] + v(t) > 0$ again forcing motion toward the line $\sigma(x,t) = x = 0$. For a positive initial condition, this is illustrated in Figure 57.8(b). The rate of convergence to the line depends on the disturbance. Nevertheless, a finite time t_1 exists for which $x(t) = 0$ for all $t \geq t_1$. If the disturbance magnitude exceeds 1, then the gain can always be adjusted to compensate. Hence, this VSC law is robust in the face of bounded disturbances, illustrating the simplicity and advantage of the VSC technique.

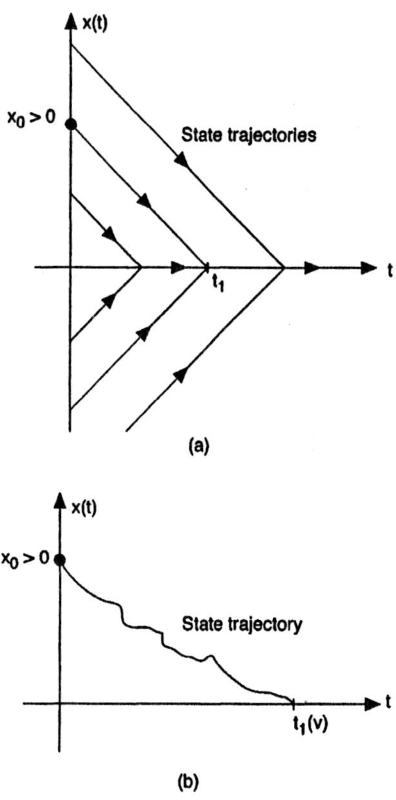

Figure 57.8 (a) State trajectories for the system $\dot{x} = -\text{sgn}(x)$; (b) State trajectory for the system $\dot{x}(t) = -\text{sgn}[x(t)] + v(t)$.

From the above example, one can see that VSC can provide a robust means of controlling (nonlinear) plants with disturbances and parameter uncertainties. Further, the advances in computer technology and high-speed switching circuitry have made the practical implementation of VSC quite feasible and of increasing interest. Indeed, pulse-width modulation control and switched dc-to-dc power converters [12] can be viewed in a VSC framework.

57.5.2 System Model, Control Structure, and Sliding Modes

System Model

This chapter investigates a class of systems with a state model nonlinear in the state vector $x(\cdot)$ and linear in the control vector $u(\cdot)$ of the form,

$$\dot{x}(t) = F(x,t,u) = f(x,t) + B(x,t)u(x,t), \quad (57.143)$$

where $x(t) \in R^n, u(t) \in R^m$, and $B(x,t) \in R^{n\times m}$; further, each entry in $f(x,t)$ and $B(x,t)$ is assumed continuous with a bounded continuous derivative with respect to x. In the linear time-invariant case, Equation 57.143 becomes

$$\dot{x} = Ax + Bu \quad (57.144)$$

with A $n\times n$ and B $n\times m$ constant matrices.

Associated with the system is a $(n-m)$-dimensional switching surface (also called a *discontinuity* or *equilibrium manifold*),

$$\mathcal{S} = \left\{(x,t) \in R^{n+1} | \sigma(x,t) = 0\right\}, \quad (57.145)$$

where

$$\sigma(x,t) = [\sigma_1(x,t), \ldots, \sigma_m(x,t)]^T = 0. \quad (57.146)$$

(We will often refer to $\mathcal{S}$ as $\sigma(x,t) = 0$.) When there is no t-dependence, this $(n-m)$-dimensional manifold in the state space R^n is determined as the intersection of m $(n-1)$-dimensional surfaces $\sigma_i(x,t) = 0$. These surfaces are designed so that the system state trajectory, restricted to $\sigma(x,t) = 0$, has a desired behavior such as stability or tracking. Although general nonlinear time-varying surfaces as in Equation 57.145 are possible, linear ones are more prevalent in design [2], [8], [11], [13], [14]. Linear surface design is taken up in Section 57.5.4.

Control Structure

After proper design of the surface, a switched controller, $u(x,t) = [u_1(x,t), \ldots, u_m(x,t)]^T$ is constructed of the form,

$$u_i(x,t) = \begin{cases} u_i^+(x,t), & \text{when} \quad \sigma_i(x,t) > 0, \\ u_i^-(x,t), & \text{when} \quad \sigma_i(x,t) < 0. \end{cases} \quad (57.147)$$

Equation 57.147 indicates that the control changes value depending on the sign of the switching surface at x and t. Thus $\sigma(x,t) = 0$ is called a switching surface. The control is undefined on the switching surface. Off the switching surface, the control values $u_i^\pm$ are chosen so that the tangent vectors of the state trajectory point towards the surface such that the state is driven to and maintained on $\sigma(x,t) = 0$. Such controllers result in discontinuous closed-loop systems.

Sliding Modes

The control $u(x,t)$ is designed so that the system state trajectory is attracted to the switching surface and, once having intercepted it, remains on the switching surface for all subsequent time. The state trajectory can be viewed as sliding along the switching surface and thus the system is in a sliding mode. A sliding mode exists if, in the vicinity of the switching surface, $\mathcal{S}$, the tangent or velocity vectors of the state trajectory point toward the switching surface. If the state trajectory intersects the sliding surface, the value of the state trajectory or "representative point" remains within an ε neighborhood of $\mathcal{S}$. If a sliding mode exists on $\mathcal{S}$, then $\mathcal{S}$, or more commonly $\sigma(x,t) = 0$, is also termed a sliding surface. Note that interception of the surface $\sigma_i(x,t) = 0$ does not guarantee sliding on the surface for all subsequent time as illustrated in Figure 57.9, although this is possible.

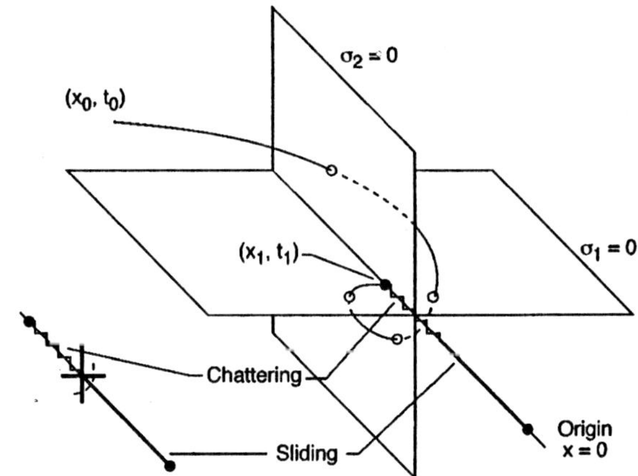

Figure 57.9 A situation in which a sliding mode exists on the intersection of the two indicated surfaces for $t \geq t_1$.

An *ideal sliding mode* exists only when the state trajectory $x(t)$ of the controlled plant satisfies $\sigma(x(t),t) = 0$ at every $t \geq t_1$ for some t_1. This may require infinitely fast switching. In real systems, a switched controller has imperfections, such as delay, hysteresis, etc., which limit switching to a finite frequency. The representative point then oscillates within a neighborhood of the switching surface. This oscillation, called *chattering* (discussed in a later section), is also illustrated in Figure 57.9. If the frequency of the switching is very high relative to the dynamic response of the system, the imperfections and the finite switching frequencies are often but not always negligible. The subsequent development focuses primarily on ideal sliding modes.

Conditions for the Existence of a Sliding Mode

The existence of a sliding mode [2], [13], [14] requires stability of the state trajectory to the switching surface $\sigma(x,t) = 0$, i.e., after some finite time t_1, the system representative point, $x(t)$, must be in some suitable neighborhood, $\{x | \|\sigma(x,t)\| < \varepsilon\}$, of $\mathcal{S}$ for suitable $\varepsilon > 0$. A domain, $\mathcal{D}$, of dimension $n-m$ in the manifold, $\mathcal{S}$, is a sliding-mode domain if, for each $\varepsilon > 0$, there is a $\delta > 0$, so that any motion starting within a n-dimensional δ-vicinity of $\mathcal{D}$ may leave the n-dimensional ε-vicinity of $\mathcal{D}$ only through the n-dimensional ε-vicinity of the boundary of $\mathcal{D}$ as illustrated in Figure 57.10.

The *region of attraction* is the largest subset of the state space from which sliding is achievable. A sliding mode is globally reachable if the domain of attraction is the entire state space. The

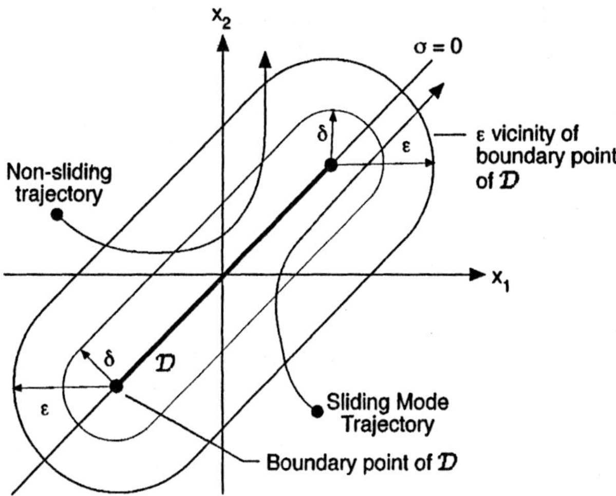

Figure 57.10 Two-dimensional illustration of the domain of a sliding mode.

second method of Lyapunov provides the natural setting for a controller design leading to a sliding mode. In this effort one uses a generalized Lyapunov function, $V(t, x, \sigma)$, that is positive definite with a negative time derivative in the region of attraction.

THEOREM 57.5 *[13, p. 83]: For the $(n - m)$-dimensional domain $\mathcal{D}$ to be the domain of a sliding mode, it is sufficient that in some n-dimensional domain $\Omega \supset \mathcal{D}$, a function $V(t, x, \sigma)$ exists, continuously differentiable with respect to all of its arguments and satisfying the following conditions:*

(1) $V(t, x, \sigma)$ is positive definite with respect to σ, i.e., for arbitrary t and x, $V(t, x, \sigma) > 0$ when $\sigma \neq 0$ and $V(t, x, 0) = 0$; on the sphere $\|\sigma\| \leq \rho > 0$, for all $x \in \Omega$ and any t, the relationships

$$\inf_{\|\sigma\|=\rho} V(t,x,\sigma) = h_\rho, \qquad h_\rho > 0$$

and

$$\sup_{\|\sigma\|=\rho} V(t,x,\sigma) = H_\rho, \qquad H_\rho > 0$$

hold where h_ρ and H_ρ depend only on ρ with $h_\rho \neq 0$ if $\rho \neq 0$.

(2) The total time derivative of $V(t, x, \sigma)$ on the trajectories of the system of Equation 57.143 has a negative supremum for all $x \in \Omega$ except for x on the switching surface where the control inputs are undefined and the derivative of $V(t, x, \sigma)$ does not exist.

In summary, two phases underly VSC design. The first phase is to construct a switching surface $\sigma(x, t) = 0$ so that the system restricted to the surface has a desired global behavior, such as stability, tracking, regulation, etc. The second phase is to design a (switched) controller $u(x, t)$ so that away from the surface $\sigma(x, t) = 0$, the tangent vectors of the state trajectories point toward the surface, i.e., so that there is stability to the switching surface. This second phase is achieved by defining an appropriate Lyapunov function $V(t, x, \sigma)$, differentiating this function so that the control $u(x, t)$ becomes explicit, and adjusting controller gains so that the derivative is negative definite. The choice of $V(t, x, \sigma)$ determines the allowable controller structures. Conversely, a workable control structure has a set of possible Lyapunov functions to verify its viability. A later section will take up the discussion of general control structures.

An Illustrative Example

To conclude this section we present an illustrative example for a single pendulum system,

$$\begin{aligned} \dot{x} &= A(x)x + Bu(x) \\ &= \begin{bmatrix} 0 & 1 \\ -\frac{\sin(x_1)}{x_1} & 0 \end{bmatrix} \begin{bmatrix} x_1 \\ x_2 \end{bmatrix} + \begin{bmatrix} 0 \\ 1 \end{bmatrix} u(x), \end{aligned}$$

with a standard feedback control structure, $u(x) = k_1(x)x_1 + k_2(x)x_2$, having nonlinear feedback gains switched according to the rule

$$k_i(x) = \begin{cases} \alpha_i(x), & \text{if} \quad \sigma(x)x_i > 0, \\ \beta_i(x), & \text{if} \quad \sigma(x)x_i < 0, \end{cases}$$

that depends on the linear switching surface $\sigma(x) = [s_1 \quad s_2]x$. For such single-input systems it is ordinarily convenient to choose a Lyapunov function of the form $V(t, x, \sigma) = 0.5\sigma^2(x)$. To determine the gains necessary to drive the system state to the surface $\sigma(x) = 0$, they may be chosen so that

$$\begin{aligned} \dot{V}(t,x,\sigma) &= 0.5\frac{d\sigma^2}{dt} = \sigma(x)\frac{d\sigma(x)}{dt} = \sigma(x)[s_1 \quad s_2]\dot{x} \\ &= \sigma(x)x_1\left[s_2\left(k_1(x) - \frac{\sin(x_1)}{x_1}\right)\right] \\ &\quad + \sigma(x)x_2\left[s_1 + s_2k_2(x)\right] < 0. \end{aligned}$$

This is satisfied whenever

$$\begin{aligned} \alpha_1(x) &= \alpha_1 < \min_{x_1}\left[\frac{\sin(x_1)}{x_1}\right] = -1, \\ \beta_1(x) &= \beta_1 > \max_{x_1}\left[\frac{\sin(x_1)}{x_1}\right] = 1, \end{aligned}$$

$\alpha_2 < -(s_1/s_2)$, and $\beta_2 > -(s_1/s_2)$. Thus, for properly chosen s_1 and s_2, the controller gains are readily computed.

This example proposed no methodology for choosing s_1 and s_2, i.e., for designing the switching surface. Section 57.5.4 takes up this topic. Further this example was only single input. For the multi-input case, ease of computation of the control gains depends on a properly chosen Lyapunov function. For most cases, a quadratic Lyapunov function is adequate. This topic is taken up in Section 57.5.5.

57.5.3 Existence and Uniqueness of Solutions to VSC Systems

VSC produces system dynamics with discontinuous right-hand sides due to the switching action of the controller. Thus they fail to satisfy conventional existence and uniqueness results of differential equation theory. Nevertheless, an important aspect of VSC is the presumption that the plant behaves uniquely when restricted to $\sigma(x, t) = 0$. One of the earliest and conceptually straightforward approaches addressing existence and uniqueness is the method of Filippov [7]. The following briefly reviews this method in the two-dimensional, single-input case.

From Equation 57.143, $\dot{x}(t) = F(x,t,u)$ and the control $u(x,t)$ satisfy Equation 57.147. Filippov's work shows that the state trajectories of Equation 57.143 with control Equation 57.147 on the switching manifold Equation 57.145 solve the equation

$$\dot{x}(t) = \alpha F^{+} + (1-\alpha)F^{-} = F^{0}, \quad 0 \le \alpha \le 1. \quad (57.148)$$

This is illustrated in Figure 57.11 where $F^{+} = F(x,t,u^{+})$, $F^{-} = F(x,t,u^{-})$, and F^{0} is the resulting velocity vector of the state trajectory in a sliding mode.

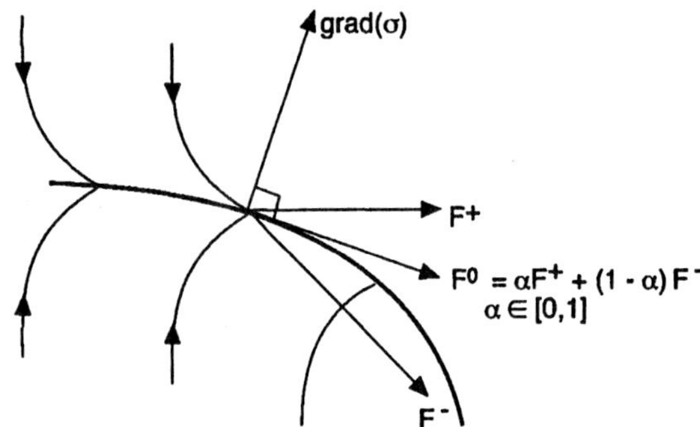

Figure 57.11 Illustration of the Filippov method of determining the desired velocity vector F^{0} for the motion of the state trajectory on the sliding surface as per Equation 57.148.

The problem is to determine α, which follows from solving the equation $<\text{grad}(\sigma), F^{0}> = 0$, i.e.,

$$\alpha = \frac{\langle \text{grad}(\sigma), F^{-} \rangle}{\langle \text{grad}(\sigma), (F^{-} - F^{+}) \rangle}$$

provided (1) $<\text{grad}(\sigma), (F^{-} - F^{+})> \neq 0$, (2) $<\text{grad}(\sigma), F^{+}> \le 0$, and (3) $<\text{grad}(\sigma), F^{-}> \ge 0$, where the notation $< a, b >$ denotes the inner product of a and b. Here, computation of F^{0} represents the "average" velocity, $\dot{x}(t)$, of the state trajectory restricted to $\sigma(x,t) = 0$. On average, the solution to Equation 57.143 with control Equation 57.147 exists and is uniquely defined on $\mathcal{S} = \{(x,t) \in R^{n+1} | \sigma(x,t) = 0\}$. This technique can also be used to determine the plant behavior in a sliding mode.

57.5.4 Switching-Surface Design

Switching-surface design is predicated upon knowledge of the system behavior in a sliding mode. This behavior depends on the parameters of the switching surface. Nonlinear switching surfaces are nontrivial to design. In the linear case, the switching-surface design problem can be converted into an equivalent state feedback design problem. In any case, achieving a switching-surface design requires analytically specifying the motion of the state trajectory in a sliding mode. The so-called method of equivalent control is essential to this specification.

Equivalent Control

Equivalent control constitutes an equivalent input which, when exciting the system, produces the motion of the system on the sliding surface whenever the initial state is on the surface. Suppose at t_1 the plant's state trajectory intercepts the switching surface, and a sliding mode exists. The existence of a sliding mode implies that, for all $t \ge t_1, \sigma(x(t),t) = 0$, and hence $\dot{\sigma}(x(t),t) = 0$. Using the chain rule, we define the equivalent control u_{eq} for systems of the form of Equation 57.143 as the input satisfying

$$\dot{\sigma} = \frac{\partial \sigma}{\partial t} + \frac{\partial \sigma}{\partial x}\dot{x} = \frac{\partial \sigma}{\partial t} + \frac{\partial \sigma}{\partial x} f(x,t) + \frac{\partial \sigma}{\partial x} B(x,t) u_{eq} = 0.$$

Assuming that the matrix product $\frac{\partial \sigma}{\partial x} B(x,t)$ is nonsingular for all t and x, one can compute u_{eq} as

$$u_{eq} = -\left[\tfrac{\partial \sigma}{\partial x} B(x,t)\right]^{-1} \left(\tfrac{\partial \sigma}{\partial t} + \tfrac{\partial \sigma}{\partial x} f(x,t)\right). \quad (57.149)$$

Therefore, given that $\sigma(x(t_1),t_1) = 0$, then, for all $t \ge t_1$, the dynamics of the system on the switching surface will satisfy

$$\begin{aligned} \dot{x}(t) &= \left[I - B(x,t)\left[\frac{\partial \sigma}{\partial x} B(x,t)\right]^{-1} \frac{\partial \sigma}{\partial x} \right] f(x,t) \\ &\quad - B(x,t)\left[\frac{\partial \sigma}{\partial x} B(x,t)\right]^{-1} \frac{\partial \sigma}{\partial t}. \end{aligned} \quad (57.150)$$

This equation represents the *equivalent system dynamics* on the sliding surface. The driving term is present when some form of tracking or regulation is required of the controlled system, e.g., when

$$\sigma(x,t) = Sx + r(t) = 0$$

with $r(t)$ serving as a "reference" signal [11].

The $(n-m)$-dimensional switching surface, $\sigma(x,t) = 0$, imposes m constraints on the plant dynamics in a sliding mode. Hence, m of the state variables can be eliminated, resulting in an equivalent reduced-order system whose dynamics represent the motion of the state trajectory in a sliding mode. Unfortunately, the structure of Equation 57.150 does not allow conveniently exploiting this fact in switching-surface design. To set forth a clearer switching-surface design algorithm, we first convert the plant dynamics to the so-called regular form.

Regular Form of the Plant Dynamics

The *regular form* of the dynamics of Equation 57.143 is

$$\begin{aligned} \dot{z}_1 &= \hat{f}_1(z,t), \\ \dot{z}_2 &= \hat{f}_2(z,t) + \hat{B}_2(z,t)u(z,t), \end{aligned} \quad (57.151)$$

where $z_1 \in R^{n-m}$, $z_2 \in R^{m}$. This form can often be constructed through a linear state transformation, $z(t) = Tx(t)$ where T has the property

$$TB(x,t) = TB(T^{-1}z,t) = \begin{bmatrix} 0 \\ \hat{B}_2(z,t) \end{bmatrix},$$

and $\hat{B}_2(z,t)$ is an $(m \times m)$ nonsingular mapping for all t and z. In general, computing the regular form requires the nonlinear transformation,

$$z(t) = T(x,t) = \begin{bmatrix} T_1(x,t) \\ T_2(x,t) \end{bmatrix},$$

where

(*i*) $T(x,t)$ is a diffeomorphic transformation, i.e., a continuous differentiable inverse mapping $\tilde{T}(z,t) = x$ exists, satisfying $\tilde{T}(0,t) = 0$ for all t,

(*ii*) $T_1(\cdot,\cdot) : R^n x R \to R^{n-m}$ and $T_2(\cdot,\cdot) : R^n x R \to R^m$, and

(*iii*) $T(x,t)$ has the property that

$$\frac{\partial T}{\partial x}B(x,t) = \begin{bmatrix} \frac{\partial T_1}{\partial x} \\ \frac{\partial T_2}{\partial x} \end{bmatrix} B(\tilde{T}(z,t),t) = \begin{bmatrix} 0 \\ \hat{B}_2(z,t) \end{bmatrix},$$

This partial differential equation has a solution only if the so-called Frobenius condition is satisfied [9]. The resulting nonlinear regular form of the plant dynamics has the structure,

$$\begin{aligned} \dot{z}_1 &= \frac{\partial T_1}{\partial x} f(\tilde{T}(z,t),t) + \frac{\partial T_1}{\partial t} \equiv \hat{f}_1(z,t), \\ \dot{z}_2 &= \frac{\partial T_2}{\partial x} f(\tilde{T}(z,t),t) + \frac{\partial T_2}{\partial t} + \frac{\partial T_2}{\partial x} B(\tilde{T}(z,t),t) \\ &\equiv \hat{f}_2(z,t) + \hat{B}_2(z,t)u \end{aligned} \quad (57.152)$$

Sometimes all nonlinearities in the plant model can be moved to $\hat{f}_2(z,t)$ so that

$$\dot{z}_1 = \hat{f}_1(z,t) = [A_{11} \quad A_{12}] \begin{bmatrix} z_1 \\ z_2 \end{bmatrix} \quad (57.153)$$

which solves the sliding-surface design problem with linear techniques (to be shown). If the original system model is linear, the regular form is given by

$$\begin{bmatrix} \dot{z}_1 \\ \dot{z}_2 \end{bmatrix} = \begin{bmatrix} A_{11} & A_{12} \\ A_{21} & A_{22} \end{bmatrix} \begin{bmatrix} z_1 \\ z_2 \end{bmatrix} + \begin{bmatrix} 0 \\ B_2 \end{bmatrix} u, \quad (57.154)$$

where $z_1 \in R^{n-m}$ and $z_2 \in R^m$ are as above.

Equivalent System Dynamics via Regular Form

The regular form of the equivalent state dynamics is convenient for analysis and switching-surface design. To simplify the development we make three assumptions: 1. The sliding surface is given in terms of the states of the regular form. 2. The surface has the linear time varying structure,

$$\sigma(z,t) = Sz + r(t) = [S_1 \quad S_2] \begin{bmatrix} z_1 \\ z_2 \end{bmatrix} + r(t) = 0,$$

where the matrix S_2 is chosen to be nonsingular. 3. The system is in a sliding mode, i.e., for some t_1, $\sigma(x(t),t) = 0$ for all $t \geq t_1$. With these three assumptions, one can solve for $z_2(t)$ as

$$z_2(t) = -S_2^{-1} S_1 z_1(t) - S_2^{-1} r(t). \quad (57.155)$$

Substituting the nonlinear regular form of Equation 57.152 yields

$$\dot{z}_1 = \hat{f}_1(z_1, z_2, t) = \hat{f}_1\left(z_1, -S_2^{-1}S_1 z_1 - S_2^{-1} r(t), t\right).$$

The goal then is to choose S_1 and S_2 to achieve a desired behavior of this nonlinear system.

If this system is linear, i.e., if Equation 57.153 is satisfied, then, using Equation 57.155, the reduced order dynamics are

$$\dot{z}_1 = \left(A_{11} - A_{12} S_2^{-1} S_1\right) z_1 - A_{12} S_2^{-1} r(t). \quad (57.156)$$

Analysis of the State Feedback Structure of Reduced-Order Linear Dynamics

The equivalent reduced-order dynamics of Equation 57.156 exhibit a state feedback structure in which $F = S_2^{-1} S_1$ is a state feedback map and A_{12} represents an "input" matrix. Under the conditions that the original (linear) system is controllable, the following well-known theorem applies:

THEOREM 57.6 *[16]: If the linear regular form of the state model (Equation 57.154) is controllable, then the pair (A_{11}, A_{12}) of the reduced-order equivalent system of Equation 57.156 is controllable.*

This theorem leads to a wealth of switching-surface design mechanisms. First, it permits setting the poles of $A_{11} - A_{12}S_2^{-1}S_1$, for stabilizing the state trajectory to zero when $r(t) = 0$ or to a prescribed rate of tracking, otherwise. Alternatively, one can determine S_1 and S_2 to solve the LQR problem when $r(t) = 0$.

As an example, suppose a system has the regular form of Equation 57.154 except that A_{21} and A_{22} are time varying and nonlinear,

$$\begin{bmatrix} \dot{z}_1 \\ \dot{z}_2 \end{bmatrix} = \left[\begin{array}{ccc|cc} 0 & 1 & 0 & 0 & 0 \\ 0 & 0 & 0 & 1 & 0 \\ 0 & 0 & 0 & 0 & 1 \\ \hline a_{11} & a_{12} & a_{13} & a_{14} & a_{15} \\ a_{21} & a_{22} & a_{23} & a_{24} & a_{25} \end{array}\right] \begin{bmatrix} z_1 \\ z_2 \end{bmatrix} + \left[\begin{array}{cc} 0 & 0 \\ 0 & 0 \\ 0 & 0 \\ \hline 1 & 0 \\ 0 & 1 \end{array}\right] u,$$

where $a_{ij} = a_{ij}(t,x)$ and $a_{ij}^{\min} \leq a_{ij}(t,x) \leq a_{ij}^{\max}$. Let the switching surface be given by

$$\sigma(z) = [S_1 \quad S_2] \begin{bmatrix} z_1 \\ z_2 \end{bmatrix} = 0.$$

The pertinent matrices of the reduced order equivalent system matrices are

$$A_{11} = \begin{bmatrix} 0 & 1 & 0 \\ 0 & 0 & 0 \\ 0 & 0 & 0 \end{bmatrix} \quad \text{and} \quad A_{12} = \begin{bmatrix} 0 & 0 \\ 1 & 0 \\ 0 & 1 \end{bmatrix}.$$

To stabilize the system, suppose the goal is to find F so that the equivalent system has eigenvalues at $\{-1, -2, -3\}$. Using Matlab's Control System's Toolbox yields,

$$F = \begin{bmatrix} 2 & 3 & 0 \\ 0 & 0 & 3 \end{bmatrix} = S_2^{-1} S_1.$$

Choosing $S_2 = I$, leaves $S_1 = F$. This then specifies the switching surface matrix $S = [F \quad I]$.

Alternatively, suppose the objective is to find the control that minimizes the performance index

$$J = \int_0^\infty \left(z_1^T Q z_1 + \hat{u}^T R \hat{u} \right) dt,$$

where the lower limit of integration refers to the initiation of sliding. This is associated with the equivalent reduced-order system

$$\dot{z}_1 = A_{11} z_1 - A_{12} \hat{u}$$

where

$$\hat{u} = S_2^{-1} S_1 z_1 \equiv F z_1.$$

Suppose weighting matrices are taken as

$$Q = \begin{bmatrix} 1.0 & 0.5 & 1.0 \\ 0.5 & 2.0 & 1.0 \\ 1.0 & 1.0 & 3.0 \end{bmatrix}$$

$$R = \begin{bmatrix} 2 & 0 \\ 0 & 1 \end{bmatrix}.$$

Using Matlab's Control Systems Toolbox, the optimal feedback is

$$F = \begin{bmatrix} 0.6420 & 1.4780 & 0.2230 \\ 0.4190 & 0.4461 & 1.7031 \end{bmatrix}.$$

Again, choosing $S_2 = I$, the switching surface matrix is given by $S = [F \quad I]$. Here, the poles of the system in sliding are $\{-1.7742, -0.7034 \pm j0.2623\}$.

57.5.5 Controller Design

Stability to Equilibrium Manifold

As mentioned, in VSC a Lyapunov approach is used for deriving conditions on the control $u(x,t)$ that will drive the state trajectory to the equilibrium manifold. Ordinarily, it is sufficient to consider only quadratic Lyapunov functions of the form,

$$V(t, x, \sigma) = \sigma^T(x,t) W \sigma(x,t), \tag{57.157}$$

where W is a symmetric positive definite matrix. The control $u(x,t)$ must be chosen so that the time derivative of $V(t,x,\sigma)$ is negative definite for $\sigma \neq 0$. To this end, consider

$$\dot{V}(t,x,\sigma) = \dot{\sigma}^T W \sigma + \sigma^T W \dot{\sigma} = 2\sigma^T W \dot{\sigma}, \tag{57.158}$$

where we have suppressed specific x and t dependencies. Recalling Equation 57.143, it follows that

$$\dot{\sigma} = \frac{\partial \sigma}{\partial t} + \frac{\partial \sigma}{\partial x}\dot{x} = \frac{\partial \sigma}{\partial t} + \frac{\partial \sigma}{\partial x} f + \frac{\partial \sigma}{\partial x} B u. \tag{57.159}$$

Substituting Equation 57.159 in Equation 57.158 leads to a Lyapunov-like theorem.

THEOREM 57.7 *A sufficient condition for the equilibrium manifold (Equation 57.145) to be globally attractive is that the control $u(x,t)$ be chosen so that*

$$\dot{V} = 2\sigma^T W \frac{\partial \sigma}{\partial t} + 2\sigma^T W \frac{\partial \sigma}{\partial x} f + 2\sigma^T W \frac{\partial \sigma}{\partial x} B u < 0 \tag{57.160}$$

for $\sigma \neq 0$, i.e., $\dot{V}(t,x,\sigma)$ is negative definite.

Observe that Equation 57.160 is linear in the control. Virtually all control structures for VSC are chosen so that this inequality is satisfied for appropriate W. Some control laws utilize an x- and t-dependent W requiring that the derivation above be generalized.

Various Control Structures

To make the needed control structures more transparent, recall the equivalent control of Equation 57.149,

$$u_{eq}(x,t) = -\left[\frac{\partial \sigma}{\partial x} B(x,t)\right]^{-1} \left(\frac{\partial \sigma}{\partial t} + \frac{\partial \sigma}{\partial x} f(x,t)\right)$$

computed assuming that the matrix product $\frac{\partial \sigma}{\partial x} B(x,t)$ is nonsingular for all t and x. We can now decompose the general control structure as

$$u(x,t) = u_{eq}(x,t) + u_N(x,t) \tag{57.161}$$

where $u_N(x,t)$ is as yet an unspecified substructure. Substituting in Equation 57.160 produces the following sufficient condition for stability to the switching surface: Choose $u_N(x,t)$ so that

$$\dot{V}(t,x,\sigma) = 2\sigma^T W \frac{\partial \sigma}{\partial x} B(x,t) u_N(x,t) < 0. \tag{57.162}$$

Because $\frac{\partial \sigma}{\partial x} B(x,t)$ is assumed nonsingular for all t and x, it is convenient to set

$$u_N(x,t) = \left[\frac{\partial \sigma}{\partial x} B(x,t)\right]^{-1} \hat{u}_N(x,t). \tag{57.163}$$

Often a switching surface $\sigma(x,t)$ can be designed to achieve a desired system behavior in sliding and, at the same time, to satisfy the constraint $\frac{\partial \sigma}{\partial x} B = I$ in which case $u_N = \hat{u}_N$. Without losing generality, we make one last simplifying assumption, $W = I$, because $W > 0$, W is nonsingular, and can be compensated for in the control structure. Hence, stability on the surface reduces to finding $\hat{u}_N(x,t)$ such that

$$\dot{V} = 2\sigma^T W \left[\frac{\partial \sigma}{\partial x} B\right] \left[\frac{\partial \sigma}{\partial x} B\right]^{-1} \hat{u}_N = 2\sigma^T \hat{u}_N < 0. \tag{57.164}$$

These simplifications allow us to specify five common controller structures:

1. Relays with constant gains: $\hat{u}_N(x,t)$ is chosen so that

$$\hat{u}_N = \alpha \operatorname{sgn}(\sigma(x,t))$$

with $\alpha = [\alpha_{ij}]$ an $m \times m$ matrix, and $\text{sgn}(\sigma(x,t))$ is defined componentwise. Stability to the surface is achieved if $\alpha = [\alpha_{ij}]$ is chosen diagonally dominant with negative diagonal entries [13]. Alternatively, if α is chosen to be diagonal with negative diagonal entries, then the control can be represented as

$$\hat{u}_{Ni} = \alpha_{ii}\text{sgn}(\sigma_i(x,t))$$

and, for $\sigma_i \neq 0$,

$$2\sigma_i \hat{u}_{Ni} = 2\alpha_{ii}\sigma_i \text{sgn}(\sigma_i) = 2\alpha_{ii}|\sigma_i| < 0$$

which guarantees stability to the surface given the Lyapunov function, $V(t,x,\sigma) = \sigma^T(x,t)\sigma(x,t)$.

2. Relays with state dependent gains: Each entry of $\hat{u}_N(x,t)$ is chosen so that

$$\hat{u}_{Ni} = \alpha_{ii}(x,t)\text{sgn}(\sigma_i(x,t)), \qquad \alpha_{ii}(x,t) < 0.$$

The condition for stability to the surface is that

$$\begin{aligned} 2\sigma_i \hat{u}_{Ni} &= 2\alpha_{ii}(x,t)\sigma_i \text{sgn}(\sigma_i) \\ &= 2\alpha_{ii}(x,t)|\sigma_i| < 0 \qquad \text{for} \qquad \sigma_i \neq 0. \end{aligned}$$

An adequate choice for $\alpha_{ii}(x,t)$ is to choose $\beta_i < 0$, $\gamma_i > 0$, and k a natural number with

$$\alpha_{ii}(x) = \beta_i \left(\sigma_i^{2k}(x,t) + \gamma_i\right).$$

3. Linear state feedback with switched gains: Here $\hat{u}_N(x,t)$ is chosen so that

$$\hat{u}_N = \Psi x; \quad \Psi = \left[\Psi_{ij}\right]; \quad \Psi_{ij} = \begin{cases} \alpha_{ij} < 0, & \sigma_i x_j > 0, \\ \beta_{ij} > 0, & \sigma_i x_j < 0. \end{cases}$$

To guarantee stability, it is sufficient to choose α_{ij} and β_{ij} so that

$$\begin{aligned} \sigma_i \hat{u}_{Ni} &= \sigma_i\left(\Psi_{i1}x_1 + \Psi_{i2}x_2 + \cdots + \Psi_{in}x_n\right) \\ &= \Psi_{i1}\sigma_i x_1 + \Psi_{i2}\sigma_i x_2 + \cdots + \Psi_{in}\sigma_i x_n < 0. \end{aligned}$$

4. Linear continuous feedback: Choose

$$\hat{u}_N = -P\sigma(x,t), \qquad P = P^T > 0,$$

i.e., $P \in R^{m\times m}$ is a symmetric positive definite constant matrix. Stability is achieved because

$$\sigma^T \hat{u}_N = -\sigma^T P \sigma < 0,$$

where P is often chosen as a diagonal matrix with positive diagonal entries.

5. Univector nonlinearity with scale factor: In this case, choose

$$\hat{u}_N = \begin{cases} \frac{\sigma(x,t)}{\|\sigma(x,t)\|}\rho, & \rho < 0 \text{ and } \sigma \neq 0, \\ 0, & \sigma = 0. \end{cases}$$

Stability to the surface is guaranteed because, for $\sigma \neq 0$,

$$\sigma^T \hat{u}_N = \frac{\sigma^T \sigma}{\|\sigma\|}\rho = \|\sigma\|\rho < 0.$$

Of course, it is possible to make ρ time dependent, if necessary, for certain tracking problems. This concludes our discussion of control structures to achieve stability to the sliding surface.

57.5.6 Design Examples

This section presents two design examples illustrating typical VSC strategies.

DESIGN EXAMPLE 1: In this example, we illustrate a constant gain relay control with nonlinear sliding surface design for a single-link robotic manipulator driven by a *dc* armature-controlled motor modeled by the normalized (i.e., scaled) simplified equations,

$$\begin{bmatrix} \dot{x}_1 \\ \dot{x}_2 \\ \dot{x}_3 \end{bmatrix} = \begin{bmatrix} x_2 \\ \sin(x_1) + x_3 \\ x_2 + x_3 \end{bmatrix} + \begin{bmatrix} 0 \\ 0 \\ 1 \end{bmatrix} u \equiv f(x) + Bu,$$

in the regular form.

To determine the structure of an appropriate sliding surface, recall the assumption that $\frac{\partial \sigma}{\partial x}B(x,t)$ is nonsingular. Because $B = [0 \quad 0 \quad 1]^T$, it follows that $\frac{\partial \sigma}{\partial x_3}$ must be nonzero. Without losing generality, we set $\frac{\partial \sigma}{\partial x_3} = 1$. Hence, it is sufficient to consider sliding surfaces of the form

$$\sigma(x) = \sigma(x_1, x_2, x_3) = \sigma_1(x_1, x_2) + x_3 = 0. \quad (57.165)$$

Our design presumes that the reduced-order dynamics have a second-order response represented by the reduced-order state dynamics,

$$\begin{bmatrix} \dot{x}_1 \\ \dot{x}_2 \end{bmatrix} = \begin{bmatrix} x_2 \\ -a_1 x_1 - a_2 x_2 \end{bmatrix} = \begin{bmatrix} 0 & 1 \\ -a_1 & -a_2 \end{bmatrix}\begin{bmatrix} x_1 \\ x_2 \end{bmatrix}.$$

This form allows us to specify the characteristic polynomial of the dynamics and thus the eigenvalues, i.e., $\pi_A(\lambda) = \lambda^2 + a_2\lambda + a_1$. Proper choice of a_1 and a_2 leads to proper rise time, settling time, overshoot, gain margin, etc.

The switching-surface structure of Equation 57.165 implies that, in a sliding mode,

$$x_3 = -\sigma_1(x_1, x_2).$$

Substituting in the given system model, the reduced-order dynamics become

$$\begin{bmatrix} \dot{x}_1 \\ \dot{x}_2 \end{bmatrix} = \begin{bmatrix} x_2 \\ \sin(x_1) - \sigma_1(x_1, x_2) \end{bmatrix} = \begin{bmatrix} x_2 \\ -a_1 x_1 - a_2 x_2 \end{bmatrix}.$$

Hence the switching-surface design is completed by setting

$$\sigma_1(x_1, x_2) = \sin(x_1) + a_1 x_1 + a_2 x_2.$$

To complete the controller design, we first compute the equivalent control,

$$u_{eq} = -\begin{bmatrix} \frac{\partial \sigma_1}{\partial x_1} & \frac{\partial \sigma_1}{\partial x_2} & 1 \end{bmatrix}\begin{bmatrix} x_2 \\ \sin(x_1) + x_3 \\ x_2 + x_3 \end{bmatrix}.$$

For the constant gain relay control structure (Equation 57.161)

$$u_N = \alpha \text{sgn}(\sigma(x)).$$

Stability to the switching surface results whenever $\alpha < 0$ as

$$\sigma\dot{\sigma} = \alpha\sigma\,\mathrm{sgn}(\sigma) = \alpha|\sigma| < 0.$$

DESIGN EXAMPLE 2: Consider the fourth-order (linear) model of a mass spring system which could represent a simplified model of a flexible structure in space with two-dimensional control (Figure 57.12).

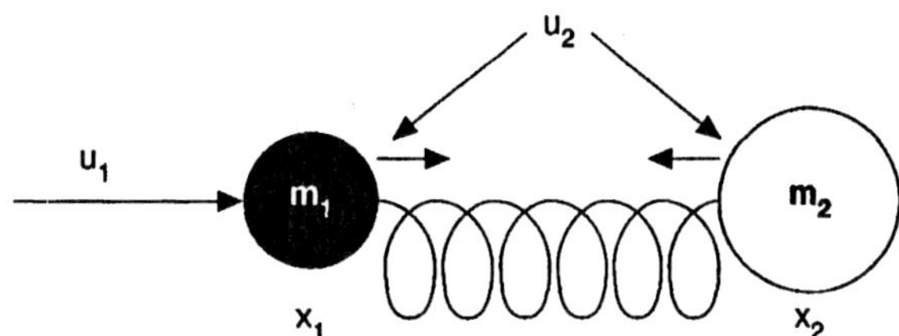

Figure 57.12 A mass spring system for Design Example 2.

Here, x_1 is the position of m_1, x_2 the position of m_2, u_1 the force applied to m_1, and u_2 the force applied between m_1 and m_2. The differential equation model has the form,

$$\begin{aligned} m_1\ddot{x}_1 + k(x_1 - x_2) &= u_1 + u_2 \\ m_2\ddot{x}_2 + k(x_2 - x_1) &= -u_2 \end{aligned}$$

where k is the spring constant. Given that $x_3 = \dot{x}_1$ and $x_4 = \dot{x}_2$, the resulting state model in regular form is

$$\begin{bmatrix} \dot{x}_1 \\ \dot{x}_2 \\ \dot{x}_3 \\ \dot{x}_4 \end{bmatrix} = \begin{bmatrix} x_3 \\ x_4 \\ -\frac{k}{m_1}x_1 + \frac{k}{m_1}x_2 \\ \frac{k}{m_2}x_1 - \frac{k}{m_2}x_2 \end{bmatrix} + \begin{bmatrix} 0 \\ 0 \\ \frac{1}{m_1}u_1 + \frac{1}{m_1}u_2 \\ -\frac{1}{m_2}u_2 \end{bmatrix}.$$

There are two simultaneous control objectives:

1. Stabilize oscillations, i.e., $x_1 = x_2$
2. Track a desired trajectory, $x_2(t) = z_{\mathrm{ref}}(t)$.

These goals are achieved if the following relationships are maintained for c_1 and $c_2 > 0$:

$$\dot{x}_1 - \dot{x}_2 + c_1(x_1 - x_2) = 0 \Rightarrow x_1 - x_2 \to 0$$

and

$$\dot{x}_2 - \dot{z}_{\mathrm{ref}} + c_2(x_2 - z_{\mathrm{ref}}) = 0 \Rightarrow x_2 - z_{\mathrm{ref}} \to 0.$$

The first step is to determine the appropriate sliding surface. To achieve the first control objective, set

$$\sigma_1(x,t) = x_3 - x_4 + c_1(x_1 - x_2) = 0,$$

and to achieve the desired tracking, set

$$\sigma_2(x,t) = x_4 - \dot{z}_{\mathrm{ref}} + c_2(x_2 - z_{\mathrm{ref}}) = 0.$$

The next step is to design a VSC law to drive the state trajectory to the intersection of these switching surfaces. In this effort we will illustrate two controller designs. The first is a hierarchical structure [2] so that, for $\sigma \neq 0$,

$$\begin{aligned} u_1 &= \alpha_1\mathrm{sgn}(\sigma_1) \\ u_2 &= \alpha_2\mathrm{sgn}(\sigma_2) \end{aligned}$$

with the sign of $\alpha_1, \alpha_2 \neq 0$ to be determined.

For stability to the surface, it is sufficient to have $\sigma_1\dot{\sigma}_1 < 0$ and $\sigma_2\dot{\sigma}_2 < 0$, as can be seen from Equation 57.158, with $W = I$. Observe that

$$\dot{\sigma}_1 = \dot{x}_3 - \dot{x}_4 + c_1(\dot{x}_1 - \dot{x}_2) = \dot{x}_3 - \dot{x}_4 + c_1(x_3 - x_4)$$

and

$$\dot{\sigma}_2 = \dot{x}_4 - \ddot{z}_{\mathrm{ref}} + c_2(\dot{x}_2 - \dot{z}_{\mathrm{ref}}) = \dot{x}_4 - \ddot{z}_{\mathrm{ref}} + c_2(x_4 - \dot{z}_{\mathrm{ref}}).$$

Substituting for the derivatives of x_3 and x_4 leads to

$$\begin{aligned} \begin{bmatrix} \dot{\sigma}_1 \\ \dot{\sigma}_2 \end{bmatrix} &= \begin{bmatrix} \frac{1}{m_1} & \frac{1}{m_1} + \frac{1}{m_2} \\ 0 & -\frac{1}{m_2} \end{bmatrix}\begin{bmatrix} u_1 \\ u_2 \end{bmatrix} + \begin{bmatrix} h_1 \\ h_2 \end{bmatrix} \\ &= \begin{bmatrix} \frac{1}{m_1} & \frac{1}{m_1} + \frac{1}{m_2} \\ 0 & -\frac{1}{m_2} \end{bmatrix}\begin{bmatrix} \alpha_1\mathrm{sgn}(\sigma_1) \\ \alpha_2\mathrm{sgn}(\sigma_2) \end{bmatrix} + \begin{bmatrix} h_1 \\ h_2 \end{bmatrix} \end{aligned} \tag{57.166}$$

where

$$h_1 = -\frac{k}{m_1}x_1 + \frac{k}{m_1}x_2 - \frac{k}{m_2}x_1 + \frac{k}{m_2}x_2 + c_1x_3 - c_1x_4$$

and

$$h_2 = \frac{k}{m_2}x_1 - \frac{k}{m_2}x_2 - \ddot{z}_{\mathrm{ref}} + c_2x_4 - c_2\dot{z}_{\mathrm{ref}}.$$

Taking a brute force approach to the computation of the control gains, stability to the switching surface is achieved provided

$$\sigma_2\dot{\sigma}_2 = \frac{-\alpha_2}{m_2}\sigma_2\mathrm{sgn}(\sigma_2) + \sigma_2h_2 < 0,$$

i.e., whenever

$$\alpha_2 > m_2|h_2|(> 0)$$

and provided

$$\sigma_1\dot{\sigma}_1 = \frac{\alpha_1}{m_1}\sigma_1\mathrm{sgn}(\sigma_1) + \sigma_1\left[\left(\frac{1}{m_1} + \frac{1}{m_1}\right)\alpha_2\mathrm{sgn}(\sigma_2) + h_1\right] < 0$$

i.e., whenever

$$\alpha_1 < -m_1|h_1| - \left(1 + \frac{m_1}{m_2}\right)\alpha_2.$$

In a second controller design, we recall Equation 57.166. For $\sigma_1 \neq 0$ and $\sigma_2 \neq 0$, it is convenient to define the controller as

$$\begin{bmatrix} u_1 \\ u_2 \end{bmatrix} = \begin{bmatrix} \frac{1}{m_1} & \frac{1}{m_1} + \frac{1}{m_2} \\ 0 & -\frac{1}{m_2} \end{bmatrix}^{-1}\begin{bmatrix} \beta_1\mathrm{sgn}(\sigma_1) \\ \beta_2\mathrm{sgn}(\sigma_2) \end{bmatrix},$$

where β_1 and β_2 are to be determined. It follows that

$$\begin{bmatrix} \dot{\sigma}_1 \\ \dot{\sigma}_2 \end{bmatrix} = \begin{bmatrix} \beta_1\mathrm{sgn}(\sigma_1) \\ \beta_2\mathrm{sgn}(\sigma_2) \end{bmatrix} + \begin{bmatrix} h_1 \\ h_2 \end{bmatrix}.$$

As in the first controller design, the state trajectory will intercept the sliding surface in finite time and sliding will occur for β_1 and β_2 sufficiently large and negative, thereby achieving the desired control objective.

57.5.7 Chattering

The VSC controllers developed earlier assure the desired behavior of the closed-loop system. These controllers, however, require an infinitely (in the ideal case) fast switching mechanism. The phenomenon of nonideal but fast switching was labeled as chattering (actually the word stems from the noise generated by the switching element). The high-frequency components of the chattering are undesirable because they may excite unmodeled high-frequency plant dynamics resulting in unforeseen instabilities. To reduce chatter, define a so-called boundary layer as

$$\{x \| \|\sigma(x)\| \leq \varepsilon, \ \ \varepsilon > 0\}, \tag{57.167}$$

whose thickness is 2ε. Now modify the control law of Equation 57.161 (suppressing t and x arguments) to

$$u = \begin{cases} u_{eq} + u_N, & \|\sigma\| \geq \varepsilon, \\ u_{eq} + p(\sigma, x), & \|\sigma\| \leq \varepsilon, \end{cases} \tag{57.168}$$

where $p(\sigma, x)$ is any continuous function satisfying $p(0, x) = 0$ and $p(\sigma, x) = u_N(x)$ when $\|\sigma\| = \varepsilon$. This control guarantees that trajectories are attracted to the boundary layer. Inside the boundary layer, Equation 57.168 offers a continuous approximation to the usual discontinuous control action. Similar to Corless and Leitmann [1], asymptotic stability is not guaranteed but ultimate boundedness of trajectories to within an ϵ-dependent neighborhood of the origin is assured.

57.5.8 Robustness to Matched Disturbances and Parameter Variations

To explore the robustness of VSC to disturbances and parameter variations, one modifies Equation 57.143 to

$$\begin{aligned} \dot{x}(t) &= [f(x,t) + \Delta f(x,t,q(t))] \\ &\quad + [B(x,t) + \Delta B(x,t,q(t))]u(x,t) + d(t) \end{aligned} \tag{57.169}$$

where $q(t)$ is a vector function representing parameter uncertainties, Δf and ΔB represent the cumulative effects of all plant uncertainties, and $d(t)$ denotes an external (deterministic) disturbance. The first critical assumption in our development is that all uncertainties and external disturbances satisfy the so-called *matching condition*, i.e., Δf, ΔB, and $d(t)$ lie in the image of $B(x, t)$ for all x and t. As such they can all be lumped into a single vector function $\xi(x, t, q, d, u)$, so that

$$\dot{x}(t) = f(x,t) + B(x,t)u(x,t) + B(x,t)\xi(x,t,q,d,u). \tag{57.170}$$

The second critical assumption is that a positive continuous bounded function $\rho(x, t)$ exists, satisfying

$$\|\xi(x,t,q,d,u)\| \leq \rho(x,t). \tag{57.171}$$

To incorporate robustness into a VSC design, we utilize the control structure of Equation 57.161, $u(x, t) = u_{eq}(x, t) + u_N(x, t)$, where $u_{eq}(x, t)$ is given by Equation 57.149. Given the plant and disturbance model of Equation 57.170, then, per Equation 57.162, it is necessary to choose $u_N(x, t)$ so that

$$\begin{aligned} \dot{V}(t,x,\sigma) &= 2\sigma^T W \frac{\partial \sigma}{\partial x} B(x,t) u_N(x,t) \\ &\quad + [B(x,t)\xi(x,t,q,d,u)] < 0. \end{aligned}$$

Choosing any one of the control structures outlined in Section 57.5.5, a choice of sufficiently "high" gains will produce a negative definite $\dot{V}(t, x, \sigma)$. Alternatively, one can use a control structure [2],

$$u_N(x,t) = \begin{cases} -\dfrac{B^T\left[\frac{\partial\sigma}{\partial x}\right]^T\sigma}{\left\|B^T\left[\frac{\partial\sigma}{\partial x}\right]^T\sigma\right\|}[\rho(x,t) + \alpha(x,t)], \\ \sigma(x,t) \neq 0 \\ 0, \text{ otherwise} \end{cases} \tag{57.172}$$

where $\alpha(x, t)$ is to be determined. Assuming $W = I$, it follows that, for $\sigma \neq 0$,

$$\begin{aligned} \dot{V}(t,x,\sigma) &= -2\sigma^T \frac{\frac{\partial\sigma}{\partial x} B B^T \left[\frac{\partial\sigma}{\partial x}\right]^T \sigma}{\left\|B^T\left[\frac{\partial\sigma}{\partial x}\right]^T\sigma\right\|} \\ &\quad \times [\rho(x,t) + \alpha(x,t)] + 2\sigma^T \frac{\partial\sigma}{\partial x} B\xi \\ &\leq -2\left\|B^T\left[\frac{\partial\sigma}{\partial x}\right]^T\sigma\right\|\alpha(x,t). \end{aligned}$$

Choosing $\alpha(x, t) = \alpha > 0$ leads to the stability of the state trajectory to the equilibrium manifold despite matched disturbances and parameter variations, demonstrating the robustness property of a VSC law.

57.5.9 Observer Design

Recall the nonlinear model of Equation 57.143 whose linear form is given in Equation 57.144. Under the condition that the state is unavailable, we postulate a linear measurement equation,

$$y = Cx, \tag{57.173}$$

where $C \in R^{p \times n}$. Our goal is to construct a dynamic system that estimates the state based on knowledge of the input and the measurements of Equation 57.173. In the linear case (Equations 57.144 and 57.173), this is referred to as the Luenberger observer,

$$\dot{\hat{x}} = A\hat{x} + Bu + L(y - C\hat{x}),$$

presupposing (C, A) is an observable pair. The context for a sliding-mode observer will be

$$\dot{\hat{x}} = A\hat{x} + Bu + L(y - C\hat{x}) + GE(y, \hat{x}) \tag{57.174}$$

where E is a user-chosen function to insure convergence in the presence of uncertainties and/or nonlinear observations; L is chosen so that A-LC is a stability matrix. Define the estimation error $e(t) = \hat{x}(t) - x(t)$ resulting in the error dynamics

$$\dot{e}(t) = (A - LC)e(t) + GE(e) - B\xi \qquad (57.175)$$

where the term $-B\xi$ represents (matched) lumped uncertainties, parameter variations, or disturbances that affect the plant. "Matched" refers to the assumption that all disturbances, etc., affect the plant dynamics through the image of $B(x, t)$ at each x and t. See Section 57.5.8.

OBSERVER DESIGN 1 [15]: A particular implementation would be $G = B$ and

$$E(e) = \eta \frac{FCe}{\|FCe\|}$$

where η is a design parameter satisfying $\eta > \|\xi\|$, and $F \in R^{m \times p}$, $p \geq m$, so that

$$FC = B^T P$$

where $P = P^T > 0$ satisfies

$$(A - LC)^T P + P(A - LC) = -Q$$

for an appropriate $Q = Q^T > 0$, if it exists. It follows that

$$\frac{d}{dt}(e^T Pe) = -e^T Qe$$

which implies $\lim_{t \to \infty} e(t) = 0$. For further analysis see [5], [8].

OBSERVER DESIGN 2 [3]: Here we set $L = 0$ and $E(y, \hat{x}) = \text{sgn}(y - C\hat{x})$. Further, we execute a nonsingular state transformation

$$\begin{bmatrix} y \\ w_1 \end{bmatrix} = \begin{bmatrix} C \\ M_1 \end{bmatrix} x$$

to obtain the equivalent plant dynamics:

$$\begin{bmatrix} \dot{y} \\ \dot{w}_1 \end{bmatrix} = \begin{bmatrix} A_{11} & A_{12} \\ A_{21} & A_{22} \end{bmatrix} \begin{bmatrix} y \\ w_1 \end{bmatrix} + \begin{bmatrix} B_1 \\ B_2 \end{bmatrix} u. \quad (57.176)$$

We will build the observer sequentially as follows:

1. Set $e_1(t) = \hat{y}(t) - y(t) = C\hat{x}(t) - y(t)$. This results in the error dynamics for the partial state "y" of the form

$$\dot{e}_1(t) = A_{11}e_1(t) + A_{12}e_2(t) - G_1\text{sgn}(e_1(t)) \quad (57.177)$$

where $e_2(t) = \hat{w}(t) - w(t)$. By choosing an appropriate nonsingular gain matrix G_1 we force Equation 57.177 into a sliding regime along the manifold $\{e_1(t) = 0\}$.

2. The "equivalent control" of Equation 57.177(i.e., along the manifold $e_1(t) = 0$) would then take the form

$$[\text{sgn}(e_1)]_{eq} = G_1^{-1} A_{12} e_2. \qquad (57.178)$$

3. Now consider the second subsystem of Equation 57.176

$$\dot{w}_1 = A_{21}y + A_{22}w_1 + B_2u. \qquad (57.179)$$

We now apply steps 1 and 2 to Equation 57.179 using a second nonsingular state transformation on the partial state w_1,

$$\begin{bmatrix} y_2 \\ w_2 \end{bmatrix} = \begin{bmatrix} G_1^{-1}A_{12} \\ M_2 \end{bmatrix} w_1$$

to obtain

$$\begin{bmatrix} \dot{y}_2 \\ \dot{w}_2 \end{bmatrix} = \begin{bmatrix} A_{11}^2 & A_{12}^2 \\ A_{21}^2 & A_{22}^2 \end{bmatrix} \begin{bmatrix} y_2 \\ w_2 \end{bmatrix} + \begin{bmatrix} B_1^2 \\ B_2^2 \end{bmatrix} u.$$

Assuming that $e_2 = \hat{y}_2 - y_2$ can be measured using Equation 57.178 one forms the estimator for the new partial state,

$$\dot{\hat{y}}_2 = A_{11}^2\hat{y}_2 + A_{11}^2 w_2 + B_1^2 u - G_2\text{sgn}(\hat{y}_2 - y_2)$$

having error dynamics

$$\dot{e}_1^2(t) = A_{11}^2 e_1^2(t) + A_{12}^2 e_2^2(t) - G_2\text{sgn}(e_1^2(t)).$$

The process is repeated until the estimator encompasses all state variables. Extensions to nonlinear plants can be found in [4].

57.5.10 Concluding Remarks

This chapter has summarized the salient results of variable structure control theory and illustrated the design procedures with various examples. A wealth of literature exists on the subject which cannot be included because of space limitations. In particular, the literature is replete with realistic applications [12], extensions to output feedback [6], extensions to decentralized control [10], as well as extensions to discrete-time systems. The reader is encouraged to search the literature for the many papers in this area.

57.5.11 Defining Terms

Chattering: The phenomenon of nonideal but fast switching. The term stems from the noise generated by a switching element.

Equilibrium (discontinuity) manifold: A specified, user-chosen manifold in the state space to which a system's trajectory is driven and maintained for all time subsequent to intersection of the manifold by a discontinuous control that is a function of the system's states, hence, discontinuity manifold. Other terms commonly used are sliding surface and switching surface.

Equivalent control: The solution to the algebraic equation involving the derivative of the equation of the switching surface and the plant's dynamic model. The equivalent control is used to determine the system's dynamics on the sliding surface.

Equivalent system dynamics: The system dynamics obtained after substituting the equivalent control into the plant's dynamic model. It characterizes state motion parallel to the sliding surface if the system's initial state is off the surface and state motion is on the sliding surface if the initial state is on the surface.

Ideal sliding mode: Motion of a system's state trajectory along a switching surface when switching in the control law is infinitely fast.

Matching condition: The condition requiring the plant's uncertainties to lie in the image of the input matrix, that is, the uncertainties can affect the plant dynamics only through the same channels as the plant's input.

Region of attraction: A set of initial states in the state space from which sliding is achievable.

Regular form: A particular form of the state-space description of a dynamic system obtained by a suitable transformation of the system's state variables.

Sliding surface: See equilibrium manifold.

Switching surface: See equilibrium manifold.

References

[1] Corless, M.J. and Leitmann, G., Continuous State Feedback Guaranteeing Uniform Ultimate Boundedness for Uncertain Dynamic Systems, *IEEE Trans. on Automat. Control*, AC-26(5),1139–1144, 1981.

[2] DeCarlo, R.A., Żak, S. H., and Matthews, G.P., Variable Structure Control of Nonlinear Multivariable Systems: A Tutorial, *Proc. IEEE*, 76(3), 212–232, 1988.

[3] Drakunov, S.V., Izosimov, D.B., Lukyanov, A. G., Utkin, V. A., and Utkin, V.I., The Block Control Principle, Part II, *Automation and Remote Control*, 51(6), 737–746, 1990.

[4] Drakunov, S.V., *Sliding-Mode Observers Based on Equivalent Control Method*, Proc. 31st IEEE Conf. Decision and Control, Tucson, Arizona, 2368–2369, 1992.

[5] Edwards, C. and Spurgeon, S.K., On the Development of Discontinuous Observers, *Int. J. Control*, 59(5), 1211–1229, 1994.

[6] El-Khazali, R. and DeCarlo, R.A., Output Feedback Variable Structure Control Design, *Automatica*, 31(5), 805–816, 1995.

[7] Filippov, A. F., *Differential Equations with Discontinuous Righthand Sides*, Kluwer Academic, Dordrecht, The Netherlands, 1988.

[8] Hui, S. and Żak, S.H., Robust Control Synthesis for Uncertain/Nonlinear Dynamical Systems, *Automatica*, 28(2), 289–298, 1992.

[9] Hunt, L.R., Su, R., and Meyer, G., Global Transformations of Nonlinear Systems, *IEEE Trans. on Automat. Control*, AC-28(1), 24–31, 1983.

[10] Matthews, G. and DeCarlo, R.A., Decentralized Variable Structure Control of Interconnected Multi-Input/Multi-Output Nonlinear Systems, *Circuits Systems & Signal Processing*, 6(3), 363–387, 1987.

[11] Matthews, G.P. and DeCarlo, R.A., Decentralized Tracking for a Class of Interconnected Nonlinear Systems Using Variable Structure Control, *Automatica*, 24(2), 187–193, 1988.

[12] Sira-Ramirez, H., Nonlinear P-I Controller Design for Switchmode dc-to-dc Power Converters, *IEEE Trans. on Circuits & Systems*, 38(4), 410–417, 1991.

[13] Utkin, V.I., *Sliding Modes and Their Application in Variable Structure Control*, Mir, Moscow, 1978.

[14] Utkin, V.I., *Sliding Modes in Control and Optimization*, Springer, Berlin, 1992.

[15] Walcott, B.L. and Żak, S.H., State Observation of Nonlinear/Uncertain Dynamical Systems, *IEEE Trans. on Automat. Control*, AC-32(2), 166–170, 1987.

[16] Young, K.-K.D., Kokotović, P.V., and Utkin, V.I., A Singular Perturbation Analysis of High-Gain Feedback Systems, *IEEE Trans. on Automat. Control*, AC-22(6), 931–938, 1977.

57.6 Control of Bifurcations and Chaos

Eyad H. Abed, Department of Electrical Engineering and the Institute for Systems Research, University of Maryland, College Park, MD

Hua O. Wang, United Technologies Research Center, East Hartford, CT

Alberto Tesi, Dipartimento di Sistemi e Informatica, Università di Firenze, Firenze, Italy

57.6.1 Introduction

This chapter deals with the control of bifurcations and chaos in nonlinear dynamical systems. This is a young subject area that is currently in a state of active development. Investigations of control system issues related to bifurcations and chaos began relatively recently, with most currently available results having been published within the past decade. Given this state of affairs, a unifying and comprehensive picture of control of bifurcations and chaos does not yet exist. Therefore, the chapter has a modest but, it is hoped, useful goal: to summarize some of the motivation, techniques, and results achieved to date on control of bifurcations and chaos. Background material on nonlinear dynamical behavior is also given, to make the chapter somewhat self-contained. However, interested readers unfamiliar with nonlinear dynamics will find it helpful to consult nonlinear dynamics texts to reinforce their understanding.

Despite its youth, the literature on control of bifurcations and chaos contains a large variety of approaches as well as interesting applications. Only a small number of approaches and applications can be touched upon here, and these reflect the background and interests of the authors. The Further Reading section provides references for those who would like to learn about alternative approaches or to learn more about the approaches discussed here.

Control system design is an enabling technology for a variety of application problems in which nonlinear dynamical behavior arises. The ability to manage this behavior can result in significant practical benefits. This might entail facilitating system operability in regimes where linear control methods break down; taking advantage of chaotic behavior to capture a desired oscillatory behavior without expending much control energy; or purposely introducing a chaotic signal in a communication system to mask a transmitted signal from an adversary while allowing perfect reception by the intended party.

The control problems addressed in this chapter are characterized by two main features:

1. nonlinear dynamical phenomena impact system behavior
2. control objectives can be met by altering nonlinear phenomena

Nonlinear dynamics concepts are clearly important in *understanding* the behavior of such systems (with or without control). Traditional linear control methods are, however, often effective in the *design* of control strategies for these systems. In other cases, such as for systems of the type discussed in Section 57.6.5, nonlinear methods are needed both for control design and performance assessment.

The chapter proceeds as follows. In section two, some basic nonlinear system terminology is recalled. This section also includes a new term, namely "candidate operating condition", which facilitates subsequent discussions on control goals and strategies. Section three contains a brief summary of basic bifurcation and chaos concepts that will be needed. Section four provides application examples for which bifurcations and/or chaotic behavior occur. Remarks on the control aspects of these applications are also given. Section five is largely a review of some basic concepts related to control of parameterized families of (nonlinear) systems. Section five also includes a discussion of what might be called "stressed operation" of a system. Section six is devoted to control problems for systems exhibiting bifurcation behavior. The subject of section seven is control of chaos. Conclusions are given in section eight. The final section gives some suggestions for further reading.

57.6.2 Operating Conditions of Nonlinear Systems

In linear system analysis and control, a blanket assumption is made that the operating condition of interest is a particular equilibrium point, which is then taken as the origin in the state space. The topic addressed here relates to applying control to alter the dynamical behavior of a system possessing multiple possible operating conditions. The control might alter these conditions in terms of their location and amplitude, and/or stabilize certain possible operating conditions, permitting them to take the place of an undesirable open-loop behavior. For the purposes of this chapter, therefore, it is important to consider a variety of possible operating conditions, in addition to the single equilibrium point focused on in linear system theory. In this section, some basic terminology regarding operating conditions for nonlinear systems is established.

Consider a finite-dimensional continuous-time system

$$\dot{x}(t) = F(x(t)) \tag{57.180}$$

where $x \in \mathbb{R}^n$ is the system state and F is smooth in x. (The terminology recalled next extends straightforwardly to discrete-time systems $x^{k+1} = F(x^k)$.) An *equilibrium point* or *fixed point* of the system (57.180) is a constant steady-state solution, i.e., a vector x^0 for which $F(x^0) = 0$. A *periodic solution* of the system is a trajectory $x(t)$ for which there is a minimum $T > 0$ such that $x(t + T) = x(t)$ for all t. An *invariant set* is a set for which any trajectory starting from an initial condition within the set remains in the set for all future and past times. An *isolated invariant set* is a bounded invariant set a neighborhood of which contains no other invariant set. Equilibrium points and periodic solutions are examples of invariant sets. A periodic solution is called a *limit cycle* if it is isolated. An *attractor* is a bounded invariant set to which trajectories starting from all sufficiently nearby initial conditions converge as $t \to \infty$. The *positive limit set* of (57.180) is the ensemble of points that some system trajectory either approaches as $t \to \infty$ or makes passes nearer and nearer to as $t \to \infty$. For example, if a system is such that all trajectories converge either to an equilibrium point or a limit cycle, then the positive limit set would be the union of points on the limit cycle and the equilibrium point. The *negative limit set* is the positive limit set of the system run with time t replaced by $-t$. Thus, the positive limit set is the set where the system ends up at $t = +\infty$, while the negative limit set is the set where the system begins at $t = -\infty$ [30]. The *limit set* of the system is the union of its positive and negative limit sets.

It is now possible to introduce a term that will facilitate the discussions on control of bifurcations and chaos. A *candidate operating condition* of a dynamical system is an equilibrium point, periodic solution or other invariant subset of its limit set. Thus, a candidate operating condition is any possible steady-state solution of the system, without regard to its stability properties. This term, though not standard, is useful since it permits discussion of bifurcations, bifurcated solutions, and chaotic motion in general terms without having to specify a particular nonlinear phenomenon. The idea behind this term is that such a solution, if stable or stabilizable using available controls, might qualify as an operating condition of the system. Whether or not it would be *acceptable* as an actual operating condition would depend on the system and on characteristics of the candidate operating condition. As an extreme example, a steady spin of an airplane *is* a candidate operating condition, but certainly *is not* acceptable as an actual operating condition!

57.6.3 Bifurcations and Chaos

This section summarizes background material on bifurcations and chaos that is needed in the remainder of the chapter.

Bifurcations

A *bifurcation* is a change in the number of candidate operating conditions of a nonlinear system that occurs as a parameter is quasistatically varied. The parameter being varied is referred to as the bifurcation parameter. A value of the bifurcation parameter at which a bifurcation occurs is called a critical value of the bifurcation parameter. Bifurcations from a nominal operating condition can only occur at parameter values for which the condition (say, an equilibrium point or limit cycle) either loses stability or ceases to exist.

To fix ideas, consider a general one-parameter family of ordinary differential equation systems

$$\dot{x} = F^{\mu}(x) \tag{57.181}$$

where $x \in \mathbb{R}^n$ is the system state, $\mu \in \mathbb{R}$ denotes the bifurcation parameter, and F is smooth in x and μ. Equation 57.181 can be viewed as resulting from a particular choice of control law in a family of nonlinear control systems (in particular, as Equation 57.188 with the control u set to a fixed feedback function $u(x, \mu)$). For any value of μ, the equilibrium points of Equation 57.181 are given by the solutions for x of the algebraic equations $F^{\mu}(x) = 0$.

Local bifurcations are those that occur in the vicinity of an equilibrium point. For example, a small-amplitude limit cycle can emerge (bifurcate) from an equilibrium as the bifurcation parameter is varied. The bifurcation is said to occur regardless of the stability or instability of the "bifurcated" limit cycle. In another local bifurcation, a pair of equilibrium points can emerge from a nominal equilibrium point. In either case, the *bifurcated solutions* are close to the original equilibrium point — hence the name local bifurcation. *Global bifurcations* are bifurcations that are not local, i.e., those that involve a domain in phase space. Thus, if a limit cycle loses stability releasing a new limit cycle, a global bifurcation is said to take place.[3]

The nominal operating condition of Equation 57.181 can be an equilibrium point or a limit cycle. In fact, depending on the coordinates used, it is possible that a limit cycle in one mathematical model corresponds to an equilibrium point in another. This is the case, for example, when a truncated Fourier series is used to approximate a limit cycle, and the amplitudes of the harmonic terms are used as state variables in the approximate model. The original limit cycle is then represented as an equilibrium in the amplitude coordinates.

If the nominal operating condition of Equation 57.181 is an equilibrium point, then bifurcations from this condition can occur only when the linearized system loses stability. Suppose, for example, that the origin is the nominal operating condition for some range of parameter values. That is, let $F^{\mu}(0) = 0$ for all values of μ for which the nominal equilibrium exists. Denote the Jacobian matrix of (57.181) evaluated at the origin by

$$A(\mu) := \frac{\partial F^{\mu}}{\partial x}(0).$$

[3]This use of the terms local bifurcation and global bifurcation is common. However, in some books, a bifurcation from a limit cycle is also referred to as a local bifurcation.

Local bifurcations from the origin can only occur at parameter values μ where $A(\mu)$ loses stability.

The scalar differential equation

$$\dot{x} = \mu x - x^3 \tag{57.182}$$

provides a simple example of a bifurcation. The origin $x^0 = 0$ is an equilibrium point for all values of the real parameter μ. The Jacobian matrix is $A(\mu) = \mu$ (a scalar). It is easy to see that the origin loses stability as μ increases through $\mu = 0$. Indeed, a bifurcation from the origin takes place at $\mu = 0$. For $\mu \leq 0$, the only equilibrium point of Equation 57.182 is the origin. For $\mu > 0$, however, there are two additional equilibrium points at $x = \pm\sqrt{\mu}$. This pair of equilibrium points is said to *bifurcate* from the origin at the *critical parameter value* $\mu_c = 0$. This is an example of a pitchfork bifurcation, which will be discussed later.

Subcritical vs. supercritical bifurcations In a very real sense, the fact that bifurcations occur when stability is lost is helpful from the perspective of control system design. To explain this, suppose that a system operating condition (the "nominal" operating condition) is not stabilizable beyond a critical parameter value. Suppose a bifurcation occurs at the critical parameter value. That is, suppose a new candidate operating condition emerges from the nominal one at the critical parameter value. Then it may be that the new operating condition is stable and occurs beyond the critical parameter value, providing an alternative operating condition near the nominal one. This is referred to as a *supercritical bifurcation*. In contrast, it may happen that the new operating condition is unstable and occurs prior to the critical parameter value. In this situation (called a *subcritical bifurcation*), the system state must leave the vicinity of the nominal operating condition for parameter values beyond the critical value. However, feedback offers the possibility of rendering such a bifurcation supercritical. This is true even if the nominal operating condition is not stabilizable. If such a feedback control can be found, then the system behavior beyond the stability boundary can remain close to its behavior at the nominal operating condition.

The foregoing discussion of bifurcations and their implications for system behavior can be gainfully viewed using graphical sketches called *bifurcation diagrams*. These are depictions of the equilibrium points and limit cycles of a system plotted against the bifurcation parameter. A bifurcation diagram is a schematic representation in which only a measure of the amplitude (or norm) of an equilibrium point or limit cycle need be plotted. In the bifurcation diagrams given in this chapter, a solid line indicates a stable solution, while a dashed line indicates an unstable solution.

Several bifurcation diagrams will now be used to further explain the meanings of supercritical and subcritical bifurcation, and to introduce some common bifurcations. It should be noted that not all bifurcations are supercritical or subcritical. For example, bifurcation can also be transcritical. In such a bifurcation, bifurcated operating conditions occur both prior to and beyond the critical parameter value. Identifying a bifurcation as supercritical, subcritical, transcritical, or otherwise is the problem of determining the *direction of the bifurcation*. A book on nonlinear dynamics should be consulted for a more extensive treatment (e.g., [10, 18, 28, 30, 32, 43, 46]). In this chapter only the basic elements of bifurcation analysis can be touched upon.

Suppose that the origin of Equation 57.181 loses stability as μ *increases* through the critical parameter value $\mu = \mu_c$. Under mild assumptions, it can be shown that a bifurcation occurs at μ_c.

Figure 57.13 serves two purposes: it depicts a subcritical bifurcation from the origin, and it shows a common consequence of subcritical bifurcation, namely hysteresis. A subcritical bifurcation occurs from the origin at the point labeled A in the figure. It leads to the unstable candidate operating condition corresponding to points on the dashed curve connecting points A and B. As the parameter μ is decreased to its value at point B, the bifurcated solution merges with another (stable) candidate operating condition and disappears. A *saddle-node bifurcation* is said to occur at point B. This is because the unstable candidate operating condition (the "saddle" lying on the dashed curve) merges with a stable candidate operating condition (the "node" lying on the solid curve). These candidate operating conditions can be equilibrium points or limit cycles — both situations are captured in the figure. Indeed, the origin can also be reinterpreted as a limit cycle and the diagram would still be meaningful. Another common name for a saddle-node bifurcation point is a *turning point.*

The physical scenario implied by Figure 57.13 can be summarized as follows. Starting from operation at the origin for small μ, increasing μ until point A is reached does not alter system behavior. If μ is increased past point A, however, the origin loses stability. The system then transitions ("jumps") to the available stable operating condition on the upper solid curve. This large transition can be intolerable in many applications. As μ is then decreased, another transition back to the origin occurs but at a lower parameter value, namely that corresponding to point B. Thus, the combination of the subcritical bifurcation at A and the saddle-node bifurcation at B can lead to a hysteresis effect.

Figure 57.14 depicts a *supercritical bifurcation* from the origin. This bifurcation is distinguished by the fact that the solution bifurcating from the origin at point A is stable, and occurs locally for parameter values μ beyond the critical value (i.e., those for which the nominal equilibrium point is unstable). In marked difference with the situation depicted in Figure 57.13, here as the critical parameter value is crossed a smooth change is observed in the system operating condition. No sudden jump occurs.

Suppose closeness of the system's operation to the nominal equilibrium point (the origin, say) is a measure of the system's performance. Then supercritical bifurcations ensure close operation to the nominal equilibrium, while subcritical bifurcations may lead to large excursions away from the nominal equilibrium point. For this reason, a supercritical bifurcation is commonly also said to be *safe* or *soft*, while a subcritical bifurcation is said to be *dangerous* or *hard* [49, 32, 52].

Given full information on the nominal equilibrium point, the occurrence of bifurcation is a consequence of the behavior of the linearized system at the equilibrium point. The manner in which the equilibrium point loses stability as the bifurcation parameter is varied determines the type of bifurcation that arises.

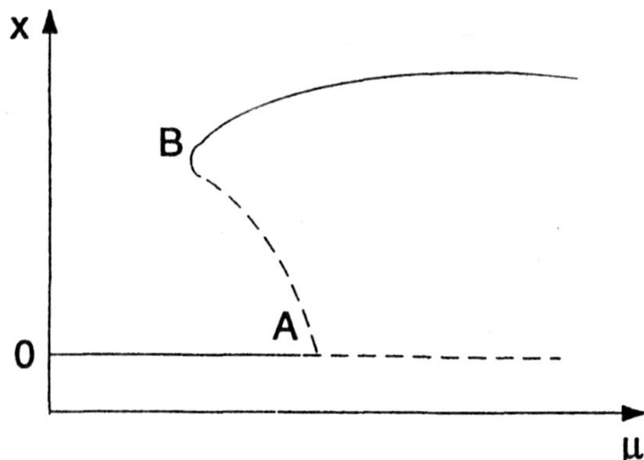

Figure 57.13 Subcritical bifurcation with hysteresis.

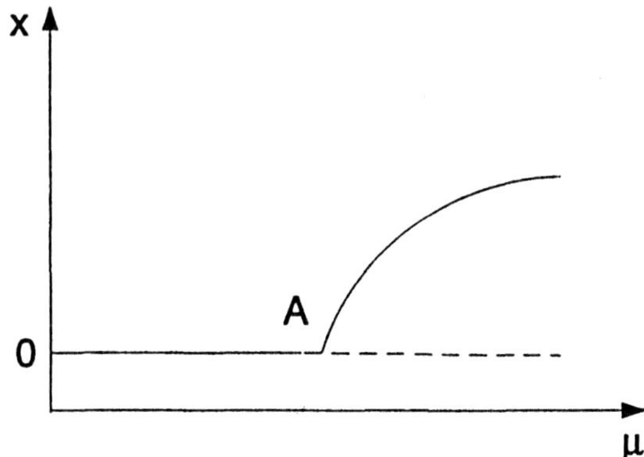

Figure 57.14 Supercritical bifurcation.

Three types of local bifurcation and a global bifurcation are discussed next. These are, respectively, the stationary bifurcation, the saddle-node bifurcation, the Andronov-Hopf bifurcation, and the period doubling bifurcation. All of these except the saddle-node bifurcation can be safe or dangerous. However, the saddle-node bifurcation is always dangerous.

There are analytical techniques for determining whether a stationary, Andronov-Hopf, or period doubling bifurcation is safe or dangerous. These techniques are not difficult to understand but involve calculations that are too lengthy to repeat here. The calculations yield formulas for so-called "bifurcation stability coefficients" [18], the meaning of which is addressed below. The references [1, 2, 4, 18, 28, 32] can be consulted for details.

What is termed here as "stationary bifurcation" is actually a special case of the usual meaning of the term. In the bifurcation theory literature [10], stationary bifurcation is any bifurcation of one or more equilibrium points from a nominal equilibrium point. When the nominal equilibrium point exists both before and after the critical parameter value, a stationary bifurcation "from a known solution" is said to occur. If the nominal solution disappears beyond the critical parameter value, a stationary bifurcation "from an unknown solution" is said to occur. To simplify the terminology, here the former type of bifurcation is called a *stationary bifurcation.* The saddle-node bifurcation is the most common example of the latter type of bifurcation.

Andronov-Hopf bifurcation also goes by other names. "Hopf bifurcation" is the traditional name in the West, but this name neglects the fundamental early contributions of Andronov and his co-workers (see, e.g., [6]). The essence of this phenomenon was also known to Poincaré, who did not develop a detailed theory but used the concept in his study of lunar orbital dynamics [37, Secs. 51–52]. The same phenomenon is sometimes called flutter bifurcation in the engineering literature. This bifurcation of a limit cycle from an equilibrium point occurs when a complex conjugate pair of eigenvalues crosses the imaginary axis into the right half of the complex plane at $\mu = \mu_c$. A small-amplitude limit cycle then emerges from the nominal equilibrium point at μ_c.

Saddle-node bifurcation and stationary bifurcation Saddle-node bifurcation occurs when the linearized system has a zero eigenvalue at $\mu = \mu_c$ but the origin doesn't persist as an equilibrium point beyond the critical parameter value. Saddle-node bifurcation was discussed briefly before, and will not be discussed in any detail in the following. Several basic remarks are, however, in order.

1. Saddle-node bifurcation of a nominal, stable equilibrium point entails the disappearance of the equilibrium upon its merging with an unstable equilibrium point at a critical parameter value.
2. The bifurcation occurring at point B in Figure 57.13 is representative of a saddle-node bifurcation.
3. The nominal equilibrium point possesses a zero eigenvalue at a saddle-node bifurcation.
4. An important feature of the saddle-node bifurcation is the *disappearance, locally, of any stable bounded solution of the system* (57.181).

Stationary bifurcation (according to the agreed upon terminology above) is guaranteed to occur when a single real eigenvalue goes from being negative to being positive as μ passes through the value μ_c. More precisely, the origin of Equation 57.181 undergoes a stationary bifurcation at the critical parameter value $\mu = 0$ if hypotheses (S1) and (S2) hold.

(S1) F of system (57.181) is sufficiently smooth in x, μ, and $F^{\mu}(0) = 0$ for all μ in a neighborhood of 0. The Jacobian $A(\mu) := \frac{\partial F^{\mu}}{\partial x}(0)$ possesses a simple real eigenvalue $\lambda(\mu)$ such that $\lambda(0) = 0$ and $\lambda'(0) \neq 0$.

(S2) All eigenvalues of the critical Jacobian $\frac{\partial F^{\mu}}{\partial x}(0)$ besides 0 have negative real parts.

Under (S1) and (S2), two new equilibrium points of Equation 57.181 emerge from the origin at $\mu = 0$. Bifurcation stability coefficients are quantities that determine the direction of bifurcation, and in particular the stability of the bifurcated solutions. Next, a brief discussion of the origin and meaning of these quantities is given.

Locally, the new equilibrium points occur for parameter values given by a smooth function of an auxiliary small parameter ϵ (ϵ can be positive or negative):

$$\mu(\epsilon) = \mu_1\epsilon + \mu_2\epsilon^2 + +\mu_3\epsilon^3 + \cdots \qquad (57.183)$$

where the ellipsis denotes higher order terms. One of the new equilibrium points occurs for $\epsilon > 0$ and the other for $\epsilon < 0$. Also, the stability of the new equilibrium points is determined by the sign of an eigenvalue $\beta(\epsilon)$ of the system linearization at the new equilibrium points. This eigenvalue is near 0 and is also given by a smooth function of the parameter ϵ:

$$\beta(\epsilon) = \beta_1\epsilon + \beta_2\epsilon^2 + \beta_3\epsilon^3 + \cdots \qquad (57.184)$$

Stability of the bifurcated equilibrium points is determined by the sign of $\beta(\epsilon)$. If $\beta(\epsilon) < 0$ the corresponding equilibrium point is stable, while if $\beta(\epsilon) > 0$ the equilibrium point is unstable. The coefficients β_i, $i = 1, 2, \ldots$ in the expansion above are the bifurcation stability coefficients mentioned earlier, for the case of stationary bifurcation. The values of these coefficients determine the local nature of the bifurcation.

Since ϵ can be positive or negative, it follows that if $\beta_1 \neq 0$ the bifurcation is neither subcritical nor supercritical. (This is equivalent to the condition $\mu_1 \neq 0$.) The bifurcation is therefore generically transcritical. In applications, however, special structures of system dynamics and inherent symmetries often result in stationary bifurcations that are not transcritical. Also, it is sometimes possible to render a stationary bifurcation supercritical using feedback control. For these reasons, a brief discussion of subcritical and supercritical pitchfork bifurcations is given next.

If $\beta_1 = 0$ and $\beta_2 \neq 0$, a stationary bifurcation is known as a *pitchfork bifurcation*. The pitchfork bifurcation is subcritical if $\beta_2 > 0$; it is supercritical if $\beta_2 < 0$. The bifurcation diagram of a subcritical pitchfork bifurcation is depicted in Figure 57.15, and that of a supercritical pitchfork bifurcation is depicted in Figure 57.16. The bifurcation discussed previously for the example system (57.182) is a supercritical pitchfork bifurcation.

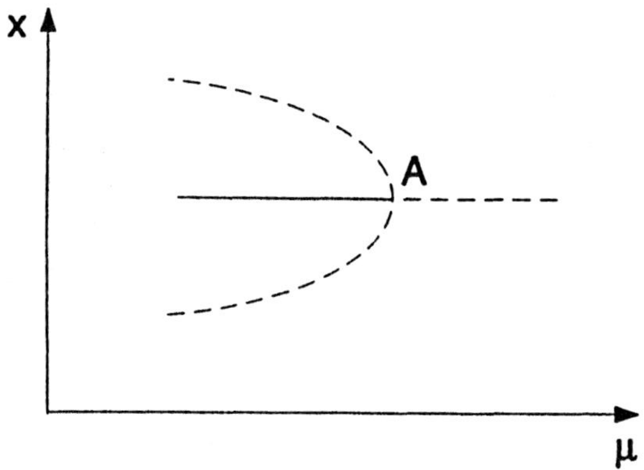

Figure 57.15 Subcritical pitchfork bifurcation.

Andronov-Hopf bifurcation Suppose that the origin of Equation 57.181 loses stability as the result of a complex conjugate pair of eigenvalues of $A(\mu)$ crossing the imaginary axis. All other eigenvalues are assumed to remain stable, i.e., their real parts are negative for all values of μ. Under this simple condition on the linearization of a nonlinear system, the nonlinear system typically undergoes a bifurcation. The word "typically" is used because there is one more condition to satisfy, but it is a

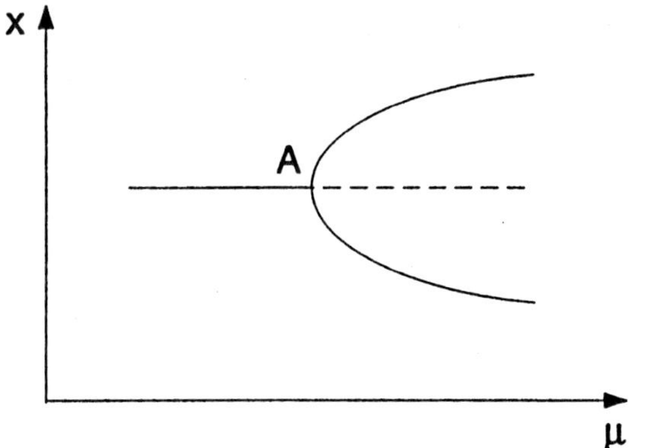

Figure 57.16 Supercritical pitchfork bifurcation.

mild condition. The type of bifurcation that occurs under these circumstances involves the emergence of a limit cycle from the origin as μ is varied through μ_c. This is the Andronov-Hopf bifurcation, a more precise description of which is given next. The following hypotheses are invoked in the Andronov-Hopf Bifurcation Theorem. The critical parameter value is taken to be $\mu_c = 0$ without loss of generality.

(AH1) F of system (57.181) is sufficiently smooth in x, μ, and $F^\mu(0) = 0$ for all μ in a neighborhood of 0. The Jacobian $\frac{\partial F^\mu}{\partial x}(0)$ possesses a complex-conjugate pair of (algebraically) simple eigenvalues $\lambda(\mu) = \alpha(\mu) + i\omega(\mu), \overline{\lambda(\mu)}$, such that $\alpha(0) = 0, \alpha'(0) \neq 0$ and $\omega_c := \omega(0) > 0$.

(AH2) All eigenvalues of the critical Jacobian $\frac{\partial F^\mu}{\partial x}(0)$ besides $\pm i\omega_c$ have negative real parts.

The Andronov-Hopf Bifurcation Theorem asserts that, under (AH1) and (AH2), a small-amplitude nonconstant limit cycle (i.e., periodic solution) of Equation 57.181 emerges from the origin at $\mu = 0$. Locally, the limit cycles occur for parameter values given by a smooth and even function of the amplitude ϵ of the limit cycles:

$$\mu(\epsilon) = \mu_2\epsilon^2 + \mu_4\epsilon^4 + \cdots \tag{57.185}$$

where the ellipsis denotes higher order terms.

Stability of an equilibrium point of the system (57.181) can be studied using eigenvalues of the system linearization evaluated at the equilibrium point. The analogous quantities for consideration of limit cycle stability for Equation 57.181 are the *characteristic multipliers* of the limit cycle. (For a definition, see for example [10, 18, 28, 30, 32, 43, 46, 48].) A limit cycle is stable (precisely: orbitally asymptotically stable) if its characteristic multipliers all have magnitude less than unity. This is analogous to the widely known fact that an equilibrium point is stable if the system eigenvalues evaluated there have negative real parts. The stability condition is sometimes stated in terms of the characteristic exponents of the limit cycle, quantities which are easily obtained from the characteristic multipliers. If the characteristic exponents have negative real parts, then the limit cycle is stable. Although it isn't possible to discuss the basic theory of limit cycle stability in any detail here, the reader is referred to almost any text on differential equations, dynamical systems, or bifurcation theory for a detailed discussion (e.g, [10, 18, 28, 30, 32, 43, 46, 48]).

The stability of the limit cycle resulting from an Andronov-Hopf bifurcation is determined by the sign of a particular characteristic exponent $\beta(\epsilon)$. This characteristic exponent is real and vanishes in the limit as the bifurcation point is approached. It is given by a smooth and even function of the amplitude ϵ of the limit cycles:

$$\beta(\epsilon) = \beta_2\epsilon^2 + \beta_4\epsilon^4 + \cdots \tag{57.186}$$

The coefficients μ_2 and β_2 in the expansions above are related by the exchange of stability formula

$$\beta_2 = -2\alpha'(0)\mu_2. \tag{57.187}$$

Generically, these coefficients do not vanish. Their signs determine the direction of bifurcation. The coefficients β_2, β_4,... in the expansion (57.186) are the bifurcation stability coefficients for the case of Andronov-Hopf bifurcation.

If $\beta_2 > 0$, then locally the bifurcated limit cycle is unstable and the bifurcation is subcritical. This case is depicted in Figure 57.17. If $\beta_2 < 0$, then locally the bifurcated limit cycle is stable (more precisely, one says that it is orbitally asymptotically stable [10]). This is the case of supercritical Andronov-Hopf bifurcation, depicted in Figure 57.18. If it happens that β_2 vanishes, then stability is determined by the first nonvanishing bifurcation stability coefficient (if one exists).

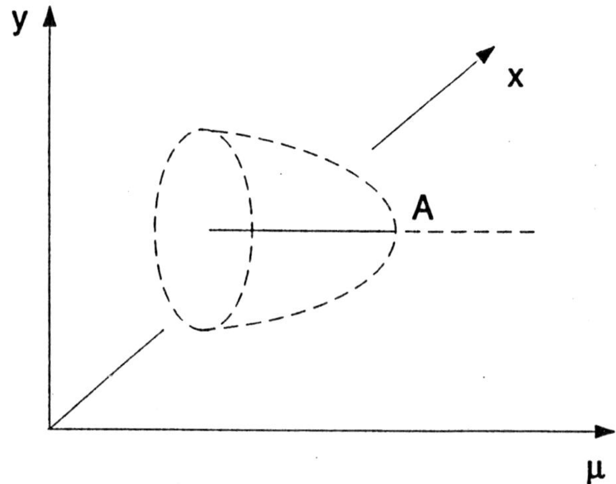

Figure 57.17 Subcritical Andronov-Hopf bifurcation.

Period doubling bifurcation The bifurcations considered above are all local bifurcations, i.e., bifurcations from an equilibrium point of the system (57.181). Solutions emerging at these bifurcation points can themselves undergo further bifurcations. A particularly important scenario involves a global bifurcation known as the *period doubling bifurcation*.

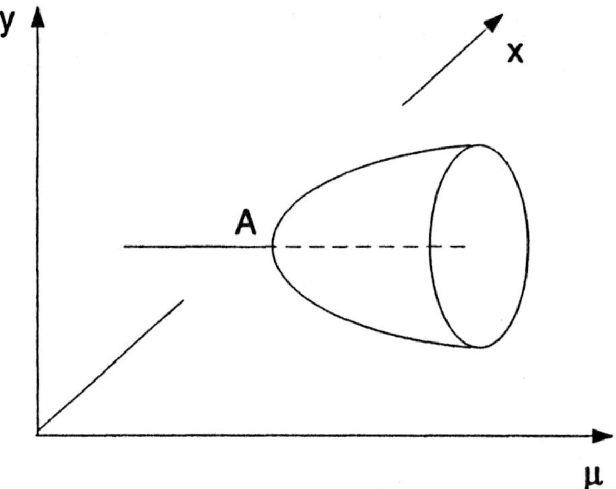

Figure 57.18 Supercritical Andronov-Hopf bifurcation.

To describe the period doubling bifurcation, consider the one-parameter family of nonlinear systems (57.181). Suppose that Equation 57.181 has a limit cycle γ^{μ} for a range of values of the real parameter μ. Moreover, suppose that for all values of μ to one side (say, less than) a critical value μ_c, all the characteristic multipliers of γ^{μ} have magnitude less than unity. If exactly one characteristic multiplier exits the unit circle at $\mu = \mu_c$, and does so at the point $(-1, 0)$, and if this crossing occurs with a nonzero rate with respect to μ, then one can show that a period doubling bifurcation from γ^{μ} occurs at $\mu = \mu_c$. (See, e.g., [4] and references therein.)

This means that another limit cycle, initially of twice the period of γ^{μ_c}, emerges from γ^{μ} at $\mu = \mu_c$. Typically, the bifurcation is either *supercritical* or *subcritical*. In the supercritical case, the period doubled limit cycle is stable and occurs for parameter values on the side of μ_c for which the limit cycle γ^{μ} is unstable. In the subcritical case, the period doubled limit cycle is unstable and occurs on the side of μ_c for which the limit cycle γ^{μ} is stable. In either case, an *exchange of stability* is said to have occurred between the nominal limit cycle γ^{μ} and the bifurcated limit cycle. Figure 57.19 depicts a supercritical period doubling bifurcation. In this figure, a solid curve represents a stable limit cycle, while a dashed curve represents an unstable limit cycle. The figure assumes that the nominal limit cycle loses stability as μ increases through μ_c.

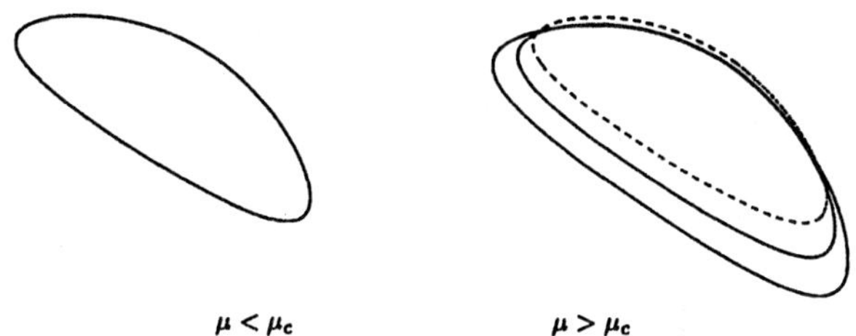

Figure 57.19 Period doubling bifurcation (supercritical case).

The direction of a period doubling bifurcation can easily be determined in discrete-time, using formulas that have been derived in the literature (see, e.g., [4]). Recently, an approximate test that applies in continuous-time has been derived using the harmonic balance approach [47].

Chaos

Bifurcations of equilibrium points and limit cycles are well understood and there is little room for alternative definitions of the main concepts. Although the notation, style, and emphasis may differ among various presentations, the main concepts and results stay the same. Unfortunately, the situation is not quite as tidy in regard to discussions of chaos. There are several distinct definitions of chaotic behavior of dynamical systems. There are also some aspects of chaotic motion that have been found to be true for many systems but have not been proved in general. The aim of this subsection is to summarize in a nonrigorous fashion some important aspects of chaos that are widely agreed upon.

The following working definition of chaos will suffice for the purposes of this chapter. The definition is motivated by [46, p. 323] and [11, p. 50]. It uses the notion of "attractor" defined in section two, and includes the definition of an additional notion, namely that of "strange attractor."

> A solution trajectory of a deterministic system (such as Equation 57.180) is *chaotic* if it converges to a strange attractor. A *strange attractor* is a bounded attractor that: (1) exhibits sensitive dependence on initial conditions, and (2) cannot be decomposed into two invariant subsets contained in disjoint open sets.

A few remarks on this working definition are in order. An aperiodic motion is one that is not periodic. Long-term behavior refers to steady-state behavior, i.e., system behavior that persists after the transient decays. Sensitive dependence on initial conditions means that for almost any initial condition lying on the strange attractor, there exists another initial condition as close as desired to the given one such that the solution trajectories starting from these two initial conditions separate by at least some prespecified amount after some time. The requirement of nondecomposability simply ensures that strange attractors are considered as being distinct if they are not connected by any system trajectories.

Often, sensitive dependence on initial conditions is defined in terms of the presence of at least one positive "Liapunov exponent" (e.g., [34, 35]). Further discussion of this viewpoint would entail technicalities that are not needed in the sequel.

From a practical point of view, a chaotic motion can be defined as a bounded invariant motion of a deterministic system that is not an equilibrium solution or a periodic solution or a quasiperiodic solution [32, p. 277]. (A quasiperiodic function is one that is composed of finitely many periodic functions with incommensurate frequencies. See [32, p. 231].)

There are two more important aspects of strange attractors and chaos that should be noted, since they play an important role in a technique for control of chaos discussed in section 57.6.7. These are:

1. A strange attractor generally has embedded within itself infinitely many unstable limit cycles. For example, Figure 57.20(a) depicts a strange attractor, and Figure 57.20(b) depicts an unstable limit cycle that is embedded in the strange attractor. Note that the shape of the limit cycle resembles that of the strange attractor. This is to be expected in general. The limit cycle chosen in the plot happens to be one of low period.
2. The trajectory starting from any point on a strange attractor will, after sufficient time, pass as close as desired to any other point of interest on the strange attractor. This follows from the indecomposability of strange attractors noted previously.

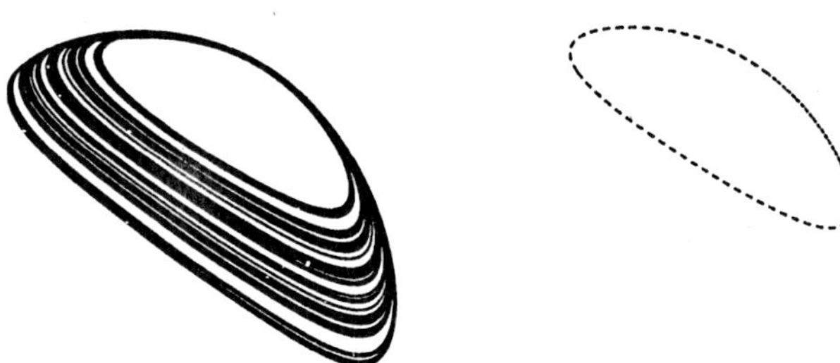

Figure 57.20 Strange attractor with embedded unstable limit cycle.

An important way in which chaotic behavior arises is through sequences of bifurcations. A well-known such mechanism is the *period doubling route to chaos*, which involves the following sequence of events:

1. A stable limit cycle loses stability, and a new stable limit cycle of double the period emerges. (The original stable limit cycle might have emerged from an equilibrium point via a supercritical Andronov-Hopf bifurcation.)
2. The new stable limit cycle loses stability, releasing another stable limit cycle of twice its period.
3. There is a cascade of such events, with the parameter separation between each two successive events decreasing geometrically.[4] This cascade culminates in a sustained chaotic motion (a strange attractor).

57.6.4 Bifurcations and Chaos in Physical Systems

In this brief section, a list of representative physical systems that exhibit bifurcations and/or chaotic behavior is given. The purpose is to provide practical motivation for study of these phenomena and their control.

[4]This is true exactly in an asymptotic sense. The ratio in the geometric sequence is a universal constant discovered by Feigenbaum. See, e.g., [11, 32, 34, 46, 43, 48].

Examples of physical systems exhibiting bifurcations and/or chaotic behavior include the following.

- An aircraft stalls for flight under a critical speed or above a critical angle-of-attack (e.g., [8, 13, 39, 51]).
- Aspects of laser operation can be viewed in terms of bifurcations. The simplest such observation is that a laser can only generate significant output if the pump energy exceeds a certain threshold (e.g., [19, 46]). More interestingly, as the pump energy increases, the laser operating condition can exhibit bifurcations leading to chaos (e.g., [17]).
- The dynamics of ships at sea can exhibit bifurcations for wave frequencies close to a natural frequency of the ship. This can lead to large-amplitude oscillations, chaotic motion, and ship capsize (e.g., [24, 40]).
- Lightweight, flexible, aircraft wings tend to experience flutter (structural oscillations) (e.g., [39]) (along with loss of control surface effectiveness (e.g., [15])).
- At peaks in customer demand for electric power (such as during extremes in weather), the stability margin of an electric power network may become negligible, and nonlinear oscillations or divergence ("voltage collapse") can occur (e.g., [12]).
- Operation of aeroengine compressors at maximum pressure rise implies reduced weight requirements but also increases the risk for compressor stall (e.g., [16, 20, 25]).
- A simple model for the weather consists of fluid in a container (the atmosphere) heated from below (the sun's rays reflected from the ground) and cooled from above (outer space). A mathematical description of this model used by Lorenz [27] exhibits bifurcations of convective and chaotic solutions. This has implications also for boiling channels relevant to heat-exchangers, refrigeration systems, and boiling water reactors [14].
- Bifurcations and chaos have been observed and studied in a variety of chemically reacting systems (e.g., [42]).
- Population models useful in the study and formulation of harvesting strategies exhibit bifurcations and chaos (e.g., [19, 31, 46]).

57.6.5 Control of Parametrized Families of Systems

Tracking of a desired trajectory (referred to as regulation when the trajectory is an equilibrium), is a standard goal in control system design [26]. In applying linear control system design to this standard problem, an evolving (nonstationary) system is modeled by a *parametrized family* of time-invariant (stationary) systems. This approach is at the heart of the gain scheduling method, for example [26].

In this section, some basic concepts in control of parameterized families of systems are reviewed, and a notion of stressed operation is introduced. Control laws for nonlinear systems usually consist of a feedforward control plus a feedback control. The feedforward part of the control is selected first, followed by the feedback part. This decomposition of control laws is discussed in the next subsection, and will be useful in the discussions of control of bifurcations and chaos. In the second subsection, a notion of "stressed operation" of a system is introduced. Stressed systems are not the only ones for which control of bifurcations and chaos are relevant, but they are an important class for which such control issues need to be evaluated.

Feedforward/Feedback Structure of Control Laws

Control designs for nonlinear system can usually be viewed as proceeding in two main steps [26, Chapters 2, 14], [45, Chapter 3]:

1. feedforward control
2. feedback control

In Step 1, the feedforward part of the control input is selected. Its purpose is to achieve a desired candidate operating condition for the system. If the system is considered as depending on one or more parameters, then the feedforward control will also depend on the parameters. The desired operating condition can often be viewed as an equilibrium point of the nonlinear system that varies as the system parameters are varied. There are many situations when this operating condition is better viewed as a limit cycle that varies with the parameters. In Step 2, an additional part of the control input is designed to achieve desired qualities of the transient response in a neighborhood of the nominal operating condition. Typically this second part of the control is selected in feedback form.

In the following, the feedforward part of the control input is taken to be designed already and reflected in the system dynamical equations. The discussion centers on design of the feedback part of the control input. Because of this, it is convenient to denote the feedback part of the control simply by u and to view the feedforward part as being determined *a priori*. It is also convenient to take u to be small (close to zero) near the nominal operating condition. (Any offset in u is considered part of the feedforward control.)

It is convenient to view the system as depending on a single exogeneous parameter, denoted by μ. For instance, μ can represent the set-point of an aircraft's angle-of-attack, or the power demanded of an electric utility by a customer. In the former example, the control u might denote actuation of the aircraft's elevator angles about the nominal settings. In the latter example, u can represent a control signal in the voltage regulator of a power generator.

Consider, then, a nonlinear system depending on a single parameter μ

$$\dot{x} = f^{\mu}(x, u). \tag{57.188}$$

Here $x \in \mathbb{R}^n$ is the n-dimensional state vector, u is the feedback part of the control input, and μ is a parameter. Both u and μ are taken to be scalar-valued for simplicity. The dependence of the system equations on x, u, and μ is assumed smooth; i.e., f is jointly continuous and several times differentiable in these variables. This system is thus actually a one-parameter *family* of nonlinear control systems. The parameter μ is viewed as being allowed to change so slowly that its variation can be taken as quasistatic.

Suppose that the nominal operating condition of the system is an equilibrium point $x^0(\mu)$ that depends on the parameter μ. For simplicity of notation, suppose that the state x has been chosen so that this nominal equilibrium point is the origin for the range of values of μ for which it exists ($x^0(\mu) \equiv 0$). Recall that the nominal equilibrium is achieved using feedforward control. Although the process of choosing a feedforward control is not elaborated on here, it is important to emphasize that in general this process aims at securing a particular desired candidate operating condition. The feedforward control thus is expected to result in an acceptable form for the nominal operating condition, but there is no reason to expect that other operating conditions will also behave in a desirable fashion. As the parameter varies, the nominal operating condition may interact with other candidate operating conditions in bifurcations. The design of the feedback part of the control should take into account the other candidate operating conditions in addition to the nominal one.

For simplicity, suppose the control u isn't allowed to introduce additional dynamics. That is, suppose u is required to be a direct state feedback (or static feedback), and denote it by $u = u(x, \mu)$. Note that in general u can depend on the parameter μ. That is, it can be *scheduled*. Since in the discussion above it was assumed that u is small near the nominal operating condition, it is assumed that $u(0, \mu) \equiv 0$ (for the parameter range in which the origin is an equilibrium).

Denote the linearization of Equation 57.188 at $x^0 = 0$, $u = 0$ by

$$\dot{x} = A(\mu)x + b(\mu)u. \tag{57.189}$$

Here,

$$A(\mu) := \frac{\partial f^{\mu}}{\partial x}(0, 0)$$

and

$$b(\mu) := \frac{\partial f^{\mu}}{\partial u}(0, 0).$$

Consider how control design for the linearized system depends on the parameter μ. Recall the terminology from linear system theory that the pair $(A(\mu), b(\mu))$ is controllable if the system (57.189) is controllable. Recall also that there are several simple tests for controllability, one of which is that the controllability matrix

$$(b(\mu), A(\mu)b(\mu), (A(\mu))^2 b(\mu), \ldots, (A(\mu))^{(n-1)} b(\mu))$$

is of full rank. (In this case this is equivalent to the matrix being nonsingular, since here the controllability matrix is square.)

If μ is such that the pair $(A(\mu), b(\mu))$ is controllable, then a standard linear systems result asserts the existence of a linear feedback $u(x, \mu) = -k(\mu)x$ stabilizing the system. (The associated closed-loop system would be $\dot{x} = (A(\mu) - b(\mu)k(\mu))x$.) Stabilizability tests not requiring controllability also exist, and these are more relevant to the problem at hand. Even more interesting from a practical perspective is the issue of output feedback stabilizability, since not all state variables are accessible for real-time measurement in many systems. As μ is varied over the desired regime of operability, the system (57.189) may lose stability and stabilizability.

Stressed Operation and Break-Down of Linear Techniques

A main motivation for the study of control of bifurcations is the need in some situations to operate a system in a condition for which the stability margin is small and linear (state or output) feedback is ineffective as a means for increasing the stability margin. Such a system is sometimes referred to as being "pushed to its limits", or "heavily loaded". In such situations, the ability of the system to function in a state of increased loading is a measure of system performance. Thus, the link between increased loading and reduced stability margin can be viewed as a performance vs. stability trade-off. Systems operating with a reduced achievable margin of stability may be viewed as being "stressed". This trade-off is not a general fact that can be proved in a rigorous fashion, but has been found to occur in a variety of applications. Examples of this trade-off are given at the end of this subsection.

Consider a system that is weakly damped and for which the available controls cannot compensate with sufficient additional damping. Such a situation may arise for a system for some ranges of parameter values and not for others. Let the *operating envelope* of a system be the possible combinations of system parameters for which system operability is being considered. Linear control system methods lose their effectiveness on that part of the operating envelope for which the operating condition of interest of system (57.189) is either:

1. not stabilizable with linear feedback using available sensors and actuators
2. linearly stabilizable using available sensors and actuators but only with unacceptably high feedback gains
3. vulnerable in the sense that small parameter changes can destroy the operating condition completely (as in a saddle-node bifurcation)

Operation in this part of the desired operating envelope can be referred to by terms such as "stressed operation".

An example of the trade-off noted previously is provided by an electric power system under conditions of heavy loading. At peaks in customer demand for electric power (such as during extremes in weather), the stability margin may become negligible, and nonlinear dynamical behaviors or divergence may arise (e.g., [12]). The divergence, known as voltage collapse, can lead to system blackout. Another example arises in operation of an aeroengine compressor at its peak pressure rise. The increased pressure rise comes at the price of nearness to instability. The unstable modes that can arise are strongly related to flow asymmetry modes that are unstabilizable by linear feedback to the compression system throttle. However, bifurcation control techniques have yielded valuable nonlinear throttle actuation techniques that facilitate operation in these circumstances with reduced risk of stall [25, 16].

57.6.6 Control of Systems Exhibiting Bifurcation Behavior

Most engineering systems are designed to operate with a comfortable margin of stability. This means that disturbances or moderate changes in system parameters are unlikely to result in loss of stability. For example, a jet airplane in straight level flight under autopilot control is designed to have a large stability margin. However, engineering systems with a usually comfortable stability margin may at times be operated at a reduced stability margin. A jet airplane being maneuvered at high angle-of-attack to gain an advantage over an enemy aircraft, for instance, may have a significantly reduced stability margin. If a system operating condition actually loses stability as a parameter (like angle-of-attack) is slowly varied, then generally it is the case that a bifurcation occurs. This means that at least one new candidate operating condition emerges from the nominal one at the point of loss of stability. The purpose of this section is to summarize some results on control of bifurcations, with an emphasis placed on control of local bifurcations. Control of a particular global bifurcation, the period doubling bifurcation, is considered in the next section on control of chaos. This is because control of a period doubling bifurcation can result in quenching of the period doubling route to chaos summarized at the end of section three.

Bifurcation control involves designing a control input for a system to result in a desired modification to the system's bifurcation behavior. In section 57.6.5, the division of control into a feedforward component and a feedback component was discussed. Both components of a control law can be viewed in terms of bifurcation control. The feedforward part of the control sets the equilibrium points of the system, and may influence the stability margin as well. The feedback part of the control has many functions, one of which is to ensure adequate stability of the desired operating condition over the desired operating envelope. Linear feedback is used to ensure an adequate margin of stability over a desired parameter range. Use of linear feedback to "delay" the onset of instability to parameter ranges outside the desired operating range is a common practice in control system design. An example is the gain scheduling technique [26]. Delaying instability modifies the bifurcation diagram of a system. Often, the available control authority does not allow stabilization of the nominal operating condition beyond some critical parameter value. At this value, instability leads to bifurcations of new candidate operating conditions. For simplicity, suppose that a single candidate operating condition is born at the bifurcation point. Another important goal in bifurcation control is to ensure that the bifurcation is *supercritical* (i.e., *safe*) and that the resulting candidate operating condition remains stable and close to the original operating condition for a range of parameter values beyond the critical value. The need for control laws that soften (stabilize) a hard (unstable) bifurcation has been discussed earlier in this chapter. This need is greatest in stressed systems, since in such systems delay of the bifurcation by linear feedback is not viable. A soft bifurcation provides the possibility of an alternative operating condition beyond the regime of operability at the nominal condition.

Both of these goals (delaying and stabilization) basically involve local considerations and can be approached analytically (if a good system model is available). Another goal might entail a reduction in amplitude of any bifurcated solutions over some prespecified parameter range. This goal is generally impossible to work with on a completely analytical basis. It requires extensive numerical study in addition to local analysis near the bifurcation(s).

In the most severe local bifurcations (saddle-node bifurcations), neither the nominal equilibrium point nor any bifurcated solution exists past the bifurcation. Even in such cases, an understanding of bifurcations provides some insight into control design for safe operation. For example, it may be possible to use this understanding to determine (or introduce via added control) a warning signal that becomes more pronounced as the severe bifurcation is approached. This signal would alert the high-level control system (possibly a human operator) that action is necessary to avoid catastrophe.

In this section, generally it is assumed that the feedforward component of the control has been pre-determined, and the goal is to design the feedback component. An exception is the following brief discussion of a real-world example of the use of feedforward control to modify a system's operating condition and its stability margin in the face of large parameter variations. In short, this is an example where feedforward controls are used to successfully *avoid* the possibility of bifurcation. During takeoff and landing of a commercial jet aircraft, one can observe the deployment of movable surfaces on the leading- and trailing-edges of the wings. These movable surfaces, called flaps and slats, or camber changers [13], result in a nonlinear change in the aerodynamics, and, in turn, in an increased lift coefficient [51, 13]. This is needed to allow takeoff and landing at reduced speeds. Use of these surfaces has the drawback of reducing the critical angle-of-attack for stall, resulting in a reduced stability margin. A common method to alleviate this effect is to incorporate other devices, called vortex generators. These are small airfoil-shaped vanes, protruding upward from the wings [13]. The incorporation of vortex generators results in a further modification to the aeodynamics, moving the stall angle-of-attack to a higher value. Use of the camber changers and the vortex generators are examples of feedforward control to modify the operating condition within a part of the aircraft's operating envelope. References [13] and [51] provide further details, as well as diagrams showing how the lift coefficient curve is affected by these devices.

Local Direct State Feedback

To give a flavor of the analytical results available in the design of the feedback component in bifurcation control, consider the nonlinear control system (57.188), repeated here for convenience:

$$\dot{x} = f^{\mu}(x, u). \tag{57.190}$$

Here, u represents the feedback part of the control law; the feedforward part is assumed to be incorporated into the function f. The technique and results of [1, 2] form the basis for the following discussion. Details are not provided since they would require introduction of considerable notation related to multivariate Taylor series. However, an illustrative example is given based on formulas available in [1, 2].

Suppose for simplicity that Equation 57.190 with $u \equiv 0$ possesses an equilibrium at the origin for a parameter range of interest (including the value $\mu = 0$). Moreover, suppose that the origin of Equation 57.190 with the control set to zero undergoes either a subcritical stationary bifurcation or a subcritical Andronov-Hopf bifurcation at the critical parameter value $\mu = 0$. Feedback control laws of the form $u = u(x)$ ("static state feedbacks") are derived in [1, 2] that render the bifurcation supercritical.

For the Andronov-Hopf bifurcation, this is achieved using a formula for the coefficient β_2 in the expansion (57.186) of the characteristic exponent for the bifurcated limit cycle. Smooth nonlinear controls rendering $\beta_2 < 0$ are derived. For the stationary bifurcation, the controlled system is desired to display a supercritical pitchfork bifurcation. This is achieved using formulas for the coefficients β_1 and β_2 in the expansion (57.184) for the eigenvalue of the bifurcated equilibrium determining stability. Supercriticality is insured by determining conditions on $u(x)$ such that $\beta_1 = 0$ and $\beta_2 < 0$.

The following example is meant to illustrate the technique of [2]. The calculations involve use of formulas from [2] for the bifurcation stability coefficients β_1 and β_2 in the analysis of stationary bifurcations. The general formulas are not repeated here.

Consider the one-parameter family of nonlinear control systems

$$\dot{x}_1 = \mu x_1 + x_2 + x_1 x_2^2 + x_1^3, \tag{57.191}$$

$$\dot{x}_2 = -x_2 - x_1 x_2^2 + \mu u + x_1 u. \tag{57.192}$$

Here x_1, x_2 are scalar state variables, and μ is a real-valued parameter. This is meant to represent a system after application of a feedforward control, so that u is to be designed in feedback form. The nominal operating condition is taken to be the origin $(x_1, x_2) = (0, 0)$, which is an equilibrium of (57.191),(57.192) when the control $u = 0$ for all values of the parameter μ.

Local stability analysis at the origin proceeds in the standard manner. The Jacobian matrix $A(\mu)$ of the right side of (57.191),(57.192) is given by

$$A(\mu) = \begin{pmatrix} \mu & 1 \\ 0 & -1 \end{pmatrix}. \tag{57.193}$$

The system eigenvalues are μ and -1. Thus, the origin is stable for $\mu < 0$ but is unstable for $\mu > 0$. The critical value of the bifurcation parameter is therefore $\mu_c = 0$. Since the origin persists as an equilibrium point past the bifurcation, and since the critical eigenvalue is 0 (not an imaginary pair), it is expected that a stationary bifurcation occurs. The stationary bifurcation that occurs for the open-loop system can be studied by solving the pair of algebraic equations

$$0 = \mu x_1 + x_2 + x_1 x_2^2 + x_1^3, \quad (57.194)$$

$$0 = -x_2 - x_1 x_2^2 \quad (57.195)$$

for a nontrivial (i.e., nonzero) equilibrium (x^1, x^2) near the origin for μ near 0. Adding Equation 57.194 to Equation 57.195 gives

$$0 = \mu x_1 + x_1^3. \quad (57.196)$$

Disregarding the origin, this gives two new equilibrium points that exist for $\mu < 0$, namely

$$x(\mu) = \begin{pmatrix} \pm\sqrt{-\mu} \\ 0 \end{pmatrix}. \quad (57.197)$$

Since these bifurcated equilibria occur for parameter values ($\mu < 0$) for which the nominal operating condition is unstable, the bifurcation is a *subcritical* pitchfork bifurcation.

The first issue addressed in the control design is the possibility of using linear feedback to delay the bifurcation to some positive value of μ. This would require stabilization of the origin at $\mu = 0$. Because of the way in which the control enters the system dynamics, however, the system eigenvalues are not affected by linear feedback at the parameter value $\mu = 0$. To see this, simply note that in Equation 57.192, the term μu vanishes when $\mu = 0$, and the remaining impact of the control is through the term $x_1 u$. This latter term would result only in the addition of *quadratic* terms to the right side of Equations 57.191 and 57.192 for any linear feedback u. Hence, the system is an example of a stressed nonlinear system for μ near 0.

Since the pitchfork bifurcation cannot be delayed by linear feedback, next consider the possibility of rendering the pitchfork bifurcation supercritical using nonlinear feedback. This rests on determining how feedback affects the bifurcation stability coefficients β_1 and β_2 for this stationary bifurcation. Once this is known, it is straightforward to seek a feedback that renders $\beta_1 = 0$ and $\beta_2 < 0$. The formulas for β_1 and β_2 derived in [2] simplify for systems with no quadratic terms in the state variables. For such systems, the coefficient β_1 always vanishes, and the calculation of β_2 also simplifies. Since the dynamical equations (57.191) and (57.192) in the example contain no quadratic terms in x, it follows that $\beta_1 = 0$ in the absence of control. Moreover, if the control contains no linear terms, then it will not introduce quadratic terms into the dynamics. Hence, for any smooth feedback $u(x)$ containing no linear terms in x, the bifurcation stability coefficient $\beta_1 = 0$.

As for the bifurcation stability coefficient β_2, the pertinent formula in [2] applied to the open-loop system yields $\beta_2 = 2$. Thus, a subcritical pitchfork bifurcation is predicted for the open-loop system, a fact that was established above using simple algebra. Now let the control consist of a quadratic function of the state and determine conditions under which $\beta_2 < 0$ for the closed-loop system.[5] Thus, consider u to be of the form

$$u(x_1, x_2) = -k_1 x_1^2 - k_2 x_1 x_2 - k_3 x_2^2. \quad (57.198)$$

The formula in [2] yields that β_2 for the closed-loop system is then given by

$$\beta_2 = 2(1 - k_1). \quad (57.199)$$

Thus, to render the pitchfork bifurcation in (57.191),(57.192) supercritical, it suffices to take u to be the simple quadratic function

$$u(x_1, x_2) = -k_1 x_1^2, \quad (57.200)$$

with any gain $k_1 > 1$. In fact, other quadratic terms, as well as any other cubic or higher order terms, can be included along with this term without changing the local result that the bifurcation is rendered supercritical. Additional terms can be useful in improving system behavior as the parameter leaves a local neighborhood of its critical value.

Local Dynamic State Feedback

Use of a static state feedback control law $u = u(x)$ has potential disadvantages in nonlinear control of systems exhibiting bifurcation behavior. To explain this, consider the case of an equilibrium $x^0(\mu)$ as the nominal operating condition. The equilibrium is not translated to the origin to illustrate how it is affected by feedback. In general, a static state feedback

$$u = u(x - x^0(\mu)) \quad (57.201)$$

designed with reference to the nominal equilibrium path $x^0(\mu)$ of Equation 57.190 will affect not only the stability of this equilibrium but also the location and stability of other equilibria. Now suppose that Equation 57.190 is only an approximate model for the physical system of interest. Then the nominal equilibrium branch will also be altered by the feedback. A main disadvantage of such an effect is the wasted control energy that is associated with the forced alteration of the system equilibrium structure. Other disadvantages are that system performance is often degraded by operating at an equilibrium that differs from the one at which the system is designed to operate. Moreover, by influencing the locations of system equilibria, the feedback control is competing with the feedforward part of the control.

For these reasons, dynamic state feedback-type bifurcation control laws have been developed that do not affect the locations

[5]Cubic terms are not included because they would result in quartic terms on the right side of Equations 57.191 and 57.192, while the formula for β_2 in [2] involves only terms up to cubic order in the state.

of system equilibria [22, 23, 50]. The method involves incorporation of filters called "washout filters" into the controller architecture. A washout filter-aided control law preserves all system equilibria, and does so without the need for an accurate system model.

A washout filter is a stable high pass filter with zero static gain [15, p. 474]. The typical transfer function for a washout filter is

$$G(s) = \frac{y_i(s)}{x_i(s)} = \frac{s}{s+d}. \tag{57.202}$$

where x_i is the input variable to the filter and y_i is the output of the filter. A washout filter produces a nonzero output only during the transient period. Thus, such a filter "washes out" sensed signals that have settled to constant steady-state values. Washout filters occur in control systems for power systems [5, p. 277], [41, Chapter 9] and aircraft [7, 15, 39, 45].

Washout filters are positioned in a control system so that a sensed signal being fed back to an actuator first passes through the washout filter. If, due to parameter drift or intentional parameter variation, the sensed signal has a steady-state value that deviates from the assumed value, the washout filter will give a zero output and the deviation will not propagate. If a direct state feedback were used instead, the steady-state deviation in the sensed signal would result in the control modifying the steady-state values of the other controlled variables. As an example, washout filters are used in aircraft yaw damping control systems to prevent these dampers from "fighting" the pilot in steady turns [39, p. 947].

From a nonlinear systems perspective, the property of washout filters described above translates to achieving equilibrium preservation, i.e., zero steady-state tracking error, in the presence of system uncertainties.

Washout filters can be incorporated into bifurcation control laws for Equation 57.190. This should be done only after the feedforward control has been designed and incorporated *and* part of the feedback control ensuring satisfactory equilibrium point structure has also been designed and incorporated into the dynamics. Otherwise, since washout filters preserve equilibrium points, there will be no possibility of modifying the equilibria. It is assumed below that these two parts of the control have been chosen and implemented.

For each system state variable x_i, $i = 1, \ldots, n$, in Equation 57.190, introduce a washout filter governed by the dynamic equation

$$\dot{z}_i = x_i - d_i z_i \tag{57.203}$$

along with output equation

$$y_i = x_i - d_i z_i. \tag{57.204}$$

Here, the d_i are positive parameters (this corresponds to using stable washout filters). Finally, require the control u to depend only on the measured variables y, and that $u(y)$ satisfy $u(0) = 0$.

In this formulation, n washout filters, one for each system state, are present. In fact, the actual number of washout filters needed, and hence also the resulting increase in system order, can usually be taken less than n.

It is straightforward to see that washout filters result in equilibrium preservation and automatic equilibrium (operating point) following. Indeed, since $u(0) = 0$, it is clear that y vanishes at steady-state. Hence, the x subvector of a closed-loop equilibrium point (x, z) agrees exactly with the open-loop equilibrium value of x. Also, since y_i may be re-expressed as

$$\begin{aligned} y_i &= x_i - d_i z_i \\ &= (x_i - x_i^0(\mu)) - d_i(z_i - z_i^0(\mu)), \end{aligned} \tag{57.205}$$

the control function $u = u(y)$ is guaranteed to center at the correct operating point.

57.6.7 Control of Chaos

Chaotic behavior of a physical system can either be desirable or undesirable, depending on the application. It can be beneficial for many reasons, such as enhanced mixing of chemical reactants, or, as proposed recently [36], as a replacement for random noise as a masking signal in a communication system. Chaos can, on the other hand, entail large amplitude motions and oscillations that might lead to system failure. The control techniques discussed in this section have as their goal the replacement of chaotic behavior by a nonchaotic steady-state behavior. The first technique discussed is that proposed by Ott, Grebogi, and Yorke [33]. The Ott-Grebogi-Yorke (OGY) method sparked significant interest and activity in control of chaos. The second technique discussed is the use of bifurcation control to delay or extinguish the appearance of chaos in a family of systems.

Exploitation of Chaos for Control

Ott, Grebogi, and Yorke [33] proposed an approach to control of chaotic systems that involves use of small controls and exploitation of chaotic behavior. To explain this method, recall from section three that a strange attractor has embedded within itself a "dense" set (infinitely many) of unstable limit cycles. The strange attractor is a candidate operating condition, according to the definition in section two. In the absence of control, it is the actual system operating condition. Suppose that system performance would be significantly improved by operation at one of the unstable limit cycles embedded in the strange attractor. The OGY method replaces the originally chaotic system operation with operation along the selected unstable limit cycle.

Figure 57.20 depicts a strange attractor along with a particular unstable limit cycle embedded in it. It is helpful to keep such a figure in mind in contemplating the OGY method. The goal is to reach a limit cycle such as the one shown in Figure 57.20, or another of some other period or amplitude. Imagine a trajectory that lies on the strange attractor in Figure 57.20, and suppose the desire is to use control to force the trajectory to reach the unstable limit cycle depicted in the figure and to remain there for all subsequent time.

Control design to result in operation at the desired limit cycle is achieved by the following reasoning. First, note that the desired unstable limit cycle is also a candidate operating condition. Next, recall from section three that a trajectory on the strange attractor will, after sufficient time, pass as close as desired to any other point of interest on the attractor. Thus, the trajectory will eventually come close (indeed, arbitrarily close) to the desired limit cycle. Thus, no control effort whatsoever is needed in order for the trajectory to reach the desired limit cycle — chaos guarantees that it will. To maintain the system state on the desired limit cycle, a small stabilizing control signal is applied once the trajectory enters a small neighborhood of the limit cycle. Since the limit cycle is rendered stable by this control, the trajectory will converge to the limit cycle. If noise drives the trajectory out of the neighborhood where control is applied, the trajectory will wander through the strange attractor again until it once again enters the neighborhood and remains there by virtue of the control.

Note that the control obtained by this method is an example of a variable structure control, since it is "on" in one region in state space and "off" in the rest of the space. Also, the particular locally stabilizing control used in the neighborhood of the desired limit cycle has not been discussed in the foregoing. This is because several approaches are possible, among them pole placement [38]. See [34, 35] for further discussion.

Two particularly significant strengths of the OGY technique are:

1. it requires only small controls
2. it can be applied to experimental systems for which no mathematical model is available

The first of these strengths is due to the assumption that operation at an unstable limit cycle embedded in the strange attractor is desirable. If none of the embedded limit cycles provides adequate performance, then a large control could possibly be used to introduce a new desirable candidate operating condition within the strange attractor. This could be followed by a small control possibly designed within the OGY framework. Note that the large control would be a feedforward control, in the terminology used previously in this chapter. For a discussion of the second main strength of the OGY control method mentioned above, see, e.g., [17, 33, 34]. It suffices to note here that a construction known as experimental delay coordinate embedding is one means to implement this control method without *a priori* knowledge of a reliable mathematical model.

Several interesting applications of the OGY control method have been performed, including control of cardiac chaos (see [35]). In [17], a multimode laser was controlled well into its usually unstable regime.

Bifurcation Control of Routes to Chaos

The bifurcation control techniques discussed in section six have direct relevance for issues of control of chaotic behavior of dynamical systems. This is because chaotic behavior often arises as a result of bifurcations, such as through the period doubling route to chaos. The bifurcation control technique discussed earlier in this chapter is model-based. Thus, the control of chaos applications of the technique also require availability of a reliable mathematical model.

Only a few comments are given here on bifurcation control of routes to chaos, since the main tool has already been discussed in section six. The cited references may be consulted for details and examples.

In [50], a thermal convection loop model is considered. The model, which is equivalent to the Lorenz equations mentioned in section four, displays a series of bifurcations leading to chaos. In [50], an Andronov-Hopf bifurcation that occurs in the model is rendered supercritical using *local* dynamic state feedback (of the type discussed in section six). The feedback is designed using local calculations at the equilibrium point of interest. It is found that this simple control law results in elimination of the chaotic behavior in the system. From a practical perspective, this allows operation of the convection loop in a *steady* convective state with a desired velocity and temperature profile.

Feedback control to render supercritical a previously subcritical period doubling bifurcation was studied in [4, 47]. In [4], a discrete-time model is assumed, whereas a continuous-time model is assumed in [47]. The discrete-time model used in [4] takes the form (k is an integer)

$$x(k+1) = f^{\mu}(x(k), u(k)) \tag{57.206}$$

where $x(k) \in \mathbb{R}^n$ is the state, $u(k)$ is a scalar control input, $\mu \in \mathbb{R}$ is the bifurcation parameter, and the mapping f^{μ} is sufficiently smooth in x, u, and μ. The continuous-time model used in [47] is identical to Equation 57.188.

For discrete-time systems, a limit cycle must have an integer period. A period-1 limit cycle sheds a period-2 limit cycle upon period doubling bifurcation. The simplicity of the discrete-time setting results in explicit formulas for bifurcation stability coefficients and feedback controls [4]. By improving the stability characteristics of a bifurcated period-2 limit cycle, an existing period doubling route to chaos can be extinguished. Moreover, the period-2 limit cycle will then remain close to the period-1 limit cycle for an increased range of parameters.

For continuous-time systems, limit cycles cannot in general be obtained analytically. Thus, [47] employs an approximate analysis technique known as *harmonic balance*. Approximate bifurcation stability coefficients are obtained, and control to delay the onset of period doubling bifurcation or stabilize such a bifurcation is discussed.

The washout filter concept discussed in section six is extended in [4] to discrete-time systems. In [47], an extension of the washout filter concept is used that allows approximate preservation of limit cycles of a certain frequency.

57.6.8 Concluding Remarks

Control of bifurcations and chaos is a developing area with many interesting avenues for research and for application. Some of the tools and ideas that have been used in this area were discussed. Connections among these concepts, and relationships to traditional control ideas, have been emphasized.

57.6.9 Acknowledgment

During the preparation of this chapter, the authors were supported in part by the Air Force Office of Scientific Research (U.S.), the Electric Power Research Institute (U.S.), the National Science Foundation (U.S.), and the Ministero della Università e della Ricerca Scientifica e Tecnologica under the National Research Plan 40% (Italy).

References

[1] Abed, E.H. and Fu, J.-H., Local feedback stabilization and bifurcation control, I. Hopf bifurcation, *Syst. Control Lett.*, 7, 11-17, 1986.

[2] Abed E.H. and Fu J.-H., Local feedback stabilization and bifurcation control, II. Stationary bifurcation, *Syst. Control Lett.*, 8, 467–473, 1987.

[3] Abed E.H. and Wang H.O., Feedback control of bifurcation and chaos in dynamical systems, in *Nonlinear Dynamics and Stochastic Mechanics*, W. Kliemann and N. Sri Namachchivaya, Eds., CRC Press, Boca Raton, 1995, 153–173.

[4] Abed E.H., Wang H.O., and Chen R.C., Stabilization of period doubling bifurcations and implications for control of chaos, *Physica D*, 70(1-2), 154–164, 1994.

[5] Anderson P.M. and Fouad A.A., *Power System Control and Stability*, Iowa State University Press, Ames, 1977.

[6] Andronov A.A., Vitt, A.A., and Khaikin S.E., *Theory of Oscillators*, Pergamon Press, Oxford, 1966 (reprinted by Dover, New York, 1987), English translation of Second Russian Edition; Original Russian edition published in 1937.

[7] Blakelock J.H., *Automatic Control of Aircraft and Missiles*, 2nd ed., John Wiley & Sons, New York, 1991.

[8] Chapman, G.T., Yates, L.A., and Szady, M.J., Atmospheric flight dynamics and chaos: Some issues in modeling and dimensionality, in *Applied Chaos*, J.H. Kim and J. Stringer, Eds., John Wiley & Sons, New York, 1992, 87–141.

[9] Chen, G. and Dong, X., From chaos to order — Perspectives and methodologies in controlling chaotic nonlinear dynamical systems, *Internat. J. Bifurcation Chaos*, 3, 1363–1409, 1993.

[10] Chow, S.N. and Hale, J.K., *Methods of Bifurcation Theory*, Springer-Verlag, New York, 1982.

[11] Devaney, R.L., *An Introduction to Chaotic Dynamical Systems*, 2nd ed., Addison-Wesley, Redwood City, CA, 1989.

[12] Dobson, I., Glavitsch, H., Liu, C.-C., Tamura, Y., and Vu, K., Voltage collapse in power systems, *IEEE Circuits and Devices Magazine*, 8(3), 40–45, 1992.

[13] Dole, C.E., *Flight Theory and Aerodynamics: A Practical Guide for Operational Safety*, John Wiley & Sons, New York, 1981.

[14] Dorning, J.J. and Kim, J.H., Bridging the gap between the science of chaos and its technological applications, in *Applied Chaos*, J.H. Kim and J. Stringer, Eds., John Wiley & Sons, New York, 1992, 3–30.

[15] Etkin, B., *Dynamics of Atmospheric Flight*, John Wiley & Sons, New York, 1972.

[16] Eveker, K.M., Gysling, D.L., Nett, C.N., and Sharma, O.P., Integrated control of rotating stall and surge in aeroengines, in *Sensing, Actuation, and Control in Aeropropulsion*, J.D. Paduano, Ed., Proc. SPIE 2494, 1995, 21–35.

[17] Gills, Z., Iwata, C., Roy, R., Schwartz, I.B., and Triandaf, I., Tracking unstable steady states: Extending the stability regime of a multimode laser system, *Phys. Rev. Lett.*, 69, 3169-3172, 1992.

[18] Hassard, B.D., Kazarinoff, N.D., and Wan, Y.H., *Theory and Applications of Hopf Bifurcation*, Cambridge University Press, Cambridge, 1981.

[19] Jackson, E.A., *Perspectives of Nonlinear Dynamics*, Vols. 1 and 2, Cambridge University Press, Cambridge, 1991.

[20] Kerrebrock, J.L., *Aircraft Engines and Gas Turbines*, 2nd ed., MIT Press, Cambridge, MA, 1992.

[21] Kim, J.H. and Stringer, J., Eds., *Applied Chaos*, John Wiley & Sons, New York, 1992.

[22] Lee, H.C., *Robust Control of Bifurcating Nonlinear Systems with Applications*, Ph.D. Dissertation, Department of Electrical Engineering, University of Maryland, College Park, 1991.

[23] Lee, H.-C. and Abed, E.H., Washout filters in the bifurcation control of high alpha flight dynamics, *Proc. 1991 Am. Control Conf.*, Boston, pp. 206–211, 1991.

[24] Liaw, C.Y. and Bishop, S.R., Nonlinear heave-roll coupling and ship rolling, *Nonlinear Dynamics*, 8, 197–211, 1995.

[25] Liaw, D.-C. and Abed, E.H., Analysis and control of rotating stall, *Proc. NOLCOS'92: Nonlinear Control System Design Symposium*, (M. Fliess, Ed.), June 1992, Bordeaux, France, pp. 88–93, Published by the International Federation of Automatic Control; See also: Active control of compressor stall inception: A bifurcation-theoretic approach, *Automatica*, 32, 1996 (to appear).

[26] Lin, C.-F., *Advanced Control Systems Design*, Prentice Hall Series in Advanced Navigation, Guidance, and Control, and their Applications, Prentice Hall, Englewood Cliffs, NJ, 1994.

[27] Lorenz, E.N., Deterministic nonperiodic flow, *J. Atmosph. Sci.*, 20, 130–141, 1963.

[28] Marsden, J.E. and McCracken, M., *The Hopf Bifurcation and Its Applications*, Springer-Verlag, New York, 1976.

[29] McRuer, D., Ashkenas, I., and Graham, D., *Aircraft Dynamics and Automatic Control*, Princeton University Press, Princeton, 1973.

[30] Mees, A.I., *Dynamics of Feedback Systems*, John Wiley & Sons, New York, 1981.

[31] Murray, J.D., *Mathematical Biology*, Springer-Verlag, New York, 1990.

[32] Nayfeh, A.H. and Balachandran, B., *Applied Nonlinear Dynamics: Analytical, Computational, and Experimental Methods*, Wiley Series in Nonlinear Science, John Wiley & Sons, New York, 1995.

[33] Ott, E., Grebogi, C., and Yorke, J.A., Controlling chaos, *Phys. Rev. Lett.*, 64, 1196–1199, 1990.

[34] Ott, E., *Chaos in Dynamical Systems*, Cambridge University Press, Cambridge, 1993.

[35] Ott, E., Sauer, T., and Yorke, J.A., Eds., *Coping with Chaos: Analysis of Chaotic Data and the Exploitation of Chaotic Systems*, Wiley Series in Nonlinear Science, John Wiley & Sons, New York, 1994.

[36] Pecora, L.M., and T.L. Carroll, Synchronization in chaotic systems, *Phys. Rev. Lett.*, 64, 821–824, 1990.

[37] Poincaré, H., *New Methods of Celestial Mechanics*, Parts 1, 2, and 3 (Edited and introduced by D.L. Goroff), Vol. 13 of the History of Modern Physics and Astronomy Series, American Institute of Physics, U.S., 1993; English translation of the French edition *Les Méthodes nouvelles de la Mécanique céleste*, originally published during 1892–1899.

[38] Romeiras, F.J., Grebogi, C., Ott, E., and Dayawansa, W.P., Controlling chaotic dynamical systems, *Physica D*, 58, 165–192, 1992.

[39] Roskam, J., *Airplane Flight Dynamics and Automatic Flight Controls (Part II)*, Roskam Aviation and Engineering Corp., Lawrence, Kansas, 1979.

[40] Sanchez, N.E. and Nayfeh, A.H., Nonlinear rolling motions of ships in longitudinal waves, *Internat. Shipbuilding Progress*, 37, 247–272, 1990.

[41] Sauer, P.W. and Pai, M.A., *Power System Dynamics and Stability*, Draft manuscript, Department of Electrical and Computer Engineering, University of Illinois at Urbana-Champaign, 1995.

[42] Scott, S.K., *Chemical Chaos*, International Series of Monographs on Chemistry, Oxford University Press, Oxford, 1991.

[43] Seydel, R., *Practical Bifurcation and Stability Analysis: From Equilibrium to Chaos*, 2nd ed., Springer-Verlag, Berlin, 1994.

[44] Shinbrot, T., Grebogi, C., Ott, E., and Yorke, J.A., Using small perturbations to control chaos, *Nature*, 363, 411–417, 1993.

[45] Stevens, B.L. and Lewis, F.L., *Aircraft Control and Simulation*, John Wiley & Sons, New York, 1992.

[46] Strogatz, S.H., *Nonlinear Dynamics and Chaos: With Applications to Physics, Biology, Chemistry, and Engineering*, Addison-Wesley, Reading, MA, 1994.

[47] Tesi, A., Abed, E.H., Genesio, R., and Wang, H.O., Harmonic balance analysis of period doubling bifurcations with implications for control of nonlinear dynamics, Report No. RT 11/95, Dipartimento di Sistemi e Informatica, Università di Firenze, Firenze, Italy, 1995 (submitted for publication).

[48] Thompson, J.M.T. and Stewart, H.B., *Nonlinear Dynamics and Chaos*, John Wiley & Sons, Chichester, U.K., 1986.

[49] Thompson, J.M.T., Stewart, H.B., and Ueda, Y., Safe, explosive, and dangerous bifurcations in dissipative dynamical systems, *Phys. Rev. E*, 49, 1019–1027, 1994.

[50] Wang, H.O. and Abed, E.H., Bifurcation control of a chaotic system, *Automatica*, 31, 1995, in press.

[51] Wegener, P.P., *What Makes Airplanes Fly?: History, Science, and Applications of Aerodynamics*, Springer-Verlag, New York, 1991.

[52] Ziegler, F., *Mechanics of Solids and Fluids*, Springer-Verlag, New York, 1991.

Further Reading

Detailed discussions of bifurcation and chaos are available in many excellent books (e.g., [6, 10, 11, 18, 19, 28, 30, 32, 34, 35, 43, 46, 48]). These books also discuss a variety of interesting applications. Many examples of bifurcations in mechanical systems are given in [52]. There are also several journals devoted to bifurcations and chaos. Of particular relevance to engineers are *Nonlinear Dynamics*, the *International Journal of Bifurcation and Chaos*, and the journal *Chaos, Solitons and Fractals*.

The book [45] discusses feedforward control in the context of "trimming" an aircraft using its nonlinear equations of motion and the available controls. Bifurcation and chaos in flight dynamics are discussed in [8]. Lucid explanations on specific uses of washout filters in aircraft control systems are given in [15, pp. 474–475], [7, pp. 144–146], [39, pp. 946–948 and pp. 1087–1095], and [45, pp. 243–246 and p. 276]. The book [31] discussed applications of nonlinear dynamics in biology and population dynamics.

Computational issues related to bifurcation analysis are addressed in [43]. Classification of bifurcations as safe or dangerous is discussed in [32, 43, 48, 49].

The edited book [14] contains interesting articles on research needs in applications of bifurcations and chaos.

The article [3] contains a large number of references on bifurcation control, related work on stabilization, and applications of these techniques. The review papers [9, 44] address control of chaos methods. In particular, [44] includes a discussion of use of sensitive dependence on initial conditions to direct trajectories to targets. The book [35] includes articles on control of chaos, detection of chaos in time series, chaotic data analysis, and potential applications of chaos in communication systems. The book [32] also contains discussions of control of bifurcations and chaos, and of analysis of chaotic data.

57.7 Open-Loop Control Using Oscillatory Inputs

J. Baillieul, *Boston University*[6]
B. Lehman, *Northeastern University*[7]

57.7.1 Introduction

The interesting discovery that the topmost equilibrium of a pendulum can be stabilized by oscillatory vertical movement of the suspension point has been attributed to Bogolyubov [11] and Kapitsa [26], who published papers on this subject in 1950 and 1951, respectively. In the intervening years, literature appeared analyzing the dynamics of systems with oscillatory forcing, e.g., [31]. Control designs based on oscillatory inputs have been proposed (for instance [8] and [9]) for a number of applications. Many classical results on the stability of operating points for systems with oscillatory inputs depend on the eigenvalues of the averaged system lying in the left half-plane. Recently, there has been interest in the stabilization of systems to which such classical results do not apply. Coron [20], for instance, has shown the existence of a time-varying feedback stabilizer for systems whose averaged versions have eigenvalues on the imaginary axis. This design is interesting because it provides smooth feedback stabilization for systems which Brockett [15] had previously shown were never stabilizable by smooth, time-invariant feedback. For conservative mechanical systems with oscillatory control inputs, Baillieul [7] has shown that stability of operating points may be assessed in terms of an energy-like quantity known as the *averaged potential.* Control designs with objectives beyond stabilization have been studied in path-planning for mobile robots [40] and in other applications where the models result in "drift-free" controlled differential equations. Work by Sussmann and Liu [36]–[38], extending earlier ideas of Haynes and Hermes, [21], has shown that, for drift-free systems satisfying a certain Lie algebra rank condition (LARC discussed in Section 57.7.2), arbitrary smooth trajectories may be interpolated to an arbitrary accuracy by appropriate choice of oscillatory controls. Leonard and Krishnaprasad [28] have reported algorithms for generating desired trajectories when certain "depth" conditions on the brackets of the defining vector fields are satisfied.

This chapter summarizes the current theory of open-loop control using oscillatory forcing. The recent literature has emphasized geometric aspects of the methods, and our discussion in Sections 57.7.2 and 57.7.3 will reflect this emphasis. Open-loop methods are quite appealing in applications in which the realtime sensor measurements needed for feedback designs are expensive or difficult to obtain. Because the methods work by virtue of the geometry of the motions in the systems, the observed effects may be quite robust. This is borne out by experiments described below. The organization of the article is as follows. In the present section we introduce oscillatory open-loop control laws in two very different ways. Example 57.9 illustrates the geometric mechanism through which oscillatory forcing produces nonlinear behavior in certain types of (drift-free) systems. Following this, the remainder of the section introduces a more classical analytical approach to control systems with oscillatory inputs. Section 57.7.2 provides a detailed exposition of open loop design methods for so-called "drift-free" systems. The principal applications are in kinematic motion control, and the section concludes with an application to grasp mechanics. Section 57.7.3 discusses some geometric results of oscillatory forcing for stabilization. Examples have been chosen to illustrate different aspects of the theory.

Noncommuting Vector Fields, Anholonomy, and the Effect of Oscillatory Inputs

We begin by describing a fundamental mathematical mechanism for synthesizing motions in a controlled dynamical system using oscillatory forcing. We shall distinguish among three *classes* of systems:

I. Drift-free systems with input entering linearly:

$$\dot{x} = \sum_{i=1}^{m} u_i g_i(x). \tag{57.207}$$

Here we assume each $g_i : \mathbf{R}^n \to \mathbf{R}^n$ is a smooth (i.e. analytic) vector field, and each "input" $u_i(\cdot)$ is a piecewise analytic function of time. Generally, we assume $m < n$.

II. Systems with drift and input entering affinely:

$$\dot{x} = f(x) + \sum_{i=1}^{m} u_i g_i(x). \tag{57.208}$$

The assumptions here are, as in the previous case, with $f : \mathbf{R}^n \to \mathbf{R}^n$ also assumed to be smooth.

III. Systems with no particular restriction on the way in which the control enters:

$$\dot{x} = f(x, u). \tag{57.209}$$

Here $u = (u_1, \ldots, u_m)^T$ is a vector of piecewise analytic inputs, as in the previous two cases, and $f : \mathbf{R}^n \times \mathbf{R}^m \to \mathbf{R}^n$ is analytic.

This is a hierarchy of types, each a special case of its successor in the list. More general systems could be considered, and indeed, in the Lagrangian models which are described in Section 57.7.3, we shall encounter systems in which the derivatives of inputs also enter the equations of motion. It will be shown that these systems can be reduced to Class III, however.

REMARK 57.10 The extra term on the right hand side of Equation 57.208 is called a "drift" because, in the absence of control

[6]The first author gratefully acknowledges support of the U.S. Air Force Office of Scientific Research under grant AFOSR-90-0226.
[7]The second author gratefully acknowledges support of an NSF Presidential Faculty Fellow Award, NSF CMS-9453473.

input, the differential equation "drifts" in the direction of the vector field f.

EXAMPLE 57.9:

Even Class I systems possess the essential features of the general mechanism (anholonomy) by which oscillatory inputs may be used to synthesize desired motions robustly. (See remarks below on the robustness of open-loop methods.) Consider the simple and widely studied "Heisenberg" system (see [16]):

$$\begin{pmatrix} \dot{x}_1 \\ \dot{x}_2 \\ \dot{x}_3 \end{pmatrix} = \begin{pmatrix} u_1(t) \\ u_2(t) \\ u_2(t)x_1 - u_1(t)x_2 \end{pmatrix}. \tag{57.210}$$

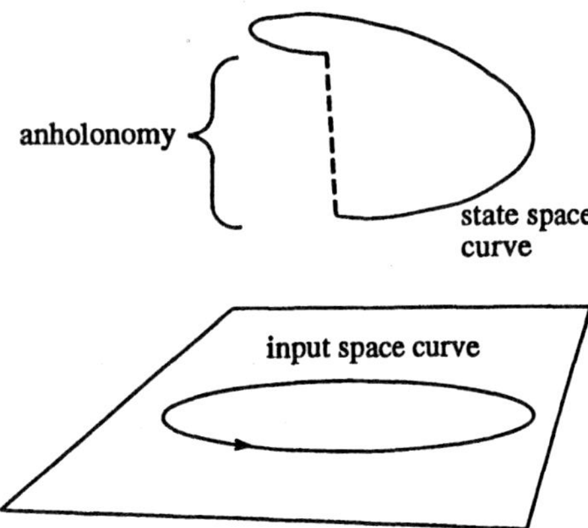

Figure 57.21 The anholonomy present in the Heisenberg system is depicted in a typical situation in which the input variables trace a closed curve and the state variables trace a curve which does not close. The distance between endpoints of the state space curve (measured as the length of the dashed vertical line) reflects the anholonomy in the system.

This system is a special case of Equation 57.207 in which $m = 2$ and

$$g_1(x) = \begin{pmatrix} 1 \\ 0 \\ -x_2 \end{pmatrix}, \quad g_2(x) = \begin{pmatrix} 0 \\ 1 \\ x_1 \end{pmatrix}.$$

If we define the *Lie bracket* of these vector fields by $[g_1, g_2] = \frac{\partial g_1}{\partial x} g_2 - \frac{\partial g_2}{\partial x} g_1$, then a simple calculation reveals $[g_1, g_2] = (0, 0, -2)^T$. Another fairly straightforward calculation shows that, from general considerations, there is a choice of inputs $(u_1(\cdot), u_2(\cdot))$ which generates a trajectory pointing approximately in the direction $(0, 0, 1)^T$, and this approximation may be made arbitrarily accurate (see, Nijmeijer and Van der Schaft, [30], p. 77 or Bishop and Crittenden, [10], p. 18.). In the present case, we can be more precise and more explicit. Starting at the origin, $(x_1, x_2, x_3) = (0, 0, 0)$, motion in any of the three coordinate directions is possible. By choosing $u_1(t) \equiv 1,\ u_2(t) \equiv 0$, for instance, motion along the x_1-axis is produced, and motion along the x_2-axis may similarly be produced by reversing the role of $u_1(\cdot)$ and $u_2(\cdot)$. Motion along the x_3-axis is more subtle. If we let the inputs, $(u_1(\cdot), u_2(\cdot))$, trace a closed curve so that the states x_1 and x_2 end with the same values with which they began, the net motion of the system is along the x_3-axis. This is illustrated in Figure 57.21. Brockett [17] has observed that the precise shape of the input curve is unimportant, but the x_3-distance is twice the (signed) area circumscribed by the (x_1, x_2)-curve. For this simple system, we thus have a way to prescribe trajectories between any two points in $\mathbf{R}^3$. Taking the case of trajectories starting at the origin, for instance, we may specify a trajectory passing through any other point (x, y, z) at time $t = 1$ by finding a circular arc initiating at the origin in (x_1, x_2)-space with appropriate length, initial direction, and curvature. This construction of inputs may be carried out somewhat more generally, as discussed below in Section 57.7.2. Ideas along this line are also treated in more depth in [28].

The geometric mechanism by which motion is produced by oscillatory forcing is fairly transparent in the case of the Heisenberg system. For systems of the form of Equation 57.208 and especially of the form of Equation 57.209, there is no comparably complete geometric theory. Indeed, much of the literature on such systems makes no mention of geometry. A brief survey/overview of some of the classical literature on such systems is given next. We shall return to the geometric point of view in Sections 57.7.2 and 57.7.3.

Oscillatory Inputs to Control Physical Systems

The idea of using oscillatory forcing to control physical processes is not new. Prompted by work in the 1960s on the periodic optimal control of chemical processes, Speyer and Evans [33] derived a sufficiency condition for a periodic process to minimize a certain integral performance criterion. This approach also led to the observation that periodic paths could be used to improve aircraft fuel economy (see [32].). Previously cited work of Bogolyubov and Kapitsa [11] and [26], led Bellman et al. [8] and [9] to investigate the systematic use of *vibrational control* as an open loop control technique in which zero average oscillations are introduced into a system's parameters to achieve a dynamic response (such as stabilizing effects). For example, it has been shown in [8],[9], [24], and [25], that the oscillation of flow rates in a continuous stirred tank reactor allows operating exothermic reactions at (average) yields which were previously unstable. Similarly, [6] and [7] describe the general geometric mechanism by which oscillations along the vertical support of an inverted pendulum stabilize the upper equilibrium point.

This section treats the basic theory of vibrational control introduced in [8] and [9]. The techniques are primarily based on [8], with additional material taken from [24] and [25]. In particular, in [24] and [25], the technique of vibrational control has been extended to delay differential equations.

Problem Statement

Consider the nonlinear differential equation (Class III)

$$\frac{dx}{dt} = f(x, u), \tag{57.211}$$

where $f : \mathbf{R}^n \times \mathbf{R}^d \to \mathbf{R}^n$ is continuously differentiable, $x \in \mathbf{R}^n$ is the state, and $u = (u_1, \ldots, u_d)^T$ is a vector of control inputs assumed to be piecewise analytic functions of time. These are the quantities which we can directly cause to vibrate.

Introduce into Equation 57.211 oscillatory inputs according to the law $u(t) = \lambda_0 + \gamma(t)$ where λ_0 is a constant vector and $\gamma(t)$ is a *periodic average zero* (**PAZ**) vector. (For simplicity, $\gamma(t)$ has been assumed periodic. However, the following discussion can be extended to the case where $\gamma(t)$ is an almost periodic zero average vector [8],[9], [24], and [25].) Then Equation 57.211 becomes

$$\frac{dx}{dt} = f(x, \lambda_0 + \gamma(t)). \tag{57.212}$$

Assume that Equation 57.211 has a fixed equilibrium point $x_s = x_s(\lambda_0)$ for fixed $u(t) = \lambda_0$.

DEFINITION 57.2 An equilibrium point $x_s(\lambda_0)$ of Equation 57.211 is said to be *vibrationally stabilizable* if, for any $\delta > 0$, there exists a PAZ vector $\gamma(t)$ such that Equation 57.212 has an asymptotically stable periodic solution, $x^*(t)$, characterized by

$$\| \bar{x}^* - x_s(\lambda_0) \| \leq \delta, \text{ where } \bar{x}^* = \frac{1}{T}\int_0^T x^*(t)dt.$$

It is often preferable that Equations 57.211 and 57.212 have the same fixed equilibrium point, $x_s(\lambda_0)$. However, this is not usually the case because the right hand side of Equation 57.212 is time varying and periodic. Therefore, the technique of vibrational stabilization is to determine vibrations $\gamma(t)$ so that the (possibly unstable) equilibrium point $x_s(\lambda_0)$ bifurcates into a stable periodic solution whose average is close to $x_s(\lambda_0)$. The engineering aspects of the problem consist of 1) finding conditions for the existence of stabilizing vibrations, 2) determining which oscillatory inputs, $u(\cdot)$, are physically realizable, and 3) determining the shape (waveform type, amplitude, phase) of the oscillations which will insure the desired response. In Section 57.7.3, we shall present an example showing how oscillatory forcing induces interesting stable motion in neighborhoods of points which are not close to equilibria of the time-varying system of Equation 57.212.

Vibrational Stabilization

It is frequently assumed that Equation 57.212 can be decomposed as

$$\frac{dx}{dt} = f_1(x(t)) + f_2(x(t), \gamma(t)), \tag{57.213}$$

where λ_0 and $\gamma(\cdot)$ are as above and where $f_1(x(t)) = f_1(\lambda_0, x(t))$ and the function $f_2(x(t), \gamma(t))$ is linear with respect to its second argument. Systems for which this assumption does not hold are discussed in Section 57.7.3. For simplicity only, assume that f_1 and f_2 are analytic functions. Additionally, assume that $\gamma(t)$, the control, is periodic of period T $(0 < T \ll 1)$ and in the form, $\gamma(t) = \omega\hat{u}(\omega t)$, where $\omega = \frac{2\pi}{T}$, and $\hat{u}(\cdot)$ is some fixed period-2π function (e.g., sin or cos). We write $\gamma(\cdot)$ in this way because, although the theory is not heavily dependent on the exact shape of the waveform of the periodic input, there is a crucial dependence on the simultaneous scaling of the frequency and amplitude. Because we are usually interested in high frequency behavior, this usually implies that the amplitude of $\gamma(t)$ is large. It is possible, however, that $\hat{u}(\cdot)$ has small amplitude, making the amplitude of $\gamma(t)$ small also.

Under these assumptions, Equation 57.213 can be rewritten as

$$\frac{dx}{dt} = f_1(x(t)) + \omega f_2(x(t), \hat{u}(\omega t)). \tag{57.214}$$

To proceed with the stability analysis, Equation 57.214 will be transformed to an ordinary differential equation in "standard" form ($\frac{dx}{dt} = \epsilon f(x, t)$) so that the method of averaging can be applied (see [11] and [24]). This allows the stability properties of the time varying system Equation 57.214 to be related to the stability properties of a simpler autonomous differential equation (the averaged equation). To make this transformation, consider the so-called "generating equation"

$$\frac{dx}{dt} = f_2(x(t), \hat{u}(t)).$$

Suppose that this generating equation has a T-periodic general solution $h(t, c)$, for some $\hat{u}(\cdot)$ and $t \geq t_0$, where $h : \mathbf{R} \times \mathbf{R}^n \to \mathbf{R}^n$ and $c \in \mathbf{R}^n$ is uniquely defined for every initial condition $x(t_0) \in \Omega \subset \mathbf{R}^n$.

Introduce into Equation 57.214 the Lyapunov substitution $x(t) = h(\omega t, q(t))$ to obtain an equation for $q(\cdot)$:

$$\frac{dq}{dt} = [\frac{\partial h(\omega t, q(t))}{\partial q}]^{-1} f_1(h(\omega t, q(t))),$$

which, in slow time $\tau = \omega t$, with $z(\tau) = q(t)$ and $\epsilon = \frac{1}{\omega}$, becomes

$$\frac{dz}{d\tau} = \epsilon[\frac{\partial h[\tau, z(\tau)]}{\partial z}]^{-1} f_1(h(\tau, z(\tau))). \tag{57.215}$$

Equation 57.215 is a periodic differential equation in "standard" form and averaging can be applied. If T denotes the period of the right hand side of Equation 57.215, then the averaged equation (autonomous) corresponding to Equation 57.215 is given as

$$\begin{aligned} \frac{dy}{d\tau} &= \epsilon\bar{Y}(y(\tau)); \text{ where} \\ \bar{Y}(c) &= \frac{1}{T}\int_0^T [\frac{\partial h(\tau, c)}{\partial z}]^{-1} f_1(h(\tau, c))\, d\tau. \end{aligned} \tag{57.216}$$

By the theory of averaging, it is known that an $\epsilon_0 > 0$ exists such that for $0 < \epsilon \leq \epsilon_0$, the hyperbolic stability properties of Equation 57.215 and Equation 57.216 are the same. Specifically, if y_s is an asymptotically stable equilibrium point of Equation 57.216, this implies that, for $0 < \epsilon \leq \epsilon_0$, a unique periodic solution, $z^*(\tau)$ of Equation 57.215 exists, in the vicinity of y_s that is asymptotically stable also. Since the transformation $x(t) = h(\omega t, q(t))$ is a homeomorphism, there will exist an asymptotically stable, periodic, solution to Equation 57.214

given by $x^*(t) = h(\omega t, z^*(\omega t))$ (converting back to fast time using the fact that $q(t) = z(\omega t)$, where $z(\cdot)$ is the solution to Equation 57.215). Using Definition 57.2, Equation 57.211 is said to be *vibrationally stabilized* provided that $\overline{x}^* = \frac{1}{T}\int_0^T x^*(t)dt$ remains in the vicinity of $x_s(\lambda_0)$. This can be formalized by the following theorem given in [8] and [24]:

THEOREM 57.8 *Assume that Equation 57.212 with $\gamma(t) = \omega\hat{u}(\omega t)$ has the form of Equation 57.214, with f_1 and f_2 analytic. Assume, also, that $h(t, c)$ is periodic and that the function $[\frac{\partial h(\tau, z(\tau))}{\partial z}]^{-1} f_1(h(\tau, z(\tau)))$ in Equation 57.214 is continuously differentiable with respect to $z \in \Omega \subset \mathbf{R}^n$.*

Then the equilibrium point $x_s(\lambda_0)$ of Equation 57.211 is vibrationally stabilizable if a $\hat{u}(t)$ exists such that Equation 57.216 has an asymptotically stable equilibrium point characterized by $\frac{1}{T}\int_0^T h(\tau, y_s)d\tau = x_s$.

The technique of vibrational control now becomes clearer. Introduce open loop oscillatory forcing into Equation 57.211, $u(t) = \lambda_0 + \omega\hat{u}(\omega t)$, such that Equation 57.211 is in the form of Equation 57.214. Transform Equation 57.214 into Equation 57.215 and study the stability properties of the corresponding average of Equation 57.215, given by Equation 57.216. Then determine parameters of $\hat{u}$ (phase, amplitude and frequency) such that Equation 57.216 has an asymptotically stable equilibrium point y_s. If $\frac{1}{T}\int_0^T h(\tau, y_s)d\tau = x_s$, then the system is vibrationally stabilizable.

The procedure of vibrational stabilization described above is trial and error. Vibrations are inserted into a system until vibrations are found which give the desired response. However, if specific classes of vibrations are analyzed, explicit algorithms are known which explain the size and location of vibration needed to give specified responses (see [8], [9], and [24]). For example, it is common to assume that the vibrations are in *linear multiplicative* form, $f_2(x(t), \gamma(t)) = B(t)x(t)$ or *vector additive* form, $f_2(x(t), \gamma(t)) = L(t)$. In each of these cases, sufficient conditions (sometimes necessary and sufficient) are known for vibrational stabilization of systems. We return to linear multiplicative vibrations in Section 57.7.3, using a more geometric framework.

Finally, it should be noted that [8] and [24] can relate the transient behavior of Equation 57.216 to the "average" transient behavior of Equation 57.213 through the estimation $x(t) \approx h(\omega t, y(\omega t))$, where $y(\tau)$ is the solution of Equation 57.216 in slow time. Hence, the technique of vibrational control has the ability to both stabilize a system and control transient response issues.

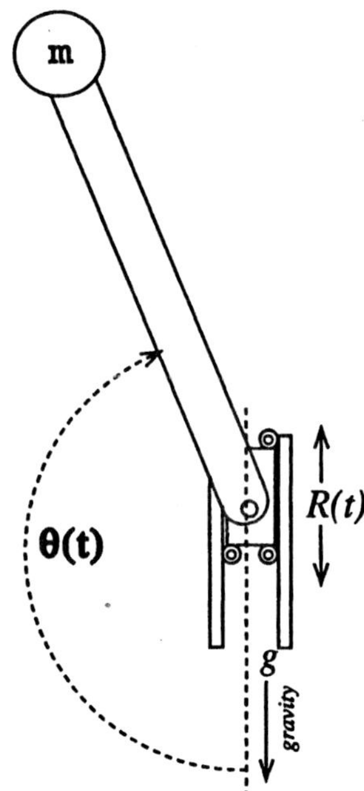

Figure 57.22 A simple pendulum whose hinge point undergoes vertical motion.

EXAMPLE 57.10: Oscillatory stabilization of a simple pendulum.

A classical example to which the theory applies involves stabilizing the upright equilibrium of a simple pendulum by the forced vertical oscillation of the pendulum's hinge point. Consider a simple pendulum consisting of a massless but rigid link of length ℓ to which a tip of mass m is attached. Suppose the hinge point of the pendulum undergoes vertical motion, and is located at time t at vertical position $R(t)$ with respect to some reference coordinate (see Figure 57.22.). Taking into account motion in this variable and the friction coefficient at the hinge ($b > 0$), the pendulum dynamics may be written as

$$\ell\ddot{\theta} + (b/m)\dot{\theta} + \ddot{R}\sin\theta + g\sin\theta = 0.$$

Consider the simple sinusoidal oscillation of the hinge-point, $R(t) = \alpha\sin\beta t$. Then $\ddot{R}(t) = -\eta\beta\sin\beta t$, where $\eta = \eta(\beta) = \alpha\beta$. Letting $x_1 = \theta$ and $x_2 = \dot{\theta}$, the system can be placed in the form of Equation 57.214. The corresponding generating equation is given as $\dot{x}_1 = 0$ and $\dot{x}_2 = (\eta/\ell)\sin t \sin x_1$, which has solution $x_1 = c_1 = h_1(t, c)$ and $x_2 = -(\eta/\ell)\cos t \sin c_1 + c_2 = h_2(t, c)$. Introducing the transformation $x_1 = z_1$ and $x_2 = -(\eta/\ell)\cos\beta t \sin z_1 + z_2$, letting $\tau = \beta t$, and letting $\epsilon = 1/\beta$, Equation 57.215 specializes to

$$\begin{aligned} \dot{z}_1 &= \epsilon[-(\eta/\ell)\cos\tau\sin z_1 + z_2] \\ \dot{z}_2 &= \epsilon[-(\eta^2/\ell^2)\cos^2\tau\cos z_1\sin z_1 \\ &\quad - (g/\ell)\sin z_1 + (\eta/\ell)z_2\cos\tau\cos z_1 \\ &\quad + (\eta b/m\ell^2)\cos\tau\sin z_1 - (b/m\ell)z_2]. \end{aligned}$$

Therefore the averaged equations are $\dot{y}_1 = \epsilon y_2$ and $\dot{y}_2 = \epsilon[-\frac{\eta^2}{2\ell^2}\cos y_1 \sin y_1 - (g/\ell)\sin y_1 - (b/m\ell)y_2]$. The upper equilibrium point in the averaged equation has been preserved and $x_s = \frac{1}{T}\int_0^T h(\tau, y_s)d\tau$. Hence, by the above theorem, if the vertical equilibrium point is asymptotically stable for the averaged equation, then for sufficiently large β the inverted pendulum with oscillatory control has an asymptotically stable periodic orbit vibrating vertically about the point $\theta = \pi$. A simple linearization of the averaged equation reveals that its upper equilibrium point is asymptotically stable when $\alpha^2\beta^2 > 2g\ell$. Under these conditions, the upper equilibrium point of the inverted pendulum is said to be *vibrationally stabilized.*

57.7.2 The Constructive Controllability of Drift-Free (Class I) Systems

Class I control systems arise naturally as kinematic models of mechanical systems. In this section, we outline the current theory of motion control for such systems, emphasizing the geometric mechanism (anholonomy) through which oscillatory inputs to Equation 57.207 produce motions of the state variables. Explicit results along the lines given in Section 57.7.1 for the Heisenberg system have been obtained in a variety of settings, and some recent work will be discussed below. The question of when such explicit constructions are possible more generally for Class I systems does not yet have a complete answer. Thus we shall also discuss computational approaches that yield useful approximate solutions. After briefly outlining the state of current theory, we conclude the section with an example involving motion specification for a ball "grasped" between two plates.

The recent literature treating control problems for such systems suggests that it is useful distinguishing between two control design problems:

P1: The *prescribed endpoint steering problem* requires that, given any pair of points $x_0, x_1 \in \mathbf{R}^n$, a vector of piecewise analytic control inputs $u(\cdot) = (u_1(\cdot), \ldots, u_m(\cdot))$ is to be determined to steer the state of Equation 57.207 from x_0 at time $t = 0$ to x_1 at time $t = T > 0$.

P2: The *trajectory approximation steering problem* requires that, given any sufficiently "regular" curve $\gamma : [0, T] \to \mathbf{R}^n$, we determine a sequence $\{u^j(\cdot)\}$ of control input vectors such that the corresponding sequence of trajectories of Equation 57.207 converges (uniformly) to γ.

A general solution to either of these problems requires that a certain *Lie algebra rank condition* (LARC) be satisfied. More specifically, with the *Lie bracket* of vector fields defined as in Section 57.7.1, define a set of vector fields

$$\mathcal{C} = \{\xi \,:\, \xi = [\xi_j, [\xi_{j-1}, [\ldots, [\xi_1, \xi_0]\ldots]]];\\ \xi_i \in \{g_1, \ldots, g_m\}, i = 1, \ldots, j;\ j = 1, \ldots, \infty\}.$$

Then $\mathcal{L} = \text{span}(\mathcal{C})$ (= the set of all linear combinations of elements of $\mathcal{C}$) is called the *Lie algebra* generated by $\{g_1, \ldots, g_m\}$. We say Equation 57.207 (or equivalently the set of vector fields $\{g_1, \ldots, g_m\}$) satisfies the *Lie algebra rank condition* on $\mathbf{R}^n$ if $\mathcal{L}$ spans $\mathbf{R}^n$ at each point of $\mathbf{R}^n$. The following result is fundamental because it characterizes those systems for which the steering problems may, in principle, be solved.

THEOREM 57.9 *A drift-free system Equation* 57.207 *is* completely controllable *in the sense that, given any $T > 0$ and any pair of points $x_0, x_1 \in \mathbf{R}^n$, there is a vector of inputs $u = (u_1, \ldots, u_m)$ which are piecewise analytic on $[0, T]$ and which steer the system from x_0 to x_1 in T units of time if, and only if, the system satisfies the Lie algebra rank condition.*

As stated, this theorem is essentially due to W.L. Chow [19], but it has been refined and tailored to control theory by others ([13] and [35]). The various versions of this theorem in the literature have all been nonconstructive. Methods for the explicit determination of optimal (open-loop) control laws for steering Class I systems between prescribed endpoints have appeared in [1], [2], [16], and [18]. The common features in all this work are illustrated by the following:

A model nonlinear optimal control problem with three states and two inputs: Find controls $u_1(\cdot), u_2(\cdot)$ which steer the system

$$\begin{pmatrix} \dot{x}_1 \\ \dot{x}_2 \\ \dot{x}_3 \end{pmatrix} = \begin{pmatrix} 0 & 0 & u_2 \\ 0 & 0 & -u_1 \\ -u_2 & u_1 & 0 \end{pmatrix} \begin{pmatrix} x_1 \\ x_2 \\ x_3 \end{pmatrix} \tag{57.217}$$

between prescribed endpoints to minimize the cost criterion

$$\eta = \int_0^1 u_1^2 + u_2^2\, dt.$$

Several comments regarding the geometry of this problem are in order. First, an appropriate version of Chow's theorem shows that Equation 57.217 is controllable on any 2-sphere, $\mathcal{S} = \{x \in \mathbf{R}^3 : \|x\| = r\}$ for some $r > 0$, centered at the origin in $\mathbf{R}^3$. Hence, the optimization problem is well-posed precisely when the prescribed endpoints $x_0, x_1 \in \mathbf{R}^3$ satisfy $\|x_0\| = \|x_1\|$. Second, the problem may be interpreted physically as seeking minimum length paths on a sphere in which only motions composed of rotations about the x-axis (associated with input u_1) and y-axis (associated with u_2) are admissible. General methods for solving this type of problem appear in [1] and [2]. Specifically, in the first author's 1975 Ph.D. thesis (see reference in [1] and [2]), it was shown that the optimal inputs have the form, $u_1(t) = \mu \sin(\omega t + \phi)$, $u_2(t) = \mu \cos(\omega t + \phi)$. The optimal inputs depend on three parameters reflecting the fact that the set (group) of rotations of the 2-sphere is three dimensional. The details for determining the values of the parameters μ, ω, and ϕ in terms of the end points x_0 and x_1 are given in the thesis cited in [1] and [2].

The general nonlinear quadratic optimal control problem of steering Equation 57.207 to minimize a cost of the form $\int_0^1 \|u\|^2\, dt$ has not yet been solved in such an explicit fashion. The general classes of problems which have been discussed in [1], [2], [16], and [18] are associated with certain details of structure in the set of vector fields $\{g_1, \ldots, g_m\}$ and the corresponding Lie algebra $\mathcal{L}$. In [16] and [18], for example, Brockett discusses various higher dimensional versions of the Lie algebraic structure characterizing the above sphere problem and the Heisenberg system.

In addition to optimal control theory having intrinsic interest, it also points to a broader approach to synthesizing control inputs. Knowing the form of optimal trajectories, we may relax the requirement that inputs be optimal and address the simpler question of whether problems P1 and P2 may be solved using inputs with the given parametric dependence. Addressing the cases where we have noted the optimal inputs are phase-shifted sinusoids, we study the effect of varying each of the parameters. For instance, consider Equation 57.210 steered by the inputs $u_1(t) = \mu \sin(\omega t + \phi)$, $u_2(t) = \mu \cos(\omega t + \phi)$, with μ and ϕ fixed and ω allowed to vary. As ω increases, the trajectories produced become increasingly tight spirals ascending about the z-axis. One consequence of this is that, although the vectorfields

$$\begin{pmatrix} 1 \\ 0 \\ -x_2 \end{pmatrix}, \text{ and } \begin{pmatrix} 0 \\ 1 \\ x_1 \end{pmatrix}$$

are both perpendicular to the x_3-axis at the origin, pure motion in the x_3-coordinate direction may nevertheless be produced to an arbitrarily high degree of approximation. This basic example can be generalized, and extensive work on the use of oscillatory inputs for approximating arbitrary motions in Class I systems has been reported by Sussmann and Liu, [36]. The general idea is that, when a curve $\gamma(t)$ is specified, a sequence of appropriate oscillatory inputs $u^j(\cdot)$ is produced so that the corresponding trajectories of Equation 57.207 converge to $\gamma(\cdot)$ *uniformly*. The interested reader is referred to [37],[38], and the earlier work of Haynes and Hermes, [21], for further details.

Progress on problem P1 has been less extensive. Although a general constructive procedure for generating motions of Equation 57.207 which begin and end exactly at specified points, x_0 and x_1, has not yet been developed, solutions in a number of special cases have been reported. For the case of arbitrary "nilpotent" systems, Lafferriere and Sussmann, [22] and [23], provide techniques for approximating solutions for general systems. Leonard and Krishnaprasad [28] have designed algorithms for synthesizing open-loop sinusoidal control inputs for point-to-point system maneuvers where up to depth-two brackets are required to satisfy the LARC.

Brockett and Dai [18] have studied a natural subclass of nilpotent systems within which the Heisenberg system Equation 57.210 is the simplest member. The underlying geometric mechanism through which a rich class of motions in $\mathbf{R}^3$ is produced by oscillatory inputs to Equation 57.210 is also present in systems with two vector fields but higher dimensional state spaces. These systems are constructed in terms of nonintegrable p-forms in the coordinate variables x_1 and x_2. We briefly describe the procedure, referring to [18] for more details.

The number of linearly independent p-forms in x_1 and x_2 is $p + 1$. (Recall that a p-form in x_1 and x_2 is a monomial of the form $x_1^k x_2^{p-k}$. The linearly independent p-forms may be listed explicitly $\{x_1^p, x_1^{p-1}x_2, \ldots, x_2^p\}$.) Thus, there are $2(p+1)$ linearly independent expressions of the form

$$\eta = \phi(x_1, x_2)\dot{x}_1 + \psi(x_1, x_2)\dot{x}_2$$

where ϕ, ψ are homogeneous polynomials of degree p in x_1 and x_2. Within the set of such expressions, there is a set of $p + 2$ linearly independent expressions of the form $\eta = \frac{d\gamma}{dt}$, where γ is a homogeneous polynomial in x_1, x_2 of degree $p + 1$ (Such expressions are called exact differentials). There is a complementary p-dimensional family ($p = 2(p+1) - (p+2)$) of η's which are not integrable.

For example, if $p = 2$, there are 2 linearly independent nonintegrable forms η, and we may take these to be $\{x_1^2\dot{x}_2, x_2^2\dot{x}_1\}$. From these, we construct a completely controllable two-input system

$$\begin{pmatrix} \dot{x}_1 \\ \dot{x}_2 \\ \dot{x}_3 \\ \dot{x}_4 \\ \dot{x}_5 \end{pmatrix} = \begin{pmatrix} u \\ v \\ x_1 v - x_2 u \\ x_1^2 v \\ x_2^2 u \end{pmatrix}. \tag{57.218}$$

More generally, for each positive integer p we could write a completely controllable two-input system whose state space has dimension $2 + p(p + 1)/2$. Brockett and Dai [18] consider the optimal control problem of steering Equation 57.218 between prescribed endpoints $x(0)$, $x(T) \in \mathbf{R}^5$ to minimize the cost functional

$$\int_0^T u^2 + v^2 \, dt.$$

It is shown that explicit solutions may also be obtained in this case, and these are given in terms of elliptic functions.

Before treating an example problem in mechanics which makes use of these ideas, we summarize some other recent work on control synthesis for Class I systems. Whereas Sussmann and his coworkers [22], [23],[36]-[38] have used the concept of a *P. Hall basis* for free Lie algebras to develop techniques applicable in complete generality to Class I systems, an approach somewhat in the opposite direction has been pursued by S. Sastry and his co-workers [29], [39], [40]. This approach sought to characterize systems controlled by sinusoidal inputs. Motivated by problems in the kinematics of wheeled vehicles and robot grasping, they have defined a class of *chained systems* in which desired motions result from inputs which are sinusoids with integrally related frequencies. While the results are no less explicit than the Heisenberg example in Section 57.7.1, the systems themselves are very special. Conditions under which a Class I system may be converted to chained form are given in [29].

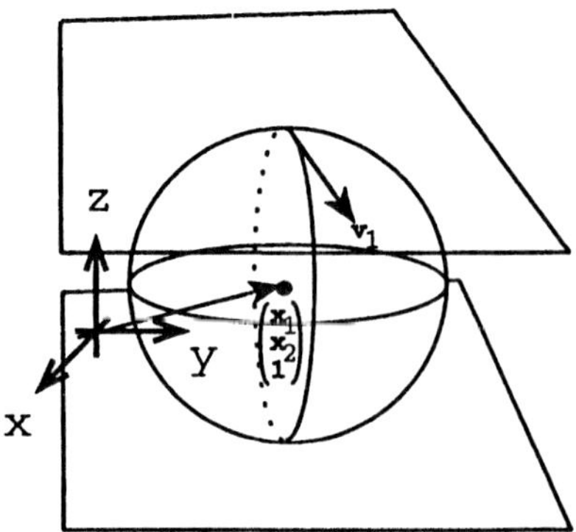

Figure 57.23 A ball rolling without slipping between two flat plates.

EXAMPLE 57.11:

The example we discuss next (due to Brockett and Dai [18]) is prototypical of applications involving the kinematics of objects in the grasp of a robotic hand. Consider a ball that rolls without slipping between two flat horizontal plates. It is convenient to assume the bottom plate is fixed. Suppose that the ball has unit radius. Fix a coordinate system whose x- and y-axes lie in the fixed bottom plate with the positive z-axis perpendicular to the plate in the direction of the ball. Call this the (bottom) "plate frame." We keep track of the ball's motion by letting $(x_1, x_2, 1)$ denote the plate-frame coordinates of the ball's center. We also fix an orthonormal frame in the ball, and we denote the plate-frame directions of the three coordinate axes by a 3×3 orthogonal matrix

$$A = \begin{pmatrix} a_{11} & a_{12} & a_{13} \\ a_{21} & a_{22} & a_{23} \\ a_{31} & a_{32} & a_{33} \end{pmatrix}.$$

The ball's position and orientation are thus specified by the 4×4 matrix

$$H = \begin{pmatrix} A & \vec{x} \\ 0 & 1 \end{pmatrix} = \begin{pmatrix} a_{11} & a_{12} & a_{13} & x_1 \\ a_{21} & a_{22} & a_{23} & x_2 \\ a_{31} & a_{32} & a_{33} & 1 \\ 0 & 0 & 0 & 1 \end{pmatrix}.$$

As the top plate moves in the plate-frame x direction with velocity v_1, the ball's center also moves in the same direction with velocity $u_1 = v_1/2$. This motion imparts a counterclockwise rotation about the y-axis, and since the ball has unit radius, the angular velocity is also u_1. Similarly, if the top plate moves in the (plate-frame) y direction with velocity v_2, the ball's center moves in the y direction with velocity $u_2 = v_2/2$, and the angular velocity imparted about the x-axis is $-u_2$. The kinematic description of this problem is obtained by differentiating H with respect to time.

$$\dot{H} = \begin{pmatrix} \Omega A & \vec{u} \\ 0 & 0 \end{pmatrix}$$

where

$$\Omega = \begin{pmatrix} 0 & 0 & u_1 \\ 0 & 0 & u_2 \\ -u_1 & -u_2 & 0 \end{pmatrix} \quad \text{and} \quad \vec{u} = \begin{pmatrix} u_1 \\ u_2 \\ 0 \end{pmatrix}.$$

This velocity relationship, at the point $(x_1, x_2) = (0, 0)$, is

$$\dot{H} = u_1 U_1 H + u_2 U_2 H, \tag{57.219}$$

where

$$U_1 = \begin{pmatrix} 0 & 0 & 1 & 1 \\ 0 & 0 & 0 & 0 \\ -1 & 0 & 0 & 0 \\ 0 & 0 & 0 & 0 \end{pmatrix} \quad \text{and} \quad U_2 = \begin{pmatrix} 0 & 0 & 0 & 0 \\ 0 & 0 & 1 & 1 \\ 0 & -1 & 0 & 0 \\ 0 & 0 & 0 & 0 \end{pmatrix}.$$

Computing the Lie bracket of the vector fields $U_1 H$ and $U_2 H$ according to the formula given in Section 57.7.1, we obtain a new vector field $U_3 H = [U_1 H, U_2 H]$, where

$$U_3 = \begin{pmatrix} 0 & -1 & 0 & 0 \\ 1 & 0 & 0 & 0 \\ 0 & 0 & 0 & 0 \\ 0 & 0 & 0 & 0 \end{pmatrix}.$$

Computing additional Lie brackets yields only quantities which may be expressed as linear combinations of $U_1 H$, $U_2 H$, and $U_3 H$. Since the number of linearly independent vector fields obtained by taking Lie brackets is three, the system is completely controllable on a three dimensional space. Comparison with the problem of motion on a sphere discussed above shows that, by having the top plate execute a high-frequency oscillatory motion—$u_1(t) = \mu \sin(\omega t + \phi)$, $u_2(t) = \mu \cos(\omega t + \phi)$, the ball may be made to rotate about its z-axis. The motion of the ball is "retrograde." If the top plate executes a small amplitude clockwise loop about the "plate-frame" z-axis, the motion of the ball is counterclockwise.

57.7.3 Systems with Drift—Stability Analysis Using Energy Methods and the Averaged Potential

Constructive design methods for the control of systems of Class II are less well developed than in the Class I case. Nevertheless, the classical results on stability described in Section 57.7.2 apply in many cases, and the special structure of Class II systems again reveals the geometric mechanisms underlying stability.

First and Second Order Stabilizing Effects of Oscillatory Inputs in Systems with Drift

Much of the published literature on systems of Class I and Class II is devoted to the case in which the vector fields are linear in the state. Consider the control system

$$\dot{x} = (A + \sum_{i=1}^{m} u_i(t) B_i) x, \tag{57.220}$$

where $A, B_1, \ldots, B_m$ are constant $n \times n$ matrices, $x(t) \in \mathbf{R}^n$ with control inputs of the type we have been considering. We shall be interested in the possibility of using high-frequency oscillatory inputs to create stable motions in a neighborhood of the origin $x = 0$.

Assume that $\hat{u}_1(\cdot), \ldots, \hat{u}_m(\cdot)$ are periodic functions and (for simplicity only) assume that each $\hat{u}_i(\cdot)$ has common fundamental period $T > 0$. To apply classical averaging theory to study the motion of Equation 57.220, we consider the effect of increasing the frequency of the forcing. Specifically, we study the dynamics of

$$\dot{x}(t) = (A + \sum \hat{u}_i(\omega t) B_i) x(t)$$

as ω becomes large. The analysis proceeds by scaling time and considering $\tau = \omega t$. Let $z(\tau) = x(t)$. This satisfies the differential equation

$$\frac{dz}{d\tau} = \frac{1}{\omega}(A + \sum \hat{u}_i(\tau) B_i) z. \tag{57.221}$$

Assuming $\omega > 0$ is large, and letting $\epsilon = \frac{1}{\omega}$, we see that Equation 57.221 is in a form to which classical averaging theory applies.

THEOREM 57.10 *Consider the system of Equation 57.220 with $u_i(\cdot) = \hat{u}_i(\cdot)$ where, for $i = 1, \ldots, m$, $\hat{u}_i(\cdot)$ is continuous on*

$0 \le t < t_f \le \infty$ and periodic of period $T \ll t_f$. Let

$$\bar{u}_i = \frac{1}{T}\int_0^T \hat{u}_i(t)\,dt,$$

and let $y(\tau)$ be a solution of the constant coefficient linear system

$$\dot{y}(\tau) = \epsilon(A + \sum \bar{u}_i B_i)y(\tau). \tag{57.222}$$

If z_0 and y_0 are respective initial conditions associated with Equations 57.221 and 57.222 such that $|z_0 - y_0| = \mathcal{O}(\epsilon)$, then $|z(\tau) - y(\tau)| = \mathcal{O}(\epsilon)$ on a time scale $\tau \sim \frac{1}{\epsilon}$. If $A + \sum \bar{u}_i B_i$ has its eigenvalues in the left half plane, then $|z(\tau) - y(\tau)| = \mathcal{O}(\epsilon)$ as $\tau \to \infty$.

This theorem relies on classical averaging theory and is discussed in [7]. A surprising feature of systems of this form is that the stability characteristics can be modified differently if both the magnitude and frequency of the oscillatory forcing are increased. A contrast to Theorem 57.10 is the following:

THEOREM 57.11 *Let $\hat{u}_1(\cdot), \ldots, \hat{u}_m(\cdot)$ be periodic functions of period $T > 0$ for $i = 1, \ldots, m$. Assume that each $\hat{u}_i(\cdot)$ has mean 0. Consider Equation 57.220, and assume that, for all $i, j = 1, \ldots, m$, $B_iB_j = 0$. Let $\epsilon = \frac{1}{\omega}$. Define for each $i = 1, \ldots, m$, the periodic function $v_i(t) = \int_0^t \hat{u}_i(s)\,ds$, and let*

$$\bar{v}_i = \tfrac{1}{T}\int_0^T v_i(s)\,ds, \quad i = 1, \ldots, m,$$

and

$$\sigma_{ij} = \tfrac{1}{T}\int_0^T v_i(s)v_j(s)\,ds, \quad i, j = 1, \ldots, m.$$

Let $y(t)$ be a solution of the constant coefficient linear system

$$\dot{y} = \Big(A + \sum_{i,j}(\bar{v}_i\bar{v}_j - \sigma_{ij})B_iAB_j\Big)y. \tag{57.223}$$

Suppose that the eigenvalues of Equation 57.223 have negative real parts. Then there is a $t_1 > 0$ such that for $\omega > 0$ sufficiently large (i.e., for $\epsilon > 0$ sufficiently small), if x_0 and y_0 are respective initial conditions for Equations 57.220 and 57.223 such that $|x_0 - y_0| = \mathcal{O}(\epsilon)$, then $|x(t) - y(t)| = \mathcal{O}(\epsilon)$ for all $t > t_1$.

The key distinction between these two theorems is that induced stability is a first order effect in Theorem 57.10 where we scale frequency alone, whereas in Theorem 57.11 where both frequency and magnitude are large, any induced stability is a second order effect (depending on the rms value of the integral of the forcing). Further details are provided in [7]. Rather than pursue these results, we describe a closely related theory which may be applied to mechanical and other physical systems.

A Stability Theory for Lagrangian and Hamiltonian Control Systems with Oscillatory Inputs

The geometric nature of emergent behavior in systems subject to oscillatory forcing is apparent in the case of conservative physical systems. In this subsection, we again discuss stabilizing effects of oscillatory forcing. The main analytical tool will be an energy-like quantity called the *averaged potential.* This is naturally defined in terms of certain types of Hamiltonian control systems of the type studied in [30]. Because we shall be principally interested in systems with symmetries most easily described by Lagrangian models, we shall start from a Lagrangian viewpoint and pass to the Hamiltonian description via the Legendre transformation. Following Brockett, [14] and Nijmeijer and Van der Schaft, [30], we define a *Lagrangian control system* on a differentiable manifold M as a dynamical system with inputs whose equations of motion are prescribed by applying the Euler-Lagrange operator to a function $L : TM \times U \to \mathbf{R}$, $L = L(q, \dot{q}; u)$, whose dependence on the configuration q, the velocity $\dot{q}$, and the control input u is smooth. U is a set of control inputs satisfying the general properties outlined in Section 57.7.1.

Lagrangian systems arising via reduction with respect to cyclic coordinates: Consider a Lagrangian control system with configuration variables $(q_1, q_2) \in \mathbf{R}^{n_1} \times \mathbf{R}^{n_2}$. The variable q_1 will be called *cyclic* if it does not enter into the Lagrangian when $u = 0$, i.e., if $\frac{\partial L}{\partial q_1}(q_1, q_2, \dot{q}_1, \dot{q}_2; 0) \equiv 0$. A symmetry is associated with the cyclic variables q_1 which manifests itself in the invariance of $L(q_1, q_2, \dot{q}_1, \dot{q}_2; 0)$ with respect to a change of coordinates $q_1 \mapsto q_1 + \alpha$ for any constant $\alpha \in \mathbf{R}^n$. We shall be interested in Lagrangian control systems with cyclic variables in which the cyclic variables may be directly controlled. In such systems, we shall show how the velocities associated with the cyclic coordinates may themselves be viewed as controls. Specifically, we shall consider systems of the form

$$L(q_1, q_2, \dot{q}_1, \dot{q}_2; u) = \mathcal{L}(q_2, \dot{q}_1, \dot{q}_2) + q_1^T u \tag{57.224}$$

where

$$\mathcal{L}(q_2, \dot{q}_1, \dot{q}_2) = \frac{1}{2}(\dot{q}_1^T, \dot{q}_2^T)\begin{pmatrix} m(q_2) & A^T(q_2) \\ A(q_2) & M(q_2)\end{pmatrix}\begin{pmatrix}\dot{q}_1 \\ \dot{q}_2\end{pmatrix} - V(q_2).$$

The matrices $m(q_2)$ and $M(q_2)$ are symmetric and positive definite of dimension $n_1 \times n_1$ and $n_2 \times n_2$ respectively, where $\dim q_1 = n_1$, $\dim q_2 = n_2$, and the matrix $A(q_2)$ is arbitrary.

To emphasize the distinguished role to be played by the velocity associated with the cyclic variable q_1, we write $v = \dot{q}_1$. Applying the usual Euler-Lagrange operator to this function leads to the equations of motion,

$$\frac{d}{dt}\frac{\partial \mathcal{L}}{\partial v} = u$$

and

$$\left(\frac{d}{dt}\frac{\partial \mathcal{L}}{\partial \dot{q}_2} - \frac{\partial \mathcal{L}}{\partial q_2}\right)\Big|_{(q_2, \dot{q}_2, v)} = 0. \tag{57.225}$$

The first of these equations may be written more explicitly as

$$\frac{d}{dt}(m(q_2)v + A(q_2)^T\dot{q}_2) = u. \tag{57.226}$$

Although, in the physical problem represented by the Lagrangian Equation 57.224, $u(\cdot)$ is clearly the input with $v(\cdot)$ determined via Equation 57.226, it is formally equivalent to take $v(\cdot)$ as the input with the corresponding $u(\cdot)$ determined by Equation 57.226. In actual practice, this may be done, provided we may command actuator inputs $u(\cdot)$ large enough to dominate the dynamics. The motion of q_2 is then determined by Equation 57.225 with the influence of the actual control input felt only through $v(\cdot)$ and $\dot{v}(\cdot)$. Viewing $v(\cdot)$ together with $\dot{v}(\cdot)$ as control input, Equation 57.225 is a Lagrangian control system in its own right. The defining Lagrangian is given by $\hat{L}(q_2, \dot{q}_2; v) = \frac{1}{2}\dot{q}_2^T M(q_2)\dot{q}_2 + \dot{q}_2^T A(q_2)v - V_a(q_2; v)$, where V_a is the *augmented potential* defined by $V_a(q; v) = V(q) - \frac{1}{2}v^T m(q_2)v$. In the remainder of our discussion, we shall confine our attention to controlled dynamical systems arising from such a *reduced Lagrangian*. Because the reduction process itself will typically not be central, we henceforth omit the subscript "2" on the generalized configuration and velocity variables which we wish to control. We write:

$$\hat{L}(q, \dot{q}; v) = \frac{1}{2}\dot{q}^T M(q)\dot{q} + \dot{q}^T A(q)v - V_a(q; v). \quad (57.227)$$

EXAMPLE 57.12: The rotating pendulum.

As in [4], we consider a mechanism consisting of a solid uniform rectangular bar fixed at one end to a universal joint as depicted in Figure 57.24. The universal joint is comprised of two single degree of freedom revolute joints with mutually orthogonal intersecting axes (labeled x and y in Figure 57.22). These joints are assumed to be frictionless. Angular displacements about the x- and y-axes are denoted ϕ_1 and ϕ_2 respectively, with $(\phi_1, \phi_2) = (0, 0)$ designating the configuration in which the pendulum hangs straight down. The pendulum also admits a controlled rotation about a spatially fixed vertical axis. Let θ denote the amount of this rotation relative to some chosen reference configuration.

To describe the dynamics of the forced pendulum, we choose a (principal axis) coordinate frame, fixed in the bar, consisting of the x- and y-axes of the universal joint together with the corresponding z-axis prescribed by the usual right-hand rule. When the pendulum is at rest, for some reference value of θ, the body frame x-, y-, and z-axes will coincide with corresponding axes x', y', and z' of an inertial frame, as in Figure 57.24, with respect to which we shall measure all motion. Let I_x, I_y, and I_z denote the principal moments of inertia with respect to the body (x, y, z)-coordinate system. Then the system has the Lagrangian

$$\begin{aligned} L(\phi_1, \dot{\phi}_1; \phi_2, \dot{\phi}_2; \dot{\theta}) &= \frac{1}{2}\Big[I_x(\dot{\theta}s_2 + \dot{\phi}_1)^2 \\ &\quad + I_y(\dot{\theta}c_2s_1 - \dot{\phi}_2c_1)^2 \\ &\quad + I_z(\dot{\theta}c_1c_2 + \dot{\phi}_2s_1)^2\Big] + c_1c_2, \end{aligned} \quad (57.228)$$

where the last term is a normalized gravitational potential, and where $s_i = \sin\phi_i$, $c_i = \cos\phi_i$. We assume that there is an actuator capable of rotating the mechanism about the inertial z-axis with any prescribed angular velocity $\dot{\theta}$. We shall study the dynamics of this system when $I_x \geq I_y \gg I_z$, and as above, we shall view $v = \dot{\theta}(\cdot)$ as a control input. The *reduced Lagrangian* Equation 57.227 takes the form

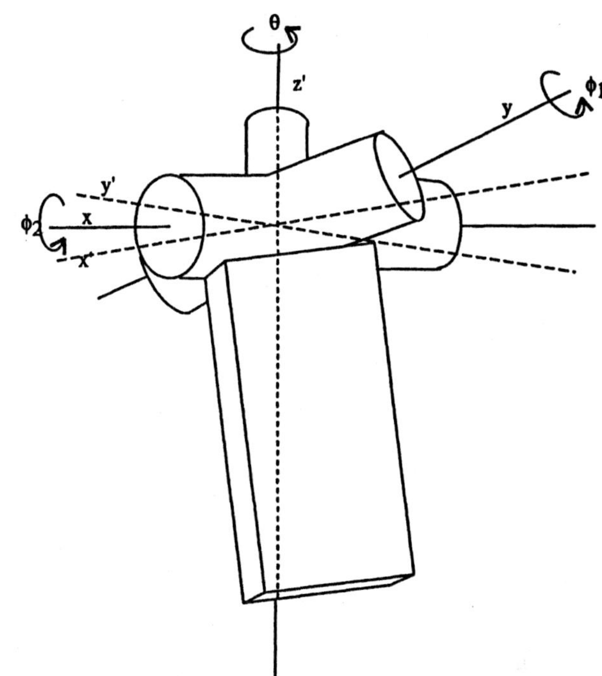

Figure 57.24 A rotating pendulum suspended from a universal joint.

$$\begin{aligned} \hat{L}(\phi, \dot{\phi}; v) &= \frac{1}{2}\left[I_x\dot{\phi}_1^2 + (I_yc_1^2 + I_zs_1^2)\dot{\phi}_2^2\right] \\ &\quad + v\left[I_xs_2\dot{\phi}_1 + (I_z - I_y)s_1c_1c_2\dot{\phi}_2\right] \\ &\quad + \left[(I_zc_1^2 + I_ys_1^2)c_2^2 + I_xs_2^2\right]v^2 + c_1c_2. \end{aligned} \quad (57.229)$$

The corresponding control system is given by applying the Euler-Lagrange operator to $\hat{L}$, and this is represented by a system of coupled second order differential equations,

$$\frac{d}{dt}\frac{\partial \hat{L}}{\partial \dot{\phi}} - \frac{\partial \hat{L}}{\partial \phi} = 0. \quad (57.230)$$

From the way in which v appears in the reduced Lagrangian, the equations of motion Equation 57.230 will have terms involving $\dot{v}$ as well as terms involving v. Though it is possible to analyze such a system directly, we shall not discuss this approach. The general analytical technique advocated here is to transform all Lagrangian models to Hamiltonian form.

The Hamiltonian viewpoint and the ***averaged potential:*** Recall that in Hamiltonian form the dynamics are represented in terms of the configuration variables q and conjugate momenta $p = \frac{\partial \hat{L}}{\partial \dot{q}}$ (see [30], p. 351). Referring to the Lagrangian in Equation 57.227 and applying the Legendre transformation $H(q, p; v) = [p\cdot\dot{q} - \hat{L}(q, \dot{q}; v)]|_{(q,p)}$, we obtain the corresponding Hamiltonian

$$H(q, p; v) = \frac{1}{2}(p - vA)^T M^{-1}(p - vA) + V_a(v, q). \quad (57.231)$$

The equations of motion are then written in the usual way in position and momentum coordinates as

$$\dot{q} = M^{-1}(p - vA) \quad (57.232)$$

$$\dot{p} = -\frac{\partial}{\partial q}[\frac{1}{2}(p - vA)^T M^{-1}(p - vA) + V_a(v, q)]. \quad (57.233)$$

We may obtain an averaged version of this system by replacing all coefficients involving the input $v(\cdot)$ by their time-averages. Assuming $v(\cdot)$ is bounded, piecewise continuous, and periodic of period $T > 0$, $v(\cdot)$ has a Fourier series representation:

$$v(t) = \sum_{k=-\infty}^{\infty} a_k e^{\frac{2\pi k}{T} it}. \quad (57.234)$$

Equations 57.232 and 57.233 contain terms of order not greater than two in v. Averaging the coefficients we obtain

PROPOSITION 57.6 Suppose $v(\cdot)$ is given by Equation 57.234. Then, if all coefficients in Equations 57.232 and 57.233 are replaced by their time averages, the resulting averaged system is Hamiltonian with corresponding Hamiltonian function

$$\begin{aligned}\bar{H}(q, p) &= \frac{1}{2}(p - A(q)\bar{v})^T M(q)^{-1}(p - A(q)\bar{v}) \\ &\quad + V_A(q), \qquad (57.235)\end{aligned}$$

where

$$\begin{aligned}V_A(q) &= V(q) + \frac{1}{2}\Big(\Sigma(q) \\ &\quad - \bar{v}^T A(q)^T M(q)^{-1} A(q)\bar{v}\Big), \qquad (57.236)\\ \bar{v} &= \frac{1}{T}\int_0^T v\,dt,\end{aligned}$$

and

$$\begin{aligned}\Sigma(q) &= \frac{1}{T}\int_0^T v(t)^T \Big(A(q)^T M(q)^{-1} A(q) \\ &\quad - m(q)\Big) v(t)\,dt.\end{aligned}$$

DEFINITION 57.3 We refer to $\bar{H}(q, p)$ in Equation 57.235 as the *averaged Hamiltonian* associated with Equation 57.231. $V_A(q)$, defined in Equation 57.236, is called the *averaged potential.*

REMARK 57.11 (Averaged kinetic and potential energies, the *averaged Lagrangian.*) Before describing the way in which the *averaged potential* may be used for stability analysis, we discuss its formal definition in more detail. The Legendre transformation used to find the Hamiltonian corresponding to Equation 57.227 makes use of the conjugate momentum

$$\begin{aligned}p &= p(q, \dot{q}, t) \\ &= \frac{\partial \hat{L}}{\partial \dot{q}} \\ &= M(q)\dot{q} + A(q)v(t).\end{aligned}$$

This explicitly depends on the input $v(t)$. Given a point in the phase space, $(q, \dot{q})$, the corresponding *averaged* momentum is

$$p = M(q)\dot{q} + A(q)\bar{v}.$$

We may think of the first term $\frac{1}{2}(p - A(q)\bar{v})^T M(q)^{-1}(p - A(q)\bar{v})$ in Equation 57.235 as an "averaged kinetic energy." It is not difficult to see that there is an "averaged Lagrangian"

$$\bar{L}(q, \dot{q}) = \frac{1}{2}\dot{q}^T M(q)\dot{q} + \dot{q}^T A(q)\bar{v} - V_A(q)$$

from which the Hamiltonian $\bar{H}$ in Equation 57.235 is obtained by means of the Legendre transformation.

The averaged potential is useful in assessing the stability of motion in Hamiltonian (or Lagrangian) control systems. The idea behind this is that strict local minima of the averaged potential will correspond to stable equilibria of the averaged Hamiltonian system. The theory describing the relationship with the stability of the forced system is discussed in [6] and [7]. The connection with classical averaging is emphasized in [6], where Rayleigh dissipation is introduced to make the critical points hyperbolically stable. In [7], dissipation is not introduced, and a purely geometric analysis is applied within the Hamiltonian framework. We state the principal stability result.

THEOREM 57.12 *Consider a Lagrangian control system prescribed by Equation 57.227 or its Hamiltonian equivalent (Equation 57.231). Suppose that the corresponding system of Equations 57.232 and 57.233 is forced by the oscillatory input given in Equation 57.234. Let q_0 be a critical point of the* averaged potential *which is independent of the period T (or frequency) of the forcing. Suppose, moreover, that, for all T sufficiently small (frequencies sufficiently large), q_0 is a strict local minimum of the* averaged potential. *Then $(q, \dot{q}) = (q_0, 0)$ is a stable equilibrium of the forced Lagrangian system, provided T is sufficiently small. If $(q, p) = (q_0, 0)$ is the corresponding equilibrium of the forced Hamiltonian system, then it is likewise stable, provided T is sufficiently small.*

This theorem is proved in [7].

We end this section with two examples to which this theorem applies, followed by a simple example which does not satisfy the hypothesis and for which the theory is currently less complete.

EXAMPLE 57.13: Oscillatory stabilization of a simple pendulum.

Consider, once again, the inverted pendulum discussed in Example 57.10. Assume now that no friction exists in the hinge and, therefore, $b = 0$. Using the classical theory of vibrational control in Example 57.14, it is not possible to draw conclusions on the stabilization of the upper equilibrium point by fast oscillating control, because the averaged equation will have purely imaginary eigenvalues when $b = 0$. The averaged potential provides a useful alternative in this case. For $b = 0$, the pendulum dynamics may be written

$$\ell\ddot{\theta} + \ddot{R}\sin\theta + g\sin\theta = 0,$$

where all the parameters have been previously defined in Example 57.10. Writing the pendulum's vertical velocity as $v(t) = \dot{R}(t)$, this is a system of the type we are considering with (reduced) Lagrangian $\hat{L}(\theta, \dot{\theta}; v) = (1/2)\ell\dot{\theta}^2 + v\dot{\theta}\sin\theta + g\cos\theta$. To find stable motions using the theory we have presented, we construct the *averaged potential* by passing to the Hamiltonian description of the system. The momentum (conjugate to θ) is $p = \frac{\partial \hat{L}}{\partial \dot{\theta}} = \ell\dot{\theta} + v\sin\theta$. Applying the Legendre transformation, we obtain the corresponding Hamiltonian

$$\begin{aligned} H(\theta, p; v) &= (p\dot{\theta} - \hat{L})|_{(\theta, p)} \\ &= \frac{1}{2\ell}(p - v\sin\theta)^2 - g\cos\theta. \end{aligned}$$

If we replace the coefficients involving $v(\cdot)$ with their time-averages over one period, we obtain the averaged Hamiltonian

$$\bar{H}(\theta, p) = \frac{1}{2\ell}(p - \bar{v}\sin\theta)^2 + \frac{1}{2\ell}(\Sigma - \bar{v}^2)\sin^2\theta - g\cos\theta,$$

where $\bar{v}$ and Σ are the time averages over one period of $v(t)$ and $v(t)^2$ respectively. The averaged potential is just $V_A(\theta) = \frac{1}{2\ell}(\Sigma - \bar{v}^2)\sin^2\theta - g\cos\theta$. Consider the simple sinusoidal oscillation of the hinge-point, $R(t) = \alpha\sin\beta t$. Then $v(t) = \alpha\beta\cos\beta t$. Carrying out the construction we have outlined, the averaged potential is given more explicitly by

$$V_A(\theta) = \frac{\alpha^2\beta^2}{4\ell}\sin^2\theta - g\cos\theta. \tag{57.237}$$

Looking at the first derivative $V_A'(\theta)$, we find that $\theta = \pi$ is a critical point for all values of the parameters. Looking at the second derivative, we find that $V_A''(\pi) > 0$ precisely when $\alpha^2\beta^2 > 2\ell g$. From Theorem 57.12 we conclude that, for sufficiently large values of the frequency β, the upright equilibrium is stable in the sense that motions of the forced system will remain nearby. This is of course completely consistent with classical results on this problem. (Cf. Example 57.10.)

EXAMPLE 57.14: Example 57.12 4, reprise: oscillatory stabilization of a rotating pendulum.

Let us return to the mechanical system treated in Example 57.12. Omitting a few details, we proceed as follows. Starting from the Lagrangian in Equation 57.229, we obtain the corresponding Hamiltonian (the general formula for which is given by Equation 57.231). The averaged potential is given by the formula in Equation 57.236. Suppose the pendulum is forced to rotate at a constant rate, perturbed by a small-amplitude sinusoid, $v(t) = \omega + \alpha\sin\beta t$. Then the coefficients in Equation 57.236 are

$$\begin{aligned} \bar{v} &= \frac{\beta}{2\pi}\int_0^{2\pi/\beta} v(t)\,dt = \omega, \text{ and} \\ \Sigma &= \frac{\beta}{2\pi}\int_0^{2\pi/\beta} v(t)^2\,dt = \omega^2 + \frac{\alpha^2}{2}, \end{aligned}$$

and some algebraic manipulation shows that the averaged potential is given in this case by

$$\begin{aligned} V_A(\phi_1, \phi_2) &= -c_1c_2 - \frac{1}{4}\frac{I_yI_zc_2^2}{I_yc_1^2 + I_zs_2^2}\alpha^2 \\ &\quad - \frac{1}{2}\left[I_xs_2^2 + (I_ys_1^2 + I_zc_1^2)c_2^2\right]\omega^2. \end{aligned}$$

Stable modes of behavior under this type of forcing correspond to local minima of the averaged potential. A brief discussion of how this analysis proceeds will illustrate the utility of the approach.

When $\alpha = 0$, the pendulum undergoes rotation at a constant rate about the vertical axis. For all rates ω, the pendulum is in equilibrium when it hangs straight down. There is a critical value, ω_{cr}, however, above which the vertical configuration is no longer stable. A critical point analysis of the averaged potential yields the relevant information and more. The partial derivatives of V_A with respect to ϕ_1 and ϕ_2 both vanish at $(\phi_1, \phi_2) = (0, 0)$ for all values of the parameters α, β, ω. To assess the stability of this critical point using Theorem 57.12, we compute the Hessian (matrix of second partial derivatives) of V_A evaluated at $(\phi_1, \phi_2) = (0, 0)$:

$$\begin{pmatrix} \frac{\partial^2 V_A}{\partial \phi_1^2}(0,0) & \frac{\partial^2 V_A}{\partial \phi_1 \partial \phi_2}(0,0) \\ \frac{\partial^2 V_A}{\partial \phi_1 \partial \phi_2}(0,0) & \frac{\partial^2 V_A}{\partial \phi_2^2}(0,0) \end{pmatrix} = \begin{pmatrix} 1 + \frac{1}{2}\frac{I_z}{I_y}(I_z - I_y)\alpha^2 - (I_y - I_z)\omega^2 & 0 \\ 0 & 1 + \frac{1}{2}I_z\alpha^2 - (I_x - I_z)\omega^2 \end{pmatrix}.$$

Let us treat the constant rotation case first. We have assumed $I_x \geq I_y \gg I_z$. When $\alpha = 0$, this means that the Hessian matrix above is positive definite for $0 \leq \omega^2 < 1/(I_x - I_z)$. This inequality gives the value of the critical rotation rate precisely as $\omega_{cr} = 1/\sqrt{I_x - I_z}$. We wish to answer the following question: Is it possible to provide a stabilizing effect by superimposing a small-amplitude, high-frequency sinusoidal oscillation on the constant-rate forced rotation? The answer emerges from Theorem 57.12 together with analysis of the Hessian. In the symmetric case, $I_x = I_y$, the answer is "no" because any nonzero value of α will decrease the (1, 1)-entry and hence the value of ω_{cr}. If $I_x > I_y$, however, there is the possibility of increasing ω_{cr} slightly, because, although the (1, 1)-entry is decreased, the more important (2, 2)-entry is increased.

Current research on oscillatory forcing to stabilize rotating systems (chains, shafts, turbines, etc.) is quite encouraging. Though only modest stabilization was possible for the rotating pendulum in the example above, more pronounced effects are generally possible with axial forcing. Because this approach to control appears to be quite robust (as seen in the next example), it merits attention in applications where feedback designs would be difficult to implement.

We conclude with an example to which Theorem 57.12 does not apply and for which the theory is currently less well developed. Methods of [6] can be used in this case.

EXAMPLE 57.15: Oscillation induced rest points in a pendulum on a cart.

We consider a slight variation on Example 57.13 wherein we consider oscillating the hinge point of the pendulum along a line which is not vertical. More specifically, consider a cart to which there is a simple pendulum (as described in Example 57.13) attached so that the cart moves along a track inclined at an angle ψ to the horizontal. Suppose the position of the cart along its track at time t is prescribed by a variable $r(t)$. Then the pendulum dynamics are expressed

$$\ell\ddot{\theta} + \ddot{r}\cos(\theta - \psi) + g\sin\theta = 0.$$

Note that when $\psi = \pi/2$, the track is aligned vertically, and we recover the problem treated in Example 57.13. In the general case, let $v(t) = \dot{r}(t)$ and write the averaged potential

$$V_A(\theta) = -g\cos\theta + \frac{1}{2\ell}\cos^2(\theta - x)\,(\Sigma - \bar{v}^2)$$

where

$$\bar{v} = \frac{1}{T}\int_0^T v(t)\,dt$$

and

$$\Sigma = \frac{1}{T}\int_0^T v(t)^2\,dt.$$

As in Example 57.13, we may take $v(t) = \alpha\beta\cos\beta t$. For sufficiently large frequencies β there are strict local minima of the averaged potential which are not equilibrium points of the forced system. Nevertheless, as noted in [6], the pendulum will execute motions in a neighborhood of such a point. To distinguish such emergent behavior from stable motions in neighborhoods of equilibria (of the nonautonomous system), we have called motions confined to neighborhoods of nonequilibrium critical points of the averaged potential *hovering motions*. For more information on such motions, the reader is referred to [41].

Remark on the robustness of open-loop methods. The last example suggests, and laboratory experiments bear out, that the stabilizing effects of oscillatory forcing of the type we have discussed are quite pronounced. Moreover, they are quite insensitive to the fine details of the mathematical models and to physical disturbances which may occur. Thus, the stabilizing effect observed in the inverted pendulum will be entirely preserved if the pendulum is perturbed or if the direction of the forcing isn't really vertical. Such robustness suggests that methods of this type are worth exploring in a wider variety of applications.

Remark on oscillatory control with feedback. There are interesting applications (e.g., laser cooling) where useful designs arise through a combination of oscillatory forcing and certain types of feedback. For the theory of time-varying feedback designs, the reader is referred to [20] and the chapters on stability by Khalil, Teel, Sontag, Praly, and Georgiou appearing in this handbook.

Defining Terms

Anholonomy: Consider the controlled differential equation 57.207, and suppose that there is a non-zero function $\phi : \mathbf{R}^n \times \mathbf{R}^n \to \mathbf{R}$ such that $\phi(x, g_i(x)) \equiv 0$ for $i = 1, \ldots, m$. This represents a constraint on the state velocities which can be commanded. Despite such a constraint, it may happen that any two specified states can be joined by a trajectory $x(t)$ generated via Equation 57.207 by an appropriate choice of inputs $u_i(\cdot)$. Any state trajectory arising from Equation 57.207 constrained in this way is said to be determined from the inputs $u_i(\cdot)$ by *anholonomy*. In principle, the notation of *anholonomy* can be extended to systems given by Equation 57.208 or 57.209. Some authors who were consulted in preparation of this chapter objected to the use of the word in this more general context.

Averaged potential: An energy-like function that describes the steady-state behavior produced by high-frequency forcing of a physical system.

Completely controllable: A system of Equation 57.209 is said to be *completely controllable* if, given any $T > 0$ and any pair of points $x_0, x_1 \in \mathbf{R}^n$, there is a control input $u(\cdot)$ producing a motion $x(\cdot)$ of Equation 57.209 such that $x(0) = x_0$ and $x(T) = x_1$.

LARC: The *Lie algebra rank condition* is the condition that the defining vector fields in systems, such as Equation 57.207 or Equation 57.208 together with their Lie brackets of all orders span $\mathbf{R}^n$.

57.7.4 Acknowledgment

The authors are indebted to many people for help in preparing this chapter. R.W. Brockett, in particular, provided useful guidance and criticism.

References

[1] Baillieul, J., Multilinear optimal control, *Proc. Conf. Geom. Control Eng.*, (NASA-Ames, Summer 1976), Brookline, MA: Math. Sci. Press, 337–359, 1977.

[2] Baillieul, J., Geometric methods for nonlinear optimal control problems, *J. Optimiz. Theory Appl.*, 25(4), 519–548, 1978.

[3] Baillieul, J., The Behavior of Super-Articulated Mechanisms Subject to Periodic Forcing, in *Analysis of Controlled Dynamical Systems*, B. Bonnard, B. Bride, J.P. Gauthier, and I. Kupka, Eds., Birkhaüser, 1991, 35–50.

[4] Baillieul, J. and Levi, M., Constrained relative motions in rotational mechanics, *Arch. Rational Mech. Anal.*, 115/2, 101–135, 1991.

[5] Baillieul, J., The behavior of single-input super-articulated mechanisms, *Proc. 1991 Am. Control Conf.*, Boston, June 26-28, pp. 1622-1626, 1991.

[6] Baillieul, J., Stable Average Motions of Mechanical Systems Subject to Periodic Forcing, *Dynamics and Control of Mechanical Systems: The Falling Cat and Related Problems*, Fields Institute Communications,

1, AMS, Providence, RI, 1–23, 1993.

[7] Baillieul, J., Energy methods for stability of bilinear systems with oscillatory inputs, *Int. J. Robust Nonlinear Control*, Special Issue on the "Control of Nonlinear Mechanical Systems," H. Nijmeijer and A.J. van der Schaft, Guest Eds., July, 285–301, 1995.

[8] Bellman, R., Bentsman, J., and Meerkov, S.M., Vibrational control of nonlinear systems: Vibrational stabilizability, *IEEE Trans. Automatic Control*, AC-31, 710–716, 1986.

[9] Bellman, R., Bentsman, J., and Meerkov, S.M., Vibrational control of nonlinear systems: Vibrational controllability and transient behavior, *IEEE Trans. Automatic Control*, AC-31, 717–724, 1986.

[10] Bishop, R.L. and Crittenden, R.J., *Geometry of Manifolds*, Academic Press, New York, 1964.

[11] Bogolyubov, N.N., Perturbation theory in nonlinear mechanics, *Sb. Stroit. Mekh. Akad. Nauk Ukr.* SSR *14*, 9–34, 1950.

[12] Bogolyubov, N.N. and Mitropolsky, Y.A., *Asymptotic Methods in the Theory of Nonlinear Oscillations*, 2nd ed., Gordon & Breach Publishers, New York, 1961.

[13] Brockett, R.W., System theory on group manifolds and coset spaces, *SIAM J. Control*, 10(2), 265–284, 1972.

[14] Brockett, R.W., Control Theory and Analytical Mechanics, in *Geometric Control Theory*, Vol. VII of Lie Groups: History, Frontiers, and Applications, C. Martin and R. Hermann, Eds., Math Sci Press, Brookline, MA, 1–46, 1977.

[15] Brockett, R.W., Asymptotic Stability and Feedback Stabilization, in *Differential Geometric Control Theory*, R.W. Brockett, R.S. Millman, and H.J. Sussmann, Eds., Birkhaüser, Basel, 1983.

[16] Brockett, R.W., Control Theory and Singular Riemannian Geometry, in *New Directions in Applied Mathematics*, Springer-Verlag, New York, 13–27, 1982.

[17] Brockett, R.W., On the rectification of vibratory motion, *Sens. Actuat.*, 20, 91–96, 1989.

[18] Brockett, R.W. and Dai, L., Nonholonomic Kinematics and the Role of Elliptic Functions in Constructive Controllability, in *Nonholonomic Motion Planning*, Kluwer Academic Publishers, 1–21, 1993.

[19] Chow, W.L., Über Systeme von Linearen Partiellen Differentialgleichungen erster Ordnung, *Math. Ann.*, 117, 98–105, 1939.

[20] Coron, J.M., Global asymptotic stabilization for controllable systems without drift, *Math. Control, Sig., Syst.*, 5, 295–312, 1992.

[21] Haynes, G.W. and Hermes, H., Nonlinear controllability via Lie theory, *SIAM J. Control*, 8(4), 450–460, 1970.

[22] Lafferriere, G. and Sussmann, H.J., Motion planning for controllable systems without drift, *Proc. IEEE Intl. Conf. Robot. Automat.*, 1148–1153, 1991.

[23] Lafferriere, G. and Sussmann, H.J., A Differential Geometric Approach to Motion Planning, in *Nonholonomic Motion Planning*, Kluwer Academic Publishers, 235–270, 1993.

[24] Lehman, B., Bentsman, J., Lunel, S.V., and Verriest, E.I., Vibrational control of nonlinear time lag systems with bounded delay: Averaging theory, stabilizability, and transient behavior, *IEEE Trans. Auto. Control*, AC-39, 898–912, 1994.

[25] Lehman, B., Vibrational Control of Time Delay Systems, in *Ordinary and Delay Equations*, J. Wiener and J. Hale, Eds., Pitman Research Notes in Mathematical Series (272), 111–115, 1992.

[26] Kapitsa, P.L., Dynamic stability of a pendulum with a vibrating point of suspension, *Zh. Ehksp. Teor. Fiz.*, 21(5), 588–598, 1951.

[27] Leonard, N.E. and Krishnaprasad, P.S., Control of Switched Electrical Networks Using Averaging on Lie Groups, *The 33rd IEEE Conference on Decision and Control*, Orlando, FL, Dec. 14-16, pp. 1919–1924, 1994.

[28] Leonard, N.E. and Krishnaprasad, P.S., Motion control of drift-free, left-invariant systems on Lie groups, *IEEE Trans. Automat. Control*, AC40(9), 1539–1554, 1995.

[29] Murray, R.M. and Sastry, S.S., Nonholonomic motion planning: steering using sinusoids, *IEEE Trans. Auto. Control*, 38(5), 700–716, 1993.

[30] Nijmeijer, H. and van der Schaft, A.J., *Nonlinear Dynamical Control Systems*, Springer-Verlag, New York, 1990.

[31] Sanders, J.A. and Verhulst, F., *Averaging Methods in Nonlinear Dynamical Systems*, Springer-Verlag, Applied Mathematical Sciences, 59, New York, 1985.

[32] Speyer, J.L., Nonoptimality of steady-state cruise for aircraft, *AIAA J.*, 14(11), 1604–1610, 1976.

[33] Speyer, J.L. and Evans, R.T., A second variational theory for optimal periodic processes, *IEEE Trans. Auto. Control*, AC-29(2), 138–148, 1984.

[34] Stoker, J.J., *Nonlinear Vibrations in Mechanical and Electrical Systems*, J. Wiley & Sons, New York, 1950. Republished 1992 in Wiley Classics Library Edition.

[35] Sussmann, H. and Jurdjevic, V., Controllability of nonlinear systems, *J. Diff. Eqs.*, 12, 95–116, 1972.

[36] Sussmann, H.J. and Liu, W., Limits of highly oscillatory controls and approximation of general paths by admissible trajectories, *30th IEEE Conf. Decision Control*, Brighton, England, 1991.

[37] Sussmann, H.J. and Liu, W., An approximation algorithm for nonholonomic systems, Rutgers University, Department of Mathematics preprint, SYCON-93-11. To appear in *SIAM J. Optimiz. Control.*

[38] Sussmann, H.J. and Liu, W., Lie Bracket Extension and Averaging: The Single Bracket Case, in *Nonholonomic Motion Planning*, Kluwer Academic Publishers, 109–147, 1994.

[39] Tilbury, D., Murray, R. and Sastry, S., Trajectory gen-

eration for the N-trailer problem using Goursat normal form, *30th IEEE Conf. Decision Control*, San Antonio, Texas, 971–977, 1993.

[40] Tilbury, D., Murray, R. and Sastry, S., Trajectory generation for the N-trailer problem using Goursat normal form, *IEEE Trans. Auto. Control*, AC40(5), 802–819, 1995.

[41] Weibel, S., Baillieul, J., and Kaper, T., Small-amplitude periodic motions of rapidly forced mechanical systems, *34th IEEE Conf. on Decision and Control*, New Orleans, 1995.

57.8 Adaptive Nonlinear Control

Miroslav Krstić, Department of Mechanical Engineering, University of Maryland, College Park, MD
Petar V. Kokotović, Department of Electrical and Computer Engineering, University of California, Santa Barbara, CA

57.8.1 Introduction: Backstepping

Realistic models of physical systems are nonlinear and usually contain parameters (masses, inductances, aerodynamic coefficients, etc.) which are either poorly known or depend on a slowly changing environment. If the parameters vary in a broad range, it is common to employ adaptation: a parameter estimator — **identifier** — continuously acquires knowledge about the plant and uses it to tune the controller "on-line".

Instabilities in nonlinear systems can be more explosive than in linear systems. During the parameter estimation transients, the state can "escape" to infinity in finite time. For this reason, adaptive nonlinear controllers cannot simply be the "adaptive versions" of standard nonlinear controllers.

Currently the most systematic methodology for adaptive nonlinear control design is **backstepping.** We introduce the idea of backstepping by carrying out a *nonadaptive* design for the system

$$\begin{aligned} \dot{x}_1 &= x_2 + \varphi(x_1)^{\mathrm{T}}\theta\,, \qquad \varphi(0) = 0 && (57.238)\\ \dot{x}_2 &= u && (57.239)\end{aligned}$$

where θ is a *known* parameter vector and $\varphi(x_1)$ is a smooth nonlinear function. Our goal is to stabilize the equilibrium $x_1 = 0$, $x_2 = -\varphi(0)^{\mathrm{T}}\theta = 0$. Backstepping design is recursive. First, the state x_2 is treated as a **virtual control** for the x_1-equation (57.238), and a **stabilizing function**

$$\alpha_1(x_1) = -c_1 x_1 - \varphi(x_1)^{\mathrm{T}}\theta\,, \qquad c_1 > 0 \qquad (57.240)$$

is designed to stabilize Equation 57.238 assuming that $x_2 = \alpha_1(x_1)$ can be implemented. Since this is not the case, we define

$$\begin{aligned} z_1 &= x_1\,, && (57.241)\\ z_2 &= x_2 - \alpha_1(x_1)\,, && (57.242)\end{aligned}$$

where z_2 is an error variable expressing the fact that x_2 is not the true control. Differentiating z_1 and z_2 with respect to time, the complete system Equations 57.238 and 57.239 is expressed in the error coordinates Equations 57.242 and 57.242:

$$\begin{aligned} \dot{z}_1 &= \dot{x}_1 = x_2 + \varphi^{\mathrm{T}}\theta\\ &= z_2 + \alpha_1 + \varphi^{\mathrm{T}}\theta = -c_1 z_1 + z_2\,, && (57.243)\end{aligned}$$

and

$$\begin{aligned} \dot{z}_2 &= \dot{x}_2 - \dot{\alpha}_1 = u - \frac{\partial \alpha_1}{\partial x_1}\dot{x}_1\\ &= u - \frac{\partial \alpha_1}{\partial x_1}\left(x_2 + \varphi^{\mathrm{T}}\theta\right). && (57.244)\end{aligned}$$

It is important to observe that the time derivative $\dot{\alpha}_1$ is implemented analytically, without a differentiator. For the system of Equations 57.243 and 57.244 we now design a control law $u = \alpha_2(x_1, x_2)$ to render the time derivative of a Lyapunov function negative definite. The design can be completed with the simplest Lyapunov function

$$V(x_1, x_2) = \frac{1}{2}z_1^2 + \frac{1}{2}z_2^2\,. \qquad (57.245)$$

Its derivative for Equations 57.243 and 57.244 is

$$\begin{aligned} \dot{V} &= z_1(-c_1 z_1 + z_2)\\ &\quad + z_2\left[u - \frac{\partial \alpha_1}{\partial x_1}\left(x_2 + \varphi^{\mathrm{T}}\theta\right)\right]\\ &= -c_1 z_1^2\\ &\quad + z_2\left[u + z_1 - \frac{\partial \alpha_1}{\partial x_1}\left(x_2 + \varphi^{\mathrm{T}}\theta\right)\right]. && (57.246)\end{aligned}$$

An obvious way to achieve negativity of $\dot{V}$ is to employ u to make the bracketed expression equal to $-c_2 z_2$ with $c_2 > 0$, namely,

$$u = \alpha_2(x_1, x_2) = -c_2 z_2 - z_1 + \frac{\partial \alpha_1}{\partial x_1}\left(x_2 + \varphi^{\mathrm{T}}\theta\right). \qquad (57.247)$$

This control may not be the best choice because it cancels some terms which may contribute to the negativity of $\dot{V}$. Backstepping design offers enough flexibility to avoid cancellation. However, for the sake of clarity, we will assume that none of the nonlinearities is useful, so that they all need to be cancelled as in the control law (57.247). This control law yields

$$\dot{V} = -c_1 z_1^2 - c_2 z_2^2\,, \qquad (57.248)$$

which means that the equilibrium $z = 0$ is globally asymptotically stable. In view of Equations 57.242 and 57.242, the same is true about $x = 0$. The resulting closed-loop system in the z-coordinates is linear:

$$\begin{bmatrix} \dot{z}_1 \\ \dot{z}_2 \end{bmatrix} = \begin{bmatrix} -c_1 & 1 \\ -1 & -c_2 \end{bmatrix}\begin{bmatrix} z_1 \\ z_2 \end{bmatrix}. \qquad (57.249)$$

In the next four sections we present adaptive nonlinear designs through examples. Summaries of general design procedures are also provided but without technical details, for which the reader is referred to the text on nonlinear and adaptive control design by Krstić, Kanellakopoulos, and Kokotović [9]. Only elementary background on Lyapunov stability is assumed, while no previous familiarity with adaptive linear control is necessary. The two main methodologies for adaptive backstepping design are the **tuning functions design**, Section 57.8.2, and the **modular design,** Section 57.8.3. These sections assume that the full state is available for feedback. Section 57.8.4 presents designs where only the output is measured. Section 57.8.5 discusses various extensions to more general classes of systems, followed by a brief literature review in Section 57.8.5.

57.8.2 Tuning Functions Design

In the tuning functions design both the controller and the parameter update law are designed recursively. At each consecutive step a **tuning function** is designed as a potential update law. The tuning functions are not implemented as update laws. Instead, the stabilizing functions use them to compensate the effects of parameter estimation transients. Only the final tuning function is used as the parameter update law.

Introductory Examples

The tuning functions design will be introduced through examples with increasing complexity:

A

$$\dot{x}_1 = u + \varphi(x_1)^{\mathrm{T}}\theta.$$

B

$$\begin{aligned} \dot{x}_1 &= x_2 + \varphi(x_1)^{\mathrm{T}}\theta, \\ \dot{x}_2 &= u. \end{aligned}$$

C

$$\begin{aligned} \dot{x}_1 &= x_2 + \varphi(x_1)^{\mathrm{T}}\theta, \\ \dot{x}_2 &= x_3, \\ \dot{x}_3 &= u. \end{aligned}$$

The adaptive problem arises because the parameter vector θ is *unknown*. The nonlinearity $\varphi(x_1)$ is known and, for simplicity, it is assumed that $\varphi(0) = 0$. The systems A, B, and C differ structurally: the number of integrators between the control u and the unknown parameter θ increases from zero at A, to two at C. Design A will be the simplest because the control u and the uncertainty $\varphi(x_1)^{\mathrm{T}}\theta$ are "matched", that is, the control does not have to overcome integrator transients to counteract the effects of the uncertainty. Design C will be the hardest because the control must act through two integrators before it reaches the uncertainty.

Design A. Let $\hat{\theta}$ be an estimate of the unknown parameter θ in the system

$$\dot{x}_1 = u + \varphi^{\mathrm{T}}\theta\,. \tag{57.250}$$

If this estimate were correct, $\hat{\theta} = \theta$, then the control law

$$u = -c_1x_1 - \varphi(x_1)^{\mathrm{T}}\hat{\theta} \tag{57.251}$$

would achieve global asymptotic stability of $x = 0$. Because $\tilde{\theta} = \theta - \hat{\theta} \neq 0$,

$$\dot{x}_1 = -c_1x_1 + \varphi(x_1)^{\mathrm{T}}\tilde{\theta}\,, \tag{57.252}$$

that is, the parameter estimation error $\tilde{\theta}$ continues to act as a disturbance which may destabilize the system. Our task is to find an update law for $\hat{\theta}(t)$ which preserves the boundedness of $x(t)$ and achieves its regulation to zero. To this end, we consider the Lyapunov function

$$V_1(x,\hat{\theta}) = \frac{1}{2}x_1^2 + \frac{1}{2}\tilde{\theta}^{\mathrm{T}}\Gamma^{-1}\tilde{\theta}\,, \tag{57.253}$$

where Γ is a positive definite symmetric matrix referred to as "adaptation gain". The derivative of V_1 is

$$\begin{aligned} \dot{V}_1 &= -c_1x_1^2 + x_1\varphi(x_1)^{\mathrm{T}}\tilde{\theta} - \tilde{\theta}^{\mathrm{T}}\Gamma^{-1}\dot{\hat{\theta}}, \\ &= -c_1x_1^2 + \tilde{\theta}^{\mathrm{T}}\Gamma^{-1}\left(\Gamma\varphi(x_1)x_1 - \dot{\hat{\theta}}\right). \end{aligned} \tag{57.254}$$

Our goal is to select an update law for $\dot{\hat{\theta}}$ to guarantee

$$\dot{V}_1 \leq 0\,. \tag{57.255}$$

The only way this can be achieved for any unknown $\tilde{\theta}$ is to choose

$$\dot{\hat{\theta}} = \Gamma\varphi(x_1)x_1\,. \tag{57.256}$$

This choice yields

$$\dot{V}_1 = -c_1x_1^2\,, \tag{57.257}$$

which guarantees global stability of the equilibrium $x_1 = 0$, $\hat{\theta} = \theta$, and hence, the boundedness of $x_1(t)$ and $\hat{\theta}(t)$. By LaSalle's invariance theorem (see the chapter by Khalil in this volume), all of the trajectories of the closed-loop adaptive system converge to the set where $\dot{V}_1 = 0$, that is, to the set where $c_1x_1^2 = 0$, implying that

$$\lim_{t\to\infty} x_1(t) = 0\,. \tag{57.258}$$

Alternatively, we can prove Equation 57.258 as follows. By integrating Equation 57.257 $\int_0^t c_1x_1(\tau)^2d\tau = V_1(x_1(0),\hat{\theta}(0)) - V_1(x_1(t),\hat{\theta}(t))$, which, due to the nonnegativity of V_1, implies that $\int_0^t c_1x_1(\tau)^2d\tau \leq V_1(x_1(0),\hat{\theta}(0)) < \infty$. Hence, x_1 is square-integrable. Due to the boundedness of $x_1(t)$ and $\hat{\theta}(t)$, from Equations 57.252 and 57.256, it follows that $\dot{x}_1(t)$ and $\dot{\hat{\theta}}$ are also bounded. By Barbalat's lemma, we conclude that $x_1(t) \to 0$.

The update law Equation 57.256 is driven by the vector $\varphi(x_1)$, called the **regressor**, and the state x_1. This is a typical form of an update law in the tuning functions design: the speed of adaptation is dictated by the nonlinearity $\varphi(x_1)$ and the state x_1.

Design B. For the system

$$\begin{aligned} \dot{x}_1 &= x_2 + \varphi(x_1)^{\mathrm{T}}\theta \\ \dot{x}_2 &= u \end{aligned} \tag{57.259}$$

we have already designed a nonadaptive controller in Section 57.8.1. To design an adaptive controller, we replace the unknown θ by its estimate $\hat{\theta}$ in the stabilizing function Equation 57.240 and in the change of coordinate Equation 57.242,

$$\begin{aligned} z_2 &= x_2 - \alpha_1(x_1,\hat{\theta}) \\ \alpha_1(x_1,\hat{\theta}) &= -c_1z_1 - \varphi^{\mathrm{T}}\hat{\theta}\,. \end{aligned} \tag{57.260}$$

Because the control input is separated from the unknown parameter by an integrator in the system (57.259), the control law Equation 57.247 will be strengthened by a term $\nu_2(x_1, x_2, \hat{\theta})$ which will compensate for the parameter estimation transients,

$$\begin{aligned} u &= \alpha_2(x_1, x_2, \hat{\theta}) \qquad (57.261) \\ &= -c_2 z_2 - z_1 + \frac{\partial \alpha_1}{\partial x_1}\left(x_2 + \varphi^{\mathrm{T}}\hat{\theta}\right) + \nu_2(x_1, x_2, \hat{\theta}). \end{aligned}$$

The resulting system in the z coordinates is

$$\begin{aligned} \dot{z}_1 &= z_2 + \alpha_1 + \varphi^{\mathrm{T}}\theta = -c_1 z_1 + z_2 + \varphi^{\mathrm{T}}\tilde{\theta}, \qquad (57.262) \\ \dot{z}_2 &= \dot{x}_2 - \dot{\alpha}_1 = u - \frac{\partial \alpha_1}{\partial x_1}\left(x_2 + \varphi^{\mathrm{T}}\theta\right) - \frac{\partial \alpha_1}{\partial \hat{\theta}}\dot{\hat{\theta}}, \\ &= -z_1 - c_2 z_2 - \frac{\partial \alpha_1}{\partial x_1}\varphi^{\mathrm{T}}\tilde{\theta} - \frac{\partial \alpha_1}{\partial \hat{\theta}}\dot{\hat{\theta}} \\ &\quad + \nu_2(x_1, x_2, \hat{\theta}), \qquad (57.263) \end{aligned}$$

or, in vector form,

$$\begin{aligned} \begin{bmatrix} \dot{z}_1 \\ \dot{z}_2 \end{bmatrix} &= \begin{bmatrix} -c_1 & 1 \\ -1 & -c_2 \end{bmatrix}\begin{bmatrix} z_1 \\ z_2 \end{bmatrix} \qquad (57.264) \\ &\quad + \begin{bmatrix} \varphi^{\mathrm{T}} \\ -\frac{\partial \alpha_1}{\partial x_1}\varphi^{\mathrm{T}} \end{bmatrix}\tilde{\theta} \\ &\quad + \begin{bmatrix} 0 \\ -\frac{\partial \alpha_1}{\partial \hat{\theta}}\dot{\hat{\theta}} + \nu_2(x_1, x_2, \hat{\theta}) \end{bmatrix}. \end{aligned}$$

The term ν_2 can now be chosen to eliminate the last brackets,

$$\nu_2(x_1, x_2, \hat{\theta}) = \frac{\partial \alpha_1}{\partial \hat{\theta}}\dot{\hat{\theta}}. \qquad (57.265)$$

This expression is implementable because $\dot{\hat{\theta}}$ will be available from the update law. Thus we obtain the **error system**

$$\begin{bmatrix} \dot{z}_1 \\ \dot{z}_2 \end{bmatrix} = \begin{bmatrix} -c_1 & 1 \\ -1 & -c_2 \end{bmatrix}\begin{bmatrix} z_1 \\ z_2 \end{bmatrix} + \begin{bmatrix} \varphi^{\mathrm{T}} \\ -\frac{\partial \alpha_1}{\partial x_1}\varphi^{\mathrm{T}} \end{bmatrix}\tilde{\theta}. \qquad (57.266)$$

When the parameter error $\tilde{\theta}$ is zero, this system becomes the linear asymptotically stable system Equation 57.249. Our remaining task is to select the update law $\dot{\hat{\theta}} = \Gamma\tau_2(x, \hat{\theta})$. Consider the Lyapunov function

$$V_2(x_1, x_2, \hat{\theta}) = V_1 + \frac{1}{2}z_2^2 = \frac{1}{2}z_1^2 + \frac{1}{2}z_2^2 + \frac{1}{2}\tilde{\theta}^{\mathrm{T}}\Gamma^{-1}\tilde{\theta}. \qquad (57.267)$$

Because $\dot{\tilde{\theta}} = -\dot{\hat{\theta}}$, the derivative of V_2 is

$$\begin{aligned} \dot{V}_2 &= -c_1 z_1^2 - c_2 z_2^2 + [z_1, z_2] \\ &\quad \begin{bmatrix} \varphi^{\mathrm{T}} \\ -\frac{\partial \alpha_1}{\partial x_1}\varphi^{\mathrm{T}} \end{bmatrix}\tilde{\theta} - \tilde{\theta}^{\mathrm{T}}\Gamma^{-1}\dot{\hat{\theta}} \\ &= -c_1 z_1^2 - c_2 z_2^2 + \tilde{\theta}^{\mathrm{T}}\Gamma^{-1} \\ &\quad \left(\Gamma\left[\varphi, -\frac{\partial \alpha_1}{\partial x_1}\varphi\right]\begin{bmatrix} z_1 \\ z_2 \end{bmatrix} - \dot{\hat{\theta}}\right). \qquad (57.268) \end{aligned}$$

The only way to eliminate the unknown parameter error $\tilde{\theta}$ is to select the update law

$$\begin{aligned} \dot{\hat{\theta}} &= \Gamma\tau_2(x, \hat{\theta}) = \Gamma\left[\varphi, -\frac{\partial \alpha_1}{\partial x_1}\varphi\right]\begin{bmatrix} z_1 \\ z_2 \end{bmatrix} \\ &= \Gamma\left(\varphi z_1 - \frac{\partial \alpha_1}{\partial x_1}\varphi z_2\right). \qquad (57.269) \end{aligned}$$

Then $\dot{V}_2$ is nonpositive,

$$\dot{V}_2 = -c_1 z_1^2 - c_2 z_2^2, \qquad (57.270)$$

which means that the global stability of $z = 0$, $\tilde{\theta} = 0$ is achieved. Moreover, by applying either the LaSalle or the Barbalat argument mentioned in Design A, we prove that $z(t) \to 0$ as $t \to \infty$. Finally, from Equation 57.260, it follows that the equilibrium $x = 0$, $\hat{\theta} = \theta$ is globally stable and $x(t) \to 0$ as $t \to \infty$.

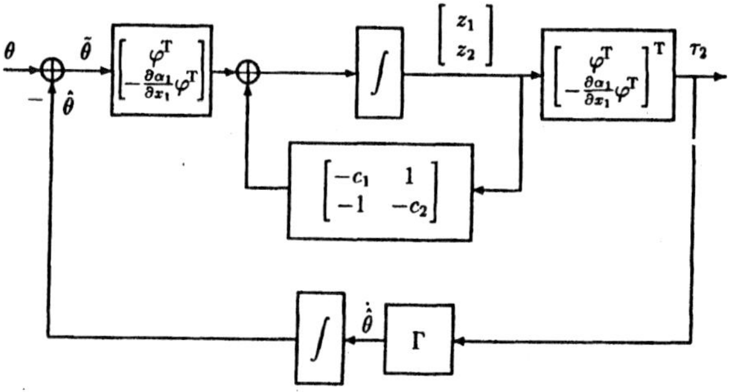

Figure 57.25 The closed-loop adaptive system (57.266), (57.269).

The crucial property of the control law in Design B is that it incorporates the ν_2-term Equation 57.265 which is proportional to $\dot{\hat{\theta}}$ and compensates for the effect of parameter estimation transients on the coordinate change Equation 57.260. It is this departure from the certainty equivalence principle that makes the adaptive stabilization possible for systems with nonlinearities of arbitrary growth.

By comparing Equations 57.269 and 57.256, we note that the first term, φz_1 is the potential update law for the z_1-system. The functions

$$\begin{aligned} \tau_1(x_1) &= \varphi z_1 \qquad (57.271) \\ \tau_2(x_1, x_2, \hat{\theta}) &= \tau_1(x_1) - \frac{\partial \alpha_1}{\partial x_1}\varphi z_2 \qquad (57.272) \end{aligned}$$

are referred to as the **tuning functions**, because of their role as potential update laws for intermediate systems in the backstepping procedure.

Design C. The system,

$$\begin{aligned} \dot{x}_1 &= x_2 + \varphi(x_1)^{\mathrm{T}}\theta, \\ \dot{x}_2 &= x_3, \qquad (57.273) \\ \dot{x}_3 &= u, \end{aligned}$$

is obtained by augmenting system Equation 57.259 with an integrator. The control law $\alpha_2(x_1, x_2, \hat{\theta})$ designed in Equation 57.261 can no longer be directly applied because x_3 is a state and not a control input. We "step back" through the integrator $\dot{x}_3 = u$ and design the control law for the actual input u. However, we keep the stabilizing function α_2 and use it to define the third error coordinate

$$z_3 = x_3 - \alpha_2(x_1, x_2, \hat{\theta}) . \quad (57.274)$$

The parameter update law Equation 57.269 will have to be modified with an additional z_3-term. Instead of $\dot{\hat{\theta}}$ in Equation 57.265, the compensating term ν_2 will now use the potential update law Equation 57.272 for the system Equation 57.266,

$$\nu_2(x_1, x_2, \hat{\theta}) = \frac{\partial \alpha_1}{\partial \hat{\theta}} \Gamma \tau_2(x_1, x_2, \hat{\theta}) . \quad (57.275)$$

Hence, the role of the tuning function τ_2 is to substitute for the actual update law in compensating for the effects of parameter estimation transients.

With Equations 57.260, 57.260, 57.274, 57.272, 57.275, and the stabilizing function α_2 in Equation 57.261,

$$\begin{aligned} \begin{bmatrix} \dot{z}_1 \\ \dot{z}_2 \end{bmatrix} &= \begin{bmatrix} -c_1 & 1 \\ -1 & -c_2 \end{bmatrix} \begin{bmatrix} z_1 \\ z_2 \end{bmatrix} + \begin{bmatrix} \varphi^{\mathrm{T}} \\ -\frac{\partial \alpha_1}{\partial x_1} \varphi^{\mathrm{T}} \end{bmatrix} \tilde{\theta} \\ &+ \begin{bmatrix} 0 \\ z_3 + \frac{\partial \alpha_1}{\partial \hat{\theta}} (\Gamma \tau_2 - \dot{\hat{\theta}}) \end{bmatrix} . \end{aligned} \quad (57.276)$$

This system differs from the error system Equation 57.266 only in its last term. Likewise, instead of the Lyapunov inequality Equation 57.270,

$$\begin{aligned} \dot{V}_2 &= -c_1 z_1^2 - c_2 z_2^2 + z_2 z_3 \\ &+ z_2 \frac{\partial \alpha_1}{\partial \hat{\theta}} (\Gamma \tau_2 - \dot{\hat{\theta}}) + \tilde{\theta}^{\mathrm{T}} (\tau_2 - \Gamma^{-1} \dot{\hat{\theta}}) . \end{aligned} \quad (57.277)$$

Differentiating Equation 57.274,

$$\begin{aligned} \dot{z}_3 &= u - \frac{\partial \alpha_2}{\partial x_1} \left(x_2 + \varphi^{\mathrm{T}} \theta \right) - \frac{\partial \alpha_2}{\partial x_2} x_3 - \frac{\partial \alpha_2}{\partial \hat{\theta}} \dot{\hat{\theta}}, \quad (57.278) \\ &= u - \frac{\partial \alpha_2}{\partial x_1} \left(x_2 + \varphi^{\mathrm{T}} \hat{\theta} \right) - \frac{\partial \alpha_2}{\partial x_2} x_3 - \frac{\partial \alpha_2}{\partial \hat{\theta}} \dot{\hat{\theta}} - \frac{\partial \alpha_2}{\partial x_1} \varphi^{\mathrm{T}} \tilde{\theta} . \end{aligned}$$

We now stabilize the (z_1, z_2, z_3)-system Equations 57.276 and 57.279 with respect to the Lyapunov function

$$\begin{aligned} V_3(x, \hat{\theta}) &= V_2 + \frac{1}{2} z_3^2 \\ &= \frac{1}{2} z_1^2 + \frac{1}{2} z_2^2 + \frac{1}{2} z_3^2 + \frac{1}{2} \tilde{\theta}^{\mathrm{T}} \Gamma^{-1} \tilde{\theta} . \end{aligned} \quad (57.279)$$

Its derivative along Equations 57.276 and 57.279 is

$$\begin{aligned} \dot{V}_3 &= -c_1 z_1^2 - c_2 z_2^2 + z_2 \frac{\partial \alpha_1}{\partial \hat{\theta}} (\Gamma \tau_2 - \dot{\hat{\theta}}) \\ &+ z_3 \left[z_2 + u - \frac{\partial \alpha_2}{\partial x_1} \left(x_2 + \varphi^{\mathrm{T}} \hat{\theta} \right) - \frac{\partial \alpha_2}{\partial x_2} x_3 - \frac{\partial \alpha_2}{\partial \hat{\theta}} \dot{\hat{\theta}} \right] \\ &+ \tilde{\theta}^{\mathrm{T}} \left(\tau_2 - \frac{\partial \alpha_2}{\partial x_1} \varphi z_3 - \Gamma^{-1} \dot{\hat{\theta}} \right) . \end{aligned} \quad (57.280)$$

Again we must eliminate the unknown parameter error $\tilde{\theta}$ from $\dot{V}_3$. For this we must choose the update law as

$$\begin{aligned} \dot{\hat{\theta}} &= \Gamma \tau_3(x_1, x_2, x_3, \hat{\theta}) = \Gamma \left(\tau_2 - \frac{\partial \alpha_2}{\partial x_1} \varphi z_3 \right) \\ &= \Gamma \left[\varphi, \ -\frac{\partial \alpha_1}{\partial x_1} \varphi, \ \frac{\partial \alpha_2}{\partial x_1} \varphi \right] \begin{bmatrix} z_1 \\ z_2 \\ z_3 \end{bmatrix} . \end{aligned} \quad (57.281)$$

Upon inspecting the bracketed terms in $\dot{V}_3$, we pick the control law,

$$\begin{aligned} u &= \alpha_3(x_1, x_2, x_3, \hat{\theta}) \quad (57.282) \\ &= -z_2 - c_3 z_3 + \frac{\partial \alpha_2}{\partial x_1} \left(x_2 + \varphi^{\mathrm{T}} \hat{\theta} \right) + \frac{\partial \alpha_2}{\partial x_2} x_3 + \nu_3 . \end{aligned}$$

The compensation term ν_3 is yet to be chosen. Substituting Equation 57.282 into Equation 57.280,

$$\begin{aligned} \dot{V}_3 &= -c_1 z_1^2 - c_2 z_2^2 - c_3 z_3^2 + z_2 \frac{\partial \alpha_1}{\partial \hat{\theta}} (\Gamma \tau_2 - \dot{\hat{\theta}}) \\ &+ z_3 \left(\nu_3 - \frac{\partial \alpha_2}{\partial \hat{\theta}} \dot{\hat{\theta}} \right) . \end{aligned} \quad (57.283)$$

From this expression it is clear that ν_3 should cancel $\frac{\partial \alpha_2}{\partial \hat{\theta}} \dot{\hat{\theta}}$. In order to cancel the cross-term $z_2 \frac{\partial \alpha_1}{\partial \hat{\theta}} (\Gamma \tau_2 - \dot{\hat{\theta}})$ with ν_3, we need to divide by z_3. However, the variable z_3 might take a zero value during the transient, and should be regulated to zero to accomplish the control objective. We resolve this difficulty by noting that

$$\begin{aligned} \dot{\hat{\theta}} - \Gamma \tau_2 &= \dot{\hat{\theta}} - \Gamma \tau_3 + \Gamma \tau_3 - \Gamma \tau_2 \\ &= \dot{\hat{\theta}} - \Gamma \tau_3 - \Gamma \frac{\partial \alpha_2}{\partial x_1} \varphi z_3 , \end{aligned} \quad (57.284)$$

so that $\dot{V}_3$ in Equation 57.283 is rewritten as

$$\begin{aligned} \dot{V}_3 &= -c_1 z_1^2 - c_2 z_2^2 - c_3 z_3^2 \quad (57.285) \\ &+ z_3 \left(\nu_3 - \frac{\partial \alpha_2}{\partial \hat{\theta}} \Gamma \tau_3 + \frac{\partial \alpha_1}{\partial \hat{\theta}} \Gamma \frac{\partial \alpha_2}{\partial x_1} \varphi z_2 \right) . \quad (57.286) \end{aligned}$$

From Equation 57.285 the choice of ν_3 is immediate:

$$\nu_3(x_1, x_2, x_3, \hat{\theta}) = \frac{\partial \alpha_2}{\partial \hat{\theta}} \Gamma \tau_3 - \frac{\partial \alpha_1}{\partial \hat{\theta}} \Gamma \frac{\partial \alpha_2}{\partial x_1} \varphi z_2 . \quad (57.287)$$

The resulting $\dot{V}_3$ is

$$\dot{V}_3 = -c_1 z_1^2 - c_2 z_2^2 - c_3 z_3^2 , \quad (57.288)$$

which guarantees that the equilibrium $x = 0$, $\hat{\theta} = \theta$ is globally stable, and $x(t) \to 0$ as $t \to \infty$. The Lyapunov design leading to Equation 57.285 is effective but does not reveal the stabilization mechanism. To provide further insight we write the (z_1, z_2, z_3)-system Equations 57.276 and 57.279 with u given in Equation 57.282 but with ν_3 yet to be selected:

$$\begin{bmatrix} \dot{z}_1 \\ \dot{z}_2 \\ \dot{z}_3 \end{bmatrix} = \begin{bmatrix} -c_1 & 1 & 0 \\ -1 & -c_2 & 1 \\ 0 & -1 & -c_3 \end{bmatrix} \begin{bmatrix} z_1 \\ z_2 \\ z_3 \end{bmatrix}$$

$$+\begin{bmatrix} \varphi^{\mathrm{T}} \\ -\frac{\partial\alpha_1}{\partial x_1}\varphi^{\mathrm{T}} \\ -\frac{\partial\alpha_2}{\partial x_1}\varphi^{\mathrm{T}} \end{bmatrix}\tilde{\theta} + \begin{bmatrix} 0 \\ \frac{\partial\alpha_1}{\partial\hat{\theta}}(\Gamma\tau_2 - \dot{\hat{\theta}}) \\ \nu_3 - \frac{\partial\alpha_2}{\partial\hat{\theta}}\Gamma\tau_3 \end{bmatrix}. \tag{57.289}$$

While ν_3 can cancel the matched term $\frac{\partial\alpha_2}{\partial\hat{\theta}}\dot{\hat{\theta}}$, it cannot cancel the term $\frac{\partial\alpha_1}{\partial\hat{\theta}}(\Gamma\tau_2 - \dot{\hat{\theta}})$ in the second equation. By substituting Equation 57.284, we note that $\frac{\partial\alpha_1}{\partial\hat{\theta}}(\Gamma\tau_2 - \dot{\hat{\theta}})$ has z_3 as a factor and absorb it into the "system matrix":

$$\begin{bmatrix} \dot{z}_1 \\ \dot{z}_2 \\ \dot{z}_3 \end{bmatrix} = \begin{bmatrix} -c_1 & 1 & 0 \\ -1 & -c_2 & 1 + \frac{\partial\alpha_1}{\partial\hat{\theta}}\Gamma\frac{\partial\alpha_2}{\partial x_1}\varphi \\ 0 & -1 & -c_3 \end{bmatrix}\begin{bmatrix} z_1 \\ z_2 \\ z_3 \end{bmatrix} + \begin{bmatrix} \varphi^{\mathrm{T}} \\ -\frac{\partial\alpha_1}{\partial x_1}\varphi^{\mathrm{T}} \\ -\frac{\partial\alpha_2}{\partial x_1}\varphi^{\mathrm{T}} \end{bmatrix}\tilde{\theta} + \begin{bmatrix} 0 \\ 0 \\ \nu_3 - \frac{\partial\alpha_2}{\partial\hat{\theta}}\Gamma\tau_3 \end{bmatrix}. \tag{57.290}$$

Now ν_3 in (57.287) yields

$$\begin{bmatrix} \dot{z}_1 \\ \dot{z}_2 \\ \dot{z}_3 \end{bmatrix} = \begin{bmatrix} -c_1 & 1 & 0 \\ -1 & -c_2 & 1 + \frac{\partial\alpha_1}{\partial\hat{\theta}}\Gamma\frac{\partial\alpha_2}{\partial x_1}\varphi \\ 0 & -1 - \frac{\partial\alpha_1}{\partial\hat{\theta}}\Gamma\frac{\partial\alpha_2}{\partial x_1}\varphi & -c_3 \end{bmatrix}\begin{bmatrix} z_1 \\ z_2 \\ z_3 \end{bmatrix} + \begin{bmatrix} \varphi^{\mathrm{T}} \\ -\frac{\partial\alpha_1}{\partial x_1}\varphi^{\mathrm{T}} \\ -\frac{\partial\alpha_2}{\partial x_1}\varphi^{\mathrm{T}} \end{bmatrix}\tilde{\theta}. \tag{57.291}$$

This choice, which places the term $-\frac{\partial\alpha_1}{\partial\hat{\theta}}\Gamma\frac{\partial\alpha_2}{\partial x_1}\varphi$ at the (2,3) position in the system matrix, achieves skew-symmetry with its positive image above the diagonal. What could not be achieved by pursuing a linearlike form, was achieved by designing a nonlinear system where the nonlinearities are 'balanced' rather than cancelled.

General Recursive Design Procedure

A systematic backstepping design with tuning functions has been developed for the class of nonlinear systems transformable into the **parametric strict-feedback form,**

$$\begin{aligned} \dot{x}_1 &= x_2 + \varphi_1(x_1)^{\mathrm{T}}\theta, \\ \dot{x}_2 &= x_3 + \varphi_2(x_1, x_2)^{\mathrm{T}}\theta, \\ &\vdots \\ \dot{x}_{n-1} &= x_n + \varphi_{n-1}(x_1, \ldots, x_{n-1})^{\mathrm{T}}\theta, \\ \dot{x}_n &= \beta(x)u + \varphi_n(x)^{\mathrm{T}}\theta, \\ y &= x_1, \end{aligned} \tag{57.292}$$

where β and

$$F(x) = [\varphi_1(x_1),\ \varphi_2(x_1, x_2),\ \cdots,\ \varphi_n(x)] \tag{57.293}$$

are smooth nonlinear functions, and $\beta(x) \neq 0$, $\forall x \in \mathbb{R}^n$. (Broader classes of systems that can be controlled by adaptive backstepping are listed in Section 57.8.5).

The general design summarized in Table 57.1 achieves asymptotic tracking, that is, the output $y = x_1$ of the system Equation 57.292 is forced to track asymptotically the reference output $y_{\mathrm{r}}(t)$ whose first n derivatives are assumed to be known, bounded and piecewise continuous.

TABLE 57.1 Summary of the Tuning Functions Design for Tracking. (For notational convenience we define $z_0 \triangleq 0, \alpha_0 \triangleq 0, \tau_0 \triangleq 0$.)

$$z_i = x_i - y_{\mathrm{r}}^{(i-1)} - \alpha_{i-1}. \tag{57.294}$$

$$\alpha_i(\bar{x}_i, \hat{\theta}, \bar{y}_{\mathrm{r}}^{(i-1)}) = -z_{i-1} - c_i z_i - w_i^{\mathrm{T}}\hat{\theta} + \sum_{k=1}^{i-1}\left(\frac{\partial\alpha_{i-1}}{\partial x_k}x_{k+1} + \frac{\partial\alpha_{i-1}}{\partial y_{\mathrm{r}}^{(k-1)}}y_{\mathrm{r}}^{(k)}\right) + \nu_i. \tag{57.295}$$

$$\nu_i(\bar{x}_i, \hat{\theta}, \bar{y}_{\mathrm{r}}^{(i-1)}) = \frac{\partial\alpha_{i-1}}{\partial\hat{\theta}}\Gamma\tau_i + \sum_{k=2}^{i-1}\frac{\partial\alpha_{k-1}}{\partial\hat{\theta}}\Gamma w_i z_k. \tag{57.296}$$

$$\tau_i(\bar{x}_i, \hat{\theta}, \bar{y}_{\mathrm{r}}^{(i-1)}) = \tau_{i-1} + w_i z_i. \tag{57.297}$$

$$w_i(\bar{x}_i, \hat{\theta}, \bar{y}_{\mathrm{r}}^{(i-2)}) = \varphi_i - \sum_{k=1}^{i-1}\frac{\partial\alpha_{i-1}}{\partial x_k}\varphi_k \tag{57.298}$$

$$i = 1, \ldots, n$$

$$\bar{x}_i = (x_1, \ldots, x_i),\ \bar{y}_{\mathrm{r}}^{(i)} = (y_{\mathrm{r}}, \dot{y}_{\mathrm{r}}, \ldots, y_{\mathrm{r}}^{(i)}).$$

Adaptive control law:

$$u = \frac{1}{\beta(x)}\left[\alpha_n(x, \hat{\theta}, \bar{y}_{\mathrm{r}}^{(n-1)}) + y_{\mathrm{r}}^{(n)}\right]. \tag{57.299}$$

Parameter update law:

$$\dot{\hat{\theta}} = \Gamma\tau_n(x, \hat{\theta}, \bar{y}_{\mathrm{r}}^{(n-1)}) = \Gamma W z. \tag{57.300}$$

The closed-loop system has the form

$$\dot{z} = A_z(z, \hat{\theta}, t)z + W(z, \hat{\theta}, t)^{\mathrm{T}}\tilde{\theta} \tag{57.301}$$

$$\dot{\hat{\theta}} = \Gamma W(z, \hat{\theta}, t)z, \tag{57.302}$$

where

$$A_z(z, \hat{\theta}, t) = \begin{bmatrix} -c_1 & 1 & 0 & \cdots & 0 \\ -1 & -c_2 & 1 + \sigma_{23} & \cdots & \sigma_{2n} \\ 0 & -1 - \sigma_{23} & \ddots & \ddots & \vdots \\ \vdots & \vdots & \ddots & \ddots & 1 + \sigma_{n-1,n} \\ 0 & -\sigma_{2n} & \cdots & -1 - \sigma_{n-1,n} & -c_n \end{bmatrix} \tag{57.303}$$

and

$$\sigma_{jk}(x,\hat{\theta}) = -\frac{\partial\alpha_{j-1}}{\partial\hat{\theta}}\Gamma w_k\,. \tag{57.304}$$

Because of the skew-symmetry of the off-diagonal part of the matrix A_z, it is easy to see that the Lyapunov function

$$V_n = \frac{1}{2}z^{\mathrm{T}}z + \frac{1}{2}\tilde{\theta}^{\mathrm{T}}\Gamma^{-1}\tilde{\theta} \tag{57.305}$$

has the derivative

$$\dot{V}_n = -\sum_{k=1}^{n} c_k z_k^2\,, \tag{57.306}$$

which guarantees that the equilibrium $z = 0, \hat{\theta} = \theta$ is globally stable, and $z(t) \to 0$ as $t \to \infty$. This means, in particular, that the system state and the control input are bounded and asymptotic tracking is achieved: $\lim_{t\to\infty}[y(t) - y_r(t)] = 0$.

To help understand how the control design of Table 57.1 leads to the closed-loop system Equations 57.301–57.304, we provide an interpretation of the matrix A_z for $n = 5$:

$$A_z = \begin{bmatrix} -c_1 & 1 & & & \\ -1 & -c_2 & 1 & & \\ & -1 & -c_3 & 1 & \\ & & -1 & -c_4 & 1 \\ & & & -1 & -c_5 \end{bmatrix} + \begin{bmatrix} & 0 & 0 & 0 & 0 \\ 0 & & \sigma_{23} & \sigma_{24} & \sigma_{25} \\ 0 & -\sigma_{23} & & \sigma_{34} & \sigma_{35} \\ 0 & -\sigma_{24} & -\sigma_{34} & & \sigma_{45} \\ 0 & -\sigma_{25} & -\sigma_{35} & -\sigma_{45} & \end{bmatrix}. \tag{57.307}$$

If the parameters were known, $\hat{\theta} \equiv \theta$, in which case we would not use adaptation, $\Gamma = 0$, the stabilizing functions Equation 57.295 would be implemented with $\nu_i \equiv 0$, and hence $\sigma_{i,j} = 0$. Then A_z would be just the above constant tridiagonal asymptotically stable matrix. When the parameters are unknown, we use $\Gamma > 0$ and, due to the change of variable $z_i = x_i - y_r^{(i-1)} - \alpha_{i-1}$, in each of the $\dot{z}_i$-equations, a term $-\frac{\partial\alpha_{i-1}}{\partial\hat{\theta}}\dot{\hat{\theta}} = \sum_{k=1}^{n}\sigma_{ik}z_k$ appears. The term $\nu_i = -\sum_{k=1}^{i}\sigma_{ik}z_k - \sum_{k=2}^{i-1}\sigma_{ki}z_k$ in the stabilizing function Equation 57.295 is crucial in compensating for the effect of $\dot{\hat{\theta}}$. The σ_{ik}-terms above the diagonal in Equation 57.307 come from $\dot{\hat{\theta}}$. Their skew-symmetric negative images come from feedback ν_i.

It can be shown that the resulting closed-loop system Equations 57.302 and 57.301, as well as each intermediate system, has a *strict passivity* property from $\tilde{\theta}$ as the input to τ_i as the output. The loop around this operator is closed (see Figure 57.25) with the vector integrator with gain Γ, which is a passive block. It follows from passivity theory that this feedback connection of one strictly passive and one passive block is globally stable.

57.8.3 Modular Design

In the tuning functions design, the controller and the identifier are derived in an interlaced fashion. This interlacing led to considerable controller complexity and inflexibility in the choice of the update law.

It is not hard to extend various standard identifiers for linear systems to nonlinear systems. It is therefore desirable to have adaptive designs where the controller can be combined with different identifiers (gradient, least-squares, passivity based, etc.). We refer to such adaptive designs as **modular.**

In nonlinear systems it is not a good idea to connect a good identifier with a controller which is good when the parameter is known (a "certainty equivalence" controller). To illustrate this, let us consider the error system

$$\dot{x} = -x + \varphi(x)\tilde{\theta} \tag{57.308}$$

obtained by applying a certainty equivalence controller $u = -x - \varphi(x)\hat{\theta}$ to the scalar system $\dot{x} = u + \varphi(x)\theta$. The parameter estimators commonly used in adaptive linear control generate bounded estimates $\hat{\theta}(t)$ with convergence rates not faster than exponential. Suppose that $\tilde{\theta}(t) = \mathrm{e}^{-t}$ and $\varphi(x) = x^3$, which, upon substitution in Equation 57.308, gives

$$\dot{x} = -x + x^3\mathrm{e}^{-t}\,. \tag{57.309}$$

For initial conditions $|x_0| > \sqrt{\frac{3}{2}}$, the system Equation 57.309 is unstable, and its solution escapes to infinity in finite time:

$$x(t) \to \infty \quad \text{as} \quad t \to \frac{1}{3}\ln\frac{x_0^2}{x_0^2 - 3/2}\,. \tag{57.310}$$

From this example we conclude that, for nonlinear systems, we need stronger controllers which prevent unbounded behavior caused by $\tilde{\theta}$.

Controller Design

We strengthen the controller for the preceding example, $u = -x - \varphi(x)\hat{\theta}$, with a **nonlinear damping** term $-\varphi(x)^2x$, that is, $u = -x - \varphi(x)\hat{\theta} - \varphi(x)^2x$. With this stronger controller, the closed-loop system is

$$\dot{x} = -x - \varphi(x)^2x + \varphi(x)\tilde{\theta}\,. \tag{57.311}$$

To see that x is bounded whenever $\tilde{\theta}$ is, we consider the Lyapunov function $V = \frac{1}{2}x^2$. Its derivative along the solutions of Equation 57.311 is

$$\begin{aligned} \dot{V} &= -x^2 - \varphi(x)^2x^2 + x\varphi(x)\tilde{\theta}, \\ &= -x^2 - \left[\varphi(x)x - \frac{1}{2}\tilde{\theta}\right]^2 + \frac{1}{4}\tilde{\theta}^2, \\ &\le -x^2 + \frac{1}{4}\tilde{\theta}^2\,. \end{aligned} \tag{57.312}$$

From this inequality it is clear that $|x(t)|$ will not grow larger than $\frac{1}{2}|\tilde{\theta}(t)|$, because then $\dot{V}$ becomes negative and $V = \frac{1}{2}x^2$ decreases. Thanks to the nonlinear damping, the boundedness of $\tilde{\theta}(t)$ guarantees that $x(t)$ is bounded.

To show how nonlinear damping is incorporated into a higher-order backstepping design, we consider the system

$$\begin{aligned} \dot{x}_1 &= x_2 + \varphi(x_1)^{\mathrm{T}}\theta, \\ \dot{x}_2 &= u\,. \end{aligned} \tag{57.313}$$

Viewing x_2 as a control input, we first design a control law $\alpha_1(x_1, \hat{\theta})$ to guarantee that the state x_1 in $\dot{x}_1 = x_2 + \varphi(x_1)^{\mathrm{T}}\theta$ is bounded whenever $\tilde{\theta}$ is bounded. In the first stabilizing function we include a nonlinear damping term[8] $-\kappa_1|\varphi(x_1)|^2 x_1$:

$$\begin{aligned} \alpha_1(x_1, \hat{\theta}) &= -c_1 x_1 - \varphi(x_1)^{\mathrm{T}}\hat{\theta} - \kappa_1|\varphi(x_1)|^2 x_1, \\ & \quad c_1, \kappa_1 > 0\,. \end{aligned} \tag{57.314}$$

Then we define the error variable $z_2 = x_2 - \alpha_1(x_1, \hat{\theta})$, and for uniformity denote $z_1 = x_1$. The first equation is now

$$\dot{z}_1 = -c_1 z_1 - \kappa_1|\varphi|^2 z_1 + \varphi^{\mathrm{T}}\tilde{\theta} + z_2\,. \tag{57.315}$$

If z_2 were zero, the Lyapunov function $V_1 = \frac{1}{2}z_1^2$ would have the derivative

$$\begin{aligned} \dot{V}_1 &= -c_1 z_1^2 - \kappa_1|\varphi|^2 z_1^2 + z_1\varphi^{\mathrm{T}}\tilde{\theta} \\ &= -c_1 z_1^2 - \kappa_1\left|\varphi z_1 - \frac{1}{2\kappa_1}\tilde{\theta}\right|^2 + \frac{1}{4\kappa_1}|\tilde{\theta}|^2 \\ &\le -c_1 z_1^2 + \frac{1}{4\kappa_1}|\tilde{\theta}|^2\,, \end{aligned} \tag{57.316}$$

so that z_1 would be bounded whenever $\tilde{\theta}$ is bounded. With $z_2 \neq 0$,

$$\dot{V}_1 \le -c_1 z_1^2 + \frac{1}{4\kappa_1}|\tilde{\theta}|^2 + z_1 z_2\,. \tag{57.317}$$

Differentiating $x_2 = z_2 + \alpha_1(x_1, \hat{\theta})$, the second expression in Equation 57.313 yields

$$\dot{z}_2 = \dot{x}_2 - \dot{\alpha}_1 = u - \frac{\partial\alpha_1}{\partial x_1}\left(x_2 + \varphi^{\mathrm{T}}\theta\right) - \frac{\partial\alpha_1}{\partial\hat{\theta}}\dot{\hat{\theta}}\,. \tag{57.318}$$

The derivative of the Lyapunov function

$$V_2 = V_1 + \frac{1}{2}z_2^2 = \frac{1}{2}|z|^2 \tag{57.319}$$

along the solutions of Equations 57.315 and 57.318 is

$$\begin{aligned} \dot{V}_2 &\le -c_1 z_1^2 + \frac{1}{4\kappa_1}|\tilde{\theta}|^2 + z_1 z_2 \\ &\quad + z_2\left[u - \frac{\partial\alpha_1}{\partial x_1}\left(x_2 + \varphi^{\mathrm{T}}\theta\right) - \frac{\partial\alpha_1}{\partial\hat{\theta}}\dot{\hat{\theta}}\right], \\ &\le -c_1 z_1^2 + \frac{1}{4\kappa_1}|\tilde{\theta}|^2 + z_2 \\ &\quad \left[u + z_1 - \frac{\partial\alpha_1}{\partial x_1}\left(x_2 + \varphi^{\mathrm{T}}\hat{\theta}\right) - \left(\frac{\partial\alpha_1}{\partial x_1}\varphi^{\mathrm{T}}\tilde{\theta} + \frac{\partial\alpha_1}{\partial\hat{\theta}}\dot{\hat{\theta}}\right)\right]. \end{aligned} \tag{57.320}$$

We note that, in addition to the $\tilde{\theta}$-dependent disturbance term $\frac{\partial\alpha_1}{\partial x_1}\varphi^{\mathrm{T}}\tilde{\theta}$, we also have a $\dot{\hat{\theta}}$-dependent disturbance $\frac{\partial\alpha_1}{\partial\hat{\theta}}\dot{\hat{\theta}}$. No such term appeared in the scalar system Equation 57.311. We now use nonlinear damping terms $-\kappa_2\left|\frac{\partial\alpha_1}{\partial x_1}\varphi\right|^2 z_2$ and $-g_2\left|\frac{\partial\alpha_1}{\partial\hat{\theta}}^{\mathrm{T}}\right|^2 z_2$ to counteract the effects of both $\tilde{\theta}$ and $\dot{\hat{\theta}}$:

$$\begin{aligned} u &= -z_1 - c_2 z_2 - \kappa_2\left|\frac{\partial\alpha_1}{\partial x_1}\varphi\right|^2 z_2 - g_2\left|\frac{\partial\alpha_1}{\partial\hat{\theta}}^{\mathrm{T}}\right|^2 z_2 \\ &\quad + \frac{\partial\alpha_1}{\partial x_1}\left(x_2 + \varphi^{\mathrm{T}}\hat{\theta}\right), \end{aligned} \tag{57.321}$$

where $c_2, \kappa_2, g_2 > 0$. Upon completing the squares, as in Equation 57.316,

$$\dot{V}_2 \le -c_1 z_1^2 - c_2 z_2^2 + \left(\frac{1}{4\kappa_1} + \frac{1}{4\kappa_2}\right)|\tilde{\theta}|^2 + \frac{1}{4g_2}|\dot{\hat{\theta}}|^2\,, \tag{57.322}$$

which means that the state of the error system,

$$\begin{aligned} \dot{z} &= \begin{bmatrix} -c_1 - \kappa_2|\varphi|^2 & 1 \\ -1 & -c_2 - \kappa_2\left|\frac{\partial\alpha_1}{\partial x_1}\varphi\right|^2 - g_2\left|\frac{\partial\alpha_1}{\partial\hat{\theta}}^{\mathrm{T}}\right|^2 \end{bmatrix} z \\ &\quad + \begin{bmatrix} \varphi^{\mathrm{T}} \\ -\frac{\partial\alpha_1}{\partial x_1}\varphi^{\mathrm{T}} \end{bmatrix}\tilde{\theta} + \begin{bmatrix} 0 \\ -\frac{\partial\alpha_1}{\partial\hat{\theta}} \end{bmatrix}\dot{\hat{\theta}}, \end{aligned} \tag{57.323}$$

is bounded whenever the disturbance inputs $\tilde{\theta}$ and $\dot{\hat{\theta}}$ are bounded. Moreover, because V_2 is quadratic in z, see Equation 57.319, we can use Equation 57.322 to show that the boundedness of z is guaranteed also when $\dot{\hat{\theta}}$ is square-integrable but not bounded. This observation is crucial for the modular design with passive identifiers where $\dot{\hat{\theta}}$ cannot be a priori guaranteed as bounded.

The recursive controller design for the parametric strict-feedback systems Equation 57.292 is summarized in Table 57.2.

Comparing the expression for the stabilizing function Equation 57.325 in the modular design with Equation 57.295 for the tuning functions design we see that the difference is in the second lines. Though the stabilization in the tuning functions design is achieved with the terms ν_i, in the modular design stabilization is accomplished with the nonlinear damping term, $s_i z_i$, where

$$s_i(\bar{x}_i, \hat{\theta}, \bar{y}_r^{(i-2)}) = \kappa_i|w_i|^2 + g_i\left|\frac{\partial\alpha_{i-1}}{\partial\hat{\theta}}^{\mathrm{T}}\right|^2. \tag{57.330}$$

The resulting error system is

$$\dot{z} = A_z(z, \hat{\theta}, t)z + W(z, \hat{\theta}, t)^{\mathrm{T}}\tilde{\theta} + Q(z, \hat{\theta}, t)^{\mathrm{T}}\dot{\hat{\theta}} \tag{57.331}$$

where A_z, W, Q are

$$A_z(z, \hat{\theta}, t) = \begin{bmatrix} -c_1 - s_1 & 1 & 0 & \cdots & 0 \\ -1 & -c_2 - s_2 & 1 & \ddots & \vdots \\ 0 & -1 & \ddots & \ddots & 0 \\ \vdots & \ddots & \ddots & \ddots & 1 \\ 0 & \cdots & 0 & -1 & -c_n - s_n \end{bmatrix}, \tag{57.332}$$

[8]The Euclidian norm of a vector v is denoted as $|v| = \sqrt{v^{\mathrm{T}}v}$.

TABLE 57.2 Summary of the Controller Design in the Modular Approach. (For notational convenience we define $z_0 \stackrel{\Delta}{=} 0, \alpha_0 \stackrel{\Delta}{=} 0$.)

$$z_i = x_i - y_r^{(i-1)} - \alpha_{i-1} \quad (57.324)$$

$$\alpha_i(\bar{x}_i, \hat{\theta}, \bar{y}_r^{(i-1)}) = -z_{i-1} - c_i z_i - w_i^T \hat{\theta} + \sum_{k=1}^{i-1}\left(\frac{\partial \alpha_{i-1}}{\partial x_k} x_{k+1} + \frac{\partial \alpha_{i-1}}{\partial y_r^{(k-1)}} y_r^{(k)}\right) - s_i z_i \quad (57.325)$$

$$w_i(\bar{x}_i, \hat{\theta}, \bar{y}_r^{(i-2)}) = \varphi_i - \sum_{k=1}^{i-1} \frac{\partial \alpha_{i-1}}{\partial x_k} \varphi_k \quad (57.326)$$

$$s_i(\bar{x}_i, \hat{\theta}, \bar{y}_r^{(i-2)}) = \kappa_i |w_i|^2 + g_i \left|\frac{\partial \alpha_{i-1}}{\partial \hat{\theta}}^T\right|^2 \quad (57.327)$$

$$i = 1, \ldots, n$$
$$\bar{x}_i = (x_1, \ldots, x_i),$$
$$\bar{y}_r^{(i)} = (y_r, \dot{y}_r, \ldots, y_r^{(i)}) \quad (57.328)$$

Adaptive control law:

$$u = \frac{1}{\beta(x)}\left[\alpha_n(x, \hat{\theta}, \bar{y}_r^{(n-1)}) + y_r^{(n)}\right] \quad (57.329)$$

Controller module guarantees:

If $\tilde{\theta} \in \mathcal{L}_\infty$ and $\dot{\hat{\theta}} \in \mathcal{L}_2$ or $\mathcal{L}_\infty$, then $x \in \mathcal{L}_\infty$

$$W(z, \hat{\theta}, t)^T = \begin{bmatrix} w_1^T \\ w_2^T \\ \vdots \\ w_n^T \end{bmatrix}$$

$$\text{and } Q(z, \hat{\theta}, t)^T = \begin{bmatrix} 0 \\ -\frac{\partial \alpha_1}{\partial \hat{\theta}} \\ \vdots \\ -\frac{\partial \alpha_{n-1}}{\partial \hat{\theta}} \end{bmatrix}. \quad (57.333)$$

Since the controller module guarantees that x is bounded whenever $\tilde{\theta}$ is bounded and $\dot{\hat{\theta}}$ is either bounded or square-integrable, then we need identifiers which independently guarantee these properties. Both the boundedness and the square-integrability requirements for $\dot{\hat{\theta}}$ are essentially conditions which limit the speed of adaptation, and only one of them needs to be satisfied. The modular design needs *slow adaptation* because the controller does not cancel the effect of $\dot{\hat{\theta}}$, as was the case in the tuning functions design.

In addition to boundedness of $x(t)$, our goal is to achieve asymptotic tracking, that is, to regulate $z(t)$ to zero. With z and $\hat{\theta}$ bounded, it is not hard to prove that $z(t) \to 0$ provided

$$W(z(t), \hat{\theta}(t), t)^T \tilde{\theta}(t) \to 0 \text{ and } \dot{\hat{\theta}}(t) \to 0.$$

Let us factor the regressor matrix W, using Equations 57.333, 57.326 and 57.293, as

$$W(z, \hat{\theta}, t)^T = \begin{bmatrix} 1 & 0 & \cdots & 0 \\ -\frac{\partial \alpha_1}{\partial x_1} & 1 & \ddots & \vdots \\ \vdots & \ddots & \ddots & 0 \\ -\frac{\partial \alpha_{n-1}}{\partial x_1} & \cdots & -\frac{\partial \alpha_{n-1}}{\partial x_{n-1}} & 1 \end{bmatrix} F(x)^T \stackrel{\Delta}{=} N(z, \hat{\theta}, t) F(x)^T. \quad (57.334)$$

Since the matrix $N(z, \hat{\theta}, t)$ is invertible, the tracking condition $W(z(t), \hat{\theta}(t), t)^T \tilde{\theta}(t) \to 0$ becomes $F(x(t))^T \tilde{\theta}(t) \to 0$.

In the next two subsections we develop identifiers for the general parametric model

$$\dot{x} = f(x, u) + F(x, u)^T \theta. \quad (57.335)$$

The parametric strict-feedback system Equation 57.292 is a special case of this model with $F(x, u)$ given by Equation 57.293 and $f(x, u) = [x_2, \ldots, x_n, \beta_0(x)u]^T$.

Before we present the design of identifiers, we summarize the properties required from the identifier module:

(i) $\tilde{\theta} \in \mathcal{L}_\infty$ and $\dot{\hat{\theta}} \in \mathcal{L}_2$ or $\mathcal{L}_\infty$,

(ii) if $x \in \mathcal{L}_\infty$, then $F(x(t))^T \tilde{\theta}(t) \to 0$ and $\dot{\hat{\theta}}(t) \to 0$.

We present two types of identifiers: the **passive** identifier and the **swapping** identifier.

Passive Identifier

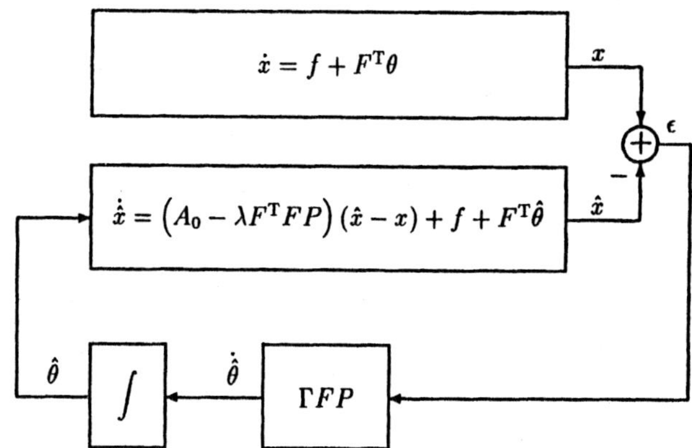

Figure 57.26 The passive identifier.

For the parametric model Equation 57.335, we implement the "observer"

$$\dot{\hat{x}} = \left[A_0 - \lambda F(x, u)^T F(x, u) P\right](\hat{x} - x)$$

$$+ f(x,u) + F(x,u)^{\mathrm{T}}\hat{\theta}\,, \tag{57.336}$$

where $\lambda > 0$ and A_0 is an arbitrary constant matrix so that

$$PA_0 + A_0^{\mathrm{T}}P = -I\,, \quad P = P^{\mathrm{T}} > 0\,. \tag{57.337}$$

By direct substitution it can be seen that the observer error

$$\epsilon = x - \hat{x} \tag{57.338}$$

is governed by

$$\dot{\epsilon} = \left[A_0 - \lambda F(x,u)^{\mathrm{T}}F(x,u)P\right]\epsilon + F(x,u)^{\mathrm{T}}\tilde{\theta}\,. \tag{57.339}$$

The observer error system Equation 57.339 has a *strict passivity* property from the input $\tilde{\theta}$ to the output $F(x,u)P\epsilon$.

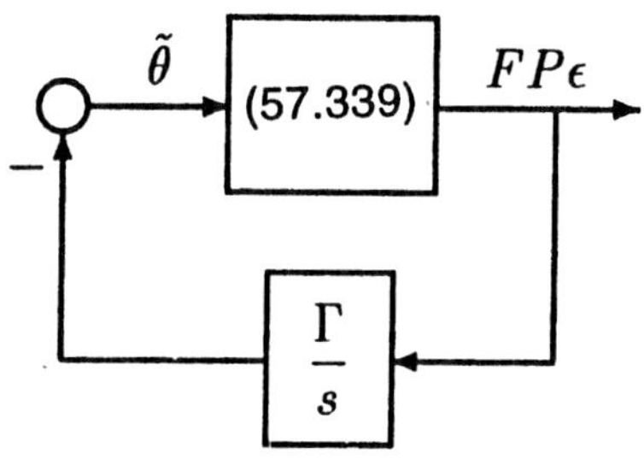

Figure 57.27 Negative feedback connection of the strictly passive system Equation 57.339 with the passive system $\frac{\Gamma}{s}$.

A standard result of passivity theory is that the equilibrium $\tilde{\theta} = 0$, $\epsilon = 0$, of the negative feedback connection of one strictly passive and one passive system is globally stable. Using integral feedback, such a connection can be formed as in Figure 57.27. This suggests the use of the following update law

$$\dot{\hat{\theta}} = \Gamma F(x,u)P\epsilon\,, \qquad \Gamma = \Gamma^{\mathrm{T}} > 0\,. \tag{57.340}$$

To analyze the stability properties of the passive identifier, we use the Lyapunov function

$$V = \tilde{\theta}^{\mathrm{T}}\Gamma^{-1}\tilde{\theta} + \epsilon^{\mathrm{T}}P\epsilon\,. \tag{57.341}$$

After uncomplicated calculations, its derivative can be shown to satisfy

$$\dot{V} \le -\epsilon^{\mathrm{T}}\epsilon - \frac{\lambda}{\bar{\lambda}(\Gamma)^2}|\dot{\hat{\theta}}|^2\,. \tag{57.342}$$

This guarantees the boundedness of $\tilde{\theta}$ and ϵ, even when $\lambda = 0$. However, $\dot{\hat{\theta}}$ cannot be shown to be bounded (unless x and u are known to be bounded). Instead, for the passive identifier one can show that $\dot{\hat{\theta}}$ is square integrable. For this we must use $\lambda > 0$, that is, we rely on the nonlinear damping term $-\lambda F(x,u)^{\mathrm{T}}F(x,u)P$ in the observer. The boundedness of $\tilde{\theta}$ and the square-integrability of $\dot{\hat{\theta}}$ imply (cf. Table 57.2) that x is bounded.

To prove the tracking, we need to show that the identifier guarantees that, whenever x is bounded, $F(x(t))^{\mathrm{T}}\tilde{\theta}(t) \to 0$ and $\dot{\hat{\theta}}(t) \to 0$. Both properties are established by Barbalat's lemma. The latter property can easily be shown to follow from the square-integrability of $\dot{\hat{\theta}}$. The regulation of $F(x)^{\mathrm{T}}\tilde{\theta}$ to zero follows upon showing that both $\epsilon(t)$ and $\dot{\epsilon}(t)$ converge to zero. Though the convergence of $\epsilon(t)$ follows by deducing its square-integrability from Equation 57.342, the convergence of $\dot{\epsilon}(t)$ follows from the fact that its integral, $\int_0^\infty \dot{\epsilon}(\tau)d\tau = \epsilon(\infty) - \epsilon(0) = -\epsilon(0)$, exists.

Swapping Identifier

For the parametric model Equation 57.335, we implement two filters,

$$\begin{aligned}\dot{\Omega}^{\mathrm{T}} &= \left[A_0 - \lambda F(x,u)^{\mathrm{T}}F(x,u)P\right]\Omega^{\mathrm{T}} \\ &\quad + F(x,u)^{\mathrm{T}}\,, \end{aligned} \tag{57.343}$$

and

$$\begin{aligned}\dot{\Omega}_0 &= \left[A_0 - \lambda F(x,u)^{\mathrm{T}}F(x,u)P\right](\Omega_0 - x) \\ &\quad - f(x,u)\,, \end{aligned} \tag{57.344}$$

where $\lambda \ge 0$ and A_0 is as defined in Equation 57.337. The **estimation error,**

$$\epsilon = x + \Omega_0 - \Omega^{\mathrm{T}}\hat{\theta}\,. \tag{57.345}$$

can be written in the form

$$\epsilon = \Omega^{\mathrm{T}}\tilde{\theta} + \tilde{\epsilon} \tag{57.346}$$

where $\tilde{\epsilon} \triangleq x + \Omega_0 - \Omega^{\mathrm{T}}\theta$ decays exponentially because it is governed by

$$\dot{\tilde{\epsilon}} = \left[A_0 - \lambda F(x,u)^{\mathrm{T}}F(x,u)P\right]\tilde{\epsilon}\,. \tag{57.347}$$

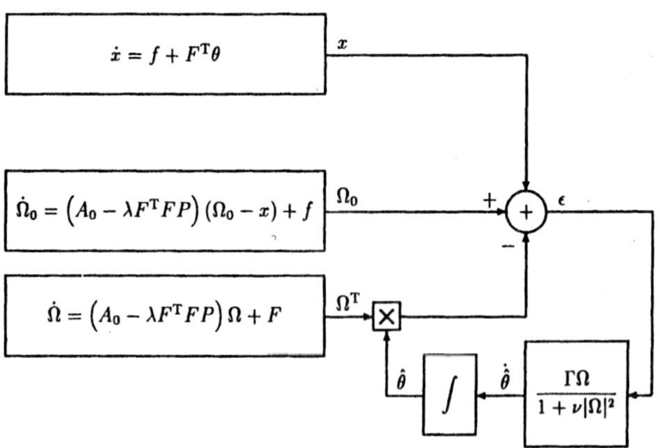

Figure 57.28 The swapping identifier.

The filters Equations 57.343 and 57.344 have converted the dynamic model Equation 57.335 into the linear static parametric model Equation 57.346 to which we can apply standard estimation algorithms. As our update law, we will employ either the gradient

$$\dot{\hat{\theta}} = \Gamma\frac{\Omega\epsilon}{1 + \nu\,\mathrm{tr}\{\Omega^{\mathrm{T}}\Omega\}}\,, \qquad \begin{array}{l}\Gamma = \Gamma^{\mathrm{T}} > 0 \\ \nu \ge 0\,,\end{array} \tag{57.348}$$

or the least squares algorithm

$$\begin{aligned}\dot{\hat{\theta}} &= \Gamma\frac{\Omega\epsilon}{1+\nu\mathrm{tr}\{\Omega^{\mathrm{T}}\Omega\}} \\ \dot{\Gamma} &= -\Gamma\frac{\Omega\Omega^{\mathrm{T}}}{1+\nu\mathrm{tr}\{\Omega^{\mathrm{T}}\Omega\}}\Gamma, \qquad \Gamma(0)=\Gamma(0)^{\mathrm{T}}>0 \\ & \qquad\qquad\qquad\qquad\qquad \nu\geq 0.\end{aligned} \tag{57.349}$$

By allowing $\nu = 0$, we encompass unnormalized gradient and least-squares. The complete swapping identifier is shown in Figure 57.28.

The update law normalization, $\nu > 0$, and the nonlinear damping, $\lambda > 0$, are two different means for slowing down the identifier in order to guarantee the boundedness and square-integrability of $\dot{\hat{\theta}}$.

For the gradient update law Equation 57.348, the identifier properties (boundedness of $\tilde{\theta}$ and $\hat{\theta}$ and regulation of $F(x)\tilde{\theta}$ and $\dot{\hat{\theta}}$) are established via the Lyapunov function

$$V = \frac{1}{2}\tilde{\theta}^{\mathrm{T}}\Gamma^{-1}\tilde{\theta} + \tilde{\epsilon}P\tilde{\epsilon} \tag{57.350}$$

whose derivative is

$$\dot{V} \leq -\frac{3}{4}\frac{\epsilon^{\mathrm{T}}\epsilon}{1+\nu\mathrm{tr}\{\Omega^{\mathrm{T}}\Omega\}}. \tag{57.351}$$

The Lyapunov function for the least-squares update law Equation 57.349 is $V = \tilde{\theta}^{\mathrm{T}}\Gamma(t)^{-1}\tilde{\theta} + \tilde{\epsilon}P\tilde{\epsilon}$.

57.8.4 Output Feedback Designs

For linear systems, a common solution to the output-feedback problem is a stabilizing state-feedback controller employing the state estimates from an exponentially converging observer. Unfortunately, this approach is not applicable to nonlinear systems. Additional difficulties arise when the nonlinear plant has unknown parameters because *adaptive* observers, in general, are not exponentially convergent.

These obstacles have been overcome for systems transformable into the **output-feedback form,**

$$\begin{aligned}\dot{x} &= Ax + \phi(y) + \Phi(y)a + \begin{bmatrix} 0 \\ b \end{bmatrix}\sigma(y)u, \quad x \in \mathbb{R}^n, \\ y &= e_1^{\mathrm{T}}x,\end{aligned} \tag{57.352}$$

where only the output y is available for measurement,

$$A = \begin{bmatrix} 0 & \\ \vdots & I_{n-1} \\ 0 & \cdots 0 \end{bmatrix}, \tag{57.353}$$

$$\phi(y) = \begin{bmatrix} \varphi_{0,1}(y) \\ \vdots \\ \varphi_{0,n}(y) \end{bmatrix},$$

$$\Phi(y) = \begin{bmatrix} \varphi_{1,1}(y) & \cdots & \varphi_{q,1}(y) \\ \vdots & & \vdots \\ \varphi_{1,n}(y) & \cdots & \varphi_{q,n}(y) \end{bmatrix}, \tag{57.354}$$

and the vectors of unknown constant parameters are

$$a = [a_1, \ldots, a_q]^{\mathrm{T}}, \; b = [b_m, \ldots, b_0]^{\mathrm{T}}. \tag{57.355}$$

We make the following assumptions: the sign of b_m is known, the polynomial $B(s) = b_m s^m + \cdots + b_1 s + b_0$ is known to be Hurwitz, and $\sigma(y) \neq 0 \, \forall y \in \mathbb{R}$. An important restriction is that the nonlinearities $\phi(y)$ and $\Phi(y)$ are allowed to depend only on the output y. Even when θ is known, this restriction is needed to achieve global stability.

TABLE 57.3 State Estimation Filters.

Filters:

$$\begin{aligned}\dot{\xi} &= A_0\xi + ky + \phi(y) && (57.356) \\ \dot{\Xi} &= A_0\Xi + \Phi(y) && (57.357) \\ \dot{\lambda} &= A_0\lambda + e_n\sigma(y)u && (57.358)\end{aligned}$$

$$\begin{aligned}v_j &= A_0^j\lambda, \quad j = 0, \ldots, m && (57.359) \\ \Omega^{\mathrm{T}} &= [v_m, \ldots, v_1, v_0, \; \Xi] && (57.360)\end{aligned}$$

We define the parameter-dependent state estimate

$$\hat{x} = \xi + \Omega^{\mathrm{T}}\theta, \tag{57.361}$$

which employs the filters given in Table 57.3, with the vector $k = [k_1, \ldots, k_n]^{\mathrm{T}}$ chosen so that the matrix $A_0 = A - ke_1^{\mathrm{T}}$ is Hurwitz, that is,

$$PA_0 + A_0^{\mathrm{T}}P = -I, \quad P = P^{\mathrm{T}} > 0. \tag{57.362}$$

The state estimation error,

$$\varepsilon = x - \hat{x}, \tag{57.363}$$

as is readily shown, satisfies

$$\dot{\varepsilon} = A_0\varepsilon. \tag{57.364}$$

The following two expressions for $\dot{y}$ are instrumental in the backstepping design:

$$\dot{y} = \omega_0 + \omega^{\mathrm{T}}\theta + \varepsilon_2, \tag{57.365}$$

and

$$\dot{y} = b_m v_{m,2} + \omega_0 + \bar{\omega}^{\mathrm{T}}\theta + \varepsilon_2, \tag{57.366}$$

where

$$\begin{aligned}\omega_0 &= \varphi_{0,1} + \xi_2, && (57.367) \\ \omega &= [v_{m,2}, v_{m-1,2}, \ldots, v_{0,2}, \; \Phi_{(1)} + \Xi_{(2)}]^{\mathrm{T}}, && (57.368)\end{aligned}$$

and

$$\bar{\omega} = [0, v_{m-1,2}, \ldots, v_{0,2}, \ \Phi_{(1)} + \Xi_{(2)}]^{\mathrm{T}}. \tag{57.369}$$

Since the states $x_2, \ldots, x_n$ are not measured, the backstepping design is applied to the system

$$\begin{aligned} \dot{y} &= b_m v_{m,2} + \omega_0 + \bar{\omega}^{\mathrm{T}}\theta + \varepsilon_2, & (57.370)\\ \dot{v}_{m,i} &= v_{m,i+1} - k_i v_{m,1}, \ i = 2, \ldots, \rho - 1, & (57.371)\\ \dot{v}_{m,\rho} &= \sigma(y)u + v_{m,\rho+1} - k_\rho v_{m,1}. & (57.372) \end{aligned}$$

The order of this system is equal to the relative degree of the plant Equation 57.352.

Output-Feedback Design with Tuning Functions

The output-feedback design with tuning functions is summarized in Table 57.4. The resulting error system is

$$\begin{aligned} \dot{z} &= A_z(z,t)z + W_\varepsilon(z,t)\varepsilon_2 + W_\theta(z,t)^{\mathrm{T}}\tilde{\theta} \\ &\quad - b_m(\dot{y}_r + \bar{\alpha}_1)e_1\tilde{\varrho} \end{aligned} \tag{57.373}$$

where

$$A_z = \tag{57.374}$$

$$\begin{bmatrix} -c_1 - d_1 & \hat{b}_m & 0 & \cdots & \cdots & 0 \\ -\hat{b}_m & -c_2 - d_2\left(\frac{\partial\alpha_1}{\partial y}\right)^2 & 1+\sigma_{23} & \sigma_{24} & \cdots & \sigma_{2,\rho} \\ 0 & -1-\sigma_{23} & \ddots & \ddots & \ddots & \vdots \\ \vdots & -\sigma_{24} & \ddots & \ddots & \ddots & \sigma_{\rho-2,\rho} \\ \vdots & \vdots & \ddots & \ddots & \ddots & 1+\sigma_{\rho-1,\rho} \\ 0 & -\sigma_{2,\rho} & \cdots & -\sigma_{\rho-2,\rho} & -1-\sigma_{\rho-1,\rho} & -c_\rho - d_\rho\left(\frac{\partial\alpha_{\rho-1}}{\partial y}\right)^2 \end{bmatrix}$$

and

$$W_\varepsilon(z,t) = \begin{bmatrix} 1 \\ -\frac{\partial\alpha_1}{\partial y} \\ \vdots \\ -\frac{\partial\alpha_{\rho-1}}{\partial y} \end{bmatrix}, \tag{57.375}$$

$$W_\theta(z,t)^{\mathrm{T}} = W_\varepsilon(z,t)\omega^{\mathrm{T}} - \hat{\varrho}(\dot{y}_r + \bar{\alpha}_1)e_1e_1^{\mathrm{T}}. \tag{57.376}$$

The nonlinear damping terms $-d_i\left(\frac{\partial\alpha_{i-1}}{\partial y}\right)^2$ in Equation 57.375 are included to counteract the exponentially decaying state estimation error ε_2. The variable $\hat{\varrho}$ is an estimate of $\varrho = 1/b_m$.

Output-Feedback Modular Design

In addition to $sgn(b_m)$, in the modular design we assume that a positive constant ς_m is known so that $|b_m| \geq \varsigma_m$.

TABLE 57.4 Output-Feedback Tuning Functions Design.

$$\begin{aligned} z_1 &= y - y_r. & (57.377)\\ z_i &= v_{m,i} - \hat{\varrho}y_r^{(i-1)} - \alpha_{i-1}, & (57.378)\\ & \quad i = 2, \ldots, \rho. \end{aligned}$$

$$\alpha_1 = \hat{\varrho}\bar{\alpha}_1, \ \bar{\alpha}_1 = -(c_1 + d_1)z_1 - \omega_0 - \bar{\omega}^{\mathrm{T}}\hat{\theta}. \tag{57.379}$$

$$\begin{aligned} \alpha_2 &= -\hat{b}_m z_1 - \left[c_2 + d_2\left(\frac{\partial\alpha_1}{\partial y}\right)^2\right]z_2 \\ &\quad + \left(\dot{y}_r + \frac{\partial\alpha_1}{\partial\hat{\varrho}}\right)\dot{\hat{\varrho}} + \frac{\partial\alpha_1}{\partial\hat{\theta}}\Gamma\tau_2 + \beta_2. \end{aligned} \tag{57.380}$$

$$\begin{aligned} \alpha_i &= -z_{i-1} - \left[c_i + d_i\left(\frac{\partial\alpha_{i-1}}{\partial y}\right)^2\right]z_i \\ &\quad + \left(y_r^{(i-1)} + \frac{\partial\alpha_{i-1}}{\partial\hat{\varrho}}\right)\dot{\hat{\varrho}} \\ &\quad + \frac{\partial\alpha_{i-1}}{\partial\hat{\theta}}\Gamma\tau_i - \sum_{j=2}^{i-1}\frac{\partial\alpha_{j-1}}{\partial\hat{\theta}}\Gamma\frac{\partial\alpha_{i-1}}{\partial y}z_j + \beta_i, \\ i &= 3, \ldots, \rho. \end{aligned} \tag{57.381}$$

$$\begin{aligned} \beta_i &= \frac{\partial\alpha_{i-1}}{\partial y}\left(\omega_0 + \omega^{\mathrm{T}}\hat{\theta}\right) + \frac{\partial\alpha_{i-1}}{\partial\xi}(A_0\xi + ky + \phi) \\ &\quad + \frac{\partial\alpha_{i-1}}{\partial\Xi}(A_0\Xi + \Phi) & (57.382)\\ &\quad + \sum_{j=1}^{i-1}\frac{\partial\alpha_{i-1}}{\partial y_r^{(j-1)}}y_r^{(j)} + k_i v_{m,1} \\ &\quad + \sum_{j=1}^{m+i-1}\frac{\partial\alpha_{i-1}}{\partial\lambda_j}(-k_j\lambda_1 + \lambda_{j+1}). & (57.383) \end{aligned}$$

$$\begin{aligned} \tau_1 &= \left(\omega - \hat{\varrho}(\dot{y}_r + \bar{\alpha}_1)e_1\right)z_1. & (57.384)\\ \tau_i &= \tau_{i-1} - \frac{\partial\alpha_{i-1}}{\partial y}\omega z_i, \ i = 2, \ldots, \rho. & (57.385) \end{aligned}$$

Adaptive control law:

$$u = \frac{1}{\sigma(y)}\left(\alpha_\rho - v_{m,\rho+1} + \hat{\varrho}y_r^{(\rho)}\right). \tag{57.386}$$

Parameter update laws:

$$\begin{aligned} \dot{\hat{\theta}} &= \Gamma\tau_\rho. & (57.387)\\ \dot{\hat{\varrho}} &= -\gamma\, sgn(b_m)(\dot{y}_r + \bar{\alpha}_1)z_1. & (57.388) \end{aligned}$$

TABLE 57.5 Output-Feedback Controller in the Modular Design.

$$z_1 = y - y_r. \quad (57.389)$$

$$z_i = v_{m,i} - \frac{1}{\hat{b}_m} y_r^{(i-1)} - \alpha_{i-1}, \quad i = 2, \ldots, \rho. \quad (57.390)$$

$$\alpha_1 = -\frac{sgn(b_m)}{\varsigma_m}(c_1 + s_1) z_1 + \frac{1}{\hat{b}_m}\bar{\alpha}_1,$$

$$\bar{\alpha}_1 = -\omega_0 - \bar{\omega}^{\mathrm{T}}\hat{\theta}. \quad (57.391)$$

$$\alpha_2 = -\hat{b}_m z_1 - (c_2 + s_2) z_2 + \beta_2. \quad (57.392)$$

$$\alpha_i = -z_{i-1} - (c_i + s_i) z_i + \beta_i, \quad i = 3, \ldots, \rho. \quad (57.393)$$

$$\beta_i = \frac{\partial \alpha_{i-1}}{\partial y}\left(\omega_0 + \omega^{\mathrm{T}}\hat{\theta}\right) + \frac{\partial \alpha_{i-1}}{\partial \xi}(A_0\xi + ky + \phi) + \frac{\partial \alpha_{i-1}}{\partial \Xi}(A_0\Xi + \Phi) + \sum_{j=1}^{i-1}\frac{\partial \alpha_{i-1}}{\partial y_r^{(j-1)}} y_r^{(j)} + k_i v_{m,1} + \sum_{j=1}^{m+i-1}\frac{\partial \alpha_{i-1}}{\partial \lambda_j}(-k_j\lambda_1 + \lambda_{j+1}). \quad (57.394)$$

$$s_1 = d_1 + \kappa_1\left|\bar{\omega} + \frac{1}{\hat{b}_m}(\dot{y}_r + \bar{\alpha}_1)e_1\right|^2. \quad (57.395)$$

$$s_2 = d_2\left(\frac{\partial \alpha_1}{\partial y}\right)^2 + \kappa_2\left|\frac{\partial \alpha_1}{\partial y}\omega - z_1 e_1\right|^2 + g_2\left|\frac{\partial \alpha_1}{\partial \hat{\theta}}^{\mathrm{T}} - \frac{1}{\hat{b}_m^2}\dot{y}_r e_1\right|^2. \quad (57.396)$$

$$s_i = d_i\left(\frac{\partial \alpha_{i-1}}{\partial y}\right)^2 + \kappa_i\left|\frac{\partial \alpha_{i-1}}{\partial y}\omega\right|^2 + g_i\left|\frac{\partial \alpha_{i-1}}{\partial \hat{\theta}}^{\mathrm{T}} - \frac{1}{\hat{b}_m^2} y_r^{(i-1)} e_1\right|^2, \quad i = 3, \ldots, \rho. \quad (57.397)$$

Adaptive control law:

$$u = \frac{1}{\sigma(y)}\left(\alpha_\rho - v_{m,\rho+1} + \frac{1}{\hat{b}_m} y_r^{(\rho)}\right). \quad (57.398)$$

The complete design of the control law is summarized in Table 57.5. The resulting error system is

$$\dot{z} = A_z^*(z,t)z + W_\varepsilon(z,t)\varepsilon_2 + W_\theta^*(z,t)^{\mathrm{T}}\tilde{\theta} + Q(z,t)^{\mathrm{T}}\dot{\hat{\theta}} \quad (57.399)$$

where

$$A_z^*(z,t) = \quad (57.400)$$

$$\begin{bmatrix} -\frac{|b_m|}{\varsigma_m}(c_1+s_1) & b_m & 0 & \cdots & 0 \\ -b_m & -(c_2+s_2) & 1 & \ddots & \vdots \\ 0 & -1 & \ddots & \ddots & 0 \\ \vdots & \ddots & \ddots & \ddots & 1 \\ 0 & \cdots & 0 & -1 & -(c_\rho+s_\rho) \end{bmatrix},$$

$$W_\varepsilon(z,t) = \begin{bmatrix} 1 \\ -\frac{\partial \alpha_1}{\partial y} \\ \vdots \\ -\frac{\partial \alpha_{\rho-1}}{\partial y} \end{bmatrix}, \quad (57.401)$$

$$W_\theta^*(z,t)^{\mathrm{T}} = \begin{bmatrix} \bar{\omega}^{\mathrm{T}} + \frac{1}{\hat{b}_m}(\dot{y}_r + \bar{\alpha}_1)e_1^{\mathrm{T}} \\ -\frac{\partial \alpha_1}{\partial y}\omega^{\mathrm{T}} + z_1 e_1^{\mathrm{T}} \\ -\frac{\partial \alpha_2}{\partial y}\omega^{\mathrm{T}} \\ \vdots \\ -\frac{\partial \alpha_{\rho-1}}{\partial y}\omega^{\mathrm{T}} \end{bmatrix},$$

and

$$Q(z,t)^{\mathrm{T}} = \begin{bmatrix} 0 \\ -\frac{\partial \alpha_1}{\partial \hat{\theta}} + \frac{1}{\hat{b}_m^2}\dot{y}_r e_1^{\mathrm{T}} \\ \vdots \\ -\frac{\partial \alpha_{\rho-1}}{\partial \hat{\theta}} + \frac{1}{\hat{b}_m^2} y_r^{(\rho-1)} e_1^{\mathrm{T}} \end{bmatrix} \quad (57.402)$$

Passive identifier

For the parametric model Equation 57.365, we introduce the scalar observer

$$\dot{\hat{y}} = -\left(c_0 + \kappa_0|\omega|^2\right)(\hat{y} - y) + \omega_0 + \omega^{\mathrm{T}}\hat{\theta}. \quad (57.403)$$

The observer error,

$$\epsilon = y - \hat{y}, \quad (57.404)$$

is governed by

$$\dot{\epsilon} = -\left(c_0 + \kappa_0|\omega|^2\right)\epsilon + \omega^{\mathrm{T}}\tilde{\theta} + \varepsilon_2. \quad (57.405)$$

The parameter update law is

$$\dot{\hat{\theta}} = \underset{\hat{b}_m}{\mathrm{Proj}}\{\Gamma\omega\epsilon\}, \quad \begin{matrix}\hat{b}_m(0)sgnb_m > \varsigma_m \\ \Gamma = \Gamma^{\mathrm{T}} > 0\end{matrix}, \quad (57.406)$$

where the projection operator is employed to guarantee that $|\hat{b}_m(t)| \geq \varsigma_m > 0, \ \forall t \geq 0$.

Swapping identifier

The estimation error

$$\epsilon = y - \xi_1 - \Omega_1^T\hat{\theta} \tag{57.407}$$

satisfies the following equation linear in the parameter error:

$$\epsilon = \Omega_1^T\tilde{\theta} + \varepsilon_1 \,. \tag{57.408}$$

The update law for $\hat{\theta}$ is either the gradient,

$$\dot{\hat{\theta}} = \underset{\hat{b}_m}{\text{Proj}}\left\{\Gamma\frac{\Omega_1\epsilon}{1+\nu|\Omega_1|^2}\right\}, \quad \begin{array}{l}\hat{b}_m(0)sgnb_m > \varsigma_m, \\ \Gamma = \Gamma^T > 0, \\ \nu > 0,\end{array} \tag{57.409}$$

or the least squares,

$$\begin{aligned}\dot{\hat{\theta}} &= \underset{\hat{b}_m}{\text{Proj}}\left\{\Gamma\frac{\Omega_1\epsilon}{1+\nu|\Omega_1|^2}\right\}, \quad \hat{b}_m(0)sgnb_m > \varsigma_m, \\ \dot{\Gamma} &= -\Gamma\frac{\Omega_1\Omega_1^T}{1+\nu|\Omega_1|^2}\Gamma, \quad \begin{array}{l}\Gamma(0) = \Gamma(0)^T > 0, \\ \nu > 0 \,.\end{array}\end{aligned} \tag{57.410}$$

57.8.5 Extensions

Adaptive nonlinear control designs presented in the preceding sections are applicable to classes of nonlinear systems broader than the parametric strict-feedback systems Equation 57.292.

Pure-feedback systems.

$$\begin{aligned}\dot{x}_i &= x_{i+1} + \varphi_i(x_1,\ldots,x_{i+1})^T\theta, \; i = 1,\ldots,n-1, \\ \dot{x}_n &= \left[\beta_0(x) + \beta(x)^T\theta\right]u + \varphi_0(x) + \varphi_n(x)^T\theta \,,\end{aligned} \tag{57.411}$$

where $\varphi_0(0) = 0,\; \varphi_1(0) = \cdots = \varphi_n(0) = 0,\; \beta_0(0) \neq 0$. Because of the dependence of φ_i on x_{i+1}, the regulation or tracking for pure-feedback systems is, in general, not global, even when θ is known.

Unknown virtual control coefficients.

$$\begin{aligned}\dot{x}_i &= b_i x_{i+1} + \varphi_i(x_1,\ldots,x_i)^T\theta, \qquad i = 1,\ldots,n-1 \\ \dot{x}_n &= b_n\beta(x)u + \varphi_n(x_1,\ldots,x_n)^T\theta \,,\end{aligned} \tag{57.412}$$

where, in addition to the unknown vector θ, the constant coefficients b_i are also unknown. The unknown b_i-coefficients are frequent in applications ranging from electric motors to flight dynamics. The signs of b_i, $i = 1,\ldots,n$, are assumed to be known. In the tuning functions design, in addition to estimating b_i, we also estimate its inverse $\varrho_i = 1/b_i$. In the modular design, we assume that, in addition to $sgnb_i$, a positive constant ς_i is known such that $|b_i| \geq \varsigma_i$. Then, instead of estimating $\varrho_i = 1/b_i$, we use the inverse of the estimate $\hat{b}_i$, i.e., $1/\hat{b}_i$, where $\hat{b}_i(t)$ is kept away from zero by using parameter projection.

Multi-input systems.

$$\begin{aligned}\dot{X}_i &= B_i(\bar{X}_i)X_{i+1} + \Phi_i(\bar{X}_i)^T\theta, \qquad i = 1,\ldots,n-1, \\ \dot{X}_n &= B_n(X)u + \Phi_n(X)^T\theta \,,\end{aligned} \tag{57.413}$$

where X_i is a ν_i-vector, $\nu_1 \leq \nu_2 \leq \cdots \leq \nu_n$, $\bar{X}_i = \left[X_1^T,\ldots,X_i^T\right]^T$, $X = \bar{X}_n$. and the matrices $B_i(\bar{X}_i)$ have full rank for all $\bar{X}_i \in \mathbb{R}^{\sum_{j=1}^{i}\nu_j}$. The input u is a ν_n-vector. The matrices B_i can be allowed to be unknown provided they are constant and positive definite.

Block strict-feedback systems.

$$\begin{aligned}\dot{x}_i &= x_{i+1} + \varphi_i(x_1,\ldots,x_i,\zeta_1,\ldots,\zeta_i)^T\theta, \; i = 1,\ldots,\rho-1, \\ \dot{x}_\rho &= \beta(x,\zeta)u + \varphi_\rho(x,\zeta)^T\theta, \\ \dot{\zeta}_i &= \Phi_{i,0}(\bar{x}_i,\bar{\zeta}_i) + \Phi_i(\bar{x}_i,\bar{\zeta}_i)^T\theta, \; i = 1,\ldots,\rho,\end{aligned} \tag{57.414}$$

with the following notation: $\bar{x}_i = [x_1,\ldots,x_i]^T$, $\bar{\zeta}_i = \left[\zeta_1^T,\ldots,\zeta_i^T\right]^T$, $x = \bar{x}_\rho$, and $\zeta = \bar{\zeta}_\rho$. Each ζ_i-subsystem of Equation 57.414 is assumed to be bounded-input bounded-state (BIBS) stable with respect to the input $(\bar{x}_i, \bar{\zeta}_{i-1})$. For this class of systems it is quite simple to modify the procedure in Tables 57.1 and 57.2. Because of the dependence of φ_i on $\bar{\zeta}_i$, the stabilizing function α_i is augmented by the term $+\sum_{k=1}^{i-1}\frac{\partial\alpha_{i-1}}{\partial\zeta_k}\Phi_{k,0}$, and the regressor w_i is augmented by $-\sum_{k=1}^{i-1}\Phi_i\left(\frac{\partial\alpha_{i-1}}{\partial\zeta_k}\right)^T$.

Partial state-feedback systems. In many physical systems there are unmeasured states as in the output-feedback form Equation 57.352, but there are also states other than the output $y = x_1$ that are measured. An example of such a system is

$$\begin{aligned}\dot{x}_1 &= x_2 + \varphi_1(x_1)^T\theta, \\ \dot{x}_2 &= \mathbf{x}_3 + \varphi_2(x_1,x_2)^T\theta, \\ \dot{\mathbf{x}}_3 &= \mathbf{x}_4 + \varphi_3(x_1,x_2)^T\theta, \\ \dot{\mathbf{x}}_4 &= x_5 + \varphi_4(x_1,x_2)^T\theta, \\ \dot{x}_5 &= u + \varphi_5(x_1,x_2,x_5)^T\theta \,.\end{aligned}$$

The states $\mathbf{x}_3$ and $\mathbf{x}_4$ are assumed not to be measured. To apply the adaptive backstepping designs presented in this chapter, we combine the state-feedback techniques with the output-feedback techniques. The subsystem $(x_2, \mathbf{x}_3, \mathbf{x}_4)$ is in the output-feedback form with x_2 as a measured output, so that we employ a state estimator for $(x_2, \mathbf{x}_3, \mathbf{x}_4)$ using the filters introduced in Section 57.8.4.

References

[1] Bastin, G., Adaptive nonlinear control of fed-batch stirred tank reactors, *Int. J. Adapt. Control Sig. Process.*, 6, 273–284, 1992.

[2] Dawson, D.M., Carroll, J.J., and Schneider, M., Integrator backstepping control of a brushed DC motor turning a robotic load, *IEEE Trans. Control Syst. Tech.*, 2, 233–244, 1994.

[3] Janković, M., Adaptive output feedback control of nonlinear feedback-linearizable systems, *Int. J. Adapt. Control Sig. Process.*, to appear.

[4] Jiang, Z.P. and Pomet, J.B., Combining backstepping and time-varying techniques for a new set of adaptive controllers, *Proc. 33rd IEEE Conf. Dec. Control*, Lake Buena Vista, FL, December 1994, pp. 2207–2212.

[5] Kanellakopoulos, I., Kokotović, P.V., and Morse, A.S., Systematic design of adaptive controllers for feedback linearizable systems, *IEEE Trans. Auto. Control*, 36, 1241–1253, 1991.

[6] Kanellakopoulos, I., Kokotović, P.V., and Morse, A.S., Adaptive nonlinear control with incomplete state information, *Int. J. Adapt. Control Sig. Process.*, 6, 367–394, 1992.

[7] Khalil, H., Adaptive output-feedback control of nonlinear systems represented by input-output models, *Proc. 33rd IEEE Conf. Dec. Control*, Lake Buena Vista, FL, December 1994, pp. 199–204; also submitted to *IEEE Trans. Auto. Control.*

[8] Krstić, M., Kanellakopoulos, I., and Kokotović, P.V., Adaptive nonlinear control without overparametrization, *Syst. Control Lett.*, 19, 177–185, 1992.

[9] Krstić, M., Kanellakopoulos, I., and Kokotović, P.V., *Nonlinear and Adaptive Control Design*, John Wiley & Sons, New York, NY, 1995.

[10] Krstić, M. and Kokotović, P.V., Adaptive nonlinear design with controller-identifier separation and swapping, *IEEE Trans. Auto. Control*, 40, 426–441, 1995.

[11] Marino, R., Peresada, S., and Valigi, P., Adaptive input-output linearizing control of induction motors, *IEEE Trans. Auto. Control*, 38, 208–221, 1993.

[12] Marino, R. and Tomei, P., Global adaptive output-feedback control of nonlinear systems, Part I: linear parametrization, *IEEE Trans. Auto. Control*, 38, 17–32, 1993.

[13] Pomet, J.B. and Praly, L., Adaptive nonlinear regulation: estimation from the Lyapunov equation, *IEEE Trans. Auto. Control*, 37, 729–740, 1992.

[14] Praly, L., Bastin, G., Pomet, J.-B., and Jiang, Z.P., Adaptive stabilization of nonlinear systems, in *Foundations of Adaptive Control*, P. V. Kokotović, Ed., Springer-Verlag, Berlin, 1991, 347–434.

[15] Sastry, S.S. and Isidori, A., Adaptive control of linearizable systems, *IEEE Trans. Auto. Control*, 34, 1123–1131, 1989.

[16] Seto, D., Annaswamy, A.M., and Baillieul, J., Adaptive control of a class of nonlinear systems with a triangular structure, *IEEE Trans. Auto. Control*, 39, 1411–1428, 1994.

[17] Teel, A.R., Adaptive tracking with robust stability, *Proc. 32nd IEEE Conf. Dec. Control*, San Antonio, TX, December 1993, pp. 570-575.

Further Reading

Here we have briefly surveyed representative results in adaptive nonlinear control, a research area that has been rapidly growing in the 1990s.

The first adaptive backstepping design was developed by Kanellakopoulos, Kokotović and Morse [5]. Its overparametrization was removed by the tuning functions design of Krstić, Kanellakopoulos and Kokotović [8]. Possibilities for extending the class of systems in [5] were studied by Seto, Annaswamy and Baillieul [16].

Among the early estimation-based results are Sastry and Isidori [15], Pomet and Praly [13], etc. They were surveyed in Praly, Bastin, Pomet and Jiang [14]. All of these designs involve some growth conditions. The modular approach of Krstić and Kokotović [10] removed the growth conditions and achieved a complete separation of the controller and the identifier.

One of the first output-feedback designs was proposed by Marino and Tomei [12]. Kanellakopoulos, Kokotović and Morse [6] presented a solution to the partial state-feedback problem. A tracking design where the regressor depends only on reference signals was given in Teel [17].

Current efforts in adaptive nonlinear control focus on broadening the class of nonlinear systems for which adaptive controllers are available. Jiang and Pomet [4] developed a design for nonholonomic systems using the tuning functions technique. Khalil [7] and Janković [3] developed semiglobal output feedback designs for a class which includes some systems not transformable into the output feedback form.

Among the applications of adaptive nonlinear control are biochemical processes (Bastin [1]) and electric motors (Marino, Peresada and Valigi [11], and Dawson, Carroll and Schneider [2]).

An important future research topic is the robustness of adaptive nonlinear designs.

For a complete and pedagogical presentation of adaptive nonlinear control the reader is referred to the text *Nonlinear and Adaptive Control Design* by Krstić, Kanellakopoulos, and Kokotović [9]. The book introduces backstepping and illustrates it with numerous applications (including jet engine, automotive suspension, aircraft wing rock, robotic manipulator, and magnetic levitation). It contains the details of methods surveyed here and their extensions. It also covers several important topics not mentioned here. Among them is the systematic improvement of *transient performance.* It also shows the advantages of applying adaptive backstepping to *linear* systems.

57.9 Intelligent Control

Kevin M. Passino, Department of Electrical Engineering, Ohio State University, Columbus, OH

57.9.1 Introduction

Intelligent control,[9] the discipline in which control algorithms are developed by emulating certain characteristics of intelligent biological systems, is an emerging area of control that is being fueled by advancements in computing technology [1], [16], [18], [8]. For instance, software development and validation tools for **expert systems** (computer programs that emulate the actions of a human who is proficient at some task) are being used to construct "expert controllers" that seek to automate the actions of a human operator who controls a system. Other knowledge-based systems such as **fuzzy systems** (rule-based systems that use fuzzy logic for knowledge representation and inference) and **planning systems** (that emulate human planning activities) are being used in a similar manner to automate the perceptual, cognitive (deductive and inductive), and action-taking characteristics of humans who perform control tasks. Artificial **neural networks** emulate biological neural networks and have been used (1) to learn how to control systems by observing the way that a human performs a control task, and (2) to learn in an on-line fashion how best to control a system by taking control actions, rating the quality of the responses achieved when these actions are used, then adjusting the recipe used for generating control actions so that the response of the system improves. **Genetic algorithms** are being used to evolve controllers via off-line computer-aided design of control systems or in an on-line fashion by maintaining a population of controllers and using "survival of the fittest" principles where "fittest" is defined by the quality of the response achieved by the controller. For fuzzy, genetic, and neural systems, in addition to software development tools there are several computer chips available that provide efficient parallel implementation so that complex reasoning processes can be implemented in real time.

Via these examples we see that computing technology is driving the development of the field of control by providing alternative strategies for the functionality and implementation of controllers for dynamical systems. In fact, there is a trend in the field of control to integrate the functions of intelligent systems, such as those listed above, with conventional control systems to form highly "autonomous" systems that have the capability to perform complex control tasks independently with a high degree of success. This trend toward the development of **intelligent autonomous control systems** is gaining momentum as control engineers have solved many problems and are naturally seeking control problems where broader issues must be taken into consideration and where the full range of capabilities of available computing technologies is used.

The development of such sophisticated controllers does, however, still fit within the conventional engineering methodology for the construction of control systems. Mathematical modeling using first principles or data from the system, along with heuristics, is used. Some intelligent control strategies rely more on the use of heuristics (e.g., direct fuzzy control) but others utilize mathematical models in the same way that they are used in conventional control, while still others use a combination of mathematical models and heuristics (see, e.g., the approaches to fuzzy adaptive control in [17]). There is a need for systematic methodologies for the construction of controllers. Some methodologies for the construction of intelligent controllers are quite *ad hoc* (e.g., for the fuzzy controller) yet often effective since they provide a method and formalism for incorporating and representing the nonlinearities that are needed to achieve high-performance control. Other methodologies for the construction of intelligent controllers are no more *ad hoc* than ones for conventional control (e.g., for neural and fuzzy adaptive controllers). There is a need for nonlinear analysis of stability, controllability, and observability properties. Although there has been significant recent progress in stability analysis of fuzzy, neural, and expert control systems, there is need for much more work in nonlinear analysis of intelligent control systems. Simulations and experimental evaluations of intelligent control systems are necessary. Comparative analysis of competing control strategies (conventional or intelligent) is, as always, important. Engineering cost-benefit analysis that involves issues of performance, stability, ease of design, lead time to implementation, complexity of implementation, cost, and other issues must be used.

Overall, while the intelligent control paradigm focuses on biologically motivated approaches, it is not clear that there are drastic differences in the behavior of the resulting controllers that are finally implemented (they are not mystical; they are simply nonlinear, often adaptive controllers). This is, however, not surprising since there seems to be an existing conventional control approach that is analogous to every new intelligent control approach that has been introduced. This is illustrated in Table 57.6 below. It is not surprising then that while there seem to be some new concepts growing from the field of intelligent control, there is a crucial role for the control engineer and control scientist to play in evaluating and developing the emerging field of intelligent control. For more detailed discussions on the relationships between conventional and intelligent control see [12] , [13].

In this chapter we briefly examine the basic techniques of intelligent control, provide an overview of intelligent autonomous control, and discuss the recent advances that have focused on comparative analysis, modeling, and nonlinear analysis of intelligent control systems. The intent is to provide only a brief introduction to the field of intelligent control and to the next two chapters; the interested reader should consult the references provided at the end of the chapter, or the chapters on fuzzy and neural control for more details.

[9] Partial support for this work came from the National Science Foundation under grants IRI-9210332 and EEC-9315257.

TABLE 57.6 Analogies Between Conventional and Intelligent Control

Intelligent Control Technique	Conventional Control Approach
Direct fuzzy control	Nonlinear control
Fuzzy adaptive/learning control	Adaptive control and identification
Fuzzy supervisory control	Gain-scheduled control, hierarchical control
Direct expert control	Controllers for automata, Petri nets, and other discrete event systems
Planning systems for control	Certain types of controllers for discrete event systems, receding horizon control of nonlinear systems
Neural control	Adaptive control and identification, optimal control
Genetic algorithms for computer-aided design (CAD) of control systems, controller tuning	CAD using heuristic optimization techniques, optimal control, receding horizon control, and stochastic adaptive control

57.9.2 Intelligent Control Techniques

The major approaches to intelligent control are outlined and references are provided in case the reader would like to learn more details about any one of these approaches. It is important to note that while each of the approaches is presented separately, in practice there is a significant amount of work being done to determine the best ways to utilize various aspects of each of the approaches in "hybrid" intelligent control techniques. For instance, neural and fuzzy control approaches are often combined. In other cases, neural networks are trained with genetic algorithms. One can imagine justification for integration of just about any permutation of the presented techniques depending on the application at hand.

Fuzzy Control

A fuzzy controller can be designed to roughly emulate the human deductive process (i.e., the process whereby we successively infer conclusions from our knowledge). As shown in Figure 57.29, the fuzzy controller consists of four main components. The rule base holds a set of "if - then" rules that are quantified via fuzzy logic and used to represent the knowledge that human experts may have about how to solve a problem in their domain of expertise. The fuzzy inference mechanism successively decides what rules are most relevant to the current situation and applies the actions indicated by these rules. The fuzzification interface converts numeric inputs into a form that the fuzzy inference mechanism can use to determine which knowledge in the rule base is most relevant at the current time. The defuzzification interface combines the conclusions reached by the fuzzy inference mechanism and provides a numeric value as an output. Overall, the fuzzy control design methodology, which primarily involves the specification of the rule base, provides a heuristic technique to construct nonlinear controllers, and this seems to be one of its main advantages. For more details on direct fuzzy control, see the next chapter.

Often it is the case that we have higher-level knowledge about how to control a process such as information on how to tune the controller while it is in operation or how to coordinate the application of different controllers based on the operating point of the system. For instance, in aircraft control, certain key variables are used in the tuning ("scheduling") of control laws, and fuzzy control provides a unique approach to the construction and implementation of such a gain scheduler. In process control, engineers or process operators often have a significant amount of heuristic expertise on how to tune proportional-integral-derivative (PID) controllers while they are in operation, and such information can be loaded into the rule base of a "fuzzy PID tuner" and used to make sure that a PID controller is tuned properly at all times. In the more general case, we may have knowledge of how to tune and coordinate the application of conventional or fuzzy controllers, and this can be used in a rule-based supervisor as it is shown in Figure 57.30. For more details on fuzzy supervisory control, see the next chapter.

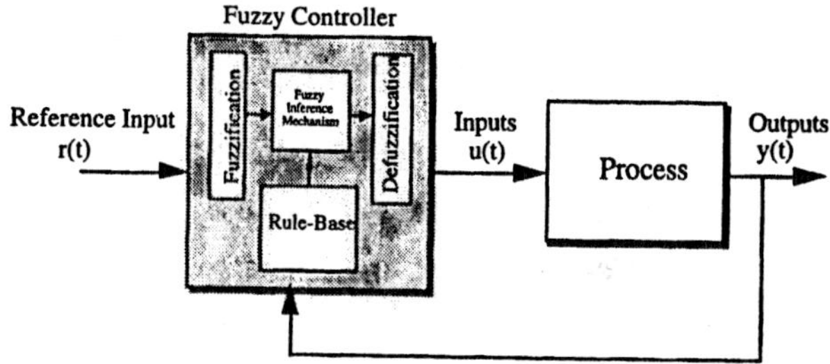

Figure 57.29 Fuzzy control system.

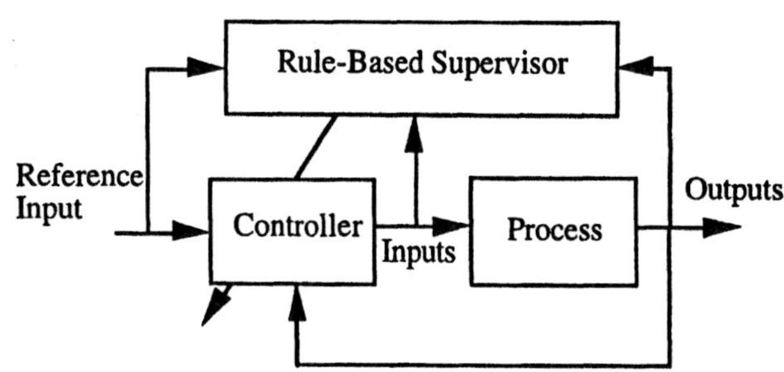

Figure 57.30 Rule based supervisory control.

In other fuzzy control approaches, rather than implementing deductive systems, the goal is to implement an "inductive system," i.e., one that can *learn* and generalize from particular examples (e.g., examples of how the system is behaving). Such approaches typically fall under the title of "fuzzy learning control" or "fuzzy adaptive control." In one approach, shown in Figure 57.31, called "fuzzy model reference learning control" (FMRLC) [10], there is a fuzzy controller with a rule base that has no knowledge about how to control the system. A "reference model" with output $y_m(t)$ is used to characterize how you would like the closed-loop system to behave (i.e., it holds the performance specifications). Then, a learning mechanism compares $y(t)$ to $y_m(t)$ (i.e., the way that the system is currently performing to how you would like it to perform) and decides how to synthesize/tune the fuzzy controller so that the difference between $y(t)$ and $y_m(t)$ goes to zero and hence the performance objectives are met.

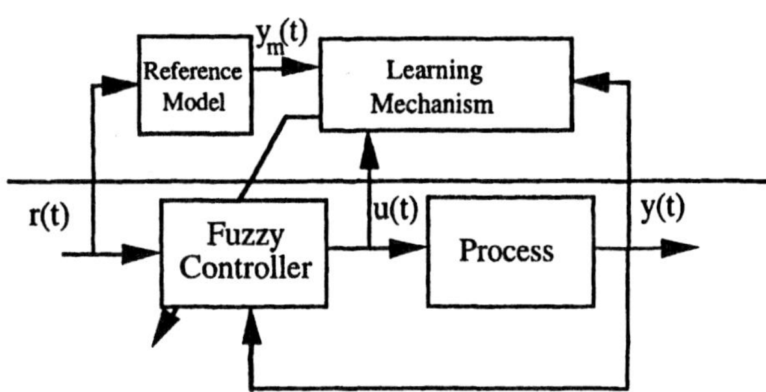

Figure 57.31 Fuzzy model reference learning control.

Overall, our experiences with the FMRLC seem to indicate that significant advantages may be obtained if one can implement a controller that can truly *learn* from its experiences (while forgetting appropriate information) so that when a similar situation is encountered repeatedly, the controller already has a good idea of how to react. This seems to represent an advance over some adaptive controllers where parameters are adapted in such a way that each time the same situation is encountered, some amount of re-adaptation must occur, no matter how often this situation is encountered (for more details on this and other fuzzy learning/adaptive control approaches, see the next chapter or [17]).

Expert and Planning Systems

While fuzzy control techniques similar to those described above have been employed in a variety of industrial control applications, more general "knowledge-based controllers" have also been used successfully. For instance, there are "expert controllers" that are being used to directly control complex processes or in a supervisory role similar to that shown in Figure 57.30 [1], [11]. Others are being used to supervise conventional control algorithms. For instance, the work by Astrom, Anton, and Arzen [5] describes the use of expert supervisory systems for conventional adaptive controllers. Expert systems are also being used as the basis for learning controllers.

In addition, there are planning systems (computer programs that emulate the way that experts plan) that have been used in path planning and high-level decisions about control tasks for robots [1], [16], [14], [6]. A generic planning system taken from Passino and Antsaklis [14], configured in the architecture of a standard control system, is shown in Figure 57.32. Here, the "problem domain" is the environment in which the planner operates, i.e., the plant. There are measured outputs y_i at step i (variables of the problem domain that can be sensed in real time), control actions u_i (the ways in which we can affect the problem domain), disturbances d_i (that represent random events that can affect the problem domain and hence the measured variable y_i), and goals g_i (what we would like to achieve in the problem domain). There are closed-loop specifications that quantify performance specifications and stability requirements.

It is the task of the planner in Figure 57.32 to monitor the measured outputs and goals and to generate control actions that will counteract the effects of the disturbances and result in the goals and the closed-loop specifications to be achieved. To do this, the planner performs "plan generation" whereby it projects into the future (usually a finite number of steps, and, often, a model of the problem domain is used) and tries to determine a set of candidate plans. Next, this set of plans is pruned to one plan that is the best one to apply at the current time. The plan is executed and, during execution, the performance resulting from the plan is monitored and evaluated. Often, due to disturbances, plans will fail and hence the planner must generate a new set of candidate plans, select one, then execute that one. While not pictured in Figure 57.32, some planning systems use "situation assessment" to try to estimate the state of the problem domain (this can be useful in execution monitoring and in plan generation); others perform "world modeling" where a model of the problem domain is developed in an on-line fashion (similar to on-line system identification), and "planner design" where information from the world modeler is used to tune the planner (so that it makes the right plans for the current problem domain). The reader will, perhaps, think of such a planning system as a general "self-tuning regulator." For more details on the use of planning systems for control see [1], [14], [6].

Neural Networks for Control

There has been a flurry of activity in the use of artificial neural networks for control [1], [18], [8], [9]. In this approach engineers are trying to emulate the low-level biological functions of the brain and to use these to solve challenging control problems. For instance, for some control problems we may train an artificial neural network to remember how to regulate a system by repeatedly providing it with examples of how to perform such a task. After the neural network has learned the task, it can be used to recall the control input for each value of the sensed output. Some other approaches to neural control, taken from Hunt et al. [9] (and the references therein), include a neural "internal model control" method and a "model reference structure" that is based on an approach to using neural networks for system identification.

Still other neural control approaches bear some similarities to the FMRLC in Figure 57.31 in the sense that they automatically learn how to control a system by observing the behavior from that system. For instance, in Figure 57.33 we show a "neural predictive control" approach from [9] where one neural network is used as an identifier (structure) for the plant and another is used

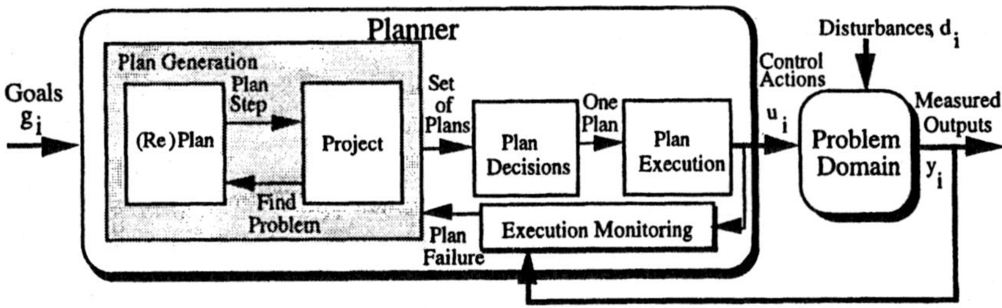

Figure 57.32 Closed-loop planning systems.

as a feedback controller for the plant that is tuned on-line. This tuning proceeds at each step by having the "optimizer" specify an input u' for the neural model of the plant over some time interval. The predicted behavior of the plant y' is obtained and used by the optimizer, along with y_m to pick the best parameters of the neural controller so that the difference between the plant and reference model outputs is as small as possible (if y' predicts y well, we would expect that the optimizer would be quite successful at tuning the controller). For more details on the multitude of techniques for using neural networks for control see the later chapter in this section or [9].

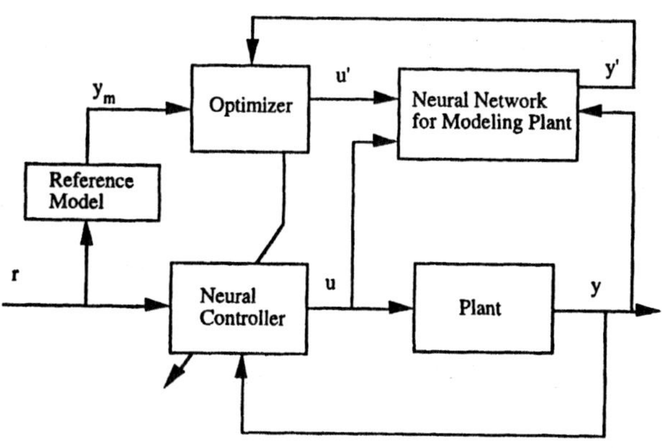

Figure 57.33 Neural predictive control.

Genetic Algorithms for Control

A genetic algorithm uses the principles of evolution, natural selection, and genetics from natural biological systems in a computer algorithm to simulate evolution [7]. Essentially, the genetic algorithm performs a parallel, stochastic, but directed search to evolve the most fit population. It has been shown that a genetic algorithm can be used effectively in the (off-line) computer-aided design of control systems because it can artificially "evolve" an appropriate controller that meets the performance specifications to the greatest extent possible. To do this, the genetic algorithm maintains a population of strings that each represent a different controller, and it uses the genetic operators of "reproduction" (which represents the "survival of the fittest" characteristic of evolution), "crossover" (which represents "mating"), and "mutation" (which represents the random introduction of new "genetic material"), coupled with a "fitness measure" (which often quantifies the performance objectives) to generate successive generations of the population. After many generations, the genetic algorithm often produces an adequate solution to a control design problem since the stochastic, but directed, search helps avoid locally optimal designs and seeks to obtain the best design possible.

Another more challenging problem is how to evolve controllers while the system is operating, rather than in off-line design. Recently, progress in this direction has been made by the introduction of the "genetic model reference adaptive controller" (GMRAC) shown in Figure 57.34 [15]. As in the FMRLC, the GMRAC uses a reference model to characterize the desired performance. For the GMRAC, a genetic algorithm maintains a population of strings that represents candidate controllers. This genetic algorithm uses a process model (e.g., a linear model of the process) and data from the process to evaluate the fitness of each controller in the population at each time step. Using this fitness evaluation, the genetic algorithm propagates controllers into the next generation via the standard genetic operators. The controller that is the most fit one in the population is used to control the system. This allows the GMRAC to automatically evolve a controller from generation to generation (i.e., from one time step to the next) and hence to tune a controller in response to changes in the process (e.g., due to temperature variations, parameter drift, etc.) or due to an on-line change of the specifications in the reference model. Early indications are that the GMRAC seems quite promising as a new technique for stochastic adaptive control since it provides a unique feature whereby alternative controllers can be applied quickly to the problem if they appear useful, and since it has some inherent capabilities to learn via evolution of its population of controllers. There is, however, a significant amount of comparative and nonlinear analysis that needs to be done to more fully evaluate this approach to control.

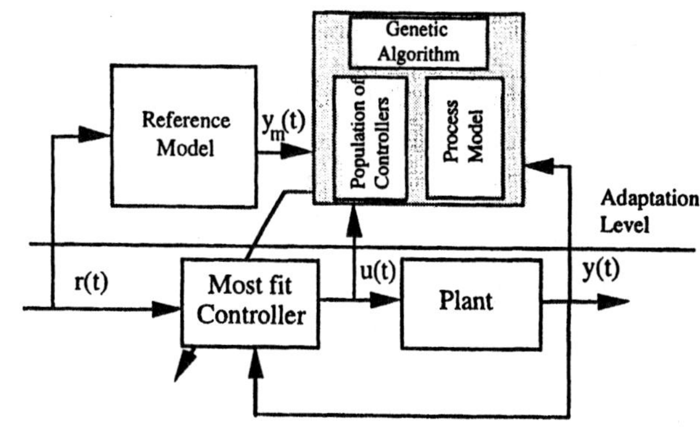

Figure 57.34 Genetic model reference adaptive controller.

57.9.3 Autonomous Control

The goal of the field of autonomous control is to design control systems that automate enough functions so that they can perform well independently under significant uncertainties for extended periods of time, even if there are significant system failures or disturbances. Next we overview some of the basic ideas from [1], [3], [2] on how to specify controllers that can in fact achieve high levels of autonomy.

The Intelligent Autonomous Controller

Figure 57.35 shows a functional architecture for an intelligent autonomous controller with an interface to the process (plant) involving sensing (e.g., via conventional sensing technology, vision, touch, smell, etc.), actuation (e.g., via hydraulics, robotics, motors, etc.), and an interface to humans (e.g., a driver, pilot, crew, etc.) and other systems. The "execution level" has low-level numeric signal processing and control algorithms [e.g., PID, optimal, or adaptive control; parameter estimators, failure detection and identification (FDI) algorithms]. The "coordination level" provides for tuning, scheduling, supervising, and redesigning the execution-level algorithms, crisis management, planning and learning capabilities for the coordination of execution-level tasks, and higher-level symbolic decision making for FDI and control algorithm management. The "management level" provides for the supervision of lower-level functions and for managing the interface to the human(s). In particular, the management level interacts with the users in generating goals for the controller and in assessing capabilities of the system. The management level also monitors performance of the lower-level systems, plans activities at the highest level (and in cooperation with the human), and performs high-level learning about the user and the lower-level algorithms. Applications that have used this type of architecture can be found in, e.g., [1], [16].

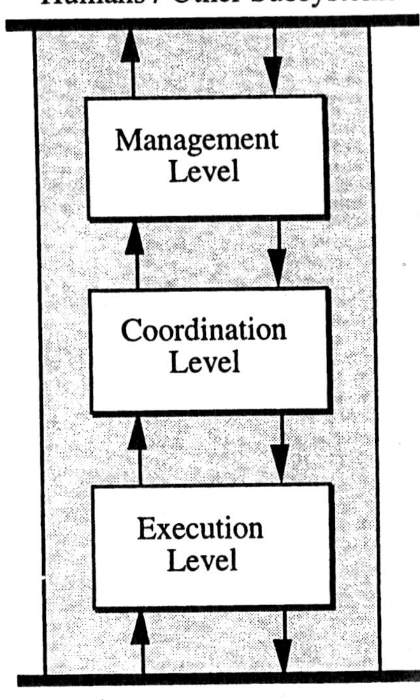

Figure 57.35 Intelligent autonomous controller.

Intelligent systems/controllers (e.g., fuzzy, neural, genetic, expert, etc.) can be employed as appropriate in the implementation of various functions at the three levels of the intelligent autonomous controller (e.g., adaptive fuzzy control may be used at the execution level; planning systems may be used at the management level for sequencing operations; and genetic algorithms may be used in the coordination level to pick an optimal coordination strategy). Hierarchical controllers composed of a hybrid mix of intelligent and conventional systems are commonly used in the intelligent control of complex dynamical systems. This is due to the fact that to achieve high levels of autonomy, we often need high levels of intelligence, which calls for incorporation of a diversity of decision-making approaches for complex dynamic reasoning.

Several fundamental characteristics have been identified for intelligent autonomous control systems (see [1], [16], [3], [2] and the references therein). For example, there is generally a successive delegation of duties from the higher to lower levels and the number of distinct tasks typically increases as we go down the hierarchy. Higher levels are often concerned with slower aspects of the system's behavior and with its larger portions, or broader aspects. There is then a smaller contextual horizon at lower levels; i.e., the control decisions are made by considering less information. Higher levels are typically concerned with longer time horizons than lower levels. It is said that there is "increasing intelligence with decreasing precision as one moves from the lower to the higher levels" (see [16] and the references therein). At the higher levels there is typically a decrease in time scale density, a decrease in bandwidth or system rate, and a decrease in the decision (control action) rate. In addition, there is typically a decrease in granularity of models used, or equivalently, an increase in model abstractness at the higher levels. Finally, we note that there is an ongoing *evolution* of the intelligent functions of an autonomous controller so that by the time one implements its functions they no longer appear intelligent, just algorithmic. It is due to this evolution principle and the fact that implemented intelligent controllers are nonlinear controllers that many researchers feel more comfortable focusing on *achieving autonomy* rather than whether the resulting controller is *intelligent.*

The Control-Theoretic View of Autonomy

Next it is explained how to incorporate the notion of autonomy into the conventional manner of thinking about control problems. Consider the general control system shown in Figure 57.36 where P is a model of the plant, C represents the controller, and T represents specifications on how we would like the closed-loop system to behave (i.e., closed-loop specifications). For some classical control problems, the scope is limited so that C and P are linear and T represents simply, for example, stability, rise time, overshoot, and steady-state tracking error specifications. In this case, intelligent control techniques may not be needed. For engineers, the simplest solution that works is the best one. We tend to need more complex controllers for more complex plants (where, for example, there is a significant amount of uncertainty) and more demanding closed-loop specifications T (see [16], [1] and the references therein).

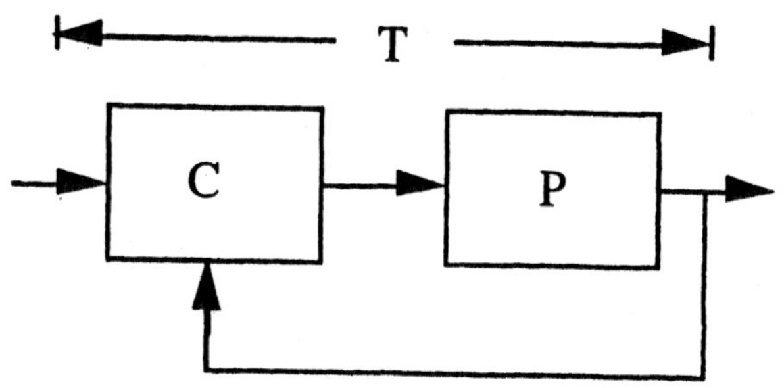

Figure 57.36 Control system.

Consider the case where

1. P is so complex that it is most convenient to represent it with ordinary differential equations and discrete-event system (DES) models (or some other "hybrid" mix of models) and for some parts of the plant the model is not known (i.e., it may be too expensive to find).
2. T is used to characterize the desire to make the system perform well and act with *high degrees of autonomy* (i.e., so that the system performs well under significant uncertainties in the system and its environment for extended periods of time and compensates for significant system failures without external intervention [1]).

The *general control problem* is how to construct C, given P, so that T holds. The intelligent autonomous controller described briefly in the previous section provides a general architecture for C to achieve highly autonomous behavior specified by T for very complex plants P.

Modeling and Analysis

Conventional approaches [10] to modeling P include the use of ordinary differential or difference equations, partial differential equations, stochastic models, models for hierarchical and distributed systems, and so on. However, often some portion of P is more easily represented with an automata model, Petri net, or some other DES model. Moreover, for analysis of the closed-loop system where a variety of intelligent and conventional controllers are working together to form C (e.g., a planning system and a conventional adaptive controller), there is the need for "hybrid" modeling formalisms that can represent dynamics with differential equations and DES models. Control engineers use a model of the plant P to aid in the construction of the model of the controller C, and this model is then implemented to control the real plant. In addition, models of C and P are used as formalisms to represent the dynamics of the closed-loop system so that analysis of the properties of the feedback system is possible before implementation (for redesign, verification, certification, and safety analysis). If the model of P is chosen to be too complex and C is very complex, it will be difficult to develop and utilize mathematical approaches for the analysis of the resulting closed-loop system. Often we want the simplest possible model P that will allow for the development of the (simplest) controller C and for it to be proven/demonstrated that the closed-loop specifications T are met (of course, a separate, more complex model P may be needed for simulation). Unfortunately, there is no clear answer to the question of how much or what type of modeling is needed for the plant P, and there is no standardization of models for intelligent control in the way that there is for many areas of conventional control. Hence, although it is not exactly clear how to proceed with the modeling task, it is clear that knowledge of many different types of models may be needed, depending on the task at hand.

Given the model of the plant P and the model of the controller C, the next task often considered by a control engineer is the use of analysis to more fully understand the behavior of P or the closed-loop system, and to show that when C and P are connected, the closed-loop specifications T are satisfied. Formal mathematical analysis can be used to verify stability, controllability, and observability properties. There is, in fact, a growing body of literature on nonlinear analysis (e.g., stability and describing function analysis) of fuzzy control systems (both direct and adaptive [17]). There is a significant amount of activity in the area of nonlinear analysis of neural control systems and results in the past on nonlinear analysis of (numerical) learning control systems (see the later chapter in this section of this book and [9]). There have already been some applications of DES theory to artificial intelligence (AI) planning systems and there have been recent results on stability analysis of expert control systems [11]. There has been recent progress in defining models and developing approaches to analysis for some hybrid systems, but there is the need for much more work in this area. Many fundamental modeling and representation issues need to be reconsidered; different design objectives and control structures need to be examined; our repertoire of approaches to analysis and design needs to be expanded; and there is the need for more work in the area of simulation and experimental evaluation for hybrid systems. The importance of the solution to the hybrid control system analysis problem is based on the importance of solving the general control problem described above; that is, hybrid system analysis techniques could provide an approach to verifying the operation of intelligent controllers that seek to obtain truly autonomous operation. Finally, we must emphasize that while formal verification of the properties of a control system is important, simulation and experimental evaluation always play an especially important role also.

[10] The reader can find references for the work on modeling and analysis of intelligent control systems discussed in [13].

57.9.4 Concluding Remarks

We have provided a brief overview of the main techniques in the field of intelligent control and have provided references for the reader who is interested in investigating the details of any one of these techniques. The intent of this chapter is to provide an overview of a quickly emerging area of control, while the intent of the next two chapters is to provide an introduction to two of the more well-developed areas in intelligent control: fuzzy and neural control. In a chapter so brief, it seems important to indicate what has been omitted. We have not discussed:

1. Failure detection and identification (FDI) methods, which are essential for a truly autonomous controller
2. Reconfigurable (fault-tolerant) control strategies that use conventional nonlinear robust control techniques and intelligent control techniques
3. Sensor fusion and integration techniques that will be needed for autonomous control
4. Architectures for intelligent and autonomous control systems (e.g., alternative ways to structure interconnections of intelligent subsystems)
5. Distributed intelligent systems
6. Applications.

Moreover, the debate over the meaning of the word "intelligent" in the field of intelligent control has been purposefully avoided. The reader interested in a detailed discussion on this topic should see [4].

57.9.5 Defining Terms:

Expert system: A computer program designed to emulate the actions of a human who is proficient at some task. Often the expert system is broken into two components: a "knowledge base" that holds information about the problem domain, and an inference mechanism (engine) that evaluates the current knowledge and decides what actions to take. An "expert controller" is an expert system that is designed to automate the actions of a human operator who controls a system.

Fuzzy system: A type of knowledge-based system that uses fuzzy logic for knowledge representation and inference. It is composed of four primary components: the fuzzification interface, the rule base (a knowledge base that is composed of rules), an inference mechanism, and a defuzzification interface. A fuzzy system that is used to control a system is called a "fuzzy controller."

Genetic algorithm: A genetic algorithm uses the principles of evolution, natural selection, and genetics from natural biological systems in a computer algorithm to simulate evolution. Essentially, the genetic algorithm performs a parallel, stochastic, but directed search to evolve the most fit population.

Intelligent autonomous control system: A control system that uses conventional and intelligent control techniques to provide enough automation so that the system can perform well independently under significant uncertainties for extended periods of time, even if there are significant system failures or disturbances.

Neural network: Artificial hardware (e.g., electrical circuits) designed to emulate biological neural networks. These may be simulated on conventional computers or on specially designed "neural processors."

Planning system: Computer program designed to emulate human planning activities. These may be a type of expert system that has a special knowledge base that has plan fragments and strategies for planning and an inference process that generates and evaluates alternative plans.

References

[1] Antsaklis, P.J., Passino, K.M., Eds., *An Introduction to Intelligent and Autonomous Control,* Kluwer Academic Publishers, Norwell, MA, 1993.

[2] Antsaklis, P.J., Passino, K.M., and Wang, S.J., An introduction to autonomous control systems, *IEEE Control Syst. Mag., Special Issue on Intelligent Control,* 11(4), 5–13, June 1991.

[3] Antsaklis, P.J., Passino, K.M., and Wang, S.J., Towards intelligent autonomous control systems: architecture and fundamental issues, *J. Intelligent Robotic Syst.,* 1(4), 315–342, 1989.

[4] Antsaklis, P.J., Defining intelligent control, report of the IEEE Control Systems Society Task Force on Intelligent Control (P.J. Antsaklis, Chair), *IEEE Control Systems Magazine,* 14(3), 4–5, 58–66, June 1994; See also Proc. IEEE Int. Conf. Intelligent Control, Columbus, OH, August 1994. (Coauthors include: J. Albus, P.J. Antsaklis, K. Baheti, J.D. Birdwell, M. Lemmon, M. Mataric, A. Meystel, K. Narendra, K. Passino, H. Rauch, G. Saridis, H. Stephanou, P. Werbos)

[5] Astrom, K.J., Anton, J.J., and Arzen, K.E., Expert control, *Automatica,* 22, 277–286, 1986.

[6] Dean, T., and Wellman, M.P., *Planning and Control,* Morgan Kaufman, CA, 1991.

[7] Goldberg, D.E., *Genetic Algorithms in Search, Optimization, and Machine Learning,* Addison Wesley, Reading, MA, 1989.

[8] Gupta, M.M. and Sinha, N.K., Eds., *Intelligent Control: Theory and Applications,* IEEE Press, Piscataway, NJ, 1996.

[9] Hunt, K.J., Sbarbaro, D., Zbikowski, R., and Gawthrop, P.J., Neural networks for control systems—a survey, in *Neuro-Control Systems: Theory and Applications,* Gupta, M.M. and Rao, D.H., Eds.,

IEEE Press, NJ, 1994.

[10] Layne, J.R. and Passino, K.M., Fuzzy model reference learning control for cargo ship steering, *IEEE Control Syst. Mag.*, 13(6), 23–34, December 1993.

[11] Passino, K.M. and Lunardhi, A.D., Qualitative analysis of expert control systems, in *Intelligent Control: Theory and Applications,* Gupta, M.M. and Sinha, N.K., Eds., IEEE Press, Piscataway, NJ, 1996.

[12] Passino, K.M., Bridging the gap between conventional and intelligent control, *Special Issue on Intelligent Control, IEEE Control Syst. Mag.*, 13(4), June 1993.

[13] Passino, K.M., Towards bridging the perceived gap between conventional and intelligent Control, in *Intelligent Control: Theory and Applications,* Gupta, M.M. and Sinha, N.K., Eds., IEEE Press, Piscataway, NJ, 1996.

[14] Passino, K.M. and Antsaklis, P.J., A system and control theoretic perspective on artificial intelligence planning systems, *Int. J. Appl. Artif. Intelligence,* 3, 1–32, 1989.

[15] Porter, L.L. and Passino, K.M., Genetic model reference adaptive control, *Proc. IEEE Int. Symp. Intelligent Control,* 219–224, Columbus, OH, August 16-18, 1994.

[16] Valavanis, K.P. and Saridis, G.N., *Intelligent Robotic Systems: Theory, Design, and Applications,* Kluwer Academic Publishers, Norwell, MA, 1992.

[17] Wang Li-Xin, *Adaptive Fuzzy Systems and Control,* Prentice Hall, Englewood Cliffs, NJ, 1994.

[18] White, D.A. and Sofge, D.A., Eds., *Handbook of Intelligent Control: Neural, Fuzzy, and Adaptive Approaches,* Van Nostrand Reinhold, NY, 1992.

Further Reading

While the books and articles referenced above (particularly [1] and [8]) should provide the reader with an introduction to the area of intelligent control, there are other sources that may also be useful. For instance, there are many relevant conferences including:

IEEE International Symposium on Intelligent Control

American Control Conference

IEEE Conference on Decision and Control

IEEE Conference on Control Applications

IEEE International Conference on Systems, Man, and Cybernetics.

In addition, there are many conferences on fuzzy systems, expert systems, genetic algorithms, and neural networks where applications to control are often studied.

There are many journals that cover the topic of intelligent control including:

IEEE Control Systems Magazine

IEEE Transactions on Control Systems Technology

IEEE Transactions on Systems, Man, and Cybernetics

IEEE Transactions on Fuzzy Systems

IEEE Transactions on Neural Networks

Engineering Applications of Artificial Intelligence

Journal of Intelligent and Robotic Systems

Applied Artificial Intelligence

Journal of Intelligent and Fuzzy Systems

Journal of Intelligent Control and Systems

There are many other journals on expert systems, neural networks, genetic algorithms, and fuzzy systems where applications to control can often be found.

The professional societies most active in intelligent control are

IEEE Control Systems Society

International Federation on Automatic Control

IEEE Systems, Man, and Cybernetics Society

57.10 Fuzzy Control[11]

Kevin M. Passino, *Department of Electrical Engineering, The Ohio State University, Columbus, OH*

Stephen Yurkovich, *Department of Electrical Engineering, The Ohio State University, Columbus, OH*

57.10.1 Introduction

When confronted with a control problem for a complicated physical process, the control engineer usually follows a predetermined design procedure which begins with the need for understanding the process and the primary control objectives. A good example of such a process is that of an automobile "cruise control," designed with the objective of providing the automobile with the capability of regulating its own speed at a driver-specified set-point (e.g., 55 mph). One solution to the automotive cruise control problem involves adding an electronic controller that can sense the speed of the vehicle via the speedometer and actuate the throttle position so as to regulate the vehicle speed at the driver-specified value even if there are road grade changes, head-winds, or variations in the number of passengers in the automobile. Control engineers typically solve the cruise control problem by (1) developing a model of the automobile dynamics (which may model vehicle and power train dynamics, road grade variations,

[11]This work was supported in part by National Science Foundation Grants IRI 9210332 and EEC 9315257.

etc.); (2) using the mathematical model to design a controller (e.g., via a linear model develop a linear controller with techniques from classical control); (3) using the mathematical model of the closed-loop system and mathematical or simulation-based analysis to study its performance (possibly leading to redesign); and (4) implementing the controller via, for example, a microprocessor, and evaluating the performance of the closed-loop system (again possibly leading to redesign). See Chapter 74 for a sophisticated example of such a system.

The difficult task of modeling and simulating complex real-world systems for control systems development, especially when implementation issues are considered, is well documented. Even if a relatively accurate model of a dynamic system can be developed it is often too complex to use in controller development, especially for many conventional control design procedures that require restrictive assumptions for the plant (e.g., linearity). It is for this reason that in practice conventional controllers are often developed via simple crude models of the plant behavior that satisfy the necessary assumptions, and via the ad hoc tuning of relatively simple linear or nonlinear controllers. Regardless, it is well understood (although sometimes forgotten) that heuristics enter the design process when the conventional control design process is used as long as one is concerned with the actual implementation of the control system. It must be acknowledged, however, that conventional control engineering approaches that use appropriate heuristics to tune the design have been relatively successful (the vast majority of all controllers currently in operation are conventional PID controllers). One may ask the following questions: How much of the success can be attributed to the use of the mathematical model and conventional control design approach, and how much should be attributed to the clever heuristic tuning that the control engineer uses upon implementation? If we exploit the use of heuristic information throughout the entire design process can we obtain higher-performance control systems?

Fuzzy control provides a formal methodology for representing, manipulating, and implementing a human's heuristic knowledge about how to control a system. Fuzzy controller design involves incorporating human expertise on how to control a system into a set of rules (a rule base). The inference mechanism in the fuzzy controller reasons over the information in the knowledge base, the process outputs, and the user-specified goals to decide what inputs to generate for the process so that the closed-loop fuzzy control system will behave properly (e.g., so that the user-specified goals are met). For the cruise control example discussed above, it is clear that anyone who has experience in driving a car can practice regulating the speed about a desired set-point and load this information into a rule base. For instance, one rule that a human driver may use is "IF speed is lower than the set-point THEN press down further on the accelerator pedal". A rule that would represent even more detailed information about how to regulate the speed would be "IF speed is lower than the set-point AND speed is approaching the set-point very fast THEN release the accelerator pedal by a small amount". This second rule characterizes our knowledge about how to make sure that we do not overshoot our desired (goal) speed. Generally speaking, if we load very detailed expertise into the rule base we enhance our chances of obtaining better performance. Overall, the focus in fuzzy control is on the use of heuristic knowledge to achieve good control, whereas in conventional control the focus is on the use of a mathematical model for control systems development and subsequent use of heuristics in implementation.

Philosophy of Fuzzy Control

Due to the substantial amount of hype and excitement about fuzzy control it is important to begin by providing a sound control engineering philosophy for this approach. First, there is a need for the control engineer to assess what (if any) advantages fuzzy control methods have over conventional methods. Time permitting, this must be done by careful comparative analyses involving modeling, mathematical analysis, simulation, implementation, and a full engineering cost-benefit analysis (which involves issues of cost, reliability, maintainability, flexibility, lead time to production, etc.). When making the assessment of what control technique to use the engineer should be cautioned that most work in fuzzy control to date has only focused on its *advantages* and has not taken a critical look at what possible *disadvantages* there could be to using it. For example, the following questions are cause for concern:

- Will the behaviors observed by a human expert include all situations that can occur due to disturbances, noise, or plant parameter variations?
- Can the human expert realistically and reliably foresee problems that could arise from closed-loop system instabilities or limit cycles?
- Will the expert be able to effectively incorporate stability criteria and performance objectives (e.g., rise-time, overshoot, and tracking specifications) into a rule base to ensure that reliable operation can be obtained?
- Can an effective and widely used synthesis procedure be devoid of mathematical modeling and subsequent use of proven mathematical analysis tools?

These questions may seem even more troublesome if: (1) the control problem involves a "critical environment" where the failure of the control system to meet performance objectives could lead to loss of human life or an environmental disaster (e.g., in aircraft or nuclear power plant control), or (2) if the human expert's knowledge implemented in the fuzzy controller is somewhat inferior to that of a very experienced specialist that we expect to have design the control system (different designers have different levels of expertise). Clearly, then, for some applications there is a need for a methodology to develop, implement, and evaluate fuzzy controllers to ensure that they are reliable in meeting their performance specifications.

As it is discussed above, the standard control engineering methodology involves repeatedly coordinating the use of modeling, controller (re)design, simulation, mathematical analysis, and experimental evaluations to develop control systems. What is the relevance of this established methodology to the development of fuzzy control systems? Engineering a fuzzy control system uses many ideas from the standard control engineering methodology, except that in fuzzy control it is often said that a formal mathematical model is assumed unavailable so that mathematical analysis is impossible. While it is often the case that it is difficult, impossible, or cost prohibitive to develop an accurate mathematical model for many processes, it is almost always possible for the control engineer to specify some type of approximate model of the process (after all, we do know what physical object we are trying to control). Indeed, it has been our experience that most often the control engineer developing a fuzzy control system does have a mathematical model available. While it may not be used directly in controller design, it is often used in simulation to evaluate the performance of the fuzzy controller before it is implemented (and it is often used for rule base redesign). Certainly there are some applications where one can design a fuzzy controller and evaluate its performance directly via an implementation. In such applications one may not be overly concerned with a high performance level of the control system (e.g., for some commercial products such as washing machines or a shaver). In such cases there may thus be no need for conducting simulation-based evaluations (requiring a mathematical model) before implementation. In other applications there is the need for a high level of confidence in the reliability of the fuzzy control system before it is implemented (e.g., in systems where there is a concern for safety).

In addition to simulation-based studies, one approach to enhancing our confidence in the reliability of fuzzy control systems is to use the mathematical model of the plant and nonlinear analysis for (1) verification of stability and performance specifications and (2) possible redesign of the fuzzy controller (for an overview of the results in this area see [1]). Some may be confident that a true expert would never need anything more than intuitive knowledge for rule-based design, and therefore never design a faulty fuzzy controller. However, a true expert will certainly use all available information to ensure the reliable operation of a control system, including approximate mathematical models, simulation, nonlinear analysis, and experimentation. We emphasize that mathematical analysis cannot alone provide the definitive answers about the reliability of the fuzzy control system because such analysis proves properties about the model of the process, not the actual physical process. It can be argued that a mathematical model is never a perfect representation of a physical process; hence, while nonlinear analysis (e.g., of stability) may appear to provide definitive statements about control system reliability, it is understood that such statements are only accurate to the extent that the mathematical model is accurate. Nonlinear analysis does not replace the use of common sense and evaluation via simulations and experimentation; it simply assists in providing a rigorous engineering evaluation of a fuzzy control system before it is implemented.

It is important to note that the advantages of fuzzy control often become most apparent for very complex problems where we have an intuitive idea about how to achieve high performance control. In such control applications an accurate mathematical model is so complex (i.e., high order, nonlinear, stochastic, with many inputs and outputs) that it is sometimes not very useful for the analysis and design of conventional control systems (since assumptions needed to utilize conventional control design approaches are often violated). The conventional control engineering approach to this problem is to use an approximate mathematical model that is accurate enough to characterize the essential plant behavior, yet simple enough so that the necessary assumptions to apply the analysis and design techniques are satisfied. However, due to the inaccuracy of the model, upon implementation the developed controllers often need to be tuned via the "expertise" of the control engineer. The fuzzy control approach, where explicit characterization and utilization of control expertise is used earlier in the design process, largely avoids the problems with model complexity that are related to design. That is, for the most part fuzzy control system design does not depend on a mathematical model unless it is needed to perform simulations to gain insight into how to choose the rule base and membership functions. However, the problems with model complexity that are related to analysis have not been solved (i.e., analysis of fuzzy control systems critically depends on the form of the mathematical model); hence, it is often difficult to apply nonlinear analysis techniques to the applications where the advantages of fuzzy control are most apparent. For instance, as shown in [1], existing results for stability analysis of fuzzy control systems typically require that the plant model be deterministic, satisfy some continuity constraints, and sometimes require the plant to be linear or "linear-analytic." The only results for analysis of steady-state tracking error of fuzzy control systems, and the existing results on the use of describing functions for analysis of limit cycles, essentially require a linear time-invariant plant (or one that has a special form so that the nonlinearities can be bundled into a separate nonlinear component in the loop).

The current status of the field, as characterized by these limitations, coupled with the importance of nonlinear analysis of fuzzy control systems, make it an open area for investigation that will help establish the necessary foundations for a bridge between the communities of fuzzy control and nonlinear analysis. Clearly fuzzy control technology is leading the theory; the practitioner will proceed with the design and implementation of many fuzzy control systems without the aid of nonlinear analysis. In the mean time, theorists will attempt to develop a mathematical theory for the verification and certification of fuzzy control systems. This theory will have a synergistic effect by driving the development of fuzzy control systems for applications where there is a need for highly reliable implementations.

Summary

The focus of this chapter is on providing a practical introduction to fuzzy control (a "users guide"); hence we omit discussions of mathematical analysis of fuzzy control systems and invite the interested reader to investigate this topic further by consulting the bibliographic references. The remainder of this chapter is arranged as follows. We begin by providing a general mathematical introduction to fuzzy systems in a tutorial fashion. Next we introduce a rotational inverted pendulum "theme problem."

Many details on control design using principles of fuzzy logic are presented via this theme problem. We perform comparative analyses for fixed (nonadaptive) fuzzy and linear controllers. Following this we introduce the area of adaptive fuzzy control and show how one adaptive fuzzy technique has proven to be particularly effective for balancing control of the inverted pendulum. In the concluding remarks we explain how the area of fuzzy control is related to other areas in the field of intelligent control and what research needs to be performed as the field of fuzzy control matures.

57.10.2 Introduction to Fuzzy Control

The functional architecture of the fuzzy system (controller)[12] is composed of a *rule base* (containing a fuzzy logic quantification of the expert's linguistic description of how to achieve good control), an *inference mechanism* (which emulates the expert's decision making in interpreting and applying knowledge about how to do good control), a *fuzzification* interface (which converts controller inputs into information that the inference mechanism can easily use to activate and apply rules), and a *defuzzification* interface (which converts the conclusions of the inference mechanism into actual inputs for the process). In this section we describe each of these four components in more detail (see Section 57.10.4 for a block diagram) [2, 3, 4].

Linguistic Rules

For our purposes, a fuzzy system is a static nonlinear mapping between its inputs and outputs (i.e., it is not a dynamic system). It is assumed that the fuzzy system has inputs $u_i \in \mathcal{U}_i$ where $i = 1, 2, \ldots, n$ and outputs $y_i \in \mathcal{Y}_i$ where $i = 1, 2, \ldots, m$. The ordinary ("crisp") sets $\mathcal{U}_i$ and $\mathcal{Y}_i$ are called the "universes of discourse" for u_i and y_i, respectively (in other words they are their domains).

To specify rules for the rule base the expert will use a "linguistic description"; hence, linguistic expressions are needed for the inputs and outputs and the characteristics of the inputs and outputs. We will use "linguistic variables" (constant symbolic descriptions of what are, in general, time-varying quantities) to describe fuzzy system inputs and outputs. For our fuzzy system, linguistic variables denoted by $\tilde{u}_i$ are used to describe the inputs u_i. Similarly, linguistic variables denoted by $\tilde{y}_i$ are used to describe outputs y_i. For instance, an input to the fuzzy system may be described as $\tilde{u}_i$ = "velocity error" and an output from the fuzzy system may be $\tilde{y}_i$ = "voltage in".

Just as u_i and y_i take on values over each universe of discourse $\mathcal{U}_i$ and $\mathcal{Y}_i$, respectively, linguistic variables $\tilde{u}_i$ and $\tilde{y}_i$ take on "linguistic values" that are used to describe characteristics of the variables. Let $\tilde{A}_i^j$ denote the jth linguistic value of the linguistic variable $\tilde{u}_i$ defined over the universe of discourse $\mathcal{U}_i$. If we assume that there exist many linguistic values defined over $\mathcal{U}_i$, then the linguistic variable $\tilde{u}_i$ takes on the elements from the set of linguistic values denoted by $\tilde{A}_i = \{\tilde{A}_i^j : j = 1, 2, \ldots, N_i\}$ (sometimes for convenience we will let the j indices take on negative integer values). Similarly, let $\tilde{B}_i^j$ denote the jth linguistic value of the linguistic variable $\tilde{y}_i$ defined over the universe of discourse $\mathcal{Y}_i$. The linguistic variable $\tilde{y}_i$ takes on elements from the set of linguistic values denoted by $\tilde{B}_i = \{\tilde{B}_i^p : p = 1, 2, \ldots, M_i\}$ (sometimes for convenience we will let the p indices take on negative integer values). Linguistic values are generally expressed by descriptive terms such as "positive large," "zero," and "negative big" (i.e., adjectives).

The mapping of the inputs to the outputs for a fuzzy system is characterized in part by a set of *condition* $\rightarrow$ *action* rules, or in modus ponens (*If*... *Then*) form,

$$\textbf{If}\ (\text{antecedent})\ \textbf{Then}\ (\text{consequent}) \tag{57.415}$$

As usual, the inputs of the fuzzy system are associated with the antecedent, and the outputs are associated with the consequent. These *If*... *Then* rules can be represented in many forms. Two standard forms, multi-input, multi-output (MIMO) and multi-input, single output (MISO), are considered here. The MISO form of a linguistic rule is

$$\begin{aligned} &\textbf{If}\ \tilde{u}_1\ \text{is}\ \tilde{A}_1^j\ \textbf{and}\ \tilde{u}_2\ \text{is}\ \tilde{A}_2^k\ \textbf{and}, \ldots, \\ &\quad \textbf{and}\ \tilde{u}_n\ \text{is}\ \tilde{A}_n^l\ \textbf{Then}\ \tilde{y}_q\ \text{is}\ \tilde{B}_q^p \end{aligned} \tag{57.416}$$

It is a whole set of linguistic rules of this form that the expert specifies on how to control the system. Note that if $\tilde{u}_1$ = "velocity error" and $\tilde{A}_1^j$ = "positive large", then "$\tilde{u}_1$ is $\tilde{A}_1^j$", a single term in the antecedent of the rule, means "velocity error is positive large." It can be shown easily that the MIMO form for a rule (i.e., one with consequents that have terms associated with each of the fuzzy controller outputs) can be decomposed into a number of MISO rules (using simple rules from logic). We assume that there are a total of R rules in the rule base numbered $1, 2, \ldots, R$. For simplicity we will use tuples $(j, k, \ldots, l; p, q)_i$ to denote the ith MISO rule of the form given in Equation 57.416. Any of the terms associated with any of the inputs for any MISO rule can be included or omitted. Finally, we naturally assume that the rules in the rule base are distinct (i.e., there are no two rules with exactly the same antecedents and consequents).

Fuzzy Sets, Fuzzy Logic, and the Rule Base

Fuzzy sets and fuzzy logic are used to heuristically quantify the meaning of linguistic variables, linguistic values, and linguistic rules that are specified by the expert. The concept of a fuzzy set is introduced by first defining a "membership function." Let $\mathcal{U}_i$ denote a universe of discourse and $\tilde{A}_i^j \in \tilde{A}_i$ denote a specific linguistic value for the linguistic variable $\tilde{u}_i$. The function $\mu(u_i)$ associated with $\tilde{A}_i^j$ that maps $\mathcal{U}_i$ to $[0, 1]$ is called a "membership function." This membership function describes the "certainty" that an element of $\mathcal{U}_i$, denoted u_i, with a linguistic description $\tilde{u}_i$, may be classified linguistically as $\tilde{A}_i^j$. Membership functions are

[12] Sometimes a fuzzy controller is called a "fuzzy logic controller" or even a "fuzzy linguistic controller" since, as we will see, it uses fuzzy logic in the quantification of linguistic descriptions.

generally subjectively specified in an ad hoc (heuristic) manner from experience or intuition. For instance, if $\mathcal{U}_i = [-150, 150]$, $\tilde{u}_i$="velocity error," and $\tilde{A}_i^j$ = "positive large," then $\mu(u_i)$ may be a bell-shaped curve that peaks at one at $u_i = 75$ and is near zero when $u_i < 50$ or $u_i > 100$. Then if $u_i = 75$, $\mu(75) = 1$ so that we are absolutely certain that u_i is "positive large." If $u_i = -25$ then $\mu(-25)$ is very near zero, which represents that we are very certain that u_i is not "positive large." Clearly many other choices for the shape of the membership function are possible (e.g., triangular and trapezoidal shapes) and these will each provide a different meaning for the linguistics that they quantify. Below, we will show how to specify membership functions for a fuzzy controller for the rotational inverted pendulum.

Given a linguistic variable $\tilde{u}_i$ with a linguistic value $\tilde{A}_i^j$ defined on the universe of discourse $\mathcal{U}_i$, and membership function $\mu_{A_i^j}(u_i)$ (membership function associated with the fuzzy set A_i^j) that maps $\mathcal{U}_i$ to $[0, 1]$, a "fuzzy set" denoted with A_i^j is defined as

$$A_i^j = \{(u_i, \mu_{A_i^j}(u_i)) : u_i \in \mathcal{U}_i\} \tag{57.417}$$

Next, we specify some set-theoretic and logical operations on fuzzy sets. Given fuzzy sets A_i^1 and A_i^2 associated with the universe of discourse $\mathcal{U}_i$ ($N_i = 2$), with membership functions denoted $\mu_{A_i^1}(u_i)$ and $\mu_{A_i^2}(u_i)$, respectively, A_i^1 is defined to be a "fuzzy subset" of A_i^2, denoted by $A_i^1 \subset A_i^2$, if $\mu_{A_i^1}(u_i) \leq \mu_{A_i^2}(u_i)$ for all $u_i \in \mathcal{U}_i$.

The intersection of fuzzy sets A_i^1 and A_i^2 which are defined on the universe of discourse $\mathcal{U}_i$ is a fuzzy set, denoted by $A_i^1 \cap A_i^2$, with a membership function defined by either:

$$\begin{aligned} \text{Minimum:} \quad \mu_{A_i^1 \cap A_i^2} &= \min\{\mu_{A_i^1}(u_i), \mu_{A_i^2}(u_i) \\ &\quad : u_i \in \mathcal{U}_i\} \\ \text{Algebraic Product:} \quad \mu_{A_i^1 \cap A_i^2} &= \{\mu_{A_i^1}(u_i)\mu_{A_i^2}(u_i) \\ &\quad : u_i \in \mathcal{U}_i\} \end{aligned} \tag{57.418}$$

Suppose that we use the notation $x * y = min\{x, y\}$ or at other times we will use it to denote the product $x * y = xy$ ($*$ is sometimes called the "triangular norm"). Then $\mu_{A_i^1}(u_i) * \mu_{A_i^2}(u_i)$ is a general representation for the intersection of two fuzzy sets. In fuzzy logic, intersection is used to represent the "and" operation.

The union of fuzzy sets A_i^1 and A_i^2, which are defined on the universe of discourse $\mathcal{U}_i$, is a fuzzy set denoted $A_i^1 \cup A_i^2$, with a membership function defined by either:

$$\begin{aligned} \text{Maximum:} \quad \mu_{A_i^1 \cup A_i^2}(u_i) &= \max\{\mu_{A_i^1}(u_i), \mu_{A_i^2}(u_i) \\ &\quad : u_i \in \mathcal{U}_i\} \\ \text{Algebraic Sum:} \quad \mu_{A_i^1 \cup A_i^2}(u_i) &= \{\mu_{A_i^1}(u_i) + \mu_{A_i^2}(u_i) \\ &\quad - \mu_{A_i^1}(u_i)\mu_{A_i^2}(u_i) \\ &\quad : u_i \in \mathcal{U}_i\} \end{aligned} \tag{57.419}$$

Suppose that we use the notation $x \oplus y = max\{x, y\}$ or at other times we will use it to denote $x \oplus y = x + y - xy$ ($\oplus$ is sometimes called the "triangular co-norm"). Then $\mu_{A_i^1}(u_i) \oplus \mu_{A_i^2}(u_i)$ is a general representation for the union of two fuzzy sets. In fuzzy logic, union is used to represent the "or" operation.

The intersection and union above are both defined for fuzzy sets that lie on the same universe of discourse. The fuzzy Cartesian product is used to quantify operations on many universes of discourse. If $A_1^j, A_2^k, \ldots, A_n^l$ are fuzzy sets defined on the universes of discourse $\mathcal{U}_1, \mathcal{U}_2, \ldots, \mathcal{U}_n$, respectively, their Cartesian product is a fuzzy set (sometimes called a "fuzzy relation"), denoted by $A_1^j \times A_2^k \times \cdots \times A_n^l$, with a membership function defined by

$$\begin{aligned} &\mu_{A_1^j \times A_2^k \times \cdots \times A_n^l}(u_1, u_2, \ldots, u_n) = \\ &\mu_{A_1^j}(u_1) * \mu_{A_2^k}(u_2) * \cdots * \mu_{A_n}(u_n). \end{aligned} \tag{57.420}$$

Next, we show how to quantify the linguistic elements in the antecedent and consequent of the linguistic *If*... *Then* rule with fuzzy sets. For example, suppose we are given the *If*... *Then* rule in MISO form in Equation 57.416. Define the fuzzy sets:

$$\begin{aligned} A_1^j &= \{(u_1, \mu_{A_1^j}(u_1)) : u_1 \in \mathcal{U}_1\} \\ A_2^k &= \{(u_2, \mu_{A_2^k}(u_2)) : u_2 \in \mathcal{U}_2\} \\ &\vdots \\ A_n^l &= \{(u_n, \mu_{A_n^l}(u_n)) : u_n \in \mathcal{U}_n\} \\ B_q^p &= \{(y_q, \mu_{B_q^p}(y_q)) : y_q \in \mathcal{Y}_q\} \end{aligned} \tag{57.421}$$

These fuzzy sets quantify the terms in the antecedent and the consequent of the given *If*... *Then* rule, to make a "fuzzy implication"

$$\textbf{If } A_1^j \textbf{ and } A_2^k \textbf{ and}, \ldots, \textbf{and } A_n^l \textbf{ Then } B_q^p \tag{57.422}$$

where the fuzzy sets $A_1^j, A_2^k, \ldots, A_n^l$, and B_q^p are defined in Equation 57.422. Therefore, the fuzzy set A_1^j is associated with, and quantifies, the meaning of the linguistic statement "$\tilde{u}_1$ is $\tilde{A}_1^j$" and B_q^p quantifies the meaning of "$\tilde{y}_q$ is $\tilde{B}_q^p$". Each rule in the rule base $(j, k, \ldots, l; p, q)_i, i = 1, 2, \ldots, R$ is represented with such a fuzzy implication (a fuzzy quantification of the linguistic rule). The reader who is interested in more mathematical details on fuzzy sets and fuzzy logic should consult [5].

Fuzzification

Fuzzy sets are used to quantify the information in the rule base, and the inference mechanism operates on fuzzy sets to produce fuzzy sets; hence, we must specify how the fuzzy system will convert its numeric inputs $u_i \in \mathcal{U}_i$ into fuzzy sets (a process called "fuzzification") so that they can be used by the fuzzy system. Let $\mathcal{U}_i^*$ denote the set of all possible fuzzy sets that can be defined on $\mathcal{U}_i$. Given $u_i \in \mathcal{U}_i$, fuzzification transforms u_i to a fuzzy set denoted by $\hat{A}_i^{\text{fuz}}$ defined[13] over the universe discourse

[13] In this section, as we introduce various fuzzy sets, we always use a hat over any fuzzy set whose membership function changes dynamically over time as the u_i change.

$\mathcal{U}_i$. This transformation is produced by the fuzzification operator $\mathcal{F}$ defined by, $\mathcal{F}: \mathcal{U}_i \to \mathcal{U}_i^*$ where $\mathcal{F}(u_i) := \hat{A}_i^{\text{fuz}}$. Quite often "singleton fuzzification" is used, which produces a fuzzy set $\hat{A}_i^{\text{fuz}} \in \mathcal{U}_i^*$ with a membership function defined by

$$\mu_{\hat{A}_i^{\text{fuz}}}(x) = \begin{cases} 1 & x = u_i \\ 0 & \text{otherwise} \end{cases} \tag{57.423}$$

(any fuzzy set with this form for its membership function is called a "singleton"). Singleton fuzzification is generally used in implementations since, without the presence of noise, we are absolutely certain that u_i takes on its measured value (and no other value) and since it provides certain savings in the computations needed to implement a fuzzy system (relative to, e.g., "Gaussian fuzzification" which would involve forming bell-shaped membership functions about input points). Throughout the remainder of this chapter we use singleton fuzzification.

The Inference Mechanism

The inference mechanism has two basic tasks: (1) determining the extent to which each rule is relevant to the current situation as characterized by the inputs u_i, $i = 1, 2, \ldots, n$ (we call this task "matching"), and (2) drawing conclusions using the current inputs u_i and the information in the rule base (we call this task an "inference step"). For matching note that $A_1^j \times A_2^k \times \cdots \times A_n^l$ is the fuzzy set representing the antecedent of the ith rule $(j, k, \ldots, l; p, q)_i$ (there may be more than one such rule with this antecedent). Suppose that at some time we get inputs u_i, $i = 1, 2, \ldots, n$, and fuzzification produces $\hat{A}_1^{\text{fuz}}, \hat{A}_2^{\text{fuz}}, \ldots, \hat{A}_n^{\text{fuz}}$, the fuzzy sets representing the inputs. The first step in matching involves finding fuzzy sets $\hat{A}_1^j, \hat{A}_2^k, \ldots, \hat{A}_n^l$ with membership functions

$$\begin{aligned} \mu_{\hat{A}_1^j}(u_1) &= \mu_{A_1^j}(u_1) * \mu_{\hat{A}_1^{\text{fuz}}}(u_1) \\ \mu_{\hat{A}_2^k}(u_2) &= \mu_{A_2^k}(u_2) * \mu_{\hat{A}_2^{\text{fuz}}}(u_2) \\ &\vdots \\ \mu_{\hat{A}_n^l}(u_n) &= \mu_{A_n^l}(u_n) * \mu_{\hat{A}_n^{\text{fuz}}}(u_n) \end{aligned}$$

(for all $j, k, \ldots, l$) that combine the fuzzy sets from fuzzification with the fuzzy sets used in each of the terms in the antecedents of the rules. If singleton fuzzification is used then each of these fuzzy sets is a singleton that is scaled by the antecedent membership function (e.g., $\mu_{\hat{A}_1^j}(\bar{u}_1) = \mu_{A_1^j}(\bar{u}_1)$ for $\bar{u}_1 = u_1$ and $\mu_{\hat{A}_1^j}(\bar{u}_1) = 0$ for $\bar{u}_1 \neq u_1$). Second, we form membership values $\mu_i(u_1, u_2, \ldots, u_n)$ for each rule that represent the overall certainty that rule i matches the current inputs. In particular we first let

$$\bar{\mu}_i(u_1, u_2, \ldots, u_n) = \mu_{\hat{A}_1^j}(u_1) * \mu_{\hat{A}_2^k}(u_2) * \cdots * \mu_{\hat{A}_n^l}(u_n) \tag{57.424}$$

be the membership function for $\hat{A}_1^j \times \hat{A}_2^k \times \cdots \times \hat{A}_n^l$. Notice that since we are using singleton fuzzification we have

$$\bar{\mu}_i(\bar{u}_1, \bar{u}_2, \ldots, \bar{u}_n) = \mu_{A_1^j}(\bar{u}_1) * \mu_{A_2^k}(\bar{u}_2) * \cdots * \mu_{A_n^l}(\bar{u}_n) \tag{57.425}$$

for $\bar{u}_i = u_i$, and $\bar{\mu}_i(\bar{u}_1, \bar{u}_2, \ldots, \bar{u}_n) := 0$ for $\bar{u}_i \neq u_i$, $i = 1, 2, \ldots, n$. Since the u_i are given, $\mu_{A_1^j}(u_1), \mu_{A_2^k}(u_2), \cdots, \mu_{A_n^l}(u_n)$ are constants. Define

$$\mu_i(u_1, u_2, \ldots, u_n) = \mu_{A_1^j}(u_1) * \mu_{A_2^k}(u_2) * \cdots * \mu_{A_n^l}(u_n) \tag{57.426}$$

which is simply a function of the inputs u_i. We use $\mu_i(u_1, u_2, \ldots, u_n)$ to represent the certainty that the antecedent of rule i matches the input information. This concludes the process of matching input information with the antecedents of the rules.

Next, the inference step is taken by computing, for the ith rule $(j, k, \ldots, l; p, q)_i$, the "implied fuzzy set" $\hat{B}_q^i$ with membership function

$$\mu_{\hat{B}_q^i}(y_q) = \mu_i(u_1, u_2, \ldots, u_n) * \mu_{B_q^p}(y_q) \tag{57.427}$$

The implied fuzzy set $\hat{B}_q^i$ specifies the certainty level that the output should be a specific crisp output y_q within the universe of discourse $\mathcal{Y}_q$, taking into consideration only rule i. Note that since $\mu_i(u_1, u_2, \ldots, u_n)$ will vary with time so will the shape of the membership functions $\mu_{\hat{B}_q^i}(y_q)$ for each rule. Alternatively, the inference mechanism could, in addition, compute the "overall implied fuzzy set" $\hat{B}_q$ with membership function

$$\mu_{\hat{B}_q}(y_q) = \mu_{\hat{B}_q^1}(y_q) \oplus \mu_{\hat{B}_q^2}(y_q) \oplus \cdots \oplus \mu_{\hat{B}_q^R}(y_q) \tag{57.428}$$

that represents the conclusion reached considering all the rules in the rule base at the same time (notice that determining $\hat{B}_q$ can, in general, require significant computational resources).

Using the mathematical terminology of fuzzy sets, the computation of $\mu_{\hat{B}_q}(y_q)$ is said to be produced by a "sup-star compositional rule of inference." The "sup" in this terminology corresponds to the $\oplus$ operation and the "star" corresponds to $*$. "Zadeh's compositional rule of inference" [6] is the special case of the sup-star compositional rule of inference when max is used for $\oplus$ and min is used for $*$. The overall justification for using the above operations to represent the inference step lies in the fact that *we can be no more certain about our conclusions than we are about our premises (antecedents)*. The operations performed in taking an inference step adhere to this principle. To see this, study Equation 57.427 and note that the scaling from $\mu_i(u_1, u_2, \ldots, u_n)$ that is produced by the antecedent matching process ensures that $\sup_{y_q}\{\mu_{\hat{B}_q^i}(y_q)\} \leq \mu_i(u_1, u_2, \ldots, u_n)$ (a similar statement holds for the overall implied fuzzy set).

Up to this point we have used fuzzy logic to quantify the rules in the rule base, fuzzification to produce fuzzy sets characterizing the inputs, and the inference mechanism to produce fuzzy sets representing the conclusions that it reaches considering the current inputs and the information in the rule base. Next, we look at how to convert this fuzzy set quantification of the conclusions to a numeric value that can be input to the plant.

Defuzzification

A number of defuzzification strategies exist. Each provides a means to choose a single output (which we denote with y_q^{crisp}) based on either the implied fuzzy sets or the overall implied fuzzy set (depending on the type of inference strategy chosen). First, we present typical defuzzification techniques for the overall implied fuzzy set $\hat{B}_q$:

- Max Criteria: A crisp output y_q^{crisp} is chosen as the point on the output universe of discourse $\mathcal{Y}_q$ for which the overall implied fuzzy set $\hat{B}_q$ achieves a maximum, i.e.

$$y_q^{\text{crisp}} \in \{\arg\sup_{\mathcal{Y}_q} \left\{\mu_{\hat{B}_q}(y_q)\right\}\} \tag{57.429}$$

  Since the supremum can occur at more than one point in $\mathcal{Y}_q$ one also needs to specify a strategy on how to pick only one point for y_q^{crisp} (e.g., choosing the smallest value). Often this defuzzification strategy is avoided due to this ambiguity.

- Mean of Maximum: A crisp output y_q^{crisp} is chosen to represent the mean value of all elements whose membership in $\hat{B}_q$ is a maximum. We define $\hat{b}_q^{\max}$ as the supremum of the membership function of $\hat{B}_q$ over the universe of discourse $\mathcal{Y}_q$. Moreover, we define a fuzzy set $\hat{B}_q^* \in \mathcal{Y}_q$ with a membership function defined as

$$\mu_{\hat{B}_q^*}(y_q) = \begin{cases} 1 & \mu_{\hat{B}_q}(y_q) = \hat{b}_q^{\max} \\ 0 & \text{otherwise} \end{cases} \tag{57.430}$$

  then a crisp output, using the mean of maximum method, is defined as

$$y_q^{\text{crisp}} = \frac{\int_{\mathcal{Y}_q} y_q \cdot \mu_{\hat{B}_q^*}(y_q) \cdot dy_q}{\int_{\mathcal{Y}_q} \mu_{\hat{B}_q^*}(y_q) \cdot dy_q} \tag{57.431}$$

  Note that the integrals in Equation 57.431 must be computed at each time instant since they depend on $\hat{B}_q$, which changes with time. This can require excessive computational resources; hence, this defuzzification technique is often avoided in practice.

- Center of Area (COA): A crisp output y_q^{crisp} is chosen as the center of area for the membership function of the overall implied fuzzy set $\hat{B}_q$. For a continuous output universe of discourse $\mathcal{Y}_q$ the center of area output is denoted by

$$y_q^{\text{crisp}} = \frac{\int_{\mathcal{Y}_q} y_q \cdot \mu_{\hat{B}_q}(y_q) dy_q}{\int_{\mathcal{Y}_q} \mu_{\hat{B}_q}(y_q) dy_q} \tag{57.432}$$

  Note that, similar to the mean of the maximum method, this defuzzification approach can be computationally expensive. Also, the fuzzy system must be defined so that $\int_{\mathcal{Y}_q} \mu_{\hat{B}_q}(y_q) dy_q \neq 0$ for all u_i.

Next, we specify typical defuzzification techniques for the implied fuzzy sets $\hat{B}_q^i$:

- Center Average: A crisp output y_q^{crisp} is chosen using the centers of each of the output membership functions and the maximum certainty of each of the conclusions represented with the implied fuzzy sets and is given by

$$y_q^{\text{crisp}} = \frac{\sum_{i=1}^{R} c_q^i \sup_{y_q}\{\mu_{\hat{B}_q^i}(y_q)\}}{\sum_{i=1}^{R} \sup_{y_q}\{\mu_{\hat{B}_q^i}(y_q)\}} \tag{57.433}$$

  where c_q^i is the center of area of the membership function of B_q^p associated with the implied fuzzy set $\hat{B}_q^i$ for the ith rule $(j, k, \ldots, l; p, q)_i$. Notice that $\sup_{y_q}\{\mu_{\hat{B}_q^i}(y_q)\}$ is often very easy to compute since if $\mu_{B_q^p}(y_q) = 1$ for at least one y_q (which is the normal way to define membership functions), then for many inference strategies $\sup_{y_q}\{\mu_{\hat{B}_q^i}(y_q)\} = \mu_i(u_1, u_2, \ldots, u_n)$, which has already been computed in the matching process. Notice that the fuzzy system must be defined so that $\sum_{i=1}^{R} \sup_{y_q}\{\mu_{\hat{B}_q^i}(y_q)\} \neq 0$ for all u_i.

- Center of Gravity (COG): A crisp output y_q^{crisp} is chosen using the center of area and area of each implied fuzzy set and is given by

$$y_q^{\text{crisp}} = \frac{\sum_{i=1}^{R} c_q^i \int_{\mathcal{Y}_q} \mu_{\hat{B}_q^i}(y_q) dy_q}{\sum_{i=1}^{R} \int_{\mathcal{Y}_q} \mu_{\hat{B}_q^i}(y_q) dy_q} \tag{57.434}$$

  where c_q^i is the center of area of the membership function of B_q^p associated with the implied fuzzy set $\hat{B}_q^i$ for the ith rule $(j, k, \ldots, l; p, q)_i$. Notice that COG can be easy to compute, since it is often easy to find closed-form expressions for $\int_{\mathcal{Y}_q} \mu_{\hat{B}_q^i}(y_q) dy_q$ which is the area under a membership function. Notice that the fuzzy system must be defined so that $\sum_{i=1}^{R} \int_{\mathcal{Y}_q} \mu_{\hat{B}_q^i}(y_q) dy_q \neq 0$ for all u_i.

Overall, we see that using the overall implied fuzzy set in defuzzification is often undesirable for two reasons: (1) the overall implied fuzzy set $\hat{B}_q$ is itself difficult to compute in general, and (2) the defuzzification techniques based on an inference mechanism that provides $\hat{B}_q$ are also difficult to compute. It is for this reason that most existing fuzzy controllers (including the ones in this chapter) use defuzzification techniques based on the implied fuzzy sets such as Centroid or COG.

57.10.3 Theme Problem: Rotational Inverted Pendulum

One of the classic problems in the study of nonlinear systems is that of the inverted pendulum. The primary control problem one considers with such a system is regulating the position of the pendulum (typically a rod with mass at the endpoint) to the vertical (up) position; i.e., "balanced." A secondary problem is that of "swinging up" the pendulum from its rest position (vertical down). Often, actuation is accomplished either via a motor at the base of the pendulum (at the hinge), or via a cart through translational motion. In this example, actuation of the pendulum is accomplished through *rotation* of a separate, attached link, referred to henceforth as the "base."

Experimental Apparatus

The test bed consists of three primary components: the plant, digital and analog interfaces, and the digital controller. The overall system is shown in Figure 57.37, where the three components can be clearly identified [7]. The plant is composed of a pendulum and a rotating base made of aluminum rods, two optical encoders as the angular position sensors with effective resolutions of 0.2 degrees for the pendulum and 0.1 degrees for the base, and a large, high-torque permanent-magnet DC motor (with rated stall torque of 5.15 N-m). As the base rotates through the angle θ_0 the pendulum is free to rotate (high precision bearings are utilized) through its angle θ_1 made with the vertical.

Interfaces between the digital controller and the plant consist of two data acquisition cards and some signal conditioning circuitry, structured for the two basic functions of sensor integration and control signal generation. The signal conditioning is accomplished via a combination of several logic gates to filter quadrature signals from the optical encoders, which are then processed through a separate data acquisition card to utilize the four 16-bit counters (accessed externally to count pulses from the circuitry itself). Another card supplies the control signal interface through its 12-bit D/A converter (to generate the actual control signal), while the board's 16-bit timer is used as a sampling clock. The computer used for control is a personal computer with an Intel 80486DX processor operating at 50 MHz. The real-time codes for control are written in C.

Mathematical Model

For brevity, and because this system is a popular example for nonlinear control, we omit details of the necessary physics and geometry for modeling. The differential equations that approximately describe the dynamics of the plant are given by

$$\ddot{\theta}_0 = -a_p\dot{\theta}_0 + K_p\, v_a \tag{57.435}$$

$$\ddot{\theta}_1 = -\frac{C_1}{J_1}\dot{\theta}_1 + \frac{m_1\, g\, \ell_1}{J_1}\sin(\theta_1) + K_1\ddot{\theta}_0 \tag{57.436}$$

where, again, θ_0 is the angular displacement of the rotating base, $\dot{\theta}_0$ is the angular speed of the rotating base, θ_1 is the angular displacement of the pendulum, $\dot{\theta}_1$ is the angular speed of the pendulum, v_a is the motor armature voltage, K_p and a_p are parameters of the DC motor with torque constant K_1, g is the acceleration due to gravity, m_1 is the pendulum mass, ℓ_1 is the pendulum length, J_1 is the pendulum inertia, and C_1 is a constant associated with friction (actual parameter values appear in [7]).

For controller synthesis (and model linearization) we require a state variable description of the system. This is easily done by defining state variables $x_1 = \theta_0$, $x_2 = \dot{\theta}_0$, $x_3 = \theta_1$, $x_4 = \dot{\theta}_1$, and control signal $u = v_a$. Linearization of these equations *about the vertical position* (i.e., $\theta_1 = 0$), and using the system physical parameters [7] results in the following linear, time-invariant state-variable description:

$$\begin{bmatrix} \dot{x}_1 \\ \dot{x}_2 \\ \dot{x}_3 \\ \dot{x}_4 \end{bmatrix} = \begin{bmatrix} 0 & 1 & 0 & 0 \\ 0 & -33.04 & 0 & 0 \\ 0 & 0 & 0 & 1 \\ 0 & 49.30 & 73.41 & -2.29 \end{bmatrix} \begin{bmatrix} x_1 \\ x_2 \\ x_3 \\ x_4 \end{bmatrix} + \begin{bmatrix} 0 \\ 74.89 \\ 0 \\ -111.74 \end{bmatrix} u \tag{57.437}$$

Swing Up Control

Because we intend to develop control laws that will be valid in regions about the vertical position ($\theta_1 = 0$), it is crucial to swing the pendulum up so that it is near-vertical at near-zero (angular) velocity. Elaborate schemes can be used for this task (such as those employing concepts from differential geometry), but for the purposes of this example we choose to use a simple heuristic procedure based on an "energy pumping strategy" proposed in [8] for a similar *under-actuated* system. The goal of this simple swing up control strategy is to "pump" energy into the pendulum link in such a way that the energy or magnitude of each swing increases until the pendulum approaches its inverted position. To apply such an approach, we simply consider how one would (intuitively) swing the pendulum from its hanging position ($\theta_1 = \pi$) to its upright position. If the rotating base is swung to the left and right continually at an appropriate frequency, the magnitude of the pendulum at each swing will increase.

The control scheme we will ultimately employ consists of two main components: the "scheduler" (which we will also call a "supervisor") observes the position of the pendulum relative to its stable equilibrium point ($\theta_1 = \pi$), then schedules the transitions between two reference positions of the rotating base (${\theta_0}^{ref} = \pm\Gamma$); and, the "positioning control" regulates the base to the desired reference point. These two components compose a closed-loop planning algorithm to command the rotating base to move in a certain direction based on the position of the pendulum. In effect, the human operator acts as the supervisor in tuning the positioning control (through trial and error on the system).

For simplicity, a proportional controller will be used as the positioning control. The gain K_p is chosen just large enough so that the actuator drives the base fast enough without saturating the control output; after several trials, K_p was set to 0.5. The parameter Γ determines how far the base is allowed to swing; larger swings transfer more energy to swinging up the pendulum. The swing-up motion of the pendulum can be approximated as an exponentially growing cosine function. The parameter Γ significantly affects the "negative damping" (i.e., exponential growth) of the swing-up motion. By tuning Γ, one can adjust the motion

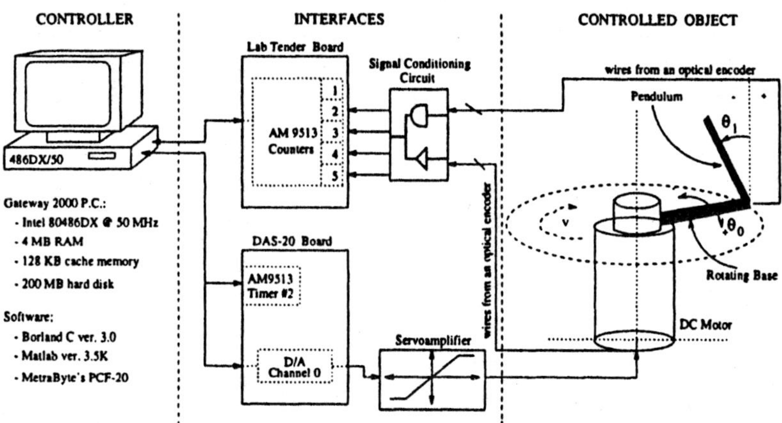

Figure 57.37 Hardware setup.

of the pendulum in such a way that the velocity of the pendulum and the control output are minimum when the pendulum reaches its inverted position (i.e., the pendulum has the largest potential energy and the lowest kinetic energy). Notice that if the dynamics of the pendulum are changed (e.g., adding extra weight to the endpoint of the pendulum), then the parameter Γ must be tuned. In [7] it is shown how a rule-based system can be used to effectively automate the swing up control by implementing fuzzy strategies in the supervisor portion of the overall scheme.

Balancing Control

Synthesis of the fuzzy controllers to follow is aided by (1) a good understanding of the pendulum dynamics (the analytical model and intuition related to the physical process), and (2) experience with performance of linear control strategies. Although numerous linear control design techniques have been applied to this particular system, here we consider the performance of only one linear strategy (the one tested) as applied to the experimental system: the linear quadratic regulator (LQR). Our purpose is twofold. First, we wish to form a baseline for comparison to fuzzy control designs to follow, and second, we wish to provide a starting point for synthesis of the fuzzy controller. It is important to note that extensive simulation results (on the nonlinear model) were carried out prior to application to the laboratory apparatus; designs were carried out on the linearized model of the system. Specifics of the design process for the LQR and other applicable linear design techniques may be found in other chapters of this volume.

Because the linearized system is completely controllable and observable, state feedback strategies, including the optimal strategies of the LQR, are applicable. Generally speaking, the system performance is prescribed via the optimal performance index

$$J = \int_0^{\infty} (x(t)^T Q x(t) + u(t)^T R u(t)) dt \quad (57.438)$$

where Q and R are the weighting matrices corresponding to the state x and input u, respectively. Given fixed Q and R, the feedback gains that optimize the function J can be uniquely determined by solving an algebraic Riccati equation. Because we are more concerned with balancing the pendulum than regulating the base, we put the highest priority in controlling θ_1 by choosing the weighting matrices $Q = diag(1, 0, 5, 0)$ and $R = [1]$. For a 10 msec sampling time, the discrete optimal feedback gains corresponding to the weighting matrices Q and R are $k_1 = -0.9$, $k_2 = -1.1$, $k_3 = -9.2$, and $k_4 = -0.9$. Although observers may be designed to estimate the states $\dot{\theta}_1$ and $\dot{\theta}_0$, we choose to use an equally effective and simple first-order approximation for each derivative.

Note that this controller is designed in simulation for the system as modeled (and subsequently linearized). When the resulting controller gains (k_1 through k_4) are implemented on the actual system, some "trial and error" tuning is required (due primarily to modeling uncertainties), which amounted to adjusting the designed gains by about 10% to obtain performance matching the predicted results from simulation. Moreover, it is critical to note that the design process (as well as the empirical tuning) has been done for the "nominal" system (i.e., the pendulum system with no additional mass on the endpoint).

Using a swing up control strategy tuned for the nominal system, the results of the LQR control design are given in Figure 57.38 for the base angle (top plot), pendulum angle (center plot), and control output (bottom plot).

57.10.4 Fuzzy Control for the Rotational Inverted Pendulum

Aside from serving to illustrate procedures for synthesizing a fuzzy controller, several reasons arise for considering the use of a nonlinear control scheme for the pendulum system. Because all linear controllers are designed based on a linearized model of the system, they are inherently valid only for a region about a specific point (in this case, the vertical, $\theta_1 = 0$ position). For this reason, such linear controllers tend to be very sensitive to parametric variations, uncertainties, and disturbances. This is indeed the case for the experimental system under study; when an extra weight or *sloshing liquid* (using a water-tight bottle) is attached at the endpoint of the pendulum, the performance of all linear controllers degrades considerably, often resulting in unstable behavior. Thus, to enhance the performance of the balancing

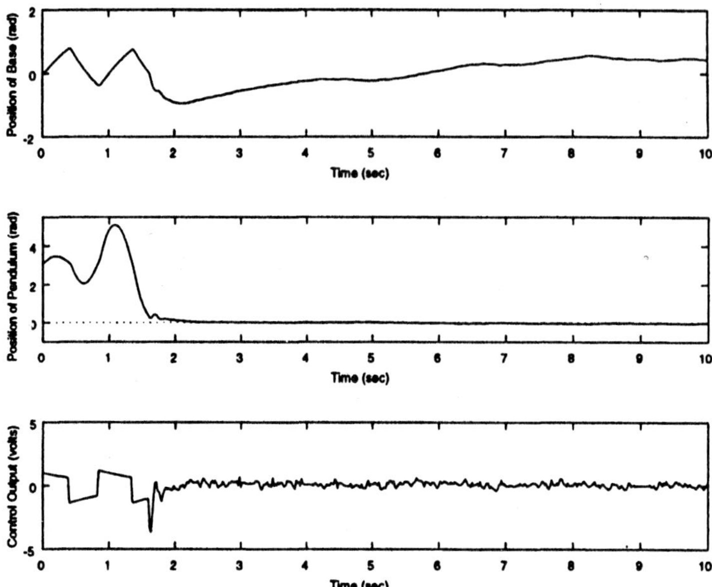

Figure 57.38 Experimental results: LQR on the *nominal system.*

control, one naturally turns to some nonlinear control scheme that is expected to exhibit improved performance in the presence of disturbances and uncertainties in modeling. Two such nonlinear controllers will be investigated here: in this section, a direct fuzzy controller is constructed and later an adaptive version of this same controller is discussed.

Controller Synthesis

For simplicity, the controller synthesis example explained next will utilize singleton fuzzification and symmetric, "triangular" membership functions on the controller inputs and output (they are, in fact, very simple to implement in real-time code). We choose to use seven membership functions for each input, uniformly distributed across their universes of discourse (over crisp values of each input e_i), as shown in Figure 57.39. The linguistic values for the ith input are denoted by $\tilde{E}_i^r$ where $r \in \{-3, -2, -1, 0, 1, 2, 3\}$. Linguistically, we would therefore define $\tilde{E}_i^{-3}$ as "negative large," $\tilde{E}_i^{-2}$ as "negative medium," $\tilde{E}_i^0$ as "zero," and so on. Note also that a "saturation nonlinearity" is built in for each input in the membership functions corresponding to the outermost regions of the universes of discourse. We use min to represent the antecedent and implication (i.e., "$*$" from Section 57.10.2 is min) and COG defuzzification.

To synthesize a fuzzy controller for our example system, we pursue the idea of seeking to "expand" the region of operation of the fixed (nonadaptive) controller. In doing so, we will utilize the results of the LQR design presented in Section 57.10.3 to lead us in the design. A block diagram of the fuzzy controller is shown in Figure 57.40. Similar to the linear quadratic regulator, the fuzzy controller for the inverted pendulum system will have four inputs and one output. The four (crisp) inputs to the fuzzy controller are the position error of the base e_1, its derivative e_2, the position error of the pendulum e_3, and its derivative e_4.

The *normalizing* gains g_i essentially serve to expand and compress the universes of discourse to some predetermined, uniform region, primarily to standardize the choice of the various parameters in synthesizing the fuzzy controller. A crude approach to choosing these gains is strictly based on intuition and does not require a mathematical model of the plant. In that case the input normalizing gains are chosen in such a way that all the desired operating regions are mapped into $[-1, +1]$. Such a simple approach in design works often for a number of systems, as witnessed by the large number of applications documented in the open literature. For complicated systems, however, such a procedure can be very difficult to implement because there are many ways to define the linguistic values and linguistic rules; indeed, it can be extremely difficult to find a viable set of linguistic values and rules just to maintain stability. Such was the case for this system.

What we propose here is an approach based on experience in designing the LQR controller for the linearized model of the plant, leading to a mechanized procedure for determining the normalizing gains, output membership functions, and rule base. Recall from our discussion in Section 57.10.2 that a fuzzy system is a static nonlinear map between its inputs and output. Certainly, therefore, a linear map such as the LQR can be easily approximated by a fuzzy system (for small values of the inputs to the fuzzy system). Two components of the LQR are the optimal gains and the summer; the optimal gains can be replaced with the normalizing gains of a fuzzy system, and the summer can essentially be incorporated into the rule base of a fuzzy system. By doing this, we can effectively utilize a fuzzy system implementation to expand the region of operation of the controller beyond the "linear region" afforded by the linearization/design process. Intuitively, this is done by making the "gain" of the fuzzy controller match that of the LQR when the fuzzy controller inputs are small, while shaping the nonlinear mapping representing the fuzzy controller for larger inputs (in regions further from zero).

As pointed out in Section 57.10.2, the rule base contains information about the relationships between the inputs and output of

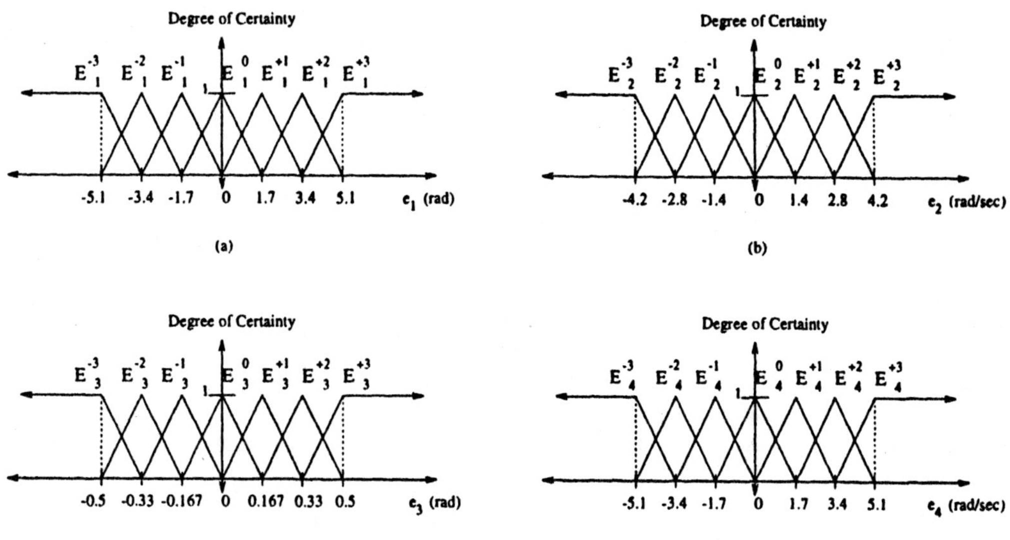

Figure 57.39 Four sets of input membership functions: (a) "base position error" ($\tilde{e}_1$), (b) "base derivative error" ($\tilde{e}_2$), (c) "pendulum position error" ($\tilde{e}_3$), and (d) "pendulum derivative error" ($\tilde{e}_4$).

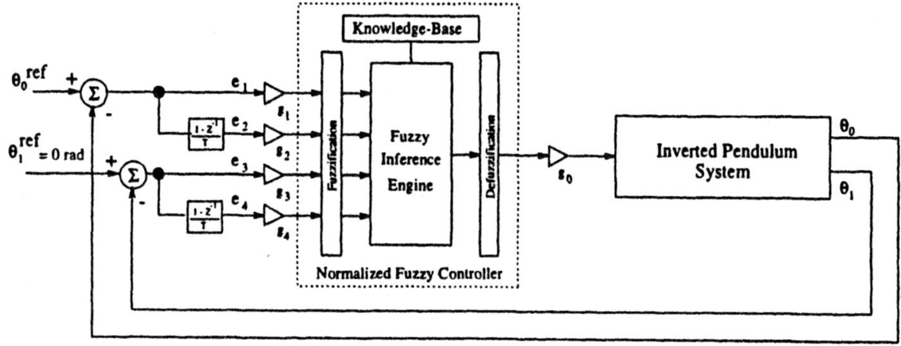

Figure 57.40 Block diagram of direct fuzzy controller.

a fuzzy system. Within the controller structure chosen, recall that we wish to construct the rule base to perform a weighted sum of the inputs. The summation operation is straightforward; prior to normalization, it is simply a matter or arranging each *If* ... *Then* rule such that the antecedent indices sum to the consequent indices. Specification of the normalizing gains is explained next.

The basic idea [9] in specifying the $g_0 - g_4$ is so that for "small" controller inputs (e_i) the local slope (about zero) of the input-output mapping representing the controller will be the same as the LQR gains (i.e., the k_i). As alluded to above, the normalizing gains $g_1 - g_4$ transform the (symmetric) universes of discourse for each input (see Figure 57.39) to $[-1, 1]$. For example, if $[-\beta_i, \beta_i]$ is the interval of interest for input i, the choice $g_i = 1/\beta_i$ would achieve this normalization, whereas the choice $g_0 = \beta_0$ would map the output of the normalized fuzzy system to the real output to achieve a corresponding interval of $[-\beta_0, \beta_0]$. Then, assuming the fuzzy system provides the summation operation, the "net gain" for the ith input-output pair is $g_i g_0$. Finally, therefore, this implies that $g_i g_0 = k_i$ is required to match local slopes of the LQR controller and the fuzzy controller (in the sense of input-output mappings).

We are now in position to summarize the gain selection procedure. Recalling (Section 57.10.3) that the optimal feedback gains based on the LQR approach are $k_1 = -0.9$, $k_2 = -1.1$, $k_3 = -9.2$, and $k_4 = -0.9$, transformation of the optimal LQR gains into the normalizing gains of the fuzzy system is achieved according to the following simple scheme:

- Choose the controller input that most greatly influences plant behavior and overall control objectives; in our case, we choose the pendulum position θ_1. Subsequently, we specify the operating range of this input (e.g., the interval $[-0.5, +0.5]$ radians, for which the corresponding normalizing input gain $g_3 = 2$).
- Given g_3, the output gain of the fuzzy controller is calculated according to $g_0 = \frac{k_3}{g_3} = -4.6$.
- Given the output gain g_0, the remaining input gains can be calculated according to $g_j = \frac{k_j}{g_0}$, where $j \in \{1, 2, 3, 4\}$, $j \neq i$ (note that $i = 3$). For $g_0 = -4.6$, the input gains g_1, g_2, g_3, and g_4 are 0.1957, 0.2391, 2, and 0.1957, respectively.

Determination of the controller output universe of discourse and corresponding normalizing gain is dependent on the structure of the rule base. A nonlinear mapping can be used to rearrange the output membership functions (in terms of their centers) for several purposes, such as to add higher gain near the center, to create a dead zone near the center, to eliminate discontinuities at the saturation points, and so on. This represents yet another area where intuition (i.e., knowledge about how to best control the process) may be incorporated into the design process. In order to preserve behavior in the "linear" region (i.e., the region near the origin) of the LQR-extended controller, but at the same time provide a smooth transition from the linear region to its extensions (e.g., regions of saturation), we choose an arctangent-type mapping to achieve this rearrangement. Because of the "flatness" of such a mapping near the origin, we expect the fuzzy controller to behave like the LQR when the states are near the process equilibrium.

It is important to note that, unlike the input membership function of Figure 57.39, the output membership functions at the outermost regions of the universe of discourse do *not* include the saturating effect; rather, they return to zero value which is required so that the fuzzy controller mapping is well defined. In general, for a fuzzy controller with n inputs and one output, the center of the controller output fuzzy set Y^s would be located at $(j+k+...+l) \times \frac{2}{(N-1)n}$, where $s = j+k+\ldots+l$ is the index of the output fuzzy set Y^s (and the output linguistic value), $\{j, k, \ldots l\}$ are the indices of the input fuzzy sets (and linguistic values), N is the number of membership functions on each input, and n is the number of inputs. Note that we must nullify the effect of divisor n by multiplying the output gain g_0 by the same factor.

Performance Evaluation

Simulation Some performance evaluation via simulation is prudent to investigate the effectiveness of the strategies employed in the controller synthesis. Using the complete nonlinear model, the simulated responses of the direct fuzzy controller with seven membership functions on each input indicate that the fuzzy controller successfully balances the pendulum, but with a slightly degraded performance as compared to that of the LQR (e.g., the fuzzy controller produces undesirable high-frequency "chattering" effects over a bandwidth that may not be realistic in implementation).

One way to increase the "resolution" of the fuzzy controller is to increase the number of membership functions. As we increase the number of membership functions on each input to 25, responses using the fuzzy controller become smoother and closer to that of the LQR. Additionally, the control surface of the fuzzy controller also becomes smoother and has a much smaller gain near the center. As a result, the control output of the fuzzy controller is significantly smoother. On the other hand, the direct fuzzy controller, with 25 membership functions on each input, comes with increased complexity in design and implementation (e.g., a four-input, one-output fuzzy system with 25 membership functions on each input has $25^4 = 390,625$ linguistic rules).

Application to Nominal System Given the experience of the simulation studies, the final step is to implement the fuzzy controller (with seven membership functions on each input) on the actual apparatus. For comparative purposes, we again consider application to the *nominal* system, that is, the pendulum alone with no added weight or disturbances. With the pendulum initialized at its hanging position ($\theta_1 = \pi$), the swing up control was tuned to give the best swing up response, as in the case of the LQR results of Section 57.10.3. The sampling time was set to 10 msec (smaller sampling times produced no significant difference in responses for any of the controllers tested on this apparatus). The only tuning required for the fuzzy control scheme (from simulation to implementation in experimentation) was in adjusting the value for g_3 upward to improve performance; recall that the gain g_3 is critical in that it essentially determines the other normalizing gains.

Figure 57.41 shows the results for the fuzzy controller on the laboratory apparatus; the top plot shows the base position (angle), the center plot shows the pendulum position (angle), and the bottom plot shows the controller output (motor voltage input). The response is comparable to that of the LQR controller (compare to Figure 57.38), in terms of the pendulum angle (ability to balance in the vertical position). However, some oscillation is noticed (particularly in the controller output, as predicted in simulation studies), but any difference in the ability to balance the pendulum is only slightly discernible in viewing the operation of the system.

Application to Perturbed System When the system experiences disturbances and changes in dynamics (by attaching additional weight to the pendulum endpoint, or by attaching a bottle half-filled with liquid), degraded responses are observed for these controllers. Such experiments are also informative for considerations of *robustness* analysis, although here we regard such perturbations on the nominal system as probing the limits of linear controllers (i.e., operating outside the linear region).

As a final evaluation of the performance of the fuzzy controller as developed above, we show results when a container half-filled with water was attached to the pendulum endpoint. This essentially gives a "sloshing liquid" effect, because the additional dynamics associated with the sloshing liquid are easily excited. In addition, the added weight shifted the pendulum center of mass away from the pivot point; as a result, the natural frequency of the pendulum decreased. Furthermore, the effect of friction becomes less dominant because the inertia of the pendulum increases. These effects obviously come to bear on the balancing controller performance, but also significantly affect the swing up controller as well. For the present article, we note that the swing up control scheme requires tuning once additional weight is added to the endpoint, and refer the interested reader to [7] for details of a *supervisory fuzzy controller* scheme where tuning of the swing up controller is carried out autonomously.

With the sloshing liquid added to the pendulum endpoint, the LQR controller (and, in fact, other linear control schemes we implemented on this system) produced an unstable response (was unable to balance the pendulum). Of course, the linear control schemes can be tuned to improve performance *for the perturbed system*, at the expense of degraded performance for the nomi-

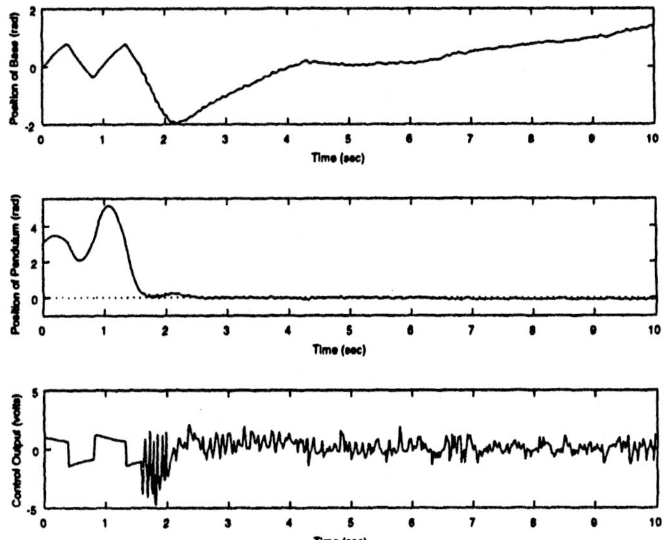

Figure 57.41 Experimental results: Direct fuzzy control on the *nominal* system.

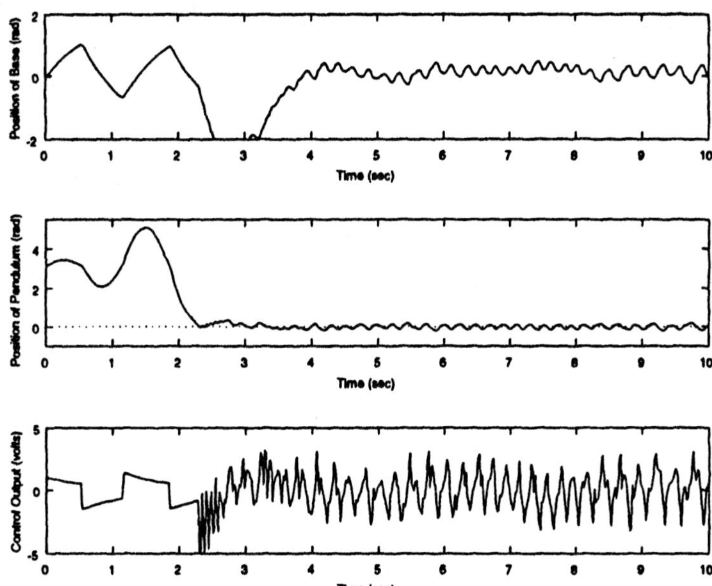

Figure 57.42 Experimental results: Direct fuzzy control on the pendulum with sloshing liquid at its endpoint.

nal system. Moreover, it is important to note that tuning of the LQR-type controller is difficult and ad hoc without additional modeling to account for the added dynamics. Such an attempt on this system produced a controller with stable but poor performance.

The fuzzy controller, on the other hand, because of its expanded region of operation (in the sense that it acts like the LQR for small inputs and induces a nonlinearity for larger signals), was able to maintain stability in the presence of the additional dynamics and disturbances caused by the sloshing liquid, *without tuning*. These results are shown in Figure 57.42, where some degradation of controller performance is apparent. Such experiments may also motivate the need for a controller that can adapt to changing dynamics during operation; this issue is discussed later when we address adaptation in fuzzy controllers.

57.10.5 Adaptive Fuzzy Control

Overview

While fuzzy control has, for some applications, emerged as a practical alternative to classical control schemes, there exist rather obvious drawbacks. We do not address all of these drawbacks here (such as stability, which is a current research direction); rather, we focus on an important emerging topical area within the realm of fuzzy control, that of adaptive fuzzy control, as it relates to some of these drawbacks.

The point that is probably most often raised in discussion of controller synthesis using fuzzy logic is that such procedures are usually performed in an ad hoc manner, where mechanized synthesis procedures, for the most part, are nonexistent (e.g., it is often not clear exactly how to justify the choices for many controller parameters, such as membership functions, defuzzification strategy, and inference strategy). On the other hand, some mechanized synthesis procedures do exist for particular applications, such as the one discussed above for the balancing control part of the inverted pendulum problem where a conventional LQR scheme was utilized in the fuzzy control design. Typically such procedures arise primarily out of necessity because of system complexity (such as when many inputs and multiple objectives must be achieved). Controller adaptation, in which a form of *automatic controller synthesis* is achieved, is one way of attacking this problem, when no other "direct" synthesis procedure is known.

Another, perhaps equally obvious, requirement in the design and operation of any controller (fuzzy or otherwise) involves questions of system robustness. For instance, we illustrated in our theme example that performance of the direct fuzzy controller constructed for the nominal plant may degrade if significant and unpredictable plant parameter variations, structural changes, or environmental disturbances occur. Clearly, controller adaptation is one way of overcoming these difficulties to achieve reliable controller performance in the presence of unmodeled parameter variations and disturbances.

Some would argue that the solution to such problems is always to incorporate more expertise into the rule base to enhance performance; however, there are several limitations to such a philosophy, including: (1) the difficulties in developing (and characterizing in a rule base) an accurate intuition about how to best compensate for the unpredictable and significant process variations that can occur for all possible process operating conditions; and (2) the complexities of constructing a fuzzy controller that potentially has a large number of membership functions and rules. Experience has shown that it is often possible to tune fuzzy controllers to perform very well if the disturbances are known. Hence, the problem does not result from a lack of basic expertise in the rule base, but from the fact that there is no facility for automatically redesigning (i.e., retuning) the fuzzy controller so that it can appropriately react to unforeseen situations as they occur.

There have been many techniques introduced for adaptive fuzzy control. For instance, one adaptive fuzzy control strategy that borrows certain ideas from conventional "model reference adaptive control" (MRAC) is called "fuzzy model reference learning control" (FMRLC) [10]. The FMRLC can automatically synthesize a fuzzy controller for the plant and later tune it if there are significant disturbances or process variations. The FMRLC has been successfully applied to an inverted pendulum, a ship-steering problem [10], antiskid brakes [11], reconfigurable control for aircraft [9], and in implementation for a flexible-link robot [12]. Modifications to the basic FMRLC approach have been studied in [13]. The work on the FMRLC and subsequent modifications to it tend to follow the main focus in fuzzy control where one seeks to employ heuristics in control. There are other techniques that take an approach that is more like conventional adaptive control in the sense that a mathematical model of the plant and a Lyapunov-type approach is used to construct the adaptation mechanism. Such work is described in [2]. There are many other "direct" and "indirect" adaptive fuzzy control approaches that have been used in a wide variety of applications (e.g., for scheduling manufacturing systems [14]). The reader should consult the references in the papers cited above for more details.

Another type of system adaptation, where a significant amount and variety of knowledge can be loaded into the rule base of a fuzzy system to achieve high-performance operation, is the *supervisory fuzzy controller*, a two-level hierarchical controller that uses a higher-level fuzzy system to supervise (coordinate or tune) a lower-level conventional or fuzzy controller. For instance, an expert may know how to control the system very well in one set of operating conditions, but if the system switches to another set of operating conditions, the controller may be required to behave differently to again achieve high-performance operation. A good example is the PID controller, which is often designed (tuned) for one set of plant operating conditions, but if the operating conditions change the controller will not be properly tuned. This is such an important problem that there is a significant amount of expertise on how to manually and automatically tune PID controllers. Such expertise may be utilized in the development of a supervisory fuzzy controller which can observe the performance of a low-level control system and automatically tune the parameters of the PID controller. Many other examples exist of applications where the control engineer may have a significant amount of knowledge about how to tune a controller. One such example is in aircraft control when controller gains are scheduled based on the operating conditions. Fuzzy supervisory controllers have been used as schedulers in such applications. In other applications we may know that conventional or fuzzy controllers need to be switched on based on the operating conditions (see the work in [15] for work on fuzzy supervision of conventional controllers for a flexible-link robot) or a supervisory fuzzy controller may be used to tune an adaptive controller (see the work in [9] where a fuzzy supervisor is used to tune an adaptive fuzzy controller that is used as a reconfigurable controller for an aircraft). It is this concept of *monitoring* and *supervising* lower-level controllers (possibly fuzzy, possibly conventional) that defines the supervisory control scheme. Indeed, in this sense supervisory and adaptive systems can be described as special cases of one another.

Autotuning for Pendulum Balancing Control

Many techniques exist for automatically tuning a fuzzy controller in order to meet the objectives mentioned above. One simple technique we present next, studied in [7, 13, 14], expands on the idea of increasing the "resolution" of the fuzzy controller in terms of the characteristics of the input membership functions. Recall from Section 57.10.4 for our theme problem that when we increased the number of membership functions on each input to 25, improved performance (and smoother control action) resulted. Likewise, we suspect that increasing the resolution would result in improved performance for the perturbed pendulum case.

To increase the resolution of the direct fuzzy controller with a limited number of membership functions (as before, we will impose a limit of seven), we propose an "autotuned fuzzy control."

To gain insight on how the autotuned fuzzy control works, consider the idea of a "fine controller," with smooth interpolation, achieved using a fuzzy system where the input and output universes of discourse are narrow (i.e., the input normalizing gains are large, and the output gain is small). In this case there are many membership functions on a small portion of the universe of discourse (i.e., "high resolution"). Intuitively, we reason that as the input gains are increased and the output gain is decreased, the fuzzy controller will have better resolution. However, we also conjecture that to get the most effective control action the input universes of discourse must also be large enough to avoid saturation. This obviously raises a question of trying to satisfy two opposing objectives. The answer is to adjust the gains based on the current operating states of the system. For example, if the states move closer to the center, then the input universe of discourse should be *compressed* to obtain better resolution, yet still cover all the active states. If the states move away from the center, then the input universe of discourse must expand to cover these states, at the expense of lowering the resolution.

The input-output gains of the fuzzy controller can be tuned periodically (e.g., every 50 samples) using the autotuning mechanism shown in Figure 57.43. Ideally, the autotuning algorithm should not alter the nominal control algorithm near the center; we therefore do not adjust each input gain independently. We can, however, tune the most significant input gain, and then adjust the rest of the gains based on this gain. For the inverted pendulum system, the most significant controller input is the position error of the pendulum, $e_3 = \theta_1$.

The input-output gains are updated every n_s samples in the following manner:

- Find the maximum e_3 over the most recent n_s samples and denote it by e_3^{max}
- Set the input gain $g_3 = \frac{1}{|e_3^{max}|}$
- Recalculate the remaining gains using the technique discussed in Section 57.10.4 so as to preserve the nominal control action near the center

We note that the larger n_s is, the slower the updating rate is, and that too fast an updating rate may cause instability. Of course, if a large enough buffer were available to store the most recent n_s samples of the input, the gains could be updated at every sample (utilizing an average); here we minimized the usage of memory and opted for the procedure mentioned above (finding the maximum value of e_3).

Simulation tests (with a 50-sample observation window and normalizing gain $g_3 = 2$) reveal that when the fuzzy controller is activated (after swing up), the input gains gradually increase while the output gain decreases, as the pendulum moves closer to its inverted position. As a result, the input and output universes of discourses contract, and the resolution of the fuzzy system increases. As g_3 reaches its maximum value[14], the control action near $\theta_1 = 0$ is smoother than that of direct fuzzy control with 25 membership functions (as investigated previously via simulation), and very good balancing performance is achieved.

When turning to actual implementation on the laboratory apparatus, some adjustments were done in order to optimize the performance of the autotuning controller. As with the direct fuzzy controller, the value of g_3 was adjusted upward, and the tuning (window) length was increased to 75 samples. The first test was to apply the scheme to the nominal system; the autotuning mechanism improved the response of the direct fuzzy controller (Figure 57.41) by varying the controller resolution online. That is, as the resolution of the fuzzy controller increased over time, the high-frequency effects diminished.

The true test of the adaptive (autotuning) mechanism is to evaluate its ability to adapt its controller parameters as the process dynamics change. Once again we investigate the performance when the "sloshing liquid" dynamics (and additional weight) are appended to the endpoint of the pendulum. As expected from simulation exercises, the tuning mechanism, which "stretches" and "compresses" the universes of discourse on the input and output, not only varied the resolution of the controller but also effectively contained and suppressed the disturbances caused by the sloshing liquid, as clearly shown in Figure 57.44.

57.10.6 Concluding Remarks

We have introduced the general fuzzy controller, shown how to design a fuzzy controller for our rotational inverted pendulum theme problem, and compared its performance to a nominal LQR controller. We overviewed supervisory and adaptive fuzzy control techniques and presented a particular adaptive scheme that worked very effectively for the balancing control of our rotational inverted pendulum theme problem. Throughout the chapter we have emphasized the importance of comparative analysis of fuzzy controllers with conventional controllers, with the hope that more attention will be given to detailed engineering cost-benefit analyses rather than the hype that has surrounded fuzzy control in the past.

There are close relationships between fuzzy control and several other intelligent control methods. For instance, expert systems are generalized fuzzy systems since they have a knowledge base (a generalized version of a rule base where information in the form of general rules or other representations may be used), an inference mechanism that utilizes more general inference strategies, and are constructed using the same general approach as fuzzy systems. There are close relationships between some types of neural networks (particularly radial basis function neural networks) and fuzzy systems and, while we did not have the space to cover it here, fuzzy systems can be trained with numerical data in the same way that neural networks can (see [2] for more details). Genetic algorithms provide for a stochastic optimization

[14]In practice, it is important to constrain the maximum value for g_3 (for our system, to a value of 10) because disturbances and inaccuracies in measurements could have adverse effects.

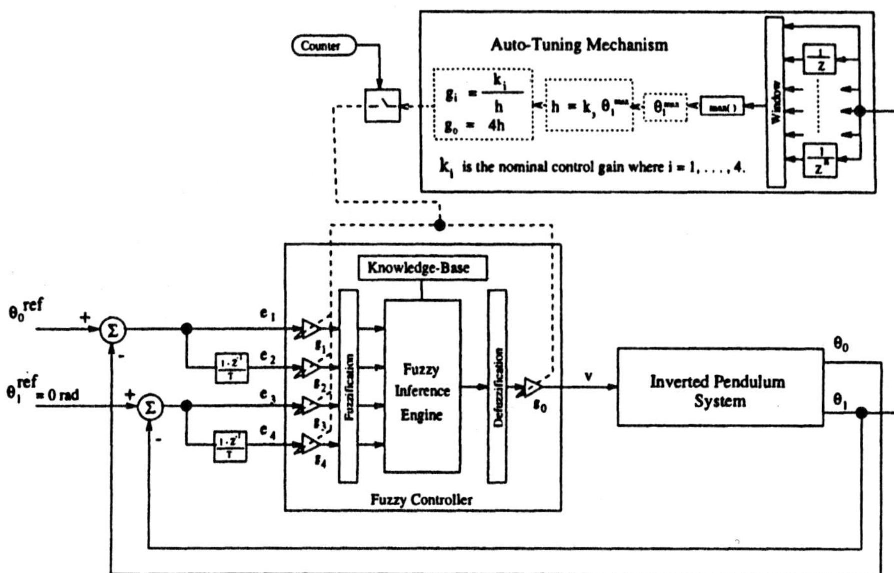

Figure 57.43 Autotuned fuzzy control.

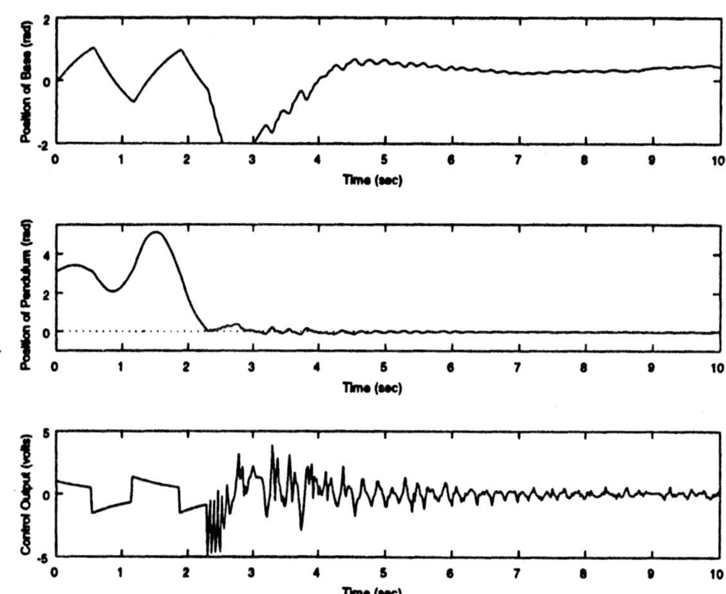

Figure 57.44 Experimental results: Autotuned fuzzy control on the pendulum with sloshing liquid at its endpoint.

technique and can be useful for computer-aided design of fuzzy controllers, training fuzzy systems for system identification, or tuning fuzzy controllers in an adaptive control setting.

While there exist such relationships between fuzzy control and other techniques in intelligent control, the exact relationships between all the techniques have not been established. Research along these lines is progressing but will take many years to complete. Another research area that is gaining increasing attention is the nonlinear analysis of fuzzy control systems and the use of comparative analysis of fuzzy control systems and conventional control systems to determine the advantages and disadvantages of fuzzy control. Such work, coupled with the application of fuzzy control to increasingly challenging problems, will help establish the technique as a viable control engineering approach.

57.10.7 Acknowledgments

A large portion of the ideas, concepts, and results reported on in this chapter have evolved over several years from the work of many students under the direction of the authors. We therefore gratefully acknowledge the contributions of: A. Angsana, E. Garcia-Benitez, S. Gyftakis, D. Jenkins, W. Kwong, E. Laukonen, J. Layne, W. Lennon, A. Magan, S. Morris, V. Moudgal, J. Spooner, and M. Widjaja. We would like, in particular, to acknowledge the work of J. Layne and E. Laukonen for their assistance in writing earlier versions of Section 57.10.2 of this chapter and M. Widjaja, who produced the experimental results for the rotational inverted pendulum.

57.10.8 Defining Terms

Rule base: A part of the fuzzy system that contains a set of If . . . Then rules that quantify a human expert's knowledge about how to best control a plant. It is a special form of a knowledge base that only contains If . . . Then rules that are quantified with fuzzy logic.

Inference Mechanism: A part of the fuzzy system that reasons over the information in the rule base and decides what actions to take. For the fuzzy system the inference mechanism is implemented with fuzzy logic.

Fuzzification: Converts standard numerical fuzzy system inputs into a fuzzy set that the inference mechanism can operate on.

Defuzzification: Converts the conclusions reached by the inference mechanism (i.e., fuzzy sets) into numeric inputs suitable for the plant.

Supervisory Fuzzy Controller: A two-level hierarchical controller which uses a higher-level fuzzy system to supervise (coordinate or tune) a lower-level conventional or fuzzy controller.

Adaptive Fuzzy Controller: An adaptive controller that uses fuzzy systems either in the adaptation mechanism or as the controller that is tuned.

References

[1] Jenkins, D. and Passino, K.M., Introduction to nonlinear analysis of fuzzy control systems, *Submitted for journal publication*, 1995.

[2] Wang, L.-X., *Adaptive Fuzzy Systems and Control: Design and Stability Analysis*, Prentice Hall, Englewood Cliffs, NJ, 1994.

[3] Driankov, D., Hellendoorn, H., and Reinfrank, M., *An Introduction to Fuzzy Control*, Springer-Verlag, New York, 1993.

[4] Pedrycz, W., *Fuzzy Control and Fuzzy Systems*, 2nd ed., Research Studies Press, New York, 1993.

[5] Klir, G.J. and Folger, T.A., *Fuzzy Sets, Uncertainty and Information*, Prentice Hall, Englewood Cliffs, NJ, 1988.

[6] Zadeh, L.A., Outline of a new approach to the analysis of complex systems and decision processes, *IEEE Trans. Syst. Man Cybern.*, 3(1), 28–44, 1973.

[7] Widjaja, M. and Yurkovich, S., Intelligent control for swing up and balancing of an inverted pendulum system, *Proc. of IEEE Conf. Control Applications*, Albany, NY, Sept. 1995.

[8] Spong, M.W., Swing up control of the acrobot, *Proc. IEEE Int. Conf. Robotics and Automation*, San Diego, May 1994, pp. 2356–2361.

[9] Kwong, W.A., Passino, K.M., Laukonen, E.G., and Yurkovich, S., Expert supervision of fuzzy learning systems for fault tolerant aircraft control, *Proc. IEEE*, 83(3), 1995.

[10] Layne, J.R. and Passino, K.M., Fuzzy model reference learning control for cargo ship steering, *IEEE Control Syst. Mag.*, 13(6), 23–34, 1993.

[11] Layne, J.R., Passino, K.M., and Yurkovich, S., Fuzzy learning control for antiskid braking systems, *IEEE Trans. Control Syst. Technol.*, 1(2), 122–129, 1993.

[12] Moudgal, V.G., Kwong, W.A., Passino, K.M., and Yurkovich, S., Fuzzy learning control for a flexible-link robot, *IEEE Trans. Fuzzy Syst.*, 3(2), 1995.

[13] Kwong, W.A. and Passino, K.M., Dynamically focused fuzzy learning control, *IEEE Trans. Syst., Man Cybern.*, in press.

[14] Angsana, A. and Passino, K.M., Distributed fuzzy control of flexible manufacturing systems, *IEEE Trans. Control Syst. Technol.*, 2(4), 423–435, 1994.

[15] Garcia-Benitez, E., Yurkovich, S., and Passino, K., Rule-based supervisory control of a two-link flexible manipulator, *J. Intelligent Robotic Syst.*, 7(2), 195–213, 1993.

Further Reading

Other introductions to fuzzy control are given in [2, 3, 4]. To study other applications of fuzzy control the reader should consult the references in the papers and books cited above. There are several conferences which have sessions or papers that study various aspects of fuzzy control. For instance, see the Proceedings of FuzzIEEE (in the World Congress on Computational Intelligence), IEEE Int. Symp. on Intelligent Control, Proceedings of the IEEE Conf. on Decision and Control, the American Control Conf., and many others. The best journals to consult are the *IEEE Transactions on Fuzzy Systems*, *Fuzzy Sets and Systems*, and the *IEEE Transactions on Systems, Man, and Cybernetics*. Also, there are many short courses that are given on the topic of fuzzy control.

57.11 Neural Control

Jay A. Farrell, College of Engineering, University of California, Riverside

57.11.1 Introduction

In control, filtering, and prediction applications the available *a priori* model information is sometimes not sufficiently accurate for a prior fixed design to satisfy all of the design specifications. In these circumstances, the designer may

1. reduce the specified level of performance,
2. expend additional efforts to reduce the level of model uncertainty, or
3. design a system to adjust itself on-line to increase its performance level as the accumulated input and

output data is used to reduce the amount of model uncertainty.

Controllers (filters and predictors can be implemented by similar techniques) which implement the last approach are the topic of this chapter. Such techniques are necessary when additional effort in modeling and validation is unacceptable because of cost and feasibility, and the desired level of performance must be maintained.

Linear adaptive control mechanisms [11] have been developed for systems that by suitable limitation of their operating domain, can be adequately modeled as linear. Such control mechanisms are well developed as described in Chapter 52. The on-line system identification and control techniques developed for nonlinear systems can be divided into two categories, parametric and nonparametric. Parametric techniques assume that the functional form of the unknown function is known, based on physical modeling principles, but that the model parameters are unknown. Nonparametric techniques are necessary when the functional form of the unknown function is unknown. In this case a general family of function approximators is selected based on the known properties of the approximation family and the characteristics of the application. In spite of the name, the objective of the nonparametric approach is still selecting a set of parameters to effect an optimal approximation to the unknown function.

Learning and neural control (and system identification) [3], [4], [5] are nonparametric techniques developed to enhance the performance of those poorly modeled nonlinear systems for which a suitable model structure is not known. This performance improvement will be achieved by exploiting experiences obtained through on-line interactions with the actual plant. Approximation-based adaptive control is an alternative name used in the literature for this class of techniques. Neural control is the name applied in the literature when a *neural network* is selected as the function approximator. To be both concise and general, learning control will be the name used throughout this chapter.

This chapter will focus on the motivation for, and implementation of, learning control systems. Topics of related interest include

- motivational applications with nonlinear dynamics and unknown model structure,
- the relationship of learning and neural control to alternative approaches, and
- the specification of the various components of learning/neural control systems.

57.11.2 Motivating Example

Although the techniques presented have wide applicability, we will cite examples from underwater vehicle control applications when useful for exemplary purposes.

At this point it is of interest to note that, based on an understanding of the physical behavior of fluids, Newton's laws, Archimede's principle, etc., a dynamic model of an underwater vehicle can be developed. However, due to both theoretical and practical limitations, the application of this model to particular vehicles may result in dynamic models too uncertain for satisfactory fixed control system design, and too complex for on-line adjustment.

In such circumstances, it is interesting to consider whether a (possibly) nonphysically based model, suitable for high performance control system design, can be constructed on-line based on performance feedback information.

57.11.3 Problem Statement

Assume that the dynamics of the system can be described as

$$\left.\begin{aligned} x(k+1) &= f(x(k), v(k)) + w(k), \\ y(k) &= x(k) + n(k) \end{aligned}\right\} \quad (57.439)$$

where $x(k)$ is the state of the system, $n(k)$ indicates noise on the measurement, and $w(k)$ is process noise. The control system design problem is to specify a control law $v(k) = K\{x(k), r(k); h[x(k), r(k)]\}$ so that the closed-loop system achieves the desired level of performance. The function $h[x(k), r(k), k]$ is any auxiliary mapping defined over a compact domain D useful in the control system design. Examples include an estimate of the system dynamics $\hat{f}[x(k), v(k)]$, and mappings from the current operating condition to either linear model or control system parameters. When the function $h[x(k), r(k)]$ is unknown or inaccurately known, it is of interest to consider whether on-line data can be used to synthesize an approximate function $\hat{h}(x(k), r(k), \theta)$ so that

$$J(\theta) = \int_D \|h(s) - \hat{h}(s; \theta)\|^2 ds \quad (57.440)$$

achieves a minimum.

Solutions to problems such as that stated above and analysis of existing solutions are still active areas of research. In the discussion following, we will see that the choice of the function approximation structure and the experimental conditions may be critical to the level of performance achieved.

57.11.4 Active and Passive Learning

Due to the fact that $h(s)$ is unknown, Equation 57.440 cannot be optimized directly. Instead, a typical approach is to optimize a cost function based on samples of the function (i.e., $y_i = h(s_i)$).[15] This results in a cost function defined as a summation of sample errors

$$J(\theta) = \frac{1}{N}\sum_{i=1}^{N} \|(y_i - \hat{h}(s_i : \theta)\|^2 \quad (57.441)$$

[15] The more general case in which the output of the approximated function cannot be directly measured is discussed in Section 57.11.7. Comments similar to those in this section still hold.

However, the solution to Equation 57.441 is usually distinct from the solution to Equation 57.440. This is true because for large N, Equation 57.441 converges to

$$J(\theta) = \int_D \|h(s) - \hat{h}(s:\theta)\|^2 p(s)ds \qquad (57.442)$$

where $p(s)$ is the experimental sample distribution. The solutions to Equations 57.440 and 57.442 will coincide only when the experimental sample distribution is uniform. Unfortunately, a uniform sample distribution is not typical for control applications. This difficulty distinguishes between active and passive learning scenarios. *Active learning* describes those situations in which the designer has the freedom to control the training sample distribution. *Passive learning* describes those situations in which the training sample density is defined by some external mechanism. On-line applications usually involve passive learning because the plant is performing some useful function.

57.11.5 Models Types: Adaptation and Learning

In the selection of a suitable model structure, it is necessary to note the domain over which the model is expected to apply. In particular, the following two definitions are of interest.

DEFINITION 57.4 *Local Approximation Structure:* A parametric model $\hat{f}(x, \hat{\theta})$ is a local approximation to $f(x)$ at x_0 if, for any ε, $\hat{\theta}$ and δ exist so that $\| f(x) - \hat{f}(x, \hat{\theta}) \| \leq \varepsilon$ for all $x \in B(x_0, \delta)$ where $B(x_0, \delta) = \{x | \| x - x_0 \| < \delta\}$.

DEFINITION 57.5 *Global Approximation Structure:* A parametric model $\hat{f}(x, \hat{\theta})$ is a global approximation to $f(x)$ over domain D if for any ε there exists $\hat{\theta}$ so that $\| f(x) - \hat{f}(x, \hat{\theta}) \| \leq \varepsilon$ for all $x \in D$.

The following items are of interest relative to the definition of local and global models:

- Physical models derived from first principles (see Chapter 7) are expected to be global models (i.e., valid over the domain of operation D).
- Whether a given approximation structure is local or global depends on the system that is being modeled and the domain D.
- The set of global models is a strict subset of the set of local models. If a set of parameters $\hat{\theta}$ exists satisfying Definition 57.5 for a particular ε, then this $\hat{\theta}$ also satisfies Definition 57.4 for the same ε at each $x_0 \in D$.
- As the desired level of accuracy increases (i.e., ε decreases), the dimension of the parameter vector or the complexity of the model structure will usually have to increase.

To maintain accuracy over domain D, a local approximation structure can either adjust its parameter vector through time, as the operating point x_0 changes or store its parameter vector as a function of the operating point. The former approach is typical of linear adaptive control methodologies in nonlinear control applications. The latter approach has been set forth [5] as an instance of learning control. The latter approach effectively constructs a global approximation structure by connecting several local approximating structures.

The conceptual differences between linear adaptive and learning control are of interest. In a nonlinear application, linear adaptive methodologies can be developed to maintain a locally accurate model or control law at the current operating point x_0. With the assumption that linearization will change over time, the adaptive mechanism will adjust the parameter estimates to maintain the accuracy of the local fit. Such an approach can provide satisfactory performance if the local linearization changes slowly, but will have to estimate the same model parameters repetitively if the system moves repetitively over a set of standard operating points. A more ambitious proposal is to develop a global model of the dynamics. One mechanism for this approach is to store the local linear models or controllers as a function of the operating point.

$$\hat{f}(x, x_0, v, v_0) = A(x_0, v_0)(x - x_0) + B(x_0, v_0)(v - v_0), \qquad (57.443)$$

where the unknown portions of the matrix functions $A(x_0, v_0)$ and $B(x_0, v_0)$ would be approximated over the domain D. The name *Learning Control* stems from the idea that past performance information is used to estimate and store information (relevant to an operating point) for use when the system operates in the vicinity of the operating point in the future. Learning Control will require more extensive use of computer memory and computational effort. The required amount of memory and computation will be determined predominantly by the selection of function approximation and parameter estimation (training) techniques.

EXAMPLE 57.16: Global underwater vehicle model

Underwater vehicle dynamics can be represented as

$$m\frac{dv_c}{dt} + m(\omega(t) \times v_c(t)) = F[v_c(t), \omega(t)] \qquad (57.444)$$

and

$$m\frac{dH}{dt} + \omega(t) \times H = M[v_c(t), \omega(t)], \qquad (57.445)$$

where v_c is the vehicle velocity, ω is the angular velocity, H is the angular momentum, m is the vehicle mass matrix, and the right hand side represents the total hydrodynamic, control, propulsion, and gravitational forces and moments acting on the vehicle.

Although this equation is intended to describe the vehicle dynamics over the operating envelope, limitations of physical measurements and approximations introduced in the specification of the right-hand side for a particular vehicle usually results in a large number (> 100) of model parameters being scheduled as a function of the operating condition. The accuracy of the global fit is improved by storing the model parameters as a function of the operating condition. Such parameter scheduling techniques are typical in aircraft and underwater vehicle applications.

EXAMPLE 57.17: Local underwater vehicle model

In the vicinity of an operating point (x_0, v_0) the vehicle dynamic model, Equations 57.444–57.445, can be approximated by a linear system,

$$\dot{x} = A_{x_0,v_0}(x - x_0) + B_{x_0,v_0}(v - v_0), \qquad (57.446)$$

where $A_{x_0,v_0} = \frac{\partial f(x,v)}{\partial x}|_{x=x_0,v=v_0}$ and $B_{x_0,v_0} = \frac{\partial f(x,v)}{\partial v}|_{x=x_0,v=v_0}$. Such models, that allow linear control system design and analysis, are only expected to hold locally. Modeling and linearization techniques are discussed in detail in various Control System textbooks (e.g., [10]) and in Chapter 5.

EXAMPLE 57.18: Connected local models of an underwater vehicle

A common practice for control system design, referred to as scheduling (see Chapter 20), is to generate a global model or controller for a nonlinear system by interpolating between various local linear models.

For example, if the dominant nonlinearities over a desired operating envelope D are due to changes in the forward velocity u, and linear models are available at integer values of the forward velocity between 5 and 15 feet per second, then

$$\hat{f}(x, v) = \sum_{i=5}^{15} (A_i(x - x_i) + B_i(v - v_i))\Gamma_i(u)$$

where $x_i = (0, \cdots, 0, u_i, 0, \cdots, 0,)$ is the ith operating point, by assumption $u_i = i$, v_i is the steady-state value of the actuation vector (i.e., propulsion force) necessary to maintain the operating condition, and

$$\Gamma_i(u) = \begin{cases} 0 & i+1 \le u < 15 \\ 1-(u-i) & i \le u < i+1 \\ 1+(u-i) & i-1 \le u < i \\ 0 & 5 \le u < i-1 \end{cases}$$

is a linearly interpolated fit to the nonlinear dynamics. Such scheduled models and controllers are commonly assumed to convert a nonlinear problem to a set of linear design problems. Such approximation structures, where each parameter has a local effect on the approximation, can have appealing properties for passive learning control applications.

57.11.6 Function Approximators

Based on the discussion of Sections 1–57.11.5, it is sometimes of interest to perform on-line function approximation to improve control performance. This is accomplished by using on-line experience (i.e., input-output data) to estimate the optimal parameters of a function approximation structure.

Examples of several classes of function approximators are given in the following. The topic of the second subsection is a discussion of several useful approximator properties. An understanding of these properties is necessary to select an appropriate class of approximation structures for a particular application.

Examples of Function Approximators

Numerous classes of approximators exist. Below are a few examples.

Sigmoidal Neural Networks A strict mathematical definition of what is or is not a neural network does not exist; however, the class of sigmoidal neural networks has become widely accepted as a function approximator.

Single hidden-layer neural networks have the form

$$\hat{h}_i(x) = \sum_{j=1}^{N} W1_{ij}\phi(W2_j^T x - B2_j) \qquad (57.447)$$

where $\hat{h} = (\hat{h}_1, \cdots, \hat{h}_n)$ is a vector of functions of the independent variable $x^T = (x_1, \cdots, x_m)$, $W1$, $B2$ and $W2$ are parameter matrices of the appropriate dimension, and ϕ, called the nodal function, is a scalar function of a single variable. The parameters in $B2$, $W1$ and $W2$ are considered unknown and must be estimated to produce an optimal approximating function. Multiple layer networks are constructed by cascading single layer networks together, with the output from one network serving as the input to the next.

Orthogonal Polynomial Series A set of polynomial functions $\phi_i(s)$, $i = 1, 2, \ldots$ is orthogonal with respect to a nonnegative weight function $w(t)$ over the interval $[a, b]$, if

$$\int_a^b w(s)\phi_i(s)\phi_j(s)ds = \begin{cases} r_i, & i = j, \\ 0, & i \ne j, \end{cases}$$

for some nonzero constants r_i. When a function $h(s)$ satisfies certain integrability conditions, it can be expanded as

$$h(s) = \sum_{n=0}^{\infty} h_n\phi_n(s)$$

where

$$h_n = \frac{1}{r_n}\int_a^b w(s)h(s)\phi_j(s)ds.$$

The mth order finite approximation of $h(s)$ with respect to the polynomial series $\phi_i(s)$ is given by

$$\hat{h}(s) = \sum_{n=0}^{m-1} h_n\phi_n(s). \qquad (57.448)$$

The integral of the error between $h(s)$ and $\hat{h}(s)$ over $[a, b]$ converges as m approaches ∞. When $h(s)$ is unknown, it may be reasonable to approximate the h_n in Equation 57.448 with on-line data.

Localized Basis Influence Functions Due to the usefulness of the interconnection of local models to generate global models, the class of *Basis Influence Functions* [1] is presented. The definition is followed by examples of several approximation architectures that satisfy the definition to demonstrate the concept and to illustrate that several approximators often discussed independently can be studied as a class within the setting of the definition. This class of models is introduced to reduce redundancy in discussing the function approximation properties in the following subsections.

DEFINITION 57.6 *Localized-basis influence functions:* A function approximator is of the BI Class if, and only if, it can be written as

$$\hat{f}(x, \hat{\theta}) = \sum_i f_i(x, \hat{\theta}; x_i)\Gamma_i(x; x_i) \tag{57.449}$$

where each $f_i(x, \hat{\theta}; x_i)$ is a local approximation to $f(x)$ for all $x \in B(x_i, \delta)$, and $\Gamma(x; x_i)$ has local support $S_i = \{x : \Gamma(x; x_i) \neq 0\}$ which is a subset of $B(x_i, \delta)$ so that $D \subseteq \bigcup_i S_i$.

BOXES One of the simplest approximation structures that satisfies Definition 57.6 is the piecewise constant function,

$$h(s; \theta) = \sum_i \theta_i \Gamma_i,$$

where

$$\Gamma_i = \begin{cases} 1, & \text{if } s \in D_i, \\ 0, & otherwise, \end{cases} \quad \text{where } \cup_i D_i = D \text{ and } D_i \cap D_j = \emptyset.$$

Often, the D_i are selected in a rectangular grid covering D. In this case, very efficient code can be written to calculate the approximation and update its parameters. Generalizations of this concept include using more complex basis elements (e.g., parameterized linear functions) in place of the constants θ_i, and interpolating between several neighboring D_i to produce a more continuous approximation.

Splinelike Functions Interpolating between regions in the BOXES approach can result in spline functions, such as (for a one dimensional domain)

$$h(s; \theta) = \lambda_i \theta_i + (1 - \lambda_i)\theta_{i+1},$$

where

$$\lambda_i = \frac{s_{i+1} - s}{s_{i+1} - s_i} \tag{57.450}$$

for $s_{i+1} > s \geq s_i$ and θ_i is the value of $h(s; \theta)$ at x_i. When the knots (i.e., x_i) are equally spaced, Equation 57.450 can be rewritten to satisfy Definition 57.6 with constant basis functions and influence functions of the form

$$\Gamma_i(s) = \begin{cases} 1 - \frac{|s - s_i|}{s_{i+1} - s_i}, & \text{if } s_{i-1} \leq s \leq s_{i+1}, \\ 0, & \text{otherwise.} \end{cases} \tag{57.451}$$

Note that such schemes can be extended to both multidimensional inputs and outputs, at the expense of increased memory and computational requirements. For example, a multidimensional output could be the state feedback gain as a function of operating condition.

It is interesting to note the similarity between spline approximations, as described above, and some fuzzy approximators. In the language of fuzzy systems, a basis function is called a *rule* and an influence function is called a *membership function.*

Radial Basis Functions An approximation structure defined as

$$\hat{h}(s, \hat{\theta}) = \sum_i \theta_i \Gamma(s; s_i),$$

where $\Gamma(s; s_i)$ is a function such as $exp[-(s-s_i)^2]$, which decays with the distance of s from s_i, is called a Radial Basis Function Approximator (RBF). In the case where $\Gamma(s; s_i)$ has only local support, the RBF also satisfies Definition 57.6, with the basis elements being constant functions. Of course, the basis functions can again be generalized to more complicated, but more inclusive, functions.

Although the exponential and inverse square nodal functions do not themselves have local support, in most applications the designer only uses nodes within a fixed distance of the current evaluation point to calculate the value of the approximating function; hence, the effective nodal function does have local support. Alternatively, if the definition of S_i were relaxed to $S_i = \{x : \Gamma(x; x_i) \leq \epsilon\}$ for some small ϵ then standard RBFs are also included in this class.

Useful Approximator Properties

The properties of families of approximators are discussed in this section to aid the designer in making an educated choice among the many available approximators.

Universal Approximation Several universal approximation results have been derived in the literature. A typical result is presented below for discussion. For less restrictive results for neural networks, see [7].

THEOREM 57.13 *[6] Let $\phi(v) : R^1 \rightarrow R^1$ be a nonconstant, bounded, and monotone increasing continuous function. Let D be a compact subset of R^m and $f(x_1, \cdots, x_m)$ be a continuous function from D to R^n. Then for any arbitrary $\varepsilon > 0$, an integer N and real constants B_2, W_1 and W_2 exist so that the approximator defined in Equation 57.447 satisfies $max_{x \in D} \|f(x) - \hat{f}(x)\|_\infty < \varepsilon$.*

This theorem indicates that if ϕ is a nodal function satisfying certain technical assumptions and ε is a specified approximation accuracy, then any given continuous function can be approximated to the desired degree of accuracy if the number of nodes N is sufficiently large. Such universal approximation results are powerful but must be interpreted with caution.

First, note that universal approximation considers increasing the number of nodes, not the number of network layers. If the nodal functions are defined as pth order polynomials, increasing the number of nodes per layer does not change the structure of the approximation from that of pth order polynomials. It is well-known that the set of pth order polynomials cannot approximate arbitrary continuous functions to a given accuracy ε. Therefore, a network of pth order polynomials cannot be a universal approximator. Note that polynomials do not meet the conditions required for the above theorem. However, it is well-known that, for any given continuous function, an integer P exists so that a polynomial of order $p > P$ can be found that approximates the function to accuracy ε over the domain D. This increase in polynomial order would correspond to an increase in the number of layers of the network. Therefore, the fact that a family of parametric approximators does not have the universal approximation property does not by itself indicate that the family of approximators should be rejected.

It should be kept in mind that universal approximation is a special case of a family of approximators being *dense* of the set of continuous functions.

Second, universal approximation requires that the number of nodes per layer be expandable. In most applications, the number of nodes per layer N_i is fixed *a priori.* Once each N_i is fixed, the network can no longer approximate an arbitrary continuous function to a specified accuracy ε; hence, approximation structures that allow criteria for the network structure specification are beneficial.

Generalization For the applications discussed, the parameters of the approximation structure will be estimated from a finite set of training data,

$$T = \{(s_1, y_1), \cdots, (s_N, y_N)\}, \tag{57.452}$$

where y_i and s_i are defined in Equations 57.439–57.440; however, the approximating function may be evaluated at any point $s \in D$. Therefore, the resulting approximation must be accurate throughout D, not only at the points specified by the training set T. The ability of parametric approximators to provide answers—good or bad—at points outside the training set is referred to in the neural network literature as *generalization.*

The fact that an approximation structure *is capable of generalizing from the training data* is not necessarily a beneficial feature. Note that the approximator will output estimates $\hat{h}(s)$ at any desired point s. The user must take appropriate steps to verify or insure the accuracy of the resulting approximation at that point. Families of approximators which allow this assessment to occur on-line are desirable.

Generalization of training data may also be described as interpolation or extrapolation of the training data. Interpolation refers to the process of filling in the holes between nearby training samples. Interpolation is a desirable form of generalization. Most approximators interpolate well. Some approximators will allow on-line analysis of the interpolation accuracy. Extrapolation refers to the process of providing answers in regions of the learning domain D that the training set T did not adequately represent. Extrapolation from the training data is risky when the functional form is unknown by assumption. Clearly, it is desirable to know whether the approximator is interpolating between the training data or extrapolating from the training data.

Linearity in Parameters Function approximators that are linear in the parameters can be represented as

$$\hat{f}(x, \upsilon, \hat{\theta}) = \phi(x, \upsilon)^T \theta \tag{57.453}$$

where θ is the unknown parameter vector, and $\phi(x, \upsilon)$ the regressor vector is a known, possibly nonlinear, vector function of x and υ. The powerful parameter estimation and performance analysis techniques that exist for approximators linear in their parameters (see [8]), make this a beneficial property.

The second major advantage of approximators that are linear in their parameters is that for square error types of cost functions, such as Equation 57.455, the cost function is a quadratic function of the parameters. Therefore, there is a unique global minimum.[16] Similar conclusions hold for more general objective functions as described in Equation 57.440. As discussed in Section 57.11.4, note that when a sample error cost function such as Equation 57.455 is used in place of an integral cost function, such as Equation 57.440, the optimal parameter estimate that results will depend on the distribution of the training samples. The fact that different parameter estimates result from different sets of training samples is *not* the result of multiple local minima; it is, instead, the result of different weightings by the sample density in the cost function.

Coverage This property can be stated formally as follows: For each f_i, and for any $x \in D$, at least one θ_j exists so that the function $|\frac{\partial f_i(x,\theta)}{\partial \theta_j}|$ is nonzero in the vicinity of x. If this property does not apply, then some point $x \in D$ exists for which the approximation cannot be changed. This is obviously an undesirable situation.

Localization This property is stated formally as follows: For all f_i and θ_j, if the function $|\frac{\partial f_i(x,\theta)}{\partial \theta_j}|$ is nonzero in the vicinity of x, then it must be zero outside some ball $B(x, \delta')$.

Under this condition the effects of changing any single parameter are limited to the local region. Thus, experience and consequent learning in one part of the input domain does not affect the knowledge accrued in other parts of the mapping domain.

For example, in the scheduling example of Section 57.11.5, the parameters of the ith linear model only affect the function output for $i - 1 < u < i + 1$; hence they should only be adjusted when u is in this range. If the gains were instead fitted with a Taylor Series, for example, then adjusting any parameter of the approximation would affect the accuracy of the approximating function over the entire domain. Localization is a desirable property because it prevents extended training in the vicinity of any given operating point from adversely affecting the approximation accuracy in distant regions of D. Therefore, localization is beneficial in passive learning applications.

The fact that a limited subset of the model parameters is relevant for a given point in the learning domain also makes it possible

[16] Possibly an equivalent connected set of parameters attaining the global minima if the approximator is overparametrized.

to reduce significantly the required amount of computation per sample interval. The trade-off is that an approximator with the localization property may require a higher number of parameters (more memory) than an approximator without the property. For example, if the domain D has dimension d, and m parameters are necessary for a given local approximator per input dimensions, then on the order of m^d parameters will generally be required to approximate the d dimensional function. This exponential increase in the memory requirements with input dimension is referred to in the literature as *the curse of dimensionality*.[17]

Discussion Function approximation structures that satisfy Definition 57.6 are of interest in learning control applications due to the fact that, when properly defined, they are dense on the set of continuous functions, and satisfy the *Coverage* and *Localization* properties. In addition, if the functions $f_i(x, \hat{\theta}; x_i)$ are each linear in their parameters, then the overall approximation is also linear in its parameters.

Additional useful properties include the ability to store *a priori* knowledge in the approximation structure, the suitability of the approximation structure to available control design methodologies, and the availability of rules for specifying the structure of the function approximator.

57.11.7 Parameter Estimation

Section 57.11.6 has presented various classes of function approximators and discussed their relevant properties. The present section discusses estimating the parameters of the approximation to optimize the objective function particular to a specific application.

Estimation Criteria

Various parameter estimation algorithms exist. Due to our interest in on-line control applications, we are interested in those techniques that minimize the amount of required calculation. Below, we will briefly consider a few parameter estimation criteria. In each case, the goal is to find the optimal parameter vector $\theta^\star$ so that $\theta^\star = argmin_\theta J(\theta)$, where $J(\theta)$ is a positive function of the parameter error. Parameter estimation is discussed in greater detail in Chapters 7 and 58 or [8]. Approximation structures that are not linear in their parameters and approximation structures with memory (i.e., their own state) will also be briefly considered.

For approximation structures that are linear in the parameters, the resulting parameter estimation algorithms will generally have the form,

$$\hat{\theta}(k+1) = \hat{\theta}(k) + W(k)\phi(k)e(k), \tag{57.454}$$

where at instant k, $\hat{\theta}(k)$ is the estimate of $\theta^\star$, $W(k)$ is a (possibly data dependent) weighting matrix, $\phi(k)$ is the regressor vector, and $e(k) = y(k) - \hat{y}(k)$ is the prediction error.

Least Squares The least-squares parameter estimate is based on minimization of

$$J(\theta) = \frac{1}{N}\sum_{i=1}^{N}\frac{1}{2}(y(i) - \hat{y}(i))^2 \tag{57.455}$$

which, for an approximator that is linear in its parameters,

$$\hat{y}(i) \quad = \quad \phi(s(i))^T\theta, \tag{57.456}$$

results in the estimate,

$$\begin{aligned} \theta^{LS} &= (\frac{1}{N}\sum_{i=1}^{N}\phi(i)\phi(i)^T)^{-1}(\frac{1}{N}\sum_{i=1}^{N}\phi(i)y(i)^T) \\ &= P(N)R(N), \end{aligned} \tag{57.457}$$

where $P(N) = [\frac{1}{N}\sum_{i=1}^{N}\phi(i)\phi(i)^T]^{-1}$ is the inverse regressor sample autocorrelation matrix and $R(N) = [\frac{1}{N}\sum_{i=1}^{N}\phi(i)y(i)^T]$ is the sample cross-correlation matrix for the regressor and the desired function output. By the matrix inversion lemma, the least-squares parameter estimate can be calculated recursively by Equation 57.454, where

$$\begin{aligned} W(k) &= \frac{P(k-1)}{1+\phi(k)^T P(k-1)\phi(k)} \quad \text{and} \\ P(k) &= P(k-1) - \frac{P(k-1)\phi(k)\phi(k)^T P(k-1)}{1+\phi(k)^T P(k-1)\phi(k)}. \end{aligned}$$

Note that the recursive least-squares solution to Equation 57.455 via Equation 57.454, with correct initialization of P, gives the same answer as the solution via Equation 57.457. The advantage of the recursive approach is that it involves inverting the p x p matrix $(I + \phi(k)^T P(k-1)\phi(k))$ where p is the dimension of h, often one. The solution by Equation 57.457 involves inverting the M x M matrix $(\frac{1}{N}\sum_{i=1}^{N}\phi(i)\phi(i)^T)$, where M is the dimension of θ which is usually very large.

The main theoretical advantages of the least-squares parameter estimate are that the estimate is unbiased and minimum variance under the assumptions that 1) the unknown function is actually in the assumed function class described by Equation 57.456, 2) the noise in the measurement of the output $y(t)$ is zero mean, and 3) there is no error in the measurement of the function input s. The convergence of the LS-based algorithms is also typically much faster than other estimation techniques.

The main disadvantages of LS-based algorithms include 1) the algorithm is computationally expensive, even in the recursive version, 2) the first and third assumptions listed in the previous paragraph do not typically hold, and 3) the algorithm is difficult to implement in the nonlinear case.

Nonlinear Least Squares The previous section discussed a special case of the problem stated in Section 57.11.3, where the unknown function was assumed linear and samples of the unknown function were measured directly. When the approximation structure is not linear or the samples of the unknown function cannot be measured directly, a nonlinear optimization

[17] *The curse of dimensionality* was originally used by Bellman to refer to the increasing complexity of the solution to dynamic programming problems with increasing input dimension.

problem results. In the first case, the parameter estimate is based on minimization of Equation 57.455 where

$$\hat{y}(i) = g(s(i), \theta) \tag{57.458}$$

and $g[s(i), \theta]$, representing the structure of the function approximator, is not linear in θ. In the second case,

$$J(\theta) = \frac{1}{N}\sum_{i=1}^{N}\frac{1}{2}\{y(i) - g[s(i), \mu(i)]\}^2, \tag{57.459}$$

and

$$\mu(i) = \phi(s(i))^T\theta, \tag{57.460}$$

where $g(s(i), \mu(i))$ is again a nonlinear function. In this case, $g(.,.)$ represents the effect of the approximator output on the plant output. There are situations in which the form of this function is not completely known.

In the general case of minimizing the two norm of the sample error $(\frac{1}{N}\sum_{i=1}^{N}\frac{1}{2}\{y(i) - g[\lambda(i), \theta]\}^2)$ as a function of θ, the iterative parameter estimation algorithm will have the form,

$$\hat{\theta}(k+1) = \hat{\theta}(k) + d\hat{\theta}(k), \tag{57.461}$$

where $d\hat{\theta}(k)$ satisfies

$$(A^T A + B)d\hat{\theta}(k) = A^T f \tag{57.462}$$

and where

$$f = \frac{1}{\sqrt{N}}\{[y(1) - g[\lambda(1), \theta(k)], \ldots, [y(N) - g[\lambda(N), \theta(k)]\}^T, \tag{57.463}$$

$$A_{ij} = -\frac{1}{\sqrt{N}}\left.\frac{\partial g(\lambda, \theta)}{\partial \theta_j}\right|_{\lambda=\lambda(i),\theta=\theta(k)}, \tag{57.464}$$

$$B_{ij} = \frac{-1}{N}\sum_{l=1}^{N}\left((y(l) - g(s(l), \theta(k))\left.\frac{\partial^2 g(\lambda, \theta)}{\partial\theta_i\partial\theta_j}\right|_{\lambda=\lambda(l),\theta=\theta(k)}\right). \tag{57.465}$$

Specialization of this algorithm to the two cases of interest illustrate some important points.

In the first special case, $A = -\frac{1}{\sqrt{N}}\left.\frac{\partial g(s,\theta)}{\partial\theta}\right|_{s=s(i),\theta=\theta(k)}$, $B = -\sum_{l=1}^{N}\left((y(l) - g(s(l), \theta(k))\left.\frac{\partial^2 g(s,\theta)}{\partial\theta_i\partial\theta_j}\right|_{s=s(l),\theta=\theta(k)}\right)$, and $g(.,.)$ is a known function because it represents the function approximator. In the special case where $g(.,.)$ is linear in the parameters, $A = \frac{-1}{\sqrt{N}}\Phi[s(i)] = \frac{-1}{\sqrt{N}}[\phi(s_1), \ldots, \phi(s_N)]^T$, $B = 0$, and the solution of Equations 57.461–57.462 reduces to the least-squares solution previously presented. In the general case, because the functional form of $g(.,.)$ is known, the derivative information required to calculate A and B could be determined. In real-time applications, simplified algorithms, such as those discussed in the next section, are often used to reduce the computational complexity.

In the second case,

$$\begin{aligned} A_i &= -\frac{1}{\sqrt{N}}\left.\frac{\partial g(\mu)}{\partial\mu}\right|_{\mu=\mu(i)}\left.\frac{\partial\mu(s,\theta)}{\partial\theta}\right|_{s=s(i),\theta=\theta(k)} \\ &= -\frac{1}{\sqrt{N}}\left.\frac{\partial g(\mu)}{\partial\mu}\right|_{\mu=\mu(i)}\phi[s(i)]^T. \end{aligned}$$

Similarly, B would involve second derivative information for $g(.,.)$. In this case, g represents the manner in which the approximating function outputs affect the plant outputs. In those situations where the functional form of g is not known, the derivative information will not be available and approximations to the algorithm will be necessary.

Standard drawbacks of the above Gauss–Newton solution to the nonlinear least-squares problem include the difficulty in calculating or estimating the second order derivative information and the expense of solving Equation 57.462. Numerous algorithms have been developed to simplify the computations and improve the estimation performance (see, for example [12], [13]).

Gradient Descent A common simplification of the above parameter estimation algorithms relies on the fact that a function's gradient with respect to a vector variable indicates the direction of the function's maximum increase with respect to the vector variable. The direction opposite the gradient indicates the direction of the function's maximum decrease with respect to the vector variable. By taking a step in the direction opposite the objective function's gradient with respect to the parameters, the objective function should be decreased. Unfortunately, the gradient does not indicate the size of the step that should be taken. The size of the step depends on higher order derivative information, which has been neglected in the simplification. Therefore, the step size is usually designed to be small. It must be small because we are using local gradient information which will change throughout the region of interest D. There is of course a tradeoff in selecting the step size. Too small a step size will lead to slow convergence, while too large a step size will cause the parameter estimate to overstep and possibly oscillate around the optimal value. Several mechanisms have been proposed for adjusting the stepsize on-line.

For the objective function and approximator structure defined in Equations 57.455–57.456, the gradient $\frac{\partial J(\theta)}{\partial\theta}$ is

$$\left.\frac{\partial J(\theta, k)^T}{\partial\theta}\right|_{\theta=\theta(k)} = \frac{-1}{N}\sum_{i=0}^{N-1}\phi(i)[y(k-i) - \hat{y}(k-i)]$$

where $\theta(k)$ is assumed constant over the previous N samples (i.e., $\theta(i) = \theta(k-N+1)$ for $i = k-N+1, \ldots, k$. This is referred to as *batch training*. The resulting parameter estimate update rule is

$$\theta(k+1) = \theta(k-N+1) - W(k)\left(\frac{1}{N}\sum_{i=1}^{N}\phi(k-i)[y(k-i) - \hat{y}(k-i)]\right)$$

where $W(k) = diag[\alpha_i(k)]$ is a diagonal matrix containing the step size for each element of the parameter vector. For convergence, $\alpha_i(k) > 0$ and because the gradient is only a *local* direction of steepest descent, usually $\alpha_i(k) << 1$. This algorithm is also in the standard form of Equation 57.454. *Incremental training* refers to the situation where $N = 1$, and the parameter vector is updated at each sample instant. In this case, the updated algorithm is

$$\theta(k+1) = \theta(k) - W(k)\left\{\phi(k-i)[y(k-i) - \hat{y}(k-i)]\right\}$$

Batch training is often used in off-line training applications, but incremental training is preferred for on-line training. Incremental training avoids the delay in updating the parameters associated with accumulating a batch of data. This is appropriate, because in passive learning applications there is no guarantee that a batch of training data would be uniformly representative of the domain D.

For the second case considered in the previous section (with N=1), the gradient algorithm is

$$\theta(k+1) = \theta(k) - W(k)\,(\phi(k-i)$$
$$\left(\left.\frac{\partial g(\mu)}{\partial \mu}\right|_{\mu=\mu(i)}\right)^T [y(k-i) - \hat{y}(k-i)]\Bigg)$$

Again, $g(.,.)$ describes the manner in which the approximated function output affects the plant output, and it may include the plant dynamics. In these cases, $\frac{\partial g(\mu)}{\partial \mu}$ will not be known, and approximate methods will again be required.

The main advantage of gradient-descent algorithms is their simplicity, which allows rapid on-line calculation. The tradeoff is that gradient-following techniques normally converge at a much slower rate than the other techniques discussed above.

Backpropagation As discussed in Section 57.11.6, some popular approximation architectures are not linear in the parameters. These approximators are referred to in Case 1 of the Nonlinear Least-Squares Section. When $g(.,)$ represents the approximator structure as in Case 1 discussed above, the gradient-descent algorithm becomes

$$\theta(k+1) = \theta(k) - W(k)$$
$$\left(\left(\left.\frac{\partial g(s,\theta)}{\partial \theta}\right|_{s=s(i),\theta=\theta(k)}\right)^T [y(k-i) - \hat{y}(k-i)]\right).$$

Care should be taken in the calculation of the partial derivative $\left.\frac{\partial g(s,\theta)}{\partial \theta}\right|_{s=s(i),\theta=\theta(k)}$, which may, depending on the approximation structure, be computationally expensive. A particularly efficient algorithm for calculating the gradient in layered approximators (e.g., sigmoidal neural networks) is the backpropagation algorithm (see, for example, [9]).

Approximation of the Gradient Due to Internal State Additional complexity will arise in situations where the approximating structure includes its own state. Consider the example of state estimation. Let an unknown plant be defined by

$$\begin{aligned} x(k+1) &= f[x(k), v(k)], \\ y(k) &= Cx(k), \end{aligned}$$

where C is assumed known and the design objective is to construct a system

$$\begin{aligned} x_e(k+1) &= g[x_e(k), v(k), \theta^\star] + L[y(k) - y_e(k), \theta^\star], \\ y_e(k) &= Cx_e(k), \end{aligned}$$

so that the state estimation error $\| x(k) - x_e(k) \|$ is minimized.

To estimate the parameter vector θ incrementally by a gradient mechanism, let

$$\Delta\theta(k) = -\alpha\frac{\partial J(\theta,k)}{\partial\theta}$$

where $J(\theta,k) = \| y(k) - y_e(k) \|$. Then,

$$\begin{aligned} &\left.\frac{\partial J(\theta,k)}{\partial\theta}\right|_{x=x(k-1),x_e=x_e(k-1),\theta=\theta(k-1)} \\ &= [y(k) - y_e(k)]\left.\frac{\partial y_e(k)}{\partial\theta}\right|_{x=x(k-1),x_e=x_e(k-1),\theta=\theta(k-1)}, \\ &= [y(k) - y_e(k)]\left.\frac{\partial y_e(k)}{\partial x_e}\frac{\partial x_e}{\partial\theta}\right|_{x=x(k-1),x_e=x_e(k-1),\theta=\theta(k-1)}, \\ &= [y(k) - y_e(k)]C\left.\frac{\partial x_e}{\partial\theta}\right|_{x=x(k-1),x_e=x_e(k-1),\theta=\theta(k-1)}, \end{aligned}$$

where

$$\begin{aligned} \frac{\partial x_e}{\partial\theta}(k) &= \left(\frac{\partial g(x_e,v,\theta)}{\partial x_e}\frac{\partial x_e}{\partial\theta}(k-1)\right. \\ &\quad \left.\left. + \frac{\partial g(x_e,v,\theta)}{\partial\theta}\right)\right|_{x_e=x_e(k-1),\theta=\theta(k-1)} \end{aligned} \tag{57.466}$$

In this case, note that the solution of the gradient-based algorithm is complicated by the solution of Equation 57.466, a nonlinear matrix difference equation. In most applications, an approximate solution to Equation 57.466 would be used.

57.11.8 Example

This example will illustrate the ideas discussed in this chapter, by means of a simplified underwater vehicle heading control example. The assumed dynamic model for vehicle heading is

$$\begin{aligned} \begin{pmatrix} \dot{h}(t) \\ \dot{r}(t) \end{pmatrix} &= \begin{pmatrix} 0 & 1 \\ 0 & -a(u(t)) \end{pmatrix}\begin{pmatrix} h(t) \\ r(t) \end{pmatrix} \\ &\quad + \begin{pmatrix} 0 \\ b(u(t)) \end{pmatrix} dr(t) \end{aligned} \tag{57.467}$$

where $h(t)$ and $r(t)$ are the heading and yaw rate, $u(t)$ is the forward velocity, $dr(t)$ is the rudder deflection, $a(u(t))$ is the velocity-dependent friction, and $b[u(t)]$ is the velocity-dependent rudder effectiveness. In reality, the model would be more complex. For example, it would include models of cross-flow friction and the effects of nonzero roll angles. The model is, however, sufficient to demonstrate the desired concepts.

The model above shows that the heading dynamics are velocity dependent. Therefore, the controller will have to react differently at different speeds. For this example, the functions $a[u(t)]$ and $b[u(t)]$ will be assumed unknown.

The equivalent discrete-time model is

$$\begin{pmatrix} h(k+1) \\ r(k+1) \end{pmatrix} = \begin{pmatrix} 1 & c(u(t)) \\ 0 & d(u(t)) \end{pmatrix} \begin{pmatrix} h(k) \\ r(k) \end{pmatrix} + \begin{pmatrix} e(u(t)) \\ f(u(t)) \end{pmatrix} dr(k) \quad (57.468)$$

where $d = e^{-aT}$, $c = \frac{1}{a}(1-d)$, $f = bc$, and $e = \frac{1}{a}(bT - f)$.

For parameter estimation, the discrete-time model will be represented as

$$\begin{aligned} \begin{pmatrix} h(k+1) \\ r(k+1) \end{pmatrix} &= d\begin{pmatrix} h(k)-h(k-1) \\ r(k)-r(k-1) \end{pmatrix} + e\begin{pmatrix} dr(k) \\ 0 \end{pmatrix} \\ &\quad + g\begin{pmatrix} dr(k-1) \\ 0 \end{pmatrix} \\ &\quad + f\begin{pmatrix} 0 \\ dr(k)-dr(k-1) \end{pmatrix}, \quad (57.469) \\ &= \Phi^T\theta, \quad (57.470) \end{aligned}$$

where $\theta^T = (d, e, g, f)$, $g = fc - de$, and

$$\Phi^T = \begin{pmatrix} (h(k)-h(k-1) & dr(k) & dr(k-1) \\ (r(k)-r(k-1) & 0 & 0 \end{pmatrix} \begin{matrix} 0 \\ dr(k)-dr(k-1) \end{matrix} \Bigg).$$

Because a and b are both functions of u, each parameter in the vector θ will also be a function of u. The objective of the example is to use on-line data to approximate the function $\theta(u)^T = (d(u), e(u), g(u), f(u))$. Because this is a simulation, the actual function is known and will be used for performance analysis. This would not be the case in most applications. For this simulation example, the continuous-time model parameters, as a function of forward velocity, were assumed to be $a(u) = 0.0461 * u$ and $b(u) = 0.0006 - 0.0050u - 0.0009u^2$ for positive values of u.

The simulation includes 18 000 input-output samples, corresponding to one hour of real-time data with the control system sample period equal to 0.2 sec. (i.e., 18 000 samples = $\frac{5\ \text{samples}}{\text{sec.}}\frac{60\ \text{sec.}}{\text{min.}}\frac{60\ \text{min.}}{\text{hr.}}$). The commanded heading during this period was

$$\begin{aligned} r(k) &= r_{rand}(k) + .2sin(0.0400k) \\ &\quad + .2cos(0.0106k) + .2sin(.0009k), \end{aligned}$$

where r_{rand} is a uniform random variable in $[-.1, .1]$ that is changed every two sec. This input signal is sufficiently and persistently exciting for the linear model with four parameters.

The forward velocity is plotted versus time in Figure 57.45. The same forward velocity trajectory will be used for both the adaptive and learning-based controllers. The commanded forward velocity is held constant for one minute intervals (i.e., 300 samples). At the end of each interval, a new commanded forward velocity is generated randomly in the interval from 2.0 to 16.0 feet per second. The actual forward velocity then changes continuously according to the dynamics $u(k+1) = u(k) + lmt\{0.1, [u_c(k) - u(k)]\}$, where u_c is the commanded velocity and $lmt(.)$ is a function that limits the magnitude of the acceleration to 0.1 feet per sec.^{2}. A short segment of Figure 57.45 is shown in Figure 57.46 to display the acceleration-limited dynamics.

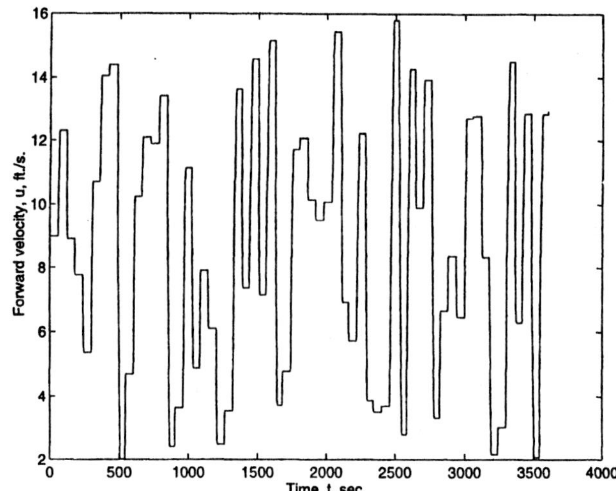

Figure 57.45 Velocity trajectory.

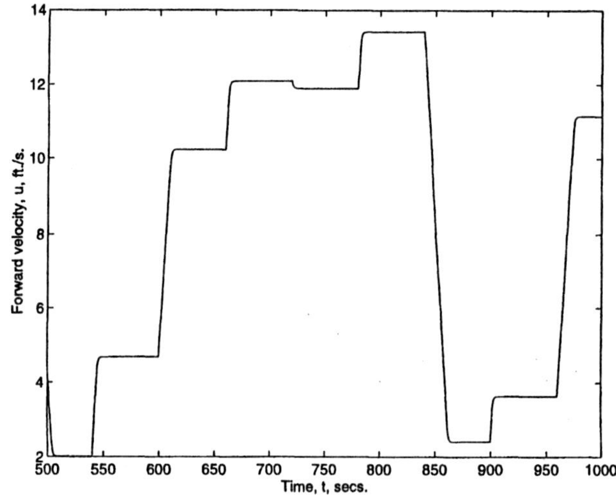

Figure 57.46 Velocity trajectory for $t \in [500, 1000]$ sec.

Using state feedback control $dr(k) = [k_h h(k) + k_r r(k)]$, with the objective of placing the discrete-time closed loop poles at $\lambda_1(u)$ and $\lambda_2(u)$, the equations for the state feedback gains become

$$k_r(u) = \frac{\lambda_1(u)\lambda_2(u)e(u) - d(u)e(u) + g(u)}{(e(u)+g(u))f(u)} \frac{[\lambda_1(u)+\lambda_2(u)-1-d(u)]}{(e(u)+g(u))f(u)}$$

$$k_h(u) = \frac{1}{e(u)}\{[\lambda_1(u) + \lambda_2(u)] - 1 - d(u) - f(u)k_r(u)\}.$$

Because the parameter vector is forward velocity dependent, the state feedback gains will be functions of the forward velocity even if the desired eigenvalues are constant. The closed-loop, discrete-time poles were selected as the constants 0.75, and 0.95. These were the fastest real poles achievable without excessive saturation (at 0.2 rads) of the rudder. Two controllers will be considered. Each controller would calculate the state feedback gains based on the estimated linear model parameters.

The first controller is adaptive. The objective of the parameter estimation is to identify the linear model parameters for the current operating condition (i.e., forward velocity). The linear model parameters will change whenever the forward velocity changes. The parameter estimation algorithm is recursive least squares with covariance resetting. The covariance is reset whenever the forward velocity changes by more than 1.0 ft. per sec. The necessity of covariance resetting or forgetting is discussed in [8]. When the covariance is reset, the last best parameter estimate, before the reset, is stored and used until the postreset parameter estimate is less than 3.00 in magnitude. Without this latching of the last best estimate, large transients may occur in the parameter estimater. Figure 57.47 displays the error (i.e., the two norm of the error) between the actual closed-loop eigenvalues and the desired eigenvalues. The actual closed-loop eigenvalues are calculated at each instant using the current linear model parameter estimates. The large errors in the closed-loop eigenvalues correspond to the times at which the forward velocity is changed. When the velocity is constant, the eigenvalue error becomes small. This is shown more clearly in Figure 57.48, which expands a portion of the time axis. For this same portion of the time axis, Figures 57.49 and 57.50 display the trajectories of the parameter estimates of d and $250e$. Although the performance of this adaptive controller could be improved with additional tuning, it is important to remember that possibly large transients will occur each time the forward velocity changes. The fact that the system may have operated at a particular forward velocity in the past will not decrease the transients at that velocity in the future, because the linear model is not capable of storing information as a function of the operating point.

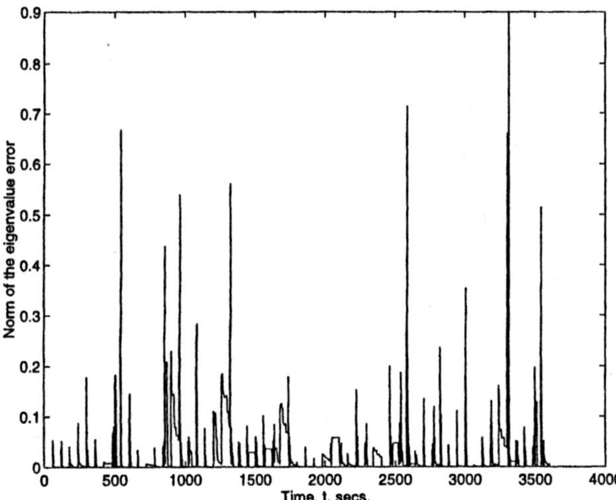

Figure 57.47 Norm of the eigenvalue error over entire training period.

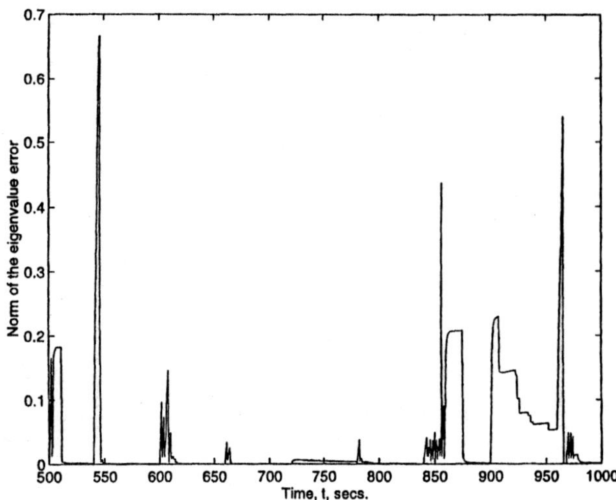

Figure 57.48 Norm of the eigenvalue error between 500 and 1000 sec.

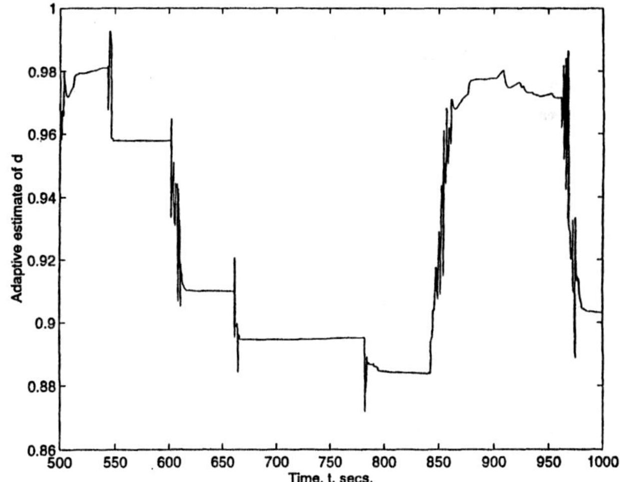

Figure 57.49 Adaptive estimate of d.

The second controller will incorporate a mechanism to synthesize a function which will store the linearized model (i.e., the θ vector) as a function of the forward velocity. Therefore, the objective function is as stated in Equation 57.440, with $h(.)$ the linearized plant dynamic model and s the forward velocity. Three approximators will be considered, but only the piecewise constant approximator will be considered in detail. The piecewise constant approximator has a single input u and four outputs, the elements of $\theta(u)$. The domain of the input variable is $D = [2, 16]$. The region D was divided into 20 regions. Each of the four outputs will be approximated by a constant value on each of the 20 regions. Because for this approximator, the regres-

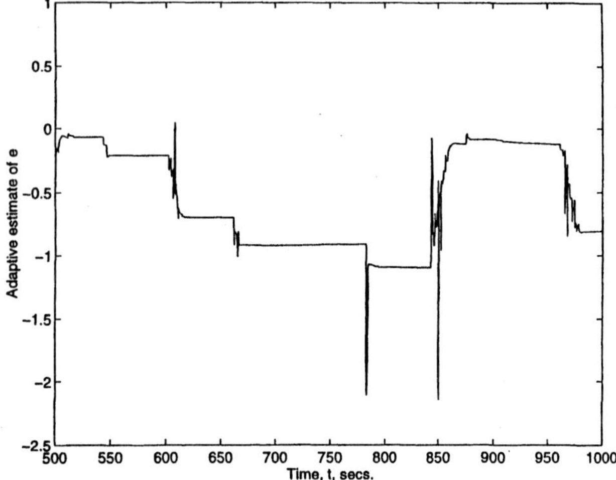

Figure 57.50 Adaptive estimate of e.

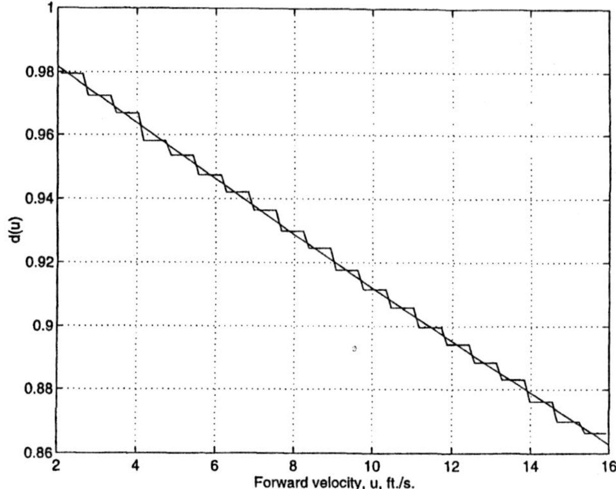

Figure 57.51 Actual and piecewise constant approximation to $d(u)$.

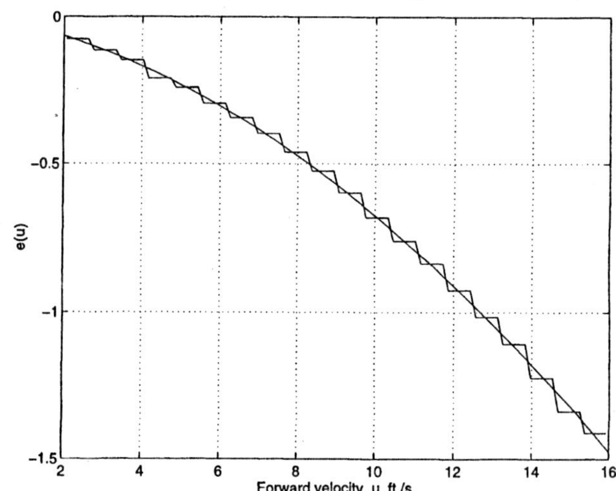

Figure 57.52 Actual and piecewise constant approximation to $e(u)$.

sor vector contains 19 elements that are zero and a single element that is one, extremely efficient algorithms can be written to solve the least-squares parameter fit. In fact, the computational load will be the same as that of the adaptive control solution described in the previous paragraphs.[18] Although there are a total of eighty parameters (4 parameters per region times 20 regions), only four parameters are relevant to any training sample. The estimates of the four parameters for a given region will not be reliable until on the order of 16 training samples are received in that region (i.e., the regressor sample autocorrelation matrix becomes sufficiently well-conditioned). Once this condition is satisfied, the approximation to the linear dynamics should be accurate over the given region despite what happens to the system when the forward velocity is outside the region. The piecewise approximation to $d(u)$, $250e(u)$, $50f(u)$, and $250g(u)$ that resulted from training over the same set of data as described for the adaptive control approach are shown in Figures 57.51–57.54.

A histogram showing the number of training samples in each region of the domain D is shown in Figure 57.55. The norm of the error between the actual and desired eigenvalues as a function of forward velocity is shown in Figure 57.56. The integral of the norm of the eigenvalue error (i.e., RMS error) over the region D is 0.016. Note that this corresponds to on-line training for one hour. At this point, if the actual system were assumed to be time invariant then the parameter update could be stopped without lossing control system performance. Repeating this example with a 20-element piecewise linear or a 5-element Legendre polynomial approximator resulted in RMS errors over the region D of $3.47e-4$ and $1.04e-7$, respectively.

Of the three approximators that were implemented, the piecewise constant approximator and the polynomial approximator had the lowest and highest computational requirements, respectively. Although the number of bins in the piecewise constant approach would have to increase significantly to achieve the same level of accuracy as the polynomial approximator, the computational complexity would not have to increase. Such an increase would not necessarily be beneficial, as the approximation could become 'spiky' due to the difficulty of getting a sufficient number of good samples in each bin. It would probably be more beneficial to move to a higher order approximator, such as the piecewise linear function. The number of parameters that must be adjusted for the piecewise linear approximator is twice that of the piecewise constant approximator, but it decreased the RMS eigenvalue error by almost two orders of magnitude.

The polynomial series approximator outperformed the piecewise linear approximator by another three orders of magnitude; however, improved performance disappears when noise and variations in time must be taken into account. Time variations are important because they require, for example, covariance resetting or forgetting in the parameter estimation routine. These techniques then require meeting a persistence of excitation condition to insure parameter convergence. The conditions for persistence of excitation are much easier to insure for local approximators.

[18]An alternative approach to this piecewise constant solution is to view it as 20 adaptive control problems, where only one adaptive control problem is valid at each sampling instant.

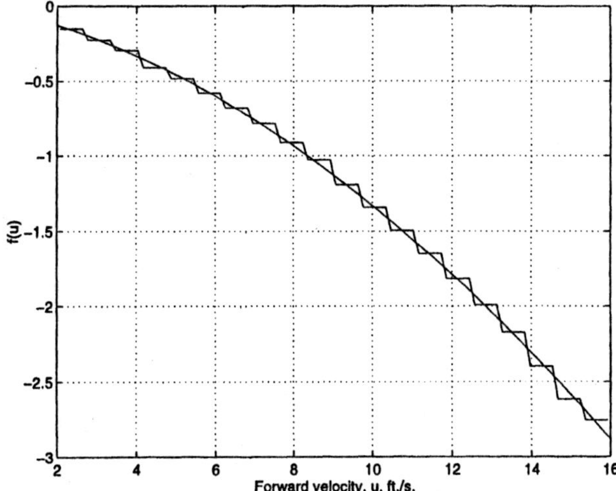

Figure 57.53 Actual and piecewise constant approximation to $f(u)$.

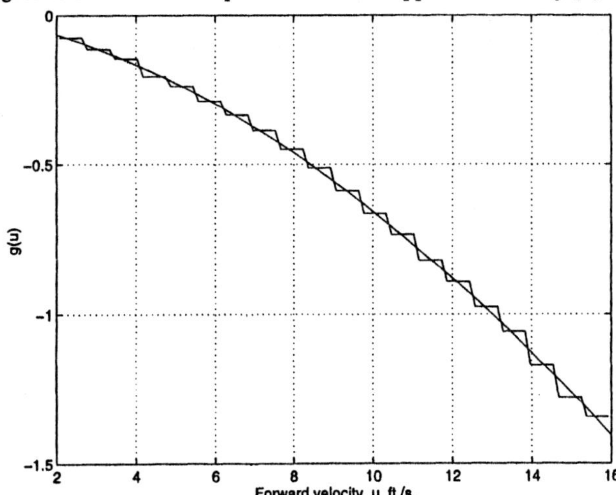

Figure 57.54 Actual and piecewise constant approximation to $g(u)$.

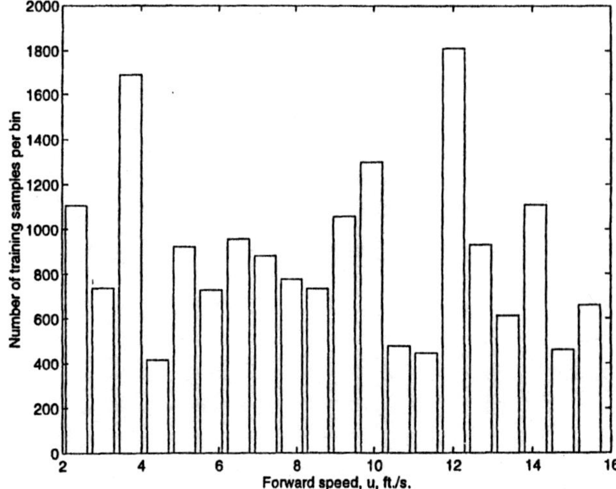

Figure 57.55 Histogram of $u(k)$.

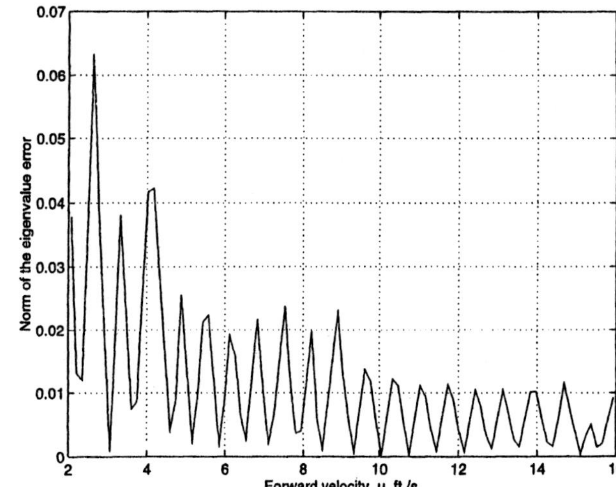

Figure 57.56 Norm of eigenvalue error versus forward velocity.

57.11.9 Conclusion

This chapter has discussed techniques for using on-line function approximation to improve control or state estimation performance. Successful implementations require proper selection of an objective function, an approximation structure, a training algorithm, and analysis of these three choices under the appropriate experimental conditions.

Several characteristics of function approximators have been discussed. In the literature, much attention is devoted to selecting certain function approximators because as a class they have the *universal approximation* property. By itself, this condition is not sufficient. First, the set of function approximators, that are dense on the set of continuous functions, is larger than the set of universal function approximators. Second, universal function approximation requires that the approximator has an arbitrarily large size (i.e., width). Once the size of the approximator is limited, a limitation is also placed on the set of functions that can be approximated to within a given ϵ. In a particular application, the designer should instead consider which approximation structure can most efficiently represent the class of expected functions. Efficiency should be measured both in terms of the number of parameters (i.e., computer memory) and the amount of computation (i.e., required computer speed). Due to the inexpensive cost of computer memory and the hard constraints on the amount of possible computation due to sampling frequency considerations, computational efficiency is often more important in applications.

Although the goal of many applications presented in the literature is to approximate a given function, the approximation accuracy is rarely evaluated directly—even in simulation examples. Instead, performance is often evaluated by analyzing graphs of the sample error, for example, a plot of the prediction error between the actual and predicted heading in the previous example. This is a dangerous approach because such graphs indicate only the accuracy of the approximation at the pertinent operating points. It is quite possible, as in the above adaptive controller example, to have a small sample error over a given time span without ever accurately approximating the desired function over the desired region. If the goal is to approximate a function over a given region, then some analysis other than monitoring the sample error during training must be performed to determine the success level. In simulation, the approximated and actual functions can be compared directly. Successful simulation performance is necessary to provide the confidence for implementation. In implementations, where the actual function is not known, a simple demonstration that the function has been approximated correctly is to show that the performance does not degrade when the parameter update law is turned off.

Several of the training algorithms discussed herein were derivative based. The main disadvantage of derivative-based parameter estimation algorithms is that they will find only the local minimum of the objective function. This is not an issue when the approximator is linear in the parameters and the objective function is convex in the approximator output because then the objective function is (quasi) convex in the parameters. It can be a major difficulty in other situations. Genetic and stochastic search techniques are not directly applicable to controller design, because a population of identical plants is not available for performance comparison. But techniques have been developed for constructing *credit assignment* functions which allow *reinforcement learning* techniques to search for globally optimal parameter estimates [1, 14].

References

[1] Millington, P., Associative Reinforcement Learning for Optimal Control, M.S. Thesis, Department of Aeronautics and Astronautics, MIT, 1991.

[2] Slotine, J. and Li, W., *Applied Nonlinear Control*, Prentice Hall, Englewood Cliffs, NJ, 1991.

[3] Fu, K., Learning Control Systems Q Review and Outlook, *IEEE Trans. Automat. Control*, AC-15(2), 1970.

[4] Tsypkin, Y., *Foundations of the Theory of Learning Systems*, Academic Press, 1971

[5] Farrell, J. and Baker, W., Learning Control Systems, In *Intelligent-Autonomous Control Systems*, Antsaklis, P. and Passino, K., Eds., Kluwer Academic, 1993.

[6] Funahashi, K., On the Approximate Realization of Continuous Mappings by Neural Networks, *Neural Networks*, 2, 183–192, 1989.

[7] Hornik, K., Stinchcombe, M., and White, H., Multilayer Feedforward Networks Are Universal Approximators, *Neural Networks*, 2, 359–366, 1989.

[8] Goodwin, G. C. and Sin, K.S., *Adaptive Filtering Prediction and Control*, Prentice Hall, Englewood Cliffs, NJ, 1984.

[9] Rumelhart, D.E., McClelland, J.L., and PDP Research Group, *Parallel Distributed Processing — Explorations in the Microstructure of Cognition, Volume 1: Foundations*, MIT Press, Cambridge, MA, 1986.

[10] Ogata, K., *Modern Control Engineering*, Prentice Hall, Englewood Cliffs, NJ, 1990.

[11] Åström, K. and Wittenmark, B., *Adaptive Control*, Addison-Wesley, 1989.

[12] *Nonlinear Optimization: Theory and Algorithms*, Dixon, L., Spedicato, E., and Szego, G., Eds., Birkhauser, Boston, 1980.

[13] Dennis, J. and Schnabel, R., *Numerical Methods for Unconstrained Optimization and Nonlinear Equations*, Prentice Hall, Englewood Cliffs, NJ, 1983.

[14] Barto, A., Sutton, R., and Anderson, C., Neuronlike Adaptive Elements that can Solve Difficult Learning Control Problems, *IEEE Trans. Syst., Man, and Cybernetics*, SMC-13(5), 1983.

SECTION XII

System Identification

58

System Identification

Lennart Ljung

Department of Electrical Engineering, Linköping University, Sweden

58.1 The Basic Ideas

58.1.1 10 Basic Questions About System Identification

1. **What is System Identification?**
System Identification allows you to build mathematical models of a dynamic system based on measured data.
2. **How is that done?**
By adjusting parameters within a given model until its output coincides as well as possible with the measured output.
3. **How do you know if the model is any good?**
A good test is to take a close look at the model's output compared to the measurements on a data set that wasn't used for the fit ("Validation Data").
4. **Can the quality of the model be tested in other ways?**
It is also valuable to look at what the model couldn't reproduce in the data ("the residuals"). There should be no correlation with other available information, such as the system's input.
5. **What models are most common?**
The techniques apply to very general models. Most common models are difference equations descriptions, such as ARX and ARMAX models, as well as all types of linear state-space models. Lately, black-box nonlinear structures, such as Artificial Neural Networks, Fuzzy models, and so on, have been much used.
6. **Do you have to assume a model of a particular type?**
For parametric models, you have to specify the structure. However, if you just assume that the system is linear, you can directly estimate its impulse or step response using Correlation Analysis or its frequency response using Spectral Analysis. This allows useful comparisons with other estimated models.
7. **How do you know what model structure to try?**
Well, you don't. For real life systems there is never any "true model", anyway. You have to be generous at this point and try out several different structures.
8. **Can nonlinear models be used in an easy fashion?**
Yes. Most common model nonlinearities are such that the measured data should be nonlinearly transformed (like squaring a voltage input if you think

0-8493-8570-9/96/$0.00+$.50

that it's the power that is the stimulus). Use physical insight about the system you are modeling and try out such transformations on models that are linear in the new variables, and you will cover a lot.

9. **What does this article cover?**
After reviewing an archetypical situation in this section, we describe the basic techniques for parameter estimation in arbitrary model structures. Section 58.3 deals with linear models of black-box structure, and Section 58.4 deals with particular estimation methods that can be used (in addition to the general ones) for such models. Physically parameterized model structures are described in Section 58.5, and nonlinear black-box models (including neural networks) are discussed in Section 58.6. Section 58.7 deals with the choices and decisions the user is faced with.

10. **Is this really all there is to System Identification?**
Actually, there is a huge amount written on the subject. Experience with real data is the driving force to further understanding. It is important to remember that any estimated model, no matter how good it looks on your screen, is a simple reflection of reality. Surprisingly often, however, this is sufficient for rational decision making.

58.1.2 Background and Literature

System Identification has its roots in standard statistical techniques and many of the basic routines have direct interpretations as well known statistical methods such as Least Squares and Maximum Likelihood. The control community took an active part in developing and applying these basic techniques to dynamic systems right after the birth of "modern control theory" in the early 1960s. Maximum likelihood estimation was applied to difference equations (ARMAX models) by [3], and thereafter a wide range of estimation techniques and model parameterizations flourished. By now, the area is mature with established and well-understood techniques. Industrial use and application of the techniques has become standard. See [13] for a common software package.

The literature on System Identification is extensive. For a practical user oriented introduction we may mention [16]. Texts that go deeper into the theory and algorithms include [14], and [26]. A classical treatment is [4].

These books all deal with the "mainstream" approach to system identification, as described in this article. In addition, there is a substantial literature on other approaches, such as "set membership" (compute all those models that reproduce the observed data within a certain given error bound), estimation of models from given frequency response measurement [24], on-line model estimation [17], nonparametric frequency domain methods [6], etc. To follow the development in the field, the IFAC series of Symposia on System Identification (Budapest, 1991, Copenhagen, 1994) is a good source.

58.1.3 An Archetypical Problem – ARX Models and the Linear Least Squares Method

The Model

We shall generally denote the system's input and output at time t by $u(t)$ and $y(t)$, respectively. The most basic relationship between the input and output is the *linear difference equation*

$$\begin{aligned} y(t) + a_1 y(t-1) + \ldots + a_n y(t-n) \\ = b_1 u(t-1) + \ldots + b_m u(t-m) \end{aligned} \tag{58.1}$$

We have chosen to represent the system in *discrete time*, primarily because observed data are always collected by sampling. It is thus more straightforward to relate observed data to discrete-time models. Nothing prevents us, however, from working with continuous-time models: we shall return to that in Section 58.5.

In Equation 58.1 we assume that the *sampling interval* is one time unit. This is not essential, but makes notation easier.

A pragmatic and useful way to see Equation 58.1 is to view it as a way of *determining the next output value* given previous observations:

$$\begin{aligned} y(t) &= -a_1 y(t-1) - \ldots - a_n y(t-n) \\ &\quad + b_1 u(t-1) + \ldots + b_m u(t-m) \end{aligned} \tag{58.2}$$

For more compact notation we introduce the vectors

$$\theta = [a_1, \ldots, a_n\, b_1, \ldots, b_m]^T, \text{ and} \tag{58.3}$$

$$\begin{aligned} \varphi(t) &= [-y(t-1) \ldots -y(t-n) \\ &\quad u(t-1) \ldots u(t-m)]^T. \end{aligned} \tag{58.4}$$

With these, Equation 58.2 can be rewritten as

$$y(t) = \varphi^T(t)\theta.$$

To emphasize that the calculation of $y(t)$ from past data in Equation 58.2 depends on the parameters in θ, we shall call this calculated value $\hat{y}(t|\theta)$ and write

$$\hat{y}(t|\theta) = \varphi^T(t)\theta. \tag{58.5}$$

The Least Squares Method

Now suppose for a given system that we do not know the values of the parameters in θ, but that we have recorded inputs and outputs over a time interval $1 \le t \le N$:

$$Z^N = \{u(1), y(1), \ldots, u(N), y(N)\}. \tag{58.6}$$

An obvious approach is then to select θ in Equation 58.1 through Equation 58.5 so as to fit the calculated values $\hat{y}(t|\theta)$ as well as possible to the measured outputs by the least squares method:

$$\min_\theta V_N(\theta, Z^N) \tag{58.7}$$

where

$$\begin{aligned} V_N(\theta, Z^N) &= \frac{1}{N}\sum_{t=1}^{N}(y(t) - \hat{y}(t|\theta))^2 = \\ &= \frac{1}{N}\sum_{t=1}^{N}(y(t) - \varphi^T(t)\theta)^2. \end{aligned} \tag{58.8}$$

We shall denote the value of θ that minimizes Equation 58.7 by $\hat{\theta}_N$:

$$\hat{\theta}_N = \arg\min_{\theta} V_N(\theta, Z^N) \tag{58.9}$$

("arg min" means the minimizing argument, i.e., that value of θ which minimizes V_N.)

Since V_N is quadratic in θ, we can find the minimum value easily by setting the derivative to zero:

$$0 = \frac{d}{d\theta} V_N(\theta, Z^N) = \frac{2}{N} \sum_{t=1}^{N} \varphi(t)(y(t) - \varphi^T(t)\theta)$$

which gives

$$\sum_{t=1}^{N} \varphi(t)y(t) = \sum_{t=1}^{N} \varphi(t)\varphi^T(t)\theta, \tag{58.10}$$

or

$$\hat{\theta}_N = \left[\sum_{t=1}^{N} \varphi(t)\varphi^T(t)\right]^{-1} \sum_{t=1}^{N} \varphi(t)y(t). \tag{58.11}$$

Once the vectors $\varphi(t)$ are defined, the solution can easily be found by modern numerical software, such as MATLAB.

EXAMPLE 58.1: First order difference equation

Consider the simple model

$$y(t) + ay(t-1) = bu(t-1).$$

This gives us the estimate according to Equations 58.3, 58.4 and 58.11

$$\begin{bmatrix} \hat{a}_N \\ \hat{b}_N \end{bmatrix} = \begin{bmatrix} \sum y^2(t-1) & -\sum y(t-1)u(t-1) \\ -\sum y(t-1)u(t-1) & \sum u^2(t-1) \end{bmatrix}^{-1} \begin{bmatrix} -\sum y(t)y(t-1) \\ \sum y(t)u(t-1) \end{bmatrix}.$$

All sums are from $t = 1$ to $t = N$. A typical convention is to set values outside the measured range at zero. In this case we would thus set $y(0) = 0$. The simple model Equation 58.1 and the well known least squares method Equation 58.11 form the archetype of System Identification. They also give the most commonly used parametric identification method and are much more versatile than perceived at first sight. In particular, one should realize that Equation 58.1 can directly be extended to several different inputs (this just calls for a redefinition of $\varphi(t)$ in Equation 58.4) and that the inputs and outputs do not have to be the raw measurements. On the contrary, it is often most important to think over the physics of the application and come up with suitable inputs and outputs for Equation 58.1 from the actual measurements.

EXAMPLE 58.2: An immersion heater

Consider a process consisting of an immersion heater in a cooling liquid. We measure

- $v(t)$: the voltage applied to the heater,
- $r(t)$: the temperature of the liquid, and
- $y(t)$: the temperature of the heater coil surface.

Suppose we need a model showing how $y(t)$ depends on $r(t)$ and $v(t)$. Some simple considerations based on common sense and high school physics ("semi-physical modeling") reveal the following:

- The change in temperature of the heater coil over one sample is proportional to the electrical power in it (the inflow power) minus the heat loss to the liquid.
- The electric power is proportional to $v^2(t)$.
- The heat loss is proportional to $y(t) - r(t)$.

This suggests the model

$$y(t) = y(t-1) + \alpha v^2(t-1) - \beta[y(t-1) - r(t-1)]$$

which fits into the form

$$y(t) + \theta_1 y(t-1) = \theta_2 v^2(t-1) + \theta_3 r(t-1))$$

This is a two input (v^2 and r) and one output model, and corresponds to choosing

$$\varphi(t) = [-y(t-1) \quad v^2(t-1) \quad r(t-1)]^T$$

in Equation 58.5.

Some Statistical Remarks

Model structures, such as Equation 58.5 that are linear in θ are known in statistics as *linear regression* and the vector $\varphi(t)$ is called the *regression vector* (its components are the *regressors*). "Regress" here alludes to the fact that we try to calculate (or describe) $y(t)$ by "going back" to $\varphi(t)$. Models such as Equation 58.1 where the regression vector, $\varphi(t)$, contains old values of the variable to be explained, $y(t)$, are then partly *autoregressions.* For that reason the model structure Equation 58.1 has the standard name ARX-model (Autoregression with extra inputs).

There is a rich statistical literature on the properties of the estimate $\hat{\theta}_N$ under varying assumptions (see, e.g., [8]). So far we have just viewed Equations 58.7 and 58.8 as "curve-fitting". In Section 58.2.2 we shall deal with a more comprehensive statistical discussion, which includes the ARX model as a special case. Some direct calculations will be done in the following subsection.

Model Quality and Experiment Design

Let us consider the simplest special case, that of a Finite Impulse Response (FIR) model, obtained from Equation 58.1 by setting $n = 0$:

$$y(t) = b_1 u(t-1) + \ldots b_m u(t-m). \tag{58.12}$$

Suppose that the observed data really have been generated by a similar mechanism

$$y(t) = b_1^0 u(t-1) + \ldots b_m^0 u(t-m) + e(t) \tag{58.13}$$

where $e(t)$ is a white noise sequence with variance λ, but otherwise unknown. (That is, $e(t)$ can be described as a sequence of independent random variables with zero mean values and variances λ). Analogous to Equation 58.5,

$$y(t) = \varphi^T(t)\theta_0 + e(t). \tag{58.14}$$

We can now replace $y(t)$ in Equation 58.11 by the above expression, and obtain

$$\begin{aligned}\hat{\theta}_N &= \left[\sum_{t=1}^{N} \varphi(t)\varphi^T(t)\right]^{-1} \sum_{t=1}^{N} \varphi(t)y(t), \\ &= \left[\sum_{t=1}^{N} \varphi(t)\varphi^T(t)\right]^{-1} \times \\ &\quad \left[\sum_{t=1}^{N} \varphi(t)\varphi^T(t)\theta_0 + \sum_{t=1}^{N} \varphi(t)e(t)\right],\end{aligned}$$

or

$$\tilde{\theta}_N = \hat{\theta}_N - \theta_0 = \left[\sum_{t=1}^{N} \varphi(t)\varphi^T(t)\right]^{-1} \sum_{t=1}^{N} \varphi(t)e(t). \tag{58.15}$$

Suppose that the input u is independent of the noise e. Then φ and e are independent in this expression, so it is easy to see that $E\tilde{\theta}_N = 0$, because e has zero mean. The estimate is consequently *unbiased*. Here E denotes *mathematical expectation*.

We can also form the expectation of $\tilde{\theta}_N\tilde{\theta}_N^T$, i.e., the covariance matrix of the parameter error. Denote the matrix within brackets by R_N. Take expectation with respect to the white noise e. Then R_N is a deterministic matrix and

$$\begin{aligned}P_N &= E\tilde{\theta}_N\tilde{\theta}_N^T = R_N^{-1} \sum_{t,s=1}^{N} \varphi(t)\varphi^T(s) \times \\ &\quad Ee(t)e(s)R_N^{-1} = \lambda R_N^{-1}\end{aligned} \tag{58.16}$$

because the double sum collapses to λR_N.

We have thus computed the covariance matrix of the estimate $\hat{\theta}_N$, determined entirely by the input properties and the noise level. Moreover define

$$\bar{R} = \lim_{N\to\infty} \frac{1}{N} R_N \tag{58.17}$$

This will be the *covariance matrix* of the input, i.e., the $i-j$-element of $\bar{R}$ is $R_{uu}(i-j)$, as defined by Equation 58.89 later on.

If the matrix $\bar{R}$ is nonsingular, the covariance matrix of the parameter estimate is approximately (and the approximation improves as $N \to \infty$)

$$P_N = \frac{\lambda}{N}\bar{R}^{-1} \tag{58.18}$$

A number of things follow from this, all typical of the general properties to be described in Section 58.2.2:

- The covariance decays like $1/N$, so that the parameters approach the limiting value at the rate $1/\sqrt{N}$.
- The covariance is proportional to the noise-to-signal ratio, that is, it is proportional to the noise variance and inversely proportional to the input power.
- The covariance does not depend on the input's or noise's signal shapes, only on their variance/covariance properties.
- Experiment design, i.e., the selection of the input u, aims at making the matrix $\bar{R}^{-1}$ "as small as possible." Note that the same $\bar{R}$ can be obtained for many different signals u.

58.1.4 The Main Ingredients

The main ingredients for the System Identification problem are as follows

- The data set Z^N,
- A class of candidate model descriptions; *a Model Structure*,
- A criterion of fit between data and models, and
- Routines to validate and accept resulting models.

We have seen in Section 58.1.3 a particular model structure, the ARX-model. In fact the major problem in system identification is to select a good model structure, and a substantial part of this article deals with various model structures. See Sections 58.3, 58.5, and 58.6, all concerning this problem. Generally speaking, a model structure is a parameterized mapping from past inputs and outputs Z^{t-1} (cf. Equation 58.6) to the space of the model outputs:

$$\hat{y}(t|\theta) = g(\theta, Z^{t-1}) \tag{58.19}$$

Here θ is the finite dimensional vector used to parameterize the mapping.

Actually, the problem of fitting a given model structure to measured data is much simpler, and can be dealt with independently of the model structure used. We shall do so in the following section.

The problem of assuring a data set with adequate information is the problem of *experiment design*, and it will be described in Section 58.7.1.

Model validation is a process to discriminate between various model structures and the final quality control station, before a model is delivered to the user. This problem is discussed in Section 58.7.2.

58.2 General Parameter Estimation Techniques

In this section we shall deal with issues that are independent of model structure. Principles and algorithms for fitting models to data, as well as the general properties of the estimated models, are all model-structure independent and equally well applicable to ARMAX models and Neural Network models.

The section is organized as follows. In the first section, the general principles for parameter estimation are outlined. Thereafter we deal with the asymptotic (in the number of observed data) properties of the models, while algorithms, both for on-line and off-line use are described in the last subsection.

58.2.1 Fitting Models to Data

In Section 58.1.3 we showed one way to parameterize descriptions of dynamical systems. There are many other possibilities and we shall spend a fair amount of this contribution discussing the different choices and approaches. *This is actually the key problem in system identification.* No matter how the problem is approached, a model parameterization leads to a predictor

$$\hat{y}(t|\theta) = g(\theta, Z^{t-1}) \tag{58.20}$$

that depends on the unknown parameter vector and past data Z^{t-1} (see Equation 58.6). This predictor can be linear in y and u. This in turn contains several special cases both in terms of black-box models and physically parameterized ones, as will be discussed in Sections 58.3 and 58.5, respectively. The predictor could also be of general, nonlinear nature, as will be discussed in Section 58.6.

In any case *we now need a method to determine a good value of* θ, based on the information in an observed, sampled data set Equation 58.6. It suggests itself that the basic least-squares approach Equations 58.7 through 58.9, still is a natural approach, even when the predictor $\hat{y}(t|\theta)$ is a more general function of θ.

A procedure with some more degrees of freedom is the following:

1. From observed data and the predictor $\hat{y}(t|\theta)$, form the sequence of prediction errors,

$$\varepsilon(t,\theta) = y(t) - \hat{y}(t|\theta), \quad t = 1, 2, \ldots N. \tag{58.21}$$

2. Filter the prediction errors through a linear filter $L(q)$,

$$\varepsilon_F(t,\theta) = L(q)\varepsilon(t,\theta) \tag{58.22}$$

(here q denotes the shift operator, $qu(t) = u(t+1)$) to enhance or depress interesting or unimportant frequency bands in the signals.

3. Choose a scalar valued, positive function $\ell(\cdot)$ to measure the "size" or "norm" of the prediction error:

$$\ell(\varepsilon_F(t,\theta)) \tag{58.23}$$

4. Minimize the sum of these norms:

$$\hat{\theta}_N = \arg\min_\theta V_N(\theta, Z^N) \tag{58.24}$$

where

$$V_N(\theta, Z^N) = \frac{1}{N}\sum_{t=1}^{N} \ell(\varepsilon_F(t,\theta)) \tag{58.25}$$

This procedure is natural and pragmatic; we can still think of it as "curve fitting" between $y(t)$ and $\hat{y}(t|\theta)$. It also has several statistical and informational theoretic interpretations. Most importantly, if the noise source in the system (as in Equation 58.62 below) is supposed to be a sequence of independent random variables $\{e(t)\}$ each having a probability density function $f_e(x)$, then Equation 58.24 becomes the Maximum Likelihood estimate (MLE) if we choose

$$L(q) = 1 \quad \text{and} \quad \ell(\varepsilon) = -\log f_e(\varepsilon). \tag{58.26}$$

The MLE has several nice statistical features and thus gives a strong "moral support" for using the outlined method. Another pleasing aspect is that the method is independent of the particular model parameterization used (although this will affect the actual minimization procedure). For example, the method of "back propagation," often used in connection with neural network parameterizations, amounts to computing $\hat{\theta}_N$ in Equation 58.24 by a recursive gradient method. We shall deal with these aspects later on in this section.

58.2.2 Model Quality

An essential question is, of course, what properties the estimate resulting from Equation 58.24 will have. These will naturally depend on the properties of the data record Z^N defined by Equation 58.6. It is, in general, a difficult problem to characterize the quality of $\hat{\theta}_N$ exactly. One normally has to be content with the asymptotic properties of $\hat{\theta}_N$ as the number of data, N, tends to infinity.

It is an important aspect of the general identification method (Equation 58.24) that the asymptotic properties of the resulting estimate can be expressed in general terms for arbitrary model parameterizations.

The first basic result is the following

$$\hat{\theta}_N \to \theta^* \quad \text{as} \quad N \to \infty \quad \text{where} \tag{58.27}$$

$$\theta^* = \arg\min_\theta E\ell[\varepsilon_F(t,\theta)], \tag{58.28}$$

that is, as more and more data become available, the estimate converges to that value θ^*, that would minimize the expected value of the "norm" of the filtered prediction errors. This is *the best possible approximation* of the true system that is available within the model structure. The expectation E in Equation 58.28 is taken with respect to all random disturbances that affect the data and it also includes averaging over the input properties. This means, in particular, that θ^* will make $\hat{y}(t|\theta^*)$ a good approximation of $y(t)$ with respect to those aspects of the system enhanced by the input signal.

The second basic result is the following one: If $\{\varepsilon(t,\theta^*)\}$ is approximately white noise, then the covariance matrix of $\hat{\theta}_N$ is approximately given by

$$E(\hat{\theta}_N - \theta^*)(\hat{\theta}_N - \theta^*)^T \sim \frac{\lambda}{N}[E\psi(t)\psi^T(t)]^{-1} \tag{58.29}$$

where

$$\lambda = E\varepsilon^2(t,\theta^*) \tag{58.30}$$

$$\text{and} \quad \psi(t) = \frac{d}{d\theta}\hat{y}(t|\theta)|_{\theta=\theta^*} \tag{58.31}$$

Think of ψ as the sensitivity derivative of the predictor with respect to the parameters. Then Equation 58.29 says that the covariance matrix for $\hat{\theta}_N$ is proportional to the inverse of the covariance matrix of this sensitivity derivative. This is a quite natural result.

Note: For all these results, the expectation operator E can, under most general conditions, be replaced by the limit of the sample mean, that is

$$E\psi(t)\psi^T(t) \leftrightarrow \lim_{N\to\infty} \frac{1}{N}\sum_{t=1}^{N} \psi(t)\psi^T(t). \tag{58.32}$$

The results Equations 58.27 through 58.31 are general and hold for all model structures, both linear and nonlinear, subject only to some regularity and smoothness conditions. They are also fairly natural and will give the guidelines for all user choices involved in the process of identification. See [14] for more details around this.

58.2.3 Measures of Model Fit

Some quite general expressions for the expected model fit, that are independent of the model structure, can also be developed.

Let us measure the (average) fit between any model Equation 58.20 and the true system as

$$\bar{V}(\theta) = E|y(t) - \hat{y}(t|\theta)|^2 \tag{58.33}$$

Here expectation E is over the data properties (i.e., expectation over "Z^∞" with the notation Equation 58.6). Recall that expectation can also be interpreted as sample means, as in Equation 58.32.

Before we continue, note that the fit $\bar{V}$ will depend, not only on the model and the true system, *but also on data properties*, like input spectra, possible feedback, etc. We shall say that the fit depends on the *experimental conditions.*

The estimated model parameter $\hat{\theta}_N$ is a random variable, because it is constructed from observed data, that can be described as random variables. To evaluate the model fit, we then take the expectation of $\bar{V}(\hat{\theta}_N)$ with respect to the estimation data. That gives our measure

$$F_N = E\bar{V}(\hat{\theta}_N). \tag{58.34}$$

In general, the measure F_N depends on a number of things:

- The model structure used,
- The number of data points N,
- The data properties for which the fit $\bar{V}$ is defined, and
- The properties of the data used to estimate $\hat{\theta}_N$.

The rather remarkable fact is that, if the two last data properties coincide, then, asymptotically in N, (see [14], Chapter 16)

$$F_N \approx \bar{V}_N(\theta^*)(1 + \frac{dim\theta}{N}) \tag{58.35}$$

Here θ^* is the value that minimizes the expected criterion Equation 58.28. The notation $dim\theta$ means the number of estimated parameters. The result also assumes that the criterion function $\ell(\varepsilon) = \|\varepsilon\|^2$, and that the model structure is successful in the sense that $\varepsilon_F(t)$ is approximately white noise.

Despite the reservations about the formal validity of Equation 58.35, it carries a most important conceptual message: If a model is evaluated on a data set with the same properties as the estimation data, then *the fit will not depend on the data properties,* and it will depend on the model structure *only in terms of the number of parameters used and of the best fit offered within the structure.*

The expression can be rewritten as follows. Let $\hat{y}_0(t|t-1)$ denote the "true" one step ahead prediction of $y(t)$, let

$$W(\theta) = E|\hat{y}_0(t|t-1) - \hat{y}(t|\theta)|^2, \tag{58.36}$$

and let

$$\lambda = E|y(t) - \hat{y}_0(t|t-1)|^2. \tag{58.37}$$

Then λ is the *innovations* variance, i.e., that part of $y(t)$ that cannot be predicted from the past. Moreover $W(\theta^*)$ is the *bias error,* i.e., the discrepancy between the true predictor and the best one available in the model structure. Under the same assumptions as above, Equation 58.35 can be rewritten as

$$F_N \approx \lambda + W(\theta^*) + \lambda\frac{dim\theta}{N}. \tag{58.38}$$

The three terms constituting the model error then have the following interpretations:

- λ is the unavoidable error, stemming from the fact that the output cannot be exactly predicted, even with perfect system knowledge.
- $W(\theta^*)$ is the bias error. It depends on the model structure, and on the experimental conditions. It will typically decrease as $dim\theta$ increases.
- The last term is the *variance error.* It is proportional to the number of estimated parameters and inversely proportional to the number of data points. It does not depend on the particular model structure or on the experimental conditions.

58.2.4 Model Structure Selection

The most difficult choice for the user is to find a suitable model structure to fit the data to. This is of course a very application-dependent problem, and it is difficult to give general guidelines. (Still, some general practical advice will be given in Section 58.7.)

The heart of the model structure selection process is handling the trade-off between bias and variance, as formalized by Equation 58.38. The "best" model structure is the one that minimizes F_N, the fit between the model and the data for a *fresh* data set, one that was not used for estimating the model. Most procedures for choosing the model structures are also aiming at finding this best choice.

Cross-Validation

A very natural and pragmatic approach is *Cross-Validation.* This means that the available data set is split into two parts:

estimation data, $Z_{\text{est}}^{N_1}$, that is used to estimate the models,

$$\hat{\theta}_{N_1} = \arg\min V_{N_1}(\theta, Z_{\text{est}}^{N_1}), \tag{58.39}$$

and *validation data*, $Z_{\text{val}}^{N_2}$, for which the criterion is evaluated,

$$\hat{F}_{N_1} = V_{N_2}(\hat{\theta}_{N_1}, Z_{\text{val}}^{N_2}). \tag{58.40}$$

Here V_N is the criterion Equation 58.25. Then $\hat{F}_N$ will be an unbiased estimate of the measure F_N, defined by Equation 58.34, discussed at length in the previous section. The procedure would be to try out a number of model structures, and choose the one that minimizes $\hat{F}_{N_1}$.

Cross-validation to find a good model structure has an immediate intuitive appeal. We simply check if the candidate model is capable of "reproducing" data it hasn't yet seen. If that works well, we have some confidence in the model, regardless of any probabilistic framework that might be imposed. Such techniques are also the most commonly used.

A few comments can be added. In the first place, one can use different splits of the original data into estimation and validation data. For example, in statistics, there is a common-cross validation technique called "leave one out." This means that the validation data set consists of one data point "at a time", but successively applied to the whole original set. In the second place, the test of the model on the validation data does not have to be in terms of the particular criterion Equation 58.40. In system identification it is common practice to simulate (or predict several steps ahead) the model using the validation data, and then visually inspect the agreement between measured and simulated (predicted) output.

Estimating the Variance Contribution – Penalizing the Model Complexity

It is clear that the criterion Equation 58.40 has to be evaluated on the validation data to be of any use. It would be strictly decreasing as a function of model flexibility if evaluated on the estimation data. In other words, the adverse effect of the dimension of θ shown in Equation 58.38 would be missed. There are a number of criteria, often derived from entirely different viewpoints, that try to capture the influence of this variance error term. The two best known are *Akaike's Information Theoretic Criterion, AIC*, which has the form (for Gaussian disturbances)

$$\tilde{V}_N(\theta, Z^N) = \left(1 + \frac{2dim\theta}{N}\right)\frac{1}{N}\sum_{t=1}^{N}\varepsilon^2(t,\theta) \tag{58.41}$$

and *Rissanen's Minimum Description Length Criterion, MDL* in which $dim\theta$ in the expression above is replaced by $\log N dim\theta$. See [1] and [23].

The criterion $\tilde{V}_N$ is then to be minimized both with respect to θ and to a family of model structures. The relationship to Equation 58.35 for F_N is obvious.

58.2.5 Algorithmic Aspects

In this section we shall discuss how to achieve the best fit between observed data and the model, i.e., how to carry out the minimization of Equation 58.24. For simplicity we here assume a quadratic criterion and set the prefilter L to unity:

$$V_N(\theta) = \frac{1}{2N}\sum_{t=1}^{N}|y(t) - \hat{y}(t|\theta)|^2. \tag{58.42}$$

No analytic solution to this problem is possible unless the model $\hat{y}(t|\theta)$ is linear in θ, so that the minimization has to be done by some numerical search procedure. A classical treatment of the problem of minimizing the sum of squares is given in [7].

Most efficient search routines are based on iterative local search in a "downhill" direction from the current point. We then have an iterative scheme of the following kind:

$$\hat{\theta}^{(i+1)} = \hat{\theta}^{(i)} - \mu_i R_i^{-1}\hat{g}_i. \tag{58.43}$$

Here $\hat{\theta}^{(i)}$ is the parameter estimate after iteration number i. The search scheme is thus made up of the three entities:

- μ_i step size.
- $\hat{g}_i$ an estimate of the gradient $V_N'(\hat{\theta}^{(i)})$, and
- R_i a matrix that modifies the search direction.

It is useful to distinguish between two different minimization situations

1. *Off-line* or *batch*: The update $\mu_i R_i^{-1}\hat{g}_i$ is based on the whole available data record Z^N.
2. *On-line* or *recursive*: The update is based only on data up to sample i (Z^i), (done so that the gradient estimate $\hat{g}_i$ is based only on data just before sample i).

We shall discuss these two modes separately below. First some general aspects will be treated.

Search Directions

The basis for the local search is the gradient

$$\begin{aligned} V_N'(\theta) &= \frac{dV_N(\theta)}{d\theta} \\ &= -\frac{1}{N}\sum_{t=1}^{N}(y(t) - \hat{y}(t|\theta))\psi(t,\theta) \end{aligned} \tag{58.44}$$

where

$$\psi(t,\theta) = \frac{\partial}{\partial\theta}\hat{y}(t|\theta). \tag{58.45}$$

The gradient ψ is, in the general case, a matrix with $dim\ \theta$ rows and $dim\ y$ columns. It is well known that gradient search for the minimum is inefficient, especially close to the minimum. There it is optimal to use the *Newton search direction*

$$R^{-1}(\theta)V_N'(\theta) \tag{58.46}$$

where

$$R(\theta) = V_N''(\theta) = \frac{d^2V_N(\theta)}{d\theta^2}$$

$$= \frac{1}{N}\sum_{t=1}^{N}\psi(t,\theta)\psi^T(t,\theta) + \frac{1}{N}\sum_{t=1}^{N}(y(t)-\hat{y}(t|\theta))\frac{\partial^2}{\partial\theta^2}\hat{y}(t|\theta). \quad (58.47)$$

The true Newton direction will thus require that the second derivative,

$$\frac{\partial^2}{\partial\theta^2}\hat{y}(t|\theta),$$

be computed. Also, far from the minimum, $R(\theta)$ need not be positive semidefinite. Therefore alternative search directions are more common in practice:

- *Gradient direction.* Simply take

$$R_i = I. \quad (58.48)$$

- *Gauss–Newton direction.* Use

$$R_i = H_i = \frac{1}{N}\sum_{t=1}^{N}\psi(t,\hat{\theta}^{(i)})\psi^T(t,\hat{\theta}^{(i)}). \quad (58.49)$$

- *Levenberg–Marquard direction.* Use

$$R_i = H_i + \delta I \quad (58.50)$$

  where H_i is defined by Equation 58.49.
- *Conjugate gradient direction.* Construct the Newton direction from a sequence of gradient estimates. Think of V_N'' as constructed by difference approximation of d gradients. The direction Equation 58.46 is, however, constructed directly without explicitly forming and inverting V''.

It is generally considered [7] that the Gauss–Newton search direction is to be preferred. For ill-conditioned problems the Levenberg–Marquard modification is recommended.

On-Line Algorithms

Equations 58.44 and 58.47 for the Gauss–Newton search clearly assume that the whole data set Z^N is available during the iterations. If the application is off-line, i.e., the model $\hat{g}_N$ is not required during the data acquisition; this is also the most natural approach.

However, many adaptive situations require on-line (or recursive) algorithms, where the data are processed as they are measured. (In Neural Network contexts such algorithms are often used off-line.) Then the measured data record is concatenated with itself several times to create a long record that is fed into the on-line algorithm. We may refer to [17] as a general reference for recursive parameter estimation algorithms. In [27] the use of such algorithms in the off-line case is discussed.

It is natural to consider the following algorithm as the basic one:

$$\hat{\theta}(t) = \hat{\theta}(t-1) + \mu_t R_t^{-1}\psi(t,\hat{\theta}(t-1))\varepsilon(t,\hat{\theta}(t-1)), \quad (58.51)$$

$$\varepsilon(t,\theta) = y(t) - \hat{y}(t|\theta), \text{ and} \quad (58.52)$$

$$R_t = R_{t-1} + \mu_t\left[\psi(t,\hat{\theta}(t-1))\psi^T(t,\hat{\theta}(t-1)) - R_{t-1}\right]. \quad (58.53)$$

The reason is that if $\hat{y}(t|\theta)$ is linear in θ, then Equations 58.51 to 58.53, with $\mu_t = 1/t$, provides the analytical solution to the minimization problem Equation 58.42. This also means that this is a natural algorithm close to the minimum, where a second order expansion of the criterion is a good approximation. In fact, it is shown in [17], that Equations 58.51 – 58.53 in general give an estimate $\hat{\theta}(t)$ with the same ("optimal") statistical, asymptotic properties as the true minimum to Equation 58.42.

The quantities $\hat{y}(t|\hat{\theta}(t-1))$ and $\psi(t,\hat{\theta}(t-1))$ would normally (except in the linear regression case) require that the whole data record be computed. This would violate the recursiveness of the algorithm. In practical implementations these quantities are therefore replaced by recursively computed approximations. The idea behind these approximations is to use the defining equation for $\hat{y}(t|\theta)$ and $\psi(t,\theta)$ (which typically are recursive equations), and replace any appearance of θ with its latest available estimate. See [17] for more details.

Some averaged variants of Equations 58.51 to 58.53) have also been discussed:

$$\hat{\hat{\theta}}(t) = \hat{\hat{\theta}}(t-1) + \mu_t R_t^{-1}\psi(t,\hat{\theta}(t-1))\varepsilon(t,\hat{\theta}(t-1)), \quad (58.54)$$

and

$$\hat{\theta}(t) = \hat{\theta}(t-1) + \rho_t[\hat{\hat{\theta}}(t) - \hat{\theta}(t-1)]. \quad (58.55)$$

The basic algorithm Equations 58.51 – 58.53 then corresponds to $\rho_t = 1$. Using $\rho_t < 1$ gives a so called "accelerated convergence" algorithm. It was introduced by [21] and has been extensively discussed by [10] and others. The remarkable thing about this averaging is that we achieve the same asymptotic statistical properties of $\hat{\theta}(t)$ by Equations 58.54 and 58.55 with $R_t = I$ (gradient search) as with Equations 58.51 – 58.53) if

$$\rho_t = 1/t,$$
$$\mu_t >> \rho_t, \quad \mu_t \to 0.$$

It is thus an alternative to Equations 58.51 to 58.53, in particular, if $\dim\theta$ is large, R_t is a big matrix.

Local Minima

A fundamental problem with minimization tasks like Equation 58.42 is that $V_N(\theta)$ may have several or many local (nonglobal) minima, where local search algorithms may get caught. There is no easy solution to this problem. It is usually well worth the effort to find a good initial value $\theta^{(0)}$ to start the iterations. Other than that, only various global search strategies are left, such as random search, random restarts, simulated annealing, and the genetic algorithm.

58.3 Linear Black-Box Systems

58.3.1 Linear System Descriptions in General

A Linear System with Additive Disturbances

A linear system with additive disturbances $v(t)$ can be described by

$$y(t) = G(q)u(t) + v(t). \tag{58.56}$$

Here $u(t)$ is the input signal, and $G(q)$ is the transfer function from input to output $y(t)$. The symbol q is the shift operator, so Equation 58.56 should be interpreted as

$$\begin{aligned} y(t) &= \sum_{k=0}^{\infty} g_k u(t-k) + v(t) \\ &= (\sum_{k=0}^{\infty} g_k q^{-k}) u(t) + v(t). \end{aligned} \tag{58.57}$$

The disturbance $v(t)$ can, in general terms, be characterized by its *spectrum*, which is a description of its frequency content. It is often more convenient to describe $v(t)$, obtained by filtering a white noise source $e(t)$ through a linear filter $H(q)$, as

$$v(t) = H(q)e(t) \tag{58.58}$$

From a linear identification perspective, this is equivalent to describing $v(t)$ as a signal with spectrum

$$\Phi_v(\omega) = \lambda |H(e^{i\omega})|^2 \tag{58.59}$$

where λ is the variance of the noise source $e(t)$. We shall assume that $H(q)$ is normalized to be monic, i.e.,

$$H(q) = 1 + \sum_{k=1}^{\infty} h_k q^{-k}. \tag{58.60}$$

Putting all of this together, we arrive at the standard linear system description

$$y(t) = G(q)u(t) + H(q)e(t). \tag{58.61}$$

Parameterized Linear Models

Now, if the transfer functions G and H in Equation 58.61 are not known, we introduce parameters θ in their description reflecting our lack of knowledge. The exact way of doing this is the topic of the present section as well as of Section 58.5.

In any case the resulting, parameterized model will be described by

$$y(t) = G(q, \theta)u(t) + H(q, \theta)e(t). \tag{58.62}$$

The parameters θ can then be estimated from data using the general procedures described in Section 58.2.

Predictors for Linear Models

Given a system description Equation 58.62 and input-output data up to time $t-1$,

$$y(s), u(s) \quad s \le t-1, \tag{58.63}$$

how shall we predict the next output value $y(t)$?

In the general case of Equation 58.62, the prediction can be deduced in the following way: Dividing Equation 58.62 by $H(q, \theta)$,

$$H^{-1}(q, \theta)y(t) = H^{-1}(q, \theta)G(q, \theta)u(t) + e(t),$$

or

$$\begin{aligned} y(t) &= [1 - H^{-1}(q, \theta)]y(t) \\ &\quad + H^{-1}(q, \theta)G(q, \theta)u(t) + e(t). \end{aligned} \tag{58.64}$$

In view of the normalization Equation 58.60,

$$1 - H^{-1}(q, \theta) = \frac{H(q, \theta) - 1}{H(q, \theta)} = \frac{1}{H(q, \theta)} \sum_{k=1}^{\infty} h_k q^{-k}$$

The expression $[1 - H^{-1}(q, \theta)]y(t)$ thus only contains old values of $y(s)$, $s \le t-1$. The right side of Equation 58.65 is thus known at time $t-1$, with the exception of $e(t)$. The prediction of $y(t)$ is obtained from Equation 58.65 by deleting $e(t)$:

$$\hat{y}(t|\theta) = [1 - H^{-1}(q, \theta)]y(t) + H^{-1}(q, \theta)G(q, \theta)u(t). \tag{58.65}$$

This generally expresses how linear models predict the next value of the output, given old values of y and u.

A Characterization of the Limiting Model in a General Class of Linear Models

Let us apply the general limit result Equations 58.27 - 58.28 to the linear model structure Equation 58.62 (or Equation 58.65). If we choose a quadratic criterion $\ell(\varepsilon) = \varepsilon^2$.(in the scalar output case), then this tells us, in the time domain, that the limiting parameter estimate is the one minimizing the filtered prediction error variance (for the input used during the experiment.) Suppose that the data actually have been generated by

$$y(t) = G_0(q)u(t) + v(t). \tag{58.66}$$

Let $\Phi_u(\omega)$ be the input spectrum and $\Phi_v(\omega)$ be the spectrum for the additive disturbance v. Then the filtered prediction error is

$$\begin{aligned} \varepsilon_F(t, \theta) &= \frac{L(q)}{H(q, \theta)}[y(t) - G(q, \theta)u(t)] \\ &= \frac{L(q)}{H(q, \theta)}[(G_0(q) \\ &\quad - G(q, \theta))u(t) + v(t)]. \end{aligned} \tag{58.67}$$

By Parseval's relationship, the prediction error variance can also be an integral over the spectrum of the prediction error. This spectrum, in turn, is directly obtained from Equation 58.67, so that the limit estimate θ^* in Equation 58.28 can also be defined as

$$\begin{aligned} \theta^* &= \arg\min_\theta \Bigg[\int_{-\pi}^{\pi} |G_0(e^{i\omega}) \\ &\quad - G(e^{i\omega}, \theta)|^2 \frac{\Phi_u(\omega)|L(e^{i\omega})|^2}{|H(e^{i\omega}, \theta)|^2} d\omega \\ &\quad + \int_{-\pi}^{\pi} \Phi_v(\omega)|L(e^{i\omega})|^2 / |H(e^{i\omega}, \theta)|^2 d\omega \Bigg]. \end{aligned} \tag{58.68}$$

If the noise model $H(q,\theta) = H_*(q)$ does not depend on θ (as in the output error model Equation 58.75), Equation 58.68 shows that the resulting model $G(e^{i\omega},\theta^*)$ will give that frequency function in the model set, that is closest to the true one, in a quadratic frequency norm with weighting function

$$Q(\omega) = \Phi_u(\omega)|L(e^{i\omega})|^2/|H_*(e^{i\omega})|^2. \tag{58.69}$$

This shows that the fit can be affected by the choice of prefilter L, the input spectrum Φ_u, and the noise model H_*.

58.3.2 Linear, Ready-Made Models

Sometimes systems or subsystems cannot be modeled based on physical insights, because the function of the system or its construction is unknown or it would be too complicated to sort out the physical relationships. Then standard models can be used, which handle a wide range of different system dynamics. Linear systems constitute the most common class of such standard models. From a modeling point of view, these models serve as *ready-made models*: tell us the size (model order), and it should be possible to find something that fits (to data).

A Family of Transfer Function Models

A very natural approach is to describe G and H in Equation 58.62 as rational transfer functions in the shift (delay) operator with unknown numerator and denominator polynomials:

$$\begin{aligned} G(q,\theta) &= \frac{B(q)}{F(q)} \qquad (58.70)\\ &= \frac{b_1 q^{-nk} + b_2 q^{-nk-1} + \cdots + b_{nb} q^{-nk-nb+1}}{1 + f_1 q^{-1} + \cdots + f_{nf} q^{-nf}}. \end{aligned}$$

Then,

$$\eta(t) = G(q,\theta)u(t) \tag{58.71}$$

is a shorthand notation for the relationship

$$\begin{aligned} \eta(t) &+ f_1\eta(t-1) + \cdots + f_{nf}\eta(t-nf) \qquad (58.72)\\ &= b_1 u(t-nk) + \cdots + b_{nb}(t-(nb+nk-1)). \end{aligned}$$

There is also a time delay of nk samples. We assume, for simplicity, that the sampling interval T is one time unit.

In the same way, the disturbance transfer function is

$$\begin{aligned} H(q,\theta) &= \frac{C(q)}{D(q)} \qquad (58.73)\\ &= \frac{1 + c_1 q^{-1} + \cdots + c_{nc} q^{-nc}}{1 + d_1 q^{-1} + \cdots + d_{nd} q^{-nd}} \end{aligned}$$

The parameter vector θ contains the coefficients b_i, c_i, d_i, and f_i of the transfer functions. This ready-made model is described by five structural parameters: nb, nc, nd, nf, and nk. When these have been chosen, it remains to adjust the parameters b_i, c_i, d_i, and f_i to data with the methods of Section 58.2. The ready-made model Equations 58.71 – 58.74 gives

$$y(t) = \frac{B(q)}{F(q)}u(t) + \frac{C(q)}{D(q)}e(t) \tag{58.74}$$

named the *Box–Jenkins (BJ) model*, after statisticians G. E. P. Box and G. M. Jenkins.

An important special case occurs when the properties of the disturbance signals are not modeled, and the noise model $H(q)$ is chosen to be $H(q) \equiv 1$, that is, $nc = nd = 0$. This special case is known as an *output error (OE) model* since the noise source $e(t)$ will then be the difference (error) between the actual output and the noise-free output:

$$y(t) = \frac{B(q)}{F(q)}u(t) + e(t). \tag{58.75}$$

A common variant is using the same denominator for G and H:

$$F(q) = D(q) = A(q) = 1 + a_1 q^{-1} + \cdots + a_{na} a^{-na}. \tag{58.76}$$

Multiplying both sides of Equation 58.74 by $A(q)$, gives

$$A(q)y(t) = B(q)u(t) + C(q)e(t). \tag{58.77}$$

This ready-made model is known as the *ARMAX model.* The name derives from the fact that $A(q)y(t)$ represents an AutoRegression, $C(q)e(t)$ is a Moving Average of white noise, and $B(q)u(t)$ represents an eXtra input (or with econometric terminology, an eXogenous variable).

The physical difference between ARMAX and BJ models is that the noise and input are subjected to the same dynamics (same poles) in the ARMAX case. This is reasonable if the dominating disturbances enter early in the process (together with the input). Consider for example an airplane where the disturbances from wind gusts give rise to the same type of forces on the airplane as the deflections of the control surfaces.

Finally, we have the special case of Equation 58.77 that, when $C(q) \equiv 1$, that is, $nc = 0$,

$$A(q)y(t) = B(q)u(t) + e(t) \tag{58.78}$$

which, with the same terminology, is called an *ARX model* and which we discussed at length in Section 58.1.3.

Figure 58.1 shows the most common model structures.

To use these ready-made models, decide on the orders na, nb, nc, nd, nf, and nk, and let the computer pick the best model in the class defined. The model obtained is then scrutinized, and it might be found that other orders must also be tested.

A relevant question is how to use the freedom that the different model structures give. Each of the BJ, OE, ARMAX, and ARX structures offer their own advantages, and we will discuss them in Section 58.7.2.

Prediction

Starting with model Equation 58.74, it is possible to predict what the output $y(t)$ will be, based on measurements of $u(s)$, $y(s)$ $s \le t-1$, using the general formula Equation 58.65. It is easiest to calculate the prediction for the OE-case, $H(q,\theta) \equiv 1$, when we obtain the model

$$y(t) = G(q,\theta)u(t) + e(t)$$

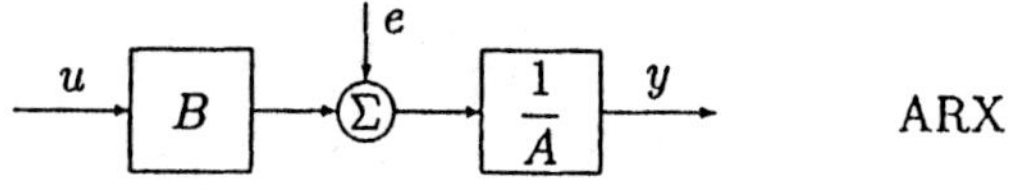

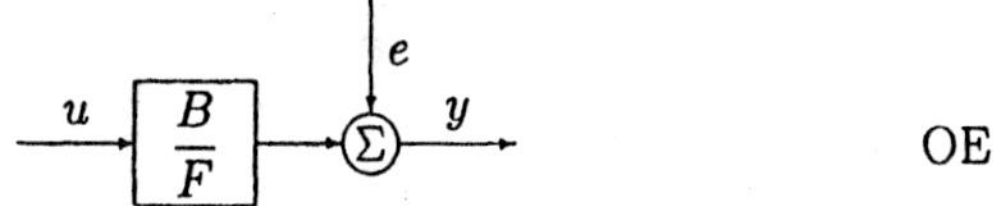

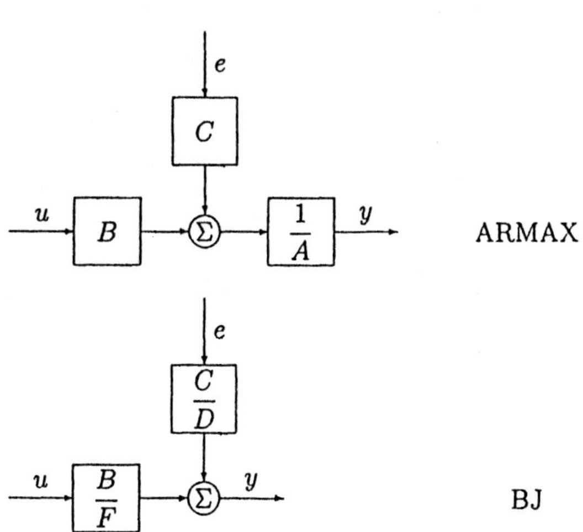

Figure 58.1 Model structures.

with the natural prediction $(1 - H^{-1} = 0)$

$$\hat{y}(t|\theta) = G(q, \theta)u(t). \tag{58.79}$$

From the ARX case Equation 58.78,

$$\begin{aligned} y(t) &= -a_1 y(t-1) - \cdots - a_{na} y(t-na) \\ &\quad + b_1 u(t-nk) + \cdots \\ &\quad + b_{nb} u(t-nk-nb+1) + e(t), \end{aligned} \tag{58.80}$$

and the prediction (delete $e(t)$!)

$$\begin{aligned} \hat{y}(t|\theta) = & -a_1 y(t-1) - \cdots - a_{na} y(t-na) \\ & + b_1 u(t-nk) + \cdots + b_{nb} u(t-nk-nb+1). \end{aligned} \tag{58.81}$$

Note the difference between Equations 58.79 and 58.81. In the OE model the prediction is based entirely on the input $\{u(t)\}$, whereas the ARX model also uses old values of the output.

Linear Regression

Both tailor-made and ready-made models describe how the predicted value of $y(t)$ depends on old values of y and u and on the parameters θ. We denote this prediction by

$$\hat{y}(t|\theta)$$

(see Equation 58.65). In general, this can be a rather complicated function of θ. The estimation work is considerably easier if the prediction is a linear function of θ,

$$\hat{y}(t|\theta) = \theta^T \varphi(t). \tag{58.82}$$

Here θ is a column vector that contains the unknown parameters, while $\varphi(t)$ is a column vector formed by old inputs and outputs. Such a model structure is called a *linear regression*. We discussed such models in Section 58.1.3 and noted that the ARX model Equation 58.78 is one common model of the linear regression type. Linear regression models can also be obtained in several other ways (see Example 58.2).

58.4 Special Estimation Techniques for Linear Black-Box Models

An important feature of a linear, time-invariant system is that it is entirely characterized by its *impulse response*. If we know the system's response to an impulse, we will also know its response to any input. Equivalently, we could study the *frequency response*, which is the Fourier transform of the impulse response.

In this section we shall consider estimation methods for linear systems that do not use particular model parameterizations. First, in Section 58.4.1, we shall consider direct methods to determine the impulse response and the frequency response by applying the definitions of these concepts.

In Section 58.4.2 methods for estimating the impulse response by correlation analysis will be described, and in Section 58.4.3, spectral analysis for frequency function estimation will be discussed. Finally, in Section 58.4.4 a recent method of estimating general linear systems (of given order, but unspecified structure) will be described.

58.4.1 Transient and Frequency Analysis

Transient Analysis

The first step in modeling is to decide which quantities and variables are important to describe what happens in the system. A simple and common experiment that shows how and in what time span various variables affect each other is called *step-response analysis* or *transient analysis*. In these experiments the inputs are varied (typically one at a time) as a step: $u(t) = u_0,\ t < t_0$; $u(t) = u_1, t \geq t_0$. The other measurable variables in the system are recorded during this time. We thus study the *step response* of the system. An alternative would be to study the impulse response of the system by letting the input be a pulse of short duration. From such measurements, the following information can be found:

1. The variables affected by the input in question. This makes it easier to draw block diagrams for the system and to decide which influences can be neglected.
2. The time constants of the system. This also allows us to decide which relationships in the model can be described as static (that is, they have significantly faster time constants than the time scale we are working with).
3. The characteristic (oscillatory, poorly damped, monotone, and the like) of the step responses, as well as the levels of static gains. Such information is

useful when studying the behavior of the final model in simulation. Good agreement with the measured step responses should give confidence in the model.

Frequency Analysis

If a linear system has the transfer function $G(q)$ and the input is

$$u(t) = u_0 \cos \omega kT, \quad (k-1)T \leq t \leq kT, \tag{58.83}$$

then the output, after possible transients have faded away, will be

$$y(t) = y_0 \cos(\omega t + \varphi), \text{ for } t = T, 2T, 3T, \ldots \tag{58.84}$$

where

$$y_0 = |G(e^{i\omega T})| \cdot u_0 \tag{58.85}$$
$$\varphi = \arg G(e^{i\omega T}) \tag{58.86}$$

If the system is driven by the input Equation 58.83 for a certain u_0 and ω_1 and we measure y_0 and φ from the output signal, it is possible to determine the complex number $G(e^{i\omega_1 T})$ using Equations 58.85 – 58.86. By repeating this procedure for a number of different ω, we can get a good estimate of the frequency function $G(e^{i\omega T})$. This method is called *frequency analysis*. Sometimes it is possible to see or measure u_0, y_0, and φ directly from graphs of the input and output signals. Most of the time, however, noise and irregularities will make it difficult to determine φ directly. A suitable procedure is then to correlate the output with $\cos \omega t$ and $\sin \omega t$.

58.4.2 Estimating Impulse Responses by Correlation Analysis

It is not necessary to use an impulse as input to estimate the impulse response of a system directly. That can also be done by correlation techniques. To explain how these work, let us first define correlation functions.

The *cross-covariance function* between two signals y and u is defined as the covariance between the random variables $y(t)$ and $u(t-\tau)$, as a function of the time difference τ:

$$R_{yu}(\tau) = E[y(t) - Ey(t)][u(t-\tau) - Eu(t-\tau)] \tag{58.87}$$

It is implicitly assumed here that the indicated expectation does not depend on absolute time t, that is, the signals are (weakly) *stationary.*

As in the case of Equation 58.32, expectation can be replaced by sample means,

$$m_y = \lim_{N\to\infty} \frac{1}{N}\sum_{t=1}^{N} y(t), \text{ and} \tag{58.88}$$

$$R_{yu}(\tau) = \lim_{N\to\infty} \frac{1}{N}\sum_{t=1}^{N} (y(t) - m_y)(u(t-\tau) - m_u). \tag{58.89}$$

As soon as we use the term covariance function, there is always an implied assumption that the involved signals are such that either Equation 58.87 or 58.89 is well defined.

The cross-correlation signal between a signal u and itself, i.e., $R_{uu}(\tau) = R_u(\tau)$, is called the (auto) *covariance function* of the signal.

We shall say that two signals are *uncorrelated* if their cross-covariance function is identically equal to zero.

Let us consider the general linear model Equation 58.56, and assume that the input u and the noise v are uncorrelated:

$$y(t) = \sum_{k=0}^{\infty} g_k u(t-k) + v(t). \tag{58.90}$$

The cross-covariance function between u and y is then

$$\begin{aligned} R_{yu}(\tau) &= Ey(t)u(t-\tau) = \sum_{k=0}^{\infty} g_k Eu(t-k)u(t-\tau) \\ &+ Ev(t)u(t-\tau) = \sum_{k=0}^{\infty} g_k R_u(\tau - k). \end{aligned} \tag{58.91}$$

If the input is white noise,

$$R_u(\tau) = \begin{cases} \lambda, & \tau = 0, \\ 0, & \tau \neq 0, \end{cases}$$

$$R_{yu}(\tau) = \lambda g_\tau \tag{58.92}$$

The cross-covariance function $R_{yu}(\tau)$ will thus be proportional to the impulse response. Of course, this function is not known, but it can be estimated from observed inputs and outputs as the corresponding sample mean:

$$\hat{R}_{yu}^N(\tau) = \frac{1}{N}\sum_{t=1}^{N} y(t)u(t-\tau). \tag{58.93}$$

In this way we also obtain an estimate of the impulse response:

$$\hat{g}_\tau^N = \frac{1}{\lambda}\hat{R}_{yu}^N(\tau). \tag{58.94}$$

If we cannot choose the input ourselves, and it is nonwhite, we can estimate its covariance function as $\hat{R}_u^N(\tau)$, analogous to Equation 58.93, and then solve for g_k from Equation 58.91 where R_u and R_{uy} have been replaced by the corresponding estimates. However, a better and more common way is the following: First note that, if both input and output are filtered through the same filter

$$y_F(t) = L(q)y(t) \quad u_F(t) = L(q)u(t) \tag{58.95}$$

then the filtered signals will be related by the same impulse response as in Equation 58.90:

$$y_F(t) = \sum_{k=1}^{\infty} g_k u_F(t-k) + v_F(t). \tag{58.96}$$

Now, for any given input $u(t)$ in the process, we can choose the filter L so that the signal $\{u_F(t)\}$ will be as white as possible. Such a filter is called a *whitening filter*. It is computed by describing $u(t)$ as an AR process (This is an ARX model without an input, cf Section 58.1.3): $A(q)u(t) = e(t)$. The polynomial $A(q) = L(q)$ can then be estimated using the least squares method. (see Section 58.1.3). We can now apply the estimate Equation 58.94 to the filtered signals.

58.4.3 Estimating the Frequency Response by Spectral Analysis

Definitions

The *cross spectrum* between two (stationary) signals $u(t)$ and $y(t)$ is defined as the Fourier transform of their cross-covariance function, provided this exists,

$$\Phi_{yu}(\omega) = \sum_{\tau=-\infty}^{\infty} R_{yu}(\tau)e^{-i\omega\tau} \tag{58.97}$$

where $R_{yu}(\tau)$ is defined by Equation 58.87 or Equation 58.89. The (auto) *spectrum* $\Phi_u(\omega)$ of a signal u is defined as $\Phi_{uu}(\omega)$, i.e., its cross spectrum with itself.

The spectrum describes the frequency content of the signal. The connection to more explicit Fourier techniques is evident from the relationship

$$\Phi_u(\omega) = \lim_{N\to\infty} \frac{1}{N}|U_N(\omega)|^2 \tag{58.98}$$

("weak limit") where U_N is the discrete time Fourier transform

$$U_N(\omega) = \sum_{t=1}^{N} u(t)e^{i\omega t}. \tag{58.99}$$

Equation 58.98 is shown, in [16].

Consider now the general linear model Equation 58.56,

$$y(t) = G(q)u(t) + v(t). \tag{58.100}$$

It is straightforward to show that the relationships between the spectra and cross spectra of y and u (provided u and v are uncorrelated) are

$$\Phi_{yu}(\omega) = G(e^{i\omega})\Phi_u(\omega) \tag{58.101}$$

$$\Phi_y(\omega) = |G(e^{i\omega})|^2\Phi_u(\omega) + \Phi_v(\omega). \tag{58.102}$$

The transfer function $G(e^{i\omega})$ and the noise spectrum $\phi_v(\omega)$ can be estimated with these expressions, if only we have a method to estimate cross spectra.

Estimation of Spectra

The spectrum is defined as the Fourier transform of the correlation function. A natural idea would then be to take the transform of the estimate $\hat{R}^N_{yu}(\tau)$ in Equation 58.93 but that will not work in most cases, because the estimate $\hat{R}^N_{yu}(\tau)$ based on only a few observations is not reliable for large τ. These "bad" estimates are mixed with good ones in the Fourier transform, creating an overall bad estimate. It is better to introduce a weighting, so that correlation estimates for large lags τ carry a smaller weights:

$$\hat{\Phi}^N_{yu}(\omega) = \sum_{\ell=-\gamma}^{\gamma} \hat{R}^N_{yu}(\ell) \cdot w_\gamma(\ell)e^{-i\ell\omega}. \tag{58.103}$$

This spectral estimation method is known as the *Blackman–Tukey approach*. Here $w_\gamma(\ell)$ is a window function that decreases with $|\tau|$. This function controls the trade-off between *frequency resolution* and *variance of the estimate*. A function that gives significant weights to the correlation at large lags will be able to provide finer frequency details (a longer time span is covered). At the same time it must use "bad" estimates, so that the statistical quality (the variance) is poorer. We shall return to this trade-off in a moment. How should we choose the shape of the window function $w_\gamma(\ell)$? There is no optimal solution to this problem, but the most common window used in spectral analysis is the *Hamming window*:

$$\begin{aligned} w_\gamma(k) &= \tfrac{1}{2}(1+\cos\tfrac{\pi k}{\gamma}) & |k| &< \gamma, \\ w_\gamma(k) &= 0, & |k| &\geq \gamma. \end{aligned} \tag{58.104}$$

From the spectral estimates Φ_u, Φ_y, and Φ_{yu} obtained in this way, we can now use Equation 58.101 to obtain a natural estimate of the frequency function $G(e^{i\omega})$:

$$\hat{G}_N(e^{i\omega}) = \frac{\hat{\Phi}^N_{yu}(\omega)}{\hat{\Phi}^N_u(\omega)}. \tag{58.105}$$

Furthermore, the disturbance spectrum can be estimated from Equation 58.102 as

$$\hat{\Phi}^N_v(\omega) = \hat{\Phi}^N_y(\omega) - \frac{|\hat{\Phi}^N_{yu}(\omega)|^2}{\hat{\Phi}^N_u(\omega)}. \tag{58.106}$$

To compute these estimates, perform the following steps:

Algorithm SPA (58.107)

1. Collect data $y(k), u(k)$ $k = 1, \ldots, N$.
2. Subtract the corresponding sample means form the data. This will avoid bad estimates at very low frequencies.
3. Choose the width of the lag window $w_\gamma(k)$.
4. Compute $\hat{R}^N_y(k)$, $\hat{R}^N_u(k)$, and $\hat{R}^N_{yu}(k)$ for $|k| \leq \gamma$ according to Equation 58.93.
5. Form the spectral estimates $\hat{\Phi}^N_y(\omega)$, $\hat{\Phi}^N_u(\omega)$, and $\hat{\Phi}^N_{yu}(\omega)$ according to Equation 58.103 and analogous expressions.
6. Form Equation 58.105 and possibly also Equation 58.106.

The user only has to choose γ. A good value for systems without sharp resonances is $\gamma = 20$ to 30. Larger values of γ may be required for systems with narrow resonances.

Quality of the Estimates

The estimates $\hat{G}_N$ and $\hat{\Phi}_v^N$ are formed entirely from estimates of spectra and cross spectra. Their properties will, therefore, be inherited from the properties of the spectral estimates. For the Hamming window with width γ, the frequency resolution will be about

$$\frac{\pi}{\gamma\sqrt{2}} \quad \text{radians/time unit.} \tag{58.108}$$

This means that details in the true frequency function, that are finer than this expression, will be smeared out in the estimate. The estimate's variances satisfy

$$\text{Var } \hat{G}_N(i\omega) \approx 0.7 \cdot \tfrac{\gamma}{N} \cdot \tfrac{\Phi_v(\omega)}{\Phi_u(\omega)} \tag{58.109}$$

and

$$\text{Var } \hat{\Phi}_v^N(\omega) \approx 0.7 \cdot \tfrac{\gamma}{N} \cdot \Phi_v^2(\omega). \tag{58.110}$$

["Variance" here refers to taking expectation over the noise sequence $v(t)$.] The relative variance in Equation 58.109 typically increases dramatically as ω tends to the Nyquist frequency because $|G(\imath\omega)|$ typically decays rapidly, and the noise-to-signal ratio $\Phi_v(\omega)/\Phi_u(\omega)$ has a tendency to increase as ω increases. In a Bode diagram the estimates will show considerable fluctuations at high frequencies. Moreover, the constant frequency resolution Equation 58.108 will look thinner and thinner at higher frequencies in a Bode diagram due to the logarithmic frequency scale.

See [16] for a more detailed discussion.

Choice of Window Size

The choice of γ is a pure trade-off between frequency resolution and variance (variability). For a spectrum with narrow resonance peaks, it is necessary to choose a large value of γ and accept a higher variance. For a flatter spectrum, smaller values of γ will do well. In practice a number of different values of γ are tried out. We start with a small value of γ and increase it successively until an estimate is found that balances the trade-off between frequency resolution (true details) and variance (random fluctuations). A typical value for spectra without narrow resonances is γ= 20–30.

58.4.4 Subspace Estimation Techniques for State-Space Models

A linear system can always be represented in state space form:

$$\begin{aligned} x(t+1) &= Ax(t) + Bu(t) + w(t) \text{ and} \\ y(t) &= Cx(t) + Du(t) + e(t). \end{aligned} \tag{58.111}$$

We assume that we have no insight into the particular structure, and we estimate any matrices A, B, C, and D that describe the input-output behavior of the system. This is not without problems, because there are an infinite number of matrices that describe the same system (the similarity transforms). The coordinate basis of the state-space realization thus needs to be fixed.

Let us for a moment assume that not only are u and y measured, but also the sequence of state vectors x. This would fix the state-space realization coordinate basis. Now, with known u, y and x, the model (58.111) becomes a linear regression: the unknown parameters, all of the matrix entries in all the matrices, mix with measured signals in linear combinations. To see this clearly, let

$$\begin{aligned} Y(t) &= \begin{pmatrix} x(t+1) \\ y(t) \end{pmatrix}, \\ \Theta &= \begin{pmatrix} A & B \\ C & D \end{pmatrix}, \\ \Phi(t) &= \begin{pmatrix} x(t) \\ u(t) \end{pmatrix}, \\ E(t) &= \begin{pmatrix} w(t) \\ e(t) \end{pmatrix} \end{aligned}$$

Then, Equation 58.111 becomes

$$Y(t) = \Theta\Phi(t) + E(t). \tag{58.112}$$

From this equation all of the matrix elements in Θ can be estimated by the simple least-squares method, described in Section 58.1.3. The covariance matrix for $E(t)$ can also be estimated as the sample sum of the model residuals. That will give the covariance matrices for w and e, as well as the cross-covariance matrix between w and e. These matrices will allow us to compute the Kalman filter for Equation 58.111. All of the above holds without changes for multivariable systems, i.e., when the output and input signals are vectors.

The only remaining problem is where to get the state vector sequence x. It has long been known ([22] and [2]) that all state vectors $x(t)$, that can be reconstructed from input-output data, are linear combinations of the components of the n k-step ahead output predictors

$$\hat{y}(t+k|t), \quad k = \{1, 2, \ldots, n\} \tag{58.113}$$

where n is the model order (the dimension of x). See also Appendix 4.A in [14]. We could then form these predictors, and select a basis among their components:

$$x(t) = L \begin{pmatrix} \hat{y}(t+1|t) \\ \vdots \\ \hat{y}(t+n|t) \end{pmatrix}. \tag{58.114}$$

The choice of L will determine the basis for the state-space realization, and is done so that it is well conditioned. The predictor $\hat{y}(t+k|t)$ is a linear function of $u(s), y(s)$, $1 \leq s \leq t$ and can efficiently be determined by linear projections directly on the input-output data. (One complication is that $u(t+1), \ldots, u(t+k)$ should not be predicted, even if they affect $y(t+k)$.)

What we have described now is the *subspace projection* approach to estimating the matrices of the state-space model Equation 58.111, including the basis for the representation and the noise covariance matrices. For a number of variants of this approach, see [19] and [12].

The approach gives very useful algorithms for model estimation and is particularly well suited for multivariable systems.

The algorithms also allow, numerically, very reliable implementations. At present, the asymptotic properties of the methods are not fully investigated, and the general results quoted in Section 58.2.2 are not directly applicable. Experience has shown, however, that confidence intervals, computed according to the general asymptotic theory, are good approximations. One may also use the estimates obtained by a subspace method as initial conditions for minimizing the prediction error criterion Equation 58.24.

58.5 Physically Parameterized Models

So far we have treated the parameters θ only as vehicles to give reasonable flexibility to the transfer functions in the general linear model Equation 58.62. This model can also be arrived at from other considerations.

Consider a continuous time state space model

$$\dot{x}(t) = A(\theta)x(t) + B(\theta)u(t) \qquad (58.115a)$$
$$y(t) = C(\theta)x(t) + v(t). \qquad (58.115b)$$

Here $x(t)$ is the state vector consisting of physical variables (such as positions and velocities). The state-space matrices A, B and C are parameterized by the parameter vector θ, reflecting physical insight into the process. The parameters could be physical constants (resistance, heat transfer coefficients, aerodynamical derivatives) whose values are not known. They could also reflect other types of insights into the system's properties.

EXAMPLE 58.3: An electric motor

Consider an electric motor with the input u the applied voltage and the output y the angular position of the motor shaft.

A first, but reasonable, approximation of the motor's dynamics is as a first-order system from voltage to angular velocity, followed by an integrator,

$$G(s) = \frac{b}{s(s+a)}.$$

If we select the state variables

$$x(t) = \begin{pmatrix} y(t) \\ \dot{y}(t) \end{pmatrix},$$

the state space form is

$$\begin{aligned} \dot{x} &= \begin{pmatrix} 0 & 1 \\ 0 & -a \end{pmatrix} x + \begin{pmatrix} 0 \\ b \end{pmatrix} u \\ y &= \begin{pmatrix} 1 & 0 \end{pmatrix} x + v \end{aligned} \qquad (58.116)$$

where v denotes disturbances and noise. In this case,

$$\begin{aligned} \theta &= \begin{pmatrix} a \\ b \end{pmatrix}, \\ A(\theta) &= \begin{pmatrix} 0 & 1 \\ 0 & -a \end{pmatrix}, \quad B(\theta) = \begin{pmatrix} 0 \\ b \end{pmatrix}, \\ \text{and } C &= \begin{pmatrix} 1 & 0 \end{pmatrix}. \end{aligned} \qquad (58.117)$$

The parameterization reflects our insight that the system contains an integration, but is in this case not directly derived from detailed physical modeling. Basic physical laws would have shown how θ depends on physical constants, such as resistance of the wiring, amount of inertia, friction coefficients and magnetic field constants.

Now, how do we fit a continuous-time model Equation 58.115 to sampled data? If the input $u(t)$ has been piecewise constant over the sampling interval

$$u(t) = u(kT), \qquad kT \leq t < (k+1)T,$$

then the states, inputs, and outputs at the sampling instants will be represented by the discrete-time model

$$\begin{aligned} x((k+1)T) &= \bar{A}(\theta)x(kT) + \bar{B}(\theta)u(kT), \\ y(kT) &= C(\theta)x(kT) + v(kT), \end{aligned} \qquad (58.118)$$

where

$$\bar{A}(\theta) = e^{A(\theta)T}, \quad \bar{B}(\theta) = \int_0^T e^{A(\theta)\tau} B(\theta) d\tau. \qquad (58.119)$$

This follows from solving Equations 58.115 and 58.115 over one sampling period. We could also further model the added noise term $v(kT)$ and represent the system in the innovations form

$$\begin{aligned} \bar{x}((k+1)T) &= \bar{A}(\theta)\bar{x}(kT) + \bar{B}(\theta)u(kT) \\ &\quad + \bar{K}(\theta)e(kT), \\ y(kT) &= C(\theta)\bar{x}(kT) + e(kT), \end{aligned} \qquad (58.120)$$

where $\{e(kT)\}$ is white noise. The step from Equations 58.118 to 58.120 is a standard Kalman filter step: $\bar{x}$ will be the one-step ahead predicted Kalman states. A pragmatic view is as follows: In Equation 58.118 the term $v(kT)$ may not be white noise. If it is colored, we may separate out that part of $v(kT)$ that cannot be predicted from past values. Denote this part by $e(kT)$: it will be the *innovation*. The other part of $v(kT)$, that can be predicted, can then be described as a combination of earlier innovations, $e(\ell T)$ $\ell < k$. Its effect on $y(kT)$ can be described via the states, by changing them from x to $\bar{x}$, where $\bar{x}$ contains additional states associated with getting $v(kT)$ from $e(\ell T)$, $k \leq \ell$.

Now Equation 58.120 can be written in input – output form (let $T = 1$) as

$$y(t) = G(q,\theta)u(t) + H(q,\theta)e(t) \qquad (58.121)$$

with

$$\begin{aligned} G(q,\theta) &= C(\theta)(qI - \bar{A}(\theta))^{-1}\bar{B}(\theta) \\ H(q,\theta) &= I + C(\theta)(qI - \bar{A}(\theta))^{-1}\bar{K}(\theta). \end{aligned} \qquad (58.122)$$

We thus return to the basic linear model Equation 58.62. The parameterization of G and H in terms of θ is however more complicated than those discussed in Section 58.3.2.

The general estimation techniques, model properties (including the frequency domain characterization Equation 58.68), algorithms, etc., apply exactly as described in Section 58.2.

From these examples, it is also quite clear that nonlinear models with unknown parameters can be approached in the same way. We then arrive at a structure,

$$\begin{aligned} \dot{x}(t) &= f(x(t), u(t), \theta) \\ y(t) &= h(x(t), u(t), \theta) + v(t) \end{aligned} \qquad (58.123)$$

In this model, all noise effects are collected as additive output disturbances $v(t)$ which is a restriction, but also a simplification. If we define $\hat{y}(t|\theta)$ as the simulated output response to Equation 58.123, for a given input, ignoring the noise $v(t)$, everything that was said in Section 58.2 about parameter estimation, model properties, etc. still applies.

58.6 Nonlinear Black-Box Models

In this section we shall describe the basic ideas behind model structures that can cover any nonlinear mapping from past data to the predicted value of $y(t)$. Recall that we defined a general model structure as a parameterized mapping in Equation 58.19:

$$\hat{y}(t|\theta) = g(\theta, Z^{t-1}) \tag{58.124}$$

We shall consequently allow quite general nonlinear mappings g. This section will deal with some general principles for constructing such mappings, and will cover Artificial Neural Networks as a special case. See [25] and [9] for recent and more comprehensive surveys.

58.6.1 Nonlinear Black-Box Structures

The model structure family Equation 58.124 is really too general, and it is useful to write g as a concatenation of two mappings: one that takes the increasing number of past observations Z^{t-1} and maps them into a finite dimensional vector $\varphi(t)$ of fixed dimension and one that takes this vector to the space of the outputs,

$$\hat{y}(t|\theta) = g(\theta, Z^{t-1}) = g(\varphi(t), \theta), \tag{58.125}$$

where

$$\varphi(t) = \varphi(Z^{t-1}). \tag{58.126}$$

Let the dimension of φ be d. As before, we shall call this vector the *regression vector* and its components will be referred to as the *regressors*. We also allow the more general case that the formation of the regressors is itself parameterized:

$$\varphi(t) = \varphi(Z^{t-1}, \eta), \tag{58.127}$$

for short, $\varphi(t, \eta)$. For simplicity, the extra argument η will, however, be used explicitly only when essential.

The choice of the nonlinear mapping in Equation 58.124 has thus, been reduced to two partial problems for dynamical systems:

1. How to choose the nonlinear mapping $g(\varphi)$ from the regressor space to the output space (i.e., from R^d to R^p).
2. How to choose the regressors $\varphi(t)$ from past inputs and outputs.

The second problem is the same for all dynamical systems. Most regression vectors are chosen to let them contain past inputs and outputs, and also past predicted/simulated outputs. The regression vector will thus be of the character Equation 58.4. We now turn to the first problem.

58.6.2 Nonlinear Mappings: Possibilities

Function Expansions and Basis Functions

The nonlinear mapping

$$g(\varphi, \theta) \tag{58.128}$$

goes from R^d to R^p for any θ. At this point it does not matter how the regression vector φ is constructed. It is just a vector that lives in R^d.

It is natural to think of the parameterized function family as function expansions,

$$g(\varphi, \theta) = \sum \theta(k) g_k(\varphi), \tag{58.129}$$

where g_k are the *basis functions* and the coefficients $\theta(k)$ are the "coordinates" of g in the chosen basis.

Now, the only remaining question is: How to choose the basis functions g_k? Depending on the support of g_k (i.e., the area in R^d for which $g_k(\varphi)$ is (practically) nonzero) we shall distinguish between three types of basis functions:

- Global basis functions,
- Semiglobal or ridge-type basis functions,
- Local basis functions

A typical and classical global basis function expansion is the Taylor series, or polynomial expansion, where g_k contains multinomials in the components of φ of total degree k. Fourier series are also examples. We shall not, however, discuss global basis functions here any further. Experience has indicated that they are inferior to semiglobal and local functions in practical applications.

Local Basis Functions

Local basis functions have their support only in some neighborhood of a given point. Think (in the case of p=1) of the indicator function for the unit cube,

$$\kappa(\varphi) = 1 \text{ if } |\varphi_k| \leq 1 \ \forall k \text{, and 0 otherwise.} \tag{58.130}$$

By scaling the cube and placing it at different locations we obtain the functions

$$g_k(\varphi) = \kappa[\alpha_k * (\varphi - \beta_k)]. \tag{58.131}$$

By allowing α to be a vector of the same dimension as φ and interpreting the multiplication $*$ as componentwise multiplication (like ".*" in MATLAB) we may also reshape the cube to be any parallelepiped. The parameters α are *scaling* or *dilation* parameters and β determines *location* or *translation*. For notational convenience,

$$g_k(\varphi) = \kappa(\alpha_k * (\varphi - \beta_k)) = \kappa(\rho_k \cdot \varphi) \tag{58.132}$$

where

$$\rho_k = [\alpha_k, \alpha_k * \beta_k].$$

In the last equality, with some abuse of notation, we expanded the regression vector φ to contain some "1"'s. This stresses the point that the argument of the basic function κ is bilinear in the scale and location parameters ρ_k and in the regression vector φ indicated by the notation $\rho_k \cdot \varphi$.

This choice of g_k in Equation 58.129 gives functions that are piecewise constant over areas in R^d that can be chosen arbitrarily small by proper choice of the scaling parameters. It should be fairly obvious that such functions g_k can approximate any reasonable function arbitrarily well.

Now it is also reasonable that the same will be true for any other localized function, such as the Gaussian bell function,

$$\kappa(\varphi) = e^{-|\varphi|^2}. \tag{58.133}$$

Ridge-Type Basis Functions

A useful alternative is to let the basis functions be local in one direction of the φ-space and global in the others. This is achieved quite analogously to Equation 58.131 as follows: Let $\sigma(x)$ be a local function from R to R. Then form

$$g_k(\varphi) = \sigma(\alpha_k^T(\varphi - \beta_k)) = \sigma(\alpha_k^T\varphi + \gamma_k) = \sigma(\rho_k \cdot \varphi) \tag{58.134}$$

where the scalar $\gamma_k = -\alpha_k^T\beta_k$, and

$$\rho_k = [\alpha_k, \gamma_k].$$

Note the difference with Equation 58.131! The scalar product $\alpha_k^T\varphi$ is constant in the subspace of R^d that is perpendicular to the scaling vector α_k. Hence the function $g_k(\varphi)$ varies like σ in a direction parallel to α_k and is constant across this direction. This leads to the term *semiglobal* or *ridge-type* for this choice of functions.

As in Equation 58.131 we expanded, in the last equality in Equation 58.134, the vector φ with the value "1", again just to emphasize that the argument of the fundamental basis function σ is bilinear in ρ and φ.

Connection to "Named Structures"

Here we briefly review some popular structures. Other structures related to interpolation techniques are discussed in [9] and [25].

Wavelets The local approach corresponding to Equations 58.129 and 58.131 has direct connections to wavelet networks and wavelet transforms. The exact relationships are discussed in [25]. Via the dilation parameters in ρ_k, we can work with different scales simultaneously to pick up both local and not-so-local variations. With appropriate translations and dilations of a single suitably chosen function κ (the "mother wavelet"), we can make the expansion (58.129) orthonormal. This is discussed extensively in [9].

Wavelet and radial basis networks The choice of Equation 58.133 without any orthogonalization is found in wavelet networks [28] and radial basis neural networks [20].

Neural networks The ridge choice Equation 58.134 with

$$\sigma(x) = \frac{1}{1 + e^{-x}}$$

gives a much used neural network structure, viz., the *one hidden layer feed-forward sigmoidal net.*

Hinging hyperplanes Instead of using the sigmoid σ function, if we choose "V-shaped" functions (in the form of a higher-dimensional "open book"), Breiman's *hinging hyperplane* structure is obtained [5]. Hinging hyperplanes model structures have the form

$$\begin{aligned} g(x) &= \max\{\beta^+x + \gamma^+ , \beta^-x + \gamma^-\} \quad \text{or} \\ g(x) &= \min\{\beta^+x + \gamma^+ , \beta^-x + \gamma^-\} . \end{aligned}$$

It can be written differently as

$$\begin{aligned} g(x) &= \frac{1}{2}[(\beta^+ + \beta^-)x + \gamma^+ + \gamma^-] \\ &\quad \pm \frac{1}{2}|(\beta^+ - \beta^-)x + \gamma^+ - \gamma^-| . \end{aligned}$$

A hinge is the superposition of a linear map and a semi global function. Therefore, we consider *hinge* functions as semiglobal or ridge-type, though not in strict accordance with our definition.

Nearest neighbors or interpolation By selecting κ as in Equation 58.130 and the location and scale vector ρ_k in the structure Equation 58.131, so that exactly one observation falls into each "cube", the nearest neighbor model is obtained: Just load the input-output record into a table, and, for a given φ, pick the pair $(\widehat{y}, \widehat{\varphi})$ for $\widehat{\varphi}$ closest to the given φ; $\widehat{y}$ is the desired output estimate. If one replaces Equation 58.130 by a smoother function and allow some overlapping of the basis functions, we get interpolation type techniques such as kernel estimators.

Fuzzy Models So-called *fuzzy models* based on fuzzy set membership belong to the model structures of the class Equation 58.129. The basis functions g_k then are constructed from the fuzzy set membership functions and the inference rules. The exact relationship is described in [25].

58.6.3 Estimating Nonlinear Black-Box Models

The model structure is determined by the following choices:

- The regression vector (typically built up from past inputs and outputs),
- The basic function κ (local) or σ (ridge), and
- The number of elements (nodes) in the expansion Equation 58.129.

Once these choices have been made, $\hat{y}(t|\theta) = g(\varphi(t), \theta)$ is a well defined function of past data and the parameters θ. The parameters are made up of coordinates in the expansion Equation 58.129 and from location and scale parameters in the different basis functions.

All of the algorithms and analytical results of Section 58.2 can thus be applied. For Neural Network applications these are also the typical estimation algorithms used, often complemented with *regularization*, which means that a term is added to the criterion Equation 58.24, that penalizes the norm of θ. This will reduce the variance of the model, in that "spurious" parameters are not allowed to take on large, and mostly random, values (see [25]).

For wavelet applications it is common to distinguish between those parameters that enter linearly in $\hat{y}(t|\theta)$ (i.e., the coordinates in the function expansion) and those that enter nonlinearly (i.e., the location and scale parameters). Often the latter are seeded to fixed values, and the coordinates are estimated by the linear least-squares method. Basis functions, that give a small contribution to the fit (corresponding to nonuseful values of the scale and location parameters), can then be trimmed away ("pruning" or "shrinking").

58.7 User's Issues

58.7.1 Experiment Design

It is desirable to affect the conditions under which the data are collected. The objective with such *experiment design* is to make the collected data set Z^N as informative as possible for the models being built from the data. A considerable amount of theory around this topic can be developed, and here we shall just review some basic points.

The first and most important point is the following:

The input signal u must expose all of the relevant properties of the system. It must not be too "simple". For example, a pure sinusoid

$$u(t) = A \cos \omega t$$

will only give information about the system's frequency response at frequency ω. This can also be seen from Equation 58.68. The rule is that

- the input must contain at least as many different frequencies as the order of the linear model being built.

Another case where the input is too simple is when it is generated by feedback such as

$$u(t) = -Ky(t). \tag{58.135}$$

If we would like to build a first order ARX model

$$y(t) + ay(t-1) = bu(t-1) + e(t),$$

we find that, for any given α, all models with

$$a + bK = \alpha$$

will give identical input-output data. We can thus not distinguish between these models using an experiment with Equation 58.135, that is, we can not distinguish between any combinations of "a" and "b" if they satisfy the above condition for a given "α". The rule is as follows:

- If closed-loop experiments have to be performed, the feedback law must not be too simple. It is preferred that a set point in the regulator is changed in a random fashion.

The second main point in experimental design is as follows:

Allocate the input power to those frequency bands where a good model is particularly important.

This is also seen from Equation 58.68.

If we let the input be filtered white noise, this gives information for choosing the filter. In the time domain, these suggestions are useful:

- Use binary (two-level) inputs if linear models are being built. This gives maximal variance for amplitude-constrained inputs.
- Check the changes between the levels so that the input occasionally stays on one level long enough for a step response from the system to settle, more or less. There is no need to let the input signal switch quickly back and forth that no response in the output is clearly visible.

Note that the second point is a reformulation in the time domain of the basic frequency domain advice: let the input energy be concentrated in the important frequency bands.

A third basic piece of advice about experiment design concerns the choice of sampling interval.

A typical good sampling frequency is 10 times the bandwidth of the system. That roughly corresponds to 5–7 samples along the rise time of a step response.

58.7.2 Model Validation and Model Selection

The system identification process has, as we have seen, these basic ingredients:

- the set of models
- the data
- the selection criterion

Once these have been decided upon, we have, at least implicitly, defined a model: The one in the set that best describes the data according to the criterion. It is thus the best available model in the chosen set. But is it good enough? It is the objective of *model validation* to answer that question. Often the answer turns out to be "no", and we then have to review the choice of model set or modify the data set (see Figure 58.2).

How do we check the quality of a model? The prime method is to investigate how well it reproduces the behavior of a new set of data *(the validation data)* that was not used to fit the model, that is, we simulate the obtained model with a new input and compare this simulated output. One may then use one's eyes or numerical measurements of fit to decide if the fit is good enough. Suppose we have obtained several different models in different model structures (say a 4th order ARX model, a 2nd order BJ model, a physically parameterized one and so on) and would like to know which one is best. The simplest and most pragmatic approach is then to simulate each one of them with validation

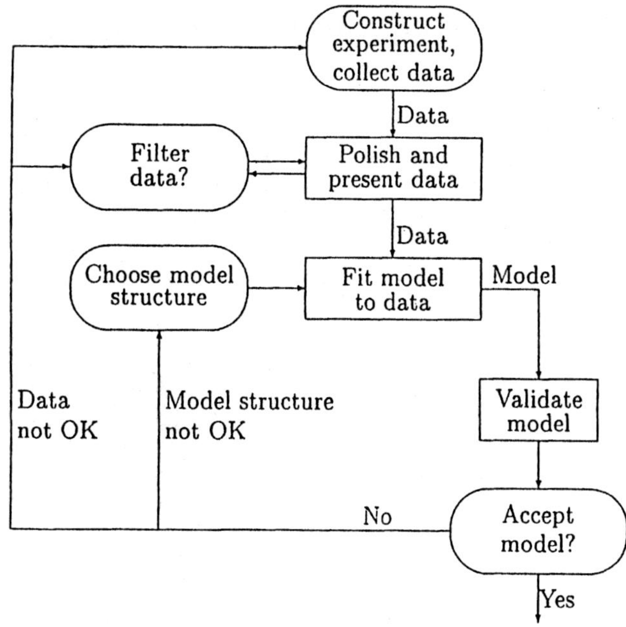

Figure 58.2 The identification loop.

data, evaluate their performance, and pick the one with the best fit to measured data. (This could be a subjective criterion!)

The second basic method for model validation is to examine the residuals ("the leftovers") from the identification process. These are the prediction errors,

$$\varepsilon(t) = \varepsilon(t, \hat{\theta}_N) = y(t) - \hat{y}(t|\hat{\theta}_N),$$

i.e., what the model could not "explain". Ideally these should be independent of information that was at hand at time $t-1$. For example, if $\varepsilon(t)$ and $u(t-\tau)$ are correlated, then elements in $y(t)$ that originate from $u(t-\tau)$ have not been properly accounted for by $\hat{y}(t|\hat{\theta}_N)$. The model has then not extracted all relevant information about the system from the data.

It is always good practice to check the residuals for such (and other) dependencies. This is known as *residual analysis.*

58.7.3 Software for System Identification

In practice, System Identification is characterized by lengthy numerical calculations to determine the best model in each given class of models. This is due to multiple user choices, trying different model structures, filtering data and so on. In practical applications, we need good software support. There are now many different commercial packages available, such as Mathwork's System Identification Toolbox [13], Matrix$_x$'s System Identification Module [18] and PIM [11]. In common they offer the following routines:

A *Handling of data, plotting, etc.*
Filtering of data, removal of drift, choice of data segments, etc.

B *Nonparametric identification methods*
Estimation of covariances, Fourier transforms, correlation and spectral analysis, etc.

C *Parametric estimation methods*
Calculation of parametric estimates in different model structures.

D *Presentation of models*
Simulation of models, estimation and plotting of poles and zeros, computation of frequency functions, and plotting Bode diagrams, etc.

E *Model validation*
Computation and analysis of residuals ($\varepsilon(t, \hat{\theta}_N)$). Comparison between different models' properties, etc.

The existing program packages differ mainly in various user interfaces and in options regarding the choice of model structure according to C above. For example, MATLAB's Identification Toolbox [13] covers all linear model structures discussed here, including arbitrarily parameterized linear models in continuous time.

Regarding the user interface, there is now a trend toward graphical orientation. This avoids syntax problems, relies more on "click and move", and avoids tedious menu-labyrinths. More aspects of CAD tools for system identification are treated in [15].

58.7.4 The Practical Side of System Identification

It follows from our discussion that, once the data have been recorded, the most essential steps in the process of identification are to try out various model structures, compute the best model in the structures, using Equation 58.24, and then validate this model. Typically this has to be repeated with quite a few different structures before a satisfactory model can be found.

The difficulties of this process should not be underestimated, and it will require substantial experience to master it. The following procedure may prove useful:

Step 1: Looking at the Data

Plot the data. Look at them carefully. Try to grasp the dynamics. Can you see the effects in the outputs of the input changes? Can you see nonlinear effects, like different responses at different levels, or different responses to a step up and a step down? Are there portions of the data that are "messy" or carry no information. Use this insight to select portions of the data for estimation and validation purposes.

Do physical levels play a role in your model? If not, detrend the data by removing their mean values. The models will then describe how changes in the input give changes in output, but will not explain the actual levels of the signals. This is the normal situation. The default situation, with good data, is that you detrend by removing means, and then select the first two-thirds or so of the data record for estimation purposes, using the remaining data for validation. (All of this corresponds to the "Data Quickstart" in the MATLAB Identification Toolbox.)

Step 2: Getting a Feel for the Difficulties
Compute and display the spectral analysis frequency response estimate, the correlation analysis impulse response estimate, and a fourth order ARX model with a delay estimated from the correlation analysis and a default order state-space model computed by a subspace method. (All of this corresponds to the "Estimate Quickstart" in the MATLAB Identification Toolbox.) This gives three plots. Look at the agreement between

- the spectral analysis estimate and the ARX and state-space models' frequency functions,
- the correlation analysis estimate and the ARX and state-space models' transient responses
- the measured validation data output and the ARX and state-space models' simulated outputs. We call this the *Model Output Plot.*

If these agreements are reasonable, the problem is not so difficult, and a relatively simple linear model will serve. Some fine tuning of model orders and noise models have to be made, and you can proceed to Step 4. Otherwise go to Step 3.

Step 3: Examining the Difficulties
There may be several reasons why the comparisons in Step 2 did not look good. This section discusses the most common ones, and how they can be handled:

- **Model Unstable:** The ARX or state-space model may be unstable, but could still be useful for control purposes. Then change to a 5- or 10-step ahead prediction instead of simulation in the Model Output Plot.
- **Feedback in Data:** If there is feedback from the output to the input, due to some regulator, then the spectral and correlation analysis estimates are not reliable. Discrepancies between these estimates and the ARX and state-space models can, therefore, be disregarded. In residual analysis of the parametric models, feedback in data can also be visible as correlation between residuals and input for negative lags.
- **Noise Model:** If the state-space model is clearly better than the ARX model at reproducing the measured output, this is an indication that the disturbances have a substantial influence, and it will be necessary to model them carefully.
- **Model Order:** If a fourth order model does not give a good Model Output Plot, try eighth order. If the fit improves, it follows that higher order models will be required, but that linear models could be sufficient.
- **Additional Inputs:** If the Model Output fit has not significantly improved by the tests so far, review the physics of the application. Are there more signals that have been, or could be, measured that might influence the output? If so, include these among the inputs and again try a fourth order ARX model from all the inputs (note that the inputs need not at all be control signals; anything measurable, including disturbances, should be treated as inputs).
- **Nonlinear Effects:** If the fit between measured and model output is still bad, consider the physics of the application. Are there nonlinear effects in the system? In that case, form the nonlinearities from the measured data. This could be as simple as forming the product of voltage and current measurements, if you realize that it is the electrical power that is the driving stimulus in, say, a heating process, and temperature is the output. This is, of course, application dependent. It does not cost very much work, however, to form a number of additional inputs by reasonable nonlinear transformations of the measured ones, and test if including them improves the fit (see Example 58.2).
- **Still Problems?** If none of these tests leads to a model that reproduces the Validation Data reasonably well, the conclusion might be that for many reasons a sufficiently good model cannot be produced from the data. The most important is that the data simply do not contain sufficient information, e.g., due to bad signal-to-noise ratios, large and nonstationary disturbances, varying system properties, etc. The system may also have some complicated nonlinearities, which cannot be realized on physical grounds. In such cases, nonlinear, black box models could be a solution. Among the most used models of this character are the Artificial Neural Networks (ANN) (see Section 58.6).

Otherwise, use the insights on which inputs to use and which model orders to expect and proceed to Step 4.

Step 4: Fine Tuning Orders and Noise Structures
For real data there is no such thing as a "correct model structure." However, different structures can give quite different model quality. The only way to determine this is to try a number of different structures and compare the properties of the models obtained. There are a few things to look for in these comparisons:

- **Fit between simulated and measured output**

  Look at the fit between the model's simulated output and the measured one for the Validation Data. Formally, you could pick that model, for which this number is the lowest. In practice, it is better to be more pragmatic, to take into account the model complexity, and to judge whether the important features of the output response are captured.

- **Residual analysis test**

  You should require of a good model that the cross-correlation function between residuals and input does not go significantly outside the confidence region. A clear peak at lag k shows that the effect from input $u(t-k)$ on $y(t)$ is not properly described. A rule of thumb is that a slowly varying cross-correlation function outside the confidence region is an indication of too few poles, while sharper peaks indicate too few zeros or incorrect delays.

- **Pole-Zero cancellations**

  If the pole-zero plot (including confidence intervals) indicates pole-zero cancellations in the dynamics, this suggests that lower order models can be used. In particular, if the order of ARX models has to be increased to get a good fit, but that pole-zero cancellations are indicated, then the extra poles are just introduced to describe the noise. Then try ARMAX, OE, or BJ model structures with an A or F polynomial of an order equal to that of the number of noncancelled poles.

What Model Structures Should be Tested?

You can spend any amount of time to check out a very large number of structures. It often takes only a few seconds to compute and evaluate a model in a certain structure, so that you should have a generous attitude to the testing. However, experience shows that when the basic properties of the system's behavior have been picked up, it is not useful to fine tune orders ad infinitum to improve the fit by fractions of a percent. For ARX models and state-space models estimated by subspace methods, there are also efficient algorithms for handling many model structures in parallel.

Multivariable Systems

Systems with many input signals and/or many output signals are called multivariable. Such systems are often more challenging to model. In particular, systems with several outputs can be difficult because the couplings between several inputs and outputs lead to more complex models. The structures involved are richer, and more parameters are required to obtain a good fit.

Generally speaking, it is preferable to work with state-space models in the multivariable case, since the model structure complexity is easier to deal with. It is just a matter of choosing the model order.

Working with Subsets of the Input–Output Channels: In the process of identifying good system models, it is useful to select subsets of the input and output channels. Partial models of the system's behavior will then be constructed. For example it might not, for example, be clear if all measured inputs have a significant influence on the outputs. That is most easily tested by removing an input channel from the data, building a model for the output(s) dependence on the remaining input channels, and checking if there is a significant deterioration in the fit of the model's output to the measured output. See also the discussion under Step 3 above. Generally speaking, the fit gets better when more inputs are included and worse when more outputs are included. To understand the latter fact, you should realize that a model required to explain the behavior of several outputs has a tougher job than one accounting for only a single output. If you have difficulties in obtaining good models for a multi-output system, model one output at a time, to find out which are the difficult ones to handle. Models to be used for simulations could very well be built up from single-output models, one output at a time. However, models for prediction and control can produce better results if constructed for all outputs simultaneously. This follows from the fact that knowing the set of all previous output channels is a better basis for prediction, than knowing only the past outputs in one channel.

Step 5: Accepting the model

The final step is to accept, at least for the time being, the model for its intended application. Recall the answer to question 10 in the introduction: *No matter how good an estimated model looks on your screen, it is only a simple reflection of reality. Surprisingly often, however, this is sufficient for rational decision making.*

References

[1] Akaike, H., A new look at the statistical model identification. *IEEE Trans. Auto. Control*, AC-19, 716–723, 1974.

[2] Akaike, H., Stochastic theory of minimal realization. *IEEE Trans. Auto. Control*, AC-19, 667–674, 1974.

[3] Åström, K. J. and Bohlin, T., Numerical identification of linear dynamic systems from normal operating records. In *IFAC Symposium on Self-Adaptive Systems*, Teddington, England, 1965.

[4] Box, G. E. P. and Jenkins, D. R., *Time Series Analysis, Forcasting and Control.* Holden-Day, San Francisco, 1970.

[5] Breiman, L., Hinging hyperplanes for regression, classification and function approximation. *IEEE Trans. Info. Theory*, 39, 999–1013, 1993.

[6] Brillinger, D., *Time Series: Data Analysis and Theory*. Holden-Day, San Francisco, 1981.

[7] Dennis, J. E. and Schnabel, R. B., *Numerical methods for unconstrained optimization and nonlinear equations*. Prentice Hall, Englewood Cliffs, NJ, 1983.

[8] Draper, N. and Smith, H., *Applied Regression Analysis*, 2nd ed., John Wiley & Sons, New York, 1981.

[9] Juditsky, A., Hjalmarsson, H., Benveniste, A., Delyon, B., Ljung, L., Sjöberg, J., and Zhang, Q., Nonlinear black-box modeling in system identification: Mathematical foundations. *Automatica*, 31, 1995.

[10] Kushner, H. J. and Yang, J., Stochastic approximation with averaging of the iterates: Optimal asymptotic rate of convergence for general processes. *SIAM J. Control Optim.*, 31(4), 1045–1062, 1993.

[11] Landau, I. D., *System Identification and Control Design Using P.I.M. + Software*. Prentice Hall, Englewood Cliffs, NJ, 1990.

[12] Larimore, W., System identification, reduced order filtering and modeling via canonical variate analysis. In *Proc. American Control Conference*, San Francisco, 1983.

[13] Ljung, L., *The System Identification Toolbox: The Manual*. The MathWorks Inc. 1st ed., (4th ed. 1994), Natick, MA, 1986.

[14] Ljung, L., *System Identification - Theory for the User*. Prentice Hall, Englewood Cliffs, NJ, 1987.

[15] Ljung, L., Identification of linear systems. In *CAD for Control Systems*, Linkens, D. A., Ed., Marcel Dekker, New York, 1993.

[16] Ljung, L. and Glad, T., *Modeling of Dynamic Systems*. Prentice Hall, Englewood Cliffs, NJ, 1994.

[17] Ljung, L. and Söderström, T., *Theory and Practice of Recursive Identification*. MIT Press, Cambridge, MA, 1983.

[18] $MATRIX_x$ users guide. *Integrated Systems Inc.*, Santa Clara, CA, 1991.

[19] Overschee, P. V. and DeMoor, B., N4SID: Subspace algorithms for the identification of combined deterministic-stochastic systems. *Automatica*, 30, 75–93, 1994.

[20] Poggio, T. and Girosi, F., Networks for approximation and learning. *Proc. of the IEEE*, 78, 1481–1497, 1990.

[21] Polyak, B. T. and Juditsky, A. B., Acceleration of stochastic approximation by averaging. *SIAM J. Control Optim.*, 30, 838–855, 1992.

[22] Rissanen, J., Basis of invariants and canonical forms for linear dynamic systems. *Automatica*, 10, 175–182, 1974.

[23] Rissanen, J., Modeling by shortest data description. *Automatica*, 14, 465–471, 1978.

[24] Schoukens, J. and Pintelon, R., *Identification of Linear Systems: A Practical Guideline to Accurate Modeling*. Pergamon, London, 1991.

[25] Sjöberg, J., Zhang, Q., Ljung, L., Benveniste, A., Delyon, B., Glorennec, P., Hjalmarsson, H., and Juditsky, A., Nonlinear black-box modeling in system identification: A unified overview. *Automatica*, 31, 1995.

[26] Söderström, T. and Stoica, P., *System Identification*. Prentice Hall Int., London, 1989.

[27] Solbrand, G., Ahlen, A., and Ljung, L., Recursive methods for off-line identification. *Int. J. Control*, 41(1), 177–191, 1985.

[28] Zhang, Q. and Benveniste, A., Wavelet networks. *IEEE Trans. Neural Networks*, 3, 889–898, 1992.

SECTION XIII

Stochastic Control

59

Discrete Time Markov Processes

Adam Shwartz
Electrical Engineering, Technion—Israel Institute of Technology, Haifa, Israel

59.1 Caveat

What follows is a quick survey of the main ingredients in the theory of discrete-time Markov processes. It is a bird's view, rather than the definitive "state of the art". To maximize accessibility, the nomenclature of mathematical probability is avoided, although rigor is not sacrificed. To compensate, examples (and counterexamples) abound and the bibliography is annotated. Relevance to control is discussed in Section 59.9.

59.2 Introduction

Discrete time Markov processes, or Markov chains, are a powerful tool for modeling and analysis of discrete time systems, whose behavior is influenced by randomness. A Markov chain is probably the simplest object which incorporates both dynamics (that is, notions of "state" and time) and randomness. Let us illustrate the idea through a gambling example.

EXAMPLE 59.1:

A gambler bets one dollar on "red" at every turn of a (fair) game of roulette. Then, at every turn he gains either one dollar (win, with probability 1/2) or (-1) (lose). In an ideal game, the gains form a sequence of independent, identically distributed (i.i.d.) random variables. We cannot predict the outcome of each bet, although we do know the odds. Denote by X_t the total fortune the gambler has at time t. If we know X_s, then we can calculate the distribution of X_{s+1} (that is, the probability that $X_{s+1} = y$, for all possible values of y), and even of X_{s+k} for any $k > 0$. The variable X_t serves as a "state" in the following sense: given X_s, knowledge of X_t for $t < s$ is irrelevant to the calculation of the distribution of future values of the state.

This notion of a state is similar to classical "state space" descriptions. Consider the standard linear model of a dynamical system

$$x_{t+1} = Ax_t + v_t, \tag{59.1}$$

or the more general nonlinear model

$$x_{t+1} = f(x_t, v_t). \tag{59.2}$$

It is intuitively clear that, given the present state, the past does not provide additional information about the future, as long as v_t is not predictable (this is why deterministic state-space models require that v_t be allowed to change arbitrarily at each t). This may remain true even when v_t is random: for example, when they are i.i.d., for then the past does not provide additional information about future v_s's. As we shall see in Theorem 59.1, in this case (59.1) and (59.2) define Markov chains.

In the next section we give the basic definitions and describe the dynamics of Markov chains. We assume that the state space S is countable. We restrict our attention to time-homogeneous dynamics (see comment following Theorem 59.1), and discuss the limiting properties of the Markov chain. Finally, we shall discuss extensions to continuous time and to more general state spaces. We conclude this section with a brief review of standard notation. All of our random variables and events are defined on a probability space Ω with a collection $\mathcal{F}$ of events (subsets of Ω) and probability $\mathbb{P}$. The "probability triple" $(\Omega, \mathcal{F}, \mathbb{P})$ is fixed, and we denote expectation by $\mathbb{E}$. For events A and B with $\mathbb{P}(B) > 0$, the basic definition of a conditional probability (the multiplication rule) is

$$\mathbb{P}(A|B) \stackrel{\triangle}{=} \frac{\mathbb{P}(A \bigcap B)}{\mathbb{P}(B)}. \tag{59.3}$$

0-8493-8570-9/96/$0.00+$.50

The abbreviation i.i.d. stands for *independent, identically distributed*, and random variables are denoted by capital letters. The identity matrix is denoted by I and $\mathbf{1}_A$ is the indicator function of A, that is, $\mathbf{1}_A(\omega) = 1$ if $\omega \in A$ and $= 0$ otherwise.

59.3 Definitions and Construction

Let $X_0, X_1, \ldots$ be a sequence of random variables, with values in a state space S. We assume that S is finite or countable, and for convenience we usually set $S = \{1, 2, \ldots\}$.

DEFINITION 59.1 A sequence $X_0, X_1, \ldots$ on S is a Markov chain if it possesses the Markov property, that is, if for all $t > 0$,

$$\begin{aligned} &\mathbb{P}(X_t = j | X_{t-1} = i_{t-1}, X_{t-2} = i_{t-2}, \ldots, X_0 = i_0) \\ &= \mathbb{P}(X_t = j | X_{t-1} = i_{t-1}). \end{aligned}$$

A Markov chain is called homogeneous if $\mathbb{P}(X_t = j | X_{t-1} = i)$ does not depend on t. In this case we denote the transition probability from i to j by

$$p_{ij} \stackrel{\Delta}{=} \mathbb{P}(X_t = j | X_{t-1} = i).$$

The matrix $P \stackrel{\Delta}{=} \{p_{ij},\ i, j = 1, \ldots\}$ is called the transition matrix.

Henceforth, we restrict our attention to homogeneous chains.

Using the Markov property, a little algebra with the definition of conditional probability gives the more general Markov property: if $t_1 \le t_2 \le \cdots \le t_k \le \cdots \le t_\ell$ then

$$\begin{aligned} &\mathbb{P}\left(X_{t_\ell} = j_\ell, X_{t_{\ell-1}} = j_{\ell-1}, \ldots, X_{t_k} = j_k \right.\\ &\qquad \left.| X_{t_{k-1}} = j_{k-1}, \ldots, X_{t_1} = j_1\right) \\ &= \mathbb{P}\left(X_{t_\ell} = j_\ell, X_{t_{\ell-1}} = j_{\ell-1}, \ldots, X_{t_k} = j_k \right.\\ &\qquad \left.| X_{t_{k-1}} = j_{k-1}\right). \end{aligned} \tag{59.4}$$

This is a precise statement of the intuitive idea given in the introduction: given the present state $X_{t_{k-1}} = j_{k-1}$, the past does not provide additional information.

A chain is called finite if S is a finite set. An alternative name for "homogeneous Markov chain" is "Markov chain with stationary transition probabilities". There are those who call a homogeneous Markov chain a "stationary Markov chain." However, since "stationary process" means something entirely different, we shall avoid such usage (see Example 59.6).

Suppose that a process is a Markov chain according to Definition 59.1, and that its initial distribution is given by the row vector $\mu(0)$, that is

$$\mathbb{P}(X_0 = i) = \mu_i(0), \quad i = 1, 2, \ldots.$$

Then we can calculate the joint probability distribution at times 0 and 1 from the definition of conditional probability,

$$\begin{aligned} \mathbb{P}(X_0 = i, X_1 = j) &= \mathbb{P}(X_1 = j | X_0 = i) \cdot \mathbb{P}(X_0 = i) \\ &= \mu_i(0) \cdot p_{ij}. \end{aligned}$$

More generally, using the Markov property:

$$\begin{aligned} &\mathbb{P}(X_0 = j_0, X_1 = j_1, \ldots, X_t = j_t) \\ &= \mu_{j_0}(0) \cdot \prod_{s=1}^{t} p_{j_{s-1} j_s}. \end{aligned}$$

So, the probability distribution of the whole process can be calculated from the two quantities: the initial probability distribution, and the transition probabilities. In particular, we can calculate the probability distribution at any time t

$$\begin{aligned} \mu_j(t) &= \mathbb{P}(X_t = j) \\ &= \sum_{j_0, j_1, \ldots, j_{t-1}} \mu_{j_0}(0) \prod_{s=1}^{t} p_{j_{s-1} j_s} \end{aligned} \tag{59.5}$$

where the sum is over all states in S, that is, each index is summed over the values 1, 2, ... In vector notation,

$$\mu(t) = \mu(t-1)P = \ldots = \mu(0)P^t. \tag{59.6}$$

(If S is countable then, of course, the matrix is infinite, but it is manipulated in exactly the same way as a finite matrix.) Thus, the probability *distribution* of a Markov chain evolves as a linear dynamical system, even when its evolution equation (59.2) is nonlinear. The one-dimensional probability distribution and the transition probabilities clearly satisfy

$$\left\{ \begin{array}{l} \mu_j(t) \ge 0 \\ \sum_{j \in S} \mu_j(t) = 1 \end{array} \right. \qquad \left\{ \begin{array}{l} p_{ij} \ge 0 \\ \sum_{j \in S} p_{ij} = 1. \end{array} \right. \tag{59.7}$$

That is, the rows of the transition matrix P sum to one. Thus, P is a *stochastic matrix*: its elements are nonnegative and its rows sum to one.

If we denote by $p_{ij}^{(n)} \stackrel{\Delta}{=} \mathbb{P}(X_{m+n} = j | X_m = i)$ the *n step transition probability* from i to j, then we obtain from the definition of conditional probability and the Markov property (59.4) the Chapman-Kolmogorov equations

$$p_{ij}^{(n+m)} = \sum_{k \in S} p_{ik}^{(n)} p_{kj}^{(m)}$$

or

$$P^{n+m} = P^n P^m. \tag{59.8}$$

Therefore, the $p_{ij}^{(n)}$ are the elements of the matrix P^n. This matrix notation yields a compact expression for expectations of functions of the state. To compute $\mathbb{E}g(X_t)$ for some function g on $S = \{1, 2, \ldots\}$, we represent g by a column vector $\underline{g} \stackrel{\Delta}{=} \{g(1), g(2), \ldots\}^T$ (where T denotes transpose). Then

$$\mathbb{E}g(X_t) = \mu(0) \cdot P^t \cdot \underline{g}.$$

Note that this expression does not depend on the particular S: since the state space is countable we can, by definition, relable the states so that the state space becomes $\{1, 2, \ldots\}$. Let us now summarize the connection between the representations (59.1)–(59.2) and Markov chains.

THEOREM 59.1 *Let $V_0, V_1, \ldots$ be a sequence of i.i.d. random variables, independent of X_0. Then for any (measurable) function f, the sequence $X_0, X_1, \ldots$ defined through (59.2) is a Markov chain. Conversely, let $\tilde{X}_0, \tilde{X}_1, \ldots$ be a Markov chain with values in S. Then there is a (probability triple and a measurable) function f and a sequence $V_0, V_1, \ldots$ of i.i.d. random variables so that the process $X_0, X_1, \ldots$ defined by (59.2) with $X_0 = \tilde{X}_0$ has the same probability distribution as $\tilde{X}_0, \tilde{X}_1, \ldots$, that is, for all t and $j_0, j_1, \ldots, j_t$,*

$$\begin{aligned}&\mathbb{P}(X_0 = j_0, X_1 = j_1, \ldots, X_t = j_t)\\&= \mathbb{P}\left(\tilde{X}_0 = j_0, \tilde{X}_1 = j_1, \ldots, \tilde{X}_t = j_t\right).\end{aligned}$$

Note that, whether the system (59.2) is linear or not, the evolution of the probability distribution (59.5)–(59.6) is always linear.

We have seen that a Markov chain defines a set of transition probabilities. The converse is also true: given a set of transition probabilities and an initial probability distribution, it is possible to construct a stochastic process which is a Markov chain with the specified transitions and probability distribution.

THEOREM 59.2 *If $X_0, X_1, \ldots$ is a homogeneous Markov chain then its probability distribution and transition probabilities satisfy (59.7) and (59.8). Conversely, given $\mu(0)$ and a matrix P that satisfy (59.7), there exists a (probability triple and a) Markov chain with initial distribution $\mu(0)$ and transition matrix P.*

The restriction to homogeneous chains is not too bad: if we define a new state $\tilde{x} \stackrel{\Delta}{=} \{t, x\}$, then it is not hard to see that we can incorporate explicit time dependence, and the new state space is still countable.

EXAMPLE 59.2:

Let $V_0, V_1, \ldots$ be i.i.d. and independent of X_0. Assume both V_t and X_0 have integer values. Then

$$X_{t+1} \stackrel{\Delta}{=} X_t + V_t = X_0 + \sum_{s=0}^{t} V_s \tag{59.9}$$

defines a Markov chain called a chain with *stationary independent increments*, with state space $\ldots, -1, 0, 1, \ldots$. The transition probability p_{ij} depends only on the difference $j - i$. It turns out [1] that the converse is also true: if the transition probabilities of a Markov chain depend only on the difference $j - i$ then the process can be obtained via (59.9) with i.i.d. V_t.

A *random walk* is a process defined through Equation 59.9, but where the V_t are not necessarily integers—they are real valued.

59.4 Properties and Classification

Given two states i and j, it may or may not be possible to reach j from i. This leads to the notion of classes.

DEFINITION 59.2 We say a state i leads to j if

$$\mathbb{P}(x_t = j \text{ for some } t | x_0 = i) > 0.$$

This holds if and only if $p_{ij}^{(t)} > 0$ for some t. We say states i and j communicate, denoted by $i \leftrightarrow j$, if i leads to j and j leads to i.

Communication is a property of pairs: it is obviously symmetric ($i \leftrightarrow j$ if and only if $j \leftrightarrow i$) and is transitive ($i \leftrightarrow j$ and $j \leftrightarrow k$ implies $i \leftrightarrow k$) by the Chapman-Kolmogorov equations (59.8). By convention, $i \leftrightarrow i$. By these three properties, $\leftrightarrow$ defines an equivalence relation. We can therefore partition S into non-empty *communicating classes* $S_0, S_1, \ldots$ with the properties

1. every state i belongs to exactly one class,
2. if i and j belong to the same class then $i \leftrightarrow j$, and
3. if i and j belong to different classes then i and j do not communicate.

We denote the class containing state i by $S(i)$. Note that if i leads to j but j does not lead to i and j *do not* communicate.

DEFINITION 59.3 A set of states C is closed, or absorbing, if $p_{ij} = 0$ whenever $i \in C$ and $j \notin C$. Equivalently,

$$\sum_{j \in C} p_{ij} = 1 \quad \text{for all } i \in C.$$

If a set C is not closed, then it is called open. A Markov chain is irreducible if all states communicate. In this case its partition contains exactly one class. A Markov chain is indecomposable if its partition contains at most one closed set.

Define the *incidence matrix* $\mathcal{I}$ as follows:

$$\mathcal{I}_{ij} = \begin{cases} 1 & \text{if } p_{ij} > 0 \\ 0 & \text{otherwise}. \end{cases}$$

We can also define a directed graph whose nodes are the states, with a directed arc between any two states for which $\mathcal{I}_{ij} = 1$. The communication properties can obviously be extracted from $\mathcal{I}$ or from the directed graph. The chain is irreducible if and only if the directed graph is connected in the sense that, going in the direction of the arcs, we can reach any node from any other node. Closed classes can also be defined in terms of the incidence matrix or the graph. The classification leads to the following maximal decomposition.

THEOREM 59.3 *By renumbering the states, if necessary, we can put the matrix P into the block form*

$$P = \begin{bmatrix} P_1 & 0 & \ldots & 0 & 0 \\ 0 & P_2 & \ldots & 0 & 0 \\ \vdots & \vdots & & \vdots & \vdots \\ 0 & 0 & \ldots & P_m & 0 \\ R_1 & R_2 & \ldots & R_m & Q \end{bmatrix} \tag{59.10}$$

where the blocks P_i correspond to closed irreducible classes. It is maximal in the sense that smaller classes will not be closed, and no subset of states corresponding to Q is a closed irreducible class.

If S is countable, then the number of classes may be infinite. Note that all the definitions in this section apply when we replace S with any closed class.

Much like other dynamical systems, Markov chains can have cyclic behavior, and can be unstable. The relevant definitions are

DEFINITION 59.4 A state i has period d if d is the greatest common divisor of the set $\{t : p_{ii}^{(t)} > 0\}$. If d is finite and $d > 1$ then the state is called periodic; otherwise it is aperiodic.

DEFINITION 59.5 A state i is called recurrent if the probability of starting at i and returning to i in finite time is 1. Formally, if

$$\mathbb{P}(X_t = i \text{ for some } t > 1 | x_1 = i) = 1.$$

Otherwise it is called transient.

EXAMPLE 59.3:

In the chain on $S = \{1, 2\}$ with $p_{ij} = 1$ if and only if $i \neq j$, both states are periodic with period $d = 2$ and both states are recurrent. The states communicate, and so S contains exactly one class, which is therefore closed. Consequently the chain is irreducible and indecomposable. However, if $p_{12} = p_{22} = 1$ then state 1 does not lead to itself, the states are not periodic, state 2 is recurrent and state 1 is transient. In this case, the partition contains two sets: the closed set $\{2\}$, and the open set $\{1\}$. Consequently, the chain is not irreducible, but it is indecomposable.

When S is finite, then either it is irreducible or it contains a closed proper subset.

EXAMPLE 59.4:

Let $S = \{1, 2, \ldots\}$. Suppose $p_{ij} = 1$ if and only if $j = i + 1$. Then all states are transient, and S is indecomposable but not irreducible. Every set of the form $\{i : i \geq k\}$ is closed, but in the partition of the state space each state is the only member in its class. Suppose now $p_{11} = 1$ and for $i > 1$, $p_{ij} = 1$ if and only if $j = i - 1$. Then state 1 is the only recurrent state, and again each state is alone in its class.

THEOREM 59.4 *Let S_k be a class. Then either all states in S_k are recurrent, or all are transient. Moreover, all states in S_k have the same period d.*

59.5 Algebraic View and Stationarity

The matrix P is positive, in the sense that its entries are positive. When S is finite, the Perron-Frobenius theorem implies [8]

THEOREM 59.5 *Let S be finite. Then P has a nonzero left eigenvector π whose entries are nonnegative, and $\pi \cdot P = \pi$, that is, the corresponding eigenvalue is 1. Moreover, $|\lambda| \leq 1$ for all other eigenvalues λ. The multiplicity of the eigenvalue 1 is equal to the number of irreducible closed subsets of the chain. In particular, if the entries of P^n are all positive for some n, then the eigenvalue 1 has multiplicity 1. In this case, the entries of π are all positive and $|\lambda| < 1$ for all other eigenvalues λ.*

If the entries of P^n are all positive for some n then the chain is irreducible and aperiodic, hence the second part of the theorem. If the chain is irreducible and periodic with period d, then the dth roots of unity are left eigenvalues of P, each is of multiplicity 1 and all other eigenvalues have strictly smaller modulus. The results for a general finite chain can be obtained by writing the chain in the block form (59.10).

DEFINITION 59.6 Let S be finite or countable. A probability distribution μ satisfying $\mu \cdot P = \mu$ is called invariant (under P) or stationary.

Theorem 59.5 thus implies that every finite Markov chain possesses at least one invariant probability distribution. For countable chains, Example 59.4 shows that this is not true.

EXAMPLE 59.5:

Returning to Example 59.3, in the first case $(1/2, 1/2)$ is the only invariant probability distribution, while in the second case $(0, 1)$ is the only invariant probability distribution. In Example 59.4, in the first case there is no invariant probability distribution, while in the second case $(1, 0, 0, \ldots)$ is the only invariant probability distribution. Finally, if $P = I$, the 2×2 identity matrix, then $\pi \stackrel{\Delta}{=} (p, 1 - p)$ is invariant for any $0 \leq p \leq 1$.

EXAMPLE 59.6:

Recall that a process $X_0, X_1, \ldots$ is called stationary if, for all positive t and s, the distribution of $\{X_0, X_1, \ldots, X_t\}$ is the same as the distribution of $\{X_s, X_{1+s}, \ldots, X_{t+s}\}$. From the definitions it follows that a (homogeneous) Markov chain (finite or not) is stationary if and only if $\mu(0)$ is invariant.

A very useful tool in the calculation of invariant probability distributions is the "balance equations":

$$\pi_i = \sum_{j:j \to i} \pi_j p_{ji} = \pi_i \sum_{j:i \to j} p_{ij} \tag{59.11}$$

where the first equality is just a restatement of the definition of invariant probability, and the second follows since by (59.7), the last sum equals 1. The intuition behind these equations is very useful: in steady state, the rate at which "probability mass enters" must be equal to the rate it "leaves". This is particularly useful for continuous-time chains. More generally, given any set S, the rate at which "probability mass enters" the set (under the stationary distribution) equals the rate it "leaves":

THEOREM 59.6 *Let S be a set of states. Then*

$$\sum_{i \in S} \sum_{j:j \to i} \pi_j p_{ji} = \sum_{i \in S} \pi_i \sum_{j:i \to j} p_{ij}.$$

EXAMPLE 59.7:

Random walk with a reflecting barrier. This example models a discrete-time queue where, at each instance, either arrival or departure occurs. The state space S is the set of non-negative integers (including 0), and

$$p_{00} = 1 - p, \quad p_{i(i+1)} = p, \quad p_{i(i-1)} = 1 - p \quad \text{for } i \geq 1.$$

Then all states communicate so that the chain is irreducible, the chain is aperiodic and recurrent. From (59.11) we obtain

$$\begin{aligned} \pi_0 &= \pi_0 p_{00} + \pi_1 p_{10} \\ \pi_i &= \pi_{i-1} p_{(i-1)i} + \pi_{i+1} p_{(i+1)i}, \quad i \geq 1. \end{aligned}$$

When $p < 1/2$, this and (59.7) imply that $\pi_i = \frac{p^i}{(1-p)^i}$ for $i \geq 0$.

EXAMPLE 59.8:

Birth-death process. A Markov chain on $S = \{0, 1, \ldots\}$ is a birth-death process if $p_{ij} = 0$ whenever $|i - j| \geq 2$. If X_t is the number of individuals alive at time t then, at any point in time, this number can increase by one (birth), decrease by one (death) or remain constant (simultaneous birth and death). Unlike Example 59.7, here the probability of a change in size may depend on the state.

59.6 Random Variables

In this section we shift our emphasis back from algebra to the stochastic process. We define some useful random variables associated with the Markov chain. It will be convenient to use $\mathbb{P}_j$ for the probability conditioned on the process starting at state j. That is, for an event A,

$$\mathbb{P}_j(A) \stackrel{\Delta}{=} \mathbb{P}(A | X_0 = j),$$

with a similar convention for expectation $\mathbb{E}_j$. The Markov property says that the past of a Markov chain is immaterial given the present. But suppose we observe a process until a random time, say the time a certain event occurs. Is this property preserved? The answer is positive, but only for non-anticipative times:

DEFINITION 59.7 Let S be a collection of states, that is, a subset of S. The hitting time τ_S of S is the first time the Markov chain visits a state in S. Formally,

$$\tau_S = \inf\{t > 0 : X_t \in S\}.$$

Note that by convention, if X_t never visits S then $\tau_S = \infty$. The initial time, here $t = 0$, does not qualify in testing whether the process did or did not visit S. By definition, hitting times have the following property. In order to decide whether or not $\tau_S = t$, it suffices to know the values of $X_0, \ldots, X_t$. This gives rise to the notion of Markov time or stopping time.

DEFINITION 59.8 A random variable τ with positive integer values is called a stopping time, or a Markov time (with respect to the process $X_0, X_1, \ldots$) if one of the following equivalent conditions hold. For each $t \geq 0$

1. it suffices to know the values of $X_0, X_1, \ldots, X_t$ in order to determine whether the event $\{\tau = t\}$ occurred or not,
2. there exists a function f_t so that
$$\mathbf{1}_{\tau=t}(\omega) = f_t\left(X_0(\omega), \ldots, X_t(\omega)\right).$$

An equivalent, and more standard definition is obtained by replacing $\tau = t$ by $\tau \leq t$. With respect to such times, the Markov property holds in a stronger sense.

THEOREM 59.7 Strong Markov property. *If τ is a stopping time for $X_0, X_1, \ldots$, then*

$$\begin{aligned} &\mathbb{P}(X_{\tau+1} = j_1, X_{\tau+2} = j_2, \ldots, X_{\tau+m} = j_m \\ &\quad | X_t = i_t,\ t < \tau,\ X_\tau = i^*) \\ &= \mathbb{P}\left(X_1 = j_1, X_2 = j_2, \ldots, X_m = j_m | X_0 = i^*\right). \end{aligned}$$

We can now rephrase and complement the definition of recurrence. We write τ_j when we really mean $\tau_{\{j\}}$.

DEFINITION 59.9 The state j is recurrent if $\mathbb{P}_j(\tau_j < \infty) = 1$. It is called positive recurrent if $\mathbb{E}_j \tau_j < \infty$, and null-recurrent if $\mathbb{E}_j \tau_j = \infty$.

If state j is recurrent, then the hitting time of j is finite. By the strong Markov property, when the processes hits j for the first time, it "restarts": therefore, it will hit j again! and again! So, let N_j be the number of times the process hits state j:

$$N_j \stackrel{\Delta}{=} \sum_{t=1}^{\infty} \mathbf{1}_{X_t = j}.$$

THEOREM 59.8

1. *If a state is positive recurrent, then all states in its class are positive recurrent. The same holds for null recurrence.*
2. *Suppose j is recurrent. Then $\mathbb{P}_j(N_j = \infty) = 1$, and consequently $\mathbb{E}_j N_j = \infty$. Moreover, for every state i,*
$$\begin{aligned} \mathbb{P}_i\left(N_j = \infty\right) &= \mathbb{P}_i\left(\tau_j < \infty\right) \cdot \mathbb{P}_j(N_j = \infty) \\ &= \mathbb{P}_i\left(\tau_j < \infty\right), \end{aligned}$$

and if $\mathbb{P}_i(\tau_j < \infty) > 0$ *then* $\mathbb{E}_i N_j = \infty$.

3. *Suppose* j *is transient. Then* $\mathbb{P}_j(N_j < \infty) = 1$, *and for all* i,

$$\mathbb{E}_i N_j = \frac{\mathbb{P}_i(\tau_j < \infty)}{1 - \mathbb{P}_j(\tau_j < \infty)} .$$

To see why the last relation should hold, note that by the strong Markov property,

$$\mathbb{P}_i(\tau_j < \infty \text{ and a second visit occurs}) = \mathbb{P}_i(\tau_j < \infty) \cdot \mathbb{P}_j(\tau_j < \infty),$$

and similarly for later visits. This means that the distribution of the number of visits is geometric: with every visit we get another chance, with equal probability, to revisit. Therefore,

$$\begin{aligned}\mathbb{E}_i N_j &= \mathbb{P}_i(\tau_j < \infty) + \mathbb{P}_i(\tau_j < \infty \\ &\qquad \text{and a second visit occurs}) + \ldots \\ &= \mathbb{P}_i(\tau_j < \infty) \\ &\qquad (1 + \mathbb{P}(\text{a second visit occurs} | \tau_j < \infty)) + \ldots . \\ &= \mathbb{P}_i(\tau_j < \infty)(1 + \mathbb{P}_j(\tau_j < \infty) + \ldots)\end{aligned}$$

which is what we obtain if we expand the denominator. A similar interpretation gives rise to

$$\mathbb{E}_i \tau_j = 1 + \sum_{k:k \neq j \in \mathsf{S}} \mathrm{p}_{ik} \mathbb{E}_k \tau_j .$$

We have a simple criterion for recurrence in terms of transition probabilities, since

$$\mathbb{E}_i N_j = \mathbb{E}_i \sum_{t=1}^{\infty} 1_{X_t = j} = \sum_{t=1}^{\infty} \mathbb{P}_i(X_t = j) = \sum_{t=1}^{\infty} \mathrm{p}_{ij}^{(t)} .$$

59.7 Limit Theorems: Transitions

Classical limit theorems concern the behavior of t-step transition probabilities, for large t. Limits for the random variables are discussed in Section 59.8.

THEOREM 59.9 *For every Markov chain, the limit*

$$P^* \stackrel{\Delta}{=} \lim_{t \to \infty} \frac{1}{t} \sum_{s=0}^{t-1} P^s \tag{59.12}$$

exists and satisfies

$$P^* \cdot P = P \cdot P^* = P^* \cdot P^* = P^* .$$

If S *is finite then* P^* *is a stochastic matrix.*

1. *Suppose the Markov chain is indecomposable, recurrent, and nonperiodic. Then, for all states* i, j, k,

$$\lim_{t \to \infty} \sum_{j \in \mathsf{S}} |\mathrm{p}_{ij}^{(t)} - \mathrm{p}_{kj}^{(t)}| = 0 .$$

2. *An irreducible chain is positive recurrent if and only if it has an invariant probability distribution* π, *and in this case* $\lim_{t \to \infty} \mathrm{p}_{ij}^{(t)} = \pi(j)$ *for all* i, j. *If it is null recurrent then for all* i, j, $\lim_{t \to \infty} \mathrm{p}_{ij}^{(t)} = 0$. *If state* j *is transient then* $\sum_t \mathrm{p}_{ij}^{(t)} < \infty$.

Since a finite Markov chain always contains a finite closed set of states, there always exists an invariant distribution. Moreover, if a set is recurrent, then it is positive recurrent.

EXAMPLE 59.9: Example 59.3 Continued.

For the periodic chain, $\mathrm{p}_{ij}^{(t)}$ clearly does not converge. However, $\mathrm{P}_{ij}^* = 1/2$ for all i, j, and the rows define an invariant measure.

EXAMPLE 59.10: Example 59.8 Continued.

Assume that for the birth death process $\mathrm{p}_{i(i+1)} > 0$ and $\mathrm{p}_{(i+1)i} > 0$ for all $i \geq 0$ and $\mathrm{p}_{ii} > 0$ for some i. Then the chain is obviously irreducible, and aperiodic (if $\mathrm{p}_{ii} = 0$ for all i then $d = 2$). Using (59.11) we obtain that an invariant probability distribution, if it exists, must satisfy

$$\pi_i = \frac{\mathrm{p}_{01} \cdots \mathrm{p}_{(i-1)i}}{\mathrm{p}_{10} \cdots \mathrm{p}_{i(i-1)}} \cdot \pi_0 . \tag{59.13}$$

Therefore, any invariant probability must satisfy $\pi_i > 0$ for all i, and in particular $\pi_0 > 0$. So, we can invoke (59.7) to obtain the following dichotomy. Either

$$Z \stackrel{\Delta}{=} \sum_{i \in \mathsf{S}} \frac{\mathrm{p}_{01} \cdots \mathrm{p}_{i-1i}}{\mathrm{p}_{10} \cdots \mathrm{p}_{ii-1}} < \infty, \tag{59.14}$$

in which case (59.13)–(59.14) determine the unique invariant probability, and we conclude that the Markov chain is positive recurrent. Or $Z = \infty$, in which case there is no invariant probability and the chain is not positive recurrent.

In terms of the transition matrix P, if a chain is nonperiodic, indecomposable, and recurrent then the matrix converges (uniformly over rows) to a matrix having identical rows, which are either all zeroes (null-recurrent case), or equal to the invariant probability distribution. Here are the missing cases from Theorem 59.9 . Denote the mean hitting time of state j starting at i by $m_{ij} = \mathbb{E}_i \tau_j$. Clearly m_{jj} is infinite if j is not positive recurrent, and we shall use the convention that $a/\infty = 0$ whenever a is finite.

THEOREM 59.10 *If a state* j *is transient then* $\lim_{t \to \infty} \mathrm{p}_{ij}^{(t)} = 0$. *If* j *is recurrent with period* d *then* $\lim_{t \to \infty} \mathrm{p}_{jj}^{(nd)} = \frac{d}{m_{jj}}$. *If* j *is nonperiodic, this remains true with* $d = 1$, *so that (by Theorem 59.9)* $\pi(j) \cdot m_{jj} = 1$.

The last statement should be intuitive: the steady state probability of visiting state j is a measure of how often this state is "visited", and this is inversely proportional to the mean time between visits. The rate at which convergence takes place depends on the second largest eigenvalue of P. Therefore, if the Markov chain is finite, indecomposable, and aperiodic with invariant probability distribution π, then

$$\left|p_{ij}^{(t)} - \pi_j\right| \leq R\rho^t \quad \text{for all } i, j$$

with $\rho < 1$. This of course implies that the one dimensional distributions converge geometrically fast. On the other hand, the Markov structure implies that if indeed the one dimensional distributions converge, then the distribution of the whole process converges:

THEOREM 59.11 *Suppose that for all i and j we have $\lim_{t\to\infty} p_{ij}^{(t)} = \pi_j$, for some probability distribution π. Then π is an invariant probability distribution, and for any i,*

$$\lim_{t\to\infty} \mathbb{P}_i (X_{t+1} = j_1, X_{t+2} = j_2 \ldots) = \mathbb{P}_\pi (X_1 = j_1, X_2 = j_2 \ldots)$$

where $\mathbb{P}_\pi$ is obtained by starting the process with the distribution π (in fact, the distribution of the process converges).

59.8 Ergodic Theorems

We do not expect the Markov chain to converge: since transition probabilities are homogeneous, the probability of leaving a given state does not change in time. However, in analogy with i.i.d. random variables, there are limit theorems under the right scaling. The connection to the i.i.d. case comes from the following construction. Fix an arbitrary state j and define

$$\begin{aligned} R_1 &= T_1 = \tau_j \\ R_k &= \inf\{t > R_{k-1} : X_t = j, \quad k > 1\} \\ T_k &= R_k - R_{k-1}, \quad k > 1. \end{aligned}$$

THEOREM 59.12 *If j is recurrent and the Markov chain starts at j (with probability one), then $T_1, T_2 \ldots$ is a sequence of i.i.d. random variables. Moreover, the random vectors*

$$Z_k \overset{\Delta}{=} \{X_{R_k}, X_{R_k+1}, \ldots, X_{R_{k+1}-1}\}$$

are independent and identically distributed (in the space of sequences of variable length!).

This is another manifestation of the fact that, once we know the Markov chain hits some state, future behavior is (probabilistically) determined. Fix a recurrent state i and a time t, and define

$$T^t \overset{\Delta}{=} \max_k \{T_k : T_k \leq t\}.$$

Denote by $N_s(j)$ the number of times in $1, 2, \ldots, s$ that $X_u = j$. By Theorem 59.12, the random variables $\{N_{R_k+1}(j) - N_{R_k}(j), k = 1, 2, \ldots\}$ are independent, and (except possibly for $k = 1$) are identically distributed for each j. By the law of large numbers this implies the following.

THEOREM 59.13 *Let i be recurrent. Then starting at i ($\mathbb{P}_i$ a.s.)*

$$\lim_{t\to\infty} \frac{N_t(j)}{N_t(\ell)} = \frac{\mathbb{E}_i \sum_{s=1}^{\tau_i} 1_{X_s=j}}{\mathbb{E}_i \sum_{s=1}^{\tau_i} 1_{X_s=\ell}} = \frac{\pi_j}{\pi_\ell},$$

and π is an invariant probability distribution, concentrated on the closed class containing i, that is,

$$\pi_k = \sum_{j\in S} \pi_j p_{jk}, \qquad \sum_{k:i\to k\in S} \pi_k = 1$$

so that $\pi_k = 0$ if $i \not\to k$. Moreover, if we start in some state j then

$$\lim_{t\to\infty} \frac{N_t(i)}{t} = \frac{1_{\tau_i<\infty}}{m_{ii}} \quad \text{with } \mathbb{P}_j\text{-probability 1.}$$

The last relation implies (59.12): taking $\mathbb{E}_j$ expectations, we obtain that if i is recurrent then $P^* = \mathbb{P}_j(\tau_i < \infty)/m_{ii}$. From here follows a limit theorem for functions of a Markov chain.

THEOREM 59.14 Ergodic Theorem. *Let S be a single recurrent class, and assume π is an invariant probability distribution. Let f and g be functions such that $\mathbb{E}_\pi |f(X_0)| < \infty$ and $\mathbb{E}_\pi |g(X_0)| < \infty$. Then for an arbitrary starting state i, with probability one,*

$$\lim_{t\to\infty} \frac{\sum_{s=0}^t f(X_s)}{\sum_{s=0}^t g(X_s)} = \frac{\mathbb{E}_i \sum_{s=1}^{\tau_i} f(X_s)}{\mathbb{E}_i \sum_{s=1}^{\tau_i} g(X_s)} = \frac{\mathbb{E}_\pi f(X_0)}{\mathbb{E}_\pi g(X_0)}$$

provided not both numerator and denominator of the last terms are zero.

Setting $g \equiv 1$ we obtain a law of large numbers for a function of the Markov chain. Here is a statement of a Central Limit theorem and a Law of Iterated Logarithm.

THEOREM 59.15 *Let S be a single positive recurrent class with an invariant probability distribution π. Fix a function f and suppose that for some i, $\mathbb{E}_i\left[\tau_i^2\right] < \infty$ and $\mathbb{E}_\pi f(X_0) = 0$ and*

$$\mathbb{E}_i \left[\left(\sum_{t=1}^{\tau_i} |f(x_t)|\right)^2\right] < \infty.$$

Define

$$\gamma^2 \overset{\Delta}{=} \pi_i \cdot \mathbb{E}_i \left[\left(\sum_{t=1}^{\tau_i} f(x_t)\right)^2\right].$$

Then $\gamma^2 < \infty$, and if $\gamma^2 > 0$ then

CLT $$\frac{\sum_{s=0}^t f(X_s)}{\sqrt{t \cdot \gamma^2}}$$

converges in distribution (as $t \to \infty$) to a standard Gaussian random variable. Moreover, the sum satisfies the law of iterated logarithm, that is, the lim sup of

LIL
$$\frac{\sum_{s=0}^{t} f(X_s)}{\sqrt{2\gamma^2 t \log\log t}},$$
as $t \to \infty$, *is* 1, *and the lim inf is* (-1).

59.9 Extensions and Comments

Markov processes are very general objects, of which our treatment covered just a fraction. Many of the basic ideas extend, but definitely not all, and usually some effort is required. In addition, the mathematical difficulties rise exponentially fast. There are two obvious directions to extend: more general state spaces, and continuous time.

When the state space is not countable, the probability that an arbitrary point in the state space is "visited" by the process is usually zero. Therefore, the notions of communication, hitting, recurrence, and periodicity have to be modified. The most extensive reference here is [4].

In the discrete-space continuous time setting, the Markov property implies that if $x_t = i$ then the values of $x_s,\ s < t$ are not relevant. In particular, the length of time from the last change in state until t should be irrelevant. This implies that the distribution of the time between jumps (= change in state) should be exponential. If the only possible transition is then from i to $i+1$ and if all transition times have the same distribution (that is, they are all exponential with the same parameter), then we obtain the Poisson process. More generally we can describe most discrete-state, continuous time Markov chains as follows. The process stays at state i an exponential amount of time with parameter $\lambda(i)$. It then jumps to the next state according to a transition probability p_{ij}, and the procedure is repeated (this is correct if, for example, $\lambda(i) > \lambda > 0$ for all i). This subject is covered, for example in [1]. If we observe such a process at jump times, then we recover a Markov chain. This is one of the major tools in the analysis of continuous time chains.

Semi-Markov processes are a further generalization, where the time between events is drawn from a general distribution, which depends on the state, and possibly on the next state. This is no longer a Markov process; however, if we observe the process only at jump times, then we recover a Markov chain.

Finally, in applications, the information structure, and consequently the set of events, is richer: we can measure more than the values of the Markov chain. This is often manifested in a recursion of the type (59.2), but where the V_t are not independent. Do we still get a Markov chain? and in what sense? The rough answer is that, if the Markov property (Equation 59.4) holds, but where we condition on all the available information, then we are back on track: all of our results continue to hold. For this to happen we need the "noise sequence" $V_0, V_1, \ldots$ to be nonanticipative in a probabilistic sense.

Criteria for Stability. As in the case of dynamical systems, there are criteria for stability and for recurrence, based on Lyapunov functions. This is one of the main tools in [4]. These techniques are often the easiest and the most powerful.

Relevance to Control. Many models of control systems subject to noise can be modeled as Markov processes, and the discrete-time, discrete-space models are controlled Markov chains. Here the transition probabilities are parameterized by the control: see the section on Dynamic Programming. In addition, many filtering and identification algorithms give rise to Markov chains (usually with values in $\mathbb{R}^d$). Limit theorems for Markov chains can then be used to analyze the limiting properties of these algorithms. See, for example [4], [9].

EXAMPLE 59.11:

Extending Example 59.2, consider the recursion

$$X_{t+1} \stackrel{\Delta}{=} X_t + V_t + U_t$$

where U_t is a control variable. This is a simple instance of a controlled recursion of the ARMA type. Suppose that U_t can only take the values ± 1. Of course, we require that the control depends only on past information. If the control values U_t depend of the past states, then $X_0, X_1, \ldots$ may not be a Markov chain. For example, if we choose $U_t = \operatorname{sign}(X_0)$, then the sequence $X_0, X_1, \ldots$ violates the Markov property (Definition 59.1). However, we do have a *controlled Markov chain*. This means that Definition 59.1 is replace with the relation

$$\begin{aligned}\mathbb{P}(X_t = j \mid X_{t-1} = i_{t-1}, \ldots, X_0 = i_0, U_{t-1}, \ldots, U_0 = u_0)\\ = \mathbb{P}(X_t = j \mid X_{t-1} = i_{t-1}, U_{t-1} = u_{t-1}).\end{aligned}$$

This in fact is the general definition of a controlled Markov chain. If we choose a feedback control, that is, $U_t = f(X_t)$ for some function f, then $X_0, X_1, \ldots$ is again a Markov chain; but the transitions and the limit behavior now depend on the choice of f. For more information on controlled Markov chains, see the section on Dynamic Programming.

References

Markov chains are covered by most introductory texts on stochastic processes. Here are some more specific references.

[1] Chung, K.L., *Markov Chains with Stationary Transition Probabilities*, Springer-Verlag, New York, 1967.

A classic on discrete space Markov chains, both discrete and continuous time. Very thorough and detailed, but mathematically not elementary.

[2] Çinlar, E., *Introduction to Stochastic Processes*, Prentice Hall, 1975.

A precise, elegant and accessible book, covering the basics.

[3] Kemeny, J.G., Snell, J., and Knapp, A.W., *Denumerable Markov Chains*, Van Nostrand, Princeton, NJ, 1966.

The most elementary in this list, but fairly thorough, not only in coverage of Markov chains, but also as introduction to Markov processes.

[4] Meyn, S.P., and Tweedie, R.L., *Markov Chains and Stochastic Stability*, Springer-Verlag, London 1993.

Deals with general discrete-time Markov chains, and covers the state of the art. It is therefore demanding mathematically, although not much measure theory is required. The most comprehensive book if you can handle it.

[5] Nummelin, E., *General Irreducible Markov Chains and Non-Negative Operators*, Cambridge University Press, 1984.

More general than [8], treats general state spaces. Introduced many new techniques; perhaps less encyclopedic than [4], but fairly mathematical.

[6] Orey, S., *Limit Theorems for Markov Chain Transition Probabilities*, Van Nostrand Reinhold, London, 1971.

A thin gem on limit theorems, fairly accessible.

[7] Revuz, D., *Markov Chains*, North Holland, Amsterdam, 1984.

A mathematical, thorough treatment.

[8] Seneta, E., *Non-Negative Matrices and Markov Chains*, Springer Verlag, New York, 1981.

Gives the algebraic point of view on Markov chains.

[9] Tijms, H.C., *Stochastic Modelling and Analysis: a Computational Approach*, John Wiley, New York, 1986.

Contains a basic introduction to the subject of Markov chains, both discrete and continuous time, with a wealth of examples for applications and computations.

60

Stochastic Differential Equations

John A. Gubner
University of Wisconsin–Madison

60.1 Introduction

This chapter deals with nonlinear differential equations of the form

$$\frac{dX_t}{dt} = a(t, X_t) + b(t, X_t)Z_t, \quad X_{t_0} = \Xi,$$

where Z_t is a Gaussian white noise driving term that is independent of the random initial state Ξ. Since the solutions of these equations are random processes, we are also concerned with the probability distribution of the solution process $\{X_t\}$. The classical example is the Langevin equation,

$$\frac{dX_t}{dt} = -\mu X_t + \beta Z_t,$$

where μ and β are positive constants. In this linear differential equation, X_t models the velocity of a free particle subject to frictional forces and to impulsive forces due to collisions. Here μ is the coefficient of friction, and $\beta = \sqrt{2\mu kT/m}$, where m is the mass of the particle, k is Boltzmann's constant, and T is the absolute temperature [7]. As shown at the end of Section 5, with a suitable Gaussian initial condition, the solution of the Langevin equation is a Gaussian random process known as the Ornstein–Uhlenbeck process.

The subject of stochastic differential equations is highly technical. However, to make this chapter as accessible as possible, the presentation is mostly on a heuristic level. On occasion, when deeper theoretical results are needed, the reader is referred to an appropriate text for details. Suggestions for further reading are given at the end of the chapter.

60.1.1 Ordinary Differential Equations (ODEs)

Consider a deterministic nonlinear system whose state at time t is $x(t)$. In many engineering problems, it is reasonable to assume that x satisfies an ordinary differential equation (ODE) of the form

$$\frac{dx(t)}{dt} = a\big(t, x(t)\big) + b\big(t, x(t)\big)z(t), \quad x(t_0) = \xi, \qquad (60.1)$$

where $z(t)$ is a separately specified input signal. Note that if we integrate both sides from t_0 to t, we obtain

$$x(t) - x(t_0) = \int_{t_0}^{t} \big[a\big(\theta, x(\theta)\big) + b\big(\theta, x(\theta)\big)z(\theta)\big]\, d\theta.$$

Since $x(t_0) = \xi$, $x(t)$ satisfies the integral equation

$$x(t) = \xi + \int_{t_0}^{t} \big[a\big(\theta, x(\theta)\big) + b\big(\theta, x(\theta)\big)z(\theta)\big]\, d\theta.$$

Under certain technical conditions, e.g., [3], it can be shown that there exists a unique solution to Equation 60.1; this is usually accomplished by solving the corresponding integral equation.

Now suppose $x(t)$ satisfies Equation 60.1. If $x(t)$ is passed through a nonlinearity, say $y(t) := g\big(x(t)\big)$, then $y(t_0) = g(\xi)$,

0-8493-8570-9/96/$0.00+$.50

and by the chain rule, $y(t)$ satisfies the differential equation,

$$\begin{aligned}\frac{dy(t)}{dt} &= g'\big(x(t)\big)\frac{dx(t)}{dt}\\ &= g'\big(x(t)\big)a\big(t,x(t)\big)+g'\big(x(t)\big)b\big(t,x(t)\big)z(t),\end{aligned} \quad (60.2)$$

assuming g is differentiable.

60.1.2 Stochastic Differential Equations (SDEs)

If x models a mechanical system subject to significant vibration, or if x models an electronic system subject to significant thermal noise, it makes sense to regard $z(t)$ as a stochastic, or random process, which we denote by Z_t. (Our convention is to denote deterministic functions by lowercase letters with arguments in parentheses and to denote random functions by uppercase letters with subscript arguments.) Now, with a random input signal Z_t, the ODE of Equation 60.1 becomes the stochastic differential equation (SDE),

$$\frac{dX_t}{dt} = a(t,X_t)+b(t,X_t)Z_t, \quad X_{t_0}=\Xi, \quad (60.3)$$

where the initial condition Ξ is also random. As our notation indicates, the solution of an SDE is a random process. Typically, we take Z_t to be a white noise process; i.e,

$$\mathsf{E}[Z_t] = 0 \quad \text{and} \quad \mathsf{E}[Z_tZ_s] = \delta(t-s),$$

where E denotes expectation and δ is the Dirac delta function. In this discussion we further restrict attention to Gaussian white noise. The surprising thing about white noise is that it cannot exist as an ordinary random process (though it does exist as a generalized process [1]). Fortunately, there is a well-defined ordinary random process, known as the **Wiener process** (also known as **Brownian motion**), denoted by W_t, that makes a good model for integrated white noise, i.e., W_t behaves as if

$$W_t = \int_0^t Z_\theta\,d\theta,$$

or symbolically, $dW_t = Z_t dt$. Thus, if we multiply Equation 60.3 by dt, and write dW_t for $Z_t dt$, we obtain

$$dX_t = a(t,X_t)\,dt+b(t,X_t)\,dW_t, \quad X_{t_0}=\Xi. \quad (60.4)$$

To give meaning to Equation 60.4 and to solve Equation 60.4, we will always understand it as shorthand for the corresponding integral equation,

$$X_t = \Xi+\int_{t_0}^t a(\theta,X_\theta)\,d\theta+\int_{t_0}^t b(\theta,X_\theta)\,dW_\theta. \quad (60.5)$$

In order to make sense of Equation 60.5, we have to assign a meaning to integrals with respect to a Wiener process. There are two different ways to do this. One is due to Itô, and the other is due to Stratonovich. Since the Itô integral is more popular, and since the Stratonovich integral can be expressed in terms of the Itô integral [1], we restrict attention in our discussion to the Itô integral.

Now suppose X_t is a solution to the SDE of Equation 60.4, and suppose we pass X_t through a nonlinearity, say $Y_t := g(X_t)$. Of course, $Y_{t_0} = g(\Xi)$, but astonishingly, by the stochastic chain rule, the analog of Equation 60.2 is [1]

$$\begin{aligned}dY_t &= g'(X_t)\,dX_t+\tfrac{1}{2}g''(X_t)b(t,X_t)^2\,dt\\ &= g'(X_t)a(t,X_t)\,dt+g'(X_t)b(t,X_t)\,dW_t\\ &\quad+\tfrac{1}{2}g''(X_t)b(t,X_t)^2\,dt,\end{aligned} \quad (60.6)$$

assuming g is twice continuously differentiable. Equation 60.6 is known as **Itô's rule**, and the last term in Equation 60.6 is called the **Itô correction term**. In addition to explaining its presence, the remainder of our discussion is as follows. Section 2 introduces the Wiener process as a model for integrated white noise. In Section 3, integration with respect to the Wiener process is defined, and a simple form of Itô's rule is derived. Section 4 focuses on SDEs. Itô's rule is derived for time-invariant nonlinearities, and its extension to time-varying nonlinearities is also given. In Section 5, Itô's rule is used to solve special forms of linear SDEs. ODEs are derived for the mean and variance of the solution in this case. When the initial condition is also Gaussian, the solution to the linear SDE is a Gaussian process, and its distribution is completely determined by its mean and variance. Nonlinear SDEs are considered in Section 6. The solutions of nonlinear SDEs are non-Gaussian Markov processes. In this case, we characterize their transition distribution in terms of the **Kolmogorov forward (Fokker–Planck)** and backward partial differential equations.

60.2 White Noise and the Wiener Process

A random process $\{Z_t\}$ is said to be a **white noise process** if

$$\mathsf{E}[Z_t] = 0 \quad \text{and} \quad \mathsf{E}[Z_tZ_s] = \delta(t-s), \quad (60.7)$$

where $\delta(t)$ is the Dirac delta, which is characterized by the two properties $\delta(t)=0$ for $t\neq 0$ and $\int_{-\infty}^{\infty}\delta(t)\,dt = 1$.

Consider the **integrated white noise**

$$W_t = \int_0^t Z_u\,du. \quad (60.8)$$

We show that integrated white noise satisfies the following five properties. For $0\le\theta\le\tau\le s\le t$,

$$\begin{aligned}W_0 &= 0, && (60.9)\\ \mathsf{E}[W_t] &= 0, && (60.10)\\ \mathsf{E}\big[(W_t-W_s)^2\big] &= t-s, && (60.11)\\ \mathsf{E}\big[(W_t-W_s)(W_\tau-W_\theta)\big] &= 0, && (60.12)\\ \mathsf{E}[W_tW_s] &= \min\{t,s\}. && (60.13)\end{aligned}$$

In other words:

- W_0 is a constant random variable with value zero.

- W_t has zero mean.
- $W_t - W_s$ has variance $t - s$.
- If $(\theta, \tau]$ and $(s, t]$ are nonoverlapping time intervals, then the increments $W_\tau - W_\theta$ and $W_t - W_s$ are uncorrelated.
- The correlation between W_t and W_s is $\mathsf{E}[W_t W_s] = \min\{t, s\}$.

(A process that satisfies Equation 60.12 is said to have orthogonal increments. A very accessible introduction to orthogonal increments processes can be found in [4].)

The property defined in Equation 60.9 is immediate from Equation 60.8. To establish Equation 60.10, write $\mathsf{E}[W_t] = \int_0^t \mathsf{E}[Z_u]\,du = 0$. To derive Equation 60.11, write

$$\begin{aligned}
\mathsf{E}[(W_t - W_s)^2] &= \mathsf{E}\left[\left(\int_s^t Z_u\,du\right)\left(\int_s^t Z_v\,dv\right)\right] \\
&= \int_s^t \left(\int_s^t \mathsf{E}[Z_u Z_v]\,du\right)dv \\
&= \int_s^t \left(\int_s^t \delta(u - v)\,du\right)dv \\
&= \int_s^t 1\,dv \\
&= t - s.
\end{aligned}$$

To obtain Equation 60.12, write

$$\begin{aligned}
\mathsf{E}[(W_t - W_s)(W_\tau - W_\theta)] &= \mathsf{E}\left[\left(\int_s^t Z_u\,du\right)\left(\int_\theta^\tau Z_v\,dv\right)\right] \\
&= \int_\theta^\tau \left(\int_s^t \delta(u - v)\,du\right)dv.
\end{aligned}$$

Because the ranges of integration, which are understood as $(\theta, \tau]$ and $(s, t]$, do not intersect, $\delta(u - v) = 0$, and the inner integral is zero. Finally, the properties defined in Equations 60.9, 60.11, and 60.12 yield Equation 60.13 by writing, when $t > s$,

$$\begin{aligned}
\mathsf{E}[W_t W_s] &= \mathsf{E}[(W_t - W_s)W_s] + \mathsf{E}[W_s^2] \\
&= \mathsf{E}[(W_t - W_s)W_s] + s \\
&= \mathsf{E}[(W_t - W_s)(W_s - W_0)] + s \\
&= s. \qquad (60.14)
\end{aligned}$$

If $t < s$, a symmetric argument yields $\mathsf{E}[W_t W_s] = t$. Hence, we can in general write $\mathsf{E}[W_t W_s] = \min\{t, s\}$.

It is well known that no process $\{Z_t\}$ satisfying Equation 60.7 can exist in the usual sense [1]. Hence, defining W_t by Equation 60.8 does not make sense. Fortunately, it is possible to define a random process $\{W_t, t \geq 0\}$ satisfying Equations 60.9 to 60.13 (as well as additional properties).

DEFINITION 60.1 The standard **Wiener process**, or **Brownian motion**, denoted by $\{W_t, t \geq 0\}$, is characterized by the following four properties:

W-1 $W_0 = 0$.

W-2 For $0 \leq s < t$, the increment $W_t - W_s$ is a Gaussian random variable with zero mean and variance $t - s$, i.e.,

$$\Pr(W_t - W_s \leq w) = \int_{-\infty}^{w} \frac{\exp\left(-\frac{x^2}{2(t-s)}\right)}{\sqrt{2\pi(t-s)}}\,dx.$$

W-3 $\{W_t, t \geq 0\}$ has independent increments; i.e., if $0 \leq t_1 \leq \cdots \leq t_n$, then the increments

$$(W_{t_2} - W_{t_1}), (W_{t_3} - W_{t_2}), \ldots, (W_{t_n} - W_{t_{n-1}})$$

are statistically independent random variables.

W-4 $\{W_t, t \geq 0\}$ has continuous sample paths with probability 1.

A proof of the existence of the Wiener process is given, for example, in [2] and in [4]. A sample path of a standard Wiener process is shown in Figure 60.1.

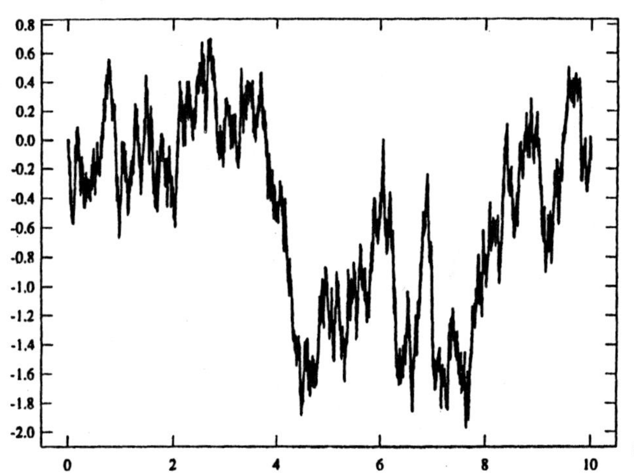

Figure 60.1 Sample path of a standard Wiener process.

We now show that properties W-1, W-2, and W-3 are sufficient to prove that the Wiener process satisfies Equations 60.9 to 60.13. Clearly, property W-1 and Equation 60.9 are the same. To establish Equation 60.10, put $s = 0$ in property W-2 and use property W-1. It is clear that property W-2 implies Equation 60.11. Also, Equation 60.12 is an immediate consequence of properties W-3 and W-2. Finally, since the Wiener process satisfies Equations 60.9, 60.11, and 60.12, the derivation in Equation 60.14 holds for the Wiener process, and thus Equation 60.13 also holds for the Wiener process.

REMARK 60.1 From Figure 60.1, we see that the Wiener process has very jagged sample paths. In fact, if we zoom in on any subinterval, say [2, 4] as shown in Figure 60.2, the sample path looks just as jagged. In other words, the Wiener process is continuous, but seems to have corners everywhere. In fact, it can be shown mathematically [2] that the sample paths of the Wiener process are nowhere differentiable. In other words, W_t cannot be the integral of any reasonable function, which is consistent with our earlier claim that continuous-time white noise cannot exist as an ordinary random process.

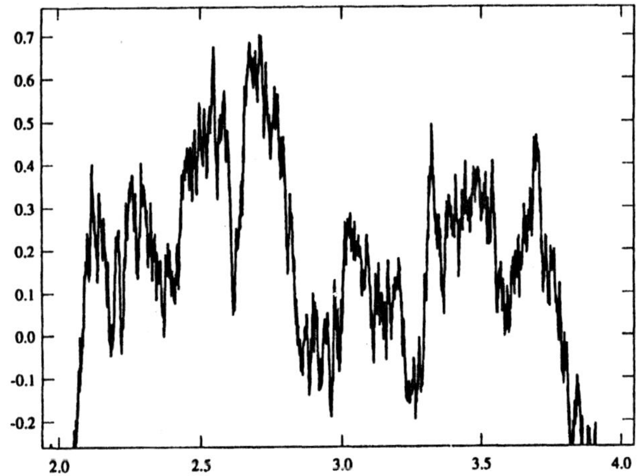

Figure 60.2 Closeup of sample path in Figure 60.1.

60.3 The Itô Integral

Let $\{W_t, t \geq 0\}$ be a standard Wiener process. The **history** of the process up to (and including) time t is denoted by $\mathcal{F}_t := \sigma(W_\theta, 0 \leq \theta \leq t)$. For our purposes, we say that a random variable X is **$\mathcal{F}_t$-measurable** if it is a function of the history up to time t; i.e., if there is a deterministic function h such that $X = h(W_\theta, 0 \leq \theta \leq t)$. For example, if $X = W_t - W_{t/2}$, then X is $\mathcal{F}_t$-measurable. Another example would be $X = \int_0^t W_\theta \, d\theta$; in this case, X depends on all of the variables $W_\theta, 0 \leq \theta \leq t$ and is $\mathcal{F}_t$-measurable. We also borrow the following notation from probability theory. If Z is any random variable, we write

$$\mathsf{E}[Z|\mathcal{F}_t] \quad \text{instead of} \quad \mathsf{E}[Z|W_\theta, 0 \leq \theta \leq t].$$

If Z is arbitrary but X is $\mathcal{F}_t$-measurable, then [2]

$$\mathsf{E}[XZ|\mathcal{F}_t] = X\,\mathsf{E}[Z|\mathcal{F}_t]. \tag{60.15}$$

A special case of Equation 60.15 is obtained if $Z = 1$; then $\mathsf{E}[Z|\mathcal{F}_t] = \mathsf{E}[1|\mathcal{F}_t] = 1$, and hence, if X is $\mathcal{F}_t$-measurable,

$$\mathsf{E}[X|\mathcal{F}_t] = X. \tag{60.16}$$

We also require the following results from probability theory [2], which we refer to as the **smoothing properties of conditional expectation:**

$$\mathsf{E}[Z] = \mathsf{E}\big[\mathsf{E}[Z|\mathcal{F}_t]\big], \quad t \geq 0, \tag{60.17}$$

and

$$\mathsf{E}[Z|\mathcal{F}_s] = \mathsf{E}\big[\mathsf{E}[Z|\mathcal{F}_t]\big|\mathcal{F}_s\big], \quad t \geq s \geq 0. \tag{60.18}$$

DEFINITION 60.2 A random process $\{H_t, t \geq 0\}$ is $\{\mathcal{F}_t\}$-**adapted** (or simply adapted if $\{\mathcal{F}_t\}$ is understood), if for each t, H_t is an $\mathcal{F}_t$-measurable random variable.

Obviously, $\{W_t, t \geq 0\}$ is $\{\mathcal{F}_t\}$-adapted. We now show that W_t is a **martingale**, i.e.,

$$\mathsf{E}[W_t|\mathcal{F}_s] = W_s, \quad t \geq s. \tag{60.19}$$

To see this, first note that since $W_0 = 0$,

$$\mathcal{F}_s := \sigma(W_\theta, 0 \leq \theta \leq s) = \sigma(W_\theta - W_0, 0 \leq \theta \leq s).$$

It follows, on account of the independent increments of the Wiener process, that $W_t - W_s$ is independent of the history $\mathcal{F}_s$ for $t \geq s$; i.e., for any function f,

$$\mathsf{E}[f(W_t - W_s)|\mathcal{F}_s] = \mathsf{E}[f(W_t - W_s)].$$

Then, since $W_t - W_s$ has zero mean,

$$\mathsf{E}[W_t - W_s|\mathcal{F}_s] = \mathsf{E}[W_t - W_s] = 0, \quad t \geq s, \tag{60.20}$$

which is equivalent to Equation 60.19.

Having shown that W_t is a martingale, we now show that $W_t^2 - t$ is also a martingale. To do this, we need the following three facts:

1. $W_t - W_s$ is independent of $\mathcal{F}_s$ with $\mathsf{E}[(W_t - W_s)^2] = t - s$.
2. W_t is a martingale (cf. Equation 60.19).
3. Properties defined in Equations 60.15 and 60.16.

For $t > s$, write

$$\begin{aligned}
\mathsf{E}[W_t^2|\mathcal{F}_s] &= \mathsf{E}[(W_t - W_s)^2 + 2W_tW_s - W_s^2|\mathcal{F}_s] \\
&= \mathsf{E}[(W_t - W_s)^2|\mathcal{F}_s] + 2W_s\mathsf{E}[W_t|\mathcal{F}_s] - W_s^2 \\
&= t - s + 2W_s^2 - W_s^2 \\
&= t + W_s^2 - s.
\end{aligned}$$

Rearranging, we have

$$\mathsf{E}[W_t^2 - t|\mathcal{F}_s] = W_s^2 - s, \quad t > s,$$

i.e., $W_t^2 - t$ is a martingale.

60.3.1 Definition of Itô's Stochastic Integral

We now define the Itô integral. As in the development of the Riemann integral, we begin by defining the integral for functions that are piecewise constant in time; i.e., we consider integrands $\{H_t, t \geq 0\}$ that are $\{\mathcal{F}_t\}$-adapted processes satisfying

$$H_t = H_{t_i} \quad \text{for} \quad t_i \leq t < t_{i+1}, \tag{60.21}$$

for some breakpoints $t_i < t_{i+1}$. Thus, while $H_t = H_{t_i}$ on $[t_i, t_{i+1})$, the value H_{t_i} is an $\mathcal{F}_{t_i}$-measurable random variable. Without loss of generality, given any $0 \le s < t$, we may assume $s = t_0 < \cdots < t_n = t$. Then the Itô integral is defined to be

$$\int_s^t H_\theta \, dW_\theta := \sum_{i=0}^{n-1} H_{t_i}(W_{t_{i+1}} - W_{t_i}). \tag{60.22}$$

To handle the general case, suppose H_t is a process for which there exists a sequence of processes H_t^k of the form of Equation 60.21 (where now n and the breakpoints $\{t_i\}$ depend on k) such that

$$\lim_{k\to\infty} \int_s^t \mathsf{E}[|H_\theta^k - H_\theta|^2] \, d\theta = 0.$$

Then we take $\int_s^t H_\theta \, dW_\theta$ to be the mean-square limit (which exists) of $\int_s^t H_\theta^k \, dW_\theta$; i.e., there exists a random variable, denoted by $\int_s^t H_\theta \, dW_\theta$, such that

$$\lim_{k\to\infty} \mathsf{E}\left[\left|\int_s^t H_\theta^k \, dW_\theta - \int_s^t H_\theta \, dW_\theta\right|^2\right] = 0. \tag{60.23}$$

See [10] for details. It should also be noted that when $H_t = h(t)$ is a deterministic function, the right-hand side of Equation 60.22 is a Gaussian random variable, and since mean-square limits of such quantities are also Gaussian [4], $\left\{\int_s^t h(\theta)\, dW_\theta, t \ge s\right\}$ is a Gaussian random process. When the integrand of an Itô integral is deterministic, the integral is sometimes called a **Wiener integral**.

60.3.2 Properties of the Itô Integral

The Itô integral satisfies the following three properties. First, the Itô integral is a zero-mean random variable, i.e.,

$$\mathsf{E}\left[\int_s^t H_\theta \, dW_\theta\right] = 0. \tag{60.24}$$

Second, the variance of $\int_s^t H_\theta \, dW_\theta$ is

$$\mathsf{E}\left[\left(\int_s^t H_\theta \, dW_\theta\right)^2\right] = \int_s^t \mathsf{E}[H_\theta^2] \, d\theta. \tag{60.25}$$

Third, if $X_t := \int_0^t H_\theta \, dW_\theta$, then X_t is a martingale, i.e., $\mathsf{E}[X_t|\mathcal{F}_s]$
$= X_s$, or in terms of Itô integrals,

$$\mathsf{E}\left[\int_0^t H_\theta \, dW_\theta \middle| \mathcal{F}_s\right] = \int_0^s H_\theta \, dW_\theta, \quad t \ge s.$$

We verify these properties when H_t is of the form of Equation 60.21. To prove Equation 60.24, take the expectations of Equation 60.22 to obtain

$$\mathsf{E}\left[\int_s^t H_\theta \, dW_\theta\right] = \sum_{i=0}^{n-1} \mathsf{E}[H_{t_i}(W_{t_{i+1}} - W_{t_i})].$$

By the first smoothing property of conditional expectation, the fact that H_{t_i} is $\mathcal{F}_{t_i}$-measurable, and the properties defined in Equations 60.15 and 60.20,

$$\begin{aligned} \mathsf{E}[H_{t_i}(W_{t_{i+1}} - W_{t_i})] &= \mathsf{E}\big[\mathsf{E}[H_{t_i}(W_{t_{i+1}} - W_{t_i})|\mathcal{F}_{t_i}]\big] \\ &= \mathsf{E}\big[H_{t_i}\mathsf{E}[W_{t_{i+1}} - W_{t_i}|\mathcal{F}_{t_i}]\big] \\ &= 0. \end{aligned}$$

To establish Equation 60.25, use Equation 60.22 to write

$$\mathsf{E}\left[\left(\int_s^t H_\theta \, dW_\theta\right)^2\right] = \sum_{i=0}^{n-1}\sum_{j=0}^{n-1} \mathsf{E}[H_{t_i}H_{t_j}(W_{t_{i+1}} - W_{t_i})(W_{t_{j+1}} - W_{t_j})].$$

First consider a typical term in the double sum for which $j = i$. Then by the first smoothing property and Equation 60.15,

$$\begin{aligned} \mathsf{E}[H_{t_i}^2(W_{t_{i+1}} - W_{t_i})^2] &= \mathsf{E}\big[\mathsf{E}[H_{t_i}^2(W_{t_{i+1}} - W_{t_i})^2|\mathcal{F}_{t_i}]\big] \\ &= \mathsf{E}\big[H_{t_i}^2\mathsf{E}[(W_{t_{i+1}} - W_{t_i})^2|\mathcal{F}_{t_i}]\big]. \end{aligned}$$

Since $W_{t_{i+1}} - W_{t_i}$ is independent of $\mathcal{F}_{t_i}$,

$$\mathsf{E}[(W_{t_{i+1}} - W_{t_i})^2|\mathcal{F}_{t_i}] = \mathsf{E}[(W_{t_{i+1}} - W_{t_i})^2] = t_{i+1} - t_i.$$

Hence,

$$\mathsf{E}[H_{t_i}^2(W_{t_{i+1}} - W_{t_i})^2] = \mathsf{E}[H_{t_i}^2](t_{i+1} - t_i).$$

For the terms with $j \ne i$, we can, without loss of generality, take $j < i$. In this case,

$$\begin{aligned} &\mathsf{E}[H_{t_i}H_{t_j}(W_{t_{i+1}} - W_{t_i})(W_{t_{j+1}} - W_{t_j})] \\ &\quad= \mathsf{E}\big[\mathsf{E}[H_{t_i}H_{t_j}(W_{t_{i+1}} - W_{t_i})(W_{t_{j+1}} - W_{t_j})|\mathcal{F}_{t_i}]\big] \\ &\quad= \mathsf{E}\big[H_{t_i}H_{t_j}(W_{t_{j+1}} - W_{t_j})\mathsf{E}[W_{t_{i+1}} - W_{t_i}|\mathcal{F}_{t_i}]\big] \\ &\quad= 0, \qquad \text{by Equation 60.20.} \end{aligned}$$

Thus,

$$\begin{aligned} \mathsf{E}\left[\left(\int_s^t H_\theta \, dW_\theta\right)^2\right] &= \sum_{i=0}^{n-1} \mathsf{E}[H_{t_i}^2](t_{i+1} - t_i) \\ &= \int_s^t \mathsf{E}[H_\theta^2] \, d\theta. \end{aligned}$$

To show that $X_t := \int_0^t H_\theta \, dW_\theta$ is a martingale, it suffices to prove that $\mathsf{E}[X_t - X_s|\mathcal{F}_s] = 0$. Write

$$\begin{aligned} \mathsf{E}[X_t - X_s|\mathcal{F}_s] &= \mathsf{E}\left[\int_s^t H_\theta \, dW_\theta \middle| \mathcal{F}_s\right] \\ &= \mathsf{E}\left[\sum_{i=0}^{n-1} H_{t_i}(W_{t_{i+1}} - W_{t_i}) \middle| \mathcal{F}_s\right] \\ &= \sum_{i=0}^{n-1} \mathsf{E}[H_{t_i}(W_{t_{i+1}} - W_{t_i})|\mathcal{F}_s]. \end{aligned}$$

Then by the second smoothing property of conditional expectation and the properties defined in Equations 60.15 and 60.20,

$$\mathsf{E}[X_t - X_s|\mathcal{F}_s] = \sum_{i=0}^{n-1} \mathsf{E}\big[H_{t_i}\mathsf{E}[W_{t_{i+1}} - W_{t_i}|\mathcal{F}_{t_i}]\big|\mathcal{F}_s\big] = 0.$$

60.3.3 A Simple Form of Itô's Rule

The following result is essential to derive Itô's rule [1].

LEMMA 60.1 For any partition of $[s, t]$, say $s = t_0 < \cdots < t_n = t$, put $\Delta := \max_{0 \le i \le n-1} |t_{i+1} - t_i|$. If

$$V := \sum_{i=0}^{n-1} (W_{t_{i+1}} - W_{t_i})^2, \qquad (60.26)$$

then

$$\mathsf{E}[|V - (t-s)|^2] \le 2\Delta(t-s).$$

The importance of the lemma is that it implies that as the Δ of the partition becomes small, the sum of squared increments V converges in mean square to the length of the interval, $t - s$.

PROOF 60.1 The first step is to note that $t - s = \sum_{i=0}^{n-1}(t_{i+1} - t_i)$. Next, let D denote the difference

$$D := V - (t-s) = \sum_{i=0}^{n-1} \left\{ (W_{t_{i+1}} - W_{t_i})^2 - (t_{i+1} - t_i) \right\}.$$

Thus, D is a sum of independent, zero-mean random variables. It follows that the expectation of the cross terms in D^2 vanishes, thus leaving

$$\mathsf{E}[D^2] = \sum_{i=0}^{n-1} \mathsf{E}\left[\left\{ (W_{t_{i+1}} - W_{t_i})^2 - (t_{i+1} - t_i) \right\}^2 \right].$$

Put $Z_i := (W_{t_{i+1}} - W_{t_i})/\sqrt{t_{i+1} - t_i}$. Then Z_i is a zero-mean Gaussian random variable with variance 1 (which implies $\mathsf{E}[Z_i^4] = 3$). We can now write

$$\begin{aligned}
\mathsf{E}[D^2] &= \sum_{i=0}^{n-1} \mathsf{E}\left[\left\{ (Z_i^2 - 1)(t_{i+1} - t_i) \right\}^2 \right] \\
&= \sum_{i=0}^{n-1} \mathsf{E}[(Z_i^2 - 1)^2](t_{i+1} - t_i)^2 \\
&= \sum_{i=0}^{n-1} 2(t_{i+1} - t_i)^2 \\
&\le \sum_{i=0}^{n-1} 2\Delta(t_{i+1} - t_i) \\
&= 2\Delta(t-s).
\end{aligned}$$

Let $\{H_t, t \ge 0\}$ be a continuous $\{\mathcal{F}_t\}$-adapted process. It can be shown [1] that as the partition becomes finer,

$$\sum_{i=0}^{n-1} H_{t_i}(W_{t_{i+1}} - W_{t_i})^2 \to \sum_{i=0}^{n-1} H_{t_i}(t_{i+1} - t_i) \to \int_s^t H_\theta \, d\theta.$$

We now derive a special case of Itô's rule. Let $g(w)$ be a twice continuously differentiable function of w. If $Y_t = g(W_t)$, our intuition about the chain rule might lead us to write $dY_t = g'(W_t)\,dW_t$. As we now show, the correct answer is $dY_t = g'(W_t)\,dW_t + \frac{1}{2}g''(W_t)\,dt$.

Consider the Taylor expansion,

$$g(w_2) - g(w_1) \approx g'(w_1)(w_2 - w_1) + \tfrac{1}{2}g''(w_1)(w_2 - w_1)^2.$$

Suppose $Y_t = g(W_t)$. For $s = t_0 < \cdots < t_n = t$, write

$$\begin{aligned}
Y_t - Y_s &= \sum_{i=0}^{n-1} g(W_{t_{i+1}}) - g(W_{t_i}) \\
&\approx \sum_{i=0}^{n-1} \left\{ g'(W_{t_i})(W_{t_{i+1}} - W_{t_i}) + \tfrac{1}{2}g''(W_{t_i})(W_{t_{i+1}} - W_{t_i})^2 \right\}.
\end{aligned}$$

Note that $g'(W_{t_i})$ and $g''(W_{t_i})$ are $\mathcal{F}_{t_i}$-measurable. Hence, as the partition becomes finer, we obtain

$$Y_t - Y_s = \int_s^t g'(W_\theta)\,dW_\theta + \tfrac{1}{2}\int_s^t g''(W_\theta)\,d\theta.$$

Writing this in differential form, we have a special case of Itô's rule: If $Y_t = g(W_t)$, then

$$dY_t = g'(W_t)\,dW_t + \tfrac{1}{2}g''(W_t)\,dt.$$

EXAMPLE 60.1:

As a simple application of this result, we show that

$$\int_0^t W_\theta\,dW_\theta = (W_t^2 - t)/2.$$

Take $g(w) = w^2$. Then $g'(w) = 2w$, and $g''(w) = 2$. The special case of Itô's rule gives

$$dY_t = 2W_t\,dW_t + 1\,dt.$$

Converting this to integral form and noting that $Y_0 = W_0^2 = 0$,

$$\begin{aligned}
Y_t &= Y_0 + \int_0^t 2W_\theta\,dW_\theta + \int_0^t 1\,d\theta \\
&= 2\int_0^t W_\theta\,dW_\theta + t.
\end{aligned}$$

Since $Y_t = g(W_t) = W_t^2$, the result follows. As noted earlier, integrals with respect to the Wiener process are martingales. Since $\int_0^t W_\theta\,dW_\theta = (W_t^2 - t)/2$, we now have an alternative proof to the one following Equation 60.20 that $W_t^2 - t$ is a martingale.

60.4 Stochastic Differential Equations and Itô's Rule

Suppose X_t satisfies the SDE

$$dX_t = a(t, X_t)\,dt + b(t, X_t)\,dW_t, \qquad (60.27)$$

or equivalently, the integral equation,

$$X_t = X_s + \int_s^t a(\theta, X_\theta)\,d\theta + \int_s^t b(\theta, X_\theta)\,dW_\theta. \qquad (60.28)$$

If $Y_t = g(X_t)$, where g is twice continuously differentiable, we show that Y_t satisfies the SDE

$$dY_t = g'(X_t)\,dX_t + \tfrac{1}{2}g''(X_t)b(t, X_t)^2\,dt. \qquad (60.29)$$

Using the Taylor expansion of g as we did in the preceding section, write

$$Y_t - Y_s \approx \sum_{i=0}^{n-1}\Big\{g'(X_{t_i})(X_{t_{i+1}} - X_{t_i}) + \tfrac{1}{2}g''(X_{t_i})(X_{t_{i+1}} - X_{t_i})^2\Big\}.$$

From Equation 60.28 we have the approximation

$$X_{t_{i+1}} - X_{t_i} \approx a(t_i, X_{t_i})(t_{i+1} - t_i) + b(t_i, X_{t_i})(W_{t_{i+1}} - W_{t_i}).$$

Hence, as the partition becomes finer,

$$\sum_{i=0}^{n-1} g'(X_{t_i})(X_{t_{i+1}} - X_{t_i})$$

converges to

$$\int_s^t g'(X_\theta)a(\theta, X_\theta)\,d\theta + \int_s^t g'(X_\theta)b(\theta, X_\theta)\,dW_\theta.$$

It remains to consider sums of the form (cf. Equation 60.26)

$$\sum_{i=0}^{n-1} g''(X_{t_i})(X_{t_{i+1}} - X_{t_i})^2.$$

The ith term in the sum is approximately

$$\begin{aligned} g''(X_{t_i})\Big\{&a(t_i, X_{t_i})^2(t_{i+1} - t_i)^2 \\ &+ 2a(t_i, X_{t_i})b(t_i, X_{t_i})(t_{i+1} - t_i)(W_{t_{i+1}} - W_{t_i}) \\ &+ b(t_i, X_{t_i})^2(W_{t_{i+1}} - W_{t_i})^2\Big\}. \end{aligned}$$

Now with $\Delta = \max_{0 \le i \le n-1} |t_{i+1} - t_i|$,

$$\begin{aligned} &\left|\sum_{i=0}^{n-1} g''(X_{t_i})a(t_i, X_{t_i})^2(t_{i+1} - t_i)^2\right| \\ &\qquad \le \Delta \sum_{i=0}^{n-1} |g''(X_{t_i})| a(t_i, X_{t_i})^2(t_{i+1} - t_i), \end{aligned}$$

which converges to zero as $\Delta \to 0$. Also,

$$\sum_{i=0}^{n-1} g''(X_{t_i})a(t_i, X_{t_i})(t_{i+1} - t_i)(W_{t_{i+1}} - W_{t_i})$$

converges to 0. Finally, note that

$$\sum_{i=0}^{n-1} g''(X_{t_i})b(t_i, X_{t_i})^2(W_{t_{i+1}} - W_{t_i})^2$$

converges to $\int_s^t g''(X_\theta)b(\theta, X_\theta)^2\,d\theta$. Putting this all together, as the partition becomes finer, we have

$$\begin{aligned} Y_t - Y_s &= \int_s^t g'(X_\theta)\big[a(\theta, X_\theta)\,dt + b(\theta, X_\theta)\,dW_\theta\big] \\ &\quad + \tfrac{1}{2}\int_s^t g''(X_\theta)b(\theta, X_\theta)^2\,d\theta, \end{aligned}$$

which is indeed the integral form of Itô's rule in Equation 60.29.

Itô's rule can be extended to handle a time-varying nonlinearity $g(t, x)$ whose partial derivatives

$$g_t := \frac{\partial g}{\partial t}, \quad g_x := \frac{\partial g}{\partial x}, \quad \text{and} \quad g_{xx} := \frac{\partial^2 g}{\partial x^2}$$

are continuous [1]. If $Y_t = g(t, X_t)$, where X_t satisfies Equation 60.27, then

$$\begin{aligned} dY_t &= g_t(t, X_t)\,dt + g_x(t, X_t)\,dX_t \\ &\quad + \tfrac{1}{2}g_{xx}(t, X_t)b(t, X_t)^2 dt \\ &= g_t(t, X_t)\,dt + g_x(t, X_t)a(t, X_t)\,dt \\ &\quad + g_x(t, X_t)b(t, X_t)\,dW_t \\ &\quad + \tfrac{1}{2}g_{xx}(t, X_t)b(t, X_t)^2\,dt. \end{aligned} \qquad (60.30)$$

EXAMPLE 60.2:

Consider the Langevin equation

$$dX_t = -3X_t\,dt + 5\,dW_t.$$

Suppose $Y_t = \sin(tX_t)$. Then with $g(t, x) = \sin(tx)$, $g_t(t, x) = x\cos(tx)$, $g_x(t, x) = t\cos(tx)$, and $g_{xx}(t, x) = -t^2\sin(tx)$. By the extended Itô's rule,

$$\begin{aligned} dY_t &= [(1 - 3t)X_t\cos(tX_t) - \tfrac{25}{2}t^2\sin(tX_t)]\,dt \\ &\quad + 5t\cos(tX_t)\,dW_t. \end{aligned}$$

60.5 Applications of Itô's Rule to Linear SDEs

Using Itô's rule, we can verify explicit solutions to linear SDEs. By a linear SDE we mean an equation of the form

$$dX_t = [\alpha(t) + c(t)X_t]\,dt + [\beta(t) + \gamma(t)X_t]\,dW_t, \quad X_{t_0} = \Xi. \qquad (60.31)$$

If Ξ is independent of $\{W_t - W_{t_0}, t \ge t_0\}$, and if c, β, and γ are bounded on a finite interval $[t_0, t_f]$, then a unique continuous solution exists on $[t_0, t_f]$; if α, c, β, and γ are bounded on $[t_0, t_f]$ for every finite $t_f > t_0$, then a unique solution exists on $[t_0, \infty)$ [1].

60.5.1 Homogeneous Equations

A linear SDE of the form

$$dX_t = c(t)X_t\,dt + \gamma(t)X_t\,dW_t, \quad X_{t_0} = \Xi, \qquad (60.32)$$

is said to be **homogeneous**. In this case, we claim that the solution is $X_t = \Xi\exp(Y_t)$, where

$$Y_t := \int_{t_0}^t [c(\theta) - \gamma(\theta)^2/2]\,d\theta + \int_{t_0}^t \gamma(\theta)\,dW_\theta,$$

or in differential form,

$$dY_t = [c(t) - \gamma(t)^2/2]\,dt + \gamma(t)\,dW_t.$$

To verify our claim, we follow [1] and simply apply Itô's rule of Equation 60.29:

$$\begin{aligned} dX_t &= \Xi\exp(Y_t)\,dY_t + \tfrac{1}{2}\Xi\exp(Y_t)\gamma(t)^2\,dt \\ &= X_t[c(t) - \gamma(t)^2/2]\,dt + X_t\gamma(t)\,dW_t + \tfrac{1}{2}X_t\gamma(t)^2\,dt \\ &= c(t)X_t\,dt + \gamma(t)X_t\,dW_t. \end{aligned}$$

Since $X_{t_0} = \Xi\exp(Y_{t_0}) = \Xi\exp(0) = \Xi$, we have indeed solved Equation 60.32.

EXAMPLE 60.3:

Consider the homogeneous SDE

$$dX_t = \cos(t)X_t\,dt + 2X_t\,dW_t, \quad X_0 = 1.$$

For this problem, $Y_t = \sin(t) - 2t + 2W_t$, and the solution is $X_t = e^{Y_t} = e^{\sin(t)-2(t-W_t)}$.

60.5.2 Linear SDEs in the Narrow Sense

A linear SDE of the form

$$dX_t = [\alpha(t) + c(t)X_t]\,dt + \beta(t)\,dW_t, \quad X_{t_0} = \Xi, \tag{60.33}$$

is said to be linear in the **narrow sense** because it is obtained by setting $\gamma(t) = 0$ in the general linear SDE in Equation 60.31. The solution of this equation is obtained as follows. First put

$$\Phi(t, t_0) := \exp\left(\int_{t_0}^t c(\theta)\,d\theta\right).$$

Observe that

$$\frac{\partial\Phi(t, t_0)}{\partial t} = c(t)\Phi(t, t_0), \tag{60.34}$$

$\Phi(t_0, t_0) = 1$, and $\Phi(t, t_0)\Phi(t_0, \theta) = \Phi(t, \theta)$. Next, let

$$Y_t := \Xi + \int_{t_0}^t \Phi(t_0, \theta)\alpha(\theta)\,d\theta + \int_{t_0}^t \Phi(t_0, \theta)\beta(\theta)\,dW_\theta,$$

or in differential form,

$$dY_t = \Phi(t_0, t)\alpha(t)\,dt + \Phi(t_0, t)\beta(t)\,dW_t.$$

Now put

$$\begin{aligned} X_t &:= \Phi(t, t_0)Y_t \\ &= \Phi(t, t_0)\Xi + \int_{t_0}^t \Phi(t, \theta)\alpha(\theta)\,d\theta \\ &\quad + \int_{t_0}^t \Phi(t, \theta)\beta(\theta)\,dW_\theta. \end{aligned} \tag{60.35}$$

In other words, $X_t = g(t, Y_t)$, where $g(t, y) = \Phi(t, t_0)y$. Using Equation 60.34, $g_t(t, y) = c(t)\Phi(t, t_0)y$. We also have $g_y(t, y) = \Phi(t, t_0)$, and $g_{yy}(t, y) = 0$. By the extended Itô's rule of Equation 60.30,

$$\begin{aligned} dX_t &= g_t(t, Y_t)\,dt + g_y(t, Y_t)\,dY_t + \tfrac{1}{2}g_{yy}(t, Y_t)\,dt \\ &= c(t)\Phi(t, t_0)Y_t\,dt \\ &\quad + \Phi(t, t_0)[\Phi(t_0, t)\alpha(t)\,dt + \Phi(t_0, t)\beta(t)\,dW_t] \\ &= c(t)X_t\,dt + \alpha(t)\,dt + \beta(t)\,dW_t, \end{aligned}$$

which is exactly Equation 60.33.

Recalling the text following Equation 60.23, and noting the form of the solution in Equation 60.35, we see that $\{X_t\}$ is a Gaussian process if and only if Ξ is a Gaussian random variable. In any case, we can always use Equation 60.35 to derive differential equations for the mean and variance of X_t. For example, put $m(t) := \mathsf{E}[X_t]$. Since Itô integrals have zero mean (recall Equation 60.24), we obtain from Equation 60.35,

$$m(t) = \Phi(t, t_0)\mathsf{E}[\Xi] + \int_{t_0}^t \Phi(t, \theta)\alpha(\theta)\,d\theta,$$

and thus

$$\begin{aligned} \frac{dm(t)}{dt} &= c(t)\Phi(t, t_0)\mathsf{E}[\Xi] + \Phi(t, t)\alpha(t) \\ &\quad + \int_{t_0}^t c(t)\Phi(t, \theta)\alpha(\theta)\,d\theta \\ &= c(t)\left\{\Phi(t, t_0)\mathsf{E}[\Xi] + \int_{t_0}^t \Phi(t, \theta)\alpha(\theta)\,d\theta\right\} + \alpha(t) \\ &= c(t)m(t) + \alpha(t), \end{aligned}$$

and $m(t_0) = \mathsf{E}[\Xi]$. We now turn to the covariance function of X_t, $r(t, s) := \mathsf{E}[(X_t - m(t))(X_s - m(s))]$. We assume that the initial condition Ξ is independent of $\{W_t - W_{t_0}, t \geq t_0\}$. Write

$$X_t - m(t) = \Phi(t, t_0)(\Xi - \mathsf{E}[\Xi]) + \int_{t_0}^t \Phi(t, \theta)\beta(\theta)\,dW_\theta.$$

For $s < t$, write

$$\int_{t_0}^t \Phi(t, \theta)\beta(\theta)\,dW_\theta = \int_{t_0}^s \Phi(t, \theta)\beta(\theta)\,dW_\theta + \int_s^t \Phi(t, \theta)\beta(\theta)\,dW_\theta.$$

Then

$$\begin{aligned} r(t, s) &= \Phi(t, t_0)\mathsf{E}[(\Xi - \mathsf{E}[\Xi])^2]\Phi(s, t_0) \\ &\quad + \int_{t_0}^s \Phi(t, \theta)\beta(\theta)^2\Phi(s, \theta)\,d\theta. \end{aligned}$$

Letting $\mathsf{var}(\Xi) := \mathsf{E}[(\Xi - \mathsf{E}[\Xi])^2]$, for arbitrary s and t, we can write

$$\begin{aligned} r(t, s) &= \Phi(t, t_0)\,\mathsf{var}(\Xi)\Phi(s, t_0) \\ &\quad + \int_{t_0}^{\min\{s,t\}} \Phi(t, \theta)\beta(\theta)^2\Phi(s, \theta)\,d\theta. \end{aligned}$$

In particular, if we put $v(t) := r(t,t) = \mathsf{E}[(X_t - m(t))^2]$, then a simple calculation shows

$$\frac{dv(t)}{dt} = 2c(t)v(t) + \beta(t)^2, \quad v(t_0) = \mathsf{var}(\Xi).$$

60.5.3 The Langevin Equation

If $\alpha(t) = 0$ and $c(t)$ and $\beta(t)$ do not depend on t, then the narrow-sense linear SDE in Equation 60.33 becomes

$$dX_t = cX_t\,dt + \beta dW_t, \quad X_{t_0} = \Xi,$$

which is the Langevin equation when $c < 0$ and $\beta > 0$. Now, since

$$\Phi(t, t_0) = \exp\left(\int_{t_0}^t c\,d\theta\right) = e^{c[t-t_0]},$$

the solution in Equation 60.35 simplifies to

$$X_t = e^{c[t-t_0]}\Xi + \int_{t_0}^t e^{c[t-\theta]}\beta\,dW_\theta.$$

Then the mean is

$$m(t) := \mathsf{E}[X_t] = e^{c[t-t_0]}\mathsf{E}[\Xi],$$

and the covariance is

$$\begin{aligned} r(t,s) &= e^{c[t-t_0+s-t_0]}\,\mathsf{var}(\Xi) + \int_{t_0}^{\min\{t,s\}} e^{c[t-\theta+s-\theta]}\beta^2\,d\theta \\ &= e^{c[t+s-2t_0]}\left[\mathsf{var}(\Xi) + \frac{\beta^2}{2c}\right] + e^{c|t-s|}\left(\frac{-\beta^2}{2c}\right). \end{aligned}$$

Now assume $c < 0$, and suppose that $\mathsf{E}[\Xi] = 0$ and $\mathsf{var}(\Xi) = -\beta^2/(2c)$. Then

$$\mathsf{E}[X_t] = 0 \quad \text{and} \quad r(t,s) = e^{c|t-s|}\left(\frac{-\beta^2}{2c}\right). \tag{60.36}$$

Since $\mathsf{E}[X_t]$ does not depend on t, and since $r(t,s)$ depends only on $|t-s|$, $\{X_t, t \geq t_0\}$ is said to be wide-sense stationary. If Ξ is also Gaussian, then $\{X_t, t \geq t_0\}$ is a Gaussian process known as the **Ornstein–Uhlenbeck process.**

60.6 Transition Probabilities for General SDEs

The **transition function** for a process $\{X_t, t \geq t_0\}$ is defined to be the conditional cumulative distribution

$$P(t, y|s, x) := \Pr(X_t \leq y | X_s = x).$$

When X_t is the solution of an SDE, we can write down a partial differential equation that uniquely determines the transition function if certain hypotheses are satisfied. Under these hypotheses, the conditional cumulative distribution $P(t, y|s, x)$ has a density $p(t, y|s, x)$ that is uniquely determined by another partial differential equation.

To motivate the general results below, we first consider the narrow-sense linear SDE in Equation 60.33, whose solution is given in Equation 60.35. Since $\Phi(t,\theta) = \Phi(t,s)\Phi(s,\theta)$, Equation 60.35 can be rewritten as

$$\begin{aligned} X_t &= \Phi(t,s)\left[\Phi(s,t_0)\Xi + \int_{t_0}^t \Phi(s,\theta)\alpha(\theta)\,d\theta \right.\\ &\quad \left. + \int_{t_0}^t \Phi(s,\theta)\beta(\theta)\,dW_\theta\right] \\ &= \Phi(t,s)\left[X_s + \int_s^t \Phi(s,\theta)\alpha(\theta)\,d\theta \right.\\ &\quad \left. + \int_s^t \Phi(s,\theta)\beta(\theta)\,dW_\theta\right] \\ &= \Phi(t,s)X_s + \int_s^t \Phi(t,\theta)\alpha(\theta)\,d\theta \\ &\quad + \int_s^t \Phi(t,\theta)\beta(\theta)\,dW_\theta. \end{aligned}$$

Now consider $\Pr(X_t \leq y | X_s = x)$. Since we are conditioning on $X_s = x$, we can replace X_s in the preceding equation by x. For notational convenience, let

$$Z := \Phi(t,s)x + \int_s^t \Phi(t,\theta)\alpha(\theta)\,d\theta + \int_s^t \Phi(t,\theta)\beta(\theta)\,dW_\theta.$$

Then $\Pr(X_t \leq y | X_s = x) = \Pr(Z \leq y | X_s = x)$. Now, in the definition of Z, the only randomness comes from the Itô integral with deterministic integrand over $[s,t]$. The randomness in this integral comes only from the increments of the Wiener process on $[s,t]$. From Equation 60.35 with t replaced by s, we see that the only randomness in X_s comes from Ξ and from the increments of the Wiener process on $[t_0, s]$. Hence, Z and X_s are independent, and we can write $\Pr(Z \leq y | X_s = x) = \Pr(Z \leq y)$. Next, from the development of the Itô integral in Section 3, we see that Z is a Gaussian random variable with mean

$$m(t|s,x) := \Phi(t,s)x + \int_s^t \Phi(t,\theta)\alpha(\theta)\,d\theta$$

and variance

$$v(t|s) := \int_s^t [\Phi(t,\theta)\beta(\theta)]^2\,d\theta.$$

Hence, the transition function is

$$P(t, y|s, x) = \int_{-\infty}^y \frac{\exp\left(\frac{1}{2}[z - m(t|s,x)]^2/v(t|s)\right)}{\sqrt{2\pi\,v(t|s)}}\,dz,$$

and the transition density is

$$p(t, y|s, x) = \frac{\exp\left(\frac{1}{2}[y - m(t|s,x)]^2/v(t|s)\right)}{\sqrt{2\pi\,v(t|s)}}.$$

EXAMPLE 60.4:

Consider the SDE $dX_t = -3X_t\,dt + 5\,dW_t$. Then $m(t|s,x) = e^{-3(t-s)}x$, and $v(t|s) = \frac{25}{6}[1 - e^{-6(t-s)}]$.

We now return to the general SDE,

$$dX_t = a(t, X_t)\,dt + b(t, X_t)\,dW_t, \quad X_{t_0} = \Xi, \tag{60.37}$$

where Ξ is independent of $\{W_t - W_{t_0}, t \geq t_0\}$. To guarantee a unique continuous solution on a finite interval, say $[t_0, t_f]$, we assume [1] that there exists a finite constant $K > 0$ such that for all $t \in [t_0, t_f]$ and all x and y, a and b satisfy the Lipschitz conditions

$$\begin{aligned} |a(t,x) - a(t,y)| &\leq K|x-y|, \\ |b(t,x) - b(t,y)| &\leq K|x-y|, \end{aligned}$$

and the growth restrictions

$$\begin{aligned} |a(t,x)|^2 &\leq K^2(1+|x|^2), \\ |b(t,x)|^2 &\leq K^2(1+|x|^2). \end{aligned}$$

If such a K exists for every finite $t_f > t_0$, then a unique solution exists on $[t_0, \infty)$ [1]. Under the above conditions for the general SDE in Equation 60.37, a unique solution exists, although one cannot usually give an explicit formula for it, and so one cannot find the transition function and density as we did in the narrow-sense linear case. However, if for some $b_0 > 0$, $b(t,x) \geq b_0$ for all t and all x, then [10] the transition function P is the unique solution of **Kolmogorov's backward equation**

$$\begin{aligned} \tfrac{1}{2}b(s,x)^2\frac{\partial^2 P(t,y|s,x)}{\partial x^2} + a(s,x)\frac{\partial P(t,y|s,x)}{\partial x} &= \\ -\frac{\partial P(t,y|s,x)}{\partial s}, \quad t_0 < s < t < t_f, \end{aligned} \tag{60.38}$$

satisfying

$$\lim_{s \uparrow t} P(t,y|s,x) = \begin{cases} 1, & y > x, \\ 0, & y < x. \end{cases}$$

Furthermore, P has a density,

$$p(t,y|s,x) = \frac{\partial P(t,y|s,x)}{\partial y}, \quad t_0 < s < t < t_f.$$

If $\partial a/\partial x$, $\partial b/\partial x$, and $\partial^2 b/\partial x^2$ also satisfy the Lipschitz and growth conditions above, and if $\partial b/\partial x \geq b_0 > 0$ and $\partial^2 b/\partial x^2 \geq b_0 > 0$, then the transition density p is the unique fundamental solution of **Kolmogorov's forward equation**

$$\begin{aligned} \frac{1}{2}\frac{\partial^2[b(t,y)^2 p(t,y|s,x)]}{\partial y^2} - \frac{\partial[a(t,y)p(t,y|s,x)]}{\partial y} &= \\ \frac{\partial p(t,y|s,x)}{\partial t}, \quad t_0 < s < t < t_f, \end{aligned} \tag{60.39}$$

satisfying

$$p(s,y|s,x) = \delta(y-x).$$

The forward partial differential equation is also known as the **Fokker–Planck equation.** Equation 60.39 is called the forward equation because it evolves forward in time starting at s; note also that x is fixed and y varies. In the backward equation, t and y are fixed, and x varies as s evolves backward in time starting at t.

REMARK 60.2 If the necessary partial derivatives are continuous, then we can differentiate the backward equation with respect to y and obtain the following equation for the transition density p:

$$\begin{aligned} \tfrac{1}{2}b(s,x)^2\frac{\partial^2 p(t,y|s,x)}{\partial x^2} + a(s,x)\frac{\partial p(t,y|s,x)}{\partial x} &= \\ -\frac{\partial p(t,y|s,x)}{\partial s}, \quad t_0 < s < t < t_f. \end{aligned}$$

EXAMPLE 60.5:

In general, it is very hard to obtain explicit solutions to the forward or backward equations. However, when $a(t,x) = a(x)$ and $b(t,x) = b(x)$ do not depend on t, it is sometimes possible to obtain a limiting density $p(y) = \lim_{t\to\infty} p(t,y|s,x)$ that is independent of s and x. The existence of this limit suggests that for large t, $p(t,y|s,x)$ settles down to a constant as a function of t. Hence, the partial derivative with respect to t on the right-hand side of the forward equation in Equation 60.39 should be zero. This results in the ODE

$$\frac{1}{2}\frac{d^2[b(y)^2 p(y)]}{dy^2} - \frac{d[a(y)p(y)]}{dy} = 0. \tag{60.40}$$

For example, let μ and λ be positive constants, and suppose that

$$dX_t = -\mu X_t\,dt + \sqrt{\lambda(1+X_t^2)}\,dW_t.$$

(The case $\mu = 1$ and $\lambda = 2$ was considered by [10].) Then Equation 60.40 becomes

$$\frac{1}{2}\frac{d^2[\lambda(1+y^2)p(y)]}{dy^2} - \frac{d[-\mu y p(y)]}{dy} = 0.$$

Integrating both sides, we obtain

$$\frac{\lambda}{2}\frac{d[(1+y^2)p(y)]}{dy} + \mu y p(y) = \kappa$$

for some constant κ. Now, the left-hand side of this equation is $(\lambda+\mu)yp(y) + \lambda(1+y^2)p'(y)/2$. If we assume that this goes to zero as $|y| \to \infty$, then $\kappa = 0$. In this case,

$$\frac{p'(y)}{p(y)} = -(1+\mu/\lambda)\frac{2y}{1+y^2}.$$

Integrating from 0 to y yields

$$\ln\frac{p(y)}{p(0)} = -(1+\mu/\lambda)\ln(1+y^2),$$

or

$$p(y) = \frac{p(0)}{(1+y^2)^{1+\mu/\lambda}}.$$

Of course, $p(0)$ is determined by the requirement that $\int_{-\infty}^{\infty} p(y)\,dy = 1$. For example, if $\mu/\lambda = 1/2$, then $p(0) = 1/2$, which can be found directly after noting that the antiderivative of $1/(1+y^2)^{3/2}$ is $y/\sqrt{1+y^2}$. As a second example, suppose $\mu/\lambda = 1$. Then $p(y)$ has the form $p(0)f(y)^2$, where $f(y) = 1/(1+y^2)$. If we let $F(\omega)$ denote the Fourier transform of $f(y)$, then Parseval's equation yields $\int_{-\infty}^{\infty} |f(y)|^2\,dy = \int_{-\infty}^{\infty} |F(\omega)|^2\,d\omega/2\pi$. Since $F(\omega) = \pi e^{-|\omega|}$, this last integral can be computed in closed form, and we find that $p(0) = 2/\pi$.

60.7 Defining Terms

Adapted: A random process $\{H_t\}$ is $\{\mathcal{F}_t\}$-adapted, where $\mathcal{F}_t = \sigma(W_\theta, 0 \le \theta \le t)$, if for each t, H_t is $\mathcal{F}_t$-measurable. See **measurable.**

Brownian motion: Synonym for **Wiener process.**

Fokker–Planck equation: Another name for **Kolmogorov's forward equation.**

History: The history of a process $\{W_\theta, \theta \ge 0\}$ up to and including time t is denoted by $\mathcal{F}_t = \sigma(W_\theta, 0 \le \theta \le t)$. See also **measurable.**

Homogeneous: Linear stochastic differential equations (SDEs) of the form in Equation 60.32 are homogeneous.

Integrated white noise: A random process W_t that behaves as if it had the representation $W_t = \int_0^t Z_\theta\,d\theta$, where Z_θ is a white noise process. The Wiener process is an example of integrated white noise.

Itô correction term: The last term in Itô's rule in Equations 60.6 and 60.30. This term accounts for the fact that the Wiener process is not differentiable.

Itô's rule: A stochastic version of the chain rule. The general form is given in Equation 60.30.

Kolmogorov's backward equation: The partial differential Equation 60.38 satisfied by the transition function of the solution of an SDE.

Kolmogorov's forward equation: The partial differential Equation 60.39 satisfied by the transition density of the solution of an SDE.

Martingale: $\{W_t\}$ is an $\{\mathcal{F}_t\}$-martingale if $\{W_t\}$ is $\{\mathcal{F}_t\}$-adapted and if for all $t \ge s \ge 0$, $\mathsf{E}[W_t|\mathcal{F}_s] = W_s$, or equivalently, $\mathsf{E}[W_t - W_s|\mathcal{F}_s] = 0$.

Measurable: See also **history**. Let $\mathcal{F}_t = \sigma(W_\theta, 0 \le \theta \le t)$. A random variable X is $\mathcal{F}_t$-measurable if it is a deterministic function of the random variables $\{W_\theta, 0 \le \theta \le t\}$.

Narrow sense: An SDE is linear in the narrow sense if it has the form of Equation 60.33.

Ornstein–Uhlenbeck process: A Gaussian process with zero mean and covariance function in Equation 60.36.

Smoothing properties of conditional expectation: See Equations 60.17 and 60.18.

Transition function: For a process $\{X_t\}$, the transition function is $P(t, y|s, x) := \Pr(X_t \le y|X_s = x)$.

White noise process: A random process with zero mean and covariance $\mathsf{E}[Z_t Z_s] = \delta(t-s)$.

Wiener integral: An Itô integral with deterministic integrand. Always yields a Gaussian process.

Wiener process: A random process satisfying properties W-1 through W-4 in Section 2. It serves as a model for integrated white noise.

Acknowledgments

The author is grateful to Bob Barmish, Wei-Bin Chang, Majeed Hayat, Bill Levine, Raúl Sequeira, and Rajesh Sharma for reading the first draft of this chapter and for their suggestions for improving it.

References

[1] Arnold, L., *Stochastic Differential Equations: Theory and Applications*, Wiley, New York, 1974, 48–56, 91, 98, 105, 113, 128, 137, 168.

[2] Billingsley, P., *Probability and Measure*, 2nd ed., Wiley, New York, 1986, sect. 37, pp 469–470 (Equations 34.2, 34.5, 34.6).

[3] Coddington, E. A. and Levinson, N., *Theory of Ordinary Differential Equations*, McGraw-Hill, New York, 1955, chap. 1.

[4] Davis, M. H. A., *Linear Estimation and Stochastic Control*, Chapman and Hall, London, 1977, chap. 3, 52-53.

[5] Elliott, R. J., *Stochastic Calculus*, Springer-Verlag, New York, 1982.

[6] Ethier, S. N. and Kurtz, T. G., *Markov Processes: Characterization and Convergence*, Wiley, New York, 1986.

[7] Karatzas, I. and Shreve, S. E., *Brownian Motion and Stochastic Calculus*, 2nd ed., Springer-Verlag, New York, 1991, 397.

[8] Protter, P., *Stochastic Integration and Differential Equations*, Springer-Verlag, Berlin, 1990.

[9] Segall, A., Stochastic processes in estimation theory, *IEEE Trans. Inf. Theory*, 22(3), 275–286, 1976.

[10] Wong, E. and Hajek, B., *Stochastic Processes in Engineering Systems*, Springer-Verlag, New York, 1985, chap. 4, pp 173-174.

Further Reading

For background on probability theory, especially conditional expectation, we recommend [2].

For linear SDEs driven by orthogonal-increments processes, we recommend the very readable text by Davis [4].

For SDEs driven by continuous martingales, there is the more advanced book by Karatzas and Shreve [7].

For SDEs driven by right-continuous martingales, the theory becomes considerably more complicated. However, the tutorial paper by Segall [9], which compares discrete-time and continuous-time results, is very readable. Also, Chapter 6 of [10] is accessible.

Highly technical books on SDEs driven by right-continuous martingales include [5] and [8].

For the reader interested in Markov processes there is the advanced text of Ethier and Kurtz [6].

61

Linear Stochastic Input—Output Models

Torsten Söderström
Systems and Control Group, Uppsala University, Uppsala, Sweden

61.1 Introduction

Stationary stochastic processes are good ways of modeling random disturbances. The treatment here is basically for linear discrete-time input-output models. Most modern systems for control and signal processing work with sampled data; hence, discrete-time models are of primary interest.

Properties of models, ways to calculate variances, and other second-order moments are treated. The chapter is organized as follows. Autoregressive moving average (ARMA) and ARMA with exogenous input (ARMAX) models are introduced in Section 61.2, while Section 61.3 deals with the effect of linear filtering. Spectral factorization, which has a key role when finding appropriate model representations for optimal estimation and control, is described in Section 61.4. Some ways to analyze stochastic systems by covariance calculations are presented in Section 61.5, while Section 61.6 gives a summary of results for continuous-time processes.

In stochastic control, it is a fundamental ingredient to predict future values of the process. A more general situation is the problem of estimating unmeasurable variables. Mean square optimal prediction is dealt with in Section 61.7, with minimal output variance control as a special application. In Section 61.8, a more general estimation problem is treated (covering optimal prediction, filtering and smoothing), using Wiener filters, which are described in some detail.

The chapter is based on [6], where proofs and derivations can be found, as well as several extensions to the multivariable case and to complex-valued signal processing problems.

0-8493-8570-9/96/$0.00+$.50

61.2 ARMA and ARMAX Models

Wide-sense stationary random processes are often characterized by their first- and second-order moments, that is, by the mean value

$$m = Ex(t) \tag{61.1}$$

and the covariance function

$$r(\tau) \triangleq E[x(t+\tau) - m][x(t) - m], \tag{61.2}$$

where t, τ take integer values $0, \pm 1 \pm 2, \ldots$. For a wide-sense stationary process, the expected values in Equations 61.1 and 61.2 are independent of t. As an alternative to the covariance function, one can use its discrete Fourier transform, that is, the spectrum,

$$\phi(z) = \sum_{n=-\infty}^{\infty} r(n) z^{-n}. \tag{61.3}$$

Evaluated on the unit circle it is called the spectral density,

$$\phi(e^{i\omega}) = \sum_{n=-\infty}^{\infty} r(n) e^{-in\omega}. \tag{61.4}$$

As

$$r(\tau) = \frac{1}{2\pi} \int_{-\pi}^{\pi} \phi(e^{i\omega}) e^{i\tau\omega} d\omega = \frac{1}{2\pi i} \oint \phi(z) z^{\tau} \frac{dz}{z} \tag{61.5}$$

(where the last integration is counterclockwise around the unit circle), the spectral density describes how the energy of the signal is distributed over different frequency bands (set $\tau = 0$ in Equation 61.5).

Similarly, the cross-covariance function between two wide-sense stationary processes $y(t)$ and $x(t)$ is defined as

$$r_{yx}(\tau) \triangleq E[y(t+\tau) - m_y][x(t) - m_x], \tag{61.6}$$

and its associated spectrum is

$$\phi_{yx}(z) = \sum_{n=-\infty}^{\infty} r_{yx}(n) z^{-n}. \quad (61.7)$$

A sequence of independent identically distributed (i.i.d.) random variables is called *white noise*. A white noise has

$$r(\tau) = 0 \text{ for } \tau \neq 0. \quad (61.8)$$

Equivalently, its spectrum is constant for all z. Hence, its energy is distributed evenly over all frequencies.

In order to simplify the development here, it is generally assumed that signals have zero mean. This is equivalent to considering only deviations of the signals from an operating point (given by the mean values).

Next, an important class of random processes, obtained by linear filtering of white noise, is introduced. Consider $y(t)$ given as the solution to the difference equation

$$\begin{aligned} y(t) \quad + \quad & a_1 y(t-1) + \ldots + a_n y(t-n) = \\ & e(t) + c_1 e(t-1) + \ldots + c_m e(t-m), \quad (61.9) \end{aligned}$$

where $e(t)$ is white noise. Such a process is called an autoregressive moving average (ARMA) process. If $m = 0$, it is called an autoregressive (AR) process; and if $n = 0$, a moving average (MA) process.

Introduce the polynomials

$$\begin{aligned} A(z) &= z^n + a_1 z^{n-1} + \ldots + a_n, \\ C(z) &= z^m + c_1 z^{m-1} + \ldots + c_m, \end{aligned} \quad (61.10)$$

and the shift operator q, $qx(t) = x(t+1)$. The ARMA model of Equation 61.9 can then be written compactly as

$$A(q)y(t-n) = C(q)e(t-m). \quad (61.11)$$

As the white noise can be "relabelled" without changing the statistical properties of $y(t)$, Equation 61.9 is much more frequently written in the form

$$A(q)y(t) = C(q)e(t). \quad (61.12)$$

Some illustrations of ARMA processes are given next.

EXAMPLE 61.1:

Consider an ARMA process

$$A(q)y(t) = C(q)e(t).$$

The coefficients of the $A(q)$ and $C(q)$ polynomials determine the properties of the process. In particular, the roots of $A(z)$, which are called the poles of the process, determine the frequency contents of the process. The closer the poles are located towards the unit circle, the slower or more oscillating the process is. Figure 61.1 illustrates the connections between the A and C coefficients, realizations of processes and their second-order moments as expressed by covariance function and spectral density.

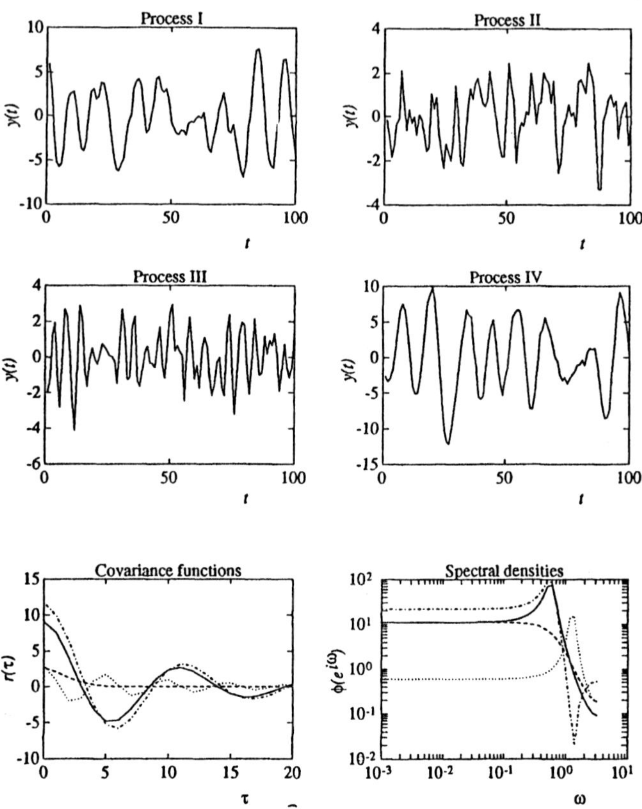

Figure 61.1 Illustration of some ARMA processes: (a) Process I: $A(q) = q^2 - 1.5q + 0.8$, $C(q) = 1$. Pole locations: $0.75 \pm i0.49$; (b) Process II: $A(q) = q^2 - 1.0q + 0.3$, $C(q) = 1$. Pole locations: $0.50 \pm i0.22$; (c) Process III: $A(q) = q^2 - 0.5q + 0.8$, $C(q) = 1$. Pole locations: $0.25 \pm i0.86$; (d) Process IV: $A(q) = q^2 - 1.0q + 0.8$, $C(q) = q^2 - 0.5q + 0.9$. Pole locations: $0.50 \pm i0.74$; (e) Covariance functions; (f) Spectral densities [Processes I (solid lines), II (dashed lines), III (dotted lines), IV (dash–dotted lines)].

As an alternative to q, one can use the *backward* shift operator, q^{-1}. An ARMA model would then be written as

$$\overline{A}(q^{-1})y(t) = \overline{C}(q^{-1})e(t),$$

where the polynomials are

$$\begin{aligned} \overline{A}(q^{-1}) &= 1 + a_1 q^{-1} + \ldots + a_n q^{-n}, \\ \overline{C}(q^{-1}) &= 1 + c_1 q^{-1} + \ldots + c_m q^{-m}. \end{aligned}$$

The advantage of using the q-formalism is that stability corresponds to the "natural" condition $|z| < 1$. An advantage of the alternative q^{-1} formalism is that causality considerations (see below) become easier to handle. The q-formalism is used here.

In some situations, such as modeling a drifting disturbance, it is appropriate to allow $A(z)$ to have zeros on or even outside the unit circle. Then the process is not wide-sense stationary, but drift away. It should hence only be considered for a finite period of time. The special simple case $A(z) = z - 1$, $C(z) = z$ is known as a random walk.

If an input signal term is added to Equation 61.10, we obtain an ARMA model with exogenous input (ARMAX):

$$A(q)y(t) = B(q)u(t) + C(q)e(t). \qquad (61.13)$$

The system must be *causal*, which means that $y(t)$ is not allowed to depend on *future* values of the input $u(t+\tau)$, $\tau > 0$. Hence, it is required that $\deg A \geq \deg B$. Otherwise, $y(t)$ would depend on *future* input values.

Sometimes higher-order moments (that is, moments of order higher than two) are useful. Such spectra are useful tools for the following:

- Extracting information due to deviations from a Gaussian distribution
- Estimating the phase of a non-Gaussian process
- Detecting and characterizing nonlinear mechanisms in time series

To exemplify, we consider the bispectrum, which is the simplest form of higher-order spectrum. Bispectra are useful only for signals that do *not* have a probability density function that is symmetric around its mean value. For signals *with* such a symmetry, spectra of order at least four are needed.

Let $x(t)$ be a scalar stationary process of zero mean. Its *third moment sequence*, $R(m,n)$ is defined as

$$R(m,n) = Ex(t)x(t+m)x(t+n) \qquad (61.14)$$

and satisfies a number of symmetry relations. The *bispectrum* is

$$B(z_1, z_2) \triangleq \sum_{m=-\infty}^{\infty} \sum_{n=-\infty}^{\infty} R(m,n) z_1^{-m} z_2^{-n}. \qquad (61.15)$$

Let us consider two special cases:

- Let $x(t)$ be zero mean Gaussian. Then

$$R(m,n) = 0, \qquad B(z_1, z_2) = 0. \qquad (61.16)$$

- Let $x(t)$ be non-Gaussian *white* noise, so that $x(t)$ and $x(s)$ are independent for $t \neq s$, $Ex(t) = 0$, $Ex^2(t) = \sigma^2$, $Ex^3(t) = \beta$. Then

$$R(m,n) = \begin{cases} \beta & \text{if } m = n = 0, \\ 0 & \text{elsewhere.} \end{cases} \qquad (61.17)$$

The bispectrum becomes a constant,

$$B(z_1, z_2) = \beta. \qquad (61.18)$$

61.3 Linear Filtering

Let $u(t)$ be a stationary process with mean m_u and spectrum $\phi_u(z)$, and consider

$$y(t) = G(q)u(t) = \sum_{k=0}^{\infty} g_k u(t-k), \qquad (61.19)$$

where $G(q) = \sum_{k=0}^{\infty} g_k q^{-k}$ is an asymptotically stable filter (that is, it has all poles strictly inside the unit circle). Then $y(t)$ is a stationary process with mean

$$m_y = G(1) m_u, \qquad (61.20)$$

and spectrum $\phi_y(z)$ and cross-spectrum $\phi_{yu}(z)$ given by, respectively,

$$\begin{aligned} \phi_y(z) &= G(z)G(z^{-1})\phi_u(z), & (61.21) \\ \phi_{yu}(z) &= G(z)\phi_u(z). & (61.22) \end{aligned}$$

The interpretation of Equation 61.20 is that the mean value m_u is multiplied by the static gain of the filter to get the output mean value m_y.

The following corollary is a form of Parseval's relation. Assume that $u(t)$ is a white noise sequence with variance λ^2. Then

$$Ey^2(t) = \lambda^2 \sum_{k=0}^{\infty} g_k^2 = \frac{\lambda^2}{2\pi i} \oint G(z)G(z^{-1}) \frac{dz}{z}. \qquad (61.23)$$

Let $u(t)$ further have bispectrum $B_u(z_1, z_2)$. Then $B_y(z_1, z_2)$ can be found after straightforward calculation,

$$B_y(z_1, z_2) = G(z_1^{-1} z_2^{-1}) G(z_1) G(z_2) B_u(z_1, z_2) \quad . \qquad (61.24)$$

This is a generalization of Equation 61.21. Note that the spectral density (the power spectrum) does not carry information about the phase properties of a filter. In contrast to this, phase properties can be recovered from the bispectrum when it exists. This point is indirectly illustrated in Example 61.2.

61.4 Spectral Factorization

Let $\phi(z)$ be a scalar spectrum that is rational in z, that is, it can be written as

$$\phi(z) = \frac{\sum_{|k| \leq m} \beta_k z^k}{\sum_{|k| \leq n} \alpha_k z^k}. \qquad (61.25)$$

Then there are two polynomials

$$\begin{aligned} A(z) &= z^n + a_1 z^{n-1} + \ldots + a_n, \\ C(z) &= z^m + c_1 z^{m-1} + \ldots + c_m, \end{aligned} \qquad (61.26)$$

and a positive real number λ^2 so that (1) $A(z)$ has all zeros inside the unit circle, (2) $C(z)$ has all zeros inside or on the unit circle, and (3)

$$\phi(z) = \lambda^2 \frac{C(z)}{A(z)} \frac{C(z^{-1})}{A(z^{-1})}. \qquad (61.27)$$

In the case where $\phi(e^{i\omega}) > 0$ for all ω, $C(z)$ will have no zeros on the circle.

Note that any continuous spectral density can be approximated arbitrarily well by a rational function in $z = e^{i\omega}$ as in Equations 61.25 to 61.27, provided that m and n are appropriately chosen. Hence, the assumptions imposed are not restrictive. Instead, the results are applicable, at least with a small approximation error, to a very wide class of stochastic processes.

It is an important implication that (as far as second-order moments are concerned) the underlying stochastic process can be regarded as generated by filtering white noise, that is, as an ARMA process

$$\begin{aligned} A(q)y(t) &= C(q)e(t), \\ Ee(t)e^*(t) &= \lambda^2. \end{aligned} \tag{61.28}$$

Hence, for describing stochastic processes (as long as they have rational spectral densities), it is no restriction to assume the input signals are white noise. In Equation 61.28, the sequence $\{e(t)\}$ is called the *output innovations.*

Spectral factorization can also be viewed as a form of *aggregation of noise sources.* Assume, for example, that an ARMA process

$$A(q)x(t) = C(q)v(t) \tag{61.29a}$$

is observed, but the observations include measurement noise

$$y(t) = x(t) + e(t), \tag{61.29b}$$

and that $v(t)$ and $e(t)$ are uncorrelated white noise sequences with variances λ_v^2 and λ_e^2, respectively. As far as the second-order properties (such as the spectrum or the covariance function) are concerned, $y(t)$ can be viewed as generated from one single noise source:

$$A(q)y(t) = D(q)\varepsilon(t). \tag{61.29c}$$

The polynomial $D(q)$ and the noise variance λ_ε^2 are derived as follows. The spectrum is, according to Equations 61.29a, 61.29b and 61.29c:

$$\phi_y(z) = \lambda_v^2 \frac{C(z)C(z^{-1})}{A(z)A(z^{-1})} + \lambda_e^2 = \lambda_\varepsilon^2 \frac{D(z)D(z^{-1})}{A(z)A(z^{-1})}.$$

Equating these two expressions gives

$$\lambda_\varepsilon^2 D(z)D(z^{-1}) \equiv \lambda_v^2 C(z)C(z^{-1}) + \lambda_e^2 A(z)A(z^{-1}). \tag{61.29d}$$

The two representations of Equations 61.29a and 61.29b, and Equation 61.29c of the process $y(t)$, are displayed schematically in Figure 61.2.

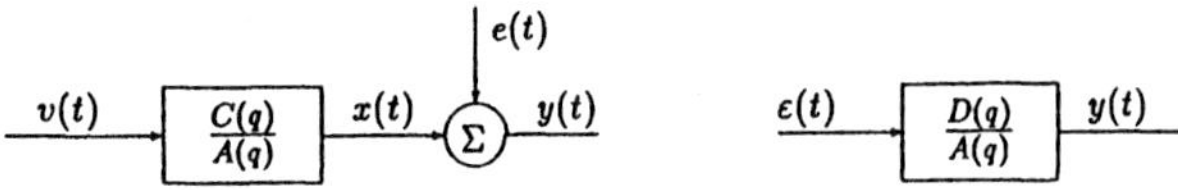

Figure 61.2 Two representations of an ARMA process with noisy observations.

Spectral factorization can also be performed using a state-space formalism. Then an algebraic Riccati equation (ARE) has to be solved. Its different solutions correspond to different polynomials $C(z)$ satisfying Equation 61.27. The positive definite solution of the ARE corresponds to the $C(z)$ polynomial with all zeros inside the unit circle.

One would expect, for a given process, the same type of filter representation to appear for the power spectrum and for the bispectrum. This is *not* so in general, as illustrated by the following example.

EXAMPLE 61.2:

Consider a process consisting of the sum of two independent AR processes

$$y(t) = \frac{1}{A(q)}e(t) + \frac{1}{C(q)}v(t), \tag{61.30}$$

$e(t)$ being Gaussian white noise and $v(t)$ non-Gaussian white noise. Both sequences are assumed to have unit variance, and $Ev^3(t) = 1$.

The Gaussian process will not contribute to the bispectrum. Further, $R_v(z_1, z_2) \equiv 1$, and according to Equation 61.24 the bispectrum will be

$$B_y(z_1, z_2) = \frac{1}{C(z_1^{-1}z_2^{-1})} \frac{1}{C(z_1)} \frac{1}{C(z_2)}, \tag{61.31}$$

so

$$H(z) = 1/C(z) \tag{61.32}$$

is the relevant filter representation as far as the bispectrum is concerned. However, the power spectrum becomes

$$\phi(z) = \frac{1}{A(z)A(z^{-1})} + \frac{1}{C(z)C(z^{-1})}, \tag{61.33}$$

and in this case it will have a spectral factor of the form

$$H(z) = \frac{B(z)}{A(z)C(z)}, \tag{61.34}$$

where

$$B(z)B(z^{-1}) \equiv A(z)A(z^{-1}) + C(z)C(z^{-1}) \tag{61.35}$$

due to the spectral factorization. Clearly, the two filter representations of Equations 61.32 and 61.34 differ.

61.5 Yule-Walker Equations and Other Algorithms for Covariance Calculations

When analyzing stochastic systems it is often important to compute variances and covariances between inputs, outputs and other variables. This can mostly be reduced to the problem of computing the covariance function of an ARMA process. Some ways to do this are presented in this section.

Consider first the case of an autoregressive (AR) process

$$\begin{aligned} y(t) + a_1 y(t-1) + \ldots + a_n y(t-n) &= e(t), \\ Ee^2(t) &= \lambda^2. \end{aligned} \tag{61.36}$$

Note that $y(t)$ can be viewed as a linear combination of all the old values of the noise, that is, $\{e(s)\}_{s=-\infty}^{t}$. By multiplying $y(t)$ by a delayed value of the process, say $y(t-\tau)$, $\tau \geq 0$, and applying the expectation operator, one obtains

$$Ey(t-\tau)[y(t)+a_1y(t-1)+\ldots+a_ny(t-n)] = Ey(t-\tau)e(t),$$

or

$$r(\tau)+a_1r(\tau-1)+\ldots+a_nr(\tau-n) = \begin{cases} 0 & \tau > 0, \\ \lambda^2 & \tau = 0, \end{cases} \qquad (61.37)$$

which is called a Yule–Walker equation. By using Equation 61.37 for $\tau = 0, \ldots, n$, one can construct the following system of equations for determining the covariance elements $r(0), r(1), \ldots, r(n)$:

$$\begin{pmatrix} 1 & a_1 & \cdots & a_n \\ a_1 & 1+a_2 & a_n & 0 \\ \vdots & & \ddots & \vdots \\ a_n & a_{n-1} & \cdots & 1 \end{pmatrix} \begin{pmatrix} r(0) \\ \vdots \\ r(n) \end{pmatrix} = \begin{pmatrix} \lambda^2 \\ 0 \\ \vdots \\ 0 \end{pmatrix}. \qquad (61.38)$$

Once $r(0), \ldots, r(n)$ are known, Equation 61.37 can be iterated (for $\tau = n+1, n+2, \ldots$) to find further covariance elements.

Consider next a full ARMA process

$$\begin{aligned} y(t) + a_1y(t-1) + \ldots + a_ny(t-n) \\ = e(t) + c_1e(t-1) + \ldots + c_me(t-m), \\ Ee^2(t) = \lambda^2. \end{aligned} \qquad (61.39)$$

Now computing the cross-covariance function between $y(t)$ and $e(t)$ must involve an intermediate step. Multiplying Equation 61.39 by $y(t-\tau)$, $\tau \geq 0$, and applying the expectation operator gives

$$\begin{aligned} r_y(\tau) + a_1r_y(\tau-1) + \ldots + a_nr_y(\tau-n) \\ = r_{ey}(\tau) + c_1r_{ey}(\tau-1) + \ldots \\ + c_mr_{ey}(\tau-m). \end{aligned} \qquad (61.40)$$

In order to obtain the output covariance function $r_y(\tau)$, the cross-covariance function $r_{ey}(\tau)$ must first be found. This is done by multiplying Equation 61.39 by $e(t-\tau)$ and applying the expectation operator, which leads to

$$\begin{aligned} r_{ey}(-\tau) + a_1r_{ey}(-\tau+1) + \ldots + a_nr_{ey}(-\tau+n) \\ = \lambda^2[\delta_{\tau,0} + c_1\delta_{\tau-1,0} + \ldots + c_m\delta_{\tau-m,0}], \end{aligned} \qquad (61.41)$$

where $\delta_{t,s}$ is the Kronecker delta ($\delta_{t,s} = 1$ if $t = s$, and 0 elsewhere). As $y(t)$ is a linear combination of $\{e(s)\}_{s=-\infty}^{t}$, it is found that $r_{ey}(\tau) = 0$ for $\tau > 0$. Hence, Equation 61.40 gives

$$\begin{aligned} r_y(\tau) + a_1r_y(\tau-1) + \ldots + a_nr_y(\tau-n) = 0, \\ \tau > m. \end{aligned} \qquad (61.42)$$

The use of Equations 61.40 to 61.42 to derive the autocovariance function is illustrated next by applying them to a first-order ARMA process.

EXAMPLE 61.3:

Consider the ARMA process

$$y(t) + ay(t-1) = e(t) + ce(t-1), \quad Ee^2(t) = \lambda^2.$$

In this case $n = 1$, $m = 1$. Equation 61.42 gives

$$r_y(\tau) + ar_y(\tau-1) = 0, \quad \tau > 1.$$

Using Equation 61.40 for $\tau = 0$ and 1 gives

$$\begin{pmatrix} 1 & a \\ a & 1 \end{pmatrix} \begin{pmatrix} r_y(0) \\ r_y(1) \end{pmatrix} = \begin{pmatrix} 1 & c \\ c & 0 \end{pmatrix} \begin{pmatrix} r_{ey}(0) \\ r_{ey}(-1) \end{pmatrix}.$$

Consider Equation 61.41 for $\tau = 0$ and 1, which gives

$$\begin{pmatrix} 1 & 0 \\ a & 1 \end{pmatrix} \begin{pmatrix} r_{ey}(0) \\ r_{ey}(-1) \end{pmatrix} = \lambda^2 \begin{pmatrix} 1 \\ c \end{pmatrix}.$$

By straightforward calculations, it is found that

$$\begin{aligned} r_{ey}(0) &= \lambda^2, \\ r_{ey}(-1) &= \lambda^2(c-a), \\ r_y(0) &= \frac{\lambda^2}{1-a^2}(1+c^2-2ac), \\ r_y(1) &= \frac{\lambda^2}{1-a^2}(c-a)(1-ac), \end{aligned}$$

and finally,

$$r_y(\tau) = \frac{\lambda^2}{1-a^2}(c-a)(1-ac)(-a)^{\tau-1}, \qquad \tau \geq 1.$$

As an example of alternative approaches for covariance calculations, consider the following situation. Assume that two ARMA processes are given:

$$\begin{aligned} A(q)y(t) &= B(q)e(t), \\ A(q) &= q^n + a_1q^{n-1} + \ldots + a_n, \\ B(q) &= b_oq^n + b_1q^{n-1} + \ldots + b_n, \end{aligned} \qquad (61.43)$$

and

$$\begin{aligned} C(q)w(t) &= D(q)e(t), \\ C(q) &= q^m + c_1q^{m-1} + \ldots + c_m, \\ D(q) &= d_oq^m + d_1q^{m-1} + \ldots + d_m. \end{aligned} \qquad (61.44)$$

Assume that $e(t)$ is the same in Equations 61.43 and 61.44 and that it is white noise of zero mean and unit variance. The problem is to find the cross-covariance elements

$$r(k) = Ey(t+k)w(t) \qquad (61.45)$$

for a number of arguments k. The cross-covariance function $r(k)$ is related to the cross-spectrum $\phi_{yw}(z)$ as (see Equations 61.7 and 61.22)

$$\begin{aligned} \phi_{yw}(z) &= \sum_{k=-\infty}^{\infty} r(k)z^{-k} \\ &= \frac{B(z)}{A(z)} \frac{D(z^{-1})}{C(z^{-1})}. \end{aligned} \qquad (61.46)$$

Introduce the two polynomials

$$\begin{aligned} F(z) &= f_o z^n + f_1 z^{n-1} + \ldots + f_n, \\ G(z^{-1}) &= g_o z^{-m} + g_1 z^{-(m-1)} + \ldots + g_{m-1} z^{-1}, \end{aligned} \tag{61.47}$$

through

$$\frac{B(z)}{A(z)} \frac{D(z^{-1})}{C(z^{-1})} \equiv \frac{F(z)}{A(z)} + z\frac{G(z^{-1})}{C(z^{-1})}, \tag{61.48}$$

or, equivalently,

$$B(z)D(z^{-1}) \equiv F(z)C(z^{-1}) + zA(z)G(z^{-1}). \tag{61.49}$$

Since $zA(z)$ and $C(z^{-1})$ are coprime (that is, these two polynomials have no common factor), Equation 61.49 has a unique solution. Note that as a linear system of equations, Equation 61.49 has $n + m + 1$ equations and the same number of unknowns. The coprimeness condition ensures that a unique solution exists. Equations 61.46 and 61.48 now give

$$\sum_{k=-\infty}^{\infty} r(k)z^{-k} = \frac{F(z)}{A(z)} + z\frac{G(z^{-1})}{C(z^{-1})}. \tag{61.50}$$

The two terms in the right-hand side of Equation 61.50 can be identified with two parts of the sum. In fact,

$$\frac{F(z)}{A(z)} = \sum_{k=0}^{\infty} r(k)z^{-k}, \quad \frac{zG(z^{-1})}{C(z^{-1})} = \sum_{k=-\infty}^{-1} r(k)z^{-k}. \tag{61.51}$$

Equating the powers of z gives

$$\begin{cases} r_{yw}(0) &= f_o, \\ r_{yw}(1) &= f_1 - a_1 r(0), \\ & \vdots \\ r_{yw}(k) &= f_k - \sum_{j=1}^{k} a_j r(k-j), \quad (2 \le k \le n), \\ r_{yw}(k) &= -\sum_{j=1}^{n} a_j r(k-j), \quad (k > n). \end{cases} \tag{61.52}$$

Note that the last part of Equation 61.52 is nothing but a Yule–Walker type of equation.

Similarly,

$$\begin{cases} r_{yw}(-1) &= g_o, \\ r_{yw}(-k) &= g_{k-1} - \sum_{j=1}^{k-1} c_j r(-k+j), \\ & \quad (2 \le k \le m), \\ r_{yw}(-k) &= -\sum_{j=1}^{m} c_j r(-k+j), \\ & \quad (k > m). \end{cases} \tag{61.53}$$

EXAMPLE 61.4:

Consider again a first-order ARMA process

$$y(t) + ay(t-1) = e(t) + ce(t-1), \quad Ee^2(t) = 1.$$

In this case, the autocovariance function is sought. Hence, choose $z(t) \equiv y(t)$ and thus $A(q) = q+a$, $B(q) = q+c$, $C(q) = q+a$, $D(q) = q + c$. Equation 61.49 becomes

$$(z+c)(z^{-1}+c) \equiv (f_o z + f_1)(z^{-1}+a) + z(z+a)(g_o z^{-1}).$$

Equating the powers of z leads to

$$f_o = \frac{1+c^2-2ac}{1-a^2}, \quad f_1 = c, \quad g_o = \frac{(c-a)(1-ac)}{1-a^2}.$$

Hence, Equation 61.52 implies that

$$\begin{aligned} r(0) &= f_o = \frac{1+c^2-2ac}{1-a^2}, \\ r(1) &= f_1 - ar(0) = \frac{(c-a)(1-ac)}{1-a^2}, \\ r(k) &= (-a)^{k-1} r(1), \quad k \ge 1 \end{aligned}$$

while Equation 61.53 gives

$$\begin{aligned} r(-1) &= g_o = \frac{(c-a)(1-ac)}{1-a^2}, \\ r(-k) &= (-a)^{k-1} r(-1), \quad k \ge 1. \end{aligned}$$

Needless to say, these expressions for the covariance function are the same as those derived in the previous example.

61.6 Continuous-Time Processes

This section illustrates how some of the properties of discrete-time stochastic systems appear in an analog form for continuous-time models. However, white noise in continuous time leads to considerable mathematical difficulties, which must be solved in a rigorous way. See Chapter 60 for more details on this aspect.

The covariance function of a process $y(t)$ is still defined as (compare Equation 61.2)

$$r(\tau) = Ey(t+\tau)y(t), \tag{61.54}$$

assuming for simplicity that $y(t)$ has zero mean. The spectrum will now be

$$\phi(s) = \int_{-\infty}^{\infty} r(\tau)e^{-s\tau}d\tau, \tag{61.55}$$

and the spectral density is

$$\phi(i\omega) = \int_{-\infty}^{\infty} r(\tau)e^{i\omega\tau}d\tau. \tag{61.56}$$

The inverse relation to Equation 61.55 is

$$r(\tau) = \frac{1}{2\pi i}\int \phi(s)e^{s\tau}ds, \tag{61.57}$$

where integration is along the whole imaginary axis.

Consider a stationary stochastic process described by a spectral density $\phi(i\omega)$ that is a rational function of $i\omega$. By pure analogy with the discrete-time case it is found that

$$\phi(i\omega) = \frac{B(i\omega)B(-i\omega)}{A(i\omega)A(-i\omega)}, \tag{61.58}$$

where the polynomials

$$\begin{aligned} A(p) &= p^n + a_1 p^{n-1} + \ldots + a_n, \\ B(p) &= b_1 p^{n-1} + \ldots + b_n, \end{aligned} \tag{61.59}$$

have all their roots in the left half-plane (i.e., in the stability area). Here p is an arbitrary polynomial argument, but can be interpreted as the differentiation operator $[py(t) = \dot{y}(t)]$.

The effect of filtering a stationary process, say $u(t)$, with an asymptotically stable filter, say $H(p)$, can be easily phrased using the spectra. Let the filtering be described by

$$y(t) = H(p)u(t). \tag{61.60}$$

Then

$$\phi_y(s) = H(s)H(-s)\phi_u(s), \tag{61.61}$$

again paralleling the discrete-time case. As a consequence, one can interpret any process with a rational spectral density Equation 61.58 as having been generated by filtering as in Equation 61.60 by using

$$H(p) = \frac{B(p)}{A(p)}. \tag{61.62}$$

The signal $u(t)$ would then have a *constant* spectral density, $\phi_u(i\omega) \equiv 1$. As for the discrete-time case, such a process is called *white noise*. It will have a covariance function $r(\tau) = \delta(\tau)$ and hence, in particular, an infinite variance. This indicates difficulties to treat it with mathematical rigor.

61.7 Optimal Prediction

Consider an ARMA process

$$A(q)y(t) = C(q)e(t), \quad Ee^2(t) = \lambda^2, \tag{61.63}$$

where $A(q)$ and $C(q)$ are of degree n and have all their roots inside the unit circle. We seek a k-step predictor, that is, a function of available data $y(t)$, $y(t-1), \ldots,$ that will be close to the future value $y(t+k)$. In particular, we seek the predictor that is optimal in a mean square sense. The clue to finding this predictor is to rewrite $y(t+k)$ into two terms. The first term is a weighted sum of future noise values, $\{e(t+j)\}_{j=1}^{k}$. As this term is uncorrelated to all available data, it cannot be reconstructed in any way. The second term is a weighted sum of past noise values $\{e(t-s)\}_{s=0}^{\infty}$. By inverting the process model, the second term can be written as a weighted sum of output values, $\{y(t-s)\}_{s=0}^{\infty}$. Hence, it can be computed exactly from data.

In order to proceed, introduce the *predictor identity*

$$z^{k-1}C(z) \equiv A(z)F(z) + L(z), \tag{61.64}$$

where

$$\begin{aligned} F(z) &= z^{k-1} + f_1 z^{k-2} + \ldots + f_{k-1}, & (61.65)\\ L(z) &= \ell_o z^{n-1} + \ell_1 z^{n-2} + \ldots + \ell_{n-1}. & (61.66) \end{aligned}$$

Equation 61.64 is a special case of a Diophantine equation for polynomials. A solution is always possible. This is analogous to the Diophantine equation

$$n = qm + r$$

for integers (for given integers n and m, there exist unique integers q and r). Now

$$\begin{aligned} y(t+k) &= \frac{C(q)}{A(q)}e(t+k)\\ &= \frac{q^{k-1}C(q)}{A(q)}e(t+1)\\ &= \frac{A(q)F(q)+L(q)}{A(q)}e(t+1)\\ &= F(q)e(t+1) + \frac{qL(q)}{A(q)}e(t)\\ &= F(q)e(t+1) + \frac{qL(q)}{A(q)}\frac{A(q)}{C(q)}y(t)\\ &= F(q)e(t+1) + \frac{qL(q)}{C(q)}y(t). \end{aligned} \tag{61.67}$$

This is the decomposition mentioned previously. The term $F(q)e(t+1)$ is a weighted sum of *future* noise values, while $qL(q)/C(q)y(t)$ is a weighted sum of *available* measurements Y^t. Note that it is crucial for stability that $C(q)$ has all zeros strictly inside the unit circle (but that this is not restrictive due to spectral factorization). As the future values of the noise are unpredictable, the *mean square optimal predictor* is given by

$$\hat{y}(t+k|t) = \frac{qL(q)}{C(q)}y(t), \tag{61.68}$$

while the associated prediction error is

$$\begin{aligned} \tilde{y}(t+k) &= F(q)e(t+1)\\ &= e(t+k) + f_1 e(t+k-1)\\ &\quad + \ldots + f_{k-1}e(t+1), \end{aligned} \tag{61.69}$$

and has variance

$$E\tilde{y}^2(t+k) = \lambda^2(1 + f_1^2 + \ldots + f_{k-1}^2). \tag{61.70}$$

As a more general case, consider prediction of $y(t)$ in the ARMAX model

$$A(q)y(t) = B(q)u(t) + C(q)e(t). \tag{61.71}$$

In this case, proceeding as in Equation 61.67,

$$\begin{aligned} y(t+k) &= \frac{B(q)}{A(q)}u(t+k) + F(q)e(t+1)\\ &\quad + \frac{qL(q)}{A(q)}\left[\frac{A(q)}{C(q)}y(t) - \frac{B(q)}{C(q)}u(t)\right]\\ &= F(q)e(t+1) + \frac{qL(q)}{C(q)}y(t)\\ &\quad + \frac{qB(q)F(q)}{C(q)}u(t). \end{aligned} \tag{61.72}$$

We find that the prediction error is still given by Equation 61.69, while the optimal predictor is

$$\hat{y}(t+k|t) = \frac{qL(q)}{C(q)}y(t) + \frac{qB(q)F(q)}{C(q)}u(t). \tag{61.73}$$

This result can also be used to derive a minimum output variance regulator. That is, let us seek a feedback control for the process Equation 61.71 that minimizes $Ey^2(t)$. Let $k = \deg A - \deg B$ denote the delay in the system. As $\tilde{y}(t+k)$ and $\hat{y}(t+k|t)$ are independent,

$$Ey^2(t+k|t) = E\hat{y}^2(t+k|t) + E\tilde{y}^2(t+k) \geq E\tilde{y}^2(t+k), \tag{61.74}$$

with equality if and only if $\hat{y}(t+k|t) = 0$, the regulator is

$$u(t) = -\frac{L(q)}{B(q)F(q)}y(t). \tag{61.75}$$

Optimal prediction can also be carried out using a state-space formalism. It will then involve computing the Kalman filter, and a Riccati equation has to be solved (which corresponds to the spectral factorization). See Chapter 39 for a treatment of linear quadratic stochastic control using state-space techniques.

61.8 Wiener Filters

The steady-state linear least mean square estimate is considered in this section. It can be computed using a state-space formalism (like Kalman filters and smoothers), but here a polynomial formalism for an input-output approach is utilized. In case time-varying or transient situations have to be handled, a state-space approach must be used. See Chapter 25 for a parallel treatment of Kalman filters.

Let $y(t)$ and $s(t)$ be two correlated and stationary stochastic processes. Assume that $y(t)$ is measured, and find a causal, asymptotically stable filter $G(q)$ such that $G(q)y(t)$ is the optimal linear mean square estimator of $s(t)$, that is, it minimizes the criterion

$$V = E[s(t) - G(q)y(t)]^2. \tag{61.76}$$

This problem is best treated in the frequency domain. This implies in particular that data are assumed to be available since the infinite past $t = -\infty$. Introduce the estimation error

$$\tilde{s}(t) = s(t) - G(q)y(t). \tag{61.77}$$

The criterion V (Equation 61.76) can be rewritten as

$$V = \frac{1}{2\pi i}\oint \phi_{\tilde{s}}(z)\frac{dz}{z}. \tag{61.78}$$

Next note that

$$\begin{aligned}\phi_{\tilde{s}}(z) &= \phi_s(z) - G(z)\phi_{ys}(z) \\ &\quad - \phi_{sy}(z)G(z^{-1}) + G(z)G(z^{-1})\phi_y(z).\end{aligned} \tag{61.79}$$

Now let $G(q)$ be the optimal filter and $G_1(q)$ any causal filter. Replace $G(q)$ in Equation 61.76 by $G(q)+\varepsilon G_1(q)$. As a function of ε, V can then be written as $V = V_0 + \varepsilon V_1 + \varepsilon^2 V_2$. For $G(q)$ to be the *optimal* filter it is required that $V \geq V_0$ for all ε, which leads to $V_1 = 0$, giving

$$0 = \operatorname{tr}\frac{1}{2\pi i}\oint [G(z)\phi_y(z) - \phi_{sy}(z)]G_1(z^{-1})\frac{dz}{z}. \tag{61.80}$$

It is possible to give an interpretation and alternative view of Equation 61.80. For the optimal filter, the estimation error, $\tilde{s}(t)$, should be uncorrelated with all past measurements, $\{y(t-j)\}_{j=0}^{\infty}$. Otherwise, there would be another linear combination of the past measurements giving smaller estimation error variance. Hence,

$$E\tilde{s}(t)y(t-j) = 0, \quad \text{all } j \geq 0, \tag{61.81}$$

or

$$E\tilde{s}(t)[G_1(q)y(t)] = 0 \quad \text{for any stable and causal } G_1(q). \tag{61.82}$$

This can be rewritten as

$$\begin{aligned}0 &= E[s(t) - G(q)y(t)][G_1(q)y(t)] \\ &= \frac{1}{2\pi i}\oint [\phi_{sy}(z) - G(z)\phi_y(z)]G_1(z^{-1})\frac{dz}{z},\end{aligned} \tag{61.83}$$

which is precisely Equation 61.80.

From Equation 61.80, one easily finds the *unrealizable Wiener filter*. Setting the integrand to zero gives $G(z)\phi_y(z) = \phi_{sy}(z)$, and

$$G(z) = \phi_{sy}(z)\phi_y^{-1}(z). \tag{61.84}$$

The filter is not realizable since it relies (except in very degenerate cases) on all *future* data points of $y(t)$. Note, though, that when "deriving" Equation 61.84 from Equation 61.80, it was effectively required that Equation 61.80 holds for *any* $G_1(z)$. However, it is required only that Equation 61.80 holds for any *causal and stable* $G_1(z)$. Such an observation will eventually lead to the optimal *realizable* filter.

To proceed, let the process $y(t)$ have the *innovations representation* (remember that this is always possible by Section 61.4)

$$\begin{aligned}y(t) &= H(q)e(t), \\ Ee(t)e(s) &= \lambda^2\delta_{t,s}, \\ H(0) &= 1,\end{aligned} \tag{61.85}$$

$$H(q),\ H^{-1}(q) \ \text{asymptotically stable.}$$

Then $\phi_y(z) = H(z)H(z^{-1})\lambda^2$. Further, introduce the *causal part* of an analytical function. Let

$$G(z) = \sum_{j=-\infty}^{\infty} g_j z^{-j}, \tag{61.86}$$

where it is required that the series converges in a strip that includes the unit circle. The *causal part* of $G(z)$ is defined as

$$[G(z)]_+ = \sum_{j=0}^{\infty} g_j z^{-j}, \tag{61.87}$$

and the *anticausal part* is the complementary part of the sum:

$$[G(z)]_- = \sum_{j=-\infty}^{-1} g_j z^{-j} = G(z) - [G(z)]_+. \tag{61.88}$$

It is important to note that the term $g_0 z^{-0}$ in Equation 61.86 appears in the causal part, $[G(z)]_+$. Note that the anticausal part $[G(z)]_-$ of a transfer function $G(z)$ has no poles inside or on the unit circle, and that a filter $G(z)$ is causal if and only if $G(z) = [G(z)]_+$. Using the conventions of Equations 61.87 and 61.88, the optimality condition of Equation 61.80 can be formulated as

$$\begin{aligned} 0 &= \frac{1}{2\pi i}\oint \{G(z)H(z) - \phi_{sy}(z)\{H(z^{-1})\}^{-1}\lambda^{-2}\} \\ &\quad \lambda^2 H(z^{-1})G_1(z^{-1})\frac{dz}{z} \qquad (61.89) \\ &= \frac{1}{2\pi i}\oint \{G(z)H(z) - [\phi_{sy}(z)\{H(z^{-1})\}^{-1}\lambda^{-2}]_+ \\ &\quad -[\phi_{sy}(z)\{H(z^{-1})\}^{-1}\lambda^{-2}]_-\}\lambda^2 H(z^{-1})G_1(z^{-1})\frac{dz}{z}. \end{aligned}$$

The stability requirements imply that the function $H(z^{-1})G_1(z^{-1})$ does not have any poles inside the unit circle. The same is true for $[\phi_{sy}(z)\{H(z^{-1})\}^{-1}\lambda^{-2}]_-$, by construction. The latter function has a zero at $z = 0$. Hence, by the residue theorem,

$$\frac{1}{2\pi i}\oint [\phi_{sy}(z)\{H(z^{-1})\}^{-1}]_- H(z^{-1})G_1(z^{-1})\frac{dz}{z} = 0. \qquad (61.90)$$

The optimal condition of Equation 61.90 is therefore satisfied if

$$G(z) = \frac{1}{\lambda^2}[\phi_{sy}(z)\{H(z^{-1})\}^{-1}]_+ H^{-1}(z). \qquad (61.91)$$

This is the *realizable Wiener filter*. It is clear from its construction that it is a causal and asymptotically stable filter.

The Wiener filter is illustrated by two examples.

EXAMPLE 61.5:

Consider the ARMA process

$$\begin{aligned} A(q)y(t) &= C(q)e(t), \\ Ee^2(t) &= \lambda^2. \end{aligned}$$

Treat the prediction problem

$$s(t) = y(t+k), \qquad k > 0.$$

In this case

$$\begin{aligned} H(z) &= \frac{C(z)}{A(z)}, \\ \phi_{sy}(z) &= z^k \phi_y(z). \end{aligned}$$

The *unrealizable* filter of Equation 61.84 becomes, as before,

$$G(z) = z^k \phi_y(z)\phi_y^{-1}(z) = z^k,$$

meaning that

$$\hat{s}(t) = y(t+k).$$

Note that it is noncausal, but it is otherwise a perfect estimate since it is without error! Next, the *realizable* filter is calculated:

$$\begin{aligned} G(z) &= \frac{1}{\lambda^2}\left[z^k \lambda^2 \frac{C(z)}{A(z)}\frac{C(z^{-1})}{A(z^{-1})}\frac{A(z^{-1})}{C(z^{-1})}\right]_+ \frac{A(z)}{C(z)} \\ &= \left[z^k \frac{C(z)}{A(z)}\right]_+ \frac{A(z)}{C(z)}. \end{aligned}$$

To proceed, let $A(z)$ and $C(z)$ have degree n, and introduce the polynomial $F(z)$ of degree $k-1$ and the polynomial $L(z)$ of degree $n-1$ by the *predictor identity* (Equation 61.64). This gives

$$\begin{aligned} G(z) &= \left[\frac{zA(z)F(z) + zL(z)}{A(z)}\right]_+ \frac{A(z)}{C(z)} \\ &= \frac{zL(z)}{A(z)}\frac{A(z)}{C(z)} = \frac{zL(z)}{C(z)}. \end{aligned}$$

The optimal predictor therefore has the form

$$\hat{s}(t) = \hat{y}(t+k|t) = \frac{qL(q)}{C(q)}y(t),$$

in agreement with Equation 61.68.

EXAMPLE 61.6:

Consider the measurement of a random signal $s(t)$ in additive noise $w(t)$. This noise source need not be white, but it is assumed to be uncorrelated with the signal $s(t)$. Model $s(t)$ and $w(t)$ as ARMA processes, (see Figure 61.3). Thus,

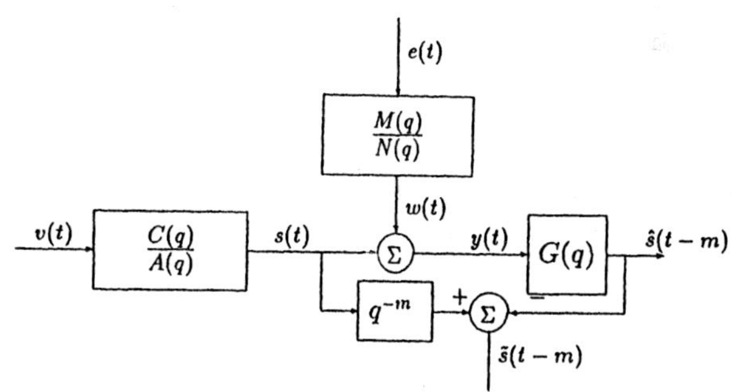

Figure 61.3 Setup for a polynomial estimation problem.

$$\begin{aligned} y(t) &= s(t) + w(t), \quad s(t) = \frac{C(q)}{A(q)}v(t), \\ w(t) &= \frac{M(q)}{N(q)}e(t), \quad Ev(t)v(s) = \lambda_v^2\delta_{t,s}, \\ Ee(t)e(s) &= \lambda_e^2\delta_{t,s}, \quad Ee(t)v(s) = 0. \end{aligned}$$

The polynomials in the model are

$$\begin{aligned} C(q) &= q^n + c_1 q^{n-1} + \ldots + c_n, \\ A(q) &= q^n + a_1 q^{n-1} + \ldots + a_n, \\ M(q) &= q^r + m_1 q^{r-1} + \ldots + m_r, \\ N(q) &= q^r + n_1 q^{r-1} + \ldots + n_r. \end{aligned}$$

Assume that all four polynomials have all their roots inside the unit circle. The problem to be treated is to estimate $s(t-m)$ from $\{y(t-j)\}_{j=0}^{\infty}$, where m is an integer. (By definition, $m = 0$ gives filtering, $m > 0$ smoothing, and $m < 0$ prediction.)

To solve the estimation problem, first perform a spectral factorization of the output spectrum

$$\begin{aligned} \phi_y(z) &= \lambda_v^2 \frac{C(z)C(z^{-1})}{A(z)A(z^{-1})} + \lambda_e^2 \frac{M(z)M(z^{-1})}{N(z)N(z^{-1})} \\ &\equiv \lambda_\varepsilon^2 \frac{B(z)B(z^{-1})}{A(z)N(z)A(z^{-1})N(z^{-1})}, \end{aligned}$$

requiring that $B(z)$ is a monic polynomial (that is, it has a leading coefficient equal to one) of degree $n+r$, and that it has all zeros inside the unit circle. The polynomial $B(z)$ is therefore uniquely given by the identity

$$\begin{aligned} \lambda_\varepsilon^2 B(z)B(z^{-1}) \equiv \lambda_v^2 C(z)C(z^{-1})N(z)N(z^{-1}) \\ + \lambda_e^2 A(z)A(z^{-1})M(z)M(z^{-1}). \end{aligned}$$

Hence,

$$H(z) = \frac{B(z)}{A(z)N(z)}, \qquad \lambda^2 = \lambda_\varepsilon^2,$$

$$\phi_{sy}(z) = z^{-m} \lambda_v^2 \frac{C(z)C(z^{-1})}{A(z)A(z^{-1})}.$$

According to Equation 61.91, the optimal filter becomes

$$\begin{aligned} G(z) &= \frac{1}{\lambda_\varepsilon^2} \left[z^{-m} \lambda_v^2 \frac{C(z)C(z^{-1})}{A(z)A(z^{-1})} \frac{A(z^{-1})N(z^{-1})}{B(z^{-1})} \right]_+ \\ &\quad \times \frac{A(z)N(z)}{B(z)} \\ &= \frac{\lambda_v^2}{\lambda_\varepsilon^2} \left[z^{-m} \frac{C(z)C(z^{-1})N(z^{-1})}{A(z)B(z^{-1})} \right]_+ \frac{A(z)N(z)}{B(z)}. \end{aligned} \tag{61.92}$$

The causal part $[\ \ ]_+$ can be found by solving the Diophantine equation

$$\begin{aligned} z^{-m} C(z)C(z^{-1})N(z^{-1}) \equiv z^{\min(0,-m)} B(z^{-1})R(z) \\ + z^{\max(0,-m)} A(z)L(z^{-1}), \end{aligned}$$

where the unknown polynomials have degrees

$$\begin{aligned} \deg R &= n - \min(0,-m), \\ \deg L &= n + r - 1 + \max(0,-m). \end{aligned}$$

Note that the "-1" that appears in $\deg L$ has no direct correspondence in $\deg R$. The reason is that the direct term $g_0 z^{-0}$ in Equation 61.86 is associated with the causal part of $G(z)$.

The optimal filter is readily found:

$$\begin{aligned} G(z) &= \frac{\lambda_v^2}{\lambda_\varepsilon^2} \left(\left[\frac{z^{\min(0,-m)} B(z^{-1})R(z)}{A(z)B(z^{-1})} \right]_+ \right. \\ &\quad \left. + \left[\frac{z^{\max(0,-m)} A(z)L(z^{-1})}{A(z)B(z^{-1})} \right]_+ \right) \frac{A(z)N(z)}{B(z)} \\ &= \frac{\lambda_v^2}{\lambda_\varepsilon^2} \left[\frac{z^{\min(0,-m)} R(z)}{A(z)} \right]_+ \frac{A(z)N(z)}{B(z)} \\ &= \frac{\lambda_v^2}{\lambda_\varepsilon^2} \frac{z^{\min(0,-m)} R(z)N(z)}{B(z)}. \end{aligned}$$

References

[1] Åström, K.J., *Introduction to Stochastic Control*, Academic Press, New York, 1970.

[2] Åström, K.J. and Wittenmark, B., *Computer Controlled Systems*, Prentice Hall, Englewood Cliffs, NJ, 1990.

[3] Grimble, M.J. and Johnson, M.A., *Optimal Control and Stochastic Estimation*, John Wiley & Sons, Chichester, UK, 1988.

[4] Hunt, K.J. Ed., *Polynomial Methods in Optimal Control and Filtering*, Peter Peregrinus Ltd, Stevenage, UK, 1993 (in particular, Chapter 6: A. Ahlén and M. Sternad: Optimal Filtering Problems).

[5] Kučera, V., *Discrete Linear Control*, John Wiley & Sons, Chichester, UK, 1979.

[6] Söderström, T., *Discrete-Time Stochastic Systems: Estimation and Control*, Prentice Hall International, Hemel Hempstead, UK, 1994.

62

Minimum Variance Control

M. R. Katebi and A. W. Ordys
Industrial Control Centre, Strathclyde University, Glasgow, Scotland

62.1 Introduction

Historically, Minimum-Variance Control (MVC) has been a practical control strategy for applying of linear stochastic control theory. The control algorithm was first formulated in [1]. Åström used the MVC technique to minimize the variance of the output signal for control of paper thickness in a paper machine. The control objective was to achieve the lowest possible variation of paper thickness with stochastic disturbances acting on the process. Since then, the MVC technique has attracted many practical applications (e.g. [8], [12]) and has gained significant theoretical development leading to a wide range of control algorithms known as Model-Based Predictive Control techniques.

The reasons for MVC's popularity lie in its simplicity, ease of interpretation and implementation, and its relatively low requirements for model accuracy and complexity. In fact, the models can be obtained from straightforward identification schemes. The MVC technique was also used for self-tuning algorithms. When MVC was developed, its low computational overhead was important. MVC is still used as a simple and efficient algorithm in certain types of control problems [10]. It is particularly well suited for situations where [1]

- the control task is to keep variables close to the operating points,
- the process can be modeled as linear, time invariant but with significant time delay, and
- disturbances can be described by their stochastic characteristics.

MVC, as it was originally presented, provides a stable control action for only a limited range of processes. One of its main assumptions is so-called *"minimum-phase"* behavior of the system. Later, extensions to MVC were introduced in [3], [5], [7] to relax the most restrictive assumptions. The penalty for this extension was increased complexity of the algorithms, eventually approaching or sometimes even exceeding the complexity of the standard *Linear Quadratic Gaussian (LQG)* solution. However, the formulation of MVC was a cornerstone in stimulating development of the more general class of Model-Based Predictive Control, of which MVC is an example.

This article is organized as follows. The model and control design requirements are described in Section 62.2.1. The formulation of an MV predictor is given in Section 62.3. The basic MVC and its variations are developed in Section 62.4. The state-space formulation is given in Section 62.5. An example of the application of MVC is given in Section 62.6. Finally, conclusions are drawn in Section 62.7.

0-8493-8570-9/96/$0.00+$.50

62.2 Basic Requirements

62.2.1 The Process Model

In its basic form, the minimum-variance control assumes a single-input, single-output (SISO), linear, time-invariant stationary stochastic process described by

$$A\left(z^{-1}\right)\cdot y(t)=z^{-d}\cdot B\left(z^{-1}\right)\cdot u(t)+C\left(z^{-1}\right)\cdot w(t) \quad (62.1)$$

where $y(t)$ represents variation of the output signal around a given steady-state operating value, $u(t)$ is the control signal, and $w(t)$ denotes a disturbance assumed to be a zero mean, Gaussian *white noise* of variance σ. $A\left(z^{-1}\right)$, $B\left(z^{-1}, C\right)\left(z^{-1}\right)$ are nth order polynomials,

$$\begin{aligned} A\left(z^{-1}\right) &= 1+a_1\cdot z^{-1}+a_2\cdot z^{-2}+\cdots \\ &\quad +a_n\cdot z^{-n}, & (62.2)\\ B\left(z^{-1}\right) &= b_0+b_1\cdot z^{-1}+b_2\cdot z^{-2}+\cdots \\ &\quad +b_n\cdot z^{-n}, \quad b_0\neq 0, & (62.3)\\ C\left(z^{-1}\right) &= 1+c_1\cdot z^{-1}+c_2\cdot z^{-2}+\cdots \\ &\quad +c_n\cdot z^{-n}, & (62.4)\end{aligned}$$

and z^{-1} is the backward shift operator, i.e.,

$$z^{-1}\cdot x(t) = x(t-1). \quad (62.5)$$

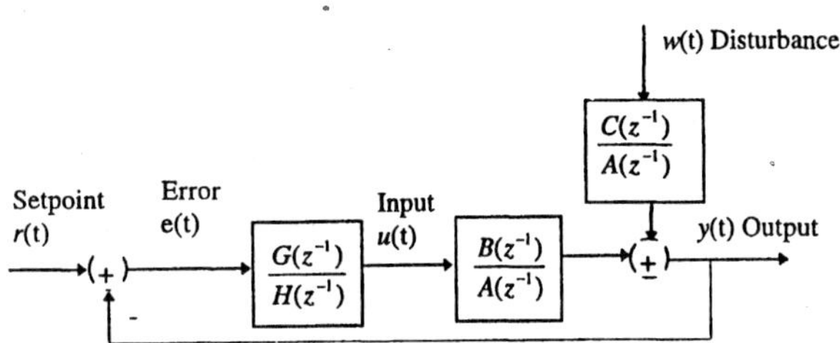

Figure 62.1 The process and the feedback system structure.

Thus z^{-d} in Equation 62.1 represents a d-step delay in the control signal. This means that the control signal starts to act on the system after d time increments. The coefficients of polynomials $A\left(z^{-1}\right)$, $B\left(z^{-1}\right)$, and $C\left(z^{-1}\right)$ are selected so that the linear dynamics of the system are accurately represented about an operating point. The coefficient values may result from considering physical laws governing the process or, more likely, from system identification schemes. Some of these coefficients may be set to zero. In particular, if certain coefficients with index n (e.g., c_n) are set to zero then the polynomials will have different orders. Also, note that the first coefficients associated with zero power of z^{-1} in $A\left(z^{-1}\right)$ and $C\left(z^{-1}\right)$ are unity. This does not impose any restrictions on the system description. For example, if the coefficient in polynomial $A\left(z^{-1}\right)$ is a_0 then both sides of Equation 62.1 may be divided by a_0 and the remaining coefficients adjusted accordingly. Similarly, if the coefficient in polynomial $C\left(z^{-1}\right)$ is c_0 then its value may be included in the disturbance description,

$$w(t)=c_0\cdot w(t). \quad (62.6)$$

The new disturbance is still a zero mean, Gaussian white noise with variance $\sigma'=c_0^2\cdot\sigma$.

62.2.2 The Disturbance Model

Consider a discrete, time-variant system of transfer function $G\left(z^{-1}\right)$: $y(t) = G\left(z^{-1}\right)w(t)$. Let the input signal be white noise of zero mean and unity variance. The spectral density of the signal may be written as

$$\phi_{yy}(\omega)=G(e^{-j\omega})G(e^{j\omega})\phi_{ww}(\omega)=|G(e^{-j\omega})|^2\phi_{ww}(\omega). \quad (62.7)$$

For the special class of a stationary stochastic process with rational spectral density, the process can be represented by the rational transfer function, $G\left(z^{-1}\right)=C\left(z^{-1}\right)/A\left(z^{-1}\right)$ driven by white noise so that G has all its poles outside the unit circle and zeros outside or on the unit circle. This process is known as spectral factorization and may be used to model the disturbance part of the model given in Equation 62.1. This implies that the output spectrum should first be calculated using the output measurement. The transfer function $G\left(z^{-1}\right)$ is then fitted to the spectrum using a least-squares algorithm.

62.2.3 The Performance Index

The performance index to be minimized is the variance of the output at $t+d$, given all the information up to time t, expressed by

$$J(t)=E\left\{y(t+d)^2\right\}, \quad (62.8)$$

where $E\{\cdot\}$ denotes the *expectation operator*. It is assumed that the control selected must be a function of information available at time t, i.e., all past control signals, all past outputs, and the present output is

$$Y(t)=[u(t-d-1), u(t-d-2), \cdots, y(t), y(t-1)\cdots]. \quad (62.9)$$

Therefore, the feasibility of the solution requires rewriting the performance index Equation 62.8 in the form ,

$$J(t)=E\left\{E\left\{y(t+k)^2\middle| Y(t)\right\}\right\}, \quad (62.10)$$

where $E\{\cdot|\cdot\}$ is the conditional expectation. Then, the optimization task is choosing control signal $u(t)$ so that $J(t)$ is minimized with respect to $u(t)$,

$$J(t)=E\left\{\min_{u(t)}\left(E\left\{y(t+d)^2\middle| Y(t)\right\}\right)\right\}. \quad (62.11)$$

62.2.4 The Control Design Specification

From a practical point of view, after applying the MVC it is important that the closed-loop system meets the following requirements :

- The closed-loop system is stable.
- The *steady-state offset* is zero.
- The system is robust, i.e., the behavior of the system does not change with respect to parameter changes.

It will be shown in later sections how the minimum-variance design fulfills these requirements.

62.3 Optimal Prediction

The control objective of the feedback system shown in Figure 62.1 is to compensate for the effect of the disturbance $w(t)$ at any time t. Because of the process time delay, z^{-d}, the control input, $u(t)$, at time t, influences the output, $y(t)$, at time $t+d$. This implies that the value of the output at time $t+d$ is needed to calculate the control input $u(t)$ at time t. Because the output is not known at the future time $t+d$, we predict its value using the process model given in Equation 62.1. Thus, we first develop a predictor in the next section.

62.3.1 The d-Step Ahead Predictor

Our objective is to predict the value of the output at time $t+d$, defined by $\hat{y}(t+d|t)$, given all the information, $Y(t) = [u(t-d-1), u(t-d-2)\cdots y(t), y(t-1), \cdots]$ up to time t. We also want to predict the output so that the variance of the prediction error, $e_p(t+d) = [y(t+d) - \hat{y}(t+d|t)]$, is minimized, i.e., find $\hat{y}(t+d|t)$ so that

$$J(t+d) = E\left[y(t+d) - \hat{y}(t+d|t)\right]^2 \tag{62.12}$$

is minimized subject to the process dynamics given in Equation 62.1.

62.3.2 Predictor Solution

We first calculate the output at time $t+d$ using the process model,

$$y(t+d) = \frac{B}{A}u(t) + \frac{C}{A}w(t+d), \tag{62.13}$$

where the argument z^{-1} is deleted from the polynomials A, B, and C for convenience.

The control input is known up to time $t-1$. The disturbance input can be split into two sets of signals, namely, $[w(t), w(t-1)\cdots,]$ and $[w(t+d-1), w(t+d-2), \cdots, w(t+1)]$. The latter are future disturbances, and so their values are unknown. The former are past disturbances and their values may be calculated from the known past inputs and outputs by

$$w(t) = \frac{A}{C}y(t) - \frac{z^{-d}B}{C}u(t). \tag{62.14}$$

The random variables $[w(t+d-1), w(t+d-2), \ldots]$ are assumed independent of the process output $[y(t), y(t-1), \ldots]$, and all future control inputs are assumed to be zero. We can therefore separate the disturbance signal into casual and noncasual parts [3].

$$y(t+d) = \frac{B}{A}u(t) + \left[E + z^{-d}\frac{F}{A}\right]w(t+d), \tag{62.15}$$

where E and F are polynomials defined as

$$E(z^{-1}) = 1 + \sum_{i=1}^{d} e_i z^{-i},$$

$$F(z^{-1}) = f_0 + \sum_{i=1}^{n-1} f_i z^{-i}, \tag{62.16}$$

and the following polynomial relationship should be satisfied:

$$A(z^{-1})E(z^{-1}) + z^{-d}F(z^{-1}) = C(z^{-1}). \tag{62.17}$$

Substituting for $w(t)$ in Equation 62.13 using Equation 62.14 gives

$$y(t+d) = \frac{B}{A}u(t) + \left(\frac{F}{A}\right)\left(\frac{A}{C}y(t) - z^{-d}\frac{B}{C}u(t)\right) + Ew(t+d) \tag{62.18}$$

which can be simplified to

$$y(t+d) = \frac{B}{A}u(t) + \left(\frac{F}{C}\right)y(t) - \left(z^{-d}\frac{FB}{AC}u(t)\right) + Ew(t+d). \tag{62.19}$$

Replacing $(z^{-d}F)$ from Equation 62.17,

$$y(t+d) = \frac{BE}{C}u(t) + \frac{F}{C}y(t) + Ew(t+d). \tag{62.20}$$

Substituting this equation in the minimum prediction cost $J(t+d)$ leads to

$$\begin{aligned} J(t+d) &= E[y(t+d) - \hat{y}(t+d|t)]^2 \\ &= E\left\{\left[Ew(t+d) + \frac{BE}{C}u(t)\frac{F}{C}y(t) - \hat{y}\right]^2\right\}, \\ &= E\left\{\left[\frac{BE}{C}u(t) + \frac{F}{C}y(t) - \hat{y}\right]^2\right\} \\ &\quad + E\left\{[Ew(t+d)]^2\right\} \\ &\quad + 2E\left\{\left[\frac{BE}{C}u(t) + \frac{F}{C}y(t) - \hat{y}\right][Ew(t+d)]\right\} \end{aligned} \tag{62.21}$$

The last term on the right-hand side is zero because $[w(t+d), w(t+d-1), \ldots, w(t+1)]$ are independent of $[y(t), y(t-1), \ldots]$ and $E\{w(t)\} = 0$. The variance of the error is therefore the sum of two positive terms. The second term represents the future disturbance input at $t+d$ and can take any value. To minimize $J(t+d)$, we make the first term zero to find the minimum variance error predictor,

$$\hat{y}(t+d|t) = \frac{BE}{C}u(t) + \frac{F}{C}y(t). \tag{62.22}$$

This can be implemented in time domain as

$$C\hat{y}(t+d|t) = EBu(t) + Fy(t). \qquad (62.23)$$

For the predictor to be stable, the polynomials C and B should have all of their roots outside the unit circle. The implementation of the above optimal predictors requires solving the polynomial Equation 62.16 to find $E(z^{-1})$ and $F(z^{-1})$. There are a number of efficient techniques to solve this equation. A simple solution is to equate the coefficients of different power of z^{-1} and solve the resulting set of linear algebraic equations. The minimum variance of the error can also be calculated as

$$\begin{aligned} J(t+d) &= E[y(t+d) - \hat{y}]^2 \\ &\geq 1 + \sum_{i=1}^{k-1} e_i^2. \end{aligned} \qquad (62.24)$$

62.4 Minimum-Variance Control

Given the dynamical process model in Equation 62.1, the minimum-variance control problem can be stated as finding a control law which minimizes the variance of the output described in the performance index of Equation 62.11.

62.4.1 Minimum-Variance Control Law

Assuming that all of the outputs $[y(t), y(t-1), \ldots]$ and the past control inputs $[u(t-1), u(t-2), \ldots]$ are known at time t, the problem is to determine $u(t)$ so that the variance of the output is as small as possible. The control signal $u(t)$ will affect only $y(t+d)$ but not any earlier outputs. Use Equation 62.17 for the output,

$$y(t+d) = \frac{EB}{C}u(t) + \frac{F}{C}y(t) + Ew(t+d). \qquad (62.25)$$

Substitute this equation in the minimum variance performance index to obtain

$$E\left\{y^2(t+d)|Y(t)\right\} = E\left\{\left[\frac{EB}{C}u(t) + \frac{F}{C}y(t)\right]^2\right\} + E\left\{[Ew(t+d)]^2\right\} \qquad (62.26)$$

The expected value of the cross-product term is zero because $[w(t+d), w(t+d-1), \ldots, w(t+1)]$ are assumed to be independent of $\{y(t), y(t-1), \ldots\}$. The performance index is minimum if the first term is zero:

$$\tfrac{BE}{C}u(t) + \tfrac{F}{C}y(t) = 0. \qquad (62.27)$$

The minimum variance control law is then given by

$$u(t) = \tfrac{G}{H}y(t) = -\tfrac{F}{BE}y(t). \qquad (62.28)$$

Note that the controller is stable only if $B(z^{-1})$ has all of its roots outside the unit circle. This implies that the process should be minimum phase for the closed-loop system to be stable.

62.4.2 Closed-Loop Stability

The closed-loop characteristic equation for the system of Figure 62.1 may be written as

$$HA + z^{-d}BG = 0 \qquad (62.29)$$

Replacing for G and H from Equation 62.27 and using the polynomial identity Equation 62.17 gives

$$CB = 0. \qquad (62.30)$$

The closed loop is asymptotically stable if the polynomials C and B have all of their roots outside the unit circle. This implies that the basic version of MVC is applicable to minimum-phase systems with disturbances which have a stable and rational power spectrum.

62.4.3 The Tracking Problem

The basic MVC regulates the variance of the output about the operating point. It is often necessary in process control to change the level of the operating point while the system is under closed-loop control. The control will then be required to minimize the variance of the error between the desired process set point and the actual output. Assuming the set point , $r(t+d)$, is known, the error at time $t+d$ is defined as

$$e(t+d) = r(t+d) - y(t+d). \qquad (62.31)$$

The performance index to be minimized may now be written

$$J(t) = E\left\{\min_{u(t)}\left(E\left\{e(t+d)^2 \middle| Y(t)\right\}\right)\right\} \qquad (62.32)$$

Following the same procedure as in the previous section, the control law may be written

$$u(t) = -\frac{F}{BE}y(t) + \frac{C}{BE}r(t+d) \qquad (62.33)$$

The feedback controller is similar to the basic MVC and hence the stability properties are the same. The set point is introduced directly in the controller and variance of the error rather than the variance of the output signal is minimized.

62.4.4 The Weighted Minimum Variance

The performance index given in Equation 62.10 does not penalize the control input $u(t)$. This can lead to excessive input changes and hence actuator saturation. Moreover, the original form of MVC developed by Åström [1] stabilizes only the minimum-phase systems. Clarke and Hastings-James [7] proposed that a weighting, R, of the manipulated variable be included in the performance index,

$$J(t) = E\left\{\min_{u(t)}\left(E\left\{e(t+d)^2 \mid Y(t) + Ru^2(t)\right\}\right)\right\}. \qquad (62.34)$$

Minimizing this performance index with respect to the control input and subject to the process dynamic will lead to the weighted MV control law,

$$u(t) = -\frac{F}{BE + \frac{R}{b_0}C}y(t) + \frac{C}{BE + \frac{R}{b_0}C}r(k+d). \quad (62.35)$$

The closed-loop characteristic equation for the weighted minimum variance is calculated from Equation 62.24 as

$$C\left[B + \frac{R}{b_0}A\right] = 0. \quad (62.36)$$

The closed-loop system is now asymptotically stable only if the roots of C are outside the unit circle. The system can now be nonminimum phase if an appropriate value of R is chosen to move the unstable zero outside the unit circle. The disadvantage is that the variance of error is not directly minimized. It should be pointed out that MVC without the control input weighting can also be designed for nonminimum-phase systems as shown by Åström and Wittenmark [3].

62.4.5 The Generalized Minimum Variance Control

Clarke and Gawthrop [5], [6] developed the Generalized Minimum-Variance Controller (GMVC) for self-tuning control application by introducing the reference signal and auxiliary variables (weighting functions) into the performance index. GMVC minimizes the variance of an auxiliary output of the form,

$$\begin{aligned}\phi(t+d) &= Q(z^{-1})y(t+d) \\ &\quad + R(z^{-1})u(t) - P(z^{-1})r(t+d), \quad (62.37)\end{aligned}$$

where $Q(z^{-1}) = Q_n(z^{-1})/Q_d(z^{-1})$, $R(z^{-1}) = R_n(z^{-1})/R_d(z^{-1})$, $P(z^{-1}) = P_n(z^{-1})/P_d(z^{-1})$ are stable weighting functions. The performance index minimized subject to the process dynamics is

$$J(t) = E[\phi^2(t+d)|t]. \quad (62.38)$$

The signals $u(t)$ and $r(t+d)$ are known at time t. The prediction of $\phi(t+d|t)$ will reduce to the prediction of $Qy(t+d)$. Multiplying the process model by Q,

$$Q_dA \cdot [Qy(t)] = z^{-d} \cdot Q_nBu(t) + Q_nCw(t). \quad (62.39)$$

Replacing A, B, and C by Q_dA, Q_nB, and Q_nC in Equation 62.23 leads to the predictor equation for GMVC,

$$Q_dC[Q\hat{y}(t+d|t)] = Q_dEBu(t) + Fy(t), \quad (62.40)$$

and the polynomial identity,

$$\begin{aligned}&Q_d(z^{-1})A(z^{-1}) + z^{-d}F(z^{-1}) \\ &= Q_n(z^{-1})C(z^{-1}). \quad (62.41)\end{aligned}$$

Adding the contribution from $u(t)$ and $r(t)$ to Equation 62.41, the predictor for the auxiliary output $\phi(t)$ is

$$\phi(t+d/t) = \frac{F}{Q_dC}y(t) + \left(\frac{EB}{C} + R\right)u(t) - Pr(t+d). \quad (62.42)$$

The control law is chosen so that the d-step ahead prediction $\phi(t+d|t)$ is zero:

$$u(t) = \frac{R_dCPr(t) - R_dFy(t)}{Q_d(R_dEB + CR_n)}. \quad (62.43)$$

Note that by setting $P_d = Q_d = 1$ and $R = P_n = 0$, the basic MVC is obtained. Q_d may be used to influence the magnitude of control input signal. In particular, choosing $Q_d = (1-z^{-1})Q_d$ introduces integral action in the controller. The weighting $R(z^{-1})$ adjusts the speed of response of the controller and hence prevents actuator saturation.

The closed-loop transfer function, derived by substituting for $u(t)$ in Equation 62.1 is

$$y(t) = \frac{EB + RC}{QB + RA}w(t) + \frac{z^{-d}BR}{QB + RA}r(t). \quad (62.44)$$

Solving the characteristic equation,

$$R_dQ_nB + R_nQ_dA = 0 \quad (62.45)$$

gives the closed loop poles. The weightings Q and R may be selected so that the poles are located in a desired position on the complex plane.

62.5 State-Space Minimum-Variance Controller

The state-space description of a linear, time-invariant discrete-time system has the form,

$$\begin{aligned}x(t+1) &= Ax(t) + Bu(t) + Gw(t); \\ y(t) &= Cx(t) + v(t), \quad x(t_0) = x_0, \quad (62.46)\end{aligned}$$

where $x(t)$, $u(t)$, $y(t)$ are state, input, and output vectors of size n, m, and r, respectively, and A, B, and C are system matrices of appropriate dimensions. Also $[w(t)]$ and $[v(t)]$ are white noise sequences. The states and outputs of the system may be estimated using a Kalman filter as discussed in other chapters of the Handbook,

$$\begin{aligned}\hat{x}(t+1) &= A\hat{x}(t) + Bu(t) + K(t)e(t), \\ \hat{y}(t) &= C\hat{x}(t), \\ e(t) &= [y(t) - C\hat{x}(t)], \quad (62.47)\end{aligned}$$

where $e(t)$ is the innovation signal with property, $E[e(t)|t] = 0$ and $K(t)$ is the Kalman filter gain.

The optimal d-step ahead predictor of $[\hat{x}(t)]$ is the conditional mean of $x(t+d)$. From Equation 62.47,

$$\begin{aligned}\hat{y}(t+d|t) &= CA^{d-1}\hat{x}(t+1) \\ &\quad + \sum_{j=t+1}^{t+d-1} CA^{t+d-j-1}B(j)u(j) \\ &\quad + \sum_{j=t+1}^{t+d-1} CA^{t+d-j-1}K(j)e(j) \quad (62.48)\end{aligned}$$

Taking the conditional expectation and using the property of innovation signal,

$$\begin{aligned}\hat{y}(t+d|t) &= CA^{d-1}\hat{x}(t+1) \qquad (62.49)\\ &= CA^{d-1}\{[A-K(t)C]\hat{x}(t) \\ &\quad + CBu(t) + K(t)y(t)\}.\end{aligned}$$

For the performance index given by Equation 62.11, the optimum control input will be achieved if the d-step prediction of output signal is zero. Therefore, the state-space version of minimum-variance controller takes the form,

$$u(t) = \left[CA^{d-1}B\right]^{-1} CA^{d-1}\{[A-K(t)C]\hat{x}(t) + K(t)y(t)\}. \qquad (62.50)$$

It can be shown [4], [13] that this controller is equivalent to the minimum-variance regulator described by Equation 62.28.

The multivariable formulation of the MVC algorithm in state-space form is similar to the one described above. The polynomial MVC for multi-input, multioutput is given in [9].

62.6 Example

As a simple example illustrating some properties of minimum-variance control, consider a plant described by the following input-output relationship:

$$\begin{aligned}y(t) &= \frac{z^{-2}(1.0-0.995z^{-1})}{1-1.81z^{-1}+0.819z^{-2}}u(t) \\ &+ \frac{1-1.8z^{-1}+0.85z^{-2}}{1-1.81z^{-1}+0.819z^{-2}}w(t).\end{aligned}$$

Note that the delay of control signal is 2. The open-loop response of the system is shown in Figure 62.2 where the variance of the output is 2.8. To illustrate the performance of the predictor, the polynomial identity is solved to give $E(z^{-1}) = (1-0.01z^{-1})$ and $F(z^{-1}) = (0.049-0.0082z^{-1})$. A sine wave, $u(t) = \sin(0.2t)$, is applied to the input and the prediction of the output is shown in Figure 62.3. The theoretical minimum prediction error variance of 1 should be obtained but the simulated error is calculated as 1.49.

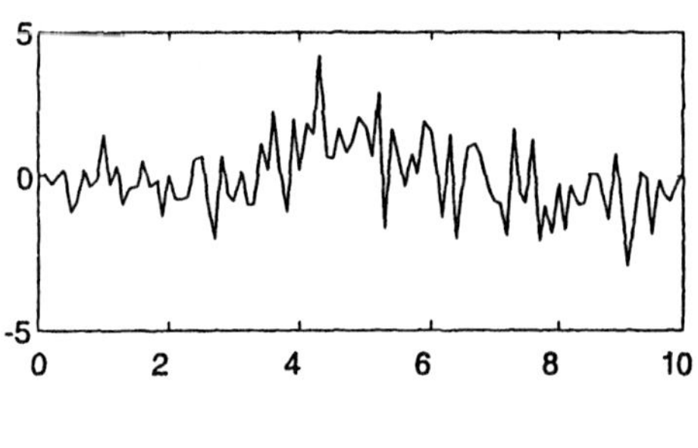

Figure 62.2 The open-loop response.

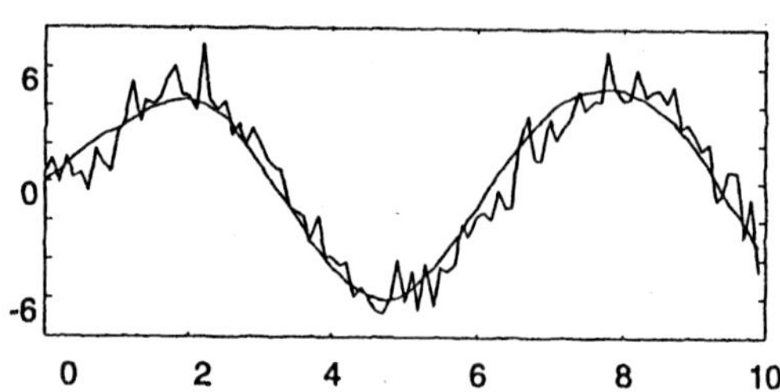

Figure 62.3 The prediction of a sine wave.

The basic MV controller Equation 62.28 is then applied to the system. The input and output are shown in Figure 62.4. The variance is now reduced to 1.26.

To examine the set point tracking of the basic MVC, a reference of magnitude 10 is used.

The results are shown in Figure 62.5. The tracking error is small, but the magnitude of the control signal is large and this is not realistic in many practical applications. To reduce the size of the control input, WMV may be used.

The result of introducing a weighting of $R = 0.8$ is shown in Figure 62.6. The set point is changed to 20 to illustrate the effect more clearly. The WMVC will reduce the control activity but it will also increase the tracking error. To keep the control input within a realistic range and to obtain zero tracking error, the GMVC may be used.

Selecting $Q_d = (1-z^{-1})$ will introduce integral action in the controller. The solution to the identity polynomial may then be obtained as $E(z^{-1}) = (1-0.01z^{-1})$ and $F(z^{-1}) = (1.06 - 1.84z^{-1} + 0.083z^{-2})$. Note that the order of the controller has increased by one due to integral action. The response of the system in Figure 62.7 shows the reduction in the tracking error and the control activity.

If the plant description changes slightly,

$$\begin{aligned}y'(t) &= \frac{z^{-2}(1.0-1.001z^{-1})}{1-1.81z^{-1}+0.819z^{-2}}u(t) \\ &+ \frac{1-1.8z^{-1}+0.85z^{-2}}{1-1.81z^{-1}+0.819z^{-2}}w(t).\end{aligned}$$

The basic MVC produces an unstable closed-loop system due to an unstable zero. The GMVC will however stabilize the system as shown in Figure 62.8.

62.7 Defining Terms

62.7.1 Expectation Operator

If $g(\zeta)$ is a function of a random variable ζ described by the probability density $\varphi(\zeta)$, the expectation operator of g is defined

$$E[g(\zeta)] = \int_{-\infty}^{\infty} [g(\zeta)\varphi(\zeta)]d\zeta.$$

The conditional expectation will be obtained if the probability density function $\varphi(\zeta)$ is replaced by conditional probability

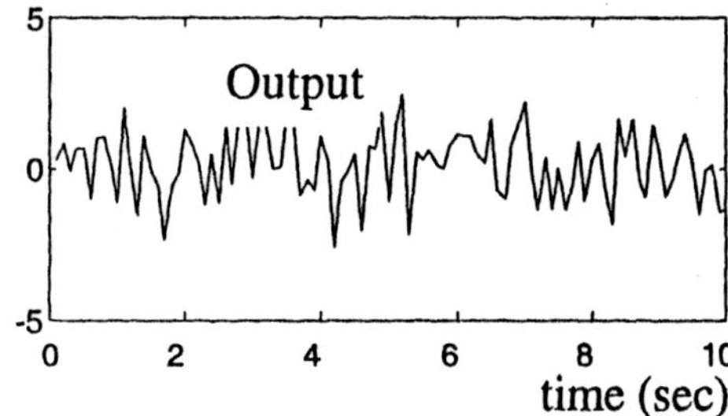

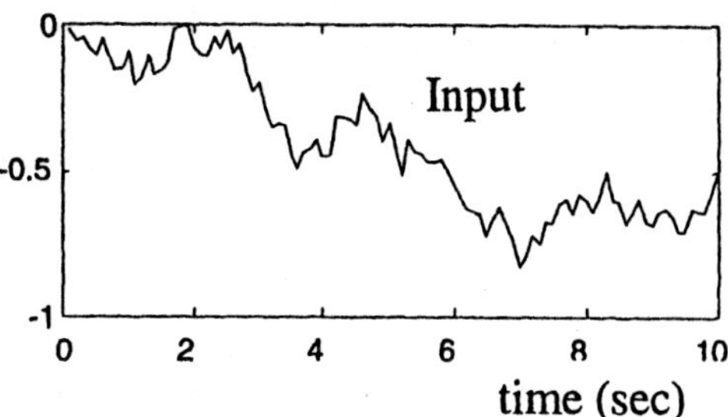

Figure 62.4 Basic MVC. (a) (output), (b) (input).

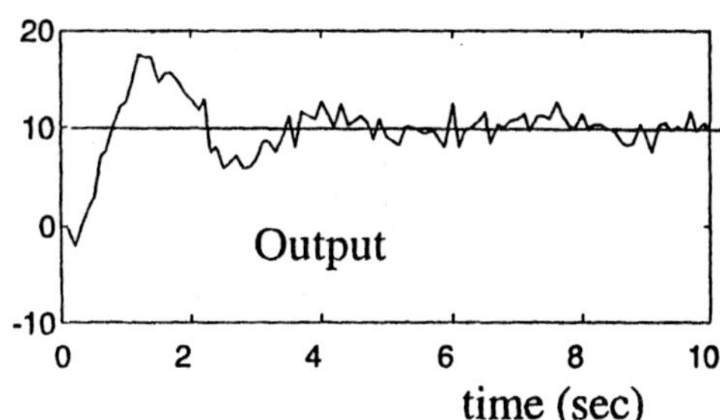

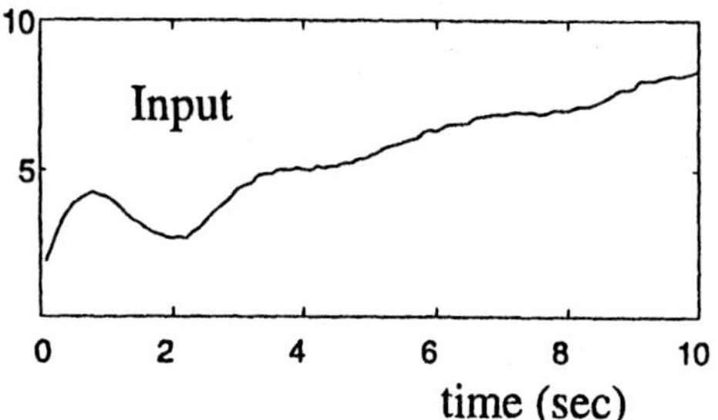

Figure 62.5 Basic MVC with set point.

$\varphi(\zeta\,|\eta)$, where η is another random variable,

$$E\left[g(\zeta\,|\eta)\right] = \int_{-\infty}^{\infty} [g(\zeta)\varphi(\zeta\,|\eta)]d\zeta.$$

The conditional probability can be obtained from Bayes equation,

$$\varphi(\zeta\,|\eta) = \frac{\varphi(\zeta,\eta)}{\varphi(\eta)}.$$

62.7.2 Linear Quadratic Gaussian (LQG)

A dynamic optimization problem can be described by a linear model of the system (either continuous-time or discrete-time) with additive disturbances which are stochastic processes with Gaussian probability density functions. The performance index to be minimized is a finite- or infinite-horizon sum involving quadratic terms on the state of the system and/or on the control.

62.7.3 Minimum Phase

A system is minimum phase if the open-loop transfer function does not include right half-plane zeros (in the continuous-time case). In the discrete-time case this corresponds to all zeros lying outside the unit circle. Practically, minimum-phase behavior of the system implies that the step response of the system starts in a "right direction". The right-hand side limit of the time derivative of the system's step response has the same sign as the steady-state gain.

62.7.4 Steady-State Offset

The difference between the set point (desired) value of the output and the actual steady-state value of the output is called steady-state offset. It is desirable to reduce the steady-state offset (to zero, if possible). It is well-known that the systems with integral action have zero steady-state offset.

62.7.5 White Noise

A stochastic process $v(t)$ is called white noise if its realizations in time are uncorrelated, i.e., the autocorrellation function, $E[v(t)v(t+\delta)]$ has nonzero value only for $\delta = 0$. The process is called Gaussian white noise if, in addition to the above assumption, the probability density function is Gaussian at each instant.

References

[1] Åström, K.J., Computer Control of a Paper Machine — An application of linear stochastic control theory, *IBM J. Res. Dev.*, 11, 389–404, 1967.

[2] Åström, K.J., *Introduction to stochastic control theory*, Academic, 1970.

[3] Åström, K.J. and Wittemnark, *Computer Controlled Systems*, Prentice Hall, 1984.

[4] Blachuta, M. and Ordys, A.W., Optimal and asymptotically optimal linear regulators resulting from a one-stage performance index, *Int. J. Syst. Sci.*, 18(7), 1377–1385, 1987.

[5] Clarke, D. W. and Gawthrop, P.J., Self tuning controller, *Proc. IEEE*, 122(9), 929–934, 1975.

[6] Clarke, D.W. and Gawthrop, P.J., Self tuning control, *Proc. IEEE*, 126(6), 1979.

[7] Clarke, D. W. and Hastings-James, R., Design of digital controllers for randomly distributed systems, *Proc. IEEE*, 118, 1503–1506, 1971.

[8] Flunkert, H.U. and Unbehauen, H., *A nonlinear adaptive Minimum-Variance Controller with application to*

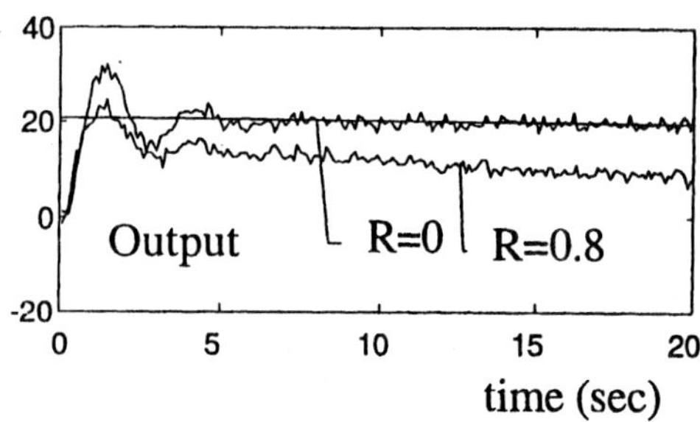

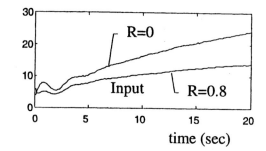

Figure 62.6 WMVC with set point.

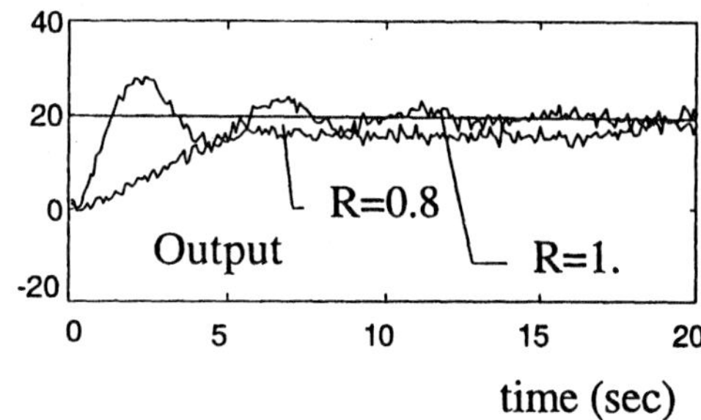

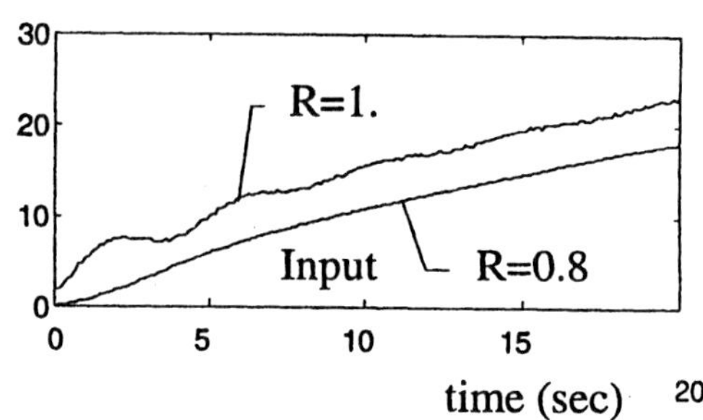

Figure 62.7 GMVC.

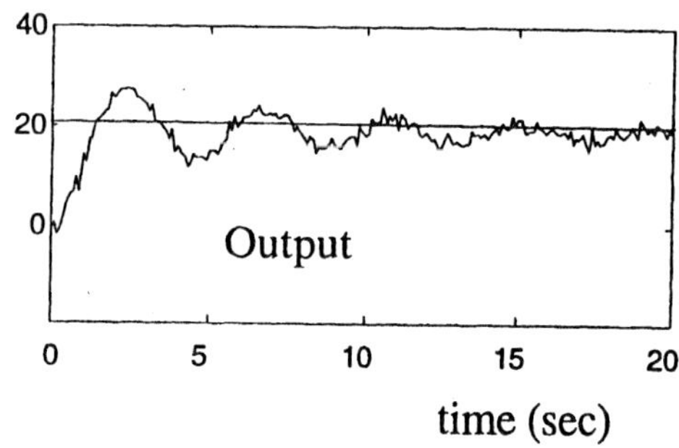

Figure 62.8 Nonminimum-phase system.

a turbo-generator set, Proc. 32nd Control and Decision Conf., San Antonio, Texas, 1993, 1597–1601.

[9] Goodwin, G.C. and Sin, K.S., *Adaptive filtering and prediction and control,* Prentice Hall Int., 1984.

[10] Grimble, M.J., Generalized minimum variance control law revisited, *Optimal Control Appl. Methods,* 9, 63–77, 1988.

[11] Isermann, R., Lachmann, K.-H., and Matko, D., *Adaptive Control Systems,* Prentice Hall, 1992.

[12] Kaminskas, V., Janickiene, D., and Vitkute, D., *Self-tuning constrained control of a power plant,* Proc. IFAC Control of Power Plants and Power Systems, Munich, Germany, 1992, pp 87–92.

[13] Warwick, K., Relationship between Åström control and the Kalman linear regulator — Caines revisited, *Optimal Control Appl. Methods,* 11, 223–232, 1990.

Further Reading

The standard minimum-variance controller was introduced in [1] where a detailed derivation can be found. The weighted minimum variance and the generalized minimum variance are discussed in [7] and [6], respectively. More general discussion of MVC and its relations to the LQG problem can be found in excellent books [2] and [9]. Also the book [11] surveys all minimum-variance control techniques.

63

Dynamic Programming

P. R. Kumar
Department of Electrical and Computer Engineering and Coordinated Science Laboratory, University of Illinois, Urbana, IL

63.1 Introduction

Dynamic programming is a recursive method for obtaining the optimal control as a function of the state in multistage systems. The procedure first determines the optimal control when there is only one stage left in the life of the system. Then it determines the optimal control where there are two stages left, etc. The recursion proceeds backward in time.

This procedure can be generalized to continuous-time systems, stochastic systems, and infinite horizon control. The cost criterion can be a total cost over several stages, a discounted sum of costs, or the average cost over an infinite horizon.

A simple example illustrates the main idea.

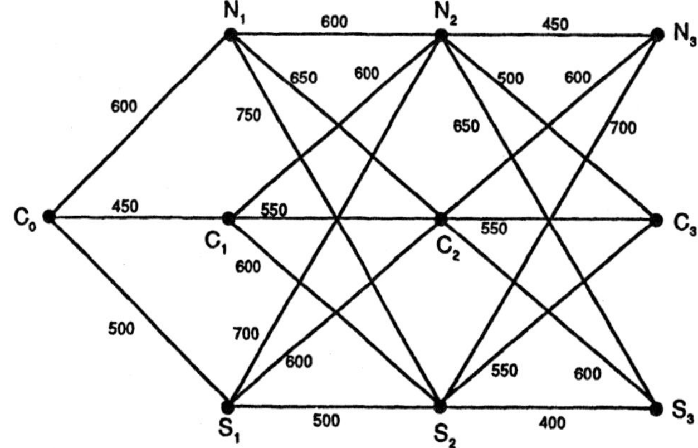

Figure 63.1 A shortest path problem.

63.1.1 Example: The Shortest Path Problem

A bicyclist wishes to determine the shortest path from Bombay to the Indian East Coast. The journey can end at any one of the cities N_3, C_3, or S_3. The journey is to be done in 3 stages.

Stage zero is the starting stage. The bicyclist is in Bombay, labelled C_0 in Figure 63.1. Three stages of travel remain. The bicyclist has to decide whether to go north, center, or south. If the bicyclist goes north, she reaches the city N_1, after travelling 600 kms. If the bicyclist goes to the center, she reaches C_1, after travelling 450 kms. If she goes south, she reaches S_1 after travelling 500 kms.

At stage one, she will therefore be in one of the cities N_1, C_1, or S_1. From wherever she is, she has to decide which city from among N_2, C_2, or S_2 she will travel to next. The distances between cities N_1, C_1, and S_1 and N_2, C_2, S_2 are shown in Figure 63.1. At stage two, she will be in one of the cities N_2, C_2, or S_2, and she will then have to decide which of the cities N_3, C_3, or S_3 to travel to. The journey ends in stage three with the bicyclist in one of the cities N_3, C_3 or S_3.

63.1.2 The Dynamic Programming Method

We first determine the optimal decision when there is only *one* stage remaining.

If the bicyclist is in N_2, she has three choices—north, center, or south, leading respectively to N_3, C_3 or S_3. The corresponding distances are 450, 500 and 650 kms. The best choice is to go

0-8493-8570-9/96/$0.00+$.50

north, and the shortest distance from N_2 to the East Coast is 450 kms.

Similarly, if the bicyclist is in C_2, the best choice is to go center, and the shortest distance from C_2 to the East Coast is 550 kms. From S_2, the best choice is to go south, and the shortest distance to the East Coast is 400 kms. We summarize the optimal decisions and the optimal costs, when only one stage remains, in Figure 63.2.

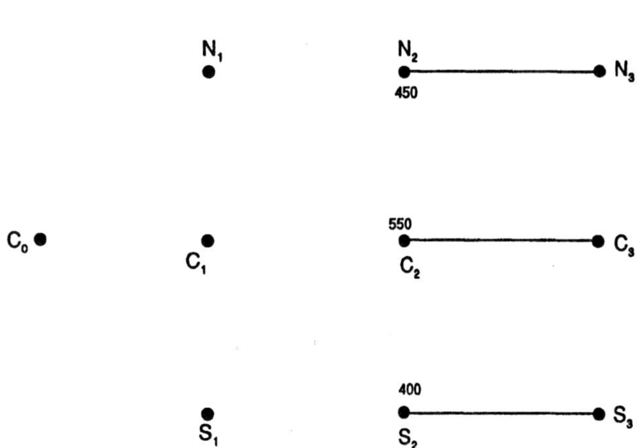

Figure 63.2 Optimal solution when one stage remains.

Now we determine the optimal decision when *two* stages remain.

Suppose the bicyclist is in N_1. If she goes north, she will reach N_2 after cycling 600 kms. Moreover from N_2, she will have to travel a *further* 450 kms to reach the East Coast, as seen from Figure 63.2. From N_1, if she goes north, she will therefore have to travel a total of 1050 kms to reach the East Coast.

If instead she goes to the center from N_1, then she will have to travel 650 kms to reach C_2, and from C_2 she will have to travel a further 550 kms to reach the East Coast. Thus, she will have to travel a total of 1200 kms to reach the East Coast.

The only remaining choice from N_1 is to go south. Then she will travel 750 kms to reach S_2, and from there she has a further 400 kms to reach the East Coast. Thus, she will have to travel 1150 kms to reach the East Coast.

The consequences of each of these three choices are shown in Figure 63.3. Thus the optimal decision from N_1 is to go north and travel to N_2. Moreover, the shortest distance from N_1 to the East Coast is 1050 kms.

Similarly, we determine the optimal paths from C_1 and S_1, also, as well as the shortest distances from them to the East Coast. The optimal solution, when two stages remain, is shown in Figure 63.4.

Now we determine the optimal decision when there are *three* stages remaining.

The bicyclist is in C_0. If she goes north, she travels 600 kms to reach N_1, and then a further 1050 kms is the minimum needed to reach the East Coast. Thus the total distance she will have to travel is 1650 kms. Similarly, if she goes to the center, she will travel a total distance of 1450 kms. On the other hand, if she goes

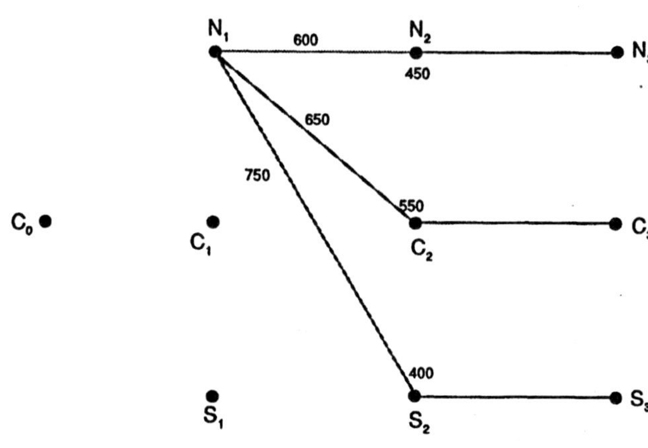

Figure 63.3 Making a decision in city N_1.

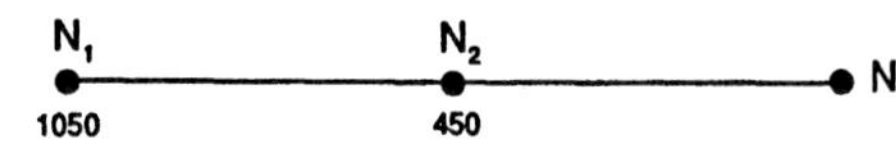

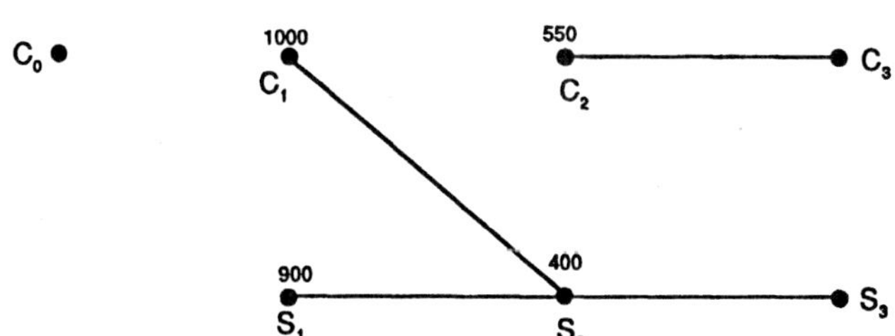

Figure 63.4 Optimal solution when two stages remain.

south, she will travel a total distance of 1400 kms to reach the East Coast. Thus the best choice from C_0 is to travel south to S_1, and the shortest distance from C_0 to the East Coast is 1400 kms. This final result is shown in Figure 63.5.

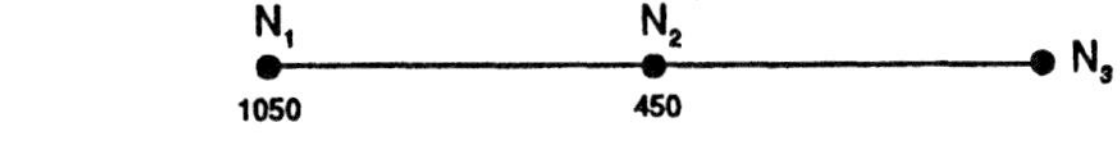

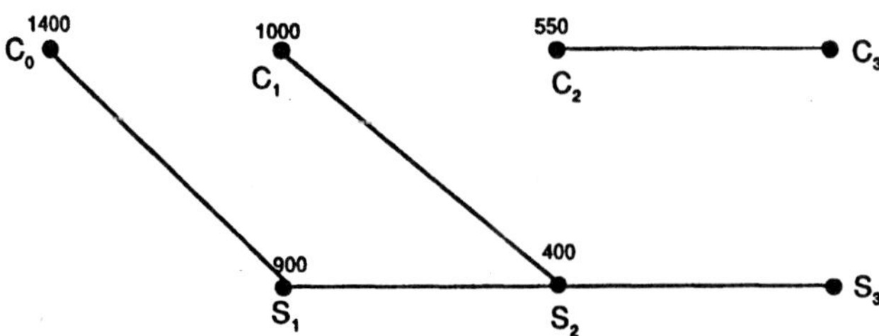

Figure 63.5 Optimal solution when three stages remain.

63.1.3 Observations on the Dynamic Programming Method

We observe a number of important properties.

1. In order to solve the shortest path from just one city C_0 to the East Coast, we actually solved the shortest path from *all* of the cities $C_0, N_1, C_1, S_1, N_2, C_2, S_2$ to the East Coast.
2. Hence, dynamic programming can be *computationally intractable* if there are many stages and many possible decisions at each stage.
3. Dynamic programming determines the optimal decision as a function of the state and the number of stages remaining. Thus, it gives an optimal *closed-loop* control policy.
4. Dynamic programming proceeds *backward* in time. After determining the optimal solution when there are s stages remaining, it determines the optimal solution when there are $(s+1)$ stages remaining.
5. Fundamental use is made of the following relationship in obtaining the backward recursion:

Optimal distance from state x to the end $= \min_i$ {Distance travelled on current stage when decision d_i is made + *optimal* distance from the state x' to the end, where x' is the state to which decision d_i takes you from state x}

Hence, if one passes through state x' from x, an optimal continuation is to take the shortest path from state x' to the end. This can be summarized by saying "segments of optimal paths are optimal in themselves." This is called the *Principle of Optimality.*

63.2 Deterministic Systems with a Finite Horizon

We can generalize the solution method to deterministic, time-varying control systems. Consider a system whose state at time t is denoted by $x(t) \in \mathcal{X}$. If a control input $u(t) \in \mathcal{U}$ is chosen at time t, then the next state $x(t+1)$ is determined as,

$$x(t+1) = f(x(t), u(t), t).$$

The system starts in some state $x(t_0) = x_0$ at time t_0. For simplicity, suppose that there are only a *finite* number of possible control actions, i.e., $\mathcal{U}$ is finite.

The cost incurred by the choice of control input $u(t)$ at time t is $c(x(t), u(t))$. We have a *finite time horizon* T. The goal is to determine the controls to minimize the cost function,

$$\sum_{t=t_0}^{T-1} c(x(t), u(t), t) \quad + \quad d(x(T)). \tag{63.1}$$

THEOREM 63.1 **Deterministic system, finite horizon, finite control set.** *Define*

$$V(x,T) := d(x), \quad \textit{and} \tag{63.2}$$

$$V(x,t-1) := \min_{u\in\mathcal{U}} \{c(x,u,t-1) + V(f(x,u,t-1),t)\}. \tag{63.3}$$

Let $u^*(x, t-1)$ be a value of u which minimizes the right-hand side (RHS) above. Then,

1. *$V(x,t) = \min_{u(\cdot)} \sum_{n=t}^{T-1} c[x(n), u(n), n] + d[x(T)]$, when $x(t) = x$, that is, it is the* optimal cost-to-go *from state x at time t to the end.*
2. *$u^*(x,t)$ is an optimal control when one is in state x at time t.*

PROOF 63.1 *Define* $V(x,t)$ as the optimal cost-to-go from state x at time t to the end. $V(x,T) = d(x)$, because there is no decision to make when one is at the end. Now we obtain the *backward recursion* of dynamic programming. Suppose we are in state x at time $t-1$. If we apply control input u, we will incur an immediate cost $c(x,u,t-1)$, and move to the state $f(x,u,t-1)$, from which the optimal cost-to-go is $V(f(x,u,t-1),t)$. Because some action u has to be taken at time $t-1$,

$$V(x,t-1) \quad \le \quad \min_{u\in\mathcal{U}}\left\{c(x,u,t-1) + V(f(x,u,t-1),t)\right\}.$$

Moreover, if $V(x,t-1)$ was strictly smaller than the RHS above, then one would obtain a contradiction to the minimality of $V(\cdot,t)$. Hence equality holds, and (63.3) is satisfied.

By evaluating the cost of the control policy given by $u^*(x,t)$, one can verify that it gives the minimal cost $V(x_0, t_0)$.

We have proven the Principle of Optimality. We have also proven that, for the cost criterion Equation 63.1, there is an optimal control policy that takes actions based only on the value of current state and current time and does not depend on past states or actions. Such a control policy is called a *Markovian policy.*

63.2.1 Infinite Control Set $\mathcal{U}$

When the control set $\mathcal{U}$ is infinite, the RHS in Equation 63.3 may not have a minimizing u. Thus the recursion requires an "inf" in place of the "min." There may be no optimal control law, only nearly optimal ones.

63.2.2 Continuous-Time Systems

Suppose one has a continuous-time system

$$\dot{x}(t) = f(x(t), u(t), t), \tag{63.4}$$

and the cost criterion is

$$d(x(T)) + \int_{t_0}^{T} c(x(t), u(t), t)dt.$$

Several technical issues arise. The differential Equation 63.4 may not have a solution for a given $u(t)$, or it may have a solution but only until some finite time. Ignoring such problems, we examine what type of equations replace the discrete-time recursions. Clearly, $V(x,T) = d(x)$. Considering a control input u which is held constant over an interval $[t, t+\Delta)$, and approximating the resulting state (all possible only under more assumptions),

$$\begin{aligned} V(x,t) &= \inf_{u\in\mathcal{U}} \{c(x,u,t)\Delta \\ &\quad + V(x+f(x,u,t)\Delta, t+\Delta) + o(\Delta)\}, \\ &= \inf_{u\in\mathcal{U}} \{c(x,u,t)\Delta + V(x,t) \\ &\quad + \frac{\partial V}{\partial x}(x,t) f(x,u,t)\Delta \\ &\quad + \frac{\partial V}{\partial t}(x,t)\Delta + o(\Delta)\Big\}. \end{aligned}$$

Thus, one looks for the partial differential equation

$$\frac{\partial V}{\partial t}(x,t) + \inf_{u\in\mathcal{U}} \left\{ \frac{\partial V}{\partial x}(x,t) f(x,u,t) + c(x,u,t) \right\} = 0, \tag{63.5}$$

to be satisfied by the optimal cost–to–go. This is called the *Hamilton–Jacobi–Bellman equation.* To prove that the optimal cost–to–go satisfies such an equation requires many technical assumptions.

However, if a smooth V exists that satisfies such an equation, then it is a lower bound on the cost, at least for Markovian control laws. To see this, consider any control law $u(x,t)$, and let $x(t)$ be the trajectory resulting from an initial condition $x(t_0) = x_0$ at time t_0. Then,

$$\begin{aligned} \frac{\partial V}{\partial t}(x(t),t) \quad + \quad & \frac{\partial V}{\partial x}(x(t),t) f(x(t),u(x(t),t),t) \\ & + c(x(t),u(x(t),t) \geq 0. \end{aligned} \tag{63.6}$$

Hence $\frac{d}{dt}[V(x(t),t)] + c(x(t),u(x(t),t),t) \geq 0$, and so,

$$\begin{aligned} V(x(t_0),t_0) &\leq V(x(T),T) + \int_{t_0}^{T} c(x(t),u(t),t)dt \\ &= d(x(T)) + \int_{t_0}^{T} c(x(t),u(t),t)dt. \end{aligned}$$

Moreover, if $u^*(x,t)$ is a control policy which attains the "inf" in Equation 63.5, then it attains equality above and is, hence, optimal.

In many problems, the optimal cost–to–go $V(x,t)$ may not be sufficiently smooth, and one resorts to the Pontryagin Minimum Principle rather than the HJB Equation 63.5.

63.3 Stochastic Systems

Consider the simplest case where time is discrete, and both the state-space $\mathcal{X}$ and the control set $\mathcal{U}$ are finite. Suppose that the system is time invariant and evolves probabilistically as a *controlled Markov chain,* i.e.,

$$\text{Prob}\,[x(t+1) = j | x(t) = i, u(t) = u] = p_{ij}(u). \tag{63.7}$$

The elements of the matrix $P(u) = [p_{ij}(u)]$ are the *transition probabilities.*

Given a starting state $x(t_0) = x_0$ at time t_0, one wishes to minimize the *expected* cost

$$E\left[\sum_{t=t_0}^{T-1} c(x(t),u(t),t) + d(x(T))\right].$$

A *nonanticipative control policy* γ is a mapping from past information to control inputs, i.e., $\gamma = (\gamma_{t_0}, \gamma_{t_0+1}, \ldots, \gamma_{T-1})$ where $\gamma_t : (x(t_0), u(t_0), x(t_0+1), u(t_0+1), \ldots, x(t)) \mapsto u(t)$. Define the conditionally expected cost–to–go when policy γ is applied, as

$$\begin{aligned} V_t^\gamma &:= E\left[\sum_{n=t}^{T-1} c(x(n),u(n),n) \right. \\ &\quad \left. + d(x(T)) | x(t_0), u(t_0), \ldots, x(t)\right]. \end{aligned}$$

Analogously to the deterministic case, set Equation 63.2, and recursively define

$$V(x,t-1) = \min_{u\in\mathcal{U}} \left\{ c(x,u,t-1) + \sum_j p_{xj}(u) V(j,t) \right\}. \tag{63.8}$$

Now let us compare V_t^γ with $V(x(t),t)$ when policy γ is used. Clearly $V_T^\gamma = V(x(T),T) = d(x(T))$ a.s. Now suppose, by induction, that $V_t^\gamma \geq V(x(t),t)$ a.s. Then

$$\begin{aligned} V_{t-1}^\gamma &= c(x(t-1),u(t-1),t-1) \\ &\quad + E\left\{ \textstyle\sum_{n=t}^{T} c[x(n),u(n),n) + d(x(T)] \right. \\ &\quad \left. |x(t_0), u(t_0), \ldots, x(t-1) \right\} \\ &= c(x(t-1),u(t-1),t-1) \\ &\quad + E\left\{ E\left[\textstyle\sum_{n=t}^{T} c[x(n),u(n),n) + d(x(T)] \right. \right. \\ &\quad |x(t_0), u(t_0), \ldots, x(t)] \\ &\quad \left. |x(t_0), u(t_0), \ldots, x(t-1) \right\} \\ &= c(x(t-1),u(t-1),t-1) \\ &\quad + E\left[V_t^\gamma | x(t_0), u(t_0), \ldots, x(t-1) \right] \\ &\geq c(x(t-1),u(t-1),t-1) \\ &\quad + E\left[V(x(t),t) | x(t_0), u(t_0), \ldots, x(t-1) \right] \\ &= c(x(t-1),u(t-1),t-1) \\ &\quad + \textstyle\sum_j p_{x(t-1)j}[u(t-1)] V(j,t) \\ &\geq \min_{u\in\mathcal{U}} \left\{ c(x(t-1),u,t-1) \right. \\ &\quad \left. + \textstyle\sum_j p_{x(t-1)j}(u) V(j,t) \right\} \\ &= V(x(t),t) a.s. \end{aligned} \tag{63.9}$$

Thus $V(x,t)$ is a lower bound on the expected cost of any non anticipative control policy. Suppose, moreover, that $u^*(x,t)$ attains the minimum on the RHS of Equation 63.8 above. Consider the Markovian control policy $\gamma^* = (\gamma_{t_0}^*, \gamma_{t_1}^*, \ldots, \gamma_{T-1}^*)$ with

$$\gamma_t^*[x(t_0), u(t_0), \ldots, x(t)] = u^*[x(t), t]. \tag{63.10}$$

For γ^* it is easy to verify that the inequalities in Equation 63.9 are equalities, and so it is optimal.

THEOREM 63.2 **Stochastic finite state, finite control system, finite horizon.** *Recursively define $V(x,t)$ from Equations 63.2 and 63.8. Let $u^*(x,t-1)$ attain the minimum on the RHS of (63.8), and consider the Markovian control policy γ^* defined in Equation 63.10. Then,*

(i) *$V(x_0, t_0)$ is the optimal cost.*

(ii) *The Markovian control policy γ^* is optimal over the class of all nonanticipative control policies.*

63.3.1 Countably Infinite State and Control Spaces

The result (i) can be extended to countably infinite state and control policies by replacing the "min" above with an "inf". However, the "inf" need not be attained, and an optimal control policy may not exist.

If one considers uncountably infinite state and control spaces, then further highly technical issues arise. A policy needs to be a *measurable* map. Moreover, $V(x, t)$ will also need to be a measurable function because one must take its expected value. One must impose appropriate conditions to insure that the "inf" over an uncountable set still gives an appropriately measurable function, and further that one can synthesize a minimizing $u(x, t)$ that is measurable.

63.3.2 Stochastic Differential Systems

Consider a system described by a *stochastic differential equation,*

$$dx(t) = f(x(t), u(t), t)dt + dw(t),$$

where w is a standard Brownian motion, with a cost criterion

$$E\{d[x(T)] + \int_{t_0}^{T} c[x(t), u(t), t]\}.$$

If $V(x, t)$ denotes the optimal cost–to–go from a starting state x when there are t time units remaining, one expects from Ito's differentiation rule that it satisfies the stochastic version of the Hamilton-Jacobi-Bellman equation:

$$\frac{\partial V(x,t)}{\partial t} + \frac{1}{2}\frac{\partial^2 V}{\partial x^2}(x,t) + \inf_{u \in \mathcal{U}} \left\{ \frac{\partial V}{\partial x}(x,t) f(x,u,t) + c(x,u,t) \right\} = 0.$$

The existence of a solution to such partial differential equations is studied using the notion of *viscosity* solutions.

63.4 Infinite Horizon Stochastic Systems

We will consider finite state, finite control, controlled Markov chains, as in Equation 63.7. We study the infinite horizon case.

63.4.1 The Discounted Cost Problem

Consider an *infinite horizon total discounted cost* of the form

$$E \sum_{t=0}^{+\infty} \beta^t c(x(t), u(t)), \tag{63.11}$$

where $0 < \beta < 1$ is a *discount factor.* The discounting guarantees that the summation is finite. We are assuming that the one-stage cost function c does not change with time. Let $W(x, \infty)$ denote the optimal value of this cost, starting in state x. The "∞" denotes that there is an infinity of stages remaining.

Let $W(x, N)$ denote the optimal value of the finite horizon cost $E \sum_{t=0}^{N-1} \beta^t c[x(t), u(t)]$, when starting in state x at time 0. Note that the index N refers to the number of *remaining* stages. From the finite horizon case, we see that

$$W(x, N) = \min_{u \in \mathcal{U}} \left\{ c(x, u) + \sum_j p_{xj}(u) \beta W(j, N-1) \right\}. \tag{63.12}$$

Note the presence of the factor β multiplying $W(j, N-1)$ above.

Denote by $R^{\mathcal{X}}$ the class of all real valued functions of the state, i.e., $V \in R^{\mathcal{X}}$ is a function of the form $V : \mathcal{X} \to R$. Define an operator $T : R^{\mathcal{X}} \to R^{\mathcal{X}}$ by its action on V as

$$TV(x) := \min_{u \in \mathcal{U}} \left\{ c(x, u) + \beta \sum_j p_{xj}(u) V(j) \right\} \quad \text{for all } x \in \mathcal{X}. \tag{63.13}$$

Using the operator T, one can rewrite Equation 63.12 as $W(\cdot, N) = TW(\cdot, N-1)$. Also, $W(\cdot, 0) = O$, where O is the identically zero function, i.e., $O(x) \equiv 0$ for all x. Note that T is a monotone operator, i.e., if $V(x) \leq \bar{V}(x)$ for all x, then $TV(x) \leq T\bar{V}(x)$ for all x. Let us suppose now that $\bar{c} \geq c(x, u) \geq 0$ for all x, u. (This can be achieved by simply adding a large enough constant to each original one-step cost.) Due to this assumption, $TO \geq 0$. Hence by monotonicity, $T^{(N)}O \geq T^{(N-1)}O \geq \cdots \geq TO \geq 0$. Thus $T^{(N)}O(x)$ converges, for every x, to a finite number (since $T^{(N)}O \leq \frac{\bar{c}}{1-\beta}$). Let

$$W(x) := \lim_{N \to \infty} T^{(N)}O(x).$$

Now $W(x, \infty) \geq W(x, N)$ for all N, because $c(x, u) \geq 0$. Hence $W(x, \infty) \geq \lim_{N\to\infty} W(x, N) = W(x)$. Moreover $W(x, \infty) \leq W(x, N) + \frac{\bar{c}\beta^N}{1-\beta}$, because one can employ an optimal policy for an N-stage problem. Thus $W(x, \infty) \leq \lim_{N\to\infty} \left[W(x, N) + \frac{\bar{c}\beta^N}{1-\beta} \right] = W(x)$. Hence $W(x) = W(x, \infty)$, and is, therefore, the *optimal infinite horizon cost.*

Hence, we would like to characterize $W(x)$. Since T is continuous, $TW(x) = \lim_{N\to\infty} T^{N+1}O(x) = W(x)$. Hence W is a *fixed point* of T, i.e.,

$$W(x) = \min_{u \in \mathcal{U}} \left\{ c(x, u) + \beta \sum_j p_{xj}(u) W(j) \right\}. \tag{63.14}$$

Simple calculations show that T is a contraction with respect to the $\|\cdot\|_\infty$ norm, i.e., $\max_x |TV(x) - T\bar{V}(x)| \leq \beta \max_x |V(x) - \bar{V}(x)|$. Thus T has a unique fixed point. Since T is a contraction, we also know that $\lim_{N\to\infty} T^{(N)}V = W$ for *any* V.

Suppose now that $u^*(x)$ attains the minimum in Equation 63.14 above. Consider the policy γ^* which always chooses control $u^*(x)$ whenever the state is x. Such a control policy is called *stationary*. If one applies the stationary control policy γ^*, then the expected cost over N days, starting in state x, is $T^{(N)}O(x)$. Thus the infinite horizon cost of γ^* is $\lim_{N\to\infty} T^N O = W$.

THEOREM 63.3 Stochastic finite state, finite control system, discounted cost criterion. *Let $T : R^{\mathcal{X}} \to R^{\mathcal{X}}$ denote the operator (Equation 63.13).*

1. *T is a contraction.*
2. *Let $W = \lim_{N\to\infty} T^{(N)}V$ for any V. Then W is the unique solution of Equation 63.14.*
3. *$W(x)$ is the optimal cost when starting in state x.*
4. *Let $\gamma^*(x)$ be the value of u which attains the minimum on the RHS in Equation 63.14. Then γ^* is a stationary control policy which is optimal in the class of all nonanticipative policies.*

The procedure for determining $W(x)$ as $\lim_{N\to\infty} T^{(N)}O$ is called *value iteration*. Another procedure which determines the optimal policy in a finite number of steps is *policy iteration*.

Policy Iteration Procedure

For a stationary policy γ, define $T_\gamma : R^{\mathcal{X}} \to R^{\mathcal{X}}$ by, $T_\gamma V(x) = c(x, \gamma(x)) + \beta \sum_j p_{xj}(\gamma(x))V(j)$.

1. Let γ_0 be any stationary policy.
2. Solve the linear system of equations $T_{\gamma_0} W_{\gamma_0} = W_{\gamma_0}$ to determine its cost W_{γ_0}.
3. If $TW_{\gamma_0} \neq W_{\gamma_0}$, then let $\gamma_1(x)$ be the value of u which attains the minimum in $\min_{u\in\mathcal{U}}\Big\{c(x,u) + \beta \sum_j p_{xj}(u)W_{\gamma_0}(j)\Big\}$.
4. Then γ_1 is a strict improvement of γ_0 (since $W_{\gamma_1} = \lim_{N\to\infty} T_{\gamma_1}^{(N)} W_{\gamma_0} \lneq W_{\gamma_0}$).
5. By repeating this procedure, one obtains a sequence of strictly improving policies, that must terminate in a finite number of steps, because the total number of stationary policies is finite.

63.4.2 The Average Cost Problem

We consider the *average cost per unit time* over an infinite horizon,

$$\limsup_{N\to\infty} \frac{1}{N} E \sum_{t=0}^{N-1} c[x(t), u(t)].$$

Then the dynamic programming equation needs to be modified slightly.

THEOREM 63.4 *Suppose a constant J^* exists, and a function $V : \mathcal{X} \to R$ exists so that*

$$J^* + V(x) = \min_{u\in\mathcal{U}}\Big\{c(x,u) + \sum_j p_{xj}(u)V(j)\Big\}. \qquad (63.15)$$

1. *Then J^* is the optimal value of the average cost criterion, starting from* any *state x.*
2. *Let $\gamma^*(x)$ be a value of u which minimizes the RHS in Equation 63.15 above. Then γ^* is a stationary policy which is optimal in the class of all nonanticipative policies.*

PROOF 63.2 Consider any nonanticipative policy. Then

$$J^* + V(x(t)) \le c(x(t), u(t)) + \sum_j p_{x(t)j}(u(t))V(j).$$

Noting that $E\sum_j p_{x(t)j}(u(t))V(j) = E\{V(x(t+1))\}$, we see that

$$E(c(x(t), u(t)) \ge J^* + E(V(x(t)) - E\{V(x(t+1))\}.$$

Hence

$$\limsup_{N\to\infty} \frac{1}{N} E \sum_{t=0}^{N-1} c(x(t), u(t)) \ge J^*$$
$$+ \limsup_{N} \frac{V(x(0)) - E\{V(x(N))\}}{N} = J^*.$$

Thus J^* is a lower bound on the average cost, starting from *any* state x. Moreover, if γ^* is the policy under consideration, then equality holds above, and so it is indeed optimal.

The question, still a topic of active research, is when does a solution $[J^*, V(\cdot)]$ exist for Equation 63.15 ? Let us consider the following simplifying assumption. For *every* stationary policy γ, the Markov chain $P_\gamma = [p_{ij}(\gamma(i))]$ is *irreducible*. By this is meant that there exists a *unique* steady state distribution π_γ, satisfying $\pi_\gamma P_\gamma = \pi_\gamma$, $\pi_\gamma(i) \ge 0$, $\sum_i \pi_\gamma(i) = 1$, which is *strictly positive*, i.e., $\pi_\gamma(i) > 0$ for all i. Then the average cost J_γ starting from any state is a constant and satisfies, $J_\gamma = \sum_i \pi_\gamma(i)c[i, \gamma(i)] = \pi_\gamma c_\gamma$ (where $c_\gamma := [c_\gamma(1), \dots, c_\gamma(1x)]$). Hence if $e = (1, \dots, 1)^T$, $\pi_\gamma(J_\gamma e - c_\gamma) = 0$. Hence $(J_\gamma e - c_\gamma)$ is orthogonal to the null space of $(P_\gamma - I)^T$, and is, therefore, in the range space of $(P_\gamma - I)$. Thus a V_γ exists so that $J_\gamma e - c_\gamma = (P_\gamma - I)V_\gamma$, which simply means that

$$J_\gamma + V_\gamma(x) = c[x, \gamma(x)] + \sum_j p_{xj}[\gamma(x)]V_\gamma(j). \qquad (63.16)$$

Note that $V_\gamma(\cdot)$ is *not* unique because $V_\gamma(\cdot) + a$ is also a solution for any a. Let us therefore fix $V_\gamma(\bar{x}) = 0$ for some $\bar{x}$. One can obtain a policy iteration algorithm, as well as prove the existence of J^* and $V(\cdot)$, as shown below.

Policy Iteration Procedure

1. Let γ_0 be any stationary policy.
2. Solve Equation 63.16, with γ replaced by γ_0, to obtain $(J_{\gamma_0}, V_{\gamma_0})$. If $(J_{\gamma_0}, V_{\gamma_0})$ does *not* satisfy Equation 63.15, then let $\gamma_1(x)$ attain the minimum in $\min_{u\in\mathcal{U}}\Big\{c(x,u) + \sum_j p_{xj}(u)V_{\gamma_0}(x)\Big\}$.

Then γ_1 is a strict improvement of γ_0. (This follows because $\pi_{\gamma_1}(i) > 0$ for all i, and so

$$\begin{aligned} J_{\gamma_0} + \pi_{\gamma_1} V_{\gamma_0} &= \pi_{\gamma_1}(J_{\gamma_0} e + V_{\gamma_0}) > \pi_{\gamma_1} c_{\gamma_1} + \pi_{\gamma_1} P_{\gamma_1} V_{\gamma_0} \\ &= J_{\gamma_1} + \pi_{\gamma_1} V_{\gamma_0}, \text{ and so } J_{\gamma_0} > J_{\gamma_1}). \end{aligned}$$

Because the policy space is finite, this procedure terminates in a finite number of steps. At termination, Equation 63.15 is satisfied.

63.4.3 Connections of Average Cost Problem with Discounted Cost Problems and Recurrence Conditions

The average cost problem can be regarded as a limit of discounted cost problems when $\beta \nearrow 1$. We illustrate this for systems with *countable state space* and finite control set.

THEOREM 63.5 **Connection between discounted and average cost.** *Let $W_\beta(x)$ denote the optimal discounted cost $E\sum_{t=0}^{+\infty}\beta^t c(x(t), u(t))$ when starting in the state x. Suppose that $|W_\beta(x) - W_\beta(x')| \le M$ all x, x', and all $\beta \in (1-\epsilon, 1)$, for some $\epsilon > 0$. For an arbitrary state $\bar{x} \in \mathcal{X}$, let $\beta_n \nearrow 1$ be a sub-sequence so that the following limits exist:*

$$\begin{aligned} &\lim_{n\to\infty}(1-\beta_n)W_{\beta_n}(x) =: J^*, \text{ and} \\ &\lim_{n\to\infty}(W_{\beta_n}(x) - W_{\beta_n}(\bar{x})) =: V(x). \end{aligned}$$

Then,

1. $J^* + V(x) = \min_{u\in\mathcal{U}}\left\{c(x,u) + \sum_j p_{xj}(j)V(j)\right\}$.
2. *If a stationary policy γ^* is optimal for a sequence of discount factors β_n with $\beta_n \nearrow 1$, then γ^* is optimal for the average cost problem.*

PROOF 63.3 The dynamic programming Equation 63.14 for the discounted cost problem can be rewritten as,

$$(1-\beta)W_\beta(x) = \min_{u\in\mathcal{U}}\left\{c(x,u) + \beta\sum_j p_{xj}[W_\beta(j) - W_\beta(x)]\right\}.$$

Taking limits along $\beta_n \nearrow 1$ yields the results.

The existence of $(J^*, V(\cdot))$ satisfying the average cost dynamic programming Equation 63.15 is guaranteed under certain uniform recurrence conditions on the controlled Markov chain.

THEOREM 63.6 **Uniformly bounded mean first passage times.** *Let τ denote the first time after time 1 that the system enters some fixed state $\bar{x}$, i.e.,*

$$\tau = \min\left\{t \ge 1 : \quad x(t) = \bar{x}\right\}.$$

Suppose that the mean first passage times are uniformly bounded, i.e.,

$$E(\tau \mid x(0) = x \text{ and } \gamma \text{ is used}) \le M < +\infty,$$

for all states x and all stationary policies γ. Then a solution $(J^*, V(\cdot))$ exists to Equation 63.15.

PROOF 63.4 Under a stationary policy γ_β which is optimal for the discount factor β,

$$\begin{aligned} W_\beta(x) &= E\left[\sum_{t=0}^{\tau-1}\beta^t c(x(t), u(t)) + \beta^\tau W_\beta(\bar{x}) \mid x(0) = x\right] \\ &\le \bar{c}E[\tau \mid x(0) = x] + W_\beta(\bar{x}) \le \bar{c}M + W_\beta(\bar{x}). \end{aligned}$$

Moreover, by Jensen's inequality,

$$\begin{aligned} W_\beta(x) &\ge E[\beta^\tau W_\beta(\bar{x}) \mid x(0) = x] \\ &= W_\beta(\bar{x})E[\beta^\tau \mid x(0) = x] \ge W_\beta(\bar{x})\beta^M. \end{aligned}$$

Hence, $-\bar{c}M \le \frac{\bar{c}(\beta^M - 1)}{1-\beta} \le W_\beta(\bar{x})(\beta^M - 1) \le W_\beta(x) - W_\beta(\bar{x}) \le \bar{c}M$, and the result follows from the preceding Theorem.

63.4.4 Total Undiscounted Cost Criterion

Consider a *total* infinite horizon cost criterion of the form

$$E\sum_{t=0}^{+\infty} c[x(t), u(t)].$$

In order for the infinite summation to exist, one often assumes that either

$$\begin{aligned} c(x,u) &\ge 0 \text{ for all } x, u \quad \text{(the positive cost case), or} \\ c(x,u) &\le 0 \text{ for all } x, u \quad \text{(the negative cost case).} \end{aligned}$$

These two cases are rather different. In both cases, one exploits the monotonicity of the operator T.

References

[1] Blackwell, D., Discounted dynamic programming. *Ann. of Math. Statist.* 36, 226–335, 1965.

[2] Strauch, R., Negative dynamic programming. *Ann. Math. Statist.*, 37, 871–890, 1966.

[3] Blackwell, D., Positive dynamic programming. *Proceedings of the 5th Berkeley Symposium on Mathematical Statistics and Probability*, 415–418, 1965.

[4] Blackwell, D., On stationary policies. *J. Roy. Statist. Soc. Ser. A*, 133, 33–38, 1970.

[5] Ornstein, D., On the existence of stationary optimal strategies. *Proc. Amer. Math. Soc.*, 20, 563–569, 1969.

[6] Blackwell, D., Discrete dynamic programming. *Ann. of Math. Statist.*, 33, 719–726, 1962.

[7] Hernandez-Lerma, O. and Lasserre, J. B., Weak conditions for average optimality in Markov control processes. *Syst. and Contr. Lett.*, 22, 287–291, 1994.

[8] Bertsekas, D. P. and Shreve, S. E., *Stochastic Optimal Control: The Discrete Time Case*, Academic, New York, 1978.

[9] Lions, P. L., Optimal control of diffusion processes and Hamilton-Jacobi equations, Part I–the dynamic programming principle and applications. *Comm. Partial Differential Equations*, 10, 1101–1174, 1983.

[10] Crandall, M., Ishii, H. and Lions, P. L., User's guide to viscosity solutions of second order partial differential equations. *Bull. Amer. Math. Soc.* 27, 1–67, 1990.

[11] Bellman, R., *Dynamic Programming*. Princeton University Press, Princeton, NJ, 1957.

[12] Bertsekas, D. P., *Dynamic Programming: Deterministic and Stochastic Models*. Prentice Hall, Englewood Cliffs, NJ, 1987.

[13] Kumar, P. R. and Varaiya, P. P., *Stochastic Systems: Estimation, Identification and Adaptive Control*. Prentice Hall, Englewood Cliffs, NJ, 1986.

[14] Ross, S. M., *Applied Probability Models with Optimization Applications*. Holden-Day, San Francisco, 1970.

Further Reading

For discounted dynamic programming, the classic reference is Blackwell [1]. For the positive cost case, we refer the reader to Strauch [2]. For the negative cost case, we refer the reader to Blackwell [3, 4] and Ornstein [5]. For the average cost case, we refer the reader to Blackwell [6] for early fundamental work, and to Hernandez-Lerma and Lasserre [7] and the references contained there for recent developments. For a study of the measurability issues which arise when considering uncountable sets, we refer the reader to Bertsekas and Shreve [8]. For continuous time stochastic control of diffusions we refer the reader to Lions [9], and to Crandall, Ishii and Lions [10] for a guide to viscosity solutions.

For the several ways in which dynamic programming can be employed, see Bellman [11].

Some recent textbooks which cover dynamic programming are Bertsekas [12], Kumar and Varaiya [13], and Ross [14].

64

Stability of Stochastic Systems

Kenneth A. Loparo and Xiangbo Feng
Department of Systems Engineering, Case Western Reserve University, Cleveland, OH

64.1 Introduction

In many applications where dynamical system models are used to capture the behavior of real world systems, stochastic components and random noises are included in the model. The stochastic aspects of the model are used to capture the uncertainty about the environment in which the system is operating and the structure and parameters of the model of the physical process being studied. The analysis and control of such systems then involves evaluating the stability properties of a random dynamical system. Stability is a qualitative property of the system and is often the first characteristic of a dynamical system (or model) studied. We know from our study of classical control theory that, before we can consider the design of a regulatory or tracking control system, we need to make sure that the system is stable from input to output. Therefore, the study of the stability properties of stochastic dynamical systems is important and considerable effort has been devoted to the study of stochastic stability. Significant results have been reported in the literature with applications to physical and engineering systems.

A comprehensive survey on the topic of stochastic stability was given by [27]. Since that time there have been many significant developments of the theory and its applications in science and engineering. In this chapter, we present some basic results on the study of stability of stochastic systems. Because of limited space, only selected topics are presented and discussed. We begin with a discussion of the basic definitions of stochastic stability and the relationships among them. Kozin's survey provides an excellent introduction to the subject and a good explanation of how the various notions of stochastic stability are related.

It is not necessary that readers have an extensive background in stochastic processes or other related mathematical topics. In this chapter, the various results and methods will be stated as simply as possible and no proofs of the theorems are given. When necessary, important mathematical concepts, which are the foundation of some of the results, will be discussed briefly. This will provide a better appreciation of the results, the application of the results, and the key steps required to develop the theory further. Those readers interested in a particular topic or result discussed in this chapter are encouraged to go to the original paper and the references therein for more detailed information.

There are at least three times as many definitions for the stability of stochastic systems as there are for deterministic systems. This is because in a stochastic setting there are three basic types of convergence: convergence in probability, convergence in mean (or moment), and convergence in an almost sure (sample path, probability one) sense. Kozin [27] presented several definitions for stochastic stability, and these definitions have been extended in subsequent research works. Readers are cautioned to examine carefully the definition of stochastic stability used when interpreting any stochastic stability results.

We begin with some preliminary definitions of stability concepts for a deterministic system. Let $x(t; x_0, t_0)$ denote the trajectory of a dynamic system initial from x_0 at time t_0.

DEFINITION 64.1 Lyapunov Stability
The equilibrium solution, assumed to be 0 unless stated otherwise, is said to be *stable* if, given $\varepsilon > 0$, $\delta(\varepsilon, t_0) > 0$ exists so that, for all $\|x_0\| < \delta$,

$$\sup_{t \geq t_0} \|x(t; x_0, t_0)\| < \varepsilon.$$

DEFINITION 64.2 Asymptotic Lyapunov Stability
The equilibrium solution is said to be *asymptotically stable* if it is stable and if $\delta' > 0$ exists so that $\|x_0\| < \delta'$, guarantees that

$$\lim_{t \to \infty} \|x(t; x_0, t_0)\| = 0.$$

If the convergence holds for all initial times, t_0, it is referred to as *uniform asymptotic stability*.

DEFINITION 64.3 Exponential Lyapunov Stability

0-8493-8570-9/96/$0.00+$.50

The equilibrium solution is said to be *exponentially stable* if it is asymptotically stable and if there exists a $\delta > 0$, an $\alpha > 0$, and a $\beta > 0$ so that $\|x_0\| < \delta$ guarantees that

$$\|x(t; x_0 t_0)\| \leq \beta \|x_0\| \exp^{-\alpha(t-t_0)}.$$

If the convergence holds for all initial times, t_0, it is referred to as *uniform exponential stability.*

These deterministic stability definitions can be translated into a stochastic setting by properly interpreting the notion of convergence, i.e., in probability, in moment, or almost surely. For example, in Definition 64.1 for Lyapunov stability, the variable of interest is $sup_{t\geq t_0} \|x(t; x_0, t_0)\|$, and we have to study the various ways in which this (now random) variable can converge. We denote the fact that the variable is random by including the variable ω, i.e., $x(t; x_0, t_0, \omega)$, and we will make this more precise later. Then,

DEFINITION 64.4 $\mathbf{I_p}$: Lyapunov Stability in Probability
The equilibrium solution is said to be *stable in probability* if, given ε, $\varepsilon' > 0$, $\delta(\varepsilon, \varepsilon', t_0) > 0$ exists so that for all $\|x_0\| < \delta$,

$$P\{\sup_{t\geq t_0} \|x(t; x_0, t_0, \omega)\| > \varepsilon'\} < \epsilon.$$

Here, P denotes probability.

DEFINITION 64.5 $\mathbf{I_m}$: Lyapunov Stability in the $\mathbf{p^{th}}$ Moment
The equilibrium solution is said to be *stable in p^{th} moment,* $p > 0$ if, given $\varepsilon > 0$, $\delta(\varepsilon, t_0) > 0$ exists so that $\|x_0\| < \delta$ guarantees that

$$E\{\sup_{t\geq t_0} \|x(t; x_0, t_0, \omega)\|^p\} < \epsilon.$$

Here, E denotes expectation.

DEFINITION 64.6 $\mathbf{I_{a.s.}}$: Almost Sure Lyapunov Stability
The equilibrium solution is said to be *almost surely stable* if

$$P\{\lim_{\|x_0\|\to 0} \sup_{t\geq t_0} \|x(t; x_0, t_0, \omega)\| = 0\} = 1.$$

Note, almost sure stability is equivalent to saying that, with probability one, all sample solutions are Lyapunov stable.

Similar statements can be made for asymptotic stability and for exponential stability. These definitions are introduced next for completeness and because they are often used in applications.

DEFINITION 64.7 $\mathbf{II_p}$: Asymptotic Lyapunov Stability in Probability
The equilibrium solution is said to be *asymptotically stable in probability* if it is stable in probability and if $\delta' > 0$ exists so that $\|x_0\| < \delta'$ guarantees that

$$\lim_{\delta\to\infty} P\{\sup_{t\geq\delta} \|x(t; x_0, t_0, \omega)\| > \varepsilon\} = 0.$$

If the convergence holds for all initial times, t_0, it is referred to as *uniform asymptotic stability in probability.*

DEFINITION 64.8 $\mathbf{II_m}$: Asymptotic Lyapunov Stability in the $\mathbf{p^{th}}$ Moment
The equilibrium solution is said to be *asymptotically p^{th} moment stable* if it is stable in the pth moment and if $\delta' > 0$ exists so that $\|x_0\| < \delta'$ guarantees that

$$\lim_{\delta\to\infty} E\{\sup_{t\geq\delta} \|x(t; x_0, t_0, \omega)\|\} = 0.$$

DEFINITION 64.9 $\mathbf{II_{a.s.}}$: Almost Sure Asymptotic Lyapunov Stability
The equilibrium solution is said to be *almost surely asymptotically stable* if it is almost surely stable and if $\delta' > 0$ exists so that $\|x_0\| < \delta'$ guarantees that, for any $\varepsilon > 0$,

$$\lim_{\delta\to\infty} \{\sup_{t\geq\delta} \|x(t; x_0, t_0, \omega)\| > \varepsilon\} = 0.$$

Weaker versions of the stability definitions are common in the stochastic stability literature. In these versions the stochastic stability properties of the system are given in terms of particular instants, t, rather than the seminfinite time interval $[t_0, \infty)$ as given in the majority of the definitions given above. Most noteworthy, are the concepts of the pth moment and almost sure exponential stability:

DEFINITION 64.10 $\mathbf{III_m}$: $\mathbf{p^{th}}$ Moment Exponential Lyapunov Stability
The equilibrium solution is said to be *p^{th} moment exponentially stable* if there exists a $\delta > 0$, an $\alpha > 0$, and a $\beta > 0$ so that $\|x_0\| < \delta$ guarantees that

$$E\{\|x(t; x_0, t_0, \omega)\|\} \leq \beta \|x_0\| \exp^{-\alpha(t-t_0)}.$$

DEFINITION 64.11 $\mathbf{III_{a.s.}}$: Almost Sure Exponential Lyapunov Stability
The equilibrium solution is said to be *almost surely exponentially stable* if there exist a $\delta > 0$, an $\alpha > 0$, and a $\beta > 0$ so that $\|x_0\| < \delta$ guarantees that

$$P\{\{\|x(t; x_0, t_0, \omega)\|\} \leq \beta \|x_0\| \exp^{-\alpha(t-t_0)}\} = 1$$

EXAMPLE 64.1:

Consider the scalar Ito equation

$$dx = axdt + \sigma xdw \tag{64.1}$$

where w is a standard Wiener process. The infinitesimal generator for the system is given by

$$\mathcal{L} = \tfrac{1}{2}\sigma^2 x^2 \tfrac{d^2}{dx^2} + ax\tfrac{d}{dx} \quad . \tag{64.2}$$

The solution process x_t for $t \geq 0$ is given by

$$x_t = e^{(a-\frac{1}{2}\sigma^2)t} e^{\sigma \int_0^t dw} x_0.$$

Hence,

$$\log \frac{x_t}{x_0} = (a - \frac{1}{2}\sigma^2)t + \sigma \int_0^t dw$$

and the asymptotic exponential growth rate of the process is

$$\lambda = \lim_{t\to\infty} \frac{1}{t} \log \frac{x_t}{x_0} = (a - \frac{1}{2}\sigma^2) + \lim_{t\to\infty} \frac{\sigma}{t} \int_0^t dw = a - \frac{1}{2}\sigma^2$$

The last equality follows from the fact that the Wiener process, with increment dw, is a zero-mean ergodic process. We then conclude that the system is almost surely exponentially stable in the sense that

$$P_{x_0}\left\{\lim_{t\to\infty} x_t = 0, \text{ at an exponential rate a.s.}\right\} = 1,$$

if, and only if, $a < \frac{1}{2}\sigma^2$.

Next we compare this with the second-moment stability result; see also example 64.2 in the next section where the same conclusion follows from a Lyapunov analysis of the system. From our previous calculation, x_t^2 for $t \geq 0$ is given by

$$x_t^2 = e^{2(a-\frac{1}{2}\sigma^2)t} e^{2\sigma \int_0^t dw} x_0^2.$$

Then,

$$E\{x_t^2\} = e^{(2a+\sigma^2)t} E\{x_0^2\},$$

and we conclude that the system is exponentially second-moment stable if, and only if, $(a + \frac{1}{2}\sigma^2) < 0$, or $a < -\frac{1}{2}\sigma^2$. Therefore, unlike deterministic systems, even though moment stability implies almost sure stability, almost sure (sample path) stability need not imply moment stability of the system. For the system Equation 64.1, the pth moment is exponentially stable if and only if $a < \frac{1}{2}\sigma^2(1-p)$ where $p = 1, 2, 3, \ldots$.

In the early stages (1940s to 1960s) of the study of the stability of stochastic systems, investigators were primarily concerned with moment stability and the stability in probability; the mathematical theory for the study of almost sure (sample path) stability was not yet fully developed. During the initial development of the theory and methods of stochastic stability, some confusion about the stability concepts, their usefulness in applications, and the relationship among the different concepts of stability existed. Kozin's survey clarified some of the confusion and provided a good foundation for further work. During the past twenty years, almost sure (sample path) stability studies have attracted increasing attention of researchers. This is not surprising because the sample paths rather than moments or probabilities associated with trajectories are observed in real systems and the stability properties of the sample paths can be most closely related to their deterministic counterpart, as argued by Kozin [27]. Practically speaking, moment stability criteria, when used to infer sample path stability, are often too conservative to be useful in applications.

One of the most fruitful and important advances in the study of stochastic stability is the development of the Lyapunov exponent theory for stochastic systems. This is the stochastic counterpart of the notion of characteristic exponents introduced in Lyapunov's work on asymptotic (exponential) stability. This approach provides necessary and sufficient conditions for almost sure asymptotic (exponential) stability, but significant computational problems must be solved. The Lyapunov exponent method uses sophisticated tools from stochastic process theory and other related branches of mathematics, and it is not a mature subject area. Because this method has the potential of providing testable conditions for almost sure asymptotic (exponential) stability for stochastic systems, this chapter will focus attention on this approach for analyzing the stability of stochastic systems.

In this chapter, we divide the results into two categories, the Lyapunov function method and the Lyapunov exponent method.

64.2 Lyapunov Function Method

The Lyapunov function method, i.e., Lyapunov's second (direct) method provides a powerful tool for the study of stability properties of dynamic systems because the technique does not require solving the system equations explicitly. For deterministic systems, this method can be interpreted briefly as described in the following paragraph.

Consider a nonnegative continuous function $V(x)$ on $\mathbb{R}^n$ with $V(0) = 0$ and $V(x) > 0$ for $x \neq 0$. Suppose for some $m \in \mathbb{R}$, the set $Q_m = \{x \in \mathbb{R}^n : V(x) < m\}$ is bounded and $V(x)$ has continuous first partial derivatives in Q_m. Let the initial time $t_0 = 0$ and let $x_t = x(t, x_0)$ be the unique solution of the initial value problem:

$$\begin{cases} \dot{x}(t) = f[x(t)], & t \geq 0, \\ x(0) = x_0 \in \mathbb{R}^n, & f(0) = 0, \end{cases} \tag{64.3}$$

for $x_0 \in Q_m$. Because $V(x)$ is continuous, the open set Q_r for $r \in (0, m]$ defined by $Q_r = \{x \in \mathbb{R}^n : V(x) < r\}$, contains the origin and monotonically decreases to the singleton set $\{0\}$ as $r \to 0^+$. If the total derivative $\dot{V}(x)$ of $V(x)$ (along the solution trajectory $x(t, x_0)$), which is given by

$$\dot{V}(x) = \frac{dV(x)}{dt} = f^T(x) \cdot \frac{\partial V}{\partial x} \stackrel{\text{def}}{=} -k(x), \tag{64.4}$$

satisfies $-k(x) \leq 0$ for all $x \in Q_m$, where $k(x)$ is continuous, then $V(x_t)$ is a nonincreasing function of t, i.e., $V(x_0) < m$ implies $V(x_t) < m$ for all $t \geq 0$. Equivalently, $x_0 \in Q_m$ implies that $x_t \in Q_m$, for all $t \geq 0$. This establishes the stability of the zero solution of Equation 64.3 in the sense of Lyapunov, and $V(x)$ is called a Lyapunov function for Equation 64.3. Let us further assume that $k(x) > 0$ for $x \in Q_m \backslash \{0\}$. Then $V(x_t)$, as a function of t, is strictly monotone decreasing. In this case, $V(x_t) \to 0$ as $t \to +\infty$ from Equation 64.4. This implies that $x_t \to 0$ as $t \to +\infty$. This fact can also be seen through an integration of

Equation 64.4, i.e.,

$$0 < V(x_0) - V(x_t) = \int_0^t k(x_s)ds < +\infty \text{ for } t \in [0, +\infty). \quad (64.5)$$

It is evident from Equation 64.5 that $x_t \to \{0\} = \{x \in Q_m : k(x) = 0\}$ as $t \to +\infty$. This establishes the asymptotic stability for system Equation 64.3.

The Lyapunov function $V(x)$ may be regarded as a generalized energy function of the system Equation 64.3. The above argument illustrates the physical intuition that if the energy of a physical system is always decreasing near an equilibrium state, then the equilibrium state is stable.

Since Lyapunov's original work, this direct method for stability study has been extensively investigated. The main advantage of the method is that one can obtain considerable information about the stability of a given system without explicitly solving the system equation. One major drawback of this method is that for general classes of nonlinear systems a systematic way does not exist to construct or generate a suitable Lyapunov function, and the stability criteria with the method, which usually provides only a sufficient condition for stability, depends critically on the Lyapunov function chosen.

The first attempts to generalize the Lyapunov function method to stochastic stability studies were made by [8] and [24]. A systematic treatment of this topic was later obtained by [30], [31] and [22] (primarily for white noise stochastic systems). The key idea of the Lyapunov function approach for a stochastic system is the following:

Consider the stochastic system defined on a probability space $(\Omega, \mathcal{F}, P)$, where Ω is the set of elementary events (sample space), $\mathcal{F}$ is the σ field which consists of all subsets of Ω that are measurable, and P is a probability measure:

$$\begin{cases} \dot{x}(t) = f(x(t), \omega), & t \geq 0, \\ x(0) = x_0. \end{cases} \quad (64.6)$$

It is not reasonable to require that $\dot{V}(x_t) \leq 0$ for all ω, where $x_t = x(t, x_0, \omega)$ is a sample solution of Equation 64.6 initial from x_0. What can be expected to insure "stability" is that the time derivative of the expectation of $V(x_t)$, denoted by $\mathcal{L}V(x_t)$, is nonpositive. Here, $\mathcal{L}$ is the infinitesimal generator of the process x_t. Suppose that the system is Markovian so that the solution process is a strong, time homogeneous Markov process. Then $\mathcal{L}$ is defined by

$$\mathcal{L}V(x_0) = \lim_{\Delta t \to 0} \frac{E_{x_0}(V(x_{\Delta t})) - V(x_0)}{\Delta t} \quad (64.7)$$

where the domain of $\mathcal{L}$ is defined as the space of functions $V(x)$ for which Equation 64.7 is well-defined. This is a natural analog of the total derivative of $V(x)$ along the solution trajectory x_t in the deterministic case. Now suppose that, for a Lyapunov function $V(x)$ which satisfies the conditions stated above, $\mathcal{L}V(x) \leq -k(x) \leq 0$. It follows that

$$\begin{aligned} 0 \leq V(x_0) - E_{x_0}V(x_t) &= E_{x_0}\int_0^t k(x_s)ds \\ &= -E_{x_0}\int_0^t \mathcal{L}V(x_s)ds < +\infty \end{aligned} \quad (64.8)$$

and for $t, s > 0$

$$E_{x_s}\left(V(x_{t+s})\right) - V(x_s) \leq 0 \quad \text{a.s.} \quad (64.9)$$

Equation 64.9 means that $V(x_t)$ is a supermartingale, and, by the martingale convergence theorem, we expect that $V(x_t) \to 0$ a.s. (almost surely) as $t \to +\infty$. This means that $x_t \to 0$ a.s. as $t \to +\infty$. A similar argument can be obtained by using Equation 64.8 an analog of Equation 64.5. It is reasonable to expect from Equation 64.8 that $x_t \to \{x \in \mathbb{R}^n : k(x) = 0\}$ almost surely. These are the key ideas behind the Lyapunov function approach to the stability analysis of stochastic systems.

Kushner [30], [31], [32] used the properties of strong Markov processes and the martingale convergence theorem to study the Lyapunov function method for stochastic systems with solution processes which are strong Markov processes with right continuous sample paths. A number of stability theorems are developed in these works and the references therein. Kushner also presented various definitions of stochastic stability. The reader should note that his definition of stability "with probability one" is equivalent to the concept of stability in probability introduced earlier here. Also, asymptotic stability "with probability one" means stability "with probability one" and sample path stability, i.e., $x_t \to 0$ a.s. as $t \to +\infty$. The key results of Kushner are based on the following supermartingale inequality which follows directly from Equation 64.8 where x_0 is given:

$$P_{x_0}\left\{\sup_{0 \leq t < +\infty} V(x_t) \geq \epsilon\right\} \leq \frac{V(x_0)}{\epsilon} \quad . \quad (64.10)$$

From this, the following typical results were obtained by Kushner. For simplicity, we assume that $x_t \in Q_m$ almost surely for some $m > 0$ and state these results in a simpler way.

THEOREM 64.1 **(Kushner)**

1. *stability "with probability one":*
   If $\mathcal{L}V(x) \leq 0$, $V(x) > 0$ for $x \in Q_m \backslash \{0\}$, then the origin is stable "with probability one".
2. *asymptotic stability "with probability one":*
   If $\mathcal{L}V(x) = -k(x) \leq 0$ with $k(x) > 0$ for $x \in Q_m \backslash \{0\}$ and $k(0) = 0$, and if for any $d > 0$ small, $\epsilon_d > 0$ exists so that $k(x) \geq d$ for $x \in \{Q_m : \|x\| \geq \epsilon_d\}$, then the origin is stable "with probability one" with

   $$P_{x_0}\{x_t \to 0 \text{ as } t \to +\infty\} \geq 1 - \frac{V(x_0)}{m}.$$

In particular, if the conditions are satisfied for arbitrarily large m, then the origin is asymptotically stable "with probability one".

3. *exponential asymptotic stability "with probability one":*
If $V(x) \geq 0$, $V(0) = 0$ and $\mathcal{L}V(x) \leq -\alpha V(x)$ on Q_m for some $\alpha > 0$, then the origin is stable "with probability one", and

$$P_{x_0}\left\{\sup_{T\leq t<+\infty} V(x_t) \geq \lambda\right\} \leq \frac{V(x_0)}{m} + \frac{V(x_0)e^{-\alpha T}}{\lambda}, \quad \forall T \geq 0.$$

In particular, if the conditions are satisfied for arbitrarily large m, then the origin is asymptotically stable "with probability one", and

$$P_{x_0}\left\{\sup_{T\leq t<+\infty} V(x_t) \geq \lambda\right\} \leq \frac{V(x_0)e^{-\alpha T}}{\lambda}.$$

Many interesting examples are also developed in Kushner's work to demonstrate the application of the stability theorems. These examples also illustrated some construction procedures for Lyapunov functions for typical systems.

EXAMPLE 64.2: (Kushner)

Consider the scalar Ito equation

$$dx = axdt + \sigma xdw \tag{64.11}$$

where w is a standard Wiener process. The infinitesimal generator for the system is given by

$$\mathcal{L} = \tfrac{1}{2}\sigma^2 x^2 \tfrac{d^2}{dx^2} + ax\tfrac{d}{dx} \quad . \tag{64.12}$$

If the Lyapunov function $V(x) = x^2$, then

$$\mathcal{L}V(x) = (\sigma^2 + 2a)x^2 \quad . \tag{64.13}$$

If $\sigma^2 + 2a < 0$, then with $Q_m = \{x : x^2 < m^2\}$, from 1) of the previous theorem, the zero solution is stable "with probability one". Let $m \to +\infty$. By 2) of the theorem,

$$\lim_{t\to\infty} x_t = 0 \qquad \text{a.s.} \tag{64.14}$$

where x_t is solution process of Equation 64.11. By 3) of the theorem,

$$P_{x_0}\left\{\sup_{T\leq t<+\infty} x_t^2 \geq \lambda\right\} \leq \frac{V(x_0)e^{-\alpha T}}{\lambda} \tag{64.15}$$

for some $\alpha > 0$.

REMARK 64.1 As calculated in the previous section, the asymptotic exponential growth rate of the process is

$$\begin{aligned}\lambda &= \lim_{t\to\infty}\frac{1}{t}\log\frac{x_t}{x_0}\\ &= (a - \frac{1}{2}\sigma^2) + \lim_{t\to\infty}\frac{\sigma}{t}\int_0^t dw = a - \frac{1}{2}\sigma^2.\end{aligned}$$

We conclude that the system is almost surely exponentially stable in the sense that

$$P_{x_0}\left\{\lim_{t\to\infty} x_t = 0, \text{ at an exponential rate a.s.}\right\} = 1,$$

if, and only if, $a < \frac{1}{2}\sigma^2$. Compare this with the stability result $a < -\frac{1}{2}\sigma^2$, given above, using the Lyapunov function method. Note that $a < -\frac{1}{2}\sigma^2$ is actually the stability criterion for second-moment stability, which is a conservative estimate of the almost sure stability condition $a < \frac{1}{2}\sigma^2$.

Has'minskii [22] and the references cited therein, provide a comprehensive study of the stability of diffusion processes interpreted as the solution process of a stochastic system governed by an Ito differential equation of the form,

$$\begin{cases} dx(t) = b(t,x)dt + \sum_{r=1}^{k}\sigma_r(t,x)d\xi_r(t)\ , \ t \geq s \\ x(s) = x_s, \end{cases} \tag{64.16}$$

where $\xi_r(t)$ are independent standard Wiener processes and the coefficients $b(t,x)$ and $\sigma_r(t,x)$ satisfy Lipschitz and growth conditions. In this case, the infinitesimal generator $\mathcal{L}$ of the system (associated with the solution processes) is a second-order partial differential operator on functions $V(t,x)$ which are twice continuously differentiable with respect to x and continuously differentiable with respect to t. $\mathcal{L}$ is given by

$$\begin{aligned}\mathcal{L}V(t,x) &= \frac{\partial V}{\partial t} + \sum_{i=1}^{n} b_i(t,x)\frac{\partial V}{\partial x_i}\\ &+ \frac{1}{2}\sum_{i,j=1}^{n} a_{ij}(t,x)\frac{\partial^2 V}{\partial x_i \partial x_j}.\end{aligned} \tag{64.17}$$

The key idea of Has'minskii's approach is to establish an inequality like Equation 64.10 developed in Kushner's work. Below are some typical results obtained by Has'minskii; the reader is referred to [22] for a more detailed development.

Let U be a neighborhood of 0 and $U_1 = \{t > 0\} \times U$. The collection of functions $V(t,x)$ defined in U_1, which are twice continuously differentiable in x except at the point $x = 0$ and continuously differentiable in t, are denoted by $C_2^0(U_1)$. A function $V(t,x)$ is said to be positive definite in the Lyapunov sense if $V(t,0) = 0$ for all $t \geq 0$ and $V(t,x) \geq \omega(x) > 0$ for $x \neq 0$ and some continuous function $\omega(x)$.

THEOREM 64.2 **(Has'minskii)**

1. *The trivial solution of Equation 64.16 is stable in probability (same as our definition) if there exists $V(t,x) \in C_2^0(U_1)$, positive definite in the Lyapunov sense, so that $\mathcal{L}V(t,x) \leq 0$, for $x \neq 0$.*

2. *If the system Equation 64.16 is time homogeneous, i.e.,* $b(t,x) = b(x)$ *and* $\sigma_r(t,x) = \sigma_r(x)$ *and if the nondegeneracy condition,*

$$\sum_{i,j=1}^{n} a_{ij}(x)\lambda_i\lambda_j > m(x)\sum_{i=1}^{n}\lambda_i^2, \quad \text{for } \lambda = (\lambda_1 \dots \lambda_n)^T \in \mathbb{R}^n, \tag{64.18}$$

is satisfied with continuous $m(x) > 0$ *for* $x \neq 0$, *then, a necessary and sufficient condition for the trivial solution to be stable in probability is that a twice continuously differentiable function* $V(x)$ *exists, except perhaps at* $x = 0$, *so that*

$$\mathcal{L}_0 V(x) = \sum_{i=1}^{n} b_i(x)\frac{\partial V}{\partial x_i} + \sum_{i,j=1}^{n} a_{ij}(x)\frac{\partial^2 V(x)}{\partial x_i \partial x_j} \leq 0$$

where $\mathcal{L}_0$ *is the generator of the time homogeneous system.*

3. *If the system Equation 64.16 is linear, i.e.,* $b(t,x) = b(t)x$ *and* $\sigma_r(t,x) = \sigma_r(t)x$, *then the system is exponentially p-stable (the* p^{th} *moment is exponentially stable), i.e.,*

$$E_{x_0}\{\|x(t,x_0,s)\|^p\} \leq A \cdot \|x\|^p \exp\{-\alpha(t-s)\}, \quad p > 0,$$

for some constant $\alpha > 0$ *if, and only if, a function* $V(t,x)$ *exists, homogeneous of degree p in x, so that for some constants* $k_i > 0$, $i = 1,2,3,4$,

$$k_1\|x\|^p \leq V(t,x) \leq k_2\|x\|^p, \quad \mathcal{L}V(t,x) \leq -k_3\|x\|^p,$$

and

$$\|\frac{\partial V}{\partial x}\| \leq k_4\|x\|^{p-1}, \quad \|\frac{\partial^2 V}{\partial x^2}\| \leq k_4\|x\|^{p-2}.$$

Besides the many stability theorems, Has'minskii also studied other asymptotic properties of stochastic systems and presented many interesting examples to illustrate the stabilizing and destabilizing effects of random noise in stochastic systems.

Just as in the case of deterministic systems, the Lyapunov function approach has the advantage that one may obtain considerable information about the qualitative (asymptotic) behavior of trajectories of the system, in particular, stability properties of the system which are of interest, without solving the system equation. However, no general systematic procedure exists to construct a candidate Lyapunov function. Even though the theorems, like Has'minskii's, provide necessary and sufficient conditions, one may never find the "appropriate" Lyapunov function in practice. Usually the nature of the stability condition obtained via a Lyapunov function critically depends on the choice of this function. Various techniques have been proposed by investigators to construct a suitable family of Lyapunov functions to obtain the "best" stability results. In the following, we summarize some of these efforts.

There are several works related to the stability of linear systems of the form,

$$\begin{cases} \dot{x}(t) = [A + F(t)]x(t);\ t \geq 0, \\ x(0) = x_0 \in \mathbb{R}^n, \end{cases} \tag{64.19}$$

with A a stable (Hurwitz) matrix and $F(t)$ a stationary and ergodic matrix-valued random process. This problem was first considered by Kozin [26] by using the Gronwall–Bellman inequality rather than a Lyapunov function technique. Kozin's results were found to be too conservative. Caughey and Gray [12] were able to obtain better results through the use of a very special type of quadratic form Lyapunov function. Later, Infante [23] extended these stability theorems by using the extremal property of the so-called regular pencil of a quadratic form. The basic idea behind Infante's work is the following:

Consider the quadratic form Lyapunov function $V(x) = x'Px$. Then the time derivative along the sample paths of the system Equation 64.19 is

$$\dot{V}(x_t) = x_t'(F'(t)P + PF(t))x_t - x_t'Qx_t \tag{64.20}$$

where Q is any positive definite matrix and P is the unique solution of the Lyapunov equation

$$A'P + PA = -Q \quad . \tag{64.21}$$

If $\lambda(t) = \dot{V}(x_t)/V(x_t)$, then

$$V(x_t) = V(x_0)\exp\{\int_0^t \lambda(s)ds\} \quad . \tag{64.22}$$

From the extremal properties of matrix pencils,

$$\lambda(t) \leq \lambda_{\max}[(A + F(t))' + Q(A + F(t))Q^{-1}]. \tag{64.23}$$

Here, $\lambda_{\max}(K)$ denotes the largest magnitude of the eigenvalues of the matrix K. By the ergodicity property of the random matrix process $F(t)$, we have the almost sure stability condition

$$\lim_{t\to\infty}\frac{1}{t}\int_0^t \lambda(\tau)d\tau = E\{\lambda(t)\} \leq E\{\lambda_{\max}[(A+F)' + Q(A+F)Q^{-1}]\} < 0.$$

Man [39] tried to generalize the results of Infante. However, several obvious mistakes can be observed in the derivation of the theorem and the two examples given in this work.

Following Infante, Kozin and Wu [29] used the distributional property of the random coefficient matrix $F(t)$ to obtain improved results for two specific second-order systems in the form,

$$\begin{cases} \ddot{x} + 2\beta\dot{x} + [c + f(t)]x = 0, \\ \ddot{x} + [2\beta + g(t)]\dot{x} + cx = 0, \end{cases} \tag{64.24}$$

where $f(t)$ and $g(t)$ are stationary and ergodic random processes.

Parthasarthy and Evan–Zwanoskii [48] presented an effective computational procedure, using an optimization technique, to apply the Lyapunov type procedure to higher order systems in the form of Equation 64.19 with $F(t) = k(t)G$. Here G is a constant matrix and $k(t)$ is a scalar (real-valued) stationary and ergodic random process. After proper parameterization of the quadratic form Lyapunov function, the Fletcher–Powell–Davidson optimization algorithm was used to optimize the stability region which depends on the system data and $k(t)$. A fourth-order system was studied using the suggested procedure, and various simulation results were obtained to show that this procedure yielded a stability region that was not unduly conservative. Because an optimization procedure was used and the solution of a Lyapunov equation was required, this procedure required an extensive computational effort.

Wiens and Sihna [56] proposed a more direct method for higher order systems defined as an interconnection of a set of second-order subsystems. Consider the system,

$$M\ddot{x} + [C_0 + C(t)]\dot{x} + [K_0 + K(t)]x = 0, \quad (64.25)$$

where M, C_0, and K_0 are nonsingular $n \times n$ matrices and $C(t)$, $K(t)$ are $n \times n$ stationary and ergodic matrix-valued, random processes. The technique for constructing a Lyapunov function suggested by Walker [55] for deterministic systems was used to construct a quadratic form Lyapunov function $V(\tilde{x})$ for the deterministic counterpart of Equation 64.25,

$$\begin{pmatrix} \dot{x}_1 \\ \dot{x}_2 \end{pmatrix} = \begin{pmatrix} 0 & I \\ -M^{-1}K_0 & -M^{-1}C_0 \end{pmatrix} \begin{pmatrix} x_1 \\ x_2 \end{pmatrix};$$

$$\tilde{x} \triangleq \begin{pmatrix} x_1 \\ x_2 \end{pmatrix} \in \mathbb{R}^{2n\times 2n}, \quad (64.26)$$

that is

$$V(\tilde{x}) = \tilde{x}' \begin{bmatrix} P_1 & \frac{1}{2}P_3' \\ \frac{1}{2}P_3 & P_2 \end{bmatrix}; \; \tilde{x} \triangleq \tilde{x}'P\tilde{x},$$

$$P_i' = P_i > 0, \text{ for } i = 1, 2,$$

with a time derivative along sample paths,

$$\dot{V}(\tilde{x}) = \tilde{x}' A_0 \tilde{x},$$

where A_0 is properly defined and the following choices are taken:

$$P_3 = P_2 M^{-1} C_0,$$

$$P_1 = P_2 M^{-1} K_0 + \frac{1}{2}(M^{-1}C_0)' P_2 (M^{-1}C_0).$$

Then, Infante's approach was used to obtain the following theorem:

THEOREM 64.3 **(Wiens and Sihna)**

The system Equation 64.25 is almost surely asymptotically stable (Definition $II_{a.s.}$ in Section 64.1) in the large if a positive definite matrix P_2 exists so that

1. *$P_2 M^{-1} K_0$ is positive definite,*
2. *the symmetric parts of $P_2 M^{-1} C_0$ and $(M^{-1}C_0)' P_2 (M^{-1}K_0)$ are positive definite, and*
3. *$E\{\lambda \max[(A_0 + C_t + K_t)P^{-1}]\} < 0$,*

where C_t and K_t are given by

$$C_t = \begin{bmatrix} 0 & \frac{1}{2}(M^{-1}C_0)' P_2 M^{-1} C(t) \\ \frac{1}{2}(M^{-1}C_0)' P_2 M^{-1} C(t) & \{P_2 M^{-1}(H)\} + \{P_2 M^{-1} C(t)\}' \end{bmatrix}$$

and

$$K_t = \begin{bmatrix} \frac{1}{2}\{(M^{-1}C_0)' P_2 M^{-1} K(t)\} + \frac{1}{2}\{(M^{-1}C_0)' P_2 M^{-1} K(t)\}' & \{P_2 M^{-1} K(t)\}' \\ P_2 M^{-1} K(t) & 0 \end{bmatrix}.$$

When applying their results to the second order system Equation 64.24, Wiens and Sihna [56] obtained the stability criteria,

$$E\{f^2(t)\} < 4\beta^2 c, \quad \text{and} \quad (64.27)$$

$$E\{g^2(t)\} < 4\beta^2 c/(c + 2\beta^2) \quad . \quad (64.28)$$

These results are similar to those obtained by Caughey and Gray [12]. Equation 64.27 is the "optimal" result of Infante, but Equation 64.28 is not. This is not surprising, because no optimization procedure was used in deriving the stability criteria. The usefulness of the theorem was demonstrated by Wiens and Sihna by applying their method to higher order systems ($n = 4$ and 6) which yielded stability regions of practical significance.

Another research direction for applying of the Lyapunov function method is the study of stochastic feedback systems for which the forward path is a time-invariant linear system and the random noise appears in the feedback path as a multiplicative feedback gain. Lyapunov functions are constructed by analogous methods for deterministic systems, e.g., the Lyapunov equation, the path-integral technique and the Kalman–Yacubovich–Popov method. However, most of the results obtained can be derived by directly using a quadratic form Lyapunov function together with the associated Lyapunov equation.

Kleinman [25] considered a stochastic system in the form,

$$dx = Axdt + Bxd\xi, \quad (64.29)$$

where ξ is a scalar Wiener process with $E\{[\xi(t) - \xi(\tau)]^2\} = \sigma^2 \mid t - \tau \mid$. By using the moment equation and a quadratic Lyapunov function, he showed that a necessary and sufficient condition for zero to be "stable with probability one" (this is the same as Kushner's definition) is that

$$I \otimes A + A \otimes I + \sigma^2 B \otimes B$$

is a stable matrix where "$\otimes$" denotes the Kronecker product of matrices. However, this result should be carefully interpreted as discussed by Willems [58].

Willems [58], [59] studied the feedback system in the form,

$$\begin{cases} dx = (A_0 x - kbcx)dt - bcxd\xi, \\ x(0) = x_0 \in \mathbb{R}^n, \end{cases} \tag{64.30}$$

with ξ a scalar-valued Wiener process with $E\{[\xi(t) - \xi(\tau)]^2\} = \sigma^2 \mid t - \tau \mid$, i.e., a system consisting of a linear time-invariant plant in the forward path with rational transfer function $H(s)$ and a minimal realization (A_0, b, c), and a multiplicative feedback gain which is the sum of a deterministic constant k and a stochastic noise component. Suppose the system is written in a companion form with state $x = (y, Dy, \ldots D^{n-1}y)$ where $D = d/dt$ and $y = cx$ is the output. The closed loop transfer function is

$$G(s) = c(sI - A_0 - kbc)^{-1}b = \frac{q(s)}{p(s)} \tag{64.31}$$

with $p(s)$ and $q(s)$ relatively prime because the realization is assumed to be minimal. Then, the input-output relation can be written in the form,

$$p(D)ydt + q(D)yd\xi = 0 \quad . \tag{64.32}$$

Willems observed that if a positive-definite quadratic form $V(x)$ was to be used as a Lyapunov function, then, following Kushner, $\mathcal{L}V(x)$ is the sum of a "deterministic part" obtained by setting the noise equal to zero and the term $\frac{1}{2}\sigma^2(q(D)y)^2(\partial^2 V/\partial x_n^2)$ $(x_n = D^{n-1}y)$. Here, $\mathcal{L}$ is the infinitesimal generator of the system. To guarantee the negativity of $\mathcal{L}V(x)$ to assure stability, using Equation 64.32, Willems used a construction technique for a Lyapunov function that was originally developed by Brockett [10], that is

$$V(x) = \int_{t(0)}^{t(x)} [p(D)yh(D)y - (q(D)y)^2]dt$$

where the polynomial $h(s) = s^n + h_{n-1}s^{n-1} + \ldots + h_0$ is the unique solution (assume $p(s)$ is strictly Hurwitz) of

$$\frac{1}{2}[h(s)p(-s) + h(-s)p(s)] = q(s)q(-s)$$

so that

$$\mathcal{L}V(x) = -(q(D)y)^2 + \frac{1}{2}h_{n-1}\sigma^2(q(D)y)^2.$$

By applying Kushner's results, Willems obtained the following theorem:

THEOREM 64.4 **(Willems)**

1. *The origin is mean-square stable in the large (in the sense that $R(t) = E\{x(t)x'(t)\} < M < +\infty$ for $t \geq 0$, $x_0 \in \mathbb{R}^n$ and $\sup_{t\geq 0} \|R(t)\| \to 0$ as $\|R(0)\| \to 0$) and stable "with probability one" (in Kushner's sense) in the large, if $p(s)$ is strictly Hurwitz and*

$$\frac{\sigma^2 h_{n-1}}{2} \leq 1 \quad . \tag{64.33}$$

Moreover, the following identity holds:

$$\frac{\sigma^2 h_{n-1}}{2} = \sigma^2 \int_0^\infty [g(t)]^2 dt = \frac{\sigma^2}{2\pi} \int_{-\infty}^{+\infty} \mid G(i\omega) \mid^2 dw$$

where $g(t)$ is the impulse response of the closed-loop (stable) deterministic system and $G(s)$ is the Laplace transform of $g(t)$.

2. *If the inequality Equation 64.33 holds, the stability of the origin as indicated in condition 1 above is asymptotic.*

Brockett and Willems [11] studied the linear feedback system given by

$$\dot{x}(t) = Ax(t) - BK(t)Cx(t) \tag{64.34}$$

where (A, B, C) is a completely symmetric realization, i.e., $A^T = A \in \mathbb{R}^{n\times n}$, $B = C' \in \mathbb{R}^{n\times m}$, and $K(t) \in \mathbb{R}^{m\times n}$ is a stationary ergodic matrix process. By using a simple quadratic form as a Lyapunov function, the specific properties of a completely symmetric system, and the well-known Kalman–Yacubovich–Popov Lemma, they obtained the following stability theorem:

THEOREM 64.5 **(Brockett and Willems)**
For the system Equation 64.34,

1. *If $K(t) = K'(t)$ almost surely and*

$$\overline{\lambda}_{\max} = E\{\lambda_{\max}(A - BK(t)C)\} < 0,$$

then the origin is almost surely asymptotically stable (in the sense that $\lim_{t\to+\infty} x(t) = 0$ a.s.). In particular, if $m = 1$, which is analogous to a single-input, single-output system, then

$$\overline{\lambda}_{\max} = \int_{Z_1}^{\infty} \sigma p\left(-\frac{1}{g(\sigma)}\right) \left|\frac{\partial g(\sigma)/\partial\sigma}{g^2(\sigma)}\right| d\sigma$$

where Z_1 is the largest zero of the open-loop transfer function $g(s) = C(sI - A)^{-1}B$ and $p(\cdot)$ is the density function of $K(0)$.

2. *If $m = 1$ and*

$$g(s) = C(sI-A)^{-1}B = \frac{q_{n-1}s^{n-1} + \ldots + q_0}{s^n + p_{n-1}s^{n-1} + \ldots + p_0},$$

the origin is almost surely asymptotically stable if a constant β exists so that

- *(a) $E\{\min(\beta, K(t))\} > 0$,*
- *(b) the poles of $g(s)$ lie in $Re\{s\} < -q_{n-1}\beta$, and*
- *(c) the locus of $G(i\omega - q_{n-1}\beta)$, $-\infty < \omega < +\infty$, does not encircle or intersect the closed disc centered at $(-1/2\beta, 0)$ with radius $1/2\beta$ in the complex plane.*

Mahalanabis and Purkayastha [38] as well as Socha [53], applied techniques similar to Willems and Willems and Brockett to study nonlinear stochastic systems. Readers should consult their papers and the references cited therein for more details on this approach.

An extension of Kushner's and Has'minskii's work on the Lyapunov function approach uses the Lyapunov function technique to study large-scale stochastic systems. The development of large-scale system theory in the past two decades is certainly a major impetus for these studies. The Lyapunov function technique of Kushner and Has'minskii, based upon the use of scalar, positive-definite functions, is not effective for large-scale systems. As in the deterministic case, the difficulties are often overcome, if a vector positive-definite function is used. Michel and Rasmussen [40], [41], [51] used vector valued positive-definite functions for studying various stability properties of stochastic large-scale systems. Their approach was to construct a Lyapunov function for the complete system from those of the subsystems. Stability properties were studied by investigating the stability properties of the lower order subsystems and the interconnection structure. Ladde and Siljak in [33] and Siljak [52] established quadratic mean stability criteria by using a vector positive-definite Lyapunov function. This is an extension of the comparison principle developed by Ladde for deterministic systems to the stochastic case. In these works, a linear comparison system was used. Bitsris [9] extended their work by considering a nonlinear comparison system. White noise systems were studied in all of the above works. Socha [54] investigated a real noise system where the noise satisfied the law of large numbers. Has'minskii's result [22] was extended to a large-scale system in this work. Interested readers are referred to the original papers and references cited therein.

64.3 The Lyapunov Exponent Method and the Stability of Linear Stochastic Systems

One of the major advanced contributions to the study of stochastic stability during the past two decades is the application of the Lyapunov exponent concept to stochastic systems. This method uses sophisticated mathematical techniques to study the sample behavior of stochastic systems and often yields necessary and sufficient conditions for almost sure (sample) stability in the sense that

$$\lim_{t\to\infty} \|x(t, x_0, \omega)\| = 0 \quad \text{a.s.} \tag{64.35}$$

This method can potentially be applied in science and engineering for the development of new theoretical results, and we expect it to be the focus of much of the future work. In this section we present a summary of this method and selected results which have been obtained to date.

After the introduction of the concept of Lyapunov exponents by A.M. Lyapunov [37] at the end of the last century, this concept has formed the foundation for many investigations into the stability properties of deterministic dynamical systems. However, it is only recently that the Lyapunov exponent method has been used to study almost sure stability of stochastic systems. The setup is as follows:

Consider the linear stochastic system in continuous time defined by

$$\begin{cases} \dot{x}(t) = A(t, \omega)x(t), & t \geq 0, \\ x(0) = x_0 \in \mathbb{R}^d. \end{cases} \tag{Σ_c}$$

Let $x(t, x_0, \omega)$ denote the unique sample solution of (Σ_c) initial from x_0 for almost all $\omega \in \Omega$. (We always denote the underlying probability space by $(\Omega, \mathcal{F}, P)$). The Lyapunov exponent $\bar{\lambda}_\omega(x_0)$ determined by x_0 is defined by the random variable,

$$\bar{\lambda}_\omega(x_0) = \overline{\lim}_{t\to+\infty} \tfrac{1}{t} \log \|x(t, x_0, \omega)\|, \tag{64.36}$$

for (Σ_c). Here, x_0 can be a random variable. For simplicity, we will concern ourselves primarily with the case where x_0 is nonrandom and fixed.

In the case $A(t, \omega) = A(t)$, a deterministic $\mathbb{R}^{d\times d}$-valued continuous bounded function, Lyapunov [37] proved the following fundamental results for the exponent $\bar{\lambda}(x_0)$ for the system,

$$\dot{x}(t) = A(t)x(t): \tag{Σ}$$

1. $\bar{\lambda}(x_0)$ is finite for all $x_0 \in \mathbb{R}^d\backslash\{0\}$.
2. The set of real numbers which are Lyapunov exponents for some $x_0 \in \mathbb{R}^d\backslash\{0\}$ is finite with cardinality p, $1 \leq p \leq d$;

$$-\infty < \lambda_1 < \lambda_2 < \ldots < \lambda p < +\infty, \quad \lambda_i \in \mathbb{R} \ \forall i.$$

3. $\bar{\lambda}(cx_0) = \bar{\lambda}(x_0)$ for $x_0 \in \mathbb{R}^d\backslash\{0\}$ and $c \in \mathbb{R}\backslash\{0\}$. $\bar{\lambda}(\alpha x_0 + \beta y_0) \leq \max\{\bar{\lambda}(x_0), \bar{\lambda}(y_0)\}$ for $x_0, y_0 \in \mathbb{R}^d\backslash\{0\}$ and $\alpha, \beta \in \mathbb{R}$ with equality if $\bar{\lambda}(x_0) < \bar{\lambda}(y_0)$ and $\beta \neq 0$. The sets $\mathcal{L}_i = \{x \in \mathbb{R}^d\backslash\{0\}: \bar{\lambda}(x) = \lambda_i\}$, $i = 1, 2 \ldots p$, are linear subspaces of $\mathbb{R}^d$, and $\{\mathcal{L}_i\}_{i=0}^p$ is a filtration of $\mathbb{R}^d$, i.e.,

$$\{0\} \triangleq \mathcal{L}_0 \subset \mathcal{L}_1 \subset \ldots \subset \mathcal{L}_p = \mathbb{R}^d$$

where $d_i \triangleq \dim(\mathcal{L}_i) - \dim(\mathcal{L}_{i-1})$ is called the *multiplicity* of the exponent λ_i for $i = 1, 2 \ldots p$ and the collection $\{(\lambda_i, d_i)\}_{i=1}^p$ is referred to as the *Lyapunov spectrum* of the system (Σ). We have the relation

$$\begin{aligned} \sum_{i=1}^{p} d_i\lambda_i &\leq \underline{\lim}_{t\to+\infty} \frac{1}{t} \log \|\Phi(t)\| \\ &\leq \overline{\lim}_{t\to+\infty} \frac{1}{t} \log \|\Phi(t)\| \end{aligned} \tag{64.37}$$

where $\Phi(t)$ is the transition matrix of (Σ). The system is said to be (forward) regular if the two inequalities in Equation 64.37 are equalities. For a forward regular system the $\underline{\lim}$ and $\overline{\lim}$ can be replaced by $\lim$.

For the stochastic system (Σ_c) with $\omega \in \Omega$ fixed, the relationship Equation 64.36 implies that if $\overline{\lambda}_\omega(x_0) < 0$, then the sample solution $x(t, x_0, \omega)$ will converge to zero at the exponential rate $|\overline{\lambda}_\omega(x_0)|$ and, if $\overline{\lambda}_\omega(x_0) > 0$, then the sample solution $x(t, x_0, \omega)$ cannot remain in any bounded region of $\mathbb{R}^d$ indefinitely. From this we see that $\overline{\lambda}_\omega(x_0)$ contains information about the sample stability of the system. As we will see later, in many cases a necessary and sufficient condition for sample stability is obtained.

Arnold and Wihstutz [5] have recently given a detailed survey of research work on Lyapunov exponents. The survey is mathematically oriented and presents a summary of general properties and results on the topic. Readers are encouraged to refer to Arnold and Wihstutz [5] for more details. Here, we are interested in the application of the Lyapunov exponents to stability studies of stochastic systems.

The fundamental studies of the Lyapunov exponent method when applied to stochastic systems are by Furstenberg, Oseledec, and Has'minskii. We will briefly review here Oseledec's and Furstenberg's work and focus attention on the work of Has'minskii, because there the idea and methodology of the Lyapunov exponent method for the study of stochastic stability are best illustrated.

The random variables $\overline{\lambda}_\omega(x_0)$ defined by Equation 64.36 are simple, nonrandom constants under certain conditions, for example, stationarity and ergodicity conditions. This is a major consequence of the multiplicative ergodic theorem of Oseledec [46]. Oseledec's theorem deals with establishing conditions for the regularity property of stochastic systems. Lyapunov used the regularity of (Σ) to determine the stability of a perturbed version of the system (Σ) from the stability of (Σ). Regularity is usually very difficult to verify for a particular system. However, Oseledec proved an almost sure statement about regularity. Because of our interest in the stability of stochastic systems, the theorem can be stated in the following special form for (Σ_c), see Arnold et al. [2].

THEOREM 64.6 **Multiplicative ergodic theorem: (Oseledec)**

Suppose $A(t, \omega)$ is stationary with finite mean, i.e., $E\{A(0, \omega)\} < \infty$. Then for (Σ_c), we have

1. *State space decomposition:*
   For almost all $\omega \in \Omega$, an integer $r = r(\omega)$ exists with $1 \le r(\omega) \le d$, real numbers $\lambda_1(\omega) < \lambda_2(\omega) < \ldots < \lambda_r(\omega)$, and linear subspaces (Oseledec spaces) $E_1(\omega), \ldots E_r(\omega)$ with dimension $d_i(\omega) = \dim[E_i(\omega)]$ so that
   $$\mathbb{R}^d = \bigoplus_{i=1}^{r} E_i(\omega)$$
   and
   $$\lim_{t\to+\infty} \frac{1}{t} \log \|\Phi(t, \omega)x_0\| = \lambda_i(\omega), \text{ if } x_0 \in E_i(\omega)$$
   where $\Phi(t, \omega)$ is the transition matrix of (Σ_c) and "$\bigoplus$" denotes the direct sum of subspaces.
2. *Domain of attraction of $E_i(\omega)$:*
   $$\lim_{t\to+\infty} \frac{1}{t} \log \|\Phi(t, \omega)x_0\| = \lambda_i(\omega), \quad \text{iff } x_0 \in \mathcal{L}_i(\omega)\backslash\mathcal{L}_{i-1}(\omega),$$
   where $\mathcal{L}_i(\omega) = \bigoplus_{j=1}^{i} E_j(\omega)$.
3. *Center of gravity of exponents:*
   $$\sum_{i=1}^{r(\omega)} d_i(\omega)\lambda_i(\omega) = \lim_{t\to+\infty} \frac{1}{t} \log \left| \det \Phi(t, \omega) \right| = tr E\{A(0, \omega)|\widetilde{\mathcal{F}}\}$$
   where $\widetilde{\mathcal{F}}$ is the σ algebra generated by the invariant sets of $A(t, \omega)$.
4. *Invariance property: If $A(t, \omega)$ is ergodic as well, then the random variables $r(\omega)$, $\lambda_i(\omega)$, and $d_i(\omega)$, $i = 1, 2 \ldots r$, are independent of ω and are non-random constants.*

Note that under the current assumptions, all $\overline{\lim}$ are actually lim and (3) is equivalent to the almost sure regularity of sample systems of the form (Σ_c). Oseledec's theorem is a very general result and the above is a special version for the system (Σ_c). A detailed statement of the theorem is beyond the scope of this chapter. As far as stability is concerned, the sign of the top exponent λ_r is of interest. We present a simple example to illustrate the application of the theorem.

EXAMPLE 64.3:

Consider the randomly switched linear system

$$\begin{cases} \dot{x}(t) = \frac{1}{2}(1 - y(t))A_{-1}x(t) + \frac{1}{2}(1 + y(t))A_1x(t), \quad t \ge 0, \\ x(0) = x_0, \end{cases} \tag{64.38}$$

where

$$A_{-1} = \begin{bmatrix} -a & 0 \\ 0 & -b \end{bmatrix}, \quad A_1 = \begin{bmatrix} c & 1 \\ 0 & c \end{bmatrix}, \quad 0 < b < a < +\infty, \quad 0 < c < +\infty, \tag{64.39}$$

and $y(t) \in \{-1, +1\}$ is the random telegraph process with mean time between jumps $\alpha^{-1} > 0$. Then,

$$e^{A_{-1}t} = \begin{bmatrix} e^{-at} & 0 \\ 0 & e^{-bt} \end{bmatrix}, \tag{64.40}$$

and

$$e^{A_1t} = \begin{bmatrix} e^{ct} & te^{ct} \\ 0 & e^{ct} \end{bmatrix}. \tag{64.41}$$

The phase curves of these two linear systems are

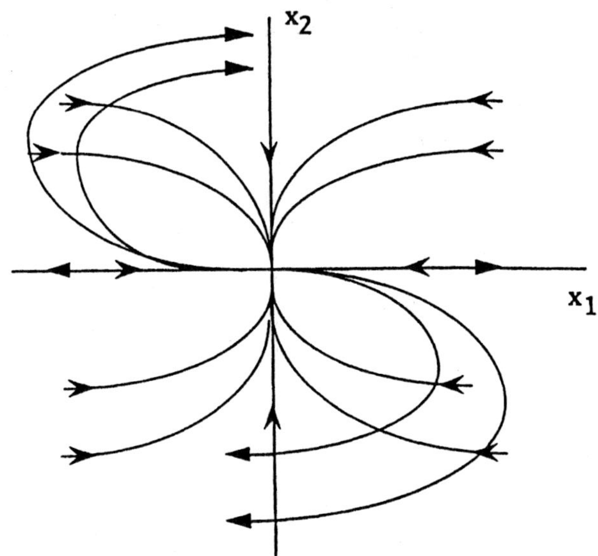

Figure 64.1 Phase curves.

It is easy to verify that, for $x_0 \in \mathbb{R}^2\backslash\{0\}$,

$$\begin{aligned}\overline{\lambda}_\omega(x_0) &= \lim_{t\to+\infty}\frac{1}{t}\log\|x(t,x_0,\omega)\| \\ &= \alpha\cdot\lim_{k\to\infty}\frac{1}{k}\log\|e^{A_{y_k}\tau^{(k)}}\ldots e^{A_{y_1}\tau^{(1)}}x_0\| \text{ a.s.}\end{aligned} \tag{64.42}$$

and

$$\begin{aligned}\overline{\lambda}_\omega(\mathbb{R}^2) &\stackrel{\text{def}}{=} \lim_{t\to+\infty}\frac{1}{t}\log\|\det\Phi(t,\omega)\| \\ &= \alpha\lim_{k\to\infty}\frac{1}{k}\log\|\det\left(e^{A_{y_k}\tau^{(k)}}\ldots e^{A_{y_1}\tau^{(1)}}\right)\| \text{ a.s.}\end{aligned} \tag{64.43}$$

where $\{y_k : k \geq 1\}$ is the embedded Markov chain of $y(t)$ and $\{\tau^{(k)} : k \geq 1\}$ is a sequence of i.i.d. random variables exponentially distributed with parameter α, which is also independent of $\{y_k : k \geq 1\}$. Because $y(t)$ is stationary and ergodic with unique stationary (invariant) distribution $P\{y(t) = 1\} = P\{y(t) = -1\} = \frac{1}{2}$, Oseledec's theorem applies. The theorem says that an $\mathcal{F}$-set N_0 exists with $P(N_0) = 0$ and two real numbers λ_1 and λ_2 (maybe $\lambda_1 = \lambda_2$) so that

$$\overline{\lambda}_\omega(x_0) \in \{\lambda_1, \lambda_2\} \quad \text{for } x_0 \in \mathbb{R}^2\backslash\{0\} \text{ and } \omega \in \Omega\backslash N_0 \tag{64.44}$$

By Equations 64.40, 64.41 and 64.43 and the law of large numbers

$$\begin{aligned}\overline{\lambda}_\omega(\mathbb{R}^2) &= \alpha\cdot\lim_{k\to\infty}\frac{1}{k}\log e^{-(a+b)(\tau^{(1)}+\ldots+\tau^{(i_k)})} \\ &\quad\cdot e^{2c(\tau_{(1)}+\ldots+\tau_{(k-i_k)})}, \\ &= -(a+b)\alpha\lim_{k\to\infty}\frac{1}{k}(\tau^{(1)}+\ldots+\tau^{(i_k)}) \\ &\quad + 2c\alpha\lim_{k\to\infty}\frac{1}{k}(\tau_{(1)}+\ldots+\tau_{(k-i_k)}), \\ &= -(a+b)/2 + c \quad \text{a.s.}\end{aligned} \tag{64.45}$$

Here i_k is the number of times that y_j takes the value -1 in k steps and $\{\tau^{(i)}, \tau_{(j)};\ i, j \geq 1\}$ is a sequence of i.i.d. random variables exponentially distributed with parameter α.

Because $E = \text{span}\{e_1 = (1,0)^T\}$ is a common eigenspace of $e^{A_{-1}t}$ and $e^{A_1 t}$, by the law of large numbers, it is clear that for any $x_0 \in E\backslash\{0\}$,

$$\overline{\lambda}_\omega(x_0) = -\tfrac{1}{2}a + \tfrac{1}{2}c \quad \text{a.s. for } x_0 \in E\backslash\{0\}. \tag{64.46}$$

By (3) of the theorem, Equations 64.45 and 64.46, we obtain

$$\lambda_1 = \tfrac{1}{2}c - \tfrac{1}{2}a \text{ and } \lambda_2 = \tfrac{1}{2}c - \tfrac{1}{2}b \tag{64.47}$$

(with $\lambda_1 < \lambda_2$). We can identify the following objects in Oseledec's theorem:

$$\left\{\begin{array}{lcl} r &=& 2,\quad E_1(\omega) = E,\ \text{for } \omega \in \Omega\backslash N_1, \\ d_1 &=& 1,\quad E_2(\omega) = E^\perp = \text{span}\{e_2 = (0,1)^T\}, \\ & & \quad\ \text{for } \omega \in \Omega\backslash N_1, \\ d_2 &=& 1,\quad \mathcal{L}_1(\omega) = E,\ \mathcal{L}_2(\omega) = \mathbb{R}^2,\ \text{for } \omega \in \Omega\backslash N_1, \end{array}\right. \tag{64.48}$$

where N_1 is an $\mathcal{F}$-set with $P(N_1) = 0$ and

$$\left\{\begin{array}{ll} \overline{\lambda}_\omega(x_0) = \lambda_1, & \text{for } x_0 \in E\backslash\{0\} \text{ and } \omega \in \Omega\backslash N_1, \\ \overline{\lambda}_\omega(x_0) = \lambda_2, & \text{for } x_0 \in \mathbb{R}^2\backslash E \text{ and } \omega \in \Omega\backslash N_1. \end{array}\right. \tag{64.49}$$

From Equation 64.49 it follows that the system Equation 64.38 is almost surely exponentially stable in the sense that

$$P\{\|x(t,\omega,x_0)\| \to 0 \text{ as } t \to +\infty, \text{ at an exponential rate }\} = 1$$

iff $\lambda_2 < 0$ (iff $c < b$). Note that the system is almost surely exponentially unstable iff $\lambda_1 > 0$(iff $c > 0$).

We remark here that we can compute λ_1 and λ_2 directly using Equation 64.46 and the center of gravity of the exponent relation for this simple system. An interesting question is what happens if $b > a$. In this case, the top exponent is $\frac{1}{2}c - \frac{1}{2}a$ and only the top exponent in Oseledec's theorem is physically "realizable" in the sense that an $x_0 \in \mathbb{R}^2$ exists so that

$$P\{\overline{\lambda}_\omega(x_0) = \frac{1}{2}c - \frac{1}{2}a\} > 0.$$

Actually, it follows from Feng and Loparo [14], [15] that $\overline{\lambda}_\omega(x_0) = \frac{1}{2}c - \frac{1}{2}a$ a.s. for any $x_0 \in \mathbb{R}^2\backslash\{0\}$ in this case.

Motivated by Bellman [7], Furstenberg and Kesten [16], as a starting point, studied the product of random matrices and generalized the classical law of large numbers of an i.i.d. sequence of random variables. They showed that under some positivity conditions of the entries of the matrices, the random variable $n^{-1}\log\|X_n\ldots X_1\|$ tended to a constant almost surely for a sequence of i.i.d. random matrices $\{X_k\}_{k=1}^\infty$. Furstenberg [17], [18] went one step further and showed that, if $X_i \in G$, for all i, where G is a noncompact semisimple connected Lie group with finite center, there existed a finite-dimensional vector space of functions ψs (rather than just $\log\|\cdot\|$) so that $n^{-1}\psi(X_n\ldots X_1)$ tended to a constant α_ψ (known as Furstenberg's constant) almost surely as $n \to +\infty$. These studies also considered the action of the group G on certain manifolds, for example, $\mathbb{P}^{d-1} \subset \mathbb{R}^d$ (the projective sphere in $\mathbb{R}^d$, i.e., $\mathbb{P}^{d-1}$ is obtained by identifying s and $-s$ on the unit sphere S^{d-1} in $\mathbb{R}^d$). Furstenberg showed that under a certain transitivity or irreducibility condition, there exists a unique invariant probability measure ν for the Markov chain $Z_n = X_n * \ldots * X_1 * Z_0$, which evolves on $\mathbb{P}^{d-1}$, with respect to μ the common probability measure of $X_i \in G$ (here, " $*$ " denotes the action of G on $\mathbb{P}^{d-1}$) and for any $x_0 \in \mathbb{R}^d \backslash \{0\}$, the Lyapunov exponent

$$\begin{aligned}\overline{\lambda}_\omega(x_0) &= \overline{\lim_{n\to\infty}} \frac{1}{n} \log \|X_n * \ldots * X_1 * x_0\| \\ &= \int\int_{G\times\mathbb{P}^{d-1}} \log \|g * z\| \mu(dg)\nu(dz). \quad (64.50)\end{aligned}$$

Furstenberg's works developed deep and significant results and Equation 64.50 is a prototype for computing the Lyapunov exponent in most of the later work which followed. The following simple discrete-time example due to Has'minskii best illustrates Furstenberg's idea.

EXAMPLE 64.4: (Has'minskii)

Consider a discrete time, $x_{n+1} = A_n(\omega)x_n$, system with $\{A_n(\omega)\}_{n=1}^{\infty}$, a sequence of i.i.d. random matrices. Let

$$\begin{cases} \rho_n = \log \|x_n\|, \quad \text{and} \\ \varphi_n = \|x_n\|^{-1} x_n \in S^{d-1}. \end{cases}$$

Then,

$$\begin{aligned}\rho_n &= \log \|x_n\| = \rho_{n-1} + \log \|A_n \rho_{n-1}\| \\ &= \rho_0 + \sum_{k=1}^{n} \log \|A_k \varphi_{k-1}\|.\end{aligned}$$

Because $\{A_n\}_{n=1}^{\infty}$ are independent, $\{A_n, \varphi_{n-1}\}_{n=1}^{\infty}$ forms a Markov chain in $\mathbb{R}^{d\times d} \times S^{d-1}$. If $\{A_n \varphi_{n-1}\}_{n=1}^{\infty}$ is ergodic, the law of large number gives

$$\begin{aligned}\overline{\lambda}_\omega(x_0) &= \overline{\lim_{n\to\infty}} \frac{1}{n} \rho_n = \lim_{n\to\infty} \frac{1}{n} \sum_{i=1}^{n} \log \|A_i \varphi_{i-1}\| \\ &= E\{\log \|A\varphi\|\} \\ &= \int\int_{\mathbb{R}^{d\times d}\times S^{d-1}} \log \|A\varphi\| \mu(dA)\nu(d\varphi) \quad \text{a.s.}\end{aligned}$$

where μ is the common probability measure of $\{A_i\}_{i=1}^{\infty}$ and ν is the unique invariant probability measure of φ_n with respect to μ, (i.e., ν is such that if φ_0 is distributed according to ν on S^{d-1}, then $\varphi_1 = \|A_1\varphi_0\|^{-1} A_1 \varphi_0$ has the same distribution ν on S^{d-1}).

Has'minskii [20] generalized Furstenberg's idea to a linear system of Ito stochastic differential equations and obtained a necessary and sufficient condition for the almost sure stability of the system:

$$\begin{cases} dx(t) = Ax(t)dt + \sum_{i=1}^{m} B_i x(t) d\xi_i(t), \quad t \geq 0 \\ x(0) = x_0 \end{cases} \quad (64.51)$$

where $\xi_i(t)$ are independent standard Wiener processes. The process $x(t)$ is then a Markov diffusion process with the infinitesimal generator

$$\mathcal{L}u = \langle Ax, u_x \rangle + \frac{1}{2} \sum_{i,j=1}^{d} \sigma_{ij}(x) \frac{\partial^2 u}{\partial x_i \partial x_j} \quad (64.52)$$

where u is a real-valued twice continuously differentiable function, $u_x = \partial u/\partial x$, $\langle \cdot, \cdot \rangle$ is the standard inner product on $\mathbb{R}^d$, and

$$\Sigma(x) = (\sigma_{ij}(x))_{d\times d} = \sum_{i=1}^{m} B_i x x' B_i \quad (64.53)$$

By introducing the polar coordinates, $\rho = \log \|x\|$ and $\varphi = \|x\|^{-1}x$ for $x \neq 0$, and applying Ito's differential rule,

$$d\rho(t) = Q(\varphi)dt + \sum_{i=1}^{m} \varphi'(t) B_i \varphi(t) d\xi_i(t) \quad (64.54)$$

$$d\varphi(t) = H_0(\varphi)dt + \sum_{j=1}^{m} H_j(\varphi) d\xi_j(t) \quad (64.55)$$

where

$$Q(\varphi) = \varphi' A \varphi + \frac{1}{2} tr \Sigma(\varphi) - \varphi' \Sigma(\varphi) \varphi$$

and $H_0(\varphi)$, $H_i(\varphi)$ are the projections onto S^{d-1} of the linear vector fields Ax and $B_i x$ on $\mathbb{R}^d \backslash \{0\}$. Then from Equation 64.55 we see that $\varphi(t)$ is independent of $\rho(t)$ and $\varphi(t)$ is a Markov diffusion process on S^{d-1}. Has'minskii assumed a nondegeneracy condition of the form,

$$(H) \quad \alpha' \Sigma(x) \alpha \geq m \|\alpha\| \cdot \|x\|, \quad \forall \alpha \in \mathbb{R}^m \text{ and } x \in \mathbb{R}^d$$

which guaranteed that the angular process $\varphi(t)$ is an ergodic process on S^{d-1} with a unique invariant probability measure ν. Then the time average of the right hand side of Equation 64.54 is equal to the ensemble average, i.e.,

$$\begin{aligned}\overline{\lambda}_\omega(x_0) &= \overline{\lim_{t\to+\infty}} \frac{1}{t} \int_0^t Q(\varphi(t))dt = E\{\varphi\} \\ &= \int_{S^{d-1}} Q(\varphi)\nu(d\varphi) = J \quad \text{a.s.} \quad (64.56)\end{aligned}$$

independent of $x_0 \in \mathbb{R}^d \backslash \{0\}$.

THEOREM 64.7 **(Has'minskii)**

Under condition (H), $J < 0$ implies that the system Equation 64.51 is almost surely stable. If $J > 0$, then the system Equation 64.51 is almost surely unstable and

$$P\{\lim_{t\to+\infty} \|x(t, x_0, \omega)\| = +\infty\} = 1.$$

If $J = 0$, then

$$P\{\overline{\lim_{t\to+\infty}} \|x(t, x_0, \omega)\| = +\infty\} > 0$$

$$P\{\lim_{t\to+\infty} \|x(t, x_0, \omega)\| < +\infty\} > 0.$$

To determine the sign of J, the invariant probability measure ν must be found. This can be done, in principle, by solving the so-called Fokker–Planck (forward) equation for the density μ of ν. For simplicity, we consider the two dimensional case ($d = 2$). It can be shown from Equation 64.55 that the angular process $\varphi(t)$ satisfies the Ito equation

$$d\varphi(t) = \Phi(\varphi)dt + \Psi(\varphi)d\xi(t) \tag{64.57}$$

with $\xi(t)$ a standard Wiener process, where Φ and Ψ are appropriate functions of φ. $\varphi(t)$ has generator

$$\mathcal{L}_\varphi = \Phi(\varphi)\frac{d}{d\varphi} + \frac{1}{2}\Psi^2(\varphi)\frac{d^2}{d\varphi^2} \quad . \tag{64.58}$$

Then, the Fokker–Planck equation for μ is

$$\mathcal{L}^*_\varphi\mu = \frac{1}{2}\frac{d^2}{d\varphi^2}(\Psi^2(\varphi)\mu) - \frac{d}{d\varphi}(\Phi(\varphi)\mu) = 0 \tag{64.59}$$

with normalization and consistency constraints

$$\int_0^{2\pi} \mu(\varphi)d\varphi = 1 \ , \ \mu(0) = \mu(2\pi). \tag{64.60}$$

Condition (H) guarantees that Equation 64.59 is nonsingular in the sense that $\Psi(\varphi) \neq 0$ for any $\varphi \in S^{d-1}$, and thus admits a unique solution satisfying Equation 64.60. However, even for simple but nontrivial systems, an analytic solution can never usually be obtained. The problem is even more complicated if (H) is not satisfied. This is more often than not the case in practice.

In this situation, singularity on S^1 is a result of $\Psi(\varphi) = 0$ for some $\varphi \in S^1$ and the Markov diffusion process $\varphi(t)$ may not be ergodic on the entire sphere S^1. One must examine each ergodic component of S^1 to determine a complete description of the invariant probability measures. The law of large numbers is then used to compute the exponent for each x_0 belonging to the ergodic components. Has'minskii presented two second-order examples to illustrate the treatment of ergodic components and the usefulness of the theorem. In one of the examples it was shown that an unstable linear deterministic system could be stabilized by introducing white noise into the system. This settled a long standing conjecture about the possibility of stabilizing an unstable linear system by noise. Has'minskii's work was fundamental in the later developments of the Lyapunov exponent method for studying stochastic stability.

Following Has'minskii, Kozin and his co-workers studied the case when the nondegeneracy condition (H) fails. This work made extensive use of one-dimensional diffusion theory to obtain analytic and numerical results for second-order linear white noise systems. Kozin and Prodromou [28] applied Has'minskii's idea to the random harmonic oscillator system in the form,

$$\frac{d^2u}{dt^2} + (1 + \sigma\dot{B})u = 0, \tag{64.61}$$

and

$$\frac{d^2u}{dt^2} + 2\xi \cdot \frac{du}{dt} + (1 + \sigma\dot{B}) = 0, \tag{64.62}$$

where $\dot{B}$ represents a Gaussian white noise process. After transforming the systems into an Ito representation, they observed that condition (H) is not satisfied with singularities $\Psi(\pm\frac{\pi}{2}) = 0$. Kozin and Prodromou then used one-dimensional diffusion theory to study the singularities and sample path behavior of the angular process $\varphi(t)$ near the singularities. The angular process traversed the circle in a clockwise direction and was never trapped at any point, nor did it remain at a singularity $+\frac{\pi}{2}$ or $-\frac{\pi}{2}$ for any positive time interval. Thus, the angular process $\varphi(t)$ was ergodic on the entire circle S^1. From the law of large numbers, they obtained

$$\begin{aligned}\overline{\lambda}_\omega(x_0) &= J = E\{Q(\varphi)\} \\ &= \frac{E\{\int_0^{\mathcal{R}} Q(\varphi(s))ds\} + E\{\int_0^{\mathcal{L}} Q(\varphi(s))ds\}}{E\{\mathcal{R}\} + E\{\mathcal{L}\}} \quad \text{a.s.}\end{aligned}$$

independent of x_0. $\mathcal{R}$ and $\mathcal{L}$ are Markov times for φ to travel the right and left circles, respectively. Using the so-called speed measure and scale measure, they were able to show the positivity of J for Equation 64.61 for $\sigma^2 \in (0, +\infty)$. This proved the almost sure instability of the random harmonic oscillator with white noise parametric excitation. For the damped oscillator Equation 64.62, they failed to obtain analytic results. The almost sure stability region in terms of the parameters (ξ, σ) was determined numerically.

Mitchell [42] studied two second-order Ito differential equations where the condition (H) is satisfied. By solving the Fokker–Planck equation, he obtained necessary and sufficient conditions for almost sure stability in terms of Bessel functions. Mitchell and Kozin [43] analyzed the linear second-order stochastic system,

$$\begin{cases} \dot{x}_1 = x_2, \quad \text{and} \\ \dot{x}_2 = -\omega^2 x_1 - 2\xi\omega x_2 - f_1 x_1 - f_2 x_2. \end{cases} \tag{64.63}$$

Here the f_i are stationary Gaussian random processes with wide bandwidth power spectral density. After introducing a correction term due to Wong and Zakai, Equation 64.63 was written as an Ito equation and Has'minskii's idea was again applied. Because the angular process $\varphi(t)$ is singular, one-dimensional diffusion theory was used to give a careful classification of the singularities on S^1 and the behavior of the sample path $\varphi(t)$ near the singularities. The ergodic components were also examined in detail. The difficulty of the analytic determination of the invariant measures corresponding to the ergodic components was again experienced. An extensive numerical simulation was conducted. The almost sure stability regions were represented graphically for different parameter sets of the system. In this work, the authors provided a good explanation of the ergodic properties of the angular process $\varphi(t)$ and also illustrated various concepts, such as stabilization and destabilization of a linear system by noise, by using examples corresponding to different parameter sets. Interestingly, they provided an example where the sample solutions of the system behaved just like a deterministic one. Examples were also used to show that the regions for second moment stability may be small when compared with the regions for a.s. stability, i.e., moment stability may be too conservative to be practically useful. This is one of the reasons why a.s. stability criteria are important.

A more detailed study of the case when the nondegeneracy condition is not satisfied was presented by Nishioka [45] for the second-order white noise system given by

$$dx(t) = Ax(t)dt + B_1x(t)d\xi_1(t) + B_2x(t)d\xi_2(t) \quad (64.64)$$

with $\xi_i(t), i = 1, 2$, being independent standard Wiener processes. Results similar to Kozin and his co-workers were obtained.

Has'minskii's results were later refined by Pinsky [49] using the so-called Fredholm alternative and the stochastic Lyapunov function $f(\rho, \varphi) = \rho + h(\varphi)$ in the following manner. From the system Equation 64.51 with the generator Equation 64.52, Pinsky observed that

$$\mathcal{L}f(\rho, \varphi) = \mathcal{L}_\rho(o) + \mathcal{L}_\varphi[h(\varphi)] \quad (64.65)$$

i.e., the action of $\mathcal{L}$ on $f(\rho, \varphi)$ separates $\mathcal{L}$ into the radial part $\mathcal{L}_\rho$ and the angular part $\mathcal{L}_\varphi$. Furthermore, $\mathcal{L}_\rho(\rho) = v(\varphi)$ is a function of φ only, and $\mathcal{L}_\varphi$ is a second-order partial differential operator which generates a Markov diffusion process on S^{d-1}. If Has'minskii's nondegeneracy condition (H) is satisfied, then $\mathcal{L}_\varphi$ is ergodic and satisfies a Fredholm alternative, i.e., the equation

$$\mathcal{L}_\varphi h(\varphi) = g(\varphi) \quad (64.66)$$

has a unique solution, up to an additive constant, provided that

$$\int_{S^{d-1}} g(\varphi)m(d\varphi) = 0 \quad (64.67)$$

with m being the unique invariant probability measure of the process $\varphi(t)$ on S^{d-1}. Define

$$q = \int_{S^{d-1}} v(\varphi)m(d\varphi). \quad (64.68)$$

Choose $h(\varphi)$ as a solution of $\mathcal{L}_\varphi h(\varphi) = q - v(\varphi)$. From Ito's formula, Pinsky obtained

$$\begin{aligned}\rho(t) + h(\varphi(t)) &= \rho(0) + h(\varphi(0)) \\ &\quad + \int_0^t \mathcal{L}f(\rho(s), \varphi(s))ds + M_t \\ &= \rho(0) + h(\varphi(0)) + qt + M_t \quad (64.69)\end{aligned}$$

where

$$M_t = \int_0^t H(\varphi(s))d\xi(s)$$

for an appropriate function H on S^{d-1}. M_t is a zero-mean martingale with respect to the σ algebra generated by the process $\{\varphi(t),\ t \geq 0\}$ and $\lim_{t\to+\infty} \frac{1}{t}M(t) = 0$ a.s. Thus, upon dividing both sides of Equation 64.69 by t and taking the limit as $t \to +\infty$,

$$\overline{\lambda}_\omega(x_0) = \lim_{t\to+\infty} \tfrac{1}{t}\rho(t) = q \quad \text{a.s.} \quad (64.70)$$

Loparo and Blankenship [35], motivated by Pinsky, used an averaging method to study a class of linear systems with general jump process coefficients

$$\dot{x}(t) = Ax(t) + \sum_{j=1}^{m} y_j(t)B_jx(t) \quad (64.71)$$

with $A' = -A$ and $(y_1 \ldots y_m)'$ a jump process with bounded generator Q. Using the transformation $z(t) = e^{-At}x(t)$, $\rho(t) = \log\|z(t)\| = \log\|x(t)\|$, and $\varphi(t) = \|z(t)\|^{-1}z(t)$, the system was transformed into a system of equations involving (ρ, φ) which are homogeneous in $y = (y_1 \ldots y_m)'$ but inhomogeneous in time. Thus an artificial process $\tau(t)$ was introduced so that $[\rho(t), \varphi(t), \tau(t), y(t)]$ was a time-homogeneous Markov process. Using the averaging method and the Fredholm alternative, a second-order partial differential operator $\overline{\mathcal{L}}$ was constructed. $\overline{\mathcal{L}}$ is the generator of a diffusion process on $\mathbb{R} \times S^{d-1}$. Applying Pinsky's idea they obtained

$$q = \int_{S^{d-1}} \overline{\mathcal{L}}_\rho(\rho)m(d\theta) = \int_{S^{d-1}} v(\theta)m(d\theta) \quad (64.72)$$

where m is the unique invariant probability measure on S^{d-1} for the diffusion $\varphi(t)$ generated by $\overline{\mathcal{L}}_\varphi$. Unfortunately, q is not the top Lyapunov exponent for the system and cannot be used to determine the a.s. stability properties of the system. This can be easily verified by considering the random harmonic oscillator

$$\ddot{u}(t) + k^2(1 + by(t))u(t) = 0 \quad (64.73)$$

with $y(t) \in \{-1, +1\}$ a telegraph process with mean time between jumps $\lambda^{-1} > 0$. The formula for q is

$$q = \frac{k^2\lambda b^2}{8(k^2+\lambda^2)} \quad \text{a.s.} \quad (64.74)$$

independent of $x_0 \in \mathbb{R}^2\backslash\{0\}$. This is only the first term in an expansion for the Lyapunov exponent $\overline{\lambda}_\omega(x_0)$; see the discussion below. Equations 64.72 and 64.74 do not provide sufficient conditions for almost sure stability or instability.

Following Loparo and Blankenship's idea, Feng and Loparo [13] concentrated on the random harmonic oscillator system Equation 64.73. By integration of the Fokker–Planck equation, the positivity of $\overline{\lambda}_\omega(x_0) = \overline{\lambda}$ was established for any $k, \lambda \in (0, +\infty)$ and $b \in (-1, 1)\backslash\{0\}$. Thus, the system Equation 64.73 is almost surely unstable for any k, λ, and b as constrained above. To compute the Lyapunov exponent $\overline{\lambda}$ explicitly, (it is known that $\overline{\lambda}_\omega(x_0) = \overline{\lambda} =$ constant a.s. for any $x_0 \in \mathbb{R}^2\backslash\{0\}$), the following procedure was used. As usual, introducing polar coordinates $\rho = \log\|x\|$, $\varphi = \|x\|^{-1}x$, where $x = (ku, \dot{u})'$, the system becomes

$$\begin{cases} \dot{\rho}(t) = yg_0(\varphi) \quad \text{and} \\ \dot{\varphi}(t) = h(\varphi, y) \end{cases} \quad (64.75)$$

for some smooth functions g_0 and h on S^1. The Markov process $(\rho(t), \varphi(t), y(t))$ has generator

$$\mathcal{L} = Q + A\varphi \cdot \frac{\partial}{\partial\varphi} + y[g_0(\varphi)\frac{\partial}{\partial\rho} + h(\varphi)\frac{\partial}{\partial\varphi}] \triangleq \mathcal{L}_1 + \mathcal{L}_2$$

where $\mathcal{L}_1 = Q + A\varphi \cdot \partial/\partial\varphi$ satisfies a Fredholm alternative. When $\mathcal{L}$ acts on the function, $F = \rho + h(\varphi) + f_1(y, \varphi)$, where $f_1(y, \varphi)$ is a correction term introduced in the work of Blankenship and Papanicolaou,

$$\begin{cases} \mathcal{L}F = q_0 + \mathcal{L}_2 h(\varphi) \quad \text{and} \\ q_0 = \overline{\pi}\mathcal{L}_2 f_1(y, \varphi) = \overline{\pi}[v(\varphi)]. \end{cases} \tag{64.76}$$

In Equation 64.76 the correction term is chosen as the solution of $\mathcal{L}_1 f_1 = -yg_0(\varphi)$, and $\mathcal{L}_2 f_1 = v(\varphi)$ is a function of φ only. Here, $\overline{\pi}$ is the uniform measure on S^1, and $h(\varphi)$ is the solution of $A\varphi \cdot \partial h/\partial\varphi = [v(\varphi) - \overline{\pi}v(\varphi)]$. It follows that $\mathcal{L}_2 h(\varphi) = yg_1(\varphi)$ and a martingale convergence argument gives

$$\begin{aligned} \overline{\lambda} &= \lim_{t\to+\infty} \frac{1}{t}\rho(t), \\ &= q_0 + \lim_{t\to+\infty} \frac{1}{t}\int_0^t (\mathcal{L}_2 h)_s(\varphi)ds, \\ &= q_0 + \overline{\lambda}_1, \end{aligned}$$

where $\overline{\lambda}_1$ is the Lyapunov exponent of the system,

$$\begin{cases} \dot{\rho}(t) = \mathcal{L}_2 h(\varphi) = yg_1(\varphi) \quad \text{and} \\ \dot{\varphi}(t) = h(\varphi, y). \end{cases} \tag{64.77}$$

Noting the similarity between the systems given by Equations 64.75 and 64.77, the above procedure can be repeated for Equation 64.77. Hence

$$\overline{\lambda} = q_0 + \overline{\lambda}_1 = q_0 + (q_1 + \overline{\lambda}_2) = \sum_{k=0}^{\infty} q_k$$

The absolute convergence of the above series for any $k, \lambda \in (0, +\infty)$, and $b \in (-1, 1)$ was proved by using Fourier series methods, and a general term q_k was obtained. To third-order (b^6), Feng and Loparo obtained

$$\begin{aligned} \overline{\lambda} = {} & \frac{k^2\lambda b^2}{8(k^2+\lambda^2)} + \frac{5k^4\lambda b^4}{64(k^2+\lambda^2)^2} \\ & + \frac{k^4\lambda b^6}{8 \times 16^2(\lambda^2+k^2)^2}\left[k^2\left(\frac{75}{k^2+\lambda^2} + \right.\right. \\ & \left. \frac{160}{\lambda^2+rk^2} + \frac{21}{\lambda^2+9k^2}\right) \\ & \left. - \lambda^2\left(\frac{25}{k^2+\lambda^2} + \frac{32}{\lambda^2+4k^2} + \frac{3}{\lambda^2+9k^2}\right)\right] \\ & + \sum_{n=3}^{\infty} q_n. \end{aligned} \tag{64.78}$$

In the formulas for the exponent obtained by most researchers, e.g., Has'minskii with Equation 64.56, an invariant probability measure was always involved in determining and evaluating the exponent. This is a very difficult problem which was overcome in the work of Feng and Loparo by sequentially applying the Fredholm alternative for the simple, but nontrivial, random harmonic oscillator problem. Due to the difficulty involved with analytic determination of the invariant probability measure, several researchers have attacked the problem by using an analytic expansion and perturbation technique; some of these efforts have been surveyed by Wihstutz [57].

Auslender and Mil'shtein [6] studied a second-order system perturbed by a small white noise disturbance, as given by

$$dx^\epsilon(t) = Bx^\epsilon(t) + \epsilon \sum_{j=1}^{k} \sigma_j x^\epsilon(t) d\xi_j(t) \tag{64.79}$$

where $\xi_j(t)$ are independent, standard Wiener processes and $0 < \epsilon < 1$ models the small noise. Assuming that Has'minskii's nondegeneracy condition (H) is satisfied, the angular process $\varphi^\epsilon(t) = \|x^\epsilon(t)\|^{-1}x^\epsilon(t)$ is ergodic on S^1, and the invariant density μ^ϵ satisfies the Fokker–Planck expression Equation 64.58. The exponent $\lambda(\epsilon)$ is given by Equation 64.56. In this case, a small parameter ϵ is involved. Auslender and Mil'shtein computed $\lambda(\epsilon)$ to second order (ϵ^2) and estimated the remainder term to obtain an expansion for $\lambda(\epsilon)$ to order ϵ^2 as $\epsilon \downarrow 0^+$. For different eigenstructures of B, they obtained the following results:

1. $B = \begin{bmatrix} a & 0 \\ 0 & b \end{bmatrix}$, $a > b$, two distinct real eigenvalues.

$$\lambda(\epsilon) = a - \frac{\epsilon^2}{2}\sum_{r=1}^{k}(\sigma_r^{11})^2 + \epsilon^4\rho(\epsilon) + \rho_0(\epsilon)$$

where $\sigma_r = (\sigma_r^{ij})2 \times 2$, $|\rho(\epsilon)| \le m < \infty$ and $|\rho_0(\epsilon)| \le ce^{-c_1/\epsilon^2}$ for some constants c and $c_1 > 0$.

2. $B = \begin{bmatrix} a & b \\ -b & a \end{bmatrix}$, $a, b > 0$, a complex conjugate pair of eigenvalues.

$$\begin{aligned} \lambda(\epsilon) = {} & a + \frac{\epsilon^2}{8}\sum_{r=1}^{k}\left[\left(\sigma_r^{12} - \sigma_r^{21}\right)^2 \right. \\ & \left. + \left(\sigma_r^{11} + \sigma_r^{22}\right)^2\right] + \epsilon^4\mathcal{R}(\epsilon) \end{aligned}$$

where $|\mathcal{R}(\epsilon)| \le m < +\infty$.

3. $B = \begin{bmatrix} a & 0 \\ 0 & a \end{bmatrix}$, one real eigenvalue of geometric multiplicity 2.

$$\lambda(\epsilon) = a + \epsilon^2 \int_0^{2\pi} \mathcal{L}(\varphi)\mu(\varphi)d\varphi$$

where

$$\mathcal{L}(\varphi) = \frac{1}{2}\sum_{r=1}^{k}\langle\sigma_r\lambda(\varphi), \sigma_r\lambda(\varphi)\rangle - \sum_{r=1}^{k}\langle\sigma_r\lambda(\varphi), \lambda(\varphi)\rangle^2,$$

$\lambda(\varphi) = (\cos\varphi, \sin\varphi)'$, and $\mu(\varphi)$ is the density determined by a Fokker–Planck equation and is independent of ϵ.

4. $B = \begin{bmatrix} a & 1 \\ 0 & a \end{bmatrix}$, one real eigenvalue of geometric multiplicity 1.

$$\lambda(\epsilon) = a + \epsilon^{2/3} \frac{\pi^{1/2}}{\Gamma(1/6)} \left[\frac{3}{4} \sum_{r=1}^{k} \left(\sigma_r^{21} \right)^2 \right]^{1/3} + O(\epsilon)$$

where $\Gamma(x)$ is the gamma function and $O(\epsilon)$ denotes the quantity of same order of ϵ (i.e., $\lim_{\epsilon \to 0} O(\epsilon)/\epsilon =$ constant).

In the above work, an intricate computation is required to obtain $\lambda(\epsilon)$ as an expansion of powers of ϵ. A much easier way to obtain an expansion is direct use of perturbation analysis of the linear operator associated with the forward equation $\mathcal{L}_\varphi^* \mu = 0$. Pinsky [50] applied this technique to the random oscillator problem,

$$\ddot{u}(t) + \{\gamma + \sigma F[\xi(t)]\} u(t) = 0 \qquad (64.80)$$

where γ is a positive constant which determines the natural frequency of the noise-free system, σ is a small parameter which signifies the magnitude of the disturbance, and $\xi(t)$ is a finite-state Markov process with state space $M = \{1, 2 \ldots N\}$. $F(\cdot)$ is a function satisfying $E\{F(\xi)\} = 0$. After introducing polar coordinates in the form $u\sqrt{\gamma} = \rho \cos\varphi$ and $\dot{u} = \rho \sin\varphi$, Pinsky obtained

$$\begin{cases} \dot{\varphi}(t) = h(\varphi, \xi) = -\sqrt{\gamma} + \frac{\sigma F(\xi)}{\sqrt{\gamma}} \cos^2\varphi \\ \dot{\rho}/\rho = q(\varphi, \xi) = \frac{\sigma F(\xi)}{2\sqrt{\gamma}} \sin 2\varphi \end{cases} \qquad (64.81)$$

where (φ, ξ) is a time-homogeneous Markov process with generator,

$$\mathcal{L} = Q + h \frac{\partial}{\partial \varphi},$$

and Q is the generator of the process $\xi(t)$. The Fokker–Planck equation of the density $p_\sigma(\varphi, \xi)$ for the invariant probability measure of (φ, ξ) is given by

$$0 = \mathcal{L}^* p_\sigma$$

where $\mathcal{L}^*$ is the formal adjoint of $\mathcal{L}$. The exponent is computed by

$$\lambda(\sigma) = \int_{S^1 \times M} q(\varphi, \xi) p_\sigma(\varphi, \xi) d\varphi d\xi \quad . \qquad (64.82)$$

Note that

$$\mathcal{L} = Q + h \frac{\partial}{\partial \varphi} = Q - \sqrt{\gamma} \frac{\partial}{\partial \varphi} + \sigma \frac{F(\xi)}{\sqrt{\gamma}} \cos^2\varphi \frac{\partial}{\partial \varphi} \triangleq \mathcal{L}_0 + \sigma \mathcal{L}_1 \quad .$$

Assume an approximation of p_σ in the form

$$p_\sigma = p_0 + \sigma_1 p_1 + \ldots + \sigma^n p_n \quad .$$

Then, it follows from

$$0 = \mathcal{L}^* p_\sigma = (\mathcal{L}_0^* + \sigma \mathcal{L}_1^*)(p_0 + \sigma_1 p_1 + \ldots + \sigma_n p_n) \qquad (64.83)$$

that

$$\mathcal{L}_0^* p_i + \mathcal{L}_1^* p_{i-1} = 0 \quad , \text{for } i = 1, 2 \ldots n, \qquad (64.84)$$

by setting the coefficients of the term σ^i in Equation 64.83 equal to zero. Pinsky showed that the expansion satisfying Equation 64.84 can be obtained by taking

$$p_0 = \frac{1}{2\pi} \quad \text{and} \quad \int_{S^1 \times M} p_n = 0 \qquad (64.85)$$

and proved the convergence of the series expansion by using the properties of the finite-state Markov process to show that

$$|p_\sigma - (p_0 + \sigma p_1 + \ldots + \sigma^n p_n)| \le c\sigma^{n+1} \max |\mathcal{L}_1^* p_n|.$$

By evaluating p_1 from Equations 64.84 and 64.85, Pinsky obtained

$$\lambda(\sigma) = \frac{\sigma^2}{4\gamma} \sum_{k=2}^{N} \frac{\lambda_k < F_1, \psi_k >^2}{\lambda_k^2 + 4\gamma} + O(\sigma^3), \quad \sigma \downarrow 0^+ \qquad (64.86)$$

where λ_k^{-1} is the mean sojourn time of $\xi(t)$ in state k and $\{\psi_i; i = 1, 2 \ldots N\}$ are normalized eigenfunctions of the linear operator Q. For the case when $F(\xi) = \xi$ is a telegraph process, Pinsky obtained a refined expansion of Equation 64.86 consisting of the first two terms in Equation 64.78 when $\sigma \downarrow 0^+$.

Contemporaneous with Pinsky, Arnold, Papanicolaou, and Wihstutz [4] used a similar technique to study the random harmonic oscillator in the form,

$$-\ddot{y}(t) + \sigma F[\xi(\tfrac{t}{\rho})] y(t) = \gamma y(t), \qquad (64.87)$$

where γ, σ and ρ are parameters modeling the natural frequency, the magnitude of noise, and the time scaling of the noise. $\xi(t)$ is only assumed to be an ergodic Markov process on a smooth connected Riemannian manifold M, and $F(\cdot)$ is a function satisfying $E\{F(\xi)\} = 0$. A summary of their results appears next.

1. small noise ($\sigma \downarrow 0^+, \ \gamma > 0, \ \rho = 1$) :

$$\lambda(\sigma) = \sigma^2 \frac{\pi}{4\gamma} \hat{f}(2\sqrt{\gamma}) + O(\sigma^3), \quad \sigma \downarrow 0^+$$

where $\hat{f}(\omega)$ is the power spectral density of $F(\xi(t))$.

2. large noise ($\sigma \uparrow +\infty, \ \gamma = \gamma_0 + \sigma\gamma_1, \ \rho = 1$) :

$$\lambda(\sigma) = \frac{\sqrt{\gamma_1}}{4\pi} \int_0^{2\pi} d\varphi \int_M \nu(d\xi) \frac{\sqrt{\gamma_1 - F(\xi)}}{\gamma_1 - F(\xi)\cos^2\varphi} Q(\log(\gamma_1 - F(\xi)\cos^2\varphi)) + O\left(\frac{1}{\sqrt{\sigma}}\right), \quad \sigma \uparrow +\infty$$

where Q is the generator of ξ and ν is the unique invariant probability measure of ξ on M.

3. fast noise ($\rho \downarrow 0^+$, σ and ν fixed): if $\gamma > 0$,

$$\lambda(\rho) = \rho \cdot \frac{\sigma^2 \pi}{4\gamma} \hat{f}(0) + O(\rho^2), \quad \rho \downarrow 0^+, \quad \text{and}$$

if $\gamma < 0$,

$$\lambda(\rho) = \sqrt{-\gamma} + \rho \frac{\sigma^2 \pi}{4\gamma} \hat{f}(0) + O(\rho^2), \quad \rho \downarrow 0^+.$$

4. slow noise ($\rho \uparrow +\infty$, σ and γ fixed): if $\gamma > \sigma \max(F)$,

$$\begin{aligned} \lambda(\rho) &= \frac{1}{\rho}\frac{\sqrt{\gamma}}{4\pi} \int_0^{2\pi} d\varphi \int_M \nu(d\xi) \frac{\sqrt{\gamma - \sigma F(\xi)}}{\gamma - \sigma F(\xi)\cos^2\varphi} \\ & \quad Q[\log(\gamma - \sigma F(\xi)\cos^2\varphi)] \\ & \quad + O\left(\frac{1}{\rho^2}\right), \quad \rho \uparrow +\infty \end{aligned}$$

if $\gamma < \sigma \min(F)$,

$$\begin{aligned} \lambda(\rho) &= \int_M \nu(d\xi)\sqrt{\sigma F(\xi) - \gamma} \\ & \quad + \frac{1}{\rho} \int_M \nu(d\xi) Q\big[(\log \sin(\varphi + \psi(\xi))\big]\Big|_{\varphi = \psi(\xi)} \\ & \quad + O\left(\frac{1}{\rho^2}\right), \quad \rho \uparrow +\infty, \end{aligned}$$

where $\psi(\xi) = \tan^{-1}\left((\sigma F(\xi) - \gamma)/\sqrt{-\gamma}\right)$.

Pardoux and Wihstutz [47] studied the two-dimensional linear white noise system Equation 64.79 by using similar perturbation techniques. Instead of using the Fokker–Planck equation, perturbation analysis was applied to the backward (adjoint) equation and a Fredholm alternative was used. Pardoux and Wihstutz were able to obtain a general scheme for computing the coefficients of the series expansion of the exponent for all powers of ϵ. Their results are the same as 1), 2), and 3) of Auslender and Mil'shtein, if only the first two terms in the expansion are considered.

There is another approach for studying the problem based on a differential geometric concept and nonlinear control theory, which been suggested by Arnold and his co-workers. For a linear system, after introducing polar coordinates, the angular process $\varphi(t)$ is separated from the radial process and is governed by the nonlinear differential equation

$$\dot{\varphi}(t) = h(\xi(t), \varphi(t)), \ \varphi \in S^{d-1} \tag{64.88}$$

where $\xi(t)$ is the noise process and h is a smooth function of ξ and φ. To determine the Lyapunov exponent $\overline{\lambda}_\omega(x_0)$, the ergodicity of the joint process $[\xi(t), \varphi(t)]$ needs to be determined. This question is naturally related to the concept of reachable sets of $\varphi(t)$ and invariant sets of Equation 64.88 on S^{d-1} for certain "controls" $\xi(t)$. From this perspective the geometric nonlinear control theory is applied.

Arnold, Kliemann, and Oeljeklaus [3] considered the system given by

$$\begin{cases} \dot{x}(t) = A[\xi(t)]x(t) \quad \text{and} \\ x(0) = x_0 \in \mathbb{R}^d \end{cases} \tag{64.89}$$

where $A : M \to \mathbb{R}^{d\times d}$ is an analytic function with domain M, an analytic connected Riemannian manifold, which is the state space of a stationary ergodic diffusion process $\xi(t)$ satisfying the Stratonovich equation,

$$d\xi(t) = X_0[\xi(t)]dt + \sum_{j=1}^{r} X_j[\xi(t)]d\xi_j(t) \tag{64.90}$$

with $\xi_j(t)$ independent, standard Wiener processes. The following Lie algebraic conditions on the vector fields were posed by Arnold, etc.:

A) $\dim LA(X_1, \dots X_r)(\xi) = \dim M, \ \ \forall \xi \in M$.

B) $\dim LA(h(\xi, \cdot), \xi \in M)(\varphi) = d - 1, \ \ \forall \varphi \in \mathbb{P}^{d-1}$.

Here $h(\xi, \psi)$ is the smooth function given in Equation 64.89 for the angular process, and $\mathbb{P}^{d-1}$ is the projective sphere in $\mathbb{R}^d$. Condition A) guarantees that a unique invariant probability density ρ of ξ on M exists solving the Fokker–Planck equation $Q^*\rho = 0$, with Q the generator of $\xi(t)$. Condition B) is equivalent to the accessibility of the angular process φ governed by Equation 64.88 or the fact that the system group,

$$G = \left\{\Pi_{i=1}^n e^{t_i A(\xi_i)};\ t_i \in \mathbb{R}, \xi_i \in M, \ \text{n finite}\right\},$$

acts transitively on S^{d-1}, i.e., for any $x, y \in S^{d-1}$, $g \in G$ exists so that $g * x = y$. Under these conditions, they showed that a unique invariant control set C in $\mathbb{P}^{d-1}$ exists which is an ergodic set and that all trajectories $\varphi(t)$ entering C remain there forever with probability one. Henceforth, a unique invariant probability measure μ of (ξ, φ) on $M \times \mathbb{P}^{d-1}$ exists with support $M \times C$. The following theorem was obtained.

THEOREM 64.8 **(Arnold, et al.)**

For the system defined by Equations 64.89 *and* 64.90, *suppose A) and B) above hold. Then*

1.

$$\begin{aligned} \lambda &= \int_{M\times C} q(\xi,\varphi)\mu(d\xi, d\varphi), \ q(\xi,\varphi) \\ &= \varphi' A(\xi)\varphi \end{aligned} \tag{64.91}$$

is the top exponent in Oseledec's theorem.

2. *For each*

$$x_0 \neq 0, \ \overline{\lambda}_\omega(x_0) = \lambda \text{ a.s.} \tag{64.92}$$

3. *For the fundamental matrix* $\Phi(t)$ *of Equation* 64.89,

$$\lim_{t\to+\infty} \tfrac{1}{t} \log \|\Phi(t)\| = \lambda \quad a.s. \tag{64.93}$$

Equation 64.91 is in the same form as Has'minskii's. The fact that $\overline{\lambda}_\omega(x_0) = \lambda = \lambda_{max}$ in Oseledec's theorem says that only the top exponent λ_{max} is realizable, i.e., observing the sample solutions of the system with probability one, one can only observe the top exponent. The difficulty is still how to determine μ.

Arnold [3], motivated by the results on the undamped random oscillator by Molchanov [44] studied the relationship between sample and moment stability properties of systems defined by Equations 64.89 and 64.90 and obtained very interesting results. Besides the conditions (A) and (B) above, another Lie algebraic condition was needed to guarantee that the generator $\mathcal{L}$ of (ξ, φ) is elliptic. Define the Lyapunov exponent for the pth moment, for $p \in \mathbb{R}$, as

$$g(p, x_0) = \overline{\lim}_{t\to+\infty} \frac{1}{t} \log E(\|x(t, x_0, \omega)\|^p), \quad x_0 \in \mathbb{R}^d \backslash \{0\} \quad . \tag{64.94}$$

By Jenson's inequality $g(p, x_0)$ is a finite convex function of p with the following properties:

1. $| g(p, x_0) | \leq k | p |, \ k = \max_{\xi\in M} \|A(\xi)\|$.
2. $g(p, x_0) \geq \lambda p \ , \ \overline{\lambda}_\omega(x_0) = \lambda$ a.s.
3. $g(p, x_0)/p$ is increasing as a function of p with $x_0 \in \mathbb{R}^d \backslash \{0\}$ fixed.
4. $g'(0^-, x_0) \leq \lambda \leq g'(0^+, x_0)$, here, g' denotes the derivative of g with respect to p.

Then linear operator theory was used to study the strongly continuous semigroup $T_t(p)$ of the generator $\mathcal{L}(p) \triangleq \mathcal{L} + pq(\xi, \varphi)$. Here, $\mathcal{L}$ is the generator of (ξ, φ) and $q(\xi, \varphi) = \varphi' A(\xi)\varphi$. Arnold showed that $g(p, x_0)$ is actually independent of x_0, i.e., $g(p, x_0) = g(p)$, and $g(p)$ is an analytic function of p with $g'(0) = \lambda$ and $g(0) = 0$. $g(p)$ can be characterized by the three figures shown in Figure 64.2 (excluding the trivial case, $g(p) \equiv 0$).

If $g(p) \equiv 0$, then $\lambda = 0$. If $g(p) \not\equiv 0$, besides 0, $g(p)$ has at most one zero $p_0 \neq 0$, and it follows that

1. $\lambda = 0$ iff $g(p) > 0, \quad \forall p \neq 0$.
2. $\lambda > 0$ iff $g(p) < 0, \quad$ for some $p < 0$.
3. $\lambda < 0$ iff $g(p) < 0, \quad$ for some $p > 0$

In the case $tr A(\xi) \equiv 0$, more information can be obtained for the second zero $p_0 \neq 0$ of $g(p)$. Under conditions that $\xi(t)$ is a reversible process and the reachability of $\varphi(t)$ on $\mathbb{P}^{d-1}$, Arnold showed that $p_0 = -d$, d is the system dimension. However, if $tr A(\xi) \not\equiv 0$, then the system Equation 64.89 is equivalent to

$$\dot{x}(t) = d^{-1}[tr A(\xi)]x(t) + A_0(\xi)x(t) \tag{64.95}$$

where $tr A_0(\xi) \equiv 0$. Observing that $d^{-1} tr A(\xi) I$ commutes with $A_0(\xi)$, it follows that

$$g(p) = \alpha_0 p + g_0(p) \tag{64.96}$$

where $\alpha_0 = d^{-1} tr\{EA[\xi(0)]\}$ and $g_0(p)$ is the exponent for the pth moment of $\dot{y}(t) = A_0(\xi)y(t)$. Therefore $g_0(-d) = 0$.

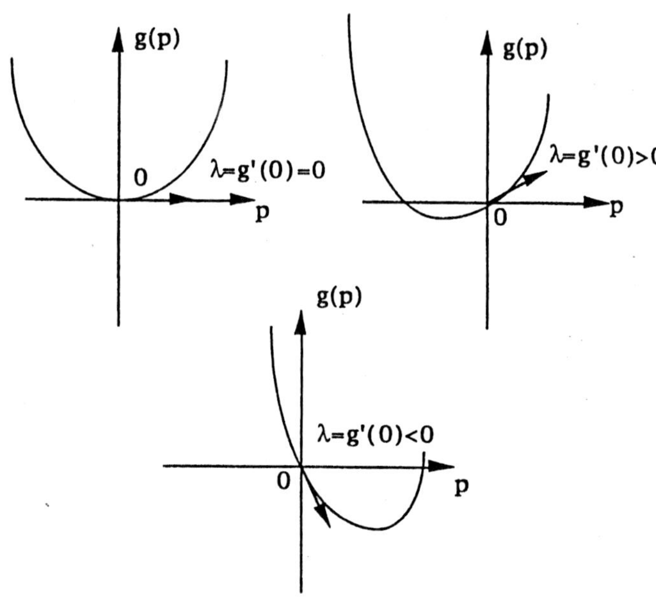

Figure 64.2

Applying the results to the damped harmonic oscillator,

$$\ddot{y}(t) + 2\beta\dot{y}(t) + [1 + \xi(t)]y(t) = 0, \tag{64.97}$$

with $x = (y, \dot{y})' \in \mathbb{R}^2$, Arnold obtained

$$g(p) = -\beta p + g_0(p) \tag{64.98}$$

where $g_0(p)$ is the moment exponent for the undamped oscillator with $g_0(-2) = 0$. Thus, it follows that

1. For the undamped oscillator Equation 64.97, if $\beta = 0$, then $\lambda > 0$ and $g(p) > 0$ for $p > 0$. This implies almost sure sample instability and p^{th} moment instability for $p > 0$.
2. We can stabilize the undamped oscillator by introducing positive damping so that with $\lambda > 0$ and $g(p) < 0$ for $p \in (0, p_1)$, p_1 is some positive real number.

We remark here that the random oscillator problem has attracted considerable attention from researchers during the past thirty years. Equation 64.97 occurs in many science and engineering applications, such as mechanical or electrical circuits, solid state theory, wave propagation in random media, and electric power systems. It is also of particular interest from a theoretical viewpoint because of the simple structure of the model. From the result 3) above, we see that p^{th} moment stability for some $p > 0$ will imply almost sure sample stability. One natural question to ask is the converse question, i.e., does almost sure sample stability have any implication for the pth moment stability? The result 2) above for the random oscillator gives a partial answer to this question. One may ask whether p_1 in 2) can be arbitrarily large or, equivalently, when does the sample stability ($\lambda < 0$) imply p^{th} moment stability for all $p > 0$? if

$$0 < \gamma \triangleq \lim_{p\to+\infty} \frac{1}{p} g(p) < +\infty,$$

then we may choose $\beta > \gamma$ sufficiently large so that $g(p) < 0$ for all $p > 0$. Hence, γ is the quantity which characterizes the questioned property.

An interesting problem in the stability study of stochastic systems is when can an unstable linear system be stabilized by introducing noise into the system. This question has attracted the attention of researchers for a long time. Has'minskii [20] presented an example giving a positive answer to the question. Arnold, Crauel and Wihstutz [2] presented a necessary and sufficient condition for stabilization of a linear system $\dot{x}(t) = A(t)x(t)$ by stationary ergodic noise $F(t) \in \mathbb{R}^{d\times d}$ in the sense that the new system $\dot{x}(t) = [A(t) + F(t)]x(t)$ will have a negative top exponent and hence is almost surely stable. This result is presented next.

THEOREM 64.9 **(Arnold, Crauel and Wihstutz)**
Given the system $\dot{x}(t) = Ax(t)$ with $A \in \mathbb{R}^{d\times d}$ a constant matrix, the perturbed system is given by $\dot{x}(t) = [A + F(t)]x(t)$ where $F(t)$ is a stationary, ergodic, and measurable stochastic process in $\mathbb{R}^{d\times d}$ with finite mean. Then

1. *For any choice of $F(t)$ with $E\{F\} = 0$, the top Lyapunov exponent $\lambda_{\max}[A + F(t)]$ of the perturbed system satisfies*

$$d^{-1}tr(A) \leq \lambda_{\max}(A + F(t)).$$

2. *For any $\epsilon > 0$ fixed, an $F(t)$ exists with $E\{F\} = 0$ so that*

$$d^{-1}tr(A) \leq \lambda_{\max}[A + F(t)] \leq d^{-1}tr(A) + \epsilon.$$

In particular, a linear system $\dot{x}(t) = Ax(t)$ can be stabilized by a zero-mean stochastic process if, and only if, $tr(A) < 0$.

Note that 2) above implies that the undamped, random harmonic oscillator cannot be stabilized by a random linear feedback control of the ergodic type.

Li and Blankenship [34] generalized Furstenberg's results on the product of random matrices and studied linear stochastic system with Poisson process coefficients of the form,

$$dx(t) = Ax(t)dt + \sum_{i=1}^{m} B_i x(t) dN_i(t), \qquad (64.99)$$

where $N_i(t)$ are independent Poisson processes (counting processes). The solution of Equation 64.99 can be written as a product of i.i.d. matrices on $\mathbb{R}^{d\times d}$ acting on the initial point $x_0 \in \mathbb{R}^d\backslash\{0\}$:

$$x(t) = \exp\left[A(t - t_{N(t)})\right] D_{\mu_{N(t)}} \cdots D_{\mu_1} \exp(A\tau_1)x_0,$$

for $t \in [t_{N(t)}, t_{N(t)+1})$, where $D_i = I + B_i$, for $i = 1, 2 \ldots m$, and $N(t) = N_1(t) + \ldots + N_m(t)$ is also a Poisson process with mean interarrival time λ^{-1}, $\{\tau_j;\ j \geq 1\}$ is the interarrival times of $N(t)$, $t_j \triangleq \tau_1 + \ldots + \tau_j$ is the occurrence times of $N(t)$, and $\mu_j(t)$ is an indicator process which encodes which $N_i(t)$ undergoes an increment at t_j.

Using the fact that $\{X_j(\omega) = D_{\mu_j} \exp(A\tau_j)\}_{j=1}^{\infty}$ is an i.i.d. sequence, they generalized Furstenberg's results from semisimple Lie groups to general semigroups (note that D_i may be singular) in the following sense. Let $M = S^{d-1} \cup \{0\}$, and let μ be the probability distribution of X_1 induced by

$$\mu(\Gamma) = P\{D_{\mu_1} e^{A\tau_1} \in \Gamma :\ \Gamma \in B(\mathbb{R}^{d\times d})\}$$

where $B(\mathbb{R}^{d\times d})$ is the Borel σ algebra on $\mathbb{R}^{d\times d}$. Let ν be an invariant probability on M with respect to μ, i.e., if $x_0 \in M$, it is distributed according to ν, and then $x_1 = \|X_1 x_0\|^{-1} X_1 x_0$ has the same distribution ν. If Q_0 denotes the collection of the so-called extremal invariant probabilities on M with respect to μ, then Li and Blankenship obtained the following result:

THEOREM 64.10 **(Li and Blankenship)**
For all $\nu \in Q_0$,

$$r_\nu \triangleq \sum_{i=1}^{m} \lambda_i \int_M \int_0^\infty \log \|D_i \exp(At)s\| e^{-\lambda t} dt \nu(ds) < +\infty$$

and

$$\overline{\lambda}_\omega(x_0) = \lim_{t\to+\infty} \frac{1}{t} \log \frac{\|x(t, x_0)\|}{\|x_0\|} = \lambda \cdot r_\nu \qquad \text{a.s.}$$

for all $x_0 \in E_\nu^0$, where λ_i^{-1} is the mean interarrival time of $N_i(t)$, λ^{-1} is the mean interarrival time of $N(t)$, and E_ν^0 is an ergodic component corresponding to $\nu \in Q_0$. There are only a finite number of different values of r_ν, say, $r_1 < r_2 < \ldots < r_\ell$, $\ell \leq d$. Furthermore, if $\bigcup_{\nu\in Q_0} E_\nu^0$ contains a basis for $\mathbb{R}^d$, then the system Equation 64.99 is asymptotically almost surely stable if $r_\ell < 0$, and Equation 64.99 is almost surely unstable if $r_1 > 0$. In the case where $r_1 < 0$ and $r_\ell > 0$, then the stability of the system depends on the initial state $x_0 \in \mathbb{R}^d\backslash\{0\}$.

Here, the difficulty is still determining the extremal invariant probabilities. Li and Blankenship also discussed large deviations and the stabilization of the system by linear state feedback.

Motivated by Oseledec and Furstenberg's work, Feng and Loparo [14] and [15] studied the linear system given by

$$\begin{cases} \dot{x}(t) = A(y(t))x(t), t \geq 0, & \text{and} \\ x(0) = x_0 \in \mathbb{R}^d, \end{cases} \qquad (64.100)$$

where $y(t) \in N = \{1, 2, \ldots n\}$ is a finite-state Markov process with the infinitesimal generator

$$\Lambda = (\lambda_{ij})_{n\times n}$$

and initial probabilities $(p_1, \ldots p_n)$, and $A(i) = A_i \in \mathbb{R}^{d\times d}$. By extensively using properties of the finite-state Markov processes and a sojourn time description of the process $y(t)$, they related the Oseledec spaces of the system to invariant subspaces of the constituent systems $\dot{x} = A_i x$ and obtained a spectrum theorem for the Lyapunov exponents of the stochastic system Equation 64.100. The exponents in the spectrum theorem given below are exactly those exponents in Oseledec's theorem which are physically realizable. They also showed that the spectrum obtained was actually independent of the choice of the initial probabilities $(p_1, \dots p_n)$ of $y(t)$. Thus, stationarity was not required. The theorem is stated next.

THEOREM 64.11 **(Feng and Loparo)**
For the system Equation 64.100, suppose $p_i > 0$ and $\lambda_{ij} > 0$ for all $i, j \in N = \{1, 2 \dots n\}$. Then there exists k reals, $k \leq d$.

$$-\infty < \lambda_1 < \lambda_2 < \dots < \lambda_k < +\infty$$

and an orthonormal basis for $\mathbb{R}^d$

$$\{e_1^{(1)} \dots e_{i_1}^{(1)} | e_1^{(2)} \dots e_{i_2}^{(2)} | \dots | e_1^{(k)} \dots e_{i_k}^{(k)}\}$$

with $i_1 + \dots + i_k = d$ and $i_j \geq 1$ for $j = 1, 2 \dots k$ so that, if

$$E^{i_j} = span\,\{e_1^{(j)} \dots e_{i_j}^{(j)}\}$$

is an i_j-dimensional subspace of $\mathbb{R}^d$ and

$$\begin{cases} \mathcal{L}_0 = \{0\} \\ \mathcal{L}_j = \oplus_{\ell=1}^{j} E^{i_\ell} \\ \quad j = 1, 2 \dots k \end{cases}$$

which is a filtration of $\mathbb{R}^d$, i.e.,

$$\{0\} \triangleq \mathcal{L}_0 \subset \mathcal{L}_1 \subset \dots \subset \mathcal{L}_k = \mathbb{R}^d$$

where "$\oplus$" denotes the direct sum of subspaces, then

1. *$\mathcal{L}_j = \{x \in \mathbb{R}^d \backslash \{0\} : \bar{\lambda}_\omega(x) \leq \lambda_j$ a.s. $\}$ and $\bar{\lambda}_\omega(x) = \lambda_j$ a.s. iff $x \in \mathcal{L}_j \backslash \mathcal{L}_{j-1}$ for $j = 1, 2 \dots k$.*
2. *$\mathcal{L}_j$ is an A_i-invariant subspace of $\mathbb{R}^d$ for all $i = 1, 2 \dots n$ and $j = 1, 2 \dots k$.*
3. *All of the results above are independent of the initial probabilities $(p_1, \dots p_n)$ chosen.*
4. *If $y(t)$ is stationary, λ_k is the top exponent in Oseledec's theorem.*

Stability results can be obtained directly from the theorem, e.g., if $\lambda_i < 0$ and $\lambda_{i+1} > 0$, then

$$P\{\overline{\lim_{t \to +\infty}} \|x(t, x_0, \omega)\| = 0\} = 1, \quad \text{if } x_0 \in \mathcal{L}_i,$$

and

$$P\{\overline{\lim_{t \to +\infty}} \|x(t, x_0, \omega)\| = +\infty\} = 1, \quad \text{if } x_0 \in \mathbb{R}^d \backslash \mathcal{L}_i.$$

64.4 Conclusions

In this chapter we have introduced the concept of stability for stochastic systems and presented two techniques for stability analysis. The first extend Lyapunov's direct method of stability analysis for deterministic systems to stochastic systems. The main ingredient is a Lyapunov function and, under certain technical conditions, the stability properties of the stochastic system are determined by the "derivative" of the Lyapunov function along sample solutions. As in the deterministic theory of Lyapunov stability, a major difficulty in applications is constructing a proper Lyapunov function for the system under study.

It is well known that, although moment stability criteria for a stochastic system may be easier to determine, often these criteria are too conservative to be practically useful. Therefore, it is important to determine the sample path (almost sure) stability properties of stochastic systems. It is the sample paths, not the moments, that are observed in applications. Focusing primarily on linear stochastic systems, we presented the concepts and theory of the Lyapunov exponent method for the sample stability of a stochastic system. Computational difficulties in computing the top Lyapunov exponent, or its algebraic sign, must still be resolved before this method will be of practical value.

References

[1] Arnold, L., A formula connecting sample and moment stability of linear stochastic systems, *SIAM J. Appl. Math.*, 44, 793–802, 1984.

[2] Arnold, L., Crauel, H., and Wihstutz, V., Stabilization of linear system by noise. *SIAM J. Control Optimiz.*, 21(3), 451–461, 1983.

[3] Arnold, L., Kliemann, W., and Oeljeklaus, E., Lyapunov exponents of linear stochastic systems. In *Lecture Notes in Math.*, No. 1186, Springer, Berlin-Heidelberg-New York-Tokyo, 1985.

[4] Arnold, L., Papanicolaou, G., and Wihstutz, V., Asymptotic analysis of the Lyapunov exponent and rotation number of the random oscillator and application. *SIAM J. App. Math.*, 46(3), 427–449, 1986.

[5] Arnold, L. and Wihstutz, V., Eds., Lyapunov exponents, In *Lecture Notes in Math.*, No. 1186, Springer, Berlin-Heidelberg-New York-Tokyo, 1985.

[6] Auslender, E.I. and Mil'shtein, G.N., Asymptotic expansion of Lyapunov index for linear stochastic system with small noise. *Prob. Math. Mech. USSR*, 46, 277–283, 1983.

[7] Bellman, R., Limit theorem for non-commutative Operator I. *Duke Math. J.*, 491–500, 1954.

[8] Bertram, J.E. and Sarachik, P.E., Stability of circuits with randomly time-varying parameters. *Trans. IRE*, PGIT-5, Special Supplement, p 260, 1959.

[9] Bitsris, G., On the stability in the quadratic mean of stochastic dynamical systems. *Int. J. Control*, 41 (4), 1061–1075, 1985.

[10] Brockett, R.W., *Finite dimensional linear systems*, John Wiley & Sons, New York, 1970.

[11] Brockett, R.W. and Willems, J.C., Average value criteria for stochastic stability. In *Lecture Notes in Math.*, No. 294, Curtain, R.F., Ed., Springer, New York, 1972.

[12] Caughey, T.K. and Gray, A.H., Jr., On the almost sure

stability of linear dynamic systems with stochastic coefficients. *J. Appl.Mech.*, 32, 365, 1965.

[13] Feng, X. and Loparo, K.A., Almost sure instability of the random harmonic oscillator. *SIAM J. Appl. Math.*, 50 (3), 744–759, 1990.

[14] Feng, X. and Loparo, K.A., A nonrandom spectrum for Lyapunov exponents of linear stochastic systems. *Stochastic Analysis and Appl.*, 9(1), 25–40, 1991.

[15] Feng, X. and Loparo, K.A., A nonrandom spectrum theorem for products of random matrices and linear stochastic systems. *J. Math. Syst., Estimation Control*, 2(3), 323–338, 1992.

[16] Furstenberg, H. and Kesten, H., Products of random matrices. *Ann. Math. Statist.*, 31, 457–469, 1960.

[17] Furstenberg, H., Noncommuting random products. *Trans. Amer. Math. Soc.*, 108, 377–428, 1963.

[18] Furstenberg, H., A Poisson formula for semi-simple Lie group. *Ann. of Math.*, 77, 335–386, 1963.

[19] Has'minskii, R.Z., A limit theorem for solutions of differential equations with random right-hand sides. *Theory Probl. Appl.*, 11, 390–406, 1966.

[20] Has'minskii, R.Z., Necessary and sufficient condition for the asymptotic stability of linear stochastic systems. *Theory Prob. Appl.*, 12, 144–147, 1967.

[21] Has'minskii, R.Z., Stability of systems of differential equations under random perturbation of their parameters. M., "Nauka", 1969.

[22] Has'minskii, R.Z., *Stochastic stability of differential equations*, Sijthoff and Noordhoff, Maryland, 1980.

[23] Infante, E.F., On the stability of some linear nonautonomous random system. *J. Appl. Mech.*, 35, 7–12, 1968.

[24] Kats, I.I. and Krasovskii, N.N., On the stability of systems with random parameters. *Prkil. Met. Mek.*, 24, 809, 1960.

[25] Kleinman, D.L., On the stability of linear stochastic systems. *IEEE Trans. on Automat. Control*, AC-14, 429–430, 1969.

[26] Kozin, F., On almost sure stability of linear systems with random coefficients. *M.I.T. Math. Phys.*, 43, 59, 1963.

[27] Kozin, F., A survey of stability of stochastic systems. *Automatica*, 5, 95–112, 1969.

[28] Kozin, F. and Prodromou, S., Necessary and sufficient condition for almost sure sample stability of linear Ito equations. *SIAM J. Appl. Math.*, 21(3), 413–424, 1971.

[29] Kozin, F. and Wu, C.M., On the stability of linear stochastic differential equations. *J. Appl. Mech.*, 40, 87–92, 1973.

[30] Kushner, H.J., *Stochastic stability and control*, Academic, New York, 1967.

[31] Kushner, H.J., *Introduction to stochastic control theory*, Holt, Rinehart and Winston, New York, 1971.

[32] Kushner, H.J., Stochastic stability. In *Lecture Notes in Math.*, No. 249, Curtain, R.F., Ed., Springer, New York, 97–124, 1972.

[33] Ladde, G.S. and Siljak, D.D., Connective stability of large scale systems. *Int. J. Syst. Sci.*, 6 (8), 713–721, 1975.

[34] Li, C.W. and Blankenship, G.L., Almost sure stability of linear stochastic system with Poisson process coefficients. *SIAM J. Appl. Math.*, 46 (5), 875–911, 1986.

[35] Loparo, K.A. and Blankenship, G.L., Almost sure instability of a class of linear stochastic system with jump parameter coefficients. In *Lyapunov Exponents, Lecture Notes in Math.*, No.1186, Springer, Berlin-Heidelberg-New York-Tokyo, 1985.

[36] Loparo, K.A. and Feng, X., Lyapunov exponent and rotation number of two-dimensional linear stochastic systems with telegraphic noise. *SIAM J. Appl. Math.*, 53 (1), 283–300, 1992.

[37] Lyapunov, A.M., Probléme générale de la stabilité du muvement. *Comm. Soc. Math. Kharkov*, *2*, 1892,*3*, 1893. Reprint *Ann. of Math. Studies*, 17, Princeton Univ. Press, Princeton, 1949.

[38] Mahalanabis, A.K. and Parkayastha, S., Frequency domain criteria for stability of a class of nonlinear stochastic systems. *IEEE Trans. Automat. Control*, AC-18 (3), 266–270, 1973.

[39] Man, F.T., On the almost sure stability of linear stochastic systems. *J. Appl. Mech.*, 37 (2), 541, 1970.

[40] Michel, A.N., Stability analysis of stochastic large scale systems. *Z. Angew. Math. Mech.*, 55, 93–105, 1975.

[41] Michel, A.N. and Rasmussen, R.D., Stability of stochastic composite systems. *IEEE Trans. Automat. Control*, AC-21, 89–94, 1976.

[42] Mitchell, R.R., Sample stability of second order stochastic differential equation with nonsingular phase diffusion. *IEEE Trans. Automat. Control*, AC-17, 706–707, 1972.

[43] Mitchell, R.R. and Kozin, F., Sample stability of second order linear differential equation with wide band noise coefficients. *SIAM J. Appl. Math.*, 27, 571–605, 1974.

[44] Molchanov, S.A., The structure of eigenfunctions of one-dimensional unordered structures. *Math. USSR Izvestija*, 12, 69–101, 1978.

[45] Nishioka, K., On the stability of two-dimensional linear stochastic systems. *Kodai Math. Sem. Rep.*, 27, 221–230, 1976.

[46] Oseledec, V.Z., A multiplicative ergodic theorem Lyapunov characteristic number for dynamical systems. *Trans. Moscow Math. Soc.*, 19, 197–231, 1969.

[47] Pardoux, E. and Wihstutz, V., Two-dimensional linear stochastic systems with small diffusion. *SIAM J. Appl. Math.*, 48, 442–457, 1988.

[48] Parthasarathy, A. and Evan-Zwanowskii, R.M., On the almost sure stability of linear stochastic systems. *SIAM J. Appl. Math.*, 34 (4), 643–656, 1978.

[49] Pinsky, M.A., Stochastic stability and the Dirichlet problem. *Comm. Pure Appl. Math.*, 27, 311–350, 1974.

[50] Pinsky, M.A., Instability of the harmonic oscillator

with small noise. *SIAM J. Appl. Math.*, 46(3), 451–463, 1986.

[51] Rasmussen, R.D. and Michel, A.N., On vector Lyapunov functions for stochastic dynamic systems. *IEEE Trans. Automat. Control,* AC-21, 250–254, 1976.

[52] Siljak, D.D., *Large-scale dynamic systems.* Amsterdam, North Holland, 1978.

[53] Socha, L., Application of Yakubovich criteria for stability of nonlinear stochastic systems. *IEEE Trans. Automat. Control,*, AC-25(2), 1980.

[54] Socha, L., The asymptotic stochastic stability in large of the composite systems. *Automatica,* 22(5), 605–610, 1986.

[55] Walker, J.A., On the application of Lyapunov's direct method to linear lumped-parameter elastic systems. *J. Appl. Mech.*, 41, 278–284, 1974.

[56] Wiens, G.J. and Sinha, S.C., On the application of Lyapunov direct method to discrete dynamic systems with stochastic parameters. *J. Sound. Vib.*, 94(1), 19–31, 1984.

[57] V. Wihstutz, V. Parameter dependence of the Lyapunov exponent for linear stochastic system. A survey. In *Lyapunov Exponents, Lecture Notes in Math.*, No. 1186, Springer, Berlin-Heidelberg-New York-Tokyo, 1985.

[58] Willems, J.L., Lyapunov functions and global frequency domain stability criteria for a class of stochastic feedback systems. In *Lecture Notes in Math.*, No. 294, Curtain, R.F., Ed., Springer, New York, 1972.

[59] Willems, J.L., Mean square stability criteria for stochastic feedback systems. *Int. J. Syst. Sci.*, 4(4), 545–564, 1973.

65

Stochastic Adaptive Control

T.E. Duncan and B. Pasik-Duncan
Department of Mathematics, University of Kansas, Lawrence, KS

65.1 Introduction

Stochastic adaptive control[1] has become a very important area of activity in control theory. This activity has developed especially during the past 25 years or so because many physical phenomena for which control is required are best modeled by stochastic systems that contain unknown parameters.

Stochastic adaptive control is the control problem of an unknown stochastic system. Typical stochastic systems are described by Markov chains, discrete time linear systems, and continuous time linear and nonlinear systems. In each case a complete solution of a stochastic adaptive control problem means that a family of strongly consistent estimators of the unknown parameter is given and an adaptive control that achieves the optimal ergodic cost for the unknown system is given. The problem of determining the unknown parameter is an identification problem. Two well-known identification schemes are least squares and maximum likelihood. In some cases these two schemes give the same family of estimates. Adaptive controls are often formed by the certainty equivalence principle, that is, a control is determined by computing the optimal control assuming that the parameter estimate is the true parameter value. Each type of stochastic system typically has special features which require special procedures for the construction of a good family of estimates or a good adaptive control.

A brief review of a few notions from stochastic analysis is given to facilitate or elucidate the subsequent discussion. Let $(\Omega, \mathcal{F}, P)$ be a complete probability space, that is, Ω is the sample space and $\mathcal{F}$ is a σ-algebra of subsets of Ω that contains all of the subsets of P-measure zero. Let $(\mathcal{F}_t, t \in \mathbb{R}_+)$ be an increasing family of sub-σ-algebras of $\mathcal{F}$ that contain all sets of P-measure zero. Such a family is often called a *filtration*. A Markov process is a family of random variables $(X_t, t \in \mathbb{R}_+)$ on $(\Omega, \mathcal{F}, P)$ such that if $t > s$ $P(X_t \in A|X_u, u \leq s) = P(X_t \in A|X_s)$ almost surely (a.s.) A standard n-dimensional Brownian motion (or Wiener process), $(B_t, t \in \mathbb{R}_+)$ on $(\Omega, \mathcal{F}, P)$ is a family of $\mathbb{R}^n$-valued random variables such that $B_0 \equiv 0$, for $t > s$ $B_t - B_s$ is Gaussian with zero mean and covariance $(t-s)I$ and this process has independent increments. The process $(Y_t; \mathcal{F}_t; t \in \mathbb{R}_+)$ is an $(\mathcal{F}_t, t \in \mathbb{R}_+)$ martingale if Y_t is integrable for all $t \in \mathbb{R}_+$ and if $t > s$ $E[Y_t|\mathcal{F}_s] = Y_s$ a.s.

[1]Supported in part by NSF Grant DMS 9305936.

0-8493-8570-9/96/\$0.00+\$.50

65.2 Adaptive Control of Markov Chains

In this section the problem of adaptive control of Markov chains is considered. A Markov chain is described by its transition probabilities. It is assumed that the transition probabilities from the state i to the state j depend on a parameter which is unknown. The adaptive control procedure consists of two stages. The first one is the estimation of the unknown parameter. The Markov chain is observed at each time t and on the basis of this observation the unknown parameter is estimated. The second stage is the construction of the control as if the estimate were the true value of the parameter. The control should be optimal in the sense that an appropriate cost function is minimized. It is important to know when the sequence of estimates converges to the true value of the parameter and when the sequence of controls as functions of the estimates converges to the control that is a function of the true value of the parameter. Is it possible to achieve the optimal cost? Some results in adaptive control of Markov chains that were obtained by Mandl, Varaiya, Borkar, Kumar, Becker, and many others are surveyed.

Let S denote a finite state space of the Markov chain and U the finite set of available controls. For each parameter $\alpha \in I$, $p(i, j, u, \alpha)$ is the probability of going from state i to state j using the control u. Precisely speaking the transition probability at time t depends upon the control action u_t taken at t and upon a parameter α

$$\text{Prob}(x_{t+1} = j | x_t = i) = p(i, j, u_t, \alpha).$$

At each time t, x_t is observed and based upon its value, u_t is selected from the set U. The parameter α has the constant value α_0 which is not known in advance. However, it is known that α_0 belongs to a fixed set I. For each parameter $\alpha \in I$ $\phi(\alpha, \cdot) : S \to U$ is a feedback control law. At time t we apply the action $u_t = \phi(\hat{\alpha}_t, x_t)$ where x_t is the state of the system at time t and $\hat{\alpha}_t$ is the maximum likelihood estimate of the unknown parameter. The estimate $\hat{\alpha}_t$ thus satisfies:

$$\begin{aligned} & P\{x_0, \ldots, x_t | x_0, u_0 \ldots, u_{t-1}, \hat{\alpha}_t\} = \\ & \prod_{s=0}^{t-1} p(x_s, x_{s+1}, u_s, \hat{\alpha}_t) \geq \\ & \geq \prod_{s=0}^{t-1} p(x_s, x_{s+1} u_s, \alpha) = \\ & = P\{x_0, \ldots, x_t | x_0, u_0 \ldots, u_{t-1}, \alpha\} \\ & \text{for all } \alpha \in I. \end{aligned} \tag{65.1}$$

If the likelihood function is maximized at more than one value of α, then a unique value is assumed to be chosen according to some fixed priority ordering. Having determined $\hat{\alpha}_t$ according to Equation 65.1, the control action is selected to be $u_t = \phi(\hat{\alpha}_t, x_t)$. The assumptions are the following:

1. S and U are finite sets.
2. I is a finite set [1] or a compact set [13], [9].
3. There is an $\varepsilon > 0$ such that for every i, j either $p(i, j, u, \alpha) > \varepsilon$ for all u, α or $p(i, j, u, \alpha) = 0$ for all u, α.
4. For every i, j there is a sequence $i_0, i_1, \ldots, i_r$ such that for all u and α, $p(i_{s-1}, i_s, u, \alpha) > 0$, $s = 1, \ldots, r+1$ where $i_0 = i, i_{r+1} = j$.

The assumption 1 is motivated by the fact that often a computer is used which can read only a finite number of (possibly quantized) states and can guarantee only a finite number of (possibly quantized) actions. The assumption 3 guarantees that the probability measures $\text{Prob}\{x_o, \ldots, x_n | x_0, u_0, \ldots, u_{t-1}, \alpha\}$, $\alpha \in I$ are mutually absolutely continuous. Since the estimation procedure will, in finite time, eliminate from future consideration those parameter values which do not yield a measure with respect to which the measure induced by α_0 is absolutely continuous, this assumption is not restrictive. The assumption 4 guarantees that the Markov chain generated by the transition probabilities $p(i, j, \phi(\alpha, i), i)$ has a single ergodic class. In other words, assumption 4 guarantees that all states communicate with each other. Some such condition is clearly needed for identification.

Many questions arise in the problems of adaptive control of Markov chains.

1. Does the sequence $\{\hat{\alpha}_t\}_{t=1}^{\infty}$ converge almost surely?
2. Does the sequence $\{\hat{\alpha}_t\}_{t=1}^{\infty}$ converge to the true value α_0 almost surely?
3. Does the sequence $\{u_t\}_{t=1}^{\infty}$, where $u_t = \phi(\hat{\alpha}_t, \cdot)$ converge almost surely?
4. Does the sequence $\{u_t\}_{t=1}^{\infty}$, where $u_t = \phi(\hat{\alpha}_t, \cdot)$ converge to $u_0 = \phi(\alpha_0, \cdot)$ almost surely?
5. Does the average cost $C_t = \frac{1}{t}\sum_{k=0}^{t-1} c(x_k, u_k)$ converge almost surely?
6. Does the average cost $C_t = \frac{1}{t}\sum_{k=0}^{t-1} c(x_k, u_k)$ converge to $J(\alpha_0)$ almost surely? Here $J(\alpha_0)$ is the optimal cost achievable for the parameter α_0, which is assumed not to depend on the initial state x_0.
7. At what rate do these quantities converge, if they do?

Some examples are given to show that 2, 4, and 6 may be negative.

The answer to the most important question: "Does the sequence of the parameter estimates $\{\hat{\alpha}_t\}$ converge to the true parameter value, α_0?" can be "no". The following is an example that illustrates such a situation.

65.2.1 A Counterexample to Convergence

Consider the two state system with the unknown parameter $\alpha \in \{0.01, 0.02, 0.03\}$ with the true value $\alpha = 0.02$ [1]. The feedback law is $u = \phi(0.01) = \phi(0.03) = 2$ and $\phi(0.02) = 1$. The transition probabilities are the following:

$$p(1, 1, u, \alpha) = 0.5 - 2\alpha + \alpha u$$

$$p(1, 2, u, \alpha) = 0.5 + 2\alpha - \alpha u$$

$$p(2, 1, u, \alpha) = 1$$

$$p(2, 2, u, \alpha) = 0.$$

The initial state is $x_0 = 1$. Suppose $u_0 = 1$. Then at $t = 1$ we have the following probabilities:

1. $$x_1 = 1, \qquad \begin{aligned} p(1, 1, u_0, 0.01) &= 0.49 \\ p(1, 1, u_0, 0.02) &= 0.48 \\ p(1, 1, u_0, 0.03) &= 0.47 \end{aligned}$$
   so that the estimate is $\hat{\alpha}_1 = 0.01$; or
2. $$x_1 = 2, \qquad \begin{aligned} p(1, 2, u_0, 0.01) &= 0.51 \\ p(1, 2, u_0, 0.02) &= 0.52 \\ p(1, 2, u_0, 0.03) &= 0.53 \end{aligned}$$
   so that the estimate is $\hat{\alpha}_1 = 0.03$.

In either case $u_1 = 2$. Since $p(i, j, 2, \alpha)$ does not depend on α, it follows that the estimate will remain unchanged. Thus, $\hat{\alpha}_t \equiv 0.01$ if $x_1 = 1$ or $\hat{\alpha}_t \equiv 0.03$ if $x_1 = 2$ and so α_0 cannot be a limit point of $\{\hat{\alpha}_t\}$.

The same situation occurs when questions 4 and 5 are asked. The example shows that the answers to the questions:

4. Does the sequence $\{\Phi(\hat{\alpha}_t, \cdot)\}$ converge to $\Phi(\alpha_0, \cdot)$ almost surely?
 and
5. Does $C_t = \frac{1}{t}\sum_{s=0}^{t-1} c(x_s, u_s)$ converge to $J(\alpha_0)$ almost surely? (Recall: $J(\alpha_0)$ denotes the optimal cost for α_0.)

can be negative.

Kumar and Becker [11] also give an example that illustrates the situation that the sequence of parameter estimates does not need to converge to the true value.

65.2.2 A Counterexample to Convergence

Consider the two state system $S = \{1, 2\}$ with the unknown parameter $\alpha = \{1, 2, 3\}$ with the true value $\alpha_0 = 1$ [11]. The transition probabilities are the following

$$\begin{aligned} p(1,1,1,1) &= p(1,1,1,3) = 0.5 \\ p(1,1,1,2) &= 0.9 \\ p(1,1,2,1) &= p(1,1,2,2) = 0.8 \\ p(1,1,2,3) &= 0.2 \\ p(2,1,u,\alpha) &= 1 \text{ for all } u \text{ and } \alpha. \end{aligned}$$

The cost function is the following:

$$c(x_t, x_{t+1}, u_t) = c(i, j, u) = 3 + (2 - i)[7.8 - 0.3u - 6j].$$

It is easy to calculate that the optimal feedback law is $\Phi(1, i) = 1$ for $i = 1, 2$ and $\Phi(2, i) = \Phi(3, i) = 2$ for $i = 1, 2$.

Assume that $x_0 = 1$ and $u_0 = 1$. With probability 0.5, x_0 goes to the state $x_1 = 1$, but then $\hat{\alpha}_1 = 2$. Hence, $u_1 = \phi(1, 2) = 2$. It is easy to verify that $\hat{\alpha}_t = 2$ for all $t \geq 1$, so the probability that $\lim_{t\to\infty} \hat{\alpha}_t \neq \alpha_0$ is at least 0.5. Borkar and Varaiya [1] proved that with probability 1 the sequence $\{\hat{\alpha}_t\}$ converges to the random variable α^* such that

$$p(i, j, \Phi(\alpha^*, i), \alpha^*) = p(i, j, \Phi(\alpha^*, i), \alpha_0) \text{ for all } i, j \in S.$$

It means that the transition probabilities are the same for the limit of the parameter estimates as for the true value, so that these two parameters α^* and α_0 are indistinguishable.

THEOREM 65.1 *(Borkar and Varaiya). If $p(i, j, u, \alpha) \geq \varepsilon > 0$ for all $i, j \in S$, $u \in U$, $\alpha \in I$ then there is a set N of zero measure, a random variable α^* and a finite random time T such that for $\omega \notin N$, $t \geq T(\omega)$:*

1. $\lim_{t\to\infty} \hat{\alpha}_t(\omega) = \alpha^*(\omega)$
2. $p(i, j, \Phi(\alpha^*(\omega), i), \alpha^*(\omega)) = p(i, j, \Phi(\alpha^*(\omega), i), \alpha_0)$ *for all* $i, j, \in S$.

The result obtained by Borkar and Varaiya (Theorem 65.1) was extended to the case where the set of parameters is compact. Mandl [13] assumed that the set of parameters is compact but he considered this situation under a condition that can be restrictive in some situations. Kumar [9] extended the result of Borkar and Varaiya assuming that the set of unknown parameters is compact. All the other assumptions are the same as those made by Borkar and Varaiya (Theorem 65.1).

THEOREM 65.2 *(P. R. Kumar). There exists a subset $N \subset \Omega$ with probability measure zero such that: $\omega \notin N$ and $\{\hat{\alpha}_t(\omega)\}$ converges to $\alpha^*(\omega) \Longrightarrow p(i, j, \Phi(\alpha^*(\omega), i), \alpha^*(\omega)) = p(i, j, \Phi(\alpha^*(\omega), i), \alpha_0)$ for every $i, j \in S$.*

The results of Borkar and Varaiya hold for every ω for which $\{\hat{\alpha}_t(\omega)\}$ converges.

Kumar showed that there is an example where with probability 1, the sequence $\{\hat{\alpha}_t\}$ diverges even though assumptions 1, 2, 3, and 4 in 65.2 are satisfied. It is easy to see that Theorem 65.2 requires the convergence of the parameter estimates. So, the result of Kumar is that the conclusion of Borkar and Varaiya holds for almost every ω for which the parameter estimates converge. Even if the parameter estimates do not converge, then another result of his is that if the regulator converges to a limiting feedback regulator (see definitions in [9]), then the true parameter and every limit point of the parameter estimates are indistinguishable, in the sense that under the limiting feedback law their closed-loop transition probabilities coincide.

As was mentioned before, Mandl was the first who considered the problem of adaptive control of Markov chains. He assumed that the set of controls U and the set of unknown parameters I is compact, $p(i, j, u, \alpha) > 0$ and $c(i, j, u)$ are continuous functions, where $c(i, j, u)$, $i, j \in S$ is the reward from a transition from i to j, when the control parameter equals u.

The parameter estimates are obtained by minimizing

$$L_t(\alpha) = \sum_{s=0}^{t-1} f(x_s, x_{s+1}, u_s, \alpha) \quad \alpha \in I,\ t = 1, 2, \ldots$$

This is called a minimum contrast estimator of α_0. The function $f(i, j, u, \alpha)$ $u \in U$, $\alpha \in I$, $i, j, \in S$ is the so-called contrast function. For this function to be a contrast function the following conditions have to be satisfied:

1. $\sum_{j=1}^{s} p(i, j, u, \alpha')(f(i, j, u, \alpha) - f(i, j, u, \alpha')) \geq 0$ for $u \in U, i \in S, \alpha', \alpha \in I$,
2. if $\alpha, \alpha' \in I, \alpha \neq \alpha'$, then there exists an $i \in S$ such that

$$\sum_{j=1}^{s} p(i, j, u, \alpha')(f(i, j, u, \alpha) - f(i, j, u, \alpha')) > 0 \quad u \in U.$$

EXAMPLE 65.1:

Assume:

a) for $i, j \in S$ either $p(i, j, u, \alpha) > 0$ for $u \in U, \alpha \in I$ or $p(i, j, u, \alpha) = 0$ for $u \in U, \alpha \in I$,

b) if $\alpha, \alpha' \in I, \alpha \neq \alpha'$, then there exists an $i \in S$ for which

$$[p(i, 1, u, \alpha), \ldots, p(i, r, u, \alpha)] \neq [p(i, 1, u, \alpha'), \ldots, p(i, r, u, \alpha')]$$

for $u \in U$ (i.e., the transition vectors are different). The contrast function leading to the maximum likelihood estimates is

$$\begin{aligned} f(i, j, u, \alpha) &= -\ln p(i, j, u, \alpha) \text{ whenever } p(i, j, u, \alpha) \neq 0, \\ f(i, j, u, \alpha) &= 1 \text{ whenever } p(i, j, u, \alpha) = 0. \end{aligned}$$

The condition b) of this example is the so-called Identifiability Condition.

THEOREM 65.3 *(Mandl). If U and I are compact, $p(i, j, u, \alpha)$, $c(i, j, u)$, $\phi(\alpha, i)$ and $f(i, u, \alpha)$ are continuous functions, $p(i, j, u, \alpha) > 0$ and the Identifiability Condition is satisfied, then*

1. *for any control*

$$\lim_{t\to\infty} \hat{\alpha}_t = \alpha_0 \quad \textit{almost surely},$$

2. *if $\Phi(\hat{\alpha}_t, x_t) = \Pi_t(x_0, u_0, \ldots, x_t)$ then*

$$\lim_{t\to\infty} \frac{1}{t}\sum_{s=0}^{t-1} c(x_s, x_{s+1}, u_s) = J(\alpha_0) \quad \textit{almost surely}.$$

The remarkable result 1 of Mandl [13] suggests that the Identifiability Condition might be too restrictive in some practical situations. To see this, Borkar and Varaiya consider the familiar Markovian system

$$x_{t+1} = ax_t + bu_t + \xi_t, \quad \alpha = (a, b), \quad t = 0, 1, \ldots$$

where x_t is a real-valued variable and $\{\xi_t\}$ is a sequence of i.i.d. random variables. The unknown parameter is $\alpha = (a, b)$. Then, for the linear control law $u_t = -gx_t$ and two parameter values $\alpha = (a, b)$ and $\alpha' = (a', b')$ such that $a/b = a'/b' = g$, $p(x_t, x_{t+1}, u_t = -gx_t, \alpha) = p(x_t, x_{t+1}, u_t = -gx_t, \alpha')$ for all x_t, x_{t+1} and so the Identifiability Condition cannot hold. However, in some problems, for example, in the control of queueing systems, such a condition is satisfied.

Others have considered special applications such as renewal processes and queueing systems and more general state spaces. These results and improvements of the previous theorems can be found in the references that are given in the survey of Kumar [10] and the book by Chen and Guo [3].

Kumar and Becker [11] introduced a new family of optimal adaptive controllers that consists of an estimator and a controller and simultaneously performs the following three tasks:

1. The successive estimates of the unknown parameter made by the adaptive controller converge, in a Cesaro sense, to an estimate of the true but unknown parameter.

2. The adaptive control law converges, in a Cesaro sense, to a control law that is optimal for the true but unknown parameter, i.e.,

$$\lim_{t\to\infty} \frac{1}{t}\sum_{s=0}^{t-1} \ell\left(u_s = \Phi\left(\alpha^*, x_s\right)\right) = 1.$$

3. The long-term average cost achieved is, almost surely, exactly equal to the optimal long-term average cost achievable if the true parameter were known.

THEOREM 65.4 *(Kumar and Becker). Let*

1. *$p(i, j, u, \alpha) > 0$ for all $i, j \in S, u \in U, \alpha \in I$.*
2. *$c(i, j, u) > 0$ for $i, j \in S, u \in U$.*
3. *I is a finite set.*
4. *$o(t)$ is a function for which $\lim_{t\to\infty} o(t) = +\infty$ and $\lim_{t\to\infty} \frac{1}{t} o(t) = 0$.*
The estimate $\hat{\alpha}_t$ and the control law u_t are chosen so that

$$\hat{\alpha}_t = \begin{cases} \underset{\alpha}{\arg\max} \; J(\alpha)^{-o(t)} \prod_{s=0}^{t-1} p(x_s, x_{s+1}, u_s, \alpha) & \text{for } t = 0, 2, 4, \ldots \\ \hat{\alpha}_{t-1} & \text{for } t = 1, 3, 5, \ldots \end{cases}$$

and $u_t = \Phi(\hat{\alpha}_t, x_t)$.
Then

5. *$\lim_{t\to\infty} \sum_{s=0}^{t-1} c(x_s, x_{s+1}, u_s) = J(\alpha_0)$ almost surely.*
6. *$\lim_{t\to\infty} \sum_{s=0}^{t-1} \ell(\hat{\alpha}_t = \alpha^*) = 1$ for some α^* almost surely.*
7. *$p(i, j, \Phi(\alpha^*, i), \alpha^*) = p(i, j, \Phi(\alpha^*, i), \alpha_0)$ almost surely.*
8. *Φ^*_α is optimal for α_0 almost surely.*

Generalizations of this result are contained in the references given in [3] and [11].

65.3 Continuous Time Linear Systems

Another important and commonly used class of systems is the class of continuous time linear systems. The models are assumed to evolve in continuous time rather than discrete time because this assumption is natural for many models and it is important for the study of discrete time models when the sampling rates are large and for the analysis of numerical round-off errors. The stochastic systems are described by linear stochastic differential equations. It is assumed that there are observations of the complete state.

The general approach to adaptive control that is described here exhibits a splitting or separation of the problems of identification of the unknown parameters and adaptive control. Maximum likelihood (or equivalently least squares) estimates are used for the identification of the unknown constant parameters. These estimates are given recursively and are shown to be strongly consistent. The adaptive control is usually constructed by the so-called certainty equivalence principle, that is, the optimal stationary controls are computed by replacing the unknown true parameter values by the current estimates of these values. Since the optimal stationary controls can be shown to be continuous functions of the unknown parameters, the self-tuning property is verified. It is shown that the family of average costs using the control from the certainty equivalence principle converges to the optimal average cost. This verifies the self-optimizing property.

A model for the adaptive control of continuous time linear stochastic systems with complete observations of the state can be described by the following stochastic differential equation

$$dX(t) = (A(\alpha)X(t) + BU(t))dt + dW(t) \tag{65.2}$$

where $X(t) \in \mathbb{R}^n$, $U(t) \in \mathbb{R}^m$.

$$A(\alpha) = A_0 + \sum_{i=1}^{p} \alpha^i A_i \tag{65.3}$$

$A_i \in \mathcal{L}(\mathbb{R}^n)$ $i = 0, \ldots, p$, $B \in \mathcal{L}(\mathbb{R}^m, \mathbb{R}^n)$, $(W(t), t \in \mathbb{R}_+)$ is a standard $\mathbb{R}^n$-valued Wiener process and $X_0 \equiv a \in \mathbb{R}^n$. It is assumed that

(3.A1) $\mathcal{A} \subset \mathbb{R}^p$ is compact and $\alpha \in \mathcal{A}$.

(3.A2) $(A(\alpha), B)$ is reachable for each $\alpha \in \mathcal{A}$.

(3.A3) The family $(A_i,\ i = 1, \ldots, p)$ is linearly independent.

Let $(\mathcal{F}_t, t \in \mathbb{R}_+)$ be a filtration such that X_t is measurable with respect to $\mathcal{F}_t$ for all $t \in \mathbb{R}_+$ and $(W(t), \mathcal{F}_t, t \in \mathbb{R}_+)$ is a Brownian martingale. The ergodic, quadratic control problem for Equation 65.2 is to minimize the ergodic cost functional

$$\limsup_{t\to\infty} \frac{1}{t} J(X_0, U, \alpha, t) \tag{65.4}$$

where

$$J(X_0, U, \alpha, t) = \int_0^t [\langle QX(s), X(s)\rangle + \langle PU(s), U(s)\rangle]ds \tag{65.5}$$

and $t \in (0, \infty]$, $X(0) = X_0$, $Q \in \mathcal{L}(\mathbb{R}^n)$ and $P \in \mathcal{L}(\mathbb{R}^m)$ are self-adjoint and P^{-1} exists, $(X(t), t \in \mathbb{R}_+)$ satisfies Equation 65.2 and $(U(t), t \in \mathbb{R}_+)$ is adapted to $(\mathcal{F}_t, t \in \mathbb{R}_+)$. It is well known [8] that if α is known then there is an optimal linear feedback control such that

$$U^*(t) = KX(t) \tag{65.6}$$

where $K = -P^{-1}B^*V$ and V is the unique, symmetric, nonnegative definite solution of the algebraic Riccati equation

$$VA + A^*V - VB^*P^{-1}BV + Q = 0. \tag{65.7}$$

For an unknown α the admissible adaptive control policies $(U(t), t \in \mathbb{R}_+)$ are linear feedback controls

$$U(t) = K(t)X(t) = \tilde{K}(t, X(u), u \le t - \Delta)X(t) \tag{65.8}$$

where $(K(t), t \ge 0)$ is an $\mathcal{L}(\mathbb{R}^n, \mathbb{R}^m)$-valued process that is uniformly bounded and there is a fixed $\Delta > 0$ such that $(K(t), t \ge 0)$ is measurable with respect to $\sigma(X_u, u \le t - \Delta)$ for each $t \ge \Delta$ and $(K(t), t \in [0, \Delta))$ is a deterministic function. For such an adaptive control, it is elementary to verify that there is a unique strong solution of Equation 65.2. The delay $\Delta > 0$ accounts for some time that is required to compute the adaptive control law from the observation of the solution of Equation 65.2.

Let $(U(t), t \ge 0)$ be an admissible adaptive control and let $(X(t), t \ge 0)$ be the associated solution of Equation 65.2. Let $\mathcal{A}(t) = (a_{ij}(t))$ and $\tilde{\mathcal{A}}(t) = (\tilde{a}_{ij}(t))$ be $\mathcal{L}(\mathbb{R}^p)$-valued processes such that

$$\begin{aligned} a_{ij}(t) &= \int_0^t \langle A_i X(s), A_j X(s)\rangle ds \\ \tilde{a}_{ij}(t) &= \frac{a_{ij}(t)}{a_{ii}(t)}. \end{aligned}$$

To verify the strong consistency of a family of least squares estimates it is assumed that

(3.A4) $\liminf_{t\to\infty} |\det \tilde{\mathcal{A}}(t)| > 0$ a.s.

The estimate of the unknown parameter vector at time t, $\hat{\alpha}(t)$, for $t > 0$ is the minimizer for the quadratic functional of α, $L(t, \alpha)$, given by

$$\begin{aligned} L(t, \alpha) = &- \int_0^t \langle (A(\alpha) + BK(s))X(s), dX(s)\rangle \\ &+ \frac{1}{2}\int_0^t |(A(\alpha) + BK(s))X(s)|^2 ds \end{aligned} \tag{65.9}$$

where $U(s) = K(s)X(s)$ is an admissible adaptive control. The following result [5] gives the strong consistency of these least squares estimators.

THEOREM 65.5 *Let $(K(t), t \ge 0)$ be an admissible adaptive feedback control law. If (3.A1–3.A4) are satisfied and $\alpha_0 \in \mathcal{A}^\circ$, the interior of $\mathcal{A}$, then the family of least squares estimates $(\hat{\alpha}(t), t > 0)$ where $\hat{\alpha}(t)$ is the minimizer of Equation 65.9, is strongly consistent, that is,*

$$P_{\alpha_0}\left(\lim_{t\to\infty} \hat{\alpha}(t) = \alpha_0\right) = 1 \tag{65.10}$$

where α_0 is the true parameter vector.

The family of estimates $(\hat{\alpha}(t), t > 0)$ can be computed recursively because this process satisfies the following equation

$$d\hat{\alpha}(t) = \mathcal{A}^{-1}(t)\langle \mathbb{A}(t)X(t), dX(t) - A(\hat{\alpha}(t))X(t)dt - BU(t)dt\rangle \tag{65.11}$$

where $\langle \mathbb{A}(t)x, y\rangle = (\langle A_i x, y\rangle)$ $i = 1, \ldots, p$.

Now the performance of some admissible adaptive controls is described.

PROPOSITION 65.1 Assume that (3.A1–3.A4) are satisfied and that

$$\lim_{t\to\infty} \tfrac{1}{t}\langle VX(t), X(t)\rangle = 0 \quad \text{a.s.} \tag{65.12}$$

$$\limsup_{t\to\infty} \tfrac{1}{t}\int_0^t |X(t)|^2 ds < \infty \quad \text{a.s.} \tag{65.13}$$

where $(X(t), t \ge 0)$ is the solution of Equation 65.2 with the admissible adaptive control $(U(t), t \ge 0)$ and $\alpha = \alpha_0 \in \mathcal{K}$

and V is the solution of the algebraic Riccati equation 65.7 with $\alpha = \alpha_0$. Then

$$\liminf_{T\to\infty} \frac{1}{T} J(X_0, U, \alpha_0, T) \geq \text{tr } V \quad \text{a.s.} \tag{65.14}$$

If U is an admissible adaptive control $U(t) = K(t)X(t)$ such that

$$\lim_{t\to\infty} K(t) = k_0 \qquad \text{a.s.} \tag{65.15}$$

where $k_0 = -P^{-1}B^*V$ then

$$\lim_{T\to\infty} \frac{1}{T} J(X_0, U, \alpha_0, T) = \text{tr } V \quad \text{a.s.} \tag{65.16}$$

COROLLARY 65.1 Under the assumptions of Proposition 65.1, if Equation 65.15 is satisfied, then Equations 65.13 and 65.13 are satisfied.

The previous results can be combined for a complete solution to the stochastic adaptive control problem (Equations 65.2 and 65.4) [5].

THEOREM 65.6 *Assume that 3.A1–3.A4 are satisfied. Let $(\hat{\alpha}(t), t > 0)$ be the family of least squares estimates where $\hat{\alpha}(t)$ is the minimizer of Equation 65.9. Let $(K(t), t \geq 0)$ be an admissible adaptive control law such that*

$$K(t) = -P^{-1}B^*V(\hat{\alpha}(t-\Delta))$$

where $V(\alpha)$ is the solution of Equation 65.7 for $\alpha \in \mathcal{A}$. Then the family of estimates $(\hat{\alpha}(t), t > 0)$ is strongly consistent,

$$\lim_{t\to\infty} K(t) = k_0 \quad a.s. \tag{65.17}$$

where $k_0 = -P^{-1}B^*V(\alpha_0)$ and

$$\lim_{T\to\infty} \frac{1}{T} J(X_0, U, \alpha_0, T) = tr\ V \quad a.s. \tag{65.18}$$

Now another model for the adaptive control of an unknown linear stochastic system is given. Let $(X(t), t \geq 0)$ be a controlled diffusion that is a solution of the stochastic differential equation

$$\begin{aligned} dX(t) &= AX(t)dt + BU(t)dt + CW(t) \\ X(0) &= X_0 \end{aligned} \tag{65.19}$$

where $X(t) \in \mathbb{R}^n$, $U(t) \in \mathbb{R}^m$ and $(W(t), t \geq 0)$ is a standard p-dimensional Wiener process. The probability space is $(\Omega, \mathcal{F}, P)$ and $(\mathcal{F}_t, t \geq 0)$ is an increasing family of sub-σ-algebras of $\mathcal{F}$ such that $\mathcal{F}_0$ contains all P-null sets, $(W(t), \mathcal{F}_t, t \geq 0)$ is a continuous martingale and $X(t) \in \mathcal{F}_t$ for all $t \geq 0$. The linear transformations A, B, C are assumed to be unknown. Since the adaptive control does not depend on C it suffices to estimate the pair (A, B). For notational simplicity let $\theta^T = [A, B]$.

For the adaptive control problem, it is required to minimize the ergodic cost functional

$$\begin{aligned} \limsup_{t\to\infty} J(t, U) &= \limsup_{t\to\infty} \frac{1}{t}\int_0^t (X^T(s)Q_1X(s) \\ &\quad + U^T(s)Q_2U(s))ds \end{aligned} \tag{65.20}$$

where $Q_1 \geq 0$ and $Q_2 > 0$ and U is an admissible control.

Since both A and B are unknown, it is necessary to ensure sufficient excitation in the control to obtain consistency of a family of estimates. This is accomplished by a diminishing excitation control (dither) that is asymptotically negligible for the ergodic cost functional Equation 65.20. Let $(v_n, n \in \mathbb{N})$ be a sequence of $\mathbb{R}^m$-valued independent, identically distributed random variables that is independent of the Wiener process $(W(t), t \geq 0)$. It is assumed that $E[v_n] = 0$, $E\left[v_n v_n^T\right] = I$ for all $n \in \mathbb{N}$ and there is a $\sigma > 0$ such that $\|v_n\|^2 \leq \sigma$ a.s. for all $n \in \mathbb{N}$. Let $\varepsilon \in \left(0, \frac{1}{2}\right)$ and fix it. Define the $\mathbb{R}^m$-valued process $(V(t), t \geq 0)$ as

$$V(t) = \sum_{n=0}^{\left[\frac{t}{\Delta}\right]} \frac{v_n}{n^{\varepsilon/2}} 1_{[n\Delta,(n+1)\Delta)}(t). \tag{65.21}$$

A family of least squares estimates $(\theta(t), t \geq 0)$ is used to estimate the unknown $\theta = [A, B]^T$. The estimate $\theta(t)$ is given by

$$\theta(t) = \Gamma(t)\textstyle\int_0^t \varphi(s)dX^T(s) + \Gamma(t)\Gamma^{-1}(0)\theta(0) \tag{65.22}$$

$$\Gamma(t) = \left(\textstyle\int_0^t \varphi(s)\varphi^T(s)ds + aI\right)^{-1} \tag{65.23}$$

$$\varphi(s) = [X^T(s) \quad U^T(s)]^T \tag{65.24}$$

where $\theta(0)$ and $a > 0$ are arbitrary. The diminishingly excited control is

$$U(t) = U^{\mathrm{d}}(t) + V(t) \tag{65.25}$$

where U^{d} is a "desired" control.

For A stable, (A, C) controllable and some measurability and asymptotic boundedness of the control, the family of estimates $(\theta(t), t \geq 0)$ is strongly consistent [2].

THEOREM 65.7 *Let $\varepsilon \in \left(0, \frac{1}{2}\right)$ be given in Equation 65.21. For Equation 65.19, if A is stable, (A, C) is controllable and the control $(U(t), t \geq 0)$ is given by Equation 65.25 where $U^d(t) \in \mathcal{F}_{(t-\Delta)\vee 0}$ for $t \geq 0$ and $\Delta > 0$ is fixed and*

$$\int_0^t \|U^d(s)\|^2 ds = O(t^{1+\delta}) \quad a.s. \tag{65.26}$$

as $t \to \infty$ for some $\delta \in [0, 1-2\varepsilon)$ then

$$\|\theta - \theta(t)\|^2 = O\left(\frac{\log t}{t^\alpha}\right) \quad a.s. \tag{65.27}$$

as $t \to \infty$ for each $\alpha \in \left(\frac{1+\delta}{2}, 1-\varepsilon\right)$ where $\theta = [A \quad B]^T$ and $\theta(t)$ satisfies Equation 65.22.

Now a self-optimizing adaptive control is constructed for the unknown linear stochastic system (65.19) with the quadratic ergodic cost functional (65.20). The adaptive control switches between a certainty equivalence control and the zero control. The family of admissible controls $\mathcal{U}(\Delta)$ is defined as follows

$$\mathcal{U}(\Delta) = \Big\{U : U(t) = U^{\mathrm{d}}(t) + U^1(t),$$

$$U^{\mathrm{d}}(t) \in \mathcal{F}_{(t-\Delta)\vee 0}$$
and
$$U^1(t) \in \sigma(V(s), (t-\Delta) \vee 0 \le s \le t)$$
for all $t \ge 0$
$$\|X(t)\|^2 = \mathrm{o}(t)$$
a.s. and $\int_0^t (\|U(s)\|^2 + \|X(s)\|^2)\mathrm{d}s$
$$= \mathrm{O}(t) \text{ a.s. as } t \to \infty\Big\}. \tag{65.28}$$

Define the $\mathbb{R}^m$-valued process $(U^0(t), t \ge \Delta)$ by the equation

$$\begin{aligned} U^0(t) &= -Q_2^{-1}B^T(t-\Delta)P(t-\Delta) \\ &\quad \left(e^{\Delta A(t)}X(t-\Delta)\right. \qquad (65.29) \\ &\quad \left. + \int_{t-\Delta}^t e^{(t-s)A(t-\Delta)}B(t-\Delta)U^{\mathrm{d}}(s)\mathrm{d}s\right) \end{aligned}$$

where $A(t)$ and $B(t)$ are the least squares estimates of A and B given by Equation 65.22 and $P(t)$ is the minimal solution of the algebraic Riccati equation

$$A^T(t)P(t) + P(t)A(t) - P(t)B(t)Q_2^{-1}B^T(t)P(t) + Q_1 = 0 \tag{65.30}$$

if $A(t)$ is stable and otherwise $P(t) = 0$.

To define the switching in the adaptive control, the following two sequences of stopping times $(\sigma_n, n = 1, 2, \ldots)$ and $(\tau_n, n = 1, 2, \ldots)$ are given as follows:

$$\begin{aligned} \sigma_0 &= 0 \\ \sigma_n &= \sup\left\{t \ge \tau_n : \int_0^t \|U^0(r)\|^2 \mathrm{d}r \le s\tau_n^\delta,\right. \\ &\quad A(s-\Delta) \text{ is stable for all} s \in [\tau_n, t)\} \qquad (65.31) \\ \tau_n &= \inf\left\{t \ge \tau_{n-1} + 1 : \int_0^t \|U^0(r)\|^2 \mathrm{d}r \le \frac{1}{2}t^{1+\delta},\right. \\ &\quad A(t-\Delta) \text{ is stable and} \\ &\quad \left.\|x(t-\Delta)\|^2 \le t^{1+\delta/2}\right\}. \qquad (65.32) \end{aligned}$$

The adaptive control $(U^*(t), t \ge 0)$ is given by

$$U^*(t) = U^{\mathrm{d}}(t) + V(t) \tag{65.33}$$

where

$$U^{\mathrm{d}}(t) = \begin{cases} 0 & \text{if } t \in [\sigma_n, \sigma_{n+1}) \quad \text{for some } n \ge 0 \\ U^0(t) & \text{if } t \in [\tau_n, \sigma_n) \quad \text{for some } n \ge 1. \end{cases} \tag{65.34}$$

The adaptive control U^* is self-optimizing [2].

THEOREM 65.8 *If A is stable and (A, C) is controllable, then the adaptive control $(U^*(t), t \ge 0)$ given by Equation 65.23 belongs to $\mathcal{U}(\Delta)$ and is self-optimizing for Equations 65.19 and 65.20, that is,*

$$\inf_{U \in \mathcal{U}(\Delta)} \limsup_{t\to\infty} J(t, U) = \lim_{t\to\infty} J(t, U^*) = tr(C^T PC) + tr(B^T PR(\Delta)PBQ_2^{-1}) \text{ a.s.} \tag{65.35}$$

where P is the minimal solution of the algebraic Riccati equation 65.30 using A and B and

$$R(\Delta) = \int_0^\Delta e^{tA}CC^T e^{tA^T} dt.$$

65.4 Continuous Time Nonlinear Systems

The stochastic adaptive control of nonlinear systems has become one of the most important areas of activity in control theory and is significantly more difficult than for linear systems. Some reasons for these difficulties are that the processes are typically not Gaussian so it is difficult to establish the existence and uniqueness of invariant measures and typically the optimal control (if it exists) is not known explicitly. To overcome these difficulties, two classes of problems are considered. The first is a financial model that describes a portfolio allocation where transaction costs are included and the second is a fairly general controlled diffusion where it is only required to find an almost self-optimizing adaptive control, that is, a control that is less than a given $\varepsilon > 0$ from the optimal ergodic cost for the known system.

MODEL I. A well-known model for optimal portfolio selection allows the investor to choose to invest in two assets: a bond B with a fixed rate of growth r and a stock S whose growth is governed by a Brownian motion with drift μ and variance σ^2. The investor controls his assets by transferring money between the stock and the bond. While it is possible to buy stocks without any cost, there is a brokerage or transaction fee that must be paid if stocks are sold such that if an amount of stock with value x is sold, then the bond only increases by an amount λx where $\lambda \in (0, 1)$ is fixed. For such a formulation it is usually not reasonable to assume that the average rate of return of the stock, μ, is known. Let $U(t)$ and $Z(t)$ denote the total amount of money transferred from S to B and B to S respectively at time t $(U(0) = Z(0) = 0)$. The processes S and B can be described by the following stochastic differential equations

$$\begin{aligned} \mathrm{d}B(t) &= rB(t)\mathrm{d}t + \lambda \mathrm{d}U(t) - \mathrm{d}Z(t) \qquad (65.36) \\ \mathrm{d}S(t) &= S(t)\left(\mu + \frac{\sigma^2}{2}\right)\mathrm{d}t + \sigma \mathrm{d}S(t)\mathrm{d}W(t) \\ &\quad + \mathrm{d}Z(t) - \mathrm{d}U(t) \qquad (65.37) \end{aligned}$$

where $(W(t), t \ge 0)$ is a real-valued standard Brownian motion. Let

$$Y(t) = S(t) + B(t) \tag{65.38}$$

be the total wealth of the investor at time t. The control problem is to find a pair of optimal controls (U^*, Z^*) such that the expected rate of growth

$$J_Y(U, Z) = E\left[\liminf_{t\to\infty}\left[\ln \frac{Y(t)}{t}\right]\right] \tag{65.39}$$

is maximized. It is shown in [14] that an optimal policy exists for the known system and for this policy the limit inferior in

Equation 65.39 can be replaced by the limit and

$$E\left[\liminf_{t\to\infty} \frac{1}{t} \ln Y(t)\right] = \lim_{t\to\infty} \frac{1}{t} \ln Y(t) \quad \text{a.s.} \tag{65.40}$$

This control problem is solved by considering the process

$$X(t) = \frac{S(t)}{B(t)} \tag{65.41}$$

and two new control processes

$$L(t) = \int_0^t S^{-1}(s)\mathrm{d}U(s) \tag{65.42}$$

and

$$R(t) = \int_0^t B^{-1}(s)\mathrm{d}Z(s) \tag{65.43}$$

which represent the cumulative percentages of stocks and bonds withdrawn until time t. By the Itô formula the process $(X(t), t \geq 0)$ satisfies the following stochastic differential equation

$$\begin{aligned} \mathrm{d}X(t) &= \sigma X(t)\mathrm{d}W(t) + aX(t)\mathrm{d}t \\ &\quad - j(X(t))\mathrm{d}L(t) + k(X(t))\mathrm{d}R(t) \end{aligned} \tag{65.44}$$

where $j(x) = x + \lambda x^2$, $k(x) = 1 + x$ and

$$a = \left(\mu + \frac{\sigma^2}{2}\right) - r. \tag{65.45}$$

The maximization of Equation 65.40 is equivalent to the minimization of

$$J(R, L) = \limsup_{t\to\infty} \frac{1}{t} E\left[\int_0^t h(X(s))\mathrm{d}s + \int_0^t g(X(t))\mathrm{d}L(s)\right] \tag{65.46}$$

where $h(x) = \frac{\sigma^2}{2}\frac{x^2}{(x+1)^2} - \frac{\alpha x}{x+1}$ and $g(x) = (1-\lambda)x(x+1)$. It is shown in [14] that the optimal policy for this control problem is a double-bound control policy that keeps the process $(X(t), t > 0)$ in the interval $[A,\ B]$ where A and B are determined by the equations

$$B^{-\alpha}\int_A^B 2\sigma^{-2}[h(A) - h(y)]y^{\alpha-2}\mathrm{d}y = \frac{g(B)}{j(B)} \tag{65.47}$$

and

$$B = \lambda^{-1}\frac{(\alpha-1)A + \alpha}{(2-\alpha)A + 1 - \alpha} \tag{65.48}$$

and $\alpha = \frac{2a}{\sigma^2}$. The optimal controls R^* and L^* are nonzero only on the boundary of the interval (local times) and are given explicitly as

$$R^* = \lim_{\varepsilon\to 0} R_\varepsilon \tag{65.49}$$

$$L^* = \lim_{\varepsilon\to 0} L_\varepsilon \tag{65.50}$$

where the limits exist in the sense of weak$*$ convergence of measures and

$$\mathrm{d}R_\varepsilon = \frac{X(s)}{k(X(s))}1_{(-\infty,\ A+\varepsilon]}(X(s))\frac{\sigma^2(B-A)}{2\varepsilon}\mathrm{d}s \tag{65.51}$$

$$\mathrm{d}L_\varepsilon = \frac{X(s)}{j(X(s))}1_{[B-\varepsilon,\ \infty)}(X(s))\frac{\sigma^2(B-A)}{2\varepsilon}\mathrm{d}s. \tag{65.52}$$

For this optimal control pair, the limit superior in Equation 65.46 is actually a limit and the optimal cost is

$$J(R^*, L^*) = h(A). \tag{65.53}$$

Since it is assumed that the average rate of return of the stock is unknown, then a in Equation 65.45 is unknown. A family of estimates $(\hat{a}(t), t \geq 0)$ based on the observations is defined and an adaptive control is defined based on the certainty equivalence principle by using a family of adaptive double bounds as follows:

$$\hat{a}(t) = \frac{1}{t}\int_0^t \frac{\mathrm{d}X(s) + j(X(s))\mathrm{d}L(s) - k(X(s))\mathrm{d}R(s)}{X(s)} \tag{65.54}$$

and

$$\hat{A}(t) = \begin{cases} \sum_{k=1}^{\infty} 1_{(k,\ k+1]}(t)A(\hat{a}(k)) & t \geq 1 \\ A_0 & 0 \leq t \leq 1 \end{cases} \tag{65.55}$$

$$\hat{B}(t) = \begin{cases} \sum_{k=0}^{\infty} 1_{(k,\ k+1]}(t)B(\hat{a}(k)) & t > 1 \\ B_0 & 0 \leq t \leq 1 \end{cases} \tag{65.56}$$

where $A(\hat{a}(k))$ and $B(\hat{a}(k))$ are the solutions of Equations 65.47 and 65.48 where $\alpha = \frac{2\hat{a}(k)}{\sigma^2}$ and A_0 and B_0 are arbitrary constants such that $0 < A_0 < B_0$. This family of estimates and the adaptive control solve the stochastic adaptive control problem [4].

THEOREM 65.9 *The family of estimates $(\hat{a}(t), t \geq 0)$ given by Equation 65.54 is strongly consistent and the adaptive control that uses the adaptive bounds given in Equations 65.55 and 65.56 is self-optimizing, that is, it achieves the optimal cost (65.53).*

MODEL II. The second unknown nonlinear system that is considered is a controlled diffusion described by a stochastic differential equation. Since the stochastic differential equation is fairly general, it is not known that a suitable optimal control exists. Furthermore, since the family of controls for the ergodic control problem is suitably measurable functions, only weak solutions of the stochastic differential equation can be defined. For the adaptive control problem only almost optimal adaptive controls are required. A family of estimates is given but no claim of consistency is made.

Let $(X(t; \alpha, u), t \geq 0)$ be a controlled diffusion process that satisfies the following stochastic differential equation

$$\begin{aligned} \mathrm{d}X(t;\alpha,u) &= f(X(t;\alpha,u))\mathrm{d}t \\ &\quad + h(X(t;\alpha,u),\alpha,u)\mathrm{d}t \\ &\quad + \sigma(X(t;\alpha,u))\mathrm{d}W(t) \\ X(0;\alpha,u) &= x \end{aligned} \tag{65.57}$$

where $X(t;\alpha,u) \in \mathbb{R}^n$, $(W(t), t \geq 0)$ is a standard $\mathbb{R}^n$-valued Wiener process, $u(t) \in U \subset \mathbb{R}^m$, U is a compact set, $\alpha \in \mathcal{A} \subset \mathbb{R}^q$ and $\mathcal{A}$ is a compact set. The functions f and σ satisfy a global Lipschitz condition $\sigma(x)\sigma^*(x) \geq cI > 0$ for all $x \in \mathbb{R}^n$ and h is a bounded, Borel function on $\mathbb{R}^n \times \mathcal{A} \times U$. The family $\mathcal{U}$ of admissible controls is

$$\mathcal{U} = \{u : u : \mathbb{R}^n \to U \text{is Borel measurable}\}.$$

The probability space for the controlled diffusion is denoted $(\Omega, \mathcal{F}, P)$. The solution of the stochastic differential equation is a weak solution that can be obtained by absolutely continuous transformation of the measure of the solution of

$$dY(t) = f(Y(t))dt + \sigma(Y(t))dW(t) \qquad (65.58)$$
$$Y(0) = x$$

which has one and only one strong solution by the Lipschitz continuity of f and σ.

For a Borel set A, let T_A be the first hitting time of A. Let Γ_1 and Γ_2 be two spheres in $\mathbb{R}^n$ with centers at 0 and radii $0 < r_1 < r_2$, respectively. Let τ be the random time defined by the equation

$$\tau = T_{\Gamma_1} + T_{\Gamma_2} \circ \theta_{T_{\Gamma_2}} \qquad (65.59)$$

where $\theta_{T_{\Gamma_2}}$ is the positive time shift by T_{Γ_2} that acts on $C(\mathbb{R}_+, \mathbb{R}^n)$. The random variable τ is the first time that the process $(X(t), t \geq 0)$ hits Γ_1 after hitting Γ_2. The following assumptions are used subsequently

(4.A1)

$$\sup_{\alpha\in\mathcal{A}} \sup_{u\in\mathcal{U}} \sup_{x\in\Gamma_1} E_x^{\alpha,u}[\tau^2] < \infty.$$

(4.A2) There is an L_0 such that for all $\alpha, \beta \in \mathcal{A}$

$$\sup_{x\in\mathbb{R}^n} \sup_{v\in\mathcal{U}} |h(x,\alpha,v) - h(x,\beta,v)| \leq L_0|\alpha - \beta|.$$

(4.A3) For each $(x, \alpha, u) \in \mathbb{R}^n \times \mathcal{A} \times \mathcal{U}$, $E_x^{\alpha,u}[T_{\Gamma_1}] < \infty$.

A family of measures $(m_x(\cdot; \alpha, u); x \in \mathbb{R}^n, \alpha \in \mathcal{A}, u \in \mathcal{U})$ on the Borel σ-algebra of $\mathbb{R}^n$, $\mathcal{B}(\mathbb{R}^n)$, is defined by the equation

$$m_x(D; \alpha, u) = E_x^{\alpha,u}\left[\int_0^\tau 1_D(X(s))ds\right] \qquad (65.60)$$

where 1_D is the indicator function of $D \in \mathcal{B}(\mathbb{R}^n)$ and τ is given by Equation 65.59 assuming 4.A1. The measure $m_x(\cdot; \alpha, u)$ is well defined for each $(\alpha, u) \in \mathcal{A} \times \mathcal{U}$.

If 4.A1 is satisfied, then there is an invariant measure, $\mu(\cdot; \alpha, u)$ on $\mathcal{B}(\mathbb{R}^n)$ for the process $(X(t; \alpha, u), t \geq 0)$ that is given by the equation

$$\mu(D; \alpha, u) = \int_{\Gamma_1} m_x(D; \alpha, u)\eta(dx; \alpha, u) \left(\int_{\Gamma_1} E_x^{\alpha,u}(\tau)\eta(dx; \alpha, u)\right)^{-1} \qquad (65.61)$$

where $D \in \mathcal{B}(\mathbb{R}^n)$ and $\eta(\cdot; \alpha, u)$ is an invariant measure for the embedded Markov chain $(x_0 \in \Gamma_1, X(\tau_n; \alpha, u), n \in \mathbb{N})$ where

$$\tau_{n+1} = \tau_n + \tau \circ \theta_{\tau_n} \qquad (65.62)$$

where $n \geq 1$ and $\tau_0 = \tau$.

It is important to verify a continuity property of the invariant measures as a function on $\mathcal{A}$. Here these measures are shown to be uniformly equicontinuous on $\mathcal{A}$ [7].

THEOREM 65.10 *If 4.A1 and 4.A2 are satisfied, then for each $\varepsilon > 0$ there is a $\delta > 0$ such that if $\alpha, \beta \in \mathcal{A}$ and $|\alpha - \beta| < \delta$*

$$\sup_{u\in\mathcal{U}} \|\mu(\cdot; \alpha, u) - \mu(\cdot; \beta, u)\| < \varepsilon$$

where $\|\cdot\|$ is the variation norm and μ is given by Equation 65.61.

The control problem for Equation 65.57 is to find an admissible control that minimizes the ergodic cost functional

$$J(u; x, \alpha) = \limsup_{t\to\infty} \frac{1}{t} E_x^{\alpha,u}\left[\int_0^t k(X(s), u(X(s)))ds\right] \qquad (65.63)$$

where $k : \mathbb{R}^n \times U \to \mathbb{R}$ is a fixed, bounded Borel function. An adaptive control is constructed that is almost optimal with respect to this cost functional (65.63).

An identification procedure is defined by a biased maximum likelihood method [11] where the estimates are changed at random times. The parameter set $\mathcal{A}$ is covered by a finite, disjoint family of sets. The adaptive control is obtained by choosing from a finite family of controls each of which is an almost optimal control for a distinguished point in one of the sets of the finite cover of $\mathcal{A}$.

Fix $\varepsilon > 0$ and choose $\delta(\varepsilon) > 0$ so that a control that is $\varepsilon/2$ optimal for $\alpha \in \mathcal{A}$ is ε optimal for $\beta \in \mathcal{A}$ where $|\alpha - \beta| < \delta(\varepsilon)$. It can be verified that this is possible. By the compactness of $\mathcal{A}$ there is a finite cover of $\mathcal{A}$, $(B(\alpha_i, \delta), i = 1, 2, \ldots, r)$ where $\alpha_i \notin B(\alpha_j, \delta)$ for $i \neq j$ and $B(\alpha, \delta)$ is the open ball with the center α and the radius $\delta > 0$. Define $(A_i(\varepsilon), i = 1, \ldots, r)$ by the equations

$$A_i(\varepsilon) = \left(B(\alpha_i, \delta) \setminus \bigcup_{j=1}^{i-1} A_i(\varepsilon)\right) \cap \mathcal{A} \qquad (65.64)$$

where $i = 2, \ldots, r$ and $A_1(\varepsilon) = B(\alpha_1, \delta) \cap \mathcal{A}$. Let $e : \mathcal{A} \to \{\alpha_1, \ldots, \alpha_r\}$ be defined by

$$e(\alpha) = \alpha_j \qquad \text{if } \alpha \in A_j(\varepsilon) \qquad (65.65)$$

and let $\lambda : \mathcal{A} \to \mathbb{R}$ be defined by

$$\lambda(\alpha) = J^*(e(\alpha)) \qquad (65.66)$$

where

$$J^*(\alpha) = \inf_{u\in\mathcal{U}} J(u; x, \alpha). \qquad (65.67)$$

It can be assumed that λ is lower semicountinuous by modifying $(A_i(\varepsilon), i = 1, \ldots, r)$ if necessary on their boundaries.

Given $\varepsilon > 0$ choose $N \in \mathbb{N}$ such that

$$2 \sup_{\alpha\in\mathcal{A}} \sup_{u\in\mathcal{U}} \|w(\cdot; \alpha, u)\| \leq \frac{\varepsilon}{4} mN \qquad (65.68)$$

where

$$w(x; \alpha, u) = \lim_{n\to\infty} E_x^{\alpha,u}\left[\int_0^{\tau_n} k(X(s), u(X(s)))ds - \int_{\mathbb{R}^n} k(z, u(z))\mu(dz; \alpha, u)ds\right]$$

and $m > 0$ satisfies

$$m \leq \inf_{x \in \Gamma_1} \inf_{u \in \mathcal{U}} \inf_{\alpha \in \mathcal{A}} E_x^{\alpha,u}(\tau).$$

Define a sequence of stopping times $(\sigma_n, n \in \mathbb{N})$ as

$$\sigma_{n+1} = \sigma_n + \tau_N \circ \theta_{\sigma_n} \qquad (65.69)$$

where $n \geq 1$, $\sigma_1 = \tau_N$ and N is given in Equation 65.67.

The unknown parameter α_0 is estimated at the random times $(\sigma_n, n \in \mathbb{N})$ by a biased maximum likelihood method [11], that is, $\hat{\alpha}(\sigma_n)$ is a maximizer of

$$L_n(\sigma) = \ln M(\sigma_n; \alpha, \alpha_0, \eta) + z(\sigma_n) \ln \left(\frac{\lambda(\alpha_0)}{\lambda(\alpha)} \right) \qquad (65.70)$$

where

$$M(\sigma; \alpha, \alpha_0, \eta) = \frac{\mathrm{d}P^{\alpha}}{\mathrm{d}P^{\alpha_0}}$$

is the likelihood function evaluated at σ_n with the control η, $z : \mathbb{R} \to \mathbb{R}_+$ satisfies $\frac{z(t)}{t} \to 0$ and $\frac{z(t)}{t^\beta} \to \infty$ for some $\beta \in \left(\frac{1}{2}, 1\right)$ as $t \to \infty$.

The family of estimates $(\hat{\alpha}(t), t \geq 0)$ is defined as follows: choose $\bar{\alpha} \in \mathcal{A}$ and let

$$\hat{\alpha}(t) = \hat{\alpha}(0) = \bar{\alpha} \qquad \text{for } 0 \leq t < \sigma_1 \qquad (65.71)$$

and for $n \geq 1$ let

$$\begin{aligned} \hat{\alpha}(\sigma_n) &= \arg\max L_n(\alpha) \\ \hat{\alpha}(t) &= \hat{\alpha}(\sigma_n) \qquad \text{for } \sigma_n \leq t < \sigma_{n+1}. \end{aligned} \qquad (65.72)$$

Using the family of estimates $(\hat{\alpha}(t), t \geq 0)$ and an approximate certainty equivalence principle, the following adaptive control is defined

$$\eta(s; \varepsilon) = u_{e(\hat{\alpha}(s))}(X(s)) \qquad (65.73)$$

where u_{α_i} for $i = 1, \ldots, r$ is a fixed $\varepsilon/2$ optimal control corresponding to the value α_i and $e(\cdot)$ is given by Equation 65.65. This adaptive control achieves ε-optimality [7].

THEOREM 65.11 *If* (A1–A3) *are satisfied then for each* $\varepsilon > 0$

$$\limsup_{t \to \infty} \frac{1}{t} \int_0^t k(X(s), \eta(s; \varepsilon)) ds \leq J^*(\alpha_0) + 2\varepsilon \quad a.s.$$

where η *is given by Equation* 65.73, $J^*(\alpha_0)$ *is given by Equation* 65.67 *and* α_0 *is the true parameter value.*

The results can be extended to a partially observed discrete time Markov process. Let us consider the following scenario. The Markov process is completely observed in a fixed recurrent domain and partially observed in the complement of this domain. Then the adaptive control problem can be formulated and solved [6].

References

[1] Borkar, V. and Varaiya, P., Adaptive control of Markov chains. I. Finite parameter set, *IEEE Trans. Autom. Control,* AC-24, 953–958, 1979.

[2] Chen, H. F., Duncan, T. E., and Pasik-Duncan, B., Stochastic adaptive control for continuous time linear systems with quadratic cost, *J. Appl. Math. Optim,* to appear.

[3] Chen, H. F. and Lei Guo., *Identification and Stochastic Adaptive Control,* Birkhäuser Verlag, 1991.

[4] Duncan, T. E., Faul, M., Pasik-Duncan, B., and Zane, O., Computational methods for the stochastic adaptive control of an investment model with transaction fees, *Proc. 33rd IEEE Conf. Decision Control, Orlando,* 2813–2816, 1994.

[5] Duncan, T. E. and Pasik-Duncan, B., Adaptive control of continuous-time linear stochastic systems, *Math. Control, Sig., Syst.,* 3, 45–60, 1990.

[6] Duncan, T. E., Pasik-Duncan, B., and Stettner, L., Some aspects of the adaptive control of a partially observed discrete time Markov processes, *Proc. 32nd Conf. Dec. Control,* 3523–3526, 1993.

[7] Duncan, T. E., Pasik-Duncan, B., and Stettner, L., Almost self-optimizing strategies for the adaptive control of diffusion processes, *J. Optim. Th. Appl.,* 81, 479–507, 1994.

[8] Fleming, W., and Rishel, R., *Deterministic and Stochastic Optimal Control,* Springer-Verlag, 1975.

[9] Kumar, P. R., Adaptive control with a compact parameter set, *SIAM J. Control Optim.,* 20, 9–13, 1982.

[10] Kumar, P. R., A survey of some results in stochastic adaptive control, *SIAM J. Control Optim.,* 23, 329–380, 1985.

[11] Kumar, P. R. and Becker, A., A new family of optimal adaptive controllers for Markov chains, *IEEE Trans. Autom. Control,* 27, 137–146, 1982.

[12] Kumar, P. R. and Varaiya, P., *Stochastic Systems: Estimation, Identification and Adaptive Control,* Prentice Hall, 1986.

[13] Mandl, P., Estimation and control in Markov chains, *Adv. Appl. Prob,* 6, 40–60, 1974.

[14] Taksar, M., Klass, M. J., and Assaf, D., A diffusion model for portfolio selection in the presence of brokerage fees, *Math. Oper. Res,* 13, 277–294, 1988.

SECTION XIV

Control of Distributed Parameter Systems

66

Controllability of Thin Elastic Beams and Plates

J. E. Lagnese
Department of Mathematics, Georgetown University, Washington, DC

G. Leugering
Fakultät für Mathematik und Physik, University of Bayreuth, Postfach Bayreuth, Germany

66.1 Dynamic Elastic Beam Models

Consider the deformation of a thin, initially curved beam of length ℓ and constant cross-section of area A, which, in its undeformed reference configuration, occupies the region

$$\begin{aligned}\Omega \quad = \quad & \{\mathbf{r} =: \mathbf{r}_0(x_1) + x_2\mathbf{e}_2(x_1) + x_3\mathbf{e}_3(x_1) |\, x_1 \in [0, \ell], \\ & (x_2, x_3) := x_2\mathbf{e}_2(x_1) + x_3\mathbf{e}_3(x_1) \in A\}\end{aligned}$$

where $\mathbf{r}_0 : [0, \ell] \to \mathbb{R}^3$ is a smooth function representing the *centerline*, or the *reference line*, of the beam at rest. The orthonormal triads $\mathbf{e}_1(\cdot)$, $\mathbf{e}_2(\cdot)$, $\mathbf{e}_3(\cdot)$ are chosen as smooth functions of x_1 so that $\mathbf{e}_1$ is the direction of the tangent vector to the centerline, i.e., $\mathbf{e}_1(x_1) = (d\mathbf{r}_0/dx_1)(x_1)$, and $\mathbf{e}_2(x_1)$, $\mathbf{e}_3(x_1)$ span the orthogonal cross section at x_1. The meanings of the variables x_i are as follows: x_1 denotes arc length along the undeformed centerline, and x_2 and x_3 denote lengths along lines orthogonal to the reference line. The set Ω can then be viewed as obtained by translating the reference curve $\mathbf{r}_0(x_1)$ to the position $x_2\mathbf{e}_2 + x_3\mathbf{e}_3$ within the cross-section perpendicular to the tangent of $\mathbf{r}_0$.

At a given time t, let $\mathbf{R}(x_1, x_2, x_3, t)$ denote the position vector after deformation to the particle which is at $\mathbf{r}(x_1, x_2, x_3)$ in the reference configuration. We introduce the displacement vector by $\mathbf{V} := \mathbf{R} - \mathbf{r}$. The position vector $\mathbf{R}(x_1, 0, 0, t)$ to the deformed reference line $x_2 = x_3 = 0$ is denoted by $\mathbf{R}_0$. Accordingly, the displacement vector of a particle on the reference line is $\mathbf{W} := \mathbf{V}(x_1, 0, 0) = \mathbf{R}_0 - \mathbf{r}_0$. The position vector $\mathbf{R}$ may be approximated to first order by

$$\mathbf{R} = \mathbf{R}_0 + x_2\mathbf{E}_2 + x_3\mathbf{E}_3,$$

where $\mathbf{E}_i$ are the tangents at $\mathbf{R}$ with respect to x_i, respectively. Note, however, that the triad $\mathbf{E}_i$ *is not necessarily orthogonal*, due to shearing.

0-8493-8570-9/96/$0.00+$.50

The deformation of $\mathbf{r}(\cdot)$ into $\mathbf{R}(\cdot, t)$ will be considered as a succession of two motions: (1) a rotation carrying the triad $\mathbf{e}_i(x_1)$ to an intermediate orthonormal triad $\hat{\mathbf{e}}_i(x_1, t)$, followed by (2) a deformation carrying $\hat{\mathbf{e}}_i(x_1, t)$ into the nonorthogonal triad $\mathbf{E}_i(x_1, t)$. The two triads $\hat{\mathbf{e}}_i$ and $\mathbf{E}_i$ then differ on account of a strain $\bar{\varepsilon}$ to be specified below. We choose to orient the intermediate (right-handed) triad $\hat{\mathbf{e}}_i$, which serves as a moving orthonormal reference frame, so that

$$\mathbf{E}_1 = (\mathbf{E}_1 \cdot \hat{\mathbf{e}}_1)\hat{\mathbf{e}}_1 = |\mathbf{E}_1|\hat{\mathbf{e}}_1, \quad \hat{\mathbf{e}}_2 \cdot \mathbf{E}_3 = \hat{\mathbf{e}}_3 \cdot \mathbf{E}_2.$$

A strain $\bar{\varepsilon}$, related to the deformation carrying the triad $\hat{\mathbf{e}}_i$ to the triad $\mathbf{E}_i$, is defined by

$$\begin{aligned}
\hat{\mathbf{e}}_1 \cdot \mathbf{E}_1 &=: 1 + \bar{\varepsilon}_{11}, \quad \hat{\mathbf{e}}_1 \cdot \mathbf{E}_2 =: 2\bar{\varepsilon}_{12}, \\
\hat{\mathbf{e}}_2 \cdot \mathbf{E}_2 &=: 1 + \bar{\varepsilon}_{22}, \quad \hat{\mathbf{e}}_1 \cdot \mathbf{E}_3 =: 2\bar{\varepsilon}_{13}, \\
\hat{\mathbf{e}}_3 \cdot \mathbf{E}_3 &=: 1 + \bar{\varepsilon}_{33}, \quad \hat{\mathbf{e}}_2 \cdot \mathbf{E}_3 =: \bar{\varepsilon}_{23}.
\end{aligned}$$

The remaining strains are defined by requiring the symmetry $\bar{\varepsilon}_{ij} = \bar{\varepsilon}_{ji}$. If distortion of the planar cross sections is neglected, then $\bar{\varepsilon}_{22} \approx \bar{\varepsilon}_{33} \approx \bar{\varepsilon}_{23} \approx 0$. The normal $\mathbf{N}$ to the cross section is then $\mathbf{N} = \mathbf{E}_2 \times \mathbf{E}_3 = \hat{\mathbf{e}}_1 - 2\bar{\varepsilon}_{21}\hat{\mathbf{e}}_2 - 2\bar{\varepsilon}_{31}\hat{\mathbf{e}}_3$.

Let Θ_i denote the angles associated with the orthogonal transformation carrying the orthonormal basis $\mathbf{e}_i$ into $\hat{\mathbf{e}}_i$, whereas the rotation of $\mathbf{e}_i$ into $\mathbf{E}_i$ is represented by the angles ϑ_i (dextral mutual rotations). Up to quadratic approximations we obtain

$$\begin{aligned}
\vartheta_1 &\doteq \Theta_1, \\
\vartheta_2 &\doteq \Theta_2 + 2\bar{\varepsilon}_{31} + 2\Theta_1\bar{\varepsilon}_{21}, \\
\vartheta_3 &\doteq \Theta_3 - 2\bar{\varepsilon}_{21} + 2\Theta_1\bar{\varepsilon}_{31}.
\end{aligned}$$

These angles are interpreted as the *global rotations*. It is obvious from these relations that the shear strains vanish if, and only if, the angles Θ_i, ϑ_i coincide, $i = 1, 2, 3$, or, what is the same, if the normal $\mathbf{N}$ to the cross section coincides with $\mathbf{E}_1$. This is what is known as the *Euler-Bernoulli hypothesis*.

To complete the representation of the reference strains in terms of the angles above and the displacements $W_i := \mathbf{W} \cdot \mathbf{e}_i$, we compute up to quadratic approximations in all rotations and linear approximations in all strains $\bar{\varepsilon}_{ij}$. To this end we introduce curvatures and twist for the undeformed reference configuration by Frénet-type formulae

$$\kappa_2 = \mathbf{e}_2 \cdot \mathbf{e}_{1,1}, \quad \kappa_3 = \mathbf{e}_3 \cdot \mathbf{e}_{1,1}, \quad \tau = \mathbf{e}_3 \cdot \mathbf{e}_{2,1}.$$

(The index separated by a comma indicates a partial derivative with respect to the corresponding variable x_i.) The reference strains are then approximated by

$$\begin{aligned}
\bar{\varepsilon}_{11} &= W_{1,1} - \kappa_2 W_2 - \kappa_3 W_3 + \frac{1}{2}((W_{3,1} + \kappa_3 W_1 + \tau W_2)^2 \\
&\quad + (W_{2,1} + \kappa_2 W_1 - \tau W_3)^2), \\
\bar{\varepsilon}_{21} &= \frac{1}{2}(W_{2,1} - \vartheta_3 + \kappa_2 W_1 - \tau W_3 \\
&\quad + \vartheta_1(\vartheta_2 + W_{3,1} + \kappa_3 W_1 + \tau W_2)), \\
\bar{\varepsilon}_{31} &= \frac{1}{2}(W_{3,1} + \vartheta_2 + \kappa_3 W_1 + \tau W_2 \\
&\quad + \vartheta_1(\vartheta_3 - W_{2,1} - \kappa_2 W_1 + \tau W_3)),
\end{aligned}$$

whereas the approximate bending strains are

$$\begin{aligned}
\tilde{\kappa}_2 &= \vartheta_{3,1} + \kappa_3\vartheta_1 + \tau\vartheta_2, \\
\tilde{\kappa}_3 &= -\vartheta_{2,1} - \kappa_2\vartheta_1 + \tau\vartheta_3, \\
\tilde{\tau} &= \vartheta_{1,1} - \kappa_2\vartheta_2 - \kappa_3\vartheta_3.
\end{aligned}$$

The approximations given above comprise the theory of rods with *infinitesimal strains and moderate rotations*; see Wempner [27].

66.2 The Equations of Motion

Under the assumptions of the previous section, the total strain (or potential) energy is given by

$$\begin{aligned}
\mathcal{U} &= \int_0^\ell \left[\frac{EA}{2}\bar{\varepsilon}_{11}^2 + 2GA(\bar{\varepsilon}_{12}^2 + \bar{\varepsilon}_{13}^2)\right. \\
&\quad \left. + \frac{EI_{22}}{2}\tilde{\kappa}_2^2 + \frac{EI_{33}}{2}\tilde{\kappa}_3^2 + \frac{GI}{2}\tilde{\tau}^2\right] dx,
\end{aligned}$$

where $E, G, I_{22}, I_{33}, I = I_{22} + I_{33}$ are Young's modulus, the shear modulus, and moments of the cross section, respectively. The kinetic energy is given by

$$\mathcal{K} = \int_0^\ell \|\dot{\mathbf{R}}\|^2 dx,$$

where $\dot{} = d/dt$. *Controls* may be introduced as distributed, pointwise, or boundary forces and couples $(\mathbf{F}, \mathbf{M}, \mathbf{f}, \mathbf{m})$ through the total work

$$\mathcal{W} = \int_0^\ell (\mathbf{F} \cdot \mathbf{W} + \mathbf{M} \cdot \vartheta)dx + \sum_{x=0,\xi,\ell} (\mathbf{f} \cdot \mathbf{W} + \mathbf{m} \cdot \vartheta).$$

Controls may also be introduced in geometric boundary conditions, but we shall not do so here. Typically, a beam will be rigidly clamped at one end (say at $x_1 = 0$) and simply supported or free at the other end, $x_1 = \ell$. If the space of test functions $V_0 := \{f \in H^1(0, \ell) | f(0) = 0\}$ is introduced, the requirement that the end $x_1 = 0$ be rigidly clamped is mathematically realized by the "geometric" constraints $W_i, \vartheta_i \in V_0$. If $x_1 = 0$ is simply supported, rather than clamped, the appropriate geometric boundary conditions are $W_i \in V_0$ for $i = 2$ and $i = 3$ only.

Let

$$\mathcal{L} = \int_0^T [\mathcal{K}(t) + \mathcal{W}(t) - \mathcal{U}(t)]dt$$

be the *Lagrangian*. Then, by *Hamilton's principle* (see, for example [26]), the dynamics of the deformation satisfy the stationarity conditions $\delta\mathcal{L}_W = 0$, $\delta\mathcal{L}_\vartheta = 0$, the variations being in V_0. The equations of motion then follow by integration by parts, collecting terms, etc., in the usual manner. Due to space limitations, neither the most general beam model nor the entirety of all meaningful boundary conditions can be described here. Rather, a partial list of beam models which, in part, have been studied in the literature and which can easily be extracted from the formulas above, will be provided. We focus on the typical situation of a beam which is clamped at $x_1 = 0$ and free (resp., controlled) at $x_1 = \ell$.

66.2.1 Initially Straight and Untwisted Linear Shearable 3-Dimensional Beams

66.2.2 Equations of motion

$$\left.\begin{aligned}
&m_0\ddot{W}_1 = [EAW_1]' && \text{(longitudinal motion)}\\
&m_0\ddot{W}_2 = [GA(W_2' - \vartheta_3)]' && \text{(lateral motion)}\\
&m_0\ddot{W}_3 = [GA(W_3 + \vartheta_2)]' && \text{(vertical motion)}\\
&m_0\ddot{\vartheta}_1 = [GI\vartheta_1']' && \text{(torsional motion)}\\
&m_0\ddot{\vartheta}_2 = [EI_{33}\vartheta_2']' - GA(W_3' + \vartheta_2), && \text{(shear around } \hat{e}_2)\\
&m_0\ddot{\vartheta}_3 = [EI_{22}\vartheta_3']' + GA(W_2' - \vartheta_3) && \text{(shear around } \hat{e}_3)
\end{aligned}\right\} \tag{66.1}$$

where $' = d/dx_1$.

66.2.3 Boundary conditions

The *geometric boundary conditions* are

$$W_i(0) = 0, \quad \vartheta_i(0) = 0, \quad i = 1, 2, 3. \tag{66.2}$$

The *dynamical boundary conditions* are

$$\left.\begin{aligned}
EAW_1'(\ell) &= f_1,\\
GA(W_2' - \vartheta_3)(\ell) &= f_2,\\
GA(W_3' + \vartheta_2)(\ell) &= f_3,\\
GI\vartheta_1'(\ell) &= m_1,\\
EI_{33}\vartheta_2'(\ell) &= m_2,\\
EI_{22}\vartheta_3'(\ell) &= m_3.
\end{aligned}\right\} \tag{66.3}$$

66.2.4 Initial conditions

$$W_i(\cdot, t = 0) = W_{i0}(\cdot), \quad \dot{W}_i(\cdot, t = 0) = W_{i1}(\cdot),$$
$$\vartheta_i(\cdot, t = 0) = 0, \quad \dot{\vartheta}_i(\cdot, t = 0) = 0, \quad i = 1, 2, 3.$$

66.2.5 Initially Straight and Untwisted, Nonshearable Nonlinear 3-Dimensional Beams

66.2.6 Equations of motion

These are comprised of four equations which describe the longitudinal, lateral, vertical, and torsional motions, respectively.

$$\left.\begin{aligned}
&m_0\ddot{W}_1 = [EA(W_1' + \tfrac{1}{2}(W_2')^2 + \tfrac{1}{2}(W_3')^2)]',\\
&m_0\ddot{W}_2 - [\rho_0 I_{22}\ddot{W}_2']' + [EI_{22}W_2'']''\\
&\qquad = [(EA(W_1' + \tfrac{1}{2}(W_2')^2 + \tfrac{1}{2}(W_3')^2)W_2']',\\
&m_0\ddot{W}_3 - [\rho_0 I_{33}\ddot{W}_3']' + [EI_{33}W_3'']''\\
&\qquad = [(EA(W_1' + \tfrac{1}{2}(W_2')^2 + \tfrac{1}{2}(W_3')^2)W_3']',\\
&\rho_0 I\ddot{\vartheta}_1 = [GI\vartheta_1']'.
\end{aligned}\right\} \tag{66.4}$$

66.2.7 Boundary conditions

The geometric boundary conditions are

$$\left.\begin{aligned}
&W_i(0) = 0, \quad i = 1, 2, 3,\\
&W_2'(0) = W_3'(0) = \vartheta_1(0) = 0,
\end{aligned}\right\} \tag{66.5}$$

while the dynamical boundary conditions are

$$\left.\begin{aligned}
&EA(W_1' + \tfrac{1}{2}((W_2')^2 + (W_3')^2))(\ell) = f_1,\\
&EI_{33}W_3''(\ell) = -m_2, \quad EI_{22}W_2''(\ell) = m_3,\\
&\left[EA(W_1' + \tfrac{1}{2}((W_2')^2 + (W_3')^2))\right]W_2'(\ell)\\
&\qquad -[EI_{22}W_2'']'(\ell) + \rho_0 I_{22}\ddot{W}_2'(\ell) = f_2,\\
&\left[EA(W_1' + \tfrac{1}{2}((W_2')^2 + (W_3')^2))\right]W_3'(\ell)\\
&\qquad -[EI_{33}W_3'']'(\ell) + \rho_0 I_{33}\ddot{W}_3'(\ell) = -f_3,\\
&GI\vartheta_1'(\ell) = m_1
\end{aligned}\right\} \tag{66.6}$$

66.2.8 Initial conditions

$$W_i(\cdot, t = 0) = W_{i0}(\cdot), \quad \dot{W}_i(\cdot, t = 0) = W_{i1}(\cdot),$$
$$\vartheta_1(\cdot, t = 0) = 0, \quad \dot{\vartheta}_1(\cdot, t = 0) = 0, \quad i = 1, 2, 3.$$

66.2.9 Nonlinear Planar, Shearable Straight Beams

66.2.10 Equations of motion

The three equations which describe the longitudinal, vertical, and shear motions, respectively, are

$$\left.\begin{aligned}
&m_0 h\ddot{W}_1 = [Eh(W_1' + \tfrac{1}{2}W_3^2)]',\\
&m_0\ddot{W}_3 = [Gh(\vartheta_2 + W_3')]' + [Eh(W_1' + \tfrac{1}{2}W_3^2)W_3']',\\
&\rho_0 I_{33}\ddot{\vartheta}_2 = EI_{33}\vartheta_2'' - Gh(\vartheta_2 + W_3').
\end{aligned}\right\} \tag{66.7}$$

66.2.11 Boundary conditions

The geometric boundary conditions are

$$W_i(0) = 0, \quad i = 1, 3, \quad \vartheta_2(0) = 0, \tag{66.8}$$

while the dynamical boundary conditions are

$$\left.\begin{aligned}
&Eh[W_1' + \tfrac{1}{2}W_3^2](\ell) = f_1,\\
&Gh[\vartheta_2 + W_3'](\ell) + [Eh(W_1' + \tfrac{1}{2}W_3^2)W_3'](\ell) = f_2,\\
&EI_{33}\vartheta_2'(\ell) = m_1.
\end{aligned}\right\} \tag{66.9}$$

66.2.12 Initial conditions

$$W_i(\cdot, t=0) = W_{i0}(\cdot), \quad \dot{W}_i(\cdot, t=0) = W_{i1}(\cdot), \; i = 1, 3,$$

$$\vartheta_2(\cdot, t=0) = \vartheta_{20}, \quad \dot{\vartheta}_2(\cdot, t=0) = \vartheta_{21}.$$

REMARK 66.1 The model in this section is attributed to Hirschhorn and Reiss [7]. If the longitudinal motion is neglected and the quadratic term in W_3 is averaged over the interval $[0, \ell]$, the model then reduces to a Woinowski–Krieger type. The corresponding partial differential equations are quasi-linear and hard to handle. If, however, one replaces Θ_2 by ϑ_2 in the expression for the strain $\bar{\varepsilon}_{11}$ (which is justified for small strains), then a semilinear partial differential equation with a cubic nonlinearity in ϑ_2 is obtained.

66.2.13 Planar, Nonshearable Nonlinear Beam

The equations of motion for the longitudinal and vertical motions, respectively, are

$$\left.\begin{aligned} m_0\ddot{W}_1 &= [EA(W_1' + \tfrac{1}{2}(W_3')^2)]', \\ m_0\ddot{W}_3 - [\rho_0 I_{33}\ddot{W}_3']' &+ [EI_{33}W_3'']'' \\ &= [(EA(W_1' + \tfrac{1}{2}(W_3')^2))W_3']'. \end{aligned}\right\} \quad (66.10)$$

We dispense with displaying the boundary and initial conditions for this and subsequent models, as those can be immediately deduced from the previous ones. This model has been derived in Lagnese and Leugering [16].

The models above easily reduce to the classical beam equations as follows. We first concentrate on the nonshearable beams.

66.2.14 The Rayleigh Beam Model

Here the longitudinal motion is not coupled to the remaining motions. The equation for vertical motion is

$$m_0\ddot{W}_3 - [\rho_0 I_{33}\ddot{W}_3']' + [EI_{33}W_3'']'' = 0. \quad (66.11)$$

66.2.15 The Euler–Bernoulli Beam Model

This is obtained by ignoring the rotational inertia of cross sections in the Rayleigh model:

$$m_0\ddot{W}_3 + [EI_{33}W_3'']'' = 0. \quad (66.12)$$

With regard to shearable beams, two systems are singled out.

66.2.16 The Bresse System

This is a model for a planar, linear shearable beam with initial curvature involving couplings of longitudinal, vertical, and shear motions.

$$\left.\begin{aligned} m_0\ddot{W}_1 &= [Eh(W_1' - \kappa_3 W_3)]' - \kappa_3 Gh(\vartheta_2 + W_3' + \kappa_3 W_1), \\ m_0\ddot{W}_3 &= [Gh(\vartheta_2 + W_3' + \kappa_3 W_1)]' + \kappa_3 Eh[W_1' - \kappa_3 W_3], \\ \rho_0 I_{33}\ddot{\vartheta}_2 &= EI_{33}\vartheta_2'' - Gh(\vartheta_2 + W_3' + \kappa_3 W_1). \end{aligned}\right\} \quad (66.13)$$

This system was first introduced by Bresse [6].

66.2.17 The Timoshenko System

This is the Bresse model for a straight beam ($\kappa_3 = 0$), so that the longitudinal motion uncouples from the other two equations, which are

$$\left.\begin{aligned} m_0\ddot{W}_3 &= [Gh(\vartheta_2 + W_3')]' \\ \rho_0 I_{33}\ddot{\vartheta}_2 &= EI_{33}\vartheta_2'' - Gh(\vartheta_2 + W_3'). \end{aligned}\right\} \quad (66.14)$$

REMARK 66.2 The models above can be taken to be the basic beam models. In applications it is also necessary to account for damping and various other (local or non-local) effects due to internal variables, such as viscoelastic damping of Boltzmann (non-local in time) or Kelvin–Voigt (local in time) types, structural damping, so-called shear-diffusion or spatial hysteresis type damping; see Russell [24] for the latter. It is also possible to impose large (and usually fast) rigid motions on the beam. We refer to [9]. A comprehensive treatment of elastic frames composed of beams of the types discussed above is given in [17]. With respect to control applications, one should also mention the modeling of beams with piezoceramic actuators; see Banks et al. [3].

66.3 Exact Controllability

66.3.1 Hyperbolic Systems

Consider the models (Equations 66.1, 66.7, 66.13, and 66.14). We concentrate on the linear equations first. All of these models can be put into the form

$$\mathbf{M}\ddot{\mathbf{z}} = [\mathbf{K}(\mathbf{z}' + \mathbf{C}\mathbf{z})]' - \mathbf{C}^T\mathbf{K}(\mathbf{z}' + \mathbf{C}\mathbf{z}) + \mathbf{f}, \quad (66.15)$$

$$\mathbf{z}(0) = 0, \quad \mathbf{K}(\mathbf{z}' + \mathbf{C}\mathbf{z})(\ell) = \mathbf{u}, \quad (66.16)$$

$$\mathbf{z}(\cdot, 0) = \mathbf{z}_0, \quad \dot{\mathbf{z}}(\cdot, 0) = \mathbf{z}_1, \quad (66.17)$$

with positive definite matrices $\mathbf{M}$,$\mathbf{K}$ depending continuously on x. In particular, for the model (Equation 66.1)

$$\mathbf{z} = (\mathrm{W}, \; \vartheta)^T, \quad \mathbf{M} = \mathrm{diag}(m_0, m_0, m_0, \rho_0 I, \rho_0 I_{33}, \rho_0 I_{22})$$

$$\mathbf{K} = \mathrm{diag}(EA, GA, GA, GI, EI_{33}, EI_{22}),$$

$$C_{20} = -1, \; C_{35} = 1, \; C_{ij} = 0 \quad \text{otherwise.}$$

In the case of the Bresse beam,

$$\mathbf{z} = (W_1, W_3, \vartheta_2)^T, \quad \mathbf{M} = \mathrm{diag}(m_0, m_0, \rho_0 I_{33}),$$

$$\mathbf{K} = \mathrm{diag}(Eh, Gh, GA, EI_{33}),$$

$$C_{12} = -\kappa_3, \; C_{21} = \kappa_3, \; C_{23} = 1, \quad C_{ij} = 0 \text{ otherwise.}$$

The Timoshenko system is obtained by setting $\kappa_3 = 0$. If $\mathbf{C} = 0$, Equation 66.15 reduces to the one-dimensional wave equation.

We introduce the spaces

$$\mathbf{H} = L^2(0, \ell, \mathbb{R}^q), \quad \mathbf{V} = \{\mathbf{z} \in H^1(0, \ell, \mathbb{R}^q) : \mathbf{z}(0) = 0\}, \tag{66.18}$$

where q is the number of state variables and $H^k(0, \ell, \mathbb{R}^q)$ denotes the Sobolev space consisting of $\mathbb{R}^q$ valued functions defined on the interval $(0, \ell)$ whose distributional derivatives up to order k are in $L^2(0, \ell; \mathbb{R}^q)$. (The reader is referred to [1] for general information about Sobolev spaces.) We further introduce the energy forms

$$(\mathbf{z}, \hat{\mathbf{z}})_{\mathbf{V}} := \frac{1}{2}\int_0^\ell \mathbf{K}(\mathbf{z}' + \mathbf{C}\mathbf{z}) \cdot (\hat{\mathbf{z}}' + \mathbf{C}\hat{\mathbf{z}})dx,$$

$$(\mathbf{z}, \hat{\mathbf{z}})_{\mathbf{H}} := \frac{1}{2}\int_0^\ell \mathbf{M}\mathbf{z} \cdot \hat{\mathbf{z}}\, dx.$$

Indeed, the norm induced by $\|\mathbf{z}\| = (\mathbf{z}, \mathbf{z})_{\mathbf{V}}^{1/2}$ is equivalent to the usual Sobolev norm of H^1 given by

$$\|\mathbf{z}\|_1 = \left(\int_0^\ell (|\mathbf{z}'|^2 + |\mathbf{z}|^2)dx\right)^{1/2}.$$

Let $\mathbf{z}_0 \in \mathbf{V}$, $\mathbf{z}_1 \in \mathbf{H}$, $\mathbf{f} \in L^2(0, T, \mathbf{H})$, $\mathbf{u} \in L^2(0, T, \mathbf{R}^q)$, $T > 0$. It can be proven that a unique function $\mathbf{z} \in C(0, T, \mathbf{V}) \cap C^1(0, T, \mathbf{H})$ exists which satisfies Equations 66.15, 66.16, and 66.17 in the following weak sense:

$$\frac{d^2}{dt^2}(\mathbf{z}(t), \phi)_{\mathbf{H}} + (\mathbf{z}(t), \phi)_{\mathbf{V}} = (\mathbf{M}^{-1}\mathbf{f}(t), \phi)_{\mathbf{H}} + \mathbf{u}(t) \cdot \phi(\ell), \quad \forall \phi \in \mathbf{V}, \tag{66.19}$$

and

$$(\mathbf{z}(0), \phi)_{\mathbf{H}} = (\mathbf{z}_0, \phi)_{\mathbf{H}}, \quad \frac{d}{dt}(\mathbf{z}(t), \phi)_{\mathbf{H}}|_{t=0} = (\mathbf{z}_1, \phi)_{\mathbf{H}}.$$

Because of space limitations, only boundary controls and constant coefficients will be considered. Distributed controls are easy to handle, while pointwise controls have much in common with boundary controls except for the liberty of their location. Thus we set $\mathbf{f} \equiv 0$ in Equation 66.15. The problem of *exact controllability* in its strongest sense can be formulated as follows: *Given initial data Equation* 66.17 *and final data* $(\mathbf{z}_{T0}, \mathbf{z}_{T1})$ *in* $\mathbf{V} \times \mathbf{H}$ *and given* $T > 0$, *find a control* $\mathbf{u} \in L^2(0, T, \mathbf{R}^q)$ *so that the solution* $\mathbf{z}$ *of Equation* 66.15 *satisfies Equations* 66.16, 66.17 *and* $\mathbf{z}(T) = \mathbf{z}_{T0}$, $\dot{\mathbf{z}}(T) = \mathbf{z}_{T1}$.

It is, in principle, possible to solve the generalized eigenvalue problem

$$\gamma^2 \frac{d^4}{dx^4}\phi = \lambda^2(I - \gamma^2\frac{d^2}{dx^2})\phi,$$
$$\phi(0) = \phi'(0) = \phi''(\ell) = (\phi''' + \lambda^2\phi')(\ell) = 0,$$

and write the solution of Equations 66.15, 66.16, 66.17 using Fourier's method of separation of variables. The solution together with its time derivative can then be evaluated at T and the control problem reduces to a trigonometric (or to a complex exponential) moment problem. The controllability requirement is then equivalent to the base properties of the underlying set of complex exponentials $[exp(i\lambda_k t)|k \in \mathbf{Z}, t \in (0, T)]$. If that set constitutes a Riesz basis in its $L^2(0, T)$-closure, then exact controllability is achieved. For conciseness, we do not pursue this approach here and, instead, refer to Krabs [12] for further reading. Rather, the approach we want to consider here, while equivalent to the former one, does not resort to the knowledge of eigenvalues and eigenelements. The controllability problem, as in finite dimensions, is a question of characterizing the image of a linear map, the control-to-state map. Unlike in finite dimensions, however, it is not sufficient here to establish a uniqueness result for the homogeneous adjoint problem, that is, to establish injectivity of the adjoint of the control-to-state map. In infinite-dimensional systems this implies only that the control-to-state map has dense range, which in turn is referred to as *approximate controllability*. Rather, we need some additional information on the adjoint system, namely, uniformity in the sense that the adjoint map is uniformly injective with respect to all finite energy initial (final) conditions. In particular, given a bounded linear map $\mathbf{L}$ between Hilbert spaces X, Y, the range of $\mathbf{L}$ is all of Y if, and only if, $\mathbf{L}^*$, the adjoint of $\mathbf{L}$, satisfies $\|\phi\| < \nu\|\mathbf{L}^*\phi\|$ for some positive ν and all $\phi \in Y$. This inequality also implies that the right inverse $\mathbf{L}^*(\mathbf{L}\mathbf{L}^*)^{-1}$ exists as a *bounded* operator (see the finite-dimensional analog, where 'bounded' is generic). It is clear that this implies *norm-minimality* of the controls constructed this way. This result extends to more general space setups. It turns out that an inequality like this is needed to assure that the set of complex exponentials is a Riesz base in its L^2-closure. As will be seen shortly, such an inequality is achieved by nonstandard energy estimates, which constitute the basis for the so-called HUM method introduced by Lions. It is thus clear that this inequality is the crucial point in the study of exact controllability.

In order to obtain such estimates, we consider smooth enough solutions ϕ of the homogeneous adjoint final value problem,

$$\mathbf{M}\ddot{\phi} = [\mathbf{K}(\phi' + \mathbf{C}\phi)]' - \mathbf{C}^T\mathbf{K}(\phi' + \mathbf{C}\phi), \tag{66.20}$$

$$\phi(0) = 0, \quad \mathbf{K}(\phi' + \mathbf{C}\phi)(\ell) = 0, \tag{66.21}$$

$$\phi(\cdot, T) = \phi_0, \quad \dot{\phi}(\cdot, T) = \phi_1. \tag{66.22}$$

Let m be a smooth, positive, increasing function of x. Multiply Equation 66.20 by $m\phi'$, where $m(\cdot)$ is a smooth function in x, and integrate by parts over (x, t). After some calculus, we obtain the following crucial identity, valid for any sufficiently smooth solution of Equation 66.20:

$$\begin{aligned} 0 \;=\; & \int_0^\ell m\mathbf{M}\dot{\phi} \cdot (\phi' + \mathbf{C}\phi)|_0^T dx \\ & - \int_0^T m(x)e(x,t)|_{x=0}^\ell dt + \int_0^T \int_0^\ell m' e\, dx dt \\ & - \int_0^T \int_0^\ell m C^T \mathbf{M}\dot{\phi} \cdot \dot{\phi}\, dx dt \\ & + \int_0^T \int_0^\ell m C^T \mathbf{K}(\phi' + \mathbf{C}\phi) \cdot (\phi' + \mathbf{C}\phi) dx dt, \end{aligned} \tag{66.23}$$

where $\rho = \int_0^\ell m\mathbf{M}\dot{\phi}\cdot(\phi'+\mathbf{C}\phi)dx$ and $e(x,t)$ denotes the energy density given by

$$e = \frac{1}{2}[\mathbf{M}\dot{\phi}\cdot\dot{\phi} + \mathbf{K}(\phi'+\mathbf{C}\phi)\cdot(\phi'+\mathbf{C}\phi)].$$

Set the total energy at time t,

$$\mathcal{E}(t) = \int_0^\ell e(x,t)\,dx, \tag{66.24}$$

and denote the norm of the matrix $\mathbf{C}$ by ν. If we choose m in Equation 66.23 so that $m'(x) - \nu m(x) \geq c_0 > 0, \forall x \in (0,\ell)$, we obtain the estimates

$$\gamma\int_0^T |\phi(\ell,t)|^2 dt \leq \mathcal{E}(0) \leq \Gamma\int_0^T |\phi(\ell,t)|^2 dt, \tag{66.25}$$

for some positive constants $\gamma(T), \Gamma(T)$, where T is sufficiently large (indeed, $T > 2\times$ "optical length of the beam" is sufficient). The second inequality in Equation 66.25 requires the multiplier above and some estimation, whereas the first requires the multiplier $m(x) = -1+2x/\ell$, and is straightforwardly proved. These inequalities can also be obtained by the method of characteristics which, in addition, yields the smallest possible control time [17]. It is then shown that the norm of the adjoint to the control-to-state map (which takes the control $\mathbf{u}$ into the final values $\mathbf{z}(T), \dot{\mathbf{z}}(T)$ (for zero initial conditions)), applied to ϕ_0, ϕ_1, is exactly equal to $\int_0^T |\phi(\ell,t)|^2 dt$. By the above argument, the original map is onto between the control space $L^2(0,T,\mathbb{R}^q)$ and the *finite energy space* $\mathbf{V}\times\mathbf{H}$.

THEOREM 66.1 *Let* $(\mathbf{z}_0, \mathbf{z}_1), (\mathbf{z}_{T0}, \mathbf{z}_{T1})$ *be in* $\mathbf{V}\times\mathbf{H}$ *and* $T > 0$ *sufficiently large. Then a unique control* $\mathbf{u} \in L^2(0,T,\mathbb{R}^q)$ *exists, with minimal norm, so that* $\mathbf{z}$ *satisfies Equations 66.15, 66.16, 66.17 and* $\mathbf{z}(T) = z_{T0}, \dot{\mathbf{z}}(T) = z_{T1}$.

REMARK 66.3 Controllability results for the fully nonlinear planar shearable beam Equation 66.7, and also for the Woinowski–Krieger-type approximation, are not known at present. The semilinear model is locally exactly controllable, using the implicit function theorem. The argument is quite similar to the one commonly used in finite-dimensional control theory.

66.3.2 Quasi-Hyperbolic Systems

In this section we discuss the (linearized) nonshearable models with rotational inertia, namely, Equations 66.4, 66.10, 66.11. We first discuss the linear subsystems. Observe that in that situation all equations decouple into wave equations governing the longitudinal and torsional motion and equations of the type of Equation 66.11. Hence it is sufficient to consider the latter. For simplicity, we restrict ourselves to constant coefficients. The system is then

$$\rho h\ddot{W} - \rho I\ddot{W}'' + EIW'''' = f,$$
$$W(0) = 0, \quad W'(0) = 0,$$

and

$$\begin{aligned} EIW''(\ell) = u_1, \quad (EIW''' - \rho I\ddot{W}')(\ell) = u_2, \\ W(\cdot,0) = W_0, \quad \dot{W}(\cdot,0) = W_1. \end{aligned} \tag{66.26}$$

Setting $\gamma^2 := I/A$ and rescaling by $t \to t\sqrt{\rho/E}$, this system can be brought into a nondimensional form. We define spaces,

$$\begin{aligned} \mathbf{H} &= [v \in H^1(0,\ell) : v(0) = 0], \\ \mathbf{V} &= [v \in H^2(0,\ell) : v(0) = v'(0) = 0], \end{aligned} \tag{66.27}$$

and forms,

$$(u,v) = \int_0^\ell uv\,dx, \quad (u,v)_{\mathbf{H}} = (u,v) + \gamma^2(u',v'),$$
$$(u,v)_{\mathbf{V}} = \gamma^2(u'',v'').$$

Let $W_0 \in \mathbf{V}$, $W_1 \in \mathbf{H}$, and $\mathbf{u} \in L^2(0,T,\mathbb{R}^2)$, $T > 0$. It may be proved that there is a unique $\mathbf{W} \in C(0,T,\mathbf{V}) \cap C^1(0,T,\mathbf{H})$ satisfying Equation 66.26 in an appropriate variational sense.

REMARK 66.4 The nonlinear models can be treated using the theory of nonlinear maximal monotone operators; see Lagnese and Leugering [16].

To produce an energy identity analogous to Equation 66.23, we multiply the first equation of Equation 66.26 by $xW' - \alpha W$, where $\alpha > 0$ is a free parameter, and then we integrate over $(0,\ell)\times(0,T)$. If we introduce the auxiliary functions $\rho_1 = \int_0^\ell \dot{W}(xW'-\alpha W)dx$ and $\rho_2 = \gamma^2\int_0^\ell \dot{W}'(xW'-\alpha W)'dx$, and $\rho = \rho_1 + \rho_2$, we find after some calculus

$$\begin{aligned} 0 = {} & \rho(T) - \rho(0) - \\ & \frac{\ell}{2}\int_0^T \{[\dot{W}(\ell,t)]^2 + \gamma^2[\dot{W}'(\ell,t)]^2\}dt \\ & + \gamma^2\int_0^T [W''(\ell,t)]^2 dt \\ & + \int_0^T (\gamma^2 W''' - \gamma^2\ddot{W}')(\ell,t)(\ell W' - \alpha W)(\ell,t))dt \\ & + \int_0^T\int_0^\ell [(\frac{1}{2}+\alpha)\dot{W}^2 + \gamma^2(\alpha - \frac{1}{2})(\dot{W}')^2 \\ & + \gamma^2(\frac{3}{2}-\alpha)(W'')^2]dxdt. \end{aligned}$$

With the total energy now defined by

$$\mathcal{E}(t) = \frac{1}{2}\{\|\dot{W}(t)\|^2_{\mathbf{H}} + \|W\|^2_{\mathbf{V}}\}, \tag{66.28}$$

this identity can now be used to derive the energy estimate,

$$\begin{aligned} \pi\int_0^T [\phi^2(\ell,t) + \gamma^2(\phi')^2(\ell,t)]dt \leq \mathcal{E}(0) \\ \leq \Pi\int_0^T [\dot{\phi}^2(\ell,t) + \gamma^2(\dot{\phi}')^2(\ell,t)]dt, \end{aligned}$$

for some positive constants $\pi(T) and \Pi(T)$, which is valid for sufficiently smooth solutions ϕ to the homogeneous system, Equation 66.26, and for sufficiently large $T > 0$ (again T is related to the "optical length," i.e., to wave velocities). The first estimate is more standard and determines the regularity of the solutions. It is again a matter of calculating the control-to-state map and its adjoint. After some routine calculation, one verifies that the norm of the adjoint, applied to the final data for the backwards running homogeneous equation, coincides with the time integral in the energy estimate. This leads to the exact controllability of the system, Equation 66.26, in the space $\mathbf{V} \times H$, with $\mathbf{V}$ and $\mathbf{H}$ as defined in Equation 66.27, using controls $\mathbf{u} = (u_1, u_2) \in L^2(0, T, \mathbb{R}^2)$.

REMARK 66.5 As in [18], for $\gamma \to 0$, the controllability results for Rayleigh beams carry over to the corresponding results for Euler-Bernoulli beams. It is however instructive and, in fact, much easier to establish controllability of the Euler-Bernoulli beam directly with control only in the shear force.

66.3.3 The Euler–Bernoulli Beam

We consider the nondimensional form of Equation 66.12, namely,

$$\begin{gathered} \ddot{W} + W'''' = 0 \\ W(0) = W'(0) = 0, \quad W''(\ell) = 0, \quad W'''(\ell) = u \qquad (66.29) \\ W(\cdot, 0) = W_0, \quad \dot{W}(\cdot, \ell) = W_1. \end{gathered}$$

We introduce the spaces

$$\mathbf{H} = L^2(0, \ell), \quad \mathbf{V} = \{v \in H^2 | v(0) = v'(0) = 0\} \qquad (66.30)$$

and the corresponding energy functional

$$\mathcal{E}(t) = \frac{1}{2}(\|\dot{W}(t)\|^2_{\mathbf{H}} + \|W(t)\|^2_{\mathbf{V}}. \qquad (66.31)$$

Again, we are going to use multipliers to establish energy identities. The usual choice is $m(x) = x$. Upon introducing $\rho = \int_0^\ell x\dot{W}W'dx$ and multiplying the first equation by mW', followed by integration by parts, we obtain

$$\begin{aligned} \rho(T) - \rho(0) + \tfrac{1}{2}\int_0^T \int_0^\ell \dot{W}^2 dxdt + \tfrac{3}{2}\int_0^T \int_0^\ell (W'')^2 dxdt \\ = \tfrac{\ell}{2}\int_0^T \dot{W}^2 dt + \ell \int_0^T W'(\ell, t)u(t)dt. \end{aligned} \qquad (66.32)$$

Using this identity for the homogeneous system solved by ϕ, we obtain the energy estimates

$$\pi \int_0^T \dot{\phi}^2(\ell, t)dt \le \mathcal{E}(0) \le \Pi \int_0^T \dot{\phi}^2(\ell, t)dt$$

where again $\pi(T)$ and $\Pi(T)$ depend on $T > 0$, with T sufficiently large.

One way to obtain the adjoint control-to-state-map is to consider

$$\frac{d}{dt}\int_0^\ell \{\dot{W}\dot{\phi} + W''\phi''\}dx = -u(t)\dot{\phi}(\ell, t),$$

for W as above and ϕ solving the backwards running adjoint equation (i.e., Equation 66.29 with final conditions ϕ_{T0} and ϕ_{T1}). Integrating with respect to time over $(0, T)$ yields

$$(L_T(u), (\phi_{T0}, \phi_{T1}))_{\mathbf{V}\times\mathbf{H}} = -\int_0^T u(t)\dot{\phi}(\ell, t)dt.$$

The same argument as above yields the conclusion of exact controllability of the system (Equation 66.29), in the space $\mathbf{V} \times \mathbf{H}$, where $\mathbf{V}$ and $\mathbf{H}$ are defined in Equation 66.30, using controls $u \in L^2(0, T, \mathbb{R})$.

REMARK 66.6 It may be shown that the control time T for the Euler-Bernoulli system can actually be taken arbitrarily small. That is typical for this kind of model (Petrovskii type systems) and is closely related to the absence of a uniform wave speed. The reader is referred to the survey article [15] for general background information on controllability and stabilizability of beams and plates.

66.4 Stabilizability

We proceed to establish uniform exponential decay for the solutions of the various beam models by *linear* feedback controls at the boundary $x = \ell$. There is much current work on nonlinear and constrained feedback laws. However, the results are usually very technical, and, therefore, do not seem suitable for reproduction in these notes. In the linear case it is known that for time reversible systems, exact controllability is equivalent to uniform exponential stabilizability. In contrast to the finite-dimensional case, however, we have to distinguish between various concepts of controllability, such as exact, spectral, or approximate controllability. Accordingly, we have to distinguish between different concepts of stabilizability, as uniform exponential decay is substantially different from nonuniform decay. Because of space limitations, we do not dwell on the relation between controllability, stabilizability, and even observability. The procedure we follow is based on Liapunov functions, and is the same in all of the models. Once again the energy identities, Equations 66.23, 66.3.2, and 66.32 are crucial. We take the hyperbolic case as an exemplar and outline the procedure in that case.

66.4.1 Hyperbolic Systems

Apply Equation 66.23 to solve Equation 66.15 (with $\mathbf{f} = 0$) and Equation 66.16 with the control $\mathbf{u}$ in Equation 66.16 replaced by a linear feedback law $\mathbf{u}(t) = -k\mathbf{z}(\ell, t)$, $k > 0$. Recall that $\rho = \int_0^\ell m\mathbf{M}\dot{\mathbf{z}} \cdot (\mathbf{z}' + \mathbf{C}\mathbf{z})dx(t)$. Then using Equation 66.23,

$$\dot{\rho} \le \gamma|\dot{\mathbf{z}}(\ell, t)|^2 - c_0\mathcal{E}(t),$$

where $\mathcal{E}$ is given by Equation 66.24. Therefore, introducing the function $\mathcal{F}_\epsilon(t) := \mathcal{E}(t) + \epsilon\rho(t)$ one finds

$$\dot{\mathcal{F}}_\epsilon \le \dot{\mathcal{E}}(t) + \gamma\epsilon|\dot{\mathbf{z}}(\ell, t)|^2 - \epsilon c_0\mathcal{E}(t).$$

However, $\dot{\mathcal{E}}(t) = -k|\dot{z}(\ell, t)|^2$, and therefore the boundary term can be compensated for by choosing ϵ sufficiently small. This results in the estimate $\dot{\mathcal{F}}_\epsilon(t) \leq -c_1\epsilon\mathcal{E}(t)$ for some $c_1 > 0$, which in turn implies

$$\mathcal{F}_\epsilon(t) + c_1\epsilon \int_0^t \mathcal{E}(s)ds \leq \mathcal{F}(0).$$

It also straightforward to see that $\mathcal{F}(t)$ satisfies

$$\pi_\epsilon \mathcal{E}(t) \leq \mathcal{F}(t) \leq \Pi_\epsilon \mathcal{E}(t).$$

The latter implies $\int_t^\infty \mathcal{E}(s)ds \leq (1/\lambda)\mathcal{E}(0)$, with $\lambda = \Pi_\epsilon/(c_1\epsilon)$ and $t \geq 0$. One defines $\eta(t) := \int_t^\infty \mathcal{E}(s)ds$ and obtains a differential inequality $\dot{\eta} + \lambda\eta \leq 0$. A standard Gronwall argument implies $\eta(t) \leq \exp(-\lambda t)\eta(0)$, that is,

$$\int_t^\infty \mathcal{E}(s)ds \leq \frac{\exp(-\lambda t)}{\lambda}\mathcal{E}(0).$$

Now, because $\mathcal{E}(t)$ is nonincreasing,

$$\tau\mathcal{E}(\tau + t) \leq \left\{\int_t^{\tau+t} + \int_{\tau+t}^\infty\right\} \mathcal{E}(s)ds \leq \frac{\exp(-\lambda t)}{\lambda}\mathcal{E}(0)$$

and this, together with the choice $\tau = 1/\lambda$, gives

$$\mathcal{E}(t) \leq e \exp(-\lambda t)\mathcal{E}(0), \quad \forall t \geq \tau. \tag{66.33}$$

Hence, we have the following result.

THEOREM 66.2 *Let* **V** *and* **H** *be given by Equation 66.18. Given initial data* $\mathbf{z}_0, \mathbf{z}_1$ *in* $\mathbf{V} \times \mathbf{H}$, *the solution to the closed-loop system, Equation 66.15, Equation 66.16, and Equation 66.17, with* $\mathbf{u}(t) = -k\dot{\mathbf{z}}(\ell, t)$, *satisfies*

$$\mathcal{E}(t) \leq M \exp(-\omega t)\mathcal{E}(0), \quad t \geq 0, \tag{66.34}$$

for some positive constants M *and* ω.

REMARK 66.7 The linear feedback law can be replaced by a monotone nonlinear feedback law with certain growth conditions. The corresponding energy estimates, however, are beyond the scope of these notes. Ultimately, the differential inequality above is to be replaced by a nonlinear one. The exponential decay has then (in general) to be replaced by an algebraic decay, see [17]. The decay rate can be optimized using "hyperbolic estimates" as in [17]. Also the dependence on the feedback parameter can be made explicit.

66.4.2 Quasi-Hyperbolic Systems

We consider the problem of Equation 66.26 with feedback controls,

$$u_1(t) = -k_1\dot{W}'(\ell, t), \quad u_2(t) = k_2\dot{W}(\ell, t), \tag{66.35}$$

with positive feedback gains k_1 and k_2. The identity Equation 66.3.2 is used to calculate the derivative of the function $\rho(t)$. By following the same procedure as above, we obtain the decay estimate Equation 66.34 for the closed-loop system, Equations 66.26 and 66.35, where the energy functional $\mathcal{E}$ is given by Equation 66.28.

REMARK 66.8 One can show algebraic decay for certain monotone nonlinear feedbacks. In addition, the nonlinear system, Equation 66.10, exhibits those decay rates as well; see Lagnese and Leugering [16].

66.4.3 The Euler–Bernoulli Beam

Here we consider the system Equation 66.29 and close the loop by setting $u(t) = -k\dot{W}(\ell, t)$, $k > 0$. By utilizing the estimate Equation 66.32 and proceeding in much the same way as above, the decay estimate Equation 66.34 can be established for the closed-loop system, where $\mathcal{E}$ is given by Equation 66.31.

66.5 Dynamic Elastic Plate Models

Let Ω be a bounded, open, connected set in $\mathbb{R}^2$ with a Lipschitz continuous boundary consisting of a finite number of smooth curves. Consider a deformable three-dimensional body which, in equilibrium, occupies the region

$$[(x_1, x_2, x_3) : (x_1, x_2) \in \overline{\Omega}, \ |x_3| \leq h/2]. \tag{66.36}$$

When the quantity h is very small compared to the diameter of Ω, the body is referred to as a *thin plate of uniform thickness h* and the planar region,

$$[(x_1, x_2, 0) : (x_1, x_2) \in \overline{\Omega}]$$

is its *reference surface.*

Two-dimensional mathematical models describing the deformation of the three-dimensional body Equation 66.36 are obtained by relating the displacement vector associated with the deformation of each point within the body to certain *state variables* defined on the reference surface. Many such models are available; three are briefly described below.

66.5.1 Linear Models

Let $\mathbf{W}(x_1, x_2, x_3, t)$ denote the displacement vector at time t of the material point located at (x_1, x_2, x_3), and let $\mathbf{w}(x_1, x_2, t)$ denote the displacement vector of the material point located at $(x_1, x_2, 0)$ in the reference surface. Further, let $\mathbf{n}(x_1, x_2, t)$ be the unit-normal vector to the deformed reference surface at the point $(x_1, x_2, 0) + \mathbf{w}(x_1, x_2, t)$. The direction of $\mathbf{n}$ is chosen so that $\mathbf{n} \cdot \mathbf{k} > 0$, where $\mathbf{i}, \mathbf{j}, \mathbf{k}$ is the natural basis for $\mathbb{R}^3$.

Kirchhoff Model

The basic kinematic assumption of this model is

$$\mathbf{W}(x_1, x_2, x_3, t) = \mathbf{w}(x_1, x_2, t) + x_3(\mathbf{n}(x_1, x_2, t) - \mathbf{k}), \tag{66.37}$$

which means that a filament in its equilibrium position, orthogonal to the reference surface, remains straight, unstretched, and orthogonal to the deformed reference surface. It is further assumed that the material is linearly elastic (Hookean), homogeneous and isotropic, that the transverse normal stress is small compared to the remaining stresses, and that the strains and the normal vector **n** are well-approximated by their linear approximations. Write $\mathbf{w} = w_1\mathbf{i} + w_2\mathbf{j} + w_3\mathbf{k}$. Under the assumptions above, there is no coupling between the in-plane displacements w_1, w_2 and the transverse displacement $w_3 := w$. The former components satisfy the partial differential equations of linear plane elasticity and the latter satisfies the equation

$$\rho h\frac{\partial^2 w}{\partial t^2} - I_\rho \Delta\frac{\partial^2 w}{\partial t^2} + D\Delta^2 w = F, \tag{66.38}$$

where $\Delta = \partial^2/\partial x_1^2 + \partial^2/\partial x_2^2$ is the harmonic operator in $\mathbb{R}^2$, ρ is the mass density per unit of reference volume, $I_\rho = \rho h^3/12$ is the polar moment of inertia, D is the modulus of flexural rigidity, and F is the transverse component of an applied force distributed over Ω. The "standard" Kirchhoff plate equation is obtained by omitting the term $I_\rho\Delta(\partial^2 w/\partial t^2)$, which accounts for the rotational inertia of cross sections, from Equation 66.38.

Reissner–Mindlin System

The basic kinematic assumption of this model is

$$\mathbf{W}(x_1, x_2, x_3, t) = \mathbf{w}(x_1, x_2, t) + x_3\mathbf{U}(x_1, x_2, t), \tag{66.39}$$

where $|\mathbf{U} + \mathbf{k}| = 1$. Equation 66.39 means that a filament in its equilibrium position, orthogonal to the reference surface, remains straight and unstretched but not necessarily orthogonal to the deformed reference surface. Write

$$\mathbf{U} = U_1\mathbf{i} + U_2\mathbf{j} + U_3\mathbf{k}, \quad \mathbf{u} = U_1\mathbf{i} + U_2\mathbf{j}.$$

In the linear approximation, $U_3 = 0$ so that the state variables of the problem are **w** and **u**. The latter variable accounts for transverse shearing of cross sections. It is further assumed that the material is homogeneous and Hookean. The stress-strain relations assume that the material is isotropic in directions parallel to the reference surface but may have different material properties in the transverse direction. As in the previous case, there is no coupling between w_1, w_2, and the remaining state variables in the linear approximations. The equations of motion satisfied by $w_3 := w$ and **u**, referred to as the Reissner or Reissner–Mindlin system, may be written

$$\left.\begin{aligned} &\rho h\frac{\partial^2 w}{\partial t^2} - Gh\operatorname{div}(\mathbf{u} + \nabla w) = F, \quad \text{and} \\ &I_\rho\frac{\partial^2 \mathbf{u}}{\partial t^2} - \frac{h^3}{12}\operatorname{div}\sigma(\mathbf{u}) + Gh(\mathbf{u} + \nabla w) = \mathbf{C}, \end{aligned}\right\} \tag{66.40}$$

where Gh is the shear modulus and $\sigma(\mathbf{u}) = (\sigma_{ij}(\mathbf{u}))$ is the stress tensor associated with **u**, i.e.,

$$\sigma_{ij}(\mathbf{u}) = 2\mu\varepsilon_{ij}(\mathbf{u}) + \frac{2\mu\lambda}{2\mu + \lambda}\delta_{ij}\sum_{k=1}^{2}\varepsilon_{kk}(\mathbf{u}), \quad i, j = 1, 2.$$

$\varepsilon_{ij}(\mathbf{u})$ denotes the linearized strain

$$\varepsilon_{ij}(\mathbf{u}) = \frac{1}{2}\left(\frac{\partial U_i}{\partial x_j} + \frac{\partial U_j}{\partial x_i}\right),$$

λ and μ are the Lamé parameters of the material,

$$\operatorname{div}(\mathbf{u} + \nabla w) = \nabla\cdot(\mathbf{u} + \nabla w), \quad \operatorname{div}\sigma(\mathbf{u}) = \sum_{j=1}^{3}\frac{\partial}{\partial x_j}\sigma_{ij}(\mathbf{u}),$$

and $\mathbf{C} = C_1\mathbf{i} + C_2\mathbf{j}$ is a distributed force couple.

66.5.2 A Nonlinear Model: The von Kármán System

Unlike the two previous models, this is a "large deflection" model. It is obtained under the same assumptions as the Kirchhoff model except for the linearization of the strain tensor. Rather, in the general strain tensor,

$$\varepsilon_{ij}(\mathbf{W}) = \frac{1}{2}\left(\frac{\partial W_i}{\partial x_j} + \frac{\partial W_j}{\partial x_i}\right) + \frac{1}{2}\sum_{k=1}^{3}\frac{\partial W_k}{\partial x_i}\frac{\partial W_k}{\partial x_j},$$

the quadratic terms involving W_3 are retained, an assumption formally justified if the planar strains are small relative to the transverse strains. The result is a nonlinear plate model in which the in-plane components of displacement w_1 and w_2 are coupled to the transverse displacement $w_3 := w$. Under some further simplifying assumptions, w_1 and w_2 may be replaced by a single function G, called an Airy stress function, related to the in-plane stresses. The resulting nonlinear equations for w and G are

$$\left.\begin{aligned} &\rho h\frac{\partial^2 w}{\partial t^2} - I_\rho\Delta\frac{\partial^2 w}{\partial t^2} + D\Delta^2 w - [w, G] = F, \\ &\Delta^2 G + \frac{Eh}{2}[w, w] = 0, \end{aligned}\right\} \tag{66.41}$$

where E is Young's modulus and where

$$[\phi, \psi] = \frac{\partial^2\phi}{\partial x_1^2}\frac{\partial^2\psi}{\partial x_2^2} + \frac{\partial^2\psi}{\partial x_1^2}\frac{\partial^2\phi}{\partial x_2^2} - 2\frac{\partial^2\phi}{\partial x_1\partial x_2}\frac{\partial^2\psi}{\partial x_1\partial x_2}.$$

One may observe that $[w, w]/2$ is the Gaussian curvature of the deformed reference surface $x_3 = w(x_1, x_2)$. The "standard" dynamic von Kármán plate system is obtained by setting $I_\rho = 0$ in Equation 66.41.

REMARK 66.9 For a derivation of various plate models, including thermoelastic plates and viscoelastic plates, see [18]. For studies of junction conditions between two or more interconnected (not necessarily co-planar) elastic plates, the reader is referred to the monographs [21] and [17] and references therein.

66.5.3 Boundary Conditions

Let Γ denote the boundary of Ω. The boundary conditions are of two types: geometric conditions, that constrain the geometry of the deformation at the boundary, and mechanical (or dynamic) conditions, that represent the balance of linear and angular momenta at the boundary.

Boundary Conditions for the Reissner–Mindlin System

Geometric conditions are given by

$$w = \bar{w}, \quad \mathbf{u} = \bar{\mathbf{u}} \text{ on } \Gamma, t > 0. \tag{66.42}$$

The case $\bar{w} = \bar{\mathbf{u}} = 0$ corresponds to a rigidly clamped boundary.

The mechanical boundary conditions are given by

$$\left.\begin{aligned} &Gh\nu \cdot (\mathbf{u} + \nabla w) = f, \\ &\frac{h^3}{12}\sigma(\mathbf{u})\nu = \mathbf{c} \text{ on } \Gamma, t > 0, \end{aligned}\right\} \tag{66.43}$$

where ν is the unit exterior pointing normal vector to the boundary of Ω, $\mathbf{c} = c_1\mathbf{i} + c_2\mathbf{j}$ is a boundary force couple and f is the transverse component of an applied force distributed over the boundary. The problem consisting of the system of Equations 66.40 and boundary conditions Equations 66.42 *or* 66.43, together with the *initial conditions*

$$\left.\begin{aligned} &w = w^0, \quad \frac{\partial w}{\partial t} = w^1, \\ &\mathbf{u} = \mathbf{u}^0, \quad \frac{\partial \mathbf{u}}{\partial t} = \mathbf{u}^1 \text{ at } t = 0, \end{aligned}\right\} \tag{66.44}$$

has a unique solution if the data of the problem is sufficiently regular. The same is true if the boundary conditions are Equation 66.42 on one part of Γ and Equation 66.43 on the remaining part, or if they consist of the first (resp., second) of the two expressions in Equation 66.42 and the second (resp., first) of the two expressions in Equation 66.43.

Boundary Conditions for the Kirchhoff Model

The geometric boundary conditions are

$$w = \bar{w}, \quad \frac{\partial w}{\partial \nu} = -\nu \cdot \bar{\mathbf{u}} \text{ on } \Gamma, t > 0, \tag{66.45}$$

and the mechanical boundary conditions may be written

$$\left.\begin{aligned} &\frac{h^3}{12}\nu \cdot \sigma(\nabla w)\nu = -\nu \cdot \mathbf{c}, \\ &\frac{\partial}{\partial \nu}\left(I_\rho \frac{\partial^2 w}{\partial t^2} - D\Delta w\right) - \frac{h^3}{12}\frac{\partial}{\partial \tau}[\tau \cdot \sigma(\nabla w)\nu] \\ &\qquad = \frac{\partial}{\partial \tau}(\tau \cdot \mathbf{c}) + f. \end{aligned}\right\} \tag{66.46}$$

Boundary Conditions for the von Kármán System

The state variables w and G are not coupled in the boundary conditions. The geometric and mechanical boundary conditions for w are those of the Kirchhoff model. The boundary conditions satisfied by G are

$$G = 0, \quad \frac{\partial G}{\partial \nu} = 0. \tag{66.47}$$

These arise if there are no in-plane applied forces along the boundary.

66.6 Controllability of Dynamic Plates

In the models discussed in the last section, some, or all, of the applied forces and moments F, $\mathbf{C}$, f, $\mathbf{c}$, and the geometric data $\bar{w}$, $\bar{\mathbf{u}}$, may be considered as *controls* which must be chosen in order to affect the transient behavior of the solution in some specified manner. These controls may either be *open loop*, or *closed loop*. Open-loop controls are usually associated with problems of controllability, which is that of steering the solution to, or nearly to, a specified state at a specified time. Closed-loop controls are usually associated with problems of stabilizability, that is, of asymptotically driving the solution towards an equilibrium state of the system.

In fact, for infinite-dimensional systems of which the above plate models are representative, there are various related but distinct concepts of controllability (spectral, approximate, exact) and of stabilizability (weak, strong, uniform), distinctions which disappear in finite-dimensional approximations of these models (see [2, Chapter 4]). Stabilizability problems will be discussed in the next section. With regard to controllability, *exact controllability* is the most stringent requirement because it requires a complete description of the configuration space (reachable set) of the solution. This is equivalent to steering any initial state of the system to any other permissible state within a specified interval of time. The notion of *spectral controllability* involves exactly controlling the span of any set of *finitely many* of the eigenmodes of the system. *Approximate controllability* involves steering an arbitrary initial state to a given, but arbitrary, neighborhood of a desired configuration within a specified time.

Among the possible controls, distinctions are made between *distributed controls* such as F and $\mathbf{C}$, which are distributed over all or a portion of the face of the plate, and *boundary controls*, such as f, $\mathbf{c}$, $\bar{w}$, $\bar{\mathbf{u}}$, which are distributed over all or a portion of the edge of the plate. Within the class of boundary controls, a further distinction is made between mechanical controls, f and $\mathbf{c}$, and geometric controls $\bar{w}$ and $\bar{\mathbf{u}}$. Because mechanical controls correspond to forces and moments, they are, in principle, physically implementable; these are the only types of controls which will be considered here. In addition, only boundary control problems will be considered in detail; however, some remarks regarding distributed control problems will be provided.

66.6.1 Controllability of Kirchhoff Plates

Assume that $\Gamma = \overline{\Gamma}_0 \cup \overline{\Gamma}_1$, where Γ_0 and Γ_1 are disjoint, relatively open subsets of Γ with $\Gamma_1 \neq \emptyset$. The problem under consideration consists of the partial differential Equation 66.38, boundary conditions Equation 66.45 on Γ_0, boundary conditions Equation 66.46 on Γ_1, and initial conditions

$$w(x, 0) = w^0(x), \quad \frac{\partial w}{\partial t}(x, 0) = w^1(x), \quad x \in \Omega. \tag{66.48}$$

In this system, the distributed force F, the geometric quantities $\bar{w}$, $\bar{\mathbf{u}}$, and the initial data (w^0, w^1) are assumed as given data, while f, $\mathbf{c}$ are the controls, chosen from a certain class C of *admissible controls*. The *configuration space*, or the *reachable set*, at time T is

$$\mathcal{R}_T = \{(w(T), \dot{w}(T)) : (f, \mathbf{c}) \in \mathcal{C}\},$$

where, for example, $w(T)$ stands for the function $[w(x,T) : x \in \Omega]$ and where $\dot{w} = \partial w/\partial t$. If z denotes the solution of the uncontrolled problem, i.e., the solution with $f = 0$ and $\mathbf{c} = 0$, then

$$\mathcal{R}_T = \mathcal{R}_T^0 \oplus \{[z(T), \dot{z}(T)]\},$$

where $\mathcal{R}_T^0$ denotes the configuration space when all of the given data are zero. Therefore, to study the reachable set it may be assumed without loss of generality that the data F, $\bar{w}$, $\bar{\mathbf{u}}$, w^0, w^1 vanish. The problem under consideration is, therefore,

$$\rho h \frac{\partial^2 w}{\partial t^2} - I_\rho \Delta \frac{\partial^2 w}{\partial t^2} + D\Delta^2 w = 0, \tag{66.49}$$

$$w = 0, \quad \frac{\partial w}{\partial \nu} = 0 \text{ on } \Gamma_0, t > 0, \tag{66.50}$$

$$\left.\begin{aligned} \frac{\partial}{\partial \nu}\left(I_\rho \frac{\partial^2 w}{\partial t^2} - D\Delta w\right) - \frac{h^3}{12}\frac{\partial}{\partial \tau}[\tau \cdot \sigma(\nabla w)\nu] \\ = \frac{\partial}{\partial \tau}(\tau \cdot \mathbf{c}) + f, \\ \frac{h^3}{12}\nu \cdot \sigma(\nabla w)\nu = -\nu \cdot \mathbf{c} \text{ on } \Gamma_1, t > 0, \end{aligned}\right\} \tag{66.51}$$

$$w(x,0) = \frac{\partial w}{\partial t}(x,0) = 0, \quad x \in \Omega. \tag{66.52}$$

If w is a solution of Equation 66.49, its *kinetic energy* at time t is

$$\mathcal{K}(t) = \frac{1}{2}\int_\Omega (\rho h \dot{w}^2 + I_\rho |\nabla \dot{w}|^2)\, d\Omega,$$

where the quantities in the integrand are evaluated at time t. The *strain energy* of this solution at time t is given by

$$\mathcal{U}(t) = \frac{1}{2}\frac{h^3}{12}\sum_{i,j=1}^{2}\int_\Omega \sigma_{ij}(\nabla w)\varepsilon_{ij}(\nabla w)\, d\Omega.$$

A pair of functions (w_0, w_1) defined on Ω is called a *finite energy pair* if

$$\sum_{i,j=1}^{2}\int_\Omega \sigma_{ij}(\nabla w_0)\varepsilon_{ij}(\nabla w_0)\, d\Omega < \infty,$$

$$\int_\Omega (\rho h w_1^2 + I_\rho |\nabla w_1|^2)\, d\Omega < \infty.$$

A solution w of Equation 66.49 is called a *finite energy solution* if $[w(t), \dot{w}(t)]$ is a finite energy pair for each $t \geq 0$ and is continuous with respect to t into the space of finite energy pairs. This means that the solution has finite kinetic and strain energies at each instant which vary continuously in time.

Many choices of the control space $\mathcal{C}$ are possible, each of which will lead to different configuration space $\mathcal{R}_T^0$. One requirement on the choice of $\mathcal{C}$ is that the solution w corresponding to given input, f, $\mathbf{c}$ be reasonably well behaved. Another is that the choice of $\mathcal{C}$ lead to a sufficiently rich configuration space. For the problem under consideration, it is very difficult to determine the precise relation between the control and configuration spaces. For example, there is no simple characterization of those inputs for which the corresponding solution has finite energy at each instant. On the other hand, when standard control spaces with simple structure, such as L^2 spaces, are utilized, the regularity properties of the solution are, in general, difficult to determine. (This is in contrast to the situation which occurs in the analogous boundary control problem for Rayleigh beams, where it is known that finite energy solutions correspond exactly to inputs which are L^2 in time.)

In order to make the ideas precise, it is necessary to introduce certain function spaces based on the energy functionals $\mathcal{K}$ and $\mathcal{U}$. Let $L^2(\Omega)$ denote the space of square integrable functions defined on Ω, and let $H^k(\Omega)$ be the Sobolev space consisting of functions in $L^2(\Omega)$ whose derivatives up to order k (in the sense of distributions) belong to $L^2(\Omega)$. Let

$$H = \{v \in H^1(\Omega) : v|_{\Gamma_0} = 0\}.$$

The quantity

$$\|v\|_H = \left(\int_\Omega (\rho h v^2 + I_\rho |\nabla v|^2)\, d\Omega\right)^{1/2}$$

defines a Hilbert norm on H which is equivalent to the standard induced $H^1(\Omega)$ norm. Similarly, define

$$V = \{v \in H : v \in H^2(\Omega), \left.\frac{\partial v}{\partial \nu}\right|_{\Gamma_0} = 0\}.$$

The quantity

$$\|v\|_V = \left(\int_\Omega \left(\frac{h^3}{12}\sigma_{ij}(\nabla v)\varepsilon_{ij}(\nabla v)\right) d\Omega\right)^{1/2}$$

defines a seminorm on V. In fact, as a consequence of Korn's lemma, $\|\cdot\|_V$ is actually a *norm* equivalent to the standard induced $H^2(\Omega)$ norm whenever $\Gamma_0 \neq \emptyset$. Such will be assumed in what follows to simplify the discussion. The Hilbert space V is dense in H and the injection $V \mapsto H$ is compact. Let H be identified with its dual space and let V^* denote the dual space of V. Then $H \subset V^*$ with compact injection. A finite energy solution of Equations 66.49–66.51 is characterized by the statements $w(t) \in V$, $\dot{w}(t) \in H$ for each t, and the mapping $t \mapsto (w(t), \dot{w}(t))$ is continuous into the space $V \times H$. The space $V \times H$ is sometimes referred to as *finite energy space.*

Write $\mathbf{c} = c_1\mathbf{i} + c_2\mathbf{j}$. In order to assure that the configuration space is sufficiently rich, the control space is chosen as

$$\mathcal{C} = \{[f, \mathbf{c}) : f \in L^2(\Gamma_1 \times (0,T)], \quad c_i \in L^2[\Gamma_1 \times (0,T)]\}. \tag{66.53}$$

The penalty for this simple choice is that the corresponding solution, which may be defined in a certain weak sense and is unique, is not necessarily a finite energy solution. In fact, it can be shown by variational methods that, if w is the solution of Equations 66.49–66.52 corresponding to an input $(f, \mathbf{c}) \in \mathcal{C}$, then $w(t) \in H$, $\dot{w}(t) \in V^*$ and the mapping $t \mapsto [w(t), \dot{w}(t)]$ is continuous into $H \times V^*$. (A more refined analysis of the regularity of the solution may be found in [20].)

Approximate Controllability

The system (66.49) - (66.52) is called *approximately controllable* at time T if $\mathcal{R}_T^0$ is dense in $H \times V^*$. To study this problem, introduce the *control-to-state map* C_T defined by

$$C_T : \mathcal{C} \mapsto H \times V^*, \quad C_T(f, c) = [w(T), \dot{w}(T)].$$

Then the system, Equations 66.49–66.52, is *approximately controllable* at time T exactly when range(C_T) is dense in $H \times V^*$. The linear operator C_T is bounded, so, therefore, is its dual operator $C_T^* : H \times V \mapsto \mathcal{C}$. Thus, proving the approximate controllability of Equations 66.49–66.52 is equivalent to showing that

$$(\phi^1, \phi^0) \in H \times V, \quad C_T^*(\phi^1, \phi^0) = 0 \Rightarrow (\phi^1, \phi^0) = 0.$$

The quantity $C_T^*(\phi^1, \phi^0)$ may be explicitly calculated (see, [14]). It is given by the trace

$$C_T^*(\phi^1, \phi^0) = (\phi, \nabla\phi)|_{\Gamma_1 \times (0,T)},$$

where ϕ is the solution of the final value problem

$$\rho h \frac{\partial^2 \phi}{\partial t^2} - I_\rho \Delta \frac{\partial^2 \phi}{\partial t^2} + D\Delta^2 \phi = 0, \tag{66.54}$$

$$\phi = 0, \quad \frac{\partial \phi}{\partial \nu} = 0 \text{ on } \Gamma_0, 0 < t < T, \tag{66.55}$$

$$\left.\begin{aligned} &\frac{\partial}{\partial \nu}\left(I_\rho \frac{\partial^2 \phi}{\partial t^2} - D\Delta\phi\right) - \frac{h^3}{12}\frac{\partial}{\partial \tau}[\tau \cdot \sigma(\nabla\phi)\nu] = 0, \\ &\frac{h^3}{12}\nu \cdot \sigma(\nabla\phi)\nu = 0 \text{ on } \Gamma_1, 0 < t < T, \end{aligned}\right\} \tag{66.56}$$

$$\phi(x, T) = \phi^0, \quad \frac{\partial \phi}{\partial t}(x, T) = \phi^1, \quad x \in \Omega. \tag{66.57}$$

Therefore, the system, Equations 66.49–66.52, is approximately controllable if the only solution of Equations 66.54–66.57, which also satisfies

$$\phi|_{\Gamma_1 \times (0,T)} = 0, \quad \nabla\phi|_{\Gamma_1 \times (0,T)} = 0, \tag{66.58}$$

is the trivial solution. However, the boundary conditions, Equations 66.56 and 66.58, together, imply that ϕ satisfies Cauchy data on $\Gamma_1 \times (0, T)$, that is, ϕ and its derivatives up to order three vanish on $\Gamma_1 \times (0, T)$. If T is large enough, a general uniqueness theorem (Holmgren's theorem) then implies that $\phi \equiv 0$ in $\Omega \times (0, T)$. This implies approximate controllability in time T.

THEOREM 66.3 *There is a $T_0 > 0$ so that the system Equations 66.49–66.52 is approximately controllable in time $T > T_0$.*

REMARK 66.10 The optimal time T_0 depends on the material parameters and the geometry of Ω and Γ_1. If Ω is convex, then $T_0 = 2\sqrt{I_\rho/D}\, d(\Omega, \Gamma_1)$, where

$$d(\Omega, \Gamma_1) = \sup_{x \in \Omega} \inf_{y \in \Gamma_1} |x - y|.$$

REMARK 66.11 (Distributed control.) Consider the problem of approximate controllability using a distributed control rather than boundary controls. Let ω be a nonempty, open subset of Ω and let the control space be

$$\mathcal{C} = [F : F \in L^2(\Omega \times (0, T)), \ F = 0 \text{ in } \Omega \backslash \omega].$$

Consider the system consisting of Equation 66.38 and (for example) the homogeneous boundary conditions,

$$w = \frac{\partial w}{\partial \nu} = 0 \text{ on } \Gamma, t > 0.$$

Assume that the initial data is zero and let

$$\mathcal{R}_T^0 = \{(w(T), \dot{w}(T)) : F \in \mathcal{C}\}.$$

For any input F taken from $\mathcal{C}$ the corresponding solution may be shown to be a finite energy solution. Let H and V be defined as above with $\Gamma_0 = \Gamma$. The control-to-state map $C_T : F \mapsto (w(T), \dot{w}(T))$ maps $\mathcal{C}$ boundedly into $V \times H$ and its dual is given by

$$C_T^*(\phi^1, \phi^0) = \phi|_{\omega \times (0,T)},$$

where ϕ is the solution of Equation 66.54 with final data, Equation 66.57, and boundary conditions

$$\phi = 0, \quad \frac{\partial \phi}{\partial \nu} = 0 \text{ on } \Gamma, 0 < t < T. \tag{66.59}$$

If $\phi|_{\omega \times (0,T)} = 0$ and T is sufficiently large, an application of Holmgren's theorem gives $\phi \equiv 0$ in $\Omega \times (0, T)$ and, therefore, the system is approximately controllable in time T. When Ω is convex, the optimal control time is $T_0 = 2\sqrt{I_\rho/D}\, d(\Omega, \omega)$.

Exact Controllability

Again consider the system Equations 66.49–66.52 with the control space given by Equation 66.53. If $\mathcal{D}$ is a subspace in $H \times V^*$, the system is *exactly controllable to* $\mathcal{D}$ at time T if $\mathcal{D} \subset \mathcal{R}_T^0$. The *exact controllability problem*, in the strictest sense, consists of explicitly identifying $\mathcal{R}_T^0$ or, in a less restricted sense, of explicitly identifying dense subspaces $\mathcal{D}$ of $H \times V^*$ contained in $\mathcal{R}_T^0$.

To obtain useful explicit information about $\mathcal{R}_T^0$ it is necessary to restrict the geometry of Γ_0 and Γ_1. It is assumed that there is a point $x_0 \in \mathbb{R}^2$ so that

$$(x - x_0) \cdot \nu \le 0, \quad x \in \Gamma_0. \tag{66.60}$$

Condition (Equation 66.60) is a "nontrapping" assumption of the sort found in early work on scattering of waves from a reflecting obstacle. Without some such restriction on Γ_0, the results described below would not be valid. It is further assumed that

$$\overline{\Gamma}_0 \cap \overline{\Gamma}_1 = \emptyset. \tag{66.61}$$

This is a technical assumption needed to assure that solutions of the uncontrolled problem have adequate regularity up to the boundary (cf. Remark 66.13 below).

THEOREM 66.4 *Under the assumptions (Equations 66.60 and 66.61), there is a $T_0 > 0$ so that*

$$\mathcal{R}_T^0 \supset V \times H \tag{66.62}$$

if $T > T_0$.

The inclusion stated in Equation 66.62 may be stated in equivalent form in terms of the control-to-state mapping C_T. In fact, let

$$\mathcal{C}_0 = \{(f, \mathbf{c}) \in \mathcal{C} : C_T(f, \mathbf{c}) \in V \times H\},$$

and consider the restriction of C_T to $\mathcal{C}_0$. This is a closed, densely defined linear operator from $\mathcal{C}_0$ into $V \times H$. The inclusion (Equation 66.62) is the same as the assertion $C_T(\mathcal{C}_0) = V \times H$, the same as proving that the dual of the operator has a bounded inverse from $V^* \times H$ to $\mathcal{C}_0$, that is,

$$\|(\phi^0, \phi^1)\|^2_{H \times V^*} \le c \int_0^T \int_{\Gamma_1} (\phi^2 + |\nabla\phi|^2)\, d\Omega, \qquad (66.63)$$

where ϕ is the solution of Equations 66.54–66.57. For large T, the "observability estimate" (Equation 66.63) was proved in [18] under the additional geometric assumption

$$(x - x_0) \cdot \nu > 0, \quad x \in \Gamma_1. \qquad (66.64)$$

However, this hypothesis may be removed by application of the results of [20].

REMARK 66.12 It is likely that the optimal control time T_0 for exact controllability is the same as that for approximate controllability, but that has not been proved.

REMARK 66.13 The hypothesis (Equation 66.61) may be replaced by the assumption that the sets $\overline{\Gamma}_0$ and $\overline{\Gamma}_1$ meet in a strictly convex angle (measured in the interior of Ω).

REMARK 66.14 The conclusion of Theorem 66.4 is false if the space of controls is restricted to finite-dimensional controllers of the form,

$$\mathbf{c}(x, t) = \sum_{i=1}^{N} \alpha_i(x)\mathbf{c}_i(t),$$

$$f(x, t) = \sum_{i=1}^{N} \beta_i(x) f_i(t), \quad x \in \Gamma_1,\ t > 0,$$

where α_i, β_i are given $L^2(\Gamma_1)$ functions and $\mathbf{c}_i$, f_i are $L^2(0, T)$ controls, $i = 1, \ldots, N$(see [25]).

REMARK 66.15 Given a desired final state $(w_0, w_1) \in V \times H$, there are many ways of constructing a control pair $(f, \mathbf{c})$ so that $w(T) = w_0$ and $\dot{w}(T) = w_1$. The *unique* control of minimum $L^2(\Gamma_1 \times (0, T))$ norm may be constructed as follows. Set $\Sigma_1 = \Gamma_1 \times (0, T)$. Let ϕ be the solution of Equations 66.54–66.57 and let w be the solution of Equations 66.49–66.52 with

$$f = \phi|_{\Sigma_1}, \quad \text{and} \quad \mathbf{c} = -\nabla\phi|_{\Sigma_1}. \qquad (66.65)$$

Then $[w(T), \dot{w}(T)]$ depends on (ϕ^0, ϕ^1). A linear mapping Λ is defined by setting

$$\Lambda(\phi^0, \phi^1) = [\dot{w}(T), -w(T)].$$

The inequality (Equation 66.63) may be used to show that, for any $(w_0, w_1) \in V \times H$, the pair (ϕ^0, ϕ^1) may be chosen so that $\Lambda(\phi^0, \phi^1) = (w_1, -w_0)$. The argument is based on the calculation (using integrations by parts)

$$\begin{aligned} 0 &= \int_0^T \int_\Omega \phi(\rho h\ddot{w} - I_\rho \Delta\ddot{w} + D\Delta^2 w)\, d\Omega dt, \\ &= \int_\Omega [\rho h\dot{w}(T)\phi^0 + I_\rho \nabla\dot{w}(T) \cdot \nabla\phi^0 \\ &\quad - \rho h w(T)\phi^1 - I_\rho \nabla w(T) \cdot \nabla\phi^1]\, d\Omega \\ &\quad - \int_{\Sigma_1} (\phi^2 + |\nabla\phi|^2)\, d\Sigma, \end{aligned}$$

that is,

$$(\Lambda(\phi^0, \phi^1), (\phi^0, \phi^1))_{H \times H} = \int_{\Sigma_1} (\phi^2 + |\nabla\phi|^2)\, d\Sigma. \quad (66.66)$$

According to Equation 66.63, for T large enough, the right-hand side of Equation 66.66 defines a Hilbert *norm* $\|(\phi^0, \phi^1)\|_F$ and a corresponding Hilbert space F which is the completion of sufficiently smooth pairs (ϕ^0, ϕ^1) with respect to $\|\cdot\|_F$. The identity (Equation 66.66) shows that Λ is exactly the Riesz isomorphism of F onto its dual space F^*. Because $(w_0, w_1) \in \mathcal{R}_T^0$ precisely when $(w_1, -w_0) \in \text{range}(\Lambda)$, it follows that

$$\mathcal{R}_T^0 = [(w_0, w_1) : (w_1, -w_0) \in F^*].$$

The inequality (Equation 66.63) implies that $H \times V^* \supset F$. Therefore $H \times V \subset F^*$, which is the conclusion of Theorem 66.4. If $(w_0, w_1) \in V \times H$, then the minimum norm control is given by Equation 66.65, where $(\phi^0, \phi^1) = \Lambda^{-1}(w_1, -w_0)$. This procedure for constructing the minimum norm control is the basis of the *Hilbert Uniqueness Method* introduced in [22], [23].

REMARK 66.16 In the situation where $I_\rho = 0$ in Equations 66.49 and 66.51, the only change is that $H = L^2(\Omega)$ rather than the space defined above and the optimal control time is known to be $T_0 = 0$, i.e., exact controllability holds in *arbitrarily* short time (cf. [28]).

REMARK 66.17 If distributed controls are used rather than boundary controls as in Remark 66.11, Equation 66.62 is not true, in general, but is valid if ω is a neighborhood of Γ.

66.6.2 Controllability of the Reissner–Mindlin System

The controllability properties of the Reissner–Mindlin system are similar to those of the Kirchhoff system. As in the last subsection, only the boundary control problem is considered in detail.

Again, we work within the context of L^2 controls and choose Equation 66.53 as the control space. As above, it may be assumed without losing generality that the data of the problem, $F, C, \bar{w}, \bar{\mathbf{u}}, w^0, w^1, \mathbf{u}^0, \mathbf{u}^1$ vanish. The problem under consideration is, therefore,

$$\left.\begin{aligned} &\rho h\frac{\partial^2 w}{\partial t^2} - Gh\operatorname{div}(\mathbf{u}+\nabla w) = 0,\\ &I_\rho\frac{\partial^2 \mathbf{u}}{\partial t^2} - \frac{h^3}{12}\operatorname{div}\sigma(\mathbf{u}) + Gh(\mathbf{u}+\nabla w) = 0, \end{aligned}\right\} \tag{66.67}$$

$$w = 0, \quad \mathbf{u} = 0 \ \text{ on } \Gamma_0, t > 0, \tag{66.68}$$

$$\left.\begin{aligned} &Gh\nu\cdot(\mathbf{u}+\nabla w) = f,\\ &\frac{h^3}{12}\sigma(\mathbf{u})\nu = \mathbf{c} \ \text{ on } \Gamma_1, t > 0, \end{aligned}\right\} \tag{66.69}$$

$$\begin{aligned} w(x,0) &= \tfrac{\partial w}{\partial t}(x,0) = 0,\\ \mathbf{u}(x,0) &= \tfrac{\partial \mathbf{u}}{\partial t}(x,0) = 0 \quad \text{in} \quad \Omega. \end{aligned} \tag{66.70}$$

For convenience, it is assumed that $\Gamma_0 \neq \emptyset$.

Set $\mathbf{w} = \mathbf{u}+w\mathbf{k}$ and introduce the configuration space (reachable set) at time T for this problem by

$$\mathcal{R}_T^0 = \{[\mathbf{w}(T), \dot{\mathbf{w}}(T)] : (f, \mathbf{c}) \in \mathcal{C}\}.$$

To describe $\mathcal{R}_T^0$, certain function spaces based on the kinetic and strain energy functionals of the above problem must be introduced. Let

$$H = [\mathbf{v} = u_1\mathbf{i} + u_2\mathbf{j} + w\mathbf{k} \stackrel{\text{def}}{=} \mathbf{u} + w\mathbf{k} : u_i, w \in L^2(\Omega)],$$

$$\|\mathbf{v}\|_H = \left(\int_\Omega (\rho h w^2 + I_\rho|\mathbf{u}|^2)\, d\Omega\right)^{1/2},$$

$$V = \{\mathbf{v} \in H : u_i, w \in H^1(\Omega),\ \mathbf{v}|_{\Sigma_1} = 0\},$$

$$\|\mathbf{v}\|_V = \left(\int_\Omega \left(\frac{h^3}{12}\sum_{i,j=1}^{2} \sigma_{ij}(\mathbf{u})\varepsilon_{ij}(\mathbf{u}) + Gh|\mathbf{u}+\nabla w|^2\right) d\Omega\right)^{1/2}.$$

It is a consequence of Korn's lemma and $\Gamma_0 \neq \emptyset$ that $\|\cdot\|_V$ is a norm equivalent to the induced $H^1(\Omega)$ norm. The space V is dense in H and the embedding $V \mapsto H$ is compact. A solution of Equations 66.67–66.70 is a *finite energy solution* if $\mathbf{w}(t) \in V$, $\dot{\mathbf{w}}(t) \in H$ for each t, and the mapping $t \mapsto (w(t), \dot{w}(t))$ is continuous into $V \times H$. This means that the solution has finite kinetic and strain energies at each instant. As with the Kirchhoff model, solutions corresponding to inputs taken from $\mathcal{C}$ are not necessarily finite energy solutions. However, it is true that $\mathbf{w}(t) \in H$, $\dot{\mathbf{w}}(t) \in V^*$ and the mapping $t \mapsto [w(t), \dot{w}(t)]$ is continuous into $H \times V^*$, where the concept of a solution is defined in an appropriate weak sense.

The system, (Equations 66.67–66.70) is called *approximately controllable* if $\mathcal{R}_T^0$ is dense in $H \times V^*$. The *exact controllability problem* consists of explicitly identifying dense subspaces of $H \times V^*$ contained in $\mathcal{R}_T^0$.

With this setup, Theorem 66.3 and a slightly weaker version of Theorem 66.4 may be proved for the Reissner-Mindlin system. The proofs again consist of an examination of the control-to-state map $C_T : \mathcal{C} \mapsto H \times V^*$ defined by $C_T(f, \mathbf{c}) = [\mathbf{w}(T), \dot{\mathbf{w}}(T)]$. The dual mapping $C_T^* : H \times V \mapsto \mathcal{C}$ is given by

$$C_T^*(\Phi^1, \Phi^0) = \Phi|_{\Gamma_1\times(0,T)},$$

where

$$\Phi = \phi + \psi\mathbf{k}, \qquad \phi = \phi_1\mathbf{i} + \phi_2\mathbf{j},$$

$$\Phi^0 = \phi^0 + \psi^0\mathbf{k}, \qquad \Phi^1 = \phi^1 + \psi^1\mathbf{k},$$

and ϕ, ψ satisfy

$$\left.\begin{aligned} &\rho h\frac{\partial^2 \psi}{\partial t^2} - Gh\operatorname{div}(\phi+\nabla\psi) = 0,\\ &I_\rho\frac{\partial^2 \phi}{\partial t^2} - \frac{h^3}{12}\operatorname{div}\sigma(\phi) + Gh(\phi+\nabla\psi) = 0, \end{aligned}\right\} \tag{66.71}$$

$$\psi = 0, \quad \phi = 0 \ \text{ on } \Gamma_0, 0 < t < T, \tag{66.72}$$

$$\left.\begin{aligned} &Gh\nu\cdot(\phi+\nabla\psi) = 0,\\ &\frac{h^3}{12}\sigma(\phi)\nu = 0 \ \text{ on } \Gamma_1, 0 < t < T, \end{aligned}\right\} \tag{66.73}$$

$$\left.\begin{aligned} &\psi(x,T) = \psi^0, \quad \frac{\partial\psi}{\partial t}(x,T) = \psi^1,\\ &\phi(x,T) = \phi^0, \quad \frac{\partial\phi}{\partial t}(x,T) = \phi^1 \ \text{ in } \Omega. \end{aligned}\right\} \tag{66.74}$$

Approximate controllability amounts to showing that $C_T^*(\Phi^1, \Phi^0) = 0$ only when $(\Phi^1, \Phi^0) = 0$. However, if Φ is the solution of Equations 66.71–66.74 and satisfies $\Phi|_{\Gamma_1\times(0,T)} = 0$, then Φ and its first derivatives vanish on $\Gamma_1 \times (0, T)$. If T is large enough, Holmgren's theorem then implies that $\Phi \equiv 0$.

With regard to the exact controllability problem, to prove the inclusion, (Equation 66.62) for the Reissner system amounts to establishing the observability estimate (cf. Equation 66.63),

$$\|(\Phi^0, \Phi^1)\|^2_{H\times V^*} \leq c\int_0^T\int_{\Gamma_1} |\Phi|^2 d\Gamma dt. \tag{66.75}$$

For sufficiently large T, this estimate has been proved in [18] under assumptions, Equations 66.60, 66.61, and 66.64. The following exact controllability result is a consequence of Equation 66.75.

THEOREM 66.5 *Under assumptions Equations 66.60, 66.61, and 66.64, there is a* $T_0 > 0$ *so that*

$$\mathcal{R}_T^0 \supset V \times H$$

if $T > T_0$.

REMARK 66.18 If Ω is convex, then the optimal control time for approximate controllability is

$$T_0 = 2\max(\sqrt{I_\rho/D}, \sqrt{\rho/G})\, d(\Omega, \Gamma_1).$$

The optimal control time for exact controllability is probably the same, but this has not been proved. See, however, [10], [11, Chapter 5]. Assumption, Equation 66.64, is probably unnecessary, but this has not been established. Remark 66.13 is valid also for the Reissner–Mindlin system. The remarks concerning approximate and exact controllability of the Kirchhoff model utilizing distributed controls remain true for the Reissner–Mindlin system.

66.6.3 Controllability of the von Kármán System

The *global* controllability results which hold for the Kirchhoff and Reissner–Mindlin models cannot be expected to hold for nonlinear partial differential equations, in general, or for the von Kármán system in particular. Rather, only *local* controllability is to be expected (although global controllability results may obtain for certain *semilinear* systems; cf. [19]), that is, in general the most that can be expected is that the reachable set (assuming zero initial data) contains *some* ball S_r centered at the origin in the appropriate energy space, with the control time depending on r.

The problem to be considered is

$$\left.\begin{aligned} &\rho h\frac{\partial^2 w}{\partial t^2} - I_\rho\Delta\frac{\partial^2 w}{\partial t^2} + D\Delta^2 w - [w, G] = 0,\\ &\Delta^2 G + \frac{Eh}{2}[w, w] = 0,\end{aligned}\right\} \quad (66.76)$$

$$w = \frac{\partial w}{\partial \nu} = 0, \quad \text{on } \Gamma_0, t > 0, \quad (66.77)$$

$$\left.\begin{aligned} &\frac{\partial}{\partial \nu}\left(I_\rho\frac{\partial^2 w}{\partial t^2} - D\Delta w\right) - \frac{h^3}{12}\frac{\partial}{\partial \tau}[\tau\cdot\sigma(\nabla w)\nu]\\ &\qquad = \frac{\partial}{\partial\tau}(\tau\cdot \mathbf{c}) + f,\\ &\frac{h^3}{12}\nu\cdot\sigma(\nabla w)\nu = -\nu\cdot\mathbf{c}, \quad \text{on } \Gamma_1, t > 0,\end{aligned}\right\} \quad (66.78)$$

$$G = 0, \quad \frac{\partial G}{\partial \nu} = 0 \ \text{ on } \Gamma, t > 0, \quad (66.79)$$

$$w(x, 0) = \frac{\partial w}{\partial t}(x, 0) = 0, \quad x \in \Omega. \quad (66.80)$$

It is assumed that $\Gamma_0 \neq \emptyset$. Note that there is no initial data for G.

The above system is usually analyzed by uncoupling w from G. This is done by solving the second equation in Equation 66.79, subject to the boundary conditions, Equation 66.79, for G in terms of w. One obtains $G = -(Eh/2)\mathcal{G}[w, w]$, where $\mathcal{G}$ is an appropriate Green's operator for the biharmonic equation. One then obtains for w the following equation with a cubic nonlinearity:

$$\rho h\frac{\partial^2 w}{\partial t^2} - I_\rho\Delta\frac{\partial^2 w}{\partial t^2} + D\Delta^2 w - \frac{Eh}{2}[w, \mathcal{G}[w, w]] = 0. \quad (66.81)$$

The problem for w consists of Equation 66.81 together with the boundary conditions, Equations 66.77, 66.78, and initial conditions Equation 66.80.

The function spaces H and V based on the kinetic and strain energy functionals, respectively, related to the transverse displacement w, are the same as those introduced in discussing controllability of the Kirchhoff model, as is the reachable set $\mathcal{R}_T^0$ corresponding to vanishing data. Let

$$S_r = \{(v, h) \in V \times H : (\|v\|_V^2 + \|h\|_H^2)^{1/2} < r\}.$$

With this notation, a local controllability result analogous to Theorem 66.4 can be established.

THEOREM 66.6 *Under assumptions, Equations 66.60, 66.61, and 66.64, there is an* $r > 0$ *and a time* $T_0(r) > 0$ *so that*

$$\mathcal{R}_T^0 \supset S_r \quad (66.82)$$

if $T > T_0$.

Curiously, a result for the von Kármán system analogous to Theorem 66.3 is not known.

Theorem 66.6 is proved by utilizing the global controllability of the linearized (i.e., Kirchhoff) problem, together with the implicit function theorem in a manner familiar in the control theory of finite-dimensional nonlinear systems.

REMARK 66.19 If the underlying dynamics are modified by introducing a dissipative term $b(x)\dot{w}$, $b(x) > 0$, into Equation 66.81, it may be proved, under assumptions, Equations 66.60 and 66.61, that the conclusion, Equation 66.82, is valid for *every* $r > 0$. However, the optimal control time T_0 will continue to depend on r, so that such a result is still local.

66.7 Stabilizability of Dynamic Plates

The problem of stabilization is concerned with the description of *feedback controls* which assure that the trajectories of the system converge asymptotically to an equilibrium state of the system. For infinite-dimensional systems in general and distributed parameter systems in particular, there are various distinct notions of asymptotic stability: weak, strong, and uniform (distinctions which, incidentally, disappear in finite-dimensional approximations of the system). The differences in the various types of stability are related to the topology in which convergence to an equilibrium takes place. The most robust notion of stability is that of uniform stability, which guarantees that all possible trajectories starting near an equilibrium of the system converge to that equilibrium *at a uniform rate.* In this concept, convergence is usually measured in the *energy norm* associated with the system. This is the classical viewpoint of stability. Strong stability, on the other hand, guarantees asymptotic convergence of each trajectory (in the energy norm) but at a rate which may become arbitrarily small, depending on the initial state of the system. The concept of weak stability is similar; however, in this case, convergence to an equilibrium takes place in a topology weaker than associated with the energy norm. In the discussion which ensues, only uniform and strong asymptotic stability will be considered.

66.7.1 Stabilizability of Kirchhoff Plates

Consider the Kirchoff system consisting of Equation 66.49, boundary conditions, Equations 66.50, and 66.51, and initial conditions

$$w(x,0)=w^0(x),\quad \frac{\partial w}{\partial t}(x,0)=w^1(x),\quad x\in\Omega. \qquad (66.83)$$

It is assumed that $\Gamma_i \neq \emptyset$, $i=0,1$. The boundary inputs $\mathbf{c}$, f are the controls. The *boundary outputs* are

$$y=\left.\frac{\partial w}{\partial t}\right|_{\Gamma_1\times(0,\infty)},\quad \mathbf{z}=\nabla\left.\left(\frac{\partial w}{\partial t}\right)\right|_{\Gamma_1\times(0,\infty)}. \qquad (66.84)$$

The problem is to determine the boundary inputs in terms of the boundary outputs to guarantee that the resulting closed-loop system is asymptotically stable in some sense.

The *total energy* of the system at time t is

$$\begin{aligned}\mathcal{E}(t) &= \mathcal{K}(t)+\mathcal{U}(t),\\ &= \frac{1}{2}\int_\Omega (\rho h\dot{w}^2+I_\rho|\nabla\dot{w}|^2)\,d\Omega\\ &\quad +\frac{1}{2}\frac{h^3}{12}\sum_{i,j=1}^{2}\int_\Omega \sigma_{ij}(\nabla w)\varepsilon_{ij}(\nabla w)\,d\Omega,\end{aligned}$$

where $\dot{w}=\partial w/\partial t$. A direct calculation shows that

$$\frac{d\mathcal{E}}{dt}=\int_{\Gamma_1}\left[(-\nu\cdot\mathbf{c})\frac{\partial\dot{w}}{\partial\nu}+\left(\frac{\partial}{\partial\tau}(\tau\cdot\mathbf{c})+f\right)\dot{w}\right]d\Gamma.$$

When the loop is closed by introducing the proportional feedback law

$$f=-k_0y,\quad \mathbf{c}=k_1\mathbf{z},\quad k_i\geq 0,\ k_0+k_1>0, \qquad (66.85)$$

it follows that

$$\begin{aligned}\frac{d\mathcal{E}}{dt} &= -\int_{\Gamma_1}\left[k_1\left(\frac{\partial\dot{w}}{\partial\nu}\right)^2+k_0(\dot{w})^2+k_1\left(\frac{\partial\dot{w}}{\partial\tau}\right)^2\right]d\Gamma,\\ &= -\int_{\Gamma_1}[k_0(\dot{w})^2+k_1|\nabla\dot{w}|^2]\,d\Gamma\leq 0.\end{aligned}$$

Thus the feedback laws Equation 66.85 are dissipative with respect to the total energy functional $\mathcal{E}$.

Let H and V be the Hilbert spaces based on the energy functionals $\mathcal{K}$ and $\mathcal{U}$, respectively, as introduced above. If $(w^0,w^1)\in V\times H$, the Kirchoff system, Equations 66.49–66.51, with initial conditions Equation 66.83, boundary outputs Equation 66.84, and feedback law Equation 66.85, is well-posed: it has a unique finite energy solution w. The system is called *uniformly asymptotically stable* if there is a positive, real-valued function $\alpha(t)$ with $\alpha(t)\to 0$ as $t\to\infty$, so that

$$\|(w(t),\dot{w}(t))\|_{V\times H}\leq\alpha(t)\|(w^0,w^1)\|_{V\times H}.$$

Therefore, $\mathcal{E}(t)\leq\alpha^2(t)\mathcal{E}(0)$. If such a function α exists, it is necessarily exponential: $\alpha(t)=Ce^{-\omega t}$ for some $\omega>0$. The system is *strongly asymptotically stable* if, for every initial state $(w^0,w^1)\in V\times H$, the corresponding solution satisfies

$$\lim_{t\to\infty}\|(w(t),\dot{w}(t))\|_{V\times H}=0$$

or, equivalently, that $\mathcal{E}(t)\to 0$ as $t\to\infty$. Uniform and strong asymptotic stability are not equivalent concepts. Strong asymptotic stability *does not* imply uniform asymptotic stability. Strong stability has the following result.

THEOREM 66.7 *Assume that $k_i>0$, $i=0,1$. Then the closed-loop Kirchhoff system is strongly asymptotically stable.*

The proof of this theorem amounts to verifying that, under the stated hypotheses, the problem has no spectrum on the imaginary axis. The latter is a consequence of the Holmgren uniqueness theorem (see [13, Chapter 4] for details).

For the closed-loop Kirchhoff system to be uniformly asymptotically stable, the geometry of Γ_i must be suitably restricted.

THEOREM 66.8 *Assume that $k_i>0$, $i=0,1$, and that Γ_i satisfy Equations 66.60 and 66.61. Then the closed-loop Kirchhoff system is uniformly asymptotically stable.*

The proof of Theorem 66.8 follows from the estimate,

$$\mathcal{E}(T)\leq C_T\int_0^T\int_{\Gamma_1}[k_0(\dot{w})^2+k_1|\nabla\dot{w}|^2]\,d\Gamma dt,\quad T\text{ large.} \qquad (66.86)$$

The proof of Equation 66.86 is highly nontrivial. From Equation 66.86 and the above calculation of $d\mathcal{E}/dt$, it follows that

$$\mathcal{E}(T)\leq\frac{1}{1+C_T}\mathcal{E}(0),$$

which implies the conclusion of the theorem.

REMARK 66.20 Theorem 66.8 was first proved in [13] under the additional assumption Equation 66.64, but the latter condition may be removed by applying the results in [20]. Assumption, Equation 66.61, may be weakened; see Remark 66.13. If $I_\rho=0$, the conclusion holds even when $k_1=0$.

REMARK 66.21 In place of the linear relationship Equation 66.85, one may consider a nonlinear feedback law

$$y = -y(\dot{w}), \quad \mathbf{z} = \mathbf{z}(\nabla \dot{w}),$$

where $y(\cdot)$ is a real-valued function, $\mathbf{z}(\cdot) : \mathbb{R}^2 \mapsto \mathbb{R}^2$ and satisfies

$$xy(x) > 0, \quad \forall x \in \mathbb{R}\backslash\{0\},$$

$$\mathbf{x} \cdot \mathbf{z}(\mathbf{x}) > 0 \quad \forall \mathbf{x} \in \mathbb{R}^2\backslash\{0\}.$$

The closed-loop system is then dissipative. In addition, suppose that both $y(\cdot)$ and $\mathbf{z}(\cdot)$ are continuous, monotone increasing functions. The closed-loop system is then well-posed in finite energy space. Under some additional assumptions on the growth of $y(\cdot)$, $\mathbf{z}(\cdot)$ at 0 and at ∞, the closed-loop system have a decay rate which, however, will be algebraic, rather than exponential, and will depend on a bound on the initial data; cf. [13, Chapter 5] and [16].

66.7.2 Stabilizability of the Reissner–Mindlin System

The system, (Equations 66.67–66.69) is considered, along with the initial conditions

$$\left.\begin{aligned} w(x,0) &= w^0(x), \quad \frac{\partial w}{\partial t}(x,0) = w^1(x), \\ \mathbf{u}(x,0) &= \mathbf{u}^0(x), \quad \frac{\partial \mathbf{u}}{\partial t}(x,0) = \mathbf{u}^1(x), \quad x \in \Omega. \end{aligned}\right\} \tag{66.87}$$

The boundary inputs $\mathbf{c}$, f are the controls. The *boundary outputs* are

$$y = \left.\frac{\partial w}{\partial t}\right|_{\Gamma_1 \times (0,\infty)}, \quad \mathbf{z} = \left.\frac{\partial \mathbf{u}}{\partial t}\right|_{\Gamma_1 \times (0,\infty)}.$$

The *total energy* of the system at time t is

$$\mathcal{E}(t) = \mathcal{K}(t) + \mathcal{U}(t) = \tfrac{1}{2}\int_\Omega (\rho h \dot{w}^2 + I_\rho |\dot{\mathbf{u}}|^2)\, d\Omega$$
$$+ \tfrac{1}{2}\int_\Omega \left(\tfrac{h^3}{12}\sum_{i,j=1}^{2} \sigma_{ij}(\mathbf{u})\varepsilon_{ij}(\mathbf{u}) + Gh|\mathbf{u} + \nabla w|^2\right) d\Omega.$$

Then

$$\frac{d\mathcal{E}}{dt} = \int_{\Gamma_1} (f\dot{w} + \mathbf{c} \cdot \dot{\mathbf{u}})\, d\Gamma,$$

which suggests that the loop be closed by introducing the proportional feedback law

$$f = -k_0 y, \quad \mathbf{c} = -k_1 \mathbf{z}, \quad k_i \geq 0, \ k_0 + k_1 > 0,$$

so that the closed-loop system is dissipative. In fact, it may be proved that the conclusions of Theorems 66.7 and 66.8 above hold for this system; see [13], where this is proved under the additional geometric assumption, Equation 66.64.

66.7.3 Stabilizability of the von Kármán System

Consider the system consisting of Equation 66.81, boundary conditions Equations 66.77 and 66.78, and initial conditions Equation 66.83. The inputs, outputs, and total energy of this system are defined as for the Kirchhoff system, and the loop is closed using the proportional feedback law Equation 66.85. The following result has been proved in [4].

THEOREM 66.9 *Assume that $k_i > 0$, $i = 0, 1$, and that Γ_i satisfy Equations 66.60, 66.61, and 66.64. Then there is an $r > 0$ so that*

$$\mathcal{E}(t) \leq Ce^{-\omega t}\mathcal{E}(0) \tag{66.88}$$

provided $\mathcal{E}(0) < r$, where $\omega > 0$ does not depend on r.

REMARK 66.22 If $I_\rho = 0$ and $\Gamma_0 = \emptyset$, the estimate Equation 66.88 was established in [13, Chapter 5] for every $r > 0$, with constants C, ω independent of r, but under a modified feedback law for $\mathbf{c}$.

REMARK 66.23 If the underlying dynamics are modified by introducing a dissipative term $b(x)\dot{w}$, $b(x) > 0$, into Equation 66.81, it is proven in [5] that, under assumptions, Equations 66.60 and 66.61, the conclusion Equation 66.88 is valid for *every* $r > 0$, where both constants C, ω depend on r. This result was later extended in [8] to the case of nonlinear feedback laws (cf. Remark 66.21).

References

[1] Adams, R.A., *Sobolev spaces*, Academic, New York, 1975.

[2] Balakrishnan, A.V., *Applied functional analysis, 2nd ed.*, Springer, New York, 1981.

[3] Banks, H.T. and Smith, R.C., Models for control in smart material structures, in *Identification and Control, in Systems Governed by Partial Differential Equations*, Banks, H.T., Fabiano, R.H., and Ito K., Eds., SIAM, 1992, pp 27–44.

[4] Bradley, M. and Lasiecka, I., Local exponential stabilization for a nonlinearly perturbed von Kármán plate, *Nonlinear Analysis: Theory, Methods and Appl.*, 18, 333–343, 1992.

[5] Bradley, M. and Lasiecka, I., Global decay rates for the solutions to a von Kármán plate without geometric constraints, *J. Math. Anal. Appl.*, 181, 254–276, 1994.

[6] Bresse, J. A. C., *Cours de mechanique applique*, Mallet Bachellier, 1859.

[7] Hirschhorn, M. and Reiss, E., Dynamic buckling of a nonlinear timoshenko beam, *SIAM J. Appl.Math.*, 37, 290–305, 1979.

[8] Horn, M. A. and Lasiecka, I., Nonlinear boundary stabilization of a von Kármán plate equation, *Differential Equations, Dynamical Systems and Control Science: A Festschrift in Honor of Lawrence Markus*, Elworthy, K. D., Everit, W. N., and Lee, E. B., Eds., Marcel Dekker, New York, 1993, pp 581–604.

[9] Kane, T.R., Ryan, R.R., and Barnerjee, A.K., Dynamics of a beam attached to a moving base, *AIAA J. Guidance, Control Dyn.*, 10, 139–151, 1987.

[10] Komornik, V., A new method of exact controllability in short time and applications, *Ann. Fac. Sci. Toulouse*, 10, 415–464, 1989.

[11] Komornik, V., *Exact controllability and stabilization*, in *The multiplier method*, Masson - John Wiley & Sons, Paris, 1994.

[12] Krabs, W., On moment theory and controllability of one-dimensional vibrating systems and heating processes, in *Lecture Notes in Control and Information Sciences*, Springer, New York, 1992, Vol. 173.

[13] Lagnese, J. E., *Boundary stabilization of thin plates*, Studies in Applied Mathematics, SIAM, Philadelphia, 1989, Vol. 10.

[14] Lagnese, J. E., The Hilbert uniqueness method: A retrospective, in *Optimal Control of Partial Differential Equations*, Hoffmann, K. H. and Krabs, W., Eds., Springer, Berlin, 1991, pp 158–181.

[15] Lagnese, J. E., Recent progress in exact boundary controllability and uniform stabilizability of thin beams and plates, in *Distributed Parameter Systems: New Trends and Applications* Chen, G., Lee, E. B., Littmann, W., and Markus, L., Eds., Marcel Dekker, New York, 1991, pp 61–112.

[16] Lagnese, J. E. and Leugering, G., Uniform stabilization of a nonlinear beam by nonlinear boundary feedback, *J. Diff. Eqns.*, 91, 355–388, 1991.

[17] Lagnese, J. E., Leugering, G., and Schmidt, E.J.P.G., *Modeling, analysis and control of dynamic elastic multi-link structures*, Birkhäuser, Boston-Basel-Berlin, 1994.

[18] Lagnese, J. E. and Lions, J. L., *Modelling, analysis and control of thin plates*, Collection RMA, Masson, Paris, 1988, Vol. 6.

[19] Lasiecka, I. and Triggiani, R., Exact controllability of semilinear abstract systems with application to waves and plates boundary control problems, *Appl. Math. Opt.*, 23, 109–154, 1991.

[20] Lasiecka, I. and Triggiani, R., Sharp trace estimates of solutions to Kirchhoff and Euler-Bernoulli equations, *Appl. Math. Opt.*, 28, 277–306, 1993.

[21] LeDret, H., *Problèmes variationnels dans les multi-domains: Modélisation des jonctions et applications*, Collection RMA, Masson, Paris, 1991, vol. 19.

[22] Lions, J. L., *Contrôlabilité exacte, perturbations et stabilisation de systèmes distribués, Tome I: Contrôlabilté exacte*, Collection RMA, Masson, Paris, 1988, Vol. 8.

[23] Lions, J. L., Exact controllability, stabilization and perturbations for distributed systems, *SIAM Review*, 30, 1–68, 1988.

[24] Russell, D. L., On mathematical models for the elastic beam with frequency-proportional damping, in *Control and Estimation in Distributed Parameter Systems*, Banks, H.T., Ed., SIAM, 1992, pp 125–169.

[25] Triggiani, R., Lack of exact controllability for wave and plate equations with finitely many boundary controls, *Diff. Int. Equations*, 4, 683–705, 1991.

[26] Washizu, K., *Variational methods in elasticity and plasticity*, 3rd ed., Pergamon, Oxford, 1982.

[27] Wempner, G., *Mechanics of solids with applications to thin bodies*, Sijthoff and Noordhoff, Rockville, MD, 1981.

[28] Zuazua, E., Contrôlabilité exacte d'un modèle de plaques vibrantes en un temps arbitrairement petit, *C. R. Acad. Sci.*, 304, 173–176, 1987.

67

Control of the Heat Equation

Thomas I. Seidman
Department of Mathematics and Statistics, University of Maryland Baltimore County, Baltimore, MD

67.1 Introduction

We must begin[1] by distinguishing between the considerations for many practical problems in controlled heat transfer and the theoretical considerations for applying the essential ideas developed for the control theory of differential equations in systems governed by partial differential equations — here, the linear *heat equation*,

$$\frac{\partial v}{\partial t} = \frac{\partial^2 v}{\partial x^2} + \frac{\partial^2 v}{\partial y^2} + \frac{\partial^2 v}{\partial z^2}$$
$$\text{for } t > 0,\ \mathbf{x} = (x, y, z) \in \Omega \subset \mathbb{R}^3. \qquad (67.1)$$

Many of the former set of problems relate to optimal design, rather than dynamic control, and many of the concerns are about fluid flow in a heat exchanger or phase changes (e.g., condensation) or other issues beyond the physical situations described by Equation 67.1. Some properties of Equation 67.1 are relevant to these problems and we shall touch on these, but the essential concerns are outside the scope of this chapter.

This chapter will focus on the distinctions between 'lumped parameter systems' (with finite-dimensional state space, governed by ordinary differential equations) and 'distributed parameter systems' governed by partial differential equations such as Equation 67.1 so that the state, for each t, is a function of position in the spatial region Ω. Though Equation 67.1 may be viewed abstractly as an ordinary differential equation[2]

$$\frac{dv}{dt} = \Delta v + \psi \quad \text{for } t > 0, \qquad (67.3)$$

abstract ordinary differential equations such as Equation 67.3 are quite different from the more familiar ordinary differential equations with finite-dimensional states. The intuition appropriate for the parabolic partial differential Equation 67.3 is quite different from that appropriate for the *wave equation*

$$\frac{d^2 w}{dt^2} = \Delta w + \psi_2, \quad \text{for } t > 0, \qquad (67.4)$$

describing a very different set of physical phenomena with very different properties (although in Section 67.5.3 we describe an interesting relationship for the corresponding theories of observation and control).

The first two sections of this chapter provide background on relevant properties of Equation 67.1, including examples and implications of these general properties for practical heat conduction problems. We then discuss the system-theoretic properties of Equation 67.1 or 67.3. We will emphasize, in particular, the considerations arising when the input/output occurs with no direct analog in the theory of lumped parameter systems — not through the equation itself, but through the boundary conditions appropriate to the partial differential Equation 67.1. This

[1]This research has been partially supported by the National Science Foundation under the grant ECS-8814788 and by the U.S. Air Force Office of Scientific Research under the grant AFOSR-91-0008.

[2]Now $v(t)$ denotes the state, as an element of an infinite-dimensional space of functions on Ω, and $\Delta = \vec{\nabla}^2$ is the *Laplace operator*, given in the three-dimensional case by

$$\Delta : v \mapsto \frac{\partial^2 v}{\partial x^2} + \frac{\partial^2 v}{\partial y^2} + \frac{\partial^2 v}{\partial z^2} \qquad (67.2)$$

with specification of the relevant boundary conditions.

0-8493-8570-9/96/$0.00+$.50

mode of interaction is quite plausible for physical implementation because it is difficult to influence or to observe directly the behavior of the system in the interior of the spatial region.

67.2 Background: Physical Derivation

Unlike situations involving ordinary differential equations with finite-dimensional state space, one cannot work with partial differential equations without developing an appreciation for the characteristic properties of the particular kind of equation. For the classical equations, such as Equation 67.1, this is closely related to physical interpretations. Thus, we begin with a discussion of some interpretations of Equation 67.1 and only then note the salient properties needed to understand its control.

Although Equation 67.1 is the *heat equation*, governing conductive heat transfer, our intuition will be aided by noting also that this same equation also governs molecular diffusion for dilute solutions and certain dispersion phenomena and the evolution of the probability distribution in the stochastic theory of Brownian motion.

For heat conduction, the fundamental notions of *heat content* Q and *temperature*, are related[3] by

$$[\text{heat content}] = [\text{heat capacity}] \cdot [\text{temperature}]. \qquad (67.6)$$

or, in symbols,

$$Q = \rho c\, T \qquad (67.7)$$

where Q is the heat density (per unit volume), ρ is the mass density, T is the temperature, and c is the 'incremental heat capacity'. The well-known physics of the situation is that *heat will flow by conduction from one body to another at a rate proportional to the difference of their temperatures.* Within a continuum one has a *heat flux* vector $\vec{q}$ describing the heat flow: $\vec{q} \cdot \vec{n}\, dA$ is the rate (per unit time) at which heat flows through any (imaginary) surface element dA, oriented by its unit normal $\vec{n}$. This is now given by *Fourier's Law*,

$$\vec{q} = -k\, grad\, T = -k\vec{\nabla}T \qquad (67.8)$$

with a (constant[4]) *coefficient of heat conduction* $k > 0$.

[3]More precisely, because the *mass density* ρ and the *incremental heat capacity* c (i.e., the amount of heat needed to raise the temperature of a unit mass of material by 1°C when it is already at temperature ϑ) are each temperature dependent, the heat content in a region $\mathcal{B}$ with temperature distribution $T(\cdot)$ is

$$Q = Q(\mathcal{B}) = \int_{\mathcal{B}} \int_0^T [\rho c](\vartheta)\, d\vartheta\, dV. \qquad (67.5)$$

For present purposes we assume that (except, perhaps, for the juxtaposition of regions with dissimilar materials) ρc is constant. This means that we assume the temperature variation is not so large as to require working with the more complicated nonlinear model of Equation 67.5. In particular, we will not treat situations involving phase changes, such as condensation or melting.

[4]This coefficient k is, in general, also temperature dependent as well as a material property. Our earlier assumption about ρc is relevant here permitting us to assume k is a constant.

For any (imaginary) region $\mathcal{B}$ in the material, the total rate of heat flow out of $\mathcal{B}$ is then $\int_{\partial\mathcal{B}} \vec{q} \cdot \vec{n}\, dA$ (where $\vec{n}$ is the outward normal to the bounding surface $\partial\mathcal{B}$) and, by the Divergence Theorem, this equals the volume integral of $\operatorname{div}\vec{q}$. Combining this with Equations 67.7 and 67.8, and using the arbitrariness of $\mathcal{B}$, gives the governing[5] heat equation,

$$\rho c \frac{\partial T}{\partial t} = \vec{\nabla} \cdot k\vec{\nabla}T + \psi \qquad (67.9)$$

where ψ is a possible source term for heat.

Let us now derive the equation governing molecular diffusion, the spread of a substance in another (e.g., a 'solute' in a 'solvent') caused by the random collisions of molecules. Assuming a solution dilute enough that one can neglect the volume fraction occupied by the solute compared with the solvent, we may present our analysis in terms of the concentration (relative density) C of the relevant chemical component. Analogous to the previous derivation, a *material flux* vector $\vec{J}$ is given by *Fick's Law*:

$$\vec{J} = -D\vec{\nabla}C \qquad (67.10)$$

where $D > 0$ is the *diffusion coefficient*[6]. As in Equation 67.9, this law for the flux immediately leads to the conservation equation

$$\frac{\partial C}{\partial t} = \vec{\nabla} \cdot D\vec{\nabla}C + \psi \qquad (67.11)$$

where ψ is now a source term for this component — say, by some chemical reaction.

A different mechanism for the spread of some substance in another depends on the effect of comparatively small relative velocity fluctuations of the medium, e.g., gusting in the atmospheric spread of the plume from a smokestack or the effect of path variation through the interstices of packed soil in the spread of a pollutant in groundwater flow. Here one again has a material flux for the concentration — given now by *Darcy's Law*, which appears identical to Equation 67.10. The situation may well be more complicated here, however, because anisotropy (D is then a matrix) and/or various forms of degeneracy may exist (e.g., D becoming 0 when $C = 0$); nevertheless, we still get Equation 67.11 with this *dispersion coefficient* D. Here, as earlier, we will assume a constant scalar $D > 0$.

Assuming constant coefficients in each case, we can simplify Equation 67.9 or 67.11 by writing these as

$$v_t = D\Delta v + \psi \qquad (67.12)$$

[5]It is essential to realize that $\vec{q}$, given in Equation 67.8, refers to heat flow *relative to the material.* If there is spatial motion of the material itself, then this argument remains valid provided the regions $\mathcal{B}$ move correspondingly, i.e., Equation 67.3 holds in material coordinates. When this is referred to stationary coordinates, the heat is transported in space by the material motion, i.e., we have convection.

[6]More detailed treatments might consider that D depends on the temperature, etc., of the solvent and also on the existing concentration, even for dilute concentrations. As earlier, these effects are insignificant for the situations considered, so that D is constant.

where v stands either for the temperature T or the concentration C, the subscript t denotes a partial derivative, and, in considering Equation 67.9, D stands for the *thermal diffusivity* $\alpha = k/\rho c$. We may, of course, always choose units to make $D = 1$ in Equation 67.12 so that it becomes Equation 67.3. It is interesting and important for applications to know the range of magnitudes of the coefficient D in Equation 67.12 in fixed units — say, cm^2/sec. For heat conduction, typical values of the coefficient $D = \alpha$ are, approximately

8.4 for diamond,
1.1 for copper,
0.2–0.6 for steam (rising with temperature),
0.17 for cast iron,
0.086 for bronze,
0.011 for ice,
7.8×10^{-3} for glass,
4×10^{-3} for soil,
1.4–1.7×10^{-3} for water,
6.2×10^{-4} for hard rubber.

For molecular diffusion, typical figures for D might be

about 0.1 for gaseous diffusion,
0.28 for the diffusion of water vapor in air,
2×10^{-5} for air dissolved in water,
2×10^{-6} for a dilute solution of water in ethanol,
8.4×10^{-6} for ethanol in water,
1.5×10^{-8} for solid diffusion of carbon in iron,
1.6×10^{-10} for hydrogen in glass.

Finally, for the dispersion of a smoke plume in mildly stable atmosphere (say, a 15 *mph* breeze) on the other hand, D might be approximately $10^6\ cm^2/sec$. As expected, dispersion is a far more effective spreading mechanism than molecular diffusion.

Assuming one knows the initial state of the physical system,

$$v(x, t = 0) = v_0(x) \qquad \text{on } \Omega, \tag{67.13}$$

where Ω is the region of $\mathbb{R}^3$ we wish to consider, we can determine the system evolution unless we also know (or can determine) the source term ψ and, unless Ω is all of $\mathbb{R}^3$, can furnish adequate information about the interaction at the boundary $\partial\Omega$. The simplest setting is no interaction at all: the physical system is *insulated* from the rest of the universe so no flux crosses the boundary. Formally, this requires that $\vec{q} \cdot \vec{n} = 0$ or, from Equation 67.11 or 67.8 with the scaling of Equation 67.12,

$$-D\frac{\partial v}{\partial n} = -D\vec{\nabla} v \cdot \vec{n} = 0. \tag{67.14}$$

More generally, the flux might be more arbitrary but known so we have the *inhomogeneous Neumann condition*:

$$-D\frac{\partial v}{\partial n} = g_1 \qquad \text{on } \Sigma = (0, T) \times \partial\Omega. \tag{67.15}$$

An alternative[7] set of data would involve knowing the temperature (concentration) at the boundary, i.e., having the *Dirichlet condition*,

$$v = g_0 \qquad \text{on } \Sigma = (0, T) \times \partial\Omega. \tag{67.17}$$

The mathematical theory supports our physical interpretation:

> *If we have Equation 67.1 on $Q = (0, T) \times \Omega$ with ψ specified on Q and the initial condition Equation 67.13 specified on Ω, then either[8] of the boundary conditions Equation 67.15 or 67.17 suffices to determine the evolution of the system on Q, i.e., for $0 < t \leq T$.*

We refer to either of these as the *direct problem*. An important property of this problem is that it is *well-posed*, i.e., a unique solution exists for each choice of the data and small changes in the data produce[9] correspondingly small changes in the solution.

67.3 Background: Significant Properties

In this section we note some of the characteristic properties of the 'direct problem' for the partial differential Equation 67.1 and, related to these, introduce the representation formulas underlying the mathematical treatment.

67.3.1 The Maximum Principle and Conservation

One characteristic property, going back to the physical derivation, is that Equation 67.1 is a *conservation equation*. In the simplest form, when heat or material is neither created nor destroyed in the interior ($\psi \equiv 0$) and if the region is insulated in Equation 67.14, then [total heat or material] $= \int_\Omega v\, dV$ is constant in time. More

[7]Slightly more plausible physically would be to assume that the ambient temperature or concentration would be known or determinable to be g 'just outside' $\partial\Omega$ and then to use the flux law (proportionality to the difference) directly,

$$-D\frac{\partial v}{\partial n} = \vec{q} \cdot \vec{n} = \lambda(v - g) \qquad \text{on } \Sigma, \tag{67.16}$$

with a *flux transfer coefficient* $\lambda > 0$. Note that, if $\lambda \approx 0$ (negligible heat or material transport), then we effectively get Equation 67.14. On the other hand, if λ is very large ($v - g = -(D/\lambda)\partial v/\partial n$ with $D/\lambda \approx 0$), then v will immediately tend to match g at $\partial\Omega$, giving Equation 67.17; see Section 67.4.1.

[8]We may also have, more generally, a partition of $\partial\Omega$ into $\Gamma_0 \cup \Gamma_1$ with data given in the form Equation 67.17 on $\Sigma_0 = (0, T) \times \Gamma_0$ and in the form Equation 67.15 on $\Sigma_1 = (0, T) \times \Gamma_1$.

[9]Making this precise — i.e., specifying the appropriate meanings of 'small' — becomes rather technical and, unlike the situation for ordinary differential equations, can be done in several ways each useful for different situations. We will see, on the other hand, that some other problems which arise in system-theoretic analysis turn out to be 'ill-posed', i.e., do not have this well-posedness property; see, e.g., Section 67.4.3.

generally,

$$\frac{d}{dt}\left[\int_{\Omega} v\, dV\right] = \int_{\partial\Omega} g_1\, dA + \int_{\Omega} \psi\, dV \tag{67.18}$$

for v satisfying Equations 67.3 – 67.15.

Another important property is the Maximum Principle:[10]

> *Let v satisfy Equation 67.3 with $\psi \geq 0$ on $\mathcal{Q}_\tau := (0, \tau) \times \Omega$. Then the minimum value of $v(t, \mathbf{x})$ on $\overline{\mathcal{Q}_\tau}$ is attained either initially ($t = 0$) or at the boundary ($\mathbf{x} \in \partial\Omega$). Unless v is a constant, this value cannot also occur in the interior of $\mathcal{Q}_\tau$; if it is a boundary minimum with $t > 0$, then one must have $\partial v/\partial\vec{n} > 0$ at that point. Similarly, if v satisfies Equation 67.3 with $\psi \leq 0$, then its maximum is attained for $t = 0$ or at $\mathbf{x} \in \partial\Omega$, etc.*

One simple argument for this rests on the observation that, at an interior minimum, one would necessarily have $v_t = \partial v/\partial t \leq 0$ and also $\Delta v \geq 0$.

The Maximum Principle shows, for example, that the mathematics of Equation 67.1 is consistent with the requirement for physically interpreting that a concentration cannot become negative and the fact that, because heat flows 'from hotter to cooler', it is impossible to develop a 'hot spot' except by providing a heat source.

67.3.2 Smoothing and Localization

The dominant feature of Equation 67.1 is that solutions rapidly smooth out, with peaks and valleys of the initial data averaging. We will see this in more mathematical detail later, but comment now on three points:

- approach to steady state,
- infinite propagation speed, and
- localization and geometric reduction.

The first simply means that if neither ψ nor the data g_0 vary in time, then the solution v of Equations 67.3–67.17 on $(0, \infty) \times \Omega$ tend, as $t \to \infty$, to the unique solution $\bar{v}$ of the (elliptic) *steady-state equation*

$$-\left[\frac{\partial^2 \bar{v}}{\partial x^2} + \frac{\partial^2 \bar{v}}{\partial y^2} + \frac{\partial^2 \bar{v}}{\partial z^2}\right] = \psi, \quad \bar{v}\Big|_{\partial\Omega} = g_0. \tag{67.19}$$

The same would hold if we were to use Equation 67.15 rather than Equation 67.17 except that, as is obvious from Equation 67.18, we must then impose a consistency condition that

$$\int_{\partial\Omega} g_1\, dA + \int_{\Omega} \psi\, dV = 0$$

for there to be a steady state at all, and then must note that the solution of the steady-state equation

$$-\left[\frac{\partial^2 \bar{v}}{\partial x^2} + \frac{\partial^2 \bar{v}}{\partial y^2} + \frac{\partial^2 \bar{v}}{\partial z^2}\right] = \psi, \quad \partial\bar{v}/\partial\vec{n} = g_1 \tag{67.20}$$

only becomes unique when one supplements Equation 67.20 by specifying, from the initial conditions Equation 67.13, the value of $\int_{\Omega} \bar{v}\, dV$.

Unlike the situation with the wave Equation 67.4, the mathematical formulation Equation 67.1, etc., implies an infinite propagation speed for disturbances — e.g., the effect of a change in the boundary data $g_0(t, \mathbf{x})$ at some point $\mathbf{x}_* \in \partial\Omega$ occurring at a time $t = t_*$ is immediately felt throughout the region, affecting the solution for every $\mathbf{x} \in \Omega$ at every $t > t_*$. One can see that this is necessary to have the Maximum Principle, for example, but it is certainly nonphysical. This phenomenon results from idealizations in our derivation and becomes consistent with our physical intuition when we note that this 'immediate influence' is extremely small: there is a noticeable delay before a perturbation has a noticeable effect at a distance.

Consistent with the last observation, we note that the behavior in any subregion will, to a great extent, be affected only very slightly (in any fixed time) by what happens at parts of the boundary which may be very far away; this is a 'localization' principle. For example, if we are only interested in what is happening close to one part of the boundary, then we may treat the far boundary as 'at infinity'. To the extent that there is little spatial variation in the data at the nearby part of the boundary, we may then approximate the solution quite well by solving the problem on a half-space with spatially constant boundary data, dependent only on time. Taking coordinates so that the boundary becomes the plane '$x = 0$', this solution will be independent of the variables y, z if the initial data and source term are. Equation 67.1 then reduces to a one-dimensional form

$$\frac{\partial v}{\partial t} = \frac{\partial^2 v}{\partial x^2} + \psi(t, x) \tag{67.21}$$

for $t > 0$ and, now, $x > 0$ with, e.g., specification of $v(t, 0) = g_0(t)$ and of $v(0, x) = v_0(x)$. Similar dimensional reductions occur in other contexts — one might get Equation 67.21 for $0 < x < L$ where L gives the thickness of a slab in appropriate units or one might get a two-dimensional form corresponding to a body which is long compared to its constant cross-section and with data relatively constant longitudinally. In any case, our equation will be Equation 67.3, with the dimensionally suitable interpretation of the Laplace operator. Even if the initial data does depend on the variables to be omitted, our first property asserts that this variation will disappear so that we may still get a good approximate after waiting through an initial transient. On the other hand, one usually cannot accept this approximation near, e.g., the ends of the body where 'end effects' due to those boundary conditions may become significant.

[10] This should not be confused with the Pontryagin Maximum Principle for optimal control.

67.3.3 Linearity

We follow Fourier in using the *linearity* of the heat equation, expressed as a 'superposition principle' for solutions, to obtain a general representation for solutions as an infinite series. Let $\{[e_k, \lambda_k] : k = 0, 1, \ldots\}$ be the pairs of *eigenfunctions* and *eigenvalues* for $-\Delta$ on Ω, i.e.,

$$-\Delta e_k = \lambda_k e_k \text{ on } \Omega \text{ (with BC) for } k = 0, 1, \ldots \quad (67.22)$$

where "BC" denotes one of the homogeneous conditions

$$e_k = 0 \text{ or } \frac{\partial e_k}{\partial \vec{n}} = 0 \quad \text{on } \partial\Omega \quad (67.23)$$

based on considering Equation 67.17 or 67.15. It is always possible to take these so that

$$\int_\Omega |e_k|^2\, dV = 1, \quad \int_\Omega e_i e_k\, dV = 0 \text{ for } i \neq k, \quad (67.24)$$

with $0 \leq \lambda_0 < \lambda_1 \leq \ldots \to \infty$. Note $\lambda_0 > 0$ for Equation 67.17 and $\lambda_0 = 0$ for Equation 67.15.

From Equation 67.22, each function $e^{-\lambda_k t} e_k(\mathbf{x})$ satisfies Equation 67.1 so, superposing, we see that

$$v(t, \mathbf{x}) = \sum_k c_k e^{-\lambda_k t} e_k(\mathbf{x}) \quad (67.25)$$

gives the "general solution" with the coefficients (c_k) obtained from Equation 67.13 by

$$c_k = \langle e_k, v_0 \rangle \qquad \text{so } v_0(\cdot) = \sum_k c_k e_k(\cdot), \quad (67.26)$$

assuming Equation 67.24. Note that $\langle \cdot, \cdot \rangle$ denotes the $L^2(\Omega)$ inner product, $\langle f, g \rangle = \int_\Omega f(\mathbf{x}) g(\mathbf{x})\, d^m\mathbf{x}$ (for m-dimensional Ω — with $m = 1, 2, 3$). The expansion Equation 67.26, and so Equation 67.25, is valid if the function v_0 is in the Hilbert space $L^2(\Omega)$, i.e., if $\int_\Omega |v_0|^2 < \infty$. The series Equation 67.26 need not converge pointwise unless one assumes more smoothness for v_0. Because it is known that, asymptotically as $k \to \infty$,

$$\lambda_k \sim C k^{2/m} \qquad \text{with } C = C(\Omega), \quad (67.27)$$

the factors $e^{-\lambda_k t}$ decrease quite rapidly for any fixed $t > 0$, and Equation 67.25 then converges nicely to a smooth function. Indeed, this is just the 'smoothing' noted above: this argument can be used to show that solutions of Equation 67.1 are analytic (representable locally by convergent power series) in the interior of Ω for any $t > 0$ and that this does not depend on having homogeneous boundary conditions.

The same approach can be used when there is a source term ψ as in Equation 67.9 but we still have homogeneous boundary conditions as, e.g., $g_0 = 0$ in Equation 67.17. The more general representation is

$$\begin{aligned} v(t, \mathbf{x}) &= \sum_k \gamma_k(t) e_k(\mathbf{x}) \quad \text{where} \\ \gamma_k(t) &= c_k e^{-\lambda_k t} + \int_0^t e^{-\lambda_k (t-s)} \psi_k(s)\, ds, \quad (67.28) \\ c_k &= \langle e_k, v_0 \rangle, \quad \psi_k(t) = \langle e_k, \psi(t, \cdot) \rangle \end{aligned}$$

for the solution of Equation 67.9. When ψ is constant in t, this reduces to

$$\gamma_k(t) = \psi_k/\lambda_k + [c_k - \psi_k/\lambda_k]\, e^{-\lambda_k t} \longrightarrow \psi_k/\lambda_k$$

which shows that $v(t, \cdot) \to \bar{v}$, as in Equation 67.19 with $g_0 = 0$, and demonstrates the exponential rate of convergence with the transient dominated by the principal terms, corresponding to the smaller eigenvalues. This last must be modified slightly when using Equation 67.15, because $\lambda_0 = 0$.

Another consequence of linearity is that the effect of a perturbation is simply additive: if $\hat{v}$ solves Equation 67.9 with data $\hat{\psi}$ and $\hat{v}_0$ and one perturbs this to obtain a new perturbed solution $\tilde{v}$ for the data $\hat{\psi} + \psi$ and $\hat{v}_0 + v_0$ (and unperturbed boundary data), then the solution perturbation $v = \tilde{v} - \hat{v}$ itself satisfies Equation 67.9 with data ψ and v_0 and homogeneous boundary conditions. If we now multiply the partial differential equation by v and integrate,

$$\frac{d}{dt}\left(\frac{1}{2}\int_\Omega |v|^2\right) + \int_\Omega |\vec{\nabla} v|^2 = \int_\Omega v\psi,$$

using the Divergence Theorem to see that $\int v\Delta v = -\int |\vec{\nabla} v|^2$ with no boundary term because the boundary conditions are homogeneous. The Cauchy–Schwartz Inequality gives $\left|\int v\psi\right| \leq \|v\|\,\|\psi\|$ where $\|\cdot\|$ is the $L^2(\Omega)$-norm: $\|v\| = \left[\int_\Omega |v|^2\right]^{1/2}$ and we can then apply the Gronwall Inequality[11] to derive, for example, the *energy inequality*

$$\|v(t)\|^2,\ 2\int_0^t \|\vec{\nabla} v\|^2\, ds \leq \left(\|v_0\|^2 + \int_0^t \|\psi\|^2\, ds\right) e^t. \quad (67.29)$$

This is one form of the well-posedness property asserted at the end of the last section.

67.3.4 Autonomy, Similarity, and Scalings

Two additional useful properties of the heat equation are *autonomy* and *causality*. The first means that the equation itself is time independent so that a time-shifted setting gives the time-shifted solution. For the pure initial-value problem, i.e., Equation 67.1 with $g = 0$ in Equation 67.17 or 67.15, 'causality' means that $v(t, \cdot)$ is determined by its 'initial data' at *any* previous time t_0, so

$$v(t, \cdot) = \mathbf{S}(t - t_0)\, v(t_0, \cdot) \quad (67.30)$$

where $\mathbf{S}(\tau)$ is the *solution operator* for Equation 67.1 for elapsed time $\tau \geq 0$. This operator $\mathbf{S}(\tau)$ is a useful linear operator in a variety of settings, e.g., $L^2(\Omega)$ or the space $C(\bar{\Omega})$ of continuous functions with the topology of uniform convergence. A comparison with Equation 67.25 shows that

$$\mathbf{S}(t) : e_k \mapsto e^{-\lambda_k t} e_k$$

---

[11] If a function $\varphi \geq 0$ satisfies $\varphi(t) \leq C + M\int_0^t \varphi(s)\, ds$ for $0 \leq t \leq T$, then it satisfies $\varphi(t) \leq Ce^{Mt}$ there.

$$\text{so } \mathbf{S}(t)\left[\sum_k c_k e_k\right] = \left[\sum_k c_k e^{-\lambda_k t} e_k\right]. \quad (67.31)$$

From Equation 67.31, the fundamental 'semigroup property'

$$\mathbf{S}(s+t) = \mathbf{S}(t) \circ \mathbf{S}(s) \quad \text{for } t, s \geq 0. \quad (67.32)$$

This means that, if one initiates Equation 67.1 with any initial data v_0 at time 0 and so obtains $v(s, \cdot) = \mathbf{S}(s)v_0$ after a time s and $v(s+t, \cdot) = \mathbf{S}(s+t)v_0$ after a longer time interval of length $s+t$, as in (67.30), 'causality' gives $v(s+t, \cdot) = \mathbf{S}(t)v(s, \cdot)$. It is possible to verify that this operator function is strongly continuous at $t = 0$:

$$\mathbf{S}(t)v_0 \to v_0 \text{ as } t \to 0 \text{ for each } v_0$$

and differentiable for $t > 0$. Equation 67.1 tells us that

$$\frac{d}{dt}\mathbf{S}(t) = \mathbf{\Delta}\mathbf{S}(t) \quad (67.33)$$

where the Laplace operator $\mathbf{\Delta}$ here includes specifying the appropriate boundary conditions; we refer to $\mathbf{\Delta}$ in (67.33) as 'the infinitesimal generator of the semigroup $\mathbf{S}(\cdot)$'.

In terms of $\mathbf{S}(\cdot)$, a new solution representation for Equation 67.9 is

$$v(t, \cdot) = \mathbf{S}(t)v_0 + \int_0^t \mathbf{S}(t-s)\psi(s, \cdot)\, ds. \quad (67.34)$$

$\mathbf{S}$ corresponds to the 'Fundamental Solution of the homogeneous equation' for ordinary differential equations and Equation 67.34 is the usual 'variation of parameters' solution for the inhomogeneous Equation 67.9; compare also with Equation 67.25. We may also treat the system with inhomogeneous boundary conditions by introducing the *Green's operator* $\mathbf{G} : g \mapsto w$, defined by solving

$$-\mathbf{\Delta} w = 0 \text{ on } \Omega, \quad \mathbf{B}w = g \text{ at } \partial\Omega, \quad (67.35)$$

with $\mathbf{B}w$ either w or $\partial w/\partial\vec{n}$, considering Equation 67.17 or 67.15. Using the fact that $u = v - w$ then satisfies $u_t = \mathbf{\Delta} u + (\psi - w_t)$ with homogeneous boundary conditions, Equations 67.34 and 67.33 yield, after an integration by parts,

$$\begin{aligned} v(t, \cdot) &= \mathbf{S}(t)v_0 + \mathbf{G}[g_0(t) - g_0(0)] \\ &\quad + \int_0^t \mathbf{S}(t-s)\psi(s, \cdot)\, ds \\ &\quad - \int_0^t \mathbf{\Delta}\mathbf{S}(t-s)\mathbf{G}g_0(s)\, ds. \end{aligned} \quad (67.36)$$

The autonomy/causality above corresponds to the invariance of Equation 67.1 under time shifting and we now note the invariance under some other transformations. For this, we temporarily ignore considerations related to the domain boundary and take Ω to be the entire three-dimensional space $\mathbb{R}^3$.

In considering Equation 67.12 (with $\psi = 0$) with constant coefficients, we have insured that we may shift solutions arbitrarily in space. Not quite as obvious mathematically is the physically obvious fact that we may rotate in space. In particular, we may consider solutions which spatially depend only on the distance from the origin so that $v = v(t, r)$ with $r = |\mathbf{x}| = \sqrt{x^2 + y^2 + z^2}$. Equation 67.12 with $\psi = \psi(t, r)$ is then equivalent to

$$\frac{\partial v}{\partial t} = D\left[\frac{\partial^2 v}{\partial r^2} + \frac{2}{r}\frac{\partial v}{\partial r}\right] + \psi \quad (67.37)$$

which involves only a single spatial variable. For the two-dimensional setting $\mathbf{x} = (x, y)$ as in Section 67.3.2, this becomes

$$\frac{\partial v}{\partial t} = D\left[\frac{\partial^2 v}{\partial r^2} + \frac{1}{r}\frac{\partial v}{\partial r}\right] + \psi. \quad (67.38)$$

More generally, for the d-dimensional case, Equations 67.37 and 67.38 can be written as

$$v_t = r^{-(d-1)}\left(r^{d-1}v_r\right)_r + \psi. \quad (67.39)$$

The apparent singularity of these equations as $r \to 0$ is, an effect of the use of polar coordinates. As in Section 67.3.3, we may seek a series representation like Equation 67.25 for solutions of Equation 67.39 with the role of the eigenfunction expansion. Equation 67.22 now played by Bessel's equation; we then obtain an expansion in Bessel functions with the exponentially decaying time dependence $e^{-\lambda_k t}$, as earlier.

Finally, we may also make a combined scaling of both time and space. If, for some constant c, we set

$$\hat{t} = c^2 Dt, \quad \hat{\mathbf{x}} = c\mathbf{x}, \quad (67.40)$$

then, for any solution v of Equation 67.12 with $\psi = 0$, the function $\hat{v}(\hat{t}, \hat{\mathbf{x}}) = v(t, \mathbf{x})$ will satisfy Equation 67.1 in the new variables. This corresponds to the earlier comment that we may make $D = 1$ by appropriately choosing units.

Closely related to the above is the observation that the function

$$k(t, \mathbf{x}) = (4\pi Dt)^{-d/2} e^{-|\mathbf{x}|^2/4Dt} \quad (67.41)$$

satisfies Equation 67.39 for $t > 0$, and a simple computation shows[12] that

$$\int_{\mathbb{R}^d} k(t, \mathbf{x})\, d_d\mathbf{x} = 1 \text{ for each } t > 0 \quad (67.42)$$

so that $k(t, \cdot)$ becomes a δ-function as $t \to 0$. Thus, $k(t-s, \mathbf{x}-\mathbf{y})$ is the *impulse response function* for an impulse at $(s, \mathbf{y})$. Taking $d = 3$,

$$v(t, \mathbf{x}) = \int_{\mathbb{R}^3} k(t, \mathbf{x} - \mathbf{y}) v_0(\mathbf{y})\, d_3\mathbf{y} \quad (67.43)$$

is a superposition of solutions (now by integration, rather than by summation) so linearity insures that v is itself a solution; also,

$$v(t, \cdot) \longrightarrow v_0 \text{ as } t \to 0, \quad (67.44)$$

[12] $k(t, \cdot)$ is a multivariate normal distribution (Gaussian) with standard deviation $\sqrt{2Dt} \to 0$ as $t \to 0$.

where the specific interpretation of this convergence depends on how smooth v_0 is assumed to be. Thus, Equation 67.43 provides another solution representation although, as noted, it ignores the effect of the boundary for a physical region which is not all of $\mathbb{R}^3$. For practical purposes, following the ideas of Section 67.3.2, the Equation 67.43 will approximate the solution as long as $\sqrt{2Dt}$ is quite small[13] compared to the distance from the point $\mathbf{x}$ to the boundary of the region.

67.4 Some Control-Theoretic Problems

In this section we provide three elementary examples to see how the considerations above apply to some control-theoretic questions. The first relates to a simplified version of a practical heat transfer problem and is treated with the use of rough approximations, to see how such heuristic treatment can be used for practical results. The second describes the problem of control to a specified terminal state, which is a standard problem in the case of ordinary differential equations but which involves some new considerations in this distributed parameter setting. The final example is a 'coefficient identification' problem: using interaction (input/output) boundary to determine the function $q = q(x)$ in an equation of the form $u_t = u_{xx} - qu$, generalizing Equation 67.21.

67.4.1 A Simple Heat Transfer Problem

Consider a slab of thickness a and diffusion coefficient D heated at constant rate ψ. On one side the slab is insulated ($v_x = 0$) and on the other it is in contact with a stream of coolant (diffusion coefficient D') moving in an adjacent duct with constant flow rate F in the y-direction. Thus, the slab occupies $\{(x, y) : 0 < x < a,\ 0 < y < L\}$ and the duct occupies $\{(x, y) : a < x < \bar{a},\ 0 < y < L\}$ with $a, \bar{a} \ll L$ and no dependence on z.

If the coolant enters the duct at $y = 0$ with input temperature u_0, how hot will the slab become. For this purpose, assume a steady state, that the coolant flow is turbulent enough to insure perfect mixing (and so constant temperature) across the duct, and that, to a first approximation, the longitudinal transfer of heat is entirely by the coolant flow so conduction in the slab is only in the transverse direction ($0 < x < a$).

The source term ψ in the slab gives heat production $a\psi$ per unit distance in y and this must be carried off by the coolant stream for a steady state. We might, as noted earlier, shift to material coordinates in the stream to obtain an equation there. More simply, we just observe that, when the coolant has reached the point y, it has absorbed the amount $a\psi y$ of heat per second, and, for a flow rate F (choosing units so that ρc in Equation 67.7 is 1), the coolant temperature increases from u_0 to $[u_0 + a\psi y/F] =: u(y)$.

Now consider the transverse conduction in the slab, where $v_t = Dv_{xx} + \psi$ with $v_t = 0$ for steady state. As $v_x = 0$ at the outer boundary $x = 0$, the solution has the form $v = v^* - (\psi/2D)x^2$ where v^* is exactly what we wish to determine. If we assume a simple temperature match of slab to coolant ($v = u(y)$ at $x = a$), then $v^*(y) - (\psi/2D)a^2 = v(a, y) = u(y) = u_0 + a\psi y/F$ so that

$$v^* = v^*(y) = u_0 + \left[\frac{y}{F} + \frac{a}{2D}\right]a\psi$$

and

$$v = u_0 + \left[\frac{y}{F} + \frac{a}{2D}\left(1 - \left[\frac{x}{a}\right]^2\right)\right]a\psi. \quad (67.46)$$

A slight correction of this derivation is worth noting: for the coolant flow we expect a boundary layer (say, of thickness δ) of 'stagnant' coolant at the duct wall and within this layer $u_x \approx \text{constant} = -\left[v(a) - u\,|_{\text{flow}}\right]/\delta$. Also, $D'u_x = \text{flux} = a\psi$ by Fourier's Law so, instead of matching $v(a) = u(y)$, $v(a) = u(y) + (\delta/D')a\psi$ which also increases v^*, v by $(\delta/D')a\psi$ as a correction to Equation 67.46; this correction notes the reduction of heat transfer by replacing the boundary conditions Equation 67.17 by Equation 67.16 with $\lambda = D'/\delta$. Much more complicated corrections are needed to consider conduction within the duct, especially with a velocity profile other than the plug flow assumed here.

In this derivation we neglected longitudinal conduction in the slab, omitting the v_{yy} term in Equation 67.12. Because Equation 67.46 gives $v_{yy} = 0$, this is consistent with the equation. It is, however, inconsistent with reasonable boundary conditions at the ends of the slab ($y = 0, L$) with 'end effects' as well as some evening out of v^*.

Although this was derived in steady state, we could use Equation 67.46 for an optimal control problem (especially if ψ varies, but slowly) with the flow rate F as control.

67.4.2 Exact Control

Consider the problem of using ψ as control to reach a specified 'target state' $\omega = \omega(x)$ at time T. Basing the discussion on the representation[14] Equation 67.28, permits treating each component independently: the condition that $v(T, \cdot) = \omega$ becomes the

[13]When $\mathbf{x}$ is too close to the boundary for this to work well, it is plausible to think of $\partial\Omega$ as 'almost flat' on the relevant spatial scale and then to extend v_0 by reflection across it, as an odd function for Equation 67.17 with $g_0 = 0$ or as an even function for Equation 67.15 with $g_1 = 0$. For (67.17) with, say, $g_0 = g_0(t)$ locally, there would then be a further correction by adding

$$\int_0^t \hat{k}(t - s, x)g_0(s)\,ds,$$

$$\hat{k}(\tau, x) = \frac{x}{\tau}k_1(\tau, x) = 2D\frac{\partial k_1}{\partial x} \quad (67.45)$$

where $k_1 = k_1(\tau, x)$ is as in Equation 67.41 for $d = 1$ and x is the distance from $\mathbf{x}$ to the boundary; compare (67.41). There are also comparable correction formulas for more complicated settings.

[14]For definiteness, one may think of the one-dimensional heat Equation 67.21 with homogeneous Dirichlet boundary conditions at

sequence of 'moment equations',

$$\begin{aligned} y_k(T) &= c_k e^{-\lambda_k T} + \int_0^T e^{-\lambda_k(T-s)} \psi_k(s)\, ds \\ &= \omega_k := \langle e_k, \omega \rangle \end{aligned} \tag{67.48}$$

for each k. This does not determine the control uniquely, when one exists, so we select by optimality, minimizing the norm of ψ in $L^2(\mathcal{Q})$ with $\mathcal{Q} = (0, T) \times \Omega$. This is equivalent to requiring that $\psi_k(t)$ is a constant times $e^{\lambda_k(T-t)}$. Noting that $\int_0^T \left|e^{-\lambda_k(T-s)}\right|^2 ds = \left[1 - e^{-2\lambda_k T}\right]/2\lambda_k$, the conditions (67.48) yield the formula

$$\psi(t, \mathbf{x}) = 2 \sum_k \lambda_k \left(\frac{\omega_k - c_k e^{-\lambda_k T}}{1 - e^{-2\lambda_k T}} \right) e^{-\lambda_k(T-t)} e_k(\mathbf{x}). \tag{67.49}$$

This formula converges if (and only if) the specified target state ω is attainable by some control in $L^2(\mathcal{Q})$.

Let us now see what happens when we attempt to use Equation 67.49. One must truncate the expansion (say, at $k = K$) and then find each coefficient $\alpha_k = \omega_k - c_k e^{-\lambda_k T}$ with an error bound ε_k by using an algorithm of numerical integration on Ω to compute the inner products $\langle e_k, \omega \rangle$. [For simplicity we assume that we know the relevant eigenvalues and eigenfunctions, as for Equation 67.21 and a variety of higher-dimensional geometries.] Denoting the optimal control by Ψ and the approximation obtained by Ψ_K, we can bound the total error by

$$\|\Psi - \Psi_K\|_{\mathcal{Q}}^2 \le 2 \sum_{k=1}^{K} \left[\frac{\lambda_k \varepsilon_k^2}{1 - e^{-2\lambda_k T}} \right] + 2 \sum_{k > K} \left[\frac{\lambda_k \alpha_k^2}{1 - e^{-2\lambda_k T}} \right]. \tag{67.50}$$

For an attainable target ω the second sum is small for large K, corresponding to convergence of Equation 67.49. A fixed error bound $|\varepsilon_k| \le \varepsilon$ for the coefficient computation makes the first sum of the order of $K^{1+(2/d)}\varepsilon$ by Equation 67.27 which becomes large as K increases. To make the total error Equation 67.50 small requires picking K and then choosing ε dependent on this choice, or using a relative error condition: $|\varepsilon_k| \le \varepsilon |\alpha_k|$. This last seems quite plausible for numerical integration with floating-point arithmetic, but one problem remains. Neglecting v_0, a plausible form of the error estimate for a method of numerical integration might be

$$|\varepsilon_k| \le C_\nu h^\nu \|\omega e_k\|_{[\nu]} \sim C'_\nu h^\nu \lambda_k^{\nu/2} \|\omega\|_{[\nu]}$$

where h characterizes a mesh size and the subscript on $\|\cdot\|_{[\nu]}$ indicates derivatives of order up to ν, with ν depending on the choice of integration method; we have noted that $\|e_k\|_{[\nu]} \sim \lambda_k^{\nu/2}$ because the differential operator $\boldsymbol{\Delta}$ is already of order 2. This means that one might have to refine the mesh progressively to get such a uniform relative error for large k.

67.4.3 System Identification

Finally, we consider a one-dimensional example governed by an equation of the form[15]

$$\frac{\partial v}{\partial t} = D \frac{\partial^2 v}{\partial x^2} - q(x) v \tag{67.51}$$

but with D and the specific coefficient function $q(\cdot)$ unknown or known with inadequate accuracy. Assume that $u = 0$ at $x = 1$, but that interaction is possible at the end $x = 0$, where one can manipulate the temperature and observe the resulting heat flux; for simplicity, we assume that $v_0 = 0$. Thus, we consider the input/output pairing: $g \mapsto f$, defined through Equation 67.51 with

$$\begin{aligned} u(t, 1) &= 0, \quad u(t, 0) = g(t), \\ f(t) &:= -D u_x(t, 0). \end{aligned} \tag{67.52}$$

By linearity, causality, and autonomy of the Equation 67.51, this pairing takes the convolution form

$$\begin{aligned} f(t) &= \int_0^t \sigma(t - s) g(s)\, ds \\ &= \int_0^t \sigma(s) g(t - s)\, ds \end{aligned} \tag{67.53}$$

where $\sigma(\cdot)$ is an impulse-response function. Much as we obtained Equation 67.25 and 67.36,

$$\begin{aligned} \sigma(t) &= \sum_k \sigma_k e^{-\lambda_k t} \\ \text{with} \quad \sigma_k &:= -D\lambda_k e'_k(0) \langle z, e_k \rangle \\ Dz'' - qz &= 0, \ z(0) = 1, \ z(1) = 0 \\ \text{and} \quad -De''_k + qe_k &= \lambda_k e_k e_k(0) = 0 = e_k(1), \end{aligned} \tag{67.54}$$

noting that z and $\{(\lambda_k, e_k)\}$ are unknown because q is unknown.

Viewing Equation 67.53 as an integral equation for σ, Equation 67.53 determines σ for appropriate choices of the input $g(\cdot)$. The simplest would be if g is a δ-function (impulse) so the observed f would be σ: otherwise we must first solve a Volterra equation of first kind, already an ill-posed problem. The function $\sigma(\cdot)$ contains all the information about the unknown q and large differences for q may produce only very small perturbations of σ. Thus, this identification problem cannot be 'well-posed', regardless of $g(\cdot)$.

$x = 0, 1$. The eigenvalues and normalized eigenfunctions are then

$$\lambda_k = k^2\pi^2, \qquad e_k(x) = \sqrt{2} \sin k\pi x, \tag{67.47}$$

so the expansions, starting at $k = 1$ for convenience, are standard Fourier sine series.

[15] For example, such an equation might arise for a rod with heat loss to the environment, appearing through a boundary condition at the surface of the rod as in Equation 67.16, with g = constant and λ varying along the rod, reduced to a simplified one-dimensional form.

None of the coefficients σ_k will be 0 so, given $\sigma(\cdot)$, Equation 67.55 uniquely determines the eigenvalues $\{\lambda_k\}$ appearing as exponents. Note that $\lambda_k \sim D\pi^2 k^2$ so $D = \lim_k \lambda_k/\pi^2 k^2$ is then determined. We can then show (by an argument involving analytic continuation, Fourier transforms, and properties of the corresponding wave equation) that $\sigma(\cdot)$ uniquely determines $q(\cdot)$, as desired.

The discussion above does not suggest how to compute D, $q(\cdot)$ from the observations. Typically, one seeks nodal values for a discretization of q. This can be done, for example, by *history matching*, an approach often used for such identification problems, in which one solves the direct problem with a guessed q to obtain a resulting '$f = f(q)$' and proceeds to find the q which makes this best match the observed f. With some further *a priori* information about the function q, say, a known bound on the derivative q', the uniqueness result, although nonconstructive, insures convergence for these computations to the correct result as the discretization is refined. The auxiliary *a priori* information converts this to a well-posed problem, although badly conditioned so that the practical difficulties do not entirely disappear.

67.5 More Advanced System Theory

In this section we consider the system-theoretic results available for the heat equation, especially regarding observability and controllability. We emphasize how, although the relevant questions are parallel to those in 'lumped parameter' control theory, new technical difficulties can occur because of the infinite-dimensional state space; this means that this section depends more heavily on the results of Functional Analysis[16] and special mathematics for the partial differential equations involved. One new consideration is that the geometry of the region Ω is relevant here. We will concentrate primarily on problems in which input/output interaction (for control and for observation) is restricted to the boundary, partly because this is physically reasonable and partly because only for a system governed by a partial differential equation can one consider 'control via the boundary conditions'.

67.5.1 The Duality of Observability/ Controllability

For the finite-dimensional case, controllability for a problem and observability for the adjoint problem are dual. One can control $\dot{x} = Ax + Bg$ ($g(\cdot) =$ control) from one arbitrary state to another if, and only if, only the trivial solution of the adjoint equation $-\dot{y} = A^* y$ can give [observation] $= B^* y(\cdot) \equiv 0$. Similar result holds for the heat equation with boundary I/O, but we must be careful in our statement.

[16][1] is a general reference for Functional Analysis, directed toward distributed parameter system theory.

We begin by computing the relevant adjoint problem, taking the boundary control problem as

$$\begin{aligned} u_t &= \Delta u \text{ on } \mathcal{Q} \text{ with} \\ \mathbf{B}u &= g \text{ on } \Sigma \text{ and } u\Big|_{t=0} = u_0 \end{aligned} \tag{67.55}$$

in which the control function g is the data for the boundary conditions, defined on $\Sigma = (0, T) \times \partial\Omega$. As for Equation 67.35, the operator $\mathbf{B}$ will correspond to either Equation 67.17 or 67.15; in specifying $\mathbf{B}$, $g(\cdot)$ must be 0 outside some fixed 'patch', i.e., a relatively open subset $\mathcal{U} \subset \Sigma$, viewed as an 'accessible' portion of Σ, and referred to as a problem of 'patch control'. Note that

$$u_T := u(T, \cdot) = \mathbf{S}(T)u_0 + \mathbf{L}g(\cdot) \tag{67.56}$$

where Equation 67.36 gives

$$\begin{aligned} \mathbf{L} : g(\cdot) \mapsto \mathbf{G}[g(T) - g(0)] \\ - \int_0^T \Delta\mathbf{S}(T-s)\mathbf{G}g(s)\,ds. \end{aligned}$$

For the adjoint problem, we consider

$$-v_t = \Delta v \qquad \mathbf{B}v = 0; \ \varphi = [\hat{\mathbf{B}}v]\Big|_{\mathcal{U}} \tag{67.57}$$

where $\hat{\mathbf{B}}$ gives the 'complementary' boundary data: $\hat{\mathbf{B}}v := \partial v/\partial\vec{n}$ if $\mathbf{B}$ corresponds to Equation 67.17 and $\hat{\mathbf{B}}v := v\Big|_{\partial\Omega}$ if $\mathbf{B}$ corresponds to Equation 67.15. Using the Divergence Theorem,

$$\begin{aligned} \left[\int_\Omega u_T v_T\right] - \left[\int_\Omega u_0 v_0\right] &= \int_{\mathcal{Q}} (uv)_t \\ &= \int_{\mathcal{Q}} [(\vec{\nabla}^2 u)v - u(\vec{\nabla}^2 v)] \\ &= \int_\Sigma [u_{\vec{n}} v - u v_{\vec{n}}] \\ &= -\int_{\mathcal{U}} g\varphi \end{aligned}$$

where we write v_T, v_0 for $v(T, \cdot)$, $v(0, \cdot)$, respectively, and set $\mathcal{Q} = (0, T) \times \Omega$. Thus, with subscripts indicating the domain for the inner product of $L^2(\cdot)$, we have the identity

$$\langle u_T, v_T\rangle_\Omega + \langle g, \varphi\rangle_{\mathcal{U}} = \langle u_0, v_0\rangle_\Omega \tag{67.58}$$

from which we wish to draw conclusions.

First, consider the *reachable set* $\mathcal{R} = \{u_T : g = \text{any} \in L^2(\mathcal{U}); u_0 = 0\}$, which is the range of the operator $\mathbf{L} : L^2(\mathcal{U}) \to L^2(\Omega)$. If this were not dense, i.e., if we did not have $\overline{\mathcal{R}} = L^2(\Omega)$, then (by the Hahn–Banach Theorem) there would be some nonzero v_T^* orthogonal to all $u_T \in \overline{\mathcal{R}}$ so Equation 67.58 would give $\langle g, \varphi^*\rangle_{\mathcal{U}} = 0$ for all g, whence $\varphi^* = 0$, violating *detectability* (i.e., that $\varphi^* = 0$ only if $v^* = 0$). Conversely, a violation of detectability would give a nonzero v^* with $v_T^* \neq 0$ orthogonal to $\overline{\mathcal{R}}$. Thus, detectability is equivalent to *approximate controllability*. This last means that one could control arbitrarily closely to any target state, even if it cannot be reached exactly. This is a meaningless distinction for finite-dimensional linear systems although significant for the heat equation, because solutions of Equation 67.1 are analytic in the interior of Ω so that only very special targets can be exactly reachable.

Detectability means that the map $v_T \mapsto v \mapsto \varphi$ is 1–1 so, inverting, $\varphi \mapsto v_T \mapsto v_0$ is well-defined: one can predict (note the time reversal in Equation 67.57) v_0 from observation of φ on $\mathcal{U}$. In the finite-dimensional case, any linear map such as $\mathbf{A} : \varphi \mapsto v_0$ would necessarily be continuous (bounded), but here this is not automatically the case; note that the natural domain of $\mathbf{A}$ is the range of $v_T \mapsto \varphi$ and, if one had continuity, this would extend to the closure $\mathcal{M} = \mathcal{M}_{\mathcal{U}} \subset L^2(\mathcal{U})$. For bounded $\mathbf{A} : \mathcal{M} \to L^2(\Omega)$, there is a bounded adjoint operator $\mathbf{A}^* : L^2(\Omega) \to \mathcal{M}$ and, if we were to set $g = \mathbf{A}^* u_0$ in Equation 67.55,

$$\begin{aligned} \langle u_T, v_T\rangle_\Omega &= \langle u_0, \mathbf{A}\varphi\rangle_\Omega - \langle \mathbf{A}^* u_0, \varphi\rangle_{\mathcal{U}} = 0 \\ &\quad \text{for every } v_T \in L^2(\Omega). \end{aligned}$$

This would imply $u_T = 0$ so that $g = \mathbf{A}^* u_0$ is a nullcontrol from u_0. Indeed, it turns out that this g is the *optimal* nullcontrol in minimizing the $L^2(\mathcal{U})$-norm. Conversely, if there is some nullcontrol $\tilde{g}$ for each u_0, there will be a minimum-norm nullcontrol g, and the linear map $\mathbf{C} : u_0 \mapsto g$ is then continuous by the Closed-Graph Theorem; further, its adjoint $\mathbf{A} = \mathbf{C}^*$ is the observation operator: $\varphi \mapsto v_T$. Thus, bounded observability for the adjoint problem is equivalent to nullcontrollability for Equation 67.55 which, from Equation 67.56, is equivalent to considering that the range of $\mathbf{L}$ contains the range of $\mathbf{S}(T)$.

Suppose we have nullcontrollability for arbitrarily small $T > 0$, always taking $\mathcal{U} = [0, T] \times U$ for some fixed patch $U \subset \partial\Omega$. A simple argument shows that the reachable set $\mathcal{R}$ must then be entirely independent of T and of the initial state u_0. No satisfactory characterization of $\mathcal{R}$ is available, although there are various known sufficient considerations for some $\omega \in \mathcal{R}$.

67.5.2 The One-Dimensional Case

From the discussion above, it is sufficient to prove that bounded observability for the one-dimensional heat equation means it has nullcontrollability also. This was not realized when these results were first proved so that each was originally proved independently. We will consider the observability problem with $\Omega = (0, 1), \mathcal{U} = (0, T) \times \{0\}$ and

$$v_t = v_{xx} \qquad v(t,0) = v(t,1) = 0; \qquad \varphi(t) := v_x(t, 0), \tag{67.59}$$

for which we know Equation 67.47. From Equation 67.25,

$$\varphi(t) = \sum_k \tilde{c}_k e^{-\lambda_k t}, \tag{67.60}$$

and

$$v(T, \cdot) = \sum_k \frac{\tilde{c}_k}{k\pi} e^{-\lambda_k T} e_k(\cdot). \tag{67.61}$$

[For convenience, we have rereversed time compared with Equation 67.57, and from Equation 67.25, $\tilde{c}_k = \sqrt{2}k\pi \int_0^1 v_0(x) \sin k\pi x\, dx$, although we won't need any information about v_0.]

The form of Equation 67.60 is a *Dirichlet series*; this becomes a power series in $\xi = e^{-\pi^2 t}$ with only the k^2 powers appearing: $e^{-\lambda_k t} = e^{-k^2\pi t} = \xi^{k^2}$. The theory of such series centers on the Müntz–Szász Theorem (extending the Weierstrass Approximation Theorem) which, for our purposes, shows that only special functions can have L^2-convergent expansions Equation 67.60 when $\Sigma 1/\lambda_k < \infty$. An estimate for Equation 67.60 of the form is

$$|\tilde{c}_k| \le \beta_k \|\varphi\|_{L^2(0,\infty)} \tag{67.62}$$

with the values of β_k explicitly computable as an infinite product

$$\beta_k = \sqrt{1 + 2\lambda_k} \prod_{i \ne k} \left| 1 + \frac{1 + 2\lambda_k}{\lambda_i - \lambda_k} \right| \tag{67.63}$$

(convergent when $\sum_k 1/\lambda_k$ is convergent); $1/\beta_k$ is the distance in $L^2(0, \infty)$ from $\exp[-\lambda_k t]$ to $\text{span}\{\exp[-\lambda_i t] : i \ne k\}$. Schwartz has further shown that for functions given as in Equation 67.60

$$\|\varphi\|_{L^2(0,\infty)} \le \Gamma_T \|\varphi\|_{L^2(0,T)}. \tag{67.64}$$

Combining these estimates

$$\|v(T, \cdot)\|_{L^2(0,1)} \le C_T \Gamma_T \|\varphi\|_{L^2(0,T)} \qquad \left(C_T^2 := \sum_k \left[\frac{\beta_k}{k\pi}\right]^2 e^{-2k^2\pi^2 T} \right). \tag{67.65}$$

The sequence β_k increases moderately rapidly as $k \to \infty$ but the exponentials $\exp[-k^2\pi^2 T]$ decay even more rapidly so that the sum giving C_T^2 is always convergent and Equation 67.65 provides a bound ($\|\mathbf{A}\| \le \Gamma_T C_T < \infty$) for the observation operator[17] $\mathbf{A} : \mathcal{M} = \mathcal{M}_{[0,T]} \to L^2(\Omega) : \varphi \mapsto v(T, \cdot)$, when $T > 0$ is arbitrarily small.

A different way of looking at this is that the linear functional, $\mathcal{M} \to \mathbb{R} : \varphi \mapsto \tilde{c}_k$, must be given by a function $g_k \in L^2(0, T)$ so that

$$\int_0^T g_k(t) e^{-\lambda_i t}\, dt = \delta_{i,k} := \begin{cases} 0 & \text{if } i \ne k, \\ 1 & \text{if } i = k. \end{cases} \tag{67.66}$$

If $g_k(\cdot)$ is defined on $\mathbb{R}$ (0 off $[0, T]$), we may take the Fourier transform and note that Equation 67.66 asserts the 'interpolation conditions'

$$\hat{g}_k(-j\lambda_i) = \sqrt{2\pi}\,\delta_{i,k}, \quad (j = \sqrt{-1}), \tag{67.67}$$

so that it suffices to construct functions $\hat{g}_k$ satisfying Equation 67.67, together with the properties required by the Paley–Wiener Theorem, to get the inverse Fourier transform in $L^2(0, T)$

[17] A recent estimate by Borwein and Erdélyi makes it possible to obtain comparable results when $\mathcal{U}$ has the form $U = \mathcal{E} \times \{0\}$ with $\mathcal{E}$ any subset of $[0, T]$ having positive measure; one consequence of this is a bang-bang principle for time-optimal constrained boundary control.

with $\|g_k\| = \beta_k$. This approach leads to the sharp asymptotic estimate

$$\ln \|\mathbf{A}\| = \mathcal{O}(1/T) \quad \text{as } T \to 0, \tag{67.68}$$

showing how much more difficult[18] observability or controllability becomes for small T, even though one does have these for every $T > 0$.

A variant on this considers the interior point observation $\varphi(t) := v(t, a)$. The observability properties now depend on number-theoretic properties of $0 < a < 1$, e.g., for rational $a = m/n$ one gets no information at all about c_k when k is a multiple of n, because then $\sin k\pi a = 0$. It can be shown that one has bounded observability (with arbitrarily small $T > 0$) for a in a set of full measure whereas the complementary set for which this fails is uncountable in each subinterval.[19] Finally, we note that an identical treatment for all of the material of this subsection would work more generally for Equation 67.21 and with other boundary conditions.

67.5.3 Higher Dimensional Geometries

For higher dimensional cases, we can obtain observability for any 'cylindrical' region $\Omega := (0, 1) \times \hat{\Omega} \subset \mathbb{R}^d$ with $\mathcal{U} = (0, T) \times [0 \times \hat{\Omega}]$ by using the method of 'separation of variables' to reduce this to a sequence of independent one-dimensional problems: noting that

$$\lambda_{k,\ell} = k^2\pi^2 + \hat{\lambda}_\ell \quad e_{k,\ell}(x, \hat{\mathbf{x}}) = [\sqrt{2} \sin k\pi x]\hat{e}_\ell(\hat{\mathbf{x}}),$$

we get

$$\mathbf{A}v_x(\cdot, 0, \cdot) = \sum_\ell [\mathbf{A}_1\varphi_\ell](x)e^{-\hat{\lambda}_\ell T}\hat{e}_\ell(\hat{\mathbf{x}}),$$

$$\text{with} \quad \varphi_\ell(t) := e^{\hat{\lambda}_\ell t}\langle v_x(t, 0, \cdot), \hat{e}_\ell\rangle_{\hat{\Omega}}$$

where $\mathbf{A}_1$ is the observability operator for Equation 67.59. It is easy to verify that this gives $\|\mathbf{A}\| \leq \|\mathbf{A}_1\| < \infty$ and we have nullcontrollability by duality.

For more general regions, when $\mathcal{U}$ is all of $\Sigma := (0, T) \times \partial\Omega$ we may shift to the context of nullcontrollability for Equation 67.55 and rely on a simple geometric observation. Suppose $\Omega \subset \tilde{\Omega} \subset \mathbb{R}^d$ where $\tilde{\Omega}$ is some conveniently chosen region (e.g., a cylinder, as above) for which we already know that we have nullcontrollability for Equation 67.55, i.e., with $\mathcal{Q}, \mathcal{U}$ replaced by $\tilde{\mathcal{Q}} = (0, T) \times \tilde{Om}$ and $\tilde{\mathcal{U}} = \tilde{\Sigma} = (0, T) \times \partial\tilde{\Omega}$, respectively. Given any initial data $u_0 \in L^2(\Omega)$ for Equation 67.55, we extend it as 0 to all of $\tilde{\Omega}$ and, as has been assumed, let $\tilde{u}$ be a (controlled) solution of Equation 67.55, vanishing on all of $\tilde{\Omega}$ at time T. The operator $\mathbf{B}$ acting (at Σ) on $\tilde{u}$ will have *some* value, which we now call 'g' and using this in Equation 67.55 necessarily (by uniqueness) gives the restriction to $\mathcal{Q}$ of $\tilde{u}$ which vanishes at T. Thus, this g is a nullcontrol for Equation 67.55. As already noted, once we have a nullcontrol g for each u_0, it follows, as noted earlier, that the nullcontrol operator $\mathbf{C}$ for this setting is well-defined and continuous; by duality, one also has bounded observability. It is unnecessary that $\mathcal{U}$ be all of Σ here: this construction works if the 'unused' part of Σ is contained in $\tilde{\Sigma} \setminus \tilde{\mathcal{U}}$.

At this point we note a deep connection between the theories for the wave and heat equations:

> *observability for the wave equation* $w_{tt} = \Delta w$ *for some* $[\Omega, U]$ *and some* $T^* > 0$ *implies observability for the heat equation* $u_t = \Delta u$ *for arbitrary* $T > 0$ *in the same geometric setting* $[\Omega, U]$.

We now have the condition $\int_{\mathcal{U}} g_k z_i \exp[-\lambda_i t] = \delta_{i,k}$ (with $z_k := \hat{\mathbf{B}}e_k$), corresponding to Equation 67.66 for this higher dimensional problem. When $\mathcal{U}$ has the form $(0, T) \times U$ for a patch $U \subset \partial\Omega$, one can take the Fourier transform (in t only) and get, as with Equation 67.67,

$$\langle \hat{g}_k(-j\lambda_i, \cdot), z_i\rangle_U = \sqrt{2\pi}\delta_{i,k}. \tag{67.69}$$

Russell observed [5] that, if h_k were the function for the boundary observability problem for the wave equation $w_{tt} = \Delta w$ on Ω corresponding to g_k, then

$$\langle \hat{h}_{\pm k}(\pm j\sqrt{\lambda_i}, \cdot), z_i\rangle_U = \sqrt{2\pi}\delta_{i,k}\delta_\pm \tag{67.70}$$

(where $\delta_\pm = 1, 0$ as the occurrences of $\pm$ on the left do or do not match), so that the spatial dependence is identical and the time dependence closely related. This suggests constructing $\hat{g}_k$ as

$$\hat{g}_k(\tau, \cdot) := [h_{+k}(\sigma) + h_{-k}(\sigma)]/2 \quad \sigma^2 := j\tau,$$

(noting that the rhs is an even analytic function of σ and so is an analytic function of σ^2), so that Equation 67.70 implies the desired Equation 67.69. This does not quite work, since it would not give $g_k \in L^2(\mathcal{U})$, but can be modified, multiplying by a suitable function $R = R(\tau)$ on the right, to get g_k as an inverse Fourier transform. We note that it is the construction of the modifier $R(\cdot)$ which gives the asymptotics Equation 67.68; it is also this multiplication which makes the implication above irreversible. The relation used here is also usable to obtain uniqueness results for a higher dimensional version of the identification problem of Section 67.4.3 from known corresponding results for the wave equation.

One gets the observability for the wave equation (hence, for the heat equation) for suitable[20] $[\Omega, U]$ by an argument [5], [7] based on Scattering Theory (existence of a uniform decay rate for the portion of the total energy remaining in Ω when this is embedded in a larger region $\tilde{\Omega}$). In particular, one can take $\tilde{\Omega}$

[18] This may be compared to the corresponding estimate $\|\mathbf{A}\| = \mathcal{O}\left(T^{-(K+1/2)}\right)$ for the finite-dimensional setting, with K the minimal index giving the rank condition there.

[19] Because the 'bad set' has measure 0, one might guess that observation using local integral averages (as a 'generalized thermometer') should always work but, somewhat surprisingly, this is false.

[20] The finite propagation speed associated with the wave equation implies the existence of a minimum time T_*, depending on Ω, for observability/nullcontrollability.

as the complement of a 'star-shaped' obstacle at which $\mathbf{B}w = 0$ provided $\partial\Omega\setminus U \subset \partial\tilde{\Omega} = \partial$ (obstacle). On the other hand, there is a significant geometric restriction on $[\Omega, U]$ for which one can observe or control the wave equation: there can be no 'trapped waves' (which continually reflect off the unused boundary $\Omega \setminus U$ without ever intersecting U) which, roughly, requires that $\partial\Omega\setminus U$ should be 'visible' from some single point outside $\overline{\Omega}$.

This argument then gives observability/controllability for the heat equation for a variety of geometric settings, but there is a price, the geometric restriction noted above. This is quite reasonable for the wave equation but seems irrelevant to the behavior of the heat equation and it was long conjectured that accessibility of an arbitrary patch $U \subset \partial\Omega$ would suffice for observation or control of Equation 67.1. Quite recently, Lebeau and Robbiano have settled this conjecture by demonstrating — using an argument based on new Carleman-type estimates for a related elliptic equation — patch nullcontrollability using an arbitrary patch[21] $\mathcal{U}$ for an arbitrary Ω.

References

[1] Curtain, R.F. and Pritchard, A.J., *Functional Analysis in Modern Applied Mathematics*, Academic, New York, 1977.

[2] Krabs, W., *On Moment Theory and Controllability of One-Dimensional Vibrating Systems and Heating Processes,* (Lecture Notes in Control and Inf. Sci. # 173), Springer, Berlin, 1992.

[3] Lasiecka, I. and Triggiani, R., *Differential and Algebraic Riccati Equations with Application to Boundary/Point Control Problems: Continuous Theory and Approximation Theory,* (Lecture Notes in Control and Inf. Sci. # 164), Springer, Berlin, 1991.

[4] Lions, J.-L., Exact controllability, stabilization, and perturbations, *SIAM Review,* 30, 1–68, 1988.

[5] Russell, D.L., A unified boundary controllability theory for hyperbolic and parabolic partial differential equations, *Stud. Appl. Math.,* LII, 189–211, 1973.

[6] Seidman, T.I., Boundary observation and control for the heat equation. In *Calculus of Variations and Control Theory,* Russell, D.L., Ed., Academic, New York, 1976, pp 321–351.

[7] Seidman, T.I., Exact boundary controllability for some evolution equations, *SIAM J. Control Opt.,* 16, 979–999, 1978.

[21]They have given this result for the case of boundary controllability described here and also for the case of distributed control — Equation 67.3 with homogeneous boundary conditions and control function ψ, vanishing outside a patch $\mathcal{U}$ open in $\mathcal{Q} = (0, T) \times \Omega$.

68

Observability of Linear Distributed-Parameter Systems

David L. Russell
Department of Mathematics, Virginia Tech, Blacksburg, VA

68.1 Comparison with the Finite Dimensional Case

The general question of *observability* in the finite dimensional setting concerns a system

$$\dot{x} = f(x, ...), x \in \mathbf{R}^n,$$

where the ellipsis indicates that the system might involve additional parameters, controls, disturbances, etc., and an *output (measurement, observation)* function

$$y = g(x, ...), y \in \mathbf{R}^m.$$

Observability theory is concerned, first of all, with the question of *distinguishability*, i.e., whether distinct system trajectories $x(t)$, $\hat{x}(t)$ necessarily give rise to distinct outputs $y(t)$, $\hat{y}(t)$ over a specified time interval. The *observability* question, properly speaking, concerns actual *identification* of the trajectory $x(t)$, equivalently, the initial state x_0, from the available observations on the system with an ultimate view to the possibility of *reconstruction* of the trajectories, or perhaps the initial or current states, which, in application, are generally not directly available from the outputs, typically consisting of a fairly small set of recorded instrument readings on the system. In the linear case, observability and distinguishability are equivalent, and both questions can be treated in the context of a vector/matrix system

$$\dot{x} = A(t)x, x \in \mathbf{R}^n, A(t) \in \mathbf{R}^{n\times n}, \tag{68.1}$$

together with an output vector y related to x via

$$y = C(t)x, y \in \mathbf{R}^m, C(t) \in \mathbf{R}^{m\times n}, \tag{68.2}$$

and it is a standard result [1] that observability obtains on an interval $[0, T]$, $T > 0$, just in case $y \equiv 0$ implies that $x \equiv 0$ on that interval. This observability property, in turn, is equivalent to the positive definiteness of the matrix integral

$$Z(T) = \int_0^T \Phi(t,0)^* C(t)^* C(t) \Phi(t,0)\, dt, \tag{68.3}$$

where $\Phi(t, s)$ is the fundamental solution matrix of the system with $\Phi(s, s) = I$. The initial state x_0 can then be reconstructed from the observation $y(t)$ by means of the *reconstruction operator* (cf. [18])

$$x_0 = Z(T)^{-1} \int_0^T \Phi(t,0)^* C(t)^* y(t)\, dt. \tag{68.4}$$

In the constant coefficient case $A(t) \equiv A$, observability (on any interval of positive length) is equivalent to the algebraic condition that no eigenvector of A should lie in the null space of C; there are many other equivalent formulations.

In the (infinite dimensional) *distributed-parameter* setting, it is not possible to provide any comparably concise description of the general system to be studied or universal definition of the terms involved, but we will make an attempt in this direction later in the chapter. To set the stage for that, let us begin with a very simple example that illustrates many of the complicating factors involved.

The term *distributed-parameter system* indicates a system whose state parameters are distributed over a spatial region rather than constituting a discrete set of dependent variables. For our example, we take the spatial region to be the interval $[0, 1]$ and represent the state by a function $w(x, t)$, $x \in [0, 1]$, $t \in [0, \infty)$. Let us think of $w(x, t)$ as representing some physical property (temperature, concentration of a dissolved chemical, etc.) in a fluid moving from left to right through a conduit whose physical extent corresponds to $0 \leq x \leq 1$ with a uniform unit velocity. If we assume no diffusion process is involved, it is straightforward to see that $w(x, t)$ will be, in some sense that we do not elaborate

0-8493-8570-9/96/$0.00+$.50

on at the moment, a solution of the first-order *partial differential equation* (PDE)

$$\frac{\partial w}{\partial t} + \frac{\partial w}{\partial x} = 0, \tag{68.5}$$

to which we need to adjoin a *boundary condition*

$$w(0, t) = v(t), t \geq 0, \tag{68.6}$$

and an *initial condition* or *initial state* given, without loss of generality, at $t = 0$,

$$w(x, 0) = w_0(x), x \in [0, 1]. \tag{68.7}$$

It can be shown (cf. [4], e.g.) that with appropriate regularity assumptions on $v(t)$ and $w_0(x)$ there is a unique solution $w(x, t)$ of Equations 68.5, 68.6, and 68.7 for $x \in [0, 1], t \in [0, \infty)$.

We will consider two different types of measurement. The first is a *point* measurement at a given location x_0,

$$y(t) \equiv w(x_0, t), \tag{68.8}$$

while the second is a *distributed* measurement, which we will suppose to have the form

$$y(t) \equiv \int_0^1 c(x)\, w(x, t)\, dx, \tag{68.9}$$

for some piecewise continuous function $c(x)$, defined and not identically equal to zero on $[0, 1]$.

Let us suppose that we have an initial state $w_0(x)$ defined on $[0, 1]$, while the boundary input (Equation 68.6) is identically equal to 0. Let us examine the simplest form of observability, *distinguishability*, for the resulting system. In this case, the solution takes the form, for $t \geq 0$,

$$w(x, t) \equiv \begin{cases} w_0(x - t), \ t \leq x \leq 1, \\ 0, x < t. \end{cases}$$

For a point observation at x_0, the output obtained is clearly

$$y(t) = w(x_0, t) = \begin{cases} w_0(x_0 - t), 0 \leq t \leq x_0, \\ 0, t > x_0. \end{cases}$$

We see that if $x_0 < 1$, the data segment consisting of the values

$$w_0(x), x_0 \leq x \leq 1$$

is lost from the data. Consequently, we do not have distinguishability in this case because initial states $w_0(x)$, $\tilde{w}_0(x)$ differing only on the indicated interval cannot be distinguished on the basis of the observation $y(t)$ on any interval $[0, T]$. On the other hand, for $x_0 = 1$ the initial state is simply rewritten, in reversed order, in the values of $y(t), 0 \leq t \leq 1$ and, thus, we have complete knowledge of the initial state $w_0(x)$ and, hence, of the solution $w(x, t)$ determined by that initial state, provided the length of the observation interval is at least unity. If the length of the interval is less than one, we are again lacking some information on the initial state. It should be noted that this is already a departure from the finite dimensional case; we have here a time-independent system for which distinguishability is dependent on the length of the interval of observation.

Now let us consider the case of a distributed observation; for definiteness, we consider the case wherein $c(x)$ is the characteristic function of the interval $[1 - \delta, 1]$ for $0 \leq \delta \leq 1$. Thus,

$$y(t) = \int_{1-\delta}^1 w(x, t)\, dx = \int_{min\{0,1-(t+\delta)\}}^{min\{0,1-t\}} w_0(x)\, dx$$

If $y(t) \equiv 0$ on the interval $[0, 1]$, then by starting with $t = 1$ and decreasing to $t = 0$ it is easy to see that $w_0(x) \equiv 0$ and thus $w(x, t) \equiv 0$; thus, we again have the property of distinguishability if the interval of observation has length ≥ 1. But now an additional feature comes into play, again not present in the finite dimensional case. If we consider trigonometric initial states

$$w_0(x) = \sin \omega x, \quad \omega > 0$$

we easily verify that $|y(t)| \leq \frac{\delta}{\omega}, \ t \geq 0$, a bound tending to zero as $\omega \to \infty$. Thus, it is not possible to bound the supremum norm of the initial state in terms of the corresponding norm of the observation, whatever the length of the observation interval. Thus, even though we have the property of distinguishability, we lack observability in a stronger sense to the extent that we cannot reconstruct the initial state from the indicated observation in a continuous (i.e., bounded) manner. This is again a departure from the finite dimensional case wherein we have noted, for linear systems, that distinguishability/observability is equivalent to the existence of a bounded reconstruction operator.

68.2 General Formulation in the Distributed-Parameter Case

To make progress toward some rigorous definitions we have to introduce a certain degree of precision into the conceptual framework. In doing this we assume that the reader has some background in the basics of functional analysis [17]. Accordingly, then, we assume that the process under study has an associated *state space*, W, which we take to be a *Banach space* with *norm* $\|w\|_W$ [in specific instances, this is often strengthened to a *Hilbert space* with *inner product* $(w, \hat{w})_W$]. The process itself is described by an operator differential equation in W,

$$\dot{w} = A\, w \quad , \tag{68.10}$$

where A is a (typically unbounded, differential) closed linear operator with domain $\mathcal{D}(A)$ constituting a dense subspace of W satisfying additional conditions (cf. [6], e.g.) so that the *semigroup of bounded operators* e^{At} is defined for $t \geq 0$, *strongly continuous* in the sense that the state trajectory associated with an initial state $w_0 \in W$,

$$w(t) = e^{At}\, w_0, \tag{68.11}$$

is continuous in W for $t \geq 0$. Many time-independent PDEs can be represented in this way, with A corresponding to the "spatial" differential operator appearing in the equation (e.g., in the case of Equation 68.1 we could take W to be $C[0,1]$, A to be the operator defined by $A\,w = -\frac{\partial w}{\partial x}$ and $\mathcal{D}(A) = C_0^1[0,1]$, the subspace of $C[0,1]$ consisting of continuously differentiable functions on $[0,1]$ with $w(0) = 0$). It should be noted that the range of e^{At}, and hence $w(t)$, is not, in general, in $\mathcal{D}(A)$. It can be shown that this is the case if $w_0 \in \mathcal{D}(A)$.

Now let Y be a second Banach space, the *output*, or *measurement* space and let $C : W \to Y$ be the observation operator; in general, unbounded but with domain including $\mathcal{D}(A)$. Further, given an observation interval $[0,T]$, $T > 0$, let $\mathcal{Y}_T$, be the *space of observations*; e.g., when Y is a Banach space we might let $\mathcal{Y}_T = C([0,T];Y)$, the space of Y continuous functions on $[0,T]$ with the supremum (Y) norm, or, in the case where Y is a Hilbert space, we might wish to take $\mathcal{Y}_T = L^2([0,T];Y)$, the space of norm square integrable functions with range in Y. This space is generally defined so that, for an initial state $w_0 \in \mathcal{D}(A)$, which results, via Equation 68.11, in a trajectory $w(t) \in \mathcal{D}(A)$, the observation function is

$$y(t) = C\,w(t) \in \mathcal{Y}_T. \tag{68.12}$$

The observation operator C is said to be *admissible* if the linear map from $\mathcal{D}(A)$ to $\mathcal{Y}_T$, $w_0 \to y(\cdot)$, has a continuous extension to a corresponding map from W to $\mathcal{Y}_T$; we will continue to describe this map via Equation 68.12 even though $C\,w(t)$ will not, in general, be defined for general $w_0 \in W$. This whole process may seem very complicated but it cannot be avoided in many, indeed, the most important, examples. In fact, it becomes necessary in the case of point observations on Equation 68.1 if that system is posed in the space $W = L^2[0,1]$ because, although the state $w(\cdot,t)$ is defined as an element of $L^2[0,1]$ for each t, the value $w(1,t)$ may not be defined for certain values of t.

With this framework in place we can introduce some definitions.

DEFINITION 68.1 The *linear observed system* of Equations 68.10 and 68.12 is *distinguishable* on an interval $[0,T]$, $T \geq 0$, if and only if $y(\cdot) = 0$ in $\mathcal{Y}_T$ implies that $w_0 = 0$ in W.

DEFINITION 68.2 Given $T \geq 0$ and $\tau \in [0,T]$, the linear observed system of Equations 68.10 and 68.12 is τ*-observable* on $[0,T]$ if and only if there exists a positive number $\gamma \geq 0$ such that, for every initial state $w_0 \in W$ and resulting state (via Equation 68.11) $w(\tau)$ in W, we have

$$\|y(\cdot)\|_{\mathcal{Y}_T} \geq \|w(\tau)\|_W, \tag{68.13}$$

$y(\cdot)$ being the observation obtained via Equations 68.11, 68.12.

REMARK 68.1 In the finite dimensional context, for $\tau_1, \tau_2 \in [0,T]$, τ_1-observability is equivalent to τ_2-observability, though the (largest) corresponding values of γ may be different. We will see, in a *heat conduction* example to be discussed later, that this need no longer be the case for distributed-parameter systems. It does remain true for *time-reversible* distributed-parameter systems of the type discussed here, corresponding, e.g., to the *wave equation* [19], or in the case of the *elastic beam equation*, which we discuss at some length later.

Let us note that, just as in the finite dimensional case, the theory of observability for distributed-parameter systems forms the basis for *observer theory* and *state estimation theory* (cf. [15], [18]) in the distributed-parameter context. Observability also enters into the question of asymptotic stability for certain linear distributed-parameter systems via its connection with the *La Salle Invariance Principle* [12], [20]. The question of observability also arises in *parameter identification* studies [3]. Distributed parameter observability plays a dual role to *controllability* for the *dual control system*, but the relation between the two is not quite as simple as it is in the corresponding finite dimensional context. The reader is referred to [5] for details of this relationship.

In the case of finite dimensional linear systems, a weaker concept than observability, *detectability*, is often introduced. The constant coefficient version of Equations 68.1, 68.2 is detectable (on the interval $[0,\infty)$ just in case $y(t) \equiv 0 \to x(t) \to 0$ as $t \to \infty$. This concept is not as useful in the distributed-parameter context because, unlike the constant coefficient linear case, those components of the solution tending to zero do not necessarily tend to zero at a uniform exponential rate (see, e.g., [20]). A more useful concept is that of γ*-detectability* for a given $\gamma \geq 0$; the system of Equations 68.10 and 68.12 enjoys this property on the interval $[0,\infty)$ just in case $y(t) \equiv 0$ implies that, for some $M \geq 0$, $\|w(\cdot,t)\|_W \leq M\,e^{\gamma t}$, $t \geq 0$.

68.3 Observation of a Heat Conduction Process

Let us consider an application involving PDEs of *parabolic type* in several space dimensions. In the steel industry, it is important that the temperature distribution in a steel ingot in preparation for rolling operations should be as uniform as possible. It is clearly difficult, if not impossible, to determine the temperature distribution in the interior of the ingot directly, but the measurement of the surface temperature is routine. We therefore encounter the problem of the observability of the temperature distribution throughout the ingot from the available surface measurements.

In order to analyze this question mathematically, we first require a model for the process. If we represent the spatial region occupied by the ingot as a region $\Omega \in \mathbf{R}^3$ with smooth boundary Γ and suppose the measurement process takes place over a time interval $0 \leq t \leq T$, we are led by the standard theory of heat conduction to consider the parabolic PDE

$$\rho\,\frac{\partial w}{\partial t} = k\left(\frac{\partial^2 w}{\partial x^2} + \frac{\partial^2 w}{\partial y^2} + \frac{\partial^2 w}{\partial z^2}\right), \tag{68.14}$$

where ρ is the specific heat of the material and k is its thermal conductivity, both assumed constant here. The rate of heat loss to the exterior environment is $k\,\frac{\partial w}{\partial \nu}$, where ν denotes the unit normal vector to Γ, external with respect to Ω, and this rate is proportional to the difference between the surface temperature w at the same point on the surface and the ambient temperature, which we will assume, for simplicity, to be zero. The system under study therefore consists of Equation 68.14 together with a Dirichlet-Neumann boundary condition

$$k\frac{\partial w}{\partial \nu} = \sigma\, w, \tag{68.15}$$

where σ is a positive constant of proportionality. Using X to stand for the triple x, y, z, the (unknown) initial condition is

$$w(X,0) \;=\; w_0(X). \tag{68.16}$$

The standard theory of parabolic PDEs [22] guarantees the existence of a unique solution $w(X,t)$, $X \in \Omega$, $t \in [0,T]$ of Equation 68.14 with the boundary/initial data of Equations 68.15 and 68.16. The available measurement data are

$$y(X,t) \;=\; w(X,t),\, X \in \Gamma,\, t \in [0,T]. \tag{68.17}$$

The question then is whether it is possible to reconstruct the temperature distribution $w(X,\tau)$, $X \in \Omega$ at a particular instant $\tau \in [0,T]$ on the basis of the measurement (Equation 68.18). This question has been extensively studied, sometimes indirectly via the dual question of boundary controllability. Brevity requirements constrain us to cite only [13], [14] here, but we will indicate below some of the mathematical issues involved.

The Laplacian operator appearing on the right-hand side of Equation 68.14, defined on a dense domain in $L^2(\Omega)$ incorporating the boundary condition of Equation 68.15, is known to be a positive self-adjoint differential operator with positive eigenvalues λ_k, $k = 1,2,3,\ldots$ and has corresponding normalized eigenfunctions ϕ_k, $k = 1,2,3,\ldots$ forming an orthonormal basis for the Hilbert space $L^2(\Omega)$. An initial state w_0 in that space has an expansion, convergent in that space,

$$w_0 = \sum_{k=1}^{\infty} c_k\, \phi_k. \tag{68.18}$$

Corresponding to this expansion, the system of Equations 68.14 to 68.16 has the solution

$$w(X,t) \;=\; \sum_{k=1}^{\infty} c_k\, e^{-\lambda_k t}\, \phi_k(X), \tag{68.19}$$

with the corresponding measurement, or observation

$$y(X,t) \;=\; \sum_{k=1}^{\infty} c_k\, e^{-\lambda_k t}\, \phi_k(X),\; X \in \Gamma,\, t \in [0,T]. \tag{68.20}$$

involving known λ_k, ϕ_k, but unknown c_k.

Let us first consider the question of distinguishability: Can two solutions $w(X,t)$, $\tilde{w}(X,t)$, corresponding to initial states w_0, $\tilde{w}_0$ produce the same observation $y(X,t)$ via Equation 68.17? From the linear homogeneous character of the system it is clear that this is equivalent to asking whether a nonzero initial state (Equation 68.18), i.e., such that not all $c_k = 0$, can give rise to an observation (Equation 68.20) that is identically zero. This immediately requires us to give attention to the boundary values $\eta_k(X,t) \;\equiv\; e^{-\lambda_k t}\,\phi_k(X)$ corresponding to $w_0(X) \;=\; \phi_k(X)$. Clearly, the boundary observation corresponding to a general initial state (Equation 68.18) is then

$$y(X,t) = \sum_{k=1}^{\infty} c_k\, \eta_k(X,t),\; X \in \Gamma,\, t \in [0,T], \tag{68.21}$$

Can $y(X,t)$, taking this form, be identically zero if the c_k are not all zero? The statement that this is not possible, hence that the system is distinguishable, is precisely the statement that the $\eta_k(X,t)$, $X \in \Gamma$, $t \in [0,T]$ are *weakly independent* in the appropriate boundary space, for simplicity, say $L^2(\Gamma \times [0,T])$. As a result of a variety of investigations [9], we can assert that this is, indeed, the case for any $T \geq 0$. In fact, these investigations show that a stronger result, *spectral observability*, is true. The functions $\eta_k(X,t)$ are actually *strongly independent* in $L^2(\Gamma \times [0,T])$, by which we mean that there exist *biorthogonal* functions, not necessarily unique, in $L^2(\Gamma \times [0,T])$ for the $\eta_k(X,t)$, i.e., functions $\zeta_k(X,t) \in L^2(\Gamma \times [0,T])$, $k = 1,2,3,\ldots$, such that

$$\int_{\Gamma\times[0,T]} \zeta_k(X,t)\,\eta_j(X,t)\,dX\,dt \;=\; \begin{cases} 0, & k \neq j, \\ 1, & k = j. \end{cases} \tag{68.22}$$

The existence of these biorthogonal functions implies that any finite number of the coefficients c_k can be constructed from the observation $y(X,t)$ via

$$c_k = \int_{\Gamma\times[0,T]} \zeta_k(X,t)\, y(X,t)\, dX\, dt. \tag{68.23}$$

Indeed, we can construct a map from the output space, here assumed to be $L^2(\Gamma \times [0,T])$, namely,

$$S_{\tau,K}\, Y = \sum_{k=1}^{K} e^{-\lambda_k \tau}\phi_k(X) \int_{\Gamma\times[0,T]} \zeta_k(X,t)\, y(X,t)\, dX\, dt, \tag{68.24}$$

carrying the observation Y into the "K-approximation" to the state $w(\cdot,\tau)$, $\tau \in [0,T]$. Since the sum (Equation 68.18) is convergent in $L^2(\Omega)$, this property of spectral observability is a form of *approximate observability* in the sense that it permits reconstruction of the initial state of Equation 68.18, or a corresponding subsequent state $w(\cdot,\tau)$, within any desired degree of accuracy. Unfortunately, there is, in general, no way to know how large K should be in Equation 68.24 in order to achieve a specified accuracy, nor is there any way to obtain a uniform estimate on the effect of errors in measurement of Y.

The question of τ-observability, for $\tau \in [0,T]$, is a more demanding one; it is the question as to whether the map $S_{\tau,K}$ defined in Equation 68.24 extends by continuity to a bounded linear map S_τ taking the observation Y into the corresponding state w_0. It turns out [21] that this is possible for any $\tau > 0$, but it is not possible for $\tau = 0$; i.e., the initial state can never be continuously reconstructed from measurements of this kind. Fortunately, it is ordinarily the terminal state $w(\cdot,T)$ that is the more relevant, and reconstruction is possible in this case. The proof of these assertions (cf. [7], [21], e.g.) relies on delicate estimates of the norms of the biorthogonal functions $\zeta_k(X, t)$ in the space $L^2(\Gamma \times [0, T])$. The boundedness property of the operator S_τ is important not only in regard to convergence of approximate reconstructions of the state $w(\cdot, \tau)$ but also in regard to understanding the effect of an error δY. If such an error is present, the estimate obtained for $w(\cdot, \tau)$ will clearly be

$$\tilde{w}(\cdot, \tau) = S_\tau (Y + \delta Y) = w(\cdot, \tau) + S_\tau \, \delta Y, \tag{68.25}$$

and the norm of the reconstruction error thus does not exceed $\| S_\tau \| \, \| \delta Y \|$. A further consequence of this clearly is the importance, since reconstruction operators are not, in general, unique, of obtaining a reconstruction operator of least possible norm. If we denote the subspace of $L^2(\Gamma \times [0, T])$ spanned by the functions $\eta_k(X, t)$, as described following Equation 68.21, by $E(\Gamma \times [0, T])$, it is easy to show that the biorthogonal functions ζ_k described via Equation 68.22 are unique and have least possible norm if we require that they should lie in $E(\Gamma \times [0, T])$. The particular reconstruction operator $\hat{S}_\tau$, constructed as the limit of operators (Equation 68.24) with the least norm biorthogonal functions $\hat{\zeta}_k$, may then be seen to have least possible norm. In applications, reconstruction operators of the type we have described here are rarely used; one normally uses a state estimator (cf. [18], e.g.) that provides only an asymptotic reconstruction of the system state, but the performance of such a state estimator is still ultimately limited by considerations of the same sort as we have discussed here.

It is possible to provide similar discussions for the "wave" counterpart of Equation 68.14, i.e.,

$$\rho \frac{\partial^2 w}{\partial t^2} = k \left(\frac{\partial^2 w}{\partial x^2} + \frac{\partial^2 w}{\partial y^2} + \frac{\partial^2 w}{\partial z^2} \right) , \tag{68.26}$$

with a variety of boundary conditions, including Equation 68.15. A very large number of such studies have been made, but they have normally been carried out in terms of the dual control system [19] rather than in terms of the linear observed system. In some cases, methods of harmonic analysis similar to those just described for the heat equation have been used [8], but the most definitive results have been obtained using methods derived from the *scattering theory* of the wave equation and from related methods such as *geometrical optics* [2] or *multiplier methods* [11]. These studies include treatment of cases wherein the observation/(dual) control process is restricted to a subset $\Gamma_1 \subset \Gamma$ having certain geometrical properties. Other contributions [21] have shown the study of the wave equation to be pivotal in the sense that results for related heat and elastic processes can be inferred, via harmonic analysis, once the wave equation results are in place.

68.4 Observability Theory for Elastic Beams

There are several different models for elastic beams, even when we restrict attention to small deformation linear models. These include the Euler-Bernoulli, Rayleigh and Timoshenko models. The oldest and most familiar of these is the Euler-Bernoulli model, consisting of the PDE

$$\rho(x)\frac{\partial^2 w}{\partial t^2} = \frac{\partial^2}{\partial x^2}\left(EI(x)\frac{\partial^2 w}{\partial x^2} \right), \ x \in [0, L], \ t \in [0, \infty), \tag{68.27}$$

wherein $\rho(x)$ denotes the mass per unit length and $EI(x)$ is the so-called *bending modulus*. We are concerned with solutions in a certain *weak* sense, which we will not elaborate upon here, corresponding to a given initial state

$$w(\cdot, 0) = w_0 \in H^2[0, L], \ \frac{\partial w}{\partial t}(x, 0) = v_0 \in L^2[0, L], \tag{68.28}$$

where $H^2[0, L]$ is the standard *Sobolev space* of functions with square integrable second derivatives on the indicated interval. Additionally, one needs to give boundary conditions at $x = 0$ and at $x = L$; these vary with the physical circumstances. For the sake of brevity, we will confine our discussion here to the *cantilever* case, where the left-hand end is assumed "clamped" while the right-hand end is "free"; the appropriate boundary conditions are then

$$w(0, t) \equiv 0, \ \frac{\partial w}{\partial x}(0, t) \equiv 0, \tag{68.29}$$

$$\frac{\partial^2 w}{\partial x^2}(L, t) \equiv 0, \ , \frac{\partial}{\partial x}\left(EI(x)\frac{\partial^2 w}{\partial x^2} \right)(L, t) \equiv 0. \tag{68.30}$$

In many applications a mechanical structure, such as a manipulator arm, is clamped to a rotating base that points the arm/beam in various directions in order to carry out particular tasks. Each "slewing" motion results in a degree of vibration of the structure, which, for most practical purposes, can be thought of as taking place within the context of the model of Equations 68.27 to 68.30. In order to attenuate the undesired vibration, it is first of all necessary to carry out an observation procedure in order to determine the oscillatory state of the system preparatory to, or in conjunction with, control operations. A number of different measurement options exist whose feasibility depends on the operational situation in hand. We will cite three of these. In the first instance, one might attach a *strain gauge* to the beam near the clamped end. The extension or compression of such a (normally piezoelectric) device provides a scaled physical realization of the mathematical measurement

$$y(t) = \frac{\partial^2 w}{\partial x^2}(0, t). \tag{68.31}$$

Alternatively, one can place an *accelerometer* near the free end of the beam to provide a measurement equivalent to

$$y(t) = \frac{\partial^2 w}{\partial t^2}(L, t), \tag{68.32}$$

or one can use a laser device, relying on the Doppler effect, to measure

$$y(t) = \frac{\partial w}{\partial t}(L, t). \tag{68.33}$$

Each of these measurement modes provides a scalar valued function $y(t)$ carrying a certain amount of information on the system state $w(x, t)$; the problem, as before, is to reconstruct the initial state, or the current state at time T, from the record $y(t), 0 \leq t \leq T$. The mathematical theory of this reconstruction is in many ways similar to the one we have just described for the heat equation, but with some significant differences. The most notable of these is immediately apparent from the equation itself; it is invariant under reversal of the time direction. This means that there is no inherent difference between initial and terminal states or, indeed, any intermediate state. We should expect this to show up in the mathematics, and it does.

The differential operator defined by

$$(A\, w) \equiv -\frac{1}{\rho}\frac{\partial^2}{\partial x^2}\left(EI(x)\frac{\partial^2 w}{\partial x^2}\right), \tag{68.34}$$

on the subspace of $H^4[0, L]$ resulting from imposition of the cantilever boundary conditions, is an unbounded positive self-adjoint operator with positive eigenvalues, listed in increasing order, λ_k, $k = 1, 2, 3, \ldots$. The corresponding normalized eigenfunctions $\phi_k(x)$ form an orthonormal basis for $L^2[0, L]$. Defining $\omega_k = \sqrt{\lambda_k}$, the solution of Equations 68.27 to 68.30 takes the form

$$w(x, t) = \sum_{k=1}^{\infty}\left(c_k\, e^{i\omega_k t} + d_k\, e^{-i\omega_k t}\right)\phi_k(x), \tag{68.35}$$

where, with w_0 and v_0 as in Equation 68.28

$$w_0(x) = \sum_{k=1}^{\infty} w_{0,k}\phi_k(x),\ v_0(x) = \sum_{k=1}^{\infty} v_{0,k}\phi_k(x), \tag{68.36}$$

with

$$w_{0,k} = c_k + d_k,\ v_{0,k} = i\omega_k(c_k - d_k). \tag{68.37}$$

The norm of the state w_0, v_0 in the state space $H^2[0, L] \otimes L^2[0, L]$ is equivalent to the norm of the double sequence $\{w_{0,k}, v_{0,k}\}$ in the Hilbert space, which we call ℓ^2_ω. That norm is

$$\|w_{0,k},\ v_{0,k}\| = \left[\sum_{k=1}^{\infty}\left(\omega_k w_{0,k}^2 + v_{0,k}^2\right)\right]^{\frac{1}{2}}.$$

Any one of the measurement modes discussed earlier now takes the form

$$y(t) = \sum_{k=1}^{\infty}\left(\gamma_k c_k\, e^{i\omega_k t} + \delta_k d_k\, e^{-i\omega_k t}\right), \tag{68.38}$$

with γ_k and δ_k depending on the particular measurement mode being employed. Thus, in the cases of Equations 68.31, 68.32 and 68.33, respectively, we have

$$\gamma_k = \delta_k = \frac{d^2\phi_k}{dx^2}(0), \tag{68.39}$$

$$\gamma_k = \delta_k = -\lambda_k\phi_k(L), \tag{68.40}$$

$$\gamma_k = i\omega_k\phi_k(L),\ \delta_k = -i\omega_k\phi_k(L). \tag{68.41}$$

Just as in the earlier example of the heat equation, but now we are concerned only with the *scalar* exponential functions

$$e^{i\omega_k t},\ e^{-i\omega_k t},\ k = 1, 2, 3, \ldots, \tag{68.42}$$

everything depends on the *independence* properties of these functions in $L^2[0, T]$, where T is the length of the observation interval, in relation to the asymptotic growth of the ω_k, $k \to \infty$, the basic character of which is that the ω_k are distinct, increasing with k, if ordered in the natural way, and

$$\omega_k = \mathcal{O}(k^2),\ k \to \infty. \tag{68.43}$$

The relationship between properties of the functions of Equation 68.42 and the asymptotic and/or separation properties of the exponents ω_k is one of the questions considered in the general topic of *nonharmonic Fourier series*, whose systematic study began in the 1930s with the work of Paley and Wiener [16]. The specific result that we make use of is due to A. E. Ingham [10]. Combined with other, more or less elementary, considerations, it implies that if the ω_k are real and satisfy a separation condition, for some positive integer K and positive number Γ,

$$\omega_{k+1} \geq \omega_k + \Gamma,\ k \geq K, \tag{68.44}$$

then the functions of Equation 68.42 are *uniformly independent* and *uniformly convergent* in $L^2[0, T]$, provided that $T \geq \frac{2\pi}{\Gamma}$, which means there are numbers b, B, $0 < b < B$, such that, for any square summable sequences of coefficients c_k, d_k, $k = 1, 2, 3, \ldots$, we have

$$\begin{aligned} b\sum_{k=1}^{\infty}\left(|c_k|^2 + |d_k|^2\right) &\leq \left\|\sum_{k=1}^{\infty}\left(c_k\, e^{i\omega_k t} + d_k\, e^{-i\omega_k t}\right)\right\|^2_{L^2[0,T]} \\ &\leq B\sum_{k=1}^{\infty}\left(|c_k|^2 + |d_k|^2\right) \end{aligned} \tag{68.45}$$

Since the asymptotic property of Equation 68.43 clearly implies, for any $\Gamma > 0$, that Equation 68.44 is satisfied if $K = K(\Gamma)$ is sufficiently large, inequalities of Equation 68.45, with $b = b(\Gamma)$, $B = B(\Gamma)$, must hold for any $T > 0$. It follows that the linear map, or operator,

$$C : \sum_{k=1}^{\infty}\left(c_k\, e^{i\omega_k t} + d_k\, e^{-i\omega_k t}\right) \to \{c_k, d_k | k = 1, 2, 3, \ldots\} \in \ell^2, \tag{68.46}$$

defined on the (necessarily closed, in view of Equation 68.45) subspace of $L^2[0, T]$ spanned by the exponential functions in question, is both bounded and boundedly invertible. Applied to the observation $y(t)$, $t \in [0, T]$, corresponding to an initial state of Equation 68.36, the boundedness of C, together with the relationship of Equation 68.37 between the $w_{0,k}$, $v_{0,k}$ and the c_k, d_k and the form of the ℓ^2_ω norm, it is not hard to see that the boundedness of the linear operator C implies the existence of a bounded reconstruction operator from the observation

(Equation 68.38) to the initial state (Equation 68.36) provided the coefficients δ_k, γ_k in Equation 68.38 satisfy an inequality of the form

$$|c_k| \geq C\,\omega_k,\ |d_k| \geq D\,\omega_k, \tag{68.47}$$

for some positive constants C and D. Using the trigonometric/exponential form of the eigenfunctions $\phi_k(x)$, one easily verifies this to be the case for each of the observation modes of Equations 68.39 to 68.41; indeed, the condition is overfulfilled in the case of the accelerometer measurement (Equation 68.40). Thus, bounded observability of elastic beam states via scalar measurements can be considered to be typical.

At any later time $t = \tau$, the system state resulting from the initial state of Equation 68.36 has the form

$$w_\tau(x) = \sum_{k=1}^{\infty}\left(\cos\omega_k t\; w_{0,k} + \frac{1}{\omega_k}\sin\omega_k t\; v_{0,k}\right)\phi_k(x),$$

$$v_\tau(x) = \sum_{k=1}^{\infty}\left(-\omega_k\sin\omega_k t\; w_{0,k} + \cos\omega_k t\; v_{0,k}\right)\phi_k(x). \tag{68.48}$$

Using Equation 68.48 with the form of the norm in ℓ^2_ω, one can see that the map from the initial state $w_0, v_0 \in H^2[0, L] \otimes L^2[0, L]$ to w_τ, v_τ in the same space is bounded and boundedly invertible; this is another way of expressing the time reversibility of the system. It follows that the state w_τ, v_τ can be continuously reconstructed from the observation $y(t)$, $t \in [0, T]$ in precisely the same circumstances as we can reconstruct the state w_0, v_0.

References

[1] Anderson, B.D.O. and Moore, J.B., *Linear Optimal Control*, Elect. Eng. Series, Prentice Hall, Englewood Cliffs, NJ, 1971, chap. 8.

[2] Bardos, C., LeBeau, G., and Rauch, J., Contrôle et stabilisation dans des problèmes hyperboliques, in *Controlabilité Exacte, Perturbations et Stabilisation de Systèmes Distribués*, Lions, J.-L., Masson, Paris, 1988, appendix 2.

[3] Beck, J.V. and Arnold, K.J., *Parameter Estimation in Engineering and Science*, John Wiley & Sons, New York, 1977.

[4] Courant, R. and Hilbert, D., *Methods of Mathematical Physics; Vol. 2: Partial Differential Equations*, Interscience Publishers, New York, 1962.

[5] Dolecki, S. and Russell, D.L., A general theory of observation and control, *SIAM J. Control Opt.*, 15, 185–220, 1977.

[6] Dunford, N. and Schwartz, J.T., *Linear Operators; Vol. 1: General Theory*, Interscience, New York, 1958.

[7] Fattorini, H.O. and Russell, D.L., Exact controllability theorems for linear parabolic equations in one space dimension, *Arch. Ration. Mech. Anal.*, 4, 272–292, 1971.

[8] Graham, K.D. and Russell, D.L., Boundary value control of the wave equation in a spherical region, *SIAM J. Control*, 13, 174–196, 1975.

[9] Ho, L.F., Observabilité frontière de l'équation des ondes, *Cah. R. Acad. Sci.*, Paris, 302, 1986.

[10] Ingham, A.E., Some trigonometrical inequalities with applications to the theory of series, *Math. Z.*, 41, 367–379, 1936.

[11] Lagnese, J., Controllability and stabilizability of thin beams and plates, in *Distributed Parameter Control Systems*, Chen, G., Lee, E.B., Littman, W., Markus, L., Eds., Lecture Notes in Pure and Applied Math., Marcel Dekker, New York, 1991, 128.

[12] LaSalle, J.P. and Lefschetz, S., *Stability by Liapunov's Direct Method, with Applications*, Academic Press, New York, 1961.

[13] Lions, J.L., *Optimal Control of Systems Governed by Partial Differential Equations*, Grund. Math. Wiss. Einz., Springer-Verlag, New York, 1971, 170.

[14] Lions, J.L., *Controlabilité Exacte, Perturbations et Stabilisation de Systèmes Distribués*, Tomes 1,2, *Recherches en Mathematiques Appliquées*, Masson, Paris, 1988, 8, 9.

[15] Luenberger, D.G., An introduction to observers, *IEEE Trans. Autom. Control*, 22, 596–602, 1971.

[16] Paley, R.E.A.C. and Wiener, N., *Fourier transforms in the complex domain, in Colloq. Pub.*, American Mathematical Society, Providence, RI, 1934, 19.

[17] Pazy, A., *Semigroups of linear operators and applications to partial differential equations, in Applied Mathematical Sciences*, Springer-Verlag, New York, 1983, 44.

[18] Russell, D.L., *Mathematics of Finite Dimensional Control Systems: Theory and Design*, Marcel Dekker, Inc., New York, 1979.

[19] Russell, D.L., *Controllability and stabilizability theory for linear partial differential equations; recent progress and open questions, SIAM Rev.*, 20, 639–739, 1978.

[20] Russell, D.L., *Decay rates for weakly damped systems in Hilbert space obtained with control-theoretic methods, J. Diff. Eq.*, 19, 344–370, 1975.

[21] Russell, D.L., *A unified boundary controllability theory for hyperbolic and parabolic partial differential equations, Stud. Appl. Math.*, LII, 189–211, 1973.

[22] Showalter, R.E., *Hilbert Space Methods for Partial Differential Equations*, Pitman Publishing Ltd., San Francisco, 1977.

PART C

APPLICATIONS OF CONTROL

SECTION XV

Process Control

69

Water Level Control for the Toilet Tank: A Historical Perspective

Bruce G. Coury
The Johns Hopkins University, Applied Physics Laboratory, Laurel, MD

69.1 Introduction

Control technologies are a ubiquitous feature of everyday life. We rely on them to perform a wide variety of tasks without giving much thought to the origins of that technology or how it became such an important part of our lives. Consider the common toilet, a device that is found in virtually every home and encountered every day. Here is a device — hidden away in its own room and rarely a major topic of conversation — that plays a significant role in our life and depends on several control systems for its effective operation. Unlike many of the typical household systems which depend on control technologies and are single purpose, self-contained devices (e.g., coffee makers), the toilet is a relatively sophisticated device that uses two types of ancient controllers. For most of us, the toilet is also the primary point of daily contact with a large distributed system for sanitation management; a system that is crucially dependent on control and has extremely important social, cultural, environmental, and political implications. In addition, we have some very clear expectations about the performance of that technology and well-defined limits for the types of tolerable errors should the technology fail. To imagine life without properly functioning modern toilets conjures up images of a lifestyle that most of us would find unacceptable and very unpleasant.

One need not look too far back in time to discover what life was *really* like without toilets linked to a major sanitation system. The toilet as we know it today is a recent technological development, coming into widespread use only at the end of the 19th century. Prior to that time, indoor toilets were relatively rare in all but the most wealthy homes, and the disposal and management of sewage was a rather haphazard affair. Only as a result of the devastating cholera epidemics of the mid-19th century in the United States and Europe were significant efforts made to control and manage waste products. Surprisingly, the technologies for sanitation had been available for quite some time. For instance, the toilet has a rich technological history with roots dating back to antiquity. The Greeks and Romans had developed the types of control technologies necessary for the operation of modern toilets, but the widespread adoption of that technology did not occur for another 1800 years.

One wonders, then, why a control technology so obviously useful took centuries to be adopted. To ponder the factors that motivate developing and adopting a technology is the essence of a historical analysis of that technology and is the first step in tracing the roots of technological development. One quickly realizes that the answers to such questions are rather complex, requiring one to explore not only the development of the technology, but also the economic, political, and social influences that shaped the development and implementation of that technology. In this respect, toilets and their controls are an ideal topic. First, it is a technology that is common, familiar, and classic in its control engineering. Although writing about the control technology of toilets is fraught with many pitfalls (one can easily fall victim to bathroom humor), toilets are a technology that people encounter every day that is part of a very large and highly distributed system.

Second, the history of the development of control technologies for toilets has many dimensions. There is the obvious chronicle of the evolution from latrine to commode to water closet to

0-8493-8570-9/96/$0.00+$.50

toilet. The history will quickly reveal that the development of toilet technology was inexplicably linked to the development of an entire water supply and sanitation system. Perhaps more important, we will see that the history of toilet technology and the entire sanitation system must be considered in the context of significant parallel social, cultural, and political trends that describe changing attitudes towards health and sanitation. Hopefully, by the end of this article, the reader will begin to grasp the complexity of technological development and the myriad forces that shape the development and use of a technology.

To consider the historical development of technology requires a specific framework of analysis. To focus only on the development of toilet technology and the adaptation of sensory and control mechanisms to the operation of toilets is to miss the larger context of the history of technological development. As Hughes [4] points out in his studies of the historical development of systems (he concentrates on electric power systems), "systems embody the physical, intellectual, and symbolic resources of the society that constructs them." As we shall see with the toilet, its development and use was driven by the threat of disease, the appalling conditions resulting from uncontrolled waste disposal, changing attitudes towards health and cleanliness, and the fashionable features of a "modern" home made possible by an abundant water supply.

The discussion begins with a description of the toilet technology relevant to this chapter, the valves for maintaining the water level in the toilet tank. What may come as a surprise to the reader is the similarity of modern toilet technology to the water closets of the late 19th century and how the same technological challenges have persisted for more than 150 years. We will then turn to the historical roots of those technologies and trace the events of the last few centuries that resulted in today's toilet and sanitation system.

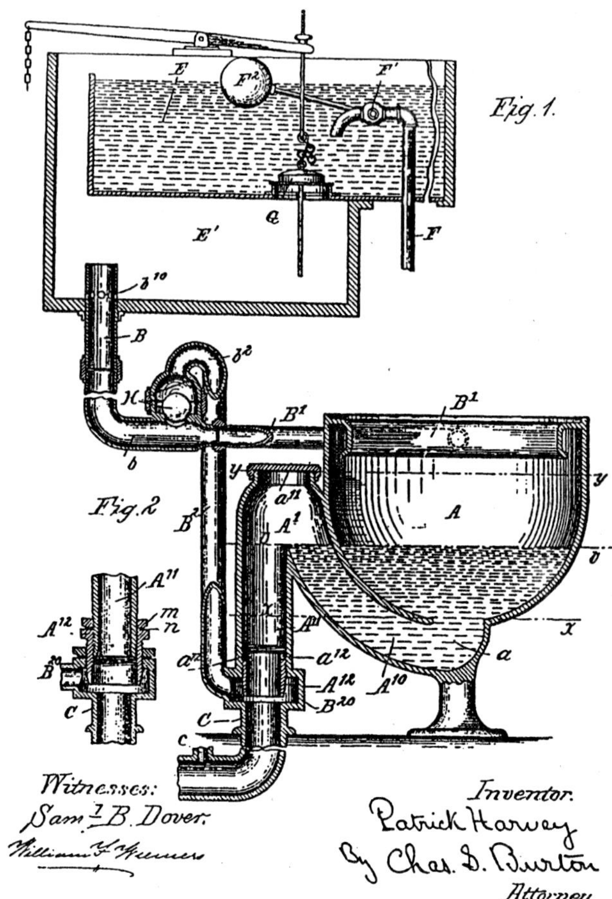

Figure 69.1 U.S. Patent drawing of the design of a water closet showing the basic components of the toilet tank, flush and float valves, and the toilet basin. *Source:* U.S. Patent No. 349,348 filed by P. Harvey, Sept. 21, 1886.

69.2 Control Technology in the Toilet

Lift the lid from the top of a toilet tank and inside is the basic control technology for the entry to a widely dispersed sanitation system. Although simple in operation (e.g., as shown in Figures 69.1, 69.2, and 69.3), the mechanisms found in the toilet tank serve one very important function: maintaining a constant water level in the tank. Modern toilets operate on the concept of flushing where a preset amount of water is used to remove waste from the toilet bowl. The tank stores water for use during flushing and controls the amount of water in storage by two valves: one controlling the amount of water entering the tank and another controlling the amount of water leaving the tank.

Push the handle on the tank to activate the flush cycle. Pushing the handle pops open the valve at the bottom of the tank (the toilet flush valve), allowing the water to rush into the toilet bowl (assuming well-maintained and nonleaking equipment). Notice that this valve (in most toilet designs) also floats, thereby keeping the valve open after the handle is released. As the water level drops, the float attached to the second valve (the float valve) also descends, opening that valve and allowing water to enter the tank. When the water level drops below a certain level, the flush valve closes, allowing the tank to refill. The water level rises, carrying the float with it until there is sufficient water in the tank to close the float valve.

Within a toilet tank we find examples of classic control devices. There are valves for controlling input and output, activation devices for initiating a control sequence, feedback mechanisms that sense water level and provide feedback to the control devices, and failure modes that minimize the cost of disruptions in control. The toilet tank is comprised of two primary, independent control mechanisms, the valve for controlling water flow into the tank and the valve for controlling water flow out of the tank, both relying on the same variable (tank water level) for feedback control. Using water level as the feedback parameter is very useful. If all works well, the water level determines when valves should be opened and closed for proper toilet operation.

The amount of water in the tank is also the measure of performance required to minimize the adverse effects of a failure. Should one of the valves fail (e.g., the float valve in the open position), the consequences of the failure (water running all over the floor) are minimized (water pours down the inside of the

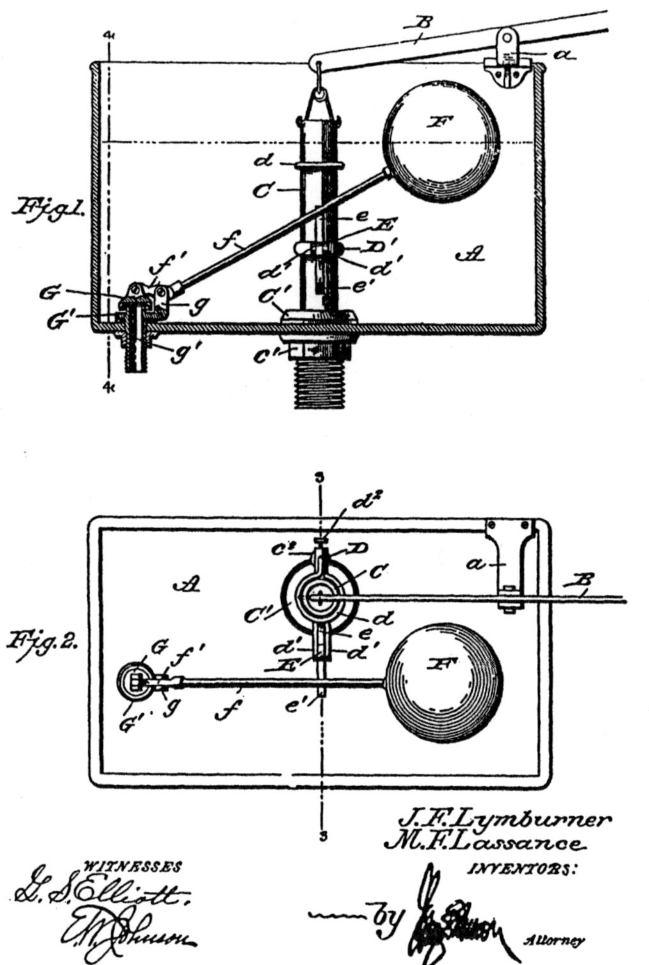

Figure 69.2 U.S. Patent drawing of the flush valve and the float valve for the toilet tank. *Source:* U.S. Patent No. 549,378 filed by J.F. Lymburner and M.F. Lassance, Nov. 5, 1895.

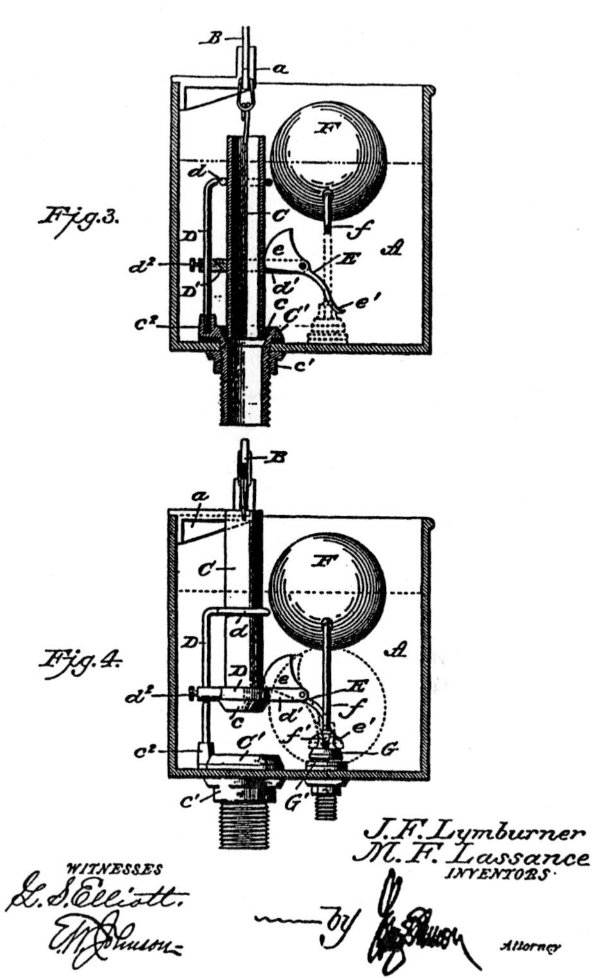

Figure 69.2 *(Continued.)* U.S. Patent drawing of the flush valve and the float valve for the toilet tank. *Source:* U.S. Patent No. 549,378 filed by J.F. Lymburner and M.F. Lassance, Nov. 5, 1895.

flush valve tube). Not all of the solutions to the failsafe maintenance of water level are the same. In the design shown in Patrick Harvey's U.S. Patent dated 1886 (Figure 69.1), a tank-within-a-tank approach is used so that the overflow from the cistern due to float valve failure goes directly to the waste pipe. The traditional approach to the same problem is shown in Figures 69.2 and 69.3; although the two designs are almost 100 years apart (Figures 69.2a and 69.2b are from the U.S. Patent filed by Joseph F. Lymburner and Mathias F. Lassance in 1895; Figure 69.3 is from the U.S. Patent filed in 1992 by Olof Olson), each uses a hollow tube set at a predetermined height for draining overflow should the float valve fail. In other words, the toilet is a control device characterized by relatively fail-safe operation.

69.3 Toilet Control

The toilet flush valve and the float valve operate on two different principles. The float valve has separate mechanisms for sensing the level of water in the tank (the float) and for controlling the input of water into the tank (the valve). The flush valve, on the other hand, combines both mechanisms for sensing and control into a single device. Both types of control technology have their origins in antiquity. The requirement for control of the level and flow of water was recognized by the Greeks in the design of water clocks. Water clocks were a rather intriguing device built on the principle that a constant flow of water could be used to measure the passage of time. In the design by Ktesibious (Figure 69.4), a mechanician serving under King Ptolemy II Philadelphus (285–247 B.C.), water flows into a container through an orifice of predetermined size in which the water level slowly rises as time passes [6]. Riding on the surface of the water is a float (labeled *P* in the diagram) attached to a mechanism that indicates the time of day; as the float rises, the pointer moves up the time scale. To assure an adequate supply of water to the clock, the orifice controlling the flow of water into the clock container was attached to a holding tank.

Ktesibious recognized that accurate time keeping requires precise control over water flow and maintenance of a constant water level in the container. Consequently, it was necessary to develop a method of control assuring a constant flow rate of water into the container. The Greeks clearly understood the relationship

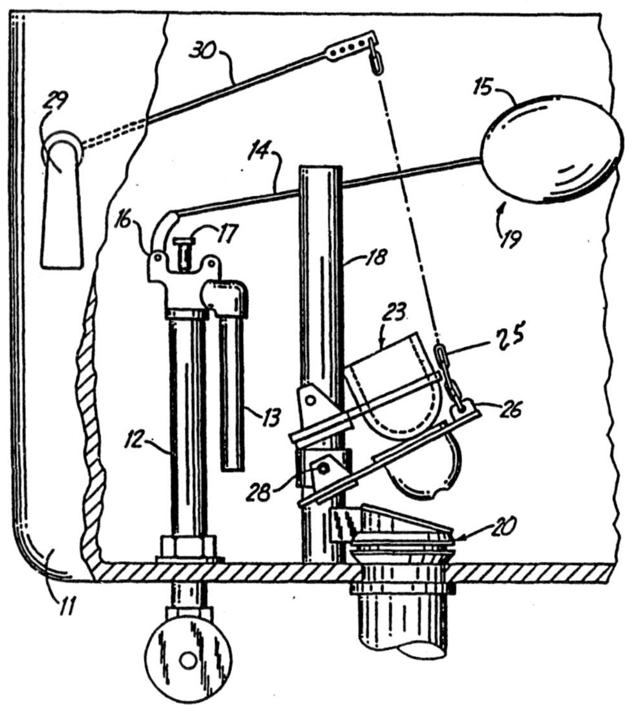

Figure 69.3 U.S. Patent drawing of a modern design for the flush valve and float valve for the toilet tank. *Source:* U.S. Patent No. 5,142,710 filed by O. Olson, Sept. 1, 1992.

between flow rate and relative pressure, and recognized that the flow rate of water into the clock would decrease as the holding tank emptied. By maintaining a constant water level in the holding tank, the problem of variations in flow rate could be solved. There are, however, a number of possible passive or active solutions to the constant water level problem. A passive approach to the problem could use either a constantly overflowing holding tank or an extremely large reservoir of water relative to the size of the water clock container. Active control solutions, on the other hand, would use some form of water level sensing device and a valve to regulate the amount of water entering the holding tank (thereby reducing the need for a large reservoir or a messy overflow management system). Ktesibious chose the active control solution by designing a float valve for the inlet to the holding tank that assured a constant water level. From the descriptions of Ktesibious' water clock [6], the valve was a solid cone (labeled *G* in Figure 69.4) that floated on the surface of the water in the holding tank serving the orifice of the water clock. The valve stopped the flow of water into the holding tank when the level of the water forced the tip of the cone into a similarly shaped valve seat. Notice again that the functions of sensing the level of water and controlling the flow of water are both contained in the float valve.

The flush valve of a modern toilet uses a similar principle. Combining both sensing and control in the same device, the flush valve is a modern variation of Ktesibious cone-shaped float that controls the amount of water flowing out of the toilet tank. Unlike the Greek original, however, the control sequence of flush

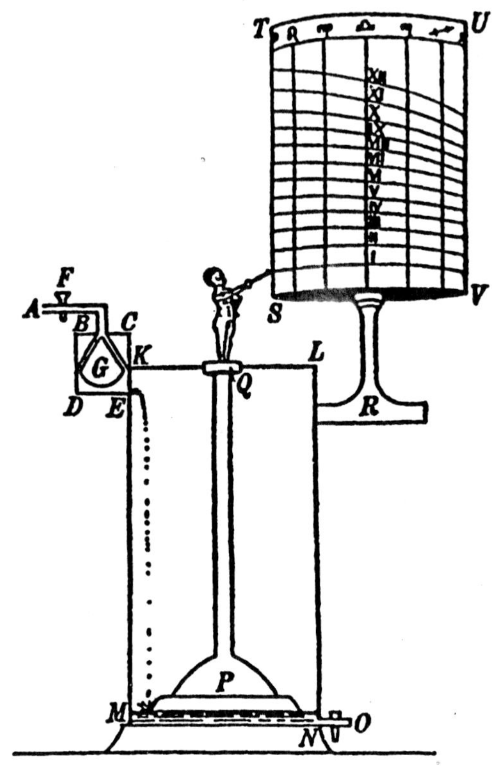

Figure 69.4 Control technology in holding vessels designed by Ktesibious (285–247 B.C.) showing the use of a float valve for regulating water flow. *Source:* Mayr, O., *The Origins of Feedback Control.* MIT Press, Cambridge, MA, 1969.

valve action is initiated by the mechanism attached to the flushing handle on the outside of the toilet tank. Thus, the toilet flush valve is a discrete control device that seeks to control the output of water after some external event has resulted in a drop in water level.

Subsequent developments of the float valve by Heron in the first century A.D. improved on the relationship between float and valve. Described in the *Pneumatica*, Heron developed float valves to maintain constant fluid levels in two separate vessels (e.g., as in Figure 69.5).

In his design, he employed a float connected to a valve by a series of jointed levers. The rise and fall of the float in one vessel would close or open a valve in another vessel. This design effectively separates the sensing of fluid level from the actions to control the flow of fluid, and most closely resembles the float valve for regulating water level in the toilet tank. Although the purpose of the toilet float valve is to maintain a constant level of water in the toilet tank, variations in that water level are usually step functions resulting from discrete, external events. The technology can be applied to continuous control, as is so audibly illustrated when the toilet flush valve leaks.

The float valve for the water clock and the holding vessels illustrates a number of important concepts critical to feedback and control. First, feedback is a fundamental component of control. Without a means for sensing the appropriate performance parameter, automatic regulation of input and output cannot be

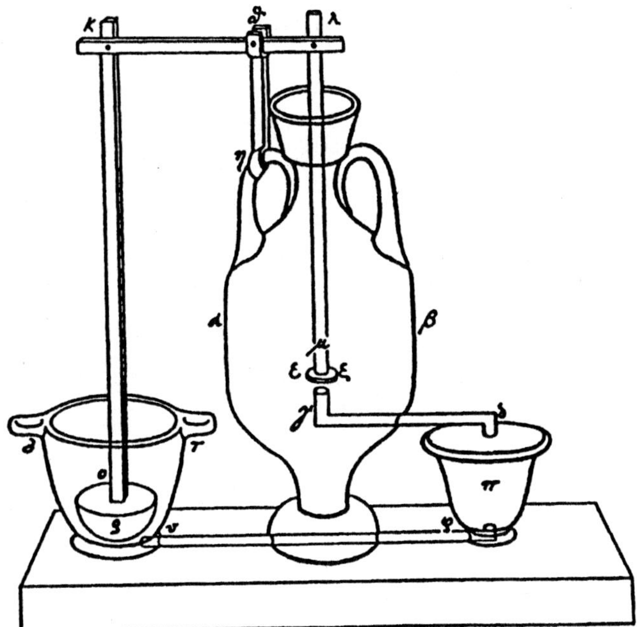

Figure 69.5 Control technology in the waterclock designed by Heron (100 A.D.) showing the use of a float connected to a valve for maintaining a constant fluid level. *Source:* Mayr, O., *The Origins of Feedback Control.* MIT Press, Cambridge, MA, 1969.

accomplished. Second, separation of the mechanisms for sensing and control provide a more sophisticated form of control technology. In the example of Heron's holding vessels, using a float to sense fluid level that was connected by adjustable levers to a valve for controlling fluid flow allowed adjustments to be made to the float level. Thus, the level of fluid required to close the valve could be varied by changing the required float level. The float valve in Ktesibious' design, on the other hand, was not adjustable; to change the water level resulting in valve closure required substituting an entirely new float valve of the proper dimensions.

69.4 The Concept of Flushing

The Greeks and Romans did not immediately recognize the relevance of this control technology to the development of toilet like devices. The lack of insight was certainly not due to the inability to understand the relevance of water to waste removal. Flowing streams have always been a source of waste removal, especially in ancient history when people tended to settle near water. By the time the Greeks and Romans built public baths, the use of water to cleanse the body and remove excrement was an integral part of the design. Even the frontier Roman forts in Britain had elaborate latrines that used surface water to flush away human wastes [8]. There was, however, no explicit mechanism for controlling the flow of water other than devices for diverting streams, pipes for routing the flow, or reservoirs to assure an ample supply of water [10].

The use of a purpose-built *flushing* mechanism (where water is stored in a holding tank until called upon for the removal of waste) was slow to develop and was relatively rare until more recent history (although Reynolds discusses the possibility for such a water closet at Knossos). For instance, 15th century evidence of such a mechanism for a latrine was found during excavation of St. Albans the Abbot in England [10]. In the most simple form of an 18th century flushing toilet, a cistern captured and stored rain water until a valve was opened (usually by pulling on a lever), allowing the water to flush away waste products. In most early applications of flushing, the amount of water used was determined by the size of the cistern, the source of water, and the patience of the user. Although such devices persisted until the latter part of the 19th century, they were considered to be a rather foul and obnoxious solution to the problem.

In general, the collection and disposal of human waste was accomplished without sophisticated technology until the mid-1800s. When a special purpose device for collecting human waste existed in a home, it was usually a commode or chamber pot (although the *close stools* built for royalty in 16th century Europe could be quite throne-like in appearance, complete with seat and arms and covered with velvet). Otherwise, residents trekked outside to a latrine or privy in the backyard or garden. The first evidence of a recognizable predecessor of the modern toilet is found in the British patents filed by Alexander Cummings in 1775 and Joseph Bramah in 1778 (although Sir John Harrington's valve closet of 1596 had some characteristics similar to 18th century water closets). Cummings proposed a valve closet that had an overhead supply cistern, a handle activated valve for the flush mechanism, and a valve that opened in the bottom of the basin to allow waste to escape. All valves and flushing mechanisms were activated and controlled by the user. Bramah's contribution was a crank activated valve for emptying the basin. No control devices that relied on water level were evident, and the valve separating the basin from the discharge pipe was not a very effective barrier against noxious odors and potentially dangerous sewer gases [10]. An early version of an American water closet of similar design is shown in Figure 69.6. Patented by Daniel Ryan and John

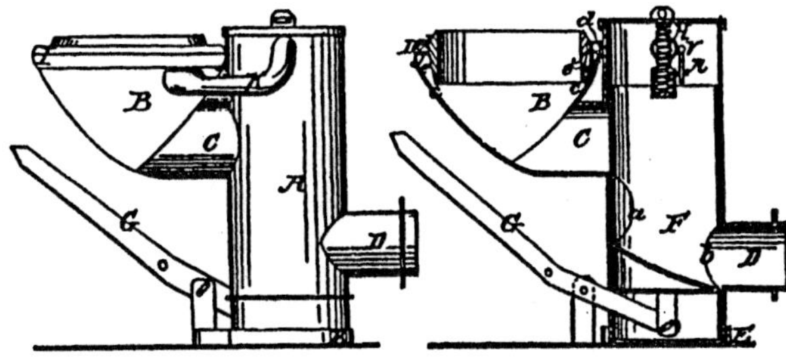

Figure 69.6 U.S. Patent drawing of an early water closet design showing manual operation of flushing cycle. *Source:* U.S. Patent No. 10,620 filed by D.Ryan and J. Flanagan, Mar. 7, 1854.

Flanagan in 1854, the basic components of a manually operated water closet are shown. The lever (labeled G) operates the water closet. When the lever is depressed, it pushes up the sliding tube

F until the opening a coincides with the pipe C from the toilet bowl. At the same time, the valve labeled V is opened to allow water to enter the basin and flush away waste.

In all situations, commode, latrine, privy, or water closet, no organized or structured system of collecting and managing the disposal of human waste products existed. Human household waste was typically dumped directly into vaults or cesspits (some of which were in the basements of homes) or heaved into ditches or drains in the street. The collection and disposal of that waste was usually handled by workers who came to the home at night with shovel and bucket to empty out the vault or cesspit and cart off the waste for disposal (usually into the most convenient ditch or source of flowing water). When a system of sewers did exist, household waste was specifically excluded. Even the Romans, who were quite advanced in building complex water and sanitation systems, used cesspits in the gardens of private homes to collect human waste products [8]. Not until the link between sanitation and health become evident and a universal concern arose in response to the health hazards of uncontrolled sewage did the need arise to develop an infrastructure to support the collection and removal of human waste.

At this point, it should be evident to the reader that the development of toilet technology did not proceed along an orderly evolutionary path from control principle to technological development. For instance, the development of critical water level control technologies by the Greeks and Romans did not immediately lead to the use of that technology in a sanitation system (despite the fact that the need for the removal of waste products had been well-established for a long time). Nor did the existence of critical technologies immediately lead to the development of more sophisticated devices for the collection and disposal of human excrement. Even where conditions surrounding the disposal of human waste products was recognized as deplorable for many centuries, little was done to develop technological solutions that remotely resembled the modern toilet until other factors came into play. Understanding those factors will lead us to a better understanding of the development of toilet control technology.

69.5 The Need for a Sanitation System

The development of sanitation systems was a direct response to health concerns. From a modern perspective, sanitation and the treatment of human waste was a haphazard affair through the early 19th century. Descriptions of conditions in British and American cities portray open sewers, waste piling up in streets and back alleys, cesspits in yards or even under living room floors, and outdoor privies shared by an entire neighborhood. Mortality rates during that period were shocking: the death rate per 1,000 children under five years of age was 240 in the English countryside and 480 in the cities [10]. In 1849, sewers delivered more than 9 million cubic feet of untreated sewage into the Thames, the same river used as a source of water for human consumption.

In his comprehensive treatment of the conditions in 19th century Newark, NJ, Galishoff [3] provides a vivid account of the situation in an American city. Streets were unpaved and lined with household and commercial wastes. Common sewers were open drains that ran down the center of streets, and many residents built drains from their home to the sewer to carry away household wastes. Prior to the construction of a sewer system in 1854, Newark's "privy and cesspool wastes not absorbed by the soil drained into the city's waterways and into ditches and other open conduits" [3]. When heavy rains came, the streets turned into filthy quagmires with the runoff filling cellars and lower levels of homes and businesses in low-lying areas. States Galishoff [3], "Despite greater and greater accumulations of street dirt (consisting mainly of garbage, rubbish, and manure), gutters and streets were cleaned only twice a year. An ordinance compelling property owners to keep streets abutting their parcels free of obstructions was ignored."

Given such horrible conditions, it is not surprising that cholera reached epidemic proportions during the period of 1832-1866; 14,000 Londoners died of cholera in 1849, another 10,000 in 1854, and more than 5,000 in the last major outbreak in 1866 [10]. Newark was struck hard in 1832, 1849, and 1866, with minor outbreaks in 1854 and 1873. Although Newark's experience was typical of many American cities (the number of deaths represented approximately 0.5 percent of the city populations), and only Boston and Charleston, SC, escaped relatively unharmed, cholera struck some cities devastating blows. Lexington, KY, for instance, lost nearly a third of its population when the disease swept through that city in 1833 [3].

Until legislative action was taken, the private sector was responsible for constructing sewage systems through most of the 19th century, especially in the United States. This arrangement resulted in highly variable service and waste treatment conditions. For example, Boston had a well developed system of sewers by the early 1700s, whereas the sewers in the city of New York were built piecemeal over a six decade period after 1700 [1]. At that time, and until the recognition of the link between disease and contaminated water in the mid-19th century, the disposal of general waste water and the disposal of human waste were treated as separate concerns. Private citizens were responsible for the disposal of the waste from their own privies, outbuildings, and cesspools. If the waste was removed (and in the more densely populated sections of American cities, such was hardly ever the case), it usually was dumped into the nearest body of water. As a result, American cities suffered the same fate as European cities during the cholera years of 1832–1873. Needless to say, lawmakers were motivated to act, and the first public health legislation was passed. London enacted its Public Health Act in 1848, and efforts were made to control drainage, close cesspits, and repair, replace, and construct sewers. Resolution of the problem was a Herculean task; not until the 1870s did the situation improve sufficiently for the death rate to decline significantly. The Newark Board of Health was created in 1857 to oversee sanitary conditions. It, too, faced an uphill battle and was still considered ineffectual in 1875.

69.6 Concerns About Disease

Fundamental to the efforts to control sewage was the realization of a direct link between contaminated water and the spread of disease. This realization did not occur until the medical profession established the basis for the transmission of disease. Prior to the mid-1800s, the most common notion of the mechanism for spreading disease was based on atmospheric poisoning (referred to as the miasmic theory of disease) caused by the release of toxic gases during the fermentation of organic matter. In this view, diseases such as cholera were spread by the toxic gases released by stagnant pools of sewage and accumulations of garbage and waste. Urban congestion and squalor appeared to confirm the theory; the highest incidence of cholera (as well as most other types of communicable diseases) occurred in the most densely populated and poorest sections of a city where sanitation was virtually nonexistent. Some believed, even in the medical community, that sewers were especially dangerous in congested, urban areas because of the large volumes of sewage that could accumulate and emit deadly concentrations of sewer gases.

The higher incidence of disease in the poorest sections of a city also contributed to the notion that certain factors predisposed specific segments of the populace to infection, a convenient way to single out immigrants and other less fortunate members of the community for discriminatory treatment in the battle against disease. Such attitudes also contributed to the slow growth in the public health movement in the United States because cleaning up the squalor in the poorest sections of a city could potentially place a significant economic burden on the city government and business community [3]. By associating the disease with a particular class of people, costly measures for improving sanitation could be ignored. Class differences in the availability of and access to water and sanitation facilities persisted in Newark from 1850–1900, with much of the resources for providing water and sanitation services directed towards the central business district and the more affluent neighborhoods [3].

The medical profession slowly realized that "atmospheric poisoning" was not the mechanism for spreading diseases. During the 19th century, medical science was evolving and notions about the spread and control of disease were being formalized into a coherent public health movement. Edwin Chadwick, as secretary of the British Poor Law Commission, reported to Parliament on the social and environmental conditions of poor health in 1848. The influence of his report was so great that Chadwick is credited with initiating the "Great Sanitary Awakening"[9]. In 1854, John Snow, a London anesthesiologist, determined that a contaminated water supply led to the deaths of 500 people when he traced the source of the infected water to a single common well. Snow had argued earlier that cholera was spread by a "poison" that attacked the intestines and was carried in human waste. His study of the 500 deaths caused by the contaminated well provided strong support for his theory. The first recorded study in the United States of the spread of disease by contaminated water was conducted by Austin Flint in 1855. In that study he established that the source of a typhoid epidemic in North Boston, NY was a contaminated well [9].

During the 1860s, the miasmic theory of disease was slowly displaced by the germ theory. Louis Pasteur established the role of microorganisms in fermentation. At about the same time, Joseph Lister introduced antiseptic methods in surgery. These changes in medicine were due to an increasing awareness of the link between microorganisms and disease and the role of sterilization and cleanliness in preventing the spread of disease. The germ theory of disease and the role of bacteria in the spread of disease were established in the 1880s by Robert Koch. His research unequivocally demonstrated that bacteria were the cause of many types of disease. By firmly establishing the link between bacteria and disease, germ theory provided the basis for understanding the link between contaminated water and health.

Once it was discovered that disease could spread as a result of ground water seepage and sewage leachate, uncontrolled dumping of waste was no longer acceptable. To eliminate water contamination, potable water had to be separated from sewage. Such an objective required that an extensive, coordinated sewer system be constructed to provide for the collection of waste products at the source. Water and drainage systems did exist in some communities and wealthy households in the mid-18th century [e.g., [10]], but, in general, widespread water and sanitation systems were slow to develop. One major force behind the development of sanitation systems was the rapid growth of cities in the years leading up to the Civil War. For example, Newark became one of the nation's largest industrial cities when its population grew from 11,000 to 246,000 during the period 1830–1900. In the three decades following 1830, Newark's population increased by more than 60,000 people. Much of the increase was due to a large influx of Irish and German immigrants. After 1900, the population of Newark almost doubled again (to 435,000 in 1918) after undergoing a new wave of immigration from eastern and southern Europe [3]. These trends paralleled urban growth in other parts of the nation; the total urban population in the United States, 322,371 in 1800, had grown to more than 6 million by 1860, and exceeded 54 million by 1920 [7].

As a consequence of such rapid growth, the demand for water for household and industrial use also increased. In the United States during the late 1700s, private companies were organized in a number of cities to provide clean water. The companies relied on a system of reservoirs, pumping stations, and aqueducts. For instance, by 1840 Philadelphia had developed one of the best water systems in the nation, delivering more than 1.6 million gallons of water per day to 4, 800 customers. The enormous population growth during and after the Civil War (and the concomitant increase in congestion) overwhelmed the capacity of these early systems and outstripped the private sector's ability to meet the demands for water and sanitation. Worsening conditions finally forced the issue, resulting in large scale public works projects to meet the demand for water. In 1842, for example, the Croton Aqueduct was completed to provide New York City with a potential capacity of 42 million gallons of water per day. That system became obsolete by the early 1880s when demand exceeded nearly 370 million gallons per day for 3.5 million people and a more extensive municipal system had to be built using public funds [1].

The abundance of water allowed for the development of indoor plumbing, thereby increasing the volumes of waste water. In the late 1800s, most of the increase in domestic water consumption was due to the installation of bathroom fixtures [3]. In Chicago, per capita water consumption increased from 33 gallons per day in 1856 to 144 in 1872. By 1880, approximately one-fourth of all Chicago households had water closets [7]. Unfortunately, early sewage systems were constructed without allowance for human and household wastes. As demand increased and both residential and industrial effluent was being diverted into storm drainage systems, it became clear that municipalities would have to build separate sewers to accommodate household wastes and adopt the "water-carriage" system of waste removal (rather than rely on cesspools and privy vaults). The first separate sewer system was built in Memphis, TN, in 1880. Because of the enormous economic requirements of such large scale sewer systems, cities assumed the responsibility for building sewer systems and embarked on some of the largest construction projects of the 19th century [1], [7]. In Newark, more than 200 miles of sewers were constructed during the period 1894–1920 at a cost exceeding several million dollars, providing service to 95 percent of the improved areas of the city [3]. The technology to treat sewage effectively, however, (other than filtering and irrigation) would not be widely available for another 10 years.

69.7 Changing Attitudes About Health and Hygiene

The major source of the increase in household water consumption was personal hygiene. Accompanying the efforts to curb the spread of disease was a renewed interest in bathing. Although a popular activity among the Greeks and Romans, the frequency of bathing and an emphasis on cleanliness has ebbed and flowed throughout history. By the Dark Ages, cleanliness had fallen out of favor, only to become acceptable once again during the time of the Crusades. Through the 16th and 17th centuries bathing was rare, except among members of the upper class and then on an infrequent basis. By the first half of the 19th century, bathing was largely a matter of appearance and daily sponge bathing was uncommon in middle-class American homes until the mid-1800s [5]. As awareness of the causes of disease increased at the end of the 19th century, cleanliness became a means to prevent the spread of disease. Bathing and personal hygiene to prevent disease, states Lupton and Miller, "was aggressively promoted by health reformers, journalists, and the manufacturers of personal care products." Thus, hygienic products and the popular press became important mechanisms for defining the importance of cleanliness for Americans in the latter half of the 19th century.

In this period of technological development, social forces significantly influenced the construction, adaptation, use, and acceptance of technologies related to hygiene. The importance of cleanliness and the appropriate solutions to the hygiene problem were defined for people by specific agents of change, namely, health professionals, journalists, and commercial interests. Thus, the meaning of and need for health care products and home sanitary systems were defined by a number of influential social groups. In the history of technology literature, this is referred to as social constructionism where the meaning of a technology (especially in terms of its utility or value) is defined through the interaction of relevant social groups [2]. In the social constructionist's view, technological development is cast in terms of the problems relevant for each social group, with progress determined by the influence exerted by a particular social group to resolve a specific problem. For instance, the health professionals were instrumental in defining the need for cleanliness and sanitation, with journalists communicating the message to the middle class in the popular press. Through marketing and advertisements, those groups concerned with producing plumbing and bathroom fixtures significantly influenced the standards (and products) for personal hygiene in the home. By increasing awareness of the need for a clean and sanitary home and defining the dominant attitudes towards health standards, these influential social forces had defined for Americans the importance of the components of a bathroom [5].

With the arrival of indoor plumbing, both hot and cold water could be piped into the home. Previous to the introduction of indoor plumbing, appliances for bathing and defecating were portable and similar in appearance to furniture (to disguise their purpose when placed in a room). The fixed nature of pipes and the use of running water required that the formerly portable equipment become stationary. To minimize the cost of installing pipes and drains, the bath, basin, and commode were placed in a single room; consequently, the "bathroom" became a central place in the home for bathing and the elimination of waste. Early designs of bathroom furniture retained the wood construction of the portable units, but the desire for easy cleaning and a nonporous surface that would not collect dirt and grime led to the use of metal and china. The bathroom became, in effect, a reflection of the modern desire for an antiseptic approach to personal hygiene. Porcelain-lined tubs, china toilets, and tiled floors and walls (typically white in color) emphasized the clinical design of the bathroom and the ease with which dirt could be found and removed. As Lupton and Miller [5] point out, the popular press and personal hygiene guides of the period compared the design of the modern bathroom to a hospital. The cleanliness of the bathroom became a measure of household standards for hygiene and sanitation. By the late 1880s, indoor plumbing and bathrooms were a standard feature in homes and a prerequisite for a middle-class lifestyle.

69.8 The Indoor Toilet

The growing market in bathroom fixtures stimulated significant technological development. Before 1860, there were very few U.S. Patents filed for water closets and related toilet components (the Ryan and Flanagan design shown in Figure 69.6 is one of only three patents for water closets filed in the period 1847–1855). Once the toilet moved into the home and became a permanent fixture in the bathroom, the technology rapidly developed. The primary concern in the development of the technology was the amount of water used by the toilet and its subsequent impact on the water supply and sewage collection and treatment systems. The fact that water could be piped directly to the cistern of the water closet (previous designs of water closets had required that they be filled by hand) necessitated some mechanism to control the flow of water to the toilet and minimize the impact of toilet use on the water supply. In addition, there was a more systemwide concern to provide some automatic way to control the amount of water used by the toilet and discharged into the sewerage system (and thereby eliminating the unacceptable continuous flow method of flushing used by the Greeks and Romans). As a consequence, much effort was devoted to developing mechanisms for controlling water flow into and out of the toilet.

During the period 1870–1920, the number of U.S. patents filed for toilet technology rapidly escalated. In the 1880s, more than 180 U.S. patents were issued for water closets. For many of the inventors concerned with the design of toilet technology, the flush and float valves became the center of attention. In 1879, William Ross of Glasgow, Scotland, filed the U.S. Patent "Improvement in Water-Closet Cisterns" shown in Figure 69.7. In the patent documents, Ross put forward an eloquent statement of the design objectives for the valves in the cistern: "The object I have in view is to provide cisterns for supplying water to water-closets, urinals, and other vessels with valves for controlling the inlet and outlet of water, the valves for admitting water to such cisterns being balanced, so that they can be operated by small floats, and also being of simple, light, and durable construction, while the valves for governing the outflow of water from such cisterns allow a certain amount of water only to escape, and thus prevent waste, and are more simple and efficient in their action than those used before for the same purpose, and are portable and self-contained, not requiring outer guiding cylinders or frames, as heretofore." It is interesting to note that the basic design put forward by Ross for the float and flush valves has changed little in the past 116 years. By 1880 the basic design of the toilet, as we know it today, was well-established (as shown in Figures 69.1 and 69.2). A cistern with flush and float valves and a basin using a water trap are clearly evident.

Some of the designs were quite complex. The patent filed in 1897 by David S. Wallace of the Denver, CO, Wallace Plumbing Improvement Co. depicts a design (shown in Figure 69.8) that would allow both manual and automatic control of the open and close cycle of the flush valve. Such an approach would allow the toilet user to adjust the amount of water to be used during the flushing cycle (a concept that would be resurrected almost 100 years later as a water saving device). Not all inventors chose to use the same approach to the flush valve. David Craig and Henry Conley proposed a quite different solution in 1895 (Figure 69.9). In operation, the valve was quite simple, using a ball to close the water outlet from the toilet tank. The flushing operation is initiated by pulling on the cord 2, which raises the ball inside the tube d', allowing the ball to roll to the f' end of the tube and the tank to empty. The ball rolls slowly back down the tube, settling into the collar c and closing the water outlet to the toilet bowl. This approach to control was not widely adopted.

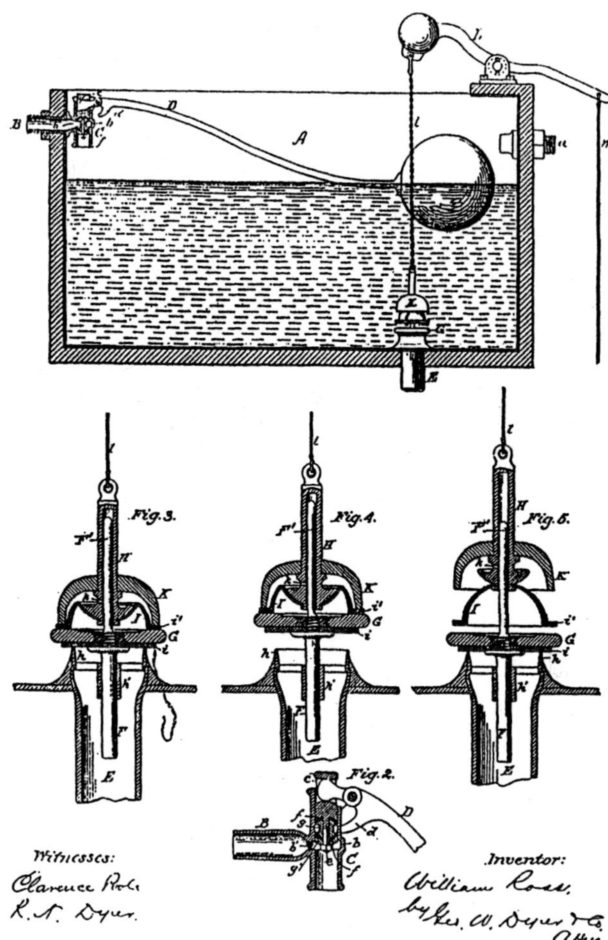

Figure 69.7 U.S. Patent drawing of the flush and float valves, including drawings of the operation of the flush valve, for a toilet tank. *Source:* U.S. Patent No. 211,260 filed by W. Ross, Jan. 7, 1879.

Throughout the history of the development of toilet control technology, much of the technological effort was directed towards improving the mechanisms for assuring correct closure of the flush valve. More recently, however, concerns about conservation of water have motivated technologists to consider new designs that increase the effectiveness of the flush valve. The design by Olson (Figure 69.3) is a good example of recent concerns in the development of control technologies for the toilet. The design allows the user to control the amount of water used during the flushing cycle. Characterized as a water saving system that uses only several pints of water rather than the several gallons of a conventional toilet, the actual amount of water used during the flushing cycle is actively controlled by the user. Recall that this is similar in concept to the approach taken by Wallace in 1897 (Figure 69.8). Notice, too, that the approach to control remains the same; once the flushing cycle has been initiated, the level of water in the toilet tank is controlled by the float valve.

69.9 Historical Approaches to Systems

It should be clear that understanding the historical development of toilet technology requires understanding the historical devel-

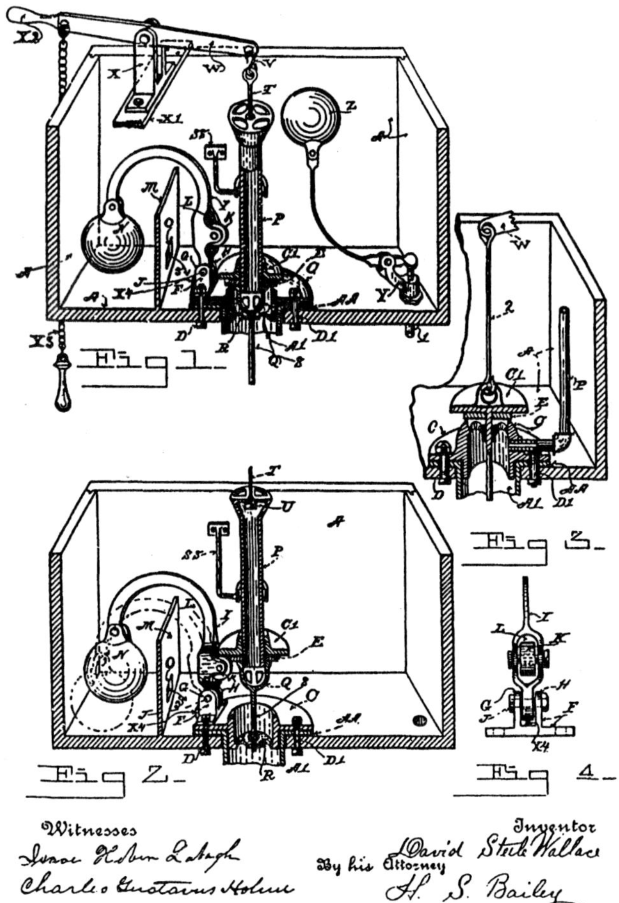

Figure 69.8 U.S. Patent drawing of the flush and float valves that allows both manual and automatic control of the open and close cycle of the flush valve. *Source:* U.S. Patent No. 577,899 filed by D.S. Wallace, Mar. 2, 1897.

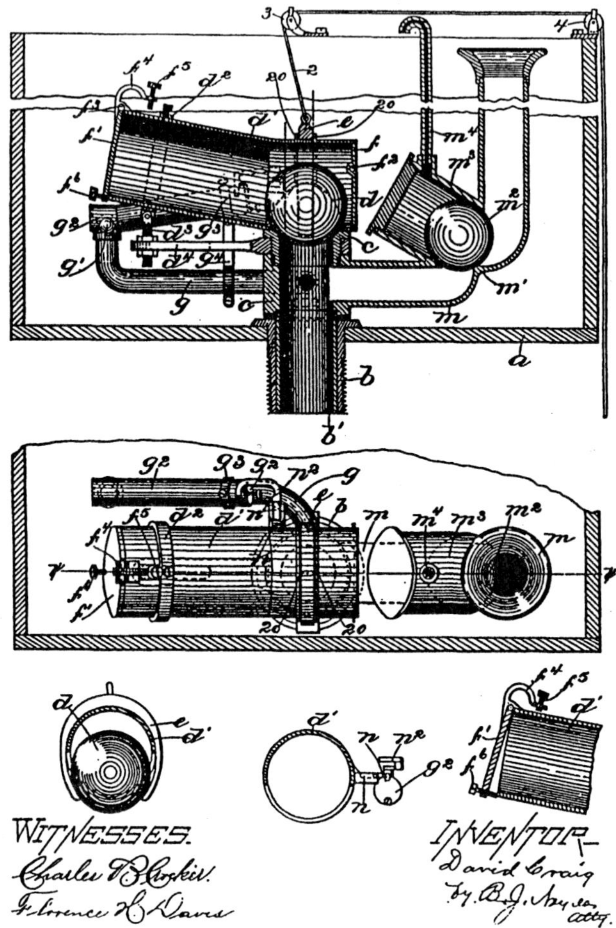

Figure 69.9 U.S. Patent drawing of a variation on the flush valve for the toilet tank. The ball labeled *d* provides the means for closing and sealing the outlet to the toilet basin during the flush cycle. *Source:* U.S. Patent No. 543,570 filed by D. Craig and H. Conley, Jul. 30, 1895.

opment of an entire system of sanitation. Historians have recognized for some time that the history of technology must be considered in terms of entire systems, and considerable recent effort has been devoted to understanding the ways in which systems and the technology required to support them develop over time.

The Thomas Hughes [4] model for the historical development of systems has been especially influential. Hughes extensively studied the development of electric power networks in the period 1880–1930 and, as a result of that research, constructed a four-phase model of the historical development of large-scale distributed systems. Each phase of development is defined by its own characteristics and a group of professionals who play a dominant role in developing the technologies employed in that phase. In the first phase, invention predominates with significant technical effort and intellectual resources devoted to developing the technologies comprising the system. The second phase is primarily concerned with the transfer of technologies from one region and society to another. Although inventors still play a predominant role in this phase, entrepreneurs and financiers become increasingly important to the growth and survival of the system.

In the third phase of the model, regional systems grow into national systems and the scalability of regional technologies becomes a major concern. In the third phase of the model, Hughes links system growth and development to reverse salients and critical problems. Borrowed from military terminology, a reverse salient describes a situation where a section of an advancing line (in this context, the development of some aspect of system technology) is slowed or halted while the remaining sections continue movement in the expected direction. As Hughes [4] states, "A reverse salient appears in an expanding system when a component of the system does not march along harmoniously with other components" and reveals "imbalances or bottlenecks within the system. The imbalances were seen as drags on the movement of the system towards its goals, especially those of lower costs or larger size."

A reverse salient has two major effects on technological development: the growth of technology slows and action must be taken for sustained growth to continue. It is the occurrence of a reverse salient and the effort expended to eliminate that bottleneck which, according to Hughes, captures the essence of technological development. Innovations in technology occur because a reverse salient and the imbalance in system development motivates inventors, industrial scientists, and engineers to concentrate their efforts in finding solutions to those problems. The imbalance defines the problem for the engineers, and the removal of that problem leads to innovations in technology, management policy, and methods of control. Hughes cites a number of examples of inventions and innovations in electrical technologies to illustrate the concept of a reverse salient. One such example is the complex information and control networks established in the 1920s to collect data, monitor plant performance, and control plant output. These networks developed in response to an increase in demand for electricity and a need for better scheduling of plant utilization.

In the fourth phase of the model, the system develops substantial momentum with "mass, velocity, and direction." Much of that momentum can be characterized by the significant investment in capital equipment and infrastructure necessary to maintain the system. At this stage of development, the system's "culture" develops, defined by the professionals, corporate entities, government agencies, investors, and workers whose existence depends upon developing and perpetuating the system.

The development of a comprehensive water and sanitation system (with the toilet as an integral part of it), fits well into Hughes model. Once the need for a system to meet the demand for potable water and provide for the collection and disposal of human waste was identified, rapid technological development occurred. The necessary components of the system were developed and employed in building regional water supply and sewerage systems during the last half of the 19th century (phases one and two of the model). During that period there was considerable transfer of knowledge about requisite sanitation methods from one community to another and among the various medical, engineering, and public health professionals concerned with sanitation policy and the construction of water supply and sewerage systems. As cities grew and demand outstripped capacity, much of the effort was directed towards transforming small-scale private systems into the large-scale municipal systems that could accommodate the requirements of rapidly expanding urban populations (phase three). Because the means for supplying water typically preceded systems of disposal, the collection and disposal of household and industrial waste products were a continual source of problems. When indoor plumbing provided the means to distribute water throughout the home, the market for sinks, baths, and toilets blossomed and the rate of domestic water consumption increased dramatically.

In general, water supply preceded waste disposal, so that an increase in the ability to supply water resulted in significantly more volumes of waste water than could be handled by existing sewer systems. The increase in waste water created a reverse salient that could not be ignored and required a massive municipal response to construct the necessary collection and disposal system. The ability to pipe water directly to bathroom fixtures also created a reverse salient that was directly related to the control problem in the toilet tank. Both the amount of water flowing into the toilet and the amount of water used by the toilet during flushing had to be controlled to minimize the impact on the water supply and sewer systems (in fact, the control technology in the toilet tank became the means for linking the two systems). As a result, significant effort was expended in the two decades following 1875 to refine the operation of the toilet flush and float valves, especially to minimize the potential for incomplete valve closure and resultant water leakage. As we have seen, many designs were proposed, but the solutions have remained fairly consistent through the years.

Once work began on the basic infrastructure for supplying water and disposing of wastes, the development of sanitation systems was well into Hughes' fourth stage. The increase in the ability to supply water created demand for water and bathroom technologies and the subsequent need for waste water disposal. A public health movement and public works facilities and services grew out of the concern for clean water and sanitary living conditions. Underlying the desire for clean water and sanitary conditions were changing attitudes towards health, cleanliness, and the causes of disease. Once urban congestion reached a point where the health and well-being of people living in cities was threatened by disease, change occurred and new norms for cleanliness and standards of living were adopted. One of the major technological developments arising from that period of change was the modern toilet.

69.10 Summary

The goal of this chapter was to place the development of a control-based technology in its historical context. In the process of attaining that goal, an attempt has been made to show how the development of such a technology, even in its simplest form, has many dimensions. The toilet and the basic control technology that resides in its water tank found its way into the home as the result of a number of important social, political, economic, and health reasons. The chronology shown in Figure 69.10 captures the multitude of events that occurred between 1830 and 1905 in the areas of disease, health, medicine, public works, and toilet technology. The chronology focuses on events in Newark, NJ during that period. Prior to 1832 and the first cholera epidemic in Newark, there was little activity in the areas of public health or public works. Until the community was motivated to do something about the appalling conditions, no significant efforts were made to supply clean water and effectively remove waste products. As far as toilet technology was concerned, there was very little development prior to 1854. The development of control technology for the toilet rapidly followed the development of water supply and sewer systems, so that, by the 1880s, considerable progress had been made in controlling water flow in the toilet tank.

There is no one single event that can be identified as the motivating factor for putting control technologies into the toilet. Clearly, the availability of an adequate water supply, indoor plumbing, and a sewer system for carrying away waste were prerequisites for developing water flow control in the toilet tank.

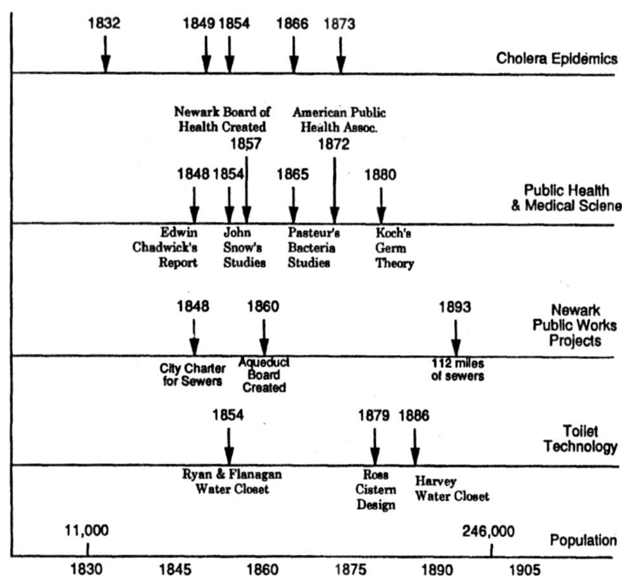

Figure 69.10 Chronology of events in Newark, NJ between 1830 and 1905 in the areas of disease, health, medicine, public works, and toilet technology.

However, those factors were not sufficient motivation for moving the toilet into the home nor adequate stimulus for the development of toilet technology. The threat of disease, changing attitudes towards health and sanitation, and a redefinition of the requirements of a modern home also contributed to the demand for bathroom technology. History has shown us how the development of a control technology, as well as the attitudes towards health, cleanliness, and standards for a middle class home, can be influenced by a number of strong, interacting forces. As Hughes' model points out, however, once development attained momentum, there was no stopping toilet technology.

References

[1] Armstrong, E.L., *History of Public Works in the United States 1776-1976,* American Public Works Association, Chicago, IL, 1976.

[2] Bijker, W.E., Hughes, T.P., and Pinch, T.J., *The Social Construction of Technological Systems,* The MIT Press, Cambridge, MA, 1987.

[3] Galishoff, S., *Newark: The Nation's Unhealthiest City, 1832-1895,* Rutgers University Press, New Brunswick, NJ, 1975.

[4] Hughes, T.P., *Networks of Power: Electrification in Western Society, 1880-1930,* The Johns Hopkins University Press, Baltimore, MD, 1983.

[5] Lupton, E. and Miller, J.A., *The Bathroom, The Kitchen, and the Aesthetics of Waste,* MIT List Visual Arts Center, Cambridge, MA, 1992.

[6] Mayr, O., *The Origins of Feedback Control,* MIT Press, Cambridge, MA, 1969.

[7] Melosi, M.V., *Pollution and Reform in American Cities, 1870-1930,* University of Texas Press, Austin, TX, 1980.

[8] Reynolds, R., *Cleanliness and Godliness,* Doubleday, New York, 1946.

[9] Winslow, C.E.A., *Man and Epidemics,* Princeton University Press, Princeton, NJ, 1952.

[10] Wright, L., *Clean and Decent,* Revised ed., Routledge & Kegan Paul, London, 1980.

70

Temperature Control in Large Buildings

Clifford C. Federspiel
Johnson Controls, Inc., Milwaukee, WI

John E. Seem
Johnson Controls, Inc., Milwaukee, WI

70.1 Introduction

Heating, ventilating, and air-conditioning (HVAC) systems in large buildings are large-scale processes consisting of interconnected electromechanical and thermo-fluid subsystems. The primary function of HVAC systems in large commercial buildings is to maintain a comfortable indoor environment and good air quality for occupants under all anticipated conditions with low operational costs and high reliability.

The importance of well-designed controls for HVAC systems becomes apparent when one considers the impact of HVAC systems on the economy, the environment, and the health of building occupants. One of the most significant operational costs of HVAC systems is energy. There is a direct cost associated with purchasing energy, and an indirect environmental cost of generating energy. It has been estimated that HVAC systems consume 18% of the total energy used in the U.S. each year [1]. Energy utilization in buildings is strongly influenced by the HVAC control systems, so the potential impact of HVAC controls on the economy and the environment is significant.

In addition to the impact on national energy consumption and the environment, HVAC systems have an impact on human health because people spend more than 90% of their lives indoors. Based on data from residential and commercial buildings, the Environmental Protection Agency (EPA) has placed indoor radon first and indoor air pollution fourth on a list of 31 environmental problems that pose cancer risks to humans [2]. Additionally, indoor air pollution other than radon was assessed as a high noncancer risk. Therefore, HVAC control systems can have a significant impact on human health because they can play a central role in controlling indoor air pollution.

In this chapter, building temperature controls and the associated flow and pressure controls are described. The emphasis of the chapter is on the current practice of control system design for large commercial buildings. Therefore, the controls described in this chapter may not reflect those for residential buildings or industrial buildings.

70.2 System and Component Description

Before one can design a control system for a process, it is necessary to understand how the process works. In this section, a prototypical HVAC system for a large building is described. Important features of subsystems that affect the control system design and performance are also described. Also, a description of the control system design is provided.

70.2.1 A Prototypical System

In large buildings, there is typically a central plant that supplies chilled and hot water or steam to a number of heating and cooling coils in air-handling units, terminal units, and perimeter heat-

0-8493-8570-9/96/$0.00+$.50

ing coils. The central plant uses one or more boilers, chillers, and cooling towers to perform this task. Primary and secondary pumping systems such as shown in Figure 70.1 may be used to lower the pumping power requirements. Chillers generally re-

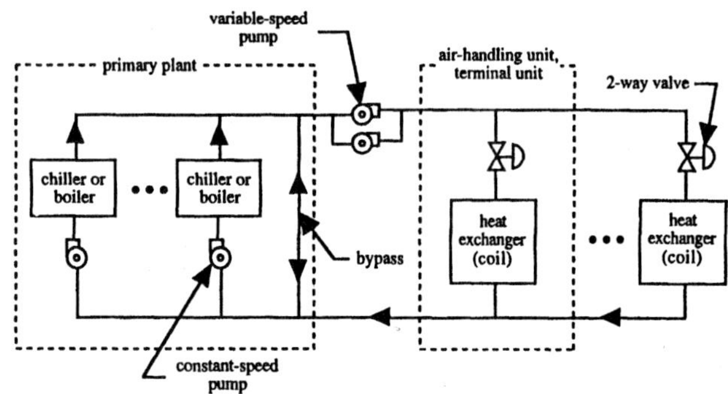

Figure 70.1 Schematic diagram of a piping distribution system. The primary system is constant-flow while the secondary systems are variable-flow.

quire a constant water flow rate, so the primary pumps are run at constant speed. The speed of the secondary pumps may be modulated to reduce pumping costs.

Air-handling units distribute conditioned air to a number of zones, which is a volume within the building that is distinct from other volumes. Often, but not always, physical partitions, walls, or floors separate zones. In large buildings, air-handling units typically supply cool air to all zones during all seasons. Core zones in buildings must be cooled even when the outdoor temperature is well below freezing because of the large number of heat sources such as lights, equipment, and people, and because core zones are insulated from weather conditions by perimeter zones. As Figure 70.2 shows, air-handling units have a number of discrete components. Not all of the components shown in the figure are

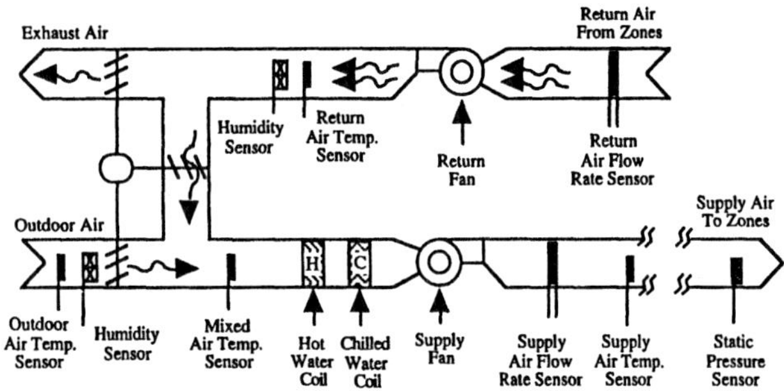

Figure 70.2 Schematic diagram of an air-handling unit.

found in every air-handling unit, and some air-handling units may contain additional components not shown in the figure. There are two basic types of air-handling units: constant-air-volume (CAV) units and variable-air-volume (VAV) units. In CAV air-handling units, the supply and return fan capacities are constant, while in VAV air-handling units, the fan capacities are modulated.

Terminal units, such as the one shown in Figure 70.3, are used to control the temperature in each zone. Terminal units are typ-

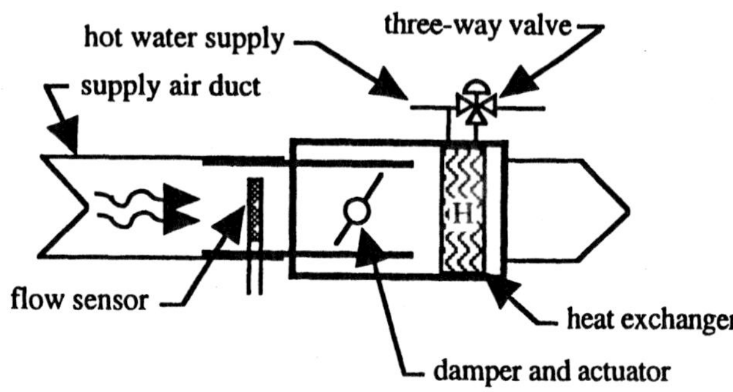

Figure 70.3 Schematic diagram of one type of VAV terminal unit. The zone is cooled by modulating the air flow rate, and the zone is heated by modulating the valve on the reheat coil.

ically installed in the supply air duct for each zone, where they are used to modulate the flow rate and/or the temperature of air supplied to a zone.

70.2.2 Components

An HVAC system consists of a number of interconnected components that each have an impact on the behavior of the system and on the ability of the control system to affect that behavior. Components that have a key influence on the control system design are described next.

Actuators

Actuators are used to modulate the flow through dampers and valves. The two most common types of actuators are electric motors and pneumatic actuators. Most of the electric motors used in HVAC controls are synchronous ac motors, some of which are stepper motors. To reduce motor power consumption and cost, gear ratios are used such that the load reflected onto the motor through the gear reduction is small.

Pneumatic actuators are usually linear motion devices consisting of a spring-loaded piston in a cylinder. Air pressure in the cylinder pushes against the spring and any other forces acting on the piston. The spring returns the piston to the "normal" position in case of a loss of pressure. Pressure is supplied to the piston from a main, typically at a pressure of 20 (psi). Typical hardware for throttling the main pressure is a dual set of solenoid valves, an electric-to-pneumatic transducer, or a pilot positioner. Pneumatic actuators are prone to have substantial friction and hysteresis nonlinearities. A detailed description of pneumatic actuators can be found in [3].

Heat Exchangers

It is important to understand the steady-state behavior of heat exchangers because it affects the process gain. There are different ways to mathematically model the steady-state behavior including the log mean temperature difference (LMTD) method and the effectiveness-number of heat transfer units (ϵ-NTU) method. These relations are used to design and size heat exchangers. For control system design, one needs to model and understand the relationship between the heat transfer rate and the controlled flow rate. The steady-state characteristics of heat exchangers can be derived from the ϵ-NTU relations, which can be found in most texts on heat transfer (e.g., [4]). For example, consider a cross-flow, water-to-air heat exchanger with unmixed flow on both sides in which the air flow rate is a constant, the water flow rate is modulated, and the water is heating the air. For such a heat exchanger, the ϵ-NTU relation is [4]

$$\epsilon = 1 - exp((\frac{1}{C_r})(NTU)^{0.22}(exp(-C_r(NTU)^{(0.78)}) - 1)) \tag{70.1}$$

where ϵ is the heat exchanger effectiveness, C_r is the ratio of the minimum to maximum heat transfer capacity rates, and NTU is the number of heat transfer units. All three of these quantities are nondimensional. Mathematically, they are defined as

$$\epsilon \equiv \frac{q}{q_{max}} \tag{70.2}$$

$$q_{max} = C_{min}(T_{w,i} - T_{a,i}) \tag{70.3}$$

$$C_{min} \equiv \min[C_a, C_w] \tag{70.4}$$

$$C_r \equiv \frac{C_{min}}{C_{max}} \tag{70.5}$$

$$C_{max} = \max[C_a, C_w] \tag{70.6}$$

$$C_a \equiv \dot{m}_a c_{p_a} \tag{70.7}$$

$$C_w \equiv \dot{m}_w c_{p_w} \tag{70.8}$$

$$NTU \equiv \frac{UA}{C_{min}} \tag{70.9}$$

where q denotes the heat transfer rate in units of energy per unit time, q_{max} denotes the maximum achievable heat transfer rate, C_{min} is the minimum heat transfer capacity rate of the two fluids in units of energy per unit time per unit temperature, $T_{w,i}$ is the temperature of the water entering the heat exchanger, $T_{a,i}$ is the temperature of the air upstream of the heat exchanger, C_{max} is the maximum heat transfer capacity rate of the two fluids, C_a is the heat transfer capacity rate of the air, C_w is the heat transfer capacity rate of the water, $\dot{m}_a$ is the mass flow rate of the air, c_{p_a} is the specific heat at constant pressure of the air in units of energy per unit mass per unit temperature, $\dot{m}_w$ is the mass flow rate of the water, c_{p_w} is the specific heat at constant pressure of the water, U is the heat transfer coefficient at the surface of the heat exchanger in units of energy per unit time per unit temperature per unit area, and A is the area of the heat exchanger (U and A must correspond to the same surface).

The steady-state relation between the water flow rate and the heat transfer is

$$q = q_{max}\epsilon \tag{70.10}$$

Figure 70.4 shows the characteristic of Equation 70.10 normalized by the heat transfer at the maximum water flow rate when $NTU_{max} = 2$, $\dot{m}_a = 5.8$ kg/s (10000 ft^3/min), and $\dot{m}_{w_{max}} = 3\dot{m}_a \frac{c_{pa}}{c_{pw}}$. The large slope at low flow and small slope at high flow is typical of heat exchanger characteristics. This fact is important when control valves are coupled with heat exchangers, and it will be discussed in more detail later.

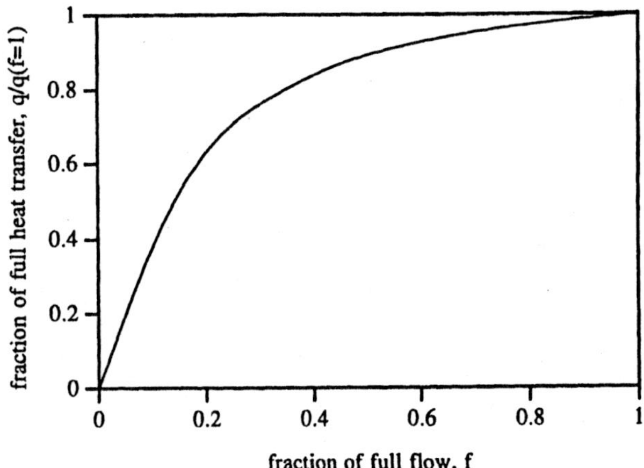

Figure 70.4 Heat exchanger characteristic for a cross-flow heat exchanger with unmixed flow in both fluids.

The dynamic behavior of heat exchangers is complex and may be difficult to model accurately because heat exchangers are nonlinear, distributed-parameter subsystems. Qualitatively, a heat exchanger introduces a phase lag and time delay into the overall system. This lag and delay vary as the flow rate through the heat exchanger is modulated.

Dampers and Valves

The flow through a damper or valve can be determined from

$$Q = fC_v\sqrt{\frac{\Delta p_v}{g_s}} \tag{70.11}$$

where Q is the flow rate, f is the flow characteristic of the valve, C_v is the flow coefficient, Δp_v is the pressure drop across the device, and g_s is the specific gravity of the fluid. The flow characteristic is the relation between the position of the damper blades or valve plug and the fraction of full flow. When the pressure difference across the damper or valve is held constant for all positions, the flow characteristic is referred to as the inherent characteristic and is denoted by f_i. The flow coefficient, C_v, is the ratio of the flow rate in the fully open position to the ratio of the square root of the pressure drop to the specific gravity. The C_v increases as the size of the device increases. For water valves in which the flow rate is measured in units of gallons per minute and pressure measured in units of pounds per square inch, typical values for the C_v are 1 for a $\frac{1}{2}$-inch valve, 37 for a $2\frac{1}{2}$-inch valve, and 150 for a 4-inch valve.

Figure 70.5 shows an example of the inherent characteristic of a damper or valve. The quantity f_i is the fraction of full flow, and L is the fraction of the fully open position of the damper blades or valve stem. The important features of a damper's inherent char-

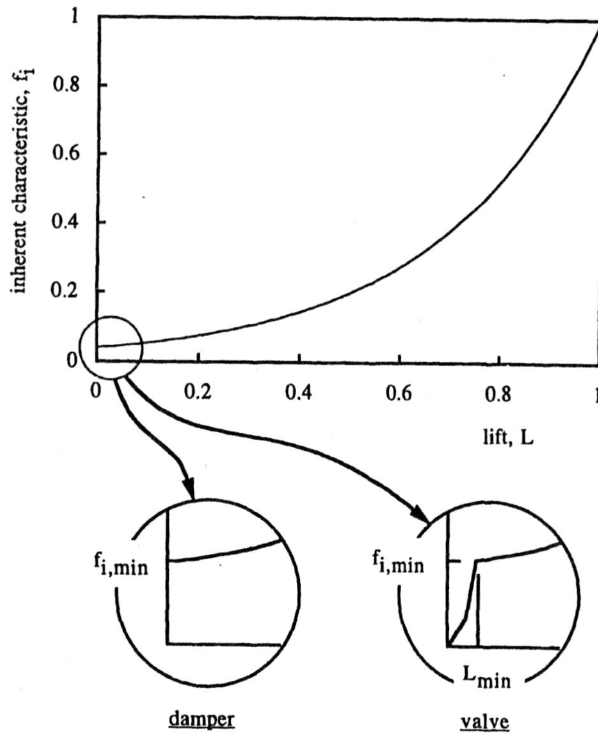

Figure 70.5 Inherent characteristic of a damper or valve.

acteristic are (1) the flow rate at the closed position (dampers leak, although seals may be used to reduce leakage) and (2) the shape of the inherent characteristic between the fully closed and fully open positions. Parallel-blade dampers have a faster-opening inherent characteristic than opposed-blade dampers.

Like dampers, one of the important features of a valve's inherent characteristic is the behavior when the lift is nearly zero. Unlike dampers, many valves do not leak when fully closed. However, when the lift is nearly zero, the characteristic changes dramatically as shown in Figure 70.5. When $L < L_{min}$, the inherent characteristic is dependent on the valve construction, seat material, manufacturing tolerances, etc. The valve may shut off quickly and completely as shown in the figure, or it may shut off quickly but not completely. If the valve seat has rubberized seals, then L_{min} will be larger than without rubberized seals.

All actuation devices have a limited resolution for positioning the valve stem. Pneumatic actuators, which are commonly used in HVAC systems, have low resolution because they are prone to stiction and hysteresis. Although the characteristic of the valve may be a continuous function of the lift for flow rates less than f_{min}, the limited resolution of the valve stem positioning implies that the characteristic may be accurately modeled as a discontinuity at zero lift. The magnitude of the discontinuity is characterized by the rangeability of the valve. The inherent rangeability of a valve is defined as the ratio of the maximum to the minimum controllable flow. Mathematically, it is

$$R_i = \frac{1}{f_{i,min}} \tag{70.12}$$

Control valves in HVAC systems typically have an inherent rangeability between 20 and 50.

The other important feature of a valve's inherent characteristic is the relationship between L and f_i for $L_{min} \leq L \leq 1$. Valves are often designed to have one of the following inherent characteristics: quick-opening, linear, or equal-percentage. Assuming the inherent characteristic is discontinuous at $L = 0$ (i.e., $L_{min} = 0$) due to limited resolution of the lift, then the inherent characteristic of a linear valve is

$$f_i = (1 - f_{i,min})L + f_{i,min} \tag{70.13}$$

and the inherent characteristic of an equal-percentage valve is

$$f_i = R_i^{L-1} \tag{70.14}$$

In most systems, the pressure drop across a damper or valve varies with the flow rate because of additional pressure losses in ducts or pipes. The authority of a damper or valve is defined to be

$$\alpha = \frac{p_v}{p_s} \tag{70.15}$$

where p_v is the pressure drop across the damper or valve at the fully open position and p_s is the total system pressure drop (e.g., the combined pressure drop across a valve, piping, and heat exchanger). The relationship between the lift and the fraction of full flow when a damper or valve is installed in a system is called the installed characteristic. The installed characteristic is dependent on both the lift and the authority. Mathematically, the installed characteristic is

$$f_s = \sqrt{\frac{1}{1 + \alpha(f_i^{-2} - 1)}} \tag{70.16}$$

Figure 70.6 shows the installed characteristics of equal-percentage valves for several different values of authority assuming that the inherent characteristic is discontinuous when the lift is zero. When the authority is unity, the installed characteristic is the same as the inherent characteristic. Note that the installed value of f_{min} is dependent on the authority. This means that the range-

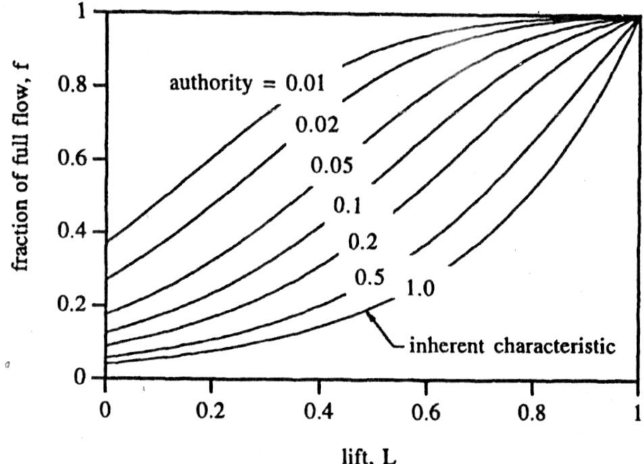

Figure 70.6 Installed characteristics of equal-percentage valves with $R_i = 25$.

ability of an installed valve or damper is not the same as the inherent rangeability. The installed rangeability is the ratio of the maximum installed flow to the minimum controllable installed flow and is mathematically defined as

$$R_s = \frac{1}{f_{s,min}} \tag{70.17}$$

where $f_{s,min}$ is the smallest controllable fraction of the maximum installed flow. For an equal percentage valve, the installed rangeability is

$$R_s = \sqrt{1 + \alpha(R_i^2 - 1)} \tag{70.18}$$

See [5] for complete details on valve characteristics and terminology.

Valves need to be properly sized to ensure proper operation. The size of the valve affects the authority of the valve; oversized valves have a low authority. Undersized valves cannot deliver sufficient flow rates to meet the maximum load conditions, but oversized valves require less pumping power. In [6] it is recommended that valves be sized so that the authority is between 25 and 33%. In the HVAC industry, oversized valves are more prevalent than undersized valves because large safety factors are used to ensure adequate capacity.

The inherent rangeability and the authority affect flow control loops at low flow rates because it is impossible to smoothly modulate the flow rate below the value of $f_{s,min}$. Low authority and/or low rangeability make this problem worse. Problems caused by low authority and low rangeability are worse when valves are coupled with heat exchangers. Consider the equal-percentage characteristic such as shown in Figure 70.6. When an equal-percentage valve is coupled with the cross-flow heat exchanger described in Section 70.2.2, the resulting steady-state characteristic of the valve and heat exchanger combination is as shown in Figure 70.7 for two different combinations of authority and inherent rangeability. One can define the heat transfer rangeability for a valve and heat exchanger combination as the ratio of the maximum to minimum controllable heat transfer at steady-state. Mathematically, the heat transfer rangeability is

$$R_q = \frac{1}{q_{min}} \tag{70.19}$$

Note that the value of R_q corresponding to $\alpha = 0.1$ and $R_i = 25$ is 2.2. The implication of the heat transfer rangeability is that a heating or cooling load less than q_{min} cannot be matched without cycling. Since the steady-state characteristic of heat exchangers has a large slope at low flow rates, the value of R_q will be very low and cycling will occur unless the inherent rangeability of the valve and the authority of the valve are high.

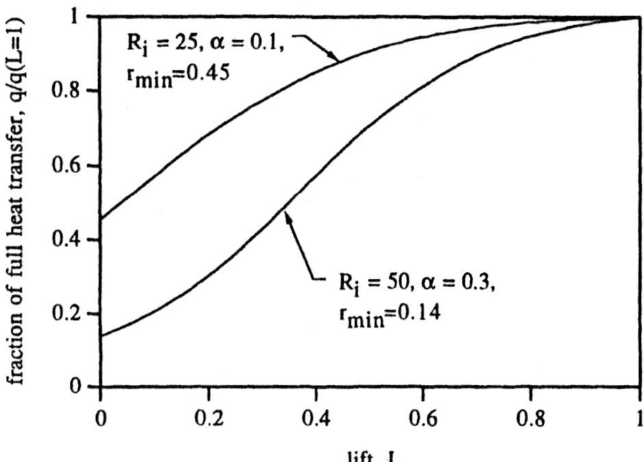

Figure 70.7 Steady-state installed heat transfer characteristic for an equal-percentage valve with $R_i = 25$ and $\alpha = 0.1$.

Fans and Pumps

Steady-state relations between head, flow rate, speed, and power consumption are nonlinear. The following are idealized similarity relations for pumps moving incompressible fluids [7]

$$\frac{Q_2}{Q_1} = \frac{\omega_1}{\omega_2}\left(\frac{D_2}{D_1}\right)^3 \tag{70.20}$$

$$\frac{H_2}{H_1} = \left(\frac{\omega_2}{\omega_1}\right)^2\left(\frac{D_2}{D_1}\right)^2 \tag{70.21}$$

$$\frac{P_2}{P_1} = \frac{\rho_2}{\rho_1}\left(\frac{\omega_2}{\omega_1}\right)^3\left(\frac{D_2}{D_1}\right)^5 \tag{70.22}$$

where Q denotes flow rate, ω denotes speed, D denotes the characteristic size, H denotes head, P denotes power consumption, and ρ denotes density.

These relations have two important consequences for control systems. The first is that the cubic relation between power and speed implies that capacity modulation by speed modulation is more efficient than by throttling; throttling results in only minor reductions in power consumption.

The second implication of the similarity relations is that the gain of pressurization loops will be nonlinear due to the quadratic relation between pressure and speed. Since the load on a modulating pump or fan varies with time due to changes in the position of control valves or dampers, the quadratic relation varies with time.

Sensors and Transmitters

One feature of sensors that affects control performance is the response time of the sensors. In [6] it is recommended that measurement and transmission time constants be less than one tenth the largest process time constant. In some cases, such as flow control loops, the process dynamics are extremely fast, so it may not be possible to meet this recommendation.

Design decisions for HVAC systems and the associated controls are strongly influenced by initial cost or installed cost. Therefore, many buildings have just the minimum number of sensors required for operation. In other words, many buildings are "sensory-starved." It is often possible to reduce operational costs when additional sensors are included in the design. Additional sensory information can be used by control systems to improve energy utilization and to increase diagnostic capabilities.

70.2.3 Controls

HVAC control systems are partially decentralized. Typically there is a separate controller for each subsystem. For example, there may be a controller for the central plant, a controller for each air-handling unit, and a controller for each terminal unit. Each subsystem may have multiple inputs and multiple outputs. The commands to the process inputs are usually generated through a logical coordination of single-input, single-output (SISO) proportional-integral-derivative (PID) algorithms. There may be multiple PID algorithms operating in a controller at one time.

Historically, pneumatic devices were used exclusively to compute and execute control commands. Today, many control systems in buildings are still partially or completely pneumatic, but the trend is toward the use of distributed digital controllers. The most modern control systems for HVAC equipment contain a network of digital controllers connected on a communication trunk or bus. There are several different methods for connecting various controllers. Figure 70.8 shows an architecture that uses peer-to-peer communications. The digital controllers and operator workstations all communicate on the same network. There may be hundreds of digital controllers on a single network. Due to the historic use of pneumatic controls, many of the digital control systems in buildings are programmed to emulate pneumatic controllers. The communications networks are used primarily for alarm reporting, monitoring, and status communications.

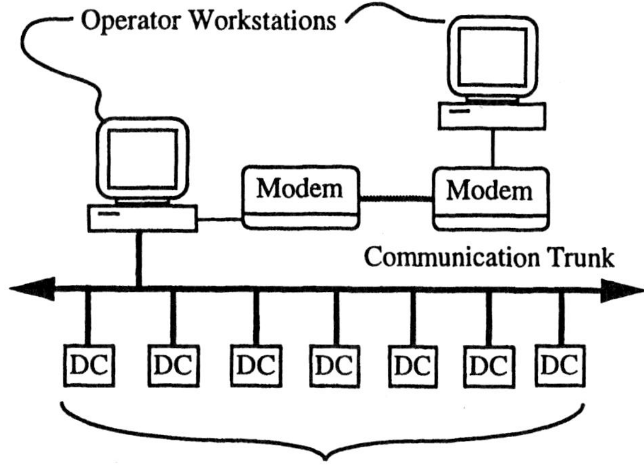

Figure 70.8 Communication architecture with peer-to-peer communications.

70.3 Control Performance Specifications

In many engineering design specifications for HVAC systems, the performance of the control system is not rigorously specified. Often there is little reference made to quantifiable measures of performance. In this section, guidelines for certain measures of performance are described. These guidelines are based on common practice and inferences from industry standards. Since most HVAC subsystems are controlled using a decentralized SISO design, this section is focused on control performance specifications for SISO systems.

70.3.1 Time-Domain Specifications

In Chapter 10.1, time-domain specifications for feedback control systems are described in detail. In this section, some time-domain specifications for certain control loops are recommended.

Zone Temperature

For zone temperature control, the steady-state error has a strong impact on the comfort of occupants. In [8] the ASHRAE standard for an acceptable range of zone temperatures is described. If temperatures remain within this range, then 80% of the occupants should be comfortable. The temperature range is 7^oF and the center of the range is dependent on the season. The implication of this standard is that the steady-state error for zone temperature control is dependent on the setpoint. If the setpoint is chosen as the center of the ASHRAE acceptable temperature range, then the maximum allowable steady-state error is 3.5^oF. It is common practice for building operators to raise the zone temperature setpoints when it is hot outside and lower the setpoints when it is cold outside to conserve energy. If the setpoint is raised or lowered to the limit of the ASHRAE standard, then the steady-state error for the zone temperature controls should be zero.

Flow Control

Flow control loops are cascaded with temperature control loops in pressure-independent VAV boxes and are used in air-handling units to regulate the outdoor air flow rate. Typically, it is expected that flow control loops will be controllable from 5 to 100% of the maximum rated flow.

Flow controllers may be operated manually during commissioning or when troubleshooting. During manual operation, the settling time of flow controllers is important because it is expensive to have an operator wait for transients to settle out. A reasonable settling time is twice the stroke time of the actuator.

For outdoor air flow control loops and cooling based on the use of cool outdoor air (economizer cooling), overshooting may be a problem when the outdoor air temperature is below freezing. If the controls overshoot, then freeze protection devices may be activated, which will shut down an air-handling unit. The maximum allowable overshoot on an outdoor air flow control or economizer control loop depends on the outdoor air temperature. When the outdoor air temperature is well below freezing, overshoot may be intolerable.

Pressurization Control

For pressure control loops, overshoot can be dangerous. Overshooting pressurization controls can trip pressure relief devices or rupture ducts or pipes. The maximum allowable percent overshoot on a pressurization control loop depends on how close the loop is being controlled to the high limit. For example, if the high limit on a duct pressurization loop is 2.5 inches of water and the setpoint is 2 inches, then the maximum allowable percent overshoot for a safety factor of 5 is 5%. However, if the setpoint is 1 inch, then the maximum allowable percent overshoot for the same safety factor is 30%.

70.3.2 Performance Indices

It is often easier to specify control performance in terms of time-domain specifications because the physical meaning of the quantities such as percent overshoot and steady-state error are clear and are particularly relevant to certain control problems. However, it is often easier to design a control system to optimize a single index of the control system performance.

Three commonly used performance indices for designing single-loop control algorithms are the integrated absolute error (IAE), the integrated squared error (ISE), and the integrated time absolute error (ITAE). These indices are defined as

$$IAE = \int_0^\infty |e(t)|dt \tag{70.23}$$

$$ISE = \int_0^\infty e^2(t)dt \tag{70.24}$$

$$ITAE = \int_0^\infty t|e(t)|dt \tag{70.25}$$

These indices are rarely used to determine parameters of control loops in the field, but are commonly used to design automated tuning algorithms or adaptive control algorithms. They are also used as benchmarks for simulation of new control algorithms. For low-order linear systems, design relations that minimize one of these indices are available for certain types of linear SISO controllers. For more information, see Chapter 10.2.

70.4 Local-Loop Control

This section describes algorithms for modulating valves, dampers, and other devices to control temperature, pressure, or flow. The emphasis is on SISO systems and PID control algorithms.

70.4.1 Mathematical Models

Despite the fact that many of the components comprising an HVAC system are most accurately modeled as nonlinear distributed-parameter systems, low-order linear models are used to determine parameters for the local-loop controllers. Many of the most commonly used process transfer function models for designing the local-loop controllers can be represented by the following transfer function:

$$G(s) = \frac{Ke^{-sT}}{(s+p_1)(s+p_2)} \tag{70.26}$$

For example, one can get a first-order delayed transfer function from Equation 70.26 by letting $p_2 \to \infty$, or one can get a second-order Type 1 system (an integrating system) by letting $p_2 \to 0$. The control systems are typically designed based on the worst-case model parameters.

70.4.2 Feedback Control Algorithms

Figure 70.9 shows the basic structure of a feedback controller. The objective of the feedback controller is to maintain the process output at the setpoint. The feedback controller uses the error to determine the control signal. The control signal is used to adjust the manipulated variable.

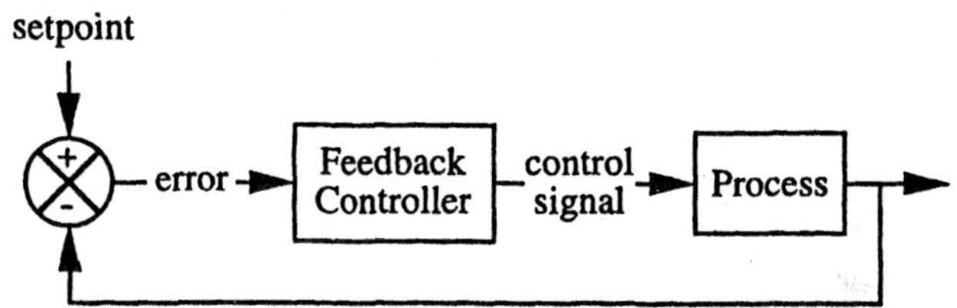

Figure 70.9 Structure of feedback controller.

According to [3], proportional-only (P) control is used in most pneumatic and older electric systems. The control signal for a P-controller is determined from

$$u(t) = \bar{u} + K_P e(t) \tag{70.27}$$

where $u(t)$ is the controller output at time t, $\bar{u}$ is a constant bias, K_P is the controller gain, and $e(t)$ is the error signal at time t.

Equation 70.27 describes the ideal behavior of a P-controller. In an actual application, the actuator usually has upper and lower limits (e.g., a valve has limits of completely open or closed). P-controllers exhibit a steady-state offset if the process is a Type 0 system (i.e., nonintegrating process) and if the controller output is not biased. The amount of steady-state offset can be reduced by setting the bias value equal to the expected nominal steady-state value of the controller output.

The steady-state offset of Type 0 systems can be eliminated by using a proportional plus integral (PI) controller. The time-domain representation of a PI controller is

$$u(t) = \bar{u} + K_P e(t) + K_I \int_0^t e(\tau)d\tau \tag{70.28}$$

where K_I is the integration gain.

There is a disadvantage of integral control action under certain conditions. If the actuator saturates and the error continues to be integrated, then the integral term will become very large and it will take a long time to bring the integral term back to a normal value. Integrating a sustained error while the actuator is saturated is commonly called integral windup or reset windup. Integral windup can occur during start-up, after a large setpoint change, or during a large disturbance in which the process output cannot be controlled to the setpoint. In [6, 9, 10], methods for reducing the effect of integral windup are described. The feature of reducing the effect of integrator windup is called anti-reset windup. An anti-reset windup strategy will significantly improve the performance of a controller with integral action.

For some systems, the performance of a PI controller can be improved by adding a derivative (D) term. The classical equation for a PID control algorithm is

$$u(t) = \bar{u} + K_P e(t) + K_I \int_0^t e(\tau)d\tau + K_D \frac{d}{dt} e(t) \quad (70.29)$$

where K_D is the derivative gain.

In commercial controllers, one does not use the classical form of the PID control algorithm because of a phenomenon known as derivative kick. If the classical PID controller is used, then the controller output goes through a large change following a step change in the controller output. This phenomenon is called derivative kick and it can be eliminated by using

$$u(t) = \bar{u} + K_P e(t) + K_I \int_0^t e(\tau)d\tau - K_D \frac{d}{dt} y(t) \quad (70.30)$$

Equation 70.30 determines the derivative contribution based on the process output rather than the error.

Derivative action generally improves the response of control loops in buildings. However, it is not commonly used in HVAC control loops because a three-parameter PID controller is more difficult to tune than a two-parameter PI controller and because derivative action makes the controller more sensitive to noise. Therefore, the additional improvement in the response time of the controller attributed to derivative action is generally outweighed by the additional cost of tuning and by the increased sensitivity to noise.

Figure 70.10 shows a block diagram for a practical feedback control algorithm in a digital controller. Next, we describe the purpose of the analog filter, noise spike filter, digital filter, and the deadzone nonlinearity.

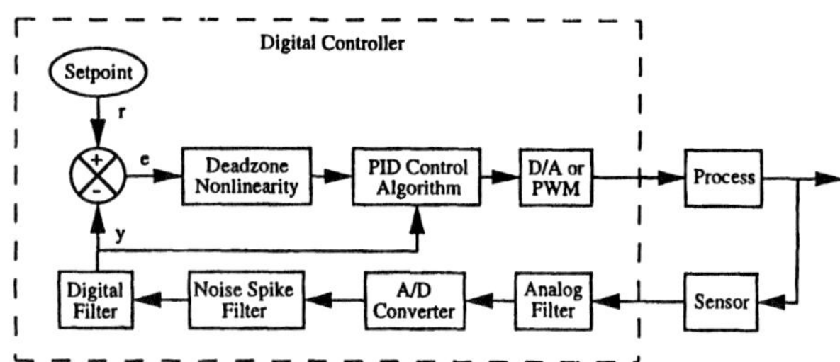

Figure 70.10 Structure of digital feedback controller.

Analog Filter

The purpose of the analog filter is to remove high frequency noise before sampling by the analog-to-digital (A/D) converter. This filter is commonly called an anti-aliasing filter or an analog prefilter. It is especially important to use an anti-aliasing filter when controlling flow or pressure in a duct or pipe because turbulence can generate a large amount of noise. The bandwidth of the analog filter should be selected based on the sampling period for the A/D converter. In the HVAC industry, a sampling period of 1 second is typical.

Noise Spike Filter

Noise spike filters improve the performance of digital control systems by removing outliers in the process output measurement. Outliers can be caused by a brief communication failure, instrument malfunction, electrical noise, or brief power failures. If noise spike filters are not used, then an outlier in the process output measurement will introduce a large disturbance into the control system. Also, if integration is used in the feedback controller, it may take a long time to recover from the disturbance. In [6, 9] simple noise spike filters are described.

Digital Filters

The sampling time for digital feedback controllers should be selected based on the dynamics of the control system. When controlling slow loops, such as the temperature in a room, a sampling period of 1 minute may be adequate for the feedback controller. However, the A/D converter may have a fixed sampling period of 1 second. For this case a combination of analog and digital filters should be used to remove aliases. The analog filter removes higher-frequency noise than the digital filter. Also, the cutoff frequency of the digital filter is adjusted in software. Thus, digital filters are especially useful for digital control algorithms that use an adjustable sampling period.

Deadzone Nonlinearity

The deadzone nonlinearity shown in Figure 70.11 is used to reduce actuator movement for small errors. The deadzone nonlinearity can be used to increase actuator life without sacrificing control performance. The input-output characteristic curve for the deadzone nonlinearity is shown in Figure 70.11. The output from the deadzone block can be determined from the following relationship:

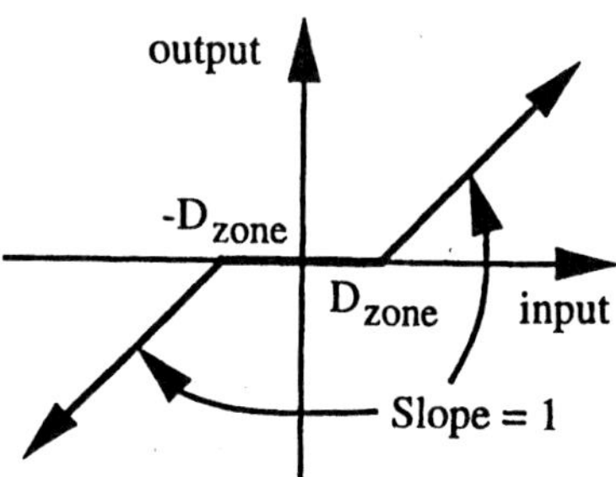

Figure 70.11 Input-output characteristic curve for a dead-zone nonlinearity.

$$\tilde{e} = \begin{cases} e - D_{zone} & \text{when} \quad e > D_{zone} \\ 0 & \text{when} \quad |e| \le D_{zone} \\ e + D_{zone} & \text{when} \quad e < -D_{zone} \end{cases} \quad (70.31)$$

where $\tilde{e}$ is the modified error, D_{zone} is the size of the dead zone, and e is the error. The dead zone nonlinearity is especially useful when controlling processes with noisy signals, such as the flow through pressure-independent VAV boxes. The size of the dead zone should be selected based on the noise in the sampled signal.

70.4.3 Tuning

Tuning methods are used to determine appropriate values of controller parameters. Since the HVAC industry is cost driven, people installing and commissioning systems do not have a long time to tune systems. Consequently, a number of systems use the default parameters shipped with the controller. For some systems, the default parameters may not be appropriate. HVAC systems are nonlinear and time-varying. Thus, fixed-gain controllers may be poorly tuned some of the time even if they were properly tuned when they were installed. Control loops for HVAC systems may require retuning during the year for the processes that have nonlinear characteristics that are dependent upon the season. If a feedback control loop is not retuned, the control response may be "poor" (e.g., the closed-loop response for the controlled variable is oscillatory). An operator would have to retune the controller to maintain "good" control. In buildings, it is important to tune loops because properly tuned loops improve indoor air quality, decrease energy consumption, and increase equipment life.

Computer control systems can automate control loop tuning. One automated method for tuning is autotuning. Autotuning involves automatic tuning of a loop on command from an operator or a computer. With autotuning, the operator or control system must issue a new command to determine new control parameters. Since HVAC processes are nonlinear and time-varying, loops may require retuning after the system load changes. Another automated method for determining control parameters is adaptive control. With adaptive control, the control parameters are continually being updated or adjusted.

In the HVAC industry, people installing and commissioning control systems often have limited analytical skills. Thus, it is necessary that autotuning and adaptive control methods be user friendly. Also, the methods should be robust and have low computational and memory requirements.

Prior to tuning a system, the operator should make sure that all elements of the control system are working properly. The controller should be reading the signal for the process output, and the actuator should be working properly. If the control system uses an electric-to-pneumatic (E-P) converter to drive the actuator, then the zero and span of the E-P converter should be properly adjusted.

Several manual tuning and autotuning methods are based on the response to a step change in the controller output. Open-loop step tests should be run in the high-gain region of the process. Running the step test in the high-gain region helps ensure that the control system remains stable for other regions of operation. (The process gain is equal to the ratio of the steady-state change in the process output to the change in the controller output following a step change in the controller output). For systems with a large amount of hysteresis in the controlled device, the user should stroke the controlled device to reduce the effect of hysteresis on the step test. Figure 70.12 shows how to reduce the effect of hysteresis for a step change in the positive direction. Initially, the controller output should move in a negative direction. Then, the controller output should move in the positive direction until it is back to the initial position. The step change in controller output should take place after the process output returns to a nearly steady-state condition. A similar procedure should be used for a step change in the negative direction.

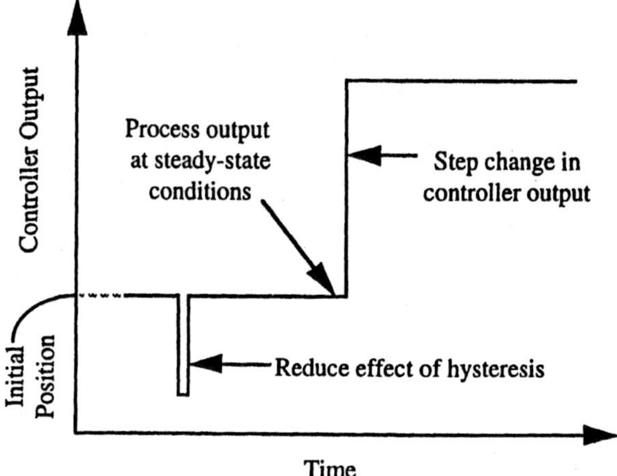

Figure 70.12 Controller output during a step test.

70.5 Logical Control

Logical operations play a central role in the control of HVAC systems. Logical operations are used to ensure the safe start-up and shutdown of subsystems such as boilers, chillers, and air-handling units. They are also used to adjust the capacity of subsystems with discrete states and to sequence modulating controls.

70.5.1 Discrete-State Devices

Discrete-state devices are common in HVAC control systems. These devices may be mechanical devices, such as a pump, or electronic or software devices, such as an alarm.

It is common to use combinations of discrete-state devices to modulate the capacity of a subsystem. For example, condenser water temperature is often controlled by sequencing fans, which may have different capacities and different discrete speeds.

When sequencing discrete-state devices to modulate capacity there is a trade-off between the switching rate and the maximum deviation from the setpoint. The faster the switching rate, the smaller the deviations. This trade-off is affected by the capacitance or inertia of the system. If the system has a large capacitance, then the switching rate required to maintain the same maximum deviation from setpoint is lower.

70.5.2 Modulating Devices

Often the capacity of a single modulating device such as a control valve is insufficient to maintain control over the entire range of operating conditions. In such cases, multiple modulating devices are operated in sequence. A common example is the control of supply air temperature in an air-handling unit or similar subsystem through the sequencing of the cooling valve, outdoor air dampers, and heating valve. When it is sufficiently cool outdoors, the desired sequence involves modulating only one device at a time and placing the outdoor air damper control between the heating valve control and the cooling valve control. This sequence guarantees that the system does not heat and cool at the same time and that the relatively inexpensive cooling with outdoor air (called economizer control) is used before resorting to the relatively expensive cooling provided with the cooling valve.

A common method of sequencing modulating devices is to span the output of a single controller over the range of the modulating devices. For example, the heating valve, economizer, and cooling valve are often controlled as shown in Figure 70.13. While

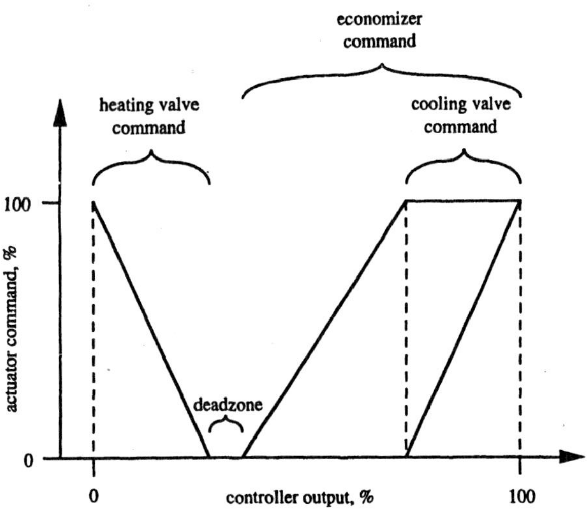

Figure 70.13 Sequencing control for heating, economizer and cooling by spanning the output of a single controller.

this method can result in a logically correct control sequence, it has a basic problem. It is difficult to tune a single fixed-gain PID controller for three different modes of operation (heating, cooling, and economizer) because the system dynamics in each mode are different. A way to get better control performance from a sequence of modulating devices is to use separate controllers for each device. While this method improves the control performance, it complicates the sequencing logic.

A problem with graphically representing sequencing logic as in Figure 70.13 is that the representation of the logic is incomplete. For example, the economizer may be disabled due to the outdoor air temperature, an override by the ventilation controls, or an override by the mixed-air temperature controls. Other methods of graphically representing sequencing logic are described in Chapter 18.

70.6 Energy-Efficient Control

Most of the local control loops in an HVAC system do not directly affect the zone temperatures. The purpose of these loops is to ensure the safe, reliable, and efficient operation of the subsystems. In this section, strategies for controlling such loops are described.

70.6.1 Economizer Cooling

The use of outdoor air for cooling is commonly referred to as economizer cooling or free cooling. In [11] simulations were used to estimate the energy savings with economizer cooling for two different building types in five different locations. Energy savings for cooling varied from 5 to 52%.

Figure 70.14 is a psychrometric chart that shows the different regions of operation for the air-handling unit as a function of the thermodynamic state of the outdoor air. The thermo-

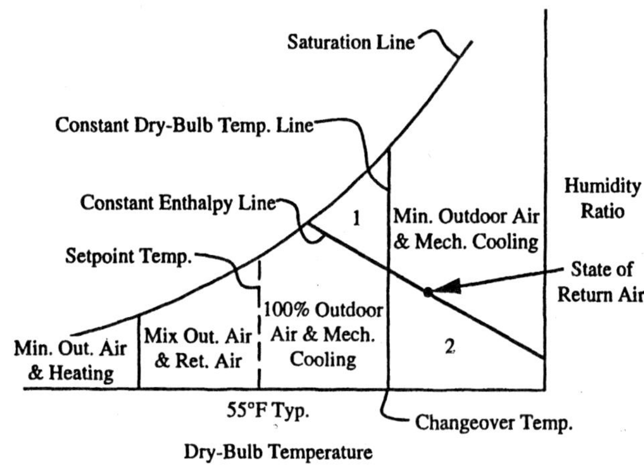

Figure 70.14 Psychrometric chart showing the decision boundaries for temperature-based and enthalpy-based economizer control.

dynamic state of the return air also is shown in Figure 70.14. There are two commonly used strategies for switching between 100% outdoor air and mechanical cooling to minimum outdoor air and mechanical cooling. One strategy compares outdoor air temperature with a changeover temperature (e.g., return air temperature), and the other strategy compares outdoor air enthalpy with a changeover enthalpy (e.g., return air enthalpy). With the temperature economizer strategy, the controls switch from minimum outdoor air to 100% outdoor air when the outdoor air temperature drops below the changeover temperature. With the enthalpy economizer cycle, the controls switch from minimum outdoor air to 100% outdoor air when the enthalpy of the outdoor air drops below the changeover enthalpy. For example, in region 1 of Figure 70.14, the temperature economizer cycle uses 100% outdoor air while the enthalpy economizer cycle uses the minimum amount of outdoor air. Also, in region 2 of Figure 70.14, the temperature economizer uses the minimum amount of outdoor air while the enthalpy economizer cycle uses 100% outdoor air.

Accurate humidity sensors should be used with the enthalpy economizer cycle because moderate sensor errors can result in significant energy penalties. Problems with low-cost humidity sensors are drift and slow recovery from saturated conditions.

It is often assumed that the energy required for cooling will always be less with the enthalpy economizer cycle. However, this is not true. In [11] it was shown that in dry climates the temperature economizer cycle may use less energy than the enthalpy economizer cycle.

70.6.2 Electrical Demand Limiting

Owners or operators of commercial buildings are commonly billed for electric power based upon energy consumed (i.e., kWh) and peak consumption over a demand interval. The rate structure of utilities varies considerably. For example, some utilities use a 30-minute demand interval and other utilities use a 5-minute demand interval.

Computer control systems can reduce the charges for peak demand by shutting off non-essential equipment during times of peak consumption. The strategy of shutting off the nonessential electric loads is commonly called demand limiting. Demand-limiting control strategies can reduce electric bills by 15 to 20% [12]. The building operator selects the desired electric demand target and enters a table of nonessential loads that can be turned off into the computer control system. Also, minimum and maximum times the load can remain off should be entered into the computer control system. A computer control system measures the electrical demand and determines the electrical loads to turn off to maintain the electrical consumption below the target.

In [13] an adaptive demand-limiting strategy with three important features is developed. First, the algorithm is easy to use because statistical methods automatically determine the characteristics of electric energy consumption for a particular building. Second, the algorithm controls the energy consumption just below the target level when there are enough nonessential loads to turn off. Third, the algorithm has low computational and memory requirements. Figure 70.15 shows the target energy consumption, actual energy consumption data with the demand-limiting algorithm, and estimated energy consumption without demand-limiting control from a manufacturing facility. Notice that the demand stays just below the target. Figure 70.16 shows the calculated load to shed from the demand-limiting algorithm and the actual load shed. To limit the electrical demand below the target level, the actual load shed is larger than the calculated load to shed from the demand limiting algorithm. Also, the actual load shed remains larger than the calculated load to shed from the demand limiting algorithm because of a minimum off time for the sheddable loads.

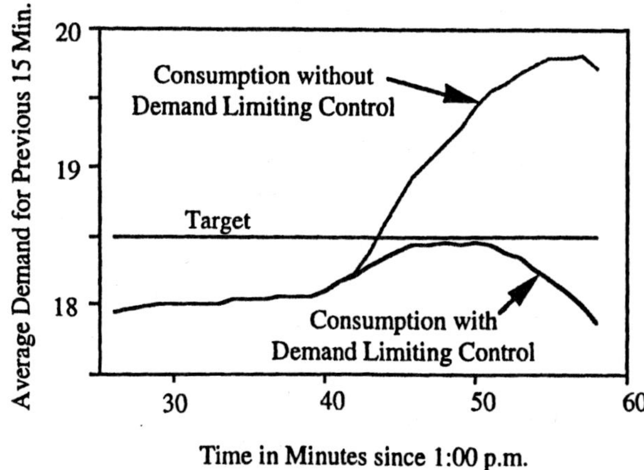

Figure 70.15 Electric consumption with and without demand limiting control.

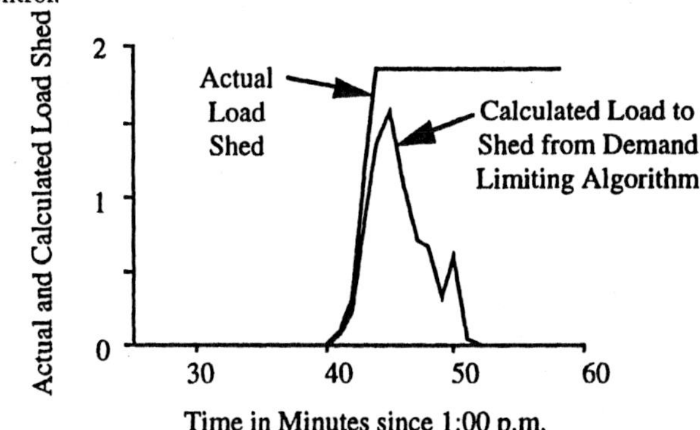

Figure 70.16 Actual amount of load shed and desired amount to shed.

70.6.3 Predicting Return Time from Night or Weekend Setback

For buildings that are not continuously occupied, energy costs can be reduced by adjusting the setpoint temperatures during unoccupied times. One common strategy for adjusting setpoint temperatures is called night setback. Night setback involves lowering the setpoint temperature for heating and raising the setpoint temperature for cooling. In [14] it is shown that night setback can result in energy savings of 12% for heavyweight buildings and 34% for lightweight buildings.

When using a night setback strategy, space conditions need to be brought to comfortable conditions prior to the time when occupants return. Ideally, the space conditions would just be in the comfortable range as the building occupants return. In [15] seven different methods for predicting return time from night or weekend setback were compared. For nights that required cooling to return, the following equation can be used to predict return time

$$\hat{\tau} = a_0 + a_1 T_{r,i} + a_2 T_{r,i}^2 \tag{70.32}$$

where $\hat{\tau}$ is the estimate of the return time from night or weekend setback, a_i is a coefficient, and $T_{r,i}$ is the initial room temperature at the beginning of the return period.

For nights that require heating, the return time can be estimated from

$$\hat{\tau} = a_0 + (1-w)(a_1 T_{r,i} + a_2 T_{r,i}^2) + w a_3 T_a \tag{70.33}$$

$$w = 1000^{-\frac{T_{r,i}-R_{h,u}}{R_{h,o}-R_{h,u}}} \tag{70.34}$$

where $R_{h,u}$ is the setpoint temperature for heating during unoccupied times, $R_{h,o}$ is the setpoint temperature for heating during occupied times, and T_a is the outdoor air temperature at the beginning of return from night setback. Recursive linear least squares with exponential forgetting is used to determine the coefficients in Equations 70.32 and 70.33.

70.6.4 Thermal Storage

Thermal storage systems use electrical energy to cool some storage media at off-peak hours, then use the cooling capacity of the media during on-peak hours. Storage systems are used for some of the same reasons as the electrical demand-limiting algorithm described previously; they allow the electrical demand at peak hours to be reduced. They also allow the chiller load to be shifted to times when electrical energy is inexpensive. Thermal storage systems may also allow one to use lower-capacity chillers.

The two most common storage media are chilled water and ice. Ice storage tanks are becoming more popular because the energy density of ice is much greater than that of water.

Water and Ice Storage Control Strategies

The simplest control strategy for water or ice storage systems is chiller priority control. With chiller priority control, the cooling load is matched by the chiller until the load exceeds the chiller capacity. Then the storage media is used to augment the chiller capacity. During the off-peak hours, the storage media is fully recharged. Chiller priority control is sometimes referred to as demand-limiting control. Chiller priority control is simple to implement, but it is substantially suboptimal except when cooling loads are high.

Storage priority control strategies make better use of the storage capacity than chiller priority control. The objective of storage priority control is to use mainly stored energy during the on-peak period when energy is expensive and to use all of the stored energy by the end of the on-peak period. An example of storage priority control is load-limiting control [16]. During the off-peak unoccupied hours, the storage media is charged as much as possible. During the on-peak unoccupied hours, the building is cooled using chiller priority control. During the on-peak hours, the chiller is run at a constant capacity such that the storage media is completely used by the end of the on-peak period. Storage priority control strategies such as load-limiting control require a forecast of the cooling load for the on-peak period.

Optimal control of thermal storage systems requires determining a sequence of control commands for charging and discharging the storage media such that the total cost of supplying chilled water is minimized subject to operating constraints. To determine the truly optimal sequence requires perfect knowledge of the behavior of the process and perfect knowledge of future cooling loads. Even with perfect knowledge of the process behavior and of future cooling loads, determining the optimal sequence is computationally intensive (e.g., requires solving a dynamic programming problem). Consequently, most of the practical work on optimal control of thermal storage systems involves the development of heuristic strategies that are nearly optimal. For example, a nearly optimal strategy for the control of ice storage systems is developed in [17].

70.6.5 Setpoint Resetting

One way to improve the operational efficiency of an HVAC system is to make adjustments to the setpoints of the local-loop controllers. This is commonly called setpoint resetting. In this section, two common setpoint resetting strategies are described.

Chiller Water Temperature

Figure 70.17 shows a schematic diagram of a chiller. The

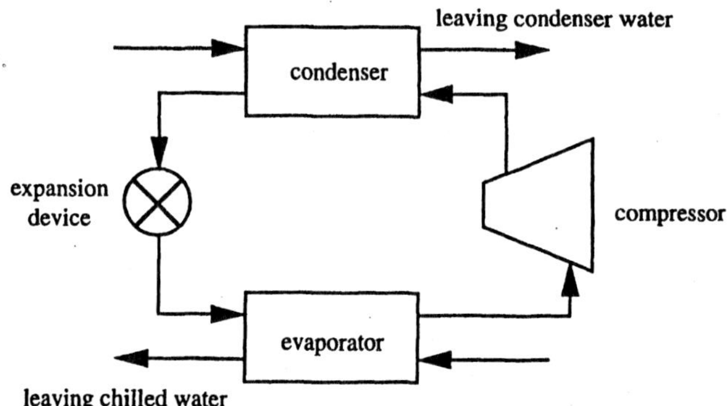

Figure 70.17 Schematic diagram of a chiller.

chiller power consumption is a function of the refrigerant flow rate and the pressure difference across the compressor. The larger the pressure difference across the compressor, the larger the chiller power consumption. The temperature difference between the water leaving the condenser and the water leaving the evaporator is strongly correlated with the pressure difference across the compressor. Therefore, the chiller power can be reduced if the leaving water temperature difference can be reduced.

One way to reduce the leaving water temperature difference is to reset the chilled water supply setpoint to a higher value whenever possible. There are numerous strategies for doing this. One is to control the position of the most-open valve in the chilled water distribution circuit to a nearly open position as shown in Figure 70.18 and described in [18].

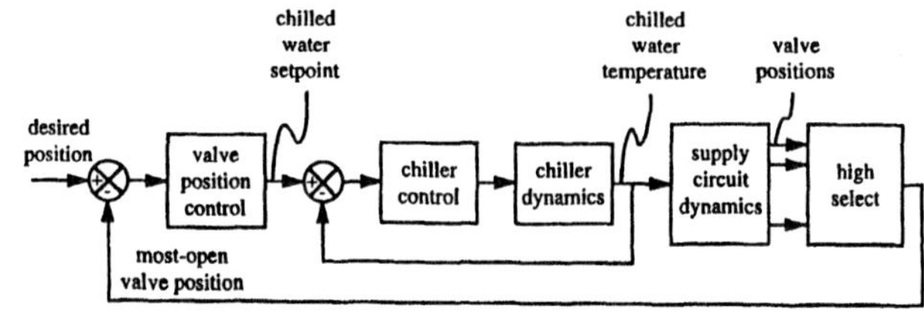

Figure 70.18 Block diagram of chilled water reset control.

An advantage of this strategy over others is that it allows the chilled water temperature to go as high as possible without forcing a cooling valve into saturation. A potential disadvantage is that, if the controller is not properly tuned, the chiller controls are sensitive to disturbances on a single cooling coil. Therefore, the controller should be tuned so that the bandwidth of the resetting loop is much smaller than the bandwidth of the slowest cooling coil control loop.

Duct Static Pressure

In VAV systems, the most common method of modulating the total supply air flow rate is to modulate the supply fan so as to regulate the static pressure at some point in the supply air duct. The operator must select a static pressure setpoint. If the setpoint is too low, one or more terminal units will open completely and still not be able to deliver the required quantity of supply air. If the setpoint is too high, the fan power consumption will be greater than necessary to supply air to the terminal units. Therefore, there is an optimal static pressure that minimizes the fan power yet supplies the necessary amount of air to all of the terminal units.

One method to control the fan efficiently is to reset the static pressure setpoint based on the position of the terminal unit that is open the most. In [19], a strategy is described in which the setpoint is increased by a fixed amount when one or more of the terminal units is saturated fully open and decreased by a fixed amount when one or none of the terminal units is saturated. It is reported that this strategy is able to reduce the required fan power by 19 to 42% over a fixed-setpoint control strategy.

70.7 Configuring and Programming

There are several different ways that digital HVAC controllers are configured or programmed. In this section, three common methods are described.

70.7.1 Textual Programming

Perhaps the oldest approach to configuring and programming controllers is textual programming. For example, BASIC, FORTRAN, or C may be used to program logical controls, local-loop controls, and supervisory controls and diagnostics. The advantage of using textual programming is that it allows the programmer great flexibility in constructing the control software. Textual programming requires substantial programming expertise. Since many controllers must be custom-programmed, the expertise required for textual programming often makes the cost of using textual programming prohibitive.

70.7.2 Graphical Programming

Control algorithms and logic may be programmed graphically through the construction of diagrams. One common type of control diagram is the ladder logic diagram. Originally, ladder logic diagrams were designed to resemble electrical relay circuits and, therefore, could perform only limited types of operations. Today, additional functions can be included in ladder logic diagrams so that the controllers can perform functions such as arithmetic and data manipulation operations.

Another common method of graphical programming involves the construction of block diagrams. Function blocks are connected by lines or wires. The blocks may perform basic logical operations or more complex functions (e.g., implement PID control algorithms). Also, the programmer may construct new blocks for a specialized purpose.

The advantage of using graphical programming is that the controller can be configured and programmed for a nonstandard system or subsystem. Specialized control functions can often be constructed quickly, and the required level of programming expertise may be lower than other programming methods.

70.7.3 Question-and-Answer Sessions

Many HVAC subsystems are designed in such a way that standardized controls can be used. In such cases, the systems are preprogrammed. To be made fully operational, only information about the size or type of subsystem and the desired control strategy need be supplied. Therefore, these types of controllers may be configured through a simple question-and-answer session. Control strategies such as those described in Section 70.6 may be preprogrammed for a variety of different equipment and can be selected to match the particular system.

The obvious disadvantage of using preprogrammed controls is that they are inflexible. The user cannot make changes to the way in which the controller operates. However, there are significant advantages to preprogrammed controls. One advantage is that it reduces the commissioning time. Furthermore, it does not require programming expertise. Another advantage is that preprogrammed controls can be tested in a laboratory so that they are guaranteed to be safe, reliable, and efficient. If the controls are custom-programmed for a specific system, rigorous testing cannot be done.

References

[1] Hirst, E., Clinton, J., Geller, H., and Kroner, W., *Energy Efficiency in Buildings: Progress and Promise,* O'Hara, F.M., Jr., Ed., American Council for an Energy Efficient Economy, 1986, 3.

[2] Environmental Protection Agency, *Unfinished Business: A Comparative Assessment of Environmental Problems*, Vol. 1, February 1987.

[3] Haines, R. W. and Hittle, D.C.,*Control Systems for Heating, Ventilating, and Air Conditioning*, Van Nostrand Reinhold, New York, 1993.

[4] Incropera, F. P. and DeWitt, David P., *Fundamentals of Heat and Mass Transfer*, Wiley, New York, 1990.

[5] Instrument Society of America, *ISA Handbook of Control Valves*, Pittsburgh, 1976.

[6] Seborg, D. E., Edgar, T.F., and Mellichamp, D.A., *Process Dynamics and Control*, Wiley, New York, 1989.

[7] White, F. M., *Fluid Mechanics*, McGraw-Hill, New York, 1979.
[8] *1993 ASHRAE Handbook: Fundamentals*, ASHRAE, Atlanta, 1993.
[9] Clarke, D. W., PID algorithms and their computer implementation, Oxford University Engineering Laboratory Report No. 1482/83, Oxford University, 1983.
[10] Shinskey, F. G., *Process Control Systems: Application, Design, and Tuning*, McGraw-Hill, New York, 1988.
[11] Spitler, J. D., Hittle, D.C., Johnson, D.L., and Pedersen, C.O., A comparative study of the performance of temperature-based and enthalpy-based economy cycles, *ASHRAE Trans.*, 93(2), 13–22, 1987.
[12] Stein, B., Reynolds, J.S., and McGuinness, W.J., *Mechanical and Electrical Equipment for Buildings*, 7th ed., Wiley, New York, 1986.
[13] Seem, J. E., Adaptive demand limiting control using load shedding, *Int. J. Heat., Vent., Air-Cond. Refrig. Res.*, 1(1), 21–34, January 1995.
[14] Bloomfield, D. P. and Fisk, D.J., The optimization of intermittent heating, *Buildings and Environment*, Vol. 12, 1997, 43–55.
[15] Seem, J. E., Armstrong, P.R., and Hancock, C.E., Algorithms for predicting recovery time from night setback, *ASHRAE Trans.*, 95(2), 1989.
[16] Braun, J. E., A comparison of chiller-priority, storage-priority, and optimal control of an ice-storage system," *ASHRAE Trans.*, 98(1), 1992.
[17] Drees, K., *Modeling and Control of Area Constrained Ice Storage Systems*, Master's thesis, Mechanical Engineering Department, Purdue University, 1994.
[18] *1991 ASHRAE Handbook: HVAC Applications*, ASHRAE, Atlanta, 1991, chap. 41.
[19] Lorenzetti, D. M., and Norford, L.K., Pressure setpoint control of adjustable speed fans, *J. Sol. Energy Eng.*, 116, 158–163, August 1994.

71

Control of pH

F. Greg Shinskey
Process Control Consultant, North Sandwich, NH

71.1 Introduction

The pH loop has been generally recognized as the most difficult single loop in process control, for many reasons. First, the response of pH to reagent addition tends to be nonlinear in the extreme. Second, the sensitivity of pH to reagent addition in the vicinity of the set point also tends to be extreme, in that a change of one pH unit can result from a fraction of a percent change in addition. Thirdly, the two relationships above are often subject to change, especially when treating wastewater. And finally, reagent flow requirements may vary over a range of 1000:1 or more, especially when treating wastewater.

As a result of these unusual characteristics, many, if not most pH control loops are unsatisfactory, either limit-cycling or slow to respond to upsets or both. Considerable effort and ingenuity have gone into designing advanced controls, nonlinear, feedforward, and adaptive, to solve these problems. Not enough has gone into designing the neutralization process to be controllable. Therefore, after describing the pH control problem and before developing control-system design guidelines, process design is covered in substantial detail. If the process is designed to be controllable, the control system can be more effective and simpler, and, therefore easier to operate and maintain.

71.2 The pH Measuring System

Most of the unique characteristics of the pH control loop center around its measuring system with its unusual combination of high sensitivity and wide range. The fundamentals of the measuring system are firm and not likely to change with time or technology. However, the methods and devices used to measure pH can be expected to continue to improve, as obstacles are overcome. Still, failures can occur, and recognizing the possible failure modes is important in anticipating the response of the controls to them.

71.2.1 The pH Scale

The pH measuring system is expressly designed to report the **activity** of hydrogen ions in an aqueous solution. (While pH measurements can be made in some nonaqueous media such as methanol, the results are not equivalent to those obtained in water and require independent study.) The true *concentration* of the hydrogen ions may differ from its measured activity at pH levels below 2, where ion mobility may be impeded. Most pH loops have control points in the pH 2–12 range, however, where activity and concentration are essentially identical. Consequently, the pH measurement will be used herein as an indication of the concentration of hydrogen ions $[H^+]$ in solution.

That concentration is expressed in terms of gram-ions of hydrogen per liter of solution. A *Normal* solution will contain one gram-ion per liter. Laboratory solutions are typically prepared using the Normal scale, e.g., 1.0 N, 0.01 N, etc., representing the number of replaceable gram-ions of hydrogen or hydroxyl ions per liter.

When a pH measuring electrode is placed in a solution with a reference electrode and a liquid junction between them, a potential difference E may be measured between the two electrodes:

0-8493-8570-9/96/$0.00+$.50

$$E = 59.16 \log[\mathrm{H}^+] - E_{ref} - E_j \qquad (71.1)$$

where E_{ref} is the potential of the reference cell and E_j that of the liquid junction. The commonly accepted method of expressing base-10 logarithms is to use p to indicate the negative power. Thus, $\mathrm{pH} = -\log[\mathrm{H}^+]$ and $[\mathrm{H}^+] = 10^{-\mathrm{pH}}$. The potential developed by the pH electrode is therefore 59.16 millivolts per pH unit.

Actually, this coefficient of 59.16 is precise only at 25°C, varying with $T/298$ where T is the absolute temperature in Kelvin. The reference cell is designed to match the internal cell of the pH electrode, buffered to pH 7. Therefore at pH 7, there is no potential difference between the measuring and reference electrodes, and temperature has no effect there, the *isopotential* point. Temperature compensation may be applied either automatically or manually, as desired.

The liquid-junction potential represents any voltage developed by diffusion of ions across the junction between the solution being measured and that of the reference electrode. Potassium chloride is selected for the reference solution, because its ions have approximately the same ionic mobility, thereby minimizing the junction potential. For most solutions, E_j is less than 1 mV, and is not considered a major source of error.

The net result of the above is a linear scale of pH vs. millivolts, with zero voltage corresponding to pH 7. The significance of pH 7 is that it represents the neutral point for water, where hydrogen and hydroxyl ions are equal. The two ions are related via the equilibrium constant, which is 10^{-14} at 25°C:

$$[\mathrm{H}^+][\mathrm{OH}^-] = 10^{-14}. \qquad (71.2)$$

The equilibrium constant is also a function of temperature, causing a given solution to change pH with temperature [1]. This effect is only significant above pH 6, and is not normally included in the temperature compensation applied to the electrodes.

71.2.2 Electrodes

The pH measuring electrode in most common use is the glass electrode. It consists of a glass bulb containing a solution buffered to pH 7, with an internal reference cell identical to that in the reference electrode. This combination produces the null voltage at pH 7. The sensitive portion of the electrode is a thin glass membrane saturated with water. The membrane must be thin enough to keep its electrical impedance in a reasonable range and yet resist breakage under normal use. The impedance of the glass can exceed 100 megohms, requiring extremely high-impedance voltage-measuring devices to produce accurate results. In view of these special requirements, ordinary voltmeters and wiring cannot be used to measure pH, as circuits are very sensitive to electrical leakage and grounding.

The reference electrode forms the other connection between the solution to be measured and the voltage measuring device. It must have a stable voltage (where that of the measuring electrode varies with solution pH) and a low impedance.

The extremely high impedance of the glass electrode makes it susceptible to *short-circuiting*. Short-circuiting reduces the measured millivolts relative to the true pH of the solution, producing pH readings closer to 7 than actually the case. This is an unfortunate failure mode, for 7 is the most common set point, especially in wastewater treatment, and will not cause alarm. Rovira [2] describes a microprocessor-based pH transmitter with the capability of checking the impedance of the glass electrode and thereby warning of this type of failure.

There is only one adjustment available to calibrate or *standardize* a pH measuring system. If the electrodes are in good working order, calibration against one buffer should produce accurate results against other buffers without further adjustment. If it does not, then the millivoltage change per pH unit is not 59.16 at 25°C, indicating that the glass electrode is damaged or that there is an electrical leak.

Coating of the pH electrode introduces a time lag between the ions in the process solution and those at the surface of the electrode. Coating can be verified by the appearance of the electrodes, but be aware that a film as thin as a millimeter could produce a time constant of several minutes. McMillan [1] recommends increasing the velocity of the process fluid past the electrodes to 7 ft/sec. to minimize fouling, although velocities exceeding 10 ft/sec. may cause excessive noise and erosion.

71.3 Process Titration Curves

What we recognize as the nonlinearity in pH control loops is the process *titration curve*, the relationship between measured pH and the amount of acid or base reagent added to a solution. The reagents are usually strong acids or bases of known concentration added to a solution that contains variable concentrations of possibly unknown substances. Laboratory titrations are conducted batchwise by adding reagent incrementally to a measured volume of process solution. In a process plant, pH control can be applied to a batch of solution, but more often is applied to a flowing stream, matching reagent flow to it. In a continuous operation, only that portion of the titration curve near the set point may be visible, but the slope of the curve in this region determines the steady-state process gain, and the amount of reagent required to reach that point determines the process load. Therefore, titration curves are as important in continuous processing as in batch.

Laboratory titrations are usually performed using standard solutions prepared to 0.01 N or similar concentration. Process reagents are usually much more concentrated, but their concentration must be known in the same terms to estimate required delivery based on laboratory titrations of process samples. The normality of 32% HCl is 10.17 N, that of 98% H_2SO_4 is 36.0 N, that of 25% NaOH is 7.93 N, and that of 10% $Ca(OH)_2$ is 2.86 N. For other solutions, normality can be calculated by multiplying the weight percent concentration of the reagent by $10n\rho/M$, where n is the number of replaceable H^+ or OH^- ions per molecule, ρ is its density in g/ml, and M is its molecular weight.

71.3.1 The Strong Acid-Strong Base Curve

A "strong " agent is one which ionizes completely in solution. Examples of strong acids are HCl and HNO_3, and of bases are the alkalies NaOH and KOH. Surprisingly, H_2SO_4 and HF are not classified as strong acids, despite their tendency to dehydrate and etch glass, respectively. A common basic reagent, $Ca(OH)_2$, similarly is not strong, and also has limited solubility.

Consider a reaction between NaOH and HCl:

$$HCl + NaOH + H_2O \rightarrow H^+ + OH^- + Na^+ + Cl^- \quad (71.3)$$

The sum of the negative and positive charges must be equal:

$$[Na^+] + [H^+] = [Cl^-] + [OH^-] \quad (71.4)$$

With both reagents being completely ionized, the concentration of the acid x_A can be substituted for $[Cl^-]$ and the concentration of the base x_B for $[Na^+]$:

$$x_B - x_A = [OH^-] - [H^+]. \quad (71.5)$$

Due to the equilibrium of hydrogen and hydroxyl ions in water, the former may be substituted for the latter, using Equation 71.2:

$$x_B - x_A = 10^{-14}/[H^+] - [H^+]. \quad (71.6)$$

Placed in terms of pH, the relationship becomes

$$x_B - x_A = 10^{pH-14} - 10^{-pH}. \quad (71.7)$$

This is the fundamental pH relationship in the absence of buffering. Although buffers are common to a limited extent in most solutions, Equation 71.7 describes the most severe titration curve that a control system must handle. It is shown for different scales of normality in Figure 71.1. Although the curves appear to have different breakpoints and therefore to be different curves, they are, in fact, a single curve shown at two different ranges of sensitivity.

Both curves have the same characteristic in that a base-acid mismatch of one-tenth of fullscale leaves the pH only one unit from the end of the curve. Similarly, a mismatch of 1/100 of full scale leaves the pH two units from the end of the curve. This logarithmic relationship holds everywhere that one of the terms on the right of Equation 71.7 is negligible compared to the other, i.e., everywhere except in the range of pH 6–8.

71.3.2 Distance from Target

The essence of the pH-control problem is the matching of acid and base at neutrality. In the strong acid-strong base system, neutrality occurs at pH 7, and therefore a perfect match will bring this result. In some process work, pH must be controlled within ± 0.1 unit, but wastewater limits are usually pH 6–9, a much broader target. Consider how difficult it can be to control wastewater pH in the absence of buffering. If the wastewater enters at pH 3, as in the inner curve of Figure 71.1, a mismatch between acid and base of 10% will leave the pH at 4 or 10; a mismatch of 1.0% will leave it at 5 or 9; a mismatch of 0.1% is required to produce a pH in the 6–8 range.

The size of the target does not change with the scale, but the size of the reagent valve does. For example, if the wastewater were to enter one pH unit farther away from neutral, ten times as much reagent would be required to neutralize it. The accuracy of delivery must then be ten times as great percentagewise to control to the same target. For example, a solution represented by the outer curve in Figure 71.1 would require a percentage accuracy in reagent delivery 100 times as great as for the inner curve.

A graphic illustration of the problem can be envisioned by comparing the size of the target to our distance from it. Think of the range of pH 6–8 as a target one foot in diameter, $x_B - x_A$ being essentially $\pm 10^{-6}$ at the edge. If the entering solution is at pH 5 or 9, the distance to the target is ten times its radius or five feet away. From this distance, the target could be easily hit by throwing darts at it. If the entering pH were 3 or 11, the same target would be 500 feet away, requiring an expert with a high-powered rifle. If the pH entering were 1 or 13, the same target would be ten miles away, requiring a guided missile.

To maximize the accuracy of reagent delivery, *dead band* in the control valves must be eliminated through the use of *valve positioners*. Dead band has been observed to cause limit-cycling in control of pH and reduction-oxidation potential, which was eliminated by installing positioners on the valves.

71.3.3 Weak Acids and Bases

Most acids and bases do not completely ionize in solution, but establish an equilibrium between the concentration of the undissociated molecule and its ions. A *monoprotic* agent is one that has a single ionizable hydrogen or hydroxyl group. An example is acetic acid, which dissociates into hydrogen and acetate ions in accordance with the following relationship:

$$[H^+][Ac^-] = K_A[HAc] \quad (71.8)$$

where K_A is the ionization constant for acetic acid at $10^{-4.75}$. A pH electrode senses only the $[H^+]$, leaving the undissociated [HAc] unmeasured. However, any adjustment to solution pH made by adding either base or another acid will shift the equilibrium. For example, increasing $[H^+]$ with a strong acid will convert some of the $[Ac^-]$ into [HAc], leaving more of it unmeasured. Conversely, reducing $[H^+]$ with a base will shift the equilibrium the other way, ionizing more of the [HAc]. This behavior tends to reduce the change in pH to additions of acids and bases, and is called buffering.

Adding acetic acid to the former solution of HCl and NaOH adds acetate ions to the charge balance:

$$[H^+] + [Na^+] = [OH^-] + [Cl^-] + [Ac^-]. \quad (71.9)$$

Let the total concentration of acetic acid be represented by

$$x_{Ac} = [HAc] + [Ac^-] \quad (71.10)$$

The [HAc] in the above expression can be replaced by its value from Equation 71.8:

$$x_{Ac} = [Ac^-](1 + [H^+]/K_A). \quad (71.11)$$

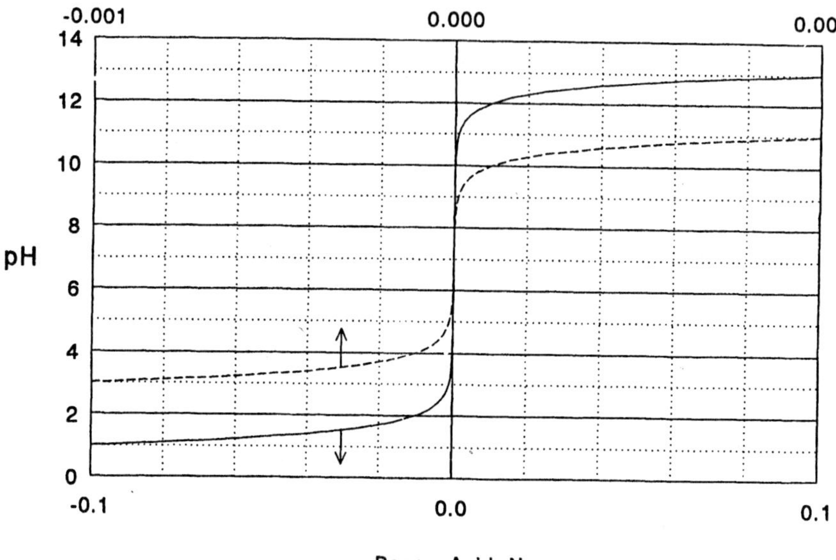

Figure 71.1 Titration curve for strong base against strong acid, two scales.

Replacing all of the ions in the charge balance with equivalent values of the three ingredients, and $[H^+]$ with pH, gives

$$x_B - x_A = 10^{\mathrm{pH}-14} - 10^{-\mathrm{pH}} + \frac{x_{Ac}}{1 + 10^{\mathrm{p}K_A - \mathrm{pH}}} \tag{71.12}$$

where $\mathrm{p}K_A$ (like pH) is the negative base-10 logarithm of the ionization constant.

The titration curve for a monoprotic weak base is derived in a fashion similar to that of the acid. As Equation 71.12 simply had the weak-acid term added to the strong acid-strong base relationship of Equation 71.7, so the following weak-base term may be added in the same way:

$$\cdots - \frac{x_{Bw}}{1 + 10^{\mathrm{p}K_B + \mathrm{pH} - 14}} \tag{71.13}$$

where x_{Bw} represents the concentration of weak base whose ionization constant has the negative logarithm $\mathrm{p}K_B$. Ammonia is a weak base having a $\mathrm{p}K_B$ of 4.75, identical to the $\mathrm{p}K_A$ value for acetic acid.

Most of the weak acids and bases encountered in industry have two ions, and are hence classified as *diprotic*. A very common diprotic weak acid is carbon dioxide, encountered in groundwater in the form of carbonate and bicarbonate ions. Dissociation of CO_2 and water into H^+ and HCO_3^- ions has a $\mathrm{p}K_{A1}$ value of 6.35, so that carbonic acid is weaker than acetic acid by a factor of about 40. A second equilibrium takes place between HCO_3^- ions and CO_3^{2-} ions with a $\mathrm{p}K_{A2}$ value of 10.25. Because this second value exceeds 7, the carbonate ion effectively acts as a weak base having a $\mathrm{p}K_B$ value of 14 – 10.25 or 3.75.

Derivation of the equations representing the titration curves for diprotic agents follows the same procedures used for monoprotic agents. The result is adding to Equation 71.7 the following for weak diprotic acids:

$$\cdots + \frac{x_{Aw}(1 + 0.5 \times 10^{\mathrm{p}K_{A2} - \mathrm{pH}})}{1 + 10^{\mathrm{p}K_{A2} - \mathrm{pH}}(1 + 10^{\mathrm{p}K_{A1} - \mathrm{pH}})} \tag{71.14}$$

where x_{Aw} represents the concentration of the weak acid.

Figure 71.2 is a titration curve for water containing 0.001 N CO_2, equivalent to a total alkalinity of 50 ppm as $CaCO_3$, the customary method of reporting water hardness. On the larger scale of ± 0.01 N, the bicarbonate shows its presence with a slight moderation of the titration curve around pH 6.35. This makes the pH of wastewaters easier to control than it would be otherwise, and most water does contain significant amounts of bicarbonate ions. The second breakpoint associated with the carbonate ionization is not visible on this scale.

On the smaller scale of ± 0.001 N, the bicarbonate buffering is even more pronounced. This characteristic makes bicarbonate useful for neutralizing stomach acids. The neutral point for this solution is pH 8, however only half as much strong base as acid is required to reach the neutral point. The reason for this is that the value of $\mathrm{p}K_{A2}$ exceeds 7, resulting in carbonates being weak bases. Na_2CO_3 is a useful reagent for neutralizing acids, because it produces a titration curve well-buffered in the region of pH 6–7; unfortunately, twice as much sodium is required per mol of acid than if NaOH were used, making it too costly for most waste treatment applications.

Carbonates are common in cleaning solutions, and therefore in wastewater, but their concentration may fluctuate considerably, as most cleaning is done batchwise. Other cleaning agents such as phosphates and silicates, ammonia and amines, and soaps and detergents of all kinds are weak agents. Consequently, the degree of buffering in industrial wastewater may be expected to be quite variable.

The term added to the pH relationship for weak diprotic bases is similar to that for acids:

$$\cdots - \frac{x_{Bw}(1 + 0.5 \times 10^{\mathrm{p}K_{B2} + \mathrm{pH} - 14})}{1 + 10^{\mathrm{p}K_{B2} + \mathrm{pH} - 14}(1 + 10^{\mathrm{p}K_{B1} + \mathrm{pH} - 14})}. \tag{71.15}$$

A list of ionization constants for weak acids and bases is given in Table 71.1. For a more complete list, consult Weast [7].

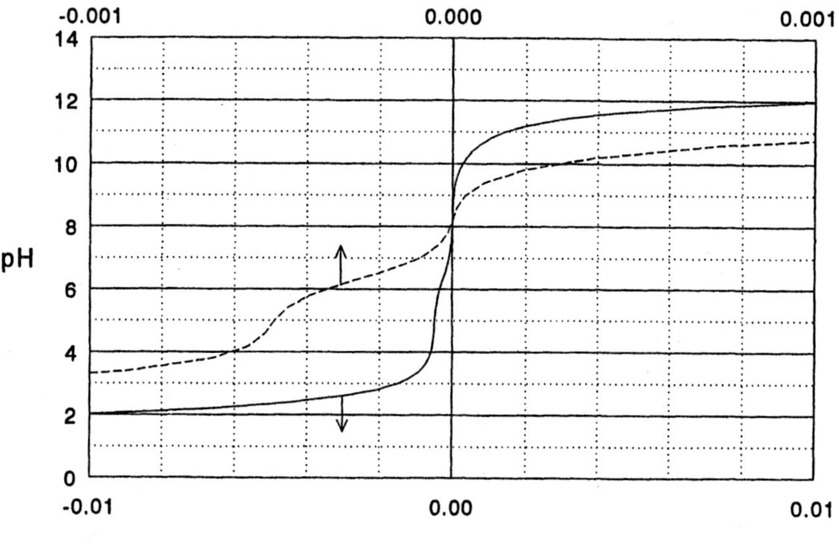

Figure 71.2 Titration of 0.001 N CO_2 by strong base on two scales.

TABLE 71.1 Ionization Constants for Common Acids and Bases.

Acid	pK_{A1}	pK_{A2}
Acetic acid	4.75	
Boric acid	4.7	9.1
Carbon dioxide	6.35	10.25
Chromic acid	0.7	6.2
Ferric ion	2.5	4.7
Ferrous ion	6.8	
Formic acid	3.65	
Hydrogen fluoride	3.17	
Hydrogen sulfide	7.0	12.9
Hypochlorous acid	7.5	
Oxalic acid	1.1	4.0
Phosphoric acid	2.23	7.21
Sulfuric acid	−3	1.99
Sulfur dioxide	1.8	6.8
Base	pK_{B1}	pK_{B2}
Aluminate ion	1.6	
Ammonia	4.75	
Barium hydroxide	0.7	
Calcium hydroxide	1.40	2.43
Cupric hydroxide	7.2	
Ethylamine	3.3	
Hydrazine	5.5	
Hydroxylamine	7.97	
Magnesium hydroxide	2.6	
Methylamine	3.28	
Phenol	4.2	
Silicon hydroxide	1.3	4.4
Urea	13.9	
Zinc hydroxide	3.02	

Any of the agents in the table could be listed either as an acid or base. For example, the ferric ion is listed as an acid with pK_A values of 2.5 and 4.7; it could have been listed as ferric hydroxide with pK_B values complementary at 9.3 and 11.5, the effect is the same. Most metal ions are weak agents, producing titration curves with multiple breakpoints like the inner curve of Figure 71.2. Therefore, effluents from metal treating operations, like pickling and plating, tend to be heavily buffered at all times, and their pH is generally easy to control.

While it is possible to create titration curves from the equations and pK values above, it is strongly recommended that several samples be taken at different times of day from the stream whose pH is to be controlled and titrated to identify how extreme the curves are and how much they may vary. This information will be useful in sizing valves and reducing the likelihood of surprises when the control system is commissioned.

71.4 Designing a Controllable Process

The extreme gain changes presented by most titration curves, and their variablilty, pose special problems for the pH control system. The typically very high gain of the titration curve in the region about set point is its most demanding feature, causing most pH loops to cycle about the set point. Uniform cycling develops when the gain of the control loop is unity. Loop gain is the product of all of the gains in the loop: process and controller, steady-state and dynamic. The high steady-state gain of the typical titration curve requires a proportionately low controller gain if cycling is to be avoided. But a low controller gain means poor response to disturbances in the flow and composition of the stream whose pH is being controlled.

To maximize the controller gain, the process gains need to be minimized. The steady-state gain terms for the process consist of the titration curve converting concentration to pH, the reagent valve converting percent controller output to reagent concentration, and the pH transmitter converting pH to percent controller input. Other than selecting a buffering reagent such as Na_2CO_3 or CO_2, which may not be feasible, we must accept the process titration curve as it exists. Similarly, the valve size is determined by process needs, as is the transmitter span.

But the process has a dynamic gain, too, and this can be minimized by proper process design. While this is always good practice, it is especially critical in pH control. Furthermore, in most processes, the design is determined by other considerations such as reaction rate, heat transfer, etc. But most acid-base reactions are zero order and generate negligible amounts of heat, so these considerations do not apply. The process designer is then free to choose vessel size, layout, mixing, piping, etc., and should choose those which favor dynamic response.

71.4.1 A Dynamic Model of Mixing

Consider an acid-base reaction performed by combining the process stream and reagent in a *static mixer* as shown in Figure 71.3 (but without the circulating pump, whose function is described later). A static mixer consists of a pipe fitted with internal baffles which alternately split and rotate the fluids, resulting in a homogeneous blend at the exit. The principal problem with the static mixer is that it presents pure deadtime to the controller. A step change in the reagent flow will produce no observable response in the pH at the exit until the new mixture emerges, at which time the full extent of the step appears at once. The dynamic gain of the mixer is 1.0 because the step at the inlet appears as a step at the exit, not diminished or spread out over time.

The time between the initiation of the step and its appearance in the response of the pH transmitter is the deadtime caused by transportation through the mixer. It is equal to the length of the mixer divided by the velocity of the mix, or the volume of the mixer divided by the flow of the mix. (The connection between the mixer exit and the pH electrodes must be included in this calculation.) In addition to the high dynamic gain of the mixer, its deadtime is also *variable,* as a function of flow. This variablilty presents an additional problem for the controller, in that its optimum integral time is contingent on the process deadtime. If tuned for the highest flow (and hence shortest deadtime), the controller will integrate too fast for the process to respond at lower flow rates, causing cycling to develop. Therefore, the controller must be tuned at the lowest expected flow (where deadtime is the longest) for stability across the flow range. However, control action will then be slower than desired at higher flow rates. And if flow should ever stop, the control loop will be open. For the above reasons, static mixers are *not* recommended for pH control, and transportation delay of effluent to the pH electrodes should be eliminated or minimized.

The undesirable properties of the static mixer can be mitigated, however, by adding the circulating pump also shown in Figure 71.3. The pump applies suction downstream of the pH electrodes, recirculating treated product at a flow rate F_a relative to the discharge rate F. This reduces the deadtime through the process from V/F to $V/(F+F_a)$, where V is the volume of the process, and also places an upper limit on the deadtime in the loop at V/F_a. But it reduces the dynamic gain of the process as well.

Consider the case where $F_a = F$, and the feed is being mixed with enough reagent introduced stepwise to change the product concentration by one unit in the steady state. As the step in reagent flow is introduced, however, the reagent is being diluted with twice as much flow as without recirculation, so that its effect on product concentration after the lapse of the deadtime is reduced by a factor of two. At this point, this half-strength product is recirculated and mixed in equal volumes with full-strength feed, causing the blend to reach 75 percent of full strength after the lapse of another deadtime. Then the 75% product is recirculated to produce 87.5% product after the next deadtime, etc. The product concentration approaches its steady-state value covering 50% of the remaining distance each deadtime. The dynamic gain of the process has effectively been reduced by 50%. Increasing recirculation F_a to $4F$ reduces deadtime to $0.2V/F$ and the dynamic gain to 0.2, as illustrated by the staircase response in Figure 71.4. The effect of recirculation is quite predictable.

Next consider a *backmixed* tank with approximately equal diameter and depth of liquid, as shown in Figure 71.5, and let the agitator be sized to recirculate liquid at a rate F_a equal to $4F$. The same step response will be obtained as with the static mixer, except that the tank itself will not produce the same plug-flow profile as the baffled static mixer. As a result, the duration of the molecule travel from the reagent valve to the pH sensor has a wider distribution. Some will take longer than the deadtime estimated as if it were a static mixer, and some will take less time, although the average will be the same. The resulting step response will be better described by the smooth curve passing through the midpoints of all of the steps in Figure 71.4. This has the advantage of reducing the deadtime by a factor of two. Therefore, for the backmixed vessel, deadtime τ_d can be estimated as

$$\tau_d = V/2(F+F_a). \tag{71.16}$$

The average time that a molecule remains in the vessel is called its *residence time,* calculated as V/F. Observe that the exponential curve of Figure 71.4 passes through 63.2% response at time V/F. This curve represents the familiar step response of a first-order lag, which is 63.2% complete when the elapsed time from the beginning of the curve is equal to the time constant of the lag. The difference between the residence time and the deadtime is therefore the time constant of the vessel's first-order lag, τ_1 :

$$\tau_1 = V/F - \tau_d. \tag{71.17}$$

A first-order lag has a phase angle ϕ_1 and a dynamic gain G_1 which are functions of its time constant and the frequency or period at which the loop is cycling. In process control, the period, the duration of a complete cycle, is easier to measure than the frequency in cycles per minute or radians per minute, which in practice is calculated from the observed period. Therefore, the phase lag and dynamic gain are given here as functions of the period of oscillation τ_o [5, p. 28]:

$$\phi_1 = -\tan^{-1}(2\pi\tau_1/\tau_o) \quad G_1 = \cos\phi_1 \tag{71.18}$$

The period of oscillation is a dependent variable, produced where the sum of the dynamic phase lags in the loop reaches 180 degrees. Process deadtime also contributes to the loop phase lag, but its dynamic gain is always unity:

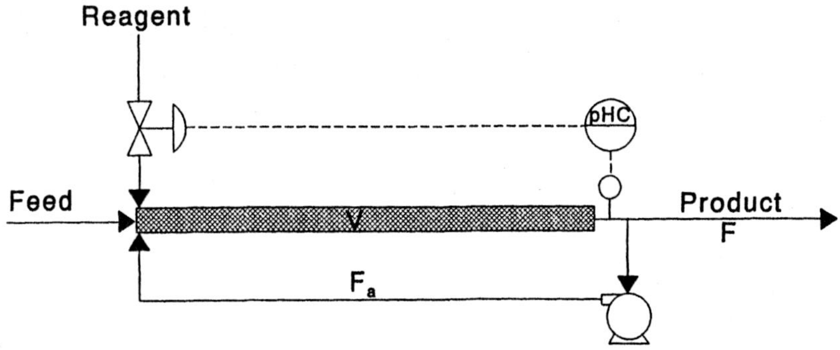

Figure 71.3 Controlling pH using a static mixer features variable deadtime and should not be used without recirculation.

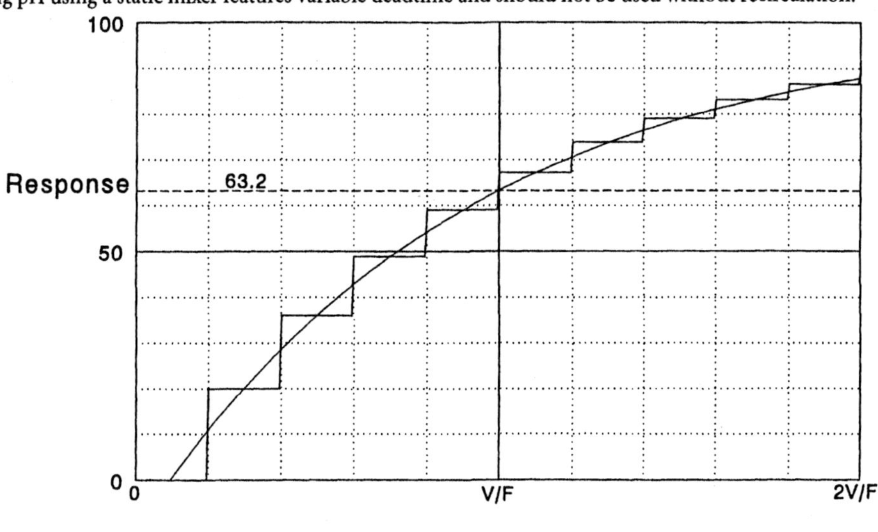

Figure 71.4 The steps are produced by recirculation through a plug-flow process; the curve is produced by a backmixed tank.

$$\phi_d = -360^\circ \tau_d/\tau_o \quad G_d = 1.0 \tag{71.19}$$

If a proportional controller is used, which has no dynamic phase lag of its own, the period of oscillation determined by the process dynamics alone is then called the *natural* period, τ_n. Solution of $\phi_1 + \phi_d = -180^\circ$ for τ_o by trial and error can give the natural period of the process; then its dynamic gain G_1 can be expressed as a function of the ratio of τ_d/τ_1 produced by a given degree of mixing. Table 71.2 displays the results of these estimates for a selected set of τ_d/τ_1 ratios.

TABLE 71.2 Natural Period and Dynamic Gain with Mixing.

			F_a/F	
τ_d/τ_1	$\tau_n/(V/F)$	G_1	Plug-Flow	Backmixed
∞	2.0	1.000	0	
2.0	1.83	0.657	0.5	
1.0	1.55	0.441	1.0	0
0.5	1.14	0.262	2.0	0.5
0.2	0.62	0.117	5.0	2.0
0.1	0.35	0.061	10.0	4.5

Reducing the τ_d/τ_1 ratio by recirculation has *two* advantages: it shortens the natural period relative to the residence time of the vessel, and it lowers its dynamic gain as well. This allows an increase in controller gain (commensurate with the reduction in process dynamic gain) and a reduction in its integral time (commensurate with the reduction in natural period). For a given upset then, the deviation from set point will be smaller and of shorter duration. The corresponding recirculation ratios F_a/F for both the plug-flow (static) mixer and the backmixed tank are included in the table.

71.4.2 Vessel Design

The principal elements to be determined in designing a vessel are its volume, its dimensions, and the location of inlet and outlet ports. Its volume is determined by the residence time needed for the reaction to go to completion at the maximum rate of feed. Since acid-base reactions are essentially zero order, the residence time on the basis of reaction rate is indeterminant. Then the limiting factor may be the response time of the pH electrode, which could be from a few seconds to a minute or more depending on the thickness of the glass and the presence of any coating [1]. Neutralization reactions have been successfully conducted within

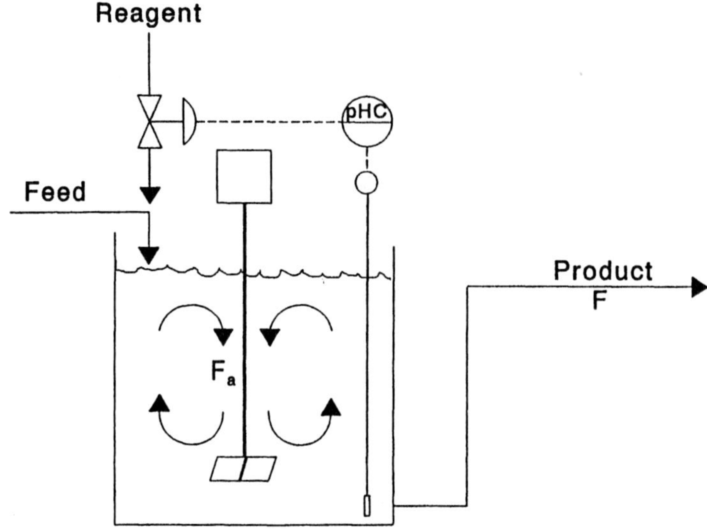

Figure 71.5 This is the preferred location of ports and of the pH measurement in a backmixed tank.

the casing of a centrifugal pump, using the impeller as a mixer.

When reacting species are not in the liquid phase, however, time must be allowed for mass transfer between phases. The most common example of this type is the use of a slurry of hydrated lime as a reagent. The $Ca(OH)_2$ already in solution reacts immediately, but must be replaced by mass transfer from the solid phase, which takes time. Neutralization of acid wastes with lime slurry has been successfully controlled in vessels having a residence time of 15 minutes. If the residence time is too short, the effluent will leave the vessel with some lime still undissolved, which will raise the pH when dissolution is complete. In the absence of mass transfer limitations such as this, a residence time of 3–5 minutes is ordinarily adequate.

If the pH of the feed stream is within the 2–12 range, neutralization can ordinarily be controlled successfully within a single vessel. Should the pH of the feed fall outside this range, it should be first pretreated in a smaller vessel, where the pH is controlled in the 2–3 range for acid feeds or 11–12 for basic feeds, and then brought to neutral in a second, larger vessel. The reason for recommending a smaller vessel for pretreatment is to separate the periods of the two pH loops. A control loop is most sensitive to cyclic disturbances of its own period, and can actually amplify those cycles rather than attenuate them. Given equal recirculation ratios in the two vessels, the periods of the two pH loops will be proportional to their residence times. The sensitivity of the second loop to cycles in the first can be reduced by a factor of ten if the ratio of their residence times is 3:1 [5, pp. 130–134].

The vessel should be of equal dimensions in all directions, to minimize deadtime between inlet and outlet. Trying to control pH in a long narrow channel is ineffective, no matter how thoroughly mixed, because it is essentially all deadtime. If the long narrow channel is already fixed, or if pH is to be controlled in a pond (e.g., a flocculation basin or an ash pond at a power plant), then a cubic section should be walled off at the inlet as shown in Figure 71.6, with pH controlled at the exit from that section. The remainder of the channel or pond can be used to attenuate fluctuations, precipitate solids, etc.

The vessel where pH is to be controlled should be either cylindrical or cubic, with inlet and outlet ports diametrically opposed as shown in Figure 71.5. This avoids short-circuiting the vessel, allowing use of its full capacity. The pH electrodes should be located directly in front of the exit port, to give accurate representation of the product quality. While faster response may be obtained by moving them closer to the inlet, the measurement no longer represents product quality. However, avoid placing the electrodes within the exit pipe itself, as that adds deadtime to the loop, which varies inversely with feed rate; it also results in an open loop whenever feed is discontinued.

Reagent should be premixed with the feed at the feed point, rather than introduced anywhere else, because reagent introduced at other locations has produced erratic results. The feed ordinarily enters through a free fall, whose turbulence affords adequate premixing if the reagent is introduced there.

Feed *could* be introduced near the bottom of the vessel and product withdrawn from the top on the other side, but this increases the deadtime between reagent addition and electrode response. Agitators are designed to pump downward (assuming aeration is not the purpose): the general pattern of flow in Figure 71.5 carries reagent down the shaft of the mixer over to the electrodes in a short path. If feed and reagent are introduced near the bottom, however, the pattern of flow takes reagent up the wall of the vessel, then down the center, and finally up the wall on the other side where the electrodes are located; this path is twice as long, doubling the deadtime in the loop. Equation 71.16 applies to the arrangement shown in Figure 71.5.

71.4.3 Effective Backmixing

For pH control, high-velocity mixing is most effective. The vessel should be fitted with a direct-drive *axial* turbine (or marine propeller in smaller sizes) to maximize circulation rate. Low-speed radial mixers intended to keep solids in suspension increase deadtime excessively.

If a vortex forms, the solution is rotating like a wheel, without

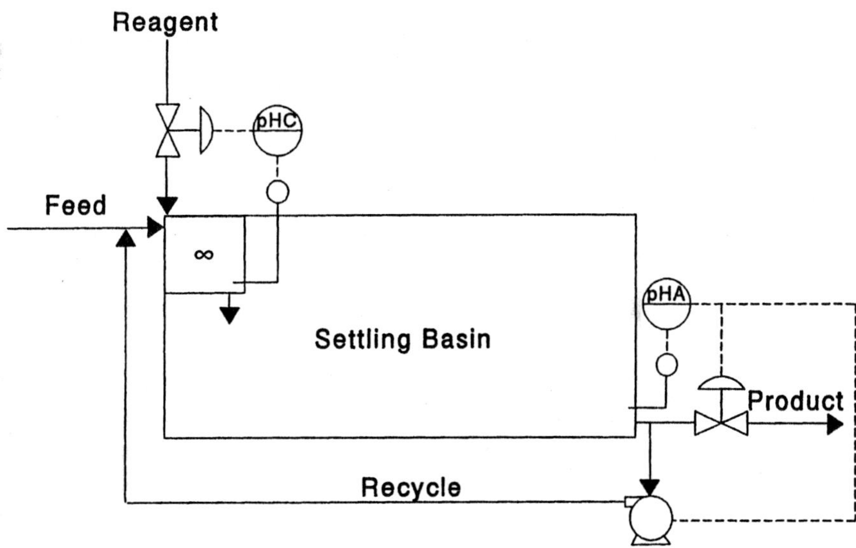

Figure 71.6 A large settling basin should be partitioned for pH control at the inlet; a pH alarm at the outlet starts recirculation of off-specification product.

distributing the reagent uniformly. In smaller cylindrical vessels, tilting the shaft of the mixer or offsetting it from center can eliminate vortex formation. In smaller rectangular vessels, the corners may be sufficient to prevent it, but in large, especially cylindrical vessels, vertical baffles are needed.

Horsepower requirements per gallon of vessel capacity decrease as the vessel capacity increases from 2.3 Hp for vessels of 1,000 gal to 0.8 Hp/1,000 gal for 100,000 gal. This level of agitation gives τ_d/τ_1 ratios of about 0.05 [3, p. 160].

71.4.4 Reagent Delivery Systems

Although Figure 71.5 shows the reagent valve located atop the vessel, it may be located in a more protected environment. However, its discharge line must not be allowed to empty when the valve is closed, as this adds deadtime to the loop. To avoid this problem, place a loop seal (pigtail) in the reagent pipe at the point of discharge, even if the discharge is below the surface of the solution.

If the reagent is a slurry such as lime, it must be kept in circulation at all times to avoid plugging. A circulation pump must convey the slurry from its agitated supply tank to the reaction vessel and back. At the reaction vessel a "Y" connection then brings slurry up to a ball control valve located at the high point in its line, from which reagent drops into the vessel [3, p. 178]. Then the ball valve will not plug on closing, as solids will settle away from it.

Often, feed is supplied by a pump driven by a level switch in a sump. In this case, feed rate is constant while it is flowing, but there are also periods of no flow. To protect the system against using reagents needlessly, the air supply to their control valves should be cut off whenever the feed pump is not running. At the same time, the pH controller should be switched to the manual mode to avoid integral *windup*, which would cause overshoot when the pump starts. When the pump is restarted, the controller is returned to the automatic mode at the same output it had when last in automatic; its set point should *not* be initialized while in manual.

71.4.5 Protection Against Failure

Failures are unavoidable. The pH electrodes could fail, reagent could plug or run out, or the feed could temporarily overload the reagent delivery system. Because of the damage which could result from off-specification product, protection is essential. A very effective strategy is shown in Figure 71.6.

A second pH transmitter is located at the outfall, downstream of the point where pH is controlled. It actuates an alarm if the product is out of specification, alerting operators to a failure. At the same time, the alarm logic closes the discharge valve and starts a recirculation pump to treat the effluent again. The process must have sufficient extra capacity to operate under these conditions for an hour or more, to allow time for diagnosis and repair of the failure.

71.5 Control System Design

If all of the above recommendations are carefully implemented, control system design is a relatively simple matter. But if they are not, even the most elaborately designed control system may not produce satisfactory results. This section considers those elements of design which will provide the simplest, most effective system.

71.5.1 Valve or Pump Selection

The first order of business is estimating the maximum reagent flow required to meet the highest demand. (This does not necessarily correspond to the highest feed rate, because in some plants, the highest feed rate is stormwater which may require no reagent at all.) Then the minimum reagent demand must also be estimated, and divided into the maximum to determine the rangeability required for both conditions. If full flow of reagent is insufficient to match the demand, the pH will violate specifications and the product will have to be impounded and retreated, as described above. If the minimum controllable flow of reagent is too high to match the lowest demand, the pH will tend to limit-cycle between that produced by the two conditions of minimum controllable flow and zero flow. The cycle will tend to be saw-toothed, with a relatively long period. Its amplitude could be within specification limits, however.

Control valves typically have a rangeability of 35–100:1, higher than metering pumps at 10–20:1. A linear valve will produce a linear relationship between controller output and reagent flow if it operates under constant pressure drop, as for example, supplied from a head tank. For higher rangeability, large and small valves must be operated in sequence, with the one not in use completely closed. Proper sequencing requires that the valves have equal-percentage (logarithmic) characteristics, yielding an overall characteristic which is also equal percentage and must be linearized by a matching characterizer. This technique is detailed by Shinskey [5, pp. 400–401]. Ball valves used to manipulate slurry flow ordinarily have equal-percentage characteristics.

If the pH of the feed could be on either side of the set point, then both acid and base valves must be sequenced, with both closed at 50% controller output. Since both valves must fail closed, the base valve should be fitted with a reverse-acting positioner, opening the valve as the controller output moves from 50% to zero.

71.5.2 Linear and Nonlinear Controllers

In the case of a well-buffered titration curve, or controlling at a set point away from the region of highest gain, a linear pH controller may be used quite satisfactorily. Where neither of these mitigating factors apply, however, a linear controller will not give satisfactory results. Many pH loops limit-cycle between pH 4–10 or 5–9 using a linear PID controller having a proportional band below 400% (proportional gain > 0.25). If the proportional band is increased (gain lowered) until cycling stops, the controller will not respond satisfactorily to load upsets, allowing large deviations to last for exceptionally long durations.

The reason for this behavior is the very low gain of the titration curve beyond pH 4–10. When a disturbance forces the pH out of the 4–10 range, the loop gain is so low that very little correction is applied by the controller, amounting to integration of the large deviation over time. Eventually, reagent flow will match the load, but it is likely to cause overshoot as shown for the step load change in Figure 71.7. The long durations during which the pH is out of specification could cause damage to the environment downstream unless protection is provided, in which case the protective system may severely curtail production.

The simplest characterizer for pH control is a three-piece function which produces a low gain within an adjustable zone around the set point, with normal gain beyond. The width of the zone and the gain within it must be adjustable to match the titration curve. (Note that this is not the same as switching the controller gain when the deviation changes zones, as this would bump the output.) Although an imperfect match for the typical curve, this compensation provides the required combination of effective response to load changes with stability at the set point. The step load response in Figure 71.8 for the same process as in Figure 71.7 shows the effectiveness of the three-piece compensator. The width of the low-gain zone is ± 2.2 pH around the set point, and its gain is 0.05 compared to a gain of 1.0 outside the zone. The proportional band of the controller is a factor of 14 lower (gain 14 times higher) than that of the linear controller of Figure 71.7. This combination actually gives more stability (lower loop gain) around the set point.

For very precise pH control of a process having a single well-defined titration curve, more accurate characterization is needed. In building such a characterizer, remember that input and output axes are reversed, pH being the input to the characterizer and equivalent concentration the output. The required characterizer simply has more points connected by straight lines than the three-piece characterizer described above. If the set point is always to remain in a fixed position, this characterizer can relate input and output in terms of deviation, as was done with the three-piece characterizer. However, if set-point adjustments are likely, then there must be two identical characterizers used, one applied to the pH measurement and the other to the set point. (In a digital controller, the same characterizer can be used for both.)

71.5.3 Controller Tuning Procedures

When a characterizer is not required, tuning of the PID controller should be carried out following the same rules recommended for other loops. Shinskey [6] gives tuning rules in terms of gain, deadtime, and time constant, or the natural period and gain at that period. For a PI controller, integral time should be set at $4.0\tau_d$ or τ_n; for a PID controller, integral time should be $2.0\tau_d$ or $0.5\tau_n$ and derivative time $0.5\tau_d$ or $0.12\tau_n$. Because pH loops tend to limit-cycle, the natural period is easy to obtain and should give better results.

Derivative action is strongly recommended for pH control. Derivative action amplifies noise, however, so that if measurement noise is excessive, derivative action may not be useful, at least without filtering. A viable alternative is the PIτ_d controller, a PI controller with deadtime compensation [6]. When properly tuned, it provides integration without phase lag and without amplifying noise. Optimum integral time is the same as for the PID controller, and optimum controller deadtime is 25% less.

Proportional gain should be adjusted for quick recovery from disturbances in the process load without excessive overshoot or cycling. Rules for fine-tuning based on observed overshoot and decay ratio in response to disturbances in the closed loop are also provided in Shinskey [6]. Most pH loops operate at the same set point all the time. If a loop need not contend with set point disturbances, it should not be tested by disturbing the set point. Instead, a simulated load change should be introduced by manually stepping the controller output from a steady state

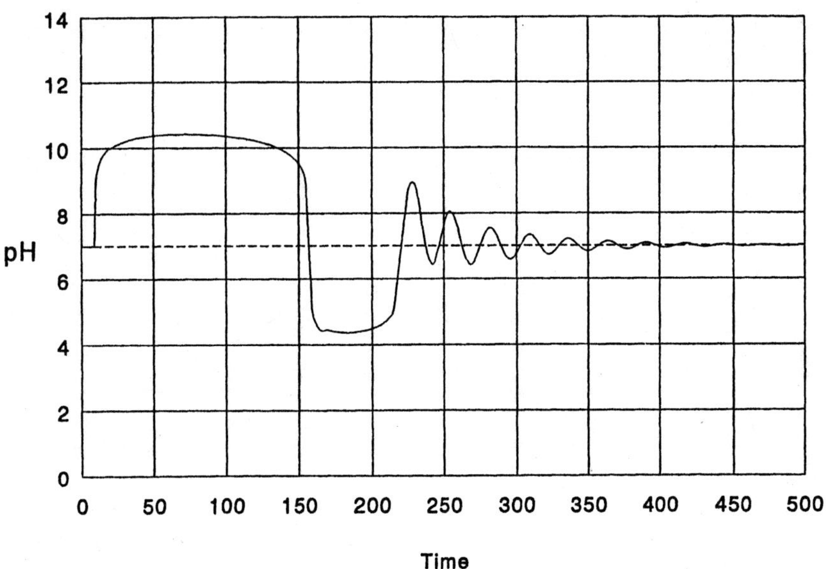

Figure 71.7 A linear PID controller tuned to avoid cycling will be slow in responding to load changes.

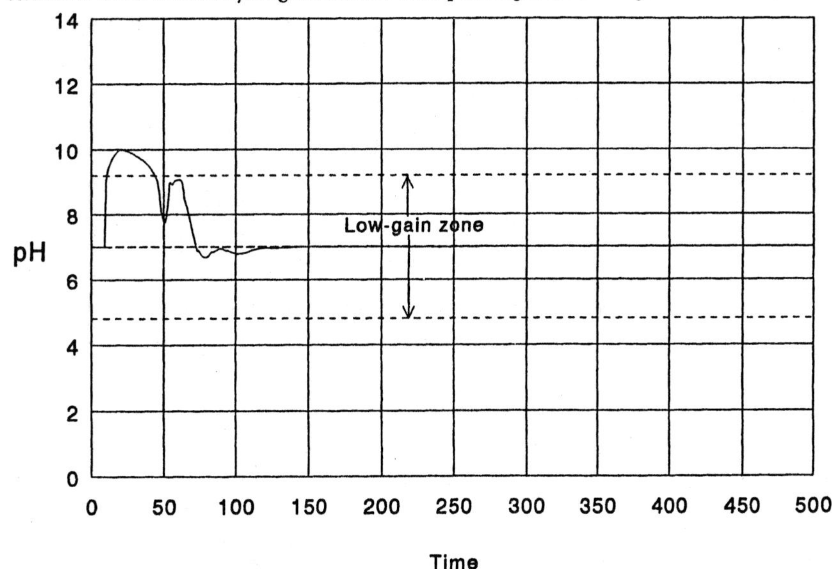

Figure 71.8 A three-piece nonlinear characterizer adds fast recovery and stability to the PID loop.

at set point, and immediately transferring the controller to the automatic mode. The effect of such a disturbance will be identical to that of the step load changes introduced in Figures 71.7 and 71.8.

If a three-piece characterizer is used to compensate the titration curve, it should be adjusted to position its breakpoints about 1 pH unit inside the knees of the curve. The slope of the characterizer between the breakpoints should be in the order of 0.05: if set $>$ 0.1, it will not provide enough gain change for effective compensation, and, if set at zero, it will cause slow cycling between the breakpoints. In the event the titration curve is unknown, the width of the low-gain zone should be initially set at zero and the proportional band decreased (gain increased) until cycling begins. The width should then be increased until the cycling abates. There is no hope of adjusting more than two breakpoints on-line.

An encounter with a particularly variable titration curve led the author to develop a self-tuning characterizer [4]. It was a three-piece characterizer with remotely adjustable width for the low-gain zone. In the presence of an oscillation, the zone would expand until the oscillation stopped; in the presence of prolonged deviation it would contract. This solved the immediate problem, but required a constant period of oscillation, because the PID settings of the controller and the dynamics of the tuner were manually set.

Now there are autotuning and self-tuning PID controllers on the market. *Autotuning* generally means that the controller tests the process on request, either with a step in the open loop or by relay-cycling in the closed loop to generate initial PID settings (which may or may not be effective). These controllers have no means for improving on the initial settings, however, or of automatically adapting them to observed variations in process parameters. A true *self-tuning* controller observes closed-loop responses to naturally occurring disturbances, adjusting the controller settings to converge to a response that is optimum or represents a desired overshoot or decay ratio. The self-tuning controller therefore is capable of automatically adapting its settings to variations in process parameters.

If a linear self-tuning controller is applied to a nonlinear pH process, it will adjust its gain low enough to dampen oscillations about set point. When a load disturbance follows, it will begin to respond like the low-gain linear controller in Figure 71.7. Observing the slow recovery of the loop, it will then increase controller gain until the set point is crossed, at which point oscillation begins, requiring that the gain be reduced to its former value. Although the resulting response is an improvement over that in Figure 71.7, it is not as effective as that in Figure 71.8. Therefore, nonlinear characterization is recommended, fit to the most severe titration curve likely to be encountered, with self-tuning used to adapt the controller gain to more moderate curves.

Self-tuning of the integral and derivative settings of the controller can adapt to a period increasing with fouling of the electrodes. While this can keep the loop stable as fouling proceeds, response to disturbances will nonetheless deteriorate due to the increasing time constants of the measurement and the controller. Much more satisfactory behavior will be achieved by keeping the electrodes clean. In the event that fouling is not continuous and cleaning is required only periodically, an alarm can be activated when the adapted integral time of the controller indicates that fouling is excessive.

71.5.4 Feedforward pH Control

Feedforward control uses a measurement of a disturbing variable to manipulate a correcting variable directly, without waiting for its effect on the controlled variable. In a pH loop, disturbing variables are the flow and composition of the feed. It is quite effective to set the flow of reagent in ratio to the feed flow, and thereby eliminate this source of upset. However, the ratio between the two flows depends on the composition of the feed, which can be quite variable. The feedback pH controller must adjust this ratio, as needed, to control pH and thereby compensate for variations in feed composition. This is done by *multiplying* the measured feed flow by the output of the pH controller, with the product being the required flow of reagent. The reagent flow must then respond linearly to this signal. This system works well when the rangeability of the reagent flow is 10:1 or less, because it depends on the rangeability of flowmeters and metering pumps.

Feedforward systems have been implemented using the pH of the feed as an indication of its composition, but with very limited success. This measurement is only representative of composition if the titration curve is fixed, which is never true for wastewater. Efforts have been made to titrate the feed to determine its composition or estimate the shape of the titration curve, but this takes time and these complications reduce the reliability of the system. Far more satisfactory performance will be realized if the process has been designed to be controllable, and a nonlinear feedback controller with self-tuning is used.

71.5.5 Batch pH Control

Some pH adjustments are carried out batchwise: this is the case for small volumes, or solutions too precious or toxic to be allowed discharge without assurance of precise control. The solution is simply impounded until its composition is satisfactory. Batch processes require a different controller than their continuous counterparts, however. In a continuous process, flow in and out can be expected to vary over time, constituting the major source of load change. In a batch process, there is no flow out, and so the load is essentially zero, while composition changes with time as the neutralization proceeds.

If a controller with integral action is used to control the endpoint on a batch process, the controller will *windup*, an expression indicating that the integral term has reached or exceeded the output limit of the controller. The cause of windup is the typically large deviation from set point experienced while reagent is being added to bring the deviation to zero. When the deviation is then finally reduced to zero, this integral term will keep the reagent valve open, causing a large overshoot. In a zero-load process, overshoot is permanent, and so must be avoided.

The recommended controller for a batch pH application is proportional-plus-derivative (PD), whose output bias is set to zero. The controller output will then be zero when the pH rests at the set point, and will be zero as the pH approaches the set point if it is moving. This controller can thereby avoid overshoot, when properly tuned. Proportional gain should be set to keep the reagent at full flow until the pH begins to approach the set point, and the derivative time set to avoid overshoot. An equal-percentage reagent valve will help to compensate for the nonlinearity of the titration curve, because its gain decreases as flow is reduced and, therefore, as the set point is approached.

71.6 Defining Terms

Activity: The effect of hydrogen ions on a hydrogen-ion electrode.

Autotuning: Tuning a controller automatically based on the application of rules to the results of a test.

Backmixing: Recirculating the contents of a vessel with its feed.

Concentration: The amount of an ingredient per unit of mixture.

Dead band: The largest change in signal which fails to cause a valve to move upon reversal of direction.

Diprotic: Denoting an acid or base which yields two hydrogen or hydroxyl ions per molecule.

Feedforward control: Conversion of a measurement of a disturbing variable into changes in the manipulated variable which will cancel its effect on the controlled variable.

Isopotential point: The pH at which changes in temperature have no effect on output voltage.

Monoprotic: Denoting an acid or base which yields one hydrogen or hydroxyl ion per molecule.

Normal: A Normal solution contains 1.0 gram-ion of hydrogen or hydroxyl ions per liter of solution.

Residence time: The average time that molecules spend in a vessel.

Self-tuning: Tuning a controller to improve the response observed to changes in set point or load.

Solution ground: A connection between the process solution and the instrument ground.

Standardize: To calibrate a pH instrument against a standard solution.

Static mixer: An in-line mixer with no moving parts.

Titration curve: A plot of pH vs. reagent added to a solution.

Valve positioner: A device which forces the position of a valve stem to match the control signal.

Windup: Saturation of the integral mode of a controller, causing overshoot of the set point.

References

[1] McMillan, G.K., Understanding Some Basic Truths of pH Measurements, *Chem.Eng.Prog.,* October, 30–37, 1991.

[2] Rovira, W.S., Microprocessors Bring New Power and Flexibility to pH Transmitters, *Contr.Eng.*, September, 24–25, 1990.

[3] Shinskey, F.G., *pH and pIon Control in Process and Waste Streams,* John Wiley & Sons, New York, 1973.

[4] Shinskey, F.G., Adaptive pH Controller Monitors Nonlinear Process, *Contr.Eng.*, February, 57–59, 1974.

[5] Shinskey, F.G., Process Control Systems, 3rd ed., McGraw-Hill, New York, 1988.

[6] Shinskey, F.G., Manual Tuning Methods, in *Feedback Controllers for the Process Industries,* McGraw-Hill, New York, 1994, pp. 143–183.

[7] Weast, R.C., *Handbook of Chemistry and Physics,* The Chemical Rubber Company, Cleveland, OH, 1970.

72

Control of the Pulp and Paper Making Process

W.L. Bialkowski
EnTech Control Engineering Inc.

72.1 Introduction

Pulp and paper products are consumed by almost every sector of modern society—a society that shows no sign of becoming "paperless", especially with the advent of the photocopier and laser printer. Paper is manufactured from wood, a naturally renewable resource, by a large industry with significant economic impact on the world economy. The manufacture of pulp and paper products represents a particularly challenging environment for the control engineer, as it is large in scale, highly nonlinear, highly stochastic, and dominated by time delays. This is not a linear-time-invariant system, but rather a "near-chaotic" environment with all of the implications that this has for process and product variability. The market for paper products, however, is increasingly quality conscious and customer driven, one in which product uniformity (the opposite of product variability) is increasingly being demanded by the customer as the prime ingredient of "quality" and the "price of admission".

This chapter reflects the experience gained [3] in the pulp and paper industry over the last decade, while attempting to improve product and process variability throughout the industry. The control engineer's challenge is not an academic one, but rather it is nothing short of enhancing the ability of his or her company to continue existing in such a competitive world. This challenge extends far beyond linear control theory and involves the control engineer as a critical member of an interdisciplinary team, who brings the essential understanding of dynamics to the manufacturing problem. Without this understanding of dynamics, it is impossible to understand variability. In the pulp and paper industry the control engineer is likely to be the only person who understands dynamics. Hence the scope of interest must extend far beyond the narrow confines of linear control theory or the closed-loop dynamic performance of single loops and must encompass the characteristics of the product required by the customer, the characteristics of the raw materials, the nonlinear and multivariable nature of the process, the effectiveness of the control strategy to remove variability, and the information and training needs of the rest of the team.

0-8493-8570-9/96/$0.00+$.50

This chapter will attempt to present the essence of this challenge by developing the subject in the following sequence: pulp and paper background, the impact of steady state design, mill variability results to date, actuator nonlinearities, linear control concepts — "Lambda Tuning", algorithms in use, control strategies for uniform manufacturing, minimizing variability by integrating process and control design, and finally conclusions. The reader is assumed to have an electrical engineering background, a strong interest in automatic control, and no prior exposure to the pulp and paper industry.

72.2 Pulp and Paper Manufacturing Background

Paper products, ever present in our daily lives, are manufactured primarily from wood in pulp and paper mills which are large, industrial complexes, each containing thousands of control loops. Let us describe some of this background before developing the control engineering perspective.

72.2.1 Wood Species and Raw Materials

Wood from trees, the primary raw material for pulp and paper products, is a renewable natural resource. In North America alone, there are nearly 1000 pulp and paper mills which depend on forest harvesting activities extending from northern Canada, most of the U.S., northern Mexico, and from Newfoundland in the east, to Alaska in the west. Some mills produce and sell pulp as a product. Others are integrated mills that produce pulp and then paper. Some paper mills buy pulp as a raw material, and make paper. Recycling mills take paper waste and make paper. The pulp and paper industry exists in many parts of the world.

The prime ingredient of wood, which makes it useful as a raw material, is **cellulose**, a long chain polymer of the sugar, glucose, which constitutes much of the wall structure of the wood cell, or fiber. Trees grow as a result of the natural process of photosynthesis, in which the energy from the sun converts atmospheric CO_2 into organic compounds, such as cellulose, with the release of O_2 gas. Growth occurs in the tree underneath the bark, as layers of cells are deposited in the form of annular rings (the age of a tree can be easily determined by counting these annular rings).

The wood cell, or **fiber**, is "cigar" shaped, roughly square in cross-section and varies in length from 0.5 mm to well over 4 mm. The center of the cell is hollow. Fiber length is a key property for paper making and is primarily determined by the wood species, which divide into two main types: softwoods and hardwoods. Softwoods are mainly coniferous trees which have needles instead of leaves, such as pines and firs. These species have long fibers: typically 3 mm long (redwoods have 7 mm long fibers) and about 30 μm wide. Hardwoods are primarily the deciduous trees with leaves, such as maple, oak, and aspen. Their fibers tend to be short and "stubby" (typically 0.5 mm).

The fiber structure consists of many layers of cellulosic "fibrils", crystalline bundles of cellulose molecules. Wood also contains **hemicellulose**, polymers of other sugars with chain lengths shorter than cellulose. The cell structure is bonded together by **lignin**, the third wood ingredient, which acts very much like a glue. Typical wood composition by fraction on a dry basis is shown in Table 72.1

TABLE 72.1 Wood Composition (Dry Basis).

Wood Species	Hardwood	Softwood
Cellulose %	50	50
Hemicellulose %	30	20
Lignin %	20	30

The process of removing fibers from solid wood is called "pulping". There are two process for pulping: mechanical and chemical. Mechanical pulping involves "ripping" the fiber out of the solid wood structure by mechanical and thermal means. Mechanical pulp fibers include all of the above ingredients, hence the pulping process has a very high yield (90%+). However, due to the nature of the pulping process, there is substantial fiber damage with a resulting loss in strength. Chemical pulping involves chemically dissolving the lignin fraction (and usually the hemicellulose fraction as well). This liberates the cellulose fraction, still in the shape of the original fiber, with relatively little damage, and results in high strength and the potential for high brightness, once all traces of lignin have been removed and the pulp has been bleached. The yield of chemical pulp is low, $<$ 50% to 60%, because the lignin and hemicellulose have been dissolved. Typically, these organic compounds are used as a fuel in a chemical pulp mill.

72.2.2 Products

Pulp and paper products cover a broad spectrum. An incomplete list includes newspaper, paper for paperback books, photocopier paper, photographic paper, paper for books, coated paper for magazines, facial tissue, toilet tissue, toweling, disposable diapers, sanitary products, corrugated boxboard, linerboard, corrugating medium, food board (e.g., cereal boxes, milk carton), Bristol board, bleached carton (cigarette carton, beer cases), molded products (paper plates), insulation board, roofing felt, fiberboard, market pulp for paper manufacture, and pulp for chemical feed stock such as the manufacture of rayon, film, or food additives.

Each of these products requires certain unique properties which must be derived from the raw material. For instance, newspaper must be inexpensive, strong enough to withstand the tension imposed by the printing press without breaking, and have good printing properties. It is often made from recycled paper with added mechanical or chemical softwood pulp for strength. Photocopier paper must have excellent brightness, a superb printing surface and must not curl or jam in the photocopier. Highly bleached chemical pulp is used. This type of paper is made by blending hardwood pulp (for a very smooth printing surface) with softwood pulp (for strength). Additives such as clay and titanium dioxide are added to enhance the printing surface. Facial tissue must be soft and absorbent. It is made from chemically bleached pulp. Market pulp is sold by pulp grade. Each grade having specifications for species, brightness, viscosity, cleanliness, etc. In most cases market pulps are made chemically and are usually bleached.

72.2.3 Manufacturing Processes

Pulp and paper mills vary greatly in design, as they reflect the product being made. The simplest division is to separate pulping from papermaking. We will discuss each of the significant areas as unit operations. The degree of control and instrumentation which exists varies widely depending on both the age and design of the plant.

72.2.4 Mechanical Pulping

Mechanical pulping can be achieved by grinding or refining. *Groundwood* mills mechanically grind whole logs against an abrasive surface. The pulping action is a combination of raising the temperature, and mechanically "ripping" the fiber from the wood surface. This is done by feeding logs to the "pockets" of grinders, which are powered by large synchronous motors (typically 5,000 to 10,000 HP). A typical groundwood mill may have 10 to 20 grinders (50 to 150 control loops). The *Thermal-Mechanical Pulping* (TMP) process (100 to 300 control loops) is the more modern way of mechanical pulping. It consists of feeding wood chips into the "gap" of rotating pressurized machines, called chip refiners, to produce pulp directly. Normally there are two refining stages. The chips disintegrate inside the refiner as they pass between the "teeth" of the refiner plates. Each refiner is powered by a synchronous motor of 10,000 to 30,000 HP. The amount of refining can be controlled by adjusting the gap between the rotating plates. Mechanical pulps can be brightened (bleached) to some degree.

72.2.5 Chemical Pulping

Chemical pulping can be divided into two main processes, sulfite and Kraft. The *sulfite process* is an acid based cooking process, whose use is in decline. The *Kraft process* is an alkali-based process in which the active chemicals are fully recycled in the *Kraft liquor cycle.* The Kraft process itself consists of eight individual unit operations. It starts with wood chips being fed to a *digester house* (unit process with 50 to 300 control loops) in which the chips are fed to a digester and impregnated with **white cooking liquor** (a solution of NaOH and Na_2S), and "cooked" at about 175°C for about an hour. In this process the lignin and hemicellulose are dissolved. The spent cooking liquor is then extracted and the pulp is "blown" into the "blow" tank. Modern digesters are continuous vertical columns, with the chips descending down the column. The impregnation, cooking, and extraction processes take about three hours or so. Some digester houses use batch digesters instead, with 6 to 20 batch digesters. From the blow tank the pulp is pumped to the *brown stock washers* (unit process with 20 to 200 control loops), in which the pulp is washed in a multistage countercurrent washing process to remove the spent cooking liquor, including the dissolved organic compounds (lignin, hemicellulose) and the spent sodium compounds which must be removed from the pulp stream and reused. The two output streams consist of: washed pulp (which goes to the bleach plant for bleaching or to the paper machines if bleaching is not required) and the spent liquor which is called *black liquor* and is returned to the Kraft liquor cycle.

The *Kraft liquor cycle* starts with pumping black liquor from the brown stock washers to the *multiple effect evaporators* (unit process with 20 to 100 control loops) where the black liquor solids are raised from about 15% to about 50% by evaporation with steam. These solids are then fed to the **recovery boiler** (unit process with 100 to 500 control loops) in which the black liquor solids are further concentrated and then the liquor is fired into the recovery boiler as a "fuel". The bottom of this boiler contains a large smoldering "char-bed" of burning black liquor solids, below which the sodium compounds form a molten pool of "smelt", containing Na_2CO_3, Na_2S and Na_2SO_4. The smelt pours from the bottom of the recovery boiler into a dissolving tank, where it is dissolved in water, and becomes *green liquor.* The upper part of the recovery boiler is conventional in design and produces about 50% of the total steam consumed by the mill. The green liquor is pumped to the **causticizing area** (unit process with 50 to 200 control loops) where "burned lime" (CaO) is reacted with the green liquor in order to re-constitute the white cooking liquor (first reaction is: $CaO+H_2O=Ca(OH)_2$; second reaction is: $Na_2CO_3+Ca(OH)_2 = 2NaOH+CaCO_3$). The resulting white liquor is sent to the digester for cooking. The calcium carbonate is precipitated and sent to the **lime kiln** (unit process with 50 to 200 control loops) where the $CaCO_3$ plus heat produces burned lime (CaO) plus carbon dioxide gas. The burned lime is used in the causticizing area to reconstitute the white cooking liquor.

72.2.6 Bleaching

After cooking, the pulp may need to be bleached. This is necessary for all products which require high brightness or the complete removal of lignin. A typical bleach plant (unit process with 50 to 300 control loops), consists of the sequential application of specific chemicals, each followed by a reaction vessel and a washing stage. A bleaching sequence which has been used often in the past involves pumping unbleached "brown stock" from the pulp mill and applying the following chemicals in turn: chlorine (to dissolve lignin), caustic (to wash out lignin by-products), chlorine dioxide (to brighten), caustic (to dissolve by-products), and finally chlorine dioxide to provide final brightening. Bleach plant technology is currently in a state of flux as a result of the environmental impact of chemicals, such as chlorine. As a result, the industry is moving towards new chemicals, chiefly oxygen and hydrogen peroxide.

72.2.7 Paper Making

Paper is made on a paper machine (unit process with 100 to 1000 control loops). Paper making involves **refining** various types of pulps (mechanical work to improve bonding tendencies) individually, and then blending them together in combination with specific additives. This "stock" is then screened and cleaned to remove any dirt, and is diluted to a very lean pulp slurry before being ejected onto a moving drainage medium, known as "the wire", Fourdrinier, or former, where a sheet of wet paper is formed. This wet paper is then pressed, dried, and calendered (pressed between smooth polished surfaces) to produce a smooth final product with a specified basis weight (weight per unit area), moisture content, caliper (thickness), smoothness, brightness, color, ash content, and many other properties. Figure 72.1 shows a diagram of a simple fine paper machine (say for photocopier paper): Hardwood and softwood stocks are stored in High Density (HD) storage tanks at about 15% consistency (mass percentage of pulp in a slurry). As the stock is pumped from the HD's it is diluted to 4% and refined. Paper machine refiners have rotating discs which fibrilate or "fluff" the fiber, thereby giving it better bonding strength. The hardwood and softwood stocks are then blended together in a desired ratio in the blend chest (subject of case study later), and sent to the machine chest, the final source of stock for the paper machine. From here the stock is diluted with "white water" (filtrate which has drained through the forming section), and is sent to the headbox (a pressurized compartment with a wide orifice or "slice lip" which allows the "stock" to be ejected as a wide jet). The stock discharges from the headbox at the required speed onto the "wire", a wide porous "conveyor belt" consisting of synthetic woven material. Most of the fiber stays on, and most of the water drains through to be recycled again as "white water".

72.2.8 The Pulp and Paper Mill

The pulp and paper mill consists of between one and some thirty unit operations, with a total of about 500 to 10,000 control loops. A typical mill produces 1000 tons/day of product (although the capacity of mills can vary from 50 to over 3000 tons). Each of the operations involving wood chips or fiber is essentially hydraulic in nature, and involves a sudden application of chemical, dilution water, mechanical energy, or thermal energy. Sometimes this is followed by a reaction vessel with long residence time. As a result, the fast hydraulic dynamics are very important in determining the final pulp or paper uniformity. A typical mill is operated some 355 days per year, 24 hours per day, by a staff of a few hundred people, often with a unionized workforce. The traditional responsibility for control loop tuning lies with the instrumentation department, which may have between 5 and 30 instrument technicians. The number of control engineers varies from a dozen in some mills to none in others.

This has been a very brief description of pulp and paper science, a well established field with a rich literature. A good introduction to this work is available in the textbook series on pulp and paper manufacture edited by Kocurek [14].

72.2.9 Competitive Marketplace

Regardless of the imperfections which might exist in the pulp and paper manufacturing environment, the marketplace now demands ever greater product uniformity. Warnings and recommendations about control performance and control engineering skill have been reported [3] in pulp and paper industry literature. Such problems are not unique to pulp and paper, and probably apply to most process industries. For instance, similar concerns have been expressed in the chemical industry [7], where the following recent prediction was made: *"In ten years the ability to produce highly consistent product will not be an issue because those companies that fail in this effort will be out of the market or possibly out of business. Among the remaining companies, the central issue ten years from now will be the efficient manufacture of products that conform to customer variability expectations"*. To a great extent the issues are not technological but human. A higher level of competency in process control is required to reduce variability [9], which requires awareness creation, training, and education. To this end, there has been extensive training of both instrument technicians and control engineers in recent years [4], with nearly 2000 industry engineers and technicians having received training in dynamics and control.

72.3 Steady-State Process Design and Product Variability

Pulp and paper mills, like most of heavy industry, have been designed only in steady state. Design starts with a steady-state flow sheet, proceeds to mass and energy balances, following which major equipment is selected and the design proceeds to the detailed stage. It is in the detailed design stage that control of the process is often considered for the first time, when sensors and valves are selected and a general purpose control system is purchased. It is generally assumed that only single PID loops will be required to control the process adequately and that these controllers can handle whatever dynamics are encountered when the process starts up. Until now, the prime focus has been on production capacity.

The last ten years have witnessed increasing concern about high product variability, as the quality "revolution" has started to engulf the pulp and paper industry. Customer loyalties can change and large long term contracts can be canceled when the product does not perform well in the customer's plant. As an example, Japanese newspaper print-rooms are now demanding paper which is guaranteed to have only one web break or less, per one thousand rolls of paper [13]. Such intense competition has not yet reached North America, where newsprint suppliers are still being rated by their customers in breaks per hundred rolls, with 3 being good — this is 30 breaks per thousand rolls! As competition intensifies, pressures such as these will create an increasing need for closer attention to variability. Such market forces are creating new demands. For instance, the consulting company, EnTech Control Engineering Inc., was formed in response to the needs for an independent variability and control auditing service with design and training expertise to help pulp

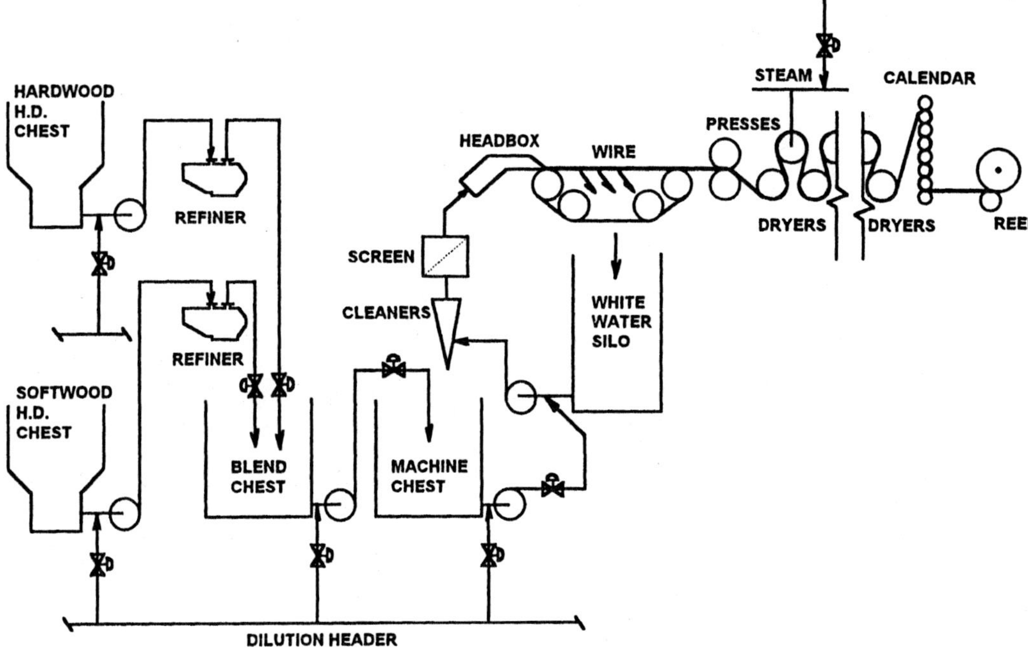

Figure 72.1 A paper machine.

and paper companies acquire the skills to become more effective competitors in the marketplace. About 200 mill variability audits have been carried out to date. The results of these process and product variability audits [3] raise serious questions about the quality of mill design, maintenance practice, control equipment (such as control valves) and control knowledge, as these issues pertain to the ability of achieving effective process dynamics and low product variability.

72.3.1 Mill Variability Audit Findings

The results shown are fairly typical of audit results which may be obtained from any part of the pulp and paper industry prior to corrective action. The blend chest area of a paper machine has been chosen as a small case study to illustrate typical mill variability problems. A design discussion of the blend chest area is continued throughout later sections of the chapter. The blend chest is included in the paper machine sketch of Figure 72.1, and is also shown in more detail in Figure 72.2 below, which illustrate a fine paper machine (fine paper refers to high quality, publication grade paper) manufacturing paper, such as photocopy paper, using 70% hardwood and 30% softwood and typically producing about 500 tons per day. The hardwood pulp is needed for surface smoothness and printability, and the softwood pulp is needed for strength. These two pulp slurries (stocks) are pumped out of their respective high density (HD) storage chests at about 5% *consistency* (mass percentage fiber slurry concentration). The hardwood stock is pumped out of the hardwood HD chest and is diluted further to 4.5% consistency at the suction of the stock pump, by the addition of dilution water. The dilution water is supplied from a common dilution header (pipe) for this part of the paper machine. In the example of Figure 72.2, the dilution water addition is modulated by the consistency control loop, NC104 (typical loop tag based on ISA terminology), regulating the consistency to a set point of about 4.5%. The consistency sensor is normally located some distance after the pump, per the manufacturer's installation instructions. A time delay of 5 seconds or longer, is typical. After consistency control, the hardwood stock is pumped into the hardwood refiner. The purpose of the refiner is to "fibrillate" the fiber, increasing its surface area and tendency to bond more effectively in the forming section. The refining process is sensitive to the mass flow of fiber and the mechanical energy being supplied to the refiner motor (specific energy). After the refiner, the stock flow is controlled by flow controller FC105. The softwood line is identical to the hardwood line. The two flow controllers FC105 and FC205 are in turn part of a cascade control structure, in which both set points are adjusted together to maintain the blend chest level at the desired set point (typically at 70%), while maintaining the desired blend ratio (70% hardwood and 30% softwood).

72.3.2 Variability Audit Results

Figure 72.3 shows the blend chest consistency NC302 on automatic control (upper plot) and on manual (lower plot) for a period of 15 minutes (900 sec). On automatic, the mean value is 3.82% consistency, and two standard deviations ("2Sig") are 0.05093% consistency. This number (which represents 95% of all readings assuming a Gaussian distribution) can also be expressed as 1.33% of mean. It is this number which is usually referred to as "variability" in the pulp and paper industry (approximately half the range of the variable expressed as a percent). For the manual data, the mean value is virtually unchanged (the loop was merely turned off control). The variability however is now 0.848%, 50% lower than on automatic. NC302 is controlled by a PI controller which has been tuned with default tuning settings

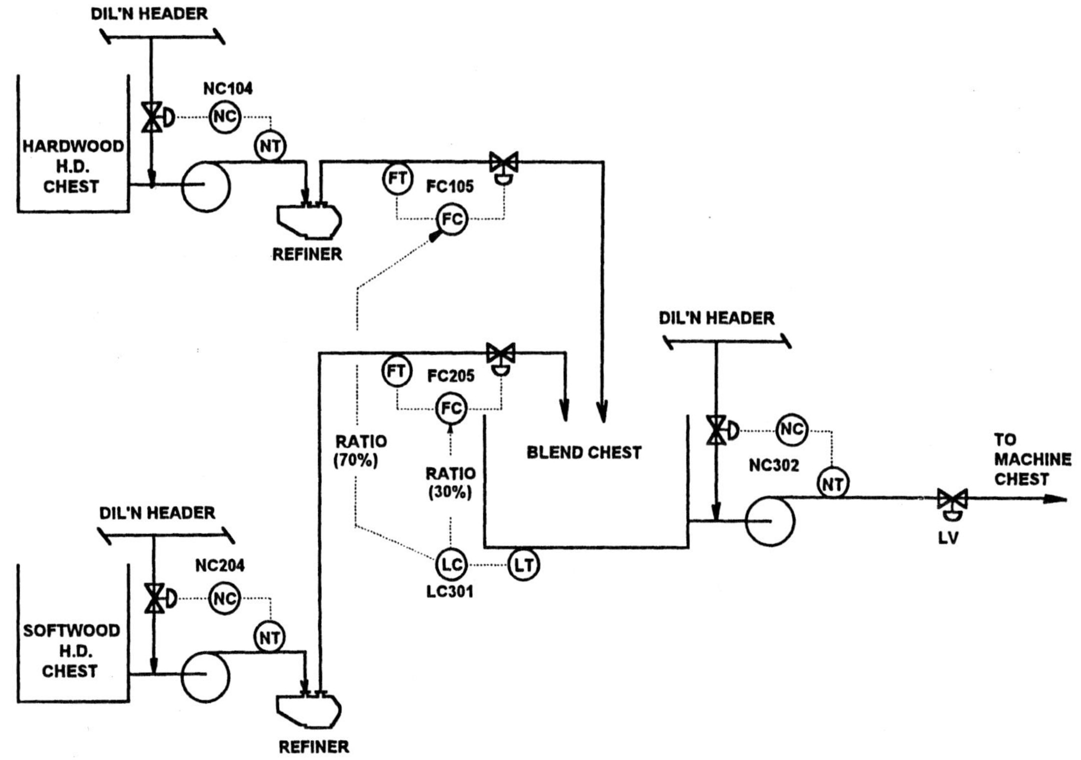

Figure 72.2 The blend chest area of a paper machine.

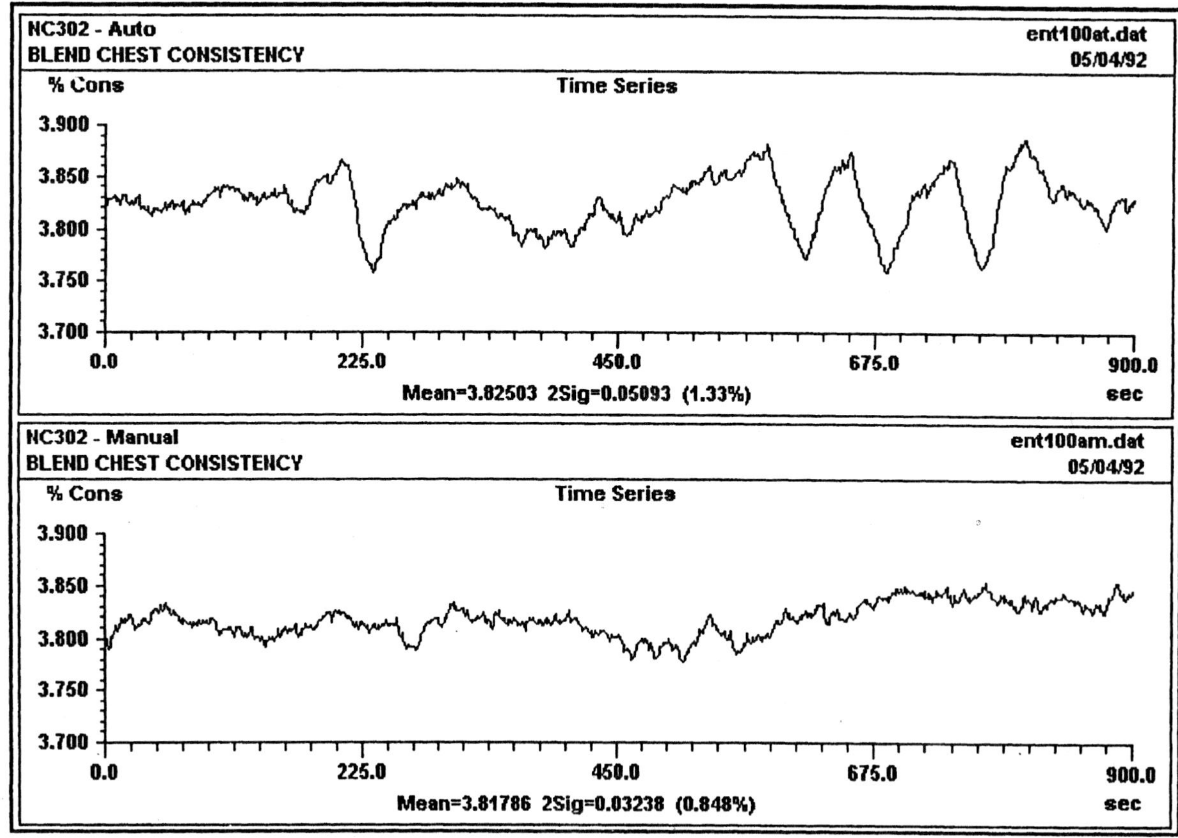

Figure 72.3 Blend chest consistency NC302 time Series — on and off control.

consisting of a controller gain of 1.0 and an integral time of 0.5 minutes/repeat (standard PI controller form). These tuning parameters occur frequently, are entered as default values prior to the shipping of controller equipment, and are "favorite numbers" for people tuning by "trial and error". However, the variabilty problems of NC302 are not all tuning related; the control valve was found to have about 1% stiction. It is this tendency which induces the limit-cycling behavior seen in Figure 72.3 at second 225 and from seconds 500 to 800.

Variability of 1.33% for a blend chest consistency is not very high, but, considering that paper customers are starting to demand paper variability approaching 1%, this represents potential for improvement, especially because the variability on manual control is 50% lower than on automatic. Figure 72.4 shows two open-loop step tests *(bump tests)* for the blend chest consistency loop. Also shown is the impact of these on the dilution header pressure control loop, PC309 (not shown in Figure 72.2). Excursions in the consistency control valve of 2% cause header pressure excursions of over 1% and pressure control valve excursions of about 1%. This example illustrates the degree of coupling that exists between these two variables and poses questions about the tendency of this dilution header design to cause interaction between all of the consistency loops which feed from it. Every time that one consistency loop demands more dilution water, it will cause a negative dilution header pressure upset and disturb the other consistency loops. Not visible in Figure 72.4 is the fact that the NC302 dilution control valve also has about 1% stiction which was mentioned as the cause of the limit cycle in Figure 72.3.

Figure 72.5 shows an expanded view of the first NC302 bump test of Figure 72.4. The top trace shows the controller output, while the lower shows the process variable. A fit of a first-order plus deadtime process transfer function model is superimposed over this data. The lower left hand of Figure 72.5 shows the resulting parameters for the transfer function form,

$$G_P(s) = \frac{K_P e^{-sT_d}}{(\tau_1 s+1)} \tag{72.1}$$

The time delay of 8.3 seconds is excessively long. It results from locating the consistency sensor too far from the pump discharge (could be less than 5 sec.), The time constant of 18.8 seconds is also excessively long (could be 3 sec.). This parameter is primarily determined by the adjustment of a low-pass filter in the consistency sensor itself. Both of these issues contribute to the sluggish behavior of NC302.

Figure 72.6 shows a series of open-loop bump tests performed on the hardwood flow FC105. There are seven steps performed on the controller output. The first five have a magnitude of about 0.7% (one negative, four positive), while the last two have a magnitude of about 1.5%. It is clear that steps 3, 4, and 5 produce no noticeable flow response. The response of step number six is 1.5% producing a flow response of about 50 gallons per minute, and the last step produces no response at all. The combined backlash and stiction in this valve assembly is about 5%. Clearly, it is impossible to control this flow to within 1%. Similar bump tests on the softwood flow FC205 indicated that the combined backlash and stiction are in the order of about 1.7%.

Figure 72.7 shows the softwood flow FC205 on automatic control. There is a limit cycle induced by the combined stiction and backlash of more than ±5%. The controller output appears like a triangular wave of amplitude 1.7%. This is a characteristic of stiction-induced limit cycles and is caused by the PI controller integral term attempting to induce valve motion. The valve, however, is stuck and will not move until the controller output has changed by about 1.7%. The valve will then release suddenly and move too far, thereby, inducing the next cycle. The period of the limit cycle is about 1.4 minutes. The period is a function of the stiction magnitude, the PI controller tuning, and the process gain. Figure 72.8 shows what happens when FC205 is taken off control. The upper plot of Figure 72.8 is the data from the last half of Figure 72.7. The bottom half of Figure 72.8 shows FC205 on manual control. It is clear that the limit cycling stops and the variability drops significantly. This is quite typical of many flow control loops.

Figure 72.9 shows the operation of the blend chest level controller, together with the two flows entering the chest. The upper left hand plot shows the blend chest level LC301. The upper right hand plot shows the LC301 power spectrum plotted on a log-log scale. There is a tendency for the level control to cycle with a period of about 17 minutes, as indicated by the time series plot and by the low-frequency lobe of the power spectrum which contains the fundamental frequency (period of 1024 seconds). Level controllers frequently cycle in industry because their tuning is not intuitive. There is a tendency for people to use tuning parameters such as the default settings of a gain of 1.0 and an integral time of 0.5 minutes. This represents far too aggressive an integral time and far too low a gain for an integrating process such as the blend chest, which requires quite different tuning. The two lower plots show the softwood FC205 (left) and hardwood FC105 (right) flows. The variability of both is about 5%. However, this is where the similarity ends. Both flows limit cycle in their own characteristic way with large amplitude swings. The softwood flow is limit cycling in a manner similar to that shown in Figure 72.7 (period now 2.6 minutes). The hardwood flow FC105, on the other hand appears to break into a violent limit cycle on a periodic basis. The amplitude of this limit cycle is more than 8% with a period of about one minute. The violent limit cycling stops periodically, followed by periods of little action lasting more than five minutes. It is interesting how such nonlinearities cause frequency shifts and multiple frequency behavior reminiscent of chaos theory.

72.3.3 Impact on Papermaking

The impact of the variability described above is quite damaging to papermaking. The variability in hardwood and softwood flows causes a number of problems. First of all, flow variability causes variable consistency, which the consistency controller cannot correct due to its time-delay dominant dynamics. In addition, the consistency controller tends to have a resonant mode due to its time-delay dominant dynamics. The net result is that the variability in the mass flow of fiber is amplified by the joint action of consistency control and stock flow control. This vari-

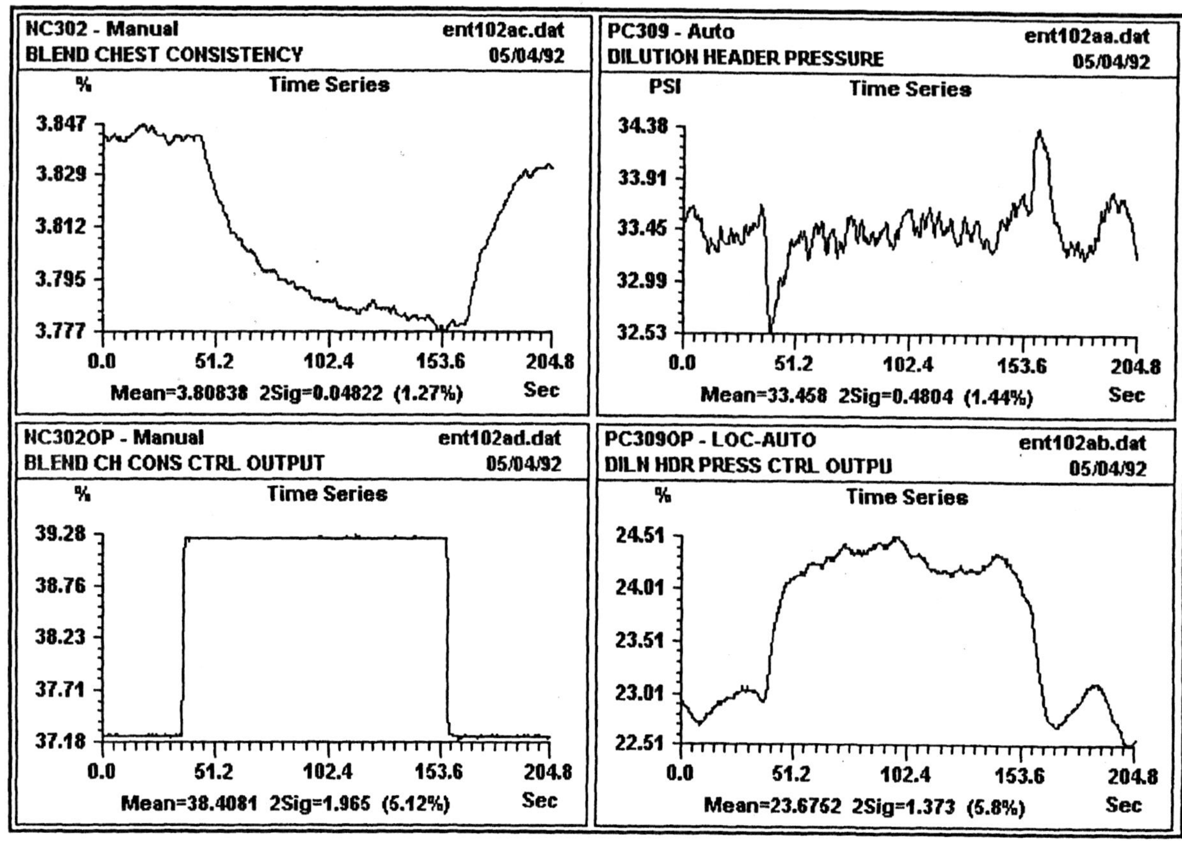

Figure 72.4 Interaction between blend chest consistency NC302 and dilution header pressure PC309.

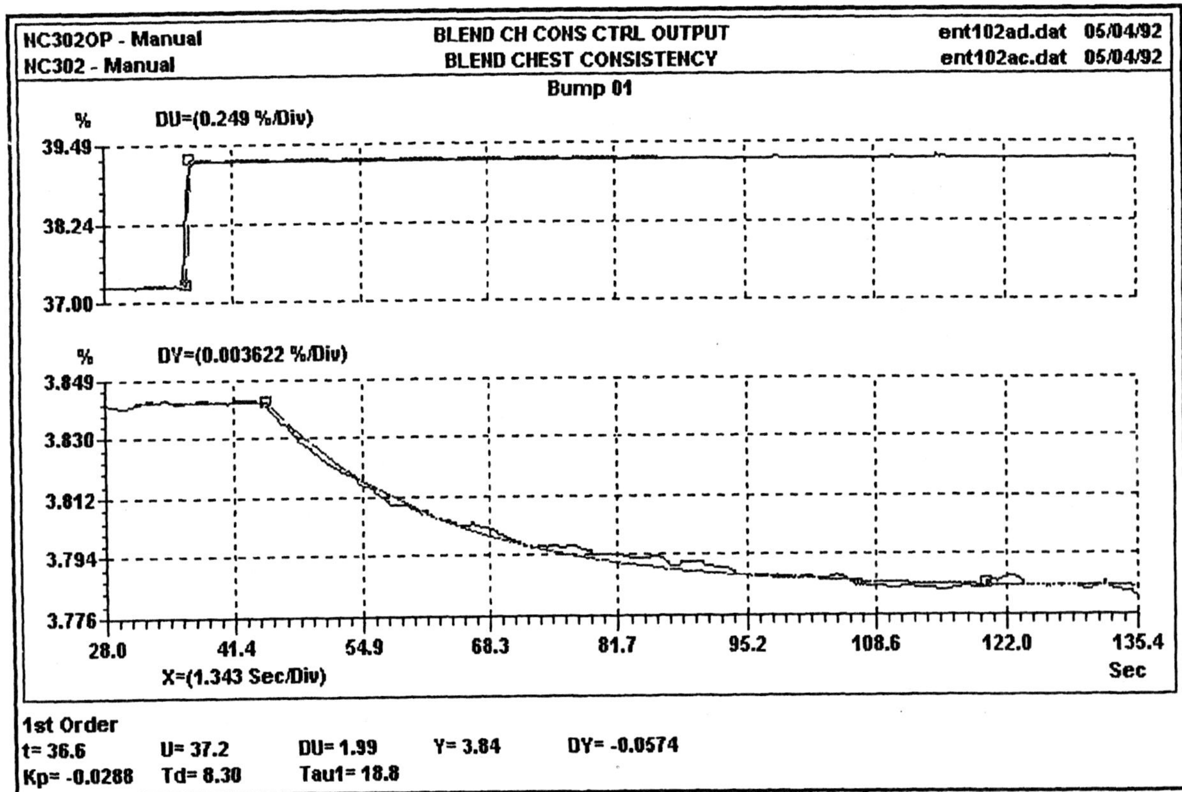

Figure 72.5 Blend chest consistency NC302 open loop bump test.

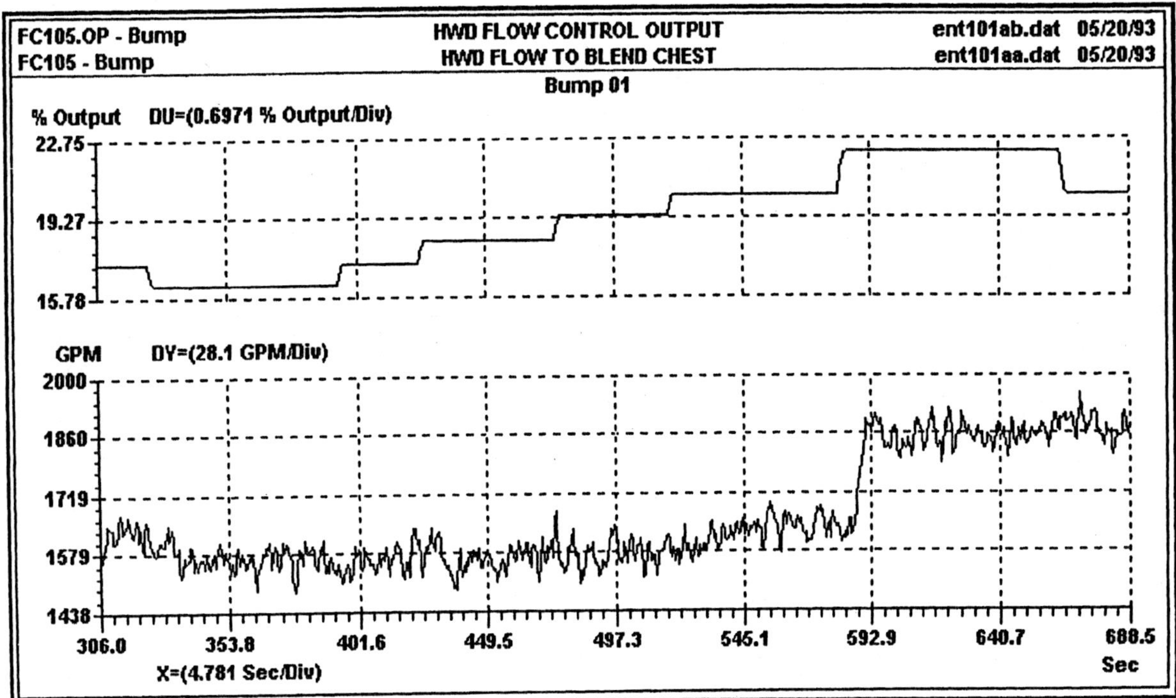

Figure 72.6 Hardwood flow FC105 bump tests showing about 5% backlash/stiction.

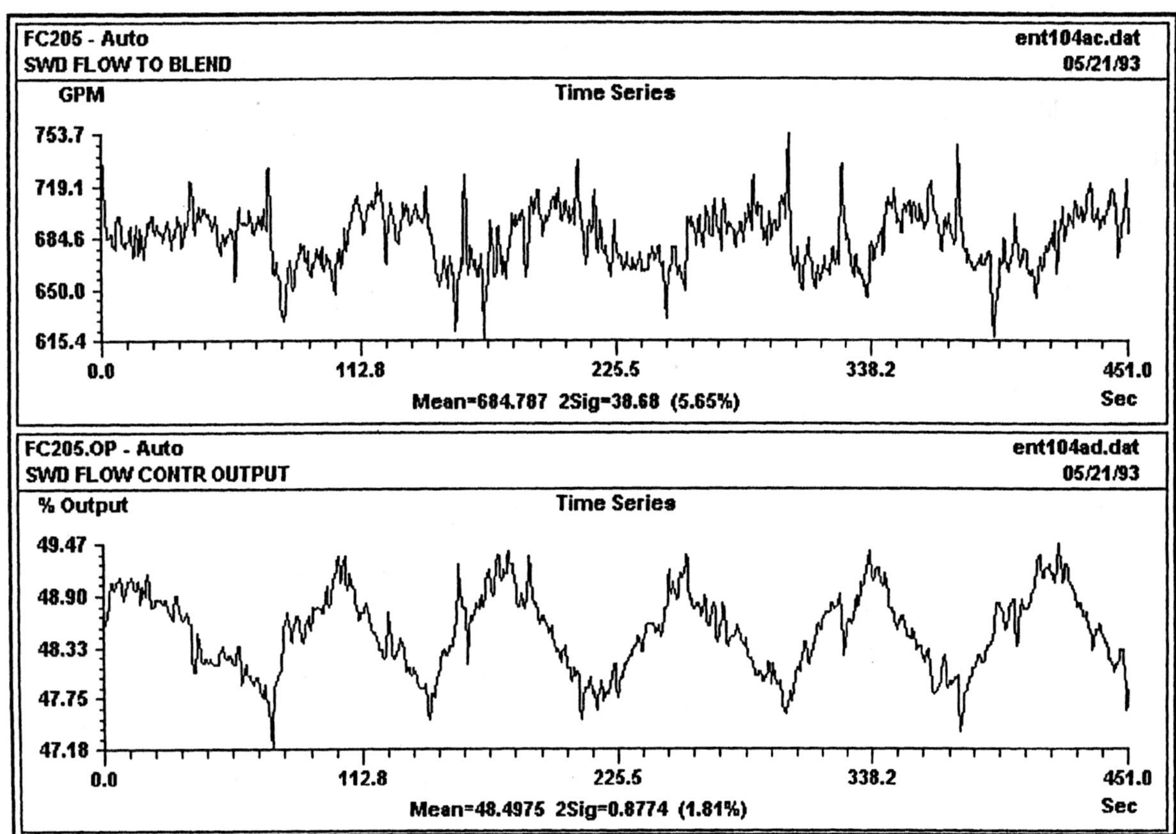

Figure 72.7 Softwood flow FC205 with 1.2% stiction showing a limit cycle.

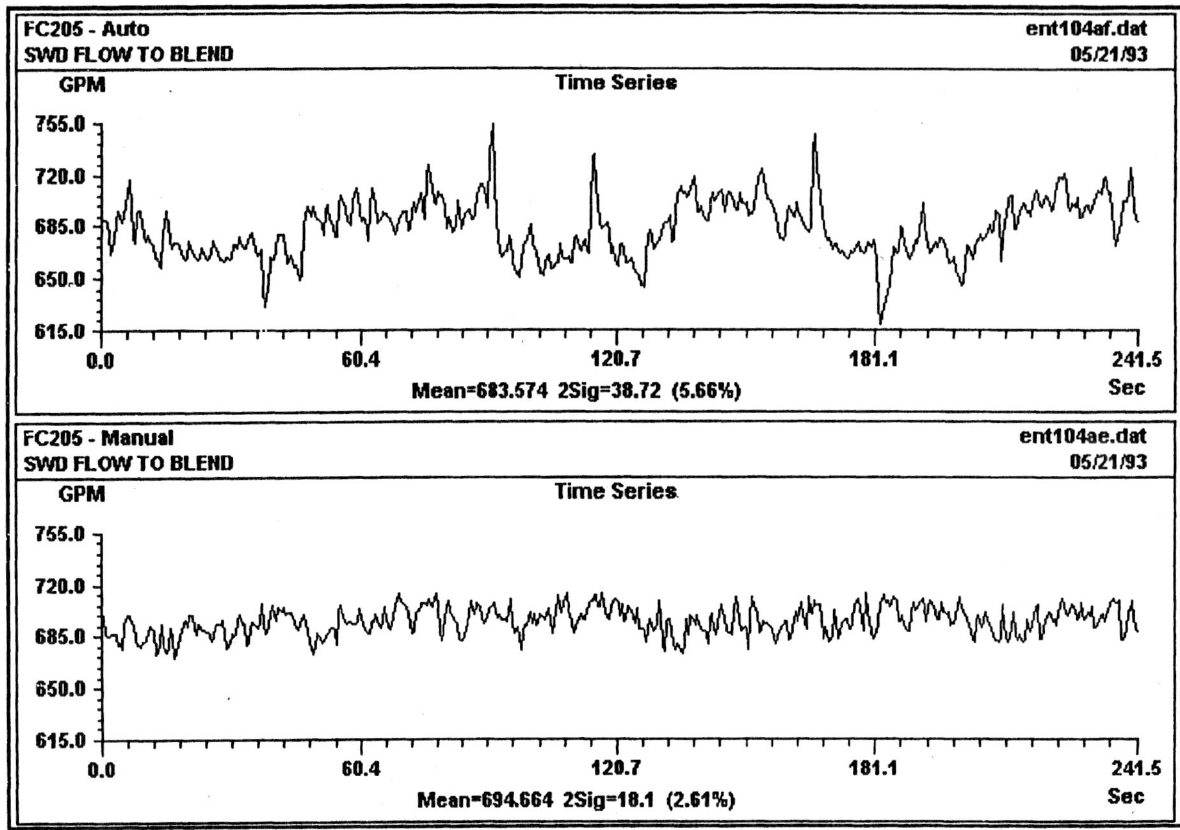

Figure 72.8 Softwood flow FC205 on and off control.

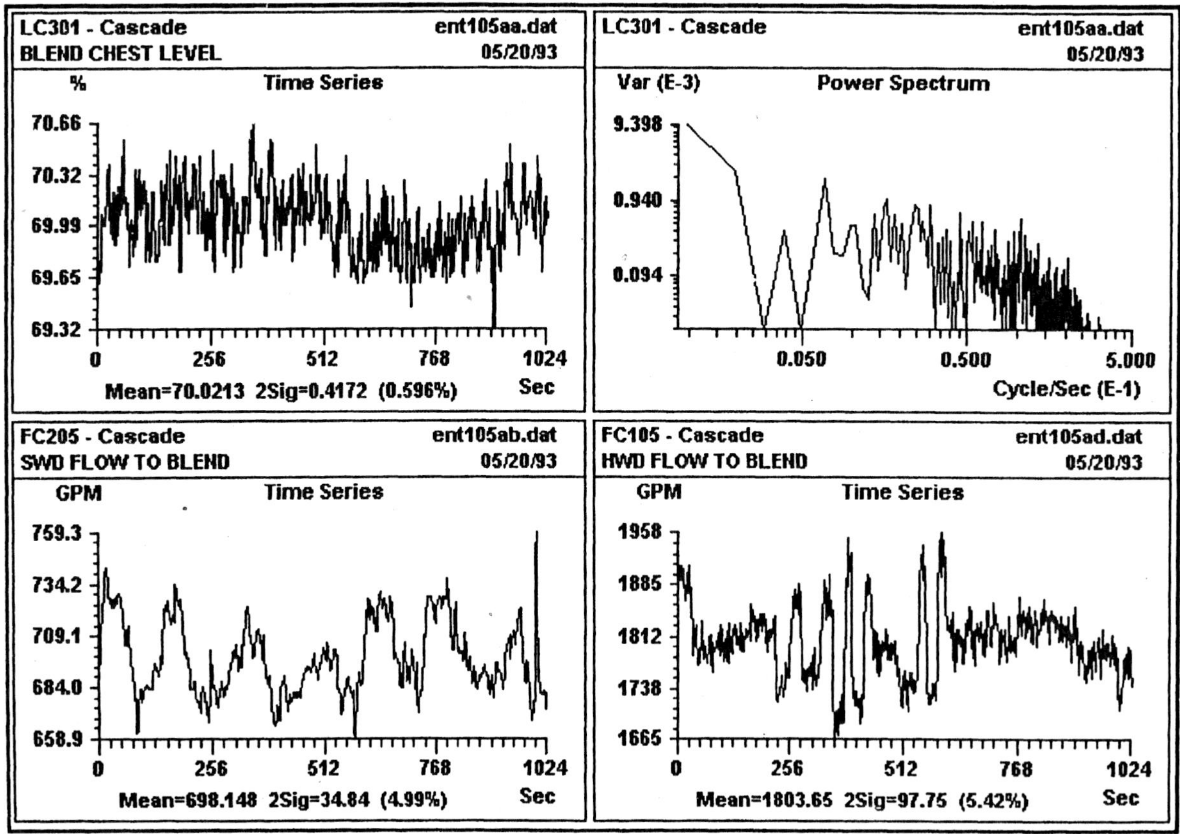

Figure 72.9 Blend chest level LC301 on automatic control.

ability is driven by the unique limit-cycle behavior of each flow controller. As a result the refining process will suffer, with the bonding strength developed by each fiber species varying in different ways. In addition, the stock blend (e.g., 70% hardwood, 30% softwood) uniformity is also compromised by the different limit cycles. At times, the hardwood/softwood ratio is off by 5% in Figure 72.8. Clearly this cannot help to deliver uniformly blended stock to the paper machine.

72.3.4 Final Product Variability

Final paper product measurements include: basis weight (mass per unit area), moisture content, caliper (thickness), smoothness, gloss, opacity, color, and brightness. Of these, basis weight is the most important and the easiest to measure, as the sensor is essentially an analog sensor. Almost all of the other sensors are digital with limited bandwidth. The final quality sensors of most paper machines are located on the scanning frame, which allows the sensors to traverse the sheet at the *dry end*. Typically each traverse takes 20 to 60 seconds. During each traverse, sensor averages are calculated to allow feedback control in the "machine-direction" (MD). At the same time, data vectors are built during the sheet traversing process to measure and control the "cross-direction" (CD) properties. To measure "fast" basis weight variability the traversing must be stopped by placing the scanning sensors in "fixed-point" (MD and CD controls are suspended for the duration). While in fixed point, it is possible to collect data from the basis weight sensor (beta gauge) at data rates limited only by the bandwidth of the data collection equipment. Figure 72.10 shows a fixed-point basis weight data collection run. The average basis weight is 55 g/m^2(gsm), with variability of 3.29%. Clearly there is a strong tendency to cycle at about 0.2 Hz, or 5 seconds per cycle. This is evident in the time series plot (upper left), in the power spectrum plot (upper right), in the autocorrelation function (bottom left), and in the cumulative spectrum (bottom right), which shows that about 25% of the variance is caused by this cycle. The cause of the cycle is not related to the blend chest data but rather to variability in the headbox area. Nevertheless, the time series data indicates that it is worthwhile to identify and eliminate this cycle. Figure 72.11 shows similar data collected at a slower rate for a longer period of time. Once again the mean basis weight is 55 gsm, and the variability is 3.7%. The time series data indicates a relatively confused behavior. From the cumulative spectrum it is clear that 50% of the variance is caused by variability slower than 0.02 Hz (50 seconds/cycle). From the power spectrum it is evident that there is significant variability from 0.003 Hz (5.6 minutes/cycle) through to about 0.0125 Hz (1.3 minutes/cycle). In addition there is considerable power at the fundamental period of 17 minutes. All of these frequencies fit the general behavior of the blend chest area, including the hardwood and softwood flows, as well as the blend chest level. Proof of cause and effect can be established by preventing the control-induced cycling (placing the control loops on manual) and collecting the basis weight data again. Figure 72.12 shows the behavior of basis weight over a period of about 2.5 hours based on scan average data (average of each traverse) collected every 40 seconds. The mean is still 55 gsm, and the variability is now 1.39%. The Nyquist frequency of 0.758 cycles/minute corresponds to a scanning time of 40 seconds. There is significant power at 0.38 cycles/minute (2.6 minutes/cycle), and at 0.057 cycles/minute (17 minutes/cycle). Once again these frequencies correspond to some of the behavior in the blend chest area.

The purpose of this small case study was only to illustrate typical behavior of automatic control loops in a fairly realistic case study. In practice, an operating unit process, such as the example paper machine, has several hundred control loops of which about a hundred would be examined carefully during a variability audit to investigate the causal interrelationships between these individual variables and the final product.

72.3.5 Variability Audit Typical Findings

The case study results are typical of audit results. As of now, our audit experience extends over approximately 200 audits, with 50 to 100 control loops analyzed in each audit. Audit recommendations frequently include 20 to 50 items covering process design, control strategy design, valve replacement, and loop tuning issues. Findings by loop are categorized into several categories as listed in Table 72.2.

TABLE 72.2 Variability Audit Findings by Category.

Loop category: Loops which...	%
reduce variablility (have adequate/good design, equipment, and tuning)	20
cycle and increase variability due to control equipment (valve backlash, stiction, etc.)	30
cycle and increase variability due to poor tuning	30
require control strategy redesign to work properly	10
requires process redesign or changes in operating practice to reduce variability	10
Total	100

These findings say that only 20% of the control loops surveyed actually reduce variability over the "short term" in their "as-found" condition. "Short term" means periods of about 15 minutes for fast variables, such as flow loops, and periods of one or two hours for loops with slow dynamics, such as basis weight. Let us examine the chief reasons for inadequate performance which are control equipment, tuning, and design.

72.3.6 Control Equipment

The control equipment category accounts for about 30% of all loops and represents a major potential for improvement. Control valve nonlinear behavior is the primary problem. However, also included are other problems, such as installation (e.g., sensor location), excessive filtering, or equipment characteristics (e.g., inappropriate controller gain adjustment ranges or control mode features). The blend chest case study discussed several problems. The most serious were the control valve backlash/stiction prob-

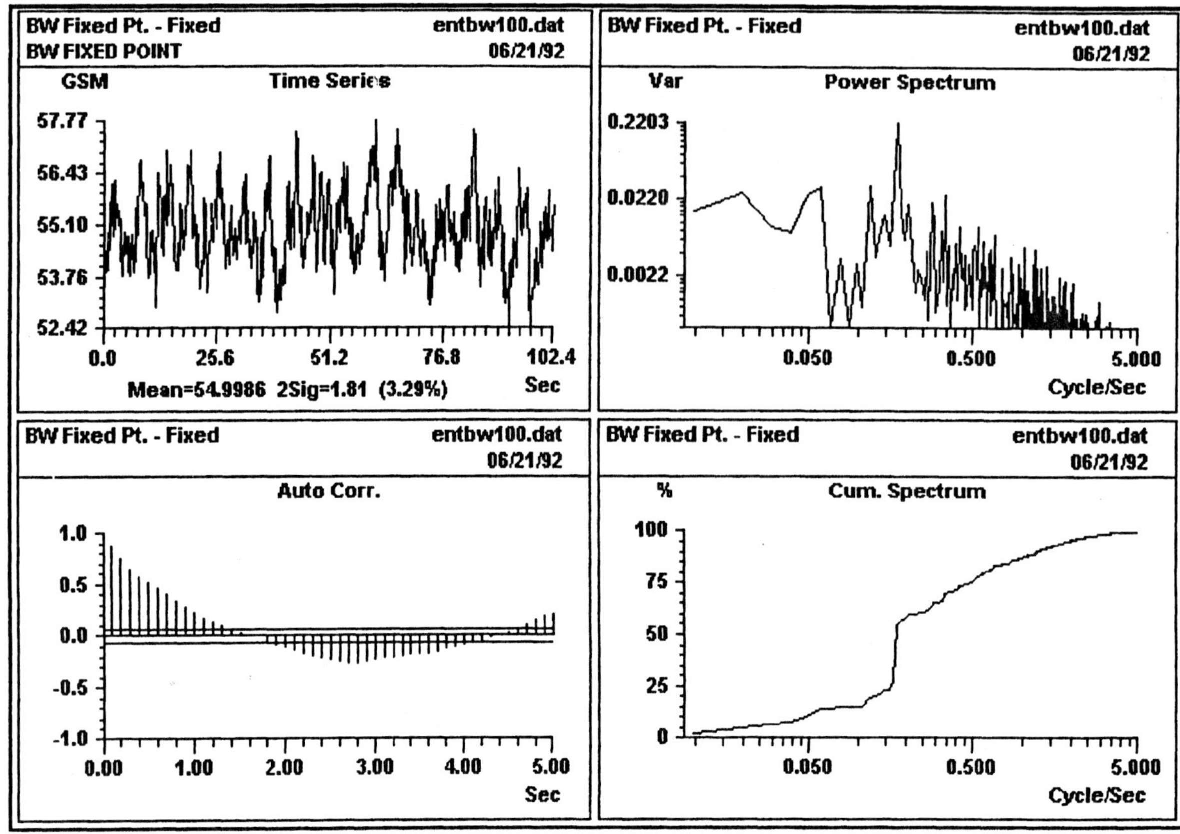

Figure 72.10 Paper final product—Basis weight (0.01 to 5Hz).

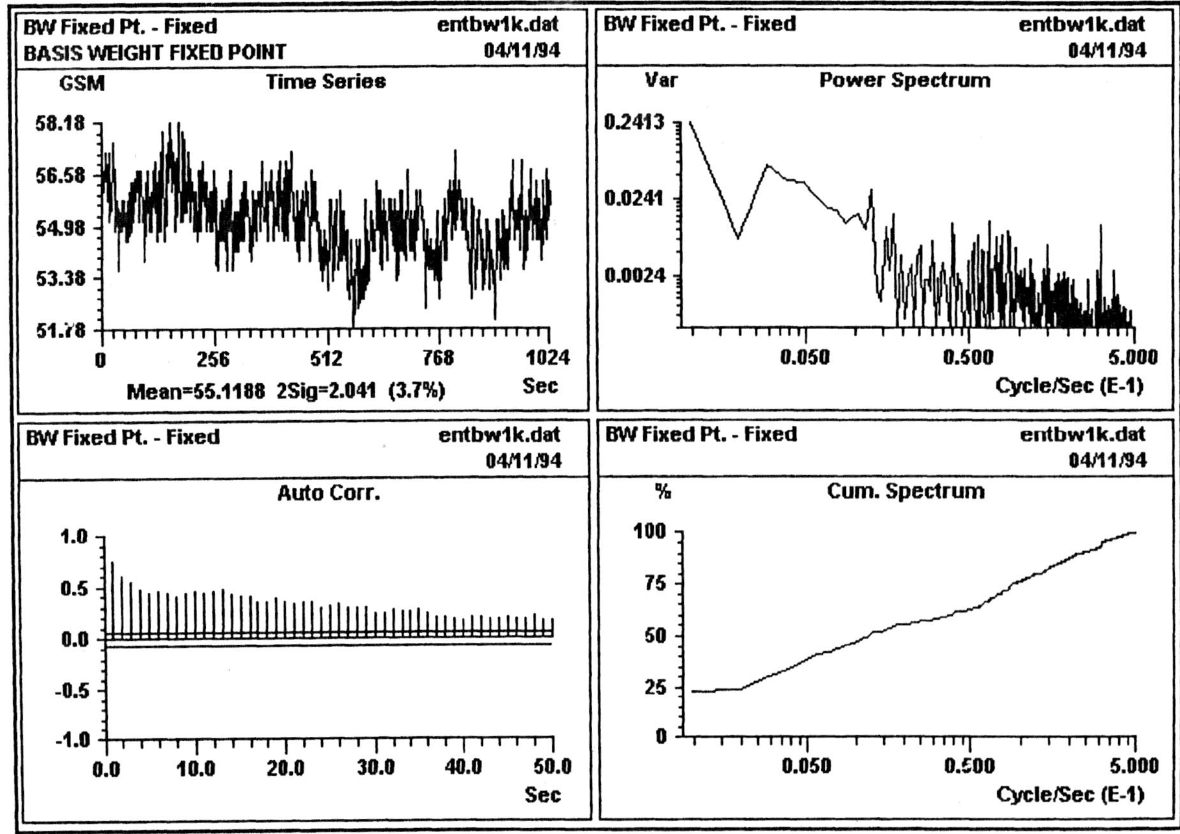

Figure 72.11 Paper final product—Basis weight (0.001 to 0.5 Hz).

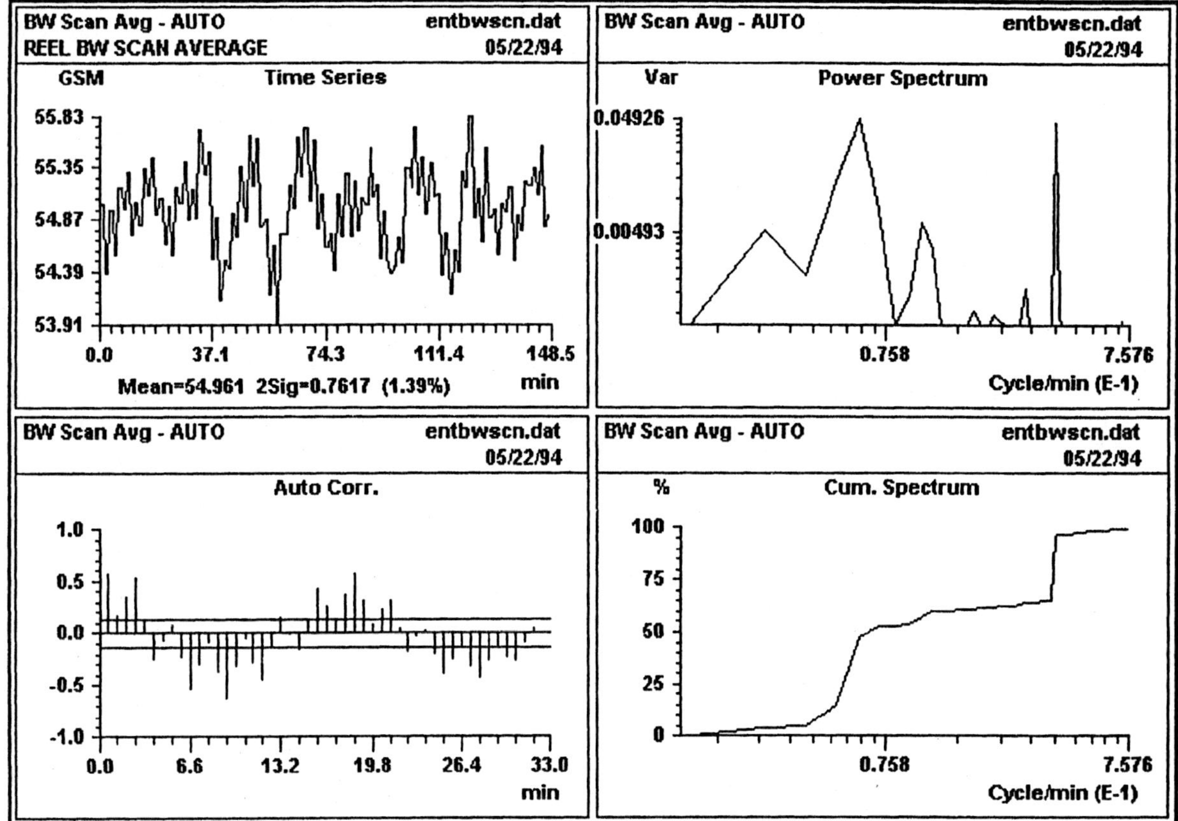

Figure 72.12 Paper final product — Basis weight (0.0001 to 0.0125 Hz).

lems exhibited by FC105 and FC205. To some extent, the blend chest consistency loop NC302 also suffered from control valve induced limit cycling. Lesser problems included the inappropriate consistency sensor location for NC302 (time delay too long) and the excessive filter time constant.

The serious problems in this category are the control valve nonlinearities such as backlash, stiction, and issues relating to control valve positioners (local pneumatic feedback servo) which react to valve stiction with time delays that are inversely proportional to the change in the valve input signal. These nonlinearites cannot be modeled through "small-signal" linearization methods and are extremely destabilizing for feedback control. The only recourse is to eliminate such equipment from the process (best way of linearizing). This is discussed further in the section on dynamic performance specifications.

72.3.7 Loop Tuning

This category accounts for 30% of all loops and represents a major potential for improvement. In the case study of the blend chest consistency, NC302 and the blend chest level, LC301 were in this category. Until recently, loop tuning has been an "art" in the process industries done by trial and error. The responsibility for loop tuning in a pulp and paper mill normally rests with the Instrumentation Shop and is carried out by Instrument Technicians. Until recently, the only training that most instrument technicians have received is the Quarter-Amplitude-Damping method of Ziegler and Nichols [21]. This method is not robust (gain margin of 2.0 and phase margin of 30°) and tends to produce excessive cycling behavior. As a result most people have taught themselves to "tune by feel". Recent training [4] throughout the industry has introduced many engineers and technicians to "Lambda Tuning", discussed in detail in a later section.

72.3.8 Control Strategy Redesign, Process Redesign, and Changes in Operating Practice

Control equipment and loop tuning problems can be corrected at relatively little cost. The remaining opportunity to gain another 20% advantage in variability reduction is spread over a number of areas including control strategy redesign, process redesign, and changes in operating practice. Issues relating to process and control design are discussed in a later section. Changing operating practice involves people and how they think about their operation. Making changes of this kind involves creating awareness and changing "culture".

72.3.9 Pulp and Paper Mill Culture and Awareness of Variability

The reader can be forgiven for the obvious question: Are people aware of the variability that they actually have? The answer is generally "No". Space does not permit a thorough discussion of this subject which determines a mill's ability to move forward (see [9]). In brief, there are many historical reasons why operating data is presented in a particular way. Even though most mills use digital "distributed control systems" (DCS) today, the psychology of the past has carried over to the present. In current DCSs, process data is manipulated in ways which tend to strip the true nature of the process variability from it. Some of these include slow input sampling rates, data aliasing, slow update rates for operator consoles and mill data archives, "report-by-exception" techniques (to minimize digital communication traffic), and display of data on a scale of 0–100% of span [11].

72.4 Control Equipment Dynamic Performance Specifications

Nothing can be done to improve control performance when the control equipment does not perform adequately. The control valve has been identified in audit findings as the biggest single cause of process variability in the pulp and paper industry. Another problem identified in audits is that some automatic controllers have inappropriate design features, such as gain adjustment ranges, which exclude legitimate yet effective settings. To help the industry move forward, dynamic performance specifications have been prepared. These include a control valve dynamic specification [10], and a controller dynamic specification [8]. Currently these specification documents are widely used in the pulp and paper industry, frequently for equipment purchase. As a result, most valve suppliers to the pulp and paper industry are actively using the valve specification as a performance guide to test their products [20]. In a number of cases new or improved control valve products have emerged. It is anticipated that future control valves will have combined backlash and stiction as low as 1% or lower, and time delays in fractions of a second. When these levels cannot be tolerated, mills are considering variable speed drive technology for pumps as an alternative to the control valve.

72.5 Linear Control Concepts — Lambda Tuning — Loop Performance Coordinated with Process Goals

In the pulp and paper industry, the term "Lambda Tuning" refers to a concept in which the designer/tuner specifies the loop performance by choosing the closed-loop time constant (called Lambda (λ)). The calculation of the parameters required by a specific control algorithm is determined by a "Lambda tuning rule," — a transformation of the closed-loop time constant, the process dynamics, and the control algorithm structure into the appropriate gains. Lambda tuning is considered a useful concept for designing and tuning control loops in continuous processes, especially when low product variability is the objective. The concept of specifying a desired performance for each loop is the key to coordinating the overall performance of a unit process with hundreds of control loops.

The idea of specifying a particular closed-loop performance is foreign to most people, who equate high bandwidth with performance, and hence would tune loops to be as fast as practically possible. The resulting closed-loop dynamics are a function only of the open-loop dynamics and the stability margins required by robustness considerations. Yet if every control loop in a paper machine application was independently tuned for maximum bandwidth, it is doubtful that the paper machine would run at all, let alone make saleable paper!

The beginnings of the Lambda tuning concept can be traced back to the work of Newton, Gould, and Kaiser [16], and the analytical design of linear feedback controls in which the concept of stochastic control systems for minimum bandwidth and minimization of mean-square error was put forward. Dahlin [6], applied the analytical design concept to controlling processes with time delay and first-order dynamics and used the notion of a user-selected bandwidth. This was done by specifying the desired closed-loop pole position, λ. Dahlin originated the expression, "Lambda Tuning," — meaning a single parametric dynamic specification for closed-loop performance. Later these ideas, generalized and extended by others [18], specified the control loop performance by choosing a closed-loop time constant, as opposed to a pole location. The evolution of the internal control concept (IMC) further extended these ideas to processes of arbitrary dynamics, while considering the associated problem of loop robustness [15]. More recently these concepts have been applied to the development of "tuning rules" for simple PI and PID controllers [5]. Finally, Lambda tuning has been adopted as the preferred controller tuning approach in the pulp and paper industry [17].

The mathematics of "Lambda tuning" is well established in the literature cited. But almost nothing is said about selecting the closed-loop time constant λ. Only the trade-off between performance and robustness [15] is recognized, and that robustness suffers as the loop performance specification is more aggressive. Yet it is precisely the ability to choose a specific closed-loop time constant for each loop that makes Lambda tuning so useful in the pulp and paper industry. This allows uniquely coordinated tuning of all of the loops which make up a unit process to enhance the manufacturing of uniform product. This idea is explored in the following blend chest design example.

72.5.1 Design Example — Blend Chest Dynamics using Lambda Tuning

For the now familiar blend chest example, let us establish the manufacturing objectives and translate these into control engineering objectives as desired closed-loop time constants for each loop.

Blend Chest Manufacturing Objectives

The manufacturing or papermaking objectives are:

1. provide a uniformly blended source of paper stock, at a desired blend ratio

2. insure that the stocks leaving both the hardwood and softwood chests are at uniform consistencies
3. insure uniform refining of both stocks, so that these fibers have uniformly developed bonding strength
4. insure that the blend chest level is maintained at the level set point, and never spills over or goes below, say 40%, while allowing the paper machine production rate to change by, say 20%

Uniform stock delivery from the hardwood and softwood chests is critically important for two reasons. First, the refiners depend on uniform stock delivery to insure that refining action is uniformly applied. Overrefined fibers provide higher bonding strength but reduce the drainage rate on the wire and cause high sheet moisture. Underrefined fibers cause loss of sheet strength, higher drainage rate, and lower moisture content in the sheet. Overrefined and underrefined fibers are discrete qualities of stock and do not "average-out".

Blend Chest Control Objectives:

1. Maintaining tight control over hardwood (NC104) and softwood (NC204) consistencies is certainly the most important objective, because it will insure constant and uniform fiber delivery. These two loops are then of prime importance. At the high density storage chests, the disturbance energy is normally very high and somewhat unpredictable. Hence, these loops should be tuned for maximum practical bandwidth. However, the consistency loops have time-delay dominant dynamics and, as a result, a high-frequency resonance (discussed later) slightly above the cutoff frequency. This high-frequency resonance typically occurs at about 0.2 radians/second (about 30 seconds per cycle, frequency depends on tuning) and should be minimized so as not to amplify process noise at this frequency and allow this resonance to propagate further down the paper machine line. The choice of the closed-loop time constant (λ), determines the extent of the resonance. For a λ two times the time delay, the resonance will be about +2 dB (**AR** = 1.26), and a λ equal to the time delay will cause a resonance of +3.6 dB (AR = 1.51). Because the noise in the resonant band is amplified by over 30%, this choice is too aggressive. The hardwood and softwood consistency loops with about 5 seconds of time delay, should be tuned for a λ of about 15 seconds.
2. The blend chest has a residence time of 20 minutes when full. At the normal operating point of 70% full, there are 14 minutes of stock residence time and an air space equivalent to 6 minutes. We are told that the paper machine production rate changes will be no more than 20%. A paper machine stoppage will cause the blend chest to overflow in 6 minutes. A 20% reduction in flow would cause the chest to overflow in 30 minutes. The process immediately down stream from the blend chest is the machine chest, which is itself level controlled. As a result the actual changes in blend chest outflow will be subject to the tuning of the machine chest level controller. The purpose of the blend chest is to provide a surge capacity. Fast tuning of the level will tightly couple disturbance in the outflow to the very sensitive upstream refiner and consistency control problems. Hence the level controller LC301, should be tuned as slowly as practically possible (minimum bandwidth), subject to the stipulation that the level must never make excursions greater than 30%. A tank level, with the inlet flows controlled by a Lambda tuned PI controller, has a load response to a step change in outflow in which the maximum excursion occurs in one closed-loop time constant. A good choice for the closed-loop time constant may then be 10 or 15 minutes. Either of these choices will maintain the level well within the required 30%. It is important to insure that the level controller never oscillates, because oscillation is damaging to all of the surrounding loops. Hence the tuning must insure that the closed-loop dominant poles remain on the negative real axis (discussed later).
3. The hardwood (FC105) and softwood (FC205) flows are the inner cascade loops for the level controller LC301, and their set points are adjusted via ratio stations which achieve the 70:30 blend ratio. To maintain a constant blend ratio under dynamic conditions, both flows must be tuned for the same absolute closed-loop time constant. Yet the hardwood and softwood lines probably differ, in pumps, pipe diameters, pipe lengths, valve sizes, valve types, flow meters, hence, open-loop dynamics. Any tuning method in which the closed-loop dynamics are a function only of the open-loop dynamics (e.g., Ziegler–Nichols, 5% overshoot, etc.) will produce different closed-loop dynamics for the hardwood and softwood flows. Only Lambda tuning can achieve the goal of identical closed-loop dynamics. The actual choice of closed-loop time constants for both flows is also critically important. Both of these flows can be tuned to closed-loop time constants under 10 seconds without too much trouble. However, too fast a time constant will disturb the consistency loops and the refiner operation through hydraulic coupling. This will be especially true when accounting for the nonlinear behaviour of the valves. Too slow a time constant will interfere with the operation of the level controller which requires the inner loops to be considerably faster than its closed-loop time constant. Analysis is required to determine the exact value of the flow time constant to insure adequate stability margins for the level control. However, for

a level λ of 15 minutes, it is likely that the flows can be tuned for λs in the two minute range. This choice will insure relatively light coupling between the flow and the consistency loops when tuned for closed-loop time constants of 15 seconds.

4. The blend chest consistency loop NC302 can be tuned in the same exact way as the other consistency loops. Alternatively, if a repeatable noise model structure for this loop can be identified by studying the noise power spectrum on manual, it may be possible to tune NC302 using minimize-variance principles. If the noise model is the fairly common integrating-moving-average (IMA) type (drifts plus noise) and if the corner frequency of the noise structure is slower than the frequency at which the consistency loop has significant resonance [12], λ can be chosen to match the corner frequency of the noise structure, thereby, approximately canceling the low-frequency drifts and producing a "white" power spectrum.

Conclusion — Design of Blend Chest Dynamics with Lambda Tuning

The blend chest example illustrates six reasons for choosing specific closed-loop time constant values: 1) maximum nonresonant bandwidth for the loops considered the most important (hardwood and softwood consistencies), 2) minimum possible bandwidth for the least important loop (level), 3) equal closed-loop time constants for loops controlling parallel processes which must be maintained at a given ratio (hardwood and softwood flows), 4) a closed-loop time constant dictated by the dynamics of an upper cascade loop (flow loops), 5) a closed-loop time constant for loops of lesser importance sufficiently slower than adjacent coupled loops of greater importance (flows versus consistencies), and, finally, 6) choosing the closed-loop time constant to minimize variance through matching the regulator sensitivity function to the inverse of an IMA type noise structure by matching the regulator cut-off frequency to the IMA corner frequency.

72.5.2 Pulp and Paper Process Dynamics

Next, let us consider the types of process dynamics present in the pulp and paper industry. Most process dynamics can be described by one of the following two general transfer functions shown below:

$$G_P(s) = \frac{K_P(\beta s+1)e^{-sT_d}}{s^a(\tau_1 s+1)(\tau_2 s+1)} \quad \text{or}$$
$$G_P(s) = \frac{K_P(\beta s+1)e^{-sT_d}}{s^a(\tau_1^2 s^2+2\varsigma\tau_1 s+1)} \tag{72.2}$$

where $a = 0$ or 1, β = lead time constant(positive or negative), τ_1, τ_2 = time constants ($\tau_1 \geq \tau_2$), ς = damping coeficient, and T_d = deadtime. Typical parameter values for pulp and paper process variables are listed in Table 72.3.

72.5.3 Lambda Tuning

Lambda tuning employs the general principles used in the Internal Model Control (IMC) concept which has the following requirements for a controller:

1. The controller should cancel the process dynamics, process poles with controller zeros, and process zeros with controller poles.
2. The controller should provide at least one free integrator (Type 1 loop) in the loop transfer function to insure that offsets from set point are canceled.
3. The controller must allow the speed of response to be specified (Lambda (λ) to be set by designer/tuner).

For the the process transfer function of Equation 72.2, these general principles call for PID control with a series filter (PID.F). For nonintegrating processes, this translates into the following controller transfer functions:

$$G_C(s) = \frac{(\tau_1 s+1)(\tau_2 s+1)}{K_P(s)(\lambda+T_d)(|\beta|s+1)} \quad \text{or}$$
$$G_C(s) = \frac{(\tau_1^2 s^2+2\varsigma\tau_1 s+1)}{K_P(s)(\lambda+T_d)(|\beta|s+1)} \tag{72.3}$$

where closed-loop time constant = λ, setpoint response bandwidth (inverse sensitivity function) $\cong \frac{1}{\lambda}$, and load response bandwidth (sensitivity function) $\cong \frac{1}{(\lambda+T_d)}$.

For integrating processes, it is normally necessary to specify control loops of Type 2 form, because a controller integrator is usually desirable to overcome offsets due to load disturbances. The controller required for this task is typically of the form,

$$G_C(s) = \frac{[(2\lambda+T_d)s+1](\tau_1 s+1)}{K_P s(\lambda+T_d)^2(|\beta|s+1)}. \tag{72.4}$$

Whereas the form of these controllers can be implemented in PID.F form, in most cases a PI controller will suffice, especially when the performance objectives are reasonably modest relative to the loop robustness limits. This gives the following advantages: the series filters need not be implemented (not a standard feature of current distributed control systems (DCSs), the resulting controller is within the training scope of the instrument technician, and the large actuation "spikes" caused by derivative control are avoided. The form of the PI controller contained in most DCSs is

$$G_C(s) = K_C\frac{(T_R s+1)}{T_R s}. \tag{72.5}$$

Equating the controller gain (K_C) and reset (or integral) time (T_R) to the general process parameter values of Equations 72.3 and 72.4 yields the following tuning rules [5]. Consider the two most important cases of process dynamics, first-order plus deadtime, and integrating plus deadtime, which together represent 85% of all process dynamics in the pulp and paper industry. The tuning rules are listed in Table 72.4 below.

These two tuning rules form the basis for the bulk of the Lambda tuning for single loops in the pulp and paper industry. Whereas the choice of closed-loop time constant λ, provides the

TABLE 72.3 Typical Dynamics of Pulp and Paper Process Variables.

Process Variable	K_P	a	β (sec.)	τ_1 (sec.)	τ_2 (sec.)	ς	T_d (sec.)
Stock flow	1.0#	0	-	3	-	-	0.2
Stock flow with air entrainment	1.0#	0	8	15	3	-	0.2
Stock flow, flexible pipe supports	1.0#	0	-	3	-	0.2	0.2
Stock consistency	−1.0#	0	-	5	-	-	5
Stock pressure	1.0#	0	-	3	-	-	0.1
Headbox total head with "hornbostle"	1.0#	0	-	3	-	0.2	-
Headbox total head with variable speed fan pump	1.0#	0	-	1	-	-	-
Paper machine basis weight	*	0	-	45	-	-	100
Paper machine moisture	*	0	-	120	-	-	120
Pulp dryer basis weight	*	0	-	10	-	-	300
Dryer steam pressure	0.005#	1	300	20	-	-	-
Chest level	0.005#	1	-	-	-	-	-
Chest level with controlled flows	0.005#	1	-	15!	-	-	-
Bleach plant chlorine gas flow	1.0#	0	-	2	-	-	0.1
Bleach plant pulp brightness	1.0#	0	-	20	-	-	20
Bleach plant steam mixer temperature	1.0#	0	-	20	5	-	-
Bleach plant oxygen gas flow	1.0#	0	-	0.05	-	-	-
Bleach plant D1 Stage 'J tube' brightness	1.0#	0	-	120	-	-	120
Digester Chip Bin Level	0.005#	1	-	-	-	-	200
Boiler feed water deaerator level	0.005#	1	-	-	-	-	20
Boiler steam drum level	0.005#	1	−30	30	-	-	-
Boiler master - steam header control with bark boiler	0.005#	1	-	-	-	-	300

* varies
# nominal value, depends on equipment sizing
! tuning dependent

mechanism for coordinating the tuning of many control loops, this must always be done with control loop stability and robustness in mind. Furthermore, variability audit experience indicates that the tendency of control loops to resonate should be avoided at all costs. Hence, closed-loop poles must always be located on the negative real axis, and time-delay-induced resonance must be limited to manageable quantities, never more than, say, +3 dB. Such design objectives should significantly limit propagation of resonances within pulp and paper processes. Let us examine these issues further.

TABLE 72.4 PI Tuning Rules for First-Order and Integrating Dynamics.

	Process Dynamics	K_C	T_R
1)	$G_P(s) = \frac{K_p e^{-sT_d}}{(\tau s+1)}$	$\frac{\tau}{K_P(\lambda+T_d)}$	τ
2)	$G_P(s) = \frac{K_P e^{-sT_d}}{s}$	$\frac{2\lambda+T_d}{K_P(\lambda+T_d)^2}$	$2\lambda + T_d$

72.5.4 Choosing the Closed-loop Time Constant for Robustness and Resonance

First-Order Plus Deadtime

Consider a consistency control loop, as an example of a first-order plus deadtime process, with the following process parameters (time in seconds, ignore negative gain): $K_P = 1$, $\tau = 3$, and $T_d = 5$. By applying the Lambda tuning rule of Table 72.4, the loop transfer function becomes

$$G_L(s) = \frac{(\tau s+1)}{K_P(\lambda+T_d)s}\,\frac{K_p e^{-sT_d}}{(\tau s+1)} = \frac{e^{-sT_d}}{(\lambda+T_d)s}. \qquad (72.6)$$

As long as pole-zero cancellation has been achieved (or nearly so), the resulting dynamics depend only on T_d and the choice of λ. Let us consider what will happen as λ is varied as a function of T_d. Figure 72.13 shows the root locus plot for the cases $\lambda/T_d = 1, 2, 3,$ and 4. The deadtime is approximated by a first-order Pade' approximation (open-loop poles are xs, zeros are os, and closed-loop poles are open squares).

Figure 72.13 shows the time-delay pole/zero pair of the Pade' approximation at ± 0.4, together with the controller integrator pole at the origin. The most aggressive of the tuning choices, $\lambda/T_d = 1, \lambda = T_d = 5$ has a damping coefficient of about 0.7 and hence is just starting to become oscillatory. Figure 72.14 shows the sensitivity functions plotted for all four cases. The resonance for the case of $\lambda = 2T_d = 10$ is fairly acceptable at +2 dB, or an amplitude ratio of 1.26, hence noise amplification by 26%. From the viewpoint of robustness, this tuning produces a gain margin of 4.7 and a phase margin of 71°, a fairly acceptable result considering that changes in loop gain by a factor of two can be expected and changes of 50% in both time constant and deadtime can also be expected. This tuning is thus recommended for general use. On the other hand, a choice of $\lambda = T_d = 5$ is barely acceptable, having a resonance of +3.6 dB, (amplitude ratio of 1.51), a gain margin of only 3.1, and a phase margin of

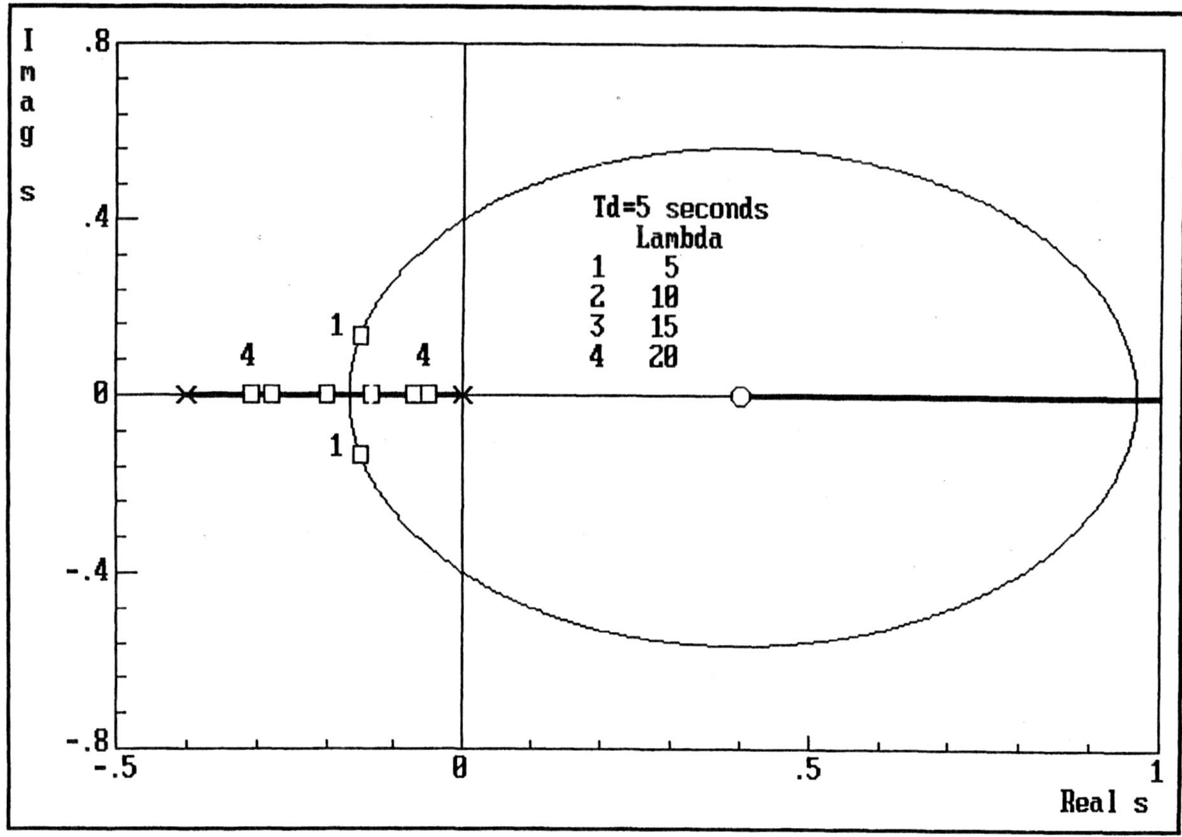

Figure 72.13 Consistency control loop with PI controller tuned for $\lambda/T_d = 1, 2, 3, 4$.

61°. A summary of these results is contained in Table 72.5. These values represent the limiting conditions for the tuning of such loops based on robustness and resonance. The actual choices of closed-loop time constants are likely to be made using process related considerations as outlined in the blend chest example.

Integrator Plus Deadtime

Now consider the integrator plus deadtime case. The loop transfer function is

$$G_L(s) = \frac{[(2\lambda + T_d)s + 1]}{K_P(\lambda + T_d)^2 s} \frac{K_P e^{-sT_d}}{s} = \frac{[(2\lambda + T_d)s + 1]\, e^{-sT_d}}{(\lambda + T_d)^2 s^2} \quad (72.7)$$

Let us use a digester chip bin level example to illustrate this case. Wood chips are transported from the chip pile via a fixed speed conveyor belt to the chip bin. The transport delay is 200 seconds. Let us approximate this as 3 minutes for the sake of this example. The chip bin level controller adjusts the speed of a variable speed feeder which deposits the chips on the belt. The process parameters are (time in minutes):$K_P = 0.005$, and $T_d = 3$. From Equation 72.7 the loop transfer function is a function only of the deadtime and the closed-loop time constant. Let us consider two cases, $\lambda/T_d = 2$ and 3.

Figure 72.15 shows the root locus plot for the case of $\lambda = 2T_d = 6$ minutes. The root locus plot shows the Pade' pole/zero pair approximation for deadtime at $\pm 2/3$, the reset zero at $-1/9$, and two poles at the origin. The closed poles are located on a root locus segment which moves quickly into the right half-plane (RHP), and the poles have a damping coefficient of only 0.69. Even more aggressive tuning speeds up the controller zero and increases the loop gain. This causes the root locus segment to bend towards the RHP earlier and reduce the damping coefficient.

Clearly this and faster tunings are far too aggressive. Figure 72.16 shows a root locus plot for $\lambda = 2T_d = 9$ minutes. The slower controller zero and lower loop gain cause the root locus to encircle the controller zero fully before breaking away again. Figure 72.16 shows an expanded plot of the region near the origin.

Even though there is still a branch containing a pair of poles which goes to the RHP, the system is quite robust to changes in loop gain. Changes in deadtime move the deadtime pole/zero pair closer to the origin as the deadtime lengthens, causing the eventual separation of the root locus branch as in Figure 72.15. From this analysis, it is clear that for reasons of robustness and resonance, the closed-loop time constant should not be made faster than three times the deadtime. Figure 72.17 shows the load response for the chip bin level control tuned for $\lambda = 3T_d = 9$ minutes, for a step change in outflow from the chip bin. The level initially sinks until the controller manages to increase the flow into the chest, to arrest the change for the first time. This occurs at 9.9 minutes in this case. For minimum-phase systems (no deadtime or RHP zeros), this point occurs exactly at λ. Otherwise the closed-loop time constant is a good approximation for

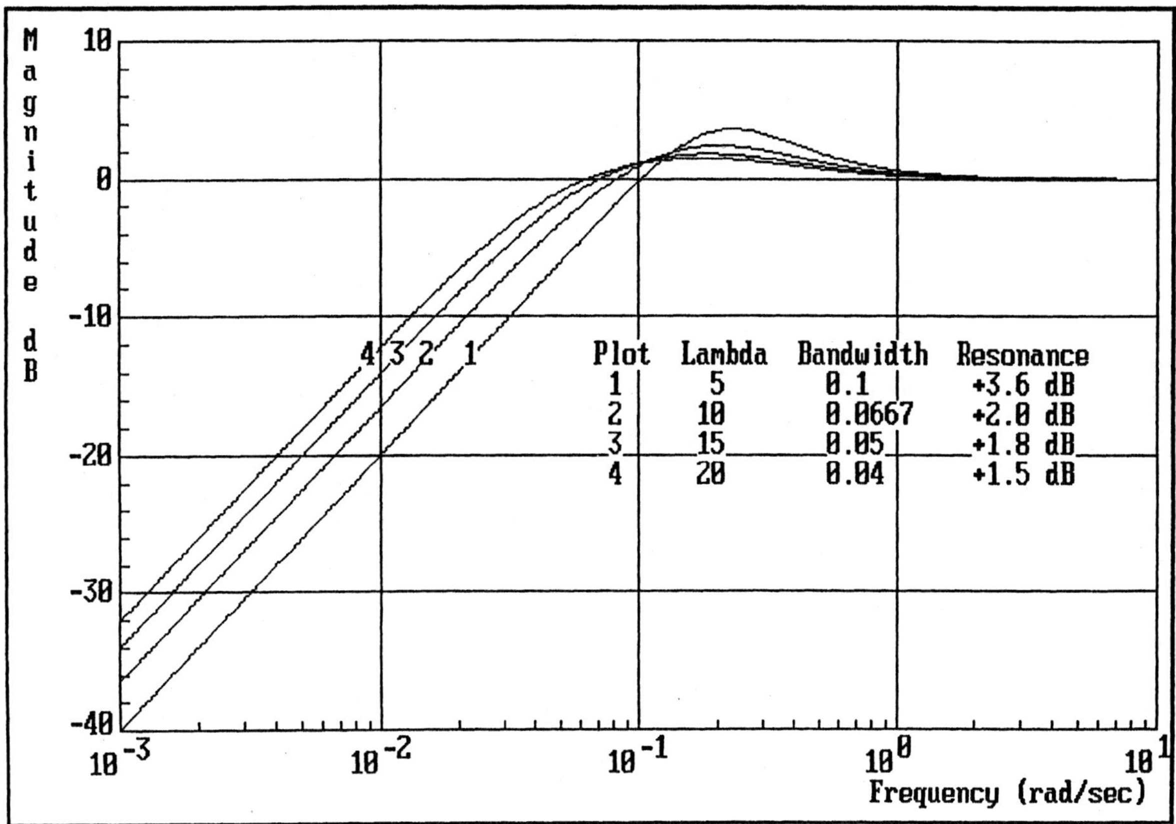

Figure 72.14 Consistency loop sensitivity functions for $\lambda/T_d = 1, 2, 3, 4$.

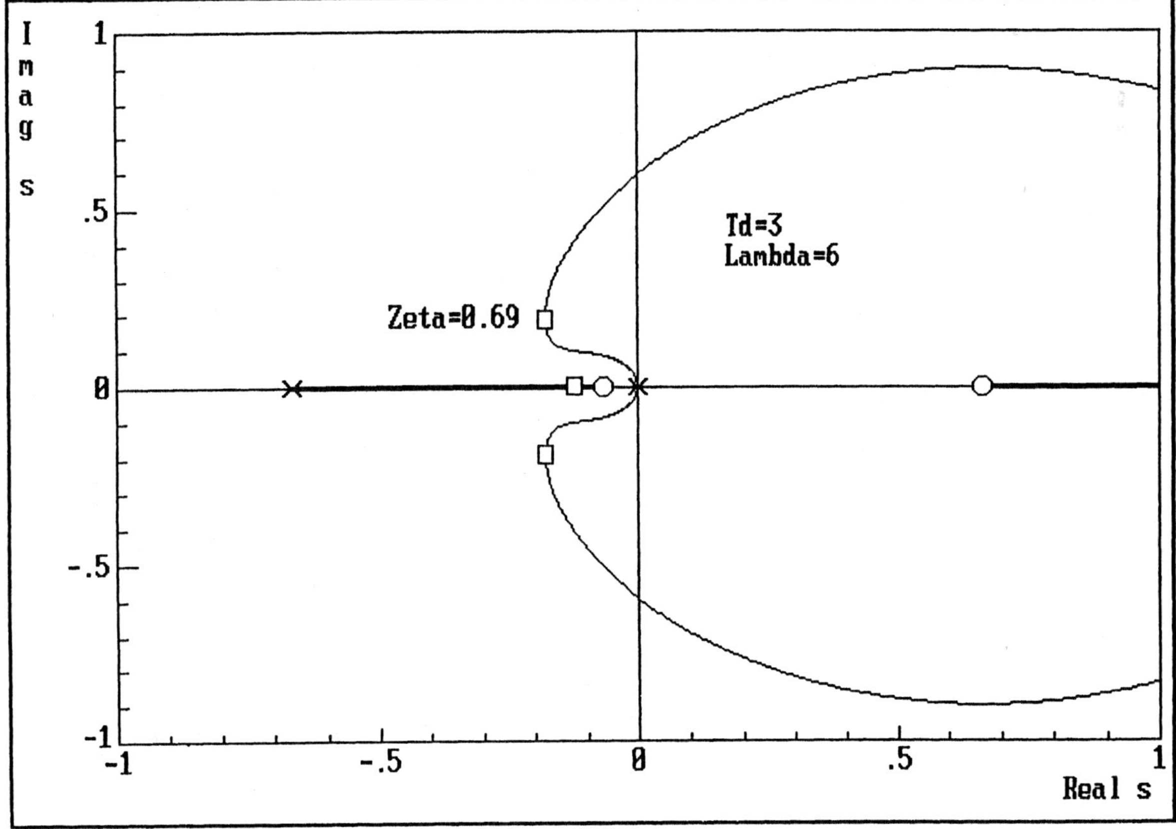

Figure 72.15 Digester chip bin level root locus for $\lambda = 2T_d$.

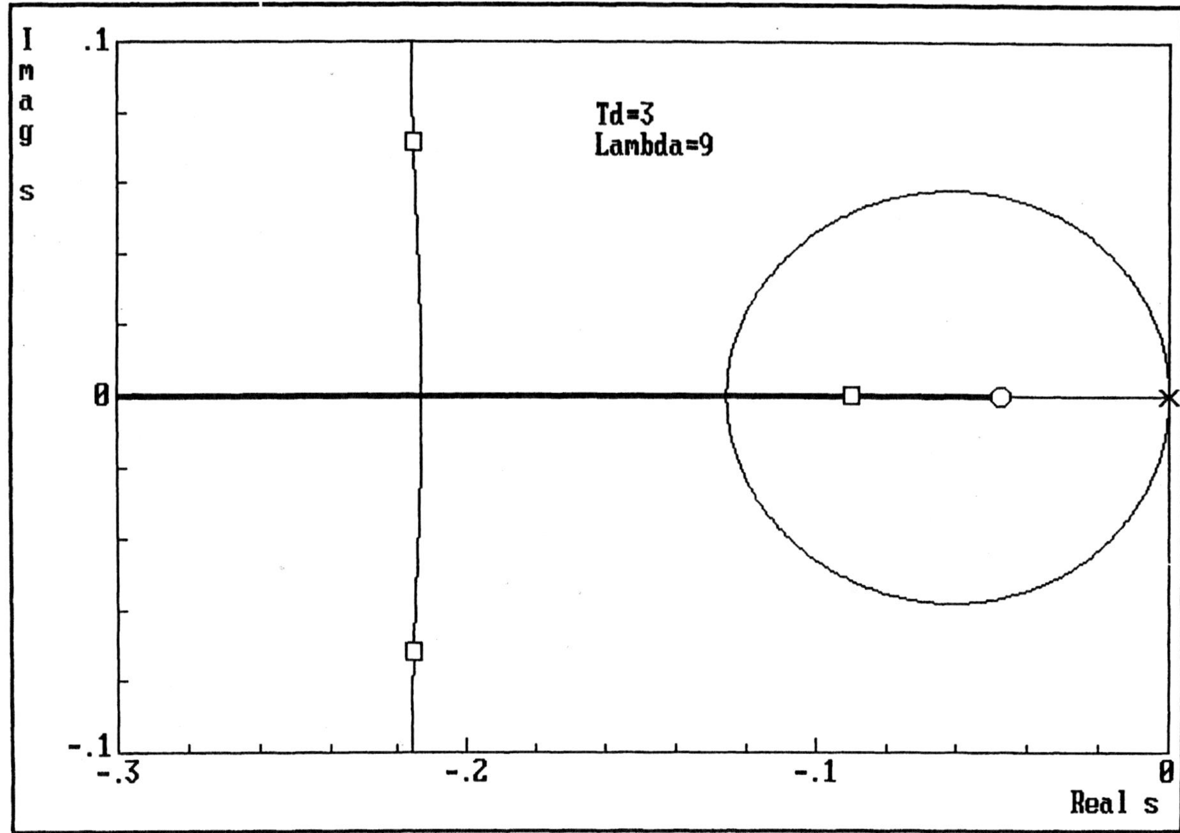

Figure 72.16 Digester chip bin level root locus for $\lambda = 3T_d$ expanded plot.

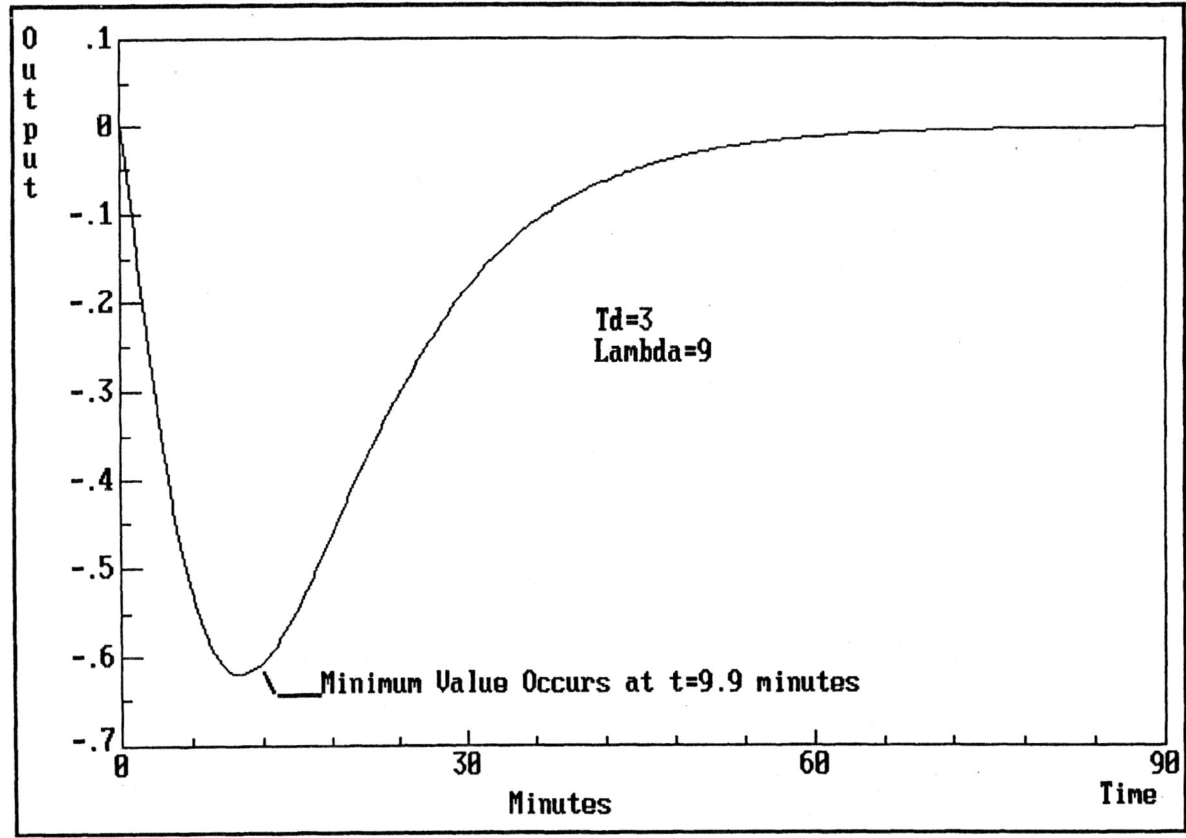

Figure 72.17 Digester chip bin level load response to step change in outlet flow.

TABLE 72.5 First-Order Plus Deadtime Lambda Tuning Performance/Robustness Trade-Off.

Tuning (λ/T_d) ratio	1	2	3	4
Closed-loop time constant λ (sec.)	5	10	15	20
Bandwidth ($1/(\lambda + T_d)$) radians/sec.	0.1	0.0667	0.05	0.04
Resonant Peak (dB)	+3.6	+2.0	+1.8	+1.5
Amplitude Ratio	1.51	1.26	1.23	1.19
Gain Margin	3.1	4.7	6.8	7.9
Phase Margin (°)	61	71	76	79

the time when the load disturbance will be arrested. The time to correct fully for the whole change is approximately six times longer.

72.5.5 Multiple Control Objectives

A control loop can have multiple control objectives, such as

- good set point tracking,
- good rejection of major load disturbances, and
- minimizing process variability.

The tuning for each of these is unlikely to be the same. The first two relate to fairly large excursions, where speed of response is of the essence and process noise does not dominate. Minimizing process variability, on the other hand, is of major importance in the pulp and paper industry, and is essentially a regulation problem around fixed set points with noise dominant. There is a strong need to be able to accommodate the different tuning requirements (different λs) of these cases, all of which apply to the same loop at different times. One way to approach this problem might be to use logic within the control system, which would gain schedule based on the magnitude of the controller error signal. For errors below some threshold, say 3%, the controller would be tuned with λs suitable for rejecting noise over a certain bandwidth. This tuning is likely to be fairly conservative. On the other hand, when the controller error is above some threshold, say 10%, the gains could correspond to an aggressive λ chosen to correct the disturbance as quickly as practically possible. In addition, it is useful to consider the use of set point feedforward control to help differentiate between set point changes and major load changes.

72.5.6 Algorithms

SISO Control Using PI, PID, PID.F, Cascade, and Feedforward

The foregoing discussion has centered on the PI controller and SISO control (98% of the control loops in pulp and paper). The reasons for such wide use of single loop control is partly historical, as this is how the industry has evolved. There is also a good technical justification given the sheer complexity of a process with hundreds of control loops. The operator may arbitrarily place any of these in manual mode for operating reasons. Hence the concept of designing a workable MIMO system is almost unworkable given the complexity (note that the pilot of an aircraft cannot put the left aileron on manual, only the entire autopilot system). Furthermore, the concept of Lambda tuning allows the PI controller to provide an effective solution in most cases. PID control is needed when the controller must be able to cancel two process poles (e.g., the control of an underdamped second-order process, the aggressive control of a second-order overdamped process). There are times when it may be advantageous to include a series filter with PID algorithm (PID.F) (e.g., the control of a second-order process with a transfer function zero). The filter is then used to cancel the zero. There is extensive use in the pulp and paper industry of cascade control (e.g., basis weight and stock flow, paper moisture and steam pressure, bleach plant brightness and bleaching chemical flow, boiler steam drum level and feedwater flow). There is also quite frequent use of feedforward control (e.g., boiler steam drum feedforward from steam demand).

Deadtime Compensation

The pulp and paper industry uses deadtime compensator algorithms extensively (e.g., basis weight control, moisture control, Kappa number control, bleach plant brightness control). The common types are the Smith predictor [19], and the Dahlin algorithm [6]. It is interesting to note that the original work of Smith focused heavily on the set point response of the algorithm, for which the technique is well suited. However, almost all of the deadtime compensators in commission are employed as regulators, operating in steady state with the goal of disturbance rejection and minimizing variability. Although not a new discovery, Haggman and Bialkowski [12] have shown that the sensitivity function of these regulators, as well as deadtime compensators using IMC structure [15], is essentially the same as that of a Lambda tuned PI controller for a given bandwidth. There is no way of escaping the time-delay phase lag induced resonance ("water-bed" effect) as long as simple process output feedback is used. These algorithms offer no advantage for low-frequency attenuation or in providing less resonance when tuned for the same bandwidth. On the other hand, they are more complex and have a greater sensitivity to model mismatch than the PI controller. One exception is the Kalman filter-based deadtime compensator [2] which uses state feedback from an upstream state variable in the deadtime model. Because the feedback control structure is identical to that of a Smith predictor, the time-delay resonance is also present. However, the Kalman filter prevents the excitation of this resonance through the low-pass dynamics of the Kalman filter update cycle which has equal bandwidth to the regulator, thereby attenuating noise in the region of the primary time-delay resonance by some −10 dB.

Adaptive and MIMO Control

With the exception of gain scheduling, adaptive control has not achieved wide use in the pulp and paper industry, because its algorithms are too complex for the current technical capability in most mills. Whereas there are occasional reports of quite advanced work being done [such as [1], on MIMO adaptive control of Kamyr digester chip level using general predictive control (GPC)], in most cases the results of such advanced work are very difficult to maintain for long in an operational state in a mill environment. As for the much simpler commercially available self-tuning controllers, most of these perform some type of identification (e.g., the relay method) and then implement a simple adaptation scheme which often takes the form of automating a Ziegler–Nichols-like tuning. This usually results in a lightly damped closed-loop system which is quite unsuitable for the control objectives previously discussed (e.g., to prevent propagation of resonances). Finally, there is fairly wide use of simple decoupling schemes which depend mainly on static decouplers (e.g., basis weight and moisture, multiply headbox control of fan pumps).

72.6 Control Strategies for Uniform Manufacturing

The relationship between algorithms, control strategy, and variability was put into a concise perspective by Downs and Doss, [7], and some of their thoughts have been put into a pulp and paper context here. The control algorithm, however simple or complex, does not eliminate variability, but it causes the variability to be shifted to another place in the process. For instance, consistency control shifts variability from the stock consistency to the dilution valve and the dilution header, from the fiber stream to the dilution stream. The control algorithm and its tuning determines the efficiency with which this is done. The control strategy determines the pathway that the variability will follow. The control strategy can be defined as

1. defining control system objectives
2. selecting sensor types and their location
3. selecting actuator types and their location
4. deciding on input/output pairing for single loops
5. designing multivariable control where needed
6. designing process changes that make control more effective

The control strategy design hence determines where the variability will attempt to go. Let us revisit the paper machine blend chest example. Both the hardwood and softwood steams have consistency control, as these stocks leave their respective stock chests. Both of these consistency controls modulate dilution valves which take dilution water from the same header. As a result, they interact (see Figure 72.4). Hence variability will move from the hardwood consistency to the dilution header, to the softwood consistency, to the blend chest consistency, and back again. When these stock flows enter the blend chest, they will undergo mixing, which will act as a low-pass filter. Hence, the consistency loops have attenuated the low-frequency content, while causing interaction between each other at high frequency. Then, the high-frequency content will be attenuated by the mixing action of the chest. However, on leaving the blend chest, the blend chest consistency controller also draws dilution water from the same header. Clearly, this design will compromise almost everything that has been gained at the high-frequency end of the spectrum.

This example illustrates also that the control engineer, acting alone in the domain of the control algorithm, cannot achieve effective control over process variability. What is needed is an integration of process and control design.

72.6.1 Minimizing Variability — Integrated Process and Control Design

To design a process which produces low variability product, a true integration of the process and control design disciplines is required. The old way of designing the process in steady state, and adding the controls later, has produced the pulp and paper mills of today, which, as variability audits have already shown, are variable far in excess of potential. Control algorithm design follows a very general methodology and is largely based on linear dynamics. When thinking about control loop performance, the engineer pays no attention to the actual behavior of the process. For instance, the most important phenomena in the pulp and paper industry concern pulp slurries and the transport of fiber in two- or three-phase flow. The physics that govern these phenomena involve the principles of Bernoulli and Reynolds and are very nonlinear. The linear transfer function is a necessary abstraction to allow the control engineer to perform linear control design, the only analysis that can be done well. Yet in the final analysis, control only moves variability from one stream to another, where hopefully it will be less harmful. Yet the process design is not fixed. What about the strategy of creating new streams?

Integrated process and control design must take a broader view of control strategy design. Control is only a high-pass attenuation mechanism for variability. Process mixing and agitation provide low-pass attenuation of variability via mixing internal process streams. Yet both of these techniques only attenuate by so many dB. But the customer demands that the absolute variability of the product be within some specified limit to meet the manufacturing needs of his process. Surely, eliminating sources of variability is the best method to insure that no variability will be present which will need attenuation. These issues are in the domain of process design and control strategy design. Control strategy design does not lend itself to elegant and general analysis. Each process is different and must be understood in its specific detail. Nonlinear dynamic simulation offers a powerful tool to allow detailed analysis of performance trade-offs and is the only available method for investigating the impact of different design decisions on variability. Such decisions must question current process design practice.

From the blend chest example, it is clear that a number of process design issues compromise variability, and that integrated process and control design could lead to the following design alternatives:

1. Eliminate all stock flow control valves, and use variable frequency pump drives. This will eliminate all of the nonlinear problems of backlash and stiction.
2. Redesign the existing blend chest with two compartments, each separately agitated. This will convert the existing agitation from a first-order low pass filter to a second-order filter and provide a high-frequency attenuation asymptote at −40 dB/decade instead of −20 dB/decade.
3. Provide a separate dilution header for NC302, and use a variable frequency pump drive instead of the control valve. This will eliminate the high-frequency noise content of the existing header from disturbing NC302, and will also provide nearly linear control of the dilution water.
4. Replace the dilution header control valve of PC309 by a variable frequency pump drive. This will allow much faster tuning of this loop, substantially reducing the interaction between NC104 and NC204.

Each of these design alternatives should be evaluated using dynamic simulation before significant funds are committed. The alterations proposed above vary in capital cost from $10,000 to $1,000,000. Hence the simulation must have high fidelity representing the phenomena of importance. In addition, network analysis techniques may determine how variability spectra propagate through a process and control strategy. Changes in process and control strategy design alter these pathways by creating new streams. The plant can be viewed as a network of connected nodes (process variables) with transmission paths (e.g., control loops, low pass process dynamics, etc.) which allow variability spectra to propagate. These ideas will need time to develop.

72.7 Conclusions

This chapter has attempted to provide a general overview of control engineering in the pulp and paper industry in the mid 1990s. There is a brief introduction to wood, pulp, paper products, and the unit processes. The results of variability audits were then presented to show how much potential exists in this industry to improve product uniformity, especially when there is an increasingly strong demand for uniform product. The concept of specifying the closed-loop performance of each control loop to match the process needs is presented as the Lambda tuning concept. The use of Lambda tuning is illustrated in a paper mill example, and the performance and robustness of tuned control loops are explored. There is a general review of various algorithms and their use in the industry. Finally, the concept of integrating control strategy and design of both the process and control is presented. This is seen as a new avenue of thought which promises to provide a design methodology for the pulp and paper mills of the future, which will be far more capable of efficiently manufacturing highly uniform product than today's mills.

72.8 Defining Terms

AR: Amplitude ratio.

Backlash: Hysteresis or lost motion in an actuator.

Basis weight: The paper property of mass per unit area (gsm, lbs/3000 sq. ft., etc.)

Bump test: Step test.

β: Process transfer function zero time constant.

C: ISA symbol for control.

Cellulose: Long chain polymer of glucose, the basic building block of wood fiber.

Chest: Tank.

Consistency: Mass percentage of solids or fiber content of a pulp or stock slurry.

CPPA: Canadian Pulp and Paper Association, 1155 Metcalfe St. Montreal, Canada, H3B 4T6.

DCS: Distributed control system.

Deadtime: Time delay.

F: ISA symbol for flow.

$G_C(s)$: Controller transfer function in the continuous (Laplace) domain.

$G_P(s)$: Process transfer function in the continuous (Laplace) domain.

gsm: Grams per square meter.

HD Chest: High Density chest, a large production capacity stock tank with consistency typically in the 10 to 15% range and with a dilution zone in the bottom.

Hemicellulose: Polymers of sugars other than glucose, a constituent of wood fiber.

IMA: Integrating moving average noise structure.

ISA: Instrument Society of America.

ISA tags: ISA tagging convention (e.g., FIC177 means *F*low *I*ndicating *C*ontroller No. 177).

K_C: Controller gain.

K_P: Process gain.

L: ISA symbol for level.

Lambda tuning: Tuning which requires the user to specify the desired closed-loop time constant, Lambda.

Lambda (λ): The desired closed-loop time constant, usually in seconds.

Lignin: Organic compound which binds the wood fiber structure together.

Limit cycle: A cycle induced in a control loop by nonlinear elements.

N: ISA symbol for consistency.

P: ISA symbol for pressure.

PI: Proportional-Integral controller.

PID: Proportional-Integral-Derivative controller.

PID.F: Proportional-Integral-Derivative controller with series filter.

Positioner: Control valve accessory which acts as a local pneumatic feedback servo.

Pulping: The process of removing individual fibers from solid wood.

Refiner: A machine with rotating plates used in pulping which disintegrates the wood chips into individual fibers through mechanical action, and in papermaking, to "fibrillate" the fibers to enhance bonding strength.

RHP: Right half-plane of the s-plane—the unstable region.

Standard PI form: $G_C(s) = K_C\left[1 + \frac{1}{T_R s}\right]$.

Stiction: Static friction in an actuator.

Stock: Pulp slurry.

$\mathbf{T_d}$: Deadtime.

$\mathbf{T_R}$: Controller reset or integral time/repeat.

τ_1, τ_2: process time constants ($\tau_1 \geq \tau_2$).

TAPPI: Tech. Asn. P & P Ind., P. O. Box 105133, Atlanta, GA, USA, 30348-5113.

References

[1] Allison, B. J., Dumont G. A., and Novak L. H., *Multi-Input Adaptive Control of Kamyr Digester Chip Level: Industrial Results and Practical Considerations*, CPPA Proc., Control Systems '90, Helsinki, Finland, 1990.

[2] Bialkowski, W. L., Application of Kalman Filters to the Regulation of Dead Time Processes, *IEEE Trans. Automat. Control*, AC-28, 3, 1983.

[3] Bialkowski, W. L., *Dreams Versus Reality: A View From Both Sides of the Gap*, Keynote Address, Control Systems '92, Whistler, British Columbia, 1992, published, *Pulp Paper Canada*, 94, 11, 1993.

[4] Bialkowski, W. L., Haggman, B. C., and Millette, S. K., Pulp and Paper Process Control Training Since 1984, *Pulp Paper Canada*, 95, 4, 1994.

[5] Chien, I-L. and Fruehauf, P. S., *Consider IMC Tuning to Improve Controller Performance, Hydrocarbon Proc.*, 1990.

[6] Dahlin, E. B., Designing and Tuning Digital Controllers, *Instrum. Control Syst.*, 41(6), 77, 1968.

[7] Downs, J. J. and Doss, J. E., *Present Status and Future Needs—a view from North American Industry*, Fourth International Conf. Chem. Proc. Control, Padre Island, Texas, 1991, 17–22.

[8] EnTech^{TM} — Automatic Controller Dynamic Specification (Version 1.0, 11/93) (EnTech Literature).

[9] EnTech^{TM} — Competency in Process Control-Industry Guidelines (Version 1.0, 3/94) (EnTech Literature).

[10] EnTech^{TM} — Control Valve Dynamic Specification (Version 2.1, 3/94) (EnTech Literature).

[11] EnTech^{TM} — Digital Measurement Dynamics - Industry Guidelines (Version 1.0, 8/94) (EnTech Literature).

[12] Haggman, B. C. and Bialkowski, W. L., Performance of Common Feedback Regulators for First-Order and Deadtime Dynamics. *Pulp Paper Canada*, 95, 4, 1994.

[13] Kaminaga, H., *One in a thousand*, Proc. CPPA Control Systems '94, Stockholm, Sweden, 1994.

[14] Kocurek, M. J., Series Ed., *Pulp and Paper Manufacture*, Joint (TAPPI, CPPA) Textbook Committee of the Pulp and Paper Industry, 1983 to 1993, Vol. 1 to 10.

[15] Morari, M. and Zafiriou, E., *Robust Process Control*, Prentice Hall, 1989.

[16] Newton, G. C., Gould, L. A., and Kaiser, J. F., *Analytical Design of Linear Feedback Controls*, John Wiley & Sons, 1957.

[17] Sell, N., Editor, Bialkowski, W. L., and Thomason, F. Y., contributors, *Process Control Fundamentals for the Pulp & Paper Industry*, TAPPI Textbook, to be published by TAPPI Press, 1995.

[18] Smith, C. A. and Corripio, A. B., *Principles and Practice of Automatic Process Control*, John Wiley & Sons, 1885.

[19] Smith, O. J. M., Closer Control of Loops with Dead Time, *Chem. Eng. Prog.*, 53(5), 217–219, 1957.

[20] Taylor, G., *The Role of Control Valves in Process Performance*, Proc. CPPA, Tech. Section, Canadian Pulp and Paper Assoc., Montreal, 1994.

[21] Ziegler, J. G. and Nichols, N. B., Optimum settings for automatic controllers, *Trans. ASME*, 759–768, 1942.

73

Control for Advanced Semiconductor Device Manufacturing: A Case History

T. Kailath, C. Schaper, Y. Cho, P. Gyugyi, S. Norman, P. Park, S. Boyd, G. Franklin, and K. Saraswat
Department of Electrical Engineering, Stanford University, Stanford, CA

M. Moslehi and C. Davis
Semiconductor Process and Design Center, Texas Instruments, Dallas, TX

73.1 Introduction

Capital[1] costs for new integrated circuit (IC) fabrication lines are growing even more rapidly than had been expected even quite recently. Figure 73.1 was prepared in 1992, but a new Mitsubishi factory in Shoji, Japan, is reported to have cost \$3 billion. Few companies can afford investments on this scale (and those that can perhaps prefer it that way). Moreover these factories are inflexible. New equipment and new standards, which account for roughly 3/4 of the total cost, are needed each time the device feature size is reduced, which has been happening about every 3 years. It takes about six years to bring a new technology on line. The very high development costs, the high operational costs (e.g., equipment down time is extremely expensive so maintenance is done on a regular schedule, whether it is needed or not), and the intense price competition compel a focus on high-volume low cost commodity lines, especially memories. Low volume, high product mix ASIC (application-specific integrated circuit) production does not fit well within the current manufacturing scenario.

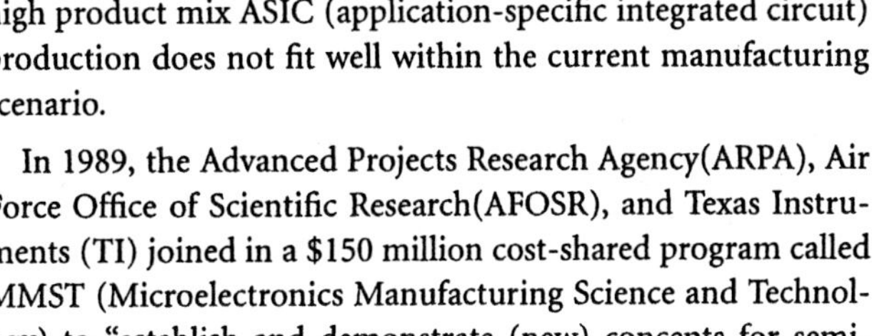

In 1989, the Advanced Projects Research Agency(ARPA), Air Force Office of Scientific Research(AFOSR), and Texas Instruments (TI) joined in a \$150 million cost-shared program called MMST (Microelectronics Manufacturing Science and Technology) to "establish and demonstrate (new) concepts for semiconductor device manufacture which will permit flexible, cost-effective manufacturing of application-specific logic integrated circuits in relatively low volume ... during the mid 1990s and beyond".

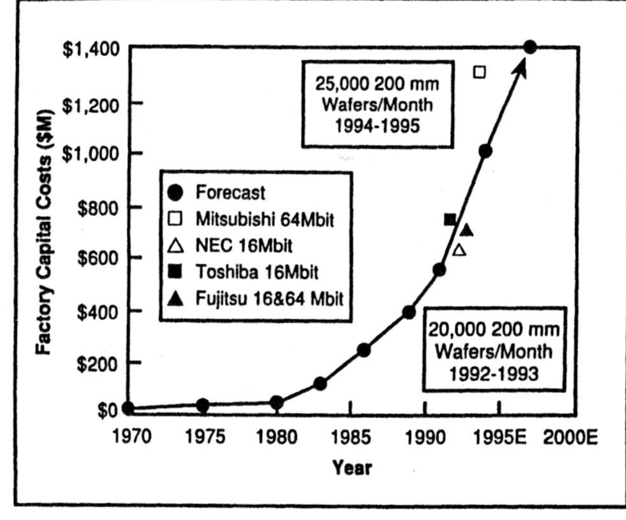

Figure 73.1 Capital cost for a new IC factory. (Source: *Texas Instruments Technical Journal*, 9(5), 8, 1992.)

The approach taken by MMST was to seek fast cycle time by performing all single-wafer processing using highly instrumented flexible equipment with advanced process controls. The goal of the equipment design and operation was to quickly adapt the equipment trajectories to a wide variety of processing specifications and to quickly reduce the effects of manufacturing disturbances associated with small lot sizes (e.g., 1, 5 or 24 wafers)

[1]This research was supported by the Advanced Research Projects Agency of the Department of Defense, under Contract F49620-93-1-0085 monitored by the Air Force Office of Scientific Research.

0-8493-8570-9/96/\$0.00+\$.50

without the need for pilot wafers. Many other novel features were associated with MMST including a factory wide CIM (computer integrated manufacturing) computer system. The immediate target was a 1000-wafer demonstration (including demonstration of "bullet wafers" with three-day cycle times) of an all single-wafer factory by May 1993.

In order to achieve the MMST objectives, a flexible manufacturing tool was needed for the thermal processing steps associated with IC manufacturing. For a typical CMOS process flow, more than 15 different thermal processing steps are used, including chemical vapor deposition (CVD), annealing, and oxidation. The MMST program decided to investigate the use of Rapid Thermal Processing (RTP) tools to achieve these objectives.

TI awarded Professor K. Saraswat of Stanford's Center for Integrated Systems (CIS) a subcontract to study various aspects of RTP. About a year later, a group of us at Stanford's Information Systems Laboratory got involved in this project. Manufacturing was much in the news at that time. Professor L. Auslander, newly arrived at ARPA's Material Science Office, soon came to feel that the ideas and techniques of control, optimization, and signal processing needed to be more widely used in materials manufacturing and processing. He suggested that we explore these possibilities, and after some investigation, we decided to work with CIS on the problems of RTP.

RTP had been in the air for more than a decade, but for various reasons, its study was still in a research laboratory phase. Though there were several small companies making equipment for RTP, the technology still suffered from various limitations. One of these was an inability to achieve adequate temperature uniformity across the wafer during the rapid heating (e.g., 20°C to 1100°C in 20 seconds), hold (e.g., at 1100°C for 1–5 minutes), and rapid cooling phases.

This chapter is a case history of how we successfully tackled this problem, using the particular "systems-way-of-thinking" very familiar to control engineers, but seemingly not known or used in semiconductor manufacturing. In a little over two years, we started with simple idealized mathematical models and ended with deployment of a control system during the May, 1993, MMST demonstration. The system was applied to eight different RTP machines conducting thirteen different thermal operations, over a temperature range of 450°C to 1100°C and pressures ranging from 10^{-3} to 1 atmosphere.

Our first step was to analyze the performance of available commercial equipment. Generally, a bank of linear lamps was used to heat the wafer (see Figure 73.2).

The conventional wisdom was that a uniform energy flux to the wafer was needed to achieve uniform wafer temperature distribution. However, experimentally it had been seen that this still resulted in substantial temperature nonuniformities, which led to crystal slip and misprocessing. To improve performance, various heuristic strategies were used by the equipment manufacturers, e.g., modification of the reactor through the addition of guard rings near the wafer edge to reflect more energy to the edge, modification of the lamp design by using multiple lamps with a fixed power ratio, and various types of reflector geometries. However, these modifications turned out to be satisfactory only for a narrow range of conditions.

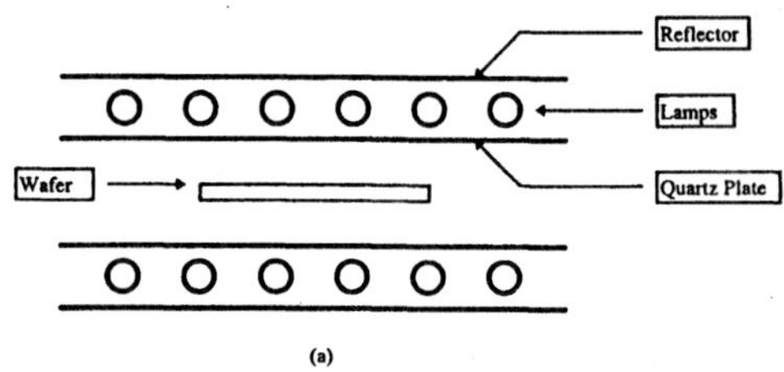

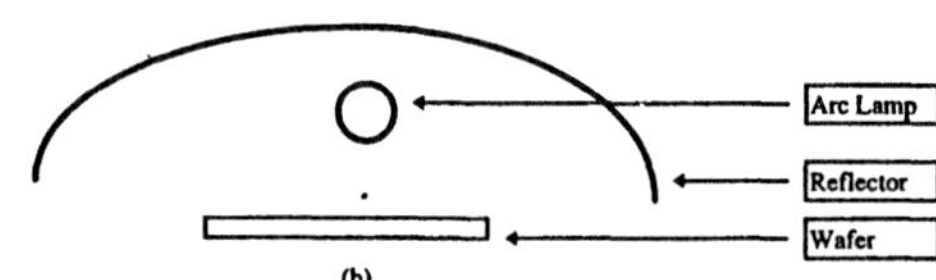

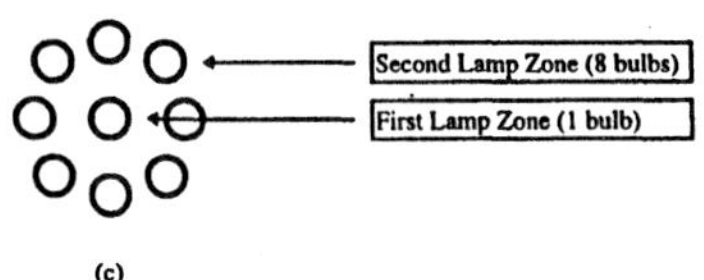

Figure 73.2 RTP lamp configurations: (a) bank of linear lamps, (b) single arc lamp, (c) two-zone lamp array.

The systems methodology suggests methods attempting to determine the performance limitations of RTP systems. To do this, we proceeded to develop a simple mathematical model, based on energy transfer relations that had been described in the literature. Computer simulations with this model indicated that conventional approaches trying to achieve uniform flux across the wafer would never work; there was always going to be a large temperature roll-off at the wafer edge (Figure 73.3). To improve performance, we decided to study the case where circularly symmetric rings of lamps were used to heat the wafer. With this configuration, two cases were considered: (1) a single power supply in a fixed power ratio, a strategy being used in the field and (2) independently controllable multiple power supplies (one for each ring of lamps). Both steady-state and dynamic studies indicated that it was necessary to use the (second) multivariable configuration to achieve wafer temperature uniformity within specifications. These modeling and analysis results are described in Sections 73.2 and 73.3, respectively.

The simulation results were presented to Texas Instruments, which had developed prototype RTP equipment for the MMST program with two concentric lamp zones, but operated in a scalar control mode using a fixed ratio between the two lamp zones. At our request, Texas Instruments modified the two zone lamp by adding a third zone and providing separate power supplies for each zone, allowing for multivariable control. The process engineers in the Center for Integrated Systems (CIS) at Stanford then evaluated the potential of multivariable control by their traditional so called "hand-tuning" methodology, which con-

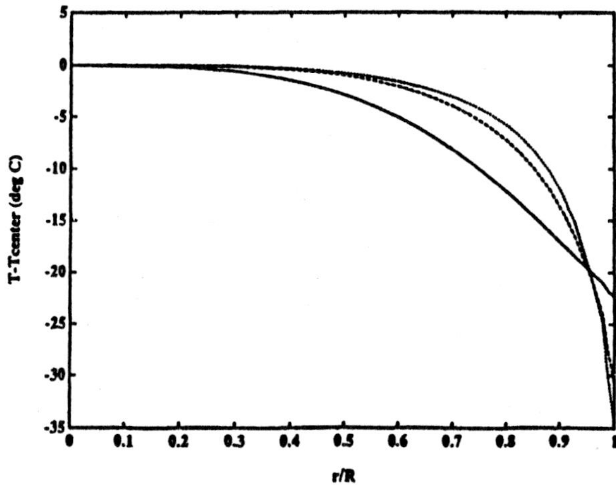

Figure 73.3 Nonuniformity in temperature induced by uniform energy flux impinging on the wafer top surface (center temperatures - solid line: 600°C; dashed line: 1000°C; dotted line: 1150°C.). R is the radius of the wafer, r is the radial distance from the center of the wafer.

sists of having experienced operators determining the settings of the three lamp powers by manual iterative adjustment based on the results of test wafers. Good results were achieved (see Figure 73.4), but it took 7–8 hours and a large number of wafers before the procedure converged. Of course, it had to be repeated the next day because of unavoidable changes in the ambient conditions or operating conditions. Clearly, an "automatic" control strategy was required.

However, the physics-based equations used to simulate the RTP were much too detailed and contained far too many uncertain parameters for control design. The two main characteristics of the simulation model were (1) the relationship between the heating zones and the wafer temperature distribution and (2) the nonlinearities (T^4) of radiant heat transfer. Two approaches were used to obtain a reduced-order model. The first used the physical relations as a basis in deriving a lower-order approximate form. The resulting model captured the important aspects of the interactions and the nonlinearities, but had a simpler structure and fewer unknown parameters. The second approach viewed the RTP system as a black box. A novel model identification procedure was developed and applied to obtain a state-space model of the RTP system. In addition to identifying the dynamics of the process, these models were also studied to assess potential difficulties in performance and control design. For example, the models demonstrated that the system gain and time constants changed by a factor of 10 over the temperature range of interest. Also, the models were used to improve the condition number of the equipment via a change in reflector design. The development of control models is described in Section 73.4.

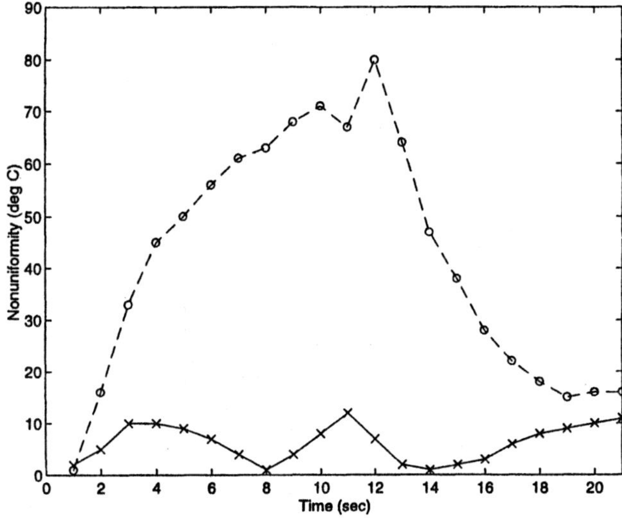

Figure 73.4 Temperature nonuniformity when the powers to the lamp were manually adjusted ("hand-tuning"). These nonuniformities correspond to a ramp and hold from nearly room temperature to 600°C at roughly 40°C/s. The upper curve (-o-) corresponds to scalar control (fixed power ratio to lamps). The lower curve (x-x) corresponds to multivariable control.

Using these models, a variety of control strategies was evaluated. The fundamental strategy was to use feedforward in combination with feedback control. Feedforward control was used to get close to the desired trajectory and feedback control was used to compensate for inevitable tracking errors. A feedback controller based on the Internal Model Control (IMC) design procedures was developed using the low-order physics-based model. An LQG feedback controller was developed using the black-box model. Gain scheduling was used to compensate for the nonlinearities. Optimization procedures were used to design the feedforward controller. Controller design is described in Section 73.5.

Our next step was to test the controller experimentally on the Stanford RTP system. After using step response and PRBS (Pseudo Random Binary Sequence) data to identify models of the process, the controllers were used to ramp up the wafer temperature from 20°C to 900°C at approximately 45°C/s, followed by a hold for 5 minutes at 900°C. For these experiments, the wafer temperature distribution was sensed by three thermocouples bonded to the wafer. The temperature nonuniformity present during the ramp was less than ±5°C from 400°C to the processing temperature and better than ±0.5°C on average during the hold. These proof-of-concept experiments are described in Section 73.6.

These results were presented to Texas Instruments, who were preparing their RTP systems for a 1000 wafer demonstration of the MMST concept. After upper level management review, it was decided that the Stanford temperature control system would be integrated within their RTP equipment. The technology transfer involved installing and testing the controller on eight different RTP machines conducting thirteen different thermal operations used in two full-flow 0.35 μm CMOS process technologies (see Figure 73.5 taken from an article appearing in a semiconductor manufacturing trade journal). More discussion concerning the

WAFER PROCESSING NEWS

Pete Singer, Senior Editor

Stanford Completes RTP Tech Transfer to TI

A real-time multivariable temperature control technique for rapid thermal processing (RTP) has been successfully transferred from Stanford University to TI for use in the Microelectronics Manufacturing Science and Technology (MMST) program. Many believe that such improved control techniques are crucial to the long-term success of RTP. RTP is a key part of the MMST program, the goal of which is to demonstrate the feasibility of 100% single wafer processing.

The transfer of Stanford's control technology, which includes hardware and software originally developed for prototype equipment at Stanford, to TI and subsequent customization was a complex process, involving:

- implementation on seven RTP machines: three machines with a four-zone TI lamp and four machines with a six-zone (G-squared) lamp,
- configurability to 0.1, 2, 3 or 4 advanced pyrometric temperature sensors,
- flexibility for arbitrary ramp and hold trajectories, including specialized trajectories for autocalibration of temperature sensors,
- usage on 13 different processes from 450°C to 1100°C (see Table),
- flexibility for process calibration (as opposed to temperature calibration),
- incorporation of software interlocks for RTP over-temperature protection,
- implementation of signal processing, strategies for noisy temperature sensors, and
- integration into the CIM MMST operational software environment.

The basic idea behind the new control concept is to manipulate the power to the lamp array to control wafer temperature, thereby achieving improved process uniformity and repeatability. This is done largely through a number of software modules for feedback, feedforward, anti-overshoot, gain scheduling and others. *PS* □

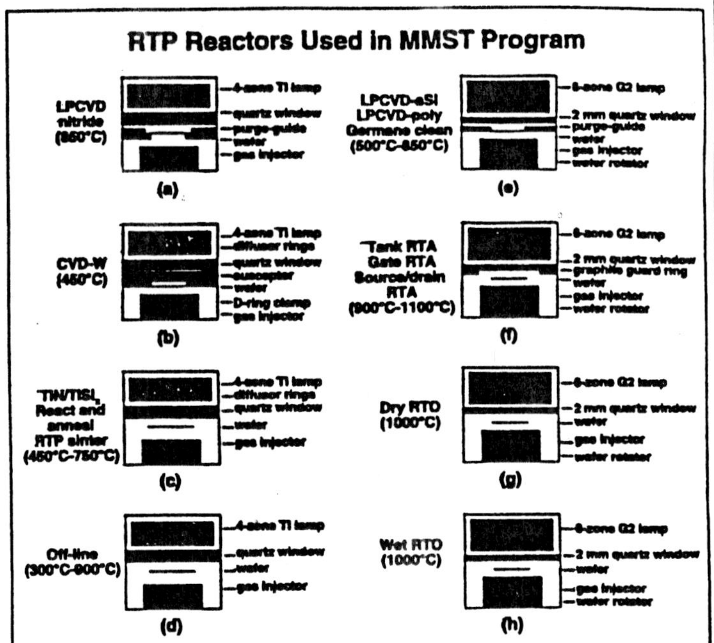

Figure 73.5 Description of technology transfer in *Semiconductor International*, 16(7), 58, 1993.

technology transfer and results of the MMST demonstration is given in Section 73.7. Finally, some overview remarks are offered in Section 73.8.

73.2 Modeling and Simulation

Three alternative lamp configurations for rapidly heating a semiconductor wafer are shown in Figure 73.2. In Figure 73.2(a), linear lamps are arranged above and below the wafer. A single arc lamp is shown in Figure 73.2(b). Concentric rings of single bulbs are presented in Figure 73.2(c). These designs can be modified with guard rings around the wafer edge, specially designed reflectors, and diffusers placed on the quartz window. These additions allowed fine-tuning of the energy flux profile to the wafer to improve temperature uniformity.

To analyze the performance of these and related equipment designs, a simulator of the heat transfer effects was developed starting from physical relations for RTP available in the literature [1], [2]. The model was derived from a set of PDE's describing the radiative, conductive and convective energy transport effects. The basic expression is

$$\frac{1}{r}\frac{\partial}{\partial r}\left(kr\frac{\partial T}{\partial r}\right)+\frac{1}{r^2}\frac{\partial}{\partial \theta}\left(k\frac{\partial T}{\partial \theta}\right)+\frac{\partial}{\partial z}\left(k\frac{\partial T}{\partial z}\right)=\rho C_p\frac{\partial T}{\partial t} \tag{73.1}$$

where T is temperature, k is thermal conductivity, ρ is density, and C_p is specific heat. Both k and C_p are temperature dependent. The boundary conditions are given by

$$\begin{aligned} k\frac{\partial T}{\partial r} &= q_{edge}(\theta, z),\ r = R, \\ k\frac{\partial T}{\partial z} &= q_{bottom}(r, \theta),\ z = 0,\ \text{and} \\ k\frac{\partial T}{\partial z} &= q_{top}(r, \theta),\ z = Z, \end{aligned}$$

where q_{edge}, q_{bottom}, and q_{top} are heat flow per unit area into the wafer edge, bottom, and top, respectively, via radiative and convective heat transfer mechanisms, Z is the thickness of the wafer, and R is the radius of the wafer. These terms coupled the effects of the lamp heating zones to the wafer.

Approximations were made to the general energy balance assuming axisymmetry and neglecting axial temperature gradients. The heating effects in RTP were developed by discretizing the wafer into concentric annular elements. Within each annular wafer element, the temperature was assumed uniform [2]. The resulting model was given by a set of nonlinear vector differential equations:

$$\begin{aligned} \mathbf{C}\dot{T} &= \mathbf{K}^{rad}T^4 + \mathbf{K}^{cond}T + \mathbf{K}^{conv}(T - T_{gas}) \\ &\quad + \mathbf{F}P + q^{wall} + q^{dist} \end{aligned} \tag{73.2}$$

where

$$\begin{aligned} T &= [T_1 \; T_2 \; \ldots \; T_N]^T \\ T^4 &= \left[T_1^4 \; T_2^4 \; \ldots \; T_N^4\right]^T \\ P &= [P_1 \; P_2 \; \ldots \; P_M]^T \end{aligned}$$

where N denotes the number of wafer elements and M denotes the number of radiant heating zones; $\mathbf{K}^{rad}$ is a full matrix describing the radiation emission characteristics of the wafer, $\mathbf{K}^{cond}$ is a tridiagonal matrix describing the conductive heat transfer effects across the wafer, $\mathbf{K}^{conv}$ is a diagonal matrix describing the convective heat transfer effects from the wafer to the surrounding gas, $\mathbf{F}$ is a full matrix quantifying the fraction of energy leaving each lamp zone that radiates onto the wafer surface, q^{dist} is a vector of disturbances, q^{wall} is a vector of energy flux leaving the chamber walls and radiating onto the wafer surface, and $\mathbf{C}$ is a diagonal matrix relating the heat flux to temperature transients. More details can be found in [2] and [3].

73.3 Performance Analysis

We first used the model to analyze the case of uniform energy flux impinging on the wafer surface. In Figure 73.3, the temperature profile induced by a uniform input energy flux is shown for the cases where the center portion of the wafer was specified to be at either 600°C, 1000°C, or 1150°C. A roll-off in temperature is seen in the plots for all cases because the edge of the wafer required a different amount of energy flux than the interior due to differences in surface area. Conduction effects within the wafer helped to smooth the temperature profile. These results qualitatively agreed with those reported in the literature where, for example, sliplines at the wafer edge were seen because of the large temperature gradients induced by the uniform energy flux conditions.

We then analyzed the multiple concentric lamp zone arrangement of Figure 73.2(c) to assess the capability of achieving uniform temperature distribution during steady-state and transients. We considered each of four lamp zones to be manipulated independently. The optimal lamp powers were determined to minimize the peak temperature difference across the wafer at a steady-state condition,

$$\max_{0 \le r \le R} \left| T^{ss}(r, P) - T^{set} \right| \tag{73.3}$$

where T^{set} is the desired wafer temperature and $T^{ss}(r, P)$ is the steady-state temperature at radius r with the constant lamp power vector P, subject to the constraint that each entry P_j of P satisfies $0 \le P_j \le P_j^{max}$. Using the finite difference model, the objective function of Equation 73.3 was approximated as

$$\max_i \left| T_i^{ss}(P) - T^{set} \right| = \left\| T^{ss}(P) - T^{set} \right\|_\infty \tag{73.4}$$

where $T_i^{ss}(P)$ is the steady-state temperature of element i with constant lamp power vector P and T^{set} is a vector with all entries equal to T^{set}. A two-step numerical optimization procedure was then employed in which two minimax error problems were solved to determine the set of lamp powers that minimize Equation 73.3 [4] and [2]. In Figure 73.6, the temperature deviation about the set points of 650°C, 1000°C, and 1150°C is shown. The deviation is less than ±1°C, much better than for the case of uniform energy flux.

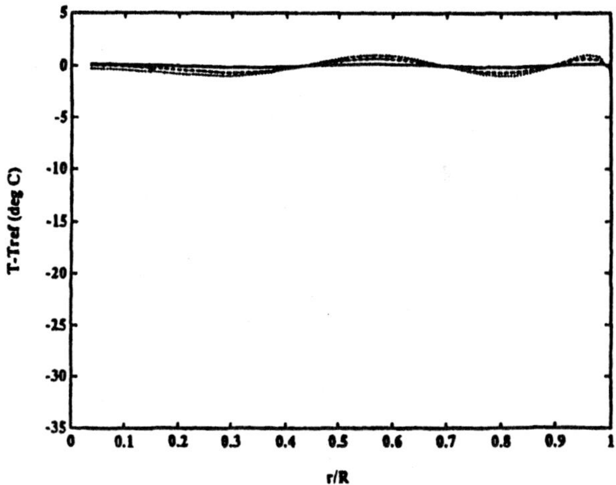

Figure 73.6 Optimal temperature profiles using a multizone RTP system (center temperatures - solid line: 600°C; dashed line: 1000°C; dotted line: 1150°C).

In addition, an analysis of the transient performance was conducted because a significant fraction of the processing and the potential for crystal slip occurs during the ramps made to wafer temperature. We compared a multivariable lamp control strategy and a scalar lamp control strategy. Industry, at that time, employed a scalar control strategy. For the scalar case, the lamps were held in a fixed ratio of power while total power was allowed to vary. We selected the optimization criterion of minimizing

$$\max_{t_o \le t \le t_f} \left\| T(t) - T^{ref}(t) \right\|_\infty \tag{73.5}$$

which denotes the largest temperature error from the specified trajectory $T^{ref}(t)$ at any point on the wafer at any time between an initial time T_o and a final time t_f. The reference temperature trajectory was selected as a ramp from 600°C to 1150°C in 5 seconds. The optimization was carried out with respect to the power to the four lamp zones, in the case of the multilamp configuration, or to the total power for a fixed ratio that was optimal only at a 1000°C steady-state condition. The temperature at the center of the wafer matched the desired temperature trajectory almost exactly for both the multivariable and scalar control cases. However, the peak temperature difference across the wafer was much less for the multivariable case compared to the scalar (fixed-ratio) case as shown in Figure 73.7.

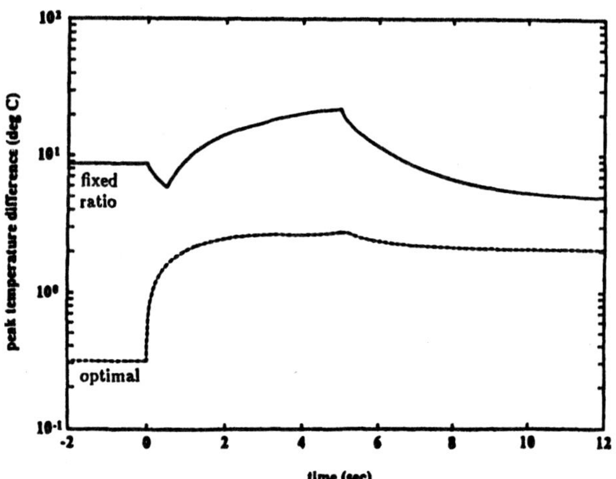

Figure 73.7 Peak temperature nonuniformity during ramp.

For the case of the fixed-ratio lamps, the peak temperature difference was more than 20°C during the transient and the multivariable case resulted in a temperature deviation of about 2°C. The simulator suggested that this nonuniformity in temperature for the fixed-ratio case would result in crystal slip as shown in Figure 73.8 which shows the normalized maximum resolved stress

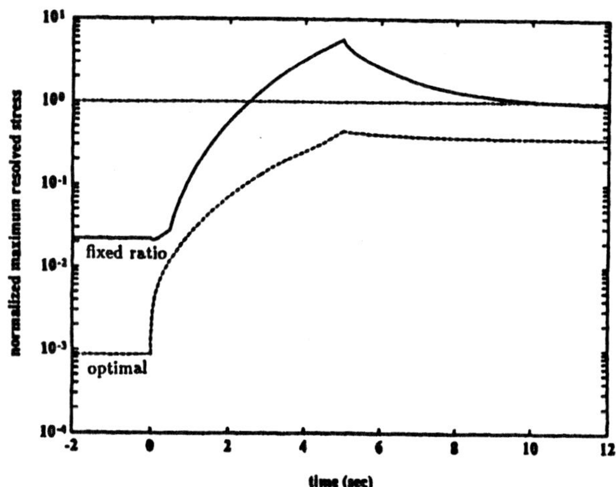

Figure 73.8 Normalized maximum resolved stress during ramp.

(based on simulation) as a function of time. No slip was present in the multivariable case. This analysis of the transient performance concluded that RTP systems configured with multiple independently controllable lamps can substantially outperform any existing scalar RTP system: for the same temperature schedule, much smaller stress and temperature variation across the wafer was achieved; and for the same specifications for stress and temperature variation across the wafer, much faster rise times can be achieved [2].

At the time of these simulations, prototype RTP equipment was being developed at Texas Instruments for implementation in the MMST program. TI had developed an RTP system with two concentric lamp zones. Their system at that time was operated in a scalar control mode with a fixed ratio between the two lamp zones. Upon presenting the above results, the two zone lamp was modified by adding a third zone and providing separate power supplies for each zone. This configuration allowed multivariable control. A resulting three-zone RTP lamp was then donated by TI to Stanford University. The chronology of this technology transfer is shown in Figure 73.9.

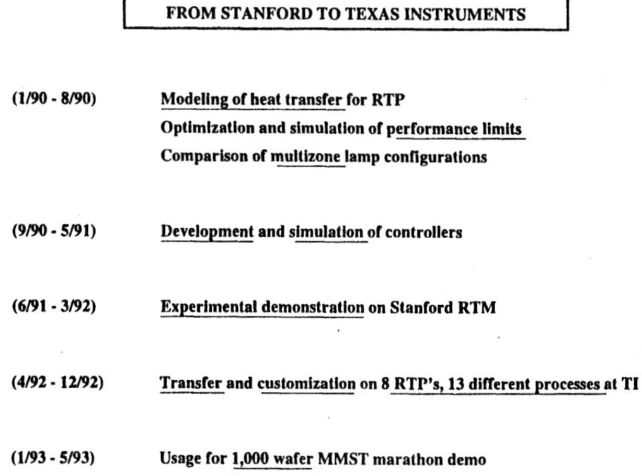

Figure 73.9 Chronology of the technology transfer to Texas Instruments.

A schematic of the Stanford RTP system and a picture of the three-zone arrangement are shown in Figures 73.10 and 73.11, respectively.

"Hand-tuning" procedures were used to evaluate the performance of the RTP equipment at Stanford quickly. In this approach, the power settings to the lamp were manually manipulated in real-time to achieve a desirable temperature response. In Figure 73.4, open-loop, hand-tuned results are shown for scalar control (i.e., fixed power ratio) and multivariable control as well as the error during the transient. Clearly, this comparison demonstrated that multivariable control was preferred to the scalar control method [5]. However, the hand-tuning approach was a trial and error procedure that was time-consuming and resulted in sub-optimal performance. An automated real-time control strategy is described in the following sections.

73.4 Models for Control

Two approaches were evaluated to develop a model for control design. In the first approach, the nonlinear physical model presented earlier was used to produce a reduced-order version. An energy balance equation on the ith annular element can be ex-

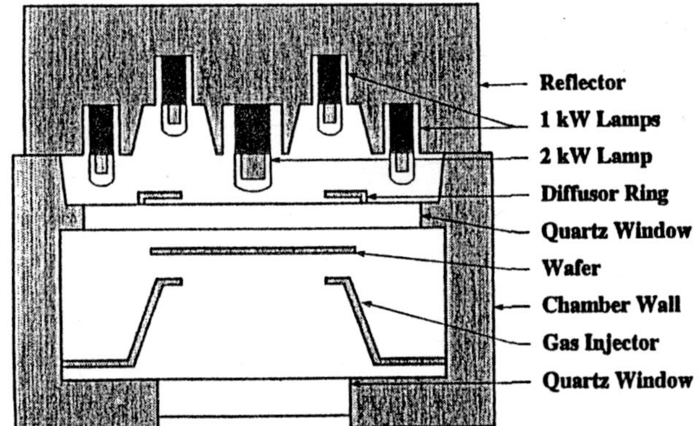

Figure 73.10 Schematic of the rapid thermal processor.

Figure 73.11 Picture of the Stanford three-zone RTM lamp.

pressed as [3] and [6]

$$\rho V_i C_p \frac{dT_i}{dt} = -\epsilon\sigma A_i \sum_{j=1}^{N} D_{i,j} T_j^4 - h_i A_i (T_i - T_{gas}) + q_i^{cond} + q_i^{wall} + q_i^{dist} + \epsilon \sum_{j=1}^{M} F_{i,j} P_j \tag{73.6}$$

where ρ is density, V_i is the volume of the annular element, C_p is heat capacity, T_i is temperature, ϵ is total emissivity, σ is the Stefan-Boltzmann constant, A_i is the surface area of the annular element, $D_{i,j}$ is a lumped parameter denoting the energy transfer due to reflections and emission, h_i is a convective heat transfer coefficient, q_i^{cond} is heat transfer due to conduction, $F_{i,j}$ is a view factor that represents the fraction of energy received by the ith annular element from the jth lamp zone, and P_j is the power from the jth lamp zone.

To develop a simpler model, the temperature distribution of the wafer was considered nearly uniform and much greater than that of the water-cooled chamber walls. With these approximations, q_i^{cond} and q_i^{wall} were negligible. In addition, the term accounting for radiative energy transport due to reflections can be simplified by analyzing the expansion,

$$\sum_{j=1}^{N} D_{i,j} T_j^4 = \sum_{j=1}^{N} D_{i,j} \left(T_i^4 + 4T_i^3\delta_{i,j} + 6T_i^2\delta_{i,j}^2 + 4T_i\delta_{i,j}^3 + \delta_{i,j}^4\right), \tag{73.7}$$

where $\delta_{i,j} = T_j - T_i$. After eliminating the terms involving $\delta_{i,j}$ (since $T_i >> \delta_{i,j}$), the resulting model was,

$$\rho V C_p \frac{dT_i}{dt} = -\epsilon\sigma A_i T_i^4 \sum_{j=1}^{N} D_{i,j} - h_i A_i (T_i - T_{ambient}) + \epsilon \sum_{j=1}^{M} F_{i,j} P_j. \tag{73.8}$$

It was noted that Equation 73.8 was interactive because each lamp zone affects the temperature of each annular element and noninteractive because the annular elements did not affect one another.

The nonlinear model given by Equation 73.8 was then linearized about an operating point $(\bar{T}_i, \bar{P}_i)$,

$$\rho V C_p \frac{d\tilde{T}_i}{dt} = -\left[4\epsilon\sigma A_i \bar{T}_i^3 \sum_{j=1}^{N} D_{i,j} + h_i A_i\right] \tilde{T}_i + \epsilon \sum_{j=1}^{M} F_{i,j} \tilde{P}_j, \tag{73.9}$$

where the deviation variables are defined as $\tilde{T}_i = T_i - \bar{T}_i$ and $\tilde{P}_i = P_i - \bar{P}_i$. This equation can be expressed more conveniently as

$$\tau_i \frac{d\tilde{T}_i}{dt} = -\tilde{T}_i + \sum_{j=1}^{M} K_{i,j} \tilde{P}_j, \tag{73.10}$$

where the gain and time-constant are given by

$$K_{i,j} = \frac{\epsilon F_{i,j}}{4\epsilon\sigma A_i \bar{T}_i^3 \sum_{j=1}^{N} D_{i,j} + h_i A_i} \tag{73.11}$$

$$\tau_i = \frac{\rho V C_p}{4\epsilon\sigma A_i \bar{T}_i^3 \sum_{j=1}^{N} D_{i,j} + h_i A_i}. \tag{73.12}$$

From Equation 73.11, the gain decreases as $\bar{T}$ was increased. Larger changes in the lamp power were required at higher $\bar{T}$ to achieve an equivalent rise in temperature. In addition, from Equation 73.12, the time constant decreases as $\bar{T}$ is increased. Thus, the wafer temperature responded faster to changes in the lamp power at higher $\bar{T}$. The nonlinearities due to temperature were substantial, as the time constant and gain vary by a factor of 10 over the temperature range associated with RTP.

The identification scheme to estimate τ_i and **K** from experimental data is described in [7], [8]. A sequence of lamp power values was sent to the RTP system. This sequence was known as a recipe. The recipe was formulated so that reasonable spatial temperature uniformity was maintained at all instants in order to satisfy the approximation used in the development of the low-order model. The eigenvalues of the system were estimated at various temperature using a procedure employing the TLS ESPRIT algorithm [9]. After the eigenvalues were estimated, the amplitude of the step response was estimated. This was difficult because of the temperature drift induced by the window heating; however, a least-squares technique can be employed. The gain of the system and view factors were then identified using a least-squares algorithm again. The results are shown in Figure 73.12 and 73.13 for the estimation of the effects of temperature on the gain and time constant, respectively.

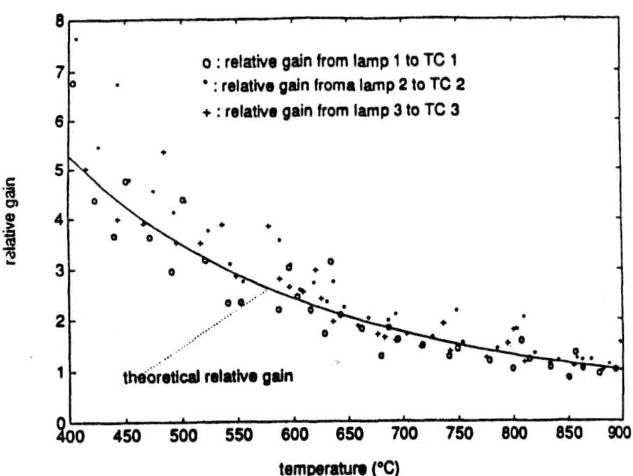

Figure 73.12 Gain of the system relative to that at 900°C and comparison with theory.

The model was expressed in discrete-time format for use in designing a control system. Using the zero-order hold to describe the input sequence, the discrete-time expression of Equation 73.10 was given by

$$\tilde{T}(z) = \Gamma(z)\mathbf{K}\tilde{P}(z) \tag{73.13}$$

where z denotes the z-transform,

$$\Gamma(z) = diag\left[\frac{(1-e^{-\Delta t/\tau_i})z^{-1}}{1-e^{-\Delta t/\tau_i}z^{-1}}\right] \tag{73.14}$$

and Δt denotes the sampling time. The system model was inherently stable since the poles lie within the unit circle for all operating temperatures.

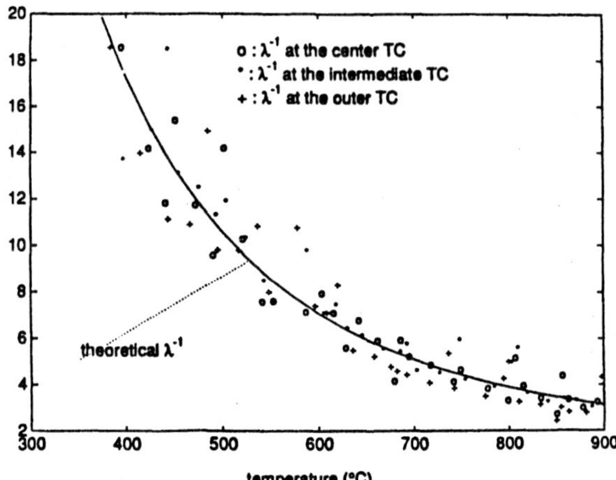

Figure 73.13 Time constant of the process model as a function of temperature and comparison with theory.

In order to obtain a complete description of the system, this relationship was combined with models describing sensor dynamics and lamp dynamics [4]. The sensor and lamp dynamics can be described by detailed models. However, for the purpose of model-based control system design, it was only necessary to approximate these dynamics as a simple time-delay relation,

$$\tilde{T}_{m,i}(t) = \tilde{T}_i(t-\theta). \tag{73.15}$$

The measured temperature at time t was denoted by $T_{m,i}(t)$, and the time delay was denoted by θ. The resulting model expressed in z-transform notation was given by

$$\tilde{T}_m(z) = z^{-d}\Gamma(z)\mathbf{K}\tilde{P}(z) \tag{73.16}$$

where $d = \theta/\Delta t$ was rounded to the nearest integer.

The power was supplied to the lamp filaments by sending a 0–10 volt signal from the computer control system to the power supplies that drive the lamps. The relation between the voltage signal applied to the supplies and the power sent to the lamps was not necessarily linear. Consequently, it was important to model that nonlinearity, if possible, so that it could be accounted for by the control system. It was possible to determine the nonlinearity with a transducer installed on the power supply to measure the average power sent to the lamps. By applying a known voltage to the power supplies and then recording the average power output, the desired relationship can be determined. This function can be described by a polynomial and then inverted to remove the nonlinearity from the loop because the model is linear with respect to radiative power from the lamps (see Equation 73.16). We noted that the average power to the lamps may not equal the radiative power from the lamps. The offset was due to heating losses within the bulb filament. However, this offset was implicitly incorporated in the model when the gain matrix, **K**, was determined from experimental data.

A second strategy that considered the RTP system as a black box was employed to identify a linear model of the process [11],

[12]. Among numerous alternatives, an ARX model was used to describe the system,

$$T(t) - T_{ss} = \sum_{k=1}^{n_a} A_k(T(t-k) - T_{ss}) + \sum_{k=1}^{n_b} B_k(P(t-k) - P_{ss}) + n(t), \quad (73.17)$$

where T is an $l \times 1$ vector describing temperature, P is an $M \times 1$ vector describing percent of maximum zone power, A_k is an $l \times l$ matrix, B_k is an $l \times M$ matrix, and l is the number of sensors on the wafer measuring temperature. The steady-state temperature and power were denoted by T_{ss} and P_{ss}, respectively. Because the steady-state temperature was difficult to determine accurately because of drift, a slight modification to the model was made. Let $T_{ss} = \hat{T} + \Delta\hat{T}$. The least squares problem for model identification can then be formulated as

$$min_{A_i,B_i} \sum_{t=1}^{N} \left\| (T(t) - \hat{T}) - \sum_{k=1}^{n_a} A_k(T(t-k) - \hat{T}) - \sum_{k=1}^{n_b} B_k(P(t-k) - P_{ss}) - T_{bias} \right\|_2^2 \quad (73.18)$$

where $T_{bias} = (I - \sum_{k=1}^{n_a} A_k)\Delta\hat{T}$.

The strategy for estimating the unknown model parameters utilized PRBS (pseudo-random binary sequence) to excite the system to obtain the necessary input-output data. The mean temperature that this excitation produced is designated as $\hat{T}$. Some other issues that were accounted for during model identification included the use of a data subsampling method so that the ARX model could span a longer time interval and observe a larger change of the temperature. Subsampling was needed because the data collection rate was 10 Hz and over that interval the temperature changed very little. Consequently, the least-squares formulation, as an identification method, may contain inherent error sources due to the effect of the measurement sensor noise (which was presumed to be Gaussian distributed) and the quantization noise (which was presumed to be uniformly distributed with quantization level of 0.5°C).

With the black box approach, the model order needed to be selected. The criterion used to determine the appropriateness of the model was to be able to make the prediction error smaller than the quantization level using as small a number of ARX model parameters as possible. For our applications, this order was three A matrices and three B matrices with subsampling selected as four.

We are now going to show the value of the identified models by using them to study an important characteristic of the RTP system, its DC gain. For the ARX model with coefficients $\{A_i, B_i\}$, the identified DC gain was given by the formula

$$\text{DC gain} = \mathbf{D}_o = (I - \sum_{i=1}^{3} A_i)^{-1} \sum_{l=1}^{3} B_i$$

Substituting in the appropriate values for 700°C (with $l = 3$, $J = 3$),

$$\mathbf{D}_o = \begin{bmatrix} 2.07 & 4.41 & 4.50 \\ 1.11 & 4.78 & 4.91 \\ 0.73 & 5.08 & 5.51 \end{bmatrix}. \quad (73.19)$$

Note that the magnitude of the first column of $\mathbf{D}_o$ is smaller than those of the second and third columns, which was due to the difference in the maximum power of each lamp: the first (center) lamp has 2 kW maximum power, the second (intermediate) lamp 12 kW maximum power, and the third (outer) lamp 24 kW maximum power. Also note the similarity of the second and third column of $\mathbf{D}_o$, which says that the second lamp will affect the wafer temperature in a manner similar to the third lamp. As a result, we have effectively two lamps rather than three (recall we have physically three lamps), which may cause difficulties in maintaining temperature uniformity in the steady state because of an inadequate number of degrees of freedom. This conclusion will be more clearly seen from an SVD (Singular Value Decomposition) analysis, about which more will be said later. To increase the independence of the control effects of the two outside lamps, a baffle was installed to redistribute the light energy from the lamps to the wafer. The same identification technique described earlier was used to identify the RTP system model, and the DC gain was computed from the identified model with the result

$$\mathbf{D}_n = \begin{bmatrix} 2.02 & 5.27 & 3.97 \\ 1.19 & 5.53 & 4.01 \\ 0.83 & 5.11 & 5.15 \end{bmatrix}. \quad (73.20)$$

We can observe that the second column of the new DC gain matrix was no longer similar to the third column, as it was in Equation 73.19. As a result, the three lamps heated the wafer in different ways. The first (center) lamp heated mostly the center of the wafer, and the third (outer) lamp heated mostly the edge of the wafer. On the other hand, the second (intermediate) lamp heated the wafer overall, acting like a bulk heater. Of course, the second lamp heated the intermediate portion of the wafer more than the center and edge of the wafer, but the difference was not so significant.

Even if the idea of installing a baffle was partly motivated by the direct investigation of the DC gain matrix, it was in fact deduced from an SVD (Singular Value Decomposition) analysis of the DC gain matrix.

The SVD of $\mathbf{D}_o$ in Equation 73.19 is given by

$$u_1 = [0.54, 0.57, 0.63], \; u_2 = [-0.80, 0.13, 0.58],$$
$$u_3 = [0.25, -0.81, 0.53] \quad (73.21)$$
$$v_1 = [0.18, 0.68, 0.71], \; v_2 = [-0.98, 0.04, 0.21],$$
$$v_3 = [0.12, -0.73, 0.67] \quad (73.22)$$
$$\sigma_1 = 12.15, \sigma_2 = 1.12, \sigma_3 = 0.11 \quad (73.23)$$

From this, we can conclude that [1, 1, 1] (u_1) is a strong output direction. Of course, u_1 is [0.54, 0.57, 0.62] and is not exactly equal to [1, 1, 1]. However, [0.54, 0.57, 0.62] was close to [1, 1, 1] in terms of direction in a 3-dimensional coordinate system and was denoted as the [1, 1, 1] direction, here. Since [1, 1, 1] was the strong output direction, we can affect the wafer temperature in the [1, 1, 1] direction by a minimal input power change. This means that if we maintain the temperature uniformity at the reference temperature (700°C), we can maintain the uniformity near 700°C (say, at 710°C) with a small change in the input lamp power. The weak output direction (the vector u_3 — approximately [1, −1, 1]) says that it is difficult to increase the temperature of the center and outer portions of the wafer while cooling down the intermediate portion of the wafer, which was, more or less, expected. The gain of the weak direction (σ_3) was two orders of magnitude smaller than that of the strong direction (σ_3). This meant that there were effectively only two lamps in the RTP system in terms of controlling the temperature of the wafer, even if there were physically three lamps. This naturally led to the idea of redesigning the RTP chamber to get a better lamp illumination pattern. Installing a baffle (see [10] for more details) into the existing RTP system improved our situation as shown in the SVD analysis of the new DC gain $\mathbf{D}_n$ in Equation 73.20. The SVD of $\mathbf{D}_n$ was given by

$$\begin{aligned} u_1 &= [0.56, 0.57, 0.60], \; u_2 = [-0.63, -0.17, 0.76], \\ u_3 &= [0.53, -0.80, 0.26] \qquad (73.24) \\ v_1 &= [0.19, -0.72, 0.67], \; v_2 = [0.76, -0.3, -0.57], \\ v_3 &= [0.63, 0.61, 0.48] \qquad (73.25) \\ \sigma_1 &= 12.14, \; \sigma_2 = 1.17, \; \sigma_3 = 0.52 \qquad (73.26) \end{aligned}$$

Compared to the SVD of the previous DC gain, the lowest singular value (σ_3) has been increased by a factor of 5, a significant improvement over the previous RTP system. In other words, only one-fifth of the power required to control the temperature in the weak direction, using the previous RTP system, was necessary for the same task with the new RTP system. As a result, we obtained three independent lamps by merely installing a baffle into the existing RTP system. Independence of the three lamps in the new RTP system was crucial in maintaining the temperature uniformity of the wafer.

73.5 Control Design

The general strategy of feedback combined with feedforward control was investigated for RTP control. In this strategy, a feedforward value of the lamp power was computed (in response to a change in the temperature set point) according to a predetermined relationship. This feedforward value was then added to a feedback value and the resultant lamp power was sent to the system. The concept behind this approach was that the feedforward power brings the temperature close to the desired temperature; the feedback value compensates for modeling errors and disturbances.

The feedback value can be determined with a variety of design techniques, two of which are described below. Several approaches were investigated to determine the feedforward value. One approach was based on replaying the lamp powers of previous runs. Another approach was based on a model-based optimization.

The physics-based model was employed to develop a controller using a variation of the classical Internal Model Control (IMC) design procedure [13], [6]. The IMC approach consisted of factoring the linearized form of the nonlinear low-order model (see Equation 73.16) as,

$$\tilde{\mathbf{G}}_p(z) = \tilde{\mathbf{G}}_p^+(z)\tilde{\mathbf{G}}_p^-(z) \qquad (73.27)$$

where $\tilde{\mathbf{G}}_p^+(z)$ contains the time delay terms, z^{-d}, all right half-plane zeros, zeros that are close to $(-1, 0)$ on the unit disk, and has unity gain. The IMC controller is then obtained by

$$\mathbf{G}_c^*(z) = \tilde{\mathbf{G}}_p^-(z)^{-1}\mathbf{F}(z) \qquad (73.28)$$

where $\mathbf{F}(z)$ is a matrix of filters used to tune the closed-loop performance and robustness and to obtain a physically realizable controller. The inversion $\tilde{\mathbf{G}}_p^-(z)^{-1}$ was relatively straightforward because the dynamics of the annular wafer elements of the linearized form of the nonlinear model were decoupled.

The tuning matrix, or IMC filter, $\mathbf{F}(z)$ was selected to satisfy several requirements of RTP. The first requirement was related to repeatability in which zero offset between the actual and desired trajectory was to be guaranteed at steady-state condition despite modeling error. The second requirement was related to uniformity in which the closed-loop dynamics of the wafer temperature should exhibit similar behavior. The third requirement was related to robustness and implementation in which the controller should be as low-order as possible. Other requirements were ease of operator usage and flexibility. One simple selection of $\mathbf{F}(z)$ that meets these requirements was given by the first-order filter

$$\begin{aligned} \mathbf{F}(z) &= f(z)\mathbf{I}, \qquad (73.29) \\ f(z) &= \frac{1-\alpha}{1-\alpha z^{-1}}, \qquad (73.30) \end{aligned}$$

where α is a tuning parameter, the speed of response. This provided us with a simple controller with parameters that could be interpreted from a physical standpoint.

In this approach to control design, the nonlinear dependency of $\mathbf{K}$ and τ_i on temperature can be parameterized explicitly in the controller. Hence, a continuous gain-scheduling procedure can be applied. It was noted that, as temperature increased, the process gain decreased. Since the controller involved the inverse of the process model, the controller gain increased as temperature was increased. Consequently, the gain-scheduling provided consistent response over the entire temperature range. Thus, control design at one temperature should also apply at other temperatures.

In addition to the IMC approach, a multivariable feedback control law was determined by an LQG design which incorporated integral control action to reduce run-to-run variations [12], [14]. The controller needed to be designed carefully, because, in a nearly singular system such as the experimental RTP, actuator saturation and integrator windup can cause problems. To solve this problem partially, integral control was applied in only the (strongly) controllable directions in temperature error space, helping to prevent the controller from trying to remove uncontrollable disturbances.

For the LQG design, the black-box model was used. It can be expressed in the time domain as follows:

$$\mathbf{y}_k^0 = \mathbf{C}\mathbf{A}^{k-1}\mathbf{B}\mathbf{u}_0^0 + \ldots + \mathbf{CAB}\mathbf{u}_{k-2}^0 + \mathbf{CB}\mathbf{u}_{k-1}^0,$$

and the resulting equations ordered from $\mathbf{y}_1^0$ to $\mathbf{y}_N^0$:

$$\begin{bmatrix} \mathbf{y}_1^0 \\ \mathbf{y}_2^0 \\ \vdots \\ \mathbf{y}_N^0 \end{bmatrix} = \begin{bmatrix} \mathbf{CB} & \cdots & \mathbf{O} \\ \mathbf{CAB} & \cdots & \mathbf{O} \\ \vdots & \ddots & \vdots \\ \mathbf{CA}^{N-1}\mathbf{B} & \cdots & \mathbf{CB} \end{bmatrix} \begin{bmatrix} \mathbf{u}_0^0 \\ \mathbf{u}_1^0 \\ \vdots \\ \mathbf{u}_{N-1}^0 \end{bmatrix}.$$

These combined equations determine a linear relationship between the input and output of the system in the form $\mathbf{Y} = \mathbf{HU}$, where the notation $\mathbf{Y}$ and $\mathbf{U}$ is used to designate the $n_o N \times 1$ and $n_i N \times 1$ stacked vectors.

The identified model of the system was augmented with new states representing the integral of the error along the m easiest to control directions, defined as $\xi \stackrel{def}{=} [\ \xi_1\ \xi_2\ \ldots\ \xi_m\ ]^T$. The new system model was, then,

$$\begin{bmatrix} \mathbf{x}_{k+1} \\ \xi_{k+1} \end{bmatrix} = \begin{bmatrix} \mathbf{A} & \mathbf{O} \\ \mathbf{U}_{1:m}^T\mathbf{C} & \mathbf{I} \end{bmatrix} \begin{bmatrix} \mathbf{x}_k \\ \xi_k \end{bmatrix} + \begin{bmatrix} \mathbf{B} \\ \mathbf{O} \end{bmatrix} \mathbf{u},$$
$$\mathbf{y}_k = \begin{bmatrix} \mathbf{C} & \mathbf{O} \\ \mathbf{O} & \mathbf{I} \end{bmatrix} \begin{bmatrix} \mathbf{x}_k \\ \xi_k \end{bmatrix},$$

where $\mathbf{U}_{1:m}$ is the first m columns of $\mathbf{U}$, the output matrix from the SVD of the open-loop transfer matrix $H(z)|_{z=1} = \mathbf{USV}^T$, and represented the (easily) controllable subspace. The output $\mathbf{y}$ then consisted of thermocouple readings and the integrator states (in practice the integrator states were computed in software from the measured temperature errors). Weights for the integrator states were chosen to provide a good transient response. A complete description of the control design can be found in [14], [15].

The goal of our feedforward control was to design, in advance, a reference input trajectory which will cause the system to follow a predetermined reference output trajectory, assuming no noise or modeling error. The approach built upon the analysis done in [16] and expressed the open-loop trajectory generation problem as a convex optimization with linear constraints and a quadratic objective function [14], [15].

The input/output relationship of the system can be described by a linear matrix equation. This relationship was used to convert convex constraints on the output trajectory into convex constraints on the input trajectory.

The RTP system imposed a number of linear constraints. These were linear constraints imposed by the RTP hardware, specifically, the actuators had both saturation effects and a maximum rate of increase.

Saturation constraints were modeled as follows: $\mathbf{P}_k^{lm}$ is defined as the $n_i \times 1$ vector of steady-state powers at the point of linearization of the above model at time k. With an outer feedback control loop running, to insure that the feedback controller has some room to work (for example ±10% leeway), the total powers should be constrained to the operating range of $10 \le \mathbf{P}^{\text{total}} \le 90$, which translates into a constraint on $\mathbf{U}$ of $(10 - \mathbf{P}^{lm}) \le \mathbf{U} \le (90 - \mathbf{P}^{lm})$.

Maximum rates of increase (or slew rate limits) for our actuators were included also. These were due to the dynamics of the halogen bulbs in our lamp. We included this constraint as $\mathbf{u}_{k+1}^0 - \mathbf{u}_k^0 \le 5$, which can be expressed in a matrix equation of the form $\mathbf{SU} \le 5$, where $\mathbf{S}$ has 1 on the main diagonal and -1 on the off-diagonal.

The quality of our optimized trajectory can be measured in two ways: minimized tracking error (following a reference trajectory) and minimized spatial nonuniformity across the wafer. Because the tracking error placed an upper bound on the nonuniformity error, we concentrate on it here. We define the desired trajectory $\mathbf{Y}^{\text{ref}}$ as relative to the same linearized starting point used for system identification. The tracking error $\mathbf{E}$ can be defined as $\mathbf{E} = \mathbf{Y} - \mathbf{Y}^{\text{ref}}$, where $\mathbf{E}$ again denotes the stacked error vector.

We define our objective function to be a quadratic constraint on $\mathbf{E}$ as $F(x) = \mathbf{E}^T\mathbf{E}$, and expand

$$F(x) = \mathbf{U}^T\mathbf{H}^T\mathbf{HU} - 2(\mathbf{Y}^{\text{ref}})^T\mathbf{HU} + (\mathbf{Y}^{\text{ref}})^T\mathbf{Y}^{\text{ref}}. \quad (73.31)$$

Software programs exist, such as the FORTRAN program LSSOL [17], which can take the convex constraints and produce a unique solution, if one exists.

After achieving successful results in simulation, the control system was implemented in a real-time computing environment linked to the actual RTP equipment. The computing environment included a VxWorks real-time operating system, SUN IPC workstation, VME I/O boards and a Motorola 68030 processor.

73.6 Proof-of-Concept Testing

The Stanford RTP system was used for multiprocessing applications in which sequential thermal process steps were performed within the same reactor. A schematic of the RTM is shown in Figure 73.10. A concentric three-zone 38-kW illuminator, constructed and donated by Texas Instruments, was used for wafer heating. The center zone consisted of a 2-kW bulb, the intermediate zone consisted of 12 1-kW bulbs and the outer zone consisted of 24 1-kW bulbs. A picture of the three-zone lamp is presented in Figure 73.11. The reflector was water and air cooled. An annular gold-plated stainless steel opaque ring was placed on the quartz window to provide improved compartmentalization between the intermediate and outer zones. This improvement was achieved by reducing the radiative energy from the outer zone impinging on the interior location of the wafer and from the intermediate zone impinging on the edge of the wafer. The RTM was used for 4-inch wafer processing. The wafer was manually loaded onto three supporting quartz pins of low thermal mass. The wafer was placed in the center of the chamber which was approximately 15 inches in diameter and 6 inches in height. Gas was injected via two jets. For the control experiments presented

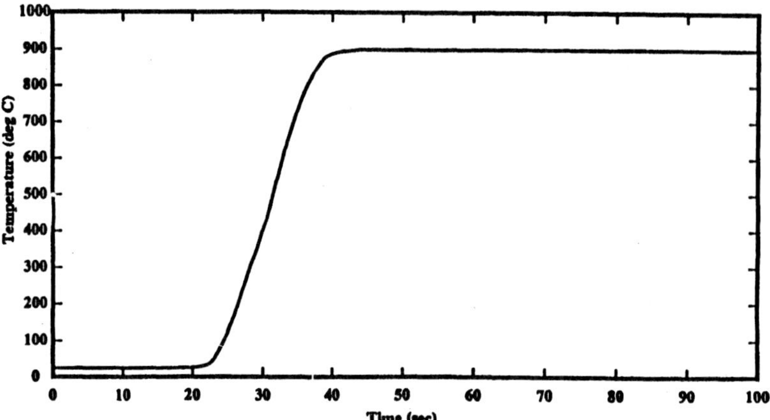

Figure 73.14 Temperature trajectory of the three sensors over the first 100 seconds and the 5 minute process hold.

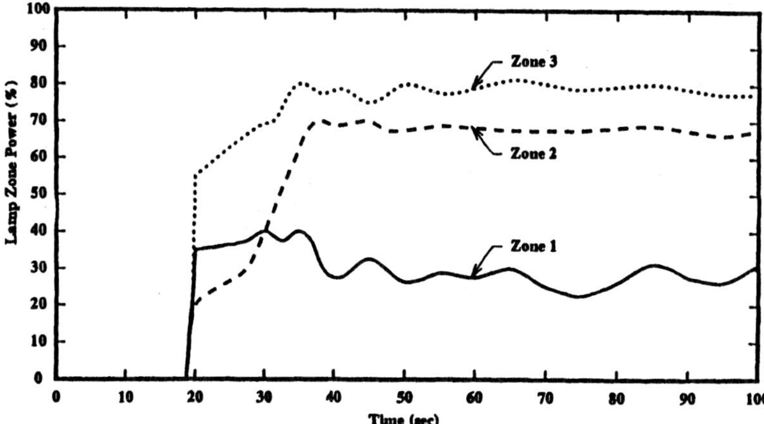

Figure 73.15 Powers of the three zones used to control temperature over the first 100 seconds.

below, temperature was measured with a thermocouple instrumented wafer. Three thermocouples were bonded to the wafer along a common diameter at radial positions of 0 inch (center), 1 inch, and 1 7/8 inches. The experiments were conducted in a N_2 environment at 1 atmosphere pressure.

The control algorithms were evaluated for control of temperature uniformity in achieving a ramp from room temperature to 900°C at a ramp rate of 45°C/s followed by a hold for 5 minutes at 1 atm pressure and 1000 sccm (cc/min gas at standard conditions) N_2 [4]. This trajectory typified low-temperature thermal oxidation or annealing operations. The ramp rate was selected to correspond to the performance limit (in terms of satisfying uniformity requirements) of the equipment. The control system utilized simultaneous IMC feedback and feedforward control. Gain scheduling was employed to compensate for the nonlinearities induced by radiative heating.

The wafer temperature for the desired trajectory over the first 100 seconds is plotted in Figure 73.14 for the center, middle, and edge locations where thermocouples are bonded to the wafer along a common diameter at radial positions of 0 inch (center), 1 inch, and 1 7/8 inches.

The ramp rate gradually increased to the specified 45°C/s and then decreased as the desired process hold temperature was approached. The corresponding lamp powers, that were manipulated by commands from the control system to achieve the desired temperature trajectory, are shown in Figure 73.15.

The time delay of the system can be seen by comparing the starting times of the lamp powers to the temperature response. Approximately, a two second delay existed in the beginning of the response. Of this delay, approximately 1.5 seconds was caused by a power surge interlock on the lamp power supplies which only functions when the lamp power is below 15% of the total power. The remaining delay was caused by the sensor and filament heating dynamics. In the power profile plot, the rate limiting of the lamp powers is seen. This rate-limiting strategy was employed as a safety precaution to prevent a large inrush current to the lamps. However, these interlocks prevented higher values of ramp rates from being achieved.

The nonuniformity of the controlled temperature trajectory was then analyzed. From the measurements of the entire five minute run (i.e., the first 100 seconds shown in Figure 73.14 along with an additional 400 second hold at 900°C not shown in the figure), the nonuniformity was computed by the peak-to-peak temperature error of the temperature measurements of the three thermocouples. The result is plotted in Figure 73.16. The maximum temperature nonuniformity of approximately 15°C occurred during the ramp around a mean temperature of 350°C. This nonuniformity occurred at a low temperature and does not effect processing or damage the wafer via slip. As the ramp progressed from this point, the nonuniformity decreased. The sig-

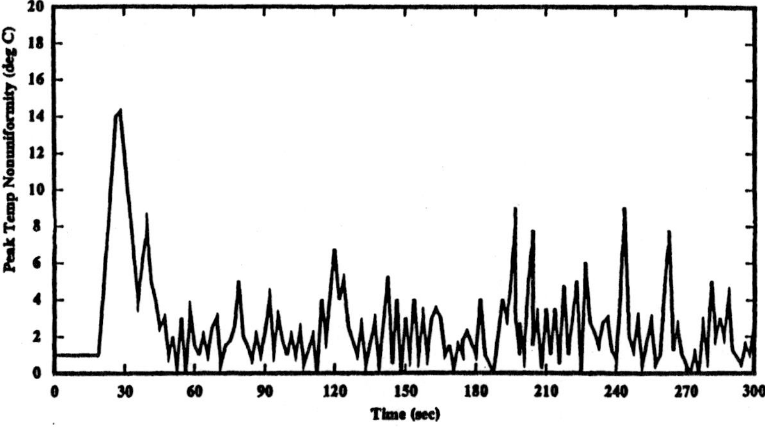

Figure 73.16 Temperature nonuniformity for the 5 minute run as measured by the three temperature sensors.

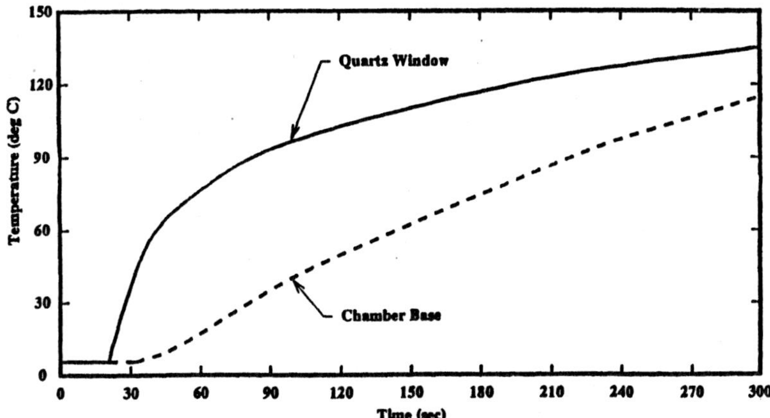

Figure 73.17 Temperatures of the quartz window and chamber base over the 5 minute run.

nificant sensor noise can be seen.

The capability of the controller to hold the wafer temperature at a desired process temperature despite the presence of dynamic heating from extraneous sources was then examined. As seen in Figure 73.14, the control system held the wafer temperature at the desired value of 900°C. Although the sensors were quite noisy and had resolution of 0.5°C, the wafer temperature averaged over the entire hold portion for the three sensors corresponded to 900.9°C, 900.7°C, and 900.8°C, respectively. This result was desired because the uniformity of the process parameters, such as film thickness and resistivity, generally depend on the integrated or averaged temperature over time. The capability of the control system to hold the wafer temperature at the desired value, albeit slightly higher, is demonstrated by plotting the dynamics of the quartz window and chamber base of the RTM in Figure 73.17.

The slow heating of these components of the RTM corresponded to slow disturbances to the wafer temperature. Because of the reduced gain of the controller to compensate for time delays, these disturbances impacted the closed-loop response by raising the temperature to a value slightly higher than the set point. However, without the feedback temperature control system, the wafer temperature would have drifted to a value more than 50°C higher than the set point as opposed to less than 1°C in the measured wafer temperature.

73.7 Technology Transfer to Industry

After demonstrating the prototype RTP equipment at Stanford, the multivariable control strategy (including hardware and software) was transferred to Texas Instruments for application in the MMST program. This transfer involved integration on eight RTP reactors: seven on-line and one off-line. These RTP systems were eventually to be used in a 1000 wafer demonstration of two full-flow sub-half-micron CMOS process technologies in the MMST program at TI [18], [19],[20],[21] and [22].

Although there were similarities between the RTP equipment at TI and the three-zone RTP system at Stanford, there were also substantial differences. Two types of illuminators were used for MMST, a four-zone system constructed at Texas Instruments for MMST applications and a six-zone system manufactured by G^2 Semiconductor Corp. Both systems utilized concentric zone heating; the TI system employed a circular arrangement of lamps, and the G^2 system used a hexagonal arrangement of lamps. A thick (roughly 15 mm) quartz window was used in the TI system to separate the lamps from the reaction chamber, and a thin (3 mm) quartz window was used in the G^2 system. Wafer rotation was not employed with the TI system but was used with the G^2 system. The rotation rate was approximately 20 rpm. Moreover, six-inch wafer processing took place using up to four pyrometers for feedback. Most reactors employed different configurations purge ring assemblies, guard rings, and susceptors (see Figure 73.5).

The on-line RTP reactors configured with the IMC controller were used for thirteen different thermal processes: LPCVD Nitride, LPCVD Tungsten, Silicide react, Silicide anneal, sinter, LPCVD polysilicon, LPCVD amorphous silicon, germane clean, dry RTO, wet RTO, source/drain anneal, gate anneal, and tank anneal. These processes ranged from 450° to 1100°C, from 1 to 650 torr pressure, and from 30 seconds to 5 minutes of processing time (see Figure 73.18).

RTP Process	Carrier Gas	T_{ph} [°C]	T_{pr} [°C]	t_{ph} [s]	t_{pr} [s]	P [Torr]
sinter	N_2	450	450	0	180	650
LPCVD-W	Ar/H_2	425-475	425-475	0	60-180	30
LPCVD-amor Si	Ar	450-500	500-560	5-15	60-180	15
LPCVD-poly	Ar	450-550	650	5-15	120-240	15
silicide react	N_2	450-500	650	5-15	180	1
silicide anneal	Ar	450-550	750	5-15	60	1
LPCVD-SiO_2	O_2	450-550	750	5-15	30-180	1-5
germane clean	H_2	450-550	650-750	5-15	120	15
LPCVD-Si_3N_4	NH_3	550-650	850	5-15	60-180	1-5
gate RTA	Ar	750-800	900	5-15	30	650
dry RTO	O_2	750-800	1000	5-15	120-180	650
wet RTO	O_2	750-800	1000	5-15	120-180	650
source/drain RTA	Ar	750-800	1000-1050	5-15	15-30	650
tank RTA	NH_3	750-800	1100	5-15	300	650

Figure 73.18 List of processes controlled during the MMST program. The preheat temperature (T_{ph}) and time (t_{ph}) and the process temperature (T_{pr}) and time (t_{pr}) are given. The carrier gases and operating pressures are also presented.

There were several challenges in customizing the temperature control system for operation in an all-RTP factory environment [13]. These challenges included substantial differences among the eight reactors and thirteen processes, operation in a prototyping development environment, ill-conditioned processing equipment, calibrated pyrometers required for temperature sensing, equipment reliability tied to control power trajectories, multiple lamp-zone/sensor configurations, detection of equipment failures, and numerous operational and communication modes.

Nonetheless, it was possible to develop a single computer control code with the flexibility of achieving all of the desired objectives. This was accomplished by developing a controller in a modular framework based on a standardized model of the process and equipment. The control structure remained the same while the model-based parameters of the controller differed from process to process and reactor to reactor. It was possible to read these parameters from a data file while holding the controller code and logic constant. Consequently, it was only necessary to maintain and modify a single computer control code for the entire RTP factory.

We present results here for an LPCVD-Nitride process that employed a TI illuminator and for an LPCVD-Poly process that employed a G^2 illuminator. Additional temperature and process control results are presented in [13] and [6].

The desired temperature trajectory for the LPCVD-Nitride process involved a ramp to 850°C and then a hold at 850°C for roughly 180 seconds in a SiH_4/NH_3 deposition environment. Temperature was measured using four radially distributed 3.3 μm InAs pyrometers. The center and edge pyrometers were actively used for real-time feedback control and the inner two pyrometers were used to monitor the temperature. The reasons for this analysis were: (1) repeatable results were possible using only two pyrometers, (2) an analysis of the benefits of using pyrometers for feedback could be assessed, and (3) fewer pyrometers were maintained during the marathon demonstration. In Figure 73.19, the center temperature measurement is shown for a 24-wafer lot process. The offsets in the plot during the ramps are merely due to differences in the starting points of the ramps.

During the hold at 850°C, the reactive gases were injected, and the deposition took place. The standard deviation (computed over the 24 runs) of the temperature measurements during the deposition time was analyzed. In Figure 73.20, the standard deviation of the four sensor measurements are shown. The controlled sensors improved repeatability over the monitored sensor locations by a factor of seven. A three-sigma interpretation shows roughly that the controlled sensors held temperature to within ±0.3°C and the monitored sensors were repeatable at ±2.0°C.

We analyzed the power trajectories to the lamp zones to evaluate the repeatability of the equipment. In Figure 73.21, the power to the center zone is presented for the 24 runs. The intermediate two zones were biased off the center and edge zones, respectively. From these results, it was clear that the lamp power decreased substantially during a nitride deposition run because the chamber and window heat more slowly than the wafer; because the chamber and window provide energy to the wafer, the necessary energy from the lamps to achieve a specified wafer temperature was less as the chamber and window heat up. In addition, we noted the chamber and window heating effect from run-to-run by observing the lowered lamp energy requirements as the lot processing progresses. These observations can be used in developing fault detection algorithms.

To study the capability of temperature control on the process parameter, we compared the thickness of the LPCVD poly process determined at the center for each wafer of a 24-wafer lot where multizone feedback temperature control was used and no real-time feedback temperature control (i.e., open-loop operation) was used. For the open-loop case, a predetermined lamp power trajectory was replayed for each wafer of the 24-wafer lot. The comparison is shown in Figure 73.22. It is clear that the multizone feedback control is much better than open-loop control. In some sense, this comparison is a worst case analysis since the lamp powers themselves for both cases had no control, not usual in industry. In our experiments, variations in line voltage

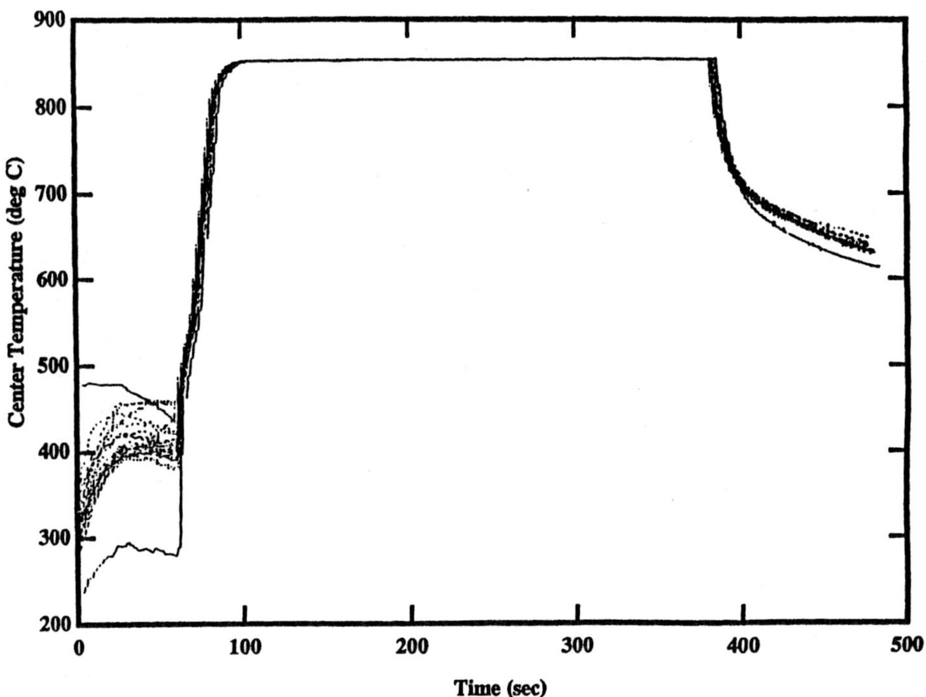

Figure 73.19 The center temperature of the LPCVD nitride process for a 24-wafer lot run.

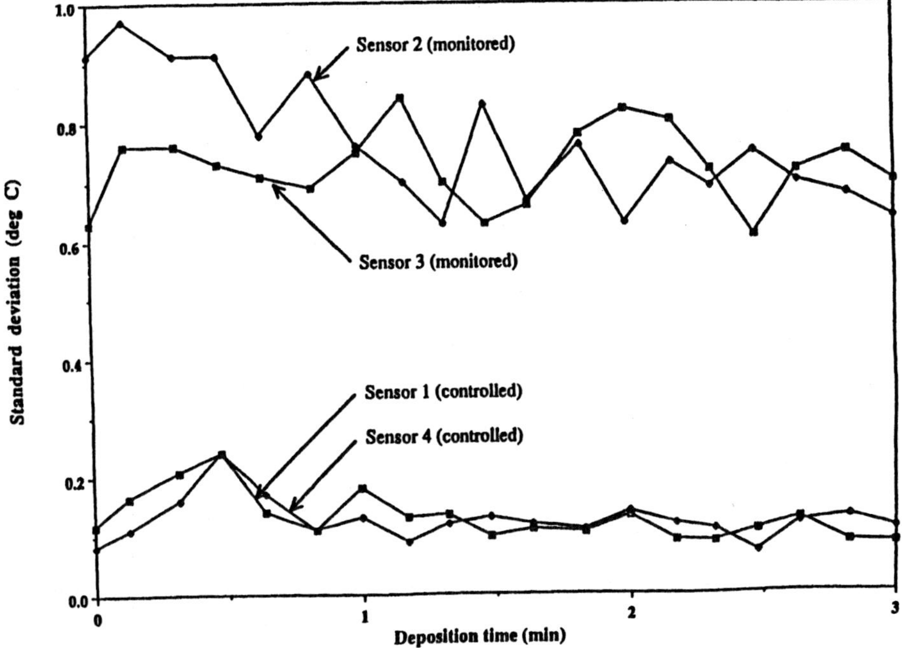

Figure 73.20 Standard deviation of temperature measurements during the three minute deposition step of the LPCVD nitride process for the 24-wafer lot run.

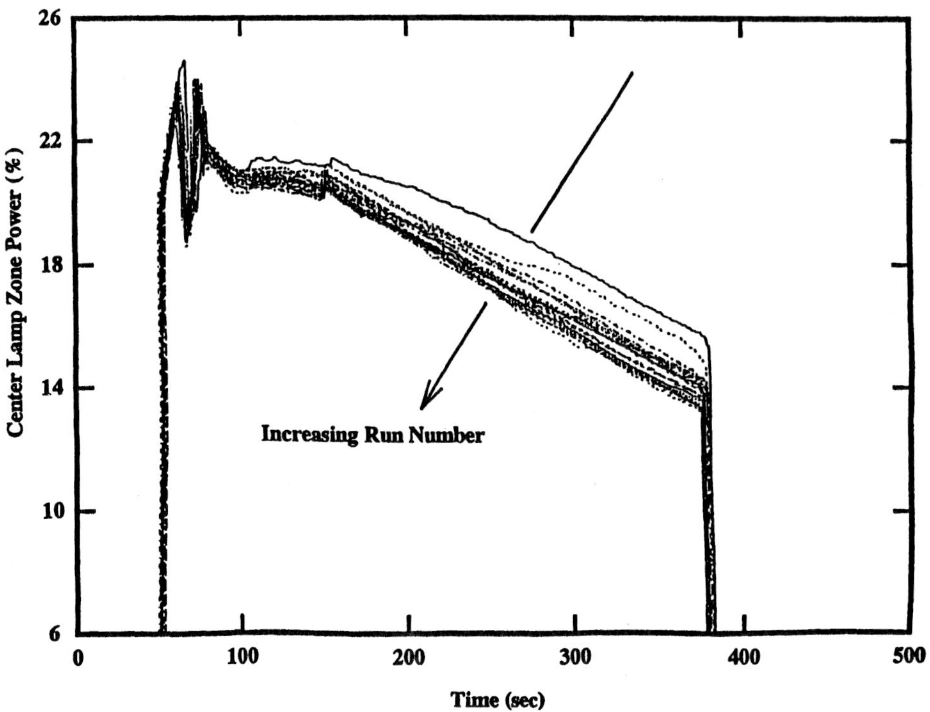

Figure 73.21 Power to the center zone for the 24-wafer lot run of the LPCVD nitride process.

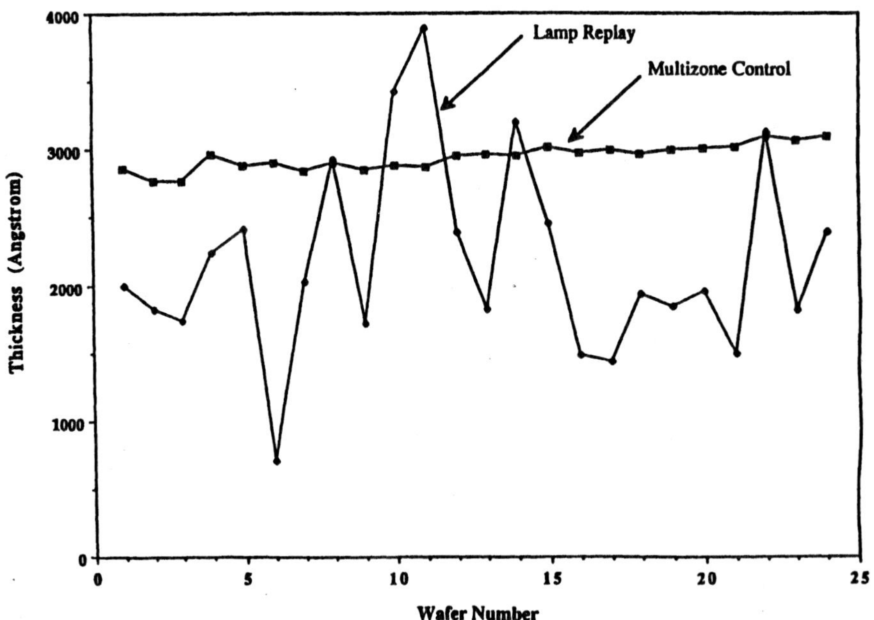

Figure 73.22 Comparison of multizone feedback control and open-loop lamp replay for LPCVD polysilicon process.

proceeded unfiltered through the reactor causing unprovoked fluctuations in the lamp power and inducing strong temperature effects. However, the feedback temperature control system can compensate somewhat for these fluctuations. For the open-loop case, these fluctuations pass on directly and result in unacceptable repeatability.

73.8 Conclusions

A systems approach has been used for a study in semiconductor manufacturing. This methodology has included developing models, analyzing alternative equipment designs from a control perspective, establishing model identification techniques to develop a model for control design, developing a real-time control system and embedding it within a control processor, proof-of-concept testing with a prototype system, and then transferring the control technology to industry. This application has shown the role that control methodologies can play in semiconductor device manufacturing.

References

[1] Lord, H., Thermal and stress analysis of semiconductor wafers in a rapid thermal processing oven, *IEEE Trans. Semicond. Manufact.*, 1, 141–150, 1988.

[2] Norman, S.A., *Wafer Temperature Control in Rapid Thermal Processing*, Ph.D. Thesis, Stanford University, 1992.

[3] Cho, Y., Schaper, C. and Kailath, T., Low order modeling and dynamic characterization of rapid thermal processing, *Appl. Phys. A: Solids and Surfaces*, A:54(4), 317–326, 1992.

[4] Norman, S.A., Schaper, C.D. and Boyd, S.P., Improvement of temperature uniformity in rapid thermal processing systems using multivariable control. In *Mater. Res. Soc. Proc.: Rapid Thermal and Integrated Processing*. Materials Research Society, 1991.

[5] Saraswat, K. and Apte, P., Rapid thermal processing uniformity using multivariable control of a circularly symmetric three-zone lamp, *IEEE Trans. on Semicond. Manufact.*, 5, 1992.

[6] Saraswat, K., Schaper, C., Moslehi, M. and Kailath, T., Modeling, identification, and control of rapid thermal processing, *J. Electrochem. Soc.*, 141(11), 3200–3209, 1994.

[7] Cho, Y., *Fast Subspace Based System Identification: Theory and Practice*, Ph.D. Thesis, Stanford University, CA, 1993.

[8] Cho, Y. and Kailath, T., Model identification in rapid thermal processing systems, *IEEE Trans. Semicond. Manufact.*, 6(3), 233–245, 1993.

[9] Roy, R., Paulraj, A. and Kailath, T., ESPRIT - a subspace rotation approach to estimation of parameters of cisoids in noise, *IEEE Trans. ASSP*, 34(5), 1340–1342, 1986.

[10] Schaper, C., Cho, Y., Park, P., Norman, S., Gyugyi, P., Hoffmann, G., Balemi, S., Boyd, S., Franklin, G., Kailath, T., and Sarawat, K., Dynamics and control of a rapid thermal multiprocessor. In *SPIE Conference on Rapid Thermal and Integrated Processing*, September 1991.

[11] Cho, Y.M., Xu, G., and Kailath, T., Fast recursive identification of state-space models via exploitation of displacement structure, *Automatica*, 30(1), 45–59, 1994.

[12] Gyugyi, P., Cho, Y., Franklin, G., and Kailath, T., Control of rapid thermal processing: A system theoretic approach. In *IFAC World Congress*, 1993.

[13] Saraswat, K., Schaper, C., Moslehi, M., and Kailath, T., Control of MMST RTP: Uniformity, repeatability, and integration for flexible manufacturing, *IEEE Trans. on Semicond. Manifact.*, 7(2), 202–219, 1994.

[14] Gyugyi, P., *Application of Model-Based Control to Rapid Thermal Processing Systems*. Ph.D. Thesis, Stanford University, 1993.

[15] Gyugyi, P.J., Cho, Y.M., Franklin, G., and Kailath, T., Convex optimization of wafer temperature trajectories for rapid thermal processing. In *The 2nd IEEE Conf. Control Appl.*, Vancouver, 1993.

[16] Norman, S.A., Optimization of transient temperature uniformity in RTP systems, *IEEE Trans. Electron Dev.*, January 1992.

[17] Gill, P.E., Hammarling, S.J., Murray, W., Saunders, M.A., and Wright, M.H., User's guide for LSSOL (Version 1.0): A FORTRAN package for constrained least-squares and convex quadratic programming, Tech. Rep. SOL 86-1, Operations Research Dept., Stanford University, Stanford, CA, 1986.

[18] Chatterjee, P. and Larrabee, G., Manufacturing for the gigabit age, *IEEE Trans. on VLSI Technology*, 1, 1993.

[19] Bowling, A., Davis, C., Moslehi, M., and Luttmer, J., Microeletronics manufacturing science and technology: Equipment and sensor technologies, *TI Technical J.*, 9, 1992.

[20] Davis, C., Moslehi, M., and Bowling, A., Microeletronics manufacturing science and technology: Single-wafer thermal processing and wafer cleaning, *TI Technical J.*, 9, 1992.

[21] Moslehi, M. et al., Single-wafer processing tools for agile semiconductor production, *Solid State Technol.*, 37(1), 35–45, 1994.

[22] Saraswat, K. et al., Rapid thermal multiprocessing for a programmable factory for adaptable manufacturing of ic's, *IEEE Trans. on Semicond. Manufact.*, 7(2), 159–175, 1994.

SECTION XVI

Mechanical Control Systems

74

Automotive Control Systems

J. A. Cook
Ford Motor Company, Scientific Research Laboratory, Control Systems Department, Dearborn, MI

J. W. Grizzle
Department of EECS, Control Systems Laboratory, University of Michigan, Ann Arbor, MI

J. Sun
Ford Motor Company, Scientific Research Laboratory, Control Systems Department, Dearborn, MI

M. K. Liubakka
Advanced Vehicle Technology, Ford Motor Company, Dearborn, MI

D.S. Rhode
Advanced Vehicle Technology, Ford Motor Company, Dearborn, MI

J. R. Winkelman
Advanced Vehicle Technology, Ford Motor Company, Dearborn, MI

P. V. Kokotović
ECE Department, University of California, Santa Barbara, CA

74.1 Engine Control

J. A. Cook, Ford Motor Company, Scientific Research Laboratory, Control Systems Department, Dearborn, MI

J. W. Grizzle, Department of EECS, Control Systems Laboratory, University of Michigan, Ann Arbor, MI

J. Sun, Ford Motor Company, Scientific Research Laboratory, Control Systems Department, Dearborn, MI

74.1.1 Introduction

Automotive engine control systems must satisfy diverse and often conflicting requirements. These include regulating exhaust emissions to meet increasingly stringent standards without sacrificing good drivability; providing increased fuel economy to satisfy customer desires and to comply with Corporate Average Fuel Economy (CAFE) regulations; and delivering these performance objectives at low cost, with the minimum set of sensors and actuators. The dramatic evolution in vehicle electronic control systems over the past two decades is substantially in response to the first of these requirements. It is the capacity and flexibility of microprocessor-based digital control systems, introduced in the 1970s to address the problem of emission control, that have resulted in the improved function and added convenience, safety, and performance features that distinguish the modern automobile [8].

Although the problem of automotive engine control may encompass a number of different power plants, the one with which this chapter is concerned is the ubiquitous four-stroke cycle, spark ignition, internal combustion gasoline engine. Mechanically, this power plant has remained essentially the same since Nikolaus Otto built the first successful example in 1876. In automotive applications, it consists most often of four, six or eight cylinders wherein reciprocating pistons transmit power via a simple connecting rod and crankshaft mechanism to the wheels. Two complete revolutions of the crankshaft comprise the following sequence of operations.

0-8493-8570-9/96/$0.00+$.50

The initial 180 degrees of crankshaft revolution is the intake stroke, where the piston travels from top-dead-center (TDC) in the cylinder to bottom-dead-center (BDC). During this time an intake valve in the top of the cylinder is opened and a combustible mixture of air and fuel is drawn in from an intake manifold. Subsequent 180-degree increments of crankshaft revolution comprise the compression stroke, where the intake valve is closed and the mixture is compressed as the piston moves back to the top of the cylinder; the combustion stroke when, after the mixture is ignited by a spark plug, torque is generated at the crankshaft by the downward motion of the piston caused by the expanding gas; and finally, the exhaust stroke, when the piston moves back up in the cylinder, expelling the products of combustion through an exhaust valve.

Three fundamental control tasks affect emissions, performance, and fuel economy in the spark ignition engine: (1) air-fuel ratio (A/F) control, that is, providing the correct ratio of air and fuel for efficient combustion to the proper cylinder at the right time; (2) ignition control, which refers to firing the appropriate spark plug at the precise instant required; and (3) control of exhaust gas recirculation to the combustion process to reduce the formation of oxide of nitrogen (NOx) emissions.

Ignition Control

The spark plug is fired near the end of the compression stroke, as the piston approaches TDC. For any engine speed, the optimal time during the compression stroke for ignition to occur is the point at which the maximum brake torque (MBT) is generated. Spark timing significantly in advance of MBT risks damage from the piston moving against the expanding gas. As the ignition event is retarded from MBT, less combustion pressure is developed and more energy is lost to the exhaust stream.

Numerous methods exist for energizing the spark plugs. For most of automotive history, cam-activated breaker points were used to develop a high voltage in the secondary windings of an induction coil connected between the battery and a distributor. Inside the distributor, a rotating switch, synchronized with the crankshaft, connected the coil to the appropriate spark plug. In the early days of motoring, the ignition system control function was accomplished by the driver, who manipulated a lever located on the steering wheel to change ignition timing. A driver that neglected to retard the spark when attempting to start a hand-cranked Model T Ford could suffer a broken arm if he experienced "kickback." Failing to advance the spark properly while driving resulted in less than optimal fuel economy and power.

Before long, elaborate centrifugal and vacuum-driven distributor systems were developed to adjust spark timing with respect to engine speed and torque. The first digital electronic engine control systems accomplished ignition timing simply by mimicking the functionality of their mechanical predecessors. Modern electronic ignition systems sense crankshaft position to provide accurate cycle-time information and may use barometric pressure, engine coolant temperature, and throttle position along with engine speed and intake manifold pressure to schedule ignition events for the best fuel economy and drivability subject to emissions and spark knock constraints. Additionally, ignition timing may be used to modulate torque to improve transmission shift quality and in a feedback loop as one control variable to regulate engine idle speed. In modern engines, the electronic control module activates the induction coil in response to the sensed timing and operating point information and, in concert with dedicated ignition electronics, routes the high voltage to the correct spark plug.

One method of providing timing information to the control system is by using a magnetic proximity pickup and a toothed wheel driven from the crankshaft to generate a square wave signal indicating TDC for successive cylinders. A signature pulse of unique duration is often used to establish a reference from which absolute timing can be determined. During the past ten years there has been substantial research and development interest in using in-cylinder piezoelectric or piezoresistive combustion pressure sensors for closed-loop feedback control of individual cylinder spark timing to MBT or to the knock limit. The advantages of combustion-pressure-based ignition control are reduced calibration and increased robustness to variability in manufacturing, environment, fuel, and component aging. The cost is in an increased sensor set and additional computing power.

Exhaust Gas Recirculation

Exhaust gas recirculation (EGR) systems were introduced as early as 1973 to control (NOx) emissions. The principle of EGR is to reduce NOx formation during the combustion process by diluting the inducted air-fuel charge with inert exhaust gas. In electronically controlled EGR systems, this is accomplished using a metering orifice in the exhaust manifold to enable a portion of the exhaust gas to flow from the exhaust manifold through a vacuum-actuated EGR control valve and into the intake manifold. Feedback based on the difference between the desired and measured pressure drop across the metering orifice is employed to duty cycle modulate a vacuum regulator controlling the EGR valve pintle position. Because manifold pressure rate and engine torque are directly influenced by EGR, the dynamics of the system can have a significant effect on engine response and, ultimately, vehicle drivability. Such dynamics are dominated by the valve chamber filling response time to changes in the EGR duty cycle command.

The system can be represented as a pure transport delay associated with the time required to build up sufficient vacuum to overcome pintle shaft friction cascaded with first-order dynamics incorporating a time constant that is a function of engine exhaust flow rate. Typically, the EGR control algorithm is a simple proportional-integral (PI) or proportional-integral-derivative (PID) loop. Nonetheless, careful control design is required to provide good emission control without sacrificing vehicle performance. An unconventional method to accomplish NOx control by exhaust recirculation is to directly manipulate the timing of the intake and exhaust valves. Variable-cam-timing (VCT) engines have demonstrated NOx control using mechanical and hydraulic actuators to adjust valve timing and to affect the amount of internal EGR remaining in the cylinder after the exhaust stroke is completed. Early exhaust valve closing has the additional advantage that unburned hydrocarbons (HC) normally emitted to the exhaust stream are recycled through a second combustion event, reducing HC emissions as well. Although VCT engines eliminate the normal EGR system dynamics, the fundamentally multivariable nature of the resulting system presages a difficult engine control problem.

Air-Fuel Ratio Control

Historically, fuel control was accomplished by a carburetor that used a venturi arrangement and a simple float-and-valve mechanism to meter the proper amount of fuel to the engine. For special operating conditions, such as idle or acceleration, additional mechanical and vacuum circuitry was required to assure satisfactory engine operation and good drivability. The demise of the carburetor was occasioned by the advent of three-way catalytic converters (TWC) for emission control. These devices simultaneously convert oxidizing [HC and carbon monoxide (CO)] and reducing (NOx) species in the exhaust, but, as shown in Figure 74.1, require precise control of A/F to the stoichiometric value to be effective. Consequently, the electronic fuel system of

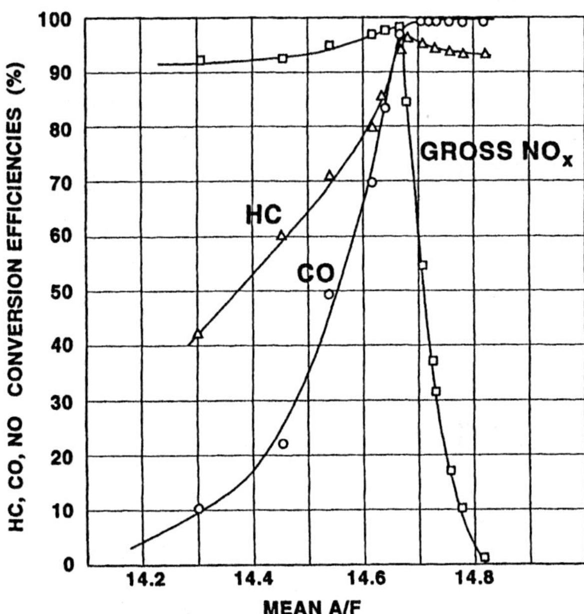

Figure 74.1 Typical TWC efficiency curves.

a modern spark ignition automobile engine employs individual fuel injectors located in the inlet manifold runners close to the intake valves to deliver accurately timed and metered fuel to all cylinders. The injectors are regulated by an A/F control system that has two primary components: a feedback portion, in which a signal related to A/F from an exhaust gas oxygen (EGO) sensor is fed back through a digital controller to regulate the pulse width command sent to the fuel injectors; and a feedforward portion, in which injector fuel flow is adjusted in response to a signal from an air flow meter.

The feedback, or closed-loop portion of the control system, is fully effective only under steady-state conditions and when the EGO sensor has attained the proper operating temperature. The feedforward, or open-loop portion of the control system, is particularly important when the engine is cold (before the closed-loop A/F control is operational) and during transient operation [when the significant delay between the injection of fuel (usually during the exhaust stroke, just before the intake valve opens) and the appearance of a signal at the EGO sensor (possibly long after the conclusion of the exhaust stroke) inhibits good control]. First, in Section 74.1.2, the open-loop A/F control problem is examined with emphasis on accounting for sensor dynamics. Then, the closed-loop problem is addressed from a modern control systems perspective, where individual cylinder control of A/F is accomplished using a single EGO sensor.

Idle Speed Control

In addition to these essential tasks of controlling ignition, A/F, and EGR, the typical on-board microprocessor performs many other diagnostic and control functions. These include electric fan control, purge control of the evaporative emissions canister, turbocharged engine wastegate control, overspeed control, electronic transmission shift scheduling and control, cruise control, and idle speed control (ISC). The ISC requirement is to maintain constant engine RPM at closed throttle while rejecting disturbances such as automatic transmission neutral-to-drive transition, air conditioner compressor engagement, and power steering lock-up.

The idle speed problem is a difficult one, especially for small engines at low speeds where marginal torque reserve is available for disturbance rejection. The problem is made more challenging by the fact that significant parameter variation can be expected over the substantial range of environmental conditions in which the engine must operate. Finally, the ISC design is subject not only to quantitative performance requirements, such as overshoot and settling time, but also to more subjective measures of performance, such as idle quality and the degree of noise and vibration communicated to the driver through the body structure. The ISC problem is addressed in Section 74.1.3.

74.1.2 Air-Fuel Ratio Control System Design

Due to the precipitous falloff of TWC efficiency away from stoichiometry, the primary objective of the A/F control system is to maintain the fuel metering in a stoichiometric proportion to the incoming air flow [the only exception to this occurs in heavy load situations where a rich mixture is required to avoid premature detonation (or knock) and to keep the TWC from overheating]. Variation in air flow commanded by the driver is treated as a disturbance to the system. A block diagram of the control structure is illustrated in Figure 74.2, and the two major subcomponents treated here are highlighted in bold outline. The first part of this section describes the development and implementation of a cylinder air charge estimator for predicting the air charge entering the cylinders downstream of the intake manifold plenum on the basis of available measurements of air mass flow rate upstream

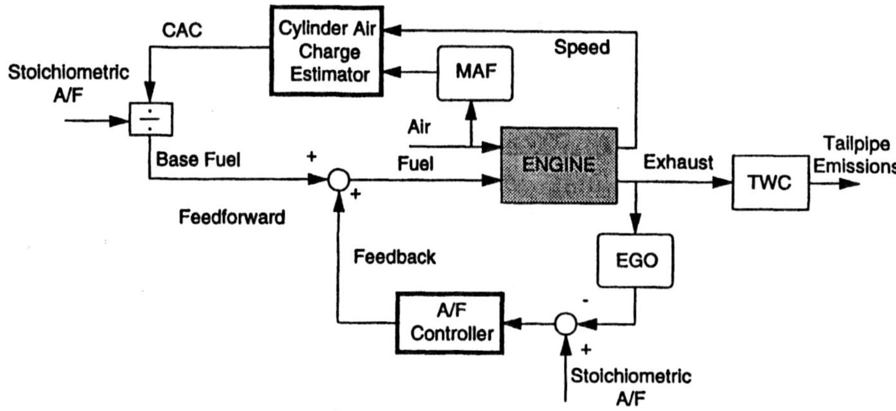

Figure 74.2 Basic A/F control loop showing major feedforward and feedback elements.

of the throttle. The air charge estimate is used to form the base fuel calculation, which is often then modified to account for any fuel-puddling dynamics and the delay associated with closed-valve fuel injection timing. Finally, a classical, time-invariant, single-input single-output (SISO) PI controller is normally used to correct for any persistent errors in the open-loop fuel calculation by adjusting the average A/F to perceived stoichiometry.

Even if the average A/F is controlled to stoichiometry, individual cylinders may be operating consistently rich or lean of the desired value. This cylinder-to-cylinder A/F maldistribution is due, in part, to injector variability. Consequently, fuel injectors are machined to close tolerances to avoid individual cylinder flow discrepancies, resulting in high cost per injector. However, even if the injectors are perfectly matched, maldistribution can arise from individual cylinders having different breathing characteristics due to a combination of factors, such as intake manifold configuration and valve characteristics. It is known that such A/F maldistribution can result in increased emissions due to shifts in the closed-loop A/F setpoint relative to the TWC [9]. The second half of this section describes the development of a nonclassical, periodically time-varying controller for tuning the A/F in each cylinder to eliminate this maldistribution.

Hardware Assumptions

The modeling and control methods presented here are applicable to just about any fuel-injected engine. For illustration purposes only, it is assumed that the engine is a port fuel-injected V8 with independent fuel control for each bank of cylinders. The cylinders are numbered one through four, starting from the front of the right bank, and five through eight, starting from the front of the left bank. The firing order of the engine is 1-3-7-2-6-5-4-8, which is not symmetric from bank to bank. Fuel injection is timed to occur on a closed valve prior to the intake stroke (induction event). For the purpose of closed-loop control, the engine is equipped with a switching type EGO sensor located at the confluence of the individual exhaust runners and just upstream of the catalytic converter. Such sensors typically incorporate a ZrO_2 ceramic thimble employing a platinum catalyst on the exterior surface to equilibrate the exhaust gas mixture. The interior surface of the sensor is exposed to the atmosphere. The output voltage is exponentially related to the ratio of O_2 partial pressures across the ceramic, and thus the sensor is essentially a switching device indicating by its state whether the exhaust gas is rich or lean of stoichiometry.

Cylinder Air Charge Computation

This section describes the development and implementation of an air charge estimator for an eight-cylinder engine. A very real practical problem is posed by the fact that the hot-wire anemometers currently used to measure mass air flow rate have relatively slow dynamics. Indeed, the time constant of this sensor is often on the order of an induction event for an engine speed of 1500 RPM and is only about four to five times faster than the dynamics of the intake manifold. Taking these dynamics into account in the air charge estimation algorithm can significantly improve the accuracy of the algorithm and have substantial benefits for reducing emissions.

Basic Model The air path of a typical engine is depicted in Figure 74.3. An associated lumped-parameter phenomenological model suitable for developing an on-line cylinder air charge estimator [2] is now described. Let P, V, T and m be the pressure in the intake manifold (psi), volume of the intake manifold and runners (liters), temperature (°R), and mass (lbm) of the air in the intake manifold, respectively. Invoking the ideal gas law, and assuming that the manifold air temperature is slowly varying, leads to

$$\frac{d}{dt}P = \frac{RT}{V}[MAF_a - Cyl(N, P, T_{EC}, T_i)], \qquad (74.1)$$

where MAF_a is the actual mass air flow metered in by the throttle, R is the molar gas constant, $Cyl(N, P, T_{EC}, T_i)$ is the average instantaneous air flow pumped out of the intake manifold by the cylinders, as a function of engine speed, N (RPM), manifold pressure, engine coolant temperature, T_{EC} (^oR), and air inlet temperature, T_i (^oR). It is assumed that both MAF_a and $Cyl(N, P, T_{EC}, T_i)$ have units of lbm/s.

The dependence of the cylinder pumping or induction function on variations of the engine coolant and air inlet temperatures

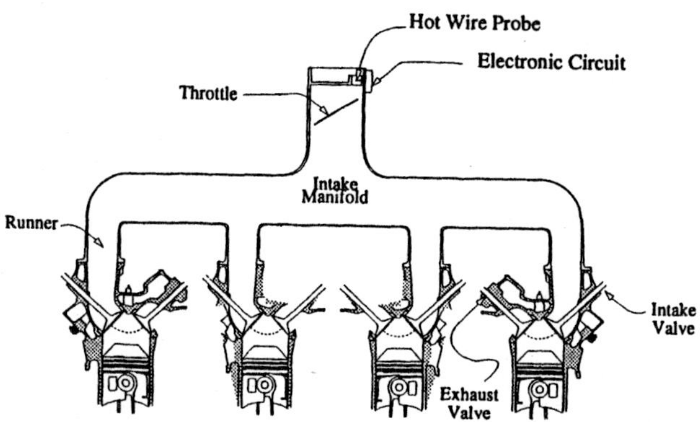

Figure 74.3 Schematic diagram of air path in engine.

is modeled empirically by [10], as

$$Cyl(N,P,T_{EC},T_i) = Cyl(N,P)\sqrt{\frac{T_i}{T_i^{\text{mapping}}}}\;\frac{T_{EC}^{\text{mapping}}+2460}{T_{EC}+2460}, \quad (74.2)$$

where the superscript "mapping" denotes the corresponding temperatures (^oR) at which the function $Cyl(N, P)$ is determined, based on engine mapping data. An explicit procedure for determining this function is explained in the next subsection.

Cylinder air charge per induction event, CAC, can be determined directly from Equation 74.1. In steady state, the integral of the mass flow rate of air pumped out of the intake manifold over two engine revolutions, divided by the number of cylinders, is the air charge per cylinder. Since engine speed is nearly constant over a single induction event, and the time in seconds for two engine revolutions is $\frac{120}{N}$, a good approximation of the inducted air charge on a per-cylinder basis is given by

$$CAC = \frac{120}{nN} Cyl(N,P,T_{EC},T_i)\ \text{lbm}, \quad (74.3)$$

where n is the number of cylinders.

The final element to be incorporated in the model is the mass air flow meter. The importance of including this was demonstrated in [2]. For the purpose of achieving rapid on-line computations, a simple first-order model is used

$$\gamma\frac{d}{dt}MAF_m + MAF_m = MAF_a\,, \quad (74.4)$$

where MAF_m is the measured mass air flow and γ is the time constant of the air meter. Substituting the left-hand side of Equation 74.4 for MAF_a in Equation 74.1 yields

$$\frac{d}{dt}P = \frac{RT}{V}\left[\gamma\frac{d}{dt}MAF_m + MAF_m - Cyl(N,P,T_{EC},T_i)\right] \quad (74.5)$$

To eliminate the derivative of MAF_m in Equation 74.5, let $x = P - \gamma\frac{RT}{V}MAF_m$. This yields

$$\frac{d}{dt}x = \frac{RT}{V}\left[MAF_m - Cyl(N, x+\gamma\frac{RT}{V}MAF_m, T_{EC}, T_i)\right]. \quad (74.6)$$

Cylinder air charge is then computed from Equation 74.3 as

$$CAC = \frac{120}{nN}\text{Cyl}(N, x+\gamma\frac{RT}{V}MAF_m, T_{EC}, T_i)\,. \quad (74.7)$$

Note that the effect of including the mass air flow meter's dynamics is to add a feedforward term involving the mass air flow rate to the cylinder air charge computation. When $\gamma = 0$, Equations 74.6 and 74.7 reduce to an estimator that ignores the air meter's dynamics or, equivalently, treats the sensor as being infinitely fast.

Determining Model Parameters The pumping function $Cyl(N, P)$ can be determined on the basis of steady-state engine mapping data. Equip the engine with a high-bandwidth manifold absolute pressure (MAP) sensor and exercise the engine over the full range of speed and load conditions while recording the steady-state value of the instantaneous mass air flow rate as a function of engine speed and manifold pressure. For this purpose, any external exhaust gas recirculation should be disabled. A typical data set should cover every 500 RPM of engine speed from 500 to 5,000 RPM and every half psi of manifold pressure from 3 psi to atmosphere. For the purpose of making these measurements, it is preferable to use a laminar air flow element as this, in addition, allows the calibration of the mass air flow meter to be verified. Record the engine coolant and air inlet temperatures for use in Equation 74.2. The function $Cyl(N, P)$ can be represented as a table lookup or as a polynomial regressed against the above mapping data. In either case, it is common to represent it in the following functional form:

$$Cyl(N,P) = \mu(N)P + \beta(N). \quad (74.8)$$

The time constant of the air meter is best determined by installing the meter on a flow bench and applying step or rapid sinusoidal variations in air flow to the meter. Methods for fitting an approximate first-order model to the data can be found in any textbook on classical control. A typical value is $\gamma = 20$ ms. Though not highly recommended, a value for the time constant can be determined by on-vehicle calibration, if an accurate determination of the pumping function has been completed. This is explained at the end of the next subsection.

Model Discretization and Validation The estimator modeled by Equations 74.6 and 74.7 must be discretized for implementation. In engine models, an event-based sampling scheme is often used [7]. For illustration purposes, the discretization is carried out here for a V8; the modifications required for other configurations will be evident. Let k be the recursion index and let Δt_k be the elapsed time in seconds per 45 degrees of crank-angle advancement, or $\frac{1}{8}$ revolution; that is, $\Delta t_k = \frac{7.5}{N_k}$ s, where N_k is the current engine speed in RPM. Then Equation 74.6 can be Euler-integrated as

$$x_k = x_{k-1} + \Delta t_k\frac{RT_{k-1}}{V}\Big[MAF_{m,k-1} - Cyl(N_{k-1}, x_{k-1} + \gamma\frac{RT_{k-1}}{V}MAF_{m,k-1}, T_{EC}, T_i)\Big] \quad (74.9)$$

The cylinder air charge is calculated by

$$CAC_k = 2\Delta t_k Cyl(N_k, x_k + \gamma \frac{RT_k}{V} MAF_{m,k}, T_{EC}, T_i) \quad (74.10)$$

and need be computed only once per 90 crank-angle degrees.

The accuracy of the cylinder air charge model can be easily validated on an engine dynamometer equipped to maintain constant engine speed. Apply very rapid throttle tip-ins and tip-outs, as in Figure 74.4, while holding the engine speed constant. If the model parameters have been properly determined, the calculated manifold pressure accurately tracks the measured manifold pressure. Figure 74.5 illustrates one such test at 1500 RPM. The dynamic responses of the measured and computed values match up quite well. There is some inaccuracy in the quasi-steady-state values at 12 psi; this corresponds to an error in the pumping function $Cyl(N, P)$ at high manifold pressures, so, in this operating condition, it should be reevaluated.

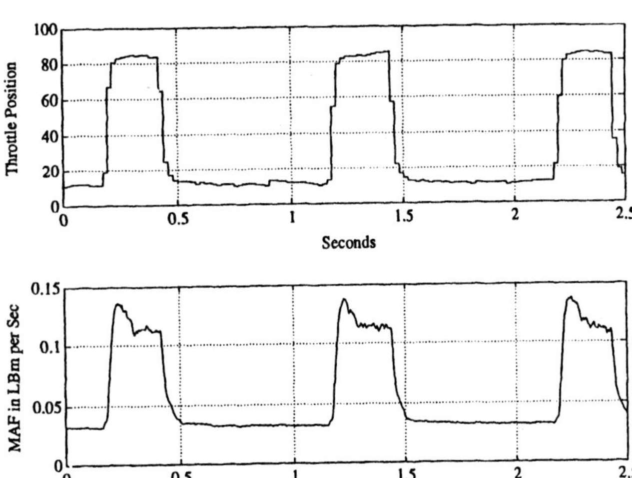

Figure 74.4 Engine operating conditions at nominal 1500 RPM.

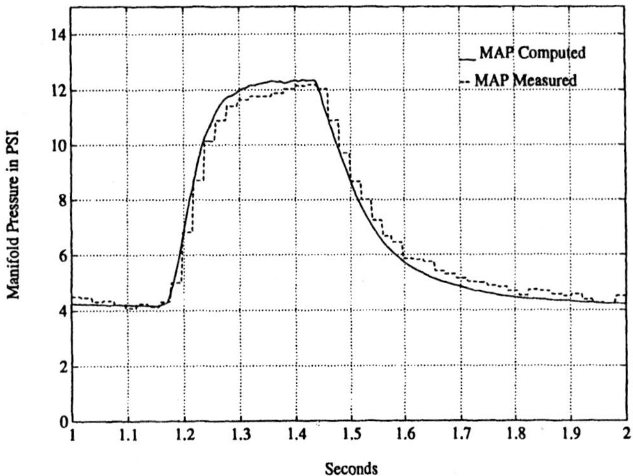

Figure 74.5 Comparison of measured and computed manifold pressure.

From Figure 74.5, it can be seen that the air meter time constant has been accurately identified in this test. If the value for γ in Equation 74.4 had been chosen too large, then the computed manifold pressure would be leading the measured manifold pressure; conversely, if γ were too small, the computed value would lag the measured value.

Eliminating A/F Maldistribution through Feedback Control

A/F maldistribution is evidenced by very rapid switching of the EGO sensor on an engine-event-by-engine-event basis. Such cylinder-to-cylinder A/F maldistribution can result in increased emissions due to shifts in the closed-loop A/F setpoint relative to the TWC [9]. The trivial control solution, which consists of placing individual EGO sensors in each exhaust runner and wrapping independent PI controllers around each injector-sensor pair, is not acceptable from an economic point of view; technically, there would also be problems due to the nonequilibrium condition of the exhaust gas immediately upon exiting the cylinder. This section details the development of a controller that eliminates A/F maldistribution on the basis of a single EGO sensor per engine bank. The controller is developed for the left bank of the engine.

Basic Model A control-oriented block diagram for the A/F system of the left bank of an eight-cylinder engine is depicted in Figure 74.6. The model evolves at an engine speed-dependent sampling interval of 90 crank-angle degrees consistent with the eight-cylinder geometry (one exhaust event occurs every 90 degrees of crankshaft rotation). The engine is represented by injector gains G1 through G8 and pure delay z^{-d}, which accounts for the number of engine events that occur from the time that an individual cylinder's fuel injector pulse width is computed until the corresponding exhaust valve opens to admit the mixture to the exhaust manifold. The transfer function H(z) represents the mixing process of the exhaust gases from the individual exhaust ports to the EGO sensor location, including any transport delay. The switching type EGO sensor is represented by a first-order transfer function followed by a preload (switch) nonlinearity.

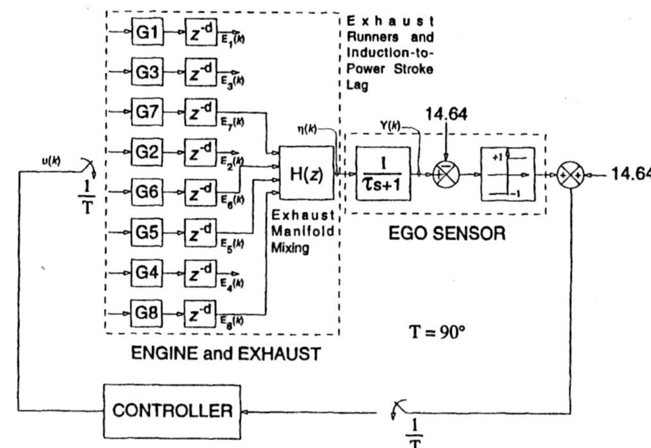

Figure 74.6 Control-oriented block diagram of A/F subsystem.

Note that only cylinders 5 through 8 are inputs to the mixing model, H(z). This is due to the fact that the separate banks of the V8 engine are controlled independently. The gains and delays for cylinders 1 through 4 correspond to the right bank of the engine and are included in the diagram only to represent the firing order. Note furthermore that cylinders 5 and 6 exhaust within 90 degrees of one another, whereas cylinders 7 and 8 exhaust 270 degrees apart. Since the exhaust stroke of cylinder 6 is not complete before the exhaust stroke of cylinder 5 commences, for any exhaust manifold configuration, there is mixing of the exhaust gases from these cylinders at the point where the EGO sensor samples the exhaust gas. We adopt the notation that the sampling index, k, is a multiple of 90 degrees, that is, $x(k)$ is the quantity x at $k\cdot 90$ degrees; moreover, if we are looking at a signal's value during a particular point of an engine cycle, then we will denote this by $x(8k + j)$, which is x at $(8k + j) \cdot 90$ degrees, or x at the j-th event of the k-th engine cycle. The initial time will be taken as $k = 0$ at TDC of the compression stroke of cylinder 1. The basic model for a V6 or a four-cylinder engine is simpler; see [3].

A dynamic model of the exhaust manifold mixing is difficult to determine with current technology. This is because a linear A/F measurement is required, and, currently, such sensors have a slow dynamic response in comparison with the time duration of an individual cylinder event. Hence, standard system identification methods break down. In [6], a model structure for the mixing dynamics and an attendant model parameter identification procedure, compatible with existing laboratory sensors, is provided. This is outlined next.

The key assumption used to develop a mathematical model of the exhaust gas mixing is that once the exhaust gas from any particular cylinder reaches the EGO sensor, the exhaust of that cylinder from the previous cycle (two revolutions) has been completely evacuated from the exhaust manifold. It is further assumed that the transport lag from the exhaust port of any cylinder to the sensor location is less than two engine cycles. With these assumptions, and with reference to the timing diagram of Figure 74.7, a model for the exhaust-mixing dynamics may be expressed as relating the A/F at the sensor over one 720-crank-angle-degree period beginning at ($8k$) as a linear combination of the A/Fs admitted to the exhaust manifold by cylinder 5 during the exhaust strokes occurring at times ($8k$+7), ($8k$-1), and ($8k$-9); by cylinder 6 at ($8k$+6), ($8k$-2), and ($8k$-10); by cylinder 7 at ($8k$+4), ($8k$-4), and ($8k$-12); and by cylinder 8 at ($8k$+1), ($8k$-7), and ($8k$-15). This relationship is given by

$$\begin{bmatrix} \eta(8k) \\ \eta(8k+1) \\ \vdots \\ \eta(8k+6) \\ \eta(8k+7) \end{bmatrix} = \begin{bmatrix} a_5(1,0) & \dots & a_8(1,0) \\ \vdots & & \\ a_5(8,0) & \dots & a_8(8,0) \end{bmatrix} \begin{bmatrix} E_5(8k+7) \\ E_6(8k+6) \\ E_7(8k+4) \\ E_8(8k+1) \end{bmatrix} + \begin{bmatrix} a_5(1,1) & \dots & a_8(1,1) \\ \vdots & & \\ a_5(8,1) & \dots & a_8(8,1) \end{bmatrix} \begin{bmatrix} E_5(8k-1) \\ E_6(8k-2) \\ E_7(8k-4) \\ E_8(8k-7) \end{bmatrix} + \begin{bmatrix} a_5(1,2) & \dots & a_8(1,2) \\ \vdots & & \\ a_5(8,2) & \dots & a_8(8,2) \end{bmatrix} \begin{bmatrix} E_5(8k-9) \\ E_6(8k-10) \\ E_7(8k-12) \\ E_8(8k-15) \end{bmatrix} \tag{74.11}$$

where η is the actual A/F at the production sensor location, E_n is the exhaust gas A/F from cylinder n ($n = 5, 6, 7, 8$), and a_n is the time-dependent fraction of the exhaust gas from cylinder n contributing to the A/F fuel ratio at the sensor. It follows from the key assumption that only 32 of the 96 coefficients in Equation 74.11 can be nonzero. Specifically, every triplet $\{a_n(k,0), a_n(k,1), a_n(k,2)\}$ has, at most, one nonzero element. This is exploited in the model parameter identification procedure.

Determining Model Parameters The pure delay z^{-d} is determined by the type of injection timing used (open or closed valve) and does not vary with engine speed or load. A typical value for closed-valve injection timing is $d = 8$. The time constant of the EGO sensor is normally provided by the manufacturer; if not, it can be estimated by installing it directly in the exhaust runner of one of the cylinders and controlling the fuel pulse width to cause a switch from rich to lean and then lean to rich. A typical average value of these two times is $\tau = 70$ ms.

The first step towards identifying the parameters in the exhaust-mixing model is to determine which one of the parameters $\{a_n(k,0), a_n(k,1), a_n(k,2)\}$ is the possibly nonzero element; this can be uniquely determined on the basis of the transport delay between the opening of the exhaust valve of each cylinder and the time of arrival of the corresponding exhaust gas pulse at the EGO sensor. The measurement of this delay is accomplished by installing the fast, switching type EGO sensor in the exhaust manifold in the production location and carefully balancing the A/F of each cylinder to the stoichiometric value. Then apply a step change in A/F to each cylinder and observe the time delay. The results of a typical test are given in [6]. The transport delays change as a function of engine speed and load and, thus, should be determined at several operating points. In addition, they may not be a multiple of 90 degrees. A practical method to account for these issues through a slightly more sophisticated sampling schedule is outlined in [3].

At steady state, Equation 74.11 reduces to

$$\begin{bmatrix} \eta(8k) \\ \eta(8k+1) \\ \vdots \\ \eta(8k+6) \\ \eta(8k+7) \end{bmatrix} = A_{mix} \begin{bmatrix} E_5(8k-1) \\ E_6(8k-2) \\ E_7(8k-4) \\ E_8(8k+1) \end{bmatrix} \tag{74.12}$$

where

$$A_{mix} = \begin{bmatrix} \sum_{j=0}^{2} a_5(1,j) & \dots & \sum_{j=0}^{2} a_8(1,j) \\ \vdots & \dots & \vdots \\ \sum_{j=0}^{2} a_5(8,j) & \dots & \sum_{j=0}^{2} a_8(1,j) \end{bmatrix} \tag{74.13}$$

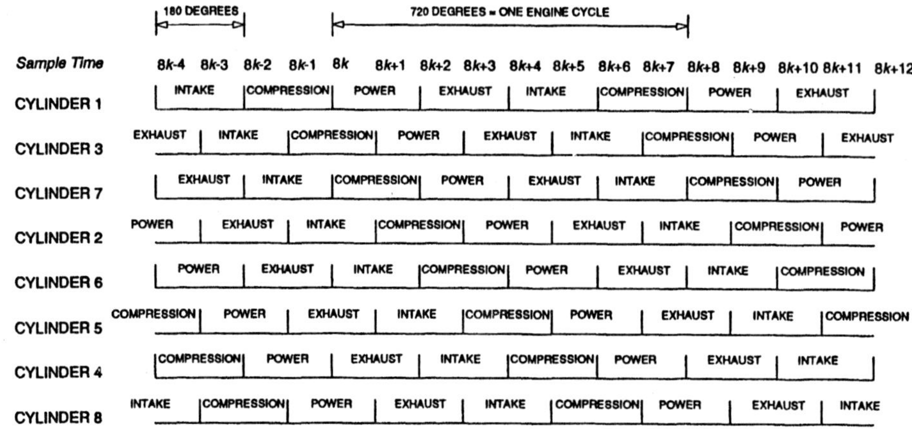

Figure 74.7 Timing diagram for eight-cylinder engine.

This leads to the second part of the parameter identification procedure in which the values of the summed coefficients of Equation 74.12 may be identified from "steady-state" experiments performed with a linear EGO sensor installed in the production location. Then, knowing which of the coefficients is nonzero, Equation 74.11 may be evaluated.

Install a linear EGO sensor in the production location. Then the measured A/F response, $y(k)$, to the sensor input, $\eta(k)$, is modeled by

$$\begin{aligned} w(k+1) &= \alpha w(k) + (1-\alpha)\eta(k) \\ y(k) &= w(k), \end{aligned} \tag{74.14}$$

where $\alpha = e^{-T/\tau_L}$, τ_L is the time constant of the linear EGO sensor, and T is the sampling interval, that is, the amount of time per 90 degrees of crank-angle advance. It follows [6] that the combined steady-state model of exhaust-mixing and linear EGO sensor dynamics is

$$Y = Q_s A_{mix} E \quad , \tag{74.15}$$

where

$$Y = \begin{bmatrix} y(8k) \\ y(8k+1) \\ \vdots \\ y(8k+6) \\ y(8k+7) \end{bmatrix}, \quad E = \begin{bmatrix} E_5(8k-1) \\ E_6(8k-2) \\ E_7(8k-4) \\ E_8(8k+1) \end{bmatrix},$$

and

$$Q_s = \frac{1-\alpha}{1-\alpha^8} \begin{bmatrix} \alpha^7 & \alpha^6 & \dots & \alpha & 1 \\ \alpha^8 & \alpha^7 & \dots & \alpha^2 & \alpha \\ \vdots & \vdots & \dots & \vdots & \vdots \\ \alpha^{13} & \alpha^{12} & \dots & \alpha^7 & \alpha^6 \\ \alpha^{14} & \alpha^{13} & \dots & \alpha^8 & \alpha^7 \end{bmatrix} + \begin{bmatrix} 0 & 0 & \dots & 0 & 0 \\ 1-\alpha & 0 & \dots & 0 & 0 \\ \vdots & \vdots & \dots & \vdots & \vdots \\ \alpha^5(1-\alpha) & \alpha^4(1-\alpha) & \dots & 0 & 0 \\ \alpha^6(1-\alpha) & \alpha^5(1-\alpha) & \dots & 1-\alpha & 0 \end{bmatrix}.$$

Next, carefully balance each of the cylinders to the stoichiometric A/F, then offset each cylinder, successively, 1 A/F rich and then 1 A/F lean to assess the effect on the A/F at the sensor location. At each condition, let the system reach steady state, then record Y and E over three to ten engine cycles, averaging the components of each vector in order to minimize the impact of noise and cycle-to-cycle variability in the combustion process. This provides input A/F vectors $\bar{E} = [\bar{E}^1 \dots \bar{E}^8]$ and output A/F vectors $\bar{Z} = [\bar{Y}^1 \dots \bar{Y}^8]$, where the overbar represents the averaged value. The least squares solution to Equation 74.15 is then given by

$$A_{mix} = Q_s^{-1} \bar{Z} \bar{E}^T (\bar{E} \bar{E}^T)^{-1} . \tag{74.16}$$

The identified coefficients of A_{mix} should satisfy two conditions: (1) the entries in the matrix lie in the interval [0, 1]; (2) the sum of the entries in any row of the matrix is unity. These conditions correspond to no chemical processes occurring in the exhaust system (which could modify the A/F) between the exhaust valve and the EGO sensor. Inherent nonlinearities in the "linear" EGO sensor or errors in the identification of its time constant often lead to violations of these conditions. In this case, the following fix is suggested. For each row of the matrix, identify the largest negative entry and subtract it from each entry so that all are non-negative; then scale the row so that its entries sum to one.

Assembling and Validating the State-Space Model A state-space model will be used for control law design. The combined dynamics of the A/F system from the fuel scheduler to the EGO sensor is shown in Figure 74.8. The coefficients $\kappa_1(k), \dots, \kappa_{16}(k)$ arise from constructing a state-space representation of Equation 74.11 and, thus, are directly related to the $a_n(k, j)$. In particular, they are periodic, with period equal to one engine cycle. Figure 74.9 provides an example of these coefficients for the model identified in [6]. Assigning state variables as indicated, the state-space model can be expressed as

$$\begin{aligned} x(k+1) &= A(k)x(k) + B(k)u(k) \\ y(k) &= C(k)x(k) . \end{aligned} \tag{74.17}$$

This is an 8-periodic SISO system.

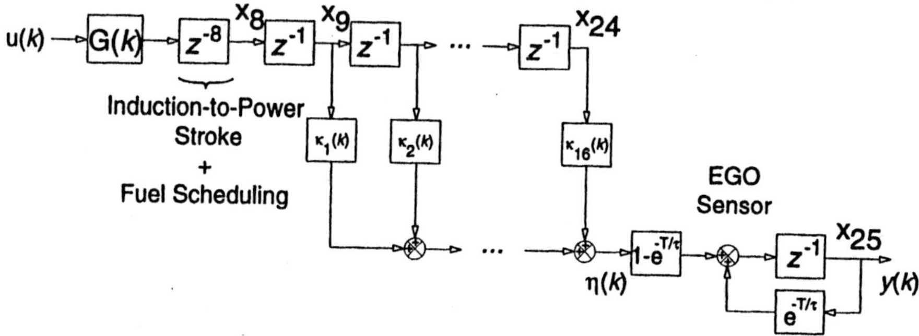

Figure 74.8 Periodically time-varying model of engine showing state variable assignments.

k	$\kappa_1(k)$	$\kappa_2(k)$	$\kappa_3(k)$	$\kappa_4(k)$	$\kappa_5(k)$	$\kappa_6(k)$	$\kappa_7(k)$	$\kappa_8(k)$	$\kappa_9(k)$	$\kappa_{10}(k)$	$\kappa_{11}(k)$	$\kappa_{12}(k)$	$\kappa_{13}(k)$	$\kappa_{14}(k)$	$\kappa_{15}(k)$	$\kappa_{16}(k)$
0	0	$a_6(1,1)$	0	$a_7(1,1)$	0	0	$a_8(1,1)$	0	$a_5(1,2)$	0	0	0	0	0	0	0
1	0	0	$a_6(2,1)$	0	$a_7(2,1)$	0	0	$a_8(2,1)$	0	$a_5(2,2)$	0	0	0	0	0	0
2	$a_8(3,0)$	0	0	$a_6(3,1)$	0	$a_7(3,1)$	0	0	0	0	$a_5(3,2)$	0	0	0	0	0
3	0	$a_8(4,0)$	0	0	$a_6(4,1)$	0	$a_7(4,1)$	0	0	0	0	$a_5(4,2)$	0	0	0	0
4	0	0	$a_8(5,0)$	0	0	$a_6(5,1)$	0	$a_7(5,1)$	0	0	0	0	$a_5(5,2)$	0	0	0
5	0	0	0	$a_8(6,0)$	0	0	$a_6(6,1)$	0	$a_7(6,1)$	0	0	0	0	$a_5(6,2)$	0	0
6	0	$a_7(7,0)$	0	0	$a_8(7,0)$	0	0	$a_6(7,1)$	0	0	0	0	0	0	$a_5(7,2)$	0
7	0	0	$a_7(8,0)$	0	0	$a_8(8,0)$	0	0	$a_6(8,1)$	0	0	0	0	0	0	$a_5(8,2)$

Figure 74.9 Time-dependent coefficients for Figure 74.8.

For control design, it is convenient to transform this system, via lifting [1], [5], to a linear, time-invariant multiple-input multiple-output (MIMO) system as follows. Let $\bar{x}(k) = x(8k)$, $Y(k) = [y(8k), \ldots, y(8k+7)]^T$, $U(k) = [u(8k), \ldots, u(8k+7)]^T$. Then

$$\begin{aligned} \bar{x}(k+1) &= \bar{A}\bar{x}(k) + \bar{B}U(k) \quad , \\ Y(k) &= \bar{C}\bar{x}(k) + \bar{D}U(k) \end{aligned} \tag{74.18}$$

where

$$\begin{aligned} \bar{A} &= A(7)A(6)\cdots A(1)A(0), \\ \bar{B} &= [A(7)A(6)\cdots A(1)B(0) : A(7)A(6)\cdots \\ &\quad \cdots A(2)B(1) ; \cdots : A(7)B(6) : B(7)] \ , \\ \bar{C} &= \begin{bmatrix} C(0) \\ C(1)A(0) \\ \vdots \\ C(6)A(5)\cdots A(0) \\ C(7)A(6)\cdots A(0) \end{bmatrix}, \end{aligned} \tag{74.19}$$

$$\bar{D} = \begin{bmatrix} 0 & \vdots & 0 & \cdots & 0 & \vdots & 0 \\ C(1)B(0) & \vdots & 0 & \cdots & 0 & \vdots & 0 \\ \vdots & \vdots & \vdots & \cdots & \vdots & \vdots & \vdots \\ C(6)A(5)\cdots A(1)B(0) & \vdots & C(6)A(5)\cdots A(2)B(1) & \cdots & 0 & \vdots & 0 \\ C(7)A(6)\cdots A(1)B(0) & \vdots & C(7)A(6)\cdots A(2)B(1) & \cdots & C(7)B(6) & \vdots & 0 \end{bmatrix}.$$

Normally, $\bar{D}$ is identically zero because the time delay separating the input from the sensor is greater than one engine cycle. Since only cylinders 5 through 8 are to be controlled, the $\bar{B}$ and $\bar{D}$ matrices may be reduced by eliminating the columns that correspond to the control variables for cylinders 1 through 4. This results in a system model with four inputs and eight outputs.

Additional data should be taken to validate the identified model of the A/F system. An example of the experimental and modeled response to a unit step input in A/F is shown in Figure 74.10.

Control Algorithm for ICAFC The first step is to check the feasibility of independently controlling the A/F in the four cylinders. This will be possible if and only if[1] the model of Equation 74.18, with all of the injector gains set to unity, has

[1]Since the model is asymptotically stable, it is automatically stabilizable and detectable.

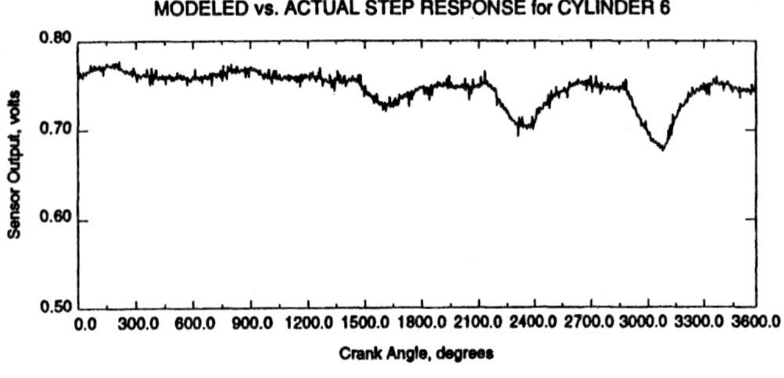

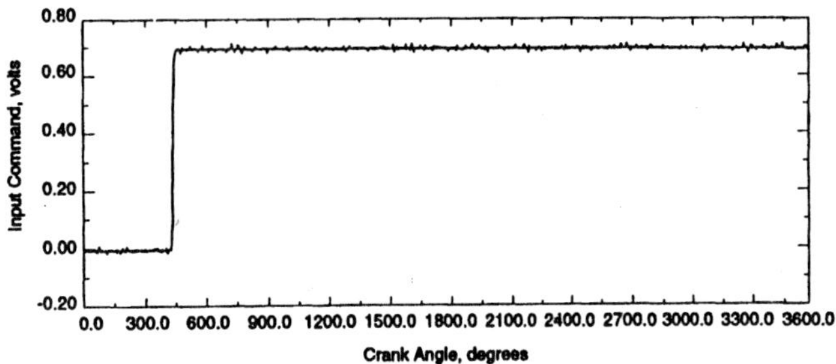

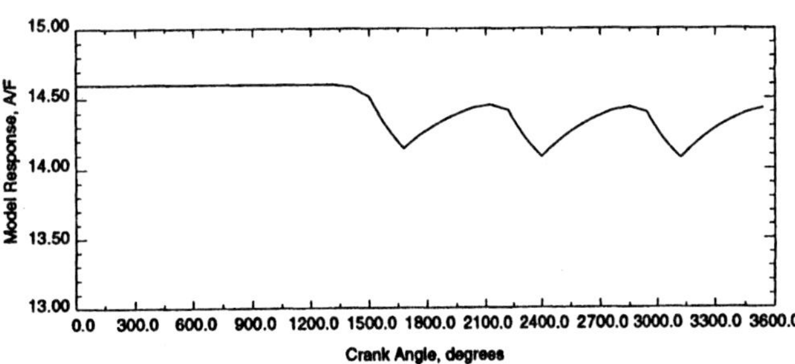

Figure 74.10 Comparison of actual and modeled step response for cylinder number 6.

"full rank[2] at dc" (no transmission zeros at 1). To evaluate this, compute the dc gain of the system

$$G_{dc} = \bar{C}(I - \bar{A})^{-1}\bar{B} + \bar{D}, \tag{74.20}$$

then compute the singular value decomposition (SVD) of G_{dc}. For the regulation problem to be feasible, the ratio of the largest to the fourth largest singular values should be no larger than 4 or 5. If the ratio is too large, then a redesign of the hardware is necessary before proceeding to the next step [6].

In order to achieve individual set-point control on all cylinders, the system model needs to be augmented with four integrators. This can be done on the input side by

$$\begin{aligned} \bar{x}(k+1) &= \bar{A}\bar{x}(k) + \bar{B}U(k) \\ U(k+1) &= U(k) + V(k) \\ Y(k) &= \bar{C}\bar{x}(k) + \bar{D}U(k), \end{aligned} \tag{74.21}$$

where $V(k)$ is the new control variable; or on the output side. To do the latter, the four components of Y that are to be regulated to stoichiometry must be selected. One way to do this is to choose four components of Y on the basis of achieving the best numerically conditioned dc gain matrix when the other four output components are deleted. Denote the resulting reduced output by $\tilde{Y}(k)$. Then integrators can be added as

$$\begin{aligned} \bar{x}(k+1) &= \bar{A}\bar{x}(k) + \bar{B}U(k) \\ W(k+1) &= W(k) + \Delta\tilde{Y}_m(k) \\ Y(k) &= \bar{C}\bar{x}(k) + \bar{D}U(k), \end{aligned} \tag{74.22}$$

where $\Delta\tilde{Y}_m$ is the error between the measured value of $\tilde{Y}$ and the stoichiometric setpoint.

In either case, it is now very easy to design a stabilizing controller by a host of techniques presented in this handbook. For implementation purposes, the order of the resulting controller can normally be significantly lowered through the use of model reduction methods. Other issues dealing with implementation are discussed in [3], such as how to incorporate the switching aspect of the sensor into the final controller and how to properly schedule the computed control signals. Specific examples of such controllers eliminating A/F maldistribution are given in [3] and [6].

74.1.3 Idle Speed Control

Engine idle is one of the most frequently encountered operating conditions for city driving. The quality of ISC affects almost every aspect of vehicle performance such as fuel economy, emissions, drivability, etc. The ISC problem has been extensively studied, and a comprehensive overview of the subject can be found in [4].

The primary objective for ISC is to maintain the engine speed at a desired setpoint in the presence of various load disturbances. The key factors to be considered in its design include:

- **Engine speed setpoint.** To maximize fuel economy, the reference engine speed is scheduled at the minimum that yields acceptable combustion quality; accessory drive requirements; and noise, vibration, and harshness (NVH) properties. As the automotive industry strives to reduce fuel consumption by lowering the idle speed, the problems associated with the idle quality (such as allowable speed droop and recovery transient, combustion quality and engine vibration, etc.) tend to be magnified and thus put more stringent requirements on the performance of the control system.
- **Accessory load disturbances.** Typical loads in today's automobile include air conditioning, power steering, power windows, neutral-to-drive shift, alternator loads, etc. Their characteristics and range of operation determine the complexity of the control design and achievable performance.
- **Control authority and actuator limitations.** The control variables for ISC are air flow (regulated by the throttle or a bypass valve) and spark timing. Other variables, such as A/F, also affect engine operation,

[2]Physically, this corresponds to being able to use constant injector inputs to arbitrarily adjust the A/F in the individual cylinders.

but A/F is not considered as a control variable for ISC because it is the primary handle on emissions. The air bypass valve (or throttle) and spark timing are subject to constraints imposed by the hardware itself as well as other engine control design considerations. For example, in order to give spark enough control authority to respond to the load disturbances, it is necessary to retard it from MBT to provide appreciable torque reserve. On the other hand, there is a fuel penalty associated with the retarded spark, which, in theory, can be compensated by the lower idle speed allowed by the increased control authority of spark. The optimal trade-off, however, differs from engine to engine and needs to be evaluated by taking into consideration combustion quality and the ignition hardware constraints (the physical time required for arming the coil and firing the next spark imposes a limitation on the allowable spark advance increase between two consecutive events).

- **Available measurement.** Typically, only engine speed is used for ISC feedback. MAP, or inferred MAP, is also used in some designs. Accessory load sensors (such as the air conditioning switch, neutral-to-drive shift switch, power steering pressure sensor, etc.) are installed in many vehicles to provide information on load disturbances for feedforward control.
- **Variations in engine characteristics over the entire operating range.** The ISC design has to consider different operational and environmental conditions such as temperature, altitude, etc. To meet the performance objectives for a large fleet of vehicles throughout their entire engine life, the control system has to be robust enough to incorporate changes in the plant dynamics due to aging and unit-to-unit variability.

The selection of desired engine setpoint and spark retard is a sophisticated design trade-off process and is beyond the scope of this chapter. The control problem addressed here is the speed tracking problem, which can be formally stated as: *For a given desired engine speed setpoint, design a controller that, based on the measured engine speed, generates commands for the air bypass valve and spark timing to minimize engine speed variations from the setpoint in the presence of load disturbances.* A schematic control system diagram is shown in Figure 74.11.

Engine Models for ISC

An engine model that encompasses the most important characteristics and dynamics of engine idle operation is given in Figure 74.12. It uses the model structure developed in [7] and consists of the actuator characteristics, manifold filling dynamics, engine pumping characteristics, intake-to-power stroke delay, torque characteristics, and engine rotational dynamics (EGR is not considered at idle). The assumption of sonic flow through the throttle, generally satisfied at idle, has led to a much simplified model where the air flow across the throttle is only a function of the throttle position. The differential equations describing the overall dynamics are given by

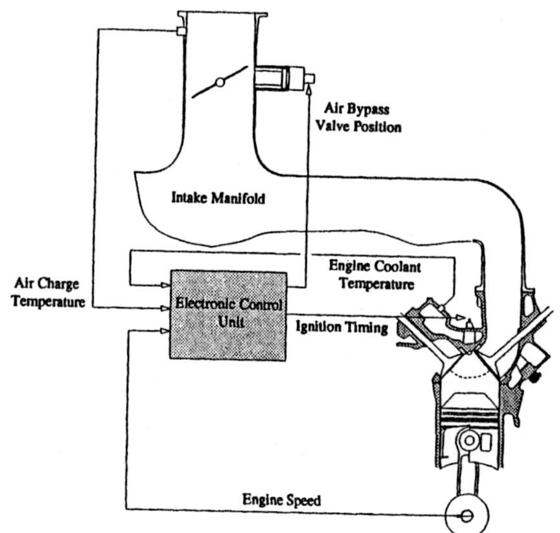

Figure 74.11 Sensor-actuator configuration for ISC.

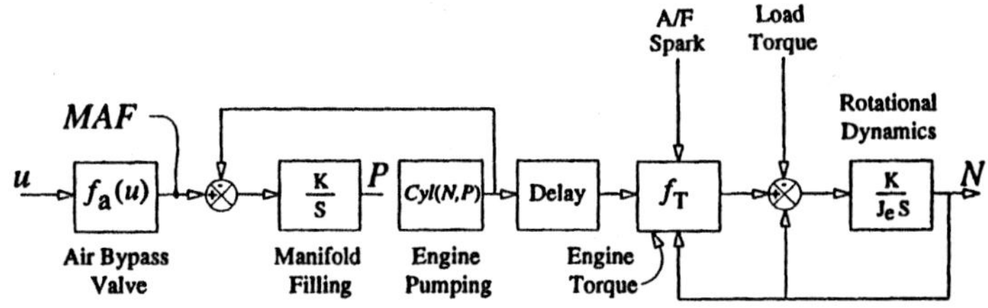

Figure 74.12 Nonlinear engine model.

$$\begin{aligned}
MAF &= f_a(u) \\
\dot{P} &= K_m(MAF - \dot{m}) \\
\dot{m} &= Cyl(N, P) \qquad\qquad (74.23) \\
J_e\dot{N} &= T_q - T_L, \\
T_q(t) &= f_T(\dot{m}(t-\sigma), N(t), r(t-\sigma), \delta(t)) \\
\text{where } u &= \text{duty cycle for the air bypass valve} \\
r &= \text{A/F} \\
\delta &= \text{spark timing in terms of crank-angle degrees before TDC} \\
T_L &= \text{load torque}
\end{aligned}$$

J_e and K_m in Equation 74.23 are two engine-dependent constants, where J_e represents the engine rotational inertia, and K_m is a function of the gas constant, air temperature, manifold volume, etc. Both J_e and K_m can be determined from engine design specifications and given nominal operating conditions. The time delay σ in the engine model equals approximately 180 degrees of crank-angle advance and, thus, is a speed-dependent parameter. This is one reason that models for ISC often use crank-angle instead of time as the independent variable. Additionally, most engine control activities are event driven and synchronized with crank position; the use of $\frac{dM}{d\theta}$, $\frac{dP}{d\theta}$ instead of $\frac{dM}{dt}$, $\frac{dP}{dt}$, respectively, tends to have a linearizing effect on the pumping and torque generation blocks.

Performing a standard linearization procedure results in the linear model shown in Figure 74.13, with inputs Δu, $\Delta\delta$, ΔT_L (the change of the bypass valve duty cycle, spark, and load torque from their nominal values, respectively) and output ΔN (the deviation of the idle speed from the setpoint). The time delay in the continuous-time feedback loop usually complicates the control design and analysis tasks. In a discrete-time representation,

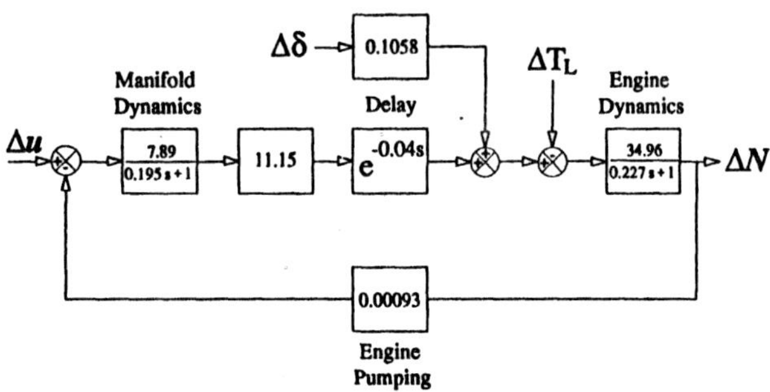

Figure 74.13 Linearized model for typical eight-cylinder engine.

however, the time delay in Figure 74.13 corresponds to a rational transfer function z^{-n} where n is an integer that depends on the sampling scheme and the number of cylinders. It is generally more convenient to accomplish the controller design using a discrete-time model.

Determining Model Parameters

In the engine model of Equation 74.23, the nonlinear algebraic functions f_a, f_T, Cyl describe characteristics of the air bypass valve, torque generation, and engine pumping blocks. These functions can be obtained by regressing engine dynamometer test data, using least squares or other curve-fitting algorithms. The torque generation function is developed on the engine dynamometer by individually sweeping ignition timing, A/F, and mass flow rate (regulated by throttle or air bypass valve position) over their expected values across the idle operating speed range. For a typical eight-cylinder engine, the torque regression is given by

$$\begin{aligned} T_q &= f_T(MAF, N, r, \delta) \\ &= -28.198 + 128.38MAF - 0.196N \\ &\quad + 13.845r - 0.306\delta - 5.669MAF^2 \\ &\quad + 7.39 \times 10^{-5}N^2 - 0.6257r^2 - 0.0257\delta^2 \\ &\quad - 0.0379MAF \cdot N + 0.2843MAF \cdot r \\ &\quad - 0.2483MAF \cdot \delta - 0.00059N \cdot r \\ &\quad + 0.00067N \cdot \delta + 0.0931r \cdot \delta \end{aligned}$$

The steady-state speed-torque relation to spark advance is illustrated in Figure 74.14.

For choked (i.e., sonic) flow, the bypass valve's static relationship is developed simply by exercising the actuator over its operating envelope and measuring either mass airflow using a

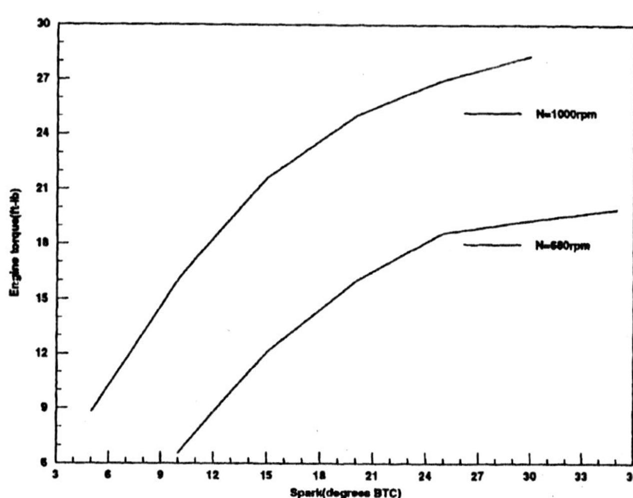

Figure 74.14 Spark-torque relation for different engine speeds (with A/F=14.64).

hot-wire anemometer or volume flow by measuring the pressure drop across a calibrated laminar flow element. The dynamic elements in the linearized idle speed model can be obtained by linearizing the model in Section 74.1.2, or they can be estimated by evaluating (small) step response data from a dynamometer. In particular, the intake manifold time constant can be validated by constant-speed, sonic-flow throttle step tests, using a sufficiently high-bandwidth sensor to measure manifold pressure.

ISC Controller Design

The ISC problem lends itself to the application of various control design techniques. Many different design methodologies, ranging from classical (such as PID) to modern (such as LQG, H_∞, adaptive, etc.) and nonconventional (such as neural networks and fuzzy logic) designs have been discussed and implemented [4]. The mathematical models described previously and commercially available software tools can be used to design different control strategies, depending on the implementor's preference and experience. A general ISC system with both feedforward and feedback components is shown in the block diagram of Figure 74.15.

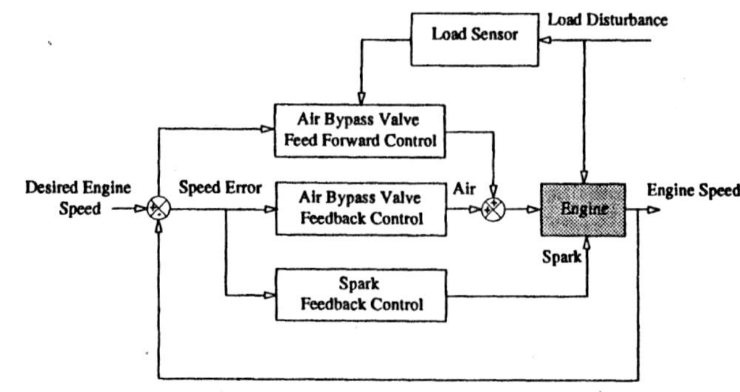

Figure 74.15 General ISC system with feedforward and feedback elements.

Feedforward Control Design Feedforward control is considered as an effective mechanism to reject load disturbances, especially for small engines. When a disturbance is measured (most disturbance sensors used in vehicles are on-off type), control signals can be generated in an attempt to counteract its effect. A typical ISC strategy has feedforward only for the air bypass valve control, and the feedforward is designed based on static engine mapping data. For example, if an air conditioning switch sensor is installed, an extra amount of air will be scheduled to prevent engine speed droop when the air conditioning compressor is engaged. The amount of feedforward control can be determined as follows. At the steady state, since $\dot{P}, \dot{N} \approx 0$, the available engine torque to balance the load torque is related to the mass air flow and engine speed through

$$T_q = f_T(MAF, N, r, \delta).$$

By estimating the load torque presented to the engine by the measured disturbance, one can calculate, for fixed A/F and spark, the amount of air that is needed to maintain the engine speed at the fixed setpoint. The feedforward control can be applied either as a multiplier or an adder to the control signal.

Feedforward control introduces extra cost due to the added sensor and software complexity; thus, it should be used only when necessary. Most importantly, it should not be used to replace the role of feedback in rejecting disturbances, since it does not address the problems of operating condition variations, miscalibration, etc.

Feedback Design Feedback design for ISC can be pursued in many different ways. Two philosophically different approaches are used in developing the control strategy. One is the SISO approach, which treats the air and spark control as separate entities and designs one loop at a time. When the SISO approach is used, the natural separation of time scale in the air and spark dynamics (spark has a fast response compared to air flow, which has a time lag due to the manifold dynamics and intake-to-power delay) suggests that the spark control be closed first as an inner loop. Then the air control, as an outer loop, is designed by including the spark feedback as part of the plant dynamics. Another approach is to treat the ISC as a multiple-input single-output (MISO) or, when the manifold pressure is to be controlled, a MIMO problem. Many control strategies, such as LQ-optimal control and H_∞ have been developed within the MIMO framework. This approach generally leads to a coordinated air and spark control strategy and improved performance.

Despite the rich literature on ISC featuring different control design methodologies, PID control in combination with static feedforward design is still viewed as the control structure of choice in the automotive industry. In many cases, controllers designed using advanced theory, such as H_∞ and LQR, are ultimately implemented in the PID format to reduce complexity and to append engineering content to design parameters. A typical production ISC feedback strategy has a PID for the air bypass valve (or throttle) control and a simple proportional feedback for the spark control. This control configuration is dictated by the following requirements: (1) at steady state, the spark should return to its nominal value independent of the load disturbances; (2) zero steady-state error has to be attained for step disturbances.

Calibration of ISC

Control system development in the automotive industry has been traditionally an empirical process with heavy reliance on manual tuning. As engine control systems have become more complex because of increased functionality, the old-fashioned trial-and-error approach has proved inadequate to achieve optimum performance for interactively connected systems. The trends in today's automotive control system development are in favor of more model-based design and systematic calibration. Tools introduced for performing systematic in-vehicle calibration include dynamic optimization packages, which are used to search for optimal parameters based on a large amount of vehicle data, and controller fine-tuning techniques. Given the reality that most ISC strategies implemented in vehicles are of PID type, we discuss two PID tuning approaches that have proved effective in ISC calibration.

The first method is based on the sensitivity functions of the engine speed with respect to the controller parameters. Let K be a generic controller parameter (possibly vector valued), and suppose that we want to minimize a performance cost function $J(\Delta N)$ [a commonly used function for J is $J = (\Delta N)^2$] by adjusting K. Viewing ΔN as a function of K and noting that $\frac{\partial \Delta N}{\partial K} = \frac{\partial N}{\partial K}$, we have

$$\begin{aligned} J(K + \Delta K) \quad &\approx \quad J(K) + 2\Delta N \frac{\partial N}{\partial K} \Delta K \\ &\quad + (\Delta K)^{\top} \left(\frac{\partial N}{\partial K}\right)^{\top} \frac{\partial N}{\partial K} \Delta K. \end{aligned}$$

According to Newton's method, ΔK, which minimizes J, is given by

$$\Delta K = -\left[\left(\frac{\partial N}{\partial K}\right)^{\top} \frac{\partial N}{\partial K}\right]^{-1} \left(\frac{\partial N}{\partial K}\right)^{\top} \Delta N. \qquad (74.24)$$

By measuring the sensitivity function $\frac{\partial N}{\partial K}$, we can use a simple gradient method or Equation 74.24 to iteratively minimize the cost function J. The controller gains for the air and spark loops can be adjusted simultaneously. The advantages of the method are that the sensitivity functions are easy to generate. For the ISC calibration, the sensitivity functions of N with respect to PID controller parameters can be obtained by measuring the signal at the sensitivity points, as illustrated in Figure 74.16.

It should be pointed out that this offline tuning principle can be used to develop an on-line adaptive PID control scheme (referred to as the M.I.T. rule in the adaptive control literature). The sensitivity function method can also be used to investigate the robustness of the ISC system with respect to key plant parameters by evaluating $\frac{\partial N}{\partial K_p}$ where K_p is the plant parameter vector.

The second method is the well-known Ziegler-Nichols PID tuning method. It gives a set of heuristic rules for selecting the optimal PID gains. For the ISC applications, modifications have

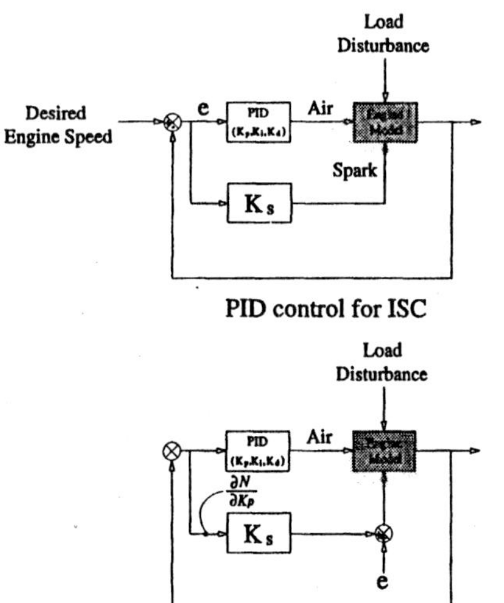

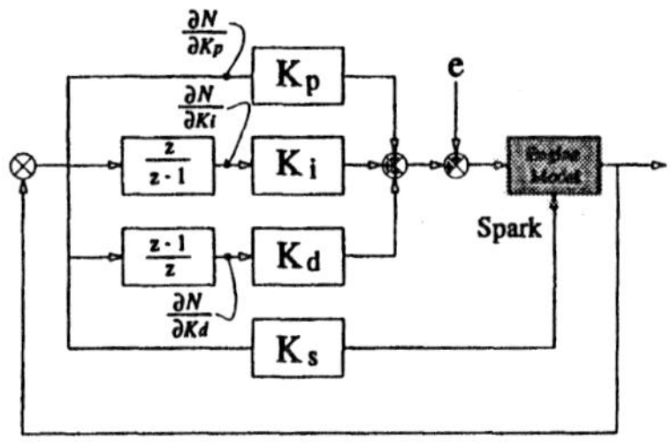

Sensitivity points for PID air-loop control

Figure 74.16 Sensitivity points for calibrating PID idle speed controller.

to be introduced to accommodate the time delay and other constraints. Generally, the Ziegler-Nichols sensitivity method is used to calibrate the PID air feedback loop after the proportional gain for the spark is fixed.

74.1.4 Acknowledgments

This work was supported in part by the National Science Foundation under contract NSF ECS-92-13551.

The authors also acknowledge their many colleagues at Ford Motor Company and the University of Michigan who contributed to the work described in this chapter, with special thanks to Dr. Paul Moraal of Ford.

References

[1] Buescher, K.L., Representation, Analysis, and Design of Multirate Discrete-Time Control Systems, Master's thesis, Department of Electrical and Computer Engineering, University of Illinois, Urbana-Champaign, 1988.

[2] Grizzle, J.W., Cook, J.A., and Milam, W.P., Improved cylinder air charge estimation for transient air fuel ratio control, in *Proc. 1994 Am. Control Conf.*, Baltimore, MD, June 1994, 1568–1573.

[3] Grizzle, J.W., Dobbins, K.L., and Cook, J.A., Individual cylinder air fuel ratio control with a single EGO sensor, *IEEE Trans. Vehicular Technol.*, 40(1), 280–286, February 1991.

[4] Hrovat, D. and Powers, W.F., Modeling and Control of Automotive Power Trains, in *Control and Dynamic Systems*, Vol. 37, Academic Press, New York, 1990, 33–64.

[5] Khargonekar, P.P., Poolla, K., and Tannenbaum, A., Robust control of linear time-invariant plants using periodic compensation, *IEEE Trans. Autom. Control*, 30(11), 1088–1096, 1985.

[6] Moraal, P.E., Cook, J.A., and Grizzle, J.W., Single sensor individual cylinder air-fuel ratio control of an eight cylinder engine with exhaust gas mixing, in *Proc. 1993 Am. Control Conf.*, San Francisco, CA, June 1993, 1761–1767.

[7] Powell, B.K. and Cook, J.A., Nonlinear low frequency phenomenological engine modeling and analysis, in *Proc. 1987 Am. Control Conf.*, Minneapolis, MN, June 1987, 332–340.

[8] Powers, W.F., Customers and controls, *IEEE Control Syst. Mag.*, 13(1), February 1993, 10–14.

[9] Shulman, M.A. and Hamburg, D.R., Non-ideal properties of $Z_r O_2$ and $T_i O_2$ exhaust gas oxygen sensors, SAE Tech. Paper Series, No. 800018, 1980.

[10] Taylor, C.F., The Internal Combustion Engine in Theory and Practice, Vol. 1: Thermodynamics, Fluid Flow, Performance, MIT Press, Cambridge, MA, 1980, 187.

74.2 Adaptive Automotive Speed Control

M. K. Liubakka, Advanced Vehicle Technology, Ford Motor Company, Dearborn, MI
D.S. Rhode, Advanced Vehicle Technology, Ford Motor Company, Dearborn, MI
J. R. Winkelman, Advanced Vehicle Technology, Ford Motor Company, Dearborn, MI
P. V. Kokotović, ECE Department, University of California, Santa Barbara, CA

74.2.1 Introduction

One of the main goals for an automobile speed control [3] (cruise control) system is to provide acceptable performance over a wide range of vehicle lines and operating conditions. Ideally, this is to be achieved with one control module, without recalibration for

[3]©1993 IEEE. Reprinted, with permission, from *IEEE Transactions on Automatic Control;* Volume 38, Number 7, Pages 1011–1020; July 1993.

different vehicle lines. For commonly used proportional feedback controllers, no single controller gain is adequate for all vehicles and all operating conditions. Such simple controllers no longer have the level of performance expected by customers.

The complexity of speed control algorithms has increased through the years to meet the more stringent performance requirements. The earliest systems simply held the throttle in a fixed position [1]. In the late 1950s speed control systems with feedback appeared [2]. These used proportional (P) feedback of the speed error, with the gain typically chosen so that 6 to 10 mph of error would pull full throttle. The next enhancement was proportional control with an integral preset or bias input (PI) [3]. This helped to minimize steady-state error as well as speed droop when the system was initialized. Only with the recent availability of inexpensive microprocessors have more sophisticated control strategies been implemented. Proportional-integral-derivative (PID) controllers, optimal LQ regulators, Kalman filters, fuzzy logic, and adaptive algorithms have all been tried [4]-[10].

Still, it is hard to beat the performance of a well-tuned PI controller for speed control. The problem is how to keep the PI controller well tuned, since both the system and operating conditions vary greatly. The optimal speed control gains are dependent on:

- Vehicle parameters (engine, transmission, weight, etc.)
- Vehicle speed
- Torque disturbances (road slope, wind, etc.)

Gain scheduling over vehicle speed is not a viable option because the vehicle parameters are not constant and torque disturbances are not measurable. Much testing and calibration work has been done to tune PI gains for a controller that works across more than one car line, but as new vehicles are added, retuning is often necessary. For example, with a PI speed control, low-power cars generally need higher gains than high-power cars. This suggests a need for adaptation to vehicle parameters. For an individual car, the best performance on flat roads is achieved with low-integral gain, while rolling hill terrain requires high-integral gain. This suggests a need for adaptation to disturbances.

Our goal was to build an adaptive controller that outperforms its fixed-gain competitors, yet retains their simplicity and robustness. This goal has been achieved with a slow-adaptation design using a sensitivity-based gradient algorithm. This algorithm, driven by the vehicle response to unmeasured load torque disturbances, adjusts the proportional and integral gains, K_p and K_i, respectively, to minimize a quadratic cost functional. Through simulations and experiments a single cost functional was found that, when minimized, resulted in satisfactory speed control performance for each vehicle and all operating conditions. Adaptive minimization of this cost functional improved the performance of every tested vehicle over varying road terrain (flat, rolling hills, steep grades, etc.). This is not possible with a fixed-gain controller.

Our optimization type adaptive design has several advantages. The slow adaption of only two adjustable parameters is simple and makes use of knowledge already acquired about the vehicle. The main requirement for slow adaptation is the existence of a fixed-gain controller that provides the desired performance when properly tuned. Since the PI control meets this requirement and is well understood, the design and implementation of the adaptive control with good robustness properties become fairly easy tasks. With only two adjustable parameters, all but perfectly flat roads provide sufficient excitation for parameter convergence and local robustness. These properties are strengthened by the sensitivity filter design and speed-dependent initialization.

The main idea of the adaptive algorithm employed here comes from a sensitivity approach proposed in the 1960s [11] but soon abandoned because of its instabilities in fast adaptation. Under the ideal model-matching conditions, such instabilities do not occur in more complex schemes developed in the 1970s to 1980s. However, the ideal model-matching requires twice as many adjustable parameters as the dynamic order of the plant. If the design is based on a reduced-order model, the resulting unmodeled dynamics may cause instability and robust redesigns are required. This difficulty motivated our renewed interest in *a sensitivity-based approach in which both the controller structure and the adjustable parameters are free to be chosen independently of the plant order.* Such an approach would be suitable for adaptive tuning of simple controllers to higher-order plants if a verifiable condition for its stability could be found. For this purpose we employ the "pseudogradient condition," recently derived by Rhode [12], [13] using the averaging results of [14]-[16]. A brief outline of this derivation is given in the appendix. From the known bounds on vehicle parameters and torque disturbances, we evaluate, in the frequency domain, a "phase-uncertainty envelope." Then we design a sensitivity filter to guarantee that the pseudogradient condition is satisfied at all points encompassed by the envelope.

74.2.2 Design Objectives

The automobile speed control is simpler than many other automotive control problems: engine efficiency and emissions, active suspension, four-wheel steering, to name only a few. It is, therefore, required that the solution to the speed control problem be simple. However, this simple solution must also satisfy a set of challenging performance and functional requirements. A typical list of these is as follows:

- *Performance requirements*
  - Speed tracking ability for low-frequency commands.
  - Torque disturbance attenuation for low frequencies, with zero steady-state error for large grades (within the capabilities of the vehicle power train).
  - Smooth and minimal throttle movement.
  - Robustness of the above properties over a wide range of operating conditions.

- *Functional requirements*
  - Universality: the same control module must meet the performance requirements for different vehicle lines without recalibration.
  - Simplicity: design concepts and diagrams should be understandable to automotive engineers with basic control background.

The dynamics that are relevant for this design problem are organized in the form of a generic vehicle model in Figure 74.17.

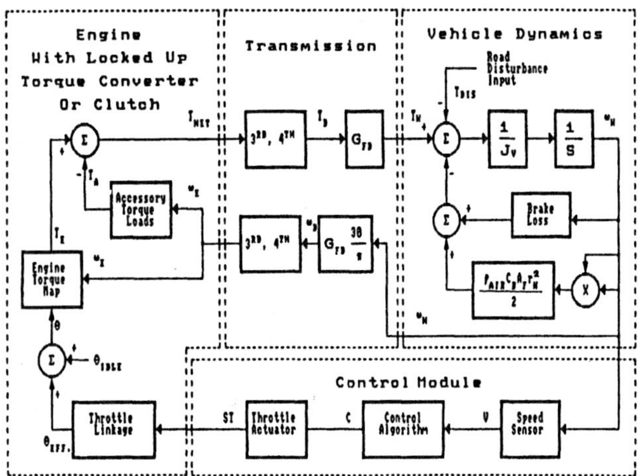

Figure 74.17 Vehicle model for speed control. Different vehicle lines are represented by different structures and parameters of individual blocks.

To represent vehicles of different lines (Escort, Mustang, Taurus, etc.) individual blocks will contain different parameters and, possibly, slightly different structures (e.g., manual or automatic transmission). Although the first-order vehicle dynamics are dominant, there are other blocks with significant higher-frequency dynamics. Nonlinearities such as dead zone, saturation, multiple gear ratios, and backlash are also present. In conjunction with large variations of static gain (e.g., low- or high-power engine) these nonlinearities may cause limit cycles, which should be suppressed, especially if noticeable to the driver.

There are two inputs: the speed set-point y_{set} and the road load disturbance torque T_{dis}. While the response to set-point changes should be as specified, the most important performance requirement is the accommodation of the torque disturbance. The steady-state error caused by a constant torque disturbance (e.g., constant-slope road) must be zero. For other types of roads (e.g., rolling hills) a low-frequency specification of disturbance accommodation is defined.

This illustrative review of design objectives, which is by no means complete, suffices to motivate the speed control design presented in Sections 74.2.3 and 74.2.4. Typical results with test vehicles presented in Section 74.3 show how the above requirements have been satisfied.

74.2.3 The Design Concept

The choice of a design concept for mass production differs substantially from an academic study of competing theories, in this case, numerous ways to design an adaptive scheme. With physical constraints, necessary safety nets, and diagnostics, the implementation of an analytically conceived algorithm may appear similar to an "expert," "fuzzy," or "intelligent" system. Innovative terminologies respond to personal tastes and market pressures, but the origins of most successful control designs are often traced to some fundamental concepts. The most enduring among these are PI control and gradient type algorithms. Recent theoretical results on conditions for stability of such algorithms reduce the necessary ad hoc fixes required to assure reliable performance. They are an excellent starting point for many practical adaptive designs and can be expanded by additional nonlinear compensators and least-square modifications of the algorithm.

For our design, a PI controller is suggested by the zero steady-state error requirement, as well as by earlier speed control designs. In the final design, a simple nonlinear compensator was added, but is not discussed in this text. The decision to adaptively tune the PI controller gains K_p and K_i was reached after it was confirmed that a controller with gain scheduling based on speed cannot satisfy performance requirements for all vehicle lines under all road load conditions. Adaptive control is chosen to eliminate the need for costly recalibration and to satisfy the universal functionality requirement.

The remaining choice was that of a parameter adaptation algorithm. Based on the data about the vehicle lines and the fact that the torque disturbance is not available for measurement, the choice was made of an optimization-based algorithm. A reference-model approach was not followed because no single model can specify the desired performance for the wide range of dynamics and disturbances. On the other hand, through simulation studies and experience with earlier designs, a *single quadratic cost functional* was constructed whose minimization led to an acceptable performance for each vehicle and each operating condition. For a given vehicle subjected to a given torque disturbance, the minimization of the cost functional generates an optimal pair of the PI controller gains K_p and K_i. In this sense, the choice of a single cost functional represents an implicit map from the set of vehicles and operating conditions to an admissible region in the parameter plane (K_p, K_i). This region was chosen to be a rectangle with preassigned bounds.

The task of parameter adaptation was to minimize the selected quadratic cost functional for each unknown vehicle and each unmeasured disturbance. A possible candidate was an indirect adaptive scheme with an estimator of the unknown vehicle and disturbance model parameters and an on-line LQ optimization algorithm. In this particular system, the frequency content in the disturbance was significantly faster than the plant dynamics. This resulted in difficulties in estimating the disturbance. After some experimentation, this scheme was abandoned in favor of a simpler sensitivity-based scheme, which more directly led to adaptive minimization of the cost functional and made better use of the knowledge acquired during its construction.

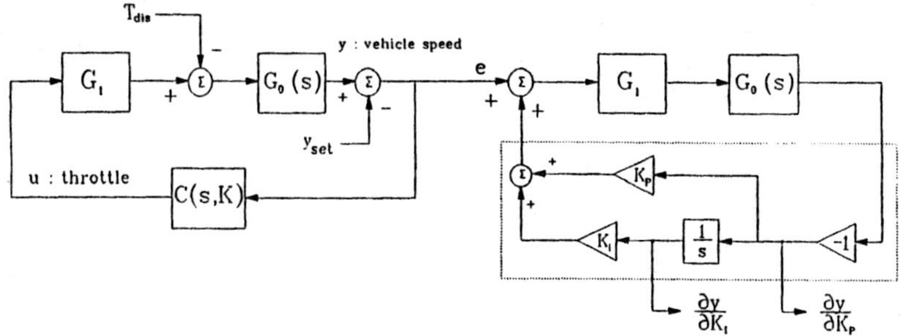

Figure 74.18 The system and its copy generate sensitivity functions for optimization of the PI controller parameters K_p and K_i.

The sensitivity-based approach to parameter optimization exploits the remarkable sensitivity property of linear systems: *the sensitivity function (i.e., partial derivative) of any signal in the system with respect to any constant system parameter can be obtained from a particular cascade connection of the system and its copy.* For a linearized version of the vehicle model in Figure 74.17, the sensitivities of the vehicle speed error $e = y - y_{set}$ with respect to the PI controller parameters K_p and K_i are obtained as in Figure 74.18, where $G_0(s)$ and $G_1(s)$ represent the vehicle and power train dynamics, respectively; $C(s, K) = -K_p - \frac{K_i}{s}$; and the control variable u is throttle position. This result can be derived by differentiation of

$$e(s, K) = \frac{1}{1 + C(s, K)G_1(s)G_0(s)} [y_{set} - G_0(s)T_{dis}] \tag{74.25}$$

with respect to $K = [K_p, K_i]$, namely,

$$\frac{\partial e}{\partial K} = \frac{\partial C}{\partial K} \frac{G_1(s)G_0(s)}{1 + C(s, K)G_1(s)G_0(s)} e(s, K) \tag{74.26}$$

where $\frac{\partial C}{\partial K} = \left(-1, -\frac{1}{s}\right)$. Expressions analogous to Equation 74.26 can be obtained for the control sensitivities $\frac{\partial u}{\partial K}$. Our cost functional also uses a high-pass filter $F(s)$ to penalize higher frequencies in u; that is, $\tilde{u}(s, K) = F(s)u(s, K)$. The sensitivities of $\tilde{u}$ are obtained simply as $\frac{\partial \tilde{u}}{\partial K} = F(s)\frac{\partial u}{\partial K}$.

When the sensitivity functions are available, a continuous-gradient algorithm for the PI controller parameters is

$$\begin{aligned} \frac{dK_p}{dt} &= -\epsilon\left(\beta_1 \tilde{u}\frac{\partial \tilde{u}}{\partial K_p} + \beta_2 e \frac{\partial e}{\partial K_p}\right) \\ \frac{dK_i}{dt} &= -\epsilon\left(\beta_1 \tilde{u}\frac{\partial \tilde{u}}{\partial K_i} + \beta_2 e \frac{\partial e}{\partial K_i}\right) \end{aligned} \tag{74.27}$$

where the adaptation speed determined by ϵ must be kept sufficiently small so that the averaging assumption (K_p and K_i are constant) is approximately satisfied. With ϵ small, the method of averaging [14]-[16] is applicable to Equation 74.27 and proves that, as $t \to \infty$, the parameters K_p and K_i converge to an ϵ neighborhood of the values that minimize the quadratic cost functional

$$J = \int_0^\infty \left(\beta_1 \tilde{u}^2 + \beta_2 e^2\right) dt. \tag{74.28}$$

With a choice of the weighting coefficients, β_1 and β_2 (to be discussed later), our cost functional is Equation 74.28. Thus, Equation 74.27 is a convergent algorithm that can be used to minimize this functional when the system is known, so that its copy can be employed to generate the sensitivities needed in Equation 74.27. In fact, our computational procedure for finding a cost functional good for all vehicle lines and operating conditions made use of this algorithm.

Unfortunately, when the vehicle parameters are unknown, the exact-gradient algorithm of Equation 74.27 cannot be used because a copy of the system is not available. In other words, an algorithm employing exact sensitivities is not suitable for adaptive control. A practical escape from this difficulty is to generate some approximations of the sensitivity functions

$$\psi_1 \approx \frac{\partial \tilde{u}}{\partial K_p}, \quad \psi_2 \approx \frac{\partial \tilde{u}}{\partial K_i}, \quad \psi_3 \approx \frac{\partial e}{\partial K_p}, \quad \psi_4 \approx \frac{\partial e}{\partial K_i} \tag{74.29}$$

and to employ them in a "pseudogradient" algorithm

$$\begin{aligned} \frac{dK_p}{dt} &= -\epsilon\,(\beta_1 \tilde{u}\psi_1 + \beta_2 e\psi_3) \\ \frac{dK_i}{dt} &= -\epsilon\,(\beta_1 \tilde{u}\psi_2 + \beta_2 e\psi_4)\ . \end{aligned} \tag{74.30}$$

A filter used to generate ψ_1, ψ_2, ψ_3, and ψ_4 is called a *pseudosensitivity filter.* The fundamental problem in the design of the pseudosensitivity filter is to guarantee not only that the algorithm of Equation 74.30 converges, but also that the values to which it converges are close to those that minimize the chosen cost functional.

74.2.4 Adaptive Controller Implementation

The adaptive speed control algorithm presented here is fairly simple and easy to implement, but care must be taken when choosing its free parameters and designing pseudosensitivity filters. This section discusses the procedure used to achieve a robust system and to provide the desired speed control performance.

Pseudosensitivity Filter

While testing the adaptive algorithm it becomes obvious that the gains K_p and K_i and the vehicle parameters vary greatly for operating conditions and vehicles. This makes it impossible to implement the exact sensitivity filters for the gradient algorithm of Equation 74.27. Our approach is to generate a "pseudogradient" approximation of $\partial J/\partial P$, satisfying the stability and convergence conditions summarized in the appendix.

In the appendix, the two main requirements for stability and convergence are: a persistently exciting (PE) input condition and a "pseudogradient condition," which, in our case, is a *phase condition on the nominal sensitivity filters.* Since we are using a reduced-order controller with only two adjustable parameters, the PE condition is easily met by the changing road loads. Road disturbances have an approximate frequency spectrum centered about zero that drops off with the square of frequency. This meets the PE condition for adapting two gains, K_p and K_i.

To satisfy the pseudogradient condition, the phase of the pseudosensitivity filter must be within $\pm 90°$ of the phase of the actual sensitivity at the dominant frequencies. To help guarantee this for a wide range of vehicles and operating conditions, we varied the system parameters in the detailed vehicle model to generate an envelope of possible exact sensitivities. Then the pseudosensitivity filter was chosen near the center of this envelope. An important element of the phase condition is the fact that it is a condition on the sum of frequencies; that is, the phase condition is most important in the range of frequencies where there are dominant dynamics. If the pseudosensitivity filters do not meet the phase conditions at frequencies where there is little dominant spectral content, the algorithm may still be convergent, provided the phase conditions are strongly met in the region of dominant dynamics. Thus, the algorithm possesses a robustness property.

Figure 74.19 shows the gain and phase plots for the pseudosensitivity filter $\partial y/\partial K_p$. The other three sensitivities are left out for brevity. Figure 74.20 shows the $\pm 90°$ phase boundary (solid lines) along with exact sensitivity phase angles (dashed lines) as vehicle inertia, engine power, and the speed control gains are varied over their full range. From this plot it is seen that the chosen pseudosensitivity filter meets the pseudogradient condition along with some safety margin to accommodate unmodeled dynamics.

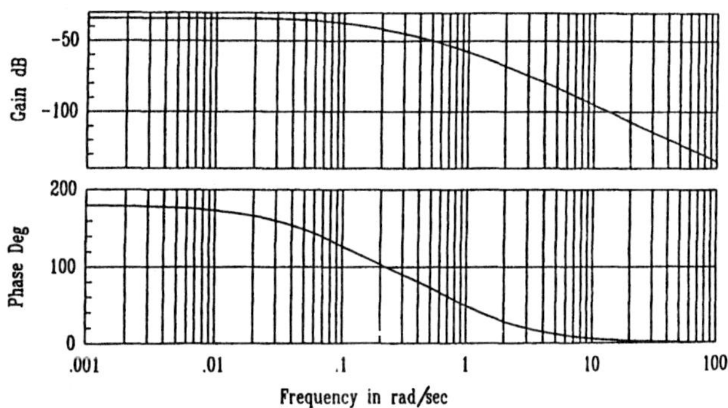

Figure 74.19 Gain and phase plot of pseudosensitivity $\partial y/\partial K_p$.

Choice of ϵ, β_1, and β_2

For implementation, the first parameters that must be chosen are the adaptation gain ϵ and the weightings in the cost functional, β_1 and β_2. The adaptation gain ϵ determines the speed of adaptation and should be chosen based on the slowest dynamics of the system. For speed control, the dominant dynamics result from the vehicle inertia and have a time constant on the order of 30 to 50 seconds. To avoid interaction between the adaptive controller states and those of the plant, the adaptation should be approximately an order of magnitude slower than the plant. As shown in [14]-[16], this allows one to use the frozen parameter system and averaging to analyze stability of the adaptive system. The adaptation law takes up to several minutes to converge, depending on initial conditions and the road load disturbances.

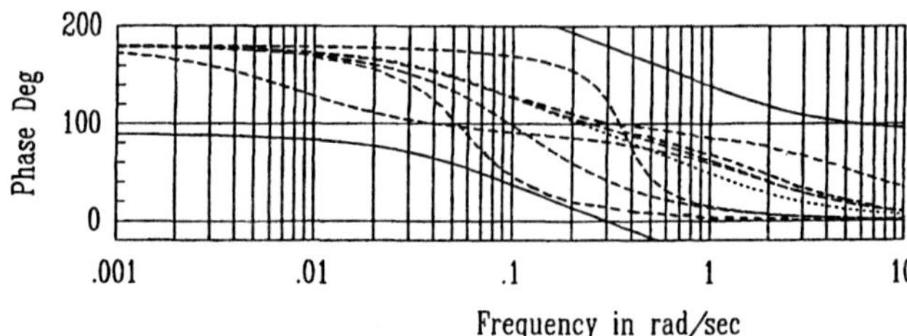

Figure 74.20 Envelope of possible exact sensitivity phase angles. The solid curves mark the $\pm 90°$ boundaries, and the dashed curves denote the limits of plant variation.

The two extreme choices of βs are (1) $\beta_1 = 0$, $\beta_2 = k$ and (2) $\beta_1 = k$, $\beta_2 = 0$ where $k > 0$. For the first extreme, $\beta_1 = 0$, we are penalizing only speed error, and the adaptation will tune to an unacceptable high-gain controller. High gain will cause too much throttle movement, resulting in nonsmooth behavior as felt by the passengers, and the system will be less robust from a stability point of view. For the second case, $\beta_2 = 0$, the adaptation will try to keep a fixed throttle angle and will allow large speed errors. Obviously, some middle values for the weightings are desired. An increase of the ratio β_1/β_2 reduces unnecessary throttle movement, while to improve tracking and transient speed errors we need to decrease this ratio.

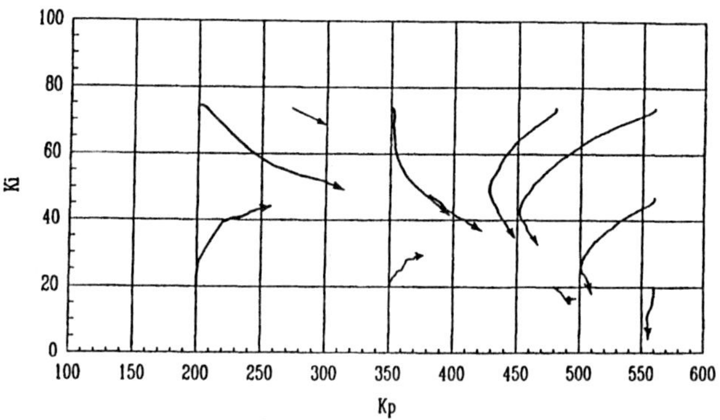

Figure 74.21 Gain trajectories on flat ground at 30 mph.

The choice of the weightings was based on experience with tuning standard PI speed controllers. Much subjective testing has been performed on vehicles to obtain the best fixed gains for the PI controller when the vehicle parameters are known. With this information, simulations were run on a detailed vehicle model with a small ϵ, and the βs were varied until the adaptation converged to approximately the gains obtained from subjective testing. The cost functional weights β_1 and β_2 are a different parameterization of the controller tuning problem. For development engineers, who may not be control system engineers, β_1

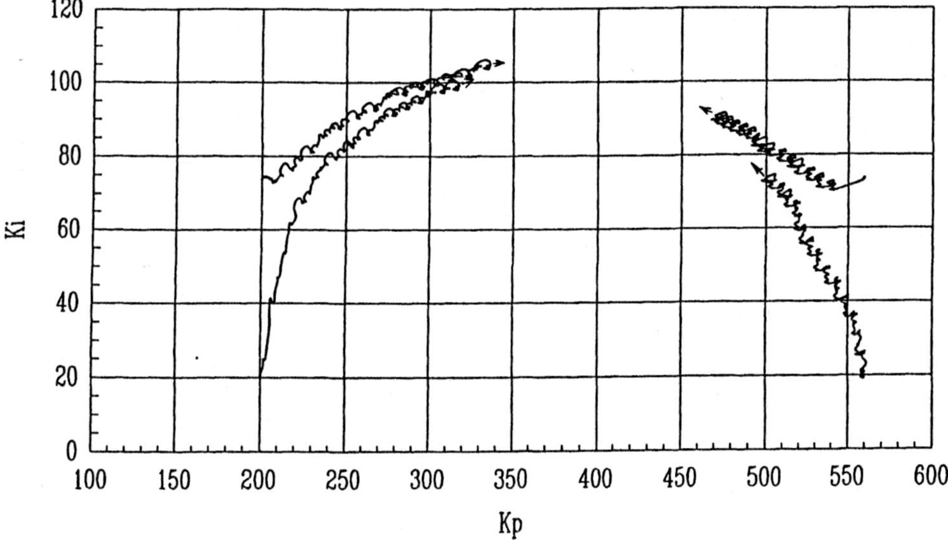

Figure 74.22 Gain trajectories on low-frequency hills at 30 mph.

and β_2 represent a pair of tunable parameters that relate directly to customer needs. This allows for a broader range of engineering inputs into the tuning process.

As examples of adaptive controller performance, simulation results in Figures 74.21 and 74.22 show the trajectories of the gains for a vehicle on two different road terrains. Much can be learned about the behavior of the adaptive algorithm from these types of simulations. In general, K_p varies proportionally with vehicle speed and K_i varies with road load. The integral gain K_i tends toward low values for small disturbances or for disturbances too fast for the vehicle to respond to. This can be seen in Figure 74.21. Conversely, K_i tends toward high values for large or slowly varying road disturbances, as can be seen in Figure 74.22.

Modifications For Improved Robustness

Additional steps have been taken to ensure robust performance in the automotive environment. First, the adaptation is turned off if the vehicle is operating in regions where the modeling assumptions are violated. These include operation at closed throttle or near wide-open throttle, during start-up transients, and when the set speed is changing. When the adaptation is turned off, the gains are frozen, but the sensitivities are still computed.

Care has been taken to avoid parameter drift due to noise and modeling imperfections. Two common ways to reduce drift are projection and a dead band on error. Projection, which limits the range over which the gains may adapt, is more attractive given the *a priori* knowledge of reasonable gains for the speed control system. To minimize computation, a simple projection is used, constraining the tuned gains to a predetermined set as shown in Figure 74.23.

There are other unmeasurable disturbances that can affect performance, such as the driver overriding speed control by use of the throttle. This condition, which can cause large gain changes, cannot be detected immediately because throttle position is not measured. To minimize these unwanted gain changes, the rate at which the gains can adapt is limited. If the adaptation adjusts more quickly than a predetermined rate, the adaptation gain ϵ is lowered, limiting the rate of change of the gains K_p and K_i.

The final parameters to choose are the initial guesses for K_p and K_i. Since the adaptation is fairly slow, a poor choice of initial gains can cause poor start-up performance. For quick convergence, the initial controller gains are scheduled with vehicle speed at the points A, B, and C. Figure 74.23 shows these initial gains along with the range of possible tuning. The proportional relationship between vehicle speed and the optimal controller gains can be seen.

74.3 Performance in Test Vehicles

The adaptive algorithm discussed in this chapter has been tried in a number of vehicles with excellent results. The adaptive control has worked well in every car tested so far and makes significant performance improvements in vehicles that have poor performance with the fixed-gain PI control. For vehicles where speed control already performs well, improvements from the adaptive algorithm are still significant at low speeds or at small throttle angles. Many vehicles with conventional speed control systems exhibit a limit cycle or surge condition at low speeds or on down slopes [7]. This is due in part to nonlinearities such as the idle stop, which limits throttle movement. In such vehicles, the adaptive controller reduces the magnitude and lengthens the period of the limit cycle, thus improving performance.

The first set of data, Figures 74.24 to 74.26, is for a vehicle that had a low-speed surge. Here the limit cycle is very noticeable in the data as well as while driving the car. The high frequency (0.2 Hz) of the limit cycle is what makes it noticeable to the driver. The adaptive controller greatly improves performance by decreasing the amplitude and frequency to the point where the driver cannot feel the limit cycle. Looking at the control gains during this test, it is seen that the gains initially decrease to reduce the limit cycle,

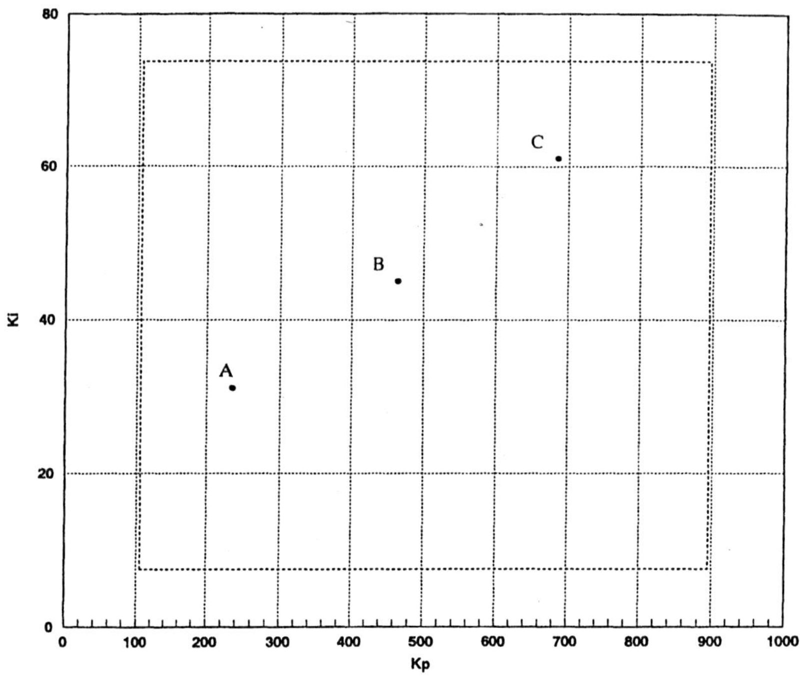

Figure 74.23 Limits and initial conditions for control gains: (A) $y_{set} < 35$ mph; (B) 35 mph $\leq y_{set} < 50$ mph; (C) $y_{set} \geq 50$ mph.

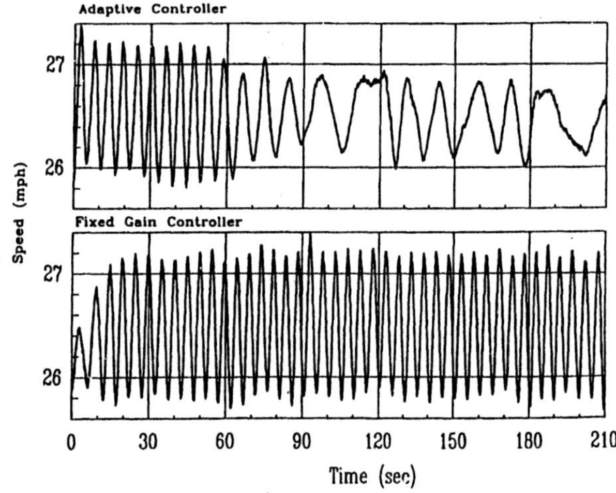

Figure 74.24 Vehicle with low speed limit cycle: adaptive controller; top, bottom, fixed gain controller.

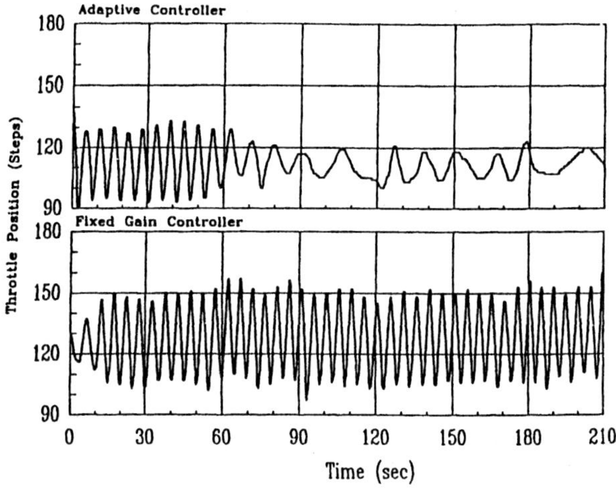

Figure 74.25 Vehicle with low-speed limit cycle: top, adaptive controller; bottom, fixed-gain controller.

then they continue to adjust in small amounts for the varying road loads as the trade-off is made between speed tracking and throttle movement. The parameter histories repeat themselves since the experiment was performed on a closed course. Only the first cycle is shown. Of course, a fixed-parameter controller could be adjusted to minimize this limit cycle. However, the controller gains that would minimize the limit cycle behavior would not produce acceptable performance on rolling hill terrain. The adaptive controller yields acceptable performance under all operating conditions.

The next three Figures show the gains for three different vehicles traveling over the same road at 40 mph. As expected, even though the initial gains are acceptable, each vehicle tunes to a different set of optimal gains for this road. One thing to notice in Figure 74.27 is the initial increase in K_i. This occurred because the integral preset of the control did not match this vehicle, causing an initial overshoot or droop in speed. To the adaptive algorithm, this offset looks like a large road disturbance. Again, since these tests were performed on a closed course, the gains eventually become periodic.

Figure 74.30 to 74.32 show a vehicle driving on a freeway which passes below the grade level to pass under surface roads. An approximate road profile for a section of this road is shown in Figure 74.33. The dips are where the surface streets cross over the freeway. Because the road disturbances are of such high frequency, the adaptive controller cannot greatly improve the tracking ability of the vehicle, but much of the high-frequency limit cycle has been removed. For this disturbance, the proportional

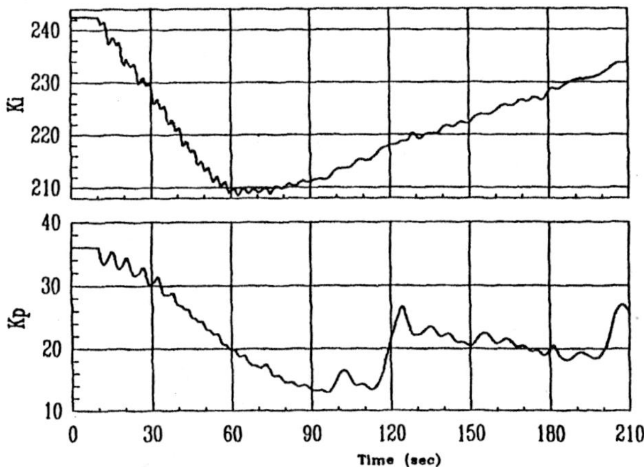

Figure 74.26 Controller gains.

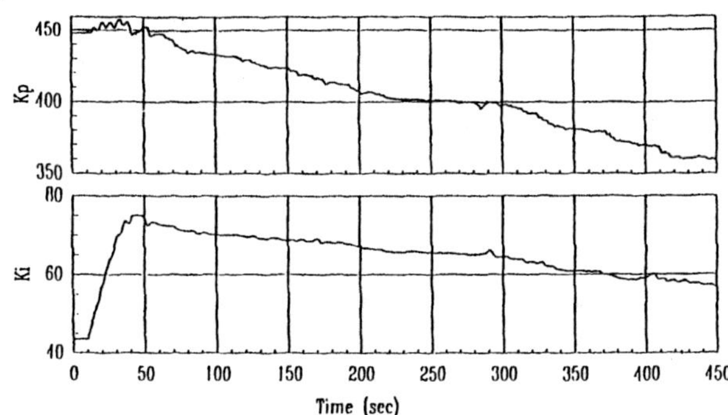

Figure 74.27 Test car A, 40 mph.

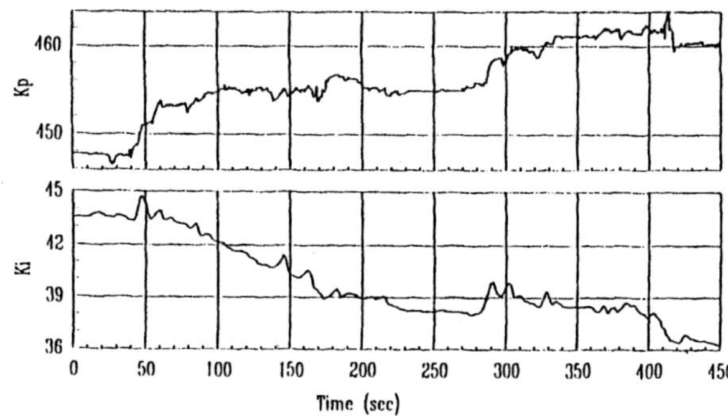

Figure 74.28 Test car B, 40 mph.

gain tunes up and the integral gain tunes down. The integral gain decreases to reduce limit cycle behavior since the road disturbance is too fast for the vehicle to track.

From the tests run in several vehicles it seems that, compared with other controllers, the adaptive algorithm is providing the best control possible as vehicle parameters change from car to car and as the road profile changes.

74.3.1 Conclusions

This chapter has presented an adaptive algorithm to adjust the gains of a vehicle speed control system. By continuously adjusting the PI control gains, speed control performance can be optimized for each vehicle and operating condition. This helps to design a single speed control module without additional calibration or

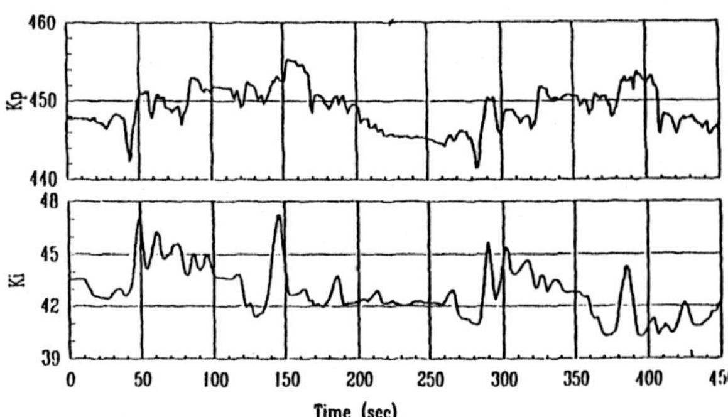

Figure 74.29 Test car C, 40 mph.

sacrifices in performance for certain car lines. It also allows improved performance for changing road conditions not possible with a fixed-gain control or other types of adaptive control.

The results of initial vehicle testing confirm the performance improvements and robustness of the adaptive controller. Vehicle speed control is not the only automotive application of adaptive control at Ford Motor Company. The adaptive control technique presented in this chapter has also been applied to solve other problems of electronically controlled automotive systems.

Appendix

The pseudogradient adaptive approach used in this application relies upon the properties of slow adaptation. Such gradient tuning algorithms can be described by

$$\dot{K}(t) = -\epsilon\Psi(t, K)e(t, K), \quad K \in \Re^m \tag{74.31}$$

where K is a vector of tunable parameters, $e(t, K)$ is an error signal, and $\Psi(t, K)$ is the regressor, which is an approximation of the sensitivity of the output with respect to the parameters $\frac{\partial y(t,K)}{\partial K}$. In this derivation, $e(t, K)$ is defined as the difference between the output $y(t, K)$ and a desired response $y_m(t)$,

$$e(t, K) \equiv y(t, K) - y_m(t), \tag{74.32}$$

so that the pseudogradient update law of Equation 74.31 aims to minimize the average of $e(t, K)^2$. However, the same procedure is readily modified to minimize a weighted sum of $e(t, K)^2$ and $u(t, K)^2$ as in this speed control application. Since slow adaptation is used only for performance improvement, and not for stabilization, we constrain the vector K of adjustable controller parameters, $k_1, ..., k_m$, to remain in a set $\bar{K}$ such that with constant K ($\epsilon = 0$), the resulting linear system is stable. Our main tool in this analysis is an integral manifold, the so-called *slow manifold* [14], that separates the fast linear plant and controller states from the slow parameter dynamics. To assure the existence of this manifold, we assume that $\Psi(t, K)$ and $e(t, K)$ are differentiable with respect to K and the input to the system, $r(t)$, is a uniformly bounded, almost periodic function of time. Then applying the averaging theorem of Bogolubov, the stability of the adaptive algorithm is determined from the response of the linear system with constant parameters and the average update law. The following analysis is restricted to initial conditions near this slow manifold.

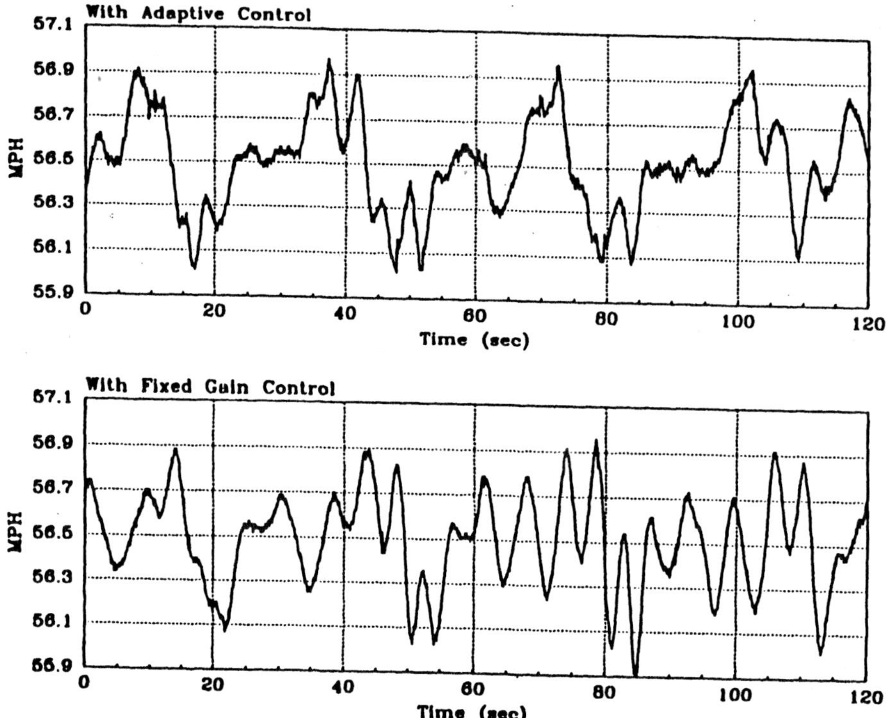

Figure 74.30 Vehicle with high-frequency road disturbance: top, with adaptive control; bottom, with fixed-gain control.

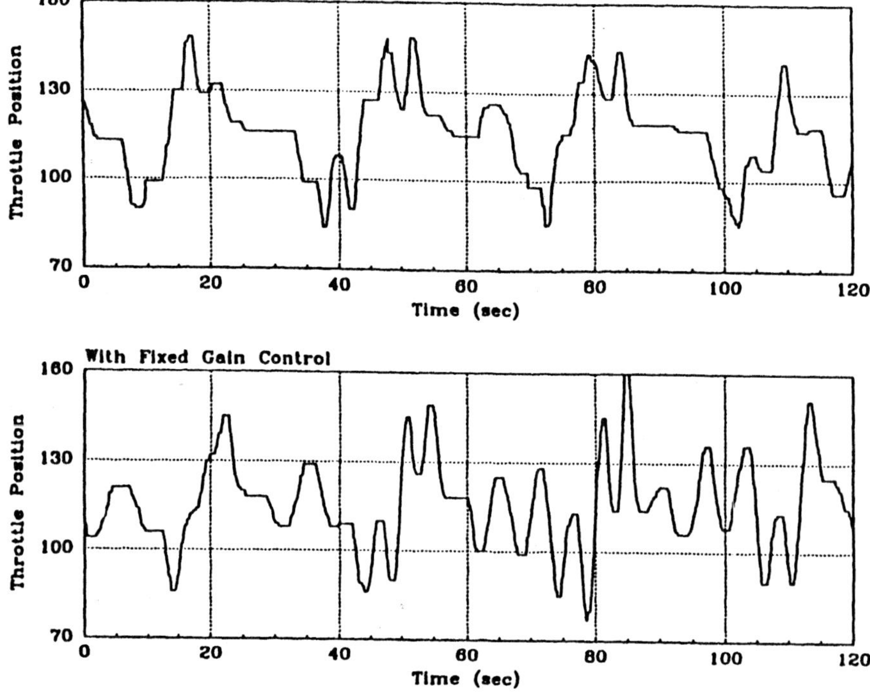

Figure 74.31 Vehicle with high-frequency road disturbance: top, with adaptive control; bottom, with fixed-gain control.

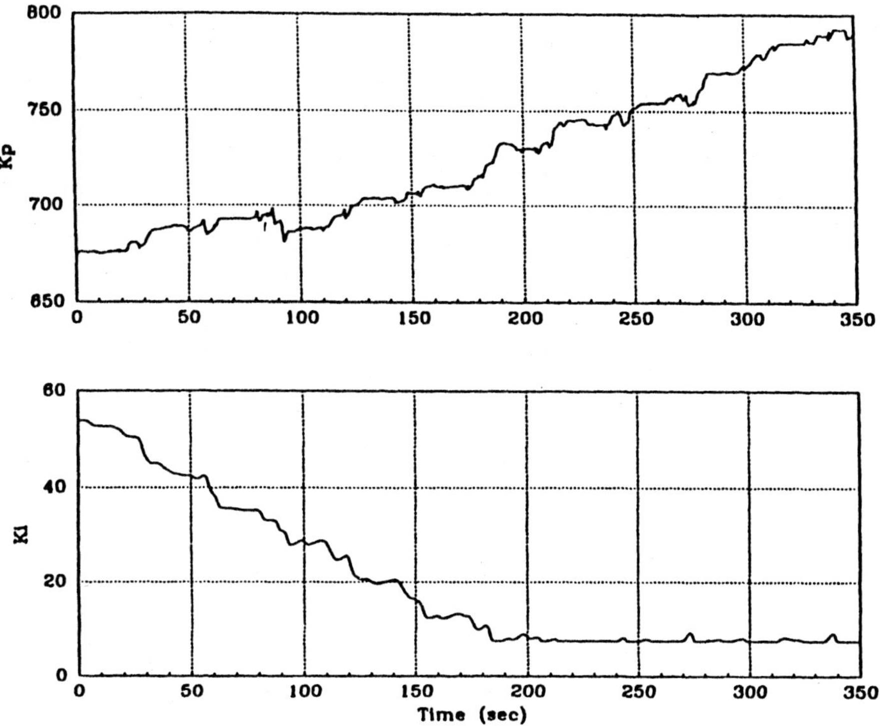

Figure 74.32 Gains with high-frequency disturbances.

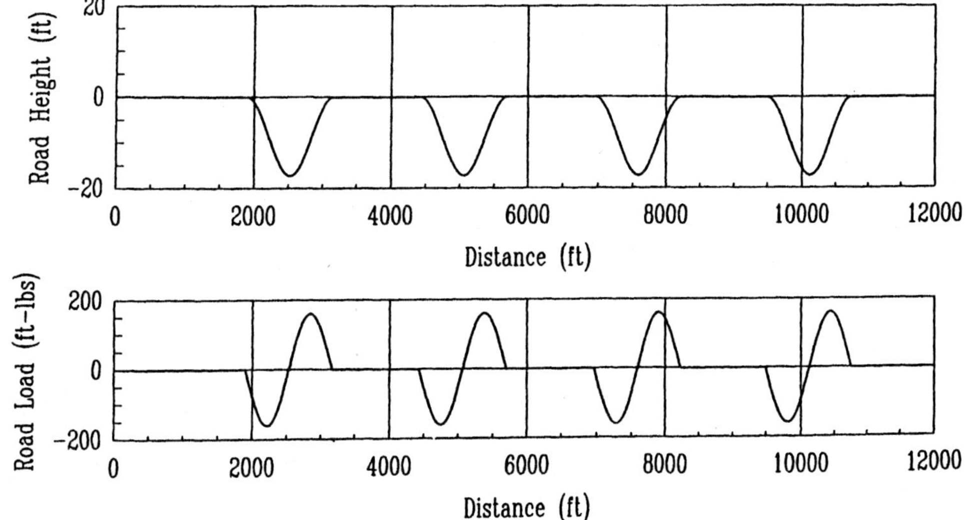

Figure 74.33 Approximate freeway profile.

In the pseudogradient approach used here, the regressor $\Psi(t, K)$ is an approximation of the sensitivities of the error with respect to parameters $\frac{\partial e(t,K)}{\partial K}$. We assume that the adjustable controller parameters K feed a common summing junction as shown in Figure 74.34. In this case, if the system in Figure 74.34 was known, these sensitivities would be obtained as in Figure 74.35, where $H_\Sigma(s, K)$ is the scalar transfer function from the summing junction input to the system output as shown in Figure 74.36. However, this transfer function is unknown, and we approximate it by the sensitivity filter $H_\Psi(s)$. The pseudogradient stability condition to be derived here specifies a bound on the allowable mismatch between $H_\Sigma(s, K^*)$ and $H_\Psi(s)$, where K^* is such that

$$\lim_{T\to\infty} \frac{1}{T}\int_t^{t+T} \Psi(\tau, K^*)e(\tau, K^*)d\tau = \qquad (74.33)$$

$$[\Psi(t, K^*)e(t, K^*)]_{\text{ave}} = 0. \qquad (74.34)$$

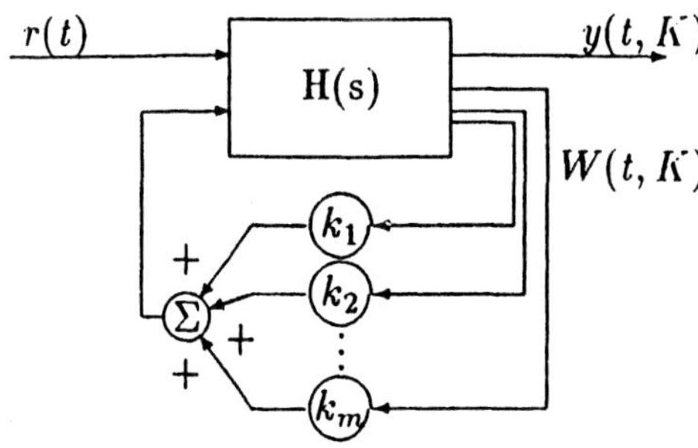

Figure 74.34 Adjustable linear system.

With reduced-order controllers, it is unrealistic to assume that the error $e(t, K)$ can be eliminated entirely for any value of controller parameters and plant variations. The remaining nonzero error is called the *tuned error*, $e^*(t) = e(t, K^*)$. The error $e(t, K)$ is comprised of $e^*(t)$ and a term caused by the parameter error $\tilde{K} = K - K^*$. Using a mixed (t, s) notation with signals as functions of time and transfer functions denoted as Laplace transforms, the error $e(t, K)$ and the regressor $\Psi(t, K)$ are both obtained from the measured signals $W(t, K)$ as follows:

$$e(t, K) = H_\Sigma(s, K^*)[\tilde{K}^T(t)W(t, K)] + e^*(t) \qquad (74.35)$$

$$\Psi(t, K) = H_\Psi(s)[W(t, K)]. \qquad (74.36)$$

The error expression illustrated in Figure 74.37 contains the exact sensitivity filter transfer function $H_\Sigma(s, K^*)$. The stability properties of this slowly adapting system are determined by examination of the average parameter update equation

$$\begin{aligned}\dot{\tilde{K}}(t) &= -\epsilon\{\Psi(t, K)V(t, K)^T\}_{\text{ave}}\tilde{K}(t) \\ &\quad - \epsilon\Delta - \epsilon\{\Psi(t, K)e^*(t)\}_{\text{ave}}. \end{aligned} \qquad (74.37)$$

where we have again used $H_\Sigma(s, K^*)$ in the definition of the signal

$$V(t, K) \equiv H_\Sigma(s, K^*)[W(t, K)]. \qquad (74.38)$$

The average swapping term Δ in Equation 74.37 is

$$\begin{aligned}\Delta &= \{H_\Sigma(s, K^*)[W(t, K)^T\tilde{K}(t)] \\ &\quad - H_\Sigma(s, K^*)[W(t, K)^T]\tilde{K}(t)\}_{\text{ave}}\end{aligned} \qquad (74.39)$$

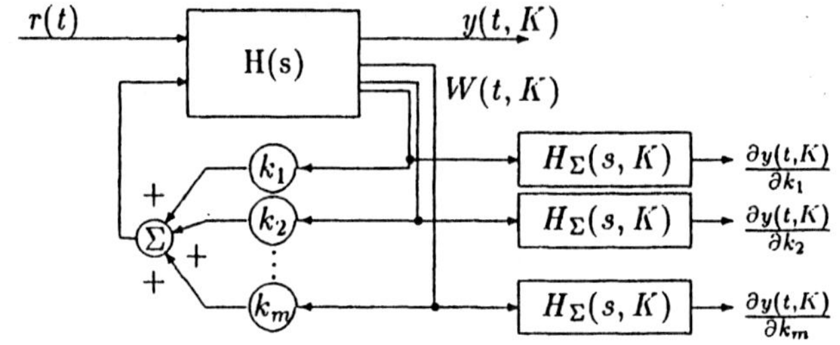

Figure 74.35 Sensitivity system.

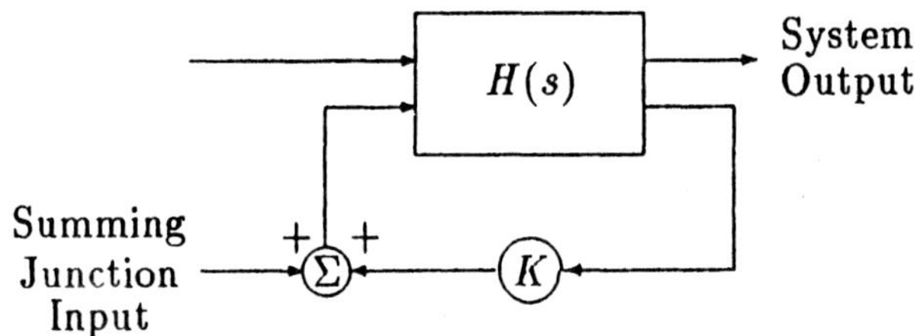

Figure 74.36 $H_\Sigma(s, K)$.

As shown in [14] and [15], for slow adaptation this term is small and does not affect the stability of Equation 74.37 in a neighborhood of the equilibrium K^*. The stability of this equilibrium, in the case when $e^*(t) = 0$, is established as follows.

LEMMA 74.1 [15] Consider the average update system (Equation 74.37) with $\Delta(t) = 0$, $e(t, K^*) = 0$, and $K = K^*$. If

$$\text{Re}\lambda\{\Psi(t, K)V^T(t, K)\}_{\text{ave}} > 0 \qquad (74.40)$$

then there exists an $\epsilon^* > 0$ such that $\forall\epsilon \in (0, \epsilon^*]$ the equilibrium $\tilde{K} = 0$ is uniformly asymptotically stable.

To interpret the condition of Equation 74.40, we represent the regressor as

$$\Psi(t, K) = H_\Psi(s)H_\Sigma^{-1}(s, K^*)[V(t, K)] \qquad (74.41)$$

so that the matrix in Equation 74.40 can be visualized as in Figure 74.38. Representing the almost periodic signal $V(t, K)$ as

$$V(t, K) = \sum_{i=-\infty}^{+\infty} v_i e^{j\omega_i t} \qquad (74.42)$$

a sufficient condition for Equation 74.40 to hold is

$$\sum_{i=-\infty}^{+\infty} \text{Re}[H_\Psi(j\omega_i)H_\Sigma^{-1}(j\omega_i, K^*)]\text{Re}[v_i\bar{v}_i^T] > 0. \qquad (74.43)$$

If the signal, $V(t)$, possesses an autocovariance, $R_V(z)$, it is PE if and only if $R_V(0) > 0$ [17]. It follows that $V(t, K)$ is PE if and only if $\sum_{i=-\infty}^{\infty} R_e[v_i\bar{v}_i^T] > 0$. Clearly, if the sensitivity filter is exact, $H_\Psi(j\omega) = H_\Sigma(j\omega, K^*)$, and $V(t, K)$ is PE, then the sufficient stability condition in Equation 74.43 holds. When $H_\Psi(s)$ cannot be made exact, this stability condition is still satisfied when the pseudosensitivity filter is chosen such that $\text{Re}[H_\Psi(j\omega)H_\Sigma^{-1}(j\omega, K^*)] > 0$ for dominant frequencies, that

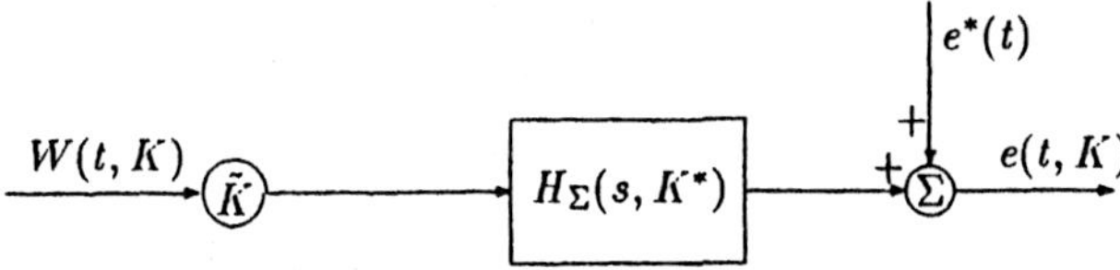

Figure 74.37 Error model.

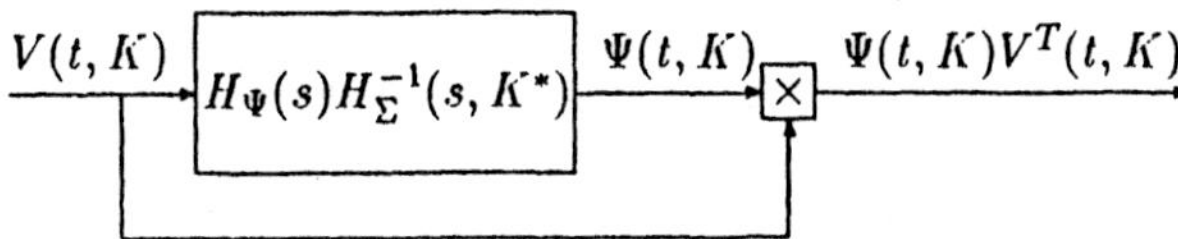

Figure 74.38 Feedback matrix.

is, the frequencies where $v_i \bar{v}_i^T$ is large. This condition serves as a guide for designing the pseudosensitivity filter $H_\Psi(s)$. As $H_\Sigma(s, K^*)$ is unknown beforehand, we use the *a priori* information about the closed-loop system to design a filter $H_\Psi(s)$ such that in the dominant frequency range the following *pseudogradient condition* is satisfied for all plant and controller parameters of interest:

$$-90^o < \angle\{H_\Psi(j\omega)H_\Sigma^{-1}(j\omega, K^*)\} < 90^o \qquad (74.44)$$

When Equation 74.44 is satisfied, then the equilibrium K^* of the average update law of Equation 74.37 with $\Delta \equiv 0$, $e^*(t) \equiv 0$ is uniformly asymptotically stable. By Bogolubov's theorem and slow manifold analysis [14], this implies the local stability property of the actual adaptive system, provided $e^*(t) \neq 0$ is sufficiently small. Since, in this approach, both the controller structure and the number of adjustable parameters are free to be chosen independently of plant order, there is no guarantee that $e^*(t)$ will be sufficiently small. Although conservative bounds for $e^*(t)$ may be calculated [12], in practice, since the design objective of the controller and pseudogradient adaptive law is to minimize the average of $e(t, K)^2$, $e^*(t)$ is typically small.

References

[1] Ball, J.T., Approaches and Trends in Automatic Speed Controls, SAE Tech. Paper #670195, 1967.

[2] Follmer, W.C., Electronic Speed Control, SAE Tech. Paper #740022, 1974.

[3] Sobolak, S.J., Simulation of the Ford Vehicle Speed Control System, SAE Tech. Paper #820777, 1982.

[4] Nakamura, K., Ochiai, T., and Tanigawa, K., Application of microprocessor to cruise control system, *Proc. IEEE Workshop Automot. Appl. Microprocessors,* 37–44, 1982.

[5] Chaudhure, B., Schwabel, R.J., and Voelkle, L.H., Speed Control Integrated into the Powertrain Computer, SAE Tech. Paper #860480, 1986.

[6] Tabe, T., Takeuchi, H., Tsujii, M., and Ohba, M., Vehicle speed control system using modern control theory, *Proc. 1986 Int. Conf. Industrial Electron., Control Instrum.,* 1, 365–370, 1986.

[7] Uriuhara, M., Hattori, T., and Morida, S., Development of Automatic Cruising Using Fuzzy Control System, *J. SAE Jpn.*, 42(2), 224–229, 1988.

[8] Abate, M. and Dosio, N., Use of Fuzzy Logic for Engine Idle Speed Control, SAE Tech. Paper #900594, 1990.

[9] Tsujii, T., Takeuchi, H., Oda, K., and Ohba, M., Application of self-tuning to automotive cruise control, *Proc. Am. Control Conf.*, 1843–1848, 1990.

[10] Hong, G. and Collings, N., Application of Self-Tuning Control, SAE Tech. Paper #900593, 1990.

[11] Kokotovic, P.V., Method of sensitivity points in the investigation and optimization of linear control systems, *Automation Remote Control,* 25, 1670–1676, 1964.

[12] Rhode, D.S., Sensitivity Methods and Slow Adaptation, Ph.D. thesis, University of Illinois at Urbana-Champaign, 1990.

[13] Rhode, D.S. and Kokotovic, P.V., Parameter Convergence conditions independent of plant order, in *Proc. Am. Control Conf.*, 981–986, 1990.

[14] Riedle, B.D. and Kokotovic, P.V., Integral manifolds of slow adaptation, *IEEE Trans. Autom. Control,* 31, 316–323, 1986.

[15] Kokotovic, P.V., Riedle, B.D., and Praly, L., On a stability criterion for continuous slow adaptation, *Sys. Control Lett.*, 6, 7–14, 1985.

[16] Anderson, B.D.O., Bitmead, R.R., Johnson, C.R., Jr., Kokotovic, P.V., Kosut, R.L., Mareels, I., Praly, L., and Riedle, B.D., *Stability of Adaptive Systems: Passivity and Averaging Analysis*, MIT Press, Cambridge, MA, 1986.

[17] Boyd S. and Sastry, S.S., Necessary and sufficient conditions for parameter convergence in adaptive control, *Automatica*, 22(6), 629–639, 1986.

75

Aerospace Controls

M. Pachter
Department of Electrical and Computer Engineering, Air Force Institute of Technology, Wright-Patterson AFB, OH

C. H. Houpis
Department of Electrical and Computer Engineering, Air Force Institute of Technology, Wright-Patterson AFB, OH

Vincent T. Coppola
Department of Aerospace Engineering, The University of Michigan, Ann Arbor, MI

N. Harris McClamroch
Department of Aerospace Engineering, The University of Michigan, Ann Arbor, MI

S. M. Joshi and A. G. Kelkar
NASA Langley Research Center

David Haessig
GEC-Marconi Systems Corporation, Wayne, NJ

75.1 Flight Control of Piloted Aircraft

M. Pachter, Department of Electrical and Computer Engineering, Air Force Institute of Technology, Wright-Patterson AFB, OH

C. H. Houpis, Department of Electrical and Computer Engineering, Air Force Institute of Technology, Wright-Patterson AFB, OH

75.1.1 Introduction

Modern Flight Control Systems (FCS) consist of (1) aerodynamic control surfaces and/or the engines' nozzles, (2) actuators, (3) sensors, (4) a sampler and ZOH device, and (5) compensators. The first four components of an FCS are hardware elements, whereas the controller (the digital implementation of the compensator) is an *algorithm* executed in real time in the on-board digital computer. In this chapter the design of the compensation/controller element/algorithm of the FCS, for a given aircraft, after the actuators, sensors and samplers have been chosen, is addressed. The way in which control theory is applied to the FCS's controller design is the main focus of this article. An advanced and comprehensive perspective on flight control is presented. The emphasis is on maneuvering flight control. Thus, attention is given to the process of setting up the flight control problem from its inception. Flight Mechanics is used to obtain a rigorous formulation of the nonlinear dynamic model of the controlled "plant;" the linearization of the latter yields Linear Time Invariant (LTI) models routinely used for controller design. Also, it is important to remember that the *pilot* will be closing an additional outer feedback loop. This transforms the FCS design problem from one of meeting flying quality specifications into one of meeting handling quality specifications.

0-8493-8570-9/96/$0.00+$.50

The essence of flight control is the design of an FCS for *maneuvering* flight. Hence, we chose not to dwell on outer-loop control associated with autopilot design and, instead, the focus is on the challenging problems of maneuvering flight control and the design of an inner-loop FCS, and pilot-in-the-loop issues. By its very nature, maneuvering flight entails large state variable excursions, which forces us to address nonlinearity and cross-coupling.

To turn, pilots will bank their aircraft and pull g's. It is thus realized that the critical control problem in maneuvering flight is the stability axis, or, velocity vector, roll maneuver. During a velocity vector roll both the lateral/directional *and* the pitch channels are controlled simultaneously. Hence, this maneuver epitomizes maneuvering flight control, for it brings into the foreground the pitch and lateral/directional channels' cross-coupling, nonlinearity, time-scale separation, tracking control design, actuator saturation concerns, and pilot-in-the-loop issues—all addressed in this chapter. Moreover, when additional simplifying assumptions apply, the velocity vector roll maneuver is general enough to serve as the starting point for derivating the classical LTI aircraft model, where the longitudinal and lateral/directional flight control channels are decoupled. Evidently, the design of a FCS for velocity vector rolls is a vehicle for exploring the important aspects of maneuvering flight control. Hence, this article's leitmotif is the high Angle Of Attack (AOA) velocity vector roll maneuver.

Since this chapter is configured around maneuvering flight, and because pilots employ a high AOA and bank in order to turn, the kinematics of maneuvering flight and high AOA velocity vector rolls are now illustrated. Thus, should the aircraft roll about its $x-$body axis, say, as a result of adverse yaw, then at the point of attainment of a bank angle of 90°, the AOA will have been totally converted into sideslip angle, as illustrated in Figure 75.1e. In this case, the aircraft's nose won't be pointed in the right direction. During a velocity vector roll the aircraft rotates about an axis aligned with its velocity vector, as illustrated in Figures 75.1a – 75.1d. The operational significance of this maneuver is obvious, for it allows the pilot to slew quickly and point the aircraft's nose using a fast roll maneuver, without pulling g's, i.e., without increasing the normal acceleration, and turning. This maneuver is also a critical element of the close air combat S maneuver, where one would like to roll the loaded airframe, rather than first unload, roll and pull gs. Thus, in this chapter, maneuvering flight control of modern fighter aircraft, akin to an F-16 derivative, is considered.

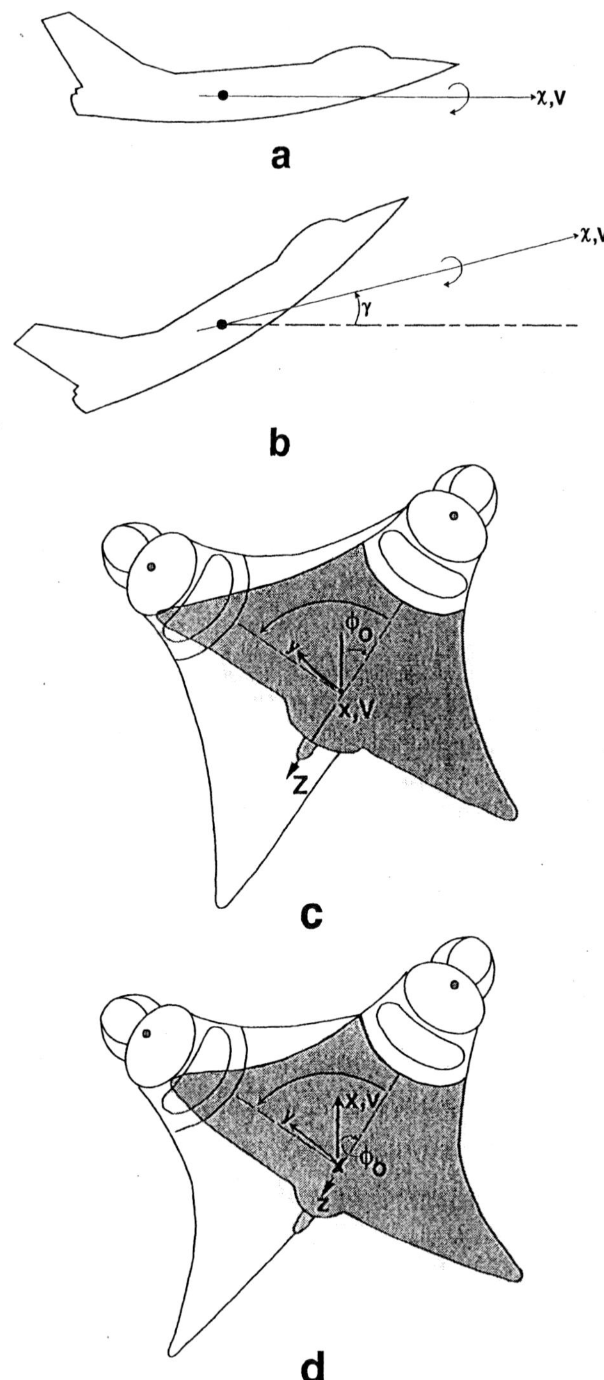

Figure 75.1 Initial and final flight configurations for high AOA maneuvers; V is always the aircraft velocity vector. a) level flight. $\psi_0 = \theta_0 = \phi_0 = 0$; $\psi_f = \theta_f = 0, \phi_f = \phi$ b) climbing flight. $\psi_0 = 0, \theta_0 = \gamma, \phi_0 = 0$; $\psi_f = 0, \theta_f = \gamma, \phi_f = \phi$ c) planar S maneuver. $\psi_0 = \theta_0 = 0, \phi_0 = -\phi_0$; $\psi_f = \theta_f = 0, \phi_f = \phi_0$ d) climbing S maneuver. $\psi_0 = 0, \theta_0 = \gamma, \phi_0 = -\phi_0$; $\psi_f = 0, \theta_f = \gamma, \phi_f = \phi_0$

75.1.2 Flight Mechanics

Proper application of the existing control theoretic methods is contingent on a thorough understanding of the "plant." Hence, a careful derivation of the aircraft, ("plant") model required in FCS design is given. To account properly for the nonlinearities affecting the control system in maneuvering flight, the plant model must be rigorously derived from the fundamental nine state equations of motion [1] and [2]. The Euler equations for a rigid body yield the following equations of motion. The *Force*

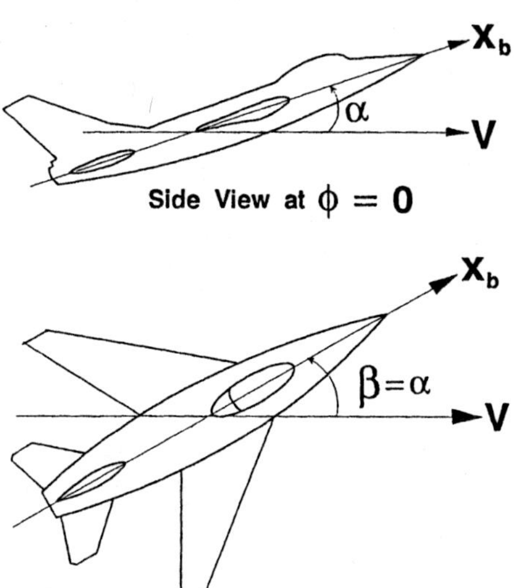

Figure 75.1 *(Cont.)*Initial and final flight configurations for high AOA maneuvers; V is always the aircraft velocity vector. e) roll about x-body axis.

Equations are

$$\dot{U} = VR - WQ + \frac{1}{m}\sum F_x, \quad (75.1)$$
$$U(0) = U_0 = \bar{U}, \quad 0 \le t \le t_f,$$

$$\dot{V} = WP - UR + \frac{1}{m}\sum F_y, \quad V(0) = V_0, \quad (75.2)$$

and

$$\dot{W} = UQ - VP + \frac{1}{m}\sum F_z, \quad W(0) = 0, \quad (75.3)$$

where m is the mass of the aircraft and U, V, and W are the components of the aircraft's velocity vector, resolved in the respective x, y, and z body axes; similarly, P, Q, and R are the components of the aircraft's rotational speed, resolved in the x, y, and z body axes (see Figure 75.2). The origin of the body axes is collocated with the aircraft's CG. The initial condition in Equation 75.3 is consistent with the use of stability axes. If body axes were used instead, then the initial condition would be $W(0) = W_0$. The *Moment Equations* are

$$\begin{aligned}\dot{P} = {} & \frac{I_{xz}}{D}(I_x - I_y + I_z)PQ \\ & + \frac{1}{D}(I_y I_z - I_z^2 - I_{xz}^2)QR \\ & + \frac{I_z}{D}\sum L + \frac{I_{xz}}{D}\sum N, \\ & P(0) = P_0, \end{aligned} \quad (75.4)$$

$$\begin{aligned}\dot{Q} = {} & \frac{I_z - I_x}{I_y}PR + \frac{I_{xz}}{I_y}(R^2 - P^2) + \frac{1}{I_y}\sum M, \\ & Q(0) = Q_0, \end{aligned} \quad (75.5)$$

and

$$\begin{aligned}\dot{R} = {} & \frac{1}{D}(I_x^2 - I_x I_y + I_{xz}^2)PQ \\ & + \frac{I_{xz}}{D}(I_y - I_x - I_z)QR \\ & + \frac{I_{xz}}{D}\sum L + \frac{I_x}{D}\sum N, \\ & R(0) = R_0, \end{aligned} \quad (75.6)$$

where I denotes the aircraft's inertia tensor and the definition $D = I_x I_z - I_{xz}^2$ is used. Also note that the modified "moment equations" above are obtained by transforming the classical Euler equations for the dynamics of a rigid body into state-space form.

The *Kinematic Equations* are concerned with the propagation of the Euler angles, which describe the attitude of the aircraft in inertial space: The Euler angles specify the orientation of the aircraft's body axes triad, and rotations are measured with reference to a right handed inertial frame whose z-axis points toward the center of the earth. The body axes are initially aligned with the inertial frame and the orientation of the body axes triad is determined by three *consecutive* rotations of ψ, θ and ϕ radians about the respective z, y, and x body axes. When the three consecutive rotations of the body axes frame is performed in the specified order above, ψ, ϕ, and θ are referred to as (3,2,1) Euler angles, the convention adhered to in this chapter. The Euler angles, which are needed to resolve the force of gravity into the body axes, are determined by the aircraft's angular rates P, Q, and R, according to the following equations:

$$\begin{aligned}\dot{\theta} = {} & Q\cos\phi - R\sin\phi, \\ & \theta(0) = \theta_0, \quad \theta(t_f) = \theta_f, \end{aligned} \quad (75.7)$$

$$\begin{aligned}\dot{\phi} = {} & P + Q\sin\phi\tan\theta + R\cos\phi\tan\theta, \\ & \phi(0) = \phi_0, \quad \phi(t_f) = \phi_f, \end{aligned} \quad (75.8)$$

$$\begin{aligned}\dot{\psi} = {} & Q\frac{\sin\phi}{\cos\theta} + R\frac{\cos\phi}{\cos\theta}, \\ & \psi(0) = \psi_0, \quad \psi(t_f) = \psi_f, \end{aligned} \quad (75.9)$$

The contribution of the force of gravity, the contributions of the aerodynamic and propulsive forces and moments acting on the aircraft, and the control input contributions, are contained in the force (F) and moment (M) summations in Equations 75.1 – 75.6. The aerodynamic forces and moments are produced by the aircraft's relative motion with respect to the air flow, and are proportional to the air density ρ and the square of the airspeed $\bar{U}$. In addition, the aerodynamic forces and moments are determined by the orientation angles with respect to the relative wind, viz., the AOA α and the sideslip angle β. The aerodynamic angles are depicted in Figure 75.2, where the aerodynamic forces and moments are also shown.

The standard [4] nondimensional aerodynamic force and moment coefficients, designated by the letter C, are introduced. The dynamic pressure $\bar{q} = \frac{1}{2}\rho(U^2 + V^2 + W^2) = \frac{1}{2}\rho\bar{U}^2$, and the wing's area S, are used to nondimensionalize the aerodynamic forces; the additional parameters, c, the wing's mean aerodynamic chord, or, b, the wing's span, are used to nondimensionalize the aerodynamic moments. Moreover, two subscript levels are used. The first subscript of C designates the pertinent aerodynamic force or moment component, and the second subscript

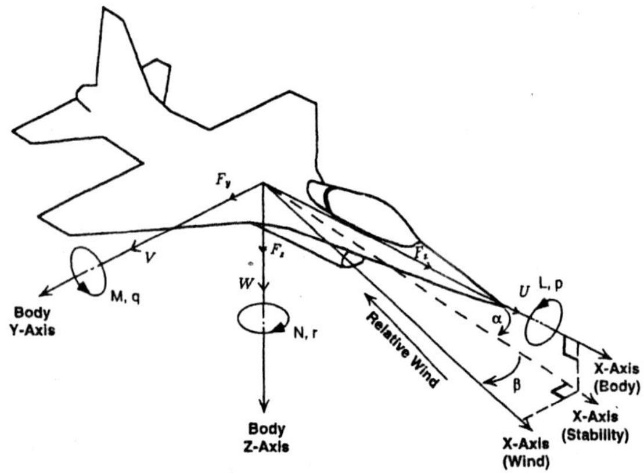

Figure 75.2 Diagram depicting axes definitions, componentes of velocity vector of the aircraft, angular rates of the aircraft, and aerodynamic forces and moments.

pertains to a specific component of the state vector. For example, the first stability derivative in Equation 75.10 is C_{l_p}, and it yields the aircraft's roll rate contribution to the nondimensional rolling moment coefficient C_l. Thus, the aerodynamic derivatives are "influence coefficients". Using three generalized control inputs, aileron, elevator, and rudder deflections, δ_a, δ_e, and δ_r, respectively, these force and moment summation equations are

$$\begin{aligned}\sum L &= \bar{q}Sb\,(\frac{b}{2\bar{U}}C_{l_p}P + \frac{b}{2\bar{U}}C_{l_r}R + C_{l_\beta}\beta + C_{l_{\delta_a}}\delta_a \\ &\quad + C_{l_{\delta_r}}\delta_r), \qquad (75.10)\end{aligned}$$

$$\begin{aligned}\sum M &= \bar{q}Sc\,(C_{m_0} + C_{m_\alpha}\alpha + \frac{c}{2\bar{U}}C_{m_q}Q \\ &\quad + \frac{c}{2\bar{U}}C_{m_{\dot\alpha}}\dot\alpha + C_{m_{\delta_e}}\delta_e), \qquad (75.11)\end{aligned}$$

$$\begin{aligned}\sum N &= \bar{q}Sb\,(\frac{b}{2\bar{U}}C_{n_p}P + \frac{b}{2\bar{U}}C_{n_r}R + C_{n_\beta}\beta \\ &\quad + C_{n_{\delta_a}}\delta_a + C_{n_{\delta_r}}\delta_r), \qquad (75.12)\end{aligned}$$

$$\begin{aligned}\sum F_y &= mg\,sin\phi\,cos\theta + \bar{q}S(\frac{b}{2\bar{U}}C_{y_p}P \\ &\quad + \frac{b}{2\bar{U}}C_{y_r}R + C_{y_\beta}\beta + C_{y_{\delta_r}}\delta_r),\end{aligned}$$

$$\begin{aligned}\sum F_z &= mg\,cos\phi\,cos\theta + T_z + \bar{q}S(-C_{z_\alpha}\alpha_{0L} \\ &\quad + C_{z_\alpha}\alpha + \frac{c}{2\bar{U}}C_{z_q}Q + \frac{c}{2\bar{U}}C_{z_{\dot\alpha}}\dot\alpha \\ &\quad + C_{z_{\delta_e}}\delta_e). \qquad (75.13)\end{aligned}$$

In Equation 75.13, T_z is the z-axis component of the thrust (usually < 0) and α_{0L} (< 0) is the zero lift Angle Of Attack (AOA) referenced from the x-stability axis. Hence, α_{0L} is determined by the choice of stability axes. The same is true for the C_{m_0} stabilty derivative in Equation 75.11. The C_{m_0} stability derivative used here is $C'_{m_0} - C_{m_\alpha}\alpha_{0L}$, where C'_{m_0} pertains to AOA measurements referenced to the aircraft's zero lift plane. Both C_{m_0} and α_{0L} are defined with reference to a nominal elevator setting ($\delta_e = 0$).

The x-axis velocity component is assumed constant throughout the short time horizon of interest in inner-loop flight control work. Hence, the thrust (control) setting is not included in inner-loop flight control work and Equation 75.1 is not used. The pertinent Equations are 75.2 – 75.13. The thrust control setting is also not included in the T_z thrust component. Furthermore, the heading angle ψ does not play a role in Equations 75.2 – 75.8 and Equations 75.10 – 75.13. Hence, one need not consider Equation 75.9 and thus the pertinent velocity vector roll dynamics are described by the seven DOF system of Equations 75.2 – 75.8 and Equations 75.10 – 75.13.

75.1.3 Nonlinear Dynamics

The seven states are P, Q, R, $\frac{V}{\bar{U}}$, $\frac{W}{\bar{U}}$, θ, and ϕ. The aerodynamic angles are defined as follows: $\alpha = tan^{-1}(\frac{W}{\bar{U}})$, $\beta = tan^{-1}(\frac{V}{\bar{U}})$. In Equations 75.2 – 75.6 and Equations 75.10 – 75.13, consolidated stability and control derivatives (which combine like terms in the equations of motion) are used, and the following nonlinear dynamics are obtained [2]:

$$\begin{aligned}\frac{d}{dt}(\frac{V}{\bar{U}}) &= (\frac{W}{\bar{U}})P + (C_{y_r} - 1)R + \frac{g}{\bar{U}}\,sin\phi\,cos\theta \\ &\quad + C_{y_p}P + C_{y_\beta}\beta + C_{y_{\delta_r}}\delta_r\,, \\ &\quad 0 \le t\,,\ V(0) = V_0 \qquad (75.14)\end{aligned}$$

$$\begin{aligned}\frac{d}{dt}(\frac{W}{\bar{U}}) &= (1 + C_{z_q})Q - (\frac{V}{\bar{U}})P + \frac{g}{\bar{U}}\,cos\phi\,cos\theta \\ &\quad + C_{T_z} - C_{z_\alpha}\alpha_{0L} + C_{z_\alpha}\alpha + C_{z_{\dot\alpha}}\dot\alpha + C_{z_{\delta_e}}\delta_e, \\ &\quad W(0) = 0 \qquad (75.15)\end{aligned}$$

$$\begin{aligned}\frac{dP}{dt} &= C_{l_{pq}}PQ + C_{l_{qr}}QR + C_{l_p}P + C_{l_r}R \\ &\quad + C_{l_\beta}\beta + C_{l_{\delta_a}}\delta_a + C_{l_{\delta_r}}\delta_r\,, \\ &\quad P(0) = P_0 \qquad (75.16)\end{aligned}$$

$$\begin{aligned}\frac{dQ}{dt} &= C_{m_{pr}}PR + C_{m_{r^2}}R^2 - C_{m_{p^2}}P^2 + C_{m_0} \\ &\quad + C_{m_\alpha}\alpha + C_{m_{\dot\alpha}}\dot\alpha + C_{m_q}Q + C_{m_\delta}\delta_e\,, \\ &\quad Q(0) = Q_0 \qquad (75.17)\end{aligned}$$

$$\begin{aligned}\frac{dR}{dt} &= C_{n_{pq}}PQ - C_{l_{pq}}QR + C_{n_p}P + C_{n_r}R \\ &\quad + C_{n_\beta}\beta + C_{n_{\delta_a}}\delta_a + C_{n_{\delta_r}}\delta_r\,, \\ &\quad R(0) = R_0 \qquad (75.18)\end{aligned}$$

It is assumed that stability axes are used. Furthermore, the particular stability axes used are chosen at time $t = 0$. Also, in Equations 75.15 and 75.17, $\dot\alpha = (\frac{\dot W}{\bar{U}})/(1 + \frac{W^2}{\bar{U}^2})$. The kinematic Equations 75.7 and 75.8 are also included in the nonlinear dynamical system.

Trim Analysis

The LHS of the seven differential Equations 75.2 – 75.6 (or Equations 75.14 – 75.18) and Equations 75.7 and 75.8, is set equal to zero to compute the so-called "trim" values of the aircraft states and controls. Trim conditions are now considered where the aircraft's angular rates P_0, Q_0, and R_0, its sideslip velocity component V_0, and the respective pitch and bank Euler angles θ_0, and ϕ_0, are constant. The pertinent seven states are V, W, P, Q, R, θ, and ϕ, and the (generalized) controls are

δ_a, δ_e, and δ_r. In the sequel, the trim controls and states are denoted with the subscript 0, with the exception of the trim value of U, which is barred. For fixed control trim settings, an initial trim condition (or nominal trajectory) is established. Also, when some of the trim states are specified, the remaining states and control trim settings can be obtained, provided that the total number of specified variables (controls and/or states) is three.

Using generalized stability and control derivatives, the following algebraic *trim equations* are obtained from Equations 75.14, 75.16, and 75.18:

$$\frac{g}{\bar{U}} cos\theta_0 \, sin\phi_0 + C_{y_p} P_0 + (C_{y_r} - 1) R_0 + C_{y\beta}\beta_0 = -C_{y\delta_r}\delta_{r_0}, \tag{75.19}$$

$$C_{l_{pq}} P_0 Q_0 + C_{l_{qr}} Q_0 R_0 + C_{l_p} P_0 + C_{l_r} R_0 + C_{l\beta}\beta_0 = -C_{l\delta_a}\delta_{a_0} - C_{l\delta_r}\delta_{r_0}, \tag{75.20}$$

and

$$C_{n_{pq}} P_0 Q_0 - C_{l_{pq}} Q_0 R_0 + C_{n_p} P_0 + C_{n_r} R_0 + C_{n\beta}\beta_0 = -C_{n\delta_a}\delta_{a_0} - C_{n\delta_r}\delta_{r_0}. \tag{75.21}$$

In Equations 75.19 – 75.21 the unknowns are P_0, Q_0, R_0, β_0, δ_{a_0}, and δ_{r_0}. In addition, and as will be shown in the sequel, Equations 75.7 and 75.8 yield $\theta_0 = \theta_0(P_0, Q_0, R_0)$ and $\phi_0 = \phi_0(P_0, Q_0, R_0)$. Hence, we have obtained three equations in six unknowns. This then requires the specification of three trim variables, whereupon the remaining three trim variables are obtained from the above three trim equations. For example, if the equilibrium angular rates P_0, Q_0, and R_0 are specified then the required control trim settings δ_{a_0} and δ_{r_0}, and the trim sideslip angle β_0, can be calculated. The calculation then entails the solution of a set of three linear equations in three unknowns. The system matrix that needs to be inverted obtained from Equations 75.19 – 75.21 is

$$M = \begin{bmatrix} C_{y\beta} & 0 & C_{y\delta_r} \\ C_{l\beta} & C_{l\delta_a} & C_{l\delta_r} \\ C_{n\beta} & C_{n\delta_a} & C_{n\delta_r} \end{bmatrix}.$$

Obviously, the M matrix must be nonsingular for a trim solution with constant angular rates to be feasible, i.e., the condition $C_{y\beta}C_{l\delta_a}C_{n\delta_r} + C_{y\delta_r}C_{l\beta}C_{n\delta_a} \neq C_{y\beta}C_{l\delta_r}C_{n\delta_a} + C_{y\delta_r}C_{l\delta_a}C_{n\beta}$ must hold.

From the remaining Equations 75.15 and 75.17, the two equations which determine the zero lift angle included between the velocity vector and the aircraft's zero lift plane, α_{0L}, and the trim elevator setting, δ_{e_0}, are obtained:

$$\frac{g}{\bar{U}} cos\theta_0 \, cos\phi_0 + C_{T_z} - C_{z_\alpha}\alpha_{0L} - P_0 tan\beta_0 + (1 + C_{z_q}) Q_0 + C_{z\delta_e}\delta_{e0} = 0, \tag{75.22}$$

and

$$C_{m_0} + C_{m\delta_e}\delta_{e0} + C_{m_q} Q_0 + C_{m_{pr}} P_0 R_0 + C_{m_{p2}} R_0^2 - C_{m_{p2}} P_0^2 = 0. \tag{75.23}$$

Once the angular rates and the sideslip angle β_0 have been established, the δ_{e_0} and α_{0L} unknowns are determined from the linear Equations 75.22 and 75.23. Obviously, δ_{e_0} must be feasible, i.e., $-\delta_{e_{max}} \leq \delta_{e_0} \leq \delta_{e_{max}}$.

An analysis of Equations 75.7 and 75.8 is now undertaken. Recall that the trim trajectory is flown at a constant pitch angle. If both Q_0 and R_0 are not 0, then the bank angle ϕ_0 must be constant. The trim pitch and bank angles are obtained from the Euler angles Equations 75.7 and 75.8. Thus, if $R_0 \neq 0$, then $\phi_0 = tan^{-1}(\frac{Q_0}{R_0})$, and $\theta_0 = -tan^{-1}(\frac{P_0}{R_0} cos\phi_0)$. Hence, if $P_0 = 0$, then $\theta_0 = 0$ and this trim condition entails a level turn, illustrated in Figure 75.1c. In the general case where $P_0 \neq 0$, a steady spiral climb or a steady corkscrew descent, as illustrated in Figure 75.1d, is being considered. Moreover, for the above flight maneuvers, Equation 75.9 yields the elegant heading rate result $\dot{\psi} = \sqrt{P_0^2 + Q_0^2 + R_0^2}$.

Obviously, not every pair of prespecified pitch and bank trim angles, θ_0 and ϕ_0, respectively, is feasible. Indeed, the solution of the equations above must satisfy $-\delta_{a_{max}} \leq \delta_{a_0} \leq \delta_{a_{max}}$ and $-\delta_{r_{max}} \leq \delta_{r_0} \leq \delta_{r_{max}}$. In the very special case where the bank angle $\phi_0 = 0$, a symmetrical flight condition ensues with $\beta_0 = \delta_{a_0} = \delta_{r_0} = 0$. A steady climb in the pitch plane is then considered, as illustrated in Figure 75.1b. Finally, α_{0L} and δ_{e_0} are then determined from Equations 75.22 and 75.23, where $\phi_0 = P_0 = Q_0 = R_0 = 0$, i.e., $\frac{g}{\bar{U}} cos\theta_0 + C_{T_z} - C_{z_\alpha}\alpha_{0L} + C_{z\delta_e}\delta_{e0} = 0$, *and* $C'_{m_0} - C_{m_\alpha}\alpha_{0L} + C_{m\delta_e}\delta_{e0} = 0$.

The trim conditions identified above represent interesting flight phases. For example, and as shown in Figure 75.1a, one can initiate a velocity vector roll in trimmed and level flight, where $P_0 = Q_0 = R_0 = \theta_0 = \phi_0 = \beta_0 = \delta_{a_0} = \delta_{r_0} = 0$, and one can subsequently arrest the velocity vector roll at a new trim flight condition which entails a level turn, shown in Figure 75.1c, and where Q_0 and R_0 are prespecified, $P_0 = \theta_0 = 0$, and $\phi_0 = tan^{-1}(\frac{Q_0}{R_0})$. The new bank angle satisfies the equation $sin\phi_0 = \frac{Q_0}{\sqrt{Q_0^2+R_0^2}} = \frac{Q_0}{\dot{\psi}}$. The required trim β_0, δ_{a_0}, and δ_{r_0} is determined by solving the linear trim equations. Since the trim bank angle changes, this will require a change in the trim elevator setting and in α_{0L}. The latter directly translates into a change in AOA. To account for the difference in the required AOA for trim before and after the maneuver is accomplished, an $\alpha = \alpha_{0L_2} - \alpha_{0L_1}$ must be commanded. To establish the new trim condition, a new trim sideslip angle $\beta = \beta_0$ must also be commanded.

Similarly, a level and symmetric S maneuver will be initiated in a level and trimmed turn, say, to the right, and the end trim condition will be a level and trimmed turn to the left. In this case, the new trim variables are $Q_0 \leftarrow Q_0$, $-R_0 \leftarrow R_0$, $P_0 = \theta_0 = 0$, $\pi - \phi_0 \leftarrow \phi_0$. Evidently, R_0 changes, hence a new trim sideslip angle must be established. A similar analysis can be conducted for climbing turns, corkscrew descents and nonplanar S maneuvers, where $\theta_0 \neq 0$.

75.1.4 Actuators

The control effectors' actuator dynamics play an important role in flight control. Typical fighter aircraft actuator Transfer Functions

(TFs) are given below.
Elevator dynamics:

$$\frac{\delta_e(s)}{\delta_{e_c}(s)} = \frac{2138\,(s^2+11.27s+6872)}{s^4+154.1s^3+1612.8s^2+495589s+14692336} \quad (75.24)$$

The Bode plot of the elevator actuator TF is shown in Figure 75.3. Note the phase lag at high frequency. High frequency operation is brought about by a high loop gain, which, in turn, is specified for robustness purposes. Thus, robust (flight) control mandates the use of high-order actuator models, which, in turn, complicates the control design problem. In addition, the following nonlinearities play an important role in actuator modeling and significantly impact the control system's design and performance. These nonlinearities are of saturation type and entail 1) Maximum elevator deflection of, e.g., $\pm 22°$ and 2) Maximum elevator deflection rate of $\pm 60°$/sec. The latter significantly impacts the aircraft's maneuverability.

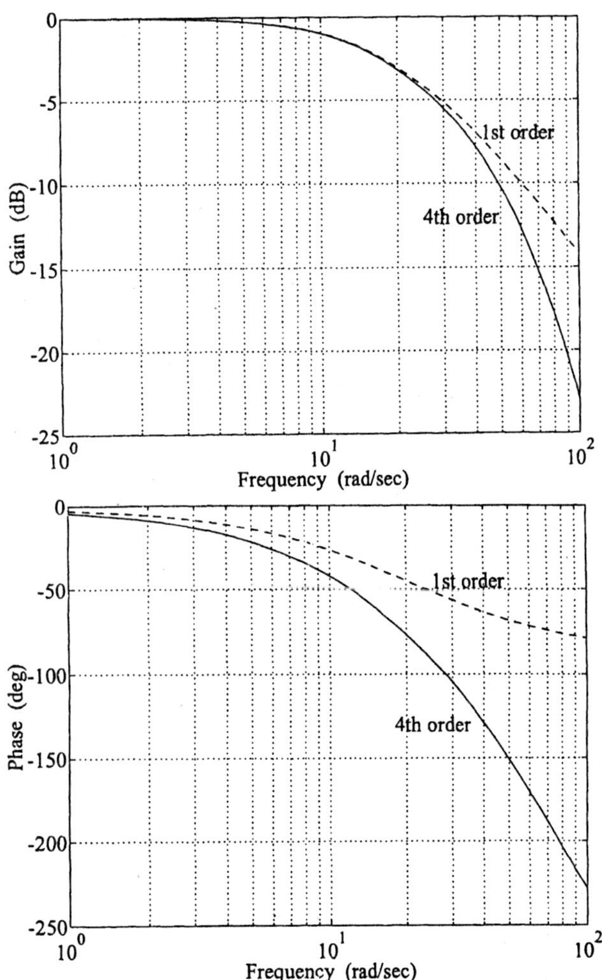

Figure 75.3 Elevator actuator transfer function.

Aileron dynamics:

$$\frac{\delta_a(s)}{\delta_{a_c}(s)} = \frac{5625}{s^2+88.5s+5625} \quad (75.25)$$

The aileron's saturation limits are similar to the elevator's.
Rudder dynamics:

$$\frac{\delta_r(s)}{\delta_{r_c}(s)} = \frac{5148}{s^2+99.36s+5148} \quad (75.26)$$

The maximum rudder deflection is $\pm 30°$ and its maximum deflection rate is $\pm 60°$/sec..
Multiaxes Thrust Vectoring (MATV) actuator dynamics:

The MATV actuator entails a thrust vectoring nozzle. Its deflection in pitch and yaw is governed by the following transfer function:

$$\frac{\delta_{TV}(s)}{\delta_{TV_c}(s)} = \frac{400}{s^2+26s+400} \quad (75.27)$$

Note the reduced bandwidth of the MATV actuator. In addition, the dynamic range of the MATV actuator is $\pm 20°$ and its maximum deflection rate is limited to $\pm 45°$/sec.

Weighting Matrix

The three generalized control inputs δ_e, δ_a, and δ_r are responsible for producing the required control moments about the aircraft's x, y, and z axes. Modern combat aircraft are, however, equipped with redundant control effectors. The latter are of aerodynamic type, and, in addition, Multiaxes Thrust Vectoring (MATV) is employed. Hence, a weighting matrix W for control authority apportionment is used to transform the three generalized command inputs into the six command signals δ_{ar}, δ_{er}, δ_{ep}, δ_{TVp}, δ_{ry}, and δ_{TVy} that control the six physical actuators. In industrial flight control circles the weighting matrix is also referred to as the "Mixer." In our flight control problem, W is a 6×3 matrix.

It is noteworthy that a weighting matrix is required even in the case of aircraft equipped with basic control effectors only, in which case the generalized controls are also the actual physical controls and the weighting matrix is 3×3. The flight control concept of an aileron rudder interconnect will mitigate some of the adverse yaw effects encountered at high AOAs. Thus, a roll command would not only command an aileron deflection, but also generate a rudder command to remove adverse yaw produced by the aileron deflection. The appropriate weighting matrix coefficient is based on the ratio of the applicable control derivatives because the latter are indicative of the yaw moment control power of the respective effector. Hence, the weighting matrix

$$W_1 = \begin{bmatrix} 1 & 0 & 0 \\ 0 & 1 & 0 \\ 0 & \frac{C_{n_{\delta_a}}}{C_{n_{\delta_r}}} & 0 \end{bmatrix} \text{ is oftentimes employed,}$$

$$\text{and} \begin{bmatrix} \delta_e \\ \delta_a \\ \delta_r \end{bmatrix} \leftarrow W_1 \begin{bmatrix} \delta_e \\ \delta_a \\ \delta_r \end{bmatrix}.$$

When the number of physical effectors exceeds the number of generalized controls, the control authority in each axis needs to be adequately apportioned among the physical effectors, as shown in Figure 75.4. Note that the six physical control effectors are partitioned into three distinct groups which correspond to the three control axes of the aircraft. There are two physical effectors in each group. The physical effectors δ_{ep} and δ_{TVp} are in the δ_e group, the physical effectors δ_{ar} and δ_{er} are in the δ_a group, and the physical effectors δ_{ry} and δ_{TVy} are in the δ_r group. Now, the underlying control concept is to have the control effectors in a group saturate simultaneously. Hence, scale factors are used in the weighting matrix to account for dissimilar maximum control deflections within the control effectors in each group. For example, a maximal aileron command should command the roll ailerons to their maximum deflection of 20 degrees, but the roll elevators to their preassigned maximum roll control authority of 5 degrees only. Thus, scaling is achieved by multiplying the generalized control command by the ratio of the respective maximum actuator limits in its group. Hence,

$$W_2 = \begin{bmatrix} 1 & 0 & 0 \\ \frac{\gamma}{\delta_{ep\max}} & 0 & 0 \\ 0 & 1 & 0 \\ 0 & \frac{\delta_{er\,max}}{\delta_{ar\max}} & 0 \\ 0 & 0 & 1 \\ 0 & 0 & \frac{\gamma}{\delta_{ry\max}} \end{bmatrix}$$

Finally, the weighting matrices W_1 and W_2 are combined to form the weighting matrix $W = W_2 W_1$.

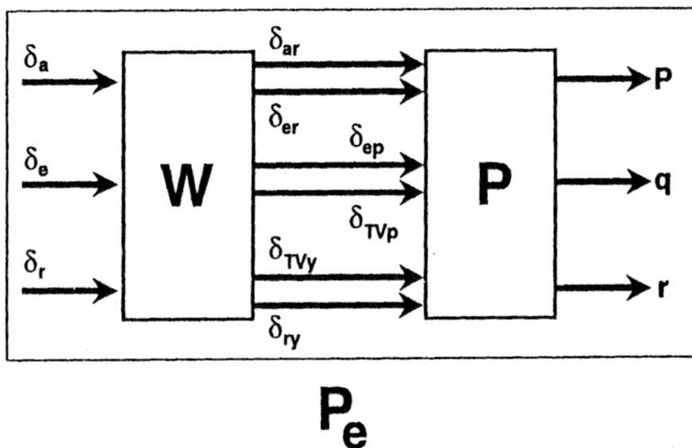

Figure 75.4 Mixer.

A more detailed analysis also considers the relative magnitudes of the control moments and control forces generated by the two effectors in each control group, which requires knowledge of the control derivatives. The control theoretic problem of optimal control authority apportionment among the various control effectors is currently a research topic in flight control.

75.1.5 Flight Control Requirements

The Flight Control System constitutes the pilot/aircraft interface and its primary mission is to enable the pilot/aircraft to accomplish a prespecified task. This entails the following:

1. Accommodate *high* α (velocity vector roll) *maneuvers.*
2. Meet handling/flying qualities specifications over the entire flight envelope and for all aircraft configurations, including tight tracking and accurate pointing for Fire Control.
3. Above and beyond item 1, obtain optimal performance. Pilot "gets what he wants."
4. Minimize number of placards, the number of limitations and restrictions the pilot must adhere to → Pilot can fly his aircraft "in abandon".

Specifications

Mil-Std 1797A [3] defines the flying quality specifications, Level 1 the best. The specifications (airworthiness criteria) are given as a mixture of time and frequency domain specifications. The longitudinal channel uses time-domain specifications based on the pitch rate q response to a step input command calculated from the two-degrees-of-freedom model given by the fast, Short Period (SP), approximation. The time-domain specifications are based on two straight lines drawn on the q response shown in Figure 75.5. To meet Level 1 flying quality specifications, the equivalent time delay (t_1) must be less than 0.12 sec, the transient peak ratio ($\frac{\Delta q_2}{\Delta q_1}$) less than 0.30, and the effective rise time ($\Delta t = t_2 - t_1$) between $9/\bar{U}$ and $500/\bar{U}$, where $\bar{U}$ is the true airspeed (ft/s).

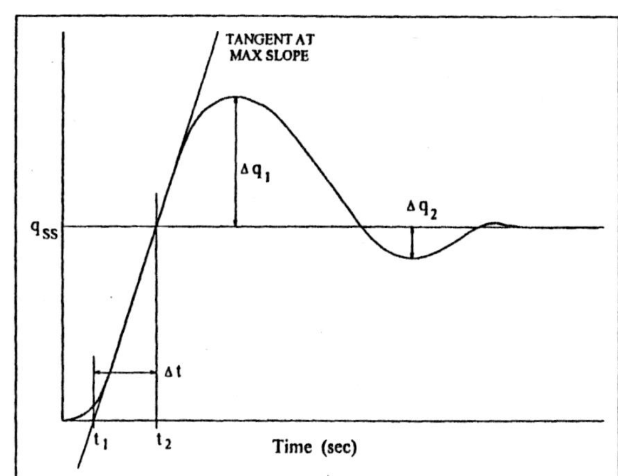

Figure 75.5 Step Elevator Response.

For the lateral/directional channel, the flying quality specifications apply to simultaneously matching the Bode plots of the final system with those of the equivalent fourth-order transfer functions given by

$$\frac{\phi(s)}{\delta_{rstk}(s)} = \frac{K_\phi(s^2 + 2\zeta_\phi\omega_\phi s + \omega_\phi^2)\, e^{-\tau_{ep} s}}{(s + \frac{1}{T_R})(s + \frac{1}{T_s})(s^2 + 2\zeta_d\omega_d s + \omega_d^2)}, \text{ and}$$

$$\frac{\beta(s)}{\delta_{rud}(s)} = \frac{(A_3 s^3 + A_2 s^2 + +A_1 s + A_0)\, e^{-\tau_{e\beta} s}}{(s + \frac{1}{T_R})(s + \frac{1}{T_s})(s^2 + 2\zeta_d \omega_d s + \omega_d^2)}.$$

To meet Level 1 flying qualities, the Roll Mode time constant (T_R) must be less than 1.0 sec, the Dutch Roll Mode damping (ζ_d) greater than 0.40, the Dutch Roll Mode natural frequency (ω_d) greater than 1 rad/sec, the roll rate time delay (τ_{ep}) less than 0.1 sec, and the time to double of the Spiral Mode ($-ln2\ T_S$) greater than 12 sec. Disturbance rejection specifications in the lateral/directional channel are based on control surface usage and are variable throughout the flight envelope.

In addition, the open-loop phase margin angle γ of both the longitudinal and lateral/directional designs must be greater than 30°, and the open-loop gain margin a must be greater than 6dB. To determine phase and gain margins, the open-loop transmissions from stick inputs to the required outputs in each loop are examined. Finally, the phase margin frequency (cutoff frequency $= \omega_\phi$) should be less than 30 rad/sec to prevent deleterious interaction with the bending modes of the aircraft.

Concerning cross-coupling disturbance bounds: One requirement is that a sustained 10° sideslip will use less than 75% of the available roll axis control power (aileron). There is also a complicated specification for the amount of sideslip β allowed for a particular roll angle. If one simplifies the requirements in a conservative fashion over the majority of the flight envelope, a roll command of 1°/sec should result in less than 0.022° of β, but, at low speeds, β is allowed to increase to 0.067°. The maximal β allowed in any roll maneuver is 6°.

Evidently, a "robust" set of specifications is not available because the flying qualities depend on flight condition throughout the flight envelope. The lack of uniform performance bounds over the entire envelope illustrates a shortcoming of the current robust control paradigms. Moreover, the specifications above must be met in the face of the following.

A. **Constraints**

Practical limitations on achievable performance (including robustness) are imposed by
(1) actuator rate limits (e.g., ca. 60°/sec), (2) actuator saturation (e.g., ca. 22°), (3) sensor noise/sensor quality, (4) low frequency elastic modes, and (5) highest possible sampling rate for digital implementation.

B. **Environment**

The FCS needs to accommodate the following:
(1) "plant" variability/uncertainty in the extended flight envelope, and aircraft configurations possibilities. Thus, in the real world, *structured* uncertainty needs to be addressed. Also, accommodation of inflight aerodynamic control surfaces failures/battle damage is required, (2) atmospheric disturbances/ turbulence suppression, (3) unstable aircraft (for reduced trim drag and improved cruise performance); guard against adverse interaction with actuator saturation, for the latter can cause loss of control, i.e., departure, (4) control authority allocation for trim, maneuverability, additional channel control, stabilization. Example: In the current F-16, the apportionment of control authority is 10, 7, 4, and 2 degrees of elevator deflection, respectively, (5) optimal control redundancy management, and (6) pilot-in-the-loop.

75.1.6 Dynamic Analysis

The linearization of the state equations of motion about trimmed flight conditions is the first step performed in classical flight control system design. The simple trim condition shown in Figure 75.1a, which entails symmetric and level, or climbing, flight, and where $P_0 = Q_0 = R_0 = \phi_0 = 0$ and $\beta_0 = \delta_{r_0} = \delta_{a_0} = 0$, is considered.

Linearization is based on the small perturbations hypothesis, which predicates that only small excursions in the state variables about a given trim condition occur [1], [2], [5]. For conventional aircraft with an (x,z) plane of symmetry, and for small roll rates, an additional benefit accrues: Linearization brings about the decoupling of the aircraft's pitch and lateral/directional dynamics, thus significantly simplifying the flight control problem. Unfortunately, the velocity vector roll is a large amplitude maneuver consisting of a high roll rate and large excursions in Euler angles. Thus, the velocity vector roll maneuver could entail a high (stability axis) roll rate P. According to the flying quality specifications in Section 75.1.5, a maximal roll rate is desired, hence P is not necessarily small. In addition, during a velocity vector roll maneuver, large excursions in the bank angle occur. Hence, for maneuvering flight, the standard linearization approach must be modified. All of the system states, other than the roll rate and the Euler angle ϕ, can still be represented as small perturbation quantities (denoted by lower case variables), and the squares and products of these state variables can be neglected as is typically done in the small perturbations based linearization process. All terms containing the roll rate P and bank angle ϕ are not linearized, however, because these variables do not represent small perturbations. Furthermore, the elevator deflection δ_e is now measured with reference to the trim elevator setting of δ_{e_0}. Therefore, α_{0L} does not feature in Equation 75.15, nor does C_{m_0} feature in Equation 75.17. Furthermore, the gravity term in Equation 75.15 is modified to account for the trim condition properly. Moreover, the FCS strives to maintain q and r small, and our choice of stabilty axes yields the velocity vector roll's end conditions $\theta_f = \theta_0$. Hence, it is assumed that, throughout the velocity vector roll, the pitch angle perturbation $\theta \approx 0$. In addition, the stabilty axes choice lets us approximate the AOA as $\alpha = tan^{-1}(\frac{w}{U}) \approx \frac{w}{U}$ and, at this particular trim condition ($\beta_0 = 0$), the sideslip angle $\beta = tan^{-1}(\frac{v}{U}) \approx \frac{v}{U}$. Thus, the "slow" dynamics in Equations 75.8, 75.15 and 75.14 are reduced to

$$\dot{\phi} = P + (q\ sin\phi + r\ cos\phi)tan\theta_0, \tag{75.28}$$

$$\begin{aligned}\dot{\alpha} &= q - P\beta + \frac{g}{U}(cos\theta_0 cos\phi - sin\theta_0\,\theta\ cos\phi - cos\theta_0) \\ &\quad + C_{z_\alpha}\alpha + C_{z_q}q + C_{z_{\dot\alpha}}\dot{\alpha} + C_{z_{\delta_e}}\delta_e,\end{aligned} \tag{75.29}$$

and

$$\dot{\beta} = P\alpha - r + \frac{g}{\bar{U}}(cos\theta_0\, sin\phi - sin\theta_0\, \theta\, sin\phi) + C_{y_p}P + C_{y_r}r + C_{y\beta}\beta + C_{y_{\delta_r}}\delta_r\,, \quad (75.30)$$

where θ now denotes the perturbation in the pitch angle. Equations 75.16, 75.17 and 75.18 yield the fast dynamics

$$\dot{P} = C_{l_p}P + C_{l_{pq}}Pq + C_{l_r}r + C_{l\beta}\beta + C_{l_{\delta_{ar}}}\delta_{ar} + C_{l_{\delta_{er}}}\delta_{er} + C_{l_{\delta_{ry}}}\delta_{ry} + C_{l_{\delta_{TVy}}}\delta_{TVy}, \quad (75.31)$$

$$\dot{q} = C_{m_{pr}}Pr - C_{m_{p2}}P^2 + C_{m_\alpha}\alpha + C_{m_q}q + C_{m_{\dot{\alpha}}}\dot{\alpha} + C_{m_{\delta_{ep}}}\delta_{ep} + C_{m_{\delta_{TVp}}}\delta_{TVp}, \quad (75.32)$$

$$\dot{r} = C_{n_{pq}}Pq + C_{n_p}P + C_{n_r}r + C_{n\beta}\beta + C_{n_{\delta_{ar}}}\delta_{ar} + C_{n_{\delta_{er}}}\delta_{er} + C_{n_{\delta_{ry}}}\delta_{ry} + C_{n_{\delta_{TVy}}}\delta_{TVy}, \quad (75.33)$$

and

$$\dot{\theta} = q\, cos\phi - r\, sin\phi. \quad (75.34)$$

Hence, the dynamics of maneuvering flight and of the velocity vector roll maneuver entail a seven DOF model which is specified by Equations 75.28 – 75.34. The three generalized control effectors' inputs δ_e, δ_a and δ_r appearing in the six DOF equations represent the six physical control effector inputs, i.e., the pitch elevator deflection δ_{e_p}, the deflection in pitch of the thrust vectoring engine nozzle δ_{TVp}, the aileron deflection δ_{ar}, the roll elevator deflection δ_{er} (we here refer to the differential tails), the rudder deflection δ_{ry}, and the deflection in yaw of the thrust vectoring engine nozzle δ_{TVy}.

When the initial trim condition is straight and level flight, $\theta_0 = 0$. Then a further simplification of the Equations of motion 75.28 – 75.30 is possible:

$$\dot{\phi} = P, \quad (75.35)$$

$$\dot{\alpha} = (1 + C_{z_q})q - P\beta + (\frac{g}{\bar{U}})(cos\phi - 1) + C_{z_{\dot{\alpha}}}\dot{\alpha} + C_{z_\alpha}\alpha + C_{z_{\delta_{ep}}}\delta_{ep} + C_{z_{\delta_{TVp}}}\delta_{TVp}, \text{ and} \quad (75.36)$$

$$\dot{\beta} = P\alpha + (\frac{g}{\bar{U}})sin\phi + C_{y_p}P + (C_{y_r} - 1)r + C_{y\beta}\beta + C_{y_{\delta_{ar}}}\delta_{ar} + C_{y_{\delta_{er}}}\delta_{er} + C_{y_{\delta_{ry}}}\delta_{ry} + C_{y_{\delta_{TVy}}}\delta_{TVy}. \quad (75.37)$$

Equations 75.35 – 75.37 are used in conjunction with the "fast" dynamics in Equations 75.31 – 75.34 to complete the description of the aircraft dynamics for an initial straight and level trim.

75.1.7 Conventional Flight Control

It is remarkable that the very same assumption of trimmed *level* flight renders in the pitch channel the classical second-order Short Period (SP) approximation. Thus, $P = \phi = 0$ yields the celebrated linear SP pitch dynamics

$$\dot{\alpha} = C_{z_\alpha}\alpha + (1 + C_{z_q})q + C_{z_{\dot{\alpha}}}\dot{\alpha} + C_{z_{\delta_e}}\delta_e \quad (75.38)$$

$$\dot{q} = C_{m_\alpha}\alpha + C_{m_q}q + C_{m_{\dot{\alpha}}}\dot{\alpha} + C_{m_{\delta_e}}\delta_e \quad (75.39)$$

In the more general case of trimmed and symmetric *climbing* flight in the vertical plane, the longitudinal SP dynamics are of third order [2], [6]:

$$\dot{\alpha} = C_{z_\alpha}\alpha + (1 + C_{z_q})q - \frac{g}{\bar{U}}sin\theta_0\, \theta + C_{z_{\dot{\alpha}}}\dot{\alpha} + C_{z_{\delta_e}}\delta_e$$

$$\dot{q} = C_{m_\alpha}\alpha + C_{m_q}q + C_{m_{\dot{\alpha}}}\dot{\alpha} + C_{m_{\delta_e}}\delta_e$$

$$\dot{\theta} = q$$

The third-order dynamics depend on the pitch angle parameter θ_0. At elevated pitch angles the SP/Phugoid time-scale separation's validity is questionable.

Furthermore, applying the small perturbations hypothesis to the additional variables P and ϕ in Equations 75.28, 75.30, 75.31, and 75.33, allows us to neglect the terms which contain perturbation products. This decouples the lateral/directional channel from the pitch channel. Thus, the conventional lateral/directional "plant" is

$$\dot{p} = C_{l_p}p + C_{l_r}r + C_{l\beta}\beta + C_{l_{\delta_a}}\delta_a + C_{l_{\delta_r}}\delta_r \quad (75.40)$$

$$\dot{r} = C_{n_p}p + C_{n_r}r + C_{n\beta}\beta + C_{n_{\delta_a}}\delta_a + C_{n_{\delta_r}}\delta_r \quad (75.41)$$

$$\dot{\beta} = C_{y_p}p + (C_{y_r} - 1)r + C_{y\beta}\beta + \frac{g}{\bar{U}}\phi + C_{y_{\delta_r}}\delta_r \quad (75.42)$$

$$\dot{\phi} = p \quad (75.43)$$

In conclusion, the horizontal flight assumption $\theta_0 = 0$ reduces the complexity of both the pitch channel and the velocity vector roll control problems.

The stability and control derivatives in Equations 75.38, 75.39, and Equations 75.40 – 75.43 depend on flight condition. Therefore, in most aircraft implementations, gain scheduling is used in the controller. Hence, low order controllers are used. A simple first-order controller whose gain and time constant are scheduled on the dynamic pressure $\bar{q}$ will do the job.

In modern approaches to flight control, robust controllers that do not need gain scheduling are sought. A full envelope controller for the F-16 VISTA that meets the flying quality specifications of Section 75.1.5 has been synthesized using the QFT robust control design method [7]. The pitch channel's inner and outer loop compensators, G_1 and G_2, and the prefilter F, are

$$G_1(s) = \frac{3.1(s + 13)(s + 17)}{s(s + 100)},$$

$$G_2(s) = \frac{30s^4 + 1356.6s^3 + 57136s^2 + 177000s + 280000}{s^4 + 109.8s^3 + 1029s^2 + 4900s},$$

$$F(s) = \frac{16}{s^2 + 6s + 16}.$$

75.1.8 Time-Scale Separation

Time-scale separation plays an important role in flight control. The fast states P, q, and r which represent the angular rates are dominant during maneuvering flight where the velocity vector roll is initiated and arrested. The perturbations α in AOA and in the sideslip angle β are maintained at near zero in the short time periods of the transition regimes. Therefore, the six DOF dynamical model obtained when $\theta_0 = 0$ can be decoupled into "slow" dynamics associated with the α, β and ϕ variables and the "fast" dynamics of the P, q, and r angular rates. Two primary regimes of the roll maneuver are identified, viz., the transition and free stream regions shown in Figure 75.6. The dynamic transition regions represent the initiation and arrest of the velocity vector roll, which occur in a relatively short amount of time (on the order of one second). In these transition regions the fast states of the aircraft, viz., the angular rates, are dominant. During these initial and/or terminal phases of the velocity vector roll maneuver a desired roll rate needs to be established, viz., the controller's modus operandi entails tracking. In the free stream regime, the established roll rate is maintained through the desired bank angle, while the perturbations in AOA and sideslip angle are regulated to zero. Thus, in the "free stream" regime, the controller's function entails regulation.

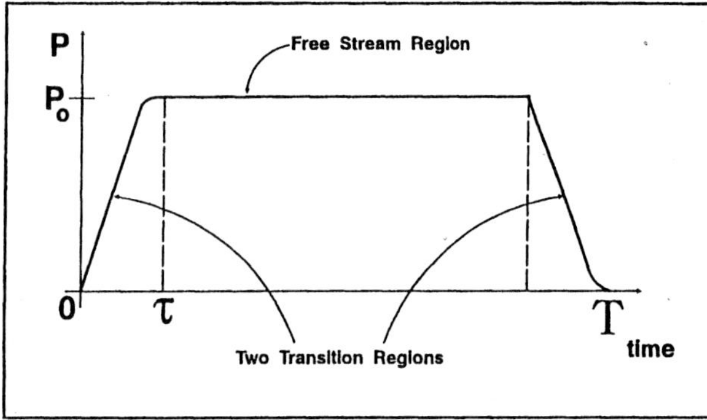

Figure 75.6 Time-scale separation.

The design of the FCS for maneuvering flight and for large amplitude velocity vector roll maneuvers at high AOAs hinges on this discernible time-scale separation, and is now outlined. The time-scale separation leads to the two-loop control structure shown in Figure 75.7. The inner loop of the FCS consists of a three-axis rate-commanded control system which controls the fast states of the system. This control function is essential in the transition regions of the velocity vector roll maneuver. The bandwidth of the inner loop is relatively high. The outer loop serves to regulate to zero the relatively slower aerodynamic angles, viz., the AOA and sideslip angle perturbations which accrue during the velocity vector roll, in particular, during the transition regions. The outer loop is therefore applicable to both the transition and free stream regimes of the velocity vector roll.

Time-Domain Approach

Maneuvering flight control entails tracking control, as opposed to operation on autopilot, where regulation is required. Tracking control in particular in the presence of actuator saturation requires a time-domain approach. Therefore, the objective is the derivation of a receding horizon/model predictive tracking control law, in the case where both fast and slow state variables feature in the plant model. An optimal control sequence for the whole planning horizon is obtained and its first element is applied to the plant, following which the control sequence is reevaluated over a one time step displaced horizon. Thus, feedback action is obtained. A procedure for synthesizing an optimal control sequence for solving a fixed horizon optimal control problem, which exploits the time-scale separation inherent in the dynamical model, is employed. The time-scale separation decomposes the receding horizon control problem into two nested, but lower dimensional, receding horizon control optimization problems. This yields an efficient control signal generation algorithm. The outer loop optimal control problem's horizon is longer than the inner loop optimal control problem's horizon, and the discretization time step in the inner loop is smaller than in the outer loop. Thus, to apply model predictive tracking control to dynamical systems with time-scale separation, the classical nested feedback control loop structure of the frequency domain is transcribed into the time domain.

The plant model for the inner loop's "fast" control system entails reduced order dynamics and hence the "fast" state x_1 consists of the angular rates P, Q, and R. In the inner loop the control signal is $[\delta_e, \delta_a, \delta_r]^T$. The reference variables are the desired angular rates, namely P_c, Q_c, and R_c. The latter are supplied by the slower outer loop. The on-line optimization performed in the inner loop yields the optimal control signals δ_e^*, δ_a^*, and δ_r^*.

The slow variables are passed into the inner loop from the outer loop. Thus, in the inner loop the slow variables $x_2 = (\alpha, \beta, \theta, \phi)$ are fixed, and if linearization is employed, their perturbations are set to 0. In the slower, outer loop, only the "slow" states α, β, θ, and ϕ are employed. In the outer loop the control variables are the "fast" angular rates P, Q, and R. Their optimal values are determined over the outer loop optimization horizon, and they are subsequently used as the reference signal in the fast inner loop. The outer loop's reference variables are commanded by the pilot. For example, at low $\bar{q}$ the pilot's commands are Q_c, P_c, β_c.

75.1.9 Actuator Saturation Mitigation in Tracking Control

The classical *regulator* paradigm applies to outer loop flight control, where autopilot design problems are addressed. More challenging maneuvering, or inner loop, flight control entails the on-line solution of *tracking* control problems. The design of an advanced pitch channel FCS which accommodates actuator saturation is presented.

Linear control law synthesis methods, e.g., LQR, H_∞, and QFT, tend to ignore "hard" control constraints, viz., actuator saturation. Unfortunately, good tracking requires the use of high gain in the linear design, and consequently actuator saturation effects are pronounced in tracking control. Hence, in this section, the basic problems of tracking control and mitigating actuator saturation effects are synergetically addressed. The emphasis is on real-time synthesis of the control law, so that implementation in a feedback flight control system is possible. Thus, a hybrid optimization approach is used for mitigating actuator saturation effects in a high gain situation. Specifically, model following control is considered. The Linear Quadratic Regulator (LQR) and

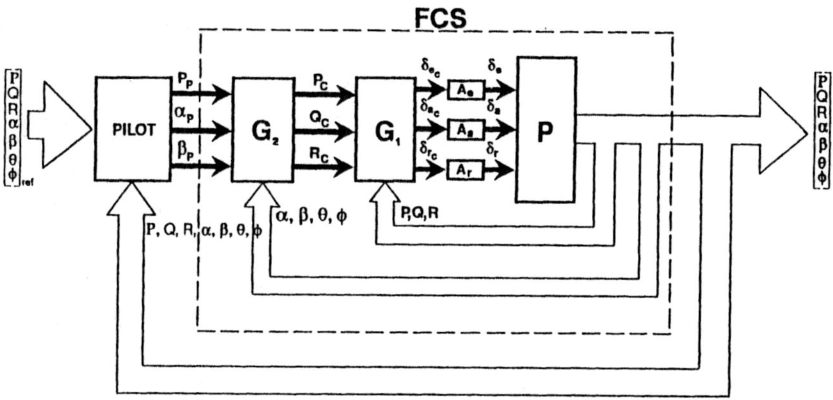

Figure 75.7 Flight control system.

Linear Programming (LP) optimization paradigms are combined [8] so that a tracking control law that does not violate the actuator constraints is synthesized in real time. Furthermore, the multiscale/multigrid approach presented in Section 75.1.8 is applied. Attention is confined to the pitch channel only and the conventionally linearized plant Equations 75.38 and 75.39 is used.

Saturation Problem

Actuator saturation reduces the benefits of feedback and degrades the tracking performance. Actuator saturation mitigation is of paramount importance in the case where feedback control is employed to stabilize open-loop unstable aircraft, e.g., the F-16: Prolonged saturation of the input signal to the plant is tantamount to opening the feedback control loop. Hence, in the event where the controller *continuously* outputs infeasible control signals which will saturate the plant[1], instability and departure will follow. At the same time, conservative approaches, which rely on small signal levels to avoid saturation, invariably yield inferior (sluggish) control performance. Hence, the following realization is crucial: The cause of saturation—precipitated instability is the continuous output by the linear controller of *infeasible* control signals. Conversely, if the controller sends only *feasible* control signals to the actuator, including controls on the boundary of the control constraint set, instability won't occur. In this case, in the event where the fed back measurement tells the controller that a reduction in control effort is called for, the latter can be instantaneously accomplished. Thus, a "windup" situation, typical in linear controllers, does not occur and no 'anti-windup" measures are required. Therefore, the feedback loop has not been opened, and the onset of instability is precluded. Obviously, a *nonlinear* control strategy is needed to generate *feasible* control signals. To command feasible control signals u, one must compromise on the tracking performance which is achievable with unconstrained controls.

[1]"Plant" here means the original plant and the actuator.

Tracking Control

The model following a state feedback control system is illustrated schematically in Figure 75.8. The pitch channel, modeled by Equations 75.38 and 75.39, is considered. The controlled variable is pitch rate q. The model M outputs the reference signal r, which is the pilot's desired pitch rate q_c. P represents the aircraft's dynamics, A the actuator dynamics, and G the controller. K_i and K_o are adaptive gains in the feedback and feedforward loops, respectively.

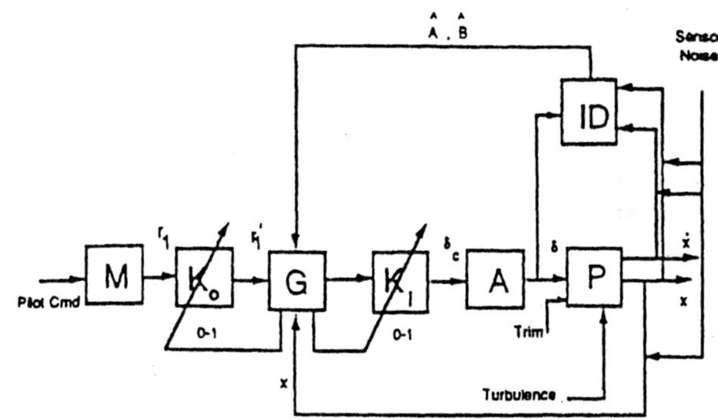

Figure 75.8 Tracking control.

A first-order actuator model with a time constant of 50 milliseconds is used. The Short Period approximation of the longitudinal dynamics of an F-16 aircraft derivative is employed. The latter is statically unstable, viz., $M_\alpha > 0$. Integral action is sought, and hence the plant is augmented to include an integrator in the forward path. The integrator's state is z, viz., $\dot{z} = r' - q$. Hence, the fed back state vector consists of the pitch rate q, the AOA α, the elevator deflection δ_e, and the integrator state z.

The command signal u= δ_{e_c} at the actuator's input consists of two parts, a fed back signal which is a linear function of the state and a signal which is a linear function of the pilot's input to the model and hence of the reference signal r. Therefore, actuator saturation will be precipitated by either a high amplitude excursion of the state, caused by a disturbance applied to the plant, or by a high amplitude reference signal when the flight control system is being driven hard by the pilot. Now, it is most desirable to track the reference model output driven directly by the pilot. However, because saturation requires the generation by the controller of feasible control signals, tracking performance must be relaxed. Hence, saturation mitigation requires that $r_1' \neq r_1$, therefore, reduce the feedforward gain K_o, or, alternatively, reduce the loop gain K_i.

Reference Signal Extrapolation

The optimal control solution of the tracking control problem (as opposed to the regulator problem) requires advance knowledge of the exogenous reference signal. Therefore, the reference signal r must be extrapolated over the optimization time horizon. However, long range extrapolation is dangerous. Hence, receding horizon optimal control provides a natural framework for tracking control.

Even though past reference signals are employed for extrapolation, some of which (r') are divorced from the pilot's reference signal input (r), a good way must be found to incorporate the pilot's extrapolated (future) reference signal demand into the extrapolated reference signal being supplied to the LQR algorithm. Hence, the extrapolation of the pilot's signal and its incorporation into the reference signal provided to the LQR algorithm provides good lead and helps enhance tracking performance. At the same time, the use of the past reference signal output by the controller (and not necessarily by the pilot-driven model) will not require abrupt changes in the current elevator position and this will ameliorate the elevator rate saturation problem.

Since the receding horizon tracking controller requires an estimate of q for the future, the latter is generated as a parabolic interpolation of r_0', the yet to be determined adjusted reference signal r_1', and $\hat{r}_{10}$; $\hat{r}_{10}$ is generated by parabolically interpolating the pilot issued commands r_{-1}, r_0, and r_1, and extrapolating ten time steps in the future, yielding $\hat{r}_{10}$. The ten point r' vector (which is the vector actually applied to the system, not the r vector) is parameterized as a function of the yet to be determined r_1'. Therefore, the objective of the controller's LP algorithm is to find the reference signal at time now, r_1', that will not cause a violation of the actuator limits over the ten-point prediction horizon.

Optimization

The discrete-time dynamical system $x_{k+1} = Ax_k + bu_k$, $x_0 \equiv x_0$, $k = 0, 1, ...N-1$, is considered over a receding ten time steps planning horizon, i.e., $N = 10$. The state vector $x \in R^4$ and the control variable is the commanded elevator deflection $\delta_{e_c} \in R^1$. In flight control, sampling rates of 100 Hz are conceivable. The quadratic cost criterion weighs the tracking error $r' - q$, the control effort (commanded elevator deflection), and control rate effort, over the ten time steps planning horizon which is 0.1 sec.

At the time instant $k = 0$, the future control sequence $u_0, u_1, ...u_{N-1}$ is planned based on the initial state information x_0 and the reference signal sequence $r_1, r_2, ...r_N$ given ahead of time. Now, the actual reference signal received at time $k = 0$ is r_1; the sequence $r_2, ...r_N$ is generated by polynomially (quadratically) extrapolating the reference signal forward in time, based on the knowledge of past reference signals. The extrapolated reference signals $r_2, ...r_N$ are linear in r_1. An open-loop optimization approach is employed. Although the complete optimal control sequence u_0, $u_1, ...u_{N-1}$ is calculated, only u_0 is exercised at time "now," thus obtaining feedback action.

An LQR tracking control law is first synthesized without due consideration of saturation. The objective here is to obtain good "small signal" tracking performance. If saturation does not occur, the LQR control law will be exercised. Saturation is addressed by using the explicit formula that relates the applied reference signal sequence $r_1', ...r_N'$ and the LQR algorithm-generated optimal control sequence $u_0, \ldots, u_N$. Thus, the dependence of the optimal control sequence on the initial state and on the reference signal r_1' is transparent:

$$\begin{bmatrix} u_0 & u_1 & . & . & . & u_{N-1} \end{bmatrix}^T = a(p)r_1' + b(p)x_0 + c(p) \tag{75.44}$$

where the vectors a, b and c $\in R^N$. The vector $p \in R^3$ represents the parameters (weights) of the LQR algorithm, viz., $p^T = [\frac{R}{Q_P}, \frac{R_R}{Q_P}, \frac{Q_I}{Q_P}]$. We are mainly concerned with actuator rate saturation and hence the important parameter is $\frac{R_R}{Q_P}$.

Next, the LP problem is formulated to enforce the nonsaturation of the command signal to the actuator and of the command rate signal. The following $2N$ (= 20) inequalities must hold:

$$\begin{array}{rcl}
-def_{\max} \leq | & u_0 & | \leq def_{\max}, \\
-defr_{\max} \leq | & \frac{u_0 - u_{-1}}{\Delta T} & | \leq defr_{\max}, \\
-def_{\max} \leq | & u_1 & | \leq def_{\max}, \\
-defr_{\max} \leq | & \frac{u_1 - u_0}{\Delta T} & | \leq defr_{\max}, \\
 & \text{and} & \\
 & \vdots & \\
-def_{\max} \leq | & u_{N-1} & | \leq def_{\max}, \\
-defr_{\max} \leq | & \frac{u_{N-1} - u_{N-2}}{\Delta T} & | \leq defr_{\max}.
\end{array}$$

The control u_{-1} is available from the previous window. The def_{max} (22°) and $defr_{max}$ (60°/sec) bounds are the maximal elevator deflection and elevator deflection rate, respectively. The sampling rate is 100 Hz and hence ΔT=0.01 sec.

The 2N inequalities above yield a set of 2N constraints in the currently applied reference signal r_1'. Thus,

$$g_i(\frac{R_R}{Q_P}) \leq r_1' \leq f_i(\frac{R_R}{Q_P}), i = 1, 2, ...2N. \tag{75.45}$$

Next, $f(\frac{R_R}{Q_P}) = min_i[f_i]$, $g(\frac{R_R}{Q_P}) = max_i[g_i]$ is defined, and the case where $g \leq f$ is considered. The set of 2N inequalities is consistent and a feasible solution r_1' exists. Then, given the pilot-injected reference signal r_1, a feasible solution is obtained, which satisfies the 2N inequalities above, and which is as close as possible to r_1. In other words,

$$r_1' = \begin{cases} r_1, & if\ g < r_1 < f \\ f, & if\ r_1 \geq f \\ g, & if\ r_1 \leq g \end{cases} \tag{75.46}$$

Finally, at time k=0, set $r_1 := r_1'$. Thus, the action of the LP algorithm is equivalent to lowering the feedforward gain K_o. In the case where the set of inequalities is inconsistent (f<g) and a solution does not exist, the loop gain K_i needs to be reduced. This is accomplished by increasing (say doubling) the LQR algorithm's R_R penalty parameter, and repeating the optimization process.

Implementation

The implementation of the advanced tracking control law developed in the preceding sections is discussed. The aircraft model represents the conventionally linearized second-order pitch plane dynamics (short period) of an F-16 aircraft derivative at the $M = 0.7$ and $h = 10,000$ ft flight condition. The bare aircraft model represented by Equations 75.38 and 75.39 is given in state-space form in Equation 75.47, where Z_α, Z_q, Z_δ, M_α, M_q, M_δ are the partial derivatives of normal force (Z) and pitching moment (M) with respect to AOA, pitch rate and stabilator deflection.

$$\begin{bmatrix} \dot{\alpha} \\ \dot{q} \end{bmatrix} = \begin{bmatrix} Z_\alpha & Z_q \\ M_\alpha & M_q \end{bmatrix} \begin{bmatrix} \alpha \\ q \end{bmatrix} + \begin{bmatrix} Z_\delta \\ M_\delta \end{bmatrix} \delta \qquad (75.47)$$

The stability and control derivatives are defined in Table 75.1. The dynamics entail the algebraic manipulation of the equations of motion to eliminate the $\dot{\alpha}$ derivatives.

TABLE 75.1 Nominal Model.

	δ	α	q
Z	−0.1770	−1.1500	0.9937
M	−19.5000	3.7240	−1.2600

Failure or damage to an aerodynamic control surface results in a loss of control authority, which, in turn, exacerbates the actuator saturation problem. A failure is modeled as a 50% loss of horizontal tail area which occurs 5 sec into the "flight." The failure affects *all* of the short period stability and control derivatives. The short period model's parameters, after failure, are shown in Table 75.2. The reference model has the same structure as the plant model. A first-order actuator with bandwidth of 20 rad/sec is included. The parameters of M are shown in Table 75.3. Only the pitch rate output of the model is tracked. The actuator output is rate limited to ±1 rad/sec and deflection limited to ±0.37 rad.

TABLE 75.2 Failure Model.

	δ	α	q
Z	−0.0885	−1.11	0.9968
M	−9.7500	5.58	−0.8400

The block diagram for the complete system is shown in Figure 75.8. The system consists of a reference model, a receding horizon LQ controller which also includes a LP algorithm, and the plant model. Integral action is included in the flight control system and is mechanized by augmenting the dynamics, viz., $z_{k+1} = z_k + (q - r')$. Thus, although the bare aircraft model has two states, with the actuator and integrator included, there are four states: α, q, δ, and z.

TABLE 75.3 Reference Model.

	δ	α	q
Z	−0.1770	−1.2693	0.9531
M	−19.5000	−9.4176	−5.7307

A parameter estimation module is also included, rendering the flight control system adaptive and reconfigurable.

The simulation is performed with both the receding horizon LQ/LP controller and the System Identification module in the loop. The pilot commands are 1.0 sec duration pitch rate pulses of 0.2 rad/sec magnitude, with polarities of +, −, +, at times 0.0, 3.0, and 6 seconds, respectively. At 5.0 seconds, a loss of one horizontal stabilator is simulated. The failure causes a change in the trim condition of the aircraft, which biases the pitch acceleration ($\dot{q}$) by $-0.21\ rad/sec^2$.

Figure 75.9 is a comparison plot of the performance of two controllers designed to prevent actuator saturation. REF is the desired pitch rate produced by a model of the desired pilot input to pitch rate response. The GAIN curve represents an inner loop gain limiter (0-1) denoted as K_i in Figure 75.10. The K_i value is computed from driving a rate and position limited actuator model and a linear actuator model with the command vector from the receding horizon controller. The ratio of the limited and unlimited value at each time step is computed; the smallest ratio is K_i (see Figure 75.10). The PROG curve is from an LP augmented solution that determined the largest magnitude of the input that can be applied without saturating the actuator. This is equivalent to outer-loop gain K_o attenuation. As can be seen, both PROG and GAIN track very well. At 5.0 seconds the failure occurs with a slightly larger perturbation for PROG than for GAIN. PROG has a slightly greater overshoot after the failure, primarily due to the reduced equivalent actuator rate.

75.1.10 Nonlinear Inner Loop Design

During high AOA maneuvers the pitch and lateral/directional channels are coupled and nonlinear effects are important. *Robust* control can be brought to bear on *nonlinear* control problems. In this Section is shown how QFT and the concept of structured plant parameter uncertainty accommodate the nonlinearities in the "fast" inner loop. The plant is specified by Equations 75.31 to 75.33 with $\alpha = \dot{\alpha} = \beta = 0$. Based on available data, a maximum bound of approximately 30°/sec can be placed on the magnitude of P for a 30 degree AOA velocity vector roll. Structured uncertainty is modeled with a set of LTI plants which describe the uncertain plant over the range of uncertainty. Therefore, intro-

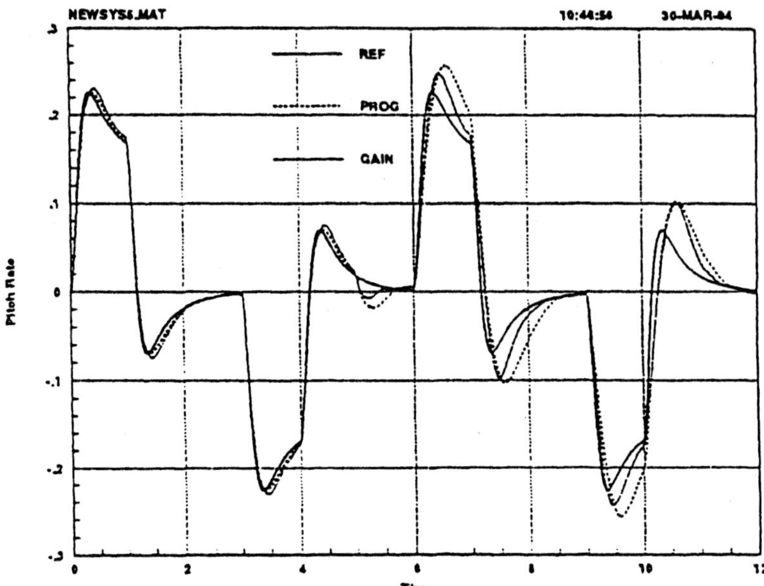

Figure 75.9 Tracking performance.

ducing a roll parameter P_p which varies over the range of 0 to 30°/sec in place of the state variable P in the nonlinear terms, allows the roll rate to be treated as structured uncertainty in the plant. Note that the roll rate parameter replaces one of the state variables in the P^2 term, viz., $P^2 \approx P_p P$. With the introduction of the roll rate parameter, a set of LTI plants with three states which model the nonlinear three DOFs plant are obtained. The generalized control vector is also three dimensional. The state and input vectors are

$$\begin{aligned} x &= \left[\begin{array}{ccc} P & q & r \end{array} \right]^T , \\ u &= \left[\begin{array}{cccccc} \delta_{ar} & \delta_{er} & \delta_{ep} & \delta_{TVp} & \delta_{TVy} & \delta_{ry} \end{array} \right]^T , \end{aligned}$$

where $\delta_{ar}, \delta_{er}, \delta_{ep}, \delta_{TVp}, \delta_{TVy}$, and δ_{ry} are the differential (roll) ailerons, differential (pitch) elevators, collective pitch elevators, pitch thrust vectoring, yaw thrust vectoring, and rudder deflections, respectively.

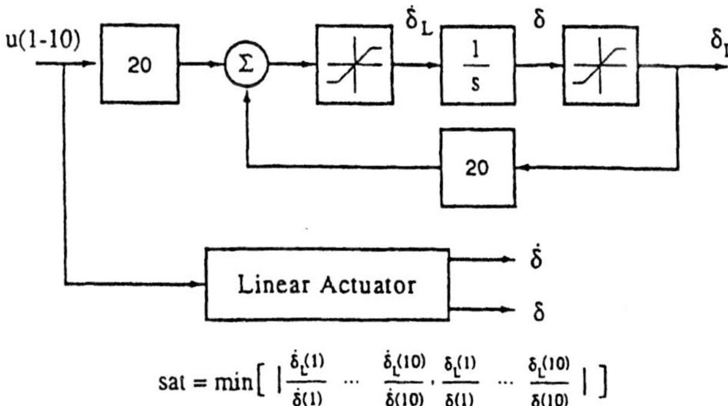

Figure 75.10 Actuator antiwindup scheme.

The linear, but parameter-dependent plant dynamics, and input matrices, are

$$A = \left[\begin{array}{ccc} C_{l_p} & C_{l_{pq}} P_p & C_{l_r} \\ -C_{m_{p^2}} P_p & C_{m_q} & C_{m_{pr}} P_p \\ C_{n_p} & C_{n_{pq}} P_p & C_{n_r} \end{array} \right] ,$$

$$B = \left[\begin{array}{cccccc} C_{l_{\delta_{ar}}} & C_{l_{\delta_{er}}} & 0 & 0 & C_{l_{\delta_{TVy}}} & C_{l_{\delta_{ry}}} \\ 0 & 0 & C_{m_{\delta_{ep}}} & C_{m_{\delta_{TVp}}} & 0 & 0 \\ C_{n_{\delta_{ar}}} & C_{n_{\delta_{er}}} & 0 & 0 & C_{n_{\delta_{TVy}}} & C_{n_{\delta_{ry}}} \end{array} \right]$$

The low $\bar{q}$ corner of the flight envelope is considered and four values of the roll rate parameter ($P_p = 0,\ 8,\ 16,\ 24$) deg/sec are used to represent the variation in roll rate during the roll onset or roll arrest phases [9], [2]. Thus, structured uncertainty is used to account for the nonlinearities introduced by the velocity vector roll maneuver. Finally, the QFT linear robust control design method is used to accommodate the structured uncertainty and design for the flying quality specifications of Section 75.1.5, thus obtaining a linear controller for a nonlinear plant. The weighting matrix

$$W = \left[\begin{array}{ccc} 0.2433 & 0 & 0 \\ 0.1396 & 0 & 0 \\ 0 & 0.0838 & 0 \\ 0 & 0.0698 & 0 \\ -0.03913 & 0 & 0.1745 \\ -0.11813 & 0 & 0.5236 \end{array} \right]$$

is used, and the respective roll, pitch and yaw channels controllers in the G_1 block, in pole-zero format, are

$$\begin{aligned} G_{1_{11}} &= \frac{205.128(-13)(-24 \pm j18)}{(0)(-50)(-60)} , \\ G_{1_{22}} &= \frac{1225(-5.5)(-17.1 \pm j22.8)(-32)}{(0)(-57 \pm j18.735)(-65)} , \\ G_{1_{33}} &= \frac{500(-6)(-14 \pm j14.2829)}{(0)(-50)(-50)} \end{aligned}$$

75.1.11 Flight Control of Piloted Aircraft

The pilot is closing an additional outer loop about the flight control system, as illustrated in Figure 75.6. This brings into the picture the human operator's weakness, namely, the pilot introduces a transport delay into the augmented flight control system. This problem is anticipated and is factored into the FCS design using the following methods.

Neal–Smith Criterion

This is a longitudinal FCS design criterion which includes control system dynamics used in preliminary design. The target tracking mission is considered and the aircraft's handling qualities are predicted.

The expected pilot workload is quantified in terms of the required generation of lead or lag by the pilot. These numbers have been correlated with qualitative Pilot Ratings (PR) of the aircraft, according to the Cooper–Harper chart [10], [3], and [4]. Hence, this design method uses a synthetic pilot model in the outer control loop. The augmented "plant" model includes the airframe and FCS dynamics. Thus, the plant model represents the transfer function from stick force F_s to pitch rate q. The augmented longitudinal FCS with the pilot in the loop is illustrated in Figure 75.11.

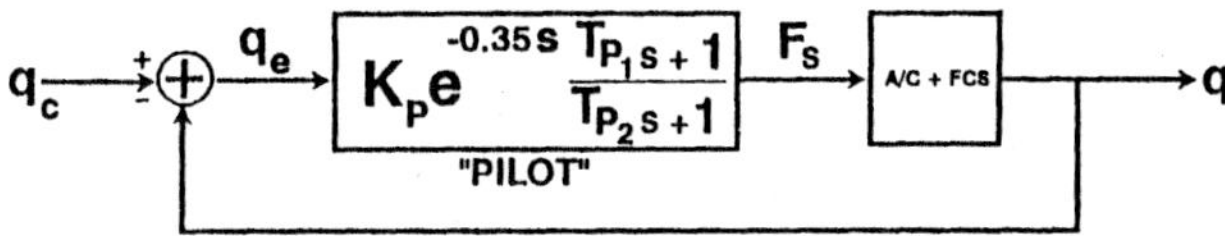

Figure 75.11 "Paper" pilot.

The "pilot" is required to solve the following parameter optimization problem: Choose a high gain K_P so that the closed-loop system's bandwidth is $\geq$ 3.5 rad/sec, and at the same time try to set the lead and lag parameters, τ_{p_1}, and τ_{p_2}, so that the peak closed-loop response $| \frac{Q(j\omega)}{Q_c(j\omega)} |_{max}$ is minimized and the droop at low frequencies is reduced. The closed-loop bandwidth is specified by the frequency where the closed-loop system's phase angle becomes $-90°$. Hence, the pilot chooses the gain and lead and lag parameters K_P, τ_{p_1}, and τ_{p_2} so that the closed-loop system's frequency response is as illustrated in Figure 75.12. Next, the angle

$$\angle_{PC} \equiv \angle \frac{\tau_{p_1}s+1}{\tau_{p_2}s+1} |_{s=j\omega, \omega=-BW}$$

is read off the chart. This angle is a measure of pilot compensation required.

If $\angle_{PC} > 0$, the model predicts that the pilot will need to apply lead compensation to obtain adequate handling qualities. If however $\angle_{PC} < 0$, the model predicts that lag compensation must be applied by the pilot. Thus, the aircraft's handling qualities in pitch are predicted from the chart shown in Figure 75.13, where PR corresponds to the predicted Cooper–Harper pilot rating.

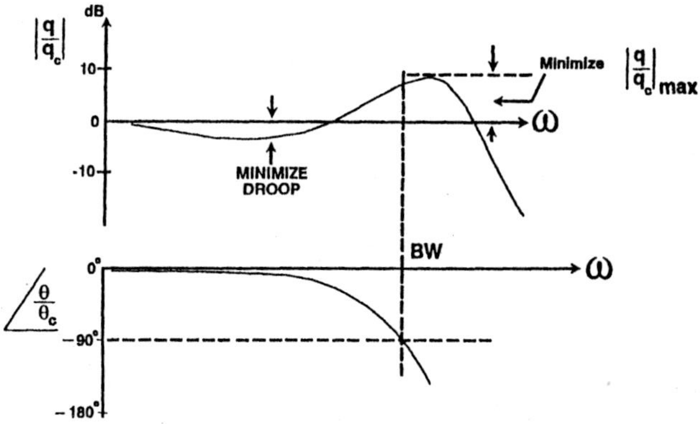

Figure 75.12 Closed-Loop frequency response.

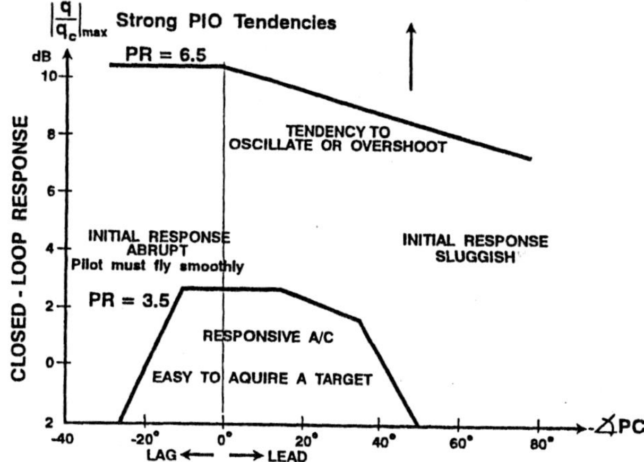

Figure 75.13 Pilot rating prediction chart.

Pilot-Induced Oscillations

Pilot-Induced Oscillations (PIOs) are a major concern in FCS design. PIOs are brought about by the following factors: (1) the RHP zero of the pilot model, (2) an airframe with high ω_{sp} (as is the case at low altitude and high $\bar{q}$ flight) and low damping ζ_{sp}, (3) excessive lag in the FCS (e.g., space shuttle in landing flare flight condition), (4) high-order compensators in digital FCSs (an effective transport delay effect is created), (5) FCS nonlinearities (interaction of nonlinearity and transport delay will cause oscillations), (6) stick force gradient too low (e.g., F-4), and/or "leading stick" (e.g., F-4, T-38 aircraft), and (7) high stress mission, where the pilot applies "high gain."

The Ralph–Smith criterion [11] has been developed for use during the FCS design cycle, to predict the possible onset of PIOs. Specifically, it is recognized that the pilot's transport delay "eats up" phase. Hence, the PIO tendency is predicted from the open-loop $\frac{n_z}{F_s}$ transfer function, and a sufficiently large Phase Margin (PM) is required for PIO prevention.

The Ralph–Smith criterion requires that the resonant frequency ω_R be determined (see Figure 75.14). The following condition must be satisfied: $PM \geq 14.3\ \omega_R$, in which case the aircraft/FCS is not PIO prone.

Pilot Role

It is of utmost importance to realize that (fighter) pilots play a crucial role when putting their aircraft through aggres-

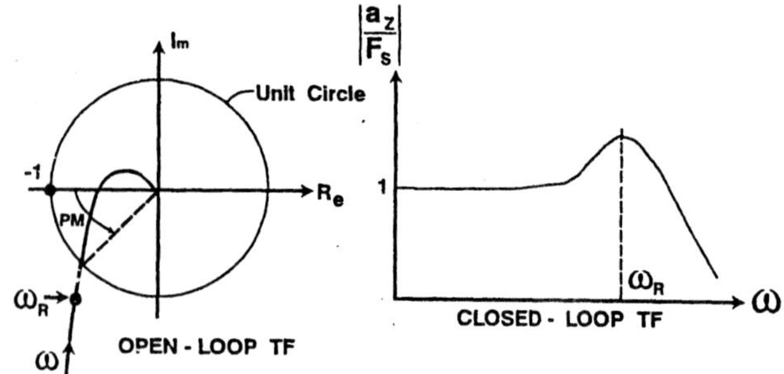

Figure 75.14 Ralph-Smith PIO criterion.

sive maneuvers. When the pilot/aircraft man/machine system is investigated, the following are most pertinent:

1. In Handling Qualities work, the problem of which variable the pilot actually controls is addressed. This is dictated by the prevailing flight condition. In the pitch channel and at low dynamic pressures $\bar{q}$, the controlled variables are α or Q; at high dynamic pressures, the controlled variables are $\dot{\alpha}$, n_z, or $C^* \equiv n_z + \bar{U}q$. In the lateral channel, the pilot commands roll rate P_C. In the directional channel and at low $\bar{q}$, the pilot controls the sideslip angle β, and, at high $\bar{q}$, the pilot controls $\dot{\beta}$ or n_y.
2. The classical analyses of "pilot in the loop" effects are usually confined to single channel flight control. The discussion centers on the deleterious effects of the $e^{-\tau s}$ transport delay introduced into the flight control system by the pilot. The transport delay-caused phase lag can destabilize the augmented FCS system, which includes the pilot in its outer loop. Hence, a linear stability analysis of the augmented flight control system is performed, as outlined in Sections 75.1.11 and 75.1.11. Thus, a very narrow and a very specialized approach is pursued. Unfortunately, this somewhat superficial investigation of control with a "pilot in the loop" only amplifies the obvious and unavoidable drawbacks of manual control of flight vehicles. It is, however, our firm belief that the above rash conclusion is seriously flawed. In simple, "one-dimensional," and highly structured scenarios it might indeed be hard to justify the insertion of a human operator into the control loop, for it is precisely in such environments that automatic machines outperform the pilot. However, in unstructured environments, (automatic) machines have a hard time beating the control prowess of humans. Human operators excel at high level tasks, where a degree of perception is required. This important facet of "pilot in the loop" operation is further amplified in items 3 and 4 in the sequel.
3. A subtle aspect of the pilot's work during maneuvering flight entails the following. The pilot compensates for deficiencies in the existing control law. These are caused by discrepancies between the simplified plant model used in control law design and the actual nonlinear dynamics of the aircraft. Obviously, some of this "modeling error" is being accommodated by the "benefits of feedback" and the balance, we believe, is being relegated to "pilot workload". However, the FCS designer's job is to strive to reduce the pilot's workload as much as possible.
4. The analysis of different trim conditions in Section 75.1.3 indicates that, in order to perform operationally meaningful maneuvers, not only the roll rate P, but also Q, R, α and β need to be controlled. Thus, control laws for the three control axes are simultaneously synthesized in the outer control loop which includes the pilot. Furthermore, since perfect decoupling is not realistically achievable, it is the pilot who synergetically works the FCS's three control channels to perform the required maneuvers. In other words, the pilot performs the nontrivial task of *multivariable control.* Although the inclusion of a pilot in the flight control loop comes not without a price—a transport time lag is being introduced—the benefits far outweigh this inherent drawback of a human operator, for the pilot brings to the table the intelligent faculty of on-line multivariable control synthesis. Indeed, high AOA and maneuvering flight entails a degree of on-line perception and pattern recognition. The latter is colloquially referred to by pilots as "seat of the pants" flying. Hence stick-and-rudder prowess is not a thing of the past, and the pilot plays a vital role in maneuvering flight.

References

[1] Blakelock, J.H., *Automatic Control of Aircraft and Missiles,* John Wiley & Sons, New York, 1991.

[2] Pachter, M., *Modern Flight Control,* AFIT Lecture Notes, 1995, obtainable from the author at AFIT/ENG, 2950 P Street, Wright Patterson AFB, OH 45433-7765.

[3] MIL-STD-1797A: Flying Qualities of Piloted Aircraft, US Air Force, Feb. 1991.

[4] Roskam, J., *Airplane Flight Dynamics, Part 1,* Roskam Aviation and Engineering Corp., Lawrence, Kansas, 1979.

[5] Etkin, B., *Dynamics of Flight: Stability and Control,* John Wiley & Sons, New York, 1982.

[6] Stevens, B.L. and Lewis, F.L., *Aircraft Control and Simulation,* John Wiley & Sons, New York, 1992.

[7] Reynolds, O.R., Pachter, M., and Houpis, C.H., *Full Envelope Flight Control System Design Using QFT,* Proceedings of the American Control Conference, pp 350–354, June 1994, Baltimore, MD; to appear in the ALAA Journal of Guidance, Control and Dynamics.

[8] Chandler, P.R., Mears, M., and Pachter, M., *A Hybrid LQR/LP Approach for Addressing Actuator Saturation*

in *Feedback Control*, Proceedings of the Conference on Decision and Control, pp 3860–3867, 1994, Orlando, FL.

[9] Boyum, K.E., Pachter, M., and Houpis, C.H., *High Angle Of Attack Velocity Vector Rolls*, Proceedings of the 13th IFAC Symposium on Automatic Control in Aerospace, pp 51–57, 1994, Palo Alto, CA, and Control Engineering Practice, 3(8), 1087–1093, 1995.

[10] Neal, T.P. and Smith, R.E., An In-Flight Investigation to Develop Control System Design Criteria for Fighter Airplanes, AFFDL-TR-70-74, Vols. 1 and 2, Air Force Flight Dynamics Laboratory, Wright Patterson AFB, 1970.

[11] Chalk, C.R., Neal, T.P., and Harris, T.M., Background Information and User Guide for MIL-F-8785 B - Military Specifications and Flying Qualities of Piloted Airplanes, AFFDL-TR-69-72, Air Force Flight Dynamics Laboratory, Wright Patterson AFB, August 1969.

Further Reading

a) Monographs

1. C. D. Perkins and R. E. Hage, "Airplane Performance, Stability and Control," Wiley, New York, 1949.
2. "Dynamics of the Airframe," Northrop Corporation, 1952.
3. W. R. Kolk, "Modern Flight Dynamics," Prentice Hall, 1961.
4. B. Etkin, "Dynamics of Atmospheric Flight," Wiley, 1972.
5. D. McRuer, I. Ahkenas and D. Graham, "Aircraft Dynamics and Automatic Control," Princeton University Press, Princeton, NJ, 1973.
6. J. Roskam, "Airplane Flight Dynamics, Part 2," Roskam Aviation, 1979.
7. A. W. Babister, "Aircraft Dynamic Stability and Response," Pergamon Press, 1980.
8. R. C. Nelson, "Flight Stability and Automatic Control," McGraw-Hill, 1989, Second Edition.
9. D. McLean, "Automatic Flight Control Systems," Prentice Hall, 1990.
10. E. H. Pallett and S. Coyle, "Automatic Flight Control," Blackwell, 1993, Fourth Edition.
11. A. E. Bryson, "Control of Spacecraft and Aircraft," Princeton University Press, Princeton, NJ, 1994.
12. "Special Issue: Aircraft Flight Control," International Journal of Control, Vol. 59, No 1, January 1994.

b) The reader is encouraged to consult the bibliography listed in References [8], [9] and [10] in the text.

75.2 Spacecraft Attitude Control

Vincent T. Coppola, *Department of Aerospace Engineering, The University of Michigan, Ann Arbor, MI*

N. Harris McClamroch, *Department of Aerospace Engineering, The University of Michigan, Ann Arbor, MI*

75.2.1 Introduction

The purpose of this chapter is to provide an introductory account of spacecraft attitude control and related control problems. Attention is given to spacecraft kinematics and dynamics, the control objectives, and the sensor and control actuation characteristics. These factors are combined to develop specific feedback control laws for achieving the control objectives. In particular, we emphasize the interplay between the spacecraft kinematics and dynamics and the control law design, since we believe that the particular attributes of the spacecraft attitude control problem should be exploited in any control law design. We do not consider specific spacecraft designs or implementations of control laws using specific sensor and actuation hardware.

Several different rotational control problems are considered. These include the problem of transferring the spacecraft to a desired, possibly time-variable, angular velocity and maintaining satisfaction of this condition without concern for the orientation. A special case is the problem of transferring the spacecraft to rest. A related but different problem is to bring the spacecraft to a desired, possibly time-variable, orientation and to maintain satisfaction of this condition. A special case is the problem of bringing the spacecraft to a constant orientation. We also consider the spin stabilization problem, where the spacecraft is desired to have a constant spin rate about a specified axis of rotation. All of these control problems are considered, and we demonstrate that there is a common framework for their study.

Our approach is to provide a careful development of the spacecraft dynamics and kinematics equations and to indicate the assumptions under which they are valid. We then provide several general control laws, stated in terms of certain gain constants. No attempt has been made to provide algorithms for selecting these gains, but standard optimal control and robust control approaches can usually be used. Only general principles are mentioned that indicate how these control laws are obtained; the key is that they result in closed-loop systems that can be studied using elementary methods to guarantee asymptotic stability or some related asymptotic property.

75.2.2 Modeling

The motion of a spacecraft consists of its orbital motion, governed by translational equations, and its attitude motion, governed by rotational equations. An inertial frame is chosen to be at the center of mass of the orbited body and nonrotating with respect to some reference (e.g., the polar axis of the earth or the fixed stars). Body axes x, y, z are chosen as a reference frame fixed in the spacecraft body with origin at its center of mass. For spacecraft in circular orbits, a local horizontal-vertical reference frame is used to measure radial pointing. The local vertical is defined to be radial from the center of mass of the orbited body with the local horizontal aligned with the spacecraft's velocity.

The spacecraft is modeled as a rigid body. The rotational inertia matrix is assumed to be constant with respect to the x, y, z axes. This rigidity assumption is rather strong considering that many real spacecraft show some degree of flexibility and/or internal motion, for example, caused by fuel slosh. These nonrigidity effects may be sometimes modeled as disturbances of the rigid spacecraft.

We assume that gravitational attraction is the dominant force experienced by the spacecraft. Since the environmental forces (e.g., solar radiation, magnetic forces) are very weak, the spacecraft's orbital motion is well modeled as an ideal two-body, purely Keplerian orbit (i.e., a circle or ellipse), at least for short time periods. Thus, the translational motion is assumed decoupled from the rotational motion.

The rotational motion of the spacecraft responds to control moments and moments arising from environmental effects such as gravity gradients. In contrast to the translational equations, the dominant moment for rotational motion is not environmental, but is usually the control moment. In such cases, all environmental influences are considered as disturbances.

75.2.3 Spacecraft Attitude Sensors and Control Actuators

There are many different attitude sensors and actuators used in controlling spacecraft. Sensors provide indirect measurements of orientation or rate; models of the sensor (and sometimes of the spacecraft itself) can be used to compute orientation and rate from available measurements. Actuators are used to control the moments applied to influence the spacecraft rotational motion in some desired way. Certain sensors and actuators applicable for one spacecraft mission may be inappropriate for another, depending on the spacecraft characteristics and the mission requirements. A discussion of these issues can be found in [5].

Several types of sensors require knowledge of the spacecraft orbital motion. These include sun sensors, horizon sensors, and star sensors, which measure orientation angles. The measurement may be taken using optical telescopes, infrared radiation, or radar. Knowledge of star (and possibly planet or moon) positions may also be required to determine inertial orientation. Magnetometers provide a measurement of orientation based upon the magnetic field. The limited ability to provide highly accurate models of the magnetosphere limit the accuracy of these devices.

Gyroscopes mounted within the spacecraft can provide measurements of both angular rate and orientation. Rate gyros measure angular rate; integrating gyros provide a measure of angular orientation. The gyroscope rotor spins about its symmetry axis at a constant rate in an inertially fixed direction. The rotor is supported by one or two gimbals to the spacecraft. The gimbals move as the spacecraft rotates about the rotor. Angular position and rate are measured by the movement of the gimbals. The model of the gyroscope often ignores the inertia of the gimbals and the friction in the bearings; these effects cause the direction of the gyro to drift over time.

An inertial measurement unit (IMU) consists of three mutually perpendicular gyroscopes mounted to a platform, either in gimbals or strapped down. Using feedback of the gyro signals, motors apply moments to the gyros to make the angular velocity of the platform zero. The platform then becomes an inertial reference.

The Global Positioning System (GPS) allows for very precise measurements of the orbital position of earth satellites. Two GPS receivers, located sufficiently far apart on the satellite, can be used to derive attitude information based on the phase shift of the signals from the GPS satellites. The orientation and angular rates are computed based upon a model of the rotational dynamics of the satellite.

Gas-jet thrusters are commonly employed as actuators for spacecraft attitude control. At least 12 thrust chambers are needed to provide three-axis rotational control. Each axis is controlled by two pairs of thrusters: one pair provides clockwise moment; the other pair, counterclockwise. A thruster pair consists of two thrusters that operate simultaneously at the same thrust level but in opposite directions. They create no net force on the spacecraft but do create a moment about an axis perpendicular to the plane containing the thrust directions. Thrusters may operate continuously or in full-on, full-off modes. Although the spacecraft loses mass during thruster firings, it is often negligible compared to the mass of the spacecraft and is ignored in the equations of motion.

Another important class of actuators used for attitude control are reaction wheel devices. Typically, balanced reaction wheels are mounted on the spacecraft so that their rotational axes are rigidly attached to the spacecraft. As a reaction wheel is spun up by an electric motor rigidly attached to the spacecraft, there is a reaction moment on the spacecraft. These three reaction moments provide the control moments on the spacecraft. In some cases, the effects of the electric motor dynamics are significant; these dynamics are ignored in this chapter.

75.2.4 Spacecraft Rotational Kinematics

The orientation or attitude of a rigid spacecraft can be expressed by a 3×3 rotation matrix R [3], [4]. Since a body-fixed x, y, z frame is rigidly attached to the spacecraft, the orientation of the spacecraft is the orientation of the body frame expressed with respect to a reference frame X, Y, Z. The columns of the rotation matrix are the components of the three standard basis vectors of the X, Y, Z frame expressed in terms of the three body-fixed standard basis vectors of the x, y, z frame. It can be shown that rotation matrices are necessarily orthogonal matrices; that is, they have the property that

$$RR^T = I \text{ and } R^T R = I, \tag{75.48}$$

where I is the 3×3 identity matrix, and $\det(R) = 1$. Upon differentiating with respect to time, we obtain

$$\frac{d}{dt}(RR^T) = \dot{R}R^T + R\dot{R}^T = 0 \tag{75.49}$$

Consequently,

$$(\dot{R}R^T) = -(\dot{R}R^T)^T \tag{75.50}$$

is a skew-symmetric matrix. It can be shown that this skew-symmetric matrix can be expressed in terms of the components of the angular velocity vector in the body-fixed coordinate frame $(\omega_x, \omega_y, \omega_z)$ as

$$\dot{R}R^T = S(\omega) \tag{75.51}$$

where

$$S(\omega) = \begin{bmatrix} 0 & \omega_z & -\omega_y \\ -\omega_z & 0 & \omega_x \\ \omega_y & -\omega_x & 0 \end{bmatrix}. \tag{75.52}$$

Consequently, the spacecraft attitude is described by the kinematics equation

$$\dot{R} = S(\omega)R. \tag{75.53}$$

This linear matrix differential equation describes the spacecraft attitude time dependence. If the angular velocity vector is a given vector function of time, and if an initial attitude is specified by a rotation matrix, then the matrix differential Equation 75.53 can be integrated using the specified initial data to obtain the subsequent spacecraft attitude. The solution of this matrix differential equation must necessarily be an orthogonal matrix at all instants of time. It should be noted that this matrix differential equation can also be written as nine scalar linear differential equations, but if these scalar equations are integrated, say, numerically, care must be taken to guarantee that the resulting solution satisfies the orthogonality property. In other words, the nine scalar entries in a rotation matrix are not independent.

The description of the spacecraft attitude in terms of a rotation matrix is conceptually natural and elegant. However, use of rotation matrices directly presents difficulties in computations and in physical interpretation. Consequently, other descriptions of attitude have been developed. These descriptions can be seen as specific parameterizations of rotation matrices using fewer than nine parameters. The most common parameterization involves the use of three Euler angle parameters.

Although there are various definitions of Euler angles, we introduce the most common definition (the 3-2-1 definition) that is widely used in spacecraft analyses. The three Euler angles are denoted by Ψ, θ, and ϕ, and are referred to as the yaw angle, the pitch angle, and the roll angle, respectively. A general spacecraft orientation defined by a rotation matrix can be achieved by a sequence of three elementary rotations, beginning with the body-fixed coordinate frame coincident with the reference coordinate frame, defined as follows:

1. A rotation of the spacecraft about the body-fixed z axis by a yaw angle Ψ
2. A rotation of the spacecraft about the body-fixed y axis by a pitch angle θ
3. A rotation of the spacecraft about the body-fixed x axis by a roll angle ϕ

It can be shown that the rotation matrix can be expressed in terms of the three Euler angle parameters Ψ, θ, ϕ according to the relationship [4]

$$R = \begin{bmatrix} \cos\theta\cos\Psi & \cos\theta\sin\Psi & -\sin\theta \\ (-\cos\phi\sin\Psi + \sin\phi\sin\theta\cos\Psi) & (\cos\phi\cos\Psi + \sin\phi\sin\theta\sin\Psi & \sin\phi\cos\theta \\ (\sin\phi\sin\Psi + \cos\phi\sin\theta\cos\Psi) & (-\sin\phi\cos\Psi + \cos\phi\sin\theta\sin\Psi) & \cos\phi\cos\theta \end{bmatrix}. \tag{75.54}$$

The components of a vector, expressed in terms of the x, y, z frame, are the product of the rotation matrix and the components of that same vector, expressed in terms of the X, Y, Z frame.

One of the deficiencies in the use of the Euler angles is that they do not provide a global parameterization of rotation matrices. In particular, the Euler angles are restricted to the range

$$\begin{aligned} -\pi &< \Psi < \pi, \\ -\tfrac{\pi}{2} &< \theta < \tfrac{\pi}{2}, \\ -\pi &< \phi < \pi, \end{aligned} \tag{75.55}$$

in order to avoid singularities in the above representation. This limitation is serious in some attitude control problems and motivates the use of other attitude representations.

Other attitude representations that are often used include axis-angle variables and quaternion or Euler parameters. The latter representations are globally defined but involve the use of four parameters rather than three. Attitude control problems can be formulated and solved using these alternative attitude parameterizations. In this chapter, we make use only of the Euler angle approach; however, we are careful to point out the difficulties that can arise by using Euler angles.

The spacecraft kinematic equations relate the components of angular velocity vector to the rates of change of the Euler angles. It can also be shown [4] that the angular velocity components in the body-fixed coordinate frame can be expressed in terms of the rates of change of the Euler angles as

$$\begin{aligned} \omega_x &= -\dot{\Psi}\sin\theta + \dot{\phi}, \\ \omega_y &= \dot{\Psi}\cos\theta\sin\phi + \dot{\theta}\cos\phi, \\ \omega_z &= \dot{\Psi}\cos\theta\cos\phi - \dot{\theta}\sin\phi. \end{aligned} \tag{75.56}$$

Conversely, the rates of change of the Euler angles can be expressed in terms of the components of the angular velocity vector as

$$\begin{aligned} \dot{\Psi} &= \omega_y\sec\theta\sin\phi + \omega_z\sec\theta\cos\phi, \\ \dot{\theta} &= \omega_y\cos\phi - \omega_z\sin\phi, \\ \dot{\phi} &= \omega_x + \omega_y\tan\theta\sin\phi + \omega_z\tan\theta\cos\phi. \end{aligned} \tag{75.57}$$

These equations are referred to as the spacecraft kinematics equations; they are used subsequently in the development of attitude control laws.

75.2.5 Spacecraft Rotational Dynamics

We consider the rotational dynamics of a rigid spacecraft. The body-fixed coordinates are chosen to be coincident with the principal axes of the spacecraft. We first consider the case where three pairs of thrusters are employed for attitude control. We next consider the case where three reaction wheels are employed for attitude control; under appropriate assumptions, the controlled spacecraft dynamics for the two cases are identical. The uncontrolled spacecraft dynamics are briefly studied.

We first present a model for the rotational dynamics of a rigid spacecraft controlled by thrusters as shown in Figure 75.15.

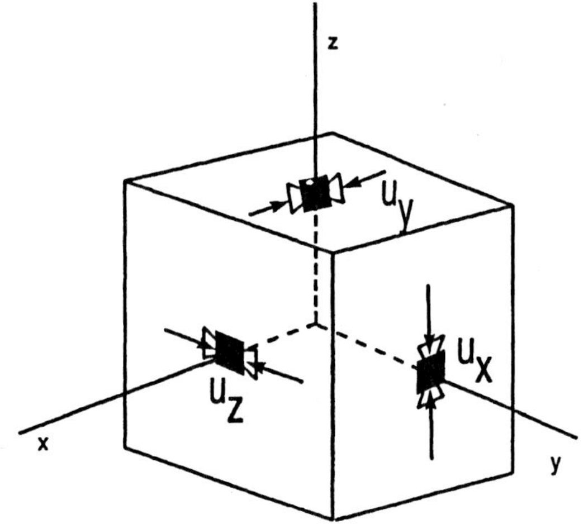

Figure 75.15 Spacecraft with gas-jet thrusters.

These dynamic equations are most naturally expressed in body-fixed coordinates. We further assume that the body-fixed coordinate frame is selected to be coincident with the spacecraft principal axes and that there are three pairs of thrusters that produce attitude control moments about these three principal axes. The resulting spacecraft dynamics are given by

$$\begin{aligned} I_{xx}\dot{\omega}_x &= \left(I_{yy} - I_{zz}\right)\omega_y\omega_z + u_x + T_x, \\ I_{yy}\dot{\omega}_y &= \left(I_{zz} - I_{xx}\right)\omega_z\omega_x + u_y + T_y, \\ I_{zz}\dot{\omega}_z &= \left(I_{xx} - I_{yy}\right)\omega_x\omega_y + u_z + T_z, \end{aligned} \tag{75.58}$$

where $(\omega_x, \omega_y, \omega_z)$ are the components of the spacecraft angular velocity vector, expressed in body-fixed coordinates; (u_x, u_y, u_z) are the components of the control moments due to the thrusters about the spacecraft principal axes; and (T_x, T_y, T_z) are the components of the external disturbance moment vector, expressed in body-fixed coordinates. The terms I_{xx}, I_{yy}, I_{zz} are the constant moments of inertia of the spacecraft with respect to its principal axes.

We now present a model for the rotational dynamics of a rigid spacecraft controlled by reaction wheels as shown in Figure 75.16.

As before, we express the rotational dynamics in body-fixed coordinates. We further assume that the body-fixed coordinate frame is selected to be coincident with the spacecraft principal axes and that there are three balanced reaction wheels that pro-

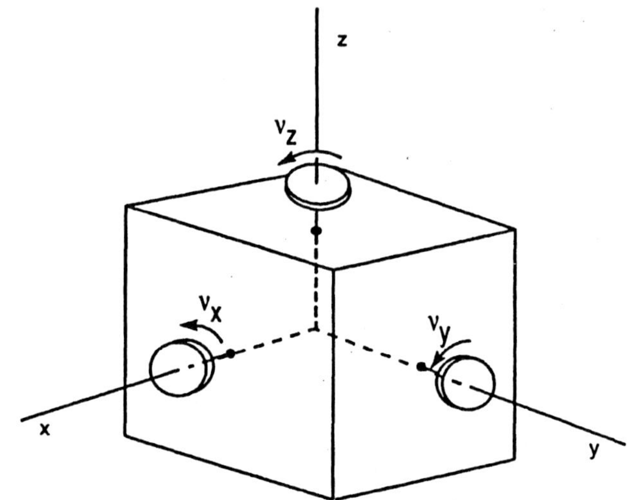

Figure 75.16 Spacecraft with three reaction wheels.

duce attitude control moments; the rotation axes of the reaction wheels are the spacecraft principal axes. The resulting spacecraft dynamics are given by [2]

$$\begin{aligned} I_{xx}\dot{\omega}_x &= \left(I_{yy} - I_{zz}\right)\omega_y\omega_z \\ &\quad + J_y\omega_z(\omega_y + \nu_y) - J_z\omega_y(\omega_z + \nu_z) \\ &\quad + u_x + T_x, \\ I_{yy}\dot{\omega}_y &= \left(I_{zz} - I_{xx}\right)\omega_z\omega_x \\ &\quad + J_z\omega_x(\omega_z + \nu_z) - J_x\omega_z(\omega_x + \nu_x) \\ &\quad + u_y + T_y, \\ I_{zz}\dot{\omega}_z &= \left(I_{xx} - I_{yy}\right)\omega_x\omega_y \\ &\quad + J_x\omega_y(\omega_x + \nu_x) - J_y\omega_x(\omega_y + \nu_y) \\ &\quad + u_z + T_z, \end{aligned} \tag{75.59}$$

and the dynamics of the reactions wheels are given by

$$\begin{aligned} J_x(\dot{\omega}_x + \dot{\nu}_x) &= -u_x, \\ J_y(\dot{\omega}_y + \dot{\nu}_y) &= -u_y, \\ J_z(\dot{\omega}_z + \dot{\nu}_z) &= -u_z, \end{aligned} \tag{75.60}$$

where $(\omega_x, \omega_y, \omega_z)$ are the components of the spacecraft angular velocity vector, expressed in body-fixed coordinates; (ν_x, ν_y, ν_z) are the relative angular velocities of the reaction wheels with respect to their respective axes of rotation (the spacecraft principal axes); (u_x, u_y, u_z) are the control moments developed by the electric motors rigidly mounted on the spacecraft with shafts aligned with the spacecraft principal axes connected to the respective reaction wheels; and (T_x, T_y, T_z) are the components of the external disturbance moment vector, expressed in body-fixed coordinates. Letting $(\mathsf{I}_{xx}, \mathsf{I}_{yy}, \mathsf{I}_{zz})$ denote the principal moments of inertia of the spacecraft, $I_{xx} = \mathsf{I}_{xx} - J_x$, $I_{yy} = \mathsf{I}_{yy} - J_y$, $I_{zz} = \mathsf{I}_{zz} - J_z$, where (J_x, J_y, J_z) are the (polar) moments of inertia of the reaction wheels.

Our interest is in attitude control of the spacecraft, so that Equation 75.60 for the reaction wheels does not play a central role. Note that if the terms in Equation 75.59 that explicitly couple the spacecraft dynamics and the reaction wheel dynamics are assumed to be small so that they can be ignored, then Equation 75.59 reduces formally to Equation 75.58.

Our subsequent control development makes use of Equation 75.58, or equivalently, the simplified form of Equation 75.59, so that the development applies to spacecraft attitude control using either thrusters or reaction wheels. In the latter case, it should be kept in mind that the coupling terms in Equation 75.59 are ignored as small. If these terms in Equation 75.59 are in fact not small, suitable modifications can be made to our subsequent development to incorporate the effect of these additional terms. One conceptually simple approach is to introduce *a priori* feedback loops that eliminate these coupling terms.

Disturbance moments have been included in the above equations, but it is usual to ignore external disturbances in the control design. Consequently, we subsequently make use of the controlled spacecraft dynamics given by

$$\begin{aligned} I_{xx}\dot{\omega}_x &= \left(I_{yy} - I_{zz}\right)\omega_y\omega_z + u_x, \\ I_{yy}\dot{\omega}_y &= \left(I_{zz} - I_{xx}\right)\omega_z\omega_x + u_y, \\ I_{zz}\dot{\omega}_z &= \left(I_{xx} - I_{yy}\right)\omega_x\omega_y + u_z, \end{aligned} \tag{75.61}$$

where it is assumed that the disturbance moment vector $(T_x, T_y, T_z) = (0, 0, 0)$.

If there are no control or disturbance moments on the spacecraft, then the rotational dynamics are described by the Euler equations

$$\begin{aligned} I_{xx}\dot{\omega}_x &= \left(I_{yy} - I_{zz}\right)\omega_y\omega_z, \\ I_{yy}\dot{\omega}_y &= \left(I_{zz} - I_{xx}\right)\omega_z\omega_x, \\ I_{zz}\dot{\omega}_z &= \left(I_{xx} - I_{yy}\right)\omega_x\omega_y. \end{aligned} \tag{75.62}$$

These equations govern the uncontrolled angular velocity of the spacecraft. Both the angular momentum H and the rotational kinetic energy T given by

$$\begin{aligned} 2T &= I_{xx}\omega_x^2 + I_{yy}\omega_y^2 + I_{zz}\omega_z^2, \\ H^2 &= (I_{xx}\omega_x)^2 + (I_{yy}\omega_y)^2 + (I_{zz}\omega_z)^2, \end{aligned} \tag{75.63}$$

are constants of the motion. These constants describe two ellipsoids in the $(\omega_x, \omega_y, \omega_z)$ space with the actual motion constrained to lie on their intersection. Since the intersection occurs as closed one-dimensional curves (or points), the angular velocity components $(\omega_x, \omega_y, \omega_z)$ are periodic time functions. Point intersections occur when two of the three components are zero: the body is said to be in a simple spin about the third body axis. Moreover, the spin axis is aligned with the (inertially fixed) angular momentum vector in the simple spin case.

Further analysis shows that simple spins about either the minor or major axis (i.e., the axes with the least and greatest moment of inertia, respectively) in Equation 75.62 are Lyapunov stable, while a simple spin about the intermediate axis is unstable. However, the minor axis spin occurs when the kinetic energy is maximum for a given magnitude of the angular momentum; thus, if dissipation effects are considered, spin about the minor axis becomes unstable. This observation leads to the major-axis rule: An energy-dissipating rigid body eventually arrives at a simple spin about its major axis. Overlooking this rule had devastating consequences for the U.S.' first orbital satellite *Explorer I.*

The orientation of the body with respect to inertial space is described in terms of the Euler angle responses Ψ, θ, ϕ [3]. This is best understood when the inertial directions are chosen so that the angular momentum vector lies along the inertial Z axis. Then the x axis is said to be precessing about the angular momentum vector at the rate $d\Psi/dt$ (which is always positive) and nutating at the rate $d\theta/dt$ (which is periodic and single signed). The angle $(\pi/2 - \theta)$ is called the coning or nutation angle. The motion of the x axis in inertial space is periodic in time; however, in general, the orientation of the body is not periodic because the spin angle ϕ is not commensurate with periodic motion of the x axis.

Although simple spins about major/minor axes are Lyapunov stable to perturbations in the angular velocities, neither is stable with respect to perturbations in orientation since the angular momentum vector cannot resist angular rate perturbations about itself [4]. However, it can resist perturbations orthogonal to itself. Hence, both the major and minor axes are said to be linearly directionally stable (when dissipation is ignored) since small orientation perturbations of the simple spin do not cause large deviations in the direction of the spin axis. That is, the direction of the spin axis is linearly stable.

75.2.6 Linearized Spacecraft Rotational Equations

The spacecraft kinematics and dynamics are described by nonlinear differential Equations 75.57 and 75.61. In this section, we develop linear differential equations that serve as good approximations for the spacecraft kinematics and dynamics in many instances.

We first consider a *linearization of the spacecraft equations near a rest solution.* Suppose that there are no external moments applied to the spacecraft; we note that the spacecraft can remain at rest in any fixed orientation. This corresponds to a constant attitude and zero angular velocity. If the reference coordinate frame defines this constant attitude, then it is easily seen that

$$R = I, \quad \omega = 0$$

satisfy the spacecraft kinematics and dynamics given by Equations 75.53 and 75.61. Equivalently, this reference attitude corresponds to the Euler angles being identically zero

$$(\Psi, \theta, \phi) = (0, 0, 0)$$

and it is easily seen that the kinematics Equation 75.56 is trivially satisfied.

Now if we assume that applied control moments do not perturb the spacecraft too far from its rest solution, then we can assume that the components of the spacecraft angular velocity vector are small; thus, we ignore the product terms to obtain the linearized approximation to the spacecraft dynamics

$$\begin{aligned} I_{xx}\dot{\omega}_x &= u_x, \\ I_{yy}\dot{\omega}_y &= u_y, \\ I_{zz}\dot{\omega}_z &= u_z. \end{aligned} \tag{75.64}$$

Similarly, by assuming that the Euler angles and their rates of change are small, we make the standard small angle assumptions to obtain the linearized approximation to the spacecraft kinematics equations

$$\begin{aligned} \dot{\phi} &= \omega_x, \\ \dot{\theta} &= \omega_y, \\ \dot{\Psi} &= \omega_z, \end{aligned} \tag{75.65}$$

Thus, the linearized dynamics and kinematics equations can be combined in a decoupled form that approximates the rolling, pitching, and yawing motions, namely,

$$\begin{aligned} I_{xx}\ddot{\phi} &= u_x, \\ I_{yy}\ddot{\theta} &= u_y, \\ I_{zz}\ddot{\Psi} &= u_z, \end{aligned} \tag{75.66}$$

We next present a *linearized model for the spacecraft equations near a constant spin solution.* Suppose that there is no control moment applied to the spacecraft. Another solution for the spacecraft motion, in addition to the rest solution, corresponds to a simple spin at constant rate of rotation about an inertially fixed axis. To be specific, we assume that the spacecraft has a constant spin rate Ω about the inertially fixed Z axis; the corresponding solution of the spacecraft kinematics Equation 75.57 and the spacecraft dynamics Equation 75.61 is given by

$$(\Psi, \theta, \phi) = (\Omega t, 0, 0), (\omega_x, \omega_y, \omega_z) = (0, 0, \Omega).$$

Now if we assume that the applied control moments do not perturb the spacecraft motion too far from this constant spin solution, then we can assume that the angular velocity components ω_x, ω_y, and $\omega_z - \Omega$ are small. Thus, we can ignore the products of small terms to obtain the linearized approximation to the spacecraft dynamics

$$\begin{aligned} I_{xx}\dot{\omega}_x &= \left(I_{yy} - I_{zz}\right)\Omega\omega_y + u_x, \\ I_{yy}\dot{\omega}_y &= \left(I_{zz} - I_{xx}\right)\Omega\omega_x + u_y, \\ I_{zz}\dot{\omega}_z &= u_z. \end{aligned} \tag{75.67}$$

Similarly, by assuming that the Euler angles θ and ϕ (but not Ψ) are small, we obtain the linearized approximation to the spacecraft kinematics

$$\begin{aligned} \dot{\Psi} &= \omega_z, \\ \dot{\theta} &= \omega_y - \Omega\phi, \\ \dot{\phi} &= \omega_x + \Omega\theta. \end{aligned} \tag{75.68}$$

Thus, the linearized dynamics and kinematics equations can be combined into

$$\begin{aligned} I_{xx}(\ddot{\phi} - \Omega\dot{\theta}) &= (I_{yy} - I_{zz})\Omega(\dot{\theta} + \Omega\phi) + u_x, \\ I_{yy}(\ddot{\theta} + \Omega\dot{\phi}) &= (I_{zz} - I_{xx})\Omega(\dot{\phi} - \Omega\theta) + u_y, \\ I_{zz}\ddot{\Psi} &= u_z, \end{aligned} \tag{75.69}$$

which makes clear that the yawing motion (corresponding to the motion about the spin axis) is decoupled from the pitching and rolling motion.

There is little loss of generality in assuming that the nominal spin axis is the inertially fixed Z axis; this is simply a matter of defining the inertial reference frame in a suitable way. This choice has been made since it leads to a simple formulation in terms of the Euler angles.

75.2.7 Control Specifications and Objectives

There are a number of possible rotational control objectives that can be formulated. In this section, several common problems are described. In all cases, our interest, for practical reasons, is to use feedback control. We use the control actuators to generate control moments that cause a desired spacecraft rotational motion; operation of these control actuators depends on feedback of the spacecraft rotational motion variables, which are obtained from the sensors. Most commonly, these sensors provide instantaneous measurements of the angular velocity vector (with respect to the body-fixed coordinate frame) and the orientation, or equivalently, the Euler angles.

The use of feedback control is natural for spacecraft rotational control applications; the same control law can be used to achieve and maintain satisfaction of the control objective in spite of certain types of model uncertainties and external disturbances. The desirable features of feedback are best understood in classical control theory, but the benefits of feedback can also be obtained for the fundamentally nonlinear rotational control problems that we consider here. Conversely, feedback, if applied unwisely, can do great harm. Consequently, we give considerable attention to a careful description of the feedback control laws and to justification for those control laws via specification of formal closed-loop stability properties. This motivates our use of mathematical models and careful analysis and design, thereby avoiding the potential difficulties associated with *ad hoc* or trial-and-error based control design.

We first consider spacecraft control problems where the control objectives are specified in terms of the spacecraft angular velocity. A general angular velocity control objective is that the spacecraft angular velocity vector exactly track a specified angular velocity vector in the sense that the angular velocity error be brought to zero and maintained at zero. If $(\omega_{xd}, \omega_{yd}, \omega_{zd})$ denotes the desired, possibly time-variable, angular velocity vector, this control objective is described by the asymptotic condition that

$$(\omega_x, \omega_y, \omega_z) \to (\omega_{xd}, \omega_{yd}, \omega_{zd}) \text{ as } t \to \infty.$$

One special case of this general control objective corresponds to bringing the spacecraft angular velocity vector to zero and maintaining it at zero. In this case, the desired angular velocity vector is zero, so the control objective is described by the asymptotic condition that

$$(\omega_x, \omega_y, \omega_z) \to (0, 0, 0) \text{ as } t \to \infty.$$

Another example of this general control objective corresponds to specifying that the spacecraft spin about a specified axis fixed to the spacecraft with a given angular velocity. For example, if $(\omega_{xd}, \omega_{yd}, \omega_{zd}) = (0, 0, \Omega)$, the control objective is that

$$(\omega_x, \omega_y, \omega_z) \to (0, 0, \Omega) \text{ as } t \to \infty.$$

so that the spacecraft asymptotically spins about its body-fixed z axis with a constant angular velocity Ω.

In all of the above cases, the spacecraft orientation is of no concern. Consequently, control laws that achieve these control objectives make use solely of the spacecraft dynamics.

We next consider spacecraft control problems where the control objectives are specified in terms of the spacecraft attitude. A more general class of control objectives involves the spacecraft attitude as well as the spacecraft angular velocity. In particular, suppose that it is desired that the spacecraft orientation exactly track a desired orientation in the sense that the attitude error is brought to zero and maintained at zero. The specified orientation may be time variable. If the desired orientation is described in terms of a rotation matrix function of time R_d, then the control objective is described by the asymptotic condition that

$$R \to R_d \text{ as } t \to \infty.$$

If we assume that R_d is parameterized by the desired Euler angle functions $(\Psi_d, \theta_d, \phi_d)$, then the control objective can be described by the asymptotic condition that

$$(\Psi, \theta, \phi) \to (\Psi_d, \theta_d, \phi_d) \text{ as } t \to \infty.$$

Since the specified matrix R_d is a rotation matrix, it follows that there is a corresponding angular velocity vector $\omega_d = (\omega_{xd}, \omega_{yd}, \omega_{zd})$ that satisfies

$$\dot{R}_d = S(\omega_d)R_d.$$

Consequently, it follows that the spacecraft angular velocity vector must also satisfy the asymptotic condition

$$(\omega_x, \omega_y, \omega_z) \to (\omega_{xd}, \omega_{yd}, \omega_{zd}) \text{ as } t \to \infty.$$

An important special case of the general attitude control objective is to bring the spacecraft attitude to a specified constant orientation and to maintain it at that orientation. Let R_d denote the desired constant spacecraft orientation; then the control objective is described by the asymptotic condition that

$$R \to R_d \text{ as } t \to \infty.$$

Letting R_d be parameterized by the desired constant Euler angles $(\Psi_d, \theta_d, \phi_d)$, the control objective can be described by the asymptotic condition that

$$(\Psi, \theta, \phi) \to (\Psi_d, \theta_d, \phi_d) \text{ as } t \to \infty.$$

Since the desired orientation is constant, it follows that the angular velocity vector must be brought to zero and maintained at zero; that is,

$$(\omega_x, \omega_y, \omega_z) \to (0, 0, 0) \text{ as } t \to \infty.$$

If the reference coordinate frame is selected to define the desired spacecraft orientation, then $R_d = I$, the 3 by 3 identity matrix, and $(\Psi_d, \theta_d, \phi_d) = (0, 0, 0)$ and the control objective is

$$R \to I \text{ as } t \to \infty,$$

or equivalently,

$$(\Psi, \theta, \phi) \to (0, 0, 0) \text{ as } t \to \infty.$$

Control laws that achieve the above control objectives should make use of both the spacecraft kinematics and the spacecraft dynamics. Special cases of such control objectives are given by the requirements that the spacecraft always point at the earth's center, a fixed star, or another spacecraft.

Another control objective corresponds to specifying that the spacecraft spin about a specified axis that is inertially fixed with a given angular velocity Ω. If the reference coordinate frame is chosen so that it is an inertial frame and its positive Z axis is the spin axis, then it follows that the control objective can be described by the asymptotic condition that

$$(\theta, \phi) \to (0, 0) \text{ as } t \to \infty,$$
$$(\omega_x, \omega_y, \omega_z) \to (0, 0, \Omega) \text{ as } t \to \infty,$$

which guarantees that the spacecraft tends to a constant angular velocity about its body-fixed z axis and that its z axis is aligned with the inertial Z axis. Control laws that achieve the above control objectives should make use of both the spacecraft kinematics and the spacecraft dynamics.

An essential part of any control design process is performance specifications. When the spacecraft control inputs are adjusted automatically according to a specified feedback control law, the resulting system is a closed-loop system. In terms of control law design, performance specifications are naturally imposed on the closed-loop system. Since the spacecraft kinematics and dynamics are necessarily nonlinear in their most general form, closedloop specifications must be given in somewhat nonstandard form. There are no uniformly accepted technical performance specifications for this class of control problems, but it is generally accepted that the closed loop should exhibit rapid transient response, good steady-state accuracy, good robustness to parameter uncertainties, a large domain of attraction, and the ability to reject certain classes of external disturbances.

These conceptual control objectives can be quantified, at least if the rotational motions of the closed loop are sufficiently small so that the dynamics and kinematics are adequately described by linear models. In such case, good transient response depends on the eigenvalues or characteristic roots. Desired closed-loop properties can be specified if there are uncertainties in the models and external disturbances of certain classes. Control design to achieve performance specifications, such as steady-state accuracy, robustness, and disturbance rejection, have been extensively treated in the theoretical literature, for both linear and nonlinear control systems, and are not explicitly studied here. We note that if there are persistent disturbances, then there may be nonzero steady-state errors for the control laws that we subsequently propose; those control laws can easily be modified to include integral error terms to improve the steady-state accuracy. Examples of relatively simple specifications of the closed loop are illustrated in the subsequent sections that deal with design of feedback control laws for several spacecraft rotational control problems.

75.2.8 Spacecraft Control Problem: Linear Control Law Based on Linearized Spacecraft Equations

In this section, we assume that the spacecraft dynamics and kinematics can be described in terms of the linearized Equations 75.64 and 75.65. We assume the control moments can be adjusted to any specified level; hence, we consider the use of linear control laws.

We first consider *control of the spacecraft angular velocity.* It is assumed that the spacecraft dynamics are described by the linearized Equation 75.64; a control law (u_x, u_y, u_z) is desired in feedback form, so that the resulting closed-loop system satisfies the asymptotic condition

$$(\omega_x, \omega_y, \omega_z) \to (\omega_{xd}, \omega_{yd}, \omega_{zd}) \text{ as } t \to \infty,$$

and has desired closed-loop properties. Since the spacecraft dynamics are linear and uncoupled first-order equations, a standard linear control approach can be used to design a linear feedback control of the form

$$\begin{aligned} u_x &= -c_x(\omega_x - \omega_{xd}) + I_{xx}\dot{\omega}_{xd}, \\ u_y &= -c_y(\omega_y - \omega_{yd}) + I_{yy}\dot{\omega}_{yd}, \\ u_z &= -c_z(\omega_z - \omega_{zd}) + I_{zz}\dot{\omega}_{zd}, \end{aligned} \tag{75.70}$$

where the control gains c_x, c_y, c_z are chosen as positive constants. Based on the linearized equations, introduce the variables

$$\delta_x = \omega_x - \omega_{xd}, \;\; \delta_y = \omega_y - \omega_{yd}, \;\; \delta_z = \omega_z - \omega_{zd}, \tag{75.71}$$

so that the closed-loop equations are

$$\begin{aligned} \dot{\delta}_x + \left(\frac{c_x}{I_{xx}}\right)\delta_x &= 0, \\ \dot{\delta}_y + \left(\frac{c_y}{I_{yy}}\right)\delta_y &= 0, \\ \dot{\delta}_z + \left(\frac{c_z}{I_{zz}}\right)\delta_z &= 0. \end{aligned} \tag{75.72}$$

Hence, the angular velocity errors in roll rate, pitch rate, and yaw rate are brought to zero asymptotically, at least for sufficiently small perturbations. The values of the control gains can be chosen to provide specified closed-loop time constants. Consequently, the control law is guaranteed to achieve exact tracking asymptotically, with speed of response that is determined by the values of the control gains.

A simple special case of the above control law corresponds to the choice that $(\omega_{xd}, \omega_{yd}, \omega_{zd}) = (0, 0, 0)$; i.e., the spacecraft is to be brought to rest. In simplified form, the control law becomes

$$\begin{aligned} u_x &= -c_x\omega_x, \\ u_y &= -c_y\omega_y, \\ u_z &= -c_z\omega_z, \end{aligned} \tag{75.73}$$

and it is guaranteed to bring the spacecraft asymptotically to rest.

It is important to note that the preceding analysis holds only for sufficiently small perturbations from rest. For large perturbations in the angular velocity of the spacecraft, the above control laws may not have the desired properties. An analysis of the closed-loop system, using the nonlinear dynamics Equation 75.61, is required to determine the domain of perturbations for which closed-loop stabilization is achieved. Such an analysis is beyond the scope of the present chapter, but we note that the stability domain depends on both the desired control objective and the control gains that are selected.

We next consider *control of the spacecraft angular attitude.* It is assumed that the spacecraft dynamics and kinematics are described by the linearized Equation 75.66 and a control law (u_x, u_y, u_z) is desired in feedback form, so that the resulting closed-loop system satisfies the attitude control conditions

$$\begin{aligned} (\Psi, \theta, \phi) &\to (\Psi_d, \theta_d, \phi_d) \text{ as } t \to \infty, \\ (\omega_x, \omega_y, \omega_z) &\to (\omega_{xd}, \omega_{yd}, \omega_{zd}) \text{ as } t \to \infty, \end{aligned}$$

where the kinematic conditions of Equation 75.56 are assumed to be satisfied, and the closed-loop system has desired closed-loop properties. Since the equations are linear and uncoupled second-order equations, a standard linear control approach can be used to obtain linear feedback control laws of the form

$$\begin{aligned} u_x &= -c_x(\omega_x - \omega_{xd}) - k_x(\phi - \phi_d) + I_{xx}\ddot{\phi}_d, \\ u_y &= -c_y(\omega_y - \omega_{yd}) - k_y(\theta - \theta_d) + I_{yy}\ddot{\theta}_d, \\ u_z &= -c_z(\omega_z - \omega_{zd}) - k_z(\Psi - \Psi_d) + I_{zz}\ddot{\Psi}_d. \end{aligned} \tag{75.74}$$

Recall that the angular velocity components are the rates of change of the Euler angles according to the linearized relations of Equation 75.65, so that the control law can equivalently be expressed in terms of the Euler angles and their rates of change. Based on the linearized equations, introduce the variables

$$\varepsilon_x = \phi - \phi_d, \;\; \varepsilon_y = \theta - \theta_d, \;\; \varepsilon_z = \Psi - \Psi_d, \tag{75.75}$$

so that the closed-loop equations are

$$\begin{aligned} I_{xx}\ddot{\varepsilon}_x + c_x\dot{\varepsilon}_x + k_x\varepsilon_x &= 0, \\ I_{yy}\ddot{\varepsilon}_y + c_y\dot{\varepsilon}_y + k_y\varepsilon_y &= 0, \\ I_{zz}\ddot{\varepsilon}_z + c_z\dot{\varepsilon}_z + k_z\varepsilon_z &= 0, \end{aligned} \tag{75.76}$$

Hence, the spacecraft attitude errors in roll angle, pitch angle, and yaw angle are brought to zero asymptotically, at least for sufficiently small perturbations. The values of the control gains can be chosen to provide specified closed-loop natural frequencies and damping ratios. Consequently, the above control laws are guaranteed to bring the spacecraft to the desired attitude asymptotically, with speed of response that is determined by the values of the control gains.

A simple special case of the above control law corresponds to the choice that $(\Psi_d, \theta_d, \phi_d)$ is a constant and $(\omega_{xd}, \omega_{yd}, \omega_{zd}) = (0, 0, 0)$. The simplified control law is of the form

$$\begin{aligned} u_x &= -c_x\omega_x - k_x(\phi - \phi_d), \\ u_y &= -c_y\omega_y - k_y(\theta - \theta_d), \\ u_z &= -c_z\omega_z - k_z(\Psi - \Psi_d), \end{aligned} \tag{75.77}$$

and it is guaranteed to bring the spacecraft asymptotically to the desired constant attitude and to maintain it in the desired attitude. The resulting control law is referred to as an attitude stabilization control law. Further details are available in [1].

Again, it is important to note that the preceding analysis holds only for sufficiently small perturbations in the spacecraft angular velocity and attitude. For large perturbations in the angular velocity and attitude of the spacecraft, the control law may not have the desired properties. An analysis of the closed-loop system, using the nonlinear dynamics Equation 75.61 and the nonlinear kinematic Equation 75.57, is required to determine the domain of perturbations for which closed-loop stabilization is achieved. Such an analysis is beyond the scope of the present chapter, but we note that the stability domain depends on the desired attitude and angular velocity and the control gains that are selected.

We now consider *spin stabilization of the spacecraft about a specified inertial axis.* A control law (u_x, u_y, u_z) is desired in feedback form, so that the resulting closed-loop system satisfies the asymptotic control objective

$$(\theta, \phi) \to (0, 0) \text{ as } t \to \infty,$$
$$(\omega_x, \omega_y, \omega_z) \to (0, 0, \Omega) \text{ as } t \to \infty,$$

corresponding to an asymptotically constant spin rate Ω about the body-fixed z axis, which is aligned with the inertial Z axis, and the closed loop has desired properties. It is assumed that the spacecraft dynamics and kinematics are described by the linearized Equations 75.67 and 75.68. Since the spacecraft equations are linear and uncoupled, a standard linear control approach can be used. Consider the linear feedback control law of the form

$$\begin{aligned} u_x &= -(I_{yy} - I_{zz})\Omega(\dot{\theta} + \Omega\phi) - I_{xx}\Omega\dot{\theta} - c_x\dot{\phi} - k_x\phi, \\ u_y &= -(I_{zz} - I_{xx})\Omega(\dot{\phi} - \Omega\theta) + I_{yy}\Omega\dot{\phi} - c_y\dot{\theta} - k_y\theta, \\ u_z &= -c_z(\omega_z - \Omega). \end{aligned} \tag{75.78}$$

Recall that the angular velocity components are related to the rates of change of the Euler angles according to the linearized relations of Equation 75.68, so that the control law can be expressed either in terms of the rates of change of the Euler angles (as in Equation 75.78) or in terms of the components of the angular velocity vector.

Based on the linearized equations, introduce the variables

$$\varepsilon_x = \phi, \quad \varepsilon_y = \theta, \quad \delta_z = \omega_z - \Omega, \tag{75.79}$$

so that the closed-loop equations are

$$\begin{aligned} I_{xx}\ddot{\varepsilon}_x + c_x\dot{\varepsilon}_x + k_x\varepsilon_x &= 0, \\ I_{yy}\ddot{\varepsilon}_y + c_y\dot{\varepsilon}_y + k_y\varepsilon_y &= 0, \\ I_{zz}\dot{\delta}_z + c_z\delta_z &= 0. \end{aligned} \tag{75.80}$$

Hence, the spacecraft attitude errors in pitch angle and roll angle are brought to zero asymptotically and the yaw rate is brought to the value Ω asymptotically, at least for sufficiently small perturbations. The values of the control gains can be chosen to provide specified closed-loop transient responses. Consequently, the above control law is guaranteed to bring the spacecraft to the desired spin rate about the specified axis of rotation, asymptotically, with speed of response that is determined by the values of the control gains. The control law of Equation 75.78 has the desirable property that it requires feedback of only the pitch and roll angles, which characterize, in this case, the errors in the instantaneous spin axis; feedback of the yaw angle is not required.

Again, it is important to note that the preceding analysis holds only for sufficiently small perturbations in the spacecraft angular velocity and attitude. For large perturbations in the angular velocity and attitude of the spacecraft, the above control law may not have the desired properties. An analysis of the closed-loop system, using the nonlinear dynamics Equation 75.61 and the nonlinear kinematic Equation 75.57, is required to determine the domain of perturbations for which closed-loop stabilization is achieved. Such an analysis is beyond the scope of the present chapter, but we note that the stability domain depends on the desired spin rate and the control gains that are selected.

75.2.9 Spacecraft Control Problem: Bang-Bang Control Law Based on Linearized Spacecraft Equations

As developed previously, it is assumed that the spacecraft kinematics and dynamics are described by linear equations obtained by linearizing about the rest solution. The linearized spacecraft dynamics are described by Equation 75.64, and the linearized spacecraft kinematics are described by Equation 75.65.

Since certain types of thrusters are most easily operated in an on-off mode, the control moments produced by each pair of thrusters may be limited to fixed values (of either sign) or to zero. We now impose these control constraints on the development of the control law by requiring that each of the control moment components can take only the values $\{-U, 0, +U\}$ at each instant of time.

We first present a simple control law that *stabilizes the spacecraft to rest*, ignoring the spacecraft attitude. The simplest control law, satisfying the imposed constraints, that stabilizes the spacecraft to rest is given by

$$\begin{aligned} u_x &= -U \operatorname{sgn}(\omega_x), \\ u_y &= -U \operatorname{sgn}(\omega_y), \\ u_z &= -U \operatorname{sgn}(\omega_z), \end{aligned} \tag{75.81}$$

where the signum function is the discontinuous function defined by

$$\operatorname{sgn}(\sigma) = \begin{bmatrix} -1 & \text{if} & \sigma < 0, \\ 0 & \text{if} & \sigma = 0, \\ 1 & \text{if} & \sigma > 0. \end{bmatrix} \tag{75.82}$$

It is easily shown that the resulting closed-loop system has the property that

$$(\omega_x, \omega_y, \omega_z) \to (0, 0, 0) \text{ as } t \to \infty,$$

at least for sufficiently small perturbations in the spacecraft angular velocity vector.

There are necessarily errors in measuring the angular velocity vector, and it can be shown that the above control law can be improved by using the modified feedback control law

$$\begin{aligned} u_x &= -U \operatorname{dez}(\omega_x, \varepsilon), \\ u_y &= -U \operatorname{dez}(\omega_y, \varepsilon), \\ u_z &= -U \operatorname{dez}(\omega_z, \varepsilon), \end{aligned} \qquad (75.83)$$

where the dead-zone function is the discontinuous function defined by

$$\operatorname{dez}(\sigma, \varepsilon) = \begin{bmatrix} -1 & \text{if} & \sigma < -\varepsilon, \\ 0 & \text{if} & -\varepsilon < \sigma < \varepsilon, \\ 1 & \text{if} & \sigma > \varepsilon, \end{bmatrix} \qquad (75.84)$$

and ε is a positive constant, the dead-zone width. The resulting closed-loop system has the property that

$$(\omega_x, \omega_y, \omega_z) \to S \text{ as } t \to \infty,$$

where S is an open set containing (0,0,0); the (maximum) diameter of S has the property that it goes to zero as ε goes to zero. Thus, the dead-zone parameter ε can be selected appropriately, so that the angular velocity vector is maintained small while the closed loop is not excessively sensitive to measurement errors in the angular velocity vector.

We now present a simple control law that *stabilizes the spacecraft to a fixed attitude,* which is given by the Euler angles being all zero. The simplest control law, satisfying the imposed constraints, is of the form

$$\begin{aligned} u_x &= -U \operatorname{sgn}(\phi + \tau_x \dot{\phi}), \\ u_y &= -U \operatorname{sgn}(\theta + \tau_y \dot{\theta}), \\ u_z &= -U \operatorname{sgn}(\Psi + \tau_z \dot{\Psi}), \end{aligned} \qquad (75.85)$$

where τ_x, τ_y, τ_z are positive constants and the linear arguments in the control laws of Equation 75.85 are the switching functions. Consequently, the closed-loop system is decoupled into independent closed loops for the roll angle, the pitch angle, and the yaw angle. By analyzing each of these closed-loop systems, it can be shown that

$$(\Psi, \theta, \phi) \to (0, 0, 0) \text{ as } t \to \infty,$$

at least for sufficiently small perturbations in the spacecraft angular velocity vector and attitude. It should be noted that a part of the solution involves a chattering solution where the switching functions are identically zero over a finite time interval.

There are necessarily errors in measuring the angular velocity vector and the attitude, and it can be shown that the above control law can be improved by using the modified feedback control law

$$\begin{aligned} u_x &= -U \operatorname{dez}(\phi + \tau_x \dot{\phi}, \varepsilon), \\ u_y &= -U \operatorname{dez}(\theta + \tau_y \dot{\theta}, \varepsilon), \\ u_z &= -U \operatorname{dez}(\Psi + \tau_z \dot{\Psi}, \varepsilon), \end{aligned} \qquad (75.86)$$

where ε is a positive constant, the dead-zone width. The resulting closed-loop system has the property that

$$(\Psi, \theta, \phi) \to S \text{ as } t \to \infty,$$

where S is an open set containing (0, 0, 0); in fact, the spacecraft attitude is asymptotically periodic with a maximum amplitude that tends to zero as ε tends to zero. Thus, the dead-zone parameter ε can be selected appropriately, so that the angular velocity vector and the attitude errors are maintained small while the closed loop is not excessively sensitive to measurement errors in the angular velocity vector or the attitude errors. Further details are available in [1].

75.2.10 Spacecraft Control Problem: Nonlinear Control Law Based on Nonlinear Spacecraft Equations

In this section, we present nonlinear control laws that guarantee that the closed-loop equations are exactly linear; this approach can be viewed as using feedback both to cancel out the nonlinear terms and then to add in linear terms that result in good closed-loop linear characteristics. This approach is often referred to as feedback linearization or dynamic inversion.

We first consider *control of the spacecraft angular velocity.* It is assumed that the spacecraft dynamics are described by the nonlinear Equation 75.61 and a control law (u_x, u_y, u_z) is desired in feedback form, so that the resulting closed-loop system is linear and satisfies the asymptotic condition

$$(\omega_x, \omega_y, \omega_z) \to (\omega_{xd}, \omega_{yd}, \omega_{zd}) \text{ as } t \to \infty.$$

Control laws that accomplish these objectives are given in the form

$$\begin{aligned} u_x &= -(I_{yy} - I_{zz})\omega_y \omega_z - c_x(\omega_x - \omega_{xd}) + I_{xx}\dot{\omega}_{xd}, \\ u_y &= -(I_{zz} - I_{xx})\omega_z \omega_x - c_y(\omega_y - \omega_{yd}) + I_{yy}\dot{\omega}_{yd}, \\ u_z &= -(I_{xx} - I_{yy})\omega_x \omega_y - c_z(\omega_z - \omega_{zd}) + I_{zz}\dot{\omega}_{zd}, \end{aligned} \qquad (75.87)$$

where the control gains c_x, c_y, c_z are chosen as positive constants. If the variables

$$\delta_x = \omega_x - \omega_{xd}, \quad \delta_y = \omega_y - \omega_{yd}, \quad \delta_z = \omega_z - \omega_{zd}, \qquad (75.88)$$

are introduced, the closed-loop equations are

$$\begin{aligned} \dot{\delta}_x + \left(\frac{c_x}{I_{xx}}\right)\delta_x &= 0, \\ \dot{\delta}_y + \left(\frac{c_y}{I_{yy}}\right)\delta_y &= 0, \\ \dot{\delta}_z + \left(\frac{c_z}{I_{zz}}\right)\delta_z &= 0. \end{aligned} \qquad (75.89)$$

Hence, the angular velocity errors in roll rate, pitch rate, and yaw rate are brought to zero asymptotically. The values of the control gains can be chosen to provide specified closed-loop time constants. Consequently, the above control laws are guaranteed to achieve exact tracking asymptotically, with speed of response that is determined by the values of the control gains.

A simple special case of the above control law corresponds to the choice that $(\omega_{xd}, \omega_{yd}, \omega_{zd}) = (0, 0, 0)$. Thus, the simplified control law is given by

$$\begin{aligned} u_x &= -(I_{yy} - I_{zz})\omega_y\omega_z - c_x\omega_x, \\ u_y &= -(I_{zz} - I_{xx})\omega_z\omega_x - c_y\omega_y, \\ u_z &= -(I_{xx} - I_{yy})\omega_x\omega_y - c_z\omega_z, \end{aligned} \tag{75.90}$$

and it is guaranteed to bring the spacecraft asymptotically to rest.

The preceding analysis holds globally, that is, for all possible perturbations in the angular velocity vector, since the closed-loop system is exactly linear and asymptotically stable. This is in sharp contrast with the development that was based on the linearized equations. Thus, the family of control laws given here provides excellent closed-loop properties. The price of such good performance is the relative complexity of the control laws and the associated difficulty in practical implementation.

We now consider *spin stabilization of the spacecraft about a specified inertial axis.* It is assumed that the spacecraft dynamics are described by the nonlinear Equation 75.61, and the kinematics are described by the nonlinear Equation 75.57. A control law (u_x, u_y, u_z) is desired in feedback form, so that the resulting closed-loop system is linear and satisfies the control objective

$$\begin{aligned} &(\theta, \phi) \rightarrow (0, 0) \text{ as } t \rightarrow \infty, \\ &(\omega_x, \omega_y, \omega_z) \rightarrow (0, 0, \Omega) \text{ as } t \rightarrow \infty. \end{aligned}$$

This corresponds to control of the spacecraft so that it asymptotically spins about its body-fixed z axis at a spin rate Ω, and this spin axis is aligned with the inertially fixed Z axis.

In order to obtain closed-loop equations that are exactly linear, differentiate each of the first two equations in Equation 75.61, substitute for the time derivatives of the angular velocities from Equation 75.61; then select the feedback control law to cancel out all nonlinear terms in the resulting equations and add in desired linear terms. After considerable algebra, this feedback linearization approach leads to the following nonlinear control law

$$\begin{aligned} u_x &= -I_{xx}f_x - c_x\dot{\phi} - k_x\phi, \\ u_y &= -[I_{yy}f_y + c_y\dot{\theta} + k_y\theta]\sec\phi, \\ u_z &= (I_{yy} - I_{xx})\omega_x\omega_y - c_z(\omega_z - \Omega), \end{aligned} \tag{75.91}$$

where

$$\begin{aligned} f_y &= \frac{c_z}{I_{zz}}(\omega_z - \Omega)\sin\phi + \frac{(I_{zz} - I_{xx})}{I_{yy}}\omega_x\omega_z\cos\phi \\ &\quad - \omega_y\dot{\phi}\sin\phi - \omega_z\dot{\phi}\cos\phi, \\ f_x &= \frac{(I_{yy} - I_{zz})}{I_{xx}}\omega_y\omega_z \\ &\quad + \frac{(I_{zz} - I_{xx})}{I_{yy}}\omega_x\omega_z\sin\phi\tan\theta \\ &\quad - [f_y + \frac{c_y}{I_{yy}}\dot{\theta} + \frac{k_y}{I_{yy}}\theta]\tan\phi\tan\theta \\ &\quad - \frac{c_z}{I_{zz}}(\omega_z - \Omega)\cos\phi\tan\theta \\ &\quad + \omega_y\frac{d}{dt}[\sin\phi\tan\theta] + \omega_z\frac{d}{dt}[\cos\phi\tan\theta]. \end{aligned} \tag{75.92}$$

Note that the control law can be expressed either in terms of the rates of change of the Euler angles or in terms of the components of the angular velocity vector; the above expression involves a mixture of both. In addition, the control law can be seen not to depend on the yaw angle.

If we introduce the variables

$$\varepsilon_x = \phi, \;\; \varepsilon_y = \theta, \;\; \delta_x = \omega_z - \Omega, \tag{75.93}$$

it can be shown (after substantial algebra) that the closed-loop equations are the linear decoupled equations

$$\begin{aligned} I_{xx}\ddot{\varepsilon}_x + c_x\dot{\varepsilon}_x + k_x\varepsilon_x &= 0, \\ I_{yy}\ddot{\varepsilon}_y + c_y\dot{\varepsilon}_y + k_y\varepsilon_y &= 0, \\ I_{zz}\dot{\delta}_z + c_z\delta_z &= 0. \end{aligned} \tag{75.94}$$

Hence, the spacecraft attitude errors in pitch angle and roll angle are brought to zero asymptotically and the yaw rate is brought to the value Ω asymptotically. The values of the control gains can be chosen to provide specified closed-loop responses. Consequently, the above control laws are guaranteed to bring the spacecraft to the desired spin rate about the specified axis of rotation with speed of response that is determined by the values of the control gains.

The preceding analysis holds nearly globally, that is, for all possible perturbations in the angular velocity vector and for all possible perturbations in the Euler angles, excepting the singular values, since the closed-loop system of Equation 75.94 is exactly linear. This is in sharp contrast with the previous development that was based on linearized equations. It can easily be seen that the control law of Equation 75.78, obtained using the linearized approximation, also results from a linearization of the nonlinear control law of Equation 75.91 obtained in this section. Thus, the control law given by Equation 75.91 provides excellent closed-loop properties. The price of such good performance is the relative complexity of the control law of Equation 75.91 and the associated difficulty in its practical implementation.

75.2.11 Spacecraft Control Problem: Attitude Control in Circular Orbit

An important case of interest is that of a spacecraft in a circular orbit. In such a case, it is natural to describe the orientation of the spacecraft not with respect to an inertially fixed coordinate frame, but rather with respect to a locally horizontal-vertical coordinate frame as reference, defined so that the X axis of this frame is tangent to the circular orbit in the direction of the orbital motion, the Z axis of this frame is directed radially at the center of attraction, and the Y axis completes a right-hand orthogonal frame. Let the constant orbital angular velocity of the locally horizontal coordinate frame be

$$\Omega = \sqrt{\frac{g}{R}} \tag{75.95}$$

where g is the local acceleration of gravity and R is the orbital radius; the direction of the orbital angular velocity vector is along the negative Y axis.

The nonlinear dynamics for a spacecraft in circular orbit can be described in terms of the body-fixed angular velocity components. However, moment terms arise from the gravitational forces on the spacecraft when the spacecraft is modeled as a finite body rather than a point mass. These gravity gradient moments can be expressed in terms of the orbital angular velocity and the Euler angles, which describe the orientation of the body-fixed frame with respect to the locally horizontal-vertical coordinate frame. The complete nonlinear dynamics equations are not presented here due to their complexity.

The linearized expressions for the spacecraft dynamics about the constant angular velocity solution $(0, -\Omega, 0)$ are obtained by introducing the perturbation variables

$$\delta_x = \omega_x, \quad \delta_y = \omega_y + \Omega, \quad \delta_z = \omega_z; \tag{75.96}$$

the resulting linearized dynamics equations are

$$\begin{aligned} I_{xx}\dot{\delta}_x &= \Omega(I_{zz} - I_{yy})\delta_z + T_x, \\ I_{yy}\dot{\delta}_y &= T_y, \\ I_{zz}\dot{\delta}_z &= \Omega(I_{yy} - I_{xx})\delta_x + T_z, \end{aligned} \tag{75.97}$$

The external moments on the spacecraft are given by

$$\begin{aligned} T_x &= 3\Omega^2(I_{zz} - I_{yy})\phi + u_x, \\ T_y &= 3\Omega^2(I_{xx} - I_{zz})\theta + u_y, \\ T_z &= u_z, \end{aligned} \tag{75.98}$$

where (u_x, u_y, u_z) denotes the control moments, and the other terms describe the linearized gravity gradient moments on the spacecraft. Thus, the linearized spacecraft dynamics are given by

$$\begin{aligned} I_{xx}\dot{\delta}_x &= \Omega(I_{zz} - I_{yy})\delta_z + 3\Omega^2(I_{zz} - I_{yy})\phi + u_x, \\ I_{yy}\dot{\delta}_y &= -3\Omega^2(I_{xx} - I_{zz})\theta + u_y, \\ I_{zz}\dot{\delta}_z &= \Omega(I_{yy} - I_{xx})\delta_x + u_z. \end{aligned} \tag{75.99}$$

The linearized kinematics equations are given by

$$\begin{aligned} \dot{\phi} &= \delta_x + \Omega\Psi, \\ \dot{\theta} &= \delta_y, \\ \dot{\Psi} &= \delta_z - \Omega\phi, \end{aligned} \tag{75.100}$$

where the extra terms arise since the locally horizontal-vertical coordinate frame has a constant angular velocity of $-\Omega$ about the Y axis. Thus, the linearized dynamics and kinematics equations can be combined in a form that makes clear that the pitching motion is decoupled from the rolling and yawing motions

$$I_{yy}\ddot{\theta} = -3\Omega^2(I_{xx} - I_{zz})\theta + u_y, \tag{75.101}$$

but the rolling and yawing motions are coupled

$$\begin{aligned} I_{xx}\ddot{\phi} &= \Omega(I_{xx} + I_{zz} - I_{yy})\dot{\Psi} \\ &\quad + 4\Omega^2(I_{zz} - I_{yy})\phi + u_x, \\ I_{zz}\ddot{\Psi} &= \Omega(I_{yy} - I_{xx} - I_{zz})\dot{\phi} - \Omega^2(I_{yy} - I_{xx})\Psi + u_z. \end{aligned} \tag{75.102}$$

A control law (u_x, u_y, u_z) is desired in feedback form, so that the resulting closed-loop system satisfies the asymptotic attitude conditions

$$\begin{aligned} (\Psi, \theta, \phi) &\to (0, 0, 0) \text{ as } t \to \infty, \\ (\delta_x, \delta_y, \delta_z) &\to (0, 0, 0) \text{ as } t \to \infty, \end{aligned}$$

which guarantees that

$$(\omega_x, \omega_y, \omega_z) \to (0, -\Omega, 0) \text{ as } t \to \infty;$$

that is, the spacecraft has an asymptotically constant spin rate consistent with the orbital angular velocity as desired.

Since the preceding spacecraft equations are linear, a standard linear control approach can be used to obtain linear feedback control laws for the pitching motion and for the rolling and yawing motion of the form

$$\begin{aligned} u_x &= -\Omega(I_{xx} + I_{zz} - I_{yy})\dot{\Psi} - c_x\dot{\phi} \\ &\quad - \left[4\Omega^2(I_{zz} - I_{yy}) + k_x\right]\phi, \\ u_y &= -c_y\dot{\theta} - [3\Omega^2(I_{xx} - I_{zz}) + k_y]\theta, \\ u_z &= -c_z\dot{\Psi} - [-\Omega^2(I_{yy} - I_{xx}) + k_z]\Psi \\ &\quad + \Omega(I_{xx} + I_{zz} - I_{yy})\dot{\phi}. \end{aligned} \tag{75.103}$$

The control law can be expressed either in terms of the rates of change of the Euler angles or in terms of the components of the angular velocity vector.

If we introduce the perturbation variables

$$\varepsilon_x = \phi, \quad \varepsilon_y = \theta, \quad \varepsilon_z = \Psi, \tag{75.104}$$

the resulting closed-loop system is described by

$$\begin{aligned} I_{xx}\ddot{\varepsilon}_x + c_x\dot{\varepsilon}_x + k_x\varepsilon_x &= 0, \\ I_{yy}\ddot{\varepsilon}_y + c_y\dot{\varepsilon}_y + k_y\varepsilon_y &= 0, \\ I_{zz}\ddot{\varepsilon}_z + c_z\dot{\varepsilon}_z + k_z\varepsilon_z &= 0. \end{aligned} \tag{75.105}$$

If the gains are chosen so that the pitching motion is asymptotically stable and the rolling and yawing motion is asymptotically stable, then the spacecraft attitude errors are automatically brought to zero asymptotically, at least for sufficiently small perturbations of the orientation from the locally horizontal-vertical reference. The values of the control gains can be chosen to provide specified closed-loop response properties. Consequently, the above control laws are guaranteed to bring the spacecraft to the desired attitude with speed of response that is determined by the values of the control gains.

We again note that the preceding analysis holds only for sufficiently small attitude perturbations from the local horizontal-vertical reference. For large perturbations, the above control laws may not have the desired properties. An analysis of the closed-loop system, using the nonlinear dynamics and kinematics equations, is required to determine the domain of perturbations for which closed-loop stabilization is achieved. Such an analysis is beyond the scope of the present chapter.

75.2.12 Other Spacecraft Control Problems and Control Methodologies

Our treatment of spacecraft attitude control has been limited, both by the class of rotational control problems considered and by the assumptions that have been made. In this section, we briefly indicate other classes of problems for which results are available in the published literature.

Throughout our development, specific control laws have been developed using orientation representations expressed in terms of Euler angles. As we have indicated, the Euler angles are not global representations for orientation. Other orientation representations, including quaternions and Euler axis-angle variables, have been studied and various control laws have been developed using these representations.

Several of our control approaches have been based on use of the linearized spacecraft kinematics and dynamics equations. We have also suggested a nonlinear control approach, feedback linearization, to develop several classes of feedback control laws for the nonlinear kinematics and dynamics equations. Other types of control approaches have been studied for spacecraft reorientation, including optimal control and pulse-width-modulated control schemes.

We should also mention other classes of spacecraft attitude control problems. Our approach has assumed use of pairs of gas-jet thrusters or reaction wheels modeled in the simplest way. Other assumptions lead to somewhat different models and, hence, somewhat different control problems. In particular, we mention that control problems in the case there are only two, rather than three, control moments have recently been studied.

A key assumption throughout our development is that the spacecraft is a rigid body. There are important spacecraft designs where this assumption is not satisfied, and the resulting attitude control problems are somewhat different from what has been considered here; usually these control problems are even more challenging. Examples of such control problems occur when nutation dampers or control moment gyros are used for attitude control. Dual-spin spacecraft and multibody spacecraft are examples where there is relative motion between spacecraft components that must be taken into account in the design of attitude control systems. Numerous modern spacecraft, due to weight constraints, consist of flexible components. Attitude control of flexible spacecraft is a very important and widely studied subject.

75.2.13 Defining Terms

Attitude: The orientation of the spacecraft with respect to some reference frame.

Body axes: A reference frame fixed in the spacecraft body and rotating with it.

Euler angles: A sequence of angle rotations that are used to parametrize a rotation matrix.

Pitch: The second angle in the 3-2-1 Euler angle sequence. For small rotation angles, the pitch angle is the rotation angle about the spacecraft y axis.

Roll: The third angle in the 3-2-1 Euler angle sequence. For small rotation angles, the roll angle is the rotation angle about the spacecraft x axis.

Rotation matrix: A matrix of direction cosines relating unit vectors of two different coordinate frames.

Simple spin: A spacecraft spinning about a body axis whose direction remains inertially fixed.

Yaw: The first rotation angle in the 3-2-1 Euler angle sequence. For small rotation angles, the yaw angle is the rotation angle about the spacecraft z axis.

References

[1] Bryson, A.E., *Control of Spacecraft and Aircraft,* Princeton University Press, Princeton, NJ, 1994.

[2] Crouch, P.E., Spacecraft attitude control and stabilization: applications of geometric control theory to rigid body models, *IEEE Trans. Autom. Control,* 29(4), 321–331, 1984.

[3] Greenwood, D.T., *Principles of Dynamics,* 2nd ed., Prentice Hall, Englewood Cliffs, NJ, 1988.

[4] Hughes, P.C., *Spacecraft Attitude Dynamics,* Wiley, New York, 1986.

[5] Wertz, J.R., Ed., *Spacecraft Attitude Determination and Control,* Kluwer, Dordrecht, Netherlands, 1978.

Further Reading

The most complete reference on the control of spacecraft is *Spacecraft Attitude Determination and Control,* a handbook edited by J. R. Wertz [5]. It has been reprinted often and is available from Kluwer Academic Publishers.

Two introductory textbooks on spacecraft dynamics are Wiesel, W.E. 1989, *Spaceflight Dynamics,* McGraw-Hill, New York, and Thomson, W.T. 1986, *Introduction to Space Dynamics,* Dover Publications (paperback), NY, 1986.

A more comprehensive treatment can be found in *Spacecraft Attitude Dynamics* by P.C. Hughes [4].

A discussion of the orbital motion of spacecraft can be found in Bate, Mueller, White, 1971, *Fundamentals of Astrodynamics,* Dover Publications, New York, and Danby, J.M.A. 1988, *Fundamentals of Celestial Mechanics,* 2nd ed., Willmann-Bell Inc., Richmond, VA.

75.3 Control of Flexible Space Structures

S. M. Joshi and A. G. Kelkar, NASA
Langley Research Center

75.3.1 Introduction[2]

A number of near-term space missions as well as future mission concepts will require flexible space structures (FSS) in low Earth and geostationary orbits. Examples of near-term missions include multipayload space platforms, such as Earth observing systems and space-based manipulators for on-orbit assembly and satellite servicing. Examples of future space mission concepts include mobile satellite communication systems, solar power satellites, and large optical reflectors, which would require large antennas, platforms, and solar arrays. The dimensions of such structures would range from 50 meters (m) to several kilometers (km). Because of their relatively light weight and, in some cases, expansive sizes, such structures tend to have low-frequency, lightly damped structural (elastic) modes. The natural frequencies of the elastic modes are generally closely spaced, and some natural frequencies may be lower than the controller bandwidth. In addition, the elastic mode characteristics are not known accurately. For these reasons, control systems design for flexible space structures is a challenging problem.

Depending on their missions, flexible spacecraft can be roughly categorized as single-body spacecraft and multibody spacecraft. Two of the most important control problems for single-body FSS are (1) fine-pointing of FSS in space with the required precision in attitude (represented by three Euler angles) and shape, and (2) large-angle maneuvering ("slewing") of the FSS to orient to a different target. The performance requirements for both of these problems are usually very high. For example, for a certain mobile communication system concept, a 122-meter diameter space antenna will have to be pointed with an accuracy of 0.03 degree root mean square (RMS). The requirements for other missions vary, but some are expected to be even more stringent, on the order of 0.01 arc-second. In some applications, it would be necessary to maneuver the FSS quickly through large angles to acquire a new target on the Earth in minimum time and with minimum fuel expenditure, while keeping the elastic motion and accompanying stresses within acceptable limits. Once the target is acquired, the FSS must point to it with the required precision.

For multibody spacecraft with articulated appendages, the main control problems are (1) fine-pointing of some of the appendages to their respective targets, (2) rotating some of the appendages to follow prescribed periodic scanning profiles, and (3) changing the orientation of some of the appendages through large angles. For example, a multipayload platform would have the first two requirements, while a multilink manipulator would have the third requirement to reach a new end-effector position.

The important feature that distinguishes FSS from conventional older generation spacecraft is their highly prominent structural flexibility which results in special dynamic characteristics. Detailed literature surveys on dynamics and control of FSS may be found in [11], [20].

The organization of this chapter is as follows. The problem of fine-pointing control of single-body spacecraft is considered in Section 75.3.2. This problem not only represents an important class of missions, but also permits analysis in the linear, time-invariant (LTI) setting. The basic linearized mathematical model of single-body FSS is presented, and the problems encountered in FSS control systems design are discussed. Two types of controller design methods, model-based controllers and passivity-based controllers, are presented. The highly desirable robustness characteristics of passivity-based controllers are summarized. Section 75.3.3 addresses another important class of missions, namely, multibody FSS. A generic nonlinear mathematical model of a multibody flexible space system is presented and passivity-based robust controllers are discussed.

75.3.2 Single-Body Flexible Spacecraft

Linearized Mathematical Model

Simple structures, such as uniform beams or plates, can be effectively modeled by infinite-dimensional systems (see [12]). In some cases, approximate infinite-dimensional models have been proposed for more complex structures such as trusses [2]. However, most of the realistic FSS are highly complex, not amenable to infinite-dimensional modeling. The standard practice is to use finite-dimensional mathematical models generated by using the finite-element method [19]. The basic approach of this method is dividing a continuous system into a number of elements using fictitious dividing lines, and applying the Lagrangian formulation to determine the forces at the points of intersection as functions of the applied forces. Suppose there are r force actuators and p torque actuators distributed throughout the structure. The ith force actuator produces the 3×1 force vector $f_i = (f_{xi}, f_{yi}, f_{zi})^T$, along the X, Y, and Z axes of a body-fixed coordinate system centered at the nominal center of mass (c.m.). Similarly, the ith torque actuator produces the torque vector $T_i = (T_{xi}, T_{yi}, T_{zi})^T$. Then the linearized equations of motion can be written as follows [12]:

rigid-body translation:

$$M\ddot{z} = \sum_{i=1}^{r} f_i, \tag{75.106}$$

rigid-body rotation:

$$J\ddot{\alpha} = \sum_{i=1}^{r} R_i \times f_i + \sum_{i=1}^{p} T_i, \quad \text{and} \tag{75.107}$$

elastic motion:

$$\ddot{q} + D\dot{q} + \Lambda q = \sum_{i=1}^{r} \Delta_i^T f_i + \sum_{i=1}^{p} \Phi_i^T T_i, \tag{75.108}$$

[2]This article is based on the work performed for the U.S. Government. The responsibility for the contents rests with the authors.

where M is the mass, z is the 3×1 position of the c.m., R_i is the location of f_i on the FSS, J is the 3×3 moment-of-inertia matrix, α is the attitude vector consisting of the three Euler rotation angles (ϕ, θ, ψ), and $q = (q_1, q_2, \ldots, q_{nq})^T$ is the $n_q \times 1$ modal amplitude vector for the n_q elastic modes. ("$\times$" in Equation 75.107 denotes the vector cross-product.) In general, the number of modes (n_q) necessary to characterize an FSS adequately is quite large, on the order of 100–1000. Δ_i^T, Φ_i^T are the $n_q \times 3$ translational and rotational mode shape matrices at the ith actuator location. The rows of Δ_i^T, Φ_i^T represent the X, Y, Z components of the translational and rotational mode shapes at the location of actuator i.

$$\Lambda = diag(\omega_1^2, \omega_2^2, \ldots, \omega_{n_q}^2) \tag{75.109}$$

where ω_k is the natural frequency of the kth elastic mode, and D is an $n_q \times n_q$ matrix representing the inherent damping in the elastic modes:

$$D = 2\,diag\,(\rho_1\omega_1, \rho_2\omega_2, \ldots, \rho_{nq}\omega_{nq}). \tag{75.110}$$

The inherent damping ratios (ρ_is) are typically on the order of 0.001–0.01. The finite-element method cannot model inherent damping. This proportional damping term is customarily added after an undamped finite-element model is obtained.

The translational and rotational positions z_p and y_p, at a location with coordinate vector R, are given by

$$z_p = z - R \times \alpha + \overline{\Delta} q \tag{75.111}$$

$$\text{and} \quad y_p = \alpha + \overline{\Phi} q, \tag{75.112}$$

where the ith columns of $\overline{\Delta}$ and $\overline{\Phi}$ represent the ith $3 \times n_q$ translational and rotational mode shapes at that location.

Figure 75.17 shows the mode-shape plots for a finite-element model of a completely free, 100 ft. $\times$ 100 ft. $\times$ 0.1 in. aluminum plate. The plots were obtained by computing the elastic displacements at many locations on the plate, resulting from nonzero values of individual modal amplitudes q_i. Figure 75.107 shows the mode-shape plots for the 122-m diameter, hoop/column antenna concept [22], which consists of a deployable central mast attached to a deployable hoop by cables held in tension.

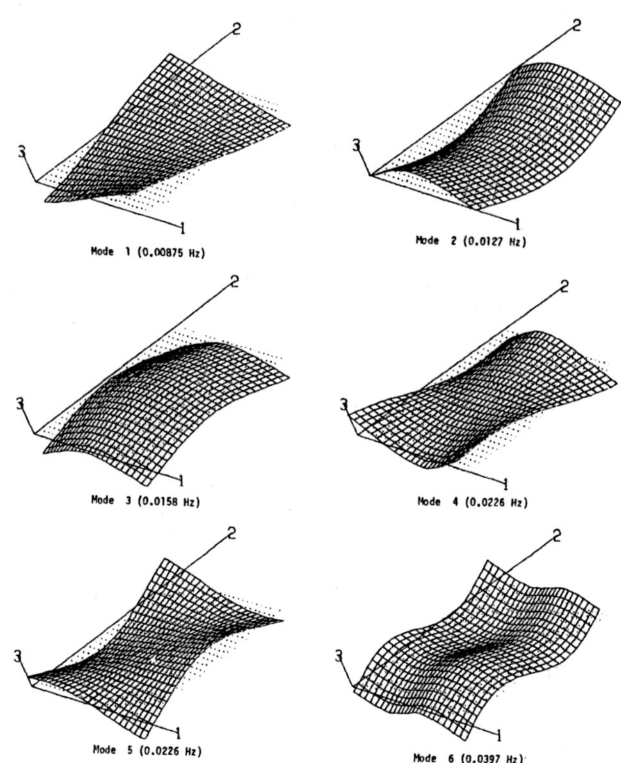

Figure 75.17 Mode-shape plots for a completely free plate.

Controllability and Observability

Precision attitude control is usually accomplished by torque actuators. Therefore, in the following material, only rotational equations of motion are considered. No force actuators are used. With $f_i = 0$ and denoting $\xi = [\alpha^T, q_1, q_2, \ldots q_{nq}]^T$, Equations 75.107 and 75.108 can be written as

$$\tilde{A}\ddot{\xi} + \tilde{B}\dot{\xi} + \tilde{C}\xi = \Gamma^T u \tag{75.113}$$

$$\tilde{A} = diag(J, I_{n_q}), \tilde{B} = diag(0_3, D), \tilde{C} = diag(0_3, \Lambda) \tag{75.114}$$

$$\Gamma^T = \begin{bmatrix} I_{3\times3} \quad I_{3\times3} \ldots I_{3\times3} \\ \Phi^T \end{bmatrix}_{n_1 \times m} \tag{75.115}$$

where $\Phi^T = \left[\Phi_1^T, \Phi_2^T, \ldots \Phi_p^T\right]$, $n_1 = n_q + 3$, and $m = 3p$. The system can be written in the state-space form as

$$\dot{x} = Ax + Bu \tag{75.116}$$

where $x = (\alpha^T, \dot{\alpha}^T, q_1, \dot{q}_1, \ldots, q_{nq}, \dot{q}_{nq})^T$ is the n-dimensional state vector ($n = 2n_1$) and u is the $m \times 1$ control vector consisting of applied torques.

$$A = diag(A_{rb}, A_1, A_2, \ldots, A_{nq}) \tag{75.117}$$

$$A_{rb} = \begin{bmatrix} 0_3 & I_3 \\ 0_3 & 0_3 \end{bmatrix}; A_i = \begin{bmatrix} 0 & 1 \\ -\omega_i^2 & -2\rho_i\omega_i \end{bmatrix} \tag{75.118}$$

(0_k and I_k denote the $k \times k$ null and identity matrices, respectively.)

$$B = \left[B_{rb}^T, : 0_{m\times1}, \phi_1 : 0_{m\times1}, \phi_2 : \ldots : 0_{m\times1}, \phi_{nq}\right]^T \tag{75.119}$$

$$B_{rb} = \begin{bmatrix} 0_{3\times m} \\ J^{-1}, J^{-1} \ldots J^{-1} \end{bmatrix}_{6\times m} \tag{75.120}$$

where ϕ_k^T represents the kth row of Φ^T.

If a three-axis attitude sensor is placed at a location where the rotational mode shapes are given by the rows of the $n_q \times 3$ matrix Ψ^T, the sensed attitude (ignoring noise) would be

$$y_p = \alpha + \Psi q = \left[I_3, \Psi^T\right]\xi. \tag{75.121}$$

The conditions for controllability are given below.

Controllability Conditions

The system given by Equation 75.116 is controllable if, and only if, (iff) the following conditions are satisfied:

1. Rows of Φ^T corresponding to each distinct (in frequency) elastic mode have at least one nonzero entry.

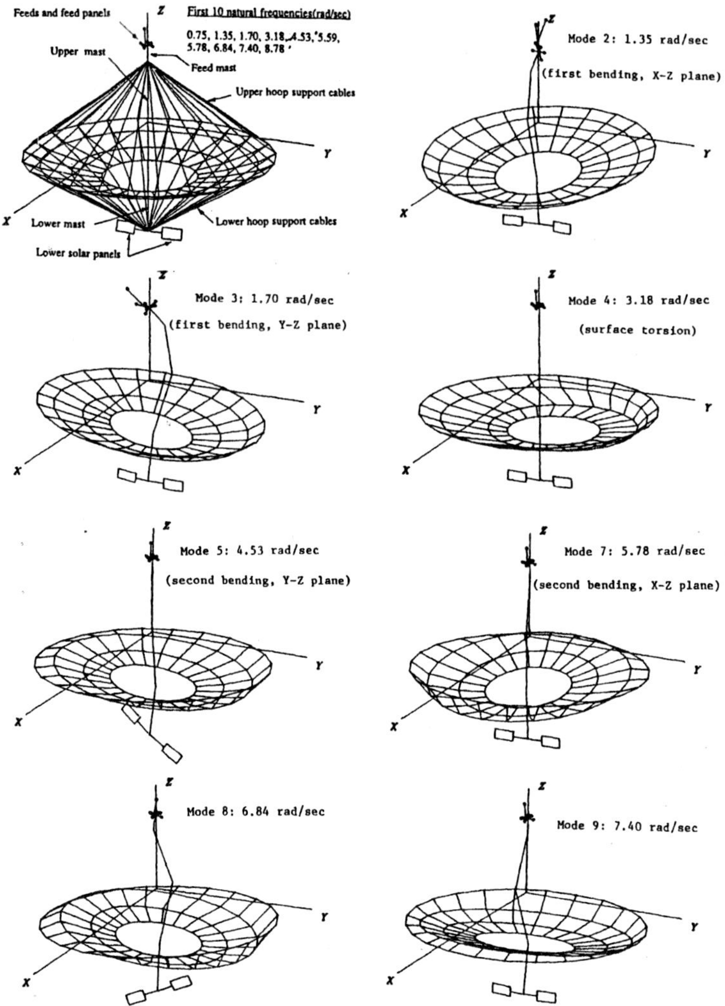

Figure 75.18 Typical mode shapes of hoop/column antenna.

2. If there are ν elastic modes with the same natural frequency $\overline{\omega}$, the corresponding rows of Φ^T form a linearly independent set.

A proof of this result can be found in [12]. Condition (1) would be satisfied iff the rotational mode shape (X, Y, or Z component) for each mode is nonzero at the location of at least one actuator. Condition (2) needs to be tested only when there is more than one elastic mode with the same natural frequency, which can typically occur in the case of symmetric structures.

Similar necessary and sufficient conditions can be obtained analogously for observability. It should be noted that the rigid-body modes are not observable using attitude-rate sensors alone without attitude sensors. However, a three-axis attitude sensor can be sufficient for observability even if no rate sensors are used.

Problems in Controller Design for FSS

Precision attitude control requires controlling the rigid rotational modes **and** suppressing the elastic vibration. The objectives of the controller are

1. **fast transient response:** Quickly damp out the pointing errors resulting from step disturbances such as thermal distortion resulting from entering or leaving Earth's shadow or nonzero initial conditions, resulting from the completion of a large-angle attitude maneuver.
2. **disturbance rejection:** Maintain the attitude as close as possible to the desired attitude in the presence of noise and disturbances.

The first objective translates into the closed-loop bandwidth requirement, and the second translates into minimizing the RMS pointing error. In addition, the elastic motion must be very small, i.e., the RMS shape distortions must be below prescribed limits. For applications such as large communications antennas, the typical bandwidth requirement is 0.1 rad/sec., with at most a 4 sec. time constant for all of the elastic modes (closed loop). Typical allowable RMS errors are 0.03 degrees pointing error, and 6-mm surface distortion.

The problems encountered in designing an attitude controller are

1. An adequate model of an FSS is of high order because it contains a large number of elastic modes; however, a practically implementable controller has to be of sufficiently low order.
2. The inherent energy dissipation (damping) is very small.
3. The elastic frequencies are low and closely spaced.
4. The parameters (frequencies, damping ratios, and mode shapes) are not known accurately.

The simplest controller design approach would be truncating the model beyond a certain number of modes and designing a reduced-order controller. This approach is routinely used for controlling relatively rigid conventional spacecraft, wherein only the rigid modes are retained in the design model. Second-order filters are included in the loop to attenuate the contribution of the elastic modes. This approach is not generally effective for FSS because the elastic modes are much more prominent. Figure 75.19 shows the effect of using a truncated design model. When constructing a control loop around the "controlled" modes, an unintentional feedback loop is also constructed around the truncated modes, which can make the closed-loop system unstable. The inadvertent excitation of the truncated modes by the input and the unwanted contribution of the truncated modes to the sensed output were aptly termed by Balas [4] as "control spillover" and "observation spillover", respectively. The spillover terms may cause performance degradation and even instability, leading to catastrophic failure.

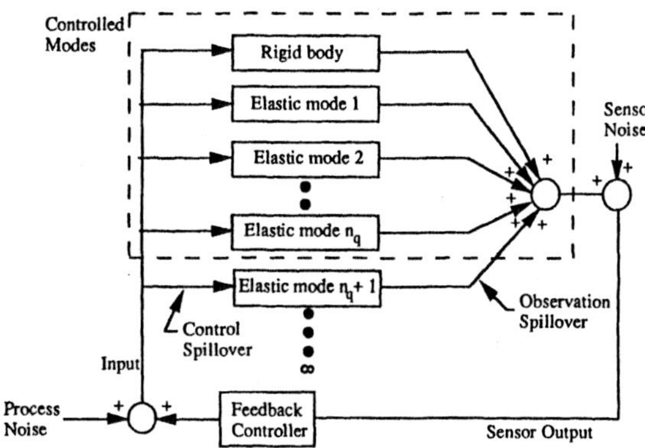

Figure 75.19 Control and observation spillover.

In addition to the truncation problem, the designer also lacks accurate knowledge of the parameters. Finite-element models give reasonably accurate estimates of the frequencies and mode shapes only for the first few modes and can provide no estimates of inherent damping ratios. Premission ground testing for parameter estimation is not generally possible because many FSS are not designed to withstand the gravitational force (while deployed), and because the test facilities required, such as a vacuum chamber, would be excessively large. Another consideration in controller design is that the actuators and sensors have nonlinearities, and finite response times. In view of these problems, the attitude controller must be a "robust" one, that is, it must at least maintain stability, and perhaps performance, despite modeling errors, uncertainties, nonlinearities, and component failures. The next two sections present linear controller design methods for the attitude control problem.

Model-Based Controllers

Consider the nth order state space model of an FSS in the presence of process noise and measurement noise,

$$\dot{x} = Ax + Bu + v, \quad y = Cx + Du + w, \tag{75.122}$$

where $v(t)$ and $w(t)$ are, respectively, the $n \times 1$ and $l \times 1$ process noise and sensor noise vectors. v and w are assumed to be mutually uncorrelated, zero-mean, Gaussian white noise processes with covariance intensity matrices V and W. A linear quadratic Gaussian (LQG) controller can be designed to minimize

$$J = \lim_{t_f \to \infty} \frac{1}{t_f} \mathcal{E} \int_0^{t_f} [x^T(t)Qx(t) + u^T(t)Ru(t)]dt \tag{75.123}$$

where "$\mathcal{E}$" denotes the expectation operator and $Q = Q^T \geq 0$, $R = R^T > 0$ are the state and control weighting matrices. The resulting nth order controller consists of a Kalman–Bucy filter (KBF) in tandem with a linear quadratic regulator (LQR) with the form,

$$\dot{\hat{x}} = (A + BG - KC)\hat{x} + Ky, \quad u = G\hat{x}, \tag{75.124}$$

where $\hat{x}$ is the controller state vector, and $G_{m\times n}$ and $K_{n\times l}$ are the LQR and KBF gain matrices. Any controller using an observer and state estimate feedback has the same mathematical structure. The order of the controller is n, the same as that of the plant. An adequate model of an FSS typically consists of several hundred elastic modes. However, to be practically implementable, the controller must be of sufficiently low order. A reduced-order controller design can be obtained in two ways, either by using a reduced-order "design model" of the plant or by obtaining a reduced-order approximation to a high-order controller. The former method is used more widely than the latter because high-order controller design relies on the knowledge of the high frequency mode parameters, which is usually inaccurate.

A number of model-order reduction methods have been developed during the past few years. The most important of these include the singular perturbation method, the balanced truncation method, and the optimal Hankel norm method (see [7] for a detailed description). In the singular perturbation method, higher frequency modes are approximated by their quasi-static representation. The balanced truncation method uses a similarity transformation that makes the controllability and observability matrices equal and diagonal. A reduced-order model is then obtained by retaining the most controllable and observable state variables. The optimal Hankel norm approximation method aims to minimize the Hankel norm of the approximation error and can yield a smaller error than the balanced truncation method. A disadvantage of the balanced truncation and the Hankel norm methods is that the resulting (transformed) state variables are mutually coupled and do not correspond to individual modes, resulting in the loss of physical insight. A disadvantage of the singular perturbation and Hankel norm methods is that they can yield non-strictly proper reduced-order models. An alternate method of overcoming these difficulties is to rank the elastic modes according their contributions to the overall transfer function, in the sense of H_2, H_∞, or $\mathcal{L}_1$ norms [9]. The highest ranked modes are then retained in the design model. This method retains the physical significance of the modes and also yields a strictly proper model. Note that the rigid-body modes must always be included in the design model, no matter which order-reduction method is used. A model-based controller can then be designed based on the reduced-order design model.

LQG Controller

An LQG controller designed for the reduced-order design model is guaranteed to stabilize the nominal design model. However, it may not stabilize the full-order plant because of the control and observation spillovers. Some time-domain methods for designing spillover-tolerant, reduced-order LQG controllers are discussed in [12]. These methods basically attempt to reduce the norms of spillover terms $\|B_t G\|$ and $\|K C_t\|$, where B_t and C_t denote the input and observation matrices corresponding to the truncated modes. Lyapunov-based sufficient conditions for stability are derived in terms of upper bounds on the spillover norms and are used as guidelines in spillover reduction. The controllers obtained by these methods are generally quite conservative and also require the knowledge of the truncated mode parameters to ensure stability.

Another approach to LQG controller design is the application of multivariable frequency-domain methods, wherein the truncated modes are represented as an additive uncertainty term $\Delta P(s)$ that appears in parallel with the design model (i.e., nominal plant) transfer function $P(s)$, as shown in Figure 75.20.

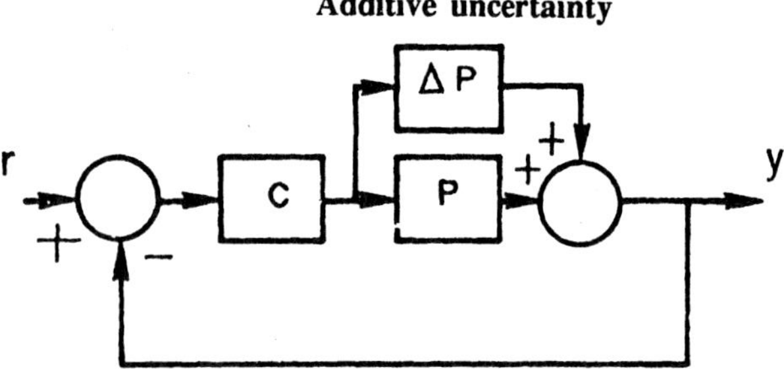

Figure 75.20 Additive uncertainty formulation of truncated dynamics.

A sufficient condition for stability is [7]:

$$\bar{\sigma}[\Delta P(j\omega)] < 1/\bar{\sigma}\{C(j\omega)[(I + P(j\omega)C(j\omega)]^{-1}\}, \quad \text{for} \quad 0 \leq \omega < \infty, \tag{75.125}$$

where $C(s)$ is the controller transfer function and $\bar{\sigma}[.]$ denotes the largest singular value. An upper bound on $\bar{\sigma}[\Delta P(j\omega)]$ can be obtained from (crude knowledge of) the truncated mode parameters to generate an "uncertainty envelope". Figure 75.21 shows the stability test Equation 75.125 for the 122-m hoop/column antenna where the design model consists of the three rigid rotational modes and the first three elastic modes.

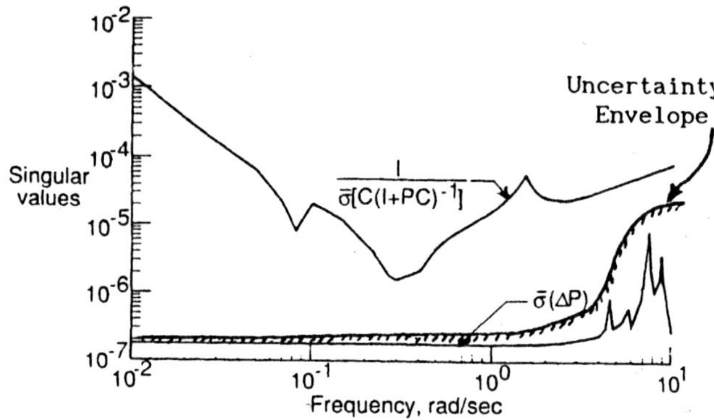

Figure 75.21 Stability test for additive uncertainty.

A measure of the nominal closed-loop performance is given by the bandwidth of the closed loop transfer function, $G_{cl} = PC(I + PC)^{-1}$, shown in Figure 75.22 for the hoop/column antenna.

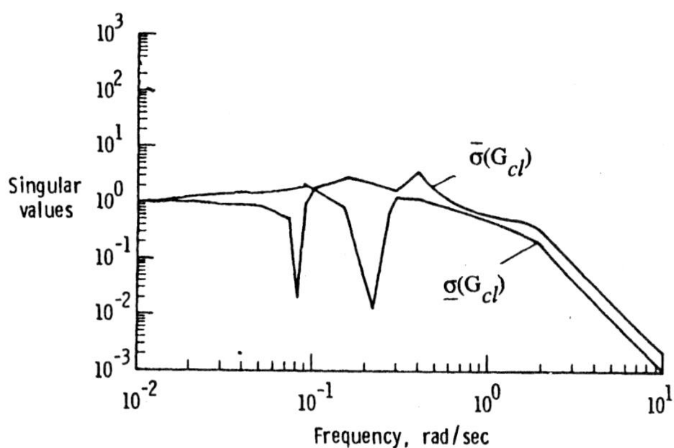

Figure 75.22 Closed-loop transfer function.

The results in Figures 75.21 and 75.22 were obtained by iteratively designing the KBF and the LQR to yield the desired closed-loop bandwidth while still satisfying Equation 75.125. The iterative method, described in [12], is loosely based on the LQG/Loop Transfer Recovery (LTR) method [23]. The resulting design is robust to any truncated mode dynamics which lie under the uncertainty envelope. However, the controller may not provide robustness to parametric uncertainties in the design model. A small uncertainty in the natural frequencies of the design model (i.e., the "controlled modes") can cause closed-loop instability because the very small open-loop damping ratios cause very sharp peaks in the frequency response, so that a small error in the natural frequency produces a large error peak in the frequency response [12].

H_∞- and μ-Synthesis Methods

The H_∞ controller design method [7] represents a systematic method for obtaining the desired performance as well as robustness to truncated mode dynamics. A typical design objective is minimizing the H_∞ norm of the frequency-weighted transfer function from the disturbance inputs (e.g., sensor and actuator noise) to the controlled variables, while insuring stability in the presence of truncated modes. An example of the application of the H_∞ method to FSS control is given in [17]. The problem also can be formulated to include parametric uncertainties represented as unstructured uncertainty. However, the resulting controller design is usually very conservative and provides inadequate performance.

The structured singular value method [6], [21] also known as the "μ-synthesis" method, can overcome the conservatism of the H_∞ method. In this method, the parametric uncertainties are individually "extracted" from the system block diagram and arranged as a diagonal block that forms a feedback connection with the nominal closed-loop system. The controller design problem is formulated as one of H_∞-norm minimization subject to a constraint on the structured singular value of an appropriate transfer function. The μ-synthesis problem can also be formulated to provide robust performance, i.e., the performance specifications must be satisfied despite model uncertainties. An application of the μ-synthesis method to FSS control is presented in [3].

The next section presents a class of controllers that can circumvent the problems due to spillovers and parametric uncertainties.

Passivity-Based Controllers

Consider the case where an attitude sensor and a rate sensor are collocated with each of the p torque actuators. Then the $m \times 1$ $(m = 3p)$ sensed attitude and rate vectors y_p and y_r are given by

$$y_p = \Gamma\xi, \quad y_r = \Gamma\dot{\xi} \tag{75.126}$$

The transfer function from $U(s)$ to $Y_p(s)$ is $G(s) = G'(s)/s$, where $G'(s)$ is given by

$$G'(s) = \frac{\mathcal{I}J^{-1}\mathcal{I}^T}{s} + \sum_{1}^{n_q} \frac{\phi_i\phi_i^T s}{s^2 + 2\rho_i\omega_i s + \omega_i^2} \tag{75.127}$$

where the $m \times 3$ matrix $\mathcal{I} = [I_3, I_3, \ldots I_3]^T$, and ϕ_i^T denotes the ith row of Φ^T. The entries of ϕ_i^T represent the rotational mode shapes for the ith mode. An important consequence of collocation of actuators and sensors is that the operator from u to y_r is passive [5], or equivalently, $G'(s)$ is Positive-Real as defined below.

DEFINITION 75.1 A rational matrix-valued function $T(s)$ of the complex variable s is said to be positive-real if all of its elements are analytic in $Re[s] > 0$, and $T(s) + T^T(s^*) \geq 0$ in $Re[s] > 0$, where $*$ denotes the complex conjugate.

Scalar positive-real (PR) functions have a relative degree (i.e., the difference between the degrees of the denominator and numerator polynomials) of –1, 0, or 1 [24]. PR matrices have no transmission zeros or poles in the open right-half of the complex plane, and the poles on the imaginary axis are simple and have nonnegative definite residues. It can be shown that it is sufficient to check for positive semidefiniteness of $T(s)$ only on the imaginary axis ($s = j\omega, 0 \leq \omega < \infty$), i.e., the condition becomes $T(j\omega) + T^*(j\omega) \geq 0$, where $*$ denotes complex conjugate transpose. Suppose (A, B, C, D) is an nth order minimal realization of $T(s)$. From [1], a necessary and sufficient condition for $T(s)$ to be positive real is that an $n \times n$ symmetric positive definite matrix P and matrices W and L exist so that

$$\begin{aligned} A^T P + PA &= -LL^T \\ C &= B^T P + W^T L \\ W^T W &= D + D^T \end{aligned} \tag{75.128}$$

This result is also generally known in the literature as the Kalman–Yakubovich lemma. Positive realness of $G'(s)$ gives rise to a large class of robustly stabilizing controllers, called dissipative controllers. Such controllers can be divided into static dissipative and dynamic dissipative controllers.

Static Dissipative Controllers

Consider the proportional-plus-derivative control law,

$$u = -G_p y_p - G_r y_r, \tag{75.129}$$

where G_p and G_r are symmetric positive-definite matrices. The closed-loop equation then becomes

$$\tilde{A}\ddot{\xi} + (\tilde{B} + \Gamma^T G_r \Gamma)\dot{\xi} + (\tilde{C} + \Gamma^T G_p \Gamma)\xi = 0. \tag{75.130}$$

It can be shown that $(\tilde{B} + \Gamma^T G_r \Gamma)$ and $(\tilde{C} + \Gamma^T G_p \Gamma)$ are positive-definite matrices, and that this control law stabilizes the plant $G(s)$, i.e., the closed-loop system of Equation 75.130 is asymptotically stable (see [12]). The closed-loop stability is not affected by the number of truncated modes or the knowledge of the parametric values, that is, the stability is ***robust***. The only requirements are that the actuators and sensors be collocated and that the feedback gains be positive definite. Furthermore, if G_p, G_r are diagonal, then the robust stability holds even when the actuators and sensors have certain types of nonlinear gains.

Stability in the Presence of Actuator and Sensor Nonlinearities

Suppose that G_p, G_r are diagonal, and that

1. the actuator nonlinearities, $\psi_{ai}(\nu)$, are monotonically nondecreasing and belong to the $(0, \infty)$ sector, i.e., $\psi_{ai}(0) = 0$, and $\nu\psi_{ai}(\nu) > 0$ for $\nu \neq 0$.
2. the attitude and rate sensor nonlinearities, $\psi_{si}(\nu)$, belong to the $(0, \infty)$ sector.

Then the closed-loop system with the static dissipative control law is globally asymptotically stable.

A proof of this result can be obtained by slightly modifying the results in [12]. Examples of permissible nonlinearities are shown in Figure 75.23. It can be seen that actuator and sensor saturation are permissible nonlinearities which will not destroy the robust stability property.

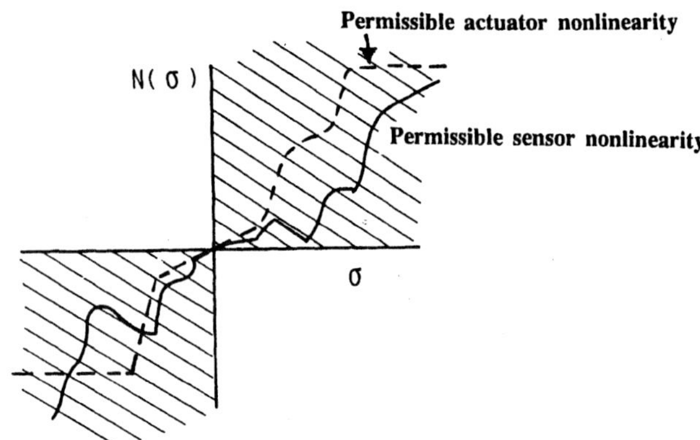

Figure 75.23 Permissible actuator and sensor nonlinearities.

Some methods for designing static dissipative controllers are discussed in [12]. In particular, the static dissipative control law minimizes the quadratic performance function,

$$\mathcal{J} = \int_0^\infty \Big[y_p^T G_p G_r^{-1} G_p y_p + y_r^T G_r y_r + 2\dot{q}^T D\dot{q} + 2y_p^T G_p G_r^{-1} u + u^T G_r^{-1} u \Big] dt.$$

This performance function can be used as a basis for controller design. Another approach for selecting gains is to minimize the norms of the differences between the actual and desired values of the closed-loop coefficient matrices, $(\tilde{B} + \Gamma^T G_r \Gamma)$ and $(\tilde{C} + \Gamma^T G_p \Gamma)$ (see [12]).

The performance of static dissipative controllers is inherently limited because of their restrictive structure. Furthermore, direct output feedback allows the addition of unfiltered sensor noise to the input. These difficulties can be overcome by using dynamic dissipative controllers, which also offer more design freedom and potentially superior performance.

Dynamic Dissipative Controllers

The positive realness of $G'(s)$ also permits robust stabilization of $G(s)$ by a class of dynamic compensators. The following definition is needed to define this class of controllers.

DEFINITION 75.2 A rational matrix-valued function $T(s)$ of the complex variable s is said to be marginally strictly positive-real (MSPR) if $T(s)$ is PR, and $T(j\omega) + T^*(j\omega) > 0$ for $\omega \in (-\infty, \infty)$.

This definition of MSPR matrices is a weaker version of the definitions of "strictly positive-real (SPR)" matrices which have appeared in the literature [18]. The main difference is that MSPR matrices can have poles on the imaginary axis. It has been shown in [13] that the negative feedback connection of a PR system and an MSPR system is stable, i.e., the composite system consisting of minimal realizations of the two systems is asymptotically stable (or equivalently, one system "stabilizes" the other system).

Consider an $m \times m$ controller transfer function matrix $\mathcal{K}(s)$ which uses the position sensor output $y_p(t)$ to generate the input $u(t)$. The following sufficient condition for stability is proved in [14].

Stability with Dynamic Dissipative Controller

Suppose that the controller transfer function $\mathcal{K}(s)$ has no transmission zeros at the origin and that $C(s) = \mathcal{K}(s)/s$ is MSPR. Then $\mathcal{K}(s)$ stabilizes the plant $G(s)$.

This condition can also be stated in terms of $[A_k, B_k, C_k, D_k]$, a minimal realization of $\mathcal{K}(s)$ [14]. The stability property depends only on the positive realness of $G'(s)$, which is a consequence of actuator-sensor collocation. Therefore, just as in the case of the static dissipative controller, the stability property holds regardless of truncated modes or the knowledge of the parameters. Controllers which satisfy the above property are said to belong to the class called "dynamic dissipative" controllers.

The condition that $\mathcal{K}(s)/s$ be MSPR is generally difficult to test. However, if $\mathcal{K}(s)$ is restricted to be diagonal, i.e., $\mathcal{K}(s) = diag[\mathcal{K}_1(s), \ldots, \mathcal{K}_m(s)]$, the condition is easier to check. For example, for the diagonal case, let

$$\mathcal{K}_i(s) = k_i \frac{s^2 + \beta_{1i}s + \beta_{0i}}{s^2 + \alpha_{1i}s + \alpha_{0i}}. \qquad (75.131)$$

It is straightforward to show that $\mathcal{K}(s)/s$ is MSPR if, and only if, (for $i = 1, \ldots, m$), $k_i, \alpha_{0i}, \alpha_{1i}, \beta_{0i}, \beta_{1i}$ are positive,

$$\alpha_{1i} - \beta_{1i} > 0 \qquad (75.132)$$

$$\alpha_{1i}\beta_{0i} - \alpha_{0i}\beta_{1i} > 0 \qquad (75.133)$$

For higher order $\mathcal{K}_i$s, the conditions on the polynomial coefficients are harder to obtain. One systematic procedure for obtaining such conditions for higher order controllers is the application of Sturm's theorem [24]. Symbolic manipulation codes can then be used to derive explicit inequalities similar to Equations 75.132 and 75.133. Using such inequalities as constraints,the controller design problem can be posed as a constrained optimization problem which minimizes a given performance function. An example design of a dynamic dissipative controller for the hoop/column antenna concept is presented in [14], wherein diagonal $\mathcal{K}(s)$ is assumed. For the case of fully populated $\mathcal{K}(s)$, however, there are no straightforward methods and it remains an area of future research.

The preceding stability result for dynamic dissipative controllers can be used to show that the robust stability property of the static dissipative controller is maintained even when the actuators have a finite bandwidth:

1. For the static dissipative controller (Equation 75.129), suppose that G_p and G_r are diagonal with positive entries (denoted by subscript i), and that actuators represented by the transfer function $G_{Ai}(s) = \frac{k_i}{(s+a_i)}$ are present in the ith control channel. Then the closed-loop system is asymptotically stable if $G_{ri} > G_{pi}/a_i$ (for $i = 1, \ldots, m$).
2. Suppose that the static dissipative controller also includes the feedback of the acceleration $y_a (= \Gamma\ddot{\xi})$, that is,

$$u = -G_p y_p - G_r y_r - G_a y_a$$

where G_p, G_r, and G_a are diagonal with positive entries. Suppose that the actuator dynamics for the ith input channel are given by $G_{Ai}(s) = k_i/(s^2 + \mu_i s + \nu_i)$, with k_i, μ_i, ν_i positive. Then the closed-loop system is asymptotically stable if

$$\frac{G_{ri}}{G_{ai}} \leq \mu_i < \frac{G_{ri}}{G_{pi}} \quad (i = 1, .., m).$$

Because of the requirement that $\mathcal{K}(s)/s$ be MSPR, the controller $\mathcal{K}(s)$ is not strictly proper. From a practical viewpoint, it is sometimes desirable to have a strictly proper controller because it attenuates sensor noise as well as high-frequency disturbances. Furthermore, the most common types of controllers, which include the LQG as well as the observer/pole placement controllers, are strictly proper (they have a first-order rolloff). It is possible to realize $\mathcal{K}(s)$ as a strictly proper controller wherein both y_p and y_r are utilized for feedback. Let $[A_k, B_k, C_k, D_k]$ be a minimal realization of $\mathcal{K}(s)$ where C_k is of full rank.

Strictly Proper Dissipative Controller

The plant with y_p and y_r as outputs is stabilized by the controller given by

$$\dot{x}_k = A_k x_k + \left[\, B_k - A_k L \quad L \,\right] \begin{bmatrix} y_p \\ y_r \end{bmatrix}, \qquad (75.134)$$

$$y_k = C_k x_k, \qquad (75.135)$$

where the $n_k \times m(n_k \geq m)$ matrix L is a solution of

$$D_k - C_k L = 0. \qquad (75.136)$$

Equation 75.136 represents m^2 equations in mn_k unknowns. If $m < n_k$ (i.e., the compensator order is greater than the number of plant inputs), there are many possible solutions for L. The solution which minimizes the Frobenius norm of L is

$$L = C_k^T (C_k C_k^T)^{-1} D_k. \qquad (75.137)$$

If $m = n_k$, Equation 75.136 gives the unique solution, $L = C_k^{-1} D_k$.

The next section addresses another important class of systems, namely, multibody flexible systems, which are described by nonlinear mathematical models.

75.3.3 Multibody Flexible Space Systems

This section considers the problem of controlling a class of nonlinear multibody flexible space systems consisting of a flexible central body with a number of articulated appendages. A complete nonlinear rotational dynamic model of a multibody flexible spacecraft is considered. It is assumed that the model configuration consists of a branched geometry, i.e., it has a central flexible body to which various flexible appendage bodies are attached (Figure 75.24). Each branch by itself can be a serial chain of structures. The actuators and sensors are assumed to be collocated. The global asymptotic stability of such systems is established using a nonlinear feedback control law. In many applications, the central body has a large mass and moments of inertia as compared to any other appendage bodies. As a result, the motion of the central body is small and can be assumed to be in the linear range. For this special case, the robust stability results are given for linear static as well as dynamic dissipative compensators. The effects of realistic nonlinearities in the actuators and sensors are also considered.

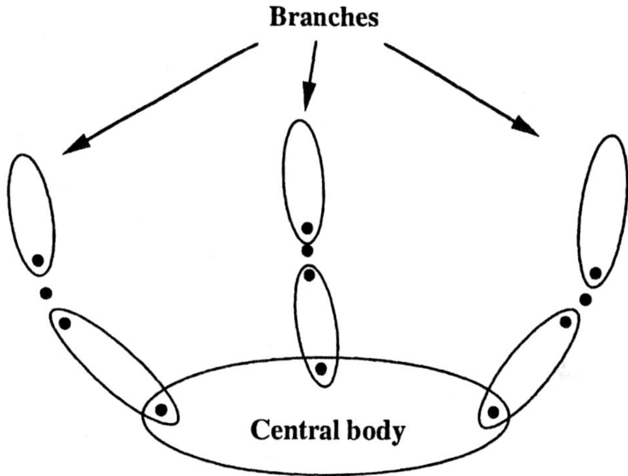

Figure 75.24 Multibody flexible system.

Mathematical Model

Equations of Motion

Consider a spacecraft (shown in Figure 75.24) consisting of a central flexible body and a chain of $(k - 3)$ flexible links. (Although a single chain is considered here, all the results are also valid when several chains are present.) Using the Lagrangian formulation, the following equations of motion can be obtained (see [15] for details):

$$M(p)\ddot{p} + C(p, \dot{p})\dot{p} + D\dot{p} + Kp = B^T u \qquad (75.138)$$

where $\dot{p} = \{\omega^T, \dot{\theta}^T, \dot{q}^T\}^T$; ω is the 3×1 inertial angular velocity vector (in body-fixed coordinates) for the central body; $\theta = (\theta_1, \theta_2, .., \theta_{(k-3)})^T$, where θ_i denotes the joint angle for the ith joint expressed in body-fixed coordinates; q is the $(n - k)$ vector of flexible degrees of freedom (modal amplitudes); $p = (\gamma^T, \theta^T, q^T)^T$, and $\dot{\gamma} = \omega$; $M(p) = M^T(p) > 0$ is the configuration-dependent, mass-inertia matrix, and $\tilde{K}$ is the symmetric positive-definite stiffness matrix related to the flexible degrees of freedom. $C(p, \dot{p})$ corresponds to Coriolis and centrifugal forces; D is the symmetric, positive semidefinite damping matrix; $B = [\, I_{k \times k} \quad 0_{k \times (n-k)} \,]$ is the control influence matrix and u is the k-vector of applied torques. The first three components of u represent the attitude control torques (about the X-, Y-, and Z- axes) applied to the central body, and the remaining components are the torques applied at the $(k - 3)$ joints. K and D are symmetric, positive-semidefinite stiffness and damping matrices,

$$K = \begin{bmatrix} 0_{k\times k} & 0_{k\times(n-k)} \\ 0_{(n-k)\times k} & \tilde{K}_{(n-k)\times(n-k)} \end{bmatrix}$$
$$D = \begin{bmatrix} 0_{k\times k} & 0_{k\times(n-k)} \\ 0_{(n-k)\times k} & \tilde{D}_{(n-k)\times(n-k)} \end{bmatrix} \qquad (75.139)$$

where $\tilde{K}$ and $\tilde{D}$ are symmetric positive definite. The angular measurements for the central body are Euler angles (not the vector γ), whereas the remaining angular measurements between bodies are relative angles. In deriving equations of motion, it is assumed that the elastic displacements are small enough to be in the linear range and that the total displacement can be obtained by the principle of superposing rigid and elastic motions. One important inherent property (which will be called "Property $\mathcal{S}$") of such systems crucial to the stability results is given next.

Property $\mathcal{S}$: For the system represented by Equation 75.138, the matrix $(\frac{1}{2}\dot{M} - C)$ is skew-symmetric.

The justification of this property can be found in [15].

The central body attitude (Euler angle) vector η is given by $E(\eta)\dot{\eta} = \omega$, where $E(\eta)$ is a 3×3 transformation matrix [8]. The sensor outputs consist of three central body Euler angles, the $(k - 3)$ joint angles, and the angular rates, i.e., the sensors are collocated with the torque actuators. The sensor outputs are then given by

$$y_p = B\hat{p} \quad \text{and} \quad y_r = B\dot{p} \qquad (75.140)$$

where $\hat{p} = (\eta^T, \theta^T, q^T)^T$ in which η is the Euler angle vector for the central body. $y_p = (\eta^T, \theta^T)^T$ and $y_r = (\omega^T, \dot{\theta}^T)^T$ are measured angular position and rate vectors, respectively. It is assumed that the body rate measurements, ω, are available from rate gyros. Using Property $\mathcal{S}$, it can be proved that the operator from u to y_r is passive [14].

Quaternion as a Measure of Attitude

The orientation of a free-floating body can be minimally represented by a three-dimensional orientation vector. However, this representation is not unique. One minimal representation commonly used to represent the attitude is Euler angles. The 3×1 Euler angle vector η is given by $E(\eta)\dot{\eta} = \omega$, where $E(\eta)$ is a 3×3 transformation matrix. $E(\eta)$ becomes singular for certain values of η. However, the limitations imposed on the allowable orientations due to this singularity are purely mathematical without physical significance. The problem of singularity in three-parameter representation of attitude has been studied in detail in the literature. An effective way of overcoming the singularity problem is to use the quaternion formulation (see [10]).

The unit quaternion α is defined as follows.

$$\alpha = \{\overline{\alpha}^T, \alpha_4\}^T, \quad \overline{\alpha} = \begin{bmatrix} \hat{\alpha}_1 \\ \hat{\alpha}_2 \\ \hat{\alpha}_3 \end{bmatrix} sin(\tfrac{\phi}{2}),$$
$$\alpha_4 = cos(\tfrac{\phi}{2}). \qquad (75.141)$$

$\hat{\alpha} = (\hat{\alpha}_1, \hat{\alpha}_2, \hat{\alpha}_3)^T$ is the unit vector along the eigen-axis of rotation and ϕ is the magnitude of rotation. The quaternion is also subjected to the norm constraint,

$$\overline{\alpha}^T\overline{\alpha} + \alpha_4^2 = 1. \qquad (75.142)$$

It can be also shown [10] that the quaternion obeys the following kinematic differential equations:

$$\dot{\overline{\alpha}} = \tfrac{1}{2}(\omega \times \overline{\alpha} + \alpha_4\omega), \quad \text{and} \qquad (75.143)$$
$$\dot{\alpha}_4 = -\tfrac{1}{2}\omega^T\overline{\alpha}. \qquad (75.144)$$

The quaternion representation can be effectively used for the central body attitude. The quaternion can be computed [10] using Euler angle measurements (Equation 75.140). The open-loop system, given by Equations 75.138, 75.143, and 75.144, has multiple equilibrium solutions: $(\overline{\alpha}_{ss}{}^T, \alpha_{4ss}, \theta_{ss}^T)^T$ where the subscript 'ss' denotes the steady-state value (the steady-state value of q is zero). Defining $\beta = (\alpha_4 - 1)$ and denoting $\dot{p} = v$, Equations 75.138, 75.143, and 75.144 can be rewritten as

$$M\dot{v} + Cv + Dv + \tilde{K}q = B^T u \qquad (75.145)$$
$$\begin{bmatrix} \dot{\theta} \\ \dot{q} \end{bmatrix} = \begin{bmatrix} 0_{(n-3)\times 3} & I_{(n-3)} \end{bmatrix} v \qquad (75.146)$$
$$\dot{\overline{\alpha}} = \tfrac{1}{2}(\omega \times \overline{\alpha} + (\beta + 1)\omega) \qquad (75.147)$$
$$\dot{\beta} = -\tfrac{1}{2}\omega^T\overline{\alpha} \qquad (75.148)$$

In Equation 75.145 the matrices M and C are functions of p, and $(p, \dot{p})$, respectively. It should be noted that the first three elements of p associated with the orientation of the central body can be fully described by the unit quaternion. The system represented by Equations 75.145–75.148 can be expressed in the state-space form as follows:

$$\dot{x} = f(x, u) \qquad (75.149)$$

where $x = (\overline{\alpha}^T, \beta, \theta^T, q^T, v^T)^T$. Note that the dimension of x is $(2n + 1)$, which is one more than the dimension of the system in Equation 75.138. However, one constraint (Equation 75.142) is now present. It can be verified from Equations 75.143 and 75.144 that the constraint (Equation 75.142) is satisfied for all $t > 0$ if it is satisfied at $t = 0$.

A Nonlinear Feedback Control Law

Consider the dissipative control law u, given by

$$u = -G_p\tilde{p} - G_r y_r \qquad (75.150)$$

where $\tilde{p} = \{\overline{\alpha}^T, \theta^T\}^T$. Matrices G_p and G_r are symmetric positive-definite $(k \times k)$ matrices and G_p is given by

$$G_p = \begin{bmatrix} (1 + \frac{(\beta+1)}{2})G_{p1} & 0_{3\times(k-3)} \\ 0_{(k-3)\times 3} & G_{p2(k-3)\times(k-3)} \end{bmatrix}. \qquad (75.151)$$

Note that Equations 75.150 and 75.151 represent a nonlinear control law. If G_p and G_r satisfy certain conditions, this control law renders, the time rate of change of the system's energy negative along all trajectories; i.e., it is a 'dissipative' control law.

The closed-loop equilibrium solution can be obtained by equating all the derivatives to zero in Equations 75.138, 75.147, and 75.148. After some algebraic manipulations, there appear to be two equilibrium points in the state space. However, it can be shown [14] that they refer to a single equilibrium point.

If the objective of the control law is to transfer the state of the system from one orientation (equilibrium) position to another without loss of generality, the target orientation can be defined as zero and the initial orientation, given by $[\overline{\alpha}(0), \alpha_4(0), \theta(0)]$, can always be defined so that $|\theta_i(0)| \leq \pi, 0 \leq \alpha_4(0) \leq 1$ (corresponding to $|\phi| \leq \pi$) and $[\overline{\alpha}(0), \alpha_4(0)]$ satisfy Equation 75.142.

The following stability result is proved in [14].

Stability Result: Suppose that $G_{p2_{(k-3)\times(k-3)}}$ and $G_{r\,(k\times k)}$ are symmetric and positive definite, and $G_{p1} = \mu I_3$, where $\mu > 0$. Then, the closed-loop system given by Equations 75.149–75.151 is globally asymptotically stable (g.a.s.).

This result states that the control law in Equation 75.150 stabilizes the nonlinear system despite unmodeled elastic mode dynamics and parametric errors, that is, a multibody spacecraft can be brought from any initial state to the desired final equilibrium state. The result also applies to a particular case, namely, single-body FSS, that is, this control law can bring a rotating spacecraft to rest or perform robustly stable, large-angle, rest-to-rest maneuvers. A generalization of the control law in Equations 75.150–75.151 to the case with fully populated G_{p1} matrix is given in [16].

The next section considers a special case of multibody systems.

Systems in Attitude-Hold Configuration

Consider a special case where the central body attitude motion is small. This can occur in many realistic situations. For example, in the case of a space-station-based or shuttle-based manipulator, the moments of inertia of the base (central body) are much larger than that of any manipulator link or payload. In such cases the rotational motion of the base can be assumed to be in the linear region, although the payloads (or links) attached to it can undergo large rotational and translational motions and nonlinear dynamic loading due to Coriolis and centripetal accelerations. For this case, the attitude of the central body is simply γ, the integral of the inertial angular velocity ω, and the use of quaternions is not necessary. The equations of motion (Equation 75.138) can now be expressed in the state-space form as

$$\dot{\overline{x}} = \begin{bmatrix} 0 & I \\ -M^{-1}K & -M^{-1}(C+D) \end{bmatrix}\overline{x} + \begin{bmatrix} 0 \\ M^{-1}B^T \end{bmatrix} u \tag{75.152}$$

where $\overline{x} = \{p^T, \dot{p}^T\}^T$, and $p = \{\gamma^T, \theta^T, q^T\}^T$. Note that M and C are functions of $\overline{x}$, and hence the system is nonlinear.

Stability with Dissipative Controllers

The static dissipative control law u is given by

$$u = -\overline{G}_p y_p - G_r y_r \tag{75.153}$$

where $\overline{G}_p$ and G_r are constant symmetric positive definite $(k \times k)$ matrices,

$$y_p = Bp \quad and \quad y_r = B\dot{p}. \tag{75.154}$$

Where y_p and y_r are measured angular position and rate vectors. The following result is proved in [14].

Stability Result: Suppose that $\overline{G}_{p_{k\times k}}$ and $G_{r_{k\times k}}$ are symmetric and positive definite. Then the closed-loop system given by Equations 75.152, 75.153, and 75.154 is globally asymptotically stable.

The significance of the two stability results presented in this section is that any nonlinear multibody system belonging to these classes can be robustly stabilized with the dissipative control laws given. In the case of manipulators, this means that one can accomplish any terminal angular position (of the links) from any initial position with guaranteed asymptotic stability. Furthermore, for the static dissipative case, the stability result also holds when the actuators and sensors have nonlinearities. In particular, the stability result of the Section on page 1321 for the linear, single-body FSS also extends to the case of nonlinear flexible multibody systems in attitude-hold configuration, that is, the closed-loop system with the static dissipative controller is globally asymptotically stable in the presence of monotonically nondecreasing $(0, \infty)$-sector actuator nonlinearities, and $(0, \infty)$-sector sensor nonlinearities. This result is proved in [14].

For the more general case where the central body motion is not in the linear range, the robust stability in the presence of actuator/sensor nonlinearities cannot be easily extended because the stabilizing control law (Equation 75.150) is nonlinear.

The robust stability with dynamic dissipative controllers (Section 75.3.2) can also be extended to the multibody case (in attitude-hold configuration). As stated previously, the advantages of using dynamic dissipative controllers include higher performance, more design freedom, and better noise attenuation.

Consider the system given by Equation 75.152 with the sensor outputs given by Equation 75.154. As in the linear case, consider a $k \times k$ controller $\mathcal{K}(s)$ which uses the angular position vector $y_p(t)$ to produce the input $u(t)$. The closed-loop system consisting of nonlinear plant (Equation 75.152) and the controller $\mathcal{K}(s)$ is shown in Figure 75.25. $\mathcal{K}(s)$ is said to stabilize the nonlinear plant if the closed-loop system is globally asymptotically stable (with $\mathcal{K}(s)$ represented by its minimal realization). The conditions under which $\mathcal{K}(s)$ stabilizes the nonlinear plant are the same as the linear, single-body case, discussed in the Section on page 1322, that is, the closed-loop system in Figure 75.25 is g.a.s. if $\mathcal{K}(s)$ has no transmission zeros at $s = 0$, and $C(s) = \frac{\mathcal{K}(s)}{s}$ is MSPR. A proof of this result is given in [14]. The proof does not make any assumptions regarding the model order or the knowledge of the parametric values. Hence, the stability is robust to modeling errors and parametric uncertainties. As shown in Section 75.3.2, this controller can also be realized as a strictly proper controller that utilizes the feedback of both y_p and y_r.

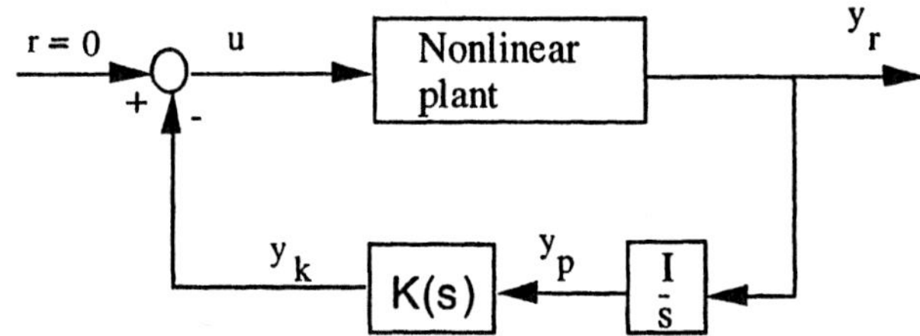

Figure 75.25 Stabilization with dynamic dissipative controller.

75.3.4 Summary

This chapter has provided an overview of various control issues for flexible space structures (FSS), which are classified into single-body FSS or multibody FSS. The first problem considered is fine attitude pointing and vibration suppression of single-body FSS, which can be formulated in the linear, time-invariant setting. For this problem, two types of controllers, model-based and passivity-based ("dissipative"), are discussed. The robust stability properties of dissipative controllers are highlighted and design techniques are discussed. For the case of multibody FSS, a generic nonlinear mathematical model is presented and is shown to have a certain passivity property. Nonlinear and linear dissipative control laws which provide robust global asymptotic stability are presented.

References

[1] Anderson, B. D. O., A System Theory Criterion for Positive-Real Matrices, *SIAM J. Control*, 5, 1967.

[2] Balakrishnan, A. V., Combined Structures-Controls Optimization of Lattice Trusses, *Computer Methods Appl. Mech. Eng.*, 94, 1992.

[3] Balas, G. J. and Doyle, J. C., Control of Lightly Damped Flexible Modes in the Controller Crossover Region, *J. Guid. Control Dyn.*, 17(2), 370–377, 1994.

[4] Balas, M. J., Trends in Large Space Structures Control Theory: Fondest Hopes, Wildest Dreams, *IEEE Trans. Automat. Control*, AC-27(3), 1982.

[5] Desoer, C. A. and Vidyasagar, M., *Feedback Systems: Input-Output Properties*, Academic, New York, 1975.

[6] Doyle, J. C., Analysis of Feedback Systems With Structured Uncertainty, *IEEE Proc.*, 129D(6), 1982.

[7] Green, M. and Limebeer, D. J. N., *Linear Robust Control*, Prentice Hall, Englewood Cliffs, NJ, 1995.

[8] Greenwood, D. T., *Principles of Dynamics*, Prentice Hall, Englewood Cliffs, NJ, 1988.

[9] Gupta, S., *State Space Characterization and Robust Stabilization of Dissipative Systems*, D.Sc. Thesis, George Washington University, 1994.

[10] Haug, E. G., *Computer-Aided Kinematics and Dynamics of Mechanical Systems*, Allyn and Bacon Series in Engineering, 1989.

[11] Hyland, D. C., Junkins, J. L., and Longman, R. W., Active Control Technology for Large Space Structures, *J. Guid. Control Dyn.*, 16(5), 1993.

[12] Joshi, S. M., *Control of Large Flexible Space Structures, Lecture Notes in Control and Information Sciences*, Springer, Berlin, 1989, Vol. 131.

[13] Joshi, S. M. and Gupta, S., Robust Stabilization of Marginally Stable Positive-Real Systems. NASA TM-109136, 1994.

[14] Joshi, S. M., Kelkar, A. G., and Maghami, P. G., A Class of Stabilizing Controllers for Flexible Multibody Systems. NASA TP-3494, 1995.

[15] Kelkar, A. G., Mathematical Modeling of a Class of Multibody Flexible Space Structures. NASA TM-109166, 1994.

[16] Kelkar, A. G. and Joshi, S. M., *Global Stabilization of Multibody Spacecraft Using Quaternion-Based Nonlinear Control Law*, Proc. Am. Control Conf., Seattle, Washington, 1995.

[17] Lim, K. B., Maghami, P. G., and Joshi, S. M., Comparison of Controller Designs for an Experimental Flexible Structure, *IEEE Control Syst. Mag.*, 3, 1992.

[18] Lozano-Leal, R. and Joshi, S. M., Strictly Positive Real Functions Revisited, *IEEE Trans. Automat. Control*, 35(11), 1243–1245, 1990.

[19] Meirovitch, L., *Methods of Analytical Dynamics*, MacGraw-Hill, New York, 1970.

[20] Nurre, G. S., Ryan, R. S., Scofield, H. N., and Sims, J. L., Dynamics and Control of Large Space Structures, *J. Guid. Control Dyn.*, 7(5), 1984.

[21] Packard, A. K., Doyle, J. C., and Balas, G. J., Linear Multivariable Robust Control With a μ-Perspective, *ASME J. Dyn. Meas. & Control*, 115(2(B)), 426–438, 1993.

[22] Russell, R. A., Campbell, T. G., and Freeland, R. E., A Technology Development Program for Large Space Antenna, NASA TM-81902, 1980.

[23] Stein, G. and Athans, M., The LQG/LTR Procedure for Multivariable Feedback Control Design, *IEEE Trans. Automat. Control*, 32(2), 105–114, 1987.

[24] Van Valkenberg, M. E., *Introduction to Modern Network Synthesis*, John Wiley & Sons, New York, 1965.

75.4 Line-of-Sight Pointing and Stabilization Control System

David Haessig, GEC-Marconi Systems Corporation, Wayne, NJ

75.4.1 Introduction

To gain some insight into the functions of line-of-sight pointing and stabilization, consider the human visual system. It is quite apparent that we can control the direction of our eyesight, but few are aware that they have also been equipped with a stabilization capability. The vestibulo-occular reflex is a physiological mechanism which acts to fix the direction of our eyes inertially when the head is bobbling about for whatever reason [1]. (Try shaking your head rapidly while viewing this text and note that you can fix your eyesight on a particular word.) Its importance becomes obvious when you imagine life without it. Catching a football while running would be next to impossible. Reading would be difficult while traveling in a car over rough road, and driving dangerous. Our vision would appear jittery and blurry. This line-of-sight pointing and stabilization mechanism clearly improves and expands our capabilities.

Similarly, many man-made systems have been improved by adding pointing and stabilization functionality. In the night vision systems used extensively during the Persian Gulf War, image clarity and resolution, and their precision strike capability, were enhanced by the pointing and stabilization devices they contained.

The Falcon Eye System [11] was developed in the late 1980s to build upon the strengths of the night vision systems developed earlier and used extensively during Desert Storm. The Falcon Eye differs from previous night vision systems developed for fighter aircraft in that it is head-steerable. It provides a high degree of night situational awareness by allowing the pilot to look in any direction including directly above the aircraft. This complicates the control system design problem because not only must the system isolate the line of sight from image blurring angular vibration, it also must simultaneously track pilot head motion. Nevertheless, this system has been successfully designed, built, extensively flight tested on a General Dynamics F-16 Fighting Falcon, and shown to markedly improve a pilot's ability to find and attack fixed or mobile targets at night.

The Falcon Eye System is pictured in Figures 75.26 and 75.27. It consists of (1) a helmet-mounted display with dual optical combiners suspended directly in front of the pilot's eyes to permit a merging of **Flir** (forward looking infrared) imagery with external light, (2) a head angular position sensor consisting of a magnetic sensor attached to the helmet and coupled to another attached to the underside of the canopy, and (3) a Flir sensor providing three-axis control of the line-of-sight orientation and including a two-axis fine stabilization assembly. The infrared image captured by this sensor is relayed to the pilot's helmet-mounted display, providing him with a realistic Flir image of the outside world. The helmet orientation serves as a commanded reference that the Flir line of sight must follow.

75.4.2 Overall System Performance Objectives

The system's primary design goal was to provide a night vision capability that works and feels as much like natural daytime vision as possible. This means the pilot must be able to turn his head to look in any direction. Also, the Flir scene cannot degrade due to the high frequency angular vibration present in the airframe where the equipment is mounted.

The Falcon Eye Flir attempts to achieve these effects by serving as a buffer between the vehicle and the outside world, which must (1) track the pilot's head with sufficient accuracy so that lag or registration errors between the Flir scene and the outside world scene are imperceptible and (2) isolate the Flir sensor from vehicle angular vibration which can cause the image to appear blurry (in a single image frame) and jittery (jumping around from frame to frame).

The pilot can select either a 1:1 or a 5.6:1 level of scene magnification. Switching to the magnified view results in a reduction in the size of the scene from 22 × 30 degrees to the very narrow 4 × 4.5 degrees, hence the name *narrow field of view.* When in narrow field-of-view mode, the scene resolution is much finer, 250 μrads, and therefore the stabilization requirements are tighter. This chapter focuses on the more difficult task of designing a controller for the stabilization assembly in narrow field-of-view mode.

Control System Structure

The Falcon Eye's field of regard is equal to that of the pilot's. Consequently, the system must precisely control the orientation of the line of sight over a very large range of angles. It is difficult to combine fine stabilization and large field-of-regard capabilities in a single mechanical control effector. Therefore, these tasks are separated and given to two different parts of the overall system (see Figure 75.28). A coarse gimbal set accomplishes the task of achieving a large field of regard. A fine stabilization assembly attached to the inner gimbal acts as a vernier which helps to track commands and acts to isolate the line of sight from angular vibration, those generated by the vehicle and those generated within the coarse gimbal set. The coarse positioning gimbal system tracks pilot head position commands, but will fall short because of its limited dynamic response and because of the effects of torque disturbances within the gimbal drive systems. These torque disturbances include stiction, motor torque ripple and cogging, bearing torque variations, and many others. The stabilization system assists in tracking the commanded head position by (1) isolating the line-of-sight from vehicle vibration and (2) reducing the residual tracking error left by the gimbal servos. The reason for sending the command rates rather than the command angles to the stabilization system will be covered in the description of the stabilization system controller.

Control System Performance Requirements

Ideally the Flir sensor's line of sight should perfectly track the pilot's line of sight. The control system must therefore limit the difference between these two pointing directions. (Recognize that the pilot's line of sight is defined here by the orientation of his head, not that of his eyes. A two-dimensional scene appears in the display and he is free to direct his eyes at any particular point in that scene, much like natural vision.)

Two quantities precisely define the pilot's line of sight in inertial space: vehicle attitude relative to inertial space θ_a and pilot head attitude relative to the vehicle ϕ_p. (θ is used for inertial variables and ϕ for relative variables.) This is depicted in one dimension in Figure 75.29, i.e., $\theta_p = \phi_p + \theta_a$.

The Flir, although connected to the aircraft, is physically separated from the pilot by a nonrigid structure which can exhibit vibrational motion relative to the pilot's location. This relative angular motion will be referred to as ϕ_v. The Flir's line of sight θ_f equals the vehicle attitude θ_a plus vehicle vibration ϕ_v, plus the Flir line of sight relative to the vehicle ϕ_f, i.e., $\theta_f = \phi_f + \phi_v + \theta_a$. The difference between the two is the error,

$$e = \theta_f - \theta_p = \phi_f - \phi_p + \phi_v. \tag{75.155}$$

As Equation 75.155 indicates, for this error to equal zero, the Flir must track pilot head motion and move antiphase to vehicle vibration. Head orientation is therefore accurately described as a commanded reference input, and vehicle vibration as a disturbance.

Figure 75.26 The Falcon Eye Flir, consisting of a helmet-mounted display, a head position sensor, a Flir optical system, and the line-of-sight position control system partially visible just forward of the canopy. (By courtesy of Texas Instruments, Defense Systems & Electronics Group.).

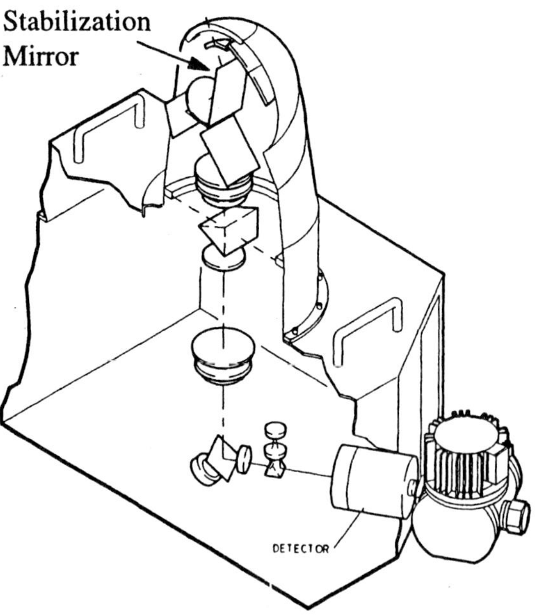

Figure 75.27 The Falcon Eye gimbals follow pilot head motion. An image stabilization assembly attached inside the ball helps to follow the pilot's head and isolates the line of sight from angular vibration.

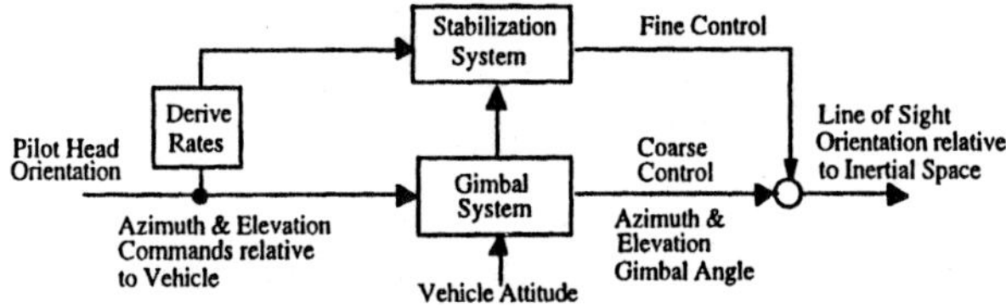

Figure 75.28 Partitioning of the control responsibility: the gimbals track commands; the stabilization system helps to track commands and stabilizes the image.

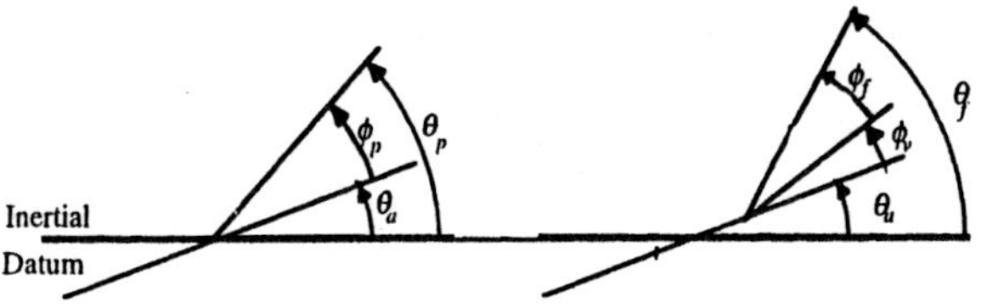

Figure 75.29 Tracking of the pilot's line of sight involves isolation from vehicle vibration ϕ_v, tracking of vehicle maneuvering θ_a, and tracking of pilot head motion relative to the vehicle ϕ_p.

The ideal response to various inputs to the system is shown in Figure 75.30. The output is the Flir line of sight angle relative to inertial space. The inputs consist of vehicle angle relative to inertial space at the Flir ($\theta_v = \theta_a + \phi_v$), pilot head commanded angle relative to the vehicle, and a third input not mentioned thus far, the tracking error due to torque disturbances generated within the gimbals. The goal is to point the Flir sensor's line of sight continuously in the same direction as the pilot's. Therefore, pilot head commands are ideally tracked perfectly, as the upper plot indicates. However, because these commands are relative to the vehicle, the vehicle maneuver motion θ_a, which is generally low frequency, must also be tracked, as the lower plot depicts. At higher frequencies, vehicle angular vibration ϕ_v injects an input that affects the Flir sensor but not the pilot. So, as the lower plot indicates, the ideal response to vehicle vibration is complete attenuation. Between these frequency regimes there is a transition region. In this application, maneuver tracking must occur for frequencies up to about 0.3 Hz. The transition occurs between 0.3 and about 3 Hz, and good isolation performance is needed in the 3 to 30 Hz range. Beyond that, because there is little vibration energy at the higher frequencies, the transfer function from vehicle motion to line-of-sight motion can increase without significant penalty.

The center curve of Figure 75.30 is the ideal system response to torque disturbances generated within the gimbals. Ideally, internally generated tracking errors should be completely attenuated by the stabilization assembly. Thus the magnitude of this transfer function is zero, as shown. This is implied by the perfect command tracking ideal response depicted in the upper plot. However, it is shown separately to distinguish and clearly identify these two characteristics.

The ideal performance requirements discussed thus far serve primarily to guide the specification of realistic, quantifiable performance requirements defining allowable responses to specific

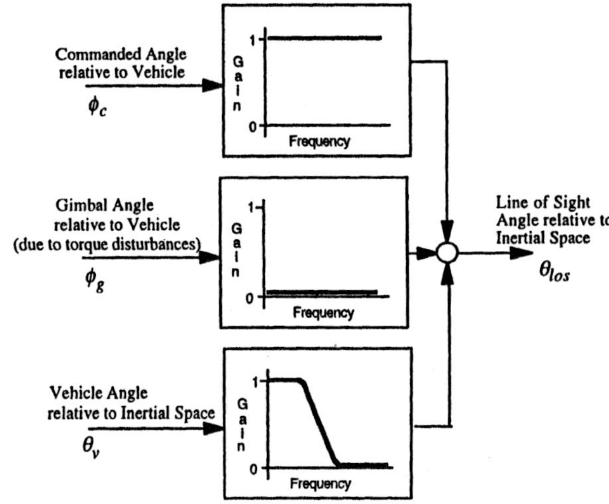

Figure 75.30 The response of an ideal line-of-sight control system to the various inputs.

input environments. For the Falcon Eye these fall into 4 areas: (1) stabilization error, the high frequency line-of-sight motion caused by vehicle vibration or other disturbances, (2) tracking smoothness, the tracking error due to internally generated torque disturbances, (3) command tracking, the error in tracking specifically defined head motion commands, and (4) maneuver tracking, the tracking error that occurs in response to specific maneuvers. In this chapter our evaluation of performance will cover the first two in detail.

Stabilization Requirement Vehicle angular vibration produces line-of-sight motion at higher frequencies (above 5 Hz) that can degrade the resolution of the image. Resolution is not impaired by vibration, but is limited by the size of the pixel associated with the infrared (IR) detector, if the line of sight *rms* motion due to vibration is less than 1/4 the size of the pixel, which in the Falcon Eye is 250 μrads in narrow field of view. The image stabilization assembly must allow no more than 62 μrads rms of line-of-sight motion when subjected to vehicle angular vibration.

Tracking Smoothness Requirement Testing has indicated that a pilot can see and is bothered by a jump or jitter in the image at intermediate frequencies (2 to 10 Hz) when they approach or exceed the size of one pixel, i.e., 250 μrads. Therefore, a smoothness of tracking requirement is applied which limits the peak-to-peak tracking error to less then 250 μrads in the 2 to 10 Hz range when following constant-rate head motion.

Command and Maneuver Tracking Requirements In the next section we will see that the stabilization system must provide command and maneuver tracking bandwidths of greater than 30 and 0.3 Hz, respectively.

75.4.3 Physical System Description and Modeling

The stabilization assembly, pictured in Figure 75.31, is rigidly attached to the inner gimbal of the coarse positioning system (see Figure 75.27). It produces small changes in the line-of-sight direction by changing the angular position of the mirror relative to the base structure. The mirror is attached to the base by a flexure hinge, a solid but flexible structure which permits angular motion only about two-axes and applies a small restoring spring torque. Rotation is limited to ± 5 μrads ($\pm 1/3°$), by hardstops under the mirror.

Between the mirror and the base there is a two-axis torquer which applies the control signals. A two-axis angular position sensor, referred to as the pickoff sensor, measures the angular position of the mirror relative to the base. A two-axis angular rate sensor attached rigidly in the base senses angular rate, including the vibration disturbance that the stabilization assembly must counter. The rate sensor is oriented to detect any rotation about axes perpendicular to the line-of-sight, because only those alter the line-of-sight direction.

While operating within the hardstops, each axis of the mirror is governed by the dynamic equation,

$$J_m\ddot{\theta}_m + \beta_m(\dot{\theta}_m - \dot{\theta}_b) + K_m(\theta_m - \theta_b) = \tau_m, \quad (75.156)$$

where θ_m and θ_b are the mirror and base inertial angular positions, respectively, and τ_m is the control torque. Other parameters are defined in Table 75.4. Unmodeled nonlinear dynamics such as centripetal, coriolis, and friction torques are small and neglected.

The variable that we are interested in controlling is the line-of-sight angle θ_{los}. This variable cannot be sensed directly but is related to other variables that are sensed. Consider the situation where the mirror is fixed to the base so that $\phi_m = \theta_m - \theta_b = 0$. Then the line of sight follows the base exactly: $\theta_{los} = \theta_b$. If the mirror rotates relative to the base and the base remains stationary, the line of sight moves λ times farther due to optical effects (λ is 2 for a typical mirror). Thus

$$\theta_{los} = \theta_b + \lambda\phi_m. \quad (75.157)$$

Perfect inertial stabilization would be achieved by driving the mirror position ϕ_m to $-\theta_b/\lambda$. However, not everything in θ_b is to be rejected, only vehicle vibration, not vehicle maneuvers.

Exogenous Input Environments

Mathematical descriptions of the input command and disturbance environments (exogenous inputs) are used in evaluating and designing the control system compensators. Models of the disturbance inputs serve as a basis for the structure of the design model and become embedded within the controller. The resulting designs were evaluated, analytically and through simulation, using these models to define or generate the inputs.

Vehicle Vibration Vehicle angular vibration at the location where the Flir equipment is mounted was expected to be equal or less than that defined by the solid line in the spectrum of Figure 75.32. It has an *rms* angular acceleration and angular position of 4.4 rad/sec^2 and 560 μrads, respectively. The bulk of its energy is concentrated around 10 Hz and it extends from 5 to 200 Hz. Angular motion contributed beyond 200 Hz is negligible in magnitude. Angular motion below 5 Hz falls in the jitter range and does not have a blurring effect.

A stationary process having the spectrum of Figure 75.32 can be generated by passing unit variance white noise through a shaping filter having the transfer function,

$$V_{\ddot{\theta}}(s) = K_v \frac{s^2(s/\alpha_1+1)(s/\alpha_2+1)}{(s/\alpha_3+1)(s/\alpha_4+1)(s/\alpha_5+1)(s^2/\Omega^2+2\zeta s/\Omega+1)}, \quad (75.158)$$

where $K_v = 2.22 \times 10^{-4}$, $\alpha_1 = 94$ rad/sec, $\alpha_2 = 377$ rad/sec, $\alpha_3 = 56.5$ rad/sec, $\alpha_4 = 188$ rad/sec, $\alpha_5 = 942$ rad/sec, $\Omega = 62.8$ rad/sec, and $\zeta = 0.5$. Twice integrating yields angular position,

$$V_\theta(s) = \frac{1}{s^2}V_{\ddot{\theta}}(s). \quad (75.159)$$

The image stabilization requirement of 65 μrads rms must be achieved when subjected to this input (see Figure 75.36).

Vehicle Maneuvering A precise mathematical definition of all input environments is not always available to the control system designer, or if one is available, it may not accurately represent the input environment that will exist when the equipment is in use. It may be necessary, as was the case here, to estimate the environment. The F-16 is capable of very high dynamic maneuvering. However, a pilot would not be doing those types of maneuvers when using this equipment in the 5.6:1 magnified, narrow field-of-view mode. (It is described to be like flying through a straw.) Under these conditions, maneuvers would be held to a minimum. Therefore, an assumption was made that the bulk of maneuvering motion energy, when in narrow field of view, would be concentrated at low frequencies, between D.C. and 0.3 Hz. This consideration will impact the design of the LQG (linear quadratic guassian) controller. By selecting the filter weighting matrices, the poles of the Kalman filter associated with the maneuver state will be positioned at locations having a magnitude of 0.3 Hz.

Pilot Head Motion Like maneuver motion, a good mathematical description of pilot head motion (when flying through the straw) was not available. One would think that the narrow field-of-view conditions would result in very little head motion, however, the pilots will glance down at their instruments, producing periods of moderately high rates of motion that must be tracked. Our assumption was that this head motion frequency content would extend out to about 3 or 4 Hz. Tight tracking of those inputs require bandwidths of 30 Hz or greater (well beyond the capabilities of the coarse positioning servo).

Self-Induced Torque Disturbances There are many sources of internally generated error present in the typical coarse line-of-sight pointing system. These include motor cogging (also called slot-lock or detent), torque ripple, stiction of seals and bearings, bearing retainer friction, bearing disturbance torques, etc. Friction or stiction in the gimbals causes a hang-off error whenever the gimbals change direction or whenever motion is initiated after a period of rest. Friction estimation and compensation techniques, described in [2], [8], as well as in this handbook in Section 77.1 (Friction Compensation), are particularly important in systems that do not include a fine stabilization assembly.

Tracking error due to cogging and torque ripple is of particular concern. Both are periodic functions of motor shaft angle

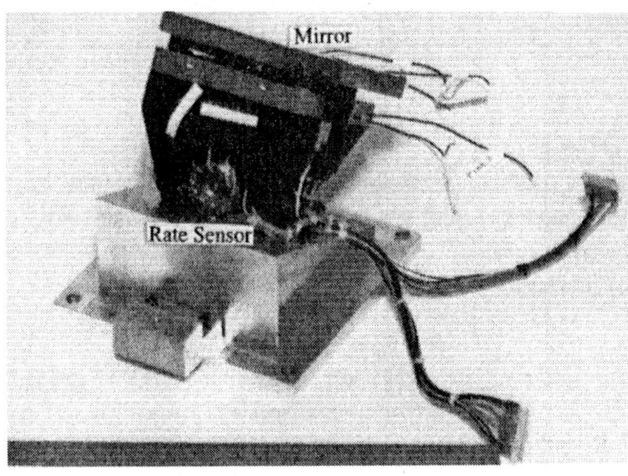

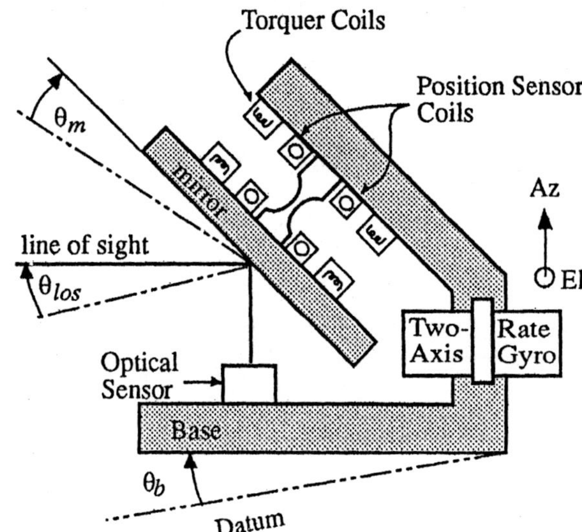

Figure 75.31 The two-axis stabilization assembly, consisting of a mirror, a two-axis angular rate Multisensor (Trademark of GEC Systems), mirror angular position sensors and torquers. Changes in line-of-sight direction are effected by tilting the mirror.

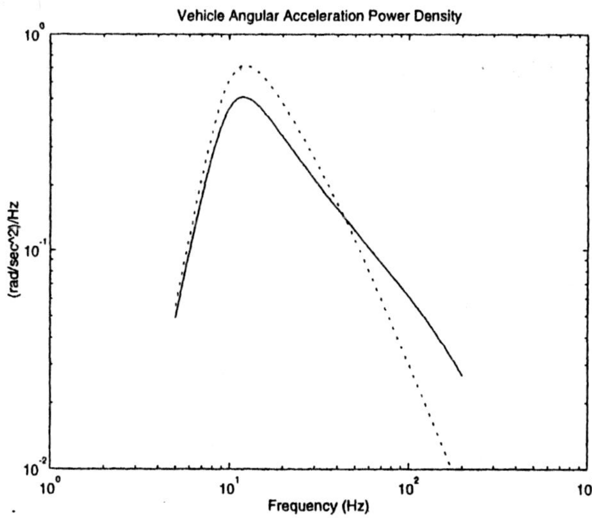

Figure 75.32 When subjected to this angular vibration environment (solid) having an amplitude of 560 μrads rms, the image stabilization assembly cannot permit more than 65 μrads rms of line-of-sight motion. An approximation (dashed) is used for controller design.

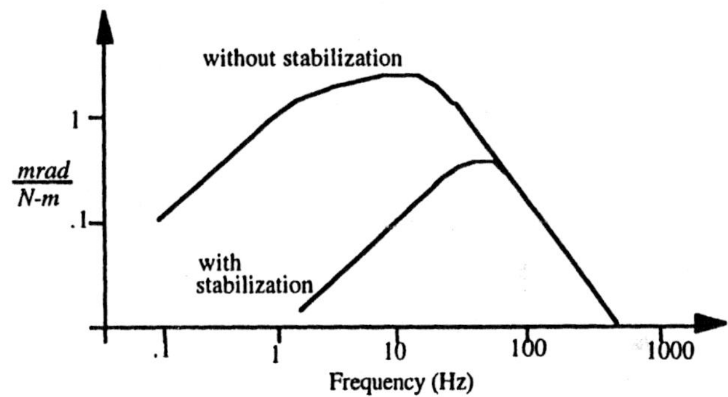

Figure 75.33 The stabilization system helps attenuate tracking error caused by internally generated torque disturbances.

which produce a noticeable jumping of the image when the pilot is steadily turning his head. The coarse positioning system will have a certain response to these disturbances as defined by their Torque Disturbance Sensitivity function, which is typically shaped like the curve in Figure 75.33. In the Falcon Eye, torque disturbances produce a worst-case gimbal tracking error of about 3 mrad peak to peak, far in excess of the 250 μrad limit imposed by the IR detector resolution. The stabilization system therefore must attenuate those tracking errors by more than a factor of 12 for frequencies up to 10 Hz.

Stabilization Control System Structure and Modeling

The architecture of the overall control system is shown in some detail in Figure 75.34. It represents a single axis of the two-axis system (coordinate transformations not shown). A discussion of the coordinate systems relating gimbals to mirror stabilization assembly is not provided because it will not help to understand this control system design effort. All that one needs to recognize is that coordinate frames used in designing the stabilization system controllers are attached locally to the base of the stabilization assembly, and that, in the actual implementation, transformations from the gimbal to the stabilization assembly coordinate frames, and back, were employed.

Linear Truth Model A linear model of one axis of the stabilization control system is given in Figure 75.35. It defines the stabilization system's response to the three inputs discussed in conjunction with Figure 75.30 and will be used in evaluating system performance with the compensators to be derived. Transfer functions from those three inputs, ϕ_g, ϕ_c, and θ_v, to

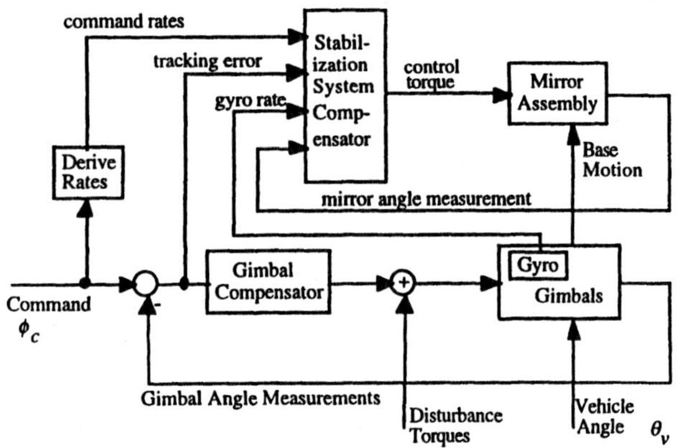

Figure 75.34 Structure controller system.

the line of sight angle θ_{los} are to be referred to as $T_g(s)$, $T_c(s)$, and $T_v(s)$, respectively. The system parameters of this model appear in Table 75.4. Units $N \cdot m$ refer to Newton-meters.

Although implemented digitally, the compensator is shown here in continuous-time form. It was possible to design and analyze the compensator in the continuous-time domain because the fastest time constants of the closed-loop system were significantly less (e.g., 4 times) than the sample frequency (2700 Hz). Nevertheless, delays due to processing and the sample and holding (S/H) of data, and other performance limiting effects were included in the model. These include a 2 kHz, three-pole, low-pass Butterworth antialias filter $A(s)$, and a 1 kHz, three-pole low-pass Butterworth reconstruction filter $R(s)$ on the control output from the D/A converter.

75.4.4 Controller Design

Design Procedure

Our objective was to develop a compensator that achieves a performance approaching the ideal performance described with Figure 75.30. This was accomplished using the tools of Linear Quadratic Gaussian (LQG) optimal control theory. The design procedure, briefly, is as follows. It begins with the development of a design model capturing the essential features of the plant and **exogenous inputs.** This model serves as a basis for the design of a steady-state LQ regulator and Kalman filter, which combine to form a linear time-invariant dynamic compensator. The designer adjusts the regulator and filter weighting matrices (Q, R, V, and W) until a design is produced which achieves satisfactory system performance. This is described in greater detail below.

1. Design Model Definition

The first step in the design process entails defining a linear state-space model of the process to be controlled. It becomes imbedded within the controller, therefore the simplest model that adequately describes the plant and exogenous disturbance inputs should be used. In this application an adequate description is one which will enable the controller, given the measured inputs, to distinguish the various exogenous inputs from one another, and thereby to effect the proper control action.

The design model will be represented in the usual state-space form:

$$\begin{aligned} \dot{x} &= Ax + Bu + v, \\ y &= Cx + w, \end{aligned} \tag{75.160}$$

where x, y, and u are the appropriately dimensioned state, measurement, and control vectors, respectively, all functions of time. The vectors v and w are independent, white Gaussian noise signals, having autocovariance matrices,

$$\begin{aligned} E\{vv'\} &= V\delta(t), \quad \text{and} \\ E\{ww'\} &= W\delta(t). \end{aligned} \tag{75.161}$$

A, B, and C are constant coefficient system matrices.

2. Linear Quadratic Regulator Design

At this stage of the design process the state and control weighting matrices, Q and R, of the quadratic performance index,

$$J = \int_0^\infty (x'Qx + u'Ru)dt, \tag{75.162}$$

are selected, and the control matrix G of the linear control law

$$u = -Gx \tag{75.163}$$

is derived in accordance with $G = R^{-1}B'M$, where M is the solution to the Control Algebraic Riccati Equation (see Chapter 36),

$$0 = MA + A'M - MBR^{-1}B'M + Q. \tag{75.164}$$

The coefficients of Q, the state weighting, are defined in accordance with the error signal to be minimized. This error will be some linear combination of state variables.

The control weighting R is adjusted to achieve some other criterion. It usually suffices to weight each control independently. Then all of the off-diagonal coefficients in R are zero. The diagonal elements are typically chosen to achieve a bandwidth criterion, as was the case in this application, described below.

3. Kalman Filter Design

The state vector x of the linear control law (Equation 75.163) is typically not completely known. Some states will not be measured directly, and those measured may be corrupted by noise. The steady-state Kalman filter

$$\begin{aligned} \dot{\hat{x}} &= A\hat{x} + B\hat{u} + K(y - C\hat{x}), \\ \hat{u} &= -G\hat{x}, \end{aligned} \tag{75.165}$$

is enlisted to provide an estimate $\hat{x}$ of the actual state x. The designer specifies the coefficients in the weighting matrices, V and W, which, along with the system matrices A and C and the Filter Algebraic Riccati Equation (also in Chapter 36),

$$0 = AP + PA' - PC'W^{-1}CP + V, \tag{75.166}$$

define the filter gain $K = PC'W^{-1}$.

It is important to recognize that we are not designing a Kalman filter for accurately estimating the system state, but we are using

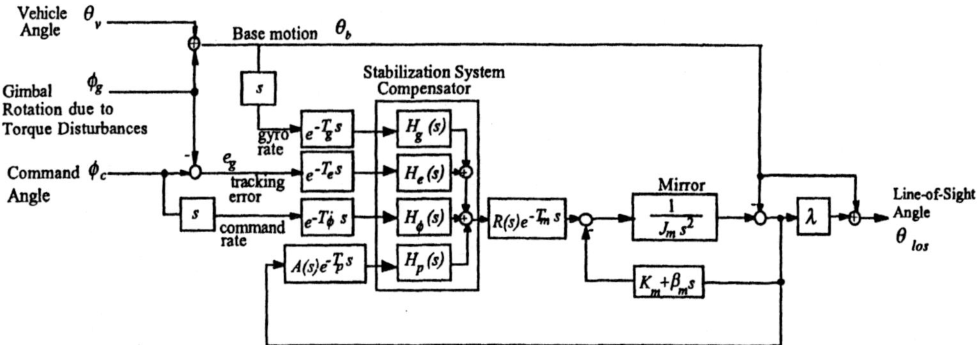

Figure 75.35 Linear model used in evaluating stabilization system performance.

TABLE 75.4 Stabilization System Actuator and Error Source Parameters

Symbol	Value	Units	Parameter name
J_m	3.5×10^{-5}	N-m-s^2	mirror inertia
β_m	4×10^{-3}	N-m-s	flexure damping
K_m	.096	N-m/rad	flexure stiffness
T_g	1.75	msec	gyro signal latency
T_e	85	μsec	tracking error processing delay
T_ϕ	40	μsec	command delay
T_p	90	μsec	pickoff measurement processing delay
T_m	240	μsec	stab. compensator processing & S/H delay
λ	1.386	-	optical scale factor

the Kalman filter as a synthesis tool for designing a compensator. The matrices, V and W, are not viewed as descriptions of actual noise present in the system, but as design parameters, i.e., knobs, that the designer adjusts to shape the dynamic response of the filter and to achieve design goals.

To simplify the task of finding an acceptable design, one would like to work with the smallest set of parameters that provides the needed design freedom. To this end we assume that the state and measurement noise vectors, v and w, are not only independent of each other, but that the individual elements of v and w are independent. Then, only the diagonal elements of V and W are nonzero and are employed as adjustable design parameters. (There are cases when the off-diagonal elements must be adjusted to achieve certain design characteristics. An example is given in [4].)

The filter and regulator combine to form a dynamic compensator having the matrix transfer function,

$$u(s) = H(s)y(s), \tag{75.167}$$

where

$$H(s) = -G(sI - A + BG + KC)^{-1}K. \tag{75.168}$$

4. Evaluation of Performance

At this stage we verify that the design goals have been satisfied. In this application, this involved evaluating the performance predicted by the linear model of the system (Figure 75.35) with the newly derived controller. If the requirements are met, a more thorough evaluation can be performed by simulating the operation of this controller in a detailed truth model of the system. This is a model that includes all of the important dynamic and nonlinear effects, the coordinate transformations, Coriolis and centripetal couplings, the gimbals, friction, backlash, vehicle and pilot head attitude data, etc. If satisfactory performance is achieved there, the design is complete. If not, a judgment is made as to what can be changed to improve performance, and the design process is repeated. This entails 1) going back to Steps 2 and 3, modifying one or more weighting matrices, Q, R, V, and W, and generating a new compensator based on the same design model, or (2) going back to Step 1, altering the design model to incorporate some effect now understood to be important, and repeating the design. This continues until satisfactory performance with an adequate controller is achieved.

In this application the stabilization performance was computed in accordance with the model of Figure 75.36, where $V_\theta(j\omega)$ is the shaping filter (see Equation 75.158) defining worst-case angular motion, and $T_v(j\omega)$ is the stabilization system's response to vehicle angular motion inputs.

The resulting line-of-sight motion due to angular vibration in the frequency range over which the stabilization requirement applies (5 to 200 Hz) is given by

$$\sigma_{los}^2 = \frac{1}{2\pi}\int_{2\pi 5}^{2\pi 200} |V_\theta(j\omega)T_v(j\omega)|^2\,\sigma_n^2 d\omega \tag{75.169}$$

where $\sigma_n^2 = 1$.

The response of the complete system to internal torque disturbances was generated in accordance with the model in Figure 75.37 where $T_g(j\omega)$ is the stabilization system's response to

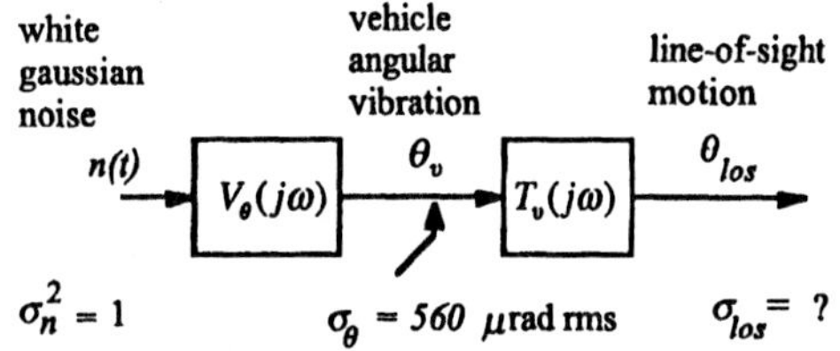

Figure 75.36 Model used to compute line-of-sight motion due to angular vehicle vibration.

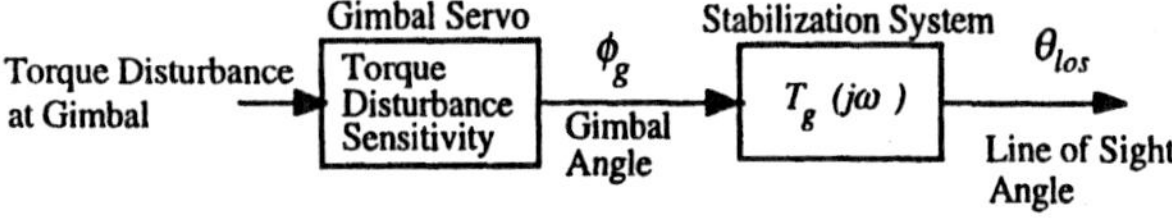

Figure 75.37 Model used to compute line of sight motion due to torque disturbances.

gimbal angular input (whose ideal value is depicted by the center curve of Figure 75.30). Because the gimbal servos are not described, this check will entail verifying that the stabilization system provides adequate isolation (i.e., > 12 in the 2 to 10 Hz range) from gimbal angular position errors due to torque disturbances.

Design Model Development

The process of developing a design model is commonly one requiring several iterations. One reason for this is that it is highly unlikely, except in simple cases, that the designer will anticipate all of the important characteristics that should be in the design model in the first attempt. Secondly, it pays to begin with a simple model which may or may not be adequate, then to increase complexity until satisfactory performance is achieved, rather than beginning immediately with what may be an unnecessarily complicated, high-order model.

The design models developed for this application all include the stabilization mirror dynamics (Equation 75.156) and some form of base motion model. (Recall that base motion is defined as motion about a local coordinate system attached to the base to which the mirror is mounted.) Since the controller must isolate the line of sight from vehicle vibration, the base motion model must include a model of that disturbance. The actual vibration model, defined by the shaping filter of Equation 75.158 is fifth-order and probably more complicated than need be. A third-order approximation was developed for the purpose of the controller design. This model, when excited by white noise, yields a signal whose second derivative has the spectrum shown as a dashed line in Figure 75.32. It has an angular position output given by the transfer function,

$$V'_\theta(s) = \frac{1}{(s/\Omega_0+1)(s^2/\Omega_1^2+2\zeta_1 s/\Omega_1+1)}, \qquad (75.170)$$

where Ω_0 = 126 rad/sec, Ω_1 = 62.8 rad/sec, and $\zeta_1 = 0.5$. This model is part of the controller design model block diagram shown in Figure 75.38, depicting the design model in its final form after several stages of development. Those stages are described below.

The first design model developed for this application involved a base motion model consisting of vehicle vibration only. The

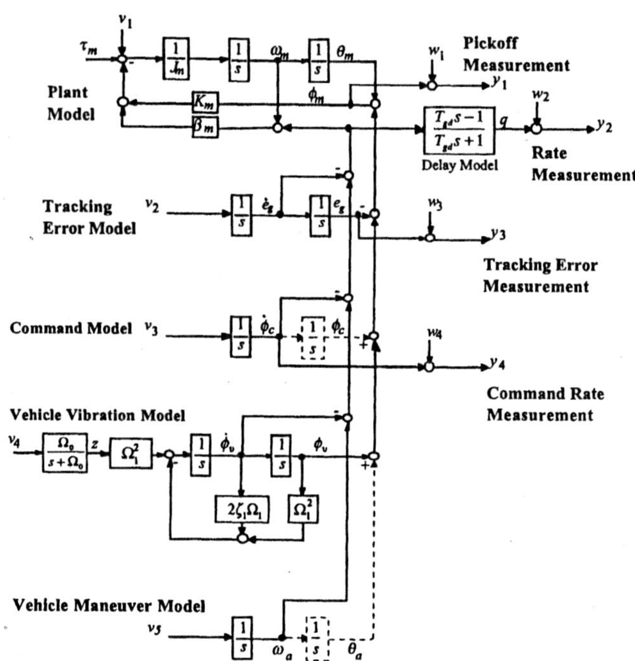

Figure 75.38 Controller design model.

resulting controller did a fine job in isolating the line of sight from vibration. However, it knew nothing of vehicle maneuvers or of gimbal motion, and interpreted those large motions, when tested in simulation, as vibration to be rejected, driving the mirror into its stops. To remedy this, models of vehicle maneuvering and commanded gimbal motion were added to the design model. Both are modeled as double integrators because white noise through a double integrator produces a signal having characteristics expected of maneuver and head motion — infinite DC gain needed to represent constant turning rates and a rolling off with increased frequency. An important distinction between the command and maneuver signals is that the command is a known input while the maneuver signal is not. The command rate is therefore included as a measured output in the design model.

A controller developed with this design model worked, in that it did not drive the mirror to the stops. However, stabilization performance was not quite adequate (85 μrads rms, relative to the required 62.) Performance was being limited by the pure delay present in the gyro signal measurement. Improved stabilization performance was achieved by adding a model of the delay to the design model. A first-order Pade' approximation to the delay incorporated in the design model increased the order of the compensator by one, but also improved stabilization performance sufficiently (to 55 μrads rms), as described below.

Although stabilization performance was now adequate, the tracking smoothness performance was not. Improved ability to isolate the line of sight from gimbal motion due to internally generated torque disturbances was achieved by adding the gimbal servo tracking error e_g, as an input to the stabilization system compensator. Two additional integrator states were added to allow including the tracking error as an input to the compensator, and therefore as an output of the design model. The resulting design model in its final form is shown in Figure 75.38.

Two of the integrators in the design model are drawn with dashed lines. The regulator gains on those states turn out to be zero. Consequently, they have no effect on the output of the controller and can be eliminated from the design model without effect. The model and the resulting compensator are therefore tenth order.

You will note that the command angle ϕ_c is no longer present as a state in the design model. And even if it were, it could not be used as a reference input to the stabilization system because the gain on that state is zero. The command rate is therefore used as the reference input defining pilot head motion.

The ten-element state vector and four-element measurement vector are ordered as follows:

$$x = \begin{bmatrix} \omega_m \\ \phi_m \\ \dot{e}_g \\ e_g \\ \dot{\phi}_c \\ z \\ \dot{\phi}_v \\ \phi_v \\ \omega_a \\ q \end{bmatrix} = \begin{bmatrix} \text{mirror inertial rate} \\ \text{mirror angular position} \\ \text{gimbal tracking error rate} \\ \text{gimbal tracking error} \\ \text{command rate} \\ \text{vehicle vibration filter state} \\ \text{vehicle vibration angular rate} \\ \text{vehicle vibration angular position} \\ \text{vehicle maneuver angular rate} \\ \text{gyro delay model variable} \end{bmatrix}$$

$$y = \begin{bmatrix} \phi_m \\ \omega_g \\ e_g \\ \dot{\phi}_c \end{bmatrix} = \begin{bmatrix} \text{mirror pickoff angle} \\ \text{gyro rate} \\ \text{gimbal tracking error} \\ \text{command rate} \end{bmatrix}$$

The system parameters contained in the design model are given in Table 75.5. The units of these parameters were changed from seconds to deciseconds (ds), adjusting the scaling of the system matrices to eliminate numerical problems which were precluding the solution of the Control and Filter Algebraic Riccati Equations. The controller gain and filter gain matrices were derived for this scaled system, and then scaled back to units of seconds prior to their usage.

The corresponding system matrices are

$$A = \begin{bmatrix} -.0133 & -.272 & .0133 & 0 & .0133 & 0 & .0133 & 0 & .0133 & 0 \\ 1 & 0 & 1 & 0 & -1 & 0 & -1 & 0 & -1 & 0 \\ 0 & 0 & \varepsilon & -.01 & 0 & 0 & 0 & 0 & 0 & 0 \\ 0 & 0 & 1 & 0 & 0 & 0 & 0 & 0 & 0 & 0 \\ 0 & 0 & 0 & 0 & \varepsilon & 0 & 0 & 0 & 0 & 0 \\ 0 & 0 & 0 & 0 & 0 & -1.26 & 0 & 0 & 0 & 0 \\ 0 & 0 & 0 & 0 & 0 & .394 & -.628 & -.394 & 0 & 0 \\ 0 & 0 & 0 & 0 & 0 & 0 & 1 & 0 & 0 & 0 \\ 0 & 0 & 0 & 0 & 0 & 0 & 0 & 0 & \varepsilon & 0 \\ 0 & 0 & -1 & 0 & 1 & 0 & 1 & 0 & 1 & -11.4 \end{bmatrix}$$

$$B = \begin{bmatrix} 2.82 & 0 & 0 & 0 & 0 & 0 & 0 & 0 & 0 & 0 \end{bmatrix}'$$

$$C = \begin{bmatrix} 0 & 1 & 0 & 0 & 0 & 0 & 0 & 0 & 0 & 0 \\ 0 & 0 & 1 & 0 & -1 & 0 & -1 & 0 & -1 & 22.8 \\ 0 & 0 & 0 & 1 & 0 & 0 & 0 & 0 & 0 & 0 \\ 0 & 0 & 0 & 0 & 1 & 0 & 0 & 0 & 0 & 0 \end{bmatrix}$$

The small negative entries ($\varepsilon = -1 \times 10^{-5}$) on the diagonal of A and the small spring term ($-.01$) at A_{34}, were also added to eliminate the numerical problems precluding the convergence of the Riccati equation solver.

LQ Regulator Design

A quadratic performance integral was formed which contains both the error e that the regulator must drive to zero and the control action τ_m :

$$J = \int_0^\infty (e^2 + \rho\tau_m^2)dt \tag{75.171}$$

Because there is only a single control variable, the mirror torque τ_m, the control weighting, ρ, is scalar.

An expression for the error e was derived from the tracking error expression given by Equation 75.155. The Flir line of sight relative to the vehicle ϕ_f is composed of the gimbal relative angle ϕ_g and the line of sight angle relative to the gimbals due to mirror rotation $\lambda\phi_m$:

$$\phi_f = \phi_g + \lambda\phi_m. \tag{75.172}$$

To include gimbal tracking error in this expression, we express the gimbal angle in terms of the command and gimbal tracking error:

$$\phi_g = \phi_c - e_g. \tag{75.173}$$

We also assume that the pilot's head attitude relative to the vehicle ϕ_p of Equation 75.155 equals the command angle ϕ_c (i.e., latency and noise in the head tracker data is ignored). Combining this and Equations 75.155, 75.172, and 75.173 yields the error equation,

$$e = \phi_v - e_g + \lambda\phi_m. \tag{75.174}$$

The state weighting matrix Q is formed to penalize this error expression. To compute Q easily, we express e in terms of the state

TABLE 75.5 Design Model Scaled System Parameters.

Parameter	Value	Units
Mirror inertia J_m	3.54×10^{-1}	N-m-ds^2
Mirror hinge damping β_m	4.0×10^{-3}	N-m-ds
Mirror hinge stiffness K_m	.096	N-m/rad
Gyro delay parameter T_{gd}	1/2(.175)	ds
Shaping filter parameter Ω_0	1.26	rad/ds
Shaping filter parameter Ω_1	.628	rad/ds
Shaping filter parameter ζ_1	0.5	—

vector $x : e = Lx$. With optical scale factor $\lambda = 1.386$ (which applies in narrow field-of-view mode) this vector L becomes

$$L = \begin{bmatrix} 0 & 1.386 & 0 & -1 & 0 & 0 & 0 & 1 & 0 & 0 \end{bmatrix}, \tag{75.175}$$

and with this the state weighting Q is computed as follows:

$$Q = L'L + \varepsilon I_{10} \tag{75.176}$$

It was necessary to add small positive terms ($\varepsilon = 1 \times 10^{-5}$) on the diagonal to makes Q slightly more positive definite, because, without them, the Ricatti Equation Solver halted when checking and mistaking Q for a negative-definite matrix.

Only a single parameter, the control weighting ρ of the matrix $R = [\rho]$, must be selected to define completely all of the matrices entering into the Control Algebraic Ricatti Equation. As ρ is decreased, the eigenvalues of the controlled plant, eig$(sI - A + BG)$, associated with the mirror dynamics, move to the left and approach asymptotes at 45° above and 45° below the negative real axis. The other eigenvalues, those associated with the exogenous inputs, do not move because those modes are uncontrollable. The coefficient ρ was adjusted to place the mirror eigenvalues at a location having a magnitude of about 100 Hz. This enables the regulator to respond adequately to exogenous inputs having lesser frequency content, and yet does not move those eigenvalues into a frequency regime where unmodeled phase lag in the system begins to have a destabilizing effect.

With $R = [0.125]$ the controller matrix G (for the scaled system), computed via solution of the Control Algebraic Ricatti Equation, is

$$\begin{aligned} G \;=\; & [1.6424 \quad 3.8249 \quad .46390 \quad -2.8286 \quad -1.6423 \\ & -.034147 \quad -.38867 \quad 2.8672 \quad -1.6423 \quad 0] \end{aligned}$$

Filter Design

There are a total of nine nonzero coefficients in V and W, all on the diagonal, which the designer must specify. To reduce the number of design parameters to a manageable number, the coefficients in W were set to fixed values proportional to the actual noise variances of the associated measurements. The tracking error and command rate signals are not actual measurements but are digital signals fed forward from the gimbal servoes, so their variances were set to values consistent with their quantization levels. The elements of W were, therefore, set to values proportional to the pickoff, gyro rate, tracking error, and command rate noise variances, respectively, which yielded

$$W = \text{diag}[1 \quad 5 \quad .1 \quad 16.7] \tag{75.177}$$

The five nonzero diagonal coefficients of V,

$$V = \text{diag}[v_1 \quad 0 \quad v_2 \quad 0 \quad v_3 \quad v_4 \quad 0 \quad 0 \quad v_5 \quad 0], \tag{75.178}$$

were selected by performing a manual search over the five-dimensional parameter space spanned by the plant noise variances v_1 through v_5. At selected points within that space, control system compensators were computed and the performance of the stabilization system with that compensator was evaluated. This included examining how closely its performance approached the ideal (Figure 75.30), and the calculations necessary to determine if the stabilization and tracking smoothness requirements were met. This manual search did not proceed blindly, however. It was possible to predict how adjustments made to specific plant noise levels should alter closed-loop performance. For example, what would you expect the compensator to do if you tell it there is more noise driving the vibration portion of the design model by increasing v_4? It should increase the degree of isolation against vehicle vibration. And in fact it does that, as shown in Figure 75.39. As v_4 is increased, the transfer function from vehicle angular motion to line-of-sight motion deepens.

This manual search settled upon

$$V = \text{diag}[500 \quad 0 \quad 100 \quad 0 \quad 100 \quad 10^5 \quad 0 \quad 0 \quad .01 \quad 0]$$

to which corresponds the Kalman gain matrix:

$$K = \begin{bmatrix} 1.9548e{+}01 & 4.5483e{+}00 & -1.2313e{+}00 & 1.0944e{-}01 \\ 7.6554e{+}00 & -1.1539e{+}00 & 7.0985e{-}01 & -5.3122e{-}02 \\ 1.0308e{+}00 & -7.1132e{-}01 & 3.1038e{+}01 & 2.1484e{-}02 \\ 7.0985e{-}02 & -9.7569e{-}02 & 7.8452e{+}00 & 4.1423e{-}03 \\ -8.8714e{-}01 & 5.9042e{-}01 & 6.9177e{-}01 & 2.4153e{+}00 \\ -9.5811e{+}01 & 7.3508e{+}01 & 9.7882e{+}01 & -5.0307e{+}00 \\ -1.1170e{+}01 & 1.2785e{+}01 & 2.2769e{+}01 & -1.5445e{+}00 \\ 1.9951e{+}00 & -2.2092e{-}01 & 1.8483e{+}00 & -5.8257e{-}01 \\ -4.3322e{-}02 & 3.8739e{-}02 & 5.9208e{-}02 & -3.8896e{-}03 \\ -8.2898e{-}01 & 1.0424e{+}00 & -5.4377e{-}01 & 4.4834e{-}02 \end{bmatrix}.$$

This matrix and the G matrix given above were unscaled, from units of deciseconds to seconds, before use.

75.4.5 Performance Achieved

Here we examine how closely the stabilization system's response approaches the ideal response defined in Figure 75.30. Transfer functions $T_v(s)$, $T_c(s)$, $T_g(s)$, from vehicle angle, command angle, and gimbal angle, respectively, to the line-of-sight angle, were

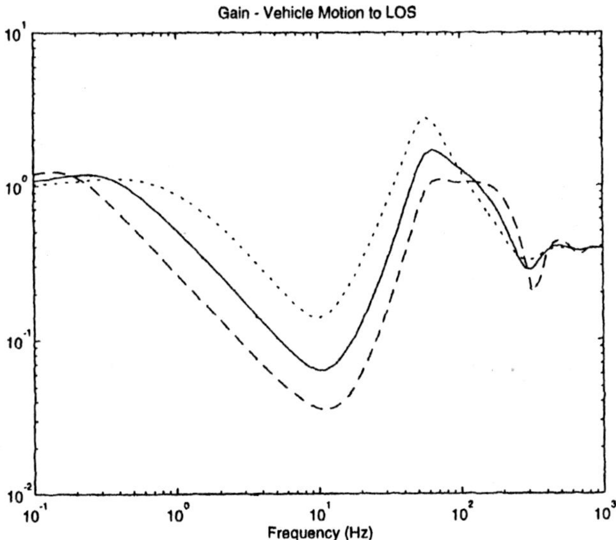

Figure 75.39 As noise variance v_4 driving the vibration part of the plant is increased, the level of line-of-sight (LOS) stabilization against vibration improves, as shown here with $v_4 = 10^4$ (dotted), 10^5 (solid), and 10^6 (dashed). All other elements in V, W, Q, and R were not changed.

derived in accordance with the linear model of Figure 75.35. The magnitudes of these transfer functions are plotted in Figure 75.40.

Clearly these plots approach the ideal response curves given in Figure 75.30, primarily in the low to middle frequency regimes. In the upper plot, command tracking is unity (the ideal) to a frequency of 10 Hz, and a command tracking bandwidth of nearly 100 Hz is provided, which exceeds the Command Tracking Requirement defined earlier. The controller contributes a measure of peaking in the response to reduce phase lag at lower frequencies where tight command tracking is needed.

As shown in the middle plot, the stabilization system achieves exceptionally good isolation from internally generated gimbal tracking errors.

Finally, in the lower plot, which concerns the system's response to vehicle angular motion, there is adequate maneuver tracking with a bandwidth of just over 0.3 Hz, and there is a transition from maneuver tracking at low frequencies to isolation from high frequency vibration.

Stabilization Performance The adequacy of the stabilization performance was verified by numerically evaluating Equation 75.169 with $T_v(j\omega)$ given by the lower plot of Figure 75.40. This results in

$$\sigma_{los} = 55 \ \mu\text{rads rms}$$

which meets the 62 μrads requirement.

Smoothness of Tracking Performance The center plot of Figure 75.40 clearly indicates that this requirement (greater than a factor of $12x$, or 22 dB of attenuation in the 2 to 10 Hz range) is met by a large margin.

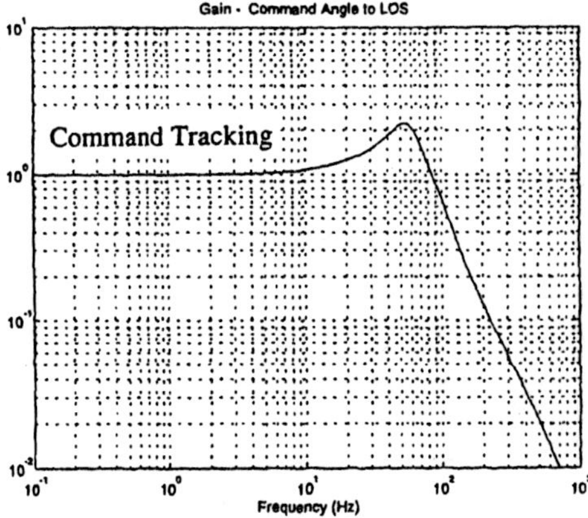

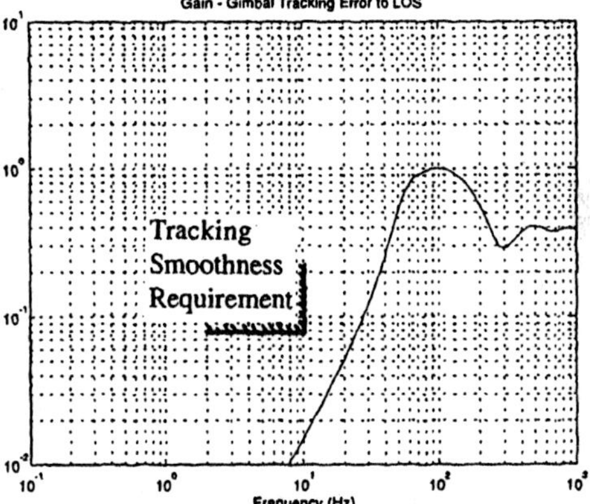

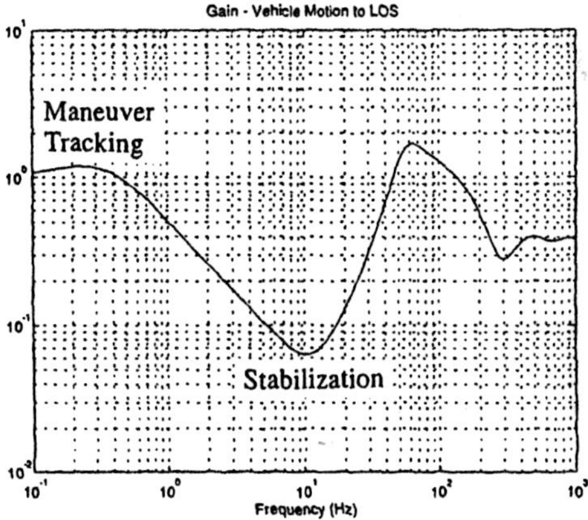

Figure 75.40 Response of the line-of-sight (LOS) stabilization system to the various inputs (compare to Figure 75.30).

75.4.6 Concluding Remarks

This chapter describes the design of a controller for the line-of-sight pointing and stabilization system contained within the Falcon Eye, a head-steered night vision system developed for the F16 Falcon. The chapter begins by discussing the purpose and performance objectives of the overall night vision system. This is followed by a description of the electro-optical subsystem used to point and stabilize the line-of-sight, i.e., the plant, and of the disturbance inputs. The chapter then discusses the performance requirements applying specifically to this subsystem. Finally, the design, using modern control techniques, of the controller that met those requirements is presented.

Although only one was described, seven distinct controllers were developed for this system. Four were developed for the two-axis stabilization assembly — one per axis with different controllers applied in wide and narrow field-of-view modes — and one for each of the three gimbal servos, azimuth, elevation, and derotation. All were extensively and successfully flight tested, as was the overall Falcon Eye night vision system. In fact, test pilots developed enough confidence in the system to use it at night when taking off, landing, and during low level flight ($\sim$ 200 ft above ground) with the Falcon Eye as their only visual source.

75.4.7 Defining Terms

Flir: An infrared imaging system typically directed to look forward in the direction of travel.

Exogenous inputs: Inputs arising from states that are unaffected by the control.

References

[1] Benson, A.J. and Barnes, G.R., Vision During Angular Oscillation: The Dynamic Interaction of Visual and Vestibular Mechanisms, *Aviation, Space, and Environmental Medicine,* 49(*1Sec.II*), 1978.

[2] Friedland, B. and Park, Y.J., On Adaptive Friction Compensation, *Proc. 30th IEEE Conf. Decision Control,* Brighton, England, 1991, pp. 2899–2903.

[3] Fuchs, C., et al., Two-axis Mirror Stabilization Assembly, U.S. Patent 4 881 800, 1989.

[4] Haessig, D., Selection of LQG/LTR Weighting Matrices through Constrained Optimization, *Proc. ACC,* Seattle, WA, 1995, pp. 458–460.

[5] Haessig, D. A., Mirror Positioning Assembly for Stabilizing the Line-of-Sight in a Two-Axis Line-of-Sight Pointing System, U.S. Patent 5 220 456, 1993.

[6] Haessig, D. A., Image Stabilization Assembly for an Optical System, U.S. Patent 5 307 206, 1994.

[7] Johnson, C.D. and Masten, M.K., Fundamental Concepts and Limitations in Precision Pointing and Tracking Problems, Acquisition, Tracking, and Pointing VII; *SPIE Proc.,* 1993, Vol. 1950, pp. 58–72.

[8] Maqueira, B. and Masten, M.K., Adaptive Friction Compensation for Line-of-Sight Pointing and Stabilization, *Proc. ACC,* 1993, pp. 1942–46.

[9] Masten, M.K.,, Electromechanical Systems for Optical Target Tracking Sensors. In *Multitarget-Multisensor Tracking: Advanced Applications,* Artech House, Norwood, MA, 1990, pp 321–360.

[10] Masten, M.K. and Sebesta, H.R., Line-of-Sight Stabilization/Tracking Systems: An Overview, *Proc. ACC,* 2, 1987, pp. 1477–1482.

[11] Scott, W.B., Falcon Eye Flir, GEC Helmet Aid F-16 Mission Flexibility, *Aviation Week,* April 17, 1989.

76

Control of Robots and Manipulators

Mark W. Spong
The Coordinated Science Laboratory, University of Illinois at Urbana-Champaign

Joris De Schutter
Katholieke Universiteit Leuven, Department of Mechanical Engineering, Leuven, Belgium

Herman Bruyninckx
Katholieke Universiteit Leuven, Department of Mechanical Engineering, Leuven, Belgium

John Ting-Yung Wen
Department of Electrical, Computer, and Systems Engineering, Rensselaer Polytechnic Institute

76.1 Motion Control of Robot Manipulators

Mark W. Spong, The Coordinated Science Laboratory, University of Illinois at Urbana-Champaign

76.1.1 Introduction

The design of intelligent, autonomous machines to perform tasks that are dull, repetitive, hazardous, or that require skill, strength, or dexterity beyond the capability of humans is the ultimate goal of robotics research. Examples of such tasks include manufacturing, excavation, construction, undersea, space, and planetary exploration, toxic waste cleanup, and robotic-assisted surgery. The field of robotics is highly interdisciplinary, requiring the integration of control theory with computer science, mechanics, and electronics.

The term *robot* has been applied to a wide variety of mechanical devices, from children's toys to guided missiles. An important class of robots are the manipulator arms, such as the PUMA robot shown in Figure 76.1 These manipulators are used primarily in materials handling, welding, assembly, spray painting, grinding, deburring, and other manufacturing applications. This chapter discusses the motion control of such manipulators.

Robot manipulators are basically multi-degree-of-freedom positioning devices. The robot, as the "plant to be controlled," is a multi-input/multi-output, highly coupled, nonlinear mechatronic system. The main challenges in the motion control problem are the complexity of the dynamics and uncertainties, both

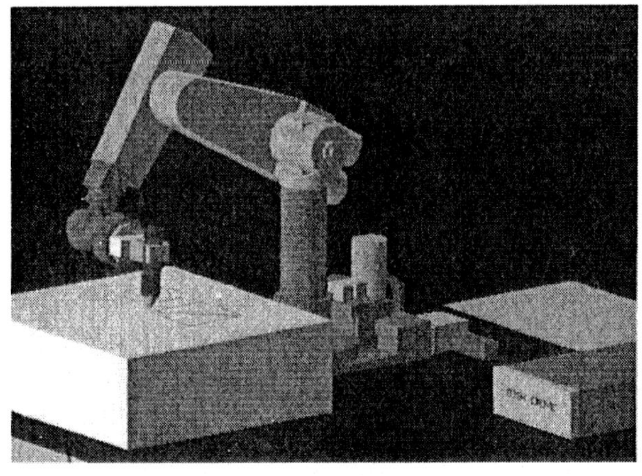

Figure 76.1 PUMA robot manipulator.

parametric and dynamic. Parametric uncertainties arise from imprecise knowledge of kinematic parameters and inertia parameters, while dynamic uncertainties arise from joint and link flexibility, actuator dynamics, friction, sensor noise, and unknown environment dynamics.

There are a number of excellent survey articles on the control of robot manipulators from an elementary viewpoint [2], [4], [6]. These articles present the basic ideas of independent joint control using linearized models and classical transfer function analysis. That material is not repeated here. Instead, we survey more recent and more advanced material that summarizes the work of many researchers from about 1985 to the present. Many of the ideas presented here are found in the papers reprinted in

0-8493-8570-9/96/$0.00+$.50

[10], which is recommended to the reader who wishes additional details. The present chapter is accessible to anyone having an elementary knowledge of Lagrangian dynamics and the state-space theory of dynamical systems, including the basics of Lyapunov stability theory. The textbooks by Asada and Slotine [1] and Spong and Vidyasagar [11] may be consulted for the background necessary to follow the material presented here.

Configuration Space and Task Space

We consider a robot manipulator with **n-links** interconnected by **joints** into an **open kinematic chain** as shown in Figure 76.2. For simplicity of exposition we assume that all joints are rotational, or **revolute**. Most of the discussion in this chapter remains valid for robots with sliding or **prismatic** joints.

The joint variables, $q_1, \ldots, q_n$, are the relative angles between the links; for example, q_i is the angle between link i and link $i-1$. A vector $q = (q_1, \ldots, q_n)^T$, with each $q_i \in [0, 2\pi)$ is called a **configuration**. The set of all possible configurations is called **configuration space** or **joint space**, which we denote as $\mathcal{C}$. The configuration space is an n-dimensional torus, $\mathcal{T}^n = S^1 \times \cdots \times S^1$, where S^1 is the unit circle.

The **task space**, or **end-effector space**, is the space of all positions and orientations (called **poses**) of the end-effector (usually a gripper or tool). If a coordinate frame, called the **base frame** or **world frame**, is established at the base of the robot and a second frame, called the **end-effector frame** or **task frame**, is attached to the end-effector, then the end-effector position is given by a vector $x \in \mathbb{R}^3$ specifying the coordinates of the origin of the task frame in the base frame, and the end-effector orientation is given by a 3×3 matrix R whose columns are the direction cosines of the coordinate axes of the task frame in the base frame. This **orientation matrix**, R, is orthogonal, i.e. $R^T R = I$, and satisfies $\det(R) = +1$. The set of all such 3×3 orientation matrices forms the **special orthogonal group**, $SO(3)$. The task space is then isomorphic to the **special Euclidean group**, $SE(3) = \mathbb{R}^3 \times SO(3)$. Elements of $SE(3)$ are called rigid motions [11].

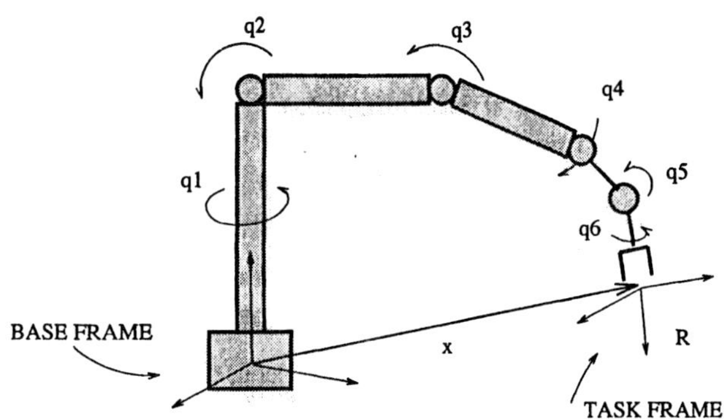

Figure 76.2 Configuration space and task space variables.

Kinematics

Kinematics refers to the geometric relationship between the motion of the robot in joint space and the motion of the end-effector in task space without consideration of the forces that produce the motion. The **forward kinematics problem** is to determine the mapping

$$X_0 = \begin{bmatrix} x(q) \\ R(q) \end{bmatrix} = f_0(q) : \mathcal{T}^n \to SE(3) \tag{76.1}$$

from configuration space to task space, which gives the end-effector pose in terms of the joint configuration. The **inverse kinematics problem** is to determine the inverse of this mapping, i.e., the joint configuration as a function of the end-effector pose. The forward kinematics map is many-to-one, so that several joint space configurations may give rise to the same end-effector pose. This means that the forward kinematics always has a unique pose for each configuration, while the inverse kinematics has multiple solutions, in general.

The kinematics problem is compounded by the difficulty of parameterizing the rotation group, $SO(3)$. It is well-known that there does not exist a minimal set of coordinates to "cover" $SO(3)$, i.e., a single set of three variables to represent all orientations in $SO(3)$ uniquely. The most common representations used are Euler angles and quaternions. Representational singularities, which are points at which the representation fails to be unique, give rise to a number of computational difficulties in motion planning and control.

Given a minimal representation for $SO(3)$, for example, a set of Euler angles ϕ, θ, ψ, the forward kinematics mapping may also be defined by a function

$$X_1 = \begin{bmatrix} x(q) \\ o(q) \end{bmatrix} = f_1(\cdot) : \mathcal{T}^n \to \mathbb{R}^6 \tag{76.2}$$

where $x(q) \in \mathbb{R}^3$ gives the Cartesian position of the end-effector and $o(q) = [\phi(q), \theta(q), \psi(q)]^T$ represents the orientation of the end-effector. The nonuniqueness of the inverse kinematics in this case includes multiplicities due to the particular representation of $SO(3)$ in addition to multiplicities intrinsic to the geometric structure of the manipulator.

The **velocity kinematics** is the relationship between the joint velocities and the end-effector velocities. If the mapping f_0 from Equation 76.1 is used to represent the forward kinematics, then the velocity kinematics is given by

$$V = \begin{pmatrix} v \\ \omega \end{pmatrix} = J_0(q)\dot{q} \tag{76.3}$$

where $J_0(q)$ is a $6 \times n$ matrix, called the **manipulator Jacobian**, and $V^T = (v^T, \omega^T)$ represents the linear and angular velocity of the end-effector. The vector $v \in \mathbb{R}^3$ is just $\frac{d}{dt}x(q)$, where $x(q)$ is the end-effector position vector from Equation 76.1. It is a little more difficult to see how the angular velocity vector ω is computed since the end-effector orientation in Equation 76.1 is specified by a matrix $R \in SO(3)$. If $\omega = (\omega_x, \omega_y, \omega_z)^T$ is a vector in $\mathbb{R}^3$, we may define a skew-symmetric matrix, $S(\omega)$, according to

$$S(\omega) = \begin{bmatrix} 0 & -\omega_z & \omega_y \\ \omega_z & 0 & -\omega_x \\ -\omega_y & \omega_x & 0 \end{bmatrix}. \tag{76.4}$$

The set of all skew-symmetric matrices is denoted by $so(3)$. Now, if $R(t)$ belongs to $SO(3)$ for all t, it can be shown that

$$\dot{R} = S(\omega(t))R, \tag{76.5}$$

for a unique vector $\omega(t)$ [11]. The vector $\omega(t)$ thus defined is the angular velocity of the end-effector frame relative to the base frame.

Singularities in the Jacobian $J_0(q)$, i.e., configurations where the Jacobian loses rank, are important configurations to identify for any manipulator. At singular configurations the manipulator loses one or more degrees-of-freedom. In many applications, it is important to place robots in work cells and to plan their motion in such a way that singular configurations are avoided.

If the mapping f_1 is used to represent the forward kinematics, then the velocity kinematics is written as

$$\dot{X}_1 = J_1(q)\dot{q} \tag{76.6}$$

where $J_1(q) = \partial f_1/\partial q$ is the 6×6 Jacobian of the function f_1. Singularities in J_1 include the representational singularities in addition to the manipulator singularities present in J_0. In the sequel, we use J to denote either the matrix J_0 or J_1.

Trajectories

Because of the complexity of both the kinematics and the dynamics of the manipulator and of the task to be carried out, the motion control problem is generally decomposed into three stages: **motion planning, trajectory generation**, and **trajectory tracking**. In the motion planning stage, desired paths are generated in the task space, $SE(3)$, without timing information, i.e., without specifying velocity or acceleration along the paths. Of primary concern is the generation of collision-free paths in the workspace. In the trajectory generation stage, the desired position, velocity, and acceleration of the manipulator along the path, as a function of time or as a function of arc length along the path, are computed. The trajectory planner may parameterize the end-effector path directly in task space, either as a curve in $SE(3)$ or as a curve in $\mathbb{R}^6$ using a particular minimal representation for $SO(3)$, or it may compute a trajectory for the individual joints of the manipulator as a curve in the configuration space $\mathcal{C}$.

In order to compute a joint space trajectory, the given end-effector path must be transformed into a joint space path via the inverse kinematics mapping. Because of the difficulty of computing this mapping on-line, the usual approach is to compute a discrete set of joint vectors along the end-effector path and to perform an interpolation in joint space among these points in order to complete the joint space trajectory. Common approaches to trajectory interpolation include polynomial spline interpolation, using trapezoidal velocity trajectories or cubic polynomial trajectories, as well as trajectories generated by reference models.

The computed reference trajectory is then presented to the controller, whose function is to cause the robot to track the given trajectory as closely as possible. A typical architecture for the robot control problem is illustrated in the block diagram of Figure 76.3. This chapter is concerned mainly with the design of the tracking controller, assuming that the path and trajectory have been precomputed. For background on the motion planning and trajectory generation problems the reader is referred to [10].

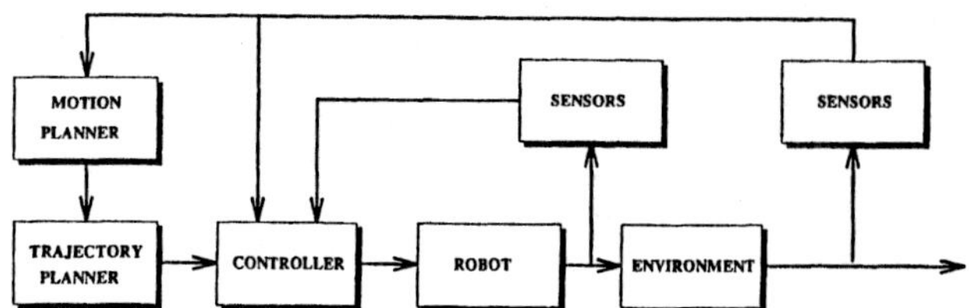

Figure 76.3 Block diagram of the robot control problem.

76.1.2 Dynamics

The dynamics of n-link manipulators are conveniently described by Lagrangian dynamics. In the Lagrangian approach, the joint variables, $q = (q_1, \ldots, q_n)^T$, serve as a suitable set of generalized coordinates. The **kinetic energy** of the manipulator is given by a symmetric, positive definite quadratic form,

$$\mathcal{K} = \frac{1}{2}\sum_{i,j=1}^{n} d_{ij}(q)\dot{q}_i\dot{q}_j = \frac{1}{2}\dot{q}^T D(q)\dot{q} \tag{76.7}$$

where $D(q)$ is the **inertia matrix** of the robot. Let $\mathcal{P} : \mathcal{C} \rightarrow \mathbb{R}$ be a continuously differentiable function (called the **potential energy**). For a rigid robot, the potential energy is due to gravity alone while for a flexible robot the potential energy also contains elastic potential energy. We define the function $\mathcal{L} = \mathcal{K} - \mathcal{P}$, which is called the **Lagrangian**. The dynamics of the manipulator are then described by Lagrange's equations [11],

$$\frac{d}{dt}\frac{\partial \mathcal{L}}{\partial \dot{q}_k} - \frac{\partial \mathcal{L}}{\partial q_k} = \tau_k, \quad k = 1, \ldots, n \tag{76.8}$$

where $\tau_1, \ldots, \tau_n$ represent input generalized forces. In local coordinates, Lagrange's equations can be written as

$$\sum_{j=1}^{n} d_{kj}(q)\ddot{q}_j + \sum_{i,j=1}^{n} \Gamma_{ijk}(q)\dot{q}_i\dot{q}_j + \phi_k(q) = \tau_k, \quad k = 1, \ldots, n \tag{76.9}$$

where

$$\Gamma_{ijk} = \frac{1}{2}\left\{\frac{\partial d_{kj}}{\partial q_i} + \frac{\partial d_{ki}}{\partial q_j} - \frac{\partial d_{ij}}{\partial q_k}\right\} \tag{76.10}$$

are known as **Christoffel symbols of the first kind**, and

$$\phi_k = \frac{\partial \mathcal{P}}{\partial q_k}. \tag{76.11}$$

In matrix form we can write Lagrange's Equation 76.9 as

$$D(q)\ddot{q} + C(q, \dot{q})\dot{q} + g(q) = \tau \tag{76.12}$$

Properties of the Lagrangian Dynamics

The Lagrangian dynamics given in Equation 76.12 possess a number of important properties that facilitate analysis and control system design. Among these are

1. The inertia matrix $D(q)$ is symmetric and positive definite, and there exist scalars, $\mu_1(q)$ and $\mu_2(q)$, such that

$$\mu_1(q)I \le D(q) \le \mu_2(q)I. \tag{76.13}$$

Moreover, if all joints are revolute, then μ_1 and μ_2 are constants.

2. The matrix $W(q, \dot{q}) = \dot{D}(q) - 2C(q, \dot{q})$ is skew symmetric. This property is easily shown by direct calculation. The kj-th component of $\dot{D}(q)$ is given by the chain rule as

$$\dot{d}_{kj} = \sum_{i=1}^{n} \frac{\partial d_{kj}}{\partial q_i}\dot{q}_i, \tag{76.14}$$

and the kj-th component of the matrix $C(q, \dot{q})$ is given as

$$\begin{aligned} c_{kj} &= \sum_{i=1}^{n} \Gamma_{ijk}\dot{q}_i \qquad (76.15)\\ &= \frac{1}{2}\sum_{i=1}^{n}\left\{\frac{\partial d_{kj}}{\partial q_i} + \frac{\partial d_{ki}}{\partial q_j} - \frac{\partial d_{ij}}{\partial q_k}\right\}\dot{q}_i \end{aligned}$$

Therefore the kj component of the matrix $W(q, \dot{q})$ is given by

$$\begin{aligned} w_{kj} &= \dot{d}_{kj} - 2c_{kj}\\ &= \sum_{i=1}^{n}\left[\frac{\partial d_{kj}}{\partial q_i} - \{\frac{\partial d_{kj}}{\partial q_i} + \frac{\partial d_{ki}}{\partial q_j} - \frac{\partial d_{ij}}{\partial q_k}\}\right]\dot{q}_i\\ &= \sum_{i=1}^{n}\left[\frac{\partial d_{ij}}{\partial q_k} - \frac{\partial d_{ki}}{\partial q_j}\right]\dot{q}_i. \end{aligned}$$

Skew symmetry of $W(q, \dot{q})$ now follows by symmetry of the inertia matrix $D(q)$. Strongly related to the skew symmetry property is the so-called **passivity property**.

3. The mapping $\tau \rightarrow \dot{q}$ is passive; i.e., there exists $\beta \ge 0$ such that

$$\int_0^T \dot{q}^T(u)\tau(u)du \ge -\beta. \tag{76.16}$$

To show this property, let H be the total energy of the system

$$H = \frac{1}{2}\dot{q}^T D(q)\dot{q} + \mathcal{P}(q). \tag{76.17}$$

Then the change in energy, $\dot{H}$, satisfies

$$\dot{H} = \frac{1}{2}\dot{q}^T \dot{D}(q)\dot{q} + \dot{q}^T[D(q)\ddot{q} + g(q)] \tag{76.18}$$

since $g(q)^T$ is the gradient of $\mathcal{P}$. Substituting Equation 76.12 into Equation 76.18 yields

$$\dot{H} = \dot{q}^T\tau + \frac{1}{2}\dot{q}^T\{\dot{D}(q) - 2C(q, \dot{q})\}\dot{q} = \dot{q}^T\tau, \tag{76.19}$$

by skew symmetry of $\dot{D} - 2C$. Integrating both sides of Equation 76.19 with respect to time gives

$$\int_0^T \dot{q}^T(u)\tau(u)du = H(T) - H(0) \ge -H(0), \tag{76.20}$$

since the total energy $H(T)$ is non-negative, and passivity follows with $\beta = H(0)$.

4. Rigid robot manipulators are fully actuated; i.e., there is an independent control input for each degree-of-freedom. By contrast, robots possessing joint or link flexibility are no longer fully actuated and the control problems are more difficult, in general.

5. The equations of motion given in Equation (76.12) are linear in the inertia parameters. In other words, there is a constant vector $\theta \in \mathbb{R}^p$ and a function $Y(q, \dot{q}, \ddot{q}) \in \mathbb{R}^{n\times p}$ such that

$$D(q)\ddot{q} + C(q, \dot{q})\dot{q} + g(q) = Y(q, \dot{q}, \ddot{q})\theta = \tau. \tag{76.21}$$

The function $Y(q, \dot{q}, \ddot{q})$ is called the **regressor**. The parameter vector θ is comprised of link masses, moments of inertia, and the like, in various combinations. The dimension of the parameter space is not unique, and the search for the parameterization that minimizes the dimension of the parameter space is an important problem. Historically, the appearance of the passivity and linear parameterization properties in the early 1980s marked watershed events in robotics research. Using these properties, researchers have been able to prove elegant global convergence and stability results for robust and adaptive control. We detail some of these results in the following.

Additional Dynamics

So far, we have discussed only rigid body dynamics. Other important contributions to the dynamic description of manipulators include the dynamics of the actuators, joint and link flexibility, friction, noise, and disturbances. In addition, whenever the manipulator is in contact with the environment, the complete dynamic description includes the dynamics of the environment and the coupling forces between the environment and the manipulator. Modeling all of these effects produces an enormously complicated model. The key in robot control system design is to model the most dominant dynamic effects for the particular manipulator under consideration and to design the controller so that it is insensitive or robust to the neglected dynamics.

Friction and joint elasticity are dominant in geared manipulators such as those equipped with harmonic drives. Actuator dynamics are important in many manipulators, while noise is present in potentiometers and tachometers used as joint position and velocity sensors. For very long or very lightweight robots, particularly in space robots, the link flexibility becomes an important consideration.

Actuator Inertia and Friction The simplest modification to the rigid robot model given in Equation 76.12 is the inclusion of the actuator inertia and joint friction. The actuator inertia is specified by an $n \times n$ diagonal matrix

$$I = \mathrm{diag}(I_1 r_1^2, \ldots, I_n r_n^2), \tag{76.22}$$

where I_i and r_i are the actuator inertia and gear ratio, respectively, of the i-th joint. The friction is specified by a vector, $f(q, \dot{q})$, and may contain only viscous friction, $B\dot{q}$, or it may include more complex models that include static friction. Defining $M(q) = D(q) + I$, we may modify the dynamics to include these additional terms as

$$M(q)\ddot{q} + C(q, \dot{q})\dot{q} + g(q) + f(q, \dot{q}) = \tau. \tag{76.23}$$

As can be seen, the inclusion of the actuator inertias and friction does not change the order of the equations. For simplicity of notation we ignore the friction terms $f(q, \dot{q})$ in the subsequent development.

Environment Forces Whenever the robot is in contact with the environment, additional forces are produced by the robot/environment interaction. Let F_e denote the force acting on the robot due to contact with the environment. It is easy to show, using the **principle of virtual work** [11], that a joint torque $\tau_e = J^T(q)F_e$ results, where $J(q)$ is the manipulator Jacobian. Thus, Equation 76.23 may be further modified as

$$M(q)\ddot{q} + C(q, \dot{q})\dot{q} + g(q) + J^T(q)F_e = \tau \tag{76.24}$$

to incorporate the external forces due to environment interaction. The problem of force control is not considered further in this chapter, but is treated in detail in Chapter 78.2.

Actuator Dynamics If the joints are actuated with permanent magnet dc motors, we may write the actuator dynamics as

$$L\frac{di}{dt} + Ri = V - K_b\dot{q}, \tag{76.25}$$

where i, V are vectors representing the armature currents and voltages, and L, R, K_b are matrices representing, respectively, the armature inductances, armature resistances, and back emf constants. Since the joint torque τ and the armature current i are related by $\tau = K_m i$, where K_m is the torque constant of the motor, we may write the complete system of Equations 76.23 to 76.25 as

$$M(q)\ddot{q} + C(q, \dot{q})\dot{q} + g(q) = K_m i \tag{76.26}$$

$$L\frac{di}{dt} + Ri = V - K_b\dot{q} \tag{76.27}$$

The inclusion of these actuator dynamics increases the dimension of the state-space from $2n$ to $3n$. Other types of actuators, such as induction motors or hydraulic actuators, may introduce more-complicated dynamics and increase the system order further.

Joint Elasticity Joint elasticity, due to elasticity in the motor shaft and gears, is an important effect to model in many robots. If the joints are elastic, then the number of degrees-of-freedom is twice the number for the rigid robot, since the joint angles and motor shaft angles are no longer simply related by the gear ratio, but are now independent generalized coordinates. If we represent the joint angles by q_1 and the motor shaft angles by q_2 and model the joint elasticity by linear springs at the joints, then we may write the dynamics as

$$D(q_1)\ddot{q}_1 + C(q_1, \dot{q}_1)\dot{q}_1 + g(q_1) + K(q_1 - q_2) = 0 \tag{76.28}$$

$$I\ddot{q}_2 + K(q_2 - q_1) = \tau \tag{76.29}$$

where I is the actuator inertia matrix and K is a diagonal matrix of joint stiffness constants. The model given in Equations 76.28 and 76.29 is derived under the assumptions that the inertia of the rotor is symmetric about its axis of rotation and that the inertia of the rotor axes, other than the motor shaft, may be neglected. For models of flexible joint robots that include these additional effects, the reader is referred to the article by DeLuca in [10, p. 98]. It is easy to show that the flexible joint model given in Equations 76.28 and 76.29 defines a passive mapping from motor torque τ to motor velocity $\dot{q}_2$, but does not define a passive mapping from τ to the link velocity $\dot{q}_1$. This is related to the classical problem of collocation of sensors and actuators and has important consequences in the control system design.

Singular Perturbation Models

In practice the armature inductances in L in Equation 76.25 are quite small, whereas the joint stiffness constants in K in Equations 76.28 and 76.29 are quite large relative to the inertia parameters in the rigid model given in Equation 76.23. This means that the rigid robot model (Equation 76.23) may be viewed as a **singular perturbation** [3] of both the system with actuator dynamics (Equations 76.23 to 76.25) and of the flexible joint system (Equations 76.28 and 76.29). In the case of actuator dynamics, suppose that all entries L_i/R_i in the diagonal matrix $R^{-1}L$ are equal to $\epsilon << 1$ for simplicity and write Equations 76.26 and 76.27 as

$$M(q)\ddot{q} + C(q, \dot{q})\dot{q} + g(q) = K_m i \tag{76.30}$$

$$\epsilon\frac{di}{dt} + i = R^{-1}(V - K_b\dot{q}) \tag{76.31}$$

In the case of joint elasticity, define $z = K(q_2 - q_1)$ and suppose that all the joint stiffness constants are equal to $1/\epsilon^2$ for simplicity. It is easy to show that Equations 76.28 and 76.29 may be written as

$$D(q_1)\ddot{q}_1 + C(q_1, \dot{q}_1)\dot{q}_1 + g(q_1) = z \tag{76.32}$$

$$\epsilon^2 I\ddot{z} + z = \tau - I\ddot{q}_1 \tag{76.33}$$

In both cases above, the rigid model given by Equation 76.23 is recovered in the limit as $\epsilon \to 0$. Singular perturbation techniques are thus of great value in designing and analyzing control laws for manipulators.

76.1.3 PD Control

In light of the complexity of the dynamics of n-link robots, it is quite remarkable to discover that very simple control laws can be used in a number of cases. For example, a simple independent joint proportional-derivative (PD) control can achieve global asymptotic stability for the set-point tracking in the absence of gravity. This fact is a fundamental consequence of the passivity property discussed earlier. To see this, consider the dynamic equations in the absence of friction

$$M(q)\ddot{q} + C(q,\dot{q})\dot{q} = u \tag{76.34}$$

where $u = \tau - g(q)$. Let

$$u = K_p(q^d - q) - K_d\dot{q} \tag{76.35}$$

be an independent joint PD control, where q^d represents a constant reference set-point, and K_p and K_d are positive, diagonal matrices of proportional and derivative gains, respectively. Consider the Lyapunov function candidate

$$V = \frac{1}{2}\dot{q}^T M(q)\dot{q} + \frac{1}{2}(q^d - q)^T K_p(q^d - q). \tag{76.36}$$

Then a simple calculation using the skew symmetry property shows that

$$\dot{V} = -\dot{q}^T K_d\dot{q} \leq 0. \tag{76.37}$$

LaSalle's invariance principle [5] can now be used to show that the equilibrium state $q = q^d$, $\dot{q} = 0$ is globally asymptotically stable. Thus, if gravity is absent or is compensated as

$$\tau = g(q) + u, \tag{76.38}$$

then a simple independent joint PD control achieves global asymptotic stability. Unfortunately, these results do not extend to the case of time-varying reference trajectories since they rely on LaSalle's Theorem.

Other researchers have investigated proportional-integral-derivative (PID) control and adaptive gravity compensation in order to overcome the requirement that the gravity parameters in Equation 76.38 be known exactly. It has also been shown that asymptotic tracking may be achieved using only the reference set-point q^d in the gravity compensation, i.e.,

$$\tau = g(q^d) + u, \tag{76.39}$$

provided the PD gains are chosen suitably, which simplifies the on-line computational requirements. These latter results require the addition of cross terms to the Lyapunov function (Equation 76.36) in order to show asymptotic stability.

In practice, however, the input given by Equation 76.35 or Equation 76.38 may result in input saturation because of the large gains that are usually required and the large initial tracking error. One of the primary reasons for trajectory generation, such as trapezoidal velocity profiles, in the first place is to help achieve proper scaling of the amplifier inputs.

Nevertheless, these simple PD results provide theoretical justification for the widespread use of such controllers in commercial industrial manipulators. In fact, it is not too difficult to show the same result for a flexible joint robot in the absence of gravity, provided the PD control is implemented using the motor variables. This is because the flexible joint robot dynamics defines a passive mapping from motor torque τ to the motor velocity $\dot{q}_2$.

76.1.4 Feedback Linearization

The notion of **feedback linearization** of nonlinear systems is a relatively recent idea in control theory, whose practical realization has been made possible by the rapid development of microprocessor technology. The basic idea of feedback linearization control is to transform a given nonlinear system into a linear system by use of a nonlinear coordinate transformation and nonlinear feedback. Feedback linearization is a useful paradigm because it allows the extensive body of knowledge from linear systems to be brought to bear to design controllers for nonlinear systems. The roots of feedback linearization in robotics predate the general theoretical development by nearly a decade, going back to the early notion of feedforward-computed torque [8].

In the robotics context, feedback linearization is known as **inverse dynamics**. The idea is to exactly compensate all of the coupling nonlinearities in the Lagrangian dynamics in a first stage so that a second-stage compensator may be designed based on a linear and decoupled plant. Any number of techniques may be used in the second stage. The feedback linearization may be accomplished with respect to the joint space coordinates or with respect to the task space coordinates. Feedback linearization may also be used as a basis for force control, such as hybrid control and impedance control.

Joint Space Inverse Dynamics

We first present the main ideas in joint space where they are easiest to understand. The control architecture we use is important as a basis for later developments. Thus, given the plant model

$$M(q)\ddot{q} + C(q,\dot{q})\dot{q} + g(q) = \tau, \tag{76.40}$$

we compute the nonlinear feedback control law

$$\tau = M(q)a_q + C(q,\dot{q})\dot{q} + g(q) \tag{76.41}$$

where $a_q \in \mathbb{R}^n$ is, as yet, undetermined. Since the inertia matrix $M(q)$ is invertible for all q, the closed-loop system reduces to the decoupled **double integrator**

$$\ddot{q} = a_q. \tag{76.42}$$

Given a joint space trajectory, $q^d(t)$, an obvious choice for the outer loop term a_q is as a **PD plus feedforward acceleration** control

$$a_q = \ddot{q}^d + K_p\left(q^d - q\right) + K_d\left(\dot{q}^d - \dot{q}\right). \tag{76.43}$$

Substituting Equation 76.43 into Equation 76.42 and defining

$$\tilde{q} = q - q^d \quad , \tag{76.44}$$

we have the linear and decoupled closed-loop system

$$\ddot{\tilde{q}} + K_d\dot{\tilde{q}} + K_p\tilde{q} = 0. \tag{76.45}$$

We can implement the joint space inverse dynamics in a so-called **inner loop/outer loop** architecture as shown in Figure 76.4.

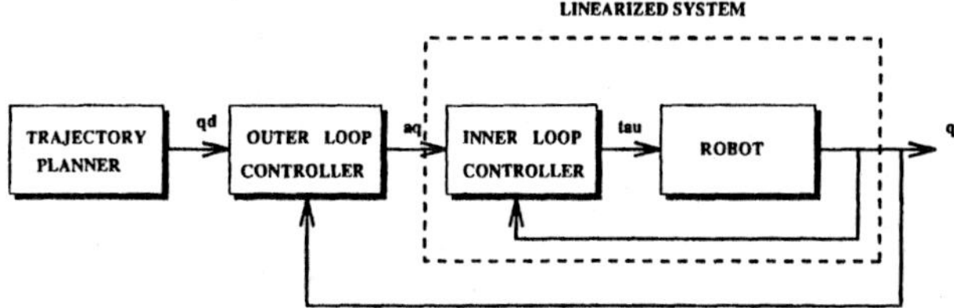

Figure 76.4 Inner loop/outer loop architecture.

The computation of the nonlinear terms in Equation 76.41 is performed in the inner loop, perhaps with a dedicated microprocessor to obtain high computation speed. The computation of the additional term a_q is performed in the outer loop. This separation of the inner loop and outer loop terms is important for several reasons. The structure of the inner loop control is fixed by Lagrange's equations. What control engineers traditionally think of as **control system design** is contained primarily in the outer loop. The outer loop control given in Equation 76.43 is merely the simplest choice of outer loop control and achieves asymptotic tracking of joint space trajectories in the ideal case of perfect knowledge of the model given by Equation 76.40. However, one has complete freedom to modify the outer loop control to achieve various other goals without the need to modify the dedicated inner loop control. For example, additional compensation terms may be included in the outer loop to enhance the robustness to parametric uncertainty, unmodeled dynamics, and external disturbances. The outer loop control may also be modified to achieve other goals, such as tracking of task space trajectories instead of joint space trajectories, regulating both motion and force, etc. The inner loop/outer loop architecture thus unifies many robot control strategies from the literature.

Task Space Inverse Dynamics

As a first illustration of the importance of the inner loop/outer loop paradigm, we show that tracking in task space can be achieved by modifying our choice of outer loop control a_q in Equation 76.42 while leaving the inner loop control unchanged. Let $X \in R^6$ represent the end-effector pose using any minimal representation of $SO(3)$. Since X is a function of the joint variables $q \in C$, we have

$$\begin{aligned} \dot{X} &= J(q)\dot{q} & (76.46)\\ \ddot{X} &= J(q)\ddot{q} + \dot{J}(q)\dot{q}. & (76.47)\end{aligned}$$

where $J = J_1$ is the Jacobian defined in Section 76.1.1. Given the double integrator (Equation 76.42) in joint space we see that if a_q is chosen as

$$a_q = J^{-1}\left\{a_X - \dot{J}\dot{q}\right\}, \tag{76.48}$$

then we have a double integrator model in task space coordinates

$$\ddot{X} = a_X. \tag{76.49}$$

Given a task space trajectory $X^d(t)$, we may choose a_X as

$$a_X = \ddot{X}^d + K_p(X^d - X) + K_d(\dot{X}^d - \dot{X}) \tag{76.50}$$

so that the Cartesian space tracking error, $\tilde{X} = X - X^d$, satisfies

$$\ddot{\tilde{X}} + K_d\dot{\tilde{X}} + K_p\tilde{X} = 0. \tag{76.51}$$

Therefore, a modification of the outer loop control achieves a linear and decoupled system directly in the task space coordinates, without the need to compute a joint trajectory and without the need to modify the nonlinear inner loop control.

Note that we have used a minimal representation for the orientation of the end-effector in order to specify a trajectory $X \in \mathbb{R}^6$. In general, if the end-effector coordinates are given in $SE(3)$, then the Jacobian J in the preceding formulation is the Jacobian J_0 defined in Section 76.1.1. In this case

$$V = \begin{pmatrix} v \\ \omega \end{pmatrix} = \begin{pmatrix} \dot{x} \\ \omega \end{pmatrix} = J(q)\dot{q} \tag{76.52}$$

and the outer loop control

$$a_q = J^{-1}(q)\{\begin{pmatrix} a_x \\ a_\omega \end{pmatrix} - \dot{J}(q)\dot{q}\} \tag{76.53}$$

applied to Equation 76.42 results in the system

$$\begin{aligned} \ddot{x} &= a_x \in \mathbb{R}^3 & (76.54)\\ \dot{\omega} &= a_\omega \in \mathbb{R}^3 & (76.55)\\ \dot{R} &= S(\omega)R, \quad R \in SO(3), \; S \in so(3). & (76.56)\end{aligned}$$

Although, in this case, the dynamics have not been linearized to a double integrator, the outer loop terms a_v and a_ω may still be used to achieve global tracking of end-effector trajectories in $SE(3)$. In both cases, we see that nonsingularity of the Jacobian is necessary to implement the outer loop control.

The inverse dynamics control approach has been proposed in a number of different guises, such as **resolved acceleration control** and **operational space control**. These seemingly distinct approaches have all been shown to be equivalent and may be incorporated into the general framework shown previously (see the article by Kreutz in [10, p. 77]).

It turns out that both the robot model including actuator dynamics given in Equations 76.26 and 76.27 and the model including joint flexibility given in Equations 76.28 and 76.29 are also feedback linearizable using a nonlinear coordinate transformation and nonlinear feedback. For the system with actuator dynamics given in Equations 76.26 and 76.27, if one chooses as state variables, the link positions, velocities, and accelerations, a nonlinear feedback control exists to reduce the system to a decoupled set of third-order integrators.

The coordinates in which the flexible joint system given in Equations 76.28 and 76.29 may be exactly linearized are the link positions, velocities, accelerations, and jerks; and a decoupled set of fourth-order integrators is achievable by suitable nonlinear feedback. See the article by Spong [p. 105] for details.

Composite Control

Although global feedback linearization in the case of actuator dynamics or joint flexibility is possible in theory, in practice it is difficult to achieve, mainly because the coordinate transformation is a function of the system parameters and, hence, sensitive to uncertainty. Also, the large differences in magnitude among the parameters, e.g., between the joint stiffness and the link inertia, may make the computation of the control ill-conditioned and the performance of the system poor. Using the singular perturbation models derived earlier, so-called **composite control** laws may produce better designs. We illustrate this idea using the flexible joint model given by Equations 76.32 and 76.33. The corresponding result for the system with actuator dynamics is similar.

Given the system of Equations 76.32 and 76.33, we define two related systems, the **quasi-steady-state system** and the **boundary layer system**, as follows. The quasi-steady-state system is the reduced-order system calculated by setting $\epsilon = 0$ in Equation 76.33,

$$\bar{z} = \bar{\tau} - I\ddot{\bar{q}}_1, \qquad (76.57)$$

where the overbar indicates quantities computed at $\epsilon = 0$, and eliminating $\bar{z}$ in Equation 76.32. It can easily be shown that the quasi-steady-state system thus derived is equal to

$$M(\bar{q}_1)\ddot{\bar{q}}_1 + C(\bar{q}_1, \dot{\bar{q}}_1)\dot{\bar{q}}_1 + g(\bar{q}_1) = \bar{\tau}. \qquad (76.58)$$

The quasi-steady-state is thus identical to the rigid robot model in terms of $\bar{q}_1$.

The boundary layer system represents the dynamics of $\eta = z - \bar{z}$, computed in the **fast time scale**, $\sigma = t/\epsilon$. This system can be shown to be

$$\frac{d^2}{d\sigma^2}\eta + (I^{-1} + D^{-1})\eta = \tau_f, \qquad (76.59)$$

where $\tau_f = \tau - \bar{\tau}$. A **composite control** for the flexible joint system (Equations 76.32 and 76.33) is a control law of the form

$$\tau = \bar{\tau}(q_1, \dot{q}_2, t) + \tau_f(\eta, \dot{\eta}) \qquad (76.60)$$

where $\bar{\tau}$ is designed for the quasi-steady-state system and τ_f is designed for the boundary layer system. The significant features of this approach are that

1. The control design is based on reduced-order systems, which is generally easier than designing a controller for the full-order system.
2. The quasi-steady-state system is just the rigid robot model, so existing controllers designed for rigid robots may be used without modification.

If τ_f is designed to render the boundary layer system asymptotically stable in the fast time scale and $\bar{\tau}$ is any control that achieves asymptotic tracking for the rigid system, it follows from Tichonov's Theorem [3] that

$$\begin{aligned} q_1(t) &= \bar{q}_1(t) + O(\epsilon) && (76.61)\\ z(t) &= \bar{z}(t) + \eta(\sigma) + O(\epsilon) && (76.62) \end{aligned}$$

uniformly on a time interval $[0, 1/\sigma]$. Stronger results showing asymptotic tracking may also be achieved, depending on the particular control chosen for the quasi-steady-state system. Roughly speaking, the response of the system with joint flexibility using the composite control given by Equation 76.60 is nearly the same as the response of the rigid robot model using only $\bar{\tau}$. An important consequence is that the inverse dynamics control for the rigid robot may be used for the flexible joint robot, provided a correction term is superimposed on it to stabilize the boundary layer system. The additional control term τ_f represents the damping of the joint oscillations. Other "rigid" controllers may also be used, such as the robust and adaptive controllers detailed in the next section.

76.1.5 Robust and Adaptive Control

The feedback linearization approach of the previous section exploits important structural properties of robot dynamics. However, the practical implementation of such controllers requires consideration of various sources of uncertainties such as modeling errors, computation errors, external disturbances, unknown loads, and noise. Robust and adaptive control are concerned with the problem of maintaining precise tracking under uncertainty. We distinguish robust from adaptive control in the sense that an adaptive algorithm typically incorporates some sort of online parameter estimation scheme while a robust, nonadaptive scheme does not.

Robust Feedback Linearization

A number of techniques from linear and nonlinear control theory have been applied to the problem of robust feedback linearization for manipulators. Chief among these are sliding modes, Lyapunov's second method, and the method of stable factorizations. Given the dynamic equations

$$M(q)\ddot{q} + C(q, \dot{q})\dot{q} + g(q) = \tau, \qquad (76.63)$$

the control input is chosen as

$$\tau = \hat{M}(q)a_q + \hat{C}(q, \dot{q})\dot{q} + \hat{g}(q) \qquad (76.64)$$

where $\hat{(\cdot)}$ represents the computed or nominal value of $(\cdot)$ and indicates that the theoretically exact feedback linearization cannot be achieved in practice due to the uncertainties in the system. The error or mismatch $\tilde{(\cdot)} = (\cdot) - \hat{(\cdot)}$ is a measure of one's knowledge of the system dynamics. The outer loop control term a_q may be used to compensate for the resulting perturbation terms.

If we set

$$a_q = \ddot{q}^d + K_d(\dot{q}^d - \dot{q}) + K_p(q^d - q) + \delta a \qquad (76.65)$$

and substitute Equations 76.64 and 76.65 into Equation 76.63 we obtain, after some algebra,

$$\ddot{\tilde{q}} + K_d\dot{\tilde{q}} + K_p\tilde{q} = \delta a + \eta(q, \dot{q}, \delta a, t) \qquad (76.66)$$

where

$$\eta = M^{-1}\{\tilde{M}(\ddot{q}^d + K_d\dot{\tilde{q}} + K_p\tilde{q} + \delta a) + \tilde{C}\dot{q} + \tilde{g}\}. \qquad (76.67)$$

In state-space we may write the system given by Equation 76.66 as

$$\dot{x} = Ax + B\{\delta a + \eta\} \tag{76.68}$$

where

$$x = \begin{pmatrix} \tilde{q} \\ \dot{\tilde{q}} \end{pmatrix}; \; A = \begin{bmatrix} 0 & I \\ -K_p & -K_d \end{bmatrix}; \; B = \begin{bmatrix} 0 \\ I \end{bmatrix}. \tag{76.69}$$

The approach is now to search for a time-varying scalar bound, $\rho(x, t) \geq 0$, on the uncertainty η, i.e.,

$$||\eta|| \leq \rho(x, t) \tag{76.70}$$

and to design the additional input term δa to guarantee asymptotic stability or, at least, ultimate boundedness of the state trajectory $x(t)$ in Equation 76.68. The bound ρ is difficult to compute, both because of the complexity of the perturbation terms in η and because the uncertainty η is itself a function of δa.

The sliding mode theory of variable structure systems has been extensively applied to the design of δa in Equation 76.68. The simplest such sliding mode control results from choosing the components δa_i of δa according to

$$\delta a_i = \rho_i(x, t)\text{sgn}(s_i), \; i = 1, \ldots, n \tag{76.71}$$

where ρ_i is a bound on the i-th component of η, $s_i = \dot{\tilde{q}}_i + \lambda_i \tilde{q}_i$ represents a sliding surface in the state-space, and sgn$(\cdot)$ is the signum function

$$\text{sgn}(s_i) = \begin{cases} +1 & \text{if} \quad s_i > 0 \\ -1 & \text{if} \quad s_i < 0 \end{cases} \tag{76.72}$$

An alternative but similar approach is the so-called theory of guaranteed stability of uncertain systems, based on Lyapunov's second method. Since K_p and K_d are chosen in Equation 76.68 so that A is a Hurwitz matrix, for any $Q > 0$ there exists a unique symmetric $P > 0$ satisfying the Lyapunov equation,

$$A^T P + PA = -Q. \tag{76.73}$$

Using the matrix P, the outer loop term δa may be chosen as

$$\delta a = \begin{cases} -\rho(x, t)\frac{B^T Px}{||B^T Px||} & ; \text{ if } \; ||B^T Px|| \neq 0 \\ 0 & ; \text{ if } \; ||B^T Px|| = 0 \end{cases} \tag{76.74}$$

The Lyapunov function $V = x^T Px$ can be used to show that $\dot{V}$ is negative definite along solution trajectories of the system given by Equation 76.68.

In both the sliding mode approach and the guaranteed stability approach, problems arise in showing the existence of solutions to the closed-loop differential equations because the control signal δa is discontinuous in the state x. In practice, a chattering control signal results due to nonzero switching delays. There have been many refinements and extensions to the above approaches to robust feedback linearization, mainly to simplify the computation of the uncertainty bounds and to smooth the chattering in the control signal [10]

The method of stable factorizations has also been applied to the robust feedback linearization problem. In this approach a linear, dynamic compensator $C(s)$ is used to generate δa to stabilize the perturbed system. Since A is a Hurwitz matrix, the Youla-parameterization may be used to generate the entire class, Ω, of stabilizing compensators for the unperturbed system, i.e., Equation 76.68 with $\eta = 0$. Given bounds on the uncertainty, the Small Gain Theorem is used to generate a sufficient condition for stability of the perturbed system, and the design problem is to determine a particular compensator, $C(s)$, from the class of stabilizing compensators Ω that satisfies this sufficient condition. The interesting feature of this problem is that the perturbation terms appearing in Equation 76.68 are finite in the L_∞ norm, but not necessarily in the L_2 norm sense. This means that standard H_∞ design methods fail for this problem. For this reason, the robust manipulator control problem was influential in the development of the L_1 optimal control field. Further details on these various outer loop designs may be found in [10].

Adaptive Feedback Linearization

Once the linear parameterization property for manipulators became widely known in the mid-1980s, the first globally convergent adaptive control results began to appear. These first results were based on the inverse dynamics or feedback linearization approach discussed earlier. Consider the plant (Equation 76.63) and control (Equation 76.64) as previously, but now suppose that the parameters appearing in Equation 76.64 are not fixed as in the robust control approach, but are time-varying estimates of the true parameters. Substituting Equation 76.64 into Equation 76.63 and setting

$$a_q = \ddot{q}^d + K_d(\dot{q}^d - \dot{q}) + K_p(q^d - q), \tag{76.75}$$

it can be shown, after some algebra, that

$$\ddot{\tilde{q}} + K_d\dot{\tilde{q}} + K_p\tilde{q} = \hat{M}^{-1}Y(q, \dot{q}, \ddot{q})\tilde{\theta} \tag{76.76}$$

where Y is the regressor function, and $\tilde{\theta} = \hat{\theta} - \theta$, and $\hat{\theta}$ is the estimate of the parameter vector θ. In state-space we write the system given by Equation 76.76 as

$$\dot{x} = Ax + B\Phi\tilde{\theta} \tag{76.77}$$

where

$$x = \begin{pmatrix} \tilde{q} \\ \dot{\tilde{q}} \end{pmatrix}; \; A = \begin{bmatrix} 0 & I \\ -K_p & -K_d \end{bmatrix};$$
$$B = \begin{bmatrix} 0 \\ I \end{bmatrix}; \; \Phi = \hat{M}^{-1}Y(q, \dot{q}, \ddot{q}) \tag{76.78}$$

with K_p and K_d chosen so that A is a Hurwitz matrix. Suppose that an output function $y = Cx$ is defined for Equation 76.77 in such a way that the transfer function $C(sI - A)^{-1}B$ is strictly positive real (SPR). It can be shown using the passivity theorem that, for $Q > 0$, there exists a symmetric, positive definite matrix P satisfying

$$A^T P + PA = -Q \tag{76.79}$$
$$PB = C^T \tag{76.80}$$

If the parameter update law is chosen as

$$\dot{\hat{\theta}} = -\Gamma^{-1}\Phi^T C x \tag{76.81}$$

where $\Gamma = \Gamma^T > 0$, then global convergence to zero of the tracking error with all internal signals remaining bounded can be shown using the Lyapunov function

$$V = x^T P x + \frac{1}{2}\tilde{\theta}^T \Gamma \tilde{\theta}. \tag{76.82}$$

Furthermore, the estimated parameters converge to the true parameters, provided the reference trajectory satisfies the condition of **persistency of excitation,**

$$\alpha I \leq \int_{t_0}^{t_0+T} Y^T(q^d, \dot{q}^d, \ddot{q}^d) Y(q^d, \dot{q}^d, \ddot{q}^d) dt \leq \beta I \tag{76.83}$$

for all t_0, where α, β, and T are positive constants.

In order to implement this adaptive feedback linearization scheme, however, one notes that the acceleration $\ddot{q}$ is needed in the parameter update law and that $\hat{M}$ must be guaranteed to be invertible, possibly by the use of projection in the parameter space. Later work was devoted to overcome these two drawbacks to this scheme, by using so-called **indirect** approaches based on a (filtered) prediction error.

Passivity-Based Approaches

By exploiting the passivity of the rigid robot dynamics it is possible to derive more elegant robust and adaptive control algorithms for manipulators, which are, at the same time, simpler to design. In the passivity-based approach we modify the inner loop control as

$$\tau = \hat{M}(q)a + \hat{C}(q,\dot{q})v + \hat{g}(q) - Kr \tag{76.84}$$

where v, a, and r are given as

$$\begin{aligned} v &= \dot{q}^d - \Lambda\tilde{q} \\ a &= \dot{v} = \ddot{q}^d - \Lambda\dot{\tilde{q}} \\ r &= \dot{q}^d - v = \dot{\tilde{q}} + \Lambda\tilde{q} \end{aligned}$$

with K, Λ diagonal matrices of positive gains. In terms of the linear parameterization of the robot dynamics, the control given in Equation 76.84 becomes

$$\tau = Y(q,\dot{q},a,v)\hat{\theta} - Kr \tag{76.85}$$

and the combination of Equation 76.84 with Equation 76.63 yields

$$M(q)\dot{r} + C(q,\dot{q})r + Kr = Y\tilde{\theta}. \tag{76.86}$$

Note that, unlike the inverse dynamics control given in Equation 76.41, the modified inner loop control of Equation 76.63 does not achieve a linear, decoupled system, even in the known parameter case, $\hat{\theta} = \theta$. However, the advantage achieved is that the regressor Y in Equation 76.86 does not contain the acceleration $\ddot{q}$, nor is the inverse of the estimated inertia matrix required.

Passivity Based Robust Control In the robust approach, the term $\hat{\theta}$ in Equation 76.85 is chosen as

$$\hat{\theta} = \theta_0 + u \tag{76.87}$$

where θ_0 is a fixed nominal parameter vector and u is an additional control term. The system given in Equation 76.86 then becomes

$$M(q)\dot{r} + C(q,\dot{q})r + Kr = Y(a,v,q,\dot{q})(\tilde{\theta} + u) \tag{76.88}$$

where $\tilde{\theta} = \theta_0 - \theta$ is a constant vector and represents the parametric uncertainty in the system. If the uncertainty can be bounded by finding a non-negative constant $\rho \geq 0$ such that

$$\|\tilde{\theta}\| = \|\theta_0 - \theta\| \leq \rho, \tag{76.89}$$

then the additional term u can be designed to guarantee stable tracking according to the expression

$$u = \begin{cases} -\rho\frac{Y^T r}{\|Y^T r\|} & ; \text{ if } \quad \|Y^T r\| \neq 0 \\ 0 & ; \text{ if } \quad \|Y^T r\| = 0 \end{cases} \tag{76.90}$$

The Lyapunov function

$$V = \frac{1}{2}r^T M(q) r + \tilde{q}\Lambda K \tilde{q} \tag{76.91}$$

can be used to show asymptotic stability of the tracking error. Note that $\tilde{\theta}$ is constant and so is not a state vector as in adaptive control. Comparing this approach with the approach in Section 76.1.5 we see that finding a constant bound ρ for the constant vector $\tilde{\theta}$ is much simpler than finding a time-varying bound for η in Equation 76.67. The bound ρ in this case depends only on the inertia parameters of the manipulator, while $\rho(x,t)$ in Equation 76.70 depends on the manipulator state vector and the reference trajectory and, in addition, requires some assumptions on the estimated inertia matrix $\hat{M}(q)$.

Various refinements of this approach are possible. By replacing the discontinuous control law with the continuous control

$$u = -\rho\frac{Y^T r}{\|Y^T r\| + \gamma_1 e^{-\gamma_2 t}} \tag{76.92}$$

not only is the problem of existence of solutions due to discontinuities in the control eliminated, but also the tracking errors can be shown to be globally exponentially stable. It is also possible to introduce an estimation algorithm to estimate the uncertainty bound ρ so that no *a priori* information of the uncertainty is needed.

Passivity-Based Adaptive Control In the adaptive approach, the vector $\hat{\theta}$ in Equation 76.86 is now taken to be a time-varying estimate of the true parameter vector θ. Instead of adding an additional control term, as in the robust approach, we introduce a parameter update law for $\hat{\theta}$. Combining the control law given by Equation 76.84 with Equation 76.63 yields

$$M(q)\dot{r} + C(q,\dot{q})r + Kr = Y\tilde{\theta}. \tag{76.93}$$

The parameter estimate $\hat{\theta}$ may be computed using standard methods, such as gradient or least squares. For example, using the gradient update law

$$\dot{\hat{\theta}} = -\Gamma^{-1}Y^T(q, \dot{q}, a, v)r \qquad (76.94)$$

together with the Lyapunov function

$$V = \frac{1}{2}r^T M(q)r + \tilde{q}^T \Lambda K \tilde{q} + \frac{1}{2}\tilde{\theta}^T \Gamma \tilde{\theta} \qquad (76.95)$$

results in global convergence of the tracking errors to zero and boundedness of the parameter estimates.

A number of important refinements to this basic result are possible. By using the reference trajectory instead of the measured joint variables in both the control and update laws, i.e.,

$$\tau = Y(q^d, \dot{q}^d, \ddot{q}^d)\hat{\theta} - K_p\dot{\tilde{q}} - K_d\tilde{q} \qquad (76.96)$$

with

$$\dot{\hat{\theta}} = -\Gamma^{-1}Y^T(q^d, \dot{q}^d, \ddot{q}^d)r, \qquad (76.97)$$

it is possible to show exponential stability of the tracking error in the known parameter case, asymptotic stability in the adaptive case, and convergence of the parameter estimation error to zero under persistence of excitation of the reference trajectory.

76.1.6 Time-Optimal Control

For many applications, such as palletizing, there is a direct correlation between the speed of the robot manipulator and cycle time. For these applications, making the robot work faster translates directly into an increase in productivity. Since the input torques to the robot are limited by the capability of the actuators as

$$\tau^{min}(t) \leq \tau(t) \leq \tau^{max}(t), \qquad (76.98)$$

it is natural to consider the problem of time-optimal control. In many applications, such as seam tracking, the geometric path of the end-effector is constrained in the task space. In such cases, it is useful to produce time-optimal trajectories, i.e., time-optimal parameterizations of the geometric path that can be presented to the feedback controller. For this reason, most of the research into the time-optimal control of manipulators has gone into the problem of generating a minimum time trajectory along a given path in task space. Several algorithms are now available to compute such time-optimal trajectories.

Formulation of the Time-Optimal Control Problem

Consider an end-effector path $p(s) \in \mathbb{R}^6$ parameterized by arc length s along the path. If the manipulator is constrained to follow this path, then

$$p(s) = f(q) \qquad (76.99)$$

where $f(q)$ is the forward kinematics map. We may also use the inverse kinematics map to write

$$q = f^{-1}(p(s)) = e(s) \in C \qquad (76.100)$$

Either the end-effector path (Equation 76.99) or the joint space path (Equation 76.100) may be used in the subsequent development. We illustrate the formulation with the joint space path (Equation 76.100) and refer the reader to [10] for further details.

The final time t_f may be specified as

$$t_f = \int_0^{t_f} dt = \int_0^{s_{max}} ds/\dot{s}, \qquad (76.101)$$

which suggests that, in order to minimize the final time, the velocity $\dot{s}$ along the path should be maximized. It turns out that the optimal solution in this case is bang-bang; i.e., the acceleration $\ddot{s}$ is either maximum or minimum along the path. Since the trajectory is parameterized by the scalar s, phase plane techniques may be used to calculate the maximum acceleration $\ddot{s}$, as we shall see. From Equation 76.100 we may compute

$$\begin{aligned} \dot{q} &= e_s(s)\dot{s} &(76.102)\\ \ddot{q} &= e_s(s)\ddot{s} + \dot{e}_s\dot{s} &(76.103) \end{aligned}$$

where e_s is the Jacobian of the mapping given by Equation 76.100. Therefore, once the optimal solution $s(t)$ is computed, the above expressions can be used to determine the optimal joint space trajectory. Substituting the expressions for $\dot{q}$ and $\ddot{q}$ into the manipulator dynamics (Equation 76.63) leads to a set of second-order equations in the scalar arc length parameter s,

$$a_i(s)\ddot{s} + b_i(s)\dot{s}^2 + c_i(s) = \tau, \; i = 1, \ldots, n. \qquad (76.104)$$

Given the bounds on the joint actuator torques from Equation 76.98 bounds on $\ddot{s}$ can be determined by substituting Equation 76.104 into Equation 76.98

$$\tau_i^{min} \leq a_i(s)\ddot{s} + b_i(s)\dot{s}^2 + c_i(s) \leq \tau_i^{max} \;\; i = 1, \ldots, n, \qquad (76.105)$$

which can be written as a set of n-constraints on $\ddot{s}$ as

$$\alpha_i(s, \dot{s}) \leq \ddot{s} \leq \beta_i(s, \dot{s}) \qquad (76.106)$$

$$\text{where } \begin{aligned} \alpha_i &= (\tau_i^\alpha - b_i\dot{s}^2 - c_i)/a_i \\ \beta_i &= (\tau_i^\beta - b_i\dot{s}^2 - c_i)/a_i \end{aligned}$$

$$\begin{aligned} &\text{with } \tau_i^\alpha = \tau_i^{min} \text{ and } \tau_i^\beta = \tau_i^{max} && \text{if} \quad a_i > 0 \\ &\tau_i^\alpha = \tau_i^{max} \text{ and } \tau_i^\beta = \tau_i^{min} && \text{if} \quad a_i < 0 \end{aligned}$$

Thus, the bounds on the scalar $\ddot{s}$ are determined as

$$\alpha(s, \dot{s}) \leq \ddot{s} \leq \beta(s, \dot{s}) \qquad (76.107)$$

where

$$\alpha(s, \dot{s}) = \max\{\alpha_i(s, \dot{s})\}\,; \;\; \beta(s, \dot{s}) = \min\{\beta_i(s, \dot{s})\} \qquad (76.108)$$

Since the solution is known to be bang-bang, the optimal control is determined by finding the times, or positions, at which $\ddot{s}$ switches between

$$\ddot{s} = \beta(s, \dot{s})$$

and

$$\ddot{s} = \alpha(s, \dot{s}).$$

These switching times may be found by constructing switching curves in the phase plane s-$\dot{s}$ corresponding to $\alpha(s, \dot{s}) = \beta(s, \dot{s})$. Various methods for constructing this switching curve are found in the references contained in [10, part 6]. A typical minimum-time solution is shown in Figure 76.5.

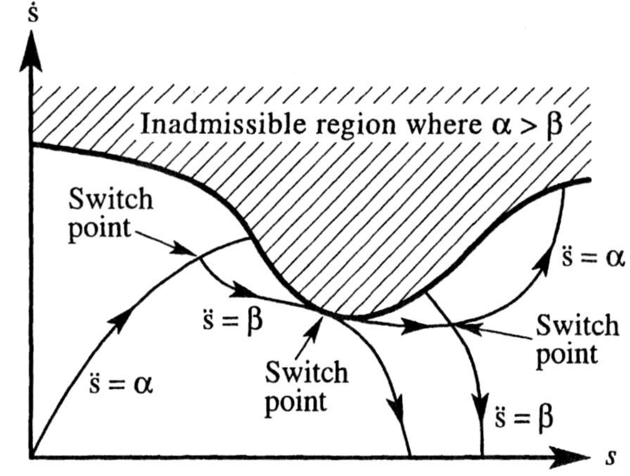

Figure 76.5 Minimum-time trajectory in the s-$\dot{s}$ plane.

76.1.7 Repetitive and Learning Control

Since many robotic applications, such as pick-and-place operations, painting, and circuit board assembly, involve repetitive motions, it is natural to consider using data gathered in previous cycles to try to improve the performance of the manipulator in subsequent cycles. This is the basic idea of **repetitive control** or **learning control.**

Consider the rigid robot model given by Equation 76.23 and suppose one is given a desired output trajectory on a finite time interval, $y_d(t), 0 \le t \le T$, which may represent a joint space trajectory or a task space trajectory. The reference trajectory $y_d(t)$ is used in repeated trials of the manipulator, assuming either that the trajectory is periodic, $y_d(T) = y_d(0)$ (repetitive control), or that the robot is reinitialized to lie on the desired trajectory at the beginning of each trial (learning control). Hereafter we use the term *learning control* to mean either repetitive or learning control.

Let $\tau_k(t)$ be the input torque during the k-th cycle, which produces an output $y_k(t), 0 \le t \le T$. The input/output pair $[\tau_k(t), y_k(t)]$ may be stored and utilized in the $k+1$-st cycle. The initial control input $\tau_0(t)$ can be any control input that produces a stable output, such as a PD control.

The learning control problem is to determine a recursive control

$$\tau_{k+1}(t) = F(\tau_k(t), \Delta y_k(t)), \quad 0 \le t \le T, \tag{76.109}$$

where $\Delta y_k(t) = y_k(t) - y_d(t)$, such that $||\Delta y_k|| \to 0$ as $k \to \infty$ in some suitably defined function space norm, $||\cdot||$. Such learning control algorithms are attractive because accurate models of the dynamics need not be known *a priori.*

Several approaches have been used to generate a suitable learning law F and to prove convergence of the output error. A P-type learning law is one of the form

$$\tau_{k+1}(t) = \tau_k(t) - \Phi \Delta y_k(t), \tag{76.110}$$

so called because the correction term to the input torque at each iteration is proportional to the error Δy_k. A D-type learning law is one of form

$$\tau_{k+1}(t) = \tau_k(t) - \Gamma \frac{d}{dt} \Delta y_k(t) \tag{76.111}$$

A more general PID type learning algorithm takes the form

$$\tau_{k+1}(t) = \tau_k(t) - \Gamma \frac{d}{dt} \Delta y_k(t) - \Phi \Delta y_k(t) - \int \psi \Delta y_k(u) du. \tag{76.112}$$

Convergence of $y_k(t)$ to $y_d(t)$ has been proved under various assumptions on the system. The earliest results considered the robot dynamics linearized around the desired trajectory and proved convergence for the linear time-varying system that results. Later results proved convergence based on the complete Lagrangian model.

Passivity has been shown to play a fundamental role in the convergence and robustness of learning algorithms. Given a joint space trajectory $q_d(t)$, let $\tau_d(t)$ be defined by the inverse dynamics, i.e.,

$$\tau_d(t) = M(q_d(t))\ddot{q}_d(t) + C(q_d(t), \dot{q}_d(t))\dot{q}_d(t) + g(q_d(t)). \tag{76.113}$$

The function $\tau_d(t)$ need not be computed; it is sufficient to know that it exists. Consider the P-type learning control law given by Equation 76.110 and subtract $\tau_d(t)$ from both sides to obtain

$$\Delta \tau_{k+1} = \Delta \tau_k - \Phi \Delta y_k \tag{76.114}$$

where $\Delta \tau_k = \tau_k - \tau_d$. It follows that

$$\Delta \tau_{k+1}^T \Phi^{-1} \Delta \tau_{k+1} = \Delta \tau_k^T \Phi^{-1} \Delta \tau_k + \Delta y_k^T \Phi \Delta y_k - 2 \Delta y_k^T \Delta \tau_k. \tag{76.115}$$

Multiplying both sides by $e^{-\lambda t}$ and integrating over $[0, T]$ it can be shown that

$$||\Delta \tau_{k+1}||^2 \le ||\Delta \tau_k||^2 - \beta ||\Phi \Delta y_k||^2 \tag{76.116}$$

provided there exist positive constants λ and β such that

$$\int_0^T e^{-\lambda t} \Delta y_k^T \Delta \tau_k(t) dt \ge \frac{1+\beta}{2} ||\Phi \Delta y_k||^2 \tag{76.117}$$

for all k. Equation 76.117 defines a passivity relationship of the exponentially weighted error dynamics. It follows from Equation 76.116 that $\Delta y_k \to 0$ in the L_2 norm. See [10, part 5] and the references therein for a complete discussion of this and other results in learning control.

References

[1] Asada, H. and Slotine, J-J. E., *Robot Analysis and Control*, John Wiley & Sons, Inc., New York, 1986.

[2] Dorf, R.C., Ed., *International Encyclopedia of Robotics: Applications and Automation*, John Wiley & Sons, Inc., 1988.

[3] Kokotović, P.V., Khalil, H.K., and O'Reilly, J., *Singular Perturbation Methods in Control: Analysis and Design*, Academic Press, Inc., London, 1986.

[4] Luh, J.Y.S., Conventional controller design for industrial robots: a tutorial, *IEEE Trans. Syst., Man, Cybern.*, 13(3), 298–316, May/June 1983.

[5] Khalil, H., *Nonlinear Systems*, Macmillan Press, New York, 1992.

[6] Nof, S.Y., Ed., *Handbook of Industrial Robotics*, John Wiley & Sons, Inc., New York, 1985.

[7] Ortega, R., and Spong, M.W., Adaptive control of rigid robots: a tutorial, in Proc. IEEE Conf. Decision Control, 1575–1584, Austin, TX, 1988.

[8] Paul, R.C., Modeling, Trajectory Calculation, and Servoing of a Computer Controlled Arm, Stanford A.I. Lab, A.I. Memo 177, Stanford, CA, November 1972.

[9] Spong, M.W., On the robust control of robot manipulators, *IEEE Trans. Autom. Control*, 37, 1782–1786, November 1992.

[10] Spong, M.W., Lewis, F., and Abdallah, C., *Robot Control: Dynamics, Motion Planning, and Analysis*, IEEE Press, 1992.

[11] Spong, M.W. and Vidyasagar, M., *Robot Dynamics and Control*, John Wiley & Sons, Inc., New York, 1989.

76.2 Force Control of Robot Manipulators

Joris De Schutter, Katholieke Universiteit Leuven, Department of Mechanical Engineering, Leuven, Belgium

Herman Bruyninckx, Katholieke Universiteit Leuven, Department of Mechanical Engineering, Leuven, Belgium

76.2.1 Introduction

Robots of the first generations were conceived as "open-loop" positioning devices; they operate with little or no feedback at all from the process in which they participate. For industrial assembly environments, this implies that all parts or subassemblies have to be prepositioned with a high accuracy, which requires expensive and rather inflexible peripheral equipment. Providing robots with sensing capabilities can reduce these accuracy requirements considerably. In particular, for industrial assembly, force feedback is extremely useful. But also for other tasks, in which a tool held by the robot has to make controlled contact with a work piece, as in deburring, polishing, or cleaning, it is not a good idea to rely fully on the positioning accuracy of the robot, and force feedback or force control becomes mandatory.

Force feedback is classified into two categories. In passive force feedback, the trajectory of the robot end-effector is modified by the interaction forces due to the inherent compliance of the robot; the compliance may be due to the structural compliance of links, joints, and end-effector or to the compliance of the position servo. In passive force feedback there is no actual force measurement, and the preprogrammed trajectory of the end-effector is never changed at execution time. On the other hand, in active force feedback, the interaction forces are measured, fed back to the controller, and used to modify, or even generate on-line, the desired trajectory of the robot end-effector.

Up till now force feedback applications in industrial environments have been mainly passive, for obvious reasons. Passive force feedback requires neither a force sensor nor a modified programming and control system and is therefore simple and cheap. In addition, it operates very fast. The Remote Center Compliance (RCC), developed by Draper Lab [1] and now widely available in various forms, is a well-known example of passive force feedback. It consists of a compliant end-effector that is designed and optimized for peg-into-hole assembly operations.

However, compared to active force feedback, passive force feedback has several disadvantages. It lacks flexibility, since for every robot task a special-purpose compliant end-effector has to be designed and mounted. Also, it can deal only with small errors of position and orientation. Finally, since no forces are measured, it can neither detect nor cope with error conditions involving excessive contact forces, and it cannot guarantee that high contact forces will never occur.

Clearly, although active force feedback has an answer to all of these issues, it is usually slower, more expensive, and more sophisticated than purely passive force feedback. Apart from a force sensor, it also requires an adapted programming and control system. In addition, it has been shown [2] that, in order to obtain a reasonable task execution speed and disturbance rejection capability, active force feedback has to be used in combination with passive force feedback.

In this chapter the design of active force controllers is discussed. More detailed reviews are given in [7] and [10].

In order to apply active force control, the following components are needed: force measurement, task specification, and control.

Force measurement. For a general force-controlled task, six force components are required to provide complete contact force information: three translational force components and three torques. Very often, a force sensor is mounted at the robot wrist, but other possibilities exist. The force signals may be obtained using strain measurements, which results in a stiff sensor, or using deformation measurements (e.g., optically), which results in a compliant sensor. The latter approach has an advantage if additional passive compliance is desired.

Task specification. For robot tasks involving constrained motion, the user has to specify more than just the desired motion as in the case of motion in free space. In addition, he has to specify how the robot has to interact with the external constraints. Basically there are two approaches, which are explained in this chapter. In the *hybrid force/position control* approach, the user specifies both desired motion and, explicitly, desired contact forces in two mutually independent subspaces. In the second approach, called *impedance control*, the user specifies how the robot has to comply with the external constraints; i.e., he specifies the dynamic relationship between contact forces and executed motion. In the hybrid approach, there is a clear separation between task specification and control; in the impedance approach, task specification and control are closely linked.

Control. The aim of the force control system is to make the actual contact forces, as measured by the sensor, equal to the desired contact forces, given by the task specification. This is called low-level or setpoint control, which is the main topic of this chapter. However, by interpreting the velocities actually executed by the robot, as well as the measured contact forces, much can be learned about the actual geometry of the constraints. This is the key for more high-level or adaptive control.

The chapter is organized as follows. Section 76.2.2 derives the system equations for control and task specification purposes. It also contains examples of hybrid force/position task specification. Section 76.2.3 describes methods for hybrid control of end-effector motion and contact force. Section 76.2.4 shows how to adapt the hybrid controller to the actual contact geometry. Section 76.2.5 briefly describes the impedance approach.

76.2.2 System Equations

Chapter 76.1 derived the general dynamic equations for a robot arm constrained by a contact with the environment:

$$M(q)\ddot{q} + C(q,\dot{q})\dot{q} + g(q) + J^T F_e = \tau. \qquad (76.118)$$

$F_e^T = \left(f^T \; m^T\right)$ is a six-vector representing Cartesian forces (f) and torques (m) occurring in the contact between robot and environment. $q = [q_1 \ldots q_n]^T$ are the n joint angles of the manipulator. $M(q)$ is the inertia matrix of the manipulator, expressed in joint space form. $C(q,\dot{q})\dot{q}$ and $g(q)$ are the velocity- and gravity-dependent terms, respectively. τ is the vector of joint torques. J is the manipulator's Jacobian matrix that transforms joint velocities $\dot{q}$ to the Cartesian linear and angular end-effector velocities represented by the six-vector $V^T = \left(v^T \; \omega^T\right)$:

$$V = J\dot{q}. \qquad (76.119)$$

The Cartesian contact force F_e is determined both by Equation 76.118 and by the dynamics of the environment. The dynamic model of the environment contains two aspects: (1) the kinematics of the contact geometry; i.e., in what directions is the manipulator's motion constrained by the environment; and (2) the relationship between force applied to the environment and the deformation of the constraint surface. We consider two cases: (1) the robot and the environment are perfectly rigid; and (2) the environment behaves as a mass-spring-damper system. The first model is most appropriate for:

- Motion specification purposes: The user has to specify the desired motion of the robot, as well as the desired contact forces. This is easier within a Cartesian, rigid, and purely geometric constraint model.
- Theoretical purposes: A perfectly rigid interaction between robot and environment is an interesting *ideal* limit case, to which all other control approaches can be compared [9].

The soft environment model corresponds better to most of the real situations. Usually, the model of the robot-environment interaction is simplified by assuming that all compliance in the system—including that of the manipulator, its servo, the force sensor, and the tool—is localized in the environment.

Rigid Environment

Two cases are considered: (1) the constraints are formulated in joint or configuration space; and (2) the constraints are formulated in Cartesian space.

Configuration Space Formulation Assume that the kinematic constraints imposed by the environment are expressed in configuration space by n_f algebraic equations

$$\psi_q(q) = 0. \qquad (76.120)$$

This assumes the constraints are rigid, bilateral, and holonomic. Assume also that these n_f constraints are mutually independent; then they take away n_f motion degrees of freedom from the manipulator.

The constrained robot dynamics are now derived from the unconstrained robot dynamics by the classical technique of incorporating the constraints into the Lagrangian function (see Chapter 76.1). For the unconstrained system, the Lagrangian is the difference between the system's kinetic energy, $\mathcal{K}$, and its potential energy, $\mathcal{P}$. Each of the constraint equations ψ_{qj}, $j = 1, \ldots, n_f$, should be identically satisfied. Lagrange's approach to satisfying both the dynamical equations of the system and the constraint requirements was to define the extended Lagrangian

$$\mathcal{L} = \mathcal{K} - \mathcal{P} - \sum_{j=1}^{n_f} \lambda_j \psi_{qj}(q).$$

$\lambda = [\lambda_1 \; \ldots \; \lambda_{n_f}]^T$ is a vector of Lagrange multipliers. The solution to the Lagrangian equations

$$\frac{d}{dt}\left(\frac{\partial \mathcal{L}}{\partial \dot{q}_k}\right) - \frac{\partial \mathcal{L}}{\partial q_k} = \tau_k, \; k = 1, \ldots, n$$

then results in

$$M(q)\ddot{q} + C(q,\dot{q})\dot{q} + g(q) + J_{\psi_q}^T(q)\lambda = \tau. \qquad (76.121)$$

$J_{\psi_q}(q)$ is the $n_f \times n$ Jacobian matrix (i.e., the matrix of partial derivatives with respect to the joint angles q) of the constraint function $\psi_q(q)$. J_{ψ_q} is of full rank, n_f, because all constraints are assumed to be independent. (If not, the constraints represent a so-called *hyperstatic* situation.) $J_{\psi_q}^T(q)\lambda$ represents the ideal contact forces, i.e., without contact friction.

Cartesian Space Formulation The kinematic constraints in Cartesian space are easily derived from the configuration space results, as long as the manipulator's Jacobian matrix J is square (i.e., $n = 6$) and nonsingular. In that case, Equation 76.121 is equivalent to [J^T denotes J transpose and $J^{-T} = (J^T)^{-1}$]

$$M(q)\ddot{q} + C(q,\dot{q})\dot{q} + g(q) + J^T\left(J^{-T}J^T_{\psi_q}(q)\right)\lambda = \tau.$$

Denote $J_{\psi_q}(q)J^{-1}$ by $J_{\psi_x}(q)$. This is an $n_f \times 6$ matrix. Then

$$M(q)\ddot{q} + C(q,\dot{q})\dot{q} + g(q) + J^T J^T_{\psi_x}(q)\lambda = \tau. \tag{76.122}$$

Comparison with Equation 76.118 shows that

$$F_e = J^T_{\psi_x}\lambda. \tag{76.123}$$

This means that the ideal Cartesian reaction forces F_e belong to an n_f dimensional vector space, spanned by the full rank matrix $J^T_{\psi_x}$. Let S_f denote a basis of this vector space, i.e., a set of n_f independent contact forces that can be generated by the constraints:

$$\forall F_e, \exists\phi = [\phi_1 \ \ldots \ \phi_{n_f}]^T \in \mathbb{R}^{n_f} : F_e = S_f\,\phi. \tag{76.124}$$

The coordinate vector ϕ contains dimensionless scalars; all physical dimensions of Cartesian force are contained in the basis S_f.

The time derivative of Equation 76.120 yields

$$0 = \frac{d\psi_q(q)}{dt} = \frac{\partial\psi_q(q)}{\partial q}\dot{q} = J_{\psi_q}\dot{q}. \tag{76.125}$$

Hence, using Equation 76.119 yields

$$\left(J_{\psi_q}J^{-1}\right)(J\dot{q}) = J_{\psi_x}V = 0. \tag{76.126}$$

Combining Equations 76.123 and 76.126 gives the kinematic *reciprocity* relationship between the ideal Cartesian reaction forces F_e (spanning the so-called *force-controlled* subspace) and the Cartesian manipulator velocities V that obey the constraints (spanning the *velocity-controlled* subspace):

$$V^T F_e = F_e^T V = 0. \tag{76.127}$$

This means that the velocity-controlled subspace is the n_x dimensional ($n_x = 6 - n_f$) reciprocal complement of the force-controlled subspace. It can be given a basis S_x, such that

$$\forall V, \exists\chi = [\chi_1 \ \ldots \ \chi_{n_x}]^T \in \mathbb{R}^{n_x} : V = S_x\,\chi. \tag{76.128}$$

Again, χ is a vector with physically dimensionless scalars; the columns of S_x have the physical dimensions of a Cartesian velocity. From Equation 76.127 it follows that

$$S_x^T S_f = 0. \tag{76.129}$$

Task Specification The task specification module specifies force and velocity setpoints, F^d and V^d, respectively. In order to be consistent with the constraints, these setpoints must lie in the force- and velocity-controlled directions, respectively. Hence, the instantaneous task description corresponds to specifying the vectors ϕ^d and χ^d:

$$F^d = S_f\,\phi^d, \quad V^d = S_x\,\chi^d. \tag{76.130}$$

These equations are invariant with respect to the choice of reference frame and with respect to a change in the physical units. However, the great majority of tasks have a set of orthogonal reference frames in which the task specification becomes very easy and intuitive. Such a frame is called a *task frame* or *compliance frame*, [6]. Figures 76.6 and 76.7 show two examples.

Inserting a round peg in a round hole. The goal of this task is to push the peg into the hole while avoiding wedging and jamming. The peg behaves as a cylindrical joint; hence, it has two degrees of motion freedom ($n_x = 2$) while the force-controlled subspace is of rank four ($n_f = 4$). Hence, the task can be achieved by the following four force setpoints and two velocity setpoints in the task frame depicted in Figure 76.6:

1. A nonzero velocity in the Z direction
2. Zero forces in the X and Y directions
3. Zero torques about the X and Y directions
4. An arbitrary angular velocity about the Z direction

The task continues until a "large" reaction force in the Z direction is measured. This indicates that the peg has hit the bottom of the hole.

Sliding a block over a planar surface. The goal of this task is to slide the block over the surface without generating too large reaction forces and without breaking the contact. There are three velocity-controlled directions and three force-controlled directions ($n_x = n_f = 3$). Hence, the task can be achieved by the following setpoints in the task frame depicted in Figure 76.7:

1. A nonzero force in the Z direction
2. A nonzero velocity in the X direction
3. A zero velocity in the Y direction
4. A zero angular velocity about the Z direction
5. Zero torques about the X and Y directions

Soft Environment

For the fully constrained case, i.e., all end-effector degrees of freedom are constrained by the environment, the dynamic equation of the robot-environment interaction is given by

$$F_e = M_e\,\Delta a + C_e\,\Delta V + K_e\,\Delta X. \tag{76.131}$$

$\Delta V = V - V_e$ is the Cartesian deformation velocity of the soft environment, with V_e the velocity of the environment. $\Delta a = (d\Delta V)/(dt)$ is the deformation acceleration, and $\Delta X = \int \Delta V\,dt$ is the deformation. Note that velocity is taken here as the basic motion input, since the difference $X - X_e$ of two-position

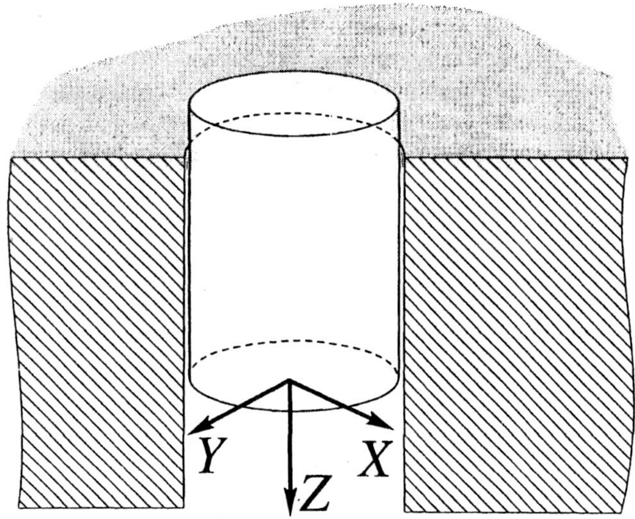

Figure 76.6 Peg-in-hole.

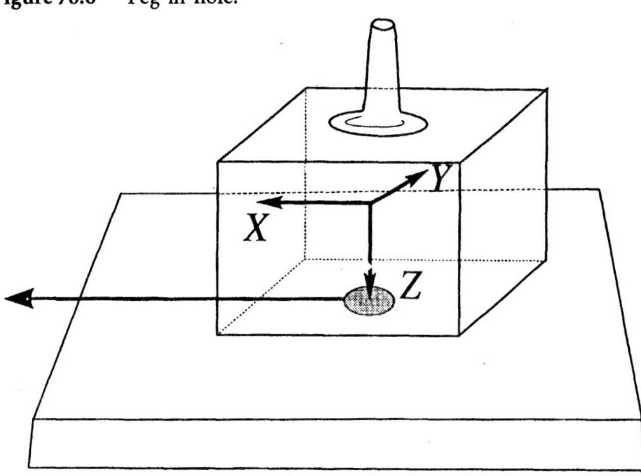

Figure 76.7 Sliding a block.

six-vectors is not well defined. Here X represents position and orientation of the robot end-effector, and X_e represents position and orientation of the environment. M_e is a positive definite inertia matrix; C_e and K_e are positive semidefinite damping and stiffness matrices. If there is sufficient passive compliance in the system, Equation 76.131 is approximated by

$$F_e = K_e \, \Delta X. \tag{76.132}$$

This case of soft environment is considered subsequently.

For partially constrained motion, the contact kinematics influence the dynamics of the robot-environment interaction. Moreover, in the case of a soft environment the measured velocity V does not completely belong to the ideal velocity subspace, defined for a rigid environment, because the environment can deform. Similarly, the measured force F_e does not completely belong to the ideal force subspace, also due to friction along the contact surface.

Hence, for control purposes the measured quantities V and F_e have to be projected onto the corresponding modeled subspaces. Algebraically, these projections are performed by projection matrices $P_x = S_x S_x^\dagger$ and $P_f = S_f S_f^\dagger$ [3]. $S_x^\dagger$ and $S_f^\dagger$ are (weighted) pseudo inverses, defined as

$$\begin{aligned} S_x^\dagger &= \left(S_x^T K_e S_x\right)^{-1} S_x^T K_e, \\ S_f^\dagger &= \left(S_f^T K_e^{-1} S_f\right)^{-1} S_f^T K_e^{-1}. \end{aligned} \tag{76.133}$$

With Equation 76.129 this yields

$$S_x^\dagger K_e^{-1} S_f = 0, \quad S_f^\dagger K_e S_x = 0. \tag{76.134}$$

The projection matrices decompose every Cartesian force F_e into a *force of constraint* $P_f F_e$, which is fully taken up by the constraint, and a *force of motion* $(I_6 - P_f) F_e$, which generates motion along the velocity-controlled directions. (I_6 is the 6×6 unity matrix.) Similarly, the Cartesian velocity V consists of a *velocity of freedom* part $P_x V$ in the velocity-controlled directions and a *velocity of constraint* part $(I_6 - P_x)V$, which deforms the environment. Within the velocity- and force-controlled subspaces, the measured velocity V and the measured force F_e correspond, respectively, to the following coordinate vectors with respect to the bases S_x and S_f:

$$\chi = S_x^\dagger \, V, \quad \phi = S_f^\dagger \, F_e. \tag{76.135}$$

Note that in the limit case of a rigid (and frictionless) environment, V and F_e do lie in the ideal velocity- and force-controlled subspaces. As a result, the projections of V and F_e onto the velocity- and force-controlled subspace, respectively, coincide with V and F_e. Hence, Equation 76.135 always gives the same results, whatever weighting matrices are chosen in Equation 76.133.

Using the projection matrices P_f and P_x, the ideal contact force of Equation 76.132 is rewritten for the partially constrained case as

$$F_e = P_f K_e (I_6 - P_x) \Delta X. \tag{76.136}$$

With Equation 76.134 this reduces to

$$F_e = P_f K_e \Delta X. \tag{76.137}$$

76.2.3 Hybrid Force/Position Control

The aim of hybrid control is to split up simultaneous control of both end-effector motion and contact force into two separate and decoupled subproblems [5]. Three different hybrid control approaches are presented here. In the first two, *acceleration-resolved* approaches, the control signals generated in the force and velocity subspaces are transformed to joint space signals, necessary to drive the robot joints, in terms of accelerations. Both the cases of a rigid and of a soft environment are considered. In the third, *velocity-resolved* approach, this transformation is performed in terms of velocities. Similarly as described in Chapter 76.1, all three approaches use a combination of an inner and an outer controller: the inner controller compensates the dynamics of the robot arm and may be model based; the outer controller is purely error driven.

We consider only the case of a constant contact geometry, so

$$\dot{S}_x = \ddot{S}_x = 0; \quad \dot{S}_f = \ddot{S}_f = 0. \tag{76.138}$$

Acceleratio-Resolved Control: Case of Rigid Environment

The system equations are given by Equation 76.118, where F_e is an ideal constraint force as in Equation 76.124. The inner loop controller is given by

$$\tau = M(q)a_q + C(q,\dot{q})\dot{q} + g(q) + J^T F^d, \qquad (76.139)$$

where F^d is the desired force, specified as in Equation 76.130. a_q is a desired joint space acceleration, which is related to a_x, the desired Cartesian space acceleration resulting from the outer loop controller:

$$a_q = J^{-1}(a_x - \dot{J}\dot{q}). \qquad (76.140)$$

The closed-loop dynamics of the system with its inner loop controller are derived as follows. Substitute Equation 76.140 in Equation 76.139, then substitute the result in Equation 76.118. Using the derivative of Equation 76.119,

$$J\ddot{q} + \dot{J}\dot{q} = A, \qquad (76.141)$$

where A is the Cartesian acceleration of the end-effector, this results in

$$A + JM^{-1}J^T F_e = a_x + JM^{-1}J^T F^d. \qquad (76.142)$$

In the case of a rigid environment, the Cartesian acceleration A is given by the derivative of Equation 76.128:

$$A = S_x \dot{\chi}. \qquad (76.143)$$

The outer loop controller generates an acceleration in the motion-controlled subspace only:

$$a_x = S_x\left(\dot{\chi}^d + K_{dx}\tilde{\chi} + K_{px}\tilde{\chi}_\Delta\right), \qquad (76.144)$$

where $\tilde{\chi} = \chi^d - \chi$; χ^d specifies the desired end-effector velocity as in Equation 76.130; χ represents the coordinates of the measured velocity, and is given by Equation 76.135; $\tilde{\chi}_\Delta$ represents the time integral of $\tilde{\chi}$; K_{dx} and K_{px} are control gain matrices with dimensions $n_x \times n_x$ and physical units $\frac{1}{time}$ and $\frac{1}{(time)^2}$, respectively.

Substituting the outer loop controller of Equation 76.144 into Equation 76.142 results in the closed-loop equation. This closed-loop equation is split up into two independent parts corresponding to the force- and velocity-controlled subspaces. First, premultiply Equation 76.142 with $S_x^\dagger K_e^{-1} J^{-T} M J^{-1}$. This eliminates the terms containing F_e and F^d, because of Equation 76.134, and results in

$$S_x^\dagger K_e^{-1} J^{-T} M J^{-1} A = S_x^\dagger K_e^{-1} J^{-T} M J^{-1} a_x. \qquad (76.145)$$

Since $S_x^\dagger K_e^{-1} J^{-T} M J^{-1} S_x$ is a $n_x \times n_x$ nonsingular matrix, and with Equations 76.143 and 76.144, this reduces to

$$\dot{\tilde{\chi}} + K_{dx}\tilde{\chi} + K_{px}\tilde{\chi}_\Delta = 0. \qquad (76.146)$$

Choosing diagonal gain matrices K_{dx} and K_{px} decouples the velocity control.

Similarly, premultiply Equation 76.142 with $S_f^\dagger K_e$. This eliminates the terms containing A and a_x, because of Equation 76.134, and results in

$$S_f^\dagger K_e J M^{-1} J^T F_e = S_f^\dagger K_e J M^{-1} J^T F^d. \qquad (76.147)$$

Since $S_f^\dagger K_e J M^{-1} J^T S_f$ is an $n_f \times n_f$ nonsingular matrix, and with Equations 76.124 and 76.130, this reduces to

$$\phi = \phi^d, \qquad (76.148)$$

or

$$\tilde{\phi} = 0, \qquad (76.149)$$

with $\tilde{\phi} = \phi^d - \phi$. This proves the complete decoupling between velocity- and force-controlled subspaces. In the velocity-controlled subspace, the dynamics are assigned by choosing appropriate matrices K_{dx} and K_{px}. The force control is very sensitive to disturbance forces, since it contains no feedback. Suppose a disturbance force F_{dist} acts on the end-effector, e.g., due to modeling errors in the inner loop controller. This changes Equation 76.149 to

$$\tilde{\phi} = \phi_{dist}, \qquad (76.150)$$

where $\phi_{dist} = S_f^\dagger F_{dist}$. This effect is compensated for by modifying F^d in Equation 76.139 to include feedback, e.g., $F^d + S_f K_{pf}(\phi^d - \phi)$, with K_{pf} a dimensionless $n_f \times n_f$ (diagonal) force control gain matrix. Using this feedback control law, Equation 76.149 results in

$$\tilde{\phi} = (I_{n_f} + K_{pf})^{-1}\phi_{dist}. \qquad (76.151)$$

Acceleration Resolved Control: Case of Soft Environment

The system equations are given by Equation 76.118 for the robot and by Equation 76.137 for the robot-environment interaction. The inner loop controller is taken as

$$\tau = M(q)a_q + C(q,\dot{q})\dot{q} + g(q) + J^T F_e, \qquad (76.152)$$

where F_e is the measured contact force, which is supposed to correspond to the real contact force. a_q is again given by Equation 76.140. The closed-loop dynamics of the system with its inner loop controller result in

$$A = a_x. \qquad (76.153)$$

The outer loop controller contains two terms:

$$a_x = a_{xx} + a_{xf}. \qquad (76.154)$$

a_{xx} corresponds to an *acceleration of freedom* and is given by Equation 76.144. On the other hand, a_{xf} corresponds to an *acceleration of constraint* and corresponds to

$$a_{xf} = K_e^{-1} S_f\left(\ddot{\phi}^d + K_{df}\dot{\tilde{\phi}} + K_{pf}\tilde{\phi}\right). \qquad (76.155)$$

K_{df} and K_{pf} are (diagonal) control gain matrices with dimensions $n_f \times n_f$ and physical units $\frac{1}{time}$ and $\frac{1}{(time)^2}$ respectively.

Substituting Equation 76.154 in Equation 76.153 leads to the closed-loop dynamic equation. The closed-loop equation is split up into two independent parts corresponding to the force- and velocity-controlled subspaces. First, premultiply Equation 76.153 with $S_x^\dagger$. This eliminates the acceleration of constraint a_{xf}, because of Equation 76.134, and results in

$$S_x^\dagger A = S_x^\dagger a_{xx}. \tag{76.156}$$

With $\dot{\chi} = S_x^\dagger A$, and with Equation 76.144, this reduces to Equation 76.146. Similarly, premultiply Equation 76.153 with $S_f^\dagger K_e$. In the case of a stationary environment, i.e., X_e is constant, the second derivative of Equation 76.137 is given by

$$\ddot{F}_e = P_f K_e A. \tag{76.157}$$

With Equation 76.157, with Equation 76.135, and Equation 76.134 this leads to:

$$\ddot{\tilde{\phi}} + K_{df}\dot{\tilde{\phi}} + K_{pf}\tilde{\phi} = 0. \tag{76.158}$$

Hence, there is complete decoupling between velocity- and force-controlled subspaces. Suppose a disturbance force F_{dist} acts on the end-effector. The influence on the force loop dynamics is derived as:

$$\ddot{\tilde{\phi}} + K_{df}\dot{\tilde{\phi}} + K_{pf}\tilde{\phi} = S_f^\dagger K_e J M^{-1} J^T F_{dist}. \tag{76.159}$$

Hence, as in the case of a rigid environment, disturbance forces directly affect the force loop; their effect is proportional to the contact stiffness. As a result, accurate force control is much easier in the case of soft contact. This is achieved by adding extra compliance in the robot end-effector.

Remarks.

1. Usually, $\dot{\phi}^d = \ddot{\phi}^d = 0$.
2. Usually, the measured force signal is rather noisy. Therefore, feedback of $\dot{\phi} = S_f^\dagger \dot{F}_e$ in the outer loop controller of Equation 76.155 is often replaced by $S_f^\dagger K_e J \dot{q}$, where the joint velocities $\dot{q}$ are measured using tachometers. In the case of a stationary environment, both signals are equivalent, and hence they result in the same closed-loop dynamics.
3. If in the inner controller of Equation 76.152 the contact force is compensated with the desired force F^d instead of the measured force F_e, the force loop dynamics become Equation 76.159, with $F_{dist} = F_e - F^d$. Hence, the dynamics of the different force coordinates are coupled.

Velocity-Resolved Control

The model-based inner loop controllers of Equations 76.139 and 76.152 are too advanced for implementation in current industrial robot controllers. The main problems are their computational complexity and the nonavailability of accurate inertial parameters of each robot link. These parameters are necessary to calculate $M(q)$, $C(q, \dot{q})$, and $g(q)$.

Instead, the state of practice consists of using a set of independent, i.e., completely decoupled, velocity controllers for each robot joint as the inner controller. Usually such a velocity controller is an analog proportional-integral (PI) type controller that controls the voltage of the joint actuator based on feedback of the joint velocity. This velocity is either obtained from a tachometer or derived by differentiating the position measurement. Such an analog velocity loop can be made very high bandwidth. Because of their high bandwidth, the independent joint controllers are able to decouple the robot dynamics in Equation 76.118 to a large extent, and they are able to suppress the effect of the environment forces to a large extent, especially if the contact is sufficiently compliant. Other disturbance forces are suppressed in the same way by the inner velocity controller before they affect the dynamics of the outer loop. This property has made this approach, in combination with a compliant end-effector, very popular for practical applications.

This practical approach is considered below. The closed-loop dynamics of the system with its high bandwidth inner velocity controller is approximated by

$$\dot{q} = \dot{q}^d, \tag{76.160}$$

or, using Equation 76.119, in the Cartesian space:

$$V = v_x. \tag{76.161}$$

v_x is the control signal generated by the outer loop controller. The outer loop controller contains two terms:

$$v_x = v_{xx} + v_{xf}. \tag{76.162}$$

v_{xx} corresponds to a *velocity of freedom* and is given by

$$v_x = S_x\left(\chi^d + K_{px}\tilde{\chi}_\Delta\right). \tag{76.163}$$

On the other hand, v_{xf} corresponds to a *velocity of constraint* and corresponds to

$$v_{xf} = K_e^{-1} S_f\left(\dot{\phi}^d + K_{pf}\tilde{\phi}\right). \tag{76.164}$$

Both K_{px} and K_{pf} have units $\frac{1}{time}$.

Substituting Equation 76.162 in Equation 76.161 leads to the closed-loop dynamic equation. The closed-loop equation is split up into two independent parts corresponding to the force- and velocity-controlled subspaces. First, premultiply Equation 76.161 with $S_x^\dagger$. This eliminates the velocity of constraint v_{xf}, because of Equation 76.134, and results in:

$$\tilde{\chi} + K_{px}\tilde{\chi}_\Delta = 0. \tag{76.165}$$

Similarly, premultiply Equation 76.161 with $S_f^\dagger K_e$. In the case of a stationary environment, i.e., X_e is constant, the derivative of Equation 76.137 is given by

$$\dot{F}_e = P_f K_e V. \tag{76.166}$$

This leads to

$$\dot{\tilde{\phi}} + K_{pf}\tilde{\phi} = 0. \tag{76.167}$$

Hence, there is complete decoupling between velocity- and force-controlled subspaces.

76.2.4 Adaptive Control

The hybrid control approach just presented explicitly relies on a decomposition of the Cartesian space into force- and velocity-controlled directions. The control laws implicitly assume that accurate models of both subspaces are available all the time. On the other hand, most practical implementations turn out to be rather robust against modeling errors. For example, for the two tasks discussed in Section 76.2.2—i.e., *peg-in-hole* and *sliding a block*—the initial relative position between the manipulated object and the environment may contain errors. As a matter of fact, to cope reliably with these situations is exactly why force control is used! The robustness of the force controller increases if it can continuously adapt its model of the force- and velocity-controlled subspaces. In this chapter, we consider only geometric parameters (i.e., the S_x and S_f subspaces), not the dynamical parameters of the manipulator and/or the environment.

The previous sections relied on the assumption that $\dot{S}_x = 0$ and $\dot{S}_f = 0$. If this assumption is not valid, the controller must follow (or *track*) the constraint's time variance by: (1) using feedforward (motion) information from the constraint equations (if known and available); (2) estimating the changes in S_x and S_f from the motion and/or force measurements. The adaptation involves two steps: (1) to identify the errors between the current constraint model and the currently measured contact situation; and (2) to feed back these identified errors to the constraint model. Figures 76.8 and 76.9 illustrate two examples of error identification.

Two-Dimensional Contour Following

The orientation of the contact normal changes if the environment is not planar. Hence, an error $\Delta\alpha$ appears. This error angle can be estimated with either the velocity or the force measurements only (Figure 76.8):

1. Velocity based: The $X_t Y_t$ frame is the modeled task frame, while $X_0 Y_0$ indicates the real task frame. Hence, the executed velocity V, which is tangential to the real contour, does not completely lie along the X_t axis, but has a small component V_{yt} along the Y_t axis. The orientation error $\Delta\alpha$ is approximated by the arc tangent of the ratio V_{yt}/V_{xt}.
2. Force based: The measured (ideal) reaction force F does not lie completely along the modeled normal direction (Y_t), but has a component F_{xt} along X_t. The orientation error $\Delta\alpha$ is approximated by the arc tangent of the ratio F_{xt}/F_{yt}.

The velocity-based approach is disturbed by mechanical compliance in the system; the force-based approach is disturbed by friction.

Sliding an Edge Over an Edge

This task has two geometric uncertainty parameters when the robot moves the object over the environment: (1) an uncertainty Δx of the position of the task frame's origin along the contacting edge of the object; and (2) an uncertainty $\Delta\alpha$ in the frame's orientation about the same edge (Figure 76.9). Identification equations for these uncertainties are

$$\begin{cases} \Delta\alpha = \arctan \frac{v_z}{v_y} \\ \Delta x = \frac{v_z}{\omega_y} \end{cases} \quad \text{(velocity based)}, \qquad (76.168)$$

$$\begin{cases} \Delta x = -\frac{f_z}{m_y} \\ \Delta\alpha = -\arctan \frac{f_z}{f_y} \end{cases} \quad \text{(force based)}. \qquad (76.169)$$

The results from the force- and/or velocity-based identification should be fed back to the model. Moreover, in the contour-following case, the identified orientation error can be converted into an error $\tilde{\chi}_\Delta$, such that the setpoint control laws of Equation 76.144 or Equation 76.163 make the robot track changes in the contact normal.

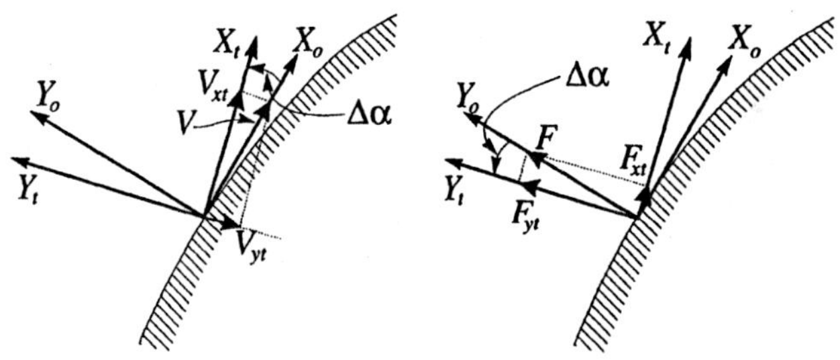

Figure 76.8 Estimation of orientation error.

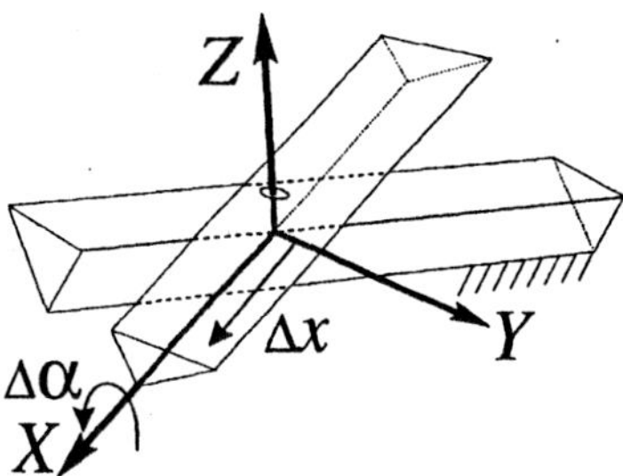

Figure 76.9 Estimation of position error.

76.2.5 Impedance Control

The previous sections focused on a *model-based* approach towards force control: the controller relies on an explicit geometric model of the force- and velocity-controlled directions. However, an alternative approach exists, called *impedance control*. It differs from the hybrid approach both in task specification and in control.

Task Specification

Hybrid control specifies desired motion *and* force trajectories; impedance control [4] specifies (1) a desired *motion* trajectory and (2) a desired *dynamic relationship* between the deviations from this desired trajectory, induced by the contact with the environment, and the forces exerted by the environment:

$$F = -M_c \dot{V} + C_c \tilde{V} + K_c \int \tilde{V} dt. \quad (76.170)$$

$\tilde{V}$ is the Cartesian error velocity, i.e., the difference between the prescribed velocity V^d and the measured velocity V; $\dot{V}$ is the Cartesian acceleration; M_c, C_c, and K_c are user-defined inertia, damping, and stiffness matrices, respectively.

Compared to hybrid force/position control, the apparent advantage of impedance control is that no explicit knowledge of the constraint kinematics is required. However, in order to obtain a satisfactory dynamic behavior, the inertia, damping, and stiffness matrices have to be tuned for a particular task. Hence, they embody implicit knowledge of the task geometry, and hence task specification and control are intimately linked.

Control

For control purposes the dynamic relationship of Equation 76.170 can be interpreted in two ways. It is the model of an *impedance*; i.e., the robot reacts to the "deformations" of its planned position and velocity trajectories by generating forces. Special cases are *stiffness control* [8], where $M_c = C_c = 0$, and *damping control*, where $M_c = K_c = 0$. However, Equation 76.170 can also be interpreted in the other way as an *admittance*; i.e., the robot reacts to the constraint forces by deviating from its planned trajectory.

Impedance Control In essence, a stiffness or damping controller is a proportional-derivative (PD) position controller, with position and velocity feedback gains adjusted in order to obtain the desired impedance. Consider a PD joint position controller (see Chapter 76.1):

$$\begin{aligned} \tau &= K_{pq}(q^d - q) + K_{dq}(\dot{q}^d - \dot{q}) && (76.171) \\ &= K_{pq}\tilde{q} + K_{dq}\dot{\tilde{q}}. && (76.172) \end{aligned}$$

This compares to Equation 76.170, with $M_c = 0$ by writing

$$\begin{aligned} \tau &= J^T F && (76.173) \\ &= J^T \left(C_c \tilde{V} + K_c \int \tilde{V} dt \right) && (76.174) \\ &= J^T \left(C_c J\dot{\tilde{q}} + K_c J\tilde{q} \right). && (76.175) \end{aligned}$$

Hence, in order to obtain the desired stiffness or damping behavior, the PD gain matrices have to be chosen as

$$K_{pq} = J^T K_c J; \quad K_{dq} = J^T C_c J. \quad (76.176)$$

Note that the gain matrices are position dependent due to the position dependence of the Jacobian J.

Admittance Control In this case, the measured constraint force F_e is used to modify the robot trajectory, given by $X^d(t)$ and $V^d(t)$. In *stiffness control* (or rather, inverse stiffness, or compliance control), a modified position X^m is commanded:

$$X^m = X^d - K_c^{-1} F_e. \quad (76.177)$$

In *damping control* (or rather, inverse damping control), a modified velocity V^m is commanded:

$$V^m = V^d - C_c^{-1} F_e. \quad (76.178)$$

Besides introducing damping and stiffness, the most general case of admittance control changes the apparent inertia of the robot manipulator. In this case a desired acceleration A^m is solved from Equation 76.170:

$$A^m = M_c^{-1} \left(-F_e + C_c \tilde{V} + K_c \int \tilde{V} dt \right). \quad (76.179)$$

X^m, V^m, and A^m are applied to the motion controller (see Chapter 76.1) in which the constraint forces may be compensated for by the measured forces as an extra term. For example, in the general admittance case, the control torques are

$$\tau = M(q)a_q + C(q, \dot{q})\dot{q} + g(q) + J^T F_e, \quad (76.180)$$

where a_q is given by Equation 76.140, in which $a_x = A^m$; and A_m is given by Equation 76.179.

References

[1] DeFazio, T. L., Seltzer, D.S., Whitney, D. E., The instrumented remote center compliance, *Industrial Robot*, 11(4), 238–242, 1984.

[2] De Schutter, J. and Van Brussel, H., Compliant robot motion, *Int. J. Robotics Res.*, 7(4), 3–33, 1988.

[3] Doty, K. L., Melchiorri, C., and Bonivento, C., A theory of generalized inverses applied to robotics, *Int. J. Robotics Res.*, 12(1), 1–19, 1993.

[4] Hogan, N., Impedance control: an approach to manipulation, I-III, *Trans. ASME, J. Dynamic Syst., Meas., Control*, 117, 1–24, 1985.

[5] Khatib, O., A Unified approach for motion and force control of robot manipulators: the operational space formulation, *IEEE J. Robotics Autom.*, 3(1), 43–53, 1987.

[6] Mason, M. T., Compliance and force control for computer controlled manipulators, *IEEE Trans. Syst., Man, Cybern.*, 11(6), 418–432, 1981.

[7] Patarinski, S. and Botev, R., Robot force control, a review, *Mechatronics*, 3(4), 377–398, 1993.

[8] Salisbury, J. K., Active stiffness control of a manipulator in cartesian coordinates, *19th IEEE Conf. Decision Control*, 95–100, 1980.

[9] Wang D. and McClamroch, N. H., Position and force control for constrained manipulator motion: Lyapunov's direct method, *IEEE Trans. Robotics Autom.*, 9(3), 308–313, 1993.

[10] Whitney, D. E., Historic perspective and state of the art in robot force control, *Int. J. Robotics Res.*, 6(1), 3–14, 1987.

76.3 Control of Nonholonomic Systems

John Ting-Yung Wen, Department of Electrical, Computer, and Systems Engineering, Rensselaer Polytechnic Institute

76.3.1 Introduction

When the generalized velocity of a mechanical system satisfies an equality condition that cannot be written as an equivalent condition on the generalized position, the system is called a nonholonomic system. Nonholonomic conditions may arise from constraints, such as pure rolling of a wheel, or from physical conservation laws, such as the conservation of angular momentum of a free floating body.

Nonholonomic systems pose a particular challenge from the control point of view, as any one who has tried to parallel park a car in a tight space can attest. The basic problem involves finding a path that connects an initial configuration to the final configuration and satisfies all the holonomic and nonholonomic conditions for the system. Both open-loop and closed-loop solutions are of interest: open loop solution is useful for off-line path generation and closed-loop solution is needed for real-time control.

Nonholonomic systems typically arise in the following classes of systems:

1. No-slip constraint
   Consider a single wheel rolling on a flat plane (see Figure 76.10). The no slippage (or pure rolling) contact condition means that the linear velocity at the contact point is zero. Let $\vec{\omega}$ and $\vec{v}$, respectively, denote the angular and linear velocity of the body frame attached to the center of the wheel. Then the no slippage condition at the contact point can be written as

$$\vec{v} - \ell\vec{\omega} \times \vec{z} = 0. \qquad (76.181)$$

   We will see later that part of this constraint is nonintegrable (i.e., not reducible to a position constraint) and, therefore, nonholonomic.

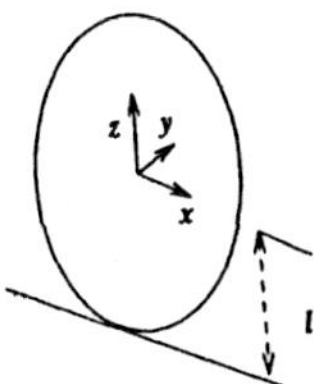

Figure 76.10 A wheel with no-slip contact.

   In modeling the grasping of an object by a robot hand, the so-called soft finger contact model is sometimes used. In this model, the finger is not allowed to rotate about the local normal, $\vec{z}\cdot\vec{\omega} = 0$, but is free to rotate about the local x and y axes. This velocity constraint is nonintegrable.
   The dynamic equations of wheeled vehicles and finger grasping are of similar forms. There are two sets of equations of motion, one for the unconstrained vehicle or finger, the other for the ground (stationary) or the payload. These two sets of equations are coupled by the constraint force/torque that keeps the vehicle on the ground with no wheel slippage or fingers on the object with no rotation about the local normal axis. These equations can be summarized in the following form:

$$\begin{aligned}
&(a) \quad M(q)\ddot{q} + C(q,\dot{q})\dot{q} + g(q) = u - J^T f,\\
&(b) \quad M_c\alpha_c + b_c + k_c = A^T f,\\
&(c) \quad Hf = 0, \qquad (76.182)\\
&(d) \quad v^+ = J\dot{\theta}, \; v^- = Av_c = v^+ + H^T W, \text{ and}\\
&(e) \quad \alpha^+ = J\ddot{\theta} + \dot{J}\dot{\phi}\, \alpha^- = A\alpha_c + a\\
&\qquad = \alpha^+ + H^T\dot{W} + \dot{H}^T W.
\end{aligned}$$

   Equations 76.182a and 76.182b are the equations of motion of the fingers and the payload, respectively, f is the constraint force related to the vehicle or fingers via the Jacobian transpose J^T, α_c denotes the payload acceleration, b_c and k_c are the Coriolis and gravity forces on the payload, H is a full row rank matrix whose null space specifies the directions where motion at the contact is allowed (these are also the directions with no constraint forces), v^+ and v^- are the velocity at the two sides of the contacts. Similarly, α^+ and α^- denote accelerations and W parameterizes the admissible velocity across the contact. The velocity constraint is specified in Equation 76.182d; premultiplying Equation 76.182d by the annihilator of H^T, denoted by $\hat{H}^T$,

$$\hat{H}^T(J\dot{\theta} - Av_c) = 0. \qquad (76.183)$$

   In the single wheel case in Figure 76.10,

$$\begin{aligned}
v_c &= 0 \;, \quad \dot{\theta} = \begin{bmatrix} \vec{\omega} \\ \vec{v} \end{bmatrix}\\
J &= \begin{bmatrix} I & 0 \\ \ell\vec{z}\times & I \end{bmatrix}, \quad H = [I \;,\; 0],\\
\hat{H}^T &= [0 \;,\; I]. \qquad (76.184)
\end{aligned}$$

   The velocity constraint Equation 76.183 is then the same as Equation 76.181.

2. Conservation of angular momentum
   In a Lagrangian system, if a subset of the generalized coordinates q_u does not appear in the mass matrix $M(q)$, they are called the cyclic coordinates. In this case, the Lagrangian equation associated with q_u is

$$\frac{d}{dt}\left(\frac{\partial L}{\partial \dot{q}_{u_i}}\right) = \frac{\partial L}{\partial q_{u_i}} = 0. \qquad (76.185)$$

After integration, we obtain the conservation of generalized momentum condition associated with the cyclic coordinates.
As an example, consider a free floating multibody system with no external torque (such as a robot attached to a floating platform in space or an astronaut unassisted by the jet pack, as shown in Figure 76.11).

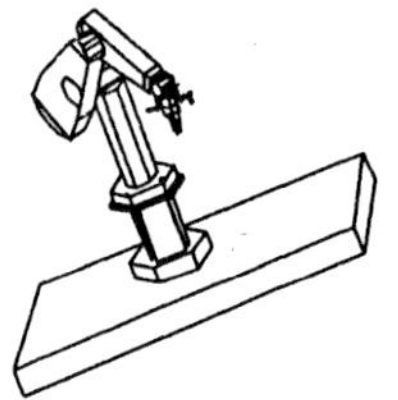

Figure 76.11 Examples of free floating multibody systems.

The equation of motion for such systems is

$$\begin{bmatrix} M(q) & M_1(q) \\ M_1^T(q) & M_b(q) \end{bmatrix} \begin{bmatrix} \ddot{q} \\ \dot{\omega} \end{bmatrix} + \begin{bmatrix} C_{11}(q,\dot{q}) & C_{12}(q,\dot{q},\omega) \\ C_{21}(q,\dot{q},\omega) & C_{22}(q,\dot{q},\omega) \end{bmatrix} \begin{bmatrix} \dot{q} \\ \omega \end{bmatrix} = \begin{bmatrix} u \\ 0 \end{bmatrix}, \tag{76.186}$$

where ω is the angular velocity of the multibody system about the center of mass. This is a special case of the situation described above with $L = \frac{1}{2}\dot{q}^T M(q)\dot{q} + \dot{q}^T M_1(q)\omega + \frac{1}{2}\omega^T M_b(q)\omega$. Identifying $\dot{q}_u$ with ω, Equation 76.185 becomes

$$M_1^T(q)\dot{q} + M_b(q)\omega = 0 \tag{76.187}$$

which is a nonintegrable condition.

3. Underactuated mechanical system
An underactuated mechanical system is one that does not have all of its degrees of freedom independently actuated. The nonintegrable condition can arise in terms of velocity, as we have seen above, or in terms of acceleration which cannot be integrated to a velocity condition. The latter case is called the *second-order nonholonomic condition* [1].
First-Order condition: Consider a rigid spacecraft with less than three independent torques.

$$I\dot{\omega} + \omega \times I\omega = Bu \tag{76.188}$$

where B is a full column rank matrix with rank less than three. Let $\hat{B}$ be the annihilator of B, i.e., $\hat{B}B = 0$. Then premultiplying Equation 76.188 by $\hat{B}$ gives $\frac{d}{dt}(\hat{B}I\omega) = 0$. Assuming the initial velocity is zero, then we arrive at a nonintegrable velocity constraint,

$$\hat{B}I\omega = 0.$$

Second-Order condition: Consider a robot with some of the joints unactuated. The general dynamic equation can be written as

$$M(q)\ddot{q} + C(q,\dot{q})\dot{q} + g(q) = \begin{bmatrix} u \\ 0 \end{bmatrix}. \tag{76.189}$$

By premultiplying by $\hat{B} = [0 \ I]$ which annihilates the input vector, we obtain a condition involving the acceleration,

$$\hat{B}(M(q)\ddot{q} + C(q,\dot{q})\dot{q} + g(q)) = 0. \tag{76.190}$$

It can be shown that this equation is integrable to a velocity condition, $h(q,\dot{q},t) = 0$, if, and only if, the following conditions hold [1]:

(a) the gravitational torque for the unactuated variables, $g_u(q) = \hat{B}g(q)$, is a constant and
(b) the mass matrix $M(q)$ does not depend on the unactuated coordinates, $q_u = \hat{B}q$.

This implies that any earthbound robots with nonplanar, articulated, underactuated degrees of freedom would satisfy a nonintegrable second-order constraint because $g_u(q)$ would not be constant.

The control problem associated with a nonholonomic system can be posed based on the kinematics alone (with an ideal dynamic controller assumed) or the full dynamical model.

In the kinematics case, nonholonomic conditions are linear in the velocity, v,

$$\Omega(q)v = 0. \tag{76.191}$$

Assuming that the rank of $\Omega(q)$ is constant over q, then Equation 76.191 can be equivalently stated as,

$$v = f(q)u \tag{76.192}$$

where the columns of $f(q)$ form a basis of the null space of $\Omega(q)$. Equation 76.192 can be regarded as a control problem with u as the control variable and the configuration variable, q, as the state if $v = \dot{q}$. If v is nonintegrable (as is the case for the angular velocity), there would be an additional kinematic equation $\dot{q} = h(q)v$ (such as the attitude kinematic equation); the control problem then becomes $\dot{q} = h(q)f(q)u$. Note that in either case, the right hand side of the differential equation does not contain a term dependent only in q. Such systems are called *driftless systems*.

Solving the control problem associated with the kinematic Equation 76.192 produces a feasible path. To actually follow the path, a real-time controller is needed to produce the required force or torque. This procedure of decomposing path planning and path following is common in industrial robot motion control. Alternatively, one can also consider the control of the full dynamical system directly. In other words, consider Equation 76.182 for the rolling constraint case, or Equations 76.186, 76.188 or 76.189 for the underactuated case, with u as the control input. In the rolling constraint case, the contact force also needs to be controlled, similar to a robot performing a contact task. Otherwise, slippage or even loss of contact may result (e.g., witness occasional truck rollovers on highway exit ramps). The dynamical equations also differ from the kinematic problem Equation 76.192 in a fundamental way: a control-independent term, called the drift term, is present in the dynamics. In contrast to driftless systems, there is no known general global controllability condition for such systems. However, the presence of the drift term sometimes simplifies the problem by rendering the linearized system locally controllable.

This chapter focus mainly on the kinematic control problem. In addition to the many research papers already published on this subject, excellent summaries of the current state of research can be found in [2, 3].

In the remainder of this chapter, we address the following aspects of the kinematic control of a nonholonomic system:

1. Determination of Nonholonomy. Given a set of constraints, how does one classify them as holonomic or nonholonomic?
2. Controllability. Given a nonholonomic system, does a path exist that connects an initial configuration to the desired final configuration?
3. Path Planning. Given a controllable nonholonomic system, how does one construct a path that connects an initial configuration to the desired final configuration?
4. Stabilizability. Given a nonholonomic system, can one construct a stabilizing feedback controller, and if it is possible, how does one do so?
5. Output stabilizability. Given a nonholonomic system, can one construct a feedback controller that drives a specified output to the desired target while maintaining the boundedness of all the states, and, if it is possible, how does one do so?

We shall use a simple example to illustrate various concepts and results throughout this section. Consider a unicycle with a fat wheel, i.e., it cannot fall (see Figure 76.12). For this system,

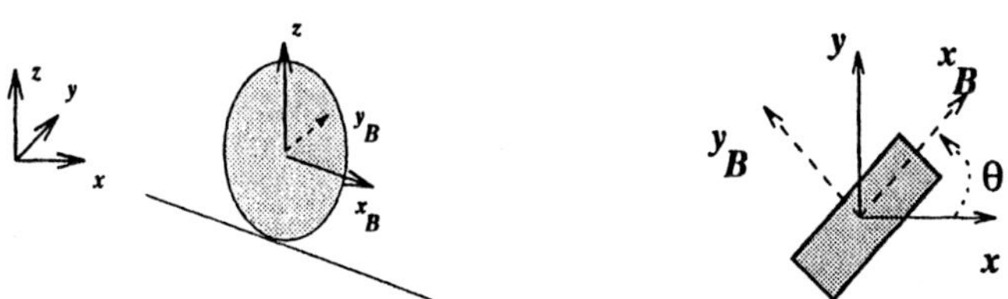

Figure 76.12 Unicycle model and coordinate definition.

there are four constraints:

$$\begin{aligned} \vec{x}_B \cdot \vec{\omega} &= 0, \\ \text{and } \vec{v} - \vec{\omega} \times \ell \vec{z} &= 0. \end{aligned} \quad (76.193)$$

The first equation specifies the no-tilt constraint and the second equation is the no-slip constraint.

76.3.2 Test of Nonholonomy

As motivated in the previous section, consider a set of constraints in the following form:

$$\Omega(q)\dot{q} = 0 \quad (76.194)$$

where $q \in \mathcal{R}^n$ is the configuration variable, $\dot{q}$ is the velocity, and $\Omega(q) \in \mathcal{R}^{\ell \times n}$ specifies the constraint directions.

The complete integrability of the velocity condition in Equation 76.194 means that $\Omega(q)$ is the Jacobian of some function, $h(q) \in \mathcal{R}^\ell$, i.e.,

$$\frac{\partial h}{\partial q} = \Omega(q). \quad (76.195)$$

In this case, Equation 76.194 can be written as an equivalent holonomic condition, $h(q) = c$, where c is some constant vector. Equation 76.194 may be only partially integrable, which means that some of the rows of $\Omega(q)$, say, $\Omega_{k+1}, \ldots, \Omega_\ell$, satisfy

$$\frac{\partial h_i}{\partial q} = \Omega_i \quad , \quad i = k+1, \ldots, \ell. \quad (76.196)$$

for some scalar functions $h_i(q)$. Substituting Equation 76.196 in Equation 76.194, we have $\ell - k$ integrable constraints

$$\frac{\partial h_i}{\partial q}\dot{q} = 0, \quad (76.197)$$

which can be equivalently written as $h_i(q) = c_i$ for some constants c_i. If $\ell - k$ is the maximum number of such $h_i(q)$ functions, the remaining k constraints are then nonholonomic.

To determine if the constraint Equation 76.194 is integrable, we can apply the Frobenius theorem. We first need some definitions:

DEFINITION 76.1

1. A *vector field* is a smooth mapping from the configuration space to the tangent space.
2. A *distribution* is the subspace generated by a collection of vector fields. The *dimension* of a distribution is the dimension of any basis of the distribution.
3. The *Lie bracket* between two vector fields, f and g, is defined as
$$[f, g] \triangleq \frac{\partial g}{\partial q} f(q) - \frac{\partial f}{\partial q} g(q).$$
4. An *involutive distribution* is a distribution that is closed with respect to the Lie bracket, that is, if f, g belong to a distribution Δ, then $[f, g]$ also belongs to Δ.
5. A distribution, Δ, with constant dimension m, consisting of vector fields in $\mathcal{R}^n$ is *integrable* if $n - m$ functions, $h_1, \ldots, h_{n-m}$ exist so that the Lie derivative of h_i along each vector field $f \in \Delta$ is zero, that is,
$$L_f h_i(q) \triangleq \frac{\partial h_i}{\partial q} \cdot f(q) = 0.$$

6. The *involutive closure* of a distribution Δ is the smallest involutive distribution that contains Δ.

The Frobenius theorem can be simply stated:

THEOREM 76.1 *A distribution is integrable if, and only if, it is involutive.*

To apply the Frobenius theorem to Equation 76.194, first observe that $\dot{q}$ must be within the null space of $\Omega(q)$ denoted by Δ. Suppose the constraints are independent throughout the configuration space, then the dimension of Δ is $n - \ell$; let a basis of Δ be $g_1(q), \ldots, g_{n-\ell}(q)$:

$$\Delta = \text{span}\{g_1(q), \ldots, g_{n-\ell}(q)\}.$$

Let $\overline{\Delta}$ be the involutive closure of Δ. Suppose the dimension of $\overline{\Delta}$ is constant, $n - \ell + k$. Since $\overline{\Delta}$ is involutive by definition, from the Frobenius theorem, $\overline{\Delta}$ is integrable. This means that functions h_i, $i = 1, \ldots, \ell - k$ exist, so that $\frac{\partial h_i}{\partial q}$ annihilates $\overline{\Delta}$:

$$\frac{\partial h_i}{\partial q} f = 0 \tag{76.198}$$

for all $f \in \overline{\Delta} \supset \Delta$. Since $\dot{q} \in \Delta$, it follows that Equation 76.196 is satisfied for all $\dot{q}$. Hence, among the ℓ constraints given by Equation 76.194, $\ell - k$ are holonomic (obtained from the annihilator of $\overline{\Delta}$) and k are nonholonomic. Geometrically, this means that the flows of the system lie on a $n - \ell + k$ dimensional manifold given by h_i =constant, $i = 1, \ldots, \ell - k$.

To illustrate the above discussion, consider the unicycle example presented at the end of Section 76.3.1. First write the constraints Equation 76.193 in the same form as Equation 76.191

$$\begin{bmatrix} \vec{x}_B \cdot & 0 \\ \ell \vec{z} \times & I \end{bmatrix} \begin{bmatrix} \vec{\omega} \\ \vec{v} \end{bmatrix} = 0.$$

This implies that

$$\begin{bmatrix} \vec{\omega} \\ \vec{v} \end{bmatrix} \in \text{span}\left\{ \begin{bmatrix} \vec{y}_B \\ \ell \vec{x}_B \end{bmatrix}, \begin{bmatrix} \vec{z} \\ 0 \end{bmatrix} \right\}.$$

Represent the top portion of each vector field in the body coordinates, $\vec{y}_B = [0, 1, 0]^T$, $\vec{z} = [0, 0, 1]^T$, and the bottom portion in the world coordinates, $\vec{x}_B = [\ell c_\theta, \ell s_\theta, 0]^T$, $c_\theta = \cos\theta$ and $s_\theta = \sin\theta$, θ is the steering angle. We have

$$\Delta = \text{span}\left\{ \begin{bmatrix} 0 \\ 1 \\ 0 \\ \ell c_\theta \\ \ell s_\theta \\ 0 \end{bmatrix}, \begin{bmatrix} 0 \\ 0 \\ 1 \\ 0 \\ 0 \\ 0 \end{bmatrix} \right\}. \tag{76.199}$$

The involutive closure of Δ can be computed by taking repeated Lie brackets:

$$\overline{\Delta} = \text{span}\left\{ \begin{bmatrix} 0 \\ 1 \\ 0 \\ \ell c_\theta \\ \ell s_\theta \\ 0 \end{bmatrix}, \begin{bmatrix} 0 \\ 0 \\ 1 \\ 0 \\ 0 \\ 0 \end{bmatrix}, \begin{bmatrix} 0 \\ 0 \\ 0 \\ -\ell s_\theta \\ \ell c_\theta \\ 0 \end{bmatrix}, \begin{bmatrix} 0 \\ 0 \\ 0 \\ \ell c_\theta \\ \ell s_\theta \\ 0 \end{bmatrix} \right\}. \tag{76.200}$$

which is of constant dimension four. The annihilator of $\overline{\Delta}$ is

$$\overline{\Delta}^{\perp} = \text{span}\{[1, 0, 0, 0, 0, 0], [0, 0, 0, 0, 0, 1]\}.$$

From the Frobenius theorem, the annihilator of $\overline{\Delta}$ is integrable. Indeed, the corresponding holonomic constraints are what one could have obtained by inspection:

$$z = \text{ constant}, \quad \psi = \text{ roll angle } = 0.$$

Eliminating the holonomic constraints results in a common form of the kinematic equation for unicycle,

$$\begin{bmatrix} \dot{x} \\ \dot{y} \\ \dot{\theta} \\ \dot{\phi} \end{bmatrix} = \begin{bmatrix} \ell c_\theta \\ \ell s_\theta \\ 0 \\ 1 \end{bmatrix} u_1 + \begin{bmatrix} 0 \\ 0 \\ 1 \\ 0 \end{bmatrix} u_2 \tag{76.201}$$

where $u_1 = \omega_{y_B}$ and $u_2 = \omega_z$. Because the exact wheel rotational angle is frequently inconsequential, the ϕ equation is often omitted. In that case, the kinematic Equation 76.201 becomes

$$\begin{bmatrix} \dot{x} \\ \dot{y} \\ \dot{\theta} \end{bmatrix} = \begin{bmatrix} \ell c_\theta \\ \ell s_\theta \\ 0 \end{bmatrix} u_1 + \begin{bmatrix} 0 \\ 0 \\ 1 \end{bmatrix} u_2. \tag{76.202}$$

We shall refer to the system described by either Equation 76.201 or Equation 76.202 as the unicycle problem.

76.3.3 Nonholonomic Path Planning Problem

The nonholonomic path planning problem, also called nonholonomic motion planning, involves finding a path connecting specified configurations that satisfies the nonholonomic condition as in Equation 76.194. As discussed in Section 76.3.1, this problem can be written as an equivalent nonlinear control problem:

Given the system

$$\dot{q} = f(q)u \ , \ q \in \mathcal{R}^n \ , u \in \mathcal{R}^m, \tag{76.203}$$

and initial and desired final configurations, $q(0) = q_0$ *and* q_f*, find* $\underline{u} = \{u(t) : t \in [0, 1]\}$ *so that the solution of Equation* 76.203 *satisfies* $q(1) = q_f$.

The terminal time has been normalized to 1. In Equation 76.203, $f(q)$ is a full rank matrix whose columns span the null space of $\Omega(q)$ in Equation 76.194, and u parameterizes the degree of freedom in the velocity space. By construction, $f(q)$ is necessarily a tall matrix, that is, an $n \times m$ matrix with $n > m$.

Controllability

For $u = 0$, every q in $\mathcal{R}^n$ is an equilibrium. The linearized system about any equilibrium $q^\star$ is

$$\frac{d}{dt}(q - q^\star) = f(q^\star)u. \qquad (76.204)$$

Because $f(q)$ is tall, this linear time-invariant system is not controllable (the controllability matrix, $[f(q^\star) : 0 : \ldots : 0]$, has maximum rank m). This is intuitively plausible; as the nonholonomic condition restricts the flows in the tangent space, the system can locally only move in directions compatible with the nonholonomic condition, contradicting the controllability requirement. However, the system may still be controllable globally.

For a driftless system, such as a nonholonomic system described by Equation 76.203, the controllability can be ascertained through the following sufficient condition (sometimes called *Chow's theorem*):

THEOREM 76.2 *The system given by Equation 76.203 is controllable if the involutive closure of the columns of $f(q)$ is of constant rank n for all q.*

The involutive closure of a set of vector fields is in general called the *Lie algebra* generated by these vector fields. In the context of control systems where the vector fields are the columns of the input matrix $f(q)$, the Lie algebra is called the *control Lie algebra*.

For systems with drift terms, the above full rank condition is only sufficient for local accessibility. For a linear time-invariant system, this condition simply reduces to the usual controllability rank condition. This theorem is nonconstructive, however. The path planning problem basically deals with finding a specific control input to steer the system from a given initial condition to a given final condition, once the controllability rank condition is satisfied.

Since the involutive closure of the null space of the constraints is just the control Lie algebra of the corresponding nonholonomic control system, the control system (with the holonomic constraints removed) is globally controllable as long as the constraints remain independent for all configurations. For the unicycle problem given by Equation 76.201, the control Lie algebra is $\overline{\Delta}$ in Equation 76.200 (with z and ψ coordinates removed). Because the dimension of $\overline{\Delta}$ and the state-space dimension are both equal to four, the system is globally controllable.

An alternate way to view Equation 76.203 is to regard it as a nonlinear mapping of the input function $\underline{u}$ to the final state $q(1)$:

$$q(1) = F(q_0, \underline{u}). \qquad (76.205)$$

Given q_0 and $\underline{u}$, denote the solution of Equation 76.203 by

$$q(t) = \phi_{\underline{u}}(t; q_0). \qquad (76.206)$$

Then, $F(q_0, \underline{u}) = \phi_{\underline{u}}(1; q_0)$. In general, the analytic expression for F is impossible to obtain.

By definition, global controllability means that $F(q_0, \cdot)$ is an onto mapping for every q_0. For a given $\underline{u}$, $\nabla_{\underline{u}} F(q_0, \underline{u})$ corresponds to the system linearized about a trajectory $\underline{q} = \{q(t) : t \in [0, 1]\}$ which is generated by $\underline{u}$:

$$\delta\dot{q} = A(t)\delta q + B(t)\delta u, \quad \delta q(0) = 0, \qquad (76.207)$$

where, $A(t) \triangleq [\frac{\partial f}{\partial q_1}(q(t))u(t) \; : \; \cdots \; : \; \frac{\partial f}{\partial q_n}(q(t))u(t)]$, and $B(t) \triangleq f(q(t))$. Since $\delta q(0) = 0$, the solution to this equation is,

$$\delta q(1) = \int_0^1 \Phi(1, s)B(s)\delta u(s)\, ds \qquad (76.208)$$

where Φ is the state transition matrix of the linearized system. It follows that

$$\left(\nabla_{\underline{u}} F(q_0, \underline{u})\right) \underline{v} = \int_0^1 \Phi(1, s)B(s)v(s)\, ds. \qquad (76.209)$$

Controllability of the system in Equation 76.207 implies that for any final state $\delta q(1) \in \mathcal{R}^n$, a control δu exists which drives the linear system from $\delta q(0) = 0$ to $\delta q(1)$. This is equivalent to the operator $\nabla_{\underline{u}} F$ being onto (equivalently, the null space of the adjoint operator, $[\nabla_{\underline{u}} F]^\star$, being zero). In the case that $\underline{u} = 0$, $\nabla_{\underline{u}} F$ reduces to the linear time-invariant system Equation 76.204. In this case, $\nabla_{\underline{u}} F$ cannot be of full rank because the linearized system is not controllable.

Path Planning Algorithms

Steering with Cyclic Input In Equation 76.203, because $f(q)$ is full rank for all q, there is a coordinate transformation so that $f(q)$ becomes $\begin{bmatrix} I \\ f_1(q) \end{bmatrix}$. In other words, the inputs are simply the velocities of m configuration variables. For example, in the unicycle problem described by Equation 76.201, u_1 and u_2 are equal to $\dot{\theta}$ and $\dot{\phi}$. The subspace corresponding to these variables is called the *base space* (also called the *shape space*). A cyclic motion in the base space returns the base variables to their starting point, but the configuration variables would have a net change (called the geometric phase) as shown in Figure 76.13. In the unicycle case, cyclic motions in θ and ϕ result in the following net changes in the x and y coordinates:

$$\begin{aligned} x(T) - x(0) &= \int_0^T \cos\theta\dot{\phi}\, dt = \oint \cos\theta\, d\phi; \\ y(T) - y(0) &= \int_0^T \sin\theta\dot{\phi}\, dt = \oint \sin\theta\, d\phi. \end{aligned} \qquad (76.210)$$

By Green's theorem, they can be written as surface integrals

$$\begin{aligned} x(T) - x(0) &= \iint_S -\sin\theta\, d\theta\, d\phi; \quad \text{and} \\ y(T) - y(0) &= \iint_S \cos\theta\, d\theta\, d\phi \end{aligned} \qquad (76.211)$$

where S is the surface enclosed by the closed contour in the (ϕ, θ) space.

A general strategy for path planning would then consist of two steps: first drive the base variables to the desired final location, then appropriately choose a closed contour in the base space to achieve the desired change in the configuration variables without affecting the base variables. This idea has served as the basis of many path planning algorithms.

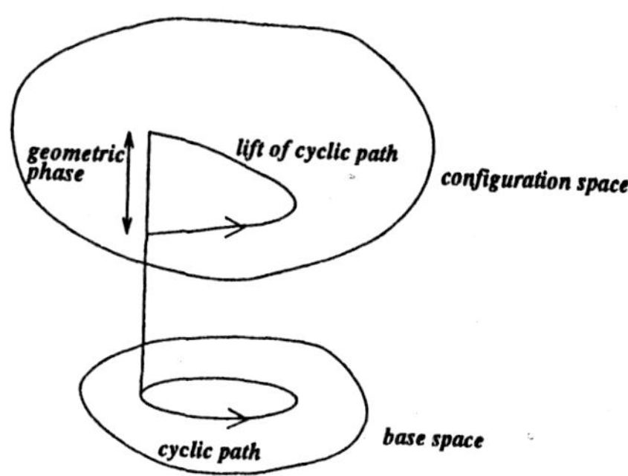

Figure 76.13 Geometric phase.

To illustrate this procedure for path planning for the unicycle example, assume that the base variables, ϕ and θ, have reached their target values. We choose them to be sinusoids with integral frequencies, so that at $t = 1$, they return to their initial values:

$$u_1 = a_1 \cos(4\pi t), \text{ and } u_2 = a_2 \cos(2\pi t). \tag{76.212}$$

By direct integration,

$$\phi = \frac{a_1}{4\pi}\sin(4\pi t), \text{ and } \theta = \frac{a_2}{2\pi}\sin(2\pi t). \tag{76.213}$$

For several values of a_1 and a_2, the closed contours in the $\phi - \theta$ plane given by Equation 76.213 are as shown in Figure 76.14. The net changes in x and y over the period $[0, 1]$ are given by the surface integrals Equation 76.211 over the area enclosed by the contours. To achieve the desired values for x and y, the two equations can be numerically solved for a_1 and a_2.

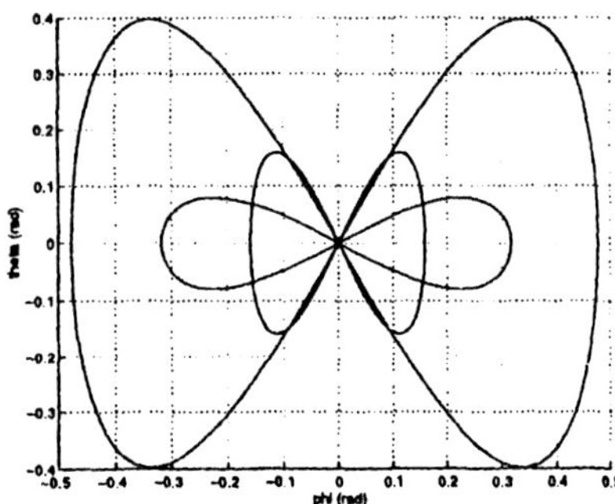

Figure 76.14 Closed contour in base space.

This procedure can also be performed directly in the time domain. For the chosen sinusoidal inputs, the changes in x and y are

$$\Delta x = \int_0^1 a_1 \cos(4\pi t)\cos\left(\frac{a_2}{2\pi}\sin(2\pi t)\right) dt \tag{76.214}$$

$$\Delta y = \int_0^1 a_1 \cos(4\pi t)\sin\left(\frac{a_2}{2\pi}\sin(2\pi t)\right) dt. \tag{76.215}$$

Using Fourier series expansion for even functions,

$$\begin{aligned} \cos\left(\frac{a_2}{2\pi}\sin(2\pi t)\right) &= \sum_{k=0}^{\infty} \alpha_k \cos(2\pi k t) \\ \sin\left(\frac{a_2}{2\pi}\sin(2\pi t)\right) &= \sum_{k=0}^{\infty} \beta_k \cos(2\pi k t). \end{aligned}$$

After the integration, we obtain

$$\Delta x = \frac{1}{2}a_1\alpha_1 \quad , \quad \Delta y = \frac{1}{2}a_1\beta_1. \tag{76.216}$$

Because α_1 and β_1 depend on a_2, given the desired motion in x and y, Equation 76.216 results in a one-dimensional line search for a_2:

$$\frac{\Delta x}{\alpha_1(a_2)} - \frac{\Delta y}{\beta_1(a_2)} = 0. \tag{76.217}$$

Once a_2 is found (there may be multiple solutions), a_1 can be found from Equation 76.216.

The above procedure of using sinusoidal inputs for path planning can be generalized to systems in the following canonical form (written for systems with two inputs), called the *chain form*:

$$\begin{bmatrix} \dot q_1 \\ \dot q_2 \\ \dot q_3 \\ \dot q_4 \\ \vdots \\ \dot q_n \end{bmatrix} = \begin{bmatrix} u_1 \\ u_2 \\ q_2 u_1 \\ q_3 u_1 \\ \vdots \\ q_{n-1} u_1. \end{bmatrix} \tag{76.218}$$

For example, the unicycle problem Equation 76.202 can be converted to the chain form by defining

$$\begin{aligned} q_1 &= \theta \\ q_2 &= c_\theta x + s_\theta y \\ q_3 &= s_\theta x - c_\theta y. \end{aligned}$$

Then

$$\begin{aligned} \dot q_1 &= u_1 \\ \dot q_2 &= -q_3 u_1 + \ell u_2 \\ \dot q_3 &= q_2 u_1. \end{aligned}$$

By defining the right hand side of the $\dot q_2$ equation as the new u_2, the system is now in the chain form.

For a general chain system, consider the sinusoidal inputs,

$$u_1 = a\sin(2\pi t), \quad u_2 = b\cos(2\pi k t) \tag{76.219}$$

It follows that, for $i < k + 2$, $q_i(t)$ consists of sinusoids with period 1; therefore, $q_i(1) = q_i(0)$. The net change in q_{k+2} can be computed as

$$q_{k+2}(1) - q_{k+2}(0) = \left(\frac{a}{4\pi}\right)^k \frac{b}{k!}. \tag{76.220}$$

The parameters a and b can then be chosen so that q_{k+2} is driven to the desired value in $[0, 1]$ without affecting all of the states preceding it. A steering algorithm will then consist of the following steps:

1. Drive q_1 and q_2 to the desired values.
2. For each q_{k+2}, $k = 1, \ldots, n-2$, drive q_{k+2} to its desired values by using the sinusoidal input Equation 76.219 with a and b determined from Equation 76.220.

Many systems can be converted to the chain form, e.g., kinematic car, space robot etc. In [4], a general procedure is provided to transform a given system to the chain form. There are also some systems that cannot be transformed to the chain form, e.g., a ball rolling on a flat plate.

Optimal Control Another approach to nonholonomic path planning is optimal control. Consider the following two-input, three-state chain system (we have shown that the unicycle can be converted to this form):

$$\dot{q} = \begin{bmatrix} u_1 \\ u_2 \\ q_2 u_1 \end{bmatrix}, \quad q(0) = q_0. \tag{76.221}$$

The inputs u_i are to be chosen to drive $q(t)$ from q_0 to $q(1) = 0$ while minimizing the input energy:

$$J = \int_0^1 \frac{1}{2} \parallel u(t) \parallel^2 dt.$$

The Hamiltonian associated with this optimal control problem is

$$H(q, u, \lambda) = \frac{1}{2} \parallel u \parallel^2 + \lambda^T f(q) u \tag{76.222}$$

where λ is the co-state vector. From the Maximum Principle, the optimal control can be found by minimizing H with respect to u:

$$u_1 = -(\lambda_1 + \lambda_3 q_2), \quad u_2 = -\lambda_2. \tag{76.223}$$

The co-state satisfies

$$\dot{\lambda} = -\frac{\partial H}{\partial q} = \begin{bmatrix} 0 \\ -\lambda_3 u_1 \\ 0 \end{bmatrix}. \tag{76.224}$$

Differentiating the optimal control in Equation 76.223,

$$\dot{u}_1 = c u_2, \text{ and } \dot{u}_2 = -c u_1 \tag{76.225}$$

where c is a constant ($c = -\lambda_3$). This implies that u_1 and u_2 are sinusoids:

$$\begin{aligned} u_1(t) &= -a \cos ct + u_1(0), \text{ and} \\ u_2(t) &= a \sin ct + u_2(0). \end{aligned} \tag{76.226}$$

Substituting in the equation of motion Equation 76.221 and choosing $c = 2\pi$,

$$\begin{aligned} q_1(1) &= u_1(0) + q_1(0) \\ q_2(1) &= u_2(0) + q_2(0) \\ q_3(1) &= \frac{a^2}{4\pi} + \frac{u_1(0)a}{2\pi} + q_2(0)u_1(0) + \frac{u_1(0)u_2(0)}{2}. \end{aligned} \tag{76.227}$$

The requirement on the zero final state can be used to solve for the constants in the control:

$$\begin{aligned} u_1(0) &= -q_1(0) \\ u_2(0) &= -q_2(0) \\ a &= q_1(0) + \sqrt{q_1^2(0) + 2\pi q_1(0) q_2(0)}. \end{aligned} \tag{76.228}$$

If the expression within the square root is negative, then the constant c should be chosen as -2π to render it positive.

The optimization approach described above can be generalized to certain higher order systems, but, in general, the optimal control for nonholonomic systems is more complicated. One can also try finding an optimal solution numerically; this would, in general, entail the solving a two-point boundary value problem. A nonlinear programming approach based on the following Ritz approximation of the input function space has also been proposed:

$$u(t) = \sum_{k=0}^{N} \alpha_k \psi_k(t) \tag{76.229}$$

where ψ_k's are chosen to be independent orthonormal functions (such as the Fourier basis) and α_k's are constant vectors parameterizing the input function. The minimum input energy criterion can then be combined with a final state penalty term, resulting in the following optimization criterion:

$$\begin{aligned} J &= \gamma \parallel q_f - q(1) \parallel^2 + \int_0^1 \parallel u(t) \parallel^2 dt \\ &= \gamma \parallel q_f - q(1) \parallel^2 + \sum_{k=0}^{N} \parallel \alpha_k \parallel^2. \end{aligned} \tag{76.230}$$

The optimal α_k's can be solved numerically by using nonlinear programming. The penalty weighting γ can be iteratively increased to enforce the final state constraint. The problem is not necessarily convex. Consequently, as in any nonlinear programming problem, only local convergence can be asserted. In next section, we will describe a similar approach without the control penalty term in J. As a result of this modification, a stronger convergence condition can be established.

Path Space Iterative Approach As shown in the beginning of Section 76.3.3, the differential equation governing the nonholonomic motion Equation 76.203 can be written as a nonlinear operator relating an input function, $\underline{u}$, to a path, $\underline{q}$. By writing the final state error as

$$y = q_f - F(q_0, \underline{u}) \tag{76.231}$$

the path planning problem can be regarded as a nonlinear least-squares problem. Global controllability means, that for any q_f, there is at least one solution $\underline{u}$. Many numerical algorithms exist for the solution of this problem. In general, the solution involves lifting a path connecting the initial y to the desired $y = 0$ to the $\underline{u}$ space (see Figure 76.15). Let $\underline{u}(0)$ be the first guess of the input function and $y(0)$ be the corresponding final state error as given by Equation 76.231. The goal is to modify $\underline{u}$ iteratively so that y converges to 0 asymptotically. To this end, choose a path in

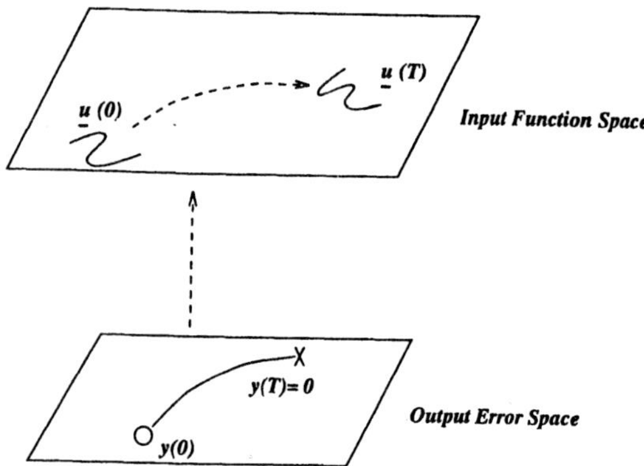

Figure 76.15 Path planning by lifting a path in output error space.

the error space connecting $y(0)$ to 0, call it $y_d(\tau)$, where τ is the iteration variable. The derivative of $y(\tau)$ is

$$\frac{dy}{d\tau} = -\nabla_{\underline{u}} F(q_0, \underline{u}) \frac{d\underline{u}}{d\tau}. \tag{76.232}$$

If $\nabla_{\underline{u}} F(q_0, \underline{u})$ is full rank, then we can choose the following update rule for $\underline{u}(\tau)$ to force y to follow y_d:

$$\frac{d\underline{u}}{d\tau} = -\left[\nabla_{\underline{u}} F(q_0, \underline{u})\right]^{+} \left[-\alpha((q_f - F(q_0, \underline{u})) - y_d) + \frac{dy_d}{d\tau}\right] \tag{76.233}$$

where $\alpha > 0$ and $\left[\nabla_{\underline{u}} F(q_0, \underline{u})\right]^{+}$ denotes the Moore-Penrose pseudo-inverse of $\nabla_{\underline{u}} F(\underline{u})$. This is essentially the continuous version of Newton's method. Equation 76.233 is an initial value problem in $\underline{u}$ with a chosen $\underline{u}(0)$. With $\underline{u}$ discretized by a finite dimensional approximation (e.g., using Fourier basis as in Equation 76.229), it can be solved numerically by an ordinary differential equation solver.

As discussed in Section 76.3.3, the gradient of F, $\nabla_{\underline{u}} F(q_0, \underline{u})$, can be computed from the system Equation 76.203 linearized about the path corresponding to $\underline{u}$. A sufficient condition for the convergence of the iterative algorithm Equation 76.233 is that $\nabla_{\underline{u}} F(q_0, \underline{u}(\tau))$ is full rank for all τ, or equivalently, the time varying linearized system Equation 76.207, generated by linearizing Equation 76.203 about $\underline{u}(\tau)$, is controllable. For controllable systems without drift, it has been shown in [5] that this full rank condition is true generically (i.e., for almost all $\underline{u}$ in the C_∞ topology).

In the cases where $\nabla_{\underline{u}} F(q_0, \underline{u})$ loses rank (possibly causing the algorithm to get stuck), a *generic loop* (see Figure 76.16) can be appended to the singular control causing the composite control to be nonsingular and thus allowing the algorithm to continue its progress toward a solution. A generic loop can be described as follows: For some small time interval $[0, T/2]$, generate a nonsingular control $\underline{v_a}$ (which can be randomly chosen, due to the genericity property). Then let $\underline{v}$ be the control on $[0, T]$ consisting of $\underline{v_a}$ on $[0, T/2]$ and $-\underline{v_a}$ on $[T/2, T]$. Because nonholonomic systems have no drift term, it follows that the system makes a "loop" starting at $q(1) = F(q_0, \underline{u})$ ending once again at the same point. Appending $\underline{v}$ to $\underline{u}$ and renormalizing the time interval to $[0, 1]$ yields a nonsingular control which does not change y. The algorithm is therefore guaranteed to converge to any arbitrary neighborhood of the desired final configuration.

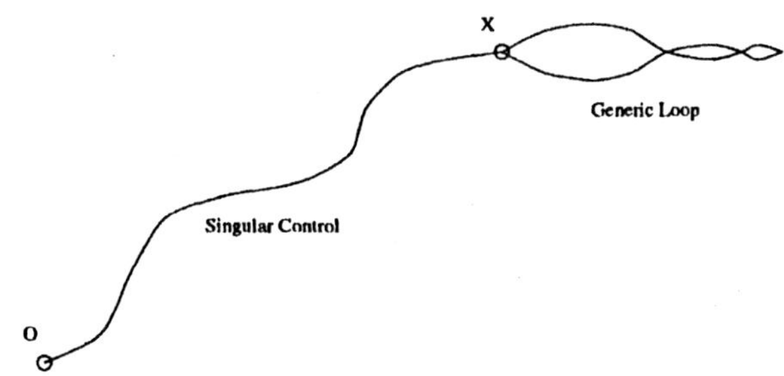

Figure 76.16 Generic loop.

This algorithm has been extended to include inequality constraints such as joint limits, collision avoidance, etc. [6], by using an exterior penalty function approach. Consider state inequality constraints given by

$$c(\underline{q}) \leq 0 \tag{76.234}$$

where $\underline{q}$ is the complete path in the configuration space, $c(\cdot)$ is a vector, and the inequality is interpreted in the componentwise sense. The state trajectory, $\underline{q}$, can be related to the input function $\underline{u}$ through a nonlinear operator (which is typically not possible to find analytically),

$$\underline{q} = \mathcal{F}(q_0, \underline{u}). \tag{76.235}$$

The inequality constraint Equation 76.234 can then be expressed in terms of $\underline{u}$:

$$c(\mathcal{F}(q_0, \underline{u})) \leq 0. \tag{76.236}$$

Inequality constraints in optimization problems are typically handled through penalty functions. There are two types, interior and exterior penalty functions. An interior penalty function sets up barriers at the boundary of the inequality constraints. As the height of the barrier increases, the corresponding path becomes closer to being feasible. If the optimization problem, in our case, the feasible path problem, can be solved for each finite barrier, then convergence to the optimal solution is assured as the barrier height tends to infinity. In the exterior penalty function approach, the ith inequality constraint is converted to an equality constraint by using an exterior penalty function,

$$z_i(\underline{u}) = \gamma_i \sum_{j=1}^{N} g(c_i(\mathcal{F}_j(q_0, \underline{u}))) \tag{76.237}$$

where $\gamma_i > 0$, c_i is the ith constraint, $\mathcal{F}_j$ denotes the jth discretized time point where the constraint is checked, and g is a continuous scalar function with the property that g is equal to zero, when c_i is less than or equal to zero, and is greater than zero and monotonic when c_i is greater than zero. The same iterative approach presented for the equality-only case can now be applied to the composite constraint vector:

$$\psi(\underline{u}) = \begin{bmatrix} y(\underline{u}) \\ z(\underline{u}) \end{bmatrix}. \quad (76.238)$$

For a certain class of convex polyhedral constraints, the generic full rank condition for the augmented problem still holds. This approach has been successfully applied to many complex examples, such as cars with multiple trailers, subject to a variety of collision avoidance and joint limits constraints [6].

76.3.4 Stabilization

State Stabilization

Stabilizability means the existence of a feedback controller that will render the closed-loop system asymptotically stable about an equilibrium point. For linear systems, controllability implies stabilizability. It would be of great value if this were true for special classes of nonlinear systems such as the nonholonomic systems considered in this article (where controllability can be checked through a rank condition on the control Lie algebra). It was shown by [7] that this assertion is not true in general. For a general nonlinear system $\dot{q} = f_0(q, u)$, with equilibrium at q_0, $f_0(q_0, 0) = 0$ and $f_0(\cdot, \cdot)$ continuous in a neighborhood of $(q_0, 0)$, a necessary condition for the existence of a continuous time-invariant control law, that renders $(q_0, 0)$ asymptotically stable, is that f maps any neighborhood of $(q_0, 0)$ to a neighborhood of 0. For a nonholonomic system described by Equation 76.203, $f_0(q, u) = f(q)u$. Then the range of $\{f_0(q, u) : (q, u)$ in a neighborhood of $(q_0, 0)\}$ is equal to the span of the columns of $f(q)$ which is of dimension m (number of inputs). Because a neighborhood about the zero state is n dimensional, the necessary condition above is not satisfied unless $m \geq n$.

There are two approaches to deal with the lack of a continuous time-invariant stabilizing feedback. The first is to relax the continuity requirement to allow piecewise smooth control laws; the second is to relax the time invariance requirement and allow a time-varying feedback.

In either approach, an obvious starting point is to begin with an initial feasible path obtained by using any one of the open-loop methods discussed in Section 76.3.3 and then to apply a feedback to stabilize the system around the path. Given an initial feasible path, if the nonlinear kinematic model linearized about the path is controllable, almost always true as mentioned in Section 76.3.3, a time-varying stabilizing controller can be constructed by using standard techniques. The resulting system will then be locally asymptotically stable.

Consider the unicycle problem Equation 76.202 as an example. Suppose an open-loop trajectory, $\{(x^\star(t), y^\star(t), \theta^\star(t), t \in [0, 1]\}$, and the corresponding input, $\{u_1^\star(t), u_2^\star(t), t \in [0, 1]\}$, are already generated by using any of the methods discussed in Section 76.3.3. The system equation can be linearized about this path:

$$\begin{aligned} \delta\dot{x} &= -(\sin(\theta^\star(t))\, u_1^\star(t))\, \delta\theta + \cos(\theta^\star(t))\, \delta u_1, \\ \delta\dot{y} &= (\cos(\theta^\star(t))u_1^\star(t))\, \delta\theta + \sin(\theta^\star(t))\, \delta u_1, \\ \text{and } \delta\dot{\theta} &= \delta u_2. \end{aligned} \quad (76.239)$$

This is a linear time-varying system. It can be easily verified that as long as $u_1^\star$ is not identically zero, the system is controllable. One can then construct a time-varying stabilizing feedback to keep the system on the planned open-loop path.

Stabilizing control laws can also be directly constructed without first finding a feasible open-loop path. In [8], it was shown that all nonholonomic systems can be feedback stabilized with a smooth periodic controller. For specific classes of systems, such as mobile robots in [9], or, more generally, the so-called power systems as in [10], explicit constructive procedures for such controllers have been demonstrated.

We will again use the unicycle example to illustrate the basic idea of constructing a time-varying stabilizing feedback by using a time-dependent coordinate transformation so that the equation of motion contains a time-varying drift term. Define a new variable z by

$$z = \theta + k(t, x, y) \quad (76.240)$$

where k is a function that will be specified later. Differentiating z,

$$\dot{z} = v \stackrel{\Delta}{=} u_2 + \frac{\partial k}{\partial t} + (\frac{\partial k}{\partial x}\cos\theta + \frac{\partial k}{\partial y}\sin\theta)u_1.$$

Consider a quadratic Lyapunov function candidate $V = \frac{1}{2}(x^2 + y^2 + z^2)$. The derivative along the solution trajectory is

$$\dot{V} = (x\cos\theta + y\sin\theta)u_1 + zv.$$

By choosing

$$\begin{aligned} u_1 &= -\alpha_1(x\cos\theta + y\sin\theta), \\ v &= -\alpha_2 z, \quad \alpha_1, \alpha_2 > 0, \end{aligned} \quad (76.241)$$

which means

$$\begin{aligned} u_2 &= -\frac{\partial k}{\partial t} - (\frac{\partial k}{\partial x}\cos\theta + \frac{\partial k}{\partial y}\sin\theta)u_1 \\ &\quad - \alpha_2(\theta + k(t, x, y)), \end{aligned}$$

we obtain a negative semidefinite $\dot{V} = -\alpha_1(x\cos\theta + y\sin\theta)^2 - \alpha_2 z^2$. This implies that, as $t \to \infty$, $z \to 0$ and $x\cos\theta + y\sin\theta \to 0$. Substituting in the definition of z, we get $\theta(t) \to -k(t, x(t), y(t))$. From the other asymptotic condition, $\theta(t)$ also converges to $-\tan^{-1}\left(\frac{x(t)}{y(t)}\right)$. As $\dot{x}$ and $\dot{y}$ asymptotically vanish, $x(t)$ and $y(t)$, and therefore, $\theta(t)$, tend to constants. Equating the two asymptotic expressions for $\theta(t)$, we conclude that $k(t, x(t), y(t))$ converges to a constant. By suitably choosing $k(t, x, y)$, e.g., $k(t, x, y) = (x^2 + y^2)\sin(t)$, the only condition under which $k(t, x, y)$ can converge to a constant is that $x^2 + y^2$ converges to zero, which in turn implies that $\theta(t) \to 0$. In contrast to the indirect approach (i.e., using a linear time varying control law to stabilize a system about a planned open-loop path), this control law is globally stabilizing.

Output Stabilization

In certain cases, it may only be necessary to control the state to a certain manifold rather than to a particular configuration.

For example, in the case of a robot manipulator on a free floating mobile base, it may only be necessary to control the tip of the manipulator so that it can perform useful tasks. In this case, a smooth output stabilizing controller can frequently be found.

Suppose the output of interest is

$$y = g(q) \ , \ \ y \in \mathcal{R}^p \tag{76.242}$$

and $p < n$. At a particular configuration, q,

$$\dot{y} = \nabla_q g(q)\, f(q)\, u. \tag{76.243}$$

Define $K(q) = \nabla_q g(q)\, f(q)$. If $K(q)$ is onto, i.e., $p \leq m$ and $K(q)$ is full rank, then the system is locally output controllable (there is a u that can move y arbitrarily within a small enough ball) though it is not locally state controllable.

The output stabilization problem involves finding a feedback controller u (possibly dependent on the full state) to drive y to a set point, y_d. Provided that $K(q)$ is of full row rank, an output stabilizing controller can be easily found:

$$u = -QK^T(q)(y - y_d) \quad , \quad Q > 0. \tag{76.244}$$

Therefore, y is governed by

$$\dot{y} = -K(q)QK^T(q)(y - y_d). \tag{76.245}$$

Under the full row rank assumption on $K(q)$, $K(q)QK^T(q)$ is positive definite, which implies that y converges to y_d asymptotically. In general, either y converges to y_d or q converges to a singular configuration of $K(q)$ (where $K(q)$ loses row rank) and $(y - y_d)$ converges to the null space of $K^T(q)$.

We will again use the unicycle problem as an illustration. Suppose the output of interest is (x, θ) and the goal is to drive (x, θ) to (x_d, θ_d) where θ_d is not a multiple of $\frac{\pi}{2}$. By choosing the control law,

$$u_1 = -\alpha_1 \cos\theta(x - x_d), \ u_2 = -\alpha_2(\theta - \theta_d). \tag{76.246}$$

The closed-loop system for the output is

$$\dot{x} = -\alpha_1(\cos^2\theta)(x - x_d), \ \dot{\theta} = -\alpha_2(\theta - \theta_d). \tag{76.247}$$

The closed-loop system contains a singularity at $\theta = \frac{\pi}{2}$, but, if $\theta_d \neq \frac{\pi}{2}$, this singularity will not be attractive. The output stabilization of (x, θ) can be concatenated with other output stabilizing controllers, with other choices of outputs, to obtain full state stabilization. For example, once x is driven to zero, θ can be independently driven to zero (with $u_1 = 0$), and, finally, y can be driven to zero without affecting x and θ. These stages can be combined together as a piecewise smooth state stabilizing feedback controller.

Consider a space robot on a platform as another example. Suppose the output of interest is the end effector coordinate, y. The singular configurations in this case are called the dynamic singularities. The output velocity is related to the joint velocity and center of mass angular velocity by the kinematic Jacobians,

$$\dot{y} = J(q)\dot{q} + J_b(q)\omega.$$

As discussed in Section 76.3.1, the nonholonomic nature of the problem follows from the conservation of the angular momentum Equation 76.187:

$$M_1^T(q)\dot{q} + M_b(q)\omega = 0.$$

Eliminating ω,

$$\dot{y} = (J(q) - J_b M_b^{-1} M_1^T(q))\dot{q}.$$

The effective Jacobian, $K(q) = J(q) - J_b M_b^{-1} M_1^T(q)$, sometimes called the dynamic Jacobian, now contains inertia parameters (hence the modifier "dynamic") in contrast to a terrestrial robot Jacobian which only depends on the kinematic parameters. If the dimension of q is at least as large as the dimension of y, the output can be effectively controlled provided that the dynamic Jacobian does not lose rank (i.e., q is away from the dynamic singularities).

References

[1] Oriolo, G. and Nakamura, Y., Control of mechanical systems with second-order nonholonomic constraints: Underactuated manipulators, *Proc. 30th IEEE Conference on Decision and Control,* 2398–2403, Brighton, England, 1991.

[2] Li, Z. and Canny, J.F., Eds., *Nonholonomic motion planning,* Kluwer Academic, Boston, MA, 1993.

[3] Sastry, S.S., Murray, R.M., and Li, Z., *A Mathematical Introduction to Robotic Manipulation,* CRC Press, Boca Raton, FL, 1993.

[4] Murray, R.M. and Sastry, S.S., Nonholonomic motion planning – steering using sinusoids, *IEEE Trans. Automat. Control,* 38, 700–716, 1993.

[5] Lin, Y. and Sontag, E.D., Universal formula for stabilization with bounded controls, *Syst. Control Lett.,* 16(6), 393–397, 1991.

[6] Divelbiss, A. and Wen, J.T., Nonholonomic motion planning with inequality constraints, *Proc. IEEE Int. Conf. Robotics Automat.,* San Diego, CA, 1994.

[7] Brockett, R.W., Asymptotic stability and feedback stabilization, in *Differential Geometric Control Theory,* Brockett, R.W., Millman, R.S., and Sussmann, J.J, Eds., Birkhauser, 1983, vol. 27, 181–208.

[8] Coron, J.-M., Global assymptotic stabilization for controllable systems without drift, *Math. Control, Signals, Syst.,* 5(3), 1992.

[9] Samson, C. and Ait-Abderrahim, K., Feedback control of a nonholonomic wheeled cart in Cartesian space, in *Proc. IEEE Robotics Automat. Conf.,* Sacramento, CA, 1991.

[10] Teel, A., Murray, R., and Walsh, G., Nonholonomic control systems: From steering to stabilization with sinusoids, *Proc. 31th IEEE Conf. Dec. Control,* Tucson, AZ, 1992.

77

Miscellaneous Mechanical Control Systems

Brian Armstrong
Department of Electrical Engineering and Computer Science, University of Wisconsin—Milwaukee, Milwaukee, WI

Carlos Canudas de Wit
Laboratoire d'Automatique de Grenoble, ENSIEG, Grenoble, France

Jacob Tal
Galil Motion Control, Inc.

Thomas R. Kurfess
The George W. Woodruff School of Mechanical Engineering, The Georgia Institute of Technology, Atlanta, GA

Hodge Jenkins
The George W. Woodruff School of Mechanical Engineering, The Georgia Institute of Technology, Atlanta, GA

Maarten Steinbuch
Philips Research Laboratories, Eindhoven, The Netherlands

Gerrit Schootstra
Philips Research Laboratories, Eindhoven, The Netherlands

Okko H. Bosgra
Mechanical Engineering Systems and Control Group, Delft University of Technology, Delft, The Netherlands

77.1 Friction Modeling and Compensation

Brian Armstrong, *Department of Electrical Engineering and Computer Science, University of Wisconsin—Milwaukee, Milwaukee, WI*
Carlos Canudas de Wit, *Laboratoire d'Automatique de Grenoble, ENSIEG, Grenoble, France*

77.1.1 Introduction

The successful implementation of friction compensation brings together aspects of servo control theory, tribology (the science of friction), machine design, and lubrication engineering. The design of friction compensation is thus intrinsically an interdisciplinary challenge. Surveys of contributions from the diverse fields that are important for friction modeling, motion analysis, and compensation are presented in [2], [10].

The challenge to good control posed by friction is often thought of as being stick-slip, which is an alternation between sliding and sticking due to static friction. Stick-slip is most common when integral control is used and can prevent a machine from ever reaching its intended goal position. However, other forms of frictional disturbance can be of equal or greater importance. The tracking error introduced by friction into multi-axis motion is an example. This error, called *quadrature glitch* is illustrated in Figure 77.1. The two-axis machine fails to accurately track the desired circular contour because as one axis goes through zero velocity, it is arrested for a moment by static friction while the other axis continues to move.

Even when frictional disturbances are eliminated, friction in mechanical servos may still impact cost and performance. As outlined in Section 77.1.5 , lubrication and hardware modification are two commonly used approaches to improving closed-loop performance. These approaches are grouped under problem avoidance, and a survey of engineers in industry [2] suggests that they are the most commonly used techniques for eliminat-

0-8493-8570-9/96/$0.00+$.50

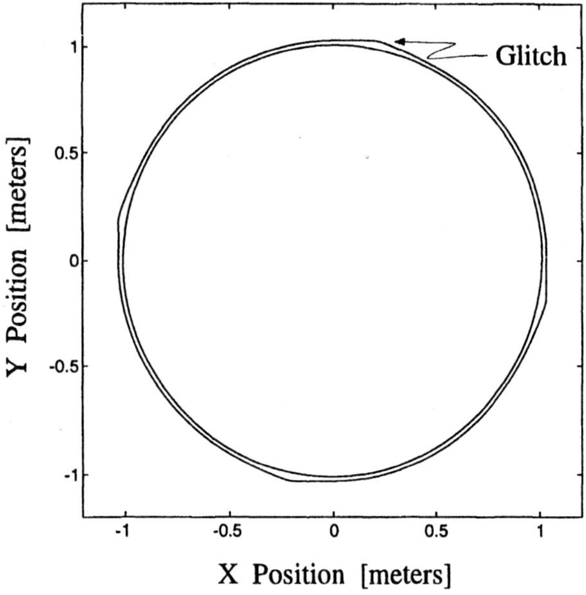

Figure 77.1 An example of quadrature glitch in 2-axis motion: X-Y position trace showing quadrature glitch. The desired trajectory is shown with 2% radial reduction to highlight the tracking error.

ing frictional disturbances. But these techniques have a cost that may be hidden if more effective servo compensation of friction is not considered. For example, special materials, called *friction materials* (see Section 77.1.5) are often used on the machine tool slideways to eliminate stick-slip. These materials have a high coulomb friction but a relatively lower static friction, and thus the machine slideway is not prone to stick-slip. But the higher coulomb friction of the slideway introduces increased power and energy requirements, which increases the initial costs for a larger drive and energy costs throughout the lifetime of the machine. More effective friction compensation by feedback control could provide higher performance at lower cost.

77.1.2 Friction Modeling

The simplest friction model has the instantaneous friction force, $F_f(t)$, expressed as a function of instantaneous sliding velocity, $v(t)$. Such a model may include coulomb, viscous, and/or static friction terms, which are described in Section 77.1.2. Typical examples are shown in Figure 77.2. The components of this simple model are given in Equations 77.1 to 77.3.

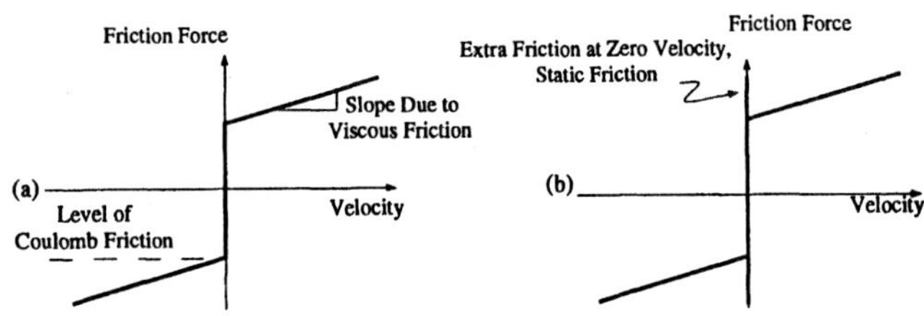

Figure 77.2 Overly simplistic, but nonetheless common, friction models: (a) coulomb + viscous friction model; (b) static + coulomb + viscous friction model.

Glossary of Terms and Effects

Coulomb friction: A force of constant magnitude, acting in the direction opposite to motion.

$$\text{When } v(t) \neq 0: \quad F_f(t) = -F_c \text{sgn}(v(t)) \tag{77.1}$$

Viscous friction: A force proportional to velocity.

$$\text{When } v(t) \neq 0: \quad F_f(t) = -F_v v(t) \tag{77.2}$$

Static friction: Is not truly a force of friction, as it is neither dissipative nor a consequence of sliding. Static friction is a force of constraint [12].

$$\begin{aligned} \text{When } v(t) &= 0: \\ F_f &= \begin{cases} u(t); & |u(t)| \leq (F_c + F_s) \\ (F_c + F_s)\text{sgn}(u(t)); & \text{otherwise} \end{cases} \end{aligned} \tag{77.3}$$

where $u(t)$ is the externally applied force.

For many servo applications, the simplest friction model–instantaneous friction as a function (any function) of instantaneous sliding velocity–is inadequate to accurately predict the interaction of control parameters and frictional disturbances to motion. In addition to coulomb, viscous, and static friction, dynamic friction must be considered. The state of the art in tribology does not yet provide a friction model derived from first principles. Four friction phenomena, however, are consistently observed in lubricated machines [10]:

1. **Stribeck friction or the Stribeck curve** is the negatively sloped and nonlinear friction-velocity characteristic occurring at low velocities for most lubricated and some dry contacts. The Stribeck curve is illustrated in Figure 77.3. The negatively sloped portion of the curve is an important contributor to stick-slip. Because of dynamic friction effects, instantaneous friction is not simply a function of instantaneous velocity. When velocity is steady, however, a steady level of friction is observed. This is friction as a function of steady-state velocity and gives the Stribeck curve.
2. **Rising static friction.** The force required for breakaway (the transition from not-sliding to sliding) varies with the time spent at zero velocity (dwell time) and with force application rate. The physics underlying rising static friction are not well understood, but experimental data and empirical models are available from the tribology literature [2] [10]. Rising static friction interacts significantly with stick-slip and is represented as a function of dwell time in the model presented in Section 77.1.2, and as a function of force application rate in the model presented in Section 77.1.2.

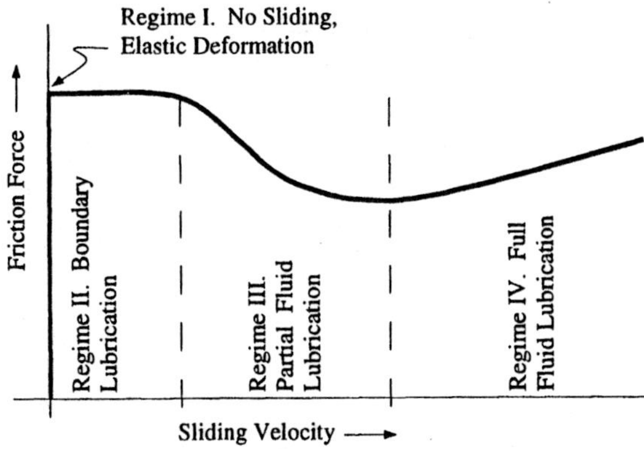

Figure 77.3 Friction as a function of steady-state velocity, the Stribeck curve. The four regimes of motion are described in section 5.1.a (from [2], courtesy of the publisher).

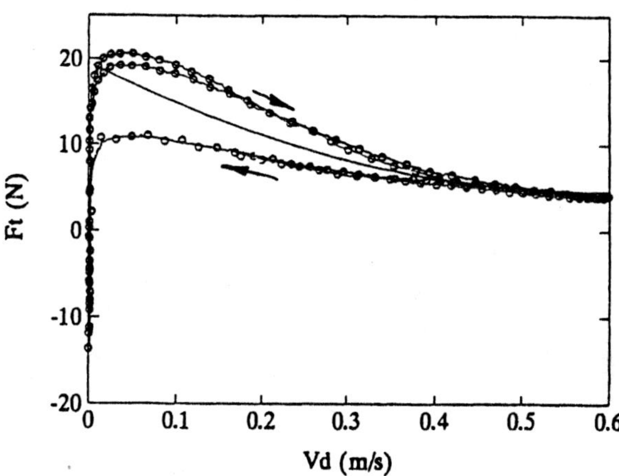

Figure 77.4 Measured friction versus sliding velocity: —, quasi-steady (equilibrium) sliding; *ooo*, intermittent sliding. (From Ibrahim, R.A. and Soom, A., Eds., *Friction-Induced Vibration, Chatter, Squeal, and Chaos, Proc. ASME Winter Annu. Meet.,* DE-Vol. 49, 139, 1992. With permission.)

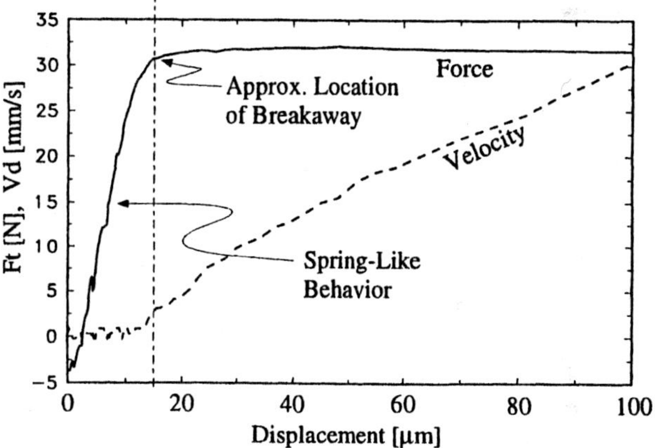

Figure 77.5 Force (—) and velocity (- -) versus displacement during the transition from static to sliding friction (breakaway). The spring-like behavior of the friction contact is seen as the linear force-displacement curve over the first 15μm of motion. (Adapted from [12].)

3. **Frictional memory** is the lag observed between changes in velocity or normal load and the corresponding change in friction force. It is most distinctly observed when there is partial fluid lubrication (partial fluid lubrication is a common condition in many machine elements, such as transmission and rolling bearings; see Section 77.1.5). As a result of frictional memory, instantaneous friction is a function of the history of sliding velocity and load as well as instantaneous velocity and load. An example of measured friction plotted against velocity is seen in Figure 77.4. The solid line indicates the friction as a function of *steady-state* sliding velocity and shows a single value for friction as a function of velocity. The solid line may be compared with Figure 77.3. The line marked *ooo* shows friction during intermittent sliding (stick-slip). The friction force is a multi-valued function of velocity: during acceleration, the friction force is higher, while during deceleration the friction force is lower. The hysteresis seen in Figure 77.4 indicates the presence of frictional memory.
4. **Presliding displacement** is the displacement of rolling or sliding contacts prior to true sliding. It arises due to elastic and/or plastic deformation of the contacting asperities and is seen in Figure 77.5. Because metal components make contact at small, isolated points called asperities, the tangential compliance of a friction contact (e.g., gear teeth in a transmission) may be substantially greater than the compliance of the bulk material. Presliding displacement in machine elements has been observed to be on the order of 1 to 5 micrometers. Because a small displacement in a bearing or between gear teeth may be amplified by a mechanism (for example, by a robot arm [1]), presliding displacement may give rise to significant output motions.

Friction models capturing these phenomena are described in Sections 77.1.2 and 77.1.2. In broad terms, these four dynamic friction phenomena can be linked to behaviors observed in servo mechanisms [2]. These connections are presented in Table 77.1.

Additional Terms

Temporal friction phenomena: This term connotes both rising static friction and frictional memory.

Stick-slip: In the literature, stick-slip is used to refer to a broad range of frictional phenomena. Here, it is used to refer to a stable limit cycle arising during motion.

Hunting: With feedback control, a type of frictional limit cycle is possible that is not possible with passive systems: a limit cycle that arises while the net motion of the system is zero. This is called hunting and is

TABLE 77.1 Observable Consequences of Dynamic Friction Phenomena

Dynamic Friction Phenomenon	Predicted/Observed Behavior
Stribeck friction	Needed to correctly predict initial conditions and system parameters leading to stick-slip
Rising static friction	Needed to correctly predict the interaction of velocity with the presence and amplitude of stick-slip
Frictional memory	Needed to correctly predict the interaction of stiffness (mechanical or feedback) with the presence of stick-slip
Presliding displacement	Needed to correctly predict motion during stick(e.g., during velocity reversal or before breakaway)

most often associated with integral control.

Standstill (or lost motion): Even where there is no stable limit cycle, the frictional disturbance to a servo system may be important. Standstill or lost motion refers to the frictional disturbance that arises when a system goes through zero velocity; it is seen in Figure 77.1. An ideal system would show continuous acceleration, but the system with friction may be arrested at zero velocity for a period of time. In machine tools, this is the frictional disturbance of greatest economic importance.

The Seven-Parameter Friction Model

The seven-parameter friction model [2] captures the four detailed friction phenomena, as well as coulomb, viscous, and static friction. The model has the advantage that the friction phenomena are explicitly represented with physically motivated model parameters. The model is presented in Equations 77.4 to 77.6, and the parameters are described in Table 77.2. The liability of the seven-parameter model is that it is not, in fact, a single integrated model, but a combination of three models: Equation 77.4 reflects the tangential force of constraint during stick; Equation 77.5 reflects frictional force during sliding; and Equation 77.6 represents a sampled process that reflects rising static friction.

Not-sliding (pre-sliding displacement):

$$F_f(x(t)) = -K_t x(t) \qquad (77.4)$$

Sliding (coulomb + viscous + Stribeck curve friction with frictional memory):

$$F_f(v(t), t) = -\Bigg(F_c + F_v\,|v(t)| + F_s(\gamma, t_2)\frac{1}{1+\left(\frac{v(t-\tau_L)}{v_s}\right)^2}\Bigg)\mathrm{sgn}[v(t)] \qquad (77.5)$$

Rising Static Friction (friction level at breakaway):

$$F_s(\gamma, t_2) = F_{s,a} + (F_{s,\infty} - F_{s,a})\frac{t_2}{t_2+\gamma} \qquad (77.6)$$

where:

$F_f(.)$	=	instantaneous friction force
F_c	=	coulomb friction force*
F_v	=	viscous friction force*
F_s	=	magnitude of the Stribeck friction [frictional force at breakaway is given by $F_f(t = t_{\text{breakaway}}) = F_c + F_s$]
$F_{s,a}$	=	magnitude of the Stribeck friction at the end of the previous sliding period
$F_{s,\infty}$	=	magnitude of the Stribeck friction after a long time at rest (or with a slow application force)*
K_t	=	tangential stiffness of the static contact*
v_s	=	characteristic velocity of the Stribeck friction*
τ_L	=	time constant of frictional memory*
γ	=	temporal parameter of the rising static friction*
t_2	=	dwell time, time at zero velocity

(*) Marks friction model parameters; other variables are state variables.

Integrated Dynamic Friction Model

Canudas et al. [6] have proposed a friction model that is conceptually based on elasticity in the contact. They introduce the variable $z(t)$ to represent the average state of deformation in the contact. The friction is given by

$$F_f(t) = \sigma_0 z(t) + \sigma_1 \frac{dz(t)}{dt} + F_v v(t) \qquad (77.7)$$

where $F_f(t)$ is the instantaneous friction and $v(t)$ is the contact sliding velocity. The state variable $z(t)$ is updated according to:

$$\frac{dz(t)}{dt} = v(t) - \frac{\sigma_0}{g(v(t))} z(t)|v(t)| \qquad (77.8)$$

In steady sliding, $\dot{z}(t) = 0$, giving

$$z(t)|_{\text{steady-state}} = \frac{g(v(t))}{\sigma_0}\mathrm{sgn}[v(t)] \qquad (77.9)$$

which then gives a steady-state friction:

$$F_f(t)\big|_{\dot{z}=0} = g(v(t)) + F_v v(t) \qquad (77.10)$$

A parameterization that has been proposed to describe the nonlinear low-velocity friction is

$$g(v(t)) = F_c + F_s e^{-[v(t)/v_s]^2} \qquad (77.11)$$

where F_c is the coulomb friction; F_s is magnitude of the Stribeck friction (the excess of static friction over coulomb friction); and v_s is the characteristic velocity of the Stribeck friction, approximately the velocity of the knee in Figure 77.3. With this description of $g(v(t))$, the model is characterized by six parameters: $\sigma_0, \sigma_1, F_v, F_c, F_s$, and v_s. For steady sliding, $[\dot{v}(t) = 0]$, the friction force (Figure 77.3) is given by:

$$\begin{aligned} F_{ss}(v(t)) &= g(v(t))\mathrm{sgn}[v(t)] + F_v v \quad (77.12) \\ &= F_c\mathrm{sgn}[v(t)] + Fse^{-[v(t)/v_s]^2}\mathrm{sgn}[v(t)] \\ &\quad + F_v v(t) \end{aligned}$$

When velocity is not constant, the dynamics of the model give rise to frictional memory, rising static friction, and presliding displacement.

The model has a number of desirable properties:

1. It captures presliding displacement, frictional memory, rising static friction, and the Stribeck curve in a single model without discontinuities.
2. The steady-state (friction–velocity) curve is captured by the function $g(v(t))$, which is chosen by the designer (e.g., Equation 77.11).
3. In simulation, the model is able to reproduce the data of a number of experimental investigations [6].

A difficulty associated with dynamic friction models lies with identifying the friction model parameters for a specific system. The static parameters involved in the function $g(v(t))$ and the viscous friction parameter F_v may be identified from experiments at constant velocity (see Section 77.1.4). The dynamic parameters σ_0 and σ_1 are more challenging to identify. These parameters are interrelated in their description of the physical phenomena, and their identification is made challenging by the fact that the state $z(t)$ is not physically measurable. A procedure for identifying these parameters is outlined in Section 77.1.4.

Magnitudes of Friction Parameters

The magnitudes of the friction model parameters naturally depend upon the mechanism and lubrication, but typical values may be offered, as seen in Table 77.2 (see [1], [2], [3], [12]). The friction force magnitudes, F_c, F_v, and $F_{s,\infty}$, are expressed as a function of normal force F_n; i.e., as coefficients of friction. Δ_x is the deflection before breakaway resulting from contact compliance.

In servo machines it is often impossible to know the magnitude of the normal force in sliding contacts. Examples of mechanisms with difficult-to-identify normal forces are motor brushes, gear teeth, and roller bearings, where the normal force is dependent on spring stiffness and wear, gear spacing, and bearing preload, respectively. For this reason, friction is often described for control design in terms of parameters with units of force, rather than as coefficients of friction.

In addition to the models described above, state variable friction models have been used to describe friction at very low velocities (μm/s). The state variable models are particularly suited to capture nonlinear low-velocity friction and frictional memory. Velocities of micrometers per second can be important in some control applications, such as wafer stepping or the machining of nonspherical optics (see [2][7]).

Friction that depends upon position has been observed in machines with gear-type transmissions. In general, mechanisms in which the normal force in sliding contacts varies during motion show position-dependent friction. Selected components of the Fourier transform and table lookup have been used to model the position-dependent friction. At least one study of friction compensation in an industrial robot has shown that it is important to model the position-dependent friction in order to accurately identify the detailed friction model parameters [1].

In practical machines, there are often many rubbing surfaces that contribute to the total friction: drive elements, seals, rotating electrical contacts, bearings, etc. In some mechanisms, a single interface may be the dominant contributor, as transmission elements often are. In other cases where there are several elements contributing at a comparable level, it may be impossible to identify their individual contributions without disassembling the machine. In these cases, it is often aggregate friction that is modeled.

The control designer faces a considerable challenge with respect to friction modeling. On the one hand, parameters of a dynamic friction model are at best difficult to identify while, on the other hand, recent theoretical investigations show that the dynamics of friction play an important role in determining frictional disturbances to motion and appropriate compensation [1], [5], [6], [7], [10], [12], [14]. Often, it is necessary to use a simplified friction model. The most common model used for control incorporates only coulomb friction, Equation 77.1. For machine tools, the Karnopp model, Equations 77.15 to 77.17, has been employed to represent static friction (e.g., [4]). And presliding displacement, Equation 77.4, has been modeled for precision control of pointing systems (e.g., [14]). Reference [1] provides an example showing how a detailed friction model can be used to achieve high-performance control, but also makes clear the technical challenges, including special sensing, associated with identifying the parameters of a detailed friction model. To date, there has been no systematic exploration of the trade-offs between model complexity and control performance.

77.1.3 Simulation

Simulation is the most widely used tool for predicting the behavior of friction compensation. The simulation of systems with friction is made challenging by the rapid change of friction as velocity goes through zero. In the simplest friction model, friction is discontinuous at zero velocity:

$$F_f(t) = -F_c\mathrm{sgn}[v(t)] \quad (77.13)$$

The discontinuity poses a difficulty for numerical integrators used in simulation. The difficulty can be addressed by "softening" the discontinuity, for example with

$$F_f(t) = -F_c\mathrm{sgn}[v(t)]\left(1 - e^{-[|v(t)|/v_0]}\right) \quad (77.14)$$

TABLE 77.2 Approximate Ranges of Detailed Friction Model Parameters for Metal-on-Metal Contacts Typical of Machines

	Parameter Range	Parameter Depends Principally Upon
F_c	$0.001 - 0.1x F_n$	Lubricant viscosity, contact geometry, and loading
F_v	0—very large	Lubricant viscosity, contact geometry, and loading
$F_{s,\infty}$	$0 - 0.1x F_n$	Boundary lubrication
k_t	$\frac{1}{\Delta_x} \times (F_s + F_c)$; $\Delta_x \simeq 1 - 50(\mu M)$	Material properties and surface finish
v_s	$0.00001 - 0.1$ (m/s)	Boundary lubrication, lubricant viscosity, material properties and surface finish, contact geometry, and loading
τ_L	$1 - 50$ (ms)	Lubricant viscosity, contact geometry, and loading
γ	$0 - 200$ (s)	Boundary lubrication

which produces a smooth curve through zero velocity. Models that are modified in this way, however, exhibit creep: small applied forces result in low but steady velocities. Some frictional contacts exhibit creep, but metal-on-metal contacts often exhibit a minimum applied force below which there is no motion.

The discontinuous friction model is a nonphysical simplification in the sense that a mechanical contact with distributed mass and compliance cannot exhibit an instantaneous change in force. Friction may be a discontinuous function of *steady-state* velocity (as are Figure 77.3, and Equation 77.11), but a system going through zero velocity is a transient event.

The integrated dynamic friction model (Section 77.1.2) uses an internal state (Equation 77.7) to represent the compliance of the contact and thereby avoids both the discontinuity in the applied forces and creep. The model has been used to reproduce, in simulation dynamic friction phenomena that have been experimentally observed. Another model that has been widely applied to simulation is the Karnopp friction model [11]. The Karnopp model solves the problems of discontinuity and creep by introducing a pseudo velocity and a finite neighborhood of zero velocity over which static friction is taken to apply. The pseudo velocity is integrated in the standard way

$$\dot{p}(t) = [F_a(t) - F_m(t)] \tag{77.15}$$

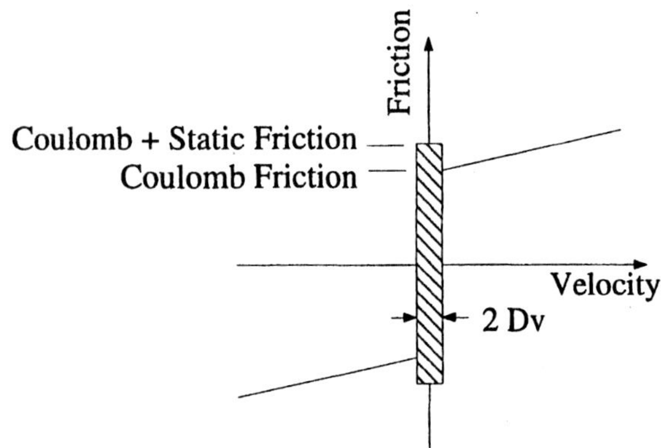

Figure 77.6 A friction-velocity representation of the Karnopp friction model (adapted from [11]).

where $p(t)$ is the pseudo velocity (identified with momentum in [11]), $F_a(t)$ is the applied force, and $F_m(t)$ is the modeled friction. The modeled friction is given by:

$$F_m(v(t), F_a(t)) = \begin{cases} -\text{sgn}[F_a(t)]\max[|F_a(t)|, (F_c + F_s)] & |v(t)| < D_V \\ -\text{sgn}[v(t)]F_c & |v(t)| \geq D_V \end{cases} \tag{77.16}$$

A small neighborhood of zero velocity is defined by D_V, as shown in Figure 77.6. Outside this neighborhood, friction is a function of velocity; coulomb friction is specified in Equation 77.16, but any friction-velocity curve could be used. Inside the neighborhood of zero velocity, friction is equal to and opposite the applied force up to the breakaway friction, and velocity is set to zero

$$v(t) = \begin{cases} 0, & |p(t)| < D_V \\ \frac{1}{M}p(t), & |p(t)| \geq D_V \end{cases} \tag{77.17}$$

where M is the mass.

The integration step must be short enough that at least one value of velocity falls in the range $|v(t)| < D_V$. At this point velocity is set to zero, according to Equation 77.17. If the applied force is less than the breakaway friction, $F_c + F_s$, then $\dot{p}(t) = 0$, and the system remains at rest. When $|F_a| > (F_c + F_s)$, $\dot{p}(t) \neq 0$ and, perhaps after some time, the condition of Equation 77.17 allows motion to begin. The Karnopp model represents static friction as applying over a region of low velocities rather than at the mathematical concept of zero velocity. The model thus eliminates the need to search for the exact point where velocity crosses zero.

While practical for simulation, and even feedback control (e.g., [4]), the Karnopp friction model is a simplification that neglects frictional memory, presliding displacement, and rising static friction. The simulation predictions are thus limited to gross motions; to accurately predict detailed motion, dynamic friction must be considered.

For any approach to simulating systems with friction, an integrator with variable time step size is important. The variable time step allows the integrator to take very short time steps near zero velocity, where friction is changing rapidly, and longer time steps elsewhere, where friction is more steady. Variable step size integration is standard in many simulation packages.

77.1.4 Off-Line Friction Parameter Identification

At this time, it is not possible to accurately predict the static and dynamic friction model parameters based on the specifications of a mechanism. Consequently, friction compensation methods that require a model require a method to identify the model parameters. The problem is one of nonlinear parameter identification, which has a large literature, including Chapter 58.

While roughly determining the coulomb and viscous friction parameters may entail only some straightforward constant force or constant velocity motions (see Section 77.1.4), determining the dynamic friction parameters often entails acceleration sensing or specialized approaches to parameter identification.

Position-Dependent Friction

Mechanisms that are spatially homogeneous, such as direct and belt-driven mechanisms, should not show a substantial position-dependent friction. But mechanisms with spatial inhomogeneities, such as gear drives, can show a large change in friction from one point to another, as seen in Figure 77.7 . The

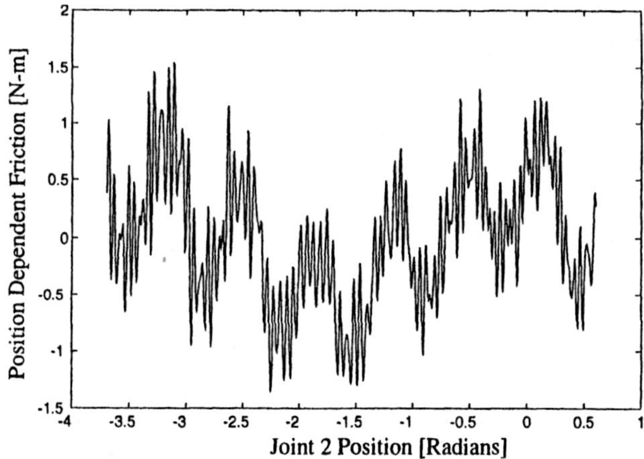

Figure 77.7 Position-dependent friction observed in joint 2 of a PUMA robot arm. The signal with a spatial period of approximately 0.7 radians corresponds to one rotation of the intermediate gear. One cycle of the predominant high-frequency signal corresponds to one rotation of the motor pinion gear. The 3 N-m peak-to-peak position dependent friction may be compared to the average Coulomb friction level of 12 N-m. (Adapted from [1].)

position-dependent friction can be measured by observing the breakaway friction, the level of force required to initiate motion, throughout the range of motion of the mechanism. To measure breakaway friction:

1. The mechanism is allowed to come to rest at the point where friction will be measured.
2. Force (torque) is applied according to a specified curve (generally a ramp).
3. The onset of motion is detected.

The applied force corresponding to the onset of motion is the breakaway friction. To achieve repeatable results, it is necessary to consider these factors:

1. In many machines, friction is observed to be higher after a period of inactivity and to decrease with the first few motions. Bringing lubricants into their steady-state condition accounts for this transient. To repeatably measure any friction parameter, the machine must be "warmed up" by typical motions prior to measurement.
2. The highest spatial frequencies present in the position-dependent friction may be quite high. Spatial sampling must be sufficient to avoid aliasing.
3. Force must be applied consistently. Because the breakaway force depends upon the dwell time and rate of force application (see Section 77.1.2) these variables must be controlled.
4. The method used to detect breakaway must be well selected. The onset of motion is not a simple matter, corresponding to the fact that motion occurs before true sliding begins (presliding displacement, see Figure 77.5). Detecting the first shaft encoder pulse after the beginning of force application, for example, is a very poor detection of sliding because the pulse can arise with presliding displacement at any point on the force application curve. Tests based on velocity or on total motion from the initial position have been used.

Gear drives can exhibit significant friction variation with motions corresponding to a tenth of the highest gear pitch. For example, in a mechanism with 15 : 1 reduction and 18 teeth on the motor pinion gear, $15x18x10 = 2700$ cycles of position-dependent friction per revolution of the mechanism. Capturing these variations during motion poses a significant sensing bandwidth challenge. Breakaway friction is a suitable tool for identifying position dependency in the sliding friction because it can be measured at many points, corresponding to a high spatial sampling.

Friction as a Function of Steady-State Velocity

The steady-state friction-velocity curve (Figure 77.3) can be observed with a series of constant velocity motions. The motions can be carried out with either open- or closed-loop control. If there is significant position-dependent friction, feed-forward compensation of the position-dependent friction will improve the accuracy of the steady-sliding friction model. An example is shown in Figure 77.8, where the data show the non-linear low-velocity friction. No measurements were taken in the region of the steep friction-velocity curve because stick-slip arises in the tested mechanism, even with acceleration feedback, and constant-velocity sliding at low velocities was not possible. The beginning of the negative velocity friction curve observed during constant-velocity sliding and the value of breakaway friction measured using the breakaway experiment (Section 77.1.4) would be sufficient to approximately identify the parameters of the Stribeck curve.

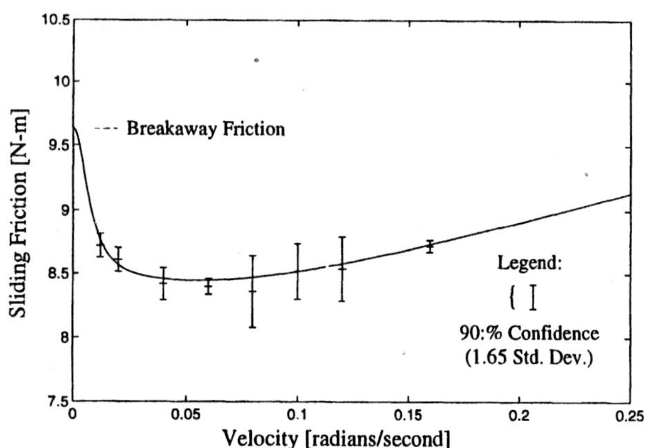

Figure 77.8 Friction as a function of velocity in joint 1 of a PUMA robot. The data points indicate the average level of friction observed in constant-velocity motions using a stiff velocity feedback. The solid curve is given by the seven parameter friction model, Equation 77.5. (Adapted from [1].)

Dynamic Friction Model Parameters

The dynamic friction phenomena–Stribeck friction, frictional lag, rising static friction, and presliding displacement–are difficult to observe directly. They operate over regions of very low velocity, in which mechanism motion may be unstable, or over short time intervals or small distances. Direct observation of the phenomena and measurement of the parameters is possible with acceleration sensing and careful elimination of sources of measurement error [1], [12].

In spite of the fact that sensitive sensing is required to directly observe the dynamic friction phenomena, their impact on motion may be substantial, motivating both their correct modeling during the design of compensation and providing a basis for identification from observable system motions. Figure 77.9 shows the simulated friction, as well as position and velocity curves, for a motor servo and two different sets of friction parameters. The integrated dynamic friction model (Section 77.1.2) was used for the simulation, with the parameters specified in Table 77.3.

TABLE 77.3 Parameters Used for the Simulations of Figure 77.9

$F_c = 0.292$(N·m)	$F_v = 0.0113$ (N·m-s/rad)
$F_s = 0.043$(N·m)	$v_s = 0.95$ (rad/s)

Case 1 :	$\sigma_0 = 2.0$	$\sigma_1 = 0.1$
Case 2 :	$\sigma_0 = 40.0$	$\sigma_1 = 2.0$

The two friction models are seen to give distinct motions, producing differences that can be used to identify the friction

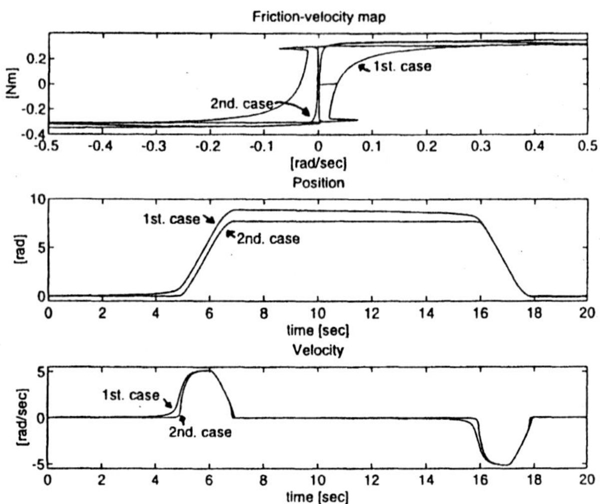

Figure 77.9 Friction and motion dependence on σ_0 and σ_1.

model parameters. The parameter identification is a nonlinear optimization problem; standard engineering software packages provide suitable routines.

The low-velocity portion of the friction-velocity curve, as well as frictional memory, presliding displacement, and rising static friction parameters, were identified in an industrial robot using open-loop motions and variable compliance. See [1] for the details of the experiment and identification.

In spite of progress in the area of modeling and demonstrations of parameter identification, there is not, at this time, a generally applicable or thoroughly tested method for identifying the parameters of a dynamic friction model. None of the methods reported to date has been evaluated in more than one application. Friction modeling thus remains, to a large extent, an application-specific exercise in which the designer identifies friction phenomena important for achieving specific design goals. Tribology and the models presented in Section 77.1.2 offer a guide to friction phenomena that will be present in a lubricated mechanism. Because of the challenges, sophisticated and successful friction compensation has been achieved in many cases without a dynamic friction model, but at the cost of entirely empirical development and tuning. One can say only that more research is needed in this area.

77.1.5 Friction Compensation

Friction compensation techniques are broken down into three categories: problem avoidance, nonmodel-based compensation techniques, and model-based compensation techniques. Problem avoidance refers to modifications to the system or its lubrication that reduce frictional disturbances. These changes often involve machine design or lubrication engineering and may not seem the domain of the control engineer. But system aspects that play a large role in the closed-loop performance, particularly the detailed chemistry of lubrication, may not have been adequately considered prior to the appearance of frictional disturbance to motion (imprecise control), and it may be up to the control engineer to suggest the use of friction modifiers in the lubricant. The division of control-based compensation techniques into model-based and nonmodel-based reflects the challenge associated with developing an accurate friction model.

Problem Avoidance

Lubricant Modification For reasons that relate to service life and performance, systematic lubrication is common in servo machines. The nature of sliding between lubricated metal contacts depends upon the sliding velocity and distance traveled. When a servo goes from standstill to rapid motion, the physics of friction transition from:

- Full solid-to-solid contact without sliding, motion by elastic deformation (presliding displacement) To solid-to-solid contact with sliding (boundary lubrication)
- A mix of fluid lubrication and solid-to-solid contact (partial fluid lubrication) To full fluid lubrication (oil or grease supports the entire load; elastohydrodynamic or hydrodynamic lubrication depending on contact geometry)

The physics of these processes are quite different from one another. The wide variety of physical processes and the difficulty of ascertaining which are active at any moment explain, in part, why a complete description of friction has been so elusive: the model must reflect all of these phenomena. Typical ranges for the friction coefficients in these different regimes are illustrated in Figure 77.10.

When sliding takes place at low velocities, actual shearing of solid material plays an important role in determining the friction. If the surfaces in contact are extraordinarily clean, the shearing takes place in the bulk material (e.g., steel) and the coefficient of friction is extremely high. More commonly, the sliding surfaces are coated with a thin boundary layer (typical thickness, 0.1 μm) of oxides or lubricants, and the shearing takes place in this layer.

Customarily, additives are present in machine lubricants, which bind to the surface and form the boundary layer, putting it under the control of the lubrication engineer. These additives constitute a small fraction of the total lubricant and are specific to the materials to which they bind. Friction modifiers are boundary lubricants that are specifically formulated to affect the coefficient of friction [9]. The control engineer should be aware of the possibilities, because lubrication is normally specified to maximize machine life, and friction modification is not always a priority of the lubrication engineer.

Hardware Modification The most common hardware modifications to reduce frictional disturbances relate to increasing stiffness or reducing mass. These modifications permit higher gains and help to increase the natural frequency of the mechanism in closed loop. The greater stiffness reduces the impact of friction directly, while a higher natural frequency interacts with frictional memory to reduce or eliminate stick-slip [7].

Other hardware modifications include special low-friction bearings and the use of "friction materials," such as Rulon^R, in machine tool slideways to eliminate stick-slip.

Nonmodel-Based Compensation Techniques

Modifications to Integral Control Integral control can reduce steady-state errors, including those introduced by friction in constant-velocity applications. When the system trajectory encounters velocity reversal, however, a simple integral control term can increase rather than reduce the frictional disturbance. A number of modifications to integral control are used reduce the impact of friction. Their application depends not only on system characteristics, but also upon the desired motions and that aspects of possible frictional disturbances that are most critical.

Position-error dead band. Perhaps the most common modification, a dead band in the input to the integrator eliminates hunting, but introduces a threshold in the precision with which a servo can be positioned. The dead band also introduces a nonlinearity, which complicates analysis. It can be modified in various ways, such as scaling the dead band by a velocity term to reduce its affect during tracking. The integrator with dead band does not reduce dynamic disturbances, such as quadrature glitch.

Lag compensation. Moving the compensator pole off the origin by the use of lag compensation with high but finite dc gain accomplishes something of the same end as a position-error dead band, without introducing a nonlinearity.

Integrator resetting. When static friction is higher than coulomb friction (a sign of ineffective boundary lubrication), the reduced friction following breakaway is overcompensated by the integral control action, and overshoot may result. In applications such as machine tools, which have little tolerance for overshoot, the error integrator can be reset when motion is detected.

Multiplying the integrator term by the sign of velocity. When there is a velocity reversal and the friction force changes direction, integral control may compound rather than compensate for coulomb friction. This behavior enlarges (but does not create) quadrature glitch. To compensate for this effect, integral control can be multiplied by the sign of velocity, a technique that depends upon coulomb friction dominating the integrated error signal. The sign of desired or reference model velocity is often used; if measured or estimated velocity is used, the modification introduces a high-gain nonlinearity into the servo loop.

High Servo Gains (Stiff Position and Velocity Control) Linearizing the Stribeck friction curve (Figure 77.3) about a point in the negatively sloped region gives a "negative viscous friction" effective during sliding at velocities in the partial fluid lubrication regime. Typically, the negative viscous friction operates over a small range of low velocities and gives a much greater destabilizing influence than can be directly compensated by velocity

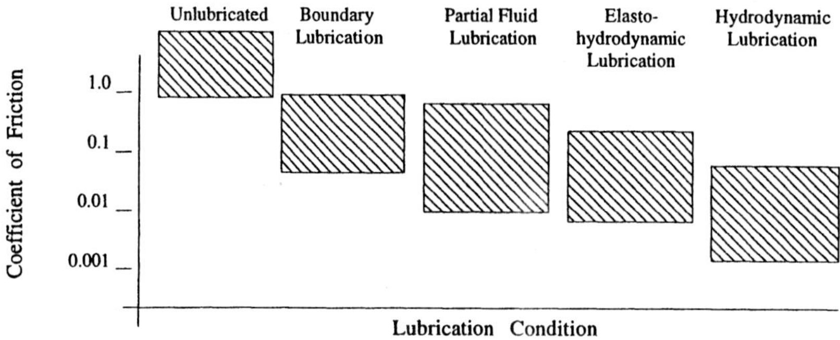

Figure 77.10 Typical ranges for friction in machines, corresponding to the friction process (adapted from [3]).

feedback. The negative viscous friction, not static friction, is often the greatest contributor to stick-slip.

As always, high servo gains reduce motion errors directly. In addition, stiff position control interacts with negative viscous friction and frictional memory to create a critical stiffness above which stick-slip is eliminated [7]. In a second-order system with proportional derivative (PD) control and frictional memory modeled as a time lag, as in Equation 77.5, a critical stiffness above which stick-slip is extinguished is given by

$$k_{cr} = M\frac{\pi^2}{\tau_L^2} \tag{77.18}$$

where τ_L is the time lag of frictional memory in seconds, and M is the mechanism mass [2].

Qualitatively, the elimination of stick-slip at high stiffness can be understood as the converse of the destabilizing effect of transport lag in a normally stable feedback loop: the *destabilizing* effect of negative viscous friction operates through the delay of frictional memory, and the system is stabilized when the delay is comparable to the natural frequency of the system. Achieving the required feedback stiffness may require a very stiff mechanical system. When increasing stiffness extinguishes stick-slip, Equation 77.18 can be used to estimate the magnitude of the frictional memory.

In some cases, variable structure control with very high damping at low velocities counters the influence of nonlinear low-velocity friction and permits smooth motion where stick-slip might otherwise be observed. Machine and lubricant characteristics determine the shape of the low-velocity friction-velocity curve (see Table 77.1), which determines the required velocity feedback for smooth motion.

Learning Control Learning control, sometimes called repetitive control, generally takes the form of a table of corrections to be added to the control signal during the execution of a specific motion. The table of corrections is developed during repetitions of the specific motion to be compensated. This type of adaptive compensation is currently available on some machine tool controllers where the typical task is repetitive and precision is at a premium. The table of corrections includes inertial as well as friction forces. This type of adaptive control is grouped with nonmodel-based control because no explicit model of friction is present in the system. For further discussion and references, see [2].

Joint Torque Control Reductions of 30:1 in apparent friction have been reported using joint torque control [2], a sensor-based technique that encloses the actuator-transmission subsystem in a feedback loop to make it behave more nearly as an ideal torque source. Disturbances due to undesirable actuator characteristics (friction, ripple, etc.) or transmission behaviors (friction, flexibility, inhomogeneities, etc.) can be significantly reduced by sensing and high-gain feedback. The basic structure is shown in Figure 77.11; an inner torque loop functions to make the applied torque, T_a, follow the command torque, T_c.

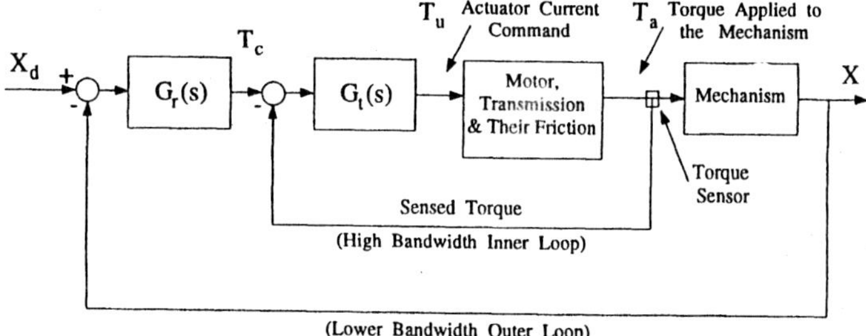

Figure 77.11 Block diagram of a joint torque control (JTC) system.

The sensor and actuator are noncollocated, separated by the compliance of the transmission and perhaps that of the sensor itself. This gives rise to the standard challenges of noncollocated sensing, including additional and possibly lightly damped modes in the servo loop.

Dither Dither is a high-frequency signal introduced into a system to improve performance by modifying nonlinearities. Dither can be introduced either into the control signal, as is often done with the valves of hydraulic servos, or by external mechanical vibrators, as is sometimes done on large pointing systems.

An important distinction arises in whether the dither force acts along a line parallel to the line of motion of the friction interface or normal to it, as shown in Figure 77.12. The effect of parallel dither is to modify the influence of friction (by averaging the nonlinearity); the effect of vibrations normal to the contact is to modify the friction itself (by reducing the friction coefficient). Dither on the control input ordinarily generates vibrations parallel to the friction interface, while dither applied by an external vibrator may be oriented in any direction.

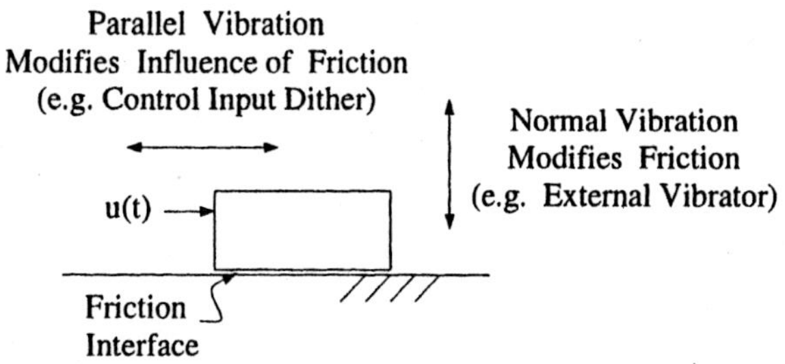

Figure 77.12 Direction and influence of dither.

Working with a simple coulomb plus viscous friction model, one would not expect friction to be reduced by normal vibrations, so long as contact is not broken; but because of contact compliance (the origin of presliding displacement), more sliding occurs during periods of reduced loading and less during periods of increased loading, which reduces the average friction. Reduction of the coefficient of friction to one third its undithered value has been reported.

Inverse Describing Function Techniques The describing function is an approximate, amplitude-dependent transfer function of a nonlinear element. The inverse describing function is a synthesis procedure that can be used when an amplitude-dependent transfer function is known and the corresponding time domain function is sought.

The inverse describing function has been used to synthesize nonlinear controllers that compensate for friction [13]. Nonlinear functions $f_P(e)$, $f_I(e)$, and $f_D(\dot{y})$ are introduced into the P, I, and D paths of proportional-integral-derivative (PID) control, as shown in Figure 77.13. The synthesis procedure is outlined in Figure 77.14. The method is included with nonmodel-based compensation because, while a friction model is required for the synthesis, it does not appear explicitly in the resulting controller.

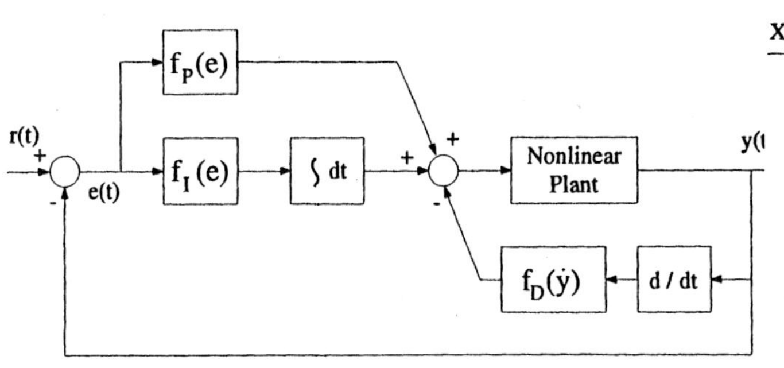

Figure 77.13 Nonlinear PI + tachometer structure.

The inverse describing function method offers the advantages that it has a systematic design procedure and it can be applied to systems with relatively complicated linear dynamics. Significant improvement in transient response is reported for a fourth-order system with flexible modes [13].

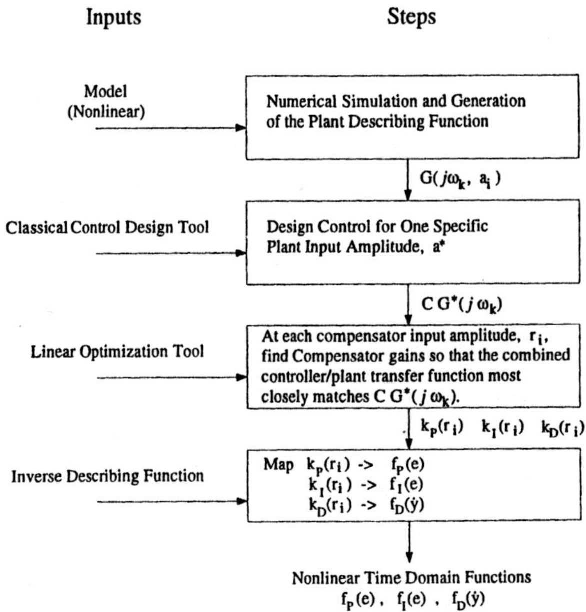

Figure 77.14 Outline of inverse describing function based synthesis of nonlinear friction compensation.

Model-Based Compensation Techniques

When friction can be predicted with sufficient accuracy, it can be compensated by feedforward application of an equal and opposite force. The basic block diagram for such a construction is shown in Figure 77.15. Examples of model-based compensation are provided in [1],[2] [4],[5],[6],[8],[14]. Compensation is addressed in the next section; questions of tuning the friction model parameters off-line and adaptively are discussed in Sections 77.1.4 and 77.1.5 respectively.

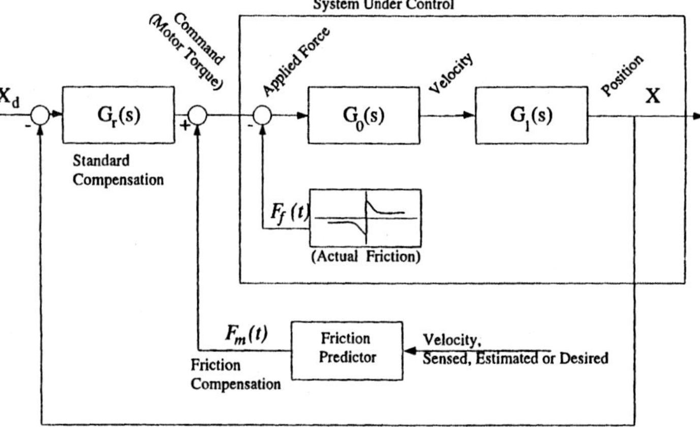

Figure 77.15 Model-based feedforward compensation for friction. (From [2], courtesy of the publisher.)

Compensation Three requirements for feedforward compensation are

1. An accurate friction model
2. Adequate control bandwidth
3. Stiff coupling between the actuator and friction element

The third item merits special consideration: significant friction often arises with the actuator itself, or perhaps with a transmission, in which case the friction may be stiffly coupled to the available control-based compensation. When the friction arises with components that are flexibly coupled to the actuator, it may not be possible to cancel friction forces with actuator commands. Some of the most successful applications of model-based friction compensation are in optical pointing and tracking systems, which can be made very stiff (e.g., [14]).

The important axes of distinction in model-based compensation are

1. The character and completeness of the friction model
2. Whether sensed, estimated, or desired velocity is used as input to the friction model

The most commonly encountered friction model used for compensation comprises only a coulomb friction term, with two friction levels, one for forward and the other for reverse motion. Industrially available machine tool controllers, for example, incorporate coulomb friction feedforward. To guard against instability introduced by overcompensation, the coulomb friction parameter is often tuned to a value less than the anticipated friction level.

The use of sensed or estimated velocity by the friction model closes a second feedback loop in Figure 77.15. If desired or reference model velocity is used, the friction compensation becomes a feedforward term. Because the coulomb friction model is discontinuous at zero velocity, and any friction model will show a rapid change in friction when passing through zero velocity, the use of sensed or estimated velocity can lead to stability problems. A survey of engineers in industry [2] indicates that the use of desired velocity is most common.

Additional friction model terms that have been used in compensation include:

1. Static friction
2. Frictional memory and presliding displacement
3. Position dependence

Static friction. Static friction may be represented in a model like the Karnopp model (see Section 77.1.3 and Figure 77.6). The most substantial demonstration of static friction compensation has been presented by Brandenburg and his co-workers [4] and is discussed in Section 77.1.5.

Frictional memory and presliding displacement. Examining Figure 77.5, it is seen that providing the full coulomb friction compensation for motions near the zero crossing of velocity would overcompensate friction. To prevent overcompensation, one industrial servo drive controller with coulomb friction compensation effectively implements the inverse of the Karnopp friction model

$$F_m(t) = \begin{cases} F_c \operatorname{sgn}[v(t)], & t_{zc} > t_{zc}^0 \\ 0, & t_{zc} \le t_{zc}^0 \end{cases} \tag{77.19}$$

where t_{zc} is the time since the last zero crossing of velocity, and t_{zc}^0 is a parameter that determines the interval over which the coulomb friction compensation is suppressed. The coulomb friction compensation could also be controlled by accumulated displacement since the last zero crossing of velocity (note that friction is plotted as a function of displacement in Figure 77.5) or as a function of time divided by acceleration (applied torque).

Friction compensation has also been demonstrated that incorporates a low-pass filter in the friction model and thus eliminates the discontinuity at zero velocity. One construction is given by

$$F_f(t) + \tau \frac{dF_f(t)}{dt} = F_c \operatorname{sgn}[v(t)] \tag{77.20}$$

where τ is a time constant. Walrath [14] reports an application of friction compensation based on Equation 77.20 that results in a 5:1 improvement in RMS pointing accuracy of a pointing telescope. Because of the accuracy requirement and the bandwidth of disturbances arising when the telescope is mounted in a vehicle, the tracking telescope pointing provides a very rigorous test of friction compensation. In the system described, τ is a function of the acceleration during the zero crossing, given by

$$\tau = \frac{1}{a_1 + a_2 \dot{v}(t)} \tag{77.21}$$

where a_1 and a_2 are empirically tuned parameters, and $\dot{v}(t)$ is estimated from the applied motor torque.

Position-dependent friction. In mechanisms with spatial inhomogeneities, such as gear teeth, friction may show substantial, systematic variation with position. Some machine tool controllers include an "anti-backlash" compensation that is implemented using table lookup. Friction is known to influence the table identification and, thus, the compensation. Explicit identification and feedforward compensation of position-dependent friction can also be done [1],[2].

Adaptive Control, Introduction Adaptive control has been proposed and demonstrated in many forms; when applied to friction compensation, adaptation offers the ability to track changes in friction. Adaptive friction compensation is indirect when an on-line identification mechanism explicitly estimates the parameters of the friction model. The advantage of indirect adaptive control is that the quality of the estimated parameters can be verified before they are used in the compensation loop. Typical examples of such verification are to test the sign of the parameters or their range of variation. By introducing an additional supervisory loop, the identified parameters can be restricted to a predetermined range.

Direct adaptive compensation rests on a different philosophy: no explicit on-line friction modeling is involved, but rather controller gains are directly adapted to minimize tracking error. Because there is no explicit friction prediction, it is more difficult to supervise the adaptive process to assure that the control reflects a physically feasible friction model or that it will give suitable transient behavior or noise rejection.

Direct Adaptive Control Model reference adaptive control (MRAC) is described elsewhere in this handbook. Perhaps the simplest implementation of model reference coulomb friction compensation is given by

$$\begin{aligned} u(t) &= \text{(PD control)} + \hat{F}_c(t)\text{sgn}[v(t)] \\ \hat{F}_c(t) &= C_1 \int_0^t \dot{e}(t)\text{sgn}[v(t)]dt \\ &\quad + C_2 \int_0^t e(t)\text{sgn}[v(t)]dt \\ \dot{e}(t) &= v_m(t) - v(t) \end{aligned} \tag{77.22}$$

where $v(t)$ is velocity; $v_m(t)$ is the reference model velocity; and $\hat{F}_c(t)$ is a coulomb friction compensation parameter. The parameters C_1 and C_2 are chosen by the designer and can be tuned to achieve the desired dynamics of the adaptive process. A Lyapunov function will show that the (idealized) process will converge for $C_1, C_2 > 0$.

Brandenburg and his co-workers [4] have carried out a thorough investigation of friction compensation in a two-mass system with backlash and flexibility (see also [2] for additional citations). They employ an MRAC structure to adapt the parameters of a coulomb friction-compensating disturbance observer. Without friction compensation, the system exhibits two stick-slip limit cycles. The MRAC design is based on a Lyapunov function; the result is a friction compensation made by applying a lag (PI) filter to $\dot{e}(t)$, the difference between model and true velocity. Combined with integrator dead band, their algorithm eliminates stick-slip and reduces lost motion during velocity reversal by a factor of 5.

Indirect Adaptive Control When the friction model is linear in the unknown parameters, it can be put in the form

$$F_f(t) = \theta^T \Phi(t) \tag{77.23}$$

where θ is the parameter vector and $\Phi(t)$ is the regressor vector. The regressor vector depends on the system state and can include nonlinear functions. For example, a model that includes the Stribeck curve is given by [5]:

$$\begin{aligned} \theta^T &= [F_c, F_s, F_v] \\ \Phi(t) &= \left[\text{sgn}(v(t)), \frac{1}{1+(v(t)/v_s)^2}, v(t)\right] \end{aligned} \tag{77.24}$$

Assuming the simplest mechanism dynamics

$$M\dot{v}(t) = u(t) - F_f(t) \tag{77.25}$$

where M is the mass and $u(t)$ is the force or torque command, and sampling at time $t = kT$, the friction prediction error is given by:

$$e_k = u_k - M\dot{v}_k - \theta^T \Phi_k \tag{77.26}$$

To avoid the explicit measurement (or calculation) of the acceleration, the friction prediction error can be based on a filtered model. By applying a stable, low-pass filter, $F(s)$, to each side of Equation 77.25, the filtered mechanism dynamics are given, and the filtered prediction error is

$$e_k = \tilde{u}_k - M\tilde{\dot{v}}_k - \theta^T \tilde{\Phi}_k \tag{77.27}$$

where $\tilde{}$ indicates a low-pass filtered signal.

An update equation for the parameter estimate is constructed

$$\hat{\theta}_k = \hat{\theta}_{k-1} + \lambda_k P_k \tilde{\Phi}_k e_k \tag{77.28}$$

where λ_k is a rate gain and P_k is the inverse of the input correlation matrix for the recursive least squares algorithm (RLS) or the identity matrix for the least mean squared algorithm (LMS).

The control law is now implemented as

$$u(t) = \text{(Standard control)} + \hat{\theta}^T \Phi(t) \tag{77.29}$$

where $\Phi(t)$ is used for compensation, rather than the filtered $\tilde{\Phi}(t)$.

During the implementation of the estimation loop the following points are important:

1. The estimation should be frozen when operating conditions are not suitable for friction identification [e.g., $v(t) = 0$; this is the persistent excitation issue].
2. The sign of the estimated parameters should be always positive.
3. The compensation can be scaled down to avoid overcompensation.
4. The sign of the reference velocity can be used in place of the sign of the measured velocity in Equation 77.24, when $v(t)$ is small.

An alternative approach has been presented by Friedland and Park [8], who justify an update law that does not depend on acceleration measurement or estimation. The friction compensation is given by

$M\dot{v}(t) = u(t) - F_f(v(t), F_c^*$	[System Dynamics, single mass]
$F_f\left(v(t), \hat{F}_c\right) = \hat{F}_c\text{sgn}[v(t)]$	[Friction Model]
$u(t) = \text{(Standard Control)} + \hat{F}_c\text{sgn}[v(t)]$	[Control Law]
$\hat{F}_c = z(t) - k\,\lvert v(t)\rvert^{\mu}$	[Friction Estimator]
$\dot{z}(t) = k\mu\,\lvert v(t)\rvert^{\mu-1}$ $\frac{1}{M}\left[u(t) - f\left(v(t), \hat{F}_c\right)\right]\text{sgn}[v(t)]$	[Friction Estimator Update Law]

where M is system mass; $u(t)$ is the control input; $z(t)$ is given by the friction estimator update law; $\hat{F}_c$ is the estimated coulomb friction; and μ and k are tunable gains. Defining the model misadjustment

$$e(t) = F_c^* - \hat{F}_c \tag{77.30}$$

one finds that

$$\dot{e}(t) = -k\mu\,\lvert v(t)\rvert^{\mu-1} e(t) \tag{77.31}$$

making $e = 0$ the stable fixed point of the process. The algorithm significantly improves dynamic response [4]; experimental results are presented in subsequent papers (see [2] for citations).

77.1.6 Conclusion

Achieving machine performance free from frictional disturbances is an interdisciplinary challenge; issues of machine design, lubricant selection, and feedback control all must be considered to cost effectively achieve smooth motion. Today, the frictional phenomena that should appear in a dynamic friction model are well understood and empirical models of dynamic friction are available, but tools for identifying friction in specific machines are lacking.

Many friction compensation techniques have been reported in both the research literature and industrial applications. Because a detailed friction model is often difficult to obtain, many of these compensation techniques have been empirically developed and tuned. Where available, reported values of performance improvement have been presented here. Reports of performance are not always available or easily compared because, to date, there has not been a controlled comparison of the effectiveness of the many demonstrated techniques for friction compensation. For many applications, it is up to the designer to seek out application-specific literature to learn what has been done before.

References

[1] Armstrong-Hélouvry, B., *Control of Machines with Friction,* Kluwer Academic Press, Boston, MA, 1991.

[2] Armstrong-Hélouvry, B., Dupont, P., and Canudas de Wit, C., A survey of models, analysis tools and compensation methods for the control of machines with friction, *Automatica,* 30(7), 1083–1138, 1994.

[3] Bowden, F.P. and Tabor, D. *Friction — An Introduction to Tribology,* Anchor Press/Doubleday, Reprinted 1982, Krieger Publishing Co., Malabar, 1973.

[4] Brandenburg, G. and Schäfer, U., Influence and partial compensation of simultaneously acting backlash and coulomb friction in a position- and speed-controlled elastic two-mass system, in *Proc. 2nd Int. Conf. Electrical Drives,* Poiana Brasov, September 1988.

[5] Canudas de Wit, C., Noel, P., Aubin, A., and Brogliato, B., Adaptive friction compensation in robot manipulators: low-velocities, *Int. J. Robotics Res.,* 10(3), 189–99, 1991.

[6] Canudas de Wit, C., Olsson, H., Aström, K.J., and Lischinsky, P., A new model for control of systems with friction, *IEEE Trans. Autom. Control,* in press, 1994.

[7] Dupont, P.E., Avoiding stick-slip through PD control, *IEEE Trans. Autom. Control,* 39(5), 1094–97, 1994.

[8] Friedland, B. and Park, Y.-J., On adaptive friction compensation, *IEEE Trans. Autom. Control,* 37(10), 1609–12, 1992.

[9] Fuller, D.D., *Theory and Practice of Lubrication for Engineers,* John Wiley & Sons, Inc., New York, 1984.

[10] Ibrahim, R.A. and Rivin, E., Eds., Special issue on friction induced vibration, *Appl. Mech. Rev.,* 47(7), 1994.

[11] Karnopp, D., Computer simulation of stick-slip friction in mechanical dynamic systems, *ASME J. Dynamic Syst., Meas., Control,* 107(1), 100–103, 1985.

[12] Polycarpou, A. and Soom, A., Transitions between sticking and slipping, in *Friction-Induced Vibration, Chatter, Squeal, and Chaos, Proc. ASME Winter Ann. Meet.,* Ibrahim, R.A. and Soom, A., Eds., DE-Vol. 49, 139–48, Anaheim: ASME; New York: ASME, 1992.

[13] Taylor, J.H. and Lu, J., Robust nonlinear control system synthesis method for electro-mechanical pointing systems with flexible modes, in *Proc. Am. Control Conf.,* AACC, San Francisco, 1993, 536–40.

[14] Walrath, C.D., Adaptive bearing friction compensation based on recent knowledge of dynamic friction, *Automatica,* 20(6), 717–27, 1984.

77.2 Motion Control Systems

Jacob Tal, *Galil Motion Control, Inc.*

77.2.1 Introduction

The motion control field has experienced significant developments in recent years. The most important one is the development of microprocessor-based digital motion controllers. Today, most motion controllers are digital in contrast to 20 years ago when most controllers were analog. The second significant development in this field is modularization, the division of motion control functions into components with well-defined functions and interfaces. Today, it is possible to use motion control components as building blocks and to integrate them into a system.

The following section describes the system components and operation including very simple mathematical models of the system. These models are adequate for the major emphasis of this chapter which discusses the features and capabilities of motion control systems. The discussion is illustrated with design examples.

77.2.2 System Elements and Operation

The elements of a typical motion control system are illustrated in the block diagram of Figure 77.16.

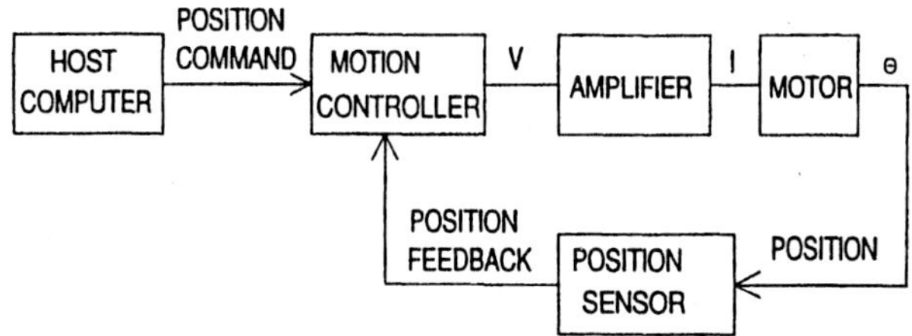

Figure 77.16 Elements of a motion control system.

The amplifier is the component that generates the current required to drive the motor. Amplifiers can be configured in different ways, thereby affecting their operation. The most common operating mode is the transconductance or current mode, where the output current, I, is proportional to the applied voltage, V. The ratio between the two signals, K_a, is known as the current gain.

$$I = K_a V \tag{77.32}$$

When current is applied to the motor, it generates a proportional torque, T_g.

$$T_g = K_t I \tag{77.33}$$

The constant K_t is called the torque constant. The effect of torque on motion is given by Newton's Second Law. Assuming negligible friction, the motor acceleration rate, α, is given by

$$\alpha = \frac{T_g}{J} \tag{77.34}$$

where J is the total moment of inertia of the motor and the load.

The motor position, Θ, is the second integral of the acceleration. This relationship is expressed by the transfer function

$$\frac{\Theta}{\alpha} = \frac{1}{s^2}. \tag{77.35}$$

The position of the motor is monitored by a position sensor. Most position sensors represent the position as a digital number with finite resolution. If the position sensor expresses the position as N units of resolution (counts) per revolution, the equivalent gain of the sensor is

$$K_f = \frac{N}{2\pi}\frac{\text{counts}}{\text{rad}}. \tag{77.36}$$

The component that closes the position loop is the motion controller. This is the "brain" of the system and performs all the computations required for closing the loop as well as providing stability and trajectory generation.

Denoting the desired motor position by R, the position feedback by C, and the position error by E, the basic equation of the closed-loop operation is

$$E = R - C. \tag{77.37}$$

The controller includes a digital filter which operates on the position error, E, and produces the output voltage, V. In order to simplify the overall modeling and analysis, it is more effective to approximate the operation of the digital filter by an equivalent continuous one. In most cases the transfer function of the filter is PID, leading to the equation,

$$H(s) = \frac{V}{E} = P + sD + \frac{I}{s}. \tag{77.38}$$

The mathematical model of the complete system can now be expressed by the block diagram of Figure 77.17. This model can be used as a basis for analysis and design of the motion control system. The analysis procedure is illustrated below.

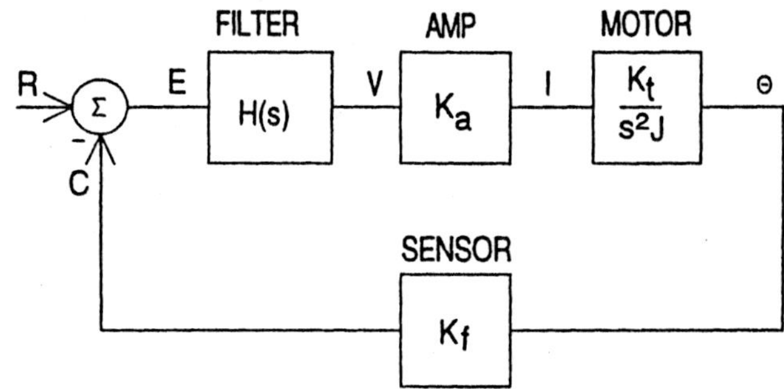

Figure 77.17 Mathematical model of a typical motion control system.

77.2.3 Stability Analysis Example

Consider a motion control system including a current source amplifier with a gain of $K_a = 2$ Amps per volt: the motor has a torque constant of $K_t = 0.1$ Nm/A and a total moment of inertia of $J = 2 \cdot 10^{-4} K_g \cdot m^2$. The position sensor is an absolute encoder with 12 bits of binary output, implying that the sensor output varies between zero and $2^n - 1$ or 4095.

This means that the sensor gain is

$$K_f = \frac{4096}{2\pi} = 652 \quad \text{counts/rad}.$$

The digital controller has a sampling period of 1 msec. It has a single input, $E(K)$, and a single output $X(K)$. The filter equations are as follows:

$$F(K) = E(K) \cdot C + F(K-1), \quad \text{and}$$
$$X(K) = E(K) \cdot A + B[E(K) - E(K-1)] + F(K).$$

The signal $X(K)$ is the filter output signal which is applied to a digital-to-analog converter (DAC) with a 14-bit resolution, and an output signal range between -10V and 10V.

A mathematical model for the DAC is developed, noting that when $X(K)$ varies over a range ± 8192 counts, the output varies over ± 10V. This implies that the gain of the DAC is

$$G = \frac{10}{8192} = 1.22 \cdot 10^{-3} \frac{\text{volts}}{\text{count}}$$

To model the equation of the filter, we note that this is a digital PID filter. This filter may be approximated by the continuous model of Equation 77.38 with the following equivalent terms:

$$\begin{aligned} P &= AG \\ I &= CG/T \\ D &= BTG \end{aligned}$$

where T is the sampling period.

For example, if filter parameters are

$$\begin{aligned} A &= 20 \\ B &= 200 \\ C &= 0.15 \end{aligned}$$

and a sampling period is

$$T = 0.001\ s$$

the resulting equivalent continuous PID parameters are

$$\begin{aligned} P &= 0.0244 \\ I &= 0.183 \\ D &= 2.44 \cdot 10^{-4} \end{aligned}$$

Assuming the filter parameters shown above, we may proceed with the stability analysis.

The open loop transfer function of the control system is

$$L(s) = \frac{K_a K_t K_f}{J s^2} \left(P + sD + \frac{I}{s} \right)$$

For the given example, $L(s)$ equals

$$L(s) = \frac{159(s^2 + 100s + 750)}{s^3}$$

Start with the crossover frequency (the frequency at which the open-loop gain equals one) to determine the stability. This equals

$$\omega_c = 181 \quad \text{rad/s}$$

The phase shift at the frequency ω_c is

$$\Theta(\omega_c) = -120^\circ$$

The system is stable with a phase margin of 60°.

The previous discussion focused on the hardware components and their operation. The system elements which form the closed-loop control system were described. In order to accomplish the required motion, it is necessary to add two more functions, motion profiling and coordination. Motion profiling generates the desired position function which becomes the input of the position loop. Coordination is the process of synchronizing various events to verify that their timing is correct. These functions are described in the following sections.

77.2.4 Motion Profiling

Consider the system in Figure 77.17 and suppose that the motor must turn 90°: The simplest way to generate this motion is by setting the reference position, R, to the final value of 90°. This results in a step command of 90°. The resulting motion is known as the step response of the system. Such a method is not practical as it provides little control on the motion parameters, such as velocity and acceleration.

An alternative approach is to generate a continuous time-dependent reference position trajectory, $R(t)$. The basic assumption here is that the control loop forces the motor to follow the reference position. Therefore, generating the reference trajectory results in the required motion.

The motion trajectory is generated outside the control loop and, therefore, has no effect on the dynamic response and the system stability. To illustrate the process of motion profiling, consider the case where the motor must turn 1 radian and come to a stop in 0.1 seconds. Simple calculation shows that if the velocity is a symmetric triangle, the required acceleration rate equals 400 rad/sec^2, and the position trajectory is

$$R(t) = \begin{cases} 200t^2 & 0 \le t \le 0.05 \\ 1 - 200(0.1 - t)^2 & 0.05 \le t \le 0.1 \end{cases}$$

The required move is accomplished by the motion controller computing the function $R(t)$ at the sample rate and applying the result as an input to the control loop.

77.2.5 Tools for Motion Coordination

Motion controllers are equipped with tools to facilitate the coordination between events. The tools may vary between specific controllers but their functions remain essentially the same. Coordination tools include

- stored programs
- control variables
- input/output interfaces
- trip points

The stored programs allow writing a set of instructions that can be executed upon command. These instructions perform a specific motion or a related function. Consider, as an example, the following stored program, written in the format of Galil controllers:

Instruction	Interpretation
#MOVE	Label
PR 5000	Relative distance of 5000 counts
SP 20000	Speed of 20,000 counts/sec
AC 100000	Acceleration rate of 100,000 counts/sec^2
BGX	Start the motion of the X-axis
EN	End the program

This program may be stored and executed in the motion controller, thus allowing independent controller operation without host intervention.

The capabilities of the motion controllers increase immensely with the use of symbolic variables. They allow the motion controller to perform mathematical functions and to use the results to modify the operation. Consider, for example, the following program which causes the X motor to follow the position of the Y motor: This is achieved by determining the difference in the motor positions, E, and by driving the X motor at a velocity, *VEL*, which is proportional to E. Both E and *VEL* are symbolic variables.

Instruction	Interpretation
#FOLLOW	Label
JG0	Set X in jog mode
AC 50000	Acceleration rate
BGX	Start X motion
#LOOP	Label
$E = \_TPY - \_TPX$	Position difference
$VEL = E * 20$	Follower velocity
JG VEL	Update velocity
JP #LOOP	Repeat the process
EN	End

The input/output interface allows motion controllers to receive additional information through the input lines. It also allows the controllers to perform additional functions through the output lines.

The input signals may be digital, representing the state of switches or digital commands from computers or programmable logic controllers (PLCs). The inputs may also be analog, representing continuous functions such as force, tension, temperature, etc.

The output signals are often digital and are aimed at performing additional functions, such as turning relays on and off. The digital output signals may also be incorporated into a method of communication between controllers or communication with PLCs and computers.

The following program illustrates the use of an input signal for tension control. An analog signal representing the tension is applied to the analog input port #1. The controller reads the signal and compares it to the desired level to form the tension error TE. To reduce the error, the motor is driven at a speed VEL that is proportional to the tension error.

Instruction	Interpretation
#TEN	Label
JG0	Zero initial velocity
AC 10000	Acceleration rate
BGX	Start motion
#LOOP	Label
TENSION = @AN [1]	Read analog input and define as tension
$TE = 6 -$ TENSION	Calculate tension error
$VEL = 3 * TE$	Calculate velocity
JG VEL	Adjust velocity
JP#LOOP	Repeat the process
EN	End

Synchronizing various activities is best achieved with trip points. The trip point mechanism specifies when a certain instruction must be executed. This can be specified in terms of motor position, input signal, or a time delay. The ability to specify the timing of events assures synchronization. Consider, for example, the following program where the motion of the X motor starts only after a start pulse is applied to input #1. After the X motor moves a relative distance of 5000 counts, the Y starts, and after the Y motor moves 3000 counts, an output signal is generated for 20 ms.

Instruction	Interpretation
#TRIP	Label
PR 10000, 10000	Motion distance for X and Y motors
SP 20000, 20000	Velocities for X and Y
AI1	Wait for start signal input 1
BGX	Start the motion of X
AD 5000	Wait until X moves 5000 counts
BGY	Start the Y motion
AD,3000	Wait until Y moves 3000 counts
SB1	Set output bit 1 high
WT20	Wait 20 ms
CB1	Clear output bit 1
EN	End

The following design example illustrates the use of the tools for programming motion.

77.2.6 Design Example – Glue Dispensing

Consider a two-axis system designed for glue dispensing, whereby an XY table moves the glue head along the trajectory of Figure 77.18. To achieve a uniform application rate of glue per unit length, it is necessary to move the table at a constant vector speed and to turn the glue valve on only when the table has reached uniform speed.

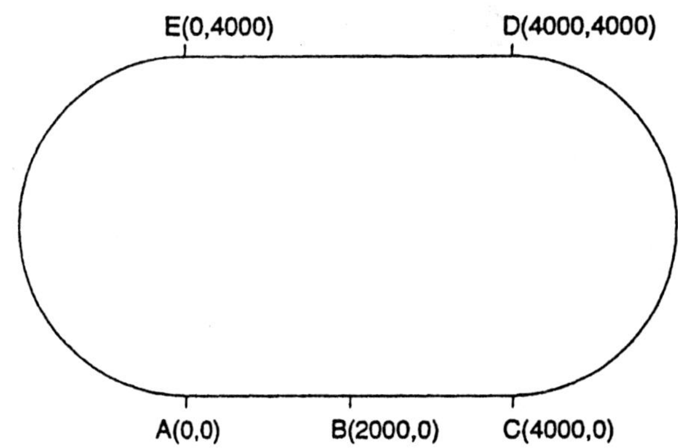

Figure 77.18 Motion path for the glue-dispensing system.

The motion starts at point A and ends at point C after a complete cycle to allow for acceleration and deceleration distances. The gluing starts and stops at point B.

The motion program is illustrated below. It is expressed in the format of Galil controllers. The instructions for straight lines and circular arcs are VP and CR, respectively. The glue valve is activated with the digital output signal #1. This signal is activated and deactivated when the X motor moves through point B. The timing of this event is specified by the trip point FM 2000 which waits until the X motor moves forward through the point $X = 2000$.

Instruction	Interpretation
#GLUE	Label
DP 0,0	Defined starting position as (0,0)
VP 4000,0	Move to point C
CR 2000,270,180	Follow the arc CD
VP 0,4000	Move to point E
CR 2000,90,180	Follow the arc EA
VP 4000,0	Move to point C
VE	End of motion
VS 20000	Vector speed
VA 100000	Vector acceleration
VD 100000	Vector deceleration
BGS	Start the motion
FM 2000	When $X = 2000$ (point B)
SB1	Set output 1–turn glue valve
WT 100	Wait 100 ms
FM 2000	When $X = 2000$ again
CB1	Turn the glue valve off
EN	End

References

[1] DC Motors, Speed Controls, Servo Systems, *Engineering Handbook,* Electrocraft Corp.

[2] Tal, J., *Step-by-Step Design of Motion Control Systems,* Galil Motion Control, 1994.

77.3 Ultra-High Precision Control

Thomas R. Kurfess, The George W. Woodruff School of Mechanical Engineering, The Georgia Institute of Technology, Atlanta, GA

Hodge Jenkins, The George W. Woodruff School of Mechanical Engineering, The Georgia Institute of Technology, Atlanta, GA

77.3.1 Introduction

Fierce international competition is placing an ever-increasing significance on precision and accuracy in ultra-high precision systems engineering. Control of dimensional accuracy, tolerance, surface finish and residual damage is necessary to achieve or inspect ultra-high precision components required for the latest technology. To achieve ultra-high precision, specialized control subsystems must be employed for modern machine tools and measurement machines. These systems are common in manufacturing today, and will become more widespread in response to continually increasing demands on manufacturing facilities. Today, ultra-high precision applications are found in the manufacture and inspection of items such as automobile bearings, specialized optics or mirrors, and platens for hard disk drives. The manufacture of x-ray optical systems for the next generation of microchip fabrication systems has also recently become internationally important.

An excellent example of an ultra-high precision machine is the large optic diamond turning machine at the Lawrence Livermore National Laboratory [8]. This machine operates with a dimensional error of less than 10^{-6} inches (1 μin) and has turned optical parts for many systems including the secondary mirror for the Keck telescope. Not only is the shape of the optics precise enough for use without further processing, but the surface finish is of high enough quality to eliminate polishing. In effect, diamond turning is capable of producing optical parts with extremely stringent optical and specular requirements.

In ultra-high precision machining, accuracies of 0.1μin and lower are targeted. To achieve these accuracies, many problems arise not commonly encountered when designing standard machine servocontrollers, such as low-velocity friction, axis-coupling, and phase lag.

Applications of Ultra-High Precision Systems

The small, but increasing, number of special applications requiring ultra-high precision control techniques include grinding, metrology, diamond turning machines, and the manufacture of semiconductors, optics and electronic mass media.

Diamond turning and grinding of optical quality surfaces on metal and glass require ultra-high precision movement. Accuracies for diamond turning of mirrors can be on the order of 10 nm. Optical grinding applications also require force control to promote ductile grinding, reducing subsurface damage of lenses [2],[13].

Another application for ultra-high precision control is metrology. Precision machined parts are either measured point by point using a vectored touch to include a sufficient amount of data to determine the part geometry, or they are measured in a continuous motion (scanning) along a part dimension using a constant force trajectory. In either case a force probe (or touch trigger) is used to make or maintain contact with the part surface. To have repeatable data, minimize the force-induced measurement errors, and extend the life of the scanning probe, the contact forces must be minimized and maintained as constant as possible.

The primary applications discussed in this chapter are diamond turning machines (DTM) and scanning coordinate measurement machines (CMMs). Both machines have many similarities including basic structure as well as velocity and position control designs. When addressing issues related to both types of machines, the term machine is used. When discussing a cutting machine (e.g., a DTM) the term machine tool is used; the term measurement machine (e.g., a CMM) is used for a machine used exclusively to measure.

History

Initial machine tool control began with simple numerical control (NC) in the 1950s and 1960s. Later (in the 1970s), computer numerical control (CNC) grew more prevalent as computer technology became less expensive and more powerful. These were either point-to-point controllers or tracking controllers, where the tool trajectory is broken up into a series of small straight-line movements. Algorithms for position feedback in most current machine tools are primarily proportional control with bandwidths of nominally 10 to 20 Hz. Increasing the accuracy and repeatability of the machine has been the driving force behind the development of NC and CNC technology.

However, not until the mid 1960s did the first diamond turning machines begin to take shape. These machines were developed at the Lawrence Livermore National Laboratory (LLNL) defense complex. DTM's differ widely from other machine tools. Two of the major differences critical for the control engineer are the mechanical design of the machine, and active analog compensators used in the machine control loop. The basic design philosophy employed by the engineers at LLNL is known as the deterministic theory [4]. The premise of this theory is that machines should be designed as well enough to eliminate the need for complicated controllers. Given such a system, the job of the control engineer is greatly simplified, however, to achieve the best possible performance, active analog compensators are still successfully employed on DTMs.

In the 1980s, other advances in machine tool and measurement system control were further developed such as parameter controllers used to servo force or feedrate. Such parameter controllers, which are fairly common today, operate at lower frequencies. For example, closed-loop force control system frequencies of less than 3 Hz are used in many applications [22].

Theoretical and experimental work has been and continues to be conducted using model reference adaptive control in machining (e.g., [24], [7], [11]. Both fixed gain and parameter adaptive controllers have been implemented for manipulating on-line feed rate to maintain a constant cutting force. Although such controllers have increased cutting efficiency, they typically reduce machine performance and may lead to instability. Although research in adaptive parameter controllers is fairly promising to date, these controllers have not been widely commercialized. Further details regarding machine tool control can be found in the literature [28].

Current Technology

The current technology for designing machines with the highest precision is based on a fundamental understanding of classical control, system identification, and precision components for the controller, drive and feedback subsystems. Control can compensate for poor equipment design only to a limited degree. A well-designed machine fitted with high precision hardware will ease the controller design task. This, of course, relates back to deterministic theory. Special hardware and actuators for ultra-high precision equipment such as laser interferometers, piezoelectric actuators, air and hydrostatic bearings are critical in achieving ultra-high precision. These subsystems and others are being used more frequently as their costs decrease and availability increases.

As stated above, the analog compensator design for precision systems is critical. For ultra-high precision surfaces, both position and velocity must be controlled at high bandwidths because air currents and floor vibrations can induce higher frequency noise. The specifications for the velocity compensator combine machine bandwidth and part requirements. Part geometry, in particular, surface gradients in conjunction with production rate requirements, dictate the velocity specifications for measurement machines and machine tools. For example, a part with high surface gradients (e.g., sharp corners) requires fast responses to velocity changes for cornering. Of course, higher production rates require higher bandwidth machines as feed rates are increased. Maximum servo velocity for a machine tool is also a function of initial surface roughness, and desired finish as well as system bandwidth. For measurement machines, only part geometry and machine bandwidth affect the system velocity specifications because no cutting occurs. One way to improve the performance of ultra-high precision machines is to use slow feed rates (on the order of 1 μin/sec), but such rates are usually unacceptable in high quantity production.

Until recently, force control was not used in many ultra-high precision machining applications, because the cutting forces are generally negligible due to the low feed rates. Furthermore, force control departs from standard precision machining approaches because force is controlled by servocontrol of feed velocity or position. Control is accomplished via position servocontrol using Hook's spring law, and damping relationships are used to control force if velocity servocontrol is used.

Unfortunately, force control in machining generates conflicting objectives between force trajectories and position trajectories (resulting in geometric errors) or velocity trajectories (resulting in undesired surface finish deviations). However, such conflicts can be avoided if the machine and its controllers are designed to decouple force from position and velocity. Such conflicts do not occur to as great an extent with measurement systems where contact forces must be kept constant between the machine measurement probe and the part being inspected.

Typical error sources encountered in precision machine control are machine nonlinearities, control law algorithms, structural response (frequency and damping), measurement errors, modeling errors, human errors, and thermal errors. Sources of machine nonlinearities include backlash, friction, actuator saturation, varying inertial loads, and machine/probe compliance. Methods for addressing these nonlinearities include high proportional gain, integral control, feedforward control, adaptive control, and learning algorithms.

Friction is a problem at the low feed rates needed to achieve high quality (optical) surface finishes. Low velocity friction (stiction) is a primary cause of nonlinearities affecting trajectory control of servo tables. At near zero velocities, stiction may cause axial motion to stop. In such a condition, the position tracking error increases and the control force builds. The axis then moves, leading to overshoot. Most industrial controllers use only proportional compensators that cannot effectively eliminate the nonlinearities of stiction [28]. Friction compensation in lead screws has been addressed by several authors concerned with coulomb friction and stiction [27]. Special techniques such as learning algorithms [25] and cross-coupling control [20] have been employed successfully in X-Y tables to reduce tracking errors caused by low velocity friction.

77.3.2 System Description

Because controller and system hardware directly affect machine performance, both must be thoroughly discussed to develop a foundation for understanding precision system control. This section describes the precision hardware components (e.g., servo drives and sensors) and controllers, in terms of their function and performance.

Basic Machine Layout

The control structure for a typical machine tool can be represented by a block diagram as shown in Figure 77.19. Typically, CNC controllers generate tool trajectories shown as the reference input in Figure 77.19. Given this reference trajectory and feedback from a position sensor, most CNC controllers (and other commercially available motion controllers) compute the position error for the system and provide an output signal (voltage) proportional to that error. A typical CNC output has a limited range of voltage (e.g., -10 to 10 V dc). The error signal is amplified to provide a power signal to the process actuators which, in turn, affect the process dynamics to change the output that is then measured by sensors and fed back to the CNC controller.

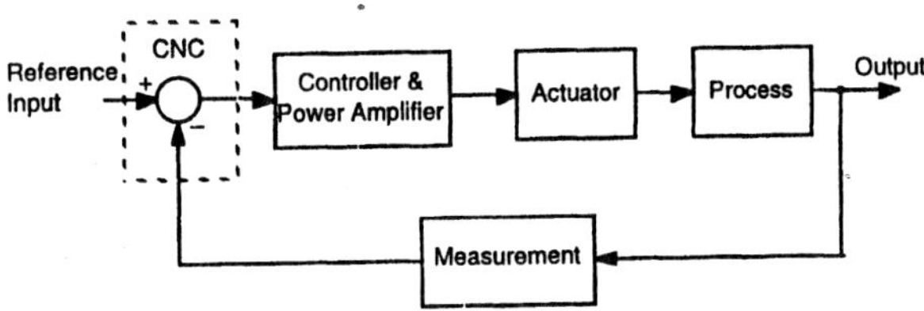

Figure 77.19 Control structure block diagram.

To improve system capability, various controllers may be placed between the position error signal and the power amplifier. This is discussed in detail in later sections. However, before beginning the controller design, it is valuable to examine the system components for adequacy. It may be desirable to replace existing motors, actuators, power amplifiers and feedback sensors, depending on performance specifications and cost constraints.

Transmissions and Drive Trains Typically, machines are based on a Cartesian or polar reference frame requiring converting the rotational displacements of actuators (i.e., servo motors) to linear displacements. Precision ball screws, friction drives or linear motors are used for this purpose. Nonlinearities, such as backlash, should be minimized or avoided in any drive system. In ball screws backlash can be greatly reduced, or eliminated in some cases, by preloading an axis. Preloading can be accomplished by suspending a weight from one side of the slide, ensuring that only one side of the screw is used to drive the slide. Backlash in gear reductions can be minimized with the use of harmonic drives. Greater precision requires friction (capstan) drives or linear motors, because position repeatability suffers from the recirculating ball screws. Capstan drives have direct contact so that no backlash occurs. However, they cannot transmit large forces, so that they are commonly used in CMMs and some DTMs where forces are low.

Rolling or antifriction elements are required to move heavy carriages along a guide way and to maintain linearity. Several configurations are used depending on the amount of precision required. Linear shafts and recirculating ball bearing pillow blocks are for less precise applications. For increased accuracy and isolation, noncontact elements are required at greater expense. Noncontact elements include air bearings, hydrostatic bearings and magnetic bearings. Air bearings are common on CMMs. Hydrostatic bearings are still a maturing technology but provide greater load carrying capability over air bearings [23].

Motors Several motor types are available for use in ultra-high precision control. Typical applications are powered by precision DC servomotors. Recent technology includes linear motors and microstepping motors.

Stepper motors provide an accurate and repeatable precise step motion, yielding a moderate holding torque. However, standard stepper motors have a fundamental step angle limitation (usually 1.8° per step). Stepper motors also have a low resonance frequency causing them to perform poorly at slow speeds, inducing vibrations on the order of 10 Hz. The step size and vibration limitations are somewhat reduced when using microstepping controllers. Finally, these motors can also dissipate large amounts of heat, warming their surroundings, and generating thermal errors due to thermal expansion of the machine.

Linear motors have the advantage of eliminating the lead screw elements of the actuator drive. They have no mechanical drive elements, but generally require rolling elements to suspend the "rotor" or slide. Heat is usually generated from a linear motor closer to the work piece and can be a significant source of thermal errors, depending on the application and machine configuration.

Power Amplifiers Power sources (amplifiers) can be either constant gain amplifiers or pulse width modulated amplifiers (for DC servo motors). Considerations for the amplifier selection are its linear range, maximum total harmonic distortion (THD) over the linear range, and maximum power output. All of these factors must be examined in designing ultra-high precision control systems. For high bandwidth systems, the amplifier dynamics may be critical. THD can be used to estimate nonlinear effects of the amplifier conservatively. The power limitations provide control limits to maintain controller linearity. (To protect the motors and amplifiers, circuit breakers or fuses should be placed appropriately. Transient currents (maximum current) and steady state current (stall current) must both be considered.)

Instrumentation: Sensors, Locations and Utilization

Instrumentation of many types is available for position, velocity and force feedback to the controller. The sensor choice depends on the particular application. For example, in diamond turning, cutting forces are generally negligible, so that only tool speed and position sensors (e.g., tachometers and encoders) are needed for control. However, in scanning metrology, the probe tip force must be controlled to a very high degree to limit probe deflections. Here force sensors must be chosen carefully to achieve the desired force response.

Because the tolerances typically held in ultra-high precision machining are on the order of 100 nm and lower, special sensors are needed for measurement and feedback. Displacement sensors used for ultra-high precision feedback include optical encoders, resolvers, precision potentiometers, linear variable differential transformers (LVDTs), eddy current sensors, capacitance gages, glass scales, magnetic encoders, and laser interferometers. Even if stepper motors are used, feedback transducers should be applied to eliminate a potential slip error. At this small measurement scale, thermal effects and floor/external vibrations not typically encountered in lower precision applications, appear more significant. Thus, the following points must be considered:

1. Allow the system to reach thermal equilibrium (attain a steady state temperature). Typically a 24-hour thermal drift test is required to assess the effect of temperature variations, [[1], Appendix C].
2. Provide environmental regulation. Once the machine has reached thermal equilibrium, it must be kept at a constant temperature to insure repeatability. The standard temperature as defined in [1] is 20°C (68°F). It is difficult, if not impossible, to keep the entire machine at a temperature of 20°C. Heat sources, such as motors and friction, will generate "hot spots" on the machine. Typically, the spatial thermal gradients are not detrimental to machine repeatability, but temporal thermal gradients are. Therefore, it is important to keep the machine temperature from varying with time.
   If laser interferometers are used, the temperature, relative humidity and pressure of the medium through which the laser beam passes (typically air) must be considered. If these quantities are not considered, the interferometer will experience a velocity of light (VOL) error, because the wavelength of light is modulated in a varying environment. It is not necessary to maintain these quantities at a fixed level (clearly, maintaining pressure is a difficult task), because interferometers can be equipped to compensate for VOL errors. In some critical applications, the laser beam is transmitted through a controlled atmosphere via separate beam ways. For the most precise machines developed, the beams are transmitted through a vacuum to eliminate all effects of temperature, humidity and pressure. This solution, however, is extremely difficult to realize.
3. The system must be isolated from floor vibrations. Granite bases, isolation damping pads, and air suspension are some of the techniques used to isolate a system from vibrations. Even the smallest disturbances, such as people conversing next to the machine, can adversely affect the surface finish.

For best results in precision control, is important to choose the appropriate feedback sensor(s) and mounting location. Several principles should be followed. Care must be taken to locate the sensor as close as possible to the line of action or contact, to avoid Abbe offset errors [3]. Of course it is always desirable to locate sensors and drives in a location that is insensitive to alignment errors and away from harm in case a crash occurs. With the appropriate design and consideration, a machine's accuracy can be improved to the level of the sensor's resolution.

It is critical to recognize that sensors have dynamics and, therefore, transfer functions. The sensor bandwidth must match or exceed the desired system frequency response. Some of these sensors will require their own compensation, such as the VOL error on laser interferometers caused by fluctuations in temperature, pressure, and humidity.

Discrete sensors, such as encoders (rotary and linear), are often used in precision machine systems. These sensors measure position by counting tick marks precisely ruled on the encoder. These discrete sensors produce two encoded phase signals for digital position indication (A and B channel quadrature). This increases their resolution by a factor of four and provides directional information. Rotary encoders and resolvers, used on the motor rotational axis, are not typically acceptable in ultra-high precision control because they are not collocated with the actual parameter being measured. Furthermore, their repeatability is affected by backlash and drive shaft (screw) compliance. However, linear versions of encoders (either magnetic or glass scales) are commonly used with good results. A limitation here is the minimum resolution size that is directly related to the spacing of the ticks ruled on the scale. Care must be taken to provide a clean environment, because dirt, smudges or other process byproducts can reduce the resolution of these sensors, limiting their utility. The requirement for cleanliness may also require that these devices be kept isolated from potentially harsh environments.

Analog sensors are also widely used for feedback. Typical analog sensors used in precision control include LVDTs, capacitance gages and inductive (eddy current) gages. The least expensive analog displacement sensor is the LVDT, which uses an inductance effect on a moving ferrous core to produce a linear voltage signal proportional to the displacement. In most cases ultra-high precision air bearing LVDTs are necessary to overcome stiction of the core. The best resolution, for analog sensors, can be obtained with capacitance gages and eddy current gages which are widely used in ultra-high precision measurements. Capacitance gages can measure 10^{-12}m displacements [14]. However, at this resolution the range is extremely limited. These sensors require a measurement from a metallic surface, so that locations are typically in-line with the metal part or offset near the equipment ways. Because of the extremely high resolution and low depth of field of these sensors, it is preferable to keep them collinear to the actuators.

The interferometer is the most preferred position sensor in precision system design, because it utilizes the wavelength of light as its standard. Thus, it is independent of any physical artifact (such as gage blocks). In general, the laser interferometer is the best obtainable long-range measurement device, with measurements accurate to 0.25 nm, nominally. Because of size limitations, interferometers are typically located with an axial offset, making measurements susceptible to offset errors, which can be minimized by good metrological practices. Similar to encoders, interferometers use A and B channel quadrature as their output format. As previously stated, care must be taken to include temperature, humidity and pressure compensation to achieve the best possible resolution for the interferometer.

Typical sensors used in force measurement include LVDTs, strain gages, and piezoelectric transducers. LVDTs are used with highly calibrated springs or flexures, allowing forces to be estimated as a function of spring deflection. Bandwidth for this type of sensor is limited. Force measurements based on strain gages are also common. However, they too suffer from limited frequency response because they operate on the same principle as sensors employing flexures (Hook's Law).

Since most ultra-high precision applications require a higher bandwidth than available from flexures and strain gage systems, piezoelectric force transducers are typically used in these applications. Piezoelectric-based dynamometers use a quartz crystal to generate a charge proportional to a force. The dynamometer may be located either behind the tool or under the work piece to measure cutting or probe forces. Although these sensors have larger bandwidth, they exhibit some hysteresis and charge leakage which results in poor capability to measure lower frequency and DC forces due to drifting. However, they are generally stiffer than strain gage force sensors. Piezoelectric force sensors also tend to be large and reduce the effective tool stiffness.

Recent work in force sensors for cutting machines has led to the development of sensors that employ a piezoelectric film. The film is a 28μm of polyvinylidene fluoride placed on tool inserts close to the cutting edge [17]. Advantages of the film are that the force measurements are close to the cutting edge and there is essentially no loss of stiffness because of its small size. Newer magnetostrictive materials, such as metglass or Terfenol^R, are currently being examined for use as force sensors. These materials emit an electromagnetic field in proportion to the force applied.

Care must be given to cabling sensors and determining the expected signal-to-noise ratio. All cables must be shielded to prevent stray magnetic fields from inducing current noise in sensor cables. A means must be provided for independent calibration of sensors (e.g., a laser interferometer can be used to independently check the capacitance gage calibration). Some sensors are also calibrated to a specific cable length impedance.

Because many sensors are digitally based, anti-aliasing filters must be provided to ensure that signal components above the system Nyquist frequency do not cause aliasing problems. Details regarding the use of antialiasing filters can be found in Section 77.3.3. It is also prudent to remember that sensors themselves are dynamic systems, which vary in magnitude and phase with frequency. Specifications are typically provided by the manufacturer. However, it is strongly recommended that sensor system responses be verified. (System identification techniques used in system response verification are presented in Section 77.3.3.) This assists in designing and trouble shooting the controller.

Compensators: Position and Velocity Loops

Analog controllers can be added to existing CNC-type controllers to improve the speed and precision of existing machines if appropriate attention is paid to understanding the system dynamics. The controller and system block diagram are depicted in Figure 77.20, highlighting the additional compensator for improved performance. The inner loop is the velocity loop, and

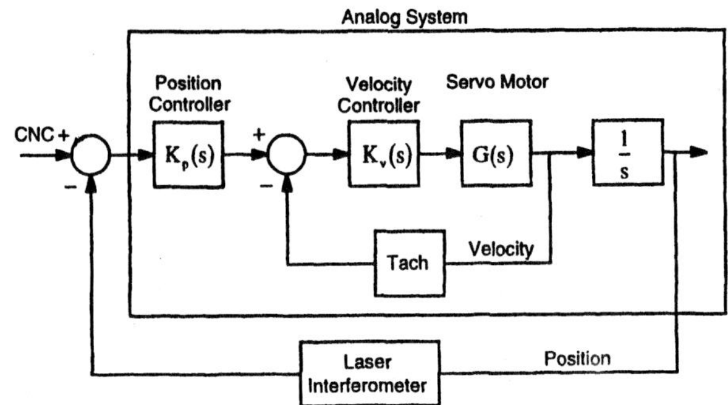

Figure 77.20 Machine with additional compensator.

the outer loop is the position loop. Both position and velocity must be well-characterized and controlled in ultra-high precision processes. Because tool position relates directly to final part geometry the position compensator, $K_p(s)$, is used to control dimensional accuracy. Similarly, since the tool velocity is directly related to surface finish, the velocity compensator, $K_v(s)$, is used to control the surface finish. The plant (or process) is depicted as $G(s)$, with velocity feedback provided by a tachometer and position feedback supplied by a laser interferometer or linear encoders. Details of the control design are provided in Section 77.3.4.

Fast Tool Servos

Recent advances have been made in the use of piezoelectric actuators for achieving greater bandwidth on precision equipment. Piezoelectric actuators are high bandwidth actuators (on the order of 1 kHz) that can produce high forces and are relatively stiff. However, the stroke of these actuators is generally limited to about 100 μm or less. Piezoelectric actuators are used with conventional actuators (lead screws, capstan drives, etc.) to yield small, high frequency displacements and the conventional actuators provide the lower speed, longer range motions. The piezoelectric actuator is powered by a high voltage supply nominally generating 1000 volts. Actuators are run in a closed-loop mode, typically using a laser interferometer or a capacitance gage for displacement feedback. When used in a closed-loop, the piezoelectric actuator is referred to as a fast tool servo (FTS).

Several ultra-high precision applications using FTS have been developed [6]. The FTS has been used to compensate for cutter runout [18] and for negating the effects of low velocity friction in a series lead screw actuator. FTS's have also been successfully implemented on diamond turning machines [13], [9].

77.3.3 System Identification and Modeling

Successful ultra-high precision control design and implementation depend on the accuracy of system models. Thus, the dynamics of each component must be well-characterized. System identification techniques provide tools for this purpose [19]. A system or component transfer function, $G(s)$, is typically represented in the frequency domain, as shown in Figure 77.21. Here $X(s)$ is defined as the input to the system and $Y(s)$ is the system output.

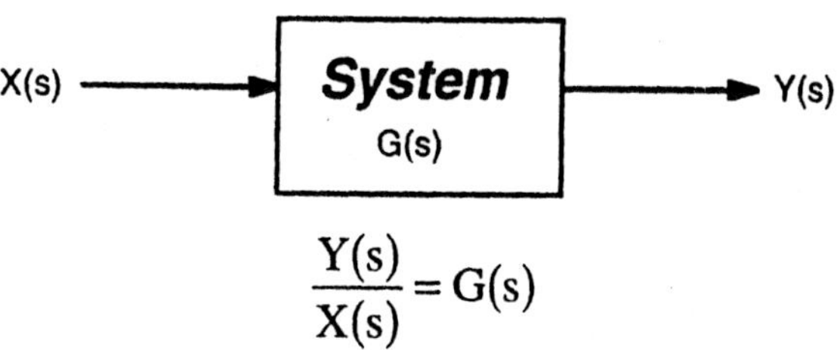

$$\frac{Y(s)}{X(s)} = G(s)$$

Figure 77.21 System input and output with resulting transfer function.

Model transfer functions, $G(s)$, are derived from time histories of the input signal, $x(t)$, and the output signal, $y(t)$, using techniques discussed in the following sections. The order of the model should be as small as possible, preferably related to physical characteristics of the machine (e.g., drive shaft compliance or carriage inertia). Several software packages provide routines that may be used to examine a range of model orders and determine the best fit for a given data set (e.g., MATLAB). However, this numerical approach should not be blindly used as a substitute for thoroughly understanding the process or physical system.

Multiple Axes – Cross-Coupling

Some machines have only a single axis or possess multiple axes that are not strongly coupled. For such machines single-axis system identification can be employed and is straightforward. The input to output transfer functions are easily found as discussed below. However, in many machining cases, such as grinding, several axes are involved in the machine control. If the dynamics of these axes are coupled, then cross-coupling effects must be considered. For example, one axis or more may be used to control force while another axis may be used to control velocity. Recent work addresses, in part, the decoupling of force and velocity loops in a grinding system by gain selection in each axis [12]. Several approaches for improved precision control with a cross-coupled system have been developed over the past 10 years. Another recent technique with superior performance is a cross-coupling controller, where the error in one axis is coupled to position control of the other axes to minimize contour errors [20].

Experimental Approaches

In the development of system models for the controllers, amplifiers and mechanical systems, several experimental approaches may be used to identify and verify component response. In this section the three most common approaches and the typical hardware configurations necessary to implement them are presented.

Stepped Sinusoidal Response Stepped sinusoidal input is the preferred method for determining transfer functions of all system components because it permits detailed system analysis at all frequencies. It is, however, the slowest of the three techniques discussed here. To use this method, sinusoids of various frequencies are fed to the system, one frequency at a time. The response of the system is measured at each frequency, recording the magnitude and phase variations between the input and output signals. The recorded data are then used to construct accurate Bode plots for the system that represent the plant transfer function, $G(s)$. Special purpose digital signal processing (DSP) equipment is available to sweep automatically through a range of frequencies and determine transfer functions in this manner. A coherence function of the data may be easily generated to assess the validity of the determined transfer function at various frequencies. The magnitude of the coherence function varies from zero to unity representing low and high confidence in results, respectively.

Step Input Response For first-order or second-order models, a step input response can be used to determine the system dynamics by applying system identification techniques on the input and output data. The sharpness of the actual step input will limit the frequency range of the identifiable response. Sharper steps have broader frequency spectrums and, therefore, may be used to generate models possessing broader frequency spectrums. Fairly sharp steps in voltage inputs can be easily generated, yielding excellent results for systems with electrical inputs, such as motors.

A fast Fourier transform (FFT) is used to determine the system model's magnitude and phase from the step input response. A frequency domain model (transfer function) or a state-space model can be derived using a least-squares parameter estimation technique. One such method is known as ARX, AutoRegressive eXternal input [19]. The ARX method minimizes the square of the error, $e(t)$, in the following set of equations

$$\begin{aligned} y(t) &= G(q)x(t) + H(q)e(t), \\ e(t) &= H^{-1}(q)[y(t) - G(q)x(t)] \equiv \text{error}, \end{aligned} \tag{77.39}$$

and

$$\left[\hat{G}_N, \hat{H}_N\right] = \min \sum_{t=1}^{N} e^2(t), \tag{77.40}$$

where $G(q)$ represents a discretized version of $G(s)$, $H(q)$ is a discretized weighting of the residual error of the model, N is the number of data points taken, and $\hat{G}_N$ and $\hat{H}_N$ are the best estimates (using the least-squares relationship given in Equation 77.40) of $G(q)$ and $H(q)$, respectively.

When using ARX procedures, several precautions must be taken. First, because ARX models tend to favor higher frequencies [26], the sampled system data should be filtered prior to identification, removing extraneous high frequency content. (The high frequency sources are typically from structural responses, disturbances, or noise, which cannot be controlled.) Also, simulation of the identified system should be compared with the actual data to verify model accuracy because the proper order model may not have been chosen. Some iteration on the model order may be required for an improved fit. As always, the lowest order model that acceptably incorporates the important system dynamics should be used.

White Noise Response Many digital signal processing boards use white noise input to determine the frequency response of a system. This method will generate Bode plots and will yield higher order systems than the step input approach. However, this technique may not be as accurate as the other techniques presented because the input signal may not be truly "white," that is, the generated input signal may not exhibit a uniform broad band spectrum. Frequency content may be lacking at certain frequencies in the input signal. Therefore, the model may be inaccurate at those frequencies.

White noise input and system output are also run through FFT algorithms to obtain Bode plots of the transfer function. Limitations on the valid frequency range of the transfer function are based on the sampling time. Windowing techniques (e.g., Hamming, Hanning, etc.) are used to decrease signal leakage in the Fourier analysis and may increase the accuracy of the attainable FFT results.

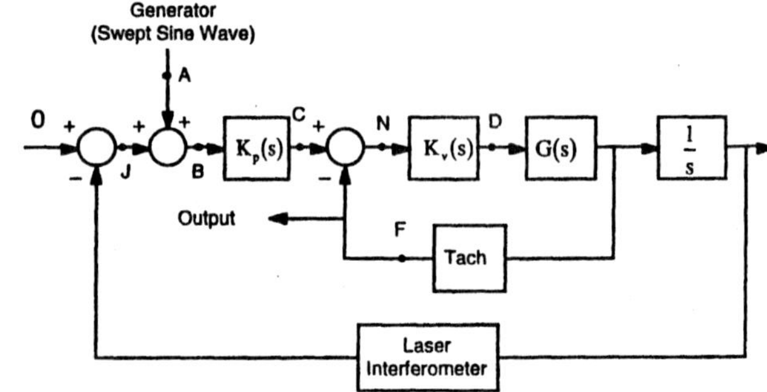

Figure 77.22 Location of input and output signals for system identification.

TABLE 77.4 Locations of Input and Output Signals for System Identification and Response

Plant/System to identify	Input location	Output location
$K_p(s)$	B	C
$K_v(s)$	N	D
$G(s)$	D	F
Closed-Loop velocity	C	F
Integrator (Laser interferometer)	F	J
Tach to position	F	J
Open-Loop position	B	J
Closed-Loop position	A	J

Experimental Set-Up

Only a few specialized hardware items are necessary to obtain transfer functions. The required equipment includes a signal generator (preferably computer based), a signal amplifier, sensors, and signal processing capability.

For the stepped-sine or white noise methods, a DSP instrument or DSP-PC board is useful for signal conditioning and transfer function analysis. The sensors required for these tests are the same as the feedback sensors listed in Section 77.3.2.

A system, shown in Figure 77.22, depicts the signal injection point from a DSP based instrument. The locations to collect input and output data for determining the component transfer functions are listed in Table 77.4. The system is connected to the CNC controller with a zero reference input (a regulator). The CNC controller is used to interpret the position from digital A and B quadrature (A quad B) signals out of the position encoder (seen in Figure 77.22 as the laser interferometer). Therefore, position measurements (point J in Figure 77.22) are obtained through the digital to analog (D/A) output of the CNC controller. This analog position value is the integral of the tachometer signal. It should be noted that the tachometer and interferometer dynamics are significantly faster than the other system components and may, therefore, be ignored.

Signal Conditioning

Because many sensor systems go through an analog to digital conversion (A/D), anti-aliasing filters are needed to prevent frequencies above the Nyquist frequency from being falsely interpreted as lower frequencies (aliasing). Although elliptic and Chebyshev filters have steeper drop-off rates, a low-pass Butterworth filter is recommended because of its flat response in the pass-band. To increase the attenuation roll-off, cascaded, higher order Butterworth filters are recommended. Because these filters have dynamics that can affect system performance, their dynamic responses must be experimentally characterized. For example, a Butterworth filter is maximally flat. However, when it is realized, it may slightly amplify or attenuate a signal and generate a slight error in a position or velocity command signal. Such slight errors may not be tolerable in ultra-high precision machines. These types of deviations, if not well understood, can reduce the performance of the machine.

77.3.4 Control Design

Good mechanical design and modeling are two important components of a precision system design. However, to achieve the highest level of performance, analog compensators must be employed to increase the bandwidth of the machine and to improve its robustness. Typically, controllers are used to improve closed-loop system performance of both the velocity and position loops, where the velocity control loop is nested inside the position loop shown in Figure 77.20. This section presents various analog controller types used in precision machines. Design methods for these controllers are presented and their implications for the actual precision servo system.

Discussion of Controllers

The controllers employed for ultra-high precision applications are typically embodied as standard analog configurations.

This approach eliminates all issues related to digital control such as sampling and quantization. The following discussion presents control elements used in both the velocity and position loops and outlines the basic the design procedures for these elements.

Velocity Loop Compensator Velocity control of a precision machine is critical in maintaining consistent and known surface finish during machining operations. Controlling velocity is also critical in maintaining consistent force trajectories while measuring or machining. The dynamic velocity loop compensator, $K_v(s)$, is a combination of a lead controller and a PI controller. The lead compensation provides additional phase margin for the system, because the integrator causes a 90° phase lag. Large amounts of phase and gain margin are usually specified for ultra-high precision systems to ensure their robustness to various outside disturbances. The integrator is used to eliminate steady-state velocity error. This is important because machines are usually servocontrolled at constant feed velocities.

Lead Compensator Design Procedure A lead compensator used to increase the phase margin for the velocity loop is depicted in Figure 77.23. Procedures for designing the controller to achieve the desired input to output voltage relationships are summarized in Equations 77.41 and 77.42 and Table 77.5. Such a compensator is designed in the following section for a diamond turning machine where actual specifications for phase and gain margin are provided.

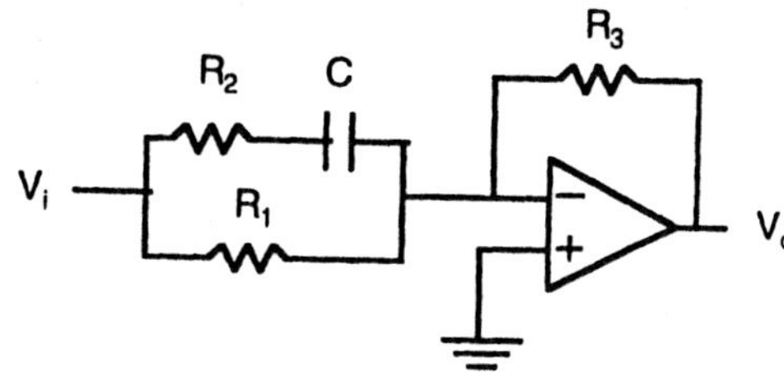

Figure 77.23 Lead compensator diagram.

$$\frac{V_0}{V_i} = -\frac{R_3}{R_1}\left(\frac{1+s(R_1+R_2)C}{1+sR_2C}\right) = -\frac{R_3}{R_1}\left(\frac{1+s\tau_1}{1+s\tau_2}\right) \tag{77.41}$$

$$\phi_{max} = \sin^{-1}\left(\frac{1-\alpha}{1+\alpha}\right); \quad \alpha = \frac{\tau_1}{\tau_2} = \frac{\omega_1}{\omega_2};$$
$$\omega_0 = \sqrt{\omega_1\omega_2} \tag{77.42}$$

TABLE 77.5 Lead Compensator Design Procedure

Determine w_0 and ϕ_{max}.
Calculate α from ϕ_{max}.
Calculate w_1 and w_2 from α and w_0.
Calculate τ_1 and τ_2 from w_1 and w_2.
Choose C.
Calculate R_1, R_2 and R_3 from τ_1, τ_2 and C.

The procedure is started by selecting the maximum phase lead angle, ϕ_{max}, and the frequency where it is to occur, w_0 (the geometric mean of the two corner frequencies). Once these two quantities have been selected, the variables α and subsequently w_1, w_2, τ_1 and τ_2 are determined. Finally, the values for C, R_1, R_2 and R_3 are determined. Although the procedure is relatively straightforward, values for the three resistors and the capacitor must be chosen so that they are commercially available. For example, a resistance value of 10 kΩ, which is commercially available, is preferred to a value of 10.3456 kΩ, which is not commercially available.

PI Compensator Design Figure 77.24 represents a PI compensator design for the velocity loop. The design procedure is summarized in Table 77.6 and Equations 77.43 and 77.44. As previously stated, the PI compensator yields zero steady-state error to a velocity trajectory that is constant, in other words, a constant feed rate.

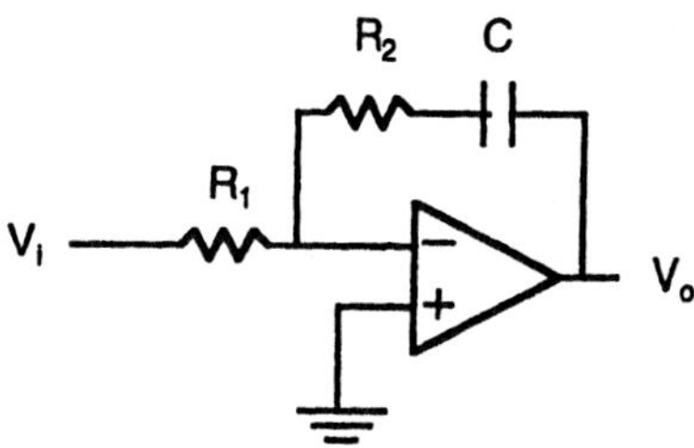

Figure 77.24 PI compensator diagram.

$$\frac{V_0}{V_i} = \frac{1+sR_2C}{sR_1C} = \frac{1+s\tau_2}{s\tau_1} \tag{77.43}$$

$$\omega_{break} = \frac{1}{\tau_2}; \left|\frac{V_0}{V_i}\right|_{\omega=\infty} = \frac{\tau_2}{\tau_1} \tag{77.44}$$

TABLE 77.6 PI Compensator Design Procedure

Determine break frequency, $\omega_{break} = 1/\tau_2$.
Choose high frequency gain.
Calculate τ_1 from τ_2.
Choose C.
Calculate R_1 and R_2.

Band Reject Filters for Structural Resonances Finally, it is worth mentioning that machine tools and measurement machines have a variety of structural modes or resonant frequencies. These are often detected during the system identification phase of the compensator design. To eliminate these modes, band reject filters are employed. These analog filters can be tuned to attenuate a specific range of frequencies and can be extremely beneficial in improving the system performance.

Input shaping is another available technique for suppressing resonance frequencies, often used in more flexible systems. The reader is referred to [21] for more detail.

Position Loop Compensator The position loop is a means of achieving the dimensional accuracy desired. For the position loop, the compensator, $K_p(s)$, is a series combination of a lead controller and a proportional control element. The lead compensator is designed in the same manner as the velocity loop. Because a natural integrator occurs in the position loop from the carriage velocity to the position sensor (see Figure 77.20), no additional integral control is necessary to achieve zero steady state to a constant command signal.

Advantages and Disadvantages

There are some advantages to this type of analog compensator scheme. Analog compensators can be used with most CNC type controllers to improve system bandwidth and precision, because most CNC controllers do not perform any dynamic compensation. Rather, they simply put out a voltage proportional to the position error of the machine tool. Another advantage of analog compensators is that their resolution is only limited to the tolerable signal-to-noise ratio, whereas digital systems are limited by quantization effects as well as noise.

There are, of course, some limitations to using analog compensators, the most significant of which is the inability to provide adaptive variable gains based on parameter estimation techniques. However, ultra-high precision plants tend to have less variability by design, eliminating the need for adaptive compensators. The other major limitation to analog compensators is that they are less flexible in design and implementation than digital systems .

Performance Criteria

System performance is typically measured in terms of bandwidth, steady-state error, rise time, overshoot, and damping. Bandwidth is important because a machine's bandwidth relates directly to its apparent stiffness. Steady-state error for both velocity and position loops must be eliminated to insure appropriate surface finishes and part geometries, respectively. Specifications for rise time assure that higher spatial frequency characteristics of the part geometry are not eliminated (i.e., the machine must be able to corner reasonably well). Damping and control of overshoot are particularly important in designing the position controller to avoid removing too much material from a part during machining.

Although the design approach presented here is nonadaptive, techniques can provide a robust compensator design. Robustness of design is necessary to assure stability and minimal performance variation of the closed-loop system in the presence of some uncertainty in the plants and environment. Robustness of the design in stability can be characterized in terms of gain and phase margins. Design techniques for obtaining compensators robust in performance include the use of root sensitivity and gain plots [16], and other parametric plots.

Root Sensitivity Gain plots are an alternate graphical representation of the Evans root locus plot [15] . They explicitly graph the eigenvalue magnitude vs. gain in a magnitude gain plot and the eigenvalue angle vs. gain in an angle gain plot. The magnitude gain plot employs a log-log scale whereas the angle gain plot uses a semilog scale (with logarithms base 10). Although gain is the variable of interest, any parameter may be used in the geometric analysis.

The root sensitivity of any system can be computed using the slopes of the gain plots [16]. In classical control theory, the root sensitivity, S_p, is defined as the relative change in a system root or eigenvalue, $\lambda_i (i = 1, ..., n)$, with respect to a system parameter, p. Most often, the parameter analyzed is the forward loop gain, k. The root sensitivity with respect to gain is given by

$$S_k = \frac{d\lambda(k)/\lambda(k)}{dk/k} = \frac{d\lambda(k)}{dk}\frac{k}{\lambda(k)}. \tag{77.45}$$

Equation 77.45 is often introduced in determining the break points of the Evans root locus plot for single-input, single-output systems. At the break points, S_k becomes infinite as at least two of the n system eigenvalues undergo a transition from the real domain to the complex domain or vice versa. This transition causes an abrupt change in the relationship between the eigenvalue angle $\angle\lambda$ and gain k, yielding an infinite eigenvalue derivative, $d\lambda/dk$.

The root sensitivity function, S_k, is a measure of the effect of parameter variations on the eigenvalues. An expression for the complex root sensitivity function is developed by employing a polar representation of the eigenvalues in the complex plane. Three assumptions are imposed: (i) the systems analyzed are lumped parameter, linear time-invariant (LTI) systems; (ii) there are no eigenvalues at the origin of the s-plane, i.e.,

$$\lambda_i \neq 0, \quad \forall i = 1, 2, \ldots, n \tag{77.46}$$

(although the eigenvalues may be arbitrarily close to the origin), and (iii) the forward scalar gain, k, is real and positive, i.e., $k \in \Re, k > 0$. Based on these assumptions, we draw the following observations: the real component of the sensitivity function is given by

$$Re\,|S_k| = \frac{d \ln|\lambda(k)|}{d\ln(k)}, \tag{77.47}$$

and the imaginary component of the sensitivity function is given by

$$Im\,|S_k| = \frac{d\angle\lambda(k)}{d\ln(k)}. \tag{77.48}$$

These observations may be proven as follows. Equation 77.45 may be rewritten in terms of the derivatives of natural logarithms as

$$S_k = \frac{d\ln[\lambda(k)]}{d\ln(k)}. \tag{77.49}$$

The natural logarithm of the complex value, λ, is equal to the sum of the logarithm of the magnitude of λ and the angle of λ multiplied by $j = \sqrt{-1}$. Thus, Equation 77.49 becomes

$$S_k = \frac{d[\ln|\lambda(k)| + j\angle\lambda(k)}{d\ln(k)}. \tag{77.50}$$

Because j is a constant, Equation 77.50 may be rewritten as

$$S_k = \frac{d\ln|\lambda(k)|}{d\ln(k)} + j\frac{d\angle\lambda(k)}{d\ln(k)}. \tag{77.51}$$

We next make the observation that the slope of the magnitude gain plot is the real component of S_k. The magnitude gain plot slope, M_m, is

$$M_m = \frac{d\log(|\lambda(k)|)}{d\log(k)}, \quad (77.52)$$

which may be rewritten as

$$M_m = \frac{d[\log(e)\ln(|\lambda(k)|)]}{d[\log(e)\ln(k)]} = \frac{d\ln(|\lambda(k)|)}{d\ln(k)} \quad (77.53)$$

corresponding to Equation 77.47.

Furthermore, the slope of the angle gain plot is the product of the imaginary component of S_k and the constant, $[log(e)]^{-1}$. The angle gain plot slope, M_a, is

$$M_a = \frac{d\angle\lambda(k)}{d\log(k)}, \quad (77.54)$$

which may be rewritten as

$$M_a = \frac{d\angle\lambda(k)}{d[\log(e)\ln(k)]} = \frac{1}{\log(e)}\frac{d\angle\lambda(k)}{d\ln(k)}, \quad (77.55)$$

and hence M_a is proportionally related to Equation 77.48 by $[log(e)]^{-1}$.

The complex root sensitivity function is now expressed with distinct real and imaginary components employing the polar form of the eigenvalues. It follows from assumption (ii) that $\ln(k)$ is real. (In general, most parameters studied are real and this proof is sufficient. If, however, the parameter analyzed is complex, as explored in [16], it is straightforward to extend the above analysis.)

The slopes of the gain plots provide a direct measure of the real and imaginary components of the root sensitivity and are available by inspection. The use of the gain plots with other traditional graphical techniques offers the control system designer important information for selecting appropriate system parameters.

Filtering and Optimal Estimation

In the presence of noisy sensor signals, special filtering techniques can be applied. Kalman filtering techniques can be used to generate a filter for optimal estimation of the machine state based on sensor data. This approach also provides an opportunity to use multiple sensor information to reduce the effects of sensor noise. However, Kalman filters are generally not necessary because much of the sensor noise can be eliminated by good precision system design and thorough control of the machine's environment.

Hardware Considerations

Selection of electronic components for the compensator is limited to commercially available resistors and capacitors. Although 1% tolerance is available on most components, the final design must be thoroughly tested to verify the response of the compensator. Off-the-shelf components may be used. However, their transfer functions must also be determined. For example if a Butterworth filter is used, it cannot be assumed that its transfer function is perfectly flat. The filter's transfer function must be mapped. Any deviation from a flat frequency response can generate positional or velocity errors that are unacceptable because of stringent requirements.

Software and Computing Considerations

Although this chapter is primarily concerned with analog control, there are some computing issues that are critical and must be addressed to assure peak performance of the closed-loop system. The tool (probe) trajectory in conjunction with the update rate must be carefully considered when designing a precision machine. If they are not an integral part of the control design, optimal system performance cannot be achieved. In many cases the CNC controller must have a fast update time corresponding to the trajectory and displacement profile desired. There are many CNC systems (including both stand-alone and PC-based controllers) on the market today, with servo update rates of 2000 Hz and higher, to perform these tasks. The controller must also be able to interface with the various sensors in the feedback design. Most of these controllers provide their own language or use g-codes, and can be controlled via PC-AT, PCMCIA, and VME bus structures. If care is not taken to interface the CNC controller to the analog portion of the machine, the resulting system will not perform to the specifications.

77.3.5 Example: Diamond turning machine

The most precise machine tools developed to date are the diamond turning machines (DTM) at the Lawrence Livermore National Laboratory (LLNL). These machine tools are single point turning machines, or lathes that are capable of turning optical finishes in a wide variety of nonferrous materials. This section presents a basic control design for such a machine. To avoid confusion, the machine vibration modes are not included in the analyses, and the experimental results have been smoothed out via filtering. The approach discussed here is the approach taken when designing analog compensators for diamond turning machines. Furthermore, the compensators developed here are in use in actual machine tools.

Figure 77.25 depicts a typical T-based diamond turning machine similar to the systems employed at LLNL. The machine consists of two axes of motion, X and Z, as well as a spindle on which the part is mounted. The machine is called a T-based machine because the cross axis (X) moves across a single column that supports it. Thus the carriage and the single column frame form a "T." In this particular T-based configuration, the entire spindle (with part) is servocontrolled in the Z direction which defines the longitudinal dimensions of the part; and the tool is servocontrolled in the X direction, defining the diametrical part dimensions.

The DTM is equipped with a CNC controller which plans the tool trajectory and determines any position error for the tool. The output of the CNC controller is a signal proportional to the tool position error, that is, the output of the summing junction for the outer (position) loop in Figure 77.20 (point J, Figure 77.22). The objective of this example is to demonstrate the design of the

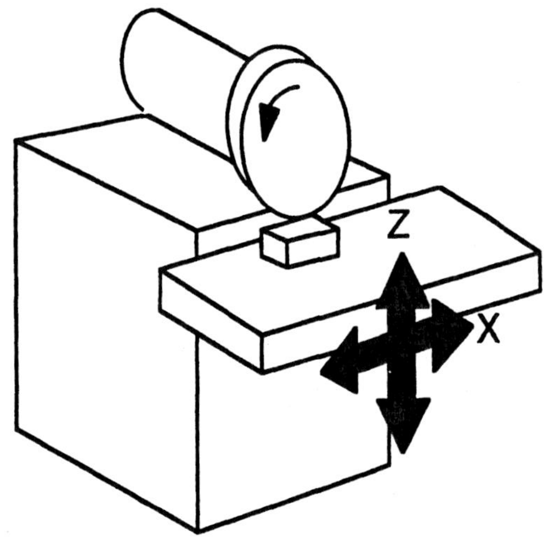

Figure 77.25 A T-Based diamond turning machine.

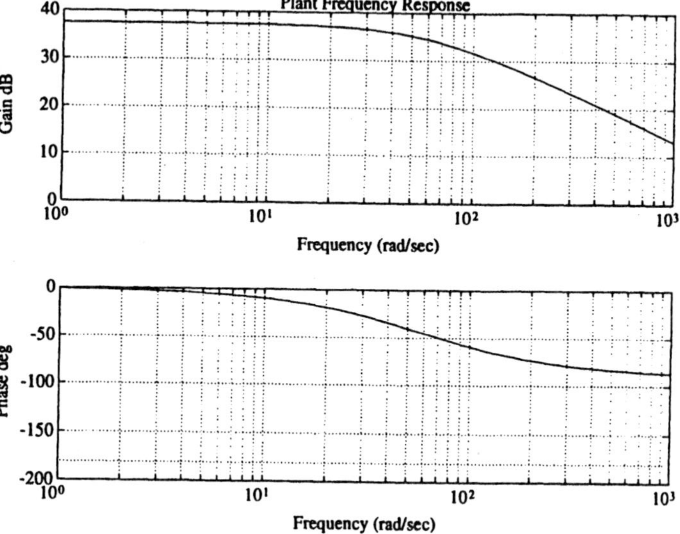

Figure 77.26 Bode plots for a diamond turning machine $G_p(s)$.

velocity and position controllers, $K_v(s)$ and $K_p(s)$, respectively, for the Z-axis. Usually, both $K_v(s)$ and $K_p(s)$ are simple gains. However, the performance required for this DTM requires active controllers in both the position and velocity loops.

Typically, the velocity compensator is designed first and, subsequently, the position compensator is designed. The velocity loop consists of the velocity compensator, a power amplifier (with unity gain that amplifies the power of the compensator signal), a unity gain tachometer mounted on the servomotor, and a summing junction. The position loop consists of the CNC controller, a position compensator, the dynamics from the closed-loop velocity system, a lead screw (with a 1 mm/rev pitch), and a (unity gain) laser interferometer providing position feedback. The lead screw acts as a natural integrator converting the rotational velocity of the servomotor into slide position. The laser interferometer provides a digital (A quad B) signal back to the CNC controller which, in turn, compares the actual location of the Z carriage to the desired location. This error is then converted to an analog signal by the CNC controller's digital to analog board and fed into the position compensator. For all practical purposes, the CNC controller may be considered a summing junction for the position loop. This assumption becomes invalid at frequencies that approach the controller's sample rate of 2 ms.

To determine the transfer function of the servo system, sinusoidal signals of various frequencies are injected into the servoamplifier, and the resulting tachometer outputs are recorded and compared to the inputs. Bode plots are then generated from the empirical data, and a transfer function is estimated from these plots. Figure 77.26 shows a typical Bode plot for the Z-axis of the DTM.

From Figure 77.26, the transfer function of the servo drive from the servo amplifier (not including the velocity compensator) input voltage, $V(s)$, to the Z-axis motor velocity, $\omega(s)$, is

$$G_p(s) = \frac{\omega(s)}{V(s)} = \frac{4500}{s+60}. \tag{77.56}$$

The high gain of the system is due to the fact that the machine moves at relatively low velocities that must generate large feedback voltage signals.

Specifications

This section presents typical specifications (and the reasons behind the specifications) for both the velocity and position loops for a DTM.

The specifications for the velocity loop are

1. Zero steady state error to a constant input in velocity, $V(s)$. This insures that the machine tool can hold constant feedrates without any steady-state error.
2. Maximum of 2 volts input into the servoamplifier. This is a limitation of the amplifier. Based on the current required to drive the motor, no more than 2 volts may be used.
3. The sensitivity to the forward loop gain should be less than 1.0. This reduces deviations in closed-loop system performance if the forward loop gain varies.
4. Damping ratio, $\zeta > 0.8$. A small amount of overshoot (2%) in velocity is acceptable. (Note that $\zeta > 0.707$ results in a closed-loop system that should not have any amplification. Thus the closed-loop Bode plot should be relatively flat below the bandwidth frequency.)
5. Maximize the closed loop bandwidth of the system. Target velocity loop bandwidth, 15 Hz $\approx$ 95 rad/s.
6. Minimum gain margin 40 dB.
7. Minimum phase margin 90°.

The specifications for the position loop are

1. Zero steady-state positioning error.
2. The sensitivity to the forward loop gain should be less than 1.5. This reduces deviations in closed-loop system performance if the forward loop gain varies.

3. Damping ratio, $\zeta > 0.9$. A small amount of overshoot (1%) in position is acceptable because the typical motion values are on the order of 0.1 μin.
4. Steady-state position error to a step should be zero.
5. Minimum gain margin 60 dB.
6. Minimum phase margin 75°.

Compensator Design: Velocity Loop

The design procedure of Section 77.3.4 is used to determine the velocity loop PI compensator. The final design of the velocity loop PI compensator is of the same form as the PI compensator of Figure 77.24, with a capacitance of 4μF and both resistance values at 1.33kΩ. These components are determined from the relationships of Equations 77.57 and 77.58. The measured frequency response of the PI compensator is shown in Figure 77.27. Note

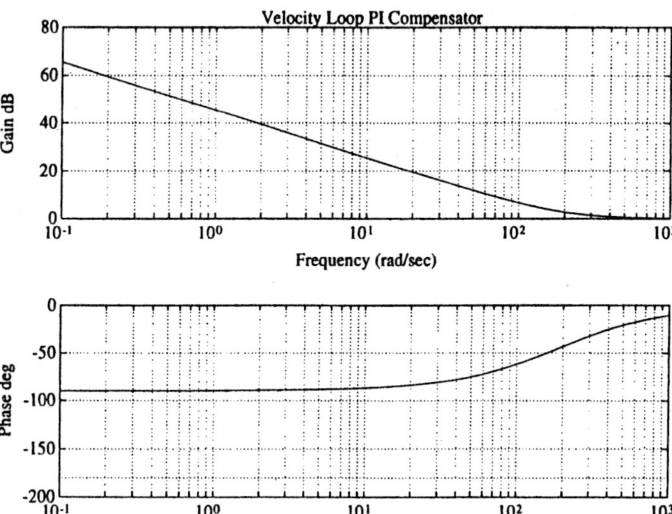

Figure 77.27 Velocity loop PI compensator response.

in Figure 77.27 that the integrator is visible with a low frequency slope of -20 dB/decade and a DC phase of $-90°$.

The lead compensator's final design is depicted in Figure 77.23, with $R_1 = 38.3k\Omega$, $R_2 = 19.6k\Omega$, $R_3 = 13.3k\Omega$, and $C = 0.0475\mu F$. The frequency response of the lead compensator is shown in Figure 77.28. Equations 77.59 and 77.60 are the governing relationships. A high impedance summing junction is placed ahead of the lead compensator. The lead compensator is designed to yield approximately 30° of lead at a frequency of 650 rad/s. As shown in Equation 77.60, this goal was accomplished within an acceptable error band. An inverting amplifier is used in the compensator design resulting in a phase shift of $-180°$ in the Bode plot of Figure 77.28. This is corrected in the complete velocity compensator by the fact that the PI and lead compensators both invert their input signals, thus canceling the effects of the inversion when they are cascaded. Figure 77.29 represents the final complete velocity compensator.

All of the components shown for this controller are commer-

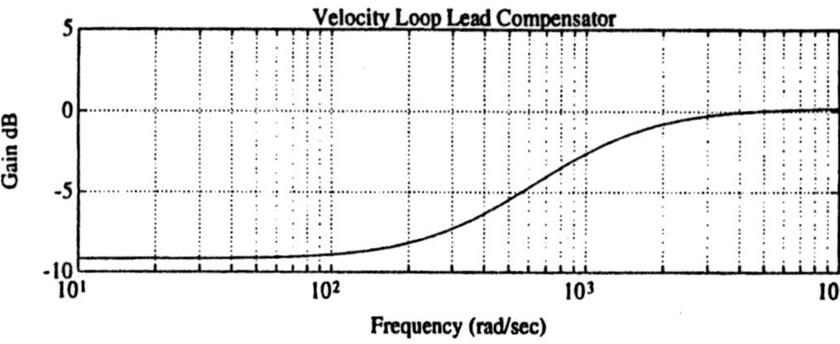

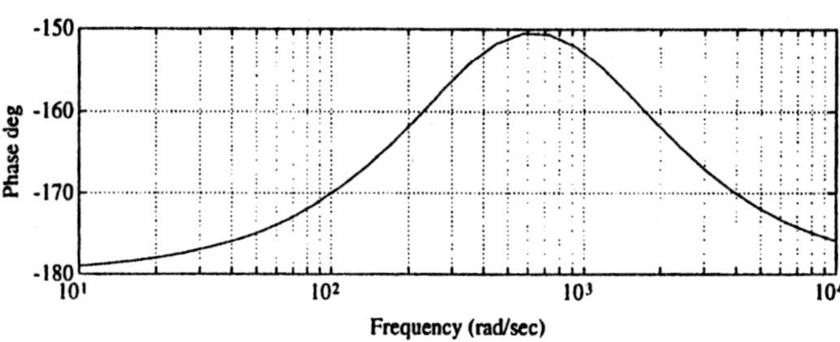

Figure 77.28 Velocity loop lead compensator frequency response.

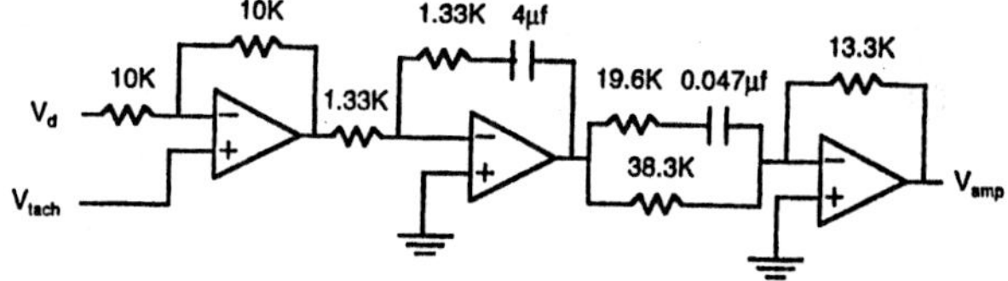

Figure 77.29 Velocity loop complete compensator.

cially available, a critical point in designing a compensator. If the components are not commercially available, special (and expensive) components would have to be fabricated. Thus, it is critical to generate designs that can be realized with commercially available resistors and capacitors.

$$\frac{V_0}{V_i} = \frac{1+0.00532s}{0.00532s} \tag{77.57}$$

$$\omega_{\text{break}} = \frac{1}{0.00532} = 188(\text{rad/s}); \left|\frac{V_0}{V_i}\right|_{\omega=\infty} = 1 \tag{77.58}$$

$$\frac{V_0}{V_i} = -0.347\frac{1+0.00272s}{1+0.000921s} \tag{77.59}$$

$$\begin{aligned} \phi_{\max} = 29.6°,\ \alpha &= \frac{0.000921}{0.00272} = \frac{368}{1085} = 0.339, \\ \omega_0 &= 632(\text{rad/s}) \end{aligned} \tag{77.60}$$

The open-loop Bode plots for the velocity loop are given in Figure 77.30. From the open-loop Bode plots, it is clear that all of the gain and phase margin specifications have been met for the system. Also, the low frequency response of the system indicates a type I system, thus there will be no steady-state error to a constant velocity input.

Now the forward loop gain is determined using the root locus and parametric plot techniques. Figure 77.31 presents the root locus of the system as the loop gain is increased. The lead compensator's final design can be assessed by using the Bode plots in conjunction with the root locus plot. The pole and zero furthest in the left hand plane are from the lead compensator. The second zero and the pole at the origin are from the PI compensator. Clearly from the plot, there is a range of gains that results

TABLE 77.7 Velocity Loop and Position Loop Controller Specification Summary

Specification	Velocity loop	Position loop
Min. closed-loop bandwidth	15 (Hz) ($\approx$95 rad/s)	150 (Hz) ($\approx$950 rad/s)
Steady-State error to a step	0	0
Min. damping ratio, ζ	0.8	0.9
Max. $\lvert S_k \rvert$	1.0	1.5
Min. phase margin, ϕ_m	90°	75°
Min. gain margin, g_m	40(dB)	60 (dB)

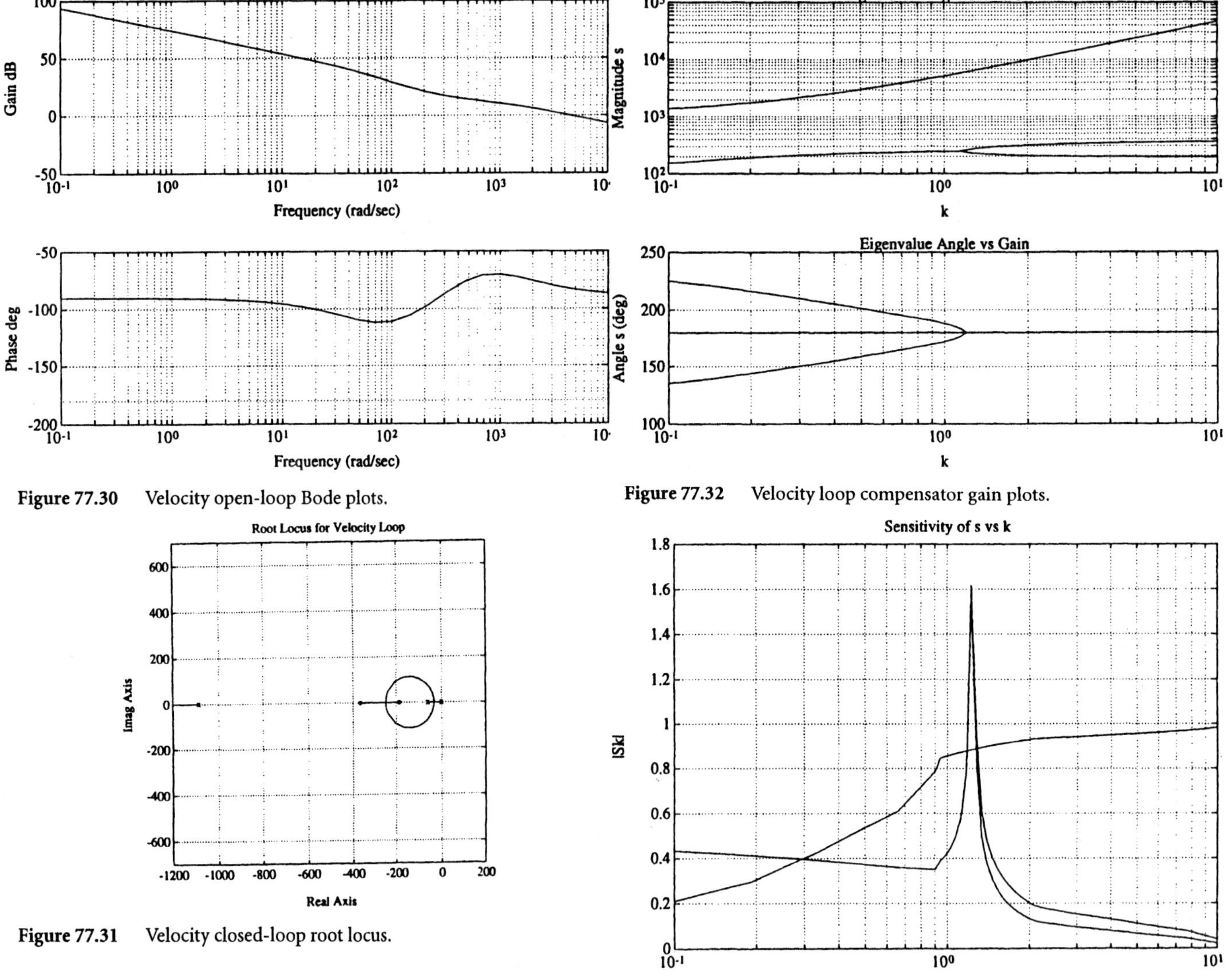

Figure 77.30 Velocity open-loop Bode plots.

Figure 77.31 Velocity closed-loop root locus.

Figure 77.32 Velocity loop compensator gain plots.

Figure 77.33 Velocity loop compensator root sensitivity.

in an underdamped system and may yield unacceptable damping for the velocity loop. To investigate further the appropriate choice of the forward loop gain, the root locus is shown as a set of gain plots, plotting pole magnitude and angle as functions of loop gain. Figure 77.32 depicts the velocity loop gain plots, showing various important aspects of the loop gain effects on the closed-loop system dynamics. In particular, low and high gain asymptotic behaviors are evident as well as the gain at the break point of the root locus. Furthermore, because the natural frequency and damping can be related to the eigenvalue (or pole) magnitude and angle, the effects of loop gain variations on the system response speed and damping may be easily visualized. Clearly, from the gain plots, gains of slightly more than unity result in an overdamped closed-loop system. This places a lower limit on the loop gain. Figure 77.33 plots the magnitude of the root sensitivity as a function of gain. The only unacceptable loop gains are those in the vicinity of the break points. Thus, the design must not use a gain in the region of $1 < K_v < 2$.

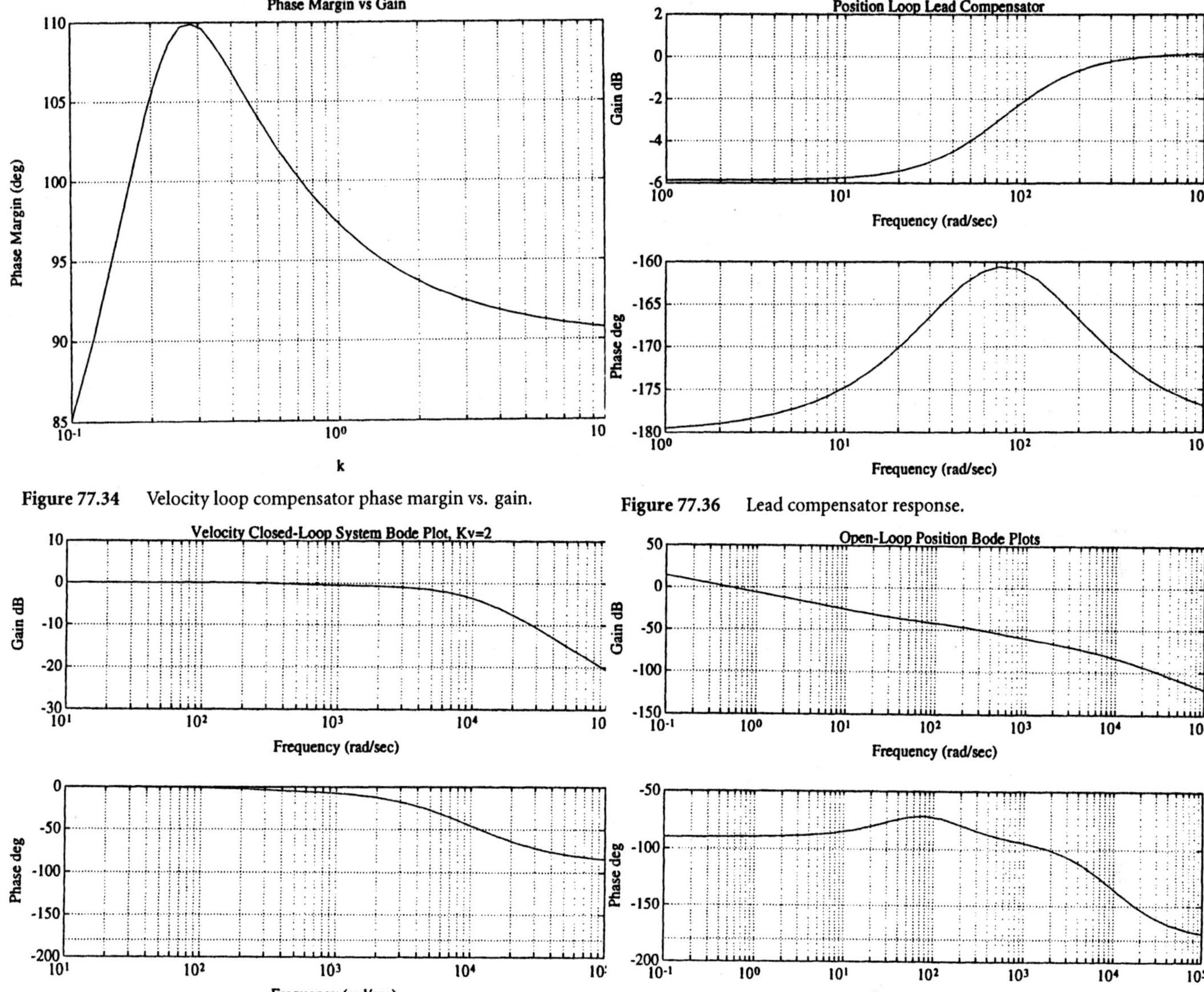

Figure 77.34 Velocity loop compensator phase margin vs. gain.

Figure 77.36 Lead compensator response.

Figure 77.35 Velocity loop compensator closed-loop response, $K_V = 2$.

Figure 77.37 Position open-loop frequency response.

Another parametric plot that can be generated is the phase margin as a function of loop gain (see Figure 77.34). Based on this plot, to achieve the appropriate phase margin the gain must be greater than 0.11. The maximum phase margin is at a gain of approximately 3. Also, at high gains, the phase margin approaches 90°, as expected, because the system has one more pole than zero. The gain margin of the system is infinite because the maximum phase shift is −90°. For this particular design, a value of 2 was chosen for the loop gain. It should be noted that the upper limit on the gain is power related, because higher gains require larger amounts of power. Thus, the target gain achieves all specifications with a minimal value. Figure 77.35 is the closed-loop Bode plot for the velocity loop.

Although gain plots and parametric plots are not necessary for the actual design of the compensators (in particular, the loop gain), they do offer insight into the various trade-offs available. For example, it is clear from the gain plots that gains much higher than 5 will not significantly affect the closed loop system response because both lower frequency poles are approaching the system transmission zeros. Higher gains only push the high frequency pole (which is dominated by the lower frequency dynamics) further out into the left hand plane. Thus higher gains, which are costly to implement, do not significantly enhance the closed-loop system dynamics beyond a certain limit and may excite higher frequency modes.

Compensator Design: Position Loop

The position loop is designed similarly to the velocity loop by determining values for the components of the lead compensator and the loop gain of the proportional controller. The final design of the lead compensator has the same form as Figure 77.23, with $R_1 = 42.2k\Omega$, $R_2 = 42.2k\Omega$, $R_3 = 21.5k\Omega$, and $C = 0.22\mu F$, and Equations 77.50 and 77.51 show its de-

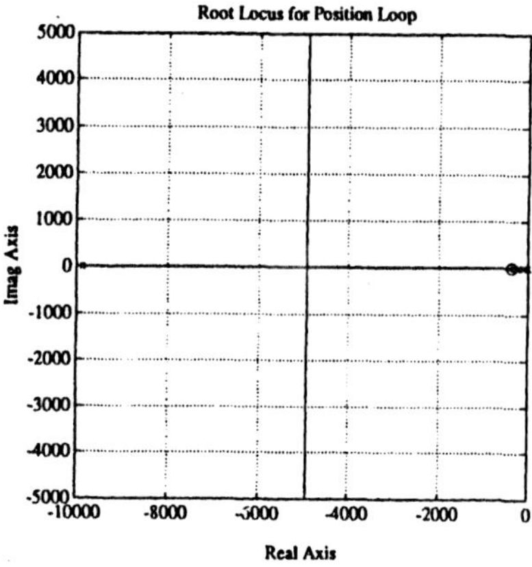

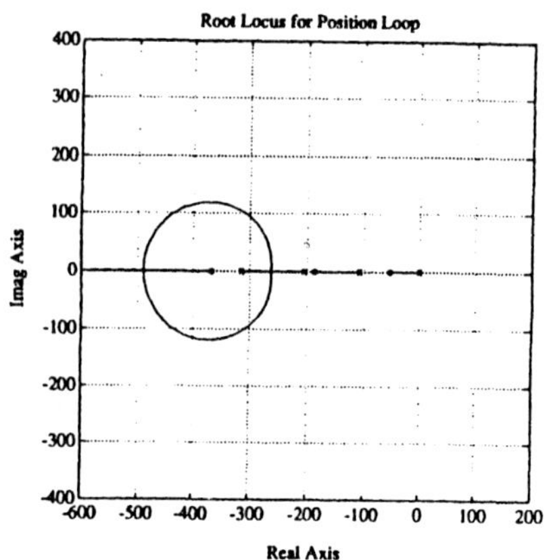

Figure 77.38 (a) and (b) - Position loop root locus.

sign values. The objective for the compensator is a maximum lead of 20° at a frequency of 75 rad/s. Figure 77.36 shows the lead compensator frequency response and Figure 77.37 plots the position open-loop response with unity gain on the proportional controller. Once again notice the low frequency slope of -20 dB/decade on the magnitude plot and $-90°$ phase in the phase plot, indicate the natural free integrator in the position loop.

$$\frac{V_0}{V_i} = -0.510\frac{1+0.0186s}{1+0.00928s}. \tag{77.61}$$

$$\begin{aligned} \phi_{\max} = 19.5°,\ \omega_0 &= 76.2(\text{rad/s}), \\ \alpha &= \frac{0.00928}{0.0186} = \frac{53.8}{108} = 0.497. \end{aligned} \tag{77.62}$$

To determine the appropriate position loop gain, K_p, for the system, a root locus plot in conjunction with gain plots and other parametric plots is used. Figure 77.38a is the root locus for the position loop. Figure 77.38b is a close-up view of the dominant poles and zeros located close to the origin. The pole at the origin is the natural integrator of the system, the other poles and zeros are from the lead compensator and the dynamics of the velocity loop. Clearly, from the root locus, there is a range of gains that yields an unacceptable damping ratio for this system. The gain plots of the position loop, shown in Figure 77.39, indicate that gains of greater than approximately 3000 are unacceptable. It is worthwhile noting that the gain plots show the two break-out points (at gain values of approximately 75 and 2600) and the break-in point at an approximate gain of 590. Due to power limitations, the gain of the position loop is limited to a maximum of 1000. Thus, the root locus shown in Figure 77.38b is more appropriate for the remainder of this design. Figure 77.40 shows the root sensitivity of the position loop. From the root sensitivity plot, it is clear that values of gain close to the break points on the root locus result in root sensitivities that do not meet specifications. Figure 77.40, also shows that the ranges $600 \le K_p \le 1000$ and $2000 \le K_p \le 3000$ are unacceptable.

Gain and phase margin can also be plotted as functions of the forward loop gain (Figures 77.41 and 77.42). The slope of the gain margin plot is -20 dB/decade as gain margin is directly related to the gain. Note that the maximum phase margin occurs at a gain of approximately 100. Based on these two plots, the gain must be lower than approximately 3000.

From the plots above, loop gains of either 300 or 1000 may be employed for the position compensator. The closed-loop frequency response is plotted in Figures 77.43 and 77.44 from the gain values of 300 and 1000, respectively. Either one of these designs results in an acceptable performance. Note that the higher value of gain yields a higher bandwidth system.

Summary Results

To summarize the results from the velocity and position compensator designs, the Tables 77.8 – 77.10 can be generated.

As can be seen from Tables 77.8 – 77.10, the performance objectives have been met for both the velocity and position loops. The only decision that remains is whether to use the lower or higher value for the position loop gain. For this particular case, it was decided to use the lower value which required a smaller and less expensive power supply.

77.3.6 Conclusions

Ultra-high precision systems are becoming more commonplace in the industrial environment as competition increases the demand for higher precision at greater speeds. The two most important factors involved in the successful implementation of an ultra-high precision system are a solid design of the open-loop system and a good model of the system and any compensator components. This chapter presented some basic concepts critical for the design and implementation of ultra-high precision control systems, including some instrumentation concerns, system identification issues and compensator design and implementation.

Clearly, this chapter is limited in its scope and there are many other details that must be considered before a machine can approach the accuracies of high performance coordinate measurement machines or diamond turning machines. Such machines have been developed over many years and continue to improve with time and technology. The interested reader is referred to the references at the end of this chapter for further details of designing and implementating ultra-high precision systems.

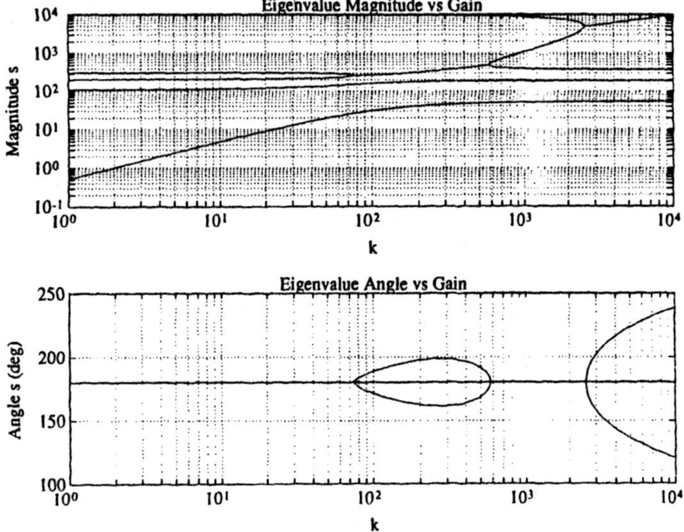

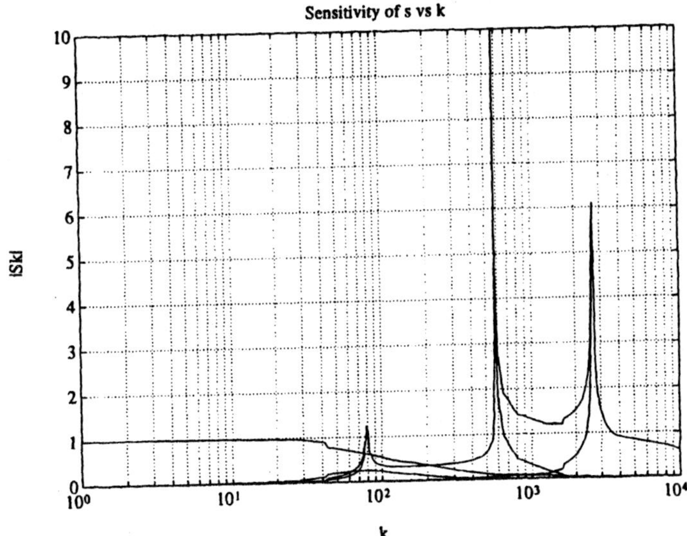

Figure 77.39 Position loop compensator gain plots.

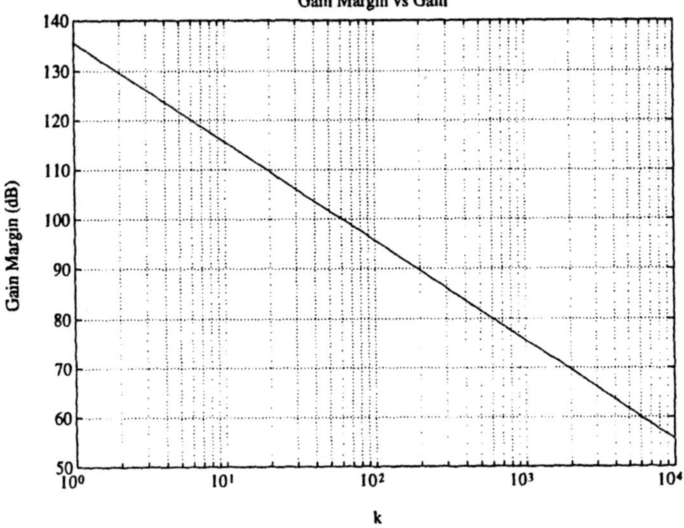

Figure 77.40 Position loop compensator root sensitivity.

Figure 77.41 Position loop gain margin.

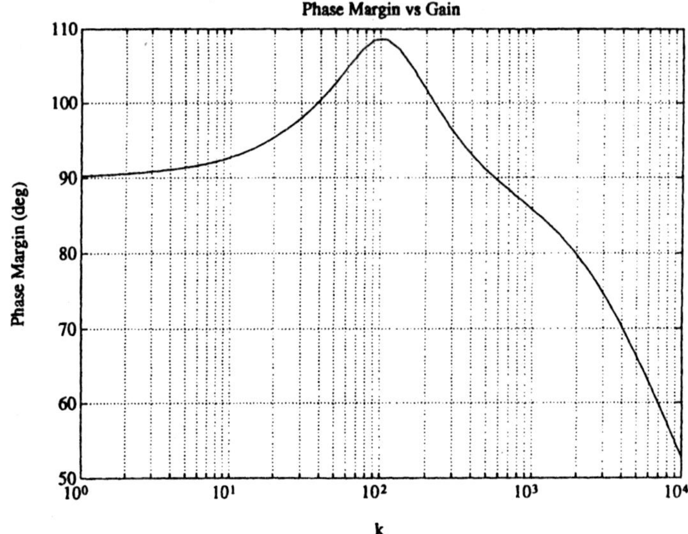

Figure 77.42 Position loop phase margin.

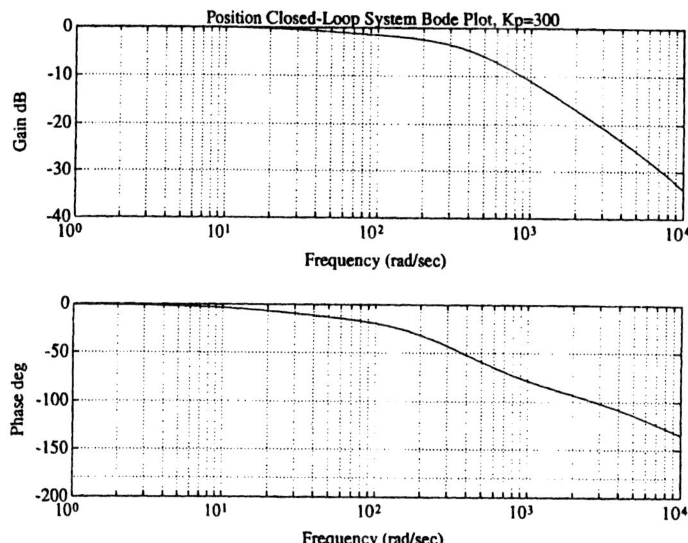

Figure 77.43 Position closed-loop frequency response, $K_p = 300$.

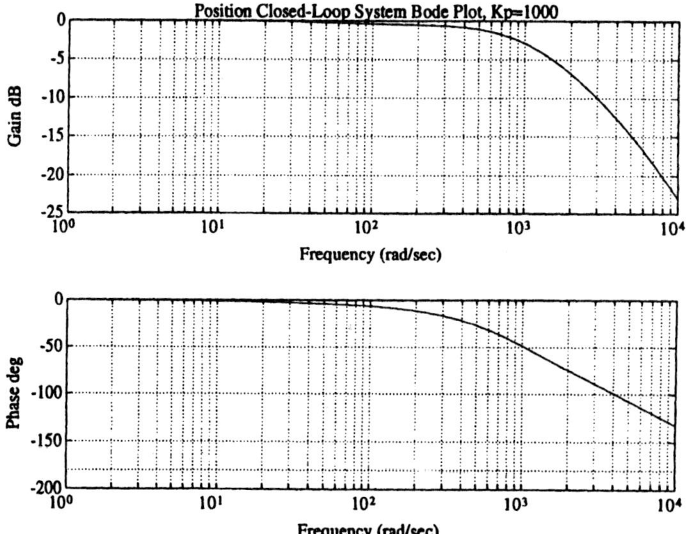

Figure 77.44 Position closed-loop frequency response, $K_p = 1000$.

TABLE 77.8 Results Velocity Loop

Specification	Design target	Value achieved
Min. closed-loop bandwidth	150 (Hz) ($\approx$950 rad/s)	637 (Hz) ($\approx$4000 rad/s)
Steady-State error to a step	0	0
Min. damping ratio, ζ	0.8	1.0
Max. $\|S_k\|$	1.0	0.92
Min. phase margin, ϕ_m	90°	93°
Min. gain margin, g_m	40 (dB)	∞ (dB)

TABLE 77.9 Position Loop $K_p = 300$

Specification	Design target	Value achieved
Min. closed-loop bandwidth	15 (Hz) ($\approx$95 rad/s)	32 (Hz) ($\approx$200 rad/s)
Steady-State error to a step	0	0
Min. damping ratio, ζ	0.9	0.93
Max. $\|S_k\|$	1.5	0.4
Min. phase margin, ϕ_m	75°	96°
Min. gain margin, g_m	60(dB)	87 (dB)

TABLE 77.10 Position Loop $K_p = 1000$

Specification	Design target	Value achieved
Min. closed-loop bandwidth	15 (Hz) ($\approx$95 rad/s)	160 (Hz) ($\approx$1000 rad/s)
Steady-State error to a step	0	0
Min. damping ratio, ζ	0.9	1.0
Max. $\|S_k\|$	1.5	1.2
Min. phase margin, ϕ_m	75°	85°
Min. gain margin, g_m	60(dB)	76 (dB)

77.3.7 Defining Terms

A quad B: Two phased (90 degrees) signals from an encoder.

A/D: Analog to digital conversion.

Abbe error: Measurement error which occurs when a gage is not collinear to the object measured.

aliasing: Identification of a higher frequency as a lower one, when the higher frequency is above the Nyquist frequency.

ARX: AutoRegressive eXternal input identification technique.

CMM: Coordinate measuring machine.

CNC: Computer numerical control.

cross-coupling: One axis affecting another.

D/A: Digital to analog conversion.

DTM: Diamond turning machine.

dynamometer: High precision and bandwidth multiaxis force sensor.

laser interferometer: Measurement instrument based on light wavelength interference.

LVDT: Linear variable differential transformer.

metglass: Magnetostrictive material (such as Terfenol^R).

metrology: The study of measurement.

piezoelectric effect: Strain-induced voltage, or voltage-induced strain.

stiction: Low velocity friction.

THD: Total harmonic distortion.

ultra-high precision: Dimensional accuracies of 0.1 μin (1 μin = 10^{-6} in).

white noise: Random signal with a uniform probability distribution.

References

[1] Methods for Performing Evaluation of Computer Numerically Controlled Machining Centers. *ANSI-ASME B5.54*, 1992.

[2] Blake, P., Bifano, T., Dow, T., and Scattergood, R., Precision Machining Of Ceramic Materials, *Am. Ceram. Soc. Bull.*, 67(6), 1038–1044, 1988.

[3] Bryan, J. B., The Abbe Principle Revisited-An Updated Interpretation, *Precision Eng.*, 1(3), 129–132, 1989.

[4] Bryan, J. B., The Power of Deterministic Thinking in Machine Tool Accuracy, UCRL-91531, 1984.

[5] Bryant, M.D. and Reeves, R.B., Precise Positioning Problems Using Piezoelectric Actuators with Force Transmission Through Mechanical Contact, *Precision Eng.*, 6, 129–134, 1984.

[6] Cetinkunt, S. and Donmez, A., CMAC Learning Controller for Servo Control of High Precision Machine Tools, *American Control Conference*, 2, 1976–80, 1993.

[7] Daneshmend, L.K. and Pak, H.A., Model Reference Adaptive Control of Feed Force in Turning, *J. Dyn. Syst. Meas. Control*, 108, 215–222, 1986.

[8] Dorf, R.C. and Kusiak, A., Eds., *Handbook of Design, Manufacturing and Automation*, John Wiley & Sons, New York, 1994.

[9] Falter, P. J. and Dow, T. A., Design And Performance Of A Small-Scale Diamond Turning Machine, *Precision Eng.*, 9(4), 185–190, 1987.

[10] Fornaro, R. J. and Dow, T. A., High-performance Machine Tool Controller, *Conf. Rec. - IEEE Ind. Appl. Soc. Annual Meeting*, 35(6), 1429–1439, 1988.

[11] Fussell, B.K. and Srinivasan, K., Model Reference Adaptive Control of Force In End Milling Operations, *Am. Control Conf.*, 2, 1189–94, 1988.

[12] Jenkins, H.E., Kurfess, T.R., and Dorf, R.C., Design of a Robust Controller for a Grinding System, *IEEE Conf. Control Appl.*, 3, 1579–84, 1994.

[13] Jeong, S. and Ro, P.I., Cutting Force-Based Feedback Control Scheme for Surface Finish Improvement in Diamond Turning, *Am. Control Conf.*, 2, 1981–1985, 1993.

[14] Jones, R. and Richardson, J., The Design and Application of Sensitive Capacitance Micrometers, *J. Phys. E: Scientific Instruments*, 6, 589, 1973.

[15] Kurfess, T. R. and Nagurka, M. L., Understanding the Root Locus Using Gain Plots, *IEEE Control Syst. Mag.*, 11(5), 37–40, 1991.

[16] Kurfess, T.R. and Nagurka, M.L., A Geometric Representation of Root Sensitivity, *J. Dyn. Syst. Meas. Control*, 116(2), 305–9, 1994.

[17] Li, C. J. and Li, S. Y., A New Sensor for Real-Time Milling Tool Condition Monitoring, *J. Dyn. Syst. Meas. Control*, 115, 285–290, 1993.

[18] Liang, S.Y. and Perry, S.A., In-Process Compensation For Milling Cutter Runout Via Chip Load Manipulation, *J. Eng. Ind. Trans. ASME*, 116(2), 153–160, 1994.

[19] Ljung, L., *System Identification: Theory for the User*, Prentice-Hall, Englewood Cliffs, NJ, 1987.

[20] Lo, C. and Koren, Y., Evaluation of Machine Tool Controllers, *Proc. Am. Control Conf.*, 1, 370–374, 1992.

[21] Meckl, P. H. and Kinceler, R., Robust Motion Control of Flexible Systems Using Feedforward Forcing Functions, *IEEE Trans. Control Syst. Technol.*, 2(3), 245–254, 1994.

[22] Pien, P.-Y. and Tomizuka, M., Adaptive Force Control of Two Dimensional Milling, *Proc. Am. Control Conf.*, 1, 399–403, 1992.

[23] Slocum, A., Scagnetti, P., and Kane, N., Ceramic Machine Tool with Self-Compensated, Water-Hydrostatic, Linear Bearings, *ASPE Proc.*, 57–60, 1994.

[24] Stute, G., Adaptive Control, *Technology of Machine Tools*, Machine Tool Controls, Lawrence Livermore Laboratory, Livermore, CA, 1980, vol. 4.

[25] Tsao, T.C. and Tomizuka, M., Adaptive and Repetitive

Digital Control Algorithms for Noncircular Machining, *Proc. Am. Control Conf.*, 1, 115–120, 1988.

[26] Tung, E. and Tomizuka, M., Feedforward Tracking Controller Design Based on the Identification of Low Frequency Dynamics, *J. Dyn. Syst. Meas. Control*, 115, 348–356, 1993.

[27] Tung, E., Anwar, G., Tomizuka, M., Low Velocity Friction Compensation and Feedforward Solution Based on Repetitive Control, *J. Dyn. Syst. Meas. Control*, 115, 279–284, 1993.

[28] Ulsoy, A.G. and Koren, Y., Control of Machine Processes, *J. Dyn. Syst. Meas. Control*, 115, 301–8, 1993.

77.4 Robust Control of a Compact Disc Mechanism

Maarten Steinbuch, Philips Research Laboratories, Eindhoven, The Netherlands

Gerrit Schootstra, Philips Research Laboratories, Eindhoven, The Netherlands

Okko H. Bosgra, Mechanical Engineering Systems and Control Group, Delft University of Technology, Delft, The Netherlands

77.4.1 Introduction

A compact disc (CD) player is an optical decoding device that reproduces high-quality audio from a digitally coded signal recorded as a spiral-shaped track on a reflective disc [2]. Apart from the audio application, other optical data systems (CD-ROM, optical data drive) and combined audio/video applications (CD-interactive, CD-video) have emerged. An important research area for these applications is the possibility of increasing the rotational frequency of the disc to obtain faster data readout and shorter access time. For higher rotational speeds, however, a higher servo bandwidth is required that approaches the resonance frequencies of bending and torsional modes of the CD mechanism. Moreover, the system behavior varies from player to player because of manufacturing tolerances of CD players in mass production, which explains the need for robustness of the controller.

Further, an increasing percentage of all CD-based applications is for portable use. Thus, additionally, power consumption and shock sensitivity play a decisive role in the performance assessment of controller design for CD systems.

In this chapter we concentrate on the possible improvements of both the track-following and focusing behavior of a CD player, using robust control design techniques.

77.4.2 Compact Disc Mechanism

A schematic view of a CD mechanism is shown in Figure 77.45. The mechanism is composed of a turntable dc motor for the rotation of the CD, and a balanced radial arm for track following. An optical element is mounted at the end of the radial arm. A diode located in this element generates a laser beam that passes through a series of optical lenses to give a spot on the information layer of the disc. An objective lens, suspended by two parallel leaf springs, can move in a vertical direction to give a focusing action.

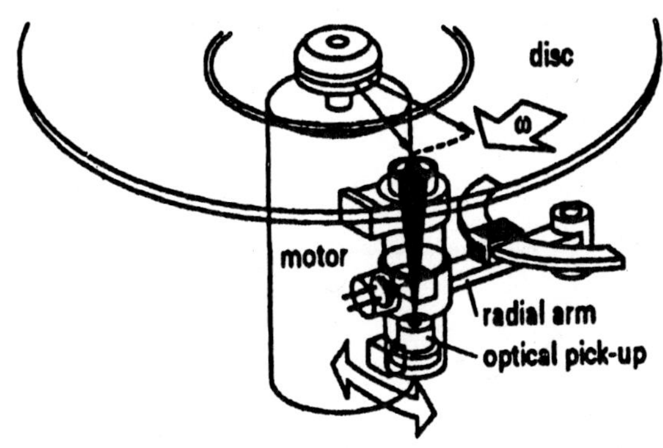

Figure 77.45 Schematic view of a rotating-arm compact disc mechanism.

Both the radial and the vertical (focus) position of the laser spot, relative to the track of the disc, have to be controlled actively. To accomplish this, the controller uses position-error information provided by four photodiodes. As input to the system, the controller generates control currents to the radial and focus actuator, which both are permanent-magnet/coil systems.

In Figure 77.46 a block diagram of the control loop is shown. The difference between the radial and vertical track position and the spot position is detected by the optical pickup; it generates a radial error signal (e_{rad}) and a focus error signal (e_{foc}) via the optical gain K_{opt}. A controller $K(s)$ feeds the system with the currents I_{rad} and I_{foc}. The transfer function from control currents to spot position is indicated by $H(s)$. Only the position-error signals after the optical gain are available for measurement. Neither the true spot position nor the track position is available as a signal.

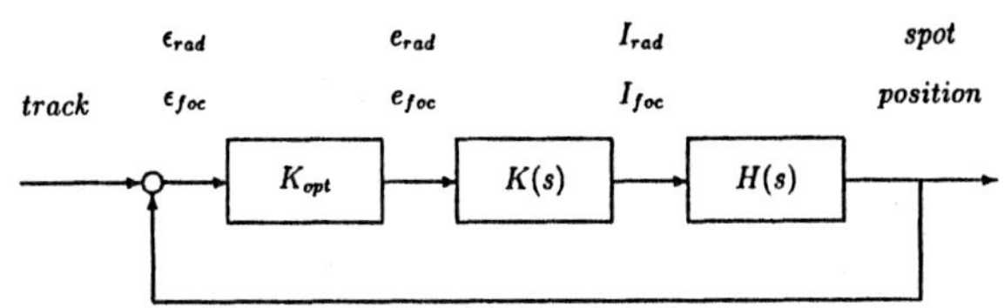

Figure 77.46 Configuration of the control loop.

In current systems, $K(s)$ is formed by two separate proportional-integral-derivative (PID) controllers [2], [6], thus, creating two single-input single-output (SISO) control loops. This is possible because the dynamic interaction between both loops is relatively low, especially from radial current to focus error. In these applications the present radial loop has a bandwidth of 500 Hz, while the bandwidth for the focus loop is 800 Hz. For more demanding applications (as discussed in the introduction) it is necessary to investigate whether improvements of the servo behavior are possible.

77.4.3 Modeling

A measured frequency response of the CD mechanism $G(s) = K_{opt}H(s)$ is given in Figure 77.47 (magnitude only). It has been determined by spectrum analysis techniques. At low frequencies, the rigid body mode of the radial arm and the lens-spring system (focus) can be easily recognized as double integrators in the 1,1 and 2,2 element of the frequency response, respectively. At higher frequencies the measurement shows parasitic dynamics, especially in the radial direction. Experimental modal analyses and finite element calculations have revealed that these phenomena are due to mechanical resonances of the radial arm, mounting plate, and disc (flexible bending and torsional modes).

With frequency-domain-based system identification, each element of the frequency response has been fitted separately using an output error model structure with a least-square criterion [10]. Frequency-dependent weighting functions have been used to improve the accuracy of the fit around the anticipated bandwidth of 1 kHz. The 2,1 element appeared to be difficult to fit because of the nonproper behavior in the frequency range of interest.

Combination of the fits of each element resulted in a 37th-order multivariable model. Using frequency-weighted balanced reduction [13], [3], this model was reduced to a 21st-order model, without significant loss in accuracy. The frequency response of the model is also shown in Figure 77.47.

Uncertainty Modeling

The most important system variations we want to account for are

1. Unstructured difference between model and measurement
2. Uncertain interaction
3. Uncertain actuator gain
4. Uncertainty in the frequencies of the parasitic resonances

The first uncertainty stems from the fact that our nominal model is only an approximation of the measured frequency response because of imperfect modeling. Further, a very-high-order nominal model is undesirable in robust control design since the design technique yields controllers with the state dimension of the nominal model plus weighting functions. For that reason, our nominal model describes only the rigid body dynamics, as well as the resonance modes that are most relevant in the controlled situation. Unmodeled high-frequency dynamics and the unstructured difference between model and measurement are modeled as a complex valued additive perturbation Δ_a, bounded by a high-pass weighting function.

The remaining uncertainty sources 2, 3 and 4 are all intended to reflect how manufacturing tolerances manifest themselves as variations in the frequency response from player to player. By so doing, we are able to appreciate the consequences of manufacturing tolerances on control design.

The uncertain interaction, item 2, is modeled using an antidiagonal output multiplicative parametric perturbation: $y = (I + \Delta_o)Cx$ where

$$\Delta_o = \begin{bmatrix} 0 & w_{o1}\delta_{o1} \\ w_{o2}\delta_{o2} & 0 \end{bmatrix}$$

The scalar weights w_{o1} and w_{o2} are chosen equal to 0.1, meaning 10% uncertainty.

Dual to the uncertain interaction, the uncertain actuator gains, item 3, are modeled as a diagonal input multiplicative parametric perturbation: $\dot{x} = Ax + B(I + \Delta_i)u$ where

$$\Delta_i = \begin{bmatrix} w_{i1}\delta_{i1} & 0 \\ 0 & w_{i2}\delta_{i2} \end{bmatrix} \tag{77.63}$$

The gain of each actuator is perturbed by 5%. With this value also non-linear gain variations in the radial loop due to the rotating-arm principle are accounted for along with further gain variations caused by variations in track shape, depth, and slope of the pits on the disc and varying quality of the transparent substrate protecting the disc [6].

Finally, we consider variations in the undamped natural frequency of parasitic resonance modes, item 4. From earlier robust control designs [11] it is known that the resonances at 0.8, 1.7, and 4.3 kHz are especially important. The modeling of the variations of the three resonance frequencies is carried out with the parametric uncertainty modeling toolbox [9]. The outcome of the toolbox is a linear fractional transformation description of the perturbed system with a normalized, block diagonal, parametric perturbation structure $\Delta_{par} = diag\{\delta_1 I_2, \delta_2 I_2, \delta_3 I_2\}$. Each frequency perturbation involves a real-repeated perturbation block of multiplicity two. The repeatedness stems from the fact that the frequencies ω_0 appear quadratically in the A-matrix. The lower and upper bounds are chosen 2.5% below and above the nominal value, respectively; see [8] for more details.

77.4.4 Performance Specification

A major disturbance source for the controlled system is track position irregularities. Based on standardization of CDs, in the radial direction the disc specifications allow a track deviation of 100 μm (eccentricity) and a track acceleration at scanning velocity of 0.4 m/s^2, while in the vertical direction these values are 1 mm and 10 m/s^2, respectively.

Apart from track position irregularities, a second important disturbance source is external mechanical shocks. Measurements show that during portable use disturbance signals occur in the frequency range from 5 up to 150 Hz, with accelerations (of the chassis) up to 50 m/s^2.

For the CD player to work properly, the maximum allowable position error is 0.1 μm in the radial direction and 1 μm in the focus direction. In the frequency domain, these performance specifications can be translated into requirements on the shape of the (output) sensitivity function $S = (I + GK)^{-1}$. Note that the track irregularities involve time-domain constraints on signals, which are hard to translate into frequency-domain specifications. To obtain the required track disturbance attenuation, the magnitude of the sensitivity at the rotational frequency should be less

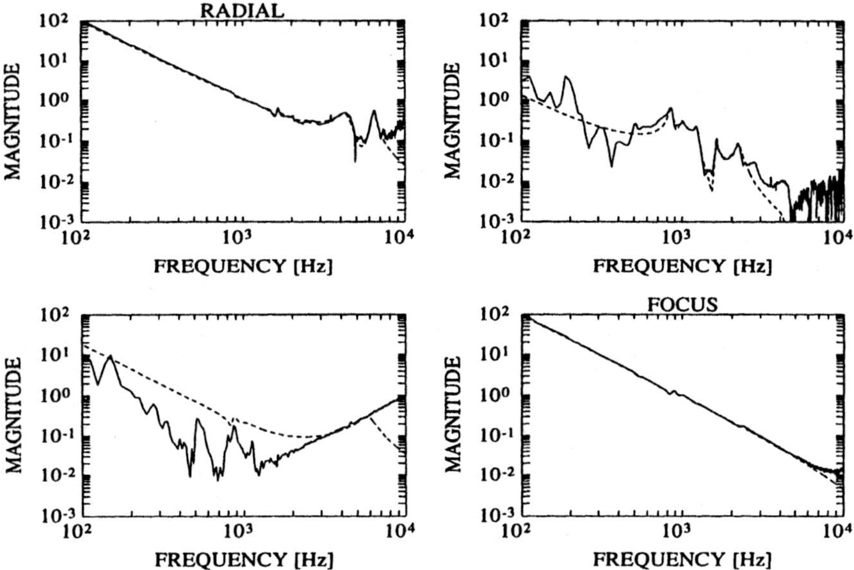

Figure 77.47 Measured frequency response of the CD mechanism (—) and of the identified 21st-order model (- -).

than 10^{-3} in both the radial and the focus direction. Further, for frequencies up to 150 Hz, the sensitivity should be as small as possible to suppress the impact of mechanical shocks and higher harmonics of the disc eccentricity.

To still satisfy these requirements for increasing rotational speed, the bandwidth has to be increased. As stated before, a higher bandwidth has implications for robustness of the design against manufacturing tolerances. But also, a higher bandwidth means a higher power consumption (very critical in portable use), generation of audible noise by the actuators, and poor playability of discs with surface scratches. Therefore, under the required disturbance rejection, we are striving towards the lowest possible bandwidth.

In addition to these conflicting bandwidth requirements, the peak magnitude of the sensitivity function should be less than 2 to create sufficient phase margin in both loops. This is important because the controller has to be discretized for implementation. Note that the Bode sensitivity integral plays an important role in the trade-off between the performance requirements.

77.4.5 μ-synthesis for the CD Player

To start with, the performance specifications on S can be combined with the transfer function KS associated with the complex valued additive uncertainty Δ_a. The performance trade-offs are then realized with a low-pass weighting function W_1 on S, while W_2 on KS reflects the size of the additive uncertainty and can also be used to force high roll-off at the input of the actuators.

In this context, the objective of achieving robust performance [1], [5] means that for all stable, normalized perturbations Δ_a the closed loop is stable and

$$\| W_1[I + (G + W_2\Delta_a)K]^{-1} \|_\infty < 1$$

This objective is exactly equal to the requirement that

$$\mu_\Delta[F_l(P, K)] < 1$$

with

$$F_l(P, K) = \begin{bmatrix} W_2KS & W_2KS \\ W_1S & W_1S \end{bmatrix} \qquad (77.64)$$

and μ_Δ is computed with respect to the structured uncertainty block $\Delta = diag(\Delta_a, \Delta_p)$, in which Δ_p represents a fictitious 2×2 performance block. Note that this problem is related to the more common H_∞ problem

$$\left\| \begin{matrix} W_2KS \\ W_1S \end{matrix} \right\|_\infty < 1$$

However, with this H_∞ design formulation it is possible to design only for nominal performance and robust stability.

An alternative that we use is to specify performance on the transfer function SG instead of S, reducing the resulting controller order by 4 [11]. Then the following standard plant results

$$F_l(P, K) = \begin{bmatrix} W_2KS & W_2KSG \\ W_1S & W_1SG \end{bmatrix} \qquad (77.65)$$

This design formulation has the additional advantage that it does not suffer from pole-zero cancellation as in the mixed sensitivity design formulations as in Equation 77.64.

Using the D-K iteration scheme [1], [5], μ-controllers are synthesized for several design problems. Starting with the robust performance problem of Equation 77.65, the standard plant is augmented step by step with the parametric perturbations 2, 3, and 4 listed earlier. In the final problem, with the complete uncertainty model, we thus arrive at a robust performance problem having nine blocks: $\Delta = diag\{\delta_1 I_2, \delta_2 I_2, \delta_3 I_2, \delta_{i1}, \delta_{i2}, \delta_{o1}, \delta_{o2}, \Delta_a, \Delta_p\}$.

The most important conclusions with respect to the use of μ-synthesis for this problem are

- Convergence of the D-K iteration scheme was fast (in most cases two steps). Although global convergence of the scheme cannot be guaranteed, in our case the resulting μ-controller did not depend on the starting controller.

- The final μ-controllers did have high order [39 for design (Equation 77.65) up to 83 for the full problem] due to the dynamic D-scales associated with the perturbations.
- Although most of the perturbations are real valued by nature, the assumption in design that all perturbations are complex valued did not introduce much conservativeness with respect to robust performance; see Figure 77.48.

  Note that the peak value of μ over frequency is 1.75, meaning that robust performance has not been achieved. This is due to the severe performance weighting W_1.
- The most difficult aspect of design appeared to be the shaping of weighting functions such that a proper trade-off is obtained between the conflicting performance and robustness specifications.

In Figure 77.49 the sensitivity transfer function is shown for the full problem μ-controller. For comparison, the result is also shown for two decentralized PID controllers achieving 800-Hz bandwidth in both loops. Clearly, the μ-controller achieves better disturbance rejection up to 150 Hz, has lower interaction, and a lower sensitivity peak value.

The controller transfer functions are given in Figure 77.50. Clearly, the μ-controller has more gain at low frequencies and actively acts upon the resonance frequencies.

77.4.6 Implementation Results

The digital implementation brings along a choice for the sampling frequency of the discretized controller. Based on experience with previous implementations of SISO radial controllers [11] and on the location of the fastest poles in the multiple-input multiple-output (MIMO) μ-controller (± 8 kHz), it is the intention to discretize the controller at 40 kHz. However, in the DSP environment used, this sampling frequency means that the order of the controller that can be implemented is, at most, 8. Higher order will lead to an overload of the DSP. This indicates a need for a dramatic controller order reduction since there is a large gap between the practically allowable controller order (8) and the controller order that have been found using the μ-synthesis methodology (83). It is, however, unlikely that the order of these controllers can be reduced to 8 without degrading the nominal and robust performance very much. To be able to implement more complex controllers there are two possibilities:

1. Designing for a sampling frequency below 40 kHz
2. Using more than one DSP system and keeping the sampling frequency at 40 kHz

In this research we chose the latter, for it is expected that the sampling frequency has to be lowered to such an extent that the performance will degrade too much (too much phase lag around the bandwidth leading to an unacceptable peak value in the sensitivity function). Although the second option introduces some additional problems, it probably gives a better indication of how much performance increase is possible with respect to PID-like control without letting this performance be influenced too much by the restrictions of implementation. To do this, model reduction of the controller is applied. The first step involves balancing and truncating of the controller states. This reduces the number of states from 83 to 53, without loss of accuracy. The next step is to split up the μ-controller into two 1-input 2-output parts: $K_{53} = \begin{bmatrix} K_{53}^1 & K_{53}^2 \end{bmatrix}$. Using a frequency-weighted closed-loop model reduction technique, the order of each of these single input multiple-output (SIMO) parts can be reduced while the loop around the other part is closed. Each of the reduced-order parts can be implemented at 40 kHz, in a separate DSP system, thus facilitating a maximum controller order of 16 (under the additional constraint that each part can be, at most, of 8th order). The resulting controller is denoted $K_8 = \begin{bmatrix} K_8^1 & K_8^2 \end{bmatrix}$.

The actual implementation of the controllers in the DSP systems has been carried out with the dSPACE Cit-Pro software [7]. Most problems occurring when implementing controllers are not essential to control theory, and most certainly not for H_∞ and μ theory, but involve problems such as scaling inputs and outputs to obtain the appropriate signal levels and to ensure that the resolution of the DSP is optimally used. These problems can be solved in a user-friendly manner using the dSPACE software.

The two DSP systems used are a Texas Instruments TMS320C30 16-bit processor with floating-point arithmetic and a Texas Instruments TMS320C25 16-bit processor with fixed-point arithmetic. The analog-to-digital converters (ADC) are also 16 bit and have a conversion time of 5 μs. The maximum input voltage is ± 10 V. The digital-to-analog converters (DAC) are 12 bit, have a conversion time of 3 μs, and also operate within the range of ± 10 V.

The first column $K_8^1(s) = [K_{11} \;\; K_{21}]^T$ has been implemented in the TMS320C30 processor at a sampling frequency of 40 kHz. The dSPACE software provides for:

- A ramp-invariant discretization
- A modal transformation
- Generation of C code that can be downloaded to the DSP

The second column $K_8^2(s) = [K_{21} \;\; K_{22}]^T$ is implemented in the TMS320C25 processor. Since this processor has fixed-point arithmetic, scaling is more involved; see also [12]. With the dSPACE software, the following steps have been carried out:

- A ramp-invariant discretization
- A modal transformation
- l_1 scaling for the nonintegrator states
- State scaling for the integrator states such that their contribution is, at most, 20%
- Generation of DSPL code that can be downloaded to the DSP

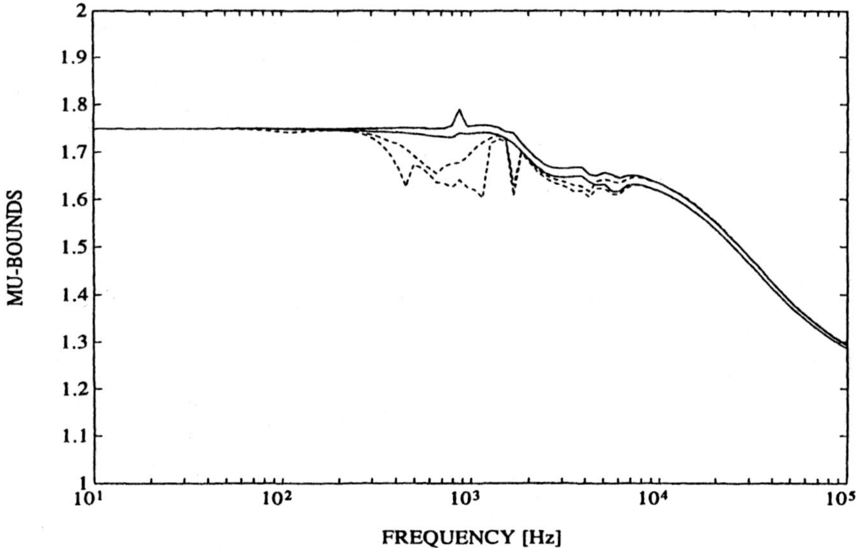

Figure 77.48 Complex (—) and real (- -) μ-bounds for the μ-controller on the standard plant, including all perturbations.

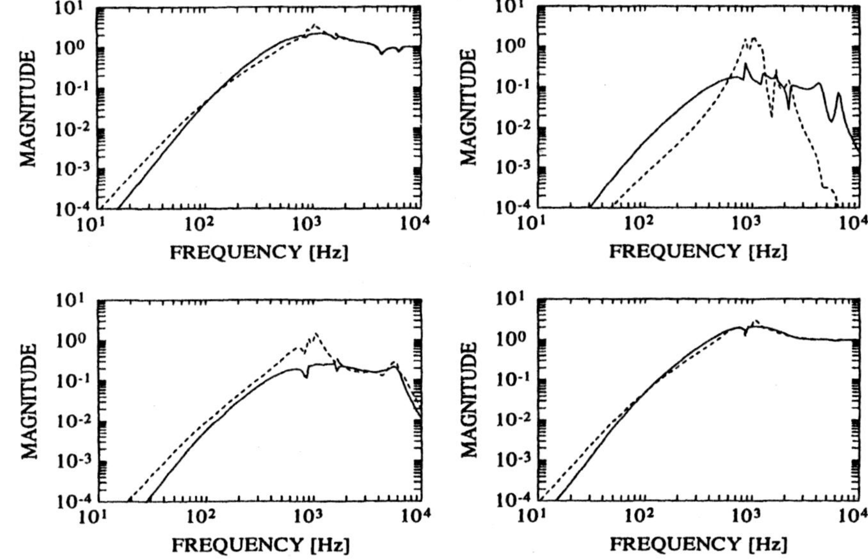

Figure 77.49 Nominal performance in terms of the sensitivity function with μ-controller (—) and two PID controllers (- -).

For the TMS320C25, a discretization at a sampling frequency of 40 kHz appeared to be too high since it resulted in a processor load of 115%. For that reason, the sampling frequency of this DSP has been lowered to 34 kHz, yielding a processor load of 98.6%.

The DSP systems have been connected to the experimental setup by means of two analog summing junctions that have been designed especially for this purpose; see Figure 77.51.

When the external 2×8 μ-controller of the complete design is connected to the experimental setup, we can measure the achieved performance in terms of the frequency response of the sensitivity function. The measurements have been carried out using a Hewlett Packard 3562 Dynamic Signal Analyzer. Because this analyzer can measure only SISO frequency responses, each of the four elements of the sensitivity function has been determined separately.

The measurements are started at the same position on the disc each time. This position is chosen approximately at the halfway point on the disc since the model is identified here and the radial gain is most constant in this region. In Figure 77.52 the measured and simulated frequency response of the sensitivity function is shown. The off-diagonal elements are not very reliable since the coherence during these measurements was very low because of small gains and nonlinearities, leading to bad signal-to-noise ratios.

The nominal performance has also been tested in terms of the possibility to increase the rotational frequency of the CD. It appeared possible to achieve an increase in speed of a factor 4 (with an open-loop bandwidth of 1 kHz).

Concluding this section, the measurements show that considerable improvements have been obtained in the suppression of track irregularities leading to the possibility of increasing the rotational frequency of the disc to a level that has not been achieved

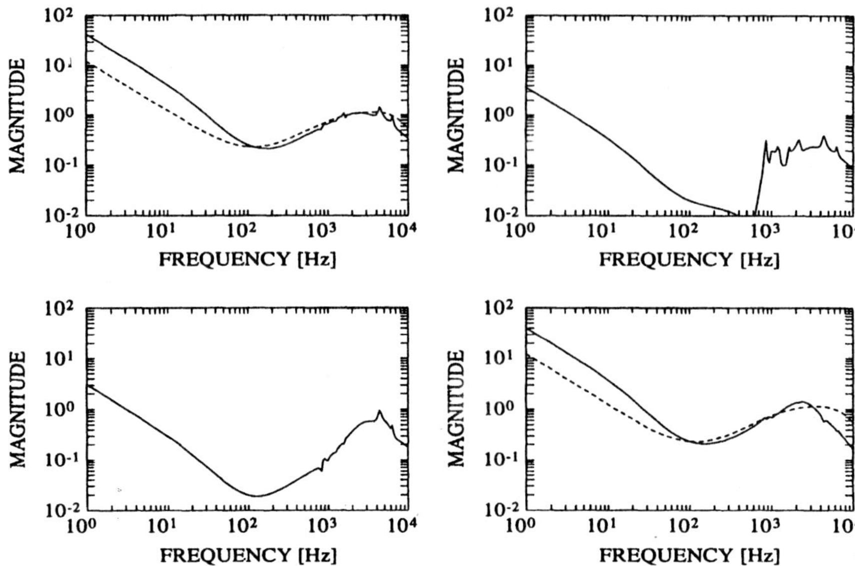

Figure 77.50 Frequency response of the μ-controller (—) and two PID controllers (- -).

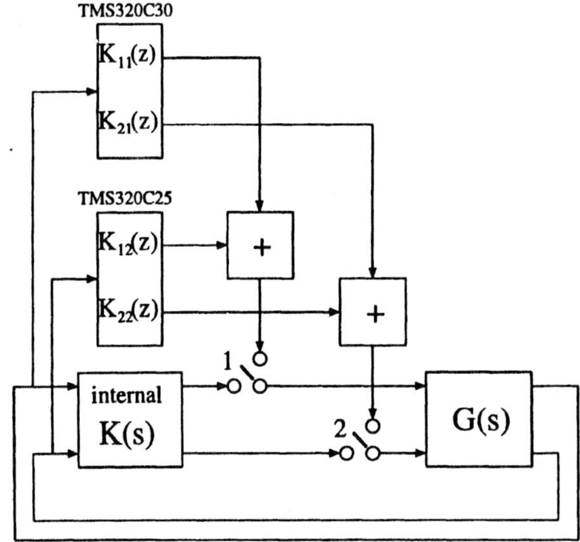

Figure 77.51 The connection of both DSP systems to the experimental setup.

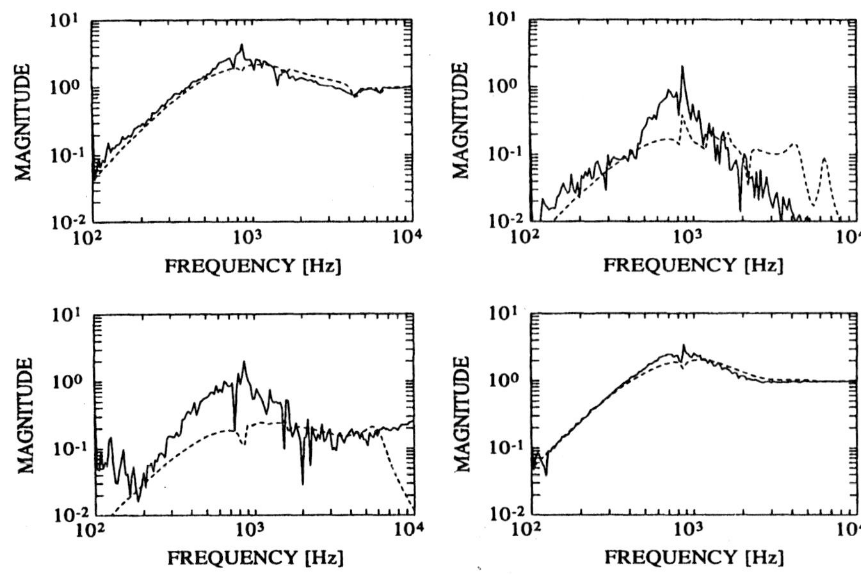

Figure 77.52 Measured frequency response of the input sensitivity function for the 2×8 reduced-order controller of the complete design (—) and the simulated output sensitivity function for the 83rd-order controller (- - -).

before. Nevertheless, the implemented performance differs on a few points from the simulated performance. It seems useful to exploit this knowledge to arrive at even better results in a next controller synthesis. Notice also the work in [4] where a controller for the radial loop has been designed using QFT, directly based on the measured frequency responses.

77.4.7 Conclusions

In this chapter μ-synthesis has been applied to a CD player. The design problem involves time-domain constraints on signals and robustness requirements for norm-bounded structured plant uncertainty. Several different uncertainty structures of increasing complexity are considered. A μ-controller has been implemented successfully in an experimental setup using two parallel DSPs connected to a CD player.

References

[1] Balas, G.J., Doyle, J.C., Glover, K., Packard, A.K., and Smith, R., *μ-Analysis and Synthesis Toolbox*, MUSYN Inc., Minneapolis, MN, 1991.

[2] Bouwhuis, G. et al., *Principles of Optical Disc Systems*, Adam Hilger Ltd., Bristol, UK, 1985.

[3] Ceton, C., Wortelboer, P., and Bosgra, O.H., Frequency weighted closed-loop balanced reduction, in *Proc. 2nd Eur. Control Conf.*, Groningen, June 26 - July 1 1993, 697–701.

[4] Chait, Y., Park, M.S., and Steinbuch, M., Design and implementation of a QFT controller for a compact disc player, *J. Syst. Eng.*, 4, 107–117, 1994.

[5] Doyle, J.C., Advances in Multivariable Control, lecture notes of the ONR/Honeywell Workshop, Honeywell, MN, 1984.

[6] Draijer, W., Steinbuch, M., and Bosgra, O.H., Adaptive control of the radial servo system of a compact disc player, *IFAC Automatica*, 28, 455–462, 1992.

[7] dSPACE GmbH, *DSPCitPro software package*, documentation of dSPACE GmbH, Paderborn, West Germany, 1989.

[8] Groos, P.J.M. van, Steinbuch, M., and Bosgra, O.H., Multivariable control of a compact disc player using μ synthesis, in Proc. 2nd Eur. Control Conf., Groningen, June 26 - July 1, 1993, 981–985.

[9] Lambrechts, P., Terlouw, J.C., Bennani, S., and Steinbuch, M., Parametric uncertainty modeling using LFTs, in Proc. 1993 Am. Control Conf., June 1993, 267–272

[10] Schrama, R., Approximate Identification and Control Design, Ph.D. thesis, Delft University of Technology, Delft, The Netherlands, 1992.

[11] Steinbuch, M., Schootstra, G., and Bosgra, O.H., Robust control of a compact disc player, IEEE 1992 Conf. Decision Control, Tucson, AZ, 2596–2600.

[12] Steinbuch, M., Schootstra, G., and Goh, H.T., Closed loop scaling in fixed-point digital control, *IEEE Trans. Control Syst. Technol.*, 2(4), 312–317, 1994.

[13] Wortelboer, P., Frequency Weighted Balanced Reduction of Closed-loop Mechanical Servo-systems, Ph.D. thesis, Delft University of Technology, Delft, The Netherlands, 1994.

SECTION XVII

Electrical and Electronic Control Systems

78

Power Electronic Controls

George C. Verghese
Massachusetts Institute of Technology

David G. Taylor
Georgia Institute of Technology, School of Electrical and Computer Engineering, Atlanta, GA

Thomas M. Jahns
GE Corporate R&D, Schenectady, NY

Rik W. De Doncker
Silicon Power Corporation, Malvern, PA

78.1 Dynamic Modeling and Control in Power Electronics

George C. Verghese, Massachusetts Institute of Technology

78.1.1 Introduction

This chapter[1] is written with the following purposes in mind:

- To describe the objectives, features, and constraints that characterize power electronic converters, emphasizing those aspects that are relevant to control.
- To outline the principles by which tractable dynamic models are obtained for power electronic converters; such models are required for application of the various control design approaches and techniques described throughout this handbook.
- To indicate how controls are typically designed and implemented in power electronics.
- To suggest to practitioners and researchers in both control and power electronics that power electronic systems constitute an interesting and important testbed for control.

Much of this chapter is distilled from the more detailed development in [3]. We begin, in this section, with an introduction to the issues that shape power electronics. The following section then describes some prototypical power electronic converters. The final two sections present more detailed discussions of dynamic modeling and control in power electronics, focusing first on a particular converter as a case study, and moving on to some extensions. The role of averaged models and sampled-data models is highlighted.

What is Power Electronics? Power electronics is concerned with high-efficiency conversion of electric power, from the form available at the input or power source, to the form required at the output or load. Most commonly, one talks of AC/DC, DC/DC, DC/AC, and AC/AC *converters*, where "AC" here typically refers to nominally *sinusoidal* voltage waveforms, while "DC" refers to nominally *constant* voltage waveforms. Small deviations from nominal are tolerable. An AC/DC converter (which has an AC power source and a DC load) is also called a *rectifier*, and a DC/AC converter is called an *inverter*. Applications of power electronics can be as diverse as high-voltage DC (HVDC) bulk-power transmission systems involving rectifiers and inverters rated in megawatts, or motor drives rated at a few kilowatts, or 50-W power supplies for electronic equipment.

The Dictates of High Efficiency High efficiency reduces energy costs, but as importantly it reduces the amount of dissipated heat that must be removed from the power converter. Efficiencies of higher than 99% can be obtained in large,

[1]The author is grateful to the following people for helpful comments: Steven Leeb, Bernard Lesieutre, Piero Maranesi, David Perreault, Seth Sanders, Aleksandar Stankovic, and Joseph Thottuvelil. For invaluable assistance with the figures, thanks are due to Deron Jackson and Steven Leeb.

0-8493-8570-9/96/$0.00+$.50

high-power systems, while small, low-power systems may have efficiencies closer to 80%. The goal of high efficiency dictates that the power processing components in the circuit be close to lossless. Switches, capacitors, inductors, and transformers are therefore the typical components in a power electronic converter.

The switches are operated cyclically, and serve to vary the circuit interconnections — or the "topological state" of the circuit — over the course of a cycle. The capacitors and inductors perform filtering actions, regulating power flows by temporarily storing or supplying energy. The transformers scale voltages and currents, and also provide electrical isolation between the source and load. *Ideal* switches, capacitors, inductors, and transformers do not dissipate power, and circuits comprising only such elements do not dissipate power either (provided that the switching operations do not result in impulsive currents or voltages, a constraint that is respected by power converters). In particular, an *ideal switch* has zero voltage across itself in its *on* (or closed, or conducting) state, zero current through itself in its *off* (or open, or blocking) state, and requires zero time to make a transition between these two states. Its power dissipation is therefore always zero. Of course, practical components depart from ideal behavior, resulting in some power dissipation.

Semiconductor Switches A switch in a power electronic converter is implemented via one or a combination of semiconductor devices. The most common such devices are

- diodes
- thyristors, which may be thought of as diodes with an additional gate terminal for control, and which are of various types (for example, silicon controlled rectifiers or SCRs, bidirectional thyristors called TRIACs that function as antiparallel SCRs, and gate turn-off thyristors or GTOs)
- bipolar junction transistors (BJTs, controlled via an appropriate drive at the base, and designed for operation in cutoff or saturation, rather than in their linear, active range), and insulated gate bipolar transistors (IGBTs, controlled via the gate)
- metal-oxide-semiconductor field effect transistors (MOSFETs, controlled via the gate)

The power loss associated with such switches comes from a nonzero voltage drop when they are closed, a nonzero leakage current when they are open, and a finite transition time from closed to open or vice versa, during which time both the voltage and current may be significant simultaneously.

A higher switching frequency generally implies a more compact converter, since smaller capacitors, inductors, and transformers can be used to meet the specified circuit characteristics. However, the higher frequency also means higher switching losses associated with the increased frequency of switch transitions, as well as other losses and limitations associated with high-frequency operation of the various components. Switching frequencies above the audible range are desirable for many applications.

Controlling the Switches Each type of semiconductor switch is amenable to a characteristic mode of control. Diodes are at one extreme, as they *cannot* be controlled; they conduct or block as a function solely of the current through them or the voltage across them, so neither their turn-on nor turn-off can be directly commanded by a control action. For SCRs and TRIACs, the turn-off happens as for a diode, but the turn-on is by command, under appropriate circuit conditions. For BJTs, IGBTs, MOSFETs and GTOs, both the turn-on and turn-off occur in response to control actions, provided circuit conditions are appropriate.

The choice of switch implementation depends on the requirements of each particular converter. For instance, the same circuit topology that is used for rectification can often be used for inversion, after appropriate modification of the switches and/or of the way they are controlled.

The only available control decisions are *when* to open and close the switches. It is by modulating the instants at which the switches are opened and closed that the dynamic behavior of a power electronic converter is regulated. Power electronic engineers have invented clever mechanisms for implementing the types of modulation schemes needed for feedback and feedforward control of power converters.

78.1.2 Prototype Converters

This section briefly describes the structure and operating principles of some basic power electronic converters. Many practical converters are directly derived from or closely related to one of the converters presented here. Also, many power electronic systems involve *combinations* of such basic converters. For instance, a high-quality power supply for electronic equipment might comprise a unity-power-factor, pulse-width-modulated (PWM) rectifier cascaded with a PWM DC/DC converter; a variable-frequency drive for an AC motor might involve a rectifier followed by a variable-frequency inverter.

High-Frequency PWM DC/DC Converters Given a DC voltage of value V (which can represent an input DC voltage, or an output DC voltage, or a DC difference between input and output voltages), we can easily use a controlled switch to "chop" the DC waveform into a *pulse waveform* that alternates between the values V and 0 at the switching frequency. This pulse waveform can then be lowpass-filtered with capacitors and/or inductors that are configured to respond to its average value, i.e., its DC component. By controlling the *duty ratio* of the switch, i.e., the fraction of time that the switch is closed in each cycle, we can control the fraction of time that the pulse waveform takes the value V, and thereby control the DC component of this waveform.

The preceding description applies directly to the simple *buck* (or voltage step-down) converter illustrated schematically in Figure 78.1. The load is taken to be just a resistor R, for simplicity. If the switch is operated periodically with a constant duty ratio D, and assuming the switch and diode to be ideal, it is easy to see that the converter settles into a periodic steady state in which the

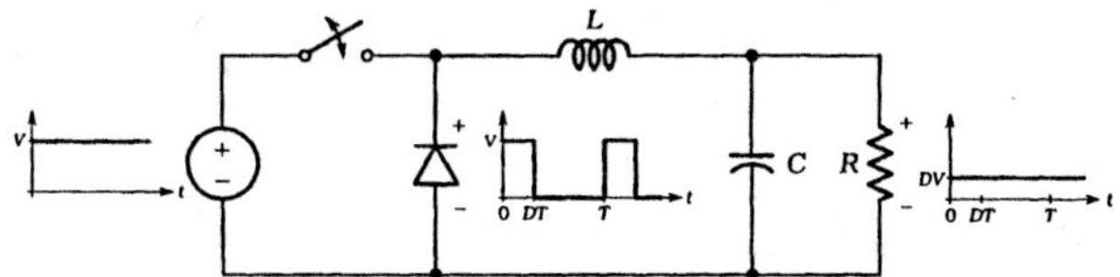

Figure 78.1 The basic principle of high-frequency PWM DC/DC conversion is illustrated here for the case of a buck converter with a resistive load. The switch operation converts the DC input voltage into a pulse waveform. The switching frequency is chosen much higher than the cutoff frequency of the lowpass LC filter, so the output voltage is essentially DC. If the switch is operated periodically with a constant duty ratio D, the converter settles into a periodic steady state in which the output voltage is DV plus a small switching-frequency ripple.

average output voltage is DV. If the switching frequency is high enough relative to the cutoff frequency of the lowpass filter, then the switching-frequency component at the output will be greatly attenuated. The output voltage will then be the DC value DV plus some small switching-frequency *ripple* superimposed on it.

The buck converter is the simplest representative of a class of DC/DC converters based on lowpass filtering of a high-frequency pulse waveform. These converters are referred to as *switching regulators* or *switched-mode* converters (to distinguish them from linear voltage regulators, which are based on transistors operating in their linear active range, and which are therefore generally less efficient.) *Pulse-width modulation* (PWM), in which the duty ratio is varied, forms the basis for the regulation of switched-mode converters, so they are also referred to as high-frequency PWM DC/DC converters. Switching frequencies in the range of 15 to 300 kHz are common.

The *boost* (or voltage step-up) converter in Figure 78.2 is a more complicated and interesting high-frequency PWM DC/DC converter, and we shall be examining it in considerably more detail. Here V_{in} denotes the voltage of a DC source, while the

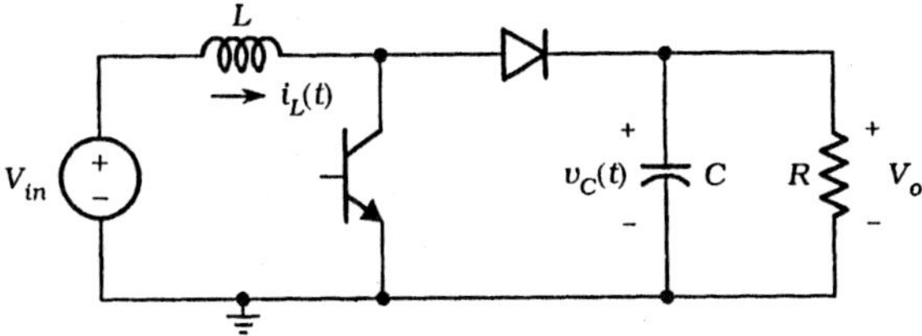

Figure 78.2 The boost converter shown here is a more complex high-frequency PWM DC/DC converter. The average voltage across the inductor must be zero in the periodic steady state that results when the transistor is switched periodically with duty ratio D. Also, if the switching frequency is high enough, the output voltage is essentially a DC voltage V_o. It follows from these facts that $V_o \approx V_{in}/(1-D)$ in the steady state, so the boost converter steps up the DC input voltage to a *higher* DC output voltage.

voltage across the load (again modeled for simplicity as being just a resistor) is essentially a DC voltage V_o , with some small switching-frequency ripple superimposed on it. The values of L and C are chosen such that the ripple in the output voltage is a suitably small percentage (typically < 5%) of the nominal load voltage. The left terminal of the inductor is held at a potential of V_{in} relative to ground, while its right terminal sees a pulse waveform that is switched between 0 (when the transistor is on, with the diode blocking) and the output voltage (when the transistor is off, with the diode conducting). In nominal operation, the transistor is switched periodically with duty ratio D, so the average potential of the inductor's right terminal is approximately $(1-D)V_o$. A periodic steady state is attained only when the inductor has zero average voltage across itself, i.e., when

$$V_o \approx \frac{V_{in}}{(1-D)} \tag{78.1}$$

Otherwise, the inductor current at the end of a switching cycle would not equal that at the beginning of the cycle, which contradicts the assumption of a periodic steady state. Since $0 < D < 1$, we see from Equation 78.1 that the output DC voltage is *higher* than the input DC voltage, which is why this converter is termed a "boost" converter.

Other High-Frequency PWM Converters Appropriate control of a high-frequency PWM DC/DC converter also enables conversion between waveforms that are not DC, but that are nevertheless slowly varying relative to the switching frequency. If, for example, the input is a slowly varying unidirectional voltage — such as the waveform obtained by rectifying a 60-Hz sinewave — while the converter is switched at a much higher rate, say 50 kHz, then we can still arrange for the output of the converter to be essentially DC.

The high-frequency *PWM rectifier* in Figure 78.3 is built around this idea, and comprises a diode bridge followed by a boost converter. The bridge circuit rectifies the AC supply, producing (for the example of the 60-Hz supply mentioned above) the unidirectional voltage $v_{in}(t) = V|\sin 120\pi t|$ of period

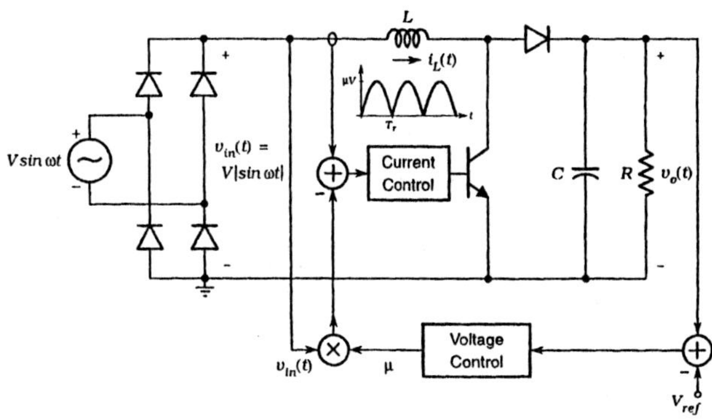

Figure 78.3 A unity-power-factor, high-frequency PWM rectifier, comprising a diode bridge followed by a boost converter. The fast, inner current-control loop of the boost converter governs the switching, and causes the inductor current to follow a reference $\mu v_{in}(t)$ that is a scaled version of the rectified input voltage $v_{in}(t)$. The scale factor μ is dynamically adjusted by the slow, outer voltage-control loop so as to obtain an essentially DC output voltage.

$T_r = 1/120$ sec from the AC voltage $V \sin 120\pi t$. The resulting voltage is applied to the boost converter. The switching of the transistor (at 50 kHz, for example) is governed by a fast, inner current-control loop which causes the inductor current to follow a reference that is a scaled replica of $v_{in}(t)$, namely the rectified sinusoid $\mu v_{in}(t)$. The current drawn by the diode bridge from the AC supply is therefore sinusoidal and in phase with the supply voltage, leading to operation at essentially unity power factor. The scale factor μ that relates the inductor current reference to $v_{in}(t)$ is dynamically adjusted by the slow, outer voltage-control loop, so as to produce a DC output voltage (corrupted by a slight 120-Hz ripple and even less 50-kHz ripple). Unity-power-factor PWM rectifiers of this sort are becoming more common as front ends in power supplies for computers and other electronic equipment, because of the desire to extract as much power as possible from the wall socket, while meeting increasingly stringent power quality standards for such equipment.

In a high-frequency *PWM inverter*, the situation is reversed. The heart of it is still a DC/DC converter, and the input to it is DC. However, the switching is controlled in such a way that the filtered output is a slowly varying rectified sinusoid at the desired frequency. This rectified sinusoid can then be "unfolded" into the desired sinusoidal AC waveform, through the action of additional controllable switches arranged in a bridge configuration. In fact, both the chopping and unfolding functions can be carried out by the bridge switches, and the resulting high-frequency PWM bridge inverter is the most common implementation, available in single-phase and three-phase versions. These inverters are often found in drives for AC servo-motors, such as the permanent-magnet synchronous motors (also called "brushless DC" motors) that are popular in robotic applications. The inductive windings of the motor perform all or part of the electrical lowpass filtering in this case, while the motor inertia provides the additional mechanical filtering that practically removes the switching-frequency component from the mechanical motion.

Other Inverters Another common approach to constructing inverters again relies on a pulse waveform created by chopping a DC voltage of value V, but with the frequency of the pulse waveform now *equal* to that of the desired AC waveform, rather than much higher. Also, the pulse waveform is now generally caused (again through controllable switches configured in a bridge arrangement) to have a mean value of zero, taking values of V, 0, and $-V$, for instance. Lowpass filtering of this pulse waveform to keep only the fundamental and reject harmonics yields an essentially sinusoidal AC waveform at the switching frequency. The amplitude of the sinusoid can be controlled by varying the duty ratio of the switches that generate the pulse waveform; this may be thought of as low-frequency PWM. It is easy to arrange for the pulse waveform to have no even harmonics, and more elaborate design of the waveform can eliminate designated low-order (e.g., third, fifth, and seventh) harmonics, in order to improve the effectiveness of the lowpass filter. This sort of inverter might be found in variable-frequency drives for large AC motors, operating at power levels where the high-frequency PWM inverters described in the previous paragraph would not be practical (because of limitations on switching frequency that become dominant at higher power levels). The lowpass filtering again involves using the inductive windings and inertia of the motor.

Resonant Converters There is an alternative approach to controlling the output amplitude of a DC/AC converter, such as that presented in the previous paragraph. Rather than varying the duty ratio of the pulse waveform, a resonant inverter uses frequency variations. In such an inverter, a resonant bandpass filter (rather than a lowpass filter) is used to extract the sinewave from the pulse waveform; the pulse waveform no longer needs to have zero mean. The amplitude of the sinewave is strongly dependent on how far the switching frequency is from resonance, so control of the switching frequency can be used to control the amplitude of the output sinewave. One difficulty, however, is that variations in the load will lead to modifications of the resonance characteristics of the filter, in turn causing significant variations in the control characteristics.

If the sinusoidal waveform produced by a resonant inverter is rectified and lowpass filtered, what is obtained is a resonant DC/DC converter, as opposed to a PWM DC/DC converter. This form of DC/DC converter can have lower switching losses and generate less electromagnetic interference (EMI) than a typical high-frequency PWM DC/DC converter operating at the same switching frequency, but these advantages come at the cost of higher peak currents and voltages, and therefore higher component stresses.

Phase-Controlled Converters We have already mentioned a diode bridge in connection with Figure 78.3, and noted that it converts an AC waveform into a unidirectional or rectified waveform. Using *controllable* switches instead of the diodes allows us to *partially* rectify a sinusoidal AC waveform, with subsequent lowpass filtering to obtain an essentially DC waveform at a specified level. This is the basis for phase-controlled rectifiers, which are used as drives for DC motors or as battery charging circuits.

A typical configuration using thyristors is shown in Figure 78.4. The associated load-voltage waveform $v_o(t)$ is drawn for the case where the load current $i_o(t)$ stays strictly positive, as would happen with an inductive load. A thyristor can be turned on ("fired") via a gate signal whenever the voltage across the thyristor is positive; the thyristor turns off when the voltage across it reverses, or the forward current through it falls to zero. The control variable is the firing angle or *delay angle*, α, which is the (electrical) angle by which the firing of a thyristor is delayed, beyond the point at which it becomes forward biased. The role of α in shaping the output voltage is evident from the waveform in Figure 78.4. The average voltage at the output of the converter is $(2V/\pi)\cos\alpha$. If the admittance of the load has a lowpass characteristic, then the steady-state current through it will be essentially DC, at a level determined by the average output voltage of the converter.

The load current in the circuit of Figure 78.4 can never become negative, because of the orientation of the thyristors. Nevertheless, the converter can operate as an inverter too. For example, if

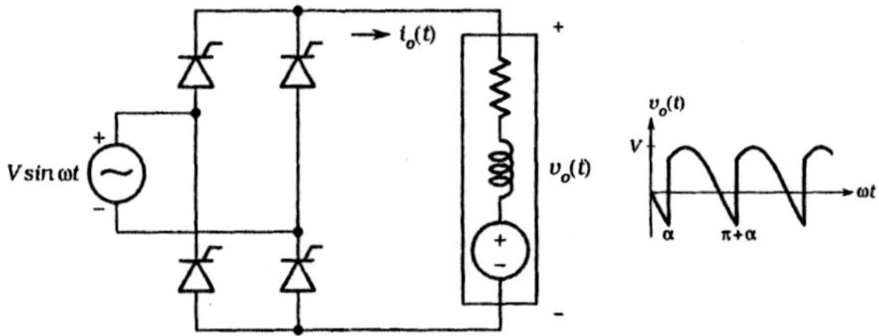

Figure 78.4 A phase-controlled converter with strictly positive load current, and the associated output voltage waveform, $v_O(t)$. The average value of $v_O(t)$ is $(2V/\pi)\cos\alpha$, where α is the firing angle or delay angle. Provided the load current does not vary significantly over a cycle, the converter functions as a rectifier when $\alpha < \pi/2$, and as an inverter when $\alpha > \pi/2$.

the load current is essentially constant over the course of a cycle (as it would be with a heavily inductive load) and if the average output voltage of the converter is made negative by setting $\alpha > \pi/2$, then the converter is actually functioning as an inverter. Such inversion might be used if the DC side of the converter is connected to a DC power source, such as a solar array or a DC generator, or a DC motor undergoing regenerative braking.

AC/AC Converters For AC/AC conversion between waveforms of the *same frequency*, we can use switches to window out sections of the source waveform, thereby reducing the fundamental component of the waveform in a controlled way; TRIACs are well suited to carrying out this operation. Subsequent filtering can be used to extract the fundamental of the windowed waveform. More intricate use of switches — in a *cycloconverter* — permits the construction of an approximately sinusoidal waveform at some specified frequency by "splicing" together appropriate segments of a set of three-phase (or multiphase) sinusoidal waveforms at a *higher* frequency; again, subsequent filtering improves the quality of the output sinusoid. While cycloconverters effect a direct AC/AC conversion, it is also common to construct an AC/AC converter as a cascade of a rectifier and an inverter (generally operating at *different* frequencies), forming a *DC-link converter.*

78.1.3 Dynamic Modeling and Control

Detailed Models Elementary circuit analysis of a power converter typically produces detailed, continuous-time, nonlinear, time-varying models in state-space form. These models have rather low order, provided one makes reasonable approximations from the viewpoint of control design: neglecting dynamics that occur at much higher frequencies than the switching frequency (for instance, dynamics due to parasitics, or to *snubber* elements that are introduced around the switches to temper the switch transitions), and focusing instead on components that are central to the power processing function of the converter. Much of this modeling phase — including the recognition of which elements are important, see [6] — is automatable. Various computer tools are also available for detailed dynamic simulation of a power converter model, specified in either circuit form or state-space form, see [9].

Simplified Models through Averaging or Sampling The continuous-time models mentioned in the preceding paragraph capture essentially all the effects that are likely to be significant for control design. However, the models are generally still too detailed and awkward to work with. The first challenge, therefore, is to extract from such a detailed model a simplified approximate model, preferably time-invariant, that is well matched to the particular control design task for the converter being considered. There are systematic ways to obtain such simplifications, notably through

- *averaging*, which blurs out the detailed switching artifacts
- *sampled-data modeling*, again to suppress the details internal to a switching cycle, focusing instead on cycle-to-cycle behavior

Both methods can produce time-invariant but still nonlinear models. Several approaches to nonlinear control design can be explored at this level. Linearization of averaged or sampled-data models around a constant operating point yields linear, time-invariant (LTI) models that are immediately amenable to a much larger range of standard control design methods.

In the remainder of this section, we illustrate the preceding comments through a more detailed examination of the boost converter that was introduced in the previous section.

A Case Study: Controlling The Boost Converter Consider the boost converter of Figure 78.2, redrawn in Figure 78.5 with some modifications. The figure includes a schematic illustration of a typical analog PWM control method that uses

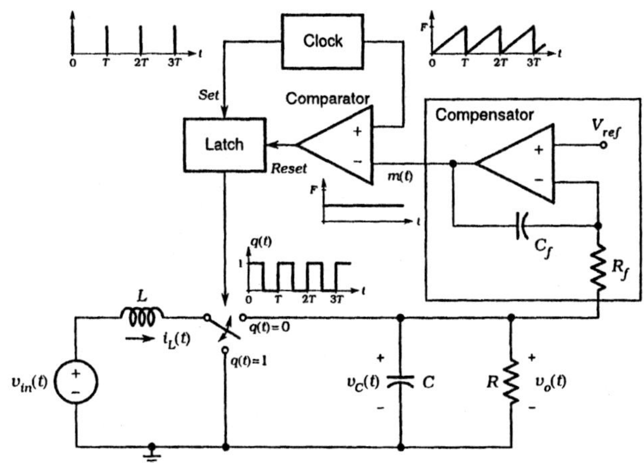

Figure 78.5 Controlling the boost converter. The operation of the switch is controlled by the latch. The switch is moved down every T seconds by a set pulse from the clock to the latch ($q(t) = 1$). The clock simultaneously initiates a ramp input of slope F/T to one terminal of the comparator. The modulating signal $m(t)$ is applied to the other terminal of the comparator. At the instant t_k in the kth cycle when the ramp crosses the level $m(t_k)$, the comparator output goes high, the latch resets ($q(t) = 0$), and the switch is moved up. The resulting duty ratio d_k in the kth cycle is $m(t_k)/F$.

output feedback. This control configuration is routinely and widely used, in a single-chip implementation; its operation will be explained shortly. We have allowed the input voltage $v_{in}(t)$ in the figure to be time varying, to allow for a source that is nominally DC at the value V_{in}, but that has some time-varying deviation or ripple around this value. Although a more realistic model of the converter for control design would also, for instance, include the equivalent series resistance — or ESR — of the output capacitor, such refinements can be ignored for our purposes here; they can easily be incorporated once the simpler case is understood. The rest of our development will therefore be for the model in Figure 78.5.

In typical operation of the boost converter under what may be called constant-frequency PWM control, the transistor in Figure 78.2 is turned on every T seconds, and turned off d_kT seconds later in the kth cycle, $0 < d_k < 1$, so d_k represents the duty ratio in the kth cycle. If we maintain a positive inductor current, $i_L(t) > 0$, then when the transistor is on, the diode is off, and vice versa. This is referred to as the *continuous conduction mode.* In the *discontinuous conduction mode*, on the other hand, the inductor current drops all the way to zero some time after the transistor is turned off, and then remains at zero, with the transistor and diode *both* off, until the transistor is turned on again. Limiting our attention here to the case of continuous conduction, the action of the transistor/diode pair in Figure 78.2 can be represented in idealized form via the double-throw switch in Figure 78.5.

We will mark the position of the switch in Figure 78.5 using a *switching function* $q(t)$. When $q(t) = 1$, the switch is down; when $q(t) = 0$, the switch is up. The switching function $q(t)$ may be thought of as (proportional to) the signal that has to be applied to the base drive of the transistor in Figure 78.2 to turn it on and off as desired. Under the constant-frequency PWM switching discipline described above, $q(t)$ jumps to 1 at the start of each cycle, every T seconds, and falls to 0 an interval d_kT later in its kth cycle. The average value of $q(t)$ over the kth cycle is therefore d_k; if the duty ratio is constant at the value $d_k = D$, then $q(t)$ is periodic, with average value D.

In Figure 78.5, $q(t)$ corresponds to the signal at the output of the latch. This signal is set to "1" every T seconds when the clock output goes high, and is reset to "0" later in the cycle when the comparator output goes high. The two input signals of the comparator are cleverly arranged so as to reset the latch at a time determined by the desired duty ratio. Specifically, the input to the "+" terminal of the comparator is a sawtooth waveform of period T that starts from 0 at the beginning of every cycle, and ramps up linearly to F by the end of the cycle. At some instant t_k in the kth cycle, this ramp crosses the level of the *modulating signal* $m(t)$ at the "−" terminal of the comparator, and the output of the comparator switches from low to high, thereby resetting the latch. The duty ratio thus ends up being $d_k = m(t_k)/F$ in the corresponding switching cycle. By varying $m(t)$ from cycle to cycle, the duty ratio can be varied.

Note that the *samples* $m(t_k)$ of $m(t)$ are what determine the duty ratios. We would therefore obtain the same sequence of duty ratios even if we added to $m(t)$ any signal that stayed negative in the first part of each cycle and crossed up through 0 in the kth cycle at the instant t_k. This fact corresponds to the familiar *aliasing* effect associated with sampling. Our standing assumption will be that $m(t)$ is not allowed to change significantly *within* a single cycle, i.e., that $m(t)$ is restricted to vary considerably more slowly than half the switching frequency. As a result, $m(t) \approx m(t_k)$ in the kth cycle, so $m(t)/F$ at any time yields the prevailing duty ratio (provided also that $0 \le m(t) \le F$, of course — outside this range, the duty ratio is 0 or 1).

The modulating signal $m(t)$ is generated by a feedback scheme. For the particular case of output feedback shown in Figure 78.5, the output voltage of the converter is compared with a reference voltage, and the difference is applied to a compensator, which produces $m(t)$. The goal of dynamic modeling, stated in the context of this example, is primarily to provide a basis for rational design of the compensator, by describing how the converter responds to variations in the modulating signal $m(t)$, or equivalently, to variations in the duty ratio $m(t)/F$. (Note that the ramp level F can also be varied in order to modulate the duty ratio, and this mechanism is often exploited to implement certain *feedforward* schemes that compensate for variations in the input voltage V_{in}.)

Switched State-Space Model for the Boost Converter

Choosing the inductor current and capacitor voltage as natural state variables, picking the resistor voltage as the output, and using the notation in Figure 78.5, it is easy to see that the following state-space model describes the idealized boost converter in that figure:

$$\begin{aligned}\frac{di_L(t)}{dt} &= \frac{1}{L}\Big[\Big(q(t)-1\Big)v_C(t) + v_{in}(t)\Big]\\ \frac{dv_C(t)}{dt} &= \frac{1}{C}\Big[\Big(1-q(t)\Big)i_L(t) - \frac{v_C(t)}{R}\Big]\\ v_o(t) &= v_C(t)\end{aligned} \tag{78.2}$$

Denoting the state vector by $\mathbf{x}(t) = [i_L(t) \;\; v_C(t)]'$ (where the prime indicates the transpose), we can rewrite the above equations as

$$\begin{aligned}\frac{d\mathbf{x}(t)}{dt} &= \Big[\Big(1-q(t)\Big)\mathbf{A}_0 + q(t)\mathbf{A}_1\Big]\mathbf{x}(t) + \mathbf{b}\,v_{in}(t)\\ v_o(t) &= \mathbf{c}\,\mathbf{x}(t)\end{aligned} \tag{78.3}$$

where the definitions of the various matrices and vectors are obvious from Equation 78.2. We refer to this model as the switched or instantaneous model, to distinguish it from the averaged and sampled-data models developed in later subsections.

If our compensator were to directly determine $q(t)$ itself, rather than determining the modulating signal $m(t)$ in Figure 78.5, then the above bilinear and time-invariant model would be the one of interest. It is indeed possible to develop control schemes directly in the setting of the switched model, Equation 78.3. In [1], for instance, a switching curve in the two-dimensional state space is used to determine when to switch $q(t)$ between its two possible values, so as to recover from a transient with a minimum number of switch transitions, eventually arriving at a periodic steady state. Drawbacks include the need for full state measurement and accurate knowledge of system parameters.

Various sliding mode schemes have also been proposed on the basis of switched models such as Equation 78.3, see for instance [13], [10], [7], and references in these papers. Sliding mode designs again specify a surface across which $q(t)$ switches, but now the (sliding) motion occurs on the surface itself, and is analyzed under the assumption of infinite-frequency switching. The requisite models are thus averaged models in effect, of the type developed in the next subsection. Any practical implementation of a sliding control must limit the switching frequency to an acceptable level, and this is often done via hysteretic control, where the switch is moved one way when the feedback signal exceeds a particular threshold, and is moved back when the signal drops below another (slightly lower) threshold. Constant-frequency implementations similar to the one in Figure 78.5 may also be used to get reasonable approximations to sliding mode behavior.

As far as the design of the compensator in Figure 78.5 is concerned, we require a model describing the converter's response to the modulating signal $m(t)$ or the duty ratio $m(t)/F$, rather than the response to the switching function $q(t)$. Augmenting the model in Equation 78.3 to represent the relation between $q(t)$ and $m(t)$ would introduce time-varying behavior and additional nonlinearity, leading to a model that is hard to work with. The models considered in the remaining sections are developed in response to this difficulty.

Nonlinear Averaged Model for the Boost Converter

To design the analog control scheme in Figure 78.5, we seek a tractable model that relates the modulating signal $m(t)$ or the duty ratio $m(t)/F$ to the output voltage. In fact, since the ripple in the instantaneous output voltage is made small by design, and since the details of this small output ripple are not of interest anyway, what we really seek is a continuous-time dynamic model that relates $m(t)$ or $m(t)/F$ to the *local average* of the output voltage (where this average is computed over the switching period). Also recall that $m(t)/F$, the duty ratio, is the local average value of $q(t)$ in the corresponding switching cycle. These facts suggest that we should look for a dynamic model that relates the local average of the switching function $q(t)$ to that of the output voltage $v_o(t)$.

Specifically, let us define the local average of $q(t)$ to be the lagged running average

$$d(t) = \frac{1}{T}\int_{t-T}^{t} q(\tau)\,d\tau \tag{78.4}$$

and call it the *continuous duty ratio* $d(t)$. Note that $d(kT) = d_k$, the actual duty ratio in the kth cycle (defined as extending from $kT - T$ to kT). If $q(t)$ is periodic with period T, then $d(t) = D$, the steady-state duty ratio. Our objective is to relate $d(t)$ in Equation 78.4 to the local average of the output voltage, defined similarly by

$$\overline{v}_o(t) = \frac{1}{T}\int_{t-T}^{t} v_o(\tau)\,d\tau \tag{78.5}$$

A natural approach to obtaining a model relating these averages is to take the local average of the state-space description in Equation 78.2. The local average of the derivative of a signal equals the derivative of its local average, because of the LTI nature of the local averaging operation we have defined. The result of averaging the model in Equation 78.2 is therefore the following set of equations:

$$\begin{aligned}\frac{d\,\overline{i}_L(t)}{dt} &= \frac{1}{L}\Big[\overline{qv_C}(t) - \overline{v}_C(t) + \overline{v}_{in}(t)\Big]\\ \frac{d\,\overline{v}_C(t)}{dt} &= \frac{1}{C}\Big[\overline{i}_L(t) - \overline{qi_L}(t) - \frac{\overline{v}_C(t)}{R}\Big]\\ \overline{v}_o(t) &= \overline{v}_C(t)\end{aligned} \tag{78.6}$$

where the overbars again denote local averages.

The terms that prevent the above description from being a state-space model are $\overline{qv_C}(t)$ and $\overline{qi_L}(t)$; the average of a product is generally *not* the product of the averages. Under reasonable assumptions, however, we can write

$$\begin{aligned}\overline{qv_C}(t) &\approx \overline{q}(t)\overline{v}_C(t) = d(t)\overline{v}_C(t)\\ \overline{qi_L}(t) &\approx \overline{q}(t)\overline{i}_L(t) = d(t)\overline{i}_L(t)\end{aligned} \tag{78.7}$$

One set of assumptions leading to the above simplification requires $v_C(\cdot)$ and $i_L(\cdot)$ over the averaging interval $[t-T, t]$ to not deviate significantly from $\overline{v}_C(t)$ and $\overline{i}_L(t)$, respectively. This condition is reasonable for a high-frequency switching converter operating with low ripple in the state variables. There are alternative assumptions that lead to the same approximations. For instance, if $i_L(\cdot)$ is essentially piecewise linear and has a slowly varying average, then the approximation in the second equation of Equation 78.7 is reasonable even if the ripple in $i_L(\cdot)$ is large; this situation is often encountered.

With the approximations in Equation 78.7, the description in Equation 78.6 becomes

$$\begin{aligned}\frac{d\,\overline{i}_L(t)}{dt} &= \frac{1}{L}\Big[\Big(d(t) - 1\Big)\overline{v}_C(t) + \overline{v}_{in}(t)\Big]\\ \frac{d\,\overline{v}_C(t)}{dt} &= \frac{1}{C}\Big[\Big(1 - d(t)\Big)\overline{i}_L(t) - \frac{\overline{v}_C(t)}{R}\Big]\\ \overline{v}_o(t) &= \overline{v}_C(t)\end{aligned} \tag{78.8}$$

What has happened, in effect, is that *all* the variables in the switched state-space model, Equation 78.2, have been replaced by their average values. In terms of the matrix notation in Equation 78.3, and with $\overline{\mathbf{x}}(t)$ defined as the local average of $\mathbf{x}(t)$, we have

$$\begin{aligned}\frac{d\,\overline{\mathbf{x}}(t)}{dt} &= \Big[\Big(1 - d(t)\Big)\mathbf{A}_0 + d(t)\mathbf{A}_1\Big]\overline{\mathbf{x}}(t) + \mathbf{b}\,\overline{v}_{in}(t)\\ \overline{v}_o(t) &= \mathbf{c}\,\overline{\mathbf{x}}(t)\end{aligned} \tag{78.9}$$

This is a nonlinear but *time-invariant* continuous-time state-space model, often referred to as the *state-space averaged* model [8]. The model is driven by the continuous-time control input $d(t)$ — with the constraint $0 \leq d(t) \leq 1$ — and by the exogenous input $\overline{v}_{in}(t)$. Note that, under our assumption of a slowly varying $m(t)$, we can take $d(t) \approx m(t)/F$; with this substitution, Equation 78.9 becomes an averaged model whose control input is the modulating signal $m(t)$, as desired.

The averaged model in Equation 78.9 leads to much more efficient simulations of converter behavior than those obtained using the switched model in Equation 78.3, provided only local averages of variables are of interest. This averaged model also forms a convenient starting point for various nonlinear control design approaches, see for instance [12], [14], [4], and references in these papers. The implementation of such nonlinear schemes would involve an arrangement similar to that in Figure 78.5, although the modulating signal $m(t)$ would be produced by some nonlinear controller rather than the simple integrator shown in the figure.

A natural circuit representation of the averaged model in Equation 78.8 is shown in Figure 78.6. This averaged circuit can actually be derived directly from the instantaneous circuit models in Figure 78.2 or Figure 78.5, using "circuit averaging" arguments [3]. All the instantaneous circuit variables are replaced by their averaged values. The approximations in Equation 78.7 then cause the transistor of the original circuit in Figure 78.2 to be replaced by a controlled voltage source of value $[1-d(t)]\bar{v}_C(t)$, while the diode of the original circuit is replaced by a controlled current source of value $[1-d(t)]\bar{i}_L(t)$. This averaged circuit representation can be used as an input to circuit simulation programs, and again leads to more efficient simulations than those obtained using the original switched model.

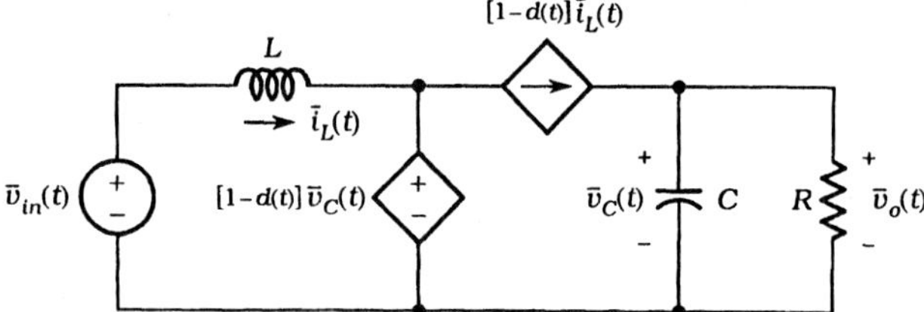

Figure 78.6 Averaged circuit model for the boost converter in Figure 78.2 and Figure 78.5, obtained by replacing instantaneous circuit variables by their local averages, and making the approximations $\overline{qv_C}(t) \approx \bar{q}(t)\bar{v}_C(t) = d(t)\bar{v}_C(t)$ and $\overline{qi_L}(t) \approx \bar{q}(t)\bar{i}_L(t) = d(t)\bar{i}_L(t)$. The transistor of the original circuit in Figure 78.2 is thereby replaced by a controlled voltage source of value $[1-d(t)]\bar{v}_C(t)$, while the diode of the original circuit is replaced by a controlled current source of value $[1-d(t)]\bar{i}_L(t)$.

Linearized Models for the Boost Converter The most common basis for control design in this class of converters is a linearization of the corresponding state-space averaged model. This linearized model approximately governs *small* perturbations of the averaged quantities from their values in some nominal (typically steady-state) operating condition. We illustrate the linearization process for the case of our boost converter operating in the vicinity of its periodic steady state.

Denote the *constant* nominal equilibrium values of the *averaged* state variables by I_L and V_C. These values can be computed from Equation 78.9 by setting the state derivative to zero, and replacing all other variables by their constant nominal values. The equilibrium state vector is thereby seen to be

$$\begin{aligned}\mathbf{X} &= -[(1-D)\mathbf{A}_0 + D\mathbf{A}_1]^{-1}\mathbf{b}\,V_{in} \\ &= \begin{pmatrix} I_L \\ V_C \end{pmatrix} = \begin{pmatrix} \frac{1}{R(1-D)^2} \\ \frac{1}{1-D} \end{pmatrix} V_{in} \qquad (78.10)\end{aligned}$$

Denote the (small) deviations of the various averaged variables from their constant equilibrium values by

$$\begin{aligned}\tilde{i}_L(t) &= \bar{i}_L(t) - I_L \\ \tilde{v}_C(t) &= \bar{v}_C(t) - V_C \\ \tilde{d}(t) &= d(t) - D \\ \tilde{v}_{in}(t) &= \bar{v}_{in}(t) - V_{in} \\ \tilde{v}_o(t) &= \bar{v}_o(t) - V_o \qquad (78.11)\end{aligned}$$

Substituting in Equation 78.8 and neglecting terms that are of second order in the deviations, we obtain the following linearized averaged model:

$$\begin{aligned}\frac{d\tilde{i}_L(t)}{dt} &= \frac{1}{L}\Big[(D-1)\tilde{v}_C(t) + V_C\tilde{d}(t) + \tilde{v}_{in}(t)\Big] \\ \frac{d\tilde{v}_C(t)}{dt} &= \frac{1}{C}\Big[(1-D)\tilde{i}_L(t) - I_L\tilde{d}(t) - \frac{\tilde{v}_C(t)}{R}\Big] \\ \tilde{v}_o(t) &= \tilde{v}_C(t) \qquad (78.12)\end{aligned}$$

This is an LTI model, with control input $\tilde{d}(t)$ and disturbance input $\tilde{v}_{in}(t)$. Rewriting this model in terms of the matrix notation in Equations 78.9 and 78.10, with

$$\tilde{\mathbf{x}}(t) = \bar{\mathbf{x}}(t) - \mathbf{X} \qquad (78.13)$$

we find

$$\begin{aligned}\frac{d\tilde{\mathbf{x}}(t)}{dt} &= \Big[(1-D)\mathbf{A}_0 + D\mathbf{A}_1\Big]\tilde{\mathbf{x}}(t) \\ &\quad + [\mathbf{A}_1 - \mathbf{A}_0]\mathbf{X}\,\tilde{d}(t) + \mathbf{b}\,\tilde{v}_{in}(t) \qquad (78.14) \\ \tilde{v}_o(t) &= \mathbf{c}\,\tilde{\mathbf{x}}(t)\end{aligned}$$

Before analyzing the above LTI model further, it is worth remarking that we could have linearized the switched model in Equation 78.3 rather than the averaged model in Equation 78.9. The only subtlety in this case is that the perturbation $\tilde{q}(t) = q(t) - Q(t)$ (where $Q(t)$ denotes the nominal value of $q(t)$ in the periodic steady state) is still a 0/1 function, so one has to reconsider what is meant by a small perturbation in $q(t)$. If we consider a small perturbation $\tilde{q}(t)$ to be one whose *area* is small in any cycle, then we still arrive at a linearized model of the form of Equation 78.14, except that each averaged variable is replaced by its instantaneous version, i.e., D is replaced by $Q(t)$, $\tilde{d}(t)$ by $\tilde{q}(t)$, and so on. For linearization around a periodic steady state, the linearized switched model is (linear and) periodically varying.

Compensator Design for the Boost Converter The LTI model in Equation 78.12 that is obtained by linearizing the state-space averaged model around its constant steady state is the standard starting point for a host of control design methods. The transfer function from $\tilde{d}(t)$ to $\tilde{v}_o(t)$ is straightforward to compute, and turns out to be

$$H(s) = -\frac{I_L}{C}\,\frac{s - \frac{V_{in}}{LI_L}}{s^2 + \frac{1}{RC}s + \frac{(1-D)^2}{LC}} \qquad (78.15)$$

(where we have also used Equation 78.1 to simplify the expression in the numerator). The denominator of $H(s)$ is the characteristic polynomial of the system. For typical parameter values, the roots of this polynomial, i.e., the poles of $H(s)$, are oscillatory and lightly damped; note that they depend on D. Also observe that the system is non-minimum phase: the zero of $H(s)$ is in the right half-plane (RHP). The RHP zero correlates with the physical fact that the initial response to a step increase in duty ratio is a *decrease* in the average output voltage (rather than the increase predicted by the steady-state expression), because the diode conducts for a smaller fraction of the time. However, the buildup in average inductor current that is caused by the increased duty ratio eventually causes the average output voltage to increase over its prior level.

The lightly damped poles and the RHP zero of $H(s)$ signify difficulties and limitations with closed-loop control, if we use only measurements of the output voltage. Nevertheless, pure integral control, which is what is shown in Figure 78.5, can generally lead to acceptable closed-loop performance and adequate low-frequency loop gain (needed to counteract the effects of parameter uncertainty and low-frequency disturbances). For closed-loop stability, the loop crossover frequency must be made smaller than the "corner" frequencies (in the Bode plot) associated with the poles and RHP zero of $H(s)$. The situation for control can be significantly improved by incorporating measurements of the inductor current as well, as is done in current-mode control, which will be described shortly.

The open-loop transfer function from $\tilde{v}_{in}(t)$ to $\tilde{v}_o(t)$ has the same poles as $H(s)$, and no (finite) zeros; its low-frequency gain is $1/(1-D)$, as expected from Equation 78.1. This transfer function is sometimes termed the open-loop *audio-susceptibility* of the converter. The control design has also to deal with rejection of the effects of input voltage disturbances on the output load voltage, i.e., with shaping the closed-loop audio-susceptibility.

In practice, a compensator designed on the basis of an LTI model would be augmented to take account of various large-signal contingencies. A *soft-start* scheme might be used to gradually increase the reference voltage V_{ref} of the controller, thereby reducing stresses in the circuit during start-up; *integrator windup* in the compensator would be prevented by placing back-to-back Zener diodes across the capacitor of the integrator, preventing large run-ups in integrator voltage during major transients; overcurrent protections would be introduced; and so on.

Current-Mode Control of the Boost Converter The attainable control performance may improve, of course, if additional measurements are taken. We have made some reference in the preceding sections to control approaches that use full state feedback. To design state feedback control for a boost converter, the natural place to start is again with the nonlinear time-invariant model (Equation 78.8 and 78.9) or the LTI model (Equations 78.12 and 78.14).

In this subsection, we examine a representative and popular state feedback scheme for high-frequency PWM converters such as the boost converter, namely current-mode control [2]. Its name comes from the fact that a fast inner loop regulates the inductor current to a reference value, while the slower outer loop adjusts the current reference to correct for deviations of the output voltage from its desired value. (Note that this is precisely what is done for the boost converter that forms the heart of the unity-power-factor PWM rectifier in Figure 78.3.) The current monitoring and limiting that are intrinsic to current-mode control are among its attractive features.

In constant-frequency peak-current-mode control, the transistor is turned on every T seconds, as before, but is turned off when the inductor current (or equivalently, the transistor current) reaches a specified reference or *peak* level, denoted by $i_P(t)$. The duty ratio, rather than being explicitly commanded via a modulating signal such as $m(t)$ in Figure 78.5, is now implicitly determined by the inductor current's relation to $i_P(t)$. Despite this modification, the averaged model in Equation 78.8 is still applicable in the case of the boost converter. (Instead of constant-frequency control, one could use hysteretic or other schemes to confine the inductor current to the vicinity of the reference current.)

A tractable and reasonably accurate continuous-time model for the dynamics of the outer loop is obtained by assuming that the average inductor current is approximately equal to the reference current:

$$\bar{i}_L(t) \approx i_P(t) \tag{78.16}$$

Making the substitution from Equation 78.16 in Equation 78.8 and using the two equations there to eliminate $d(t)$, we are left with the following first-order model:

$$\frac{d\bar{v}_C^2(t)}{dt} + \frac{2}{RC}\bar{v}_C^2(t) = -\frac{2i_P(t)}{C}\left(L\frac{di_P(t)}{dt} - \bar{v}_{in}\right) \tag{78.17}$$

This equation is simple enough that one can use it to explore various nonlinear control possibilities for adjusting $i_P(t)$ to control $\bar{v}_C(t)$ or $\bar{v}_C^2(t)$. The equation shows that, for constant $i_P(t)$ (or periodic $i_P(t)$, as in the nominal operation of the unity-power-factor PWM rectifier in Figure 78.3), $\bar{v}_C^2(t)$ approaches its constant (respectively, periodic) steady state exponentially, with time constant $RC/2$.

Linearizing the preceding equation around the equilibrium corresponding to a constant $i_P(t) = I_P$, we get

$$\begin{aligned}\frac{d\tilde{v}_C(t)}{dt} &+ \frac{2}{RC}\tilde{v}_C = \\ &- \frac{LI_P}{CV_C}\left(\frac{d\tilde{i}_P(t)}{dt} - \frac{V_{in}}{LI_P}\tilde{i}_P(t)\right) + \frac{I_P}{CV_C}\tilde{v}_{in}(t)\end{aligned} \tag{78.18}$$

It is a simple matter to obtain from Equation 78.18 the transfer functions needed for small-signal control design with disturbance rejection. We are still left with the same RHP zero as before, see Equation 78.15, in the transfer function from $\tilde{i}_P$ to $\tilde{v}_C$, but there is now only a single well-damped pole to deal with.

Current-mode control may in fact be seen as perhaps the oldest, simplest, and most common representative of a sliding mode control scheme in power electronics (we made reference earlier to more recent and more elaborate sliding mode controls). The inductor current is made to slide along the time-varying surface $i_L(t) = i_P(t)$. Equation 78.17 or 78.18 describes the system dynamics in the sliding mode, and also provides the basis for controlling the sliding surface in a way that regulates the output voltage as desired.

Sampled-Data Models for the Boost Converter Sampled-data models are naturally matched to power electronic converters, first because of the cyclic way in which power converters are operated and controlled, and second because such models are well suited to the design of digital controllers, which are used increasingly in power electronics (particularly for machine drives). Like averaged models, sampled-data models allow us to focus on cycle-to-cycle behavior, ignoring details of the intracycle behavior.

We illustrate how a sampled-data model may be obtained for our boost converter example. The state evolution of Equation 78.2 or 78.3 for each of the two possible values of $q(t)$ can be described very easily using the standard matrix exponential expressions for LTI systems. The trajectories in each segment can then be pieced together by invoking the continuity of the state variables. Under the switching discipline of constant-frequency PWM, where $q(t) = 1$ for the initial fraction d_k of the kth switching cycle, and $q(t) = 0$ thereafter, and assuming the input voltage is constant at V_{in}, we find

$$\mathbf{x}(kT + T) = e^{(1-d_k)\mathbf{A}_0 T}\left(e^{d_k \mathbf{A}_1 T}\mathbf{x}(kT) + \Gamma_1 V_{in}\right) + \Gamma_0 V_{in} \tag{78.19}$$

where

$$\begin{aligned}\Gamma_0 &= \int_0^{(1-d_k)T} e^{\mathbf{A}_0 t}\mathbf{b}\,dt \\ \Gamma_1 &= \int_0^{d_k T} e^{\mathbf{A}_1 t}\mathbf{b}\,dt\end{aligned} \tag{78.20}$$

The nonlinear, *time-invariant* sampled-data model in Equation 78.19 can directly be made the basis for control design. One interesting approach to this task is suggested in [5]. Alternatively, a linearization around the equilibrium point yields a discrete-time LTI model that can be used as the starting point for established methods of control design.

For a well-designed high-frequency PWM DC/DC converter in continuous conduction, the state trajectories in each switch configuration are close to linear, because the switching frequency is much higher than the filter cutoff frequency. What this implies is that the matrix exponentials in Equation 78.19 are well approximated by just the first two terms in their Taylor series expansions:

$$\begin{aligned}e^{(1-d_k)\mathbf{A}_0 T} &\approx \mathbf{I} + (1-d_k)\mathbf{A}_0 T \\ e^{d_k \mathbf{A}_1 T} &\approx \mathbf{I} + d_k \mathbf{A}_1 T\end{aligned} \tag{78.21}$$

If we use these approximations in Equation 78.19 and neglect terms in T^2, the result is the following approximate sampled-data model:

$$\mathbf{x}(kT + T) = \left(\mathbf{I} + (1-d_k)\mathbf{A}_0 T + d_k \mathbf{A}_1 T\right)\mathbf{x}(kT) + \mathbf{b}T V_{in} \tag{78.22}$$

This model is easily recognized as the forward-Euler approximation of the continuous-time model in Equation 78.9. Retaining the terms in T^2 leads to more refined, but still very simple, sampled-data models.

The sampled-data models in Equations 78.19 and 78.22 were derived from Equation 78.2 or 78.3, and therefore used samples of the natural state variables, $i_L(t)$ and $v_C(t)$, as state variables. However, other choices are certainly possible, and may be more appropriate for a particular implementation. For instance, we could replace $v_C(kT)$ by $\overline{v}_C(kT)$, i.e., the sampled local average of the capacitor voltage.

78.1.4 Extensions

The preceding development suggests how dynamic modeling and control in power electronics can be effected on the basis of either averaged or sampled-data models. Although a boost converter was used for illustration, the same general approaches apply to other converters, either directly or after appropriate extensions. To conclude this chapter, we outline some examples of such extensions.

Generalized Averaging It is often useful or necessary — for instance, in modeling the dynamic behavior of resonant converters — to study the *local fundamental* and *local harmonics* [11], in addition to local averages of the form shown in Equations 78.4 and 78.5. For a variable $x(t)$, the *local $\ell\omega_s$-component* may be defined as the following lagged running average:

$$<x>_\ell(t) = \frac{1}{T}\int_{t-T}^{t} x(\tau)e^{-j\ell\omega_s\tau}\,d\tau \tag{78.23}$$

In this equation, ω_s is usually chosen as the switching frequency, i.e., $2\pi/T$, and ℓ is an integer. The local averages in Equations 78.4 and 78.5 thus correspond to the choice $\ell = 0$; the choice $|\ell| = 1$ yields the local fundamental, while $|\ell| > 1$ yields the local ℓth harmonic. A key property of the local $\ell\omega_s$-component is that

$$\left\langle\frac{dx}{dt}\right\rangle_\ell = j\ell\omega_s <x>_\ell + \frac{d}{dt} <x>_\ell \tag{78.24}$$

where we have omitted the time argument t to keep the notation simple. For $\ell = 0$, we recover the result that was used in obtaining Equation 78.6 from Equation 78.2, namely that the local average of the derivative equals the derivative of the local average. More generally, we could evaluate the local $\ell\omega_s$-component of both sides of a switched state-space model such as Equation 78.3, for *several* values of ℓ. With suitable approximations, this leads to an augmented state-space model whose state vector comprises the local $\ell\omega_s$-components for *all* these values of ℓ. In the case of the boost converter, for instance, we could choose $\ell = +1$ and $\ell = -1$ in addition to the choice $\ell = 0$ that was used to get Equation 78.9 from Equation 78.3. The approximation that was used in Equation 78.7, which (with the time argument t still suppressed for notational convenience) we can rewrite as

$$\overline{q\mathbf{x}} \approx <q>_0<\mathbf{x}>_0 \tag{78.25}$$

can now be refined to

$$\overline{q\mathbf{x}} \approx <q>_0<\mathbf{x}>_0 + <q>_{-1}<\mathbf{x}>_1 + <q>_1<\mathbf{x}>_{-1} \tag{78.26}$$

The resulting state-space model will have a state vector comprising $<\mathbf{x}>_0$, $<\mathbf{x}>_1$, and $<\mathbf{x}>_{-1}$, and the solution $\mathbf{x}(t)$ will be approximated as

$$\mathbf{x}(t) \approx <\mathbf{x}>_0 + <\mathbf{x}>_1 e^{j\omega_s t} + <\mathbf{x}>_{-1} e^{-j\omega_s t} \tag{78.27}$$

Results in [11] show the significant improvements in accuracy that can be obtained this way, relative to the averaged model in Equation 78.9, while maintaining the basic simplicity and efficiency of averaged models relative to switched models. That paper also shows how generalized averaging may be applied to the analysis of resonant converters.

Generalized State-Space Models A sampled-data model for a power converter will almost invariably involve a state-space description of the form

$$\mathbf{x}(kT + T) = \mathbf{f}\Big(\mathbf{x}(kT), \mathbf{u}_k, \mathbf{T}_k, k\Big) \tag{78.28}$$

The vector $\mathbf{u}_k$ here comprises a set of parameters that govern the state evolution in the kth cycle (e.g., parameters that describe control choices and source variations during the kth cycle), and $\mathbf{T}_k$ is a vector of *switching times*, comprising the times at which switches in the converter open or close. The switching-time vector $\mathbf{T}_k$ will satisfy a set of constraints of the form

$$0 = \mathbf{c}\Big(\mathbf{x}(kT), \mathbf{u}_k, \mathbf{T}_k, k\Big) \tag{78.29}$$

If $\mathbf{T}_k$ can be solved for in Equation 78.29, then the result can be substituted in Equation 78.28 to obtain a standard sampled-data model in state-space form. However, there are many cases in which the constraint Equation 78.29 is *not* explicitly solvable for $\mathbf{T}_k$, so one is forced to take Equations 78.28 and 78.29 together as the sampled-data model. Such a pair, comprising a state evolution equation along with a side constraint, is what we refer to as a generalized state-space model. Note that the *linearized* version of Equation 78.29 will allow the switching-time *perturbations* to be calculated explicitly, provided the Jacobian matrix $\partial\mathbf{c}/\partial\mathbf{T}_k$ has full column rank. The result can be substituted in the linearized version of Equation 78.28 to obtain a linearized state-space model in standard — rather than generalized — form.

Hierarchical Modeling A hierarchical approach to modeling is mandated by the range of time scales encountered in a typical power electronic system. A single "all-purpose" model that captured all time scales of interest, from the very fast transients associated with the switch transitions and parasitics, to very slow transients spanning hundreds or thousands of switching cycles, would not be much good for any particular purpose. What is needed instead is a collection of models, focused at different time scales, suited to the distinct types of analysis that are desired at each of these scales, and capable of being linked to each other in some fashion.

Consider, for example, the high-power-factor PWM rectifier in Figure 78.3, which is examined further in [3] and [15]. The detailed evaluation of the switch stresses and the design of any snubbers requires circuit-level simulation and analysis over a short interval, comparable with the switching period T. The design of the inner current-control loop is conveniently done using a continuous-time averaged model, with averaging carried out over a window of length T; the model in Equation 78.8 is representative of this stage. The design of the outer voltage-control loop may be done using an averaged or sampled-data model computed over a window of length T_r, the period of the rectified sinusoidal input voltage; the model Equation 78.17 is representative of — or at least a precursor to — this stage. The lessons of each level have to be taken into account at all the other levels.

78.1.5 Conclusion

This has been a brief introduction to some of the major features and issues of dynamic modeling and control in power electronics. The intent was to be comprehensible rather than comprehensive; much that is interesting and relevant has inevitably been left out. Also, the emphasis here is undoubtedly skewed toward the material with which the author is most familiar. The interested reader who probes further will discover power electronics to be a vigorous, rewarding, and important domain for applications and extensions of the full range of control theory and methodology.

References

[1] Burns, W.W. and Wilson, T.G., Analytic derivation and evaluation of a state trajectory control law for DC-DC converters, in *IEEE Power Electronics Specialists Conf. Rec.*, 70–85, 1977.

[2] Hsu, S.P., Brown, A., Resnick, L., and Middlebrook, R.D., Modeling and analysis of switching DC-to-DC converters in constant-frequency current-programmed mode, in *IEEE Power Electronics Specialists Conf. Rec.*, 284–301, 1979.

[3] Kassakian, J.G., Schlecht, M.F., and Verghese, G.C., *Principles of Power Electronics*, Addison-Wesley, Reading, MA, 1991.

[4] Kawasaki, N., Nomura, H., and Masuhiro, M., The new control law of bilinear DC-DC converters developed by direct application of Lyapunov, *IEEE Trans. Power Electron.*, 10(3), 318–325, 1995.

[5] Khayatian, A. and Taylor, D.G., Multirate modeling and control design for switched-mode power converters, *IEEE Trans. Auto. Control*, 39(9), 1848–1852, 1994.

[6] Leeb, S.B., Verghese, G.C., and Kirtley, J.L., Recognition of dynamic patterns in DC-DC switching converters, *IEEE Trans. Power Electron.*, 6(2), 296–302, 1991.

[7] Malesani, L., Rossetto, L., Spiazzi, G., and Tenti, P., Performance optimization of Ćuk converters by sliding-mode control. *IEEE Trans. Power Electron.*. 10(3), 302–309, 1995.

[8] Middlebrook, R.D., and Ćuk, S., A general unified approach to modeling switching converter power stages, in *IEEE Power Electronics Specialists Conf. Rec.*, 18–34, 1976.

[9] Mohan, N., Robbins, W.P., Undeland, T.M., Nilssen, R., and Mo, O., Simulation of power electronic and motion control systems — an overview, *Proc. IEEE*, 82(8), 1287–1302, 1994.

[10] Sanders, S.R., Verghese, G.C., and Cameron, D.E., Nonlinear control of switching power converters, *Control — Theory Adv. Technol.*, 5(4), 601–627, 1989.

[11] Sanders, S.R., Noworolski, J.M., Liu, X.Z., and Verghese, G.C., Generalized averaging method for power conversion circuits. *IEEE Trans. Power Electron.*, 6(2), 251–259, 1991.

[12] Sanders, S.R. and Verghese, G.C., Lyapunov-based control for switched power converters, *IEEE Trans. Power Electron.*, 7(1), 17–24, 1992.

[13] Sira-Ramírez, H., Sliding motions in bilinear switched networks, *IEEE Trans. Circuits Syst.*, 34(8), 919–933, 1987.

[14] Sira-Ramírez, H. and Prada-Rizzo, M.T., Nonlinear feedback regulator design for the Ćuk converter, *IEEE Trans. Auto. Control*, 37(8), 1173–1180, 1992.

[15] Thottuvelil, V.J., Chin, D., and Verghese, G.C., Hierarchical approaches to modeling high-power-factor AC-DC converters, *IEEE Trans. Power Electron.*, 6(2), 179–187, 1991.

Further Reading

The list of references above suggests what are some of the journals and conferences relevant to the topic of this chapter. However, a large variety of other journals and conferences are devoted to, or occasionally contain, useful material. We limit ourselves here to providing just a few leads.

A useful perspective on the state-of-the-art in power electronics may be gleaned from the August 1994 special issue of the *Proceedings of the IEEE*, devoted to "Power Electronics and Motion Control."

Several papers on dynamic modeling and control for power electronics are presented every year at the IEEE Power Electronics Specialists Conference (PESC) and the IEEE Applied Power Electronics Conference (APEC). Many of these papers, plus others, appear in expanded form in the *IEEE Transactions on Power Electronics*. The April 1991 special issue of these *Transactions* was devoted to "Modeling in Power Electronics". In addition, the IEEE Power Electronics Society holds a biennial workshop on "Computers in Power Electronics", and a special issue of the *Transactions* with this same theme is slated to appear in May 1997.

The power electronics text of Kassakian et al. referenced above has four chapters devoted to dynamic modeling and control in power electronics. The first two of the following books consider the dynamics and control of switching regulators in some detail, while the third is a broader text that provides a view of a range of applications:

[1] Kislovski, A.S., Redl, R., and Sokal, N.O., *Dynamic Analysis of Switching-Mode DC/DC Converters*, Van Nostrand Reinhold, New York, 1991.

[2] Mitchell, D.M., *Switching Regulator Analysis*. McGraw-Hill, New York, 1988.

[3] Mohan, N., Undeland, T.M., and Robbins, W.P., *Power Electronics: Converters, Applications, and Design*, Wiley, New York, 1995.

78.2 Motion Control with Electric Motors by Input-Output Linearization[2]

David G. Taylor, Georgia Institute of Technology, School of Electrical and Computer Engineering, Atlanta, GA

78.2.1 Introduction

Due to the increasing availability of improved power electronics and digital processors at reduced costs, there has been a trend to seek higher performance from electric machine systems through the design of more sophisticated control systems software. There exist significant challenges in the search for improved control system designs, however, since the dynamics of most electric machine systems exhibit significant nonlinearities, not all state variables are necessarily measured, and the parameters of the system can vary significantly from their nominal values.

Electric machines are electromechanical energy converters, used for both motor drives and power generation. Nearly all electric power used throughout the world is generated by synchronous machines (operated as generators), and a large fraction of all this electric power is consumed by induction machines (operated as motors). The various types of electric machines in use differ with respect to construction materials and features, as well as the underlying principles of operation.

The first DC machine was constructed by Faraday around 1820, the first practical version was made by Henry in 1829, and the first commercially successful version was introduced in 1837. The three-phase induction machine was invented by Tesla around 1887. Although improved materials and manufacturing methods continue to refine electric machines, the fundamental issues relating to electromechanical energy conversion have been established for well over a century.

In such an apparently well-established field, it may come as a surprise that today there is more research and development

[2]This work was supported in part by the National Science Foundation under Grant ECS-9158037 and by the Air Force Office of Scientific Research under Grant F49620-93-1-0147.

activity than ever before. Included in a modern electric machine system is the electric machine itself, power electronic circuits, electrical and/or mechanical sensors, and digital processors equipped with various software algorithms. The recent developments in power semiconductors, digital electronics, and permanent-magnet materials have led to "enabling technology" for today's advanced electric machine systems. Perhaps more than any other factor, the increasing use of computers, for both the design of electric machines and for their real-time control, is enhancing the level of innovation in this field.

This chapter provides an overview of primarily one recent development in control systems design for electric machines operated as motor drives. The chapter takes a broad perspective in the sense that a wide variety of different machine types is considered, hopefully from a unifying point of view. On the other hand, in order to limit the scope substantially, an effort was made to focus on one more recent nonlinear control method, specifically input-output linearization, as opposed to the classical methods which have less potential for achieving high dynamic performance. An unavoidable limitation of the presentation is a lack of depth and detail beyond the specific topic of input-output linearization; however, the intention was to highlight nonlinear control technology for electric machines to a broad audience, and to guide the interested reader to a few appropriate sources for further study.

78.2.2 Background

Input-Output Linearization

The most common control designs for electric machines today, for applications requiring high dynamic performance, are based on forms of exact linearization [8]. The design concept is reflected by a two-loop structure of the controller: in the first design step, nonlinear compensation is sought which explicitly cancels the nonlinearities present in the motor (without regard to any specific control objective), and this nonlinear compensation is implemented as an inner feedback loop; in the second design step, linear compensation is derived on the basis of the resulting linear dynamics of the precompensated motor to achieve some particular control objective, and this linear compensation is implemented as an outer feedback loop. The advantage of linear closed-loop dynamics is clearly that selection of controller parameters is simplified, and the achievable transient responses are very predictable.

Not all nonlinear systems can be controlled in this fashion; the applicability of exact linearization is determined by the type and location of the model nonlinearities. Furthermore, exact linearization is not really a single methodology, but instead represents two distinct notions of linearizability, though in both cases the implementation requires full state feedback. In the first case, exact linearization of the input-output dynamics of a system is desired, with the output taken to be the controlled variables. This case, referred to as input-output linearization, is the more intuitive form of exact linearization, but can be applied in a straightforward way only to so-called minimum-phase systems (those systems with stable zero dynamics). A system can be input-output linearized if it has a well-defined relative degree (see [8] for clarification). In the second case, exact linearization of the entire state-space dynamics of a system is desired, and no output needs to be declared. This case, referred to as input-state linearization, has the advantage of eliminating any potential difficulties with internal dynamics but is less intuitively appealing and can be more difficult to apply in practice. Input-state linearization applies only to systems that are characterized by integrable, or nonsingular and involutive, distributions (see [8] for clarification).

Standard models of most electric machines are exactly linearizable, in the sense(s) described above. Prior literature has disclosed many examples of exact linearization applied to electric machines, including experimental implementation, for various machine types and various types of models. For any machine type, the most significant distinction in the application of exact linearization relates to the order of the model used. When full-order models are used, the stator voltages are considered to be the control inputs. When a reduced-order model is used, the assignment of the control inputs depends on how the order reduction has been performed: if the winding inductance is neglected, then voltage will still be the input but the mechanical subsystem model will be altered; if a high-gain current loop is employed, then current will be the input in an unaltered mechanical subsystem model. In either case, exact linearization provides a systematic method for designing the nonlinearity compensation within the inner nonlinear loop, so that the outer linear loop is concerned only with the motion control part of the total system.

Relation to Field Orientation

Prior to the development of a formal theory for exact linearization design, closely related nonlinear feedback control schemes had already been developed for the induction motor. The classical "field oriented control," introduced in [1] over 20 years ago, involves the transformation of electrical variables into a frame of reference which rotates with the rotor flux vector (the dq frame). This reference frame transformation, together with a nonlinear feedback, serves to reduce the complexity of the dynamic equations, provided that the rotor flux is not identically zero. Under this one restriction, the rotor flux amplitude dynamics are made linear and decoupled and, moreover, if the rotor flux amplitude is regulated to a constant value, the speed dynamics will also become linear and decoupled. Provided that the rotor flux amplitude may be kept constant, the field oriented control thus achieves an asymptotic linearization and decoupling, where the d-axis voltage controls rotor flux amplitude and the q-axis voltage controls speed.

Although the field oriented approach to induction motor control is widely used today and has achieved considerable success, the formal use of exact linearization design can provide alternative nonlinear control systems of comparable complexity, but achieving true (as opposed to asymptotic) linearization and decoupling of flux and torque or speed. For example, using a reduced-order electrical model (under the assumption that the rotor speed is constant), an input-state linearization static state-feedback design is reported in [6] that achieves complete decoupling of the rotor flux amplitude and torque responses. The full-order electromechanical model of an induction motor turns out not to be input-state linearizable [13]. However, as shown in [12] (see also [13]), input-output linearization methods do apply to the full-order electromechanical model.

With rotor flux amplitude and speed chosen as outputs to be controlled, simple calculations show that the system has well-defined relative degree {2, 2}, provided that the rotor flux is not identically zero. Hence, under the constraint of nonzero rotor flux, it is possible to derive a nonlinear static state-feedback that controls rotor flux amplitude and speed in a noninteracting fashion, with linear second-order transients for each controlled output (and with bounded first-order internal dynamics [13]). Although the full-order induction motor model is not input-state linearizable, the augmented system obtained by adding an integrator to one of the inputs does satisfy this property locally.

Performance Optimization

Although the difference between the classical control of [1] and the exact linearization control of [12, 13] may appear to be a minor one, the complete decoupling of speed and flux dynamics in the closed-loop system (during transients as well as in steady state) provides the opportunity to optimize performance. For example, as mentioned in [13], the flux reference will need to be reduced from nominal as the speed reference is increased above rated speed, in order to keep the required feed voltages within the inverter limits. Operation in this flux-weakening regime will excite the coupling between flux and speed in the classical field oriented control, causing undesired speed fluctuations (and perhaps instability).

There are other motivations for considering time-varying flux references as well. For instance, in [9, 10] the flux is adjusted as a function of speed in order to maximize power efficiency (i.e., only the minimum stator input power needed to operate at the desired speed is actually sourced). In [9], the flux reference is computed on the basis of predetermined relationships derived off-line and the control is implemented in a reference frame rotating with the stator excitation. In [10], the flux reference is computed on-line using a minimum power search method and the control is implemented in a fixed stator frame of reference. Yet another possibility, presented in [2], would be to vary the flux reference as a function of speed in order to achieve optimum torque (maximum for acceleration and minimum for deceleration) given limits on allowable voltage and current. In each of these references, high-gain current loops are used so that the exact linearization is performed with respect to current inputs rather than voltage inputs. Clearly, exact linearization permits optimization goals (which require variable flux references) and high dynamic performance in position, speed, or torque control, to be achieved simultaneously.

Wide Applicability

Exact linearization has been suggested for the control of many other types of electric machines as well. Both input-state linearization and input-output linearization are used to design controllers for wound-field brush-commutated DC motors, in [4, 5, 14]. For various permanent-magnet machines, there are many references illustrating the use of exact linearization. For instance, in [7], a three-phase wye-connected permanent-magnet synchronous motor with sinusoidally distributed windings is modeled with piecewise-constant parameters (which depend on current to account for reluctance variations and magnetic saturation), and an input-state linearization controller is derived from the rotor reference frame model. In [16], input-state linearization is applied to the hybrid permanent-magnet stepper motor with cogging torque accounted for, and it is further shown how constant load torques may be rejected using a nonlinear observer. This work was continued in [3], where the experimental implementation is described, including treatment of practical issues such as speed estimation from position sensors and operation at high speeds despite voltage limitations. Optimization objectives can be considered within exact linearization designs for these other types of machines too.

Review of Theory

In order to appreciate the concept of input-output linearization as it applies to electric motor drives, a brief review of the general theory is called for. See also Chapters 55 and 57.

For the present purposes, it is sufficient to consider nonlinear multivariable systems of the form

$$\dot{x} = f(x) + \sum_{i=1}^{m} g_i(x)u_i \tag{78.30}$$

$$y_j = h_j(x), \quad j = 1, \ldots, m \tag{78.31}$$

where $x \in R^n$ is the state vector, $u \in R^m$ is the input vector (i.e., the control), and $y \in R^m$ is the output vector (i.e., to be controlled). Note that this is a square system, with the same number of inputs as outputs.

Given such a nonlinear system, it is said to possess relative degree $\{r_1, \ldots, r_m\}$ at x^0 if

$$L_{g_i} L_f^k h_j(x) \equiv 0 \quad \forall\, k < r_j - 1 \quad 1 \leq i, j \leq m \tag{78.32}$$

and

$$\text{rank} \begin{bmatrix} L_{g_1} L_f^{r_1-1} h_1(x^0) & \cdots & L_{g_m} L_f^{r_1-1} h_1(x^0) \\ \vdots & \ddots & \vdots \\ L_{g_1} L_f^{r_m-1} h_m(x^0) & \cdots & L_{g_m} L_f^{r_m-1} h_m(x^0) \end{bmatrix} = m \tag{78.33}$$

The notation in Equations 78.32 and 78.33 stands for the Lie-derivative of a scalar function with respect to a vector function (see [8]). The first property implies that a chain structure exists, whereas the second property implies that input-output decoupling is possible.

For nonlinear systems possessing a well-defined relative degree, a simple diffeomorphic change of coordinates will bring the system into so-called normal form. In particular, under the change of variables

$$z_{1j} = h_j(x) \;\; z_{2j} = L_f h_j(x) \;\; \cdots \;\; z_{r_j j} = L_f^{r_j-1} h_j(x) \tag{78.34}$$

it is easy to show that $y_j = z_{1j}$ where

$$\begin{aligned}
\dot{z}_{1j} &= L_f^1 h_j(x) + \sum_{i=1}^{m} \underbrace{L_{g_i} L_f^0 h_j(x)}_{\equiv 0} u_i = z_{2j} \\
\dot{z}_{2j} &= L_f^2 h_j(x) + \sum_{i=1}^{m} \underbrace{L_{g_i} L_f^1 h_j(x)}_{\equiv 0} u_i = z_{3j} \\
&\vdots \\
\dot{z}_{r_j j} &= L_f^{r_j} h_j(x) + \sum_{i=1}^{m} \underbrace{L_{g_i} L_f^{r_j-1} h_j(x)}_{\neq 0} u_i
\end{aligned} \tag{78.35}$$

In other words, the output to be controlled, y_j, is the output of a chain of cascaded integrators, fed by a nonlinear but invertible forcing term. If $r_1 + \cdots + r_m < n$, then not all of the system's dynamics are accounted for in Equation 78.35. However, any remaining internal dynamics will not influence the input-output response, once the nonlinear feedback described below is applied.

More specifically, for nonlinear systems with well-defined relative degree, it is possible to solve for the control vector u from the system of algebraic equations

$$\begin{bmatrix} y_1^{(r_1)} \\ \vdots \\ y_m^{(r_m)} \end{bmatrix} = \underbrace{\begin{bmatrix} L_f^{r_1} h_1(x) \\ \vdots \\ L_f^{r_m} h_m(x) \end{bmatrix}}_{b(x)} \tag{78.36}$$

$$+ \underbrace{\begin{bmatrix} L_{g_1} L_f^{r_1-1} h_1(x) & \cdots & L_{g_m} L_f^{r_1-1} h_1(x) \\ \vdots & \ddots & \vdots \\ L_{g_1} L_f^{r_m-1} h_m(x) & \cdots & L_{g_m} L_f^{r_m-1} h_m(x) \end{bmatrix}}_{A(x)} \begin{bmatrix} u_1 \\ \vdots \\ u_m \end{bmatrix}$$

for all x near x^0, given some desired choice v for the term on the left-hand side where the superscript $^{(\cdot)}$ denotes time differentiation. Hence, the inner-loop nonlinearity compensation is performed by

$$u = \alpha(x) + \beta(x) v \tag{78.37}$$

where

$$\alpha(x) = -A^{-1}(x) b(x) \quad \beta(x) = A^{-1}(x) \tag{78.38}$$

which means that for each $j = 1, \ldots, m$

$$y_j^{(r_j)} = v_j \tag{78.39}$$

or, in state variable form,

$$\dot{z}_j = \tilde{A}_j z_j + \tilde{b}_j v_j \tag{78.40}$$

$$y_j = \tilde{c}_j^T z_j \tag{78.41}$$

where

$$\tilde{A}_j = \begin{bmatrix} 0 & 1 & \cdots & 0 \\ 0 & 0 & \ddots & 0 \\ \vdots & \vdots & \ddots & 1 \\ 0 & 0 & \cdots & 0 \end{bmatrix} \quad \tilde{b}_j = \begin{bmatrix} 0 \\ \vdots \\ 0 \\ 1 \end{bmatrix}$$

$$\tilde{c}_j^T = [\, 1 \;\; 0 \;\; \cdots \;\; 0 \,] \tag{78.42}$$

and

$$z_j = \begin{bmatrix} h_j(x) \\ \vdots \\ L_f^{r_j-1} h_j(x) \end{bmatrix} \tag{78.43}$$

To implement this inner-loop design, it is necessary to measure the state vector x, and to have accurate knowledge of the system model nonlinearities $f, g_1, \ldots, g_m, h_1, \ldots, h_m$.

The outer-loop design, which is problem dependent, is very straightforward due to the linearity and complete decoupling of the input-output dynamics, as given in Equation 78.39. The outer-loop feedback design will be nonlinear with respect to the original state x, but linear with respect to the computed state of the normal form z. As an example of outer-loop design, to guarantee that $y_j \to y_{jd}$ as $t \to \infty$, given a desired output trajectory y_{jd} and its first r_j time derivatives, it suffices to apply

$$v_j = -k_j^T (z_j - z_{jd}) + y_{jd}^{(r_j)} \tag{78.44}$$

where z_j is computed from x according to Equation 78.43, the desired state trajectory vector z_{jd} is

$$z_{jd} = \begin{bmatrix} y_{jd} \\ \vdots \\ y_{jd}^{(r_j-1)} \end{bmatrix} \tag{78.45}$$

and where the gain vector k_j is selected such that the matrix $\tilde{A}_j - \tilde{b}_j k_j^T$ has left-half plane eigenvalues. Implementation of this outer-loop design requires measurement of the state vector x, and accurate knowledge of the nonlinearities $f, h_1, \ldots, h_m$.

In the remainder of this chapter, the formalism outlined above on the theory of input-output linearization will be applied to a variety of DC and AC motor drive types. This review section has established the notion of relative degree as the critical feature of input-output linearization, and hence this chapter will now focus on this feature in particular and will bypass the explicit computation of the feedback, defined by Equations 78.37 and 78.44, for each motor. The objective is to provide, within a single self-contained document, a catalog of the main issues to be addressed when designing input-output linearizing controllers for electric motors. The consistent notation which will be used also adds to the clarity and usefulness of the results. The process begins with the fundamental modeling tools, such as those presented in [11], but ends with a formulation of the state variable models of various machines, and the corresponding relative degree checks. Implementation of input-output linearization for electric motors will thus be a straightforward extrapolation from the contents of this chapter.

78.2.3 DC Motors

The most logical point of departure for this catalog of results on electric motor input-output linearizability would be DC motors with separate field and armature windings, since these motors are simpler to model than AC motors yet possess a significant nonlinearity. Field coils are used to establish an air gap flux between stationary iron poles and the rotating armature. The armature has axially directed conductors which are connected to a brush commutator (a mechanical switch), and these conductors are continuously switched such that those located under a pole carry similarly directed currents. Interaction of axially directed armature currents and radially directed field flux produces a shaft torque.

Dynamic Model

The mechanical dynamic equation which models the rotor velocity ω is the same for all types of electric motors. Under the assumption that the mechanical load consists only of a constant inertia J, viscous friction with friction coefficient B, and a constant load torque T_l, the mechanical dynamics are given by

$$J\dot{\omega} = T_e - B\omega - T_l \tag{78.46}$$

Each type of electric motor has its own unique expression for electrical torque T_e, which for DC motors is

$$T_e = Mi_f i_a \tag{78.47}$$

where M designates the mutual inductance between the field and armature windings which carry currents i_f and i_a, respectively. The electrical dynamic equations describing the flow of currents in the field and armature windings are

$$v_f = R_f i_f + L_f \frac{di_f}{dt} \tag{78.48}$$

$$v_a = R_a i_a + M i_f \omega + L_a \frac{di_a}{dt} \tag{78.49}$$

where v_f and v_a are the voltages applied to the field and armature, R_f and R_a are the resistances of the field and armature, and L_f and L_a are the self inductances of the field and armature, respectively. The above model is complete for the case where separate voltage sources are used to excite the field and armature windings. For further modeling details, see [11].

When it is desired to operate the DC motor from a single voltage source v, the two windings must be connected together in parallel (shunt connection) or in series (series connection). In either of these configurations, operation of the motor at high velocities without exceeding the source limits is made possible by including an external variable resistance to limit the field flux. The electrical dynamic equations then become

$$v = (R_f + R_x) i_f + L_f \frac{di_f}{dt} \tag{78.50}$$

$$v = R_a i_a + M i_f \omega + L_a \frac{di_a}{dt} \tag{78.51}$$

for the shunt-wound motor, where external resistance R_x is in series with the field winding, and

$$0 = (R_f + R_x) i_f - R_x i_a + L_f \frac{di_f}{dt} \tag{78.52}$$

$$v = (R_a + R_x) i_a - R_x i_f + M i_f \omega + L_a \frac{di_a}{dt} \tag{78.53}$$

for the series-wound motor, where external resistance R_x is in parallel with the field winding. Operation without field weakening requires $R_x = 0$ for the shunt-wound motor, and $R_x = \infty$ for the series-wound motor. For the latter case, as $R_x \to \infty$ an order reduction occurs since the currents flowing in the field and armature windings are identical in the limit.

In order to determine the extent to which the various operating modes of DC motors are input-output linearizable, it is necessary to determine the state-variable models and to assess the relative degree of these models. For sake of brevity, only position control will be taken as a primary control objective; speed control and torque control follow in an obvious manner. Hence, all models will include a state equation for rotor position θ (although this is typically unnecessary for speed and torque control).

Separately Excited DC Motor

Consider first the separately excited DC motor, in which the field and armature windings are fed from separate voltage sources. In order to match the common notation used for the review of input-output linearization principles, the variables of the separately excited DC motor are assigned according to

$$x_1 = \theta \quad x_2 = \omega \quad x_3 = i_f \quad x_4 = i_a \quad u_1 = v_f \quad u_2 = v_a \quad y_1 = \theta \tag{78.54}$$

Note that only one output, associated with the position control objective, is specified at this point. Using Equations 78.46 to 78.49, these assignments lead to the state variable model defined by

$$f(x) = \begin{bmatrix} x_2 \\ \frac{1}{J}\left(Mx_3x_4 - Bx_2 - T_l\right) \\ -\frac{1}{L_f}\left(R_f x_3\right) \\ -\frac{1}{L_a}\left(R_a x_4 + M x_2 x_3\right) \end{bmatrix} \quad g_1(x) = \begin{bmatrix} 0 \\ 0 \\ \frac{1}{L_f} \\ 0 \end{bmatrix}$$

$$g_2(x) = \begin{bmatrix} 0 \\ 0 \\ 0 \\ \frac{1}{L_a} \end{bmatrix} \quad h_1(x) = x_1 \tag{78.55}$$

In order to check the relative degree according to the definition given in Equations 78.30 to 78.33, a second output to be controlled needs to be declared, so that the system will be square. Nevertheless, the various Lie-derivative calculations associated with the first output (rotor position) are easily verified to be

$$\begin{array}{lllllll} L_{g_1}h_1(x) & \equiv & 0 & \quad & L_{g_2}h_1(x) & \equiv & 0 \\ L_{g_1}L_f h_1(x) & \equiv & 0 & & L_{g_2}L_f h_1(x) & \equiv & 0 \\ L_{g_1}L_f^2 h_1(x) & = & \frac{M}{JL_f}x_4 & & L_{g_2}L_f^2 h_1(x) & = & \frac{M}{JL_a}x_3 \end{array} \tag{78.56}$$

If the second output is chosen to be the field current i_f, i.e., if $h_2(x) = x_3$, then the remaining Lie-derivative calculations are given by

$$L_{g_1}h_2(x) = \frac{1}{L_f} \quad L_{g_2}h_2(x) = 0 \tag{78.57}$$

In this case, the calculation

$$\det\begin{bmatrix} \frac{M}{JL_f}x_4 & \frac{M}{JL_a}x_3 \\ \frac{1}{L_f} & 0 \end{bmatrix} = -\frac{M}{JL_aL_f}x_3 \tag{78.58}$$

indicates that the decoupling matrix is nonsingular almost globally, and that the relative degree is well defined almost globally, i.e.,

$$\{r_1, r_2\} = \{3, 1\} \;\left(\text{if } i_f \neq 0\right) \tag{78.59}$$

Note that the singularity at $i_f = 0$ corresponds to a particular value of one of the controlled outputs, namely $y_2 = 0$. Consequently, this singularity is easily avoidable during operation and can be handled at start-up without difficulty.

If instead the second output is chosen to be the armature current i_a, i.e., if $h_2(x) = x_4$, then the remaining Lie-derivative calculations

$$L_{g_1}h_2(x) = 0 \quad L_{g_2}h_2(x) = \frac{1}{L_a} \tag{78.60}$$

lead to a decoupling matrix

$$\det\begin{bmatrix} \frac{M}{JL_f}x_4 & \frac{M}{JL_a}x_3 \\ 0 & \frac{1}{L_a} \end{bmatrix} = \frac{M}{JL_aL_f}x_4 \tag{78.61}$$

which again is nonsingular almost globally. This implies that the relative degree is again well defined almost globally, i.e.,

$$\{r_1, r_2\} = \{3, 1\} \;(\text{if } i_a \neq 0) \tag{78.62}$$

with a singularity when $i_a = 0$. Again the singularity corresponds to a particular value of one of the controlled outputs, namely $y_2 = 0$. Consequently, this singularity is easily avoidable during operation and can be handled at start-up without difficulty.

Provided that some alternative start-up procedure is used to establish a nonzero current in the appropriate winding and that the commanded second output is chosen to be away from zero, the nonlinearity compensation defined by $\alpha(x)$ and $\beta(x)$ in Equations 78.37 and 78.38 is well defined and easily implemented for the separately excited DC motor. Unfortunately, the value of input-output linearization for the single-source excitation strategies is significantly more limited.

Shunt Wound DC Motor

Consider now the shunt wound DC motor, in which a single source excites both field and armature windings due to the parallel connection between the two windings (with external resistance in series with the field winding to limit the field flux if desired). The variable assignments

$$x_1 = \theta \quad x_2 = \omega \quad x_3 = i_f \quad x_4 = i_a \quad u = v \quad y = \theta \tag{78.63}$$

are essentially the same as before, with the exception that just one source voltage v is present. With this variable assignment, the state variable model of the shunt wound DC motor from Equations 78.50 and 78.51 becomes

$$f(x) = \begin{bmatrix} x_2 \\ \frac{1}{J}\left(Mx_3x_4 - Bx_2 - T_l\right) \\ -\frac{1}{L_f}\left(R_f + R_x\right)x_3 \\ -\frac{1}{L_a}\left(R_ax_4 + Mx_2x_3\right) \end{bmatrix} \tag{78.64}$$

$$g(x) = \begin{bmatrix} 0 \\ 0 \\ \frac{1}{L_f} \\ \frac{1}{L_a} \end{bmatrix} \qquad h(x) = x_1$$

The evaluation of relative degree is simpler than before, since the shunt wound DC motor is a single-input system, and there is no need to consider defining a second output. The Lie-derivative calculations

$$L_gh(x) \equiv 0 \quad L_gL_fh(x) \equiv 0 \quad L_gL_f^2h(x) = \frac{M}{J}\left(\frac{x_4}{L_f} + \frac{x_3}{L_a}\right) \tag{78.65}$$

indicate that this system has a well-defined relative degree, except when the field and armature currents satisfy a particular algebraic constraint. In particular, it is clear that

$$r = 3 \quad \left(\text{if } i_f \neq -\frac{L_a}{L_f}i_a\right) \tag{78.66}$$

Note that the singularity occurs in a region of the state-space which is not directly defined by the value of the output variable (rotor position). Hence, singularity avoidance is no longer a simple matter. Most important, though, is the fact that the interconnected windings impose the constraint of unipolar torque (i.e., either positive torque or negative torque); hence, it is customary to use just a unipolar voltage source. Consequently, the shunt wound DC motor is not nearly as versatile in operation as the separately excited DC motor, despite the fact that it possesses well-defined relative degree for a large subset of the state-space.

Series Wound DC Motor

Consider now the series wound DC motor, in which a single source excites both field and armature windings due to the series connection between the two windings (with external resistance in parallel with the field winding to limit the field flux if desired). The variable assignments

$$x_1 = \theta \quad x_2 = \omega \quad x_3 = i_f \quad x_4 = i_a \quad u = v \quad y = \theta \tag{78.67}$$

are the same as for the shunt wound DC motor, with the exception that when field weakening is unnecessary ($R_x = \infty$) the order of the model effectively drops (x_3 and x_4 are not independent). With this variable assignment, the state variable model of the series wound DC motor from Equations 78.52 and 78.53 becomes

$$f(x) = \begin{bmatrix} x_2 \\ \frac{1}{J}\left(Mx_3x_4 - Bx_2 - T_l\right) \\ -\frac{1}{L_f}\left((R_f + R_x)x_3 - R_xx_4\right) \\ -\frac{1}{L_a}\left((R_a + R_x)x_4 - R_xx_3 + Mx_2x_3\right) \end{bmatrix}$$

$$g(x) = \begin{bmatrix} 0 \\ 0 \\ 0 \\ \frac{1}{L_a} \end{bmatrix} \qquad h(x) = x_1 \qquad (78.68)$$

The evaluation of relative degree for this system is completed by computing the Lie-derivatives

$$L_g h(x) \equiv 0 \quad L_g L_f h(x) \equiv 0 \quad L_g L_f^2 h(x) = \frac{M}{JL_a} x_3 \qquad (78.69)$$

and the result is that relative degree is well defined provided that the field current is nonzero, i.e.,

$$r = 3 \quad \left(\text{if } i_f \neq 0\right) \qquad (78.70)$$

Again the singularity occurs in a region of the state-space which is not directly defined by the value of the output variable (rotor position). Hence, singularity avoidance is again not so simple. Moreover, as before, the interconnected windings impose the constraint of unipolar torque (i.e., either positive torque or negative torque), so a unipolar voltage source would be used. Consequently, the series wound DC motor is also not nearly as versatile in operation as the separately excited DC motor, despite the fact that it possesses well-defined relative degree for a large subset of the state-space.

78.2.4 AC Induction Motors

The most appropriate AC motor to consider first would be the induction motor, due to its symmetry. This motor is without doubt the most commonly used motor for a wide variety of industrial applications.

Dynamic Model

The three-phase, wye-connected induction motor is constructed from a magnetically smooth stator and rotor. The stator is wound with identical sinusoidally distributed windings displaced 120°, with resistance R_s, and these windings are wye-connected. The rotor may be considered to be wound with three identical short-circuited and wye-connected sinusoidally distributed windings displaced 120°, with resistance R_r.

The voltage equations in machine variables may be expressed by

$$\begin{bmatrix} v_s \\ 0 \end{bmatrix} = \begin{bmatrix} \mathcal{R}_s & 0 \\ 0 & \mathcal{R}_r \end{bmatrix} \begin{bmatrix} i_s \\ i_r \end{bmatrix} + \begin{bmatrix} \dot{\lambda}_s \\ \dot{\lambda}_r \end{bmatrix} \qquad (78.71)$$

where v_s is the vector of stator voltages, i_s is the vector of stator currents, λ_s is the vector of stator flux linkages, i_r is the vector of induced rotor currents, and λ_r is the vector of induced rotor flux linkages. The resistance matrices for the stator and rotor windings are

$$\mathcal{R}_s = \begin{bmatrix} R_s & 0 & 0 \\ 0 & R_s & 0 \\ 0 & 0 & R_s \end{bmatrix} \quad \mathcal{R}_r = \begin{bmatrix} R_r & 0 & 0 \\ 0 & R_r & 0 \\ 0 & 0 & R_r \end{bmatrix}. \qquad (78.72)$$

Denoting any of the above stator or rotor vectors (voltage, current, flux) by generic notation f_s or f_r, respectively, the vector structure will be

$$f_s = \begin{bmatrix} f_{as} & f_{bs} & f_{cs} \end{bmatrix}^T \qquad (78.73)$$

$$f_r = \begin{bmatrix} f_{ar} & f_{br} & f_{cr} \end{bmatrix}^T \qquad (78.74)$$

where components are associated with phases a, b, c in machine variables. Assuming magnetic linearity, the flux linkages may be expressed by

$$\begin{bmatrix} \lambda_s \\ \lambda_r \end{bmatrix} = \begin{bmatrix} \mathcal{L}_s & \mathcal{L}_m(\theta) \\ \mathcal{L}_m^T(\theta) & \mathcal{L}_r \end{bmatrix} \begin{bmatrix} i_s \\ i_r \end{bmatrix} \qquad (78.75)$$

with inductance matrices

$$\mathcal{L}_s = \begin{bmatrix} L_s & M_s & M_s \\ M_s & L_s & M_s \\ M_s & M_s & L_s \end{bmatrix} \qquad (78.76)$$

$$\mathcal{L}_m(\theta) = \qquad (78.77)$$

$$M \begin{bmatrix} \cos(N\theta) & \cos(N\theta + \frac{2\pi}{3}) & \cos(N\theta - \frac{2\pi}{3}) \\ \cos(N\theta - \frac{2\pi}{3}) & \cos(N\theta) & \cos(N\theta + \frac{2\pi}{3}) \\ \cos(N\theta + \frac{2\pi}{3}) & \cos(N\theta - \frac{2\pi}{3}) & \cos(N\theta) \end{bmatrix}$$

$$\mathcal{L}_r = \begin{bmatrix} L_r & M_r & M_r \\ M_r & L_r & M_r \\ M_r & M_r & L_r \end{bmatrix} \qquad (78.78)$$

where L_s is the stator self-inductance, M_s is the stator-to-stator mutual inductance, L_r is the rotor self-inductance, M_r is the rotor-to-rotor mutual inductance, M is the magnitude of the stator-to-rotor mutual inductances which depend on rotor angle θ, and N is the number of pole pairs.

For the mechanical dynamics, the differential equation for rotor velocity ω is

$$J\dot{\omega} = T_e - B\omega - T_l \qquad (78.79)$$

where the torque of electrical origin may be determined from the inductance matrices using the general expression

$$T_e = \frac{1}{2} i^T \frac{d\mathcal{L}(\theta)}{d\theta} i = -\frac{1}{2} \lambda^T \frac{d\mathcal{L}^{-1}(\theta)}{d\theta} \lambda \qquad (78.80)$$

where $\mathcal{L}(\theta)$, i, λ denote the complete inductance matrix, current vector, and flux linkage vector appearing in Equation 78.75. For further modeling details, see [11].

Reference Frame Transformation

Because of the explicit dependence of the voltage equations on rotor angle θ, direct analysis of induction motor operation using machine variables is quite difficult. Even the determination of steady-state operating points is not straightforward. Hence, it is customary to perform a nonsingular change of variables, called a reference frame transformation, in order to effectively replace the variables associated with the physical stator and/or rotor windings with variables associated with fictitious windings oriented within the specified frame of reference.

For the symmetrical induction motor, the interaction between stator and rotor can be easily understood by considering a reference frame fixed on the stator, fixed on the rotor, rotating in synchronism with the applied stator excitation, or even rotating with an arbitrary velocity. The generality with which reference frame transformations may be applied with success to the induction motor is due to the assumed symmetry of both the stator and rotor windings.

For present purposes, it suffices to consider a stationary frame of reference located on the stator. In the new coordinates, there will be a q-axis aligned with phase a on the stator, an orthogonal d-axis located between phase a and phase c on the stator, and a 0-axis which carries only trivial information. The transformation matrix used to express rotor variables in this reference frame is

$$K_{rs}(\theta) = \tag{78.81}$$

$$\sqrt{\frac{2}{3}}\begin{bmatrix} \cos(N\theta) & \cos(N\theta+\frac{2\pi}{3}) & \cos(N\theta-\frac{2\pi}{3}) \\ -\sin(N\theta) & -\sin(N\theta+\frac{2\pi}{3}) & -\sin(N\theta-\frac{2\pi}{3}) \\ \frac{1}{\sqrt{2}} & \frac{1}{\sqrt{2}} & \frac{1}{\sqrt{2}} \end{bmatrix}$$

which is naturally θ-dependent, and the transformation matrix used to transform the stator variables is the constant matrix

$$K_{ss} = \sqrt{\frac{2}{3}}\begin{bmatrix} 1 & -\frac{1}{2} & -\frac{1}{2} \\ 0 & -\frac{\sqrt{3}}{2} & \frac{\sqrt{3}}{2} \\ \frac{1}{\sqrt{2}} & \frac{1}{\sqrt{2}} & \frac{1}{\sqrt{2}} \end{bmatrix} \tag{78.82}$$

Both matrices are orthonormal, meaning that they are constructed using orthogonal unit vectors, and hence their inverses are equal to their transposes. Formally stated, the change of variables considered is defined by

$$\tilde{f}_r = \begin{bmatrix} f_{qr} & f_{dr} & f_{0r} \end{bmatrix}^T \quad \tilde{f}_r = K_{rs}(\theta) f_r \tag{78.83}$$

$$\tilde{f}_s = \begin{bmatrix} f_{qs} & f_{ds} & f_{0s} \end{bmatrix}^T \quad \tilde{f}_s = K_{ss} f_s \tag{78.84}$$

where the tilde represents the appropriate stator or rotor variable in the new coordinates.

Since the windings on both the stator and rotor are wye-connected, the sum of the stator currents, as well as the sum of the rotor currents, must always be equal to zero. In other words, $i_{as}+i_{bs}+i_{cs}=0$ and $i_{ar}+i_{br}+i_{cr}=0$. Note that the reference frame transformation, primarily intended to eliminate θ from the voltage equations, will also satisfy the algebraic current constraint by construction.

After some tedious but straightforward algebra, it can be shown that the transformed voltage equations become

$$v_{qs} = R_s i_{qs} + \dot{\lambda}_{qs} \tag{78.85}$$

$$v_{ds} = R_s i_{ds} + \dot{\lambda}_{ds} \tag{78.86}$$

$$v_{0s} = R_s i_{0s} + \dot{\lambda}_{0s} \tag{78.87}$$

$$0 = R_r i_{qr} - N\omega\lambda_{dr} + \dot{\lambda}_{qr} \tag{78.88}$$

$$0 = R_r i_{dr} + N\omega\lambda_{qr} + \dot{\lambda}_{dr} \tag{78.89}$$

$$0 = R_r i_{0r} + \dot{\lambda}_{0r} \tag{78.90}$$

and the transformed flux linkage equations become

$$\begin{bmatrix} \lambda_{qs} \\ \lambda_{ds} \\ \lambda_{0s} \\ \lambda_{qr} \\ \lambda_{dr} \\ \lambda_{0r} \end{bmatrix} = \begin{bmatrix} L_s-M_s & 0 & 0 & \frac{3}{2}M & 0 & 0 \\ 0 & L_s-M_s & 0 & 0 & \frac{3}{2}M & 0 \\ 0 & 0 & L_s+2M_s & 0 & 0 & 0 \\ \frac{3}{2}M & 0 & 0 & L_r-M_r & 0 & 0 \\ 0 & \frac{3}{2}M & 0 & 0 & L_r-M_r & 0 \\ 0 & 0 & 0 & 0 & 0 & L_r+2M_r \end{bmatrix} \begin{bmatrix} i_{qs} \\ i_{ds} \\ i_{0s} \\ i_{qr} \\ i_{dr} \\ i_{0r} \end{bmatrix} \tag{78.91}$$

Note that all dependence on rotor angle θ has been eliminated and, hence, all analysis is substantially simplified. On the other hand, this change of variables does not eliminate nonlinearity entirely. Note also that $i_{0s}=i_{0r}=0$ so that the 0-axis equations from above can be completely ignored.

The transformed electrical model is presently expressed in terms of the two orthogonal components of current and flux linkage, on both the stator and the rotor. Any state variable description will require that half of the transformed electrical variables be eliminated. Six possible permutations are available to select from, namely (i_s, i_r), (λ_s, λ_r), (i_s, λ_r), (λ_s, i_r), (i_s, λ_s), and (i_r, λ_r). Substituting the transformed inductances of Equation 78.91 into the torque expression (Equation 78.80), these permutations result in six possible torque expressions, namely

$$T_e = \begin{cases} \frac{3}{2}MN(i_{qs}i_{dr} - i_{ds}i_{qr}) \\ \frac{3}{2}MN(\lambda_{qs}\lambda_{dr} - \lambda_{ds}\lambda_{qr}) \\ \frac{\frac{3}{2}MN}{L_r-M_r}(i_{qs}\lambda_{dr} - i_{ds}\lambda_{qr}) \\ \frac{\frac{3}{2}MN}{L_s-M_s}(\lambda_{qs}i_{dr} - \lambda_{ds}i_{qr}) \\ N(i_{qs}\lambda_{ds} - i_{ds}\lambda_{qs}) \\ N(\lambda_{qr}i_{dr} - \lambda_{dr}i_{qr}) \end{cases} \tag{78.92}$$

Any of these expressions is valid, provided that the voltage equations are rewritten in terms of the same set of electrical state variables.

Input-Output Linearizability

With the above modeling background, it is now possible to proceed with the main objective of determining the extent to which the induction motor is input-output linearizable. The only remaining modeling step is to select which permutation of electrical state variables to use, and then to construct the state variable model. Taking stator current and rotor flux as state variables

$$x_1=\theta \quad x_2=\omega \quad x_3=\lambda_{qr} \quad x_4=\lambda_{dr} \quad x_5=i_{qs} \quad x_6=i_{ds} \tag{78.93}$$

and selecting the orthogonal components of stator voltage as the two inputs, rotor angle θ as the primary output and rotor flux magnitude squared $\lambda_{qr}^2 + \lambda_{dr}^2$ as the secondary output, the remaining standard notation will be

$$u_1 = v_{qs} \quad u_2 = v_{ds} \quad y_1 = \theta \quad y_2 = \lambda_{qr}^2 + \lambda_{dr}^2 \tag{78.94}$$

Using the notations assigned above, the resulting state variable model is defined by

$$f(x) = \begin{bmatrix} x_2 \\ \frac{1}{J}\left(k(x_4x_5 - x_3x_6) - Bx_2 - T_l\right) \\ -\alpha_r x_3 + Nx_2x_4 + \beta_r x_5 \\ -\alpha_r x_4 - Nx_2x_3 + \beta_r x_6 \\ -\alpha_s x_5 - \gamma x_2x_4 + \beta_s x_3 \\ -\alpha_s x_6 + \gamma x_2x_3 + \beta_s x_4 \end{bmatrix}$$

$$g_1(x) = \begin{bmatrix} 0 \\ 0 \\ 0 \\ 0 \\ \delta \\ 0 \end{bmatrix} \qquad g_2(x) = \begin{bmatrix} 0 \\ 0 \\ 0 \\ 0 \\ 0 \\ \delta \end{bmatrix} \tag{78.95}$$

$$h_1(x) = x_1 \qquad h_2(x) = x_3^2 + x_4^2$$

with constant coefficients given by

$$\begin{aligned} k &= \frac{\frac{3}{2}MN}{L_r - M_r} \\ \alpha_r &= \frac{R_r}{L_r - M_r} \\ \beta_r &= \frac{R_r(\frac{3}{2}M)}{L_r - M_r} \\ \alpha_s &= \frac{R_s(L_r - M_r)^2 + R_r(\frac{3}{2}M)^2}{((L_s - M_s)(L_r - M_r) - (\frac{3}{2}M)^2)(L_r - M_r)} \\ \beta_s &= \frac{R_r(\frac{3}{2}M)}{((L_s - M_s)(L_r - M_r) - (\frac{3}{2}M)^2)(L_r - M_r)} \\ \gamma &= \frac{\frac{3}{2}MN}{(L_s - M_s)(L_r - M_r) - (\frac{3}{2}M)^2} \\ \delta &= \frac{L_r - M_r}{(L_s - M_s)(L_r - M_r) - (\frac{3}{2}M)^2} \end{aligned} \tag{78.96}$$

The nonlinearities of the induction motor are clearly apparent in Equation 78.95.

In order to check relative degree, the Lie-derivative calculations

$$\begin{array}{rclrcl} L_{g_1}h_1(x) & \equiv & 0 & L_{g_2}h_1(x) & \equiv & 0 \\ L_{g_1}L_f h_1(x) & \equiv & 0 & L_{g_2}L_f h_1(x) & \equiv & 0 \\ L_{g_1}L_f^2 h_1(x) & = & \frac{k\delta}{J}x_4 & L_{g_2}L_f^2 h_1(x) & = & -\frac{k\delta}{J}x_3 \end{array} \tag{78.97}$$

for the first output and

$$\begin{array}{rclrcl} L_{g_1}h_2(x) & \equiv & 0 & L_{g_2}h_2(x) & \equiv & 0 \\ L_{g_1}L_f h_2(x) & = & 2\beta_r\delta x_3 & L_{g_2}L_f h_2(x) & = & 2\beta_r\delta x_4 \end{array} \tag{78.98}$$

for the second output lead to a decoupling matrix with singularity condition

$$\det\begin{bmatrix} \frac{k\delta}{J}x_4 & -\frac{k\delta}{J}x_3 \\ 2\beta_r\delta x_3 & 2\beta_r\delta x_4 \end{bmatrix} = \frac{2k\beta_r\delta^2}{J}\left(x_3^2 + x_4^2\right) \tag{78.99}$$

and to the conclusion that

$$\{r_1, r_2\} = \{3, 2\} \text{ (if } y_2 \neq 0) \tag{78.100}$$

Hence, input-output linearization may be applied to the transformed model of the induction motor (and, hence, to the machine variable model of the induction motor via inverse reference frame transformations) provided that the rotor flux is nonzero. Since the singularity at zero rotor flux corresponds to a particular value of the secondary output variable, i.e., to $y_2 = 0$, it is easy to avoid this singularity in operation by commanding rotor fluxes away from zero and by using a start-up procedure to premagnetize the rotor prior to executing the input-output linearization calculations on-line.

78.2.5 AC Synchronous Motors

An important class of AC machines frequently used as actuators in control applications is the class of synchronous machines. Though these machines essentially share the same stator structure with the induction motor, the construction of the rotor is quite different and accounts for the asymmetry present in the modeling.

Dynamic Model

The class of synchronous machines contains several specific machines that are worth covering in this chapter, and these specific cases differ with respect to their rotor structures. These specific cases can all be considered to be special cases of a general synchronous machine. Hence, this section will begin with a presentation of the basic modeling equations for the general synchronous machine, and then will specialize these equations to the specific cases of interest prior to evaluating input-output linearizability for each specific case.

The general synchronous machine, which is commonly used as a generator of electric power, consists of a magnetically smooth stator with identical three-phase wye-connected sinusoidally distributed windings, displaced 120°. The rotor may or may not possess magnetic saliency in the form of physical poles. It may or may not possess a rotor cage (auxiliary rotor windings) for the purpose of providing line-start capability and/or to damp rotor oscillations. Finally, it may or may not possess the capability of establishing a rotor field flux, via either a rotor field winding or permanent magnets mounted on the rotor; however, if no provision for rotor field flux exists, then necessarily the rotor must have a salient pole construction.

In machine variables, the general expression for the voltage equations will involve the symmetric stator phases (designated by subscripts as, bs, and cs for phases a, b, and c) and the asymmetric rotor windings (designated by subscripts kq and kd for the q-axis and d-axis auxiliary windings and by subscript fd

for the field winding, which is assumed to be oriented along the d-axis). Hence, the voltage equations are written

$$\begin{bmatrix} v_s \\ v_r \end{bmatrix} = \begin{bmatrix} \mathcal{R}_s & 0 \\ 0 & \mathcal{R}_r \end{bmatrix} \begin{bmatrix} i_s \\ i_r \end{bmatrix} + \begin{bmatrix} \dot{\lambda}_s \\ \dot{\lambda}_r \end{bmatrix} \quad (78.101)$$

where v_s is the vector of stator voltages, i_s is the vector of stator currents, λ_s is the vector of stator flux linkages, v_r is the vector of rotor voltages (zero for the auxiliary windings), i_r is the vector of rotor currents, and λ_r is the vector of rotor flux linkages. The stator and rotor resistance matrices are given by

$$\mathcal{R}_s = \begin{bmatrix} R_s & 0 & 0 \\ 0 & R_s & 0 \\ 0 & 0 & R_s \end{bmatrix} \quad \mathcal{R}_r = \begin{bmatrix} R_{kq} & 0 & 0 \\ 0 & R_{fd} & 0 \\ 0 & 0 & R_{kd} \end{bmatrix} \quad (78.102)$$

When denoting any of the above stator or rotor vectors (voltage, current, flux) by generic notation f_s or f_r, respectively, the vector structure will be

$$f_s = \begin{bmatrix} f_{as} & f_{bs} & f_{cs} \end{bmatrix}^T \quad (78.103)$$

$$f_r = \begin{bmatrix} f_{kq} & f_{fd} & f_{kd} \end{bmatrix}^T \quad (78.104)$$

where stator components are associated with phase windings in machine variables, and rotor components are associated with the two auxiliary windings and the field winding in machine variables. Assuming magnetic linearity, the flux linkages may be expressed by

$$\begin{bmatrix} \lambda_s \\ \lambda_r \end{bmatrix} = \begin{bmatrix} \mathcal{L}_s(\theta) & \mathcal{L}_m(\theta) \\ \mathcal{L}_m^T(\theta) & \mathcal{L}_r \end{bmatrix} \begin{bmatrix} i_s \\ i_r \end{bmatrix} \quad (78.105)$$

with inductance matrices

$$\mathcal{L}_s(\theta) = \begin{bmatrix} L_s & M_s & M_s \\ M_s & L_s & M_s \\ M_s & M_s & L_s \end{bmatrix} \quad (78.106)$$

$$- L_m \begin{bmatrix} \cos(2N\theta) & \cos(2N\theta - \frac{2\pi}{3}) & \cos(2N\theta + \frac{2\pi}{3}) \\ \cos(2N\theta - \frac{2\pi}{3}) & \cos(2N\theta + \frac{2\pi}{3}) & \cos(2N\theta) \\ \cos(2N\theta + \frac{2\pi}{3}) & \cos(2N\theta) & \cos(2N\theta - \frac{2\pi}{3}) \end{bmatrix}$$

$$\mathcal{L}_m(\theta) = \quad (78.107)$$

$$\begin{bmatrix} M_q \cos(N\theta) & M_f \sin(N\theta) & M_d \sin(N\theta) \\ M_q \cos(N\theta - \frac{2\pi}{3}) & M_f \sin(N\theta - \frac{2\pi}{3}) & M_d \sin(N\theta - \frac{2\pi}{3}) \\ M_q \cos(N\theta + \frac{2\pi}{3}) & M_f \sin(N\theta + \frac{2\pi}{3}) & M_d \sin(N\theta + \frac{2\pi}{3}) \end{bmatrix}$$

$$\mathcal{L}_r = \begin{bmatrix} L_{qr} & 0 & 0 \\ 0 & L_{fr} & M_r \\ 0 & M_r & L_{dr} \end{bmatrix} \quad (78.108)$$

where L_s is (average) the stator self-inductance, M_s is the (average) stator-to-stator mutual inductance, L_m is the stator inductance coefficient that accounts for rotor saliency, L_{qr} is the self-inductance of the q-axis auxiliary winding, L_{dr} is the self-inductance of the d-axis auxiliary winding, L_{fr} is the self-inductance of the field winding, M_r is the mutual inductance between the two d-axis rotor windings, M_q, M_d, and M_f are the magnitudes of the angle-dependent mutual inductance between the stator windings and the various rotor windings, and N is the number of pole pairs. Note that this model does not account for the possibility of stator saliency (which gives rise to magnetic cogging).

For the mechanical dynamics, the differential equation for rotor velocity ω is

$$J\dot{\omega} = T_e - B\omega - T_l \quad (78.109)$$

where the torque of electrical origin may be determined from the inductance matrices using the general expression

$$T_e = \frac{1}{2} i^T \frac{d\mathcal{L}(\theta)}{d\theta} i = -\frac{1}{2} \lambda^T \frac{d\mathcal{L}^{-1}(\theta)}{d\theta} \lambda \quad (78.110)$$

where $\mathcal{L}(\theta), i, \lambda$ denote the complete inductance matrix, current vector, and flux linkage vector appearing in Equation 78.105. For further modeling details, see [11].

Reference Frame Transformation

The model derived above is not only nonlinear, it is also not in a form convenient for determining the steady-state conditions needed for achieving constant velocity operation, due to the model's periodic dependence on position. This dependence on position can be eliminated by a nonsingular change of variables, which effectively projects the stator variables onto a reference frame fixed to the rotor. Although it is possible to construct transformations to other frames of reference, these would not eliminate the position dependence due to the asymmetry present in the rotor. Since the asymmetric rotor windings are presumed to be aligned with the rotor frame of reference just one transformation matrix is necessary, which transforms circuit variables from the stator windings to fictitious windings which rotate with the rotor, and it is given by

$$K_{sr}(\theta) = \sqrt{\frac{2}{3}} \begin{bmatrix} \cos(N\theta) & \cos(N\theta - \frac{2\pi}{3}) & \cos(N\theta + \frac{2\pi}{3}) \\ \sin(N\theta) & \sin(N\theta - \frac{2\pi}{3}) & \sin(N\theta + \frac{2\pi}{3}) \\ \frac{1}{\sqrt{2}} & \frac{1}{\sqrt{2}} & \frac{1}{\sqrt{2}} \end{bmatrix} \quad (78.111)$$

This matrix is orthonormal and, hence, its inverse is equal to its transpose. Formally stated, the change of variables considered is defined by

$$\tilde{f}_s = \begin{bmatrix} f_{qs} & f_{ds} & f_{0s} \end{bmatrix}^T \quad \tilde{f}_s = K_{sr}(\theta) f_s \quad (78.112)$$

where the tilde represents the stator variables in the new coordinates.

Since the windings on the stator are wye-connected, the sum of the stator currents must always be equal to zero. In other words, $i_{as}+i_{bs}+i_{cs} = 0$. Note that the reference frame transformation, primarily intended to eliminate θ from the voltage equations, will also satisfy the algebraic current constraint by construction.

After some tedious but straightforward algebra, it can be shown that the transformed voltage equations become

$$v_{qs} = R_s i_{qs} + N\omega\lambda_{ds} + \dot{\lambda}_{qs} \quad (78.113)$$

$$v_{ds} = R_s i_{ds} - N\omega\lambda_{qs} + \dot{\lambda}_{ds} \quad (78.114)$$

$$v_{0s} = R_s i_{0s} + \dot{\lambda}_{0s} \quad (78.115)$$

$$0 = R_{kq} i_{kq} + \dot{\lambda}_{kq} \quad (78.116)$$

$$v_{fd} = R_{fd} i_{fd} + \dot{\lambda}_{fd} \quad (78.117)$$

$$0 = R_{kd} i_{kd} + \dot{\lambda}_{kd} \quad (78.118)$$

and the transformed flux linkage equations become

$$\begin{bmatrix} \lambda_{qs} \\ \lambda_{ds} \\ \lambda_{0s} \\ \lambda_{kq} \\ \lambda_{fd} \\ \lambda_{kd} \end{bmatrix} = \begin{bmatrix} L_s - M_s - \frac{3}{2}L_m & 0 & 0 & \kappa M_q & 0 & 0 \\ 0 & L_s - M_s + \frac{3}{2}L_m & 0 & 0 & \kappa M_f & \kappa M_d \\ 0 & 0 & L_s + 2M_s & 0 & 0 & 0 \\ \kappa M_q & 0 & 0 & L_{qr} & 0 & 0 \\ 0 & \kappa M_f & 0 & 0 & L_{fr} & M_r \\ 0 & \kappa M_d & 0 & 0 & M_r & L_{dr} \end{bmatrix} \begin{bmatrix} i_{qs} \\ i_{ds} \\ i_{0s} \\ i_{kq} \\ i_{fd} \\ i_{kd} \end{bmatrix} \quad (78.119)$$

where $\kappa = \sqrt{\frac{3}{2}}$. Note that all dependence on rotor angle θ has been eliminated and, hence, all analysis is substantially simplified. However, nonlinearity has not been entirely eliminated. Note also that $i_{0s} = 0$ so that the 0-axis equation from above can be completely ignored.

Due to the rotor asymmetry, the most convenient set of electrical variables for expressing the electrical torque consists of stator current and stator flux. Using these variables, the torque expression becomes

$$T_e = N(i_{qs}\lambda_{ds} - i_{ds}\lambda_{qs}) \quad (78.120)$$

This and other expressions will now be specialized to cover three common types of synchronous motors: the rotor-surface permanent magnet motor, the rotor-interior permanent magnet motor, and the reluctance motor.

SPM Synchronous Motor

The surface-magnet PM synchronous motor is obtained when the rotor field winding is replaced by permanent magnets attached to the surface of a smooth rotor, with auxiliary rotor windings removed. In other words, this is the special case where i_{fd} = constant, $i_{kq} = i_{kd} = 0$, and $L_m = 0$. Of course, incorporating the effects of auxiliary windings is straightforward, but this is not pursued here.

In order to simplify notation, the new coefficients

$$\lambda_m = \kappa M_f i_{fd} \quad L = L_s - M_s \quad (78.121)$$

concerning magnet flux and inductance are defined. Note that these new coefficients allow the torque to be expressed by

$$T_e = N\lambda_m i_{qs} \quad (78.122)$$

Variables are assigned standard notation for states, inputs, and outputs according to

$$\begin{aligned} x_1 = \theta \quad x_2 = \omega \quad x_3 = i_{qs} \quad x_4 = i_{ds} \\ u_1 = v_{qs} \quad u_2 = v_{ds} \quad y_1 = \theta \quad y_2 = i_{ds} \end{aligned} \quad (78.123)$$

Since for this motor the torque depends only on i_{qs} but not on i_{ds}, the d-axis current is the appropriate choice of the second output.

With the above variable assignments, the state variable model is

$$f(x) = \begin{bmatrix} x_2 \\ \frac{1}{J}(N\lambda_m x_3 - Bx_2 - T_l) \\ -\frac{1}{L}(R_s x_3 + N(\lambda_m + Lx_4)x_2) \\ -\frac{1}{L}(R_s x_4 - NLx_3 x_2) \end{bmatrix} \quad (78.124)$$

$$g_1(x) = \begin{bmatrix} 0 \\ 0 \\ \frac{1}{L} \\ 0 \end{bmatrix} \quad g_2(x) = \begin{bmatrix} 0 \\ 0 \\ 0 \\ \frac{1}{L} \end{bmatrix}$$

$$h_1(x) = x_1 \quad h_2(x) = x_4$$

The presence of nonlinearity in the electrical subdynamics is clear. In checking for relative degree, the Lie-derivative calculations

$$\begin{aligned} L_{g_1}h_1(x) &\equiv 0 & L_{g_2}h_1(x) &\equiv 0 \\ L_{g_1}L_f h_1(x) &\equiv 0 & L_{g_2}L_f h_1(x) &\equiv 0 \\ L_{g_1}L_f^2 h_1(x) &= \frac{N\lambda_m}{JL} & L_{g_2}L_f^2 h_1(x) &= 0 \end{aligned} \quad (78.125)$$

for the first output and

$$L_{g_1}h_2(x) = 0 \quad L_{g_2}h_2(x) = \frac{1}{L} \quad (78.126)$$

for the second output lead to a globally nonsingular decoupling matrix, as confirmed by

$$\det \begin{bmatrix} \frac{N\lambda_m}{JL} & 0 \\ 0 & \frac{1}{L} \end{bmatrix} = \frac{N\lambda_m}{JL^2} \quad (78.127)$$

Hence, the conclusion is that the relative degree is globally well defined and

$$\{r_1, r_2\} = \{3, 1\} \ (\text{globally}) \quad (78.128)$$

IPM Synchronous Motor

The interior-magnet PM synchronous motor is obtained when the rotor field winding is replaced by permanent magnets mounted inside the rotor, thus introducing rotor saliency, and with auxiliary rotor windings removed. In other words, this is the special case where i_{fd} = constant, $i_{kq} = i_{kd} = 0$, and $L_m \neq 0$. Of course, incorporating the effects of auxiliary windings is straightforward, but this is not pursued here.

In order to simplify notation, the new coefficients

$$\lambda_m = \kappa M_f i_{fd} \quad L_q = L_s - M_s - \frac{3}{2}L_m \quad L_d = L_s - M_s + \frac{3}{2}L_m \quad (78.129)$$

concerning magnet flux and inductance are defined. Note that these new coefficients allow the torque to be expressed by

$$T_e = N\lambda_m i_{qs} + N(L_d - L_q)i_{qs}i_{ds} \quad (78.130)$$

Variables are assigned standard notation for states, inputs, and outputs according to

$$x_1 = \theta \quad x_2 = \omega \quad x_3 = i_{qs} \quad x_4 = i_{ds}$$
$$u_1 = v_{qs} \quad u_2 = v_{ds} \quad y_1 = \theta \tag{78.131}$$

Since torque depends on both the q-axis and d-axis currents, the appropriate choice of the second output is nonunique and should account for the relative magnitude of the two torque production mechanisms.

With the above variable assignments, the state variable model is

$$f(x) = \begin{bmatrix} x_2 \\ \frac{1}{J}\left(N\lambda_m x_3 + N(L_d - L_q)x_3x_4 - Bx_2 - T_l\right) \\ -\frac{1}{L_q}\left(R_s x_3 + N(\lambda_m + L_d x_4)x_2\right) \\ -\frac{1}{L_d}\left(R_s x_4 - NL_q x_3 x_2\right) \end{bmatrix}$$

$$g_1(x) = \begin{bmatrix} 0 \\ 0 \\ \frac{1}{L_q} \\ 0 \end{bmatrix} \quad g_2(x) = \begin{bmatrix} 0 \\ 0 \\ 0 \\ \frac{1}{L_d} \end{bmatrix} \quad h_1(x) = x_1 \tag{78.132}$$

For this synchronous motor, nonlinearity also exists in the torque expression.

In checking for relative degree, each of the currents will be individually selected as the second output. The Lie-derivative calculations which are common to both cases are

$$\begin{aligned} L_{g_1}h_1(x) &\equiv 0 \qquad L_{g_2}h_1(x) \equiv 0 \\ L_{g_1}L_f h_1(x) &\equiv 0 \qquad L_{g_2}L_f h_1(x) \equiv 0 \\ L_{g_1}L_f^2 h_1(x) &= \frac{N}{JL_q}\left(\lambda_m + (L_d - L_q)x_4\right) \\ L_{g_2}L_f^2 h_1(x) &= \frac{N}{JL_d}(L_d - L_q)x_3 \end{aligned} \tag{78.133}$$

For the case when $h_2(x) = x_4$, the remaining Lie-derivative calculations

$$L_{g_1}h_2(x) = 0 \quad L_{g_2}h_2(x) = \frac{1}{L_d} \tag{78.134}$$

provide the decoupling matrix

$$\det\begin{bmatrix} \frac{N}{JL_q}\left(\lambda_m + (L_d - L_q)x_4\right) & \frac{N}{JL_d}(L_d - L_q)x_3 \\ 0 & \frac{1}{L_d} \end{bmatrix} = \frac{N}{JL_dL_q}\left(\lambda_m + (L_d - L_q)x_4\right) \tag{78.135}$$

which yields

$$\{r_1, r_2\} = \{3, 1\} \left(\text{if } y_2 \neq -\frac{\lambda_m}{L_d - L_q}\right) \tag{78.136}$$

For the case when $h_2(x) = x_3$, the remaining Lie-derivative calculations

$$L_{g_1}h_2(x) = \frac{1}{L_q} \quad L_{g_2}h_2(x) = 0 \tag{78.137}$$

provide a decoupling matrix

$$\det\begin{bmatrix} \frac{N}{JL_q}\left(\lambda_m + (L_d - L_q)x_4\right) & \frac{N}{JL_d}(L_d - L_q)x_3 \\ \frac{1}{L_q} & 0 \end{bmatrix} = -\frac{N}{JL_dL_q}(L_d - L_q)x_3 \tag{78.138}$$

which yields

$$\{r_1, r_2\} = \{3, 1\} \ (\text{if } y_2 \neq 0) \tag{78.139}$$

Hence, for this motor, it is clear that input-output linearization is a viable control strategy provided that isolated singularities are avoided, and this is simply achieved due to the fact that the singularities correspond to particular values of the second controlled output variable.

Synchronous Reluctance Motor

The synchronous reluctance motor is obtained when the rotor field winding and auxiliary windings are removed, but rotor saliency is introduced. In other words, this is the special case where $i_{fd} = i_{kq} = i_{kd} = 0$ and $L_m \neq 0$. Of course, incorporating the effects of auxiliary windings is straightforward, but this is not pursued here.

In order to simplify notation, the new coefficients

$$L_q = L_s - M_s - \frac{3}{2}L_m \quad L_d = L_s - M_s + \frac{3}{2}L_m \tag{78.140}$$

concerning inductance are defined. Note that these new coefficients allow the torque to be expressed by

$$T_e = N(L_d - L_q)i_{qs}i_{ds} \tag{78.141}$$

Variables are assigned standard notation for states, inputs, and outputs according to

$$x_1 = \theta \quad x_2 = \omega \quad x_3 = i_{qs} \quad x_4 = i_{ds}$$
$$u_1 = v_{qs} \quad u_2 = v_{ds} \quad y_1 = \theta \tag{78.142}$$

Since torque depends on both the q-axis and d-axis currents in a symmetric way, the appropriate choice of the second output is nonunique and either of these currents would serve the purpose equally well.

With the above variable assignments, the state variable model is

$$f(x) = \begin{bmatrix} x_2 \\ \frac{1}{J}\left(N(L_d - L_q)x_3x_4 - Bx_2 - T_l\right) \\ -\frac{1}{L_q}\left(R_s x_3 + NL_d x_4 x_2\right) \\ -\frac{1}{L_d}\left(R_s x_4 - NL_q x_3 x_2\right) \end{bmatrix}$$

$$g_1(x) = \begin{bmatrix} 0 \\ 0 \\ \frac{1}{L_q} \\ 0 \end{bmatrix} \quad g_2(x) = \begin{bmatrix} 0 \\ 0 \\ 0 \\ \frac{1}{L_d} \end{bmatrix} \quad h_1(x) = x_1 \tag{78.143}$$

Again, nonlinearity is present in both the electrical and mechanical subdynamics. In checking for relative degree, two cases are considered, corresponding to the two obvious choices for the second output. The common Lie-derivative calculations are given by

$$
\begin{aligned}
L_{g_1}h_1(x) &\equiv 0 \quad L_{g_2}h_1(x) \equiv 0 \\
L_{g_1}L_f h_1(x) &\equiv 0 \quad L_{g_2}L_f h_1(x) \equiv 0 \\
L_{g_1}L_f^2 h_1(x) &= \frac{N(L_d - L_q)}{J L_q} x_4 \\
L_{g_2}L_f^2 h_1(x) &= \frac{N(L_d - L_q)}{J L_d} x_3
\end{aligned}
\tag{78.144}
$$

For the case when $h_2(x) = x_4$, the remaining calculations

$$
L_{g_1}h_2(x) = 0 \quad L_{g_2}h_2(x) = \frac{1}{L_d} \tag{78.145}
$$

provide the decoupling matrix

$$
\det\begin{bmatrix} \frac{N(L_d-L_q)}{JL_q}x_4 & \frac{N(L_d-L_q)}{JL_d}x_3 \\ 0 & \frac{1}{L_d} \end{bmatrix} = \frac{N(L_d - L_q)}{J L_d L_q} x_4 \tag{78.146}
$$

which suggests that

$$
\{r_1, r_2\} = \{3, 1\} \ (\text{if } y_2 \neq 0) \tag{78.147}
$$

For the case when $h_2(x) = x_3$, the remaining calculations

$$
L_{g_1}h_2(x) = \frac{1}{L_q} \quad L_{g_2}h_2(x) = 0 \tag{78.148}
$$

provide the decoupling matrix

$$
\det\begin{bmatrix} \frac{N(L_d-L_q)}{JL_q}x_4 & \frac{N(L_d-L_q)}{JL_d}x_3 \\ \frac{1}{L_q} & 0 \end{bmatrix} = -\frac{N(L_d - L_q)}{J L_d L_q} x_3 \tag{78.149}
$$

which suggests that

$$
\{r_1, r_2\} = \{3, 1\} \ (\text{if } y_2 \neq 0) \tag{78.150}
$$

Hence, for this motor, it is clear that input-output linearization is a viable control strategy provided that isolated singularities are avoided, and this is simply achieved due to the fact that the singularities correspond to particular values of the second controlled output variable.

78.2.6 Concluding Remarks

This chapter has described how input-output linearization may be used to achieve motion control for a fairly wide class of motors. Although the most crucial issues of state-variable modeling and relative degree have been adequately covered, a few remaining points need to be made. In several cases, namely the shunt-connected and series-connected DC motors and the induction motor, the choice of outputs has led to first-order internal dynamics. It is not difficult to show, however, that these internal dynamics present no stability problems for input-output linearization. Also, it should be emphasized that the control possibilities for shunt-connected and series-connected DC motors are rather limited because of their unipolar operation, but that the isolated singularities found for all the other motors should not pose any real problems in practice.

The material presented in this chapter can be extended in several directions, beyond the obvious step of completely specifying the explicit feedback controls. Input-output linearization may be applied also to motors with nonsinusoidal winding distribution, to motors operating in magnetic saturation, or to motors with salient pole stators and concentrated windings (e.g., the switched reluctance motor). Moreover, there are still many supplementary issues that are important to mention, such as the augmentation of the basic controllers with algorithms to estimate unmeasured states and/or to identify unknown parameters.

Since model-based nonlinear controllers depend on parameters that may be unknown or slowly varying, an on-line parameter identification scheme can be included to achieve indirect adaptive control. Especially parameters describing the motor load are subject to significant uncertainty. There are several motives for agreeing to the additional complexity required for implementation of indirect adaptive control, including the potential for augmented performance, and improved reliability due to the use of diagnostic parameter checks. Typical examples of parameter identification schemes and adaptive control may be found in [13] for induction motors, and in [15] for permanent-magnet synchronous motors.

One of the challenges in practical nonlinear control design for electric machines is to overcome the need for full state measurement. The simplest example of this is commutation sensor elimination (e.g., elimination of Hall-effect devices from inside the motor frame). For electronically commutated motors, such as certain permanent-magnet synchronous motors, some applications do not require control of instantaneous torque, but can get by with simple commutation excitation with a variable firing angle. Even at this level of control, the need for a commutation sensor is crucial in order to maintain synchronism. The literature on schemes used to drive commutation controllers without commutation sensors is quite extensive.

A more difficult problem in the same direction is high-accuracy position estimation suitable for eliminating the high-resolution position sensor required by most of the nonlinear controls discussed earlier. In some applications, despite the need for high dynamic performance, the use of a traditional position sensor is considered undesirable due to cost, the volume and/or weight of the sensor, or the potential unreliability of the sensor in harsh environments. In this situation, the only alternative is to extract position (and/or speed) information from the available electrical terminal measurements. Of course, this is not an easy thing to do, precisely because of the nonlinearities involved. Some promising results have been reported in the literature on this topic, for permanent-magnet synchronous motors, switched reluctance motors, and for induction motors.

For the induction motor specifically, there is also a need to estimate either the rotor fluxes or rotor currents, in order to implement the input-output linearizing controller discussed earlier. Thorough treatments of this estimation problem are also available in the literature.

References

[1] Blaschke, F., The principle of field orientation applied to the new transvector closed-loop control system for rotating field machines, *Siemens Rev.*, 39, 217–220, 1972.

[2] Bodson, M., Chiasson, J., and Novotnak, R., High-performance induction motor control via input-output linearization, *IEEE Control Syst. Mag.*, 14(4), 25–33, 1994.

[3] Bodson, M., Chiasson, J.N., Novotnak, R.T., and Rekowski, R.B., High-performance nonlinear feedback control of a permanent magnet stepper motor, *IEEE Trans. Control Syst. Technol.*, 1(1), 5–14, 1993.

[4] Chiasson, J. and Bodson, M., Nonlinear control of a shunt DC motor, *IEEE Trans. Autom. Control*, 38(11), 1662–1666, 1993.

[5] Chiasson, J., Nonlinear differential-geometric techniques for control of a series DC motor, *IEEE Trans. Control Syst. Technol.*, 2(1), 35–42, 1994.

[6] De Luca, A. and Ulivi, G., Design of an exact nonlinear controller for induction motors, *IEEE Trans. Autom. Control*, 34(12), 1304–1307, 1989.

[7] Hemati, N., Thorp, J.S., and Leu, M.C., Robust nonlinear control of brushless dc motors for direct-drive robotic applications, *IEEE Trans. Ind. Electron.*, 37(6), 460–468, 1990.

[8] Isidori, A., *Nonlinear Control Systems*, 2nd ed., Springer-Verlag, New York, 1989.

[9] Kim, D.I., Ha, I.J., and Ko, M.S., Control of induction motors via feedback linearization with input-output decoupling, *Int. J. Control*, 51(4), 863–883, 1990.

[10] Kim, G.S., Ha, I.J., and Ko, M.S., Control of induction motors for both high dynamic performance and high power efficiency, *IEEE Trans. Ind. Electron.*, 39(4), 323–333, 1992.

[11] Krause, P.C., *Analysis of Electric Machinery*, McGraw-Hill, New York, 1986.

[12] Krzeminski, Z., Nonlinear control of induction motor, *Proc. 10th IFAC World Congress*, Munich, Germany, 1987, pp. 349–354.

[13] Marino, R., Peresada, S., and Valigi, P., Adaptive input-output linearizing control of induction motors, *IEEE Trans. Autom. Control*, 38(2), 208–221, 1993.

[14] Oliver, P.D., Feedback linearization of dc motors, *IEEE Trans. Ind. Electron.*, 38(6), 498–501, 1991.

[15] Sepe, R.B. and Lang, J.H., Real-time adaptive control of the permanent-magnet synchronous motor, *IEEE Trans. Ind. Appl.*, 27(4), 706–714, 1991.

[16] Zribi, M. and Chiasson, J., Position control of a PM stepper motor by exact linearization, *IEEE Trans. Autom. Control*, 36(5), 620–625, 1991.

78.3 Control of Electrical Generators

Thomas M. Jahns, *GE Corporate R&D, Schenectady, NY*

Rik W. De Doncker, *Silicon Power Corporation, Malvern, PA*

78.3.1 Introduction

Electric machines are inherently bidirectional energy converters that can be used to convert electrical energy into mechanical energy during motoring operation, or mechanical into electrical energy during generating operation. Although the underlying principles are the same, there are some significant differences between the machine control algorithms developed for motion control applications and those for electric power generation. While shaft torque, speed, and position are the controlled variables in motion control systems, machine terminal voltage and current are the standard regulated quantities in generator applications.

In this section, control principles will be reviewed for three of the most common types of electric machines used as electric power generators — DC machines, synchronous machines, and induction machines. Although abbreviated, this discussion is intended to introduce many of the fundamental control principles that apply to the use of a wide variety of alternative specialty electric machines in generating applications.

78.3.2 DC Generators

Introduction

DC generators were among the first electrical machines to be deployed reliably in military and commercial applications. Indeed, DC generators designed by the Belgian Professor F. Nollet and built by his assistant J. Van Malderen were used as early as 1859 as DC sources for arc lights illuminating the French coast [8]. During the French war in 1870, arc lights powered by the same magneto-electro machines were used to illuminate the fields around Paris to protect the city against night attacks [25]. Another Belgian entrepreneur, Zenobe Gramme, who had worked with Van Malderen, developed the renowned Gramme winding and refined the commutator design to its present state in 1869. In that year he demonstrated that the machine was capable of driving a water pump at the World Exposition held in Vienna.

Indeed, DC generators are constructed identically to DC motors, although they are controlled differently. Whereas DC motors are controlled by the armature current to regulate the torque production of the machine [23], DC generators are controlled by the field current to maintain a regulated DC terminal voltage. Typical applications for DC generators have included power supplies for electrolytic processes and variable-voltage supplies for DC servo motors, such as in rolling mill drives. Rotating machine uninteruptible power supplies (UPS), which use batteries for energy storage, require AC and DC generators, as well as AC and DC motors.

Other important historical applications include DC substations for DC railway supplies. However, since the discovery of DC rectifiers, especially silicon rectifiers (1958), rotating DC generators have been steadily replaced in new DC generator installations by diode or thyristor solid-state rectifiers. In those situations where mechanical-to-electrical energy conversion takes place, DC generators have been replaced by synchronous machines feeding DC bridge rectifiers, particularly in installations at high power, i.e., installations requiring several tens of kilowatts. Synchronous generators with DC rectifiers are also referred to as brushless DC generators.

There are multiple reasons for the steady decline in the number of DC generators. Synchronous machines with solid-state rectifiers do not require commutators or brushes that require maintenance. The commutators of DC machines also produce arcs that are not acceptable in mining operations and chemical plants. In a given mechanical frame, synchronous machines with rectifiers can be built with higher overload ratings. Furthermore, as the technology of solid-state rectifiers matured, the rectifier systems and the brushless DC generator systems became more economical to build than their DC machine counterparts.

Despite this declining trend in usage, it is still important to understand the DC generator's basic operating principles because they apply directly to other rotating machine generators and motors (especially DC motors). The following subsections describe the construction and equivalent model of the DC generator and the associated voltage control algorithms.

DC Machine Generator Fundamentals

Machine Construction and Vector Diagram Most rotating DC generators are part of a rotating machine group where AC induction machines or synchronous machines drive the DC generator at a constant speed. Figure 78.7 illustrates a typical rotating machine lineup that converts AC power to DC power. Note that this rotating machine group has bidirectional power flow capability because each machine can operate in its opposite mode. For example, the DC generator can operate as a DC motor. Meanwhile, the AC motor will feed power back into the AC grid and operate as an AC generator.

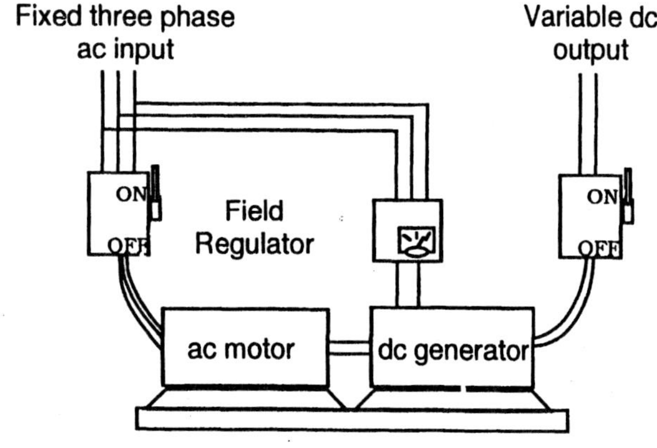

Figure 78.7 Typical line-up of DC generator driven by AC motor.

Figure 78.8(a) shows the DC generator construction, which is similar to a DC shunt motor. The space flux vector diagram of Figure 78.8(b) shows the amplitude and relative spatial position of the stator and armature fluxes that are associated with the winding currents. The moving member of the machine is called the armature, while the stationary member is called the stator. The armature current i_a is taken from the armature via a commutator and brushes. A stationary field is provided by an electromagnet or a permanent magnet and is perpendicular to the field produced by the armature. In the case of an electromagnet, the field current can be controlled by a DC power supply.

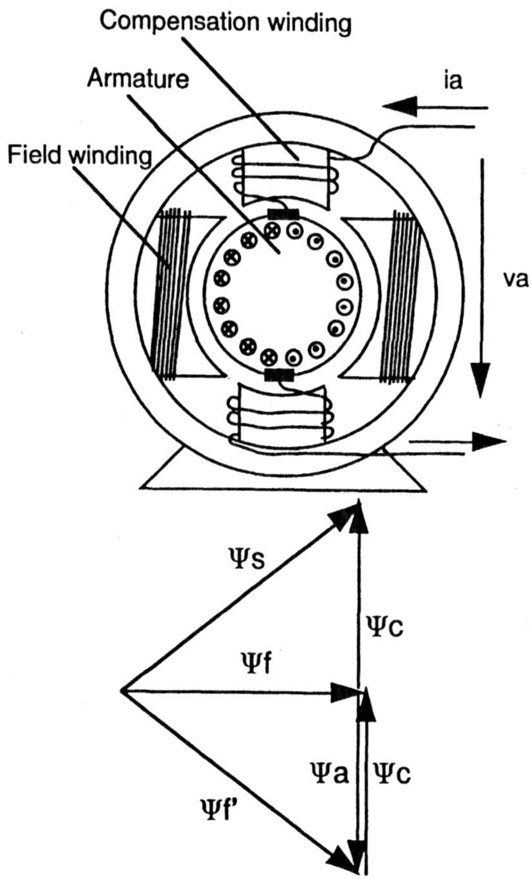

Figure 78.8 (a) Construction of DC generator, showing the stator windings and armature winding. (b) Vector diagram representing flux linked to field winding (Ψ_f), armature winding (Ψ_a), and series compensation winding (Ψ_c).

Most DC machines are constructed with interpole windings to compensate for the armature reaction. Without this compensation winding the flux inside the machine would increase to Ψ_f' when the armature current (and flux Ψ_a) increases. Due to the magnetic saturation of the pole iron, this increased flux leads to a loss in internal back-emf voltage. Hence, the output voltage would drop faster than predicted by the internal armature resistance whenever load increases. Conducting the armature current through the series compensation winding creates a compensation flux Ψ_c which adds to the flux Ψ_f' such that the total flux of the machine remains at the original level Ψ_f determined by the field current.

The vector diagram of Figure 78.8(b) shows another way to illustrate the flux linkages inside the DC generator. One can construct first the so-called stator flux Ψ_s of the DC machine. The stator flux can be defined as the flux produced solely by the stator windings, i.e., field winding and compensation winding. Adding the armature flux Ψ_a to the stator flux Ψ_s yields the field winding flux Ψ_f which represents the total flux experienced by the magnetic circuit.

Another advantage of the compensation winding is a dramatic reduction of the armature leakage inductance because the flux inside the DC machine does not change with varying armature current. In other words, the magnetic stored energy inside the machine is greatly decoupled from changes of load current, making the machine a stiffer voltage source during transient conditions.

DC Generator Equivalent Circuit and Equations The DC generator equivalent circuit is depicted in Figure 78.9. The rotating armature windings produce an induced voltage that is proportional to the speed and the field flux of the machine (from the Bvl rule [6]):

$$e_a = k\omega_m \Psi_f \tag{78.151}$$

Parasitic armature elements are the armature resistance R_a (re-

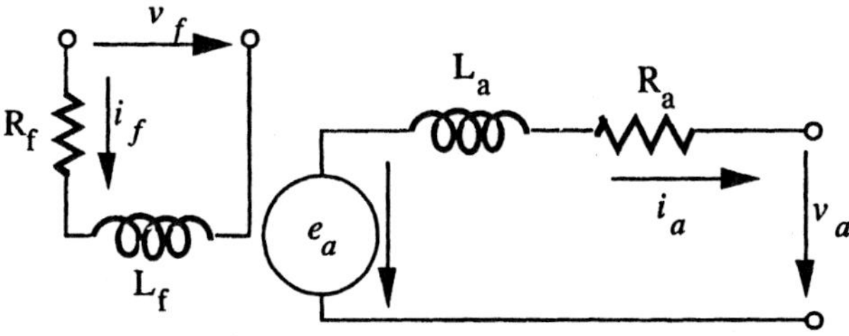

Figure 78.9 Equivalent circuit of DC generator showing the armature impedances (R_a and L_a) and armature back-emf (e_a). Also shown are the field winding impedances (R_f and L_f), the field supply voltage (v_f), and current (i_f).

sistance of armature and compensation windings, cables, and brushes) and the armature leakage inductance L_a (flux not mutual coupled between armature and compensation winding). As a result, the armature voltage loop equation is given by:

$$v_a = e_a - L_a\frac{di_a}{dt} - R_a i_a \tag{78.152}$$

The field winding is perpendicular to the armature winding and the compensation winding and experiences no induced voltages. Hence, its voltage-loop equation simplifies to:

$$v_f = L_f\frac{di_f}{dt} + R_f i_f = \frac{d\Psi_f}{dt} + \frac{R_f}{L_f}\Psi_f \tag{78.153}$$

The electromagnetic torque T_a produced by the DC generator is proportional to the field flux and the armature currents (from the Bil rule [6]) and can be expressed as:

$$T_a = k\Psi_f i_a \tag{78.154}$$

Note that this torque T_a acts as a load on the AC motor that drives the DC generator, as shown in Figure 78.7. Hence, to complete the set of system equations that describe the dynamic behavior of the DC generator, a mechanical motion equation describing the interaction between the DC generator and the AC drive motor is necessary. We assume that the mechanical motion equation of a DC generator fed by an electric motor can be characterized by the dynamic behavior of the combined inertia of the generator armature and the rotor of the motor and their speed-proportional damping. Speed-proportional damping is typically caused by a combination of friction, windage losses, and electrical induced losses in the armature and rotor:

$$T_m - T_a = J\frac{d\omega_m}{dt} + D\omega_m \tag{78.155}$$

where J is the total combined inertia of the DC generator and drive motor, ω_m is the angular velocity of the motor-generator shaft, and D is the speed-proportional damping coefficient.

The motor torque T_m is determined by the type of prime mover used and its control functions. For example, synchronous machines do not vary their average speed whenever the load torque (produced by the DC generator) varies. On the other hand, the speed of AC induction machines slips a small amount at increased load torque. The torque response of speed-controlled DC machines or AC machines depends greatly on the speed or position feedback loop used to control the machine. Assuming a proportional-integral speed control feedback loop, the torque of the driving motor can be described by the following function:

$$T_m = K_p(\omega_m^* - \omega_m) + K_i\int(\omega_m^* - \omega_m)dt \tag{78.156}$$

where ω_m is the measured speed of the drive motor, ω_m^* is the desired speed of the drive motor, K_p is the proportional gain of the drive motor feedback controller, and K_i is the integral gain of the drive motor feedback controller.

Assuming no field regulation and assuming a constant armature speed, the DC generator output voltage varies in steady state as a function of the load current according to:

$$v_a = e_a - R_a i_a \tag{78.157}$$

Figure 78.10 illustrates the DC generator steady-state load characteristic. In some applications the drop of the output voltage with increasing load current is not acceptable and additional control is required. Also, during dynamic conditions, e.g., a step change in the load current, the armature voltage may respond too slowly or with insufficient damping. The following section analyzes the dynamic response and describes a typical control algorithm to improve the DC generator performance.

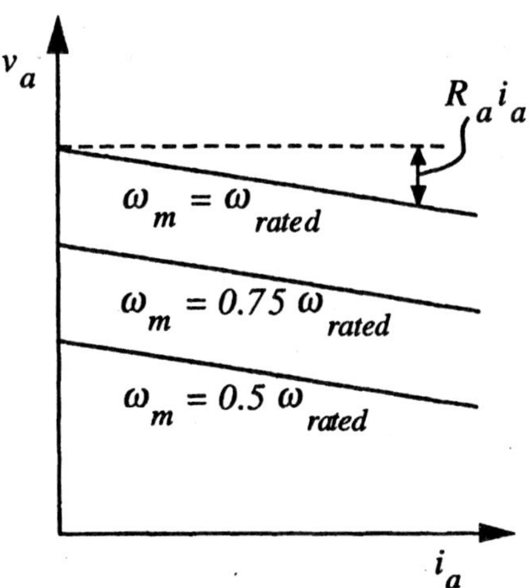

Figure 78.10 DC generator steady-state load characteristic.

DC Generator Control Principles

Basic Generator System Control Characteristics A control block diagram can be derived from the DC generator equations and is illustrated in Figure 78.11. As indicated by this block diagram, the DC generator system exhibits some complicated coupled dynamics.

For example, whenever the generator load increases quickly (e.g., due to switching of resistive-inductive loads), a rapid increase of the armature current will occur because the armature time constant is typically less than 100 msec. This leads to an equally fast increase in the generator load torque T_a. The speed regulator of the drive motor will respond to this fast torque increase with some time delay (determined by the inertia of the drive and the speed PI controller). Some speed variation can be expected as a result of this delay. Such speed variations directly influence the armature-induced voltage e_a. This causes the armature voltage and the armature current to change, thereby altering the generator load torque.

In conclusion, the DC generator drive behaves as a system of higher order. Figure 78.12 shows that a step change in load current has an impact on the armature voltage, the torque of the DC generator, and the speed of the rotating group. In this example, the output voltage settles to its new steady-state value after an oscillatory transient. Under certain circumstances (e.g., low inertia) the drive may ultimately become unstable.

Generator Voltage Regulation A field controller can be added to enhance the dynamic performance of the generator system. However, this controller acts on the proportionality factor between the back-emf voltage and the speed of the drive. Hence, this control loop makes the system nonlinear, and simple feedback control can be difficult to apply. Feedforward back-emf decoupling (from speed) combined with feedback control is therefore often used in machine controllers because of its simplicity and fast dynamic response.

Figure 78.13 shows the block diagram of a field controller which utilizes feedback and feedforward control. The feedforward loop acts to decouple the speed variations from the generator back-emf e_a. This controller is obtained by inverting the model equations of the back-emf and the field winding.

Figure 78.14 shows the response of the generator system when only feedback control is applied. The armature voltage does not have a steady-state error but still shows some oscillatory behavior at the start of the transient. To improve further the dynamic response of the generator it is essential to decouple the influence of the speed variation on the armature back-emf e_a. This can be achieved using feedforward control by measuring the drive speed and computing the desired field voltage to maintain a constant back-emf e_a as shown in Figure 78.13. A lead network compensates for the field winding inductive lag (within power limits of the field winding power supply). In practice, the field power supply has a voltage limit which can be represented by a saturation function in the field controller block diagram.

Note that the feedforward control loop relies on good estimates of the machine parameters. The feedback control loop does not require these parameters and guarantees correct steady-state operation. Figure 78.15 illustrates the response of the DC generator assuming a 10% error in the control parameters in the feedforward controller. Clearly, a faster response is obtained with less transient error.

78.3.3 Synchronous Generators

Synchronous Machine Fundamentals and Models

Introduction Synchronous machines provide the basis for the worldwide AC electric power generating and distribution system, earning them recognition as one of the most important classes of electrical generators. In fact, the installed power-generating capacity of AC synchronous generators exceeds that of all other types of generators combined by many times. The scalability of synchronous generators is truly impressive, with available machine ratings covering at least 12 orders of magnitude from milliwatts to gigawatts.

A synchronous generator is an AC machine which, in comparison to the DC generator discussed in Section 78.3.2, has its key electrical components reversed, with field excitation mounted on the rotor and the corresponding armature windings mounted in slots along the inner periphery of the stator. The armature windings are typically grouped into three phases and are specially distributed to create a smoothly rotating magnetic flux wave in the generator's airgap when balanced three-phase sinusoidal currents flow in the windings. The rotor-mounted field can be supplied by either currents flowing in a directed field winding (wound-field synchronous generator, or WFSG) or, in special cases, by permanent magnets (PMSG). Figure 78.16 includes a simplified cross-sectional view of a wound-field machine showing the physical relationship of its key elements. A wealth of technical literature is available which addresses the basic operating principles and construction of AC synchronous machines in every level of desired detail [7], [22], [24].

Synchronous Generator Model There are many alternative approaches which have been proposed for modeling synchronous machines, but one of the most powerful from the standpoints of both analytical usefulness and physical insight is

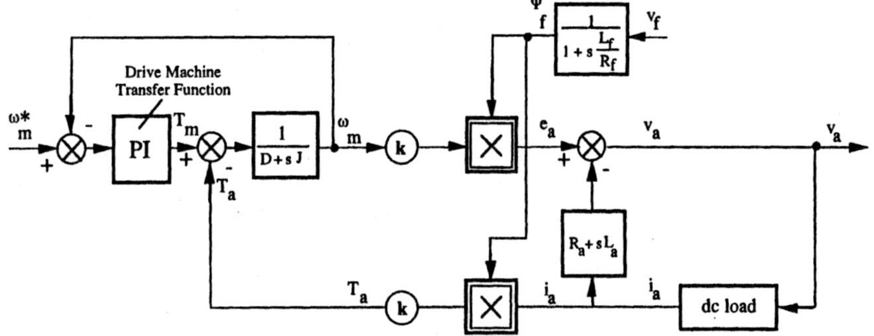

Figure 78.11 Block diagram of basic DC generator system.

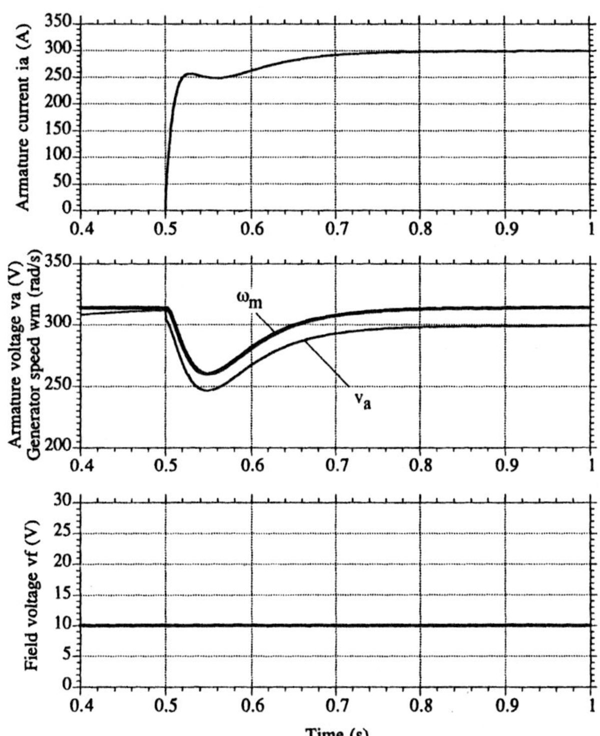

Figure 78.12 Response of DC generator with constant field v_f.

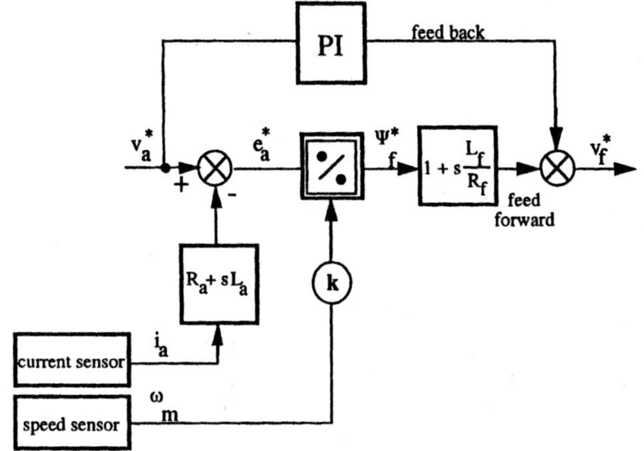

Figure 78.13 Feedforward and feedback control loops for DC generator field control.

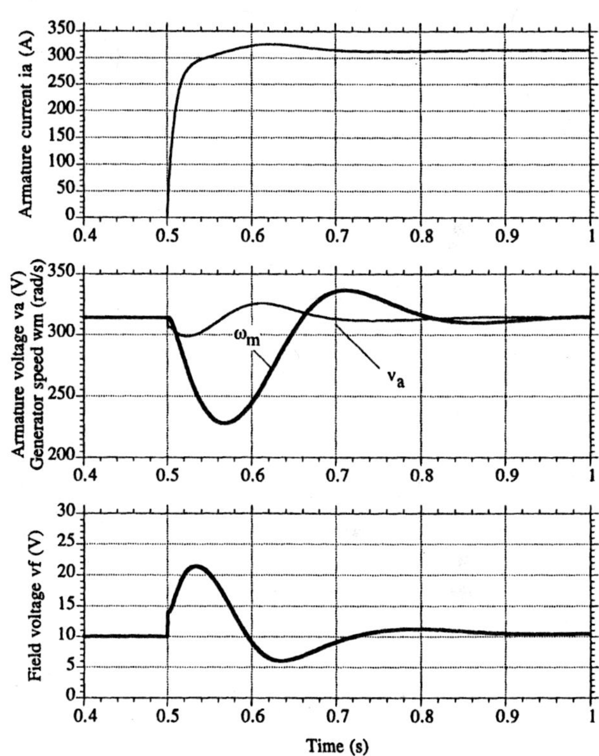

Figure 78.14 Response of DC generator with feedback control on field excitation v_f.

the dq equivalent circuit model shown in Figure 78.17. These coupled equivalent circuits result from application of the Park-Blondel dq transformation [15] which transforms the basic electrical equations of the synchronous machine from the conventional stationary reference frame locked to the stator windings into a rotating reference frame revolving in synchronism with the rotor. The direct (d) axis is purposely aligned with the magnetic flux developed by the rotor field excitation, as identified in Figure 78.16, while the quadrature (q) axis is orthogonally oriented at 90 electrical degrees of separation from the d-axis. As a result of this transformation, AC quantities in the stator-referenced equations become DC quantities during steady-state operation in the synchronously rotating reference frame.

Each of the d- and q-axis circuits in Figure 78.17 takes the form of a classic coupled-transformer equivalent circuit built around the mutual inductances L_{md} and L_{mq} representing the coupled

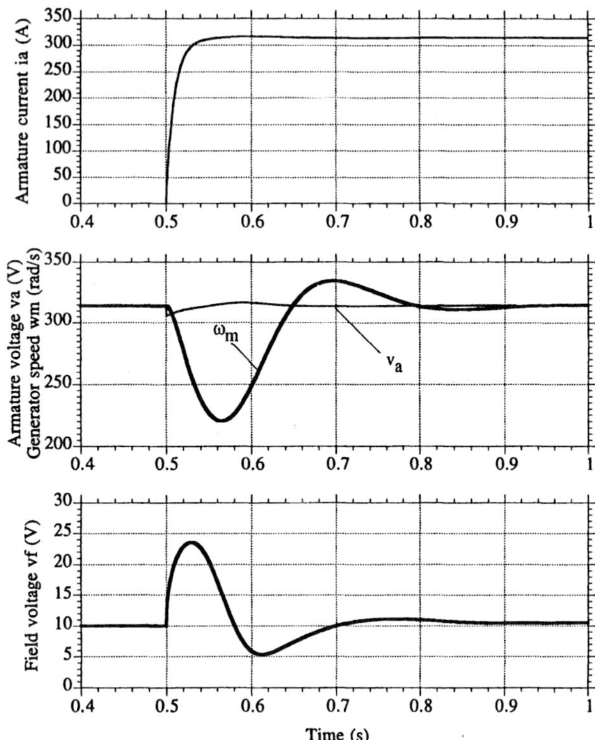

Figure 78.15 Response of DC generator with feedforward and feedback control on field excitation v_f with a 10% error in the feedforward control parameters.

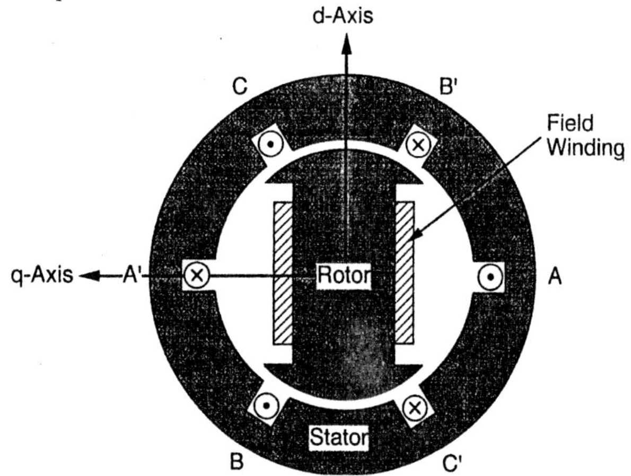

Figure 78.16 Simplified cross-sectional view of wound-field synchronous generator, including identification of d-q axes in rotor-oriented reference frame.

stator-rotor magnetic flux in each of the two axes. The principal difference between the d- and q-axis circuits is the presence of an extra set of external excitation input terminals in the d-axis rotor circuit modeling the field excitation. The externally controlled voltage source v_f which appears in this rotor excitation circuit represents the principal input port for controlling the output voltage, current, and power of a wound-field synchronous generator. The other rotor circuit legs which appear in both the d- and q-axis circuits are series resistor-inductor combinations modeling the passive damping effects of special rotor-mounted damper windings or eddy-current circuits in solid-iron rotor cores.

When analyzing the interaction of a synchronous generator (or several generators) with the associated power system, it is often very convenient to reduce the higher-order dq synchronous generator model to a much simpler Thevenin equivalent circuit model, as shown in Figure 78.18, consisting of a voltage source behind an inductive impedance [10]. The corresponding values of equivalent inductance and source voltage in this model change quite significantly depending on whether the analysis is addressing either steady-state operation or transient response. In particular, the generator's transient synchronous inductance $L_s^{'}$ is significantly smaller than the steady-state synchronous inductance L_s in most cases, while the transient model source voltage $E_s^{'}$ is, conversely, noticeably larger than the steady-state source voltage E_s.

Exciters for Wound-Field Synchronous Generators

The exciter is the functional block that regulates the output voltage and current characteristics of a wound-field synchronous generator by controlling the instantaneous voltage (and, thus, the current) applied to the generator's field winding. Exciters for large utility-class synchronous generators (> 1000 MW) must handle on the order of 0.5% of the generators' rated output power, thereby requiring high-power excitation equipment with ratings of 5 MW or higher [27].

Typically, the exciter's primary responsibility is to regulate the generator's output AC voltage amplitude for utility power system applications, leading to the simplified representation of the resulting closed-loop voltage regulation system shown in Figure 78.19. Almost all of the basic exciter control algorithms in use today for large synchronous generators are based on classical control techniques which are well suited for the machine's dynamic characteristics and limited number of system inputs and outputs. The dynamics of such large generators are typically dominated by their long field winding time constants, which are on the order of several seconds, making it possible to adequately model the generator as a very sluggish low-pass filter for small-signal excitation analyses.

While such sluggish generator dynamics may simplify the steady-state voltage regulator design, it complicates the task of fulfilling a second major responsibility of the exciter which is to improve power system stability in the event of large electrical disturbances (e.g., faults) in the generator's vicinity. This large-disturbance stabilization requires that the exciter be designed to increase its output field voltage to its maximum (ceiling) value as rapidly as possible in order to help prevent the generator from losing synchronism with the power grid (i.e., "pull-out") [21]. Newer "high initial response (HIR)" exciters are capable of increasing their output voltages (i.e., the applied field voltage V_f) from rated to ceiling values in less than 0.1 sec. [12].

Exciter Configurations A wide multitude of exciter designs have been successfully developed for utility synchronous generators and can be found in operation today. These designs vary in such key regards as the method of control signal amplifi-

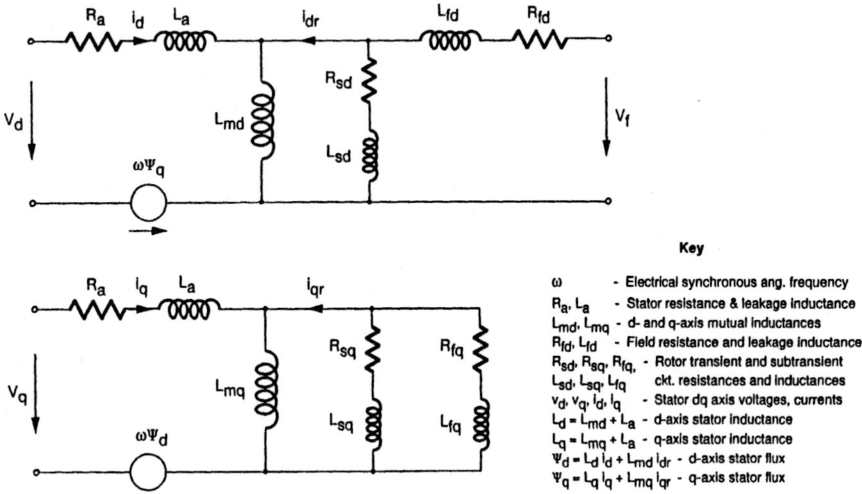

Key
ω - Electrical synchronous ang. frequency
Ra, La - Stator resistance & leakage inductance
Lmd, Lmq - d- and q-axis mutual inductances
Rfd, Lfd - Field resistance and leakage inductance
Rsd, Rsq, Rfq, - Rotor transient and subtransient
Lsd, Lsq, Lfq ckt. resistances and inductances
vd, vq, id, iq - Stator dq axis voltages, currents
Ld = Lmd + La - d-axis stator inductance
Lq = Lmq + La - q-axis stator inductance
Ψd = Ld id + Lmd idr - d-axis stator flux
Ψq = Lq iq + Lmq iqr - q-axis stator flux

Figure 78.17 Synchronous generator d- and q-axis equivalent circuits.

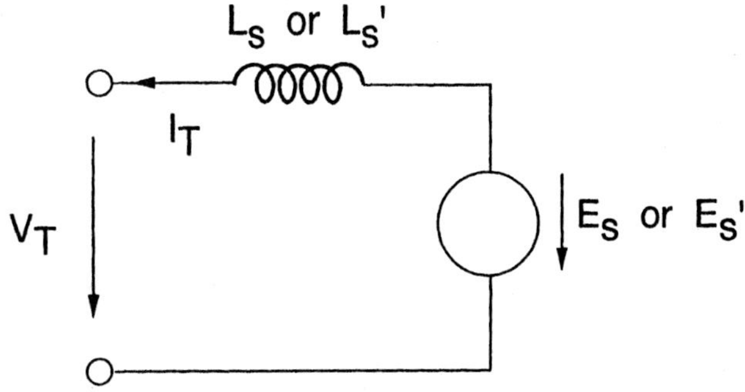

Figure 78.18 Synchronous generator Thevenin-equivalent circuit model (unprimed variables identify steady-state circuit model variables; primed variables identify transient model).

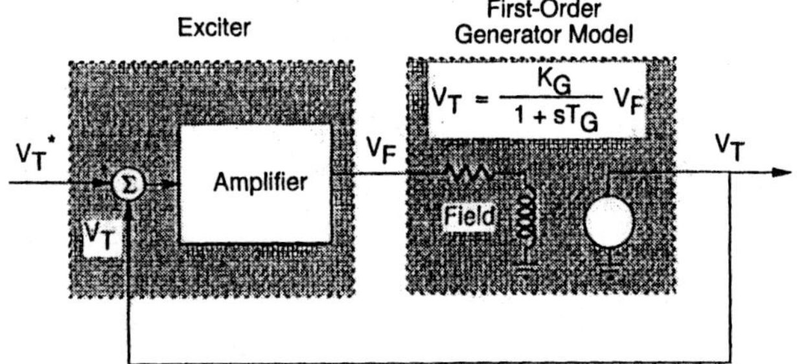

Figure 78.19 Simplified block diagram of synchronous generator closed-loop voltage regulation system.

cation to achieve the required field winding power rating, and the selected source of the excitation power (i.e., self-excitation from the generator itself vs. external excitation using an independent power source). Despite such diversity, the majority of modern exciters in use today can be classified into one of two categories based on the means of power amplification control: "rotating exciters" which use rotating machines as the control elements, and "static exciters" which use power semiconductors to perform the amplification.

Rotating Exciters Rotating exciters have been in use for many years and come in many varieties using both DC commutator machines and AC alternators as the main exciter machines [1], [4]. A diagram of one typical rotating exciter using an AC alternator and an uncontrolled rectifier to supply the generator field is shown in Figure 78.20(a). The output of the exciter alternator is controlled by regulating its field excitation, reflecting the cascade nature of the field excitation to achieve the necessary power amplification. One interesting variation of this scheme is the so-called "brushless exciter" which uses an inverted AC alternator as the main exciter with polyphase armature windings on the rotor and rotating rectifiers to eliminate the need for slip rings to supply the main generator's field winding [26].

The standard control representation developed by IEEE for the class of rotating exciters typified by the design in Figure 78.20(a) is shown in the accompanying Figure 78.20(b) [11]. The dynamics of the exciter alternator are modeled as a low-pass filter $[1/(K_E + T_E)]$ dominated by its field constant, just as in the case of the main generator discussed earlier (see Figure 78.19). The time constant (T_A) of the associated exciter amplifier is considerably shorter than the exciter's field time constant and plays a relatively minor role in determining the exciter's principal control characteristics.

The presence of the rate feedback block $[K_F, T_F]$ in the exciter control diagram is crucial to the stabilization of the overall voltage regulator. Without it, the dynamics of the voltage regulating loop in Figure 78.19 are dominated by the cascade combination of two low-frequency poles which yield an oscillatory response as the amplifier gain is increased to improve steady-state regulation. The rate feedback (generally referred to as "excitation control system stabilization") provides adjustable lag-lead compensation for the main regulating loop, making it possible to increase the regulator loop gain crossover frequency to 30 rad/sec or higher.

Static Exciters In contrast, a typical static exciter configuration such as the one shown in Figure 78.21(a) uses large silicon controller rectifiers (SCRs) rather than a rotating machine to control the amount of generator output power fed back to the field winding [17]. The voltage regulating function is performed

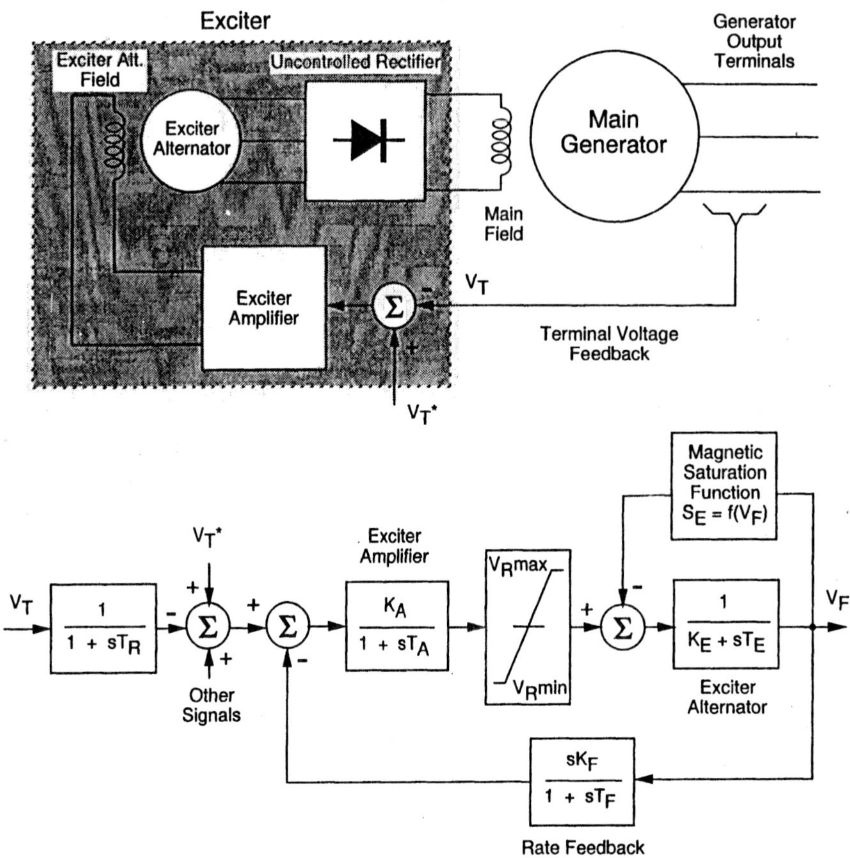

Figure 78.20 Block diagrams of a rotating exciter scheme using an uncontrolled rectifier showing (a) physical configuration, and (b) corresponding exciter standard control representation [12].

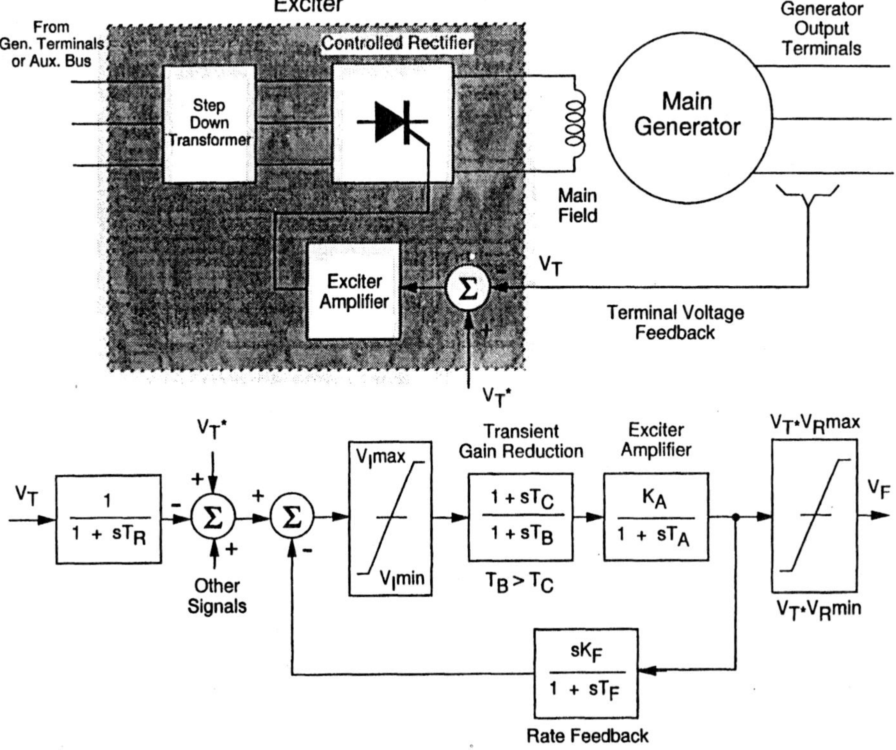

Figure 78.21 Block diagrams of a static exciter scheme using phase-controlled SCRs showing (a) physical configuration, and (b) corresponding exciter standard control representation [18]. (Note that either the "Transient Gain Reduction" block or the "Rate Feedback" block is necessary for system stabilization, but not both.)

using classic phase control principles [14] which determine when the SCRs are gated on during each 60-Hz cycle to supply the desired amount of field excitation.

Static exciters generally enjoy the advantage of much faster response times than rotating exciters, making them excellent candidates for "high initial response" exciters as discussed above. This is reflected in the standard static exciter control representation shown in Figure 78.21(b), which lacks the sluggish field time constant T_E that dominates the dynamics of the rotating exciter [13]. As a result, the task of stabilizing the main voltage regulating loop is simplified in the absence of this low-frequency pole.

Nevertheless, lag-lead compensation is often added to the exciter control in order to prevent the generator's high-frequency regulator gain from reducing the stability margin of the interconnected power system [13]. This compensation (referred to as "transient gain reduction") is provided in the Figure 78.21(b) control diagram using either the $T_B - T_C$ block in the forward gain path or the same type of $K_F - T_F$ rate feedback block introduced previously in Figure 78.20(b) (only one or the other is necessary in a given installation).

Additional Exciter Responsibilities In addition to its basic responsibilities for regulating the generator's output voltage described above, the exciter is also responsible for additional important tasks including power system stabilization, load or reactive power compensation, and exciter/generator protection. A block diagram of the complete generator-plus-exciter system identifying these supplementary functions is provided in Figure 78.22. Each of these exciter functions will be addressed briefly in this section.

Power System Stabilization Since a utility generator is typically one component in a large power grid involving a multitude of generators and loads distributed over a wide geographic area, the interactions of all these mechanical and electrical systems give rise to complex system dynamics. In some cases, the result is low-frequency dynamic instabilities in the range of 0.1 to 2 Hz involving one or more generators "swinging" against the rest of the grid across intervening transmission lines [5]. One rather common source of "local" mode instabilities is a remote generator located at the mouth of a mine that is connected to the rest of the power grid over a long transmission line with relatively high per-unit impedance. Other examples of dynamic instabilities involve the interactions of several generators, giving rise to more complicated "inter-area" modes which can be more difficult to analyze and resolve.

Unfortunately, the steps that are taken to increase exciter response for improved generator transient stability characteristics (i.e., "high initial response" exciters) tend to aggravate the power system stability problems. Weak transmission systems which result in large power angles between the generator internal voltage and the infinite bus voltage exceeding 70 degrees tend to demonstrate a particular susceptibility to this type of power system instability while under voltage regulator control.

The typical approach to resolving such dynamic instability problems is to augment the basic exciter with an additional feedback control loop [13] known as the power system stabilizer (PSS) as shown in simplified form in Figure 78.23. As indicated in this figure, alternative input signals for the PSS that are presently being successfully used in the field include changes in the generator shaft speed ($\Delta\omega_r$), generator electrical frequency ($\Delta\omega_e$), and the electrical power (ΔP_e). The primary control function performed by the PSS is to provide a phase shift using one or more adjustable lead-lag stages, which compensates for the destabilizing phase delays accumulated in the generator and exciter electrical circuits. Additional PSS signal processing in Figure 78.23 is typically added to filter out undesired torsional oscillations and to prevent the PSS control loop from interfering with the basic exciter control actions during major transients caused by sudden load changes or power system faults.

Proper settings for the primary PSS lead-lag gain parameters vary from site to site depending on the characteristics of the generator, its exciter, and the connected power system. Since the resulting dynamic characteristics can get quite complicated, a combination of system studies and field tests are typically required in order to determine the proper PSS gain settings for the best overall system performance. Effective empirical techniques for setting these PSS control gains in the field have gradually been developed on the basis of a significant experience base with successful PSS installations [16].

Load or Reactive Power Compensation A second auxiliary function provided in many excitation systems is the tailored regulation of the generator's terminal voltage to compensate for load impedance effects or to control the reactive power delivered by the generator [13]. One particularly straightforward version of this type of compensation is shown in Figure 78.24. As indicated in this figure, the compensator acts to supplement the measured generator's terminal voltage that is being fed back to the exciter's summing junction with extra terms proportional to the generator output current. The current-dependent compensation terms are added both in phase and 90° out of phase with the terminal voltage, with the associated compensation gains, R_c and X_c, having dimensions of resistive and reactive impedance.

Depending on the values and polarities of these gains, this type of compensation can be used for different purposes. For example, if R_c and X_c are negative in polarity, this block can be used to compensate for voltage drops in power system components such as step-up transformers that are downstream from the generator's terminals where the voltage is measured. Alternatively, positive values of R_c and X_c can be selected when two or more generators are bussed together with no intervening impedance in order to force the units to share the delivered reactive power more equally.

Although not discussed here, more sophisticated compensation schemes have also been developed which modify the measured terminal voltage based on calculated values of the real and reactive power rather than the corresponding measured current components. Such techniques provide means of achieving more precise control of the generator's output power characteristics and find their origins in exciter development work that was completed several decades ago [20].

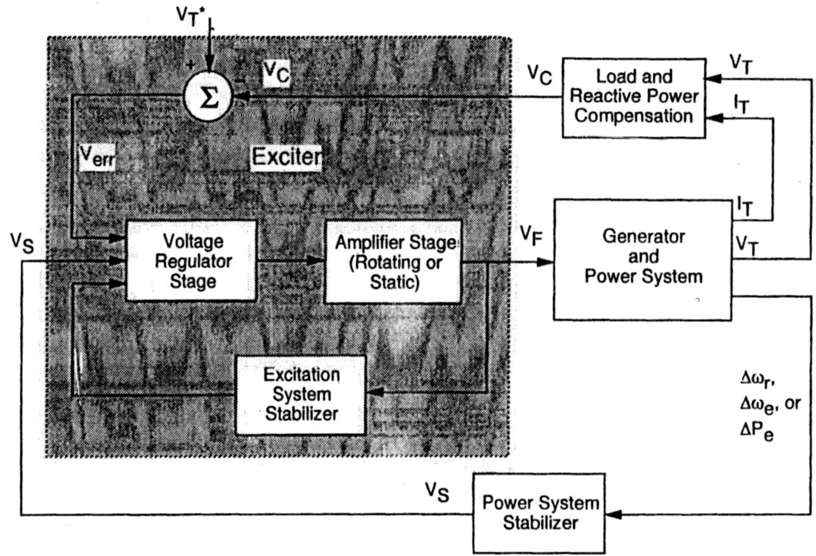

Figure 78.22 Block diagram of complete generator-plus-exciter control system, identifying supplementary control responsibilities.

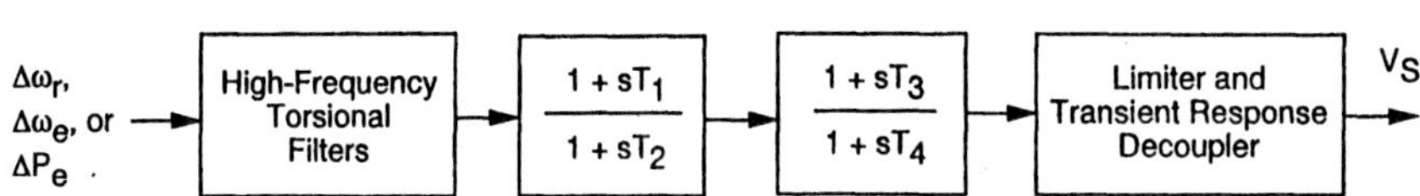

Figure 78.23 Basic control block diagram of power system stabilizer (PSS).

Exciter/Generator Protection Although a general review of the important issue of generator protection is well beyond the scope of this chapter [2], the specific role of the exciter in the generator's protection system deserves a brief discussion. Some of these key responsibilities include the following:

- Underexcited reactive ampere limit (URAL) — the minimum excitation level is limited as a function of the output reactive current since excessive underexcitation of the generator can cause dangerous overheating in the stator end turns
- Generator maximum excitation limit — at the other extreme, the maximum excitation current is limited to prevent damage to the exciter equipment and to the generator field winding due to overheating
- Volts-per-Hertz limiter — excessive magnetic flux levels in the generator iron which can cause internal overheating are prevented by using the exciter to limit the generator's output voltage as a function of output frequency (i.e., shaft speed).

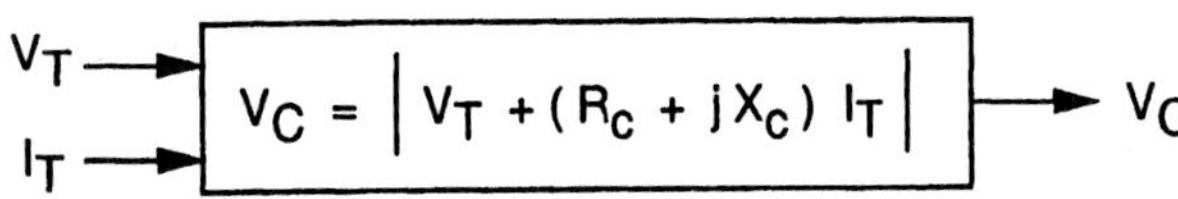

Figure 78.24 Example of generator load/reactive power compensation.

78.3.4 Induction Generators

Induction Generator Fundamentals and Models

Introduction Induction generators are induction machines that operate above synchronous speed and thereby convert mechanical power into electrical power. Induction generators have been extensively used in applications such as wind turbines and hydroelectric storage pumping stations. Frequent direct line starting is required in both of these applications. The AC induction machine offers the advantage that it can be designed for direct line-start operation, thereby avoiding additional synchronization machines and control. Furthermore, induction machines tolerate the thermal load associated with direct AC line-starting transients better than synchronous machines.

At high power levels above 1 MVA, efficiency considerations favor synchronous generators, while at very low power levels below 1 kVA, permanent magnet synchronous generators (e.g., automobile alternators) are more cost effective. One can conclude that induction generators are preferred in generator applications that require frequent starting and that are in a power range of 10 to 750 kVA.

Induction generators can operate in two distinctively different modes. In the first mode, the induction generator is connected to a fixed-frequency AC voltage source (e.g., utility line voltage) or a variable-frequency voltage-controlled AC source, such as pulse-width-modulated (PWM) inverters. The AC source provides the excitation (i.e., the magnetization current) for the induction machine. In this mode, the magnetizing flux is determined or controlled by the AC source voltage.

The second mode of operation is the so-called self-excited

mode. During self-excitation the magnetizing current for the induction generator is provided by external reactive elements (usually capacitors) or voltage-source inverters operating in six-step waveform mode. Neither of these schemes make it convenient to regulate the terminal voltage of the machine. The output voltage of the generator depends on many variables and parameters such as generator speed, load current, magnetization characteristics of the machine, and capacitor values.

The induction generator itself must deliver the necessary power to offset losses induced by the circulating reactive currents during self-excited operation, and the associated stator copper losses are typically high. As a result, self-excited induction generators are rarely used for continuous operation because they do not achieve high efficiency. Moreover, it is difficult to start the excitation process under loaded conditions. Nevertheless, self-excited operation with capacitors is sometimes used to brake induction motors to standstill in applications demanding rapid system shutdowns. In addition, six-step voltage-source inverters are often used in traction motor drive applications to apply regenerative braking.

The control principles of induction generators feeding power into a controlled AC voltage source will be discussed in the following sections. Self-excited operation of induction generators has been analyzed extensively, and interested readers are referred to the available technical literature for more details [18].

Induction Generator Model Induction generators are constructed identically to induction motors. A typical induction machine has a squirrel-cage rotor and a three-phase stator winding. Figure 78.25 illustrates the construction of a two-pole (or one pole pair), two-phase induction machine. The stator winding consists of two windings (d and q) that are magnetically perpendicular. The squirrel-cage rotor winding consists of rotor bars that are shorted at each end by rotor end rings.

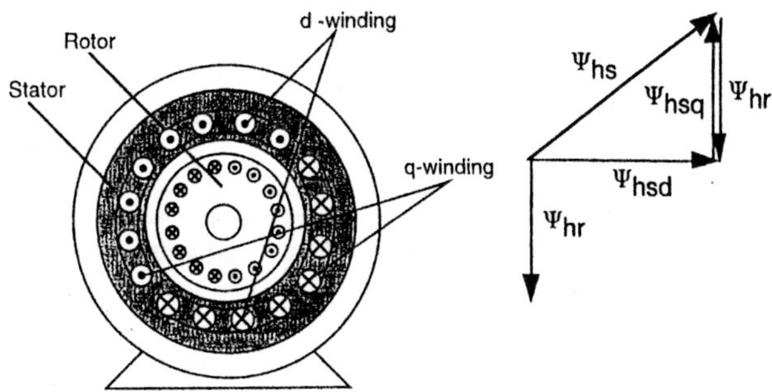

Figure 78.25 Two-phase induction machine, showing stator and rotor windings and flux diagram associated with stator and rotor currents.

The flux linkages associated with currents in each set of windings (i.e., the stator magnetizing flux $\underline{\Psi}_{hs}$, the rotor magnetizing flux $\underline{\Psi}_{hr}$, and the d- and q-axis stator magnetizing flux components $\underline{\Psi}_{hsd}$ and $\underline{\Psi}_{hsq}$,) are shown in the flux diagram. (Underlined variables designate vector quantities.) Note that each magnetizing flux vector represents a flux component produced by the current in a particular winding. These flux components are not to be confused with the total stator or rotor flux linkages $\underline{\Psi}_s$ and $\underline{\Psi}_r$ which represent the superimposed flux coupling from all windings, as discussed later in this section.

The basic equivalent circuit of an induction machine for steady-state operation is illustrated in Figure 78.26. This equivalent circuit shows that each phase winding has parasitic resistances and leakage inductances and that the stator and the rotor are magnetically coupled. However, other equivalent circuits can be derived for an induction machine, as illustrated in Figure 78.27. These equivalent circuits are obtained by transforming stator or rotor current and voltage quantities with a turns ratio "a". Figure 78.27 also specifies the different turns ratios and the corresponding flux vector diagrams which identify the flux reference vector used for each of the three equivalent circuits.

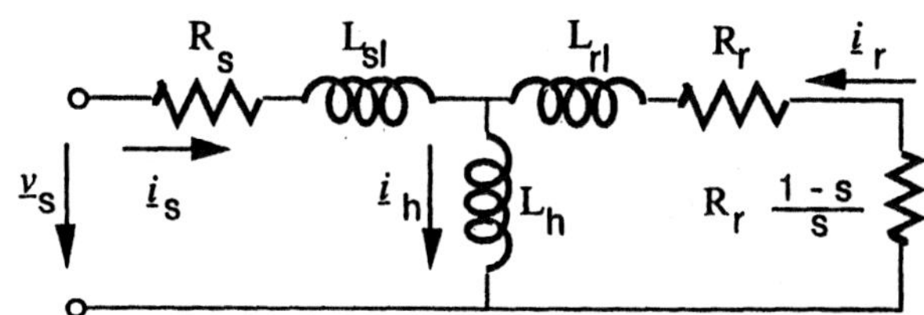

Figure 78.26 Single-phase equivalent (steady-state) circuit of induction machine. L_h, main (magnetizing) inductance; L_{sl}, stator leakage inductance; L_{rl}, rotor leakage inductance; $L_s = L_h + L_{sl}$, stator inductance; $L_r = L_h + L_{rl}$, rotor inductance; R_s, stator resistance; R_r, rotor resistance; $\underline{v}_s = v_{sd} + jv_{sq}$, stator (line-to-neutral) voltage, dq component; $\underline{i}_s = i_{sd} + ji_{sq}$, stator current, dq component; $\underline{i}_r = i_{rd} + ji_{rq}$, rotor current, dq component; $\underline{i}_h = \underline{i}_s + \underline{i}_r$, magnetizing current; s, slip of the induction machine.

Some equivalent circuits are simpler for analysis because one leakage inductance can be eliminated [3]. For example, a turns ratio $a = L_h/L_r$ transforms the equivalent circuit of Figure 78.26 into the topmost circuit of Figure 78.27, which has no leakage inductance in the rotor circuit. Hence, the rotor flux is selected here as the main flux reference vector. Also, the d-axis d_a of the dq synchronous reference frame that corresponds with this turns ratio "a" is linked to the rotor flux so that $d_a = d_r$.

Torque-Slip Characteristics The power the induction generator delivers depends on the slip frequency or, equivalently, the slip of the machine. Slip of an induction machine is defined as the relative speed difference of the rotor with respect to the synchronous speed set by the excitation frequency:

$$s = \frac{f_e - f_m}{f_e} = \frac{\omega_e - \omega_m}{\omega_e} = \frac{n_e - n_m}{n_e} \tag{78.158}$$

with s being the slip of the induction machine, f_e the stator electrical excitation frequency (Hz), f_m the rotor mechanical rotation frequency (Hz), ω_e the stator excitation angular frequency (rad/s), ω_m the rotor mechanical angular frequency (rad/s), n_e the rotational speed of excitation flux in airgap (r/min), n_m the rotor mechanical shaft speed (r/min).

In the case of machines with higher pole-pair numbers, the

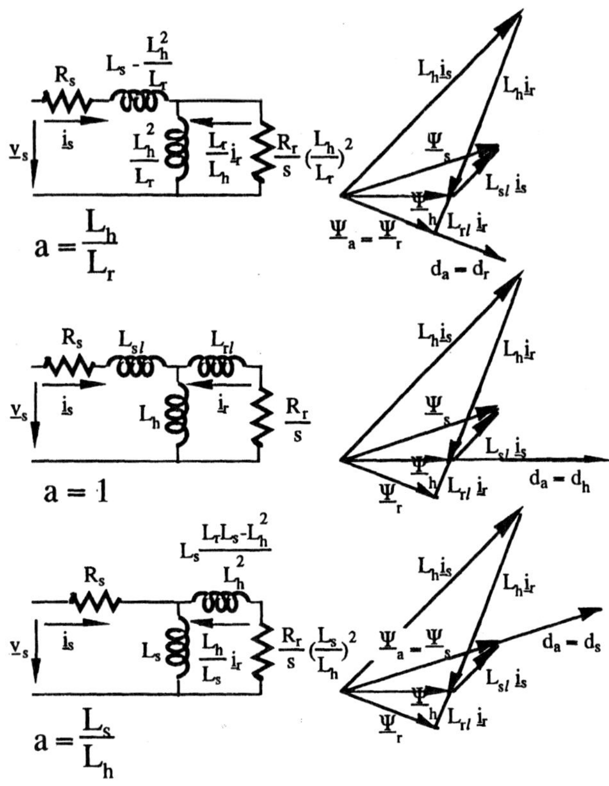

Figure 78.27 Modified equivalent circuits of induction machine, showing equivalent circuits and corresponding flux vector diagrams identifying the flux reference vector.

$$\underline{\Psi}_s = L_s \underline{i}_s + L_h \underline{i}_r$$

$$\underline{\Psi}_r = L_r \underline{i}_r + L_h \underline{i}_s$$

$$\underline{\Psi}_h = L_h \underline{i}_h = L_h \underline{i}_s + L_h \underline{i}_r = \underline{\Psi}_{hs} + \underline{\Psi}_{hr}$$

mechanical speed is usually defined in electrical degrees according to:

$$n_m = p n_{rotor} \qquad (78.159)$$

where p is the pole pair number and n_{rotor} is the rotor speed as measured by observer in mechanical degrees.

The stator resistance R_s of medium and large induction machines can usually be neglected because the designer strives to optimize the efficiency of the induction machine by packing as much copper in the stator windings as possible. Using this approximation together with the equivalent circuit of Figure 78.27 that eliminates the stator leakage inductance ($a = L_s/L_h$), the steady-state torque per phase of the AC induction machine can easily be calculated as a function of the supply voltage and the slip frequency, yielding:

$$\frac{T_{em}}{T_k} = \frac{2}{\frac{s}{s_k} + \frac{s_k}{s}} \qquad (78.160)$$

where

$$T_k = \frac{p V_s^2}{2\omega_e^2 L_l} \qquad (78.161)$$

$$s_k = \frac{R_2}{\omega_e L_l} \qquad (78.162)$$

$$L_l = L_s \frac{L_s L_r - L_h^2}{L_h^2} \qquad (78.163)$$

where T_{em} is the electromagnetic torque per phase (Nm), T_k is the per-phase pull-out (maximum) torque (Nm), s_k is the pull-out slip associated with T_k, V_s is the rms stator line-to-neutral supply voltage (V), and L_l is the leakage inductance (H).

Figure 78.28 illustrates a typical torque-slip characteristic of an induction machine. According to Equation 78.160, the torque of the induction machine at high slip values varies approximately inversely with the slip frequency. Operating an induction machine in this speed range beyond the pull-out slip is inefficient and unstable in the absence of active control, and, hence, this operating regime is of little interest. Stable operation is achieved in a narrow speed range around the synchronous speed ($s = 0$) between $-s_k$ and $+s_k$ which are identified in Figure 78.28.

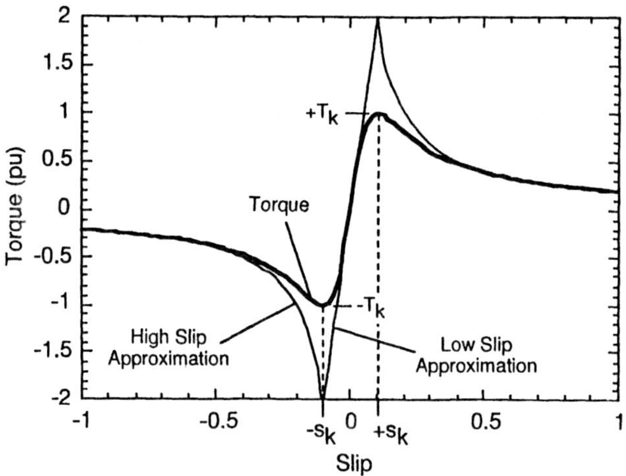

Figure 78.28 Typical slip-torque characteristic of induction machine.

Induction Generator Control Principles

Basic Slip Control Principle Whenever the slip is below the pull-out slip s_k, the torque varies approximately linearly with slip. Hence, the induction machine behaves similarly to a DC generator with constant armature voltage and constant field excitation. As soon as the speed of the generator exceeds the no-load (synchronous) speed (i.e., negative slip), mechanical power is transformed into electrical power (i.e., negative torque). Conversely, the machine operates as a motor with positive torque when the speed is below the no-load speed (i.e., positive slip).

Clearly, control of the slip frequency provides a direct means for controlling the AC induction generator's output torque and power. In applications where the electrical supply frequency is practically constant (e.g., utility systems), slip control can be realized by sensing the rotor speed and driving the generator shaft with the prime mover at the desired slip value with respect to the measured stator excitation frequency.

New induction generator systems use inverters connected to the machine's stator terminals to control the generator power, as illustrated in Figure 78.29. Inverters are power electronic devices that transform DC power to polyphase AC power or vice versa. Both the amplitude and frequency of the output waveforms delivered by the inverter are independently adjustable. In operation, the inverter provides the magnetization energy for the induction generator while the generator shaft is driven by a motor or some other type of prime mover. The AC-to-DC inverter converts the generator AC power to DC power, and this DC power can be transformed to AC power at fixed frequency (e.g., 50 or 60 Hz) using a second DC-to-AC inverter as shown in Figure 78.29. Both inverters are constructed as identical bidirectional units but are controlled in opposite power-flow modes.

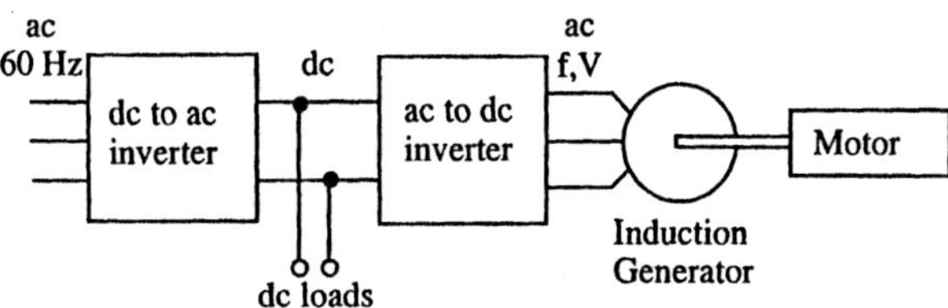

Figure 78.29 Inverter-fed induction generator system.

Induction generator applications that have a wide speed range or operate under fast varying dynamic load conditions require variable-frequency control to allow stable operation. Indeed, typical rated slip of induction machines is below 2%. Hence, a 2% speed variation around the synchronous speed changes torque from zero to 100% of its rated value. This poses significant design challenges in applications such as wind turbines operating in the presence of strong wind gusts. The high stiffness of the induction generator's torque-speed characteristic makes it very difficult to implement a speed governor to control the pitch of the turbine blades that is sufficiently fast-acting and precise to adequately regulate the machine's slip within this narrow range. On the other hand, an inverter can rapidly adjust the generator's electrical excitation frequency and slip to provide stable system operation with constant output power under all operating conditions.

Field-Oriented Control Principles The fundamental quantity that needs to be controlled in an induction generator is torque. Torque control of the inverter-fed induction machine is usually accomplished by means of field-oriented control principles to ensure stability. With field-oriented control, the torque and the flux of the induction generator are independently controlled in a similar manner to a DC generator with a separately excited field winding, as discussed earlier in this chapter.

The principles of field orientation can best be explained by recognizing from the flux vector diagrams shown in Figure 78.27 that the rotor current vector $\underline{i}_r$ is perpendicular to the rotor flux $\underline{\Psi}_r$. Furthermore, these vector diagrams illustrate that the stator current space vector $\underline{i}_s$ is composed of the rotor current $\underline{i}_r$ and the magnetizing current $\underline{i}_h$. Note that the space vectors in the vector diagrams are drawn in a reference frame that is rotating at the synchronous excitation frequency ω_e. By aligning a dq coordinate system with the rotating rotor flux vector $\underline{\Psi}_r$, one can prove that under all conditions (including transient conditions) the torque of the induction machine per phase is given by [22]:

$$T_{em} = -\frac{p}{2} i_{rq} \Psi_r = \frac{p}{2} \frac{L_h}{L_r} i_{sq} \Psi_r \tag{78.164}$$

Positive torque signifies motoring operation, while a negative torque indicates generator operation. Hence, in a synchronous dq reference frame linked to the rotor flux (top diagram in Figure 78.27), the q-axis component of the stator current i_{sq} corresponds to the torque-producing stator current component, being equal in amplitude to the rotor current component i_{rq}, with the opposite sign. The d-axis component of the stator current i_{sd} equals the magnetizing current, corresponding to the rotor flux-producing current component.

A control scheme that allows independent control of rotor flux and torque in the synchronously rotating rotor flux reference frame can now be derived, and the resulting control block diagram is shown in Figure 78.30. A Cartesian-to-polar coordinate transformation calculates the amplitude and the angle γ_{rs} of the stator current commands (in the synchronous reference frame) corresponding to the desired flux- and torque-producing components. (Controller command signals are marked with superscript $*$ in Figure 78.30, with negative torque commands corresponding to generator operation.) It is very important to note that the torque and the flux commands are decoupled (i.e., independently controlled) using this field-oriented control scheme. The controller can be seen as an open-loop disturbance feedforward controller, the disturbance signal being the variation of the flux position γ_r.

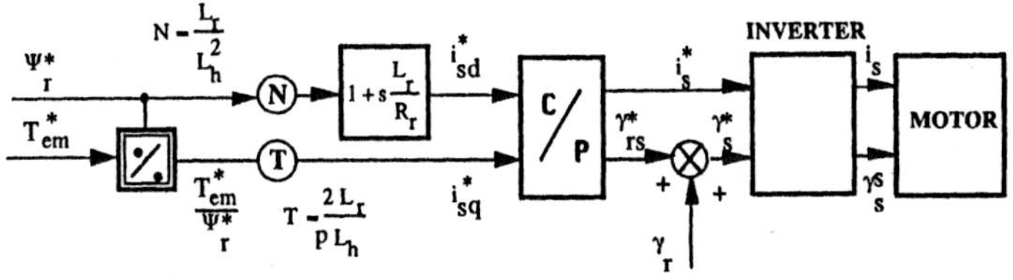

Figure 78.30 Field oriented controller allowing independent flux and torque control. γ_r, angular position of rotor flux with respect to stationary reference; γ_s, angular position of stator current with respect to stationary reference; γ_{rs}, angle between stator current and rotor flux.

The C/P block in Figure 78.30 indicates a Cartesian-to-polar coordinate transformation according to the following generalized equations:

$$x = \sqrt{x_d^2 + x_q^2} \tag{78.165}$$

$$\tan\alpha = \frac{x_q}{x_d} \tag{78.166}$$

with x_d the d-component or real component of space vector $\underline{x} = x_d + jx_q$, x_q the q-component or imaginary component of $\underline{x}$, x the amplitude of the space vector $\underline{x}$, and α the angular position of space vector $\underline{x}$.

Direct vs. Indirect Field Orientation To complete the controller calculation loops, one needs the rotor flux position γ_r to calculate the stator current vector position in the stationary reference frame that is linked to the stator of the machine. In other words, one needs to determine the orientation of the rotating field flux vector. It is for this reason that the control method illustrated in Figure 78.30 was called "field orientation".

Two field orientation strategies have been derived to detect the rotor flux position. Direct field orientation methods use sensors to directly track the flux position. Hall sensors are seldom used because of the high temperature inside the induction machine. Typical flux sensors are flux coils (sensing induced voltage) followed by integrators. The latter gives satisfactory results at frequencies above 5 to 10 Hz.

The direct field orientation control block diagram can be completed as shown in Figure 78.31. However, control problems can arise because most flux sensors are positioned on the stator and not on the rotor. As a result, these sensors monitor the stator flux Ψ_s^s in a stationary reference frame (marked with superscript "s") and not the rotor flux which is used in the decoupling network. This makes it necessary to use the flux linkage equations to derive the rotor flux from the flux sensor measurements. The required calculations introduce estimated machine parameters (leakage inductances) into the disturbance feedforward path [3] leading to detuning errors. Another approach is to decouple the machine equations in the stator flux reference frame in which the flux sensors are actually operating. This method requires a decoupling network of greater complexity but achieves high accuracy and potentially zero detuning error under steady-state conditions [9], [19].

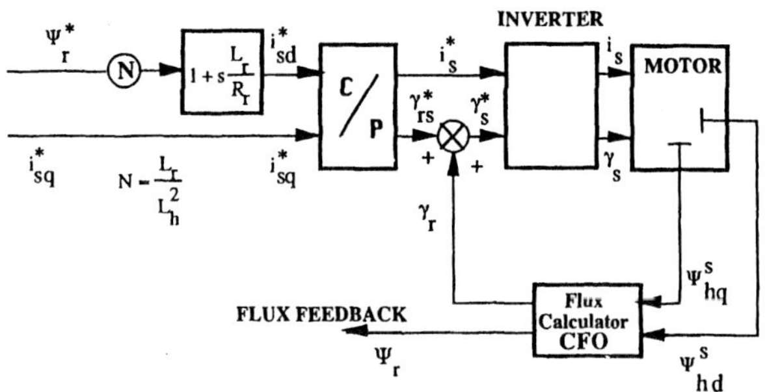

Figure 78.31 Direct field orientation method.

The second category of field orientation methods are called indirect field orientation because they derive the flux position using a calculated or estimated value of the angle γ_{mr} between the flux and the rotor position. This angle is nothing other than the rotor flux "slip" angle which varies at the slip frequency ω_{mr}. The rotor position γ_m is measured using a shaft position sensor, while the flux slip angle is derived from the slip frequency by integration. The dynamic field-oriented system equations of the induction machine are used to derive the slip frequency ω_{mr} and the slip angle γ_{mr}, as follows:

$$\omega_{mr} = \frac{L_h}{L_r}\frac{R_r}{p}\frac{i_{sq}^*}{\Psi_r^*} \tag{78.167}$$

$$\gamma_{mr} = \int \omega_{mr} dt \tag{78.168}$$

Figure 78.32 illustrates how a controller can be constructed to calculate the slip frequency command ω_{mr}^*, the slip angle command γ_{mr}^*, and the rotor flux position command γ_r^*. As Figure 78.32 shows, most indirect field-oriented controllers are constructed as open-loop disturbance feedforward controllers using the commanded current components instead of measured current quantities. This approach is justified because state-of-the-art, high-frequency, current-regulated inverters produce relatively precise current waveforms with respect to the current commands.

Note that indirect field orientation depends on the rotor time constant L_r / R_r which is composed of estimated machine parameters. As stated above, direct field orientation also needs machine parameters to calculate the flux vector position and most direct flux sensors do not operate at low frequencies. To circumvent these problems, both methods can be combined using special field-oriented algorithms to ensure greater accuracy.

Control Method Comparison Field orientation is a more advanced control technique than the slip controller discussed above. Field orientation offers greater stability during fast-changing transients because it controls torque while maintaining constant flux. Hence, the trapped magnetic energy in the machine does not change when speed or torque variations occur. This decoupled control strategy allows field orientation to control an AC induction machine exactly the same as a separately excited DC machine with a series compensation armature winding (see the section on DC generators).

The reader is invited to compare the vector diagram of Figure 78.8 in the DC generator section (illustrating the independent flux and torque control of a DC generator) and the space vector diagram of an induction generator illustrated in Figure 78.25 or Figure 78.27. The rotor flux vector $\underline{\Psi}_{hr}$ of the induction machine corresponds to the armature flux Ψ_a in the DC machine, while the stator magnetizing flux d-component $\underline{\Psi}_{hsd}$ corresponds to the field winding flux Ψ_f. The compensation winding flux Ψ_c of the DC machine relates to the q-component of the stator magnetizing flux vector $\underline{\Psi}_{hsq}$ of the AC machine. While the spatial orientation of the flux vectors in the DC machine is fixed, the space vectors of the induction machine rotate at synchronous speed. As a result, independent control of the individual space vectors in the AC machine can only be achieved by controlling the amplitude and the phase angle of the stator flux (i.e., stator voltage and current vectors).

Another approach to understanding the difference between a field-oriented controller and a slip controller is to consider the fact that field-oriented controllers control the flux slip *angle* and the stator current vector *angle* while slip controllers only regulate the *frequency* of these vectors. Controlling the angle of space vectors in electrical systems is analogous to position control

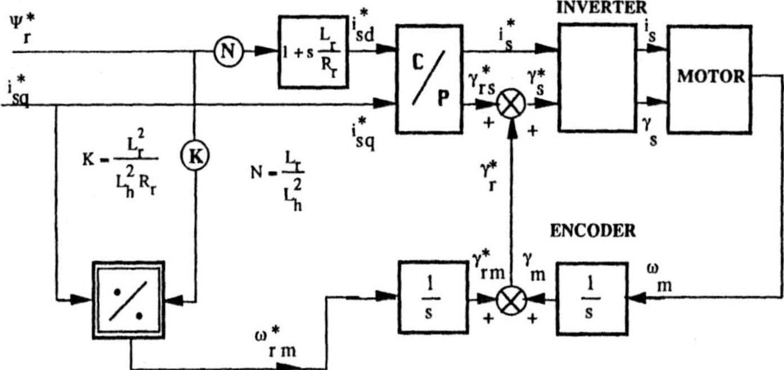

Figure 78.32 Indirect field orientation method.

in mechanical systems, while frequency control corresponds to speed control. It is immediately recognized that position control always offers greater stiffness than speed control. Hence, field orientation can offer the same stiffness improvements compared to slip-frequency controllers during transient conditions.

One disadvantage of field-oriented control is that it requires considerably more computation power than the simpler slip control algorithms. Many field-oriented controllers are implemented using digital signal processors (DSPs). Consequently, increased use of field orientation for induction generators will depend on future trends in the cost of digital controllers and sensors as well as the development of new control algorithms that decrease the controller's field installation time (e.g., machine parameter autotuning) while optimizing generator efficiency.

78.3.5 Concluding Remarks

This chapter has attempted to provide a concise overview of the major classes of electrical generators and their associated control principles. Despite the notable differences between the three types of electrical machines reviewed—DC, synchronous, and induction—there are some important underlying control aspects which they share in common. These include the single-input/single-output nature of the basic regulating control problem in each case, with dynamic response typically dominated by a long magnetic flux (i.e., field) time constant. This long time constant is responsible for the slow dynamic response which characterizes the majority of generator regulating systems now in the field.

As pointed out in the chapter, the introduction of power electronics provides access to generator control variables which can circumvent the limitations imposed by the flux time constant, leading to significant improvements in the regulator's dynamic response and other performance characteristics. Such advances have already had a significant impact in many applications, and work is continuing in many locations to extend these techniques to achieve further improvements in generator control performance and economics.

References

[1] Barnes, H. C., Oliver, J.A., Rubenstein, A.S., and Temoshok, M., Alternator-rectifier exciter for cardinal plant, *IEEE Trans. Power Appar. Syst.*, 87, 1189–1198, 1968.

[2] Berdy, J., Crenshaw, M.L., and Temoshok, M., Protection of large steam turbine generators during abnormal operating conditions, *Proc. CIGRE Int. Conf. on Large High Tension Electric Systems*, Paper No. 11-05, 1972.

[3] Blaschke, F. and Bayer, K. H., Die Stabilität der Feldorientierten Regelung von Asynchron-Maschinen, *Siemens Forsch. Entwick. Ber.*, 7(2), 77–81, 1978.

[4] Bobo, P.O., Carleton, J. T., and Horton, W.F., A new regulator and excitation system, *AIEE Trans. Power Appar. Syst.*, 72, 175–183, 1953.

[5] Bollinger, K.E. (Coordinator), *Power System Stabilization via Excitation on Control*, IEEE Tutorial Course Notes, Pub. No. 81 EHO 175-0 PWR, IEEE Press, New York, 1981.

[6] Brown, D. and Hamilton, E.P., *Electromechanical Energy Conversion*, Macmillan, New York, 1984.

[7] Concordia, C., *Synchronous Machines*, John Wiley & Sons, New York, 1951.

[8] Daumas, M., Ed., *Histoire Générale des Techniques*, Vol. 3, 1978, p. 330–335.

[9] De Doncker, R.W. and Novotny, D.W., The universal field oriented controller, *IEEE Trans. Ind. Appl.*, 30(1), 92–100, 1994.

[10] Fitzgerald, A.E., Kingsley, C., and Umans, S.D., *Electric Machinery*, McGraw-Hill, New York, 1983.

[11] IEEE Committee Report, Computer representation of excitation systems, *IEEE Trans. Power Appar. Syst.*, 87, 1460–1464, 1968.

[12] IEEE Standard 421-1972, Criteria and Definitions for Excitation Systems for Synchronous Machines, Institute of Electrical and Electronics Engineers, New York.

[13] IEEE Committee Report, Excitation system models for power system stability studies, *IEEE Trans. Power Appar. Syst.*, 100, 494–507, 1981.

[14] Kassakian, J.G., Schlecht, M.F., and Verghese, G.C., *Principles of Power Electronics*, Addison-Wesley, Reading, MA, 1991.

[15] Krause, P.C., Wasynczuk, O., and Sudhoff, S.D., *Analysis of Electric Machines*, IEEE Press, New York, NY, 1995.

[16] Larsen, E.V. and Swann, D.A., Applying Power System Stabilizers, Part II. Performance Objectives and Tuning Concepts, Paper 80 SM 559-5, presented at IEEE PES Summer Meeting, Minneapolis, 1980.

[17] McClymont, K.R., Manchur, G., Ross, R.J., and Wilson, R.J., Experience with high-speed rectifier excitation systems, *IEEE Trans. Power Appar. Syst.*, 87, 1464–1470, 1968.

[18] Novotny, D., Gritter, D., and Studtman, G., Self-excitation in inverter driven induction machines, *IEEE Trans. Power Appar. Syst.*, 96(4), 1117–1183, 1977.

[19] Profumo, F., Griva, G., Pastorelli, M., Moreira, J., and De Doncker, R., Universal field oriented controller based on air gap sensing via third harmonic stator voltage, *IEEE Trans. Ind. Appl.* 30(2), 448–455, 1994.

[20] Rubenstein, A.S. and Walkey, W.W., Control of reactive kVA with modern amplidyne voltage regulators, *AIEE Trans. Power Appar. Syst.*, 76, 961–970, 1957.

[21] Sarma, M., *Synchronous Machines (Their Theory, Stability, and Excitation Systems)*, Gordon and Breach, New York, NY, 1979.

[22] Say, M.G., *Alternating Current Machines*, 5th ed., Pitman, Bath, U.K., 1983.

[23] Sen, P., *Thyristor DC Drives*, John Wiley & Sons, New York, 1981.

[24] Slemon, G.R. and Straughen, A., *Electric Machines*, Addison-Wesley, Reading, MA, 1980.

[25] Tissandier, G., *Causeries Sur La Science*, Librairie Hachette et Cie, Paris, 1890.

[26] Whitney, E.C., Hoover, D.B., and Bobo, P.O., An electric utility brushless excitation system, *AIEE Trans. Power Appar. Syst.*, 78, 1821–1824, 1959.

[27] Wildi, T., *Electrical Machines, Drives, and Power Systems*, Prentice Hall, Englewood Cliffs, NJ, 1991.

79

Control of Electrical Power

Harry G. Kwatny
Drexel University

Claudio Maffezzoni
Politecnico Di Milano

John J. Paserba, Juan J. Sanchez-Gasca, and Einar V. Larsen
GE Power Systems Engineering, Schenectady, NY

79.1 Control of Electric Power Generating Plants

Harry G. Kwatny, *Drexel University*
Claudio Maffezzoni, *Politecnico Di Milano*

79.1.1 Introduction

This chapter provides an overview of the dynamics and control of electric power generating plants. The main goals are to characterize the essential plant physics and dynamical behavior, summarize the principle objectives of power plant control, and describe the major control structures in current use. Because of space limitations the discussion will be limited to fossil-fueled, drum-type steam generating plants. Much of it, however, is also relevant to once-through and nuclear powered plants.

The presentation is organized into four major sections. Section 79.1.2 provides a description of a typical plant configuration, explains in some detail the specific objectives of plant control, and describes the overall control system architecture. The control system is organized in a hierarchy, based on time scale separation, in which the highest level establishes set points for lower level regulators so as to meet the overall unit operating objectives.

Section 79.1.3 develops somewhat coarse linear models which qualitatively portray the small signal process behavior, characterize the essential interactions among process variables, and can be used to explain and justify the traditional regulator architectures. They are also useful for obtaining initial estimates of control system parameters which can then be fine-tuned using more detailed, nonlinear simulations of the plant.

The configurations commonly used in modern power plants for the main process variables are described in Section 79.1.4. These include controllers for pressure and generation, evaporator (drum level) temperature, and combustion control. The discussion in Section 79.1.4 is mainly qualitative, based on the understanding of plant behavior developed in Section 79.1.3.

Once a control configuration is chosen, the various compensator design parameters are established by applying analytical control design methods combined with extensive simulation studies. Because space is limited, it is not possible to provide such an analysis for each of the plant subsystems. However, in Section 79.1.5 we do so for the drum level regulator. Drum level control is chosen for illustration because it is particularly important to plant operation and because it highlights the difficulties associated with low load plant dynamics and control. There are many important and outstanding issues regarding automation at low load steam generation levels. In practice, most plants require considerable manual intervention when maneuvering at low load. The most important concerns relate to the evaporation process (the circulation loop) and to the combustion process (furnace). Section 79.1.5 revisits the circulation loop, examines the behavioral changes that take place as generation level is reduced, and explains the consequences for control.

0-8493-8570-9/96/$0.00+$.50

79.1.2 Overview of a Power Plant and its Control Systems

Overall Plant Structure

A typical power plant using fossil fuel as its energy source is organized into three main subsystems, corresponding to the three basic energy conversions taking place in the process: the steam generator (**SG**) (or boiler), the turbine (**TU**) integrated with the feed-water heater train, and the electric generator (**EG**) (or alternator). The SG converts the chemical energy available in the fuel (either oil, or natural gas or coal) into internal energy of the working fluid (the steam). The TU transforms the internal energy of steam flowing from the SG into mechanical power and makes it available at the shaft for the final conversion into electrical power in the EG.

The interactions among the principal subsystems are sketched in Figure 79.1, where only the mass and energy flows at the subsystem's boundaries are displayed.

The overall process can be described as follows: the feed-water coming from the feed-water heater train enters the SG where, due to the heat released by fuel combustion, ShS is generated and admitted into the HPT through a system of control valves (TV-hp). Here, the steam expands down to the reheat pressure, transferring power to the HPT shaft, and is discharged into a steam reheater (part of SG) which again superheats the steam (RhS). RhS is admitted into the RhT through the control valve TV-rh, normally working fully open; the steam expands successively in RhT and LPT down to the condenser pressure, releasing the rest of the available power to the turbine shaft. Condensed water is extracted from the condenser and fed to low-pressure feed-water heaters, where the feed-water is preheated using the steam extractions from RhT and LPT. Then the pressure is increased to its highest value by FwP and the feed-water gets its final preheating in the high-pressure feed-water heaters using steam extractions from HPT and RhT. The mechanical power released by the entire compound turbine is transferred to the EG, which converts that power into electrical power delivered to the grid via a three-phase line.

The control objectives in such a complex process can be synthesized as follows: transferring to the grid the demanded electrical power P_e with the maximum efficiency, with the minimum risk of plant trip, and with the minimum consumption of equipment life. As is usual in process control, such a global objective is transformed into a set of simpler control tasks, based on two principal criteria: (1) the outstanding role of certain process variables in characterizing the process efficiency and the operating constraints; (2) the weakness of a number of process interactions, which permits the decomposition of the overall process into subprocesses.

Referring to Figure 79.1, we observe that the EG, under normal operating conditions, is connected to the grid and is consequently forced to run at synchronous speed. Under those conditions it acts as a mechanical-electrical power converter with almost negligible dynamics. So, neglecting high frequency, we may assume that the EG merely implies $P_e = P_m$ (where P_m is the mechanical power delivered from the turbine). Of course, the EG is equipped with its own control, namely voltage control, which has totally negligible interactions with the control of the rest of the system (it works at much higher bandwidth).

Moreover, the turbines have very little storage capacity, so that, neglecting high frequency effects, turbines may be described by their steady-state equations:

$$P_m = P_{HP} + P_{LP} \quad (79.1)$$

$$P_{HP} = \alpha_T w_T (h_T - h_{tR}) \quad (79.2)$$

$$P_{LP} = \alpha_R w_R (h_R - h_0) \quad (79.3)$$

where P_{HP} and P_{LP} are the mechanical power released by the HPT and the RhT and LPT, respectively. w_T is the ShS mass flow-rate, h_T the corresponding enthalpy, h_{tR} the steam enthalpy at the HPT discharge, w_R is the RhS mass flow-rate, h_R the corresponding enthalpy, h_0 the fluid enthalpy at the LPT discharge, and α_T, α_R are suitable constants (≤ 1) accounting for the steam extractions from the HPT and the RhT and LPT, respectively.

With the aim of capturing the fundamental process dynamics, one may observe that the enthalpy drops $(h_T - h_{tR})$ and $(h_R - h_0)$ remain approximately unchanged as the plant load varies, because turbines are designed to work with constant pressure ratios across their stages, while the steam flow varies. This means that the output power P_m consists of two contributions, P_{HP} and P_{LP}, which are approximately proportional to the ShS flow and to the RhS flow, respectively. In turn, the flows w_T and w_R are determined by the state of the SG (i.e., pressures and temperatures) and by the hydraulic resistances that the turbines (together with their control valves) present at the SG boundaries.

Steam extractions (see Figure 79.1) mutually influence subsystems SG and TU: any variations in the principal steam flow w_T create variation in SE flow and, consequently, a change in the feed-water temperature at the inlet of the SG. Feed-water mass flow-rate, on the contrary, is essentially imposed by the FwP, which is generally equipped with a flow control system which makes the FwP act as a "flow-generator". Fortunately, the overall gain of the process loop due to the steam extractions is rather small, so that the feed-water temperature variations may be considered a small disturbance for the SG, which is, ultimately, the subprocess where the fundamental dynamics take place.

In conclusion, power plant control may be studied as a function of steam generator dynamics with the turbine flow characteristics acting as boundary conditions at the steam side, the feed-water mass flow-rate and the feed-water temperature acting as exogenous variables, and Equations 79.1, 79.2, and 79.3 determining the power output. To understand the process dynamics, it is necessary to analyze the internal structure of the SG. In the following, we will make reference to a typical drum boiler [1] ; once-through boilers are not considered for brevity.

Typical Structure of a Steam Generator

A typical scheme of a fossil-fueled steam generator, in Figure 79.2 depicts the principal components.

In Figure 79.2, the air-gas subsystem is clearly recognizable; the combustion air is sucked in by the fan (1) and conveyed

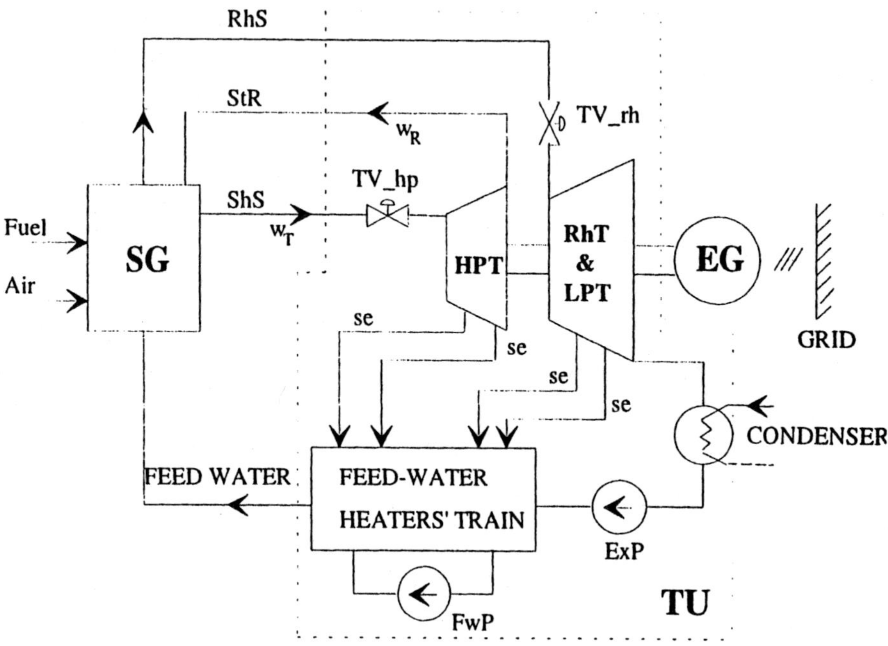

Figure 79.1 Subsystems interaction. RhS = Reheated steam; StR = Steam to reheat, ShS = Superheated steam; HPT = High pressure turbine; RhT = Reheat turbine; LPT = Low pressure turbine; se = Steam extraction; ExP = Extraction pump; FwP = Feed-water pump; TV-hp = Turbine valve, high pressure; TV-rh = Turbine valve, reheated steam.

through the air heaters (2) (using auxiliary steam) and (3) (exchanging heat counter flow with the flue gas leaving the furnace backpass) to the furnace wind box (4), where air is distributed to the burners, normally arranged in rows. Fuel and air, mixed at the burner nozzles, produce hot combustion gas in the furnace (5), where heat is released, principally by radiation, from the gas (and the luminous flame) to the furnace walls, usually made of evaporating tubes. The hot gas releases almost 50% of its available heat within the furnace and leaves it at high temperature; the rest of the internal energy of the hot gas is transferred to the steam through a cascade of heat exchangers in the back-pass of the furnace ((6) and (9) superheat the steam to high pressure, while (7) and (8) reheat steam and, at the end of the backpass, to the feed-water in the economizer (10). The gas is finally used in a special air heater (3) (called Ljungstroem) to capture the residual available energy. The flue gas is conveyed to the stack (12), possibly through induced draft fans (11), which are employed with coal-fired furnaces to keep the furnace pressure slightly below the atmospheric pressure.

The heat exchangers making up the furnace walls and the various banks arranged along the flue-gas path are connected on the steam side to generate superheated steam; this can be split into four subprocesses; water preheating, boiling, superheating, and reheating. The flow diagram of the water-steam subsystems is shown in Figure 79.3, where common components are labeled with the same numbers as in Figure 79.2.

In the scheme of Figure 79.3, the evaporator is the natural circulation type (also called drum-boiler). It consists of risers, the tubes forming the furnace walls where boiling takes place, and the steam separator (drum), where the steam-water mixture from the risers is separated into dry steam (flowing to superheating) and saturated water which, after mixing with feed-water, feeds the downcomers. There are two special devices (called spray desuperheaters), one in the high-pressure section, the other in the reheat section, which regulate superheated and reheated steam temperatures.

Process dynamics in a steam boiler is determined by the energy stored in the different sections of the steam-water system, especially in the working fluid and in the tube walls containing the fluid. Storage of energy in the combustion gas is practically negligible, because hot gas has a very low density. For those reasons, it is natural to focus on the steam-water subsystem, except for those special control issues where (fast) combustion dynamics are directly involved.

Control Objectives

A generation unit of the type described in Figures 79.1, 79.2, and 79.3 is asked to supply a certain power output P_e, that is (see Equations 79.1–79.3) certain steam flows to the turbines, while insuring that the process variables determining process efficiency and plant integrity are optimal. Because efficiency increases as the pressure and the temperature at the turbine inlet (i.e., at the throttle) increase, whereas stress on machinery goes in the opposite direction, the best trade-off between steam cycle efficiency and plant life results in prescribing certain values to throttle pressure p_T and temperature T_T and to reheat temperature T_R (reheat pressure is not specified because there is no throttling along the reheating section under normal conditions).

Moreover, proper operation of the evaporation section requires correct steam separator conditions, meaning a specified

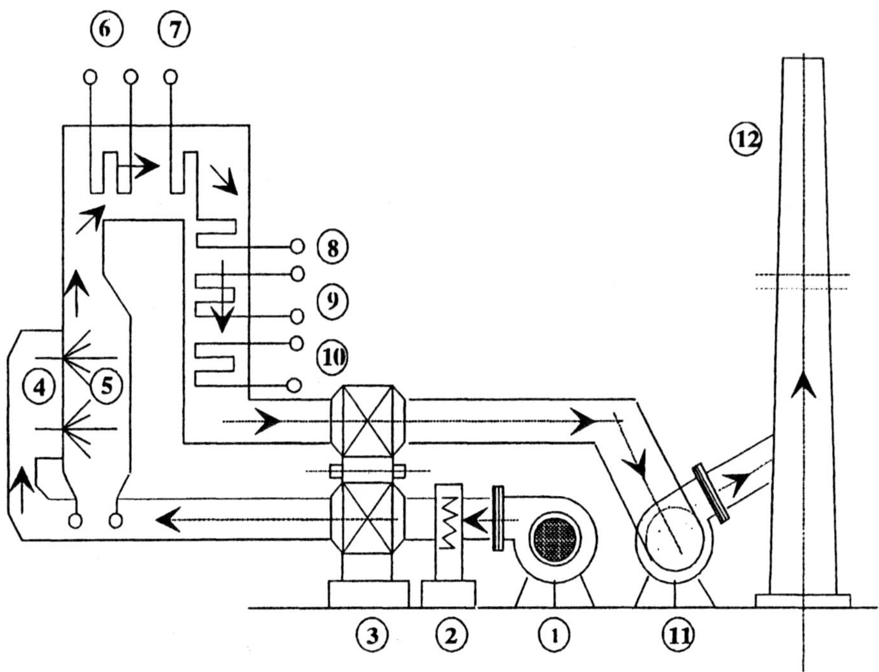

Figure 79.2 Typical scheme of the steam generator. (1) Air fan; (2) Auxiliary air heater; (3) Principal air heater; (4) Wind box; (5) Furnace (with burners); (6) High-temperature superheater; (7) High-temperature part of the reheater; (8) Low-temperature part of the reheater; (9) Low-temperature superheater; (10) Economizer ; (11) Flue-gas fan (present only with balanced draft furnace); (12) Stack.

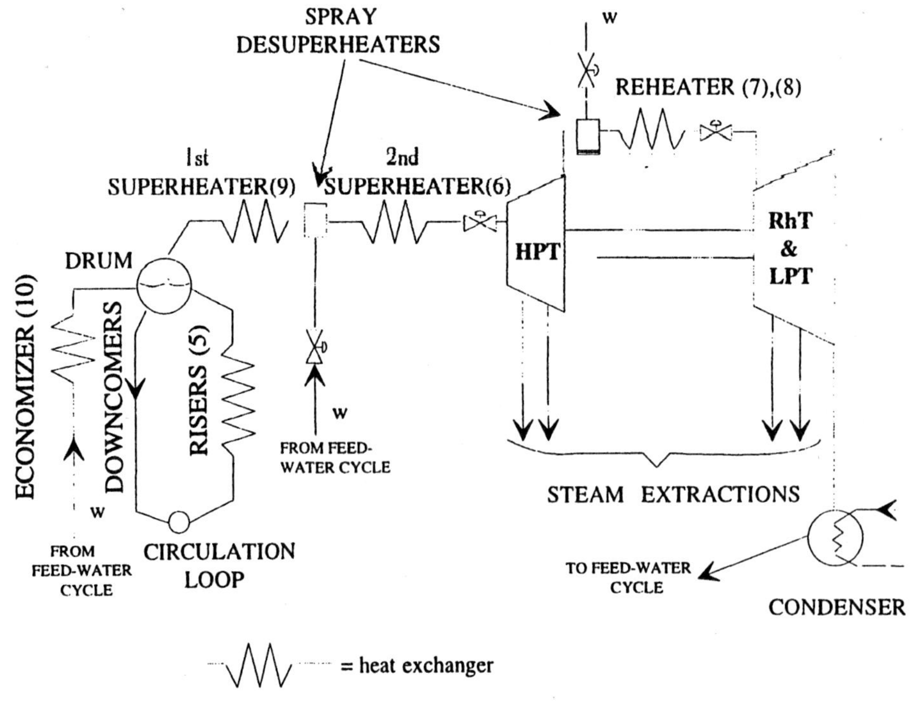

Figure 79.3 Steam-water subsystem.

water level y_D in the drum.

Overall efficiency is substantially affected by combustion quality and by the waste of energy in the flue gas. Because operating conditions also have environmental impact, they are controlled by properly selecting the air-to-fuel ratio, which depends on the condition of the firing equipment (burners etc.). In coal fired units, the furnace needs to be operated slightly below atmospheric pressure to minimize soot dispersion to the environment. Furnace pressure requires careful control, integrated with combustion control.

Early control systems gave a static interpretation of the above requirements, because the process variables and control values were set at the design stage based on the behavior expected. More recent control systems allow some adaptation to off-design conditions experienced in operation. This produces a hierarchically structured control system, whose general organization is shown in Figure 79.4.

In the scheme of Figure 79.4 are three main control levels:

- The *unit control level*, where the overall unit objective in meeting the power system demand is transformed into more specific control tasks, accounting for the actual plant status (partial unavailability of components, equipment stress, operating criteria); the decomposition into control subtasks is generally achieved by computation of set points for the main process variables.
- The *principal regulation level*, where the main process variables are controlled by a proper combination of feedforward (model based) and feedback actions. Decoupling of the overall control into independent controllers is based on the special nature of the process.
- The *dependent loop level*, where the physical devices allowing the modulation of basic process variables are controlled in a substantially independent manner with a control bandwidth much wider than the upper level regulation. These loops are the means by which the principal regulations may be conceived and designed to control process variables (like feed-water flow) rather than acting as positioning devices affected by sensitive nonlinearities (e.g., the Voigt speed control of the feed-water pump).

The tendency to decentralize control actions is quite common in process control and should be adopted generally to allow system operability. To avoid conflict with overall unit optimization, most recent control systems have extended the role of the unit coordinator, which does not interfere with the individual functionality of lower loops, but acts as a set point computer finding the optimal solution within the operation allowed by plant constraints.

When assessing control objectives, one of the crucial problems is to define system performance. For power plants, one needs to clarify the following:

- the kinds of services the unit is required to perform, usually defined in terms of maximal rate for large ramp load variations, the participation band for the power-frequency control of the power system, and the maximum amplitude and response time for the primary speed regulation in case of contingencies;
- the maximal amplitude of temperature fluctuations during load variations, to limit equipment stress due to creep or fatigue;
- maximal transient deviation of throttle pressure and drum level, to avoid potentially dangerous conditions, evaluated for the largest disturbances (e.g., in case of load rejection).

There are a few control issues still debated. The first is whether it is more convenient to operate the unit at fixed pressure (nominal constant pressure at the throttle), to let the pressure slide with a fully open turbine control valve, or to operate the unit with controlled sliding pressure. The second is the question of how much pressure variation during plant transients should be considered deleterious to some aspect of plant performance. The two questions are connected, because adopting a pure sliding pressure control strategy contradicts the concept of pressure control.

Consider the first issue. When the turbine load (i.e., the steam flow) is reduced, steam pressures at the different turbine stages are approximately proportional to the flow. Therefore, it is natural to operate the turbine at sliding pressure. On the other hand, the pressure in the steam generator is the index of the energy stored in the evaporator. Because drum boilers have a very large energy capacitance in the evaporator, boiler pressure is very difficult to change. Therefore, sliding pressure in the boiler affects load variation slowly (not the case of once-through boilers). The best condition would be to keep the pressure fixed in the boiler while sliding pressure in the turbine: this strategy would require significant throttling on the control valves with dramatic loss of efficiency in the overall steam cycle. The most popular strategy for drum boiler power plants is, therefore, controlled sliding pressure, where the boiler is operated at constant pressure above a certain load (this may be either the technical minimum or 50–60% of the MCR)[1] and pressure is reduced at lower loads. To insure high efficiency at any load, the turbine is equipped with a control stage allowing partial arc admission (i.e., with the control valves always opening in sequence to limit the amount of throttling).

The second issue is often the source of misleading design. There is no evidence that loss of performance is directly related to transient pressure deviations within safety limits, which may be very large. On the other hand, it has been demonstrated [2] that, because throttle pressure control can cause furnace overfiring, too strict pressure control can substantially disturb steam temperature.

In the following, we will consider the most common operating condition for a drum boiler, that is, with throttle pressure con-

[1] MCR = Maximum Continuous Rate

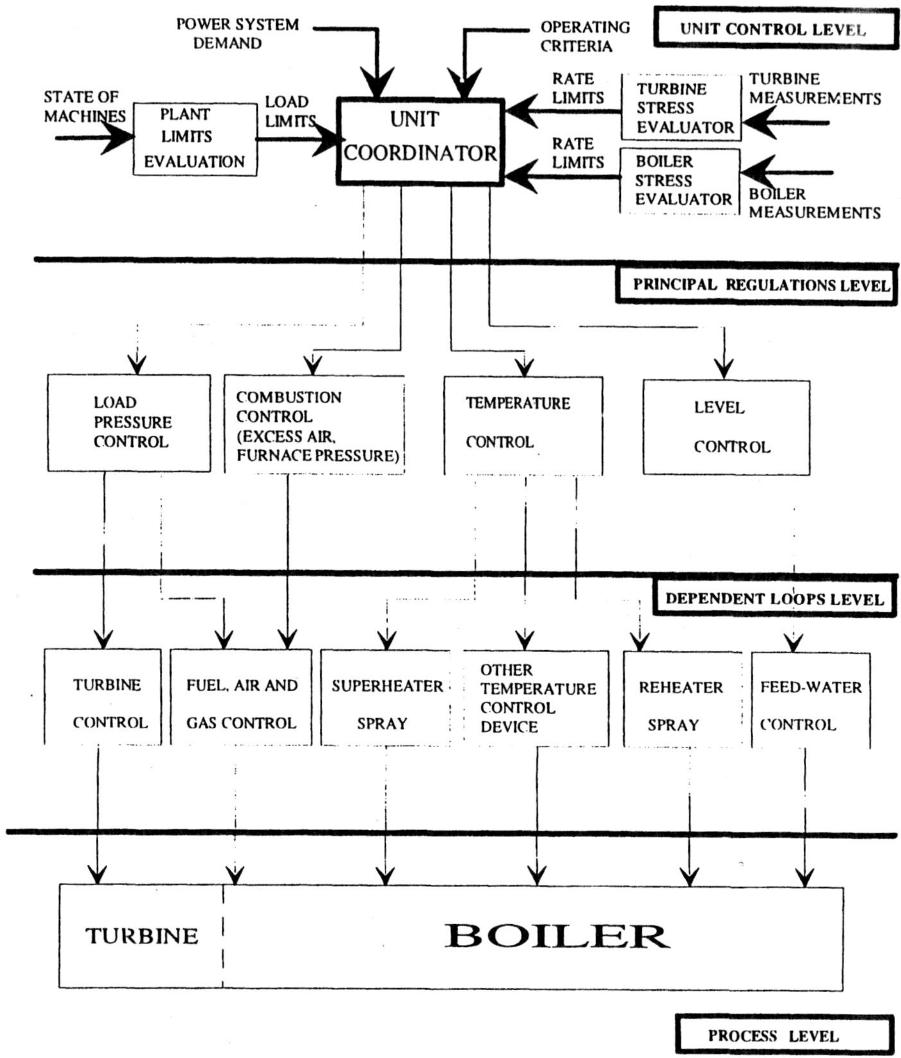

Figure 79.4 Hierarchical organization of power plant control.

trolled at a constant value during load variation, with the main objective of returning pressure to the nominal value within a reasonable time after the disturbance (e.g., load variation), while strictly insuring that it remains within safety limits (which may also depend on the amplitude of the disturbance).

79.1.3 Power Plant Modeling and Dynamical Behavior

Models for Structural Analysis

Investigating the dynamics of power plants [3] requires detailed models with precise representation of plant components. These models are generally used to build plant simulators, from which control strategies may be assessed. Large scale models are generally based on first principle equations (mass, momentum, and energy balances), combined with empirical correlations (like heat transfer correlations), and may be considered as *knowledge models*, i.e., models through which process dynamics can be thoroughly ascertained and understood. *Knowledge models* are the only reliable way, beside extensive experimentation, to learn about power plant dynamics, in particular, the many interactions among process variables and their relevance.

Today power plant simulators are broadly accepted: overall control system testing and tuning has been carried out successfully with a real-time simulator of new and novel generating unit design [4]. These detailed models are built by considering many "small" fluid or metal volumes containing matter in homogeneous conditions and writing balance equations for each volume. The resulting system can include from hundreds up to some thousands of equations.

A different kind of model has proved very helpful in establishing and justifying the basic structure of power plant control systems. Only first-cut dynamics are captured revealing the essential input-output interactions. These models, called *interpretation models*, are based on extremely coarse lumping of mass and energy balances, whose selection is guided by previous knowledge of the fundamental process dynamics. *Interpretation models* are credible because they have been derived from and compared with *knowledge models* and should be considered as useful tutorial tools to explain fundamental dynamics. Because the scope of this presentation is modeling to support control analysis, only simple *interpretation models* will be developed. However, these simple models are not useful for dynamic performance evaluation of control systems, because the dynamics they account for are only first order approximations. Nevertheless, they account for gross process interactions, into which they give good qualitative insight.

We may start developing the process model by referring to the considerations on the overall plant features presented in Section 79.1.2 and summarized in Figure 79.5. According to Section 79.1.2, the effect of the feed-water cycle on the main steam-water subsystem (SWS) variables is accounted for by including the feed-water temperature (or enthalpy) as a disturbance among the inputs of the SWS. There are some drastic simplifications in the scheme of Figure 79.5. Thermal energy released from the hot gas to the SWS walls is not totally independent of the SWS state (i.e., of wall temperatures); there is almost full independence of Q_{ev} (because heat is transferred by radiation from the hot combustion gas). Q_{SH} and Q_{RH} are more sensitive to the wall temperature because the temperature of the combustion gas is progressively decreasing; even more sensitive is Q_{ECO}, where the gas temperature is quite low. However, because the economizer definitely plays a secondary role in boiler dynamics, the scheme of Figure 79.5 is substantially correct. The dominant inputs affecting the thermal energy transferred to the SWS are the fuel and air flows and other inputs to the combustion system.

In Section 79.1.2, it was also noted that the SWS dynamics (due to massive storages of mass and energy) are far slower than combustion and air-gas (C&AG) dynamics; for that reason C&AG dynamics are negligible when the response of the main SWS variables is considered. C&AG dynamics are only relevant for specific control and dynamic problems regarding either combustion stability (relevant at low load in coal fired furnaces) or the control of the furnace pressure p_g (of great importance to protect the furnace from implosion in case of fuel trip). Then, in most control problems, we may consider the C&AG system together with its feeding system as a nondynamic process segment whose crucial role is determining energy release (and energy release partition) to different sections of the SWS. In this regard, it is important to identify how C&AG inputs may be used to influence steam generation.

Increasing the total fuel flow into the furnace will simultaneously increase all heat inputs to the different boiler sections; air flow is varied in strict relation to fuel flow so as to insure the "optimal" air-to-fuel ratio for combustion.

Because of the nonlinearity of heat transfer phenomena, Q_{EC}, Q_{EV}, Q_{SH}, and Q_{RH} do not vary in proportion to the fuel input, i.e., while varying the total heat input to the boiler, the partition of heat release is also changed. For instance, when the fuel input is increased, the heat input to the evaporator Q_{EV} (that is released in the furnace) increases less than the other heat inputs (Q_{EC}, Q_{SH}, and Q_{RH}) which are released in the upper section and backpass of the furnace. Thus, while raising the steam generation (roughly proportional to Q_{EV}), steam superheating and reheating would generally increase if proper corrective measures were not applied to rebalance heat distribution. Those measures are usually viewed as temperature control devices, because they allow superheating and reheating control in off-design conditions. One type of control measure acts on the C&AG system: the most popular approaches are 1) the recirculation of combustion gas from the backpass outlet to the furnace bottom, 2) tilting burners for tangentially fired furnaces, and 3) partitioning of the backpass by a suitable screen, equipped with gas dampers to control the gas flow partition between the two branches. The first two approaches influence the ratio between Q_{EV} and the rest of the heat release. The last varies the ratio between Q_{SH} and Q_{RH}.

The second type of temperature control measure acts on the SWS. This is the spray desuperheaters (see Figure 79.3), which balances heat input variations by injecting spray water into the superheating and the reheating path. Although superheater spray does not affect the global efficiency, reheater spray worsens efficiency so that it is only used for emergency control (when it is not possible to keep reheater temperature below a limit value by other control measures). Typical drum boiler power plants provide desuperheating sprays in the high pressure and reheat parts of the SWS and, in addition, gas recirculation as a " normal" means of controlling reheat temperature.

From this discussion, it should also be clear that modulation of heat input to the furnace and variation of recirculation gas flow simultaneously affect all process variables, because they influence heat release to all sections of the SWS. This is the principal source of interactions in the process. Because the air-to-fuel ratio is varied within a very narrow range to optimize combustion, we may assume that the boiler heat rate is proportional to w_f. Thus, to analyze the SWS dynamics, we may consider w_f and the recirculation gas mass flow-rate w_{rg} as the equivalent inputs from the gas side, because they determine the heat transfer rates Q_{EC}, Q_{EV}, Q_{SH}, and Q_{RH}.

Pressure Dynamics

A very simple *interpretation model* of evaporator dynamics can be derived with the following "drastic" assumptions:

1. The fluid in the whole circulation loop (drum, risers, and downcomers) is saturated.
2. The metal walls in the entire circulation loop are at saturation temperature.
3. The steam quality in the risers is, at any time, linearly dependent on the tube abscissa.
4. The fluid pressure variations along the circulation loop can be neglected for evaluating mass and energy storage.

The first three assumptions can be roughly considered low-frequency approximations because, excluding rapid pressure variations, water subcooling at the downcomers' inlet is very small. Moreover, because of the very high value of the heat transfer coefficient in the risers (in the order of 100 kW/m^{2}K), the metal wall very quickly follows any temperature variation in the fluid. Finally, steam quality is nearly linear at steady state because the heat flux to the furnace wall is evenly distributed. The last

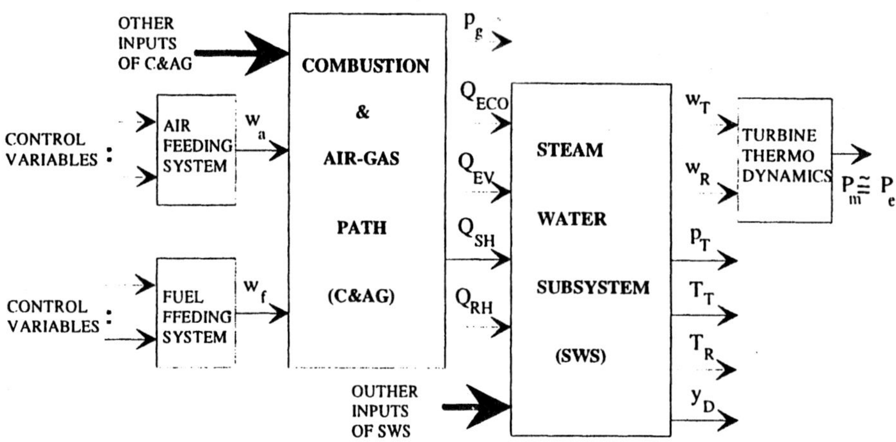

Figure 79.5 Input-output structure of the process. w_a = air mass flow-rate; w_f = fuel mass flow-rate; p_g = gas pressure in the furnace; Q_{ECO} = thermal energy to ECOnomizer; Q_{EV} = thermal energy to EVaporator; Q_{SH} = thermal energy to SuperHeaters; Q_{RH} = thermal energy to ReHeater.

assumption is based on the fact that pressure differences along the loop (which are essential for circulation) are of the order of 1% of the absolute fluid pressure in the drum so that we may identify the pressure in the evaporator with the pressure p_D in the drum. Then the global energy balance and the global mass balance in the evaporator are

$$\frac{dE_{EV}(p_D,\alpha)}{dt} = w_w h_E - w_V h_{VS}(p_D) + Q_{EV}, \quad (79.4)$$

$$\frac{dM_{EV}(p_D,\alpha)}{dt} = w_w - w_V, \quad (79.5)$$

where w_w is the feed-water mass flow-rate (mfr), w_V is the steam mfr at the drum outlet, h_E is the water enthalpy at the economizer outlet, $h_{VS}(p)$ is the steam saturation enthalpy at the pressure p, Q_{EV} is the heat-rate to the evaporator, E_{EV} is the total energy stored in the evaporator (fluid and metal of drum, downcomers and risers), and M_{EV} is the total mass stored in the evaporator. E_{EV} and M_{EV}, beside obvious geometrical parameters like volumes, depend on two process variables: the pressure p_D and the void fraction α in the evaporator, defined as the ratio of the volume occupied by steam and the total volume of the evaporator.

A better index for the mean energy level in the evaporator is obtained by subtracting the total mass multiplied by the inlet enthalpy h_E from Equation 79.4.

$$\frac{dE_{EV}(p_D,\alpha)}{dt} - h_E\frac{dM_{EV}(p_D,\alpha)}{dt} = - w_V\,[h_{VS}(p_D) - h_E] + Q_{EV}. \quad (79.6)$$

Noting that h_E is subject to limited and very slow variations (the economizer is a huge heat exchanger exploiting a limited temperature difference between flue gas and water), so that dh_E/dt is usually small, Equation 79.6 can be interpreted by introducing the net energy storage in the evaporator, $E^*_{EV} := E_{EV} - h_E M_{EV}$: the difference between the input heat transfer rate Q_{EV} and the power spent for steam generation, $P_{sg} := w_V\,[h_{VS}(p_D) - h_E]$, in transient conditions, is balanced by the storage of the net energy E^*_{EV} in the evaporator. Moreover, whereas the mass M_{EV} depends mainly on α, the net energy E^*_{EV} depends mainly on p_D, because about 50% of the energy is stored in the metal walls, which are insensitive to α.

Equation 79.6 can be rewritten approximately as

$$C_{EV}\frac{dp_D}{dt} = Q_{EV} - w_V(h_{VS}(p_D) - h_E), \quad (79.7)$$

where

$$C_{EV} = \frac{\partial E^*_{EV}(p_D,\bar{\alpha})}{\partial p_D},$$

$\bar{\alpha}$ being the nominal void fraction. Equation 79.7 yields the fundamental dynamics of drum pressure and justifies the popular claim that drum pressure is associated with the stored evaporator energy. Equation 79.7 may be usefully rewritten in normalized per unit (p.u.) form, i.e., referring energy and pressure to the nominal conditions, $Q^\circ_{EV} = w^\circ_V(h_{VS}(p^\circ_D) - h^\circ_E)$, where the superscript $^\circ$ denotes nominal value:

$$\tau_{EV}\,\dot{p}_{Dn} = Q_{EVn} - w_{Vn}\frac{h_{VS}(p_D) - h_E}{h_{VS}(p^0_D) - h^0_E} \quad (79.8)$$

with the subscript n denoting the variable expressed in p.u.. Typical values for the normalized "capacitance" τ_{EV} are 200–300 sec. It may also be observed that τ_{EV} is a function of the drum pressure p_D and is roughly inversely proportional to the pressure; thus, pressure dynamics will slow down while reducing the operating pressure.

Although for the evaporator h_E can be considered a slowly varying exogenous variable, w_V depends on the drum pressure p_D and on the total hydraulic resistance opposed by the cascade of superheaters and turbine.

Let's first characterize the turbine. For simplicity, assume that the turbine control valves are governed in full arc mode (i.e., with parallel modulation). Then the control stage of the turbine (generally of impulse type) can be viewed as the cascade of a throttle valve and a nozzle. This implies

$$w_T = C_V(x)\sqrt{\rho_T p_T}\,\chi_V(p_N/p_T), \quad (79.9)$$

$$w_T = K_N\sqrt{\rho_N p_N}\,\chi_N(p'/p_N), \quad (79.10)$$

where $C_V(x)$ is the flow coefficient of the control valve set (dependent on the valve's position χ), ρ_T is the steam density at throttle, p_N is the valve outlet pressure, $\chi_V(\beta)$ is a suitable function of the valve pressure ratio, K_N is a nozzle flow constant, ρ_N the density at the nozzle inlet, p' the pressure at the nozzle outlet, and χ_N a function similar to χ_V.

Because the HPT consists of many cascaded stages, the pressure ratio (p'/p_N) across the control stage will remain nearly constant with varying flow w_T. Then $\chi_N(p'/p_N = constant$.

Bearing in mind that superheated steam behaves like an ideal gas and that valve throttling is an isenthalpic expansion, so that $p_N/\rho_T \cong p_N/p_T$, eliminate the pressure ratio p_N/p_T by dividing Equation 79.9 by Equation 79.10. The ratio p_N/p_T is a monotonic function of $C_V(x)$. Substituting this function in Equation 79.9 results in

$$w_T = f_T\,(C_V(x))\,\sqrt{\rho_T p_T} = f_T^*(x)\sqrt{\rho_T p_T}, \qquad (79.11)$$

where $f_T(\cdot)$ is a monotonic function of its argument. Equation 79.11 says that the cascade of the turbine and its control valves behave like a choked-flow valve with a "special" opening law $f_T^*(x)$.

Even when the turbine control valves are commanded in partial-arc mode (i.e., with sequential opening), one arrives at a flow equation of the same type as Equation 79.11, but differently dependent on the valve opening command signal x. To summarize, the HPT with its control valves determines a boundary condition for the steam flow given by Equation 79.11; with typical opening strategies of full-arc and partial-arc admission, the global flow characteristic $f_T^*(x)$ looks like that in Figure 79.6. A more elaborate characterization of sequentially opened valves may be found in [7].

To obtain flow conditions for w_V instead of w_T (i.e., at the evaporator outlet), flow through superheaters must be described (see Figure 79.3). First-cut modeling of superheaters' hydrodynamics is based on the following remarks:

1. Mass storage in the superheaters is very limited (as compared with energy storage in the evaporator) because steam has low density and desuperheating spray flow w_{ds} is small compared with w_V. Thus, head losses along superheaters may be computed with the approximation $w_V \approx w_T$.
2. Head losses in the superheaters develop in turbulent flow, so that

$$p_D - p_T = k_{SH}\frac{w_T^2}{\rho_{SH}} \qquad (79.12)$$

where ρ_{SH} is a mean density of the superheated steam and a k_{SH} constant.

Equations 79.11 and 79.12 can be combined with Equation 79.7 or Equation 79.8 to build a simple model of the fundamental pressure dynamics. To this end, we derive a linearized model for small variations about a given steady state condition, identified as follows:

- assuming that the unit is operated (at least in the considered load range) at constant throttle pressure, p_T at any steady state equals the nominal pressure p_T°;
- the unit is usually operated at constant throttle temperature ($T_T + T_T^\circ$, so that the temperature profile along the superheaters does not change significantly; we may therefore assume that the mean superheating temperature T_{SH}, at any steady state, equals its value in nominal conditions T_{SH}°;
- based on Equations 79.1–79.3 and the related remarks, the load L (in p.u.) of the plant at any steady state equals the ratio between the steam flow w_T and its nominal value w_T°.

Moreover, the following assumptions are made:

(a) superheated steam behaves like an ideal gas:

$$\rho_{SH} = \frac{p_{SH}}{RT_{SH}} \approx \frac{p_T}{RT_{SH}},\ \rho_T = \frac{p_T}{RT_T}$$

where R is the gas constant;

(b) in nominal conditions, desuperheating spray mass flow-rate is zero:

$$w_{ds}^\circ = 0,$$

so that $w_{ds}^\circ = w_T^\circ$.

The model will be expressed in p.u. variables by defining

$$\delta_p = \Delta_p/p_T^\circ,$$

for any pressure p,

$$\delta_w = \Delta_w/w_T^\circ$$

for any mass flow-rate w,

$$\delta_h = \Delta_h/\left(h_{VS}(p_D^\circ) - h_E^\circ\right)$$

for any enthalpy h,

$$\delta_T = \Delta_T/T^\circ$$

for any temperature T, and

$$\delta Q_{ev} = \Delta Q_{ev}/Q_{ev}^\circ$$

($^\circ$ denotes, as usual, nominal conditions). Then Equations 79.8, 79.11, and 79.12 yield the following linearized system:

$$\begin{aligned}\tau_{EV}\,\delta p_D &= \delta Q_{ev} - \alpha\delta w_V + L\beta_1\delta p_D \\ &\quad + L\delta h_E && (79.13)\\ \delta w_T &= \delta Y + L\delta p_T, \quad \text{and} && (79.14)\\ \delta p_T &= \frac{1}{1-\gamma^\circ L^2}\delta p_D \\ &\quad - \frac{2\gamma^\circ L}{(1-\gamma^\circ L^2)}\delta w_T \\ &\quad - \gamma^\circ L^2\delta T_{SH}, && (79.15)\end{aligned}$$

where

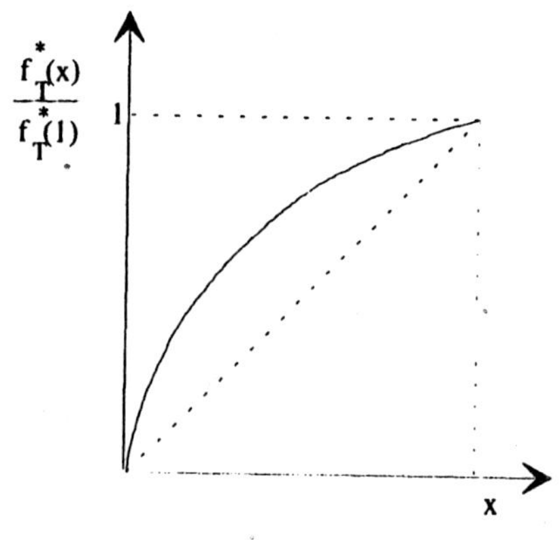

(a) Full arc command mode

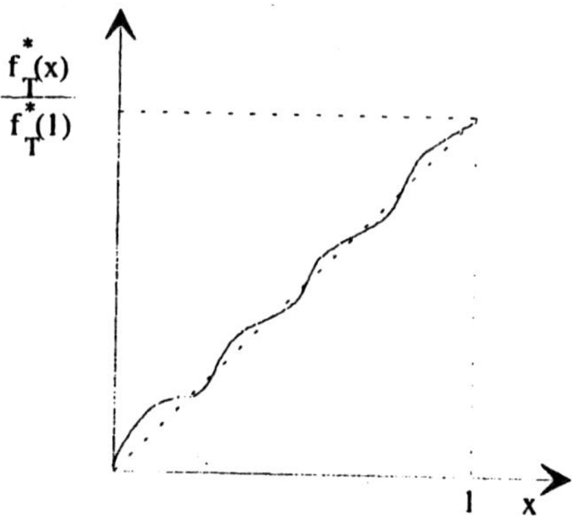

(b) Partial-arc command mode (with 4 control valves)

Figure 79.6 Flow characteristic of HPT.

$$
\begin{aligned}
\tau_{EV} &= \tau_{EV}(1+\gamma^\circ), \\
\gamma^\circ &= \frac{p_D^\circ}{p_T^\circ} - 1, \\
\alpha &= \frac{h_{VS}(p_D^\circ) - h_E}{h_{VS}(p_D^\circ) - h_E^\circ}, \\
\beta_1 &= -\left.\frac{dh_{VS}}{dp_D}\right|_\circ \frac{p_T^\circ}{(h_{VS}(p_D^\circ) - h_E^\circ)}, \\
Y &:= f_T^*(x)/\sqrt{RT_T}
\end{aligned}
$$

is the turbine "admittance", and $\bar{h}_E$ is the value of h_E at the linearization steady state.

Note that γ° is usually about 0.05 or less, $\bar{h}_E$ undergoes limited variations (α is very close to 1), β_1 is positive for $p_D^\circ > 30$ bar and is generally small because $h_{VS}(p)$ is a flat thermodynamic function. Moreover temperature variations δT_{SH} are generally slow and limited amplitude, so that $\gamma^\circ L^2 \delta T_{SH}$ is totally negligible.

Then, pressure dynamics may be approximately represented by the very simple block diagram of Figure 79.7, where

$$\mu_P = \frac{1}{1+2\gamma^\circ L^2} \quad \text{and}$$

$$\mu_Y = \frac{2\gamma^\circ L}{1+2\gamma^\circ L^2}.$$

There is a seeming inconsistency in Figure 79.7, because the variable Y is considered as an input variable, but its definition, $Y := w_T/p_T$, implies that it depends on the control variable x and also on the throttle temperature T_T.

However, it is a common practice to equip the turbine control valve with a wide band feedback loop of the type shown in Figure 79.8. Because valve servomotors today are very fast and no other lags are in the loop, at any frequency of interest for the model of Figure 79.7, $Y \cong \bar{Y}$. So the turbine admittance actually becomes a control variable and the loop of Figure 79.8 serves two complementary purposes: first, it linearizes the nonlinear characteristics of Figure 79.6 and, second, it rejects the effect of temperature fluctuations on the steam flow to the turbine, thereby decoupling pressure dynamics and temperature dynamics.

Let's analyze the scheme of Figure 79.7, bearing in mind that $0 \le \alpha < 1$, $0 < \beta_1 \le 0.1$ (for $p_D^\circ > 30$ bar, $0.9 < \mu_P < 1$, $0 < \mu_Y < 0.1$ and, of course, $L^* \le L \le 1$, where L is the minimal technical load (typically $L^* \approx 0.3$). We may observe that the pressure dynamics are characterized by a time constant τ_P, given by

$$\tau_P = \frac{\tau_{ev}}{L(\mu_P\alpha - \beta_1)} \approx \frac{\tau_{ev}}{L} \tag{79.16}$$

that is, the ratio between the evaporator energy capacity and the load. Thus, the open-loop response of the pressure to exogenous variables slows down as the load decreases.

Neglecting the effects of the small disturbances δh_E and δw_{ds}, a natural way to follow plant load demand in the fixed-pressure operating strategy is to let the turbine admittance δY vary according to the load demand ΔL_d and let the heat transfer rate δQ_{ev} vary so as to balance the power spent for steam generation, i.e., $\alpha\delta V_Y$. This means that

$$\delta Y = \Delta L_d \tag{79.17}$$

$$\delta Q_{ev} = \alpha \delta w_v. \tag{79.18}$$

As a consequence,

$$\delta p_d = 0$$

$$\delta p_T = -\mu_Y \delta Y,$$

and $$\delta w_T = \delta Y(1 - L\mu_Y).$$

Because of the head losses along the superheaters ($\mu_Y \neq 0$), the strategy expressed by Equation 79.18, keeping the energy storage (i.e., p_D) constant in the evaporator, actually determines a drop $-\mu_Y\delta Y$ of the throttle pressure p_T and, consequently, a reduced power output ($\delta w_T = \delta Y(1 - L\mu_Y) < \Delta L_D$). In other words, if one wants to keep the throttle pressure p_T constant when the load is increased, the energy storage in the evaporator also needs to be slightly increased because of the head losses:

$$
\begin{aligned}
\delta p_D \mu_P &= \delta Y \mu_Y, \qquad (79.19) \\
\delta p_d &= 2\gamma^\circ L \delta Y = 2\gamma^\circ L \Delta L_d.
\end{aligned}
$$

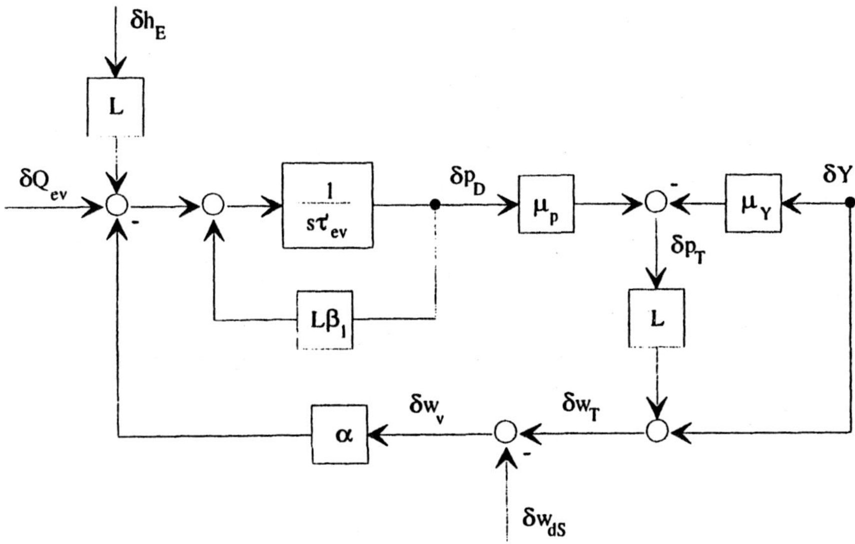

Figure 79.7 Block diagram of the linearized pressure dynamics.

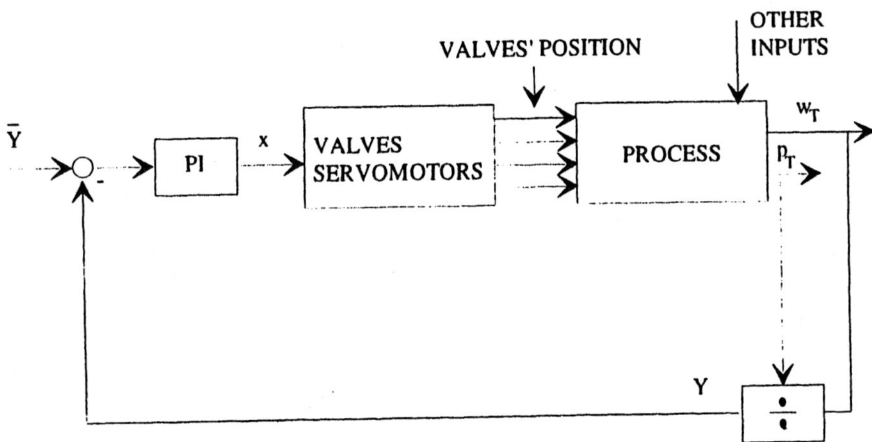

Figure 79.8 Turbine admittance feedback loop.

In Figure 79.7, observe that only Equation 79.19 implies a boiler overfiring, i.e., a transient extra fuel supply during load increase to adjust the energy stored in the evaporator.

If the feedback compensation Equation 79.18 is applied to the boiler, the pressure dynamics become slightly unstable because of the "intrinsic" positive feedback due to β_1. The same result happens if the feedback loop of Figure 79.8 is realized, as is sometime the case, not as an admittance loop but as a simple mass flow-rate loop (i.e., omitting dividing by p_T). Thus, when applying either mass flow-rate feedback, boiler stabilization must be provided.

Drum Level Dynamics

Computing drum pressure by the scheme of Figure 79.7, we return to Equation 79.5 that establishes the global mass balance in the evaporator. Equation 79.5 may be linearized as

$$\sigma_p \; \delta p_D - \sigma_\alpha \Delta \; \alpha = \delta w_w - \delta w_V, \tag{79.20}$$

where

$$\sigma_p = \frac{V p_T^\circ}{w_T^\circ}\left[(1-\bar{\alpha})\left(\frac{d\bar{\rho}_{LS}}{dp}\right) + \bar{\alpha}\left(\frac{d\bar{\rho}_{VS}}{dp}\right)\right]$$

and $\sigma_\alpha = \dfrac{V(\bar{\rho}_{LS} - \bar{\rho}_{VS})}{w_T^\circ}$.

V is total fluid volume in the evaporator, ρ_{LS} and ρ_{VS} are the liquid and vapor densities as functions of the pressure, and the upper script $^-$ denotes the steady state of linearization. If the unit is operated at constant pressure, σ_α is independent of the load, and σ_p only slightly dependent.

However, Equation 79.20 determines only the global void fraction α, while the relevant variable for the control is the level in the drum.

We may write

$$\alpha = \frac{V_r}{V}\alpha_r + \frac{V_D}{V}\alpha_D \tag{79.21}$$

where V_r and V_D are the volumes of the risers and of the drum, respectively, and α_r and α_d are the separate void fractions relative to V_r and V_D. If we assume that, at the considered steady state, the level y_D is equal to the drum radius R_D, then,

$$\Delta\alpha_D = -\frac{2}{\pi}\delta y_D, \tag{79.22}$$

with

$$\delta y_D \quad := \quad \Delta y_D / R_D.$$

Combining Equations 79.20, 79.21, and 79.22, the following equation is obtained:

$$\begin{aligned} \delta \dot{y}_D &= \frac{1}{\tau_L}(\delta w_w - \delta w_v) + k_p \delta \dot{p}_D \\ &\quad + k_r \Delta \dot{\alpha}_r \qquad (79.23) \\ \tau_L &= \frac{V_D(\bar{\rho}_{LS} - \bar{\rho}_{VS})}{w_T^{\circ}} \frac{2}{\pi}, \\ k_p &= -\frac{\sigma_p}{\sigma_\alpha} \frac{\pi}{2} \frac{V}{V_D}, \\ \text{and} \quad k_r &= \frac{\pi}{2} \frac{V_r}{V_D}. \end{aligned}$$

Equation 79.23 shows that the drum level is subject to three different kinds of variations: the first, of integral type, is due to the imbalance between feed-water flow and outlet steam flow; the second, of proportional type, arises because, even with constant stored mass, the mean fluid density depends on the evaporator pressure p_D; the third, more involved, comes from possible variations of the void fraction in the risers and might occur rapidly because any variation, $\Delta\alpha_r$ immediately reflects onto δy_D. We need to understand where $\Delta\alpha_r$ comes from. Recall the assumptions at the beginning of Section 79.1.3, write equations similar to Equations 79.4 and 79.5 but limited to the circulation tubes (i.e., to the downcomers and the risers), and derive the "net energy" stored corresponding to Equation 79.6, where, instead of enthalpy h_E, the inlet enthalpy h_{LS} of the downcomer tubes is used:

$$\begin{aligned} &\frac{dE_{ct}(p_D, \alpha_r)}{dt} - h_{LS}(p_D)\frac{dM_{ct}(p_D, \alpha_r)}{dt} \\ &= Q_{EV} - \chi_r w_r \left[h_{VS}(p_D) - h_{LS}(p_D)\right], \quad (79.24) \end{aligned}$$

where E_{ct} is the total energy (fluid + metal) stored in the circulation tubes, M_{ct} the corresponding fluid mass, h_{LS} and h_{VS} the liquid and vapor saturation enthalpies, and χ_r and w_r the steam quality and the mass flow-rate at the risers' outlet.

Then, based on assumption (3) stated at the beginning of Section 79.1.3, the following relationship is obtained:

$$\begin{aligned} \alpha_r &= (1+\beta)\left[1 - \frac{\beta}{\chi_r}\ln\left(1 + \frac{\chi_r}{\beta}\right)\right], \quad (79.25) \\ \beta &= \rho_{VS}/(\rho_{LS} - \rho_{VS}). \end{aligned}$$

To solve the model, we need to derive the circulation mass-flow rate w_r, which is obtained from the momentum equation applied to the circulation tubes:

$$\alpha_r(\rho_{LS} - \rho_{VS}) = w_r^2\left(\frac{C_{dc}}{\rho_{LS}} + \frac{C_r}{\rho_r}\right), \quad (79.26)$$

where

$$\rho_r = \rho_{LS} - \alpha_r(\rho_{LS} - \rho_{VS})$$

is the mean density in the risers and C_{dc}, C_r are suitable constants yielding the head losses in the downcomers and risers tubes, respectively.

Equations 79.25 and 79.26 may be used to eliminate w_r and x_r from Equation 79.24; through trivial but cumbersome computations, the following linearized model can be obtained for $\Delta\alpha_r$ (in $\mathcal{L}$-transform form):

$$\Delta\alpha_r = \frac{1}{1+sT_2}\left[\lambda_2(\delta Q_{EV} - \tau_{rt} s \delta p_D) + \lambda_1 \delta p_D\right], \quad (79.27)$$

where τ_{rt} is a normalized capacitance similar to τ_{EV} in Equation 79.8 but related only to the circulation tubes (typically $\tau_{rt} \approx 0.7\tau_{EV}$), T_2 is a small time-constant (a few seconds) associated with the dynamics of the void fraction within the risers, and λ_2 and λ_1 are suitable constants. The difference $\delta Q_{EV} - \tau_{rt} s \delta p_D$ is the heat transfer rate available for steam generation in the risers, given by the (algebraic) sum of the input thermal energy δQ_{EV} and the energy $-\tau_{rt}\delta \dot{p}_D$ released in the case of pressure decrease and corresponding to a reduction of the stored energy. The L-transformation of Equation 79.23 and substitution of Equation 79.27 give

$$\begin{aligned} \delta y_D &= \frac{1}{s\tau_L}(\delta w_v - \delta w_w) + \frac{k_2}{1+sT_2}\delta Q_{EV} \\ &\quad + k_1 \frac{1-sT_1}{1+sT_2}\delta p_D, \qquad (79.28) \end{aligned}$$

where

$$k_2 \;=\; k_r\lambda_2, \quad k_1 = k_p + k_r\lambda_1, \quad T_1 = \frac{k_2\tau_{rt} - k_p T_2}{k_1}.$$

The parameters of model Equation 79.28 are dimensionless, with the following typical values: $\tau_L \approx 130$ sec., $k_2 \approx 0.25$, $k_1 \approx 0.5$, $T_2 \approx 4$ sec., $\tau_{rt} \approx 0.7$, $\tau_{EV} \approx 150$ sec. (at nominal pressure), $k_p \approx 0.8$.

Since $T_2 \ll \tau_{rt}$, the time constant $T_1 \approx 70$ sec. is always positive and is essentially determined by the "capacitance effect" (τ_{rt}). The model Equation 79.28 clearly accounts for the well-known shrink and swell effect due to the nonminimum phase zero $(1 - sT_1$. Because T_2 is very small, any perturbation producing sudden pressure derivatives causes a sudden variation of the drum level in the direction opposite to the long-term trend. To this aim, referring to Figure 79.7, consider a step perturbation of δY, with, e.g., $\delta Y = \Delta/s$. Then

$$\delta p_D = -\frac{\mu'\Delta}{s}\frac{1}{1+sT_3}, \qquad \delta w_v = -\mu''\frac{\Delta}{s}\frac{1-sT_4}{1+sT_3}, \qquad (79.29)$$

with μ and T_3 suitable constants. At nominal load ($L = 1$) and with typical values, ($\mu_Y = 0.1, \alpha = 1, \mu_p = 0.9, \beta_1 = 0.05, \tau_{EV} = 200$ sec.) $\mu' \cong 1.06$, $T_3 \cong 248$ sec., $\mu'' \cong 0.054$, $T_4 \cong 4600$ sec. The effect of the step variation with $\Delta = 0.1$ is depicted in Figure 79.9. Equation 79.29 is a very useful model to conceive level control structure.

Reheat and Superheat Steamside Dynamics

Superheaters and reheaters are large heat exchangers with steam flowing into the tubes and gas crossing the tube banks

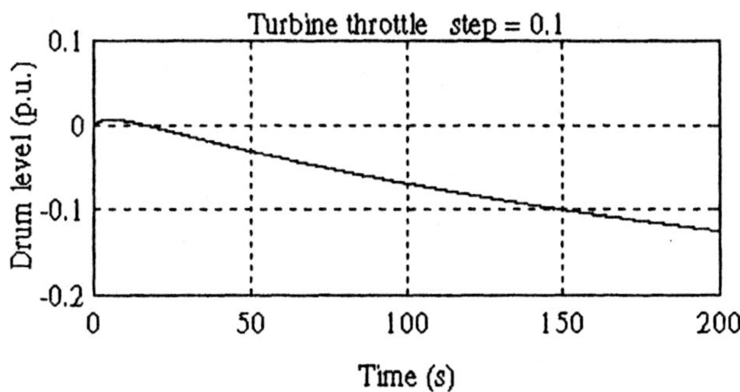

Figure 79.9 Drum level dynamic.

in cross-flow. There are some general properties that are worth recalling:

1. The heat transfer coefficient on the gas side is much smaller that the one on the steam side, so that steady state behavior is nearly independent of steamside coefficients;
2. The dynamics of these heat exchangers are essentially due to the considerable energy stored in the metal wall, because flue gas has negligible density and steam has much lower capacitance than the corresponding metal wall;
3. The mass stored creates much faster dynamics than energy stored, because only steam is involved.

Property 3 can easily be checked bearing in mind that the fundamental time constant of mass storage is,

$$\tau_{MS} = M_V / w_V,$$

where M_V is the mass of steam within the heat exchanger and w_V the mass flow-rate flowing through it, whereas the fundamental time constant of energy storage is ,

$$\tau_{ES} = \frac{M_v C_v + M_m C_m}{c_p w_v},$$

where C_v and C_p are the specific heats of steam at constant volume and pressure, M_m and C_m the mass and the specific heat of metal wall. Because $M_m C_m \gg M_v C_v$ and $C_p \approx 1.3 C_v$, $\tau_{ES} \gg \tau_{MS}$. For a typical superheater τ_{MS} is a few seconds, but τ_{ES} is more than 20 times τ_{MS}. Then, when considering temperature dynamics, which, because of property (1) are due to the metal wall capacitance, mass storage may be neglected (i.e., one can consider the steam flow independent of the tube abscissa).

Superheater or reheater outlet temperature T_{ox} is influenced by three different variables, the heat transfer rate Q_x to the external wall, the steam flow w_x, and the inlet temperature T_{ix}. Pressure fluctuations within the heat exchanger have a limited influence on T_{ox} and may be neglected. For small variations the situation is described in Figure 79.10.

It is relevant to control design to characterize the transfer functions $G_T(s)$, $G_w(s)$, and $G_Q(s)$. It is known [3] that adequate modeling of superheaters and reheaters requires a distributed parameters approach. However, reasonable lumped parameter approximation may be used for G_T, G_w, and G_Q.

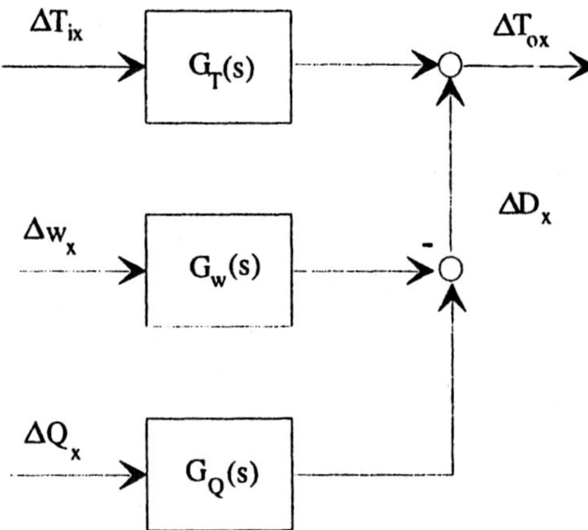

Figure 79.10 Conceptual scheme of temperature dynamics.

G_w and G_Q behave like first-order transfer functions, similar to each other, dominated by a time constant not far from τ_{ES} and with gain essentially dependent on the gas-to-wall heat transfer coefficient. The function $G_T(s)$ is, on the contrary, approximated well by:

$$G_T(s) \approx \mu_T \frac{1}{(1 + s\tau_{ES}/N)^N}, \qquad (79.30)$$

where N is the integer nearest to $S_i \gamma_i / 2 w_x c_p$ (with S_i and γ_i the steam-to-wall exchange surface and heat transfer coefficient) and μ_T is less than 1. Typically, secondary superheaters are short heat exchangers with $N \approx 2$. Larger reheaters may have $N \approx 3 - 4$.

Moreover, it appears that the process is nonlinear, because τ_{ES} is nearly proportional to the inverse of the load. N is only slightly dependent on w_x because $\gamma_i \equiv w_x^{0.8}$.

The temperature dynamics are affected by multiple lags, varying with the load. In addition, transducers for steam temperature are generally affected by a small (a few second) and a larger (some tens of seconds) time lag due to the thermal inertia of the cylinder where the sensor is placed.

Of course, multiple lags are in the loop when ΔT_{ix} is the control variable. This is the case when desuperheating spray is used to achieve mixing between the superheated steam at the outlet of the preceding component (e.g., the primary superheater) and the water spray is modulated by a suitable valve. Because the attemperator has a very small volume, storage in it is negligible, and its equations are given by steady-state mass and energy balances (see Figure 79.11; referring to the superheating section):

$$w_v + w_{ds} = w_T \quad \text{and} \qquad (79.31)$$

$$w_v h_v + w_{ds} h_{ds} = w_T h_i . \qquad (79.32)$$

In normal plant operation, the steam flow w_T in the secondary superheater is imposed (over a wide band) by the load controller, h_v is determined by the upstream superheater, and h_{ds} is nearly constant. The second superheater inlet temperature T_i is given by the following variation equation:

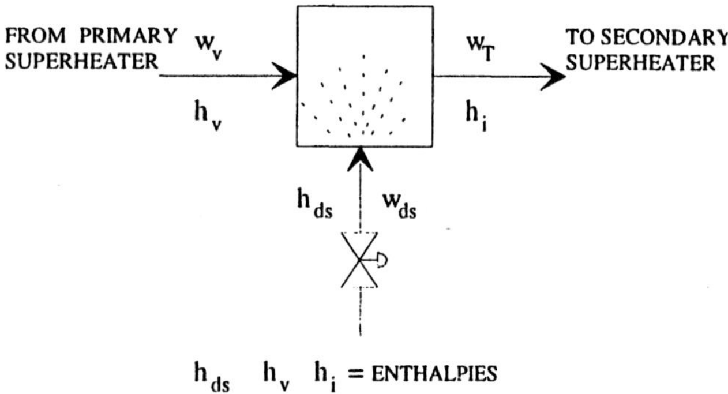

Figure 79.11 Desuperheating spray.

$$\begin{aligned}\Delta T_i &= \frac{1}{C_p}\Delta h_i = \frac{(\bar{h}_v - \bar{h}_i)}{C_p \bar{w}_T}\Delta w_T \\ &+ \frac{\bar{w}_v}{\bar{w}_T}\Delta T_v - \frac{(\bar{h}_v - \bar{h}_{ds})}{C_p \bar{w}_T}\Delta w_{ds}, \quad (79.33)\end{aligned}$$

where

$$\Delta T_i = \Delta h_v / c_p,$$

and the upper script $\bar{}$ denotes the steady-state of linearization. In Equation 79.33 Δw_{ds} is the control variable which directly modulates temperature T_i, and Δw_T and ΔT_v represent disturbances to the temperature. Because the influence of Δw_{ds} on pressure is very small, temperature control via desuperheating spray does not significantly influence boiler pressure.

In Figure 79.10, the heat transfer rate ΔQ_x may also be used to control temperature. This would be very effective because $G_Q(s)$ incorporates less phase lags than $G_T(s)$. Unfortunately (see Section 79.1.3) it is impossible to modulate the heat transfer rate to a single heat exchanger in the boiler (e.g., by varying fuel flow or recirculation gas flow) without simultaneously influencing the heat released to all of the other heat exchangers in the boiler. For instance, when recirculation gas dampers are modulated to control reheat temperature, the heat transfer rates to the evaporator and to the superheaters are also simultaneously varied. This fact generates interaction among the different process variables, to the extent that it is often necessary to introduce feedforward decoupling actions to achieve acceptable control performance. When spray is used to control superheater temperature, then Δw_x and ΔQ_x constitute disturbances for the temperature control induced, for instance, by varying fuel flow required by load-pressure control.

Fortunately (see Figure 79.7), when the heat transfer rate to the evaporator is increased, the steam generation is also increased so that Δw_x and ΔQ_x grow nearly as much. Because $G_w(s)$ and $G_Q(s)$ are similar, the global disturbance ΔD_x is much smaller than the two individual disturbances. However, as recalled in Section 79.1.3, when the heat transfer rate Q_{EV} to the evaporator is varied by the fuel flow, the heat transfer rates to the superheaters and reheater do not vary in the same percentage, so that Δw_x (nearly proportional to ΔQ_{EV}) does not exactly balance ΔQ_x. This means that $\Delta D_x \neq 0$.

Power Generation

According to Equations 79.1–79.3 and the subsequent remarks, it can be approximately assumed that

$$\delta P_e = \delta P_m \cong k_{HP}\delta w_T + k_{RH}\delta w_r, \quad (79.34)$$

where

$$\begin{aligned} k_{HP} &:= \alpha_T(h_T^\circ - h_{TR}^\circ)w_T^\circ/P_e^\circ, \\ k_{RP} &:= \alpha_R(h_R^\circ - h_0^\circ)w_R^\circ/P_e^\circ,\end{aligned}$$

with the upper script $^\circ$ denoting nominal values, and $\delta_z = \Delta z/z^\circ$, for any variable z. We know from Section 79.1.3 that w_T is given by the scheme of Figure 79.7 and is sensitive only to the high pressure part of the steam-water subsystem. To understand the factors influencing w_R, let's refer to Figure 79.12.

Because HPT has negligible storage

$$w_T + w_B = w_{se}. \quad (79.35)$$

The reheater is a large steam heat exchanger; feed-water heaters (FWH) fed by steam extractions are large tube and shell heat exchangers where steam extractions are condensed to heat feedwater. Both of these components have significant mass storage. The FWH represented in Figure 79.12 accounts (in an equivalent way) for the overall capacitance of FWHs. Neglecting the desuperheating spray (normally zero), the relevant mass balances are

$$\begin{aligned}\frac{dM_R}{dt} &= w_B - w_R, \quad (79.36)\\ \frac{dM_{FH}}{dt} &= w_{se} - w_e, \quad (79.37)\end{aligned}$$

where M_R is the steam mass in the reheater, M_{FH} is the steam mass in FWHs, and w_e is the condensation mass flow-rate in FWHs.

Pressure losses in the reheater are small and can be neglected. Because V_E is normally fully open, we may assume that the pressure in the entire reheater is the same as at the RHT inlet. Moreover, reheater temperature has much slower dynamics than mass storage (see Section 79.1.3), so that it may be considered constant while evaluating dM_R/dt.

Applying an equation similar to Equation 79.10 to RHT and considering superheated steam as an ideal gas, the RHT flow equation is

$$w_R = k'_{RHT} p_B/\sqrt{RT_{RH}}, \quad (79.38)$$

where T_{RH} is the reheater outlet temperature, R is the gas constant, and k'_{RHT} is a suitable constant. Then, taking variations of Equations 79.36, 79.37, and 79.38,

$$\delta w_R = \frac{1}{1+s\tau_R}\left(\frac{w_T^\circ}{w_R^\circ}\delta w_T - \eta\delta w_w\right), \quad (79.39)$$

where τ_R is a time constant resulting from the sum of the storage capacitance of the reheater and of FWHs, multiplied by the flow resistance of RHT, and the last term results from considering that

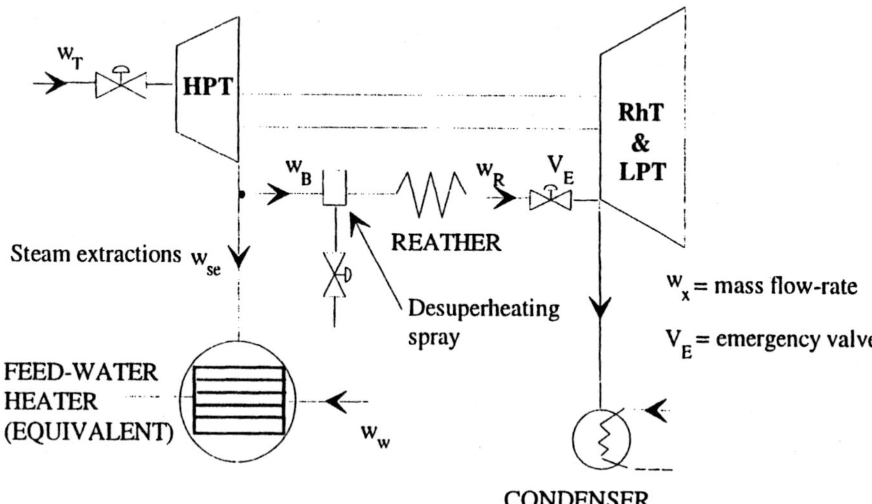

Figure 79.12 Steam reheating.

condensation flow-rate variation, Δw_c is essentially due to feed-water flow variation Δw_w. In controlled conditions δw_w strictly follows δw_T, so that Equation 79.39 becomes

$$\delta w_R = \frac{1}{1+s\tau_R}\delta w_T. \tag{79.40}$$

Equation 79.39 deserves a couple of remarks [5]: the time constant τ_R has values of about 10–12 sec., nearly 50% due to FWH's capacitance; the possibility of varying the principal steam flow δw_R by changing the feed-water flow $(\eta\delta w_w)$ has suggested one of the most recent expedients to realize quick power variations even without HPT throttle reserve [6].

When special control of feed-water is not applied, Equation 79.40 can be substituted in Equation 79.34 with the following conclusion:

$$\delta P_E \approx \left(k_{HP} + \frac{k_{RH}}{1+s\tau_R}\right)\delta w_T, \tag{79.41}$$

realizing that about 1/3 ($k_{HP} \approx 0.3 - 0.4$) of electric power output is strictly proportional to the steam flow and about 2/3 ($k_{HP} + k_{RH} = 1$) is affected by a time lag τ_R, which cannot be negligible when load varies rapidly (as in the case of load rejection incidents or turbine speed control problems). Finally, observe that the parameters of Equation 79.41 are slightly dependent on plant load.

Combustion and Air-Gas Dynamics

Referring to Figure 79.2, we see that the air-gas subsystem forms a complex circuit, where the largest storage (of mass and, thus, of energy) is the furnace. Apart from air preheaters (2) and (3), the dynamics of the air-gas subsystem are very fast, so that it is substantially decoupled from the dynamics of the steam-water subsystem. Air-gas dynamics are relevant only to two special problems:

1. control of furnace pressure in a balanced draft furnace (recognizable from the induced draft fan (11) of Figure 79.2), and
2. flame stability at low loads in coal fired plants.

The second problem requires a very complex analysis and is relevant only in very particular situations.

To analyze (1), two dynamical phenomena must be studied:

- combustion kinetics, i.e., the chemical process governing fuel oxidation and its heat release, and
- mass and energy storage in the furnace, possibly augmenting the furnace storage to account for the rest of the air-gas circuit.

Distributed parameter modeling would be required to describe these phenomena accurately. Again, a simple lumped model may be used to explain basic concepts. A simple way to account for combustion kinetics is to introduce a combustion time constant τ_e relating fuel flow-rate w_f to heat transfer rate Q_f in the furnace,

$$\Delta Q_f = \frac{1}{1+s\tau_e}H_f\Delta w_f, \tag{79.42}$$

where H_f is the heat value of the fuel. The time constant τ_e is of the order of a second, smaller for oil or gas, and larger for pulverized coal.

Mass and energy balances for the furnace can be derived from the following assumptions:

- the gas pressure p_g is uniform in the furnace,
- heat transfer from gas to wall in the furnace is computed from the mean gas temperature, T_g, and
- the combustion gas behaves as an ideal gas.

Then, without considering gas recirculation:

$$\frac{dE_g}{dt} = w_a h_a + Q_f - w_{go}h_{go} - Q_r, \tag{79.43}$$

and

$$\frac{dM_g}{dt} = w_a + w_f - w_{go}, \tag{79.44}$$

where w_a is the air-flow rate with enthalpy h_a, Q_r is the heat transfer rate radiated to the wall, w_{go} is the outlet gas flow-rate with enthalpy h_{go}, $M_g = V_f \rho_g$ is the total mass of gas in the furnace (ρ_g is the mean density), and $E_g = c_{vg} T_g M_g$ is the total energy of the gas (c_{vg} is the specific heat at constant volume). Ideal gas law and radiation law equations are

$$\frac{p_g}{\rho_g} = R_g T_g, \qquad (79.45)$$

$$Q_r = k_{rr}\left(T_g^4 - T_w^4\right) \approx k_{rr} T_g^4, \qquad (79.46)$$

where k_{rr} is a constant and T_w is the wall temperature ($T_w^4 \ll T_g^4$).

Boundary conditions are determined by head losses along the air-gas circuit and by a forced and induced draft fan. Considering the outlet boundary conditions,

$$p_g - p_o \cong -p_v + \left(k_a + k_f(z)\right) w_{go}, \qquad (79.47)$$

where p_o is the atmospheric pressure, p_v is the head at $w_{go} = 0$ of the induced draft fan, k_a is the constant yielding the head losses along the gas circuit, and $k_f(z)$ is the constant yielding the head losses of the fan depending on the control inlet vane position z.

Assuming that the forced draft fans are controlled with air flow-rate w_a and that, in the mass equation, $w_f \ll w_a$ may be approximated by Q_f/h_f, linearization of the model Equations 79.43–79.47 yields

$$\delta p_g = \frac{r_g^\circ L}{(1+s\tau_1)(1+s\tau_2)}\left\{\left(\frac{w_f}{w_{go}}\right)^*(1+s\tau_3)\delta Q_f + \left(\frac{w_a}{w_{go}}\right)^*(1+s\tau_4)\delta w_a - \mu_v L(1+s\tau_5)\delta z\right\}, \qquad (79.48)$$

where all the variables δ_y are expressed in p.u., the superscript * denotes the value at the linearization steady state, L is the plant load, and $r_g^\circ, \tau_1, \tau_2, \tau_3, \tau_4, \tau_5, \mu_v$ are suitable constants computed from design or operating data.

Omitting cumbersome computations for brevity, Equation 79.48 deserves some remarks. The time constants $\tau_j, j = 2\ldots,5$, are proportional to the furnace crossing time $t_f = (M_g/w_g)^*$. M_g^* is nearly insensitive to the load, above the technical minimum, w_g^* is nearly proportional to the load, and the time constants $\tau_j, j = 2\ldots,5$ become larger as the load decreases. The smallest time constant, on the contrary, decreases as the load decreases. And because it is only a few tenths of second at the maximum load, it may be neglected. Typical values are $t_f \approx 2-4$ sec., $\tau_1 < 0.1t_t$, $\tau_2 \approx 0.25t_f$, $\tau_3 \approx 5-6t_f$, $\tau_4 \approx 0.1t_f$, r_g° is about 0.1.

Due to the large value of τ_3, fuel trip from maximum load ($\delta Q_f = 1$) causes large pressure drops, with extreme risk of implosion. Inlet vane control is introduced to attenuate the effect of such a disturbance.

Concluding Remarks on Dynamics

To summarize the analysis of the preceding Sections, it is useful to identify the interactions in the system. Consider the input control variables, namely, the fuel flow-rate w_f, the air flow-rate w_a, the turbine admittance Y, the feed-water flow-rate w_w, the recirculation gas flow-rate w_{rg}, desuperheating spray flows w_{ds} and w_{dr}, induced fan vane position z, and the output variables to be controlled, namely, electric power P_e, throttle pressure p_T, drum level y_D, throttle temperature T_T, reheater outlet temperature T_{RH}, furnace pressure p_g, and air-to-fuel ratio λ_{af} (or combustion efficiency). From Figure 79.7 and Equation 79.41, P_e and p_T are strictly related and simultaneously affected by the heat transfer rate (i.e., w_f) and turbine admittance Y.

Flow rates w_{ds} and w_{dr} slightly affect P_e and p_T because they have only a small mass effect on steam flow. Similarly, feed-water flow w_w only marginally affects power (see Equation 79.39). Air flow is nearly proportionally to fuel flow, so that trimming actions to optimize λ_{af} have little influence on the heat transfer rate Q_{EV} to the evaporator. Only w_{rg} changes the heat release partition in the boiler; the control band width of this variable (used to control reheat temperature) is, however, rather narrow for that reason. Therefore, pressure and power form a (2x2) subsystem, tightly coupled but with limited disturbance from outside.

Drum level y_D is the only output variable markedly influenced by feed-water flow w_w; y_D is also "disturbed" (see Equation 79.28) by steam flow w_v, by drum pressure p_o, and by the evaporator heat transfer rate Q_{EV}, so that w_w does not influence the load-pressure subsystem, but y_D is considerably influenced by the control variables of that subsystem.

Superheated temperature T_T, according to Figure 79.10 and Equation 79.33, is influenced by w_{ds} and is "disturbed" by steam flow and heat transfer rate Q_{SH}, which, in turn, follows fuel flow variations. So, disturbance of the power-pressure subsystem to temperature is relevant, even though there is a natural partial compensation due to the boiler behavior (see ΔD_x in Figure 79.10).

Reheater spray follows a rule similar to superheater spray.

Furnace pressure is dynamically decoupled from all of the variables related to the steam-water side subsystems and, according to Equations 79.42 and 79.48 is affected by air flow, fuel flow, and induced draft fan inlet vane position (normally used as its control variable). A similar criterion applies to the fuel-air ratio.

Input-output relations are summarized in Figure 79.13, where only the major interactions are considered.

Finally, in coal fired units fuel is supplied by pulverizers, not influenced by the rest of the plant. The pulverizers have sluggish dynamics due to the dead time of the grinding process and to the uncertain behavior of such machines caused by coal quality variation and machine wear. For those plants, where fuel flow cannot be considered a directly manipulated variable, the slow response of coal pulverizers must be cascaded with the evaporator dynamics of Figure 79.7 often creating severe problems for system stability and control.

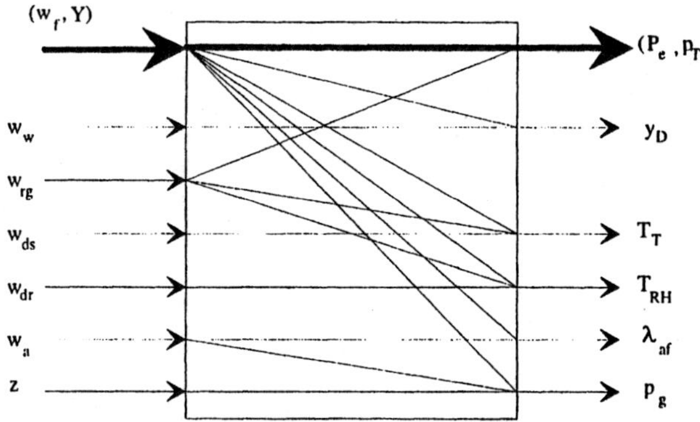

Figure 79.13 Process interactions.

79.1.4 Control Systems: Basic Architectures

Introductory Remarks on Control

This section discusses plant control at the "principal regulations level" (see Figure 79.4) which is organized around four control subsystems:

- load and pressure control: the regulation of power generation and steam pressure at the throttle,
- drum level control: the regulation of water level in the drum (steam/water separator),
- temperature control: the regulation of steam temperature at the superheater and reheater outlets, and
- combustion control: the regulation of heating rate (fuel flow), excess oxygen (air flow), and furnace pressure.

Power plant control systems have evolved over many decades. Today there are thousands of electric generating plants operating throughout the world. It would be difficult to find more than a few with identical control systems. Yet certain basic configurations are almost universally employed with relatively minor variations. These are described in the following paragraphs.

Control of Pressure and Generation

As discussed in Section 79.1.3, steam pressure and power generation are tightly coupled process variables. Both are strongly affected by energy (fuel) input and throttle valve position. This two-input, two-output system must be considered as such. Even though single-input, single-output compensation arrangements are successful, they must be designed (tuned) as a unit. Three basic architectures are commonly employed:

- turbine following: generation is paired with fuel rate and pressure with throttle valve position,
- boiler following: generation is paired with throttle valve position and pressure with fuel rate, and
- coordinated control: a true two-input, two-output configuration of which there are variations.

The turbine-following arrangement, shown in Figure 79.14 has distinctive attributes. First, the control of the energy input to the boiler is relatively slow compared with the positioning of the throttle valve. As a result, turbine-following control allows rapid regulation of throttles pressure and slow, but stable, regulation of generation. Consequently, turbine-following control is preferred for plants not used for load following.

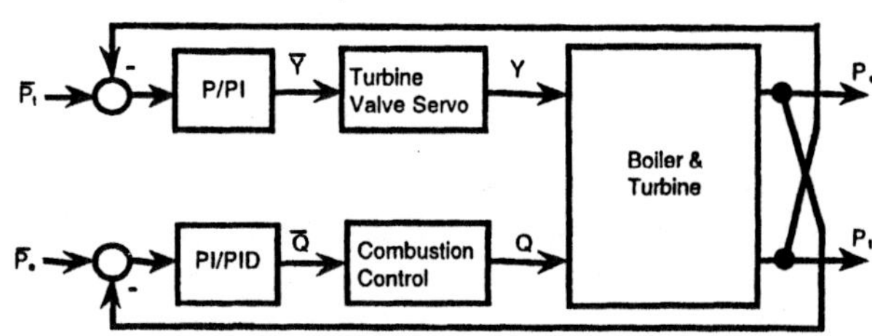

Figure 79.14 The turbine following configuration for pressure and generation control.

The boiler-following architecture is illustrated in Figure 79.15. It produces substantially more rapid responses to generation commands but they can be quite oscillatory. Moreover, pressure response is typically oscillatory.

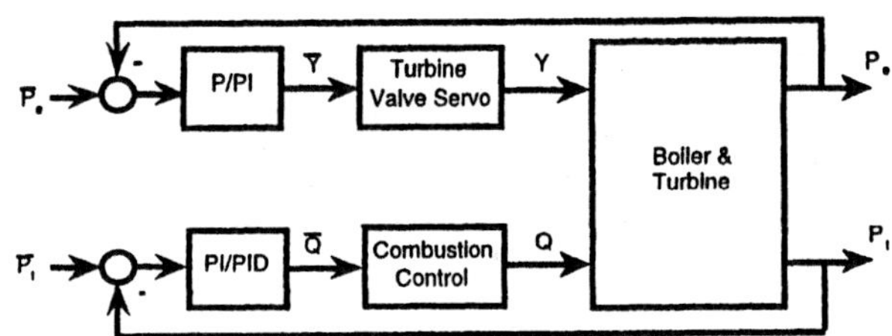

Figure 79.15 The boiler following configuration for pressure and generation control.

Modern requirements for load following have led to the widespread use of two-input, two-output pressure and generation control. There are a number of approaches to coordinated control. One configuration (commonly referred to as "coordinated control" or "integrated control") is shown in Figure 79.16. Properly designed coordinated-control systems can provide excellent response to load demand changes.

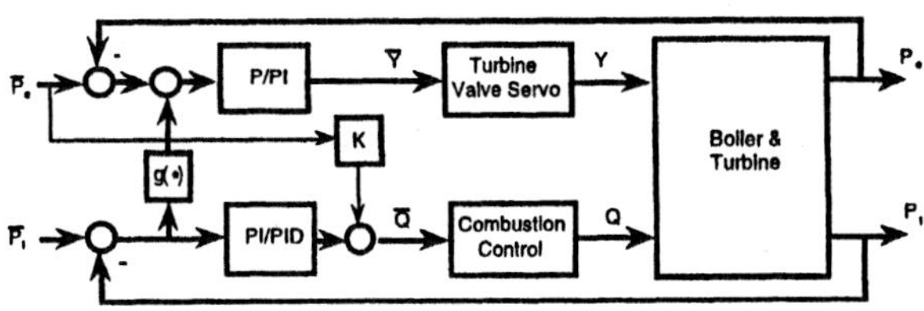

Figure 79.16 A coordinated-control configuration for pressure and generation control.

Generator speeds naturally synchronize because of their interconnection via the electrical network. Ultimately, the (steady-state) network synchronous speed is regulated by a system level controller through the assignment of generation commands to individual units. Nevertheless, speed governing on a substantial fraction of the network's generating units is essential to damping the power system's electromechanical oscillations. As a result, in many plants, the goal of turbine flow control includes speed governing as well as regulating power output. This dual requirement is almost always accomplished with the "frequency bias" arrangement shown in Figure 79.17. Here turbine speed error is fed directly through a proportional compensator to the turbine valve servo and simultaneously a frequency error correction is added to the power generation demand signal through the frequency bias constant B_f. Ideally, B_f is precisely the sensitivity of system load to synchronous frequency. The frequency bias arrangement can be incorporated in either the boiler-following or coordinated-control configurations.

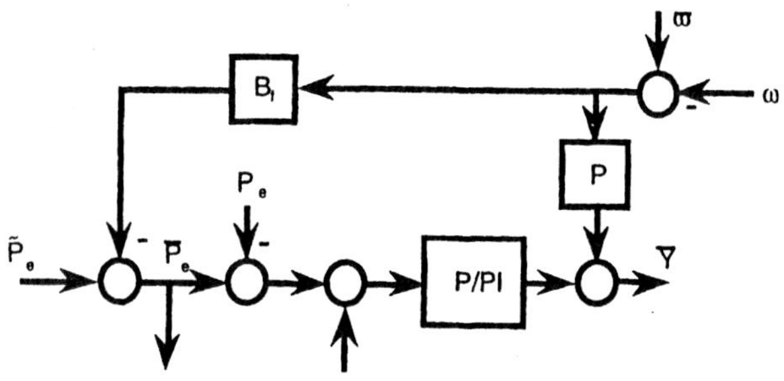

Figure 79.17 The turbine valve may be used to stabilize turbine speed and regulate generation with the "frequency bias" modification of the boiler-following or coordinated-control configurations.

Drum Level Control

The goal of the drum level controller is to manipulate the flow of feedwater into the drum so that the drum water level remains sufficiently close to a desired value. Feedwater flow is typically regulated by a flow control valve or by adjusting the speed of the feedwater pump. Drum level controllers are classified as single-, two-, or three-element. A single-element level controller utilizes feedback of a drum level measurement as illustrated in Figure 79.18(a). Two- and three-element controllers include "feedforward" measurements of steam flow and both steam and water flow, respectively, as illustrated in Figures 79.18(b) and 79.18(c).

The importance of the steam flow feedforward can be appreciated by examining the leading term in Equation 79.28 which shows that the drum level deviation is proportional to the integral of the difference between steam and water flow. Any sustained difference between steam flow and water flow can quickly empty or fill the drum. In current practice, three-element controllers are typically used during normal operation but are not suitable at very low loads, where it is common to switch to single-element configurations. Drum level dynamics are also nonminimum phase (the so-called "shrink and swell" effect) as can be observed in the last term of Equation 79.28. Drum level control

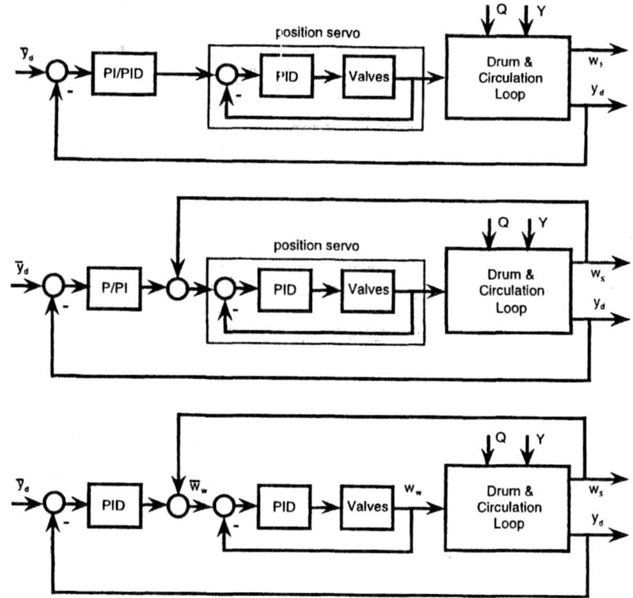

Figure 79.18 (a). A typical single-element drum level configuration. (b). A typical two-element drum level configuration. (c). A typical three-element drum level configuration.

will be examined in more detail in Section 79.1.5.

Temperature Control

An important goal of plant control is to regulate steam temperature at the turbine entry points, i.e., at the superheater and reheater outlets. There are a number of control means for accomplishing this. The most direct are attemperators which inject water at the heat exchanger inlet. By moderating the fluid temperature entering the heat exchanger, it is possible to control the outlet temperature. Other possibilities are associated with adjusting the heat transferred to the fluid as it passes through the exchanger. This can be accomplished by changing the mass flow rate of gas past the heat transfer surfaces with recirculated gas or excess air flow, or the gas temperature at the exchanger surfaces by altering the burner positions or "tilt" of the burners. Sometimes a combination of these methods is employed. In the following discussion it is assumed that control is affected by attemperators.

The dynamics of superheaters and reheaters have been discussed in Section 79.1.3. Recall that the response of the outlet temperature to a change in inlet temperature is characterized by series of first-order lags with time constants that vary inversely with the steam flow rate through the heat exchanger. Because of the significant time delay of the outlet temperature response, a cascade control arrangement, as illustrated in Figure 79.19, is typically required for temperature regulation. The attemperator outlet temperature is a convenient intermediate feedback variable although, depending on the heat exchanger construction, other intermediate steam temperatures may also be available for measurement. Because of the strong dependence of the time lag on steam flow, parameterization of the regulator parameters on steam flow is necessary for good performance over a wide load range. Some control systems incorporate disturbance feedforward.

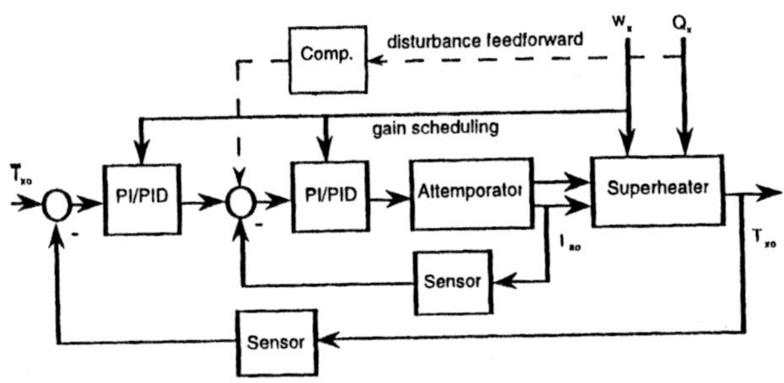

Figure 79.19 A fairly sophisticated control temperature arrangement employs a cascade arrangement, gain scheduling and disturbance feed forward.

Combustion Control

The main purpose of the combustion control system is to regulate the fuel and air inputs into the furnace to maintain the desired heat input into the steam generation process while assuring appropriate combustion conditions (excess oxygen). In most instances, regulating the furnace gas pressure is a secondary, but important, function of the combustion controller. A typical combustion control configuration is illustrated in Figure 79.20. Recall that the heating rate command signal $\bar{Q}$ is generated by the pressure-generation controller (Figure 79.14).

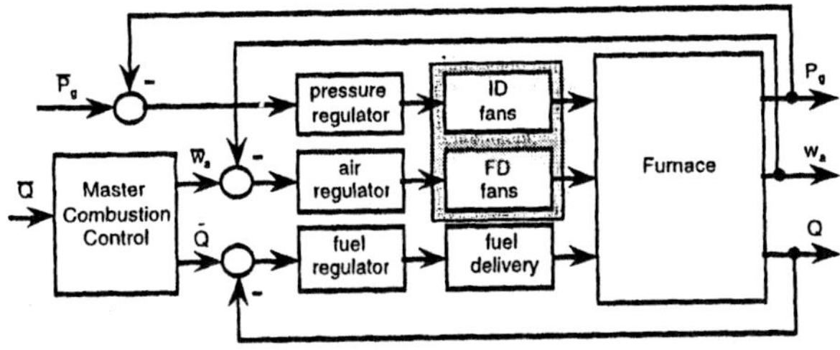

Figure 79.20 Combustion control includes regulation of fuel and air flow and furnace pressure.

The details of the combustion control system depend significantly on the type of fuel. Oil and gas are typically regulated with flow control valves. These controls are usually fast. Also, oil and gas flow and the caloric content of the fuel can be reliably measured. Pulverized coal presents a different situation. Fuel flow is regulated by adjusting the "feeder speed" (which directly changes the flow rate of coal into the pulverizer) and the primary air flow (the air flow through the pulverizer that carries the pulverized coal into the furnace). The pulverizing process is quite slow and adds a delay of 100–300 sec. in the fuel delivery process. Moreover, the flow rate of coal is difficult to estimate accurately and coal's caloric content varies.

The master combustion control proportions the fuel and air requirements and establishes set points for the lower level controllers. Calorimetry information (notably excess oxygen) is provided on a sampled or continuous basis. Minimum fuel and air (primary and secondary) flow constraints are also accommodated.

79.1.5 Design of a Drum Level Controller

Issues Involving Design Over a Wide Load Range

The control architectures described in Section 79.1.4 represent a starting point for detailed design of the plant control system. Once the configuration is chosen, it is necessary to determine compensator parameters that produce acceptable performance over a wide range of operating conditions. Doing this involves applying various analytical control design tools in conjunction with detailed simulation studies and generally terminating with field tuning during plant commissioning. This section provides an example of the analytical phase in designing a drum level controller.

Certain necessary plant operations, e.g., startup, rapid load following and controlled runback, evoke behavior that can only be attributed to the nonlinearity of the steam generation process. Deleterious nonlinear effects are particularly evident at low generation rates and when relatively large changes in generation level are made. Regulating drum level and combustion stability are particularly problematic at low loads. Nonlinear behavior can be readily observed in two ways: by comparing large and small excursions from a given equilibrium point or by comparing small signal behavior at different equilibria corresponding to distinct load levels. In the following paragraphs, we will examine the steam generation process from the latter point of view. This is particularly useful when designing control systems based on small signal (linear) behavior.

Power plant controllers are designed to provide good performance at or near rated conditions. At off-design conditions, e.g., as load is decreased, performance deteriorates because the plant behaves differently than at the design point. Typically, satisfactory performance can be achieved down to about 30% of rated generation. At some point performance becomes unsatisfactory, requiring manual intervention or a switch to a retuned or even restructured control system. At low loads the nonlinearities associated with the evaporation and combustion processes [8], in particular, are quite severe. Compounding this complexity, plant operation over a wide generating range may require configuration changes. For example, as generation is reduced, the number of burners and fuel pumps or pulverizers will be reduced (see, for example, [9]), and steam by-pass systems may be employed, particularly, in large once-through fossil plants and nuclear plants [10].

The assumptions behind the model of evaporator dynamics developed in Section 79.1.3 are not suitable for capturing dynamics at low steam generation rates. In Section 79.1.5, following the assumptions Equations 79.3 and 79.4 are relaxed to allow more general variations of steam quality and fluid pressure through the evaporator. This model is then used to investigate the small signal dynamical behavior at various load levels. Control system design is considered in Section 79.1.5. Much of the discussion herein is based on [11] which emphasizes the nonlinear behavior of the circulation loop. Earlier discussions on level control system design included [12] and [13].

Circulation Loop Dynamics Revisited

The system (see Figure 79.3) is comprised of a natural circulation loop whose main components are the drum which separates the steam and water, the downcomer piping which carries water from the drum to the bottom of the furnace, and the riser tubes which are exposed to the burning furnace gas and in which boiling takes place. A feedwater valve regulates the flow of water into the system and the throttle valve regulates the steam flow out of it. The heat absorbed by the fluid in the risers is a third input. The drum dynamical equations are considered first and then the remainder of the circulation loop.

Drum Dynamics

The dynamical equations for the drum can be formulated in a number of different ways depending on the variables chosen to characterize the thermodynamic state of the drum. Two thermodynamic variables must be selected from the four possible pairs: (T, v), (s, P), (s, v), (T, P), where T, P, v, and s denote temperature, pressure, specific volume, and specific entropy, respectively.[2] Because the fluid in the drum is in the saturated state, the pair (T, P) is not suitable but any of the other choices is valid. As an example, we give the equations for the (s, P) formulation. The following assumptions are made:

1. Drum liquid and gas are in the saturated state.
2. Pressure is uniform throughout the drum.
3. All liquid resides at the bottom of the drum and all gas at the top.

Then, the drum dynamics, derivable from mass and energy balance equations, (see [9]) are

$$\frac{dP_d}{dt} = -\frac{[w_e + (1 - x_r)w_r - w_{dc}]\frac{v_{df}^2}{v_{dg}} + [x_r w_r - w_s]\frac{1}{v_{df}}}{(V - V_w)\frac{v_{df}}{v_{dg}^2}\frac{\partial v_g}{\partial P_d} + V_w\frac{1}{v_{dg}}\frac{\partial v_f}{\partial P_d}}, \tag{79.49}$$

and

$$\frac{dV_w}{dt} = -\frac{[w_e + (1 - x_r)w_r - w_{dc}](V - V_w)}{(V - V_w)\frac{v_{df}}{v_{dg}^2}\frac{\partial v_g}{\partial P_d} + V_w\frac{1}{v_{dg}}\frac{\partial v_f}{\partial P_d}}$$

$$\frac{\left(\frac{v_{df}}{v_{dg}}\right)^2\frac{\partial v_g}{\partial P_d} - V_w[x_r w_r - w_s]\frac{\partial v_f}{\partial P_d}}{(V - V_w)\frac{v_{df}}{v_{dg}^2}\frac{\partial v_g}{\partial P_d} + V_w\frac{1}{v_{dg}}\frac{\partial v_f}{\partial P_d}}, \tag{79.50}$$

[2]In the standard convention, specific entropy is denoted by the symbol s which is also used to denote the Laplace transform variable. This will not lead to confusion in the subsequent discussion.

where the following nomenclature has been adopted:

w_r, w_{dc}, w_s	mass flow rates, riser, downcomer and turbine, respectively,
v_{df}, v_{dg}	drum specific volume, liquid and gas, respectively,
P_d	drum pressure,
T_d	drum temperature,
V	total drum volume,
V_w	volume of water in drum, and
x_d	net drum quality, $x_d = V_w/V$.

In addition to these differential equations we require the constitutive relations (coexistence curve)

$$v_f = v_f(P) \quad \text{and} \quad v_g = v_g(P) \tag{79.51}$$

and the drum level equation

$$y_D = f(V_w). \tag{79.52}$$

Under the stated assumptions, the drum thermodynamic state, entropy and pressure (s, P), is equivalent to (x_d, P) or (V_w, P) because

$$x_d = v_f(P) + x_d[v_g(P) - v_f(P)]. \tag{79.53}$$

Hence, we refer to the above equations as the (s, P) formulation even though s does not explicitly appear.

The steam flow out of the drum to the turbine is governed by the relationship

$$w_s = w_{s0}A_1\left(\frac{P_d}{P_{d0}}\right) \tag{79.54}$$

where w_{s0}, P_{d0} are the throttle flow and drum pressure at rated conditions, respectively, and A_t denotes the normalized valve position, with rated conditions corresponding to $A_t = 1$.

Circulation Loop Dynamics

The main deficiency of the model in Section 79.1.3 was the simplified treatment of the circulating fluid flow and of the complex two-phase flow dynamics of the riser loop. Consequently, it does not adequately represent the nonminimum phase characteristics associated with shrink-swell phenomenon at lower load levels. Another approach to riser modeling is based on discretizing the time-dependent, nonlinear partial differential equations of one-dimensional, two-phase flow using the method of collocation by splines, a form of finite element analysis.

The circulation loop, composed of the downcomer and riser, is assumed to be characterized by homogeneous single or two-phase flow. The conservation of mass, energy, and momentum lead to three first-order partial differential equations which are then discretized using collocation by linear splines. We use a single element for the downcomer containing fluid in a single phase (liquid) and N elements of equal length for the riser which consists of two-phase flow. In the latter case the fluid properties change significantly along the spatial coordinate. In the entropy-pressure (s, P) formulation, the downcomer equations are

$$
\begin{aligned}
\frac{dw_1}{dt} &= -A\left(\frac{P_1 - P_0}{L_{do}}\right) - \frac{2w_1v_1}{A_{do}}\left(\frac{w_1 - w_0}{L_{do}}\right) \\
&\quad - \frac{w_1^2}{A_{do}}\left(\frac{v_1 - v_0}{L_{do}}\right) \\
&\quad - A_{do}\left(-\frac{g}{v_1} + f_{do}w_1^2\right), \qquad (79.55)\\
\frac{ds_1}{dt} &= v_1\left\{\frac{1}{A_{do}T_1}\frac{\partial q}{\partial z} - \frac{w_1}{A_{do}}\left(\frac{s_1 - s_0}{L_{do}}\right) + \frac{w_1v_1}{A_{do}T_1}f_{do}w_1^2\right\}, \qquad (79.56)
\end{aligned}
$$

and

$$
\begin{aligned}
\frac{dP_0}{dt} &= \frac{1}{\gamma_A}\left\{\frac{v_0^2}{A_{do}}\left(\frac{w_1 - w_0}{L_{do}}\right) - \gamma_B v_{1-1}\right. \\
&\quad \left.\left(-\frac{w_0}{A_{do}}\left(\frac{s_1 - s_0}{L_{do}}\right) + \frac{w_0v_0}{A_{do}T_0}f_{do}w_0^2\right)\right\}. \qquad (79.57)
\end{aligned}
$$

The riser equations are

$$
\begin{aligned}
\frac{dw_i}{dt} &= -A\left(\frac{P_i - P_{i-1}}{L}\right) \\
&\quad - \frac{2w_iv_i}{A}\left(\frac{w_i - w_{i-1}}{L}\right) \\
&\quad - \frac{w_i^2}{A}\left(\frac{v_i - v_{i-1}}{L}\right) \\
&\quad - A\left(\frac{g}{v_i} + f_r w_i^2\right), \qquad (79.58a)\\
\frac{ds_i}{dt} &= v_i\left\{\frac{1}{AT_i}\frac{\partial q}{\partial z} - \frac{w_i}{A}\left(\frac{s_i - s_{i-1}}{L}\right) \right. \\
&\quad \left. + \frac{w_iv_i}{AT_i}f_r w_i^2\right\}, \qquad (79.58b)
\end{aligned}
$$

and

$$
\begin{aligned}
\frac{dP_{i-1}}{dt} &= \frac{1}{\gamma_A}\left\{\frac{v_{i-1}{}^2}{A}\left(\frac{w_i - w_{i-1}}{L}\right) - \gamma_B v_{i-1}\right. \\
&\quad \left(\frac{1}{AT_{i-1}}\frac{\partial q}{\partial z} - \frac{w_{i-1}}{A}\left(\frac{s_i - s_{i-1}}{L}\right)\right. \\
&\quad \left.\left. + \frac{w_{i-1}v_{i-1}}{AT_{i-1}}f_r w_{i-1}^2\right)\right\} \qquad (79.58c)
\end{aligned}
$$

for $i = 2, \ldots, N+1$. The following nomenclature is employed:

N	number of riser sections,
L_{do}, L	downcomer length and riser section length (total riser length/N),
A_{do}, A	downcomer, riser cross section areas,
w_i	mass flow rate at ith node,
P_i	pressure at ith node,
T_i	temperature at ith node,
s_i	aggregate entropy at ith node, and
v_i	specific volume at ith node.

Once again, we need constitutive relations to complete the model. These are required in the form

$$
v = v(s, P), T = T(s, P) \qquad (79.59a)
$$

$$
\gamma_A = \left(\frac{\partial v}{\partial P}\right)_s, \quad \text{and} \quad \gamma_B = \left(\frac{\partial v}{\partial S}\right)_P. \qquad (79.59b)
$$

Reduction of Circulation Loop Equations

The circulation loop model described above contains fast dynamics irrelevant to the control problem. In general terms, fast dynamics are associated with certain pressure-flow dynamics (hydraulic/acoustic oscillations) and slow dynamics are associated with thermal (entropy) transients. Formally, we can approach the problem of identifying and approximating fast dynamics using asymptotic analysis, because fast dynamics are associated with the fact that the parameter γ_A is small. Our analysis is based on two assumptions:

- only the slowest mode of fast pressure-flow dynamics is significant to the control problem; we shall refer to this as hydraulic dynamics.
- spatial variations in flow and pressure along the circulation loop are negligible as far as the hydraulic dynamics are concerned.

The flow and pressure equations can be written

$$
\begin{aligned}
a_i\frac{dw_i}{dt} &= -(P_i - P_{i-1}) \\
&\quad + F_i(w - i, w_{i-1}, s_i, s_{i-1}, P_i, P_{i-1}), \\
&\quad a_i := L_i/A_i \quad \text{and} \qquad (79.60a)\\
b_i\frac{dP_{i-1}}{dt} &= (w_i - w_{i-1}) + g_i(w_{i-1}, s_i, s_{i-1}, P_{i-1}), \\
&\quad b_i := \gamma_A A_i L_i / v_{i-1}^2. \qquad (79.60b)
\end{aligned}
$$

Now, we define the average circulation loop flow and pressure:

$$
w_{av} := \sum_{i=1}^{N+1}\alpha_i w_i, \alpha_i = \frac{a_i}{\sum_{i=1}^{N+1}a_i} \qquad (79.61)
$$

$$
\text{and} \quad P_{av} := \sum_{i=1}^{N+1}\beta_i P_i, \beta_i = \frac{b_i}{\sum_{i=1}^{N+1}b_i}. \qquad (79.62)
$$

Let us also define the functions

$$
f_i(w_{av}, s_i, s_{i-1}, P_{av}) := F_i(w_{av}, w_{av}, s_i, s_{i-1}, P_{av}, P_{av}). \qquad (79.63)
$$

We can state the key assumption:

Assumption 1: For w_i, s_i and $P_i, i = 0, \ldots, N+1$, the following approximations are valid in the slow time scale:

$$
f_i(w_{av}, s_i, s_{i-1}, P_{av}) \approx F_i(w_i, w_{i-1}, s_i, s_{i-1}, P_i, P_{i-1}) \qquad (79.64)
$$

and

$$
g_i(w_{av}, s_i, s_{i-1}, P_{av}) \approx g_i(w_{i-1}, s_i, s_{i-1}, P_{i-1}). \qquad (79.65)
$$

It is easy to validate this assumption in the equilibrium state. Invoking these approximations and simply adding Equations 79.60 results in

$$\frac{dw_{av}}{dt} = \left\{-(P_d - P_0) + \sum_{i=1}^{N+1} f_i(w_{av}, s_i, s_{i-1}, P_{av})\right\} \tag{79.66a}$$

and

$$\frac{dP_{av}}{dt} = \left\{-(w_r - w_0) + \sum_{i=1}^{N+1} g_i(w_{av}, s_i, s_{i-1}, P_{av})\right\}. \tag{79.66b}$$

Assumption 2: The slow time scale flow and pressure distribution through the circulation loop is adequately approximated by equilibrium conditions of Equations 79.60 and the approximations of Assumption 1, i.e.,

$$-(P_i - P_{i-1}) + f_i(w_{av}, s_i, s_{i-1}, P_{av}) = 0, \quad i = 1, \ldots, N+1, \tag{79.67a}$$

and

$$(w_i - w_{i-1}) + g_i(w_{av}, s_i, s_{i-1}, P_{av}) = 0, \quad i = 1, \ldots, N+1. \tag{79.67b}$$

These equations are important because they allow the computation of P_0 and $\omega_r = \omega_{N+1}$ which are necessary to establish the interface of the circulation loop with the drum. From Equations 79.20 and the definitions of ω_{av}, P_{av}, we can derive the following relationships:

$$P_0 = P_{av} - \sum_{i=1}^{N}\left(\sum_{j=1}^{N+1-i} \alpha_j\right) f_i(w_{av}, s_i, s_{i-1}, P_{av}) \tag{79.68a}$$

and

$$w_r = w_{av} - \sum_{i=2}^{N+1}\left(\sum_{j=1}^{i-1} \beta_j\right) g_i(w_{av}, s_i, s_{i-1}, P_{av}). \tag{79.68b}$$

The remaining interface equation is

$$w_0 = \sqrt{|P_d - P_0|}sign(P_d - P_0)/f_{de}. \tag{79.69}$$

Equilibria and Perturbation Dynamics

First, we consider the open-loop behavior. The procedure followed is

1. Trim the system at load levels ranging from near 5% to 100%.
2. Compute the linear perturbation equations.
3. Analyze the pole-zero patterns as a function of load level.

Equilibrium values are computed by specifying the desired load, drum pressure, and drum level and then computing the required control inputs and the remaining state variables. A Taylor linearization at each equilibrium point yields a linear model of the perturbation dynamics. Thus, it is possible to determine the system poles and zeros and to examine how they change as a function of load. Figure 79.21 gives a sample of the results obtained from solving the equilibrium equations.

Figure 79.22 is an eigenvalue plot showing how the plant dynamics vary with load. Table 79.2 summarizes a complete plant modal analysis at 100% load.

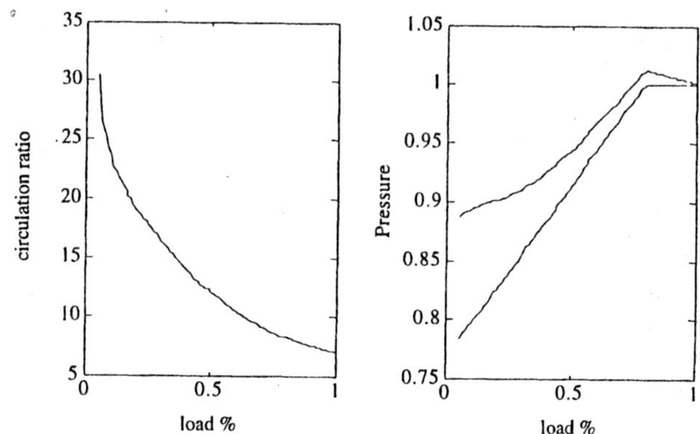

Figure 79.21 Typical equilibrium curves show the circulation ratio, average loop pressure (upper), and drum pressure (lower) as a function of load.

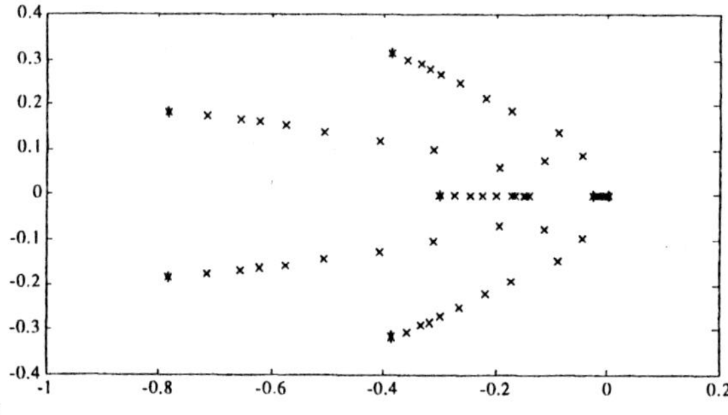

Figure 79.22 All but one of the eigenvalues of the circulation loop are illustrated. The missing eigenvalue is relatively far to the left at approximately -8 to -20 (depending on load level). There is an eigenvalue at the origin for all load levels as anticipated. The symbol (*) denotes 100% load.

Zero Dynamics

The linearized plant transmission zeros were also calculated as a function of load level using various combinations of inputs and outputs. These results are summarized in Figures 79.23 and 79.24. These figures, which characterize the relation between feedwater flow and drum level, and heat rate and drum level, re-

TABLE 79.1 Summary of Dynamical Equations

$$u_1 = q, u_2 = w_e, u_3 = A_1$$
$$\frac{dw_{av}}{dt} = f_1(w_{av}, s_1, s_2, s_3, P_{av}, P_d)$$
$$\frac{ds_1}{dt} = f_2(w_{av}, s_1, P_{av}) + g_{21}(P_{av}, s_1)u_1 + g_{22}(w_{av}, P_d)u_2$$
$$\frac{ds_2}{dt} = f_3(w_{av}, s_1, s_2, P_{av}) + g_{31}(P_{av}, s_2)u_1$$
$$\frac{ds_3}{dt} = f_4(w_{av}, s_2, s_3, P_{av}) + g_{41}(P_{av}, s_3)u_1$$
$$\frac{ds_4}{dt} = f_5(w_{av}, s_3, s_4, P_{av}) + g_{51}(P_{av}, s_4)u_1$$
$$\frac{dP_{av}}{dt} = f_6(w_{av}, s_1, s_2, s_3, s_4, P_{av}, P_d) + g_{61}(w_{av}, s_1, s_2, s_3, s_4, P_{av})u_1$$
$$\frac{dP_d}{dt} = f_7(w_{av}, s_1, s_2, s_3, s_4, P_{av}, P_d, V_w) + g_{71}(w_{av}, s_1, s_2, s_3, s_4, P_{av}, P_d, V_w)u_1 + g_{72}(P_d, V_w)u_2 + g_{73}(P_d, V_w)u_3$$
$$\frac{dV_w}{dt} = f_8(\omega_{av}, s_1, s_2, s_3, s_4, P_{av}, P_d, V_w) + g_{81}(\omega_{av}, s_1, s_2, s_3, s_4, P_{av}, P_d, V_w)u_1 + g_{82}(P_d, V_w)u_2 + g_{83}(P_d, V_w)u_3$$
$$y_1 = P_d, y_2 = y_D = f(V_w), y_3 = w_s = h_3(P_d)$$

TABLE 79.2 Eigenvalues and Eigenvectors at 100% Load

Mode	1	2&3	4&5	6	7	8
Mode description	Drum pressure-circulation flow rebalance	Drum-Riser mass balance oscilation	Riser flow-density oscilation	Circulation flow-drum level rebalance	Energy-coupled drum-level	Drum level mass balance
Eigenvalue	−7.7758	−.7817±.1835i	−.3854±.3145i	−0.3023	−0.0274	0.0000
ω_{av}	0.5288	.6866±.5787i	−.9500±.1684i	0.9692	0.4173	0.0000
s_1	−0.0066	−.0052±.0063i	.0034±.0046i	0.0045	−0.0193	0.0000
s_2	0.0005	.0025±.0148i	.0097±.0058i	0.0058	−0.0212	0.0000
s_3	0.0001	.0186±.0222i	.0029±.0207i	0.0071	−0.0230	0.0000
s_4	0.0001	−.0933±.0099i	−.0178±.0280i	0.0087	−0.0247	0.0000
P_{av}	−0.0323	.0451±.0178i	−.0107±.0270i	−0.0267	−0.1285	0.0000
P_d	0.7644	.0561±.0283i	−.0321±.0170i	0.0187	−0.1117	0.0000
V_d	0.3675	−.3925±.1526i	.0879±.2397i	0.2436	0.8916	1.0000

spectively, show the nonminimum phase behavior typical of such systems, with zeros in the right half-plane.

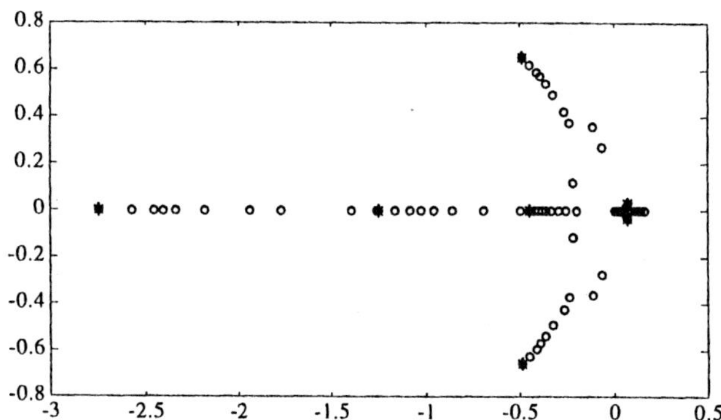

Figure 79.23 The zeros of the transfer function $\omega_e \rightarrow$ lev show the expected nonminimum phase characteristic of drum level dynamics.

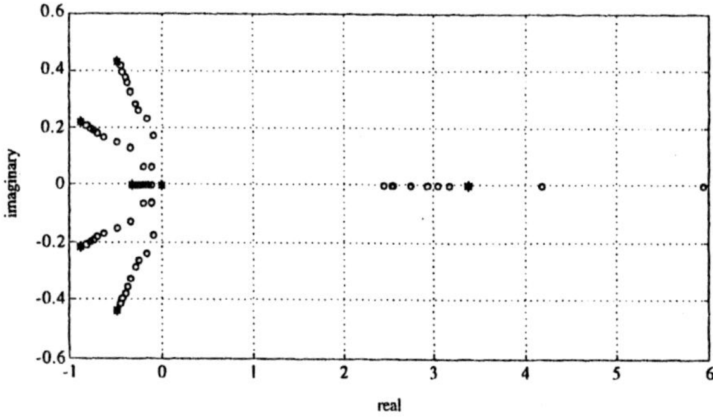

Figure 79.24 Zeros of the transfer function $Q \rightarrow$ lev. The nonminimum phase characteristic which produces the shrink/swell phenomenon is clearly evident.

Drum Level Control at High and Low Loads

The control configurations of interest in this section are shown in Figure 79.18, which illustrates typical one-, two-, and three-element feedwater regulators with proportional-integral-derivative (PID) compensation applied to the water level error signal and "feedforward" of the steam flow. This is the control structure most commonly used for feedwater regulation in drum type power plants. The unfortunate terminology "feedforward" arises from the view that steam flow is a disturbance vis-a-vis the drum level control loop, in which case such terminology would be appropriate. From the point of view here, however, because throttle valve position is the disturbance and steam flow is an output, we will refer to steam flow "feedback". This seemingly innocuous distinction is critical to understanding low load performance. If steam flow feedback is omitted, the result is a single-element controller.

With the configuration specified, the control system design problem reduces to selecting PID parameters for the compensator. It would be desirable to identify one set of parameters to provide acceptable performance at all load levels but it is well-known that this is not possible. Open-loop analysis shows that the dynamics vary dramatically over the load range and, in particular, the position of the open-loop zeros, which limit achievable performance at all loads, is particularly poorly located at low loads. Thus, it is common practice to use two different sets of parameter values, one for high loads and the other for low loads. Even so, it will be seen that the achievable low load performance is not very good and that feedback of steam flow degrades performance at the very low end.

Compensator Design

Performance Limitations There is a basic performance trade-off associated with the PID compensator for feedwater regulation. The regulator introduces a (second) pole at the origin and allows for placement of two zeros. The three design parameters may be viewed as these two zero locations and the loop gain. There are four dominant modes: the three modes 6, 7, and 8 described in Table 79.2 and the mode introduced by the compensator and corresponding to the new pole at the origin. The latter will be referred as the "drum level trim" mode. This terminology is consistent with the intent of compensator integral action, to eliminate steady-state errors from the drum level. One classical design approach is to fix the compensator zero locations and to examine the root locus with respect to loop gain. Figure 79.25 through Figure 79.27 illustrate a root locus plot at 100% load. One of the compensator zeros is placed on the real axis close to the origin because the undesirable, destabilizing phase lag introduced by the integral action is compensated for by the neighboring zero. Of course this traps a pole near the origin. This pole corresponds to the trim mode and the result is a very slow trimming of the steady-state drum level error.

The placement of the second zero is much more critical. Here it is placed on the real axis between the poles corresponding to modes 7 and 8. The logic is to pull mode 8, the critical drum mass balance mode, from the origin to the left, the farther the better, which translates into high gain. As seen in the figures, this draws modes 6 and 7 together, coupled into an oscillatory mode which approaches instability as the loop gain is increased. These poles are attracted by the nonminimum phase zeros. This represents one basic trade-off. A gain must be chosen which provides acceptable speed of response for mode 8, while retaining reasonable damping for modes 6 and 7.

Performance Under Load Variation It is important to examine the performance of this controller at lower load levels. In terms of pole location, performance degrades as load is reduced:

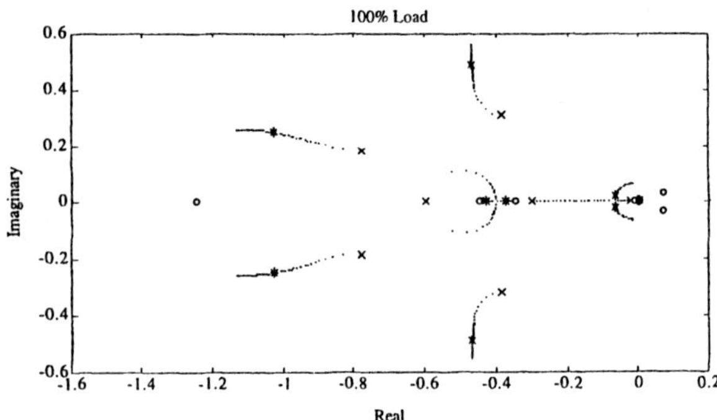

Figure 79.25 This figure provides an overview of the root locus at 100% load. The designations are: open loop poles (x), open loop zeros (o), design poles (*). All but one (the leftmost) pole is shown. There are also two additional zeros, both real and to the left. Notice the two right half-plane zeros. Two of the four dominant poles (near the origin) are not clearly visible on this scale.

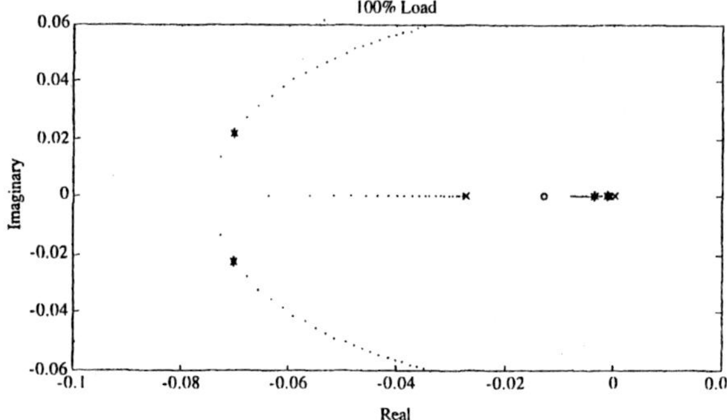

Figure 79.26 A closer view of the dominant poles is provided by this figure. Notice that the selected value of the gain provides closed-loop poles for the disturbance-coupled drum level mode, which are slightly underdamped. Further increases in gain reduce the damping, and eventually these poles move into the right half-plane attracted by the right half-plane zeros. Although some additional gain would be acceptable (in fact, desirable) at this load level, it has adverse affects at lower loads, as will be seen.

- At 60% load the disturbance-coupled drum dynamics have significantly degraded. Higher gain is called for, but once again, lower load dynamics would suffer.
- At 50% load the disturbance-coupled dynamics have further degraded, although the drum level mode response time is somewhat improved. This is a critical load which marks a transition in the qualitative closed-loop behavior. As load is reduced, the open-loop poles associated with modes 6 and 7 move to the right along the real axis. At 50% load the mode 7 pole is precisely located at one of the compensator zeros. Because of this, mode 7 is effectively unregulated and remains fixed as gain is varied. In modern terminology, this is an input decoupling zero. It is possible, of course, to place the compensator zero more to the right in order to lower the load level (50%) at which the transition takes place. This strat-

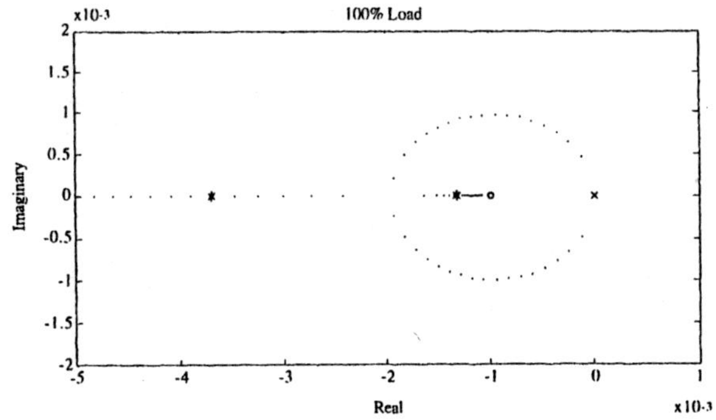

Figure 79.27 This figure provides an even closer look at the poles associated with the drum level mode and the level trim mode. These are by far the slowest modes. At the design condition, the drum level mode has a time constant of about 200 sec. This response time could be improved by increasing the gain, but at the expense of lower load stability. The trim mode is considerably slower.

egy reduces the achievable response time of mode 8. Consequently, this is a second basic trade-off which must be addressed.

- As load is further reduced to 40%, mode 7 couples with mode 8, rather than mode 6, to form a slow, but acceptably damped, oscillatory mode. The damping, however, diminishes as load is further reduced because the (open-loop) mode 7 pole migrates to the right and also because one of the right half-plane zeros approaches the origin from the right. There is a distinct change in root locus behavior in the transition from 60%–50%–40% load levels. Figure 79.28 illustrates the dominant poles at 40% load.
- At 30% load there is a fairly dramatic degradation of performance. Although the system remains stable, the dominant dynamic is a slow oscillation (a period of over 600 sec.) which is very lightly damped. A reduction in gain would improve performance, although it would be marginal at best.
- At 20% load the closed-loop system is unstable. A reduction in gain is required to stabilize it. Performance can be optimized by adjusting all three controller parameters, but even the best achievable performance is not very good. Moreover, even stable performance is attainable only with substantial degradation of high load performance.

Low Load Regulator Eventually, it is necessary to face the dilemma that stable low load operation is not achievable with the conventional control configuration tuned to provide reasonably good performance at higher loads. Thus, a new set of PID parameters is needed for low load operation. This is accomplished by drastically moving both compensator zeros to the right and choosing a new loop gain by examining the root locus. Figure 79.29 shows the root locus for a low load design at 20% load. It is important to note that the critical mode 8 has a closed-loop time constant 4 to 5 times slower than obtainable at high load. Here again a transition for this regulator, of exactly the same type as described earlier at 50% load, takes place between 10% and 5% load.

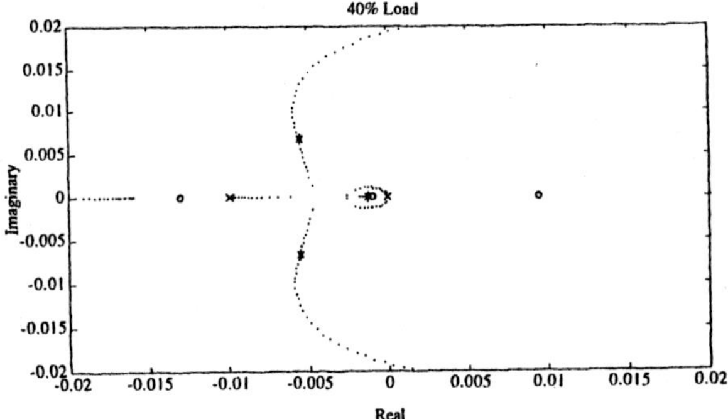

Figure 79.28 This figure provides a closer look at the dominant poles at 40% load. The damping ratio is already somewhat less than ideal so that increased gain would not be desirable.

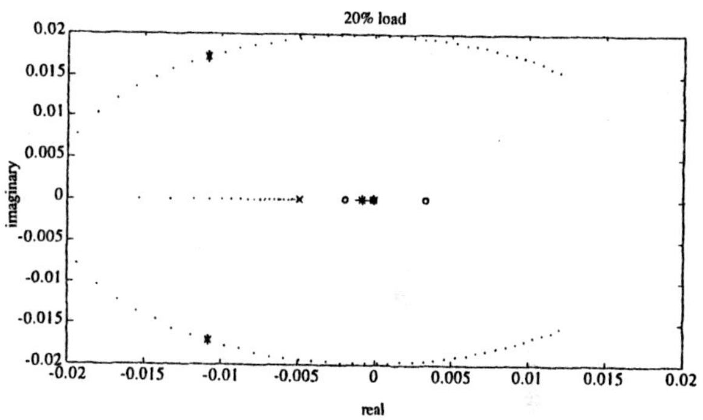

Figure 79.29 The low load regulator root locus at 20% load shows that the system is stable but has much slower response of the dominant modes.

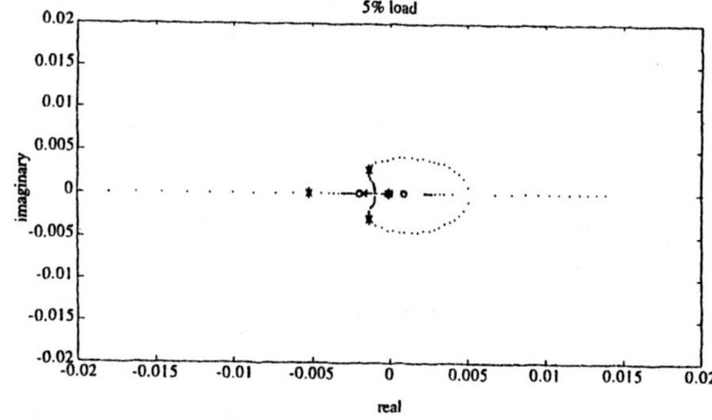

Figure 79.30 At 5% load a transition has taken place and, even though the system is still stable, the damping is substantially reduced.

Effect of Steam Flow Feedback So far the effect of steam flow feedback has not been addressed. It is possible to examine the effect of the steam flow feedback loop on the otherwise open-loop plant. Computation shows that the effect on most of the eigenvalues is negligible. Only mode 7 is significantly influenced. At all load levels the steam flow feedback moves the eigenvalues to the right. Although the largest changes are associated with the high load levels, the consequences are important only at low load levels. It is necessary to redesign the compensator with the steam flow feedback in place.

Figures 79.31 and 79.32 show the closed-loop pole location for the redesigned high load controller with steam flow feedback for load levels of 100% and 40%. Once again 50% load is a critical transition point. At 30% load stability, margins are poor and probably not acceptable. The system is unstable at 20% load. Figures 79.33 and 79.34 show the eigenvalue locations when the low load controller defined above is employed with steam flow feedforward. By comparison with Figures 79.29 and 79.30, steam flow feedback marginally degrades stability at 20% (and also at 10%, not shown) load, but the system is unstable at 5%. The choice not to use steam flow feedback with the low load controller is made to retain stability at very low loads.

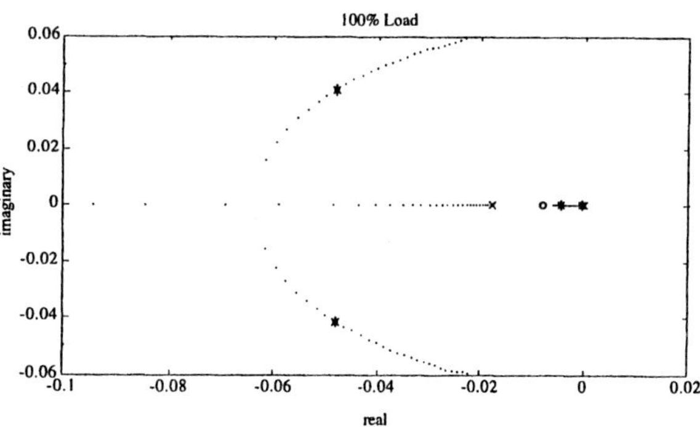

Figure 79.31 Root locus for redesigned regulator using steam flow feedback (three-element control) at 100% load.

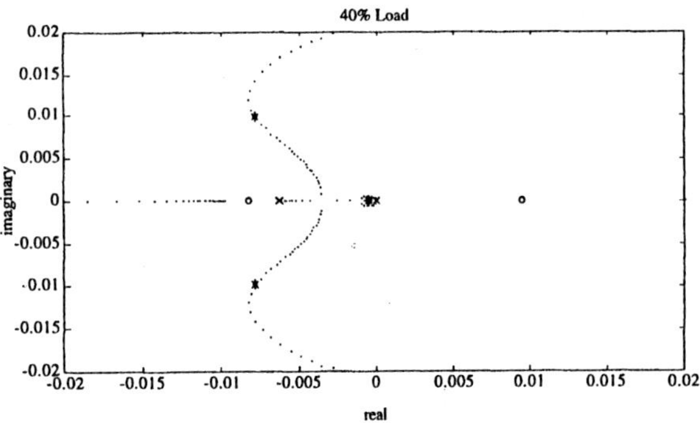

Figure 79.32 Root locus using three-element control at 40% load. Stability is somewhat degraded.

Disturbance Response It is useful to assess the relative importance of each of the closed-loop dynamical modes in terms of their appearance in the drum level response to disturbances,

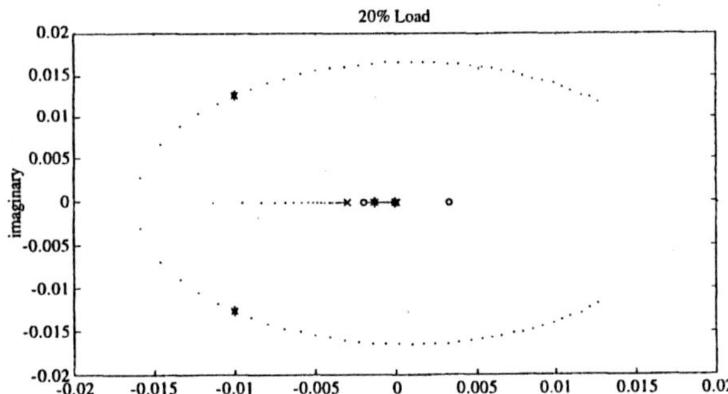

Figure 79.33 A three element-controller retuned for low load real operation provides reasonable stability at 20% load which is only marginally inferior to a single control shown above in Figure 79.29.

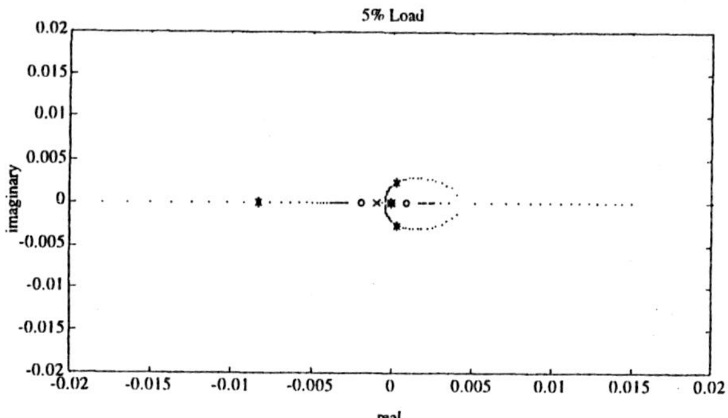

Figure 79.34 At 5% load the three-element controller results in an unstable system. The essential point is that at some load level the three-element controller will lead to an unstable response unless the controller is again retuned. Even retuning for stability, however, is not likely to result in acceptable performance.

either heat rate disturbance or throttle valve disturbance, or to drum level set point change. To do this, we look at the residues associated with each closed-loop pole when the system is subjected to a disturbance input or commanded set point change. Tables 79.3 and 79.4 summarize the closed loop eigenvalues at high and low loads, respectively. Table 79.3 corresponds to the high load controller and Table 79.4 to the low load controller. Table 79.3 contains data corresponding to the three-element high load controller whereas Table 79.4 is generated using the low load single-element controller. Thus, the 30% column in the two tables can be compared for the effect of the change in controllers on the closed-loop eigenvalues.

Notice that there are 11 closed-loop eigenvalues or poles. The open-loop modes 6, 7, and 8 correspond to closed-loop poles 8, 9, and 10. Closed loop pole 11 is the slow drum level trim mode introduced by the controller integral action. In the ramp disturbance case, there is a 12th pole at the origin which contributes a constant offset in the steady-state drum level because the controller is only of type 1. The drum level response is a weighted sum of the time responses corresponding to the simple poles, and the coefficients are the transfer function residues at the respective poles. Consequently, the relative significance of each mode in the drum level response can be identified by comparing the residue magnitudes. These computations (see Figures 79.36 and

TABLE 79.3 High Load Closed-Loop Eigenvalues

100%	85%	70%	60%	50%	40%	30%
−6.2157	−6.2375	−6.2468	−6.2386	−6.2401	−6.2794	−6.4753
−1.0655	−1.0013	−0.9477	−0.9165	−0.8764	−0.8126	−0.7230
−0.2552*i*	−0.2387*i*	−0.2216*i*	−0.2093*i*	−0.1909*i*	−0.1539*i*	−0.0577*i*
−1.0655	−1.0013	−0.9477	−0.9165	−0.8764	−0.8126	−0.7230
+0.2552*i*	+0.2387*i*	+0.2216*i*	+0.2093*i*	+0.1909*i*	+0.1539*i*	+0.0577*i*
−0.4714	−0.4362	−0.4050	−0.3851	−0.3601	−0.3221	−0.2700
−0.5156*i*	−0.4874*i*	−0.4609*i*	−0.4436*i*	−0.4190*i*	−0.3764*i*	−0.3134*i*
−0.4714	−0.4362	−0.4050	−0.3851	−0.3601	−0.3221	−0.2700
+0.5156*i*	+0.4874*i*	+0.4609*i*	+0.4436*i*	+0.4190*i*	+0.3764*i*	+0.3134*i*
−0.4357	−0.3937	+0.3844	−0.3773	−0.3680	−0.3536	−0.3309
	−0.0159*i*	−0.0370*i*	−0.0445*i*	−0.0503*i*	−0.0533*i*	−0.0483*i*
−0.3689	−0.3937	+0.3844	−0.3773	−0.3680	−0.3536	−0.3309
	+0.0159*i*	+0.0370*i*	+0.0445*i*	+0.0503*i*	+0.0533*i*	+0.0483*i*
−0.0482	+0.0441	−0.0372	−0.0304	−0.0225	−0.0236	−0.0265
−0.0413*i*	−0.0290*i*	−0.0207*i*	−0.0182*i*	−0.0140*i*		
−0.0482	−0.0441	−0.0372	−0.0304	−0.0225	−0.0078	−0.0006
+0.0413*i*	+0.0290*i*	+0.0207*i*	+0.0182*i*	+0.0140*i*	−0.0099*i*	−0.0091*i*
−0.0046	−0.0050	−0.0056	−0.0063	−0.0083	−0.0078	−0.0006
					+0.0099*i*	+0.0091*i*
−0.0005	−0.0005	−0.0005	−0.0005	−0.0005		−0.0005

TABLE 79.4 Low Load Closed-Loop Eigenvalues

30%	20%	10%	5%
−6.4923	−7.2444	−8.8627	−12.3819
−0.7237 − 0.0570*i*	−0.7765	−0.7189	−0.6922
−0.7237 + 0.0570*i*	−0.5246	−0.5187	−0.5185
−0.2710 − 0.3133*i*	−0.2269 − 0.2594*i*	−0.1247 − 0.2111*i*	−0.1734 − 0.0660*i*
−0.2710 + 0.3133*i*	−0.2269 + 0.2594*i*	−0.1247 + 0.2111*i*	−0.1734 + 0.0660*i*
−0.3300 − 0.0458*i*	−0.2970 − 0.0372*i*	−0.2612	−0.0625 − 0.1397*i*
−0.3300 + 0.0458*i*	−0.2970 + 0.0372*i*	−0.2058	−0.0625 + 0.1397*i*
−0.0130 − 0.0224*i*	−0.0088 − 0.0182*i*	−0.0045 − 0.0104*i*	−0.0009 − 0.0036*i*
−0.0130 + 0.0224*i*	−0.0088 + 0.0182*i*	−0.0045 + 0.0104*i*	−0.0009 + 0.0036*i*
−0.0008	−0.0009	−0.0012	−0.0039
−0.0001	−0.0001	−0.0001	−0.0001

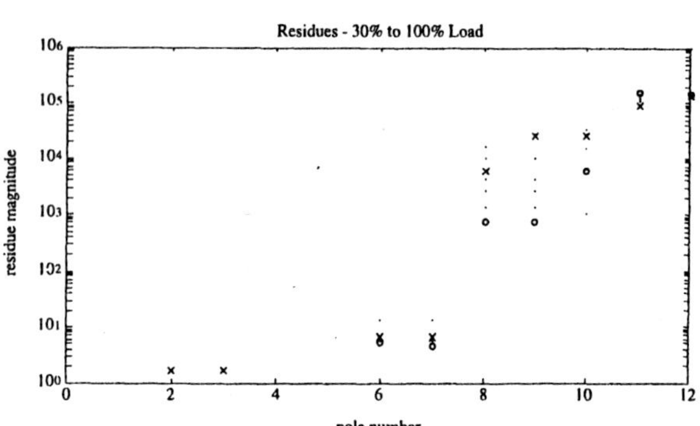

Figure 79.35 Drum level residues for the closed-loop system responding to a ramp firing rate disturbance at load levels from 30% (x) to 100% (o). Recall that the 12th pole is at the origin and is contributed by the ramp disturbance. The nonzero residue merely confirms that the type 1 controller leads to a constant drum level offset when subjected to a ramp input. Otherwise, the dominant modes in the response are the three open-loop modes 6, 7, and 8 and the trim mode. The trim mode clearly dominates at 100% load, but, at lower loads, poles 9 and 10 are quite important also. These poles correspond to open-loop modes 7 and 8 which, through the feedback loop, have been coupled into a lightly damped oscillation.

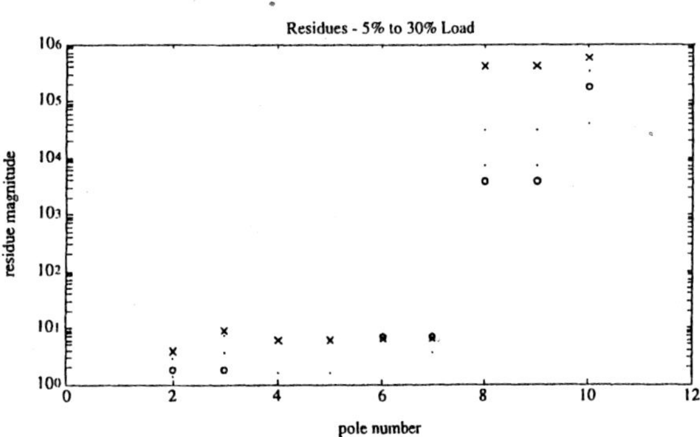

Figure 79.36 The low load controller is subjected to the same ramp firing rate disturbance. The residues are shown for the load range 5% (x) to 30% (o). Although there is a 12th pole at the origin, its residue is very small because the controller zeros are much closer to the origin and, as load is dropped, an open-loop plant zero moves toward the origin from the right. The trim mode residue is also extremely small for the same reason. The remaining dominant modes are the same as the high load case (open-loop modes 6, 7, and 8). Notice that the residue value values are considerably larger, however.

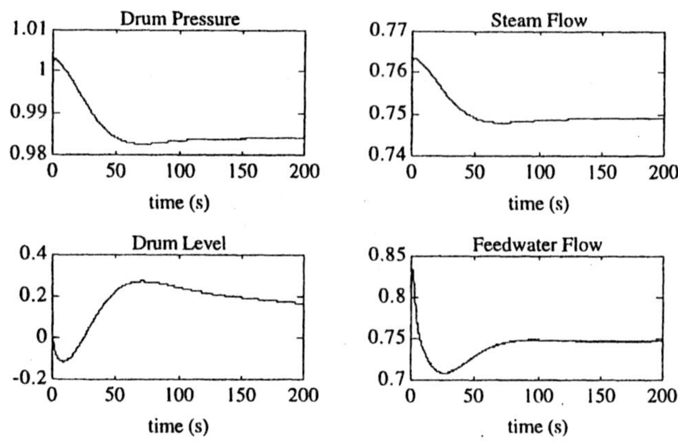

Figure 79.37 Three-element controller response to a 5% step command changing load from 80% to 75%.

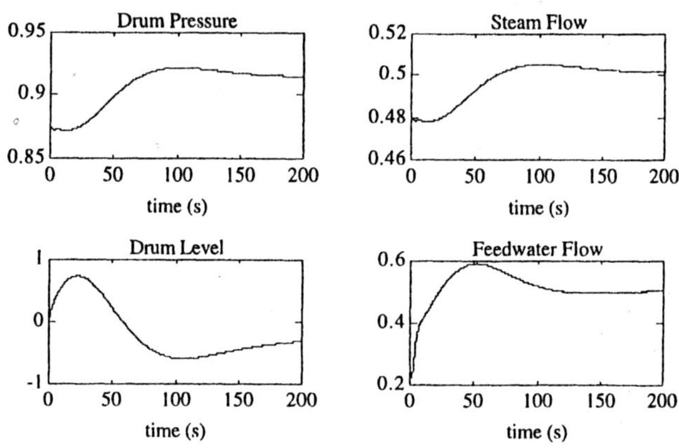

Figure 79.38 Three-element controller response to a 10% step command changing load from 40% to 50%.

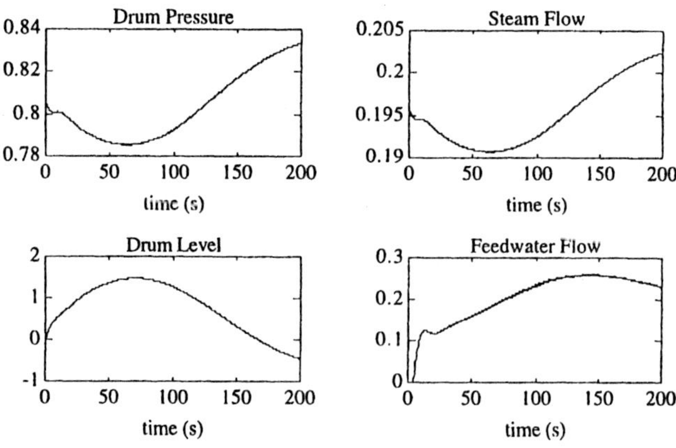

Figure 79.39 Single-element controller response to a 5% step command changing load from 15% to 20%. Below 30% load, a single-element controller is used. For this particular plant, the 15%–20% range is particularly troublesome.

79.37) confirm that the dominant dynamics are the four slowest modes identified above in Table 79.3: circulation flow-drum level rebalance (mode 6), energy coupled drum level (mode 7), drum level mass rebalance (mode 8), and the drum level trim mode. Control system design must focus on the regulation of these modes.

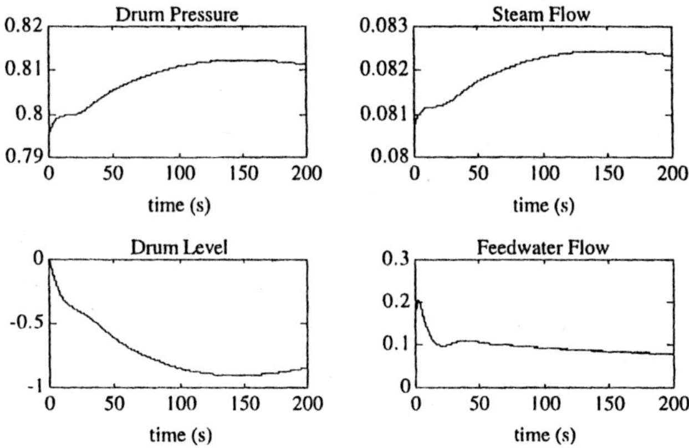

Figure 79.40 Single-element controller response to a 2% step command changing load from 10% to 8%. At this very low load, larger step changes are not possible.

The effect of steam flow feedforward can be seen in the 30% load results in Figures 79.35 and 79.36, which correspond to the three-element and single-element controllers, respectively. Consider pole number 10 which corresponds to the mass balance mode (open-loop mode 8). The 30% residue of this pole is the 'x' in Figure 79.35 and the 'o' in Figure 79.36. There is an order of magnitude improvement with the three-element controller in this important mode.

Simulation Responses

Another perspective is obtained from the closed-loop time responses. The following figures correspond to a feedback regulator structure that employs a coventional three element feedwater controller above 30% load and a single element controller below 30% load. These controllers cannot tolerate load level step changes of more than about 5%–10%, and somewhat smaller step changes at very low loads. The following figures present a sequence of simulation results of step load change commands spanning the range from high load to very low load. The three-element controller used at higher loads responds much better to load disturbances than the single element controller because the residues associated with the dominant closed-loop modes are much smaller. However, the destabilizing effect of steam flow feedback (disturbance feedforward) precludes its use at low loads.

References

[1] *Steam, its Generation and Use*, 39th Ed., *Babcock & Wilcox*, New York.

[2] Bolis, V., Maffezzoni, C., and Ferrarini, L., *Synthesis of the overall boiler-turbine control system by single loop auto-tuning technique*, Proc. 12th World IFAC Congress, 3, 409–414, 1993 (also to appear in *Contr. Eng. Practice*, 1995).

[3] Maffezzoni, C., *Issues in modeling and simulation of power plants*. Proc. IFAC Symposium on Control of Power Plants and Power Systems, 1, 19–27, 1992.

[4] Groppelli, P., Maini, M., Pedrini, G., and Radice, A., *On plant testing of control systems by a real-time simulator*, 2nd Annual IDSA/EPRI Joint Control and Instrumentation Conf., Kansas City, 1992.

[5] Colombo, F., De Marco, A., Ferrari, E., and Magnani, G., *Considerations upon the representation of turbine and boiler in the dynamic response of fossil-fired electrical units*. Proc. Cigrè-IFAC Symposium on Control Applications for Power System Security, Florence (Italy), 1983.

[6] Fuetterer, B., Lausterer, G.K., and Leibbrandt, S.R., *Improved unit dynamic response using condensate stoppage*, Proc. IFAC Symposium on Control of Power Plants and Power Systems, 1, 129–146, 1992.

[7] Kalnitsky, K.C. and Kwatny, H.G., First Principle Model for Steam Turbine Control Analysis. *J. Dyn. Syst., Meas. Contr.*, 103, 61–68, 1981.

[8] Kwatny, H.G. and Bauerle, J., *Simulation Analysis of the Stability of Coal Fired Furnaces at Low Load*, Proc. 2nd IFAC Workshop on Power Plant Dynamics and Control, Philadelphia, 1986.

[9] Kwatny, H.G., McDonald, J.P., and Spare, J.H., *Nonlinear Model for Reheat Boiler Turbine-Generator Systems, Part 1-General Description and Evaluation, Part 2-Development*. 12th Joint Automat. Contr. Conf., 1971.

[10] Kwatny, H.G. and Fink, L.H., Acoustics, Stability and Compensation in Boiling Water Reactor Pressure Control Systems. *IEEE Trans. Automat. Contr.* AC-20, 727–739, 1975.

[11] Kwatny, H.G. and Berg, J., *Drum Level Regulation at All Loads: A Study of System Dynamics and Conventional Control Structures*. Proc. 12th IFAC World Congress, Sydney, 1993.

[12] Schulz, R., The Drum Water Level in The Multivariable Control of a Steam Generator, *IEEE Trans. Ind. Electron. Contr. Instrum. IECI-20*, 164–169, 1973.

[13] Nahavandi, A.N. and Batenburg, A., Steam Generator Water Level Control, *Trans. ASME J. Basic Eng.*, 343–354, 1966.

[14] Chien, K.L., Ergin, E.I., Ling, C., and Lee, A., Dynamic Analysis of a Boiler. *Trans. ASME*, 80, 1809–1819, 1958.

[15] Profos, P., *Die Regulung von Dampenflagen*, (The Control of Thermal Power Plants), Springer, Berlin, 1962, (in German).

[16] Dolezal, R. and Varcop, L., *Process Dynamics*, Elsevier, Amsterdam, 1970.

[17] Klefenz, G., *La regulation dans les central thermiques* (The control of thermal power plants). Edition Eyrolles, Paris, 1974 (in French).

[18] Friedly, J.C., *Dynamic Behavior of Processes,* Prentice Hall, Englewood Cliffs, NJ, 1972.

[19] Kecman, V., *State-Space Models of Lumped and Distributed Systems,* Springer, New York, 1988.

[20] Dukelow, S.G., *The Control of Boilers,* ISA Press, 1986.

[21] Maffezzoni, C., *Dinamica dei generatori di vapore* (The dynamics of steam generators), Masson Ed., 1989 (also printed by Hartmann and Braun, 1988), (in Italian).

[22] Maffezzoni, C., *Controllo dei generatori di vapore* (The control of steam generators), Masson Ed., 1990 (also printed by Hartmann and Braun, 1990), (in Italian).

[23] Quazza, G. and Ferrari, E ., Role of Power Station Control in Overall System Operation. In *Real-Time Control of Electric Power Systems,* Handschin, E., Ed., Elsevier, Amsterdam, 1972

[24] Nicholson, H., *Dynamic Optimization of a Boiler,* Proc. IEEE, 111, 1478–1499, 1964.

[25] Anderson, J.H., *Dynamic Control of a Power Boiler,* Proc. IEEE, 116, 1257–1268, 1969.

[26] McDonald, J.P. and Kwatny, H.G., Design and Analysis of Boiler-Turbine-Generator Controls Using Optimal Linear Regulator Theory. *IEEE Trans. Automat. Contr.* AC-18, 202–209, 1973.

[27] Maffezzoni, C., *Concepts, practice and trends in fossil fired power plant control.* Proc. IFAC Symposium on Control of Power Plants and Power Systems, Pergamon Press, pp. 1-9, 1986.

[28] Calvaer, A., Ed., *Power systems. Modeling and control applications.* Proc. IFAC Symposium, Pergamon, 1988.

[29] Welfonder, E., Lausterer, G.K., and Weber H., Eds., *Control of power plants and power systems,* Proc. IFAC Symposium, Pergamon, 1992.

[30] *10th World Cong. Automat. Contr.*, IFAC Proceedings, Pergamon, 1987.

[31] *11th World Cong. Automat. Contr.*, IFAC Proceedings, Pergamon, 1990.

[32] *12th World Cong. Automat. Contr.*, IFAC Proceedings, Pergamon, 1993.

[33] Uchida, M. and Nakamura, H., *Optimal control of thermal power plants in Kyushu electric power company,* Proc. 1st IFAC Workshop on Modeling and Control of Electric Power Plants, Pergamon, 1984.

[34] Mann, J., *Temperature control using state feedback in a fossil-fired power plant,* Proc. IFAC Symposium on Control of Power Plants and Power Systems, Pergamon, 1992.

Further Reading

Power plant automatic control systems have evolved over many decades as have the generating plants themselves and the environment within which they operate. In the early years the main issues revolved around the control devices themselves, along with actuators and sensors. Indeed the physical design and construction of the early pneumatic, mechanical, and analog electrical control components were quite remarkable and intricate. In contrast, the overall control architectures themselves were very simple and the setting of controller parameter values was always done in the field with carefully controlled tests carried out by engineers possessing keen knowledge of process analysis and feedback concepts combined with extensive operating experience.

In this environment, most of the work done to improve the control systems of electric power plants was performed by the major suppliers of control equipment in close cooperation with the utility companies that operated them. Much of the expertise was recorded in numerous obscure corporate documents that generally did not find their way into the public domain. So, assembling a complete technical and historical record is very difficult and the present work does not attempt that task. By the late 1950s, the maturation of the field of automatic control, both theory and devices, and the demands for improved generating plant performance through automation coalesced. A serious effort at model-based control system analysis and design for power generating plants began and with it has emerged a sizable technical literature. Our goal is to provide a connection to it. While there are undoubtedly antecedents, much of the literature on boiler modeling, related to the special needs of control system design, traces back to the 1958 paper of Chien et al. [14] which deserves special mention.

In the open literature are some books, with different points of view, that deal with boiler dynamics and control providing good balance between theory and practice. Again, for its historical significance, we note the book by Profos [15]. Another important one (1970) is Dolezal and Varcop [16], where attention concentrates on boiler dynamics. The analysis is strongly based on mathematical modeling, with much analytical detail. It is very useful in explaining the origin of basic dynamical behavior and identifying the process parameters on which they depend. The book is a milestone in boiler dynamics, though, perhaps, not so "friendly" to read. A book (1974) by Klefenz [17], which is the French translation of the original German version, has a limited circulation. Most of the concepts illustrated there are still valid and applied in practice. An excellent general text for methods and examples of process dynamics, with many references, is the book by Friedly [18]. Another is by Kecman [19].

The book by Dukelow [20], is an excellent condensation of engineering skill and field experience. Without equations, boiler control concepts and schematics are plainly discussed and explained. It represents a good first reference for anyone approaching the subject without practical experience. Several books (1989-1990), published in Italian by one of the authors, propose a systematic approach to (simplified) boiler dynamics [21] based on first principles as a prerequisite to understanding the fundamentals of boiler control [22], and the most popular control concepts and their underlying motivations.

The books above cited all have a "classical" approach to power plant control, following the approach typical of process control, i.e., structuring the control system in a number of hierarchically arranged loops and trying to minimize interactions among them. Nonclassical control strategies have been studied and proposed in the last quarter century with the application of "modern" control techniques, such as optimal control, adaptive control, robust control, and variable structure control. Much of the motivation for improved dynamic performance can be attributed to the plant's importance to overall power system operation [23]. Early investigations that applied modern control methods to power plants include [24]–[26]. The interested reader is referred to the survey paper [27] summarizing that research effort up to 1986 and to the Proceedings of subsequent dedicated events (primarily [28], [29] and the target area sessions within the world IFAC Congresses [30], [31], and [32]). There are many outstanding contributions among the many interesting papers in the literature on nonclassical control concepts applied to power plants; we want to mention a couple of cases where the new concepts have been introduced in continuously operating in commercial units: the steam temperature control of once-through boilers based on optimal control theory (see [33] and related references) and an observer-based state feedback scheme applied to superheated steam temperature control [34].

Concluding this short survey of contributed work, testing and installing improved control systems in commercial power plants is very expensive, so that it is often impossible for researchers to test their findings; yet, it is very important to validate new concepts on a realistic test-bench (possibly the plant). Complex, large simulators are now available that allow credible testing [4].

79.2 Control of Power Transmission

John J. Paserba, Juan J. Sanchez-Gasca, and Einar V. Larsen, GE Power Systems Engineering, Schenectady, NY

79.2.1 Introduction

The power transmission network serves to deliver electrical energy from power plant generators to consumer loads. While the behavior of the power system is generally predictable by simulation, its nonlinear character and vast size lead to challenging demands on planning and operating engineers. The experience and intuition of these engineers is generally more important to successful operation of the system than elegant control designs.

In most developed countries, reliability of electrical supply is essential to the national economy, therefore, reliability plays a large role in all aspects of designing and operating the transmission grid. Two key aspects of reliability are adequacy and security [1]. Adequacy means the ability of the power system to meet the power transfer needs within component ratings and voltage limits. Security means being able to cope with sudden major system changes (contingencies) without uncontrolled loss of load. In this section, the focus will be mostly on the security aspect, emphasizing power system stability.

There are two main categories of power system performance problems. One category involving steady-state issues, such as thermal and voltage operating limits, is not discussed in this section. The second broad category involves power system dynamics, dealing with transient, oscillatory, and voltage stability. The following three subsections provide an overview of these stability issues.

Transient Stability

Transient stability has traditionally been the performance issue most limiting on power systems, and thus, the phenomenon is well understood by power system engineers. Transient stability refers to the ability of all machines in the system to maintain synchronism following a disturbance such as a transmission system fault, a generator or transmission line trip, or sudden loss of load. Unlike the other stability issues described here, transient stability is a relatively fast phenomenon, with loss of synchronism typically occurring within two seconds of a major disturbance.

Traditional stability analysis software programs are designed to study and identify transient stability problems. These software programs provide the ability to model electrical and mechanical dynamics of generators, excitation systems, turbine/governors, large motors, and other equipment such as static Var compensators (SVC) and high voltage direct current (HVDC) systems. Scenarios studied typically involve major events such as faults applied to a transmission line with line trips. A transient instability is identified by examining the angles (or speeds) of all machines in the system. If the angle (or speed) of any machine does not return toward an acceptable steady-state value within one to three seconds following a power system disturbance, the machine is said to have lost synchronism with the system. Although not always modeled for transient stability analysis, this loss of synchronism will typically result in the generator being tripped off-line by a protective relay.

Another way of looking at the transient stability issue is to consider voltage swings in the transmission network. For example, many utilities require that the voltage swing during a transient event remains above a pre-determined value (e.g., 80% of nominal) because relays may trip generators or loads if the voltage is too low for too long. If the voltage swing does not meet this criterion, the event causing the disturbance, while not unstable in the classical definition of transient stability, is still considered transiently unacceptable.

In many systems, discrete, open-loop action is applied to insure stable operation in the power system following a major disturbance. These so-called "Special Protection Systems" or "Remedial-Action Schemes" can be as simple as undervoltage load shedding on selected loads or as complex as wide area generator tripping and intentional network separation [2]. Such schemes have proven highly useful, and will likely always be needed. Designing such schemes and keeping them updated as systems evolve is an ongoing challenge. To date automation is not applied to developing such schemes; experience and large-scale simulation tools are relied upon.

Voltage Stability

With the evolution of modern power systems, voltage stability has emerged as the limiting consideration in many systems. Voltage stability refers to the power systems' ability to maintain acceptable voltages under normal operating conditions and after a major system disturbance. A system is said to be voltage unstable (or in voltage collapse) when an increase in load demand, a major disturbance, or a change in the system causes a progressive and uncontrollable decrease in voltage. Means of protecting and reinforcing power systems against voltage collapse depend mostly on the supply and control of reactive power in the system. The phenomenon of voltage instability (collapse) is dynamic, yet frequently evolves very slowly, from the perspective of a transient stability. A voltage collapse initiated by loss of infeed into a load center, for example, can progress over a one to five minute period. This type of phenomenon presents two challenges to the power system engineer. The first challenge is in identifying and simulating the voltage collapse. The second challenge is selecting the system reinforcements to correct the problem in the most economic manner. A recently emerging class of computer simulation software provides utility engineers with powerful new tools for analyzing long-term dynamic phenomena. The ability to perform long-term dynamic simulations either with detailed dynamic modeling or simplified quasi-steady-state modeling permits assessment of critical power system problems more accurately than with conventional power flow and stability programs. The voltage stability issue is discussed in great detail in [3].

Oscillatory Stability

Power systems contain electromechanical modes of oscillation due to synchronous machine rotor masses swinging relative to one another. A power system having several such machines will act like a set of masses (rotating inertia of machines) interconnected by a network of springs (ac transmission), and will exhibit multiple modes of oscillation. These "power-swing modes" usually occur in the frequency range of 0.1 to 2 Hz. Particularly troublesome are the so-called inter-area oscillations which usually occur in the frequency range of 0.1 to 1 Hz. The inter-area modes are usually associated with groups of machines swinging relative to other groups across a relatively weak transmission path. The higher frequency modes (1 to 2 Hz) usually involve one or two machines swinging against the rest of the power system or electrically close machines swinging against each other [4].

Because there is great incentive to minimize transmission losses in the power system, power swing modes have very little inherent damping. Damping which is present is usually due to steam or water flow against the turbine blades of generating units and to special conductors placed in the rotor surface of the generators, known as amortisseur (damper) windings. High power flows in the transmission system create conditions where swing modes can experience destabilization. The effect of these oscillations on the power system may be quite disruptive if they become too large or are underdamped. The actual power (watts) swings may themselves not be overly troublesome, but they can result in voltage oscillations in the power system adversely affecting the system's performance. Limitations may be imposed on the power transfer between areas to reduce the possibility of sustained or growing oscillations, or special controls may be added to damp these oscillations.

Once lightly damped, sustained, or growing oscillations have been observed or predicted by time simulation for a specific system condition, the most common and effective way to investigate them in detail is with linear or small-signal, frequency-domain analysis. With the linear approach to studying power systems, the engineer can view the stability problem from a different perspective than time-domain simulations. This approach enhances the overall understanding of the system's performance. The basic elements and tools of linear analysis commonly applied to power systems include eigenvalues (poles, modes, or characteristic roots), root locus, transfer functions, Bode plots, Nyquist plots, eigenvectors (mode shapes), participation factors, Prony processing, and others. While power systems are nonlinear in nature, analysis based on small-signal linearization around a specific operating point can be important in solving damping problems. When combined with practical engineering experience, time-domain simulations, and field measurements, linear analysis can assure reasonable stability margins under many operating conditions.

Considering that the majority of closed-loop supplemental controls in the power system are for damping oscillations, the remainder of this section on "Control of Power Transmission" will focus on the closed-loop controls which affect power system oscillatory stability.

79.2.2 Impact of Generator Excitation Control on the Transmission System

The main control function of the excitation system is to regulate the generator terminal voltage. This is accomplished by adjusting the field voltage in response to terminal voltage variations. A typical excitation system includes a voltage regulator, an exciter, protective circuits, limiters, and measurement transducers. The voltage regulator processes the voltage deviations from a desired set point and adjusts the required input signals to the exciter, which provides the dc voltage and current to the field windings, to take corrective action. Figure 79.41 represents a single generator with its excitation system, supplying power through a transmission line to an "infinite bus" (a power grid having significantly more capacity than the generator under study).

Early excitation systems were manually controlled, i.e., an operator manually adjusted the excitation system current with a rheostat to obtain a desired voltage. Research and development in the 1930s and 1940s showed that applying a continuously acting proportional control in the voltage regulator significantly increased the generator steady-state stability limits. Beginning in the late 1950s and early 1960s most of the new generating units were equipped with continuously acting voltage regulators. As these units became a larger percentage of the generating capacity, it became apparent that the voltage regulator action could have a detrimental impact on the overall stability of the power system. Low frequency oscillations often persisted for long periods of time and in some cases presented limitations to the system's power transfer capability. In the 1960s, power oscillations were observed following the interconnection of the Northwest and Southwest U.S. power grids. Power oscillations were also detected between Saskatchewan, Manitoba, and Ontario systems in the Canadian power system. It was found that reducing the voltage regulator gain of excitation systems improved the system stability.

Great effort has been directed toward understanding the effect of the excitation system on the dynamic performance of power systems. The single generator infinite bus system shown in Figure 79.41 provides a good vehicle to develop concepts related to the stability of a synchronous generator equipped with an excitation system. Figure 79.42 is the block diagram associated with the linear representation of the system shown in Figure 79.41. This diagram has become a standard tool to gain insight into the stability characteristics of the system [5]. The stability of the linearized system can be studied by considering the torques that comprise the accelerating torque T_a: the synchronizing torque T_s, which is in phase with the rotor angle deviations, the damping torque T_d which is in phase with the rotor speed deviations, and the torque T_{ex} due to the excitation system acting through the field winding of the machine. The mechanical torque, T_m, is assumed constant. In Figure 79.42 the phase relations are indicated for conditions of positive synchronizing torque which tends to restore the rotor to steady-state conditions by accelerating or decelerating the rotor inertia. The damping torque is also positive and tends to damp rotor oscillations. T_{ex} is shown as contributing positive synchronizing torque and negative damping torque. The primary synchronizing effect is exhibited in the term K_1 of Figure 79.42, which represents the change in torque mostly in phase with angle. The damping parameter D represents turbine-generator friction, windage, and the impact of steam or water flow through the turbine. Typically, the torque-angle loop is stable due to the inherent damping and restoring forces in the power system. However, unstable oscillations can result from the introduction of negative damping torques added through the excitation system loop [5], [6].

Phase lags are introduced by the voltage regulator and the generator field dynamics so that the resulting torque is out of phase with both rotor angle and speed deviations. The phase characteristics of this path depend on the generator and voltage regulator characteristics, and the parameters K_4, K_5, and K_6. K_5 plays a dominant role in the phase relationships with respect to the damping torque. For weak systems and heavy loads, K_5 is typically negative. Coupled with the phase lags due to the voltage regulator and generator field winding, the path via K_5 can produce a torque T_{ex} which contributes positive synchronizing torque and a negative damping torque. This negative damping component can cancel the small inherent positive damping due to the damping torque and lead to an unstable system. A detailed description of these parameters is given in [5].

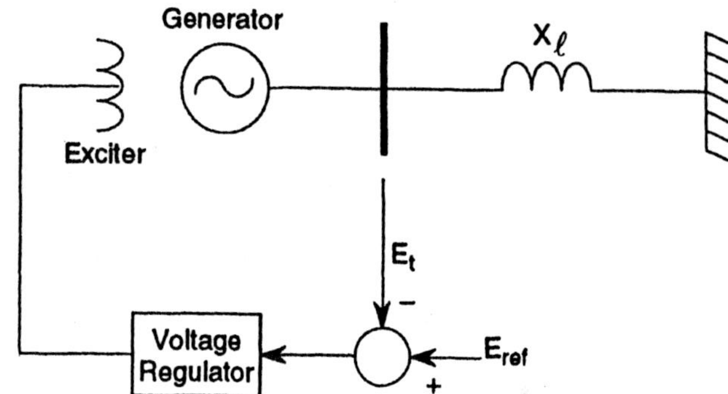

Figure 79.41 Single generator and excitation system.

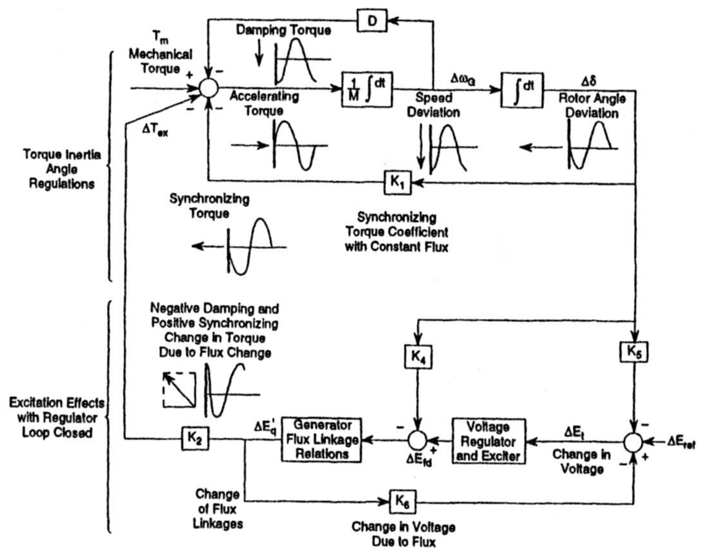

Figure 79.42 Phase relationships for simplified model of single machine to infinite bus.

79.2.3 Power System Stabilizer (PSS) Modulation on Generator Excitation Systems

Because a major source of negative damping has been introduced to the system by the application of high-response excitation systems, an effective way to increase damping is to modify the action of these excitation systems. A Power System Stabilizer (PSS) is often used to provide a supplementary signal to the excitation system. The basic function of the PSS is to extend stability limits by modulating generator excitation to provide positive damping torque to power swing modes.

To provide damping, a PSS must produce a component of electrical torque on the rotor in phase with speed deviations. The implementation details differ, depending upon the stabilizer input signal employed. PSS input signals which have been used include generator speed, frequency, and power [6]. However, for any input signal, the transfer function of the PSS must compensate for the gain and phase characteristics of the excitation system, the generator, and the power system. These collectively determine the transfer function from the stabilizer output to the component of electrical torque which can be modulated via excitation control.

Figure 79.43 is an extension of Figure 79.42 to include a speed-input PSS (PSS(s)). The transfer function $G(s)$ represents the characteristics of the generator, the excitation system, and the power system. A PSS utilizing shaft speed as an input must compensate for the lags in $G(s)$ to produce a component of torque in phase with speed changes so as to increase the damping of the rotor oscillations. An ideal PSS characteristic would therefore be inversely proportional to $G(s)$. Such a stabilizer would be impractical because perfect compensation for the lags of $G(s)$ requires pure differentiation with its associated high gain at high frequencies. A practical speed PSS must utilize lead/lag stages to compensate for phase lags in $G(s)$ over the frequency range of interest. The gain must be attenuated at high frequencies to limit the impact of noise and minimize interaction with torsional modes of turbine-generator shaft vibration. Low-pass and possibly band reject filters are required. A washout stage is included to prevent steady-state voltage offsets as system frequency changes. A typical transfer function of a practical PSS which meets the above criteria is given by

$$PSS(s) = K_s \frac{T_W s(1+sT_1)(1+sT_3)}{(1+T_W s)(1+sT_2)(1+sT_4)} F(s)$$

where $F(s)$ represents a filter designed to eliminate torsional frequencies.

A PSS must be tuned to provide the desired system performance under the power system condition which requires stabilization, typically weak systems with heavy power transfer, while at the same time being robust in that undesirable interactions are avoided for all system conditions. To develop such a design, it is important to understand the path $G(s)$ through which the PSS operates:

1. The phase characteristics of $G(s)$ are nearly identical to the phase characteristics of the closed-loop voltage regulator.
2. The gain of $G(s)$ increases with generator load.
3. The gain of $G(s)$ increases as the ac system strength increases. This effect is amplified with high-gain voltage regulators.
4. The phase lag of $G(s)$ increases as the ac system becomes stronger. This has the greatest influence with high-gain exciters, because the voltage regulator crossover frequency approaches that of the frequency of the swing modes.

$G(s)$ has the highest gain and greatest phase lag under conditions of full load on the generator and the strongest transmission system (i.e., higher short circuit strength). These conditions therefore represent the limiting case for achievable gain with a speed-input PSS and constitute the base condition for stabilizer design. Because the gain of the plant decreases as the system becomes weaker, the damping contribution for the strong system should be maximized to insure best performance with a weakened system.

In general, the highest compensation center frequency which provides adequate local mode damping will yield the greatest contribution to intertie modes of oscillation. The following are guidelines for setting the lead/lag stages to achieve adequate local mode damping with maximum contribution to intertie modes of oscillation. Two basic criteria in terms of phase compensation are

1. It is most important to maximize the bandwidth within which the phase lag remains less than 90°. This is true even though less than perfect phase compensation results at the local model frequency.
2. The phase lag at the local mode frequency should be less than about 45°. This can be improved somewhat by decreasing the washout time constant, but too low a washout time constant will add phase lead and an associated desynchronizing effect to the intertie oscillations. In general, it is best to keep the washout time constant greater than one second.

The gain and frequency at which an instability occurs also provide an indication of appropriate lead/lag settings. The relationship of these parameters to performance is useful in root locus analysis and in field testing. The following observations hold:

3. The frequency at which an instability occurs is highest for the best lead/lag settings. This is related to maximizing the bandwidth within which the phase lag remains less than 90°.
4. The optimum gain for a particular lead/lag setting is consistently about one-third of the instability gain.

79.2.4 Practical Issues for Supplemental Damping Controls Applied to Power Transmission Equipment

In many power systems, equipment is installed in the transmission network to improve various performance issues such as transient, oscillatory, or voltage stability. Often this equipment is based on power electronics which generally means that the device can be rapidly and continuously controlled. Examples of such equipment include high voltage dc systems (HVDC), static Var compensators (SVC), and thyristor-controlled series compensation (TCSC). To improve damping in a power system, a supplemental damping controller can be applied to the primary regulator of a device. The supplemental control action should

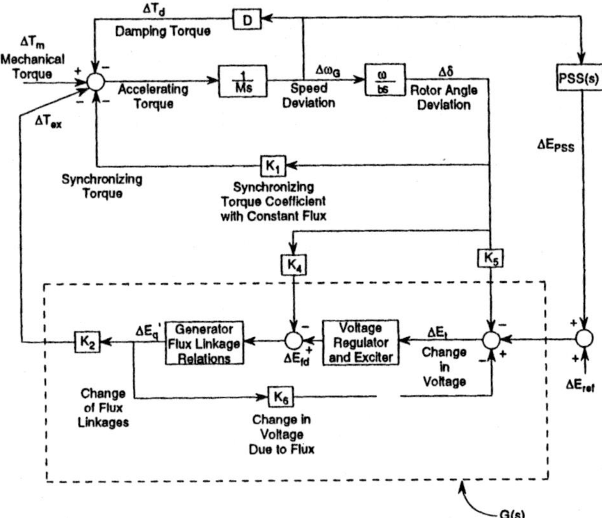

Figure 79.43 Simplified model of single machine to infinite bus.

modulate the output of a device, in turn affecting power transfer and adding damping to the power system swing modes. This section provides an overview of some practical issues affecting the ability of damping controls to improve power system dynamic performance [7]. Section 79.2.5 on "Examples" illustrates the application of these concepts [8].

Siting

Siting plays an important role in the ability of a device to stabilize a swing mode. Many controllable power system devices are sited based on issues unrelated to stabilizing the network (e.g., HVDC transmission and generators). In other situations (e.g., SVC or TCSC), the equipment is installed primarily to help support the transmission system, and siting will be heavily influenced by its stabilizing potential. Device cost represents an important driving force in selecting a location. In general, there will be one location which makes optimum use of the controllability of a device. If the device is located at a different location, a larger sized device would be needed to achieve a desired stabilization objective. In some cases, overall costs may be minimized with non-optimum locations of individual devices, because other considerations must also be taken into account, such as land price and availability, environmental regulations, etc.

The inherent ability of a device to achieve a desired stabilization objective in a robust manner, while minimizing the risk of adverse interactions, is another consideration which can influence the siting decision. Most often, these other issues can be overcome by appropriate selection of input signals, signal filtering, and control design discussed later in this section. This is not always possible, however, so these issues should be included in the decision-making process for choosing a site. For many applications, it will be desirable to apply the devices in a distributed manner. This approach helps to maintain a more uniform voltage profile across the network, during both steady-state operation and after transient events. Greater security may also be possible with distributed devices because the overall system can more likely tolerate the loss of one of the devices.

Objectives for Control Design and Operation

Several aspects of control design and operation must be satisfied during both transient and steady-state operation of the power system before and after a major disturbance. These aspects suggest that controls applied to the power system should meet these requirements:

1. Survive the first few swings after a major system disturbance with some degree of safety. The safety factor is usually built into a Reliability Council's criteria (e.g., keeping voltages above some threshold during the swings).
2. Provide some minimum level of damping in the steady-state condition after a major disturbance (post-contingent operation) because, in addition to providing security for contingencies, some applications will require "ambient" damping to prevent spontaneous growth of oscillations in steady-state operation.
3. Minimize the potential for adverse side effects, which can be classified as follows:
   (a) Interactions with high-frequency phenomena on the power system, such as turbine-generator torsional vibrations and resonances in the ac transmission network.
   (b) Local instabilities within the bandwidth of the desired control action.
4. Be robust. This means that the control will meet its objectives for a wide range of operating conditions encountered in power system applications. The control should have minimum sensitivity to system operating conditions and component parameters because power systems operate over a wide range of operating conditions and there is often uncertainty in the simulation models used for evaluating performance. The control should also have minimum communication requirements.
5. Be highly dependable. This means that the control has a high probability of operating as expected when needed to help the power system. This suggests that the control should be testable in the field to ascertain that the device will act as expected in a contingency. The control response should be predictable. The security of system operations depends on knowing, with a reasonable certainty, what the various control elements will do in a contingency.

Closed-Loop Control Design

Closed-loop control is utilized in many power system components. Voltage regulators are commonplace in generator excitation systems, capacitor and reactor banks, tap-changing transformers, and SVCs. Modulation controls to enhance power system stability have been applied extensively to generator exciters and to HVDC and SVC systems. A notable advantage of closed-loop control is that stabilization objectives can often be met with less equipment and impact on the steady-state power flows than with open-loop controls.

Typically, a closed-loop controller is always active. One benefit of such a continuing response to low-level motion in the system is that it is easy to test continuously for proper operation. Another benefit is that, once the system is designed for the worst-case contingency, the chance of a less severe contingency causing a system breakup is lower than if only open-loop controls are applied. Disadvantages of closed-loop control involve mainly the potential for adverse interactions. Another possible drawback is the need for small step sizes, or vernier control in the equipment, which will have some impact on cost. If communication is needed, this could also be a problem. However, experience suggests that adequate performance should be attainable using only locally measurable signals.

One of the most critical steps in control design is to select an appropriate input signal. The other issues are determining the input filtering and control algorithm to assure attainment of the stabilization objectives in a robust manner with minimal risk of adverse side effects. The following paragraphs discuss design approaches for closed-loop stability controls, so that the potential benefits of such control can be realized in the power system.

Input-Signal Selection The choice of local signals as inputs to a stabilizing control function is based on several considerations:

1. The input signal must be sensitive to the swings on the machines and lines. In other words, the swing modes must be "observable" in the input signal selected. This is mandatory for the controller to provide a stabilizing influence.
2. The input signal should have as little sensitivity as possible to other swing modes in the power system. For example, in a transmission line device, the control action will benefit only those modes which involve power swings on that line. Should the input signal respond to local swings within an area at one end of the line, then valuable control range would be wasted in responding to an oscillation over which the damping device has little or no control.
3. The input signal should have little or no sensitivity to its own output in the absence of power swings. Similarly, there should be as little sensitivity as possible to the action of other stabilizing controller outputs. This decoupling minimizes the potential for local instabilities within the controller bandwidth.

These considerations have been applied to a number of modulation control designs, which have proven themselves in many actual applications. The application of PSS controls on generator excitation systems was the first such study. The study concluded that speed or power were best and that frequency of the generator substation voltage was an acceptable choice as well [6]. When applied to SVCs, the magnitude of line current flowing past the SVC was the best choice [9]. For torsional damping controllers on HVDC systems, it was found that the frequency of a synthesized voltage close to the internal voltage of the nearby generator (calculated with locally measured voltages and currents) was best [10]. In the case of a series device in a transmission line, these considerations lead to the conclusion that frequency of a synthesized remote voltage to "find" the center of an area involved in a swing mode is a good choice. Synthesizing voltages at either end of the line allows determining the frequency difference across the line at the device location, which can then be used to make the line behave like a damper across the inherent "spring" nature of the ac line. Synthesizing input signals is discussed further in the section on "Examples".

Input-Signal Filtering To prevent interactions with phenomena outside the desired control bandwidth, low-pass and high-pass filtering is used for the input signal. In certain applications, notch filtering is needed to prevent interactions with specific lightly damped resonances. This has been the case with SVCs interacting with ac network resonances and modulation controls interacting with generator torsional vibrations. On the low-frequency end, the high-pass filter must have enough attenuation to prevent excessive response during slow ramps of power or during the long-term settling following a loss of generation or load. This filtering must be considered while designing the overall control, as it will strongly affect performance and the potential for local instabilities within the control bandwidth. However, finalizing such filtering usually must wait until the design for performance is completed, after which the attenuation needed at specific frequencies can be determined. During the control design work, a reasonable approximation of these filters needs to be included. Experience suggests a high-pass break near 0.05 Hz (three-second washout time constant), and a double low-pass break near 4 Hz (40-msec. time constant) as shown in Figure 79.44, is suitable as a starting point. A control design which adequately stabilizes the power system with these settings for the input filtering will probably be adequate after the input filtering parameters are finalized.

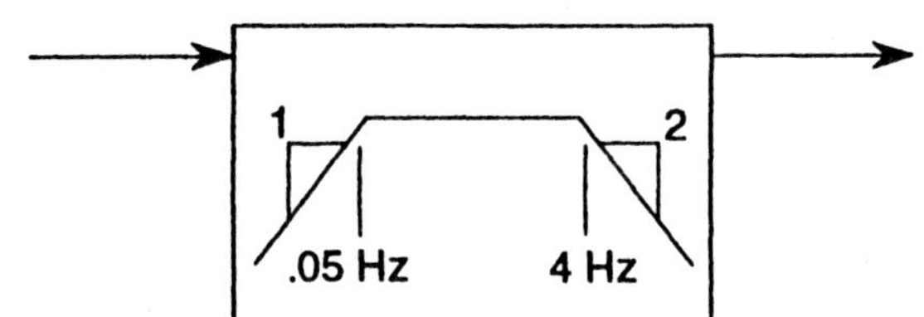

Figure 79.44 Initial input-signal filtering.

Control Algorithm Typically, the control algorithm for damping controllers leads to a transfer function which relates an input signal(s) and a device output. When the input is a speed or frequency type and the output affects the real power, the control algorithm should approach a proportional gain to provide a pure damping influence. This is the starting point for understanding how deviations in the control algorithm affect system performance.

In general, the transfer function of the control and input-signal filtering is most readily discussed in terms of its gain and phase relationship versus frequency. A phase shift of 0° in the transfer function means that the output is simply proportional to the input, and, for discussion, is assumed to represent a pure damping effect on a lightly damped power swing mode. Phase lag in the transfer function (up to 90°), translates to a positive synchronizing effect, tending to increase the frequency of the swing mode when the control loop is closed. The damping effect will decrease with the sine of the phase lag. Beyond 90°, the damping effect will become negative. Conversely, phase lead is a desynchronizing influence and will decrease the frequency of the swing mode when the control loop is closed. Generally, the desynchronizing effect should be avoided, so the preferred transfer function is one which lags between 0° and 45° in the frequency range of the swing modes the control is designed to damp.

After the shape of the transfer function meeting the desired control phase characteristics is designed, the gain of the transfer function is selected to obtain the desired level of damping. The gain should be high enough to insure full utilization of the device for the critical disturbances, but no greater, so that risks of adverse effects are minimized. Typically, the gain selection is done with root locus or Nyquist analysis. This handbook presents many other control design methods that can be utilized to design supplemental controls for the power transmission system.

Performance Evaluation Good simulation tools are essential for applying damping controls to power transmission equipment for system stabilization. The controls must be designed and tested for robustness with these simulation tools. For many operating conditions, the only feasible means of testing the system is by simulation, so that confidence in the power system model is crucial. A typical large-scale power system model may contain up to 15,000 state variables or more. For design purposes, a reduced-order model of the power system is often desirable. If the size of the study system is excessive, the large number of system variations and parametric studies required becomes tedious and prohibitively resource limited for many linear analysis techniques in general use. Good understanding of the system performance can be obtained with a model of only the relevant dynamics for the problem under study. The key conditions which establish that controller performance is adequate and robust can be identified from the reduced-order model, and then tested with the full-scale model. Recent improvements have been made in deriving meaningful reduced-order equivalent power system models [11].

Field testing is also essential for applying supplemental controls to power systems. Testing needs to be performed with the controller open loop, comparing the measured response at its own input and the inputs of other planned controllers against the simulation models. Once these comparisons are acceptable, the system can be tested with the control loop closed. Again, the test results should correlate reasonably with the simulation program. Methods have been developed for performing testing of the overall power system to provide benchmarks for validating the full-system model. Testing can also be done on the simulation program to obtain the reduced-order models for the advanced control design methods [12]. Methods have also been developed to improve the modeling of individual components.

Adverse Side Effects Historically in the power industry, each major advance in improving system performance has had adverse side effects. For example, adding high-speed excitation systems more than 40 years ago caused the destabilization known as the "hunting" mode of the generators. The solution was power system stabilizers, but it took more than 10 years to learn to tune them properly, and there were some unpleasant surprises involving interactions with torsional vibrations on the turbine-generator shaft [6].

HVDC systems also interacted adversely with torsional vibrations (the so-called subsynchronous torsional interaction (SSTI) problem), especially when augmented with supplemental modulation controls to damp power swings. Similar SSTI phenomena exist with SVCs, although to a lesser degree than with HVDC. Detailed study methods have since been established which permit designing systems with confidence that these effects will not disturb normal operation. Protective relaying exists to cover unexpected contingencies [10], [13].

Another potentially adverse side effect is with SVC systems which can interact unfavorably with network resonances. This has caused problems in the initial application of SVCs to transmission systems. Design methods now exist to deal with this phenomenon, and protective functions exist within SVC controls to prevent continuing exacerbation of an unstable condition [9].

As technologies continue to evolve, such as the current industry focus on Flexible AC Transmission Systems (FACTS), new opportunities arise for improving power system performance. FACTS devices introduce capabilities that may be an order of magnitude greater than existing equipment for stability improvement. Therefore there may be much more serious consequences if new devices fail to operate properly. Robust, non-interacting controls for FACTS devices are critically important for stability of the power system [14].

79.2.5 Examples

Large power-electronic devices have been applied to power transmission grids for many years. One of the most common and well-established uses of large-scale power electronics is high voltage dc transmission systems (HVDC). As noted above, current industry activity is focusing on new types of equipment which can affect ac transmission systems, using the acronym "FACTS". This acronym arises from the concept of "Flexible AC Transmission Systems," where power-electronic devices can improve flexibility beyond what is possible with conventional switched compensating equipment. Because these devices are thyristor-controlled, their operating point can be adjusted within a few milliseconds, and there is no mechanical wear associated with rapid cycling of the operating point. These characteristics are the primary features which made power-electronic devices attractive for stabilization functions [8].

In this section, three types of power-electronic systems are described and two examples are presented, emphasizing their ability to add damping to power system modes of oscillation.

High-Voltage DC (HVDC)

High-voltage dc (HVDC) transmission is used to interconnect asynchronous ac grids for power transfer over very long distances. The system includes line-commutated converters at both the sending and receiving ac terminals with a dc link between. In long-distance transmission schemes, this dc link is a transmission line or cable. In back-to-back schemes, both converters are in the same building and the dc link is simply a short length of busbar.

HVDC provides a means to control real power transfer directly between ac networks, independent of ac phase angle. This feature is sometimes utilized to provide a modulating function to the controls for damping power-swing oscillations. Because the converters are line commutated, they draw a reactive power load from the ac networks. This reactive power is related to the real power and both ac and dc voltages, which tends to be a detriment to adding modulation controls. Considerable study has been directed at supplemental controls for HVDC systems. Some schemes have benefited from modulation [15].

Static Var Compensator (SVC)

The SVC provides rapid control of an effective shunt susceptance on the transmission grid and is typically used to regulate voltage at a bus. This action is similar to conventional switched reactors or shunt capacitors which have been used for many years, except that with the SVC, the control can be accomplished quickly and continuously. The voltage-regulating function of an SVC is sometimes augmented by a supplemental control to modulate the voltage set point to add damping to power-swing modes.

Generally, the best location for shunt compensation is the point where the voltage swings are greatest for the dominant swing mode. In a simple remote generation system, this is at the electrical midpoint between the internal voltage of the generator and that of the receiving system. In an interconnected system, the siting decision is complicated by the fact that there are multiple modes of oscillation. Knowing the general characteristics of the dominant swing modes, (based on which generators swing relative to the others), siting decisions may be made using the same basic concept as for the simple remote generation system just described. In addition, properly selecting input signals can allow the damping benefits for a specific mode, while not risking instability of others. These important issues are described in detail in [9].

When an SVC is located near a load area, modulation may be ineffective or even detrimental to power-swing damping. The difficulty lies in the fact that the change in power of the generators in the load area is affected in two ways by a variation in voltage. One is due to the change in power flow across the transmission lines from the sending system, and the other is due to the change in local load. These effects counteract each other, and the leverage for controlling the swing mode is very small. Worse than being small, however, is that the sign of the effect can change with operating point, so that a control designed for a beneficial effect in some operating conditions may have a detrimental effect in others. This inherent characteristic implies that a robust damping control is difficult to achieve on an SVC sited in a load area for swing modes which involve that area. Note that there is usually value in simply stiffening the voltage in a load area, however, so applying an SVC with only voltage control may help stability.

As an example of the effect of SVC on damping power system oscillations, consider the three-area system in Figure 79.45. This system has two modes of oscillation, one near 0.4 Hz and the other near 0.9 Hz. The lower frequency mode involves Area 1 swinging against Areas 2 and 3, while the higher frequency mode involves Area 2 swinging against Area 3. With the SVC located midway along the major intertie (Bus 1 to Bus 5 is the intertie, Bus 4 is midway as shown in Figure 79.45), the leverage of the SVC on damping (known as controllability) the two modes varies with intertie power transfer. The controllability is greatest at higher power transfer and is different for each of the modes, based on the participation of generators in the specific swing mode. The best input signal for the SVC power-swing damping control, based on work presented in [9], is line current magnitude. Figure 79.46 shows the input-signal filtering and controller configuration, and Figure 79.47 shows the corresponding transfer function.

Figure 79.48 shows the power system response following a disturbance in Area 1, while Area 1 is exporting power. The thin curves are for the system with no damping controller on the SVC, and the thicker curves are the case with a damping control. The swings of Area 1 angle reflect primarily the 0.4-Hz mode, while the Area 2 angle contains some of both modes. Figure 79.49 shows the simulation results for a disturbance in Area 2 with Area 1 importing power. Here, Area 2 shows primarily the 0.9-Hz mode, and Area 1 shows primarily the 0.4-Hz mode. In both cases, the simulation results show that significant damping is added to the system with a control added to modulate the SVC voltage set point, even for greatly varying power system operating conditions. These results suggest that the control design is robust.

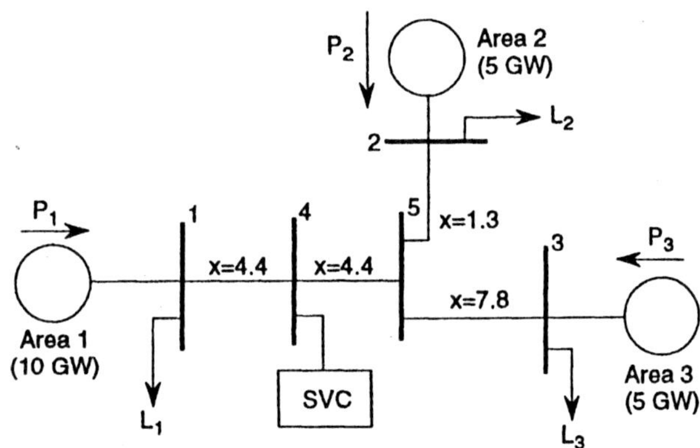

Figure 79.45 Example of three-area system (SVC is rated at 7.5% of total system generation).

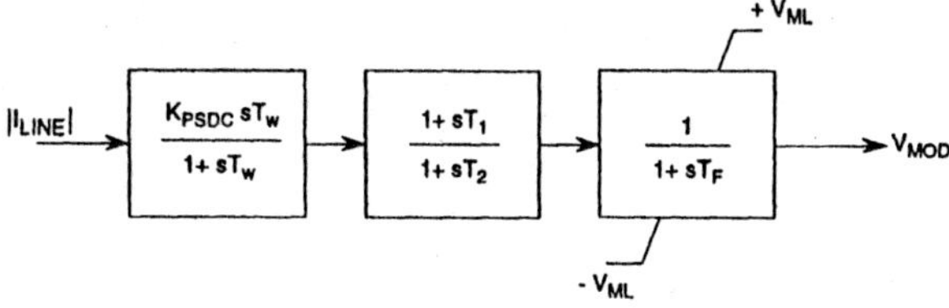

Figure 79.46 Simple power-swing damping control structure for an SVC.

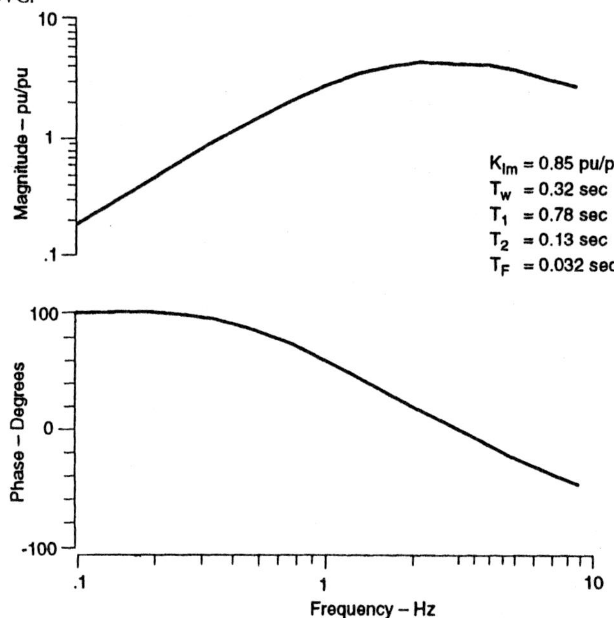

Figure 79.47 Sample transfer function of a simple SVC damping control.

Thyristor-Controlled Series Compensation (TCSC)

Series capacitors have long been applied to transmission systems for reducing the effective impedance between load and source to increase power transfer capability. Adding thyristor control provides significant leverage in directing steady-state power flow along desired paths. In addition, this controllability can be highly effective for damping power swings. The thyristor control also mitigates the adverse side effects from the resonance between the series capacitor and the inductance of the transmission lines, transformers, and generators (known as subsynchronous resonance or SSR) [16].

A TCSC module consists of a series capacitor and a parallel path with a thyristor switch and a surge inductor. Also in parallel, as is typical with series capacitor applications, is a metal-oxide varistor (MOV) for overvoltage protection. A complete compensation system may be made up of several of these modules in series and may also include a conventional (fixed) series capacitor bank as part of the overall scheme, as shown in Figure 79.50. The TCSC can be rapidly and continuously controlled in a wide band of operating points that can make it appear as a net capacitive or net inductive impedance in a transmission line [17].

The following is an example of a damping control for an actual TCSC installation. The objective of the control was to con-

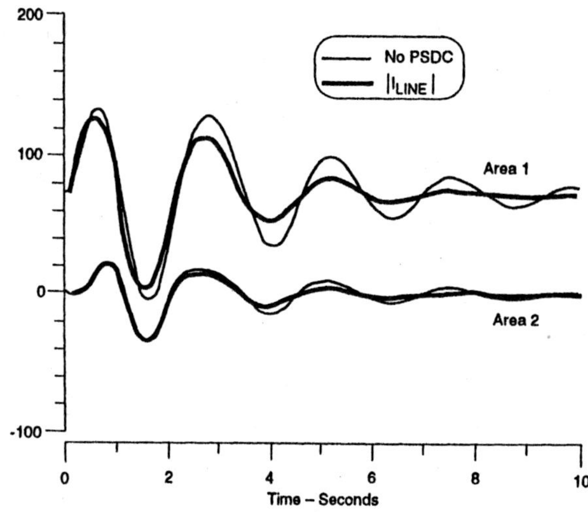

Figure 79.48 Angle swings for a disturbance in Area 1 with Area 1 exporting power (Area 3 is reference).

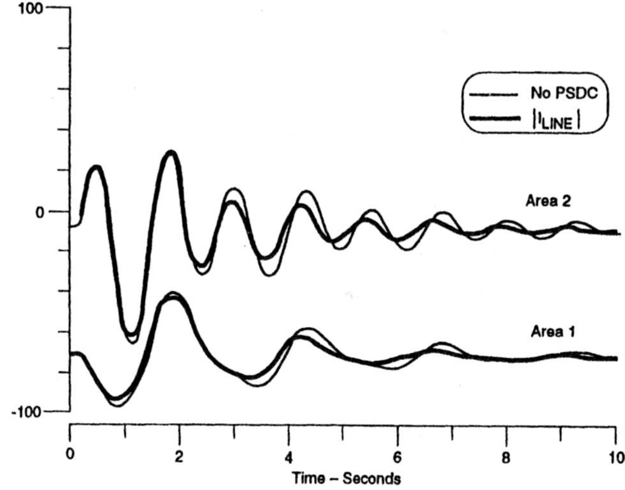

Figure 79.49 Angle swings for a disturbance in Area 2 with Area 1 importing power (Area 3 is reference).

tinuously adjust the series reactance (X_C) of the TCSC to add damping to critical electromechanical modes of oscillation. An appropriate input signal with the proper characteristics to meet the stabilization objective of this design was selected. The signal that proved most appropriate for this application is the frequency difference across the system obtained by synthesizing voltages behind reactances in both directions away from the TCSC. This concept is illustrated in Figure 79.51 and described in [14] and [18]. This signal is synthesized from *local* voltage and current measurements and eliminating the need for long-distance communication of signals. To prevent interaction with phenomena outside the desired control bandwidth, low-pass, high-pass, and torsional filtering is included in the modulation control path, as well as some phase compensation. The control structure for one specific installation [16] is shown in Figure 79.52. The transfer function of this control and input signal filtering are shown in Figure 79.53. Note that the 6-Hz and 13-Hz notch filters are in the input-signal filtering path to minimize the potential for site-specific torsional frequency interaction.

The system examined here is a radial machine swinging against the rest of a power system at a mode near 1 Hz. This configuration is significant because actual TCSC controls were factory tested and field tested on a system with a topology such as this [16]. The TCSC for this system has an *rms* line-to-line voltage rating of 500 kV, *rms* line current rating of 2900 *Amps*, and a nominal reactance rating of 8Ω. Figure 79.54 shows analog simulator results for a severe fault applied on one of the system buses. At approximately one-half second, a fault is placed on the system and the frequency shows significant oscillations without the supplemental damping control. At six seconds into this test, the TCSC power swing damping control is engaged and the oscillations are eliminated within three cycles of the 1-Hz oscillation. The remainder of the simulator test results show the ambient damping of the system. The control deadband prohibits further reduction at this point. These results clearly demonstrate the potential leverage of a TCSC for damping power system oscillations.

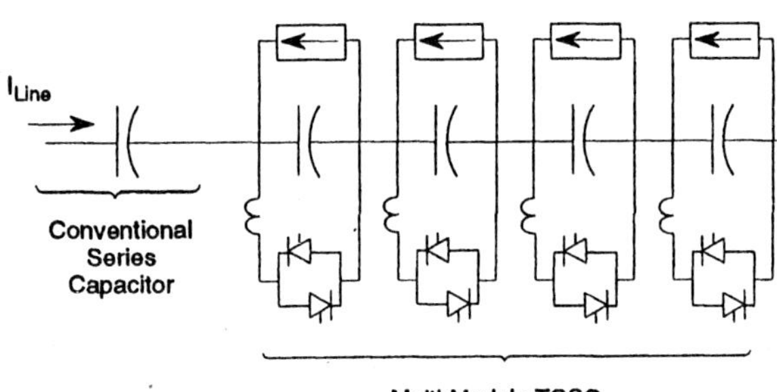

Figure 79.50 TCSC compensation scheme including a multimodule TCSC.

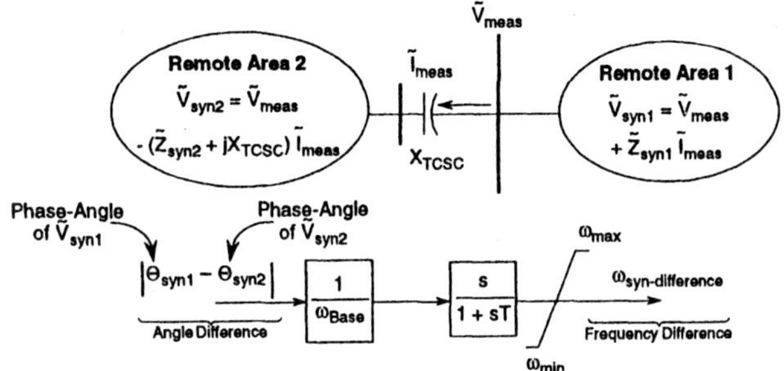

Figure 79.51 Synthesis of remote voltage phasers, angle difference, and frequency difference.

79.2.6 Recent Developments in Control Design

Traditional approaches to aid the damping of power swings include the use of Power System Stabilizers (PSS) to modulate the

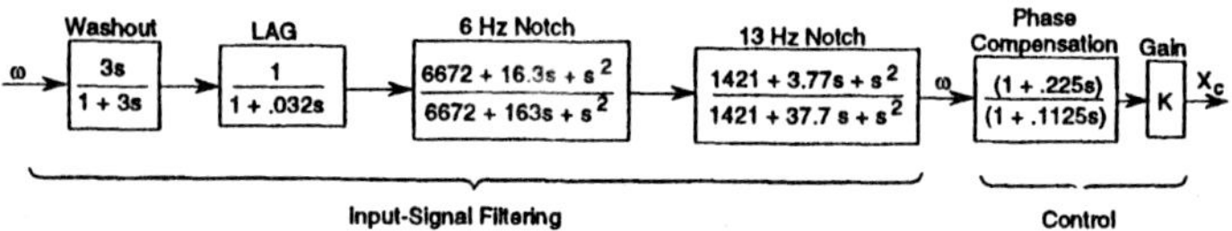

Figure 79.52 Input-Signal filtering and control structure for a specific TCSC installation.

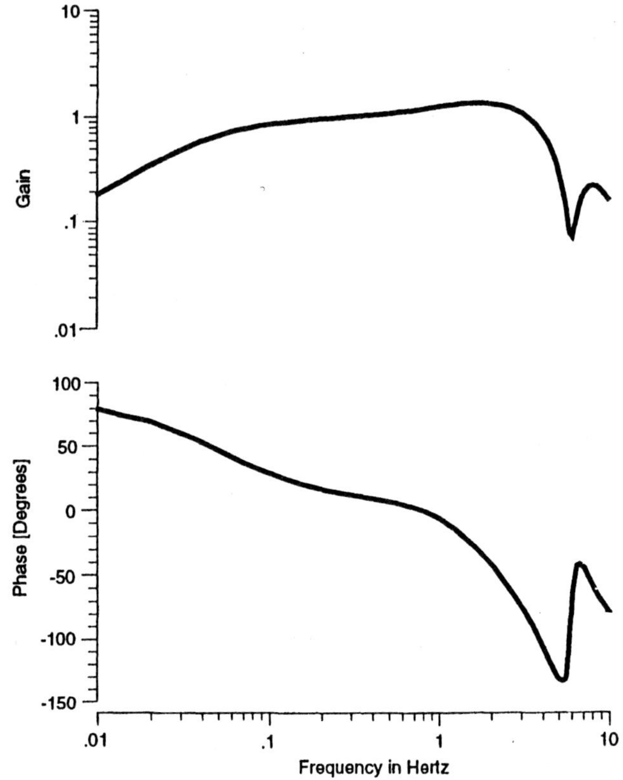

Figure 79.53 Transfer function of a TCSC damping control.

generator excitation control, for which much experience and insight exist in the industry. Unlike PSS control at a generator location, the speed deviations of the machines of interest are not readily available to a controller sited in the transmission path. Further, since the intent is to damp complex swings involving large numbers of generators, speed signals themselves are not necessarily the best choice for an input signal. It is desired to extract an input signal from the locally measurable quantities at the controller location. Finding an appropriate combination of measurements is the most important aspect of control design.

Approximate Multimodal Decomposition

The approach outlined in this section and described in [18] makes a few approximations to develop engineering insight in the control design process. One key approximation is that the modes of interest exhibit light damping. Another is that the impact of any individual control on the frequency and mode shape of the power swing is small. These assumptions permit breaking the system apart based upon the approximate mode shape information, and determining the incremental effect of controllers on each mode separately. The effect of the controllers upon themselves also becomes apparent, and the design can proceed using

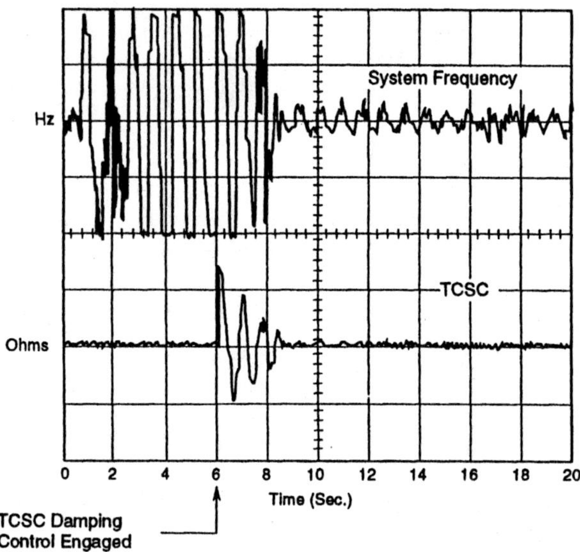

Figure 79.54 Damping benefits of a TCSC as shown on an analog simulator.

this information to assure minimum self-interaction effects on the final response.

Formulation

In the single-machine model described earlier, the mechanical swing mode is represented in terms of a synchronizing and a damping torque with control loops built around it. The same type of representation for each of the swing modes can be obtained by transforming the linearized representation of the power system into the equivalent form [18]:

$$\begin{bmatrix} \Delta\dot{\delta}_{mi} \\ \Delta\dot{\omega}_{mi} \\ \dot{Z}_{mi} \end{bmatrix} = \begin{bmatrix} 0 & \omega_b & 0 \\ -k_{mi} & -d_{mi} & -A_{d23} \\ A_{d31} & A_{d32} & A_{d33} \end{bmatrix} \begin{bmatrix} \Delta\delta_{mi} \\ \Delta\omega_{mi} \\ Z_{mi} \end{bmatrix} + \begin{bmatrix} 0 \\ -B_{d2} \\ B_{d3} \end{bmatrix} u,$$

$$y = \begin{bmatrix} C_{d1} & C_{d2} & C_{d3} \end{bmatrix} \begin{bmatrix} \Delta\delta_{mi} \\ \Delta\omega_{mi} \\ Z_{mi} \end{bmatrix} + Du,$$

where $\Delta\delta_{mi}$ and $\Delta\omega_{mi}$ represent the modal angle and modal speed, respectively, associated with a swing mode λ_i, k_{mi} and d_{mi} represent modal synchronizing and damping coefficients, and Z_{mi} consists of all the other state variables. Based on this representation, a block diagram, similar to the one developed in [5], can be constructed for mode λ_i. Such a block diagram is shown in Figure 79.55. In this figure, $K_{mi}(s)$, $K_{ci}(s)$, $K_{oi}(s)$, and $K_{ILi}(s)$ denote *modal, controllability, observability,* and *inner loop* transfer functions, respectively. These transfer functions evaluated at $s = j\omega$ are complex, providing both gain and phase information which can be used to select effective transfer functions for feedback control.

From Figure 79.55, the effective control action can be described

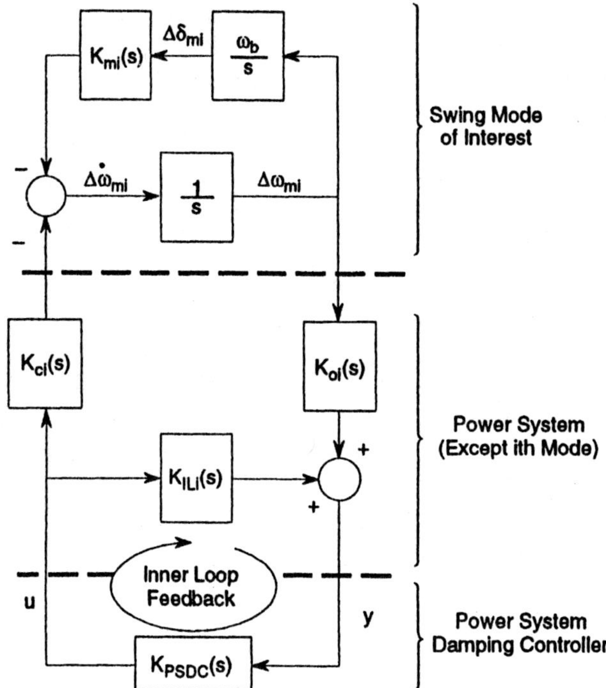

Figure 79.55 Multimodal decomposition block diagram.

by the transfer function:

$$K_{ei}(s) = K_{ci}(s)\frac{K_{PSDC}(s)}{1 - K_{ILi}(s)K_{PSDC}(s)}K_{oi}(s)$$

This relationship describes the impact of a given damping controller, $K_{PSDC}(s)$, on the modal system and is useful in estimating the eigenvalue sensitivity of the ith swing mode. This relationship shows that the effective control action $K_{ei}(s)$ directly relates to the controllability $K_{ci}(s)$ and observability $K_{oi}(s)$ functions. Assuming the perturbation of the complex pair of eigenvalues corresponding to the ith swing mode as $\Delta\lambda_i = -\Delta\sigma \pm j\Delta\omega_i$, the following relationships can be shown [18]:

$$\begin{aligned} \Delta\sigma_i &\approx \text{Real}\{\{K_{ei}(j\omega_i)/2\}, \\ \Delta\omega_i &\approx -\text{Imag}\{\{K_{ei}(j\omega_i)/2\} \end{aligned}$$

Thus, for an ideal damping controller design, $K_{ei}(s)$ is desired to be real for the frequency range of interest. For a practical controller design, some phase lag at the swing mode frequency is acceptable, since it tends to increase synchronizing torque. Also, a good bandwidth is desired, such that high-frequency interactions can be limited by the controller design, as described earlier in this section.

This direct relationship of modal sensitivity to the controller and power system characteristics provides the key to achieving the desired insight into control design. Subsequent discussion will focus on each of the terms in the above equations, to show how they relate to control performance and how they can be used to select effective measurements and controls.

Interpretations

Controllability The effect of a controller on a given swing mode is defined as the controllability function $K_{ci}(s)$, and $K_{ei}(s)$ is directly proportional to $K_{ci}(s)$. In PSS design, $K_{ci}(s)$ depends mostly on the excitation system, generator flux dynamics and network impedances. For network control devices such as TCSC, $K_{ci}(s)$ depends mostly on the network structure and loads. When evaluated at $s = j\omega_i$, $K_{ci}(j\omega_i)$ provides a measure of how controllable the ith mode is by the control signal u. If $K_{ci}(j\omega_i)$ is zero, then mode i is not affected by u. When more than one damping controller is used, $K_{ci}(s)$ is a vector transfer function.

In general, $K_{ci}(j\omega_i)$ is different for each swing mode. For example, a TCSC sited on a tie line would have significant controllability on the associated inter-area mode, but much smaller controllability over local modes. In cases with multiple inter-area modes, the controllability may be nearly 180° out of phase from one mode to the next. Such a condition would mean that, if the machine speeds were averaged and transmitted to the controller, then the action of improving the damping on one mode would simultaneously decrease the damping of another mode. This situation must be compensated for by using an appropriate set of measurements so that this inherent adverse impact will be minimized or eliminated.

The quantity $K_{ci}(j\omega_i)$ is a good indicator for evaluating effective locations to apply damping controllers, because the larger this is, the greater the leverage on the swing mode. For example, an SVC on a bus needing voltage support will be more effective for damping control than one close to a generator terminal bus.

Observability The effective control action $K_{ei}(s)$ is also directly proportional to the observability function $K_{oi}(s)$. The function $K_{oi}(s)$ relates the measured signal y to the ith modal speed $\Delta\omega_{mi}$. For a PSS using the machine speed as the input signal, $K_{oi}(s) = 1$ for the local mode of that machine. When evaluated at $s = j\omega_i$, $K_{oi}(j\omega_i)$ gives an indication of the modal content of the ith swing mode in the measured signal y. Its magnitude can be used to assess the effectiveness of measurements y for damping control applications. In a multimodal system, because $K_{oi}(s)$ is defined with respect to the modal speed, measurements directly related to machine speeds will have observability gains that are predominantly real. Signals more closely related to angular separation, such as power flow, will have an integral characteristic, i.e., nearly 90° of lag with respect to the speed. If $K_{oi}(j\omega_i)$ is small, then the ith mode is weakly observable from the measurement y. Thus having large $K_{oi}(j\omega_i)$ for the dominant modes of interest is one of the criteria in selecting an input signal for a damping controller.

Inner Loop The control design must also consider the effect of the controller output on its input (i.e., the component of the measured signal y due to the control u), other than via the swing mode of interest. This effect may be considered a "feed-forward" term, but here we have called it the "inner-loop" effect, symbolized by the transfer function $K_{ILi}(s)$.

The inner loop transfer function $K_{ILi}(s)$ is extremely important in damping controller design using input signals other than generator speeds. In [9] and [18], simple analysis based on the constraint imposed by $K_{ILi}(s)$, are developed to aid the selection of an appropriate measurement signal.

79.2.7 Summary

This section provides an overview of some key issues for control design, implementation, and operation. Basic stability issues for power systems, such as transient, voltage, and oscillatory stability were introduced. Given the focus of this book, the remainder of this section was on closed-loop controls that affect power system oscillatory stability.

Basic concepts and a historic overview of generator excitation controls as they affect power system stability were introduced and described. Supplemental controls typically applied to generator excitation systems were also discussed (power system stabilizers (PSS)). Furthermore, practical issues for applying supplemental controls to power system transmission equipment (such as HVDC, SVC, and TCSC) were presented. The issues addressed included siting, objectives for control design and operation, closed-loop control design, input-signal selection, input-signal filtering, control algorithms, performance evaluation, and a discussion on potentially adverse side effects due to the application of supplemental controls. Two detailed examples of power electronic devices (SVC and TCSC) were provided illustrating these basic concepts. Finally, a presentation on recent developments in control design was included. Several references were provided for further reading on the issues presented in this section.

This information shows that supplemental control can be beneficial in increasing the stability and utilization of electric power systems.

References

[1] Bertoldi, O. and CIGRE WG 37.08, *Adequacy and Security of Power Systems at the Planning Stage: Main Concepts and Issues,* Paper 1A-05, Symposium on Electric Power System Reliability, Montreal, Canada, 1991.

[2] Anderson, P.M. and LeReverend, B.K., Industry Experience with Special Protection Schemes, *IEEE Power Eng. Soc.,* (PES) Paper 94-WM184-2 PWRS, New York, 1994.

[3] Voltage Stability of Power Systems: Concepts, Analytical Tools and Industry Experiences, *IEEE PES* Special Publication 90TH0358-2-PWR, 1990.

[4] Kundur, P., *Power System Stability and Control,* McGraw-Hill, New York, 1994.

[5] deMello, F.P. and Concordia, C., Concepts of Synchronous Machine Stability as Affected by Excitation Control, *IEEE Trans. Power Apparatus Syst.,* PAS-88, 316–329, 1969.

[6] Larsen, E.V. and Swann, D.A., Applying Power System Stabilizers, Parts I, II, and III, *IEEE Trans. Power Apparatus Syst.,* PAS-100, 3017–3046, 1981.

[7] Paserba, J.J., Larsen, E.V., Grund, C.E., and Murdoch, A., Mitigation of Inter-Area Oscillations by Control, *IEEE PES* Special Publication 95-TP-101 on Inter-Area Oscillations in Power Systems, 1995.

[8] Paserba, J.J., et al., *Opportunities for Damping Oscillations by Applying Power Electronics in Electric Power Systems,* CIGRE Symp. Power Electron. Electr. Power Systems, Tokyo, Japan, 1995.

[9] Larsen, E.V. and Chow, J.H., SVC Control Design Concepts for System Dynamic Performance, in *Application of Static Var Systems for System Dynamic Performance,* IEEE PES Special Publication No. 87TH1087-5-PWR, 1987, 36–53.

[10] Piwko, R.J. and Larsen, E.V., HVDC System Control for Damping Subsynchronous Oscillations, *IEEE Trans. Power Apparatus Syst.,* PAS-101(7), 2203–2211, 1982.

[11] Chow, J.H., Date, R.A., Othman, H.A., and Price, W.W., *Slow Coherency Aggregation of Large Power Systems,* IEEE Symp. Eigenanalysis and Frequency Domain Methods for System Dynamic Performance, IEEE PES Special Publication 90TH0292-3-PWR, 1990, 50–60.

[12] Hauer, J.F., Application of Prony Analysis to the Determination of Model Content and Equivalent Models for Measured Power Systems Response, *IEEE Trans. Power Syst.,* 1062–1068, 1991.

[13] Bahrman, M.P., Larsen, E.V., Piwko, R.J., and Patel, H.S., Experience with HVDC Turbine-Generator Torsional Interaction at Square Butte, *IEEE Trans. Power Apparatus Syst.,* PAS-99, 966–975, 1980.

[14] Clark, K., Fardanesh, B., and Adapa, R., *Thyristor-Controlled Series Compensation Application Study—Control Interaction Considerations,* IEEE PES Paper 94-SM-478-8-PWRD, San Francisco, CA, 1994.

[15] Cresap, R.L., Mittelstadt, W.A., Scott, D.N., and Taylor, C.W., Operating Experience with Modulation of the Pacific HVDC Intertie, *IEEE Trans. Power Apparatus Syst.,* PAS-9, 1053–1059, 1978.

[16] Piwko, R.J., Wegner, C.A., Damsky, B.L., Furumasu, B.C., and Eden, J.D., *The Slatt Thyristor-Controlled Series Capacitor Project — Design, Installation, Commissioning, and System Testing,* CIGRE Paper 14-104, Paris, France, 1994.

[17] Larsen, E.V., Clark, K., Miske, S.A., and Urbanek, J., Characteristics and Rating Considerations of Thyristor Controlled Series Compensation, *IEEE Trans. Power Delivery,* 992–1000, 1994.

[18] Larsen, E.V., Sanchez-Gasca, J.J., and Chow, J.H., *Concepts for Design of FACTS Controllers to Damp Power Swings,* IEEE PES Paper 94-SM-532-1-PWRS, San Francisco, CA, 1994.

SECTION XVIII

Control Systems Including Humans

80

Human-in-the-Loop Control

R. A. Hess
University of California, Davis

80.1 Introduction

Interest in modeling the behavior of a human as an active feedback control device began during World War II, when engineers and psychologists attempted to improve the performance of pilots, gunners, and bombardiers. To design satisfactory manually controlled systems these researchers began analyzing the neuromuscular characteristics of the human operator. Their approach, e.g., [19] was to consider the human as an inanimate servomechanism with a well-defined input and output. Figure 80.1(a) is a schematic representation of a tracking task in which the human is attempting to keep a moving target within the reticle of a gun sight. Figure 80.1(b) is a feedback block diagram of the gunnery task of Figure 80.1(a) in which the human has been represented as an error-activated compensation element.

The input to the human in Figure 80.1 is a visual "signal" indicating the error between desired and actual system output, the gun azimuth. The output of the human is a command to the device which drives the gun in azimuth. This device might be a simple gearing mechanism linked to a wheel which the human turns or an electric motor which transforms a joystick input into a proportional rate of change of the gun barrel's azimuth angle. What mathematical representation should be used for the block labeled "human controller"? The early researchers in the discipline known as "manual control" hypothesized this answer to the question: the same types of mathematical equations used to describe linear servomechanisms, namely, sets of linear, constant-coefficient differential equations. The hypothesis

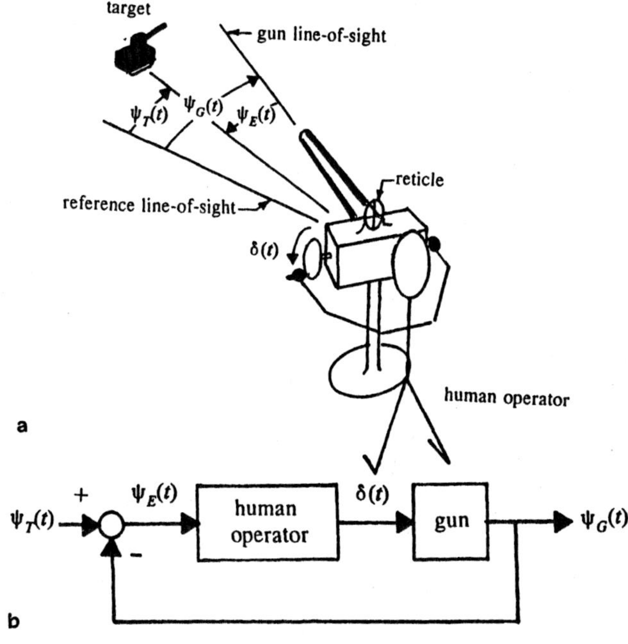

Figure 80.1 (a) A human-in-the-loop control problem. (b) A block diagram representation of the task of Figure (a).

of these early researchers turned out to be true, and the *control-theory paradigm* for quantifying human control behavior was born. This paradigm has become fundamental for manual control engineers [13].

The evolution of the control-theory paradigm for the human

0-8493-8570-9/96/$0.00+$.50

controller or operator paralleled the development of new synthesis techniques in feedback control. Thus, "optimal control models" (OCMs) of the human operator [8] appeared as linear quadratic Gaussian (LQR) control system design techniques were being developed. "Fuzzy controller" models [12] and "H-infinity" models [1] of the human operator closely followed the appearance of these design techniques.

The models discussed here will primarily be those which have successfully solved human-in-the-loop control problems. This approach will neglect some of the more recent human operator models, which are incompletely tested.

80.2 Frequency-Domain Modeling of the Human Operator

Modeling the human operator as a set of linear, constant-coefficient differential equations suggests representing the human as a transfer function. This approach, generalized to describing function descriptions, captured the attention of some of the earliest and most influential manual control engineers [14]. Figure 80.2 shows a describing function representing the human operator or controller in a single-input, single-output (SISO) tracking task, such as an aircraft pitch-attitude tracking task. Here the transfer function representation of the human has been generalized as a quasi-linear describing function [4] by the addition of an additive "remnant" signal, $n_e(t)$. This signal represents the portion of the system error signal $e(t)$ unexplainable by linear operator behavior, and not linearly correlated with the system input $c(t)$. The spectral measurements of remnant coalesced best when the remnant was assumed to be injected into the displayed error $e(t)$ rather than the operator's output $\delta(t)$. For this reason, the remnant portion of the quasi-linear describing function is almost universally shown with error-injected remnant.

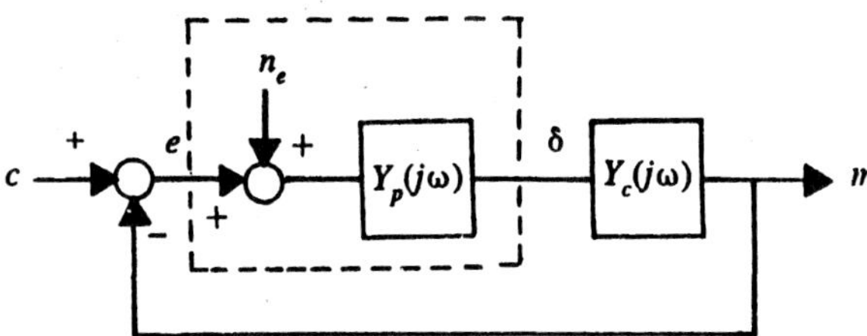

Figure 80.2 A quasi-linear describing function representation of the human operator.

Frequency-domain identification of human operator describing functions in simple laboratory tracking tasks has been actively studied over the past three decades [15]. In these experiments, the describing functions identified were $Y_p(j\omega)$ (or$Y_p(j\omega)Y_c(j\omega)$) and $\Phi_{n_e n_e}(\omega)$, where the latter quantity is defined as the power spectral density of the remnant signal $n_e(t)$. The controlled element or plant was a member of a set of "stereotypical" controlled-element dynamics, i.e., $Y_c(s) = \frac{K_c}{s^k}$, $k = 0, 1, 2$, and the inputs or disturbances were random-appearing signals, often generated as sums of sinusoids. The results led to one of the first true engineering models of the human operator, referred to as the **crossover model.**

80.2.1 The Crossover Model of the Human Operator

The crossover model is based on the following experimentally verifiable fact: In a Bode diagram representing the loop transmission $Y_p(j\omega) \cdot Y_c(j\omega)$ of the system, such as shown in Figure 80.2, the human adopts dynamic characteristics $Y_p(j\omega)$ so that

$$Y_p(j\omega) \cdot Y_c(j\omega) \approx \frac{\omega_c e^{-\tau_e \omega}}{j\omega}, \quad \text{for} \quad \omega \quad \text{near} \quad \omega_c. \tag{80.1}$$

The crossover frequency, ω_c, is defined as the frequency where $|Y_p Y_c(j\omega)| = 1.0$. Equation 80.1 is valid in a broad frequency range (1 to 1.5 decades) around the crossover frequency ω_c. The factor τ_e, referred to as an effective time delay, represents the cumulative effect of actual time delays in the human information processing system (e.g., visual detection times, neural conduction times, etc.), the low-frequency effects of higher frequency human operator dynamics (e.g., muscle actuation dynamics), and higher frequency dynamics in the controlled element, itself. Here, "higher frequency" refers to frequencies well above ω_c.

Associated with Equation 80.1 is a model of $\Phi_{n_e n_{e(\omega)}}$, the power spectral density of the error-injected remnant. Again, extensive experimental evidence suggests the following form:

$$\Phi_{n_e n_e}(\omega) \approx \frac{R\bar{e}^2}{\omega^2 + \omega_R^2} \tag{80.2}$$

where $\bar{e}^2$ represents the mean-square value of the error signal $e(t)$ in Figure 80.2. Table 80.1 shows approximate parameter values for the crossover and remnant models with related equations. In applying the crossover model, plant dynamics as simple as those in Table 80.1 will be rare. In these cases, one should interpret the stereotypical dynamics in the table to reflect the actual plant characteristics in the crossover region.

A detailed summary of empirically derived rules for selecting crossover model parameters, given the controlled-element dynamics and the bandwidth of the input signal, are given by [5]. In addition, simplified techniques are given for estimating human-machine performance (e.g., root-mean-square tracking error $\sqrt{\bar{e}^2}$) in continuous tasks with random-appearing inputs.

The reason for beginning this discussion with the crossover model is that it is basic for manual control modeling. Any valid model of the human operator in continuous tasks with random-appearing inputs *must* exhibit the characteristics of Equation 80.1.

The loop transmission prescribed by Equation 80.1 is similar to that which an experienced control system designer would select in a frequency-domain synthesis of a control system with an inanimate compensation element and performance requirements similar to the manually controlled system [16].

TABLE 80.1 Parameters and Relations for Crossover Model

Y_c (around crossover)	τ_0 (s)	ω_{c_0} (rad/s)	R	ω_R (rad/s)
K	0.30	5.0	0.1 to 0.5	3.0
$\frac{K}{s}$	0.35	4.5	0.1 to 0.5	3.0
$\frac{K}{s^2}$	0.50	3.0	0.1 to 0.5	1.0

$$\begin{aligned} \omega_c &\approx \omega_{c_0} + 0.18\omega_{BW_c} \\ \tau_e &\approx \tau_0 - 0.08\omega_{BW_c} \\ \tau_0 &= \frac{\pi}{2\omega_{c_0}} \end{aligned}$$

80.2.2 A Structural Model of the Human Operator

A model of the human operator which follows the crossover model, but provides a more detailed representation of human operator dynamics is offered in [7] and is referred to as a **structural model of the human operator**. The model is shown in Figure 80.3. Table 80.2 lists the model parameter values which depend on the order of the controlled-element dynamics in the crossover region. The parameter "k" in Table 80.2 refers to the controlled-element order around crossover, i.e., $k = 0$ for zero*th* order, $k = 1$ for first order, etc.

TABLE 80.2 Nominal Parameters for Structural Model.

k	K_e	K_1	K_2	T_1 (s)	T_2 (s)	τ (s)	ζ_n	ω_n (rad/s)
0	1.0^a	1.0	2.0	5.0	^{b}	0.15	0.707	10.0
1	1.0	1.0	2.0	5.0	^{c}	0.15	0.707	10.0
2	1.0	1.0	10.0	2.5	^{b}	0.15	0.707	10.0

[a] K_e chosen to provide desired crossover frequency
[b] selected to achieve K/s–like crossover characteristics
[c] Parameter not applicable

The structural model of the human operator is called an "isomorphic model", because the model's internal feedback structure reflects hypothesized proprioceptive feedback activity in the human.

If the control engineer can reasonably estimate the crossover frequency for the manual control task at hand, Figure 80.3 and Table 80.2 can provide a model of the linear human operator dynamics ($Y_p(j\omega)$) adequate for many engineering applications. The empirically derived rule for determining crossover frequency for the crossover model can be used to define the crossover frequency for the structural model. Remnant parameter estimates can also be obtained in a similar fashion.

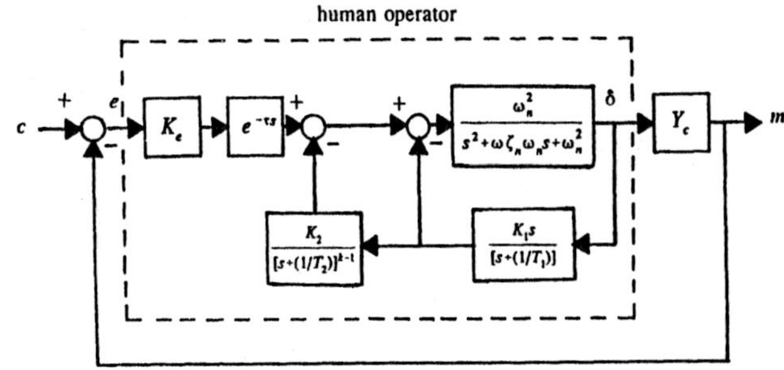

Figure 80.3 The structural model of the human operator.

80.3 Time-Domain Modeling of the Human Operator

80.3.1 An Optimal Control Model of the Human Operator

The advent of a time-domain control synthesis technique in the mid-1960s, referred to as linear quadratic Gaussian (LQG) design led to a powerful model of the human operator called the **Optimal Control Model** (OCM). This model differs from those defined in the preceding because it is algorithmic and is based on a time-domain optimization procedure. The model is algorithmic because the quantitative specification of certain human operator information processing limitations, such as signal-to-noise ratios on observed and control variables and sensory-motor time delays, together with an objective function which the human is assumed to be minimizing in the task at hand, can lead to direct computation of the linear human operator dynamics and remnant. In addition, this algorithmic capability is not limited to SISO systems but also can be extended to human control of multi-input, multioutput (MIMO) systems.

Figure 80.4 shows the basic OCM structure. Focusing for the moment on a SISO system, the elements in Figure 80.4 from time delay to neuromuscular dynamics form the human operator dynamics. Strictly, speaking, the OCM is never a SISO model, because a fundamental hypothesis in the model formulation is that, if a variable $d(t)$ is displayed to the operator, its time derivative $\dot{d}(t)$ is also sensed, both signals corrupted with white observation noise. Thus, in its simplest form, the $y(t)$ in Figure 80.4 is a column vector whose two elements are the displayed signal and

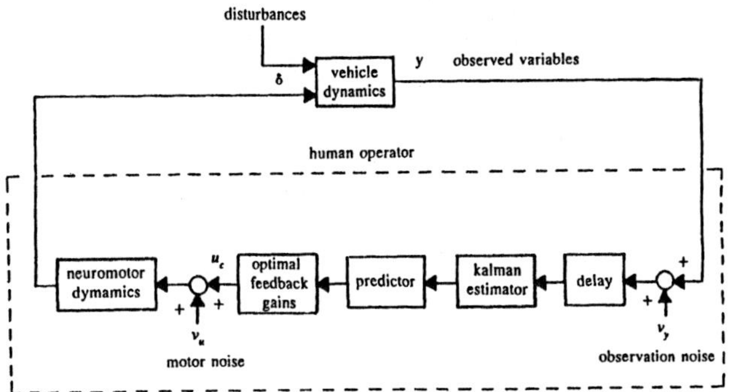

Figure 80.4 The optimal control model of the human operator.

its time derivative. Likewise, the $v_y(t)$ in Figure 80.4 is also a column vector whose elements are the observation noise signals. The covariances of these observation noise signals are assumed to scale with the covariances of the displayed/observed signals $y(t)$ and $\dot{y}(t)$. In addition to observation noise, the signal $u_c(t)$ is also assumed to be corrupted with "motor" noise. This noise provides performance predictions which are more realistic than those produced when motor noise is absent.

Table 80.3 shows "nominal" values of the noise-signal ratios and time delay for typical applications of the OCM to tracking or disturbance regulation tasks with random-appearing inputs or disturbances. The remaining user-defined element in application of the OCM is the quadratic index of performance which the operator is assumed to be minimizing. In many SISO applications, the following index of performance has been employed:

$$J = \int_0^{\infty} \left[y^2(t) + \rho\dot{\delta}^2(t)\right] dt. \tag{80.3}$$

TABLE 80.3 Nominal Parameter Values for Optimal Control Model (OCM).

$\frac{V_{y_i}}{\bar{e}^2_{y_i}}$ [a]	$\frac{V_u}{u_c^2}$ [b]	τ (s)
0.01π	0.001π	0.15

[a] noise-to-signal ratio for observation noise $v_{y_i}(t)$.
[b] noise-to-signal ratio for motor noise $v_u(t)$.

Inclusion of control rate in the index of performance will produce first-order lag dynamics in the OCM [8]. These constitute the neuromotor dynamics shown in Figure 80.4. As in any LQG synthesis application, selection of the index of performance weighting coefficients is nontrivial. One procedure for selecting ρ in Equation 80.3 is a trial and error technique in which ρ is varied until the time constant of the first-order lag (neuromotor dynamics) of approximately 0.1 s. Another procedure for choosing ρ which does not require trial and error iteration has been used successfully [6] and is referred to as an "effective time constant" method.

The power of the OCM lies in its ability to provide accurate representations of human control behavior given "standard" values for quantifying the information processing limitations mentioned in the preceding.

Finally, it should be noted that other human operator models have been developed which are progeny of the OCM. The biomorphic model of the human pilot [3], is one example, as is the OCM extension and refinement described by [21], and the more recent H-infinity model of [1].

80.4 Alternate Modeling Approaches

80.4.1 Fuzzy Control Models of the Human Operator

Fuzzy set theory leads to a description of cause and effect relationships which differ considerably from the control theoretic approaches discussed in the preceding [20]. In classical set theory, there is a distinct difference between elements which belong to a set and those which do not. Fuzzy set theory allows elements to belong to more than one set and assigns each element a membership value, M, between 0 and 1 for each set of which it is a member. Consider, for example, the use of fuzzy sets to model the manner in which a pilot might control the pitch attitude of an aircraft, creating what can be termed a **fuzzy control model** of the human. Assume that the aircraft's pitch attitude, $\theta(t)$, can vary within $-20° \leq \theta(t) \leq +20°$. This range is often referred to as the "universe of discourse" in fuzzy set theory. On can define a number of membership functions over this range of discourse as in Figure 80.5(a) where five functions are shown. The shape defining each function as well as the number of functions spanning the universe of discourse is entirely up to the analyst. The five triangular functions of Figure 80.5(a) have been chosen for simplicity. The functions have also been given linguistic definitions, i.e., "$\theta(t)$ very negative, $\theta(t)$ negative", etc. Now, as in Figure 80.5(a), an aircraft pitch attitude of $-12.5°$ is a member of five sets with membership values of: $M(-12.5) = 0.25$ (very negative), 0.75 (negative), 0 (around zero), 0 (positive), and 0 (very positive).

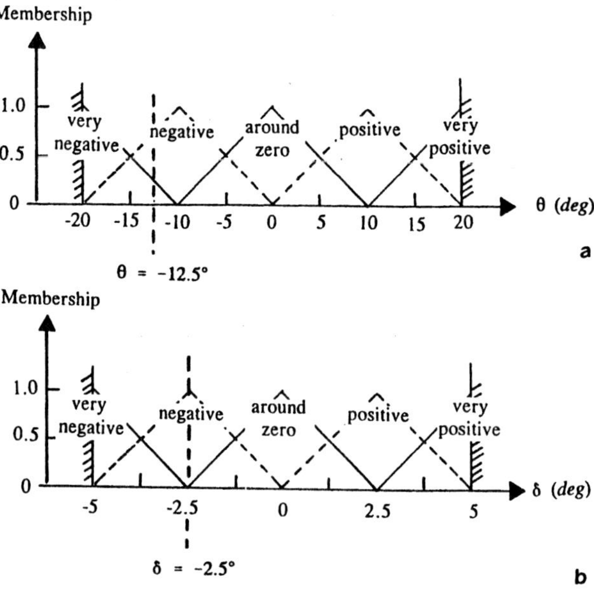

Figure 80.5 (a) Defining input membership functions for a fuzzy control model of the human operator. (b) Defining output membership functions for a fuzzy control model of the human operator.

Modeling the human operator with fuzzy sets involves using fuzzy relations between input and output values. Let us assume the output set here consists of elevator deflections which the pilot commands in response to an observed pitch attitude deviation from zero. Output fuzzy sets can be defined over the universe of discourse for the elevator deflections, here defined as $-5^\circ \leq \delta(t) \leq +5^\circ$, using membership functions shown in Figure 80.5(b). Again, linguistic definitions have been employed to describe these membership functions. Perhaps the pitch attitude of -12.5° produced a pilot commanded elevator angle of -2.5°, as shown in Figure 80.5(b). Now, as will be discussed in a later section, by observing the pitch-attitude change and resultant control actions of an experienced pilot, or group of experienced pilots in a flight simulator, the control engineer can create a "relational matrix" R between $\theta(t)$ and $\delta(t)$ as $R = M(\theta)XM(\delta)$, where "X" denotes a Cartesian product, or minimum for each element pairing of $M(\theta)$ and $M(\delta)$. Assume that, on the basis of simulator observations, the relational matrix below is obtained:

$$R = \begin{bmatrix} 0.6 & 0.5 & 0 & 0 & 0 \\ 0.2 & 0.75 & 0.5 & 0 & 0 \\ 0 & 0.4 & 1 & 0.4 & 0 \\ 0 & 0 & 0.5 & 0.75 & 0.0 \\ 0 & 0 & 0 & 0.4 & 0.6 \end{bmatrix}. \tag{80.4}$$

The R of Equation 80.4 forms the basis of a "fuzzy control" model of the pilot. In exercising the model, one can assume that the pilot makes observations every ΔT sec. For example, at some observation instant assume $\theta(n\Delta T) = -15.0^\circ$. As shown in Figure 80.5(a), the memberships of the input are

> $M(\theta)$ = 0.5 (very negative pitch), 0.5 (negative pitch), 0 (around zero pitch), 0 (positive pitch), and 0 (very positive pitch)

Using R and $M(\theta)$, the memberships of the fuzzy output can be obtained [17] as

> $M(\delta)$ = max of {min[0.6, 0.5],min[0.2, 0.5], min[0, 0], min[0, 0],min[0, 0]} for very negative elev.
>
> = max of {min[0.5, 0.5],min[0.75, 0.5], min[0.4, 0], min[0, 0],min[0, 0]} for negative elev.
>
> = max of{min[0, 0.5],min[0.5, 0.5],min[1, 0], min[0.5, 0],min[0, 0]} for around zero elev.
>
> = max of{min[0, 0.5],min[0, 0.5],min[0.4, 0], min[0.75, 0],min[0.4, 0]} for positive elev.
>
> = max of{min[0, 0.5],min[0, 0.5],min[0., 0], min[0, 0],min[0.6, 0]} for very positive elev.

The right hand sides of each of the five equations above constitute the membership values for $\delta(t)$ at this instant. These values have been obtained by pairing the elements of R with elements of $M(\theta)$ as would be done in the matrix multiplication $M\Theta R$. The minimum value of each pairing was found, and finally, the maximum of the j elements selected. This yields

> $M(\delta)$ = 0.5 (very negative elev.), 0.5 (negative elev.), 0.5 (around zero elev.), 0 (positive elev.), and 0 (very positive elev.)

Notice that the output of this fuzzy relation or fuzzy mapping is not a definite value of $\delta(n\Delta t)$, but a collection of membership values. In order to implement this fuzzy pilot model, say, in a simulation of an aircraft, one needs to "defuzzify" model outputs such as the $M(\delta)$ above into nonfuzzy $\delta(n\Delta T)$, which can then serve as useful inputs to the simulation.

It is worth emphasizing that determining fuzzy models of human operator behavior is rarely as simple as described. For example, human operator actions are often predicated upon system error and upon error rate. This obviously complicates the definition of the human's input membership functions. Despite these complications, fuzzy models of the human operator have been successful in modeling human-in-the-loop control in a variety of tasks. Examples include modeling motorcycle drivers [12], ships helmsmen [18], and automobile drivers [9].

80.5 Modeling Higher Levels of Skill Development

In the models described briefly in the preceding paragraphs, it has been tacitly assumed that the human was operating upon displayed system error (and error rate in case of the OCM). Such tasks are referred to as **compensatory tracking tasks.** Often displays for human operators contain error information and output information as well. Such displays are referred to as "pursuit" displays, and the resulting tasks are referred to as **pursuit tracking tasks.** A discussion of such tasks and their associated human operator models can be found in [5]. However, another, more subtle modeling issue can arise in which the human can actually develop pursuit tracking behavior with only a compensatory display. Such human behavior has been the subject of the "Successive Organizations of Perception" (SOP) theory, [10]. The highest level of skill development is referred to as "precognitive" behavior and implies human execution of preprogrammed responses to certain stimuli without the necessity of continuous feedback information. Modeling any of these higher levels of skill development is beyond the scope of this chapter.

80.6 Applications

The preceding sections have outlined approaches for modeling the human operator in SISO control systems in which the human is sensing system error and where the system input is random or random appearing. The various modeling approaches have met with success in engineering analyses. It is useful at this juncture to consider applying a subset of these models to a very simple but illustrative human-in-the-loop control example.

80.6.1 An Input Tracking Problem

Consider again the block diagram of Figure 80.2 representing a human-in-the-loop SISO control problem. Here, the human's task is to null a displayed error signal (difference between command input and system output) by manipulating a control device producing an output $\delta(t)$. The plant dynamics are very simple, consisting of an integrator, i.e., the plant transfer function is $Y_c(s) = \frac{1}{s}$. Such dynamics are often referred to as rate-command, because a constant control input $\delta(t)$ produces a constant output rate, $\dot{m}(t)$. The command input here is filtered white noise, wherein the filter transfer function is $F(s) = \frac{1}{(s+1)^2}$. Three models of the human will be generated here: 1) a crossover model, 2) a structural model, and 3) an Optimal Control Model. In addition, a brief discussion of how a fuzzy control model might be developed is also included.

Crossover Model

Development of the crossover model of Equation 80.1 begins by estimating the system crossover frequency, ω_c, and the effective time delay, τ_e. These parameters have been found empirically to depend upon the plant dynamics and input bandwidth [15], via the equations shown in Table 80.1, i.e.,

$$\begin{aligned} \omega_c &\approx \omega_{c_0} + 0.18\omega_{BW_c}, \\ \tau_e &\approx \tau_0 - 0.08\omega_{BW_c}, \end{aligned}$$

and

$$\tau_0 = \frac{\pi}{2\omega_{c_0}}, \tag{80.5}$$

where ω_{BW_c} refers to the input bandwidth, here approximated by the break frequency of the input filter $F(s)$ as 1 rad/s. Thus, Equations 80.5 yield

$$\begin{aligned} \omega_c &= 4.5 + 0.18(1.0) \quad \text{rad/s} \\ &= 4.68 \quad \text{rad/s} \\ \tau_e &= 0.35 - 0.08(1.0) \quad \text{s} \\ &= 0.27 \quad \text{s} \end{aligned} \tag{80.6}$$

Equation 80.1 becomes

$$Y_p Y_c(j\omega) \approx \frac{4.68e^{-0.27s}}{(j\omega)} \tag{80.7}$$

Now the power spectral density of the error-injected remnant is given by Equation 80.2. Using Table 80.1, the remnant power spectral density becomes

$$\Phi_{n_e n_e}(\omega) = \frac{0.4\overline{e}^2}{\omega^2+(0.3)^2}. \tag{80.8}$$

In Equation 80.8 an intermediate value of R has been used. Simple block diagram algebra, fundamental spectral analysis techniques and the "$\frac{1}{3}$ power law" can be used to derive the following equation for estimating human-in-the-loop tracking performance [5]:

$$\overline{e}^2 \approx \frac{\frac{1}{3}\overline{c}^2\left(\frac{\omega_{BW_c}}{\omega_c}\right)^2}{1-\frac{R}{\omega_R^2\tau_e}I_1} \tag{80.9}$$

where $\overline{c}^2$ refers to the mean-square value of the system input, here assumed to be unity. The I_1 in Equation 80.9 can be obtained from Figure 80.6 [15]. In this case, $I_1 \approx 3.6$ and Equation 80.9 gives an estimate of tracking performance as $\overline{e}^2 = 0.037$ with root-mean-square (RMS) tracking error $\sqrt{\overline{e}^2} = 0.19$. The one caveat that should accompany the use of Equation 80.9 is that it assumes a rectangular input spectra. In the task at hand, the input was formed by passing white noise through a second-order filter, and thus was not rectangular. However, to obtain a preliminary estimate of tracking performance, one can use the cutoff frequency of the continuous-input spectrum as the cutoff of the rectangular spectrum. Tracking performance can be more accurately predicted by a computer simulation of the crossover model, with injected remnant. Some iteration will be required, because the magnitude of the injected remnant has to scale with $\overline{e}^2$, the quantity one is trying to obtain from the simulation. Nonetheless, the author has found that the iterative process is fairly brief.

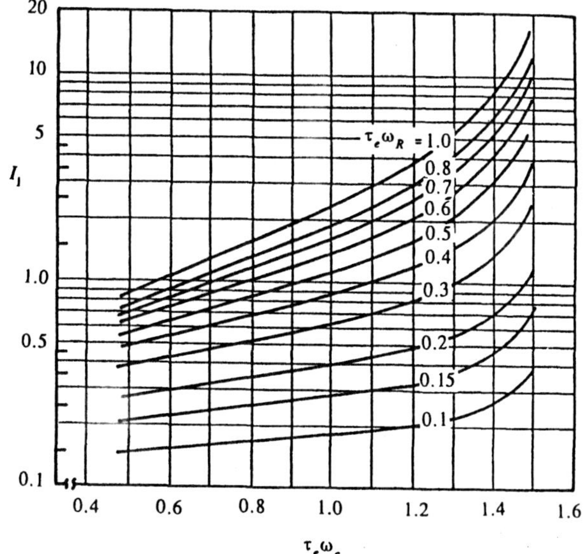

Figure 80.6 Determining I_1 for Equation 80.9.

Note that the crossover model implies a very simple model of the human operator, i.e.,

$$Y_p(j\omega) = \omega_c e^{-0.27j\omega}, \quad \text{for } \omega \text{ near } \omega_c. \tag{80.10}$$

Structural Model

Because the structural model follows the crossover model, one can use the first of Equations 80.5 to determine the crossover frequency as $\omega_c = 4.68$ rad/s. Referring to Table 80.2, and recalling that the plant exhibits first-order dynamics at all frequencies (i.e., the slope of the magnitude portion of its Bode diagram is -20 dB/dec at all frequencies), one uses the structural model parameters corresponding to $k = 1$. Thus, all of the structural model parameters can be chosen except K_e which can be selected by requiring $|Y_p(j\omega_c)Y_c(j\omega_c)| = 1.0$. This yields $K_e = 18.1$.

As opposed to Equation 80.10, the structural model offers a more detailed representation of human operator dynamics. In this application

$$Y_p(j\omega) = \frac{1810(j\omega+.2)e^{-0.15j\omega}}{(j\omega+.0497)[(j\omega)^2+14.29(j\omega)+402.1]}. \tag{80.11}$$

One can also utilize the same remnant model, as discussed, with the crossover model and approximate tracking performance with Equation 80.9 and Figure 80.6.

Optimal Control Model

Given the nominal OCM parameters from Table 80.3, only the selection of the weighting coefficient ρ in Equation 80.3 remains before the LQG synthesis technique produces a pilot model. One approach to selecting ρ has been mentioned in the preceding as an "effective time constant" method [6]. However, a simpler approach will be adopted which again calls on the empirical relation in the first of Equations 80.5 to define the crossover frequency ω_C. It can be shown that a relatively simple relationship exists between the plant dynamics and the coefficient ρ in an optimal regulator problem [11], namely,

$$\omega_{BW} \approx \left[\left(\frac{1}{\rho}\right)^{\frac{1}{2}}\right]^{\frac{1}{n-m+1}} \tag{80.12}$$

where ω_{BW} here is the bandwidth of the closed-loop, human-vehicle system and the parameters m and n are obtained from the plant dynamics when expressed as

$$Y_c(s) = \frac{K(s^m+a_{m-1}s^{m-1}+\cdots+a_1s+a_0)}{s^n+b_{n-1}s^{n-1}+\cdots+b_1+b_0}. \tag{80.13}$$

In more precise terms, ω_{BW} is the magnitude of that closed-loop pole closest to the frequency where the magnitude of the closed-loop system transfer function is 6 dB below its zero-frequency value. Now one can approximate the crossover frequency, ω_C, given ω_{BW} from Equation 80.12 by

$$\omega_C \approx 0.56\omega_{BW}. \tag{80.14}$$

Using the nominal parameter values given in Table 80.3, the OCM can now be applied to the tracking task. With $Y_c = \frac{1}{s}$, $K = 1, m = 0$, and $n = 1$ in Equation 80.13. Equations 80.12 and 80.14 and $\omega_C = 4.68$ rad/s. used in the previous models, yield $\rho = \left(\frac{0.56}{\omega_C}\right)^4 = \left(\frac{0.56}{4.68}\right)^4 = 2.05 \cdot 10^{-4}$.

The OCM computer program utilized in this study is described in [2]. As opposed to the crossover and structural models, error-injected remnant and RMS tracking error are obtained as part of the OCM model solution. The RMS tracking error predicted by the OCM was $\sqrt{\bar{e}^2} = 0.21$, which compares favorably with the 0.19 value obtained with the $\frac{1}{3}$ power law in the crossover and structural models. Finally, Figure 80.7 compares the Bode diagrams for the loop transmission $Y_pY_c(j\omega)$ obtained with each of the three modeling approaches. Again, the comparison is favorable.

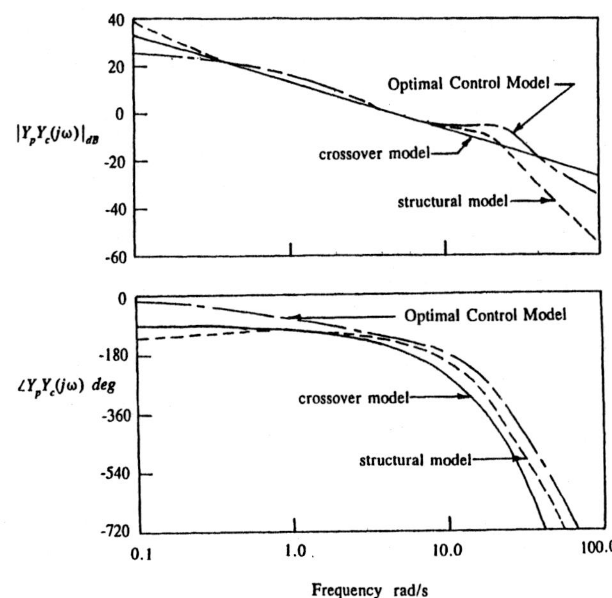

Figure 80.7 Bode diagrams of loop transmissions ($Y_pY_c(j\omega)$) for application example.

Fuzzy Control Model Formulation

Developing a fuzzy control model for the human operator in the input tracking task defined in the preceding would require actual simulator tracking data or a less attractive alternative involving detailed discussions with trained human operators who might describe how they select their control actions given certain displayed errors. This procedure assumes that the control engineer is unwilling or unable to employ any of the models just discussed. This might be the case, for example, if the plant dynamics were unknown, were suspected of being highly nonlinear, or if the task were sufficiently removed from that of tracking random-appearing input signals, e.g., tracking transient signals.

Space does not permit a detailed discussion of the fuzzy control approach to this problem, other than a very general, rudimentary outline.

Assume that a set of human operator input and output time histories ($e(t)$ and $\delta(t)$ respectively in Figure 80.2) are available from a human-in-the-loop control simulation of the task at hand. Let us further assume that, on the basis of discussions with the operator, the manual control engineer has been convinced that only error and not some combination of error and error rate or error and integral error are used by the operator in generating appropriate control inputs, $\delta(t)$.

Next, the engineer would create a set of membership functions for the error and output signals, similar to those shown in Figure 80.5(a). Choosing the number and shapes of the functions are part of the art in applying fuzzy control. Next, human operator time delay, τ, would be estimated. This delay would be considerably smaller than the effective delay, τ_e, discussed in the crossover model as it is intended only to approximate the delay between a visual stimulus and the initiation of a control response by the operator. Values on the order of 0.1 s. would be appropriate. The engineer would then tabulate pairs of error and delayed output values $e(n\Delta T)$ and $\delta(n\Delta T + \tau)$.

Using the paired input and output values, a relational matrix such as Equation 80.4 might be created as follows: For each input/output pair, form a relational matrix R_i as $R_i = M(e_i)XM(\delta_i)$, i.e., the Cartesian product of $M(e_i)$ and $M(\delta_i)$, or minimum for each element pairing of $M(e_i)$ and $M(\delta_i)$. After this has been done for all of the input-output pairs, the final relational matrix R is obtained by using, as each element $R(i, j)$, the largest corresponding element found in any R_i. This last operation is sometimes referred to as finding the "union" of the relational matrices.

With the relational matrix R thus obtained, one has the basis of a fuzzy control model which can produce membership numbers for a $\delta(n\Delta T + \tau)$ for any error $e(n\Delta T)$. This result is still fuzzy (recall the $M(\delta)$ obtained after Equation 80.4) and needs to be "defuzzified" for appropriate use in any simulation employing the fuzzy control model.

80.7 Closure

This treatment of human-in-the-loop control has been brief, and many important topics have gone untouched. Two examples of such unexplored topics are the modeling of the human operator in tasks where motion cues are important, and the modeling of human control of MIMO systems. It is hoped that the foregoing material has provided the reader with sufficient introductory background to explore these omitted topics.

80.8 Defining Terms

Compensatory tracking task: A tracking task in which the human is presented with a display of system error only.

Crossover model: A model of the human operator/plant combination (loop transmission) for compensatory tasks stating that the loop transmission in manually controlled SISO systems can be approximated by an integrator and time delay around the crossover frequency.

Fuzzy control model: A model of the human operator derived from fuzzy set theory. In its simplest form, this model is defined by input and output membership functions and a relational matrix describing mappings between these functions.

Optimal control model: An algorithmic, time-domain based model of the human operator based upon linear quadratic Gaussian control system design.

Pursuit tracking task: A tracking task in which the human is presented with a display of system error and system output.

Structural model: A model of the human operator which follows the crossover model, but provides a more detailed representation of human operator dynamics.

References

[1] Anderson, M., Standard optimal pilot models, *AIAA*, 94–3627, 1994.

[2] Curry, R. E., Hoffman, W. C., and Young, L. R., Pilot modeling for manned simulation, Air Force Flight Dynamics Lab., AFFDL-TR-76-124, 1976, Vol. 1.

[3] Gerlach, O. H., The biomorphic model of the human pilot, Delft University of Technology, Dept. of Aerospace Engineering, Report LR-310, 1980.

[4] Graham, D. and McRuer, D. T., *Analysis of nonlinear control systems,* Dover, New York, 1971, Chapter 10.

[5] Hess, R. A., Feedback control models, in *Handbook of Human Factors,* Salvendy, G., Ed., John Wiley & Sons, New York, 1212–1242, 1987.

[6] Hess, R. A. and Kalteis, R., Technique for Predicting Longitudinal Pilot-Induced Oscillations, *J. Guidance, Control, Dyn.,* 14(1), 198–204, 1991.

[7] Hess, R. A., Methodology for the Analytical Assessment of Aircraft Handling Qualities, in *Control and Dynamic Systems, Vol. 33, Advances in Aerospace Systems Dynamics and Control Systems,* Part 3, Leondes C.T., Ed., Academic, San Diego, 1990, 129–150.

[8] Kleinman, D. L., Levison, W. H., and Baron, S., An optimal control model of human response, part I: theory and validation, *Automatica,* 6(3), 357–369, 1970.

[9] Kramer, U., On the application of fuzzy sets to the analysis of the system-driver-vehicle environment, *Automatica,* 21(1), 101–107, 1985.

[10] Krendel, E. S. and McRuer, D. T., A servomechanism approach to skill development, *J. Franklin Inst.,* 269(1), 24–42, 1960.

[11] Kwakernaak, J. and Sivan, R., *Linear optimal control systems,* John Wiley & Sons, New York, 1972.

[12] Liu, T. S. and Wu, J. C., A model for a rider-motorcycle system using fuzzy control, *IEEE Trans. Syst., Man, and Cybernetics,* 23(1), 267–276, 1993.

[13] McRuer, D. T., Human dynamics in man-machine systems, *Automatica,* 16(3), 237–253, 1980.

[14] McRuer, D. T. and Krendel, E. S., Dynamic Response of Human Operators, Wright Air Development Center, WADC TR 56-524, 1957.

[15] McRuer, D. T., Graham, D., Krendel, E. S., and Reisener, W., Jr., Human pilot dynamics in compensatory systems, Air Force Flight Dynamics Lab., AFFDL-65-15, 1965.

[16] Nise, N., *Control Systems Engineering,* 1st ed., Benjamin/Cummings, Redwood City, CA, 1992, Chapter 11.

[17] Sheridan, T. B., *Telerobotics, Automation, and Human Supervisory Control,* MIT, Cambridge, MA, 1992, Chapter 1.

[18] Sutton, R. and Towill, D. R., *Modelling the Helmsman in a Ship Steering System Using Fuzzy Sets,* Proc. of the IFAC Conference on Man-Machine systems: Analysis, Design and Evaluation, Oulu, Finland, 1988.

[19] Tustin, A., The nature of the operator's response in manual control and its implication for controller design, *J. IEE,* 94, IIa(2), 1947.

[20] Zadeh, L., Outline of a new approach to the analysis of

complex systems and decision processes, *IEEE Trans. Syst., Man, Cybernetics,* SMC-3(1), 28–44, 1973.

[21] Wewerinke, P. H., Models of the human observer and controller of a dynamic system, Ph.D. Thesis, University of Twente, the Netherlands, 1989.

Further Reading

An excellent summary of application of human operator models to modeling human pilot behavior, circa 1974, is available in the AGARD report referenced herein, *Mathematical Models of Human Pilot Behavior,* by McRuer and Krendel.

The excellent text by Thomas Sheridan and William Ferrell, *Man-Machine Systems,* published by MIT Press in 1974, provides a broad view of the topics involved with human-machine interaction.

The 1987 Wiley publication *Handbook of Human Factors,* [5] contains a chapter by the author entitled Feedback Control Models. This chapter covers some MIMO applications of the classical and modern (OCM) human operator models. A new edition of this Handbook is currently in press.

The book, *Modelling Human Operators in Control System Design,* by Robert Sutton and published by Wiley in 1990 offers a very readable treatment of human-in-the-loop control systems. This book includes chapters devoted to modeling the human operator using fuzzy sets.

Index